EIGHT PEAK INDEX OF MASS SPECTRA

The eight most abundant ions in 81,123 mass spectra, indexed by molecular weight, elemental composition and most abundant ions

Compiled by

The Mass Spectrometry Data Centre

Fourth Edition 1991

Volume 1 Part 2

Published by

THE MASS SPECTROMETRY DATA CENTRE
The Royal Society of Chemistry
Thomas Graham House, Science Park
Milton Road, Cambridge, CB4 4WF, UK

ISBN 0-85186-417-1

Computer set by Computaprint
Printed and bound by Staples Printers Rochester Limited

CONTENTS

INTRODUCTION

The success of the previous three editions of the *Eight Peak Index of Mass Spectra* has confirmed the value of abbreviated mass spectra as aids for the identification of unknown compounds. The popularity of the previous editions has also been attributed to the easy to read format of the data and to the different ways in which the spectra are sorted.

The rapid growth in the number of available mass spectra in the 1970's and 1980's is exemplified by the number of spectra contained in previous editions of the Index:

1970	1st Edition	17,124 mass spectra
1974	2nd Edition	31,101 mass spectra
1983	3rd Edition	66,720 mass spectra

This 4th Edition contains 81,123 mass spectra covering 65,600 compounds. The format of the previous editions has been retained so that the Index may be used as an aid for compound identification and as a molecular weight and formula index.

Those users familiar with the previous editions will notice that in this edition iodine has been specified in the elemental formula. Also the collection code letters and numbers have been reassigned allowing for a more simple identification of the source of the data.

ORIGIN OF DATA

Each spectrum is identified by a collection letter and number, the letter denoting the source collection. For convenience, letters and sources are tabulated on page vi and on the inside back cover of each part.

The MSDC spectra originate from the MSDC collection of full spectra which have been collected on a worldwide basis since 1967. This collection includes the American Petroleum Institute (standard and matrix), the Thermodynamics Research Center (standard and matrix), the American Society for Testing and Materials, the DOW Chemical Company and Downstream Data collections.

For the third edition the NIH-EPA Chemical Information Service made available the complete archive tapes used to prepare the database for the Mass Spectral Search System. Since that time this spectra collection has developed into the *NIST/EPA/MSDC Mass Spectral Database*, compiled by the National Institute of Standards and Technology (NIST). For this Fourth Edition NIST made available all spectra added to the database subsequent to the compilation of the 3rd Edition.

The literature spectra have been extracted from published papers by Imperial Chemical Industries PLC and the MSDC, using the Mass Spectrometry Bulletin as a guide.

The tabulated data include a CAS Registry Number of the compound where this was readily available. A maximum of three spectra for any one compound has been set. These have been selected, where possible, to reflect differing experimental conditions.

Whilst great effort has been made to ensure that the information, and in particular the m/z values, in this Index is correct, this cannot be guaranteed. Difficulties occur, in particular in literature spectra, where the mode of presentation of the spectrum (such as a small line diagram or a brief tabulation) is not conclusive as to the major peaks in the original.

EDITORIAL CHECKING OF DATA

To ensure that the information included in the Index is both consistent and accurate, every record has been inspected manually and validated against many checks by computer. Even with the aid of a purpose built database system, the editorial task on a collection of this size has been enormous. This has been considered essential to maintain the high quality of the data that appear in the Index.

Computer checks on the data have alerted the editors to spectra that have low mass peaks (including air), high mass peaks significantly greater than the parent ion and invalid multiply charged ions. Further computer checks confirmed the consistency of the molecular weight with the molecular formula and the validity of the CAS Registry Number. Compound names have been edited for consistency and, as far as possible, given the most structurally informative name rather than the trivial or trade name. Duplicate spectra (identical compounds with identical m/z and relative intensity ± 1%) have not been included in this compilation.

Compounds whose spectra are considered suspect or which contain obvious impurity peaks have not been included in this collection. Spectra that have been reported with less than eight peaks have been included only when they are unique or considered to be of value. The quality of the data has been considered to be more significant than the quantity of spectra that is included in the *Eight Peak Index of Mass Spectra*.

THE MASS SPECTROMETRY DATA CENTRE

The Mass Spectrometry Data Centre (MSDC) has been serving the needs of mass spectrometrists since 1965. Its primary function has been to provide a comprehensive information and data service to all users of mass spectrometry. The *Eight Peak Index of Mass Spectra* is one of these valuable services.

The MSDC also publishes the *Mass Spectrometry Bulletin* (MSB), the leading current awareness abstracts journal in the field of mass spectrometry. This respected publication offers the most comprehensive abstracting service available to mass spectrometrists. The MSB has been available in printed form since 1966 and is also available online via ESA-IRS and in a personal computer (PC) version. Each issue contains items on approximately 800 references to the latest published literature, each item containing full bibliographical details and specialist indexing information.

The MSDC is also involved in the collection and dissemination, in collaboration with the National Institute of Standards and Technology (NIST), of mass spectra. Scientists from around the world contribute spectra to the MSDC collection, and these spectra are incorporated in the *NIST/EPA/MSDC Mass Spectral Database* and, in abbreviated form, in the *Eight Peak Index of Mass Spectra*.

DATA ELEMENTS

The three different ways in which the data are sorted in the three volumes are explained on the Contents page.

The example entry below is annotated to give a brief explanation of the various elements of the data. In more detail the information listed is:

(a) The mass-to-charge ratios of the eight most abundant peaks in the spectrum (integral values except for multiply charged ions). The relevant m/z in Volume 3 is indicated by heavy type. In a few cases spectra with less than eight available peaks have been included when it was considered that the information was of value.

(b) The molecular weight of the compound (calculated from integral values of the atomic weights of the most abundant isotope of each element present). Polymers have been given an arbitrary molecular weight of 9999.

(c) The relative intensities of the eight most abundant peaks (integral values and proportional to 100 for the most abundant ion).

(d) The relative intensity of the parent (molecular) ion (blank if already in the eight most abundant ions, zero if absent or unknown).

(e) The molecular formula (the number of atoms for C, H, O, N, S, F, Cl, Br, I, Si, P, B. The total number of atoms of any other elements present is shown under X. Polymers are not given a formula (except 1 in the X element column).

(f) The compound name.

(g) The CAS Registry Number, where readily available.

(h) The collection code letters and number (see Spectra Collection Identification).

The relevant molecular weight is given in square brackets at the head or foot of each page in Volumes 1 and 2; the relevant m/z value in Volume 3 appears in the square brackets.

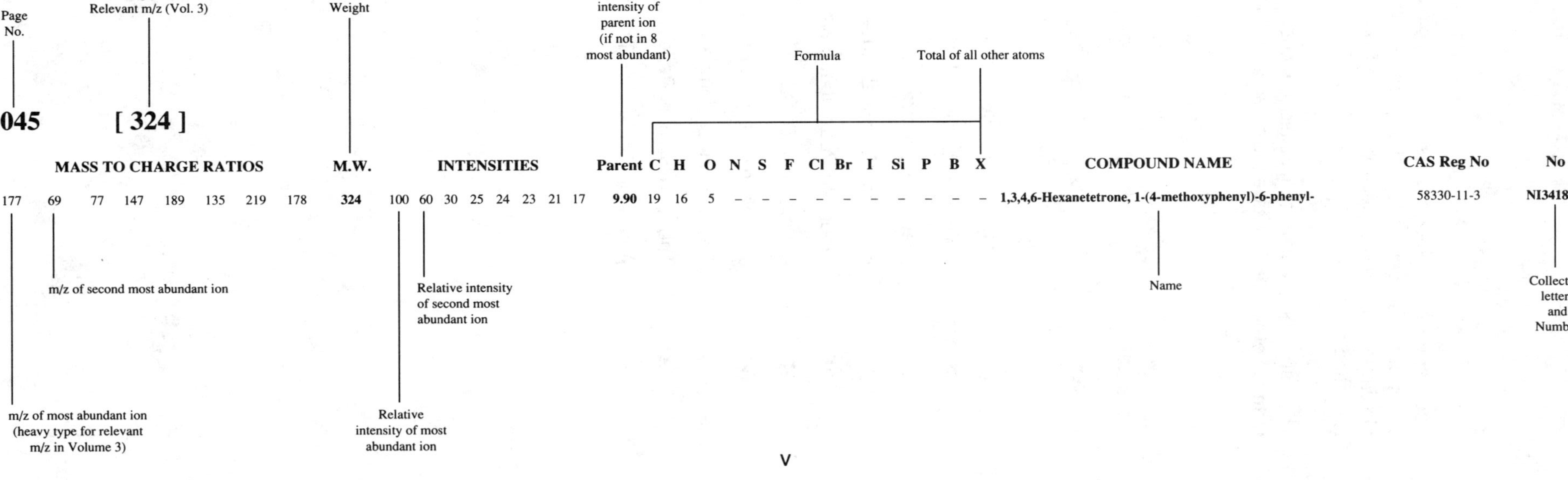

SPECTRA COLLECTION IDENTIFICATION

Each entry in the Index is assigned a collection code letter which denotes the source of the mass spectrum. The collection code letter is followed by a five digit number ranging from 1 up to the maximum number of spectra selected from the particular collection. Thus every spectrum is uniquely identified.

The table below lists the collection code letters and corresponding source along with the total number of spectra included from each of the collections.

Collection Code Letters	*Collection*	*No. of Spectra*
NI	National Institute of Standards and Technology	47,737
MS	Mass Spectrometry Data Centre	11,118
LI	Literature Spectra	10,172
IC	Imperical Chemical Industries PLC	4,654
AI	American Petroleum Institute	2,416
DD	Downstream Data	1,837
WI	John Wiley & Sons	1,602
DO	Dow Chemical Company	1,123
TR	Thermodynamics Research Center	305
AM	American Society for Testing Materials	159
		81,123

ACKNOWLEDGEMENTS

The Mass Spectrometry Data Centre wishes to acknowledge with thanks all the organisations and individuals who have contributed spectra to the *Eight Peak Index of Mass Spectra*, either directly or through the various spectra collections. Special thanks are also due to Mr S Down for his assistance in the editing of this edition.

MASS TO CHARGE RATIOS	M.W.	INTENSITIES	Parent	C	H	O	N	S	F	Cl	Br	I	Si	P	B	X	COMPOUND NAME	CAS Reg No	No
73 228 107 75 199 150 106 56	**243**	100 73 39 19 13 11 10 10	**0.00**	7	12	3	1	–	3	–	–	–	1	–	–	–	**Glycine, N-(trifluoroacetyl)-, trimethylsilyl ester**	52558-79-9	**NI23457**
86 57 28 155 140 45 43 29	**243**	100 64 42 40 33 28 24 23	**0.00**	8	12	4	1	–	3	–	–	–	–	–	–	–	**α-Daunosamine, N-trifluoroacetyl-**		**NI23458**
169 243 75 64 111 137 170 46	**243**	100 77 67 47 46 46 41 39		9	6	1	1	2	–	1	–	–	–	–	–	–	**4-Thiazolidinone, 3-(4-chlorophenyl)-2-thioxo-**	13037-55-3	**NI23459**
170 134 199 243 172 198 201 200	**243**	100 64 56 51 44 36 20 20		9	6	3	1	1	–	1	–	–	–	–	–	–	**3(2H)-Benzothiazoleacetic acid, 4-chloro-2-oxo-**	3813-05-6	**NI23461**
170 243 134 172 199 198 245 200	**243**	100 46 40 36 33 33 18 16		9	6	3	1	1	–	1	–	–	–	–	–	–	**3(2H)-Benzothiazoleacetic acid, 4-chloro-2-oxo-**	3813-05-6	**NI23460**
243 228 214 44 42 131 208 52	**243**	100 84 79 57 33 26 20 19		9	8	–	5	–	3	–	–	–	–	–	–	–	**2-Pteridinamine, N,N-dimethyl-4-(trifluoromethyl)-**	32706-25-5	**NI23462**
243 228 214 44 42 131 208 181	**243**	100 87 80 57 34 27 22 18		9	8	–	5	–	3	–	–	–	–	–	–	–	**2-Pteridinamine, N,N-dimethyl-4-(trifluoromethyl)-**	32706-25-5	**LI04802**
57 44 29 43 73 28 31 42	**243**	100 41 35 29 27 23 22 16	**0.00**	9	13	5	3	–	–	–	–	–	–	–	–	–	**Cytidine**	65-46-3	**NI23463**
112 140 151 111 44 69 43 42	**243**	100 43 23 23 18 10 8 8	**2.00**	9	13	5	3	–	–	–	–	–	–	–	–	–	**Cytidine**	65-46-3	**NI23464**
244 112 245 133 111 243 227 113	**243**	100 57 30 28 22 20 9 8		9	13	5	3	–	–	–	–	–	–	–	–	–	**Cytidine**	65-46-3	**NI23465**
208 71 45 243 245 210 138 75	**243**	100 96 81 73 67 48 28 23		10	7	–	1	1	–	2	–	–	–	–	–	–	**Thiazole, 4-(chloromethyl)-2-(4-chlorophenyl)-**	17969-22-1	**NI23466**
184 59 186 123 199 114 201 125	**243**	100 86 66 63 38 37 24 21	**1.90**	10	7	2	1	–	–	2	–	–	–	–	–	–	**2H-Azirine-3-carboxylic acid, 2-(2,6-dichlorophenyl)-, methyl ester**	98081-82-4	**NI23467**
52 159 243 80 106 107 187 215	**243**	100 84 28 25 23 16 15 6		10	9	3	1	–	–	–	–	–	–	–	–	1	**Chromium, 2-aminotolyltricarbonyl-**		**LI04803**
52 159 187 80 243 107 106 215	**243**	100 87 30 20 19 19 17 1		10	9	3	1	–	–	–	–	–	–	–	–	1	**Chromium, 3-aminotolyltricarbonyl-**		**LI04804**
159 129 44 158 36 130 89 102	**243**	100 95 71 54 44 37 32 25	**0.00**	10	10	4	1	–	–	1	–	–	–	–	–	–	**Isoquinolinium, 2-methyl-, perchlorate**	4185-65-3	**NI23468**
129 159 44 36 130 131 102 128	**243**	100 59 55 41 38 31 28 24	**0.00**	10	10	4	1	–	–	1	–	–	–	–	–	–	**Quinolinium, 1-methyl-, perchlorate**	4185-64-2	**NI23469**
120 108 107 164 243 65 121 93	**243**	100 60 56 29 28 28 20 18		10	13	4	1	1	–	–	–	–	–	–	–	–	**1,5-Benzoxazepine, 3-hydroxy-5-methylsulphonyl-2,3,4,5-tetrahydro-**		**LI04805**
91 108 155 179 43 197 65 55	**243**	100 94 66 55 35 32 28 15	**5.50**	10	13	4	1	1	–	–	–	–	–	–	–	–	**Carbamic acid, [(4-methylphenyl)sulphonyl]-, ethyl ester**	5577-13-9	**NI23471**
179 108 197 155 91 180 198 171	**243**	100 74 71 44 16 11 6 6	**0.00**	10	13	4	1	1	–	–	–	–	–	–	–	–	**Carbamic acid, [(4-methylphenyl)sulphonyl]-, ethyl ester**	5577-13-9	**NI23470**
91 108 155 179 197 43 65 171	**243**	100 93 65 54 33 33 28 5	**1.00**	10	13	4	1	1	–	–	–	–	–	–	–	–	**Carbamic acid, [(4-methylphenyl)sulphonyl]-, ethyl ester**	5577-13-9	**LI04806**
91 42 155 28 92 65 88 58	**243**	100 72 49 35 26 21 20 19	**10.00**	10	13	4	1	1	–	–	–	–	–	–	–	–	**1,3,5-Dioxazine, 5-(p-tolylsulphonyl)dihydro-**	56221-08-0	**NI23472**
214 228 243 200 43 172 216 58	**243**	100 91 77 52 44 41 34 33		10	18	–	5	–	–	1	–	–	–	–	–	–	**1,3,5-Triazine-2,4-diamine, 6-chloro-N,N-diethyl-N′-isopropyl-**	1912-25-0	**NI23473**
243 245 202 208 64 244 167 175	**243**	100 33 25 16 15 14 13 10		11	6	–	5	–	–	1	–	–	–	–	–	–	**1,3,6-Triazacyclo[3.3.3]azine, 9-chloro-4-cyano-2-methyl-**		**NI23474**
243 77 44 27 46 245 166 244	**243**	100 84 27 23 19 14 13 11		11	9	–	5	1	–	–	–	–	–	–	–	–	**5H-Pyrazolo[3,4-e]-1,2,4-triazine-3-thiol, 7-methyl-5-phenyl-**	27165-57-7	**NI23475**
56 121 129 212 147 94 81 128	**243**	100 83 52 46 28 25 21 20	**0.00**	11	12	2	–	–	–	–	–	–	–	–	1	1	**Boronic acid, ferrocenyl-, methyl ester**		**LI04807**
106 79 28 107 77 104 29 51	**243**	100 11 11 8 7 5 3 2	**0.31**	11	18	3	1	–	–	–	–	–	–	1	–	–	**Phosphonic acid, 1-amino-1-phenylmethyl-, diethyl ester**		**MS05195**
171 107 243 65 18 134 39 29	**243**	100 90 83 77 75 70 59 53		11	18	3	1	–	–	–	–	–	–	1	–	–	**Phosphonic acid, (2,6-dimethyl-4-pyridinyl)-, diethyl ester**	26384-85-0	**MS05196**
107 18 135 106 79 170 77 29	**243**	100 51 32 24 24 17 14 14	**10.00**	11	18	3	1	–	–	–	–	–	–	1	–	–	**Phosphonic acid, (3,4-dimethyl-2-pyridyl)-, diethyl ester**	26384-82-7	**MS05197**
43 170 74 41 32 57 68 42	**243**	100 52 52 45 41 30 29 27	**21.70**	11	21	3	3	–	–	–	–	–	–	–	–	–	**3-Imidazoline, 1-hydroxy-4-isopropylcarbamoyl-2,2,5,5-tetramethyl-, 3-oxide**		**NI23476**
243 242 174 59 51 102 83 244	**243**	100 95 33 30 23 19 19 18		12	9	1	3	1	–	–	–	–	–	–	–	–	**5H-1,3,4-Thiadiazolo[3,2-a]pyrimidin-5-one, 2-methyl-7-phenyl-**	42484-76-4	**NI23477**
214 167 30 180 168 77 215 169	**243**	100 69 36 24 18 15 14 14	**0.00**	12	9	3	3	–	–	–	–	–	–	–	–	–	**Aniline, 2-nitro-N-nitroso-N-phenyl-**	21565-15-1	**LI04808**
214 167 30 180 168 77 169 51	**243**	100 69 36 24 18 15 14 14	**0.00**	12	9	3	3	–	–	–	–	–	–	–	–	–	**Aniline, 2-nitro-N-nitroso-N-phenyl-**	21565-15-1	**MS05198**
214 167 30 168 166 184 215 77	**243**	100 76 50 31 17 16 14 14	**0.00**	12	9	3	3	–	–	–	–	–	–	–	–	–	**Aniline, 4-nitro-N-nitroso-N-phenyl-**	3665-70-1	**MS05199**
213 167 30 168 166 184 214 77	**243**	100 76 50 31 17 16 14 14	**0.00**	12	9	3	3	–	–	–	–	–	–	–	–	–	**Aniline, 4-nitro-N-nitroso-N-phenyl-**	3665-70-1	**LI04809**
243 136 244 166 137 242 227 138	**243**	100 77 13 10 8 5 5 4		12	9	3	3	–	–	–	–	–	–	–	–	–	**Diazene, (4-nitrophenyl)phenyl-, 1-oxide**	13921-68-1	**LI04810**
77 51 243 105 63 91 50 64	**243**	100 55 38 35 27 23 23 19		12	9	3	3	–	–	–	–	–	–	–	–	–	**Diazene, (4-nitrophenyl)phenyl-, 1-oxide**	13921-68-1	**NI23478**
243 185 178 105 121 56 160 137	**243**	100 69 25 24 23 21 19 18		12	13	1	1	–	–	–	–	–	–	–	–	1	**Ferrocene oxime, acetyl-, anti-**		**MS05200**
243 185 178 105 160 137 121 138	**243**	100 30 21 18 15 15 15 14		12	13	1	1	–	–	–	–	–	–	–	–	1	**Ferrocene oxime, acetyl-, syn-**		**MS05201**
91 164 92 42 65 120 212 41	**243**	100 99 45 36 35 24 16 16	**0.00**	12	18	2	1	–	–	1	–	–	–	–	–	–	**Propanamine, N-benzyl-3-chloro-N-(2-hydroxyethyl)-2-hydroxy-**		**IC03211**
158 98 186 43 41 71 70 42	**243**	100 48 17 12 11 8 7 7	**0.00**	12	21	4	1	–	–	–	–	–	–	–	–	–	**4-Oxazolidinecarboxylic acid, 4-ethyl-3-formyl-2-tert-butyl-, methyl ester, (2R,4S)-**		**DD00634**
121 243 214 39 77 90 89 118	**243**	100 66 29 29 26 24 19 14		13	9	2	1	1	–	–	–	–	–	–	–	–	**Thiazolo[3,2-a]pyridinium, 3,8-dihydroxy-2-phenyl-, hydroxide, inner salt**	35143-57-8	**NI23479**
121 122 138 243 139 78 77 46	**243**	100 8 2 1 1 1 1 1		13	9	4	1	–	–	–	–	–	–	–	–	–	**Benzoic acid, 2-nitro-, phenyl ester**		**LI04811**
150 151 106 93 243 107 94 46	**243**	100 8 4 3 1 1 1 1		13	9	4	1	–	–	–	–	–	–	–	–	–	**Benzoic acid, 2-nitro-, phenyl ester**		**LI04812**
243 244 213 227 93 245 166 122	**243**	100 16 5 3 3 2 1 1		13	9	4	1	–	–	–	–	–	–	–	–	–	**Benzoic acid, 4-nitro-, phenyl ester**		**LI04813**
138 243 139 227 244 121 46	**243**	100 28 8 3 1 1 1		13	9	4	1	–	–	–	–	–	–	–	–	–	**Benzoic acid, 4-nitro-, phenyl ester**		**LI04814**
105 77 166 242 51 41 43 120	**243**	100 72 56 48 25 24 22 21	**0.00**	13	9	4	1	–	–	–	–	–	–	–	–	–	**Benzophenone, 4-hydroxy-3-nitro-**		**IC03212**
121 243 65 93 122 76 39 104	**243**	100 29 16 12 8 6 6 5		13	9	4	1	–	–	–	–	–	–	–	–	–	**Benzophenone, 4′-hydroxy-4-nitro-**		**IC03213**
243 228 200 62 244 114 172 53	**243**	100 69 25 16 15 14 13 11		13	9	4	1	–	–	–	–	–	–	–	–	–	**1,3-Dioxolo[4,5-g]furo[2,3-b]quinoline, 9-methoxy-**	524-89-0	**NI23480**
243 186 198 214 242 215	**243**	100 37 7 6 2 1		13	9	4	1	–	–	–	–	–	–	–	–	–	**2(3H)-Furanone, dihydro-4-methyl-3-phthalimido-**		**LI04815**
91 77 78 51 64 39 50 92	**243**	100 33 23 23 14 13 10 9	**8.00**	13	11	3	2	–	–	–	–	–	–	–	–	–	**Aniline, N-(3-nitrobenzylidene)-, N-oxide**		**LI04816**
243 242 228 213 241 214 200 157	**243**	100 76 30 24 14 13 13 13		13	13	2	3	–	–	–	–	–	–	–	–	–	**3H-Pyrazolo[3,4-b]quinolin-9-one, 1,3-dimethyl-4,9-dihydro-7-methoxy-**		**MS05202**
55 41 77 133 91 39 118 172	**243**	100 98 89 82 62 57 49 48	**20.00**	13	17	–	5	–	–	–	–	–	–	–	–	–	**1H-Tetrazole, 5-(cyclohexylamino)-1-phenyl-**		**NI23481**
55 83 41 133 93 77 214 243	**243**	100 77 75 67 62 59 55 55		13	17	–	5	–	–	–	–	–	–	–	–	–	**1H-Tetrazole, 1-cyclohexyl-5-(phenylamino)-**		**NI23482**
85 99 100 159 117 142 84 112	**243**	100 70 55 37 31 25 23 16	**1.00**	13	25	3	1	–	–	–	–	–	–	–	–	–	**Pentanoic acid, 3-[(1-oxopentyl)amino]propyl ester**	55044-82-1	**NI23483**

MASS TO CHARGE RATIOS								M.W.	INTENSITIES								Parent	C	H	O	N	S	F	Cl	Br	I	Si	P	B	X	COMPOUND NAME	CAS Reg No	No
97	132	126	112	173	130	115	84	243	100	79	71	71	63	56	56	34	16.00	13	26	1	1	–	–	–	–	–	–	1	–	–	Phosphetan, 2,2,3,4,4-pentamethyl-1-(piperidino)-, 1-oxide		NI23484
243	141	166	242	104	70	103	244	243	100	33	17	14	14	11	10	9		14	13	1	1	1	–	–	–	–	–	–	–	–	Thiazolo[3,2-a]pyridinium, 2,3-dihydro-8-hydroxy-5-methyl-2-phenyl-, hydroxide, inner salt	39137-97-8	NI23485
107	137	79	77	90	51	78	89	243	100	89	72	52	22	13	12	11	0.10	14	13	3	1	–	–	–	–	–	–	–	–	–	Benzeneethanol, 4-nitro-α-phenyl-	20273-74-9	NI23486
55	128	243	42	187	41	156	226	243	100	99	85	80	75	71	69	62		14	13	3	1	–	–	–	–	–	–	–	–	–	4H-1,2-Benzoxazine-3,5-dione, 2,3,5,6,7,8-hexahydro-4-phenyl-		LI04817
243	228	165	154	152	168	167	153	243	100	71	43	36	32	29	23	21		14	13	3	1	–	–	–	–	–	–	–	–	–	Furan, 2,3-dihydro-3-[2-methyl-3-(4-nitrophenyl)-1-propenylidene]-	55256-12-7	NI23487
226	228	198	183	115	213	227	51	243	100	96	57	16	15	14	13	13	13.00	14	13	3	1	–	–	–	–	–	–	–	–	–	7H-Furo[3,2-g]benzopyran, 2-(1-amino-isopropyl)-7-oxo-		LI04818
81	91	96	120	243	82	92	118	243	100	81	34	14	11	7	6	5		14	13	3	1	–	–	–	–	–	–	–	–	–	Glycidamide, N-furfuryl-3-phenyl-, trans-	19464-99-4	NI23488
215	129	55	101	157	75	51	103	243	100	70	54	38	31	31	28	23	3.00	14	13	3	1	–	–	–	–	–	–	–	–	–	1,2-Naphthalenedione, 4-(4-morpholinyl)-	4569-83-9	NI23489
243	242	184	77	212	51	53	104	243	100	77	46	42	23	18	14	12		14	13	3	1	–	–	–	–	–	–	–	–	–	5-Pyridinecarboxylic acid, 1,2-dihydro-3-methyl-2-oxo-1-phenyl-, methyl ester		DD00635
243	77	228	160	118	51	144	200	243	100	78	71	57	40	35	28	23		14	13	3	1	–	–	–	–	–	–	–	–	–	2(1H)-Pyridinone, 3-acetyl-4-hydroxy-6-methyl-1-phenyl-	13959-06-3	LI04819
243	77	228	160	118	51	144	200	243	100	83	72	60	40	35	28	20		14	13	3	1	–	–	–	–	–	–	–	–	–	2(1H)-Pyridinone, 3-acetyl-4-hydroxy-6-methyl-1-phenyl-	13959-06-3	NI23490
200	77	243	173	201	160	51	69	243	100	92	56	50	39	35	35	29		14	13	3	1	–	–	–	–	–	–	–	–	–	2(1H)-Pyridone, 4-hydroxy-6-methyl-1-phenyl-, acetate	13959-05-2	NI23491
200	77	243	173	201	160	51	69	243	100	88	56	47	38	34	34	28		14	13	3	1	–	–	–	–	–	–	–	–	–	2(1H)-Pyridone, 4-hydroxy-6-methyl-1-phenyl-, acetate	13959-05-2	LI04820
228	243	200	121	120	214	160	186	243	100	55	18	9	4	2	2	1		14	13	3	1	–	–	–	–	–	–	–	–	–	Pyrrolo[2,1-b][1,3]benzoxazin-10-one, 4-acetyl-1,2,3,10-tetrahydro-		LI04821
123	183	199	184	182	200	91	44	243	100	80	67	43	34	33	24	24	10.00	14	17	1	3	–	–	–	–	–	–	–	–	–	Cyclohexanone, 2-benzylidene-, semicarbazone	17026-13-0	NI23492
42	58	243	43	41	183	201	200	243	100	78	60	44	39	36	35	24		14	17	1	3	–	–	–	–	–	–	–	–	–	1,3-Diazaadamantan-6-one, 5-phenyl-, oxime		MS05203
170	57	77	41	29	201	226	243	243	100	75	50	50	40	35	25	24		14	17	1	3	–	–	–	–	–	–	–	–	–	4-Pyrazolyl tert-butyl ketone, oxime, anti-		LI04822
187	243	120	41	188	244	186	57	243	100	42	36	25	14	7	4	4		14	17	1	3	–	–	–	–	–	–	–	–	–	5-Pyrimidinamine, 2-(4-butoxyphenyl)-	84609-99-4	NI23493
85	98	41	18	44	99	43	28	243	100	43	15	14	10	9	9	7	0.00	14	29	2	1	–	–	–	–	–	–	–	–	–	Dodecanamide, N-(2-hydroxyethyl)-		MS05204
214	182	215	58	212	55	41	43	243	100	41	15	10	8	8	7	5	0.80	14	29	2	1	–	–	–	–	–	–	–	–	–	Tridecanoic acid, 11-amino-, methyl ester	56817-91-5	NI23494
152	125	154	127	89	153	91	77	243	100	62	33	21	14	10	9	9	5.30	15	14	–	1	–	–	1	–	–	–	–	–	–	Benzeneethanamine, N-(4'-chlorobenzylidene)-		MS05205
91	106	138	65	152	121	78	45	243	100	58	57	9	7	7	5	4	3.00	15	17	–	1	1	–	–	–	–	–	–	–	–	Benzylamine, N-benzyl-3-(methylthio)-		MS05206
187	186	84	57	86	115	188	243	243	100	90	40	35	25	23	17	14		15	17	–	1	1	–	–	–	–	–	–	–	–	Pyridine, 4-phenyl-2-tert-butylmercapto-		MS05207
84	86	32	187	47	186	49	57	243	100	62	55	24	22	21	17	12	3.00	15	17	–	1	1	–	–	–	–	–	–	–	–	Pyridine, 4-phenyl-3-tert-butylmercapto-		MS05208
57	169	187	143	59	114	243	115	243	100	74	68	53	42	32	29	29		15	17	2	1	–	–	–	–	–	–	–	–	–	Carbamic acid, (1H-inden-1-ylidenemethyl)-, tert-butyl ester	55760-22-0	NI23495
57	41	116	143	115	44	56	187	243	100	90	88	80	80	60	37	35	7.00	15	17	2	1	–	–	–	–	–	–	–	–	–	Cycloprop[a]inden-1,6-imine-7-carboxylic acid, 1,1a,6,6a-tetrahydro-, tert-butyl ester	34813-09-7	NI23496
243	81	162	134	120	202	77	244	243	100	95	47	28	27	19	19	17		15	17	2	1	–	–	–	–	–	–	–	–	–	4H-Isoindoline, 5,7a-epoxy-N-(2-methoxyphenyl)-		NI23497
243	162	81	134	244	120	130	53	243	100	68	60	32	17	16	13	13		15	17	2	1	–	–	–	–	–	–	–	–	–	4H-Isoindoline, 5,7a-epoxy-N-(4-methoxyphenyl)-		NI23498
105	77	106	122	215	243	81	41	243	100	36	10	6	5	5	4	3		15	17	2	1	–	–	–	–	–	–	–	–	–	1,3-Oxazin-6-one, 4,5-dihydro-4,4-pentamethylene-2-phenyl-		NI23499
188	200	243	134	214	132	104	105	243	100	93	71	11	10	6	6	5		15	17	2	1	–	–	–	–	–	–	–	–	–	5H-Pyrano[2,3-b]quinolin-5-one, 2,3,4,10-tetrahydro-1,2,2-trimethyl-		LI04823
213	170	171	131	91	104	214	115	243	100	93	56	35	33	24	21	19	5.70	15	19	1	2	–	–	–	–	–	–	–	–	–	Spiro[1.2]pentane-3-imidazoline, 5,5-dimethyl-4-phenyl-, 1-oxile		NI23500
74	44	128	212	43	30	72	41	243	100	24	19	18	13	12	10	8	1.60	15	33	1	1	–	–	–	–	–	–	–	–	–	Dodecanamine, N-(2-hydroxyethyl)-2-methyl-		IC03214
74	212	44	30	43	41	72	55	243	100	90	79	43	41	30	26	21	3.80	15	33	1	1	–	–	–	–	–	–	–	–	–	Tridecanamine, N-(2-hydroxyethyl)-		IC03215
46	84	56	92	134	91	67	39	243	100	51	38	34	30	29	26	23	0.00	16	21	1	1	–	–	–	–	–	–	–	–	–	Formimidoic acid, N-cyclohexyl-, cinnamyl ester		NI23501
159	174	243	41	134	228	121	84	243	100	74	39	25	24	18	16	13		16	21	1	1	–	–	–	–	–	–	–	–	–	Indolizidine, 6,7-dehydro-6-(4-methoxyphenyl)-7-methyl-		MS05209
136	228	129	118	200	108	124	243	243	100	74	74	59	30	30	23	22		16	21	1	1	–	–	–	–	–	–	–	–	–	Isoxazolidine, 4-(2-methyl-1-propenylidene)-3-phenyl-2,5,5-trimethyl-		DD00636
123	200	243	81	104	105	95	96	243	100	97	47	37	26	26	24	18		16	21	1	1	–	–	–	–	–	–	–	–	–	Octadienamide, N-(2-phenylethyl)-, (2Z,4E)-		NI23502
122	137	105	79	55	243	77	136	243	100	98	97	92	88	86	75	68		16	21	1	1	–	–	–	–	–	–	–	–	–	1,2-Oxazine, 3,6-dihydro-3,4-dimethyl-6-isobutenyl-2-phenyl-		LI04824
105	122	77	43	243	41	107	55	243	100	71	36	29	27	25	16	16		16	21	1	1	–	–	–	–	–	–	–	–	–	4H-1,3-Oxazine, 5,6-dihydro-6,6-dimethyl-4,5-tetramethylene-2-phenyl-	102869-93-2	NI23503
162	107	69	93	79	77	104	91	243	100	99	93	86	71	66	63	61		16	21	1	1	–	–	–	–	–	–	–	–	–	1,2-Oxazine, 3,6-dihydro-4-(4-methyl-3-pentenyl)-2-phenyl-		LI04825
77	243	93	73	69	107	182	168	243	100	98	94	92	92	58	51	50		16	21	1	1	–	–	–	–	–	–	–	–	–	1,2-Oxazine, 3,6-dihydro-5-(4-methyl-3-pentenyl)-2-phenyl-		LI04826
105	200	77	243	137	106	51	41	243	100	52	42	27	12	8	8	8		16	21	1	1	–	–	–	–	–	–	–	–	–	Quinoline, 1-benzoyldecahydro-, cis-	5710-04-3	NI23504
105	77	200	243	137	51	106	41	243	100	42	32	27	13	9	8	8		16	21	1	1	–	–	–	–	–	–	–	–	–	Quinoline, 1-benzoyldecahydro-, cis-	5710-04-3	LI04827
105	77	200	243	137	106	51	201	243	100	42	31	25	13	8	8	7		16	21	1	1	–	–	–	–	–	–	–	–	–	Quinoline, 1-benzoyldecahydro-, trans-	22218-33-3	NI23505
105	77	200	243	137	106	51	214	243	100	42	32	26	13	9	9	3		16	21	1	1	–	–	–	–	–	–	–	–	–	Quinoline, 1-benzoyldecahydro-, trans-	22218-33-3	LI04828
157	57	43	55	71	41	69	83	243	100	73	70	61	54	49	48	40	35.76	16	21	1	1	–	–	–	–	–	–	–	–	–	(2E,4E)-Undecadiene-8,10-diynamide, N-(2-methylbutyl)-		NI23506
128	157	43	242	243	129	127	41	243	100	63	63	58	47	42	38	34		16	21	1	1	–	–	–	–	–	–	–	–	–	(2E,4Z)-Undecadiene-8,10-diynamide, N-(2-methylbutyl)-		NI23507
243	187	228	188	131	242	229	226	243	100	41	26	14	9	6	6	3		17	25	–	1	–	–	–	–	–	–	–	–	–	Acridine, 1,2,3,4,5,6,7,8-octahydro-3,3,6,6-tetramethyl-	56798-25-5	NI23508
200	91	243	84	242	186	166	201	243	100	50	29	24	23	20	19	16		17	25	–	1	–	–	–	–	–	–	–	–	–	Piperidine, 1-(1-phenylcyclohexyl)-	77-10-1	NI23509
200	91	84	28	242	243	115	129	243	100	66	47	47	37	32	31	29		17	25	–	1	–	–	–	–	–	–	–	–	–	Piperidine, 1-(1-phenylcyclohexyl)-	77-10-1	NI23510
200	91	243	84	242	186	166	201	243	100	48	26	25	23	21	18	17		17	25	–	1	–	–	–	–	–	–	–	–	–	Piperidine, 1-(1-phenylcyclohexyl)-	77-10-1	MS05210
243	242	244	241	121.5	120.5	240	108.5	243	100	26	20	16	12	9	8	8		18	13	–	1	–	–	–	–	–	–	–	–	–	Benz[a]acridine, 10-methyl-		MS05211
243	242	244	241	108.5	120.5	240	121.5	243	100	25	20	17	12	9	7	7		18	13	–	1	–	–	–	–	–	–	–	–	–	Benz[c]acridine, 5-methyl-		MS05212

MASS TO CHARGE RATIOS								M.W.	INTENSITIES								Parent	C	H	O	N	S	F	Cl	Br	I	Si	P	B	X	COMPOUND NAME	CAS Reg No	No
243	51	77	140	244	50	166	39	243	100	46	31	21	20	16	15	15		18	13	–	1	–	–	–	–	–	–	–	–	–	9H-Carbazole, 9-phenyl-	1150-62-5	AI01742
243	244	241	242	240	166	140	51	243	100	18	18	12	4	3	3	3		18	13	–	1	–	–	–	–	–	–	–	–	–	9H-Carbazole, 9-phenyl-	1150-62-5	NI23511
243	215	244	73	213	242	241	216	243	100	23	11	9	3	2	2	1		18	13	–	1	–	–	–	–	–	–	–	–	–	6-Chrysenamine	2642-98-0	NI23512
42	244	44	43	166	69	229	209	244	100	77	76	72	26	26	19	17		2	6	–	2	1	3	1	–	–	–	2	–	–	Phosphoramidimidic difluoride, N′-(chlorofluorophosphinothioyl)-N,N-dimethyl-	27352-03-0	LI04829
42	244	44	43	166	69	177	229	244	100	77	76	72	26	26	25	19		2	6	–	2	1	3	1	–	–	–	2	–	–	Phosphoramidimidic difluoride, N′-(chlorofluorophosphinothioyl)-N,N-dimethyl-	27352-03-0	MS05213
85	161	165	167	69	159	80	87	244	100	72	65	46	44	40	32	24	8.00	3	2	–	–	–	3	2	1	–	–	–	–	–	Propane, 2-bromo-1,2-dichloro-1,1,2-trifluoro-		IC03216
45	103	244	122	123	179	28	76	244	100	76	59	58	24	21	20	16		4	4	–	–	6	–	–	–	–	–	–	–	–	2,4,6,8,9,10-Hexathiatricyclo[3.3.1.1^{3,7}]decane	281-40-3	NI23513
57	85	119	121	55	147	149	246	244	100	57	44	33	30	27	26	16	6.23	4	6	2	–	–	–	–	2	–	–	–	–	–	2-Butene-1,4-diol, 2,3-dibromo-	3234-02-4	NI23514
29	57	39	85	119	31	121	55	244	100	95	82	54	51	50	42	30	7.10	4	6	2	–	–	–	–	2	–	–	–	–	–	2-Butene-1,4-diol, 2,3-dibromo-	3234-02-4	NI23515
60	185	187	244	246	189	143	63	244	100	99	90	74	65	21	19	16		4	6	4	–	–	–	–	–	–	–	–	–	2	Bis(copper acetate)		LI04830
117	131	103	143	145	118	119	157	244	100	51	47	18	10	10	8	4	0.00	4	9	–	2	1	–	–	–	1	–	–	–	–	1H-Imidazole, 4,5-dihydro-2-(methylthio)-, monohydriodide	5464-11-9	NI23516
73	203	45	43	74	205	201	109	244	100	14	14	10	8	7	7	5	0.29	4	10	–	–	–	–	–	2	–	1	–	–	–	Silane, (dibromomethyl)trimethyl-	2612-42-2	NI23517
165	229	135	150	244	120	199	214	244	100	85	36	31	23	21	16	5		4	12	2	–	1	–	–	–	–	–	–	–	1	Tin, trimethyl(methylsulphinyl)-		LI04831
246	110	176	178	199	197	31	227	244	100	75	64	63	49	37	37	34	0.00	5	–	–	–	–	6	2	–	–	–	–	–	–	1-Cyclopentene, 1,2-dichlorohexafluoro-		IC03217
244	109	246	175	195	177	197	226	244	100	73	64	53	50	34	32	31		5	–	–	–	–	6	2	–	–	–	–	–	–	1-Cyclopentene, 1,2-dichlorohexafluoro-		DO01030
109	244	175	194	246	177	196	209	244	100	83	83	62	57	53	39	33		5	–	–	–	–	6	2	–	–	–	–	–	–	1-Cyclopentene, 1,2,3,3,4,5-hexafluoro-4,5-dichloro-		IC03218
127	109	165	43	79	95	166	93	244	100	65	63	26	13	11	5	3	2.07	5	10	4	–	–	–	–	1	–	–	1	–	–	Phosphoric acid, 2-bromo-1-isopropenyl-, dimethyl ester	22752-26-7	NI23518
185	229	73	45	101	43	75	59	244	100	78	31	16	10	10	8	7	0.30	5	13	1	–	–	–	–	–	1	1	–	–	–	Silane, (2-iodoethoxy)trimethyl-	26305-99-7	NI23519
87	244	209	188	94	181	216	153	244	100	60	58	23	22	21	18	15		6	–	2	–	–	–	4	–	–	–	–	–	–	2,5-Cyclohexadiene-1,4-dione, 2,3,5,6-tetrachloro-	118-75-2	IC03219
87	246	244	209	211	89	248	28	244	100	53	41	37	36	32	26	26		6	–	2	–	–	–	4	–	–	–	–	–	–	2,5-Cyclohexadiene-1,4-dione, 2,3,5,6-tetrachloro-	118-75-2	NI23520
87	112	77	246	209	244	211	89	244	100	87	62	50	42	37	36	36		6	–	2	–	–	–	4	–	–	–	–	–	–	2,5-Cyclohexadiene-1,4-dione, 2,3,5,6-tetrachloro-	118-75-2	IC03220
218	118	190	216	153	155	188	120	244	100	94	84	81	68	66	65	63	8.77	6	–	2	–	–	–	4	–	–	–	–	–	–	3,5-Cyclohexadiene-1,2-dione, 3,4,5,6-tetrachloro-	2435-53-2	NI23521
218	190	118	216	83	188	153	155	244	100	85	83	79	72	66	66	65	11.04	6	–	2	–	–	–	4	–	–	–	–	–	–	3,5-Cyclohexadiene-1,2-dione, 3,4,5,6-tetrachloro-	2435-53-2	NI23522
218	190	216	153	188	155	181	183	244	100	83	80	70	68	68	63	59	15.00	6	–	2	–	–	–	4	–	–	–	–	–	–	3,5-Cyclohexadiene-1,2-dione, 3,4,5,6-tetrachloro-	2435-53-2	NI23523
211	209	213	193	247	245	195	79	244	100	84	41	39	37	37	33	23	0.00	6	3	2	–	1	–	3	–	–	–	–	–	–	Benzenesulphonyl chloride, 2,5-dichloro-	5402-73-3	NI23524
93	94	135	110	65	63	244	95	244	100	63	52	24	24	19	14	13		6	18	–	2	2	–	–	–	–	–	2	–	–	1,2-Ethanediamine, N,N′-bis(dimethylthiophosphinyl)-		MS05214
246	244	203	248	201	231	229	131	244	100	91	70	57	56	52	43	37		7	4	1	–	–	–	4	–	–	–	–	–	–	Benzene, 1,2,3,4-tetrachloro-5-methoxy-	938-86-3	NI23525
231	246	229	244	203	233	248	201	244	100	94	77	72	51	49	46	40		7	4	1	–	–	–	4	–	–	–	–	–	–	Benzene, 1,2,3,5-tetrachloro-4-methoxy-	938-22-7	MS05215
231	246	229	203	233	244	201	205	244	100	90	87	79	71	58	55	39		7	4	1	–	–	–	4	–	–	–	–	–	–	Benzene, 1,2,3,5-tetrachloro-4-methoxy-	938-22-7	NI23526
244	117	90	63	122	62	39	116	244	100	32	16	14	10	8	6	5		7	5	–	2	–	–	–	–	1	–	–	–	–	1H-Pyrrolo[2,3-b]pyridine, 3-iodo-	23616-57-1	NI23527
69	59	117	58	61	107	57	87	244	100	81	70	47	32	23	21	20	5.86	7	7	4	–	1	3	–	–	–	–	–	–	–	2-Propenoic acid, 3-[(trifluoroacetoxy)methylthio]-, methyl ester		NI23528
28	73	59	139	137	43	66		244	100	27	15	10	10	9	5		0.00	7	9	4	–	–	–	–	–	–	1	–	–	1	Cobalt, (trimethylsilyl)tetracarbonyl-		LI04832
181	161	182	31	244	162	155	69	244	100	15	10	10	9	5	5	5		8	2	1	–	–	5	1	–	–	–	–	–	–	Acetophenone, 2-chloro-2′,3′,4′,5′,6′-pentafluoro-	22600-29-9	NI23529
181	161	31	244	162	155	69	131	244	100	15	10	9	5	5	5	4		8	2	1	–	–	5	1	–	–	–	–	–	–	Benzeneacetic acid, pentafluoro-, chloride		LI04833
112	95	139	39	66	38	65	156	244	100	88	54	46	39	33	24	22	3.30	8	4	9	–	–	–	–	–	–	–	–	–	–	Furantetracarboxylic acid	20416-04-0	NI23530
137	109	75	101	74	89	51	50	244	100	82	49	37	35	32	32	28	21.40	8	5	2	–	1	–	–	1	–	–	–	–	–	Benzo[b]thiophene, 2-bromo-, 1,1-dioxide		MS05216
244	246	229	227	69	166	149	75	244	100	88	76	68	52	31	30	28		8	5	4	–	–	–	–	1	–	–	–	–	–	1,4-Benzenedicarboxylic acid, 2-bromo-	586-35-6	NI23531
229	244	77	69	63	81	59	67	244	100	54	45	39	33	28	17	14		8	6	1	–	–	5	–	–	–	–	1	–	–	Phosphine oxide, dimethyl(pentafluorophenyl)-	5525-96-2	NI23532
70	244	69	216	42	201	44	189	244	100	27	27	18	17	13	13	8		8	7	–	6	–	3	–	–	–	–	–	–	–	Pyrimido[5,4-e]-1,2,4-triazin-7-amine, N,N-dimethyl-5-(trifluoromethyl)-	32709-26-5	LI04834
70	244	69	216	42	201	44	55	244	100	27	27	19	16	11	11	9		8	7	–	6	–	3	–	–	–	–	–	–	–	Pyrimido[5,4-e]-1,2,4-triazin-7-amine, N,N-dimethyl-5-(trifluoromethyl)-	32709-26-5	NI23533
213	201	142	170	59	44	84	78	244	100	35	29	26	26	24	20	18	4.00	8	8	5	2	1	–	–	–	–	–	–	–	–	4,5-Thiazoledicarboxylic acid, 2-carbamoyl-, dimethyl ester	22061-97-8	LI04835
213	244	201	142	170	59	44	84	244	100	42	35	29	26	26	24	20		8	8	5	2	1	–	–	–	–	–	–	–	–	4,5-Thiazoledicarboxylic acid, 2-carbamoyl-, dimethyl ester	22061-97-8	NI23534
29	41	27	39	108	57	55	42	244	100	95	64	62	47	47	41	36	0.81	8	12	–	–	–	3	–	1	–	–	–	–	–	2-Octene, 1-bromo-1,1,2-trifluoro-	74810-71-2	NI23535
147	244	200	131	104	216	69	93	244	100	80	80	54	47	42	33	27		9	7	1	4	–	3	–	–	–	–	–	–	–	Pteridine, 2-ethoxy-4-(trifluoromethyl)-	32706-08-4	NI23536
147	200	244	131	104	216	69	229	244	100	82	80	55	48	43	34	27		9	7	1	4	–	3	–	–	–	–	–	–	–	Pteridine, 2-ethoxy-4-(trifluoromethyl)-	32706-08-4	LI04836
244	246	50	94	245	243	51	65	244	100	93	65	65	53	47	47	25		9	9	3	–	–	–	–	1	–	–	–	–	–	Benzaldehyde, 2-bromo-4,5-dimethoxy-	5392-10-9	NI23537
163	108	244	135	134	242	164	109	244	100	50	44	35	33	22	21	21		9	12	1	2	–	–	–	–	–	–	–	–	1	Urea, 4-ethoxyphenylseleno-		MS05217
44	77	71	103	110	141	51	180	244	100	59	51	26	22	19	18	14	2.02	9	12	2	2	2	–	–	–	–	–	–	–	–	Urea, 1-ethyl-3-(phenylsulphonyl)-2-thio-	6171-12-6	NI23538
44	77	71	103	110	141	51	43	244	100	59	52	27	23	19	19	13	2.00	9	12	2	2	2	–	–	–	–	–	–	–	–	Urea, 1-ethyl-3-(phenylsulphonyl)-2-thio-	6171-12-6	NI23539
202	244	43	138	186	92	107	79	244	100	75	74	42	34	25	24	12		9	12	4	2	1	–	–	–	–	–	–	–	–	Acetamide, N-[4-(aminosulfonyl)-2-methoxyphenyl]-		NI23540
201	72	88	102	160	129	96	70	244	100	30	28	18	16	16	13	13	4.00	9	12	4	2	1	–	–	–	–	–	–	–	–	1-Imidazolidinecarboxylic acid, 3-[1-(acetyloxy)vinyl]-2-thioxo-, methyl ester	67845-08-3	NI23541

MASS TO CHARGE RATIOS								M.W.	INTENSITIES								Parent	C	H	O	N	S	F	Cl	Br	I	Si	P	B	X	COMPOUND NAME	CAS Reg No	No
160	102	244	72	88	84	201	69	**244**	100	47	35	35	32	30	25	25		9	12	4	2	1	–	–	–	–	–	–	–	–	**1-Imidazolidinecarboxylic acid, 3-(1,3-dioxobutyl)-2-thioxo-, methyl ester**	67845-10-7	**NI23542**
113	73	133	57	43	55	142	69	**244**	100	86	72	55	34	29	28	28	**5.00**	9	12	6	2	–	–	–	–	–	–	–	–	–	**2,4(1H,3H)-Pyrimidinedione, 1-β-D-ribofuranosyl-**	58-96-8	**NI23543**
112	113	69	73	57	45	43	42	**244**	100	81	74	72	57	57	47	47	**1.66**	9	12	6	2	–	–	–	–	–	–	–	–	–	**2,4(1H,3H)-Pyrimidinedione, 1-β-D-ribofuranosyl-**	58-96-8	**NI23544**
112	57	69	42	73	28	29	43	**244**	100	70	60	50	36	36	29	28	**0.00**	9	12	6	2	–	–	–	–	–	–	–	–	–	**2,4(1H,3H)-Pyrimidinedione, 1-β-D-ribofuranosyl-**	58-96-8	**NI23545**
141	98	82	125	55	208	170	60	**244**	100	38	30	28	26	23	22	22	**2.20**	9	12	6	2	–	–	–	–	–	–	–	–	–	**2,4(1H,3H)-Pyrimidinedione, 5-β-D-ribofuranosyl-**		**LI04837**
61	60	208	82	141	125	137	55	**244**	100	97	58	50	44	33	31	27	**0.33**	9	12	6	2	–	–	–	–	–	–	–	–	–	**2,4(1H,3H)-Pyrimidinedione, 5-β-D-ribofuranosyl-**	1445-07-4	**NI23546**
231	229	149	139	137	244	246	232	**244**	100	95	79	78	74	24	23	20		9	13	1	–	–	–	–	1	–	1	–	–	–	**Silane, (2-bromophenoxy)trimethyl-**	36601-47-5	**NI23547**
231	229	137	139	149	150	244	246	**244**	100	95	73	69	64	45	42	41		9	13	1	–	–	–	–	1	–	1	–	–	–	**Silane, (3-bromophenoxy)trimethyl-**	36971-28-5	**NI23548**
231	229	139	149	137	150	244	246	**244**	100	96	70	68	67	64	52	51		9	13	1	–	–	–	–	1	–	1	–	–	–	**Silane, (4-bromophenoxy)trimethyl-**	17878-44-3	**NI23549**
27	41	29	42	28	39	31	229	**244**	100	91	91	87	85	85	80	69	**0.00**	9	21	4	–	–	–	–	–	–	–	–	–	1	**Vanadium, Oxotri(isopropoxo)-**	5538-84-1	**NI23550**
144	143	244	129	145	127	245	109	**244**	100	58	47	40	14	13	9	9		10	10	–	–	–	6	–	–	–	–	–	–	–	**Bicyclo[2.2.2]oct-2-ene, 2,3-dimethyl-1,4,5,5,6,6-hexafluoro-**		**NI23551**
244	246	173	245	243	174	175	159	**244**	100	64	26	23	19	19	17	16		10	10	1	2	–	–	2	–	–	–	–	–	–	**4-Imidazolidin-2-one, 1-(3,4-dichlorophenyl)-3-methyl-**		**NI23552**
150	135	77	121	39	43	108	78	**244**	100	84	31	16	15	14	13	13	**5.00**	10	12	2	–	–	–	–	–	–	–	–	–	1	**3-Selenetanol, 3-(4-methoxyphenyl)-**		**MS05218**
97	112	100	185	98	87	85	186	**244**	100	85	80	75	75	75	75	60	**45.00**	10	16	3	2	1	–	–	–	–	–	–	–	–	**Biotin**	58-85-5	**LI04838**
97	112	100	183	85	87	98	184	**244**	100	85	80	78	76	75	73	58	**45.63**	10	16	3	2	1	–	–	–	–	–	–	–	–	**Biotin**	58-85-5	**NI23553**
43	117	88	185	176	143	134	60	**244**	100	97	87	44	27	27	22	22	**18.00**	10	16	3	2	1	–	–	–	–	–	–	–	–	**L-Cysteine, N-acetyl-S-(1-methyl-2-cyanoethyl)-, methyl ester**	82971-44-6	**NI23554**
56	197	97	41	244	69	61	55	**244**	100	93	67	56	55	52	47	46		10	16	3	2	1	–	–	–	–	–	–	–	–	**2,4,6(1H,3H,5H)-Pyrimidinetrione, 5,5-diethyl-1-(methylthiomethyl)-**		**MS05219**
43	111	99	142	124	98	69	28	**244**	100	7	7	4	4	3	3	3	**0.00**	10	17	6	–	–	–	–	–	–	–	–	1	–	**1,3,2-Dioxaborinane-4-methanol, 5-(acetyloxy)-2-ethyl-, acetate, (4S-cis)-**	74841-62-6	**NI23555**
43	111	99	142	124	113	171	112	**244**	100	8	6	4	4	4	3	3	**0.00**	10	17	6	–	–	–	–	–	–	–	–	1	–	**1,3,2-Dioxaborinane-4-methanol, 5-(acetyloxy)-2-ethyl-, acetate, trans-**	74793-15-0	**NI23556**
43	111	124	110	112	29	99	73	**244**	100	35	14	9	4	4	3	3	**0.00**	10	17	6	–	–	–	–	–	–	–	–	1	–	**1,3,2-Dioxaborolane-4,5-dimethanol, 2-ethyl-, diacetate, cis-**	74793-17-2	**NI23557**
43	111	124	110	123	112	29	28	**244**	100	48	15	13	4	4	4	4	**0.00**	10	17	6	–	–	–	–	–	–	–	–	1	–	**1,3,2-Dioxaborolane-4,5-dimethanol, 2-ethyl-, diacetate, (4S-trans)-**	74841-61-5	**NI23558**
43	103	99	98	124	111	86	44	**244**	100	24	24	7	4	4	4	3	**0.00**	10	17	6	–	–	–	–	–	–	–	–	1	–	**1,2-Ethanediol, 1-(2-ethyl-1,3,2-dioxaborolan-4-yl)-, diacetate, (R*,S*)-**	74793-16-1	**NI23559**
43	103	99	111	98	124	86	28	**244**	100	16	16	5	5	4	3	3	**0.00**	10	17	6	–	–	–	–	–	–	–	–	1	–	**1,2-Ethanediol, 1-(2-ethyl-1,3,2-dioxaborolan-4-yl)-, diacetate, [S-(R*,R*)]-**	74793-27-4	**NI23560**
67	39	68	53	41	55	27	42	**244**	100	82	73	65	59	43	39	36	**8.80**	10	18	–	–	–	–	–	–	–	–	–	–	1	**Palladium, bis[(1,2,3-η)-2-pentenyl]-**	74630-72-1	**NI23561**
147	73	244	45	148	173	133	201	**244**	100	87	34	27	17	14	13	11		10	20	3	–	–	–	–	–	–	2	–	–	–	**Succinic anhydride, bis(trimethylsilyl)-**	77220-06-5	**NI23562**
229	73	114	100	57	129	128	147	**244**	100	57	30	30	28	20	19	15	**8.00**	10	24	1	2	–	–	–	–	–	2	–	–	–	**2-Pyrrolidinone, 3-amino-, N,N′-bis(trimethylsilyl)-**		**MS05220**
56	121	160	134	216	188	244	95	**244**	100	85	79	77	62	50	42	30		11	8	3	–	–	–	–	–	–	–	–	–	1	**Iron, π-cyclopentadienyl-(2-furanyl)dicarbonyl-**	59307-98-1	**NI23563**
104	134	78	103	56	160	77	216	**244**	100	63	61	59	54	49	35	29	**1.98**	11	8	3	–	–	–	–	–	–	–	–	–	1	**Iron, tricarbonyl(π-cyclooctatetraene)-**		**MS05221**
77	104	103	78	105	160	63	56	**244**	100	87	53	51	38	37	35	35	**3.00**	11	8	3	–	–	–	–	–	–	–	–	–	1	**Iron, tricarbonyl(η⁴-2-methylene-4,6-cycloheptadiene-1,3-diyl)-**	41576-77-6	**NI23564**
28	77	51	40	104	103	78	50	**244**	100	11	11	11	10	6	6	6	**0.34**	11	8	3	–	–	–	–	–	–	–	–	–	1	**Iron, tricarbonyl(7-methylene-1,3,5-cycloheptatriene)-**	33506-46-6	**MS05222**
160	55	105	120	188	56	158	79	**244**	100	79	69	63	40	39	28	22	**19.00**	11	9	3	–	–	–	–	–	–	–	–	–	1	**Manganese, tricarbonyl[(1,2,3,4,5-η)-2,4,6-cyclooctatrien-1-yl]-**	49626-38-2	**NI23565**
244	209	193	246	191	165	179	229	**244**	100	96	39	30	30	30	24	20		11	13	4	–	–	–	1	–	–	–	–	–	–	**Benzaldehyde, 6-(chloromethyl)-2,3,4-trimethoxy-**	75676-80-1	**NI23566**
43	87	125	55	93	59	116	169	**244**	100	16	15	15	13	13	12	10	**0.00**	11	16	6	–	–	–	–	–	–	–	–	–	–	**1,2-Cyclopentanedicarboxylic acid, 4-hydroxy-, acetate, dimethyl ester, (1α,2α,4β)-**		**DD00637**
158	126	55	99	157	185	114	143	**244**	100	84	80	78	68	64	62	60	**0.00**	11	16	6	–	–	–	–	–	–	–	–	–	–	**1,4-Dioxaspiro[4.4]nonanedicarboxylic acid, dimethyl ester, cis-**		**DD00638**
111	82	124	70	55	156	54	83	**244**	100	47	36	33	26	21	20	19	**7.60**	11	16	6	–	–	–	–	–	–	–	–	–	–	**1,8-Dioxaspiro[4.4]nonane-2,7-dione, 4,6-dimethyl-6-hydroxy-3-(methoxymethyl)-**	54370-16-0	**NI23567**
100	43	28	111	71	142	129	69	**244**	100	76	34	26	16	15	13	10	**0.10**	11	16	6	–	–	–	–	–	–	–	–	–	–	**β-D-Erythro-hex-2-enopyranoside, methyl 2,3-dideoxy-, diacetate**	20701-48-8	**MS05223**
100	111	141	129	81	69	97	71	**244**	100	34	28	22	15	15	14	14	**0.00**	11	16	6	–	–	–	–	–	–	–	–	–	–	**β-D-Erythro-hex-2-enopyranoside, methyl 2,3-dideoxy-, diacetate**	20701-48-8	**MS05224**
165	83	123	91	41	121	93	79	**244**	100	90	71	48	45	42	42	42	**7.00**	11	17	1	–	–	–	–	1	–	–	–	–	–	**2-Cyclopenten-1-ol, 1-(1-bromo-2-ethyl-3-methylcycloprop-1-yl)-**		**MS05225**
81	55	67	41	95	165	244	109	**244**	100	61	59	59	55	46	45	45		11	17	1	–	–	–	–	1	–	–	–	–	–	**2(1H)-Naphthalenone, 3-bromooctahydro-4a-methyl-, (3α,4aβ,8aα)-**	21794-03-6	**NI23568**
44	73	30	143	43	72	100	29	**244**	100	40	36	19	19	16	12	10	**0.47**	11	20	4	2	–	–	–	–	–	–	–	–	–	**L-Alanine, N-(N-acetylglycyl)-, butyl ester**	62167-72-0	**NI23569**
114	72	46	186	44	172	115	86	**244**	100	51	13	12	8	7	7	6	**3.14**	11	20	4	2	–	–	–	–	–	–	–	–	–	**1,3-Dioxolane-4,5-dicarboxamide, N,N,N′,N′,2,2-hexamethyl-**	74752-95-7	**NI23570**
44	86	87	43	41	29	114	30	**244**	100	62	41	23	12	12	10	9	**0.18**	11	20	4	2	–	–	–	–	–	–	–	–	–	**Glycine, N-(N-acetyl-L-alanyl)-, butyl ester**	55712-36-2	**NI23571**
70	30	229	245	200	211	199	227	**244**	100	88	83	64	59	39	39	21	**0.90**	11	20	4	2	–	–	–	–	–	–	–	–	–	**2,4-Imidazolidinedione, 1,3-bis(2-hydroxypropyl)-5,5-dimethyl-**		**MS05226**
61	168	59	76	58	114	28	112	**244**	100	81	71	67	58	49	47	42	**0.00**	11	23	–	–	–	–	–	–	–	–	1	–	1	**Nickel, (2-methyl-2-propenyl)[(1,2,3-η)-2-methyl-2-propenyl](trimethylphosphine)-**	61233-22-5	**NI23572**
214	167	168	110	244	198	166	215	**244**	100	81	48	37	33	25	25	15		12	8	2	2	1	–	–	–	–	–	–	–	–	**10H-Phenothiazine, 2-nitro-**	1628-76-8	**NI23573**
198	139	115	168	63	141	116	140	**244**	100	61	43	43	21	17	15	15	**0.00**	12	8	4	2	–	–	–	–	–	–	–	–	–	**1,1′-Biphenyl, 2,2′-dinitro-**	2436-96-6	**NI23574**
198	139	115	168	51	63	141	116	**244**	100	78	61	52	32	30	24	21	**0.00**	12	8	4	2	–	–	–	–	–	–	–	–	–	**1,1′-Biphenyl, 2,2′-dinitro-**	2436-96-6	**NI23575**
152	151	244	30	139	63	150	50	**244**	100	91	54	42	37	33	31	25		12	8	4	2	–	–	–	–	–	–	–	–	–	**1,1′-Biphenyl, 4,4′-dinitro-**	1528-74-1	**NI23576**
155	244	183	154	199	182	212	127	**244**	100	95	89	89	83	79	75	58		12	8	4	2	–	–	–	–	–	–	–	–	–	**9,10-Diazaanthracene, 1,6-dihydroxy-, 5,10-dioxide**		**LI04839**
244	216	77	142	131	103	76	102	**244**	100	43	33	29	29	23	23	20		12	8	4	2	–	–	–	–	–	–	–	–	–	**9,10-Diazaanthracene, 1,2,3,4-tetrahydroxy-**		**MS05227**
244	168	89	246	203	63	51	115	**244**	100	67	49	33	29	25	23	22		12	9	–	4	–	–	1	–	–	–	–	–	–	**Imidazo[1,2-b]-1,2,4-triazine, 3-chloro-2-methyl-6-phenyl-**		**NI23577**

MASS TO CHARGE RATIOS	M.W.	INTENSITIES	Parent	C	H	O	N	S	F	Cl	Br	I	Si	P	B	X	COMPOUND NAME	CAS Reg No	No
52 96 244 148 103 90 71 77	**244**	100 62 38 38 31 23 23 15		12	10	–	–	–	2	–	–	–	–	–	–	1	**Chromium, bis(η6-fluorobenzene)-**	42087-90-1	**NI23578**
103 197 244 128 127 144 104 77	**244**	100 99 92 42 21 15 15 15		12	12	–	4	1	–	–	–	–	–	–	–	–	**1H-Imidazo[1,2-b][1,2,4]triazole, 1-methyl-2-(methylthio)-6-phenyl-**		**MS05228**
244 174 229 146 162 119 91 133	**244**	100 75 70 58 48 29 27 26		12	16	–	6	–	–	–	–	–	–	–	–	–	**Aniline, 4-(1H-1,2,4-triazol-3-ylazo)-, N,N-diethyl-**		**IC03221**
145 41 106 244 97 132 120 67	**244**	100 81 64 59 40 38 30 30		12	20	1	–	2	–	–	–	–	–	–	–	–	**6,10-Dithiaspiro[4.5]decan-1-ol, 2-allyl-2-methyl-**	92640-73-8	**NI23579**
132 41 244 99 58 74 71 216	**244**	100 25 9 9 9 5 5 3		12	20	1	–	2	–	–	–	–	–	–	–	–	**6,10-Dithiaspiro[4.5]decan-1-one, 2-methyl-2-(n-propyl)-**	59579-80-5	**NI23580**
143 111 116 83 171 101 181 195	**244**	100 88 43 39 17 17 12 8	**0.00**	12	20	5	–	–	–	–	–	–	–	–	–	–	**Decanedioic acid, 3,6-epoxy-, dimethyl ester**		**NI23581**
55 69 41 29 101 59 129 125	**244**	100 98 97 90 65 65 61 58	**0.00**	12	20	5	–	–	–	–	–	–	–	–	–	–	**Propanoic acid, 2-(6-methoxycarbonyl-5-methyl-2-tetrahydropyranyl)-, methyl ester**		**NI23582**
114 141 159 57 71 85 55 127	**244**	100 61 55 48 46 45 40 29	**0.00**	12	20	5	–	–	–	–	–	–	–	–	–	–	**Propanoic acid, 3-(2-tert-butyl-6-oxo-1,3-dioxan-4-yl)-, methyl ester, (2R,4R)-**		**DD00639**
43 59 101 171 111 85 229 55	**244**	100 36 28 22 20 20 19 17	**0.00**	12	20	5	–	–	–	–	–	–	–	–	–	–	**D-Xylohexofuranoside, 3-deoxy-1,2:5,6-di-O-isopropylidene-**		**LI04840**
44 57 107 67 91 82 81 109	**244**	100 65 51 25 20 19 19 18	**1.40**	12	21	–	–	–	–	–	1	–	–	–	–	–	**3-Dodecyne, 1-bromo-**		**NI23583**
196 76 152 136 50 244 168 180	**244**	100 40 37 31 28 25 24 17		13	8	3	–	1	–	–	–	–	–	–	–	–	**9H-Thioxanthen-9-one, 10,10-dioxide**	3166-15-2	**NI23584**
196 76 152 136 244 168 180 151	**244**	100 40 37 31 25 24 16 16		13	8	3	–	1	–	–	–	–	–	–	–	–	**9H-Thioxanthen-9-one, 10,10-dioxide**	3166-15-2	**MS05229**
244 245 108 69 216 215 51 187	**244**	100 14 9 6 5 4 4 3		13	8	5	–	–	–	–	–	–	–	–	–	–	**9-Xanthenone, 1,3,8-trihydroxy-**		**MS05230**
165 244 246 163 166 164 83 81	**244**	100 25 25 19 12 12 11 6		13	9	–	–	–	–	–	1	–	–	–	–	–	**9H-Fluorene, 2-bromo-**	1133-80-8	**NI23585**
165 246 244 82.5 163 166 164 81.5	**244**	100 24 24 23 19 16 11 11		13	9	–	–	–	–	–	1	–	–	–	–	–	**9H-Fluorene, 2-bromo-**	1133-80-8	**AI01743**
165 166 163 82 164 139 200 115	**244**	100 15 12 11 8 5 4 3	**0.11**	13	9	–	–	–	–	–	1	–	–	–	–	–	**9H-Fluorene, 9-bromo-**	1940-57-4	**NI23586**
244 215 246 217 153 245 91 216	**244**	100 45 33 15 15 14 10 6		13	9	1	2	–	–	1	–	–	–	–	–	–	**2H-Benzimidazol-2-one, 6-chloro-1,3-dihydro-1-phenyl-**	54986-47-9	**NI23587**
244 246 181 111 77 65 209 104	**244**	100 33 20 20 20 20 14 12		13	9	1	2	–	–	1	–	–	–	–	–	–	**1H-Indazole, 1-(4-chlorophenyl)-3-hydroxy-**		**LI04841**
244 215 246 153 91	**244**	100 45 33 15 10		13	9	1	2	–	–	1	–	–	–	–	–	–	**3-Indazolinone, 5-chloro-2-phenyl-**		**LI04842**
244 201 246 203 214 215 63 202	**244**	100 55 36 20 15 8 8 7		13	9	1	2	–	–	1	–	–	–	–	–	–	**Phenazine, 2-chloro-8-methoxy-**	18450-08-3	**LI04843**
244 201 246 203 245 214 215 63	**244**	100 55 36 20 15 15 8 8		13	9	1	2	–	–	1	–	–	–	–	–	–	**Phenazine, 2-chloro-8-methoxy-**	18450-08-3	**NI23588**
123 201 227 95 244 44 78 149	**244**	100 90 77 60 54 14 12 11		13	9	2	2	–	1	–	–	–	–	–	–	–	**2-Pyridinecarboxamide, 5-(4-fluorobenzoyl)-**	82465-64-3	**NI23589**
180 244 216 152 76 50 151 181	**244**	100 81 64 51 44 23 20 18		13	9	3	–	–	–	–	–	–	–	1	–	–	**9-Phosphaanthracen-9,10-dione, 9,10-dihydro-9-hydroxy-**		**MS05231**
227 109 93 51 77 136 150 244	**244**	100 61 56 48 45 28 27 25		13	12	1	2	1	–	–	–	–	–	–	–	–	**Benzamide, 2-phenylthio-, oxime**		**MS05232**
129 29 244 89 128 130 155 216	**244**	100 38 35 34 24 23 20 19		13	12	1	2	1	–	–	–	–	–	–	–	–	**2(1H)-Isoquinolinecarboxylic acid, 1-cyanothio-, O-ethyl ester**	19821-41-1	**NI23590**
93 210 28 77 151 109 135 66	**244**	100 87 65 53 44 42 35 35	**3.54**	13	12	1	2	1	–	–	–	–	–	–	–	–	**Thiourea, N-(2-hydroxyphenyl)-N'-phenyl-**	17073-34-6	**NI23591**
215 41 104 77 132 128 39 244	**244**	100 49 44 25 21 20 20 19		13	12	3	2	–	–	–	–	–	–	–	–	–	**2,4,6(1H,3H,5H)-Pyrimidinetrione, 5-phenyl-5-allyl-**	115-43-5	**NI23592**
215 41 244 104 28 77 51 129	**244**	100 82 60 58 39 35 27 24		13	12	3	2	–	–	–	–	–	–	–	–	–	**2,4,6(1H,3H,5H)-Pyrimidinetrione, 5-phenyl-5-allyl-**	115-43-5	**NI23593**
144 129 91 145 116 215 77 65	**244**	100 64 53 38 24 11 11 11	**7.00**	13	15	1	–	–	3	–	–	–	–	–	–	–	**5H-Benzocycloheptene, 6,7,8,9-tetrahydro-5-(2,2,2-trifluoroethoxy)-**		**MS05233**
244 106 226 138 161 179 245 91	**244**	100 38 37 35 32 17 15 15		13	16	1	–	–	–	–	–	–	–	–	–	1	**Ferrocene, 1-(1-hydroxyethyl)-3-methyl-**	12289-11-1	**NI23594**
244 121 199 134 56 178 160 135	**244**	100 22 12 12 11 9 6 6		13	16	1	–	–	–	–	–	–	–	–	–	1	**Ferrocene, (3-hydroxypropyl)-**	12093-88-8	**NI23595**
115 171 41 57 114 87 116 43	**244**	100 99 94 80 79 77 56 53	**0.00**	13	24	4	–	–	–	–	–	–	–	–	–	–	**Butanedioic acid, methyl-, dibutyl ester**	18447-89-7	**NI23596**
115 57 41 43 88 73 87 42	**244**	100 46 38 22 18 18 16 16	**0.00**	13	24	4	–	–	–	–	–	–	–	–	–	–	**Butanedioic acid, methyl-, di-sec-butyl ester**	57983-31-0	**NI23597**
129 187 99 199 85	**244**	100 83 54 17 9	**0.00**	13	24	4	–	–	–	–	–	–	–	–	–	–	**1,3-Dioxolane-2-butanoic acid, 2-butyl-, ethyl ester**		**LI04844**
115 201 199 155 99 71	**244**	100 56 10 10 10 8	**0.00**	13	24	4	–	–	–	–	–	–	–	–	–	–	**1,3-Dioxolane-2-pentanoic acid, 2-propyl-, ethyl ester**	4066-85-7	**LI04845**
115 201 43 71 155 111 99 109	**244**	100 57 20 16 11 10 10 7	**0.00**	13	24	4	–	–	–	–	–	–	–	–	–	–	**1,3-Dioxolane-2-pentanoic acid, 2-propyl-, ethyl ester**	4066-85-7	**NI23598**
143 97 69 18 153 199 85 115	**244**	100 80 77 52 48 40 33 28	**0.00**	13	24	4	–	–	–	–	–	–	–	–	–	–	**Heptanedioic acid, 4,4-dimethyl-, diethyl ester**	55134-05-9	**WI01254**
55 199 152 83 157 111 41 88	**244**	100 72 68 67 64 48 44 40	**0.00**	13	24	4	–	–	–	–	–	–	–	–	–	–	**Nonanedioic acid, diethyl ester**		**IC03222**
88 69 55 125 157 83 41 59	**244**	100 48 48 38 33 33 33 31	**0.00**	13	24	4	–	–	–	–	–	–	–	–	–	–	**Nonanedioic acid, 2,4-dimethyl-, dimethyl ester, (2S,4S)-**	85547-46-2	**NI23599**
54 88 15 41 69 59 29 83	**244**	100 87 82 71 53 51 46 41	**0.00**	13	24	4	–	–	–	–	–	–	–	–	–	–	**Nonanedioic acid, 2,6-dimethyl-, dimethyl ester**	5487-98-9	**NI23600**
85 41 55 56 43 101 57 67	**244**	100 26 25 24 21 17 17 12	**0.00**	13	24	4	–	–	–	–	–	–	–	–	–	–	**Octanoic acid, 8-[(tetrahydro-2H-pyran-2-yl)oxy]-**	54699-43-3	**NI23601**
115 171 44 43 85 142 57 41	**244**	100 44 21 20 16 14 13 13	**0.00**	13	24	4	–	–	–	–	–	–	–	–	–	–	**Pentanedioic acid, dibutyl ester**	6624-57-3	**NI23602**
115 42 57 41 114 87 43 55	**244**	100 57 47 44 43 38 35 32	**0.00**	13	24	4	–	–	–	–	–	–	–	–	–	–	**Pentanedioic acid, di-sec-butyl ester**	57983-33-2	**NI23603**
171 88 115 57 41 189 59 87	**244**	100 99 51 50 43 25 24 20	**0.00**	13	24	4	–	–	–	–	–	–	–	–	–	–	**Propanedioic acid, dimethyl-, dibutyl ester**	2917-71-7	**NI23604**
57 88 115 41 87 171 70 42	**244**	100 93 90 54 45 42 21 18	**0.00**	13	24	4	–	–	–	–	–	–	–	–	–	–	**Propanedioic acid, dimethyl-, di-sec-butyl ester**	57983-08-1	**NI23605**
173 29 41 18 55 143 127 97	**244**	100 69 34 33 32 32 31 30	**0.00**	13	24	4	–	–	–	–	–	–	–	–	–	–	**Propanedioic acid, dipropyl-, diethyl ester**	6065-63-0	**NI23606**
115 171 57 133 41 56 88 189	**244**	100 92 54 37 29 21 20 16	**0.00**	13	24	4	–	–	–	–	–	–	–	–	–	–	**Propanedioic acid, ethyl-, dibutyl ester**	1113-92-4	**NI23607**
115 57 41 133 88 171 73 43	**244**	100 69 32 28 24 23 16 14	**0.00**	13	24	4	–	–	–	–	–	–	–	–	–	–	**Propanedioic acid, ethyl-, di-sec-butyl ester**	57983-52-5	**NI23608**
174 128 115 55 97 87 41 199	**244**	100 38 25 21 19 19 17 13	**0.00**	13	24	4	–	–	–	–	–	–	–	–	–	–	**Propanedioic acid, 2-methyl-2-pentyl-, diethyl ester**	98061-07-5	**NI23609**
98 41 55 73 84 60 43 69	**244**	100 97 92 83 69 65 54 42	**0.00**	13	24	4	–	–	–	–	–	–	–	–	–	–	**Tridecanedioic acid**	505-52-2	**NI23610**
55 74 59 41 98 43 69 139	**244**	100 87 77 77 64 63 57 43	**0.00**	13	24	4	–	–	–	–	–	–	–	–	–	–	**Undecanedioic acid, dimethyl ester**	4567-98-0	**NI23611**
74 98 55 43 84 69 41 59	**244**	100 99 92 64 61 61 57 47	**0.00**	13	24	4	–	–	–	–	–	–	–	–	–	–	**Undecanedioic acid, dimethyl ester**	4567-98-0	**IC03223**
98 74 213 171 55 84 69 41	**244**	100 83 70 69 68 50 49 43	**1.00**	13	24	4	–	–	–	–	–	–	–	–	–	–	**Undecanedioic acid, dimethyl ester**	4567-98-0	**LI04846**
86 244 29 41 72 30 57 44	**244**	100 90 85 82 75 75 73 48		13	28	–	2	1	–	–	–	–	–	–	–	–	**Thiourea, tributyl-**	2422-88-0	**NI23612**

MASS TO CHARGE RATIOS	M.W.	INTENSITIES	Parent	C	H	O	N	S	F	Cl	Br	I	Si	P	B	X	COMPOUND NAME	CAS Reg No	No
86 244 29 41 72 57 30 44	**244**	100 90 87 82 74 73 73 51		13	28	–	2	1	–	–	–	–	–	–	–	–	**Thiourea, tributyl-**	2422-88-0	**LI04847**
73 75 117 229 132 129 41 55	**244**	100 86 56 43 24 19 18 17	**0.00**	13	28	2	–	–	–	–	–	–	1	–	–	–	**Decanoic acid, trimethylsilyl ester**	55494-15-0	**NI23613**
229 117 73 132 75 145 230 131	**244**	100 71 65 37 37 21 18 14	**3.16**	13	28	2	–	–	–	–	–	–	1	–	–	–	**Decanoic acid, trimethylsilyl ester**	55494-15-0	**NI23615**
75 73 229 117 41 129 43 132	**244**	100 94 53 45 31 30 30 26	**2.60**	13	28	2	–	–	–	–	–	–	1	–	–	–	**Decanoic acid, trimethylsilyl ester**	55494-15-0	**NI23614**
187 159 215 131 59 65 188 79	**244**	100 75 61 46 32 29 20 19	**0.00**	13	32	–	–	–	–	–	–	–	2	–	–	–	**Methane, bis(tetraethylsilyl)-**		**MS05234**
91 244 102 77 103 119 125 179	**244**	100 92 92 68 66 34 19 15		14	12	2	–	1	–	–	–	–	–	–	–	–	**Benzene, [2-(phenylethenyl)sulphonyl]-**	5418-11-1	**NI23616**
244 243 228 229 227 171 211 184	**244**	100 77 30 25 22 22 17 16		14	12	2	–	1	–	–	–	–	–	–	–	–	**4H-Benzo[4,5]cyclohepta[1,2-b]thiophen-4-ol, 10-methoxy-**		**MS05235**
105 75 106 122 51 64 76 93	**244**	100 86 55 44 36 25 16 15	**0.00**	14	12	4	–	–	–	–	–	–	–	–	–	–	**Benzaldehyde diperoxide**		**LI04848**
201 216 109 244 229 214 137 144	**244**	100 90 68 66 40 40 23 20		14	12	4	–	–	–	–	–	–	–	–	–	–	**Benzo[1,2-b:5,4-b′]difuran-4,8-dione, dihydro-2-isopropenyl-5-methyl-**		**LI04849**
229 189 244 188 91 230 160 76	**244**	100 62 54 35 21 15 13 10		14	12	4	–	–	–	–	–	–	–	–	–	–	**2H,6H-Benzo[1,2-b:5,4-b′]dipyran-2,6-dione, 7,8-dihydro-8,8-dimethyl-**	16499-05-1	**NI23617**
201 244 216 173 102 175 146 202	**244**	100 61 39 36 33 21 17 15		14	12	4	–	–	–	–	–	–	–	–	–	–	**Benzo[1,2-b:4,3-b′]dipyran-3,6(2H,8H)-dione, 2,2-dimethyl-**	55256-11-6	**NI23618**
212 184 92 213 244 128 76 50	**244**	100 70 56 29 28 23 23 17		14	12	4	–	–	–	–	–	–	–	–	–	–	**Benzoic acid, 2-(2-hydroxyphenoxy)-, methyl ester**	3487-84-1	**NI23619**
244 213 124 121 39 245 77 92	**244**	100 65 29 22 16 15 14 11		14	12	4	–	–	–	–	–	–	–	–	–	–	**Benzoic acid, 2-(4-hydroxyphenoxy)-, methyl ester**	21905-66-8	**NI23620**
121 244 124 65 122 51 39 77	**244**	100 34 18 11 9 9 9 8		14	12	4	–	–	–	–	–	–	–	–	–	–	**Benzoic acid, 2-(2-methoxyphenoxy)-**	21905-73-7	**NI23621**
121 244 124 92 64 199 77 63	**244**	100 67 28 20 18 14 14 14		14	12	4	–	–	–	–	–	–	–	–	–	–	**Benzoic acid, 2-(3-methoxyphenoxy)-**	21905-75-9	**NI23622**
121 124 244 227 65 151 39 95	**244**	100 44 42 36 33 28 22 19		14	12	4	–	–	–	–	–	–	–	–	–	–	**Benzophenone, 2,2′-dihydroxy-4-methoxy-**	131-53-3	**NI23623**
121 124 227 151 244 65 39 122	**244**	100 49 33 29 26 21 14 12		14	12	4	–	–	–	–	–	–	–	–	–	–	**Benzophenone, 2,2′-dihydroxy-4-methoxy-**	131-53-3	**NI23624**
137 243 229 244 227 135 108 107	**244**	100 85 82 71 41 36 28 22		14	12	4	–	–	–	–	–	–	–	–	–	–	**Benzophenone, 2,4,4′-trihydroxy-2′-methyl-**	4520-99-4	**NI23625**
134 162 244 55 83 105 77 51	**244**	100 93 53 47 28 23 22 22		14	12	4	–	–	–	–	–	–	–	–	–	–	**2-Butenal, 2-methyl-4-[(2-oxo-2H-1-benzopyran-7-yl)oxy]-**	38971-86-7	**MS05236**
244 77 214 124 200 51 157 44	**244**	100 69 39 34 32 31 29 29		14	12	4	–	–	–	–	–	–	–	–	–	–	**Carbonic acid, 3-methoxyphenyl phenyl ester**	17145-96-9	**NI23626**
77 244 123 124 91 185 51 200	**244**	100 90 85 80 48 39 23 21		14	12	4	–	–	–	–	–	–	–	–	–	–	**Carbonic acid, 4-methoxyphenyl phenyl ester**	17145-95-8	**NI23627**
77 123 124 240 185 51 65 95	**244**	100 94 69 68 41 39 35 25	**10.89**	14	12	4	–	–	–	–	–	–	–	–	–	–	**Carbonic acid, 4-methoxyphenyl phenyl ester**	17145-95-8	**NI23628**
244 128 212 156 184	**244**	100 81 55 37 24		14	12	4	–	–	–	–	–	–	–	–	–	–	**Cyclopenta[c]pyran-1(7H)-one-4-carboxylic acid, 7-crotonylidene-, methyl ester**		**LI04850**
213 244 185 126 154 214 245 77	**244**	100 75 38 26 17 15 12 11		14	12	4	–	–	–	–	–	–	–	–	–	–	**2,6-Naphthalenedicarboxylic acid, dimethyl ester**	840-65-3	**NI23629**
213 244 185 126 154 77 214 63	**244**	100 80 35 23 20 15 13 12		14	12	4	–	–	–	–	–	–	–	–	–	–	**2,6-Naphthalenedicarboxylic acid, dimethyl ester**	840-65-3	**IC03224**
213 244 185 126 77 154 214 63	**244**	100 66 39 20 17 16 14 12		14	12	4	–	–	–	–	–	–	–	–	–	–	**2,6-Naphthalenedicarboxylic acid, dimethyl ester**	840-65-3	**IC03225**
160 43 161 202 244 159 131 77	**244**	100 30 14 12 8 7 7 6		14	12	4	–	–	–	–	–	–	–	–	–	–	**1,4-Naphthalenediol, diacetate**		**MS05237**
160 43 131 202 161 244 159 77	**244**	100 28 14 13 13 8 5 4		14	12	4	–	–	–	–	–	–	–	–	–	–	**1,5-Naphthalenediol, diacetate**		**MS05238**
244 200 158 172 102 76 201 45	**244**	100 82 72 70 65 52 40 33		14	12	4	–	–	–	–	–	–	–	–	–	–	**1,4-Naphthoquinone, 2-(1,4-dioxan-2-yl)-**	24161-37-3	**NI23630**
208 229 244 165 231 104 178	**244**	100 86 75 72 31 22 14		14	13	–	–	–	–	1	–	–	1	–	–	–	**9-Silanthracene, 9-chloro-9,10-dihydro-9-methyl-**		**LI04851**
208 229 244 165 231 246 166 209	**244**	100 85 76 71 32 26 24 23		14	13	–	–	–	–	1	–	–	1	–	–	–	**9-Silanthracene, 9-chloro-9,10-dihydro-9-methyl-**		**LI04852**
208 131 207 209 180 91 77 130	**244**	100 67 38 16 16 14 14 11	**0.00**	14	13	–	2	–	–	1	–	–	–	–	–	–	**Benzeneacetonitrile, 4-amino-α-phenyl-, monohydrochloride**	69833-17-6	**NI23631**
182 198 227 183 209 131 208 130	**244**	100 19 17 16 13 7 4 2	**0.00**	14	13	–	2	–	–	1	–	–	–	–	–	–	**Benzeneacetonitrile, 4-amino-α-phenyl-, monohydrochloride**	69833-17-6	**NI23632**
227 228 213 243 244 195 229 165	**244**	100 82 62 32 24 19 17 14		14	13	2	–	–	–	–	–	–	–	1	–	–	**Phenoxaphosphine, 2,8-dimethyl-, 10-oxide**		**LI04853**
212 244 165 229 213 166 245 183	**244**	100 89 69 48 21 20 14 13		14	13	2	–	–	–	–	–	–	–	1	–	–	**9-Phosphaanthracene, 9,10-dihydro-9-methoxy-, 9-oxide**		**MS05239**
244 190 243 157 77 135 79 93	**244**	100 90 88 64 64 45 39 32		14	16	–	2	1	–	–	–	–	–	–	–	–	**1H,4H,4′H,5H,8H,8′H-3,1-Benzothiazine, 2-phenylimino-, cis-**		**MS05240**
244 243 77 190 79 157 93 91	**244**	100 73 50 46 45 33 32 24		14	16	–	2	1	–	–	–	–	–	–	–	–	**1H,4H,4′H,5H,8H,8′H-3,1-Benzothiazine, 2-phenylimino-, trans-**		**MS05241**
167 244 154 107 137 166 91 77	**244**	100 30 30 25 21 20 20 8		14	16	2	–	–	–	–	–	–	1	–	–	–	**Silane, dimethoxydiphenyl-**	6843-66-9	**NI23633**
211 244 229 77 151 91 75 51	**244**	100 69 51 47 40 33 29 22		14	16	2	–	–	–	–	–	–	1	–	–	–	**Silane, dimethyldiphenoxy-**	3440-02-6	**NI23634**
201 158 244 159 202 143 115 91	**244**	100 79 50 29 21 10 10 10		14	16	2	2	–	–	–	–	–	–	–	–	–	**Benzene, 3,5-bis(isocyanatomethyl)-1,2,4,5-tetramethyl-**		**IC03226**
244 229 198 212 245 214 213 197	**244**	100 35 27 20 16 13 12 12		14	16	2	2	–	–	–	–	–	–	–	–	–	**[1,1′-Biphenyl]-2,2′-diamine, 5,5′-dimethoxy-**		**LI04854**
244 201 229 186 158 79 245 202	**244**	100 72 28 27 22 17 16 10		14	16	2	2	–	–	–	–	–	–	–	–	–	**[1,1′-Biphenyl]-4,4′-diamine, 3,3′-dimethoxy-**	119-90-4	**NI23635**
244 201 158 79 186 91 229 245	**244**	100 85 32 29 23 22 13 13		14	16	2	2	–	–	–	–	–	–	–	–	–	**[1,1′-Biphenyl]-4,4′-diamine, 3,3′-dimethoxy-**	119-90-4	**NI23636**
144 173 143 244 115 130 142 174	**244**	100 99 99 60 60 48 45 41		14	16	2	2	–	–	–	–	–	–	–	–	–	**1H-Oxazolo[3,4-b]-β-carboline, 9-hydroxymethyl-3,3a,4,10-tetrahydro-**		**LI04855**
244 215 216 199 171 245 243 172	**244**	100 51 35 20 17 16 11 8		14	16	2	2	–	–	–	–	–	–	–	–	–	**2-Propenoic acid, 2-cyano-3-(4-dimethylaminophenyl)-, ethyl ester**	1886-52-8	**NI23637**
201 244 77 202 216 229 92 91	**244**	100 52 20 15 11 10 10 9		14	16	2	2	–	–	–	–	–	–	–	–	–	**3H-Pyrazol-3-one, 2,4-dihydro-5-methyl-4-(1-oxobutyl)-2-phenyl-**	22616-35-9	**MS05242**
201 244 77 202 216 229 245 92	**244**	100 53 20 13 10 8 7 7		14	16	2	2	–	–	–	–	–	–	–	–	–	**3H-Pyrazol-3-one, 2,4-dihydro-5-methyl-4-(1-oxobutyl)-2-phenyl-**	22616-35-9	**LI04856**
157 169 244 229 156 185 183 115	**244**	100 66 57 53 44 40 21 20		14	16	2	2	–	–	–	–	–	–	–	–	–	**1H-Pyrido[3,4-b]indole-3-carboxylic acid, 2,3,4,9-tetrahydro-1-methyl-, methyl ester**	16108-10-4	**NI23638**
244 188 146 171 160 187 133 41	**244**	100 94 75 38 36 34 28 24		14	16	2	2	–	–	–	–	–	–	–	–	–	**2-Pyrimidinone, 5-(4-butoxyphenyl)-1,2-dihydro-**		**NI23639**
167 244 243 229 187 215	**244**	100 84 48 47 44 23		14	16	2	2	–	–	–	–	–	–	–	–	–	**2-Pyrimidinone, 4-methyl-6-phenyl-5-propanoyl-1,2,3,4-tetrahydro-**		**LI04857**
93 78 66 244 65 51 41 42	**244**	100 36 27 17 15 15 9 8		14	16	2	2	–	–	–	–	–	–	–	–	–	**2-Quinuclidinecarboxanilide, 3-oxo-**	34291-65-1	**LI04858**
93 78 66 65 244 94 52 51	**244**	100 35 25 16 14 10 10 10		14	16	2	2	–	–	–	–	–	–	–	–	–	**2-Quinuclidinecarboxanilide, 3-oxo-**	34291-65-1	**NI23640**
118 90 200 83 117 55 91 89	**244**	100 72 45 36 33 27 26 25	**4.50**	14	17	3	–	–	–	–	–	–	–	–	1	–	**Benzeneacetic acid, α-hydroxy-, cyclohexyl boronate**		**MS05243**
71 43 41 55 57 140 72 173	**244**	100 41 11 8 7 5 5 4	**2.56**	14	28	1	–	1	–	–	–	–	–	–	–	–	**Butanethioic acid, S-decyl ester**	2432-55-5	**NI23641**

MASS TO CHARGE RATIOS								M.W.	INTENSITIES								Parent	C	H	O	N	S	F	Cl	Br	I	Si	P	B	X	COMPOUND NAME	CAS Reg No	No
57	127	55	41	159	43	109	128	**244**	100	95	20	18	15	15	10	8	**0.40**	14	28	1	–	1	–	–	–	–	–	–	–	–	Octanethioic acid, S-hexyl ester	55590-85-7	NI23642
57	187	43	55	29	154	41	69	**244**	100	10	10	9	9	8	7	5	**3.18**	14	28	1	–	1	–	–	–	–	–	–	–	–	Propanethioic acid, S-undecyl ester	2432-46-4	NI23643
113	103	85	75	47	199	183	216	**244**	100	70	65	41	36	2	2	1	**0.00**	14	28	3	–	–	–	–	–	–	–	–	–	–	4-Hexene, 2,4-dimethyl-1,1,3-triethoxy-		LI04859
89	43	41	55	71	29	57	42	**244**	100	73	61	45	39	36	29	23	**0.00**	14	28	3	–	–	–	–	–	–	–	–	–	–	Tetradecanoic acid, 2-hydroxy-	2507-55-3	NI23644
45	55	73	41	60	69	200	43	**244**	100	61	55	47	37	35	30	29	**1.00**	14	28	3	–	–	–	–	–	–	–	–	–	–	Tetradecanoic acid, 13-hydroxy-	17278-73-8	NI23645
75	229	73	83	103	97	69	55	**244**	100	55	45	30	29	25	25	25	**0.00**	14	32	1	–	–	–	–	–	–	1	–	–	–	Undecanol, [(trimethylsilyl)oxy]-	17957-64-1	NI23646
244	181	165	152	229	188	153	216	**244**	100	40	36	27	25	25	25	24		15	13	1	–	–	–	1	–	–	–	–	–	–	5H-Dibenzo[a,d]cyclohepten-5-one, 7-chloro-1,2,3,4-tetrahydro-		LI04860
93	244	77	91	148	65	105	41	**244**	100	99	90	86	80	65	55	50		15	16	3	–	–	–	–	–	–	–	–	–	–	Benzo[1,2-b:3,4-b′]difuran-2(3H)-one, 4a,8-dimethyl-3-methylene-4-vinyl-3a,4,5,8b-tetrahydro-, [3aS-(3aα,4α,8bα)]-	957-66-4	NI23647
93	77	91	148	65	244	105	53	**244**	100	88	86	80	66	60	55	47		15	16	3	–	–	–	–	–	–	–	–	–	–	Benzo[1,2-b:3,4-b′]difuran-2(3H)-one, 4a,8-dimethyl-3-methylene-4-vinyl-3a,4,5,8b-tetrahydro-, [3aS-(3aα,4α,8bα)]-		MS05244
229	244	189	230	245	41	159	77	**244**	100	75	23	14	11	10	8	7		15	16	3	–	–	–	–	–	–	–	–	–	–	2H-1-Benzopyran-2-one, 7-methoxy-6-(3-methyl-2-butenyl)-	581-31-7	NI23648
229	244	189	230	245	41	159	131	**244**	100	75	25	15	13	11	8	6		15	16	3	–	–	–	–	–	–	–	–	–	–	2H-1-Benzopyran-2-one, 7-methoxy-6-(3-methyl-2-butenyl)-		LI04861
244	229	201	189	131	213			**244**	100	85	65	53	44	42				15	16	3	–	–	–	–	–	–	–	–	–	–	2H-1-Benzopyran-2-one, 7-methoxy-8-(3-methyl-2-butenyl)-		LI04862
244	229	189	201	213	131	187	159	**244**	100	94	70	63	40	35	30	26		15	16	3	–	–	–	–	–	–	–	–	–	–	2H-1-Benzopyran-2-one, 7-methoxy-8-(3-methyl-2-butenyl)-	484-12-8	NI23649
244	214	229	198	245	199	183	115	**244**	100	46	40	27	18	13	12	12		15	16	3	–	–	–	–	–	–	–	–	–	–	1,1′-Biphenyl, 2,2′,5-trimethoxy-	19718-53-7	NI23650
141	158	140	157	129	244	128	139	**244**	100	44	33	29	25	19	15	14		15	16	3	–	–	–	–	–	–	–	–	–	–	Butanoic acid, 4-hydroxy-, 1-naphthylmethyl ester	86328-51-0	NI23651
244	161	199	135	148	245	91	105	**244**	100	24	23	20	17	16	16	14		15	16	3	–	–	–	–	–	–	–	–	–	–	6H-4,7-Methenofuro[3,2-c]oxacycloundecin-6-one, 3,11-dimethyl-4,8,9,12-tetrahydro-, [R-(E)]-		LI04863
244	199	161	133	148	245	91	105	**244**	100	25	25	20	17	16	15	14		15	16	3	–	–	–	–	–	–	–	–	–	–	6H-4,7-Methenofuro[3,2-c]oxacycloundecin-6-one, 3,11-dimethyl-4,8,9,12-tetrahydro-, [R-(E)]-	728-61-0	NI23652
185	244	186	170	141	245	154	153	**244**	100	35	17	10	7	6	6	6		15	16	3	–	–	–	–	–	–	–	–	–	–	2-Naphthaleneacetic acid, 6-methoxy-α-methyl-, methyl ester	30012-51-2	NI23653
185	244	170	141	186	115	153	154	**244**	100	26	14	14	13	11	8	7		15	16	3	–	–	–	–	–	–	–	–	–	–	2-Naphthaleneacetic acid, 6-methoxy-α-methyl-, methyl ester, (+)-	26159-35-3	NI23654
244	226	227	211	243	183	115	198	**244**	100	65	28	24	23	23	18	18		15	16	3	–	–	–	–	–	–	–	–	–	–	1-Naphthalenecarboxaldehyde, 2,8-dihydroxy-4-isopropyl-6-methyl-	55824-28-7	NI23655
201	216	202	128	115	217	158	77	**244**	100	53	16	10	10	9	7	7	**4.20**	15	16	3	–	–	–	–	–	–	–	–	–	–	1,2-Naphthalenedione, 6-hydroxy-3,8-dimethyl-5-isopropyl-	7715-96-0	MS05245
188	244	77	41	189	160	29	105	**244**	100	44	23	20	18	18	18	16		15	16	3	–	–	–	–	–	–	–	–	–	–	1,4-Naphthalenedione, 2-hydroxy-3-isopentyl-	3343-38-2	NI23656
28	77	243	107	94	118	149	117	**244**	100	63	57	55	49	30	26	26	**16.25**	15	16	3	–	–	–	–	–	–	–	–	–	–	2-Propanol, 1,3-diphenoxy-		DO01031
186	187	244	77	185	158	245	141	**244**	100	67	53	17	10	10	9	9		15	16	3	–	–	–	–	–	–	–	–	–	–	2-Propanol, 1-[(4-phenoxy)phenoxy]-		DO01032
202	45	91	92	201	59	134	77	**244**	100	98	81	65	64	52	50	40	**27.05**	15	17	1	–	–	–	–	–	–	–	1	–	–	Phosphine, diphenylisopropoxy-		IC03227
157	106	107	77	185	59	115	167	**244**	100	73	60	54	44	21	15	11	**6.00**	15	20	1	2	–	–	–	–	–	–	–	–	–	Acetic acid, 1-(1-cyclohexen-1-yl)-2-methyl-2-phenylhydrazide	67134-49-0	NI23657
119	44	244	93	42	243	43	91	**244**	100	68	55	31	21	20	20	17		15	20	1	2	–	–	–	–	–	–	–	–	–	1,3-Benzoxazine, 3-methyl-2-phenylimino-perhydro-, cis-		MS05246
244	243	106	119	161	149	118	44	**244**	100	54	49	36	35	31	24	20		15	20	1	2	–	–	–	–	–	–	–	–	–	1,3-Benzoxazine, 3-methyl-2-phenylimino-perhydro-, trans-		MS05247
244	109	229	136	135	108	107	245	**244**	100	84	58	55	47	47	30	18		15	20	1	2	–	–	–	–	–	–	–	–	–	2,2′-Dipyrrolyl ketone, 3,4,5,3′,4′,5′-hexamethyl-		LI04864
244	96	118	99	226	146	126	168	**244**	100	80	39	39	19	19	19	9		15	20	1	2	–	–	–	–	–	–	–	–	–	2-Indolemethanol, α-(1-methylpiperidin-4-yl)-		LI04865
243	244	149	245	148	146	150	136	**244**	100	99	37	19	13	13	12	12		15	20	1	2	–	–	–	–	–	–	–	–	–	Matridin-15-one, 11,12,13,14-tetradehydro-, (6β)-	6838-34-2	NI23658
98	244	146	97	96	55	160	136	**244**	100	44	16	16	15	13	11	10		15	20	1	2	–	–	–	–	–	–	–	–	–	7,14-Methano-4H,6H-dipyrido[1,2-a:1′,2′-e][1,5]diazocin-4-one, 7,7a,8,9,10,11,13,14-octahydro-, [7R-(7α,7aα,14α)]-	486-90-8	NI23659
98	146	147	190	160	244	148	134	**244**	100	39	33	22	17	15	15	14		15	20	1	2	–	–	–	–	–	–	–	–	–	7,14-Methano-4H,6H-dipyrido[1,2-a:1′,2′-e][1,5]diazocin-4-one, 7,7a,8,9,10,11,13,14-octahydro-, [7R-(7α,7aβ,11α)]-		LI04866
98	146	147	190	160	148	242	97	**244**	100	39	32	20	15	14	13	12	**0.00**	15	20	1	2	–	–	–	–	–	–	–	–	–	7,14-Methano-4H,6H-dipyrido[1,2-a:1′,2′-e][1,5]diazocin-4-one, 7,7a,8,9,10,11,13,14-octahydro-, [7R-(7α,7aβ,14α)]-	486-89-5	NI23660
98	244	97	243	146	245	136	96	**244**	100	60	13	12	12	11	9	9		15	20	1	2	–	–	–	–	–	–	–	–	–	7,14-Methano-4H,6H-dipyrido[1,2-a:1′,2′-e][1,5]diazocin-4-one, 7,7a,8,9,10,11,13,14-octahydro-, [7R-(7α,7aβ,14α)]-	486-89-5	NI23661
58	98	146	55	160	203	244	147	**244**	100	32	14	14	13	11	9	7		15	20	1	2	–	–	–	–	–	–	–	–	–	1,5-Methano-8H-pyrido[1,2-a][1,5]diazocin-8-one, 3-(3-butenyl)-1,2,3,4,5,6-hexahydro-, (1R)-	529-78-2	NI23662
203	58	160	146	98	244			**244**	100	74	16	10	10	4				15	20	1	2	–	–	–	–	–	–	–	–	–	1,5-Methano-8H-pyrido[1,2-a][1,5]diazocin-8-one, 3-(3-butenyl)-1,2,3,4,5,6-hexahydro-, (1R)-	529-78-2	NI23663
203	96	160	82	97	244	146	98	**244**	100	43	28	26	22	18	17	17		15	20	1	2	–	–	–	–	–	–	–	–	–	3H-2,5-Methano-1H-pyrrolo[3,4-i]quinolizin-8(1H)-one,3a,4,5,6-tetrahydro-1-(2-propenyl)-, [1R-(1α,2α,3aβ,5α,11aR*)]-	73575-31-2	NI23664
244	96	203	97	82	160	98	146	**244**	100	79	73	73	61	40	31	25		15	20	1	2	–	–	–	–	–	–	–	–	–	3H-2,5-Methano-1H-pyrrolo[3,4-i]quinolizin-8(9H)-one,3a,4,5,6-tetrahydro-1-(2-propenyl)-, [1R-(1α,2α,3aβ,5α,11aR*)]-	70711-82-9	NI23665
96	244	203	97	82	160	146	98	**244**	100	92	87	81	67	61	36	36		15	20	1	2	–	–	–	–	–	–	–	–	–	3H-2,5-Methano-1H-pyrrolo[3,4-i]quinolizin-8(9H)-one,3a,4,5,6-tetrahydro-1-(2-propenyl)-, [1S-(1α,2β,3aα,5β,11aS*)]-	73610-25-0	NI23666
84	131	58	244	91	41	125	42	**244**	100	47	22	21	14	10	9	9		15	20	1	2	–	–	–	–	–	–	–	–	–	Spiro[1.2]pentane-3-imidazoline, 5,5-dimethyl-4-phenyl-, 3-oxide		NI23667
57	43	56	41	71	55	69	45	**244**	100	70	64	52	50	50	45	41	**0.00**	15	32	2	–	–	–	–	–	–	–	–	–	–	Dodecane, 1-(2-hydroxyethoxy)-2-methyl-		IC03228
75	213	85	83	41	214	87	67	**244**	100	38	20	19	18	10	9	9	**0.00**	15	32	2	–	–	–	–	–	–	–	–	–	–	Tridecane, 1,1-dimethoxy-		IC03229

MASS TO CHARGE RATIOS	M.W.	INTENSITIES	Parent	C	H	O	N	S	F	Cl	Br	I	Si	P	B	X	COMPOUND NAME	CAS Reg No	No
57 43 71 55 41 85 69 63	**244**	100 85 62 56 54 39 39 37	**0.00**	15	32	2	–	–	–	–	–	–	–	–	–	–	**Tridecane, 1-(2-hydroxyethoxy)-**		**IC03230**
244 91 167 135 77 67 147 105	**244**	100 90 68 67 56 46 27 25		16	20	–	–	1	–	–	–	–	–	–	–	–	**1,3-Butadiene, 1-cyclohexyl-2-(phenylthio)-, (E)-**	83877-76-3	**DD00640**
244 91 135 77 67 129 167 147	**244**	100 88 74 53 45 39 36 30		16	20	–	–	1	–	–	–	–	–	–	–	–	**1,3-Butadiene, 1-cyclohexyl-2-(phenylthio)-, (Z)-**	83877-78-5	**DD00641**
160 39 127 116 244 55 28 129	**244**	100 83 79 70 65 64 62 52		16	20	–	–	1	–	–	–	–	–	–	–	–	**Naphthalene, 1-hexylthio-**		**AI01744**
160 116 129 39 55 244 162 126	**244**	100 82 81 79 65 61 56 46		16	20	–	–	1	–	–	–	–	–	–	–	–	**Naphthalene, 2-hexylthio-**		**AI01745**
67 135 91 77 79 244 93 41	**244**	100 91 91 77 76 71 71 67		16	20	–	–	1	–	–	–	–	–	–	–	–	**Tricyclo[6.1.1.0^{2,7}]decane, 9-endo-(phenylthio)-exo-**		**NI23668**
134 244 112 159 133 91 131 93	**244**	100 81 73 57 56 54 33 22		16	20	2	–	–	–	–	–	–	–	–	–	–	**Benz[e]-as-indacene-7,10-dione, 1,2,3,4,5,6,6a,6b,8,9,10a,10b-dodecahydro-**	31991-61-4	**NI23669**
161 244 176 107 147 123 201 91	**244**	100 27 23 8 5 5 4 4		16	20	2	–	–	–	–	–	–	–	–	–	–	**1,3-Benzenediol, 2-[(Z)-p-menthadien-1,8-yl]-**		**LI04867**
161 176 244 123 115 91 229 147	**244**	100 60 26 8 7 7 6 6		16	20	2	–	–	–	–	–	–	–	–	–	–	**1,3-Benzenediol, 4-[(Z)-p-menthadien-1,8-yl]-**		**LI04868**
146 202 128 131 141 145 129 132	**244**	100 72 40 35 29 28 26 24	**10.00**	16	20	2	–	–	–	–	–	–	–	–	–	–	**5-Benzocyclooctenol, 5,6,7,8-tetrahydro-8,8-dimethyl-, acetate**	69576-86-9	**NI23670**
244 130 18 43 115 158 129 201	**244**	100 82 77 68 52 51 51 40		16	20	2	–	–	–	–	–	–	–	–	–	–	**Benzo[9,10]-2,5-dioxacyclo[6.4.0]dodec-1(8)-ene, 6,6-dimethyl-**		**NI23671**
184 156 43 155 132 91 244 130	**244**	100 74 54 48 46 46 36 29		16	20	2	–	–	–	–	–	–	–	–	–	–	**Bicyclo[2.2.2]octan-1-ol, 4-phenyl-, acetate**	54986-35-5	**NI23672**
188 143 128 173 189 144 105 91	**244**	100 98 26 20 15 13 13 13	**2.00**	16	20	2	–	–	–	–	–	–	–	–	–	–	**1-Cyclopropene-3-carboxylic acid, 2-ethyl-1-phenyl-, isobutyl ester**		**MS05248**
143 188 128 173 105 187 91 189	**244**	100 97 29 20 19 17 16 14	**8.00**	16	20	2	–	–	–	–	–	–	–	–	–	–	**1-Cyclopropene-3-carboxylic acid, 2-ethyl-1-phenyl-, n-butyl ester**		**MS05249**
143 188 128 173 144 189 115 105	**244**	100 75 22 15 13 11 9 9	**2.00**	16	20	2	–	–	–	–	–	–	–	–	–	–	**1-Cyclopropene-3-carboxylic acid, 2-ethyl-1-phenyl-, sec-butyl ester**		**MS05250**
161 201 244 123 176 107 162 121	**244**	100 75 64 25 24 14 13 13		16	20	2	–	–	–	–	–	–	–	–	–	–	**Dibenzo[b,d]pyran, 1-hydroxy-6,6,9-trimethyl-6a,7,10,10a-tetrahydro-**		**LI04869**
169 143 184 129 244 142 141 128	**244**	100 67 56 35 30 25 22 22		16	20	2	–	–	–	–	–	–	–	–	–	–	**De-A-Estra-5,7,9-triene-17β-ol, acetate**		**NI23673**
188 105 173 244 189 77 187 143	**244**	100 25 16 15 15 13 8 7		16	20	2	–	–	–	–	–	–	–	–	–	–	**Furan, 2-isobutoxy-4-ethyl-5-phenyl-**		**MS05251**
173 188 57 244 102 174 189 41	**244**	100 72 21 17 14 13 11 9		16	20	2	–	–	–	–	–	–	–	–	–	–	**Furan, 2-isobutoxy-5-ethyl-4-phenyl-**		**MS05252**
188 105 244 77 187 189 173 143	**244**	100 58 24 19 18 17 17 7		16	20	2	–	–	–	–	–	–	–	–	–	–	**Furan, 2-n-butoxy-4-ethyl-5-phenyl-**		**MS05253**
173 188 244 57 174 102 187 189	**244**	100 65 29 16 13 13 12 10		16	20	2	–	–	–	–	–	–	–	–	–	–	**Furan, 2-n-butoxy-5-ethyl-4-phenyl-**		**MS05254**
188 105 173 189 77 244 143 187	**244**	100 24 18 14 14 10 8 7		16	20	2	–	–	–	–	–	–	–	–	–	–	**Furan, 2-sec-butoxy-4-ethyl-5-phenyl-**		**MS05255**
173 188 174 102 244 57 189 187	**244**	100 78 14 13 12 12 11 7		16	20	2	–	–	–	–	–	–	–	–	–	–	**Furan, 2-sec-butoxy-5-ethyl-4-phenyl-**		**MS05256**
244 215 171 57 91 143 128 115	**244**	100 97 48 48 46 31 30 29		16	20	2	–	–	–	–	–	–	–	–	–	–	**2,3-Hexadienoic acid, 2-ethyl-4-phenyl-, ethyl ester**		**LI04870**
244 215 28 171 57 91 143 128	**244**	100 98 49 48 48 47 33 32		16	20	2	–	–	–	–	–	–	–	–	–	–	**2,3-Hexadienoic acid, 2-ethyl-4-phenyl-, ethyl ester**	38701-10-9	**NI23674**
229 169 59 244 141 230 170 245	**244**	100 78 44 34 12 10 9 5		16	20	2	–	–	–	–	–	–	–	–	–	–	**Naphthalene, 2,6-bis(2-hydroxy-2-propyl)-**	24157-82-2	**NI23675**
244 136 229 175 189 161 69 81	**244**	100 42 35 29 19 19 17 13		16	20	2	–	–	–	–	–	–	–	–	–	–	**1,4-Naphthalenediol, 7-isopropenyl-6-methyl-5,6,7,8-tetrahydro-6-vinyl-**		**NI23676**
121 41 109 39 44 55 67 244	**244**	100 79 62 56 55 51 47 45		16	20	2	–	–	–	–	–	–	–	–	–	–	**1(2H)-Naphthalenone, 2-(2-furanylmethylene)octahydro-8a-methyl-, (4aR-trans)-**	54482-61-0	**NI23677**
121 41 78 81 216 91 55 106	**244**	100 84 46 44 43 39 39 37	**32.32**	16	20	2	–	–	–	–	–	–	–	–	–	–	**1(2H)-Naphthalenone, 2-(2-furanylmethylene)octahydro-8a-methyl-, cis-**	56192-67-7	**NI23678**
104 117 244 185 105 91 118 184	**244**	100 75 38 37 36 36 27 26		16	20	2	–	–	–	–	–	–	–	–	–	–	**[8]Paracycloph-4-ene, 3-acetoxy-, cis-**		**NI23679**
104 117 185 91 105 244 184 118	**244**	100 96 50 49 40 30 29 29		16	20	2	–	–	–	–	–	–	–	–	–	–	**[8]Paracycloph-4-ene, 3-acetoxy-, trans-**		**NI23680**
96 131 81 55 103 67 77 150	**244**	100 87 63 54 51 40 37 31	**3.40**	16	20	2	–	–	–	–	–	–	–	–	–	–	**2-Propenoic acid, 3-phenyl-, cyclohexylmethyl ester**	6918-90-7	**NI23681**
85 128 115 143 116 129 55 41	**244**	100 92 87 59 47 38 36 35	**1.00**	16	20	2	–	–	–	–	–	–	–	–	–	–	**2H-Pyran, tetrahydro-2-[(1-methyl-4-phenyl-2-butynyl)oxy]-**	54966-32-4	**NI23682**
72 58 71 73 42 229 56 44	**244**	100 42 6 5 5 4 3 3	**2.41**	16	24	–	2	–	–	–	–	–	–	–	–	–	**Benzeneacetonitrile, α-[2-(dimethylamino)propyl]-α-isopropyl-**	77-51-0	**NI23683**
54 79 80 67 93 122 53 66	**244**	100 87 77 60 33 21 21 18	**17.00**	16	24	–	2	–	–	–	–	–	–	–	–	–	**Hydrazine, N,N′-bis[(5Z)-cycloocten-1,2-ylidenyl]-**		**NI23684**
55 97 83 41 29 39 27 56	**244**	100 64 52 28 10 8 6 5	**0.05**	16	24	–	2	–	–	–	–	–	–	–	–	–	**Propanedinitrile, cyclohexyl(2-methylcyclohexyl)-**	74764-55-9	**NI23685**
115 117 91 129 139 128 159 141	**244**	100 92 92 52 51 48 42 26	**10.00**	17	24	1	–	–	–	–	–	–	–	–	–	–	**2,9-Heptadecadien-4,6-diyn-1-ol, (E,Z)-**		**LI04871**
91 159 41 43 117 55 145 131	**244**	100 83 83 76 74 72 64 64	**4.00**	17	24	1	–	–	–	–	–	–	–	–	–	–	**2,9-Heptadecadien-4,6-diyn-8-ol, (E,Z)-**		**NI23686**
57 159 215 173 187 201	**244**	100 35 10 5 3 1	**0.00**	17	24	1	–	–	–	–	–	–	–	–	–	–	**1,9-Heptadecadiene-4,6-diyn-3-one, 1,2-dihydro-8-hydroxy-, (Z)-**		**LI04872**
91 41 93 43 55 80 69 92	**244**	100 56 40 24 23 22 22 21	**2.15**	17	24	1	–	–	–	–	–	–	–	–	–	–	**2-Heptene, 6-benzyloxy-2-methyl-6-vinyl-**		**NI23687**
159 187 244 128 115 171 141 91	**244**	100 63 29 17 15 11 10 8		17	24	1	–	–	–	–	–	–	–	–	–	–	**Indene, 3,3-dimethyl-1-isopropoxy-2-isopropyl-**		**LI04873**
244 28 176 41 91 67 79 148	**244**	100 91 66 52 49 41 36 35		17	24	1	–	–	–	–	–	–	–	–	–	–	**Gon-9(11)-en-12-one**	55029-38-4	**NI23688**
91 105 41 55 148 130 104 121	**244**	100 53 49 46 34 31 31 24	**0.97**	17	24	1	–	–	–	–	–	–	–	–	–	–	**3-Pentanone, 1-cyclohexyl-5-phenyl-**		**AI01746**
131 117 229 211 143 129 141 128	**244**	100 77 75 57 50 47 43 34	**25.60**	17	24	1	–	–	–	–	–	–	–	–	–	–	**1-Phenanthrenemethanol, 1,2,3,4,4a,9,10,10a-octahydro-1,4a-dimethyl-, [1S-(1α,4aα,10aβ)]-**	727-20-8	**NI23689**
211 226 143 141 41 157 129 144	**244**	100 63 55 50 42 38 33 32	**15.00**	17	24	1	–	–	–	–	–	–	–	–	–	–	**2-Phenanthrenol, 1,2,3,4,4a,9,10,10a-octahydro-1,1,4a-trimethyl-**	61141-20-6	**NI23690**
211 226 143 141 157 129 144 117	**244**	100 63 56 50 38 33 32 32	**15.14**	17	24	1	–	–	–	–	–	–	–	–	–	–	**2-Phenanthrenol, 1,2,3,4,4a,9,10,10a-octahydro-1,1,4a-trimethyl-, [2S-(2α,4aα,10aβ)]-**	55902-87-9	**NI23691**
244 147 159 133 229 69 44 160	**244**	100 64 47 47 42 40 27 25		17	24	1	–	–	–	–	–	–	–	–	–	–	**3-Phenanthrenol, 4b,5,6,7,8,8a,9,10-octahydro-4b,8,8-trimethyl-, (4bS-trans)-**	15340-76-8	**NI23692**
244 147 229 133 159 69 44 160	**244**	100 65 53 49 48 44 28 26		17	24	1	–	–	–	–	–	–	–	–	–	–	**3-Phenanthrenol, 4b,5,6,7,8,8a,9,10-octahydro-4b,8,8-trimethyl-, (4bS-trans)-**	15340-76-8	**NI23693**
163 91 134 119 105 77 79 67	**244**	100 53 48 47 30 29 25 25	**0.00**	17	24	1	–	–	–	–	–	–	–	–	–	–	**6(7H)-Phenanthrenone, 1,4,4aβ,8,8a,9,10,10aβ-octahydro-2,3,8aβ-trimethyl-**		**DD00642**

MASS TO CHARGE RATIOS								M.W.	INTENSITIES								Parent	C	H	O	N	S	F	Cl	Br	I	Si	P	B	X	COMPOUND NAME	CAS Reg No	No
201	229	244	200	100	230	202	101	**244**	100	95	56	55	30	17	17	11		18	12	1	–	–	–	–	–	–	–	–	–	–	Pyrene, 1-acetyl-	3264-21-9	NI23694
41	79	67	55	229	81	91	93	**244**	100	66	59	59	50	49	48	39	**39.00**	18	28	–	–	–	–	–	–	–	–	–	–	–	13,16-Seco-D-nor-5α-androsta-13(18),15-diene	31239-26-6	NI23695
118	91	131	117	146	115	129	244	**244**	100	90	81	79	31	27	26	24		18	28	–	–	–	–	–	–	–	–	–	–	–	Benzene, (1-propyl-1-nonenyl)-	54986-39-9	NI23696
145	117	29	115	146	129	41	244	**244**	100	42	15	13	12	12	12	10		18	28	–	–	–	–	–	–	–	–	–	–	–	1H-Indene, 2,3-dihydro-1-ethyl-3-heptyl-, cis-		AI01747
145	117	29	28	115	129	41	43	**244**	100	41	18	16	15	14	14	13	**10.66**	18	28	–	–	–	–	–	–	–	–	–	–	–	1H-Indene, 2,3-dihydro-1-ethyl-3-heptyl-, trans-		AI01748
131	132	91	29	244	41	115	27	**244**	100	11	8	8	7	7	6	6		18	28	–	–	–	–	–	–	–	–	–	–	–	1H-Indene, 2,3-dihydro-1-methyl-3-octyl-	29138-84-9	WI01255
173	174	244	57	157	143	142	131	**244**	100	14	7	5	3	3	3	3		18	28	–	–	–	–	–	–	–	–	–	–	–	1H-Indene, 2,3-dihydro-1-neopentyl-1,1,3,5-tetramethyl-		AI01749
173	174	57	143	131	41	244	157	**244**	100	16	8	4	4	4	3	3		18	28	–	–	–	–	–	–	–	–	–	–	–	1H-Indene, 2,3-dihydro-3-neopentyl-1,1,3,5-tetramethyl-		AI01750
173	174	57	41	29	143	157	128	**244**	100	15	12	9	8	7	6	6	**1.96**	18	28	–	–	–	–	–	–	–	–	–	–	–	1H-Indene, 2,3-dihydro-3-neopentyl-1,1,3,5-tetramethyl-		AI01751
117	29	41	118	115	91	43	27	**244**	100	14	13	11	9	9	9	9	**4.50**	18	28	–	–	–	–	–	–	–	–	–	–	–	1H-Indene, 2,3-dihydro-1-nonyl-	29138-83-8	WI01256
229	57	28	230	244	41	29	173	**244**	100	28	24	18	14	12	12	8		18	28	–	–	–	–	–	–	–	–	–	–	–	1H-Indene, 2,3-dihydro-1,1,3,3,4-pentamethyl-6-tert-butyl-		AI01752
229	244	230	57	245	173	231	107	**244**	100	18	18	10	5	5	2	2		18	28	–	–	–	–	–	–	–	–	–	–	–	1H-Indene, 2,3-dihydro-1,1,3,3,4-pentamethyl-6-tert-butyl-		AI01753
187	117	131	29	129	41	130	128	**244**	100	95	83	46	39	29	24	24	**20.42**	18	28	–	–	–	–	–	–	–	–	–	–	–	Naphthalene, 1,4-dibutyl-1,2,3,4-tetrahydro-	29138-92-9	WI01257
229	230	244	29	187	185	41	155	**244**	100	18	14	10	6	6	6	5		18	28	–	–	–	–	–	–	–	–	–	–	–	Naphthalene, 6,7-diethyl-1,2,3,4-tetrahydro-1,1,4,4-tetramethyl-		AI01754
229	57	41	29	244	230	187	131	**244**	100	64	36	26	19	19	13	12		18	28	–	–	–	–	–	–	–	–	–	–	–	Naphthalene, 2,6-di-tert-butyl-1,2,3,4-tetrahydro-		AI01755
145	146	117	29	244	91	41	129	**244**	100	12	8	8	7	7	7	6		18	28	–	–	–	–	–	–	–	–	–	–	–	Naphthalene, 4-heptyl-1-methyl-1,2,3,4-tetrahydro-		AI01756
201	159	202	145	173	57	43	171	**244**	100	33	18	12	10	10	8	7	**5.83**	18	28	–	–	–	–	–	–	–	–	–	–	–	Naphthalene, 1-isopropyl-1,2,4,4,7-pentamethyl-1,2,3,4-tetrahydro-		AI01757
131	132	91	244	129	115	41	29	**244**	100	18	13	10	7	7	7	7		18	28	–	–	–	–	–	–	–	–	–	–	–	Naphthalene, 1-octyl-1,2,3,4-tetrahydro-	29138-91-8	WI01258
229	244	173	215	201	245	230	185	**244**	100	95	41	32	22	20	19	14		18	28	–	–	–	–	–	–	–	–	–	–	–	Styrene, 2,3,4,5,6-pentaethyl-	2715-34-6	NI23697
244	243	167	165	245	168	229	228	**244**	100	40	26	24	20	16	14	13		19	16	–	–	–	–	–	–	–	–	–	–	–	1,4-Cyclohexadiene, 3,3-diphenyl-6-methylene-	18636-59-4	NI23698
244	167	165	243	166	245	152	164	**244**	100	78	69	64	39	19	16	11		19	16	–	–	–	–	–	–	–	–	–	–	–	Methane, triphenyl-		IC03231
244	167	165	166	243	245	153	168	**244**	100	61	50	32	26	21	13	9		19	16	–	–	–	–	–	–	–	–	–	–	–	Methane, triphenyl-		IC03232
244	165	167	166	243	32	229	239	**244**	100	99	96	45	33	32	22	21		19	16	–	–	–	–	–	–	–	–	–	–	–	Methane, triphenyl-		IC03233
229	244	228	128	115	129	202	215	**244**	100	53	42	26	22	21	13	12		19	16	–	–	–	–	–	–	–	–	–	–	–	1,4-Methanonaphthalene, 1,4-dihydro-9-(1-phenylethylidene)-	116971-99-4	DD00643
244	229	243	245	230	165	114	166	**244**	100	90	38	21	18	16	16	11		19	16	–	–	–	–	–	–	–	–	–	–	–	Naphthalene, 2-methyl-1-styryl-		LI04874
82	61	63	31	50	35	44	47	**245**	100	83	72	16	9	9	8	3	**0.60**	2	–	–	1	2	2	3	–	–	–	–	–	–	Methanimine, dichloro-N-(chlorodifluoromethyl)dithio-		NI23699
69	245	138	157	226	107	31	50	**245**	100	18	13	8	7	5	4	1		5	–	–	1	–	9	–	–	–	–	–	–	–	1-Propynamine, 3,3,3-trifluoro-N-bis(trifluoromethyl)-		LI04875
73	47	66	230	92	87	172	85	**245**	100	95	75	20	20	15	10	10	**3.00**	6	18	1	1	–	2	–	–	–	2	1	–	–	Phosphoramidic difluoride, bis(trimethylsilyl)-	41309-92-6	MS05257
43	44	82	80	55	69	41	68	**245**	100	94	47	46	45	36	35	33	**11.98**	7	12	–	5	–	–	–	1	–	–	–	–	–	1,3,5-Triazine-2,4-diamine, 6-bromo-N,N′-diethyl-	3084-94-4	NI23700
91	245	118	65	89	119	39	63	**245**	100	43	43	30	21	17	15	11		8	8	–	1	1	5	–	–	–	–	–	–	–	(4-Methylbenzylidenimino)sulphur pentafluoride	90598-13-3	NI23701
166	51	90	77	89	136	78	120	**245**	100	25	22	13	11	11	10	9	**2.00**	8	8	3	1	–	–	–	1	–	–	–	–	–	Benzene, 2-(bromomethyl)-1-methoxy-4-nitro-	3913-23-3	NI23702
245	247	228	230	200	198	216	52	**245**	100	96	63	58	42	36	22	20		8	8	3	1	–	–	–	1	–	–	–	–	–	Benzoic acid, 4-amino-5-bromo-2-methoxy-		MS05258
228	170	119	118	245	42	198	200	**245**	100	81	81	60	43	40	34	31		8	11	4	3	1	–	–	–	–	–	–	–	–	Benzenesulphamide, 3-dimethylamino-4-nitro-		IC03234
42	41	75	44	111	43	51	74	**245**	100	63	55	46	33	32	17	13	**0.00**	9	8	3	1	1	–	1	–	–	–	–	–	–	3-Azetidinone, N-(4-chlorobenzenesulphonyl)-	82380-61-8	NI23703
164	245	243	163	121	122	165	241	**245**	100	56	28	16	16	15	13	11		9	11	2	1	–	–	–	–	–	–	–	–	1	Benzeneselenocarboxamide, 3,5-dimethoxy-		MS05259
100	73	130	72	43	173	157	113	**245**	100	82	63	59	47	31	28	20	**12.01**	9	15	5	3	–	–	–	–	–	–	–	–	–	Glycine, N-[N-(N-acetylglycyl)glycyl]-, methyl ester	55255-89-5	NI23704
245	139	96	71	230	168	247	210	**245**	100	88	42	39	35	34	30	30		9	16	1	5	–	–	1	–	–	–	–	–	–	1,3,5-Triazine-2,4-diamine, 6-(2-chloroethoxy)-N,N,N′,N′-tetramethyl-		NI23705
245	39	30	246	54	64	43	55	**245**	100	40	26	25	25	22	15	12		10	7	3	5	–	–	–	–	–	–	–	–	–	2,1,3-Benzoxadiazole, 4-[1-(2-methylimidazyl)]-7-nitro-	91485-30-2	NI23706
245	30	169	42	168	246	55	43	**245**	100	55	54	36	29	25	24	24		10	7	3	5	–	–	–	–	–	–	–	–	–	2,1,3-Benzoxadiazole, 4-[1-(4-methylimidazyl)]-7-nitro-	91485-29-9	NI23707
189	41	57	56	245	191	159	63	**245**	100	99	90	57	47	37	25	21		10	12	2	1	1	–	1	–	–	–	–	–	–	Benzene, 4-butylthio-3-chloro-1-nitro		IC03235
128	118	130	75	57	43	44	120	**245**	100	72	37	29	26	23	21	20	**4.46**	10	12	4	1	–	–	1	–	–	–	–	–	–	1,2-Propanediol, 3-(4-chlorophenoxy)-, 1-carbamate	886-74-8	NI23708
245	59	172	41	158	43	200	173	**245**	100	74	59	35	32	30	29	29		10	15	2	1	2	–	–	–	–	–	–	–	–	Δ(4)-Thiazoline-2-thione, 3-(1-carboxyethyl)-4-isopropyl-5-methyl-, L-		NI23709
245	41	158	45	200	29	39	27	**245**	100	55	51	35	34	24	23	23		10	15	2	1	2	–	–	–	–	–	–	–	–	Δ(4)-Thiazoline-2-thione, 3-(1-carboxyethyl)-4-tert-butyl-, DL-		NI23710
43	185	140	172	112	45	44	87	**245**	100	90	90	54	36	36	36	27	**3.20**	10	15	6	1	–	–	–	–	–	–	–	–	–	Butenedioic acid, [(2-acetoxyethyl)amino]-, dimethyl ester		NI23711
246	248	68	166	82	67	124	95	**245**	100	91	66	58	35	28	23	22	**5.18**	10	16	1	1	–	–	–	1	–	–	–	–	–	2-Azetidinone, 1-(4-bromobutyl)-4-methyl-4-vinyl-		NI23712
171	70	99	245	126	157	85		**245**	100	73	39	31	27	18	10			10	19	4	3	–	–	–	–	–	–	–	–	–	Pyrrolidine, 1-[1,2-bis(ethoxycarbonyl)hydrazino]-		LI04876
199	172	41	66	169	145	144	170	**245**	100	45	43	41	38	38	34	31	**5.00**	11	11	2	5	–	–	–	–	–	–	–	–	–	1-Cyclopentene, 2-amino-4-ethyl-3-methyl-3-nitro-1,5,5-tricyano-		MS05260
100	57	59	42	101	43	98	70	**245**	100	53	50	46	45	45	43	41	**3.00**	11	16	3	1	–	–	1	–	–	–	–	–	–	2-Furanone, 3-chloro-2,5-dihydro-4,5-dimethyl-5-[(4-morpholino)methyl]-	104502-46-7	NI23713
166	165	41	167	245	108	55	79	**245**	100	43	38	37	30	24	24	23		11	19	3	1	1	–	–	–	–	–	–	–	–	7-Thiabicyclo[4.2.0]octane, 1-(4-morpholino)-, 7,7-dioxide	34314-98-2	NI23714
43	101	86	28	95	143	100	128	**245**	100	69	36	34	29	27	24	23	**0.50**	11	19	5	1	–	–	–	–	–	–	–	–	–	α-Daunosaminide, methyl N,O-diacetyl-		NI23715
131	45	101	89	88	115	75	85	**245**	100	60	60	47	44	40	29	26	**0.00**	11	19	5	1	–	–	–	–	–	–	–	–	–	D-Galactonitrile, 5-O-acetyl-2,3,4-tri-O-methyl-6-deoxy-		NI23716
84	43	130	172	126	56	112	57	**245**	100	53	47	45	32	25	23	13	**2.00**	11	19	5	1	–	–	–	–	–	–	–	–	–	L-Glutamic acid, N-acetyl-, diethyl ester	1446-19-1	NI23717
57	41	70	120	44	82	74	102	**245**	100	45	18	15	15	11	9	7	**0.00**	11	19	5	1	–	–	–	–	–	–	–	–	–	Glycine, N-(tert-butoxycarbonyl)-, (Z)-4-hydroxy-2-butenyl ester		DD00644
43	57	139	41	138	61	44	153	**245**	100	50	41	25	17	12	9	7	**0.00**	11	19	5	1	–	–	–	–	–	–	–	–	–	L-Serine, N-acetyl-, sec-butyl ester, acetate	57983-72-9	NI23718

MASS TO CHARGE RATIOS								M.W.	INTENSITIES								Parent	C	H	O	N	S	F	Cl	Br	I	Si	P	B	X	COMPOUND NAME	CAS Reg No	No
86	128	73	43	75	230	172	30	**245**	100	73	45	38	35	17	17	17	**0.00**	11	23	3	1	–	–	–	–	–	1	–	–	–	D-Alloisoleucine, N-acetyl-, trimethylsilyl ester	55887-48-4	NI23719
86	128	73	44	75	172	230	30	**245**	100	73	45	38	35	17	17	17	**0.42**	11	23	3	1	–	–	–	–	–	1	–	–	–	D-Alloisoleucine, N-acetyl-, trimethylsilyl ester	55887-48-4	NI23720
73	158	43	75	99	230	172	71	**245**	100	55	55	44	31	28	25	23	**0.00**	11	23	3	1	–	–	–	–	–	1	–	–	–	Glycine, N-(1-oxohexyl)-, trimethylsilyl ester	55494-05-8	NI23721
86	128	44	73	75	230	172	30	**245**	100	73	46	45	35	17	17	17	**0.42**	11	23	3	1	–	–	–	–	–	1	–	–	–	D-Isoleucine, N-acetyl-, trimethylsilyl ester	72361-03-6	NI23722
147	73	148	188	45	75	59	114	**245**	100	60	17	14	14	9	9	9	**0.63**	11	27	1	1	–	–	–	–	–	2	–	–	–	Acetamide, tert-butyldimethylsilyl-		NI23723
245	109	137	41	53	124	81	246	**245**	100	86	51	34	33	30	29	26		12	11	3	3	–	–	–	–	–	–	–	–	–	1,3-Benzenediol, 4-(2-hydroxy-5-methyl-3-pyridylazo)-		MS05261
119	245	103	85	105	104	142	77	**245**	100	36	32	32	24	22	20	20		12	11	3	3	–	–	–	–	–	–	–	–	–	Isoxazolo[5,4-d]pyrimidine-4,6(5H,7H)-dione, 3a,7a-dihydro-5-methyl-3-phenyl-, cis-	35053-73-7	NI23724
176	147	245	91	148	132	43	77	**245**	100	39	22	22	18	16	16	15		12	14	1	1	–	3	–	–	–	–	–	–	–	Aniline, 2,6-diethyl-N-trifluoroacetyl-		NI23725
160	114	101	126	161	128	162	145	**245**	100	37	11	10	9	7	6	6	**0.04**	12	23	2	1	1	–	–	–	–	–	–	–	–	4-Thiazolidinecarboxylic acid, 5,5-dimethyl-2-hexyl-		NI23726
28	88	59	200	43	100	146	75	**245**	100	90	82	62	50	43	42	39	**0.00**	12	27	2	1	–	–	–	–	–	1	–	–	–	Diethylamine, 2,2′-dimethoxy-, N-[(allyldimethylsilyl)methyl]-		DD00645
93	111	126	245	83	97	152	153	**245**	100	91	65	37	12	6	4	3		13	11	2	1	1	–	–	–	–	–	–	–	–	Acetamide, 2-(2-thienoyl)-N-phenyl-		LI04877
93	111	126	245	83	97	153	119	**245**	100	91	45	37	19	8	3	3		13	11	2	1	1	–	–	–	–	–	–	–	–	Acetamide, 2-(3-thienoyl)-N-phenyl-		LI04878
245	175	217	189	69	173	246	188	**245**	100	47	27	23	21	17	15	13		13	11	2	1	1	–	–	–	–	–	–	–	–	1H-Benzo[d]pyrido[2,1-b]thiazole, 3-ethoxy-1-oxo-		LI04879
126	200	115	245	154	114	171	125	**245**	100	96	66	65	50	35	31	22		13	11	4	1	–	–	–	–	–	–	–	–	–	1-Naphthalenecarboxylic acid, 5-nitro-, ethyl ester	91901-43-8	NI23727
108	186	245	211	60	185	77		**245**	100	95	90	60	40	35	30			13	12	–	1	1	–	–	–	–	–	1	–	–	Phosphine, diphenyl(thiocarbamoyl)-		LI04880
245	244	229	77	247	231	228	209	**245**	100	67	55	41	38	19	18	18		13	12	–	3	–	–	1	–	–	–	–	–	–	5-Chloro-2-aminobenzophenone, hydrazone		MS05262
80	146	65	115	91	216	184	202	**245**	100	78	53	38	28	12	11	4	**0.00**	13	15	–	3	1	–	–	–	–	–	–	–	–	1,3-Heptadiene, 7-azido-3-(phenylthio)-, (E)-	108782-19-0	DD00646
80	146	77	71	115	91	216	217	**245**	100	77	56	50	33	27	11	4	**0.00**	13	15	–	3	1	–	–	–	–	–	–	–	–	1,3-Heptadiene, 7-azido-3-(phenylthio)-, (Z)-	108782-20-3	DD00647
56	84	245	203	57	83	43	42	**245**	100	51	49	27	27	26	20	11		13	15	2	3	–	–	–	–	–	–	–	–	–	Acetamide, N-(2,3-dihydro-1,5-dimethyl-3-oxo-2-phenyl-1H-pyrazol-4-yl)-	83-15-8	NI23728
72	15	77	18	39	51	40	42	**245**	100	41	37	29	21	20	16	15	**8.10**	13	15	2	3	–	–	–	–	–	–	–	–	–	Carbamic acid, dimethyl-, 3-methyl-1-phenyl-1H-pyrazol-5-yl ester	87-47-8	NI23729
158	159	102	116	117	82	81	87	**245**	100	98	85	76	72	70	70	66	**0.00**	13	27	3	1	–	–	–	–	–	–	–	–	–	L-Threonine, N,O-bis(isopropyl)-, isopropyl ester	56771-56-3	MS05263
77	245	244	105	228	246	247	51	**245**	100	97	79	46	42	39	36	33		14	12	1	1	–	–	1	–	–	–	–	–	–	Benzophenone, 5-chloro-2-(methylamino)-	1022-13-5	NI23732
246	105	245	244	248	247	168	210	**245**	100	67	37	32	29	26	20	16		14	12	1	1	–	–	1	–	–	–	–	–	–	Benzophenone, 5-chloro-2-(methylamino)-	1022-13-5	NI23731
245	244	77	246	228	105	247	193	**245**	100	75	44	38	37	33	32	21		14	12	1	1	–	–	1	–	–	–	–	–	–	Benzophenone, 5-chloro-2-(methylamino)-	1022-13-5	NI23730
105	77	104	31	51	32	52	29	**245**	100	89	85	30	29	22	15	15	**1.36**	14	12	1	1	–	–	1	–	–	–	–	–	–	2H-1,3-Benzoxazine, 6-chloro-3,4-dihydro-3-phenyl-	51892-04-7	NI23733
105	104	245	78	77	247	125	106	**245**	100	50	43	27	17	12	12	11		14	12	1	1	–	–	1	–	–	–	–	–	–	4H-3,1-Benzoxazine, 1,2-dihydro-2-(p-chlorophenyl)-	82085-90-3	NI23734
28	124	213	214	105	76	84	41	**245**	100	62	30	29	29	28	25	18	**6.00**	14	15	1	1	1	–	–	–	–	–	–	–	–	1-Thianaphthalen-4-one, 1,4-dihydro-3-piperidino-		MS05264
217	144	188	145	218	117	130	170	**245**	100	86	62	19	18	14	10	8	**0.60**	14	15	3	1	–	–	–	–	–	–	–	–	–	2H-1,4-Ethanoquinoline-2-carboxylic acid, 3,4-dihydro-3-oxo-, ethyl ester		LI04881
217	144	188	145	218	117	143	130	**245**	100	95	61	20	17	15	14	10	**0.70**	14	15	3	1	–	–	–	–	–	–	–	–	–	2H-1,4-Ethanoquinoline-2-carboxylic acid, 3,4-dihydro-3-oxo-, ethyl ester	34291-56-0	NI23735
245	214	159	158	130	103	144	107	**245**	100	98	97	97	93	88	60	50		14	15	3	1	–	–	–	–	–	–	–	–	–	1H-Indole-3-butanoic acid, 1-methyl-γ-oxo-, methyl ester	40547-43-1	NI23736
245	185	198	170	115	212	156	142	**245**	100	47	25	22	18	10	6	6		14	15	3	1	–	–	–	–	–	–	–	–	–	2-Indolecarboxylic acid, 7-allyl-6-methoxy-, methyl ester		MS05265
144	245	159	145	116	89	214	117	**245**	100	21	12	12	10	10	8	5		14	15	3	1	–	–	–	–	–	–	–	–	–	1H-Indole-3-pentanoic acid, δ-oxo-, methyl ester		LI04882
202	130	217	91	245	35	29	173	**245**	100	47	46	35	33	33	25	22		14	15	3	1	–	–	–	–	–	–	–	–	–	1,3-Isobenzofurandione, 4-(cyclohexylamino)-		LI04883
148	78	90	105	79	103	67	76	**245**	100	98	28	26	24	23	20	14	**0.00**	14	15	3	1	–	–	–	–	–	–	–	–	–	Tricyclo[4.2.0.0^{2,4}]octane, 5-(4-nitrophenoxy)-	56772-08-8	NI23737
73	147	107	244	106	77	111	144	**245**	100	83	81	56	53	43	26	23	**17.00**	14	19	1	1	–	–	–	–	–	1	–	–	–	2,3-Butadienamide, 2-(trimethylsilyl)-N-methyl-N-phenyl-		NI23738
230	213	171	156	83	245	231	187	**245**	100	42	31	19	19	18	17	13		14	19	1	3	–	–	–	–	–	–	–	–	–	3-Buten-2-one, 4-mesityl-, semicarbazone	17014-28-7	NI23739
189	57	118	245	188	82	187	130	**245**	100	56	49	17	17	13	9	8		14	19	1	3	–	–	–	–	–	–	–	–	–	1,3,8-Triazaspiro[4.5]decan-4-one, 3-methyl-1-phenyl-		MS05266
123	108	43	80	165	125	140	65	**245**	100	95	41	37	30	30	16	13	**1.00**	15	19	2	1	–	–	–	–	–	–	–	–	–	Aniline, 2-methoxy-, N-(2-acetylcyclohexylidene)-		NI23740
105	77	99	55	41	51	27	98	**245**	100	82	60	32	29	26	23	19	**3.40**	15	19	2	1	–	–	–	–	–	–	–	–	–	2-Azacyclodecanone, 3-benzoyl-	102222-14-0	NI23741
198	91	129	128	115	141	171	199	**245**	100	68	43	39	37	29	22	17	**0.00**	15	19	2	1	–	–	–	–	–	–	–	–	–	Benzene, 1-(1-nitroethyl)-2-heptynyl-		DD00648
158	245	244	184	186	214	230		**245**	100	50	45	40	35	30	20			15	19	2	1	–	–	–	–	–	–	–	–	–	3,4-Benzolupinic acid, methyl ester		LI04884
116	59	115	146	117	41	145	56	**245**	100	88	82	77	68	59	36	29	**0.50**	15	19	2	1	–	–	–	–	–	–	–	–	–	Carbamic acid, [(1,3-dihydro-2H-inden-2-ylidene)methyl]-, tert-butyl ester	34813-10-0	NI23742
109	93	77	91	60	245	79	67	**245**	100	54	51	47	44	44	42	40		15	19	2	1	–	–	–	–	–	–	–	–	–	Carbamic acid, N-phenyl-, 2-exo-methyl-2-endo-norbornyl ester		MS05267
134	245	135	96	97	244	160	119	**245**	100	63	12	9	9	9	9	7		15	19	2	1	–	–	–	–	–	–	–	–	–	7-Indolizidinone, 6-(4-methoxyphenyl)-		MS05268
106	245	67	173	105	117	104	200	**245**	100	57	40	32	30	26	25	24		15	19	2	1	–	–	–	–	–	–	–	–	–	1,4-Oxazepin-3-one, 2-methyl-5-phenyl-6,7-trimethyleneperhydro-, cis-		MS05269
104	245	55	110	91	41	164	158	**245**	100	68	59	48	46	42	38	33		15	19	2	1	–	–	–	–	–	–	–	–	–	2-Oxazolidinone, 3-cyclohexyl-5-phenyl-	56805-21-1	MS05270
70	230	104	229	43	231	91	42	**245**	100	99	29	25	20	16	15	14	**3.73**	15	24	1	1	–	–	–	–	–	–	–	1	–	2H-1,3,2-Oxazaborine, 2-ethyltetrahydro-4-methyl-3-isopropyl-6-phenyl-	74685-88-4	NI23743
245	205	217	191	44	230	190	189	**245**	100	68	62	36	30	25	25	25		16	11	–	3	–	–	–	–	–	–	–	–	–	Benz[c,d]indole, 1,2-dihydro-2-dicyanomethylene-1-ethyl-		IC03236
42	44	41	39	55	43	77	51	**245**	100	77	73	48	45	42	40	32	**0.00**	16	23	1	1	–	–	–	–	–	–	–	–	–	8-Azabicyclo[4.3.1]decan-10α-ol, 8-methyl-10-phenyl-, syn-	13493-45-3	NI23744
97	112	91	55	98	113	92	56	**245**	100	43	29	11	8	6	3	1	**1.00**	16	23	1	1	–	–	–	–	–	–	–	–	–	2-Azetidinone, 1-benzyl-3-tert-butyl-4,4-dimethyl-	22607-04-1	NI23745
97	112	91	55	98	113	246	245	**245**	100	43	29	11	8	6	1	1		16	23	1	1	–	–	–	–	–	–	–	–	–	2-Azetidinone, 1-benzyl-3-tert-butyl-4,4-dimethyl-		LI04885
54	76	56	105	162	163	202	78	**245**	100	93	83	72	62	54	46	44	**37.00**	16	23	1	1	–	–	–	–	–	–	–	–	–	Benzamide, N-cyclohexyl-N-isopropyl-	52812-84-7	NI23746

MASS TO CHARGE RATIOS								M.W.	INTENSITIES								Parent	C	H	O	N	S	F	Cl	Br	I	Si	P	B	X	COMPOUND NAME	CAS Reg No	No
245	130	131	42	129	41	159	28	**245**	100	96	49	42	38	37	35	35		16	23	1	1	–	–	–	–	–	–	–	–	–	Morpholine, 3,5-dimethyl-4-(1,2,3,4-tetrahydro-2-naphthyl)-	55134-10-6	WI01259
203	188	91	245	131	230	83	57	**245**	100	63	52	43	36	33	33	33		16	23	1	1	–	–	–	–	–	–	–	–	–	2H-Pyrrole, 5-tert-butyl-3,4-dihydro-2,2-dimethyl-3-phenyl-, 1-oxide	50455-41-9	NI23747
121	77	105	91	122	230	180	124	**245**	100	21	20	11	10	8	8	7	**2.00**	16	23	1	1	–	–	–	–	–	–	–	–	–	Pyrrolidine, 3-(2,2-dimethyl-3-methoxy-3-phenylcyclopropan-1-yl)-		NI23748
245	216	246	217	109	108	79	69	**245**	100	17	13	10	5	5	5	5		17	11	1	1	–	–	–	–	–	–	–	–	–	Benz[c]acridin-7(12H)-one		NI23749
245	217	246	189	85	94.5	216	218	**245**	100	26	19	13	9	8	6	5		17	11	1	1	–	–	–	–	–	–	–	–	–	7H-Benz[de]anthracen-7-one, 6-amino-		IC03237
245	246	217	216	51	142	77	244	**245**	100	19	11	9	7	6	6	5		17	11	1	1	–	–	–	–	–	–	–	–	–	Naphth[2,3-b]azet-2(1H)-one, 1-phenyl-	19275-01-5	NI23750
245	246	217	216	100	51	142	77	**245**	100	18	10	9	9	7	6	6		17	11	1	1	–	–	–	–	–	–	–	–	–	Naphth[2,3-b]azet-2(1H)-one, 1-phenyl-		LI04886
245	201	229	100	200	100.5	246	202	**245**	100	58	51	32	31	26	19	12		17	11	1	1	–	–	–	–	–	–	–	–	–	2-Pyrenecarboxamide		IC03238
160	91	118	132	119	174	104	120	**245**	100	48	45	31	25	18	17	15	**4.00**	17	27	–	1	–	–	–	–	–	–	–	–	–	1-Decanamine, N-benzylene-	20172-41-2	NI23751
160	91	118	132	119	41	174	104	**245**	100	49	46	32	26	21	19	17	**4.95**	17	27	–	1	–	–	–	–	–	–	–	–	–	1-Decanamine, N-benzylene-	20172-41-2	NI23752
202	245	244	41	216	55	134	203	**245**	100	84	80	46	28	25	24	23		17	27	–	1	–	–	–	–	–	–	–	–	–	9-Aza[d]homogona-13(14)-ene, (±)-		LI04887
244	245	167	165	166	51	243	246	**245**	100	67	31	26	18	13	12	11		18	15	–	1	–	–	–	–	–	–	–	–	–	Pyridine, 2-(diphenylmethyl)-	3678-70-4	NI23753
167	245	165	244	168	166	51	246	**245**	100	91	37	34	25	23	19	17		18	15	–	1	–	–	–	–	–	–	–	–	–	Pyridine, 4-(diphenylmethyl)-	3678-72-6	NI23754
213	215	211	119	117	141	71	143	**246**	100	65	63	34	34	32	32	31	**2.90**	3	–	–	–	–	–	6	–	–	–	–	–	–	1-Propene, 1,1,2,3,3,3-hexachloro-	1888-71-7	NI23755
69	145	76	246	32	64	44	147	**246**	100	90	76	26	16	11	9	8		3	–	–	–	3	6	–	–	–	–	–	–	–	Carbonotrithioic acid, bis(trifluoromethyl) ester	461-08-5	NI23756
39	89	167	121	119	59	38	28	**246**	100	50	25	19	19	19	19	19	**0.00**	3	4	–	–	1	5	–	1	–	–	–	–	–	Sulphur, (2-bromo-1-propenyl)pentafluoro-		MS05271
131	31	28	44	69	36	85	147	**246**	100	98	92	86	81	51	34	23	**0.00**	4	1	2	–	–	5	2	–	–	–	–	–	–	Butanoic acid, 3,4-dichloropentafluoro-		TR00280
89	69	116	88	70	178	90	140	**246**	100	92	19	11	11	7	5	4	**0.00**	4	3	–	–	–	7	–	–	–	–	2	–	–	2-Butene, 2,3-bis(difluorophosphino)-1,1,1-trifluoro-		DD00649
75	42	169	89	61	103	77	49	**246**	100	93	78	70	65	57	52	35	**0.00**	4	4	2	–	1	5	1	–	–	–	–	–	–	2-Oxetanone, 4-chloro-4-(pentafluorosulphur)methyl-		NI23757
57	121	149	119	151	41	31	169	**246**	100	65	64	64	62	47	40	30	**0.00**	4	8	2	–	–	–	–	2	–	–	–	–	–	1,4-Butanediol, 2,3-dibromo-	20163-90-0	MS05272
102	130	70	161	160	142	189	104	**246**	100	53	23	12	10	9	7	6	**0.00**	4	14	–	4	4	–	–	–	–	–	–	–	–	Carbamodithioic acid, 1,2-ethanediylbis-, diammonium salt	3566-10-7	NI23758
161	31	44	102	60	29	59	46	**246**	100	57	30	25	24	18	15	15	**0.00**	4	14	–	4	4	–	–	–	–	–	–	–	–	Carbamodithioic acid, 1,2-ethanediylbis-, diammonium salt	3566-10-7	NI23759
183	211	185	29	213	140	108	142	**246**	100	89	67	62	61	49	32	32	**7.70**	5	5	1	2	1	–	3	–	–	–	–	–	–	1,2,4-Thiadiazole, 5-ethoxy-3-(trichloromethyl)-	2593-15-9	NI23760
117	31	167	246	248	98	79	93	**246**	100	70	65	49	48	32	31	25		6	–	–	–	–	5	–	1	–	–	–	–	–	Benzene, bromopentafluoro-	344-04-7	NI23762
246	248	167	117	31	98	93	79	**246**	100	98	75	75	20	18	18	13		6	–	–	–	–	5	–	1	–	–	–	–	–	Benzene, bromopentafluoro-	344-04-7	NI23761
117	167	31	246	248	98	93	148	**246**	100	68	68	49	47	33	25	15		6	–	–	–	–	5	–	1	–	–	–	–	–	Benzene, bromopentafluoro-	344-04-7	NI23763
248	246	250	147	111	87	210	252	**246**	100	89	45	33	30	22	11	9		6	2	2	–	–	–	4	–	–	–	–	–	–	1,2-Benzenediol, 3,4,5,6-tetrachloro-	1198-55-6	NI23764
79	71	55	42	54	41	43	175	**246**	100	89	57	53	52	36	31	24	**0.00**	6	14	6	–	2	–	–	–	–	–	–	–	–	1,4-Butanediol, dimethanesulfonate	55-98-1	NI23765
79	71	55	54	42	109	175	41	**246**	100	68	51	39	34	29	28	28	**0.00**	6	14	6	–	2	–	–	–	–	–	–	–	–	1,4-Butanediol, dimethanesulphonate	55-98-1	NI23766
88	28	60	125	69	89	18	61	**246**	100	31	26	25	22	21	17	16	**6.23**	6	15	2	–	3	–	–	–	–	–	1	–	–	Phosphorodithioic acid, S-[2-(ethylthio)ethyl] O,O-dimethyl ester	640-15-3	NI23767
88	60	29	61	89	125	93	47	**246**	100	42	35	34	33	29	27	26	**3.90**	6	15	2	–	3	–	–	–	–	–	1	–	–	Phosphorodithioic acid, S-[2-(ethylthio)ethyl] O,O-dimethyl ester	640-15-3	NI23768
30	246	74	109	75	63	44	137	**246**	100	69	59	40	38	32	26	25		7	3	6	2	–	–	1	–	–	–	–	–	–	Benzoic acid, 4-chloro-3,5-dinitro-	118-97-8	NI23769
168	167	97	43	169	41	124	153	**246**	100	81	26	26	17	17	14	13	**0.00**	7	7	3	2	–	–	–	1	–	–	–	–	–	2,4,6(1H,3H,5H)-Pyrimidinetrione, 5-(2-bromo-2-propenyl)-	56282-33-8	NI23770
56	87	96	162	91	190	127	97	**246**	100	76	72	58	58	40	40	40	**1.00**	7	7	4	–	–	–	1	–	–	–	–	–	1	Iron, (2-methoxy-π-allyl)tricarbonylchloro-		MS05273
246	248	218	220	147	86	250	62	**246**	100	85	74	68	56	38	28	25		8	1	1	2	–	–	3	–	–	–	–	–	–	1,3-Benzenedicarbonitrile, 5-hydroxy-2,4,6-trichloro-		DD00650
86	246	248	62	218	220	147	36	**246**	100	98	88	71	68	64	54	51		8	1	1	2	–	–	3	–	–	–	–	–	–	1,3-Benzenedicarbonitrile, 6-hydroxy-2,4,5-trichloro-	28343-61-5	NI23771
246	104	131	69	52	226	45	27	**246**	100	62	59	54	52	43	37	33		8	5	–	4	1	3	–	–	–	–	–	–	–	Pteridine, 2-(methylthio)-4-(trifluoromethyl)-		LI04888
246	104	131	69	52	226	45	46	**246**	100	60	59	54	50	41	37	32		8	5	–	4	1	3	–	–	–	–	–	–	–	Pteridine, 2-(methylthio)-4-(trifluoromethyl)-	32710-61-5	NI23772
105	232	77	91	51	65	120		**246**	100	35	21	18	13	4	1		**0.00**	8	7	1	–	–	–	–	–	1	–	–	–	–	Benzophenone, α-iodo-	4636-16-2	NI23773
166	167	80	82	39	137	81	79	**246**	100	50	50	49	25	21	20	20	**4.00**	8	7	4	–	–	–	–	1	–	–	–	–	–	Benzaldehyde, 6-(bromomethyl)-2,3,4-trihydroxy-	75676-82-3	NI23774
43	97	206	204	175	177	131	159	**246**	100	33	25	25	16	15	14	13	**0.00**	8	7	4	–	–	–	–	1	–	–	–	–	–	2-Cyclohexen-1-one, 4-acetoxy-2-bromo-5,6-epoxy-, (4S,5R,6R)-		DD00651
43	69	53	85	125	97	55	93	**246**	100	26	10	7	3	3	3	2	**0.00**	8	7	4	–	–	–	–	1	–	–	–	–	–	2H-Pyran-2,4-dione, 3-acetyl-3-bromo-3,4-dihydro-6-methyl-		LI04889
167	43	248	246	125	69	178	176	**246**	100	68	26	26	18	14	12	12		8	7	4	–	–	–	–	1	–	–	–	–	–	2H-Pyran-2-one, 3-acetyl-5-bromo-4-hydroxy-6-methyl-		LI04890
167	43	69	111	153	139	97	55	**246**	100	63	25	21	20	20	20	20	**0.00**	8	7	4	–	–	–	–	1	–	–	–	–	–	2H-Pyran-2-one, 3-acetyl-6-bromomethyl-4-hydroxy-		LI04891
167	153	43	85	69	151	93	168	**246**	100	99	99	62	29	18	17	16	**10.00**	8	7	4	–	–	–	–	1	–	–	–	–	–	2H-Pyran-2-one, 3-bromoacetyl-4-hydroxy-6-methyl-		LI04892
204	206	205	163	161	207	70	120	**246**	100	96	45	42	41	38	36	36	**18.77**	8	11	2	2	–	–	–	1	–	–	–	–	–	Uracil, 5-bromo-3-isopropyl-6-methyl-	314-42-1	NI23775
80	139	246	65	98	109	108	218	**246**	100	35	33	29	23	22	22	21		8	11	5	2	–	–	–	–	–	–	1	–	–	Phosphoramidic acid, ethyl p-nitrophenyl diester		MS05274
127	177	246	227	128	178	107	207	**246**	100	73	37	8	7	6	6	4		9	5	–	–	–	7	–	–	–	–	–	–	–	Benzene, heptafluoroisopropyl-	378-34-7	NI23776
145	146	127	96	125	119	246	147	**246**	100	88	36	19	13	10	9	9		9	5	–	–	–	7	–	–	–	–	–	–	–	Bicyclo[2.2.2]oct-5-ene, 1,4,7,7,8,8-hexafluoro-2-fluoromethylene-, (Z)-		NI23777
145	146	127	125	119	246	207	126	**246**	100	81	36	12	10	8	5	5		9	5	–	–	–	7	–	–	–	–	–	–	–	Bicyclo[2.2.2]oct-5-ene, 1,4,7,7,8,8-hexafluoro-2-fluoromethylene-, (Z)-		NI23778
177	42	228	187	160	91	146	53	**246**	100	68	53	46	24	24	18	15	**1.00**	9	9	1	4	–	3	–	–	–	–	–	–	–	4-Pteridinol, 3,4-dihydro-6,7-dimethyl-4-(trifluoromethyl)-	23658-22-2	NI23779
79	51	167	111	246	95	77	68	**246**	100	57	55	37	28	19	18	18		9	10	6	–	1	–	–	–	–	–	–	–	–	Benzoic acid, 3-methoxy-4-[(methylsulphonyl)oxy]-	67258-10-0	NI23780
91	119	246	65	92	41	117	51	**246**	100	23	22	12	10	10	5	5		9	11	–	–	–	–	–	–	1	–	–	–	–	Benzene, (3-iodopropyl)-	4119-41-9	NI23781
91	246	119	92	65	41	117	39	**246**	100	44	20	9	9	6	5	5		9	11	–	–	–	–	–	–	1	–	–	–	–	Benzene, (3-iodopropyl)-	4119-41-9	LI04893

MASS TO CHARGE RATIOS								M.W.	INTENSITIES								Parent	C	H	O	N	S	F	Cl	Br	I	Si	P	B	X	COMPOUND NAME	CAS Reg No	No
189	105	231	246	191	141	233	188	**246**	100	71	69	39	34	30	27	20		9	11	2	2	1	–	1	–	–	–	–	–	–	Acetamide, 4-amino-3-chloro-5-(methylsulphonyl)-N-phenyl-		NI23782
117	73	45	130	99	43	87	71	**246**	100	57	42	40	25	23	18	14	**2.00**	9	11	5	2	–	1	–	–	–	–	–	–	–	Uridine, 2′-deoxy-5-fluoro-	50-91-9	NI23783
117	73	45	130	99	44	87	131	**246**	100	55	42	40	25	24	18	13	**2.00**	9	11	5	2	–	1	–	–	–	–	–	–	–	Uridine, 2′-deoxy-5-fluoro-	50-91-9	LI04894
28	73	143	27	42	43	55	29	**246**	100	60	57	30	27	26	23	23	**0.00**	9	14	6	2	–	–	–	–	–	–	–	–	–	Uridine, 5,6-dihydro-	5627-05-4	NI23784
139	184	246	140	211	71	69	141	**246**	100	78	70	70	64	60	50	40		9	15	2	4	–	–	1	–	–	–	–	–	–	1,3,5-Triazin-2-amine, 4-(2-chloroethoxy)-N,N-dimethyl-6-ethoxy-	69825-86-1	NI23785
140	246	69	70	71	83	184	182	**246**	100	62	58	54	41	40	32	24		9	15	2	4	–	–	1	–	–	–	–	–	–	1,3,5-Triazin-2(1H)-one, 1-(2-chloroethyl)-4-dimethylamino-6-ethoxy-	69825-97-4	NI23786
75	73	116	131	147	26	231	45	**246**	100	91	63	43	42	29	26	22	**7.54**	9	20	3	1	–	–	–	–	–	2	–	–	–	Malonyldiamide, bis(trimethylsilyl)-		NI23787
246	166	186	199	140	102	76		**246**	100	54	50	39	37	27	24			10	6	2	4	1	–	–	–	–	–	–	–	–	Naphtho[2,3-c][1,2,5]thiadiazol-4,9-dione, dioxime		NI23788
70	162	55	190	93	246	119	163	**246**	100	85	68	18	14	12	10	8		10	7	4	–	–	–	–	–	–	–	–	–	1	Manganese, acetylcyclopentadienyltricarbonyl-		LI04895
148	183	213	248	122	157	118	48	**246**	100	93	58	39	29	6	6	6	**0.00**	10	8	–	–	–	–	2	–	–	–	–	–	1	Titanium, di-(π-cyclopentadienyl)dichloro-		LI04896
106	107	70	246	77	69	79	51	**246**	100	70	43	27	25	24	17	11		10	9	2	2	–	3	–	–	–	–	–	–	–	Urea, N-methyl-N-phenyl-N′-(trifluoroacetyl)-		DD00652
161	218	58	162	187	160	114	199	**246**	100	17	12	9	6	6	6	5	**0.00**	10	9	2	2	–	3	–	–	–	–	–	–	–	Urea, N-methyl-N′-[4-(trifluoroacetyl)phenyl]-	57346-65-3	NI23789
162	68	121	67	53	94	93	54	**246**	100	63	56	31	30	28	28	28	**12.00**	10	10	2	6	–	–	–	–	–	–	–	–	–	Pteridine, 2,4-diacetylamino-		LI04897
100	146	59	29	101	72	246	28	**246**	100	55	36	28	25	22	20	19		10	14	3	–	2	–	–	–	–	–	–	–	–	4H-1,3-Dithiin-4-one, 2-[1-(acetyloxy)ethyl]-2,6-dimethyl-	55044-92-3	NI23790
142	246	186	143	113	109	247	112	**246**	100	55	30	30	13	10	8	8		10	14	3	–	2	–	–	–	–	–	–	–	–	Ethanol, 2-[4-hydroxy-2-(2-mercaptoethylthio)phenoxy]-	93885-82-6	NI23791
114	96	101	111	156	157	202	246	**246**	100	76	49	39	28	20	6	3		10	14	7	–	–	–	–	–	–	–	–	–	–	β-D-Glucopyranose, 1,6-anhydro-2,4-di-O-acetyl-	23740-49-0	NI23792
43	103	85	145	127	187	101	102	**246**	100	15	14	9	6	5	5	4	**0.00**	10	14	7	–	–	–	–	–	–	–	–	–	–	D-Glycero-tetrulose, 1,3,4-tri-O-acetyl-	78215-66-4	NI23793
43	115	143	55	41	56	42	44	**246**	100	23	16	5	4	3	3	2	**0.00**	10	14	7	–	–	–	–	–	–	–	–	–	–	Propanoic acid, 3-(acetyloxy)-, anhydride	55656-58-1	NI23794
109	137	246	110	28	81	18	63	**246**	100	60	46	23	19	14	14	12		10	15	1	–	2	–	–	–	–	–	1	–	–	Phosphonodithioic acid, ethyl-, O-ethyl S-phenyl ester	944-22-9	NI23795
109	137	246	110	63	81	77	65	**246**	100	49	34	20	14	12	12	11		10	15	1	–	2	–	–	–	–	–	1	–	–	Phosphonodithioic acid, ethyl-, O-ethyl S-phenyl ester	944-22-9	NI23796
47	91	93	92	75	77	121	179	**246**	100	89	63	52	37	28	25	24	**0.00**	10	15	2	–	–	–	–	1	–	–	–	–	–	2-Cyclopenten-1-ol, 1-(1-bromo-2-ethoxycycloprop-1-yl)-		MS05275
87	43	79	109	153	99	81	80	**246**	100	67	21	14	11	11	11	11	**0.00**	10	15	2	–	–	–	–	1	–	–	–	–	–	1,3-Dioxolane, 2-methyl-2-[(1E,3Z-E)-5-bromo-4-methyl-1,3-pentadienyl]-	80283-28-9	NI23797
99	55	27	18	167	39	41	67	**246**	100	30	21	20	16	15	14	13	**10.91**	10	15	2	–	–	–	–	1	–	–	–	–	–	Spiro[bicyclo[2.2.2]octane-2,2′-[1,3]dioxolane], 3-bromo-	55956-34-8	NI23798
161	126	90	163	120	92	127	211	**246**	100	44	40	34	18	16	15	14	**3.78**	10	15	3	2	–	–	1	–	–	–	–	–	–	Uracil, 5-chloro-6-cyclopentyloxy-5,6-dihydro-5-methyl-		NI23799
55	89	73	231	175	59	75	45	**246**	100	58	45	44	25	23	18	15	**1.50**	10	18	5	–	–	–	–	–	–	1	–	–	–	2-Pentenedioic acid, 2-[(trimethylsilyl)oxy]-, dimethyl ester	55590-96-0	NI23800
147	73	231	45	148	149	75	133	**246**	100	45	21	16	15	8	8	5	**0.80**	10	22	3	–	–	–	–	–	–	2	–	–	–	2-Butenoic acid, 2-[(trimethylsilyl)oxy]-, trimethylsilyl ester	55590-70-0	NI23801
147	73	231	148	45	149	66	75	**246**	100	45	28	18	13	9	7	5	**0.00**	10	22	3	–	–	–	–	–	–	2	–	–	–	2-Butenoic acid, 2-[(trimethylsilyl)oxy]-, trimethylsilyl ester		MS05276
73	147	231	157	45	148	75	232	**246**	100	97	73	42	32	23	23	22	**4.00**	10	22	3	–	–	–	–	–	–	2	–	–	–	3-Butenoic acid, 3-[(trimethylsilyl)oxy]-, trimethylsilyl ester		MS05277
147	73	231	157	45	75	148	232	**246**	100	91	73	28	27	18	17	15	**2.74**	10	22	3	–	–	–	–	–	–	2	–	–	–	3-Butenoic acid, 3-[(trimethylsilyl)oxy]-, trimethylsilyl ester	56272-64-1	NI23802
147	73	231	45	157	148	28	75	**246**	100	79	65	26	21	16	16	16	**2.52**	10	22	3	–	–	–	–	–	–	2	–	–	–	3-Butenoic acid, 3-[(trimethylsilyl)oxy]-, trimethylsilyl ester	56272-64-1	NI23803
177	246	135	231	115	85	91	147	**246**	100	99	55	45	45	38	23	19		11	9	1	–	1	3	–	–	–	–	–	–	–	3-Buten-2-one, 1,1,1-trifluoro-4-mercapto-4-(4-methylphenyl)-	55674-02-7	NI23804
218	97	134	184	56	128	152	100	**246**	100	72	59	48	44	41	20	20	**20.00**	11	10	1	–	1	–	–	–	–	–	–	–	1	Iron, carbonyl(π-3-thienyl)-		LI04898
121	56	55	66	136	190	134	65	**246**	100	75	64	58	51	44	44	42	**1.99**	11	10	3	–	–	–	–	–	–	–	–	–	1	Iron, (3-oxobut-1-enyl)dicarbonyl-π-cyclopentadienyl-	12288-63-0	NI23805
246	216	230	145	131	90	200	147	**246**	100	71	34	30	30	28	26	19		11	10	3	4	–	–	–	–	–	–	–	–	–	2-Pyrazolin-5-one, 4-aminomethylene-3-methyl-1-(4-nitrophenyl)-		NI23806
162	79	107	56	160	246	218	190	**246**	100	55	42	40	37	23	21	21		11	11	3	–	–	–	–	–	–	–	–	–	1	Manganese, tricarbonyl[(1,2,3,4,5-η)-2,4-cyclooctadien-1-yl]-	49626-35-9	NI23807
162	56	160	79	107	218	134	190	**246**	100	48	42	42	39	35	20	18	**17.00**	11	11	3	–	–	–	–	–	–	–	–	–	1	Manganese, tricarbonyl[(1,2,3,5,6-η)-2,5-cyclooctadien-1-yl]-	49626-36-0	NI23808
71	43	111	75	55	44	50	41	**246**	100	84	76	75	72	64	61	58	**6.60**	11	12	2	–	–	–	2	–	–	–	–	–	–	2-Butanone, 1-chloro-1-(4-chlorophenoxy)-3-methyl-		NI23809
41	69	43	77	70	111	141	75	**246**	100	55	28	26	26	25	22	22	**3.00**	11	12	2	–	–	–	2	–	–	–	–	–	–	2-Butanone, 3-chloro-1-(4-chlorophenoxy)-3-methyl-		NI23810
43	100	59	85	129	246	231	101	**246**	100	99	63	50	38	24	24	20		11	18	4	–	1	–	–	–	–	–	–	–	–	α-D-Xylofuranose, 1,2:3,5-di-O-isopropylidene-4-thio-	32848-96-7	NI23811
231	59	173	83	87	115	85	155	**246**	100	69	45	35	33	26	21	11	**2.40**	11	18	6	–	–	–	–	–	–	–	–	–	–	Diethyl (2,3-O-isopropylidene)-L-tartrate		MS05278
173	127	29	128	201	55	100	99	**246**	100	63	49	47	36	36	35	28	**0.34**	11	18	6	–	–	–	–	–	–	–	–	–	–	1,1,2-Ethanetricarboxylic acid, triethyl ester		DO01033
100	43	74	71	45	84	58	126	**246**	100	100	45	30	30	23	23	22	**0.00**	11	18	6	–	–	–	–	–	–	–	–	–	–	D-Glucofuranoside, methyl 2,4-di-O-acetyl-3,6-dideoxy-		NI23812
99	43	45	159	84	55	71	83	**246**	100	100	40	25	12	12	10	9	**0.00**	11	18	6	–	–	–	–	–	–	–	–	–	–	D-Glucofuranoside, methyl 2,5-di-O-acetyl-3,6-dideoxy-		NI23813
100	43	111	142	153	71	101	143	**246**	100	50	33	24	14	8	7	4	**0.00**	11	18	6	–	–	–	–	–	–	–	–	–	–	α-D-erythro-Hexopyranoside, methyl 2,3-dideoxy-, diacetate	64551-84-4	NI23814
43	111	56	139	72	154	41	57	**246**	100	28	20	19	17	16	16	15	**0.00**	11	18	6	–	–	–	–	–	–	–	–	–	–	2H-Pyran-2-one, 5,6-dihydro-4,5,6-trimethyl-, (5R,6R)-		NI23815
73	115	75	128	88	86	101	143	**246**	100	85	47	27	23	22	19	11	**0.60**	11	20	5	1	–	–	–	–	–	–	–	–	–	α-D-Glucosaminide, methyl N-acetyl-		LI04899
73	69	75	41	146	45	171	55	**246**	100	62	58	48	46	32	30	23	**0.00**	11	22	4	–	–	–	–	–	–	1	–	–	–	1,4-Butanedicarboxylic acid, 2,2-dimethyl-, methyl trimethylsilyl diester		MS05279
173	175	73	75	55	57	174	157	**246**	100	94	57	54	20	13	13	13	**0.52**	11	22	4	–	–	–	–	–	–	1	–	–	–	1,4-Butanedioic acid, sec-butyl trimethylsilyl diester		NI23816
73	75	69	231	29	84	143	45	**246**	100	59	41	20	19	18	15	15	**0.00**	11	22	4	–	–	–	–	–	–	1	–	–	–	1,3-Propanedioic acid, 2-ethyl-2-methyl-, ethyl trimethylsilyl diester		NI23817
121	91	61	41	107	59	77	39	**246**	100	53	15	15	14	14	9	9	**0.00**	11	22	4	–	–	–	–	–	–	1	–	–	–	Silane, trimethoxy[2-(7-oxabicyclo[4.1.0]hept-3-yl)ethyl]-	3388-04-3	NI23818
102	56	45	41	74	84	43	55	**246**	100	30	27	20	19	18	18	14	**5.60**	11	22	4	2	–	–	–	–	–	–	–	–	–	Carbamic acid, 1,5-pentanediylbis-, diethyl ester	55148-17-9	NI23819
102	58	57	101	56	115	73	72	**246**	100	62	53	46	32	31	26	26	**7.78**	11	22	4	2	–	–	–	–	–	–	–	–	–	1,4,7-Trioxa-10,12-diazacyclotetradecan-11-one, 10,12-dimethyl-	83540-66-3	NI23820
73	143	147	103	75	45	144	67	**246**	100	44	36	26	22	14	11	11	**1.17**	11	26	2	–	–	–	–	–	–	2	–	–	–	1,4-Butanediol, 1,4-bis(trimethylsilyl)-3-methylene-		NI23821
246	201	247	123	69	45	93	214	**246**	100	23	16	16	16	15	14	9		12	6	–	–	3	–	–	–	–	–	–	–	–	Benzo[1,2-b:3,4-b′:5,6-b″]trithiophene		DD00653
246	45	93	201	69	247	170	169	**246**	100	31	27	26	23	16	13	12		12	6	–	–	3	–	–	–	–	–	–	–	–	Benzo[1,2-b:3,4-b′:6,5-b″]trithiophene		DD00654
246	199	200	124	125	216	218	184	**246**	100	55	54	49	47	40	40	38		12	10	2	2	1	–	–	–	–	–	–	–	–	Aniline, 4-[(4-nitrophenyl)thio]-	101-59-7	NI23822

MASS TO CHARGE RATIOS	M.W.	INTENSITIES	Parent	C	H	O	N	S	F	Cl	Br	I	Si	P	B	X	COMPOUND NAME	CAS Reg No	No
246 124 200 199 247 80 167 99	**246**	100 49 32 25 20 18 14 12		12	10	2	2	1	–	–	–	–	–	–	–	–	Aniline, 4-[(4-nitrophenyl)thio]-	101-59-7	NI23823
218 201 246 29 127 245 128 50	**246**	100 99 71 51 42 29 29 25		12	10	4	2	–	–	–	–	–	–	–	–	–	2-Propenoic acid, 2-cyano-3-(3-nitrophenyl)-, ethyl ester	18925-00-3	NI23824
218 201 246 199 29 155 30 127	**246**	100 76 70 61 49 33 25 24		12	10	4	2	–	–	–	–	–	–	–	–	–	2-Propenoic acid, 2-cyano-3-(4-nitrophenyl)-, ethyl ester	2286-33-1	NI23825
217 77 218 199 51 78 94 47	**246**	100 69 58 58 42 24 19 16		12	11	2	–	–	–	–	–	–	1	1	–	–	Phosphinic acid, diphenyl-		IC03239
231 117 246 115 91 232 77 133	**246**	100 47 29 24 24 12 10 9		12	13	2	–	–	3	–	–	–	–	–	–	–	p-Cymen-2-ol, trifluoroacetate		LI04900
231 117 91 246 115 105 133 77	**246**	100 52 34 32 30 18 13 13		12	13	2	–	–	3	–	–	–	–	–	–	–	p-Cymen-3-ol, trifluoroacetate		LI04901
231 117 246 133 91 132 118 232	**246**	100 97 52 34 32 29 22 20		12	13	2	–	–	3	–	–	–	–	–	–	–	p-Cymen-7-ol, trifluoroacetate		LI04902
78 246 150 247 168 73 121 56	**246**	100 80 20 12 10 10 8 8		12	14	2	–	–	–	–	–	–	–	–	–	1	Ferrocene, 1,1′-bis(hydroxymethyl)-	1291-48-1	NI23826
200 246 143 201 145 144 247 197	**246**	100 72 19 17 12 12 9 6		12	14	2	4	–	–	–	–	–	–	–	–	–	4-Pyrazolecarboxylic acid, 3(5)-amino-5(3)-phenylamino-, ethyl ester	61485-23-2	NI23827
43 41 121 42 39 18 28 27	**246**	100 52 44 32 32 31 29 29	**0.00**	12	16	1	–	–	–	2	–	–	–	–	–	–	Phenol, 6-butyl-3,5-dichloro-2,4-dimethyl-		IC03240
231 233 169 246 171 141 41 248	**246**	100 63 43 20 16 16 15 13		12	16	1	–	–	–	2	–	–	–	–	–	–	Phenol, 4-tert-butyl-β,2-dichloro-4-ethoxy-		DO01034
145 41 69 97 55 106 143 246	**246**	100 53 50 39 39 31 25 23		12	22	1	–	2	–	–	–	–	–	–	–	–	6,10-Dithiaspiro[4.5]decan-1-ol, 2-methyl-2-(n-propyl)-	92640-85-2	NI23828
89 145 57 117 173 90 41 71	**246**	100 57 33 29 20 19 19 17	**0.00**	12	22	5	–	–	–	–	–	–	–	–	–	–	Butanedioic acid, hydroxy-, dibutyl ester	1587-18-4	NI23829
173 145 117 247 191 135 219 174	**246**	100 84 73 50 29 18 13 9	**0.00**	12	22	5	–	–	–	–	–	–	–	–	–	–	Butanedioic acid, hydroxy-, dibutyl ester	1587-18-4	NI23830
127 113 81 73 131 145 67 55	**246**	100 85 39 38 35 25 25 25	**0.00**	12	22	5	–	–	–	–	–	–	–	–	–	–	Dodecanedioic acid, 6-hydroxy-	50870-54-7	MS05280
73 59 133 99 103 246 191 231	**246**	100 26 19 18 13 7 7 6		12	26	3	–	–	–	–	–	–	1	–	–	–	1-Butene, 1-[dimethyl[2-(2-ethoxyethoxy)ethoxy]silyl]-, (E)-		DD00655
131 73 75 55 69 231 113 29	**246**	100 97 52 35 20 19 16 14	**0.00**	12	26	3	–	–	–	–	–	–	1	–	–	–	Hexanoic acid, 5-methyl-5-[(trimethylsilyl)oxy]-, ethyl ester		NI23831
73 131 75 103 217 188 231 29	**246**	100 69 14 13 13 11 10 10	**0.00**	12	26	3	–	–	–	–	–	–	1	–	–	–	Pentanoic acid, 2-ethyl-3-[(trimethylsilyl)oxy]-, ethyl ester		NI23832
73 131 75 231 103 188 133 217	**246**	100 80 28 22 16 15 12 11	**0.00**	12	26	3	–	–	–	–	–	–	1	–	–	–	Pentanoic acid, 2-ethyl-3-[(trimethylsilyl)oxy]-, ethyl ester		NI23833
43 57 41 118 119 162 105 56	**246**	100 73 53 45 38 32 25 25	**6.67**	12	27	–	–	–	–	–	–	–	–	–	–	1	Arsine, tributyl-	5852-58-4	NI23835
43 55 57 29 28 27 41 105	**246**	100 93 87 86 82 64 42 37	**15.60**	12	27	–	–	–	–	–	–	–	–	–	–	1	Arsine, tributyl-	5852-58-4	NI23834
161 203 119 91 89 41 27 77	**246**	100 51 33 13 9 4 4 3	**0.00**	12	28	–	–	–	–	–	–	–	–	–	–	1	Germane, tetrapropyl-		MS05281
217 189 161 66 80 218 105 133	**246**	100 80 46 24 22 21 20 16	**0.00**	12	30	1	–	–	–	–	–	–	2	–	–	–	Disiloxane, hexaethyl-		NI23836
246 218 248 139 91 220 69 183	**246**	100 46 39 28 21 17 15 11		13	7	1	–	1	–	1	–	–	–	–	–	–	9H-Thioxanthen-9-one, 2-chloro-	86-39-5	NI23837
77 153 218 51 109 18 65 28	**246**	100 71 24 24 21 20 13 13	**2.00**	13	10	1	–	2	–	–	–	–	–	–	–	–	Carbonodithioic acid, O,S-diphenyl ester	13509-35-8	NI23838
109 18 218 28 64 65 110 77	**246**	100 54 45 35 27 26 19 18	**7.00**	13	10	1	–	2	–	–	–	–	–	–	–	–	Carbonodithioic acid, S,S-diphenyl ester	13509-36-9	NI23839
134 40 58 45 90 89 246 141	**246**	100 14 8 8 7 7 3 3		13	10	3	–	1	–	–	–	–	–	–	–	–	Benzo[b]cyclobuta[d]thiophene-1,2-dicarboxylic anhydride, 1,2,2a,7b-tetrahydro-1-methyl-		LI04903
134 40 135 63 46 136 90 89	**246**	100 12 10 7 7 6 6 6	**3.00**	13	10	3	–	1	–	–	–	–	–	–	–	–	Benzo[b]cyclobuta[d]thiophene-1,2-dicarboxylic anhydride, 1,2,2a,7b-tetrahydro-1-methyl-	34002-19-2	NI23840
137 246 229 81 110 69 108 53	**246**	100 37 24 17 12 12 8 8		13	10	5	–	–	–	–	–	–	–	–	–	–	Benzophenone, 2,4,2′,4′-tetrahydroxy-		MS05282
246 231 147 174 247 160 175 232	**246**	100 84 21 19 16 15 14 13		13	10	5	–	–	–	–	–	–	–	–	–	–	2H-Furo[2,3-h][1]benzopyran-2-one, 5,6-dimethoxy-	131-12-4	NI23841
246 230 247 188 231 203 160 175	**246**	100 80 8 7 6 6 6 5		13	10	5	–	–	–	–	–	–	–	–	–	–	7H-Furo[3,2-g][1]benzopyran-7-one, 4,9-dimethoxy-	482-27-9	NI23842
231 246 247 188 160 203 175 232	**246**	100 91 18 17 15 14 14 13		13	10	5	–	–	–	–	–	–	–	–	–	–	7H-Furo[3,2-g][1]benzopyran-7-one, 4,9-dimethoxy-	482-27-9	NI23843
231 246 175 188 203 160 147 232	**246**	100 81 30 23 22 18 17 14		13	10	5	–	–	–	–	–	–	–	–	–	–	7H-Furo[3,2-g][1]benzopyran-7-one, 4,9-dimethoxy-	482-27-9	NI23844
203 246 160 76 231 204 247 51	**246**	100 80 62 21 20 14 13 11		13	10	5	–	–	–	–	–	–	–	–	–	–	7H-Furo[3,2-g][1]benzopyran-7-one, 5,6-dimethoxy-	27272-64-6	NI23845
203 246 160 76 231 204 132 104	**246**	100 82 63 22 19 14 13 11		13	10	5	–	–	–	–	–	–	–	–	–	–	7H-Furo[3,2-g][1]benzopyran-7-one, 5,6-dimethoxy-		LI04904
231 246 163 232 191 247 38 96	**246**	100 37 18 13 8 5 4 3		13	10	5	–	–	–	–	–	–	–	–	–	–	5H-Furo[3,2-g][1]benzopyran-5-one, 4-hydroxy-9-methoxy-7-methyl-		MS05283
231 246 163 232 191 247 96 133	**246**	100 38 17 13 8 6 4 2		13	10	5	–	–	–	–	–	–	–	–	–	–	5H-Furo[3,2-g][1]benzopyran-5-one, 4-hydroxy-9-methoxy-7-methyl-		LI04905
246 217 245 171 200 160 247 187	**246**	100 58 25 23 20 17 14 14		13	10	5	–	–	–	–	–	–	–	–	–	–	5H-Furo[3,2-g][1]benzopyran-5-one, 7-(hydroxymethyl)-4-methoxy-		MS05284
58 217 246 189 146 218 247 59	**246**	100 65 56 21 12 11 10 9		13	10	5	–	–	–	–	–	–	–	–	–	–	2-Naphthalenecarboxylic acid, 4,6-dihydroxy-3-hydroxymethyl-7-methoxy-, γ-lactone		MS05285
127 93 129 28 119 246 65 91	**246**	100 61 32 28 26 18 16 13		13	11	1	2	–	–	1	–	–	–	–	–	–	Urea, N-(2-chlorophenyl)-N′-phenyl-		MS05286
61 93 127 246 129 119 153 65	**246**	100 51 50 16 16 13 11 11		13	11	1	2	–	–	1	–	–	–	–	–	–	Urea, N-(3-chlorophenyl)-N′-phenyl-		MS05287
136 246 108 84 137 69 247 217	**246**	100 45 25 17 15 8 7 7		13	14	1	2	1	–	–	–	–	–	–	–	–	4H-5,6-Benzo-1,3-thiazin-4-one, 2-piperidinyl-		NI23846
246 245 43 231 203 184 247 189	**246**	100 31 28 27 22 22 19 16		13	14	3	2	–	–	–	–	–	–	–	–	–	Aniline, N-(2-acetylcyclopentylidene)-4-nitro-		NI23847
97 107 79 77 203 105 91 110	**246**	100 52 19 18 9 7 6 5	**0.06**	13	14	3	2	–	–	–	–	–	–	–	–	–	1,3-Dioxan, 5-hydroxy-2-phenyl-4-(pyrazol-3(5)-yl)-(2R,4S,5R)-		NI23848
200 91 65 39 228 218 246	**246**	100 96 56 25 14 10 3		13	14	3	2	–	–	–	–	–	–	–	–	–	Formamide, N-benzyl-N-[(5-hydroxy-2-oxo-3-pyrrolidinylidene)methyl]-		LI04906
200 91 246 201 132 77 67 51	**246**	100 47 21 20 20 20 14 8		13	14	3	2	–	–	–	–	–	–	–	–	–	1H-Pyrazole-4-carboxylic acid, 4,5-dihydro-3-methyl-5-oxo-1-phenyl-, ethyl ester	29711-06-6	LI04907
200 91 246 201 132 77 67 51	**246**	100 46 18 18 18 18 11 6		13	14	3	2	–	–	–	–	–	–	–	–	–	1H-Pyrazole-4-carboxylic acid, 4,5-dihydro-3-methyl-5-oxo-1-phenyl-, ethyl ester	29711-06-6	MS05288
173 246 92 172 145 174 200 65	**246**	100 31 30 25 19 12 11 10		13	14	3	2	–	–	–	–	–	–	–	–	–	4H-Pyrido[1,2-a]pyrimidine-3-acetic acid, 6-methyl-4-oxo-, ethyl ester	50609-57-9	NI23849
173 246 92 174 65 55 201 119	**246**	100 27 16 13 7 5 4 4		13	14	3	2	–	–	–	–	–	–	–	–	–	4H-Pyrido[1,2-a]pyrimidine-3-acetic acid, 7-methyl-4-oxo-, ethyl ester	64399-34-4	NI23850
173 246 92 174 65 55 247 119	**246**	100 23 16 13 6 5 4 4		13	14	3	2	–	–	–	–	–	–	–	–	–	4H-Pyrido[1,2-a]pyrimidine-3-acetic acid, 8-methyl-4-oxo-, ethyl ester	50609-56-8	NI23851
173 246 92 174 65 145 55 39	**246**	100 25 21 13 9 5 5 5		13	14	3	2	–	–	–	–	–	–	–	–	–	4H-Pyrido[1,2-a]pyrimidine-3-acetic acid, 9-methyl-4-oxo-, ethyl ester	50834-58-7	NI23852

MASS TO CHARGE RATIOS								M.W.	INTENSITIES								Parent	C	H	O	N	S	F	Cl	Br	I	Si	P	B	X	COMPOUND NAME	CAS Reg No	No
172	200	106	246	201	173	146	79	**246**	100	45	31	24	16	16	16	11		13	14	3	2	–	–	–	–	–	–	–	–	–	**4H-Pyrido[1,2-a]pyrimidine-3-carboxylic acid, 6,8-dimethyl-4-oxo-, ethyl ester**	16867-54-2	**NI23853**
172	159	78	173	246	201	200	144	**246**	100	83	47	37	36	21	12	12		13	14	3	2	–	–	–	–	–	–	–	–	–	**4H-Pyrido[1,2-a]pyrimidine-3-propanoic acid, 4-oxo-, ethyl ester**	64399-37-7	**NI23854**
217	117	118	146	77	103	91	115	**246**	100	38	32	27	20	19	16	15	**11.50**	13	14	3	2	–	–	–	–	–	–	–	–	–	**2,4,6(1H,3H,5H)-Pyrimidinetrione, 5-ethyl-1-methyl-5-phenyl-**	115-38-8	**NI23855**
218	117	146	118	219	28	161	39	**246**	100	20	19	17	13	13	10	10	**7.71**	13	14	3	2	–	–	–	–	–	–	–	–	–	**2,4,6(1H,3H,5H)-Pyrimidinetrione, 5-ethyl-1-methyl-5-phenyl-**	115-38-8	**NI23857**
247	248	171	219	249	220	172		**246**	100	16	16	14	2	2	1		**0.00**	13	14	3	2	–	–	–	–	–	–	–	–	–	**2,4,6(1H,3H,5H)-Pyrimidinetrione, 5-ethyl-1-methyl-5-phenyl-**	115-38-8	**NI23856**
173	145	174	246	200	201	92	172	**246**	100	47	44	36	27	15	15	13		13	14	3	2	–	–	–	–	–	–	–	–	–	**2(1H)-Quinoxalinone, 3-(1-ethoxycarbonylethyl)-**		**MS05289**
173	200	145	172	201	144	90	92	**246**	100	64	59	55	47	28	27	24	**23.00**	13	14	3	2	–	–	–	–	–	–	–	–	–	**2(1H)-Quinoxalinone, 3-(2-ethoxycarbonylethyl)-**		**MS05290**
200	145	173	246	144	172	201	146	**246**	100	55	43	41	36	35	26	26		13	14	3	2	–	–	–	–	–	–	–	–	–	**2(1H)-Quinoxalinone, 3-(ethoxycarbonylmethylene)-6-methyl-**		**MS05291**
130	43	187	28	131	77	44	246	**246**	100	24	16	15	11	10	9	8		13	14	3	2	–	–	–	–	–	–	–	–	–	**L-Tryprophan, Nα-acetyl-**		**MS05292**
130	187	131	77	43	103	117	129	**246**	100	11	11	9	6	6	4	4	**2.50**	13	14	3	2	–	–	–	–	–	–	–	–	–	**DL-Tryptophan, N-acetyl-**	87-32-1	**NI23858**
172	45	30	57	246	75	200	74	**246**	100	16	8	7	4	3	2	2		13	14	3	2	–	–	–	–	–	–	–	–	–	**L-Tryptophan, N-acetyl-**	1218-34-4	**MS05293**
81	231	41	93	69	43	95	137	**246**	100	70	57	38	38	34	33	30	**19.07**	13	24	–	–	–	–	1	–	–	–	1	–	–	**Phosphinous chloride, isopropyl(1,7,7-trimethylbicyclo[2.2.1]hept-2-yl)-**	74630-17-4	**NI23859**
83	165	55	41	149	246	99	164	**246**	100	50	41	21	7	5	4	3		13	26	2	–	1	–	–	–	–	–	–	–	–	**Cyclohexane, heptylsulphonyl-**		**LI04908**
43	41	60	73	55	57	155	29	**246**	100	91	77	72	70	67	62	60	**0.00**	13	26	4	–	–	–	–	–	–	–	–	–	–	**Decanoic acid, 2,3-dihydroxypropyl ester**	2277-23-8	**WI01260**
98	60	129	73	57	71	155	112	**246**	100	89	82	80	61	47	45	45	**0.00**	13	26	4	–	–	–	–	–	–	–	–	–	–	**Decanoic acid, 2,3-dihydroxypropyl ester**	2277-23-8	**WI01261**
155	98	215	134	74	173	112	172	**246**	100	85	62	60	49	44	40	37	**0.00**	13	26	4	–	–	–	–	–	–	–	–	–	–	**Decanoic acid, 2-hydroxy-1-(hydroxymethyl)ethyl ester**	3376-48-5	**WI01262**
43	155	41	57	55	29	71	27	**246**	100	94	69	62	55	52	48	29	**1.00**	13	26	4	–	–	–	–	–	–	–	–	–	–	**Decanoic acid, 2-hydroxy-1-(hydroxymethyl)ethyl ester**	3376-48-5	**WI01263**
101	69	231	131	73	189	157	87	**246**	100	45	18	7	6	1	1	1	**0.00**	13	26	4	–	–	–	–	–	–	–	–	–	–	**Glycerol, 1-O-(2-methoxyhexyl)-2,3-O-isopropylidene-**		**NI23860**
246	248	176	87	123	88	174	175	**246**	100	62	56	24	23	23	17	16		14	8	–	–	–	–	2	–	–	–	–	–	–	**Anthracene, 9,10-dichloro-**	605-48-1	**NI23861**
246	176	248	88	87	175	247	174	**246**	100	69	68	28	22	20	18	17		14	8	–	–	–	–	2	–	–	–	–	–	–	**9H-Fluorene, 9-(dichloromethylene)-**		**MS05294**
246	176	248	88	247	175	123	105	**246**	100	68	67	28	19	15	15	14		14	8	–	–	–	–	2	–	–	–	–	–	–	**9H-Fluorene, 9-(dichloromethylene)-**	835-17-6	**NI23862**
77	245	246	169	105	51	247	168	**246**	100	85	62	52	49	40	34	22		14	11	2	–	–	–	1	–	–	–	–	–	–	**Acetophenone, 5-chloro-2-hydroxy-4-methyl-**	68751-90-6	**NI23863**
139	91	111	246	90	65	75	107	**246**	100	64	17	13	11	11	10	9		14	11	2	–	–	–	1	–	–	–	–	–	–	**Benzoic acid, 4-chloro-, benzyl ester**		**IC03241**
204	206	246	105	77	205	122	51	**246**	100	33	20	17	15	14	13	9		14	11	2	–	–	–	1	–	–	–	–	–	–	**[1,1′-Biphenyl]-4-ol, 4′-chloro-, acetate**	57396-87-9	**NI23864**
211	246	212	153	152	182	155	127	**246**	100	26	16	13	9	8	8	7		14	11	2	–	–	–	1	–	–	–	–	–	–	**Naphtho[2,1-b]furan-4-carboxylic acid, 1,2-dihydro-2-methyl-, chloride**		**NI23865**
91	121	214	122	92	182	200	123	**246**	100	21	20	16	13	12	10	10	**4.00**	14	14	–	–	2	–	–	–	–	–	–	–	–	**Disulphide, dibenzyl-**	150-60-7	**LI04909**
91	121	214	122	92	65	182	123	**246**	100	22	20	17	14	13	12	10	**3.50**	14	14	–	–	2	–	–	–	–	–	–	–	–	**Disulphide, dibenzyl-**	150-60-7	**NI23866**
123	246	124	247	91	182	108	213	**246**	100	76	17	13	13	8	6	4		14	14	–	–	2	–	–	–	–	–	–	–	–	**Disulphide, di-p-tolyl-**	103-19-5	**LI04910**
123	246	124	91	247	122	121	182	**246**	100	75	15	12	11	8	8	7		14	14	–	–	2	–	–	–	–	–	–	–	–	**Disulphide, di-p-tolyl-**	103-19-5	**NI23867**
123	246	79	124	39	122	121	91	**246**	100	39	20	13	12	9	8	6		14	14	–	–	2	–	–	–	–	–	–	–	–	**Disulphide, di-p-tolyl-**	103-19-5	**DD00656**
137	108	246	65	45	135	123	138	**246**	100	42	13	9	9	8	8	7		14	14	–	–	2	–	–	–	–	–	–	–	–	**Ethane, 1,2-bis(phenylthio)-**		**IC03242**
171	246	127	108	204	203	172	103	**246**	100	29	27	21	16	15	13	7		14	14	–	–	2	–	–	–	–	–	–	–	–	**1-Naphthalenecarbodithioic acid, propyl ester**	35972-83-9	**NI23868**
228	180	165	151	77	246	166	103	**246**	100	91	75	65	62	61	59	55		14	14	2	–	1	–	–	–	–	–	–	–	–	**Benzene, 1,2-dimethyl-3-(phenylsulphonyl)-**	28523-22-0	**NI23869**
180	246	77	78	151	91	165	51	**246**	100	81	75	62	58	51	46	42		14	14	2	–	1	–	–	–	–	–	–	–	–	**Benzene, 1,2-dimethyl-3-(phenylsulphonyl)-**	28523-22-0	**NI23870**
228	246	180	165	151	181	166	211	**246**	100	98	92	79	71	66	66	64		14	14	2	–	1	–	–	–	–	–	–	–	–	**Benzene, 1,3-dimethyl-4-(phenylsulphonyl)-**	4212-74-2	**NI23871**
125	246	153	77	78	105	121	91	**246**	100	94	54	45	32	25	24	24		14	14	2	–	1	–	–	–	–	–	–	–	–	**Benzene, 1,3-dimethyl-5-(phenylsulphonyl)-**		**NI23872**
246	228	180	151	77	165	103	78	**246**	100	99	95	78	70	70	60	59		14	14	2	–	1	–	–	–	–	–	–	–	–	**Benzene, 1,4-dimethyl-2-(phenylsulphonyl)-**	2548-26-7	**NI23873**
180	65	246	91	167	77	165	137	**246**	100	98	84	81	61	57	53	50		14	14	2	–	1	–	–	–	–	–	–	–	–	**Benzene, 1,1′-sulphonylbis[2-methyl-**	5097-12-1	**NI23874**
139	246	91	65	140	107	247	141	**246**	100	35	25	20	12	12	7	7		14	14	2	–	1	–	–	–	–	–	–	–	–	**Benzene, 1,1′-sulphonylbis[4-methyl-**		**IC03243**
128	155	186	127	187	143	115	129	**246**	100	70	58	45	43	24	18	17	**5.00**	14	14	4	–	–	–	–	–	–	–	–	–	–	**Benzene, 1,2-bis[(E)-(2-methoxycarbonylethenyl)]-**		**MS05295**
41	149	189	104	76	132	50	188	**246**	100	39	15	11	10	9	7	5	**0.00**	14	14	4	–	–	–	–	–	–	–	–	–	–	**1,2-Benzenedicarboxylic acid, diallyl ester**		**MS05296**
41	149	189	39	132	104	98	76	**246**	100	34	14	11	9	8	7	6	**0.05**	14	14	4	–	–	–	–	–	–	–	–	–	–	**1,2-Benzenedicarboxylic acid, diallyl ester**		**IC03244**
41	149	28	189	104	132	39	76	**246**	100	82	35	26	25	24	19	11	**0.00**	14	14	4	–	–	–	–	–	–	–	–	–	–	**1,2-Benzenedicarboxylic acid, diallyl ester**	131-17-9	**NI23875**
189	190	41	150	39	246	104	103	**246**	100	12	12	7	7	6	5	5		14	14	4	–	–	–	–	–	–	–	–	–	–	**1,4-Benzenedicarboxylic acid, diallyl ester**		**IC03245**
191	213	246	190	162	189	161	214	**246**	100	99	89	85	35	24	16	15		14	14	4	–	–	–	–	–	–	–	–	–	–	**2H,6H-Benzo[1,2-b:5,4-b′]dipyran-2-one, 7,8-dihydro-6-hydroxy-8,8-dimethyl-**	16498-71-8	**NI23876**
203	246	173	218	102	63	69	76	**246**	100	48	34	29	29	24	21	17		14	14	4	–	–	–	–	–	–	–	–	–	–	**2H,8H-Benzo[1,2-b:3,4-b′]dipyran-2-one, 9,10-dihydro-9-hydroxy-8,8-dimethyl-, (R)-**		**LI04911**
176	175	246	91	77	188	177	147	**246**	100	69	36	17	13	12	12	12		14	14	4	–	–	–	–	–	–	–	–	–	–	**2H,8H-Benzo[1,2-b:3,4-b′]dipyran-2-one, 9,10-dihydro-9-hydroxy-8,8-dimethyl-, (R)-**	19380-05-3	**NI23877**
176	175	246	91	77	213	188	177	**246**	100	70	36	18	12	10	10	10		14	14	4	–	–	–	–	–	–	–	–	–	–	**2H,8H-Benzo[1,2-b:3,4-b′]dipyran-2-one, 9,10-dihydro-9-hydroxy-8,8-dimethyl-, (R)-**		**LI04912**
191	233	246	190	162	189	161	234	**246**	100	99	88	84	34	23	16	14		14	14	4	–	–	–	–	–	–	–	–	–	–	**2H,8H-Benzo[1,2-b:5,4-b′]dipyran-2-one, 6,7-dihydro-6-hydroxy-8,8-dimethyl-**		**LI04913**
246	28	43	55	57	41	69	203	**246**	100	76	56	55	45	43	36	33		14	14	4	–	–	–	–	–	–	–	–	–	–	**[1,1′-Biphenyl]-4,4′-diol, 3,3′-dimethoxy-**	4433-09-4	**NI23878**

MASS TO CHARGE RATIOS								M.W.	INTENSITIES								Parent	C	H	O	N	S	F	Cl	Br	I	Si	P	B	X	COMPOUND NAME	CAS Reg No	No
246	203	231	247	123	232	188	160	**246**	100	29	23	16	8	8	7	6		14	14	4	–	–	–	–	–	–	–	–	–	–	[1,1'-Biphenyl]-4,4'-diol, 3,3'-dimethoxy-	4433-09-4	NI23879
187	188	246	160	59	213	131	189	**246**	100	87	50	30	29	17	16	15		14	14	4	–	–	–	–	–	–	–	–	–	–	2H-Furo[2,3-h]-1-benzopyran-2-one, 8,9-dihydro-8-(1-hydroxy-1-isopropyl)-, (S)-	3804-70-4	LI04914
187	188	213	160	246	159	228	185	**246**	100	68	42	30	28	16	13	10		14	14	4	–	–	–	–	–	–	–	–	–	–	2H-Furo[2,3-h]-1-benzopyran-2-one, 8,9-dihydro-8-(1-hydroxy-1-isopropyl)-, (S)-	3804-70-4	NI23880
187	188	246	59	229	160	213	43	**246**	100	84	53	40	31	27	22	20		14	14	4	–	–	–	–	–	–	–	–	–	–	7H-Furo[3,2-g][1]benzopyran-7-one, 2,3-dihydro-2-(1-hydroxy-1-isopropyl)-, (S)-	13849-08-6	NI23881
246	211	205	231	115				**246**	100	29	22	21	9					14	14	4	–	–	–	–	–	–	–	–	–	–	2-Naphthaldehyde, 4,5,6-trimethoxy-		LI04915
246	231	115	203					**246**	100	69	21	11						14	14	4	–	–	–	–	–	–	–	–	–	–	2-Naphthaldehyde, 4,5,8-trimethoxy-		LI04916
246	231	203	185	247	129	89	173	**246**	100	95	33	30	16	16	16	15		14	14	4	–	–	–	–	–	–	–	–	–	–	1-Naphthalenol, 2-acetyl-3,4-dimethoxy-	55044-93-4	NI23882
114	56	133	86	115	77	69	103	**246**	100	85	53	53	18	13	11	8	**4.20**	14	14	4	–	–	–	–	–	–	–	–	–	–	2H-Pyran-2-one, 5,6-dihydro-5-hydroxy-4-methoxy-6-(2-phenylvinyl)-, 5R-(5α,6αE)-	52247-79-7	NI23883
114	86	133	115	91	77	103	69	**246**	100	40	16	11	7	7	6	6	**5.00**	14	14	4	–	–	–	–	–	–	–	–	–	–	2H-Pyran-2-one, 5,6-dihydro-5-hydroxy-4-methoxy-6-(2-phenylvinyl)-, 5R-(5α,6βE)-	52247-80-0	NI23884
231	213	91	65	246	165	232	227	**246**	100	51	36	25	20	20	16	14		14	15	2	–	–	–	–	–	–	–	1	–	–	Phosphinic acid, di-2-tolyl-		MS05297
217	77	104	218	202	141	51	199	**246**	100	88	55	38	35	33	33	29	**9.00**	14	15	2	–	–	–	–	–	–	–	1	–	–	Phosphinic acid, ethyldiphenyl-		MS05298
183	185	108	218	246	213	199	77	**246**	100	94	86	63	59	41	26	21		14	16	–	–	–	–	–	–	–	–	2	–	–	Phosphine, diphenyl(2-phosphinoethyl)-	34664-50-1	MS05299
132	131	246	113	128	99	77	115	**246**	100	31	26	26	22	21	21	10		14	18	–	2	1	–	–	–	–	–	–	–	–	4H-1,3-Thiazine, 3,4-dihydro-2-phenylimino-3,4,4,6-tetramethyl-	37489-69-3	NI23885
91	217	65	39	149	92	51	218	**246**	100	46	15	10	9	8	6	6	**0.79**	14	18	–	2	1	–	–	–	–	–	–	–	–	1,3-Thiazolidine, 2-benzylimino-4-ethyl-4-methyl-5-methylene-	71224-25-4	NI23886
203	100	41	77	204	39	42	246	**246**	100	19	18	17	14	11	11	10		14	18	–	2	1	–	–	–	–	–	–	–	–	1,3-Thiazolidine, 4-methyl-5-methylene-2-phenylimino-4-propyl-		NI23887
134	133	246	105	104	132	176	177	**246**	100	67	62	61	45	33	26	20		14	18	2	2	–	–	–	–	–	–	–	–	–	1H-1,8-Benzodiazacycloundecine, 2,3,4,5,6,7,8,9-octahydro-1-methyl-2,9-dioxo-		LI04917
151	95	246	152	107	231	106		**246**	100	64	43	21	7	6	6			14	18	2	2	–	–	–	–	–	–	–	–	–	1H-Imidazole, 4(5)-(2-(3,4-dimethoxyphenyl)ethyl)-5(4)-methyl-		LI04918
58	43	42	59	30	130	15	32	**246**	100	9	7	6	6	4	4	3	**0.00**	14	18	2	2	–	–	–	–	–	–	–	–	–	1H-Indole-3-ethanamine, 1-(acetyloxy)-N,N-dimethyl-	55191-09-8	NI23888
150	189	246	151	57	188	174	190	**246**	100	63	19	13	13	10	10	8		14	18	2	2	–	–	–	–	–	–	–	–	–	Phenol, 3,5-dimethyl-4-(methyl-2-propynylamino)-, methylcarbamate (ester)	23623-49-6	NI23889
150	189	246	151	57	188	174	190	**246**	100	62	17	12	12	10	10	8		14	18	2	2	–	–	–	–	–	–	–	–	–	Phenol, 3,5-dimethyl-4-(methyl-2-propynylamino)-, methylcarbamate (ester)	23623-49-6	NI23890
116	130	187	117	42	56	144	246	**246**	100	72	48	39	38	36	21	20		14	18	2	2	–	–	–	–	–	–	–	–	–	L-Tryptophan, N,N-dimethyl-, methyl ester	35214-77-8	NI23891
116	130	187	149	117	42	115	45	**246**	100	39	16	13	13	12	10	8	**7.60**	14	18	2	2	–	–	–	–	–	–	–	–	–	L-Tryptophan, N,N-dimethyl-, methyl ester	35214-77-8	NI23892
218	246	190	203	219	189	109	187	**246**	100	65	36	31	19	19	15	13		15	10	–	4	–	–	–	–	–	–	–	–	–	1H-Tetrazole, 5-(9-anthryl)-		NI23893
218	246	115	190	114	219	247	217	**246**	100	62	43	19	17	15	9	9		15	10	–	4	–	–	–	–	–	–	–	–	–	[1,2,3]Triazolo[1,5-c]quinazoline, 5-phenyl-	67824-30-0	NI23894
91	65	118	120	92	63	51	89	**246**	100	16	14	12	12	11	10	9	**1.64**	15	15	1	–	–	–	1	–	–	–	–	–	–	Benzene, 1-(2-chloroethyl)-2-benzyloxy-	56052-45-0	NI23895
91	92	121	126	103	125	128	65	**246**	100	82	71	69	37	31	23	20	**4.70**	15	15	1	–	–	–	1	–	–	–	–	–	–	Benzenepropanol, β-(4-chlorophenyl)-	55133-86-3	NI23896
231	246	233	103	77	98	248	41	**246**	100	44	44	23	14	13	12	10		15	15	1	–	–	–	1	–	–	–	–	–	–	Phenol, 2-chloro-4-(isopropylphenyl)-		IC03246
190	41	40	191	133	105	39	76	**246**	100	77	61	47	46	36	35	35	**22.20**	15	15	2	–	–	1	–	–	–	–	–	–	–	1,4-Naphthoquinone, 2-fluoro-3-pentyl-	78758-19-7	NI23897
91	123	245	18	105	137	246	228	**246**	100	84	63	56	48	38	37	35		15	18	1	–	1	–	–	–	–	–	–	–	–	1,4-Cyclohexadiene, (–)-(R(S))-4,5-dimethyl-1-(p-tolylsulphinyl)-		MS05300
91	190	41	71	57	191	92	58	**246**	100	67	50	21	21	14	14	14	**1.40**	15	18	1	–	1	–	–	–	–	–	–	–	–	Thiophene, 2-tert-butoxy-5-benzyl-	55059-20-6	NI23898
246	91	173	217	172	39	41	77	**246**	100	31	28	23	23	19	16	15		15	18	3	–	–	–	–	–	–	–	–	–	–	Azuleno[4,5-b]furan-2,7-dione, 3,3a,4,5,9a,9b-hexahydro-3,6,9-trimethyl-, [3R-(3α,3aβ,9aβ,9bα)]-	5956-04-7	MS05301
214	128	41	171	45	143	43	231	**246**	100	95	81	70	64	52	48	42	**18.00**	15	18	3	–	–	–	–	–	–	–	–	–	–	Benzofuran, 2-isopropenyl-5-(1-methoxyethyl)-6-methoxy-		NI23899
189	246	190	77	159	131	247	103	**246**	100	43	14	14	9	7	6	4		15	18	3	–	–	–	–	–	–	–	–	–	–	2H-1-Benzopyran-2-one, 7-methoxy-6-(3-methylbutyl)-	38409-36-8	NI23900
189	246	131	190	176	191	177	103	**246**	100	39	36	22	22	10	8	8		15	18	3	–	–	–	–	–	–	–	–	–	–	2H-1-Benzopyran-2-one, 7-methoxy-8-(3-methylbutyl)-	6619-22-3	NI23901
151	43	95	135	109	41	81	39	**246**	100	95	20	8	8	8	7	7	**4.39**	15	18	3	–	–	–	–	–	–	–	–	–	–	2,3'-Bifuran, 2,3,4,5-tetrahydro-5-methyl-5-[(4-methyl-2-furanyl)methyl]-	55721-94-3	NI23902
104	130	78	77	91	204	117		**246**	100	60	26	24	16	13	13		**0.00**	15	18	3	–	–	–	–	–	–	–	–	–	–	Cyclobutanebutanoic acid, 2-oxo-4-phenyl-, methyl ester		LI04919
91	104	92	246	105	137	64	41	**246**	100	27	24	19	12	11	10	10		15	18	3	–	–	–	–	–	–	–	–	–	–	2H-Cyclopenta[b]furan-2-one, hexahydro-5-hydroxy-3a-(2-phenylethyl)-	69687-97-4	NI23903
213	91	231	53	185	162	157	77	**246**	100	78	47	40	39	39	36	36	**28.67**	15	18	3	–	–	–	–	–	–	–	–	–	–	Cycloprop[2,3]indeno[5,6-b]furan-2(4H)-one, 3,6b-dimethyl-4a,5,5a,6,6a,6b-hexahydro-5-(hydroxymethyl)-, [4aS-(4aα,5α,5aα,6aα,6bβ)]-		NI23904
246	41	159	91	145	77	53	105	**246**	100	89	69	69	64	64	53	42		15	18	3	–	–	–	–	–	–	–	–	–	–	9,14-Dioxatricyclo[9.3.0.1^{7,10}]pentadeca-1(11),3,12-trien-8-one, 3,12-dimethyl-		LI04920
205	109	108	99	43	44	246	98	**246**	100	78	74	65	64	63	48	48		15	18	3	–	–	–	–	–	–	–	–	–	–	Furanoeremophila-3,7,11-trien-1-one, 10-hydroxy-		NI23905
246	105	96	93	228	148	122	200	**246**	100	99	98	89	88	84	80	77		15	18	3	–	–	–	–	–	–	–	–	–	–	1β-Guaia-4(15),10(14),11(13)-trien-12-oic acid, 3β,6α-dihydroxy-, γ-lactone	16838-87-2	DD00658
91	228	96	131	107	87	122	117	**246**	100	84	82	77	71	70	68	68	**0.00**	15	18	3	–	–	–	–	–	–	–	–	–	–	1β-Guaia-4(15),10(14),11(13)-trien-12-oic acid, 3β,6β-dihydroxy-, γ-lactone	67667-64-5	DD00657

MASS TO CHARGE RATIOS								M.W.	INTENSITIES								Parent	C	H	O	N	S	F	Cl	Br	I	Si	P	B	X	COMPOUND NAME	CAS Reg No	No
246	228	131	83	123	105	43	95	246	100	60	46	38	37	36	36	35		15	18	3	–	–	–	–	–	–	–	–	–	–	Guaia-1(10),4(15),11(13)-trieno-12,6α-lactone, 3α-hydroxy-	119273-14-2	DD00659
150	122	246	105	96	79	161	131	246	100	47	25	25	25	25	22	22		15	18	3	–	–	–	–	–	–	–	–	–	–	Guaia-4(15),9,11(13)-trieno-12,6α-lactone, 3α-hydroxy-	119273-15-3	DD00660
246	41	159	91	77	145	53	65	246	100	90	70	70	65	64	53	43		15	18	3	–	–	–	–	–	–	–	–	–	–	4,7-Methanofuro[3,2-c]oxacycloundecin-6(4H)-one, 7,8,9,12-tetrahydro-3,11-dimethyl-	728-60-9	NI23906
246	173	41	135	91	172	39	231	246	100	91	74	65	49	42	33	31		15	18	3	–	–	–	–	–	–	–	–	–	–	Naphtho[1,2-b]furan-2,8(3H,4H)-dione, 3a,5,5a,9b-tetrahydro-3,5a,9-trimethyl-, [3R-(3α,3aβ,5aα,9bα)]-	481-07-2	MS05302
246	173	83	85	41	108	91	145	246	100	63	58	54	50	41	37	32		15	18	3	–	–	–	–	–	–	–	–	–	–	Naphtho[1,2-b]furan-2,8(3H,4H)-dione, 3a,5,5a,9b-tetrahydro-3,5a,9-trimethyl-, [3S-(3α,3aα,5aβ,9bα)]-	1618-78-6	MS05304
246	173	41	91	135	172	77	39	246	100	85	61	47	45	41	33	31		15	18	3	–	–	–	–	–	–	–	–	–	–	Naphtho[1,2-b]furan-2,8(3H,4H)-dione, 3a,5,5a,9b-tetrahydro-3,5a,9-trimethyl-, [3S-(3α,3aα,5aβ,9bβ)]-		MS05303
91	173	246	135	231	175	121	172	246	100	92	71	57	38	35	35	33		15	18	3	–	–	–	–	–	–	–	–	–	–	Naphtho[1,2-b]furan-2,8(3H,4H)-dione, 3a,5,5a,9b-tetrahydro-3,5a,9-trimethyl-, [3S-(3α,3aα,5aβ,9bβ)]-	481-06-1	NI23907
91	173	246	135	175	231	121	172	246	100	90	70	57	36	35	35	32		15	18	3	–	–	–	–	–	–	–	–	–	–	Naphtho[1,2-b]furan-2,8(3H,4H)-dione, 3a,5,5a,9b-tetrahydro-3,5a,9-trimethyl-, [3S-(3α,3aα,5aβ,9bβ)]-		LI04921
246	110	136						246	100	85	17							15	18	3	–	–	–	–	–	–	–	–	–	–	Naphtho[2,3-b]furan-4,9-dione, 4a,5,6,7,8,8a-hexahydro-3,4a,5-trimethyl-, 4aR-(4aα,5α,8aα)-	18452-51-2	NI23908
173	246	187	174	135	158	172	91	246	100	50	17	17	13	10	9	9		15	18	3	–	–	–	–	–	–	–	–	–	–	Naphtho[1,2-b]furan-2(3H)-one, 3a,4,5,9b-tetrahydro-8-hydroxy-3,6,9-tr	5951-46-2	NI23909
145	105	246	231	228	173	185	169	246	100	76	65	59	35	34	32	29		15	18	3	–	–	–	–	–	–	–	–	–	–	2-Oxabicyclo[3.3.0]octan-6-one, 3,5-dimethyl-1-hydroxy-3-phenyl-		DD00661
178	147	69	41	179	119	118	148	246	100	64	35	20	14	8	7	6	3.50	15	18	3	–	–	–	–	–	–	–	–	–	–	2-Propenoic acid, 3-[4-[(3-methyl-1-butenyl)oxy]phenyl]-, methyl ester	55256-10-5	LI04922
178	147	69	41	179	119	118	148	246	100	63	34	19	14	8	7	6	3.35	15	18	3	–	–	–	–	–	–	–	–	–	–	2-Propenoic acid, 3-[4-[(3-methyl-1-butenyl)oxy]phenyl]-, methyl ester	55256-10-5	NI23910
191	149	246	231	190	175	161	204	246	100	75	51	50	30	25	23	22		15	18	3	–	–	–	–	–	–	–	–	–	–	8H-[1]Pyrano[2,3-h]-1-Benzopyran-2(2H)-one, 9,10-dihydro-4,8,8-trimethyl-		NI23911
135	95	53	162	91	77	246	79	246	100	80	63	52	47	44	42	34		15	18	3	–	–	–	–	–	–	–	–	–	–	Smyrnicordiolide	97983-42-1	NI23912
98	246	97	245	96	163	109	136	246	100	70	60	32	27	14	14	13		15	22	1	2	–	–	–	–	–	–	–	–	–	Aphyllidine	643-32-3	LI04923
91	57	41	147	146	28	42	56	246	100	80	73	70	67	53	47	46	2.00	15	22	1	2	–	–	–	–	–	–	–	–	–	Diaziridinone, tert-butyl(α,α-dimethylphenylethyl)-	19656-67-8	NI23913
231	246	85	161	189	232	174	115	246	100	75	72	28	26	18	18	18		15	22	1	2	–	–	–	–	–	–	–	–	–	1H-2,6-Methano-2,3-benzodiazocine, 3,4,5,6-tetrahydro-8-methoxy-3,6,11-trimethyl-	51578-78-0	NI23914
246	134	245	148	247	231	136	248	246	100	46	24	24	21	12	11	9		15	22	1	2	–	–	–	–	–	–	–	–	–	7,14-Methano-4H,6H-dipyrido[1,2-a:1′,2′-e][1,5]diazocin-4-one, 1,2,3,7,9,10,11,13,14,14a-decahydro-, [7S-(7α,14α,14aα)]-	35611-60-0	NI23915
98	97	246	245	121	96	84	85	246	100	45	34	10	10	9	9	8		15	22	1	2	–	–	–	–	–	–	–	–	–	7,14-Methano-4H,6H-dipyrido[1,2-a:1′,2′-e][1,5]diazocin-4-one, 2,3,7,7a,8,9,10,11,13,14-decahydro-, [7S-(7α,7aβ,14α,)]-	32101-29-4	NI23916
121	148	120	245	231	91	246	84	246	100	85	65	59	58	41	37	35		15	22	1	2	–	–	–	–	–	–	–	–	–	4H-1,3-Oxazine, 5,6-dihydro-2-(N-methyl-m-tolylamino)-4,4,6-trimethyl-	53004-35-6	NI23917
121	245	120	148	231	246	91	84	246	100	74	73	68	54	44	32	32		15	22	1	2	–	–	–	–	–	–	–	–	–	4H-1,3-Oxazine, 5,6-dihydro-2-(N-methyl-p-tolylamino)-4,4,6-trimethyl-	53004-34-5	NI23918
56	246	120	69	41	133	163	245	246	100	72	70	53	45	43	43	42		15	22	1	2	–	–	–	–	–	–	–	–	–	1,3-Oxazine, 3,4,4,6-tetramethyl-2-m-tolyliminotetrahydro-	53004-28-7	NI23919
56	120	41	69	246	133	163	55	246	100	70	46	45	44	42	34	25		15	22	1	2	–	–	–	–	–	–	–	–	–	1,3-Oxazine, 3,4,4,6-tetramethyl-2-p-tolyliminotetrahydro-	53004-27-6	NI23920
98	247	245						246	100	77	20						0.00	15	22	1	2	–	–	–	–	–	–	–	–	–	2-Piperidinecarboxamide, N-(2,6-dimethylphenyl)-1-methyl-	96-88-8	NI23921
84	57	69	91	107	203	28	118	246	100	72	54	48	46	40	37	31	22.63	15	23	2	–	–	–	–	–	–	–	–	1	–	1,3,2-Dioxaborinane, 2-ethyl-5-methyl-5-phenyl-4-propyl-	74421-13-9	NI23922
246	230	229	247	218	220	217	231	246	100	62	24	20	20	18	14	11		16	10	1	2	–	–	–	–	–	–	–	–	–	Benzo[a]phenazine, 7-oxide		MS05305
230	218	246	217	229	231	219	75	246	100	97	86	31	27	20	16	16		16	10	1	2	–	–	–	–	–	–	–	–	–	Benzo[a]phenazine, 12-oxide	18636-87-8	NI23923
133	149	132	96	97	105	81	150	246	100	97	76	53	42	32	26	23	5.80	16	22	2	–	–	–	–	–	–	–	–	–	–	Benzoic acid, 2,6-dimethyl-, cyclohexanemethyl ester		NI23924
119	246	134	105	91	187	94	171	246	100	96	87	83	65	60	58	51		16	22	2	–	–	–	–	–	–	–	–	–	–	3,9,11-Guaiatrien-12-oic acid, methyl ester		MS05306
203	41	69	151	91	246	231	77	246	100	89	82	69	69	54	53	47		16	22	2	–	–	–	–	–	–	–	–	–	–	1,5-Heptadiene, 2-(2,6-dimethoxyphenyl)-6-methyl-		LI04924
246	83	164	163	55	91	119	105	246	100	99	60	60	50	35	25	25		16	22	2	–	–	–	–	–	–	–	–	–	–	2-Hexen-3-one, 2-methyl-4-(3-hydroxy-2,4,6-trimethylphenyl)-		LI04925
178	69	41	109	177	246	110	231	246	100	77	52	39	32	27	26	25		16	23	–	–	–	–	–	–	–	–	1	–	–	Phosphine, bis(2,2-dimethylcyclopropyl)phenyl-	62337-91-1	NI23925
73	135	204	115	91	231	128	172	246	100	90	30	22	20	13	12	6	0.00	16	26	–	–	–	–	–	–	–	1	–	–	–	2-Heptene, 3-phenyl-2-(trimethylsilyl)-	118298-75-2	DD00662
122	135	41	246	39	136	123	80	246	100	66	34	15	15	11	11	10		16	26	–	2	–	–	–	–	–	–	–	–	–	Pyrazine, 2,3-dimethyl-5-(3,7-dimethyl-6-octen-1-yl)-		NI23926
122	135	41	123	69	136	121	57	246	100	57	19	12	12	10	10	10	9.00	16	26	–	2	–	–	–	–	–	–	–	–	–	Pyrazine, 2,5-dimethyl-3-(3,7-dimethyl-6-octen-1-yl)-		NI23927
122	135	41	246	136	121	123	55	246	100	63	16	12	12	9	8	7		16	26	–	2	–	–	–	–	–	–	–	–	–	Pyrazine, 2,6-dimethyl-3-(3,7-dimethyl-6-octen-1-yl)-		NI23928
246	189	218	247	94.5	187	190	109	246	100	30	25	23	14	9	8	8		17	10	2	–	–	–	–	–	–	–	–	–	–	7H-Benz[de]anthracen-7-one, 4-hydroxy-		IC03247
246	189	247	245	109	190	187	123	246	100	30	19	12	9	6	6	5		17	10	2	–	–	–	–	–	–	–	–	–	–	11H-Benzo[a]fluoren-11-one, 5-hydroxy-		IC03248
246	215	245	244	216	247	243	217	246	100	63	52	43	43	28	26	12		17	14	–	–	–	–	–	–	–	1	–	–	–	7-Silapleiadene, 7,7-dichloro-7,12-dihydro-	51986-73-3	NI23929
246	245	77	115	142	116	219	117	246	100	39	33	30	29	28	25	23		17	14	–	2	–	–	–	–	–	–	–	–	–	1H-Imidazole, 2-phenyl-1-(2-phenylvinyl)-, (E)-	58275-54-0	NI23930
43	161	41	69	119	147	105	231	246	100	32	28	23	20	16	15	14	10.00	17	26	1	–	–	–	–	–	–	–	–	–	–	Cedr-8-ene, 9-acetyl-		MS05307
231	246	189	203	57	41	232	190	246	100	46	42	29	29	15	10	8		17	26	1	–	–	–	–	–	–	–	–	–	–	2,4-Cyclohexadienone, 2,6-di-tert-butyl-4-isopropenyl-		IC03249
91	41	43	133	55	105	147	29	246	100	83	76	70	69	59	56	56	2.00	17	26	1	–	–	–	–	–	–	–	–	–	–	Heptadeca-9-en-4,6-diyn-8-ol, (Z)-		NI23931
121	41	91	203	217	185	199	231	246	100	60	40	6	5	4	3	1	0.70	17	26	1	–	–	–	–	–	–	–	–	–	–	Heptadec-16-en-4,6-diyn-8-ol		LI04926

MASS TO CHARGE RATIOS								M.W.	INTENSITIES								Parent	C	H	O	N	S	F	Cl	Br	I	Si	P	B	X	COMPOUND NAME	CAS Reg No	No
203	246	43	189	109	57	231		246	100	83	57	29	27	17	8			17	26	1	-	-	-	-	-	-	-	-	-	-	1,3-Hexadiene, 4-isopropyl-1-(2-methylfuranyl)-1,5,5-trimethyl-		LI04927
95	109	138	123	96	41	108	246	246	100	73	54	54	48	47	46	44		17	26	1	-	-	-	-	-	-	-	-	-	-	1(4H)-Phenanthrenone, 4a,4b,5,6,7,8,8a,9,10,10a-decahydro-4b,8,8-trimethyl-, 4aS-(4aα,4bβ,8aα,10aβ)-	57684-15-8	NI23932
109	123	246	69	107	81	122	55	246	100	77	74	60	59	57	54	54		17	26	1	-	-	-	-	-	-	-	-	-	-	3(4H)-Phenanthrenone, 4a,4b,5,6,7,8,8a,9,10,10a-decahydro-4b,8,8-trimethyl-, 4aS-(4aα,4bβ,8aα,10aβ)-	57684-12-5	NI23933
123	246	109	110	55	136	231	163	246	100	68	39	37	32	29	26	14		17	26	1	-	-	-	-	-	-	-	-	-	-	3(2H)-Phenanthrenone, 1,4b,5,6,7,8,8a,9,10,10a-decahydro-4b,8,8-trimethyl-, 4bS-(4bα,8aβ,10aα)-	57684-11-4	NI23934
43	203	161	246	41	218	119	105	246	100	67	50	45	38	31	26	25		17	26	1	-	-	-	-	-	-	-	-	-	-	Tricyclo[4.4.0.2^{1,4}]dec-6-ene, 7-acetyl-4,10,10-trimethyl-		MS05308
43	231	41	203	246	133	119	95	246	100	44	33	32	22	22	21	21		17	26	1	-	-	-	-	-	-	-	-	-	-	Tricyclo[5.2.0.0^{1,6}]undec-3-ene, 4-acetyl-2,2,3,7-tetramethyl-		MS05309
246	190	189	95	218	247	191	215	246	100	69	65	53	29	19	11	10		18	14	1	-	-	-	-	-	-	-	-	-	-	Benz[a]anthracene-1-(2H)-one, 3,4-dihydro-	57652-74-1	NI23935
246	247	245	215	202	244	227	217	246	100	21	15	7	6	5	5	5		18	14	1	-	-	-	-	-	-	-	-	-	-	Bicyclo[3.1.0]hex-3-en-2-one, 6,6-diphenyl-	13304-07-9	NI23936
246	152	245	170	247	77	51	229	246	100	35	21	21	20	14	14	13		18	14	1	-	-	-	-	-	-	-	-	-	-	1,1'-Biphenyl, 2-phenoxy-		IC03250
246	152	245	247	229	227	217	169	246	100	32	24	19	16	11	8	8		18	14	1	-	-	-	-	-	-	-	-	-	-	1,1'-Biphenyl, 2-phenoxy-		DO01035
246	152	247	51	115	77	229	151	246	100	39	20	12	11	10	9	9		18	14	1	-	-	-	-	-	-	-	-	-	-	1,1'-Biphenyl, 2-phenoxy-		MS05310
246	247	152	18	141	115	169	77	246	100	18	13	12	9	7	5	5		18	14	1	-	-	-	-	-	-	-	-	-	-	1,1'-Biphenyl, 4-phenoxy-		MS05311
246	218	217	202	115	245	215	203	246	100	68	40	34	28	27	26	21		18	14	1	-	-	-	-	-	-	-	-	-	-	2,5-Cyclohexadien-1-one, 4,4-diphenyl-	13304-12-6	NI23938
246	218	245	247	215	202	227	231	246	100	24	22	20	17	14	12	11		18	14	1	-	-	-	-	-	-	-	-	-	-	2,5-Cyclohexadien-1-one, 4,4-diphenyl-	13304-12-6	NI23937
246	215	131	245	231	115	202	116	246	100	43	42	41	31	30	24	23		18	14	1	-	-	-	-	-	-	-	-	-	-	1H-Inden-1-one, 2-(2,3-dihydro-1H-inden-1-ylidene)-2,3-dihydro-	5706-06-9	NI23939
246	247	245	202	215	217	123	227	246	100	20	10	8	7	5	5	3		18	14	1	-	-	-	-	-	-	-	-	-	-	Phenol, 2,4-diphenyl-		DO01036
246	247	245	202	215	217	227	115	246	100	20	18	10	9	8	5	4		18	14	1	-	-	-	-	-	-	-	-	-	-	Phenol, 2,6-diphenyl-		DO01037
91	147	189	105	92	246	104	41	246	100	23	12	11	8	7	7	6		18	30	-	-	-	-	-	-	-	-	-	-	-	Benzene, (1-butyloctyl)-	2719-63-3	AI01758
91	147	105	92	189	119	41	43	246	100	18	14	8	7	6	6	5	2.46	18	30	-	-	-	-	-	-	-	-	-	-	-	Benzene, (1-butyloctyl)-	2719-63-3	NI23940
91	147	41	105	189	92	43	104	246	100	20	13	12	9	9	9	6	5.00	18	30	-	-	-	-	-	-	-	-	-	-	-	Benzene, (1-butyloctyl)-	2719-63-3	AI01759
119	91	41	120	29	105	43	27	246	100	20	17	10	8	7	7	5	2.30	18	30	-	-	-	-	-	-	-	-	-	-	-	Benzene, (1,1-dimethyldecyl)-	27854-40-6	WI01264
119	91	120	44	41	105	43	118	246	100	19	10	9	8	7	5	4	2.10	18	30	-	-	-	-	-	-	-	-	-	-	-	Benzene, (1,1-dimethyldecyl)-	27854-40-6	WI01265
57	71	43	91	92	85	41	99	246	100	56	56	52	47	41	25	14	7.21	18	30	-	-	-	-	-	-	-	-	-	-	-	Benzene, (2,3-dimethyldecyl)-	55134-08-2	WI01266
91	57	43	105	71	85	147	41	246	100	35	27	24	24	22	20	14	12.01	18	30	-	-	-	-	-	-	-	-	-	-	-	Benzene, (3,3-dimethyldecyl)-	55134-09-3	WI01267
57	91	92	246	41	190	43	56	246	100	49	39	16	16	13	13	10		18	30	-	-	-	-	-	-	-	-	-	-	-	Benzene, (9,9-dimethyldecyl)-	54986-45-7	WI01268
120	119	43	78	246	105	121	91	246	100	58	21	14	12	11	10	7		18	30	-	-	-	-	-	-	-	-	-	-	-	Benzene, 1,4-dimethyl-2-(3,7-dimethyloctyl)-	19550-60-8	AI01760
120	119	43	78	246	105	121	91	246	100	61	21	16	13	12	10	8		18	30	-	-	-	-	-	-	-	-	-	-	-	Benzene, 1,4-dimethyl-2-(3,7-dimethyloctyl)-	19550-60-8	NI23941
91	92	246	43	41	105	104	29	246	100	95	26	16	14	11	10	10		18	30	-	-	-	-	-	-	-	-	-	-	-	Benzene, dodecyl-	123-01-3	AI01762
92	91	43	41	29	57	246	105	246	100	77	28	25	19	14	12	9		18	30	-	-	-	-	-	-	-	-	-	-	-	Benzene, dodecyl-	123-01-3	AI01761
91	92	246	43	41	105	57	29	246	100	94	24	20	19	12	12	12		18	30	-	-	-	-	-	-	-	-	-	-	-	Benzene, dodecyl-	123-01-3	NI23942
91	119	41	43	105	217	92	120	246	100	50	21	13	12	10	10	7	6.25	18	30	-	-	-	-	-	-	-	-	-	-	-	Benzene, (1-ethyldecyl)-		AI01763
91	119	217	41	246	105	92	29	246	100	53	15	11	10	10	8	7		18	30	-	-	-	-	-	-	-	-	-	-	-	Benzene, (1-ethyldecyl)-		AI01764
53	247	231	67	41	246	217	55	246	100	67	64	49	44	43	28	24		18	30	-	-	-	-	-	-	-	-	-	-	-	Benzene, hexaethyl-		AM00131
53	247	231	67	246	217	55	233	246	100	67	64	49	43	28	24	22		18	30	-	-	-	-	-	-	-	-	-	-	-	Benzene, hexaethyl-	604-88-6	NI23943
231	246	232	217	247	175	147	161	246	100	48	19	16	10	7	7	6		18	30	-	-	-	-	-	-	-	-	-	-	-	Benzene, hexaethyl-	604-88-6	NI23944
105	106	91	246	41	104	77	29	246	100	13	12	7	5	4	4	4		18	30	-	-	-	-	-	-	-	-	-	-	-	Benzene, (1-methylundecyl)-		AI01765
105	91	106	41	43	246	104	77	246	100	15	13	11	8	6	5	4		18	30	-	-	-	-	-	-	-	-	-	-	-	Benzene, (1-methylundecyl)-		AI01766
119	91	120	57	41	105	118	43	246	100	16	10	10	10	8	5	5	2.40	18	30	-	-	-	-	-	-	-	-	-	-	-	Benzene, (1,1,4,6,6-pentamethylheptyl)-	55134-07-1	WI01269
91	161	105	175	41	92	43	104	246	100	13	12	10	10	8	7	6	4.79	18	30	-	-	-	-	-	-	-	-	-	-	-	Benzene, (1-pentylheptyl)-		AI01767
91	161	175	105	104	92	246	29	246	100	18	15	12	10	8	7	7		18	30	-	-	-	-	-	-	-	-	-	-	-	Benzene, (1-pentylheptyl)-		AI01768
91	133	41	92	203	105	43	246	246	100	26	12	9	8	8	8	5		18	30	-	-	-	-	-	-	-	-	-	-	-	Benzene, (1-propylnonyl)-		AI01769
91	133	203	92	246	105	41	29	246	100	26	10	8	7	7	6	6		18	30	-	-	-	-	-	-	-	-	-	-	-	Benzene, (1-propylnonyl)-		AI01770
231	43	246	232	147	119	41	203	246	100	23	21	19	13	9	9	8		18	30	-	-	-	-	-	-	-	-	-	-	-	Benzene, 1,2,4,5-tetraisopropyl-		MS05312
231	246	232	203	91	247	147	119	246	100	27	19	8	7	5	5	5		18	30	-	-	-	-	-	-	-	-	-	-	-	Benzene, 1,2,4,5-tetraisopropyl-		AI01771
161	246	203	147	105	162	119	204	246	100	33	33	19	16	14	13	10		18	30	-	-	-	-	-	-	-	-	-	-	-	Benzene, 1,2,4-tributyl-	14800-16-9	NI23945
204	147	246	105	161	148	203	119	246	100	45	35	31	30	30	28	22		18	30	-	-	-	-	-	-	-	-	-	-	-	Benzene, 1,3,5-tributyl-	841-07-6	NI23946
161	204	147	246	203	105	148	119	246	100	81	53	47	45	43	32	29		18	30	-	-	-	-	-	-	-	-	-	-	-	Benzene, tributyl-	61142-83-4	NI23947
105	57	71	43	85	106	41	91	246	100	20	19	18	13	10	9	7	6.91	18	30	-	-	-	-	-	-	-	-	-	-	-	Benzene, (1,3,3-trimethylnonyl)-	54986-44-6	WI01270
91	85	43	57	84	161	162	41	246	100	76	69	35	30	22	21	20	1.50	18	30	-	-	-	-	-	-	-	-	-	-	-	Benzene, 1-(1,2,2-trimethylpropyl)hexyl-	55134-06-0	WI01271
217	28	29	57	41	27	218	43	246	100	93	63	32	27	19	10	10	10.20	18	30	-	-	-	-	-	-	-	-	-	-	-	Benzene, 1,3,5-tri-sec-butyl-		AI01772
57	231	41	175	246	232	43	189	246	100	53	34	17	15	10	10	8		18	30	-	-	-	-	-	-	-	-	-	-	-	Benzene, 1,2,4-tri-tert-butyl-		AM00132
57	231	41	175	246	232	43	91	246	100	53	34	17	15	10	10	8		18	30	-	-	-	-	-	-	-	-	-	-	-	Benzene, 1,3,5-tri-tert-butyl-	1460-02-2	WI01272
231	57	29	41	232	28	246	27	246	100	64	32	29	20	16	12	10		18	30	-	-	-	-	-	-	-	-	-	-	-	Benzene, 1,3,5-tri-tert-butyl-		AI01773

MASS TO CHARGE RATIOS								M.W.	INTENSITIES								Parent	C	H	O	N	S	F	Cl	Br	I	Si	P	B	X	COMPOUND NAME	CAS Reg No	No
67	246	135	81	55	95	79	136	**246**	100	94	94	80	69	62	41	31		18	30	–	–	–	–	–	–	–	–	–	–	–	Chrysene, octadecahydro-		AI01775
246	135	81	95	121	41	67	79	**246**	100	88	68	48	46	45	44	40		18	30	–	–	–	–	–	–	–	–	–	–	–	Chrysene, octadecahydro-		AI01774
246	135	81	95	121	67	79	93	**246**	100	88	68	49	47	44	40	29		18	30	–	–	–	–	–	–	–	–	–	–	–	Chrysene, octadecahydro-	2090-14-4	NI23948
133	189	57	41	246	205	175	147	**246**	100	85	84	78	38	34	31	25		18	30	–	–	–	–	–	–	–	–	–	–	–	Cyclopentane, 1,3-Bis[(E)-2,2-dimethylpropylidene]-2-isopropylidene-		MS05313
123	81	41	57	67	69	109	55	**246**	100	84	44	42	41	34	32	32	**5.00**	18	30	–	–	–	–	–	–	–	–	–	–	–	3,7-Decadiyne, 2,2,5,5,6,6,9,9-octamethyl-	17553-35-4	MS05314
246	142	155	247	91	141	218	167	**246**	100	61	27	22	21	17	16	16		19	18	–	–	–	–	–	–	–	–	–	–	–	Bicyclo[3.1.0]hexane, 4-methylene-1,6-diphenyl-	56052-57-4	NI23949
130	247	69	58	149	101	60	249	**247**	100	83	75	52	45	41	40	33		1	3	–	1	2	3	1	–	–	–	2	–	–	Dithioimidodiphosphoric acid, P-chloro-P,P′,P′-trifluoro-N-methyl-	40427-73-4	MS05315
69	130	101	47	28	110	247	117	**247**	100	99	59	49	45	26	19	14		1	3	1	1	1	2	2	–	–	–	2	–	–	Thioimidodiphosphoric acid, P,P-dichloro-P′,P′-difluoro-P-oxo-, N-methyl-	39564-20-0	MS05316
114	91	69	63	105	130	163	128	**247**	100	55	52	37	36	33	24	22	**15.00**	1	3	1	1	1	2	2	–	–	–	2	–	–	Thioimidodiphosphoric acid, P,P-dichloro-P′,P′-difluoro-P′-oxo-, N-methyl-	41006-37-5	NI23950
69	247	228	96	140	159	158	132	**247**	100	19	15	13	10	9	5	4		5	2	–	1	–	9	–	–	–	–	–	–	–	1-Propene, 3,3,3-trifluoro-1-bis(trifluoromethyl)amino-		LI04928
69	247	228	140	166	159	158	95	**247**	100	24	15	15	7	7	5	5		5	2	–	1	–	9	–	–	–	–	–	–	–	1-Propene, 3,3,3-trifluoro-1-bis(trifluoromethyl)amino-, (Z)-		LI04929
44	69	36	70	71	43	115	53	**247**	100	70	50	38	32	30	14	9	**0.00**	5	13	–	3	–	–	3	–	–	–	–	–	1	Aluminium, trichloro(1,1,3,3-tetramethylguanidine)-	26769-35-7	NI23951
44	69	36	70	71	42	115	96	**247**	100	70	50	38	32	30	14	9	**0.00**	5	13	–	3	–	–	3	–	–	–	–	–	1	Aluminium, trichloro(1,1,3,3-tetramethylguanidine)-		LI04930
109	30	74	247	111	73	97	62	**247**	100	95	92	62	35	34	28	25		6	2	6	3	–	–	1	–	–	–	–	–	–	Benzene, 2-chloro-1,3,5-trinitro	88-88-0	LI04931
74	30	109	247	111	73	249	97	**247**	100	83	82	56	26	21	20	19		6	2	6	3	–	–	1	–	–	–	–	–	–	Benzene, 2-chloro-1,3,5-trinitro-	88-88-0	NI23952
249	247	168	234	232	218	220	125	**247**	100	96	61	38	33	20	17	17		7	10	–	3	1	–	–	1	–	–	–	–	–	Pyrimidine, 5-bromo-2-dimethylamino-4-methylthio-		NI23953
247	69	42	58	227	197	248	53	**247**	100	37	28	19	17	17	11	10		8	8	1	5	–	3	–	–	–	–	–	–	–	Pyrimido[5,4-e]-1,2,4-triazine, 1,2-dihydro-7-methoxy-3-methyl-5-(trifluoromethyl)-	32709-21-0	NI23954
109	96	15	79	30	63	230	135	**247**	100	76	76	35	33	26	20	17	**15.81**	8	10	6	1	–	–	–	–	–	–	1	–	–	Phosphoric acid, dimethyl 4-nitrophenyl ester	950-35-6	NI23955
109	96	247	79	230	63	200	135	**247**	100	73	37	35	32	23	20	20		8	10	6	1	–	–	–	–	–	–	1	–	–	Phosphoric acid, dimethyl 4-nitrophenyl ester	950-35-6	NI23956
211	82	69	154	166	83	138	71	**247**	100	86	86	80	67	52	49	48	**22.99**	8	14	2	5	–	–	1	–	–	–	–	–	–	1,3,5-Triazine-2,4-diamine, 6-(2-chloroethoxy)-N-methoxy-N′,N′-dimethyl-	80197-64-4	NI23957
28	79	52	51	135	56	50	26	**247**	100	24	20	8	5	5	4	4	**0.40**	9	5	4	1	–	–	–	–	–	–	–	–	1	Iron, tetracarbonyl(pyridinyl)-	53317-88-7	NI23958
201	114	232	85	160	88	113	143	**247**	100	83	76	65	56	47	45	39	**0.00**	9	13	7	1	–	–	–	–	–	–	–	–	–	α-D-Glucofuranose, 3,6-anhydro-1,2-O-isopropylidene-3-C-nitro-	42776-32-9	NI23959
81	43	69	99	159	57	53	45	**247**	100	97	81	55	25	20	20	19	**0.00**	9	13	7	1	–	–	–	–	–	–	–	–	–	2H-Pyran-3,5-diol, 4-nitro-, diacetate, (3R,4S,5S)-		DD00663
177	43	176	41	56	55	128	119	**247**	100	96	85	64	62	47	42	32	**0.00**	9	14	1	1	–	5	–	–	–	–	–	–	–	Hexanamine, pentafluoropropanoyl ester		MS05317
58	42	43	59	57	67	44	124	**247**	100	42	40	36	32	25	22	15	**3.00**	9	14	2	1	–	–	–	1	–	–	–	–	–	2-Furanone, 3-bromo-2,5-dihydro-4,5-dimethyl-5-(dimethylaminomethyl)-		NI23960
73	75	147	130	45	74	28	102	**247**	100	30	28	22	19	10	10	9	**0.00**	9	21	3	1	–	–	–	–	–	2	–	–	–	Glycine, N-formyl-N-(trimethylsilyl)-, trimethylsilyl ester	55517-31-2	NI23961
73	147	130	75	45	232	204	74	**247**	100	33	24	21	19	10	10	9	**0.00**	9	21	3	1	–	–	–	–	–	2	–	–	–	Glycine, N-formyl-N-(trimethylsilyl)-, trimethylsilyl ester	55517-31-2	NI23962
73	147	45	232	59	204	130	74	**247**	100	42	25	25	20	16	15	13	**9.91**	9	21	3	1	–	–	–	–	–	2	–	–	–	Propanoic acid, 2-[(trimethylsilyl)oximo]-, trimethylsilyl ester		NI23963
158	91	74	72	232	248	90	93	**247**	100	88	85	67	44	33	28	26	**0.00**	9	21	3	1	–	–	–	–	–	2	–	–	–	Propanoic acid, 3-[[(trimethylsilyl)oxy]imino]-, trimethylsilyl ester	75365-78-5	NI23964
71	43	60	45	61	173	57	44	**247**	100	99	97	66	46	42	41	32	**0.00**	10	8	3	1	–	3	–	–	–	–	–	–	–	Hippuric acid, 3-(trifluoromethyl)-		MS05318
146	55	71	163	91	147	65	56	**247**	100	99	30	29	13	9	9	9	**1.79**	10	10	3	1	–	–	–	–	–	–	–	–	1	Manganese, tricarbonyl-π-(1-aminoethyl)cyclopentadienyl-	12214-72-1	NI23965
72	212	86	58	44	42	43	214	**247**	100	53	50	37	30	26	16	14	**0.00**	10	11	2	1	–	–	2	–	–	–	–	–	–	Acetamide, 2-(2,4-dichlorophenoxy)-N,N-dimethyl-	6333-39-7	NI23966
31	45	44	28	29	63	42	162	**247**	100	90	74	68	58	57	52	49	**0.00**	10	11	2	1	–	–	2	–	–	–	–	–	–	Acetamide, 2-(2,4-dichlorophenoxy)-N,N-dimethyl-	6333-39-7	NI23967
184	77	141	40	51	185	78	142	**247**	100	90	65	21	18	10	10	5	**3.00**	10	14	2	1	1	–	1	–	–	–	–	–	–	Benzenesulphonamide, N-(3-chloropropyl)-N-methyl-	74810-82-5	NI23968
137	36	38	35	138	122	28	211	**247**	100	79	27	12	12	11	7	6	**0.00**	10	14	4	1	–	–	1	–	–	–	–	–	–	L-Alanine, 3-(3-hydroxy-4-methoxyphenyl)-, hydrochloride	35296-56-1	NI23969
88	57	132	144	188	176	100	117	**247**	100	92	85	52	43	43	43	34	**18.00**	10	17	4	1	1	–	–	–	–	–	–	–	–	L-Cysteine, N-acetyl-S-(2-oxobutyl)-, methyl ester	92079-03-3	NI23970
188	117	88	71	113	76	145	118	**247**	100	66	63	56	41	25	23	23	**0.00**	10	17	4	1	1	–	–	–	–	–	–	–	–	L-Cysteine, N-acetyl-S-(3-oxobutyl)-, methyl ester	92079-04-4	NI23971
43	67	95	101	96	45	73	69	**247**	100	77	57	41	40	35	31	24	**0.00**	10	17	6	1	–	–	–	–	–	–	–	–	–	1,3-Cyclohexanediol, 5-ethoxy-2-nitro, 1-acetate, (1S,2S,3R,5R)-		DD00664
29	42	74	102	174	43	116	188	**247**	100	90	55	52	50	28	26	22	**5.60**	10	17	6	1	–	–	–	–	–	–	–	–	–	Glycine, N-(ethoxycarbonyl)methyl-N-(methoxycarbonyl)methylethyl-		IC03251
81	55	41	28	27	39	62	111	**247**	100	58	55	34	32	30	29	27	**0.00**	10	18	2	3	–	–	1	–	–	–	–	–	–	Urea, N-(2-chloroethyl)-N′-(4-methylcyclohexyl)-N-nitroso-	13909-09-6	NI23972
174	73	147	232	175	86	176	204	**247**	100	36	24	23	19	19	13	10	**0.00**	10	21	4	1	–	–	–	–	–	1	–	–	–	1,3-Propanedioic acid, 2-[(trimethylsilyl)amino]-, diethyl ester		NI23973
163	119	79	56	104	160	202	63	**247**	100	99	81	78	76	71	57	40	**4.40**	10	21	4	1	–	–	–	–	–	1	–	–	–	Silane, triethoxy(3-isocyanatopropyl)-	24801-88-5	NI23974
130	73	45	131	75	43	147	74	**247**	100	93	24	12	12	12	9	9	**0.00**	10	25	2	1	–	–	–	–	–	2	–	–	–	Alanine, 2-methyl-N-(trimethylsilyl)-, trimethylsilyl ester		MS05319
130	73	45	147	75	41	131	43	**247**	100	92	30	15	15	15	14	12	**0.00**	10	25	2	1	–	–	–	–	–	2	–	–	–	Alanine, 2-methyl-N-(trimethylsilyl)-, trimethylsilyl ester	54745-23-2	MS05320
73	188	84	89	59	45	56	75	**247**	100	75	55	30	30	23	23	18	**6.63**	10	25	2	1	–	–	–	–	–	2	–	–	–	Alanine, 2-methyl-N-(trimethylsilyl)-, trimethylsilyl ester	54745-23-2	NI23975
102	147	73	232	115	75	142	45	**247**	100	55	47	24	24	22	17	9	**4.00**	10	25	2	1	–	–	–	–	–	2	–	–	–	Butanoic acid, 2-[(trimethylsilyl)amino]-, trimethylsilyl ester		MS05321
73	130	45	59	75	131	43	147	**247**	100	84	27	18	17	13	13	12	**0.00**	10	25	2	1	–	–	–	–	–	2	–	–	–	Butanoic acid, 2-[(trimethylsilyl)amino]-, trimethylsilyl ester	55133-91-0	NI23976
73	116	45	75	100	59	43	147	**247**	100	94	30	25	22	19	17	16	**0.00**	10	25	2	1	–	–	–	–	–	2	–	–	–	Butanoic acid, 3-[(trimethylsilyl)amino]-, trimethylsilyl ester		MS05322
102	147	73	232	115	75	142	45	**247**	100	55	47	24	24	22	17	9	**4.00**	10	25	2	1	–	–	–	–	–	2	–	–	–	Butanoic acid, 4-[(trimethylsilyl)amino]-, trimethylsilyl ester		MS05323
130	73	45	131	59	147	43	75	**247**	100	89	25	13	13	12	12	11	**0.00**	10	25	2	1	–	–	–	–	–	2	–	–	–	L-Norvaline, N-(trimethylsilyl)-, trimethylsilyl ester		MS05324

MASS TO CHARGE RATIOS								M.W.	INTENSITIES								Parent	C	H	O	N	S	F	Cl	Br	I	Si	P	B	X	COMPOUND NAME	CAS Reg No	No
73	102	45	75	59	147	43	74	**247**	100	99	33	28	23	17	17	13	**0.00**	10	25	2	1	–	–	–	–	–	2	–	–	–	**Propanoic acid, 2-methyl-3-[(trimethylsilyl)amino]-, trimethylsilyl ester**		**MS05325**
102	73	45	147	75	59	103	43	**247**	100	67	16	13	13	10	9	9	**0.00**	10	25	2	1	–	–	–	–	–	2	–	–	–	**Propanoic acid, 2-methyl-3-[(trimethylsilyl)amino]-, trimethylsilyl ester**		**MS05326**
247	174	129	77	214	51	147	248	**247**	100	64	58	44	33	19	17	16		11	9	–	3	2	–	–	–	–	–	–	–	–	**Thiazolo[2,3-c]-s-triazole, 3-methylthio-4-phenyl-**		**LI04932**
247	124	96	52	41	95	94	55	**247**	100	77	44	28	28	27	25	22		11	9	4	3	–	–	–	–	–	–	–	–	–	**1,3-Benzenediol, 4-(2,6-dihydroxy-3-pyridylazo)-**		**MS05327**
174	216	247	130	175	217	161	103	**247**	100	80	21	19	11	10	10	10		11	13	2	5	–	–	–	–	–	–	–	–	–	**Alanine, N-(1-oxo-1,2-dihydrophthalazin-4-yl)-, hydrazide**		**NI23977**
98	69	114	96	127	168	164	140	**247**	100	57	41	41	27	5	5	5	**0.40**	11	21	5	1	–	–	–	–	–	–	–	–	–	**β-D-Galactosylamine, N-cyclopentyl-**		**NI23978**
98	56	69	60	114	127	216	193	**247**	100	44	37	34	17	12	3	3	**0.20**	11	21	5	1	–	–	–	–	–	–	–	–	–	**β-D-Mannosylamine, N-cyclopentyl-**		**NI23979**
204	41	166	178	69	56	247	110	**247**	100	54	21	20	20	20	18	11		11	22	3	1	–	–	–	–	–	–	1	–	–	**1,3,2-Dioxaphosphorinane, 2-(cyclohexylamino)-5,5-dimethyl-, 2-oxide**		**NI23980**
121	219	247	92	70	128	134	187	**247**	100	59	38	31	31	29	26	26		12	9	3	1	1	–	–	–	–	–	–	–	–	**Benzopyrano[3,4-d]oxazol-4(H)-one, 2-(ethylthio)-**		**NI23981**
43	205	74	104	45	206	77	51	**247**	100	69	35	31	25	9	9	7	**5.00**	12	9	3	1	1	–	–	–	–	–	–	–	–	**4-Isothiazolol, 5-formyl-3-phenyl-, acetate**		**LI04933**
247	219	218	248	77	220	161	103	**247**	100	72	29	16	13	11	9	9		12	9	3	1	1	–	–	–	–	–	–	–	–	**Pyrano[3,2-f]benzoxazol-6(H)-one, 4,8-dimethyl-2-mercapto-**	84737-01-9	**NI23982**
204	172	43	77	247	205	91	51	**247**	100	96	60	50	43	33	26	15		12	13	1	3	1	–	–	–	–	–	–	–	–	**1,2,3-Triazole, 4-(acetylthiomethyl)-5-methyl-2-phenyl-**		**NI23983**
120	52	79	78	80	51	106	200	**247**	100	61	56	49	44	41	34	27	**27.00**	12	13	3	3	–	–	–	–	–	–	–	–	–	**4,5-Isoxazoledione, 3-methyl-, 4-[(2-ethoxyphenyl)hydrazone]**	5669-79-4	**NI23984**
120	52	79	78	80	51	106	247	**247**	100	61	56	49	44	41	34	27		12	13	3	3	–	–	–	–	–	–	–	–	–	**4,5-Isoxazoledione, 3-methyl-, 4-[(2-ethoxyphenyl)hydrazone]**		**LI04934**
135	247	161	206	222	119	134	77	**247**	100	76	28	25	21	17	16	16		12	13	3	3	–	–	–	–	–	–	–	–	–	**1,2,3-Triazole-4-carboxylic acid, 5-(4-methoxyphenyl)-, ethyl ester**		**MS05328**
91	120	42	202	83	92	84	111	**247**	100	57	48	25	25	18	16	15	**0.00**	12	17	1	5	–	–	–	–	–	–	–	–	–	**2,6-Piperazinediimine, 4-benzyl-, formamide**		**MS05329**
172	41	43	116	205	29	247	60	**247**	100	54	51	48	35	32	30	26		12	25	–	1	2	–	–	–	–	–	–	–	–	**Carbamodithioic acid, dibutyl-, propyl ester**	19047-84-8	**MS05330**
158	160	116	247	173	87	74	159	**247**	100	98	79	24	17	15	15	13		12	25	2	1	1	–	–	–	–	–	–	–	–	**L-Alanine, N-isopropyl-3-(isopropylthio)-, isopropyl ester**	31552-14-4	**MS05331**
125	111	230	75	231	127	247	121	**247**	100	58	54	54	52	43	35	32		13	10	2	1	–	–	1	–	–	–	–	–	–	**Aniline, 4-chloro-N-(2-hydroxybenzylidene)-, N-oxide**		**MS05332**
139	247	141	111	249	140	75	248	**247**	100	35	33	20	12	8	8	7		13	10	2	1	–	–	1	–	–	–	–	–	–	**Benzamide, 2-chloro-N-(4-hydroxyphenyl)-**	35607-02-4	**NI23985**
171	170	115	156	130	129	172	74	**247**	100	56	51	28	22	20	14	14	**0.00**	13	13	–	1	2	–	–	–	–	–	–	–	–	**1,3-Thiazole-5(4H)-thione, 4,4-dimethyl-2-styryl-**		**DD00665**
91	247	92	155	65	182	39	168	**247**	100	66	59	54	43	17	14	11		13	13	2	1	1	–	–	–	–	–	–	–	–	**Benzenesulphonamide, 4-methyl-N-phenyl-**	68-34-8	**NI23986**
106	77	247	79	51	107	104	78	**247**	100	34	33	20	18	12	12	11		13	13	2	1	1	–	–	–	–	–	–	–	–	**Benzenesulphonamide, N-(2-methylphenyl)-**	18457-86-8	**NI23987**
43	119	247	118	44	204	87	59	**247**	100	80	61	52	44	32	30	30		13	13	2	1	1	–	–	–	–	–	–	–	–	**Thiazole, 5-acetyl-4-(hydroxymethyl)-2-(4-methylphenyl)-**	67387-03-5	**NI23988**
91	43	148	106	133	132	247	65	**247**	100	50	40	34	28	28	17	15		13	13	4	1	–	–	–	–	–	–	–	–	–	**2,5-Furandione, tetrahydro-3-(N-benzylacetamido)-**		**IC03252**
157	118	90	115	55	128	91	89	**247**	100	89	69	44	44	43	40	28	**0.00**	13	13	4	1	–	–	–	–	–	–	–	–	–	**1-Naphthalone, 2-nitro-2-(2-formylethyl)-1,2,3,4-tetrahydro-**		**DD00666**
247	230	202	248	187	232	203	231	**247**	100	49	29	21	16	8	8	8		13	13	4	1	–	–	–	–	–	–	–	–	–	**1-Naphthol, 5,8-dimethoxy-6-methyl-2-nitroso-**		**NI23989**
150	247	232	104	120	76	151	92	**247**	100	45	24	24	14	12	9	6		13	13	4	1	–	–	–	–	–	–	–	–	–	**1,2-Pentadien-1-ol, 3-methyl-, 4-nitrobenzoate**	56009-22-4	**NI23990**
160	202	247	205	130	148	187	104	**247**	100	58	12	12	12	10	8	6		13	13	4	1	–	–	–	–	–	–	–	–	–	**Pentanoic acid, 2-amino-, N-phthaloyl-**		**LI04935**
160	173	229	161	130	247	188	174	**247**	100	28	25	20	20	15	14	12		13	13	4	1	–	–	–	–	–	–	–	–	–	**Pentanoic acid, 5-phthalimido-**		**LI04936**
150	104	79	81	76	219	80	53	**247**	100	19	17	16	15	14	14	12	**2.00**	13	13	4	1	–	–	–	–	–	–	–	–	–	**1-Pentyn-3-ol, 3-methyl-, 4-nitrobenzoate**	56009-23-5	**NI23991**
170	229	187	201	159	132	215	188	**247**	100	93	79	69	61	52	51	42	**40.00**	13	13	4	1	–	–	–	–	–	–	–	–	–	**1H-Pyrrolo[1,2-a]indole-9-carboxylic acid, 9a-hydroxy-3-oxo-2,3,9α,9aα-tetrahydro-, methyl ester, (±)-**		**NI23992**
202	148	247	160	130	203	104	187	**247**	100	62	42	42	37	36	33	32		13	13	4	1	–	–	–	–	–	–	–	–	–	**Valine, phthaloyl-**	19506-85-5	**NI23993**
130	73	247	77	202	131	75	103	**247**	100	64	32	14	13	12	10	9		13	17	2	1	–	–	–	–	–	1	–	–	–	**1H-Indole-1-acetic acid, trimethylsilyl ester**		**NI23995**
105	206	73	77	207	236	75	51	**247**	100	99	56	32	27	18	13	7	**0.00**	13	17	2	1	–	–	–	–	–	1	–	–	–	**1H-Indole-1-acetic acid, trimethylsilyl ester**	74367-53-6	**NI23994**
188	247	89	158	232	189	248	73	**247**	100	77	47	28	17	17	15	12		13	17	2	1	–	–	–	–	–	1	–	–	–	**1H-Indole-2-carboxylic acid, 1-methyl-, trimethylsilyl ester**		**NI23996**
247	69	41	55	138	248	54	122	**247**	100	65	28	26	18	14	13	12		13	17	2	3	–	–	–	–	–	–	–	–	–	**Cyclohexanone, 4-methyl-, 4-nitrophenylhydrazone**		**MS05333**
55	149	41	57	43	70	118	28	**247**	100	98	95	91	80	79	63	52	**0.00**	13	17	2	3	–	–	–	–	–	–	–	–	–	**5,9-Methano-5H-[1,4,2,3]dioxadiazolo[2,3-a][1,2]diazepin-2-amine, tetrahydro-N-phenyl-**	56909-14-9	**NI23997**
219	102	218	91	247	120	90	220	**247**	100	68	46	32	26	25	24	19		14	9	–	5	–	–	–	–	–	–	–	–	–	**1,2,3-Benzotriazine, 4-(4-cyananilino)-**		**LI04937**
118	102	221	91	219	103	90	77	**247**	100	-52	46	46	25	25	18	15	**2.70**	14	9	–	5	–	–	–	–	–	–	–	–	–	**Triazene, bis(2-cyanophenyl)-**		**LI04938**
247	105	166	106	104	171	91	248	**247**	100	57	55	45	36	33	29	19		14	17	1	1	1	–	–	–	–	–	–	–	–	**1,3-Benzoxazine, perhydro-4-phenyl-2-thioxo-, cis-**		**MS05334**
247	105	106	104	166	91	28	171	**247**	100	79	74	70	61	45	37	31		14	17	1	1	1	–	–	–	–	–	–	–	–	**1,3-Oxazine-2-thione, 4-phenyl-5,6-tetramethylenetetrahydro-**		**NI23998**
175	190	89	145	247	176	52	117	**247**	100	57	41	40	37	27	26	24		14	17	3	1	–	–	–	–	–	–	–	–	–	**Cinnamamide, N-isobutyl-3,4-(methylenedioxy)-**		**LI04939**
39	247	43	172	51	150	53	41	**247**	100	91	76	69	67	50	41	40		14	17	3	1	–	–	–	–	–	–	–	–	–	**[1,3]Dioxepino[5,6-c]pyridine, 1,5-dihydro-3,3,8-trimethyl-9-(2-propynyloxy)-**	69022-74-8	**NI23999**
105	120	173	172	131	145	130	132	**247**	100	31	13	11	11	7	6	5	**0.00**	14	17	3	1	–	–	–	–	–	–	–	–	–	**Ethaneperoxoic acid, 1-cyano-1-phenylpentyl ester**	58422-73-4	**NI24000**
247	188	160	63	215	43	216	106	**247**	100	59	38	31	30	26	20	20		14	17	3	1	–	–	–	–	–	–	–	–	–	**1H-Indole-3-acetic acid, 5-methoxy-1,2-dimethyl-, methyl ester**	55133-88-5	**NI24001**
247	202	218	175	131	130	101	248	**247**	100	74	49	43	34	31	20	18		14	17	3	1	–	–	–	–	–	–	–	–	–	**1H-Indole-3-carboxylic acid, 1,2-dimethyl-5-methoxy-, ethyl ester**		**DD00667**
188	43	105	104	77	28	85	189	**247**	100	74	30	29	26	24	21	14	**0.00**	14	17	3	1	–	–	–	–	–	–	–	–	–	**Oxazole-4-carboxylic acid, 4,5-dihydro-4-ethyl-5-methyl-2-phenyl-, methyl ester, (4S,5S)-**		**DD00668**
247	105	77	248	51	106	42	41	**247**	100	99	92	84	65	50	45	33		14	17	3	1	–	–	–	–	–	–	–	–	–	**4-Piperidineacetic acid, 1-benzoyl-**	56772-11-3	**NI24002**
83	145	173	85	146	191	190	103	**247**	100	98	96	65	64	60	44	40	**0.40**	14	17	3	1	–	–	–	–	–	–	–	–	–	**2-Pyridinecarboxylic acid, 5-(3-hydroxy-3,4,4-trimethyl-1-pentyn-1-yl)-**		**NI24003**
130	29	129	202	158	27	128	102	**247**	100	65	55	54	29	21	18	16	**3.35**	14	17	3	1	–	–	–	–	–	–	–	–	–	**1(2H)-Quinolinecarboxylic acid, 2-ethoxy-, ethyl ester**	16357-59-8	**NI24004**
106	134	247	114	78	56	82	55	**247**	100	95	92	90	90	88	84	68		14	17	3	1	–	–	–	–	–	–	–	–	–	**Securinan-11-one, 4β-methoxy-**	20072-02-0	**NI24005**

MASS TO CHARGE RATIOS								M.W.	INTENSITIES								Parent	C	H	O	N	S	F	Cl	Br	I	Si	P	B	X	COMPOUND NAME	CAS Reg No	No
56	114	78	106	82	134	247	55	247	100	78	47	44	34	22	16	16		14	17	3	1	–	–	–	–	–	–	–	–	–	Securitinine		LI04940
38	248	233	231	217	215	249	232	247	100	88	87	86	86	85	83	82	37.37	14	19	2	2	–	–	–	–	–	–	–	–	–	1H-Imidazol-1-yloxy, 2,5-dihydro-2,2,5,5-tetramethyl-4-(4-methylphenyl)-,3-oxide-	81889-92-1	NI24006
191	247	75	232	73	246	158	205	247	100	95	92	91	85	77	57	41		14	21	1	1	–	–	–	–	–	1	–	–	–	1H-Indole, 2-methoxy-3,5-dimethyl-1-(trimethylsilyl)-		NI24007
133	72	132	119	148	106	177	147	247	100	99	83	63	56	46	26	25	0.00	14	21	1	3	–	–	–	–	–	–	–	–	–	2H-Indol-2-one, 3-[[1-(aminomethyl)propyl]ethylamino]-1,3-dihydro-	55570-81-5	NI24008
56	41	36	30	43	122	57	55	247	100	16	16	16	14	12	10	10	0.00	14	30	–	1	–	–	1	–	–	–	–	–	–	Cyclotetradecanamine hydrochloride		MS05335
231	77	39	247	50	232	76	179	247	100	44	23	22	20	18	18	17		15	9	1	3	–	–	–	–	–	–	–	–	–	2-Quinoxalinecarbonitrile, 3-phenyl-, 1-oxide	18916-51-3	NI24009
231	247	230	76	232	179	102	50	247	100	25	21	21	18	18	18	17		15	9	1	3	–	–	–	–	–	–	–	–	–	2-Quinoxalinecarbonitrile, 3-phenyl-, 4-oxide	18457-82-4	NI24010
105	120	106	81	77	41	172	18	247	100	64	61	48	41	41	39	38	21.00	15	21	2	1	–	–	–	–	–	–	–	–	–	Cyclobutanemethanol, 3-methyl-1-[(1-phenylethyl)carbamoyl]-	85318-01-0	NI24011
248	249	247	246	250				247	100	17	6	6	2					15	21	2	1	–	–	–	–	–	–	–	–	–	4-Piperidinecarboxylic acid, 1-methyl-4-phenyl-, ethyl ester	57-42-1	NI24012
71	70	42	28	57	43	172	25	247	100	54	49	47	33	31	30	22	22.02	15	21	2	1	–	–	–	–	–	–	–	–	–	4-Piperidinecarboxylic acid, 1-methyl-4-phenyl-, ethyl ester	57-42-1	NI24013
71	70	103	91	247	172	96	77	247	100	61	26	23	22	17	16	16		15	21	2	1	–	–	–	–	–	–	–	–	–	4-Piperidinecarboxylic acid, 1-methyl-4-phenyl-, ethyl ester	57-42-1	WI01273
105	77	42	132	128	56	51	131	247	100	86	40	29	26	26	24	21	0.00	15	21	2	1	–	–	–	–	–	–	–	–	–	4-Piperidinol, 4-methyl-N-(3-oxo-3-phenylpropyl)-		MS05336
247	96	151	232	246	84	216	218	247	100	90	80	70	60	46	40	38		15	21	2	1	–	–	–	–	–	–	–	–	–	Pyrrolidine, 2-[2-(3,4-dimethoxyphenyl)vinyl]-1-methyl-, (E)-(±)-	67257-62-9	NI24014
247	201	200	100	189	248	202	99	247	100	100	69	59	36	18	16	14		16	9	2	1	–	–	–	–	–	–	–	–	–	Fluoranthene, 1-nitro-	13177-28-1	NI24015
247	201	200	100	189	248	202	199	247	100	98	66	36	28	18	17	14		16	9	2	1	–	–	–	–	–	–	–	–	–	Fluoranthene, 2-nitro-		NI24016
247	200	201	189	217	100	248	191	247	100	71	70	45	31	30	19	18		16	9	2	1	–	–	–	–	–	–	–	–	–	Fluoranthene, 3-nitro-		NI24017
201	247	200	189	217	100	248	202	247	100	97	63	55	43	27	18	18		16	9	2	1	–	–	–	–	–	–	–	–	–	Pyrene, 1-nitro-		IC03253
201	247	100	200	189	217	87	99	247	100	85	82	76	60	38	21	19		16	9	2	1	–	–	–	–	–	–	–	–	–	Pyrene, 1-nitro-	5522-43-0	NI24018
207	247	194	208	206	205	103	152	247	100	54	49	23	23	20	19	12		16	13	–	3	–	–	–	–	–	–	–	–	–	1H-Pyrrolo[2,3-b]pyridine-1-propanenitrile, 2-phenyl-	23616-60-6	NI24019
190	232	247	57	191	43	41	134	247	100	80	32	29	28	28	26	18		16	25	1	1	–	–	–	–	–	–	–	–	–	Acetanilide, 2,4-di-tert-butyl-		MS05337
135	57	41	43	39	58	79	42	247	100	70	58	56	33	25	19	18	2.20	16	25	1	1	–	–	–	–	–	–	–	–	–	Aziridinone, 1-(1-adamantyl)-3-tert-butyl-	16664-32-7	NI24020
135	57	93	219	107	204	162	77	247	100	35	20	18	18	17	16	15	1.00	16	25	1	1	–	–	–	–	–	–	–	–	–	Aziridinone, 1-(1-adamantyl)-3-tert-butyl-		LI04941
81	167	57	41	79	66	53	115	247	100	96	73	66	38	33	33	31	11.07	16	25	1	1	–	–	–	–	–	–	–	–	–	(2E,4Z,8Z,10E)-Dodecatetraenamide, N-isobutyl-		NI24021
190	191	162	136	134	204	176	135	247	100	33	21	21	17	15	15	14	11.86	16	25	1	1	–	–	–	–	–	–	–	–	–	Lycopodan-5-one, 16-methyl-, (13α,16R)-	6883-69-8	NI24023
190	191	247	188	162	134	137	136	247	100	10	5	4	3	3	2	2		16	25	1	1	–	–	–	–	–	–	–	–	–	Lycopodan-5-one, 16-methyl-, (16R)-	466-61-5	NI24022
156	72	85	56	44	145	91	41	247	100	70	35	35	20	15	13	13	0.20	16	25	1	1	–	–	–	–	–	–	–	–	–	4-Octanone, 2-(dimethylamino)-1-phenyl-	27820-10-6	NI24024
247	219	217	218	108.5	248	245	109.5	247	100	63	30	27	26	16	12	11		17	13	1	1	–	–	–	–	–	–	–	–	–	Benzo[a,f]quinolizin-6-one, 12,13-dihydro-		MS05338
247	176	248	218	246	177	189	217	247	100	22	20	20	12	12	8	7		17	13	1	1	–	–	–	–	–	–	–	–	–	3H-Naphth[1,2-e]indol-10-ol, 3-methyl-	98033-21-7	NI24025
247	246	18	218	217	248	39	189	247	100	67	35	31	27	21	20	17		17	13	1	1	–	–	–	–	–	–	–	–	–	Quinoline, 4-(4-hydroxystyryl)-	789-76-4	NI24026
120	107	106	121	41	77	134	43	247	100	94	69	31	30	25	21	16	12.50	17	29	–	1	–	–	–	–	–	–	–	–	–	Pyridine, 2-methyl-5-undecyl-	52535-38-3	NI24027
205	232	246	247	191	41	233	216	247	100	97	51	41	32	19	17	17		17	29	–	1	–	–	–	–	–	–	–	–	–	Pyridine, 2,4,6-tri-tert-butyl-	20336-15-6	NI24028
248	69	179	169	129	160	229	119	248	100	60	50	33	23	21	9	3		3	–	–	–	–	7	–	1	–	–	–	–	–	Propane, 2-bromo-1,1,1,2,3,3,3-heptafluoro-	422-77-5	NI24029
69	100	178	70	31	159	51	28	248	100	60	40	39	22	20	12	6	0.40	3	–	–	2	1	8	–	–	–	–	–	–	–	Sulphur difluorodimide, N-hexafluoroisopropylidine-		LI04942
63	78	45	248	61	15	169	48	248	100	66	62	43	20	18	17	10		3	6	1	–	1	–	–	2	–	–	–	–	–	Dimethylsulphoxonium dibromomethylide		MS05339
115	117	63	250	248	150	116	252	248	100	35	20	17	13	12	10	9		4	4	–	–	–	–	4	–	–	2	–	–	–	1,4-Disilacyclohexa-2,5-diene, 1,1,4,4-tetrachloro-	33432-30-3	NI24030
250	248	252	213	215	108	217	254	248	100	70	62	31	27	23	17	13		6	1	–	–	–	–	5	–	–	–	–	–	–	Benzene, pentachloro-	608-93-5	NI24031
250	252	248	254	215	213	108	217	248	100	65	63	21	17	13	12	8		6	1	–	–	–	–	5	–	–	–	–	–	–	Benzene, pentachloro-	608-93-5	NI24032
216	250	214	252	248	218	108	179	248	100	91	82	57	54	50	31	24		6	1	–	–	–	–	5	–	–	–	–	–	–	Benzene, pentachloro-	608-93-5	NI24033
248	246	231	79	233	164	69	63	248	100	76	49	42	40	37	35	35		7	3	1	–	–	6	–	–	–	–	1	–	–	Phosphine oxide, fluoromethyl(pentafluorophenyl)-	53381-00-3	NI24034
248	231	65	76	203	50	127	74	248	100	60	17	17	17	16	10	9		7	5	2	–	–	–	–	–	1	–	–	–	–	Benzoic acid, 2-iodo-	88-67-5	NI24035
248	65	76	50	249	39	93	51	248	100	12	12	10	8	5	5	4		7	5	2	–	–	–	–	–	1	–	–	–	–	Benzoic acid, 2-iodo-	88-67-5	NI24036
169	171	90	89	63	250	62	50	248	100	96	51	38	30	14	14	11	7.24	7	6	–	–	–	–	–	2	–	–	–	–	–	Benzene, 1-bromo-3-(bromomethyl)-	823-78-9	NI24037
169	171	90	89	63	62	50	51	248	100	94	51	41	33	15	12	10	1.00	7	6	–	–	–	–	–	2	–	–	–	–	–	Benzene, (dibromomethyl)-	618-31-5	NI24038
89	63	90	250	169	171	62	248	248	100	79	76	71	71	69	40	38		7	6	–	–	–	–	–	2	–	–	–	–	–	Toluene, 2,5-dibromo-	615-59-8	NI24039
250	248	252	169	171	89	88	63	248	100	51	49	47	46	24	24	15		7	6	–	–	–	–	–	2	–	–	–	–	–	Toluene, 3,5-dibromo-	1611-92-3	NI24040
169	171	90	89	63	39	250	62	248	100	97	46	32	24	15	14	11	7.28	7	6	–	–	–	–	–	2	–	–	–	–	–	Toluene, α,2-dibromo-	3433-80-5	NI24041
179	248	198	148	229	117	98	141	248	100	80	59	23	21	17	16	12		8	–	–	–	–	8	–	–	–	–	–	–	–	Styrene, octafluoro-	652-23-3	NI24042
179	248	198	148	229	117	98	141	248	100	81	63	24	20	17	17	15		8	–	–	–	–	8	–	–	–	–	–	–	–	Styrene, octafluoro-	652-23-3	LI04943
143	94	109	205	105	82	248	233	248	100	75	67	58	58	58	50	50		8	6	2	–	–	6	–	–	–	–	–	–	–	Bicyclo[2.2.0]hex-2-ene, 2,3-dimethoxyhexafluoro-		NI24043
248	205	143	233	109	81	93	229	248	100	51	51	40	36	36	24	20		8	6	2	–	–	6	–	–	–	–	–	–	–	1,3-Cyclohexadiene, 2,3-dimethoxyhexafluoro-		NI24044
164	163	220	248	191	219	192	247	248	100	47	46	43	19	16	15	3		8	6	3	–	–	–	–	–	–	–	–	–	1	Molybdenum, π-cyclopentadienylhydridotricarbonyl-		LI04944
179	120	131	230	69	200	119	93	248	100	43	32	28	28	18	18	18	4.12	8	7	2	4	–	3	–	–	–	–	–	–	–	4-Pteridinol, 3,4-dihydro-2-methoxy-4-(trifluoromethyl)-	32710-63-7	NI24045
248	220	93	65	63	29	64	39	248	100	78	75	62	40	33	31	31		8	9	1	–	–	–	–	–	1	–	–	–	–	Phenetole, 3-iodo-	29052-00-4	NI24046

MASS TO CHARGE RATIOS								M.W.	INTENSITIES								Parent	C	H	O	N	S	F	Cl	Br	I	Si	P	B	X	COMPOUND NAME	CAS Reg No	No
248	220	65	93	39	63	64	29	**248**	100	90	65	51	37	36	30	28		8	9	1	–	–	–	–	–	1	–	–	–	–	**Phenetole, 4-iodo-**	699-08-1	**NI24047**
248	155	45	110	47	46	103	58	**248**	100	95	81	54	52	48	46	43		8	12	1	2	3	–	–	–	–	–	–	–	–	**4-Isothiazolecarboxamide, N-ethyl-3,5-bis(methylthio)-**	37572-35-3	**NI24048**
248	155	45	110	47	46	103	233	**248**	100	95	81	54	52	48	46	43		8	12	1	2	3	–	–	–	–	–	–	–	–	**4-Isothiazolecarboxamide, N,N-dimethyl-3,5-bis(methylthio)-**	4886-19-5	**LI04945**
204	248	72	45	131	46	44	76	**248**	100	42	37	31	26	22	22	15		8	12	1	2	3	–	–	–	–	–	–	–	–	**4-Isothiazolecarboxamide, N,N-dimethyl-3,5-bis(methylthio)-**	4886-19-5	**NI24049**
201	248	139	233	71	69	250	165	**248**	100	71	50	45	28	28	27	24		8	13	1	4	1	–	1	–	–	–	–	–	–	**1,3,5-Triazin-2-amine, 4-(2-chloroethoxy)-6-methylthio-N,N-dimethyl-**	65248-19-3	**NI24050**
201	248	139	203	233	250	235	202	**248**	100	52	48	46	36	29	22	16		8	13	1	4	1	–	1	–	–	–	–	–	–	**1,3,5-Triazin-2(1H)-one, 1-(2-chloroethyl)-4-dimethylamino-6-methylthio-**	65248-21-7	**NI24051**
107	96	106	97	143	248	140	79	**248**	100	98	73	64	53	43	41	40		8	13	3	2	1	–	–	–	–	–	1	–	–	**Phosphorothioic acid, O,O-diethyl O-pyrazinyl ester**	297-97-2	**NI24052**
103	86	128	60	248	147	69	102	**248**	100	64	60	56	45	41	37	30		8	16	1	4	2	–	–	–	–	–	–	–	–	**1-Oxa-4,6,9,11-tetraazacyclotridecane-5,10-dithione**	111039-12-4	**NI24053**
115	248	250	113	88	114	141	89	**248**	100	95	94	83	44	24	22	13		9	5	–	4	–	–	–	1	–	–	–	–	–	**s-Triazolo[3,4-a]phthalazine**		**LI04946**
164	72	55	248	192	165	93	80	**248**	100	52	36	22	14	7	5	4		9	5	5	–	–	–	–	–	–	–	–	–	1	**Manganese, tricarbonylcarboxycyclopentadienyl-**		**LI04947**
61	32	46	248	160	60	133	45	**248**	100	50	21	7	7	7	6	6		9	10	2	2	–	–	2	–	–	–	–	–	–	**Urea, N′-(3,4-dichlorophenyl)-N-methoxy-N-methyl-**	330-55-2	**NI24054**
187	124	189	161	159	61	126	88	**248**	100	79	63	43	34	29	28	15	**6.00**	9	10	2	2	–	–	2	–	–	–	–	–	–	**Urea, N′-(3,4-dichlorophenyl)-N-methoxy-N-methyl-**	330-55-2	**NI24055**
107	105	169	79	63	77	91	90	**248**	100	64	46	35	26	16	9	9	**1.20**	9	12	4	–	2	–	–	–	–	–	–	–	–	**Benzene, [bis(methylsulphonyl)methyl]-**	58751-73-8	**NI24056**
249	139	155	95	181	75	157	151	**248**	100	45	28	25	22	20	13	13	**0.00**	9	12	4	–	2	–	–	–	–	–	–	–	–	**Benzene, 1-methyl-4-[[(methylsulphonyl)methyl]sulphonyl]-**	64399-25-3	**NI24057**
91	155	121	65	39	63	105	92	**248**	100	80	68	60	31	29	24	23	**15.29**	9	12	4	–	2	–	–	–	–	–	–	–	–	**Benzene, 1-methyl-4-[[(methylsulphonyl)methyl]sulphonyl]-**	64399-25-3	**NI24058**
91	105	65	92	104	248	63	39	**248**	100	10	10	8	7	2	2	2		9	12	4	–	2	–	–	–	–	–	–	–	–	**Benzene, [[[(methylsulphonyl)methyl]sulphonyl]methyl]-**	56620-03-2	**NI24059**
43	57	164	163	162	165	120	86	**248**	100	99	63	63	61	60	28	28	**0.00**	9	13	3	–	–	–	–	1	–	–	–	–	–	**2H,4H-1,3-Dioxin-4-one, 5-bromo-2-tert-butyl-6-methyl-, (2R)-**		**DD00669**
153	90	132	61	189	117	170	104	**248**	100	78	77	62	58	42	42	39	**1.61**	9	13	4	2	–	–	1	–	–	–	–	–	–	**Uracil, 6-acetoxy-5-chloro-5,6-dihydro-1,3,5-trimethyl-**		**NI24060**
102	130	217	186	103	230	101	171	**248**	100	58	54	35	35	22	22	12	**0.00**	9	16	6	2	–	–	–	–	–	–	–	–	–	**L-Serine, N-(N-carboxyethyl)-, N-glycyl-, methyl ester**		**LI04948**
165	151	233	135	248	121	150	120	**248**	100	53	21	20	9	8	7	5		9	20	–	–	–	–	–	–	–	–	–	–	1	**Stannane, (cyclohexyl)trimethyl-**		**LI04949**
165	229	163	151	161	149	164	147	**248**	100	93	74	53	45	42	29	26	**8.80**	9	20	–	–	–	–	–	–	–	–	–	–	1	**Stannane, (cyclohexyl)trimethyl-**		**MS05340**
219	217	149	147	215	145	218	191	**248**	100	76	76	57	46	35	32	29	**2.10**	9	20	–	–	–	–	–	–	–	–	–	–	1	**Stannane, (cyclopropyl)triethyl-**		**MS05341**
147	75	73	45	148	44	47	66	**248**	100	63	54	19	15	14	10	9	**0.00**	9	20	4	–	–	–	–	–	–	2	–	–	–	**Propanedioic acid, bis(trimethylsilyl) ester**	18457-04-0	**MS05342**
147	73	75	148	45	66	149	233	**248**	100	67	20	17	16	11	9	8	**1.30**	9	20	4	–	–	–	–	–	–	2	–	–	–	**Propanedioic acid, bis(trimethylsilyl) ester**	18457-04-0	**NI24061**
147	73	75	148	45	66	149	233	**248**	100	67	20	17	16	12	9	8	**1.30**	9	20	4	–	–	–	–	–	–	2	–	–	–	**Propanedioic acid, bis(trimethylsilyl) ester**	18457-04-0	**NI24062**
90	63	64	42	39	136	91	248	**248**	100	62	47	41	38	37	31	23		10	8	4	4	–	–	–	–	–	–	–	–	–	**4,5-Isoxazoledione, 3-methyl-, 4-[(2-nitrophenyl)hydrazone]**	5669-80-7	**NI24063**
59	28	30	89	115	88	130	103	**248**	100	46	39	33	23	19	16	14	**3.32**	10	8	4	4	–	–	–	–	–	–	–	–	–	**2-Propenoic acid, 2-azido-3-(4-nitrophenyl)-, methyl ester**	106053-85-4	**NI24064**
183	185	39	148	65	248	83	250	**248**	100	67	55	50	33	31	24	20		10	10	–	–	–	–	2	–	–	–	–	–	1	**Titanium, dichlorobis(η^{5}-2,4-cyclopentadien-1-yl)-**	1271-19-8	**NI24065**
175	59	177	185	111	248	145	109	**248**	100	69	63	62	56	50	50	43		10	10	3	–	–	–	2	–	–	–	–	–	–	**Acetic acid, (2,4-dichlorophenoxy)-, ethyl ester**	533-23-3	**NI24066**
43	162	45	87	41	164	63	39	**248**	100	99	96	82	73	71	43	30	**4.41**	10	10	3	–	–	–	2	–	–	–	–	–	–	**Butanoic acid, 4-(2,4-dichlorophenoxy)-**	94-82-6	**NI24067**
162	164	45	87	43	41	27	63	**248**	100	87	76	64	57	54	40	29	**13.41**	10	10	3	–	–	–	2	–	–	–	–	–	–	**Butanoic acid, 4-(2,4-dichlorophenoxy)-**	94-82-6	**NI24068**
162	164	87	43	45	41	166	63	**248**	100	66	33	29	27	16	11	10	**2.86**	10	10	3	–	–	–	2	–	–	–	–	–	–	**Butanoic acid, 4-(2,4-dichlorophenoxy)-**	94-82-6	**NI24069**
117	59	43	248	131	183	132	40	**248**	100	41	33	27	25	16	16	8		10	16	1	–	3	–	–	–	–	–	–	–	–	**2-Oxa-4,6,8-trithiaadamantane, 1,3,5,7-tetramethyl-, (±)-**	16680-00-5	**NI24070**
56	117	248	55	41	200	88	84	**248**	100	39	32	19	19	13	13	11		10	16	3	–	2	–	–	–	–	–	–	–	–	**8,10-Dioxa-4,9-dithiabicyclo[5.3.0]dec-1(7)-ene, 2,2,6,6-tetramethyl-, 9-oxide**		**MS05343**
103	41	129	29	59	27	28	31	**248**	100	68	26	22	21	16	15	14	**0.00**	10	16	7	–	–	–	–	–	–	–	–	–	–	**Methyl allyl diglycolcarbonate**		**NI24071**
89	45	117	87	161	72	205	42	**248**	100	82	48	38	23	10	7	7	**0.10**	10	16	7	–	–	–	–	–	–	–	–	–	–	**1,4,7,10,13-Pentaoxacyclopentadecane-2,3-dione**		**MS05344**
69	74	73	114	102	87	115	86	**248**	100	50	43	41	39	39	37	37	**0.00**	10	16	7	–	–	–	–	–	–	–	–	–	–	**β-D-Xylopyranoside, methyl 2,3-di-O-acetyl-**	70003-50-8	**NI24072**
68	86	57	69	115	85	60	74	**248**	100	95	84	76	73	69	69	64	**0.00**	10	16	7	–	–	–	–	–	–	–	–	–	–	**β-D-Xylopyranoside, methyl 2,4-di-O-acetyl-**	74162-08-6	**NI24073**
69	68	128	86	115	74	61	57	**248**	100	76	66	42	28	22	20	20	**0.00**	10	16	7	–	–	–	–	–	–	–	–	–	–	**β-D-Xylopyranoside, methyl 3,4-di-O-acetyl-**	63629-70-9	**NI24074**
29	41	57	155	43	39	95	27	**248**	100	79	66	54	44	43	38	38	**12.00**	10	17	2	–	–	–	–	1	–	–	–	–	–	**2-Pentenoic acid, 3-(bromomethyl)-4,4-dimethyl-, ethyl ester, (Z)-**	36976-65-5	**NI24075**
70	54	116	101	72	55	68	56	**248**	100	93	73	63	63	60	45	41	**0.00**	10	17	3	2	–	–	1	–	–	–	–	–	–	**L-Valine, N-[2-(chloroimino)-3-methyl-1-oxobutyl]-**	55570-83-7	**NI24076**
98	84	71	248	169	233	219	205	**248**	100	60	40	38	37	10	5	3		10	20	3	2	1	–	–	–	–	–	–	–	–	**Aziridine, 2-isopropyl-3-methyl- N-[N-(methylsulphonyl)ethoxycarbimidoyl]-, cis-**		**DD00670**
73	189	85	173	129	89	59	71	**248**	100	92	70	69	69	69	35	33	**0.00**	10	20	5	–	–	–	–	–	–	1	–	–	–	**Pentanedioic acid, 2-[(trimethylsilyl)oxy]-, dimethyl ester**	55590-88-0	**NI24077**
89	127	73	59	75	45	233	99	**248**	100	92	82	58	40	34	30	30	**0.00**	10	20	5	–	–	–	–	–	–	1	–	–	–	**Pentanedioic acid, 3-[(trimethylsilyl)oxy]-, dimethyl ester**	55590-89-1	**NI24078**
73	131	147	75	45	132	148	41	**248**	100	94	70	14	14	12	11	11	**0.00**	10	24	3	–	–	–	–	–	–	2	–	–	–	**Butanoic acid, 2-[(trimethylsilyl)oxy]-, trimethylsilyl ester**	55133-93-2	**NI24079**
73	131	147	75	45	132	74	59	**248**	100	78	61	16	16	9	9	9	**0.00**	10	24	3	–	–	–	–	–	–	2	–	–	–	**Butanoic acid, 2-[(trimethylsilyl)oxy]-, trimethylsilyl ester**	55133-93-2	**MS05345**
147	73	117	75	191	45	148	88	**248**	100	98	61	30	29	22	16	14	**0.00**	10	24	3	–	–	–	–	–	–	2	–	–	–	**Butanoic acid, 3-[(trimethylsilyl)oxy]-, trimethylsilyl ester**	55133-94-3	**NI24082**
147	73	117	191	75	233	45	88	**248**	100	78	51	30	21	17	16	15	**0.00**	10	24	3	–	–	–	–	–	–	2	–	–	–	**Butanoic acid, 3-[(trimethylsilyl)oxy]-, trimethylsilyl ester**	55133-94-3	**NI24081**
147	73	117	191	75	233	45	148	**248**	100	88	59	31	26	21	17	16	**0.09**	10	24	3	–	–	–	–	–	–	2	–	–	–	**Butanoic acid, 3-[(trimethylsilyl)oxy]-, trimethylsilyl ester**	55133-94-3	**NI24080**
73	75	147	45	41	59	117	43	**248**	100	74	64	42	23	21	20	17	**0.00**	10	24	3	–	–	–	–	–	–	2	–	–	–	**Butanoic acid, 4-[(trimethylsilyl)oxy]-, trimethylsilyl ester**	55133-95-4	**MS05346**
147	73	75	233	117	148	45	149	**248**	100	52	25	24	21	16	12	9	**0.00**	10	24	3	–	–	–	–	–	–	2	–	–	–	**Butanoic acid, 4-[(trimethylsilyl)oxy]-, trimethylsilyl ester**	55133-95-4	**NI24083**
147	73	233	75	117	148	149	143	**248**	100	46	30	23	23	16	10	7	**0.30**	10	24	3	–	–	–	–	–	–	2	–	–	–	**Butanoic acid, 4-[(trimethylsilyl)oxy]-, trimethylsilyl ester**	55133-95-4	**NI24084**
73	131	147	75	45	205	132	74	**248**	100	98	56	18	16	14	12	9	**0.00**	10	24	3	–	–	–	–	–	–	2	–	–	–	**Propanoic acid, 2-methyl-2-[(trimethylsilyl)oxy]-, trimethylsilyl ester**	55133-92-1	**NI24085**
73	131	147	45	75	43	205	74	**248**	100	56	42	24	19	14	12	9	**0.00**	10	24	3	–	–	–	–	–	–	2	–	–	–	**Propanoic acid, 2-methyl-2-[(trimethylsilyl)oxy]-, trimethylsilyl ester**	55133-92-1	**NI24086**

MASS TO CHARGE RATIOS								M.W.	INTENSITIES								Parent	C	H	O	N	S	F	Cl	Br	I	Si	P	B	X	COMPOUND NAME	CAS Reg No	No
73	131	147	75	45	205	132	74	**248**	100	98	56	18	16	14	12	9	**0.00**	10	24	3	–	–	–	–	–	–	2	–	–	–	Propanoic acid, 2-methyl-2-[(trimethylsilyl)oxy]-, trimethylsilyl ester	55133-92-1	NI24087
147	73	28	75	233	148	66	103	**248**	100	71	23	20	20	16	16	15	**0.00**	10	24	3	–	–	–	–	–	–	2	–	–	–	Propanoic acid, 2-methyl-3-[(trimethylsilyl)oxy]-, trimethylsilyl ester	55530-42-2	NI24088
147	73	28	233	75	148	103	45	**248**	100	71	23	20	20	16	15	15	**0.00**	10	24	3	–	–	–	–	–	–	2	–	–	–	Propanoic acid, 2-methyl-3-[(trimethylsilyl)oxy]-, trimethylsilyl ester	55530-42-2	NI24089
248	169	142	114	43	87	63	57	**248**	100	63	29	26	24	21	21	20		11	8	3	2	1	–	–	–	–	–	–	–	–	1H-Naphtho[2,3-c][1,2,6]thiadiazin-4(3H)-one, 2,2-dioxide	40467-22-9	NI24090
92	248	156	183	64	91	213	155	**248**	100	57	53	26	13	10	9	9		11	9	1	–	–	–	1	–	–	–	–	–	1	Ferrocene, (chlorocarbonyl)-	1293-79-4	NI24091
189	105	119	77	190	139	127	91	**248**	100	61	39	37	31	24	19	19	**18.61**	11	11	3	–	–	3	–	–	–	–	–	–	–	Propanoic acid, 2-methoxy-2-phenyl-3,3,3-trifluoro-, methyl ester		NI24092
170	205	171	104	102	130	128	77	**248**	100	71	41	38	36	33	33	33	**10.00**	11	12	1	4	1	–	–	–	–	–	–	–	–	4-Amino-5-carbamoyl-6-phenyl-2-thioxo-1,2,3,6-tetrahydropyrimidine		NI24093
28	79	67	80	93	39	108	41	**248**	100	39	19	18	17	15	14	12	**0.00**	11	12	3	–	–	–	–	–	–	–	–	–	1	Iron, tricarbonyl[(1,2,3,4-η)-1,3-cyclooctadienyl]-	33270-50-7	NI24094
67	54	79	39	41	80	27	57	**248**	100	96	89	85	68	61	55	50	**0.00**	11	12	3	–	–	–	–	–	–	–	–	–	1	Iron, tricarbonyl[(1,2,5,6-η)-1,5-cyclooctadienyl]-	12093-20-8	NI24095
162	220	56	190	192	221	163	191	**248**	100	99	42	40	13	10	9	4	**3.00**	11	12	3	–	–	–	–	–	–	–	–	–	1	Iron, tricarbonyl[(1,2,3,4-η)-2,3-dimethyl-1,3-cyclohexadienyl]-	55090-53-4	NI24096
148	56	220	149	192	84	91	176	**248**	100	28	21	8	7	7	6	2	**1.00**	11	12	3	–	–	–	–	–	–	–	–	–	1	Iron, tricarbonyl[(1,2,3,4-η)-5,5-dimethyl-1,3-cyclohexadienyl]-	32732-61-9	NI24097
162	160	164	192	220	56	134	163	**248**	100	93	77	71	62	44	20	9	**8.01**	11	12	3	–	–	–	–	–	–	–	–	–	1	Iron, tricarbonyl(η(^{4})-1-vinylcyclohexenyl)-	32732-60-8	NI24098
162	56	220	91	190	105	160	248	**248**	100	64	28	16	15	12	10	4		11	12	3	–	–	–	–	–	–	–	–	–	1	Tricarbonyl[(1,2,3,4-η)-1,3-dimethyl-1,3-cyclohexadiene]iron		DD00671
204	104	103	58	44	77	76	59	**248**	100	94	81	48	45	38	33	30	**2.00**	11	12	3	4	–	–	–	–	–	–	–	–	–	4-Isoxazolecarboxamide, 5-[(aminocarbonyl)amino]-4,5-dihydro-3-phenyl-, cis-	35053-69-1	NI24099
105	77	231	106	248	42	78	41	**248**	100	85	40	19	13	9	7	6		11	12	3	4	–	–	–	–	–	–	–	–	–	Piperazine, 2,6-bis(hydroxyimino)-4-benzoyl-		MS05347
125	117	118	43	119	55	82	83	**248**	100	74	72	42	24	20	18	16	**0.00**	11	20	4	–	1	–	–	–	–	–	–	–	–	2(3H)-Furanone, dihydro-3-(methylsulphonyl)-5,5-dipropyl-	67472-96-2	NI24100
81	125	55	124	95	67	123	109	**248**	100	77	60	52	51	39	31	26	**0.00**	11	20	4	–	1	–	–	–	–	–	–	–	–	4-Heptenoic acid, 2-(methylsulphonyl)-4-propyl-, (E)-	67428-09-5	NI24101
81	44	125	41	55	124	95	67	**248**	100	82	77	62	60	52	51	39	**0.00**	11	20	4	–	1	–	–	–	–	–	–	–	–	4-Heptenoic acid, 2-(methylsulphonyl)-4-propyl-, (E)-	67428-09-5	NI24102
115	161	101	83	55	59	203	87	**248**	100	38	33	31	30	26	19	15	**10.00**	11	20	4	–	1	–	–	–	–	–	–	–	–	Pentanoic acid, 4-[(2-ethoxy-2-oxoethyl)thio]-, ethyl ester	54751-71-2	NI24103
45	43	101	74	58	85	75	44	**248**	100	93	34	29	21	17	14	13	**0.00**	11	20	6	–	–	–	–	–	–	–	–	–	–	Acetic acid, methoxy-, tetrahydro-4,6-dihydroxy-2,4,5-trimethyl-2H-pyran-3-yl ester, ($2\alpha,3\alpha,4\alpha,5\alpha,6\beta$)-	55044-44-5	NI24104
71	101	102	87	111	72	147	143	**248**	100	64	36	31	28	19	19	19	**0.00**	11	20	6	–	–	–	–	–	–	–	–	–	–	D-Galactitol, 1,5-anhydro-2-O-acetyl-3,4,6-tri-O-methyl-	102850-62-4	NI24105
101	74	75	71	171	129	114	143	**248**	100	77	75	45	39	37	30	22	**0.00**	11	20	6	–	–	–	–	–	–	–	–	–	–	D-Galactitol, 1,5-anhydro-3-O-acetyl-2,4,6-tri-O-methyl-	102850-63-5	NI24106
97	87	129	74	71	171	103	85	**248**	100	80	71	64	63	59	48	40	**0.00**	11	20	6	–	–	–	–	–	–	–	–	–	–	D-Galactitol, 1,5-anhydro-4-O-acetyl-2,3,6-tri-O-methyl-	102850-64-6	NI24107
87	101	75	88	71	130	73	188	**248**	100	87	68	62	53	27	23	22	**0.00**	11	20	6	–	–	–	–	–	–	–	–	–	–	D-Galactitol, 1,5-anhydro-6-O-acetyl-2,3,4-tri-O-methyl-	102850-65-7	NI24108
71	75	87	101	102	131	129	72	**248**	100	91	52	41	38	30	27	22	**0.00**	11	20	6	–	–	–	–	–	–	–	–	–	–	D-Glucitol, 1,4-anhydro-5-O-acetyl-2,3,6-tri-O-methyl-	90405-87-1	NI24109
101	71	102	87	143	111	147	88	**248**	100	97	73	34	32	28	22	22	**0.00**	11	20	6	–	–	–	–	–	–	–	–	–	–	D-Glucitol, 1,5-anhydro-2-O-acetyl-3,4,6-tri-O-methyl-	102850-60-2	NI24110
75	74	101	71	143	115	114	83	**248**	100	68	61	50	44	24	21	18	**0.00**	11	20	6	–	–	–	–	–	–	–	–	–	–	D-Glucitol, 1,5-anhydro-3-O-acetyl-2,4,6-tri-O-methyl-	97275-51-9	NI24111
129	171	97	143	87	71	203	75	**248**	100	95	90	78	67	55	54	46	**1.06**	11	20	6	–	–	–	–	–	–	–	–	–	–	D-Glucitol, 1,5-anhydro-4-O-acetyl-2,3,6-tri-O-methyl-	81847-57-6	NI24112
87	101	75	88	71	130	73	129	**248**	100	85	62	56	39	34	20	16	**0.00**	11	20	6	–	–	–	–	–	–	–	–	–	–	D-Glucitol, 1,5-anhydro-6-O-acetyl-2,3,4-tri-O-methyl-	97275-52-0	NI24113
129	111	203	71	87	143	97	103	**248**	100	32	25	24	21	19	14	12	**0.00**	11	20	6	–	–	–	–	–	–	–	–	–	–	D-Glucitol, 2,5-anhydro-3-O-acetyl-1,4,6-tri-O-methyl-	102850-67-9	NI24114
90	43	41	99	33	81	59	29	**248**	100	34	24	20	17	15	11	11	**0.00**	11	20	6	–	–	–	–	–	–	–	–	–	–	Heptanoic acid, 3-carbomethoxy-2,3-dihydroxy-6-methyl-, methyl ester		MS05348
43	59	145	69	167	113	55	129	**248**	100	92	58	54	47	45	44	43	**0.00**	11	20	6	–	–	–	–	–	–	–	–	–	–	Heptanoic acid, 3-carbomethoxy-3,6-dihydroxy-6-methyl-, methyl ester		MS05349
71	101	87	102	111	72	85	88	**248**	100	71	46	41	41	24	20	19	**0.00**	11	20	6	–	–	–	–	–	–	–	–	–	–	D-Mannitol, 1,5-anhydro-2-O-acetyl-3,4,6-tri-O-methyl-	93635-82-6	NI24115
75	101	74	71	143	85	83	69	**248**	100	64	55	26	24	16	15	13	**0.00**	11	20	6	–	–	–	–	–	–	–	–	–	–	D-Mannitol, 1,5-anhydro-3-O-acetyl-2,4,6-tri-O-methyl-	93635-89-3	NI24116
129	97	171	87	143	85	71	111	**248**	100	92	83	65	53	43	43	39	**0.00**	11	20	6	–	–	–	–	–	–	–	–	–	–	D-Mannitol, 1,5-anhydro-4-O-acetyl-2,3,6-tri-O-methyl-	102850-61-3	NI24117
87	75	101	88	71	130	129	73	**248**	100	76	76	49	37	33	21	18	**0.00**	11	20	6	–	–	–	–	–	–	–	–	–	–	D-Mannitol, 1,5-anhydro-6-O-acetyl-2,3,4-tri-O-methyl-	93635-96-2	NI24118
129	87	111	143	71	203	101	85	**248**	100	51	34	34	26	24	23	14	**0.00**	11	20	6	–	–	–	–	–	–	–	–	–	–	D-Mannitol, 2,5-anhydro-3-O-acetyl-1,4,6-tri-O-methyl-		NI24119
88	75	74	43	72	116	101	73	**248**	100	67	46	43	38	22	16	16	**0.00**	11	20	6	–	–	–	–	–	–	–	–	–	–	α-D-Mannopyranoside, methyl 2-O-acetyl-6-deoxy-3,4-di-O-methyl-	72922-26-0	NI24120
74	88	116	101	43	75	72	45	**248**	100	90	78	66	61	47	47	23	**0.00**	11	20	6	–	–	–	–	–	–	–	–	–	–	α-L-Mannopyranoside, methyl 3-O-acetyl-6-deoxy-2,4-di-O-methyl-	55821-15-3	NI24121
88	43	75	73	74	85	45	101	**248**	100	34	32	13	12	12	10	10	**0.00**	11	20	6	–	–	–	–	–	–	–	–	–	–	α-L-Mannopyranoside, methyl 4-O-acetyl-6-deoxy-2,4-di-O-methyl-	72945-56-3	NI24122
43	31	74	41	116	55	101	88	**248**	100	30	23	23	18	18	15	15	**0.00**	11	20	6	–	–	–	–	–	–	–	–	–	–	Methyl 3-O-acetyl-2,4-di-O-methyl-rhamnopyranoside		LI04950
73	179	160	105	248	201	89	233	**248**	100	43	26	13	12	12	12	3		11	24	–	–	2	–	–	–	–	1	–	–	–	1-Butene, 2-[3-(methylthio)propylthio]-1-(trimethylsilyl)-	80402-96-6	NI24123
73	128	99	127	147	105	248	233	**248**	100	48	37	19	18	18	16	2		11	24	–	–	2	–	–	–	–	1	–	–	–	1-Butene, 2-[3-methylthio-3-(trimethylsilyl)propylthio]-		NI24124
73	205	174	165	91	93	233	248	**248**	100	59	27	14	13	10	9	5		11	24	–	–	2	–	–	–	–	1	–	–	–	1,3-Dithiane, 2-isopropyl-2-(trimethylsilylmethyl)-	80402-87-5	NI24125
73	174	146	248	93	165	205	233	**248**	100	50	50	25	17	16	15	10		11	24	–	–	2	–	–	–	–	1	–	–	–	1,3-Dithiane, 2-propyl-2-(trimethylsilylmethyl)-	80402-85-3	NI24126
44	30	58	185	55	190	117	43	**248**	100	83	67	58	58	57	53	45	**14.00**	11	24	2	2	1	–	–	–	–	–	–	–	–	Urea, N-isopropyl-N′-(6-methylsulphinylhexyl)-		MS05350
73	145	147	75	45	146	74	219	**248**	100	96	43	28	13	13	9	7	**0.00**	11	28	2	–	–	–	–	–	–	2	–	–	–	Butane, 1,2-bis(trimethylsilyl)-2-methyl-		NI24127
73	145	147	75	146	45	74	59	**248**	100	94	41	17	13	12	9	7	**0.00**	11	28	2	–	–	–	–	–	–	2	–	–	–	Butane, 1,2-bis(trimethylsilyl)-3-methyl-		NI24128
117	73	147	75	143	158	45	145	**248**	100	97	50	23	19	18	13	11	**0.00**	11	28	2	–	–	–	–	–	–	2	–	–	–	Butane, 1,3-bis(trimethylsilyl)-2-methyl-		NI24129
131	73	147	75	103	143	45	132	**248**	100	85	51	33	23	19	14	12	**0.00**	11	28	2	–	–	–	–	–	–	2	–	–	–	Butane, 1,3-bis(trimethylsilyl)-3-methyl-		NI24130
73	147	130	116	75	103	69	45	**248**	100	99	46	42	32	26	23	20	**0.00**	11	28	2	–	–	–	–	–	–	2	–	–	–	Butane, 1,4-bis(trimethylsilyl)-2-methyl-		NI24131
147	73	130	116	103	75	69	148	**248**	100	81	47	42	25	25	17	16	**0.00**	11	28	2	–	–	–	–	–	–	2	–	–	–	Butane, 1,4-bis(trimethylsilyl)-2-methyl-		NI24132
73	131	147	103	75	143	45	132	**248**	100	96	30	29	21	14	12	12	**0.00**	11	28	2	–	–	–	–	–	–	2	–	–	–	Pentane, 1,3-bis[(trimethylsilyl)oxy]-		NI24133
131	73	147	103	75	132	55	143	**248**	100	89	31	31	18	12	10	9	**0.00**	11	28	2	–	–	–	–	–	–	2	–	–	–	Pentane, 1,3-bis[(trimethylsilyl)oxy]-		NI24134

MASS TO CHARGE RATIOS								M.W.	INTENSITIES								Parent	C	H	O	N	S	F	Cl	Br	I	Si	P	B	X	COMPOUND NAME	CAS Reg No	No
147	73	143	75	103	45	177	158	**248**	100	79	77	74	60	36	21	15	**0.00**	11	28	2	–	–	–	–	–	–	2	–	–	–	**Pentane, 1,5-bis[(trimethylsilyl)oxy]-**	54494-06-3	**MS05351**
147	69	73	143	75	41	103	148	**248**	100	83	46	34	25	22	18	16	**0.00**	11	28	2	–	–	–	–	–	–	2	–	–	–	**Pentane, 1,5-bis[(trimethylsilyl)oxy]-**	54494-06-3	**NI24135**
147	69	73	75	41	143	103	148	**248**	100	88	58	35	32	31	18	16	**0.00**	11	28	2	–	–	–	–	–	–	2	–	–	–	**Pentane, 1,5-bis[(trimethylsilyl)oxy]-**	54494-06-3	**NI24136**
147	73	129	75	45	158	148	103	**248**	100	89	69	27	19	16	16	12	**0.00**	11	28	2	–	–	–	–	–	–	2	–	–	–	**Propane, 1,3-bis(trimethylsilyl)-2-ethyl-**		**NI24137**
248	45	249	203	250	69	171	39	**248**	100	20	16	16	14	13	6	6		12	8	–	–	3	–	–	–	–	–	–	–	–	**2,2′-4′,2″-Terthienyl**		**AI01776**
248	45	249	250	69	203	127	124	**248**	100	20	16	14	14	11	11	7		12	8	–	–	3	–	–	–	–	–	–	–	–	**2,2′-5′,2″-Terthienyl**		**AI01777**
168	248	139	246	169	63	250	245	**248**	100	57	37	23	18	13	12	11		12	8	1	–	–	–	–	–	–	–	–	–	1	**Phenoxaselenin**	262-22-6	**LI04951**
168	248	139	246	169	64	245	244	**248**	100	56	35	23	16	11	10	10		12	8	1	–	–	–	–	–	–	–	–	–	1	**Phenoxaselenin**	262-22-6	**NI24138**
248	200	249	250	58	184	139	171	**248**	100	36	25	20	20	17	15	14		12	8	2	–	2	–	–	–	–	–	–	–	–	**Thianthrene, 5,5-dioxide**		**MS05352**
248	184	200	171	172	69	232	139	**248**	100	85	72	59	42	40	33	26		12	8	2	–	2	–	–	–	–	–	–	–	–	**Thianthrene, 5,10-dioxide**		**LI04952**
248	184	200	171	172	69	232	50	**248**	100	85	72	59	42	41	33	27		12	8	2	–	2	–	–	–	–	–	–	–	–	**Thianthrene, 5,10-dioxide**		**MS05353**
176	51	204	132	52	104	103	78	**248**	100	91	82	75	63	55	48	46	**25.40**	12	8	6	–	–	–	–	–	–	–	–	–	–	**1,1,3,4-Cyclohexanetetracarboxylic acid, 1-vinyl ester, dianhydride**		**NI24139**
248	205	249	151	69	77	51	41	**248**	100	19	15	9	4	3	3	3		12	8	6	–	–	–	–	–	–	–	–	–	–	**1,4-Naphthaquinone, 3-acetyl-2,5,7-trihydroxy-**	54725-01-8	**NI24140**
233	248	205	43	69	249	234	206	**248**	100	99	33	20	18	14	13	6		12	8	6	–	–	–	–	–	–	–	–	–	–	**1,4-Naphthoquinone, 6-acetyl-2,5,7-trihydroxy-**	13378-89-7	**NI24141**
233	248	43	205	69	44	249	234	**248**	100	94	41	38	30	19	15	15		12	8	6	–	–	–	–	–	–	–	–	–	–	**1,4-Naphthoquinone, 6-acetyl-2,5,7-trihydroxy-**	13378-89-7	**NI24142**
248	233	205	43	249	69	53	234	**248**	100	50	34	34	15	13	13	11		12	8	6	–	–	–	–	–	–	–	–	–	–	**1,4-Naphthoquinone, 6-acetyl-2,5,8-trihydroxy-**	13379-24-3	**NI24143**
248	233	205	43	249	69	53	234	**248**	100	71	34	34	15	13	13	10		12	8	6	–	–	–	–	–	–	–	–	–	–	**1,4-Naphthoquinone, 6-acetyl-2,5,8-trihydroxy-**	13379-24-3	**NI24144**
51	77	248	250	50	39	63	38	**248**	100	94	81	80	66	56	48	31		12	9	1	–	–	–	–	1	–	–	–	–	–	**Benzene, 1-bromo-4-phenoxy-**	101-55-3	**NI24145**
248	250	141	77	169	51	249	168	**248**	100	99	45	32	24	19	16	14		12	9	1	–	–	–	–	1	–	–	–	–	–	**Benzene, bromophenoxy-**	36563-47-0	**NI24146**
155	127	126	141	77	156	115	63	**248**	100	66	18	16	13	12	12	10	**8.83**	12	9	1	–	–	–	–	1	–	–	–	–	–	**Naphthalene, 2-(bromoacetyl)-**	613-54-7	**NI24147**
248	168	249	84	139	169	63	51	**248**	100	18	15	15	8	3	2	2		12	9	4	–	–	–	–	–	–	–	1	–	–	**Phosphoric acid, 2,2′-biphenylylene-**	35227-84-0	**NI24148**
125	80	93	124	97	207	81	73	**248**	100	28	24	21	16	15	14	9	**0.00**	12	12	–	2	2	–	–	–	–	–	–	–	–	**Aniline, 2,2′-dithiodi-**	1141-88-4	**NI24149**
124	248	125	80	249	126	53	250	**248**	100	43	36	20	7	7	5	4		12	12	–	2	2	–	–	–	–	–	–	–	–	**Aniline, 4,4′-dithiodi-**	722-27-0	**NI24150**
248	215	45	158	128	89	148	77	**248**	100	16	16	12	11	10	5	5		12	12	–	2	2	–	–	–	–	–	–	–	–	**1,2-Diazine, 3,6-bis(methylthio)-4-phenyl-**		**MS05354**
248	140	92	65	184	249	109	183	**248**	100	94	88	71	27	27	21	21		12	12	2	2	1	–	–	–	–	–	–	–	–	**Aniline, 2,2′-sulphonylbis-**	53347-49-2	**NI24151**
248	183	65	184	182	167	249	92	**248**	100	89	35	31	28	24	19	18		12	12	2	2	1	–	–	–	–	–	–	–	–	**Aniline, 2,3′-sulphonylbis-**	34262-29-8	**NI24152**
248	183	182	249	123	125	77	80	**248**	100	31	21	21	18	15	12	11		12	12	2	2	1	–	–	–	–	–	–	–	–	**Aniline, 2,4-sulphonylbis-**		**NI24153**
127	155	157	141	249	112	95	126	**248**	100	28	22	19	18	12	12	10	**0.00**	12	12	2	2	1	–	–	–	–	–	–	–	–	**Aniline, 3,3′-sulphonylbis-**	599-61-1	**NI24154**
248	108	140	184	92	65	249	57	**248**	100	63	41	36	28	27	18	17		12	12	2	2	1	–	–	–	–	–	–	–	–	**Aniline, 3,4′-sulphonylbis-**	34262-32-3	**NI24155**
248	108	140	65	92	141	249	39	**248**	100	74	47	35	29	19	16	16		12	12	2	2	1	–	–	–	–	–	–	–	–	**Aniline, 4,4′-sulphonylbis-**	80-08-0	**NI24156**
248	65	140	92	184	249	66	64	**248**	100	82	55	50	21	14	11	10		12	12	2	2	1	–	–	–	–	–	–	–	–	**Benzenamine, 3,3′-sulfonylbis-**	599-61-1	**NI24157**
248	140	65	92	184	125	249	183	**248**	100	38	37	29	22	21	17	10		12	12	2	2	1	–	–	–	–	–	–	–	–	**Benzenamine, 3,3′-sulfonylbis-**	599-61-1	**NI24158**
248	232	108	140	92	45	65	249	**248**	100	52	49	25	20	19	18	17		12	12	2	2	1	–	–	–	–	–	–	–	–	**Benzenamine, 4,4′-sulfonylbis-**	80-08-0	**NI24159**
136	248	108	137	191	218	220	205	**248**	100	24	18	11	10	7	5	5		12	12	2	2	1	–	–	–	–	–	–	–	–	**4H-5,6-Benzothiazin-4-one, 2-morpholino-**	24069-10-1	**NI24160**
162	148	190	163	149	106	248	89	**248**	100	32	17	10	9	9	9	8		12	12	2	2	1	–	–	–	–	–	–	–	–	**2-Propenoic acid, 2-cyano-3-(2-thiocyanatocyclopent-1-en-1-yl)-, ethyl ester**		**MS05355**
80	105	93	248	184	79	141	104	**248**	100	67	61	56	50	50	22	22		12	12	2	2	1	–	–	–	–	–	–	–	–	**Pyridinium, 1-[[(4-methylphenyl)sulphonyl]amino]-, hydroxide, inner salt**	40949-56-2	**LI04953**
80	105	93	248	65	184	79	66	**248**	100	66	59	54	52	50	49	47		12	12	2	2	1	–	–	–	–	–	–	–	–	**Pyridinium, 1-[[(4-methylphenyl)sulphonyl]amino]-, hydroxide, inner salt**	40949-56-2	**NI24161**
107	78	80	93	168	184	248	92	**248**	100	80	28	23	18	15	13	11		12	12	2	2	1	–	–	–	–	–	–	–	–	**Pyridinium, 2-methyl-1-[(phenylsulphonyl)amino]-, hydroxide, inner salt**	34456-58-1	**LI04954**
107	78	80	77	51	93	168	184	**248**	100	80	28	28	24	23	18	15	**13.00**	12	12	2	2	1	–	–	–	–	–	–	–	–	**Pyridinium, 2-methyl-1-[(phenylsulphonyl)amino]-, hydroxide, inner salt**	34456-58-1	**NI24162**
248	176	203	89	104	133	29	134	**248**	100	88	66	52	38	34	30	23		12	12	2	2	1	–	–	–	–	–	–	–	–	**5-Thiazolecarboxylic acid, 2-amino-4-phenyl-, ethyl ester**	64399-23-1	**NI24163**
248	188	130	144	158	103	117	145	**248**	100	97	41	38	29	25	22	21		12	12	2	2	1	–	–	–	–	–	–	–	–	**Uracil, 1-(2-ethylthio)-5-phenyl-**		**LI04955**
189	144	89	117	116	90	190	118	**248**	100	16	15	14	14	12	11	9	**8.00**	12	12	2	2	1	–	–	–	–	–	–	–	–	**Uracil, 3-(2-ethylthio)-5-phenyl-**		**LI04956**
189	104	47	172	146	77	190	103	**248**	100	21	17	16	15	14	13	10	**4.44**	12	12	2	2	1	–	–	–	–	–	–	–	–	**Uracil, 3-(2-ethylthio)-6-phenyl-**	29558-49-4	**NI24164**
189	104	172	47	146	67	190	103	**248**	100	20	15	15	14	13	12	10	**4.00**	12	12	2	2	1	–	–	–	–	–	–	–	–	**Uracil, 3-(2-ethylthio)-6-phenyl-**	29558-49-4	**LI04957**
248	52	151	111	138	150	110	28	**248**	100	60	52	52	45	43	40	31		12	12	4	2	–	–	–	–	–	–	–	–	–	**2,5-Butanocyclobuta[1,2-c:3,4-c′]dipyrrole-1,3,4,6-tetrone, tetrahydro-**	23213-90-3	**NI24165**
148	91	120	118	65	147	51	89	**248**	100	95	79	51	37	27	27	23	**7.00**	12	12	4	2	–	–	–	–	–	–	–	–	–	**Formamide, N-[2-(2,5-dihydro-3,4-dihydroxy-5-imino-2-furanyl)phenyl]-N-methyl-**	52381-18-7	**NI24166**
151	70	110	82	54	26	56	41	**248**	100	91	72	63	48	21	20	17	**0.10**	12	12	4	2	–	–	–	–	–	–	–	–	–	**2(5H)-Furanone, 5,5′-(1,4-butanediyldinitrilo)bis-**	6120-42-9	**NI24167**
151	70	110	82	54	26	56	55	**248**	100	91	72	63	48	21	20	17	**0.05**	12	12	4	2	–	–	–	–	–	–	–	–	–	**2(5H)-Furanone, 5,5′-(1,4-butanediyldinitrilo)bis-**	6120-42-9	**MS05356**
110	151	82	54	26	70	56	111	**248**	100	65	31	29	19	15	15	14	**12.00**	12	12	4	2	–	–	–	–	–	–	–	–	–	**2(5H)-Furanone, 5,5′-(1,4-butanediyldinitrilo)bis-**	6120-42-9	**MS05357**
248	175	174	202	39	203	94	55	**248**	100	70	58	48	24	21	18	18		12	12	4	2	–	–	–	–	–	–	–	–	–	**4H-Pyrido[1,2-a]pyrimidine-3-acetic acid, 6-hydroxy-4-oxo-, ethyl ester**	64399-36-6	**NI24168**
175	248	94	176	147	121	39	55	**248**	100	38	13	11	8	8	8	6		12	12	4	2	–	–	–	–	–	–	–	–	–	**4H-Pyrido[1,2-a]pyrimidine-3-acetic acid, 9-hydroxy-4-oxo-, ethyl ester**	50609-61-5	**NI24169**
219	248	57	148	120	220	58	71	**248**	100	71	55	45	44	43	40	27		12	12	4	2	–	–	–	–	–	–	–	–	–	**2,4,6(1H,3H,5H)-Pyrimidinetrione, 5-ethyl-6,6-dihydroxy-5-phenyl-**		**LI04958**
110	151	82	54	26	70	56	111	**248**	100	65	31	29	19	15	15	14	**12.01**	12	12	4	2	–	–	–	–	–	–	–	–	–	**1H-Pyrrole-2,5-dione, 1,1′-(1,4-butanediyl)bis-**	28537-70-4	**NI24170**
247	248	249	250	219	212	211	106	**248**	100	50	36	15	12	10	5	5		12	13	–	4	–	–	1	–	–	–	–	–	–	**2,4-Pyrimidinediamine, 5-(4-chlorophenyl)-6-ethyl-**	58-14-0	**NI24171**

MASS TO CHARGE RATIOS								M.W.	INTENSITIES								Parent	C	H	O	N	S	F	Cl	Br	I	Si	P	B	X	COMPOUND NAME	CAS Reg No	No
249	251	250	248	252	247			248	100	33	16	7	5	1				12	13	–	4	–	–	1	–	–	–	–	–	–	2,4-Pyrimidinediamine, 5-(4-chlorophenyl)-6-ethyl-	58-14-0	NI24172
247	248	249	250	219	212	232	221	248	100	50	36	15	12	10	5	5		12	13	–	4	–	–	1	–	–	–	–	–	–	2,4-Pyrimidinediamine, 5-(4-chlorophenyl)-6-ethyl-	58-14-0	MS05358
173	30	172	77	205	44	105	45	248	100	53	45	40	22	19	18	17	10.70	12	16	–	4	1	–	–	–	–	–	–	–	–	1,2,3-Triazole, 2-phenyl-4-[(2-aminoethyl)thiomethyl]-5-methyl-		NI24173
83	55	167	166	41	168	81	169	248	100	45	37	37	33	15	13	12	8.42	12	22	1	–	–	–	1	–	–	–	1	–	–	Phosphinic chloride, dicyclohexyl-	15873-72-0	NI24174
57	43	135	137	55	41	71	69	248	100	97	94	92	53	53	49	40	0.56	12	25	–	–	–	–	–	1	–	–	–	–	–	Decane, 1-bromo-		DO01038
45	57	46	41	135	69	71	137	248	100	41	41	28	27	21	20	18	0.13	12	25	–	–	–	–	–	1	–	–	–	–	–	Decane, 1-bromo-		IC03254
135	57	43	55	71	41	69	85	248	100	92	84	54	49	44	39	29	0.40	12	25	–	–	–	–	–	1	–	–	–	–	–	Decane, 1-bromo-		LI04959
131	122	97	192	55	41	111	57	248	100	89	75	47	42	34	30	30	9.00	12	25	1	–	1	–	–	–	–	–	1	–	–	Phosphetane, 1-(butylthio)-2,2,3,4,4-pentamethyl-, 1-oxide		NI24175
122	97	57	192	55	41	111	69	248	100	60	53	45	42	36	34	32	3.00	12	25	1	–	1	–	–	–	–	–	1	–	–	Phosphetane, 1-(isobutylthio)-2,2,3,4,4-pentamethyl-, 1-oxide		NI24176
122	97	41	192	55	193	69	57	248	100	63	45	40	40	36	36	36	12.00	12	25	1	–	1	–	–	–	–	–	1	–	–	Phosphetane, 1-(sec-butylthio)-2,2,3,4,4-pentamethyl-, 1-oxide		NI24177
122	192	41	57	193	70	97	178	248	100	75	53	50	45	42	34	32	5.00	12	25	1	–	1	–	–	–	–	–	1	–	–	Phosphetane, 1-(tert-butylthio)-2,2,3,4,4-pentamethyl-, 1-oxide		NI24178
248	44	150	220	43	75	250	74	248	100	94	85	73	73	70	60	59		13	6	1	–	–	–	2	–	–	–	–	–	–	9H-Fluoren-9-one, 2,7-dichloro-	6297-11-6	NI24179
248	250	150	220	185	75	222	249	248	100	64	37	22	20	20	15	14		13	6	1	–	–	–	2	–	–	–	–	–	–	9H-Fluoren-9-one, 3,6-dichloro-	34223-82-0	NI24180
77	141	248	94	224	65	51	44	248	100	35	32	29	27	24	23	17		13	9	3	–	–	–	1	–	–	–	–	–	–	Carbonic acid, 4-chlorophenyl phenyl ester	27087-46-3	NI24181
248	220	157	205	250	222	127	185	248	100	72	50	38	32	28	28	24		13	9	3	–	–	–	1	–	–	–	–	–	–	2H-Pyrano[3,2-d]-1-benzoxepin-2-one, 3-chloro-5,6-dihydro-	86896-95-9	NI24182
123	232	149	139	125	218	124	168	248	100	43	35	15	14	13	13	8	0.00	13	12	1	–	2	–	–	–	–	–	–	–	–	Benzenesulphinic acid, p-tolyl ester		LI04960
248	105	171	77	231	185	59	51	248	100	63	55	54	38	30	21	20		13	12	1	–	2	–	–	–	–	–	–	–	–	Benzophenone, α-(4,5-dimethyl-3H-1,2-dithiol-3-ylidene)-	38489-99-5	NI24183
91	65	92	248	94	63	66	39	248	100	16	9	7	7	4	3	3		13	12	3	–	1	–	–	–	–	–	–	–	–	Phenol, 2-benzenesulphonyl-		MS05359
202	129	248	203	249	45	204	146	248	100	75	47	33	20	12	11	11		13	12	3	–	1	–	–	–	–	–	–	–	–	3-Thiophenecarboxylic acid, 4-hydroxy-2-phenyl-, ethyl ester	1022-06-6	NI24184
56	193	175	75	41	103	148	55	248	100	80	60	31	27	23	14	12	0.00	13	12	5	–	–	–	–	–	–	–	–	–	–	1,2,4-Benzenetricarboxylic acid, cyclic 1,2-anhydride, butyl ester	1528-45-6	NI24185
248	150	149	135	163	134	164	247	248	100	62	50	45	28	28	21	16		13	12	5	–	–	–	–	–	–	–	–	–	–	1H,3H-Furo[3,4-c]furan-1-one, tetrahydro-4-[3,4-(methylenedioxy)phenyl]-		LI04961
190	146	118	43	191	89	134	28	248	100	67	54	42	19	18	13	11	4.62	13	12	5	–	–	–	–	–	–	–	–	–	–	Malonic acid, (3-hydroxybenzylidene)-, cyclic isopropylidene ester	15795-58-1	NI24186
146	190	118	43	162	134	191	89	248	100	66	65	43	40	31	25	24	19.54	13	12	5	–	–	–	–	–	–	–	–	–	–	Malonic acid, (4-hydroxybenzylidene)-, cyclic isopropylidene ester	17474-27-0	NI24187
136	248	203	164	202	75	249	175	248	100	96	67	50	27	23	16	14		13	12	5	–	–	–	–	–	–	–	–	–	–	Naphtho[2,3-c]furan-1,4-dione, 3,3a,9,9a-tetrahydro-6-hydroxy-7-methoxy-		MS05360
136	248	203	164	202	204	176	137	248	100	96	67	50	27	13	13	13		13	12	5	–	–	–	–	–	–	–	–	–	–	Naphtho[2,3-c]furan-1,4-dione, 3,3a,9,9a-tetrahydro-6-hydroxy-7-methoxy-		LI04962
248	233	250	235	249	234	216	169	248	100	96	77	74	64	55	49	35		13	13	1	2	–	–	1	–	–	–	–	–	–	1,2-Benzenediamine, 4-chloro-N'-(4-methoxyphenyl)-	1750-94-3	MS05361
93	77	155	233	201	247	139	248	248	100	99	7	3	3	2	2	1		13	13	3	–	–	–	–	–	–	–	1	–	–	Phosphonic acid, phenyl-, methyl phenyl diester		MS05362
136	108	41	248	191	69	55	137	248	100	55	24	20	20	20	20	18		13	16	1	2	1	–	–	–	–	–	–	–	–	4H-1,3-Benzothiazin-4-one, 2,3-dihydro-2-N-pentylimino-		NI24188
135	192	77	248	235	136	41		248	100	88	85	63	60	48	43			13	16	1	2	1	–	–	–	–	–	–	–	–	Isoleucine, 3-phenyl-2-thiohydantoyl-		LI04963
192	248	135	77	136	41	219	191	248	100	98	70	48	30	25	20	18		13	16	1	2	1	–	–	–	–	–	–	–	–	Isoleucine, 3-phenyl-2-thiohydantoyl-		LI04964
135	248	77	193	136	43	41	205	248	100	83	56	49	27	23	22	20		13	16	1	2	1	–	–	–	–	–	–	–	–	L-Leucine, 3-phenyl-2-thiohydantoyl-	29588-99-6	NI24189
248	135	192	77	205	136	249	43	248	100	92	47	40	22	22	18	17		13	16	1	2	1	–	–	–	–	–	–	–	–	Leucine, 3-phenyl-2-thiohydantoyl-	4399-40-0	NI24190
135	77	248	192	136				248	100	80	75	65	38					13	16	1	2	1	–	–	–	–	–	–	–	–	Leucine, 3-phenyl-2-thiohydantoyl-	4399-40-0	LI04965
135	248	77	192	136	205	43	219	248	100	98	54	52	26	25	20	18		13	16	1	2	1	–	–	–	–	–	–	–	–	Leucine, 3-phenyl-2-thiohydantoyl-	4399-40-0	LI04966
119	91	82	64	69	55	63	77	248	100	93	77	62	53	38	29	28	2.00	13	16	1	2	1	–	–	–	–	–	–	–	–	Pyrrolo[1,2-b][1,2,4]oxadiazole-2(1H)-thione, tetrahydro-5,5-dimethyl-1-phenyl-	50455-64-6	NI24191
217	218	203	248	233	219	215	201	248	100	38	10	9	7	7	5	4		13	16	1	2	1	–	–	–	–	–	–	–	–	Quinazoline, 2-tert-butyl-4-(methylsulphinyl)-	55133-87-4	NI24192
233	75	77	105	73	205	45	51	248	100	89	63	52	50	44	23	22	3.68	13	16	3	–	–	–	–	–	–	1	–	–	–	2-Propenoic acid, 3-benzoyl-, trimethylsilyl ester	25432-36-4	NI24193
189	176	248	161	174	30	190	43	248	100	79	49	39	23	23	15	15		13	16	3	2	–	–	–	–	–	–	–	–	–	Acetamide, N-[2-(6-hydroxy-5-methoxy-1H-indol-3-yl)ethyl]-	2208-41-5	NI24194
176	189	248	161	174	190	177	249	248	100	92	53	29	18	13	13	8		13	16	3	2	–	–	–	–	–	–	–	–	–	Acetamide, N-[2-(6-hydroxy-5-methoxy-1H-indol-3-yl)ethyl]-	2208-41-5	NI24195
176	189	161	248	174	43	177	190	248	100	86	42	32	23	20	18	14		13	16	3	2	–	–	–	–	–	–	–	–	–	Acetamide, N-[2-(6-hydroxy-5-methoxy-1H-indol-3-yl)ethyl]-	2208-41-5	NI24196
204	77	203	105	118	104	134	191	248	100	84	80	80	74	70	52	44	0.00	13	16	3	2	–	–	–	–	–	–	–	–	–	2,4-Imidazolidinedione, 1,3-dimethyl-5-ethoxy-5-phenyl-	116927-97-0	DD00672
233	148	248	56	234	28	133	63	248	100	75	28	18	15	13	11	10		13	16	3	2	–	–	–	–	–	–	–	–	–	2,4-Imidazolidinedione, 5-(4-methoxyphenyl)-1,3,5-trimethyl-	55822-91-8	NI24197
105	87	77	70	86	114	134	57	248	100	80	32	19	16	15	12	9	2.50	13	16	3	2	–	–	–	–	–	–	–	–	–	Morpholine, 4-[(N-benzamido)acetyl]-		NI24198
219	175	76	190	103	71	149	104	248	100	86	70	65	55	51	35	35	1.50	13	16	3	2	–	–	–	–	–	–	–	–	–	2-Oxazoline, 4,5-diethyl-2-(p-nitrophenyl)-, cis-	25943-08-2	NI24199
219	175	76	71	190	103	149	104	248	100	50	45	38	37	31	22	18	1.50	13	16	3	2	–	–	–	–	–	–	–	–	–	2-Oxazoline, 4,5-diethyl-2-(p-nitrophenyl)-, trans-	25943-15-1	NI24200
248	69	41	55	138	122	110	152	248	100	64	28	24	18	10	10	4		13	18	2	3	–	–	–	–	–	–	–	–	–	Cyclohexanone, 4-methyl-, 4-nitrophenylhydrazone		LI04967
205	149	248	163	150	206	164	136	248	100	66	26	26	23	23	20	18		13	20	1	4	–	–	–	–	–	–	–	–	–	1H-Pyrazolo[3,4-d]pyrimidin-4-ol, 1,5-dibutyl-		NI24201
83	55	69	57	43	97	41	139	248	100	45	41	27	22	19	19	18	8.82	13	26	–	–	–	–	1	–	–	–	1	–	–	Phosphinous chloride, isopropyl(methylisopropylcyclohexyl)-	74710-02-4	NI24202
178	139	201	75	166	165	248	77	248	100	62	57	57	37	35	33	33		14	10	–	–	–	–	2	–	–	–	–	–	–	Ethylene, 1,1-bis(4-chlorophenyl)-	2642-81-1	NI24204
178	88	248	75	213	250	177	176	248	100	43	38	35	30	28	27	27		14	10	–	–	–	–	2	–	–	–	–	–	–	Ethylene, 1,1-bis(4-chlorophenyl)-	2642-81-1	NI24203
137	139	138	140	117				248	100	36	9	3	2				0.00	14	10	–	–	–	–	2	–	–	–	–	–	–	Ethylene, 1,1-bis(4-chlorophenyl)-	2642-81-1	NI24205
165	178	166	88	176	248	212	177	248	100	21	18	16	15	14	13	10		14	10	–	–	–	–	2	–	–	–	–	–	–	9H-Fluorene, 9-(dichloromethyl)-	31859-82-2	NI24206

MASS TO CHARGE RATIOS								M.W.	INTENSITIES								Parent	C	H	O	N	S	F	Cl	Br	I	Si	P	B	X	COMPOUND NAME	CAS Reg No	No
248	233	203	174	218	217			248	100	86	83	55	35	23				14	16	4	–	–	–	–	–	–	–	–	–	–	3(2H)-Benzofuranone, 5-acetyl-6-hydroxy-2-isopropenyl-7-methoxy-		LI04968
233	203	248	234	218	191	69	77	248	100	22	17	13	7	7	7	5		14	16	4	–	–	–	–	–	–	–	–	–	–	2H-1-Benzopyran, 6-acetyl-7-hydroxy-5-methoxy-2,2-dimethyl-	529-70-4	NI24207
233	215	248	234	200	216	77	69	248	100	25	20	15	13	7	6	6		14	16	4	–	–	–	–	–	–	–	–	–	–	2H-1-Benzopyran, 8-acetyl-7-hydroxy-5-methoxy-2,2-dimethyl-	484-18-4	NI24209
233	43	234	215	200	69	109	77	248	100	20	19	16	9	9	8	8	2.00	14	16	4	–	–	–	–	–	–	–	–	–	–	2H-1-Benzopyran, 8-acetyl-7-hydroxy-5-methoxy-2,2-dimethyl-	484-18-4	NI24208
43	149	164	163	85	77	248		248	100	87	36	26	20	20	9			14	16	4	–	–	–	–	–	–	–	–	–	–	4H-1-Benzopyran-4-one, 8-acetyl-4a,8a-dihydro-7-hydroxy-2,5,8a-trimethyl-		LI04969
104	129	188	184	128	189	105	217	248	100	27	25	12	11	10	10	9	8.50	14	16	4	–	–	–	–	–	–	–	–	–	–	1,1-Cyclobutanedicarboxylic acid, 3-phenyl-, dimethyl ester		NI24210
102	203	158	130	29	248	103	135	248	100	99	86	75	59	58	46	44		14	16	4	–	–	–	–	–	–	–	–	–	–	1,1-Ethenedicarboxylic acid, 2-phenyl-, diethyl ester		NI24211
205	43	151	233	54	53	39	149	248	100	65	37	36	36	34	30	29	26.27	14	16	4	–	–	–	–	–	–	–	–	–	–	2(5H)-Furanone, 5-[(3,4-dimethyl-5-oxo-2(5H)-furanylidene)methyl]-3,4,5-trimethyl-	56248-63-6	NI24212
105	206	77	43	106	51	163	207	248	100	27	22	13	11	8	5	3	2.00	14	16	4	–	–	–	–	–	–	–	–	–	–	2,4-Pentanedione, 3-benzoyloxy-3-ethyl-		MS05363
102	203	158	130	248	103	29	131	248	100	75	59	57	46	46	43	38		14	16	4	–	–	–	–	–	–	–	–	–	–	Propanedioic acid, benzyl-, diethyl ester	5292-53-5	NI24213
114	91	56	248	85	117	104	143	248	100	51	46	40	36	31	28	20		14	16	4	–	–	–	–	–	–	–	–	–	–	2H-Pyran-2-one, 5,6-dihydro-5-hydroxy-4-methoxy-6-(2-phenylethyl)-		LI04970
114	56	91	143	117	86	104	57	248	100	57	55	28	28	26	24	19	13.00	14	16	4	–	–	–	–	–	–	–	–	–	–	2H-Pyran-2-one, 5,6-dihydro-5-hydroxy-4-methoxy-6-(2-phenylethyl)-, (5R-cis)-	52247-81-1	NI24214
114	56	91	86	157	77	139	43	248	100	94	73	66	47	47	39	27	11.00	14	16	4	–	–	–	–	–	–	–	–	–	–	2H-Pyran-2-one, 5,6-dihydro-5-hydroxy-4-methoxy-6-(2-phenylethyl)-, (5S-trans)-	56083-47-7	NI24215
248	193	249	205	141	77	128	247	248	100	20	18	14	14	14	12	9		14	17	2	–	–	–	–	–	–	–	1	–	–	Phosphinic acid, phenyl(cyclohexaneethylidene)-		LI04971
41	107	116	43	132	55	39	42	248	100	75	68	52	50	50	48	45	35.49	14	20	–	2	1	–	–	–	–	–	–	–	–	4H-1,3-Thiazine, 5,6-dihydro-2-[m-tolylamino(imino)]-4,4,6-trimethyl-	53004-47-0	NI24216
107	97	41	247	106	77	100	39	248	100	98	92	91	86	85	71	45	41.99	14	20	–	2	1	–	–	–	–	–	–	–	–	4H-1,3-Thiazine, 5,6-dihydro-2-(N-methylphenylamino)-4,4,6-trimethyl-	53004-54-9	NI24217
132	248	56	61	41	77	133	106	248	100	56	51	38	36	35	28	26		14	20	–	2	1	–	–	–	–	–	–	–	–	1,3-Thiazine, 2-phenylimino-3,4,4,6-tetramethyltetrahydro-	62642-86-8	NI24218
248	205	43	206	164	249	134	135	248	100	76	46	45	18	16	16	14		14	20	2	2	–	–	–	–	–	–	–	–	–	2,3′-Bipyridine, 1,1′-diacetyl-1,1′,4,4′,5,5′,6,6′-octahydro-	52195-95-6	NI24219
145	175	118	130	156	146	91	117	248	100	17	16	13	12	11	10	10	3.94	14	20	2	2	–	–	–	–	–	–	–	–	–	3-Imidazolin-1-ol, 4-(p-tolyl)-2,2,5,5-tetramethyl-, 3-oxide	78163-38-9	NI24220
58	131	130	77	132	103	102	59	248	100	57	37	10	7	6	4	4	0.00	14	20	2	2	–	–	–	–	–	–	–	–	–	1H-Indole-3-ethanamine, α-ethyl-, monoacetate	118-68-3	NI24221
30	72	133	56	43	116	41	104	248	100	91	73	16	15	13	12	11	6.50	14	20	2	2	–	–	–	–	–	–	–	–	–	2-Propanol, 1-(1H-indol-4-yloxy)-3-[(1-methylethyl)amino]-	13523-86-9	NI24222
93	43	57	94	55	41	29	56	248	100	18	16	15	13	13	12	9	9.00	14	20	2	2	–	–	–	–	–	–	–	–	–	Urea, N-(4-hydroxy-2-methylcyclohexyl)-N′-phenyl-	25546-05-8	NI24223
93	43	57	55	29	92	119	248	248	100	18	16	12	12	9	7	6		14	20	2	2	–	–	–	–	–	–	–	–	–	Urea, N-(4-hydroxy-2-methylcyclohexyl)-N′-phenyl-	25546-05-8	LI04972
93	42	56	94	54	40	29	55	248	100	18	17	16	13	13	12	10	7.01	14	20	2	2	–	–	–	–	–	–	–	–	–	Urea, N-(4-hydroxy-3-methylcyclohexyl)-N′-phenyl-	55521-13-6	NI24224
109	110	44	248	56	55	135	108	248	100	16	15	12	9	9	4	4		14	20	2	2	–	–	–	–	–	–	–	–	–	Urea, N-(4-hydroxyphenyl)-N′-(2-methylcyclohexyl)-	25546-03-6	NI24225
109	110	43	248	56	54	108		248	100	14	14	9	8	8	4			14	20	2	2	–	–	–	–	–	–	–	–	–	Urea, N-(4-hydroxyphenyl)-N′-(2-methylcyclohexyl)-	25546-03-6	LI04973
79	152	28	94	67	124	91	95	248	100	82	51	50	47	44	39	38	26.06	14	22	–	–	–	–	–	–	–	–	–	–	1	Nickel, bis[(1,2,3-η)-2-cyclohepten-1-yl]-	38816-29-4	NI24226
107	63	109	41	43	29	55	45	248	100	41	31	30	28	22	21	17	0.00	14	29	1	–	–	–	1	–	–	–	–	–	–	Dodecane, 2-(2-chloroethoxy)-		MS05364
248	220	18	138	192	164	249	247	248	100	25	23	14	13	13	12	10		15	8	2	2	–	–	–	–	–	–	–	–	–	Anthraquinone imidazole		IC03255
248	176	108	134	247	249	107	106	248	100	45	27	25	18	16	14	12		15	12	–	4	–	–	–	–	–	–	–	–	–	Benzimidazole, 2,2′-methylbis-		IC03256
171	77	248	115	249	172	51	65	248	100	88	77	61	20	20	19	10		15	12	–	4	–	–	–	–	–	–	–	–	–	1H-Pyrazole, 1-phenyl-4-(phenylazo)-	25503-12-2	NI24227
178	152	151	177	179	150	139	113	248	100	11	9	7	6	6	5	5	1.00	15	12	–	4	–	–	–	–	–	–	–	–	–	1,2,4-Triazin-3-amine, 5,6-diphenyl-	4511-99-3	LI04974
248	178	152	151	249	177	179	150	248	100	43	5	4	3	3	2	2		15	12	–	4	–	–	–	–	–	–	–	–	–	1,2,4-Triazin-3-amine, 5,6-diphenyl-	4511-99-3	NI24229
178	248	152	151	177	249	179	150	248	100	23	11	10	7	6	5	5		15	12	–	4	–	–	–	–	–	–	–	–	–	1,2,4-Triazin-3-amine, 5,6-diphenyl-	4511-99-3	NI24228
105	77	192	106	202	120	146	51	248	100	24	17	9	7	7	6	6	1.00	15	20	3	–	–	–	–	–	–	–	–	–	–	Acetic acid, 2-butylbenzoyl-, ethyl ester		MS05365
105	77	192	106	42	51	120	146	248	100	25	14	9	7	6	5	4	0.00	15	20	3	–	–	–	–	–	–	–	–	–	–	Acetic acid, 2-tert-butylbenzoyl-, ethyl ester		MS05366
108	43	109	41	248	79	95	91	248	100	98	55	40	36	24	22	22		15	20	3	–	–	–	–	–	–	–	–	–	–	Alexandrofuran		NI24230
83	55	91	151	248	150	149	193	248	100	48	4	3	2	2	2	1		15	20	3	–	–	–	–	–	–	–	–	–	–	Anthemis lactone		NI24231
248	81	119	41	121	91	53	93	248	100	94	85	73	71	67	64	57		15	20	3	–	–	–	–	–	–	–	–	–	–	Azuleno[6,5-b]furan-2,5-dione, decahydro-4a,8-dimethyl-3-methylene-, [3aR-(3aα,4aβ,7aα,8β,9aα)]-	19908-69-1	NI24232
133	248	175	41	55	79	206	135	248	100	81	40	35	26	22	21	21		15	20	3	–	–	–	–	–	–	–	–	–	–	Azuleno[6,5-b]furan-2-one, 3,3a,4,7,7a,8,9,9a-octahydro-4-hydroxy-3,5,8-trimethyl-, [3R-(3α,3aα,4α,7aβ,8,9aα)]-		MS05367
107	55	77	79	81	105	67	96	248	100	97	51	49	33	31	27	24	2.00	15	20	3	–	–	–	–	–	–	–	–	–	–	Benzeneacetic acid, α-hydroxy-, cyclohexanemethyl ester	86328-77-0	NI24233
152	135	153	81	107	55	77	248	248	100	42	13	12	10	9	9	8		15	20	3	–	–	–	–	–	–	–	–	–	–	Benzoic acid, 3-methoxy-, cyclohexanemethyl ester	86338-71-8	NI24234
233	248	234	174	173	187	31	203	248	100	90	83	73	46	43	27	20		15	20	3	–	–	–	–	–	–	–	–	–	–	2H-1-Benzopyran, 2,6-dimethyl-8-methoxy-7-(3-hydroxypropyl)-		NI24235
43	81	107	229	95	248	109	105	248	100	75	33	32	30	28	28	22		15	20	3	–	–	–	–	–	–	–	–	–	–	Cyclodeca[b]furan-2(3H)-one, 3a,4,5,8,9,11a-hexahydro-6,10-dimethyl-3-methylene-, [3aS-(3aR*,6E,10E,11aS*)]-, 1,10-epoxide		NI24236
43	190	79	145	105	107	93	91	248	100	52	42	35	35	32	31	29	2.83	15	20	3	–	–	–	–	–	–	–	–	–	–	2H-Cyclohepta[b]furan-2-one, 3,3a,4,7,8,8a-hexahydro-7-methyl-3-methylene-6-(3-oxobutyl)-, [3aR-(3aα,7β,8aα)]-	33649-15-9	NI24237
83	166	121	55	105	84	39	77	248	100	20	9	9	8	6	6	5	2.50	15	20	3	–	–	–	–	–	–	–	–	–	–	2,5-Cyclohexadiene, 2-formyl-4-(3-methyl-2-butenoyloxy)-1,1,5-trimethyl-		LI04975
41	79	43	77	57	233	191	95	248	100	80	65	57	57	56	55	52	26.00	15	20	3	–	–	–	–	–	–	–	–	–	–	3-Cyclopenten-1-one, 3-hydroxy-2-(1-hydroxy-3-methylbutylidene)-5-(3-methyl-2-butenylidene)-	38574-23-1	NI24238

MASS TO CHARGE RATIOS								M.W.	INTENSITIES								Parent	C	H	O	N	S	F	Cl	Br	I	Si	P	B	X	COMPOUND NAME	CAS Reg No	No
248	233	191	95	249	234	79	215	248	100	81	72	25	16	16	15	14		15	20	3	–	–	–	–	–	–	–	–	–	–	Dehydrohumilic acid		LI04976
68	67	107	53	121	91	93	135	248	100	47	44	42	35	35	29	28	4.00	15	20	3	–	–	–	–	–	–	–	–	–	–	8,9-Epoxy-8,9-seco-glechomanolide		NI24239
248	192	41	55	124	85	233	136	248	100	70	36	34	27	25	23	22		15	20	3	–	–	–	–	–	–	–	–	–	–	Eudesma-1,4-dien-12-oic acid, 1,2-dihydro-6a-hydroxy-3-oxo-, γ-lactone, (11S)-		MS05368
69	70	57	91	55	53	107	123	248	100	70	11	9	8	8	7	6	2.00	15	20	3	–	–	–	–	–	–	–	–	–	–	Seco-germacrane anhydride		NI24240
83	174	175	85	248	43	157	95	248	100	87	71	66	60	46	45	44		15	20	3	–	–	–	–	–	–	–	–	–	–	Guaia-4(15),10(14)-dieno-12,6α-lactone, 3α-hydroxy-, (11S)-	82206-87-9	DD00673
83	231	85	185	213	151	95	248	248	100	74	64	39	20	18	15	14		15	20	3	–	–	–	–	–	–	–	–	–	–	2-Hydroxy-α-alantolactone		NI24241
121	135	150	41	159	43	91	105	248	100	93	92	56	38	35	29	20	0.30	15	20	3	–	–	–	–	–	–	–	–	–	–	Laureacetal-C	84744-66-1	NI24242
132	107	91	118	105	248	119	95	248	100	68	47	46	45	41	36	33		15	20	3	–	–	–	–	–	–	–	–	–	–	Marasmen-3-one		MS05369
93	136	165	41	91	121	108	107	248	100	42	37	34	22	20	20	19	16.00	15	20	3	–	–	–	–	–	–	–	–	–	–	1,4-Methano-3-benzoxepin-5,10(4H)-dione, 1,2,7,8,9,9a-hexahydro-2,2,9,9a-tetramethyl-	41988-45-8	NI24243
55	91	41	105	127	248	39	120	248	100	92	90	73	63	58	57	53		15	20	3	–	–	–	–	–	–	–	–	–	–	7-Naphthaleneone, 3,4,4a,5,6,7-hexahydro-1-(2-methoxycarbonylethyl)-2-methyl-		LI04977
105	146	117	135	119	148	147	192	248	100	18	15	10	8	7	7	5	0.07	15	20	3	–	–	–	–	–	–	–	–	–	–	Oxiranecarboxylic acid, 2-ethyl-2-phenyl-, tert-butyl ester		NI24244
83	43	151	55	29	248	108	39	248	100	52	50	20	11	10	10	10		15	20	3	–	–	–	–	–	–	–	–	–	–	3-Penten-2-one, 1-[5-(3-furanyl)tetrahydro-2-methyl-2-furanyl]-4-methyl-	36238-02-5	NI24245
173	188	147	145	107	146	91	187	248	100	78	66	65	41	35	33	27	10.00	15	20	3	–	–	–	–	–	–	–	–	–	–	Phenol, 2-(2-hydroxy-1,2-dimethylcyclopentyl)-, 1-acetate, trans-	39877-96-8	NI24246
41	91	79	43	77	55	39	93	248	100	95	83	79	76	55	49	46	2.00	15	20	3	–	–	–	–	–	–	–	–	–	–	Spiro-11-oxabicyclo[8.1.0]undec-6-ene-2,2′-oxiran-3-one, 8-methylene-5-isopropyl-, 1R-(1R*,2R*,5S*,6E,10R*)-		MS05370
111	187	206	41	149	220	55	77	248	100	33	30	4	4	4	3	3	1.46	15	20	3	–	–	–	–	–	–	–	–	–	–	Tetracyclo[6.4.2.$0^{1,9}$.$0^{4,14}$]tetradeca-3,13-dione, 4-hydroxy-8-methyl-		NI24247
97	41	55	79	39	77	67	53	248	100	62	36	36	33	29	28	27	2.54	15	20	3	–	–	–	–	–	–	–	–	–	–	Tetracyclo[6.4.2.$0^{1,9}$.$0^{4,14}$]tetradeca-3,13-dione, 4-hydroxy-9-methyl-		NI24248
86	58	85	116	87	57	56	72	248	100	15	9	7	7	7	7	6	0.18	15	24	1	2	–	–	–	–	–	–	–	–	–	Acetamide, 2-(diethylamino)-N-(2,4,6-trimethylphenyl)-	616-68-2	NI24249
138	248	95	205	249	220	219	82	248	100	86	69	17	15	12	12	11		15	24	1	2	–	–	–	–	–	–	–	–	–	Bis(8-azabicyclo[3.2.1]oct-3-yl) ketone, (endo,endo)-	56847-09-7	NI24250
86	87	30	58	248	72	29	28	248	100	13	13	11	9	9	5	4		15	24	1	2	–	–	–	–	–	–	–	–	–	Benzene, 1-ethyl-3-methyl-2-(diethylaminoacetylamino)-		NI24251
248	96	247	150	98	137	55	205	248	100	95	63	55	48	47	47	45		15	24	1	2	–	–	–	–	–	–	–	–	–	1H,5H,10H-Dipyrido[2,1-f:3′,2′,1′-ij][1,6]naphthyridin-15-one, dodecahydro-	519-02-8	NI24252
249	247	248	250	136	149	98	148	248	100	56	44	17	7	6	6	3		15	24	1	2	–	–	–	–	–	–	–	–	–	7,14-Methano-4H,6H-dipyrido[1,2-a:1′,2′-e][1,5]diazocin-4-one, dodecahydro-, (6α,7β,9β,11β)-	486-88-4	NI24253
136	248	97	137	247	220	96	98	248	100	70	53	47	46	45	45	44		15	24	1	2	–	–	–	–	–	–	–	–	–	7,14-Methano-2H,6H-dipyrido[1,2-a:1′,2′-e][1,5]diazocin-6-one, dodecahydro-, [7R-(7α,7aα,14α,14aα)]-	1218-51-5	NI24254
97	136	98	110	84	137	96	220	248	100	99	90	72	59	56	55	45	41.00	15	24	1	2	–	–	–	–	–	–	–	–	–	7,14-Methano-2H,6H-dipyrido[1,2-a:1′,2′-e][1,5]diazocin-6-one, dodecahydro-, [7R-(7α,7aα,14α,14aα)]-	1218-51-5	LI04978
97	248	98	110	220	136	55	123	248	100	97	91	80	58	48	34	33		15	24	1	2	–	–	–	–	–	–	–	–	–	7,14-Methano-2H,6H-dipyrido[1,2-a:1′,2′-e][1,5]diazocin-6-one, dodecahydro-, [7R-(7α,7aα,14α,14aβ)]-	489-72-5	NI24255
136	248	137	97	247	98	220	138	248	100	70	52	47	41	35	31	28		15	24	1	2	–	–	–	–	–	–	–	–	–	7,14-Methano-2H,6H-dipyrido[1,2-a:1′,2′-e][1,5]diazocin-6-one, dodecahydro-, [7R-(7α,7aβ,14α,14aα)]-	577-37-7	LI04979
136	248	137	97	247	96	98	220	248	100	80	49	47	45	37	34	33		15	24	1	2	–	–	–	–	–	–	–	–	–	7,14-Methano-2H,6H-dipyrido[1,2-a:1′,2′-e][1,5]diazocin-6-one, dodecahydro-, [7R-(7α,7aβ,14α,14aα)]-	577-37-7	NI24256
98	150	97	96	151	55	248	110	248	100	76	64	41	33	28	26	21		15	24	1	2	–	–	–	–	–	–	–	–	–	7,14-Methano-2H,6H-dipyrido[1,2-a:1′,2′-e][1,5]diazocin-15-one, dodecahydro-, [7S-(7α,7aα,14α,14aβ)]-	33984-02-0	NI24257
97	98	248	110	220	136	123	111	248	100	78	63	58	36	35	25	22		15	24	1	2	–	–	–	–	–	–	–	–	–	7,14-Methano-2H,6H-dipyrido[1,2-a:1′,2′-e][1,5]diazocin-15-one, dodecahydro-, [7S-(7α,7aα,14α,14aβ)]-		LI04980
136	248	149	247	150	98	97	55	248	100	89	56	46	28	28	22	20		15	24	1	2	–	–	–	–	–	–	–	–	–	7,14-Methano-4H,6H-dipyrido[1,2-a:1′,2′-e][1,5]diazocin-4-one, dodecahydro-, [7S-(7α,7aα,14α,14aα)]-	486-87-3	NI24258
136	41	55	149	248	150	27	42	248	100	72	63	56	49	43	43	39		15	24	1	2	–	–	–	–	–	–	–	–	–	7,14-Methano-4H,6H-dipyrido[1,2-a:1′,2′-e][1,5]diazocin-4-one, dodecahydro-, [7S-(7α,7aβ,14α,14aα)]-	550-90-3	NI24259
136	248	149	150	247	98	97	55	248	100	58	57	41	40	32	32	27		15	24	1	2	–	–	–	–	–	–	–	–	–	7,14-Methano-4H,6H-dipyrido[1,2-a:1′,2′-e][1,5]diazocin-4-one, dodecahydro-, [7S-(7α,7aβ,14α,14aα)]-	550-90-3	MS05371
29	57	86	156	28	248	41	100	248	100	55	25	18	14	13	12	8		15	24	1	2	–	–	–	–	–	–	–	–	–	Urea, N,N-dibutyl-N′-phenyl-	2589-21-1	NI24260
30	57	86	156	248	29	41	100	248	100	56	25	18	13	13	11	8		15	24	1	2	–	–	–	–	–	–	–	–	–	Urea, N,N-dibutyl-N′-phenyl-		LI04981
176	248	28	18	204	88	44	177	248	100	84	44	42	31	29	19	15		16	8	3	–	–	–	–	–	–	–	–	–	–	Phenanthrene-9,10-dicarboxylic acid, anhydride		IC03257
105	77	248	51	106	28	78	50	248	100	94	93	85	51	39	27	14		16	12	1	2	–	–	–	–	–	–	–	–	–	Benzamide, N-(β-cyanostyryl)-		LI04982
105	77	248	106	28	78	249	89	248	100	95	92	51	38	27	23	20		16	12	1	2	–	–	–	–	–	–	–	–	–	Benzamide, N-(β-cyanostyryl)-	23228-06-0	NI24261
77	105	171	248	247	128	219	51	248	100	91	79	78	38	31	27	17		16	12	1	2	–	–	–	–	–	–	–	–	–	Benzamide, N-1-isoquinolinyl-	33357-47-0	NI24262
127	248	141	77	219	126	65	51	248	100	31	26	26	14	12	12	12		16	12	1	2	–	–	–	–	–	–	–	–	–	Diazene, 2-naphthalenylphenyl-, 1-oxide	16914-55-9	NI24263
77	116	248	91	115	127	51	140	248	100	40	28	28	23	21	21	19		16	12	1	2	–	–	–	–	–	–	–	–	–	Diazene, 2-naphthalenylphenyl-, 2-oxide	16954-76-0	NI24264
104	248	117	77	78	51	247	249	248	100	99	97	48	25	24	19	17		16	12	1	2	–	–	–	–	–	–	–	–	–	4H-Imidazol-4-one, 3,5-dihydro-2-phenyl-5-benzylene-	18511-00-7	NI24265

MASS TO CHARGE RATIOS								M.W.	INTENSITIES								Parent	C	H	O	N	S	F	Cl	Br	I	Si	P	B	X	COMPOUND NAME	CAS Reg No	No
248	104	117	77	78	51	247	90	**248**	100	99	97	48	25	24	19	16		16	12	1	2	–	–	–	–	–	–	–	–	–	4H-Imidazol-4-one, 3,5-dihydro-2-phenyl-5-benzylene-	18511-00-7	LI04983
248	168	81	219	205	109.5	139	140	**248**	100	21	18	15	12	10	6	5		16	12	1	2	–	–	–	–	–	–	–	–	–	4H-Indolo[3,2,1-de][1,5]naphthyridin-4-one, 6-ethyl-		LI04984
247	248	171	129	77	249	128	51	**248**	100	92	63	56	36	34	34	31		16	12	1	2	–	–	–	–	–	–	–	–	–	Isoquinolinium, 2-(benzoylamino)-, hydroxide, inner salt	31382-98-6	NI24266
143	115	248	77	51	171	65	114	**248**	100	69	60	57	26	18	14	11		16	12	1	2	–	–	–	–	–	–	–	–	–	2-Naphthalenol, 1-(phenylazo)-	842-07-9	NI24267
248	143	115	77	171	249	247	144	**248**	100	76	27	27	20	18	10	9		16	12	1	2	–	–	–	–	–	–	–	–	–	2-Naphthalenol, 1-phenylazo-	842-07-9	IC03258
248	249	179	178	152	151	150	195	**248**	100	18	17	17	17	17	12	11		16	12	1	2	–	–	–	–	–	–	–	–	–	3H-Phenanthro[3,4-d]imidazol-10-ol, 2-methyl-	98033-26-2	NI24268
248	220	85	84	117	249	77	116	**248**	100	97	54	53	35	33	27	25		16	12	1	2	–	–	–	–	–	–	–	–	–	2-Pyrazinone, 1,2-dihydro-3,6-diphenyl-		MS05372
248	77	105	249	171	51	220		**248**	100	53	51	20	15	15	12			16	12	1	2	–	–	–	–	–	–	–	–	–	Pyrazole, 3-benzoyl-5-phenyl-	21111-32-0	NI24269
248	247	102	191	115	249	219	90	**248**	100	77	39	20	18	17	17	17		16	12	1	2	–	–	–	–	–	–	–	–	–	Pyridazine, 3,6-diphenyl-, 1-oxide	21111-33-1	NI24270
247	248	77	249	104	102	76	141	**248**	100	69	17	12	9	9	8	7		16	12	1	2	–	–	–	–	–	–	–	–	–	2-Pyrimidinol, 4,6-diphenyl-		LI04985
248	218	128	217	77	104	102	191	**248**	100	55	55	44	35	34	33	20		16	12	1	2	–	–	–	–	–	–	–	–	–	Pyrrole, 2,5-diphenyl-3-nitroso-	75096-67-2	NI24271
156	248	128	129	157	249	101	102	**248**	100	52	30	14	12	9	9	8		16	12	1	2	–	–	–	–	–	–	–	–	–	8-Quinolinecarboxamide, N-phenyl-	22765-51-1	NI24272
246	97	41	79	189	147	203	136	**248**	100	80	80	75	30	30	25	20	**0.00**	16	24	–	–	1	–	–	–	–	–	–	–	–	3-Decene, 4-(phenylthio)-		DD00674
57	121	246	137	191	231	248	217	**248**	100	90	60	15	11	8	7	7		16	24	–	–	1	–	–	–	–	–	–	–	–	3-Octene, 7,7-dimethyl-4-(phenylthio)-		DD00675
95	94	93	79	188	189	121	133	**248**	100	52	40	40	37	35	25	23	**20.00**	16	24	2	–	–	–	–	–	–	–	–	–	–	5-Azuleneacetic acid, 1,2,4,5,6,7,8,8a-octahydro-3,8-dimethyl-α-methylene-, methyl ester, [5S-(5α,8β,8aα)]-	57110-50-6	NI24273
121	93	107	95	189	81	248	122	**248**	100	95	90	84	75	72	65	65		16	24	2	–	–	–	–	–	–	–	–	–	–	5-Azuleneacetic acid, 1,2,3,4,5,6,7,8-octahydro-3,8-dimethyl-α-methyl ene-, methyl ester, [3S-(3α,5α,8α)]-		NI24274
119	91	137	41	65	70	55	43	**248**	100	47	41	30	21	17	17	17	**0.52**	16	24	2	–	–	–	–	–	–	–	–	–	–	Benzoic acid, 4-methyl-, 2-octyl ester		IC03259
119	137	136	91	41	65	29	55	**248**	100	99	81	56	22	19	18	17	**2.10**	16	24	2	–	–	–	–	–	–	–	–	–	–	Benzoic acid, 4-methyl-, octyl ester		IC03260
124	41	55	67	81	125	79	95	**248**	100	59	55	44	43	39	37	33	**8.70**	16	24	2	–	–	–	–	–	–	–	–	–	–	Cyclobuta[1,2:3,4]dicyclooctene-1,7(2H,6bH)-dione, dodecahydro-, (6aα,6bα,12aα,12bβ)-	69926-83-6	NI24275
41	29	177	43	27	248	191	55	**248**	100	83	79	72	68	67	55	52		16	24	2	–	–	–	–	–	–	–	–	–	–	2,5-Cyclohexadiene-1,4-dione, 2,5-bis(1,1-dimethylpropyl)-	4584-63-8	WI01274
109	41	124	55	125	39	248		**248**	100	36	25	23	22	16	11			16	24	2	–	–	–	–	–	–	–	–	–	–	1-Cyclohexen-3-one, 5,5-dimethyl-4-(5,5-dimethyl-3-oxocyclohex-1-yl)-		LI04986
41	28	55	67	27	81	79	29	**248**	100	95	82	74	70	66	61	57	**33.66**	16	24	2	–	–	–	–	–	–	–	–	–	–	Cyclooctenone, dimer	61219-51-0	NI24277
41	67	55	79	28	81	136	39	**248**	100	79	71	67	60	57	53	37	**32.06**	16	24	2	–	–	–	–	–	–	–	–	–	–	Cyclooctenone, dimer	61219-51-0	NI24276
41	136	67	79	55	81	248	137	**248**	100	98	88	79	74	59	54	49		16	24	2	–	–	–	–	–	–	–	–	–	–	Cyclooctenone, dimer	61219-51-0	NI24278
165	91	151	77	69	79	248	65	**248**	100	22	14	14	14	10	7	7		16	24	2	–	–	–	–	–	–	–	–	–	–	5-Heptene, 2-(2,6-dimethoxyphenyl)-6-methyl-		LI04987
119	145	188	41	91	233	79	56	**248**	100	97	85	70	67	50	44	44	**29.85**	16	24	2	–	–	–	–	–	–	–	–	–	–	1H-3a,6-Methanoazulene-3-carboxylic acid, 2,3β,4,5,6β,7,8aα-octahydro-7,7-dimethyl-8-methylene-, methyl ester		MS05373
145	233	119	131	91	41	248	173	**248**	100	65	36	31	30	28	24	22		16	24	2	–	–	–	–	–	–	–	–	–	–	1H-3a,6-Methanoazulene-3-carboxylic acid, 2,3β,4,5,6,7-hexahydro-7,7,8-trimethyl-, methyl ester		MS05374
91	108	57	92	43	71	41	55	**248**	100	77	34	31	27	21	21	18	**1.60**	16	24	2	–	–	–	–	–	–	–	–	–	–	Nonanoic acid, benzyl ester		NI24279
91	150	105	104	43	41	151	92	**248**	100	85	17	13	12	11	10	10	**6.01**	16	24	2	–	–	–	–	–	–	–	–	–	–	Nonanoic acid, 2-phenyl-, methyl ester	30368-25-3	NI24280
121	91	174	104	118	175	248	117	**248**	100	92	80	50	47	23	22	20		16	24	2	–	–	–	–	–	–	–	–	–	–	Nonanoic acid, 3-phenyl-, methyl ester	30368-26-4	NI24281
91	117	118	31	32	105	177	174	**248**	100	65	50	48	33	32	25	25	**21.02**	16	24	2	–	–	–	–	–	–	–	–	–	–	Nonanoic acid, 4-phenyl-, methyl ester	30368-27-5	NI24282
91	117	147	191	160	104	248	159	**248**	100	47	25	20	20	14	13	13		16	24	2	–	–	–	–	–	–	–	–	–	–	Nonanoic acid, 5-phenyl-, methyl ester	30368-28-6	NI24283
55	91	44	59	70	41	42	39	**248**	100	66	57	44	36	33	25	25	**7.01**	16	24	2	–	–	–	–	–	–	–	–	–	–	Nonanoic acid, 6-phenyl-, methyl ester	30368-29-7	NI24284
91	119	187	248	169	219	41	92	**248**	100	72	49	23	20	15	12	10		16	24	2	–	–	–	–	–	–	–	–	–	–	Nonanoic acid, 7-phenyl-, methyl ester	30368-30-0	NI24285
105	91	106	248	104	41	79	77	**248**	100	23	21	17	8	8	7	7		16	24	2	–	–	–	–	–	–	–	–	–	–	Nonanoic acid, 8-phenyl-, methyl ester	30368-31-1	NI24286
91	92	74	87	31	248	105	55	**248**	100	56	42	21	19	17	17	16		16	24	2	–	–	–	–	–	–	–	–	–	–	Nonanoic acid, 9-phenyl-, methyl ester	24197-55-5	NI24287
57	149	41	164	121	105	85	91	**248**	100	42	22	17	6	6	6	4	**3.03**	16	24	2	–	–	–	–	–	–	–	–	–	–	Propanoic acid, 2,2-dimethyl-, 2-tert-butyl-4-methylphenyl ester	54644-42-7	NI24288
57	149	41	85	164	105	121	77	**248**	100	51	32	20	14	10	9	7	**4.00**	16	24	2	–	–	–	–	–	–	–	–	–	–	Propanoic acid, 2,2-dimethyl-, 2-tert-butyl-6-methylphenyl ester	54644-45-0	NI24289
77	51	85	55	78	41	39	65	**248**	100	62	59	33	28	25	25	21	**0.00**	16	24	2	–	–	–	–	–	–	–	–	–	–	2H-Pyran, tetrahydro-2-(2,5-undecadiynyloxy)-	55947-04-1	NI24290
54	67	55	82	124	68	81	69	**248**	100	99	90	78	58	56	43	31	**16.00**	16	28	–	2	–	–	–	–	–	–	–	–	–	9,9′-Bi-9-azabicyclo[6.1.0]nonane	75442-62-5	NI24291
158	58	203	159	248	204	160	143	**248**	100	30	17	17	15	8	8	7		16	28	–	2	–	–	–	–	–	–	–	–	–	Benzene, 1,2,4,5-tetramethyl-3,5-bis(N,N-dimethylaminomethyl)-		IC03261
192	135	136	193	45	205	53	41	**248**	100	51	17	14	12	10	9	7	**3.23**	16	28	–	2	–	–	–	–	–	–	–	–	–	Pyrazine, 2,5-dimethyl-3,6-dipentyl-	54600-85-0	NI24292
248	233	41	43	220	55	67	79	**248**	100	90	53	36	35	31	26	24		16	28	–	2	–	–	–	–	–	–	–	–	–	Pyrazine, tetraisopropyl-	72119-29-0	NI24293
41	57	67	248	233	206	177	39	**248**	100	67	56	50	50	44	31	25		16	28	–	2	–	–	–	–	–	–	–	–	–	Pyrazine, 2,3,5-tri-tert-butyl-	39950-98-6	NI24294
248	247	249	124	110.5	245	123.5	246	**248**	100	29	20	12	12	11	7	6		17	12	–	–	1	–	–	–	–	–	–	–	–	Naphtho[2,3-b]benzothiophene, 7-methyl-		AI01778
248	247	171	249	245	172	122	170	**248**	100	95	44	28	17	13	11	9		17	12	–	–	1	–	–	–	–	–	–	–	–	2H-Naphtho[1,8-bc]thiophene, 2-phenyl-		LI04988
248	247	249	245	124	110.5	221	246	**248**	100	25	20	10	10	9	7	6		17	12	–	–	1	–	–	–	–	–	–	–	–	11-Thiabenzo[b]fluorene, 8-methyl-		AI01779
248	247	249	124	245	246	250	122.5	**248**	100	39	20	11	10	7	6	6		17	12	–	–	1	–	–	–	–	–	–	–	–	11-Thiabenzo[b]fluorene, 10-methyl-		AI01780
248	220	115	191	231	77	127	51	**248**	100	58	33	26	25	25	23	20		17	12	2	–	–	–	–	–	–	–	–	–	–	7H-Benzocyclohepten-7-one, 2-phenoxy-		LI04989
105	248	77	51	115	220	143	124	**248**	100	44	40	8	5	2	2	2		17	12	2	–	–	–	–	–	–	–	–	–	–	2(3H)-Furanone, 5-phenyl-3-benzylene-	4361-96-0	LI04990
105	248	77	249	106	51	115	78	**248**	100	44	40	8	8	8	5	3		17	12	2	–	–	–	–	–	–	–	–	–	–	2(3H)-Furanone, 5-phenyl-3-benzylene-	4361-96-0	NI24295

MASS TO CHARGE RATIOS	M.W.	INTENSITIES	Parent	C	H	O	N	S	F	Cl	Br	I	Si	P	B	X	COMPOUND NAME	CAS Reg No	No
155 127 156 126 77 128 101 51	**248**	100 59 11 8 7 6 4 3	**1.50**	17	12	2	–	–	–	–	–	–	–	–	–	–	**1-Naphthalenecarboxylic acid, phenyl ester**		**IC03262**
220 102 77 51 248 105 32 110	**248**	100 84 73 53 44 37 33 29		17	12	2	–	–	–	–	–	–	–	–	–	–	**4H-Pyran-4-one, 2,6-diphenyl-**		**MS05375**
220 248 191 77 105 115 51 102	**248**	100 80 70 60 50 40 25 20		17	12	2	–	–	–	–	–	–	–	–	–	–	**2H-Pyrone, 4,6-diphenyl-**		**NI24296**
120 248 170 105 155 233 53 43	**248**	100 88 68 63 54 47 32 23		17	16	–	–	–	–	–	–	–	1	–	–	–	**Naphthalene, methylphenylsilyl-**	7427-12-5	**MS05376**
248 247 249 204 203 88 123 102	**248**	100 55 18 12 9 6 6 6		17	16	–	2	–	–	–	–	–	–	–	–	–	**Benzene, 1-dimethylamino-4-(2-cyano-2-phenylethenyl)-**	1222-61-3	**NI24297**
248 247 77 249 104 180 144 130	**248**	100 51 32 30 12 8 8 6		17	16	–	2	–	–	–	–	–	–	–	–	–	**Imidazole, 4,5-dimethyl-1,2-diphenyl-**		**LI04991**
98 135 43 150 41 55 248 233	**248**	100 67 54 51 30 22 14 13		17	28	1	–	–	–	–	–	–	–	–	–	–	**8,13-Oxido-14,15,16-trisnor-labd-5(6)-ene**		**MS05377**
248 95 69 55 123 81 109 91	**248**	100 45 40 38 36 36 33 24		17	28	1	–	–	–	–	–	–	–	–	–	–	**1-Phenanthrenol, 1,4,4a,4b,5,6,7,8,8a,9,10,10a-dodecahydro-4b,8,8-trimethyl-, [1R-(1α,4aβ,4bα,8aβ,10aα)]-**	57684-14-7	**NI24298**
248 69 95 55 109 123 81 91	**248**	100 39 37 37 34 33 33 30		17	28	1	–	–	–	–	–	–	–	–	–	–	**1-Phenanthrenol, 1,4,4a,4b,5,6,7,8,8a,9,10,10a-dodecahydro-4b,8,8-trimethyl-, [1S-(1α,4aα,4bβ,8aα,10aβ)]-**	57684-13-6	**NI24299**
248 123 233 41 55 109 69 163	**248**	100 67 64 60 57 46 40 34		17	28	1	–	–	–	–	–	–	–	–	–	–	**3(2H)-Phenanthrenone, dodecahydro-4b,8,8-trimethyl-, [4aS-(4aα,4bβ,8aα,10aβ)]-**	57793-22-3	**NI24300**
107 248 41 43 77 55 79 39	**248**	100 20 17 12 11 9 8 8		17	28	1	–	–	–	–	–	–	–	–	–	–	**Phenol, 2-undecyl-**	20056-71-7	**NI24301**
107 91 108 248 41 43 77 55	**248**	100 35 26 21 16 12 10 9		17	28	1	–	–	–	–	–	–	–	–	–	–	**Phenol, 2-undecyl-**	20056-71-7	**NI24302**
108 107 109 41 248 43 121 77	**248**	100 29 18 12 12 10 9 8		17	28	1	–	–	–	–	–	–	–	–	–	–	**Phenol, 3-undecyl-**	20056-72-8	**NI24303**
108 107 109 248 41 121 43 54	**248**	100 30 18 12 11 10 10 7		17	28	1	–	–	–	–	–	–	–	–	–	–	**Phenol, 3-undecyl-**	20056-72-8	**NI24304**
108 107 109 248 40 120 42 77	**248**	100 28 18 11 11 9 9 7		17	28	1	–	–	–	–	–	–	–	–	–	–	**Phenol, 3-undecyl-**	20056-72-8	**LI04992**
107 108 248 41 133 77 43 39	**248**	100 17 12 7 5 5 5 4		17	28	1	–	–	–	–	–	–	–	–	–	–	**Phenol, 4-undecyl-**	20056-73-9	**NI24305**
107 108 248 41 77 43 134 39	**248**	100 16 12 7 5 5 4 4		17	28	1	–	–	–	–	–	–	–	–	–	–	**Phenol, 4-undecyl-**	20056-73-9	**LI04993**
107 108 248 41 77 43 133 55	**248**	100 17 12 8 6 6 5 4		17	28	1	–	–	–	–	–	–	–	–	–	–	**Phenol, 4-undecyl-**	20056-73-9	**NI24306**
123 138 248 55 95 81 69 82	**248**	100 78 74 72 64 60 58 48		17	28	1	–	–	–	–	–	–	–	–	–	–	**8β-Podocarpan-14-one**	5720-73-0	**NI24307**
206 248 205 204 192 191 220 165	**248**	100 85 45 36 36 28 21 21		18	16	1	–	–	–	–	–	–	–	–	–	–	**Bicyclo[3.1.0]hexan-2-one, 6,6-diphenyl-**	22524-16-9	**NI24308**
178 89 43 176 179 76 88 177	**248**	100 30 30 27 25 25 20 15	**3.00**	18	16	1	–	–	–	–	–	–	–	–	–	–	**1-Butanone, 1-(2-anthryl)-**	32352-86-6	**NI24309**
77 105 128 247 51 248 115 171	**248**	100 85 41 40 34 30 30 25		18	16	1	–	–	–	–	–	–	–	–	–	–	**1-Cyclopentene, 1-benzoyl-2-phenyl-**		**LI04994**
247 248 105 77 171 128 143 115	**248**	100 78 37 36 20 19 14 14		18	16	1	–	–	–	–	–	–	–	–	–	–	**1-Cyclopentene, 1-benzoyl-2-phenyl-**		**LI04995**
105 248 128 143 142 106 115 141	**248**	100 18 10 9 8 7 5 4		18	16	1	–	–	–	–	–	–	–	–	–	–	**2-Cyclopentene, 1-benzoyl-2-phenyl-**		**LI04996**
248 233 219 215 118 231 249 90	**248**	100 89 65 52 28 27 20 20		18	16	1	–	–	–	–	–	–	–	–	–	–	**2,2′-Spirobiindan-1-one, 1′-methyl-, (1R)-**	82309-97-5	**NI24310**
95 191 41 109 67 81 55 96	**248**	100 96 69 59 56 52 51 33	**19.00**	18	32	–	–	–	–	–	–	–	–	–	–	–	**Anthracene, 9-butyl-1,2,3,4,5,6,7,8,9,10-tetradecahydro-**		**AI01781**
191 95 41 67 109 81 55 248	**248**	100 87 70 54 53 48 45 26		18	32	–	–	–	–	–	–	–	–	–	–	–	**Anthracene, 9-butyl-1,2,3,4,5,6,7,8,9,10-tetradecahydro-**		**AI01782**
83 165 164 55 109 82 41 67	**248**	100 62 42 40 34 24 24 21	**3.00**	18	32	–	–	–	–	–	–	–	–	–	–	–	**Cyclohexane, 1,2-dicyclohexyl-**		**AI01784**
83 41 55 165 164 67 109 82	**248**	100 69 66 57 37 35 27 26	**2.02**	18	32	–	–	–	–	–	–	–	–	–	–	–	**Cyclohexane, 1,2-dicyclohexyl-**		**AI01783**
83 82 55 165 109 164 41 67	**248**	100 55 45 42 30 29 29 26	**13.00**	18	32	–	–	–	–	–	–	–	–	–	–	–	**Cyclohexane, 1,3-dicyclohexyl-**	1706-50-9	**AI01785**
83 82 55 165 109 164 67 41	**248**	100 52 44 39 35 26 24 22	**7.71**	18	32	–	–	–	–	–	–	–	–	–	–	–	**Cyclohexane, 1,3-dicyclohexyl-**	1706-50-9	**NI24311**
83 55 41 82 67 164 165 81	**248**	100 77 72 61 41 33 29 24	**12.74**	18	32	–	–	–	–	–	–	–	–	–	–	–	**Cyclohexane, 1,4-dicyclohexyl-**		**AI01786**
83 82 55 164 109 67 41 81	**248**	100 65 47 41 30 29 26 21	**18.80**	18	32	–	–	–	–	–	–	–	–	–	–	–	**Cyclohexane, 1,4-dicyclohexyl-**		**AI01787**
81 95 55 41 67 137 43 69	**248**	100 76 62 59 44 41 39 32	**2.00**	18	32	–	–	–	–	–	–	–	–	–	–	–	**1H-Indene, 1-(1,5-dimethyl-2-hexenyl)octahydro-7a-methyl-, [1R-[1α(1R*2Z),3aβ,7aα]]-**	54411-95-9	**NI24312**
67 41 81 55 82 54 68 95	**248**	100 74 66 44 33 32 31 29	**1.12**	18	32	–	–	–	–	–	–	–	–	–	–	–	**(1,11E,13Z)-Octadecatriene**	80625-36-1	**NI24313**
81 123 43 55 124 67 69 41	**248**	100 70 44 39 38 38 33 26	**2.69**	18	32	–	–	–	–	–	–	–	–	–	–	–	**2,5,9-Tetradecatriene, 3,12-diethyl-**	74685-87-3	**NI24314**
118 119 117 248 91 115 129 249	**248**	100 52 32 31 13 11 8 6		19	20	–	–	–	–	–	–	–	–	–	–	–	**2,2′-Spirobiindan, 1,1′-dimethyl-, (S,S)-**		**NI24316**
118 119 117 248 91 115 129 249	**248**	100 61 43 33 14 12 11 7		19	20	–	–	–	–	–	–	–	–	–	–	–	**2,2′-Spirobiindan, 1,1′-dimethyl-, (S,S)-**		**NI24315**
118 119 248 117 91 115 129 249	**248**	100 64 45 43 14 13 11 9		19	20	–	–	–	–	–	–	–	–	–	–	–	**2,2′-Spirobiindan, 1,1′-dimethyl-, (1R,1′S,2S)-**		**NI24317**
251 253 249 216 214 255 218 93	**249**	100 65 61 31 26 22 15 11		5	–	–	1	–	–	5	–	–	–	–	–	–	**Pyridine, pentachloro-**	2176-62-7	**NI24318**
251 253 249 216 214 218 109	**249**	100 67 63 33 27 17 12		5	–	–	1	–	–	5	–	–	–	–	–	–	**Pyridine, pentachloro-**	2176-62-7	**LI04997**
166 69 78 134 96 77 97 33	**249**	100 63 47 11 7 7 6 6	**0.00**	5	4	–	1	–	9	–	–	–	–	–	–	–	**Propane, 3-[bis(trifluoromethyl)amino]-1,1,1-trifluoro-**		**LI04998**
249 203 76 50 127 92 74 75	**249**	100 48 41 18 16 14 13 9		6	4	2	1	–	–	–	–	1	–	–	–	–	**Benzene, 1-iodo-2-nitro-**	609-73-4	**NI24319**
249 203 76 50 74 127 75 30	**249**	100 48 43 16 13 11 10 7		6	4	2	1	–	–	–	–	1	–	–	–	–	**Benzene, 1-iodo-3-nitro-**	645-00-1	**NI24320**
249 76 203 219 50 74 191 75	**249**	100 44 33 24 18 16 15 12		6	4	2	1	–	–	–	–	1	–	–	–	–	**Benzene, 1-iodo-4-nitro-**	636-98-6	**NI24321**
249 205 78 51 50 77 204 76	**249**	100 98 95 88 83 81 73 65		6	4	2	1	–	–	–	–	1	–	–	–	–	**2-Pyridinecarboxylic acid, 6-iodo-**	55044-68-3	**NI24322**
251 90 249 253 91 170 172 143	**249**	100 60 57 56 41 32 31 16		6	5	–	1	–	–	–	2	–	–	–	–	–	**Aniline, 2,4-dibromo-**	615-57-6	**NI24323**
251 90 63 253 249 52 91 45	**249**	100 72 52 52 50 35 31 26		6	5	–	1	–	–	–	2	–	–	–	–	–	**Aniline, 2,5-dibromo-**	3638-73-1	**NI24324**
251 90 253 249 63 91 52 45	**249**	100 88 51 51 47 28 27 20		6	5	–	1	–	–	–	2	–	–	–	–	–	**Aniline, 2,6-dibromo-**	608-30-0	**NI24325**

MASS TO CHARGE RATIOS								M.W.	INTENSITIES								Parent	C	H	O	N	S	F	Cl	Br	I	Si	P	B	X	COMPOUND NAME	CAS Reg No	No
249	229	109	42	230	250	179	83	249	100	90	63	25	24	14	12	11		7	6	–	5	1	3	–	–	–	–	–	–	–	Pyrimido[5,4-e]-1,2,4-triazine-7(1H)-thione, 2,6-dihydro-3-methyl-5-(trifluoromethyl)-	32709-35-6	NI24326
249	229	109	230	42	250	179	171	249	100	90	62	25	25	13	12	10		7	6	–	5	1	3	–	–	–	–	–	–	–	Pyrimido[5,4-e]-1,2,4-triazine-7(6H)-thione, 1,2-dihydro-3-methyl-5-(trifluoromethyl)-		LI04999
63	75	122	76	65	186	187	185	249	100	31	25	25	25	21	17	16	14.20	8	8	4	1	1	–	1	–	–	–	–	–	–	Benzene, 1-[(2-chloroethyl)sulphonyl]-4-nitro-	6461-63-8	NI24327
63	122	76	65	186	75	187	185	249	100	26	26	26	21	21	18	17	16.00	8	8	4	1	1	–	1	–	–	–	–	–	–	Benzene, 1-[(2-chloroethyl)sulphonyl]-4-nitro-	6461-63-8	LI05000
136	30	78	106	89	90	28	63	249	100	62	52	39	39	37	24	17	0.00	8	8	4	1	1	–	1	–	–	–	–	–	–	Benzene, 1-[[(chloromethyl)sulphonyl]methyl]-4-nitro-	69688-53-5	NI24328
136	79	90	43	89	106	137	187	249	100	32	25	22	21	19	13	10	1.20	8	8	4	1	1	–	1	–	–	–	–	–	–	Benzene, 1-[[(chloromethyl)sulphonyl]methyl]-4-nitro-	69688-53-5	MS05378
249	234	70	179	56	140	43	208	249	100	40	34	29	27	22	19	18		8	9	1	1	–	6	–	–	–	–	–	–	–	Azetidine, 2-methyl-1-[3,3,3-trifluoro-1-oxo-2-(trifluoromethyl)propyl]-	50837-79-1	NI24329
249	234	70	179	56	140	43	55	249	100	24	20	18	16	12	12	10		8	9	1	1	–	6	–	–	–	–	–	–	–	Azetidine, 2-methyl-1-[3,3,3-trifluoro-1-oxo-2-(trifluoromethyl)propyl]-	50837-79-1	NI24330
249	234	70	179	56	140	43	208	249	100	24	19	17	16	13	11	10		8	9	1	1	–	6	–	–	–	–	–	–	–	Azetidine, 2-methyl-1-[3,3,3-trifluoro-1-oxo-2-(trifluoromethyl)propyl]-	50837-79-1	LI05001
164	192	249	218	204	219	69	43	249	100	77	68	56	52	46	41	37		8	10	1	5	–	3	–	–	–	–	–	–	–	1,3,5-Triazine, 2-amino-4-morpholino-6-(trifluoromethyl)-		MS05379
249	203	204	202	251	131	250	190	249	100	54	36	17	14	13	12	11		8	11	2	1	3	–	–	–	–	–	–	–	–	4-Isothiazolecarboxylic acid, 3,5-bis(methylthio)-, ethyl ester	2273-03-2	NI24331
142	106	107	127	79	51	52	39	249	100	97	59	55	41	35	25	24	0.00	8	12	–	1	–	–	–	–	1	–	–	–	–	Pyridinium, 2-ethyl-1-methyl-, iodide	52806-01-6	NI24332
92	142	107	65	127	39	79	51	249	100	85	85	35	25	19	18	14	0.00	8	12	–	1	–	–	–	–	1	–	–	–	–	Pyridinium, 3-ethyl-1-methyl-, iodide	42493-47-0	NI24333
107	142	106	127	36	121	66	39	249	100	64	40	36	34	26	24	24	0.00	8	12	–	1	–	–	–	–	1	–	–	–	–	Pyridinium, 1,2,6-trimethyl-, iodide	2525-19-1	NI24334
39	125	200	83	154	103	67	252	249	100	7	5	5	4	4	4	4	3.54	8	14	2	2	–	–	–	1	–	–	–	–	–	4-Bromomethyl-2,2,5,5-tetramethyl-3-imidazoline-3-oxide-1-oxile	51973-27-4	NI24335
204	249	233	232	211	184	205	157	249	100	68	40	39	12	12	10	10		9	6	2	1	1	3	–	–	–	–	–	–	–	2H-1,4-Benzothiazin-3(4H)-one, 4-hydroxy-6-(trifluoromethyl)-	4875-01-8	NI24336
175	159	176	141	123	111	231	77	249	100	59	38	30	27	20	18	17	0.90	9	9	3	1	–	–	2	–	–	–	–	–	–	Carbamic acid, N-(3,4-Dichlorobenzyl)-N-(hydroxymethyl)-		LI05002
73	132	75	116	77	45	69	93	249	100	68	57	45	29	22	21	18	0.00	9	23	3	1	–	–	–	–	–	2	–	–	–	Serine, N-(trimethylsilyl)-, trimethylsilyl ester		MS05380
73	133	146	119	249	59	147	86	249	100	47	29	22	19	18	18	13		9	27	1	1	–	–	–	–	–	3	–	–	–	Hydroxylamine, tris(trimethylsilyl)-	21023-20-1	NI24337
73	147	133	146	249	45	59	119	249	100	67	34	26	24	20	17	15		9	27	1	1	–	–	–	–	–	3	–	–	–	Hydroxylamine, tris(trimethylsilyl)-	21023-20-1	NI24338
73	133	146	119	249	59	130	147	249	100	63	33	26	25	17	15	13		9	27	1	1	–	–	–	–	–	3	–	–	–	Hydroxylamine, tris(trimethylsilyl)-	21023-20-1	NI24339
249	90	164	134	63	106	64	118	249	100	95	79	75	38	29	23	21		10	7	5	3	–	–	–	–	–	–	–	–	–	2,4,5-Imidazolidinetrione, 1-methyl-3-(4-nitrophenyl)-		MS05381
135	28	44	108	43	31	149	54	249	100	67	46	37	32	27	25	22	9.22	10	11	3	5	–	–	–	–	–	–	–	–	–	7,10-Epoxy-6H-azepino[1,2-e]purine-8,9-diol, 4-amino-7,8,9,10-tetrahydro-, [7R-(7α,8α,9α,10α)]-	3415-89-2	NI24340
132	100	88	117	172	60	191	55	249	100	31	26	20	16	13	11	10	0.00	10	19	4	1	1	–	–	–	–	–	–	–	–	L-Cysteine, N-Acetyl-S-(2-hydroxybutyl)-, methyl ester	92079-06-6	NI24341
113	119	61	190	88	71	104	172	249	100	41	41	38	36	25	23	15	1.00	10	19	4	1	1	–	–	–	–	–	–	–	–	L-Cysteine, N-Acetyl-S-(3-hydroxybutyl)-, methyl ester	92079-05-5	NI24342
71	140	88	190	60	113	172	131	249	100	93	89	70	38	34	28	28	0.00	10	19	4	1	1	–	–	–	–	–	–	–	–	L-Cysteine, N-Acetyl-S-(4-hydroxybutyl)-, methyl ester	86560-44-3	NI24343
73	43	115	74	58	139	75	45	249	100	55	43	24	21	19	17	17	0.00	10	19	6	1	–	–	–	–	–	–	–	–	–	α-D-Galactopyranoside, methyl 2-(acetylamino)-2-deoxy-3-O-methyl-	17296-08-1	NI24344
59	43	45	101	60	73	139	114	249	100	80	52	43	36	32	23	22	0.00	10	19	6	1	–	–	–	–	–	–	–	–	–	α-D-Galactopyranoside, methyl 2-(acetylamino)-2-deoxy-6-O-methyl-	17296-07-0	NI24345
101	88	75	126	102	157	73	186	249	100	20	16	5	5	4	3	2	0.00	10	19	6	1	–	–	–	–	–	–	–	–	–	α-D-Glucopyranosiduronamide, methyl 2,3,4-tri-O-methyl-		NI24346
100	112	150	152	126	169	113	154	249	100	85	72	71	53	50	46	42	5.00	10	20	1	1	–	–	–	1	–	–	–	–	–	Acetamide, N-(3-bromopropyl)-N-pentyl-	55044-64-9	NI24347
174	73	59	86	45	175	61	100	249	100	84	36	33	22	19	15	14	7.20	10	27	–	1	1	–	–	–	–	2	–	–	–	Propanamine, 3-(methylthio)-N,N-bis(trimethylsilyl)-		MS05382
146	130	73	75	144	56	147	59	249	100	46	39	27	18	18	13	12	0.59	10	27	2	1	–	–	–	–	–	2	–	–	–	Diethanolamine, bis[(trimethylsilyl)oxy]-		IC03263
118	249	43	91	73	41	65	119	249	100	52	30	21	17	13	10	10		11	11	–	3	2	–	–	–	–	–	–	–	–	1,3,5-Thiadiazine-4-thione, 6-methylamino-2-(m-tolyl)-		NI24348
184	185	92	65	108	186	183	66	249	100	83	60	40	26	11	9	9	1.00	11	11	2	3	1	–	–	–	–	–	–	–	–	Benzenesulphonamide, 4-amino-N-2-pyridinyl-	144-83-2	NI24349
249	77	205	135	204	51	93	203	249	100	80	54	53	39	30	29	28		11	11	2	3	1	–	–	–	–	–	–	–	–	4-Imidazolidinone, 5-carbamoylmethyl-3-phenyl-2-thioxo-		LI05003
77	94	158	51	103	50	32	65	249	100	86	74	56	28	25	24	22	0.00	11	11	4	3	–	–	–	–	–	–	–	–	–	4-Isoxazolecarboxylic acid, 5-[(aminocarbonyl)amino]-4,5-dihydro-3-phenyl-, cis-	35053-68-0	NI24350
213	108	81	77	54	136	51	52	249	100	97	81	42	42	28	20	11	0.00	11	12	–	5	–	–	1	–	–	–	–	–	–	Pyridine, 2,6-diamino-3-(phenylazo)-, monohydrochloride	136-40-3	NI24351
186	42	66	121	71	56	65	44	249	100	48	47	44	42	32	26	26	0.57	11	15	2	1	–	–	–	–	–	–	–	–	1	Iron, π-cyclopentadienyl-π-(dimethylaminopropionyl)carbonyl-		LI05004
186	42	66	121	71	56	44	65	249	100	48	46	43	42	31	25	23	0.56	11	15	2	1	–	–	–	–	–	–	–	–	1	Iron, π-cyclopentadienyl-π-(dimethylaminopropionyl)carbonyl-		MS05383
234	249	58	192	207	250	235	43	249	100	95	36	30	28	16	15	14		11	19	–	7	–	–	–	–	–	–	–	–	–	1,3,5-Triazine-2,4,6-triamine, N-cyano-N′,N″-di-isopropyl-N-methyl-	67704-68-1	NI24352
178	110	41	69	56	179	249	166	249	100	98	79	43	35	23	14	14		11	24	3	1	–	–	–	–	–	–	1	–	–	1,3,2-Dioxaphosphorinane, 2-(hexylamino)-5,5-dimethyl-, 2-oxide		NI24353
249	251	214	250	217	216	188	248	249	100	39	20	19	14	14	13	10		12	8	1	1	1	–	1	–	–	–	–	–	–	10H-Phenothiazin-2-ol, 8-chloro-	5909-61-5	NI24354
249	251	214	250	217	124	188	216	249	100	38	20	18	15	14	13	12		12	8	1	1	1	–	1	–	–	–	–	–	–	10H-Phenothiazin-2-ol, 8-chloro-	5909-61-5	LI05005
249	251	214	250	125	217	216	188	249	100	37	19	17	14	13	12	12		12	8	1	1	1	–	1	–	–	–	–	–	–	10H-Phenothiazin-2-ol, 8-chloro-	5909-61-5	NI24355
249	251	250	124	214	217	248	186	249	100	39	20	15	14	13	12	10		12	8	1	1	1	–	1	–	–	–	–	–	–	10H-Phenothiazin-3-ol, 2-chloro-	16770-99-3	LI05006
249	251	250	214	217	124	248	188	249	100	40	21	18	17	17	15	10		12	8	1	1	1	–	1	–	–	–	–	–	–	10H-Phenothiazin-3-ol, 2-chloro-	16770-99-3	NI24356
249	251	250	214	217	248	216	186	249	100	38	20	15	14	12	9	8		12	8	1	1	1	–	1	–	–	–	–	–	–	10H-Phenothiazin-3-ol, 2-chloro-	16770-99-3	NI24357
249	251	250	214	124	217	248	216	249	100	40	20	17	16	15	14	12		12	8	1	1	1	–	1	–	–	–	–	–	–	10H-Phenothiazin-3-ol, 8-chloro-	2002-32-6	LI05007
249	251	250	214	248	217	216	188	249	100	41	21	17	15	15	12	9		12	8	1	1	1	–	1	–	–	–	–	–	–	10H-Phenothiazin-3-ol, 8-chloro-	2002-32-6	NI24358
249	251	250	188	214	217	216	248	249	100	38	17	15	14	9	9	8		12	8	1	1	1	–	1	–	–	–	–	–	–	10H-Phenothiazin-4-ol, 8-chloro-	19807-55-7	NI24359
77	108	81	51	249	184	80	53	249	100	89	74	63	50	45	39	36		12	11	3	1	1	–	–	–	–	–	–	–	–	Benzenesulphonamide, N-(3-hydroxyphenyl)-	59149-19-8	NI24360

MASS TO CHARGE RATIOS	M.W.	INTENSITIES	Parent	C	H	O	N	S	F	Cl	Br	I	Si	P	B	X	COMPOUND NAME	CAS Reg No	No
249 221 250 222 193 120 148 122	**249**	100 89 16 12 11 11 8 8		12	11	5	1	–	–	–	–	–	–	–	–	–	**4H-1-Benzopyran-2-carboxylic acid, 5-amino-6-hydroxy-4-oxo-, ethyl ester**	32142-43-1	**NI24361**
249 221 250 120 151 222 193 95	**249**	100 72 16 10 9 8 7 7		12	11	5	1	–	–	–	–	–	–	–	–	–	**4H-1-Benzopyran-2-carboxylic acid, 6-amino-7-hydroxy-4-oxo-, ethyl ester**	30192-11-1	**NI24362**
249 221 151 250 120 222 220 123	**249**	100 61 16 14 9 7 7 7		12	11	5	1	–	–	–	–	–	–	–	–	–	**4H-1-Benzopyran-2-carboxylic acid, 8-amino-7-hydroxy-4-oxo-, ethyl ester**	30192-51-9	**NI24363**
233 205 247 217 189 75 104 79	**249**	100 81 53 28 25 18 17 17	**3.00**	12	11	5	1	–	–	–	–	–	–	–	–	–	**4H-1-Benzopyran-2-carboxylic acid, 6-(hydroxyamino)-4-oxo-, ethyl ester**	30095-80-8	**NI24364**
77 249 170 94 248 51 156 65	**249**	100 64 58 49 43 26 23 19		12	12	3	1	–	–	–	–	–	–	1	–	–	**Phosphoramidic acid, diphenyl ester**	2015-56-7	**LI05008**
77 249 170 248 94 51 156 65	**249**	100 65 59 43 29 26 23 20		12	12	3	1	–	–	–	–	–	–	1	–	–	**Phosphoramidic acid, diphenyl ester**	2015-56-7	**NI24365**
249 165 122 207 147 164 123 189	**249**	100 83 69 61 61 44 42 22		12	15	3	3	–	–	–	–	–	–	–	–	–	**1,2,4-Benzenetriacetamide**		**IC03264**
83 28 249 82 41 56 208 84	**249**	100 98 73 58 48 37 25 23		12	15	3	3	–	–	–	–	–	–	–	–	–	**1,3,5-Triazine-2,4,6(1H,3H,5H)-trione, 1,3,5-triallyl-**	1025-15-6	**IC03265**
83 249 82 70 56 208 84 41	**249**	100 83 68 50 37 30 23 23		12	15	3	3	–	–	–	–	–	–	–	–	–	**1,3,5-Triazine-2,4,6(1H,3H,5H)-trione, 1,3,5-triallyl-**	1025-15-6	**NI24366**
193 202 134 249 207 194 192 135	**249**	100 69 61 35 26 22 22 17		13	15	–	1	2	–	–	–	–	–	–	–	–	**Thiazole, 2-(butylthio)-4-phenyl-**	69390-10-9	**NI24367**
249 162 135 134 250 190 91 160	**249**	100 75 54 34 18 18 18 16		13	15	4	1	–	–	–	–	–	–	–	–	–	**1H-3-Benzazepine-2-carboxylic acid, 2,3,4,5-tetrahydro-8-methoxy-4-oxo-, methyl ester**	17639-49-5	**NI24368**
41 91 87 151 39 118 176 65	**249**	100 88 77 69 68 41 39 32	**7.00**	13	15	4	1	–	–	–	–	–	–	–	–	–	**2,9-Benzofurandione, 4,7-dihydro-4-(5-oxo-3,4-dihydro-2H-pyrrolyl)-6-methyl-**		**MS05384**
76 41 50 99 39 54 78 105	**249**	100 72 67 61 50 47 32 31	**4.00**	13	15	4	1	–	–	–	–	–	–	–	–	–	**2,9-Benzofurandione, 4,7-dihydro-4-(2-oxo-3,4,5,6-tetrahydropyridyl)-**		**MS05385**
100 30 107 72 43 207 108 249	**249**	100 97 92 52 49 24 23 22		13	15	4	1	–	–	–	–	–	–	–	–	–	**Benzoic acid, 4-(3-acetylamino-2-oxopropyl)-, methyl ester**		**NI24369**
118 157 129 90 115 91 128 185	**249**	100 72 70 67 61 55 43 39	**0.00**	13	15	4	1	–	–	–	–	–	–	–	–	–	**1-Naphthalenone, 2-(3-hydroxypropyl)-2-nitro-1,2,3,4-tetrahydro-**		**DD00676**
91 114 160 70 65 92 40 204	**249**	100 32 16 11 11 10 8 7	**2.08**	13	15	4	1	–	–	–	–	–	–	–	–	–	**1,2-Pyrrolidinedicarboxylic acid, 1-benzyl ester, (S)-**	1148-11-4	**NI24370**
77 55 43 41 104 161 249 52	**249**	100 68 63 61 41 40 38 29		13	19	2	3	–	–	–	–	–	–	–	–	–	**Hexanal, 4-methyl-, (2-nitrophenyl)hydrazone**	55320-77-9	**NI24371**
70 121 120 94 84 122 71 250	**249**	100 90 52 43 43 38 21 3	**0.00**	13	19	2	3	–	–	–	–	–	–	–	–	–	**1H-Pyrazole, 1-acetyl-4,5-dihydro-3-[[4′-(pyrrolidin-1-yl)-2-butynyl]oxy]-**		**DD00677**
71 70 121 122 120 123 108 84	**249**	100 99 94 80 80 72 68 52	**0.00**	13	19	2	3	–	–	–	–	–	–	–	–	–	**3-Pyrazolidinone, 1-acetyl-2-[4-(pyrrolidin-1-yl)-2-butynyl]-**		**DD00678**
127 129 125 43 65 140 64 92	**249**	100 28 23 21 19 16 13 12	**0.00**	14	16	1	1	–	–	1	–	–	–	–	–	–	**Aniline, N-(2-acetylcyclohexylidene)-4-chloro-**		**NI24372**
42 43 118 57 91 117 41 249	**249**	100 72 68 68 66 63 57 54		14	19	1	1	1	–	–	–	–	–	–	–	–	**1,3-Oxazine-2-thione, 6-benzyl-4,4,6-trimethyltetrahydro-**		**NI24373**
58 100 43 56 101 42 15 59	**249**	100 56 21 15 6 6 6 5	**0.70**	14	19	3	1	–	–	–	–	–	–	–	–	–	**Acetamide, N-[2-(acetyloxy)-1-methyl-2-phenylethyl]-N-methyl-**	55133-90-9	**NI24374**
58 100 43 134 77 190 148 101	**249**	100 69 23 8 8 7 5 5	**0.00**	14	19	3	1	–	–	–	–	–	–	–	–	–	**Acetamide, N-[2-(acetyloxy)-1-methyl-2-phenylethyl]-N-methyl-**	55133-90-9	**MS05386**
249 163 218 190 161 182 148 191	**249**	100 75 61 60 60 55 50 40		14	19	3	1	–	–	–	–	–	–	–	–	–	**3H-3-Benzazocine, 3-methyl-5-methoxy-8,9-methylenedioxy-1,2,4,5,6-pentahydro-**		**LI05009**
249 163 134 94 135 250 133 107	**249**	100 68 38 23 22 16 15 14		14	19	3	1	–	–	–	–	–	–	–	–	–	**Benzofuran, 2,3-dihydro-3,3-dimethyl-2-morpholino-**		**IC03266**
109 179 43 71 27 40 110 108	**249**	100 33 30 10 7 7 6 5	**2.40**	14	19	3	1	–	–	–	–	–	–	–	–	–	**Butanoic acid, 4-[1-(oxobutyl)amino]phenyl ester**		**NI24375**
105 77 249 134 106 135 162 144	**249**	100 35 13 9 9 8 7 7		14	19	3	1	–	–	–	–	–	–	–	–	–	**Hexanoic acid, 6-(benzoylamino)-, methyl ester**		**LI05010**
100 115 143 249 234	**249**	100 84 66 26 18		14	23	1	1	–	–	–	–	–	1	–	–	–	**Morpholine, 3-methyl-2-phenyl-N-trimethylsilyl-**		**LI05011**
100 73 115 143 114 59 45 249	**249**	100 64 62 37 24 24 16 11		14	23	1	1	–	–	–	–	–	1	–	–	–	**Morpholine, 3-methyl-2-phenyl-N-trimethylsilyl-**		**NI24376**
93 41 192 91 79 107 177 55	**249**	100 90 80 75 75 69 65 65	**0.00**	14	23	1	3	–	–	–	–	–	–	–	–	–	**Bicyclo[3.3.1]nonan-2-one, 9-isopropylidene-5-methyl-**		**LI05012**
234 216 83 249 41 235 160 91	**249**	100 67 28 23 19 15 14 12		14	23	1	3	–	–	–	–	–	–	–	–	–	**3-Buten-2-one, 4-(2,6,6-trimethyl-1-cyclohexen-1-yl)-, semicarbazone**	2302-89-8	**NI24377**
129 171 248 249 128 102 51 172	**249**	100 81 68 50 22 22 17 12		15	11	1	3	–	–	–	–	–	–	–	–	–	**Isoquinolinium, 2-[(4-pyridinylcarbonyl)amino]-, hydroxide, inner salt**	32339-25-6	**NI24378**
220 249 143 116 221 248 79 51	**249**	100 83 74 70 66 43 27 25		15	11	1	3	–	–	–	–	–	–	–	–	–	**1-Naphthalenol, 2-(2-pyridinylazo)-**		**LI05013**
115 143 28 78 51 114 89 79	**249**	100 69 37 25 25 22 16 15	**0.00**	15	11	1	3	–	–	–	–	–	–	–	–	–	**2-Naphthalenol, 1-(2-pyridinylazo)-**	85-85-8	**NI24379**
221 116 143 222 51 79 38 115	**249**	100 73 50 35 27 22 22 17	**9.00**	15	11	1	3	–	–	–	–	–	–	–	–	–	**2-Naphthalenol, 1-(2-pyridinylazo)-**	85-85-8	**LI05014**
220 115 143 221 249 78 114 51	**249**	100 48 47 41 21 13 11 11		15	11	1	3	–	–	–	–	–	–	–	–	–	**2-Naphthalenol, 1-(2-pyridinylazo)-**	85-85-8	**IC03267**
118 249 89 91 77 63 221 192	**249**	100 84 53 36 25 25 24 21		15	11	1	3	–	–	–	–	–	–	–	–	–	**1,2,4-Triazine, 3,5-diphenyl-6-hydroxy-**		**MS05387**
104 146 249 103 130 77 118 89	**249**	100 70 64 53 49 48 33 25		15	11	1	3	–	–	–	–	–	–	–	–	–	**1,2,4-Triazine, 3,6-diphenyl-5-hydroxy-**		**MS05388**
77 104 103 76 249 51 248 130	**249**	100 97 97 73 49 47 42 28		15	11	1	3	–	–	–	–	–	–	–	–	–	**1,2,4-Triazine, 4,6-diphenyl-2-hydroxy-**		**MS05389**
249 178 165 248 89 250 77 76	**249**	100 87 45 41 20 16 14 14		15	11	1	3	–	–	–	–	–	–	–	–	–	**1,2,4-Triazin-3(2H)-one, 5,6-diphenyl-**	4512-00-9	**NI24380**
178 165 104 105 249 191 163 250	**249**	100 58 46 42 35 27 25 23		15	11	1	3	–	–	–	–	–	–	–	–	–	**1,2,4-Triazin-3(2H)-one, 5,6-diphenyl-**	4512-00-9	**NI24381**
118 249 89 91 63 77 221 192	**249**	100 83 54 36 26 26 23 21		15	11	1	3	–	–	–	–	–	–	–	–	–	**1,2,4-Triazin-6(1H)-one, 3,5-diphenyl-**	31633-54-2	**NI24382**
158 162 116 102 120 91 77 163	**249**	100 98 96 92 90 90 90 87	**17.00**	15	23	2	1	–	–	–	–	–	–	–	–	–	**L-Alanine, N-isopropyl-3-phenyl-, isopropyl ester,**	31552-15-5	**MS05390**
87 174 75 104 55 77 43 172	**249**	100 31 19 18 17 14 12 6	**1.00**	15	23	2	1	–	–	–	–	–	–	–	–	–	**Aniline, N-2-(1,1-dimethoxy-3,4-dimethylpentylidene)-**		**NI24383**
164 107 94 178 248 249 234	**249**	100 54 34 6 3 2 2		15	23	2	1	–	–	–	–	–	–	–	–	–	**2-Butenol, 2-methyl-4-[6-(3-furanyl)-3-methyl-2-piperidinyl]-, 3-epi-**		**LI05015**
164 249 234 147 132 106 165 150	**249**	100 33 31 28 26 18 11 11		15	23	2	1	–	–	–	–	–	–	–	–	–	**Carbamic acid, (α-methylbenzyl)-, hexyl ester**	32589-35-8	**NI24384**
164 106 43 105 120 132 77 42	**249**	100 82 67 62 40 38 31 31	**20.02**	15	23	2	1	–	–	–	–	–	–	–	–	–	**Carbamic acid, (α-methylbenzyl)-, hexyl ester**	32589-35-8	**NI24385**
105 43 165 106 150 57 164 41	**249**	100 49 43 41 35 34 31 23	**2.97**	15	23	2	1	–	–	–	–	–	–	–	–	–	**Carbamic acid, (α-methylbenzyl)-, 1,1,2-trimethylpropyl ester**	32589-46-1	**NI24386**
105 43 106 165 150 164 57 85	**249**	100 48 38 36 28 24 23 22	**2.00**	15	23	2	1	–	–	–	–	–	–	–	–	–	**Carbamic acid, (α-methylbenzyl)-, 1,2,2-trimethylpropyl ester**	32589-42-7	**NI24387**

MASS TO CHARGE RATIOS	M.W.	INTENSITIES	Parent	C	H	O	N	S	F	Cl	Br	I	Si	P	B	X	COMPOUND NAME	CAS Reg No	No
133 249 165 134 179 160 57 147	**249**	100 50 50 37 26 26 26 24		15	23	2	1	–	–	–	–	–	–	–	–	–	**1-Octanone, 1-(2-hydroxy-5-methylphenyl)-, oxime, (E)-**		**DD00679**
72 56 134 133 115 249 117 73	**249**	100 17 11 10 8 7 7 6		15	23	2	1	–	–	–	–	–	–	–	–	–	**2-Propanol, 1-(indan-4-yl)-3-(isopropylamino)**		**MS05391**
72 30 73 115 102 56 43 41	**249**	100 21 5 4 4 4 4 4	**3.42**	15	23	2	1	–	–	–	–	–	–	–	–	–	**2-Propanol, 1-(isopropylamino)-3-[2-(2-propenyl)phenoxy]-**	13655-52-2	**NI24388**
72 41 56 43 44 39 91 73	**249**	100 13 12 12 11 7 6 6	**1.70**	15	23	2	1	–	–	–	–	–	–	–	–	–	**2-Propanol, 1-(isopropylamino)-3-[2-(2-propenyl)phenoxy]-**	13655-52-2	**MS05392**
114 94 249 136	**249**	100 78 70 61		15	23	2	1	–	–	–	–	–	–	–	–	–	**2H-Quinolizine-3-methanol, 6-(3-furanyl)octahydro-9-methyl-, [3S-(3α,6α,9β,9aβ)]-**	6874-86-8	**LI05016**
94 114 136 41 249 55 82 81	**249**	100 96 50 42 32 29 22 22		15	23	2	1	–	–	–	–	–	–	–	–	–	**2H-Quinolizine-3-methanol, 6-(3-furanyl)octahydro-9-methyl-, [3S-(3α,6α,9β,9aβ)]-**	6874-86-8	**NI24389**
114 249 94 136 178 96 71 231	**249**	100 96 96 85 42 34 23 20		15	23	2	1	–	–	–	–	–	–	–	–	–	**2H-Quinolizin-1-ol, 6-(3-furanyl)octahydro-3,9-dimethyl-**		**LI05017**
114 94 136 249 178 107 96 43	**249**	100 87 82 43 39 24 24 22		15	23	2	1	–	–	–	–	–	–	–	–	–	**2H-Quinolizin-3-ol, 6-(3-furanyl)octahydro-3,9-dimethyl-, [3R-(3α,6β,9α,9aα)]-**	38681-18-4	**NI24390**
55 83 68 28 54 27 82 39	**249**	100 47 40 31 23 20 13 13	**0.00**	15	27	–	3	–	–	–	–	–	–	–	–	–	**1H,6H,11H-Tripyrido[1,2-a:1′,2′-c:1″,2″-e][1,3,5]triazine, dodecahydro-**	522-33-8	**LI05018**
249 217 250 248 216 247 218 204	**249**	100 39 21 19 13 11 9 9		16	11	–	1	1	–	–	–	–	–	–	–	–	**12H-Benzo[a]phenothiazine**	225-83-2	**NI24391**
249 104 163 208 248 164 91 165	**249**	100 17 14 13 9 9 5 4		16	11	–	1	1	–	–	–	–	–	–	–	–	**Phenanthro[9,10-d]thiazole, 2-methyl-**	21639-90-7	**LI05019**
249 250 104 163 208 248 164 91	**249**	100 17 17 14 13 9 9 5		16	11	–	1	1	–	–	–	–	–	–	–	–	**Phenanthro[9,10-d]thiazole, 2-methyl-**	21639-90-7	**NI24392**
202 101 203 249 100 200 88 201	**249**	100 32 29 21 14 14 12 10		16	11	2	1	–	–	–	–	–	–	–	–	–	**Anthracene, 9-(2-nitroethenyl)-**	58349-77-2	**NI24393**
249 206 177 151 248 207 178 190	**249**	100 23 12 6 5 4 4 3		16	11	2	1	–	–	–	–	–	–	–	–	–	**10-Anthracen-ol, 9-cyano-2-methoxy-**		**LI05020**
102 205 51 77 191 189 105 63	**249**	100 21 21 20 17 16 13 12	**12.01**	16	11	2	1	–	–	–	–	–	–	–	–	–	**5(4H)-Isoxazolone, 3-phenyl-4-benzylidene-, (Z)-**	36298-61-0	**NI24394**
102 249 105 103 248 104 118 106	**249**	100 76 48 42 38 26 13 7		16	11	2	1	–	–	–	–	–	–	–	–	–	**5(4H)-Isoxazolone, 3-phenyl-4-benzylidene-, (Z)-**	36298-61-0	**LI05021**
105 77 249 51 106 50 250 89	**249**	100 46 19 13 8 4 4 3		16	11	2	1	–	–	–	–	–	–	–	–	–	**2-Oxazolin-5-one, 4-benzylidene-2-phenyl-**	842-74-0	**NI24395**
105 77 249 51 50 116	**249**	100 41 23 11 4 2		16	11	2	1	–	–	–	–	–	–	–	–	–	**5(4H)-Oxazolone, 2-phenyl-4-benzylidene-**	842-74-0	**LI05022**
105 77 249 51 106 250 50 78	**249**	100 41 23 11 8 4 4 3		16	11	2	1	–	–	–	–	–	–	–	–	–	**5(4H)-Oxazolone, 2-phenyl-4-benzylidene-**	842-74-0	**NI24396**
249 221 218 220 204 205	**249**	100 30 28 20 10 7		16	11	2	1	–	–	–	–	–	–	–	–	–	**1,8-Oxyiminomethanooxymethanophenanthrene**		**LI05023**
249 151 250 152 150 196 180 220	**249**	100 21 18 18 16 14 10 6		16	11	2	1	–	–	–	–	–	–	–	–	–	**Phenanthro[3,4-d][1,3]oxazol-10-ol, 2-methyl-**	98033-24-0	**NI24397**
250 232 206	**249**	100 5 2	**0.00**	16	11	2	1	–	–	–	–	–	–	–	–	–	**4-Quinolinecarboxylic acid, 2-phenyl-**	132-60-5	**NI24398**
234 249 233 248 77 120 106 103	**249**	100 40 21 20 9 7 7 7		16	15	–	3	–	–	–	–	–	–	–	–	–	**Quinazoline, 4-[(2,6-dimethylphenyl)amino]-**		**MS05393**
84 247 98 124 206 248 218 136	**249**	100 71 70 66 62 18 18 16	**1.00**	16	27	1	1	–	–	–	–	–	–	–	–	–	**Bicyclo[2.2.1]heptan-3-endo-ol, 2-exo-(3′-exo-dimethylaminobicyclo[2.2.1]-hept-2′-endo-yl)-**		**NI24399**
69 41 166 98 192 249 235 110	**249**	100 52 50 48 46 38 38 36		16	27	1	1	–	–	–	–	–	–	–	–	–	**2-Butenamide, N,N-dicyclohexyl-**	55955-94-7	**NI24400**
234 165 195 249 167 152 57 56	**249**	100 63 46 42 41 31 30 30		17	15	1	1	–	–	–	–	–	–	–	–	–	**Benz[c,e]azepin-5-one, 6-allyl-6,7-dihydro-**		**MS05394**
120 105 134 136 132 121 146 116	**249**	100 96 18 11 11 11 8 7	**3.57**	17	15	1	1	–	–	–	–	–	–	–	–	–	**Benzenepentanenitrile, δ-oxo-α-phenyl-**	22228-41-7	**NI24401**
143 43 115 249 77 116 128 178	**249**	100 26 23 16 11 10 9 8		17	15	1	1	–	–	–	–	–	–	–	–	–	**Isoquinoline, 1-(2-hydroxyphenethyl)-**	78744-81-7	**NI24402**
249 248 143 107 129 156 115 233	**249**	100 73 71 51 31 23 22 11		17	15	1	1	–	–	–	–	–	–	–	–	–	**Isoquinoline, 1-(4-hydroxyphenethyl)-**	78744-89-5	**NI24403**
249 248 233 234 219 250 232 191	**249**	100 53 25 22 19 17 16 13		17	15	1	1	–	–	–	–	–	–	–	–	–	**3H-Napth[1,2-e]indol-10-ol, 1,2-dihydro-3-methyl-**	98033-22-8	**NI24404**
165 249 77 206 166 117 103 51	**249**	100 85 58 43 40 33 32 20		17	15	1	1	–	–	–	–	–	–	–	–	–	**Oxazole, 2-ethyl-4,5-diphenyl-**	20662-94-6	**NI24405**
249 191 203 193 204 178 205 101	**249**	100 80 63 55 47 37 30 25		17	15	1	1	–	–	–	–	–	–	–	–	–	**Phenanthro[10,1-bc]azepin-6(7H)-one, 4,5,7a,8-tetrahydro-**		**MS05395**
172 167 165 249 82 152 91 41	**249**	100 88 39 22 22 20 19 11		18	19	–	1	–	–	–	–	–	–	–	–	–	**Azetidine, 1-(diphenylmethyl)-3-ethylidene-**		**DD00680**
206 193 249 207 180 192 204 248	**249**	100 84 81 36 36 27 22 18		18	19	–	1	–	–	–	–	–	–	–	–	–	**Spiro[acridine-9(10H),1′-cyclohexane]**	14458-75-4	**NI24406**
173 171 175 93 91 79 81 92	**250**	100 50 49 22 22 18 17 13	**3.90**	1	1	–	–	–	–	–	3	–	–	–	–	–	**Methane, tribromo-**	75-25-2	**NI24407**
101 85 117 70 116 63 66 48	**250**	100 97 75 42 37 35 33 31	**0.00**	2	–	2	–	2	5	1	–	–	–	–	–	–	**1,3-Dithietane-3,3-dione, 2-chloro-1,1,2,4,4-pentafluoro-**		**MS05396**
47 66 28 44 31 50 48 32	**250**	100 68 24 4 4 2 1 1	**0.00**	2	–	5	–	1	6	–	–	–	–	–	–	–	**Bis(trifluoromethyl) peroxymonosulphate**		**MS05397**
83 85 132 134 87 167 169 165	**250**	100 77 27 17 16 9 6 5	**0.00**	3	1	–	–	–	2	5	–	–	–	–	–	–	**Propane, 3,3-difluoro-1,1,2,2,3-pentachloro-**		**DO01039**
44 137 57 46 45 43 31 26	**250**	100 1 1 1 1 1 1 1	**0.00**	3	2	–	–	–	1	1	2	–	–	–	–	–	**Cyclopropane, 1,1-dibromo-2-chloro-2-fluoro-**	24071-57-6	**NI24408**
69 96 51 166 184 146 46 181	**250**	100 95 43 13 10 10 10 9	**0.10**	3	2	–	2	1	8	–	–	–	–	–	–	–	**Sulphurous imide difluoride, N-[2-(2-aminohexafluoropropyl)]-**		**MS05398**
117 63 214 216 119 116 118 252	**250**	100 62 58 53 35 32 28 24	**18.00**	4	6	–	–	–	–	4	–	–	2	–	–	–	**4-Hexene, 1,1,3,3-tetrachloro-1,3-disila-**	84154-20-1	**NI24409**
131 100 31 69 81 93 50 231	**250**	100 24 23 16 14 10 5 4	**0.36**	5	–	–	–	–	10	–	–	–	–	–	–	–	**Cyclopentane, decafluoro-**		**AI01788**
69 111 139 92 250 70 47 42	**250**	100 43 40 31 9 9 9 4		5	–	3	2	–	6	–	–	–	–	–	–	–	**Imidazolidinetrione, 1,3-bis(trifluoromethyl)-**		**NI24410**
101 51 69 133 131 31 83 33	**250**	100 85 63 56 50 22 17 13	**0.00**	5	3	1	–	–	9	–	–	–	–	–	–	–	**Propane, 2,2,3,3,3-pentafluoro-(1,1,2,2-tetrafluoroethoxy)-**		**IC03268**
252 92 254 250 65 64 171 173	**250**	100 86 52 51 35 33 27 26		5	4	–	2	–	–	–	2	–	–	–	–	–	**Pyridine, 2-amino-3,5-dibromo-**	35486-42-1	**NI24411**
252 250 254 215 217 180 126 145	**250**	100 77 54 19 18 12 12 9		6	–	–	–	–	2	4	–	–	–	–	–	–	**Benzene, 1,2,3,5-tetrachloro-4,6-difluoro-**	1198-56-7	**LI05024**
252 250 254 215 217 256 180 126	**250**	100 77 51 18 17 12 11 11		6	–	–	–	–	2	4	–	–	–	–	–	–	**Benzene, 1,2,3,5-tetrachloro-4,6-difluoro-**	1198-56-7	**NI24412**
252 63 250 254 143 62 145 172	**250**	100 40 26 25 13 13 12 10		6	4	1	–	–	–	–	2	–	–	–	–	–	**Phenol, 2,4-dibromo-**	615-58-7	**NI24413**
222 220 252 124 250 90 206 125	**250**	100 80 62 56 49 38 34 34		6	4	2	2	–	–	1	1	–	–	–	–	–	**Aniline, 2-bromo-6-chloro-4-nitro-**	99-29-6	**NI24414**
252 250 222 124 220 206 90 125	**250**	100 76 56 52 45 41 35 33		6	4	2	2	–	–	1	1	–	–	–	–	–	**Aniline, 2-bromo-6-chloro-4-nitro-**	99-29-6	**IC03269**

MASS TO CHARGE RATIOS								M.W.	INTENSITIES								Parent	C	H	O	N	S	F	Cl	Br	I	Si	P	B	X	COMPOUND NAME	CAS Reg No	No
252	250	171	125	237	235	219	217	250	100	96	29	29	27	26	26	26		6	7	–	2	2	–	–	1	–	–	–	–	–	Pyrimidine, 5-bromo-2,4-bis(methylthio)-		NI24415
31	75	76	157	77	155	158	156	250	100	50	38	33	33	32	23	23	10.00	6	7	2	2	1	–	–	1	–	–	–	–	–	Benzenesulphonic acid, 4-bromo-, hydrazide	2297-64-5	NI24416
36	89	54	179	90	56	181	214	250	100	92	86	68	68	54	46	41	0.00	6	10	–	2	–	–	4	–	–	–	–	–	–	Propanimidamide, 2-chloro-N-(1,2-dichloro-1-propenyl)-, monohydrochloride	40645-80-5	NI24417
42	56	44	181	71	29	28	69	250	100	99	99	98	98	98	89	51	0.00	6	10	2	2	1	4	–	–	–	–	–	–	–	Diethylaminosulphur oxide monofluoride trifluoroacetylimide		LI05025
102	215	137	169	75	78	46	76	250	100	98	98	79	79	73	44	35	0.00	7	4	–	2	2	–	2	–	–	–	–	–	–	4-(4-Chlorophenyl)-1,2,3,5-dithiadiazolium chloride	87512-51-4	NI24418
250	181	136	164	94	52	69	67	250	100	80	63	40	40	20	18	18		7	5	1	4	1	3	–	–	–	–	–	–	–	2(1H)-Pteridinethione, 3,4-dihydro-4-hydroxy-4-(trifluoromethyl)-	32706-28-8	LI05026
250	181	136	164	94	40	52	69	250	100	80	64	40	40	20	19	18		7	5	1	4	1	3	–	–	–	–	–	–	–	2(1H)-Pteridinethione, 3,4-dihydro-4-hydroxy-4-(trifluoromethyl)-	32706-28-8	NI24419
250	249	89	91	50	63	117	169	250	100	90	33	21	21	20	19	18		7	6	–	–	–	–	–	–	–	–	–	–	2	1,3-Benzodiselenole		DD00681
50	157	155	75	76	252	250	221	250	100	92	92	85	80	59	58	53		7	7	3	–	1	–	–	1	–	–	–	–	–	Benzenesulphonic acid, 4-bromo-, methyl ester	6213-85-0	WI01275
89	87	119	116	117	85	191	189	250	100	66	65	55	53	48	41	38	18.00	7	18	–	–	–	–	–	–	–	–	–	–	2	1,2-Digermacyclopentane, 1,1,2,2-tetramethyl-	35839-71-5	LI05027
89	87	119	116	117	85	191	193	250	100	66	64	54	52	47	41	30	13.46	7	18	–	–	–	–	–	–	–	–	–	–	2	1,2-Digermacyclopentane, 1,1,2,2-tetramethyl-	35839-71-5	NI24420
87	51	143	55	115	39	89	69	250	100	93	86	81	64	54	36	34	14.00	8	4	5	–	–	–	2	–	–	–	–	–	–	Bicyclo[2.2.1]heptane-2,3-dione, 5,6-carbonyldioxy-5,6-dichloro-		NI24421
153	135	199	185	167	251	215	139	250	100	64	25	20	19	16	14	14	0.76	8	7	5	–	1	–	1	–	–	–	–	–	–	Benzoic acid, 5-(chlorosulphonyl)-2-methoxy-	51904-91-7	NI24422
215	151	63	136	167	95	64	76	250	100	77	51	47	32	31	31	27	24.05	8	7	5	–	1	–	1	–	–	–	–	–	–	Benzoic acid, 5-(chlorosulphonyl)-2-methoxy-	51904-91-7	NI24423
170	169	155	59	82	80	171	79	250	100	99	41	30	27	27	20	15	0.00	8	11	–	–	2	–	–	1	–	–	–	–	–	1,2-Benzodithiol-1-ium, 4,5,6,7-tetrahydro-3-methyl-, bromide	55836-83-4	NI24424
170	169	80	82	171	137	81	79	250	100	97	47	46	21	20	17	16	0.00	8	11	–	–	2	–	–	1	–	–	–	–	–	4H-Cyclopenta-1,2-dithiol-1-ium, 3-ethyl-5,6-dihydro-, bromide	55836-82-3	NI24425
122	128	43	127	79	77	107	121	250	100	56	55	31	28	20	18	12	0.00	8	11	1	–	–	–	–	–	1	–	–	–	–	Pyrylium, 2,4,6-trimethyl-, iodide	32339-27-8	NI24426
130	131	57	29	41	87	56	58	250	100	65	54	49	47	43	32	22	5.50	8	11	4	2	1	1	–	–	–	–	–	–	–	Uracil, 1-butanesulphonyl-5-fluoro-	69207-34-7	NI24427
165	207	235	135	222	177	150	43	250	100	70	65	45	33	14	12	10	0.10	8	18	1	–	–	–	–	–	–	–	–	–	1	Stannane, (3-acetylpropyl)trimethyl-		LI05028
116	89	195	197	252	250	63	50	250	100	23	21	20	15	13	10	9		9	7	–	4	–	–	–	1	–	–	–	–	–	s-Tetrazine, 1-(4-bromobenzyl)-		MS05399
91	139	47	111	77	41	42	55	250	100	93	68	30	21	12	10	9	4.00	9	12	1	–	–	6	–	–	–	–	–	–	–	5-Nonanone, 1,1,1,9,9,9-hexafluoro-	24322-11-0	NI24428
250	59	217	117	185	131	252	125	250	100	69	38	35	27	21	18	14		9	14	–	–	4	–	–	–	–	–	–	–	–	2,4,6,8-Tetrathiatricyclo[3.3.1.1^{3,7}]decane, 1,3,5-trimethyl-	13787-69-4	NI24429
250	99	117	45	163	131	252	87	250	100	45	34	25	23	23	18	16		9	14	–	–	4	–	–	–	–	–	–	–	–	2,4,6,8-Tetrathiatricyclo[3.3.1.1^{3,7}]decane, 9,9,10-trimethyl-	17749-64-3	NI24430
123	81	39	27	67	127	79	41	250	100	73	48	35	33	32	31	30	4.60	9	15	–	–	–	–	–	–	1	–	–	–	–	Bicyclo[2.2.2]octane, 1-iodo-4-methyl-	55044-63-8	NI24431
41	81	67	55	123	39	27	69	250	100	95	86	73	61	56	46	44	5.54	9	15	–	–	–	–	–	–	1	–	–	–	–	Cyclohexane, (2-iodocyclopropyl)-	55682-67-2	MS05400
81	41	67	55	39	54	69	53	250	100	72	54	43	25	24	22	16	0.00	9	15	–	–	–	–	–	–	1	–	–	–	–	Cyclopropane, 1-(5-hexenyl)-2-iodo-	74685-40-8	NI24432
250	93	41	43	125	39	95	81	250	100	71	71	63	60	57	54	54		9	15	–	–	–	–	–	–	1	–	–	–	–	3-Heptyne, 7-iodo-2,2-dimethyl-	55402-07-8	NI24433
55	41	113	83	57	15	18	58	250	100	89	86	64	64	54	53	52	0.00	9	18	4	2	1	–	–	–	–	–	–	–	–	2-Butanone, 3,3-dimethyl-1-(methylsulphonyl)-, O-[(methylamino)carbonyl]oxime	39184-59-3	NI24434
151	149	207	29	135	147	205	43	250	100	75	58	57	52	48	44	42	7.00	9	22	–	–	–	–	–	–	–	–	–	–	1	Tin, dimethylbutylisopropyl-		MS05401
179	149	207	177	151	205	147	175	250	100	93	86	78	77	65	65	46	5.30	9	22	–	–	–	–	–	–	–	–	–	–	1	Tin, isopropyltriethyl-		MS05402
179	207	151	149	121	221	193	120	250	100	89	72	44	43	40	34	21	5.40	9	22	–	–	–	–	–	–	–	–	–	–	1	Tin, isopropyltriethyl-		LI05029
179	177	221	151	219	175	147	121	250	100	75	69	56	52	45	42	33	2.20	9	22	–	–	–	–	–	–	–	–	–	–	1	Tin, propyltriethyl-		MS05403
179	221	151	121	149	207	120	193	250	100	68	55	32	25	17	12	10	2.10	9	22	–	–	–	–	–	–	–	–	–	–	1	Tin, propyltriethyl-		LI05030
149	192	163	250	235	147	102	207	250	100	87	77	54	37	37	35	30		9	27	–	4	–	–	–	–	–	1	1	–	–	Phosphorimidic triamide, N,N,N′,N′,N″,N″-hexamethyl-N‴-(trimethylsilyl)-	53167-50-3	NI24435
181	139	250	121	120	152	85	69	250	100	42	41	22	22	21	20	15		10	6	1	–	1	4	–	–	–	–	–	–	–	3-Buten-2-one, 1,1,1-trifluoro-4-(4-fluorophenyl)-4-mercapto-	55674-01-6	NI24436
235	208	250	236	217	237	193	45	250	100	22	15	14	12	10	10	10		10	10	–	4	2	–	–	–	–	–	–	–	–	4,5′-Bipyrimidine, 4′,6-bis(methylthio)-		NI24437
92	186	185	28	65	108	187	93	250	100	71	51	48	21	15	10	8	0.00	10	10	2	4	1	–	–	–	–	–	–	–	–	Benzenesulphonamide, 4-amino-N-2-pyrimidinyl-	68-35-9	NI24438
92	186	185	28	65	108	187	156	250	100	72	52	49	22	15	12	5	2.00	10	10	2	4	1	–	–	–	–	–	–	–	–	Benzenesulphonamide, 4-amino-N-2-pyrimidinyl-	68-35-9	LI05031
91	65	39	92	77	79	78	63	250	100	20	17	16	15	8	8	7	0.03	10	10	4	–	–	–	–	–	–	–	–	–	1	Iron, (π-2,4-cycloheptadien-1-ol)tricarbonyl-		MS05404
194	186	122	121	152	222	56	42	250	100	99	92	75	74	35	31	22	1.51	10	10	4	–	–	–	–	–	–	–	–	–	1	Iron, (π-cyclopentadienyl)(methoxycarbonylmethyl)dicarbonyl-		MS05405
39	30	202	41	203	250	63	169	250	100	90	85	85	78	64	58	55		10	10	4	4	–	–	–	–	–	–	–	–	–	2-Butenal, (2,4-dinitrophenyl)hydrazone	1527-96-4	MS05406
39	202	41	203	250	63	169	28	250	100	84	84	78	63	57	55	53		10	10	4	4	–	–	–	–	–	–	–	–	–	2-Butenal, (2,4-dinitrophenyl)hydrazone	1527-96-4	NI24439
250	187	218	219	186	251	160	188	250	100	59	50	25	22	12	7	6		10	10	4	4	–	–	–	–	–	–	–	–	–	5,5′-Dipyrazole-4,4′-dicarboxylic acid, dimethyl ester		MS05407
208	43	178	79	15	134	124	56	250	100	83	24	14	14	13	11	11	10.00	10	10	4	4	–	–	–	–	–	–	–	–	–	1H-1,2,4-Triazole, 1-acetyl-5-ethyl-3-(5-nitro-2-furanyl)-	35732-76-4	NI24440
57	29	194	195	164	28	56	27	250	100	82	81	57	25	24	23	19	15.00	10	10	4	4	–	–	–	–	–	–	–	–	–	1H-1,2,4-Triazole, 5-methyl-3-(5-nitro-2-furanyl)-1-propionyl-	35732-75-3	NI24441
147	148	43	49	58	84	45	149	250	100	90	90	38	34	21	17	15	0.00	10	11	–	–	1	4	–	–	–	–	–	1	–	Benzo[b]thiophenium, 1,2-dimethyl-, tetrafluoroborate		LI05032
134	47	135	89	67	90	136	69	250	100	16	10	9	9	8	5	5	0.00	10	11	–	–	1	4	–	–	–	–	–	1	–	Benzo[b]thiophenium, 1-ethyl-, tetrafluoroborate		LI05033
235	250	96	41	210	71	155	69	250	100	75	55	45	42	27	25	25		10	14	–	6	1	–	–	–	–	–	–	–	–	1,3,5-Triazine-2,4-diamine, N-allyl-N-cyano-N′,N′-dimethyl-6-methylthio-	93286-78-3	NI24442
194	73	56	178	121	71	250	222	250	100	95	76	66	61	53	47	37		10	14	2	–	–	–	–	–	–	1	–	–	1	Iron, dicarbonyl(η^5-2,4-cyclopentadien-1-yl)(trimethylsilyl)-	31811-63-9	LI05034
194	73	56	178	121	93	250	128	250	100	95	76	66	61	53	47	45		10	14	2	–	–	–	–	–	–	1	–	–	1	Iron, dicarbonyl(η^5-2,4-cyclopentadien-1-yl)(trimethylsilyl)-	31811-63-9	MS05408
55	201	91	41	39	203	49	93	250	100	88	71	60	48	26	24	21	1.00	10	16	1	2	–	–	2	–	–	–	–	–	–	1,3,4-Oxadiazole, 2,5-bis(2-chloro-1,1-dimethylethyl)-	93289-72-6	NI24443
75	88	45	73	101	132	71	85	250	100	64	26	18	15	14	12	11	0.00	10	18	7	–	–	–	–	–	–	–	–	–	–	α-D-Glucopyranosiduronic acid, methyl 2,3-di-O-methyl-, methyl ester	51433-21-7	NI24444
75	74	85	88	45	87	101	117	250	100	39	37	25	21	19	16	13	0.00	10	18	7	–	–	–	–	–	–	–	–	–	–	α-D-Glucopyranosiduronic acid, methyl 3,4-di-O-methyl-, methyl ester	51433-28-4	NI24445

MASS TO CHARGE RATIOS	M.W.	INTENSITIES	Parent	C	H	O	N	S	F	Cl	Br	I	Si	P	B	X	COMPOUND NAME	CAS Reg No	No
75 88 73 45 132 101 71 85	**250**	100 62 22 22 15 14 9 9	**0.00**	10	18	7	–	–	–	–	–	–	–	–	–	–	**Methyl(methyl 2,3-di-O-methyl-α-D-mannopyranoside)uronate**	89177-20-8	**NI24446**
101 85 74 75 117 88 45 89	**250**	100 39 39 19 17 17 17 10	**0.00**	10	18	7	–	–	–	–	–	–	–	–	–	–	**Methyl(methyl 2,4-di-O-methyl-α-D-mannopyranoside)uronate**	89177-24-2	**NI24447**
75 85 101 74 88 45 87 117	**250**	100 28 23 23 21 12 11 10	**0.00**	10	18	7	–	–	–	–	–	–	–	–	–	–	**Methyl(methyl 3,4-di-O-methyl-α-D-mannopyranoside)uronate**	89177-28-6	**NI24448**
165 109 137 55 81 91 178 155	**250**	100 92 44 42 39 31 28 25	**0.00**	10	19	5	–	–	–	–	–	–	–	1	–	–	**2H-Azepin-2-one, 3-(diethoxyphosphinyl)hexahydro-**		**DD00682**
191 81 111 163 140 65 59 99	**250**	100 79 24 22 15 8 8 7	**0.00**	10	19	5	–	–	–	–	–	–	–	1	–	–	**Phosphonic acid, ethyl 1-(methoxycarbonyl)cyclohexyl diester**		**DD00683**
73 117 116 66 103 147 75 74	**250**	100 50 23 14 12 12 9 9	**0.00**	10	26	3	–	–	–	–	–	–	2	–	–	–	**Ethanol, 2,2′-oxybis-, bis(trimethylsilyl ether)**	16654-74-3	**NI24449**
73 117 116 147 103 75 74 44	**250**	100 43 17 11 11 11 8 8	**0.10**	10	26	3	–	–	–	–	–	–	2	–	–	–	**Ethanol, 2,2′-oxybis-, bis(trimethylsilyl ether)**	16654-74-3	**NI24450**
73 117 116 147 103 75 66 74	**250**	100 42 18 10 10 9 9 8	**0.00**	10	26	3	–	–	–	–	–	–	2	–	–	–	**Ethanol, 2,2′-oxybis-, bis(trimethylsilyl ether)**	16654-74-3	**NI24451**
73 147 205 45 89 117 148 59	**250**	100 78 33 25 19 15 13 9	**0.00**	10	26	3	–	–	–	–	–	–	2	–	–	–	**1,3-Propanediol, 1-methoxy-, bis(trimethylsilyl ether)**	62185-57-3	**NI24452**
250 249 231 223 200 229 205 100	**250**	100 73 33 20 19 14 8 7		11	4	–	–	–	6	–	–	–	–	–	–	–	**Naphthalene, 1,2-dihydrohexafluoro-3-methyl-**		**NI24453**
250 77 135 43 95 204 119 136	**250**	100 80 76 70 35 28 27 23		11	10	3	2	1	–	–	–	–	–	–	–	–	**5-Imidazolidinecarboxylic acid, 4-oxo-3-phenyl-2-thioxo-, methyl ester**		**LI05035**
166 194 222 56 250 167 195 223	**250**	100 70 57 49 14 9 6 5		11	14	3	–	–	–	–	–	–	–	–	–	1	**Iron, tricarbonyl[(2,3,4,5-η)-2,5-dimethyl-2,4-hexadiene]-**	33010-48-9	**NI24454**
180 194 196 182 67 91 179 53	**250**	100 97 92 86 86 83 82 79	**0.00**	11	20	2	–	–	–	1	–	–	–	1	–	–	**1,2-Oxaphosphole, 2-chloro-3,5-di-tert-butyl-2,5-dihydro-, 2-oxide**	42087-74-1	**NI24455**
101 205 119 75 141 145 70 45	**250**	100 62 39 30 17 13 13 13	**0.00**	11	22	6	–	–	–	–	–	–	–	–	–	–	**β-D-Fructofuranoside, methyl 1,3,4,6-tetra-O-methyl-**	23259-20-3	**LI05036**
101 205 119 75 141 145 71 45	**250**	100 62 40 30 17 13 13 13	**0.00**	11	22	6	–	–	–	–	–	–	–	–	–	–	**β-D-Fructofuranoside, methyl 1,3,4,6-tetra-O-methyl-**	23259-20-3	**NI24456**
205 119 88 101 75 141 115 45	**250**	100 99 99 70 42 35 28 27	**0.00**	11	22	6	–	–	–	–	–	–	–	–	–	–	**β-D-Fructopyranoside, methyl 1,3,4,5-tetra-O-methyl-**	4451-14-3	**NI24457**
101 75 44 45 88 71 102 73	**250**	100 17 14 13 10 10 8 5	**0.00**	11	22	6	–	–	–	–	–	–	–	–	–	–	**α-D-Galactofuranoside, methyl 2,3,4,5-tetra-O-methyl-**		**MS05409**
88 75 101 45 149 73 71 89	**250**	100 64 55 33 15 14 14 8	**0.00**	11	22	6	–	–	–	–	–	–	–	–	–	–	**α-D-Galactofuranoside, methyl 2,3,4,6-tetra-O-methyl-**		**MS05410**
101 45 75 88 59 89 71 29	**250**	100 49 35 17 14 13 13 12	**0.00**	11	22	6	–	–	–	–	–	–	–	–	–	–	**α-D-Galactofuranoside, methyl 2,3,5,6-tetra-O-methyl-**	10225-58-8	**MS05411**
101 75 45 59 41 161 88 71	**250**	100 38 37 11 11 10 10 10	**0.00**	11	22	6	–	–	–	–	–	–	–	–	–	–	**α-D-Galactofuranoside, methyl 2,3,5,6-tetra-O-methyl-**	10225-58-8	**MS05412**
101 45 75 89 88 71 29 41	**250**	100 57 48 18 17 17 14 13	**0.00**	11	22	6	–	–	–	–	–	–	–	–	–	–	**β-D-Galactofuranoside, methyl 2,3,5,6-tetra-O-methyl-**	17152-57-7	**MS05413**
88 75 101 45 73 149 71 187	**250**	100 75 50 31 18 12 9 6	**0.00**	11	22	6	–	–	–	–	–	–	–	–	–	–	**α-D-Galactopyranoside, methyl 2,3,4,6-tetra-O-methyl-**	3149-64-2	**NI24459**
88 75 101 45 73 149 74 71	**250**	100 64 55 32 18 16 12 10	**0.00**	11	22	6	–	–	–	–	–	–	–	–	–	–	**α-D-Galactopyranoside, methyl 2,3,4,6-tetra-O-methyl-**	3149-64-2	**NI24458**
88 75 101 45 73 71 127 131	**250**	100 82 73 27 26 17 5 4	**0.00**	11	22	6	–	–	–	–	–	–	–	–	–	–	**β-D-Galactopyranoside, methyl 2,3,4,6-tetra-O-methyl-**	2296-47-1	**NI24460**
101 75 45 88 71 102 73 59	**250**	100 18 14 10 10 8 6 5	**0.00**	11	22	6	–	–	–	–	–	–	–	–	–	–	**α-D-Galactoseptanoside, methyl 2,3,4,5-tetra-O-methyl-**	17152-59-9	**NI24461**
88 45 101 75 71 41 73 44	**250**	100 59 58 55 23 23 17 17	**0.00**	11	22	6	–	–	–	–	–	–	–	–	–	–	**α-D-Glucopyranoside, methyl 2,3,4,6-tetra-O-methyl-**	605-81-2	**NI24462**
88 75 101 73 45 71 149 111	**250**	100 57 47 11 11 9 6 3	**0.00**	11	22	6	–	–	–	–	–	–	–	–	–	–	**α-D-Glucopyranoside, methyl 2,3,4,6-tetra-O-methyl-**	605-81-2	**NI24463**
88 101 75 45 73 71 149 89	**250**	100 53 53 17 15 12 10 9	**0.00**	11	22	6	–	–	–	–	–	–	–	–	–	–	**α-D-Glucopyranoside, methyl 2,3,4,6-tetra-O-methyl-**	605-81-2	**NI24464**
88 101 75 45 71 73 127 149	**250**	100 46 30 21 12 11 9 7	**0.00**	11	22	6	–	–	–	–	–	–	–	–	–	–	**β-D-Glucopyranoside, methyl 2,3,4,6-tetra-O-methyl-**	3149-65-3	**NI24466**
88 149 75 101 45 71 89 73	**250**	100 58 55 53 22 15 11 11	**0.00**	11	22	6	–	–	–	–	–	–	–	–	–	–	**β-D-Glucopyranoside, methyl 2,3,4,6-tetra-O-methyl-**	3149-65-3	**NI24465**
88 75 101 45 73 71 149 145	**250**	100 62 51 22 18 13 7 3	**0.00**	11	22	6	–	–	–	–	–	–	–	–	–	–	**α-D-Mannopyranoside, methyl 2,3,4,6-tetra-O-methyl-**	3149-62-0	**NI24467**
88 75 101 45 73 149 71 145	**250**	100 73 47 18 13 5 4 2	**0.00**	11	22	6	–	–	–	–	–	–	–	–	–	–	**β-D-Mannopyranoside, methyl 2,3,4,6-tetra-O-methyl-**	3445-71-4	**NI24468**
250 251 75 219 235 190 101 45	**250**	100 56 22 21 15 10 8 6		11	22	6	–	–	–	–	–	–	–	–	–	–	**β-D-Mannopyranoside, methyl 2,3,4,6-tetra-O-methyl-**	3445-71-4	**LI05037**
96 73 138 82 141 109 68 221	**250**	100 92 70 70 31 31 30 25	**0.00**	11	23	1	–	–	–	–	1	–	–	–	–	–	**6-Bromo-1-ethyloctyl methyl ether**	20599-95-5	**NI24469**
73 41 109 55 67 45 82 221	**250**	100 23 21 18 16 16 15 9	**0.00**	11	23	1	–	–	–	–	1	–	–	–	–	–	**6-Bromo-1-ethyloctyl methyl ether**	20599-95-5	**NI24471**
221 223 73 72 222 224 139 83	**250**	100 98 45 30 10 8 3 3	**2.70**	11	23	1	–	–	–	–	1	–	–	–	–	–	**6-Bromo-1-ethyloctyl methyl ether**	20599-95-5	**NI24470**
55 41 69 83 43 68 56 97	**250**	100 86 69 45 41 26 24 23	**0.00**	11	23	1	–	–	–	–	1	–	–	–	–	–	**1-Undecanol, 11-bromo-**	1611-56-9	**NI24472**
179 123 208 151 138 81 69 43	**250**	100 34 31 28 12 8 6 6	**0.00**	11	23	4	–	–	–	–	–	–	–	1	–	–	**Phosphonic acid, (1-methyl-1-propyl-2-oxopropyl)-, diethyl ester**		**DD00684**
99 138 139 41 57 195 250 56	**250**	100 37 30 17 13 11 9 6		11	23	4	–	–	–	–	–	–	–	1	–	–	**Phosphoric acid, dibutyl isopropenyl ester**	5954-40-5	**NI24473**
99 138 139 41 57 195 250 55	**250**	100 38 30 17 15 12 10 6		11	23	4	–	–	–	–	–	–	–	1	–	–	**Phosphoric acid, dibutyl isopropenyl ester**	5954-40-5	**NI24474**
138 41 56 99 57 83 139 43	**250**	100 49 44 31 28 17 15 13	**9.79**	11	23	4	–	–	–	–	–	–	–	1	–	–	**Phosphoric acid, dibutyl 1-propenyl ester**	55822-87-2	**NI24475**
138 41 56 99 57 83 139 250	**250**	100 48 42 30 30 22 17 10		11	23	4	–	–	–	–	–	–	–	1	–	–	**Phosphoric acid, dibutyl 1-propenyl ester**	55822-87-2	**NI24476**
250 252 215 251 254 253 152 217	**250**	100 68 30 18 13 12 11 10		12	8	–	–	–	–	2	–	–	1	–	–	–	**9H-Fluorene, 9,9-dichloro-9-sila-**	18030-58-5	**NI24477**
250 252 215 152 217 251 253 254	**250**	100 68 57 39 20 19 13 12		12	8	–	–	–	–	2	–	–	1	–	–	–	**1H-Phenalene, 1,1-dichloro-1-sila-**	20470-89-7	**NI24478**
111 75 139 113 50 51 76 38	**250**	100 60 35 32 30 18 13 13	**11.70**	12	8	–	2	–	–	2	–	–	–	–	–	–	**Azobenzene, 4,4′-dichloro-**		**IC03270**
111 75 139 113 250 141 50 76	**250**	100 65 64 35 34 14 13 10		12	8	–	2	–	–	2	–	–	–	–	–	–	**Azobenzene, 4,4′-dichloro-**		**LI05038**
215 250 168 252 139 216 251 84	**250**	100 87 42 33 21 14 13 10		12	8	2	–	–	–	1	–	–	–	1	–	–	**Phosphorous acid, chloro, 2,2′-biphenylylene-**	16611-68-0	**NI24479**
250 141 184 69 217 108 65 45	**250**	100 37 34 34 29 27 27 22		12	10	–	–	3	–	–	–	–	–	–	–	–	**4,4′-Dimercaptodiphenyl sulphide**		**IC03271**
77 125 109 250 141 97 185 152	**250**	100 83 48 39 33 13 8 4		12	10	2	–	2	–	–	–	–	–	–	–	–	**Benzenesulphonothioic acid, S-phenyl ester**	1212-08-4	**LI05039**
77 125 109 250 141 97 78 185	**250**	100 81 48 39 33 11 8 7		12	10	2	–	2	–	–	–	–	–	–	–	–	**Benzenesulphonothioic acid, S-phenyl ester**	1212-08-4	**NI24480**
235 250 163 43 236 237 251 44	**250**	100 66 22 22 14 11 10 10		12	10	2	–	2	–	–	–	–	–	–	–	–	**2,2′-Bithienyl, 5,5′-diacetyl-**		**IC03272**
110 81 82 53 39 69 63 54	**250**	100 32 31 24 15 13 12 10	**0.00**	12	10	4	–	1	–	–	–	–	–	–	–	–	**1,3-Benzenediol, 4,4′-thiobis-**	97-29-0	**NI24481**
250 94 44 157 109 65 93 28	**250**	100 67 47 46 37 37 36 35		12	10	4	–	1	–	–	–	–	–	–	–	–	**Phenol, 2,4′-sulfonyldi-**	5397-34-2	**NI24482**
250 110 141 142 93 39 65 94	**250**	100 71 68 44 31 31 27 22		12	10	4	–	1	–	–	–	–	–	–	–	–	**Phenol, 4,4′-sulfonyldi-**	80-09-1	**IC03273**
250 141 110 39 142 65 93 94	**250**	100 97 97 59 54 54 46 32		12	10	4	–	1	–	–	–	–	–	–	–	–	**Phenol, 4,4′-sulfonyldi-**	80-09-1	**IC03274**
250 141 110 142 91 94 92 93	**250**	100 83 79 72 65 58 55 52		12	10	4	–	1	–	–	–	–	–	–	–	–	**Phenol, 4,4′-sulfonyldi-**	80-09-1	**NI24483**

MASS TO CHARGE RATIOS								M.W.	INTENSITIES								Parent	C	H	O	N	S	F	Cl	Br	I	Si	P	B	X	COMPOUND NAME	CAS Reg No	No
157	91	93	92	250	65	94	109	**250**	100	83	80	75	74	69	60	58		12	10	4	–	1	–	–	–	–	–	–	–	–	Phenoxy-4-hydroxyphenylsulphonate		NI24484
250	219	163	251	192	76	220	94	**250**	100	47	30	15	10	8	7	7		12	10	6	–	–	–	–	–	–	–	–	–	–	2,2′-Bifuran-5′,5-dicarboxylic acid, dimethyl ester		MS05414
69	53	250	39	66	51	151	95	**250**	100	77	38	37	30	30	29	27		12	10	6	–	–	–	–	–	–	–	–	–	–	1,4-Naphthalenedione, 5,8-dihydroxy-2,7-dimethoxy-	2808-46-0	NI24485
250	207	222	251	235	208	179	53	**250**	100	51	18	13	7	7	5	5		12	10	6	–	–	–	–	–	–	–	–	–	–	1,4-Naphthalenone, 6-ethyl-2,3,5,7-tetrahydroxy-	13378-99-9	NI24486
217	154	139	183	185	107	152	215	**250**	100	65	43	28	17	13	6	2	**0.65**	12	11	–	–	2	–	–	–	–	–	1	–	–	Phosphinodithioic acid, diphenyl-		MS05415
201	18	250	77	28	202	91	92	**250**	100	29	29	25	15	13	12	11		12	11	2	2	–	–	1	–	–	–	–	–	–	2-Pyrazolin-5-one, 4-chloroacetyl-3-methyl-1-phenyl-	31197-05-4	NI24487
177	250	164	136	191	163	135	109	**250**	100	96	72	58	50	26	21	20		12	14	2	2	1	–	–	–	–	–	–	–	–	Benzothiazol-2-amine, N-methyl-N-(2-carbomethoxyethyl)-		MS05416
177	136	191	250	178	149	109	164	**250**	100	35	28	19	16	10	10	6		12	14	2	2	1	–	–	–	–	–	–	–	–	Benzothiazoline, 2-[(2-carbomethoxyethyl)imino]-3-methyl-		MS05417
164	136	135	191	250	109	165	150	**250**	100	50	29	26	24	18	17	12		12	14	2	2	1	–	–	–	–	–	–	–	–	Benzothiazoline, 3-(2-carbomethoxyethyl)-2-(methylimino)-		MS05418
164	192	77	165	118	93	146	78	**250**	100	17	10	9	8	6	5	5	**4.70**	12	14	2	2	1	–	–	–	–	–	–	–	–	2,4-Pentadienoic acid, 2-cyano-5-ethyl-4-methyl-5-thiocyanato-, ethyl ester		MS05419
123	151	45	79	77	42	91	124	**250**	100	67	25	24	22	18	16	13	**4.00**	12	14	2	2	1	–	–	–	–	–	–	–	–	Sydnone, 4-methyl-3-[2-(4-methylphenylthio)ethyl]-	53492-47-0	NI24488
160	159	129	131	102	76	130	103	**250**	100	41	20	18	16	16	15	15	**0.00**	12	14	4	2	–	–	–	–	–	–	–	–	–	1,2,3,4-Butanetetrol, 1-(2-quinoxalinyl)-, [1R-(1R*,2S*,3R*)]-	4711-06-2	NI24489
105	77	41	78	43	28	18	40	**250**	100	38	37	22	21	18	18	16	**4.37**	12	14	4	2	–	–	–	–	–	–	–	–	–	2-Butanol, 2-methyl-3-nitroimino-, benzoate		MS05420
175	30	105	91	130	103	77	187	**250**	100	99	96	48	44	44	40	36	**0.00**	12	14	4	2	–	–	–	–	–	–	–	–	–	2H-Isoxazolo[2,3-b]isoxazole, tetrahydro-2-methyl-4-phenyl-3a-nitro-		LI05040
30	175	105	91	130	103	104	174	**250**	100	99	97	51	46	46	43	41	**0.00**	12	14	4	2	–	–	–	–	–	–	–	–	–	2H-Isoxazolo[2,3-b]isoxazole, tetrahydro-2-methyl-4-phenyl-3a-nitro-		LI05041
105	77	106	99	30	186	161	144	**250**	100	80	68	60	48	20	20	20	**0.00**	12	14	4	2	–	–	–	–	–	–	–	–	–	2H-Isoxazolo[2,3-b]isoxazole, tetrahydro-2-methyl-5-phenyl-3a-nitro-		LI05043
105	77	106	99	30	44	186	29	**250**	100	79	69	60	50	22	20	20	**0.00**	12	14	4	2	–	–	–	–	–	–	–	–	–	2H-Isoxazolo[2,3-b]isoxazole, tetrahydro-2-methyl-5-phenyl-3a-nitro-		LI05042
105	30	77	106	104	107	186	187	**250**	100	99	92	80	64	36	32	28	**0.00**	12	14	4	2	–	–	–	–	–	–	–	–	–	2H-Isoxazolo[2,3-b]isoxazole, tetrahydro-4-methyl-2-phenyl-3a-nitro-		LI05045
30	105	77	106	104	99	186	107	**250**	100	99	92	82	65	65	34	34	**0.00**	12	14	4	2	–	–	–	–	–	–	–	–	–	2H-Isoxazolo[2,3-b]isoxazole, tetrahydro-4-methyl-2-phenyl-3a-nitro-		LI05044
165	250	151	150	166	77	91	121	**250**	100	34	20	17	12	12	11	9		12	14	4	2	–	–	–	–	–	–	–	–	–	1,2,3-Oxadiazolium, 3-[2-(2,4-dimethoxyphenyl)ethyl]-5-hydroxy-, hydroxide, inner salt	55590-68-6	NI24490
95	93	41	136	121	55	110	69	**250**	100	43	37	35	27	24	20	19	**1.00**	12	17	2	–	–	3	–	–	–	–	–	–	–	Bornyl trifluoroacetate		LI05046
93	121	95	136	41	55	81	79	**250**	100	79	76	65	64	33	29	29	**2.00**	12	17	2	–	–	3	–	–	–	–	–	–	–	Epiisobornyl trifluoroacetate		LI05047
81	41	67	55	79	93	122	69	**250**	100	41	31	31	27	22	19	19	**1.00**	12	17	2	–	–	3	–	–	–	–	–	–	–	α-Fenchyl trifluoroacetate		LI05048
81	41	93	67	55	121	136	82	**250**	100	24	22	20	17	16	13	12	**1.00**	12	17	2	–	–	3	–	–	–	–	–	–	–	β-Fenchyl trifluoroacetate		LI05049
95	55	93	121	136	41	67	79	**250**	100	99	73	60	52	45	22	21	**0.70**	12	17	2	–	–	3	–	–	–	–	–	–	–	Isobornyl trifluoroacetate		LI05050
81	93	121	136	79	41	107	55	**250**	100	84	69	68	61	61	58	42	**1.00**	12	17	2	–	–	3	–	–	–	–	–	–	–	Isopulegyl trifluoroacetate		LI05051
69	41	93	39	136	70	67	53	**250**	100	41	19	9	7	6	6	6	**1.00**	12	17	2	–	–	3	–	–	–	–	–	–	–	Lavandulyl trifluoroacetate		LI05052
81	41	67	55	93	107	79	69	**250**	100	26	20	19	18	16	16	14	**1.00**	12	17	2	–	–	3	–	–	–	–	–	–	–	Methyl-camphenilyl trifluoroacetate		LI05053
68	91	67	41	93	92	134	79	**250**	100	68	34	30	26	26	22	22	**0.00**	12	17	2	–	–	3	–	–	–	–	–	–	–	Perillyl trifluoroacetate		LI05054
43	69	45	81	71	84	108	154	**250**	100	60	53	51	45	44	43	42	**0.00**	12	17	2	–	–	3	–	–	–	–	–	–	–	α-Terpinyl trifluoroacetate		LI05055
108	80	81	107	52	133	132	235	**250**	100	46	24	22	14	12	11	9	**3.00**	12	18	2	4	–	–	–	–	–	–	–	–	–	Benzene, 1,2-bis(2-amino-1-methoxyethylimino)-		NI24491
250	164	222	194	235	251	207	150	**250**	100	46	35	31	21	13	12	11		12	18	2	4	–	–	–	–	–	–	–	–	–	1H-Purine-2,6-dione, 1,3,7-triethyl-3,7-dihydro-8-methyl-	57396-62-0	NI24492
45	28	57	29	41	89	43	59	**250**	100	60	46	37	28	22	21	15		12	26	5	–	–	–	–	–	–	–	–	–	–	Tetraethylene glycol, monobutyl ether		IC03275
45	57	89	29	41	59	44	56	**250**	100	42	27	19	16	15	15	13	**0.00**	12	26	5	–	–	–	–	–	–	–	–	–	–	3,6,9,12-Tetraoxahexadecan-1-ol	1559-34-8	NI24493
83	41	29	57	27	56	39	55	**250**	100	35	31	26	16	14	8	7	**0.17**	12	27	3	–	–	–	–	–	–	–	1	–	–	Phosphorous acid, tributyl ester	102-85-2	IC03276
83	57	41	29	139	56	28	27	**250**	100	29	20	20	17	9	7	5	**0.27**	12	27	3	–	–	–	–	–	–	–	1	–	–	Phosphorous acid, tributyl ester	102-85-2	NI24494
83	57	41	29	139	134	56	28	**250**	100	33	33	25	21	15	15	11	**0.00**	12	27	3	–	–	–	–	–	–	–	1	–	–	Phosphorous acid, tributyl ester	102-85-2	IC03277
139	141	111	75	250	252	105	50	**250**	100	40	37	30	22	15	15	13		13	8	1	–	–	–	2	–	–	–	–	–	–	Benzophenone, 2,2′-dichloro-	5293-97-0	NI24495
105	173	250	77	175	252	51	106	**250**	100	38	35	33	26	23	11	8		13	8	1	–	–	–	2	–	–	–	–	–	–	Benzophenone, 2,4-dichloro-	19811-05-3	NI24496
105	173	250	77	175	252	51	145	**250**	100	38	34	33	26	23	13	9		13	8	1	–	–	–	2	–	–	–	–	–	–	Benzophenone, 2,4-dichloro-	19811-05-3	LI05056
137	39	18	180	147	27	43	41	**250**	100	34	29	27	24	21	20	20	**0.00**	13	8	1	–	–	–	2	–	–	–	–	–	–	Benzophenone, 2,4′-dichloro-	85-29-0	NI24497
139	111	75	141	250	50	113	252	**250**	100	40	35	33	17	14	13	11		13	8	1	–	–	–	2	–	–	–	–	–	–	Benzophenone, 2,4′-dichloro-	85-29-0	NI24498
105	77	250	173	252	175	51	145	**250**	100	33	29	25	19	16	12	10		13	8	1	–	–	–	2	–	–	–	–	–	–	Benzophenone, 3,4-dichloro-	6284-79-3	NI24499
139	111	141	75	250	252	113	50	**250**	100	43	36	22	17	11	11	8		13	8	1	–	–	–	2	–	–	–	–	–	–	Benzophenone, 4,4′-dichloro-	90-98-2	IC03278
139	111	141	250	252	76	140	113	**250**	100	33	32	30	18	17	10	10		13	8	1	–	–	–	2	–	–	–	–	–	–	Benzophenone, 4,4′-dichloro-	90-98-2	NI24500
139	141	250	111	252	75	113	140	**250**	100	33	30	30	19	17	9	8		13	8	1	–	–	–	2	–	–	–	–	–	–	Benzophenone, 4,4′-dichloro-	90-98-2	NI24501
139	107	143	250	91	65	39	79	**250**	100	96	95	94	89	71	45	38		13	11	2	–	1	1	–	–	–	–	–	–	–	Benzene, 1-fluoro-4-[(4-methylphenyl)sulphonyl]-	1643-97-6	NI24502
65	91	67	39	139	41	92	95	**250**	100	89	81	78	63	54	53	52	**0.00**	13	14	3	–	1	–	–	–	–	–	–	–	–	2,5-Hexadienal, 3-(p-tolylsulphonyl)-, (E)-	118356-35-7	DD00685
153	98	97	28	136	70	41	107	**250**	100	62	48	45	42	35	31	19	**1.32**	13	14	5	–	–	–	–	–	–	–	–	–	–	Allamcin	93452-23-4	NI24503
191	190	250	159	175	103	131	55	**250**	100	70	52	45	37	26	25	18		13	14	5	–	–	–	–	–	–	–	–	–	–	Benzeneacetic acid, 3-hydroxy-4-methoxy-2-(1-oxo-2-propenyl)-, methyl ester	40992-07-2	NI24504
163	208	148	77	149	164	71	70	**250**	100	64	61	54	50	45	40	40	**0.00**	13	14	5	–	–	–	–	–	–	–	–	–	–	1,2-Benzenedicarboxylic acid, methyl 3-oxobutyl diester		MS05421
135	177	69	217	43	190	104	77	**250**	100	65	65	57	55	44	32	29	**21.00**	13	14	5	–	–	–	–	–	–	–	–	–	–	Benzenepentanoic acid, 4-methoxy-β,δ-dioxo-, methyl ester	36568-14-6	NI24505

MASS TO CHARGE RATIOS								M.W.	INTENSITIES								Parent	C	H	O	N	S	F	Cl	Br	I	Si	P	B	X	COMPOUND NAME	CAS Reg No	No
250	206	217	204	232	203	91	77	**250**	100	86	77	52	51	41	32	31		13	14	5	–	–	–	–	–	–	–	–	–	–	**3H-2-Benzopyran-7-carboxylic acid, 4,6-dihydro-8-hydroxy-3,4,5-trimethyl-6-oxo-, (3R-trans)-**	518-75-2	**NI24506**
206	44	250	217	28	43	191	204	**250**	100	98	61	49	43	36	35	32		13	14	5	–	–	–	–	–	–	–	–	–	–	**3H-2-Benzopyran-7-carboxylic acid, 4,6-dihydro-8-hydroxy-3,4,5-trimethyl-6-oxo-, (3R-trans)-**	518-75-2	**MS05422**
207	235	250	136	164	192	221	83	**250**	100	94	70	44	15	13	13	12		13	14	5	–	–	–	–	–	–	–	–	–	–	**4H-1-Benzopyran-4-one, 2,8-dimethyl-6,7-dimethoxy-5-hydroxy-**	84782-35-4	**NI24507**
208	177	18	145	43	209	17	150	**250**	100	67	56	28	22	21	17	10	**9.64**	13	14	5	–	–	–	–	–	–	–	–	–	–	**Cinnamic acid, 4-hydroxy-3-methoxy-, acetate, methyl ester**		**MS05423**
208	43	110	135	250	108	209	82	**250**	100	96	35	18	16	16	13	8		13	14	5	–	–	–	–	–	–	–	–	–	–	**4,7-Ethanoisobenzofuran-1,3-dione, 5-(acetyloxy)-3a,4,7,7a-tetrahydro-6-methyl-**	52918-79-3	**NI24508**
43	110	180	208	152	166	108	28	**250**	100	42	17	15	12	11	10	9	**5.85**	13	14	5	–	–	–	–	–	–	–	–	–	–	**4,7-Ethanoisobenzofuran-1,3-dione, 5-(acetyloxy)-3a,4,7,7a-tetrahydro-8-methyl-**	52918-80-6	**NI24509**
105	43	166	77	58	85	218	51	**250**	100	75	62	43	28	26	24	14	**0.00**	13	14	5	–	–	–	–	–	–	–	–	–	–	**5-Furancarboxylic acid, 4-hydroxy-4-methyl-2-oxo-5-phenyltetrahydro-, methyl ester**		**DD00686**
219	131	89	132	151	63	77	45	**250**	100	79	48	43	39	39	38	36	**34.00**	13	14	5	–	–	–	–	–	–	–	–	–	–	**Propanedioic acid, 2-methoxybenzylidene-, dimethyl ester**		**LI05057**
250	190	151	219	132	159	251	59	**250**	100	66	49	36	17	16	15	13		13	14	5	–	–	–	–	–	–	–	–	–	–	**Propanedioic acid, 3-methoxybenzylidene-, dimethyl ester**		**LI05058**
250	190	219	151	132	59	89	63	**250**	100	57	45	45	32	21	16	15		13	14	5	–	–	–	–	–	–	–	–	–	–	**Propanedioic acid, 4-methoxybenzylidene-, dimethyl ester**		**LI05059**
250	190	151	219	132	89	251	160	**250**	100	71	57	49	36	16	14	12		13	14	5	–	–	–	–	–	–	–	–	–	–	**Propanedioic acid, 4-methoxybenzylidene-, dimethyl ester**		**IC03279**
235	161	191	250	75	236	175	118	**250**	100	76	71	50	33	18	16	16		13	18	3	–	–	–	–	–	–	1	–	–	–	**Cinnamic acid, m-methoxy-, trimethylsilyl ester**		**MS05424**
250	235	73	203	251	75	236	219	**250**	100	87	60	50	22	20	17	17		13	18	3	–	–	–	–	–	–	1	–	–	–	**Cinnamic acid, m-(trimethylsiloxy)-, methyl ester**	32342-00-0	**NI24510**
89	73	250	235	59	219	118	90	**250**	100	78	75	46	43	32	24	23		13	18	3	–	–	–	–	–	–	1	–	–	–	**Cinnamic acid, o-(trimethylsiloxy)-, methyl ester**	27927-16-8	**NI24511**
89	250	235	73	219	146	118	90	**250**	100	39	33	19	18	15	14	11		13	18	3	–	–	–	–	–	–	1	–	–	–	**Cinnamic acid, o-(trimethylsiloxy)-, methyl ester**	27927-16-8	**NI24512**
161	235	250	191	75	133	77	73	**250**	100	68	45	41	26	19	18	14		13	18	3	–	–	–	–	–	–	1	–	–	–	**Cinnamic acid, p-methoxy-, trimethylsilyl ester**		**MS05425**
250	235	251	219	203	236	252	73	**250**	100	36	21	16	13	7	6	6		13	18	3	–	–	–	–	–	–	1	–	–	–	**Cinnamic acid, p-(trimethylsiloxy)-, methyl ester**	27798-69-2	**NI24514**
250	235	73	203	219	89	45	251	**250**	100	74	66	45	39	26	23	21		13	18	3	–	–	–	–	–	–	1	–	–	–	**Cinnamic acid, p-(trimethylsiloxy)-, methyl ester**	27798-69-2	**NI24513**
250	235	73	219	203	251	236	102	**250**	100	55	33	30	22	20	12	12		13	18	3	–	–	–	–	–	–	1	–	–	–	**Cinnamic acid, p-(trimethylsiloxy)-, methyl ester**	27798-69-2	**NI24515**
58	120	59	205	65	165	91	64	**250**	100	33	23	15	13	11	11	10	**0.00**	13	18	3	2	–	–	–	–	–	–	–	–	–	**Benzoic acid, 2-methylalanyl-4-amino-, ethyl ester**		**DD00687**
141	221	81	79	41	91	93	77	**250**	100	97	54	42	36	29	28	28	**0.00**	13	18	3	2	–	–	–	–	–	–	–	–	–	**Heptabarbital**	509-86-4	**NI24516**
221	141	81	222	79	67	41	93	**250**	100	22	16	14	14	9	9	8	**0.00**	13	18	3	2	–	–	–	–	–	–	–	–	–	**Heptabarbital**	509-86-4	**NI24517**
251	252	221	157	253	135			**250**	100	15	6	3	2	1			**0.00**	13	18	3	2	–	–	–	–	–	–	–	–	–	**Heptabarbital**	509-86-4	**NI24518**
235	81	169	179	171	79	170	41	**250**	100	43	23	20	20	18	16	15	**7.00**	13	18	3	2	–	–	–	–	–	–	–	–	–	**Hexobarbital, N-methyl-**	726-79-4	**NI24520**
169	81	171	79	235	170	236	77	**250**	100	73	53	45	44	42	34	24	**7.81**	13	18	3	2	–	–	–	–	–	–	–	–	–	**Hexobarbital, N-methyl-**	726-79-4	**NI24521**
235	81	169	171	79	236	170	91	**250**	100	40	26	23	19	13	12	11	**6.01**	13	18	3	2	–	–	–	–	–	–	–	–	–	**Hexobarbital, N-methyl-**	726-79-4	**NI24519**
110	59	70	71	163	36	250	64	**250**	100	72	66	52	28	27	20	20		13	18	3	2	–	–	–	–	–	–	–	–	–	**2-Propenamide, N-(4-aminobutyl)-3-(3,4-dihydroxyphenyl)-, (E)-**	26148-06-1	**LI05060**
110	59	70	71	163	250	64	56	**250**	100	70	65	50	28	20	20	20		13	18	3	2	–	–	–	–	–	–	–	–	–	**2-Propenamide, N-(4-aminobutyl)-3-(3,4-dihydroxyphenyl)-, (E)-**	26148-06-1	**NI24522**
177	250	176	204	55	205	178	123	**250**	100	43	35	28	27	17	14	14		13	18	3	2	–	–	–	–	–	–	–	–	–	**4H-Pyrido[1,2-a]pyrimidine-3-acetic acid, 6,7,8,9-tetrahydro-6-methyl-4-oxo-, ethyl ester**	54804-24-9	**NI24523**
177	178	250	135	39	204	55	41	**250**	100	21	20	11	11	10	10	9		13	18	3	2	–	–	–	–	–	–	–	–	–	**4H-Pyrido[1,2-a]pyrimidine-3-acetic acid, 6,7,8,9-tetrahydro-7-methyl-4-oxo-, ethyl ester**	54504-54-0	**NI24524**
250	249	235	82	149	148	251	163	**250**	100	32	25	22	19	19	16	15		13	18	3	2	–	–	–	–	–	–	–	–	–	**4H-Pyrido[1,2-a]pyrimidine-3-carboxylic acid, 1,6,7,8-tetrahydro-1,6-dimethyl-4-oxo-, ethyl ester**	35569-43-8	**NI24525**
235	165	150	236	79	137	250	164	**250**	100	25	25	21	19	17	14	12		13	18	3	2	–	–	–	–	–	–	–	–	–	**Pyrimidinedione, methoxydimethyl(1-cyclohexen-1-yl)-**	55334-00-4	**NI24526**
102	148	92	250	65	76	51	130	**250**	100	71	42	31	31	23	22	16		14	10	1	4	–	–	–	–	–	–	–	–	–	**Bis(2-cyanoanilino) ether**		**IC03280**
250	77	103	94	224	51	222	93	**250**	100	40	31	28	26	24	23	20		14	10	1	4	–	–	–	–	–	–	–	–	–	**Pyrazolo[4,5-c]pyrrolo[1,2-a]-1H-pyrazin-8-one, N-phenyl-**	94692-14-5	**NI24527**
250	180	165	105	252	215	235	200	**250**	100	98	79	76	65	55	31	25		14	12	–	–	–	–	2	–	–	–	–	–	–	**1,1′-Biphenyl, 2,4-dichloro-2′,5′-dimethyl-**	69299-48-5	**NI24528**
139	141	140	142					**250**	100	31	9	3					**0.00**	14	12	–	–	–	–	2	–	–	–	–	–	–	**Ethane, 1,1-bis(4-chlorophenyl)-**	3547-04-4	**NI24529**
166	250	121	165	167	45	168	134	**250**	100	39	23	14	12	11	10	10		14	18	–	–	2	–	–	–	–	–	–	–	–	**Benzo[b]thiophene, 2-(hexylthio)-**		**AI01789**
166	250	121	45	165	167	134	168	**250**	100	27	24	17	14	13	11	10		14	18	–	–	2	–	–	–	–	–	–	–	–	**Benzo[b]thiophene, 3-(hexylthio)-**		**AI01790**
43	250	86	79	163	208	95	107	**250**	100	88	74	65	61	56	54	52		14	18	4	–	–	–	–	–	–	–	–	–	–	**2-Adamantanecarboxylic acid, 4,8-dioxo-, isopropyl ester**	15782-81-7	**MS05426**
149	43	41	150	76	105	104	65	**250**	100	22	13	11	10	7	7	6	**0.00**	14	18	4	–	–	–	–	–	–	–	–	–	–	**1,2-Benzenedicarboxylic acid, diisopropyl ester**	605-45-8	**MS05427**
149	150	105	43	209	122	41	59	**250**	100	11	11	8	6	6	5	4	**0.00**	14	18	4	–	–	–	–	–	–	–	–	–	–	**1,2-Benzenedicarboxylic acid, diisopropyl ester**	605-45-8	**WI01276**
149	43	41	150	76	104	209	65	**250**	100	12	12	9	9	8	7	6	**0.00**	14	18	4	–	–	–	–	–	–	–	–	–	–	**1,2-Benzenedicarboxylic acid, dipropyl ester**	131-16-8	**MS05428**
149	150	209	191	43	41	104	76	**250**	100	9	8	8	7	6	5	4	**0.00**	14	18	4	–	–	–	–	–	–	–	–	–	–	**1,2-Benzenedicarboxylic acid, dipropyl ester**	131-16-8	**WI01277**
149	43	41	150	76	104	209	57	**250**	100	12	12	9	9	8	7	7	**0.28**	14	18	4	–	–	–	–	–	–	–	–	–	–	**1,2-Benzenedicarboxylic acid, dipropyl ester**	131-16-8	**NI24530**
219	235	15	250	191	91	203	220	**250**	100	66	47	33	22	17	16	14		14	18	4	–	–	–	–	–	–	–	–	–	–	**1,2-Benzenedicarboxylic acid, 5-isopropyl-4-methyl-, dimethyl ester**		**MS05429**
235	207	115	91	117	145	219	59	**250**	100	27	26	20	19	16	14	14	**4.13**	14	18	4	–	–	–	–	–	–	–	–	–	–	**1,3-Benzenedicarboxylic acid, 5-tert-butyl-, dimethyl ester**	16308-65-9	**NI24531**
180	43	27	71	250	194	163	179	**250**	100	99	90	37	18	16	12	10		14	18	4	–	–	–	–	–	–	–	–	–	–	**Benzene, 1-(1-hydroxyallyl)-4-isobutoxy-3-methoxy-**		**LI05061**

MASS TO CHARGE RATIOS								M.W.	INTENSITIES								Parent	C	H	O	N	S	F	Cl	Br	I	Si	P	B	X	COMPOUND NAME	CAS Reg No	No
195	236	250	179	177	69	233	196	**250**	100	75	51	30	21	20	13	10		14	18	4	–	–	–	–	–	–	–	–	–	–	2H-1-Benzopyran, 6-acetyl-3,4-dihydro-7-hydroxy-5-methoxy-2,2-dimethyl-	27364-68-7	NI24532
195	250	179	43	235	177	207	196	**250**	100	33	22	19	13	13	9	8		14	18	4	–	–	–	–	–	–	–	–	–	–	2H-1-Benzopyran, 8-acetyl-3,4-dihydro-7-hydroxy-5-methoxy-2,2-dimethyl-		MS05430
235	69	97	81	59	123	203	250	**250**	100	70	63	35	35	34	13	10		14	18	4	–	–	–	–	–	–	–	–	–	–	1,4-Benzoquinone, 2,3-dimethoxy-6-methyl-5-(3-methyl-but-2-enyl)-		LI05062
250	191	190	193	163	162	222	207	**250**	100	92	52	50	43	39	34	16		14	18	4	–	–	–	–	–	–	–	–	–	–	Bicyclo[3.3.0]octa-4,6-diene-2-carboxylic acid, 2,4-dimethyl-7-ethoxy-3-oxo-, methyl ester, trans-		MS05431
125	43	126	189	109	53	135	111	**250**	100	74	20	13	8	6	5	5	**5.00**	14	18	4	–	–	–	–	–	–	–	–	–	–	2,3-Butanediol, 2,3-bis(5-methyl-2-furanyl)-	56335-95-6	NI24533
73	250	191	173	119	187	189		**250**	100	38	30	25	25	24	14			14	18	4	–	–	–	–	–	–	–	–	–	–	1,3-Cyclohexadiene-1,6-dicarboxylic acid, 4,5-dimethyl-2-isobutenyl-		LI05063
126	137	98	55	139	113	250	70	**250**	100	88	81	56	51	42	21	18		14	18	4	–	–	–	–	–	–	–	–	–	–	Ethylene, 2,2′-bis(3-hydroxy-2-cyclohexen-1-one)-	22381-60-8	NI24534
107	179	105	79	77	95	101	106	**250**	100	72	27	26	21	11	10	8	**0.40**	14	18	4	–	–	–	–	–	–	–	–	–	–	1-Hexenitol, 4,6-O-benzylidene-2,3-dideoxy-1-O-methyl-, (E)-		NI24535
101	69	105	77	106	91	59	250	**250**	100	30	20	15	11	8	7	6		14	18	4	–	–	–	–	–	–	–	–	–	–	α-D-erythro-Hexopyranoside, methyl 4,6-O-benzylidene-2,3-dideoxy-		NI24536
101	69	105	91	77	59	207	250	**250**	100	28	11	8	7	7	6	4		14	18	4	–	–	–	–	–	–	–	–	–	–	β-D-erythro-Hexopyranoside, methyl 4,6-O-benzylidene-2,3-dideoxy-		NI24537
201	250	149	202	232	217	43	159	**250**	100	26	21	18	17	15	11	10		14	18	4	–	–	–	–	–	–	–	–	–	–	2-Hydroxypterosin c		NI24538
250	219	201	217	232	189	173	203	**250**	100	70	63	47	43	42	40	32		14	18	4	–	–	–	–	–	–	–	–	–	–	11-Hydroxypterosin c		NI24539
202	250	203	205	219	187	119	232	**250**	100	44	23	21	17	16	16	14		14	18	4	–	–	–	–	–	–	–	–	–	–	1H-Inden-1-one, 2,3-dihydro-3-hydroxy-6-(2-hydroxyethyl)-5-(hydroxymethyl)-2,7-dimethyl-, (2S-trans)-	56227-01-1	NI24540
105	166	77	107	106	79	85	67	**250**	100	97	59	43	39	28	23	16	**0.00**	14	18	4	–	–	–	–	–	–	–	–	–	–	Methyl α-hydroxy-α-(1-hydroxy-1-cyclopentyl)benzeneacetate		DD00688
159	232	144	160	158	91	115	203	**250**	100	18	18	15	13	8	7	5	**0.00**	14	18	4	–	–	–	–	–	–	–	–	–	–	2-Naphthalenecarboxylic acid, 3-hydroxy-6-methoxy-1,2,3,4-tetrahydro-, ethyl ester, (2R,3S)-		DD00689
159	232	144	160	134	115	91	29	**250**	100	22	16	14	11	8	7	5	**0.00**	14	18	4	–	–	–	–	–	–	–	–	–	–	2-Naphthalenecarboxylic acid, 3-hydroxy-7-methoxy-1,2,3,4-tetrahydro-, ethyl ester, (2R,3S)-		DD00690
159	158	144	232	160	115	91	104	**250**	100	26	23	17	14	11	10	9	**0.00**	14	18	4	–	–	–	–	–	–	–	–	–	–	2-Naphthalenecarboxylic acid, 3-hydroxy-8-methoxy-1,2,3,4-tetrahydro-, ethyl ester, (2R,3S)-		DD00691
131	91	176	159	148	103	250	177	**250**	100	54	45	31	31	29	24	24		14	18	4	–	–	–	–	–	–	–	–	–	–	Propanedioic acid, benzyl-, diethyl ester	607-81-8	NI24541
250	150	177	145	222	89	77	205	**250**	100	85	70	59	52	31	30	29		14	18	4	–	–	–	–	–	–	–	–	–	–	2-Propenoic acid, 3-(4-ethoxy-3-methoxyphenyl)-, ethyl ester, (E)-	75332-48-8	NI24542
150	250	177	145	222	77	51	89	**250**	100	89	76	72	56	41	41	37		14	18	4	–	–	–	–	–	–	–	–	–	–	2-Propenoic acid, 3-(4-ethoxy-3-methoxyphenyl)-, ethyl ester, (Z)-	75332-47-7	NI24543
161	178	133	250	134	162	179	90	**250**	100	97	33	21	21	14	13	11		14	18	4	–	–	–	–	–	–	–	–	–	–	2-Propenoic acid, 3-(4-methoxyphenyl)-, ethoxyethyl ester		NI24544
43	68	69	94	96	39	55	66	**250**	100	87	66	52	29	24	23	20	**0.22**	14	18	4	–	–	–	–	–	–	–	–	–	–	Pyran, 2-(acetoxymethyl)-6-phenoxytetrahydro-		IC03281
193	235	73	208	147	75	89	151	**250**	100	47	37	27	20	12	7	5	**2.00**	14	22	2	–	–	–	–	–	–	1	–	–	–	Valerophenone, 2′-(trimethylsiloxy)-	33342-91-5	LI05064
121	193	235	41	83	43	73	65	**250**	100	80	67	45	40	38	30	22	**1.98**	14	22	2	–	–	–	–	–	–	1	–	–	–	Valerophenone, 2′-(trimethylsiloxy)-	33342-91-5	NI24545
193	235	73	208	121	147	194	75	**250**	100	47	37	27	26	20	17	11	**3.00**	14	22	2	–	–	–	–	–	–	1	–	–	–	Valerophenone, 2′-(trimethylsiloxy)-	33342-91-5	NI24546
193	208	73	194	147	209	57	89	**250**	100	50	28	19	8	6	6	5	**4.00**	14	22	2	–	–	–	–	–	–	1	–	–	–	Valerophenone, 4′-(trimethylsiloxy)-	33342-92-6	NI24547
193	208	73	194	209	147	57	89	**250**	100	49	27	17	10	8	6	5	**4.00**	14	22	2	–	–	–	–	–	–	1	–	–	–	Valerophenone, 4′-(trimethylsiloxy)-	33342-92-6	NI24548
193	208	73	147	235	89	250	75	**250**	100	50	27	8	5	5	4	3		14	22	2	–	–	–	–	–	–	1	–	–	–	Valerophenone, 4′-(trimethylsiloxy)-	33342-92-6	LI05065
99	56	100	44	70	40	42	121	**250**	100	35	20	17	16	15	13	9	**1.00**	14	22	2	2	–	–	–	–	–	–	–	–	–	Piperazine, 1-(2-hydroxy-3-methoxy-3-phenylpropyl)-		NI24549
232	43	189	207	44	122	165	84	**250**	100	88	81	71	45	42	40	34	**12.87**	14	22	2	2	–	–	–	–	–	–	–	–	–	Pyridine, 1-acetyl-5-(1-acetyl-2-piperidinyl)-1,2,3,4-tetrahydro-	54966-21-1	NI24550
219	235	73	250	220	45	163	236	**250**	100	94	52	24	24	22	20	19		14	26	–	–	–	–	–	–	–	2	–	–	–	Benzene, 1,2-dimethyl-4,5-bis(trimethylsilyl)-		NI24551
73	161	162	147	135	45	145	59	**250**	100	52	34	32	20	17	15	15	**2.18**	14	26	–	–	–	–	–	–	–	2	–	–	–	2,4-Disilapentane, 3-benzyl-2,2,4,4-tetramethyl-		MS05432
250	121	249	102	90	129	218	103	**250**	100	30	18	14	13	9	8	5		15	10	–	2	1	–	–	–	–	–	–	–	–	Thiazolo[3,2-a]benzimidazole, 2-phenyl-		LI05066
250	249	102	205	90	218	217	134	**250**	100	19	10	9	5	3	2	2		15	10	–	2	1	–	–	–	–	–	–	–	–	Thiazolo[3,2-a]benzimidazole, 3-phenyl-		LI05067
193	222	250	194	168	223	140	75	**250**	100	87	69	45	33	16	14	14		15	10	2	2	–	–	–	–	–	–	–	–	–	6H-Indolo[3,2,1-de][1,5]naphthyridin-2,6-dione, 3-methyl-	82652-21-9	NI24552
250	221	249	192	179	152	207	251	**250**	100	71	52	30	25	25	24	21		15	10	2	2	–	–	–	–	–	–	–	–	–	6H-Indolo[3,2,1-de][1,5]naphthyridin-6-one, 5-methoxy-	15071-56-4	NI24553
250	208	221	132	77	180	28	251	**250**	100	56	35	26	21	19	18	17		15	10	2	2	–	–	–	–	–	–	–	–	–	Isocyanic acid, methylenedi-p-phenylene ester	101-68-8	IC03282
250	208	221	132	51	39	251	249	**250**	100	59	31	28	19	18	17	16		15	10	2	2	–	–	–	–	–	–	–	–	–	Isocyanic acid, methylenedi-p-phenylene ester	101-68-8	WI01278
203	250	176	88	204	177	202	205	**250**	100	83	48	41	40	32	23	21		15	10	2	2	–	–	–	–	–	–	–	–	–	Nitrobenzene, 3-(2-cyano-2-phenylethenyl)-	6720-37-2	NI24554
250	203	204	176	88	177	251	165	**250**	100	62	38	28	25	22	17	15		15	10	2	2	–	–	–	–	–	–	–	–	–	Nitrobenzene, 4-(2-cyano-2-phenylethenyl)-	7431-35-8	NI24555
105	77	222	250	145	103	51	50	**250**	100	99	36	28	27	25	25	13		15	10	2	2	–	–	–	–	–	–	–	–	–	1,3,4-Oxadiazole, 5-benzoyl-2-phenyl-		MS05433
105	77	51	222	106	165	78	50	**250**	100	52	24	16	9	6	4	4	**2.00**	15	10	2	2	–	–	–	–	–	–	–	–	–	1,3-Propanedione, 2-diazo-1,3-diphenyl-	2085-31-6	MS05434
206	234	205	77	179	76	75	103	**250**	100	77	60	47	43	42	23	20	**0.00**	15	10	2	2	–	–	–	–	–	–	–	–	–	2-Quinoxalinecarboxaldehyde, 3-phenyl-, 1-oxide	55030-22-3	NI24556
120	250	42	77	105	102	104	79	**250**	100	48	35	28	23	18	15	15		15	14	–	4	–	–	–	–	–	–	–	–	–	Aniline, 4-(4-cyanophenylazo)-N,N-dimethyl-		LI05068
207	250	221	180	194	193	178	103	**250**	100	81	47	47	46	18	18	13		15	14	–	4	–	–	–	–	–	–	–	–	–	1,2,3-Benzotriazine, 3,4-dihydro-3-ethyl-4-phenylimino-		LI05069
235	250	39	41	77	193	52	51	**250**	100	90	26	23	20	16	16	15		15	14	–	4	–	–	–	–	–	–	–	–	–	Ethenetricarbonitrile, (4-amino-3,5-diethylphenyl)-	23957-76-8	MS05435
165	145	215	180	101	167	137	102	**250**	100	56	48	45	45	38	34	25	**7.81**	15	19	1	–	–	–	1	–	–	–	–	–	–	1-Nonen-3-one, 1-(2-chlorophenyl)-	36383-91-2	NI24557
165	41	30	43	28	145	180	39	**250**	100	91	62	60	57	55	41	35	**15.54**	15	19	1	–	–	–	1	–	–	–	–	–	–	1-Nonen-3-one, 1-(4-chlorophenyl)-	36383-92-3	NI24558
92	91	79	93	80	107	108	77	**250**	100	95	55	52	43	40	35	31	**0.00**	15	19	1	–	–	–	1	–	–	–	–	–	–	Norbornane, exo-2-chloro-syn-7-(benzyloxymethyl)-		DD00692

MASS TO CHARGE RATIOS	M.W.	INTENSITIES	Parent	C	H	O	N	S	F	Cl	Br	I	Si	P	B	X	COMPOUND NAME	CAS Reg No	No
121 214 250 185 159 94 91 215	**250**	100 19 11 8 6 6 6 5		15	19	1	–	–	–	1	–	–	–	–	–	–	Norbornane, exo-2-chloro-syn-7-(4-methoxybenzyl)-		DD00693
41 126 95 81 55 109 43 69	**250**	100 83 71 71 71 69 67 44	**36.00**	15	22	3	–	–	–	–	–	–	–	–	–	–	Azuleno[5,6-c]furan-1(3H)-one, 3a,4,4a,5,6,7,7a,8-octahydro-4-hydroxy-6,6,8-trimethyl-, (3aα,4α,4aα,7aα,8α)-	62860-40-6	NI24559
235 32 41 57 250 236 55 44	**250**	100 50 49 25 20 17 16 16		15	22	3	–	–	–	–	–	–	–	–	–	–	Benzoic acid, 3,5-di-tert-butyl-4-hydroxy-	1421-49-4	IC03283
235 250 41 57 236 191 44 207	**250**	100 20 20 18 16 12 11 8		15	22	3	–	–	–	–	–	–	–	–	–	–	Benzoic acid, 3,5-di-tert-butyl-4-hydroxy-	1421-49-4	NI24560
152 135 153 250 107 77 136 92	**250**	100 39 22 21 11 7 6 5		15	22	3	–	–	–	–	–	–	–	–	–	–	Benzoic acid, 3-methoxy-, heptyl ester	69833-40-5	NI24561
152 135 153 136 77 250 107 92	**250**	100 65 24 7 7 5 5 5		15	22	3	–	–	–	–	–	–	–	–	–	–	Benzoic acid, 4-methoxy-, heptyl ester	5452-08-4	NI24562
235 203 219 250	**250**	100 55 7 2		15	22	3	–	–	–	–	–	–	–	–	–	–	1,4-Benzopyran-3-carboxylic acid, 2,4,4,7-tetramethyl-5,6,7,8-tetrahydro-, methyl ester		LI05070
235 43 203 91 67 219 250 191	**250**	100 55 15 8 8 7 2 1		15	22	3	–	–	–	–	–	–	–	–	–	–	1,4-Benzopyran-3-carboxylic acid, 2,4,4,7-tetramethyl-5,6,7,8-tetrahydro-, methyl ester		LI05071
87 43 147 41 149 91 219 105	**250**	100 48 36 18 16 15 14 14	**0.00**	15	22	3	–	–	–	–	–	–	–	–	–	–	1,4-Dioxane, 2-[3-[3,3-dimethyl-2-isopropyloxiranyl]-1-methylene-2-propynyl]-	63922-60-1	NI24563
43 41 67 39 125 97 68 55	**250**	100 40 18 16 15 15 14 14	**1.00**	15	22	3	–	–	–	–	–	–	–	–	–	–	5,12-Dioxatricyclo[9.1.0.0^{4,6}]dodecan-8-one, 1,6-dimethyl-9-isopropylidene-		MS05436
123 232 217 135 203 81 107 121	**250**	100 64 50 50 47 47 44 43	**32.00**	15	22	3	–	–	–	–	–	–	–	–	–	–	1,3-Dioxolane, 2,2-dimethyl-4-hydroxymethyl-5-methoxymethyl-, p-toluenesulphonate, (+)-(4S,5S)-	83481-29-2	NI24564
149 86 162 73 250 91 148 119	**250**	100 78 77 54 37 18 17 13		15	22	3	–	–	–	–	–	–	–	–	–	–	1,3-Dioxolane, 2-[3-(2-methoxy-4-methylphenyl)butyl]-	63646-83-3	NI24565
126 81 97 250 179 165 111 207	**250**	100 71 58 56 50 45 42 29		15	22	3	–	–	–	–	–	–	–	–	–	–	4,8-Epidioxyazulen-5-one, 1,4-dimethyl-7-isopropyl-1,2,3,3a,4,5,8,8a-octahydro-		DD00694
95 55 41 122 82 69 123 53	**250**	100 95 85 79 79 67 45 40	**20.02**	15	22	3	–	–	–	–	–	–	–	–	–	–	4ξ-Germacr-9-en-12-oic acid, 6α-hydroxy-1-oxo-, γ-lactone, (11S)-	19906-69-5	NI24566
108 93 124 123 91 109 107 55	**250**	100 94 78 51 50 50 43 38	**15.20**	15	22	3	–	–	–	–	–	–	–	–	–	–	Isotrichodermol		NI24567
96 105 132 91 119 95 173 204	**250**	100 92 88 88 78 77 73 72	**42.00**	15	22	3	–	–	–	–	–	–	–	–	–	–	Marasmene, 3β-hydroxy-		MS05437
105 173 91 159 55 131 79 77	**250**	100 91 90 72 59 51 48 46	**15.00**	15	22	3	–	–	–	–	–	–	–	–	–	–	Marasmene, 13-hydroxy-		MS05438
105 55 91 204 57 173 79 59	**250**	100 98 97 74 73 72 66 66	**9.00**	15	22	3	–	–	–	–	–	–	–	–	–	–	Marasmene, 14-hydroxy-		MS05439
41 248 43 55 39 91 83 95	**250**	100 95 95 63 57 56 46 41	**0.00**	15	22	3	–	–	–	–	–	–	–	–	–	–	4a(2H)-Naphthalenecarboxaldehyde, octahydro-4-methyl-6-isopropyl-3,5-dioxo-, [4S-(4α,4aα,6α,8aα)]-	56298-89-6	NI24568
59 43 41 93 163 91 77 79	**250**	100 66 56 55 42 33 26 25	**0.00**	15	22	3	–	–	–	–	–	–	–	–	–	–	1(2H)-Naphthalenone, 2,3-epoxy-4a,5-dimethyl-3,4,4a,5,6,7-hexahydro-4-(1-hydroxy-1-methylethyl)-, (+)-	27062-01-7	LI05072
250 153 95 109 121 123 97 93	**250**	100 60 48 45 40 33 33 31		15	22	3	–	–	–	–	–	–	–	–	–	–	Naphtho[1,2-b]furan-2,8(3H,4H)-dione, octahydro-3,5a,9-trimethyl-, [3S-(3α,3aα,5aβ,9α,9aα,9bβ)]-	13902-54-0	NI24569
250 153 95 109 121 93 123 97	**250**	100 79 71 66 54 52 47 46		15	22	3	–	–	–	–	–	–	–	–	–	–	Naphtho[1,2-b]furan-2,8(3H,4H)-dione, octahydro-3,5a,9-trimethyl-, [3S-(3α,3aα,5aβ,9α,9aα,9bβ)]-	13902-54-0	NI24570
121 97 95 109 123 250 107 93	**250**	100 79 73 65 58 55 52 52		15	22	3	–	–	–	–	–	–	–	–	–	–	Naphtho[1,2-b]furan-2,8(3H,4H)-dione, octahydro-3,5a,9-trimethyl-, [3S-(3α,3aα,5aβ,9β,9aβ,9bβ)]-	17956-11-5	NI24571
43 85 57 151 41 39 29 83	**250**	100 97 89 77 73 47 45 39	**10.48**	15	22	3	–	–	–	–	–	–	–	–	–	–	2-Pentanone, 4-methyl-1-(2,3,4,5-tetrahydro-5-methyl[2,3′-bifuran]-5-yl)-	20007-82-3	NI24572
43 93 41 250 109 108 55 235	**250**	100 81 70 53 50 50 47 42		15	22	3	–	–	–	–	–	–	–	–	–	–	Sambucoin		NI24573
201 89 91 75 117 118 47 202	**250**	100 71 59 39 35 28 22 19	**0.00**	15	26	1	–	–	–	–	–	–	1	–	–	–	Silane, triethyl(3-phenylpropoxy)-	2290-40-6	NI24574
83 250 210 153 69 126 122 95	**250**	100 50 43 22 20 19 19 19		15	26	1	2	–	–	–	–	–	–	–	–	–	Azacyclotridecane-1-propanenitrile, 2-oxo-	67171-80-6	NI24575
72 58 73 44 42 71 70 56	**250**	100 14 9 8 6 5 3 3	**0.06**	15	26	1	2	–	–	–	–	–	–	–	–	–	2-Propanamine, N,N-dimethyl-1-[4-[2-(dimethylamino)ethoxy]phenyl]-		MS05440
250 221 222 193 205 176 194	**250**	100 45 35 8 7 7 6		16	10	3	–	–	–	–	–	–	–	–	–	–	1,8-Epoxycarbonyloxymethanophenanthrene		LI05073
28 44 178 250 14 179 40 32	**250**	100 22 5 3 3 2 2 2		16	10	3	–	–	–	–	–	–	–	–	–	–	Maleic anhydride, diphenyl-		IC03284
205 222 176 177 88 250 221 206	**250**	100 89 54 51 51 49 49 20		16	10	3	–	–	–	–	–	–	–	–	–	–	Phenanthro[4,5-cde]oxepin-6-(4H)-one, 4-hydroxy-		MS05441
222 250 249 221 223 91 251 194	**250**	100 88 77 75 18 15 11 11		16	14	1	2	–	–	–	–	–	–	–	–	–	1H-1,4-Benzodiazepin-2-one, 2,3-dihydro-7-methyl-5-phenyl-	5571-63-1	NI24576
208 250 207 77 209 104 78 51	**250**	100 48 32 17 13 12 7 6		16	14	1	2	–	–	–	–	–	–	–	–	–	1H-1,5-Benzodiazepin-2-one, 2,3-dihydro-7-methyl-5-phenyl-	64376-02-9	NI24577
208 250 207 77 209 104 78 51	**250**	100 47 32 19 12 12 8 5		16	14	1	2	–	–	–	–	–	–	–	–	–	1H-1,5-Benzodiazepin-2-one, 2,3-dihydro-8-methyl-4-phenyl-	64376-00-7	NI24578
222 249 250 223 221 77 91 207	**250**	100 90 70 40 33 20 17 16		16	14	1	2	–	–	–	–	–	–	–	–	–	2H-1,4-Benzodiazepin-2-one, 1,3-dihydro-1-methyl-5-phenyl-		MS05442
250 249 251 131 221 178 179 124	**250**	100 30 17 15 12 12 8 8		16	14	1	2	–	–	–	–	–	–	–	–	–	3-Cinnolinol, 4-(4-methylbenzyl)-		AI01791
250 180 221 104 207 165 42 166	**250**	100 82 80 76 75 72 66 60		16	14	1	2	–	–	–	–	–	–	–	–	–	2-Imidazolin-4-one, 5,5-diphenyl-2-methyl-	24133-90-2	NI24579
107 144 250 89 116 106 143 115	**250**	100 74 71 42 15 14 13 12		16	14	1	2	–	–	–	–	–	–	–	–	–	1H-Indole-2-carboxamide, N-(3-tolyl)-		LI05074
107 250 144 89 106 116 115 143	**250**	100 52 51 35 15 12 12 11		16	14	1	2	–	–	–	–	–	–	–	–	–	1H-Indole-2-carboxamide, N-(4-tolyl)-		LI05075
107 144 250 116 89 106 143 115	**250**	100 83 35 19 16 9 5 5		16	14	1	2	–	–	–	–	–	–	–	–	–	1H-Indole-3-carboxamide, N-(3-tolyl)-		LI05076
107 144 250 116 106 89 115 143	**250**	100 54 20 16 15 15 5 4		16	14	1	2	–	–	–	–	–	–	–	–	–	1H-Indole-3-carboxamide, N-(4-tolyl)-		LI05077
133 250 132 105 104 251 221 77	**250**	100 89 42 28 18 16 15 12		16	14	1	2	–	–	–	–	–	–	–	–	–	2-Indolinone, 1-(4-methylbenzylideneamino)-		AI01792
250 132 77 118 221 251 119 103.5	**250**	100 70 53 52 22 19 19 19		16	14	1	2	–	–	–	–	–	–	–	–	–	2-Indolinone, 1-(α-methylbenzylideneamino)-		AI01793
77 105 104 249 119 250 130 78	**250**	100 84 84 68 63 62 45 30		16	14	1	2	–	–	–	–	–	–	–	–	–	2-Pyrimidinone, 3,6-diphenyl-1,2,3,4-tetrahydro-		LI05078

MASS TO CHARGE RATIOS	M.W.	INTENSITIES	Parent	C	H	O	N	S	F	Cl	Br	I	Si	P	B	X	COMPOUND NAME	CAS Reg No	No
173 249 77 104 250	250	100 38 35 23 22		16	14	1	2	–	–	–	–	–	–	–	–	–	2-Pyrimidinone, 4,6-diphenyl-1,2,3,4-tetrahydro-		LI05079
235 250 91 233 65 132 236 76	250	100 50 41 32 23 19 18 16		16	14	1	2	–	–	–	–	–	–	–	–	–	4(3H)-Quinazolinone, 2-methyl-3-(2-methylphenyl)-	72-44-6	MS05443
235 250 91 233 65 236 76 40	250	100 40 37 29 22 20 18 18		16	14	1	2	–	–	–	–	–	–	–	–	–	4(3H)-Quinazolinone, 2-methyl-3-(2-methylphenyl)-	72-44-6	NI24580
235 91 250 233 236 76 65 132	250	100 47 43 29 17 17 17 15		16	14	1	2	–	–	–	–	–	–	–	–	–	4(3H)-Quinazolinone, 2-methyl-3-(2-methylphenyl)-	72-44-6	NI24581
250 91 249 173 119 65 92 251	250	100 85 63 63 47 37 23 18		16	14	1	2	–	–	–	–	–	–	–	–	–	4(3H)-Quinazolinone, 2-phenylethyl-		LI05080
233 250 249 221 234 77 76 232	250	100 93 88 64 56 45 35 24		16	14	1	2	–	–	–	–	–	–	–	–	–	Quinoxaline, 2-ethyl-3-phenyl-, 4-oxide	16007-76-4	NI24582
163 107 164 105 29 41 235 147	250	100 29 14 11 9 8 7 7	6.00	16	26	2	–	–	–	–	–	–	–	–	–	–	1-Adamantaneacetic acid, ethyl ester		IC03285
208 151 167 43 209 123 41 250	250	100 36 32 29 15 15 15 9		16	26	2	–	–	–	–	–	–	–	–	–	–	4-Azulenol, 6-tert-butyl-1,2,3,5,6,7,8,8a-octahydro-, acetate, anti-		NI24583
208 151 167 43 41 209 123 250	250	100 39 26 26 17 16 15 9		16	26	2	–	–	–	–	–	–	–	–	–	–	4-Azulenol, 6-tert-butyl-1,2,3,5,6,7,8,8a-octahydro-, acetate, syn-		NI24584
221 29 43 41 27 250 55 222	250	100 43 38 35 29 18 18 16		16	26	2	–	–	–	–	–	–	–	–	–	–	1,4-Benzenediol, 2,5-bis(1,1-dimethylpropyl)-	79-74-3	WI01279
175 193 77 43 135 41 121 161	250	100 87 56 54 53 53 51 50	0.00	16	26	2	–	–	–	–	–	–	–	–	–	–	Benzenemethanol, α,α-di-tert-butyl-2-methoxy-		MS05444
81 127 123 99 126 124 167 250	250	100 91 69 37 33 31 18 10		16	26	2	–	–	–	–	–	–	–	–	–	–	Cyclobuta[1,2:3,4]dicyclopenten-1(2H)-one, 3,3a,3b,4,5,6,6a,6b-octahydro-6-isopropyl-3b-methyl-, (3aα,3bβ,6α,6β)-, ethylene ketal		LI05081
141 109 107 250 81 108 121 82	250	100 97 77 53 43 41 38 37		16	26	2	–	–	–	–	–	–	–	–	–	–	2-Cyclohexene-1-carboxylic acid, 2,6-dimethyl-6-(4-methyl-3-pentenyl)-, methyl ester	38142-31-3	NI24585
107 135 167 109 165 250 68 69	250	100 78 75 71 57 53 52 50		16	26	2	–	–	–	–	–	–	–	–	–	–	2-Cyclohexene-1-carboxylic acid, 2,6-dimethyl-6-(4-methyl-4-pentenyl)-, methyl ester	38142-32-4	NI24586
43 123 177 41 69 149 93 67	250	100 38 30 25 15 13 12 12	0.00	16	26	2	–	–	–	–	–	–	–	–	–	–	Cyclopentanemethanol, α-[[3,3-dimethyl-2-isopropyloxiranyl]ethynyl]-α-methyl-	63954-68-7	NI24587
69 41 114 81 121 82 55 53	250	100 39 29 19 13 10 8 8	2.00	16	26	2	–	–	–	–	–	–	–	–	–	–	2,6,10-Dodecatrienoic acid, 3,7,11-trimethyl-, methyl ester, (E,E)-	3675-00-1	NI24588
69 41 114 81 121 82 67 53	250	100 62 40 28 24 14 13 12	2.89	16	26	2	–	–	–	–	–	–	–	–	–	–	2,6,10-Dodecatrienoic acid, 3,7,11-trimethyl-, methyl ester, (E,E)-	3675-00-1	NI24589
69 41 114 81 82 121 55 53	250	100 41 30 29 12 11 9 9	2.00	16	26	2	–	–	–	–	–	–	–	–	–	–	2,6,10-Dodecatrienoic acid, 3,7,11-trimethyl-, methyl ester, (E,Z)-	4176-79-8	NI24590
43 69 147 121 109 93 81 175	250	100 65 20 18 17 17 15 8	0.00	16	26	2	–	–	–	–	–	–	–	–	–	–	1,6,10-Dodecatrien-1-ol, 7,11-dimethyl-, acetate, (6E)-		DD00695
43 69 147 95 121 81 109 175	250	100 65 25 18 16 16 15 7	0.00	16	26	2	–	–	–	–	–	–	–	–	–	–	1,6,10-Dodecatrien-1-ol, 7,11-dimethyl-, acetate, (6Z)-		DD00696
43 192 59 177 149 41 235 93	250	100 53 42 30 30 20 16 14	0.00	16	26	2	–	–	–	–	–	–	–	–	–	–	3,4-Heptadien-2-one, 3-cyclopentyl-6-hydroxy-6-methyl-5-isopropyl-	63922-48-5	NI24591
95 55 41 138 81 67 123 69	250	100 51 33 28 27 23 17 17	1.00	16	26	2	–	–	–	–	–	–	–	–	–	–	3-Heptene, 1-(1,2-epoxycyclopentylcarbonyl)-4-propyl-		LI05082
149 43 192 123 59 177 164 41	250	100 68 59 50 50 27 25 18	0.00	16	26	2	–	–	–	–	–	–	–	–	–	–	3-Heptyn-2-one, 5-cyclopentyl-6-hydroxy-6-methyl-5-isopropyl-	63922-46-3	NI24592
43 138 123 139 181 41 95 69	250	100 74 40 34 20 20 16 16	0.00	16	26	2	–	–	–	–	–	–	–	–	–	–	4-Heptyn-2-one, 6-cyclopentyl-6-hydroxy-3-methyl-3-isopropyl-	63922-47-4	NI24593
137 123 114 69 81 95 109 41	250	100 99 70 59 58 56 55 51	14.00	16	26	2	–	–	–	–	–	–	–	–	–	–	1-Naphthalenecarboxylic acid, decahydro-5,5,8a-trimethyl-2-methylene-, methyl ester, (1β,4aα,8aβ)-	81752-89-8	NI24594
250 111 207 96 41 235 222 251	250	100 75 62 39 23 22 22 18		16	26	2	–	–	–	–	–	–	–	–	–	–	Naphtho[8a,1-b]furan-2-one, 3a,4,7,7-tetramethylperhydro-, (3aR,4S,6aR,10aS)-		NI24595
43 41 123 82 69 95 55 81	250	100 78 66 59 54 51 50 47	14.00	16	26	2	–	–	–	–	–	–	–	–	–	–	Naphtho[2,1-b]furan-2(1H)-one, decahydro-3a,6,6,9a-tetramethyl-, [3aR-(3aα,5aβ,9aα,9bβ)]-	564-20-5	NI24596
43 41 123 82 81 54 95 125	250	100 80 67 50 50 50 47 42	15.00	16	26	2	–	–	–	–	–	–	–	–	–	–	Naphtho[2,1-b]furan-2(1H)-one, decahydro-3a,6,6,9a-tetramethyl-, [3aR-(3aα,5aβ,9aα,9bβ)]-	564-20-5	LI05083
43 41 123 82 81 95 54 125	250	100 81 68 51 51 50 50 42	13.01	16	26	2	–	–	–	–	–	–	–	–	–	–	Naphtho[2,1-b]furan-2(1H)-one, decahydro-3a,6,6,9a-tetramethyl-, [3aR-(3aα,5aβ,9aα,9bβ)]-	564-20-5	NI24597
136 235 81 123 121 69 95 109	250	100 39 22 19 19 18 13 11	10.00	16	26	2	–	–	–	–	–	–	–	–	–	–	Naphtho[2,1-b]furan-2(1H)-one, decahydro-3a,6,6,9a-tetramethyl-, [3aS-(3aα,5aα,9aβ,9bα)]-	30450-17-0	LI05084
136 235 137 81 123 121 69 124	250	100 39 23 23 22 21 20 17	11.00	16	26	2	–	–	–	–	–	–	–	–	–	–	Naphtho[2,1-b]furan-2(1H)-one, decahydro-3a,6,6,9a-tetramethyl-, [3aS-(3aα,5aα,9aβ,9bα)]-	30450-17-0	NI24598
136 235 81 123 121 69 109 95	250	100 38 22 20 20 18 12 12	10.01	16	26	2	–	–	–	–	–	–	–	–	–	–	Naphtho[2,1-b]furan-2(1H)-one, decahydro-3a,6,6,9a-tetramethyl-, [3aS-(3aα,5aα,9aβ,9bα)]-	30450-17-0	NI24599
235 45 250 57 236 41 161 147	250	100 32 25 16 15 11 10 10		16	26	2	–	–	–	–	–	–	–	–	–	–	Phenol, 2,6-di-tert-butyl-4-(methoxymethyl)-	87-97-8	IC03287
235 250 45 219 236 57 41 193	250	100 35 26 20 18 16 13 9		16	26	2	–	–	–	–	–	–	–	–	–	–	Phenol, 2,6-di-tert-butyl-4-(methoxymethyl)-	87-97-8	IC03286
235 45 250 57 219 161 91 236	250	100 56 29 28 22 20 18 16		16	26	2	–	–	–	–	–	–	–	–	–	–	Phenol, 2,6-di-tert-butyl-4-(methoxymethyl)-	87-97-8	NI24600
55 69 140 205 250	250	100 75 58 37 22		16	26	2	–	–	–	–	–	–	–	–	–	–	Tetradecatrienoic acid, ethyl ester, (2E,4E,8Z)-		LI05085
83 41 55 109 165 95 69 81	250	100 78 59 51 47 43 42 33	1.21	16	31	1	–	–	–	–	–	–	–	–	1	–	Borinic acid, diethyl-, 1-cyclododecen-1-yl ester	61142-73-2	NI24601
250 235 234 251 117 236 249 248	250	100 82 45 20 17 16 11 11		17	14	–	–	1	–	–	–	–	–	–	–	–	Naphtho[2,3-b]benzothiopene, 9,10-dihydro-7-methyl-		AI01794
250 121 221 249 223 149 165 202	250	100 12 11 10 10 8 6 5		17	14	2	–	–	–	–	–	–	–	–	–	–	Benzofuran, 2-(4-hydroxyphenyl)-5-propenyl-, (5E)-		MS05445
250 178 251 207 179 249 222 221	250	100 55 38 32 32 28 21 19		17	14	2	–	–	–	–	–	–	–	–	–	–	5,10-Methano-5H-dibenzo[a,d]cyclohepten-11-one, 10,11-dihydro-2-methoxy-		NI24602
250 178 179 207 219 251 249 222	250	100 49 35 25 19 18 18 17		17	14	2	–	–	–	–	–	–	–	–	–	–	5,10-Methano-5H-dibenzo[a,d]cyclohepten-11-one, 10,11-dihydro-7-methoxy-		NI24603

MASS TO CHARGE RATIOS								M.W.	INTENSITIES								Parent	C	H	O	N	S	F	Cl	Br	I	Si	P	B	X	COMPOUND NAME	CAS Reg No	No
250	178	251	207	179	222	191	249	**250**	100	30	20	20	16	11	11	9		17	14	2	–	–	–	–	–	–	–	–	–	–	5,11-Methano-5H-dibenzo[a,d]cyclohepten-10-one, 10,11-dihydro-2-methoxy-		NI24604
250	249	251	207	179	219	189	191	**250**	100	79	67	60	47	41	26	21		17	14	2	–	–	–	–	–	–	–	–	–	–	5,11-Methano-5H-dibenzo[a,d]cyclohepten-10-one, 10,11-dihydro-7-methoxy-		NI24605
105	250	77	207	43	251	51	179	**250**	100	70	57	47	22	13	13	12		17	14	2	–	–	–	–	–	–	–	–	–	–	4-Pentene-1,3-dione, 1,5-diphenyl-		LI05086
250	191	208	189	179	190	43	178	**250**	100	95	89	59	43	38	21	20		17	14	2	–	–	–	–	–	–	–	–	–	–	9-Phenanthrenemethanol, acetate	53440-11-2	NI24606
43	77	117	182	132	44	104	57	**250**	100	79	79	77	51	45	43	43	**19.94**	17	18	–	2	–	–	–	–	–	–	–	–	–	2-Azetidinimine, 3,3-dimethyl-1,4-diphenyl-	62834-02-0	NI24607
132	105	91	133	77	117	90	65	**250**	100	13	12	11	5	4	4	4	**0.00**	17	18	–	2	–	–	–	–	–	–	–	–	–	Benzeneacetonitrile, α-[(1-methyl-2-phenylethyl)amino]-	17590-01-1	NI24608
132	105	91	77	65	89	79	90	**250**	100	50	42	23	21	16	15	13	**0.00**	17	18	–	2	–	–	–	–	–	–	–	–	–	Benzeneacetonitrile, α-[(1-methyl-2-phenylethyl)amino]-	17590-01-1	NI24609
250	249	77	120	251	235	117	131	**250**	100	38	25	21	19	16	16	14		17	18	–	2	–	–	–	–	–	–	–	–	–	6H,12H-5,11-Methanodibenzo[b,f][1,5]diazocine, 2,8-dimethyl-		IC03288
55	41	81	250	80	95	82	69	**250**	100	90	59	47	44	40	36	36		17	30	1	–	–	–	–	–	–	–	–	–	–	9-Cycloheptadecen-1-one, (Z)-	542-46-1	WI01280
79	67	80	41	91	93	81	55	**250**	100	97	80	68	65	57	57	55	**9.04**	17	30	1	–	–	–	–	–	–	–	–	–	–	5,8,11-Heptadecatrien-1-ol	22117-09-5	NI24610
250	41	111	55	69	81	123	95	**250**	100	96	92	90	89	83	80	80		17	30	1	–	–	–	–	–	–	–	–	–	–	1-Phenanthrenol, tetradecahydro-4b,8,8-trimethyl-, [1R-(1α,4aβ,4bα,8aβ,10aα)]-	5720-71-8	NI24611
232	123	217	41	147	55	250	81	**250**	100	90	89	89	84	77	74	73		17	30	1	–	–	–	–	–	–	–	–	–	–	3-Phenanthrenol, tetradecahydro-4b,8,8-trimethyl-, [3R-(3α,4aα,4bβ,8aα,10aβ)]-	54155-12-3	NI24612
217	41	121	55	148	147	250	69	**250**	100	97	93	89	88	86	82	80		17	30	1	–	–	–	–	–	–	–	–	–	–	1-Phenanthrenol, tetradecahydro-4b,8,8-trimethyl-, [1S-(1α,4aα,4bβ,8aα,10aβ)]-	57760-27-7	NI24613
123	232	41	217	55	69	81	95	**250**	100	79	77	66	64	59	58	55	**37.00**	17	30	1	–	–	–	–	–	–	–	–	–	–	3-Phenanthrenol, tetradecahydro-4b,8,8-trimethyl-, [3S-(3α,4aβ,4bα,8aβ,10aα)]-	54155-13-4	NI24614
104	146	250	145	103	77	78	251	**250**	100	96	69	51	38	29	25	15		18	18	1	–	–	–	–	–	–	–	–	–	–	4-Acetyl[2.2]paracyclophane		MS05446
206	193	180	115	165	91	179	167	**250**	100	98	66	57	51	40	38	37	**30.00**	18	18	1	–	–	–	–	–	–	–	–	–	–	Cyclohexanone, 2,2-diphenyl-	22612-62-0	NI24615
209	131	103	210	181	178	91	165	**250**	100	28	24	19	18	17	17	16	**3.00**	18	18	1	–	–	–	–	–	–	–	–	–	–	5H-Dibenzo[a,d]cyclohepten-5-ol, 5-cyclopropyl-10,11-dihydro-	3241-97-2	NI24616
130	129	131	77	250	115	105	91	**250**	100	34	24	24	23	19	19	9		18	18	1	–	–	–	–	–	–	–	–	–	–	5-Hexenophenone, 6-phenyl-	32363-56-7	NI24617
91	131	159	250	115	65	77	129	**250**	100	50	22	15	15	14	8	7		18	18	1	–	–	–	–	–	–	–	–	–	–	3-Penten-2-one, 1,5-diphenyl-4-methyl-		NI24618
159	91	144	141	131	129	65	115	**250**	100	43	26	20	20	12	12	11	**0.00**	18	18	1	–	–	–	–	–	–	–	–	–	–	3-Penten-2-one, 3,5-diphenyl-4-methyl-	81826-04-2	NI24619
91	131	115	159	65	250	129	77	**250**	100	39	15	14	14	12	7	7		18	18	1	–	–	–	–	–	–	–	–	–	–	4-Penten-2-one, 1,5-diphenyl-4-methyl-	81826-05-3	NI24620
105	28	77	250	235	43	249	91	**250**	100	86	80	78	75	28	27	25		18	18	1	–	–	–	–	–	–	–	–	–	–	2-Penten-1-one, 2-methyl-1,3-diphenyl-	53190-02-6	NI24621
105	235	250	77	28	249	43	91	**250**	100	85	80	79	50	33	29	27		18	18	1	–	–	–	–	–	–	–	–	–	–	2-Penten-1-one, 2-methyl-1,4-diphenyl-	61142-08-3	NI24622
207	77	105	235	250	249	131	208	**250**	100	33	32	26	25	24	18	17		18	18	1	–	–	–	–	–	–	–	–	–	–	2-Propen-1-one, 3-[4-isopropylphenyl]-1-phenyl-	36336-80-8	NI24623
55	83	41	97	166	69	82	111	**250**	100	93	87	72	44	41	37	34	**0.69**	18	34	–	–	–	–	–	–	–	–	–	–	–	Hexane, 1,1-dicyclohexyl-		AI01795
83	82	55	41	67	96	69	43	**250**	100	74	61	44	17	11	11	8	**5.53**	18	34	–	–	–	–	–	–	–	–	–	–	–	Hexane, 1,6-dicyclohexyl-		AI01796
67	41	81	82	68	43	95	55	**250**	100	62	61	57	54	54	52	50	**0.29**	18	34	–	–	–	–	–	–	–	–	–	–	–	3-Octadecyne	61886-64-4	NI24624
81	67	54	55	95	41	82	43	**250**	100	86	75	67	59	57	47	47	**0.31**	18	34	–	–	–	–	–	–	–	–	–	–	–	5-Octadecyne	71899-42-8	NI24625
81	67	55	95	82	41	54	43	**250**	100	98	84	67	63	63	59	53	**0.41**	18	34	–	–	–	–	–	–	–	–	–	–	–	9-Octadecyne	35365-59-4	NI24626
55	83	41	97	69	111	207	43	**250**	100	90	77	67	34	30	25	20	**1.23**	18	34	–	–	–	–	–	–	–	–	–	–	–	Propane, 1,3-dicyclohexyl-2-propyl-		AI01797
193	115	91	250	180	178	165	167	**250**	100	64	42	36	35	26	24	23		19	22	–	–	–	–	–	–	–	–	–	–	–	1-Heptene, 1,1-diphenyl-		AI01798
193	115	250	91	180	178	165	194	**250**	100	60	40	40	35	25	22	18		19	22	–	–	–	–	–	–	–	–	–	–	–	1-Heptene, 1,1-diphenyl-		AI01799
216	184	251	69	115	63	77	36	**251**	100	50	43	30	16	13	8	6		–	–	–	1	1	2	3	–	–	–	2	–	–	Phosphoramidothioic dichloride, N-(chlorodifluorophosphine)-		LI05087
251	101	231	42	53	253	233	69	**251**	100	49	41	41	37	20	17	17		7	5	–	5	–	3	1	–	–	–	–	–	–	Pyrimido[5,4-e]-1,2,4-triazine, 7-chloro-1,2-dihydro-3-methyl-5-(trifluoromethyl)-	32709-20-9	NI24627
251	101	231	42	53	253	233	69	**251**	100	54	45	45	41	22	18	18		7	5	–	5	–	3	1	–	–	–	–	–	–	Pyrimido[5,4-e]-1,2,4-triazine, 7-chloro-1,2-dihydro-3-methyl-5-(trifluoromethyl)-	32709-20-9	NI24628
124	42	41	44	27	39	67	28	**251**	100	65	32	21	21	20	15	15	**7.90**	8	14	–	1	–	–	–	–	1	–	–	–	–	1-Azabicyclo[2.2.2]octane, 4-(iodomethyl)-		NI24629
60	73	127	57	55	43	41	85	**251**	100	70	66	59	43	42	35	19	**0.54**	8	16	2	–	–	–	–	–	–	–	–	–	1	Octanoic acid, silver(1+) salt	24927-67-1	NI24630
205	142	169	170	143	140	207	105	**251**	100	60	46	36	36	33	26	25	**1.00**	9	6	2	5	–	–	1	–	–	–	–	–	–	1-Cyclopenten-2-amine, 3-chloro-4-methyl-3-nitro-1,5,5-tricyano-		MS05447
134	91	106	77	65	253	251	135	**251**	100	53	20	13	11	10	10	9		9	8	1	1	–	–	3	–	–	–	–	–	–	Acetamide, 2,2,2-trichloro-N-(2-methylphenyl)-	4257-87-8	NI24631
134	91	106	77	65	135	78	51	**251**	100	67	29	23	18	10	9	9	**9.00**	9	8	1	1	–	–	3	–	–	–	–	–	–	Acetamide, 2,2,2-trichloro-N-(3-methylphenyl)-	2563-96-4	NI24632
134	106	91	77	251	253	65	79	**251**	100	55	30	17	14	13	11	10		9	8	1	1	–	–	3	–	–	–	–	–	–	Acetamide, 2,2,2-trichloro-N-(4-methylphenyl)-	2564-09-2	NI24633
43	28	126	84	176	72	27	29	**251**	100	99	80	70	37	35	35	31	**0.00**	9	17	5	1	1	–	–	–	–	–	–	–	–	α-D-Arabinofuranoside, ethyl 2-deoxy-2-[(methoxycarbonyl)amino]-1-thio-	56978-64-4	NI24634
43	133	59	60	206	71	144	72	**251**	100	88	88	57	46	30	24	21	**0.00**	9	17	5	1	1	–	–	–	–	–	–	–	–	D-Galactopyranoside, methyl 2-(acetylamino)-2-deoxy-1-thio-	56847-07-5	NI24635
168	126	84	169	201	98	60	109	**251**	100	46	20	13	11	9	9	8	**0.00**	9	17	5	1	1	–	–	–	–	–	–	–	–	1-Methylsulphonyl-2-acetamido-4-dihydroxyethylarabinose		NI24636
126	43	84	42	41	72	176	29	**251**	100	90	75	60	50	47	44	35	**0.00**	9	17	5	1	1	–	–	–	–	–	–	–	–	β-D-Xylofuranoside, ethyl 2-deoxy-2-[(methoxycarbonyl)amino]-1-thio-	56909-08-1	NI24637

MASS TO CHARGE RATIOS								M.W.	INTENSITIES								Parent	C	H	O	N	S	F	Cl	Br	I	Si	P	B	X	COMPOUND NAME	CAS Reg No	No
126	251	253	114	207	63	75	74	251	100	55	54	30	20	20	18	16		10	6	2	1	–	–	–	1	–	–	–	–	–	Naphthalene, 1-bromo-5-nitro-		IC03289
126	252	250	114	63	207	205	195	251	100	56	51	28	21	20	20	18	7.84	10	6	2	1	–	–	–	1	–	–	–	–	–	Naphthalene, 1-bromo-5-nitro-		MS05448
172	77	103	91	173	51	64	251	251	100	19	16	12	12	10	10	7		10	10	–	3	–	–	–	1	–	–	–	–	–	1,2,3-Triazole, 4-(bromomethyl)-5-methyl-2-phenyl-		NI24638
202	204	140	27	63	138	111	77	251	100	98	53	41	40	32	25	24	15.00	10	12	–	1	–	–	3	–	–	–	–	–	–	Aniline, N,N-bis(2-chloroethyl)-2-chloro-		IC03290
222	251	69	198	95	236	97	208	251	100	64	46	37	29	26	26	22		10	13	1	5	1	–	–	–	–	–	–	–	–	4H-Thiazolo[3,2-a]-1,3,5-triazine-6-carbonitrile, 6,7-dihydro-2-diethylamino-4-oxo-	81287-24-3	NI24639
178	148	149	135	43	220	108	136	251	100	74	69	66	45	37	36	29	0.50	10	13	3	5	–	–	–	–	–	–	–	–	–	Adenine, 9-[2-hydroxy-1-(2-hydroxy-1-methyleneethoxy)ethyl]-		NI24640
135	162	136	108	43	164	251	45	251	100	33	21	14	14	10	9	9		10	13	3	5	–	–	–	–	–	–	–	–	–	Adenosine, 2′-deoxy-	958-09-8	NI24641
135	162	136	108	164	73	45	43	251	100	29	27	17	10	9	9	8	3.18	10	13	3	5	–	–	–	–	–	–	–	–	–	Adenosine, 2′-deoxy-	958-09-8	NI24642
135	162	136	108	178	164	45	43	251	100	37	30	20	13	12	12	11	3.80	10	13	3	5	–	–	–	–	–	–	–	–	–	Adenosine, 2′-deoxy-	958-09-8	NI24643
135	164	44	43	56	136	108	41	251	100	49	41	39	36	34	28	28	7.94	10	13	3	5	–	–	–	–	–	–	–	–	–	Adenosine, 3′-deoxy-	73-03-0	NI24644
135	164	136	108	57	73	43	178	251	100	97	96	23	21	20	19	18	8.01	10	13	3	5	–	–	–	–	–	–	–	–	–	Adenosine, 5′-deoxy-	4754-39-6	NI24645
135	162	136	108	45	73	43	191	251	100	37	30	20	15	11	10	7	4.20	10	13	3	5	–	–	–	–	–	–	–	–	–	9H-Purin-6-amine, 9-(2-deoxy-α-D-erythro-pentofuranosyl)-	3413-66-9	NI24646
135	178	164	136	108	57	43	41	251	100	77	75	53	36	36	16	13	5.00	10	13	3	5	–	–	–	–	–	–	–	–	–	9H-Purin-6-amine, 9-(5-deoxy-α-D-xylofuranosyl)-	55670-08-1	NI24647
251	45	236	44	89	195	58	182	251	100	31	28	24	18	12	12	11		10	17	1	7	–	–	–	–	–	–	–	–	–	1,3,5-Triazine-2,4-diamine, 6-(2-imino-3-oxazolidinyl)-N,N,N′,N′-tetramethyl-	87166-33-4	NI24648
251	208	96	45	206	236	163	166	251	100	60	50	35	34	28	26	25		10	17	1	7	–	–	–	–	–	–	–	–	–	1,3,5-Triazine-2,4,6-triamine, N-cyano-N-methoxymethyl-N′,N′,N″,N″-tetramethyl-	87166-09-4	NI24649
218	98	251	79	56	41	176	128	251	100	88	39	20	20	18	16	14		10	22	2	1	1	–	–	–	–	–	1	–	–	Phosphoramidothioic acid, N-cyclohexyl-, methyl isopropyl diester		LI05088
251	206	123	29	139	223	151	252	251	100	97	53	53	48	44	43	29		11	10	2	3	–	–	1	–	–	–	–	–	–	1,2,3-Triazole-4-carboxylic acid, 5-(4-chlorophenyl)-, ethyl ester		NI24650
178	193	251	177	179	145	221	146	251	100	67	36	19	17	15	12	12		11	13	2	3	1	–	–	–	–	–	–	–	–	Sydnone, 3-[4-(dimethylamino)phenyl]-4-(methylthio)-	69978-08-1	NI24651
172	77	103	173	91	51	64	251	251	100	18	15	12	9	7	6	5		11	13	2	3	1	–	–	–	–	–	–	–	–	1,2,3-Triazole, 4-(methylsulphonylmethyl)-5-methyl-2-phenyl-		NI24652
177	251	205	165	121	206	80	178	251	100	52	27	27	14	13	12	11		11	13	4	3	–	–	–	–	–	–	–	–	–	3H-Imidazo[1,5-b]pyridazine-5,7-dione, 2,6-dimethyl-4-(ethoxycarbonyl)-		NI24653
251	179	206	223	237	149	252	166	251	100	51	35	30	23	19	16	12		11	13	4	3	–	–	–	–	–	–	–	–	–	Pyrrolo[1,2-a]-1,3,5-triazine-7-carboxylic acid, 1,2,3,4-tetrahydro-1,3-dimethyl-2,4-dioxo-, ethyl ester	54449-29-5	NI24654
90	178	42	234	190	106	65	91	251	100	40	26	21	17	11	10	9	1.10	11	17	2	5	–	–	–	–	–	–	–	–	–	Triazane, 1,3-bis[1-(hydroxyimino)ethyl]-2-benzyl-	56700-85-7	MS05449
252	128	127	250	236	129	206	114	251	100	62	19	12	8	7	2	2	0.00	11	26	3	1	–	–	–	–	–	–	1	–	–	Phosphonic acid, methyl-, ethyl diisopropylaminoethyl diester		NI24655
114	70	192	72	112	84	127	43	251	100	57	34	34	32	31	22	16	0.00	11	26	3	1	–	–	–	–	–	–	1	–	–	Phosphonic acid, methyl-, ethyl diisopropylaminoethyl diester		MS05450
43	251	70	236	104	105	44	202	251	100	29	23	22	22	16	15	14		12	10	1	1	1	–	1	–	–	–	–	–	–	Thiazole, 1-acetyl-4-(chloromethyl)-2-phenyl-	41981-05-9	NI24656
159	251	131	59	103	71	161	252	251	100	61	53	28	22	11	10	10		12	13	1	1	2	–	–	–	–	–	–	–	–	1,4-Dithiine-3-carboxamide, 5,6-dihydro-2-methyl-N-phenyl-		MS05451
104	130	43	147	76	41	57	50	251	100	92	92	88	88	77	50	46	1.00	12	13	3	1	1	–	–	–	–	–	–	–	–	1H-Isoindole-1,3(2H)-dione, 2-(butylsulphinyl)-	40318-14-7	NI24657
43	39	131	28	65	159	27	93	251	100	29	26	26	25	18	18	13	4.70	12	13	3	1	1	–	–	–	–	–	–	–	–	1,4-Oxathiin-3-carboxamide, 5,6-dihydro-2-methyl-N-phenyl-, 4-oxide	17757-70-9	NI24658
131	143	159	93	87	65	77	175	251	100	98	87	79	61	56	50	39	33.10	12	13	3	1	1	–	–	–	–	–	–	–	–	1,4-Oxathiin-3-carboxamide, 5,6-dihydro-2-methyl-N-phenyl-, 4-oxide	17757-70-9	NI24659
179	29	28	43	18	44	132	27	251	100	72	40	39	32	18	16	16	4.96	12	13	5	1	–	–	–	–	–	–	–	–	–	Butanoic acid, 3-oxo-2-(p-nitrophenyl)-, ethyl ester		MS05452
192	193	132	104	251	92	220	133	251	100	13	11	9	8	8	7	7		12	13	5	1	–	–	–	–	–	–	–	–	–	2-Butenedioic acid, (6-methyl-2-pyridon-1-yl)-, dimethyl ester		LI05089
251	206	91	77	221	115	103	162	251	100	48	34	31	30	28	26	24		12	13	5	1	–	–	–	–	–	–	–	–	–	1H-Indene-1-one, 2,3-dihydro-5,6-dimethoxy-3-methyl-7-nitro-		NI24660
91	73	45	92	77	65	74	51	251	100	11	10	8	8	6	5	5	0.00	12	17	3	1	–	–	–	–	–	1	–	–	–	Acetic acid, [(benzyloxy)imino]-, trimethylsilyl ester	55494-08-1	NI24661
206	75	105	93	77	103	207	147	251	100	78	72	32	31	25	22	16	0.74	12	17	3	1	–	–	–	–	–	1	–	–	–	Glycine, N-benzoyl-, trimethylsilyl ester	2078-24-2	NI24662
105	73	77	75	45	51	206	47	251	100	93	83	43	40	38	37	17	0.10	12	17	3	1	–	–	–	–	–	1	–	–	–	Glycine, N-benzoyl-, trimethylsilyl ester	2078-24-2	NI24663
105	73	77	206	75	51	207	45	251	100	73	57	53	22	20	16	16	0.00	12	17	3	1	–	–	–	–	–	1	–	–	–	Glycine, N-benzoyl-, trimethylsilyl ester	2078-24-2	NI24664
208	163	209	251	138	71	113	46	251	100	39	38	25	25	25	20	17		12	21	1	5	–	–	–	–	–	–	–	–	–	Guanidine, N,N-dimethyl-N′-(4-hydroxy-5-butyl-6-methylpyrimid-2-yl)-		IC03291
251	173	145	182	78	51	106	231	251	100	81	63	62	41	33	30	18		13	8	1	1	–	3	–	–	–	–	–	–	–	Pyridine, 3-(3-trifluoromethylbenzoyl)-	59190-60-2	NI24665
78	145	51	173	182	106	251	232	251	100	92	92	89	81	78	11	11		13	8	1	1	–	3	–	–	–	–	–	–	–	Pyridine, 3-(4-trifluoromethylbenzoyl)-	21221-92-1	NI24666
122	149	56	163	109	251	42	58	251	100	72	67	66	49	44	28	28		13	16	2	2	–	1	–	–	–	–	–	–	–	3-Imidazoline, 2,2,5,5-tetramethyl-4-(p-fluorophenyl)-, 3-oxide-1-oxile	72353-18-5	NI24667
91	96	155	41	68	65	94	56	251	100	55	47	47	28	26	21	21	14.89	13	17	2	1	1	–	–	–	–	–	–	–	–	Benzenesulphonamide, N,N-diallyl-4-methyl-		IC03292
150	43	137	30	151	135	15	209	251	100	44	30	30	12	11	10	8	2.20	13	17	4	1	–	–	–	–	–	–	–	–	–	Acetamide, N-[2-[4-(acetyloxy)-3-methoxyphenyl]ethyl]-	55044-58-1	NI24668
150	43	137	30	151	135	28	15	251	100	44	30	30	12	11	11	10	2.20	13	17	4	1	–	–	–	–	–	–	–	–	–	Acetamide, N-[2-[4-(acetyloxy)-3-methoxyphenyl]ethyl]-	55044-58-1	MS05453
150	137	18	30	43	28	209	151	251	100	24	21	19	18	14	11	11	4.39	13	17	4	1	–	–	–	–	–	–	–	–	–	Acetamide, N-[2-[4-(acetyloxy)-3-methoxyphenyl]ethyl]-	55044-58-1	MS05454
204	43	91	41	119	39	65	77	251	100	37	33	33	25	25	23	22	14.00	13	17	4	1	–	–	–	–	–	–	–	–	–	1,2-Benzenedicarboxylic acid, 3-(N-isopropylamino)-, dimethyl ester		LI05090
188	219	56	204	187	159	160	220	251	100	97	38	37	30	29	28	23	22.00	13	17	4	1	–	–	–	–	–	–	–	–	–	1,3-Cyclopentadiene-1,3-dicarboxylic acid, 2-methyl-5-[1-(methylamino)ethylidene]-, dimethyl ester	13061-79-5	NI24669
113	43	85	41	57	114	55	69	251	100	82	20	13	12	9	8	4	2.97	13	17	4	1	–	–	–	–	–	–	–	–	–	Heptanoic acid, 3-nitrophenyl ester	56052-18-7	NI24670
43	71	58	100	28	77	151	41	251	100	58	46	29	28	21	20	18	0.12	13	17	4	1	–	–	–	–	–	–	–	–	–	2-Hexanone, 3-[hydroxy(2-nitrophenyl)methyl]-	62238-21-5	NI24671
105	122	77	28	146	177	106	148	251	100	30	21	17	11	7	7	6	0.20	13	17	4	1	–	–	–	–	–	–	–	–	–	lyxo-Hexose, 3-amino-2,3,6-trideoxy-, N-benzoyl-		NI24672
91	108	42	79	107	28	77	27	251	100	73	45	42	38	32	24	19	3.53	13	17	4	1	–	–	–	–	–	–	–	–	–	DL-Norvaline, N-carboxy-, N-benzyl ester,	21691-43-0	NI24673
164	151	93	192	132	220	117	104	251	100	79	71	68	40	22	18	16	0.90	13	17	4	1	–	–	–	–	–	–	–	–	–	2-Pyridinebutanoic acid, 3-(methoxycarbonyl)-α-methyl-, methyl ester	41173-81-3	NI24675

MASS TO CHARGE RATIOS								M.W.	INTENSITIES								Parent	C	H	O	N	S	F	Cl	Br	I	Si	P	B	X	COMPOUND NAME	CAS Reg No	No
204	165	192	236	79	132	117	104	**251**	100	97	63	49	49	48	47	45	**9.00**	13	17	4	1	–	–	–	–	–	–	–	–	–	2-Pyridinebutanoic acid, 3-(methoxycarbonyl)-α-methyl-, methyl ester	41173-81-3	NI24674
206	251	178	205	151	195	150	177	**251**	100	51	49	30	18	17	16	16		13	17	4	1	–	–	–	–	–	–	–	–	–	3,5-Pyridinedicarboxylic acid, 2,6-dimethyl-, diethyl ester	1149-24-2	NI24676
204	165	132	150	192	236	160	79	**251**	100	85	76	67	59	54	53	47	**9.00**	13	17	4	1	–	–	–	–	–	–	–	–	–	2-Pyridinepropanoic acid, 3-(methoxycarbonyl)-α,β-dimethyl-, methyl ester, [S-(R*,R*)]-	35721-84-7	NI24677
107	178	43	192	18	147	136	44	**251**	100	46	41	37	31	25	19	19	**0.80**	13	17	4	1	–	–	–	–	–	–	–	–	–	L-Tyrosine, N-acetyl-, ethyl ester	840-97-1	NI24678
91	108	101	79	107	42	28	77	**251**	100	64	42	31	30	24	21	18	**2.67**	13	17	4	1	–	–	–	–	–	–	–	–	–	DL-Valine, N-[(benzyloxy)carbonyl]-	3588-63-4	NI24679
73	160	192	45	91	161	59	89	**251**	100	99	21	16	13	11	11	10	**0.00**	13	21	2	1	–	–	–	–	–	1	–	–	–	Phenylalanine, N-trimethylsilyl-, methyl ester		NI24680
91	73	236	130	162	59	45	135	**251**	100	85	84	57	45	38	37	30	**12.70**	13	25	–	1	–	–	–	–	–	2	–	–	–	Benzylamine, N-trimethylsilyl-	18406-59-2	MS05455
236	73	251	162	237	174	135	91	**251**	100	35	27	27	25	20	20	20		13	25	–	1	–	–	–	–	–	2	–	–	–	Benzylamine, N-trimethylsilyl-	18406-59-2	NI24681
236	73	251	162	174	135	91	237	**251**	100	35	27	27	20	20	20	19		13	25	–	1	–	–	–	–	–	2	–	–	–	Benzylamine, N-trimethylsilyl-	18406-59-2	NI24682
105	77	223	90	195	179	146	167	**251**	100	37	14	5	4	3	2	1	**0.60**	14	9	2	3	–	–	–	–	–	–	–	–	–	1,2,3-Benzotriazin-4-one, 3-benzoyl-		MS05456
55	82	56	41	83	68	67	28	**251**	100	79	78	63	59	53	20	19	**0.00**	14	18	1	1	–	–	1	–	–	–	–	–	–	2H-1,3-Benzoxazine, 6-chloro-3-cyclohexyl-3,4-dihydro-	35183-44-9	NI24683
195	162	150	180	251	177	196	252	**251**	100	65	62	42	26	23	10	3		14	21	3	1	–	–	–	–	–	–	–	–	–	Benzeneacetic acid, 2-amino-4-methyl-5-methoxy-, tert-butyl ester		NI24684
163	164	88	82	149	42	18	147	**251**	100	37	36	15	11	10	10	9	**0.56**	14	21	3	1	–	–	–	–	–	–	–	–	–	Benzenepropanoic acid, α-amino-α-methyl-3-tert-butyl-4-hydroxy-		MS05457
136	251	148	208	219	137	69	121	**251**	100	30	29	20	15	12	11	10		14	21	3	1	–	–	–	–	–	–	–	–	–	Butanoic acid, 3-oxo-2-(2-pyrrolidinocyclopentenyl)-, methyl ester		MS05458
41	91	93	77	39	119	79	121	**251**	100	95	83	65	65	48	45	43	**3.00**	14	21	3	1	–	–	–	–	–	–	–	–	–	2-Cyclohexene-6-carboxylic acid, 3-methyl-1-(5-oxo-3,4-dihydro-2H-pyrrolyl)-, ethyl ester		MS05459
151	41	86	84	93	78	91	79	**251**	100	82	78	76	74	48	44	39	**4.00**	14	21	3	1	–	–	–	–	–	–	–	–	–	2-Cyclohexene-6-carboxylic acid, 4-methyl-1-(5-oxo-3,4-dihydro-2H-pyrrolyl)-, ethyl ester		MS05460
79	77	55	108	41	106	100	39	**251**	100	63	38	36	34	26	26	25	**6.00**	14	21	3	1	–	–	–	–	–	–	–	–	–	2-Cyclohexene-6-carboxylic acid, 1-(2-oxo-3,4,5,6-tetrahydropyridyl)-, ethyl ester		MS05461
236	206	237	220	179	190	192	176	**251**	100	28	17	8	8	7	6	6	**0.00**	14	21	3	1	–	–	–	–	–	–	–	–	–	8-Isoquinolol, 1,2,3,4-tetrahydro-6,7,8-trimethoxy-1,2-dimethyl-, (S)-		LI05091
57	43	41	71	70	69	55	29	**251**	100	96	60	57	54	43	42	38	**6.52**	14	21	3	1	–	–	–	–	–	–	–	–	–	Octyl 2-nitrophenyl ether		NI24685
152	251	208	164	150	86	153	151	**251**	100	47	33	18	13	12	11	11		14	21	3	1	–	–	–	–	–	–	–	–	–	2,4-Pentanedione, 3-(2-morpholinocyclopentenyl)-		MS05462
166	122	153	109	251	207	150	135	**251**	100	68	52	48	34	32	30	23		14	21	3	1	–	–	–	–	–	–	–	–	–	2-Pyridinecarboxylic acid, 5-octyl-, 1-oxide		NI24686
97	123	134	251	112	55	181	140	**251**	100	84	71	47	45	40	38	27		14	22	1	1	–	–	–	–	–	–	1	–	–	Phosphetan, 1-anilino-2,2,3,4,4-pentamethyl-, 1-oxide		NI24687
72	73	42	44	56	45	163	75	**251**	100	18	9	7	6	6	4	4	**0.00**	14	25	1	1	–	–	–	–	–	1	–	–	–	Benzenepropanamine, α,N,N-trimethyl-β-[(trimethylsilyl)oxy]-		MS05463
223	251	28	167	139	224	195	168	**251**	100	65	25	23	19	17	16	12		15	9	3	1	–	–	–	–	–	–	–	–	–	Anthraquinone-2-carboxaldehyde, 1-amino-	6363-87-7	NI24688
209	210	251	252	211				**251**	100	15	12	1	1					15	13	1	3	–	–	–	–	–	–	–	–	–	Acetamide, N-(8-methyl-2-phenazinyl)-	20297-56-7	NI24689
209	208	251	207	104	210	77	43	**251**	100	92	60	27	20	17	17	13		15	13	1	3	–	–	–	–	–	–	–	–	–	Acetamide, N-(2-phenyl-1H-pyrrolo[2,3-b]pyridin-3-yl)-	23616-67-3	NI24690
146	132	77	105	147	78	251	131	**251**	100	50	25	20	14	14	9	8		15	13	1	3	–	–	–	–	–	–	–	–	–	Acetophenone, 2-(3,4-dihydropyrido[2,3-d]pyrimidin-4-yl)-	28732-76-5	MS05464
146	83	77	119	85	251	105	57	**251**	100	38	36	24	24	20	17	14		15	13	1	3	–	–	–	–	–	–	–	–	–	Benzimidazole, [2-(N-benzoyl)methylamino]-		NI24691
251	222	223	250	252	69	224	152	**251**	100	50	49	21	11	5	4	3		15	13	1	3	–	–	–	–	–	–	–	–	–	1H-1,4-Benzodiazepin-2-one, 7-amino-2,3-dihydro-5-phenyl-	4928-02-3	NI24692
251	208	209	77	250	252	180	223	**251**	100	52	24	22	20	18	16	10		15	13	1	3	–	–	–	–	–	–	–	–	–	3H-1,3,4-Benzothiazepin-2-one, 1,2-dihydro-3-methyl-5-phenyl-		MS05465
91	251	106	147	65	130	103	92	**251**	100	43	14	12	12	11	11	8		15	13	1	3	–	–	–	–	–	–	–	–	–	1-Phthalazinone, 4-benzylamino-1,2-dihydro-		NI24693
236	105	77	237	66	202	200	185	**251**	100	48	33	18	15	14	11	10	**4.00**	15	13	1	3	–	–	–	–	–	–	–	–	–	4H-Pyran-2-amine, 3,5-dicyano-4,4-dimethyl-6-phenyl-		NI24694
236	251	204	250	207	127	180	219	**251**	100	52	40	32	24	22	20	12		15	13	1	3	–	–	–	–	–	–	–	–	–	4H-Pyran-3,5-dicarbonitrile, 2-amino-4,4-dimethyl-6-phenyl-	89809-95-0	DD00697
251	118	252	250	117	159	91	77	**251**	100	25	18	7	7	6	6	6		15	13	1	3	–	–	–	–	–	–	–	–	–	5-Pyrazolol, 3-methyl-1-phenyl-4-(2-pyridyl)-		IC03293
91	77	65	251	30	51	92	63	**251**	100	10	7	7	7	7	7	7		15	13	1	3	–	–	–	–	–	–	–	–	–	1H-1,2,4-Triazole, 1-benzyl-3-phenyl-, 4-oxide	19081-69-7	NI24695
91	251	132	119	252	92	133	120	**251**	100	80	22	18	12	8	3	2		15	13	1	3	–	–	–	–	–	–	–	–	–	3H-1,2,4-Triazol-3-one, 2,4-dihydro-5-methyl-2,4-diphenyl-	19796-34-0	NI24696
209	218	251	236	57	29	222	27	**251**	100	76	75	43	37	33	32	30		15	25	–	1	1	–	–	–	–	–	–	–	–	Pyridine, 2,4-dibutyl-6-(ethylthio)-	74630-33-4	NI24697
222	218	209	251	176	190	106	166	**251**	100	75	51	38	31	28	25	19		15	25	–	1	1	–	–	–	–	–	–	–	–	Pyridine, 3,6-dibutyl-2-(ethylthio)-	74630-32-3	NI24698
123	80	82	44	251	45	81	79	**251**	100	79	78	77	69	36	27	27		15	25	2	1	–	–	–	–	–	–	–	–	–	Benzenenonanamine, 3,4-dihydroxy-	105318-75-0	NI24699
251	210	165	252	208	211	181	152	**251**	100	74	25	20	13	12	10	10		16	13	–	1	1	–	–	–	–	–	–	–	–	Thiazole, 4-[1,1′-biphenyl]-4-yl-2-methyl-	24864-19-5	NI24700
104	251	103	105	77	78	252	51	**251**	100	36	10	9	8	7	6	4		16	13	2	1	–	–	–	–	–	–	–	–	–	2,4-Azetidinedione, 3-benzyl-1-phenyl-		IC03294
159	131	93	103	251	92	77	209	**251**	100	90	37	32	28	20	13	8		16	13	2	1	–	–	–	–	–	–	–	–	–	Benzamide, 2-(2-propynyloxy)-N-phenyl-		NI24701
130	131	77	251	104	91	92	120	**251**	100	78	41	31	18	13	10	8		16	13	2	1	–	–	–	–	–	–	–	–	–	1,3-Benzoxazin-4-one, 2,3-dihydro-3-phenyl-2-vinyl-		NI24702
135	251	103	77	105	132	92	136	**251**	100	59	53	41	16	12	12	9		16	13	2	1	–	–	–	–	–	–	–	–	–	Isoxazole, 3-(3-methoxyphenyl)-5-phenyl-	25856-12-6	LI05092
105	251	77	174	135	252	149	106	**251**	100	76	47	13	12	11	11	8		16	13	2	1	–	–	–	–	–	–	–	–	–	Isoxazole, 3-(3-methoxyphenyl)-5-phenyl-	25856-12-6	NI24703
105	251	133	77	102	103	90	174	**251**	100	77	57	55	30	29	28	17		16	13	2	1	–	–	–	–	–	–	–	–	–	Isoxazole, 3-(4-methoxyphenyl)-5-phenyl-	3672-52-4	NI24704
251	105	135	77	252	125.5	174	146	**251**	100	51	48	25	22	10	9	4		16	13	2	1	–	–	–	–	–	–	–	–	–	Isoxazole, 3-(4-methoxyphenyl)-5-phenyl-	3672-52-4	LI05093
135	251	77	252	136	92	125.5	107	**251**	100	63	22	19	11	7	6	6		16	13	2	1	–	–	–	–	–	–	–	–	–	Isoxazole, 5-(4-methoxyphenyl)-3-phenyl-	3672-51-3	LI05094
135	251	103	77	105	132	92	252	**251**	100	59	53	41	16	12	12	10		16	13	2	1	–	–	–	–	–	–	–	–	–	Isoxazole, 5-(4-methoxyphenyl)-3-phenyl-	3672-51-3	NI24705
105	251	133	77	102	103	90	174	**251**	100	77	57	55	30	29	28	17		16	13	2	1	–	–	–	–	–	–	–	–	–	Isoxazole, 5-(4-methoxyphenyl)-3-phenyl-	3672-51-3	LI05095
147	91	251	105	69	146	104	106	**251**	100	98	48	40	36	35	33	19		16	13	2	1	–	–	–	–	–	–	–	–	–	4-Isoxazolin-3-one, 2-benzyl-5-phenyl-		LI05096
236	151	29	179	150	251	27	208	**251**	100	84	69	47	42	38	34	29		16	17	–	3	–	–	–	–	–	–	–	–	–	Benzo[c]cinnoline, 2-(diethylamino)-	16371-76-9	NI24706

MASS TO CHARGE RATIOS								M.W.	INTENSITIES								Parent	C	H	O	N	S	F	Cl	Br	I	Si	P	B	X	COMPOUND NAME	CAS Reg No	No
222	207	251	208	223	180	152	151	**251**	100	53	27	24	20	19	19	15		16	17	–	3	–	–	–	–	–	–	–	–	–	Benzo[c]cinnoline, 4-(diethylamino)-	19174-76-6	NI24707
236	235	91	251	237	132	89	250	**251**	100	27	20	18	17	17	15	13		16	18	1	1	–	–	–	–	–	–	–	1	–	1,3,2-Oxazaborolidine, 3,4-dimethyl-2,5-diphenyl-	26574-29-8	NI24708
127	140	152	28	251	31	41	55	**251**	100	43	36	36	29	20	19	14		16	29	1	1	–	–	–	–	–	–	–	–	–	1-Cyclododecene, 1-morpholino-		MS05466
132	181	117	180	77	251	182	91	**251**	100	76	62	48	29	23	14	14		17	17	1	1	–	–	–	–	–	–	–	–	–	2-Azetidinone, 3,3-dimethyl-1,4-diphenyl-	5438-81-3	NI24709
132	181	117	251	180	77	182	133	**251**	100	78	44	39	34	25	12	12		17	17	1	1	–	–	–	–	–	–	–	–	–	2-Azetidinone, 3,4-dimethyl-1,4-diphenyl-		NI24710
105	146	77	174	251	234	91	118	**251**	100	78	70	67	65	38	26	19		17	17	1	1	–	–	–	–	–	–	–	–	–	Aziridine, cis-2-benzoyl-1-ethyl-3-phenyl-	38751-79-0	NI24711
105	77	146	251	174	234	91	118	**251**	100	71	52	47	37	33	24	22		17	17	1	1	–	–	–	–	–	–	–	–	–	Aziridine, trans-2-benzoyl-1-ethyl-3-phenyl-	38751-78-9	NI24712
131	103	251	104	77	91	146	132	**251**	100	30	27	20	20	12	11	10		17	17	1	1	–	–	–	–	–	–	–	–	–	Benzenepropenamide, N-phenylethyl-		TR00281
211	251	43	212	165	196	179	152	**251**	100	32	23	17	13	12	11	9		17	17	1	1	–	–	–	–	–	–	–	–	–	Butanenitrile, 3-(4'-methoxy-1,1'-biphenyl-4-yl)-	86799-93-1	NI24713
131	104	77	251	63	58	91	105	**251**	100	66	29	25	22	19	18	13		17	17	1	1	–	–	–	–	–	–	–	–	–	Non-2-ene-6,8-diynamide, N-(2-phenylethyl)-, (2E)-		NI24714
145	144	143	115	43	251	132	77	**251**	100	67	36	30	30	26	21	19		17	17	1	1	–	–	–	–	–	–	–	–	–	Isoquinoline, 3,4-dihydro-1-(2-hydroxyphenylethyl)-	78744-80-6	NI24715
118	251	117	91	131	129	234	221	**251**	100	70	63	48	21	20	18	8		17	17	1	1	–	–	–	–	–	–	–	–	–	Isoxazoline, 3,5-ditolyl-		NI24716
118	105	117	251	77	104	29	43	**251**	100	47	40	31	24	22	16	12		17	17	1	1	–	–	–	–	–	–	–	–	–	4H-1,3-Oxazine, 5,6-dihydro-2,6-diphenyl-5-methyl-	23665-17-0	NI24717
118	77	105	28	103	51	117	43	**251**	100	73	70	51	46	35	31	31	**22.44**	17	17	1	1	–	–	–	–	–	–	–	–	–	4H-1,3-Oxazine, 5,6-dihydro-2,6-diphenyl-6-methyl-		NI24718
148	104	145	90	105	147	149	91	**251**	100	21	17	17	15	14	13	13	**5.00**	17	17	1	1	–	–	–	–	–	–	–	–	–	Oxazole, 2,5-dihydro-2,2-dimethyl-4,5-diphenyl-	51304-30-4	NI24719
179	251	180	43	181	152	178	151	**251**	100	91	80	52	42	20	15	14		17	17	1	1	–	–	–	–	–	–	–	–	–	Phenanthridine, 5,6-dihydro-5-isobutanoyl-		LI05097
104	106	251	132	103	194	77		**251**	100	40	33	9	8	7	7			17	17	1	1	–	–	–	–	–	–	–	–	–	4-Piperidone, 2,6-diphenyl-		IC03295
251	250	178	179	165				**251**	100	30	11	4	3					17	17	1	1	–	–	–	–	–	–	–	–	–	Stilbene, 4-dimethylamino-4'-formyl-		LI05098
57	43	97	41	55	110	83	96	**251**	100	92	84	68	61	54	48	46	**1.00**	17	33	–	1	–	–	–	–	–	–	–	–	–	Heptadecanenitrile	5399-02-0	NI24720
43	41	57	29	55	97	27	110	**251**	100	89	75	58	53	46	28	27	**1.50**	17	33	–	1	–	–	–	–	–	–	–	–	–	Heptadecanenitrile	5399-02-0	WI01281
91	92	117	194	132	195	65	251	**251**	100	32	30	26	16	14	12	11		18	21	–	1	–	–	–	–	–	–	–	–	–	Azetidine, 1-benzyl-3,3-dimethyl-2-phenyl-	22606-97-9	NI24721
251	250	236	252	248	222	41	39	**251**	100	93	37	19	8	8	8	8		18	21	–	1	–	–	–	–	–	–	–	–	–	Benz[b]acridine, 1,2,3,4,7,8,9,10-octahydro-12-methyl-	55044-74-1	WI01282
251	250	236	252	222	248	237	39	**251**	100	92	36	19	8	7	7	7		18	21	–	1	–	–	–	–	–	–	–	–	–	Benz[c]acridine, 1,2,3,4,8,9,10,11-octahydro-7-methyl-	55044-73-0	WI01283
69	216	101	146	114	253	251	115	**252**	100	69	54	35	35	29	29	29		–	–	–	1	2	2	3	–	–	–	1	–	–	N-Trichlorophosphoranylidenethiophosphoryl difluoride		LI05099
202	200	252	199	217	215	250	201	**252**	100	88	80	71	66	62	57	57		1	3	–	–	–	–	1	–	–	–	–	–	1	Mercury, chloromethyl-	115-09-3	NI24722
252	250	202	217	215	200	251	214	**252**	100	74	72	62	57	56	52	49		1	3	–	–	–	–	1	–	–	–	–	–	1	Mercury, chloromethyl-	115-09-3	NI24723
85	87	169	167	69	147	219	217	**252**	100	32	9	9	9	8	7	7	**0.00**	3	–	–	–	–	4	4	–	–	–	–	–	–	Propane, 1,2,2,3-tetrachloro-1,1,3,3-tetrafluoro-		DO01040
83	85	69	135	87	98	129	100	**252**	100	77	38	16	15	11	9	9	**0.50**	4	1	–	–	–	7	2	–	–	–	–	–	–	Butane, 1,1-dichloro-2,2,3,3,4,4,4-heptafluoro-		IC03296
113	112	42	94	238	250	249	252	**252**	100	80	32	15	10	9	4	1		4	16	–	2	–	4	–	–	–	–	2	2	–	Diborane, bis(dimethylaminodifluorophosphinyl)-		LI05100
69	70	156	140	97	112	183	209	**252**	100	66	18	14	11	8	3	1	**0.00**	5	2	3	2	–	6	–	–	–	–	–	–	–	Urea, N,N-bis(trifluoroacetyl)-	81464-55-3	NI24724
28	140	59	87	74	47	168	43.5	**252**	100	49	45	32	21	17	16	16	**5.60**	5	3	4	–	–	2	–	–	–	1	–	–	1	Cobalt, (difluorosilyl)methyltetracarbonyl-		LI05101
252	82	28	27	52	54	253	53	**252**	100	41	28	15	14	10	9	8		5	5	2	2	–	–	–	–	1	–	–	–	–	Thymine, 6-iodo-	31458-38-5	MS05467
69	252	117	155	253	93	183	98	**252**	100	92	33	14	13	11	8	8		6	–	1	–	–	7	–	–	–	–	1	–	–	Phosphonic difluoride, (pentafluorophenyl)-	22474-68-6	NI24725
30	252	132	102	74	86	72	36	**252**	100	55	26	20	20	7	6	6		6	–	6	6	–	–	–	–	–	–	–	–	–	Benzotrifuroxan	3470-17-5	IC03297
252	30	132	74	102	70	192	85	**252**	100	40	33	13	10	7	5	5		6	–	6	6	–	–	–	–	–	–	–	–	–	Benzotrifuroxan	3470-17-5	LI05102
252	74	102	70	86	192	253	60	**252**	100	59	44	29	20	19	18	11		6	–	6	6	–	–	–	–	–	–	–	–	–	Benzotrifuroxan	3470-17-5	NI24726
147	185	183	149	111	109	187	75	**252**	100	89	87	78	70	56	45	43	**0.00**	6	5	–	–	–	–	5	–	–	–	–	–	–	Cyclohexene, 1a,2e,3e,4e,5e-pentachloro-		NI24727
181	183	147	111	146	145	185	75	**252**	100	95	36	35	35	32	29	27	**11.30**	6	5	–	–	–	–	5	–	–	–	–	–	–	Cyclohexene, 1,2,3,4,5-pentachloro-	28903-24-4	NI24728
181	183	219	111	217	75	37	146	**252**	100	94	48	41	36	36	36	35	**2.49**	6	5	–	–	–	–	5	–	–	–	–	–	–	Cyclohexene, pentachloro-		NI24729
43	39	41	127	69	142	29	58	**252**	100	96	78	65	61	57	45	43	**1.00**	6	5	3	–	–	–	–	–	1	–	–	–	–	4H-Pyran-4-one, 5-hydroxy-2-(iodomethyl)-	16065-34-2	NI24730
133	135	175	177	139	137	41	141	**252**	100	100	61	58	45	34	27	27	**0.00**	6	12	–	–	–	–	4	–	–	1	–	–	–	Silane, dichlorobis(3-chloropropyl)-	33317-65-6	NI24731
58	36	42	38	157	199	69	35	**252**	100	95	77	33	30	15	15	14	**0.00**	6	13	3	6	–	–	1	–	–	–	–	–	–	Ethanimidamide, 2-[3,5-bis(hydroxyimino)-1-piperazinyl]-N-hydroxy-, monohydrochloride	56700-83-5	MS05468
30	18	63	81	53	17	62	69	**252**	100	71	23	22	21	19	18	17	**16.11**	7	3	5	2	–	3	–	–	–	–	–	–	–	Phenol, 2,6-dinitro-4-(trifluoromethyl)-	393-77-1	NI24732
223	195	137	224	104	165	193	225	**252**	100	24	22	20	17	16	15	13	**0.39**	7	20	4	–	–	–	–	–	–	3	–	–	–	Cyclotrisiloxane, 1-methoxy-1,3,5-triethyl-	75221-52-2	NI24733
152	151	252	101	102	134	205	187	**252**	100	28	21	18	15	13	8	8		8	4	–	–	–	8	–	–	–	–	–	–	–	Bicyclo[2.2.2]oct-2-ene, 1,2,3,4,5,5,6,6-octafluoro-		NI24734
183	126	133	113	252	69	163	120	**252**	100	89	87	59	50	44	41	37		8	4	–	–	–	8	–	–	–	–	–	–	–	Tricyclo[3.3.0.0^{2,6}]octane, 3,3,4,4,7,7,8,8-octafluoro-		NI24735
120	121	64	77	48	51	42	106	**252**	100	81	54	39	26	22	19	17	**0.00**	8	11	3	3	1	–	–	–	–	–	–	–	1	Diazenesulphonic acid, [4-(dimethylamino)phenyl]-, sodium salt	140-56-7	NI24736
252	114	250	209	195	115	237	43	**252**	100	61	47	43	43	40	28	24		8	12	–	–	2	–	–	–	–	–	–	–	1	Thiophene, 2-(methylselenyl)-5-(propylthio)-	31053-59-5	NI24737
252	130	209	250	207	254	114	210	**252**	100	80	70	46	38	24	23	20		8	12	–	–	2	–	–	–	–	–	–	–	1	Thiophene, 2-(methylthio)-5-(propylselenyl)-	31053-60-8	NI24738
252	130	209	250	207	114	210	43	**252**	100	85	76	48	40	25	22	19		8	12	–	–	2	–	–	–	–	–	–	–	1	Thiophene, 2-(methylthio)-5-(propylselenyl)-	31053-60-8	LI05103
130	133	132	192	134	129	128	252	**252**	100	99	99	96	90	90	84	63		8	12	4	–	–	–	–	–	–	–	–	–	1	Selenophene, 3,4-diacetoxytetrahydro-		LI05104
43	152	154	180	182	134	136	73	**252**	100	93	91	44	43	35	34	31	**0.00**	8	13	4	–	–	–	–	1	–	–	–	–	–	Propanedioic acid, bromomethyl-, diethyl ester	29263-94-3	NI24739
165	135	150	179	237	219	194	167	**252**	100	48	24	18	17	16	16	15	**0.00**	8	20	1	–	–	–	–	–	–	–	–	–	1	2-Butanol, 3-methyl-4-(trimethylstannyl)-	53044-16-9	NI24740

MASS TO CHARGE RATIOS								M.W.	INTENSITIES								Parent	C	H	O	N	S	F	Cl	Br	I	Si	P	B	X	COMPOUND NAME	CAS Reg No	No
165	237	167	151	135	191	150	195	**252**	100	50	33	30	30	18	10	8	**0.00**	8	20	1	–	–	–	–	–	–	–	–	–	1	1-Pentanol, 5-(trimethylstannyl)-	53044-13-6	NI24741
217	193	195	59	219	197	123	221	**252**	100	84	79	63	58	22	19	13	**11.45**	9	7	2	–	–	–	3	–	–	–	–	–	–	Benzeneacetic acid, 2,3,6-trichloro-, methyl ester		NI24742
105	77	51	106	252	254	78	50	**252**	100	28	11	9	8	7	4	4		9	7	2	–	–	–	3	–	–	–	–	–	–	Ethanol, 2,2,2-trichloro-, benzoate	37934-99-9	NI24743
159	173	252	254	133	172	153	27	**252**	100	53	21	20	18	17	16	14		9	8	–	–	–	3	–	1	–	–	–	–	–	Benzene, 1-(2-bromoethyl)-3-(trifluoromethyl)-	1997-80-4	NI24744
210	63	30	149	51	79	122	43	**252**	100	43	43	36	35	27	25	25	**3.50**	9	8	5	4	–	–	–	–	–	–	–	–	–	Ethanol, 1-[(2,4-dinitrophenyl)amino]-1-isocyano-	56667-05-1	NI24745
161	59	101	133	129	69	193	91	**252**	100	56	38	23	20	20	12	11	**0.00**	9	13	7	–	–	1	–	–	–	–	–	–	–	1,5-Pentanedioic acid, 3-carboxy-3-hydroxy-fluoro-, trimethyl ester, (1R,2S)-		NI24746
161	59	101	175	133	129	69	63	**252**	100	68	36	27	22	19	16	13	**0.00**	9	13	7	–	–	1	–	–	–	–	–	–	–	1,5-Pentanedioic acid, 3-carboxy-3-hydroxy-fluoro-, trimethyl ester, (1S,2S)-		NI24747
121	77	55	141	47	73	91	41	**252**	100	90	88	66	37	35	20	19	**0.00**	9	14	1	–	–	6	–	–	–	–	–	–	–	5-Nonanol, 1,1,1,9,9,9-hexafluoro-	54699-52-4	NI24748
97	84	98	96	126	179	83	43	**252**	100	47	32	23	16	15	15	13	**9.20**	9	15	6	–	–	–	–	–	–	–	–	3	–	allo-Inositol tris(methylboronate)		NI24749
97	84	98	96	83	113	126	43	**252**	100	43	29	27	17	13	11	11	**6.12**	9	15	6	–	–	–	–	–	–	–	–	3	–	cis-Inositol tris(methylboronate)		NI24750
97	96	84	98	113	43	83	42	**252**	100	23	14	12	9	7	4	4	**1.62**	9	15	6	–	–	–	–	–	–	–	–	3	–	myo-Inositol tris(methylboronate)		NI24751
97	84	96	98	113	43	83	126	**252**	100	29	25	21	14	11	9	7	**4.60**	9	15	6	–	–	–	–	–	–	–	–	3	–	muco-Inositol tris(methylboronate)		NI24752
132	178	131	206	179	55	79	252	**252**	100	83	80	76	63	56	54	48		9	16	6	–	1	–	–	–	–	–	–	–	–	Propandioic acid, (methylsulphonyl)-, diethyl ester		NI24753
83	55	73	103	41	75	167	139	**252**	100	72	30	18	15	13	11	11	**0.00**	9	21	1	–	–	–	–	1	–	1	–	–	–	1-Hexanol, 6-bromo-1-trimethylsilyl-	26306-00-3	NI24754
193	109	151	43	41	194	91	57	**252**	100	55	32	25	21	11	9	9	**0.39**	9	21	3	–	–	–	–	–	–	–	–	–	1	Arsenous acid, tripropyl ester	15606-91-4	NI24755
193	109	151	43	41	194	116	91	**252**	100	53	31	25	23	9	9	8	**1.00**	9	21	3	–	–	–	–	–	–	–	–	–	1	Arsenous acid, tripropyl ester	26306-00-3	LI05105
61	43	57	55	44	36	41	69	**252**	100	84	68	64	59	55	52	49	**30.00**	10	5	4	2	–	–	1	–	–	–	–	–	–	Naphthalene, 2-chloro-1,6-dinitro-		IC03298
216	170	218	115	217	129	251	114	**252**	100	76	45	28	15	13	11	11	**0.00**	10	6	–	4	–	–	2	–	–	–	–	–	–	s-Triazolo[3,4-a]phthalazine, 3-dichloromethyl-		LI05106
111	252	151	83	45				**252**	100	60	16	13	4					10	8	2	2	2	–	–	–	–	–	–	–	–	Hydrazine, N,N′-dithien-2-oyl-		LI05107
167	195	162	30	222	223	32	119	**252**	100	92	79	77	62	54	49	40	**1.00**	10	8	2	2	2	–	–	–	–	–	–	–	–	Spiro[1,3-dithiolane-2,2′-[2H]indol]-3′(1′H)-one, 1′-nitroso-	38168-20-6	MS05469
210	212	252	254	211	209	131	213	**252**	100	94	26	25	23	18	18	16		10	9	1	2	–	–	–	1	–	–	–	–	–	1H-Benzodiazepin-2-one, 7-bromo-2,3-dihydro-4-methyl-	39623-52-4	NI24756
210	212	211	252	131	254	209	213	**252**	100	93	31	22	22	21	21	17		10	9	1	2	–	–	–	1	–	–	–	–	–	1H-Benzodiazepin-2-one, 8-bromo-2,3-dihydro-4-methyl-	39623-48-8	NI24757
112	139	207	180	76	91	126	64	**252**	100	95	88	88	88	86	69	56	**40.04**	10	9	2	4	–	–	1	–	–	–	–	–	–	2H-Tetrazole-5-carboxylic acid, 2-(3-chlorophenyl)-, ethyl ester	55044-76-3	NI24758
165	43	44	163	54	55	234	138	**252**	100	95	87	61	42	41	37	33	**4.38**	10	12	4	4	–	–	–	–	–	–	–	–	–	[4,4′-Bipyrimidine]-2,2′,6(1H,1′H,3H)-trione, 4,5-dihydro-5-hydroxy-5,5′-dimethyl-	20545-93-1	NI24759
28	30	41	27	29	79	39	78	**252**	100	66	54	53	51	49	36	35	**26.60**	10	12	4	4	–	–	–	–	–	–	–	–	–	Butanal, (2,4-dinitrophenyl)hydrazone	1527-98-6	IC03299
55	79	41	27	78	43	30	252	**252**	100	99	97	90	76	76	74	72		10	12	4	4	–	–	–	–	–	–	–	–	–	Butanal, (2,4-dinitrophenyl)hydrazone	1527-98-6	MS05470
252	79	55	41	78	43	152	122	**252**	100	38	36	33	31	30	25	24		10	12	4	4	–	–	–	–	–	–	–	–	–	Butanal, (2,4-dinitrophenyl)hydrazone	1527-98-6	NI24760
30	41	43	252	79	78	55	122	**252**	100	81	75	74	70	55	54	52		10	12	4	4	–	–	–	–	–	–	–	–	–	2-Butanone, (2,4-dinitrophenyl)hydrazone		MS05471
42	79	73	55	252	41	78	31	**252**	100	53	50	50	40	37	35	32		10	12	4	4	–	–	–	–	–	–	–	–	–	2-Butanone, (2,4-dinitrophenyl)hydrazone	958-60-1	NI24761
252	42	73	55	181	43	41	152	**252**	100	66	46	35	30	29	26	21		10	12	4	4	–	–	–	–	–	–	–	–	–	2-Butanone, (2,4-dinitrophenyl)hydrazone		IC03300
136	81	28	117	54	43	73	44	**252**	100	44	26	25	25	18	17	17	**0.00**	10	12	4	4	–	–	–	–	–	–	–	–	–	Inosine, 2′-deoxy-	890-38-0	NI24762
235	41	205	181	55	252	164	78	**252**	100	43	39	38	38	29	26	18		10	12	4	4	–	–	–	–	–	–	–	–	–	Pyridine, 3,5-dinitro-2-(1-piperidino)-	90871-11-7	NI24763
127	126	55	83	54	110	39	27	**252**	100	76	40	17	10	6	5	5	**0.00**	10	12	4	4	–	–	–	–	–	–	–	–	–	Thymine dimer	28806-14-6	NI24764
127	126	55	125	54	110	84	83	**252**	100	91	84	77	49	48	48	43	**0.00**	10	12	4	4	–	–	–	–	–	–	–	–	–	Thymine dimer, cis-syn	3660-32-0	NI24765
127	126	55	125	110	84	54	83	**252**	100	91	85	61	48	48	48	43	**0.10**	10	12	4	4	–	–	–	–	–	–	–	–	–	Thymine dimer, cis-syn	3660-32-0	LI05108
43	44	165	163	55	56	234	138	**252**	100	89	83	63	43	41	39	35	**4.00**	10	12	4	4	–	–	–	–	–	–	–	–	–	Thymine, 5-hydroxy-6-(4-(5-methylpyrimidin-2-one))-dihydro-		LI05109
41	28	39	68	27	42	56	15	**252**	100	81	62	47	36	35	25	25	**6.20**	10	13	–	6	–	–	1	–	–	–	–	–	–	Propanenitrile, 2-[[4-chloro-6-(cyclopropylamino)-1,3,5-triazin-2-yl]amino]-2-methyl-	32889-48-8	NI24766
237	73	252	45	238	84	223	43	**252**	100	76	57	26	15	15	13	13		10	16	2	4	–	–	–	–	–	1	–	–	–	1H-Purine-2,6-dione, 3,7-dihydro-1,3-dimethyl-7-(trimethylsilyl)-	62374-32-7	NI24767
165	29	152	151	123	28	179	41	**252**	100	92	88	82	71	64	57	56	**2.00**	10	21	5	–	–	–	–	–	–	–	1	–	–	Butanoic acid, 2-(diethoxyphosphinyl)-, ethyl ester	17145-91-4	LI05110
28	207	224	179	155	123	138	109	**252**	100	88	84	79	76	69	67	67	**0.00**	10	21	5	–	–	–	–	–	–	–	1	–	–	Butanoic acid, 2-(diethoxyphosphinyl)-, ethyl ester	17145-91-4	NI24768
252	77	132	28	165	91	210	42	**252**	100	76	42	32	31	18	16	16		11	8	–	8	–	–	–	–	–	–	–	–	–	8H-Pyrazolo[4,3-e][1,2,3,4]tetrazolo[1,5-a][1,2,4]triazine, 6-methyl-8-phenyl-	95571-16-7	NI24769
159	105	77	160	131	84	86	51	**252**	100	15	13	11	11	10	7	7	**3.10**	11	9	2	–	–	–	–	1	–	–	–	–	–	2(5H)-Furanone, 5-(bromomethyl)-5-phenyl-	53774-22-4	NI24770
252	254	209	100	91	194	78	211	**252**	100	93	77	56	41	31	29	25		11	9	3	2	–	–	1	–	–	–	–	–	–	1,3-Benzenedicarbonitrile, 5-chloro-2,4,6-trimethoxy-		DD00698
216	115	188	142	170	187	162	149	**252**	100	30	29	29	10	1	1	1		11	9	3	2	–	–	1	–	–	–	–	–	–	Pyridine, N-(4-nitrophenyl)-, 3-oxide		LI05111
105	77	147	51	191	29	43	27	**252**	100	63	24	23	21	21	20	18	**0.06**	11	12	3	2	1	–	–	–	–	–	–	–	–	Benzamide, N-(aminocarbonyl)-N-(1-thioxopropoxy)-	74779-60-5	NI24771
42	223	70	224	183	71	225	120	**252**	100	74	35	8	7	4	3	3	**1.00**	11	12	3	2	1	–	–	–	–	–	–	–	–	2-Butanone, O-1,2-benzisothiazol-3-yloxime 1,1-dioxide	13604-36-9	NI24772
160	159	129	45	145	64	161	117	**252**	100	43	21	21	17	17	13	12	**0.00**	11	12	3	2	1	–	–	–	–	–	–	–	–	6-Thia-2,3-diazabicyclo[3.2.0]hept-2-ene, 4-(4-methoxyphenyl)-, 6,6-dioxide		LI05112
108	109	137	136	121	135	203	252	**252**	100	35	34	30	30	22	13	11		11	12	5	2	–	–	–	–	–	–	–	–	–	Propanedioic acid, 2-(4-ethoxyphenyl)hydrazino-		NI24773
141	113	95	59	41	67	43	111	**252**	100	78	55	39	38	36	31	27	**0.00**	11	15	3	–	–	3	–	–	–	–	–	–	–	Cyclopropanecarboxylic acid, 2,2-dimethyl-3-(2,3,3-trifluoro-1-hydroxy-2-propenyl)-, ethyl ester	106908-95-6	DD00699

MASS TO CHARGE RATIOS								M.W.	INTENSITIES								Parent	C	H	O	N	S	F	Cl	Br	I	Si	P	B	X	COMPOUND NAME	CAS Reg No	No
72	18	15	42	28	152	27	224	252	100	47	45	28	21	17	14	11	1.10	11	16	3	4	–	–	–	–	–	–	–	–	–	Carbamic acid, dimethyl-, 2-(formylmethylamino)-5,6-dimethyl-4-pyrimidinyl ester	27218-04-8	NI24774
126	55	42	127	58	44	41	43	252	100	14	11	10	10	10	10	7	0.89	11	19	1	2	–	3	–	–	–	–	–	–	–	Acetamide, 2,2,2-trifluoro-N-[2-(hexahydro-1(2H)-azocinyl)ethyl]-	55649-50-8	NI24775
237	43	225	210	205	58	195	57	252	100	34	30	30	30	30	26	26	24.99	11	20	1	6	–	–	–	–	–	–	–	–	–	1,3,5-Triazine-2-carboximidic acid, 4,6-bis(isopropylamino)-, methyl ester		NI24776
45	59	89	58	103	43	87	29	252	100	71	27	21	17	17	14	14	0.00	11	24	6	–	–	–	–	–	–	–	–	–	–	2,5,8,11,14-Pentaoxahexadecan-16-ol	23778-52-1	NI24777
45	59	29	44	58	43	31	89	252	100	80	41	36	33	25	25	22	0.00	11	24	6	–	–	–	–	–	–	–	–	–	–	2,5,8,11,14-Pentaoxahexadecan-16-ol	23778-52-1	IC03301
217	252	254	126	253	256	127	189	252	100	96	82	40	38	33	28	27		12	6	2	–	–	–	2	–	–	–	–	–	–	Dibenzo[b,e][1,4]dioxin, 1,6-dichloro-	38178-38-0	NI24778
252	254	126	189	253	256	127	50	252	100	64	24	22	15	11	10	9		12	6	2	–	–	–	2	–	–	–	–	–	–	Dibenzo[b,e][1,4]dioxin, 2,3-dichloro-	29446-15-9	AI01800
252	254	189	126	28	50	253	63	252	100	84	40	40	33	23	19	16		12	6	2	–	–	–	2	–	–	–	–	–	–	Dibenzo[b,e][1,4]dioxin, 2,3-dichloro-	29446-15-9	NI24779
252	254	188	126	256	253	190	217	252	100	62	22	17	10	10	10	8		12	6	2	–	–	–	2	–	–	–	–	–	–	Dibenzo[b,e][1,4]dioxin, 2,7-dichloro-	33857-26-0	NI24780
252	254	189	126	191	217	253	161	252	100	65	50	37	16	14	14	12		12	6	2	–	–	–	2	–	–	–	–	–	–	Dibenzo[b,e][1,4]dioxin, 2,7-dichloro-	33857-26-0	NI24781
252	254	189	217	126	219	75	253	252	100	92	61	47	25	22	21	15		12	6	2	–	–	–	2	–	–	–	–	–	–	Dibenzo[b,e][1,4]dioxin, 2,7-dichloro-	33857-26-0	NI24782
252	254	189	126	253	63	191	256	252	100	66	35	15	13	13	12	11		12	6	2	–	–	–	2	–	–	–	–	–	–	Dibenzo[b,e][1,4]dioxin, 2,8-dichloro-		IC03302
252	251	205	157	204	253	175	235	252	100	80	24	20	16	14	12	8		12	7	2	–	–	2	–	–	–	–	1	–	–	Phenoxaphosphine, 2,8-difluoro-, 10-oxide		LI05113
125	252	159	77	51	28	75	254	252	100	65	63	59	47	38	26	24		12	9	2	–	1	–	1	–	–	–	–	–	–	Benzene, 1-chloro-4-(phenylsulfonyl)-	80-00-2	NI24783
125	159	77	51	252	75	50	161	252	100	44	44	38	30	21	19	16		12	9	2	–	1	–	1	–	–	–	–	–	–	Benzene, 1-chloro-4-(phenylsulphonyl)-	80-00-2	NI24784
252	136	224	254	209	138	205	226	252	100	75	57	35	35	25	20	19		12	9	2	–	1	–	1	–	–	–	–	–	–	4H-Pyran-4-one, 2-(4-chlorophenyl)-6-methylthio-		MS05472
252	144	141	109	39	65	93	95	252	100	83	81	81	77	66	47	45		12	9	3	–	1	1	–	–	–	–	–	–	–	Phenol, 4-(4-fluorophenylsulphonyl)-		IC03303
154	77	175	252	219	215	152	153	252	100	74	73	70	19	17	7	5		12	10	–	–	–	–	2	–	–	1	–	–	–	Silane, dichlorodiphenyl-	80-10-4	NI24785
252	143	217	145	139	107	63	183	252	100	84	38	33	29	29	20	19		12	10	–	–	1	–	1	–	–	–	1	–	–	Diphenylphosphinothioic chloride		MS05473
252	254	217	216	253	218	182	256	252	100	94	78	72	68	52	51	47		12	10	–	2	–	–	2	–	–	–	–	–	–	1,2-Benzenediamine, 4-chloro-N[1]-(4-chlorophenyl)-	23648-87-5	MS05474
252	254	253	126	127	91	182	144	252	100	66	18	13	12	11	10	10		12	10	–	2	–	–	2	–	–	–	–	–	–	[1,1′-Biphenyl]-4,4′-diamine, 3,3′-dichloro-	91-94-1	NI24786
252	254	154	126	127	91	181	182	252	100	72	28	23	22	22	16	14		12	10	–	2	–	–	2	–	–	–	–	–	–	[1,1′-Biphenyl]-4,4′-diamine, 3,3′-dichloro-	91-94-1	NI24787
252	254	91	253	126	127	77	154	252	100	64	16	15	15	11	11	10		12	10	–	2	–	–	2	–	–	–	–	–	–	[1,1′-Biphenyl]-4,4′-diamine, 3,3′-dichloro-	91-94-1	IC03304
168	167	97	205	252	110	206	142	252	100	86	62	50	45	45	34	31		12	12	–	–	3	–	–	–	–	–	–	–	–	Thiophene, 2,4-di-2-thienyltetrahydro-, trans-		LI05114
178	210	119	28	179	43	16	15	252	100	56	50	44	38	21	20	18	0.37	12	12	6	–	–	–	–	–	–	–	–	–	–	1,4-Benzenedicarboxylic acid, 6-hydroxy-, acetate		MS05475
221	222	252	162	103	161	193	119	252	100	23	12	8	8	7	6	6		12	12	6	–	–	–	–	–	–	–	–	–	–	1,2,4-Benzenetricarboxylic acid, trimethyl ester	2459-10-1	IC03305
221	83	85	222	75	252	103	47	252	100	20	14	10	8	7	7	6		12	12	6	–	–	–	–	–	–	–	–	–	–	1,2,4-Benzenetricarboxylic acid, trimethyl ester	2459-10-1	NI24788
221	62	27	49	64	63	252	98	252	100	68	53	39	23	23	17	16		12	12	6	–	–	–	–	–	–	–	–	–	–	1,3,5-Benzenetricarboxylic acid, trimethyl ester		DO01041
221	252	193	222	161	163	147	133	252	100	17	16	13	12	11	6	6		12	12	6	–	–	–	–	–	–	–	–	–	–	1,3,5-Benzenetricarboxylic acid, trimethyl ester		LI05115
221	252	193	222	161	147	75	133	252	100	21	19	13	10	7	7	5		12	12	6	–	–	–	–	–	–	–	–	–	–	1,3,5-Benzenetricarboxylic acid, trimethyl ester		IC03306
126	43	168	210	127	125	97	15	252	100	71	31	13	7	5	4	4	2.20	12	12	6	–	–	–	–	–	–	–	–	–	–	1,2,4-Benzenetriol, triacetate	613-03-6	NI24789
135	73	77	237	165	136	92	75	252	100	57	30	26	15	15	13	13	1.20	12	16	4	–	–	–	–	–	–	1	–	–	–	Benzeneacetic acid, 2-methoxy-α-oxo-, trimethylsilyl ester		MS05478
135	73	77	75	179	136	92	45	252	100	36	31	24	8	6	6	6	0.00	12	16	4	–	–	–	–	–	–	1	–	–	–	Benzeneacetic acid, 2-methoxy-α-oxo-, trimethylsilyl ester		MS05477
135	73	75	77	179	136	92	45	252	100	32	25	19	8	8	6	6	1.20	12	16	4	–	–	–	–	–	–	1	–	–	–	Benzeneacetic acid, 2-methoxy-α-oxo-, trimethylsilyl ester		MS05476
73	135	165	77	237	208	177	107	252	100	64	35	31	26	20	20	18	2.40	12	16	4	–	–	–	–	–	–	1	–	–	–	Benzeneacetic acid, 3-methoxy-α-oxo-, trimethylsilyl ester		MS05479
73	75	252	135	237	77	207	45	252	100	27	24	18	15	14	11	10		12	16	4	–	–	–	–	–	–	1	–	–	–	Benzeneacetic acid, 3,4-methylenedioxy-, trimethylsilyl ester		MS05480
195	120	43	210	135	75	92	73	252	100	52	46	33	24	20	16	16	0.10	12	16	4	–	–	–	–	–	–	1	–	–	–	Salicylic acid, trimethylsilyl ester, acetate	25436-28-6	NI24790
41	97	55	157	156	69	237	83	252	100	98	60	58	50	33	30	30	28.00	12	16	4	2	–	–	–	–	–	–	–	–	–	Barbituric acid, 5-(1-cyclohexen-1-yl)-1,5-dimethyl-1′,2′-epoxy-	66403-25-6	NI24791
176	235	161	150	194	160	175	133	252	100	63	46	37	19	19	17	17	0.00	12	16	4	2	–	–	–	–	–	–	–	–	–	Benzenebutanoic acid, α,2-diamino-3-methoxy-γ-oxo-, methyl ester	55090-41-0	NI24792
203	218	44	219	107	209	204	177	252	100	90	78	47	21	19	19	18	0.01	12	16	4	2	–	–	–	–	–	–	–	–	–	1,3-Benzodioxole, 5-(hydroxyiminomethyl)-4-methoxy-6-[(2-methylamino)ethyl]-		NI24793
70	252	97	41	55	156	152	195	252	100	90	81	42	39	22	17	13		12	16	4	2	–	–	–	–	–	–	–	–	–	2,4(3H,4aH)-Benzofuro[2,3-d]pyrimidinedione, 3,4a-dimethyl-4b-hydroxy-5,6,7,8,8a-pentahydro-	77715-61-8	NI24794
43	70	55	57	41	69	56	71	252	100	79	77	63	55	54	50	35	0.00	12	19	2	–	–	3	–	–	–	–	–	–	–	Citronellyl trifluoroacetate		LI05116
81	95	138	41	55	43	123	96	252	100	99	84	81	71	71	59	59	1.00	12	19	2	–	–	3	–	–	–	–	–	–	–	Isomenthyl trifluoroacetate		LI05117
81	41	95	55	43	96	138	123	252	100	75	68	41	39	33	32	30	0.00	12	19	2	–	–	3	–	–	–	–	–	–	–	Menthyl trifluoroacetate		LI05118
81	95	43	96	41	55	138	67	252	100	97	52	51	51	46	41	30	0.40	12	19	2	–	–	3	–	–	–	–	–	–	–	Neoisomenthyl trifluoroacetate		LI05119
95	41	138	55	43	96	67	82	252	100	57	55	50	50	45	31	25	0.50	12	19	2	–	–	3	–	–	–	–	–	–	–	Neomenthyl trifluoroacetate		LI05120
180	72	58	42	252	237	181	57	252	100	60	40	34	31	15	11	10		12	20	2	4	–	–	–	–	–	–	–	–	–	Pyrimidine, 4-[(dimethylamino)carbonyloxy]-5,6-dimethyl-2-ethylmethylamino-		IC03307
166	58	86	29	252	152	42	44	252	100	51	40	30	22	20	19	15		12	20	2	4	–	–	–	–	–	–	–	–	–	Pyrimidine, 2-dimethylamino-5,6-dimethyl-4-[(methylethylamino)carbonyloxy]-		IC03308
237	73	60	179	97	238	83	177	252	100	75	40	30	28	26	25	22	11.00	12	24	–	–	–	–	–	–	–	3	–	–	–	Silane, (1-methyl-1,2-pentadien-4-yne-1,3,5-triyl)tris[dimethyl-	61255-22-9	NI24795
237	73	59	179	238	177	97	83	252	100	76	45	33	29	22	22	21	12.00	12	24	–	–	–	–	–	–	–	3	–	–	–	Silane, (5-methyl-1,3-pentadiyn-1-yl-5-ylidene)tris[dimethyl-	61227-82-5	NI24796
56	70	57	43	41	237	111	69	252	100	10	8	7	7	6	5	5	1.00	12	27	3	–	–	–	–	–	–	–	–	3	–	Boroxin, tri-tert-butyl-	13155-00-5	NI24797

MASS TO CHARGE RATIOS								M.W.	INTENSITIES								Parent	C	H	O	N	S	F	Cl	Br	I	Si	P	B	X	COMPOUND NAME	CAS Reg No	No
237	239	252	139	209	254	211	150	252	100	82	81	69	63	58	40	33		13	10	1	–	–	–	2	–	–	–	–	–	–	1,1'-Biphenyl, 3,5-dichloro-4-methoxy-	74298-90-1	NI24798
41	39	65	91	77	139	155	92	252	100	77	76	72	55	53	43	41	0.00	13	16	3	–	1	–	–	–	–	–	–	–	–	2,5-Hexadien-1-ol, 3-(p-tolylsulphonyl)-, (E)-	118356-20-0	DD00700
105	77	45	51	106	27	193	252	252	100	29	12	11	10	8	6	3		13	16	3	–	1	–	–	–	–	–	–	–	–	Propanoic acid, 2-(benzoylthio)-, isopropyl ester		MS05481
43	116	74	216	101	15	117	59	252	100	45	21	8	8	5	4	4	0.00	13	16	3	–	1	–	–	–	–	–	–	–	–	2-Thiatricyclo[3.3.1.1^{3,7}]decane-4,6,8-trione, 1,3,5,7-tetramethyl-, (±)-	57289-20-0	NI24799
163	149	181	77	164	105	104	135	252	100	84	62	58	52	52	41	39	0.00	13	16	5	–	–	–	–	–	–	–	–	–	–	1,2-Benzenedicarboxylic acid, methyl 3-hydroxybutyl diester		MS05482
163	149	104	76	164	71	181	105	252	100	88	43	32	30	30	26	24	0.00	13	16	5	–	–	–	–	–	–	–	–	–	–	1,2-Benzenedicarboxylic acid, methyl 4-hydroxybutyl diester		MS05483
252	237	206	179	222	253	190	225	252	100	25	23	19	18	16	14	10		13	16	5	–	–	–	–	–	–	–	–	–	–	1,3-Benzodioxole, 4,5,7-trimethoxy-6-(2-propenyl)-		LI05121
91	101	92	161	234	119	59	193	252	100	45	43	32	28	21	20	18	0.00	13	16	5	–	–	–	–	–	–	–	–	–	–	Butanedioic acid, 2-benzyl-2-hydroxy-, dimethyl ester		LI05122
91	105	133	165	150	193	220	252	252	100	92	74	54	47	30	5	2		13	16	5	–	–	–	–	–	–	–	–	–	–	2,6-Cycloheptadiene-1-carboxylic acid, 1-(2-carboxymethyl-2-oxoethyl)-, methyl ester		MS05484
91	118	119	151	105	133	92	220	252	100	18	16	15	13	13	13	9	1.40	13	16	5	–	–	–	–	–	–	–	–	–	–	1,6-Cycloheptadiene-1-carboxylic acid, 3-[(1-methoxycarbonyl)ethenyloxy]-, methyl ester		MS05485
252	220	221	237	205	192	191	163	252	100	84	78	62	58	53	42	34		13	16	5	–	–	–	–	–	–	–	–	–	–	1(3H)-Isobenzofuranone, 5,7-dimethoxy-4-(methoxymethyl)-6-methyl-		NI24800
91	105	150	220	118	119	133	151	252	100	54	41	27	26	26	22	13	11.80	13	16	5	–	–	–	–	–	–	–	–	–	–	Oxatricyclodecenedicarboxylic acid, dimethyl ester		MS05486
91	105	84	161	133	149	77	195	252	100	97	88	66	56	47	46	44	26.40	13	16	5	–	–	–	–	–	–	–	–	–	–	Oxatricyclodecenedicarboxylic acid, dimethyl ester		MS05487
43	210	151	150	168	135	91	252	252	100	32	32	22	10	10	10	4		13	16	5	–	–	–	–	–	–	–	–	–	–	Phenol, 2-methoxy-4-(1-hydroxyethyl)-, diacetate	35960-49-7	NI24801
252	130	196	198	105	183	131	211	252	100	94	63	62	49	42	34	20		13	17	–	–	–	–	–	1	–	–	–	–	–	Benzene, 1-bromo-5-cyclohexyl-2-methyl-		LI05123
42	84	127	237	153	41	252	56	252	100	85	56	42	32	31	21	20		13	17	1	2	–	–	1	–	–	–	–	–	–	4H-1,3-Oxazine, 5,6-dihydro-2-[m-chlorophenylamino(imino)]-4,4,6-trimethyl-	37719-89-4	NI24802
149	122	121	109	162	179	163	42	252	100	46	42	34	26	24	20	20	8.80	13	17	2	2	–	1	–	–	–	–	–	–	–	3-Imidazoline, 1-hydroxy-4-(p-fluorophenyl)-2,2,5,5-tetramethyl-, 3-oxide	72342-93-9	NI24803
73	181	237	209	75	163	252	179	252	100	58	48	48	48	41	37	36		13	20	3	–	–	–	–	–	–	1	–	–	–	Benzeneacetic acid, 2-[(trimethylsilyl)oxy]-, ethyl ester		MS05488
252	179	165	163	237	191	181	75	252	100	50	47	47	45	41	28	25		13	20	3	–	–	–	–	–	–	1	–	–	–	Benzeneacetic acid, 3-[(trimethylsilyl)oxy]-, ethyl ester		MS05489
179	73	252	163	180	75	237	45	252	100	51	43	26	25	16	14	11		13	20	3	–	–	–	–	–	–	1	–	–	–	Benzeneacetic acid, 4-[(trimethylsilyl)oxy]-, ethyl ester		MS05490
73	179	75	131	180	77	45	74	252	100	99	52	34	32	26	21	17	0.00	13	20	3	–	–	–	–	–	–	1	–	–	–	Benzeneacetic acid, 4-[(trimethylsilyl)oxy]-, ethyl ester		MS05491
73	193	75	251	176	91	147	194	252	100	86	64	22	21	19	15	14	4.40	13	20	3	–	–	–	–	–	–	1	–	–	–	Benzeneacetic acid, 4-[(trimethylsilyl)oxy]-, ethyl ester		MS05492
73	103	89	118	222	59	237	104	252	100	99	96	88	60	34	29	26	2.00	13	20	3	–	–	–	–	–	–	1	–	–	–	Benzeneacetic acid, α-[[(trimethylsilyl)oxy]methyl]-, methyl ester	57397-46-3	NI24804
134	121	91	73	75	237	119	252	252	100	67	42	41	38	30	28	27		13	20	3	–	–	–	–	–	–	1	–	–	–	Benzenepropanoic acid, 2-methoxy-, trimethylsilyl ester		MS05493
134	75	237	73	252	135	121	91	252	100	59	38	35	31	28	15	11		13	20	3	–	–	–	–	–	–	1	–	–	–	Benzenepropanoic acid, 3-methoxy-, trimethylsilyl ester		MS05494
121	134	75	73	252	237	135	122	252	100	83	19	19	18	10	10	8		13	20	3	–	–	–	–	–	–	1	–	–	–	Benzenepropanoic acid, 4-methoxy-, trimethylsilyl ester		MS05495
179	252	180	253	192	131	73	89	252	100	61	16	12	11	10	10	9		13	20	3	–	–	–	–	–	–	1	–	–	–	Benzenepropanoic acid, 4-[(trimethylsilyl)oxy]-, methyl ester	27798-74-9	NI24805
73	193	237	89	162	161	194	177	252	100	87	68	57	47	43	15	14	0.00	13	20	3	–	–	–	–	–	–	1	–	–	–	Benzenepropanoic acid, α-[(trimethylsilyl)oxy]-, methyl ester	27854-49-5	NI24806
73	193	43	195	210	237	75	252	252	100	95	85	74	61	41	37	36		13	20	3	–	–	–	–	–	–	1	–	–	–	Benzoic acid, 4-[(trimethylsilyl)oxy]-, propyl ester	27739-19-1	NI24807
73	159	151	75	41	77	69	45	252	100	98	77	49	24	21	21	21	1.98	13	20	3	–	–	–	–	–	–	1	–	–	–	Butanoic acid, 4-phenoxy-, trimethylsilyl ester	21273-10-9	NI24808
73	159	252	151	237	69	166	117	252	100	98	78	67	53	48	37	17		13	20	3	–	–	–	–	–	–	1	–	–	–	Butanoic acid, 4-phenoxy-, trimethylsilyl ester	21273-10-9	NI24809
73	159	151	75	69	166	41	160	252	100	98	67	18	17	14	14	13	3.00	13	20	3	–	–	–	–	–	–	1	–	–	–	Butanoic acid, 4-phenoxy-, trimethylsilyl ester	21273-10-9	NI24810
58	181	124	137	71	205	234	252	252	100	34	29	25	23	10	8	7		13	20	3	2	–	–	–	–	–	–	–	–	–	Acetamide, N-(2-dimethylaminoethyl)-(4-methoxyphenoxy)-		NI24811
252	123	72	70	136	223	36	55	252	100	55	52	47	46	23	23	20		13	20	3	2	–	–	–	–	–	–	–	–	–	Benzenepropanamide, N-(4-aminobutyl)-3,4-dihydroxy-	55760-27-5	LI05124
252	123	72	70	136	223	55	59	252	100	55	52	47	46	22	20	17		13	20	3	2	–	–	–	–	–	–	–	–	–	Benzenepropanamide, N-(4-aminobutyl)-3,4-dihydroxy-	55760-27-5	NI24812
72	56	43	137	73	41	208	44	252	100	7	6	4	4	4	3	3	0.00	13	20	3	2	–	–	–	–	–	–	–	–	–	2-Propanol, 1-(2-formamidophenoxy)-3-isopropylamino-		MS05496
183	252	42	137	69	110	98	97	252	100	80	80	65	65	60	50	40		13	20	3	2	–	–	–	–	–	–	–	–	–	4H-Pyrido[1,2-a]pyrimidine-3-carboxylic acid, 1,6,7,8,9,9a-hexahydro-1,6-dimethyl-4-oxo-, ethyl ester	30238-40-5	NI24813
196	195	41	138	181	111	209	169	252	100	66	48	21	19	16	15	13	0.60	13	20	3	2	–	–	–	–	–	–	–	–	–	2,4,6(1H,3H,5H)-Pyrimidinetrione, 1,3-dimethyl-5-(2-methylpropyl)-5-(2-propenyl)-	55045-04-0	NI24814
252	253	224	220	90	76	63	136	252	100	19	13	13	10	9	9	8		14	8	1	2	1	–	–	–	–	–	–	–	–	12H-Benzimidazo[2,1-b]-1,3-benzothiazin-12-one		NI24815
252	77	169	224	51	67	104	196	252	100	99	55	54	46	37	30	29		14	12	1	4	–	–	–	–	–	–	–	–	–	4-Pyrazoleformamide, 1-phenyl-5-(1-pyrryl)-	94692-10-1	NI24816
105	77	252	106	253	78			252	100	30	20	8	3	2				14	12	1	4	–	–	–	–	–	–	–	–	–	Pyrazolo[5,1-c]-as-triazine, 6-benzoyl-4,6-dihydro-3-methyl-4-methylene-	6763-76-4	NI24817
252	163	181	135	207	149	219	189	252	100	73	64	44	40	40	18	18		14	20	4	–	–	–	–	–	–	–	–	–	–	Benzopyran, 3,4-dihydro-3,5-dihydroxy-2,2-dimethyl-6-(3-hydroxypropyl)-		LI05125
43	55	139	70	39	27			252	100	73	25	22	15	12			0.00	14	20	4	–	–	–	–	–	–	–	–	–	–	Bicyclo[3.3.1]nonan-6,9-dione, 2-hydroxy-2-methyl-5-(3'-oxobutyl)-		LI05126
71	56	127	126	111	82	41	43	252	100	56	48	42	39	37	37	26	0.16	14	20	4	–	–	–	–	–	–	–	–	–	–	Cyclobuta[1,2-b:4,3-b']dipyran-4,5-dione, octahydro-2,2,7,7-tetramethyl-	54637-09-1	NI24818
161	133	91	220	126	111	105	193	252	100	99	91	89	86	83	83	77	15.00	14	20	4	–	–	–	–	–	–	–	–	–	–	1-Cyclohexene-1,4-dicarboxylic acid, 3-methyl-4-(1-propenyl)-, dimethyl ester		MS05497
252	179	111	123	41	221	95	79	252	100	88	73	47	33	30	29	19		14	20	4	–	–	–	–	–	–	–	–	–	–	Cyclopentaneacetic acid, 4-(hydroxymethylene)-3-oxo-2-(2-pentenyl)-, methyl ester	66971-95-7	NI24819
183	196	43	252	41	209	178	177	252	100	83	62	60	48	36	35	34		14	20	4	–	–	–	–	–	–	–	–	–	–	1,2,4-Cyclopentanetrione, 3-isobutyl-5-isopentyl-		LI05127

MASS TO CHARGE RATIOS								M.W.	INTENSITIES								Parent	C	H	O	N	S	F	Cl	Br	I	Si	P	B	X	COMPOUND NAME	CAS Reg No	No
69	41	43	252	166	183	71	57	252	100	81	65	57	46	35	32	25		14	20	4	–	–	–	–	–	–	–	–	–	–	2-Cyclopenten-1-one, 3,4-dihydroxy-5-(3-methyl-2-butenyl)-2-(2-methyl-1-oxopropyl)-	38602-14-1	NI24820
43	109	41	55	39	29	107	176	252	100	59	58	54	51	47	45	39	0.29	14	20	4	–	–	–	–	–	–	–	–	–	–	1,3-Dioxane-4,6-dione, 2,2-dimethyl-5,5-bis(2-methyl-2-propenyl)-	74793-71-8	NI24821
137	43	192	123	124	81	121	41	252	100	98	85	55	30	29	23	21	18.00	14	20	4	–	–	–	–	–	–	–	–	–	–	2-Furanheptanoic acid, 5-acetyl-, methyl ester		NI24822
91	92	164	107	65	203	146	133	252	100	20	15	7	7	3	3	3	0.00	14	20	4	–	–	–	–	–	–	–	–	–	–	6-Heptene-1,2,4-triol, 3-(benzyloxy)-, (2S,3R,4R)-	120418-02-2	DD00701
91	92	164	107	65	203	146	133	252	100	20	15	7	7	3	3	3	0.00	14	20	4	–	–	–	–	–	–	–	–	–	–	6-Heptene-1,2,4-triol, 3-(benzyloxy)-, (2S,3S,4R)-	120520-95-8	DD00702
105	77	43	166	45	106	79	107	252	100	56	55	48	34	33	27	25	0.00	14	20	4	–	–	–	–	–	–	–	–	–	–	Hexanoic acid, 2,3-dihydroxy-3-methyl-2-phenyl-, methyl ester		DD00703
234	91	92	120	206	93	71	119	252	100	64	59	47	35	35	34	22	2.00	14	20	4	–	–	–	–	–	–	–	–	–	–	9H-Naphtho[1,2-c]furan-9-one, 7α-hydroxy-1,3,5,5a,6,7,8,9a-octahydro-		MS05498
166	105	107	77	57	79	106	87	252	100	74	57	47	45	35	29	21	0.00	14	20	4	–	–	–	–	–	–	–	–	–	–	Pentanoic acid, 2,3-dihydroxy-3-ethyl-2-phenyl-, methyl ester		DD00704
120	91	92	105	132	189	133	252	252	100	86	73	48	22	18	18	11		14	20	4	–	–	–	–	–	–	–	–	–	–	Propanedioic acid, [(1,7-cyclooctadien-1-yl)methyl]-, dimethyl ester		DD00705
237	97	252	124	55	167	77	69	252	100	89	45	40	40	38	32	32		14	21	2	–	–	–	–	–	–	–	1	–	–	Phosphetan, 2,2,3,4,4-pentamethyl-1-phenoxy-, 1-oxide		NI24823
179	151	223	123	45	59	77	121	252	100	79	69	40	29	23	22	20	0.39	14	24	2	–	–	–	–	–	–	1	–	–	–	Silane, triethyl(2-phenoxyethoxy)-	16654-60-7	NI24824
69	73	159	94	151	75	103	41	252	100	90	76	69	48	46	39	37	12.00	14	24	2	–	–	–	–	–	–	1	–	–	–	Silane, trimethyl[(5-phenoxypentyl)oxy]-	16654-53-8	NI24825
84	127	126	43	128	85	86	44	252	100	80	60	27	21	20	12	12	4.00	14	24	2	2	–	–	–	–	–	–	–	–	–	2,3′-Bipiperidine, 1,1′-diacetyl-	53508-18-2	NI24826
123	183	71	42	122	144	121	165	252	100	71	67	61	58	54	54	41	4.76	14	24	2	2	–	–	–	–	–	–	–	–	–	Butanoic acid, 4-[methyl[4-(1-pyrrolidinyl)-2-butynyl]amino]-, methyl ester	56196-54-4	NI24827
84	96	179	252	55	140	41	69	252	100	38	26	21	17	12	11	10		14	24	2	2	–	–	–	–	–	–	–	–	–	1-(1-Cyclohexenyl)-1-pentamethylene-2-ethoxycarbonylhydrazinium, innersalt	83487-77-8	NI24828
72	180	123	57	55	41	151	43	252	100	90	32	32	26	25	24	24	12.00	14	24	2	2	–	–	–	–	–	–	–	–	–	Propanediamide, 2-(1-cyclohexen-1-yl)-N,N,N′,N′,2-pentamethyl-	42948-59-4	NI24829
72	180	123	151	252	135	237	207	252	100	90	32	24	12	10	7	7		14	24	2	2	–	–	–	–	–	–	–	–	–	Propanediamide, 2-(1-cyclohexen-1-yl)-N,N,N′,N′,2-pentamethyl-	42948-59-4	LI05128
57	86	71	94	74	122	77	130	252	100	83	51	50	48	37	33	32	14.00	14	24	2	2	–	–	–	–	–	–	–	–	–	2-Propanol, 1-(2-amino-3-methylphenoxy)-3-(tert-butylamino)-		MS05499
44	29	59	43	99	41	79	80	252	100	81	74	66	61	48	39	25	11.00	14	24	2	2	–	–	–	–	–	–	–	–	–	9-Aza[3.3.2]propellane-10-carboxylic acid, 9-dimethylamino-, ethyl ester		NI24830
73	135	82	163	155	45	67	74	252	100	55	19	19	18	16	12	10	3.00	14	28	–	–	–	–	–	–	–	2	–	–	–	1,4-Cyclohexadiene, 1,2-bis(trimethylsilyl)-3,6-dimethyl-		NI24831
73	163	149	164	155	45	82	59	252	100	67	31	26	18	18	16	14	7.70	14	28	–	–	–	–	–	–	–	2	–	–	–	1,4-Cyclohexadiene, 1,2-bis(trimethylsilyl)-4,5-dimethyl-		NI24832
127	111	126	69	85	83	68	112	252	100	99	85	71	42	39	19	16	4.00	14	28	–	4	–	–	–	–	–	–	–	–	–	1,3-Diazetidine, 1,3-diisopropyl-2,4-diisopropylimino-		MS05500
252	253							252	100	16								15	8	4	–	–	–	–	–	–	–	–	–	–	1-Anthracenecarboxylic acid, 9,10-dihydro-9,10-dioxo-	602-69-7	NI24833
253	252							252	100	6								15	8	4	–	–	–	–	–	–	–	–	–	–	2-Anthracenecarboxylic acid, 9,10-dihydro-9,10-dioxo-	117-78-2	NI24834
105	252	77	75	51	106	103	208	252	100	99	92	80	74	59	55	54		15	8	4	–	–	–	–	–	–	–	–	–	–	1H,3H-Isobenzofuran-1,3-dione, 5-benzoyl-	3886-01-9	NI24835
252	251	225	253	218	77	224	91	252	100	38	21	15	14	13	13	13		15	12	–	2	1	–	–	–	–	–	–	–	–	1H-1,4-Benzodiazepine-2-thione, 2,3-dihydro-5-phenyl-	34099-69-9	NI24836
252	224	117	64	118	90	63	145	252	100	47	38	23	21	21	20	13		15	12	2	2	–	–	–	–	–	–	–	–	–	Aniline, 2-methoxy-N-(2-oxo-3-indolinylidine)-		LI05129
224	252	64	63	223	118	107	90	252	100	76	35	20	14	8	8	6		15	12	2	2	–	–	–	–	–	–	–	–	–	Aniline, 3-methoxy-N-(2-oxo-3-indolinylidine)-		LI05130
252	151	179	196	150	75	152	207	252	100	73	57	47	39	24	22	20		15	12	2	2	–	–	–	–	–	–	–	–	–	Benzo[c]cinnoline-2-carboxylic acid, ethyl ester	19174-79-9	NI24837
180	151	150	181	252	152	75	208	252	100	35	21	17	14	12	11	10		15	12	2	2	–	–	–	–	–	–	–	–	–	Benzo[c]cinnoline-4-carboxylic acid, ethyl ester	19174-80-2	NI24838
180	179	223	252	178	181	152	177	252	100	35	28	25	15	12	9	7		15	12	2	2	–	–	–	–	–	–	–	–	–	6H-Dibenzo[b,f]oxireno[d]azepine-6-carboxamide, 1a,10b-dihydro-	36507-30-9	NI24839
207	179	178	180	177	208	152	151	252	100	74	36	22	18	16	14	12	0.00	15	12	2	2	–	–	–	–	–	–	–	–	–	6H-Dibenzo[b,f]oxireno[d]azepine-6-carboxamide, 1a,10b-dihydro-	36507-30-9	MS05501
180	104	77	209	252	223	51	165	252	100	68	58	54	51	40	32	24		15	12	2	2	–	–	–	–	–	–	–	–	–	2,4-Imidazolidinedione, 5,5-diphenyl-	57-41-0	NI24840
180	77	209	104	223	252	51	181	252	100	57	56	56	44	34	28	25		15	12	2	2	–	–	–	–	–	–	–	–	–	2,4-Imidazolidinedione, 5,5-diphenyl-	57-41-0	NI24841
180	104	252	223	209	77	181	165	252	100	65	60	55	50	50	30	30		15	12	2	2	–	–	–	–	–	–	–	–	–	2,4-Imidazolidinedione, 5,5-diphenyl-	57-41-0	NI24842
133	252	132	105	104	77	253	223	252	100	82	44	42	26	17	14	14		15	12	2	2	–	–	–	–	–	–	–	–	–	2-Indolinone, 1-(4-hydroxybenzylideneamino)-		AI01801
91	104	105	148	106	130	78	76	252	100	99	96	94	70	66	66	50	23.00	15	12	2	2	–	–	–	–	–	–	–	–	–	1H-Isoindole-1,3(2H)-dione, 2-(3-azatricyclo[3.2.1.0^{2,4}]oct-6-en-3-yl)-	75442-61-4	NI24843
105	104	76	106	91	77	50	79	252	100	91	83	56	52	42	39	36	32.00	15	12	2	2	–	–	–	–	–	–	–	–	–	1H-Isoindole-1,3(2H)-dione, 2-(3-azatricyclo[3.2.1.0^{2,4}]oct-6-en-3-yl)-, (1α,2β,4β,5α)-	24948-29-6	NI24844
252	106	78	174	237	253	146	126	252	100	33	26	24	21	18	11	10		15	12	2	2	–	–	–	–	–	–	–	–	–	Isoxazole, 3-(4-methoxyphenyl)-5-pyridyl-		LI05131
252	133	135	119	103	77	64	91	252	100	81	80	52	23	22	22	18		15	12	2	2	–	–	–	–	–	–	–	–	–	1,2,4-Oxadiazole, 3-phenyl-5-(3-methoxyphenyl)-		NI24845
133	252	135	119	103	64	77	253	252	100	90	58	37	19	17	16	15		15	12	2	2	–	–	–	–	–	–	–	–	–	1,2,4-Oxadiazole, 3-phenyl-5-(4-methoxyphenyl)-		NI24846
183	252	77	18	184	105	91	51	252	100	94	94	43	35	28	20	16		15	12	2	2	–	–	–	–	–	–	–	–	–	3,5-Pyrazolidinedione, 1,2-diphenyl-	2652-77-9	MS05502
77	252	183	105	184	51	91	64	252	100	74	74	28	27	24	21	16		15	12	2	2	–	–	–	–	–	–	–	–	–	3,5-Pyrazolidinedione, 1,2-diphenyl-	2652-77-9	NI24847
252	235	251	119	77	76	50	59	252	100	76	61	52	30	27	25	24		15	12	2	2	–	–	–	–	–	–	–	–	–	4(3H)-Quinazolinone, 3-(2-hydroxy-6-methylphenyl)-		NI24848
236	90	252	221	77	235	223	251	252	100	76	66	59	54	49	43	39		15	12	2	2	–	–	–	–	–	–	–	–	–	Quinoxaline, 2-methoxy-3-phenyl-, 4-oxide	18457-83-5	NI24849
43	149	121	152	93	107	119	41	252	100	58	44	39	32	31	23	22	0.00	15	24	3	–	–	–	–	–	–	–	–	–	–	Acetic acid, 9-hydroxy-1,5,5,9-tetramethylbicyclo[4.3.0]non-6-en-2-yl ester, (1RS,2RS,9RS)-		DD00706
159	43	174	119	118	91	234	105	252	100	38	23	17	17	14	10	10	0.00	15	24	3	–	–	–	–	–	–	–	–	–	–	Acetic acid, 9-hydroxy-1,5,5,9-tetramethylbicyclo[4.3.0]non-6-en-2-yl ester, (1RS,2SR,9RS)-		DD00707
43	139	210	179	153	193	192	252	252	100	52	19	13	6	3	2	1		15	24	3	–	–	–	–	–	–	–	–	–	–	Acetic acid, 1-methoxy-2-pentylbicyclo[2.2.1]hept-2-en-7-yl ester, cis-		MS05503
43	152	121	93	149	119	107	41	252	100	57	42	33	32	28	28	28	0.00	15	24	3	–	–	–	–	–	–	–	–	–	–	Acetic acid, 1,2,6,6-tetramethyl-10-oxatricyclo[5.2.1.0^{2,7}]dec-3-yl ester, (1RS,2RS,3RS)-		DD00708

MASS TO CHARGE RATIOS	M.W.	INTENSITIES	Parent	C	H	O	N	S	F	Cl	Br	I	Si	P	B	X	COMPOUND NAME	CAS Reg No	No
43 119 121 107 136 93 192 91	**252**	100 84 76 72 52 48 39 38	**0.00**	15	24	3	–	–	–	–	–	–	–	–	–	–	**Acetic acid, 2,4,4-trimethyl-3-(3-oxobutyl)-2-cyclohexen-1-yl ester**		**DD00709**
43 41 55 224 123 71 95 69	**252**	100 51 32 29 26 26 25 24	**0.27**	15	24	3	–	–	–	–	–	–	–	–	–	–	**Cyclohexanecarboxaldehyde, 6-methyl-3-isopropyl-2-oxo-1-(3-oxobutyl)-**	22339-30-6	**NI24850**
163 179 252 41 95 29 123 55	**252**	100 68 36 26 24 24 18 17		15	24	3	–	–	–	–	–	–	–	–	–	–	**3-Cyclohexene-1-one, 4-(2-Carboethoxy-2,2-dimethylethyl)-5,5-dimethyl-**		**IC03309**
192 163 121 135 43 137 177 107	**252**	100 93 60 49 48 40 34 24	**1.00**	15	24	3	–	–	–	–	–	–	–	–	–	–	**2-Cyclohexen-1-one, 3-[2-(acetyloxy)butyl]-2,4,4-trimethyl-**	75332-27-3	**NI24851**
183 196 252 42 209 178 177 195	**252**	100 83 60 60 35 35 35 32		15	24	3	–	–	–	–	–	–	–	–	–	–	**1,3-Cyclopentanedione, 4-isopentyl-2-isovaleryl-**	957-61-9	**NI24852**
182 177 195 41 57 140 237 55	**252**	100 43 32 24 21 18 17 16	**14.00**	15	24	3	–	–	–	–	–	–	–	–	–	–	**1,3-Cyclopentanedione, 2-(1-oxopentyl)-4-pentyl-**	69796-08-3	**NI24853**
81 163 109 123 149 234 191 135	**252**	100 98 90 79 71 65 65 45	**0.00**	15	24	3	–	–	–	–	–	–	–	–	–	–	**4,8-Epidioxyazulen-5-ol, 1,2,3,3a,4,5,8,8a-octahydro-1,4-dimethyl-7-isopropyl-, [1R-(1α,3aβ,4β,5β,8β,8aβ)]-**	94388-64-4	**DD00710**
137 95 109 43 194 97 176 252	**252**	100 33 27 27 25 25 22 2		15	24	3	–	–	–	–	–	–	–	–	–	–	**1,4-Ethano-1H-cyclopent[c]oxepin-2-one, octahydro-1,3,3,6-tetramethyl-7-hydroxy-, [1S-(1α,4α,5aβ,6α,8aα)]-**		**LI05132**
43 41 59 205 39 149 109 107	**252**	100 69 51 27 26 24 22 22	**5.00**	15	24	3	–	–	–	–	–	–	–	–	–	–	**3aH-Furo[3,2-c]isobenzofuran-8-methanol, 5,5a,6,7,8,9-hexahydro-α,α,3a,5a-tetramethyl-, [3aR-(3aα,5aβ,8β,9aR*)]-**		**NI24854**
125 55 41 167 81 69 252 43	**252**	100 86 52 40 32 25 20 20		15	24	3	–	–	–	–	–	–	–	–	–	–	**4ξ,10ξ-Germacran-12-oic acid, 6α-hydroxy-1-oxo-, γ-lactone, (11S)-**	19906-67-3	**NI24855**
166 43 167 151 41 71	**252**	100 31 24 21 16 7	**4.00**	15	24	3	–	–	–	–	–	–	–	–	–	–	**Illudin M, tetrahydro-**		**MS05504**
85 150 132 151 234 106 86 108	**252**	100 80 14 11 10 9 7 6	**0.00**	15	24	3	–	–	–	–	–	–	–	–	–	–	**1H-Inden-5-ol, 2,3,5,6,7,7a-hexahydro-7a-methyl-1-[(tetrahydro-2H-pyran-2-yl)oxy]-, (1α,5α,7aα)-**	67884-35-9	**NI24856**
82 109 43 67 153 107 91 119	**252**	100 90 69 33 28 23 22 21	**0.00**	15	24	3	–	–	–	–	–	–	–	–	–	–	**Indicumenone**		**NI24857**
113 95 55 81 110 69 252 67	**252**	100 56 47 42 39 31 30 28		15	24	3	–	–	–	–	–	–	–	–	–	–	**1,4-Methano-3-benzoxepin-5,10-diol, 1,2,4,5,7,8,9,9a-octahydro-2,2,9,9a-tetramethyl-**	41988-43-6	**MS05505**
113 43 41 95 121 109 55 69	**252**	100 35 31 24 23 20 20 18	**0.00**	15	24	3	–	–	–	–	–	–	–	–	–	–	**1,4-Methano-3-benzoxepin-5,10-diol, 1,2,4,5,7,8,9,9a-octahydro-2,2,9,9a-tetramethyl-**	41988-43-6	**NI24858**
105 216 183 201 237 198 219 234	**252**	100 32 28 24 18 17 10 8	**1.00**	15	24	3	–	–	–	–	–	–	–	–	–	–	**1,3,13-Naphthalenetriol, 1,2,3,4,4a,5,6,7-octahydro-4,4a-dimethyl-6-isopropenyl-, [1R-(1α,3β,4β,4aα,6α)]-**		**NI24859**
43 177 192 41 121 55 69 149	**252**	100 74 52 39 34 32 31 26	**3.90**	15	24	3	–	–	–	–	–	–	–	–	–	–	**2(1H)-Naphthalenone, 3,4,4a,5,6,7,8,8aα-octahydro-5α-hydroxy-4aα,7,7-trimethyl-, acetate**	24795-35-5	**NI24860**
192 177 43 137 69 97 121 41	**252**	100 71 61 51 28 27 23 23	**6.10**	15	24	3	–	–	–	–	–	–	–	–	–	–	**2(1H)-Naphthalenone, 3,4,4a,5,6,7,8,8aα-octahydro-5β-hydroxy-4aα,7,7-trimethyl-, acetate**	28684-99-3	**NI24861**
113 41 93 95 91 67 250 55	**252**	100 21 17 16 12 10 4 3	**0.00**	15	24	3	–	–	–	–	–	–	–	–	–	–	**1(2H)-Naphthalenone, 3,4,4a,5,6,7-hexahydro-3-hydroxy-4-(1-hydroxyisopropyl)-4a,5-dimethyl-**	41988-40-3	**NI24862**
59 43 41 134 91 161 119 107	**252**	100 86 50 49 37 31 31 25	**1.00**	15	24	3	–	–	–	–	–	–	–	–	–	–	**1(2H)-Naphthalenone, 3,4,4a,5,6,7-hexahydro-3-hydroxy-4-(1-hydroxyisopropyl)-4a,5-dimethyl-, [3R-(3α,4α,4aβ,5β)]-**		**LI05133**
175 163 109 105 119 55 91 159	**252**	100 76 76 76 73 68 65 61	**9.00**	15	24	3	–	–	–	–	–	–	–	–	–	–	**Naphtho[1,2-c]furan-1-ol, 6,6,9a-trimethyl-1,3,5,5a,6,7,8,9,9a,10-decahydro-**		**MS05506**
107 122 43 91 93 79 41 77	**252**	100 56 13 9 5 5 4 3	**0.00**	15	24	3	–	–	–	–	–	–	–	–	–	–	**6-Oxabicyclo[3.2.1]oct-2-ene-7-ethanol, α,2,8,8-tetramethyl-, acetate**	66465-83-6	**NI24863**
125 140 152 98 69 252 41 112	**252**	100 73 62 32 28 26 17 16		15	24	3	–	–	–	–	–	–	–	–	–	–	**2H-Pyran-2-one, 4-methoxy-6-nonyl-**		**MS05507**
119 134 139 135 109 149 137 177	**252**	100 76 74 70 65 62 56 48	**0.80**	15	24	3	–	–	–	–	–	–	–	–	–	–	**Spiro[cyclobutane-1,2′-[7]oxabicyclo[4.1.0]heptan]-2-ol, 2,4′,4′,6′-tetramethyl-, acetate**	67884-48-4	**NI24864**
43 137 75 41 221 161 107 203	**252**	100 69 43 39 32 28 27 26	**4.00**	15	24	3	–	–	–	–	–	–	–	–	–	–	**Spiro[4.5]dec-6-en-8-one, 2-(1,2-dihydroxy-1-methylethyl)-6,10-dimethyl-**	94444-27-6	**NI24865**
42 139 149 135 134 192 177 210	**252**	100 26 17 15 15 9 7 6	**0.00**	15	24	3	–	–	–	–	–	–	–	–	–	–	**Spiro[3.5]nonan-5-one, 1-(acetyloxy)-1,6,8,8-tetramethyl-**	37643-31-5	**NI24866**
252 237 233 253 238 234 235 251	**252**	100 82 20 19 17 16 7 3		16	12	3	–	–	–	–	–	–	–	–	–	–	**9,10-Anthracenedione, 1-ethoxy-**	22924-20-5	**NI24867**
237 252 233 253 238 234 235 251	**252**	100 90 24 23 20 19 9 4		16	12	3	–	–	–	–	–	–	–	–	–	–	**9,10-Anthracenedione, 1-ethoxy-**	22924-20-5	**LI05134**
251 252 236 208 105 209 152 122	**252**	100 56 13 8 7 6 6 6		16	12	3	–	–	–	–	–	–	–	–	–	–	**3(2H)-Benzofuranone, 2-benzylidene-6-methoxy-**	4940-52-7	**NI24868**
132 131 252 121 77 43 253 221	**252**	100 84 69 58 15 15 11 9		16	12	3	–	–	–	–	–	–	–	–	–	–	**[1]Benzopyrano[3,4-b][1]benzopyran-12-one, 6,6a,12,12a-tetrahydro-**		**LI05135**
132 252 89 117 251 63 92 221	**252**	100 81 41 35 31 30 28 17		16	12	3	–	–	–	–	–	–	–	–	–	–	**4H-1-Benzopyran-4-one, 4′-methoxy-2-phenyl-**	4143-74-2	**NI24869**
209 252 253 77 210 105 152 254	**252**	100 98 25 20 16 14 12 9		16	12	3	–	–	–	–	–	–	–	–	–	–	**7H-1-Benzopyran-7-one, 5-methoxy-2-phenyl-**		**MS05508**
208 252 75 180 179 178 207 151	**252**	100 90 86 84 84 76 74 41		16	12	3	–	–	–	–	–	–	–	–	–	–	**Δ^1,α-Acenaphtheneacetic acid, 2-oxo-, ethyl ester, (Z)-**	22561-90-6	**NI24870**
208 252 75 180 179 178 207 89.5	**252**	100 92 87 85 85 76 75 66		16	12	3	–	–	–	–	–	–	–	–	–	–	**Δ^1,α-Acenaphtheneacetic acid, 2-oxo-, ethyl ester, (Z)-**	22561-90-6	**LI05136**
252 221 163 253 64 192 155 68	**252**	100 92 20 19 9 8 7 7		16	12	3	–	–	–	–	–	–	–	–	–	–	**9H-Fluorene-2-carboxylic acid, 6-methyl-9-oxo-, methyl ester**		**NI24871**
141 95 252 140 39 139 157 128	**252**	100 57 45 19 11 10 8 6		16	12	3	–	–	–	–	–	–	–	–	–	–	**2-Furancarboxylic acid, 1-naphthaleneylmethyl ester**	86328-53-2	**NI24872**
118 147 252 105 119 89 77 90	**252**	100 90 70 56 48 47 42 40		16	12	3	–	–	–	–	–	–	–	–	–	–	**2(5H)-Furanone, 4-hydroxy-3,5-diphenyl-**	5435-01-8	**NI24873**
252 237 170 67 251 234 114 253	**252**	100 93 38 37 33 32 21 18		16	12	3	–	–	–	–	–	–	–	–	–	–	**1H-Naphtho[2,1-b]pyran-1-one, 2-acetyl-3-methyl-**	5891-83-8	**NI24874**
237 252 67 115 39 171 234 170	**252**	100 99 68 61 56 50 45 44		16	12	3	–	–	–	–	–	–	–	–	–	–	**4H-Naphtho[1,2-b]pyran-4-one, 3-acetyl-2-methyl-**	69796-59-4	**NI24875**
221 220 252 235 163 165 177 222	**252**	100 99 67 32 29 24 18 16		16	12	3	–	–	–	–	–	–	–	–	–	–	**4-Phenanthrenecarboxylic acid, 5-methoxy-**		**MS05509**
179 178 180 153 224 154 152 89	**252**	100 38 28 25 23 19 19 18	**11.25**	16	12	3	–	–	–	–	–	–	–	–	–	–	**Phenanthro[1,2-c]furan-1,3-dione, 3a,3b,11,11a-tetrahydro-**	63084-72-0	**NI24876**
252 251 77 196 90 118 89 105	**252**	100 51 36 35 35 32 21 20		16	12	3	–	–	–	–	–	–	–	–	–	–	**Spiro[benzofuran-2(3H),2′-oxiran]-3-one, 5-methyl-3′-phenyl-**	35405-30-2	**NI24877**
252 251 195 90 77 118 135 105	**252**	100 53 36 36 36 34 20 20		16	12	3	–	–	–	–	–	–	–	–	–	–	**Spiro[benzofuran-2(3H),2′-oxiran]-3-one, 5-methyl-3′-phenyl-**	35405-30-2	**LI05137**

MASS TO CHARGE RATIOS	M.W.	INTENSITIES	Parent	C	H	O	N	S	F	Cl	Br	I	Si	P	B	X	COMPOUND NAME	CAS Reg No	No
209 194 195 252 45 219 117 43	**252**	100 99 61 55 50 37 30 30		16	16	1	2	–	–	–	–	–	–	–	–	–	Acetamide, N-[2-[2-(2-aminophenyl)vinyl]phenyl]-	69395-26-2	NI24878
195 252 223 181	**252**	100 26 12 5		16	16	1	2	–	–	–	–	–	–	–	–	–	β-Carboline, 2-(3-oxopentyl)-		LI05138
237 145 132 159 252 133 65 92	**252**	100 83 59 53 49 45 35 32		16	16	1	2	–	–	–	–	–	–	–	–	–	11-Oxa-2,9-diazatetracyclo[8.7.1.0^{3,8}.0^{12,17}]octadeca-3,5,7,12,14,16-hexaene, 10-methyl-		MS05510
105 77 223 103 104 149 91 117	**252**	100 80 60 21 20 10 10 7	**4.00**	16	16	1	2	–	–	–	–	–	–	–	–	–	Propiophenone, 2-hydroxy-, benzoylhydrazone		LI05139
221 252 237 222 180 197 204	**252**	100 59 58 31 19 18 14		16	16	1	2	–	–	–	–	–	–	–	–	–	Xanthine, 4,5-dihydro-6-methyl-6-methoxy-		LI05140
90 252 107 89 174 117 251 118	**252**	100 67 46 40 38 29 28 27		16	17	2	–	–	–	–	–	–	–	–	1	–	1,3,2-Dioxaborolane, 2-ethyl-4,5-diphenyl-	74810-50-7	NI24879
217 203 252 237 133 117 131 105	**252**	100 55 22 21 21 14 13 13		16	25	–	–	–	–	1	–	–	–	–	–	–	Benzene, α-chloromethyl-2,4,6-tri-isopropyl-		LI05141
97 98 252 111 55 99 57 110	**252**	100 39 17 13 8 7 5 4		16	28	–	–	1	–	–	–	–	–	–	–	–	Thiophene, 2-dodecyl-		AI01802
43 57 153 99 136 237 177 41	**252**	100 95 88 69 53 39 38 37	**0.00**	16	28	2	–	–	–	–	–	–	–	–	–	–	Bicyclo[4.3.0]non-9-en-7-ol, 5-isopropoxy-2,2,6,7-tetramethyl-, (5RS,6RS,7RS)-		DD00711
43 119 46 121 107 41 93 91	**252**	100 82 61 60 51 41 39 35	**0.00**	16	28	2	–	–	–	–	–	–	–	–	–	–	2-Butanone, 4-(3-isopropoxy-2,6,6-trimethyl-1-cyclohexen-1-yl)-		DD00712
43 67 80 55 82 66 93 135	**252**	100 99 97 83 81 81 79 39	**31.00**	16	28	2	–	–	–	–	–	–	–	–	–	–	2-Cyclopentene-1-nonanol, acetate, (±)-	67886-20-8	NI24880
41 39 55 67 60 43 82 42	**252**	100 43 40 35 30 30 20 18	**1.00**	16	28	2	–	–	–	–	–	–	–	–	–	–	2-Cyclopentene-1-undecanoic acid	459-67-6	WI01284
43 41 29 55 69 27 111 67	**252**	100 93 65 63 45 40 36 34	**1.00**	16	28	2	–	–	–	–	–	–	–	–	–	–	2-Cyclopropene-1-carboxylic acid, 2,3-dihexyl-	54467-87-7	NI24881
99 97 81 151 169 168 43 252	**252**	100 85 68 67 60 53 48 40		16	28	2	–	–	–	–	–	–	–	–	–	–	2,4-Decadienoic acid, hexyl ester, (E,Z)-		LI05142
55 97 156 128 179 207 41 111	**252**	100 49 41 27 20 19 19 17	**13.58**	16	28	2	–	–	–	–	–	–	–	–	–	–	2,6-Decadienoic acid, 8-methyl-4-propyl-, ethyl ester	56771-51-8	MS05511
109 69 82 55 95 41 83 81	**252**	100 98 75 72 69 67 56 44	**4.12**	16	28	2	–	–	–	–	–	–	–	–	–	–	8-Decenoic acid, 5-vinyl-3,5,9-trimethyl-, methyl ester	56114-53-5	NI24882
150 43 87 107 147 41 190 115	**252**	100 88 71 48 29 28 22 21	**4.00**	16	28	2	–	–	–	–	–	–	–	–	–	–	1,3-Dioxolane, 2-(4,8-dimethyl-7-methylene-3-nonenyl)-2-methyl-, (Z)-	66972-06-3	NI24883
69 109 209 41 123 95 81 67	**252**	100 89 49 40 37 22 18 18	**4.26**	16	28	2	–	–	–	–	–	–	–	–	–	–	6,10-Dodecadienoic acid, 3,7,11-trimethyl-, methyl ester, (E)-(S)-	34114-96-0	NI24884
88 55 81 95 109 41 67 123	**252**	100 23 21 19 18 18 16 5	**0.00**	16	28	2	–	–	–	–	–	–	–	–	–	–	1α,5α-Guaian-12-oic acid, methyl ester		MS05512
45 159 41 174 144 86 136 43	**252**	100 94 42 37 32 30 29 29	**0.00**	16	28	2	–	–	–	–	–	–	–	–	–	–	2-Oxabicyclo[4.3.1]dec-6(10)-ene, 3-isopropoxy-3,7,7,10-tetramethyl-		DD00713
82 67 81 41 55 96 68 95	**252**	100 99 91 89 87 78 60 59	**16.01**	16	28	2	–	–	–	–	–	–	–	–	–	–	Oxacycloheptadec-8-en-2-one	123-69-3	NI24885
41 82 67 95 55 81 43 83	**252**	100 85 66 45 26 24 22 13	**0.00**	16	28	2	–	–	–	–	–	–	–	–	–	–	2-Oxacyclohexadecanone, 6-methylidene-		DD00714
193 41 55 43 137 194 83 141	**252**	100 60 59 53 49 39 38 36	**0.00**	16	28	2	–	–	–	–	–	–	–	–	–	–	1-Oxaspiro[4.5]decane, 7-isopropoxy-6-methylidene-2,10,10-trimethyl-, (2RS,7RS)-		DD00715
193 41 194 55 43 141 209 83	**252**	100 44 38 37 37 34 28 22	**0.00**	16	28	2	–	–	–	–	–	–	–	–	–	–	1-Oxaspiro[4.5]decane, 7-isopropoxy-6-methylidene-2,10,10-trimethyl-, (2RS,7SR)-		DD00716
161 120 135 43 119 176 41 159	**252**	100 71 53 51 43 40 39 32	**0.00**	16	28	2	–	–	–	–	–	–	–	–	–	–	10-Oxatricyclo[5.2.1.0^{2,7}]decane, 3-isopropoxy-1,2,6,6-tetramethyl-, (1RS,2RS,3RS,7RS)-		DD00717
124 41 81 111 43 109 91 93	**252**	100 82 75 73 67 60 57 48	**0.00**	16	28	2	–	–	–	–	–	–	–	–	–	–	3-Oxatricyclo[5.4.0.0^{1,4}]undecane, 11-isopropoxy-4,8,8-trimethyl-		DD00718
74 55 41 87 81 67 43 69	**252**	100 56 55 44 41 32 32 28	**0.00**	16	28	2	–	–	–	–	–	–	–	–	–	–	14-Pentadecynoic acid, methyl ester	56909-04-7	NI24886
73 55 141 41 114 115 83 69	**252**	100 66 64 58 57 49 46 45	**22.75**	16	29	–	–	–	–	–	–	–	–	1	–	–	Phosphine, [methylisopropylcyclohexyl]di-2-propenyl-	74710-01-3	NI24887
83 42 43 41 55 57 54 44	**252**	100 9 8 7 5 4 4 4	**0.60**	16	32	–	2	–	–	–	–	–	–	–	–	–	Dodecanamine, N-(2-cyanoethyl)-2-methyl-		IC03310
83 212 41 43 42 44 55 56	**252**	100 15 13 12 12 11 8 6	**2.00**	16	32	–	2	–	–	–	–	–	–	–	–	–	Tridecanamine, N-(2-cyanoethyl)-		IC03311
251 252 109 90 89 253 233 133	**252**	100 49 24 13 10 8 7 7		17	13	1	–	–	1	–	–	–	–	–	–	–	1-Naphthalenone, 3,4-dihydro-2-(4-fluorophenylmethylene)-	66045-89-4	NI24888
252 224 237 253 221 223 222 208	**252**	100 35 25 19 14 10 10 9		17	16	–	–	1	–	–	–	–	–	–	–	–	Naphtho[2,3-b]benzothiophene, 6-methyl-7,8,9,10-tetrahydro-		AI01803
128 143 115 65 39 252 51 77	**252**	100 93 74 73 67 60 48 27		17	16	–	–	1	–	–	–	–	–	–	–	–	1,3-Pentadiene, 5-phenyl-3-(phenylthio)-, (Z)-	110027-08-2	DD00719
252 210 253 208 221 211 197 209	**252**	100 62 19 13 12 12 11 9		17	16	–	–	1	–	–	–	–	–	–	–	–	11-Thiabenzo[b]fluorene, 8-methyl-6,7,8,9-tetrahydro-		AI01804
208 165 179 91 128 117 77 252	**252**	100 46 43 37 29 29 25 18		17	16	2	–	–	–	–	–	–	–	–	–	–	2H-Azuleno[1,8-bc]furan-2-one, 4,5,7,8,8a,8b-hexahydro-4-phenyl-, (±)-(4α,8aα,8bβ)-		DD00720
104 131 103 77 105 91 102 78	**252**	100 44 29 26 13 9 5 5	**1.42**	17	16	2	–	–	–	–	–	–	–	–	–	–	Cinnamic acid, phenylethyl ester	103-53-7	NI24889
105 219 77 220 91 51 28 224	**252**	100 90 78 36 30 27 18 16	**13.00**	17	16	2	–	–	–	–	–	–	–	–	–	–	2,3-Dioxabicyclo[2.2.1]heptane, 1,4-diphenyl-	62121-85-1	NI24890
43 165 192 193 166 163 179 164	**252**	100 99 90 21 21 13 9 7	**7.00**	17	16	2	–	–	–	–	–	–	–	–	–	–	9H-Fluorene, 1-[1-(hydroxymethyl)ethyl]-		NI24891
252 107 158 43 145 237 251 131	**252**	100 39 36 33 30 23 21 21		17	16	2	–	–	–	–	–	–	–	–	–	–	1,4-Pentadiene, 1,3-bis(4-hydroxyphenyl)-	17676-24-3	NI24892
128 77 79 107 156 51 129 105	**252**	100 78 75 62 43 35 24 23	**0.00**	17	16	2	–	–	–	–	–	–	–	–	–	–	2-Pentyne-1,5-diol, 1,5-diphenyl-, meso-		MS05513
252 237 209 166 165 253 126 135	**252**	100 42 31 25 23 21 18 11		17	16	2	–	–	–	–	–	–	–	–	–	–	Phenanthrene, 3,6-dimethoxy-9-methyl-	15638-09-2	WI01285
252 178 207 180 165 179 152 89	**252**	100 92 76 48 36 32 28 20		17	16	2	–	–	–	–	–	–	–	–	–	–	2-Propenoic acid, 3-(4-phenylphenyl)-, ethyl ester		NI24893
135 237 252 178 105 73 43 159	**252**	100 33 11 7 7 6 6 4		17	20	–	–	–	–	–	–	–	1	–	–	–	Ethene, 1,1-diphenyl-2-(trimethylsilyl)-	51318-07-1	DD00721
223 195 224 196 252 193 225 194	**252**	100 39 23 8 7 7 5 5		17	20	–	–	–	–	–	–	–	1	–	–	–	9-Silaanthracene, 9,9-diethyl-9,10-dihydro-		NI24894
121 252 105 77 126 79 251 237	**252**	100 31 14 11 8 7 5 5		17	20	–	2	–	–	–	–	–	–	–	–	–	Methanimidamide, N,N'-bis(2,4-dimethylphenyl)-	16596-04-6	NI24896
121 252 120 106 132 122 105 79	**252**	100 20 17 13 11 9 6 4		17	20	–	2	–	–	–	–	–	–	–	–	–	Methanimidamide, N,N'-bis(2,4-dimethylphenyl)-	16596-04-6	NI24895
121 132 106 120 252 105 122 77	**252**	100 29 17 13 12 11 9 8		17	20	–	2	–	–	–	–	–	–	–	–	–	Methanimidamide, N,N'-bis(2,6-dimethylphenyl)-	16596-05-7	NI24897
121 252 106 120 122 105 251 77	**252**	100 30 12 10 9 8 7 6		17	20	–	2	–	–	–	–	–	–	–	–	–	Methanimidamide, N,N'-bis(3,5-dimethylphenyl)-	64107-05-7	NI24898
44 222 223 252 206 207 208 194	**252**	100 67 63 32 31 27 26 22		17	20	–	2	–	–	–	–	–	–	–	–	–	1,5-Methano-1H-azocino[4,3-b]indole, 12-ethyl-2,3,4,5,6,7-hexahydro-6-methylene-	3620-83-5	NI24899

MASS TO CHARGE RATIOS								M.W.	INTENSITIES								Parent	C	H	O	N	S	F	Cl	Br	I	Si	P	B	X	COMPOUND NAME	CAS Reg No	No
55	41	71	58	43	69	59	98	**252**	100	89	81	77	51	40	37	34	**21.22**	17	32	1	–	–	–	–	–	–	–	–	–	–	**Cycloheptadecanone**	3661-77-6	**WI01286**
55	41	80	67	81	82	95	94	**252**	100	91	84	79	74	71	58	51	**0.80**	17	32	1	–	–	–	–	–	–	–	–	–	–	**9-Cycloheptadecen-1-ol**	17344-59-1	**WI01287**
96	81	67	55	54	28	95	41	**252**	100	47	36	31	29	28	26	24	**1.80**	17	32	1	–	–	–	–	–	–	–	–	–	–	**12-Heptadecyn-1-ol**	56554-76-8	**NI24900**
82	28	67	81	55	96	41	95	**252**	100	63	62	52	47	46	38	24	**0.00**	17	32	1	–	–	–	–	–	–	–	–	–	–	**13-Heptadecyn-1-ol**	56554-77-9	**NI24901**
81	67	55	95	96	82	41	68	**252**	100	86	86	70	57	52	52	43	**1.00**	17	32	1	–	–	–	–	–	–	–	–	–	–	**8-Hexadecyn-1-ol, 14-methyl-, (R)-(–)-**	64566-18-3	**NI24902**
57	41	55	38	29	97	85	83	**252**	100	28	25	24	19	15	13	13	**1.01**	17	32	1	–	–	–	–	–	–	–	–	–	–	**6-Undecen-3-one, 5-butyl-2,2-dimethyl-, (E)-**	55976-05-1	**NI24903**
146	117	104	91	105	77	118	147	**252**	100	90	74	42	26	23	20	14	**8.50**	18	20	1	–	–	–	–	–	–	–	–	–	–	**Oxepane, 2,7-diphenyl-**		**MS05514**
104	105	77	91	78	103	51	39	**252**	100	75	64	50	49	47	44	31	**9.30**	18	20	1	–	–	–	–	–	–	–	–	–	–	**Pyran, 2,6-diphenyl-4-methyltetrahydro-, cis-**		**MS05515**
237	252	91	107	57	165	178	152	**252**	100	68	19	14	13	12	8	8		18	20	1	–	–	–	–	–	–	–	–	–	–	**Stilbene, 2-hydroxy-4′-tert-butyl-, (E)-**		**MS05516**
252	122	106	134	117	91	133	253	**252**	100	64	52	44	42	39	26	23		18	20	1	–	–	–	–	–	–	–	–	–	–	**Tricyclo[10.2.2.2^{5,8}]octadeca-5,7,12,14,15,17-hexaen-6-ol**	24770-04-5	**NI24904**
69	83	55	43	70	71	41	97	**252**	100	97	88	88	86	83	70	64	**3.20**	18	36	–	–	–	–	–	–	–	–	–	–	–	**Cyclopropane, 1-(1,2-dimethylpropyl)-1-methyl-2-nonyl-**	41977-42-8	**NI24905**
82	83	55	57	41	69	43	111	**252**	100	80	71	46	43	42	41	28	**0.14**	18	36	–	–	–	–	–	–	–	–	–	–	–	**Dodecane, 2-cyclohexyl-**	13151-82-1	**NI24906**
83	57	55	82	69	43	41	71	**252**	100	99	88	82	68	65	62	46	**0.00**	18	36	–	–	–	–	–	–	–	–	–	–	–	**Dodecane, 3-cyclohexyl-**	13151-83-2	**NI24907**
83	82	55	57	69	43	41	71	**252**	100	85	68	66	48	45	45	36	**1.96**	18	36	–	–	–	–	–	–	–	–	–	–	–	**Dodecane, 3-cyclohexyl-**	13151-83-2	**MS05517**
83	57	55	82	43	41	71	69	**252**	100	91	86	64	61	52	51	36	**0.00**	18	36	–	–	–	–	–	–	–	–	–	–	–	**Dodecane, 4-cyclohexyl-**	13151-84-3	**NI24908**
57	55	83	43	82	71	41	97	**252**	100	85	84	66	62	59	55	51	**0.00**	18	36	–	–	–	–	–	–	–	–	–	–	–	**Dodecane, 5-cyclohexyl-**	13151-85-4	**NI24909**
57	55	83	43	82	71	41	97	**252**	100	83	78	68	58	54	53	51	**0.00**	18	36	–	–	–	–	–	–	–	–	–	–	–	**Dodecane, 6-cyclohexyl-**	13151-86-5	**NI24910**
56	55	69	57	41	43	83	111	**252**	100	99	93	90	71	62	58	42	**3.20**	18	36	–	–	–	–	–	–	–	–	–	–	–	**7-Heptadecene, 7-methyl-, (E)-**		**AI01805**
56	55	69	57	41	43	83	97	**252**	100	93	82	75	61	54	51	39	**11.00**	18	36	–	–	–	–	–	–	–	–	–	–	–	**8-Heptadecene, 8-methyl-, (E)-**		**AI01806**
41	43	55	57	83	97	69	29	**252**	100	99	91	75	73	72	61	51	**6.41**	18	36	–	–	–	–	–	–	–	–	–	–	–	**1-Octadecene**	112-88-9	**WI01288**
43	41	55	57	69	83	97	29	**252**	100	83	76	72	51	49	40	39	**5.42**	18	36	–	–	–	–	–	–	–	–	–	–	–	**1-Octadecene**	112-88-9	**IC03312**
43	41	55	57	83	69	97	29	**252**	100	95	88	71	70	59	53	49	**5.07**	18	36	–	–	–	–	–	–	–	–	–	–	–	**1-Octadecene**	112-88-9	**AI01807**
167	165	168	166	152	252	91	29	**252**	100	15	14	7	7	6	5	5		19	24	–	–	–	–	–	–	–	–	–	–	–	**Heptane, 1,1-diphenyl-**		**AI01808**
167	168	165	152	91	252	166	41	**252**	100	15	11	6	6	5	4	4		19	24	–	–	–	–	–	–	–	–	–	–	–	**Heptane, 1,1-diphenyl-**		**AI01809**
119	91	120	105	41	77	79	118	**252**	100	58	37	33	18	16	13	8	**0.00**	19	24	–	–	–	–	–	–	–	–	–	–	–	**Pentane, 2,4-dimethyl-2,4-diphenyl-**	30387-24-7	**NI24911**
148	147	133	91	252	134	41	105	**252**	100	70	57	39	31	31	14	13		19	24	–	–	–	–	–	–	–	–	–	–	–	**Propane, 1-phenyl-3-(2,3,5,6-tetramethylphenyl)-**		**AI01810**
252	253	126	250	125	113	124	112	**252**	100	23	23	19	15	8	7	7		20	12	–	–	–	–	–	–	–	–	–	–	–	**Benz[e]acephenanthrylene**		**MS05518**
252	253	126	250	125	113	124	112	**252**	100	23	22	21	19	9	8	8		20	12	–	–	–	–	–	–	–	–	–	–	–	**Benzo[j]fluoranthene**	205-82-3	**NI24912**
252	126	253	250	125	113	112	124	**252**	100	27	23	20	16	8	8	7		20	12	–	–	–	–	–	–	–	–	–	–	–	**Benzo[k]fluoranthene**	207-08-9	**NI24913**
252	253	126	250	125	113	112	124	**252**	100	22	21	19	13	8	7	6		20	12	–	–	–	–	–	–	–	–	–	–	–	**Benzo[k]fluoranthene**	207-08-9	**MS05519**
252	253	126	254	250	125	113	112	**252**	100	78	40	39	28	20	15	12		20	12	–	–	–	–	–	–	–	–	–	–	–	**Benzo[k]fluoranthene, 7,12-dihydro-**		**NI24914**
252	126	253	250	125	113	124	112	**252**	100	23	21	16	15	9	7	7		20	12	–	–	–	–	–	–	–	–	–	–	–	**Benzo[a]pyrene**	50-32-8	**NI24916**
252	250	253	126	125	124	113	112	**252**	100	3	2	1	1	1	1	1		20	12	–	–	–	–	–	–	–	–	–	–	–	**Benzo[a]pyrene**	50-32-8	**NI24915**
252	253	250	126	125	228	113	251	**252**	100	23	21	20	15	13	12	9		20	12	–	–	–	–	–	–	–	–	–	–	–	**Benzo[a]pyrene**	50-32-8	**NI24917**
252	250	253	126	125	124	113	112	**252**	100	3	2	2	1	1	1	1		20	12	–	–	–	–	–	–	–	–	–	–	–	**Benzo[e]pyrene**	192-97-2	**NI24918**
252	250	253	125	248	126	124	113	**252**	100	3	2	2	1	1	1	1		20	12	–	–	–	–	–	–	–	–	–	–	–	**Benzo[e]pyrene**	192-97-2	**NI24919**
252	250	253	125	126	113	251	124	**252**	100	24	21	16	14	8	7	7		20	12	–	–	–	–	–	–	–	–	–	–	–	**Benzo[e]pyrene**	192-97-2	**NI24920**
252	132	130	253	250	128	251	91	**252**	100	33	29	27	27	11	9	9		20	12	–	–	–	–	–	–	–	–	–	–	–	**Perylene**	198-55-0	**IC03313**
252	253	250	126	125	251	124	248	**252**	100	22	22	22	17	7	7	6		20	12	–	–	–	–	–	–	–	–	–	–	–	**Perylene**	198-55-0	**NI24921**
252	253	250	126	125	251	124	112	**252**	100	22	21	21	17	7	7	5		20	12	–	–	–	–	–	–	–	–	–	–	–	**Perylene**	198-55-0	**WI01289**
69	70	184	134	46	51	146	31	**253**	100	44	42	38	24	21	18	13	**0.03**	3	–	–	1	1	9	–	–	–	–	–	–	–	**Sulphur difluoride imide, N-(heptafluoroisopropyl)-**		**LI05143**
253	166	44	177	146	45	42	209	**253**	100	99	90	89	89	89	77	76		4	12	–	3	1	3	–	–	–	–	2	–	–	**Phosphoramidothioic difluoride, bis[(dimethylamino)fluorophosphoranylidene]**		**LI05144**
253	126	255	99	63	128	90	127	**253**	100	34	33	16	13	12	12	7		6	5	–	1	–	–	1	–	1	–	–	–	–	**Aniline, 3-chloro-4-iodo-**		**NI24922**
77	73	28	159	105	45	43	147	**253**	100	90	45	26	13	10	8	7	**0.00**	7	13	–	1	–	6	–	–	–	1	–	–	–	**Ethylamine, N,N-bis(trifluoromethyl)-2-(trimethylsilyl)-**	28871-56-9	**NI24923**
135	107	77	92	253	78	76	120	**253**	100	82	65	43	30	21	14	12		9	7	2	3	2	–	–	–	–	–	–	–	–	**Benzamide, N-(5H-1,3,2,4-dithia(3-SIV)diazol-5-ylidene)-4-methoxy-**	57726-55-3	**NI24924**
253	255	171	169	256	254	212	210	**253**	100	99	95	95	10	10	10	10		9	8	1	3	–	–	–	1	–	–	–	–	–	**Δ^2-1,2,4-Triazolin-5-one, 1-(p-bromophenyl)-3-methyl-**	32579-49-0	**NI24925**
91	130	65	92	39	51	40	76	**253**	100	11	9	8	5	4	4	3	**1.50**	9	8	1	3	–	–	–	1	–	–	–	–	–	**1,2,3-Triazole, 3-benzyl-4-bromo-, 1-oxide**		**NI24926**
253	255	42	169	90	171	70	63	**253**	100	98	41	35	34	34	25	21		9	8	1	3	–	–	–	1	–	–	–	–	–	**1,2,3-Triazole, 4-bromo-2-(4-hydroxyphenyl)-5-methyl-**		**NI24927**
208	72	74	30	44	43	60	45	**253**	100	83	78	38	31	30	23	23	**0.60**	9	19	5	1	1	–	–	–	–	–	–	–	–	**D-erythro-α-D-Galacto-octopyranoside, methyl 6-amino-6,8-dideoxy-1-thio-**	14810-93-6	**NI24929**
208	72	74	160	30	44	43	60	**253**	100	83	77	39	37	31	30	23	**5.97**	9	19	5	1	1	–	–	–	–	–	–	–	–	**D-erythro-α-D-Galacto-octopyranoside, methyl 6-amino-6,8-dideoxy-1-thio-**	14810-93-6	**NI24928**
255	253	133	89	51	134	63	77	**253**	100	98	80	66	40	33	33	32		10	8	–	1	1	–	–	1	–	–	–	–	–	**Isothiazole, 4-bromo-3-methyl-5-phenyl-**		**MS05520**

MASS TO CHARGE RATIOS								M.W.	INTENSITIES								Parent	C	H	O	N	S	F	Cl	Br	I	Si	P	B	X	COMPOUND NAME	CAS Reg No	No
253	255	174	71	51	77	45	39	**253**	100	99	55	54	47	42	35	25		10	8	–	1	1	–	–	1	–	–	–	–	–	Isothiazole, 4-bromo-5-methyl-3-phenyl-		LI05145
91	253	43	28	134	119	79	65	**253**	100	43	21	15	14	10	10	10		10	8	1	1	–	5	–	–	–	–	–	–	–	Propanamide, 2,2,3,3,3-pentafluoro-N-benzyl-		NI24930
160	76	50	104	77	161	105	130	**253**	100	21	16	11	11	10	9	8	**6.17**	10	8	2	1	–	–	–	1	–	–	–	–	–	1H-Isoindole-1,3(2H)-dione, 2-(2-bromoethyl)-	574-98-1	NI24931
160	76	50	104	77	255	105	130	**253**	100	29	15	13	10	8	8	7	**7.41**	10	8	2	1	–	–	–	1	–	–	–	–	–	1H-Isoindole-1,3(2H)-dione, 2-(2-bromoethyl)-	574-98-1	IC03314
185	183	197	195	76	89	157	155	**253**	100	99	74	71	49	45	44	44	**16.89**	10	8	2	1	–	–	–	1	–	–	–	–	–	4H-1,3-Oxazin-5-one, 2-(4-bromophenyl)-5,6-dihydro-	77580-71-3	NI24932
211	165	252	213	181	97	153	76	**253**	100	26	25	22	17	14	12	11	**3.29**	10	8	3	3	–	–	1	–	–	–	–	–	–	1H-1,5-Benzodiazepin-2-one, 8-chloro-2,3-dihydro-4-methyl-7-nitro-		NI24933
236	76	253	118	103	177	78	77	**253**	100	73	70	58	58	51	51	51		10	11	5	3	–	–	–	–	–	–	–	–	–	Morpholine, N-(2,4-dinitrophenyl)-	39242-76-7	NI24934
223	104	139	121	181				**253**	100	91	79	67	21				**0.00**	10	11	5	3	–	–	–	–	–	–	–	–	–	3,4-Pyridinol, 1-nitroso-2-cyano-1,2,5,6-tetrahydro-, diacetate		LI05146
154	253	238	43	54	70	211	155	**253**	100	58	42	27	19	15	6	6		10	14	4	–	–	–	–	–	–	–	–	–	1	Manganese, bis(2,4-pentanedionato-O,O′)-	14024-58-9	NI24935
41	27	39	69	56	58	68	18	**253**	100	75	46	45	34	32	30	29	**0.00**	10	15	3	5	–	–	–	–	–	–	–	–	–	Alanine, N-[6-(cyclopropylamino)-1,4-dihydro-4-oxo-1,3,5-triazin-2-yl]-2-methyl-	62059-49-8	NI24936
253	57	225	86	56	45	43	210	**253**	100	90	50	50	50	40	35	33		10	19	1	7	–	–	–	–	–	–	–	–	–	1,3,5-Triazine-2,4-diamine, 6-(2-formylhydrazino)-N,N′-bis(isopropyl)-	84305-82-8	NI24937
57	41	196	140	84	113	27	63	**253**	100	88	64	60	38	27	18	8	**0.00**	10	21	–	1	–	–	2	–	–	1	–	–	–	Silanamine, 1,1-dichloro-1-methyl-N-[3-methyl-1-(2-methylpropyl)butylidene]-	55836-78-7	NI24938
99	253	252	98	28	44	56	45	**253**	100	31	27	27	27	21	18	18		10	30	–	5	–	–	–	–	–	–	–	3	–	Triborane, pentakis(dimethylamino)-		LI05147
209	71	45	211	138	72	210	140	**253**	100	57	51	36	32	17	15	14	**13.40**	11	8	2	1	1	–	1	–	–	–	–	–	–	4-Thiazoleacetic acid, 2-(4-chlorophenyl)-	17969-20-9	NI24939
209	71	211	45	138	253	210	72	**253**	100	54	50	35	34	26	20	15		11	8	2	1	1	–	1	–	–	–	–	–	–	4-Thiazoleacetic acid, 2-(4-chlorophenyl)-	17969-20-9	IC03315
253	159	206	160	173	89	121	205	**253**	100	21	20	20	18	15	14	11		11	11	–	1	3	–	–	–	–	–	–	–	–	Isothiazole, 3,5-bis(methylthio)-4-phenyl-	35537-36-1	NI24940
238	253	118	145	131	239	149	91	**253**	100	43	26	22	19	16	15	14		11	15	4	3	–	–	–	–	–	–	–	–	–	Aniline, 4-tert-butyl-N-methyl-2,6-dinitro-	33629-43-5	NI24941
127	195	253	197	153	255	111	152	**253**	100	83	62	29	25	23	22	21		11	16	–	5	–	–	1	–	–	–	–	–	–	Biguanide, 1-(4-chlorophenyl)-5-isopropyl-	500-92-5	NI24942
253	159	206	160	173	89	121	205	**253**	100	21	20	20	18	15	14	11		12	11	–	1	–	–	–	–	–	3	–	–	–	Isothiazole, 3,5-bis(methylthio)-4-phenyl-		LI05148
253	252	43	254	104	235	57	251	**253**	100	90	21	17	15	13	9	8		12	11	–	7	–	–	–	–	–	–	–	–	–	2,4,7-Pteridinetriamine, 6-phenyl-	396-01-0	NI24943
254	255	250	240	256	252			**253**	100	16	6	2	1	1			**0.00**	12	11	–	7	–	–	–	–	–	–	–	–	–	2,4,7-Pteridinetriamine, 6-phenyl-	396-01-0	NI24944
70	91	65	155	52	64	51	69	**253**	100	96	64	17	15	14	14	14	**0.00**	12	15	3	1	1	–	–	–	–	–	–	–	–	3-Azetidinone, 2-ethyl-N-(4-methylphenylsulphonyl)-	82380-63-0	NI24945
181	253	182	72	254	148	255	151	**253**	100	44	12	9	7	6	3	3		12	15	3	1	1	–	–	–	–	–	–	–	–	Benzene, 5-(2-isothiocyanatoethyl)-1,2,3-trimethoxy-	35424-93-2	NI24946
91	43	194	162	92	86	65	60	**253**	100	22	21	12	10	7	7	7	**4.55**	12	15	3	1	1	–	–	–	–	–	–	–	–	Cysteine, N-acetyl-S-benzyl-		MS05521
223	73	253	191	45	224	59	254	**253**	100	81	78	30	26	18	18	15		12	19	3	1	–	–	–	–	–	1	–	–	–	Benzaldehyde, 3-methoxy-4-[(trimethylsilyl)oxy]-, O-methyloxime		LI05149
223	73	253	191	45	59	224	206	**253**	100	81	76	30	26	20	18	15		12	19	3	1	–	–	–	–	–	1	–	–	–	Benzaldehyde, 3-methoxy-4-[(trimethylsilyl)oxy]-, O-methyloxime		MS05522
238	223	239	73	208	191	253	45	**253**	100	37	22	20	19	15	14	11		12	19	3	1	–	–	–	–	–	1	–	–	–	Benzoic acid, 2-(dimethylamino)-3-[(trimethylsilyl)oxy]-	55723-97-2	NI24947
176	44	91	41	167	42	104	120	**253**	100	92	66	58	47	47	38	37	**7.00**	12	19	3	1	–	–	–	–	–	1	–	–	–	1,3-Dioxa-6-aza-2-silacyclooctane, 6-methyl-2-methoxy-2-phenyl-		MS05523
184	43	41	127	55	69	126	57	**253**	100	98	94	90	83	60	57	43	**0.80**	12	22	1	1	–	3	–	–	–	–	–	–	–	Decanamine, N-trifluoroacetyl-		MS05524
91	73	147	253					**253**	100	25	14	3						12	23	1	1	–	–	–	–	–	2	–	–	–	Silanamine, 1,1,1-trimethyl-N-phenyl-N-[(trimethylsilyl)oxy]-	53783-47-4	NI24948
253	254	136	108	225	209	255	130	**253**	100	18	11	10	9	8	7	6		13	7	1	3	1	–	–	–	–	–	–	–	–	11H-Pyrido[3′,2′:4,5]imidazo[2,1-b]benzothiazin-11-one		NI24949
125	218	208	127	253	146	42	180	**253**	100	49	38	34	32	30	27	25		13	16	2	1	–	–	1	–	–	–	–	–	–	Acetic acid, (acetonylidene-o-chlorobenzylamine)-, ethyl ester		NI24950
166	167	236	168	125	253	139	91	**253**	100	96	91	91	91	83	78	57		13	16	2	1	–	–	1	–	–	–	–	–	–	4-Piperidone, 6-(4-chlorophenyl)-3e-hydroxy-1,3-dimethyl-	36106-53-3	NI24951
95	93	41	121	136	43	55	67	**253**	100	85	82	64	53	46	41	32	**0.00**	13	19	2	1	1	–	–	–	–	–	–	–	–	Acetic acid, thiocyanato-, 1,7,7-trimethylbicyclo[2.2.1]hept-2-yl ester, exo-	115-31-1	NI24952
210	91	155	253	98	65	211	41	**253**	100	90	80	36	35	21	13	12		13	19	2	1	1	–	–	–	–	–	–	–	–	Benzenesulphonamide, N-cyclohexyl-4-methyl-	80-30-8	NI24953
182	77	156	92	197	65	183	106	**253**	100	19	16	13	12	12	10	9	**0.00**	13	19	2	1	1	–	–	–	–	–	–	–	–	Sulphoximine, S-methyl-S-(1-oxohexyl)-S-phenyl-	54090-94-7	NI24954
152	80	43	253	120	108	193	138	**253**	100	82	40	31	28	25	22	19		13	19	4	1	–	–	–	–	–	–	–	–	–	9-Azabicyclo[3.3.1]non-2-ene-9-carboxylic acid, 6-(acetyloxy)-, ethyl ester, endo-	49690-31-5	NI24955
224	196	208	206	252	180	152	29	**253**	100	57	46	29	25	21	20	20	**19.00**	13	19	4	1	–	–	–	–	–	–	–	–	–	3,5-Pyridinedicarboxylic acid, 1,4-dihydro-2,6-dimethyl-, diethyl ester	1149-23-1	IC03316
224	196	208	252	253	152	42	180	**253**	100	48	44	25	22	15	14	13		13	19	4	1	–	–	–	–	–	–	–	–	–	3,5-Pyridinedicarboxylic acid, 1,4-dihydro-2,6-dimethyl-, diethyl ester	1149-23-1	NI24956
224	196	208	42	29	252	152	180	**253**	100	71	38	22	22	19	18	16	**15.50**	13	19	4	1	–	–	–	–	–	–	–	–	–	3,5-Pyridinedicarboxylic acid, 1,4-dihydro-2,6-dimethyl-, diethyl ester	1149-23-1	NI24957
224	178	253	162	208	207	160	206	**253**	100	97	84	48	44	44	37	23		13	19	4	1	–	–	–	–	–	–	–	–	–	Pyrrole, 2,4-diethoxy-5-ethyl-3-methyl-	69687-84-9	NI24958
117	77	105	253	134	148	135	121	**253**	100	28	26	22	8	7	7	7		13	19	4	1	–	–	–	–	–	–	–	–	–	Retronecine, 9-angelyl-, N-oxide		LI05150
156	83	113	55	98	58	99	41	**253**	100	31	20	15	14	13	12	9	**1.00**	13	23	2	3	–	–	–	–	–	–	–	–	–	1H-1,4-Diazepine-5,7(2H,6H)-dione, 3-(dimethylamino)-6,6-diethyl-2,2-dimethyl-	69315-92-0	NI24959
113	43	126	41	28	29	55	127	**253**	100	50	40	37	27	25	15	14	**5.49**	13	27	–	5	–	–	–	–	–	–	–	–	–	4H-1,2,4-Triazole-3,4-diamine, 5-undecyl-		NI24960
113	43	41	126	28	29	55	42	**253**	100	48	43	35	28	22	16	15	**6.49**	13	27	–	5	–	–	–	–	–	–	–	–	–	4H-1,2,4-Triazole-3,4-diamine, 5-undecyl-		NI24961
190	191	225	188	227	163	224	95	**253**	100	47	44	18	15	15	12	12	**0.00**	14	8	–	3	–	–	1	–	–	–	–	–	–	3H-Indazole-3-carbonitrile, 3-(4-chlorophenyl)-	56630-97-8	NI24962
253	118	77	150	121	51	252	254	**253**	100	73	49	46	26	25	22	20		14	11	–	3	1	–	–	–	–	–	–	–	–	1,3,4-Thiadiazole, 2-anilino-5-phenyl-		IC03317
253	252	91	254	255	194	103	92	**253**	100	66	23	20	6	5	3	3		14	11	–	3	1	–	–	–	–	–	–	–	–	2H-1,2,4-Triazoline-3-thione, 1,3-diphenyl-		NI24963
253	150	254	91	255	104	92	77	**253**	100	19	17	11	6	6	4	4		14	11	–	3	1	–	–	–	–	–	–	–	–	2H-1,2,4-Triazoline-3-thione, 1,5-diphenyl-	5055-74-3	NI24964
253	146	206	131	145	205	207	252	**253**	100	95	93	86	85	80	79	75		14	11	2	3	–	–	–	–	–	–	–	–	–	1H-Benzimidazole, 1-methyl-2-(4-nitrophenyl)-		NI24965
253	224	196	225	197	169	29	195	**253**	100	40	39	31	30	21	15	11		14	11	2	3	–	–	–	–	–	–	–	–	–	9H-Carbazole, 3,6-bis(N-formamido)-	98805-10-8	NI24966

MASS TO CHARGE RATIOS								M.W.	INTENSITIES								Parent	C	H	O	N	S	F	Cl	Br	I	Si	P	B	X	COMPOUND NAME	CAS Reg No	No
253	221	194	166	102	220	167	193	**253**	100	60	50	30	10	9	9	8		14	11	2	3	–	–	–	–	–	–	–	–	–	2-Phenazinecarboxylic acid, 3-amino-, methyl ester	18450-09-4	LI05151
253	221	194	166	254	102	220	167	**253**	100	60	50	30	15	10	9	9		14	11	2	3	–	–	–	–	–	–	–	–	–	2-Phenazinecarboxylic acid, 3-amino-, methyl ester	18450-09-4	NI24967
102	253	220	238	204				**253**	100	51	28	15	12					14	23	1	1	1	–	–	–	–	–	–	–	–	Bicyclo[5.3.0]decan-8-one, 2,7-dimethyl-5-(N-methylthiocarbamoyl)-		LI05152
110	168	85	152	169	150	153	123	**253**	100	88	70	66	53	41	31	23	**3.20**	14	23	3	1	–	–	–	–	–	–	–	–	–	Pyridine, 2,5-dihydro-6-methoxy-2-[3-[(tetrahydro-2H-pyran-2-yl)oxy]propyl]-		NI24968
253	77	165	105	147	122	104	51	**253**	100	65	56	54	33	31	29	27		15	11	1	1	1	–	–	–	–	–	–	–	–	2(3H)-Oxazolethione, 4,5-diphenyl-	6670-13-9	NI24969
253	205	204	224	254	210	237	206	**253**	100	85	79	35	18	18	17	15		15	11	1	1	1	–	–	–	–	–	–	–	–	Phenothiazine, 10-(1-propynyl)-, 5-oxide		NI24970
236	204	198	205	253	237	186	39	**253**	100	59	39	26	22	20	20	13		15	11	1	1	1	–	–	–	–	–	–	–	–	Phenothiazine, 10-(2-propynyl)-, 5-oxide		NI24971
180	86	78	253	181	212	77	110	**253**	100	30	30	25	22	20	15	12		15	11	1	1	1	–	–	–	–	–	–	–	–	Thiazolium, 4-hydroxy-2,3-diphenyl-, hydroxide, inner salt	13288-67-0	NI24972
253	235	224	254	76	152	236	151	**253**	100	38	22	20	19	15	14	14		15	11	3	1	–	–	–	–	–	–	–	–	–	9,10-Anthracenedione, 1-amino-2-(hydroxymethyl)-	24094-44-8	IC03318
253	44	235	224	180	152	207	208	**253**	100	54	41	30	27	22	19	19		15	11	3	1	–	–	–	–	–	–	–	–	–	9,10-Anthracenedione, 1-amino-2-(hydroxymethyl)-	24094-44-8	NI24973
253	254	102	146	123				**253**	100	16	2	1	1					15	11	3	1	–	–	–	–	–	–	–	–	–	Benzoic acid, 4-cyano-, 3-methoxyphenyl ester	53327-11-0	NI24974
91	180	104	77	225	253	181	39	**253**	100	45	27	24	15	12	12	7		15	11	3	1	–	–	–	–	–	–	–	–	–	1H,3H-Isobenzofuran-1,3-dione, 4-anilino-		LI05153
252	253	103	77	131	76	104	206	**253**	100	85	52	43	40	28	24	20		15	11	3	1	–	–	–	–	–	–	–	–	–	2-Propen-1-one, 1-(4-nitrophenyl)-3-phenyl-	1152-48-3	NI24975
77	105	51	253	76	102	236	50	**253**	100	60	44	42	35	29	26	25		15	11	3	1	–	–	–	–	–	–	–	–	–	2-Propen-1-one, 3-(4-nitrophenyl)-1-phenyl-	1222-98-6	NI24976
253	105	77	236	206	176	207	178	**253**	100	67	56	34	26	26	19	16		15	11	3	1	–	–	–	–	–	–	–	–	–	2-Propen-1-one, 3-(4-nitrophenyl)-1-phenyl-	1222-98-6	NI24977
253	91	96	225	254	130	131	196	**253**	100	80	58	26	20	18	16	14		15	11	3	1	–	–	–	–	–	–	–	–	–	2H-Pyrido[2,1-b][1,3]-oxazin-5-ium, 3-benzyl-2-oxo-, 4-olate		NI24978
253	252	196	254	237	76	206	168	**253**	100	80	21	20	20	17	13	13		15	11	3	1	–	–	–	–	–	–	–	–	–	2(1H)-Quinolinone, 3-hydroxy-4-(3-hydroxyphenyl)-	14484-44-7	NI24979
253	252	196	236	254	76	206	166	**253**	100	80	21	18	17	17	13	13		15	11	3	1	–	–	–	–	–	–	–	–	–	2(1H)-Quinolinone, 3-hydroxy-4-(3-hydroxyphenyl)-	14484-44-7	LI05154
169	168	141	97	98	126	43	41	**253**	100	29	29	26	24	20	20	19	**12.01**	15	27	2	1	–	–	–	–	–	–	–	–	–	2H-Pyran-2-one, 4-(diethylamino)-5,6-dihydro-5-methyl-6-pentyl-	55090-40-9	NI24980
218	253	141	217	219	191	77	38	**253**	100	60	60	45	40	33	30	26		16	12	–	1	–	–	1	–	–	–	–	–	–	Cyclopropanecarbonitrile, 1-(p-chlorophenyl)-2-phenyl-	32589-55-2	LI05155
218	253	140	217	191	219	77	39	**253**	100	60	60	45	32	30	29	25		16	12	–	1	–	–	1	–	–	–	–	–	–	Cyclopropanecarbonitrile, 1-(p-chlorophenyl)-2-phenyl-	32589-55-2	NI24981
253	218	255	217	254	175	252	216	**253**	100	51	35	23	21	15	10	10		16	12	–	1	–	–	1	–	–	–	–	–	–	Quinoline, 6-chloro-2-methyl-4-phenyl-		LI05156
211	253	108	212	210	254	69	109	**253**	100	76	44	38	26	18	18	12		16	15	–	1	1	–	–	–	–	–	–	–	–	1,5-Benzothiazepine, 2,3-dihydro-2-methyl-4-phenyl-	74148-61-1	NI24982
130	91	253	131	77	129	103	102	**253**	100	28	13	12	12	9	8	8		16	15	–	1	1	–	–	–	–	–	–	–	–	1H-Indole, 3-[[benzylthio]methyl]-	21903-71-9	NI24983
252	253	206	178	237	77	102	165	**253**	100	74	40	22	15	10	9	8		16	15	–	1	1	–	–	–	–	–	–	–	–	Isoquinoline, 3,4-dihydro-6-methylthio-1-phenyl-		MS05525
180	181	210	179	211	43	253	165	**253**	100	63	51	39	38	37	36	33		16	15	2	1	–	–	–	–	–	–	–	–	–	Acetamide, N-(9H-fluoren-2-yl)-N-methoxy-		NI24984
105	77	253	211	131	78	106	132	**253**	100	50	25	18	14	14	13	12		16	15	2	1	–	–	–	–	–	–	–	–	–	Benzamide, N-[2-(2-oxopropyl)phenyl]-	19051-04-8	NI24985
148	210	194	211	77	134	253	195	**253**	100	86	48	32	20	16	10	10		16	15	2	1	–	–	–	–	–	–	–	–	–	Benzophenone, 2-(N-methylacetamido)-	84297-09-6	NI24986
238	180	153	126	152	239	198	91	**253**	100	78	37	36	30	22	22	22	**18.80**	16	15	2	1	–	–	–	–	–	–	–	–	–	N-tert-Butylnaphthalisomide		MS05526
131	91	146	193	145	144	236	109	**253**	100	62	58	34	32	32	28	17	**10.00**	16	15	2	1	–	–	–	–	–	–	–	–	–	Cinnamamide, N-(benzyloxy)-	22472-17-9	LI05157
91	131	77	103	146	105	106	147	**253**	100	83	50	42	29	26	23	17	**10.19**	16	15	2	1	–	–	–	–	–	–	–	–	–	Cinnamamide, N-(benzyloxy)-	22472-17-9	NI24987
131	123	253	103	77	132	254	108	**253**	100	99	51	39	23	10	9	9		16	15	2	1	–	–	–	–	–	–	–	–	–	Cinnamamide, N-(2-methoxyphenyl)-		TR00282
131	123	103	77	253	51	132	52	**253**	100	55	48	31	24	12	10	7		16	15	2	1	–	–	–	–	–	–	–	–	–	Cinnamamide, N-(3-methoxyphenyl)-		TR00283
131	123	253	103	77	254	108	132	**253**	100	96	64	37	18	11	11	10		16	15	2	1	–	–	–	–	–	–	–	–	–	Cinnamamide, N-(4-methoxyphenyl)-		TR00284
162	91	253	134	119	163	116	65	**253**	100	48	39	29	19	12	10	10		16	15	2	1	–	–	–	–	–	–	–	–	–	1H-Indole, 5-methoxy-6-(phenylmethoxy)-	2426-59-7	NI24988
162	91	253	134	119	163	116	106	**253**	100	40	39	25	13	10	6	6		16	15	2	1	–	–	–	–	–	–	–	–	–	1H-Indole, 5-methoxy-6-(phenylmethoxy)-	2426-59-7	NI24989
162	253	91	134	119	163	116	65	**253**	100	38	30	24	18	13	7	6		16	15	2	1	–	–	–	–	–	–	–	–	–	1H-Indole, 6-methoxy-5-(phenylmethoxy)-		NI24990
105	43	77	235	68	58	83	41	**253**	100	55	50	29	22	22	18	14	**14.00**	16	15	2	1	–	–	–	–	–	–	–	–	–	2-Indolinol, 1-benzoyl-3-methyl-	22397-26-8	LI05158
105	43	77	235	69	58	83	253	**253**	100	56	50	30	23	23	19	17		16	15	2	1	–	–	–	–	–	–	–	–	–	2-Indolinol, 1-benzoyl-3-methyl-	22397-26-8	NI24991
208	76	130	77	207	209	104	102	**253**	100	38	35	27	22	18	18	16	**0.00**	16	15	2	1	–	–	–	–	–	–	–	–	–	1H-Isoindol-1-one, 2,3-dihydro-2-ethoxy-3-phenyl-		LI05159
209	253	176	130	77	105	152	236	**253**	100	66	66	31	29	22	18	9		16	15	2	1	–	–	–	–	–	–	–	–	–	1H-Isoindol-1-one, 2,3-dihydro-2-ethyl-3-hydroxy-3-phenyl-		LI05160
118	119	147	253	77	120	91	165	**253**	100	57	46	31	14	8	8	4		16	15	2	1	–	–	–	–	–	–	–	–	–	2-Oxazolidinone, 3-methyl-4,5-diphenyl-, cis-	32461-34-0	LI05161
118	119	147	253	77	41	91	78	**253**	100	57	45	31	15	10	8	8		16	15	2	1	–	–	–	–	–	–	–	–	–	2-Oxazolidinone, 3-methyl-4,5-diphenyl-, cis-	32461-34-0	NI24992
118	119	147	253	41	77	91	120	**253**	100	60	41	18	17	16	11	9		16	15	2	1	–	–	–	–	–	–	–	–	–	2-Oxazolidinone, 3-methyl-4,5-diphenyl-, trans-	32461-35-1	NI24993
121	104	91	93	77	122	119	105	**253**	100	16	16	13	12	9	9	8	**2.80**	16	15	2	1	–	–	–	–	–	–	–	–	–	2-Oxazolidinone, 4-phenyl-5-p-tolyl-, trans-	32461-31-7	NI24994
253	148	77	238	133	254	105	51	**253**	100	65	51	34	34	23	21	20		16	19	–	3	–	–	–	–	–	–	–	–	–	Azobenzene, 4-(diethylamino)-		IC03319
222	207	223	208	237	91	253	206	**253**	100	46	30	20	17	13	11	10		16	19	–	3	–	–	–	–	–	–	–	–	–	5-Methyl-2,N,N-dimethylaminobenzophenone hydrazone		MS05527
102	57	95	253	238	196			**253**	100	43	11	3	2	2				16	31	1	1	–	–	–	–	–	–	–	–	–	Cyclohexanamine, N-(2,2-dimethylpropanoyl)-1-methyl-4-tert-butyl-, cis-		LI05162
102	57	154	95	253	238	196		**253**	100	42	9	9	3	3	2			16	31	1	1	–	–	–	–	–	–	–	–	–	Cyclohexanamine, N-(2,2-dimethylpropanoyl)-1-methyl-4-tert-butyl-, trans-		LI05163
132	211	194	253	196	43	107	195	**253**	100	88	66	51	41	39	33	27		17	19	1	1	–	–	–	–	–	–	–	–	–	Acetamide, N-(2,6-dimethylphenyl)-N-(2-methylphenyl)-	68014-51-7	NI24995
253	209	175	91	148	160	115	236	**253**	100	64	43	34	32	30	29	23		17	19	1	1	–	–	–	–	–	–	–	–	–	1H-2-Benzazepine, 6-methoxy-5-phenyl-2,3,4,5-tetrahydro-		MS05528
253	209	149	223	236	91	178	115	**253**	100	63	51	43	32	28	27	25		17	19	1	1	–	–	–	–	–	–	–	–	–	1H-2-Benzazepine, 8-methoxy-5-phenyl-2,3,4,5-tetrahydro-		MS05529
181	197	152	253	153	182	238	151	**253**	100	36	33	27	21	15	9	9		17	19	1	1	–	–	–	–	–	–	–	–	–	[1,1′-Biphenyl]-2-carboxamide, N-tert-butyl-		MS05530
181	152	153	197	253	182	151	41	**253**	100	53	42	31	30	15	14	10		17	19	1	1	–	–	–	–	–	–	–	–	–	[1,1′-Biphenyl]-3-carboxamide, N-tert-butyl-	42498-35-1	NI24996

MASS TO CHARGE RATIOS								M.W.	INTENSITIES								Parent	C	H	O	N	S	F	Cl	Br	I	Si	P	B	X	COMPOUND NAME	CAS Reg No	No
181	197	253	152	153	182	198	77	**253**	100	50	43	41	28	17	11	10		17	19	1	1	–	–	–	–	–	–	–	–	–	[1,1′-Biphenyl]-4-carboxamide, N-tert-butyl-	42498-34-0	NI24997
132	121	162	117	91	105	253	77	**253**	100	93	63	31	28	22	20	10		17	19	1	1	–	–	–	–	–	–	–	–	–	Cinnamylamine, N-(3-methoxybenzyl)-		MS05531
253	238	224	252	255	254	239	210	**253**	100	92	28	27	20	20	19	19		17	19	1	1	–	–	–	–	–	–	–	–	–	2,5-Cyclohexadien-1-one, 2,5-dimethyl-4-[(2,4,5-trimethylphenyl)imino]-	54245-93-1	NI24998
141	128	127	156	54	129	41	155	**253**	100	26	26	25	24	22	17	16	**3.33**	17	19	1	1	–	–	–	–	–	–	–	–	–	Cyclohexanone oxime, O-(naphthylmethyl)-		MS05532
253	148	77	105	254	236	149	176	**253**	100	60	38	35	18	17	6	4		17	19	1	1	–	–	–	–	–	–	–	–	–	4-Hexen-2-yn-1-one, 1-phenyl-5-(1-piperidinyl)-	29743-43-9	NI24999
132	133	43	117	130	105	72	36	**253**	100	83	49	46	45	37	32	25	**0.00**	17	19	1	1	–	–	–	–	–	–	–	–	–	1-Isoquinolinemethanol, α-methyl-α-phenyl-u-1,2,3,4-tetrahydro-		DD00722
146	147	44	119	57	43	41	132	**253**	100	41	29	25	16	12	11	9	**0.00**	17	19	1	1	–	–	–	–	–	–	–	–	–	1,3-Oxazolidine, 4-methyl-2-(4-methylphenyl)-5c-phenyl-		MS05533
146	147	119	132	91	77	105	117	**253**	100	55	27	16	11	9	7	7	**0.00**	17	19	1	1	–	–	–	–	–	–	–	–	–	1,3-Oxazolidine, 4-methyl-2-(4-methylphenyl)-5t-phenyl-		MS05534
84	94	110	95	85	55	82	108	**253**	100	50	14	13	8	8	7	5	**0.10**	17	35	–	1	–	–	–	–	–	–	–	–	–	Pyrrolidine, 2-dodecyl-1-methyl-	3447-08-3	NI25000
84	94	110	95	55	82	108	57	**253**	100	50	14	13	8	7	5	3	**0.05**	17	35	–	1	–	–	–	–	–	–	–	–	–	Pyrrolidine, 2-dodecyl-1-methyl-	3447-08-3	MS05535
84	85	41	43	55	44	238	82	**253**	100	10	7	6	5	3	3	2	**1.39**	17	35	–	1	–	–	–	–	–	–	–	–	–	Pyrrolidine, 2-dodecyl-5-methyl-		NI25001
148	105	162	119	91	44	42	77	**253**	100	69	61	50	29	28	23	21	**1.00**	18	23	–	1	–	–	–	–	–	–	–	–	–	Benzeneethanamine, N,3-dimethyl-N-(2-phenylethyl)-	52059-46-8	NI25002
148	105	162	119	91	42	77	79	**253**	100	64	48	44	23	21	18	12	**1.00**	18	23	–	1	–	–	–	–	–	–	–	–	–	Benzeneethanamine, N,4-dimethyl-N-(2-phenylethyl)-	52059-45-7	NI25003
72	167	165	42	152	73	44	77	**253**	100	9	6	5	4	4	4	3	**2.06**	18	23	–	1	–	–	–	–	–	–	–	–	–	Benzenepropanamine, N,N,α-trimethyl-γ-phenyl-	13957-55-6	NI25004
69	253	41	169	170	238	168	184	**253**	100	66	49	48	47	46	21	15		18	23	–	1	–	–	–	–	–	–	–	–	–	Indole, 2-(1,1-dimethylprop-2-enyl)-6-(3,3-dimethylprop-2-enyl)-		MS05536
227	228	229	114	100	226	225	200	**253**	100	28	11	10	9	9	8	7	**1.50**	19	11	–	1	–	–	–	–	–	–	–	–	–	Azabenzo[bjk]fluoranthene		NI25005
127	254	63.5						**254**	100	50	22							–	–	–	–	–	–	–	–	2	–	–	–	–	Iodine		AM00133
127	129	48	131	254	175	47	50	**254**	100	78	26	23	14	14	14	10		1	1	–	–	–	–	1	1	1	–	–	–	–	Methane, bromochloroiodo-	34970-00-8	NI25006
177	175	179	100	96	61	129	26	**254**	100	62	45	23	22	19	18	15	**0.70**	2	2	–	–	–	–	2	2	–	–	–	–	–	Ethane, 1,2-dibromo-1,2-dichloro-	683-68-1	NI25007
65	175	177	111	113	95	191	31	**254**	100	88	85	22	21	12	11	10	**2.20**	3	3	–	–	–	3	–	2	–	–	–	–	–	Propane, 1,1-dibromo-1,2,2-trifluoro-		IC03320
209	239	224	254	223	240	222	238	**254**	100	83	68	44	8	3	3	2		3	9	–	–	–	–	–	–	–	–	–	–	1	Bismuthine, trimethyl-	593-91-9	NI25008
127	77	51	254	108	27	57	177	**254**	100	97	33	25	19	11	8	7		4	3	–	–	–	4	–	–	1	–	–	–	–	1-Butene, 3,3,4,4-tetrafluoro-4-iodo-	33831-83-3	MS05537
161	163	64	63	77	48	79	83	**254**	100	67	61	60	54	21	18	16	**0.00**	4	8	4	–	2	–	2	–	–	–	–	–	–	2,4-Dithiahexane, 3,3-dichloro-, 2,2,4,4-tetroxide		MS05538
254	44	46	73	70	176	207	224	**254**	100	99	82	76	64	63	59	39		6	2	2	6	2	–	–	–	–	–	–	–	–	4H,8H-Benzo[1,2-c:4,5-c′]bis[1,2,5]thiadiazol-4,8-dion-dioxin	83274-17-3	NI25009
50	75	157	155	76	221	219	74	**254**	100	94	89	85	70	58	58	54	**15.82**	6	4	2	–	1	–	1	1	–	–	–	–	–	Benzenesulphonyl chloride, 4-bromo-	98-58-8	NI25011
221	219	155	157	256	76	75	254	**254**	100	99	88	87	46	46	44	40		6	4	2	–	1	–	1	1	–	–	–	–	–	Benzenesulphonyl chloride, 4-bromo-	98-58-8	NI25010
221	219	155	157	256	76	75	50	**254**	100	99	88	87	46	46	44	40	**40.00**	6	4	2	–	1	–	1	1	–	–	–	–	–	Benzenesulphonyl chloride, 4-bromo-	98-58-8	NI25012
69	141	99	113	140	127	97	43	**254**	100	17	15	14	9	7	6	5	**0.00**	6	4	4	–	–	6	–	–	–	–	–	–	–	Acetic acid, trifluoro-, 1,2-ethanediyl ester	2613-44-7	NI25013
147	183	149	185	111	75	109	77	**254**	100	65	65	64	50	43	40	31	**0.10**	6	7	–	–	–	–	5	–	–	–	–	–	–	Cyclohexane, 1a,2e,3e,4e,5e-pentachloro-	22138-39-2	NI25014
91	93	105	169	79	123	171	187	**254**	100	33	27	24	24	16	15	11	**0.00**	6	10	2	–	–	–	4	–	–	–	–	–	–	2,5-Hexanediol, 1,3,4,6-tetrachloro-	36120-61-3	LI05164
91	93	105	169	79	123	171	187	**254**	100	44	36	31	31	21	20	14	**0.00**	6	10	2	–	–	–	4	–	–	–	–	–	–	2,5-Hexanediol, 1,3,4,6-tetrachloro-	36120-61-3	NI25015
83	29	81	68	125	39	129	89	**254**	100	80	75	70	65	65	55	55	**0.00**	6	11	1	2	1	5	–	–	–	–	–	–	–	(Piperidinoformamido)sulphur pentafluoride	90598-09-7	NI25016
200	198	256	254	191	163	93	75	**254**	100	99	96	95	68	66	57	46		7	1	2	–	1	–	3	–	–	–	–	–	–	1,3-Benzoxathiol-2-one, 4,5,7-trichloro-		LI05165
99	155	219	75	74	101	221	157	**254**	100	81	72	69	49	35	28	27	**14.10**	7	4	4	–	1	–	2	–	–	–	–	–	–	Benzoic acid, 2-chloro-5-(chlorosulphonyl)-	137-64-4	NI25017
198	196	197	195	200	168	40	59	**254**	100	96	39	36	34	22	22	21	**18.00**	7	5	2	2	–	–	3	–	–	–	–	–	–	2-Pyridinecarboxylic acid, 4-amino-3,5,6-trichloro-, methyl ester	14143-55-6	NI25019
198	196	86	36	59	197	125	62	**254**	100	94	52	48	45	36	36	36	**16.01**	7	5	2	2	–	–	3	–	–	–	–	–	–	2-Pyridinecarboxylic acid, 4-amino-3,5,6-trichloro-, methyl ester	14143-55-6	NI25018
196	198	197	195	200	86	59	125	**254**	100	95	32	28	28	20	18	16	**11.60**	7	5	2	2	–	–	3	–	–	–	–	–	–	2-Pyridinecarboxylic acid, 4-amino-3,5,6-trichloro-, methyl ester	14143-55-6	NI25020
127	99	111	75	50	254	31		**254**	100	53	45	32	10	9	6			7	7	3	–	–	–	1	–	–	–	–	–	1	Benzeneselenonic acid, 4-chloro-, methyl ester	25633-04-9	LI05166
127	99	111	129	75	128	101	112	**254**	100	53	44	35	30	26	19	17	**8.87**	7	7	3	–	–	–	1	–	–	–	–	–	1	Benzeneselenonic acid, 4-chloro-, methyl ester	25633-04-9	NI25021
59	254	112	113	99	176	145	131	**254**	100	74	61	36	35	34	34	34		7	10	–	–	5	–	–	–	–	–	–	–	–	2,4,6,8,9-Pentathiatricyclo[3.3.1.1^{3,7}]decane, 3,7-dimethyl-	57304-96-8	NI25022
56	181	199	41	127	29	153	198	**254**	100	87	52	25	21	20	19	17	**3.20**	7	11	2	–	–	–	–	–	1	–	–	–	–	Propanoic acid, 3-iodo-, butyl ester, (E)-		IC03321
77	39	51	65	69	158	121	160	**254**	100	45	43	35	22	20	14	13	**2.00**	8	3	–	2	–	3	2	–	–	–	–	–	–	1H-Benzimidazole, 5,6-dichloro-2-(trifluoromethyl)-	2338-25-2	NI25023
196	198	200	254	197	167	97	256	**254**	100	94	31	26	24	24	24	23		8	5	3	–	–	–	3	–	–	–	–	–	–	Acetic acid, (2,4,5-trichlorophenoxy)-	93-76-5	NI25024
196	198	254	256	200	167	149	97	**254**	100	99	77	69	44	39	34	33		8	5	3	–	–	–	3	–	–	–	–	–	–	Acetic acid, (2,4,5-trichlorophenoxy)-	93-76-5	NI25025
196	198	254	256	200	209	197	167	**254**	100	97	50	49	31	23	23	22		8	5	3	–	–	–	3	–	–	–	–	–	–	Acetic acid, (2,4,5-trichlorophenoxy)-	93-76-5	NI25026
255	257	209	207	239	259	224	211	**254**	100	96	58	55	36	31	30	30	**0.00**	8	5	3	–	–	–	3	–	–	–	–	–	–	Benzoic acid, 2,3,5-trichloro-6-methoxy-	2307-49-5	NI25027
77	113	64	51	27	185	108	78	**254**	100	50	32	9	7	4	4	3	**0.40**	8	6	–	–	–	8	–	–	–	–	–	–	–	Cyclobutane, 1,1,2,2-tetrafluoro-3-(1,1,2,2-tetrafluoro-3-butenyl)-	35207-93-3	MS05539
87	43	45	41	44	39	27	254	**254**	100	59	46	39	30	28	22	21		8	14	4	–	–	–	–	–	–	–	–	–	1	Butanoic acid, 4,4′-selenobis-	20846-67-7	NI25028
87	45	43	41	44	39	27	254	**254**	100	46	45	39	31	28	23	22		8	14	4	–	–	–	–	–	–	–	–	–	1	Butanoic acid, 4,4′-selenobis-	20846-67-7	LI05167
97	69	55	41	70	105	111	57	**254**	100	89	89	67	56	45	42	40	**32.00**	8	16	–	–	1	–	–	1	–	–	1	–	–	Phosphetane, 1-bromo-2,2,3,4,4-pentamethyl-, 1-sulphide		NI25029
225	31	195	121	91	165	40	135	**254**	100	78	78	75	57	50	25	17	**0.00**	8	22	5	–	–	–	–	–	–	2	–	–	–	Disiloxane, 1,3-diethyl-1,1,3,3-tetramethoxy-	18420-17-2	NI25030
239	73	226	254	195	211	135	147	**254**	100	27	22	21	21	20	17	16		8	23	3	–	–	–	–	–	–	2	1	–	–	Phosphonic acid, ethyl-, bis(trimethylsilyl) ester	1641-57-2	NI25031
239	240	73	226	254	195	211	135	**254**	100	32	27	22	21	21	20	17		8	23	3	–	–	–	–	–	–	2	1	–	–	Phosphonic acid, ethyl-, bis(trimethylsilyl) ester	1641-57-2	NI25032
58	254	117	136	184	197	155	168	**254**	100	67	55	48	44	33	26	24		9	3	3	–	–	5	–	–	–	–	–	–	–	1,3-Propanedial, 2-(pentafluorophenoxy)-		NI25033

MASS TO CHARGE RATIOS								M.W.	INTENSITIES								Parent	C	H	O	N	S	F	Cl	Br	I	Si	P	B	X	COMPOUND NAME	CAS Reg No	No
91	155	65	254	69	63	89	83	**254**	100	63	41	29	21	18	16	15		9	9	3	–	1	3	–	–	–	–	–	–	–	Ethanol, 2,2,2-trifluoro-, 4-methylbenzenesulphonate	433-06-7	NI25034
181	164	136	163	254	52	104	236	**254**	100	54	36	27	23	18	15	14		9	10	3	4	1	–	–	–	–	–	–	–	–	4-Pteridinecarboxylic acid, 1,2,3,4-tetrahydro-4-hydroxy-2-thioxo-, ethyl ester	18204-27-8	NI25035
181	164	136	163	254	52	165	104	**254**	100	54	36	27	23	18	14	14		9	10	3	4	1	–	–	–	–	–	–	–	–	4-Pteridinecarboxylic acid, 1,2,3,4-tetrahydro-4-hydroxy-2-thioxo-, ethyl ester	18204-27-8	NI25036
164	163	181	136	104	52	236	165	**254**	100	90	81	48	45	45	36	27	**6.00**	9	10	3	4	1	–	–	–	–	–	–	–	–	4-Pteridinecarboxylic acid, 1,2,3,4-tetrahydro-4-hydroxy-2-thioxo-, ethyl ester	18204-27-8	NI25037
254	183	71	156	96	211	199	255	**254**	100	97	80	46	40	30	24	23		9	14	1	6	1	–	–	–	–	–	–	–	–	4H-Triazolo[3,2-a]-1,3,5-triazin-2-one, 6,7-dihydro-2-dimethylamino-, acetylhydrazone	83809-56-7	NI25038
102	250	57	41	83	103	37	56	**254**	100	99	74	73	72	60	60	52	**5.00**	9	18	–	–	4	–	–	–	–	–	–	–	–	1,4,7,10-Tetrathiacyclotridecane	25423-54-5	NI25039
254	106	107	87	61	60	133	73	**254**	100	92	60	58	53	48	42	36		9	18	–	–	4	–	–	–	–	–	–	–	–	1,4,7,10-Tetrathiacyclotridecane	25423-54-5	NI25040
43	57	71	41	85	55	29	27	**254**	100	78	74	42	38	20	20	13	**7.00**	9	19	–	–	–	–	–	–	1	–	–	–	–	Nonane, 1-iodo-		LI05168
226	254	209	89	252	224	207	181	**254**	100	99	85	67	60	50	45	44		10	6	3	–	–	–	–	–	–	–	–	–	1	Benzo[b]selenophene-3-carboxylic acid, 2-formyl-	26526-49-8	NI25041
254	127	128	77	76	63	51	50	**254**	100	53	5	4	2	2	2	2		10	7	–	–	–	–	–	–	1	–	–	–	–	Naphthalene, 1-iodo-		LI05169
127	254	126	74	77	75	50	51	**254**	100	70	21	13	13	11	9	9		10	7	–	–	–	–	–	–	1	–	–	–	–	Naphthalene, 1-iodo-	90-14-2	NI25042
254	256	214	216	107	226	79	228	**254**	100	97	42	39	17	14	14	13		10	7	3	–	–	–	–	1	–	–	–	–	–	Chromone, 5-bromo-6-hydroxy-2-methyl-	30095-75-1	NI25043
254	128	256	127	39	129	64	102	**254**	100	80	58	32	20	18	17	14		10	8	–	–	–	–	2	–	–	–	–	–	1	Ferrocene, 1,1′-dichloro-	1293-67-0	NI25044
254	128	256	127	39	64	102	155	**254**	100	83	58	33	19	18	14	10		10	8	–	–	–	–	2	–	–	–	–	–	1	Ferrocene, 1,1′-dichloro-	1293-67-0	LI05170
254	128	256	182	127	255	63	258	**254**	100	80	63	27	22	13	13	10		10	8	–	–	–	–	2	–	–	–	–	–	1	Ferrocene, 1,2-dichloro-	12287-95-5	LI05171
254	128	256	127	182	255	63	45	**254**	100	78	62	21	14	12	11	10		10	8	–	–	–	–	2	–	–	–	–	–	1	Ferrocene, 1,2-dichloro-	12287-95-5	NI25045
59	77	75	119	145	63	43	28	**254**	100	99	79	73	60	54	50	50	**17.00**	10	11	–	–	–	5	–	–	–	1	–	–	–	Silane, dimethyl[2-(pentafluorophenyl)ethyl]-	58751-81-8	NI25046
241	239	256	254	159	115	160	145	**254**	100	97	82	80	78	15	12	10		10	11	1	–	–	–	–	1	–	1	–	–	–	Naphthalene, 2,2-dimethyl-3-bromo-1-oxa-2-sila-1,2-dihydro-	80269-48-3	NI25047
184	254	207	185	43	142	41	29	**254**	100	33	23	23	23	22	16	12		10	14	–	4	2	–	–	–	–	–	–	–	–	Thiazolo[5,4-d]pyrimidine, 5-amino-2-(pentylthio)-	19844-41-8	NI25048
155	254	43	239	71	85	56	255	**254**	100	83	69	63	28	18	14	13		10	14	4	–	–	–	–	–	–	–	–	–	1	Iron, bis(2,4-pentanedionato-O,O′)-	14024-17-0	NI25049
146	119	118	174	91	218	147	145	**254**	100	39	39	26	22	19	14	14	**1.00**	10	14	4	4	–	–	–	–	–	–	–	–	–	4-Pteridinecarboxylic acid, 5,6,7,8-tetrahydro-6,7-dihydroxy-7-methyl-, ethyl ester	17445-69-1	NI25050
174	219	218	220	182	254	146	175	**254**	100	75	45	27	27	18	18	15		10	14	4	4	–	–	–	–	–	–	–	–	–	4-Pteridinecarboxylic acid, 5,6,7,8-tetrahydro-6,7-dihydroxy-7-methyl-, ethyl ester	17445-69-1	NI25051
95	68	53	109	67	180	123	81	**254**	100	73	58	34	31	30	24	20	**14.00**	10	14	4	4	–	–	–	–	–	–	–	–	–	Xanthine, 7-(2,3-dihydroxypropyl)-1,3-dimethyl-		NI25052
99	98	111	43	70	100	83	156	**254**	100	33	16	12	9	6	6	5	**0.03**	10	16	6	–	–	–	–	–	–	–	–	2	–	D-Gulonic acid, γ-lactone, cyclic 2,3:5,6-bis(ethylboronate)	74128-60-2	NI25053
57	29	71	31	42	43	70	137	**254**	100	98	97	87	85	72	60	43	**0.00**	10	22	7	–	–	–	–	–	–	–	–	–	–	1,3-Propanediol, 2,2′-[oxybis(methylene)]bis[2-(hydroxymethyl)-	126-58-9	NI25054
254	115	223	252	221	222	220	195	**254**	100	79	76	49	35	28	25	21		11	10	2	–	–	–	–	–	–	–	–	–	1	Benzo[b]selenophene-2-carboxylic acid, 3-methyl-, methyl ester	20984-13-8	NI25055
254	115	223	252	222	195	194	220	**254**	100	86	57	49	46	41	31	30		11	10	2	–	–	–	–	–	–	–	–	–	1	Benzo[b]selenophene-3-carboxylic acid, 2-methyl-, methyl ester	26526-43-2	NI25056
153	43	195	69	85	212	111	254	**254**	100	50	39	38	23	15	11	9		11	10	7	–	–	–	–	–	–	–	–	–	–	2-Butenoic acid, 2-hydroxy-4-oxo-4-(4-hydroxy-6-methyl-2-pyrone-3-yl)-, methyl ester	86488-04-2	NI25057
169	173	171	257	255	209	211	237	**254**	100	99	96	63	60	53	51	23	**7.00**	11	11	2	–	–	–	–	1	–	–	–	–	–	Benzeneacetic acid, 4-bromo-, 2-propenyl ester	53603-13-7	NI25058
211	209	102	256	254	51	75	228	**254**	100	96	89	39	34	29	28	25		11	11	2	–	–	–	–	1	–	–	–	–	–	2-Propenoic acid, 3-(3-bromophenyl)-, ethyl ester	24398-80-9	NI25059
102	209	211	256	254	75	226	76	**254**	100	87	85	34	30	29	27	26		11	11	2	–	–	–	–	1	–	–	–	–	–	2-Propenoic acid, 3-(3-bromophenyl)-, ethyl ester	24398-80-9	NI25060
102	209	211	256	75	254	76	226	**254**	100	92	88	38	34	32	30	28		11	11	2	–	–	–	–	1	–	–	–	–	–	2-Propenoic acid, 3-(4-bromophenyl)-, ethyl ester	24393-53-1	NI25061
104	42	77	51	147	41	149	69	**254**	100	80	68	65	58	55	51	50	**9.00**	11	11	2	–	–	–	–	1	–	–	–	–	–	2H-Pyran-2-one, 5-bromo-6-phenyltetrahydro-	54966-45-9	NI25062
211	209	105	254	131	207	213	208	**254**	100	52	46	33	23	20	19	18		11	14	–	2	–	–	–	–	–	–	–	–	1	Selenocyanic acid, 4-(butylamino)phenyl ester	22037-10-1	NI25063
239	237	254	211	105	29	252	184	**254**	100	50	45	24	24	24	21	20		11	14	–	2	–	–	–	–	–	–	–	–	1	Selenocyanic acid, 4-(diethylamino)phenyl ester		LI05172
225	223	118	119	29	254	91	41	**254**	100	51	50	48	29	27	22	22		11	14	–	2	–	–	–	–	–	–	–	–	1	Selenocyanic acid, 4-(sec-butylamino)phenyl ester	22037-11-2	NI25064
118	132	91	254	119	110	81	164	**254**	100	65	43	38	38	30	28	27		11	14	3	2	1	–	–	–	–	–	–	–	–	1H-2,1,3-Benzothiadiazin-4(3H)-one, 1,3-diethyl-, 2,2-dioxide	40864-18-4	NI25065
118	132	91	254	119	110	81	185	**254**	100	65	43	38	38	30	28	27		11	14	3	2	1	–	–	–	–	–	–	–	–	1H-2,1,3-Benzothiadiazin-4(3H)-one, 1,3-diethyl-, 2,2-dioxide	40864-18-4	LI05173
44	137	150	149	131	123	43	42	**254**	100	50	27	16	16	11	10	10	**5.80**	11	14	3	2	1	–	–	–	–	–	–	–	–	Glycine, N-[2-(4-methylphenylthio)ethyl]-N-nitroso-		NI25066
30	29	225	27	77	39	15	107	**254**	100	99	66	66	57	42	37	35	**14.71**	11	14	5	2	–	–	–	–	–	–	–	–	–	Anisole, 2-sec-butyl-4,6-dinitro-	6099-79-2	NI25067
225	77	195	254	107	91	118	89	**254**	100	37	24	22	21	17	17	17		11	14	5	2	–	–	–	–	–	–	–	–	–	Anisole, 2-sec-butyl-4,6-dinitro-	6099-79-2	NI25068
27	43	41	29	211	39	30	225	**254**	100	96	84	80	67	59	54	50	**11.21**	11	14	5	2	–	–	–	–	–	–	–	–	–	Phenol, 2-(1-methylbutyl)-4,6-dinitro-	4097-36-3	NI25069
73	159	45	59	83	139	137	74	**254**	100	17	17	14	13	11	11	10	**3.00**	11	15	–	–	–	–	–	1	–	1	–	–	–	1,3-Cyclopentadiene, bromo-5-[2-(trimethylsilyl)cycloprop-1-ylidene]-		MS05540
152	71	224	56	254	169	73	153	**254**	100	47	36	33	30	15	15	14		11	15	2	–	–	–	–	–	–	–	–	–	1	1,3,2-Dioxarsenane, 5,5-dimethyl-2-phenyl-	34715-47-4	NI25070
152	71	224	56	254	169	153	73	**254**	100	46	39	34	29	18	18	18		11	15	2	–	–	–	–	–	–	–	–	–	1	1,3,2-Dioxarsenane, 5,5-dimethyl-2-phenyl-	34715-47-4	LI05174
171	213	169	143	168	39	41	144	**254**	100	49	25	20	19	19	18	13	**11.37**	11	19	–	–	–	–	–	–	–	–	–	–	1	Rhodium, [(1,2,3-η)-3-methyl-2-butenyl]bis(η^3-2-propenyl)-	74779-88-7	NI25071
170	55	198	115	254	226	89	80	**254**	100	49	19	14	13	10	6	6		12	7	3	–	–	–	–	–	–	–	–	–	1	Manganese, tricarbonyl[(1,2,3,3a,7a-η)-1H-inden-1-yl]-	12203-33-7	NI25072
254	256	184	75	255	258	218	108	**254**	100	83	65	36	35	32	31	31		12	8	–	–	1	–	2	–	–	–	–	–	–	Benzene, 1,1′-thiobis[4-chloro-	5181-10-2	NI25073
254	184	256	108	75	218	143	255	**254**	100	77	67	23	19	16	16	15		12	8	–	–	1	–	2	–	–	–	–	–	–	Benzene, 1,1′-thiobis[4-chloro-	5181-10-2	NI25074

MASS TO CHARGE RATIOS	M.W.	INTENSITIES	Parent	C	H	O	N	S	F	Cl	Br	I	Si	P	B	X	COMPOUND NAME	CAS Reg No	No
254 256 155 255 258 127 63 257	**254**	100 63 11 8 5 5 4 4		12	8	2	–	–	–	2	–	–	–	–	–	–	**[1,1′-Biphenyl]-3,3′-diol, 4,4′-dichloro-**	53905-37-6	**NI25075**
143 95 75 254 159 115 83 188	**254**	100 55 30 24 12 12 12 11		12	8	2	–	1	2	–	–	–	–	–	–	–	**Benzene, 1,1′-sulphonylbis[4-fluoro-**	383-29-9	**NI25076**
93 135 66 160 119 108 65 39	**254**	100 28 26 25 16 9 9 8	**5.67**	12	10	1	6	–	–	–	–	–	–	–	–	–	**Urea, N-phenyl-N′-(1H-purin-6-yl)-**	26325-06-4	**NI25077**
193 177 76 104 137 103 194 149	**254**	100 31 26 23 21 14 13 8	**1.37**	12	14	2	–	2	–	–	–	–	–	–	–	–	**1,3-Benzenedicarboxylic acid, bis(2-mercaptoethyl) ester**		**IC03322**
193 211 149 104 65 76 121 194	**254**	100 62 59 20 18 16 14 13	**0.00**	12	14	6	–	–	–	–	–	–	–	–	–	–	**1,4-Benzenedicarboxylic acid, bis(hydroxyethyl) ester**		**IC03323**
170 155 43 212 254 171 69 109	**254**	100 31 27 9 8 7 7 5		12	14	6	–	–	–	–	–	–	–	–	–	–	**1,4-Benzenediol, 2,5-dimethoxy-, diacetate**	14786-37-9	**NI25078**
43 180 212 222 152 137 197 165	**254**	100 88 47 40 27 23 21 19	**9.00**	12	14	6	–	–	–	–	–	–	–	–	–	–	**3-Furancarboxylic acid, 2-methyl-5-(1-(methoxycarbonyl)-2-oxopropyl)-, methyl ester**		**MS05541**
43 180 222 152 223 212 137 165	**254**	100 73 46 15 13 12 11 10	**4.00**	12	14	6	–	–	–	–	–	–	–	–	–	–	**Furo[3,2-b]furan-3,6-dicarboxylic acid, 3a,6a-dihydro-2,5-dimethyl-, dimethyl ester, cis-**		**MS05542**
253 111 254 152 226 75 138 41	**254**	100 62 57 56 53 51 50 48		12	15	–	2	1	–	1	–	–	–	–	–	–	**2-Thiazoline, 5-methyl-2-(N-ethyl-p-chlorophenylamino)-**	75220-53-0	**NI25079**
255 256 213 257 229 227 185 184	**254**	100 15 8 6 3 2 2 2	**0.00**	12	18	2	2	1	–	–	–	–	–	–	–	–	**4,6(1H,5H)-Pyrimidinedione, dihydro-5-sec-amyl-5-(2-propenyl)-2-thioxo-**	77-27-0	**NI25080**
43 41 184 168 167 97 55 53	**254**	100 86 84 58 50 36 35 32	**3.00**	12	18	2	2	1	–	–	–	–	–	–	–	–	**4,6(1H,5H)-Pyrimidinedione, dihydro-5-sec-amyl-5-(2-propenyl)-2-thioxo-**	77-27-0	**NI25081**
73 254 181 121 75 166 77 41	**254**	100 39 22 22 20 14 14 12		12	18	4	–	–	–	–	–	–	1	–	–	–	**Acetic acid, (o-methoxyphenoxy)-, trimethylsilyl ester**		**MS05543**
224 165 164 254 239 135 209 73	**254**	100 86 45 34 30 27 26 26		12	18	4	–	–	–	–	–	–	1	–	–	–	**Benzoic acid, 2,3-dimethoxy-, trimethylsilyl ester**		**MS05544**
165 239 254 166 164 122 195 107	**254**	100 26 13 10 7 7 6 6		12	18	4	–	–	–	–	–	–	1	–	–	–	**Benzoic acid, 2,4-dimethoxy-, trimethylsilyl ester**		**MS05545**
224 254 165 209 239 135 225 136	**254**	100 50 50 30 22 17 16 16		12	18	4	–	–	–	–	–	–	1	–	–	–	**Benzoic acid, 2,5-dimethoxy-, trimethylsilyl ester**		**MS05546**
195 239 165 254 196 240 77 73	**254**	100 85 82 49 14 13 12 10		12	18	4	–	–	–	–	–	–	1	–	–	–	**Benzoic acid, 3,4-dimethoxy-, trimethylsilyl ester**		**MS05547**
195 239 165 254 240 196 137 73	**254**	100 98 93 50 17 15 9 9		12	18	4	–	–	–	–	–	–	1	–	–	–	**Benzoic acid, 3,4-dimethoxy-, trimethylsilyl ester**		**MS05548**
224 239 254 193 225 255 240 223	**254**	100 65 61 18 17 12 12 11		12	18	4	–	–	–	–	–	–	1	–	–	–	**Benzoic acid, 3-methoxy-4-[(trimethylsilyl)oxy]-, methyl ester**	27798-53-4	**NI25082**
224 239 193 254 73 225 223 59	**254**	100 44 43 37 25 18 18 9		12	18	4	–	–	–	–	–	–	1	–	–	–	**Benzoic acid, 4-methoxy-3-[(trimethylsilyl)oxy]-, methyl ester**	55590-91-5	**NI25083**
58 210 44 136 254 196 239 123	**254**	100 70 68 40 35 14 12 12		12	18	4	2	–	–	–	–	–	–	–	–	–	**2-Propanol, 1-(ethylamino)-3-(3-methyl-2-nitrophenoxy)-**		**MS05549**
212 183 156 169 58 41 182 126	**254**	100 91 78 72 39 32 31 30	**1.00**	12	18	4	2	–	–	–	–	–	–	–	–	–	**2,4,6(1H,3H,5H)-Pyrimidinetrione, 1,3-dimethyl-5-isopropyl-5-(oxiranylmethyl)-**	57397-01-0	**NI25084**
43 55 70 57 41 69 56 71	**254**	100 74 72 64 55 51 45 34	**0.00**	12	21	2	–	–	3	–	–	–	–	–	–	–	**Acetic acid, trifluoro-, 3,7-dimethyloctyl ester**		**LI05175**
73 40 75 45 254 74 44 77	**254**	100 52 17 14 12 8 8 7		12	22	2	–	–	–	–	–	–	2	–	–	–	**1,2-Benzenediol bis(trimethylsilyl) ether**	5075-52-5	**NI25085**
73 254 45 74 239 75 255 151	**254**	100 19 11 9 5 5 5 3		12	22	2	–	–	–	–	–	–	2	–	–	–	**1,2-Benzenediol bis(trimethylsilyl) ether**	5075-52-5	**NI25086**
239 254 73 240 255 112 147 45	**254**	100 58 30 29 13 13 8 8		12	22	2	–	–	–	–	–	–	2	–	–	–	**1,3-Benzenediol bis(trimethylsilyl) ether**	4520-29-0	**LI05176**
239 254 240 73 255 112 241 147	**254**	100 69 23 22 17 12 9 7		12	22	2	–	–	–	–	–	–	2	–	–	–	**1,3-Benzenediol bis(trimethylsilyl) ether**	4520-29-0	**NI25087**
254 239 255 240 73 256 241 223	**254**	100 99 23 21 12 9 8 2		12	22	2	–	–	–	–	–	–	2	–	–	–	**1,4-Benzenediol bis(trimethylsilyl) ether**	2117-24-0	**NI25088**
239 254 73 240 255 75 45 43	**254**	100 66 40 22 15 14 13 10		12	22	2	–	–	–	–	–	–	2	–	–	–	**1,4-Benzenediol bis(trimethylsilyl) ether**	2117-24-0	**MS05550**
73 254 45 74 255 239 75 151	**254**	100 35 14 13 8 8 7 5		12	22	2	–	–	–	–	–	–	2	–	–	–	**Silane, [1,2-phenylenebis(oxy)]bis[trimethyl-**	5075-52-5	**NI25089**
127 126 139 154 43 71 55 29	**254**	100 45 32 24 24 10 10 8	**0.36**	12	24	4	–	–	–	–	–	–	–	–	2	–	**Allitol, 1,6-dideoxy-3,4-di-C-methyl-, cyclic 2,4:3,5-bis(ethylboronate)**	74793-75-2	**NI25090**
127 126 43 139 71 154 57 55	**254**	100 34 17 16 16 8 6 5	**0.00**	12	24	4	–	–	–	–	–	–	–	–	2	–	**DL-Altritol, 1,6-dideoxy-3,4-di-C-methyl-, cyclic 2,4:3,5-bis(ethylboronate)**	74841-66-0	**NI25091**
99 98 43 41 57 138 83 29	**254**	100 33 27 18 17 13 13 12	**0.00**	12	24	4	–	–	–	–	–	–	–	–	2	–	**1,3,2-Dioxaborolane, 4,4′-(1,4-butanediyl)bis[2-ethyl-**	74793-62-7	**NI25092**
210 56 254 121 211 152 208 153	**254**	100 23 22 17 16 12 8 8		13	10	2	–	–	–	–	–	–	–	–	–	1	**Ferrocenylpropynoic acid**		**MS05551**
210 56 254 121 211 152 153 89	**254**	100 23 18 17 16 12 7 7		13	10	2	–	–	–	–	–	–	–	–	–	1	**Ferrocenylpropynoic acid**		**LI05177**
198 142 141 56 196 226 78 121	**254**	100 86 72 71 57 53 46 36	**16.72**	13	10	2	–	–	–	–	–	–	–	–	–	1	**Iron, dicarbonyl(η^5-2,4-cyclopentadien-1-yl)phenyl-**	12126-73-7	**MS05552**
198 56 196 142 226 141 172 121	**254**	100 69 63 49 46 43 35 32	**9.99**	13	10	2	–	–	–	–	–	–	–	–	–	1	**Iron, dicarbonyl(η^5-2,4-cyclopentadien-1-yl)phenyl-**	12126-73-7	**NI25093**
78 179 254 225 106 90 136 77	**254**	100 82 74 74 61 58 42 40		13	10	2	4	–	–	–	–	–	–	–	–	–	**1H-Benzotriazole, 1-[(4-nitrophenyl)methyl]-**	27799-81-1	**NI25094**
141 125 198 143 89 124 154 127	**254**	100 44 33 32 21 18 17 16	**0.26**	13	15	3	–	–	–	1	–	–	–	–	–	–	**Propanoic acid, 2-(4-chlorophenyl)-2,3-epoxy-, tert-butyl ester**	89058-50-4	**NI25095**
226 181 254 198 225 154 211 182	**254**	100 97 74 47 21 21 20 19		13	18	5	–	–	–	–	–	–	–	–	–	–	**Benzoic acid, 4-ethoxy-3,5-dimethoxy-, ethyl ester**	75332-45-5	**NI25096**
166 105 43 107 77 89 45 106	**254**	100 65 40 35 30 25 25 24	**0.00**	13	18	5	–	–	–	–	–	–	–	–	–	–	**Butanoic acid, 2,3-dihydroxy-3-(methoxymethyl)-2-phenyl-, methyl ester**		**DD00723**
79 107 163 80 181 135 77 153	**254**	100 74 57 52 50 45 33 32	**0.00**	13	18	5	–	–	–	–	–	–	–	–	–	–	**2-Oxabicyclo[2.2.2]oct-5-ene-3,3-dicarboxylic acid, diethyl ester**		**MS05553**
225 43 197 202 179 153 254 181	**254**	100 91 41 20 9 9 8 8		13	18	5	–	–	–	–	–	–	–	–	–	–	**4H-Pyran-3,5-dicarboxylic acid, 2,6-dimethyl-, diethyl ester**		**LI05178**
116 114 222 86 163 151 69 73	**254**	100 45 29 25 18 14 14 13	**5.74**	13	18	5	–	–	–	–	–	–	–	–	–	–	**α-D-Xylopyranoside, methyl 3-O-benzyl-**	80973-63-3	**NI25097**
163 103 73 133 114 87 45 84	**254**	100 32 19 10 9 8 7 7	**0.42**	13	18	5	–	–	–	–	–	–	–	–	–	–	**β-D-Xylopyranoside, methyl 2-O-benzyl-**	51755-03-4	**NI25098**
114 163 151 116 86 73 69 74	**254**	100 50 33 24 24 24 22 21	**4.80**	13	18	5	–	–	–	–	–	–	–	–	–	–	**β-D-Xylopyranoside, methyl 3-O-benzyl-**	72521-39-2	**NI25099**
163 87 73 254 133 60 176 70	**254**	100 22 19 9 9 9 7 6		13	18	5	–	–	–	–	–	–	–	–	–	–	**β-D-Xylopyranoside, methyl 4-O-benzyl-**	69984-20-9	**NI25100**
73 103 75 101 45 104 74 129	**254**	100 84 19 12 11 8 8 7	**6.00**	13	19	1	–	–	–	1	–	–	1	–	–	–	**1-Butene, 2-(4-chlorophenyl)-4-[(trimethylsilyl)oxy]-**	55724-01-1	**NI25101**
43 72 128 30 153 86 41 254	**254**	100 68 48 42 38 26 20 18		13	19	1	2	–	–	1	–	–	–	–	–	–	**Urea, N-(3-chlorophenyl)-N′,N′-dipropyl-**		**MS05554**
43 57 56 41 55 70 69 71	**254**	100 54 47 46 34 31 30 23	**0.00**	13	22	–	–	–	4	–	–	–	–	–	–	–	**2-Tridecene, 1,1,1,2-tetrafluoro-, (Z)-**	108377-18-0	**DD00724**
184 169 41 43 185 183 112 69	**254**	100 92 16 13 12 12 10 8	**6.00**	13	22	3	2	–	–	–	–	–	–	–	–	–	**2,4,6(1H,3H,5H)-Pyrimidinetrione, 5-ethyl-1,3-dimethyl-5-(1-methylbutyl)-**	28239-47-6	**NI25102**

MASS TO CHARGE RATIOS								M.W.	INTENSITIES								Parent	C	H	O	N	S	F	Cl	Br	I	Si	P	B	X	COMPOUND NAME	CAS Reg No	No
169	184	183	185	112	170	41	43	**254**	100	91	14	12	12	11	10	8	**0.00**	13	22	3	2	–	–	–	–	–	–	–	–	–	**2,4,6(1H,3H,5H)-Pyrimidinetrione, 5-ethyl-1,3-dimethyl-5-(1-methylbutyl)-**	28239-47-6	**NI25104**
169	184	41	43	183	69	112	55	**254**	100	85	27	18	15	14	13	12	**0.20**	13	22	3	2	–	–	–	–	–	–	–	–	–	**2,4,6(1H,3H,5H)-Pyrimidinetrione, 5-ethyl-1,3-dimethyl-5-(1-methylbutyl)-**	28239-47-6	**NI25103**
184	169	170	185	112	41	55	183	**254**	100	96	95	77	56	44	37	36	**0.00**	13	22	3	2	–	–	–	–	–	–	–	–	–	**2,4,6(1H,3H,5H)-Pyrimidinetrione, 5-ethyl-1,3-dimethyl-5-(3-methylbutyl)-**	28239-46-5	**NI25106**
169	184	185	170	112	55	183	69	**254**	100	87	17	15	12	12	10	10	**0.20**	13	22	3	2	–	–	–	–	–	–	–	–	–	**2,4,6(1H,3H,5H)-Pyrimidinetrione, 5-ethyl-1,3-dimethyl-5-(3-methylbutyl)-**	28239-46-5	**NI25105**
169	184	41	170	56	112	185	58	**254**	100	81	33	22	22	19	18	15	**0.00**	13	22	3	2	–	–	–	–	–	–	–	–	–	**2,4,6(1H,3H,5H)-Pyrimidinetrione, 5-ethyl-1,3-dimethyl-5-(3-methylbutyl)-**	28239-46-5	**LI05179**
127	77	128	51	254	75	50	39	**254**	100	16	8	7	3	2	2	2		14	10	–	–	–	4	–	–	–	–	–	–	–	**Ethane, 1,2-diphenyltetrafluoro-**	425-32-1	**NI25107**
136	137	254	108	118	69	77	51	**254**	100	80	75	65	40	33	21	21		14	10	1	2	1	–	–	–	–	–	–	–	–	**4H-1,3-Benzothiazin-4-one, 2,3-dihydro-2-phenylimino-**		**NI25108**
105	77	45	57	55	69	46	31	**254**	100	56	55	39	34	30	29	29	**6.30**	14	10	3	2	–	–	–	–	–	–	–	–	–	**Aniline, 4-nitro-N-(benzoylmethylene)-**		**MS05555**
236	208	237	254	76	118	179	77	**254**	100	19	16	11	10	9	7	7		14	10	3	2	–	–	–	–	–	–	–	–	–	**3-Indazolone, 2-(2-carboxyphenyl)-**		**MS05556**
254	128	195	168	226	223	129	69	**254**	100	72	31	28	24	22	20	20		14	10	3	2	–	–	–	–	–	–	–	–	–	**1H-Pyrimido[1,2-a]quinoline-2-carboxylic acid, 1-oxo-, methyl ester**	64399-32-2	**NI25109**
254	119	97	83	76	71	130	81	**254**	100	99	80	75	74	74	73	73		14	10	3	2	–	–	–	–	–	–	–	–	–	**4(3H)-Quinazolinone, 3-(2,6-dihydroxyphenyl)-**		**NI25110**
253	254	239	196	255	211	240	119	**254**	100	79	48	43	17	13	10	10		14	14	1	2	–	–	–	–	–	1	–	–	–	**5-Sila-5H,10H,11H-benzo[b]pyrido[4,3-e]azepin-11-one, 5,5-dimethyl-**	89052-47-1	**NI25111**
253	254	239	196	211	252	255	240	**254**	100	49	35	23	18	15	9	8		14	14	1	2	–	–	–	–	–	1	–	–	–	**5-Sila-5H,10H,11H-benzo[e]pyrido[3,4-b]azepin-10-one, 5,5-dimethyl-**	89052-45-9	**NI25112**
126	69	78	43	55	41	29	27	**254**	100	96	95	39	36	35	11	11	**0.00**	14	22	2	–	1	–	–	–	–	–	–	–	–	**Octane, 1-phenylsulphonyl-2-hydroxy-**		**NI25113**
41	123	195	135	55	93	67	81	**254**	100	70	42	40	39	36	33	31	**0.00**	14	22	4	–	–	–	–	–	–	–	–	–	–	**Acetic acid, acetoxy(3,3-dimethyl-2-norbornyl)-, methyl ester**	78571-68-3	**NI25114**
93	57	199	153	122	154	121	94	**254**	100	90	39	37	25	22	20	20	**0.00**	14	22	4	–	–	–	–	–	–	–	–	–	–	**Bicyclo[3.1.1]heptane-1,5-dicarboxylic acid, 1-tert-butyl 5-methyl diester**		**DD00725**
194	135	195	93	254	41	91	79	**254**	100	41	36	32	25	25	24	24		14	22	4	–	–	–	–	–	–	–	–	–	–	**3-Cyclobutene-1,2-dicarboxylic acid, 3,4-dipropyl-, dimethyl ester, cis-**	55673-93-3	**NI25115**
222	194	135	93	41	195	79	91	**254**	100	70	56	54	47	42	42	40	**30.20**	14	22	4	–	–	–	–	–	–	–	–	–	–	**3-Cyclobutene-1,2-dicarboxylic acid, 3,4-dipropyl-, dimethyl ester, trans-**	55673-96-6	**NI25116**
223	79	41	78	55	67	91	77	**254**	100	75	71	57	57	52	46	45	**5.40**	14	22	4	–	–	–	–	–	–	–	–	–	–	**Cyclohexanone, 3-(1,3-heptadienyl)-5,6-dihydroxy-2-hydroxymethyl-**		**IC03324**
41	55	109	111	67	98	83	165	**254**	100	97	72	64	63	55	51	45	**19.00**	14	22	4	–	–	–	–	–	–	–	–	–	–	**1-Cylooctanecarboxylic acid, 1-(2-formylethyl)-2-oxo-, ethyl ester**		**NI25117**
195	162	222	223	41	91	59	77	**254**	100	53	49	23	23	21	17	15	**8.50**	14	22	4	–	–	–	–	–	–	–	–	–	–	**2,4-Hexadienedioic acid, 3,4-dipropyl-, dimethyl ester, (E,E)-**	58367-40-1	**NI25118**
195	196	254	223	91	163	162	41	**254**	100	17	7	7	5	4	4	4		14	22	4	–	–	–	–	–	–	–	–	–	–	**2,4-Hexadienedioic acid, 3,4-dipropyl-, dimethyl ester, (E,Z)-**	58367-41-2	**NI25119**
79	180	160	93	135	230	209	254	**254**	100	51	49	49	44	36	30	29		14	22	4	–	–	–	–	–	–	–	–	–	–	**1,3-Propanedioic acid, 2-(1,1-dimethyl-2-methylidene-3-butenyl)-, diethyl ester**		**MS05557**
95	254	163	79	107	67	160	181	**254**	100	45	36	36	33	31	28	26		14	22	4	–	–	–	–	–	–	–	–	–	–	**1,3-Propanedioic acid, 2-(2-isopropylidene-3-butenyl)-, diethyl ester**		**MS05558**
198	183	254	255	199	239	197		**254**	100	40	33	21	19	9	8			14	23	2	–	–	–	–	–	–	–	1	–	–	**Phosphinic acid, (p-tert-butylphenyl)tert-butyl-**		**LI05180**
198	183	254	78	255	199	91	55	**254**	100	5	4	4	2	2	1	1		14	23	2	–	–	–	–	–	–	–	1	–	–	**Phosphinic acid, (p-tert-butylphenyl)tert-butyl-**		**MS05559**
143	142	29	199	141	41	77	78	**254**	100	24	24	20	19	19	18	17	**1.98**	14	23	2	–	–	–	–	–	–	–	1	–	–	**Phosphonous acid, phenyl-, dibutyl ester**	3030-90-8	**NI25120**
75	57	73	197	69	55	149	71	**254**	100	80	75	56	50	48	45	40	**0.00**	14	26	2	–	–	–	–	–	–	1	–	–	–	**2,4-Pentadienoic acid, 4-methyl-5-tert-butyldimethylsilyl-, ethyl ester, (E,E)-**		**MS05560**
68	41	54	96	69	44	55	43	**254**	100	79	54	53	36	30	28	25	**0.00**	14	26	2	2	–	–	–	–	–	–	–	–	–	**Diazene, bis(3-methylcyclohexyl)-, 1,2-dioxide**	56052-56-3	**NI25121**
225	197	141	169	113	226	111	85	**254**	100	44	43	32	27	26	16	13	**0.00**	14	30	–	–	–	–	–	–	–	2	–	–	–	**Acetylene, bis(triethylsilyl)-**	17947-98-7	**DD00726**
73	59	151	45	74	85	180	43	**254**	100	26	18	9	8	7	5	5	**1.05**	14	30	–	–	–	–	–	–	–	2	–	–	–	**1,7-Octadiene, 1,8-bis[(trimethylsilyl)oxy]-**	74807-81-1	**NI25122**
58	196	72	100	116	154	140	210	**254**	100	77	46	34	15	11	10	9	**0.00**	14	30	–	2	–	–	–	–	–	1	–	–	–	**1,5-Hexadiene, 1-[dimethyl[[[2-(N,N-dimethylamino)ethyl]-N-methylamino]methyl]silyl]-, (E)-**		**DD00727**
58	196	102	72	127	95	154	140	**254**	100	53	48	33	11	11	10	10	**0.00**	14	30	–	2	–	–	–	–	–	1	–	–	–	**1,5-Hexadiene, 1-[dimethyl[[[2-(N,N-dimethylamino)ethyl]-N-methylamino]methyl]silyl]-, (Z)-**		**DD00728**
58	196	72	102	116	173	154	140	**254**	100	53	28	16	11	10	7	7	**0.00**	14	30	–	2	–	–	–	–	–	1	–	–	–	**1,5-Hexadiene, 3-[dimethyl[[[2-(N,N-dimethylamino)ethyl]-N-methylamino]methyl]silyl]-**		**DD00729**
227	89	255	63	90	229	165	228	**254**	100	62	39	37	35	33	28	18	**0.00**	15	9	1	1	–	–	1	–	–	–	–	–	–	**Benz[d][1,3]oxazepine, 2-(4-chlorophenyl)-**		**LI05181**
254	255	152	226	197	115	198	63	**254**	100	16	15	14	13	12	10	9		15	10	4	–	–	–	–	–	–	–	–	–	–	**9,10-Anthracenedione, 1,8-dihydroxy-3-methyl-**	481-74-3	**NI25123**
254	63	255	76	226	152	197	115	**254**	100	16	15	14	13	13	11	11		15	10	4	–	–	–	–	–	–	–	–	–	–	**9,10-Anthracenedione, 1,8-dihydroxy-3-methyl-**	481-74-3	**NI25124**
254	225	208	151	236	255	76	226	**254**	100	95	20	19	18	17	17	16		15	10	4	–	–	–	–	–	–	–	–	–	–	**9,10-Anthracenedione, 1-hydroxy-2-(hydroxymethyl)-**	24094-45-9	**NI25126**
225	254	152	28	226	208	151	236	**254**	100	93	35	33	26	23	22	20		15	10	4	–	–	–	–	–	–	–	–	–	–	**9,10-Anthracenedione, 1-hydroxy-2-(hydroxymethyl)-**	24094-45-9	**NI25125**
254	225	152	76	151	226	255	208	**254**	100	88	34	26	21	20	18	17		15	10	4	–	–	–	–	–	–	–	–	–	–	**9,10-Anthracenedione, 1-hydroxy-2-(hydroxymethyl)-**	24094-45-9	**IC03325**
254	225	211	127	208	183	255	126	**254**	100	54	36	27	22	22	21	17		15	10	4	–	–	–	–	–	–	–	–	–	–	**9,10-Anthracenedione, 1-hydroxy-2-methoxy-**	6003-11-8	**NI25127**
254	225	211	127	183	255	208	126	**254**	100	53	35	26	22	21	20	17		15	10	4	–	–	–	–	–	–	–	–	–	–	**9,10-Anthracenedione, 1-hydroxy-2-methoxy-**	6003-11-8	**LI05182**
208	254	236	127	139	77	209	126	**254**	100	77	30	22	20	18	17	17		15	10	4	–	–	–	–	–	–	–	–	–	–	**9,10-Anthracenedione, 1-hydroxy-2-methoxy-**		**LI05183**
254	57	43	208	41	55	236	69	**254**	100	49	45	43	42	41	40	36		15	10	4	–	–	–	–	–	–	–	–	–	–	**9,10-Anthracenedione, 1-hydroxy-8-methoxy-**		**IC03326**
254	226	255	197	253	227	69	60	**254**	100	90	47	42	38	36	33	31		15	10	4	–	–	–	–	–	–	–	–	–	–	**2H-1-Benzopyran-2-one, 6,8-dihydroxy-4-phenyl-**		**NI25128**
254	81	152	73	71	255	83	95	**254**	100	27	20	19	19	18	18	17		15	10	4	–	–	–	–	–	–	–	–	–	–	**4H-1-Benzopyran-4-one, 5,7-dihydroxy-2-phenyl-**	480-40-0	**NI25130**

MASS TO CHARGE RATIOS	M.W.	INTENSITIES	Parent	C	H	O	N	S	F	Cl	Br	I	Si	P	B	X	COMPOUND NAME	CAS Reg No	No
228 226 229 227 114 113 224 101	**254**	100 22 20 8 8 7 5 5	**0.00**	15	10	4	–	–	–	–	–	–	–	–	–	–	4H-1-Benzopyran-4-one, 5,7-dihydroxy-2-phenyl-	480-40-0	NI25129
254 152 226 124 255 113 253 227	**254**	100 27 25 18 16 12 10 10		15	10	4	–	–	–	–	–	–	–	–	–	–	4H-1-Benzopyran-4-one, 5,7-dihydroxy-2-phenyl-	480-40-0	NI25131
254 253 152 124 69 43 102 96	**254**	100 50 45 20 7 7 6 6		15	10	4	–	–	–	–	–	–	–	–	–	–	4H-1-Benzopyran-4-one, 5,7-dihydroxy-3-phenyl-	4044-00-2	NI25132
254 253 150 124 102 96 69 43	**254**	100 50 47 19 6 6 6 6		15	10	4	–	–	–	–	–	–	–	–	–	–	4H-1-Benzopyran-4-one, 5,7-dihydroxy-3-phenyl-	4044-00-2	LI05184
237 254 121 238 255 253 127 197	**254**	100 82 24 16 13 8 8 5		15	10	4	–	–	–	–	–	–	–	–	–	–	4H-1-Benzopyran-4-one, 3-hydroxy-2-(2-hydroxyphenyl)-		LI05185
238 210 136 108 254 237 239 137	**254**	100 68 44 33 28 21 20 20		15	10	4	–	–	–	–	–	–	–	–	–	–	4H-1-Benzopyran-4-one, 7-hydroxy-2-(4-hydroxyphenyl)-	2196-14-7	NI25133
252 141 132 77 158 156 51 89	**254**	100 78 68 66 37 37 28 22	**2.57**	15	10	4	–	–	–	–	–	–	–	–	–	–	4H-1-Benzopyran-4-one, 7-hydroxy-2-(4-hydroxyphenyl)-	2196-14-7	NI25134
254 39 45 226 44 73 94 197	**254**	100 77 65 53 34 24 17 13		15	10	4	–	–	–	–	–	–	–	–	–	–	2-Isobenzofuranone, 2-benzylidene-1,3-dihydro-4′,5-dihydroxy-		LI05186
254 44 226 43 41 57	**254**	100 64 50 34 34 26		15	10	4	–	–	–	–	–	–	–	–	–	–	1(3H)-Isobenzofuranone, 5-hydroxy-3-[(4-hydroxyphenyl)methylene]-	56783-95-0	MS05561
254 253 256 255 77 51 218 127	**254**	100 29 28 27 27 13 8 8		15	11	–	2	–	–	1	–	–	–	–	–	–	Pyrazole, 3-(4-chlorophenyl)-1-phenyl-		MS05562
253 254 51 255 39 77 40 90	**254**	100 77 64 51 43 38 29 25		15	11	–	2	–	–	1	–	–	–	–	–	–	Quinoline, 2-anilino-4-chloro-		NI25135
120 118 177 77 254 51 253 90	**254**	100 98 46 43 38 21 14 11		15	14	–	2	1	–	–	–	–	–	–	–	–	Benzoic acid, thio-, benzylidenemethylhydrazide	20185-01-7	NI25137
103 76 151 121 50 51 150 77	**254**	100 46 43 38 37 35 31 31	**0.00**	15	14	–	2	1	–	–	–	–	–	–	–	–	Benzoic acid, thio-, benzylidenemethylhydrazide	20185-01-7	NI25138
254 177 121 118 253 77 255 43	**254**	100 77 69 64 46 33 23 20		15	14	–	2	1	–	–	–	–	–	–	–	–	Benzoic acid, thio-, benzylidenemethylhydrazide	20185-01-7	NI25136
105 77 139 121 103 133 152 135	**254**	100 35 18 16 13 11 10 9	**8.00**	15	14	2	2	–	–	–	–	–	–	–	–	–	Acetophenone, 2-hydroxy-, benzoylhydrazone		LI05187
77 121 105 254 51 152 65 76	**254**	100 36 28 25 20 13 13 11		15	14	2	2	–	–	–	–	–	–	–	–	–	2-Azobenzenecarboxylic acid, ethyl ester		MS05563
105 133 77 254 134 121 149 122	**254**	100 81 52 14 10 4 3 2		15	14	2	2	–	–	–	–	–	–	–	–	–	Benzaldehyde, 4-methoxy-, benzoylhydrazone		LI05188
105 77 43 107 51 254 106 39	**254**	100 43 20 15 14 13 10 6		15	14	2	2	–	–	–	–	–	–	–	–	–	Benzamide, N-acetyl-N-(4-pyridinylmethyl)-	35854-38-7	MS05564
91 254 163 92 75 136 255 176	**254**	100 50 30 17 17 9 8 5		15	14	2	2	–	–	–	–	–	–	–	–	–	2-Benzimidazolinone, 1-benzyl-5-methoxy-	2509-28-6	NI25139
91 254 255 176 92 177 163 65	**254**	100 50 8 7 7 6 6 6		15	14	2	2	–	–	–	–	–	–	–	–	–	1H-Benzimidazol-4-ol, 1-benzyl-5-methoxy-	54986-61-7	NI25140
67 171 114 191 254 77 51 209	**254**	100 97 53 36 30 16 16 12		15	14	2	2	–	–	–	–	–	–	–	–	–	3-Butenoic acid, 2-diazo-4-phenyl-, (2E)-2,4-pentadienyl ester		DD00730
163 117 90 91 164 136 89 254	**254**	100 33 19 15 10 8 8 7		15	14	2	2	–	–	–	–	–	–	–	–	–	Ethanamine, N-(4-nitrobenzylidene)-2-phenyl-		MS05565
91 254 176 92 177 163 65	**254**	100 50 7 7 6 6 6		15	14	2	2	–	–	–	–	–	–	–	–	–	3-Indazolol, 1-benzyl-5-methoxy-		LI05189
254 211 239 108 118 225 195 135	**254**	100 32 30 20 17 15 15 15		15	14	2	2	–	–	–	–	–	–	–	–	–	3H-Indazol-3-one, 1,2-dihydro-2-(4-methoxyphenyl)-5-methyl-	28561-69-5	LI05190
254 211 239 108 118 255 135 107	**254**	100 32 30 20 17 16 15 15		15	14	2	2	–	–	–	–	–	–	–	–	–	3H-Indazol-3-one, 1,2-dihydro-2-(4-methoxyphenyl)-5-methyl-	28561-69-5	NI25141
91 254 194 92 77 255 195 221	**254**	100 87 65 47 18 14 14 11		15	14	2	2	–	–	–	–	–	–	–	–	–	2-Methoxy-2-phenylhydrazonoacetophenone		AI01811
254 224 239 253 255 211 225	**254**	100 34 31 24 22 16 11		15	14	2	2	–	–	–	–	–	–	–	–	–	9H-Pyrido[3,4-b]indole, 4,8-dimethoxy-1-vinyl-	65236-62-6	NI25142
170 254 197 114 142 253 141 171	**254**	100 70 25 25 23 21 21 15		15	15	3	–	–	–	–	–	–	–	–	1	–	1-Naphthalenecarboxylic acid, 2-hydroxy-, n-butyl boronate		MS05566
170 254 142 114 141 253 197 171	**254**	100 62 41 22 19 16 14 14		15	15	3	–	–	–	–	–	–	–	–	1	–	3-Naphthalenecarboxylic acid, 2-hydroxy-, n-butyl boronate		MS05567
142 121 178 161 123 120 97 109	**254**	100 42 39 30 30 30 29 27	**1.00**	15	26	3	–	–	–	–	–	–	–	–	–	–	2β,4βH,5β,10βH-Guaian-7-ol, 10,11-epoxy-		LI05191
123 81 194 55 43 151 109 45	**254**	100 80 50 30 30 28 22 20	**0.00**	15	26	3	–	–	–	–	–	–	–	–	–	–	1-Butanol, 1-[(2,6,6-trimethyl-1-cyclohexen-1-yl)oxy]-, acetate	54345-66-3	NI25143
127 168 108 85 57 109 136 211	**254**	100 99 94 77 71 58 53 50	**0.00**	15	26	3	–	–	–	–	–	–	–	–	–	–	Cyclohexanecarboxylic acid, 4-(1,5-dimethyl-3-oxohexyl)-, 4(R)-cis-	38963-91-6	MS05568
138 43 119 121 96 41 55 109	**254**	100 70 50 37 32 32 26 24	**0.00**	15	26	3	–	–	–	–	–	–	–	–	–	–	2-Cyclohexen-1-ol, 3-(3-hydroxybutyl)-2,4,4-trimethyl-, 1-acetate		DD00731
239 110 43 41 81 95 139 55	**254**	100 57 47 45 42 40 33 32	**1.00**	15	26	3	–	–	–	–	–	–	–	–	–	–	1,4-Methano-3-benzoxepin-5,10-diol, decahydro-2,2,9,9a-tetramethyl-	41988-44-7	NI25144
109 59 43 41 55 85 81 67	**254**	100 85 76 63 36 27 27 26	**3.00**	15	26	3	–	–	–	–	–	–	–	–	–	–	1(2H)-Naphthalenone, 4a,5-dimethyl-3-hydroxy-4-(1-hydroxy-1-methylethyl)-3,4,4a,5,6,7,8,8a-octahydro-		LI05192
59 43 109 41 55 95 81 178	**254**	100 71 58 54 36 32 26 17	**1.00**	15	26	3	–	–	–	–	–	–	–	–	–	–	1(2H)-Naphthalenone, 4a,5-dimethyl-3-hydroxy-4-(1-hydroxy-1-methylethyl)-3,4,4a,5,6,7,8,8a-octahydro-		LI05193
84 97 41 98 55 85 112 111	**254**	100 49 48 46 40 37 28 25	**0.00**	15	26	3	–	–	–	–	–	–	–	–	–	–	2-Oxacyclohexadecane-1,6-dione		DD00732
188 96 173 105 119 107 122 55	**254**	100 63 48 28 25 25 24 24	**10.00**	15	26	3	–	–	–	–	–	–	–	–	–	–	3,11,12-Trihydroxydrimene		MS05569
109 55 173 81 119 188 163 93	**254**	100 26 24 23 22 21 21 21	**1.00**	15	26	3	–	–	–	–	–	–	–	–	–	–	11,12,14-Trihydroxydrimene		MS05570
87 225 169 137 81 89 71 61	**254**	100 65 34 21 21 20 19 17	**1.62**	15	30	1	–	–	–	–	–	–	1	–	–	–	Silane, vinylethylmethyl[[5-methyl-2-isopropylcyclohexyl]oxy]-	74685-34-0	NI25145
254 70 87 224 113 211 198 196	**254**	100 60 48 44 39 38 34 34		15	30	1	2	–	–	–	–	–	–	–	–	–	Azacyclotridecan-2-one, 1-(3-aminopropyl)-	64414-61-5	NI25146
254 70 141 84 87 112 100 211	**254**	100 90 58 45 42 40 40 38		15	30	1	2	–	–	–	–	–	–	–	–	–	1,5-Diazacycloheptadecan-6-one	64414-58-0	NI25147
254 91 173 108 151 101 127 76	**254**	100 87 74 56 20 19 17 12		16	6	–	4	–	–	–	–	–	–	–	–	–	Acenaphtho[1,2-b]pyrazine, 8,9-dicyano-		MS05571
105 121 254 77 51 122 106 165	**254**	100 85 43 42 9 8 8 6		16	14	1	–	1	–	–	–	–	–	–	–	–	1,4-Oxathiine, 5,6-dihydro-2,3-diphenyl-		MS05572
105 107 77 36 43 149 51 106	**254**	100 31 21 18 13 12 9 8	**0.00**	16	14	3	–	–	–	–	–	–	–	–	–	–	Acetophenone, α-hydroxy-α-phenyl-, acetate		DO01042
239 254 253 91 163 240 119 77	**254**	100 60 51 37 23 20 19 17		16	14	3	–	–	–	–	–	–	–	–	–	–	1,3-Benzodioxolane, 5-(p-methylbenzoyl)-6-methyl-	52806-36-7	NI25148
239 254 240 255 211 141 139 115	**254**	100 81 10 9 4 4 4 4		16	14	3	–	–	–	–	–	–	–	–	–	–	5-Benzofuranol, 6-methoxy-2-methyl-3-phenyl-	78134-83-5	NI25149
239 254 32 240 255 127 115 77	**254**	100 78 45 21 14 9 8 8		16	14	3	–	–	–	–	–	–	–	–	–	–	5-Benzofuranol, 6-methoxy-3-methyl-2-phenyl-	69470-93-5	NI25150
163 222 194 221 91 195 223 119	**254**	100 68 64 43 43 41 32 31	**13.01**	16	14	3	–	–	–	–	–	–	–	–	–	–	Benzoic acid, 2-(2-methylbenzoyl)-, methyl ester	21147-26-2	NI25151
163 119 77 76 91 254 104 50	**254**	100 44 22 19 18 16 14 14		16	14	3	–	–	–	–	–	–	–	–	–	–	Benzoic acid, 2-(4-methylbenzoyl)-, methyl ester	6424-25-5	NI25152
105 119 118 91 77 65 254 51	**254**	100 96 57 49 32 22 19 17		16	14	3	–	–	–	–	–	–	–	–	–	–	Benzoic acid, 2-methyl-, 2-oxo-2-phenylethyl ester	55153-21-4	NI25153
105 119 91 118 65 106 51 77	**254**	100 46 23 16 9 8 7 6	**4.00**	16	14	3	–	–	–	–	–	–	–	–	–	–	Benzoic acid, 3-methyl-, 2-oxo-2-phenylethyl ester	55153-20-3	NI25154
105 119 91 77 118 106 65 51	**254**	100 74 25 16 9 9 9 7	**4.00**	16	14	3	–	–	–	–	–	–	–	–	–	–	Benzoic acid, 4-methyl-, 2-oxo-2-phenylethyl ester	54797-44-3	NI25155
105 77 254 222 223 195 194 165	**254**	100 28 14 12 8 5 4 4		16	14	3	–	–	–	–	–	–	–	–	–	–	Benzoic acid, 2-(2-oxo-2-phenylethyl)-, methyl ester		LI05194

MASS TO CHARGE RATIOS								M.W.	INTENSITIES								Parent	C	H	O	N	S	F	Cl	Br	I	Si	P	B	X	COMPOUND NAME	CAS Reg No	No
91	119	254	103	120	195	236	107	**254**	100	90	88	62	56	48	46	43		16	14	3	–	–	–	–	–	–	–	–	–	–	Benzo[b]naphtho[2,1-e][1,4]dioxin-6a(6H)-ol, 5,12a-dihydro-	55320-08-6	NI25156
134	254	253	121	119	91	135	108	**254**	100	43	30	30	13	12	11	10		16	14	3	–	–	–	–	–	–	–	–	–	–	4H-1-Benzopyran-4-one, 2,3-dihydro-2-(4-methoxyphenyl)-	3034-08-0	NI25157
134	254	121	253	119	91	135	108	**254**	100	85	46	43	29	21	20	17		16	14	3	–	–	–	–	–	–	–	–	–	–	4H-1-Benzopyran-4-one, 2,3-dihydro-2-(4-methoxyphenyl)-	3034-08-0	NI25158
150	254	122	151	104	107	255	79	**254**	100	29	27	22	17	13	6	6		16	14	3	–	–	–	–	–	–	–	–	–	–	4H-1-Benzopyran-4-one, 2,3-dihydro-7-methoxy-3-phenyl-	5128-69-8	NI25160
136	240	104	108	78	149	91	77	**254**	100	82	35	25	25	20	9	9	**0.00**	16	14	3	–	–	–	–	–	–	–	–	–	–	4H-1-Benzopyran-4-one, 2,3-dihydro-7-methoxy-3-phenyl-	5128-69-8	NI25159
181	76	152	153	182	77	151	254	**254**	100	93	82	63	32	31	23	12		16	14	3	–	–	–	–	–	–	–	–	–	–	[1,1′-Biphenyl]-4-acetic acid, α-oxo-, ethyl ester	6244-53-7	NI25161
212	69	254	121	81	165	57	105	**254**	100	71	59	38	32	29	29	25		16	14	3	–	–	–	–	–	–	–	–	–	–	2(5H)-Furanone, 4-(2-acetylethenyl)-3-(2-phenylethenyl)-		DD00733
254	239	199	225	155	171	183	226	**254**	100	75	58	55	31	21	14	12		16	14	3	–	–	–	–	–	–	–	–	–	–	7H-Furo[3,2-g][1]benzopyran-7-one, 3-(1,1-dimethylpropen-2-yl)-		NI25162
91	119	254	77	103	28	120	195	**254**	100	90	88	64	62	60	55	47		16	14	3	–	–	–	–	–	–	–	–	–	–	Isochromano[4,3:2,3]chroman, 3(3)-hydroxy-		MS05573
134	254	121	253	65	147	108	39	**254**	100	58	55	35	24	16	16	14		16	14	3	–	–	–	–	–	–	–	–	–	–	2-Propen-1-one, 1-(2-hydroxyphenyl)-3-(4-methoxyphenyl)-	3327-24-0	NI25163
254	165	223	194	195	166	167	152	**254**	100	46	31	29	26	20	19	18		16	14	3	–	–	–	–	–	–	–	–	–	–	Stilbene-4′-carboxylic acid, 2-hydroxy-, methyl ester, (E)-		MS05574
212	254	211	165	183	153	152	118	**254**	100	31	17	17	12	11	11	11		16	14	3	–	–	–	–	–	–	–	–	–	–	2,4′-Stilbenediol, 4′-acetate, (E)-		MS05575
91	65	39	120	63	51	29	28	**254**	100	45	27	17	14	14	13	13	**0.00**	16	14	3	–	–	–	–	–	–	–	–	–	–	2,3,5-Trioxabicyclo[2.1.0]pentane, 1,4-bis(phenylmethyl)-	56247-48-4	NI25164
239	161	240	254	120	40	119	93	**254**	100	22	20	16	13	8	7	7		16	18	1	2	–	–	–	–	–	–	–	–	–	1,2-Benzenediamine, 4,5-dimethyl-, 2-hydroxyacetophenonylidene-	103133-80-8	NI25165
134	149	254	148	135	77	255	253	**254**	100	61	40	38	13	11	8	7		16	18	1	2	–	–	–	–	–	–	–	–	–	4H-3,1-Benzoxazine, 1,2-dihydro-2-(p-N,N-dimethylanilino)-		NI25166
254	85	83	42	234	154	170	130	**254**	100	83	56	52	50	50	38	29		16	18	1	2	–	–	–	–	–	–	–	–	–	Ergolin-10-ol, 8,9-didehydro-6,8-dimethyl-		LI05195
170	169	254	253	171	142	143		**254**	100	62	46	18	16	13	12			16	18	1	2	–	–	–	–	–	–	–	–	–	Indolo[2,3-a]quinolizine, 1,4,5,6,7,12b-hexahydro-3-hydroxymethyl-		LI05196
169	197	184	254	211	196	157	130	**254**	100	97	65	54	46	39	39	39		16	18	1	2	–	–	–	–	–	–	–	–	–	1,5-Methano-6H-azocino[4,3-b]indol-6-one, 12-ethyl-1,2,3,4,5,7-hexahydro-	3620-81-3	NI25167
127	109	82	55	110	69	59	67	**254**	100	84	59	49	43	34	30	29	**2.00**	16	30	2	–	–	–	–	–	–	–	–	–	–	[1,1′-Bicyclohexyl]-1,1′-diol, 2,2′,3,3′-tetramethyl-	69576-81-4	NI25168
41	236	55	123	97	81	69	67	**254**	100	98	79	72	72	66	63	53	**3.00**	16	30	2	–	–	–	–	–	–	–	–	–	–	Bicyclohomofarnesane-12,16-diol, (8R,9R)-		MS05576
123	69	55	81	41	236	95	109	**254**	100	62	55	54	50	48	48	40	**3.00**	16	30	2	–	–	–	–	–	–	–	–	–	–	Bicyclohomofarnesane-12,16-diol, (8R,9S)-		MS05577
138	139	45	43	119	41	121	109	**254**	100	36	34	28	27	25	19	19	**0.00**	16	30	2	–	–	–	–	–	–	–	–	–	–	2-Butanol, 4-(3-isopropoxy-2,6,6-trimethylcyclohex-1-en-1-yl)-		DD00734
142	41	55	29	83	43	81	97	**254**	100	73	53	48	45	43	37	33	**5.80**	16	30	2	–	–	–	–	–	–	–	–	–	–	Cyclohexanecarboxylic acid, 1-octyl-, methyl ester	62338-22-1	NI25169
76	87	55	41	83	143	254	69	**254**	100	76	64	38	34	18	16	15		16	30	2	–	–	–	–	–	–	–	–	–	–	Cyclohexanenonanoic acid, methyl ester		LI05197
41	55	27	29	39	43	69	67	**254**	100	62	43	41	38	35	34	34	**1.00**	16	30	2	–	–	–	–	–	–	–	–	–	–	Cyclopentaneundecanoic acid	6053-49-2	WI01290
88	101	69	55	41	83	111	97	**254**	100	56	49	41	38	23	20	17	**0.00**	16	30	2	–	–	–	–	–	–	–	–	–	–	11-Dodecenoic acid, 2,4,6-trimethyl-, methyl ester, (R,R,R)-(–)-	30459-92-8	WI01291
88	101	69	55	83	41	111	97	**254**	100	59	58	47	42	40	21	17	**1.31**	16	30	2	–	–	–	–	–	–	–	–	–	–	11-Dodecenoic acid, 2,4,6-trimethyl-, methyl ester, (R,R,R)-(–)-	30459-92-8	NI25170
88	101	69	55	41	83	111	97	**254**	100	59	59	48	41	26	22	18	**0.00**	16	30	2	–	–	–	–	–	–	–	–	–	–	11-Dodecenoic acid, 2,4,6-trimethyl-, methyl ester, (2S,4R,6R)-(+)-	30459-93-9	WI01292
85	43	55	18	69	83	57	97	**254**	100	45	44	43	37	34	34	30	**4.60**	16	30	2	–	–	–	–	–	–	–	–	–	–	2(3H)-Furanone, 5-dodecyldihydro-	730-46-1	NI25171
55	41	69	43	83	29	28	56	**254**	100	84	69	54	44	34	33	32	**3.80**	16	30	2	–	–	–	–	–	–	–	–	–	–	9-Hexadecenoic acid	2091-29-4	NI25172
55	41	69	43	83	97	82	29	**254**	100	91	58	48	46	34	34	34	**10.01**	16	30	2	–	–	–	–	–	–	–	–	–	–	2-Oxacycloheptadecanone	109-29-5	NI25173
55	41	69	84	56	43	83	70	**254**	100	93	84	78	58	52	47	46	**0.00**	16	30	2	–	–	–	–	–	–	–	–	–	–	2-Oxacyclohexadecanone, 6-methyl-		DD00735
55	41	69	83	43	72	56	97	**254**	100	71	66	44	43	38	38	36	**3.00**	16	30	2	–	–	–	–	–	–	–	–	–	–	2-Oxacyclohexadecanone, 16-methyl-	4459-57-8	NI25174
57	69	55	41	111	127	143	87	**254**	100	94	51	46	45	42	30	27	**9.20**	16	30	2	–	–	–	–	–	–	–	–	–	–	2-Pentene, 2,4,4-trimethyl-3-oxy[bis-		IC03327
69	41	87	70	56	43	55	57	**254**	100	93	77	53	47	44	40	35	**0.00**	16	30	2	–	–	–	–	–	–	–	–	–	–	2-Propenoic acid, 2-methyl-, dodecyl ester		IC03328
69	87	70	83	97	84	71	88	**254**	100	60	38	31	22	19	16	14	**0.10**	16	30	2	–	–	–	–	–	–	–	–	–	–	2-Propenoic acid, 2-methyl-, dodecyl ester	142-90-5	WI01293
43	82	67	81	68	55	54	95	**254**	100	81	80	66	65	57	55	54	**0.00**	16	30	2	–	–	–	–	–	–	–	–	–	–	5-Tetradecen-1-ol, acetate, (Z)-	35153-13-0	NI25175
43	41	55	67	82	68	81	29	**254**	100	51	43	31	30	25	23	19	**0.00**	16	30	2	–	–	–	–	–	–	–	–	–	–	11-Tetradecen-1-ol, acetate, (Z)-	20711-10-8	NI25176
55	56	126	70	140	159	128	98	**254**	100	90	73	67	57	56	50	50	**33.30**	16	34	–	2	–	–	–	–	–	–	–	–	–	1,10-Diazacyclooctadecane		MS05578
107	77	149	79	105	51	108	43	**254**	100	35	34	32	30	10	8	6	**0.00**	17	18	2	–	–	–	–	–	–	–	–	–	–	Acetophenone, α-isopropoxy-α-phenyl-	6652-28-4	NI25177
193	194	223	224	208	207	239	178	**254**	100	91	62	52	41	28	26	22	**0.00**	17	18	2	–	–	–	–	–	–	–	–	–	–	Anthracene, 9,10-dihydro-9,10-dimethoxy-9-methyl-		LI05198
239	254	105	41	91	240	77	117	**254**	100	29	23	20	18	17	16	14		17	18	2	–	–	–	–	–	–	–	–	–	–	Benzaldehyde, 3-(4-tert-butylphenoxy)-	69770-23-6	NI25178
91	163	254	107	65	164	103	92	**254**	100	81	28	28	13	10	9	9		17	18	2	–	–	–	–	–	–	–	–	–	–	Benzene, 2-methoxy-1-(phenylmethoxy)-4-(1-propenyl)-	120-11-6	NI25179
239	121	254	240	28	135	93	65	**254**	100	45	40	18	17	10	10	10		17	18	2	–	–	–	–	–	–	–	–	–	–	Benzophenone, 4-hydroxy-4′-tert-butyl-		DO01043
209	254	181	165	166	255	226	210	**254**	100	78	52	43	39	15	15	15		17	18	2	–	–	–	–	–	–	–	–	–	–	[1,1′-Biphenyl]-4-carboxylic acid, 2′,4′-dimethyl-, ethyl ester	69299-53-2	NI25180
44	41	195	57	77	155	167	91	**254**	100	81	68	31	26	25	21	19	**5.00**	17	18	2	–	–	–	–	–	–	–	–	–	–	[1,1′-Biphenyl]-4-carboxylic acid, 4′-tert-butyl-		WI01294
181	254	194	165	182	166	152	197	**254**	100	55	25	24	22	18	13	12		17	18	2	–	–	–	–	–	–	–	–	–	–	[1,1′-Biphenyl]-4-propanoic acid, β-methyl-, methyl ester	24254-67-9	NI25181
141	69	158	97	140	41	115	254	**254**	100	56	22	22	19	18	13	9		17	18	2	–	–	–	–	–	–	–	–	–	–	Cyclopentanecarboxylic acid, 1-naphthalenemethyl ester	86328-52-1	NI25182
197	198	115	141	77	41	51	57	**254**	100	14	12	12	8	8	5	5	**2.47**	17	18	2	–	–	–	–	–	–	–	–	–	–	1-Propanone, 2,2-dimethyl-1-(4-phenoxyphenyl)-	55814-54-5	NI25183
254	255	135	239	165	127	224	196	**254**	100	19	18	17	9	8	7	7		17	18	2	–	–	–	–	–	–	–	–	–	–	1-Propene, 2,3-bis(4-methoxyphenyl)-		MS05579
45	209	254	181	165	166	148	105	**254**	100	62	60	44	33	25	22	21		17	18	2	–	–	–	–	–	–	–	–	–	–	Stilbene, 2-methoxymethoxy-4′-methyl-, (E)-		MS05580
254	209	45	181	165	166	148	105	**254**	100	98	94	58	42	33	32	24		17	18	2	–	–	–	–	–	–	–	–	–	–	Stilbene, 2-methoxymethoxy-4′-methyl-, (Z)-		MS05581
107	254	163	79	91	135	255	55	**254**	100	97	90	77	64	25	19	19		17	18	2	–	–	–	–	–	–	–	–	–	–	Tricyclo[3.3.1.1^{3,7}]decane-2,6-dione, 4-benzyl-	56781-96-5	MS05582
78	52	77	51	39	50	186	79	**254**	100	17	16	16	13	12	6	6	**1.72**	17	19	–	–	–	–	–	–	–	–	1	–	–	Phosphine, (2,2-dimethylcyclopropyl)diphenyl-	61142-51-6	NI25184
254	253	134	210	255	118	237	126	**254**	100	62	29	26	20	15	14	14		17	22	–	2	–	–	–	–	–	–	–	–	–	Aniline, 4,4′-methylenebis[N,N-dimethyl-	101-61-1	IC03329

MASS TO CHARGE RATIOS								M.W.	INTENSITIES								Parent	C	H	O	N	S	F	Cl	Br	I	Si	P	B	X	COMPOUND NAME	CAS Reg No	No
254	253	134	126	210	118	127	237	**254**	100	70	44	38	35	27	22	15		17	22	–	2	–	–	–	–	–	–	–	–	–	**Aniline, 4,4'-methylenebis[N,N-dimethyl-**	101-61-1	NI25185
211	167	209	254	168	183	212	169	**254**	100	34	30	26	23	20	18	17		17	22	–	2	–	–	–	–	–	–	–	–	–	**Aniline, N-phenyl-4-(1,2-dimethylpropylamino)-**		IC03330
225	54	167	254	223	195	77	226	**254**	100	87	47	44	36	27	15	13		17	22	–	2	–	–	–	–	–	–	–	–	–	**Aniline, N-phenyl-4-(1-ethylpropylamino)-**		IC03331
148	163	105	120	149	44	77	42	**254**	100	61	51	37	13	13	11	11	**1.00**	17	22	–	2	–	–	–	–	–	–	–	–	–	**Benzeneethanamine, 3-amino-N-methyl-N-(2-phenylethyl)-**	52059-44-6	NI25186
148	105	120	106	163	42	77	149	**254**	100	50	27	18	17	17	15	13	**1.00**	17	22	–	2	–	–	–	–	–	–	–	–	–	**Benzeneethanamine, 4-amino-N-methyl-N-(2-phenylethyl)-**	52059-41-3	NI25187
134	120	91	77	254	42	118	105	**254**	100	61	17	14	12	9	8	8		17	22	–	2	–	–	–	–	–	–	–	–	–	**Ethylenediamine, N,N'-dimethyl-N-phenyl-N'-3-tolyl-**	32869-60-6	LI05199
134	120	91	77	254	135	42	105	**254**	100	61	17	14	12	11	9	8		17	22	–	2	–	–	–	–	–	–	–	–	–	**Ethylenediamine, N,N'-dimethyl-N-phenyl-N'-3-tolyl-**	32869-60-6	NI25188
134	120	254	135	91	77	119	118	**254**	100	44	15	10	8	6	5	5		17	22	–	2	–	–	–	–	–	–	–	–	–	**Ethylenediamine, N,N'-dimethyl-N-phenyl-N'-4-tolyl-**	32869-58-2	NI25189
120	91	42	104	107	147	105	132	**254**	100	83	75	62	61	56	50	49	**10.00**	17	22	–	2	–	–	–	–	–	–	–	–	–	**1,3-Propanediamine, N,N'-dimethyl-N,N'diphenyl-**		IC03332
93	162	106	120	107	163	94	41	**254**	100	68	39	13	9	8	8	6	**1.02**	17	22	–	2	–	–	–	–	–	–	–	–	–	**Pyridine, 2,2'-(1,7-heptanediyl)bis-**	57476-56-9	NI25190
93	162	106	120	107	163	94	92	**254**	100	68	39	13	9	8	8	6	**1.03**	17	22	–	2	–	–	–	–	–	–	–	–	–	**Pyridine, 2,2'-(1,7-heptanediyl)bis-**	57476-56-9	NI25191
43	41	82	96	68	127	110	124	**254**	100	55	21	10	7	4	3	2	**0.00**	17	34	1	–	–	–	–	–	–	–	–	–	–	**Cyclododecane, 3-methylbutoxy-**		DD00736
55	82	57	41	96	43	69	68	**254**	100	97	83	83	70	63	57	53	**0.80**	17	34	1	–	–	–	–	–	–	–	–	–	–	**Cycloheptadecanol**	4429-77-0	WI01295
71	85	69	83	82	96	254	97	**254**	100	26	25	14	14	12	11	11		17	34	1	–	–	–	–	–	–	–	–	–	–	**2-Heptadecanone**	2922-51-2	WI01296
43	58	59	71	41	29	55	57	**254**	100	99	58	40	32	21	20	18	**5.71**	17	34	1	–	–	–	–	–	–	–	–	–	–	**2-Heptadecanone**	2922-51-2	WI01297
127	155	170	142	254				**254**	100	66	20	18	13					17	34	1	–	–	–	–	–	–	–	–	–	–	**8-Heptadecanone**		LI05200
71	43	57	141	58	41	29	55	**254**	100	98	91	75	74	73	62	51	**3.40**	17	34	1	–	–	–	–	–	–	–	–	–	–	**9-Heptadecanone**	540-08-9	WI01298
141	57	58	71	156	96	41	157	**254**	100	62	56	52	37	32	24	22	**11.10**	17	34	1	–	–	–	–	–	–	–	–	–	–	**9-Heptadecanone**	540-08-9	IC03333
141	71	57	43	58	41	55	156	**254**	100	98	92	92	73	58	46	37	**9.00**	17	34	1	–	–	–	–	–	–	–	–	–	–	**9-Heptadecanone**	540-08-9	LI05201
58	43	41	57	55	71	69	81	**254**	100	29	23	22	19	17	9	7	**0.84**	17	34	1	–	–	–	–	–	–	–	–	–	–	**Hexadecanal, 2-methyl-**	55019-46-0	NI25192
71	41	82	43	96	68	29	67	**254**	100	33	27	17	15	14	14	12	**1.24**	17	34	1	–	–	–	–	–	–	–	–	–	–	**1-Hexadecene, 1-methoxy-**	15519-14-9	NI25193
55	70	67	82	81	69	41	83	**254**	100	80	69	62	61	59	56	54	**0.50**	17	34	1	–	–	–	–	–	–	–	–	–	–	**8-Hexadecen-1-ol, 14-methyl-, (R)-(–)-(Z)-**	30689-78-2	NI25194
55	70	67	57	81	69	41	83	**254**	100	90	70	69	67	66	65	61	**1.00**	17	34	1	–	–	–	–	–	–	–	–	–	–	**Quinoxaline, 2-phenyl-3-[(trimethylsilyl)oxy]-**		NI25195
57	83	98	58	41	99	55	141	**254**	100	45	37	36	25	17	17	16	**2.00**	17	34	1	–	–	–	–	–	–	–	–	–	–	**6-Undecanone, 2,2,4,8,10,10-hexamethyl-**		IC03334
239	254	105	240	98	255	197	41	**254**	100	41	27	18	14	8	8	8		18	22	1	–	–	–	–	–	–	–	–	–	–	**Benzene, 2-[(3-tert-butyl-4-hydroxyphenyl)ethyl]-**		IC03335
254	44	225	173	91	186	239	115	**254**	100	93	58	51	48	47	45	45		18	22	1	–	–	–	–	–	–	–	–	–	–	**Cyclopropa[3,4]cyclohepta[1,2-a]naphthalene, 1,1a,1b,2,3,7b,8,9,10,10a-decahydro-5-methoxy-10-methylene-**	57983-83-2	NI25196
254	159	160	133	146	172	239	157	**254**	100	57	41	38	32	28	26	24		18	22	1	–	–	–	–	–	–	–	–	–	–	**Estra-1,3,5(10),16-tetraen-3-ol**	1150-90-9	NI25197
254	91	183	155	141	197	239		**254**	100	58	41	39	36	34	31			18	22	1	–	–	–	–	–	–	–	–	–	–	**2(1H)-Naphthalenone, 3,4,4a,5,6,7-hexahydro-1-benzyl-1-methyl-**		LI05202
115	226	109	91	67	131	55	211	**254**	100	97	84	78	67	58	56	50	**31.00**	18	22	1	–	–	–	–	–	–	–	–	–	–	**1(2H)-Naphthalenone, octahydro-8a-methyl-2-benzylene-, cis-**	17429-44-6	NI25198
226	109	115	254	211	91	131	116	**254**	100	71	59	57	49	48	47	45		18	22	1	–	–	–	–	–	–	–	–	–	–	**1(2H)-Naphthalenone, octahydro-8a-methyl-2-benzylene-, cis-**	17429-44-6	NI25199
109	226	115	254	116	67	91	41	**254**	100	68	57	52	40	39	38	35		18	22	1	–	–	–	–	–	–	–	–	–	–	**1(2H)-Naphthalenone, octahydro-8a-methyl-2-benzylene-, trans-**	17622-48-9	NI25200
197	212	41	141	129	183	254	128	**254**	100	58	46	43	38	35	34	33		18	22	1	–	–	–	–	–	–	–	–	–	–	**Spiro[cyclobutane-1,1'(2'H)-phenanthren]-2-one, 3',4',4'a,9',10',10'a-hexahydro-, (1'α,4'aβ,10'aα)-**	41487-68-7	NI25201
57	43	71	41	85	211	55	98	**254**	100	51	39	36	28	24	23	19	**0.05**	18	38	–	–	–	–	–	–	–	–	–	–	–	**Dodecane, 4,9-dipropyl-**		AI01812
57	43	41	71	55	85	211	98	**254**	100	58	42	31	27	21	15	15	**0.02**	18	38	–	–	–	–	–	–	–	–	–	–	–	**Dodecane, 4,9-dipropyl-**		AI01813
57	43	41	71	55	85	98	69	**254**	100	60	43	34	25	23	14	12	**0.02**	18	38	–	–	–	–	–	–	–	–	–	–	–	**Dodecane, 4,9-dipropyl-**		AI01814
56	41	43	55	71	69	85	58	**254**	100	85	55	36	31	18	16	16	**0.09**	18	38	–	–	–	–	–	–	–	–	–	–	–	**Dodecane, 2,2,4,9,11,11-hexamethyl-**		AI01815
43	57	41	71	85	55	56	42	**254**	100	75	45	40	27	24	16	15	**1.12**	18	38	–	–	–	–	–	–	–	–	–	–	–	**Heptadecane, 2-methyl-**		AI01817
43	57	41	71	85	55	56	42	**254**	100	80	46	44	29	24	18	16	**0.87**	18	38	–	–	–	–	–	–	–	–	–	–	–	**Heptadecane, 2-methyl-**		AI01816
43	71	57	41	70	55	85	211	**254**	100	72	67	50	40	37	33	25	**2.86**	18	38	–	–	–	–	–	–	–	–	–	–	–	**Heptadecane, 4-methyl-**	26429-11-8	NI25202
57	71	43	41	85	112	55	56	**254**	100	70	68	33	23	21	21	16	**1.60**	18	38	–	–	–	–	–	–	–	–	–	–	–	**Heptadecane, 7-methyl-**		AI01818
57	43	71	85	41	126	55	154	**254**	100	87	78	51	39	34	27	23	**7.60**	18	38	–	–	–	–	–	–	–	–	–	–	–	**Heptadecane, 8-methyl-**		AI01819
57	71	43	85	41	55	56	69	**254**	100	78	77	44	36	25	17	12	**0.20**	18	38	–	–	–	–	–	–	–	–	–	–	–	**Hexadecane, 7,9-dimethyl-**		AI01820
57	41	71	58	56	43	85	69	**254**	100	7	6	4	3	3	2	2	**0.00**	18	38	–	–	–	–	–	–	–	–	–	–	–	**Hexane, 3,4-di-tert-butyl-2,2,5,5-tetramethyl-**	62850-21-9	NI25203
57	43	71	41	85	55	42	69	**254**	100	94	60	45	39	23	12	10	**1.71**	18	38	–	–	–	–	–	–	–	–	–	–	–	**Octadecane**		AI01821
57	43	71	85	41	55	29	56	**254**	100	80	59	38	33	23	20	16	**7.58**	18	38	–	–	–	–	–	–	–	–	–	–	–	**Octadecane**		IC03336
57	43	71	41	85	29	55	56	**254**	100	90	58	44	40	31	25	18	**9.89**	18	38	–	–	–	–	–	–	–	–	–	–	–	**Octadecane**		AI01822
57	71	85	113	99	98	112	169	**254**	100	70	40	28	25	22	20	18	**0.00**	18	38	–	–	–	–	–	–	–	–	–	–	–	**Pentadecane, 2,6,10-trimethyl-**		LI05204
57	71	43	85	55	99	98	41	**254**	100	67	45	30	23	19	17	11	**1.00**	18	38	–	–	–	–	–	–	–	–	–	–	–	**Pentadecane, 2,6,10-trimethyl-**		LI05203
169	141	155	183	170	43	254	128	**254**	100	44	35	19	17	13	11	11		19	26	–	–	–	–	–	–	–	–	–	–	–	**Naphthalene, nonyl-**	27193-93-7	WI01299
126	254	113	127	125				**254**	100	91	61	58	54					20	14	–	–	–	–	–	–	–	–	–	–	–	**6,7-Acechrysene**		LI05205
254	253	252	126	255	113	250	125	**254**	100	52	41	28	20	16	15	12		20	14	–	–	–	–	–	–	–	–	–	–	–	**Anthracene, 9-phenyl-**	602-55-1	NI25204
253	252	254	250	255	251	126	248	**254**	100	85	71	26	13	9	6	5		20	14	–	–	–	–	–	–	–	–	–	–	–	**Benzo[b]fluoranthene, 1,2-dihydro-**		NI25205
254	253	252	126	250	255	113	125	**254**	100	76	73	25	22	19	15	10		20	14	–	–	–	–	–	–	–	–	–	–	–	**Benzo[a]pyrene, 4,5-dihydro-**	57652-66-1	NI25206
253	254	252	250	255	251	239	248	**254**	100	91	90	33	19	13	10	7		20	14	–	–	–	–	–	–	–	–	–	–	–	**Benzo[e]pyrene, 4,5-dihydro-**	95676-42-9	NI25207

MASS TO CHARGE RATIOS								M.W.	INTENSITIES								Parent	C	H	O	N	S	F	Cl	Br	I	Si	P	B	X	COMPOUND NAME	CAS Reg No	No
253	252	254	126	250	113	57	255	**254**	100	85	78	36	27	20	18	15		20	14	–	–	–	–	–	–	–	–	–	–	–	4,6′-Biazulenyl	94154-49-1	NI25208
253	252	254	126	250	251	113	255	**254**	100	94	80	34	31	18	18	17		20	14	–	–	–	–	–	–	–	–	–	–	–	4,8′-Biazulenyl		NI25209
254	253	252	126	255	127	125	113	**254**	100	69	61	46	20	18	18	18		20	14	–	–	–	–	–	–	–	–	–	–	–	1,1′-Binaphthalene	604-53-5	WI01300
254	253	252	127	255	250	239	125	**254**	100	91	72	28	21	20	15	15		20	14	–	–	–	–	–	–	–	–	–	–	–	1,1′-Binaphthalene	604-53-5	IC03337
254	253	252	126	255	113	125	127	**254**	100	75	52	42	20	17	15	14		20	14	–	–	–	–	–	–	–	–	–	–	–	1,2′-Binaphthalene	4325-74-0	WI01301
254	253	252	126	255	250	113	125	**254**	100	90	61	29	19	18	13	10		20	14	–	–	–	–	–	–	–	–	–	–	–	1,2′-Binaphthalene	4325-74-0	IC03338
254	253	252	126	255	250	127	125	**254**	100	72	53	27	19	13	10	10		20	14	–	–	–	–	–	–	–	–	–	–	–	1,2′-Binaphthalene	4325-74-0	AI01823
254	252	255	125	253	127	250	113	**254**	100	37	23	17	15	15	7	7		20	14	–	–	–	–	–	–	–	–	–	–	–	2,2′-Binaphthalene		IC03339
254	252	255	126	253	127	113	125	**254**	100	29	21	15	13	13	6	5		20	14	–	–	–	–	–	–	–	–	–	–	–	2,2′-Binaphthalene		AI01824
254	253	252	126	255	250	125	113	**254**	100	98	67	36	25	18	17	16		20	14	–	–	–	–	–	–	–	–	–	–	–	9H-Fluorene, 9-benzylene-	1836-87-9	NI25210
253	254	126	252	113	125	250	112	**254**	100	93	66	60	46	35	23	19		20	14	–	–	–	–	–	–	–	–	–	–	–	9H-Fluorene, 9-benzylene-	1836-87-9	NI25211
253	254	252	126	113	125	250	255	**254**	100	85	83	68	43	31	29	20		20	14	–	–	–	–	–	–	–	–	–	–	–	1H-Indene, 1,1′-(1,2-ethanediylidene)bis-	72088-04-1	NI25212
254	253	252	239	226	250	126	113	**254**	100	50	42	19	19	13	13	8		20	14	–	–	–	–	–	–	–	–	–	–	–	Pleiadene, 7,12-dihydro-7,12-dimethylene-		LI05206
254	253	252	126	255	113	125	127	**254**	100	92	46	37	20	18	15	12		20	14	–	–	–	–	–	–	–	–	–	–	–	Triptycene	477-75-8	NI25213
253	254	252	126	255	250	126.5	125	**254**	100	91	54	32	20	16	15	15		20	14	–	–	–	–	–	–	–	–	–	–	–	Triptycene	477-75-8	LI05207
257	255	210	208	242	240	45	129	**255**	100	94	35	33	29	27	24	22		5	6	–	1	3	–	–	1	–	–	–	–	–	Isothiazole, 4-bromo-3,5-bis(methylthio)-	37572-44-4	NI25214
257	255	210	208	242	240	45	163	**255**	100	94	35	33	29	27	24	22		5	6	–	1	3	–	–	1	–	–	–	–	–	Isothiazole, 4-bromo-3,5-bis(methylthio)-	37572-44-4	LI05208
46	30	76	43	29	31	28	41	**255**	100	15	11	11	11	7	7	5	**0.00**	5	9	9	3	–	–	–	–	–	–	–	–	–	2-Methyl-2-nitro-oxymethyltrimethylene dinitrate		MS05583
257	255	194	133	167	98			**255**	100	70	44	22	13	13				6	1	–	3	–	–	4	–	–	–	–	–	–	s-Triazolo[4,3-a]pyridine, 3,6,7,8-tetrachloro-		LI05209
220	222	255	75	257	74	63	110	**255**	100	40	25	24	18	15	15	13		6	3	4	1	1	–	2	–	–	–	–	–	–	Benzenesulphonyl chloride, 4-chloro-2-nitro-	4533-96-4	NI25215
220	222	255	257	110	221	145	128	**255**	100	41	26	18	15	8	7	7		6	3	4	1	1	–	2	–	–	–	–	–	–	Benzenesulphonyl chloride, 4-chloro-2-nitro-	4533-96-4	LI05210
220	75	222	110	74	156	109	30	**255**	100	58	38	35	34	32	25	21	**17.51**	6	3	4	1	1	–	2	–	–	–	–	–	–	Benzenesulphonyl chloride, 4-chloro-3-nitro-	97-08-5	NI25216
158	256	91	220	204	222	160	258	**255**	100	76	60	57	44	43	40	37	**0.00**	6	3	4	1	1	–	2	–	–	–	–	–	–	Benzenesulphonyl chloride, 4-chloro-3-nitro-	97-08-5	NI25217
186	50	122	76	75	92	85	64	**255**	100	87	87	84	81	45	43	26	**0.00**	7	4	4	1	1	3	–	–	–	–	–	–	–	Nitrobenzene, 4-trifluoromethylsulphonyl-	432-87-1	NI25218
92	140	196	60	168	81	29	255	**255**	100	61	51	49	34	29	29	28		7	14	3	1	2	–	–	–	–	–	1	–	–	Phosphoramidic acid, 1,3-dithiolan-2-ylidene-, diethyl ester	947-02-4	NI25219
211	73	226	195	75	135	147	45	**255**	100	48	46	36	29	27	25	22	**16.01**	7	22	3	1	–	–	–	–	–	2	1	–	–	Phosphonic acid, (aminomethyl)-, bis(trimethylsilyl) ester	53044-29-4	NI25221
211	73	226	75	135	147	133	240	**255**	100	48	46	29	27	25	19	16	**16.00**	7	22	3	1	–	–	–	–	–	2	1	–	–	Phosphonic acid, (aminomethyl)-, bis(trimethylsilyl) ester	53044-29-4	NI25220
192	165	193	209	162	146	255	237	**255**	100	74	66	58	56	56	47	43		8	9	5	5	–	–	–	–	–	–	–	–	–	Pyrazolo[5,1-c][1,2,4]triazine-3-carboxylic acid, 4,7-dihydro-6-nitro-7-hydroxy-, ethyl ester		NI25222
98	142	156	42	127	99	43	113	**255**	100	42	27	23	22	13	13	12	**0.00**	8	18	–	1	–	–	–	–	1	–	–	–	–	Piperidinium, 1-ethyl-1-methyl-, iodide	4186-71-4	NI25223
242	240	257	255	133	75	74	182	**255**	100	97	56	54	47	47	20	17		9	6	1	1	1	–	–	1	–	–	–	–	–	3H-Indol-3-one, 6-bromo-2-(methylthio)-	50630-70-1	NI25224
92	156	191	108	65	140	255	55	**255**	100	96	89	55	52	18	17	15		9	9	2	3	2	–	–	–	–	–	–	–	–	Benzenesulphonamide, 4-amino-N-2-thiazolyl-		LI05211
92	191	156	108	65	255	93	55	**255**	100	91	82	60	51	20	20	20		9	9	2	3	2	–	–	–	–	–	–	–	–	Benzenesulphonamide, 4-amino-N-2-thiazolyl-		NI25225
91	48	255	47	60	150	76	45	**255**	100	90	64	64	51	44	43	39		9	9	2	3	2	–	–	–	–	–	–	–	–	Hydrazinecarbodithioic acid, [(4-nitrophenyl)methylene]-, methyl ester	20184-97-8	NI25226
255	149	91	207	208	103	76	50	**255**	100	38	33	32	27	25	18	18		9	9	2	3	2	–	–	–	–	–	–	–	–	Hydrazinecarbodithioic acid, [(4-nitrophenyl)methylene]-, methyl ester	20184-97-8	NI25227
196	191	176	255	104	150	219		**255**	100	85	79	48	37	26	12			9	9	6	3	–	–	–	–	–	–	–	–	–	β-Alanine, 2,4-dinitrophenyl-		LI05212
210	164	118	255	134	91			**255**	100	26	22	19	18	17				9	9	6	3	–	–	–	–	–	–	–	–	–	Alanine, 2,4-dinitrophenyl-		LI05213
147	73	183	255	77	81	113	75	**255**	100	94	75	57	47	28	25	25		9	10	–	1	–	5	–	–	–	1	–	–	–	Silanamine, 1,1,1-trimethyl-N-(pentafluorophenyl)-	22529-97-1	NI25228
86	118	56	42	224	45	31	178	**255**	100	65	29	29	28	26	25	22	**1.10**	9	22	5	1	–	–	–	–	–	–	1	–	–	Phosphonic acid, N,N-bis(2-hydroxyethyl)aminomethyl-, diethyl ester		IC03340
91	77	255	51	186	104			**255**	100	43	36	13	10	10				10	7	–	1	–	6	–	–	–	–	–	–	–	Aziridine, N-phenyl-2,2-bis(trifluoromethyl)-		LI05214
210	255	208	253	182	227	206	207	**255**	100	61	47	30	23	19	19	18		10	9	2	1	–	–	–	–	–	–	–	–	1	Selenolo[2,3-b]pyridine-2-carboxylic acid, ethyl ester	41323-19-7	NI25229
171	173	43	213	215	134	92	65	**255**	100	96	84	78	77	31	10	8	**6.87**	10	10	2	1	–	–	–	1	–	–	–	–	–	Aniline, 2-bromo-N,N-diacetyl-		IC03341
237	43	93	220	58	239			**255**	100	84	52	39	37	32			**6.00**	10	10	3	3	–	–	1	–	–	–	–	–	–	3(2H)-Pyridazinone, 5-amino-4-chloro-2-(2,3-cis-dihydroxycyclohexa-4,6-diene-1-yl)-		NI25230
154	41	42	59	29	69	57	27	**255**	100	48	35	30	29	17	15	15	**0.00**	10	16	3	1	–	3	–	–	–	–	–	–	–	Alanine, 2-methyl-N-(trifluoroacetyl)-, butyl ester	34815-07-1	NI25231
154	41	59	57	42	155	69	114	**255**	100	60	45	45	43	35	20	18	**0.00**	10	16	3	1	–	3	–	–	–	–	–	–	–	Alanine, 2-methyl-N-(trifluoroacetyl)-, sec-butyl ester	57983-07-0	NI25232
154	57	59	155	114	182	69	84	**255**	100	23	18	16	10	8	7	4	**0.00**	10	16	3	1	–	3	–	–	–	–	–	–	–	Alanine, 2-methyl-N-(trifluoroacetyl)-, sec-butyl ester, (R)-	55887-90-6	NI25233
154	41	57	59	155	42	114	69	**255**	100	25	24	19	15	14	7	6	**0.00**	10	16	3	1	–	3	–	–	–	–	–	–	–	Alanine, 2-methyl-N-(trifluoroacetyl)-, sec-butyl ester, (S)-	54986-70-8	NI25234
140	153	181	55	182	42	102	158	**255**	100	84	47	41	39	25	23	21	**8.10**	10	16	3	1	–	3	–	–	–	–	–	–	–	β-Alanine, N-methyl-N-(trifluoroacetyl)-, butyl ester	34815-13-9	NI25235
154	42	110	155	56	41	69	57	**255**	100	31	21	10	10	10	9	8	**0.45**	10	16	3	1	–	3	–	–	–	–	–	–	–	L-Alanine, N-methyl-N-(trifluoroacetyl)-, butyl ester	34815-08-2	NI25236
154	110	42	41	155	69	57	56	**255**	100	17	15	10	6	6	5	5	**1.97**	10	16	3	1	–	3	–	–	–	–	–	–	–	L-Alanine, N-methyl-N-(trifluoroacetyl)-, butyl ester	34815-08-2	NI25237
140	182	199	55	153	57	42	102	**255**	100	79	58	47	32	20	18	17	**2.00**	10	16	3	1	–	3	–	–	–	–	–	–	–	β-Alanine, N-methyl-N-(trifluoroacetyl)-, sec-butyl ester, (S)-	54986-71-9	NI25238
154	42	57	110	155	182	41	56	**255**	100	23	20	19	17	10	10	9	**0.00**	10	16	3	1	–	3	–	–	–	–	–	–	–	L-Alanine, N-methyl-N-(trifluoroacetyl)-, sec-butyl ester, (S)-	54986-65-1	NI25239
55	168	43	70	42	41	71	69	**255**	100	70	57	25	16	16	14	13	**0.00**	10	16	3	1	–	3	–	–	–	–	–	–	–	β-Alanine, N-(trifluoroacetyl)-, 1-methylbutyl ester	57983-45-6	NI25240

MASS TO CHARGE RATIOS								M.W.	INTENSITIES								Parent	C	H	O	N	S	F	Cl	Br	I	Si	P	B	X	COMPOUND NAME	CAS Reg No	No
43	140	71	141	69	41	55	42	**255**	100	72	41	36	23	22	14	13	**0.00**	10	16	3	1	–	3	–	–	–	–	–	–	–	L-Alanine, N-(trifluoroacetyl)-, 1-methylbutyl ester	57983-11-6	NI25241
41	154	29	57	27	69	155	39	**255**	100	82	54	32	24	21	20	15	**0.00**	10	16	3	1	–	3	–	–	–	–	–	–	–	Butanoic acid, 2-[(trifluoroacetyl)amino]-, butyl ester	34815-09-3	NI25242
140	69	182	56	57	153	41	42	**255**	100	98	80	63	53	52	44	19	**2.89**	10	16	3	1	–	3	–	–	–	–	–	–	–	Butanoic acid, 3-[(trifluoroacetyl)amino]-, butyl ester	34815-12-8	NI25243
41	56	29	182	69	57	126	27	**255**	100	61	59	46	44	41	36	30	**0.60**	10	16	3	1	–	3	–	–	–	–	–	–	–	Butanoic acid, 4-[(trifluoroacetyl)amino]-, butyl ester	34815-14-0	NI25244
154	57	41	155	182	69	140	126	**255**	100	70	64	47	17	11	10	10	**0.00**	10	16	3	1	–	3	–	–	–	–	–	–	–	Butanoic acid, 2-[(trifluoroacetyl)amino]-, sec-butyl ester	54986-68-4	NI25245
140	69	182	41	42	57	56	43	**255**	100	98	82	66	65	56	54	47	**0.00**	10	16	3	1	–	3	–	–	–	–	–	–	–	Butanoic acid, 3-[(trifluoroacetyl)amino]-, sec-butyl ester	54986-67-3	NI25248
69	57	41	182	56	74	42	126	**255**	100	71	70	67	36	29	20	17	**0.00**	10	16	3	1	–	3	–	–	–	–	–	–	–	Butanoic acid, 3-[(trifluoroacetyl)amino]-, sec-butyl ester	54986-67-3	NI25246
69	182	140	57	56	41	153	200	**255**	100	90	81	32	26	26	20	15	**0.00**	10	16	3	1	–	3	–	–	–	–	–	–	–	Butanoic acid, 3-[(trifluoroacetyl)amino]-, sec-butyl ester	54986-67-3	NI25247
182	41	154	57	126	181	56	140	**255**	100	31	24	24	20	15	15	12	**0.00**	10	16	3	1	–	3	–	–	–	–	–	–	–	Butanoic acid, 4-[(trifluoroacetyl)amino]-, sec-butyl ester, (S)-	54986-66-2	NI25249
154	126	56	57	41	140	155	42	**255**	100	52	39	26	20	13	9	9	**3.14**	10	16	3	1	–	3	–	–	–	–	–	–	–	Glycine, N-ethyl-N-(trifluoroacetyl)-, butyl ester	34815-10-6	NI25250
154	57	126	155	41	42	199	56	**255**	100	74	73	57	51	47	45	39	**4.08**	10	16	3	1	–	3	–	–	–	–	–	–	–	Glycine, N-ethyl-N-(trifluoroacetyl)-, sec-butyl ester	57983-21-8	NI25251
154	57	199	126	155	41	182	140	**255**	100	53	39	32	26	25	24	22	**19.02**	10	16	3	1	–	3	–	–	–	–	–	–	–	Glycine, N-ethyl-N-(trifluoroacetyl)-, sec-butyl ester, (S)-	54986-51-5	NI25252
43	126	85	41	127	154	69	57	**255**	100	37	37	28	21	19	15	15	**0.00**	10	16	3	1	–	3	–	–	–	–	–	–	–	Glycine, N-(trifluoroacetyl)-, 1-methylpentyl ester	57983-14-9	NI25253
101	55	73	69	126	142	166	182	**255**	100	52	49	22	17	10	6	5	**0.00**	10	16	3	1	–	3	–	–	–	–	–	–	–	Propanamine, N-(trifluoroacetyl)-3-(4,5-dimethyl-1,3-dioxolan-2-yl)-	85236-74-4	NI25254
182	153	69	74	41	126	56	154	**255**	100	47	40	28	26	16	16	12	**0.00**	10	16	3	1	–	3	–	–	–	–	–	–	–	Propanoic acid, 2-methyl-3-[(trifluoroacetyl)amino]-, butyl ester	54986-69-5	NI25255
55	41	29	168	56	27	69	57	**255**	100	71	68	49	49	44	42	37	**0.00**	10	16	3	1	–	3	–	–	–	–	–	–	–	Propanoic acid, 2-methyl-3-[(trifluoroacetyl)amino]-, butyl ester	54986-69-5	NI25256
69	57	41	182	56	74	42	126	**255**	100	71	70	67	36	29	20	17	**0.00**	10	16	3	1	–	3	–	–	–	–	–	–	–	Propanoic acid, 2-methyl-3-[(trifluoroacetyl)amino]-, sec-butyl ester	54986-50-4	NI25257
182	69	57	41	74	56	153	126	**255**	100	95	46	41	30	19	17	15	**0.00**	10	16	3	1	–	3	–	–	–	–	–	–	–	Propanoic acid, 2-methyl-3-[(trifluoroacetyl)amino]-, sec-butyl ester	54986-50-4	NI25258
210	195	96	167	71	152	181	166	**255**	100	81	54	40	28	26	22	16	**0.00**	10	17	3	5	–	–	–	–	–	–	–	–	–	Propanoic acid, (4,6-bis(dimethylamino)-1,3,5-triazinyloxy)-	63617-39-0	NI25259
240	254	255	73	98	74	241	170	**255**	100	82	50	50	28	22	21	20		10	21	1	3	–	–	–	–	–	2	–	–	–	4-Pyrimidinamine, N-(trimethylsilyl)-2-[(trimethylsilyl)oxy]-	18037-10-0	NI25260
255	83	133	257	176	85	91	87	**255**	100	68	65	62	54	46	43	41		11	7	2	1	–	–	2	–	–	–	–	–	–	2(5H)-Furanone, 3,4-dichloro-5-(4-methylphenylimino)-		MS05584
255	133	257	87	176	106	78	77	**255**	100	86	74	43	37	18	18	17		11	7	2	1	–	–	2	–	–	–	–	–	–	1H-Pyrrole-2,5-dione, 3,4-dichloro-1-(4-methylphenyl)-	29244-55-1	NI25261
196	133	197	76	77	104	43	105	**255**	100	70	36	30	27	26	26	19	**0.00**	11	13	4	1	1	–	–	–	–	–	–	–	–	1,2-Benzisothiazolin-3-one, 2-(propoxymethoxy)-, 1,1-dioxide	27148-08-9	NI25262
42	171	107	77	121	56	92	64	**255**	100	73	41	39	30	26	23	23	**10.48**	11	13	4	1	1	–	–	–	–	–	–	–	–	3-Pyrrolidinone, 1-(4-methoxybenzenesulfonyl)-	83132-97-2	NI25263
183	185	157	155	76	257	255	200	**255**	100	97	31	31	30	29	29	27		11	14	1	1	–	–	–	1	–	–	–	–	–	Benzamide, 3-bromo-N-tert-butyl-	42498-39-5	NI25264
183	185	202	200	157	155	76	255	**255**	100	97	28	28	27	27	26	25		11	14	1	1	–	–	–	1	–	–	–	–	–	Benzamide, 4-bromo-N-tert-butyl-	42498-38-4	NI25265
43	154	71	126	42	41	155	199	**255**	100	59	38	24	21	18	14	10	**0.00**	11	20	2	1	–	3	–	–	–	–	–	–	–	Glycine, N-ethyl-N-(2,2,2-trifluoroethyl)-, 1-methylbutyl ester	57983-68-3	NI25266
43	58	68	27	18	41	29	44	**255**	100	75	43	43	42	38	36	30	**28.82**	11	21	–	5	1	–	–	–	–	–	–	–	–	1,3,5-Triazine-2,4-diamine, 6-(ethylthio)-N,N′-diisopropyl-	4147-51-7	NI25267
154	73	45	102	155	59	43	74	**255**	100	98	19	14	13	12	11	9	**0.00**	11	25	–	3	–	–	–	–	–	2	–	–	–	Histamine, bis(trimethylsilyl)-		MS05585
173	175	145	254	109	147	75	39	**255**	100	65	27	19	19	17	17	16	**12.09**	12	11	1	1	–	–	2	–	–	–	–	–	–	Benzamide, 3,5-dichloro-N-(1,1-dimethyl-2-propynyl)-	23950-58-5	NI25268
173	175	255	145	257	254	147	256	**255**	100	85	37	31	25	24	23	22		12	11	1	1	–	–	2	–	–	–	–	–	–	Benzamide, 3,5-dichloro-N-(1,1-dimethyl-2-propynyl)-	23950-58-5	NI25269
56	154	102	139	74	70	45	42	**255**	100	86	78	59	37	18	16	15	**6.50**	12	17	3	1	1	–	–	–	–	–	–	–	–	Alanine, N-[2-(4-methoxyphenylthio)ethyl]-	64276-43-3	NI25270
119	91	255	118	211	120	225	105	**255**	100	29	22	20	16	13	9	6		12	17	3	1	1	–	–	–	–	–	–	–	–	4H-1,3,2,6-Dioxathiazocine, 4-methyl-6-(p-tolyl)-5,6,7,8-tetrahydro-, 2-oxide	84868-74-6	NI25271
196	126	41	98	42	197	185	55	**255**	100	35	29	28	19	16	16	14	**4.50**	12	17	5	1	–	–	–	–	–	–	–	–	–	2-Butenedioic acid, (hexahydro-2-oxoazepin-1-yl)-, dimethyl ester, (Z)-		NI25272
164	196	83	59	55	136	113	39	**255**	100	80	50	40	40	30	30	30	**2.00**	12	17	5	1	–	–	–	–	–	–	–	–	–	2-Butenedioic acid, (1-methoxy-3-methyl-2-butenyl)imino-, dimethyl ester, (Z)-		NI25273
88	43	74	134	45	71	75	150	**255**	100	97	60	25	25	19	17	16	**0.13**	12	17	5	1	–	–	–	–	–	–	–	–	–	1,3-Cyclohexanediol, 2-(cyanomethylene)-4,5-dimethoxy-, 1-acetate, (E)-		NI25274
240	18	211	43	41	140	55	42	**255**	100	46	38	31	29	23	19	19	**15.59**	12	21	3	3	–	–	–	–	–	–	–	–	–	Isoxazolo[2,3-a]pyridine, 2,7-dicarbaldehyde dioxime-2,4,7-trimethyl-perhydro- (β form)	72589-85-6	NI25275
211	238	240	43	255	41	124	42	**255**	100	38	26	22	19	19	15	15		12	21	3	3	–	–	–	–	–	–	–	–	–	Isoxazolo[2,3-a]pyridine, 2,7-dicarbaldehyde dioxime-2,4,7-trimethyl-perhydro- (β form)	72589-85-6	NI25276
226	115	114	143	255	128	101	75	**255**	100	99	74	60	44	37	31	28		13	9	1	3	1	–	–	–	–	–	–	–	–	2-Naphthalenol, 1-(2-thiazolylazo)-	1147-56-4	NI25277
155	91	226	184	65	41	227	42	**255**	100	97	85	78	18	12	11	10	**2.00**	13	21	2	1	1	–	–	–	–	–	–	–	–	Benzenesulphonamide, 4-methyl-N,N-dipropyl-	723-42-2	NI25278
81	134	67	41	44	94	68	109	**255**	100	98	94	94	81	78	70	64	**17.00**	13	21	2	1	1	–	–	–	–	–	–	–	–	(+)-Thujan-3-spiro-2′-thiazolidine-4-carboxylic acid		NI25279
94	43	212	196	255	151	108	82	**255**	100	44	26	21	13	13	11	11		13	21	4	1	–	–	–	–	–	–	–	–	–	9-Azabicyclo[3.3.1]nonane-2,6-diol, 9-methyl-, diacetate (ester), (endo,endo)-	49656-42-0	NI25280
94	43	196	82	42	212	152	255	**255**	100	96	92	69	51	45	25	22		13	21	4	1	–	–	–	–	–	–	–	–	–	9-Azabicyclo[4.2.1]nonane-2,5-diol, 9-methyl-, diacetate (ester), (endo,endo)-	49656-41-9	NI25281
136	124	68	182	96	170	55	41	**255**	100	93	64	50	50	41	41	38	**33.00**	13	21	4	1	–	–	–	–	–	–	–	–	–	1-Cyclohexene-1-carboxylic acid, 2-[[(ethoxycarbonyl)methyl]amino]-, ethyl ester		MS05586
58	59	42	222	57	237	44	196	**255**	100	23	8	7	4	4	4	4	**2.60**	13	21	4	1	–	–	–	–	–	–	–	–	–	Phenylethanolamine, N,N-dimethyl-3,4,5-trimethoxy-		NI25282
255	223	256	111	191	196			**255**	100	64	27	20	10	4				14	9	–	1	2	–	–	–	–	–	–	–	–	5H-1,4-Benzodithieno[2,3-b]indole		LI05215
136	108	255	69	91	63	137	64	**255**	100	33	23	17	12	12	10	9		14	9	2	1	1	–	–	–	–	–	–	–	–	4H-5,6-Benzo-1,3-thiazin-4-one, 2-(2-hydroxyphenyl)-		NI25283
121	255	77	51	63	64	92	102	**255**	100	75	41	20	13	7	6	2		14	9	2	1	1	–	–	–	–	–	–	–	–	5-Thiazolone, 4-furfurylidene-2-phenyl-		LI05216
220	254	255	92	75	102	65	257	**255**	100	10	8	8	8	7	7	3		14	10	–	3	–	–	1	–	–	–	–	–	–	Quinazoline, 4-[N-(2-chlorophenyl)amino]-		MS05587

MASS TO CHARGE RATIOS								M.W.	INTENSITIES								Parent	C	H	O	N	S	F	Cl	Br	I	Si	P	B	X	COMPOUND NAME	CAS Reg No	No
150	255	106	51	77	151	39	65	255	100	76	72	20	19	18	17	16		14	13	–	3	1	–	–	–	–	–	–	–	–	2H-Benzimidazole-2-thione, 1,3-dihydro-1-[2-(4-pyridinyl)ethyl]-	13083-37-9	NI25284
93	119	255	136	91	77	65	135	255	100	40	30	23	19	19	19	18		14	13	2	3	–	–	–	–	–	–	–	–	–	Imidodicarbonic diamide, N,N′-diphenyl-	28584-90-9	NI25285
157	214	255	130	185	82	76	54	255	100	95	44	32	31	24	24	22		14	13	2	3	–	–	–	–	–	–	–	–	–	4(3H)-Quinazolinone, 2-ethyl-3-(3-methyl-5-isoxazolyl)-		NI25286
121	255	94	256	149	108	226	122	255	100	69	29	12	9	9	8	8		14	17	–	5	–	–	–	–	–	–	–	–	–	Azobenzene, 5,4′-dimethyl-2,3′,4-triamino-		IC03342
58	128	43	44	72	42			255	100	53	30	28	23	15			**1.24**	14	22	1	1	–	–	1	–	–	–	–	–	–	Benzeneethanol, 4-chloro-α-[2-(dimethylamino)-isopropyl]-α-methyl-	14860-49-2	NI25287
113	55	126	152	41	74	185	83	255	100	43	36	31	31	23	20	20	**6.16**	14	25	3	1	–	–	–	–	–	–	–	–	–	1-Aziridinenonanoic acid, 2,2-dimethyl-θ-oxo-, methyl ester	42550-29-8	MS05588
71	113	55	70	72	56	41	224	255	100	48	48	46	36	25	20	19	**17.12**	14	25	3	1	–	–	–	–	–	–	–	–	–	2-Oxazoleoctanoic acid, 4,5-dihydro-5,5-dimethyl-, methyl ester	42550-30-1	MS05589
198	57	170	197	76	41	43	69	255	100	82	52	29	29	29	27	25	**25.25**	14	29	1	3	–	–	–	–	–	–	–	–	–	5-Nonanone, 2,2,8,8-tetramethyl-, semicarbazone	5709-96-6	NI25288
227	89	255	63	90	229	165	228	255	100	62	38	37	35	32	28	17		15	10	1	1	–	–	1	–	–	–	–	–	–	3,1-Benzoxazepine, 2-(4-chlorophenyl)-	23246-32-4	NI25289
105	77	255	51	257	180	220	103	255	100	62	31	18	11	11	9	9		15	10	1	1	–	–	1	–	–	–	–	–	–	Isoxazole, 4-chloro-3,5-diphenyl-		LI05217
105	77	255	137	257	139	103	89	255	100	89	48	28	16	12	9	9		15	10	1	1	–	–	1	–	–	–	–	–	–	Isoxazole, 3-(4-chlorophenyl)-5-phenyl-	24097-17-4	NI25290
105	255	77	257	256	106	178	127.5	255	100	63	27	22	11	8	6	6		15	10	1	1	–	–	1	–	–	–	–	–	–	Isoxazole, 3-(4-chlorophenyl)-5-phenyl-	24097-17-4	LI05218
139	255	141	257	256	111	69	144	255	100	91	33	30	20	20	18	16		15	10	1	1	–	–	1	–	–	–	–	–	–	Isoxazole, 5-(4-chlorophenyl)-3-phenyl-		LI05219
139	255	103	141	111	77	257	144	255	100	43	32	30	30	20	16	13		15	10	1	1	–	–	1	–	–	–	–	–	–	Isoxazole, 5-(4-chlorophenyl)-3-phenyl-	1148-87-4	NI25291
105	77	255	137	257	139	178	103	255	100	89	48	28	16	12	9	9		15	10	1	1	–	–	1	–	–	–	–	–	–	Isoxazole, 5-(4-chlorophenyl)-3-phenyl-		LI05220
255	226	225	256	254	227	212	184	255	100	54	30	29	20	19	19	13		15	13	3	1	–	–	–	–	–	–	–	–	–	9-Acridanone, 1-hydroxy-10-methyl-3-methoxy-		MS05590
255	165	166	256	178	209	194	28	255	100	46	25	17	13	8	8	7		15	13	3	1	–	–	–	–	–	–	–	–	–	Benzene, 1-methoxy-4-[2-(4-nitrophenyl)vinyl]-, (E)-	4648-33-3	NI25292
255	227	210	182	154				255	100	60	50	30	10					15	13	3	1	–	–	–	–	–	–	–	–	–	10-Benzindolizinecarboxylic acid, 3-hydroxy-, ethyl ester		LI05221
107	106	76	104	50	18	77	39	255	100	77	65	61	50	31	26	23	**1.50**	15	13	3	1	–	–	–	–	–	–	–	–	–	Benzoic acid, 2-[[(3-methylphenyl)amino]carbonyl]-	85-72-3	NI25293
105	136	135	77	119	64	91	51	255	100	84	81	42	25	20	16	16	**10.00**	15	13	3	1	–	–	–	–	–	–	–	–	–	1,3,4-Dioxazole, 2-(4-methoxyphenyl)-5-phenyl-		NI25294
149	106	134	77	255	51	105	78	255	100	47	40	38	37	23	21	21		15	13	3	1	–	–	–	–	–	–	–	–	–	1,3,4-Dioxazole, 5-(4-methoxyphenyl)-2-phenyl-		NI25295
149	106	134	77	255	51	105	78	255	100	47	40	38	37	23	21	21		15	13	3	1	–	–	–	–	–	–	–	–	–	1,4,2-Dioxazole, 3-(4-methoxyphenyl)-5-phenyl-	55076-22-7	NI25296
105	136	135	77	119	64	91	52	255	100	84	81	42	25	20	16	16	**10.00**	15	13	3	1	–	–	–	–	–	–	–	–	–	1,4,2-Dioxazole, 5-(4-methoxyphenyl)-3-phenyl-	55076-25-0	NI25297
255	240	57	224	239	71	55	69	255	100	98	68	62	43	43	43	37		15	13	3	1	–	–	–	–	–	–	–	–	–	9-Oximo-2,7-dimethoxyfluorene		MS05591
255	46	208	256	238	225	207	209	255	100	60	22	19	18	11	10	4		15	13	3	1	–	–	–	–	–	–	–	–	–	1-Propanone, 2-(2-nitrophenyl)-1-phenyl-	58751-67-0	NI25298
255	46	256	136	225				255	100	36	16	10	2					15	13	3	1	–	–	–	–	–	–	–	–	–	1-Propanone, 3-(4-nitrophenyl)-1-phenyl-	1669-48-3	NI25299
57	128	155	102	85	142	170	43	255	100	87	60	44	44	43	41	33	**0.00**	15	29	2	1	–	–	–	–	–	–	–	–	–	Propanamide, N-(4-oxodecyl)-2,2-dimethyl-	116437-31-1	DD00737
42	43	86	84	44	57	41	127	255	100	74	66	66	58	57	52	45	**13.20**	15	33	–	3	–	–	–	–	–	–	–	–	–	Triazine, N,N,N-tributylhexahydro-		MS05592
136	255	212	213	109	131	91	65	255	100	63	56	44	25	21	15	12		16	17	–	1	1	–	–	–	–	–	–	–	–	1,5-Benzothiazepine, 2-methyl-4-phenyl-2,3,4,5-tetrahydro-	13338-10-8	NI25300
178	240	211	255	254	91	165	136	255	100	28	28	26	8	8	7	7		16	17	–	1	1	–	–	–	–	–	–	–	–	Isoquinoline, 5-methylthio-4-phenyl-1,2,3,4-tetrahydro-		MS05593
178	254	255	91	77	163	51	208	255	100	31	23	15	13	10	10	7		16	17	–	1	1	–	–	–	–	–	–	–	–	Isoquinoline, 6-methylthio-1-phenyl-1,2,3,4-tetrahydro-		MS05594
179	226	255	31	89	165	151	105	255	100	41	40	11	10	8	8	5		16	17	–	1	1	–	–	–	–	–	–	–	–	Isoquinoline, 7-methylthio-4-phenyl-1,2,3,4-tetrahydro-		MS05595
91	84	65	55	77	97	164	254	255	100	77	74	56	44	40	38	34	**0.00**	16	17	–	1	1	–	–	–	–	–	–	–	–	5H-2-Pyrindine, 4-(benzylthio)-6,7-dihydro-3-methyl-	99702-57-5	DD00738
118	91	255	77	65	105	51	178	255	100	89	19	13	8	7	7	4		16	17	–	1	1	–	–	–	–	–	–	–	–	2-Thiopheneethanamine, N-benzylidene-3-methyl-		MS05596
91	182	255	65	180	92	79	78	255	100	40	31	29	19	17	16	16		16	17	2	1	–	–	–	–	–	–	–	–	–	Acetic acid, [1-benzyl-2(1H)-pyridinylidene]-, ethyl ester	14759-28-5	NI25301
135	151	77	255	92	136	91	256	255	100	34	34	32	29	21	14	6		16	17	2	1	–	–	–	–	–	–	–	–	–	Benzeneethanamine, 2-(4-methoxybenzoyl)-		MS05597
121	255	122	77	161	135	78	91	255	100	11	10	7	6	5	5	4		16	17	2	1	–	–	–	–	–	–	–	–	–	Benzenemethanamine, 4-methoxy-N-[(4-methoxyphenyl)methylene]-	3261-60-7	MS05598
223	208	255	222	224	180	194	77	255	100	64	47	32	23	22	18	14		16	17	2	1	–	–	–	–	–	–	–	–	–	Benzoic acid, 2-[(2,3-dimethylphenyl)amino]-, methyl ester	1222-42-0	NI25302
148	77	51	105	42	104	78	149	255	100	44	23	20	18	10	10	9	**0.00**	16	17	2	1	–	–	–	–	–	–	–	–	–	Benzoin, 4-dimethylamino-	6317-85-7	NI25303
255	254	212	200	226	213	227		255	100	50	25	20	17	10	8			16	17	2	1	–	–	–	–	–	–	–	–	–	Crinan, 1,2-didehydro-	6902-20-1	LI05222
255	254	226	227	240	200			255	100	39	21	16	15	15				16	17	2	1	–	–	–	–	–	–	–	–	–	Crinan, 2,3-didehydro-	6793-60-8	LI05223
255	212	254	179	197	151	211	240	255	100	83	42	31	28	28	20	13		16	17	2	1	–	–	–	–	–	–	–	–	–	Dibenz[c,e]azocine-1,2-diol, 5,6,7,8-tetrahydro-6-methyl-		LI05224
110	68	43	146	255	212	90	145	255	100	64	58	46	45	29	25	24		16	17	2	1	–	–	–	–	–	–	–	–	–	2(1H)-Naphthalenone, 1-[(3,5-dimethyl-4-isoxazolyl)methyl]-3,4-dihydro-	54986-56-0	NI25304
184	196	185	255	183	156	141	115	255	100	48	44	30	24	20	16	16		16	17	2	1	–	–	–	–	–	–	–	–	–	DL-7-Oxohasubanalactam	34087-44-0	NI25305
58	91	72	71	197	185	65	42	255	100	65	20	19	13	9	9	9	**3.80**	16	21	–	3	–	–	–	–	–	–	–	–	–	1,2-Ethanediamine, N,N-dimethyl-N′-benzyl-N′-2-pyridinyl-	91-81-6	NI25306
211	256	254	185	239				255	100	51	41	27	13				**0.00**	16	21	–	3	–	–	–	–	–	–	–	–	–	1,2-Ethanediamine, N,N-dimethyl-N′-benzyl-N′-2-pyridinyl-	91-81-6	NI25307
58	91	197	72	71	185	184	65	255	100	93	30	26	25	16	13	8	**3.00**	16	21	–	3	–	–	–	–	–	–	–	–	–	1,2-Ethanediamine, N,N-dimethyl-N′-benzyl-N′-2-pyridinyl-	91-81-6	NI25308
115	58	100	128	72	43	44	29	255	100	37	33	29	17	12	11	10	**4.00**	16	33	1	1	–	–	–	–	–	–	–	–	–	Dodecanamide, N,N-diethyl-		MS05599
85	98	44	41	43	29	55	27	255	100	37	21	20	18	18	15	14	**0.00**	16	33	1	1	–	–	–	–	–	–	–	–	–	Hexadecanamide		IC03343
59	72	28	43	57	41	55	60	255	100	44	40	21	16	14	13	12	**3.00**	16	33	1	1	–	–	–	–	–	–	–	–	–	Hexadecanamide		MS05600
100	114	43	128	142	44	71	184	255	100	50	35	24	16	15	13	11	**4.00**	16	33	1	1	–	–	–	–	–	–	–	–	–	Hexanamide, N,N-dipentyl-		MS05601
100	114	43	128	142	44	71	41	255	100	50	35	24	16	15	13	11	**4.00**	16	33	1	1	–	–	–	–	–	–	–	–	–	Hexanamide, N,N-dipentyl-		LI05225
170	156	128	43	198	55	60	41	255	100	67	66	51	30	28	27	25	**0.10**	16	33	1	1	–	–	–	–	–	–	–	–	–	6-Undecanamine, N-acetyl-5-propyl-		MS05602
91	212	255	92	122	65	256	213	255	100	59	40	16	13	13	9	9		17	21	1	1	–	–	–	–	–	–	–	–	–	Aniline, 3-benzyloxy-4-butyl-		MS05603
58	73	45	167	165	44	42	59	255	100	89	8	5	5	4	4	3	**0.70**	17	21	1	1	–	–	–	–	–	–	–	–	–	Ethanamine, 2-(diphenylmethoxy)-N,N-dimethyl-	58-73-1	MS05604
58	73	167	165	45	168	44	42	255	100	25	10	9	7	5	5	4	**0.10**	17	21	1	1	–	–	–	–	–	–	–	–	–	Ethanamine, 2-(diphenylmethoxy)-N,N-dimethyl-	58-73-1	NI25309

MASS TO CHARGE RATIOS								M.W.	INTENSITIES								Parent	C	H	O	N	S	F	Cl	Br	I	Si	P	B	X	COMPOUND NAME	CAS Reg No	No
58	73	45	44	59	42	165	167	**255**	100	80	49	22	16	14	14	11	**0.00**	17	21	1	1	–	–	–	–	–	–	–	–	–	Ethanamine, 2-(diphenylmethoxy)-N,N-dimethyl-	58-73-1	NI25310
58	255	40	72	71	59	42	91	**255**	100	10	7	5	4	4	4	3		17	21	1	1	–	–	–	–	–	–	–	–	–	Ethanamine, N,N-dimethyl-2-[2-benzylphenoxy]-	92-12-6	NI25311
148	164	105	121	91	77	43	44	**255**	100	96	69	40	39	32	32	29	**1.00**	17	21	1	1	–	–	–	–	–	–	–	–	–	Phenol, 3-[2-[methyl(2-phenylethyl)amino]ethyl]-	52059-51-5	NI25312
148	105	121	164	77	42	91	149	**255**	100	58	38	29	26	25	22	13	**1.00**	17	21	1	1	–	–	–	–	–	–	–	–	–	Phenol, 4-[2-[methyl(2-phenylethyl)amino]ethyl]-	52059-47-9	NI25313
170	184	212	55	43	41	85	57	**255**	100	96	78	23	21	21	15	15	**0.20**	17	37	–	1	–	–	–	–	–	–	–	–	–	6-Dodecanamine, N-ethyl-6-propyl-		MS05605
30	57	43	71	41	44	55	85	**255**	100	39	37	22	20	18	17	14	**2.00**	17	37	–	1	–	–	–	–	–	–	–	–	–	1-Heptadecanamine	4200-95-7	NI25314
156	58	44	157	57	43	71	41	**255**	100	19	14	11	9	7	5	5	**1.00**	17	37	–	1	–	–	–	–	–	–	–	–	–	1-Hexanamine, 2-ethyl-N-(2-ethylhexyl)-N-methyl-	39168-98-4	NI25315
44	142	43	155	41	57	55	56	**255**	100	83	76	71	56	26	23	18	**3.33**	17	37	–	1	–	–	–	–	–	–	–	–	–	Nonanamine, N-octyl-	24552-01-0	NI25316
156	58	157	44	43	41	57	42	**255**	100	33	12	11	10	8	7	5	**1.10**	17	37	–	1	–	–	–	–	–	–	–	–	–	1-Octanamine, N-methyl-N-octyl-	4455-26-9	NI25317
156	170	43	41	154	55	168	57	**255**	100	96	16	16	14	14	13	13	**0.10**	17	37	–	1	–	–	–	–	–	–	–	–	–	7-Tetradecanamine, N-ethyl-7-methyl-		MS05606
105	164	93	79	77	162	41	91	**255**	100	54	18	16	16	14	13	13	**4.60**	18	25	–	1	–	–	–	–	–	–	–	–	–	Benzeneethanamine, N-(–)-thujylidene		NI25318
148	106	149	91	79	41	77	28	**255**	100	12	11	8	8	6	4	4	**0.00**	18	25	–	1	–	–	–	–	–	–	–	–	–	Benzenemethanamine, α-2,5,7-octatrienyl-N-propyl-		NI25319
213	212	77	255	196	224	198	168	**255**	100	58	49	45	39	24	22	22		18	25	–	1	–	–	–	–	–	–	–	–	–	Pyridine, 1,2-dihydro-3,5-diethyl-2-propyl-1-phenyl-		IC03344
64	256	128	160	192	258	96	66	**256**	100	42	31	28	17	15	9	8		–	–	–	–	8	–	–	–	–	–	–	–	–	Cyclooctasulphur		IC03345
69	149	50	130	31	80	99	168	**256**	100	25	23	17	15	12	10	9	**0.00**	2	–	–	–	–	8	–	–	–	–	–	–	1	Selenium, bis(trifluoromethyl)difluoro-		NI25320
101	117	85	70	69	63	51	82	**256**	100	83	48	44	36	35	20	14	**0.00**	2	–	–	–	2	7	1	–	–	–	–	–	–	1,3-Dithietane, 2-chloro-1,1,2,3,3,4,4-heptafluoro-		MS05607
179	177	98	181	129	127	100	142	**256**	100	77	28	24	16	11	9	8	**2.50**	2	1	–	–	–	2	1	2	–	–	–	–	–	Ethane, 1,2-dibromo-1,1-difluoro-		IC03346
15	257	259	109	79	29	93	47	**256**	100	55	53	50	38	29	27	26	**0.00**	4	8	4	–	–	–	3	–	–	–	1	–	–	Phosphonic acid, (2,2,2-trichloro-1-hydroxyethyl)-, dimethyl ester	52-68-6	NI25321
79	109	110	139	145	80	112	95	**256**	100	96	72	55	53	32	27	25	**0.00**	4	8	4	–	–	–	3	–	–	–	1	–	–	Phosphonic acid, (2,2,2-trichloro-1-hydroxyethyl)-, dimethyl ester	52-68-6	NI25322
186	221	171	256	101	85	258	69	**256**	100	77	59	56	46	41	36	36		6	–	–	–	–	6	2	–	–	–	–	–	–	1,3,5-Hexatriene, 1,6-dichloroperfluoro-		DO01044
72	58	129	127	86	81	56	55	**256**	100	62	26	23	22	21	18	14	**0.10**	6	13	1	2	1	5	–	–	–	–	–	–	–	[(Diethylamino)methoxymethylenimino]sulphur pentafluoride	81439-15-8	NI25323
82	97	131	75	81	163	101	113	**256**	100	83	77	31	23	20	17	11	**0.00**	7	4	1	–	–	8	–	–	–	–	–	–	–	Bicyclo[2.2.0]hexane, 2-methoxy-1,2,3,4,5,5,6,6-octafluoro-		NI25324
122	177	123	70	45	95	69	51	**256**	100	32	18	14	13	11	8	7	**5.00**	7	5	–	4	1	–	–	1	–	–	–	–	–	s-Tetrazine, 1-[5-(bromomethyl)-2-thienyl]-		MS05608
200	121	256	135	228	177	56	174	**256**	100	92	47	46	42	18	9	6		7	5	2	–	–	–	–	1	–	–	–	–	1	Iron, dicarbonyl-π-cyclopentadienylbromo-		MS05609
193	53	166	167	67	68	184	180	**256**	100	60	58	52	36	29	26	24	**10.00**	7	8	5	6	–	–	–	–	–	–	–	–	–	1,2,4-Triazolo[5,1-c][1,2,4]triazine-2-carboxylic acid, 4,7-dihydro-7-hydroxy-6-nitro-, ethyl ester		NI25325
83	41	55	256	43	67	82	39	**256**	100	98	94	70	62	56	50	41		7	13	3	–	–	–	–	–	–	–	1	–	1	1-Phospha-2,6,7-trioxabicyclo[2.2.2]octane, 4-isopropyl-, 1-selenide	77414-33-6	NI25326
55	97	41	39	27	69	43	135	**256**	100	86	62	29	24	23	17	15	**0.00**	7	14	–	–	–	–	–	2	–	–	–	–	–	Heptane, 1,7-dibromo-	4549-31-9	NI25327
241	73	133	242	45	163	256	59	**256**	100	34	18	18	15	14	13	12		7	21	4	–	–	–	–	–	–	2	1	–	–	Phosphoric acid, bis(trimethylsilyl)-, monomethyl ester		NI25328
79	64	113	77	51	59	177	110	**256**	100	83	50	22	21	14	7	7	**0.10**	8	8	–	–	–	8	–	–	–	–	–	–	–	Cyclobutane, 1,1,2,2-tetrafluoro-3-(1,1,2,2-tetrafluorobutyl)-	35207-97-7	MS05610
105	106	61	147	73	45	107	43	**256**	100	65	46	41	18	16	14	14	**0.14**	8	16	5	–	2	–	–	–	–	–	–	–	–	D-Glucose, cyclic 1,2-ethanediyl mercaptal	3650-65-5	NI25329
55	69	43	41	129	171	27	97	**256**	100	87	59	59	29	26	25	23	**2.54**	8	17	1	–	–	–	–	–	1	–	–	–	–	2-Octanol, 1-iodo-		IC03347
42	41	50	75	76	122	92	64	**256**	100	63	59	33	33	30	18	16	**0.00**	9	8	5	2	1	–	–	–	–	–	–	–	–	3-Azetidinone, N-(4-nitrobenzenesulphonyl)-	82380-62-9	NI25330
241	93	145	63	256	81	147	77	**256**	100	77	44	39	28	23	22	21		9	9	1	–	–	5	–	–	–	1	–	–	–	Silane, methoxydimethyl(pentafluorophenyl)-	23761-74-2	NI25331
77	73	256	163	81	113	47	49	**256**	100	87	52	37	29	28	24	17		9	9	1	–	–	5	–	–	–	1	–	–	–	Silane, trimethyl(pentafluorophenoxy)-	22577-86-2	NI25332
221	194	256	139	241	96	165	83	**256**	100	84	72	46	33	30	25	25		9	13	1	6	–	–	1	–	–	–	–	–	–	1,3,5-Triazine-2,4-diamine, 6-(2-chloroethyloxy)-N-cyano-N,N′,N′-trimethyl-	77455-08-4	NI25333
55	77	69	149	56	49	41	73	**256**	100	97	63	51	43	27	27	21	**0.00**	9	14	4	–	–	–	2	–	–	–	–	–	–	1,3-Propanediol, 2,2-dimethyl-, dichloroacetate		IC03348
135	105	256	107	136	75	133	45	**256**	100	25	20	20	19	17	15	11		9	20	4	–	2	–	–	–	–	–	–	–	–	D-Arabinose, diethyl mercaptal	1941-50-0	NI25334
186	142	121	97	153	187	93	125	**256**	100	47	47	39	33	31	22	21	**15.34**	9	21	2	–	2	–	–	–	–	–	1	–	–	Phosphorodithioic acid, O,O-diethyl S-pentyl ester	55955-86-7	NI25335
256	258	221	260	257	223	158	259	**256**	100	72	31	18	16	13	13	11		10	6	–	–	1	–	2	–	–	1	–	–	–	1-Sila-2,3-benzo-4,5-thiophenocyclopentadiene, 1,1-dichloro-	60203-42-1	NI25336
68	119	69	188	139	169	138	127	**256**	100	37	12	8	8	7	4	4	**0.00**	10	6	1	–	–	6	–	–	–	–	–	–	–	11-Oxa-3,4,4,5,5,6-hexafluorotetracyclo[6.2.1.0^{2,7}.0^{3,6}]undec-9-ene		NI25337
89	213	132	161	147	169	241	63	**256**	100	57	54	46	45	41	38	35	**34.00**	10	8	1	–	1	–	–	–	–	–	–	–	1	2-Benzo[b]thiophenecarboxaldehyde, 3-methylseleno-	39857-04-0	NI25338
89	161	132	213	241	169	256	121	**256**	100	82	67	60	49	49	48	44		10	8	1	–	1	–	–	–	–	–	–	–	1	3-Benzo[b]thiophenecarboxaldehyde, 2-methylseleno-	39857-07-3	NI25339
43	185	183	75	50	76	157	155	**256**	100	33	33	25	23	20	10	9	**1.00**	10	9	3	–	–	–	–	1	–	–	–	–	–	Acetic acid, 4-bromophenacyl ester		NI25340
135	214	52	107	18	134	27	41	**256**	100	86	80	56	51	39	38	33	**17.01**	10	12	4	2	1	–	–	–	–	–	–	–	–	1H-2,1,3-Benzothiadiazin-4(3H)-one, 6-hydroxy-3-isopropyl-, 2,2-dioxide	60374-42-7	NI25341
107	18	52	135	17	214	78	51	**256**	100	90	50	41	36	34	27	26	**9.70**	10	12	4	2	1	–	–	–	–	–	–	–	–	1H-2,1,3-Benzothiadiazin-4(3H)-one, 8-hydroxy-3-isopropyl-, 2,2-dioxide	60374-43-8	NI25342
43	109	151	65	92	156	45	108	**256**	100	51	38	20	17	15	15	13	**5.20**	10	12	4	2	1	–	–	–	–	–	–	–	–	Sulphanilamide, N^1,N^4-diacetyl-		NI25343
110	143	69	142	256	168	114	140	**256**	100	97	84	63	51	35	20	17		10	12	6	2	–	–	–	–	–	–	–	–	–	3,6-Epoxy-2H,8H-pyrimido[6,1-b][1,3]oxazocine-8,10(9H)-dione, 3,4,5,6-tetrahydro-4,5-dihydroxy-9-methyl-, [3R-(3α,4β,5β,6α)]-	41547-66-4	NI25344
221	256	67	193	258	68	81	194	**256**	100	71	65	59	29	29	28	18		10	13	2	4	–	–	1	–	–	–	–	–	–	Theophylline, 8-(chloromethyl)-7-ethyl-	27042-88-2	NI25345
43	111	214	29	139	98	83	44	**256**	100	17	14	10	8	5	5	4	**0.04**	10	13	7	–	–	–	–	–	–	–	–	1	–	α-D-Glucofuranuronic acid, γ-lactone, cyclic 1,2-(ethylboronate) 5-acetate	74779-82-1	NI25346
157	256	241	43	158	258	243	159	**256**	100	99	96	68	48	40	40	40		10	14	4	–	–	–	–	–	–	–	–	–	1	Nickel, bis(2,4-pentanedionato-O,O′)-	3264-82-2	NI25347

MASS TO CHARGE RATIOS								M.W.	INTENSITIES								Parent	C	H	O	N	S	F	Cl	Br	I	Si	P	B	X	COMPOUND NAME	CAS Reg No	No
241	91	239	237	240	211	238	209	**256**	100	76	75	44	33	28	27	22	**3.00**	10	16	–	–	–	–	–	–	–	–	–	–	1	Tin, m-tolyltrimethyl-		MS05611
241	239	237	240	211	238	91	245	**256**	100	70	44	32	30	25	19	16	**2.30**	10	16	–	–	–	–	–	–	–	–	–	–	1	Tin, o-tolyltrimethyl-		MS05612
241	239	237	240	211	92	238	209	**256**	100	73	44	33	30	29	27	24	**2.90**	10	16	–	–	–	–	–	–	–	–	–	–	1	Tin, p-tolyltrimethyl-		MS05613
184	44	168	41	42	57	43	256	**256**	100	95	54	45	35	31	28	19		10	16	2	4	1	–	–	–	–	–	–	–	–	2-Imidazolidinone, 3-[5-(1,1-dimethylethyl)-1,3,4-thiadiazol-2-yl]-4-hydroxy-1-methyl-	55511-98-3	NI25348
71	99	112	101	72	73	84	83	**256**	100	21	13	8	6	3	1	1	**0.00**	10	16	4	4	–	–	–	–	–	–	–	–	–	Propanenitrile, 3-amino-, (E)-2-butenedioate (2:1)	2079-89-2	NI25349
100	214	113	170	157	55	87	84	**256**	100	79	71	56	50	46	42	40	**1.99**	10	16	4	4	–	–	–	–	–	–	–	–	–	1,4,7,10-Tetraazacyclotetradecane-3,8,11,14-tetrone	92456-55-8	NI25350
225	111	224	125	169	57	124	99	**256**	100	58	49	49	30	29	24	24	**0.00**	10	18	6	–	–	–	–	–	–	–	–	2	–	β-D-Fructopyranose, cyclic 2,3:4,5-bis(ethylboronate)	64780-37-6	NI25351
111	28	57	225	125	110	99	69	**256**	100	47	34	30	28	27	25	20	**0.20**	10	18	6	–	–	–	–	–	–	–	–	2	–	α-L-Sorbofuranose, cyclic 2,3:4,6-bis(ethylboronate)	64780-40-1	NI25352
113	257	241	141	119	114	185	112	**256**	100	36	33	18	18	15	11	10	**2.00**	10	20	2	2	–	–	–	–	–	2	–	–	–	Pyrimidine, 2,4-bis[(trimethylsilyl)oxy]-		NI25353
73	99	241	45	147	256	43	100	**256**	100	85	70	57	47	31	26	21		10	20	2	2	–	–	–	–	–	2	–	–	–	Pyrimidine, 2,4-bis[(trimethylsilyl)oxy]-		MS05614
73	99	241	45	147	256	43	100	**256**	100	85	70	57	47	31	26	21		10	20	2	2	–	–	–	–	–	2	–	–	–	2,4(1H,3H)-Pyrimidinedione, 1,3-bis(trimethylsilyl)-	3442-82-8	NI25354
140	42	55	83	82	73	56	54	**256**	100	63	53	47	27	10	10	10	**3.50**	10	20	2	2	–	–	–	–	–	2	–	–	–	2,4(1H,3H)-Pyrimidinedione, 1,3-bis(trimethylsilyl)-	3442-82-8	NI25355
149	177	77	256	150	91	121	79	**256**	100	21	18	14	11	10	8	7		11	12	5	–	1	–	–	–	–	–	–	–	–	2(3H)-Benzofuranone, 3,3-dimethyl-5-[(methylsulfonyl)oxy]-	26244-33-7	NI25356
149	15	77	43	41	177	39	91	**256**	100	67	43	34	32	30	29	23	**14.01**	11	12	5	–	1	–	–	–	–	–	–	–	–	2(3H)-Benzofuranone, 3,3-dimethyl-5-[(methylsulfonyl)oxy]-	26244-33-7	NI25357
163	119	18	91	164	28	17	65	**256**	100	48	37	29	17	15	15	12	**0.00**	11	13	2	–	–	–	–	1	–	–	–	–	–	1,3-Dioxolane, 2-bromomethyl-2-(2-methylphenyl)-		NI25358
256	169	154	258	170	211	56	171	**256**	100	68	42	39	29	24	23	21		11	13	3	2	–	–	1	–	–	–	–	–	–	Imidazolidinone, 1-(3-chloro-4-methoxyphenyl)-3-methoxy-		NI25359
194	58	45	166	167	225	80	59	**256**	100	68	59	57	55	53	32	26	**5.30**	11	16	5	2	–	–	–	–	–	–	–	–	–	3,4-Pyrazoledicarboxylic acid, 1-(2-methoxyethyl)-5-methyl-, dimethyl ester	82581-95-1	NI25360
141	140	167	73	45	125	117	256	**256**	100	87	24	18	15	12	12	10		11	16	5	2	–	–	–	–	–	–	–	–	–	2(1H)-Pyrimidinone, 1-(2-deoxy-β-D-erythro-pentofuranosyl)-4-methoxy-5-methyl-	50591-13-4	NI25361
117	140	141	126	73	99	45	125	**256**	100	53	37	34	30	25	20	18	**17.50**	11	16	5	2	–	–	–	–	–	–	–	–	–	Uridine, 2′-deoxy-5-ethyl-	15176-29-1	MS05615
117	140	99	141	73	125	45	43	**256**	100	43	40	37	37	26	23	23	**8.70**	11	16	5	2	–	–	–	–	–	–	–	–	–	Uridine, 2′-deoxy-5-ethyl-	15176-29-1	NI25362
119	117	115	94	77	118	121	212	**256**	100	75	56	42	26	25	24	23	**1.00**	11	18	2	–	–	–	–	–	–	–	–	–	1	Germane, trimethyl(2-phenoxyethoxy)-	16654-61-8	NI25363
119	117	115	94	77	118	121	65	**256**	100	74	55	43	26	24	23	23	**0.32**	11	18	2	–	–	–	–	–	–	–	–	–	1	Germane, trimethyl(2-phenoxyethoxy)-	16654-61-8	NI25364
188	32	88	42	44	56	132	144	**256**	100	93	51	28	24	22	19	12	**6.60**	11	20	3	2	–	–	–	–	–	1	–	–	–	2,9,10-Trioxa-6-aza-1-silatricyclo[4.3.3.0^{1,6}]dodecane, 1-(3-cyanopropyl)-		MS05616
186	150	256	258	151	188	221	110	**256**	100	42	36	33	33	32	30	30		12	7	–	–	–	–	3	–	–	–	–	–	–	1,1′-Biphenyl, 2,2′,5-trichloro-	37680-65-2	NI25365
186	256	258	150	110	151	188	111	**256**	100	63	59	52	40	37	34	31		12	7	–	–	–	–	3	–	–	–	–	–	–	1,1′-Biphenyl, 2,2′,6-trichloro-	38444-73-4	NI25366
186	150	256	258	110	151	188	111	**256**	100	55	46	44	44	40	32	31		12	7	–	–	–	–	3	–	–	–	–	–	–	1,1′-Biphenyl, 2,3,4-trichloro-	55702-46-0	NI25367
256	258	186	150	260	188	151	257	**256**	100	99	89	36	35	15	14	11		12	7	–	–	–	–	3	–	–	–	–	–	–	1,1′-Biphenyl, 2,3,4-trichloro-	55702-46-0	NI25368
186	256	258	75	150	93	188	151	**256**	100	99	91	50	39	37	33	30		12	7	–	–	–	–	3	–	–	–	–	–	–	1,1′-Biphenyl, 2,3,4-trichloro-	55702-46-0	NI25369
186	256	258	93	150	75	188	260	**256**	100	99	96	41	39	38	32	31		12	7	–	–	–	–	3	–	–	–	–	–	–	1,1′-Biphenyl, 2,3,4′-Trichloro-	38444-85-8	NI25370
256	186	258	150	260	188	151	75	**256**	100	99	96	45	33	29	25	21		12	7	–	–	–	–	3	–	–	–	–	–	–	1,1′-Biphenyl, 2,3,6-trichloro-	55702-45-9	NI25371
186	150	110	256	151	258	188	111	**256**	100	55	42	37	37	36	31	30		12	7	–	–	–	–	3	–	–	–	–	–	–	1,1′-Biphenyl, 2,3,6-trichloro-	55702-45-9	NI25372
256	258	186	150	260	188	151	257	**256**	100	94	87	19	16	15	15	11		12	7	–	–	–	–	3	–	–	–	–	–	–	1,1′-Biphenyl, 2,3′,4-trichloro-	55712-37-3	NI25373
256	258	186	260	257	259	188	150	**256**	100	99	42	35	16	15	15	12		12	7	–	–	–	–	3	–	–	–	–	–	–	1,1′-Biphenyl, 2,3′,5-trichloro-	38444-81-4	NI25374
186	150	256	258	151	110	111	187	**256**	100	57	43	41	40	39	28	16		12	7	–	–	–	–	3	–	–	–	–	–	–	1,1′-Biphenyl, 2,3′,5-trichloro-	38444-81-4	NI25375
256	186	258	75	93	150	188	260	**256**	100	98	96	57	43	39	33	30		12	7	–	–	–	–	3	–	–	–	–	–	–	1,1′-Biphenyl, 2,3′,5-trichloro-	38444-81-4	NI25376
186	256	258	75	93	150	151	188	**256**	100	82	74	47	38	35	32	31		12	7	–	–	–	–	3	–	–	–	–	–	–	1,1′-Biphenyl, 2,4,4′-trichloro-	7012-37-5	NI25377
256	258	186	93	150	260	75	188	**256**	100	97	85	36	33	32	30	28		12	7	–	–	–	–	3	–	–	–	–	–	–	1,1′-Biphenyl, 2,4,5-trichloro-	15862-07-4	NI25378
256	258	186	150	260	188	151	100	**256**	100	95	93	40	31	31	31	12		12	7	–	–	–	–	3	–	–	–	–	–	–	1,1′-Biphenyl, 2,4,5-trichloro-	15862-07-4	NI25379
186	256	258	150	110	151	188	111	**256**	100	61	59	54	42	35	33	30		12	7	–	–	–	–	3	–	–	–	–	–	–	1,1′-Biphenyl, 2,4,6-trichloro-	35693-92-6	NI25380
256	186	258	150	188	260	151	110	**256**	100	98	93	44	35	16	15	12		12	7	–	–	–	–	3	–	–	–	–	–	–	1,1′-Biphenyl, 2,4,6-trichloro-	35693-92-6	NI25381
256	258	186	150	260	188	75	151	**256**	100	95	93	35	29	28	25	22		12	7	–	–	–	–	3	–	–	–	–	–	–	1,1′-Biphenyl, 2,4,6-trichloro-	35693-92-6	NI25382
186	256	258	150	151	110	188	111	**256**	100	65	61	53	42	36	34	26		12	7	–	–	–	–	3	–	–	–	–	–	–	1,1′-Biphenyl, 2,4′,5-trichloro-	16606-02-3	NI25383
256	258	186	150	260	151	188	257	**256**	100	94	85	37	34	30	29	11		12	7	–	–	–	–	3	–	–	–	–	–	–	1,1′-Biphenyl, 2,4′,5-trichloro-	16606-02-3	NI25384
186	256	258	75	150	93	188	151	**256**	100	83	76	42	38	34	31	30		12	7	–	–	–	–	3	–	–	–	–	–	–	1,1′-Biphenyl, 2,4′,5-trichloro-	16606-02-3	NI25385
256	258	186	150	260	188	151	257	**256**	100	95	94	41	32	32	29	11		12	7	–	–	–	–	3	–	–	–	–	–	–	1,1′-Biphenyl, 2′,3,4-trichloro-	38444-86-9	NI25386
256	258	186	75	93	260	150	188	**256**	100	95	76	41	33	31	28	24		12	7	–	–	–	–	3	–	–	–	–	–	–	1,1′-Biphenyl, 3,4,4′-Trichloro-	38444-90-5	NI25387
230	232	186	256	258	160	221	234	**256**	100	97	66	62	61	35	35	32		12	7	–	–	–	–	3	–	–	–	–	–	–	1,1′-Biphenyl, trichloro-	25323-68-6	NI25388
30	77	51	256	76	179	180	28	**256**	100	66	55	46	41	37	36	29		12	8	3	4	–	–	–	–	–	–	–	–	–	4-Benzofurazanamine, 7-nitro-N-phenyl-	18378-15-9	NI25389
256	77	179	51	193	30	192	180	**256**	100	54	51	40	36	28	18	17		12	8	3	4	–	–	–	–	–	–	–	–	–	Benzofurazan, 5-anilino-4-nitro-	18378-16-0	NI25390
256	241	213	258	243	215	170	198	**256**	100	78	70	63	51	43	28	23		12	10	2	–	–	–	2	–	–	–	–	–	–	Naphthalene, 1,5-dichloro-2,6-dimethoxy-	25315-05-3	NI25391
256	241	213	258	243	215	170	198	**256**	100	71	70	63	51	44	28	23		12	10	2	–	–	–	2	–	–	–	–	–	–	Naphthalene, 1,5-dichloro-2,6-dimethoxy-	25315-05-3	LI05226
172	52	256	173	200	53	257	120	**256**	100	80	27	23	13	10	7	6		12	12	3	–	–	–	–	–	–	–	–	–	1	Chromium, mesitylenetricarbonyl-		LI05227
241	243	200	256	202	201	203	258	**256**	100	65	50	40	35	31	22	16		12	14	–	2	–	–	2	–	–	–	–	–	–	1H-1,5-Benzodiazepine, 7,8-dichloro-2,3-dihydro-2,2,4-trimethyl-		LI05228

MASS TO CHARGE RATIOS								M.W.	INTENSITIES								Parent	C	H	O	N	S	F	Cl	Br	I	Si	P	B	X	COMPOUND NAME	CAS Reg No	No
184	36	38	92	185	183	91	35	**256**	100	54	18	17	14	11	10	8	**0.00**	12	14	–	2	–	–	2	–	–	–	–	–	–	[1,1′-Biphenyl]-4,4′-diamine, dihydrochloride	531-85-1	NI25392
156	50	155	52	128	51	157	76	**256**	100	57	44	19	15	15	14	13	**0.00**	12	14	–	2	–	–	2	–	–	–	–	–	–	4,4′-Bipyridinium, 1,1′-dimethyl-, dichloride	1910-42-5	NI25393
106	28	45	78	27	29	135	26	**256**	100	93	92	87	85	52	48	45	**0.00**	12	16	2	–	2	–	–	–	–	–	–	–	–	Benzoic acid, bis[(methylthio)methyl] ester		LI05229
112	111	97	44	39	59	53	54	**256**	100	47	40	40	25	20	14	12	**0.70**	12	16	4	–	1	–	–	–	–	–	–	–	–	2-Thiabicyclo[3.2.0]hept-3-ene-6,7-dicarboxylic acid, 1,3,6,7-tetramethyl-	34002-18-1	NI25394
238	164	256	166	220	150	149	178	**256**	100	76	52	48	36	36	30	22		12	16	4	–	1	–	–	–	–	–	–	–	–	α-D-Xylopyranose, 4-thio-4-S-benzyl-		LI05230
142	137	210	29	256	55	136	27	**256**	100	90	79	79	69	60	59	54		12	16	6	–	–	–	–	–	–	–	–	–	–	1,4-Cyclohexanedicarboxylic acid, 2,5-dioxo-, diethyl ester		NI25395
137	210	142	29	55	256	136	27	**256**	100	88	81	73	72	63	61	57		12	16	6	–	–	–	–	–	–	–	–	–	–	1,4-Cyclohexanedicarboxylic acid, 2,5-dioxo-, diethyl ester		NI25396
137	96	59	79	151	165	98	105	**256**	100	88	41	39	34	33	27	25	**1.00**	12	16	6	–	–	–	–	–	–	–	–	–	–	1-Cyclohexene-1,5-dicarboxylic acid, 3-[(hydroxycarbonyl)methyl]-, dimethyl ester		MS05617
43	164	136	183	105	137	44	59	**256**	100	82	62	24	20	17	17	16	**0.00**	12	16	6	–	–	–	–	–	–	–	–	–	–	2-Cyclopentene-1-acetic acid, 5-(acetyloxy)-2-(methoxycarbonyl)-, methyl ester	69745-78-4	NI25397
54	53	70	71	99	155	225	169	**256**	100	40	34	21	20	16	14	14	**0.00**	12	16	6	–	–	–	–	–	–	–	–	–	–	6,11-Dioxaspiro[6.4]undec-8-ene-2,3-dicarboxylic acid, monomethyl ester, (2R-cis)-		DD00739
31	45	142	27	137	29	210	46	**256**	100	54	35	32	30	28	26	26	**22.01**	12	16	6	–	–	–	–	–	–	–	–	–	–	Di-p-tolyl ether		IC03349
105	77	123	165	51	106	61	43	**256**	100	29	20	10	9	8	8	8	**0.00**	12	16	6	–	–	–	–	–	–	–	–	–	–	DL-Xylitol, 1-benzoate	74793-68-3	NI25398
67	41	44	55	82	54	94	133	**256**	100	71	66	55	38	38	37	27	**0.00**	12	17	1	–	–	–	–	1	–	–	–	–	–	Benzene, [(6-bromohexyl)oxy]-	51795-97-2	NI25399
59	74	67	41	94	45	82	43	**256**	100	69	69	69	65	62	54	46	**11.53**	12	17	4	–	–	–	–	–	–	–	1	–	–	1,3,2-Dioxaphosphorinane, 4,4,6-trimethyl-2-phenoxy-, 2-oxide	19219-96-6	MS05618
199	209	256	41	186	241	153	69	**256**	100	76	47	35	27	26	9	7		12	20	–	2	2	–	–	–	–	–	–	–	–	1,2-Diazine, 3,6-bis(methylthio)-4-hexyl-		MS05619
83	29	111	81	155	27	256	183	**256**	100	76	57	44	27	21	15	14		12	20	4	2	–	–	–	–	–	–	–	–	–	2,3-Diazabicyclo[2.2.1]heptane-2,3-dicarboxylic acid, 1-methyl-, diethyl ester	74793-69-4	NI25400
69	29	141	97	27	42	43	41	**256**	100	63	22	22	15	14	11	11	**5.37**	12	20	4	2	–	–	–	–	–	–	–	–	–	2,3-Diazabicyclo[2.2.1]heptane-2,3-dicarboxylic acid, 5-methyl-, diethyl ester	74793-70-7	NI25401
45	43	169	58	41	55	29	39	**256**	100	84	82	72	68	64	60	52	**0.00**	12	20	4	2	–	–	–	–	–	–	–	–	–	2,4,6(1H,3H,5H)-Pyrimidinetrione, 5-butyl-5-ethyl-3-hydroxy-N,N′-dimethyl-		LI05231
57	41	58	69	42	56	153	68	**256**	100	66	55	37	35	34	33	31	**15.30**	12	22	3	3	–	–	–	–	–	–	–	–	–	3-Imidazoline, 4-tert-butylcarbamoyl-2,2,5,5-tetramethyl-, 3-oxide-1-oxile		NI25402
57	41	58	153	69	42	56	226	**256**	100	55	45	41	40	32	30	30	**20.50**	12	22	3	3	–	–	–	–	–	–	–	–	–	3-Imidazoline, 4-tert-butylcarbamoyl-2,2,5,5-tetramethyl-, 3-oxide-1-oxile		NI25403
163	256	164	152	210	239	209	151	**256**	100	77	66	47	37	37	35	32		13	8	4	2	–	–	–	–	–	–	–	–	–	9H-Fluorene, 2,5-dinitro-	15110-74-4	NI25404
163	256	164	210	152	209	239	151	**256**	100	90	65	54	45	38	35	20		13	8	4	2	–	–	–	–	–	–	–	–	–	9H-Fluorene, 2,5-dinitro-	15110-74-4	NI25405
163	256	164	152	210	46	239	151	**256**	100	69	65	47	46	28	18	17		13	8	4	2	–	–	–	–	–	–	–	–	–	9H-Fluorene, 2,7-dinitro-	5405-53-8	NI25406
256	144	172	89	90	117	228	116	**256**	100	99	44	44	31	24	19	19		13	8	4	2	–	–	–	–	–	–	–	–	–	1,4-Phenazinedione, 1,4-dihydro-2,3-dihydroxy-7-methyl-		MS05620
144	142	117	89	170	115	90	116	**256**	100	63	45	45	37	36	31	30	**18.00**	13	8	4	2	–	–	–	–	–	–	–	–	–	1,2,3,4-Phenazinetetrone, 7-methyl-1,2,3,4-terahydro-		MS05621
256	163	164	210	152	239	209	151	**256**	100	92	67	50	35	27	24	12		13	8	4	2	–	–	–	–	–	–	–	–	–	Toluene, 2,7-dinitro-		NI25407
228	193	166	256	192	90	230	201	**256**	100	74	45	44	40	36	34	28		13	9	–	4	–	–	1	–	–	–	–	–	–	1,2,3-Benzotriazine, 4-(2-chloroanilino)-		LI05232
193	228	166	192	256	90	230	111	**256**	100	96	61	57	41	32	30	20		13	9	–	4	–	–	1	–	–	–	–	–	–	1,2,3-Benzotriazine, 3-(2-chlorophenyl)-3,4-dihydro-4-imino-		LI05233
165	256	89	90	199	258			**256**	100	22	20	19	8	6				13	9	–	4	–	–	1	–	–	–	–	–	–	1H-Benzotriazole, 1-(benzylidenamino)-5-chloro-		LI05234
165	89	256	91	199	111	164	103	**256**	100	30	26	9	6	5	4	4		13	9	–	4	–	–	1	–	–	–	–	–	–	1H-Benzotriazole, 1-[(2-chlorobenzylidene)amino]-	33682-98-3	NI25408
165	89	256	111	91	199	164	163	**256**	100	27	25	8	6	5	5	4		13	9	–	4	–	–	1	–	–	–	–	–	–	1H-Benzotriazole, 1-[(3-chlorobenzylidene)amino]-	33682-99-4	NI25409
165	89	256	258	91	199			**256**	100	30	26	7	6	5				13	9	–	4	–	–	1	–	–	–	–	–	–	1H-Benzotriazole, 1-[(4-chlorobenzylidene)amino]-	33683-00-0	LI05235
165	89	256	111	91	199	124	164	**256**	100	31	25	8	8	6	6	4		13	9	–	4	–	–	1	–	–	–	–	–	–	1H-Benzotriazole, 1-[(4-chlorobenzylidene)amino]-	33683-00-0	NI25410
256	191	89	56	145	257	95	173	**256**	100	60	21	20	18	16	16	15		13	12	2	–	–	–	–	–	–	–	–	–	1	3-Ferrocenepropanoic acid		LI05236
256	191	89	56	145	257	173	118	**256**	100	60	21	20	18	16	16	16		13	12	2	–	–	–	–	–	–	–	–	–	1	3-Ferrocenepropanoic acid		LI05237
256	213	44	257	43	185	171	58	**256**	100	56	42	40	35	31	23	19		13	12	2	4	–	–	–	–	–	–	–	–	–	Benzo[g]pteridine-2,4(3H,10H)-dione, 7,8,10-trimethyl-	1088-56-8	NI25411
256	213	44	257	43	185	171	158	**256**	100	55	43	40	34	31	25	20		13	12	2	4	–	–	–	–	–	–	–	–	–	Benzo[g]pteridine-2,4(3H,10H)-dione, 7,8,10-trimethyl-	1088-56-8	LI05238
256	255	194	77	209	180	91	239	**256**	100	57	54	49	43	26	26	23		13	12	2	4	–	–	–	–	–	–	–	–	–	5-Nitro-2-aminobenzophenone hydrazone		MS05622
184	256	211	143	28	27	29	18	**256**	100	80	34	31	23	20	15	14		13	12	2	4	–	–	–	–	–	–	–	–	–	1,3,6,9b-Tetraazaphenalene-4-carboxylic acid, 2-methyl-, ethyl ester	37550-67-7	NI25412
256	241	227	255	226	128	186	172	**256**	100	91	71	27	20	18	9	9		13	16	–	6	–	–	–	–	–	–	–	–	–	3H-Cyclohepta[1,2-d:4,5-d′]diimidazole-2,6-diamine, N6,N6,3,4-tetramethyl-	40451-47-6	MS05623
69	41	57	123	179	109	77	65	**256**	100	52	42	40	34	24	23	21	**3.00**	13	17	1	–	1	–	1	–	–	–	–	–	–	2-Propanone, 3-chloro-1-phenylthio-1-tert-butyl-		MS05624
193	109	224	197	192	225	196	256	**256**	100	94	43	38	31	27	22	11		13	20	5	–	–	–	–	–	–	–	–	–	–	Bicyclo[3.3.0]octane-2-carboxylic acid, 7,7-dimethoxy-2-methyl-3-oxo-, methyl ester, $(1\alpha,2\beta,5\alpha)$-		MS05625
66	85	79	94	67	81	77	95	**256**	100	84	71	50	48	31	30	28	**0.00**	13	20	5	–	–	–	–	–	–	–	–	–	–	1H-Cyclopenta[c]furan-1-one, hexahydro-4-(hydroxymethyl)-5-[(tetrahydro-2H-pyran-2-yl)oxy]-, $[3aR-[3a\alpha,4\alpha,5\beta,6a\alpha]]$-		NI25413
55	168	193	151	136	59	196	164	**256**	100	78	52	52	44	34	30	30	**14.00**	13	20	5	–	–	–	–	–	–	–	–	–	–	1,3-Cyclopentanedipropanoic acid, 2-oxo-, dimethyl ester		MS05626
96	109	136	41	81	165	67	183	**256**	100	88	34	33	32	30	24	23	**0.00**	13	20	5	–	–	–	–	–	–	–	–	–	–	2H-Pyran-6,6-dicarboxylic acid, 5,6-dihydro-3,4-dimethyl-, diethyl ester		MS05627
108	167	54	136	81	80	182	30	**256**	100	60	60	39	37	33	30	27	**3.33**	13	20	5	–	–	–	–	–	–	–	–	–	–	2H-Pyran-2,2-dicarboxylic acid, 3,6-dihydro-3,6-dimethyl-, diethyl ester, cis-	36736-30-8	NI25414

MASS TO CHARGE RATIOS								M.W.	INTENSITIES								Parent	C	H	O	N	S	F	Cl	Br	I	Si	P	B	X	COMPOUND NAME	CAS Reg No	No
167	93	54	136	108	182	81	80	**256**	100	60	59	53	53	46	42	41	**2.89**	13	20	5	–	–	–	–	–	–	–	–	–	–	2H-Pyran-2,2-dicarboxylic acid, 3,6-dihydro-3,6-dimethyl-, diethyl ester, trans-	36736-31-9	NI25415
256	228	102	126	257	115	77	130	**256**	100	39	14	14	13	12	11	9		14	8	5	–	–	–	–	–	–	–	–	–	–	9,10-Anthracenedione, 1,2,4-trihydroxy-	81-54-9	NI25416
256	240	257	63	228	115	212	77	**256**	100	60	17	13	12	10	9	9		14	8	5	–	–	–	–	–	–	–	–	–	–	9,10-Anthracenedione, 1,2,6-trihydroxy-		MS05628
200	199	115	256	92	120	228	116	**256**	100	75	75	69	65	24	20	20		14	8	5	–	–	–	–	–	–	–	–	–	–	Naphtho[2,3-b]pyran-2,5,10-trione, 9-hydroxy-3-methyl-		LI05239
256	258	176	177	88	151	257	150	**256**	100	99	65	41	38	17	16	16		14	9	–	–	–	–	–	1	–	–	–	–	–	Anthracene, 9-bromo-	1564-64-3	NI25417
256	258	176	88	177	151	150	75	**256**	100	98	89	76	60	29	27	25		14	9	–	–	–	–	–	1	–	–	–	–	–	Anthracene, 9-bromo-	1564-64-3	NI25418
258	256	15	30	175	76	176	177	**256**	100	80	76	57	46	45	36	35		14	9	–	–	–	–	–	1	–	–	–	–	–	Phenanthrene, 4-bromo-	19462-79-4	NI25419
256	258	88	176	177	74	151	89	**256**	100	96	60	55	44	19	18	18		14	9	–	–	–	–	–	1	–	–	–	–	–	Phenanthrene, 9-bromo-	573-17-1	LI05240
88	256	258	176	177	75	151	150	**256**	100	71	67	62	49	28	18	17		14	9	–	–	–	–	–	1	–	–	–	–	–	Phenanthrene, 9-bromo-	573-17-1	NI25420
153	256	155	77	258	90	51	125	**256**	100	63	35	28	23	22	19	15		14	9	1	2	–	–	1	–	–	–	–	–	–	1,2,4-Oxadiazole, 3-(4-chlorophenyl)-5-phenyl-		NI25421
255	256	257	221	258	77	214	192	**256**	100	74	43	26	24	13	10	9		14	9	1	2	–	–	1	–	–	–	–	–	–	2(1H)-Quinazolinone, 6-chloro-4-phenyl-		MS05629
105	77	135	51	93	256	50	65	**256**	100	88	42	35	30	19	13	11		14	12	1	2	1	–	–	–	–	–	–	–	–	Benzamide, N-[(phenylamino)thioxomethyl]-	4921-82-8	NI25422
223	238	135	105	121	77	51	58	**256**	100	93	62	60	55	48	29	24	**22.77**	14	12	1	2	1	–	–	–	–	–	–	–	–	Benzoic acid, 2-(phenylthioxomethyl)hydrazide	43089-09-4	NI25423
119	91	118	210	120	256	77	51	**256**	100	81	42	25	17	13	13	11		14	12	3	2	–	–	–	–	–	–	–	–	–	Aniline, N-(2-methoxybenzylidene)-4-nitro-	24588-82-7	NI25424
119	91	118	210	120	256	51	77	**256**	100	81	40	24	17	12	12	11		14	12	3	2	–	–	–	–	–	–	–	–	–	Aniline, N-(2-methoxybenzylidene)-4-nitro-	24588-82-7	LI05241
93	118	77	256	164	210	104	94	**256**	100	46	36	30	21	12	10	10		14	12	3	2	–	–	–	–	–	–	–	–	–	Aniline, N-(2-methoxy-5-nitrobenzylidene)-	19652-31-4	NI25425
93	118	77	256	164	210	94	51	**256**	100	45	36	29	21	12	9	9		14	12	3	2	–	–	–	–	–	–	–	–	–	Aniline, N-(2-methoxy-5-nitrobenzylidene)-	19652-31-4	LI05242
150	256	104	76	77	163	120	50	**256**	100	49	28	27	21	20	15	14		14	12	3	2	–	–	–	–	–	–	–	–	–	Benzamide, N-methyl-4-nitro-N-phenyl-	961-61-5	NI25426
105	104	256	78	77	106	134	257	**256**	100	52	45	31	13	11	8	7		14	12	3	2	–	–	–	–	–	–	–	–	–	4H-3,1-Benzoxazine, 1,2-dihydro-2-(p-nitrophenyl)-	82085-89-0	NI25427
121	105	77	28	51	238	79	65	**256**	100	85	48	32	28	23	18	17	**4.90**	14	12	3	2	–	–	–	–	–	–	–	–	–	Hydrazine, N-benzoyl-N-salicyl-		IC03350
121	136	120	137	65	256	93	39	**256**	100	29	29	23	23	21	16	15		14	12	3	2	–	–	–	–	–	–	–	–	–	Hydrazine, N-salicyl-N′-salicylidene-		IC03351
256	225	130	198	77	257	103	128	**256**	100	62	58	42	17	16	16	12		14	12	3	2	–	–	–	–	–	–	–	–	–	1H-Pyrimido[1,2-a]quinoline-2-carboxylic acid, 5,6-dihydro-1-oxo-, methyl ester	64399-33-3	NI25428
256	118	117	171	115	138	257	56	**256**	100	69	36	30	23	20	18	17		14	16	1	–	–	–	–	–	–	–	–	–	1	Iron, (η^5-2,4-cyclopentadien-1-yl) (1,2,3,3a,7a-η)-4,5,6,7-tetrahydro-4-hydroxy-1H-inden-1-yl-	1271-95-0	NI25430
256	257	188	115	254	237	121	117	**256**	100	17	12	9	8	8	8	8		14	16	1	–	–	–	–	–	–	–	–	–	1	Iron, (η^5-2,4-cyclopentadien-1-yl) (1,2,3,3a,7a-η)-4,5,6,7-tetrahydro-4-hydroxy-1H-inden-1-yl-	1271-96-1	NI25429
256	136	92	121	255	93	257	65	**256**	100	51	49	25	19	17	16	16		14	16	1	4	–	–	–	–	–	–	–	–	–	Azobenzene, 4,4′-diamino-3-methoxy-6-methyl-		IC03352
116	43	41	71	256	57	29	55	**256**	100	82	62	54	49	42	40	28		14	24	–	–	2	–	–	–	–	–	–	–	–	Thiophene, 2-(decylthio)-	54986-42-4	NI25431
116	43	41	256	29	71	39	45	**256**	100	82	62	49	40	34	22	21		14	24	–	–	2	–	–	–	–	–	–	–	–	Thiophene, 2-(decylthio)-	54986-42-4	AI01825
41	43	57	144	129	71	55	45	**256**	100	99	58	54	36	33	31	29	**14.36**	14	24	–	–	2	–	–	–	–	–	–	–	–	Thiophene, 2-ethyl-5-(octylthio)-		AI01826
131	73	101	111	241	189	157	79	**256**	100	67	64	22	17	6	3	3	**0.10**	14	24	4	–	–	–	–	–	–	–	–	–	–	Glycerol, 1-O-(2-methoxy-4-heptynyl)-2,3-O-isopropylidene-		NI25432
143	97	69	115	41	55	98	83	**256**	100	95	75	51	49	36	30	25	**3.00**	14	24	4	–	–	–	–	–	–	–	–	–	–	Hexanoic acid, 7-(2-methyl-5-oxocyclopentyloxy)-, ethyl ester, trans-		NI25433
173	160	55	133	29	101	127	41	**256**	100	74	32	23	22	20	18	18	**0.00**	14	24	4	–	–	–	–	–	–	–	–	–	–	Malonic acid, cyclohexylmethyl-, diethyl ester		DO01045
155	41	57	183	99	70	43	81	**256**	100	98	96	69	55	55	55	44	**0.00**	14	24	4	–	–	–	–	–	–	–	–	–	–	α-Pyran-6-carboxylic acid, 2-butoxy-5,6-dihydro-, butyl ester		LI05243
155	41	57	183	43	99	70	81	**256**	100	99	96	70	57	56	56	45	**0.50**	14	24	4	–	–	–	–	–	–	–	–	–	–	α-Pyran-6-carboxylic acid, 2-butoxy-5,6-dihydro-, butyl ester		MS05630
43	136	41	95	81	67	164	68	**256**	100	96	92	69	65	53	48	45	**1.00**	14	24	4	–	–	–	–	–	–	–	–	–	–	10-Undecenoic acid, 2-(acetyloxy)-, methyl ester	54986-41-3	WI01302
241	75	143	73	129	242	55	41	**256**	100	47	34	32	31	22	15	13	**1.53**	14	28	2	–	–	–	–	–	–	1	–	–	–	2-Hendecenoic acid, trimethylsilyl ester		NI25434
73	75	117	55	41	241	129	29	**256**	100	99	50	46	36	31	28	17	**1.95**	14	28	2	–	–	–	–	–	–	1	–	–	–	Undecenoic acid, trimethylsilyl ester	24338-09-8	NI25435
84	58	126	213	114	128	100	72	**256**	100	95	87	79	67	55	52	49	**19.08**	14	28	2	2	–	–	–	–	–	–	–	–	–	Acetamide, N-[3-(acetylethylamino)propyl]-N-pentyl-	39160-03-7	NI25436
30	43	73	72	41	86	44	184	**256**	100	49	23	20	16	14	14	13	**4.00**	14	28	2	2	–	–	–	–	–	–	–	–	–	1,10-Decanediamine, N,N′-diacetyl-		LI05244
72	184	142	198	212	114	97	213	**256**	100	33	22	13	12	10	8	4	**0.20**	14	28	2	2	–	–	–	–	–	–	–	–	–	Propanediamide, 2-ethyl-N,N,N′,N′-tetramethyl-2-(1-methylbutyl)-	42948-60-7	LI05245
72	184	142	186	212	114	97	41	**256**	100	33	22	13	12	10	8	8	**0.10**	14	28	2	2	–	–	–	–	–	–	–	–	–	Propanediamide, 2-ethyl-N,N,N′,N′-tetramethyl-2-(1-methylbutyl)-	42948-60-7	NI25437
256	165	258	228	193	163	257	221	**256**	100	43	34	32	30	23	20	20		15	9	2	–	–	–	1	–	–	–	–	–	–	9,10-Anthracenedione, 1-chloro-2-methyl-		IC03353
256	228	165	193	82	258	230	163	**256**	100	87	74	69	35	34	31	27		15	9	2	–	–	–	1	–	–	–	–	–	–	Isocoumarin, 4-(4-chlorophenyl)-	28773-76-4	NI25438
105	191	178	208	207	256	165	77	**256**	100	54	51	50	40	27	21	18		15	12	2	–	1	–	–	–	–	–	–	–	–	1,2-Diphenylprop-1-ene (1-3)sultine	84735-17-1	NI25439
211	165	212	256	121	166	178	77	**256**	100	99	70	59	37	30	28	18		15	12	2	–	1	–	–	–	–	–	–	–	–	1,3-Oxathiolan-4-one, 5,5-diphenyl-	54986-52-6	NI25440
165	211	256	212	121	77	178	138	**256**	100	99	88	58	42	22	19	18		15	12	2	–	1	–	–	–	–	–	–	–	–	1,3-Oxathiolan-5-one, 4,4-diphenyl-	19962-73-3	NI25441
211	165	212	256	121	166	178	77	**256**	100	98	70	58	38	30	28	20		15	12	2	–	1	–	–	–	–	–	–	–	–	1,3-Oxathiolan-5-one, 4,4-diphenyl-	19962-73-3	LI05246
197	256	152	43	196	139	184	145	**256**	100	70	23	23	19	14	9	7		15	12	2	–	1	–	–	–	–	–	–	–	–	Thiophene, 5-acetoxymethyl-2-phenylethynyl-		LI05247
256	196	197	43	213	185	163	152	**256**	100	87	73	61	41	36	35	29		15	12	2	–	1	–	–	–	–	–	–	–	–	Thiophene, 5-(3-acetoxypropen-1-yl)-2-phenyl-		LI05248
256	241	149	128	152	257	135	111	**256**	100	68	31	19	17	16	15	12		15	12	4	–	–	–	–	–	–	–	–	–	–	1,3-Benzenediol, 4-(6-methoxy-2-benzofuranyl)-	67685-32-9	NI25442
197	227	256	228	213	198	257	128	**256**	100	83	75	47	29	14	12	11		15	12	4	–	–	–	–	–	–	–	–	–	–	2-Benzofuranone, 2,3-dihydro-5-hydroxy-3-(4-methoxyphenyl)-		IC03354
224	223	196	121	65	256	77	168	**256**	100	76	76	42	40	36	36	34		15	12	4	–	–	–	–	–	–	–	–	–	–	Benzoic acid, 2-(2-hydroxybenzoyl)-, methyl ester	21147-22-8	NI25443
224	223	196	121	65	77	225	168	**256**	100	76	76	42	40	36	34	34	**0.00**	15	12	4	–	–	–	–	–	–	–	–	–	–	Benzoic acid, 2-(2-hydroxybenzoyl)-, methyl ester	21147-22-8	LI05249

MASS TO CHARGE RATIOS								M.W.	INTENSITIES								Parent	C	H	O	N	S	F	Cl	Br	I	Si	P	B	X	COMPOUND NAME	CAS Reg No	No
121	256	163	225	93	65	122	39	**256**	100	30	15	11	11	11	8	7		15	12	4	–	–	–	–	–	–	–	–	–	–	Benzoic acid, 2-(4-hydroxybenzoyl)-, methyl ester	21147-23-9	NI25444
121	256	163	225	93	65	122	77	**256**	100	30	15	11	11	11	8	7		15	12	4	–	–	–	–	–	–	–	–	–	–	Benzoic acid, 2-(4-hydroxybenzoyl)-, methyl ester	21147-23-9	LI05250
105	121	77	256	120	65	39	51	**256**	100	39	26	19	19	15	13	11		15	12	4	–	–	–	–	–	–	–	–	–	–	Benzoic acid, 2-hydroxy-, 2-oxo-2-phenylethyl ester	55153-16-7	NI25445
105	121	256	77	65	93	39	122	**256**	100	95	24	23	16	15	12	10		15	12	4	–	–	–	–	–	–	–	–	–	–	Benzoic acid, 4-hydroxy-, 2-oxo-2-phenylethyl ester	55153-15-6	NI25446
135	77	181	149	121	92	180	65	**256**	100	32	27	25	20	16	13	13	**11.01**	15	12	4	–	–	–	–	–	–	–	–	–	–	Benzoic acid, 2-(2-methoxybenzoyl)-	1151-04-8	NI25447
135	256	77	92	149	136	212	65	**256**	100	23	16	14	9	9	7	7		15	12	4	–	–	–	–	–	–	–	–	–	–	Benzoic acid, 2-(4-methoxybenzoyl)-	1151-15-1	NI25448
256	255	179	152	43	124	257	104	**256**	100	69	53	47	47	25	17	16		15	12	4	–	–	–	–	–	–	–	–	–	–	2H-1-Benzopyran-4-one, 3,4-dihydro-5,7-dihydroxy-2-phenyl-		LI05251
256	255	179	152	43	124	257	41	**256**	100	69	53	47	47	25	17	16		15	12	4	–	–	–	–	–	–	–	–	–	–	4H-1-Benzopyran-4-one, 2,3-dihydro-5,7-dihydroxy-2-phenyl-, (S)-	480-39-7	NI25449
256	137	163	255	120	57	43	71	**256**	100	91	66	49	38	29	21	19		15	12	4	–	–	–	–	–	–	–	–	–	–	4H-1-Benzopyran-4-one, 2,3-dihydro-7-hydroxy-2-(3-hydroxyphenyl)-	62252-06-6	NI25450
137	256	120	163	255	42	60	57	**256**	100	91	61	33	24	21	18	16		15	12	4	–	–	–	–	–	–	–	–	–	–	4H-1-Benzopyran-4-one, 2,3-dihydro-7-hydroxy-2-(4-hydroxyphenyl)-	69097-97-8	NI25451
158	256	183	159	182	169	181	115	**256**	100	36	17	12	9	9	7	7		15	12	4	–	–	–	–	–	–	–	–	–	–	1,2-Dibenzofurandicarboxylic acid, 4-methyl-1,2,3,11-tetrahydro-, anhydride		LI05252
256	137	255	120	163	257	107	91	**256**	100	61	45	40	28	21	12	12		15	12	4	–	–	–	–	–	–	–	–	–	–	2-Propen-1-one, 1-(2,4-dihydroxyphenyl)-3-(4-hydroxyphenyl)-, (E)-	13745-20-5	NI25452
137	120	91	256	69	53	51	65	**256**	100	85	30	28	28	28	27	26		15	12	4	–	–	–	–	–	–	–	–	–	–	2-Propen-1-one, 1-(2,4-dihydroxyphenyl)-3-(4-hydroxyphenyl)-, (E)-	13745-20-5	NI25453
51	69	77	55	57	53	62	76	**256**	100	98	91	75	73	66	63	55	**0.00**	15	12	4	–	–	–	–	–	–	–	–	–	–	9H-Xanthen-9-one, 3,6-dihydroxy-1,8-dimethyl-		MS05631
256	258	126	257	255	118	112	77	**256**	100	33	29	26	21	20	13	13		15	13	–	2	–	–	1	–	–	–	–	–	–	2-Pyrazoline, 1-(4-chlorophenyl)-3-phenyl-		MS05632
256	91	77	258	255	257	51	64	**256**	100	55	39	33	29	28	16	15		15	13	–	2	–	–	1	–	–	–	–	–	–	2-Pyrazoline, 3-(4-chlorophenyl)-1-phenyl-		MS05633
256	255	51	257	77	258	75	39	**256**	100	98	65	50	48	36	22	20		15	13	–	2	–	–	1	–	–	–	–	–	–	Quinoxaline, 6-chloro-1,4-dihydro-1-methyl-4-phenyl-		NI25454
256	147	255	257	164	136	106	120	**256**	100	70	24	20	19	19	19	18		15	16	–	2	1	–	–	–	–	–	–	–	–	Thiourea, N-ethyl-N,N′-diphenyl-	15093-51-3	NI25455
91	106	256	65	28	79	92	39	**256**	100	91	39	35	26	25	16	16		15	16	–	2	1	–	–	–	–	–	–	–	–	Thiourea, N,N′-dibenzyl-	1424-14-2	NI25456
107	91	106	222	149	65	39	77	**256**	100	90	89	54	43	42	37	30	**20.01**	15	16	–	2	1	–	–	–	–	–	–	–	–	Thiourea, N,N′-di-p-tolyl-	621-01-2	NI25457
256	105	93	120	58	43	151	77	**256**	100	44	35	27	25	21	20	20		15	16	–	2	1	–	–	–	–	–	–	–	–	Thiourea, N-phenyl-N′-(1-phenylethyl)-	15093-41-1	NI25458
256	151	93	152	257	105	77	119	**256**	100	41	29	27	18	16	15	14		15	16	–	2	1	–	–	–	–	–	–	–	–	Thiourea, N-phenyl-N′-(2-phenylethyl)-	15093-42-2	NI25459
224	225	256	194	165	209	226	195	**256**	100	27	23	13	11	9	8	8		15	16	2	–	–	–	–	–	–	1	–	–	–	9-Silaanthracene, 9,10-dihydro-9,9-dimethoxy-	76470-25-2	NI25460
256	255	77	108	92	240	135	123	**256**	100	32	28	24	20	18	18	18		15	16	2	2	–	–	–	–	–	–	–	–	–	Benzophenone, 3,3′-methoxy-, hydrazone	67437-15-4	NI25461
181	180	256	221	167	166	182	193	**256**	100	95	75	58	58	55	52	40		15	16	2	2	–	–	–	–	–	–	–	–	–	5H-Dibenzo[d,f][1,3]diazepine, 6-ethyl-6,7-dihydro-5,7-dihydroxy-	38224-74-7	NI25462
213	256	184	185	255	200	214	199	**256**	100	68	41	40	32	30	20	20		15	16	2	2	–	–	–	–	–	–	–	–	–	6H-Indolo[3,2,1-de][1,5]naphthyridin-6-one, 1,2,3,3a,4,5-hexahydro-8-hydroxy-3-methyl-	50630-67-6	NI25463
59	44	55	117	90	89	74	111	**256**	100	72	68	56	52	37	31	27	**0.00**	15	16	2	2	–	–	–	–	–	–	–	–	–	Naphthalene-2,3-dicarboximide, 9-ethyl-1,4-imino-1,2,3,4-tetrahydro-N-methyl-, endo-	123565-80-0	DD00740
256	255	241	257	226	227	198	213	**256**	100	68	45	20	20	17	13	9		15	16	2	2	–	–	–	–	–	–	–	–	–	9H-Pyrido[3,4-b]indole, 4,8-dimethoxy-1-ethyl-		NI25464
107	15	39	228	121	41	79	53	**256**	100	88	73	71	64	61	42	41	**5.30**	15	16	2	2	–	–	–	–	–	–	–	–	–	5-Pyrimidinemethanol, α-cyclopropyl-α-(4-methoxyphenyl)-	12771-68-5	NI25465
256	121	107	77	91	257	92	78	**256**	100	99	33	29	23	20	19	18		15	16	2	2	–	–	–	–	–	–	–	–	–	Toluene, 2-(4-methoxyphenyldiazo)-5-methoxy-		NI25466
241	256	242	183	227	257	243	198	**256**	100	34	28	20	13	9	9	8		15	20	–	–	–	–	–	–	–	2	–	–	–	1,3(2H)-Disilaphenalene, 1,1,3,3-tetramethyl-		NI25467
98	111	103	104	77	159	70	144	**256**	100	81	73	64	52	44	44	42	**0.00**	15	20	–	4	–	–	–	–	–	–	–	–	–	1H-1,2,3-Triazole, 4,5-dihydro-4-methyl-1-(1-phenylvinyl)-5-(1-pyrrolidinyl)-	77726-82-0	NI25468
99	114	72	84	71	70	96	73	**256**	100	71	49	41	29	29	19	15	**0.00**	15	28	3	–	–	–	–	–	–	–	–	–	–	Furan, 2-(2,2,3,3-cyclopropoxy)-4-cis-hydroxy-3,3,5,5-tetramethyltetrahydro-		LI05253
99	71	72	70	114	73	84		**256**	100	68	61	58	47	29	18		**0.00**	15	28	3	–	–	–	–	–	–	–	–	–	–	Furan, 2-(2,2,3,3-cyclopropoxy)-4-trans-hydroxy-3,3,5,5-tetramethyltetrahydro-		LI05254
96	69	97	55	115	143	41	81	**256**	100	70	67	62	61	53	46	45	**0.00**	15	28	3	–	–	–	–	–	–	–	–	–	–	Hexanoic acid, 7-(2,2-dimethylcyclopentyloxy)-, ethyl ester		NI25469
143	69	55	111	45	155	207	157	**256**	100	80	25	22	22	2	1	1	**0.10**	15	28	3	–	–	–	–	–	–	–	–	–	–	Tetradecanoic acid, 7-methoxy-, (7S)-, (E)-		MS05634
57	43	41	71	55	85	197	95	**256**	100	75	51	40	39	23	20	18	**0.00**	15	28	3	–	–	–	–	–	–	–	–	–	–	Tetradecanoic acid, 2-oxo-, methyl ester	55682-82-1	NI25470
43	55	41	58	83	167	71	69	**256**	100	40	39	30	27	23	23	23	**0.00**	15	28	3	–	–	–	–	–	–	–	–	–	–	Tetradecanoic acid, 13-oxo-, methyl ester	18993-10-7	NI25471
41	55	67	81	74	43	82	96	**256**	100	98	86	81	81	62	48	47	**0.00**	15	28	3	–	–	–	–	–	–	–	–	–	–	5-Tetradecenoic acid, 14-hydroxy-, methyl ester, (Z)-	69494-18-4	NI25472
104	103	152	91	256	135	136	240	**256**	100	100	86	86	57	57	29	14		16	16	1	–	1	–	–	–	–	–	–	–	–	Ethene, 1-(2-phenyl)ethylsulfinyl-2-phenyl-, (E)-	86394-27-6	NI25473
152	104	135	103	256	136	240	134	**256**	100	100	88	88	63	38	25	25		16	16	1	–	1	–	–	–	–	–	–	–	–	Ethene, 1-(2-phenyl)ethylsulfinyl-2-phenyl-, (Z)-	81562-56-3	NI25474
151	105	77	51	91	152			**256**	100	43	41	14	13	11			**0.00**	16	16	3	–	–	–	–	–	–	–	–	–	–	Acetophenone, α-phenyl-α,α-dimethoxy-	24650-42-8	NI25475
91	256	92	65	39	257	41	77	**256**	100	16	8	8	4	3	3	2		16	16	3	–	–	–	–	–	–	–	–	–	–	Benzaldehyde, 4-benzyloxy-3,6-dimethyl-2-hydroxy-		MS05635
105	183	211	77	184	51	212	106	**256**	100	84	54	51	12	9	8	7	**0.84**	16	16	3	–	–	–	–	–	–	–	–	–	–	Benzeneacetic acid, α-hydroxy-α-phenyl-, ethyl ester	52182-15-7	MS05636
197	77	91	51	256	65	103	104	**256**	100	44	43	30	29	26	23	21		16	16	3	–	–	–	–	–	–	–	–	–	–	Benzeneacetic acid, α-methyl-3-phenoxy-, methyl ester, (±)-	74793-79-6	NI25476
134	43	91	195	167	77	256	165	**256**	100	69	62	36	26	26	24	19		16	16	3	–	–	–	–	–	–	–	–	–	–	2H-Benzopyran-3,4-diol, 3,4-dihydro-3-methyl-2-phenyl-		MS05637
238	107	104	91	119	131	132	77	**256**	100	80	80	70	62	58	54	50	**26.00**	16	16	3	–	–	–	–	–	–	–	–	–	–	2H-Benzopyran-3-ol, 3,4-dihydro-2-[2-(hydroxymethyl)phenyl]-, (+)-trans-	14894-98-5	MS05638
69	152	91	39	124	43	77	51	**256**	100	59	52	48	46	45	40	37	**25.00**	16	16	3	–	–	–	–	–	–	–	–	–	–	2H-Benzopyran-7-ol, 3,4-dihydro-5-methoxy-		MS05639
134	121	256	135	91	119	238	237	**256**	100	33	22	21	20	18	14	12		16	16	3	–	–	–	–	–	–	–	–	–	–	2H-1-Benzopyran-4-ol, 3,4-dihydro-2-(4-methoxyphenyl)-	5162-64-1	NI25478

MASS TO CHARGE RATIOS	M.W.	INTENSITIES	Parent	C	H	O	N	S	F	Cl	Br	I	Si	P	B	X	COMPOUND NAME	CAS Reg No	No
134 119 121 238 237 135 122 256	**256**	100 23 22 15 15 11 6 4		16	16	3	–	–	–	–	–	–	–	–	–	–	**2H-1-Benzopyran-4-ol, 3,4-dihydro-2-(4-methoxyphenyl)-**	5162-64-1	**NI25477**
241 182 167 181 209 256 152 242	**256**	100 81 69 55 46 26 18 17		16	16	3	–	–	–	–	–	–	–	–	–	–	**1-Dibenzofurancarboxylic acid, 3,9b-dihydro-4,9b-dimethyl-, methyl ester**	40801-38-5	**NI25479**
241 182 197 181 256 209 152 183	**256**	100 75 56 55 54 41 20 16		16	16	3	–	–	–	–	–	–	–	–	–	–	**1-Dibenzofurancarboxylic acid, 1,9b-dihydro-4,9b-dimethyl-, methyl ester, cis-**	40801-40-9	**NI25480**
183 241 197 256 182 209 152 184	**256**	100 96 91 74 69 32 22 18		16	16	3	–	–	–	–	–	–	–	–	–	–	**1-Dibenzofurancarboxylic acid, 1,9b-dihydro-4,9b-dimethyl-, methyl ester, trans-**	40801-41-0	**NI25481**
255 256 105 179 257 225 107 151	**256**	100 92 42 25 17 12 11 10		16	16	3	–	–	–	–	–	–	–	–	–	–	**4,5-Dimethoxy-2-methylphenyl phenyl ketone**	52806-39-0	**NI25482**
225 256 77 105 177 257 151 107	**256**	100 92 63 41 25 17 10 10		16	16	3	–	–	–	–	–	–	–	–	–	–	**4,5-Dimethoxy-2-methylphenyl phenyl ketone**	52806-39-0	**NI25483**
159 256 174 157 77 202 43 197	**256**	100 43 41 40 36 35 32 31		16	16	3	–	–	–	–	–	–	–	–	–	–	**2,5-Hexadienoic acid, 5-(3-methyl-2-benzofuranyl)-, methyl ester**	55937-83-2	**NI25484**
256 45 211 183 165 155 150 181	**256**	100 70 66 51 26 24 22 16		16	16	3	–	–	–	–	–	–	–	–	–	–	**Stilbene, 4′-hydroxy-2-methoxymethoxy-, (E)-**		**MS05640**
141 71 115 256 41 139 140 128	**256**	100 65 16 13 13 8 5 5		16	16	3	–	–	–	–	–	–	–	–	–	–	**2-Tetrahydrofurancarboxylic acid, 1-naphthalenemethyl ester**	86338-68-3	**NI25485**
167 73 179 165 256 178 45 241	**256**	100 68 44 40 38 24 24 22		16	20	1	–	–	–	–	–	–	1	–	–	–	**Silane, (diphenylmethoxy)trimethyl-**	14629-59-5	**NI25486**
136 120 121 77 256 78 137 91	**256**	100 79 16 16 15 15 10 10		16	20	1	2	–	–	–	–	–	–	–	–	–	**Ethylene diamine, N,N′-dimethyl-N-(3-hydroxyphenyl)-N′-phenyl-**		**LI05255**
136 120 121 77 256 78 137 91	**256**	100 79 16 16 15 15 10 10		16	20	1	2	–	–	–	–	–	–	–	–	–	**Ethylene diamine, N,N′-dimethyl-N-(4-hydroxyphenyl)-N′-phenyl-**	32857-40-2	**NI25487**
42 57 144 256 89 84 241 85	**256**	100 98 94 75 73 57 52 51		16	20	1	2	–	–	–	–	–	–	–	–	–	**3H-Indol-3-one, 2-[(1,2-dimethyl-3-piperidinyl)methyl]-**	54986-55-9	**NI25488**
255 256 197 170 169 100 156 239	**256**	100 96 35 31 31 27 21 19		16	20	1	2	–	–	–	–	–	–	–	–	–	**Indolo[2,3-a]quinolizin-2-ol, 1,2,3,4,6,7,12,12b-octahydro-2-methyl-, cis-**	51598-42-6	**NI25489**
256 255 197 100 170 169 156 239	**256**	100 88 28 28 22 22 20 18		16	20	1	2	–	–	–	–	–	–	–	–	–	**Indolo[2,3-a]quinolizin-2-ol, 1,2,3,4,6,7,12,12b-octahydro-2-methyl-, trans-**	51598-41-5	**NI25490**
238 237 96 195 194 132 256	**256**	100 69 34 27 25 25 20		16	20	1	2	–	–	–	–	–	–	–	–	–	**1-Methyl-indol-2-yl-(1,2,5,6-tetrahydro-1-methyl-4-pyridyl)-methanol**		**LI05256**
183 256 155 237 182 154 46 168	**256**	100 92 74 55 55 53 48 45		16	20	1	2	–	–	–	–	–	–	–	–	–	**2-Propen-1-ol, 2-methyl-3-[1,3,4,5-tetrahydro-4-(methylamino)benz[cd]indol-5-yl]-, [R-(4α,5β(E))]-**	2390-99-0	**NI25491**
158 157 256 130 159 118 211 241	**256**	100 72 53 41 21 18 17 13		16	20	1	2	–	–	–	–	–	–	–	–	–	**Pyridine, 5-(1-hydroxyethyl)-2-indolyl-1-methyl-1,2,3,6-tetrahydro-**		**MS05641**
82 55 173 67 83 41 43 81	**256**	100 97 82 81 72 56 31 24	**20.50**	16	32	–	–	1	–	–	–	–	–	–	–	–	**Cyclohexylsulfide, decyl-**	41346-30-9	**NI25492**
101 41 43 55 87 67 85 102	**256**	100 21 19 15 11 11 9 8	**3.90**	16	32	–	–	1	–	–	–	–	–	–	–	–	**Thiacyclohexane, 2-undecyl-**	1801-98-5	**NI25493**
57 43 97 55 71 41 171 85	**256**	100 86 79 78 75 72 69 56	**0.15**	16	32	2	–	–	–	–	–	–	–	–	–	–	**Decanoic acid, 5-ethyl-3,5,9-trimethyl-, methyl ester**	56247-63-3	**NI25494**
87 29 45 43 116 41 55 31	**256**	100 69 65 62 59 50 39 34	**4.57**	16	32	2	–	–	–	–	–	–	–	–	–	–	**Dodecanoic acid, 2-propyl-, methyl ester**		**IC03355**
88 101 256 41 43 55 87 57	**256**	100 88 84 64 55 50 42 41		16	32	2	–	–	–	–	–	–	–	–	–	–	**Dodecanoic acid, 2,4,6-trimethyl-, methyl ester**	57289-31-3	**NI25495**
101 74 256 55 41 75 43 199	**256**	100 71 40 32 27 24 24 23		16	32	2	–	–	–	–	–	–	–	–	–	–	**Dodecanoic acid, 2,6,10-trimethyl-, methyl ester**		**LI05257**
101 74 171 143 87 256 139 129	**256**	100 81 5 5 5 3 3 3		16	32	2	–	–	–	–	–	–	–	–	–	–	**Dodecanoic acid, 3,7,11-trimethyl-, methyl ester**		**MS05642**
98 55 41 84 57 43 83 69	**256**	100 90 71 31 31 29 24 24	**1.30**	16	32	2	–	–	–	–	–	–	–	–	–	–	**Hexadecanoic acid**	57-10-3	**WI01303**
73 71 83 129 256 98 85 97	**256**	100 34 25 23 22 21 20 17		16	32	2	–	–	–	–	–	–	–	–	–	–	**Hexadecanoic acid**	57-10-3	**NI25496**
43 73 60 41 57 55 256 29	**256**	100 97 90 77 76 65 44 44		16	32	2	–	–	–	–	–	–	–	–	–	–	**Hexadecanoic acid**	57-10-3	**NI25497**
57 70 71 43 41 127 112 88	**256**	100 57 27 27 24 20 20 20	**0.00**	16	32	2	–	–	–	–	–	–	–	–	–	–	**Hexanoic acid, 2-ethyl-, 2-ethylhexyl ester**	7425-14-1	**NI25498**
145 112 157 43 70 41 84 55	**256**	100 89 74 58 46 43 42 41	**3.60**	16	32	2	–	–	–	–	–	–	–	–	–	–	**Octanoic acid, octyl ester**		**IC03356**
74 87 127 43 55 41 57 88	**256**	100 58 52 50 31 28 23 22	**1.47**	16	32	2	–	–	–	–	–	–	–	–	–	–	**Pentadecanoic acid, methyl ester**		**IC03357**
74 87 43 41 55 256 75 57	**256**	100 61 38 31 27 22 19 16		16	32	2	–	–	–	–	–	–	–	–	–	–	**Pentadecanoic acid, methyl ester**		**IC03358**
74 87 43 41 55 57 69 143	**256**	100 62 49 38 34 31 20 16	**0.00**	16	32	2	–	–	–	–	–	–	–	–	–	–	**Pentadecanoic acid, methyl ester**		**MS05643**
88 101 43 41 55 73 57 29	**256**	100 47 26 18 17 14 14 14	**11.91**	16	32	2	–	–	–	–	–	–	–	–	–	–	**Tetradecanoic acid, ethyl ester**	124-06-1	**WI01304**
88 101 43 55 41 73 57 29	**256**	100 56 30 23 23 21 17 17	**6.40**	16	32	2	–	–	–	–	–	–	–	–	–	–	**Tetradecanoic acid, ethyl ester**	124-06-1	**NI25499**
88 101 43 41 29 256 55 57	**256**	100 43 33 25 23 21 21 16		16	32	2	–	–	–	–	–	–	–	–	–	–	**Tetradecanoic acid, ethyl ester**	124-06-1	**IC03359**
88 41 43 55 101 57 45 69	**256**	100 37 34 33 31 19 19 18	**6.09**	16	32	2	–	–	–	–	–	–	–	–	–	–	**Tetradecanoic acid, 2-methyl-, methyl ester**	55554-09-1	**IC03360**
88 101 256 57 43 55 41 143	**256**	100 31 17 13 13 12 12 9		16	32	2	–	–	–	–	–	–	–	–	–	–	**Tetradecanoic acid, 2-methyl-, methyl ester**	55554-09-1	**NI25500**
88 101 213 57 43 55 41 69	**256**	100 73 32 30 25 24 23 16	**6.00**	16	32	2	–	–	–	–	–	–	–	–	–	–	**Tetradecanoic acid, 2-methyl-, methyl ester**	55554-09-1	**NI25501**
74 101 43 41 75 65 67 69	**256**	100 67 27 25 21 20 16 13	**0.00**	16	32	2	–	–	–	–	–	–	–	–	–	–	**Tetradecanoic acid, 3-methyl-, methyl ester**		**LI05258**
213 256 185 163 181 225 187 214	**256**	100 72 51 42 33 25 20 18		16	32	2	–	–	–	–	–	–	–	–	–	–	**Tetradecanoic acid, 11-methyl-, methyl ester**		**LI05259**
74 41 55 87 43 29 57 69	**256**	100 88 75 73 69 58 55 36	**7.31**	16	32	2	–	–	–	–	–	–	–	–	–	–	**Tetradecanoic acid, 12-methyl-, methyl ester**	5129-66-8	**NI25502**
74 87 43 55 41 57 69 75	**256**	100 83 40 39 37 36 24 21	**9.00**	16	32	2	–	–	–	–	–	–	–	–	–	–	**Tetradecanoic acid, 12-methyl-, methyl ester**	5129-66-8	**MS05644**
74 87 55 57 43 41 69 83	**256**	100 72 44 41 32 27 20 15	**13.00**	16	32	2	–	–	–	–	–	–	–	–	–	–	**Tetradecanoic acid, 12-methyl-, methyl ester, (S)-**	62691-05-8	**NI25503**
74 87 143 256 101 129 225 199	**256**	100 63 13 10 7 5 2 2		16	32	2	–	–	–	–	–	–	–	–	–	–	**Tetradecanoic acid, 13-methyl-, methyl ester**		**LI05260**
55 69 83 41 61 57 56 97	**256**	100 89 88 85 80 73 66 65	**0.00**	16	32	2	–	–	–	–	–	–	–	–	–	–	**1-Tetradecanol, acetate**	638-59-5	**NI25504**
88 101 185 57 43 41 69 55	**256**	100 50 15 14 14 12 11 11	**1.20**	16	32	2	–	–	–	–	–	–	–	–	–	–	**Tridecanoic acid, 2,4-dimethyl-, methyl ester**	73105-69-8	**NI25505**
88 101 43 55 41 57 157 69	**256**	100 44 19 18 15 15 12 12	**3.40**	16	32	2	–	–	–	–	–	–	–	–	–	–	**Tridecanoic acid, 2,6-dimethyl-, methyl ester**	73105-76-7	**NI25506**
88 101 157 57 69 55 43 41	**256**	100 64 50 30 21 20 20 19	**2.90**	16	32	2	–	–	–	–	–	–	–	–	–	–	**Tridecanoic acid, 2,8-dimethyl-, methyl ester**	73105-83-6	**NI25507**
102 29 45 87 41 55 43 31	**256**	100 82 79 76 68 67 67 39	**5.56**	16	32	2	–	–	–	–	–	–	–	–	–	–	**Tridecanoic acid, 2-ethyl-, methyl ester**		**IC03361**
227 256 177 195 213 185 225 187	**256**	100 91 87 84 53 50 29 25		16	32	2	–	–	–	–	–	–	–	–	–	–	**Tridecanoic acid, 11-ethyl-, methyl ester**		**LI05261**

MASS TO CHARGE RATIOS								M.W.	INTENSITIES								Parent	C	H	O	N	S	F	Cl	Br	I	Si	P	B	X	COMPOUND NAME	CAS Reg No	No
44	87	30	43	57	41	56	55	256	100	77	37	29	26	24	22	18	0.30	16	36	–	2	–	–	–	–	–	–	–	–	–	Dodecanamine, N-(3-aminopropyl)-2-methyl-		IC03362
44	30	57	87	43	41	70	210	256	100	59	39	38	38	37	29	26	1.30	16	36	–	2	–	–	–	–	–	–	–	–	–	Tridecanamine, N-(3-aminopropyl)-		IC03363
112	135	256	257	91	134	136	123	256	100	90	63	12	12	11	9	9		17	20	2	–	–	–	–	–	–	–	–	–	–	Benzene, 3,5-dimethyl-2-hydroxy-[1,1′-methylenebis-		IC03364
91	107	165	92	79	65	77	166	256	100	64	58	15	15	12	8	6	0.00	17	20	2	–	–	–	–	–	–	–	–	–	–	Benzene, 1,1′-[1,3-propanediylbis(oxymethylene)]bis-	53088-81-6	NI25508
71	141	167	115	256	128	129	127	256	100	71	50	45	39	35	30	26		17	20	2	–	–	–	–	–	–	–	–	–	–	2,8,10-Heptadecatrien-4,6-diyn-14-one, 1-hydroxy-, (E,E,E)-		LI05262
241	256	255	242	239	135	257	134	256	100	80	23	19	18	18	16	14		17	20	2	–	–	–	–	–	–	–	–	–	–	Methane, 4,4′-dihydroxy-3,3′,5,5′-tetramethyldiphenyl-		IC03365
256	241	255	257	242	239	135	134	256	100	99	25	19	18	18	18	13		17	20	2	–	–	–	–	–	–	–	–	–	–	Methane, 4,4′-dihydroxy-3,3′,5,5′-tetramethyldiphenyl-		MS05645
155	141	256	41	57	115	56	43	256	100	97	55	21	20	17	17	17		17	20	2	–	–	–	–	–	–	–	–	–	–	2-Naphthalenepentanoic acid, β,β-dimethyl-	64184-13-0	NI25509
43	256	115	71	255	128	141	129	256	100	69	55	55	46	35	33	23		17	20	2	–	–	–	–	–	–	–	–	–	–	6-Nonenoic acid, 9-(3-isopropylphenyl)-2,6-dimethyl-, methyl ester, (Z)-		LI05263
41	55	39	117	67	43	69	57	256	100	80	55	45	45	40	38	35	3.50	17	20	2	–	–	–	–	–	–	–	–	–	–	Oxirane, 2-(6-heptenyl)-3-(oct-6(Z)-en-2,4-diynoyl)-		LI05264
111	256	72	98	255	144	143	84	256	100	65	60	35	30	21	18	17		17	24	–	2	–	–	–	–	–	–	–	–	–	1,2-Seco-1,2-butano-β-carboline, 2-ethyl-1,2,3,4-tetrahydro-		LI05265
241	212	256	128	197	242	113	105	256	100	50	47	19	18	17	16	12		17	24	–	2	–	–	–	–	–	–	–	–	–	1H-Pyrrole, 3-ethyl-5-[(4-ethyl-3,5-dimethyl-2H-pyrrol-2-ylidene)methyl]-2,4-dimethyl-	2407-83-2	NI25510
241	212	256	197	242	128	113	211	256	100	50	44	18	17	16	14	12		17	24	–	2	–	–	–	–	–	–	–	–	–	1H-Pyrrole, 3-ethyl-5-[(4-ethyl-3,5-dimethyl-2H-pyrrol-2-ylidene)methyl]-2,4-dimethyl-	2407-83-2	LI05266
43	55	41	57	69	83	29	56	256	100	90	84	75	69	67	47	44	0.00	17	36	1	–	–	–	–	–	–	–	–	–	–	1-Heptadecanol	1454-85-9	MS05646
83	57	69	43	55	97	70	56	256	100	96	89	83	82	74	57	54	0.00	17	36	1	–	–	–	–	–	–	–	–	–	–	1-Heptadecanol	1454-85-9	IC03366
71	70	84	69	112	83	85	97	256	100	85	77	68	48	44	36	30	0.00	17	36	1	–	–	–	–	–	–	–	–	–	–	1-Heptadecanol	1454-85-9	NI25511
45	43	57	55	41	83	71	69	256	100	36	30	25	22	18	18	18	0.14	17	36	1	–	–	–	–	–	–	–	–	–	–	2-Heptadecanol	16813-18-6	NI25512
43	57	71	41	58	29	141	55	256	100	91	88	79	69	61	59	57	0.12	17	36	1	–	–	–	–	–	–	–	–	–	–	9-Heptadecanol		AI01827
69	143	83	57	55	43	238	41	256	100	62	52	32	32	25	22	20	0.00	17	36	1	–	–	–	–	–	–	–	–	–	–	9-Heptadecanol		LI05267
57	43	28	55	41	69	56	71	256	100	89	68	67	63	58	57	51	0.34	17	36	1	–	–	–	–	–	–	–	–	–	–	1-Hexadecanol, 2-methyl-	2490-48-4	NI25513
239	57	171	43	155	69	141	85	256	100	45	39	37	18	14	12	12	0.00	17	36	1	–	–	–	–	–	–	–	–	–	–	5-Nonanol, 2,2-dimethyl-5-(3-hexyl)-		DD00741
70	45	83	57	55	69	97	71	256	100	92	74	72	67	58	55	52	0.00	17	36	1	–	–	–	–	–	–	–	–	–	–	Pentadecane, 1-methoxy-13-methyl-	56196-09-9	NI25514
69	83	55	57	43	111	97	41	256	100	75	54	53	42	41	35	28	0.00	17	36	1	–	–	–	–	–	–	–	–	–	–	6-Tridecanol, 3,9-diethyl-		DO01046
255	256	128	257	127	126	254		256	100	99	20	18	13	10	5			18	12	–	2	–	–	–	–	–	–	–	–	–	7H-[4,5]benzobenzimidazo[2,1-a]isoquinoline		LI05268
256	255	154	227	257	101	254	75	256	100	78	23	22	20	19	13	12		18	12	–	2	–	–	–	–	–	–	–	–	–	Benzonitrile, 4-[2-(4-quinolinyl)vinyl]-, (E)-	40317-06-4	NI25515
256	255	128	257	102	75	127	77	256	100	49	25	20	4	4	3	3		18	12	–	2	–	–	–	–	–	–	–	–	–	2,2′-Biquinoline	119-91-5	AI01828
256	255	257	128	101	73	75	253	256	100	66	18	12	5	4	3	3		18	12	–	2	–	–	–	–	–	–	–	–	–	2,2′-Biquinoline	119-91-5	NI25516
256	255	128	101	257	75	114	51	256	100	62	48	17	16	10	8	8		18	12	–	2	–	–	–	–	–	–	–	–	–	2,2′-Biquinoline	119-91-5	NI25517
256	255	257	41	128	43	55	27	256	100	33	21	14	13	13	12	9		18	12	–	2	–	–	–	–	–	–	–	–	–	6,6′-Biquinoline		LI05269
256	255	257	128	127	126			256	100	55	19	16	15	15				18	12	–	2	–	–	–	–	–	–	–	–	–	12H-Isoindolo[2,1-a]perimidine		LI05270
247	246	18	218	217	248	39	189	256	100	67	35	31	27	22	20	17		18	12	–	2	–	–	–	–	–	–	–	–	–	Quinoline, 4-(4-cyanostyryl)-, (E)-		LI05271
91	105	107	79	41	65	77	92	256	100	18	17	15	14	10	10	10	0.26	18	24	1	–	–	–	–	–	–	–	–	–	–	Bicyclo[3.1.1]hept-2-ene, 6,6-dimethyl-2-[2-(phenylmethoxy)ethyl]-, (1S)-	74851-17-5	NI25518
256	159	257	146	185	160	172	157	256	100	25	20	19	15	15	11	11		18	24	1	–	–	–	–	–	–	–	–	–	–	Estra-1,3,5(10)-trien-3-ol	53-63-4	NI25519
241	43	256	242	185	143	183	165	256	100	89	59	58	45	29	22	20		18	24	1	–	–	–	–	–	–	–	–	–	–	s-Indacene, 4-acetyl-1,2,3,5,6,7-hexahydro-1,1,5,5-tetramethyl-		TR00285
241	256	185	242	199	142	257	141	256	100	30	26	19	11	7	6	6		18	24	1	–	–	–	–	–	–	–	–	–	–	s-Indacen-1(2H)-one, 3,5,6,7-tetrahydro-3,3,4,5,5,8-hexamethyl-	38754-94-8	NI25520
241	256	143	113	185	43	41	29	256	100	23	23	12	11	10	10	10		19	28	–	–	–	–	–	–	–	–	–	–	–	as-Indacene, 1,2,3,6,7,8-hexahydro-4-isopropyl-1,1,6,6-tetramethyl-		AI01829
129	130	143	41	43	128	256	213	256	100	88	76	58	52	49	39	39		19	28	–	–	–	–	–	–	–	–	–	–	–	1H-Indene, 2-butyl-3-hexyl-		AI01830
241	143	171	157	185	129	128	41	256	100	28	24	24	22	22	21	20	15.74	19	28	–	–	–	–	–	–	–	–	–	–	–	Naphthalene, 6-ethyl-1,2,3,4-tetrahydro-1,1,4,4-tetramethyl-7-isopropenyl-		NI25521
91	41	256	93	79	241	105	77	256	100	86	85	81	73	71	68	58		19	28	–	–	–	–	–	–	–	–	–	–	–	18-Norisopimara-4(19),7,15-triene		MS05647
256	241	239	119.5	240	120	257	242	256	100	75	59	38	36	29	21	15		20	16	–	–	–	–	–	–	–	–	–	–	–	Benz[a]anthracene, 1,12-dimethyl-		AI01831
256	241	257	240	119.5	120	128	113	256	100	70	48	46	37	36	22	21		20	16	–	–	–	–	–	–	–	–	–	–	–	Benz[a]anthracene, 7,12-dimethyl-		MS05648
256	241	239	240	119	257	120	242	256	100	40	37	24	23	21	20	18		20	16	–	–	–	–	–	–	–	–	–	–	–	Benz[a]anthracene, 9,10-dimethyl-		AI01832
241	256	239	242	257	114	253	128	256	100	76	26	22	16	10	9	9		20	16	–	–	–	–	–	–	–	–	–	–	–	Benz[a]anthracene, 7-ethyl-	3697-30-1	WI01305
241	256	239	242	257	120.5	119.5	114	256	100	76	26	22	16	12	12	10		20	16	–	–	–	–	–	–	–	–	–	–	–	Benz[a]anthracene, 7-ethyl-	3697-30-1	AM00134
104	256	152	255	165	103	78	257	256	100	80	33	27	24	21	21	19		20	16	–	–	–	–	–	–	–	–	–	–	–	Benzo[k]fluoranthene, 6b,7,12,12a-tetrahydro-	19638-45-0	NI25522
228	256	252	114	255	229	253	226	256	100	73	24	22	20	19	18	16		20	16	–	–	–	–	–	–	–	–	–	–	–	Benzo[k]fluoranthene, 1,2,3,12b-tetrahydro-	95785-04-9	NI25523
256	228	255	252	226	227	239	257	256	100	97	56	45	37	34	25	21		20	16	–	–	–	–	–	–	–	–	–	–	–	Benzo[b]fluoranthene, 9,10,11,12-tetrahydro-		NI25524
256	228	255	226	257	239	227	113	256	100	61	27	24	19	16	16	15		20	16	–	–	–	–	–	–	–	–	–	–	–	Benzo[k]fluoranthene, 8,9,10,11-tetrahydro-	33942-81-3	NI25525
256	241	239	119.5	240	226	120	257	256	100	86	78	37	28	26	25	22		20	16	–	–	–	–	–	–	–	–	–	–	–	Benzo[c]phenanthrene, 1,12-dimethyl-	4076-43-1	AM00135
256	241	239	240	226	120	257	242	256	100	86	78	28	26	25	22	17		20	16	–	–	–	–	–	–	–	–	–	–	–	Benzo[c]phenanthrene, 1,12-dimethyl-	4076-43-1	WI01306
256	255	257	128	119.5				256	100	32	28	21	19					20	16	–	–	–	–	–	–	–	–	–	–	–	Benzo[c]phenanthrene, 5,8-dimethyl-	54986-63-9	LI05272
256	257	239	241	240	120	113	255	256	100	22	21	15	11	9	9	8		20	16	–	–	–	–	–	–	–	–	–	–	–	Benzo[c]phenanthrene, 5,8-dimethyl-	54986-63-9	WI01307
256	257	239	241	119.5	240	120	113	256	100	22	21	15	12	11	9	9		20	16	–	–	–	–	–	–	–	–	–	–	–	Benzo[c]phenanthrene, 5,8-dimethyl-	54986-63-9	AM00136

MASS TO CHARGE RATIOS								M.W.	INTENSITIES								Parent	C	H	O	N	S	F	Cl	Br	I	Si	P	B	X	COMPOUND NAME	CAS Reg No	No
256	255	252	253	257	239	254	240	**256**	100	44	26	22	21	16	11	10		20	16	–	–	–	–	–	–	–	–	–	–	–	Benzo[a]pyrene, 4,5,11,12-tetrahydro-	109292-58-2	NI25526
256	228	252	227	239	226	255	257	**256**	100	56	36	35	31	29	25	22		20	16	–	–	–	–	–	–	–	–	–	–	–	Benzo[a]pyrene, 7,8,9,10-tetrahydro-	17750-93-5	NI25527
256	227	228	252	226	239	257	241	**256**	100	45	39	37	32	29	23	19		20	16	–	–	–	–	–	–	–	–	–	–	–	Benzo[e]pyrene, 9,10,11,12-tetrahydro-	67099-81-4	NI25528
256	241	239	240	257	120	242	226	**256**	100	76	53	25	23	22	17	13		20	16	–	–	–	–	–	–	–	–	–	–	–	Chrysene, 5-ethyl-	54986-62-8	NI25529
256	230	154	115	178	91	257	77	**256**	100	54	32	30	26	26	22	18		20	16	–	–	–	–	–	–	–	–	–	–	–	Cyclooctatetraene, 1,5-diphenyl-	119770-22-8	DD00742
256	178	179	255	241	165	239	257	**256**	100	50	38	30	23	23	22	21		20	16	–	–	–	–	–	–	–	–	–	–	–	Ethylene, triphenyl-		LI05273
69	107	109	64	78	65	48	80	**257**	100	64	62	54	29	25	25	21	**0.00**	1	1	3	–	1	3	–	–	–	–	–	–	1	Methanesulphonic acid, trifluoro-, silver(1+) salt	2923-28-6	NI25530
127	69	89	135	101	103	188	129	**257**	100	41	33	16	14	13	6	5	**0.00**	2	–	–	1	1	8	1	–	–	–	–	–	–	Ethylidenimine, 1-chloro-2,2,2-trifluoro-N-pentafluorosulphanyl-	2375-40-8	NI25531
70	46	127	89	130	47	48	67	**257**	100	70	60	25	22	10	7	7	**0.00**	2	–	2	3	2	5	–	–	–	–	–	–	–	Sulphilimine, S,S-diisocyanato-N-pentafluorosulphanyl-	80044-07-1	NI25532
178	69	128	33	159	109	259	28	**257**	100	68	63	37	30	30	11	5	**0.00**	4	2	–	1	–	6	–	1	–	–	–	–	–	Aziridine, 1-bromo-2,2-bis(trifluoromethyl)-		LI05274
69	178	107	105	90	259	257	40	**257**	100	95	48	48	26	17	17	7		4	2	–	1	–	6	–	1	–	–	–	–	–	Ethylene, 1-bis(trifluoromethyl)amino-1-bromo-		LI05275
259	257	69	171	169	107	105	90	**257**	100	99	90	31	31	21	21	14		4	2	–	1	–	6	–	1	–	–	–	–	–	Ethylene, 1-bis(trifluoromethyl)amino-2-bromo-, cis-		LI05276
259	257	69	171	169	107	105	96	**257**	100	99	84	34	34	27	27	25		4	2	–	1	–	6	–	1	–	–	–	–	–	Ethylene, 1-bis(trifluoromethyl)amino-2-bromo-, trans-		LI05277
93	125	126	79	87	172	47	63	**257**	100	83	47	34	32	23	21	20	**6.97**	6	12	4	1	2	–	–	–	–	–	1	–	–	Phosphorodithioic acid, S-[2-(formylmethylamino)-2-oxoethyl] O,O-dimethyl ester	2540-82-1	NI25533
139	111	75	78	76	211	257	137	**257**	100	83	30	22	16	14	12	11		8	4	1	3	2	–	1	–	–	–	–	–	–	Benzamide, 3-chloro-N-5H-1,3,2,4-dithia(3-SIV)diazol-5-ylidene-	57726-58-6	NI25534
139	111	75	211	78	257	137	76	**257**	100	31	25	11	10	9	6	5		8	4	1	3	2	–	1	–	–	–	–	–	–	Benzamide, 4-chloro-N-5H-1,3,2,4-dithia(3-SIV)diazol-5-ylidene-	57726-53-1	NI25535
120	257	241	162	104	79	78	15	**257**	100	76	36	28	25	24	23	19		8	7	3	3	2	–	–	–	–	–	–	–	–	1,3,4-Thiadiazole, 2-methylsulphonyl-5-(1-oxido-4-pyridyl)-	90045-18-4	NI25536
197	15	156	59	166	69	85	43	**257**	100	64	56	31	27	22	11	11	**0.80**	8	10	5	1	–	3	–	–	–	–	–	–	–	L-Aspartic acid, N-(trifluoroacetyl)-, dimethyl ester	688-09-5	NI25537
110	41	36	42	39	227	229	69	**257**	100	32	16	14	14	13	13	10	**1.10**	8	12	1	2	–	–	3	–	–	–	–	–	–	3-Imidazoline, 2,2,5,5-tetramethyl-4-trichloromethyl-, 1-oxile	84359-77-3	NI25538
73	242	121	213	70	75	164	243	**257**	100	67	53	31	13	10	9	8	**0.00**	8	14	3	1	–	3	–	–	–	1	–	–	–	L-Alanine, N-(trifluoroacetyl)-, trimethylsilyl ester	52558-80-2	NI25539
73	188	192	45	77	74	100	72	**257**	100	15	14	14	13	9	8	7	**1.80**	8	18	1	1	–	3	–	–	–	2	–	–	–	Acetamide, 2,2,2-trifluoro-N,N-bis(trimethylsilyl)-	21149-38-2	NI25540
73	188	45	77	192	74	100	43	**257**	100	14	14	13	12	8	6	6	**1.44**	8	18	1	1	–	3	–	–	–	2	–	–	–	Acetamide, 2,2,2-trifluoro-N,N-bis(trimethylsilyl)-	21149-38-2	NI25541
73	188	192	45	74	77	43	75	**257**	100	17	15	10	8	7	4	4	**2.86**	8	18	1	1	–	3	–	–	–	2	–	–	–	Acetamide, 2,2,2-trifluoro-N,N-bis(trimethylsilyl)-	21149-38-2	NI25542
147	159	148	170	120	149	44	42	**257**	100	18	15	11	9	7	7	5	**0.00**	8	18	1	1	–	3	–	–	–	2	–	–	–	Acetamidic acid, trifluoro-N,O-bis(trimethylsilyl)-		LI05278
73	29	192	188	77	75	45	74	**257**	100	26	24	18	14	14	12	8	**4.10**	8	18	1	1	–	3	–	–	–	2	–	–	–	Acetamidic acid, trifluoro-N,O-bis(trimethylsilyl)-		NI25543
86	29	156	27	30	58	101	28	**257**	100	77	50	46	40	35	18	16	**0.00**	8	20	–	1	–	–	–	–	1	–	–	–	–	Ammonium, tetraethyl-, iodide	68-05-3	NI25544
29	86	58	127	156	56	101	72	**257**	100	99	74	73	51	24	11	10	**0.00**	8	20	–	1	–	–	–	–	1	–	–	–	–	Ammonium, tetraethyl-, iodide	68-05-3	NI25545
259	257	150	230	228	178	231	229	**257**	100	95	85	60	59	46	32	32		9	8	1	1	1	–	–	1	–	–	–	–	–	Thiazolo[3,2-a]pyridinium, 7-bromo-8-hydroxy-3,5-dimethyl-, hydroxide, inner salt	30277-16-8	NI25546
91	104	150	105	78	77	103	51	**257**	100	65	31	30	22	22	18	16	**0.11**	9	10	2	–	–	–	–	–	–	–	–	–	1	Benzenepropanoic acid, silver(1+) salt	75112-79-7	NI25547
196	198	44	97	45	200	132	43	**257**	100	92	48	40	40	38	36	32	**5.00**	9	11	2	3	2	–	–	–	–	–	–	–	–	5H-1,3,4-Thiadiazolo[3,2-a]pyrimidin-5-one, 2-(butylthio)-7-hydroxy-		MS05649
242	257	188	170	44	97	69	83	**257**	100	78	65	58	50	23	22	18		9	11	2	3	2	–	–	–	–	–	–	–	–	5H-1,3,4-Thiadiazolo[3,2-a]pyrimidin-5-one, 6-ethyl-2-(ethylthio)-7-hydroxy-		MS05650
136	184	44	42	43	30	73	137	**257**	100	99	98	62	45	38	31	29	**7.00**	9	11	6	3	–	–	–	–	–	–	–	–	–	1,2-Benzenediol, 4-[1-hydroxy-2-(methylnitrosoamino)ethyl]-5-nitro-, (–)-	25775-86-4	LI05279
136	182	44	42	43	137	108	51	**257**	100	99	99	62	44	30	18	14	**5.61**	9	11	6	3	–	–	–	–	–	–	–	–	–	1,2-Benzenediol, 4-[1-hydroxy-2-(methylnitrosoamino)ethyl]-5-nitro-, (–)-	25775-86-4	NI25548
59	155	58	28	113	86	140	43	**257**	100	48	43	38	23	23	13	13	**0.00**	9	14	4	1	–	3	–	–	–	–	–	–	–	lyxo-Hexose, methyl 3-amino-2,3,6-trideoxy-N-trifluoroacetyl-		NI25549
57	139	138	69	56	110	55	43	**257**	100	87	55	51	19	13	13	13	**0.00**	9	14	4	1	–	3	–	–	–	–	–	–	–	L-Serine, N-(trifluoroacetyl)-, butyl ester	23403-43-2	NI25550
36	44	206	38	149	191	28	164	**257**	100	66	65	35	31	30	25	21	**0.00**	10	10	2	2	1	–	1	–	–	–	–	–	–	2(3H)-Thiazolimine, 3-hydroxy-4-(4-methoxyphenyl)-, monohydrochloride	58275-66-4	NI25551
158	43	242	257	28	155	85	59	**257**	100	86	68	66	39	12	12	10		10	14	4	–	–	–	–	–	–	–	–	–	1	Cobalt, bis(acetylacetonato)-	14024-48-7	NI25552
158	257	241	43	143	258	159	242	**257**	100	50	50	13	8	6	6	5		10	14	4	–	–	–	–	–	–	–	–	–	1	Cobalt, bis(acetylacetonato)-	14024-48-7	NI25553
111	73	212	39	75	213	45	242	**257**	100	77	44	23	21	19	18	9	**1.37**	10	15	3	1	1	–	–	–	–	1	–	–	–	Glycine, N-(2-thienylcarbonyl)-, trimethylsilyl ester	55887-50-8	NI25554
125	126	28	55	27	43	29	54	**257**	100	71	70	50	49	44	34	33	**0.83**	10	15	5	3	–	–	–	–	–	–	–	–	–	Cytidine, 5-methyl-	2140-61-6	NI25555
112	44	151	45	146	43	87	29	**257**	100	91	86	84	42	42	38	32	**2.01**	10	15	5	3	–	–	–	–	–	–	–	–	–	Cytidine, 2′-O-methyl-	2140-72-9	NI25556
112	44	151	45	146	43	87	28	**257**	100	91	85	85	43	41	37	31	**3.00**	10	15	5	3	–	–	–	–	–	–	–	–	–	Cytidine, 2′-O-methyl-	2140-72-9	LI05280
112	170	140	111	87	71	69	151	**257**	100	24	24	13	11	8	8	7	**0.00**	10	15	5	3	–	–	–	–	–	–	–	–	–	Cytidine, 3′-O-methyl-	20594-00-7	NI25557
112	170	140	111	87	71	69	32	**257**	100	24	23	13	11	9	7	6	**0.00**	10	15	5	3	–	–	–	–	–	–	–	–	–	Cytidine, 3′-O-methyl-	20594-00-7	LI05281
155	112	126	125	185	111	222	197	**257**	100	99	92	67	30	21	15	15	**0.00**	10	15	5	3	–	–	–	–	–	–	–	–	–	D-Mannitol, 1-(4-amino-2-oxo-1(2H)-pyrimidinyl)-3,6-anhydro-1-deoxy-	35823-84-8	NI25558
43	68	58	41	110	85	42	69	**257**	100	83	47	36	24	23	23	22	**4.20**	10	19	1	5	1	–	–	–	–	–	–	–	–	1,3,5-Triazine-2,4-diamine, N,N′-diisopropyl-6-(methylsulphinyl)-	55702-48-2	NI25559
51	87	222	143	89	52	224	104	**257**	100	84	83	37	30	28	27	26	**14.33**	11	9	2	1	–	–	2	–	–	–	–	–	–	Carbamic acid, (3-chlorophenyl)-, 4-chloro-2-butynyl ester	101-27-9	NI25560
222	51	87	143	153	104	224	69	**257**	100	74	67	40	39	39	34	29	**23.06**	11	9	2	1	–	–	2	–	–	–	–	–	–	Carbamic acid, (3-chlorophenyl)-, 4-chloro-2-butynyl ester	101-27-9	NI25562
51	222	87	143	52	224	104	99	**257**	100	90	78	39	31	29	28	26	**20.90**	11	9	2	1	–	–	2	–	–	–	–	–	–	Carbamic acid, (3-chlorophenyl)-, 4-chloro-2-butynyl ester	101-27-9	NI25561
157	115	50	36	156	44	142	128	**257**	100	45	35	34	31	30	28	25	**0.00**	11	12	4	1	–	–	1	–	–	–	–	–	–	Quinolinium, 1,2-dimethyl-, perchlorate	20729-86-6	NI25563

MASS TO CHARGE RATIOS								M.W.	INTENSITIES								Parent	C	H	O	N	S	F	Cl	Br	I	Si	P	B	X	COMPOUND NAME	CAS Reg No	No
176	257	94	148	255	202	122	65	**257**	100	58	41	32	30	30	26	24		11	15	1	1	–	–	–	–	–	–	–	–	1	Propanamide, seleno-, N-(4-ethoxyphenyl)-		MS05651
194	132	119	146	76	91	74	120	**257**	100	41	33	31	28	24	24	22	**4.10**	11	15	2	1	2	–	–	–	–	–	–	–	–	Ethanamine, 2-methylsulphinyl-2-methylthio-N-salicylidene-		NI25564
120	108	134	178	257	65	121	93	**257**	100	50	37	31	19	19	16	14		11	15	4	1	1	–	–	–	–	–	–	–	–	1,5-Benzoxazepin-3-ol, 5-methylsulphonyl-3-methyl-2,3,4,5-tetrahydro-		LI05282
121	200	137	15	91	58	77	28	**257**	100	66	61	60	50	39	34	31	**0.70**	11	15	4	1	1	–	–	–	–	–	–	–	–	Carbamic acid, methyl-, 4-(methylsulphonyl)-3,5-xylyl ester	2179-25-1	NI25565
193	91	155	108	58	65	136	86	**257**	100	80	33	27	22	21	14	14	**0.00**	11	15	4	1	1	–	–	–	–	–	–	–	–	Carbamic acid, methyl(p-tolylsulphonyl)-, ethyl ester	56805-37-9	NI25566
91	108	45	155	43	65	151	107	**257**	100	99	55	50	49	20	13	12	**2.00**	11	15	4	1	1	–	–	–	–	–	–	–	–	Carbamic acid, (p-tolylsulphonyl)-, isopropyl ester	18303-02-1	NI25567
91	108	45	155	43	65	151	107	**257**	100	99	58	54	53	27	17	16	**2.00**	11	15	4	1	1	–	–	–	–	–	–	–	–	Carbamic acid, (p-tolylsulphonyl)-, isopropyl ester	18303-02-1	LI05283
91	155	43	108	197	193	57	65	**257**	100	76	54	44	30	28	28	23	**0.00**	11	15	4	1	1	–	–	–	–	–	–	–	–	Carbamic acid, (p-tolylsulphonyl)-, propyl ester	18303-01-0	NI25568
91	155	108	197	193	107	257		**257**	100	76	44	30	28	10	3			11	15	4	1	1	–	–	–	–	–	–	–	–	Carbamic acid, (p-tolylsulphonyl)-, propyl ester	18303-01-0	LI05284
211	257	183	212	29	166	66	139	**257**	100	78	49	43	37	35	30	23		11	15	4	1	1	–	–	–	–	–	–	–	–	2,4-Thiophenedicarboxylic acid, 5-amino-3-methyl-, diethyl ester	4815-30-9	NI25569
222	223	165	250	224	193	166	87	**257**	100	11	7	4	1	1	1	1	**0.00**	11	16	2	3	–	–	1	–	–	–	–	–	–	Methanimidamide, N,N-dimethyl- N′-[3-[[(methylamino)carbonyl]oxy]phenyl]-, monohydrochloride	23422-53-9	NI25571
44	163	164	221	149	122	42	36	**257**	100	91	81	67	59	59	42	39	**0.00**	11	16	2	3	–	–	1	–	–	–	–	–	–	Methanimidamide, N,N-dimethyl- N′-[3-[[(methylamino)carbonyl]oxy]phenyl]-, monohydrochloride	23422-53-9	NI25570
44	163	164	221	36	122	149	42	**257**	100	77	76	62	47	46	42	31	**0.00**	11	16	2	3	–	–	1	–	–	–	–	–	–	Methanimidamide, N,N-dimethyl- N′-[3-[[(methylamino)carbonyl]oxy]phenyl]-, monohydrochloride	23422-53-9	NI25572
242	228	257	72	83	93	45	166	**257**	100	36	30	19	16	10	10	9		11	19	–	3	2	–	–	–	–	–	–	–	–	1,2-Diazine, 4-diethylamino-3,6-bis(methylthio)-5-methyl-		MS05652
72	100	29	44	27	28	45	42	**257**	100	96	91	83	56	53	52	51	**14.91**	12	16	1	1	1	–	1	–	–	–	–	–	–	Carbamothioic acid, diethyl-, S-[(4-chlorophenyl)methyl] ester	28249-77-6	NI25573
81	176	198	86	257	134	91	60	**257**	100	29	23	23	20	19	19	18		12	19	3	1	1	–	–	–	–	–	–	–	–	L-Cysteine, N-acetyl-S-(2-cyclohexenyl)-, methyl ester	77549-16-7	NI25574
70	71	187	72	257	28	188	118	**257**	100	27	25	12	11	8	7	5		12	24	1	3	–	–	–	–	–	–	1	–	–	Phosphine, tripyrrolidino-, oxide	6415-07-2	NI25575
257	256	174	77	83	258	229	147	**257**	100	97	39	25	24	19	19	18		13	11	1	3	1	–	–	–	–	–	–	–	–	5H-1,3,4-Thiadiazolo[3,2-a]pyrimidin-5-one, 2-ethyl-7-phenyl-	41837-57-4	NI25576
242	167	257	243	166	127.5	241	121	**257**	100	25	15	15	7	6	4	4		13	12	–	1	–	–	–	–	–	–	–	–	1	Phenarsazine, 5,10-dihydro-10-methyl-		LI05285
257	259	31	258	229	201	139	36	**257**	100	34	25	18	17	12	11	11		14	8	2	1	–	–	1	–	–	–	–	–	–	9,10-Anthracenedione, 1-amino-4-chloro-		IC03367
127	257	102	146	129	259	258	128	**257**	100	76	74	36	35	25	8	6		14	8	2	1	–	–	1	–	–	–	–	–	–	Benzoic acid, 4-cyano-, 3-chlorophenyl ester	53327-09-6	NI25577
257	127	146	259	102	129	258	128	**257**	100	88	48	38	36	30	15	10		14	8	2	1	–	–	1	–	–	–	–	–	–	Benzoic acid, 4-cyano-, 4-chlorophenyl ester	32792-59-9	NI25578
222	76	223	104	50	75	257	194	**257**	100	18	13	6	6	4	3	3		14	8	2	1	–	–	1	–	–	–	–	–	–	Phthalimide, N-(2-chlorophenyl)-		MS05653
257	76	213	259	104	178	258	50	**257**	100	43	37	33	23	18	15	13		14	8	2	1	–	–	1	–	–	–	–	–	–	Phthalimide, N-(3-chlorophenyl)-		MS05654
257	76	213	259	104	178	258	215	**257**	100	42	41	33	21	18	15	13		14	8	2	1	–	–	1	–	–	–	–	–	–	Phthalimide, N-(4-chlorophenyl)-		MS05655
119	256	91	258	64	120	89	63	**257**	100	41	17	14	10	9	9	8	**7.00**	14	10	1	2	–	–	1	–	–	–	–	–	–	1,2,4-Oxadiazole, 5-(2-chlorophenyl)-3-phenyl-		NI25579
119	256	91	258	64	120	111	75	**257**	100	47	18	16	12	9	9	9	**8.00**	14	10	1	2	–	–	1	–	–	–	–	–	–	1,2,4-Oxadiazole, 5-(3-chlorophenyl)-3-phenyl-		NI25580
119	256	91	258	64	75	111	63	**257**	100	47	18	17	12	10	9	9	**8.00**	14	10	1	2	–	–	1	–	–	–	–	–	–	1,2,4-Oxadiazole, 5-(4-chlorophenyl)-3-phenyl-		NI25581
198	43	199	186	167	215	140	257	**257**	100	27	18	16	14	9	9	2		14	11	2	1	1	–	–	–	–	–	–	–	–	10H-Phenothiazine, 10-acetyl-, 5-oxide	1217-37-4	LI05286
198	43	199	186	167	140	215	166	**257**	100	30	20	18	15	10	9	9	**3.00**	14	11	2	1	1	–	–	–	–	–	–	–	–	10H-Phenothiazine, 10-acetyl-, 5-oxide	1217-37-4	NI25582
198	43	199	215	186	167	166	140	**257**	100	27	18	15	15	14	8	8	**1.81**	14	11	2	1	1	–	–	–	–	–	–	–	–	10H-Phenothiazine, 10-acetyl-, 5-oxide	1217-37-4	MS05656
121	255	77	51	63	256	122	64	**257**	100	75	41	20	13	11	10	7	**3.00**	14	11	2	1	1	–	–	–	–	–	–	–	–	5-Thiazolidinone, 4-(2-furanylmethylene)-3-phenyl-	54986-57-1	NI25583
257	163	46	93	179	258	136		**257**	100	60	22	19	17	16	8			14	11	4	1	–	–	–	–	–	–	–	–	–	Benzeneacetic acid, 4-nitro-, phenyl ester	6335-82-6	NI25584
166	167	257	122	46				**257**	100	8	3	2	2					14	11	4	1	–	–	–	–	–	–	–	–	–	Benzoic acid, 4-nitro-, benzyl ester	14786-27-7	NI25585
199	43	127	155	143	59	200	171	**257**	100	79	54	40	32	32	27	21	**1.06**	14	11	4	1	–	–	–	–	–	–	–	–	–	Benzonitrile, 4-[(2,2-dimethyl-4,6-dioxo-1,3-dioxan-5-ylidene)methyl]-	53942-70-4	NI25586
91	83	92	65	257	56	41	39	**257**	100	12	11	10	6	6	6	5		14	15	2	3	–	–	–	–	–	–	–	–	–	Propaneitrile, 3-(3-benzyl-2,5-dioxo-6-piperazinyl)-, (3S,6S)-		MS05657
170	110	171	96	98	111	70	86	**257**	100	96	54	47	43	42	41	38	**15.00**	14	27	3	1	–	–	–	–	–	–	–	–	–	L-Proline, 4-isopropoxy-1-isopropyl-, isopropyl ester, trans-	56771-71-2	MS05658
157	57	85	74	257	105	128	101	**257**	100	45	35	12	10	10	9	9		14	27	3	1	–	–	–	–	–	–	–	–	–	Sylvamide	83905-60-6	NI25587
30	43	73	72	41	55	184	60	**257**	100	50	37	35	32	26	18	18	**4.00**	14	27	3	1	–	–	–	–	–	–	–	–	–	Undecanoic acid, 11-(N-acetylamino)-, methyl ester		LI05287
131	103	257	77	127	132	259	51	**257**	100	26	23	15	11	9	8	5		15	12	1	1	–	–	1	–	–	–	–	–	–	Propenanilide, 4′-chloro-3-phenyl-		TR00286
61	77	194	51	180	165	211	257	**257**	100	26	15	15	14	7	5	3		15	15	1	1	1	–	–	–	–	–	–	–	–	Benzophenone, O-[(methylthio)methyl]oxime	25056-49-9	NI25588
61	194	91	77	165	180	105	257	**257**	100	21	18	13	11	8	7	6		15	15	1	1	1	–	–	–	–	–	–	–	–	Methylamine, N-(diphenylmethylene)-1-(methylthio)-, N-oxide		LI05288
77	136	257	107	108	78	123	94	**257**	100	73	40	23	20	20	15	4		15	15	3	1	–	–	–	–	–	–	–	–	–	Acetamide, N-(4-methoxyphenyl)-2-phenoxy-		MS05659
200	239	229	242	257				**257**	100	62	43	16	13					15	15	3	1	–	–	–	–	–	–	–	–	–	(2,4-Dioxo-1,2,3,4-tetrahydroquinoline)-3-spiro-2′-(4′,5′-dimethyl-3′,6′-dihydro-2′H-pyran)		LI05289
189	188	82	257	41	70	53	67	**257**	100	64	63	56	45	30	25	24		15	15	3	1	–	–	–	–	–	–	–	–	–	4,7-Epoxy-1H-isoindole-1,3(2H)-dione, hexahydro-3a-methyl-2-phenyl-, [3aR-(3aα,4β,7β,7aα)]-	54382-61-5	NI25589
110	95	257	81	77	67	53	91	**257**	100	54	50	29	21	17	16	14		15	15	3	1	–	–	–	–	–	–	–	–	–	1H-Isoindole-1,3(2H)-dione, 3a,4,5,7a-tetrahydro-4-hydroxy-3a-methyl-2-phenyl-, [3aR-(3aα,4β,7aα)]-	54346-13-3	NI25590
95	110	257	77	96	93	91	67	**257**	100	77	24	18	15	13	13	12		15	15	3	1	–	–	–	–	–	–	–	–	–	1H-Isoindole-1,3(2H)-dione, 3a,6,7,7a-tetrahydro-7-hydroxy-3a-methyl-2-phenyl-, [3aS-(3aα,7β,7aα)]-	54346-14-4	NI25591
78	56	77	65	174	55	57	115	**257**	100	27	25	19	18	17	16	14	**9.61**	15	15	3	1	–	–	–	–	–	–	–	–	–	Noroxocrinine		NI25592
212	213	122	198	91	53	44	65	**257**	100	72	36	36	23	22	18	15	**1.08**	15	15	3	1	–	–	–	–	–	–	–	–	–	1H-Pyrrole-2-acetic acid, 1-methyl-5-(4-methylbenzoyl)-	26171-23-3	NI25593

MASS TO CHARGE RATIOS								M.W.	INTENSITIES								Parent	C	H	O	N	S	F	Cl	Br	I	Si	P	B	X	COMPOUND NAME	CAS Reg No	No
137	213	214	197	115	198	91	41	**257**	100	62	43	28	22	17	17	15	**5.00**	15	19	1	3	–	–	–	–	–	–	–	–	–	Cycloheptanone, 2-benzylidene-, semicarbazone	17026-14-1	NI25594
178	80	152	106	79	257	91	179	**257**	100	63	35	29	24	18	17	15		16	19	2	1	–	–	–	–	–	–	–	–	–	2-Azabicyclo[2.2.2]oct-5-ene-2-carboxylic acid, 3-phenyl-, ethyl ester	3693-55-8	MS05660
105	77	70	42	257	51	41	106	**257**	100	93	33	24	21	13	12	10		16	19	2	1	–	–	–	–	–	–	–	–	–	1-Azaspiro[3.5]nonan-2-one, 1-benzoyl-5-methyl-	77064-43-8	NI25595
105	77	229	257	122	214	108	106	**257**	100	38	18	18	17	15	10	10		16	19	2	1	–	–	–	–	–	–	–	–	–	1-Azaspiro[3.5]nonan-2-one, 1-benzoyl-6-methyl-	77064-44-9	NI25596
121	150	77	106	51	91	42	30	**257**	100	74	50	29	24	23	19	17	**1.00**	16	19	2	1	–	–	–	–	–	–	–	–	–	Benzeneethanamine, β-hydroxy-N-(3-methoxybenzyl)-		MS05661
256	228	229	201	257	255	202	185	**257**	100	42	32	19	18	17	16	15		16	19	2	1	–	–	–	–	–	–	–	–	–	Crinan	510-70-3	NI25597
257	256	201	202	185	228	214	200	**257**	100	22	21	17	16	15	13	10		16	19	2	1	–	–	–	–	–	–	–	–	–	Crinan	510-70-3	NI25598
97	257	240	122	256	83	96	82	**257**	100	96	83	54	44	41	34	30		16	19	2	1	–	–	–	–	–	–	–	–	–	Elaeocarpine		LI05290
95	163	257	120	96	134	77	135	**257**	100	21	19	8	7	6	6	5		16	19	2	1	–	–	–	–	–	–	–	–	–	4H-5,7a-Epoxyisoindoline, N-(4-methoxyphenyl)-5-methyl-		NI25599
143	161	55	89	144	257	115	116	**257**	100	67	51	37	30	26	23	11		16	19	2	1	–	–	–	–	–	–	–	–	–	2-Indolecarboxylic acid, cyclohexanemethyl ester	86328-79-2	NI25600
97	257	81	122	240	83	96	256	**257**	100	80	50	43	38	30	27	22		16	19	2	1	–	–	–	–	–	–	–	–	–	(+)-Isoelaeocarpine		LI05291
105	77	229	122	201	106	186	200	**257**	100	47	31	14	10	10	9	9	**3.09**	16	19	2	1	–	–	–	–	–	–	–	–	–	1,3-Oxazin-6-one, 4,5-dihydro-4,4-hexamethylene-2-phenyl-		NI25601
105	77	56	40	55	106	103	78	**257**	100	78	52	38	31	9	5	5	**3.19**	16	19	2	1	–	–	–	–	–	–	–	–	–	1,3-Oxazin-6-one, 4,5-dihydro-5-methyl-4,4-pentamethylene-2-phenyl-		NI25602
107	104	91	105	149	150	77	257	**257**	100	77	77	48	32	26	18	11		16	19	2	1	–	–	–	–	–	–	–	–	–	Phenylpropanoic acid, 6-cyanohexanyl ester		LI05292
58	108	183	214	199	134	186	107	**257**	100	15	11	9	8	8	7	5	**3.40**	16	20	–	1	–	–	–	–	–	–	1	–	–	Phosphine, 2-[(dimethylamino)ethyl]diphenyl-		IC03368
147	257	146	67	256	41	78	80	**257**	100	99	66	53	47	40	39	34		16	29	–	1	–	–	–	–	–	–	–	2	–	9-Borabicyclo[3.3.1]nonan-9-amine, N-9-borabicyclo[3.3.1]non-9-yl-	74421-04-8	NI25603
256	257	229	76	84	86	153	103	**257**	100	64	15	15	14	7	5	5		17	11	–	3	–	–	–	–	–	–	–	–	–	Quinoxaline, 2-[2-(4-cyanophenyl)ethenyl]-, (E)-		MS05662
166	148	105	44	123	167	103	57	**257**	100	63	44	31	30	13	9	9	**3.75**	17	20	–	1	–	1	–	–	–	–	–	–	–	Benzeneethanamine, 3-fluoro-N-methyl-N-(2-phenylethyl)-	52059-54-8	NI25604
148	166	105	123	42	103	77	44	**257**	100	84	67	52	35	24	22	19	**3.46**	17	20	–	1	–	1	–	–	–	–	–	–	–	Benzeneethanamine, 4-fluoro-N-methyl-N-(2-phenylethyl)-	52059-52-6	NI25605
105	186	257	77	124	82	149	242	**257**	100	76	36	35	23	17	16	15		17	23	1	1	–	–	–	–	–	–	–	–	–	Bicyclo[2.2.1]heptane, 1-benzamido-2,3,3-trimethyl-		LI05293
124	229	110	125	41	82	80	98	**257**	100	38	38	23	21	16	16	14	**5.80**	17	23	1	1	–	–	–	–	–	–	–	–	–	Cyclohexanone, 2-(piperidin-1-yl)-6-phenyl-		IC03369
45	136	128	115	44	41	257	46	**257**	100	11	8	8	8	8	7	7		17	23	1	1	–	–	–	–	–	–	–	–	–	Morphinan, 3-methoxy-	1531-25-5	NI25606
59	150	157	200	57	55	198	160	**257**	100	46	20	18	16	12	11	10	**0.00**	17	23	1	1	–	–	–	–	–	–	–	–	–	Morphinan-3-ol, 17-methyl-	77-07-6	NI25607
257	150	256	59	157	200	42	44	**257**	100	98	67	53	48	44	39	25		17	23	1	1	–	–	–	–	–	–	–	–	–	Morphinan-3-ol, 17-methyl-	77-07-6	NI25608
59	42	150	157	44	256	200	58	**257**	100	59	32	22	17	14	14	13	**12.10**	17	23	1	1	–	–	–	–	–	–	–	–	–	Morphinan-3-ol, 17-methyl-, (9α,13α,14α)-	125-73-5	NI25609
72	141	257	115	73	43	56	41	**257**	100	14	10	6	6	6	4	3		17	23	1	1	–	–	–	–	–	–	–	–	–	Naphthalenebutylamine, α-hydroxy-N-isopropyl-		MS05663
105	136	77	107	29	122	257	28	**257**	100	66	30	28	25	19	17	17		17	23	1	1	–	–	–	–	–	–	–	–	–	4H-1,3-Oxazine, 5,6-dihydro-6-ethyl-6-methyl-2-phenyl-4,5-tetramethylene-		NI25610
180	77	257	165	51	256	258	181	**257**	100	98	79	39	36	29	17	15		19	15	–	1	–	–	–	–	–	–	–	–	–	Aniline, N-benzylidene-	574-45-8	NI25611
257	256	258	241	255	254	128.5	127	**257**	100	23	20	12	10	9	8	8		19	15	–	1	–	–	–	–	–	–	–	–	–	Benza[a]acridine, 1,5-dimethyl-		MS05664
257	256	258	255	240	239	254	128.5	**257**	100	29	20	16	16	15	12	10		19	15	–	1	–	–	–	–	–	–	–	–	–	Benza[a]acridine, 5,9-dimethyl-		MS05665
257	258	256	128.5	241	113.5	254	126	**257**	100	20	10	8	7	7	6	6		19	15	–	1	–	–	–	–	–	–	–	–	–	Benza[c]acridine, 5,10-dimethyl-		MS05666
257	258	128.5	256	240	127	120.5	127.5	**257**	100	21	21	14	14	14	13	12		19	15	–	1	–	–	–	–	–	–	–	–	–	Benza[c]acridine, 7,8-dimethyl-		MS05667
257	242	258	256	121	241	254	129	**257**	100	23	21	20	17	14	13	10		19	15	–	1	–	–	–	–	–	–	–	–	–	Benza[c]acridine, 7,9-dimethyl-	963-89-3	NI25612
257	258	256	242	241	254	128.5	255	**257**	100	21	13	11	9	8	8	7		19	15	–	1	–	–	–	–	–	–	–	–	–	Benza[c]acridine, 7,10-dimethyl-		MS05668
179	258	129	189	239	158	69		**258**	100	16	16	6	2	2	2			2	–	–	–	–	4	–	2	–	–	–	–	–	Ethane, 1,1-dibromo-1,2,2,2-tetrafluoro-	27336-23-8	NI25613
179	181	129	131	100	31	260	50	**258**	100	97	34	33	17	13	12	8	**6.42**	2	–	–	–	–	4	–	2	–	–	–	–	–	Ethane, 1,2-dibromo-1,1,2,2-tetrafluoro-	124-73-2	DO01047
179	181	31	129	131	100	50	69	**258**	100	96	92	61	59	37	32	18	**1.90**	2	–	–	–	–	4	–	2	–	–	–	–	–	Ethane, 1,2-dibromo-1,1,2,2-tetrafluoro-	124-73-2	NI25614
179	181	69	31	100	131	129	81	**258**	100	99	95	76	52	36	36	26	**0.00**	2	–	–	–	–	4	–	2	–	–	–	–	–	Ethane, 1,2-dibromo-1,1,2,2-tetrafluoro-	124-73-2	IC03370
180	182	117	144	222	119	224	184	**258**	100	68	36	29	24	24	15	12	**2.50**	2	6	2	2	1	–	3	–	–	–	1	–	–	Phosphoryl chloride, (dimethylamino)imidosulphuroxychloro-		LI05294
181	179	143	225	183	227	223	141	**258**	100	91	79	63	63	53	39	34	**0.00**	3	3	–	–	–	–	4	1	–	–	–	–	–	Butane, 1-bromo-2,4,4-tetrachloro-		LI05295
99	57	180	178	82	80	55	72	**258**	100	88	77	76	63	62	60	36	**0.00**	3	4	–	2	1	–	–	2	–	–	–	–	–	2-Thiazolamine, 5-bromo-, monohydrobromide	61296-22-8	NI25615
101	129	181	179	209	207	141	221	**258**	100	47	33	33	25	23	12	8	**0.00**	3	4	–	2	1	–	–	2	–	–	–	–	–	2-Thiazolamine, 5-bromo-, monohydrobromide	61296-22-8	NI25616
225	227	223	190	260	188	118	141	**258**	100	65	63	43	36	32	30	29	**19.22**	4	–	–	–	–	–	6	–	–	–	–	–	–	1,3-Butadiene, 1,1,2,3,4,4-hexachloro-	87-68-3	NI25617
225	227	223	260	190	262	188	141	**258**	100	64	63	43	39	34	30	26	**22.40**	4	–	–	–	–	–	6	–	–	–	–	–	–	1,3-Butadiene, 1,1,2,3,4,4-hexachloro-	87-68-3	IC03371
225	227	223	190	260	188	118	262	**258**	100	57	50	38	36	33	30	28	**19.30**	4	–	–	–	–	–	6	–	–	–	–	–	–	1,3-Butadiene, 1,1,2,3,4,4-hexachloro-	87-68-3	NI25618
59	164	117	157	153	129	199	188	**258**	100	38	17	14	11	9	5	4	**0.00**	4	3	2	–	–	–	5	–	–	–	–	–	–	Propanoic acid, pentachloro-, methyl ester	813-46-7	NI25619
243	105	91	121	213	107	195	109	**258**	100	61	59	53	36	34	21	20	**0.03**	4	12	3	–	–	–	–	–	–	–	–	–	2	Arsinic acid, dimethyl-, anhydride	56298-86-3	NI25620
43	81	53	260	258	179	262	181	**258**	100	50	43	40	24	24	20	14		5	5	1	–	–	–	2	1	–	1	–	–	–	1-Oxa-2-silacyclo-3,5-hexadiene, 3-bromo-2,2-dichloro-6-methyl-		NI25621
59	45	122	258	193	136	117	103	**258**	100	46	21	19	19	19	15	15		5	6	–	–	6	–	–	–	–	–	–	–	–	2,4,6,8,9,10-Hexathiaadamantane, 1-methyl-	13847-89-7	NI25622
106	108	55	29	181	179	187	27	**258**	100	97	84	66	44	43	37	33	**0.00**	5	8	2	–	–	–	–	2	–	–	–	–	–	Propanoic acid, 2,3-dibromo-, ethyl ester	3674-13-3	NI25623
75	243	73	116	45	117	43	76	**258**	100	77	53	23	18	9	9	8	**6.01**	5	11	2	–	–	–	–	–	1	1	–	–	–	Acetic acid, iodo-, trimethylsilyl ester	55030-41-6	NI25624
223	159	92	75	50	76	28	64	**258**	100	91	66	59	51	31	30	28	**11.13**	6	4	4	–	2	1	1	–	–	–	–	–	–	Benzenesulphonyl fluoride, 4-(chlorosulphonyl)-	30672-72-1	NI25625
258	228	67	91	119	92	30	52	**258**	100	43	28	21	18	17	15	13		6	6	6	6	–	–	–	–	–	–	–	–	–	1,3,5-Benzenetriamine, 2,4,6-trinitro-	67539-61-1	NI25626

MASS TO CHARGE RATIOS								M.W.	INTENSITIES								Parent	C	H	O	N	S	F	Cl	Br	I	Si	P	B	X	COMPOUND NAME	CAS Reg No	No
97	98	43	28	111	99	45	58	**258**	100	55	15	15	8	8	8	7	**0.00**	6	9	–	–	1	6	–	–	–	–	1	–	–	Thiophen-1-ium, 1,3-dimethyl-, hexafluorophosphate		LI05296
197	107	89	62	137	258	66	196	**258**	100	80	50	47	40	28	26	24		6	15	–	–	3	–	–	–	–	–	–	–	1	Arsenotrithious acid, triethyl ester	34666-79-0	NI25627
260	179	258	74	75	144	181	69	**258**	100	88	78	48	47	36	29	27		7	3	–	–	–	3	1	1	–	–	–	–	–	Benzene, 4-bromo-1-chloro-2-(trifluoromethyl)-	445-01-2	NI25628
133	260	258	202	200	106	39	94	**258**	100	84	81	64	60	34	33	32		7	3	2	2	1	–	–	1	–	–	–	–	–	Thieno[2,3-b]pyridine, 3-bromo-2-nitro-	53399-40-9	NI25629
260	258	133	106	202	200	214	212	**258**	100	95	49	35	27	27	18	17		7	3	2	2	1	–	–	1	–	–	–	–	–	Thieno[3,2-c]pyridine, 3-bromo-2-nitro-	28783-15-5	NI25630
77	76	65	30	166	120	52	28	**258**	100	95	69	67	37	27	27	27	**24.00**	7	6	7	4	–	–	–	–	–	–	–	–	–	o-Anisidine, 3,4,5-trinitro-		IC03372
258	230	172	202	144	103	131	28	**258**	100	75	72	70	64	64	60	56		7	7	4	–	–	–	–	–	–	–	–	–	1	Rhodium, dicarbonyl(2,4-pentanedionato-O,O')-	14874-82-9	NI25631
61	62	123	159	125	89	187	161	**258**	100	93	92	90	87	87	83	83	**16.79**	8	6	1	–	–	–	4	–	–	–	–	–	–	Benzene, 1-(chloromethyl)-2-methoxy-trichloro-	74312-97-3	NI25632
232	230	234	260	29	258	262	27	**258**	100	80	48	17	14	13	8	8		8	6	1	–	–	–	4	–	–	–	–	–	–	Benzene, 1,2,3,5-tetrachloro-4-ethoxy-	56818-02-1	MS05669
258	131	63	90	259	104	129	62	**258**	100	40	20	15	10	10	9	9		8	7	–	2	–	–	–	–	1	–	–	–	–	Benzimidazole, 5-iodo-2-methyl-	2818-70-4	NI25633
197	139	171	199	111	121	143	125	**258**	100	81	63	53	41	35	28	26	**0.00**	8	19	3	–	2	–	–	–	–	–	1	–	–	Phosphorothioic acid, O,O-diethyl O-[2-(ethylthio)ethyl] ester	298-03-3	NI25634
88	89	60	61	171	29	97	115	**258**	100	62	35	25	14	14	12	11	**2.28**	8	19	3	–	2	–	–	–	–	–	1	–	–	Phosphorothioic acid, O,O-diethyl O-[2-(ethylthio)ethyl] ester	298-03-3	NI25636
88	60	89	126	115	61	114	170	**258**	100	38	21	14	12	12	11	10	**3.84**	8	19	3	–	2	–	–	–	–	–	1	–	–	Phosphorothioic acid, O,O-diethyl O-[2-(ethylthio)ethyl] ester	298-03-3	NI25635
88	89	60	61	171	115	97	28	**258**	100	57	35	25	17	10	10	10	**2.04**	8	19	3	–	2	–	–	–	–	–	1	–	–	Phosphorothioic acid, O,O-diethyl S-[2-(ethylthio)ethyl] ester	126-75-0	NI25637
89	88	91	90	171	143	139	199	**258**	100	33	15	10	8	5	5	4	**0.00**	8	19	3	–	2	–	–	–	–	–	1	–	–	Phosphorothioic acid, O,O-diethyl S-[2-(ethylthio)ethyl] ester	126-75-0	NI25638
69	93	121	185	214	111	82	47	**258**	100	78	68	28	26	12	7	7	**1.00**	8	20	5	–	–	–	–	–	–	–	2	–	–	Diphosphorous acid, tetraethyl ester	21646-99-1	NI25639
258	161	230	180	148	179	58	79	**258**	100	94	70	63	39	29	29	19		9	1	1	–	–	7	–	–	–	–	–	–	–	2-Indanone, 1,1,3,4,5,6,7-heptafluoro-		NI25640
148	258	260	149	53	52	83	67	**258**	100	39	25	17	15	14	7	7		9	8	1	4	–	–	2	–	–	–	–	–	–	Pyrazolo[5,1-c]-as-triazine, 6-(dichloroacetyl)-4,6-dihydro-3-methyl-4-methylene-	6763-75-3	NI25641
70	230	69	215	42	258	43	41	**258**	100	36	23	18	15	8	7	7		9	9	–	6	–	3	–	–	–	–	–	–	–	Pyrimido[5,4-e]-1,2,4-triazin-7-amine, N,N,3-trimethyl-5-(trifluoromethyl)-	32709-27-6	NI25642
61	46	258	91	260	32	172	170	**258**	100	22	14	14	13	13	10	10		9	11	2	2	–	–	–	1	–	–	–	–	–	Urea, N'-(4-bromophenyl)-N-methoxy-N-methyl-	3060-89-7	NI25643
259	261	146	85	260	127	171	73	**258**	100	95	25	16	12	10	9	9	**1.40**	9	11	2	2	–	–	–	1	–	–	–	–	–	Urea, N'-(4-bromophenyl)-N-methoxy-N-methyl-	3060-89-7	NI25644
58	41	90	61	170	28	172	56	**258**	100	23	10	10	8	8	7	7	**6.88**	9	11	2	2	–	–	–	1	–	–	–	–	–	Urea, N'-(4-bromophenyl)-N-methoxy-N-methyl-	3060-89-7	NI25645
95	84	69	44	43	141	41	40	**258**	100	51	50	39	34	33	32	31	**8.00**	9	13	2	–	–	–	3	–	–	–	–	–	–	1,3-Dioxolane, 5,5-dimethyl-4-isopropylidene-2-trichloromethyl-		LI05297
126	109	258	110	73	57	133	127	**258**	100	29	22	12	9	9	8	8		9	14	5	4	–	–	–	–	–	–	–	–	–	5-Imidazolecarboxamide, 4-amino-, riboside		NI25646
202	230	258	28	172	146	144	43	**258**	100	76	56	55	48	41	41	35		9	15	2	–	–	–	–	–	–	–	–	–	1	Rhodium, bis(η^2-ethene)(2,4-pentanedionato-O,O')-	12082-47-2	NI25647
86	43	216	58	129	44	130	59	**258**	100	83	68	47	40	39	34	33	**14.10**	9	18	3	6	–	–	–	–	–	–	–	–	–	2,4,6-Triazinetrione, 1,2,3,4,5,6-hexahydro-1,3,5-tris(dimethylamino)-	67500-31-6	NI25648
173	223	175	225	101	137	102	75	**258**	100	91	65	29	26	22	21	15	**7.01**	10	8	2	2	–	–	2	–	–	–	–	–	–	Sydnone, 3-(2,4-dichlorophenethyl)-	26574-35-6	NI25649
66	193	65	39	129	130	128	115	**258**	100	69	68	37	29	26	19	19	**16.55**	10	10	–	–	–	2	–	–	–	–	–	–	1	Zirconium, bis(cyclopentadienyl)difluoro-		NI25650
131	91	258	129	115	103	77	51	**258**	100	30	25	20	20	17	15	15		10	11	–	–	–	–	–	–	1	–	–	–	–	Naphthalene, 1,2,3,4-tetrahydro-5-iodo-	56804-95-6	NI25651
131	258	91	129	128	115	103	77	**258**	100	93	36	28	25	22	22	19		10	11	–	–	–	–	–	–	1	–	–	–	–	Naphthalene, 1,2,3,4-tetrahydro-6-iodo-	56804-94-5	NI25652
165	77	121	166	63	51	122	91	**258**	100	11	8	8	7	6	5	5	**2.41**	10	11	3	–	–	–	–	1	–	–	–	–	–	Acetophenone, α-bromo-2,4-dimethoxy-	60965-26-6	NI25653
216	174	258	109	218	217	198	229	**258**	100	75	53	14	13	12	12	11		10	14	–	2	3	–	–	–	–	–	–	–	–	4-Isothiazolecarbonitrile, 3,5-bis(propylthio)-	24135-14-6	NI25654
258	214	130	85	72	58	129	187	**258**	100	99	80	74	20	19	15	14		10	14	2	2	2	–	–	–	–	–	–	–	–	1,3-Dithietane-2,4-dicarboxamide, N,N'-diethyl-2,4-dimethylene-	55030-39-2	LI05298
214	258	130	85	72	58	259	129	**258**	100	99	83	75	21	20	15	15		10	14	2	2	2	–	–	–	–	–	–	–	–	1,3-Dithietane-2,4-dicarboxamide, N,N'-diethyl-2,4-dimethylene-	55030-39-2	NI25655
258	214	130	85	187	186			**258**	100	95	83	75	9	8				10	14	2	2	2	–	–	–	–	–	–	–	–	1,3-Dithietane-2,4-dicarboxamide, N,N'-diethyl-2,4-dimethylene-	55030-39-2	LI05299
258	214	187	143	85	71	116	130	**258**	100	90	80	46	32	22	20	19		10	14	2	2	2	–	–	–	–	–	–	–	–	1,3-Dithiole-4,5-dicarboxamide, N,N-diethyl-2-methylene-		LI05300
258	214	187	143	85	128	71	116	**258**	100	89	77	45	30	25	23	20		10	14	2	2	2	–	–	–	–	–	–	–	–	1,3-Dithiole-4,5-dicarboxamide, N,N'-diethyl-2-methylene-	55030-40-5	NI25656
58	77	85	117	141	41	43	110	**258**	100	69	40	31	30	21	19	18	**3.00**	10	14	2	2	2	–	–	–	–	–	–	–	–	Urea, 1-(phenylsulphonyl)-3-propyl-2-thio-	24539-89-7	NI25657
84	186	126	228	144	96	114	258	**258**	100	88	68	52	44	38	28	17		10	14	6	2	–	–	–	–	–	–	–	–	–	3,4-Piperidinediol, 1-nitroso-2,5-methylenedioxy-, diacetate		LI05301
126	69	127	28	73	42	57	44	**258**	100	75	56	55	35	35	32	24	**4.09**	10	14	6	2	–	–	–	–	–	–	–	–	–	Uridine, 3-methyl-	2140-69-4	NI25658
126	127	55	73	57	133	31	28	**258**	100	57	44	35	35	31	16	16	**12.71**	10	14	6	2	–	–	–	–	–	–	–	–	–	Uridine, 5-methyl-	1463-10-1	NI25659
87	146	147	45	27	71	57	113	**258**	100	21	20	17	16	14	13	11	**0.00**	10	14	6	2	–	–	–	–	–	–	–	–	–	Uridine, 2'-O-methyl-	2140-76-3	LI05302
87	146	147	45	28	71	57	100	**258**	100	22	19	16	16	13	12	11	**1.46**	10	14	6	2	–	–	–	–	–	–	–	–	–	Uridine, 2'-O-methyl-	2140-76-3	NI25660
87	113	71	147	88	85	69	45	**258**	100	34	34	27	24	11	11	11	**0.00**	10	14	6	2	–	–	–	–	–	–	–	–	–	Uridine, 3'-O-methyl-	6038-59-1	NI25661
169	171	245	243	179	73	90	89	**258**	100	97	78	77	65	62	46	35	**7.73**	10	15	1	–	–	–	–	1	–	1	–	–	–	Toluene, 4-bromo-α-[(trimethylsilyl)oxy]-	86605-93-8	NI25662
93	243	165	91	77	135	215	185	**258**	100	52	40	22	18	16	11	9	**8.75**	10	18	–	–	–	–	–	–	–	–	–	–	1	Stannane, bicyclo[2.2.1]hept-2-en-7-yltrimethyl-, anti-	34208-81-6	NI25663
165	185	215	135	230	150	120	243	**258**	100	49	20	19	16	13	13	8	**0.00**	10	18	–	–	–	–	–	–	–	–	–	–	1	Stannane, bicyclo[2.2.1]hept-2-en-7-yltrimethyl-, syn-	38573-93-2	NI25664
73	45	74	102	75	243	59	43	**258**	100	18	11	10	10	8	8	6	**0.00**	10	22	2	2	–	–	–	–	–	2	–	–	–	3-Isoxazolol, 5-(aminomethyl)-1,3-bis(trimethylsilyl)-		MS05670
258	73	243	114	143	259	116	86	**258**	100	79	56	27	24	23	22	19		10	22	2	2	–	–	–	–	–	2	–	–	–	2,5-Piperazinedione, 1,3-bis(trimethylsilyl)-	3553-95-5	NI25665
147	73	243	100	45	93	59	72	**258**	100	90	51	51	39	25	24	18	**6.62**	10	22	2	2	–	–	–	–	–	2	–	–	–	2,4(1H,3H)-Pyrimidinedione, dihydro-1,3-bis(trimethylsilyl)-	74810-47-2	NI25666
172	213	228	243	156	202			**258**	100	90	34	20	10	5			**0.00**	10	24	2	2	–	–	–	–	–	–	–	–	2	Aluminium, dimethylacetoximatodimethyldi-		LI05303
61	77	147	105	194	151	51	63	**258**	100	23	20	18	17	14	10	5	**0.00**	11	14	3	–	2	–	–	–	–	–	–	–	–	Propanophenone, β-[(methylthio)methyl]sulphonyl-	55030-38-1	NI25667
61	77	147	105	51	63	62	45	**258**	100	23	20	18	10	5	3	3	**0.00**	11	14	3	–	2	–	–	–	–	–	–	–	–	Propanophenone, β-[(methylthio)methyl]sulphonyl-	55030-38-1	LI05304
43	15	179	79	137	77	105	161	**258**	100	98	73	56	46	40	36	34	**16.61**	11	14	5	–	1	–	–	–	–	–	–	–	–	2,5-Benzofurandiol, 2,3-dihydro-3,3-dimethyl-, 5-methanesulfonate, (±)-	62059-52-3	NI25668

MASS TO CHARGE RATIOS								M.W.	INTENSITIES								Parent	C	H	O	N	S	F	Cl	Br	I	Si	P	B	X	COMPOUND NAME	CAS Reg No	No
149	161	77	147	79	179	91	105	**258**	100	63	46	42	36	33	30	18	**5.88**	11	14	5	–	1	–	–	–	–	–	–	–	–	2,5-Benzofurandiol, 2,3-dihydro-3,3-dimethyl-, 5-methanesulfonate, (±)-	62059-52-3	NI25669
68	85	43	156	97	86	69	157	**258**	100	80	62	30	19	16	16	9	**0.00**	11	14	7	–	–	–	–	–	–	–	–	–	–	2H-Pyran-2,3,5-triol, 5,6-dihydro-, triacetate	56248-06-7	NI25670
258	69	41	191	149	56	67	133	**258**	100	94	71	34	18	18	7	5		11	15	3	–	1	–	–	–	–	–	1	–	–	1,3,2-Dioxaphosphorinane, 5,5-dimethyl-2-phenoxy-, 2-sulphide	31951-84-5	NI25671
122	72	186	258	123	108	243	44	**258**	100	56	49	36	26	12	11	8		11	18	3	2	1	–	–	–	–	–	–	–	–	Benzenesulphonamide, 3-amino-N,N-diethyl-4-methoxy-		IC03373
84	18	140	30	55	83	41	28	**258**	100	55	29	22	17	15	14	14	**5.00**	11	18	5	2	–	–	–	–	–	–	–	–	–	5-Pyrrolidinecarboxylic acid, 1-(5-amino-5-carboxypentyl)-2-oxo-		NI25672
75	73	128	115	141	243	100	77	**258**	100	65	57	29	25	24	20	20	**5.00**	11	26	1	2	–	–	–	–	–	2	–	–	–	2-Piperidinone, 3-amino-, 1,3-bis(trimethylsilyl)-		MS05671
128	73	115	243	141	169	100	147	**258**	100	75	55	53	38	33	30	21	**10.00**	11	26	1	2	–	–	–	–	–	2	–	–	–	2-Piperidinone, 3-amino-, 1,3-bis(trimethylsilyl)-		MS05672
128	73	115	243	141	169	147	70	**258**	100	76	55	52	42	26	21	18	**8.00**	11	26	1	2	–	–	–	–	–	2	–	–	–	2-Piperidinone, 3-amino-, 1,3-bis(trimethylsilyl)-		LI05305
242	171	156	170	243	172	130	116	**258**	100	71	43	23	18	9	7	7	**0.00**	12	10	3	4	–	–	–	–	–	–	–	–	–	Benzo[g]pteridine-2,4(1H,3H)-dione, 1,3-dimethyl-, 5-oxide	2962-89-2	NI25673
43	57	258	41	56	69	42	215	**258**	100	94	88	81	79	56	47	42		12	10	3	4	–	–	–	–	–	–	–	–	–	Benzo[g]pteridine-2,4(3H,10H)-dione, 8-hydroxy-7,10-dimethyl-	37163-35-2	NI25674
105	166	43	77	107	106	79	51	**258**	100	98	75	50	40	35	18	17	**0.00**	12	15	4	–	–	–	1	–	–	–	–	–	–	Butanoic acid, 3-(chloromethyl)-2,3-dihydroxy-2-phenyl-, methyl ester		DD00743
145	258	59	167	225	147	146	118	**258**	100	23	23	20	14	10	9	9		12	18	–	–	3	–	–	–	–	–	–	–	–	2,6,9-Trithiabicyclo[3.3.1]nona-3,7-diene, 1,3,4,5,7,8-hexamethyl-, DL-	17879-07-1	NI25675
171	243	59	229	85	69	113	143	**258**	100	89	89	82	73	71	47	45	**0.00**	12	18	6	–	–	–	–	–	–	–	–	–	–	β-D-Arabino-hexos-2-ulo-2,6-pyranose, 2,3:4,5-bis-O-isopropylidene-	32786-02-0	MS05673
213	185	29	84	112	140	39	157	**258**	100	99	87	35	28	27	27	25	**25.00**	12	18	6	–	–	–	–	–	–	–	–	–	–	1,2,3-Cyclopropanetricarboxylic acid, triethyl ester		IC03374
43	142	101	143	116	59	111	69	**258**	100	85	57	49	45	45	36	25	**0.50**	12	18	6	–	–	–	–	–	–	–	–	–	–	Decanedioic acid, 3,8-dioxo-, dimethyl ester	55030-37-0	WI01308
215	55	229	81	258	69	59	101	**258**	100	75	19	16	10	10	10	9		12	18	6	–	–	–	–	–	–	–	–	–	–	1,4-Dioxaspiro[4.5]decane-2,3-dicarboxylic acid, dimethyl ester, (2R,3R)-		DD00744
113	172	171	157	71	140	55	199	**258**	100	99	76	45	43	41	39	38	**0.00**	12	18	6	–	–	–	–	–	–	–	–	–	–	6,10-Dioxaspiro[5.4]decane-2,3-dicarboxylic acid, dimethyl ester, cis-		DD00745
114	43	44	111	156	153	86	57	**258**	100	40	39	29	14	10	10	7	**0.00**	12	18	6	–	–	–	–	–	–	–	–	–	–	Hex-2-enopyranoside, ethyl 2,3-dideoxy-, diacetate	56196-39-5	NI25676
29	112	139	157	212	213	84	27	**258**	100	71	57	50	33	30	28	27	**1.86**	12	18	6	–	–	–	–	–	–	–	–	–	–	1-Propene-1,2,3-tricarboxylic acid, triethyl ester		IC03375
111	185	153	166	41	81	84	69	**258**	100	90	66	21	20	19	17	16	**1.00**	12	18	6	–	–	–	–	–	–	–	–	–	–	2H-Pyran-6,6-dicarboxylic acid, 5,6-dihydro-2-methoxy-, diethyl ester		LI05306
73	75	225	101	125	226	183	153	**258**	100	56	51	26	11	10	9	9	**0.00**	12	22	4	–	–	–	–	–	–	1	–	–	–	2,3-Dioxabicyclo[4.3.0]nonan-7-one, 1-hydroxy-6-methyl-4-[(trimethylsilyl)methyl]-		DD00746
44	87	86	43	114	72	29	41	**258**	100	43	35	10	8	6	6	5	**0.00**	12	22	4	2	–	–	–	–	–	–	–	–	–	L-Alanine, N-(N-acetyl-L-alanyl)-, butyl ester	55712-39-5	NI25677
72	129	69	59	45	73	130	44	**258**	100	54	42	42	42	27	20	19	**0.50**	12	22	4	2	–	–	–	–	–	–	–	–	–	Butanoic acid, 2-[[[(dimethylamino)carbonyl]methylamino]carbonyl]-2-ethyl-, methyl ester	42948-63-0	NI25678
142	143	202	170	227	144	127	113	**258**	100	83	44	36	22	16	15	13	**9.80**	12	22	4	2	–	–	–	–	–	–	–	–	–	Glycine, N-(2-amino-1,4-dioxononyl)-, methyl ester, (S)-	55030-36-9	NI25679
43	99	41	39	27	29	15	56	**258**	100	12	7	6	3	3	3	3	**0.00**	12	22	4	2	–	–	–	–	–	–	–	–	–	Hydrazine, N,N'-bis(2-methyl-4-oxopent-4-yl)-N,N'-dihydroxy-	94514-30-4	NI25680
72	129	97	101	157	214	227	154	**258**	100	54	17	13	10	9	4	4	**0.50**	12	22	4	2	–	–	–	–	–	–	–	–	–	Malonuric acid, N,N',N'-trimethyl-2,2-diethyl-, methyl ester		LI05307
111	42	44	43	110	96	129	113	**258**	100	69	56	45	38	33	31	31	**10.86**	12	22	4	2	–	–	–	–	–	–	–	–	–	Di-α-L-Xylofuranose 1,2':2,1'-dianhydride, 4,4',5,5'-tetradeoxy-4,4'-bis(methylamino)-	37679-76-8	NI25681
258	147	73	259	148	45	75	149	**258**	100	90	71	25	15	15	12	10		12	26	2	–	–	–	–	–	–	2	–	–	–	1-Cyclohexene, 1,2-bis[(trimethylsilyl)oxy]-	6838-67-1	NI25682
229	201	72	100	230	86	128	173	**258**	100	76	46	30	21	21	14	12	**0.00**	12	32	–	2	–	–	–	–	–	–	–	–	2	Aluminium, diethyl(N,N-dimethylamido)-, dimer		LI05308
229	201	72	100	230	86	128	158	**258**	100	76	46	30	21	21	14	12	**0.00**	12	32	–	2	–	–	–	–	–	–	–	–	2	Aluminium, tetraethylbis[μ-(N-methylmethanaminato)]di-	23129-29-5	NI25683
105	77	94	212	106	258	51	228	**258**	100	50	23	20	10	7	5	3		13	10	4	2	–	–	–	–	–	–	–	–	–	Aniline, N-benzoyl-3-hydroxy-4-nitro-		IC03376
166	167	152	165	77	51	139	210	**258**	100	54	51	42	39	37	36	35	**3.10**	13	10	4	2	–	–	–	–	–	–	–	–	–	Benzene, 1,1'-methylenebis[2-nitro-		IC03377
258	165	212	166	241	106	153	89	**258**	100	91	34	30	15	12	10	9		13	10	4	2	–	–	–	–	–	–	–	–	–	Benzene, 1,1'-methylenebis[4-nitro-	1817-74-9	NI25684
258	164	259	165	136	46			**258**	100	60	16	6	3	3				13	10	4	2	–	–	–	–	–	–	–	–	–	Benzoic acid, 2-amino-4-nitro-, phenyl ester	55076-31-8	NI25685
258	259	46	93	164	136			**258**	100	17	9	4	2	1				13	10	4	2	–	–	–	–	–	–	–	–	–	Benzoic acid, 3-amino-5-nitro-, phenyl ester	55076-34-1	NI25686
242	225	39	241	258	196	155	169	**258**	100	56	44	38	36	30	26	25		13	10	4	2	–	–	–	–	–	–	–	–	–	Phenazine, 2-hydroxy-7-methoxy-, 5,10-dioxide		IC03378
258	77	230	78	144	89	259	156	**258**	100	33	27	24	23	18	16	14		13	10	4	2	–	–	–	–	–	–	–	–	–	Phenazine, 1,2,3,4-tetrahydroxy-7-methyl-		MS05674
103	214	104	240	196	168	77	44	**258**	100	76	46	38	25	19	15	10	**6.00**	13	10	4	2	–	–	–	–	–	–	–	–	–	4,5-Pyrimidinedicarboxylic acid, 6-methyl-2-phenyl-	61416-98-6	NI25687
57	101	153	99	152	98	74	102	**258**	100	54	33	31	29	25	25	24	**0.00**	13	22	5	–	–	–	–	–	–	–	–	–	–	Hexanoic acid, 3-hydroxy-2-methylene-6-methoxy-, tert-butyl ester		MS05675
101	129	147	185	157	203	259	175	**258**	100	64	58	48	48	37	32	13	**0.50**	13	22	5	–	–	–	–	–	–	–	–	–	–	1,5-Pentanedioic acid, 2-oxo-, dibutyl ester		NI25688
57	41	68	124	58	43	42	56	**258**	100	28	9	7	7	6	5	4	**0.00**	13	23	1	2	–	–	1	–	–	–	–	–	–	Oxetane, 2,3-di-tert-butylimino-4-methyl-4-chloromethyl-		MS05676
88	89	144	71	145	185	116	70	**258**	100	46	34	28	20	19	15	11	**0.00**	13	26	3	2	–	–	–	–	–	–	–	–	–	Carbamic acid, (2-amino-1-oxopropyl)isobutyl-, tert-butyl ester, (S)-	56771-90-5	NI25689
179	178	166	165	260	180	258	89	**258**	100	56	56	46	45	33	22	18		14	10	–	–	–	–	–	–	–	–	–	–	1	Dibenzo[c,e]selenepin	220-00-8	NI25690
165	258	77	51	94	44	135	166	**258**	100	44	23	21	17	17	14	12		14	10	5	–	–	–	–	–	–	–	–	–	–	1,4-Benzenedicarboxylic acid, 2-phenoxy-		NI25691
137	240	213	110	239	76	104	198	**258**	100	69	56	46	40	40	37	33	**30.00**	14	10	5	–	–	–	–	–	–	–	–	–	–	Benzoic acid, 2-(2,4-dihydroxybenzoyl)-		LI05309
137	240	213	110	239	76	104	198	**258**	100	69	56	46	40	40	37	33	**30.03**	14	10	5	–	–	–	–	–	–	–	–	–	–	Benzoic acid, 2-(2,5-dihydroxybenzoyl)-	6272-39-5	NI25692
258	165	241	257	121	92	256	164	**258**	100	94	89	79	71	32	14	14		14	10	5	–	–	–	–	–	–	–	–	–	–	Benzoic acid, 3-(3-hydroxybenzoyl)-4-hydroxy-		NI25693
258	230	259	115	229	69	51	187	**258**	100	16	15	15	13	12	11	10		14	10	5	–	–	–	–	–	–	–	–	–	–	6H-Dibenzo[b,d]pyran-6-one, 3,7,9-trihydroxy-1-methyl-	641-38-3	NI25694
258	259	129	115	69	229	230	187	**258**	100	35	32	25	24	22	18	17		14	10	5	–	–	–	–	–	–	–	–	–	–	6H-Dibenzo[b,d]pyran-6-one, 3,7,9-trihydroxy-1-methyl-	641-38-3	NI25695
258	241	65	141	259	167	39	50	**258**	100	40	28	20	16	15	15	14		14	10	5	–	–	–	–	–	–	–	–	–	–	3,4'-Dicarboxydiphenyl ether		NI25696
258	215	259	187	243	131	129	216	**258**	100	19	16	12	10	4	4	3		14	10	5	–	–	–	–	–	–	–	–	–	–	9H-Xanthen-9-one, 1,5-dihydroxy-6-methoxy-		NI25697
258	215	187	243	259	131	103	102	**258**	100	59	44	28	22	14	12	12		14	10	5	–	–	–	–	–	–	–	–	–	–	9H-Xanthen-9-one, 1,5-dihydroxy-6-methoxy-		LI05310
258	243	215	187	259	228	131	103	**258**	100	54	53	37	19	19	14	12		14	10	5	–	–	–	–	–	–	–	–	–	–	9H-Xanthen-9-one, 1,6-dihydroxy-5-methoxy-		LI05311

MASS TO CHARGE RATIOS								M.W.	INTENSITIES								Parent	C	H	O	N	S	F	Cl	Br	I	Si	P	B	X	COMPOUND NAME	CAS Reg No	No
258	215	243	121	131	228			**258**	100	88	73	12	9	6				14	10	5	–	–	–	–	–	–	–	–	–	–	9H-Xanthen-9-one, 2,3-dihydroxy-1-methoxy-		LI05312
258	212	229	244	228	240	257	259	**258**	100	36	35	31	26	25	19	15		14	10	5	–	–	–	–	–	–	–	–	–	–	9H-Xanthen-9-one, 3,8-dihydroxy-1-methoxy-		MS05677
258	229	212	257	228	259	241	240	**258**	100	44	40	36	19	15	13	10		14	10	5	–	–	–	–	–	–	–	–	–	–	9H-Xanthen-9-one, 5,6-dihydroxy-1-methoxy-		NI25698
258	259	229	257	69	230	115	77	**258**	100	20	18	13	9	6	6	5		14	10	5	–	–	–	–	–	–	–	–	–	–	9H-Xanthen-9-one, 1,3,6-trihydroxy-8-methyl-	20716-98-7	NI25699
258	229	259	257	69	115	201	77	**258**	100	28	20	13	10	6	5	5		14	10	5	–	–	–	–	–	–	–	–	–	–	9H-Xanthen-9-one, 1,3,6-trihydroxy-8-methyl-	20716-98-7	LI05313
178	179	258	260	89	76	176	88	**258**	100	57	55	53	35	28	15	15		14	11	–	–	–	–	–	1	–	–	–	–	–	Stilbene, 3-bromo-, (E)-	14064-45-0	NI25700
178	179	258	260	89	76	177	88	**258**	100	78	75	72	34	29	15	14		14	11	–	–	–	–	–	1	–	–	–	–	–	Stilbene, 4-bromo-	4714-24-3	NI25701
178	258	260	179	89	76	259	88	**258**	100	96	94	53	42	32	17	17		14	11	–	–	–	–	–	1	–	–	–	–	–	Stilbene, 4-bromo-, (E)-	13041-70-8	NI25702
179	178	180	258	260	89	76	177	**258**	100	69	46	25	24	24	20	13		14	11	–	–	–	–	–	1	–	–	–	–	–	Stilbene, α-bromo-	40389-50-2	NI25703
105	77	121	89	258	104	103	90	**258**	100	38	28	6	2	2	2	2		14	11	1	2	–	–	1	–	–	–	–	–	–	Benzaldehyde, 2-chloro-, benzoylhydrazone		LI05314
105	77	121	258	89	136	90	137	**258**	100	32	31	6	5	2	2	1		14	11	1	2	–	–	1	–	–	–	–	–	–	Benzaldehyde, 4-chloro-, benzoylhydrazone		LI05315
258	260	91	215	229	259	138	65	**258**	100	33	33	32	23	15	12	10		14	11	1	2	–	–	1	–	–	–	–	–	–	1H-Benzimidazole, 5-chloro-2-(4-methylphenyl)-, 3-oxide	55030-34-7	NI25704
91	258	260	92	65	180	182	181	**258**	100	37	12	12	12	10	8	8		14	11	1	2	–	–	1	–	–	–	–	–	–	1H-Benzimidazol-4-ol, 1-benzyl-5-chloro-	55030-35-8	NI25705
91	258	75	74	73	71	69	92	**258**	100	30	15	15	15	15	15	12		14	11	1	2	–	–	1	–	–	–	–	–	–	2H-Benzimidazol-2-one, 1-benzyl-5-chloro-1,3-dihydro-	2215-54-5	NI25706
258	40	77	51	259	257	256	242	**258**	100	67	49	36	25	21	18	13		14	11	1	2	–	–	1	–	–	–	–	–	–	2H-Benzimidazol-2-one, 5-chloro-1-methyl-3-phenyl-2,3-dihydro-		NI25707
258	260	91	215	229	138	195	65	**258**	100	33	33	32	23	12	10	10		14	11	1	2	–	–	1	–	–	–	–	–	–	3-Indazolinone, 5-chloro-2-(4-tolyl)-		LI05316
91	258	260	92	65	180	182	181	**258**	100	37	12	12	12	10	8	8		14	11	1	2	–	–	1	–	–	–	–	–	–	1H-Indazol-3-ol, 1-benzyl-5-chloro-		LI05317
228	181	230	199	152	258	180	200	**258**	100	44	30	27	27	26	23	12		14	11	3	–	–	–	–	–	–	–	1	–	–	9-Phosphaanthracene, 9,10-dihydro-9-methoxy-9,10-dioxo-		MS05678
143	155	27	258	215	29	183	115	**258**	100	45	37	36	35	31	23	19		14	14	1	2	1	–	–	–	–	–	–	–	–	2(1H)-Isoquinolinecarbothioic acid, 1-cyano-1-methyl-, S-ethyl ester	17954-39-1	NI25708
258	227	259	197	228	196	260	229	**258**	100	33	21	18	13	5	4	3		14	14	3	–	–	–	–	–	–	1	–	–	–	9-Oxa-10-sila-phenanthrene, 9,10-dihydro-10,10-dimethoxy-		NI25709
258	121	107	77	92	106	78	64	**258**	100	84	62	44	40	30	30	28		14	14	3	2	–	–	–	–	–	–	–	–	–	Diazene, bis(4-methoxyphenyl)-, 1-oxide	1562-94-3	NI25710
243	258	227	94	215	185	173	183	**258**	100	98	83	41	30	29	25	23		14	14	3	2	–	–	–	–	–	–	–	–	–	4H-Pyrazolo[5,1-c][1,4]benzoxazine, 1-acetyl-2-methoxymethyl-		MS05679
130	143	258	77	115				**258**	100	43	21	7	3					14	14	3	2	–	–	–	–	–	–	–	–	–	Succinimide, N-(2-indol-3-yl)ethyl-		LI05318
129	185	101	59	57	41	43	56	**258**	100	59	59	32	30	27	26	18	**0.00**	14	26	4	–	–	–	–	–	–	–	–	–	–	Butanedioic acid, 2,2-dimethyl-, dibutyl ester	56051-56-0	NI25711
129	57	41	56	185	101	130	74	**258**	100	50	42	35	33	33	27	25	**0.00**	14	26	4	–	–	–	–	–	–	–	–	–	–	Butanedioic acid, 2,3-dimethyl-, dibutyl ester	56051-57-1	NI25712
129	101	57	59	41	102	43	56	**258**	100	65	65	49	40	34	32	26	**0.16**	14	26	4	–	–	–	–	–	–	–	–	–	–	Butanedioic acid, 2,2-dimethyl-, diisopropyl ester	57983-28-5	NI25713
129	57	41	56	102	101	130	73	**258**	100	58	32	25	23	21	15	15	**0.10**	14	26	4	–	–	–	–	–	–	–	–	–	–	Butanedioic acid, 2,3-dimethyl-, diisopropyl ester	57983-29-6	NI25714
125	55	97	98	88	138	83	101	**258**	100	85	81	66	62	56	53	48	**0.00**	14	26	4	–	–	–	–	–	–	–	–	–	–	Decanedioic acid, diethyl ester	110-40-7	NI25716
55	29	41	69	97	43	88	83	**258**	100	90	57	40	36	31	30	30	**0.00**	14	26	4	–	–	–	–	–	–	–	–	–	–	Decanedioic acid, diethyl ester	110-40-7	NI25715
88	55	69	171	41	139	97	59	**258**	100	53	44	43	42	33	25	24	**0.00**	14	26	4	–	–	–	–	–	–	–	–	–	–	Decanedioic acid, 2,4-dimethyl-, dimethyl ester, (2S,4S)-	85547-47-3	NI25717
55	41	69	43	59	74	98	185	**258**	100	90	80	67	65	56	43	40	**0.00**	14	26	4	–	–	–	–	–	–	–	–	–	–	Decanedioic acid, 3,6-dimethyl-, dimethyl ester	5024-28-2	NI25718
55	69	41	74	185	59	153	43	**258**	100	76	76	61	51	51	47	38	**0.00**	14	26	4	–	–	–	–	–	–	–	–	–	–	Decanedioic acid, 3,7-dimethyl-, dimethyl ester	5131-46-4	NI25719
55	41	69	185	59	43	87	74	**258**	100	58	50	40	40	40	38	31	**0.00**	14	26	4	–	–	–	–	–	–	–	–	–	–	Decanedioic acid, 4,7-dimethyl-, dimethyl ester	5024-27-1	NI25720
102	55	87	41	69	44	43	83	**258**	100	62	55	45	43	43	36	33	**0.00**	14	26	4	–	–	–	–	–	–	–	–	–	–	Decanedioic acid, 2-ethyl-, dimethyl ester		IC03379
185	153	69	74	83	96	111	227	**258**	100	55	42	39	28	27	24	23	**0.00**	14	26	4	–	–	–	–	–	–	–	–	–	–	Decanoic acid, 4-(methoxycarbonylmethyl)-, methyl ester		NI25721
98	74	185	227	55	84	69	112	**258**	100	70	58	54	51	49	44	38	**0.00**	14	26	4	–	–	–	–	–	–	–	–	–	–	Dodecane-1,2-dioic acid, dimethyl ester		LI05319
98	74	55	84	69	185	41	227	**258**	100	84	72	54	51	49	49	41	**1.92**	14	26	4	–	–	–	–	–	–	–	–	–	–	Dodecanedioic acid, dimethyl ester	1731-79-9	IC03380
98	74	153	84	185	227	112	69	**258**	100	79	67	66	61	48	47	45	**1.00**	14	26	4	–	–	–	–	–	–	–	–	–	–	Dodecanedioic acid, dimethyl ester	1731-79-9	NI25722
98	74	55	84	69	41	59	43	**258**	100	91	83	60	59	55	41	40	**0.00**	14	26	4	–	–	–	–	–	–	–	–	–	–	Dodecanedioic acid, dimethyl ester	1731-79-9	IC03381
131	73	101	113	99	81	243	226	**258**	100	49	29	22	17	17	11	9	**0.10**	14	26	4	–	–	–	–	–	–	–	–	–	–	Glycerol, 1-O-(2-methoxy-4-heptenyl)-2,3-O-isopropylidene		NI25723
185	130	112	55	56	87	41	57	**258**	100	86	47	33	29	28	28	27	**0.89**	14	26	4	–	–	–	–	–	–	–	–	–	–	Hexanedioic acid, dibutyl ester	105-99-7	DO01048
185	129	111	41	55	44	156	56	**258**	100	76	44	25	23	23	20	20	**0.00**	14	26	4	–	–	–	–	–	–	–	–	–	–	Hexanedioic acid, dibutyl ester	105-99-7	NI25724
129	185	57	111	56	55	41	29	**258**	100	58	46	26	22	22	22	14	**0.47**	14	26	4	–	–	–	–	–	–	–	–	–	–	Hexanedioic acid, diisobutyl ester		DO01049
129	185	111	57	55	128	100	41	**258**	100	20	19	14	14	12	11	11	**0.17**	14	26	4	–	–	–	–	–	–	–	–	–	–	Hexanedioic acid, di-sec-butyl ester	38447-22-2	DO01050
129	55	41	57	185	56	101	111	**258**	100	82	77	55	43	40	30	29	**0.74**	14	26	4	–	–	–	–	–	–	–	–	–	–	Hexanedioic acid, di-sec-butyl ester	38447-22-2	NI25725
87	55	41	69	43	116	45	82	**258**	100	99	80	68	68	67	61	49	**0.00**	14	26	4	–	–	–	–	–	–	–	–	–	–	Nonanedioic acid, 2-propyl-, dimethyl ester		IC03382
185	152	73	111	69	179	115	83	**258**	100	51	42	38	28	25	25	22	**0.00**	14	26	4	–	–	–	–	–	–	–	–	–	–	Octanedioic acid, tetramethyl-, dimethyl ester	72361-22-9	NI25726
129	57	41	101	56	55	128	100	**258**	100	43	34	31	24	22	19	13	**0.00**	14	26	4	–	–	–	–	–	–	–	–	–	–	Pentanedioic acid, 2-methyl-, bis(1-methylpropyl) ester	57983-49-0	NI25727
129	41	57	56	128	69	100	42	**258**	100	71	55	52	48	39	36	35	**1.12**	14	26	4	–	–	–	–	–	–	–	–	–	–	Pentanedioic acid, 2-methyl-, bis(1-methylpropyl) ester	57983-37-6	NI25728
129	41	56	101	100	57	185	87	**258**	100	61	42	35	35	35	27	27	**0.00**	14	26	4	–	–	–	–	–	–	–	–	–	–	Pentanedioic acid, 3-methyl-, dibutyl ester	56051-60-6	NI25729
188	173	142	99	185	69	213	41	**258**	100	61	61	30	25	23	18	17	**0.90**	14	26	4	–	–	–	–	–	–	–	–	–	–	Propanedioic acid, ethylisopropyl-, diethyl ester	76-72-2	NI25730
41	55	98	73	39	60	43	84	**258**	100	88	78	74	70	62	62	59	**0.00**	14	26	4	–	–	–	–	–	–	–	–	–	–	Tetradecanedioic acid	821-38-5	NI25731
88	55	112	69	41	98	59	74	**258**	100	44	33	33	33	28	24	22	**0.00**	14	26	4	–	–	–	–	–	–	–	–	–	–	Undecanedioic acid, 2-methyl-, dimethyl ester		IC03383
115	87	103	75	225	109	59	154	**258**	100	74	66	54	52	45	41	27	**0.00**	14	30	2	–	–	–	–	–	–	1	–	–	–	3-Hexene-2,5-diol, 2,5-dimethyl-3-(triethylsilyl)-, (Z)-		NI25732
133	75	89	59	243	199	103	117	**258**	100	73	69	63	29	28	10	8	**0.00**	14	30	2	–	–	–	–	–	–	1	–	–	–	1-Nonene, 1-[dimethyl(2-methoxyethoxy)silyl]-, (E)-		DD00747
43	41	42	56	44	57	55	39	**258**	100	73	68	54	44	33	32	26	**0.00**	14	30	2	2	–	–	–	–	–	–	–	–	–	Nitramine, diheptyl-		MS05680

MASS TO CHARGE RATIOS								M.W.	INTENSITIES								Parent	C	H	O	N	S	F	Cl	Br	I	Si	P	B	X	COMPOUND NAME	CAS Reg No	No
141	126	98	112	70	183	111	96	258	100	96	95	91	75	53	50	37	20.00	14	30	2	2	–	–	–	–	–	–	–	–	–	L-Ornithine, N^2,N^5-diisopropyl-, isopropyl ester	56771-70-1	MS05681
178	165	179	89	258	63	166	163	258	100	61	28	14	12	11	10	9		15	11	2	–	–	–	1	–	–	–	–	–	–	Carbonochloridic acid, 9H-fluoren-9-ylmethyl ester	28920-43-6	NI25733
118	44	90	88	76	119	179	178	258	100	19	17	17	11	8	2	2	0.09	15	11	2	–	–	–	1	–	–	–	–	–	–	Propanoic acid, 2-hydroxy-2-phenyl-3-(4-chlorophenyl)-, lactone		LI05320
147	120	258	121	257	65	101	259	258	100	93	87	69	65	43	37	36		15	11	2	–	–	–	1	–	–	–	–	–	–	2-Propen-1-one, 3-(4-chlorophenyl)-1-(2-hydroxyphenyl)-	3033-96-3	NI25734
105	77	123	118	95	106	51	75	258	100	22	20	16	11	9	9	5	1.00	15	11	3	–	–	1	–	–	–	–	–	–	–	Benzoic acid, 2-fluoro-, 2-oxo-2-phenylethyl ester	55153-24-7	NI25735
105	77	123	95	106	118	51	75	258	100	20	15	11	10	6	6	4	1.00	15	11	3	–	–	1	–	–	–	–	–	–	–	Benzoic acid, 3-fluoro-, 2-oxo-2-phenylethyl ester	55153-23-6	NI25736
105	123	77	95	106	118	51	75	258	100	21	21	13	9	8	6	4	1.00	15	11	3	–	–	1	–	–	–	–	–	–	–	Benzoic acid, 4-fluoro-, 2-oxo-2-phenylethyl ester	55153-22-5	NI25737
135	225	258	151	77	92	242	136	258	100	86	72	71	43	34	33	28		15	14	2	–	1	–	–	–	–	–	–	–	–	Benzophenone, 4,4′-dimethoxythio-	958-80-5	NI25738
258	151	225	227	189	57	210	108	258	100	98	97	37	32	31	28	28		15	14	2	–	1	–	–	–	–	–	–	–	–	Benzophenone, 4,4′-dimethoxythio-	958-80-5	NI25739
178	179	258	167	177	152	51	176	258	100	87	71	43	23	23	20	18		15	14	2	–	1	–	–	–	–	–	–	–	–	Styrene, 1-methylsulphonyl-1-phenyl-		LI05321
136	123	135	258	149	111	109	134	258	100	81	63	61	38	36	31	26		15	14	4	–	–	–	–	–	–	–	–	–	–	1,3-Benzenediol, 4-(3,4-dihydro-7-hydroxy-2H-1-benzopyran-3-yl)-	65332-45-8	NI25740
152	107	151	258	78	153	77	52	258	100	69	47	36	17	8	7	7		15	14	4	–	–	–	–	–	–	–	–	–	–	1,3-Benzodioxan, 2-(3-hydroxy-4-methoxyphenyl)-		MS05682
107	108	214	258	134	240	165	151	258	100	75	70	58	33	15	13	13		15	14	4	–	–	–	–	–	–	–	–	–	–	Benzoic acid, 2,3-dihydroxy-6-(2-phenylethyl)-	75112-81-1	NI25741
226	258	183	127	198	155	115	139	258	100	40	20	16	15	13	12	8		15	14	4	–	–	–	–	–	–	–	–	–	–	Benzoic acid, 2-hydroxy-4-methoxy-6-phenyl-, methyl ester		LI05322
258	121	77	65	164	165	120	94	258	100	99	91	42	39	38	33	28		15	14	4	–	–	–	–	–	–	–	–	–	–	Benzoic acid, 4-(2-phenoxyethoxy)-		IC03384
135	77	181	149	121	92	180	65	258	100	32	27	25	20	16	13	13	0.00	15	14	4	–	–	–	–	–	–	–	–	–	–	Benzophenone, 2-carboxy-2′-methoxy-		LI05323
229	258	214	259	189	131	129	159	258	100	33	18	15	15	15	12	11		15	14	4	–	–	–	–	–	–	–	–	–	–	2H-1-Benzopyran-2-one, 7-methoxy-6-(3-isopropenyloxiranyl)-, trans-	23631-16-5	NI25742
257	258	203	227	83	55	243	89	258	100	91	71	56	43	33	21	14		15	14	4	–	–	–	–	–	–	–	–	–	–	2H-1-Benzopyran-2-one, 7-methoxy-6-(3-methyl-1-oxo-2-butenyl)-	16850-91-2	NI25743
258	257	227	203	82	243	175	56	258	100	90	49	40	24	23	21	16		15	14	4	–	–	–	–	–	–	–	–	–	–	2H-1-Benzopyran-2-one, 7-methoxy-6-(3-methyl-1-oxo-2-butenyl)-	16850-91-2	NI25744
226	211	240	186	39	31	18	102	258	100	47	34	32	32	27	21	20	2.00	15	14	4	–	–	–	–	–	–	–	–	–	–	1,3-Butanedione, 1-(3-hydroxy-1-methoxy-2-naphthalenyl)-	55030-42-7	NI25745
243	258	185	200	101	129	183	215	258	100	99	90	53	47	27	22	18		15	14	4	–	–	–	–	–	–	–	–	–	–	Dibenzofuran, 1,2,3-trimethoxy-	88256-08-0	NI25746
243	258	200	129	215	142	259	244	258	100	91	33	19	19	17	17	16		15	14	4	–	–	–	–	–	–	–	–	–	–	Dibenzofuran, 1,2,4-trimethoxy-	88256-09-1	NI25747
258	243	200	129	142	114	215	259	258	100	89	46	28	24	21	21	17		15	14	4	–	–	–	–	–	–	–	–	–	–	Dibenzofuran, 1,3,4-trimethoxy-	88256-10-4	NI25748
258	243	185	200	101	129	183	259	258	100	96	47	40	35	35	30	16		15	14	4	–	–	–	–	–	–	–	–	–	–	Dibenzofuran, 2,3,4-trimethoxy-		NI25749
243	227	258	185	213				258	100	90	40	25	18					15	14	4	–	–	–	–	–	–	–	–	–	–	2H-Furo[2,3-h]-1-benzopyran-2-one, 8-(1-hydroxymethyl-1-isopropyl)-		LI05324
174	105	175	115	173	76	216	77	258	100	17	13	13	11	10	10	9	4.40	15	14	4	–	–	–	–	–	–	–	–	–	–	1,4-Naphthalenediol, 2-methyl-, diacetate	573-20-6	NI25750
213	258	155	198	183	139	128	127	258	100	75	36	20	20	20	20	20		15	14	4	–	–	–	–	–	–	–	–	–	–	Naphthalenepropenoic acid, 2,7-dimethoxy-	89229-14-1	NI25751
159	160	258	72	43	197	187	172	258	100	44	36	27	24	23	23	23		15	14	4	–	–	–	–	–	–	–	–	–	–	Naphtho[1,2-b]furan-4,5-dione, 2,3-dihydro-2-(1-hydroxy-1-isopropyl)-, (±)-	41019-53-8	NI25752
200	172	59	115	43	144	199	66	258	100	40	35	16	16	13	12	9	4.00	15	14	4	–	–	–	–	–	–	–	–	–	–	Naphtho[2,3-b]furan-4,9-dione, 2,3-dihydro-2-(1-hydroxy-1-isopropyl)-	41192-67-0	NI25753
159	43	72	160	71	230	258	188	258	100	95	83	76	76	62	55	38		15	14	4	–	–	–	–	–	–	–	–	–	–	2H-Naphtho[1,2-b]pyran-5,6-dione, 3,4-dihydro-3-hydroxy-2,2-dimethyl-, (R)-	41135-46-0	NI25754
230	215	231	201	159	158	115	243	258	100	33	17	16	15	15	15	14	10.00	15	14	4	–	–	–	–	–	–	–	–	–	–	Naphtho[1,8-bc]pyran-7,8-dione, 2,3-dihydro-4-hydroxy-3,6,9-trimethyl-	13383-59-0	NI25755
222	91	221	223	65	77	131	104	258	100	39	24	17	14	6	5	5	0.00	15	15	–	2	–	–	1	–	–	–	–	–	–	Formamidine, α-chloro-N,N′-bis(4-tolyl)-		IC03385
258	133	91	132	135	77	101	42	258	100	62	55	45	41	39	31	27		15	18	–	2	1	–	–	–	–	–	–	–	–	1H,4H,4′H,8H,8′H-3,1-Benzothiazine, 1-methyl-2-phenylimino-, cis-		MS05683
258	91	132	133	77	167	257	259	258	100	62	47	35	32	30	23	21		15	18	–	2	1	–	–	–	–	–	–	–	–	1H,4H,4′H,8H,8′H-3,1-Benzothiazine, 1-methyl-2-phenylimino-, trans-		MS05684
84	111	96	55	54	110	56	82	258	100	72	45	40	20	16	13	12	3.00	15	18	2	2	–	–	–	–	–	–	–	–	–	2,5-Pyrrolidinedione, 1-phenyl-2-(1-piperidino)-		LI05325
216	188	146	43	130	144	69	217	258	100	91	89	73	40	36	30	26	20.83	15	18	2	2	–	–	–	–	–	–	–	–	–	Pyrrolo[2,3-b]indole, 1,8-diacetyl-1,2,3,3a,8,8a-hexahydro-3a-methyl-, (3aS-cis)-	55030-11-0	MS05685
216	188	146	43	130	217	144	69	258	100	88	88	71	40	34	34	29	21.02	15	18	2	2	–	–	–	–	–	–	–	–	–	Pyrrolo[2,3-b]indole, 1,8-diacetyl-1,2,3,3a,8,8a-hexahydro-3a-methyl-, (3aS-cis)-	55030-11-0	NI25756
85	55	41	56	83	69	101	29	258	100	47	41	26	24	23	20	20	0.34	15	30	3	–	–	–	–	–	–	–	–	–	–	1-Decanol, 10-[(tetrahydro-2H-pyran-2-yl)oxy]-	43047-93-4	NI25757
69	41	87	99	71	43	70	105	258	100	32	22	20	18	13	12	11	1.50	15	30	3	–	–	–	–	–	–	–	–	–	–	4,8-Dioxa-11-dodecen-1-ol, 3,3,7,7,11-pentamethyl-		MS05686
199	69	83	55	97	43	57	41	258	100	87	79	77	75	69	61	55	23.00	15	30	3	–	–	–	–	–	–	–	–	–	–	Tetradecanoic acid, 2-hydroxy-, methyl ester	56009-40-6	NI25758
57	55	41	43	69	29	83	97	258	100	99	94	72	67	67	66	32	0.00	15	30	3	–	–	–	–	–	–	–	–	–	–	Tetradecanoic acid, 2-hydroxy-, methyl ester	56009-40-6	NI25759
103	43	71	41	74	55	57	61	258	100	70	33	33	32	23	19	18	0.00	15	30	3	–	–	–	–	–	–	–	–	–	–	Tetradecanoic acid, 3-hydroxy-, methyl ester	55682-83-2	NI25760
103	43	41	71	55	29	74	61	258	100	88	60	55	52	51	30	21	0.00	15	30	3	–	–	–	–	–	–	–	–	–	–	Tetradecanoic acid, 3-hydroxy-, methyl ester	55682-83-2	NI25761
75	103	73	243	55	43	41	69	258	100	75	70	26	20	18	18	17	0.00	15	34	1	–	–	–	–	–	–	1	–	–	–	Decanol, 2-ethyl-, trimethylsilyl ether	74421-16-2	NI25762
117	73	75	41	55	43	118	32	258	100	50	49	11	10	10	9	6	0.00	15	34	1	–	–	–	–	–	–	1	–	–	–	2-Dodecanol, trimethylsilyl ether	39789-27-0	NI25763
73	131	75	83	229	41	55	69	258	100	92	63	18	15	15	14	13	0.00	15	34	1	–	–	–	–	–	–	1	–	–	–	3-Dodecanol, trimethylsilyl ether	39789-26-9	NI25764
73	145	75	69	103	215	41	55	258	100	90	65	27	25	23	15	13	0.00	15	34	1	–	–	–	–	–	–	1	–	–	–	4-Dodecanol, trimethylsilyl ether	39789-28-1	NI25765
73	159	75	69	201	103	55	41	258	100	77	64	38	37	29	17	15	0.00	15	34	1	–	–	–	–	–	–	1	–	–	–	5-Dodecanol, trimethylsilyl ether	39789-29-2	NI25766
73	75	173	187	55	103	41	43	258	100	68	64	48	28	25	19	13	0.00	15	34	1	–	–	–	–	–	–	1	–	–	–	6-Dodecanol, trimethylsilyl ether	39789-30-5	NI25767
75	243	73	103	83	69	55	97	258	100	65	45	32	29	29	25	24	0.00	15	34	1	–	–	–	–	–	–	1	–	–	–	Dodecanol, trimethylsilyl ether	6221-88-1	NI25769
75	73	43	41	243	29	55	103	258	100	75	68	68	53	46	35	26	0.44	15	34	1	–	–	–	–	–	–	1	–	–	–	Dodecanol, trimethylsilyl ether	6221-88-1	NI25768
75	103	73	243	41	55	43	69	258	100	73	64	26	19	18	18	14	0.00	15	34	1	–	–	–	–	–	–	1	–	–	–	Octanol, 2-butyl-, trimethylsilyl ether	74421-15-1	NI25770

MASS TO CHARGE RATIOS								M.W.	INTENSITIES								Parent	C	H	O	N	S	F	Cl	Br	I	Si	P	B	X	COMPOUND NAME	CAS Reg No	No
258	200	202	201	259	75	88	100	258	100	35	31	25	20	19	16	12		16	10	–	4	–	–	–	–	–	–	–	–	–	4,4′-Dicinnolyl		LI05326
45	258	213	185	183	109	152	165	258	100	43	42	17	16	14	11	10		16	15	2	–	–	1	–	–	–	–	–	–	–	Stilbene, 4′-fluoro-2-methoxymethoxy-, (E)-		MS05687
45	213	258	185	183	109	152	165	258	100	29	27	14	14	11	9	8		16	15	2	–	–	1	–	–	–	–	–	–	–	Stilbene, 4′-fluoro-2-methoxymethoxy-, (Z)-		MS05688
227	135	151	228	77	197	152	153	258	100	69	26	24	21	15	12	10	5.00	16	18	3	–	–	–	–	–	–	–	–	–	–	Di-p-anisylmethyl ether		LI05327
91	258	92	65	39	240	77	242	258	100	12	9	9	6	5	3	2		16	18	3	–	–	–	–	–	–	–	–	–	–	Benzenemethanol, 4-benzyloxy-3,6-dimethyl-2-hydroxy-		MS05689
43	28	55	229	258				258	100	50	46	34	7					16	18	3	–	–	–	–	–	–	–	–	–	–	1,3-Cyclohexanedione, 2-(4-propanoylbenzyl)-		LI05328
57	29	202	91	258				258	100	58	17	12	1					16	18	3	–	–	–	–	–	–	–	–	–	–	1,3-Cyclohexanedione, 2-(4-propanoylbenzyl)-		LI05329
169	155	198	141	215	115	128	91	258	100	78	26	25	19	18	14	14	0.00	16	18	3	–	–	–	–	–	–	–	–	–	–	2-Oxatricyclo[7.4.0.0^{1,5}]tridec-8-en-3-one, 11-acetyl-6,7-dimethylidene-, (1RS,5RS,11RS)-		DD00748
216	174	43	115	159	145	77	55	258	100	37	28	26	17	15	15	14	2.50	16	18	3	–	–	–	–	–	–	–	–	–	–	Phenalenone, 9-acetoxy-2,3,3a,4,5,6-hexahydro-6-methyl-		LI05330
195	155	240	258	243	153	152	196	258	100	71	46	34	33	18	18	16		16	18	3	–	–	–	–	–	–	–	–	–	–	2-Propenoic acid, 2-[6-(1-hydroxy-1-methylethyl)naphthalen-2-yl]-		NI25771
120	138	258	121	77	105	42	91	258	100	18	11	10	6	4	4	3		16	19	–	2	–	1	–	–	–	–	–	–	–	Ethylenediamine, N-(m-fluorophenyl)-N,N′-dimethyl-N′-phenyl-	32857-43-5	NI25772
120	138	258	77	105	104	95	91	258	100	36	12	11	6	6	5	3		16	19	–	2	–	1	–	–	–	–	–	–	–	Ethylenediamine, N-(p-fluorophenyl)-N,N′-dimethyl-N′-phenyl-	32857-41-3	LI05331
120	138	258	77	121	42	104	95	258	100	35	12	10	9	7	5	5		16	19	–	2	–	1	–	–	–	–	–	–	–	Ethylenediamine, N-(p-fluorophenyl)-N,N′-dimethyl-N′-phenyl-	32857-41-3	NI25773
136	135	121	120	60	68	93	137	258	100	58	35	22	21	18	15	12	0.00	16	22	1	2	–	–	–	–	–	–	–	–	–	Aniline, N-(2-acetylcyclohexylidene)-4-(N,N-dimethylamino)-		NI25774
100	57	56	42	43	41	77	103	258	100	48	47	43	40	35	34	30	22.00	16	22	1	2	–	–	–	–	–	–	–	–	–	1-Butanimine, 1-morpholino-N-styryl-	85572-32-3	NI25775
258	241	84	70	198	153	147	110	258	100	63	24	17	16	12	12	11		16	22	1	2	–	–	–	–	–	–	–	–	–	1,3-Diazaadamantan-6-ol, 5,7-dimethyl-6-phenyl-		MS05690
106	258	91	163	259	105	67	257	258	100	85	52	41	18	16	16	15		16	22	1	2	–	–	–	–	–	–	–	–	–	1,3-Oxazine, 3-benzyl-2-methylimino-4,5-tetramethylenetetrahydro-, (E)-		NI25776
106	258	91	163	58	259	67	105	258	100	85	59	50	27	19	15	14		16	22	1	2	–	–	–	–	–	–	–	–	–	1,3-Oxazine, 3-benzyl-2-methylimino-4,5-tetramethylenetetrahydro-, (Z)-		NI25777
128	130	83	85	41	96	47	129	258	100	21	17	11	10	9	9	8	4.00	16	22	1	2	–	–	–	–	–	–	–	–	–	Piperidine, 1-[2-(indol-3-yl)ethyl]-3-methoxy-		LI05332
72	143	181	102	44	156	60	157	258	100	96	62	59	59	55	44	40	29.60	16	22	1	2	–	–	–	–	–	–	–	–	–	2-Propanol, 1-isopropylamino-3-(1-naphthylamino)-		MS05691
107	135	108	258	94	150	106	243	258	100	87	67	49	39	37	33	26		16	22	1	2	–	–	–	–	–	–	–	–	–	Pyrrole, 3,5-dimethyl-4-ethyl-2-(3-ethyl-4-methyl-2-pyrroloyl)-		LI05333
187	172	258	71	157	131	77	144	258	100	25	11	7	6	6	6	5		16	22	1	2	–	–	–	–	–	–	–	–	–	Quinoxaline, 2-tert-butyl-3,4-dihydro-3-(2′-tetrafuranyl)-		LI05334
98	131	84	28	91	258	41	56	258	100	52	20	20	16	16	11	10		16	22	1	2	–	–	–	–	–	–	–	–	–	Spiro[1.2]hexane-3-imidazoline, 5,5-dimethyl-4-phenyl-, 3-oxide		NI25778
79	91	92	108	94	258	107	80	258	100	97	64	56	53	43	43	30		16	22	1	2	–	–	–	–	–	–	–	–	–	Urea, N-(exo-bicyclo[4.1.0]hept-3-en-7-yl)-N′-(endo-bicyclo[4.1.0]hept-3-en-7-methyl)-	90013-15-3	NI25779
258	257	120	93	67	121	79	147	258	100	46	32	25	23	22	20	19		16	28	1	–	–	–	–	–	–	–	–	2	–	9-Borabicyclo[3.3.1]nonane, 9,9′-oxybis-	74744-62-0	NI25780
145	69	43	55	41	56	70	57	258	100	74	74	73	62	61	59	52	20.54	16	34	–	–	1	–	–	–	–	–	–	–	–	Dioctyl sulphide		MS05692
41	43	55	29	57	69	83	97	258	100	82	71	48	47	43	36	27	4.70	16	34	–	–	1	–	–	–	–	–	–	–	–	1-Hexadecanethiol	2917-26-2	NI25781
57	43	71	55	83	69	97	41	258	100	76	62	62	59	54	46	45	0.14	16	34	2	–	–	–	–	–	–	–	–	–	–	Ethanol, 2-(tetradecyloxy)-	2136-70-1	DO01051
57	43	71	55	41	45	63	85	258	100	79	63	51	51	43	42	39	0.14	16	34	2	–	–	–	–	–	–	–	–	–	–	Ethanol, 2-(tetradecyloxy)-	2136-70-1	NI25782
57	71	43	85	63	55	45	83	258	100	69	58	42	41	31	31	29	0.04	16	34	2	–	–	–	–	–	–	–	–	–	–	Ethanol, 2-(tetradecyloxy)-	2136-70-1	IC03386
83	69	55	97	43	57	41	111	258	100	98	89	88	78	75	65	48	0.00	16	34	2	–	–	–	–	–	–	–	–	–	–	1,2-Hexadecanediol		MS05693
71	41	43	82	29	75	55	31	258	100	42	20	19	18	15	13	13	0.00	16	34	2	–	–	–	–	–	–	–	–	–	–	Tetradecane, 1,1-dimethoxy-	14620-53-2	NI25783
71	82	75	68	96	236	110	138	258	100	19	15	11	8	2	2	1	0.00	16	34	2	–	–	–	–	–	–	–	–	–	–	Tetradecane, 1,1-dimethoxy-	14620-53-2	LI05335
160	28	39	127	258	116	129	162	258	100	94	83	79	75	65	54	50		17	22	–	–	1	–	–	–	–	–	–	–	–	Heptyl 1-naphthyl sulphide		AI01833
105	93	136	77	68	94	121	81	258	100	84	65	64	57	44	39	36	0.03	17	22	2	–	–	–	–	–	–	–	–	–	–	Benzoic acid, β-terpinyl ester, (E)-		NI25784
185	258	186	212	199	217	216	230	258	100	79	29	20	16	9	9	8		17	22	2	–	–	–	–	–	–	–	–	–	–	Cyclohepta[de]naphthalen-3-propanoic acid, 1,2,3,7,8,9,10,10a-octahydro-		LI05336
130	117	129	115	91	184	113	185	258	100	80	53	51	41	36	36	32	7.81	17	22	2	–	–	–	–	–	–	–	–	–	–	Cyclopropanecarboxylic acid, 2,2-dimethyl-3-(1-phenyl-2-propenyl)-, ethyl ester	74810-40-5	NI25785
258	175	243	215	259	137	244	161	258	100	80	78	62	21	19	15	14		17	22	2	–	–	–	–	–	–	–	–	–	–	Dibenzo[b,d]pyran-1-ol, 6a,7,8,10a-tetrahydro-3,6,6,6-tetramethyl-		LI05337
175	258	137	190	259	215	176	91	258	100	97	29	28	19	17	15	15		17	22	2	–	–	–	–	–	–	–	–	–	–	Dibenzo[b,d]pyran-1-ol, 6a,7,10,10a-tetrahydro-3,6,6,9-tetramethyl-		LI05338
43	71	115	128	89	187	258	159	258	100	43	39	31	26	23	19	18		17	22	2	–	–	–	–	–	–	–	–	–	–	4,6-Heptadecadi-8,10-enediyn-1-ol, 12-oxo-, (E,E)-		LI05339
41	71	55	67	89	39	90	54	258	100	84	76	72	65	62	51	39	0.30	17	22	2	–	–	–	–	–	–	–	–	–	–	11,13-Heptadeca-1,15-dien-10-ol, cis-8,9-epoxy-		LI05340
81	41	258	229	201	91	53	55	258	100	62	61	57	55	43	35	34		17	22	2	–	–	–	–	–	–	–	–	–	–	1(2H)-Naphthalenone, 8aβ-ethyl-2-furfurylideneoctahydro-	17429-51-5	NI25786
147	258	133	159	226	157	145	107	258	100	88	62	43	41	36	27	20		17	22	2	–	–	–	–	–	–	–	–	–	–	1-Phenanthrenecarboxaldehyde, 1,2,3,4,4a,9,10,10a-octahydro-6-hydroxy-1,4a-dimethyl-, [1S-(1α,4aα,10aβ)]-	24035-48-1	NI25787
197	243	141	129	258	128	143	198	258	100	72	27	26	24	24	20	17		17	22	2	–	–	–	–	–	–	–	–	–	–	1-Phenanthrenecarboxylic acid, 1,2,3,4,4a,9,10,10a-octahydro-1,4a-dimethyl-, [1S-(1α,4aα,10aα)]-	57345-30-9	NI25789
197	243	141	129	128	143	117	115	258	100	66	30	30	25	24	22	20	16.80	17	22	2	–	–	–	–	–	–	–	–	–	–	1-Phenanthrenecarboxylic acid, 1,2,3,4,4a,9,10,10a-octahydro-1,4a-dimethyl-, [1S-(1α,4aα,10aα)]-	57345-30-9	NI25788
197	243	141	117	129	258	143	115	258	100	64	48	34	25	24	23	22		17	22	2	–	–	–	–	–	–	–	–	–	–	1-Phenanthrenecarboxylic acid, 1,2,3,4,4a,9,10,10a-octahydro-1,4a-dimethyl-, [1S-(1α,4aα,10aβ)]-	4586-72-5	NI25790
129	101	144	198	128	258	130	170	258	100	61	60	59	50	49	47	41		17	22	2	–	–	–	–	–	–	–	–	–	–	1-Phenanthrenecarboxylic acid, 1,2,3,4,4a,9,10,10a-octahydro-1-methyl-, methyl ester, [1S-(1α,4aα,10aβ)]-	57378-56-0	NI25791

MASS TO CHARGE RATIOS								M.W.	INTENSITIES								Parent	C	H	O	N	S	F	Cl	Br	I	Si	P	B	X	COMPOUND NAME	CAS Reg No	No
148	175	201	159	190	121	215	160	**258**	100	77	43	33	30	30	25	25	**25.00**	17	22	2	–	–	–	–	–	–	–	–	–	–	**Phenol, 2-(1,2-dimethylbicyclo[3.1.0]hex-2-yl)-5-methyl-, acetate, (1α,2α,5α)-(±)-**	39892-69-8	**NI25792**
201	175	190	148	159	121	160	215	**258**	100	94	91	91	74	51	50	45	**0.00**	17	22	2	–	–	–	–	–	–	–	–	–	–	**Phenol, 2-(1,2-dimethylbicyclo[3.1.0]hex-2-yl)-5-methyl-, acetate, (1α,2α,5α)-(±)-**	39892-69-8	**LI05341**
148	117	119	201	258	159	121	215	**258**	100	42	40	38	30	30	30	22		17	22	2	–	–	–	–	–	–	–	–	–	–	**Phenol, 6-(1,2-dimethylbicyclo[3.1.0]hex-2-yl)-3-methyl-, acetate, (1S,2R)-(–)-**	13002-87-4	**NI25793**
258	185	121	77	91	79	134	115	**258**	100	92	22	12	11	10	7	7		17	22	2	–	–	–	–	–	–	–	–	–	–	**Tricyclo[3.3.1.1^{3,7}]decan-2-ol, 1-(4-methoxyphenyl)-**	40571-16-2	**NI25794**
73	142	216	59	129	258	115	243	**258**	100	52	37	25	20	12	11	4		17	26	–	–	–	–	–	–	–	1	–	–	–	**Naphthalene, 1-butyl-3,4-dihydro-2-(trimethylsilyl)-**	118298-82-1	**DD00749**
91	79	162	94	119	105	133	93	**258**	100	87	77	59	57	53	52	42	**0.00**	17	26	–	2	–	–	–	–	–	–	–	–	–	**Indazole-3-spirocyclohexane, 4,5,6,7-tetrahydro-**		**NI25795**
139	70	258	42	91	43	28	41	**258**	100	48	37	37	36	36	28	22		17	26	–	2	–	–	–	–	–	–	–	–	–	**Piperazine, 1-methyl-4-(4-phenylcyclohexyl)-**		**TR00287**
258	70	58	42	43	139	56	143	**258**	100	82	81	59	57	43	43	37		17	26	–	2	–	–	–	–	–	–	–	–	–	**Piperazine, 1-methyl-4-(1,2,3,4-tetrahydro-3,7-dimethyl-2-naphthalenyl)-**	55591-14-5	**NI25796**
258	257	259	230	202	231	201	200	**258**	100	28	27	13	9	6	4	4		18	10	2	–	–	–	–	–	–	–	–	–	–	**Benz[a]anthracene-7,12-dione**	2498-66-0	**MS05694**
258	101	202	100	230	200	201	88	**258**	100	55	48	40	36	32	29	27		18	10	2	–	–	–	–	–	–	–	–	–	–	**Benz[a]anthracene-7,12-dione**	2498-66-0	**NI25797**
258	259	229	200	202	201	260	75	**258**	100	21	8	8	5	4	3	3		18	10	2	–	–	–	–	–	–	–	–	–	–	**Benzo[1,2-b:3,4-b′]bisbenzofuran**		**NI25798**
258	259	200	229	202	201	176	75	**258**	100	21	10	7	7	6	5	3		18	10	2	–	–	–	–	–	–	–	–	–	–	**Benzo[1,2-b:4,3-b′]bisbenzofuran**		**NI25799**
258	259	229	200	202	201	260	75	**258**	100	21	9	8	5	4	3	3		18	10	2	–	–	–	–	–	–	–	–	–	–	**Benzo[1,2-b:4,5-b′]bisbenzofuran**		**NI25800**
258	259	229	200	202	201	260	75	**258**	100	21	7	7	5	4	3	3		18	10	2	–	–	–	–	–	–	–	–	–	–	**Benzo[1,2-b:6,5-b′]bisbenzofuran**		**NI25801**
258	176	202	88	101	257	230	259	**258**	100	53	50	33	18	17	15	15		18	10	2	–	–	–	–	–	–	–	–	–	–	**1,4(1H,4H)-Chrysenedione**	100900-16-1	**NI25802**
258	180	257	181	259	182	255	260	**258**	100	100	87	43	28	27	8	7		18	14	–	–	–	–	–	–	–	1	–	–	–	**9-Silafluorene, 9-phenyl-**	18557-54-5	**NI25803**
180	77	258	155	181	257	51	259	**258**	100	85	64	40	23	22	16	12		18	14	–	2	–	–	–	–	–	–	–	–	–	**Aniline, N-(phenyl-2-pyridinylmethylene)-**	30894-24-7	**NI25804**
257	212	258	242	230				**258**	100	45	20	11	9					18	14	–	2	–	–	–	–	–	–	–	–	–	**1H-Benzo[de]cinnoline, 1-methyl-3-phenyl-**		**LI05342**
258	243	154	229	204	182	149		**258**	100	95	73	46	36	36	31			18	14	–	2	–	–	–	–	–	–	–	–	–	**Benzo[1,2:4,5]dicycloheptene, 6,12-dicyano-1,4,5,9-tetrahydro-**		**LI05343**
153	77	152	258	154	151	51	105	**258**	100	63	51	40	13	13	13	10		18	14	–	2	–	–	–	–	–	–	–	–	–	**1,1′-Biphenyl, 4-phenylazo-**		**MS05695**
258	241	243	242	257	240	189	259	**258**	100	51	37	36	32	30	26	20		18	14	–	2	–	–	–	–	–	–	–	–	–	**2-Butene, 3-amino-2-cyano-1-(9-fluorenylidene)-, (E)-**		**NI25805**
258	259	257	242	129	241	127	255	**258**	100	33	23	17	15	9	7	6		18	14	–	2	–	–	–	–	–	–	–	–	–	**Dibenzo[b,h][1,6]naphthyridine, 9,10-dimethyl-**	4240-85-1	**NI25806**
258	259	257	243	129	242	127	121.5	**258**	100	32	22	17	14	8	7	5		18	14	–	2	–	–	–	–	–	–	–	–	–	**Dibenzo[b,h][1,6]naphthyridine, 9,10-dimethyl-**	4240-85-1	**LI05344**
258	180	129	230	203	229	154	259	**258**	100	44	36	30	24	22	20	19		18	14	–	2	–	–	–	–	–	–	–	–	–	**Naphthalene, 1,4-dicyano-1-phenyl-1,2,3,4-tetrahydro-**	25679-87-2	**NI25807**
41	110	149	79	55	91	67	77	**258**	100	86	52	52	52	48	45	38	**34.36**	18	26	1	–	–	–	–	–	–	–	–	–	–	**Estr-4-en-3-one**	4811-77-2	**NI25808**
110	258	149	135	79	40	94	67	**258**	100	47	45	33	28	27	26	25		18	26	1	–	–	–	–	–	–	–	–	–	–	**Estr-4-en-3-one**	4811-77-2	**NI25809**
43	109	91	107	105	77	95	119	**258**	100	96	60	52	52	41	40	38	**14.00**	18	26	1	–	–	–	–	–	–	–	–	–	–	**3,5,7-Octatrien-2-one, 6-methyl-8-(2,6,6-trimethyl-1-cyclohexenyl)-, (3E,5E,7E)-**	17974-57-1	**NI25810**
133	258	55	107	83	41	189	254	**258**	100	60	56	53	44	36	27	26		18	26	1	–	–	–	–	–	–	–	–	–	–	**Phenol, 2,4-dicyclohexyl-**		**IC03387**
133	107	176	258	55	254	120	41	**258**	100	56	30	29	24	22	22	20		18	26	1	–	–	–	–	–	–	–	–	–	–	**Phenol, 2,6-dicyclohexyl-**		**IC03388**
243	43	258	244	201	159	259	128	**258**	100	53	28	20	8	6	6	5		18	26	1	–	–	–	–	–	–	–	–	–	–	**Tetralin, 7-acetyl-6-ethyl-1,1,4,4-tetramethyl-**	88-29-9	**NI25811**
258	215	259	242	228	257	229	226	**258**	100	27	22	22	18	17	15	12		19	14	1	–	–	–	–	–	–	–	–	–	–	**Benz[a]anthracene, 5,6-dihydro-5,6-epoxy-7-methyl-**		**NI25812**
181	258	152	77	105	153	51	182	**258**	100	50	44	32	22	15	13	11		19	14	1	–	–	–	–	–	–	–	–	–	–	**Benzophenone, 4-phenyl-**	2128-93-0	**NI25813**
230	258	43	41	56	39	229	231	**258**	100	49	40	32	27	23	22	20		19	14	1	–	–	–	–	–	–	–	–	–	–	**1,4-Pentadiyn-3-one, 1,5-bis(4-methylphenyl)-**	34793-64-1	**MS05696**
230	258	177	40	231	229	117	91	**258**	100	67	23	22	21	18	17	17		19	14	1	–	–	–	–	–	–	–	–	–	–	**2,4-Pentadiyn-1-one, 1,5-bis(4-methylphenyl)-**	29372-68-7	**MS05697**
156	127	128	129	155	158	157	77	**258**	100	82	77	70	60	50	23	21	**2.20**	19	14	1	–	–	–	–	–	–	–	–	–	–	**2-Propen-1-one, 3-(1-naphthalenyl)-1-phenyl-**	42299-49-0	**NI25814**
181	182	258	257	178	259	152	255	**258**	100	30	15	14	12	6	6	4		19	14	1	–	–	–	–	–	–	–	–	–	–	**9H-Xanthene, 9-phenyl-**		**IC03389**
189	204	258	95	120	79	41	67	**258**	100	94	40	36	33	33	32	30		19	30	–	–	–	–	–	–	–	–	–	–	–	**Androst-2-ene**	40061-86-7	**NI25815**
243	258	149	148	147	95	109	41	**258**	100	73	54	52	35	34	27	27		19	30	–	–	–	–	–	–	–	–	–	–	–	**Androst-7-ene, (5α)-**	54411-76-6	**NI25816**
117	173	29	131	41	27	129	115	**258**	100	98	32	31	28	22	17	17	**12.57**	19	30	–	–	–	–	–	–	–	–	–	–	–	**Indan, 2-butyl-1-hexyl-**		**AI01834**
173	117	131	29	41	115	129	43	**258**	100	96	30	26	19	17	15	15	**12.99**	19	30	–	–	–	–	–	–	–	–	–	–	–	**Indan, 2-butyl-1-hexyl-**		**AI01835**
173	117	41	43	131	115	91	129	**258**	100	60	27	22	17	17	16	15	**12.43**	19	30	–	–	–	–	–	–	–	–	–	–	–	**Indan, 2-butyl-1-hexyl-**		**AI01836**
187	258	117	43	131	188	129	41	**258**	100	47	19	16	15	14	13	9		19	30	–	–	–	–	–	–	–	–	–	–	–	**Indan, 2-butyl-5-hexyl-**		**AI01837**
187	258	188	43	29	129	117	41	**258**	100	56	30	28	25	20	20	18		19	30	–	–	–	–	–	–	–	–	–	–	–	**Indan, 2-butyl-5-hexyl-**		**AI01838**
187	258	188	43	41	117	129	131	**258**	100	37	34	34	25	24	18	17		19	30	–	–	–	–	–	–	–	–	–	–	–	**Indan, 2-butyl-5-hexyl-**		**AI01839**
145	258	187	131	29	146	129	41	**258**	100	32	16	15	15	12	10	10		19	30	–	–	–	–	–	–	–	–	–	–	–	**Indan, 5-butyl-6-hexyl-**		**AI01842**
145	258	131	187	146	41	43	129	**258**	100	26	21	20	16	16	13	11		19	30	–	–	–	–	–	–	–	–	–	–	–	**Indan, 5-butyl-6-hexyl-**		**AI01840**
145	258	187	129	215	128	259	131	**258**	100	29	19	7	6	6	5	5		19	30	–	–	–	–	–	–	–	–	–	–	–	**Indan, 5-butyl-6-hexyl-**		**AI01841**
117	258	104	131	116	118	43	41	**258**	100	82	65	42	35	32	32	32		19	30	–	–	–	–	–	–	–	–	–	–	–	**Indan, 2-decyl-**		**AI01844**
117	104	41	131	43	118	116	258	**258**	100	64	39	38	38	35	30	29		19	30	–	–	–	–	–	–	–	–	–	–	–	**Indan, 2-decyl-**		**AI01843**
131	132	258	117	41	29	91	43	**258**	100	30	29	15	12	12	9	9		19	30	–	–	–	–	–	–	–	–	–	–	–	**Indan, 5-decyl-**		**AI01846**
131	132	258	117	41	43	91	115	**258**	100	34	21	16	16	13	10	8		19	30	–	–	–	–	–	–	–	–	–	–	–	**Indan, 5-decyl-**		**AI01845**
131	132	91	41	29	258	43	115	**258**	100	11	8	8	8	6	6	5		19	30	–	–	–	–	–	–	–	–	–	–	–	**Indan, 1-methyl-3-nonyl-**	29138-85-0	**WI01309**

MASS TO CHARGE RATIOS								M.W.	INTENSITIES								Parent	C	H	O	N	S	F	Cl	Br	I	Si	P	B	X	COMPOUND NAME	CAS Reg No	No
117	131	187	201	29	129	41	43	**258**	100	97	61	46	45	34	30	26	**20.22**	19	30	–	–	–	–	–	–	–	–	–	–	–	**Naphthalene, 1-butyl-4-pentyl-1,2,3,4-tetrahydro-**	29138-93-0	**WI01310**
131	132	91	29	258	41	129	43	**258**	100	11	9	7	6	6	5	5		19	30	–	–	–	–	–	–	–	–	–	–	–	**Naphthalene, 1-nonyl-1,2,3,4-tetrahydro-**	33425-49-9	**WI01311**
118	117	91	41	131	173	43	115	**258**	100	72	55	33	23	22	21	15	**7.85**	19	30	–	–	–	–	–	–	–	–	–	–	–	**6-Tridecene, 7-phenyl-**		**AI01848**
118	117	91	41	173	131	43	129	**258**	100	54	50	28	23	20	19	13	**11.40**	19	30	–	–	–	–	–	–	–	–	–	–	–	**6-Tridecene, 7-phenyl-**		**AI01847**
215	258	216	202	229	203	259	228	**258**	100	82	40	36	32	22	18	16		20	18	–	–	–	–	–	–	–	–	–	–	–	**Benzo[b]fluoranthene, 3b,4,5,6,7,7a-hexahydro-**		**NI25817**
215	258	202	216	203	259	201	200	**258**	100	64	47	42	17	16	16	14		20	18	–	–	–	–	–	–	–	–	–	–	–	**Benzo[e]pyrene, 8b,9,10,11,12,12a-hexahydro-**	86439-17-0	**NI25818**
258	215	259	257	230	229	202	216	**258**	100	24	22	21	21	20	15	14		20	18	–	–	–	–	–	–	–	–	–	–	–	**Benzo[a]pyrene, 4,5,7,8,9,10-hexahydro-**	73712-75-1	**NI25819**
258	215	230	229	259	257	202	216	**258**	100	30	28	23	22	22	19	15		20	18	–	–	–	–	–	–	–	–	–	–	–	**Benzo[a]pyrene, 4,5,7,8,9,10-hexahydro-**	73712-75-1	**NI25820**
258	215	257	259	216	202	229	230	**258**	100	41	23	22	22	19	17	15		20	18	–	–	–	–	–	–	–	–	–	–	–	**Benzo[a]pyrene, 7,8,9,10,11,12-hexahydro-**	73712-71-7	**NI25821**
258	230	229	215	217	252	259	226	**258**	100	40	38	34	23	22	21	20		20	18	–	–	–	–	–	–	–	–	–	–	–	**Benzo[e]pyrene, 1,2,3,6,7,8-hexahydro-**		**NI25822**
230	215	258	229	252	228	216	226	**258**	100	95	73	39	35	22	20	18		20	18	–	–	–	–	–	–	–	–	–	–	–	**Benzo[e]pyrene, 4,5,9,10,11,12-hexahydro-**		**NI25823**
104	258	130	153	128	154	129	115	**258**	100	73	47	43	34	31	21	21		20	18	–	–	–	–	–	–	–	–	–	–	–	**1,2′-Binaphthyl, 1′,2′,3′,4′-tetrahydro-**		**AI01849**
243	165	244	181	166	258	77	239	**258**	100	35	21	11	9	7	6	4		20	18	–	–	–	–	–	–	–	–	–	–	–	**Ethane, 1,1,1-triphenyl-**	5271-39-6	**NI25824**
167	165	168	152	91	166	39	51	**258**	100	65	53	32	26	25	11	9	**1.12**	20	18	–	–	–	–	–	–	–	–	–	–	–	**Ethane, 1,1,2-triphenyl-**		**AI01850**
258	228	243	165	115	215	242	91	**258**	100	26	21	20	19	16	15	15		20	18	–	–	–	–	–	–	–	–	–	–	–	**2,6-Octadien-4-yne, 3,6-diphenyl-**		**NI25825**
258	117	167	154	91	115	165	259	**258**	100	74	59	46	37	30	28	22		20	18	–	–	–	–	–	–	–	–	–	–	–	**1,3,5,7-Octatetraene, 1,8-diphenyl-, (all-E)-**	22828-29-1	**NI25826**
258	215	257	259	252	229	239	230	**258**	100	30	27	22	21	21	17	15		20	18	–	–	–	–	–	–	–	–	–	–	–	**Perylene, 1,2,3,7,8,9-hexahydro-**	68912-99-2	**NI25827**
258	215	257	259	252	229	239	230	**258**	100	30	27	22	21	21	17	15		20	18	–	–	–	–	–	–	–	–	–	–	–	**Perylene, 1,2,3,7,8,9-hexahydro-**	68912-99-2	**NI25828**
258	215	230	229	257	259	239	228	**258**	100	30	23	23	21	20	13	12		20	18	–	–	–	–	–	–	–	–	–	–	–	**Perylene, 1,2,3,10,11,12-hexahydro-**	7350-92-7	**NI25829**
258	215	230	229	257	259	239	252	**258**	100	30	23	23	21	20	13	9		20	18	–	–	–	–	–	–	–	–	–	–	–	**Perylene, 1,2,3,10,11,12-hexahydro-**	7350-92-7	**NI25830**
215	203	108	201	261	213	143	259	**259**	100	99	86	81	77	75	60	59		6	1	2	1	–	–	4	–	–	–	–	–	–	**Benzene, 1,2,3,4-tetrachloro-5-nitro-**	879-39-0	**NI25831**
203	201	215	261	213	108	259	30	**259**	100	88	87	73	71	70	58	51		6	1	2	1	–	–	4	–	–	–	–	–	–	**Benzene, 1,2,4,5-tetrachloro-3-nitro-**	117-18-0	**NI25832**
261	215	259	213	203	201	263	217	**259**	100	89	80	72	71	57	51	45		6	1	2	1	–	–	4	–	–	–	–	–	–	**Benzene, 1,2,4,5-tetrachloro-3-nitro-**	117-18-0	**NI25833**
215	261	203	213	201	259	205	217	**259**	100	96	91	74	73	57	53	47		6	1	2	1	–	–	4	–	–	–	–	–	–	**Benzene, 1,2,4,5-tetrachloro-3-nitro-**	117-18-0	**NI25834**
246	244	248	261	218	259	216	263	**259**	100	71	43	34	34	31	29	21		7	5	1	1	–	–	4	–	–	–	–	–	–	**Aniline, 2,3,4,5-tetrachloro-6-methoxy-**	42138-72-7	**NI25835**
246	244	248	261	86	259	218	216	**259**	100	80	52	36	33	28	23	19		7	5	1	1	–	–	4	–	–	–	–	–	–	**Aniline, 2,3,5,6-tetrachloro-4-methoxy-**	70439-96-2	**NI25836**
261	218	259	216	263	220	146	110	**259**	100	90	81	74	53	46	25	17		7	5	1	1	–	–	4	–	–	–	–	–	–	**Aniline, 2,4,5,6-tetrachloro-3-methoxy-**	941-79-7	**NI25837**
126	90	127	190	63	128	259	52	**259**	100	84	69	44	39	36	29	24		9	10	2	3	1	–	1	–	–	–	–	–	–	**Benzenesulphonamide, 3-amino-4-chloro-N-(2-cyanoethyl)-**	64415-13-0	**NI25838**
156	139	259	140	152	44	123	59	**259**	100	71	40	27	20	17	13	13		9	13	6	3	–	–	–	–	–	–	–	–	–	**1H-Pyrazole-5-carboxamide, 4-hydroxy-3-β-D-ribofuranosyl-**	30868-30-5	**NI25839**
259	258	212	213	52	198	185	56	**259**	100	98	50	36	16	14	14	13		10	8	2	3	–	3	–	–	–	–	–	–	–	**1H-Benzimidazole, 2-ethyl-4-nitro-6-(trifluoromethyl)-**	51026-15-4	**NI25840**
42	111	175	75	56	84	113	177	**259**	100	22	16	11	9	7	7	6	**2.19**	10	10	3	1	1	–	1	–	–	–	–	–	–	**3-Pyrrolidinone, 1-(4-chlorobenzenesulphonyl)-**	83132-99-4	**NI25841**
200	172	119	103	259	228			**259**	100	36	15	15	4	3				10	13	7	1	–	–	–	–	–	–	–	–	–	**3,4,5-Isoxazoletricarboxylic acid, 2,3-dihydro-2-methyl-, trimethyl ester**		**LI05345**
44	87	86	30	18	43	114	28	**259**	100	62	51	22	20	17	16	7	**1.30**	10	17	5	3	–	–	–	–	–	–	–	–	–	**Glycine, N-[N-(N-acetyl-L-alanyl)glycyl]-, methyl ester**	33105-22-5	**NI25842**
259	190	162	128	146	130	69	77	**259**	100	62	41	38	34	33	25	23		11	8	1	1	1	3	–	–	–	–	–	–	–	**2(1H)-Quinolinone, 4-methyl-3-[(trifluoromethyl)thio]-**		**DD00750**
69	50	77	146	76	63	51	118	**259**	100	75	56	47	38	31	31	21	**0.00**	11	8	1	1	1	3	–	–	–	–	–	–	–	**4(1H)-Quinolinone, 2-methyl-3-[(trifluoromethyl)thio]-**		**DD00751**
168	30	244	81	214	259	52	39	**259**	100	73	67	57	50	44	43	40		11	9	3	5	–	–	–	–	–	–	–	–	–	**1,3,2-Benzodioxazole, 4-[1-(2-ethylimidazolyl)]-7-nitro-**	91485-31-3	**NI25843**
59	101	41	140	45	112	108	69	**259**	100	43	43	31	28	26	23	23	**0.20**	11	17	6	1	–	–	–	–	–	–	–	–	–	**2-Butenedioic acid, [[3-(methoxycarbonyl)propyl]amino]-, dimethyl ester, (Z)-**		**NI25844**
68	67	180	69	260	82	262	95	**259**	100	52	35	35	34	34	32	26	**0.00**	11	18	1	1	–	–	–	1	–	–	–	–	–	**2-Azetidinone, 1-(5-bromopentyl)-4-methyl-4-vinyl-**		**NI25845**
186	84	140	114	171	259	99		**259**	100	92	40	33	16	13	5			11	21	4	3	–	–	–	–	–	–	–	–	–	**1,2-Hydrazinedicarboxylic acid, 1-piperidino-, diethyl ester**		**LI05346**
142	73	45	143	43	59	147	75	**259**	100	75	20	14	10	8	7	7	**0.00**	11	25	2	1	–	–	–	–	–	2	–	–	–	**L-Proline, 1-(trimethylsilyl)-, trimethylsilyl ester**	7364-47-8	**MS05698**
142	73	143	147	216	74	144	75	**259**	100	50	14	6	6	4	4	4	**0.24**	11	25	2	1	–	–	–	–	–	2	–	–	–	**L-Proline, 1-(trimethylsilyl)-, trimethylsilyl ester**	7364-47-8	**NI25846**
73	142	45	43	75	59	143	72	**259**	100	79	32	15	13	13	10	10	**0.00**	11	25	2	1	–	–	–	–	–	2	–	–	–	**2-Pyrrolidineformic acid, 1-(trimethylsilyl)-, trimethylsilyl ester**		**MS05699**
259	196	167	168	166	260	50	92	**259**	100	27	26	24	20	15	13	12		12	9	4	3	–	–	–	–	–	–	–	–	–	**Diphenylamine, 2,2′-dinitro-**	18264-71-6	**LI05347**
259	196	167	168	166	260	50	39	**259**	100	27	26	24	20	15	13	12		12	9	4	3	–	–	–	–	–	–	–	–	–	**Diphenylamine, 2,2′-dinitro-**	18264-71-6	**MS05700**
259	167	166	77	260	168	139	51	**259**	100	31	25	16	15	14	13	13		12	9	4	3	–	–	–	–	–	–	–	–	–	**Diphenylamine, 2,4-dinitro-**	961-68-2	**MS05701**
259	167	166	260	139	154	50	76	**259**	100	19	19	15	12	11	10	9		12	9	4	3	–	–	–	–	–	–	–	–	–	**Diphenylamine, 2,4′-dinitro-**	612-36-2	**MS05702**
259	167	166	260	139	154	50	229	**259**	100	19	19	15	12	11	10	9		12	9	4	3	–	–	–	–	–	–	–	–	–	**Diphenylamine, 2,4′-dinitro-**	612-36-2	**LI05348**
259	167	229	166	183	260	28	139	**259**	100	46	25	22	16	13	11	10		12	9	4	3	–	–	–	–	–	–	–	–	–	**Diphenylamine, 4,4′-dinitro-**	1821-27-8	**MS05703**
43	93	41	28	39	27	137	65	**259**	100	67	29	28	24	22	18	18	**0.00**	12	10	–	5	–	–	1	–	–	–	–	–	–	**1H-Purin-6-amine, [(3-chlorophenyl)methyl]-**	74421-48-0	**NI25847**
227	212	43	28	58	68	69	170	**259**	100	75	75	75	64	47	33	31	**0.00**	12	10	–	5	–	–	1	–	–	–	–	–	–	**1H-Purin-6-amine, [(4-chlorophenyl)methyl]-**	74421-47-9	**NI25848**
186	159	109	90	259	214	89	117	**259**	100	97	65	37	28	26	26	24		12	12	2	1	–	3	–	–	–	–	–	–	–	**Aziridinecarboxylic acid, 3-phenyl-2-(trifluoromethyl)-, ethyl ester, cis-**		**DD00752**
117	187	259	118	186	166	90	91	**259**	100	96	79	46	42	32	29	27		12	12	2	1	–	3	–	–	–	–	–	–	–	**1-Propene, 2-[(ethoxycarbonyl)amino]-1-phenyl-3,3,3-trifluoro-, (Z)-**		**DD00753**
172	144	43	170	171	145	41	213	**259**	100	99	97	65	62	40	37	21	**1.00**	12	13	2	5	–	–	–	–	–	–	–	–	–	**1-Cyclopenten-2-amine, 3-methyl-3-nitro-4-propyl-1,5,5-tricyano-**		**MS05704**

MASS TO CHARGE RATIOS								M.W.	INTENSITIES								Parent	C	H	O	N	S	F	Cl	Br	I	Si	P	B	X	COMPOUND NAME	CAS Reg No	No
259	260	44	231	203	55	109	43	**259**	100	15	13	9	9	8	7	6		12	13	2	5	–	–	–	–	–	–	–	–	–	Imidazo[1,2-a]pyrimidin-7-one, 1-(5-methyl-4-oxo-3,4-dihydro-2-pyrimidyl)-6-methyl-1,2,3,7-tetrahydro-		NI25849
161	163	259	261	217	165	219	162	**259**	100	65	35	22	13	11	9	7		12	15	1	1	–	–	2	–	–	–	–	–	–	Pentanamide, N-(3,4-dichlorophenyl)-2-methyl-	2533-89-3	NI25850
42	43	41	140	141	44	96	56	**259**	100	86	76	70	60	57	52	44	**3.00**	12	21	5	1	–	–	–	–	–	–	–	–	–	1-Azepanecarboxylic acid, 4-ethoxycarbonyloxy-, ethyl ester		MS05705
113	43	59	85	30	100	114	229	**259**	100	72	53	27	20	18	17	15	**1.50**	12	21	5	1	–	–	–	–	–	–	–	–	–	α-D-Glucofuranose, 6-amino-6-deoxy-1,2:3,5-bis-O-(isopropylidene)-	34322-93-5	NI25851
73	142	244	189	99	41	75	29	**259**	100	29	27	26	26	26	22	21	**0.00**	12	25	3	1	–	–	–	–	–	1	–	–	–	Octanoic acid, 2-(methoxyimino)-, trimethylsilyl ester	55494-11-6	NI25852
73	189	142	99	244	75	89	41	**259**	100	45	43	26	23	17	14	14	**0.00**	12	25	3	1	–	–	–	–	–	1	–	–	–	Octanoic acid, 2-(methoxyimino)-, trimethylsilyl ester	55494-11-6	NI25853
259	77	257	258	178	256	69	71	**259**	100	64	55	54	50	38	33	30		13	9	–	1	–	–	–	–	–	–	–	–	1	1,2-Benzisoselenazole, 3-phenyl-	40193-45-1	NI25854
259	77	257	258	178	256	69	255	**259**	100	65	55	54	50	38	33	30		13	9	–	1	–	–	–	–	–	–	–	–	1	1,2-Benzisoselenazole, 3-phenyl-	40193-45-1	LI05349
259	156	257	154	256	255	117	260	**259**	100	58	55	29	23	21	19	18		13	9	–	1	–	–	–	–	–	–	–	–	1	Benzoselenazole, 2-phenyl-	32586-68-8	NI25855
165	137	182	109	46	259			**259**	100	65	21	3	2	1				13	9	5	1	–	–	–	–	–	–	–	–	–	Benzoic acid, 2-hydroxy-5-nitro-, phenyl ester	616-71-7	NI25856
165	166	121	93	122	258	167	46	**259**	100	93	39	16	15	11	8	6	**4.00**	13	9	5	1	–	–	–	–	–	–	–	–	–	Benzoic acid, 2-hydroxy-6-nitro-, phenyl ester	55076-36-3	NI25857
259	258	138	260	46	137	165	93	**259**	100	35	21	15	15	8	3	3		13	9	5	1	–	–	–	–	–	–	–	–	–	Benzoic acid, 3-hydroxy-5-nitro-, phenyl ester	55076-33-0	NI25858
77	259	44	40	214	94	141	43	**259**	100	70	70	61	47	38	35	34		13	9	5	1	–	–	–	–	–	–	–	–	–	Carbonic acid, 4-nitrophenyl phenyl ester	17175-11-0	NI25859
77	200	65	51	94	44	259	50	**259**	100	99	24	22	20	17	15	14		13	9	5	1	–	–	–	–	–	–	–	–	–	Carbonic acid, 4-nitrophenyl phenyl ester	17175-11-0	NI25860
259	261	76	260	50	258	51	75	**259**	100	98	87	81	73	69	61	59		13	10	–	1	–	–	–	1	–	–	–	–	–	Aniline, 4-bromo-N-benzylidene-	780-20-1	NI25861
83	31	171	42	32	69	84	70	**259**	100	72	66	49	48	40	38	34	**2.00**	13	25	4	1	–	–	–	–	–	–	–	–	–	Acetic acid, 2-(2-sec-butyl-2-methylaminoacetoxy)-2-isopropyl-, methyl ester		LI05350
59	114	146	214	73	182	228	130	**259**	100	79	68	56	20	13	10	10	**0.00**	13	29	2	1	–	–	–	–	–	1	–	–	–	1-Butene, 1-[[[bis(2-methoxyethyl)amino]methyl]dimethylsilyl]-, (Z)-		DD00754
214	146	204	59	114	100	260	69	**259**	100	76	57	56	37	17	16	12	**0.00**	13	29	2	1	–	–	–	–	–	1	–	–	–	1-Butene, 3-[[[bis(2-methoxyethyl)amino]methyl]dimethylsilyl]-		DD00755
105	77	103	51	106	76	224	89	**259**	100	87	9	9	8	5	4	4	**2.00**	14	10	2	1	–	–	1	–	–	–	–	–	–	Benzenohydroximidoyl chloride, o-benzoyl-		NI25862
119	105	259	91	77	120	64	51	**259**	100	34	23	22	17	12	12	11		14	10	2	1	–	–	1	–	–	–	–	–	–	1,3,4-Dioxazole, 2-(4-chlorophenyl)-5-phenyl-		NI25863
153	155	259	139	105	77	90	125	**259**	100	34	32	28	25	20	18	17		14	10	2	1	–	–	1	–	–	–	–	–	–	1,3,4-Dioxazole, 5-(4-chlorophenyl)-2-phenyl-		NI25864
153	155	259	139	105	77	90	125	**259**	100	34	32	28	25	20	18	17		14	10	2	1	–	–	1	–	–	–	–	–	–	1,4,2-Dioxazole, 3-(4-chlorophenyl)-5-phenyl-	55076-20-5	NI25865
119	105	259	91	139	77	120	64	**259**	100	34	23	22	18	17	12	12		14	10	2	1	–	–	1	–	–	–	–	–	–	1,4,2-Dioxazole, 5-(4-chlorophenyl)-3-phenyl-	55076-24-9	NI25866
178	213	176	259	215	105	151	261	**259**	100	49	24	17	17	15	12	6		14	10	2	1	–	–	1	–	–	–	–	–	–	Ethene, 1-chloro-1,2-diphenyl-2-nitro-, (E)-	57337-95-8	DD00756
156	77	94	104	182	105	65	51	**259**	100	96	81	58	55	52	34	34	**0.00**	14	13	2	1	1	–	–	–	–	–	–	–	–	Sulphoximine, N-benzoyl-S-methyl-S-phenyl-	54090-90-3	NI25867
123	95	110	259	149	136	81	67	**259**	100	77	56	29	15	13	10	10		14	13	4	1	–	–	–	–	–	–	–	–	–	Acetamide, N-(3-anisolyl)-2-(2-furoyl)-		LI05351
95	123	81	149	110	259	67	136	**259**	100	72	45	37	37	35	25	14		14	13	4	1	–	–	–	–	–	–	–	–	–	Acetamide, N-(3-anisolyl)-2-(3-furoyl)-		LI05352
149	95	123	259	110	67	81	136	**259**	100	96	64	48	14	8	7	6		14	13	4	1	–	–	–	–	–	–	–	–	–	Acetamide, N-(4-anisolyl)-2-(3-furoyl)-		LI05353
259	244	216	186	260	201	173	53	**259**	100	45	19	17	14	14	13	7		14	13	4	1	–	–	–	–	–	–	–	–	–	Furo[2,3-b]quinoline, 4,6,7-trimethoxy-	484-08-2	NI25868
259	258	229	228	244	213	199	173	**259**	100	52	46	25	22	21	18	15		14	13	4	1	–	–	–	–	–	–	–	–	–	Furo[2,3-b]quinoline, 4,6,8-trimethoxy-	522-19-0	NI25869
244	159	230	216	201	258	229	213	**259**	100	70	52	31	28	26	26	25	**0.00**	14	13	4	1	–	–	–	–	–	–	–	–	–	Furo[2,3-b]quinoline, 4,7,8-trimethoxy-	83-95-4	NI25870
259	244	258	260	214	201	231	217	**259**	100	33	21	15	10	10	9	8		14	13	4	1	–	–	–	–	–	–	–	–	–	Furo[2,3-b]quinolin-4(9H)-one, 6,7-dimethoxy-9-methyl-	523-15-9	NI25871
259	244	258	260	214	201	231	230	**259**	100	33	21	15	10	10	9	8		14	13	4	1	–	–	–	–	–	–	–	–	–	Furo[2,3-b]quinolin-4(9H)-one, 6,7-dimethoxy-9-methyl-	523-15-9	LI05354
259	244	260	201	245	229	129.5	216	**259**	100	68	15	13	10	9	8	6		14	13	4	1	–	–	–	–	–	–	–	–	–	Furo[2,3-b]quinolin-4(9H)-one, 7,8-dimethoxy-9-methyl-	436-14-6	LI05355
259	244	260	201	245	229	216	173	**259**	100	68	15	13	10	9	6	4		14	13	4	1	–	–	–	–	–	–	–	–	–	Furo[2,3-b]quinolin-4(9H)-one, 7,8-dimethoxy-9-methyl-	436-14-6	NI25872
120	259	42	105	77	75	261	104	**259**	100	80	30	28	23	20	18	18		14	14	–	3	–	–	1	–	–	–	–	–	–	Aniline, N,N-dimethyl-4-(3-chlorophenylazo)-		LI05356
120	259	42	261	105	77	148	111	**259**	100	68	34	23	20	20	15	15		14	14	–	3	–	–	1	–	–	–	–	–	–	Aniline, N,N-dimethyl-4-(4-chlorophenylazo)-		LI05357
120	259	42	105	111	77	261	75	**259**	100	46	25	18	17	17	15	13		14	14	–	3	–	–	1	–	–	–	–	–	–	Benzene, 4-chloro-4′-(dimethylaminoazo)-		NI25873
41	91	71	51	160	109	122	198	**259**	100	56	50	43	42	42	40	37	**0.00**	14	17	–	3	1	–	–	–	–	–	–	–	–	1,3-Octadiene, 8-azido-3-(phenylthio)-, (E)-	123289-43-0	DD00757
194	222	36	223	165	38	224	97	**259**	100	99	82	46	37	28	22	20		15	12	2	–	–	–	1	–	–	–	–	–	–	4′-Hydroxyflavylium hydrochloride		MS05706
36	223	194	222	165	38	224	35	**259**	100	56	56	53	43	34	33	23		15	12	2	–	–	–	1	–	–	–	–	–	–	7-Hydroxyflavylium hydrochloride		MS05707
91	118	119	92	65	77	42	51	**259**	100	20	19	14	13	11	8	7	**0.32**	15	14	1	1	–	–	1	–	–	–	–	–	–	2H-1,3-Benzoxazine, 3-benzyl-6-chloro-3,4-dihydro-	55955-92-5	NI25874
207	242	259	77	91	244	261	182	**259**	100	93	86	57	43	39	27	25		15	14	1	1	–	–	1	–	–	–	–	–	–	5-Chloro-2-N,N-dimethylaminobenzophenone		MS05708
105	231	110	126	77	55	42	68	**259**	100	70	49	40	20	20	20	16	**2.00**	15	17	3	1	–	–	–	–	–	–	–	–	–	1-Azabicyclo[2.2.2]octan-3-one, 6-[(benzoyloxy)methyl]-, cis-	36074-41-6	NI25875
105	231	110	126	55	42	77	41	**259**	100	70	42	36	23	22	20	17	**2.00**	15	17	3	1	–	–	–	–	–	–	–	–	–	1-Azabicyclo[2.2.2]octan-3-one, 6-[(benzoyloxy)methyl]-, trans-	36074-39-2	NI25876
215	44	91	259	216	45	92	64	**259**	100	90	61	34	31	29	28	21		15	17	3	1	–	–	–	–	–	–	–	–	–	1,2-Benzenediol, 4-[1-(4-hydroxyphenyl)-2-(methylamino)ethyl]-		LI05358
187	72	159	131	103	259	215	200	**259**	100	99	80	60	50	40	40	20		15	17	3	1	–	–	–	–	–	–	–	–	–	1,3-Indandione, 2-(diethylcarbamoyl)-2-methyl-	36692-53-2	DD00758
172	259	203	184	212	115	199	226	**259**	100	82	73	19	18	15	9	8		15	17	3	1	–	–	–	–	–	–	–	–	–	2-Indolecarboxylic acid, 6-hydroxy-7-(3-methylbut-2-enyl)-, methyl ester		MS05709
148	259	135	151	146	136	190	133	**259**	100	56	20	10	7	3	3	2		15	17	3	1	–	–	–	–	–	–	–	–	–	5,7(5H,7H)-Indolizidinedione, 6-(4-methoxyphenyl)-		MS05710
188	259	189	202	77	55	93	43	**259**	100	48	44	35	31	26	24	20		15	17	3	1	–	–	–	–	–	–	–	–	–	1H-Isoindole-1,3(2H)-dione, hexahydro-4-hydroxy-3a-methyl-2-phenyl-, [3aR-(3aα,4β,7aα)]-	54346-16-6	NI25877
93	189	188	259	70	97	77	55	**259**	100	80	76	70	70	57	41	31		15	17	3	1	–	–	–	–	–	–	–	–	–	1H-Isoindole-1,3(2H)-dione, hexahydro-4-hydroxy-7a-methyl-2-phenyl-, [3aS-(3aα,4β,7aα)]-	54355-48-5	NI25878

MASS TO CHARGE RATIOS	M.W.	INTENSITIES	Parent	C	H	O	N	S	F	Cl	Br	I	Si	P	B	X	COMPOUND NAME	CAS Reg No	No
218 176 43 200 77 105 97 104	**259**	100 65 64 44 40 39 30 18	**0.00**	15	17	3	1	–	–	–	–	–	–	–	–	–	4-Oxazolecarboxylic acid, 4,5-dihydro-5-methyl-2-phenyl-4-(2-propenyl)-, methyl ester, (4S,5S)-		DD00759
175 259 176 145 84 89 117 63	**259**	100 70 52 48 48 39 25 24		15	17	3	1	–	–	–	–	–	–	–	–	–	Piperidine, N-(3,4-methylenedioxycinnamoyl)-		LI05359
187 103 131 159 215 214 259 204	**259**	100 62 60 42 24 20 16 12		15	17	3	1	–	–	–	–	–	–	–	–	–	Propanamide, N,N-diethyl-2-phthalidylidene-	36692-55-4	DD00760
201 259 173 186 130 260 202 158	**259**	100 94 67 48 47 44 37 36		15	17	3	1	–	–	–	–	–	–	–	–	–	2H-Pyrano[4,3-b]indol-1-one, 1,4-dihydro-3,3-dimethyl-8-methoxy-5-methyl-		DD00761
138 121 139 77 91 105 110 82	**259**	100 63 30 18 12 10 7 7	**3.00**	15	17	3	1	–	–	–	–	–	–	–	–	–	2-Pyrrolidone, 1-(4-methoxy-4-phenyl-3-oxo-1-butenyl)-		NI25879
105 231 110 126 55 42 77 81	**259**	100 70 42 37 23 23 20 17	**2.00**	15	17	3	1	–	–	–	–	–	–	–	–	–	5-Quinuclidone, 2-benzoyloxymethyl-, cis-		LI05360
105 231 110 126 77 55 42 96	**259**	100 70 47 38 21 21 21 15	**2.00**	15	17	3	1	–	–	–	–	–	–	–	–	–	5-Quinuclidone, 2-benzoyloxymethyl-, trans-		LI05361
97 259 140 69 119 98 260 125	**259**	100 37 25 13 8 7 5 5		15	21	1	3	–	–	–	–	–	–	–	–	–	1H-Pyrazole-1-carboxamide, 4-ethyl-4,5-dihydro-N-phenyl-5-propyl-	55030-12-1	NI25880
71 259 189 118 96 201 130 109	**259**	100 45 34 17 15 12 7 7		15	21	1	3	–	–	–	–	–	–	–	–	–	1,3,8-Triazaspiro[4.5]decan-4-one, 3,8-dimethyl-1-phenyl-		MS05711
155 259 51 129 116 153 156 76	**259**	100 37 34 24 16 13 9 7		16	16	–	–	–	–	–	–	–	–	–	–	1	Vanadium, bis(η^8-1,3,5,7-cyclooctatetraene)-	35796-61-3	NI25881
79 91 107 92 80 106 153 174	**259**	100 62 33 26 19 17 14 2	**1.00**	16	21	2	1	–	–	–	–	–	–	–	–	–	exo-Bicyclo[4.1.0]hept-3-en-7-carbamic acid, (endo-bicyclo[4.1.0]hept-3-en-7-methyl) ester	90013-18-6	NI25882
79 91 107 92 153 109 108 80	**259**	100 70 67 36 24 24 23 23	**7.00**	16	21	2	1	–	–	–	–	–	–	–	–	–	exo-Bicyclo[4.1.0]hept-3-en-7-carbamic acid, (exo-bicyclo[4.1.0]hept-3-en-7-methyl) ester	94730-34-4	NI25883
83 68 84 103 77 91 157 82	**259**	100 95 31 20 17 12 11 11	**4.50**	16	21	2	1	–	–	–	–	–	–	–	–	–	3-Cyclohexene-1-carboxylic acid, 2-(methylamino)-1-phenyl-, ethyl ester		MS05712
123 97 122 150 231 37 216 124	**259**	100 42 29 26 19 11 10 10	**10.00**	16	21	2	1	–	–	–	–	–	–	–	–	–	13-Dehydro-15(14)-anhydroserratinine		LI05362
161 259 162 145 91 188 130 99	**259**	100 55 39 34 34 24 24 20		16	21	2	1	–	–	–	–	–	–	–	–	–	1H-Dibenz[e,g]isoindole-1,3(2H)-dione, 3a,3b,4,5,6,7,8,9,10,11,11a,11b-dodecahydro-, (3aα,3bα,11aα,11bα)-	51624-45-4	NI25884
123 259 244 122 258 242 150 124	**259**	100 95 63 60 40 16 16 16		16	21	2	1	–	–	–	–	–	–	–	–	–	Elaeocarpine, 15,16-dihydro-, (+)-		LI05363
123 122 259 244 97 69 82 55	**259**	100 72 32 28 28 27 24 24		16	21	2	1	–	–	–	–	–	–	–	–	–	Isoelaeocarpine, 15,16-dihydro-, (–)-		LI05364
105 77 216 106 51 138 188 217	**259**	100 40 32 9 8 7 6 5	**1.19**	16	21	2	1	–	–	–	–	–	–	–	–	–	1,3-Oxazin-6-one, 4,5-dihydro-4,4-diisopropyl-2-phenyl-	77074-13-6	NI25885
176 259 160 175 177 174 244 188	**259**	100 76 22 18 16 16 15 12		16	21	2	1	–	–	–	–	–	–	–	–	–	1,9b-Propano-5H-indeno[7,1-bc]azepin-5-one, 2,3,6,6a,7,8-hexahydro-7-hydroxy-8-methyl-		LI05365
72 30 28 115 144 116 73 43	**259**	100 37 12 8 7 6 6 6	**3.81**	16	21	2	1	–	–	–	–	–	–	–	–	–	2-Propanol, 1-(isopropylamino)-3-(1-naphthalenyloxy)-	525-66-6	NI25886
72 115 144 56 43 73 116 41	**259**	100 11 10 8 8 6 5 5	**2.00**	16	21	2	1	–	–	–	–	–	–	–	–	–	2-Propanol, 1-(isopropylamino)-3-(1-naphthalenyloxy)-	525-66-6	IC03390
72 144 115 73 43 116 56 127	**259**	100 8 8 8 8 7 7 4	**2.90**	16	21	2	1	–	–	–	–	–	–	–	–	–	2-Propanol, 1-(isopropylamino)-3-(1-naphthalenyloxy)-	525-66-6	MS05713
143 170 129 154 142 59 43 185	**259**	100 79 71 52 51 45 39 36	**0.00**	16	21	2	1	–	–	–	–	–	–	–	–	–	Quinolinium, 1-methyl-8-tert-butyl-, acetate		NI25887
156 200 171 214 215 128 259 185	**259**	100 90 87 77 35 33 27 25		16	25	–	3	–	–	–	–	–	–	–	–	–	Tricyclo[5.2.1.0^{4,10}]deca-2,5,8-triene, 1,4,7-tris(dimethylamino)-		NI25888
230 70 43 84 104 229 41 103	**259**	100 82 65 48 39 29 28 19	**0.62**	16	26	1	1	–	–	–	–	–	–	–	1	–	2H-1,3,2-Oxazaborine, 2,4-diethyl-3-isopropyl-6-phenyltetrahydro-	74630-86-7	NI25889
118 70 244 117 142 86 91 43	**259**	100 76 37 31 22 20 19 16	**2.88**	16	26	1	1	–	–	–	–	–	–	–	1	–	2H-1,3,2-Oxazaborine, 2-ethyl-4,5-dimethyl-3-isopropyl-6-phenyltetrahydro-	74630-85-6	NI25890
181 259 258 78 77 155 51 156	**259**	100 94 78 77 77 75 55 26		17	13	–	3	–	–	–	–	–	–	–	–	–	Aniline, 2,2′-dipyridylketonylidene-	100288-61-7	NI25891
258 259 134 216 41 135 136 55	**259**	100 65 48 38 34 30 24 22		17	25	1	1	–	–	–	–	–	–	–	–	–	9-Aza-D-homogona-13(14)-en-6-one, (±)-		LI05366
174 259 146 189 147 134 188 145	**259**	100 54 31 26 20 15 14 13		17	25	1	1	–	–	–	–	–	–	–	–	–	1-Nonen-3-one, 1-[4-(dimethylamino)phenyl]	36383-60-5	NI25892
105 104 77 202 130 203 259 214	**259**	100 73 71 65 55 12 7 4		17	25	1	1	–	–	–	–	–	–	–	–	–	2-Oxazoline, 4,5-dibutyl-2-phenyl-, cis-		LI05367
105 104 77 202 130 203 259 106	**259**	100 73 71 65 55 12 7 7		17	25	1	1	–	–	–	–	–	–	–	–	–	2-Oxazoline, 4,5-dibutyl-3-phenyl-, cis-	32161-19-6	NI25893
202 105 104 77 130 203 259 106	**259**	100 79 72 70 45 18 7 5		17	25	1	1	–	–	–	–	–	–	–	–	–	2-Oxazoline, 4,5-dibutyl-2-phenyl-, trans-	32161-20-9	NI25894
202 105 104 77 130 203 259 214	**259**	100 79 72 70 45 18 7 1		17	25	1	1	–	–	–	–	–	–	–	–	–	2-Oxazoline, 4,5-dibutyl-2-phenyl-, trans-	32161-20-9	LI05368
105 146 77 91 104 161 216 122	**259**	100 72 60 40 35 25 16 15	**10.01**	17	25	1	1	–	–	–	–	–	–	–	–	–	2-Oxazoline, 4-octyl-2-phenyl-	32014-89-4	NI25895
117 105 77 122 91 146 104 259	**259**	100 95 45 25 21 15 12 8		17	25	1	1	–	–	–	–	–	–	–	–	–	2-Oxazoline, 5-octyl-2-phenyl-	32014-89-4	LI05369
117 105 77 122 91 146 104 118	**259**	100 95 45 25 21 15 12 8	**8.01**	17	25	1	1	–	–	–	–	–	–	–	–	–	2-Oxazoline, 5-octyl-2-phenyl-	32014-90-7	NI25896
131 103 259 161 146 188 132 77	**259**	100 27 24 23 18 13 13 13		17	25	1	1	–	–	–	–	–	–	–	–	–	Propenamide, N-octyl-3-phenyl-		TR00288
259 230 260 115 258 114 100 77	**259**	100 15 14 11 9 6 4 4		18	13	1	1	–	–	–	–	–	–	–	–	–	Benz[c]acridin-7(12H)-one, 11-methyl-		NI25897
217 189 259 216 218 187 214 188	**259**	100 51 39 26 22 12 10 9		18	13	1	1	–	–	–	–	–	–	–	–	–	1-Pyreneacetamide	22755-15-3	NI25898
259 258 260 115 129 91 65 128	**259**	100 28 19 18 15 10 7 6		19	17	–	1	–	–	–	–	–	–	–	–	–	Pyridine, 2,6-di-p-tolyl-	14435-88-2	NI25899
44 76 45 46 47 60 64 48	**260**	100 83 48 45 17 15 13 12	**0.00**	2	2	–	2	4	–	–	–	–	–	–	–	2	1,2-Hydrazinedicarbodithioic acid, dipotassium salt		NI25900
43 42 44 69 260 159 217 28	**260**	100 93 91 55 21 21 17 14		2	6	–	2	2	4	–	–	–	–	2	–	–	Phosphorodifluoridothioic hydrazide, 1-(difluorophosphinothioyl)-2,2-dimethyl-	36267-51-3	MS05714
260 69 191 64 127 95 141 51	**260**	100 31 24 23 17 15 11 9		3	2	–	–	–	5	–	–	1	–	–	–	–	Propane, 1-iodo-2,2,3,3,3-pentafluoro-		IC03391
179 155 157 177 119 181 192 190	**260**	100 88 83 79 57 49 47 37	**2.40**	4	2	–	–	–	–	6	–	–	–	–	–	–	1-Butene, 1,1,3,3,4,4-hexachloro-		MS05715
29 27 28 231 229 260 202 26	**260**	100 29 11 8 7 6 6 6		4	10	–	–	–	–	–	–	–	–	–	–	1	Mercury, diethyl-		AI01851
225 227 175 109 85 177 262 260	**260**	100 68 45 31 27 25 20 20		5	–	–	–	–	5	3	–	–	–	–	–	–	Cyclopentene, 3,4,5-trichloropentafluoro-		IC03392

MASS TO CHARGE RATIOS								M.W.	INTENSITIES								Parent	C	H	O	N	S	F	Cl	Br	I	Si	P	B	X	COMPOUND NAME	CAS Reg No	No
90	127	133	225	84	92	169	135	**260**	100	96	95	47	42	38	33	32	**5.20**	5	10	–	2	1	5	1	–	–	–	–	–	–	[Chloro(diethylamino)methylenimino]sulphur pentafluoride	81439-13-6	NI25901
89	231	152	229	233	247	245	173	**260**	100	10	10	6	4	2	1	1	**0.00**	5	10	2	–	–	–	–	2	–	–	–	–	–	Propane, 1,1-dibromo-2,2-dimethoxy-		NI25902
73	179	177	89	231	87	105	88	**260**	100	94	94	66	54	54	44	44	**0.00**	5	10	2	–	–	–	–	2	–	–	–	–	–	Propane, 1,3-dibromo-2,2-dimethoxy-		NI25903
214	71	41	43	31	39	53	212	**260**	100	78	77	65	63	57	56	52	**0.00**	5	10	2	–	–	–	–	2	–	–	–	–	–	1,3-Propanediol, 2,2-bis(bromomethyl)-	3296-90-0	NI25904
56	96	97	135	125	95	84	176	**260**	100	60	55	49	37	28	27	24	**1.00**	6	5	3	–	–	–	–	1	–	–	–	–	1	Iron, (π-allyl)bromotricarbonyl-		MS05716
109	93	135	127	15	79	173	229	**260**	100	96	56	53	39	33	31	30	**8.54**	6	14	7	–	–	–	–	–	–	–	2	–	–	Phosphonic acid, (1-hydroxyvinyl)-, dimethyl ester, dimethyl phosphate	3328-33-4	NI25905
181	31	161	81	182	93	69	75	**260**	100	29	18	11	10	10	10	8	**2.60**	7	2	–	–	–	5	–	1	–	–	–	–	–	Benzene, (bromomethyl)pentafluoro-	1765-40-8	NI25906
181	31	161	81	93	69	75	131	**260**	100	29	18	11	10	10	8	7	**2.55**	7	2	–	–	–	5	–	1	–	–	–	–	–	Benzene, (bromomethyl)pentafluoro-	1765-40-8	LI05370
262	260	264	47	227	229	212	210	**260**	100	83	65	53	38	35	33	30		7	4	–	–	1	–	4	–	–	–	–	–	–	Benzene, 1,2,4,5-tetrachloro-3-(methylthio)-	68671-90-9	NI25907
247	245	262	260	249	87	264	219	**260**	100	82	62	52	50	39	30	15		7	4	2	–	–	–	4	–	–	–	–	–	–	Phenol, 2,3,5,6-tetrachloro-4-methoxy-	484-67-3	NI25908
211	56	120	92	213	175	27	42	**260**	100	74	66	41	33	20	14	14	**0.06**	7	15	2	2	–	–	2	–	–	–	1	–	–	2H-1,3,2-Oxazaphosphorine, 2-[bis(2-chloroethylamino)]tetrahydro-, 2-oxide	50-18-0	NI25909
79	69	68	55	15	67	56	29	**260**	100	82	75	42	38	30	27	25	**0.00**	7	16	6	–	2	–	–	–	–	–	–	–	–	1,5-Pentanediol, dimethanesulphonate	2374-22-3	NI25910
68	79	41	67	85	56	69	55	**260**	100	55	48	38	36	34	32	32	**0.00**	7	16	6	–	2	–	–	–	–	–	–	–	–	1,5-Pentanediol, dimethanesulphonate	2374-22-3	NI25911
75	121	260	97	28	16	47	65	**260**	100	32	18	18	17	17	16	14		7	17	2	–	3	–	–	–	–	–	1	–	–	Phosphorodithioic acid, O,O-diethyl S-[(ethylthio)methyl] ester	298-02-2	NI25912
75	97	47	65	121	45	93	41	**260**	100	24	24	23	22	12	11	11	**6.40**	7	17	2	–	3	–	–	–	–	–	1	–	–	Phosphorodithioic acid, O,O-diethyl S-[(ethylthio)methyl] ester	298-02-2	NI25913
29	109	18	81	27	45	47	183	**260**	100	89	57	55	54	30	26	25	**0.00**	7	17	4	–	2	–	–	–	–	–	1	–	–	Phosphorothioic acid, O,O-diethyl S-[(ethylsulphinyl)methyl] ester	2588-05-8	NI25914
260	63	69	245	93	181	61	199	**260**	100	73	63	60	56	44	43	31		8	6	–	–	1	5	–	–	–	–	1	–	–	Phosphine sulphide, dimethyl(pentafluorophenyl)-	19100-54-0	NI25915
89	260	57	130	131	73	83	98	**260**	100	38	34	28	23	16	13	8		8	8	–	2	4	–	–	–	–	–	–	–	–	Bis(3-methylisothiazol-5-yl) disulphide		LI05371
133	92	106	77	51	79	65	127	**260**	100	80	68	62	31	28	24	21	**9.70**	8	9	–	2	1	5	–	–	–	–	–	–	–	[(Methylphenylamino)methylenimino]sulphur pentafluoride	90598-17-7	NI25916
241	260	210	141	160	117	191	93	**260**	100	64	44	20	9	8	7	7		9	–	–	–	–	8	–	–	–	–	–	–	–	Indene, perfluoro-		NI25917
124	159	161	28	260	126	18	262	**260**	100	63	45	39	30	30	23	21		9	6	3	2	–	–	2	–	–	–	–	–	–	1,2,4-Oxadiazolidine-3,5-dione, 2-(3,4-dichlorophenyl)-4-methyl-	20354-26-1	NI25918
159	124	260	161	262	126	163	88	**260**	100	97	71	71	50	41	16	15		9	6	3	2	–	–	2	–	–	–	–	–	–	1,2,4-Oxadiazolidine-3,5-dione, 2-(3,4-dichlorophenyl)-4-methyl-	20354-26-1	NI25919
245	139	135	243	215	213	137	133	**260**	100	93	93	78	76	58	58	53	**6.67**	9	13	–	–	–	1	–	–	–	–	–	–	1	Stannane, (4-fluorophenyl)trimethyl-	14101-14-5	NI25920
260	262	206	202	208	204	55	56	**260**	100	99	79	79	16	16	8	4		9	13	2	2	–	–	–	1	–	–	–	–	–	Uracil, 5-bromo-1-sec-butyl-6-methyl-		LI05372
205	207	40	39	70	206	190	162	**260**	100	94	42	20	19	17	17	15	**3.90**	9	13	2	2	–	–	–	1	–	–	–	–	–	Uracil, 5-bromo-3-sec-butyl-6-methyl-	314-40-9	NI25921
205	207	206	231	233	204	162	164	**260**	100	96	19	15	13	13	13	12	**8.31**	9	13	2	2	–	–	–	1	–	–	–	–	–	Uracil, 5-bromo-3-sec-butyl-6-methyl-	314-40-9	AI01852
205	207	262	260	206	204	233	231	**260**	100	99	52	52	13	8	7	7		9	13	2	2	–	–	–	1	–	–	–	–	–	Uracil, 5-bromo-3-sec-butyl-6-methyl-	314-40-9	NI25922
188	88	32	42	44	56	144	132	**260**	100	33	30	24	20	16	13	12	**0.00**	9	16	3	2	1	–	–	–	–	1	–	–	–	2,9,10-Trioxa-6-aza-1-silatricyclo[4.3.3.0^{1,6}]dodecane, 1-(thiocyanatomethyl)-		MS05717
160	159	260	191	119	241	158	141	**260**	100	29	8	8	8	5	5	5		10	7	–	–	–	7	–	–	–	–	–	–	–	Bicyclo[2.2.2]octa-2,5-diene, 2,3-dimethyl-1,4,6,7,7,8,8-heptafluoro-		NI25923
183	185	155	198	157	200	76	50	**260**	100	88	34	34	31	30	29	28	**1.49**	10	10	1	–	–	–	1	1	–	–	–	–	–	Butyrophenone, 4′-bromo-4-chloro-	4559-96-0	NI25924
200	230	258	202	174	96	135	161	**260**	100	29	27	27	18	16	15	13	**0.00**	10	10	2	–	–	–	–	–	–	–	–	–	1	Molybdenum, (π-allyl)dicarbonylcyclopentadienyl-		NI25925
260	174	262	42	176	186	230	189	**260**	100	87	53	53	43	37	33	27		10	10	2	2	–	–	2	–	–	–	–	–	–	Imidazolidinone, 1-(2,4-dichlorophenyl)-3-methoxy-		NI25926
174	42	260	230	176	56	262	232	**260**	100	96	81	73	62	62	58	54		10	10	2	2	–	–	2	–	–	–	–	–	–	Imidazolidinone, 1-(2,5-dichlorophenyl)-3-methoxy-		NI25927
260	174	262	42	176	187	230	189	**260**	100	87	53	53	43	37	33	27		10	10	2	2	–	–	2	–	–	–	–	–	–	Imidazolidinone, 1-(3,4-dichlorophenyl)-3-methoxy-		NI25928
104	132	118	91	76	131	130	77	**260**	100	94	90	86	63	57	56	55	**44.20**	10	12	4	–	2	–	–	–	–	–	–	–	–	1,3-Dithiane, 5-phenyl-, 1,1,3,3-tetraoxide		NI25929
113	42	86	131	43	69	87	44	**260**	100	89	70	41	38	23	19	14	**0.00**	10	12	8	–	–	–	–	–	–	–	–	–	–	1,4,8,11-Tetraoxacyclotetradecane-5,7,12,14-tetrone	21620-29-1	NI25930
69	41	137	39	28	67	55	53	**260**	100	89	70	52	39	31	30	28	**0.00**	10	13	3	–	–	–	–	1	–	–	–	–	–	3-Oxabicyclo[3.2.1]octane-2,4-dione, 1-bromo-5,8,8-trimethyl-		NI25931
97	182	98	85	123	242	84	111	**260**	100	92	92	85	69	54	54	38	**15.00**	10	16	4	2	1	–	–	–	–	–	–	–	–	1H-Thieno[3,4-d]imidazole-4-pentanoic acid, hexahydro-2-oxo-, 5-oxide, [3aS-(3aα,4β,6aα)]-		LI05373
102	101	217	72	85	100	83	160	**260**	100	97	85	58	56	55	48	45	**1.44**	10	16	6	2	–	–	–	–	–	–	–	–	–	1,4,10-Trioxa-7,13-diazacyclopentadecane-2,6,14-trione	74229-36-0	NI25932
245	73	147	75	246	143	45	247	**260**	100	61	37	27	19	16	16	9	**0.09**	10	20	4	–	–	–	–	–	–	2	–	–	–	2-Butenedioic acid (E)-, bis(trimethylsilyl) ester	17962-03-7	NI25933
245	73	147	41	75	40	246	45	**260**	100	60	38	38	28	27	20	16	**0.00**	10	20	4	–	–	–	–	–	–	2	–	–	–	2-Butenedioic acid (E)-, bis(trimethylsilyl) ester	17962-03-7	NI25934
245	73	147	75	45	246	143	74	**260**	100	99	58	40	27	21	17	10	**0.00**	10	20	4	–	–	–	–	–	–	2	–	–	–	2-Butenedioic acid (E)-, bis(trimethylsilyl) ester	17962-03-7	NI25935
73	147	45	75	245	83	148	43	**260**	100	93	40	37	27	24	14	13	**0.00**	10	20	4	–	–	–	–	–	–	2	–	–	–	2-Butenedioic acid (Z)-, bis(trimethylsilyl) ester	23508-82-9	NI25936
147	73	245	45	148	75	149	74	**260**	100	83	21	20	17	16	9	8	**0.00**	10	20	4	–	–	–	–	–	–	2	–	–	–	2-Butenedioic acid (Z)-, bis(trimethylsilyl) ester	23508-82-9	NI25937
219	245	217	247	147	260	262	97	**260**	100	95	94	89	85	62	57	53		11	7	1	–	–	–	3	–	–	–	–	–	–	Naphthalene, 2,4,5-trichloro-1-methoxy-	74298-68-3	NI25938
211	183	213	185	113	62	148	74	**260**	100	99	94	84	83	82	58	56	**40.93**	11	7	1	–	–	–	3	–	–	–	–	–	–	Naphthalene, 2,4,8-trichloro-1-methoxy-	74298-67-2	NI25939
91	122	66	148	56	176	204	65	**260**	100	82	58	50	44	43	41	37	**2.09**	11	8	4	–	–	–	–	–	–	–	–	–	1	Iron, (2-formylnorbornadienyl)tricarbonyl-		NI25940
55	120	204	80	66	65	172	260	**260**	100	87	46	28	22	16	13	7		11	9	4	–	–	–	–	–	–	–	–	–	1	Manganese, cyclopentadienyl(methoxycarbonylvinylidene)dicarbonyl-	73766-80-0	NI25941
100	72	187	29	225	58	189	30	**260**	100	60	34	27	25	23	22	18	**5.20**	11	14	1	2	–	–	2	–	–	–	–	–	–	Urea, N-(2,5-dichlorophenyl)-N′,N′-diethyl-		MS05718
100	72	29	187	44	260	58	30	**260**	100	54	29	18	16	14	14	12		11	14	1	2	–	–	2	–	–	–	–	–	–	Urea, N-(3,4-dichlorophenyl)-N′,N′-diethyl-		MS05719
147	73	245	148	45	149	75	40	**260**	100	52	17	17	12	10	9	6	**1.60**	11	24	3	–	–	–	–	–	–	2	–	–	–	2-Butenoic acid, 3-methyl-2-[(trimethylsilyl)oxy]-, trimethylsilyl ester	55044-79-6	NI25942
73	147	103	75	232	45	148	245	**260**	100	82	34	25	19	16	14	11	**0.00**	11	24	3	–	–	–	–	–	–	2	–	–	–	2-Butenoic acid, 3-methyl-2-[(trimethylsilyl)oxy]-, trimethylsilyl ester	55044-79-6	NI25943
73	147	45	75	245	148	55	43	**260**	100	99	38	24	13	12	12	12	**1.10**	11	24	3	–	–	–	–	–	–	2	–	–	–	2-Pentenoic acid, 2-[(trimethylsilyl)oxy]-, trimethylsilyl ester	55045-17-5	NI25944

MASS TO CHARGE RATIOS	M.W.	INTENSITIES	Parent	C	H	O	N	S	F	Cl	Br	I	Si	P	B	X	COMPOUND NAME	CAS Reg No	No
147 73 245 148 45 149 75 217	260	100 52 20 17 15 10 9 5	2.44	11	24	3	–	–	–	–	–	–	2	–	–	–	2-Pentenoic acid, 2-[(trimethylsilyl)oxy]-, trimethylsilyl ester	55045-17-5	NI25945
73 147 245 45 75 171 148 246	260	100 86 61 29 28 18 14 13	3.44	11	24	3	–	–	–	–	–	–	2	–	–	–	2-Pentenoic acid, 2-[(trimethylsilyl)oxy]-, trimethylsilyl ester	55045-17-5	NI25946
73 147 245 231 45 75 171 170	260	100 66 30 25 25 20 16 13	5.97	11	24	3	–	–	–	–	–	–	2	–	–	–	2-Pentenoic acid, 3-[(trimethylsilyl)oxy]-, trimethylsilyl ester	98779-01-2	NI25947
73 147 245 45 75 231 148 171	260	100 76 28 28 22 20 12 12	4.96	11	24	3	–	–	–	–	–	–	2	–	–	–	2-Pentenoic acid, 3-[(trimethylsilyl)oxy]-, trimethylsilyl ester	98779-01-2	NI25948
73 147 245 231 45 171 148 75	260	100 95 45 30 20 15 15 15	8.00	11	24	3	–	–	–	–	–	–	2	–	–	–	2-Pentenoic acid, 3-[(trimethylsilyl)oxy]-, trimethylsilyl ester	98779-01-2	NI25949
73 143 147 75 45 144 74 260	260	100 79 29 20 16 11 9 8		11	24	3	–	–	–	–	–	–	2	–	–	–	2-Pentenoic acid, 4-[(trimethylsilyl)oxy]-, trimethylsilyl ester		MS05720
93 65 167 139 77 260 79 51	260	100 65 63 36 34 29 23 18		12	8	5	2	–	–	–	–	–	–	–	–	–	Benzene, 2,4-dinitro-1-phenoxy-	2486-07-9	NI25950
260 139 168 63 214 113 30 197	260	100 42 16 7 5 5 5 3		12	8	5	2	–	–	–	–	–	–	–	–	–	1,1′-Biphenyl, 3,3′-dinitro-4-hydroxy-		NI25951
243 260 139 213 167 77 105 63	260	100 84 81 50 33 28 20 20		12	8	5	2	–	–	–	–	–	–	–	–	–	1,1′-Biphenyl, 3,5-dinitro-2-hydroxy-	731-92-0	NI25952
260 139 105 77 53 127 155 115	260	100 44 22 17 14 13 12 12		12	8	5	2	–	–	–	–	–	–	–	–	–	1,1′-Biphenyl, 3,5-dinitro-4-hydroxy-	4097-53-4	NI25953
77 51 157 155 260 262 76 105	260	100 41 40 38 33 32 31 30		12	9	–	2	–	–	–	1	–	–	–	–	–	Azobenzene, 4-bromo-		MS05721
260 163 107 91 103 261 77 69	260	100 82 30 25 18 13 13 13		12	11	3	–	–	3	–	–	–	–	–	–	–	Phenol, 2-methoxy-4-(1-propenyl)-, trifluoroacetate		LI05374
260 163 103 91 107 147 41 261	260	100 58 26 22 17 16 15 14		12	11	3	–	–	3	–	–	–	–	–	–	–	Phenol, 2-methoxy-4-(2-propenyl)-, trifluoroacetate		LI05375
168 260 259 233 194 180 142 103	260	100 95 91 37 28 25 24 17		12	13	–	–	–	–	–	–	–	–	–	–	1	Rhodium, (η^4-norbornadiene)(η^5-cyclopentadienyl)-	12295-14-6	NI25954
202 200 204 206 203 150 201 198	260	100 97 86 64 54 51 49 48	27.29	12	18	–	–	–	–	–	–	–	–	–	–	1	Molybdenum, tris(η^4-1,3-butadiene)-	51733-17-6	NI25955
69 41 67 43 111 29 141 97	260	100 52 50 34 33 33 22 22	0.00	12	20	2	–	2	–	–	–	–	–	–	–	–	Dithiocarbonic acid, O-[ethoxy(cyclopentylidene)methyl] S-propyl diester		DD00762
186 125 141 218 185 81 157 58	260	100 99 95 93 92 90 71 59	0.00	12	20	2	–	2	–	–	–	–	–	–	–	–	Dithiocarbonic acid, O-[methoxy(cyclohexylidene)methyl] S-propyl diester		DD00763
108 79 121 180 92 149 109 67	260	100 58 46 40 38 28 28 27	6.00	12	20	4	–	1	–	–	–	–	–	–	–	–	1-Cyclohexene-1-propanoic acid, β-methyl-α-(methylsulphonyl)-, methyl ester	67428-18-6	NI25956
101 127 43 59 159 55 73 74	260	100 83 76 72 50 39 37 35	0.00	12	20	6	–	–	–	–	–	–	–	–	–	–	α-D-Allofuranose, 1,2:5,6-di-O-isopropylidene-	2595-05-3	NI25957
101 245 43 73 59 55 127 41	260	100 96 84 78 72 61 42 39	0.00	12	20	6	–	–	–	–	–	–	–	–	–	–	β-D-Allofuranose, 2,3:5,6-di-O-isopropylidene-	27108-13-0	NI25958
85 69 117 72 73 100 71 101	260	100 90 86 82 56 46 44 36	0.00	12	20	6	–	–	–	–	–	–	–	–	–	–	α-D-Fructopyranose, 1,2:4,5-di-O-isopropylidene-	20880-93-7	NI25959
69 85 127 245 171 68 115 113	260	100 45 40 39 25 18 16 16	0.00	12	20	6	–	–	–	–	–	–	–	–	–	–	β-D-Fructopyranose, 2,3:4,5-di-O-isopropylidene-	20880-92-6	NI25960
43 59 69 100 85 245 71 41	260	100 35 19 17 17 16 15 15	0.00	12	20	6	–	–	–	–	–	–	–	–	–	–	α-D-Galactopyranose, 1,2:3,4-di-O-isopropylidene-	4064-06-6	NI25962
69 85 245 100 71 81 113 127	260	100 92 90 90 81 68 52 40	0.00	12	20	6	–	–	–	–	–	–	–	–	–	–	α-D-Galactopyranose, 1,2:3,4-di-O-isopropylidene-	4064-06-6	NI25961
113 59 43 142 129 114 85 245	260	100 60 60 23 23 22 22 18	0.00	12	20	6	–	–	–	–	–	–	–	–	–	–	α-D-Glucofuranose, 1,2:3,5-di-O-isopropylidene-	28528-94-1	NI25963
101 43 245 59 127 187 85 55	260	100 85 70 60 25 20 18 16	0.00	12	20	6	–	–	–	–	–	–	–	–	–	–	α-D-Glucofuranose, 1,2:5,6-di-O-isopropylidene-		NI25964
85 73 69 127 72 71 60 245	260	100 91 85 62 62 58 47 45	0.00	12	20	6	–	–	–	–	–	–	–	–	–	–	α-D-Glucoturanose, 1,2:5,6-di-O-isopropylidene-		NI25965
43 101 59 245 85 73 55 41	260	100 62 43 37 16 16 16 13	0.00	12	20	6	–	–	–	–	–	–	–	–	–	–	β-D-Mannopyranose, 1,2:3,4-di-O-isopropylidene-	72361-04-7	NI25966
57 29 173 71 129 117 130 58	260	100 15 8 8 4 4 3 3	0.00	12	20	6	–	–	–	–	–	–	–	–	–	–	1,2,3-Propanetriol, tripropanoate	139-45-7	DO01052
57 29 173 129 117 58 28 27	260	100 21 7 4 3 3 3 3	0.00	12	20	6	–	–	–	–	–	–	–	–	–	–	1,2,3-Propanetriol, tripropanoate	139-45-7	NI25967
245 69 113 171 97 101 165 85	260	100 84 65 51 50 47 42 41	0.00	12	20	6	–	–	–	–	–	–	–	–	–	–	L-Sorbopyranose, 1,2:4,6-di-O-isopropylidene-		NI25968
43 245 101 59 246 41 55 127	260	100 97 38 22 18 17 14 13	0.00	12	20	6	–	–	–	–	–	–	–	–	–	–	β-D-Talofuranose, 1,2:5,6-bis-O-isopropylidene-	23262-79-5	LI05376
43 245 101 59 41 55 127 73	260	100 48 37 22 17 14 13 12	0.00	12	20	6	–	–	–	–	–	–	–	–	–	–	β-D-Talofuranose, 1,2:5,6-bis-O-isopropylidene-	23262-79-5	NI25969
101 43 59 245 73 187 55 82	260	100 99 76 54 25 20 18 17	0.00	12	20	6	–	–	–	–	–	–	–	–	–	–	α-D-Talofuranose, 2,3:5,6-di-O-isopropylidene-	23262-78-4	LI05377
43 101 59 245 73 187 55 72	260	100 99 75 26 26 19 17 12	0.00	12	20	6	–	–	–	–	–	–	–	–	–	–	α-D-Talofuranose, 2,3:5,6-di-O-isopropylidene-	23262-78-4	NI25970
43 74 75 116 85 101 100 59	260	100 99 98 80 49 25 21 13	0.00	12	20	6	–	–	–	–	–	–	–	–	–	–	L-Talopyranoside, methyl 2,4-di-O-acetyl-3-O-methyl-6-deoxy-		LI05378
49 75 84 86 185 93 51 203	260	100 75 57 37 34 30 30 23	0.00	12	21	2	–	–	–	1	–	–	1	–	–	–	7-Oxabicyclo[4.1.0]hept-3-ene, 2-[(tert-butyldimethylsilyl)oxy]-4-chloro-, (1β,2β,6β)-		DD00764
55 98 97 99 56 28 41 44	260	100 96 62 52 36 28 25 24	2.20	12	24	–	2	2	–	–	–	–	–	–	–	–	Piperidine, 4,4′-dimethyl-1,1′-dithiabis-		IC03393
56 28 83 55 41 98 135 97	260	100 45 42 40 40 34 22 19	1.67	12	24	2	2	1	–	–	–	–	–	–	–	–	Sulphamide, dicyclohexyl-		LI05379
73 83 75 29 245 45 201 103	260	100 53 49 25 16 16 14 13	0.00	12	24	4	–	–	–	–	–	–	1	–	–	–	Propanedioic acid, diethyl-, ethyl trimethylsilyl diester		NI25971
73 75 43 175 56 131 129 173	260	100 75 63 40 29 29 27 26	0.00	12	24	4	–	–	–	–	–	–	1	–	–	–	Propanedioic acid, (2-methylbutyl)-, methyl trimethylsilyl diester		NI25972
55 58 43 104 56 41 97 28	260	100 80 71 62 54 50 46 44	1.81	12	24	4	2	–	–	–	–	–	–	–	–	–	Carbamic acid, isopropyl-, 2-[[(aminocarbonyl)oxy]methyl]-2-methylpentyl ester	78-44-4	NI25973
176 200 158 97 261	260	100 32 32 23 8	0.00	12	24	4	2	–	–	–	–	–	–	–	–	–	Carbamic acid, isopropyl-, 2-[[(aminocarbonyl)oxy]methyl]-2-methylpentyl ester	78-44-4	NI25974
58 55 97 43 104 158 62 56	260	100 77 55 53 44 41 39 38	2.80	12	24	4	2	–	–	–	–	–	–	–	–	–	Carbamic acid, isopropyl-, 2-[[(aminocarbonyl)oxy]methyl]-2-methylpentyl ester	78-44-4	NI25975
73 170 75 147 157 45 155 171	260	100 67 32 27 23 15 13 10	0.00	12	28	2	–	–	–	–	–	–	2	–	–	–	1,4-Butanediol, 2,3-dimethyl-1,4-bis(trimethylsilyl)-		NI25976
245 83 58 43 260 246 173 147	260	100 60 39 36 21 21 19 19		12	28	2	–	–	–	–	–	–	2	–	–	–	tert-Butylketene bis(trimethylsilyl) acetal		MS05722
81 147 73 170 116 75 217 101	260	100 61 51 44 32 25 23 14	0.00	12	28	2	–	–	–	–	–	–	2	–	–	–	1,3-Cyclohexanediol, bis(trimethylsilyl) ether	55320-10-0	NI25977
73 131 75 81 170 169 45 147	260	100 89 66 54 41 38 27 24	0.10	12	28	2	–	–	–	–	–	–	2	–	–	–	1,4-Cyclohexanediol, bis(trimethylsilyl) ether	55724-30-6	NI25978
73 147 142 81 245 129 75 133	260	100 92 62 39 38 26 24 15	0.00	12	28	2	–	–	–	–	–	–	2	–	–	–	1,2-Cyclohexanediol, bis(trimethylsilyl) ether, cis-	39789-20-3	NI25979

MASS TO CHARGE RATIOS								M.W.	INTENSITIES								Parent	C	H	O	N	S	F	Cl	Br	I	Si	P	B	X	COMPOUND NAME	CAS Reg No	No
81	147	73	170	116	75	217	149	260	100	61	50	44	32	25	23	16	1.00	12	28	2	–	–	–	–	–	–	2	–	–	–	1,3-Cyclohexanediol, bis(trimethylsilyl) ether, cis-	32282-77-2	NI25980
131	81	73	75	147	170	169	149	260	100	94	86	52	32	29	13	13	0.00	12	28	2	–	–	–	–	–	–	2	–	–	–	1,4-Cyclohexanediol, bis(trimethylsilyl) ether, cis-	29753-63-7	NI25981
131	81	73	147	170	75	169	132	260	100	63	58	37	35	27	24	13	0.00	12	28	2	–	–	–	–	–	–	2	–	–	–	1,4-Cyclohexanediol, bis(trimethylsilyl) ether, cis-	29753-63-7	NI25982
131	81	73	170	75	169	147	132	260	100	44	44	39	29	22	22	11	0.00	12	28	2	–	–	–	–	–	–	2	–	–	–	1,4-Cyclohexanediol, bis(trimethylsilyl) ether, cis-	29753-63-7	NI25983
73	147	142	245	81	75	129	148	260	100	78	56	37	37	32	30	16	0.00	12	28	2	–	–	–	–	–	–	2	–	–	–	1,2-Cyclohexanediol, bis(trimethylsilyl) ether, trans-	39789-21-4	NI25984
73	170	81	75	147	169	116	56	260	100	93	73	55	42	40	36	27	2.00	12	28	2	–	–	–	–	–	–	2	–	–	–	1,3-Cyclohexanediol, bis(trimethylsilyl) ether, trans-	29753-64-8	NI25986
170	73	75	169	116	155	171	115	260	100	77	51	39	20	17	14	14	0.00	12	28	2	–	–	–	–	–	–	2	–	–	–	1,3-Cyclohexanediol, bis(trimethylsilyl) ether, trans-	29753-64-8	NI25985
131	73	74	169	170	81	245	45	260	100	76	61	54	53	22	14	13	0.10	12	28	2	–	–	–	–	–	–	2	–	–	–	1,4-Cyclohexanediol, bis(trimethylsilyl) ether, trans-	29753-62-6	NI25987
73	131	75	170	169	81	155	132	260	100	91	60	45	45	16	12	11	0.00	12	28	2	–	–	–	–	–	–	2	–	–	–	1,4-Cyclohexanediol, bis(trimethylsilyl) ether, trans-	29753-62-6	NI25988
147	73	142	245	129	81	75	148	260	100	68	60	51	33	32	17	15	0.00	12	28	2	–	–	–	–	–	–	2	–	–	–	Cyclohexanediol, bis(trimethylsilyl) ether, trans-		NI25989
73	147	142	245	129	81	75	148	260	100	70	70	43	41	34	33	26	0.00	12	28	2	–	–	–	–	–	–	2	–	–	–	Cyclohexanediol, bis(trimethylsilyl) ether, trans-		NI25990
81	73	147	170	75	116	217	149	260	100	65	54	40	33	27	17	15	0.00	12	28	2	–	–	–	–	–	–	2	–	–	–	Cyclohexanediol, bis(trimethylsilyl) ether, trans-		NI25991
231	203	175	28	147	27	26	173	260	100	76	51	29	27	19	19	15	0.65	12	30	2	–	–	–	–	–	–	–	–	–	2	Aluminium, di-μ-ethoxytetraethyldi-	15629-79-5	NI25992
260	152	232	258	180	262	256	230	260	100	65	55	50	47	21	20	19		13	8	1	–	–	–	–	–	–	–	–	–	1	9H-Selenoxanthen-9-one	4734-58-1	NI25993
260	152	232	180	114	130			260	100	66	48	46	14	10				13	8	1	–	–	–	–	–	–	–	–	–	1	9H-Selenoxanthen-9-one	4734-58-1	LI05380
260	232	261	190	233	231	162	43	260	100	54	26	14	13	11	9	9		13	8	6	–	–	–	–	–	–	–	–	–	–	Naphtho[2,3-b]furan-5,8-dione, 2-methyl-4,7,9-trihydroxy-		NI25994
260	261	69	116	231	232	203	262	260	100	24	12	8	7	6	6	5		13	8	6	–	–	–	–	–	–	–	–	–	–	9H-Xanthen-9-one, 1,3,6,8-tetrahydroxy-		MS05723
105	183	185	260	77	262	76	155	260	100	54	53	44	44	43	21	17		13	9	1	–	–	–	–	1	–	–	–	–	–	Benzophenone, 4-bromo-		DO01053
111	139	75	260	65	121	113	93	260	100	60	50	42	42	37	37	25		13	9	2	2	–	–	1	–	–	–	–	–	–	Azobenzene-2′-carboxylic acid, 4-chloro-		MS05724
90	119	118	92	77	91	65	51	260	100	98	88	75	58	53	40	35	17.52	13	12	2	2	1	–	–	–	–	–	–	–	–	Benzenesulphonic acid, benzylidenehydrazide	1666-16-6	NI25995
182	245	77	78	183	125	65	93	260	100	42	19	11	10	9	9	7	0.00	13	12	2	2	1	–	–	–	–	–	–	–	–	Sulphoximine, S-methyl-S-phenyl-N-(2-pyridinylcarbonyl)-	54090-93-6	NI25996
182	245	77	78	183	125	51	65	260	100	43	19	11	10	10	10	9	0.00	13	12	2	2	1	–	–	–	–	–	–	–	–	Sulphoximine, S-methyl-S-phenyl-N-(2-pyridinylcarbonyl)-	54090-93-6	NI25997
260	141	116	143	59	115	117	142	260	100	81	57	54	45	35	33	32		13	12	4	2	–	–	–	–	–	–	–	–	–	4a,8a-Methanophthalazine-1,4-dicarboxylic acid, dimethyl ester		DD00765
130	131	202	129	187	113	201	117	260	100	10	5	3	2	2	1	1	0.10	13	20	–	1	–	–	2	–	–	–	–	–	–	Diethyl(indol-3-ylmethyl)ammonium chloride, hydrochloride		NI25998
115	101	59	245	71	43	187	95	260	100	70	26	23	19	19	16	16	0.50	13	24	5	–	–	–	–	–	–	–	–	–	–	D-Glucitol, 1-deoxy-2,3:5,6-di-O-isopropylidene-4-O-methyl-		MS05725
73	145	75	231	43	113	103	146	260	100	81	36	28	24	23	22	11	0.00	13	28	3	–	–	–	–	–	–	1	–	–	–	Pentanoic acid, 2-ethyl-3-methyl-3-[(trimethylsilyl)oxy]-, ethyl ester		NI25999
58	101	147	72	84	90	176	70	260	100	63	38	27	25	17	14	13	13.00	13	29	1	2	–	–	–	–	–	–	1	–	–	Phosphetane, 1-[1-amino-3-(dimethylamino)propane]-2,2,3,4,4-pentamethyl-, 1-oxide		NI26000
139	149	181	111	75	141	65	76	260	100	70	60	51	38	31	21	20	12.70	14	9	3	–	–	–	1	–	–	–	–	–	–	Benzoic acid, 2-(4-chlorobenzoyl)-	85-56-3	NI26001
215	181	152	216	214	44	151	217	260	100	86	80	52	47	44	36	35	19.00	14	9	3	–	–	–	1	–	–	–	–	–	–	Fluorene-9-carboxylic acid, 2-chloro-9-hydroxy-		MS05726
179	178	166	165	260	180	152	167	260	100	56	56	46	45	33	12	10		14	12	–	–	–	–	–	–	–	–	–	–	1	Dibenzo[c,e]selenepin, 5,7-dihydro-	33948-83-3	NI26002
105	77	51	196	91	94	65	50	260	100	75	35	26	16	11	10	10	0.00	14	12	3	–	1	–	–	–	–	–	–	–	–	Acetophenone, 2-(phenylsulphonyl)-	3406-03-9	NI26003
105	77	51	78	154	126	106	50	260	100	87	59	52	42	30	17	15	0.00	14	12	3	–	1	–	–	–	–	–	–	–	–	1,3,2-Dioxathiolane, 4,5-diphenyl-, 2-oxide	4464-79-3	NI26004
134	135	39	44	90	89	59	69	260	100	12	9	8	7	7	6	5	4.00	14	12	3	–	1	–	–	–	–	–	–	–	–	4-Thia-2,3-benzobicyclo[3.2.0]hept-2-ene-6,7-dicarboxylic anhydride, 6,7-dimethyl-		LI05381
134	135	40	45	90	89	61	69	260	100	10	8	7	5	5	5	4	4.00	14	12	3	–	1	–	–	–	–	–	–	–	–	Thiepino[3,2-e]isobenzofuran-1,3-dione, 3a,10b-dihydro-3a,10b-dimethyl-	55044-57-0	NI26005
162	134	260	242	163	99	51	77	260	100	65	13	13	10	7	7	6		14	12	5	–	–	–	–	–	–	–	–	–	–	2-Butenoic acid, 2-methyl-4-[(2-oxo-2H-1-benzopyran-7-yl)oxy]-	16851-00-6	MS05727
260	245	216	217	231	189	177	261	260	100	97	77	44	37	27	24	15		14	12	5	–	–	–	–	–	–	–	–	–	–	5H-Furo[3,2-g][1]benzopyran-5-one, 4,9-dimethoxy-7-methyl-	82-02-0	MS05728
203	204	175	105	157	146	173	129	260	100	50	48	48	38	32	24	24	4.00	14	12	5	–	–	–	–	–	–	–	–	–	–	2H-Naphtho[2,3-b]furan-5,10-dione, 3,4-dihydro-3,4-dihydroxy-2-methyl-, [2R-(2α,3β,4β)]-	75442-57-8	NI26006
203	175	105	157	146	173	129	204	260	100	48	48	38	32	24	24	15	1.00	14	12	5	–	–	–	–	–	–	–	–	–	–	2H-Naphtho[2,3-b]furan-5,10-dione, 3,4-dihydro-3,4-dihydroxy-2-methyl-, [2S-(2α,3α,4α)]-	75442-58-9	NI26007
260	245	261	231	43	40	57	44	260	100	59	16	15	11	11	9	9		14	12	5	–	–	–	–	–	–	–	–	–	–	3H-Naphtho[2,3-c]furan-3-one, 6,8-dimethoxy-7-hydroxy-		MS05729
159	39	54	78	51	27	131	77	260	100	45	39	21	20	19	18	18	1.00	14	12	5	–	–	–	–	–	–	–	–	–	–	4-Oxatricyclo[5.2.2.0^{2,6}]undec-10-ene-3,5,9-trione-8-spiro[2-(4-methyleneoxolane)]		MS05730
91	169	171	262	260	90	65	167	260	100	42	39	17	17	16	13	12		14	13	–	–	–	–	–	1	–	–	–	–	–	Bibenzyl, 4-bromo-	14310-24-8	NI26008
141	93	106	77	143	225	119	140	260	100	99	35	35	32	31	25	20	15.90	14	13	1	2	–	–	1	–	–	–	–	–	–	Urea, N-(2-chloro-6-methylphenyl)-N′-phenyl-		MS05731
127	107	129	133	106	28	260	101	260	100	37	33	21	21	20	18	12		14	13	1	2	–	–	1	–	–	–	–	–	–	Urea, N-(2-chlorophenyl)-N′-(2-methylphenyl)-		MS05732
127	107	260	129	106	91	133	28	260	100	69	33	32	32	14	12	12		14	13	1	2	–	–	1	–	–	–	–	–	–	Urea, N-(3-chlorophenyl)-N′-(2-methylphenyl)-		MS05733
260	195	261	196	259	242	97	152	260	100	15	14	10	6	6	6	5		14	13	3	–	–	–	–	–	–	–	1	–	–	Phenoxaphosphinic acid, 2,6-dimethyl-		NI26009
136	260	245	108	137	261	98	217	260	100	66	26	20	14	12	11	5		14	16	1	2	1	–	–	–	–	–	–	–	–	4H-5,6-Benzothiazin-4-one, 2-(3-methylpiperidino)-		NI26010
105	77	231	51	99	114	41	45	260	100	60	56	21	12	9	9	9	0.99	14	16	1	2	1	–	–	–	–	–	–	–	–	1,3-Thiazolidine, 2-benzoylimino-4-ethyl-4-methyl-5-methylene-		NI26011
105	77	260	51	245	56	183	41	260	100	62	27	26	25	22	14	10		14	16	1	2	1	–	–	–	–	–	–	–	–	1,3-Thiazolidine, 2-benzoylimino-5-methylene-3,4,4-trimethyl-	59000-10-1	NI26012
177	260	44	95	67	41	130	55	260	100	98	86	73	50	50	47	47		14	16	3	2	–	–	–	–	–	–	–	–	–	Cycloheptanone, 7-[(4-nitropenylamino)methylene]-		MS05734
133	260	134	132	104	56	105	77	260	100	50	13	13	7	6	5	4		14	16	3	2	–	–	–	–	–	–	–	–	–	Imidazolidine-2,4,5-trione, 1-butyl-3-(p-methylphenyl)-		MS05735
69	41	191	260	192	261	70	124	260	100	55	47	30	8	7	7	6		14	16	3	2	–	–	–	–	–	–	–	–	–	Phenol, 2,4-bis[(2-methyl-2-propenoyl)amino]-		IC03394

MASS TO CHARGE RATIOS								M.W.	INTENSITIES								Parent	C	H	O	N	S	F	Cl	Br	I	Si	P	B	X	COMPOUND NAME	CAS Reg No	No
187	260	106	188	186	159	122	79	**260**	100	30	24	15	14	13	11	8		14	16	3	2	–	–	–	–	–	–	–	–	–	**4H-Pyrido[1,2-a]pyrimidine-3-acetic acid, 6,8-dimethyl-4-oxo-, ethyl ester**	50609-65-9	**NI26013**
173	92	260	186	172	145	65	187	**260**	100	56	54	44	31	29	26	20		14	16	3	2	–	–	–	–	–	–	–	–	–	**4H-Pyrido[1,2-a]pyrimidine-3-propanoic acid, 6-methyl-4-oxo-, ethyl ester**	50609-73-9	**NI26014**
203	188	232	117	204	115	40	70	**260**	100	53	23	16	13	11	11	9	**1.50**	14	16	3	2	–	–	–	–	–	–	–	–	–	**4,6(1H,5H)-Pyrimidinedione, 5-ethyl-2-methoxy-1-methyl-5-phenyl-**	55030-33-6	**NI26015**
232	146	175	117	118	233	188	103	**260**	100	45	41	41	40	22	18	18	**13.02**	14	16	3	2	–	–	–	–	–	–	–	–	–	**2,4,6(1H,3H,5H)-Pyrimidinetrione, 5-ethyl-1,3-dimethyl-5-phenyl-**	730-66-5	**NI26017**
117	175	118	146	232	233	188	103	**260**	100	98	85	81	68	54	43	35	**8.12**	14	16	3	2	–	–	–	–	–	–	–	–	–	**2,4,6(1H,3H,5H)-Pyrimidinetrione, 5-ethyl-1,3-dimethyl-5-phenyl-**	730-66-5	**NI26016**
232	118	175	117	146	44	233	103	**260**	100	32	30	28	27	17	16	12	**0.00**	14	16	3	2	–	–	–	–	–	–	–	–	–	**2,4,6(1H,3H,5H)-Pyrimidinetrione, 5-ethyl-1,3-dimethyl-5-phenyl-**	730-66-5	**NI26018**
177	203	105	141	104	176	260	133	**260**	100	97	66	53	53	37	32	28		14	16	3	2	–	–	–	–	–	–	–	–	–	**2,4(2H,4H)-Quinazolinedione, 1-methyl-3-(4-oxopentyl)-**		**LI05382**
130	201	131	260	77	41	170	103	**260**	100	30	13	10	8	8	7	7		14	16	3	2	–	–	–	–	–	–	–	–	–	**L-Tryptophan, N-acetyl-, methyl ester**	2824-57-9	**NI26020**
130	43	44	201	41	57	55	77	**260**	100	46	30	23	23	17	15	13	**4.30**	14	16	3	2	–	–	–	–	–	–	–	–	–	**L-Tryptophan, N-acetyl-, methyl ester**	2824-57-9	**NI26019**
171	73	170	245	79	260	103	117	**260**	100	83	46	29	25	25	23	15		14	20	1	2	–	–	–	–	–	1	–	–	–	**Quinoxaline, 2-methyl-3-[(2-trimethylsilyloxy)ethyl]-**		**NI26021**
57	81	41	137	95	136	29	69	**260**	100	49	34	26	16	15	15	14	**8.39**	14	26	–	–	–	–	1	–	–	–	1	–	–	**Phosphinous chloride, tert-butyl(1,7,7-trimethylbicyclo[2.2.1]hept-2-yl)-**	74630-34-5	**NI26022**
57	41	103	59	15	29	159	113	**260**	100	28	21	21	21	11	4	2	**0.00**	14	28	4	–	–	–	–	–	–	–	–	–	–	**Acetic acid, di-tert-butoxy-, tert-butyl ester**		**LI05383**
83	115	101	245	131	73	189	82	**260**	100	63	31	10	7	6	3	1	**0.00**	14	28	4	–	–	–	–	–	–	–	–	–	–	**Glycerol, 1-O-(2-methoxy-5-methylhexyl)-2,3-O-isopropylidene-**		**NI26023**
91	260	225	210	77	181	134	65	**260**	100	99	83	69	65	63	62	54		15	13	–	–	1	–	1	–	–	–	–	–	–	**Ethylene, 1-chloro-1-[(4-methylphenyl)thio]-2-phenyl-**		**DD00766**
201	165	166	203	202	260	164	82	**260**	100	69	50	34	14	9	8	8		15	13	2	–	–	–	1	–	–	–	–	–	–	**Acetic acid, (p-chlorophenyl)phenyl-, methyl ester**	5359-46-6	**NI26024**
171	260	127	204	203	172	261	173	**260**	100	26	21	20	13	13	5	5		15	16	–	–	2	–	–	–	–	–	–	–	–	**1-Naphthalenecarbodithioic acid, butyl ester**	35972-84-0	**NI26025**
260	194	181	242	179	91	151	195	**260**	100	91	84	81	68	61	59	56		15	16	2	–	1	–	–	–	–	–	–	–	–	**Benzene, 2,4-dimethyl-1-[(3-methylphenyl)sulfonyl]-**	28523-18-4	**NI26026**
194	242	260	179	151	91	77	195	**260**	100	97	92	80	77	71	63	59		15	16	2	–	1	–	–	–	–	–	–	–	–	**Benzene, 2,4-dimethyl-1-[(4-methylphenyl)sulfonyl]-**	13249-97-3	**NI26027**
260	194	181	242	179	151	195	165	**260**	100	98	96	90	79	76	71	67		15	16	2	–	1	–	–	–	–	–	–	–	–	**Benzene, 2,5-dimethyl-1-[(3-methylphenyl)sulfonyl]-**	28523-19-5	**NI26028**
260	242	194	151	179	243	104	91	**260**	100	96	87	74	74	70	63	62		15	16	2	–	1	–	–	–	–	–	–	–	–	**Benzene, 2,6-dimethyl-1-[(4-methylphenyl)sulfonyl]-**		**NI26029**
242	165	194	179	91	180	77	105	**260**	100	77	77	67	65	58	50	50	**34.08**	15	16	2	–	1	–	–	–	–	–	–	–	–	**Benzene, 2,4,6-trimethyl-1-(phenylsulfonyl)-**	3112-82-1	**NI26030**
173	77	65	86	109	42	53	260	**260**	100	69	58	56	54	40	29	28		15	16	2	–	1	–	–	–	–	–	–	–	–	**Bicyclo[2.2.1]hept-5-en-2-one, 6-(phenylthio)-, ethylene ketal**	119595-42-5	**DD00767**
260	118	117	216	91	187	215	119	**260**	100	59	48	45	40	34	32	30		15	16	4	–	–	–	–	–	–	–	–	–	–	**4H-Azuleno[1,8-bc:4,5-b′]difuran-4,8(2aH)-dione, 1,7-dimethyl-6,6a,7,9a,9b,9c-hexahydro-, [2aR-(2aα,6aα,7α,9aβ,9bα,9cα)]-**	79396-08-0	**NI26031**
260	77	245	91	69	261	105	131	**260**	100	47	35	33	18	18	15	15		15	16	4	–	–	–	–	–	–	–	–	–	–	**Benzene, 2,3,4-trimethoxy-1,1′-oxybis-**	88037-83-6	**NI26032**
260	245	217	77	69	125	114	261	**260**	100	93	40	36	29	22	17	16		15	16	4	–	–	–	–	–	–	–	–	–	–	**Benzene, 2,3,5-trimethoxy-1,1′-oxybis-**	88037-84-7	**NI26033**
260	245	77	217	69	124	261	125	**260**	100	53	47	33	29	29	17	15		15	16	4	–	–	–	–	–	–	–	–	–	–	**Benzene, 2,4,5-trimethoxy-1,1′-oxybis-**	88037-85-8	**NI26034**
245	260	77	217	187	69	261	246	**260**	100	86	51	36	20	20	17	16		15	16	4	–	–	–	–	–	–	–	–	–	–	**Benzene, 3,4,5-trimethoxy-1,1′-oxybis-**	88037-86-9	**NI26035**
205	260	217	204	206	41	44	43	**260**	100	52	37	25	23	16	14	14		15	16	4	–	–	–	–	–	–	–	–	–	–	**2H,6H-Benzo[1,2-b:5,4-b′]dipyran-6-one, 3,4-dihydro-5-hydroxy-2,2,8-trimethyl-**	13475-11-1	**MS05737**
205	260	217	245	176	192	123	231	**260**	100	52	37	13	10	6	6	3		15	16	4	–	–	–	–	–	–	–	–	–	–	**2H,6H-Benzo[1,2-b:5,4-b′]dipyran-6-one, 3,4-dihydro-5-hydroxy-2,2,8-trimethyl-**	13475-11-1	**MS05736**
205	260	217	43	206	41	245	39	**260**	100	31	29	19	17	13	12	12		15	16	4	–	–	–	–	–	–	–	–	–	–	**2H,10H-Benzo[1,2-b:3,4-b′]dipyran-10-one, 3,4-dihydro-5-hydroxy-2,2,8-trimethyl-**	13475-12-2	**MS05738**
205	260	217	245	176	123	231	192	**260**	100	32	19	13	6	6	3	3		15	16	4	–	–	–	–	–	–	–	–	–	–	**2H,10H-Benzo[1,2-b:3,4-b′]dipyran-10-one, 3,4-dihydro-5-hydroxy-2,2,8-trimethyl-**	13475-12-2	**MS05739**
205	217	260	245	206	69	243	43	**260**	100	82	41	25	13	10	8	8		15	16	4	–	–	–	–	–	–	–	–	–	–	**4H-1-Benzopyran-4-one, 5,7-dihydroxy-2-methyl-6-(3-methyl-2-butenyl)-**	578-72-3	**MS05741**
205	217	260	245	192	123	231	165	**260**	100	86	42	26	8	7	5	5		15	16	4	–	–	–	–	–	–	–	–	–	–	**4H-1-Benzopyran-4-one, 5,7-dihydroxy-2-methyl-6-(3-methyl-2-butenyl)-**	578-72-3	**MS05740**
260	245	217	205	69	41	77	51	**260**	100	98	92	59	52	46	28	20		15	16	4	–	–	–	–	–	–	–	–	–	–	**2H-1-Benzopyran-2-one, 3-(1,1-dimethylallyl)-7-hydroxy-6-methoxy-**		**MS05742**
245	260	205	217	246	189	261		**260**	100	53	50	43	22	20	10			15	16	4	–	–	–	–	–	–	–	–	–	–	**2H-1-Benzopyran-2-one, 8-(1,1-dimethylallyl)-7-hydroxy-5-methoxy-**		**LI05384**
217	189	260	159	131	218	190	187	**260**	100	84	67	32	23	19	19	14		15	16	4	–	–	–	–	–	–	–	–	–	–	**2H-1-Benzopyran-2-one, 6-[(3,3-dimethyloxiranyl)methyl]-7-methoxy-**	38409-28-8	**NI26036**
189	245	260	203	43	242	227	190	**260**	100	62	45	42	37	20	14	11		15	16	4	–	–	–	–	–	–	–	–	–	–	**2H-1-Benzopyran-2-one, 6-(3-hydroxy-3-methyl-1-butenyl)-7-methoxy-, (E)-**	18529-47-0	**NI26037**
190	189	131	175	161	160	77	103	**260**	100	31	21	13	12	10	6	5	**0.00**	15	16	4	–	–	–	–	–	–	–	–	–	–	**2H-1-Benzopyran-2-one, 8-(2-hydroxy-3-methyl-2-butenyl)-7-methoxy-**	61235-25-4	**NI26038**
192	164	69	41	193	205	260	245	**260**	100	47	35	31	12	6	5	5		15	16	4	–	–	–	–	–	–	–	–	–	–	**2H-1-Benzopyran-2-one, 5-methoxy-7-(3,3-dimethylallyloxy)-**		**LI05385**
203	260	204	218	245	89	261	160	**260**	100	20	14	10	6	5	4	4		15	16	4	–	–	–	–	–	–	–	–	–	–	**2H-1-Benzopyran-2-one, 7-methoxy-6-(3-methyl-1-oxobutyl)-**		**LI05386**
189	190	43	260	71	131	191	44	**260**	100	76	54	24	22	10	9	9		15	16	4	–	–	–	–	–	–	–	–	–	–	**2H-1-Benzopyran-2-one, 7-methoxy-6-(3-methyl-2-oxobutyl)-**	38409-25-5	**NI26039**
53	55	81	136	105	91	137	203	**260**	100	83	67	53	37	30	25	24	**17.00**	15	16	4	–	–	–	–	–	–	–	–	–	–	**Cyclodeca[b]furan-10-carboxylic acid, 7α-hydroxy-6-methyl-3-methylene-2,3,3a,4,7,8,11,11a-octahydro-2-oxo-, 7,10-lactone, [3aR-(3aR*,5E,9Z,11aS*)]-**	82003-39-2	**NI26040**
203	260	204	218	245	89	150	76	**260**	100	20	14	11	6	5	4	4		15	16	4	–	–	–	–	–	–	–	–	–	–	**Cyclohexene, 4-vinyl-4-methyl-3-isopropenyl-**	450-16-8	**NI26041**
217	260	192	205	190	231	245	176	**260**	100	91	43	31	23	20	8	7		15	16	4	–	–	–	–	–	–	–	–	–	–	**Dihydrodeoxykarenin**		**MS05743**
260	159	145	91	171	135	41	187	**260**	100	52	40	28	22	18	16	14		15	16	4	–	–	–	–	–	–	–	–	–	–	**2H,10,1a-(Epoxymethano)oxireno[4,5]cyclodeca[1,2-b]furan-12-one,5,9-dimethyl-3,6,10,10a-tetrahydro-, [1aS-(1aR*,4E,10R*,10aR*)]-**	13476-25-0	**LI05387**

MASS TO CHARGE RATIOS								M.W.	INTENSITIES								Parent	C	H	O	N	S	F	Cl	Br	I	Si	P	B	X	COMPOUND NAME	CAS Reg No	No
260	159	145	161	91	173	261	135	**260**	100	53	41	40	27	22	18	17		15	16	4	–	–	–	–	–	–	–	–	–	–	2H,10,1a-(Epoxymethano)oxireno[4,5]cyclodeca[1,2-b]furan-12-one,5,9-dimethyl-3,6,10,10a-tetrahydro-, [1aS-(1aR*,4E,10R*,10aR*)]-	13476-25-0	MS05744
189	190	43	260	71	159	131	191	**260**	100	75	54	24	21	12	8	7		15	16	4	–	–	–	–	–	–	–	–	–	–	Isogeijerin		LI05388
161	260	145	217	43	133	91	189	**260**	100	76	65	58	39	38	33	31		15	16	4	–	–	–	–	–	–	–	–	–	–	5H-7,4-Methenofuro[3,2-c]oxireno[f]oxacycloundecin-5-one, 8,11a-dimethyl-1a,2,3,7,11,11a-hexahydro-, [1aR-(1aR*,7R*,11aR*)]-	15004-50-9	MS05745
260	227	245	242	115	128	199	259	**260**	100	48	33	25	15	13	12	9		15	16	4	–	–	–	–	–	–	–	–	–	–	1-Naphthalenecarboxaldehyde, 4-isopropyl-6-methyl-2,3,8-trihydroxy-	40817-07-0	NI26042
202	59	192	227	43	159	77	105	**260**	100	91	77	56	53	50	40	33	**7.01**	15	16	4	–	–	–	–	–	–	–	–	–	–	1,4-Naphthalenedione, 2-hydroxy-3-(1-hydroxy-3-methylbutyl)-	55030-44-9	NI26043
202	59	192	227	43	159	77	105	**260**	100	90	75	54	51	49	40	32	**6.00**	15	16	4	–	–	–	–	–	–	–	–	–	–	1,4-Naphthalenedione, 2-hydroxy-3-(3-hydroxy-3-methylbutyl)-	15298-01-8	MS05746
129	128	174	201	130	115	260	127	**260**	100	76	72	68	56	48	36	36		15	16	4	–	–	–	–	–	–	–	–	–	–	1-Naphthalenepropanoic acid, 2,7-dimethoxy-	32834-63-2	NI26044
136	260	124						**260**	100	25	8							15	16	4	–	–	–	–	–	–	–	–	–	–	Naphtho[2,3-b]furan-4,6,9(5H)-trione, 4a,7,8,8a-tetrahydro-3,4a,5-trimethyl-, [4aR-(4aα,5α,8aα)]-	61176-53-2	NI26045
260	115	202	200	228	201	116	144	**260**	100	86	76	71	61	61	57	48		15	16	4	–	–	–	–	–	–	–	–	–	–	Piperonal, 6-(4-methoxy-1-cyclohexen-1-yl)-	16831-73-5	NI26046
129	91	29	128	77	157	39	51	**260**	100	89	72	58	56	54	54	46	**36.80**	15	16	4	–	–	–	–	–	–	–	–	–	–	2,4,6,8,10-Undecapentaenoic acid, 11,11-diformyl-, ethyl ester		NI26047
245	108	217	183	246	75	186	121	**260**	100	36	32	28	17	15	14	14	**5.40**	15	17	–	–	1	–	–	–	–	–	1	–	–	Ethane, 2-(diphenylphosphino)-1-(methylthio)-		IC03395
260	212	245	152	130	151	227	259	**260**	100	99	56	15	14	13	12	10		15	20	–	2	1	–	–	–	–	–	–	–	–	2,2′-Dipyrrylthione, 3,4,5,3′,4′,5′-hexamethyl-		LI05389
261	97	202	216	259				**260**	100	64	47	28	24				**0.00**	15	20	–	2	1	–	–	–	–	–	–	–	–	1,2-Ethanediamine, N,N-dimethyl-N′-phenyl-N′-(2-thienylmethyl)-	493-78-7	NI26048
176	148	177	260	91	120	146	65	**260**	100	14	13	10	7	4	3	3		15	20	2	2	–	–	–	–	–	–	–	–	–	Aniline, N-(2-cyanoethoxyethyl)-N-ethyl-4-formyl-3-methyl-		IC03396
229	245	134	259	190	175	132	260	**260**	100	51	38	33	30	24	23	22		15	20	2	2	–	–	–	–	–	–	–	–	–	2,3-Benzo-1,5-diazacycloundec-4-en-11-one, 1-methyl-4-methoxy-		LI05390
260	114	146	261	243	160	159	152	**260**	100	46	16	15	10	10	10	10		15	20	2	2	–	–	–	–	–	–	–	–	–	7,14-Methano-4H,6H-dipyrido[1,2-a:1′,2′-e][1,5]diazicin-4-one, 7,7a,8,9,10,11,13,14-octahydro-9-hydroxy-	27773-56-4	LI05391
91	260	141	169	43	92	86	41	**260**	100	37	36	24	20	16	16	16		15	20	2	2	–	–	–	–	–	–	–	–	–	Piperazine-2,5-dione, 3-benzyl-6-isobutyl-		LI05392
161	245	260	244	246	162	259	77	**260**	100	94	31	26	16	13	8	8		15	21	3	–	–	–	–	–	–	–	–	1	–	p-Cymene-2-carboxylic acid, 3-hydroxy-, n-butylboronate		MS05747
194	260	28	124	196	126	58	138	**260**	100	43	42	41	39	36	35	33		15	22	–	–	–	–	–	–	–	–	–	–	1	Nickel, [(1,2,3-η)-2-cyclodecen-1-yl](η⁵-2,4-cyclopentadien-1-yl)-	61233-23-6	NI26049
43	57	243	55	41	71	69	83	**260**	100	99	69	57	56	47	40	29	**0.00**	15	32	1	–	1	–	–	–	–	–	–	–	–	Tetradecane, 1-(methylsulfinyl)-	3079-31-0	WI01312
43	41	31	29	55	45	27	42	**260**	100	92	92	90	57	43	43	41	**2.40**	15	32	3	–	–	–	–	–	–	–	–	–	–	1,2,15-Pentadecanetriol	57289-60-8	NI26050
260	259	261	130	103	77	18	76	**260**	100	60	27	16	14	11	10	10		16	12	–	4	–	–	–	–	–	–	–	–	–	Dibenzo[b,h][1,4,7,10]tetraazacyclododecine	38063-20-6	NI26051
260	231	155	129	247	261	156	106	**260**	100	27	23	23	20	19	17	16		16	12	–	4	–	–	–	–	–	–	–	–	–	2-Naphthalamine, 1-(2H-benzotriaz-2-yl)-		IC03397
259	260	232	234	233	258	261	130	**260**	100	94	47	29	29	19	18	16		16	12	–	4	–	–	–	–	–	–	–	–	–	Pyrazino[1,2-a:4,3-a′]dibenzimidazole, 6,7-dihydro-		IC03398
260	231	206	179	232	205	178	166	**260**	100	88	74	51	49	41	41	26		16	12	–	4	–	–	–	–	–	–	–	–	–	p-Quinodimethane, 7,7,8,8-tetracyano-2,5-diethyl-		MS05748
108	79	77	91	55	109	39	41	**260**	100	15	14	13	11	10	10	9	**1.30**	16	20	3	–	–	–	–	–	–	–	–	–	–	5-Benzofuranacetic acid, 3,6-dimethyl-α-methylene-4,5,6,7-tetrahydro-6-vinyl-, methyl ester	19912-86-8	MS05749
108	79	77	91	39	109	41	105	**260**	100	15	14	13	11	10	9	8	**1.30**	16	20	3	–	–	–	–	–	–	–	–	–	–	5-Benzofuranacetic acid, 3,6-dimethyl-α-methylene-4,5,6,7-tetrahydro-6-vinyl-, methyl ester	19912-86-8	LI05393
245	217	215	199	260				**260**	100	16	15	6	1					16	20	3	–	–	–	–	–	–	–	–	–	–	Benzopyran-3-carboxylic acid, 2,4,4,7-tetramethyl-, ethyl ester		LI05394
245	199	217	215	43	128	100	67	**260**	100	26	16	15	9	7	5	4	**1.00**	16	20	3	–	–	–	–	–	–	–	–	–	–	Benzopyran-3-carboxylic acid, 2,4,4,7-tetramethyl-, ethyl ester		LI05395
108	260	109	192	148	161	145	133	**260**	100	16	7	6	6	4	4	3		16	20	3	–	–	–	–	–	–	–	–	–	–	Cyclodeca[b]furan-6-carboxylic acid, 3,10-dimethyl-4,7,8,11-tetrahydro-, methyl ester, (E,E)-	21678-22-8	MS05750
108	260	109	148	192	161	145	201	**260**	100	16	7	6	4	4	4	2		16	20	3	–	–	–	–	–	–	–	–	–	–	Cyclodeca[b]furan-6-carboxylic acid, 3,10-dimethyl-4,7,8,11-tetrahydro-, methyl ester, (E,E)-	21678-22-8	LI05396
178	177	260	176	245	163	139	217	**260**	100	20	11	7	6	5	5	3		16	20	3	–	–	–	–	–	–	–	–	–	–	1,5-Dioxaphenanthren-6-one, 9-hydroxy-2-(4-methyl-4-butenyl)-2-methyl-		LI05397
177	260	178	245	217	179	189		**260**	100	23	11	10	10	8	5			16	20	3	–	–	–	–	–	–	–	–	–	–	3,9-Dioxatetracyclo[6.4.2.1^{10,13}.0^{4,14}]pentadeca-4,6,8(14)-triene, 6-methoxy-2,2,10-trimethyl-		LI05398
136	260	55	175	161	227	91	77	**260**	100	96	54	51	46	40	38	37		16	20	3	–	–	–	–	–	–	–	–	–	–	Naphthalene-1,4-diol, 5,6,7,8-tetrahydro-7-(1-hydroxymethylvinyl)-6-methyl-6-vinyl-		NI26052
187	260	202	188	172	147	91		**260**	100	47	15	15	13	13	9			16	20	3	–	–	–	–	–	–	–	–	–	–	Naphtho[1,2-b]furan-2(3H)-one, 3a,4,5,9b-tetrahydro-8-methoxy-3,6,9-trimethyl-, (3α,3aα,9bα)-	55057-88-0	NI26053
187	260	188	201	172	147	91		**260**	100	46	14	13	13	12	9			16	20	3	–	–	–	–	–	–	–	–	–	–	Naphtho[1,2-b]furan-2(3H)-one, 3a,4,5,9b-tetrahydro-8-methoxy-3,6,9-trimethyl-, (3α,3aα,9bα)-		LI05399
177	245	260	217	163	139	246	178	**260**	100	80	70	50	30	30	20	20		16	20	3	–	–	–	–	–	–	–	–	–	–	10-Oxaphenanthrene-2,4-diol, 4a,7,8,8a,9,10-hexahydro-6,9,9-trimethyl-		LI05400
177	260	178	179	189	245	217	139	**260**	100	15	11	8	6	5	5	5		16	20	3	–	–	–	–	–	–	–	–	–	–	2-Oxatricyclo[6.4.0.1^{3,7}]trideca-1(8),9,11-triene-9,11-diol, 6-isopropenyl-3-methyl-		LI05401
229	169	230	260	59	170	141	107	**260**	100	50	16	12	9	8	8	4		16	20	3	–	–	–	–	–	–	–	–	–	–	1,2-Propanediol, 2-[6-(1-hydroxy-1-methylethyl)naphthalen-2-yl]-		NI26054
245	135	107	43	75	91	260	183	**260**	100	99	37	32	31	19	11	11		16	24	1	–	–	–	–	–	–	1	–	–	–	Furan, 3-(dimethylphenylsilyl)-2,2,5,5-tetramethyldihydro-		NI26055
169	110	98	105	151	140	81	91	**260**	100	55	29	27	25	21	16	10	**1.00**	16	24	1	2	–	–	–	–	–	–	–	–	–	Cyclohexane, 1-acetamido-2-(phenethylamino)-, cis-		MS05751
169	60	110	81	140	98	105	127	**260**	100	30	22	17	16	15	10	5	**0.00**	16	24	1	2	–	–	–	–	–	–	–	–	–	Cyclohexane, 1-acetamido-3-(phenethylamino)-, cis-		MS05752
169	81	140	110	60	105	91	98	**260**	100	29	28	27	23	20	17	15	**0.00**	16	24	1	2	–	–	–	–	–	–	–	–	–	Cyclohexane, 1-acetamido-4-(phenethylamino)-, cis-		MS05753

MASS TO CHARGE RATIOS								M.W.	INTENSITIES								Parent	C	H	O	N	S	F	Cl	Br	I	Si	P	B	X	COMPOUND NAME	CAS Reg No	No
110	151	169	105	81	140	98	91	**260**	100	64	56	32	32	26	26	25	**1.00**	16	24	1	2	–	–	–	–	–	–	–	–	–	Cyclohexane, 1-acetamido-2-(phenethylamino)-, trans-		MS05754
140	169	60	81	98	105	91	110	**260**	100	87	28	23	20	15	12	10	**2.00**	16	24	1	2	–	–	–	–	–	–	–	–	–	Cyclohexane, 1-acetamido-3-(phenethylamino)-, trans-		MS05755
140	169	60	81	105	98	110	91	**260**	100	82	34	21	18	14	10	7	**0.00**	16	24	1	2	–	–	–	–	–	–	–	–	–	Cyclohexane, 1-acetamido-4-(phenethylamino)-, trans-		MS05756
57	43	71	55	41	69	91	85	**260**	100	88	52	51	46	38	33	32	**0.00**	16	33	–	–	–	–	1	–	–	–	–	–	–	Hexadecane, 1-chloro-	4860-03-1	IC03399
57	43	71	91	41	55	85	69	**260**	100	82	55	44	40	39	33	31	**2.00**	16	33	–	–	–	–	1	–	–	–	–	–	–	Hexadecane, 1-chloro-	4860-03-1	LI05402
43	41	55	57	29	69	83	56	**260**	100	94	84	81	55	51	40	40	**0.20**	16	33	–	–	–	–	1	–	–	–	–	–	–	Hexadecane, 1-chloro-	4860-03-1	WI01313
135	117	119	260	107	105	77	79	**260**	100	67	64	52	51	44	32	30		17	12	1	2	–	–	–	–	–	–	–	–	–	2-Oxazoleacetonitrile, 4,5-diphenyl-	65913-22-6	NI26056
260	231	232	259	130	102	204	248	**260**	100	37	35	24	15	14	14	13		17	12	1	2	–	–	–	–	–	–	–	–	–	6H-Pyrido[4,3-b]carbazole, 5-formyl-11-methyl-	77251-57-1	NI26057
259	39	90	77	76	64	51	260	**260**	100	77	60	46	37	37	28	26		17	12	1	2	–	–	–	–	–	–	–	–	–	Quinazoline, 2-phenyl-4-(2-propynyloxy)-		LI05403
129	260	102	259	156	128	130	157	**260**	100	84	51	32	30	30	28	25		17	12	1	2	–	–	–	–	–	–	–	–	–	4(1H)-Quinazolinone, 2-phenyl-1-(2-propynyl)-	26197-90-0	LI05404
129	260	102	259	156	128	130	103	**260**	100	86	50	32	30	30	28	25		17	12	1	2	–	–	–	–	–	–	–	–	–	4(1H)-Quinazolinone, 2-phenyl-1-(2-propynyl)-	26197-90-0	NI26058
260	259	231	39	261	232	77	76	**260**	100	89	28	22	19	17	14	14		17	12	1	2	–	–	–	–	–	–	–	–	–	4(3H)-Quinazolinone, 2-phenyl-3-(2-propynyl)-		LI05405
105	77	260	51	106	129	128	50	**260**	100	40	12	11	9	6	4	4		17	12	1	2	–	–	–	–	–	–	–	–	–	2-Quinolinecarbonitrile, 1-benzoyl-1,2-dihydro-	13721-17-0	NI26059
245	57	246	260	143	41	78	29	**260**	100	32	19	16	14	14	10	10		17	24	2	–	–	–	–	–	–	–	–	–	–	1H-Indene-4-acetic acid, 6-tert-butyl-2,3-dihydro-1,1-dimethyl-	55591-05-4	NI26060
57	41	105	163	43	29	85	69	**260**	100	20	14	13	12	9	8	6	**0.00**	17	24	2	–	–	–	–	–	–	–	–	–	–	Tetracyclo[10.2.0.0]dodecane, 3,12-diacetyl-11-methyl-		NI26061
260	259	258	129	261	130	129.5	213	**260**	100	73	56	32	22	11	10	9		18	12	–	–	1	–	–	–	–	–	–	–	–	Naphtho[2,1-b]thiophene, 1-phenyl-		AI01853
130	129	130.5	104	39	40	213	116	**260**	100	41	21	18	18	17	16	15	**0.00**	18	12	–	–	1	–	–	–	–	–	–	–	–	Naphtho[2,1-b]thiophene, 2-phenyl-		AI01854
260	261	258	130	40	129	262	259	**260**	100	19	15	12	11	6	5	5		18	12	–	–	1	–	–	–	–	–	–	–	–	1-Thiaindene, 2-(2-naphthyl)-		AI01855
260	259	258	129	261	130	213	262	**260**	100	65	45	21	20	9	6	5		18	12	–	–	1	–	–	–	–	–	–	–	–	1-Thiaindene, 3-(1-naphthyl)-		AI01856
260	261	258	215	259	129	40	130	**260**	100	20	18	10	8	8	8	7		18	12	–	–	1	–	–	–	–	–	–	–	–	1-Thiaindene, 3-(2-naphthyl)-		AI01857
202	260	231	101	203	200	201	88	**260**	100	64	49	35	35	28	28	19		18	12	2	–	–	–	–	–	–	–	–	–	–	7H-Benzo[c]fluorene-6-carbaldehyde, 5-hydroxy-		MS05757
260	202	231	203	215				**260**	100	96	66	58	18					18	12	2	–	–	–	–	–	–	–	–	–	–	1,1′-Biindanylidene, 3,3′-dioxo-		LI05406
102	44	28	260	262	232	231	261	**260**	100	98	90	73	33	31	27	23		18	12	2	–	–	–	–	–	–	–	–	–	–	2,5-Cyclohexadiene-1,4-dione, 2,5-diphenyl-	844-51-9	IC03400
102	260	259	232	261	76	231	203	**260**	100	79	38	22	14	14	13	12		18	12	2	–	–	–	–	–	–	–	–	–	–	2,5-Cyclohexadiene-1,4-dione, 2,5-diphenyl-	844-51-9	NI26062
102	44	28	260	259	262	232	203	**260**	100	89	88	76	52	37	31	21		18	12	2	–	–	–	–	–	–	–	–	–	–	2,5-Cyclohexadiene-1,4-dione, 2,5-diphenyl-	844-51-9	MS05758
182	181	183	105	260	180	259	184	**260**	100	58	26	22	18	15	7	6		18	16	–	–	–	–	–	–	–	1	–	–	–	Silane, triphenyl-	789-25-3	NI26064
182	181	183	105	180	155	260	259	**260**	100	79	24	23	19	8	6	6		18	16	–	–	–	–	–	–	–	1	–	–	–	Silane, triphenyl-	789-25-3	NI26063
260	261	183	167	130	77	51	259	**260**	100	20	11	11	9	9	6	3		18	16	–	2	–	–	–	–	–	–	–	–	–	Benzene, 1,4-dianilino-		IC03401
204	203	202	232	191	215	205	165	**260**	100	61	54	49	37	18	18	16	**1.00**	18	16	–	2	–	–	–	–	–	–	–	–	–	Benzo[c]cyclopenta[f]-1,2-diazepine, 10-phenyl-1,2,3,3a-tetrahydro-	28749-06-6	NI26065
122	121	146	160	135	260	134	161	**260**	100	36	34	19	19	18	16	14		18	28	1	–	–	–	–	–	–	–	–	–	–	Cyclohexane, 1-(m-methoxyphenethyl)-2,2,6-trimethyl-, cis-		LI05407
121	260	161	135	57	245	148	203	**260**	100	82	75	40	37	18	18	14		18	28	1	–	–	–	–	–	–	–	–	–	–	Cyclohexane, 4-tert-butyl-1-(2-methoxyphenyl)-1-methyl-, cis-		LI05408
161	245	260	121	148	57	135	189	**260**	100	70	59	54	33	24	19	8		18	28	1	–	–	–	–	–	–	–	–	–	–	Cyclohexane, 4-tert-butyl-1-(4-methoxyphenyl)-1-methyl-, cis-		LI05409
161	260	121	245	148	57	135	189	**260**	100	53	45	41	29	24	20	6		18	28	1	–	–	–	–	–	–	–	–	–	–	Cyclohexane, 4-tert-butyl-1-(4-methoxyphenyl)-1-methyl-, trans-		LI05410
217	135	43	69	41	161	67	93	**260**	100	60	60	54	52	47	47	45	**5.00**	18	28	1	–	–	–	–	–	–	–	–	–	–	1-Cyclopropen-3-ol, 1,3-dicyclopentyl-2-isopropyl-, acetate	69611-52-5	NI26066
260	217	81	245	67	218	95	107	**260**	100	77	59	58	54	40	38	35		18	28	1	–	–	–	–	–	–	–	–	–	–	Estran-2-one, (5α)-	4967-97-9	NI26067
43	217	41	67	191	69	260	95	**260**	100	93	68	50	43	43	36	36		18	28	1	–	–	–	–	–	–	–	–	–	–	3,4-Heptadien-2-one, 3,5-dicyclopentyl-6-methyl-	63922-51-0	NI26068
218	148	109	175	108	203	81	41	**260**	100	56	54	49	48	44	35	31	**2.00**	18	28	1	–	–	–	–	–	–	–	–	–	–	D-Norandrostan-16-one, (5α)-	32319-06-5	NI26069
260	215	261	259	217	202	229	245	**260**	100	24	21	16	15	15	12	12		19	16	1	–	–	–	–	–	–	–	–	–	–	Benzene, 1,2-diphenyl-4-methoxy-		NI26070
183	105	77	260	165	154	182	155	**260**	100	76	57	51	24	23	21	21		19	16	1	–	–	–	–	–	–	–	–	–	–	Benzenemethanol, α,α-diphenyl-	76-84-6	MS05759
105	183	77	260	154	155	182	51	**260**	100	80	66	22	22	18	16	15		19	16	1	–	–	–	–	–	–	–	–	–	–	Benzenemethanol, α,α-diphenyl-	76-84-6	NI26071
260	167	165	77	261	51	166	152	**260**	100	95	38	33	21	21	19	18		19	16	1	–	–	–	–	–	–	–	–	–	–	1,1′-Biphenyl, 4-phenoxy-		IC03402
259	260	115	261	116	130	129	91	**260**	100	66	23	11	8	7	7	7		19	16	1	–	–	–	–	–	–	–	–	–	–	Cyclopentanone, 2,5-dibenzylidene-	895-80-7	MS05760
259	260	115	116	261	129	91	130	**260**	100	66	35	14	12	11	10	9		19	16	1	–	–	–	–	–	–	–	–	–	–	Cyclopentanone, 2,5-dibenzylidene-	895-80-7	NI26072
179	194	151	180	195	150	152	66	**260**	100	78	47	14	13	11	8	8	**6.19**	19	16	1	–	–	–	–	–	–	–	–	–	–	Fluoranthene, 3-acetyl-7,10-methano-6b,7,10,10a-tetrahydro-, endo-		MS05761
260	245	202	122.5	215	43	231	130	**260**	100	53	13	13	8	7	4	4		19	16	1	–	–	–	–	–	–	–	–	–	–	Phenanthro[1,2-b]furan, 1-ethyl-2-methyl-		LI05411
245	260	41	95	55	203	81	67	**260**	100	98	82	76	66	65	65	64		19	32	–	–	–	–	–	–	–	–	–	–	–	Androstane	24887-75-0	NI26073
82	55	81	67	217	95	79	109	**260**	100	81	58	56	43	43	38	37	**0.00**	19	32	–	–	–	–	–	–	–	–	–	–	–	Androstane, (5α)-	438-22-2	NI26074
260	245	203	94	81	67	204	135	**260**	100	83	48	37	33	30	28	28		19	32	–	–	–	–	–	–	–	–	–	–	–	Androstane, (5α)-	438-22-2	NI26075
260	245	203	135	95	149	109	107	**260**	100	97	72	61	57	54	45	42		19	32	–	–	–	–	–	–	–	–	–	–	–	Androstane, (5α)-	438-22-2	NI26076
245	203	260	135	81	95	44	107	**260**	100	81	79	75	64	60	45	44		19	32	–	–	–	–	–	–	–	–	–	–	–	Androstane, (5α,14β)-	22511-92-8	NI26078
260	203	245	95	81	67	135	55	**260**	100	86	78	75	64	51	50	47		19	32	–	–	–	–	–	–	–	–	–	–	–	Androstane, (5α,14β)-	22511-92-8	NI26077
245	260	95	81	203	135	67	149	**260**	100	62	44	42	41	40	31	30		19	32	–	–	–	–	–	–	–	–	–	–	–	Androstane, (5β)-	438-23-3	NI26080
245	260	203	95	81	135	67	55	**260**	100	81	47	47	44	43	36	33		19	32	–	–	–	–	–	–	–	–	–	–	–	Androstane, (5β)-	438-23-3	NI26079
245	260	95	81	135	203	67	41	**260**	100	74	64	57	49	45	41	35		19	32	–	–	–	–	–	–	–	–	–	–	–	Androstane, (5β)-	438-23-3	NI26081
260	245	95	203	81	135	67	55	**260**	100	82	60	53	52	43	39	37		19	32	–	–	–	–	–	–	–	–	–	–	–	Androstane, (5β,14β)-	13077-81-1	NI26082
203	147	204	28	260	148	57	41	**260**	100	26	22	18	15	14	10	9		19	32	–	–	–	–	–	–	–	–	–	–	–	Benzene, 2,6-bis(2,2-dimethylpropyl)-1,3,5-trimethyl-	33770-73-9	NI26083

MASS TO CHARGE RATIOS								M.W.	INTENSITIES								Parent	C	H	O	N	S	F	Cl	Br	I	Si	P	B	X	COMPOUND NAME	CAS Reg No	No
119	95	69	91	107	149	109	177	**260**	100	98	98	84	70	66	63	43	**7.00**	19	32	–	–	–	–	–	–	–	–	–	–	–	Friedo-15-norlabda-8-(17),9-diene, (10S)-		MS05762
92	91	260	105	133	57	43	93	**260**	100	82	17	11	9	9	9	8		19	32	–	–	–	–	–	–	–	–	–	–	–	Tridecane, 1-phenyl-	123-02-4	NI26084
105	118	131	91	106	43	41	260	**260**	100	19	18	16	14	6	6	5		19	32	–	–	–	–	–	–	–	–	–	–	–	Tridecane, 2-phenyl-	4534-53-6	MS05763
105	106	91	104	43	41	260	79	**260**	100	12	9	4	4	4	3	3		19	32	–	–	–	–	–	–	–	–	–	–	–	Tridecane, 2-phenyl-	4534-53-6	NI26085
105	106	91	260	104	79	77	103	**260**	100	16	15	7	6	5	5	3		19	32	–	–	–	–	–	–	–	–	–	–	–	Tridecane, 2-phenyl-	4534-53-6	WI01314
91	119	105	41	43	120	92	231	**260**	100	60	13	13	10	8	8	7	**3.22**	19	32	–	–	–	–	–	–	–	–	–	–	–	Tridecane, 3-phenyl-	4534-52-5	NI26086
91	119	105	41	231	43	260	104	**260**	100	57	23	20	17	15	10	10		19	32	–	–	–	–	–	–	–	–	–	–	–	Tridecane, 3-phenyl-	4534-52-5	IC03403
91	133	92	105	43	41	217	119	**260**	100	24	9	8	7	7	5	5	**2.03**	19	32	–	–	–	–	–	–	–	–	–	–	–	Tridecane, 4-phenyl-	4534-51-4	NI26087
91	147	105	41	203	92	260	43	**260**	100	22	20	12	10	10	9	9		19	32	–	–	–	–	–	–	–	–	–	–	–	Tridecane, 5-phenyl-	4534-50-3	IC03404
91	147	105	92	41	203	119	43	**260**	100	19	13	8	7	5	5	5	**2.48**	19	32	–	–	–	–	–	–	–	–	–	–	–	Tridecane, 5-phenyl-	4534-50-3	NI26088
91	105	161	119	92	189	41	43	**260**	100	19	13	9	8	7	6	5	**2.63**	19	32	–	–	–	–	–	–	–	–	–	–	–	Tridecane, 6-phenyl-	4534-49-0	NI26089
91	175	105	43	41	92	104	55	**260**	100	26	13	13	12	9	8	5	**4.78**	19	32	–	–	–	–	–	–	–	–	–	–	–	Tridecane, 7-phenyl-		AI01858
91	175	105	41	43	92	104	119	**260**	100	25	15	12	11	9	8	7	**4.95**	19	32	–	–	–	–	–	–	–	–	–	–	–	Tridecane, 7-phenyl-		AI01859
91	175	105	104	43	41	29	260	**260**	100	38	15	12	9	9	9	8		19	32	–	–	–	–	–	–	–	–	–	–	–	Tridecane, 7-phenyl-		AI01860
245	260	105	230	246	167	215	91	**260**	100	59	42	33	21	18	17	16		20	20	–	–	–	–	–	–	–	–	–	–	–	Benzene, 1,1′-(1,4-dimethyl-2,5-cyclohexadiene-1,4-diyl)bis-	74421-24-2	NI26090
260	169	232	217	43	128	261	215	**260**	100	87	40	38	28	22	21	21		20	20	–	–	–	–	–	–	–	–	–	–	–	Benzo[j]fluoranthene, 4,5,6,6a,6b,7,8,12b-octahydro-		AI01861
169	260	232	217	128	129	115	28	**260**	100	82	32	32	27	22	22	22		20	20	–	–	–	–	–	–	–	–	–	–	–	Benzo[j]fluoranthene, 4,5,6,6a,6b,7,8,12b-octahydro-, cis,anti-		AI01862
260	232	129	217	128	130	28	115	**260**	100	48	37	36	36	29	27	26		20	20	–	–	–	–	–	–	–	–	–	–	–	Benzo[j]fluoranthene, 4,5,6,6a,6b,7,8,12b-octahydro-, cis,syn-		AI01863
260	232	217	202	215	203	231	204	**260**	100	100	38	38	34	30	26	25		20	20	–	–	–	–	–	–	–	–	–	–	–	Benzo[b]fluoranthene, 1,2,3,3a,9,10,11,12-octahydro-		NI26091
232	260	217	202	203	204	215	231	**260**	100	92	66	66	64	53	52	30		20	20	–	–	–	–	–	–	–	–	–	–	–	Benzo[k]fluoranthene, 1,2,3,8,9,10,11,12b-octahydro-		NI26092
156	141	155	260	104	119	115	128	**260**	100	40	28	27	19	12	12	10		20	20	–	–	–	–	–	–	–	–	–	–	–	Benzo[d][2,2]paracyclopane, 1,2-dihydro-		LI05412
260	217	261	202	259	232	203	215	**260**	100	29	21	19	14	14	14	11		20	20	–	–	–	–	–	–	–	–	–	–	–	Benzo[a]pyrene, 4,5,7,8,9,10,11,12-octahydro-	73712-70-6	NI26093
260	245	131	215	261	115	216	230	**260**	100	43	43	23	22	22	18	15		20	20	–	–	–	–	–	–	–	–	–	–	–	1,1′-Biindanylidene, (3S,3′S)-dimethyl-, (E)-	96144-92-2	NI26094
217	215	218	202	260	216	213	203	**260**	100	22	20	20	16	14	4	4		20	20	–	–	–	–	–	–	–	–	–	–	–	15H-Cyclopenta[a]phenanthrene, 16,17-dihydro-3-isopropyl-	57325-68-5	NI26095
129	128	260	130	131	115			**260**	100	63	41	40	38	21				20	20	–	–	–	–	–	–	–	–	–	–	–	Dibenzo[a,g]biphenylene, 5,6,6a,6b,11,12,12a,12b-octahydro-		LI05413
217	203	260	114	218	101	230	107	**260**	100	28	24	21	18	16	12	11		20	20	–	–	–	–	–	–	–	–	–	–	–	9,10-Ethanophenanthrene, 11,11,12,12-tetramethyl-	108794-66-7	DD00768
156	91	115	128	141	157	155	129	**260**	100	47	20	15	14	13	12	11	**3.50**	20	20	–	–	–	–	–	–	–	–	–	–	–	Spiro[2.4]heptane, 1-phenyl-5-benzylidene-	74793-06-9	NI26096
176	177	67	152	39	54	41	151	**260**	100	75	69	65	58	57	55	48	**4.00**	20	20	–	–	–	–	–	–	–	–	–	–	–	Triptycene, 11,12,13,14,15,16-hexahydro-		LI05414
61	63	98	76	44	35	31	79	**261**	100	90	28	27	22	18	11	5	**0.00**	2	–	–	1	2	1	4	–	–	–	–	–	–	Methimine, bis(dichlorosulfanyl)-N-dichlorofluoromethyl-		NI26097
69	31	88	54	104	85	50	97	**261**	100	6	5	5	4	4	4	3	**1.30**	4	–	1	1	–	8	–	–	–	–	1	–	–	Propane, 2-cyano-2-(μ-oxodifluorophosphine)hexafluoro-		LI05415
100	69	150	261	131	114	214		**261**	100	71	39	15	12	12	11			5	–	1	1	–	9	–	–	–	–	–	–	–	2-Pyrrolidinone, hexafluoro-N-(trifluoromethyl)-		LI05416
261	263	63	62	78	52	89	169	**261**	100	99	85	71	54	49	39	38		6	4	4	3	–	–	–	1	–	–	–	–	–	Aniline, 2-bromo-4,6-dinitro-	1817-73-8	IC03405
115	113	174	264	262	246	248	198	**261**	100	64	59	49	48	47	39	37	**10.83**	6	4	4	3	–	–	–	1	–	–	–	–	–	Aniline, 2-bromo-4,6-dinitro-	1817-73-8	NI26099
263	261	90	169	171	63	231	78	**261**	100	98	87	60	56	42	40	40		6	4	4	3	–	–	–	1	–	–	–	–	–	Aniline, 2-bromo-4,6-dinitro-	1817-73-8	NI26098
261	134	262	127	107	63	90	75	**261**	100	37	10	8	8	8	6	6		7	4	–	1	1	–	–	–	1	–	–	–	–	Thieno[2,3-b]pyridine, 3-iodo-	53399-37-4	NI26100
134	261	107	77	92	133	64	50	**261**	100	75	64	54	29	25	23	22		8	8	1	1	1	5	–	–	–	–	–	–	–	(4-Methoxybenzylidenimino)sulphur pentafluoride	90598-14-4	NI26101
80	181	82	39	152	138	83	55	**261**	100	99	99	74	69	54	50	39	**0.00**	8	8	2	1	1	–	–	1	–	–	–	–	–	Thiazolo[3,2-a]pyridinium, 3,8-dihydroxy-2-methyl-, bromide	32002-90-7	NI26102
261	233	69	42	213	109	262	232	**261**	100	90	28	27	18	18	14	13		9	10	1	5	–	3	–	–	–	–	–	–	–	Pyrimido[5,4-e]-1,2,4-triazine, 7-ethoxy-1,2-dihydro-3-methyl-5-(trifluoromethyl)-	32709-22-1	NI26103
109	244	79	127	63	51	90	89	**261**	100	52	21	15	15	14	14	13	**10.31**	9	12	6	1	–	–	–	–	–	–	1	–	–	Phosphoric acid, dimethyl 3-methyl-4-nitrophenyl ester	2255-17-6	NI26104
199	96	226	155	261	156	200	184	**261**	100	82	72	54	24	23	22	20		9	16	–	5	1	–	1	–	–	–	–	–	–	1,3,5-Triazine-2,4-diamine, 6-[(2-chloroethyl)thio]-N,N,N′,N′-tetramethyl-		NI26105
231	168	230	261	139	232	233	83	**261**	100	66	64	42	40	36	30	18		9	16	2	5	–	–	1	–	–	–	–	–	–	1,3,5-Triazine-2,4-diamine, 6-(2-chloroethoxy)-N-methoxy-N,N′,N′-trimethyl-	101132-50-7	NI26106
201	203	202	261	195	139	153	83	**261**	100	34	22	18	16	16	12	12		9	16	2	5	–	–	1	–	–	–	–	–	–	1,3,5-Triazin-2(1H)-one, 1-(2-chloroethyl)-4-dimethylamino-6-(N-methoxy-N-methylamino)-		NI26107
177	159	261	55	205	133	104	122	**261**	100	66	55	47	40	32	20	19		10	8	4	1	–	–	–	–	–	–	–	–	1	Manganese, [(1,2,3,4,5-η)-1-acetyl-2-amino-2,4-cyclopentadien-1-yl]tricarbonyl-	33195-11-8	NI26108
177	261	55	159	205	133	122	94	**261**	100	49	42	40	33	20	18	17		10	8	4	1	–	–	–	–	–	–	–	–	1	Manganese, [(1,2,3,4,5-η)-1-acetyl-3-amino-2,4-cyclopentadien-1-yl]tricarbonyl-, stereoisomer	33221-15-7	NI26109
190	42	43	29	27	191	149	80	**261**	100	35	20	19	17	15	14	12	**6.80**	10	11	2	7	–	–	–	–	–	–	–	–	–	1,2,4-Triazolo[3,4-a]pyrazolo[3,4-e]pyrimidin-4(4H)-one, 5-[(ethoxymethylen)imino]-2-methyl-		NI26110
43	147	261	105	162	246	149	231	**261**	100	69	40	37	34	31	28	25		10	14	4	–	–	–	–	–	–	–	–	–	1	Copper, bis(acetylacetonato)-	13395-16-9	NI26111
261	147	246	162	263	105	231	149	**261**	100	77	54	54	46	46	42	35		10	14	4	–	–	–	–	–	–	–	–	–	1	Copper, bis(acetylacetonato)-	13395-16-9	NI26112
177	176	41	43	57	142	55	56	**261**	100	66	53	51	48	40	39	34	**0.00**	10	16	1	1	–	5	–	–	–	–	–	–	–	1-Heptanamine, N-pentafluoropropanoyl-		MS05764

MASS TO CHARGE RATIOS								M.W.	INTENSITIES								Parent	C	H	O	N	S	F	Cl	Br	I	Si	P	B	X	COMPOUND NAME	CAS Reg No	No
82	94	122	42	95	43	202	204	**261**	100	87	86	82	56	50	38	37	**10.00**	10	16	2	1	–	–	–	1	–	–	–	–	–	8-Azabicyclo[3.2.1]octan-3-ol, 6-bromo-8-methyl-, acetate (ester)	54725-45-0	NI26113
125	109	108	29	97	80	27	65	**261**	100	97	85	76	58	58	50	36	**33.73**	10	16	3	1	1	–	–	–	–	–	1	–	–	Phosphorothioic acid, O-(4-aminophenyl) O,O-diethyl ester	3735-01-1	NI26114
73	147	45	144	59	75	246	218	**261**	100	38	23	18	17	14	13	12	**7.29**	10	23	3	1	–	–	–	–	–	2	–	–	–	Butanoic acid, 2-oxo-, O-(trimethylsilyl)oxime, trimethylsilyl ester		NI26115
73	75	147	144	43	102	45	218	**261**	100	46	44	39	34	30	29	21	**3.80**	10	23	3	1	–	–	–	–	–	2	–	–	–	Glycine, N-acetyl-N-(trimethylsilyl)-, trimethylsilyl ester	55124-99-7	NI26116
73	144	102	147	45	75	43	246	**261**	100	64	52	41	25	25	19	19	**4.75**	10	23	3	1	–	–	–	–	–	2	–	–	–	Glycine, N-acetyl-N-(trimethylsilyl)-, trimethylsilyl ester	55124-99-7	NI26117
73	147	75	56	146	131	130	45	**261**	100	68	59	52	40	31	22	20	**3.18**	10	23	3	1	–	–	–	–	–	2	–	–	–	Propanedioic acid monoamide, methyl-N-(trimethylsilyl)-, trimethylsilyl ester		NI26118
66	167	51	39	94	78	79	261	**261**	100	93	72	64	56	43	35	33		11	7	5	3	–	–	–	–	–	–	–	–	–	Pyridine, 3-(2,4-dinitrophenoxy)-	40604-27-1	NI26119
66	167	94	78	261	183			**261**	100	93	56	43	33	4				11	7	5	3	–	–	–	–	–	–	–	–	–	Pyridine, 3-(2,4-dinitrophenoxy)-	40604-27-1	LI05417
78	155	233	76	51	75	50	154	**261**	100	88	75	68	68	64	64	27	**0.00**	11	8	–	3	–	–	–	1	–	–	–	–	–	Pyridine, 2-[(p-bromophenyl)azo]-		LI05418
78	157	155	235	233	76	51	50	**261**	100	88	88	75	75	68	68	64	**0.00**	11	8	–	3	–	–	–	1	–	–	–	–	–	Pyridine, 2-[(p-bromophenyl)azo]-	25117-50-4	NI26120
155	183	261	51	76	78	75	50	**261**	100	78	75	55	47	36	36	36		11	8	–	3	–	–	–	1	–	–	–	–	–	Pyridine, 4-[(p-bromophenyl)azo]-		LI05419
155	157	185	183	263	261	51	76	**261**	100	99	78	78	75	75	55	47		11	8	–	3	–	–	–	1	–	–	–	–	–	Pyridine, 4-[(p-bromophenyl)azo]-	20815-54-7	NI26121
135	261	148	121	147	52	69	80	**261**	100	68	66	34	19	19	18	17		11	10	3	1	–	3	–	–	–	–	–	–	–	Acetamide, 2,2,2-trifluoro-N-[2-(hexahydro-1(2H)-azocinyl)ethyl]-	55649-52-0	NI26122
133	177	55	99	56	205	44	261	**261**	100	38	37	13	11	10	10	9		11	12	3	1	–	–	–	–	–	–	–	–	1	Manganese, tricarbonyl[(1,2,3,4,5-η)-1-[(dimethylamino)methyl]-2,4-cyclopentadien-1-yl]-	49626-33-7	NI26123
29	154	188	39	152	41	116	190	**261**	100	58	31	30	23	23	22	21	**9.00**	11	13	2	1	–	–	2	–	–	–	–	–	–	Cyclopropane-1-carboxylic acid, 1-cyano-2,2-dimethyl-3-(2,2-dichlorovinyl)-, ethyl ester		IC03406
200	129	59	67	261	155	113	99	**261**	100	90	41	25	20	16	15	15		11	19	2	1	2	–	–	–	–	–	–	–	–	2,5-Pyrrolidinedione, 3-[1,1-bis(ethylthio)ethyl]-4-methyl-	58467-36-0	NI26124
72	142	172	114	71	84	113	60	**261**	100	99	98	86	78	58	54	52	**0.00**	11	19	6	1	–	–	–	–	–	–	–	–	–	Allopyranoside, 3-acetamido-3-deoxy-1,2-O-isopropylidene-	28297-43-0	NI26125
43	101	126	83	186	144	160	246	**261**	100	98	36	16	10	10	8	5	**1.70**	11	19	6	1	–	–	–	–	–	–	–	–	–	α-D-Glucofuranosylamine, N-acetyl-5,6-O-isopropylidene-	83937-98-8	NI26126
202	88	114	170	143	28	54	110	**261**	100	92	87	66	48	39	30	29	**1.00**	11	19	6	1	–	–	–	–	–	–	–	–	–	Piperidine-2,6-dicarboxylic acid, 4,4-dimethoxy-, cis-, dimethyl ester		NI26127
114	202	88	28	170	143	54	110	**261**	100	99	93	91	59	59	44	41	**0.00**	11	19	6	1	–	–	–	–	–	–	–	–	–	Piperidine-2,6-dicarboxylic acid, 4,4-dimethoxy-, trans-, dimethyl ester		NI26128
72	148	55	44	41	150	82	125	**261**	100	96	73	45	36	31	31	27	**0.10**	11	20	2	3	–	–	1	–	–	–	–	–	–	Hydroxylamine, N-(1-cyanoisobutyl)-N-(1-chloroacetamidoisobutyl)		NI26129
174	73	142	59	175	89	86	246	**261**	100	67	46	27	20	18	18	17	**0.35**	11	27	2	1	–	–	–	–	–	2	–	–	–	Butanoic acid, 4-aminomethyl-, N-trimethylsilyl-, trimethylsilyl ester		NI26130
144	73	45	145	75	147	43	74	**261**	100	85	20	13	11	10	9	8	**0.00**	11	27	2	1	–	–	–	–	–	2	–	–	–	Butanoic acid, 4-[(trimethylsilyl)amino]-2-methyl-, trimethylsilyl ester		MS05765
144	73	45	145	59	147	75	43	**261**	100	86	21	13	12	10	10	10	**0.00**	11	27	2	1	–	–	–	–	–	2	–	–	–	Norvaline, N-(trimethylsilyl)-, trimethylsilyl ester		MS05766
102	73	75	156	147	45	117	59	**261**	100	48	33	20	19	12	11	6	**3.00**	11	27	2	1	–	–	–	–	–	2	–	–	–	Pentanoic acid, 5-[(trimethylsilyl)amino]-, trimethylsilyl ester		MS05767
144	73	218	147	45	246	100	59	**261**	100	31	28	11	5	4	4	3	**2.00**	11	27	2	1	–	–	–	–	–	2	–	–	–	L-Valine, N-(trimethylsilyl)-, trimethylsilyl ester		MS05768
73	144	45	218	43	75	59	147	**261**	100	89	25	16	16	13	13	12	**0.00**	11	27	2	1	–	–	–	–	–	2	–	–	–	L-Valine, N-(trimethylsilyl)-, trimethylsilyl ester		MS05769
220	261	222	263	262	187	221	260	**261**	100	97	39	37	20	19	16	14		12	8	–	3	1	–	1	–	–	–	–	–	–	Pyrimidine, 6-(m-chlorophenyl)-5-cyano-3,4-dihydro-2-methyl-4-thioxo-		NI26131
182	183	185	50	75	78	51	155	**261**	100	37	35	34	30	25	25	22	**3.11**	12	8	1	1	–	–	–	1	–	–	–	–	–	Pyridine, 3-(4-bromobenzoyl)-	14548-45-9	NI26132
261	85	121	70	262	102	263	143	**261**	100	71	28	25	18	16	13	11		12	11	–	3	2	–	–	–	–	–	–	–	–	5H-Imidazo[4,5-d]thiazole-5-thione, 4,6-dihydro-4,6-dimethyl-2-phenyl-	23888-58-6	NI26133
86	83	119	85	91	78	92	261	**261**	100	69	59	46	22	22	15	11		12	11	4	3	–	–	–	–	–	–	–	–	–	Diazeto[3,4-b]dioxindicarboxylic acid, 2,3,4a,5,6,6a-hexahydro-, N-phenylimide		LI05420
228	261	230	190	218	260	227	201	**261**	100	47	41	41	31	30	30	30		12	12	1	1	2	–	–	–	–	–	–	1	–	4H-Borepino[3,2-b:6,7-b']dithiene, 4-(2-aminoethyl)-	51081-15-3	NI26134
168	167	41	43	153	124	97	195	**261**	100	99	50	39	33	29	27	17	**0.00**	12	18	3	2	–	–	–	–	–	–	–	–	1	2,4,6(1H,3H,5H)-Pyridiminetrione, 5-(1-methylbutyl)-5-(2-propenyl)-, monosodium salt	309-43-3	NI26135
168	167	41	43	97	195	124	169	**261**	100	86	37	35	27	18	18	16	**0.00**	12	18	3	2	–	–	–	–	–	–	–	–	1	2,4,6(1H,3H,5H)-Pyrimidinetrione, 5-(1-methylbutyl)-5-(2-propenyl)-, monosodium salt	309-43-3	NI26136
168	167	41	43	124	97	169	195	**261**	100	85	37	33	21	19	18	16	**0.00**	12	18	3	2	–	–	–	–	–	–	–	–	1	2,4,6(1H,3H,5H)-Pyrimidinetrione, 5-(1-methylbutyl)-5-(2-propenyl)-, monosodium salt	309-43-3	NI26137
105	77	261	51	106	184	263	138	**261**	100	51	28	13	12	11	10	8		13	8	3	1	–	–	1	–	–	–	–	–	–	Benzophenone, 2-chloro-5-nitro-	34052-37-4	NI26138
107	139	123	65	79	77	91	45	**261**	100	55	51	51	47	44	41	25	**17.43**	13	11	3	1	1	–	–	–	–	–	–	–	–	Benzene, 1-(4-methylphenylsulfinyl)-4-nitro-	22865-49-2	NI26139
202	217	261	218	189	174	133	216	**261**	100	99	70	40	31	25	22	20		13	11	5	1	–	–	–	–	–	–	–	–	–	1,3-Dioxolo[4,5-g]quinoline-7-carboxylic acid, 5-ethyl-5,8-dihydro-8-oxo-	14698-29-4	NI26140
202	77	47	201	60	227	185	108	**261**	100	95	95	56	39	28	22	17	**11.00**	13	12	1	1	1	–	–	–	–	–	1	–	–	Phosphine, diphenylthiocarbamoyl-, oxide		LI05421
230	207	29	31	30	77	75	45	**261**	100	85	66	38	35	31	26	25	**8.50**	13	19	1	5	–	–	–	–	–	–	–	–	–	2-Buten-1-ol, 2-methyl-4-(1,2,8-trimethyl-1H-purin-6-ylamino)-, (E)-		NI26141
219	104	149	133	134	105	103	116	**261**	100	40	26	25	23	22	21	18	**7.10**	14	7	1	5	–	–	–	–	–	–	–	–	–	5H-Pyrazino[2,3-b]indole, 5-acetyl-2,3-dicyano-		MS05770
140	77	261	107	108	78	94	154	**261**	100	98	60	58	42	18	10	8		14	12	2	1	–	–	1	–	–	–	–	–	–	Acetamide, N-(3-chlorophenyl)-2-phenoxy-		MS05771
140	261	77	107	108	78	94	154	**261**	100	80	60	40	23	10	4	4		14	12	2	1	–	–	1	–	–	–	–	–	–	Acetamide, N-(4-chlorophenyl)-2-phenoxy-		MS05772
111	75	119	127	91	125	77	51	**261**	100	93	90	82	81	80	76	60	**11.84**	14	12	2	1	–	–	1	–	–	–	–	–	–	Aniline, 4-chloro-N-(2-methoxybenzylidene)-, N-oxide		MS05773
106	91	155	260	77	197	107	65	**261**	100	68	43	35	29	23	19	18	**6.00**	14	15	2	1	1	–	–	–	–	–	–	–	–	Benzenesulphonamide, N,4-dimethyl-N-phenyl-	599-62-2	NI26142
61	89	81	261	184	91	77	140	**261**	100	94	81	68	62	42	23	20		14	15	2	1	1	–	–	–	–	–	–	–	–	Cyclohexanespiro-2'-[4-(phenylthio)-1,3-oxazol-5(2H)-one]		DD00769
174	141	173	175	171	129	142	172	**261**	100	56	45	19	19	16	16	13	**0.00**	14	15	2	1	1	–	–	–	–	–	–	–	–	Cysteine, S-(7-methyl-1-naphthyl)-	85896-60-2	NI26143
43	261	218	44	71	246	45	72	**261**	100	65	60	46	44	39	29	21		14	15	2	1	1	–	–	–	–	–	–	–	–	Thiazole, 5-acetyl-2-(4-ethoxyphenyl)-4-methyl-	71633-86-8	NI26144
246	128	261	202	43	127	115	87	**261**	100	51	26	23	19	18	18	18		14	15	4	1	–	–	–	–	–	–	–	–	–	1,3-Dioxolane, 2-methyl-2-[4-(4-nitrophenyl)-1,3-butadienyl]-	80283-25-6	NI26145

MASS TO CHARGE RATIOS								M.W.	INTENSITIES								Parent	C	H	O	N	S	F	Cl	Br	I	Si	P	B	X	COMPOUND NAME	CAS Reg No	No
160	148	199	161	130	261	243	105	261	100	21	15	14	12	10	10	8		14	15	4	1	–	–	–	–	–	–	–	–	–	Hexanoic acid, 6-phthalimido-		LI05422
160	187	205	216	132	148	104	130	261	100	93	70	60	59	56	47	46	20.93	14	15	4	1	–	–	–	–	–	–	–	–	–	2-Isoindolineacetic acid, α-sec-butyl-1,3-dioxo-	19506-84-4	NI26146
160	187	205	216	132	148	104	130	261	100	92	70	60	60	55	48	46	21.00	14	15	4	1	–	–	–	–	–	–	–	–	–	Isoleucine, N-phthaloyl-		LI05423
44	59	43	58	41	57	51	145	261	100	80	63	21	18	17	16	13	2.20	14	15	4	1	–	–	–	–	–	–	–	–	–	3-Isoxazolecarboperoxoic acid, 5-phenyl-, tert-butyl ester	35145-87-0	NI26147
160	216	130	148	104	205	261	187	261	100	60	21	18	14	12	11	7		14	15	4	1	–	–	–	–	–	–	–	–	–	Norleucine, N-phthaloyl-		LI05424
170	229	201	230	261	171	198	142	261	100	92	88	70	48	41	40	31		14	15	4	1	–	–	–	–	–	–	–	–	–	1H-Pyrrolo[1,2-a]indole-9-carboxylic acid, 9a-methoxy-3-oxo-2,3,9α,9aα-tetrahydro-, methyl ester, (±)-		NI26148
58	97	72	71	191	190	79	78	261	100	71	22	19	11	10	8	8	4.00	14	19	–	3	1	–	–	–	–	–	–	–	–	1,2-Ethanediamine, N,N-dimethyl-N′-2-pyridinyl-N′-(2-thienylmethyl)-	91-80-5	NI26149
217	262	260	191	245				261	100	23	16	10	5				0.00	14	19	–	3	1	–	–	–	–	–	–	–	–	1,2-Ethanediamine, N,N-dimethyl-N′-2-pyridinyl-N′-(2-thienylmethyl)-	91-80-5	NI26151
58	97	72	84	71	191	190	261	261	100	71	28	24	18	15	12	11		14	19	–	3	1	–	–	–	–	–	–	–	–	1,2-Ethanediamine, N,N-dimethyl-N′-2-pyridinyl-N′-(2-thienylmethyl)-	91-80-5	NI26150
58	97	72	71	79	78	42	40	261	100	68	24	22	15	14	14	10	1.30	14	19	–	3	1	–	–	–	–	–	–	–	–	1,2-Ethanediamine, N,N-dimethyl-N′-2-pyridinyl-N′-(3-thienylmethyl)-	91-79-2	NI26152
217	262	260	245	191				261	100	18	12	6	6				0.00	14	19	–	3	1	–	–	–	–	–	–	–	–	1,2-Ethanediamine, N,N-dimethyl-N′-2-pyridinyl-N′-(3-thienylmethyl)-	91-79-2	NI26153
202	100	41	204	261	39	45	47	261	100	73	46	34	32	32	24	24		14	19	2	1	–	–	–	–	–	1	–	–	–	1H-Indoleacetic acid, N-trimethylsilyl-, methyl ester		NI26154
171	261	156	202	73	75	246	172	261	100	53	51	30	29	27	25	23		14	19	2	1	–	–	–	–	–	1	–	–	–	1H-Indole-2-carboxylic acid, 5-ethyl-, trimethylsilyl ester		NI26155
218	143	246	261	188	115	73	174	261	100	76	62	38	24	24	22	19		14	19	2	1	–	–	–	–	–	1	–	–	–	1H-Indole-2-carboxylic acid, 1-(trimethylsilyl)-, ethyl ester		NI26156
221	261	192	191	222	205	206	135	261	100	35	34	17	14	11	8	8		14	19	2	3	–	–	–	–	–	–	–	–	–	Acetamide, N-[3-[(2-cyanoethyl)ethylamino]-4-methoxyphenyl]-	19433-94-4	NI26157
80	82	78	81	94	96	93	175	261	100	93	48	44	42	38	8	8	1.89	14	19	2	3	–	–	–	–	–	–	–	–	–	5,6-Quinolinol, 8-[(4-amino-1-methylbutyl)amino]-		NI26158
102	130	131	103	75	76	77	104	261	100	48	34	17	11	10	9	6	3.60	15	11	–	5	–	–	–	–	–	–	–	–	–	Triazine, 3-methyl-1,3-di(2-cyanophenyl)-		LI05425
95	96	261	57	138	263	167	55	261	100	9	7	7	6	5	5	5		15	16	1	1	–	–	1	–	–	–	–	–	–	4H-Isoindole, N-(4-chlorophenyl)-5,7a-epoxy-5-methyl-1,2,3,7a-tetrahydro-		NI26159
230	261	229	202	231	201	186		261	100	76	37	19	16	11	11			15	16	2	1	–	1	–	–	–	–	–	–	–	1H-Pyrrole-2-carboxylic acid, 4-fluoro-5-(2,4,6-trimethylphenyl)-, methyl ester	51504-17-7	NI26160
44	217	120	106	105	174	104	160	261	100	90	83	80	73	67	45	39	0.00	15	19	3	1	–	–	–	–	–	–	–	–	–	1-Cyclobutanecarboxylic acid, 3-methyl-1-[(1-phenylethyl)carbamoyl]-	85317-98-2	NI26161
43	39	41	131	91	132	51	42	261	100	74	71	48	38	33	29	29	24.00	15	19	3	1	–	–	–	–	–	–	–	–	–	2-Aza-6,8-dioxatetracyclo[8.2.1.0^{2,12}.0^{4,10}]tridec-3-en-11-one, 7-isopropyl-12-methyl-13-methylidene-		NI26162
161	131	132	43	91	39	130	160	261	100	95	72	63	51	51	49	46	0.00	15	19	3	1	–	–	–	–	–	–	–	–	–	2-Aza-6,8-dioxatetracyclo[8.2.1.0^{2,12}.0^{4,10}]tridec-3-en-11-one, 7,7,12-trimethyl-13-methylidene-		NI26163
172	173	218	261	39	106	51	43	261	100	17	5	4	4	3	3	3		15	19	3	1	–	–	–	–	–	–	–	–	–	[1,3]Dioxepino[5,6-c]pyridine, 1,5-dihydro-8-methyl-3-isopropyl-9-(2-propynyloxy)-	69022-73-7	NI26164
105	120	187	186	131	130	148	121	261	100	52	11	9	8	8	7	6	0.00	15	19	3	1	–	–	–	–	–	–	–	–	–	Ethaneperoxoic acid, 1-cyano-4-methyl-1-phenylpentyl ester	58422-74-5	NI26165
246	173	41	43	172	135	231	91	261	100	75	49	37	32	32	29	26	0.79	15	19	3	1	–	–	–	–	–	–	–	–	–	Naphtho[1,2-b]furan-2,8(3H,9bH)-dione, 3a,4,5,5a-tetrahydro-3,5a,9-trimethyl-, 8-oxime, [3S-(3α,3aα,5aβ,9bβ)]-	1618-82-2	NI26166
43	202	105	218	77	176	104	103	261	100	84	82	63	63	40	39	32	0.00	15	19	3	1	–	–	–	–	–	–	–	–	–	4-Oxazolecarboxylic acid, 4,5-dihydro-4-isopropyl-5-methyl-2-phenyl-, methyl ester, (4R,5R)-		DD00770
72	83	189	147	175	118	117	119	261	100	27	23	20	19	17	14	13	12.00	15	23	1	3	–	–	–	–	–	–	–	–	–	Pyrrolidine, 1-[4-(dimethylamino)-1-oxobutyl]-2-(3-pyridinyl)-, (S)-	69730-92-3	NI26167
56	122	36	43	30	41	182	55	261	100	24	23	21	20	19	17	16	0.00	15	32	–	1	–	–	1	–	–	–	–	–	–	Cyclopentadecanamine hydrochloride		MS05774
105	156	51	77	106	157	50	76	261	100	99	41	40	22	12	12	10	5.00	16	11	1	3	–	–	–	–	–	–	–	–	–	1-Phthalazinecarbonitrile, 2-benzoyl-1,2-dihydro-	13925-27-4	NI26168
134	105	135	36	91	79	77	106	261	100	95	12	12	10	10	10	8		16	20	–	1	–	–	1	–	–	–	–	–	–	Benzeneethanamine, N-(2-phenylethyl)-, hydrochloride	6332-28-1	NI26169
186	104	77	75	55	51	83	261	261	100	79	39	32	32	11	9	4		16	23	2	1	–	–	–	–	–	–	–	–	–	Aniline, N-(1-cyclohexyl-2,2-dimethoxyethylidene)-		NI26170
184	77	229	81	51	104	214	113	261	100	46	31	21	20	18	17	15	1.50	16	23	2	1	–	–	–	–	–	–	–	–	–	Aniline, N-[1-(1-methoxycyclohexyl)-2-methoxyethylidene]-		NI26171
175	41	174	42	146	219	176	53	261	100	41	40	24	21	19	17	14	3.00	16	23	2	1	–	–	–	–	–	–	–	–	–	Annotinine, 8-deoxo-	69796-60-7	NI26172
107	149	57	78	108	79	77	72	261	100	79	71	61	57	42	38	37	15.91	16	23	2	1	–	–	–	–	–	–	–	–	–	1H-Azepine-4-carboxylic acid, hexahydro-1-methyl-4-phenyl-, ethyl ester	77-15-6	NI26174
262	260	216	188					261	100	32	12	5					0.00	16	23	2	1	–	–	–	–	–	–	–	–	–	1H-Azepine-4-carboxylic acid, hexahydro-1-methyl-4-phenyl-, ethyl ester	77-15-6	NI26173
123	122	82	69	246	97	83	261	261	100	59	32	27	22	19	19	16		16	23	2	1	–	–	–	–	–	–	–	–	–	Elaeocarpine, 13,14,15,16-tetrahydro-, (+)-		LI05426
134	43	41	70	42	96	69	55	261	100	58	31	27	24	16	16	13	11.51	16	23	2	1	–	–	–	–	–	–	–	–	–	Indolizidine, 7-hydroxy-6-(4-methoxyphenyl)-7-methyl-		MS05775
123	122	82	261	246	55	97	83	261	100	85	33	31	31	28	25	19		16	23	2	1	–	–	–	–	–	–	–	–	–	Isoelaeocarpine, 13,14,15,16-tetrahydro-, (-)-		LI05427
174	191	41	190	91	77	55	42	261	100	81	50	33	26	26	26	26	16.01	16	23	2	1	–	–	–	–	–	–	–	–	–	Lycopodan-8-one, 11,12-didehydro-5-hydroxy-15-methyl-, (5β,15R)-	664-24-4	NI26175
262	140	112	260	246				261	100	40	24	21	12				0.00	16	23	2	1	–	–	–	–	–	–	–	–	–	1-Piperidinepropanol, 2-methyl-, benzoate (ester), (±)-	136-82-3	NI26176
172	42	187	57	84	43	44	77	261	100	50	37	36	34	26	23	18	5.60	16	23	2	1	–	–	–	–	–	–	–	–	–	4-Piperidinol, 1,3-dimethyl-4-phenyl-, propanoate (ester), cis-	77-20-3	NI26177
172	187	42	84	129	144	44	91	261	100	66	65	54	48	40	34	33	6.41	16	23	2	1	–	–	–	–	–	–	–	–	–	4-Piperidinol, 1,3-dimethyl-4-phenyl-, propanoate (ester), cis-	77-20-3	NI26178
188	262	189	263	187	261	230	172	261	100	30	15	5	5	3	2	2		16	23	2	1	–	–	–	–	–	–	–	–	–	4-Piperidinol, 1,3-dimethyl-4-phenyl-, propanoate (ester), cis-	77-20-3	NI26179
136	162	176	261	59	122	80	218	261	100	75	49	20	16	10	9	5		16	27	–	1	–	–	–	–	–	1	–	–	–	1-Nonene, 1-[dimethyl(2-pyridyl)silyl]-, (E)-		DD00771
136	176	137	59	162	261	190	106	261	100	10	9	7	5	4	4	3		16	27	–	1	–	–	–	–	–	1	–	–	–	1-Nonene, 3-[dimethyl(2-pyridyl)silyl]-		DD00772
179	105	261	77	204	177	51	207	261	100	91	73	45	39	32	30	27		17	11	2	1	–	–	–	–	–	–	–	–	–	2-Furancarbonitrile, 2,5-dihydro-3,4-diphenyl-5-oxo-	122949-76-2	DD00773
261	262	232	233	260	204	205	176	261	100	91	68	67	57	53	52	40		17	11	2	1	–	–	–	–	–	–	–	–	–	Naphtho[2,3-e]indole-6,11-dione, 3-methyl-		NI26180
189	215	261	216	213	244	214	202	261	100	98	88	82	78	55	49	26		17	11	2	1	–	–	–	–	–	–	–	–	–	Pyrene, 1-methyl-2-nitro-		MS05776

MASS TO CHARGE RATIOS	M.W.	INTENSITIES	Parent	C	H	O	N	S	F	Cl	Br	I	Si	P	B	X	COMPOUND NAME	CAS Reg No	No
215 261 213 214 203 216 202 189	261	100 72 56 46 26 21 21 19		17	11	2	1	–	–	–	–	–	–	–	–	–	Pyrene, 1-methyl-7-nitro-		MS05777
261 135 262 176 162 148 218 190	261	100 30 18 11 5 5 4 4		17	27	1	1	–	–	–	–	–	–	–	–	–	Benzamide, N-decyl-	53044-19-2	NI26181
81 181 71 41 43 55 57 79	261	100 67 30 24 21 15 14 14	6.35	17	27	1	1	–	–	–	–	–	–	–	–	–	(2E,4E,8Z,10E)-Dodecatetraenamide, N-(2-methylbutyl)-		NI26182
170 171 148 57 45 55 43 42	261	100 10 8 7 7 5 5 5	0.10	17	27	1	1	–	–	–	–	–	–	–	–	–	5-Nonanone, 2-(dimethylamino)-1-phenyl-	27820-12-8	NI26183
105 77 183 51 106 78 79 261	261	100 80 68 48 46 44 38 26		18	15	1	1	–	–	–	–	–	–	–	–	–	Benzenemethanol, α-phenyl-α-(3-pyridyl)-	1620-30-0	NI26184
232 233 234 102 204 203 88 63	261	100 97 24 14 13 10 7 7	0.00	18	15	1	1	–	–	–	–	–	–	–	–	–	12-Aza-6-oxabenz[a]anthracene, 6,6-dimethyl-5,6-dihydro-		NI26185
261 217 230 260 218 216 262 189	261	100 63 37 36 35 26 23 23		18	15	1	1	–	–	–	–	–	–	–	–	–	Quinoline, 4-[2-(4-methoxyphenyl)vinyl]-, (E)-	31059-68-4	NI26186
134 81 67 41 79 55 120 43	261	100 59 59 54 46 42 36 35	1.84	18	31	–	1	–	–	–	–	–	–	–	–	–	6-Octadecynenitrile	56600-19-2	NI26187
41 81 67 55 54 95 68 29	261	100 97 94 82 79 58 57 50	1.04	18	31	–	1	–	–	–	–	–	–	–	–	–	9-Octadecynenitrile	56599-96-3	NI26188
67 41 81 54 55 68 95 82	261	100 97 94 88 85 78 67 61	0.54	18	31	–	1	–	–	–	–	–	–	–	–	–	11-Octadecynenitrile	56599-92-9	NI26189
93 106 120 41 94 43 29 107	261	100 23 14 12 8 8 8 6	1.10	18	31	–	1	–	–	–	–	–	–	–	–	–	Pyridine, 2-tridecyl-	33912-24-2	WI01315
91 233 170 232 168 115 143 259	261	100 44 35 30 26 22 19 18	0.00	19	19	–	1	–	–	–	–	–	–	–	–	–	Carbazolenine, 11-benzyl-1,2,3,4-tetrahydro-		MS05778
261 91 233 170 232 168 115 143	261	100 99 44 35 30 26 22 19		19	19	–	1	–	–	–	–	–	–	–	–	–	Carbazolenine, 11-benzyl-1,2,3,4-tetrahydro-		LI05428
262 261 260 219 247	261	100 58 45 23 7		19	19	–	1	–	–	–	–	–	–	–	–	–	1H-Indeno[2,1-c]pyridine, 2,3,4,9-tetrahydro-2-methyl-9-phenyl-	82-88-2	NI26191
260 261 42 202 57 217 218 203	261	100 77 35 26 25 21 19 19		19	19	–	1	–	–	–	–	–	–	–	–	–	1H-Indeno[2,1-c]pyridine, 2,3,4,9-tetrahydro-2-methyl-9-phenyl-	82-88-2	NI26190
260 261 42 57 217 202 184 218	261	100 68 40 28 17 17 15 15		19	19	–	1	–	–	–	–	–	–	–	–	–	1H-Indeno[2,1-c]pyridine, 2,3,4,9-tetrahydro-2-methyl-9-phenyl-	82-88-2	NI26192
261 117 105 106 91 118 119 156	261	100 73 57 46 42 38 30 29		19	19	–	1	–	–	–	–	–	–	–	–	–	Tricyclo[10.2.2.2^{5,8}]octadeca-5,7,12,14,15,17-hexaene-6-carbonitrile	24770-03-4	NI26193
264 266 185 187 183 106 104 262	262	100 92 88 47 46 41 39 37		2	1	–	–	–	–	–	3	–	–	–	–	–	Ethylene, 1,1,2-tribromo-		IC03407
162 243 93 193 143 262 69 131	262	100 46 30 21 21 20 20 15		6	–	–	–	–	10	–	–	–	–	–	–	–	Cyclohexene, perfluoro-		IC03408
162 93 69 243 143 193 131 155	262	100 72 64 48 41 32 30 22	11.04	6	–	–	–	–	10	–	–	–	–	–	–	–	Cyclohexene, perfluoro-		IC03409
162 93 243 69 143 193 131 31	262	100 68 54 44 34 27 26 24	0.00	6	–	–	–	–	10	–	–	–	–	–	–	–	Cyclohexene, perfluoro-		LI05429
169 109 125 168 142 110 79 170	262	100 63 44 12 11 11 11 7	0.00	6	15	5	–	2	–	–	–	–	–	1	–	–	Phosphorothioic acid, S-[2-(ethylsulphonyl)ethyl] O,O-dimethyl ester	17040-19-6	NI26194
229 227 118 218 216 231 120 221	262	100 99 78 76 70 67 51 47	26.61	7	6	2	–	–	–	4	–	–	–	–	–	–	1,3-Cyclopentadiene, 5,5-dimethoxy-1,2,3,4-tetrachloro-	2207-27-4	NI26195
93 65 92 119 262 91 66 39	262	100 35 28 24 21 14 14 12		7	7	1	2	1	5	–	–	–	–	–	–	–	(3-Phenylureido)sulphur pentafluoride	90598-04-2	NI26196
193 83 42 55 69 41 28 84	262	100 29 25 20 16 13 12 11	0.50	7	10	2	2	1	4	–	–	–	–	–	–	–	Piperidinosulphur(VI)oxide monofluoride(trifluoroacetylimide)		LI05430
262 231 203 104 135 76 103 77	262	100 78 9 3 2 2 2 1		8	7	2	–	–	–	–	–	1	–	–	–	–	Benzoic acid, 4-iodo-, methyl ester	619-44-3	NI26197
104 183 185 51 78 77 103 39	262	100 78 77 40 36 33 29 25	5.00	8	8	–	–	–	–	–	2	–	–	–	–	–	Benzene, 1,3-bis(bromomethyl)-	626-15-3	NI26198
104 183 185 51 103 77 78 50	262	100 65 60 27 20 18 14 13	2.88	8	8	–	–	–	–	–	2	–	–	–	–	–	Benzene, 1,4-bis(bromomethyl)-	623-24-5	NI26199
51 183 185 264 104 103 77 50	262	100 85 81 72 63 62 60 40	37.71	8	8	–	–	–	–	–	2	–	–	–	–	–	Benzene, 1,4-dibromo-2,5-dimethyl-	1074-24-4	NI26200
104 183 185 103 51 77 78 50	262	100 48 47 41 41 37 32 26	0.70	8	8	–	–	–	–	–	2	–	–	–	–	–	Benzene, (1,2-dibromoethyl)-	93-52-7	NI26201
61 122 62 76 262 18 183 30	262	100 93 81 78 70 55 53 46		8	10	4	2	2	–	–	–	–	–	–	–	–	Benzenesulphonamide, N-(dimethylthio)-4-nitro-		MS05779
133 83 41 77 55 39 247 105	262	100 60 33 22 20 20 18 18	3.78	8	11	–	–	–	4	–	1	–	–	–	–	–	3-Hexene, 1-bromo-1,1,2,2-tetrafluoro-5,5-dimethyl-	74793-50-3	NI26202
56 43 41 42 27 77 55 39	262	100 54 44 36 25 19 15 14	2.89	8	11	–	–	–	4	–	1	–	–	–	–	–	3-Octene, 1-bromo-1,1,2,2-tetrafluoro-	74793-72-9	NI26203
262 167 193 166 69 108 263 105	262	100 62 40 21 20 15 12 11		9	5	–	2	2	3	–	–	–	–	–	–	–	Benzothiazole, 2-[(trifluorothioacetyl)amino]-		MS05780
36 210 168 38 212 37 133 89	262	100 60 41 35 23 21 19 18	0.00	9	8	1	2	1	–	2	–	–	–	–	–	–	2(3H)-Thiazolimine, 4-(4-chlorophenyl)-3-hydroxy-, monohydrochloride	58275-67-5	NI26204
161 163 56 63 74 124 187 90	262	100 58 36 22 22 19 16 16	4.89	9	8	3	2	–	–	2	–	–	–	–	–	–	Ureidoacetic acid, N-(3,5-dichlorophenyl)-		NI26205
134 169 133 157 135 30 143 156	262	100 77 69 62 58 55 48 38	12.40	9	8	4	2	–	1	1	–	–	–	–	–	–	Propane, 1,3-dinitro-2-(2-chloro-6-fluorophenyl)-		IC03410
87 57 147 149 41 69 29 88	262	100 98 45 44 38 23 20 5	0.00	9	11	4	–	–	–	–	1	–	–	–	–	–	2H,4H-1,3-Dioxin-6-carboxaldehyde, 5-bromo-2-tert-butyl-4-oxo-, (2R)-		DD00774
262 247 185 264 227 136 113 116	262	100 63 43 42 39 33 27 26		9	15	1	4	1	–	1	–	–	–	–	–	–	1,3,5-Triazin-2-amine, 4-(2-chloroethoxy)-N-isopropyl-6-methylthio-	73775-70-9	NI26206
215 139 227 217 262 216 247 71	262	100 50 32 32 25 12 10 10		9	15	1	4	1	–	1	–	–	–	–	–	–	1,3,5-Triazin-2(1H)-one, 1-(3-chloropropyl)-4-dimethylamino-6-methylthio-	74483-59-3	NI26207
73 30 147 75 247 102 45 129	262	100 90 39 34 19 19 18 17	1.32	9	22	3	2	–	–	–	–	–	2	–	–	–	Glycine, carbamoyl-, N-trimethylsilyl-, trimethylsilyl ester		NI26208
55 178 144 59 60 88 206 234	262	100 85 70 43 19 18 17 15	3.99	10	7	3	–	1	–	–	–	–	–	–	–	1	Cymantrene, thioacetyl-		NI26209
87 43 45 111 75 262 159 135	262	100 97 30 24 14 12 10 10		10	11	4	–	1	–	1	–	–	–	–	–	–	Acetic acid, [(4-chlorophenyl)sulphonyl]-, ethyl ester	3636-65-5	NI26210
87 43 45 111 75 262 159 135	262	100 97 30 24 14 12 10 10		10	11	4	–	1	–	1	–	–	–	–	–	–	Ethanol, 2-[(4-chlorophenyl)sulphonyl]-, acetate		LI05431
263 247 265 266 249 163 251 165	262	100 81 58 46 46 42 31 23	0.00	10	14	4	–	–	–	–	–	–	–	–	–	1	Zinc, bis(2,4-pentanedionato)-	14024-63-6	NI26211
59 43 60 262 127 57 71 55	262	100 87 30 27 22 22 18 17		10	14	6	–	1	–	–	–	–	–	–	–	–	α-D-Allofuranose, 1,2-O-isopropylidene-, cyclic 5,6-(thiocarbonate)-	30645-01-3	NI26212
60 59 85 187 43 86 69 55	262	100 99 90 65 60 42 34 25	1.00	10	14	6	–	1	–	–	–	–	–	–	–	–	α-D-Glucofuranose, 1,2-O-isopropylidene-, cyclic 5,6-(thiocarbonate)-		LI05432
102 101 103 72 61 219 73 159	262	100 64 60 57 55 53 53 49	2.06	10	14	8	–	–	–	–	–	–	–	–	–	–	1,4,7,10,13-Pentaoxacyclopentadecane-2,5,9-trione	79687-33-5	NI26213
177 179 69 57 98 41 87 126	262	100 98 75 66 58 41 37 26	0.00	10	15	3	–	–	–	–	1	–	–	–	–	–	2H,4H-1,3-Dioxin-4-one, 6-(1-bromoethyl)-2-tert-butyl-, (2R)-		DD00775
15 109 18 262 79 45 77 121	262	100 64 58 46 24 20 17 16		10	15	4	–	1	–	–	–	–	–	1	–	–	Phosphoric acid, dimethyl 3-methyl-4-(methylthio)phenyl ester	6552-12-1	NI26214
262 109 79 247 77 121 153 45	262	100 74 26 24 19 19 18 16		10	15	4	–	1	–	–	–	–	–	1	–	–	Phosphoric acid, dimethyl 3-methyl-4-(methylthio)phenyl ester	6552-12-1	NI26215
70 72 85 83 71 87 101 88	262	100 28 25 23 23 22 22 22	17.50	10	18	6	2	–	–	–	–	–	–	–	–	–	(1S,2S)-1,2-Dihydroxy-3,14-dioxo-7,10-dioxa-4,13-diazacyclotetradecane		MS05781
147 72 247 74 148 149 44 56	262	100 46 23 15 14 8 8 7	2.00	10	22	4	–	–	–	–	–	–	2	–	–	–	Butanedioic acid, bis(trimethylsilyl) ester	40309-57-7	NI26217

MASS TO CHARGE RATIOS								M.W.	INTENSITIES								Parent	C	H	O	N	S	F	Cl	Br	I	Si	P	B	X	COMPOUND NAME	CAS Reg No	No
147	73	75	148	247	45	55	149	**262**	100	73	27	16	15	15	13	9	**0.00**	10	22	4	–	–	–	–	–	–	2	–	–	–	**Butanedioic acid, bis(trimethylsilyl) ester**	40309-57-7	**NI26216**
147	73	148	75	247	149	45	55	**262**	100	91	62	56	45	32	24	20	**2.40**	10	22	4	–	–	–	–	–	–	2	–	–	–	**Butanedioic acid, bis(trimethylsilyl) ester**	40309-57-7	**MS05782**
73	147	247	103	45	101	75	66	**262**	100	54	26	25	16	15	14	13	**10.01**	10	22	4	–	–	–	–	–	–	2	–	–	–	**2(3H)-Furanone, dihydro-3,4-bis[(trimethylsilyl)oxy]-, cis-**	55220-75-2	**NI26219**
73	147	103	101	75	45	247	116	**262**	100	53	25	16	16	16	13	12	**4.59**	10	22	4	–	–	–	–	–	–	2	–	–	–	**2(3H)-Furanone, dihydro-3,4-bis[(trimethylsilyl)oxy]-, cis-**	55220-75-2	**NI26218**
73	147	247	45	101	75	103	116	**262**	100	67	21	20	15	14	13	12	**8.00**	10	22	4	–	–	–	–	–	–	2	–	–	–	**2(3H)-Furanone, dihydro-3,4-bis[(trimethylsilyl)oxy]-, trans-**	55220-79-6	**MS05783**
73	147	247	45	101	75	103	59	**262**	100	67	21	20	15	14	13	12	**8.01**	10	22	4	–	–	–	–	–	–	2	–	–	–	**2(3H)-Furanone, dihydro-3,4-bis[(trimethylsilyl)oxy]-, trans-**	55220-79-6	**NI26220**
73	147	45	56	75	66	148	74	**262**	100	65	29	26	21	11	10	8	**0.13**	10	22	4	–	–	–	–	–	–	2	–	–	–	**Propanedioic acid, methyl-, bis(trimethylsilyl) ester**	40333-07-1	**NI26221**
147	73	148	75	247	218	149	157	**262**	100	26	14	12	11	10	8	3	**0.66**	10	22	4	–	–	–	–	–	–	2	–	–	–	**Propanedioic acid, methyl-, bis(trimethylsilyl) ester**	40333-07-1	**NI26222**
147	73	75	148	45	56	149	66	**262**	100	62	27	15	15	12	9	8	**0.00**	10	22	4	–	–	–	–	–	–	2	–	–	–	**Propanedioic acid, methyl-, bis(trimethylsilyl) ester**	40333-07-1	**MS05784**
73	131	189	116	262	173	174	115	**262**	100	89	58	42	34	21	16	16		10	30	–	–	–	–	–	–	–	4	–	–	–	**Disilane, pentamethyl[bis-**		**MS05785**
148	56	206	178	176	262	234	207	**262**	100	89	46	34	23	13	6	6		11	10	4	–	–	–	–	–	–	–	–	–	1	**Iron, tricarbonyl[(1,3,4,5-η)-1-methyl-2-oxo-4-cycloheptene-1,3-diyl]-**		**DD00776**
43	95	44	45	262	169	199	64	**262**	100	55	30	22	20	18	16	16		11	10	4	4	–	–	–	–	–	–	–	–	–	**Pyrazole, 3,5-dimethyl-1-(2,4-dinitrophenyl)-**		**IC03411**
217	262	150	234	29	115	89	88	**262**	100	94	68	59	59	43	36	36		11	10	4	4	–	–	–	–	–	–	–	–	–	**1,2,3-Triazole-5-carboxylic acid, 4-(3-nitrophenyl)-, ethyl ester**		**MS05786**
262	234	217	150	115	29	190	171	**262**	100	64	62	27	23	23	22	22		11	10	4	4	–	–	–	–	–	–	–	–	–	**1,2,3-Triazole-5-carboxylic acid, 4-(4-nitrophenyl)-, ethyl ester**		**MS05787**
43	175	262	162	177	264	164	41	**262**	100	39	32	28	26	23	23	23		11	12	3	–	–	–	2	–	–	–	–	–	–	**Acetic acid, (2,4-dichlorophenoxy)-, isopropyl ester**	94-11-1	**NI26223**
43	175	18	177	41	162	27	262	**262**	100	46	38	35	25	21	19	16		11	12	3	–	–	–	2	–	–	–	–	–	–	**Acetic acid, (2,4-dichlorophenoxy)-, isopropyl ester**	94-11-1	**NI26224**
101	15	59	41	42	29	69	39	**262**	100	87	71	58	42	38	32	31	**2.50**	11	12	3	–	–	–	2	–	–	–	–	–	–	**Butanoic acid, 4-(2,4-dichlorophenoxy)-, methyl ester**	18625-12-2	**NI26225**
60	100	42	69	162	101	43	64	**262**	100	92	31	28	20	16	16	11	**0.00**	11	12	3	–	–	–	2	–	–	–	–	–	–	**Butanoic acid, 4-(2,4-dichlorophenoxy)-, methyl ester**	18625-12-2	**NI26226**
121	155	135	77	136	157	91	76	**262**	100	38	38	29	24	14	14	10	**0.30**	11	12	3	–	–	–	2	–	–	–	–	–	–	**2-Butanone, 1,3-dichloro-4-hydroxy-4-(4-methoxyphenyl)-**		**MS05788**
155	43	85	117	116	172	101	41	**262**	100	97	73	58	48	37	32	27	**0.00**	11	18	7	–	–	–	–	–	–	–	–	–	–	**2H-Furo[3,2-b]pyran-2-one, hexahydro-3a-methoxy-3,3,7-trihydroxy-5,6,7-trimethyl-, (3aα,5β,6β,7α,7aα)-**	50392-34-2	**NI26227**
43	59	87	247	173	74	113	100	**262**	100	43	30	18	14	14	10	10	**0.00**	11	18	7	–	–	–	–	–	–	–	–	–	–	**α-D-Galactopyranosiduronic acid, methyl 3,4-O-isopropylidene-, methyl ester**	25253-47-8	**LI05433**
43	59	45	87	29	31	247	85	**262**	100	44	30	29	27	24	20	20	**1.70**	11	18	7	–	–	–	–	–	–	–	–	–	–	**α-D-Galactopyranosiduronic acid, methyl 3,4-O-isopropylidene-, methyl ester**	25253-47-8	**MS05789**
43	59	45	87	247	85	74	173	**262**	100	45	30	28	20	18	18	16	**3.16**	11	18	7	–	–	–	–	–	–	–	–	–	–	**α-D-Galactopyranosiduronic acid, methyl 3,4-O-isopropylidene-, methyl ester**	25253-47-8	**NI26228**
87	42	45	69	131	175	71	89	**262**	100	90	78	20	17	13	6	5	**0.10**	11	18	7	–	–	–	–	–	–	–	–	–	–	**1,4,7,10,13-Pentaoxacyclohexadecane-14,16-dione**		**MS05790**
43	75	74	116	85	18	101	28	**262**	100	99	99	80	48	32	21	19	**0.00**	11	18	7	–	–	–	–	–	–	–	–	–	–	**L-Talopyranose, 6-deoxy-3-O-methyl-, 2,4-diacetate**	55836-48-1	**NI26229**
129	43	87	58	102	74	69	100	**262**	100	100	90	59	53	42	40	36	**0.00**	11	18	7	–	–	–	–	–	–	–	–	–	–	**α-D-Xylopyranoside, methyl 2,3-di-O-acetyl-4-O-methyl-**	53872-84-7	**NI26230**
75	74	43	71	100	99	69	45	**262**	100	100	100	59	34	23	20	18	**0.00**	11	18	7	–	–	–	–	–	–	–	–	–	–	**α-D-Xylopyranoside, methyl 2,4-di-O-acetyl-3-O-methyl-**	72922-31-7	**NI26231**
74	43	100	69	87	142	129	45	**262**	100	100	51	45	43	29	24	22	**0.00**	11	18	7	–	–	–	–	–	–	–	–	–	–	**α-D-Xylopyranoside, methyl 3,4-di-O-acetyl-2-O-methyl-**	53872-83-6	**NI26232**
71	130	104	58	74	129	174	73	**262**	100	92	58	50	39	33	32	30	**24.57**	11	22	3	2	1	–	–	–	–	–	–	–	–	**1,4,7-Trioxa-10,12-diazacyclotetradecane-11-thione, 10,12-dimethyl-**	89863-08-1	**NI26233**
205	203	201	149	41	204	147	29	**262**	100	72	52	49	49	40	38	37	**7.14**	11	24	–	–	1	–	–	–	–	–	–	–	1	**1,2-Thiagermolane, 2,2-dibutyl-**	4514-08-3	**NI26234**
205	203	201	149	41	204	147	57	**262**	100	73	54	51	50	42	38	26	**9.00**	11	24	–	–	1	–	–	–	–	–	–	–	1	**Thiophene, 2-dibutylgermanotetrahydro-**		**LI05434**
73	145	147	75	45	146	219	43	**262**	100	90	48	17	16	12	10	9	**0.00**	11	26	3	–	–	–	–	–	–	2	–	–	–	**Butanoic acid, 2-methyl-2-[(trimethylsilyl)oxy]-, trimethylsilyl ester**	55557-18-1	**NI26235**
73	117	147	75	45	218	88	148	**262**	100	82	62	25	16	12	11	10	**0.00**	11	26	3	–	–	–	–	–	–	2	–	–	–	**Butanoic acid, 2-methyl-3-[(trimethylsilyl)oxy]-, trimethylsilyl ester**	55557-17-0	**NI26236**
73	117	147	75	45	218	88	148	**262**	100	82	62	25	16	12	11	10	**0.00**	11	26	3	–	–	–	–	–	–	2	–	–	–	**Butanoic acid, 2-methyl-3-[(trimethylsilyl)oxy]-, trimethylsilyl ester**	55557-17-0	**NI26237**
147	73	75	247	148	45	56	146	**262**	100	75	26	18	16	16	14	12	**0.00**	11	26	3	–	–	–	–	–	–	2	–	–	–	**Butanoic acid, 2-methyl-4-[(trimethylsilyl)oxy]-, trimethylsilyl ester**		**NI26238**
147	73	75	247	148	146	45	56	**262**	100	73	25	25	19	15	13	13	**0.16**	11	26	3	–	–	–	–	–	–	2	–	–	–	**Butanoic acid, 2-methyl-4-[(trimethylsilyl)oxy]-, trimethylsilyl ester**		**NI26239**
73	145	147	45	75	146	74	148	**262**	100	73	45	16	15	10	9	7	**0.00**	11	26	3	–	–	–	–	–	–	2	–	–	–	**Butanoic acid, 3-methyl-2-[(trimethylsilyl)oxy]-, trimethylsilyl ester**	55124-92-0	**NI26240**
73	147	145	75	45	133	43	219	**262**	100	55	44	24	24	13	12	9	**0.00**	11	26	3	–	–	–	–	–	–	2	–	–	–	**Butanoic acid, 3-methyl-2-[(trimethylsilyl)oxy]-, trimethylsilyl ester**	55124-92-0	**NI26241**
73	145	147	45	75	146	74	148	**262**	100	73	45	16	15	10	9	7	**0.00**	11	26	3	–	–	–	–	–	–	2	–	–	–	**Butanoic acid, 3-methyl-2-[(trimethylsilyl)oxy]-, trimethylsilyl ester**	55124-92-0	**NI26242**
73	131	147	75	45	247	95	148	**262**	100	87	79	45	18	17	15	14	**0.00**	11	26	3	–	–	–	–	–	–	2	–	–	–	**Butanoic acid, 3-methyl-3-[(trimethylsilyl)oxy]-, trimethylsilyl ester**	55124-90-8	**NI26243**
131	73	147	75	247	148	95	132	**262**	100	88	86	25	22	14	13	12	**0.00**	11	26	3	–	–	–	–	–	–	2	–	–	–	**Butanoic acid, 3-methyl-3-[(trimethylsilyl)oxy]-, trimethylsilyl ester**	55124-90-8	**NI26244**
147	73	75	117	247	45	103	131	**262**	100	90	36	20	19	17	17	16	**0.00**	11	26	3	–	–	–	–	–	–	2	–	–	–	**Butanoic acid, 3-methyl-4-[(trimethylsilyl)oxy]-, trimethylsilyl ester**		**NI26245**
73	131	147	75	46	247	148	95	**262**	100	80	76	37	16	12	12	11	**0.00**	11	26	3	–	–	–	–	–	–	2	–	–	–	**Pentanoic acid, 2-[(trimethylsilyl)oxy]-, trimethylsilyl ester**	55887-46-2	**NI26246**
73	145	147	45	75	146	103	74	**262**	100	78	55	16	15	11	10	9	**0.00**	11	26	3	–	–	–	–	–	–	2	–	–	–	**Pentanoic acid, 2-[(trimethylsilyl)oxy]-, trimethylsilyl ester**	55887-46-2	**NI26247**
73	145	147	45	75	146	103	74	**262**	100	78	55	16	15	11	10	9	**0.00**	11	26	3	–	–	–	–	–	–	2	–	–	–	**Pentanoic acid, 2-[(trimethylsilyl)oxy]-, trimethylsilyl ester**	55887-46-2	**NI26248**
73	147	131	75	45	28	205	74	**262**	100	50	38	20	14	14	12	9	**0.00**	11	26	3	–	–	–	–	–	–	2	–	–	–	**Pentanoic acid, 3-[(trimethylsilyl)oxy]-, trimethylsilyl ester**	79628-62-9	**NI26249**
73	117	147	75	157	247	55	83	**262**	100	83	73	55	43	34	29	19	**0.00**	11	26	3	–	–	–	–	–	–	2	–	–	–	**Pentanoic acid, 4-[(trimethylsilyl)oxy]-, trimethylsilyl ester**		**NI26250**
147	73	75	247	148	157	149	59	**262**	100	69	41	32	16	13	12	9	**0.58**	11	26	3	–	–	–	–	–	–	2	–	–	–	**Pentanoic acid, 5-[(trimethylsilyl)oxy]-, trimethylsilyl ester**		**NI26251**
147	73	247	117	131	103	197	157	**262**	100	85	29	21	18	18	14	12	**0.00**	11	26	3	–	–	–	–	–	–	2	–	–	–	**Pentanoic acid, 5-[(trimethylsilyl)oxy]-, trimethylsilyl ester**		**MS05791**
147	73	75	103	45	247	55	148	**262**	100	82	24	18	17	17	16	16	**0.00**	11	26	3	–	–	–	–	–	–	2	–	–	–	**Propanoic acid, 2-ethyl-3-[(trimethylsilyl)oxy]-, trimethylsilyl ester**		**NI26252**
147	73	75	247	148	103	177	41	**262**	100	62	20	18	14	14	12	12	**0.00**	11	26	3	–	–	–	–	–	–	2	–	–	–	**Propanoic acid, 2-ethyl-3-[(trimethylsilyl)oxy]-, trimethylsilyl ester**		**MS05792**
139	247	83	57	231	211	73	262	**262**	100	79	72	48	25	25	19	3		11	28	–	2	–	–	1	–	–	1	–	1	–	**Borane, chloro(tert-butylamino)[tert-butyl(trimethylsilyl)amino]-**		**MS05793**

MASS TO CHARGE RATIOS								M.W.	INTENSITIES								Parent	C	H	O	N	S	F	Cl	Br	I	Si	P	B	X	COMPOUND NAME	CAS Reg No	No
262	234	160	149	205	263	235	132	**262**	100	61	17	17	14	13	10	8		12	6	7	–	–	–	–	–	–	–	–	–	–	5-Oxacyclopenta[f]naphthalene-1,4-dione, 2,3,8,9-tetrahydroxy-		LI05435
262	215	216	217	264	180	242	152	**262**	100	96	46	37	35	33	32	17		12	7	3	2	–	–	1	–	–	–	–	–	–	10H-Phenoxazine, 3-chloro-1-nitro-	39522-55-9	MS05794
77	91	105	262	157	93	65	137	**262**	100	48	43	39	24	20	20	18		12	10	3	2	1	–	–	–	–	–	–	–	–	Azobenzene-4-sulphonic acid		LI05436
262	89	117	261	160	132	202	263	**262**	100	64	38	36	18	15	15	14		12	10	3	2	1	–	–	–	–	–	–	–	–	4,6(1H,3H,5H)-Pyrimidinetrione, 5-(p-anisylidene)-2-thioxo-		NI26253
262	118	85	77	117	91	116	59	**262**	100	84	18	16	14	12	7	7		12	10	3	2	1	–	–	–	–	–	–	–	–	1,3-Thiazine-6-carboxylic acid, 2,3-dihydro-4-oxo-2-phenylimino-, methyl ester		MS05795
188	262	142	216	189	75	29	18	**262**	100	85	31	18	14	14	14	12		12	10	5	2	–	–	–	–	–	–	–	–	–	4(1H)-Quinolinone, 2-ethoxycarbonyl-6-nitro-		MS05796
262	156	78	121	56	199	185	183	**262**	100	39	28	20	18	17	16	15		12	11	1	–	–	–	1	–	–	–	–	–	1	Ferrocene, (chloroacetyl)-	51862-24-9	NI26254
232	103	198	77	247	262	213		**262**	100	72	53	53	52	40	36			12	11	1	4	–	–	1	–	–	–	–	–	–	6(1H,6H)-Pyridazinone, 5-chloro-4-hydrazino-1-(3-vinylphenyl)-		LI05437
147	59	262	115	113	41	99	148	**262**	100	89	30	14	13	11	9	7		12	22	–	–	3	–	–	–	–	–	–	–	–	2,6,9-Trithiabicyclo[3.3.1]nonane, 1,3,3,5,7,7-hexamethyl-, (±)-	57289-14-2	NI26255
122	153	182	81	55	82	95	139	**262**	100	96	56	54	50	46	40	36	**3.00**	12	22	4	–	1	–	–	–	–	–	–	–	–	4-Heptenoic acid, 2-(methylsulphonyl)-4-propyl-, methyl ester, (E)-	67428-08-4	NI26256
143	111	101	167	262	199	204	119	**262**	100	68	40	36	17	17	8	4		12	22	6	–	–	–	–	–	–	–	–	–	–	D-Arabino-hex-1-enitol, 3,4,5-tri-O-methyl-1,2-O-isopropylidene-	66851-40-9	NI26257
76	57	105	132	41	29	161	77	**262**	100	81	75	49	38	37	16	16	**0.00**	12	22	6	–	–	–	–	–	–	–	–	–	–	1,4-Butanedioic acid, 2,3-dihydroxy-, diethyl ester		NI26258
101	69	71	43	129	41	113	59	**262**	100	64	57	47	41	39	37	34	**0.00**	12	22	6	–	–	–	–	–	–	–	–	–	–	2-Furanhexanoic acid, tetrahydro-β,δ-dihydroxy-5-methoxy-, methyl ester	18142-13-7	NI26259
59	247	131	173	43	85	101	71	**262**	100	94	68	38	28	25	19	12	**0.00**	12	22	6	–	–	–	–	–	–	–	–	–	–	Galactitol, 2,3:4,5-bis-O-isopropylidene-	20581-93-5	NI26260
173	43	85	101	71	69	129	73	**262**	100	73	64	48	30	30	28	27	**0.00**	12	22	6	–	–	–	–	–	–	–	–	–	–	Galactitol, 2,3:4,5-bis-O-isopropylidene-	20581-93-5	LI05438
59	247	48	173	131	85	101	73	**262**	100	67	44	33	31	23	22	14	**0.00**	12	22	6	–	–	–	–	–	–	–	–	–	–	Galactitol, 1,2:4,5-di-O-isopropylidene-	20581-94-6	NI26261
173	247	131	48	101	73	111	70	**262**	100	99	92	66	65	42	36	36	**0.00**	12	22	6	–	–	–	–	–	–	–	–	–	–	Galactitol, 1,2:4,5-di-O-isopropylidene-	20581-94-6	LI05439
59	41	43	129	97	101	117	58	**262**	100	81	62	33	26	10	9	9	**0.00**	12	22	6	–	–	–	–	–	–	–	–	–	–	D-Glucitol, 1,4-di-O-acetyl-2,5-di-O-methyl-3,6-dideoxy-		NI26262
101	43	59	247	73	57	111	72	**262**	100	86	45	17	16	16	14	14	**0.11**	12	22	6	–	–	–	–	–	–	–	–	–	–	D-Mannitol, 1,2:5,6-di-O-isopropylidene-	1707-77-3	NI26263
59	247	43	189	111	131	101	72	**262**	100	99	88	69	65	55	54	53	**0.00**	12	22	6	–	–	–	–	–	–	–	–	–	–	D-Mannitol, 1,2:5,6-di-O-isopropylidene-	1707-77-3	LI05440
101	247	43	59	189	111	131	72	**262**	100	36	33	18	13	12	10	10	**0.00**	12	22	6	–	–	–	–	–	–	–	–	–	–	D-Mannitol, 1,2:5,6-di-O-isopropylidene-	1707-77-3	NI26264
145	185	43	69	167	59	129	113	**262**	100	99	87	72	66	66	54	47	**0.00**	12	22	6	–	–	–	–	–	–	–	–	–	–	Octanoic acid, 3,7-dihydroxy-3-methoxycarbonyl-7-methyl-, methyl ester		MS05797
73	174	179	219	89	99	121	105	**262**	100	20	17	12	10	8	7	7	**5.00**	12	26	–	–	2	–	–	–	–	1	–	–	–	1-Butene, 3-methyl-2-[3-(methylthio)propylthio]-1-trimethylsilyl-	80403-02-7	NI26265
73	99	142	147	262	141	105	247	**262**	100	52	44	19	18	16	16	3		12	26	–	–	2	–	–	–	–	1	–	–	–	1-Butene, 3-methyl-2-[3-methylthio-3-(trimethylsilyl)propylthio]-	80403-03-8	NI26266
73	93	146	188	106	262	205	247	**262**	100	88	53	36	24	11	9	6		12	26	–	–	2	–	–	–	–	1	–	–	–	1,3-Dithiane, 2-butyl-2-(trimethylsilylmethyl)-	80402-86-4	NI26267
73	146	188	93	262	165	205	247	**262**	100	75	36	29	25	18	15	13		12	26	–	–	2	–	–	–	–	1	–	–	–	1,3-Dithiane, 2-(2-methylpropyl)-2-[(trimethylsilyl)methyl]-	80402-88-6	NI26268
73	179	146	174	262	105	89	121	**262**	100	27	26	15	13	12	12	11		12	26	–	–	2	–	–	–	–	1	–	–	–	1-Pentene, 2-[3-(methylthio)propylthio]-1-trimethylsilyl-		NI26269
73	142	99	114	147	105	262	247	**262**	100	38	33	29	20	19	18	3		12	26	–	–	2	–	–	–	–	1	–	–	–	1-Pentene, 2-[3-methylthio-3-(trimethylsilyl)propylthio]-		NI26270
57	117	150	262	56	85	206	44	**262**	100	50	47	42	41	38	20	13		12	26	–	2	2	–	–	–	–	–	–	–	–	Piperazine, 1,4-bis(tert-butylthio)-	56771-87-0	MS05798
132	74	56	100	58	57	118	87	**262**	100	85	85	69	62	62	54	54	**0.45**	12	26	4	2	–	–	–	–	–	–	–	–	–	1,4,10,13-Tetraoxa-7,16-diazacyclooctadecane	23978-55-4	NI26271
55	147	83	73	75	262	103	82	**262**	100	84	71	63	49	40	30	26		12	30	2	–	–	–	–	–	–	2	–	–	–	Hexane, 1,6-bis[(trimethylsilyl)oxy]-	6222-22-6	NI26272
147	55	73	83	75	149	103	45	**262**	100	84	66	52	49	25	24	21	**0.00**	12	30	2	–	–	–	–	–	–	2	–	–	–	Hexane, 1,6-bis[(trimethylsilyl)oxy]-	6222-22-6	NI26273
159	73	103	147	75	160	45	41	**262**	100	98	71	45	15	14	12	11	**0.00**	12	30	2	–	–	–	–	–	–	2	–	–	–	Pentane, 1,2-bis[(trimethylsilyl)oxy]-4-methyl-		NI26274
131	73	103	147	75	132	45	233	**262**	100	76	31	26	12	12	10	10	**0.00**	12	30	2	–	–	–	–	–	–	2	–	–	–	Pentane, 1,3-bis[(trimethylsilyl)oxy]-2-methyl-		NI26275
199	262	264	201	163	200	164	263	**262**	100	54	38	35	31	19	16	15		13	8	–	2	–	–	2	–	–	–	–	–	–	11H-Dibenzo[c,f][1,2]diazepine, 3,8-dichloro-	1084-98-6	NI26277
199	201	262	163	264	164	200	81	**262**	100	32	31	29	20	18	16	11		13	8	–	2	–	–	2	–	–	–	–	–	–	11H-Dibenzo[c,f][1,2]diazepine, 3,8-dichloro-	1084-98-6	NI26276
77	153	18	28	51	262	110	17	**262**	100	99	65	35	23	20	16	13		13	10	–	–	3	–	–	–	–	–	–	–	–	Carbonotrithioic acid, diphenyl ester	2314-54-7	NI26278
262	159	231	69	203	115	63	39	**262**	100	94	54	54	38	35	33	33		13	10	2	–	2	–	–	–	–	–	–	–	–	2,2′-Bithienyl-5′-carboxylic acid, methyl ester		LI05441
105	77	51	262	185	76	50	106	**262**	100	42	14	12	11	11	9	7		13	10	4	–	1	–	–	–	–	–	–	–	–	Benzophenone-4-sulphonic acid		IC03412
262	191	263	219	190	176	247	147	**262**	100	53	17	17	10	10	5	5		13	10	6	–	–	–	–	–	–	–	–	–	–	7H-Furo[3,2-g][1]benzopyran-7-one, 5-hydroxy-6,9-dimethoxy-	35779-45-4	LI05442
262	191	234	163	244	263	219	176	**262**	100	52	33	28	23	15	15	9		13	10	6	–	–	–	–	–	–	–	–	–	–	7H-Furo[3,2-g][1]benzopyran-7-one, 5-hydroxy-6,9-dimethoxy-	35779-45-4	NI26279
262	219	191	176	247	263	190	51	**262**	100	46	38	38	20	16	10	7		13	10	6	–	–	–	–	–	–	–	–	–	–	7H-Furo[3,2-g][1]benzopyran-7-one, 6-hydroxy-4,5-dimethoxy-	27230-32-6	NI26280
262	219	191	176	247	263	190	68	**262**	100	48	38	38	21	17	11	8		13	10	6	–	–	–	–	–	–	–	–	–	–	7H-Furo[3,2-g][1]benzopyran-7-one, 6-hydroxy-4,5-dimethoxy-	27230-32-6	LI05443
43	247	262	219	41	55	45	204	**262**	100	92	87	62	58	47	40	38		13	10	6	–	–	–	–	–	–	–	–	–	–	1,4-Naphthoquinone, 2-acetyl-5,8-dihydroxy-3-methoxy-	14090-55-2	NI26281
262	264	77	51	155	263	265	89	**262**	100	92	41	30	20	16	15	15		13	11	1	–	–	–	–	1	–	–	–	–	–	Benzene, 1-(bromomethyl)-3-phenoxy-		IC03413
262	264	77	51	155	263	89	63	**262**	100	87	36	25	20	17	13	13		13	11	1	–	–	–	–	1	–	–	–	–	–	Benzene, 1-bromo-2-methyl-4-phenoxy-		IC03414
262	168	263	181	128	232	169	131	**262**	100	21	14	7	3	3	3	3		13	11	4	–	–	–	–	–	–	–	1	–	–	Phosphoric acid, 2,2′-biphenylylene-, methyl ester	76045-15-3	NI26282
121	92	198	94	107	91	77	65	**262**	100	99	39	38	28	28	28	28	**6.01**	13	14	2	2	1	–	–	–	–	–	–	–	–	Pyridinium, 2,6-dimethyl-1-[(phenylsulphonyl)amino]-, hydroxide, inner salt	34456-62-7	NI26283
84	218	116	129	130	115	117	126	**262**	100	59	44	41	31	29	22	21	**6.00**	13	14	4	2	–	–	–	–	–	–	–	–	–	Carbamic acid, 3-(2,3,3a,8b,tetrahydro-2-oxo-4H-indeno[2,1-d]oxazolyl)-, ethyl ester		LI05444
84	28	110	44	82	165	98	54	**262**	100	96	66	50	44	40	34	34	**0.80**	13	14	4	2	–	–	–	–	–	–	–	–	–	Crotonic acid, 4,4′-(pentamethylenedinitrilo)bis[4-hydroxy-, cyclic 1,4:1′,4′-dianhydride		MS05799

MASS TO CHARGE RATIOS								M.W.	INTENSITIES								Parent	C	H	O	N	S	F	Cl	Br	I	Si	P	B	X	COMPOUND NAME	CAS Reg No	No
84	28	110	44	82	165	54	26	**262**	100	96	66	50	44	40	34	34	**1.00**	13	14	4	2	–	–	–	–	–	–	–	–	–	Crotonic acid, 4,4′-(pentamethylenedinitrilo)bis[4-hydroxy-, cyclic 1,4:1′,4′-dianhydride, (Z)-	28537-77-1	NI26284
116	262	115	117	149	129	263	103	**262**	100	27	21	14	6	6	5	2		13	14	4	2	–	–	–	–	–	–	–	–	–	4H-Indeno[2,1-e]-1,3,4-oxadiazine-4-carboxylic acid, 4a,9b-dihydro-2-methoxy-, methyl ester		LI05445
262	68	110	195	111	52	152	151	**262**	100	72	62	61	42	41	30	25		13	14	4	2	–	–	–	–	–	–	–	–	–	2,5-Pentanocyclobuta[1,2-c:3,4-c′]dipyrrole-1,3,4,6-tetrone, tetrahydro-	23213-91-4	MS05800
110	262	165	111	82	181	152	28	**262**	100	55	39	24	22	18	18	17		13	14	4	2	–	–	–	–	–	–	–	–	–	1H-Pyrrole-2,5-dione, 1,1′-(1,5-pentanediyl)bis-	5443-21-0	NI26285
233	201	219	77	105	104	134	78	**262**	100	49	23	18	16	16	10	10	**0.00**	13	14	4	2	–	–	–	–	–	–	–	–	–	2-Quinazolinecarboxylic acid, 2-acetyl-1,2,3,4-tetrahydro-1-methyl-4-oxo-, methyl ester	54833-71-5	NI26286
43	41	57	55	135	137	71	69	**262**	100	93	76	66	40	39	36	36	**0.20**	13	27	–	–	–	–	–	1	–	–	–	–	–	Tridecane, 1-bromo-	765-09-3	NI26287
73	117	69	75	55	77	132	41	**262**	100	72	63	50	45	37	30	23	**1.30**	13	27	2	–	–	1	–	–	–	1	–	–	–	Decanoic acid, 10-fluoro-, trimethylsilyl ester	26305-85-1	NI26288
123	95	76	31	262	124	140	28	**262**	100	32	9	9	8	6	4	4		14	8	3	–	–	2	–	–	–	–	–	–	–	Benzoic acid, 3-fluoro-, anhydride	56666-54-7	NI26289
123	95	76	124	262	69	50	28	**262**	100	44	17	13	4	4	4	4		14	8	3	–	–	2	–	–	–	–	–	–	–	Benzoic acid, 4-fluoro-, anhydride	25569-77-1	NI26291
105	77	28	121	183	51	43	50	**262**	100	70	31	28	20	19	11	8	**0.00**	14	8	3	–	–	2	–	–	–	–	–	–	–	Benzoic acid, 4-fluoro-, anhydride	25569-77-1	NI26290
91	65	262	39	92	63	93	89	**262**	100	33	17	17	16	11	9	9		14	14	–	–	–	–	–	–	–	–	–	–	1	Dibenzyl selenide	1842-38-2	NI26292
123	246	124	247	139	121	182	122	**262**	100	65	14	11	11	11	8	8	**0.00**	14	14	1	–	2	–	–	–	–	–	–	–	–	p-Toluenesulphinic acid, thio-, S-p-tolyl ester	6481-73-8	NI26293
123	246	124	247	139	182	108	97	**262**	100	67	15	12	12	8	7	6	**0.00**	14	14	1	–	2	–	–	–	–	–	–	–	–	p-Toluenesulphinic acid, thio-, S-p-tolyl ester	6481-73-8	LI05446
105	77	106	79	103	104	78	65	**262**	100	10	9	9	7	6	4	3	**0.70**	14	14	3	–	1	–	–	–	–	–	–	–	–	2-Hydroxyphenyl 1-phenylethyl sulphone		MS05801
91	155	262	107	65	92	77	108	**262**	100	66	29	18	13	12	12	7		14	14	3	–	1	–	–	–	–	–	–	–	–	p-Toluenesulphonic acid, 2-methylphenyl ester		IC03415
91	155	262	65	92	39	107	108	**262**	100	59	28	14	11	8	7	5		14	14	3	–	1	–	–	–	–	–	–	–	–	p-Toluenesulphonic acid, 3-methylphenyl ester		IC03416
91	155	262	107	65	92	77	79	**262**	100	76	45	32	15	12	11	9		14	14	3	–	1	–	–	–	–	–	–	–	–	p-Toluenesulphonic acid, 4-methylphenyl ester		IC03417
246	231	228	213	203	185			**262**	100	43	8	7	7	4				14	14	5	–	–	–	–	–	–	–	–	–	–	3(2H)-Benzofuranone, 5-acetyl-6-hydroxy-2-isopropenyl-7-methoxy-		LI05447
192	43	165	71	262				**262**	100	40	39	23	11					14	14	5	–	–	–	–	–	–	–	–	–	–	2H-1-Benzopyran-2-one, 7-butanoyl-5-methoxy-		LI05448
230	262	191	43	77	29	39	263	**262**	100	94	39	33	21	21	19	18		14	14	5	–	–	–	–	–	–	–	–	–	–	4H-1-Benzopyran-4-one, 2,5-dimethyl-7-hydroxy-8-methoxycarbonyl-		NI26294
162	134	101	262	163	77	51	55	**262**	100	44	23	19	11	10	10	9		14	14	5	–	–	–	–	–	–	–	–	–	–	Butanoic acid, 2-methyl-4-[(2-oxo-2H-1-benzopyran-7-yl)oxy]-, (-)-	38971-84-5	MS05802
247	219	262	204	203				**262**	100	90	76	14	13					14	14	5	–	–	–	–	–	–	–	–	–	–	2-Naphthaldehyde, 5-hydroxy-4,6,8-trimethoxy-		LI05449
162	262	217	134	157	55	188	215	**262**	100	98	68	46	44	41	33	29		14	14	5	–	–	–	–	–	–	–	–	–	–	9H-Xanthene-1,8-dione, 2-hydroxy-1,2,3,4,5,6,7,8-octahydro-, formate		LI05450
117	118	183	105	91	39	65	79	**262**	100	60	58	48	38	33	30	24	**5.60**	14	15	–	–	–	–	–	1	–	–	–	–	–	4,7-Methano-2,3,8-methenocyclopent[a]indene, 10-bromo-1,2,3,3a,3b,4,7,7a,8,8a-decahydro-, stereoisomer	38560-40-6	NI26295
136	205	69	262	229	108	164	219	**262**	100	66	54	40	33	31	26	24		14	18	1	2	1	–	–	–	–	–	–	–	–	4H-1,3-Benzothiazin-4-one, 2,3-dihydro-3-methyl-2-(n-pentylimino)-		NI26296
42	91	55	83	119	77	64	51	**262**	100	59	52	49	45	45	42	42	**0.00**	14	18	1	2	1	–	–	–	–	–	–	–	–	Pyrrolo[1,2-b][1,2,4]oxadiazole-2(1H)-thione, tetrahydro-5,5,7a-trimethyl-1-phenyl-	50455-71-5	NI26297
83	60	57	187	44	202	98	104	**262**	100	60	56	52	52	50	36	30	**0.50**	14	18	1	2	1	–	–	–	–	–	–	–	–	Pyrrolo[1,2-b][1,2,4]oxadiazole-2(1H)-thione, tetrahydro-1,5,5-trimethyl-6-phenyl-	50455-88-4	NI26298
114	247	148	133	113	99	59	41	**262**	100	59	54	37	37	33	29	25	**22.99**	14	18	1	2	1	–	–	–	–	–	–	–	–	4H-1,3-Thiazine, 2-[p-methoxyphenylamino(imino)]-4,4,6-trimethyl-	85298-47-1	NI26299
81	41	53	221	79	39	178	77	**262**	100	100	90	84	72	59	39	37	**13.90**	14	18	3	2	–	–	–	–	–	–	–	–	–	Barbituric acid, 5-allyl-1-methyl-5-(1-methyl-2-pentynyl)-, (±)-	151-83-7	NI26300
205	174	262	160	206	204	57	161	**262**	100	43	18	18	13	13	12	8		14	18	3	2	–	–	–	–	–	–	–	–	–	2H-Furo[2,3-b]indol-5-ol,3a,8-dimethyl-3,3a,8,8a-tetrahydro-, methylcarbamate, (3aS-cis)-	6091-05-0	MS05803
119	87	91	86	70	120	65	176	**262**	100	56	20	11	9	9	7	6	**3.60**	14	18	3	2	–	–	–	–	–	–	–	–	–	Morpholine, 4-[N-acetyl-(4-methylacetamido)]-		NI26301
113	83	121	55	91	149	232	77	**262**	100	63	33	23	18	16	13	9	**0.46**	14	18	3	2	–	–	–	–	–	–	–	–	–	2-Piperidineacetic acid, 1-nitroso-α-phenyl-, methyl ester	55557-03-4	NI26302
221	67	196	181	164	41	222	107	**262**	100	96	89	53	40	33	26	22	**0.31**	14	18	3	2	–	–	–	–	–	–	–	–	–	2,4,6(1H,3H,5H)-Pyrimidinetrione, 5-allyl-5-(2-cyclopenten-1-yl)-1,3-dimethyl-	28239-48-7	NI26303
263	183	211	221	155				**262**	100	52	18	15	6				**0.00**	14	18	3	2	–	–	–	–	–	–	–	–	–	2,4,6(1H,3H,5H)-Pyrimidinetrione, 5-allyl-1-methyl-5-(1-methyl-2-pentynyl)-, (±)-	151-83-7	NI26304
57	43	41	55	29	262	164	56	**262**	100	39	32	27	19	16	14	13		14	30	–	–	2	–	–	–	–	–	–	–	–	Diheptyl disulphide	10496-16-9	MS05804
57	43	41	55	164	29	262	56	**262**	100	34	31	26	19	18	15	13		14	30	–	–	2	–	–	–	–	–	–	–	–	Diheptyl disulphide	10496-16-9	NI26305
55	69	83	97	73	107	103	43	**262**	100	93	75	60	46	32	31	30	**0.00**	14	31	1	–	–	1	–	–	–	1	–	–	–	Undecane, 11-fluoro-1-[(trimethylsilyl)oxy]-	26305-81-7	NI26306
129	262	77	130	75	263	51	101	**262**	100	46	13	11	10	9	9	6		15	10	1	4	–	–	–	–	–	–	–	–	–	3H-Pyrazol-3-one, 4-diazo-2,4-dihydro-2,5-diphenyl-	56666-62-7	MS05805
43	142	167	129	141	202	77	155	**262**	100	72	33	30	26	25	23	13	**3.00**	15	15	2	–	–	–	1	–	–	–	–	–	–	Norbornane, cis-2-acetoxy-3-chloro-endo-5,6-(o-phenylene)-		NI26307
167	43	128	129	155	145	168	141	**262**	100	59	30	25	23	15	14	13	**11.00**	15	15	2	–	–	–	1	–	–	–	–	–	–	Norbornane, cis-7-acetoxy-2-chloro-endo-5,6-(o-phenylene)-		NI26308
167	43	138	139	168	166	145	165	**262**	100	48	36	33	24	24	24	21	**15.00**	15	15	2	–	–	–	1	–	–	–	–	–	–	Norbornane, cis-7-acetoxy-2-chloro-exo-5,6-(o-phenylene)-		NI26309
262	32	18	173	91	244	69	145	**262**	100	77	72	58	58	57	54	51		15	18	4	–	–	–	–	–	–	–	–	–	–	Azuleno[4,5-b]furan-2,7-dione, 4-hydroxy-2,3,3a,4,5,7,9a,9b-octahydro-3,6,9-trimethyl-		IC03418
97	166	105	107	77	68	79	55	**262**	100	50	43	42	38	30	27	15	**0.00**	15	18	4	–	–	–	–	–	–	–	–	–	–	Benzeneacetic acid, α-hydroxy-α-(1-hydroxy-2-cyclohexen-1-yl)-, methyl ester		DD00777
262	206	177	135	163	191	263	205	**262**	100	76	73	48	25	22	21	20		15	18	4	–	–	–	–	–	–	–	–	–	–	1,2-Benzocycloheptene-3,7-dione, 3′,6′-dimethoxy-5,5-dimethyl-		LI05451
186	189	262	173	115	187	77	128	**262**	100	79	46	32	31	27	22	21		15	18	4	–	–	–	–	–	–	–	–	–	–	1,3-Benzodioxole-5-methanol, 6-(4-methoxycyclohexenyl)-		NI26310

MASS TO CHARGE RATIOS								M.W.	INTENSITIES								Parent	C	H	O	N	S	F	Cl	Br	I	Si	P	B	X	COMPOUND NAME	CAS Reg No	No
206	205	262	219	207	123	69	55	**262**	100	74	16	14	12	9	6	6		15	18	4	–	–	–	–	–	–	–	–	–	–	4H-1-Benzopyran-4-one, 5,7-dihydroxy-2-methyl-6-(3-methyl-2-butyl)-		MS05807
206	205	262	219	123	165	177	137	**262**	100	75	16	15	9	5	4	3		15	18	4	–	–	–	–	–	–	–	–	–	–	4H-1-Benzopyran-4-one, 5,7-dihydroxy-2-methyl-6-(3-methyl-2-butyl)-		MS05806
205	262	165	206	177	137	123	233	**262**	100	17	16	12	2	2	2	1		15	18	4	–	–	–	–	–	–	–	–	–	–	4H-1-Benzopyran-4-one, 5,7-dihydroxy-2-methyl-8-(3-methyl-2-butyl)-		MS05808
205	262	165	206	69	43	41	39	**262**	100	17	15	12	4	2	2	2		15	18	4	–	–	–	–	–	–	–	–	–	–	4H-1-Benzopyran-4-one, 5,7-dihydroxy-2-methyl-8-(3-methyl-2-butyl)-		MS05809
205	206	262	175	57	162	70	71	**262**	100	12	9	7	6	4	4	3		15	18	4	–	–	–	–	–	–	–	–	–	–	2H-1-Benzopyran-2-one, 6-(1-hydroxy-3-methylbutyl)-7-methoxy-	16850-96-7	NI26311
43	111	41	95	139	53	123	91	**262**	100	80	70	59	54	54	44	41	**9.60**	15	18	4	–	–	–	–	–	–	–	–	–	–	2H-Bisoxireno[2,3:8,8a]azuleno[4,5-b]furan-7(3aH)-one, octahydro-3a,8c-dimethyl-6-methylene-	36416-50-9	NI26312
68	53	91	244	105	216	96	138	**262**	100	67	63	57	55	43	37	25	**18.00**	15	18	4	–	–	–	–	–	–	–	–	–	–	Cyclodeca[b]furan-10-carboxylic acid, 2,3,3a,4,7,8,11,11a-octahydro-6-methyl-3-methylene-2-oxo-, [3aR-(3aR*,5E,9Z,11aS*)]-	87741-69-3	NI26313
105	244	111	139	77	69	245	226	**262**	100	36	23	19	14	8	7	2	**0.00**	15	18	4	–	–	–	–	–	–	–	–	–	–	2,3-Dioxabicyclo[4.4.0]decan-7-one, 1-hydroxy-6-methyl-4-phenyl-		DD00778
125	43	69	44	219	41	193	57	**262**	100	88	42	23	21	21	19	18	**12.01**	15	18	4	–	–	–	–	–	–	–	–	–	–	2,4(3H,5H)-Furandione, 5-(1,3-hexadienyl)-5-methyl-3-(1-oxo-2-butenyl)-, (E,E,E)-	55045-15-3	NI26314
43	133	70	72	41	175	115	171	**262**	100	7	7	5	5	4	4	3	**0.13**	15	18	4	–	–	–	–	–	–	–	–	–	–	2-Furanone, 4-acetoxy-3-benzyl-5,5-dimethyltetrahydro-		LI05452
135	95	53	79	136	126	91	77	**262**	100	70	46	31	27	26	26	25	**22.00**	15	18	4	–	–	–	–	–	–	–	–	–	–	Furo[2,3-h][3]benzoxepin-2(4H)-one, 8-hydroxy-4a,9a,10,10a-tetrahydro-3,5,9a-trimethyl-, [4aR-(4aα,8β,9aβ,10aβ)]-		NI26315
218	176	262	120	91	161	119	134	**262**	100	55	44	39	39	29	27	26		15	18	4	–	–	–	–	–	–	–	–	–	–	1,15-Marasmenedione		MS05810
132	262	119	91	105	120	67	77	**262**	100	44	34	31	22	21	21	20		15	18	4	–	–	–	–	–	–	–	–	–	–	3,15-Marasmenedione		MS05811
262	44	205	121	42	91	77	219	**262**	100	87	85	84	60	46	43	41		15	18	4	–	–	–	–	–	–	–	–	–	–	4,7-Methanofuro[3,2-c]oxireno[f]oxacycloundecin-5(2H)-one, 8,11a-dimethyl-1a,3,4,7,11,11a-hexahydro-, [1aR-(1aR*,4S*,7R*,11aS*)]-	20450-39-9	LI05453
262	43	205	121	91	77	219	164	**262**	100	86	85	84	46	42	41	33		15	18	4	–	–	–	–	–	–	–	–	–	–	4,7-Methanofuro[3,2-c]oxireno[f]oxacycloundecin-5(2H)-one, 8,11a-dimethyl-1a,3,4,7,11,11a-hexahydro-, [1aR-(1aR*,4S*,7R*,11aS*)]-	20450-39-9	MS05812
230	215	262	202	247	231	234		**262**	100	52	45	44	23	5	4			15	18	4	–	–	–	–	–	–	–	–	–	–	1-Naphthalenecarboxylic acid, 1,4,4a,5,8,8a-hexahydro-4,6,8a-trimethyl-5,8-dioxo-, methyl ester, (1α,4α,4aα,8aβ)-	58865-15-9	NI26316
68	230	234	262	231	202	215	247	**262**	100	25	24	17	10	10	6	5		15	18	4	–	–	–	–	–	–	–	–	–	–	1-Naphthalenecarboxylic acid, 1,4,4a,5,8,8a-hexahydro-4,6,8a-trimethyl-5,8-dioxo-, methyl ester, (1α,4α,4aβ,8aβ)-	58822-93-8	NI26317
247	229	201	248	183	262	211	173	**262**	100	36	27	18	18	13	11	9		15	18	4	–	–	–	–	–	–	–	–	–	–	Naphtho[1,2-b]furan-2,6(3H,4H)-dione, 9-hydroxy-3a,5,5a,9,9a,9b-hexahydro-5a,9-dimethyl-, [3aS-(3aα,5aβ,9α,9aα,9bβ)]-	16981-97-8	MS05813
91	135	121	262	244	149	122	171	**262**	100	84	74	54	52	50	46	38		15	18	4	–	–	–	–	–	–	–	–	–	–	Naphtho[1,2-b]furan-2,8(3H,4H)-dione, 4-hydroxy-3a,5,5a,9b-tetrahydroxy-3,5a,9-trimethyl-, [3S-(3α,3aα,4α,5aβ,9bβ)]-	481-05-0	NI26318
91	41	71	262	69	39	135	77	**262**	100	96	72	68	64	61	53	49		15	18	4	–	–	–	–	–	–	–	–	–	–	Naphtho[1,2-b]furan-2,8(3H,4H)-dione, 4-hydroxy-3a,5,5a,9b-tetrahydroxy-3,5a,9-trimethyl-, [3S-(3α,3aα,4α,5aβ,9bβ)]-	481-05-0	MS05814
161	262	162	187	175	91	99	43	**262**	100	97	57	31	31	30	26	24		15	18	4	–	–	–	–	–	–	–	–	–	–	3-Oxabicyclo[8.4.0]tetradeca-1(10),11,13-triene-2,7-dione, 4-methyl-12-methoxy-		DD00779
43	97	151	41	98	39	69	95	**262**	100	83	78	55	40	38	29	22	**9.23**	15	18	4	–	–	–	–	–	–	–	–	–	–	2-Pentenal, 2-methyl-4-oxo-5-(2,3,4,5-tetrahydro-5-methyl[2,3′-bifuran]-5-yl)-	55721-95-4	NI26319
41	97	39	121	91	120	261	100	**262**	100	84	55	50	46	44	34	34	**24.39**	15	22	–	2	1	–	–	–	–	–	–	–	–	4H-1,3-Thiazine, 5,6-dihydro-2-[(p-tolylamino)methylamino]-4,4,6-trimethyl-		NI26320
81	41	79	64	82	80	55	28	**262**	100	42	34	32	31	29	28	28	**0.55**	15	22	2	2	–	–	–	–	–	–	–	–	–	Cyclohexane, 1,1′-methylenebis[4-isocyanato-		IC03419
245	161	174	246	175	55	218	146	**262**	100	51	48	38	25	24	22	21	**10.00**	15	22	2	2	–	–	–	–	–	–	–	–	–	Matrine, 5,17-dehydro-N-oxide, (+)-		MS05815
150	262	55	110	151	112	84	97	**262**	100	64	40	33	27	27	25	22		15	22	2	2	–	–	–	–	–	–	–	–	–	7,14-Methano-4H,6H-dipyrido[1,2-a:1′,2′-e][1,5]diazocin-3,4-dione, dodecahydro-, [7S-(7α,7aβ,14α,14aα)]		LI05454
150	262	84	152	124	151	111	110	**262**	100	80	63	47	22	21	18	15		15	22	2	2	–	–	–	–	–	–	–	–	–	7,14-Methano-4H,6H-dipyrido[1,2-a:1′,2′-e][1,5]diazocin-10,13-dione, dodecahydro-, [7S-(7α,7aβ,14α,14aα)]		LI05455
262	150	151	234	110	263	55	112	**262**	100	79	32	25	25	20	20	16		15	22	2	2	–	–	–	–	–	–	–	–	–	7,14-Methano-4H,6H-dipyrido[1,2-a:1′,2′-e][1,5]diazocine-4,13(14H)-dione, decahydro-, [7S-(7α,7aβ,14α,14aα)]-	4697-83-0	NI26321
150	262	84	152	55	41	151	124	**262**	100	88	70	52	29	24	23	23		15	22	2	2	–	–	–	–	–	–	–	–	–	7,14-Methano-6H,13H-dipyrido[1,2-a:1′,2′-e][1,5]diazocine-6,13-dione, dodecahydro-, [7R-(7α,7aα,14α,14aβ)]-	52717-73-4	NI26322
221	41	44	150	42	262	112	222	**262**	100	28	25	24	24	18	16	14		15	22	2	2	–	–	–	–	–	–	–	–	–	1,5-Methano-2H-pyrido[1,2-a][1,5]diazocine-4,8(1H,3H)-dione, 3-(3-butenyl)hexahydro-, [1S-(1α,5α,11aα)]-	3382-89-6	NI26323
84	262	163	42	108	41	247	135	**262**	100	87	87	75	65	54	52	49		15	22	2	2	–	–	–	–	–	–	–	–	–	4H-1,3-Oxazine, 5,6-dihydro-2-[p-ethoxyphenylamino(imino)]-4,4,6-trimethyl-	37719-88-3	NI26324
72	117	190	262	57	162	146	43	**262**	100	22	20	19	19	18	16	16		15	22	2	2	–	–	–	–	–	–	–	–	–	Propanediamide, 2-ethyl-N,N,N′,N′-tetramethyl-2-phenyl-	42948-61-8	NI26325
72	91	262	190	162	146	177	247	**262**	100	19	13	13	7	4	2	1		15	22	2	2	–	–	–	–	–	–	–	–	–	Propanediamide, 2-ethyl-N,N,N′,N′-tetramethyl-2-phenyl-	42948-61-8	LI05456
72	117	190	262	162	146	91	177	**262**	100	22	20	19	18	16	15	14		15	22	2	2	–	–	–	–	–	–	–	–	–	Propanediamide, 2-ethyl-N,N,N′,N′-tetramethyl-2-phenyl-	42948-61-8	LI05457
43	91	30	65	42	41	120	106	**262**	100	48	43	10	9	8	6	6	**1.00**	15	22	2	2	–	–	–	–	–	–	–	–	–	Putrescine, N,N′-diacetyl-N-benzyl-		LI05458
92	91	45	59	262	155	43	93	**262**	100	26	17	15	13	13	9	8		15	26	–	–	–	–	–	–	–	2	–	–	–	Bicyclo[4.2.1]nona-2,4,7-triene, 7,8-bis(trimethylsilyl)-		NI26326

MASS TO CHARGE RATIOS								M.W.	INTENSITIES								Parent	C	H	O	N	S	F	Cl	Br	I	Si	P	B	X	COMPOUND NAME	CAS Reg No	No
73	173	59	159	45	174	74	188	262	100	31	24	19	18	16	12	8	0.50	15	26	–	–	–	–	–	–	–	2	–	–	–	Bicyclo[4.3.0]nona-2,4,8-triene, 2,3-bis(trimethylsilyl)-, cis-	101652-97-5	NI26327
246	230	218	247	229	217	220	76	262	100	84	38	19	19	18	16	16	7.00	16	10	2	2	–	–	–	–	–	–	–	–	–	Benzo[a]phenazine, 7,12-dioxide		MS05816
262	234	205	103	76	131	206	77	262	100	26	26	26	17	14	12	12		16	10	2	2	–	–	–	–	–	–	–	–	–	[$\Delta^{2,2'}$-Biindoline]-3,3'-dione	482-89-3	LI05459
105	141	77	51	114	106	50	142	262	100	39	34	21	10	6	4	3	1.25	16	10	2	2	–	–	–	–	–	–	–	–	–	Propanedinitrile, (benzoyloxy)phenyl-	5467-94-7	NI26328
131	262	77	132	263	104	51	90	262	100	56	23	20	11	8	5	3		16	14	–	4	–	–	–	–	–	–	–	–	–	Benzimidazole, 1,1'-ethylenebis-		IC03420
262	263	91	261	89	65	64	144	262	100	22	12	10	9	9	9	8		16	14	–	4	–	–	–	–	–	–	–	–	–	Pyrimidine, 2,4-bis(4-aminophenyl)-		NI26329
117	90	116	118	262	89	91	39	262	100	25	21	17	16	13	5	4		16	14	–	4	–	–	–	–	–	–	–	–	–	1,2,4,5-Tetrazine, 3,6-di-m-tolyl-		MS05817
110	95	82	67	109	55	66	41	262	100	58	49	41	35	29	28	26	0.10	16	22	1	–	1	–	–	–	–	–	–	–	–	Cyclopropanol, 1-[2-methyl-1-(phenylthio)cyclohexyl]-	74231-52-0	NI26330
136	229	244	262	175	43	173	162	262	100	75	72	62	53	46	34	32		16	22	3	–	–	–	–	–	–	–	–	–	–	1,4,8a-Anthracenetriol, 5,5-dimethyl-5,6,7,8,8a,9,10,10a-octahydro-		NI26331
161	123	121	201	43	55	107	93	262	100	92	74	59	59	37	30	19	10.00	16	22	3	–	–	–	–	–	–	–	–	–	–	1,4-Benzenediol, 2-(8-hydroxy-3,7-dimethylocta-2,6-dienyl)-		NI26332
91	205	131	101	153	104	59	41	262	100	32	21	21	15	12	11	11	9.00	16	22	3	–	–	–	–	–	–	–	–	–	–	Benzenepentanoic acid, butyl-β-oxo-, methyl ester	30414-59-6	NI26333
138	109	120	69	121	41	55	65	262	100	84	83	82	65	52	47	44	4.20	16	22	3	–	–	–	–	–	–	–	–	–	–	Benzoic acid, 2-hydroxy-, 3,3,5-trimethylcyclohexyl ester	118-56-9	NI26334
91	107	108	55	156	85	79	162	262	100	47	38	38	37	33	23	18	0.00	16	22	3	–	–	–	–	–	–	–	–	–	–	7,9-Dioxabicyclo[4.2.1]nonane, 6-[2-(benzyloxy)ethyl]-, (1S,6S)-		DD00780
131	159	91	117	176	205	132	57	262	100	72	67	54	51	49	47	15	0.00	16	22	3	–	–	–	–	–	–	–	–	–	–	1,3-Dioxan-4-one, 5-benzyl-2-tert-butyl-5-methyl-, (2S,5R)-		DD00781
159	91	176	117	132	57	158	130	262	100	99	73	70	40	28	22	15	0.00	16	22	3	–	–	–	–	–	–	–	–	–	–	1,3-Dioxan-4-one, 5-benzyl-2-tert-butyl-6-methyl-, (2R,5R,6R)-		DD00782
105	69	97	129	77	132	79	106	262	100	33	32	25	16	12	12	11	0.00	16	22	3	–	–	–	–	–	–	–	–	–	–	Hexanoic acid, 5-(2-phenylpropanoyl)-, methyl ester		NI26335
220	205	43	191	192	164	219	55	262	100	60	40	30	19	17	15	14	14.00	16	22	3	–	–	–	–	–	–	–	–	–	–	5H-Inden-5-one, 3-[1-(acetyloxy)-2-methylpropyl]-1,2,3,6,7,7a-hexahydro-7a-methyl-	66708-20-1	NI26336
177	262	179	178	139	247	192		262	100	17	13	13	8	5	5			16	22	3	–	–	–	–	–	–	–	–	–	–	2-Oxatricyclo[6.4.0.1^{3,7}]trideca-1(8),9,11-triene, 9,11-dihydroxy-6-isopropyl-		LI05460
135	57	163	206	160	145	147	205	262	100	92	88	82	80	67	66	65	0.50	16	22	3	–	–	–	–	–	–	–	–	–	–	Propanoic acid, 2,3-epoxy-2-isopropyl-2-phenyl-, tert-butyl ester		NI26337
73	75	189	116	107	41	206	190	262	100	44	35	26	21	17	15	14	0.00	16	26	1	–	–	–	–	–	–	1	–	–	–	Bicyclo[4.2.0]oct-2-ene, 7,7-dimethyl-3-[(1-oxo-3-(trimethylsilyl)-2-butenyl)]-, (E)-cis-		DD00783
134	128	262	86	77	135	105	106	262	100	44	22	17	16	12	11	10		16	26	1	2	–	–	–	–	–	–	–	–	–	Benzamide, N,N-dibutyl-2-(methylamino)-	15236-36-9	NI26338
134	128	262	77	105	86	106	135	262	100	41	27	25	15	14	13	12		16	26	1	2	–	–	–	–	–	–	–	–	–	Benzamide, N,N-dibutyl-2-(methylamino)-	15236-36-9	TR00289
262	117	263	234	188	187	94	93.5	262	100	20	18	6	5	5	4	4		17	10	3	–	–	–	–	–	–	–	–	–	–	7H-Benz[de]anthracen-7-one, 6,7-dihydroxy-		IC03421
127	77	262	92	155	79	135	143	262	100	70	31	23	22	21	18	17		17	14	1	2	–	–	–	–	–	–	–	–	–	Anisole, 2-(2-naphthylazo)-	18277-99-1	NI26339
261	262	219	220	218	77	128	263	262	100	55	52	9	8	8	7	7		17	14	1	2	–	–	–	–	–	–	–	–	–	1H-1,5-Benzodiazepin-2-one, 2,3-dihydro-4-styryl-	27185-47-3	NI26340
246	262	233	245	130	77	103	104	262	100	67	64	44	29	26	23	20		17	14	1	2	–	–	–	–	–	–	–	–	–	Imidazo[2,1-a]isoquinoline, 5,6-dihydro-2-phenyl-, 1-oxide		NI26341
118	262	117	91	201	65	90	89	262	100	84	54	22	16	14	13	13		17	14	1	2	–	–	–	–	–	–	–	–	–	2-Imidazolin-5-one, 4-benzylidene-2-p-tolyl-		LI05461
118	262	117	91	263	201	65	89	262	100	84	54	22	16	16	14	13		17	14	1	2	–	–	–	–	–	–	–	–	–	2-Imidazolin-5-one, 4-benzylidene-2-p-tolyl-	32997-17-4	NI26342
157	262	114	77	158	263	127	142	262	100	56	19	19	14	13	11	9		17	14	1	2	–	–	–	–	–	–	–	–	–	Naphthalene, 1-methoxy-4-phenylazo-		IC03422
262	261	263	220	163	151	165	247	262	100	23	20	20	17	12	10	9		17	14	1	2	–	–	–	–	–	–	–	–	–	3H-Phenanthro[3,4-d]imidazol-10-ol, 2,3-dimethyl-	98033-25-1	NI26343
261	262	77	104	115	263	130	140	262	100	48	19	18	15	9	9	8		17	14	1	2	–	–	–	–	–	–	–	–	–	2(1H)-Pyrimidone, 4,6-diphenyl-5-methyl-		LI05462
261	262	77	91	104	263	115	116	262	100	68	20	17	13	12	12	11		17	14	1	2	–	–	–	–	–	–	–	–	–	2(1H)-Pyrimidone, 4-(4-methylphenyl)-6-phenyl-	24030-10-2	NI26344
261	262	77	91	263	115	104	116	262	100	68	19	17	13	13	13	11		17	14	1	2	–	–	–	–	–	–	–	–	–	2(1H)-Pyrimidone, 6-(4-methylphenyl)-4-phenyl-		LI05463
128	262	77	104	231	263	101	51	262	100	70	32	30	20	15	15	11		17	14	1	2	–	–	–	–	–	–	–	–	–	3(H)-Pyrrol-3-one, 2,5-diphenyl-, O-methyloxime		NI26345
262	143	77	142	115	263	261	117	262	100	57	35	34	28	20	20	16		17	14	1	2	–	–	–	–	–	–	–	–	–	2(1H)-Quinazolinone, 3,4-dihydro-3-phenyl-1-(2-propynyl)-	31078-97-4	NI26346
93	30	169	143	74	31			262	100	35	26	21	18	15			3.78	17	14	1	2	–	–	–	–	–	–	–	–	–	Urea, N'-(1-naphthyl)-N-phenyl-		MS05818
95	109	260	183	217	161	262	147	262	100	80	55	16	15	14	5	5		17	26	–	–	1	–	–	–	–	–	–	–	–	3-Decene, 2-methyl-4-(phenylthio)-		DD00784
135	57	109	260	95	151	203	161	262	100	80	70	58	50	45	10	10	0.00	17	26	–	–	1	–	–	–	–	–	–	–	–	3-Octene, 4-(phenylthio)-2,7,7-trimethyl-		DD00785
152	43	67	41	95	55	81	111	262	100	92	62	54	43	37	30	28	13.49	17	26	2	–	–	–	–	–	–	–	–	–	–	11H-Benzo[a]cyclopenta[d]cycloocten-11-one, 4-acetyltetradecahydro-	18938-06-2	NI26347
189	187	131	91	159	133	145	202	262	100	42	23	21	19	19	16	15	0.00	17	26	2	–	–	–	–	–	–	–	–	–	–	Coralloidin C		DD00786
81	137	41	109	95	93	79	55	262	100	49	31	28	25	16	14	14	2.22	17	26	2	–	–	–	–	–	–	–	–	–	–	Cyclopentanecarboxylic acid, 3-methylene-, 1,7,7-trimethylbicyclo[2.2.1]hept-2-yl ester	74793-59-2	NI26348
91	108	43	92	57	41	69	71	262	100	72	36	30	20	20	19	17	1.60	17	26	2	–	–	–	–	–	–	–	–	–	–	Decanoic acid, benzyl ester		NI26349
43	91	131	117	79	105	67	145	262	100	63	45	42	41	33	30	25	1.00	17	26	2	–	–	–	–	–	–	–	–	–	–	5,10-Pentadecadiyn-1-ol, acetate	64275-45-2	NI26350
94	43	79	93	122	91	77	121	262	100	42	41	40	34	33	23	19	0.00	17	26	2	–	–	–	–	–	–	–	–	–	–	2-Penten-1-ol, 2-methyl-5-(2-methyl-3-methylene-2-norbornyl)-, (–)-(E)-	77-43-0	NI26351
94	122	43	93	79	91	77	41	262	100	35	34	27	26	17	13	12	0.00	17	26	2	–	–	–	–	–	–	–	–	–	–	2-Penten-1-ol, 2-methyl-5-(2-methyl-3-methylene-2-norbornyl)-, (–)-(Z)-		NI26352
94	43	79	93	122	91	134	77	262	100	60	35	33	27	26	20	20	0.00	17	26	2	–	–	–	–	–	–	–	–	–	–	2-Penten-1-ol, 2-methyl-5-(2-methyl-3-methylene-2-norbornyl)-, (+)-epi-	41414-75-9	NI26353
207	150	262	247	177				262	100	85	77	53	14					17	26	2	–	–	–	–	–	–	–	–	–	–	Phenol, 3-methoxy-2-(3-methyl-2-butenyl)-5-(1-pentyl)-		NI26354
57	163	41	178	85	91	161	43	262	100	22	16	12	8	5	4	4	1.00	17	26	2	–	–	–	–	–	–	–	–	–	–	Propanoic acid, 2,2-dimethyl-, 2,6-diisopropylphenyl ester	54644-44-9	NI26355
189	99	55	141	140	69	109	247	262	100	97	84	58	58	52	46	42	29.00	17	26	2	–	–	–	–	–	–	–	–	–	–	15,16,20-Trisnorlabd-6-ene-8,13-dione		MS05819
124	138	151	262	55	219	193	69	262	100	88	86	56	55	54	39	35		17	27	–	–	–	–	–	–	–	–	1	–	–	Phosphine, methyl[methyl(isopropyl)cyclohexyl]phenyl-	74709-99-2	NI26356
262	247	263	261	131	248	245	110.5	262	100	26	20	10	9	8	7	7		18	14	–	–	1	–	–	–	–	–	–	–	–	11-Thiabenzo[b]fluorene, 8,9-dimethyl-		AI01864
262	261	263	247	131	264	258	123.5	262	100	22	21	12	9	7	7	6		18	14	–	–	1	–	–	–	–	–	–	–	–	11-Thiabenzo[b]fluorene, 8,10-dimethyl-		AI01865

MASS TO CHARGE RATIOS	M.W.	INTENSITIES	Parent	C	H	O	N	S	F	Cl	Br	I	Si	P	B	X	COMPOUND NAME	CAS Reg No	No
262 18 77 263 51 169 29 168	**262**	100 45 31 20 20 18 16 13		18	14	2	–	–	–	–	–	–	–	–	–	–	**Benzene, 1,2-diphenoxy-**		**MS05820**
262 169 142 77 263 168 28 51	**262**	100 32 27 22 20 16 15 12		18	14	2	–	–	–	–	–	–	–	–	–	–	**Benzene, 1,2-diphenoxy-**		**DO01054**
262 46 29 27 141 43 77 263	**262**	100 65 63 54 49 42 25 21		18	14	2	–	–	–	–	–	–	–	–	–	–	**Benzene, 1,3-diphenoxy-**		**MS05821**
262 77 51 263 185 141 129 115	**262**	100 55 19 17 10 9 7 6		18	14	2	–	–	–	–	–	–	–	–	–	–	**Benzene, 1,4-diphenoxy-**		**MS05822**
262 77 263 192 111 190 185 141	**262**	100 28 20 18 15 14 12 12		18	14	2	–	–	–	–	–	–	–	–	–	–	**Benzene, 1,4-diphenoxy-**		**DO01055**
105 141 77 140 115 139 51 262	**262**	100 69 31 23 18 14 9 6		18	14	2	–	–	–	–	–	–	–	–	–	–	**Benzoic acid, 1-naphthalenemethyl ester**	69238-67-1	**NI26357**
231 215 262 216 78 244 232 202	**262**	100 99 78 67 56 46 29 28		18	14	2	–	–	–	–	–	–	–	–	–	–	**5,6-Chrysenediol, 5,6-dihydro-, trans-**		**NI26358**
262 105 202 217 263 215 234 179	**262**	100 66 30 28 20 17 13 10		18	14	2	–	–	–	–	–	–	–	–	–	–	**2(3H)-Furanone, 4,5-dihydro-3-(2,2-diphenylvinylidene)-**		**DD00787**
233 262 205 203 202 218 215 217	**262**	100 78 48 30 30 22 17 15		18	14	2	–	–	–	–	–	–	–	–	–	–	**2(5H)-Furanone, 3-(2,2-diphenylvinyl)-**		**DD00788**
185 261 77 157 262 105 186 114	**262**	100 98 23 22 21 21 15 11		18	14	2	–	–	–	–	–	–	–	–	–	–	**Naphthalene, 6-benzoyl-2-methoxy-**		**MS05823**
262 202 244 245 189 122 215 165	**262**	100 62 32 30 29 18 10 10		18	14	2	–	–	–	–	–	–	–	–	–	–	**Phenanthro[1,2-b]furan, 1-hydroxymethyl-2-methyl-**		**LI05464**
234 191 262 44 43 189 235 192	**262**	100 75 51 30 30 27 18 14		18	14	2	–	–	–	–	–	–	–	–	–	–	**2H-Pyran-2-one, 6-methyl-3,4-diphenyl-**	32727-38-1	**NI26359**
262 183 108 263 261 184 107 185	**262**	100 43 28 20 16 11 11 6		18	15	–	–	–	–	–	–	–	–	1	–	–	**Phosphine, triphenyl-**	603-35-0	**IC03423**
262 183 263 108 261 184 107 185	**262**	100 42 20 20 10 10 10 8		18	15	–	–	–	–	–	–	–	–	1	–	–	**Phosphine, triphenyl-**	603-35-0	**NI26360**
262 183 108 263 261 184 185 107	**262**	100 50 33 23 17 12 9 9		18	15	–	–	–	–	–	–	–	–	1	–	–	**Phosphine, triphenyl-**	603-35-0	**IC03424**
220 262 233 221 232 217 218 263	**262**	100 47 27 18 16 16 15 10		18	18	–	2	–	–	–	–	–	–	–	–	–	**Azepino[4,5-b]indole, 5-phenylhexahydro-**		**MS05824**
247 77 262 129 116 248 219 263	**262**	100 86 56 28 23 19 12 11		18	18	–	2	–	–	–	–	–	–	–	–	–	**Benzimidoyl cyanide, 4-tert-butyl-N-phenyl-**		**MS05825**
262 134 261 263 77 118 51 42	**262**	100 77 39 30 8 6 6 6		18	18	–	2	–	–	–	–	–	–	–	–	–	**Cyclopropanecarbonitrile, 2-[p-(dimethylamino)phenyl]-1-phenyl-**	6114-58-5	**NI26361**
262 134 261 263 135 76 51 43	**262**	100 75 38 29 10 8 7 6		18	18	–	2	–	–	–	–	–	–	–	–	–	**Cyclopropanecarbonitrile, 2-[p-(dimethylamino)phenyl]-1-phenyl-**	6114-58-5	**NI26362**
263 134 262 264 135 76 246 51	**262**	100 76 39 30 10 9 7 7		18	18	–	2	–	–	–	–	–	–	–	–	–	**Cyclopropanecarbonitrile, 2-[p-(dimethylamino)phenyl]-1-phenyl-**	6114-58-5	**LI05465**
121 91 262 37 122 93 36 35	**262**	100 44 25 23 15 14 14 12		18	30	1	–	–	–	–	–	–	–	–	–	–	**Benzene, 1-methoxy-2-undecyl-**	20056-62-6	**LI05466**
121 91 262 40 122 93 39 38	**262**	100 45 25 22 15 13 13 12		18	30	1	–	–	–	–	–	–	–	–	–	–	**Benzene, 1-methoxy-2-undecyl-**	20056-62-6	**NI26363**
121 91 262 41 122 93 40 43	**262**	100 45 25 23 15 14 14 13		18	30	1	–	–	–	–	–	–	–	–	–	–	**Benzene, 1-methoxy-2-undecyl-**	20056-62-6	**NI26364**
122 121 40 41 91 262 39 43	**262**	100 36 25 21 21 14 14 14		18	30	1	–	–	–	–	–	–	–	–	–	–	**Benzene, 1-methoxy-3-undecyl-**	20056-63-7	**NI26365**
122 121 39 40 91 262 42 38	**262**	100 37 26 21 20 14 13 13		18	30	1	–	–	–	–	–	–	–	–	–	–	**Benzene, 1-methoxy-3-undecyl-**	20056-63-7	**NI26366**
122 121 36 91 37 262 35 39	**262**	100 36 25 21 20 14 14 13		18	30	1	–	–	–	–	–	–	–	–	–	–	**Benzene, 1-methoxy-3-undecyl-**	20056-63-7	**LI05467**
121 262 122 41 91 40 77 43	**262**	100 13 10 9 7 6 5 4		18	30	1	–	–	–	–	–	–	–	–	–	–	**Benzene, 1-methoxy-4-undecyl-**	20056-64-8	**LI05468**
121 262 122 41 91 77 40 39	**262**	100 12 10 9 7 7 6 5		18	30	1	–	–	–	–	–	–	–	–	–	–	**Benzene, 1-methoxy-4-undecyl-**	20056-64-8	**NI26367**
121 262 122 41 91 40 77 43	**262**	100 14 11 9 7 7 6 5		18	30	1	–	–	–	–	–	–	–	–	–	–	**Benzene, 1-methoxy-4-undecyl-**	20056-64-8	**NI26368**
41 43 55 191 69 95 189 57	**262**	100 88 78 64 60 58 47 46	**34.50**	18	30	1	–	–	–	–	–	–	–	–	–	–	**1-Naphthalene, 1-(3-oxobutyl)-3,4,5,6,7,8,9,10-octahydro-**		**LI05469**
43 109 262 191 95 81 123 69	**262**	100 71 56 48 47 47 42 30		18	30	1	–	–	–	–	–	–	–	–	–	–	**1H-Naphtho[2,1-b]pyran, 4a,5,6,6a,7,8,9,10,10a,10b-decahydro-3,4a,7,7,10a-pentamethyl-, [4aR-(4aα,6aβ,10aα,10bβ)]-**	5153-92-4	**NI26369**
43 95 136 81 121 107 123 93	**262**	100 98 89 77 61 59 48 39	**2.08**	18	30	1	–	–	–	–	–	–	–	–	–	–	**1H-Naphtho[2,1-b]pyran, 4a,5,6,6a,7,8,9,10,10a,10b-decahydro-3,4a,7,7,10a-pentamethyl-, [4aR-(4aα,6aβ,10aβ,10bβ)]-**	56245-50-2	**NI26370**
218 109 148 175 108 203 81 95	**262**	100 58 57 49 48 44 36 35	**3.00**	18	30	1	–	–	–	–	–	–	–	–	–	–	**D-Norandrostan-16-ol, (5α,16β)-**	35575-61-2	**NI26371**
41 79 67 55 29 27 95 39	**262**	100 63 59 45 43 37 32 32	**3.70**	18	30	1	–	–	–	–	–	–	–	–	–	–	**9,12,15-Octadecatrienal**	26537-71-3	**NI26372**
41 42 68 82 57 96 262 110	**262**	100 27 13 8 7 5 4 2		18	30	1	–	–	–	–	–	–	–	–	–	–	**9,12,15-Octadecatrienal**	26537-71-3	**LI05470**
43 109 95 81 41 123 262 69	**262**	100 82 64 60 58 50 48 42		18	30	1	–	–	–	–	–	–	–	–	–	–	**5-Octen-2-one, 6-methyl-8-(2,6,6-trimethyl-1-cyclohexen-1-yl)-**	55638-41-0	**NI26373**
43 69 41 81 55 95 93 107	**262**	100 65 52 32 15 14 14 11	**1.00**	18	30	1	–	–	–	–	–	–	–	–	–	–	**5,9,13-Pentadecatrien-2-one, 6,10,14-trimethyl-**		**LI05471**
43 69 41 81 93 107 95 68	**262**	100 98 54 41 15 14 14 14	**3.00**	18	30	1	–	–	–	–	–	–	–	–	–	–	**5,9,13-Pentadecatrien-2-one, 6,10,14-trimethyl-, (E,E)-**	1117-52-8	**NI26374**
247 262 219 57 248 41 263 29	**262**	100 37 37 27 23 13 8 7		18	30	1	–	–	–	–	–	–	–	–	–	–	**Phenol, 2,6-di-tert-butyl-4-butyl-**		**IC03425**
135 121 107 149 41 43 55 29	**262**	100 42 40 35 15 14 13 11	**2.08**	18	30	1	–	–	–	–	–	–	–	–	–	–	**Phenol, 4-dodecyl-**		**MS05826**
135 121 149 107 41 43 29 55	**262**	100 40 36 36 32 25 22 16	**3.30**	18	30	1	–	–	–	–	–	–	–	–	–	–	**Phenol, dodecyl-**	27193-86-8	**WI01316**
191 57 192 41 247 29 262 175	**262**	100 25 14 8 5 5 4 3		18	30	1	–	–	–	–	–	–	–	–	–	–	**Phenol, 2-tert-butyl-4-(2,4,4-trimethylpent-2-yl)-**		**IC03426**
233 262 58 234 29 41 56 28	**262**	100 21 19 18 16 15 14 9		18	30	1	–	–	–	–	–	–	–	–	–	–	**Phenol, 2,4,5-tri-sec-butyl-**		**IC03427**
233 234 262 29 58 41 28 56	**262**	100 18 17 16 14 12 7 6		18	30	1	–	–	–	–	–	–	–	–	–	–	**Phenol, 2,4,6-tri-sec-butyl-**		**IC03428**
247 57 262 248 28 41 29 80	**262**	100 30 19 19 14 12 7 6		18	30	1	–	–	–	–	–	–	–	–	–	–	**Phenol, 2,4,6-tri-tert-butyl-**	732-26-3	**WI01317**
247 57 262 248 41 94 80 263	**262**	100 27 20 19 9 5 5 4		18	30	1	–	–	–	–	–	–	–	–	–	–	**Phenol, 2,4,6-tri-tert-butyl-**	732-26-3	**IC03429**
247 57 248 262 41 231 263 125	**262**	100 28 18 11 11 5 3 3		18	30	1	–	–	–	–	–	–	–	–	–	–	**Phenol, 2,4,6-tri-tert-butyl-**	732-26-3	**IC03430**
43 41 69 93 83 55 133 107	**262**	100 95 91 90 70 51 43 40	**1.00**	18	30	1	–	–	–	–	–	–	–	–	–	–	**Tricyclo[7.2.0.2^{4,5}]tridecan-12-one, 8-methylene-4,11,11,13,13-pentamethyl-**		**LI05472**
83 41 69 55 93 109 137 29	**262**	100 66 50 50 39 22 20 15	**0.10**	18	30	1	–	–	–	–	–	–	–	–	–	–	**Tricyclo[7.2.0.2^{4,5}]tridecan-13-one, 8-methylene-4,11,11,12,12-pentamethyl-**		**LI05473**
117 262 145 115 116 91 247 219	**262**	100 95 81 74 67 62 52 24		19	18	1	–	–	–	–	–	–	–	–	–	–	**1,4-Butadien-3-one, 2,4-dimethyl-1,5-diphenyl-**		**IC03431**
262 220 205 219 234 191 206 91	**262**	100 70 68 38 36 32 31 26		19	18	1	–	–	–	–	–	–	–	–	–	–	**2-Cyclohexen-1-one, 3-methyl-4,4-diphenyl-**	17429-38-8	**NI26375**
157 43 262 128 91 171 129 77	**262**	100 63 61 61 50 45 45 33		19	18	1	–	–	–	–	–	–	–	–	–	–	**1,3-Heptadien-5-one, 1,7-diphenyl-**		**MS05827**

MASS TO CHARGE RATIOS								M.W.	INTENSITIES								Parent	C	H	O	N	S	F	Cl	Br	I	Si	P	B	X	COMPOUND NAME	CAS Reg No	No
171	43	91	172	127	128	155	170	**262**	100	78	15	11	11	7	6	5	**1.17**	19	18	1	–	–	–	–	–	–	–	–	–	–	**2-Propanol, 2-(1-naphthyl)-1-phenyl-**		**AI01866**
262	247	233	229	133	263	145	115	**262**	100	82	54	27	22	21	18	18		19	18	1	–	–	–	–	–	–	–	–	–	–	**2,2′-Spirobiindan-1-one, 1′,3-dimethyl-, (1′R,2S,3R)-**		**NI26376**
247	262	233	229	248	263	232	133	**262**	100	89	46	32	19	18	18	15		19	18	1	–	–	–	–	–	–	–	–	–	–	**2,2′-Spirobiindan-1-one, 1′,3-dimethyl-, (1′R,2S,3S)-**		**NI26377**
97	83	55	41	179	178	67	96	**262**	100	94	81	64	47	31	24	18	**0.32**	19	34	–	–	–	–	–	–	–	–	–	–	–	**Methane, tricyclohexyl-**		**AI01867**
262	219	220	263	191	205	233	203	**262**	100	42	24	22	19	15	14	14		20	22	–	–	–	–	–	–	–	–	–	–	–	**Benzo[b]fluoranthene, 3b,4,5,6,7,7a,9,10,11,12-decahydro-**		**NI26378**
262	252	205	263	219	234	233	203	**262**	100	25	23	22	17	16	15	15		20	22	–	–	–	–	–	–	–	–	–	–	–	**Benzo[e]pyrene, 1,2,3,3a,4,5,5a,6,7,8-decahydro-**		**NI26379**
262	234	233	219	252	202	205	203	**262**	100	76	46	42	32	28	27	26		20	22	–	–	–	–	–	–	–	–	–	–	–	**Benzo[e]pyrene, 4,5,5a,6,7,8,9,10,11,12-decahydro-**		**NI26380**
262	205	219	263	234	203	233	202	**262**	100	30	21	20	20	19	18	18		20	22	–	–	–	–	–	–	–	–	–	–	–	**Benzo[e]pyrene, 1,2,3,6,7,8,9,10,11,12-decahydro-**	92387-50-3	**NI26381**
91	115	128	116	262	104	77	39	**262**	100	66	60	46	36	30	25	23		20	22	–	–	–	–	–	–	–	–	–	–	–	**1,2′-Binaphthyl, 1,1′,2,2′,3,3′,4,4′-octahydro-**		**AI01868**
131	262	104	130	129	91			**262**	100	42	40	39	32	29				20	22	–	–	–	–	–	–	–	–	–	–	–	**2,2′-Binaphthyl, 1,1′,2,2′,3,3′,4,4′-octahydro-**		**LI05474**
131	262	104	130	91	129	115	128	**262**	100	47	39	36	32	31	21	17		20	22	–	–	–	–	–	–	–	–	–	–	–	**2,2′-Binaphthyl, 1,1′,2,2′,3,3′,4,4′-octahydro-**		**MS05828**
104	130	262	131	129	115	128	144	**262**	100	99	50	27	23	21	17	16		20	22	–	–	–	–	–	–	–	–	–	–	–	**2,5′-Binaphthyl, 1,1′,2,2′,3,3′,4,4′-octahydro-**		**AI01869**
262	234	219	263	205	203	202	191	**262**	100	72	32	22	19	18	17	15		20	22	–	–	–	–	–	–	–	–	–	–	–	**Perylene, 4,5,6,7,8,9,9a,10,11,12-decahydro-**		**NI26382**
69	184	85	263	64	115	107	50	**263**	100	82	75	36	33	24	24	22		–	–	–	1	1	4	–	1	–	–	2	–	–	**Phosphorimidic bromide difluoride, (difluorophosphinothioyl)-**	27352-10-9	**MS05829**
130	263	164	265	60	101	69	63	**263**	100	80	62	54	47	40	34	33		1	3	–	1	2	2	2	–	–	–	2	–	–	**Thioimidodiphosphoryl chloride fluoride, methyl-**	39564-19-7	**MS05830**
134	46	70	69	50	31	131	129	**263**	100	20	18	17	8	8	7	7	**0.00**	2	–	–	1	1	6	–	1	–	–	–	–	–	**Thioimide, difluoro-, N-(2-bromo-1,1,2,2-tetrafluoroethyl)-**		**LI05475**
111	69	81	83	92	91	96	180	**263**	100	51	34	21	10	10	6	5	**0.00**	5	2	1	1	–	9	–	–	–	–	–	–	–	**Propanamide, N,N-bis(trifluoromethyl)-3,3,3-trifluoro-**		**LI05476**
35	28	86	47	265	60	95	61	**263**	100	95	90	85	84	83	73	68	**50.05**	6	2	–	1	–	–	5	–	–	–	–	–	–	**Aniline, 2,3,4,5,6-pentachloro-**	527-20-8	**NI26383**
265	32	267	263	78	98	269	194	**263**	100	73	70	67	33	21	18	17		6	2	–	1	–	–	5	–	–	–	–	–	–	**Aniline, 2,3,4,5,6-pentachloro-**	527-20-8	**NI26384**
94	150	65	127	66	39	77	93	**263**	100	22	19	17	11	11	8	8	**1.10**	7	6	2	1	1	5	–	–	–	–	–	–	–	**(Phenoxyformamido)sulphur pentafluoride**	81439-23-8	**NI26385**
203	205	221	223	69	52	265	263	**263**	100	98	94	84	31	31	24	23		7	6	3	1	1	–	–	1	–	–	–	–	–	**3-Thiophenecarboxylic acid, 2-acetamido-5-bromo-**		**IC03432**
52	53	51	104	105	184	77	186	**263**	100	87	82	81	79	75	75	74	**29.00**	7	7	–	1	–	–	–	2	–	–	–	–	–	**Aniline, 2,4-dibromo-6-methyl-**		**IC03433**
44	42	155	157	43	50	75	76	**263**	100	32	29	27	22	22	22	21	**9.74**	8	10	2	1	1	–	–	1	–	–	–	–	–	**Benzenesulfonamide, 4-bromo-N,N-dimethyl-**	707-60-8	**NI26386**
57	176	41	177	204	119	56	58	**263**	100	68	62	57	31	18	16	14	**0.00**	8	10	3	1	–	5	–	–	–	–	–	–	–	**Carbamic acid, (2,2,3,3,3-pentafluoro-1-oxopropyl)-, isopropyl ester**	57983-74-1	**NI26387**
43	176	177	41	204	119	27	58	**263**	100	71	67	30	17	15	13	10	**0.00**	8	10	3	1	–	5	–	–	–	–	–	–	–	**Glycine, N(O,S)-pentafluoropropanyl-, isopropyl ester**		**MS05831**
109	125	79	47	63	263	93	62	**263**	100	77	55	45	42	38	21	16		8	10	5	1	1	–	–	–	–	–	1	–	–	**Phosphorothioic acid, O,O-dimethyl O-(4-nitrophenyl) ester**	298-00-0	**NI26388**
264	234	140	110	111	262	109		**263**	100	63	40	24	18	14	11		**10.50**	8	10	5	1	1	–	–	–	–	–	1	–	–	**Phosphorothioic acid, O,O-dimethyl O-(4-nitrophenyl) ester**	298-00-0	**NI26389**
109	125	263	79	93	63	47	264	**263**	100	91	68	21	20	12	9	7		8	10	5	1	1	–	–	–	–	–	1	–	–	**Phosphorothioic acid, O,O-dimethyl O-(4-nitrophenyl) ester**	298-00-0	**AI01870**
42	111	175	75	113	58	28	177	**263**	100	79	56	34	27	27	25	21	**7.00**	9	10	4	1	1	–	1	–	–	–	–	–	–	**4H-1,3,5-Dioxazine, 5-(p-chlorophenylsulphonyl)-5,6-dihydro-**	56221-09-1	**NI26390**
263	203	204	221	190	265	128	45	**263**	100	46	35	26	17	14	14	14		9	13	2	1	3	–	–	–	–	–	–	–	–	**4-Isothiazolecarboxylic acid, 3,5-bis(methylthio)-, isopropyl ester**	37572-39-7	**NI26391**
263	203	204	190	265	264	202	47	**263**	100	62	50	18	16	14	14	14		9	13	2	1	3	–	–	–	–	–	–	–	–	**4-Isothiazolecarboxylic acid, 3,5-bis(methylthio)-, propyl ester**	37572-40-0	**NI26392**
97	43	157	115	45	85	98	87	**263**	100	93	62	39	22	19	12	11	**0.00**	9	13	6	1	1	–	–	–	–	–	–	–	–	**2H-Thiopyran-3,5-diol, 4-nitro-3,4,5,6-tetrahydro-, diacetate, (3R,4S,5S)-**		**DD00789**
217	113	63	263	101	75	187	87	**263**	100	98	50	30	20	20	19	19		10	5	6	3	–	–	–	–	–	–	–	–	–	**Naphthalene, 1,3,8-trinitro-**		**IC03434**
263	72	113	71	42	60	45	230	**263**	100	48	33	29	27	23	20	18		10	9	2	5	1	–	–	–	–	–	–	–	–	**Thiazole, 2-amino-4-methyl-5-(4-nitrophenylazo)-**		**IC03435**
163	205	136	162	164	137	263	190	**263**	100	63	52	39	37	25	17	14		10	9	4	5	–	–	–	–	–	–	–	–	–	**7-Pteridinecarboxylic acid, 2-acetylamino-3,4-dihydro-4-oxo-, methyl ester**		**LI05477**
75	161	163	263	127	265	162	126	**263**	100	96	40	29	18	14	11	9		10	11	3	1	–	–	2	–	–	–	–	–	–	**Propanamide, N-(3,4-dichlorophenyl)-2,3-dihydroxy-2-methyl-**	41362-84-9	**NI26393**
69	229	231	233	230	161	133	63	**263**	100	36	24	4	4	4	4	4	**0.00**	10	11	3	1	–	–	2	–	–	–	–	–	–	**Propanamide, N-(3,4-dichlorophenyl)-2,3-dihydroxy-2-methyl-**	41362-84-9	**NI26394**
189	174	191	203	176	110	205	175	**263**	100	74	65	58	56	38	34	25	**10.00**	10	11	3	1	–	–	2	–	–	–	–	–	–	**Propanoic acid, β-amino-N-(3,5-dichloro-2-hydroxybenzyl)-**		**NI26395**
73	103	70	147	60	45	100	75	**263**	100	20	18	15	11	11	9	9	**1.43**	10	25	3	1	–	–	–	–	–	2	–	–	–	**Propanal, 2,3-bis[(trimethylsilyl)oxy]-, O-methyloxime, (S)-**	56196-02-2	**NI26396**
263	265	184	264	262	143	142	115	**263**	100	98	98	47	34	22	15	15		11	10	–	3	–	–	–	1	–	–	–	–	–	**2-Pyrimidinamine, 5-bromo-6-methyl-4-phenyl-**		**LI05478**
137	60	150	43	41	87	27	39	**263**	100	75	57	53	46	35	35	33	**20.90**	11	12	3	1	–	3	–	–	–	–	–	–	–	**Acetamide, 2,2,2-trifluoro-N-[2-(4-hydroxy-2-methoxyphenyl)ethyl]-**	56805-14-2	**MS05832**
73	43	41	55	29	71	57	31	**263**	100	87	81	68	64	64	59	47	**0.20**	11	21	4	1	1	–	–	–	–	–	–	–	–	**4,5,6,7-Decanetetrol, 1-isothiocyanato-**		**NI26397**
73	43	115	45	75	74	57	84	**263**	100	57	47	45	24	22	20	11	**0.00**	11	21	6	1	–	–	–	–	–	–	–	–	–	**α-D-Galactopyranoside, methyl 2-acetamido-2-deoxy-3,6-di-O-methyl-**		**MS05833**
71	128	59	101	45	43	102	73	**263**	100	60	54	40	36	36	27	22	**0.00**	11	21	6	1	–	–	–	–	–	–	–	–	–	**α-D-Galactopyranoside, methyl 2-acetamido-2-deoxy-4,6-di-O-methyl-**	17296-06-9	**NI26398**
133	230	69	68	263	231	67	165	**263**	100	95	91	22	16	13	8	4		11	22	2	1	1	–	–	–	–	–	1	–	–	**1,3,2-Dioxaphosphorinane, 2-(cyclohexylamino)-5,5-dimethyl-, 2-sulphide**	4073-56-7	**NI26399**
207	41	151	125	152	123	165	56	**263**	100	86	82	63	60	59	55	43	**13.00**	11	22	4	1	–	–	–	–	–	–	1	–	–	**Butanamide, 4-(diethoxyphosphinoyl)-N-(2-propenyl)-**		**NI26400**
200	75	111	202	235	50	63	237	**263**	100	75	43	33	30	24	22	20	**10.01**	12	7	–	3	–	–	2	–	–	–	–	–	–	**1H-Benzotriazole, 5-chloro-1-(4-chlorophenyl)-**	29328-99-2	**NI26401**
200	75	111	235	50	237	165	113	**263**	100	75	43	30	24	20	16	16	**10.00**	12	7	–	3	–	–	2	–	–	–	–	–	–	**1H-Benzotriazole, 5-chloro-1-(4-chlorophenyl)-**	29328-99-2	**LI05479**
77	51	170	78	28	65	182	152	**263**	100	59	57	42	30	29	26	21	**9.48**	12	9	4	1	1	–	–	–	–	–	–	–	–	**Benzene, 2-nitro-1,1′-sulphonylbis-**	31515-43-2	**NI26402**
77	125	93	170	263	51	65	76	**263**	100	79	79	72	70	61	44	31		12	9	4	1	1	–	–	–	–	–	–	–	–	**Benzene, 3-nitro-1,1′-sulphonylbis-**	18513-19-4	**NI26403**
125	263	77	93	51	170	28	31	**263**	100	74	59	55	45	44	20	18		12	9	4	1	1	–	–	–	–	–	–	–	–	**Benzene, 4-nitro-1,1′-sulphonylbis-**	1146-39-0	**NI26404**
218	89	190	162	146	145	219	173	**263**	100	33	28	17	15	15	14	14	**0.00**	12	9	6	1	–	–	–	–	–	–	–	–	–	**4H-1-Benzopyran-2-carboxylic acid, 6-nitro-4-oxo-, ethyl ester**	30095-79-5	**LI05480**
218	89	190	162	105	173	146	145	**263**	100	33	28	17	16	15	15	15	**1.00**	12	9	6	1	–	–	–	–	–	–	–	–	–	**4H-1-Benzopyran-2-carboxylic acid, 6-nitro-4-oxo-, ethyl ester**	30095-79-5	**NI26405**

MASS TO CHARGE RATIOS	M.W.	INTENSITIES	Parent	C	H	O	N	S	F	Cl	Br	I	Si	P	B	X	COMPOUND NAME	CAS Reg No	No
263 28 44 125 43 123 149 57	263	100 27 23 20 19 18 18 17		12	13	2	3	1	–	–	–	–	–	–	–	–	Benzene, 3,4,4′-triamino-1,1′-sulphonylbis-	17828-44-3	NI26406
77 59 135 51 205 246 263 206	263	100 69 54 38 35 32 21 21		12	13	2	3	1	–	–	–	–	–	–	–	–	4-Imidazolidinepropanamide, 5-oxo-1-phenyl-2-thioxo-		LI05481
263 246 204 77 59 135 203 205	263	100 91 77 74 63 48 44 40		12	13	2	3	1	–	–	–	–	–	–	–	–	4-Imidazolidinepropanamide, 5-oxo-1-phenyl-2-thioxo-, (S)-	29588-05-4	NI26407
86 87 119 88 73 177 91 84	263	100 98 61 61 47 43 37 35	28.00	12	13	4	3	–	–	–	–	–	–	–	–	–	1-(1,4-Dioxan-2-yl)-4-phenyl-1,2,4-triazolidine-3,5-dione		LI05482
103 44 76 119 98 104 50 245	263	100 50 35 25 18 15 13 12	0.00	12	13	4	3	–	–	–	–	–	–	–	–	–	4-Isoxazolecarboxylic acid, 4,5-dihydro-5-[[(methylamino)carbonyl]amino]-3-phenyl-, cis-	35053-70-4	NI26408
263 219 136 264 220 164 137 265	263	100 72 45 13 8 5 5 1		12	13	4	3	–	–	–	–	–	–	–	–	–	Δ^2-1,3,4-Oxadiazolin-5-one, 2-isobutyl-4-(p-nitrophenyl)-	28741-00-6	NI26409
101 86 45 114 115 58 42 84	263	100 86 69 39 24 22 21 20	4.00	12	25	5	1	–	–	–	–	–	–	–	–	–	α-D-Glucopyranoside, methyl 2-deoxy-2-(dimethylamino)-3,4,6-tri-O-methyl-	92414-15-8	NI26410
56 57 86 88 87 100 74 72	263	100 90 80 70 70 50 50 50	1.59	12	25	5	1	–	–	–	–	–	–	–	–	–	1,4,7,10,13-Pentaoxa-16-azacyclooctadecane	33941-15-0	NI26411
72 191 58 73 71 44 70 263	263	100 82 34 30 25 16 15 13		12	30	1	3	–	–	–	–	–	–	1	–	–	Phosphoric triamide, N,N,N′,N′,N″,N″-hexaethyl-	2622-07-3	NI26412
263 248 265 220 250 264 185 222	263	100 75 38 33 27 18 14 12		13	10	1	1	1	–	1	–	–	–	–	–	–	10H-Phenothiazine, 2-chloro-3-methoxy-	17800-08-7	NI26413
263 140 248 265 220 250 264 185	263	100 84 75 38 34 28 19 15		13	10	1	1	1	–	1	–	–	–	–	–	–	10H-Phenothiazine, 2-chloro-3-methoxy-	17800-08-7	NI26414
263 41 59 57 43 39 55 74	263	100 62 61 58 49 46 43 39		13	10	1	1	1	–	1	–	–	–	–	–	–	10H-Phenothiazine, 2-chloro-6-methoxy-	56438-22-3	NI26415
263 248 265 220 264 250 185 222	263	100 42 39 20 19 16 9 7		13	10	1	1	1	–	1	–	–	–	–	–	–	10H-Phenothiazine, 2-chloro-7-methoxy-	1730-44-5	NI26416
263 248 265 220 264 250 131 185	263	100 43 40 20 19 17 12 10		13	10	1	1	1	–	1	–	–	–	–	–	–	10H-Phenothiazine, 2-chloro-7-methoxy-	1730-44-5	LI05483
263 265 220 248 264 228 185 222	263	100 39 25 20 19 10 10 9		13	10	1	1	1	–	1	–	–	–	–	–	–	10H-Phenothiazine, 2-chloro-8-methoxy-	17800-09-8	NI26417
263 220 265 248 264 185 219 45	263	100 40 37 27 19 18 14 14		13	10	1	1	1	–	1	–	–	–	–	–	–	10H-Phenothiazine, 2-chloro-8-methoxy-	17800-09-8	NI26418
263 265 220 248 264 131 185 228	263	100 38 25 20 19 13 10 9		13	10	1	1	1	–	1	–	–	–	–	–	–	10H-Phenothiazine, 2-chloro-8-methoxy-	17800-09-8	LI05484
127 95 110 129 263 137 265 136	263	100 97 78 31 19 7 6 6		13	10	3	1	–	–	1	–	–	–	–	–	–	Acetamide, N-(3-chlorophenyl)-2-(2-furoyl)-		LI05485
127 95 110 129 263 136 265 137	263	100 92 75 33 23 9 7 7		13	10	3	1	–	–	1	–	–	–	–	–	–	Acetamide, N-(4-chlorophenyl)-2-(2-furoyl)-		LI05486
95 110 263 127 265 67 136 81	263	100 41 19 8 7 7 6 5		13	10	3	1	–	–	1	–	–	–	–	–	–	Acetamide, N-(4-chlorophenyl)-2-(3-furoyl)-		LI05487
95 127 110 129 263 265 67 136	263	100 81 56 26 21 7 6 4		13	10	3	1	–	–	1	–	–	–	–	–	–	Acetamide, N-(4-chlorophenyl)-2-(3-furoyl)-		LI05488
162 221 263 144 77 115 176 172	263	100 70 34 24 21 19 16 16		13	13	3	1	1	–	–	–	–	–	–	–	–	1H-1-Benzazepine-5-carboxylic acid, 2,3-dihydro-4-methyl-2-oxo-, S-methyl ester		IC03436
176 144 263 77 59 204 172 51	263	100 85 62 51 31 27 23 22		13	13	5	1	–	–	–	–	–	–	–	–	–	4,5-Isoxazoledicarboxylic acid, 4,5-dihydro-3-phenyl-, dimethyl ester, cis-	17669-30-6	NI26419
57 41 85 207 150 40 206 151	263	100 37 14 11 6 5 4 3	3.00	13	17	3	3	–	–	–	–	–	–	–	–	–	1,2,4-Oxadiazole, 3-methyl-5-(4-nitrophenyl)-4-tert-butyl-		LI05489
207 263 216 148 147 221 234 115	263	100 63 59 41 36 32 18 14		14	17	–	1	2	–	–	–	–	–	–	–	–	Thiazole, 2-(butylthio)-4-(4-methylphenyl)-	69390-11-0	NI26420
144 29 77 263 117 190 31 27	263	100 52 50 45 30 28 28 28		14	17	4	1	–	–	–	–	–	–	–	–	–	Butenedioic acid, 2-(anilino)-, diethyl ester, (E)-		MS05834
95 83 55 70 69 81 43 68	263	100 93 91 87 87 80 79 78	38.00	14	17	4	1	–	–	–	–	–	–	–	–	–	2,4-Dioxabicyclo[3.2.1]octane, 6,7-dimethyl-3-(4-nitrophenyl)-	55821-20-0	NI26421
69 55 95 41 263 83 179 59	263	100 74 70 45 35 27 21 19		14	17	4	1	–	–	–	–	–	–	–	–	–	2,4-Dioxabicyclo[3.2.1]octane, 6,7-dimethyl-3-(4-nitrophenyl)-	55821-20-0	NI26422
59 77 73 57 44 105 43 41	263	100 94 91 86 83 77 69 57	0.00	14	17	4	1	–	–	–	–	–	–	–	–	–	3-Isoxazolecarboperoxoic acid, 4,5-dihydro-5-phenyl-, tert-butyl ester	35145-84-7	NI26423
55 157 45 118 200 43 129 199	263	100 87 72 68 64 59 56 52	0.00	14	17	4	1	–	–	–	–	–	–	–	–	–	1(2H)-Naphthalenone, 2-(3-hydroxybutyl)-2-nitro-1,2,3,4-tetrahydro-		DD00790
133 85 135 131 91 90 55 101	263	100 98 46 44 42 36 36 35	0.00	14	17	4	1	–	–	–	–	–	–	–	–	–	3-Oxabicyclo[8.4.0]tetradeca-1(10),11,13-trien-2-one, 4-methyl-7-nitro-		DD00791
118 39 41 91 43 90 89 63	263	100 98 96 95 89 88 67 55	1.00	14	17	4	1	–	–	–	–	–	–	–	–	–	2-Oxaindan-1,3-dione, hexahydro-6-methyl-4-(2-oxo-3,4,5,6-tetrahydropyridyl)-		MS05835
43 55 41 77 79 91 39 56	263	100 51 44 41 39 35 29 28	1.00	14	17	4	1	–	–	–	–	–	–	–	–	–	2-Oxaindan-1,3-dione, hexahydro-4-(7-Oxo-1,3,4,5,6,7-hexahydro-2H-azepin-1-yl)-		MS05836
28 56 158 75 162 40 263 42	263	100 84 78 34 30 18 16 13		14	17	4	1	–	–	–	–	–	–	–	–	–	Propanedioic acid, 2-[1-(methylamino)ethylidene]-, ethyl phenyl diester		MS05837
70 135 263 264 232 128 85 84	263	100 48 35 21 21 16 12 12		14	17	4	1	–	–	–	–	–	–	–	–	–	2-Pyrrolidinemethanol, 1-(1,3-benzodioxol-5-ylacetyl)-	38847-94-8	MS05838
161 56 175 73 263 58 149 41	263	100 82 57 50 34 31 28 28		14	19	3	2	–	–	–	–	–	–	–	–	–	3-Imidazoline, 4-(p-methoxyphenyl)-2,2,5,5-tetramethyl-, 3-oxide-1-oxile		NI26424
152 109 79 53 81 67 136 111	263	100 91 80 74 71 71 56 52	3.00	14	21	2	3	–	–	–	–	–	–	–	–	–	Benzimidazole, 2-methyl-1-(2-oximinocyclohexyl)-4,5,6,7-tetrahydro-, 1-oxide		LI05490
263 262 221 56 31 264 103 41	263	100 88 31 27 19 17 17 17		15	9	2	3	–	–	–	–	–	–	–	–	–	Imidazo[1,2-a]pyridin-2-ol, 3-(1-oxo-1H-isoindol-3-yl)-		IC03437
178 263 264 179 176 77 76 177	263	100 97 16 16 8 8 8 1		15	13	–	5	–	–	–	–	–	–	–	–	–	1,2,4-Triazin-3(2H)-one, 5,6-diphenyl-, hydrazone	21383-24-4	NI26425
57 120 105 146 77 162 41 190	263	100 86 79 65 47 29 24 11	0.00	15	21	3	1	–	–	–	–	–	–	–	–	–	Carbamic acid, (4-phenyl-4-oxobutyl)-, tert-butyl ester	116437-41-3	DD00792
91 172 263 42 98 218 146 134	263	100 65 18 11 10 9 9 9		15	21	3	1	–	–	–	–	–	–	–	–	–	4-Piperidinecarboxylic acid, 1-benzyl-3-hydroxy-, ethyl ester, (3R,4R)-		DD00793
58 213 50 91 45 69 71 105	263	100 21 9 4 3 2 2 1	0.00	15	34	–	1	–	–	1	–	–	–	–	–	–	Dodecyltrimethylammonium chloride	112-00-5	NI26426
235 263 178 236 179 152 151 164	263	100 75 29 25 25 25 23 21		16	9	3	1	–	–	–	–	–	–	–	–	–	7H-Dibenzo[de,g]quinolin-7-one, 1,2-dihydroxy-	65400-36-4	NI26427
263 188 216 187 189 246 217 233	263	100 55 32 31 28 20 20 3		16	9	3	1	–	–	–	–	–	–	–	–	–	1-Pyrenol, 2-nitro-		MS05839
263 105 264 130 77 235 91 102	263	100 25 19 18 16 14 12 11		16	13	1	3	–	–	–	–	–	–	–	–	–	Imidazo[2,1-a]phthalazin-6-one, 5-phenyl-2,3,5,6-tetrahydro-		NI26428
144 263 77 51 246 145 89 27	263	100 74 70 37 17 9 7 7		16	13	1	3	–	–	–	–	–	–	–	–	–	1H-Pyrazole, 4-benzoyl-1-phenyl-, oxime, anti-		LI05491
91 132 263 77 264 51 64 103	263	100 87 65 26 12 11 10 8		16	13	1	3	–	–	–	–	–	–	–	–	–	2-Pyrazolin-5-one, 3-methyl-1-phenyl-4-phenylimino-		MS05840
91 132 263 77 264 51 64 103	263	100 87 65 26 13 12 10 8		16	13	1	3	–	–	–	–	–	–	–	–	–	3H-Pyrazol-3-one, 2,4-dihydro-5-methyl-2-phenyl-4-(phenylimino)-	27807-95-0	NI26429
178 179 176 152 177 151 113 165	263	100 16 10 7 6 6 6 5	2.00	16	13	1	3	–	–	–	–	–	–	–	–	–	1,2,4-Triazine, 3-methoxy-5,6-diphenyl-		LI05492

MASS TO CHARGE RATIOS								M.W.	INTENSITIES								Parent	C	H	O	N	S	F	Cl	Br	I	Si	P	B	X	COMPOUND NAME	CAS Reg No	No
214	77	229	158	57	41	191	263	**263**	100	18	17	16	14	14	10	1		16	25	–	1	1	–	–	–	–	–	–	–	–	Thioacetamide, di-tert-butyl-N-phenyl-		LI05493
164	146	263	148	147	264	91	165	**263**	100	77	69	16	13	12	12	10		16	25	2	1	–	–	–	–	–	–	–	–	–	Acetophenone, 2-hydroxy-5-octyl-, oxime, (E)-		DD00794
172	173	148	161	189	174	162	147	**263**	100	12	10	3	2	2	2	2	**0.50**	16	25	2	1	–	–	–	–	–	–	–	–	–	Butanoic acid, 3-(dimethylamino)-4-phenyl-, butyl ester	27820-09-3	NI26430
172	116	41	148	42	57	71	72	**263**	100	40	30	27	24	23	20	18	**0.60**	16	25	2	1	–	–	–	–	–	–	–	–	–	Butanoic acid, 3-(dimethylamino)-4-phenyl-, butyl ester	27820-09-3	NI26431
106	105	41	57	69	55	44	43	**263**	100	96	79	78	75	53	45	37	**0.00**	16	25	2	1	–	–	–	–	–	–	–	–	–	Carbamic acid, (α-methylbenzyl)-, 1-ethyl-1-methylbutyl ester	32692-89-0	NI26432
150	164	105	57	106	165	43	120	**263**	100	85	83	62	61	52	25	20	**8.01**	16	25	2	1	–	–	–	–	–	–	–	–	–	Carbamic acid, (α-methylbenzyl)-, 1-methylhexyl ester	32589-43-8	NI26433
94	112	138	28	126	45	44	77	**263**	100	78	73	65	42	39	39	32	**0.98**	16	25	2	1	–	–	–	–	–	–	–	–	–	Cyclohexanamine, N-[2-(2-phenoxyethoxy)ethyl]-	55955-93-6	NI26434
96	220	263	109	108	136	28	178	**263**	100	88	36	18	18	16	16	14		16	25	2	1	–	–	–	–	–	–	–	–	–	Dendroban-12-one	2115-91-5	NI26435
203	263	202	204	188	160	262	245	**263**	100	31	26	18	13	10	9	8		16	25	2	1	–	–	–	–	–	–	–	–	–	Lycopodane-5,6-diol, 11,12-didehydro-15-methyl-, (5β,6α,15R)-	22594-91-8	NI26436
263	190	246	152	151	192	124	123	**263**	100	82	72	64	50	43	43	39		16	25	2	1	–	–	–	–	–	–	–	–	–	Lycopodan-5-one, 12-hydroxy-15-methyl-, (12α,15R)-	21061-90-5	NI26437
263	190	246	152	151	123	192	124	**263**	100	92	88	65	60	60	51	50		16	25	2	1	–	–	–	–	–	–	–	–	–	Lycopodan-5-one, 12-hydroxy-15-methyl-, (15R)-	6900-92-1	NI26438
263	122	190	150	40	55	151	246	**263**	100	70	62	57	55	54	53	50		16	25	2	1	–	–	–	–	–	–	–	–	–	Lycopodan-5-one, 12-hydroxy-15-methyl-, (15R)-	6900-92-1	LI05494
133	132	43	59	146	60	147	44	**263**	100	95	44	41	37	37	34	34	**0.00**	16	25	2	1	–	–	–	–	–	–	–	–	–	Quinolinium, 1-methyl-8-tert-butyl-5,6,7,8-tetrahydro-, acetate		NI26439
84	69	60	122	194	109	83	263	**263**	100	84	72	24	22	20	19	18		16	29	–	3	–	–	–	–	–	–	–	–	–	Guanidine, N^1,N^2,N^3-triisopentenyl-		LI05495
170	115	171	263	142	93	264	89	**263**	100	98	93	86	57	25	16	12		17	13	2	1	–	–	–	–	–	–	–	–	–	Aniline, N-(3-hydroxynaphthoyl)-		IC03438
179	235	76	234	206	50	77	178	**263**	100	86	49	47	33	30	23	21	**2.00**	17	13	2	1	–	–	–	–	–	–	–	–	–	1-Benzazocine-3,6-dione, 4,5-dihydro-2-phenyl-	90732-27-7	NI26440
105	141	77	115	263	142	140	51	**263**	100	44	38	32	28	24	21	17		17	13	2	1	–	–	–	–	–	–	–	–	–	3-Butenenitrile, 2-(benzoyloxy)-4-phenyl-		LI05496
105	141	77	115	263	142	140	51	**263**	100	22	20	17	14	12	10	8		17	13	2	1	–	–	–	–	–	–	–	–	–	3-Butenenitrile, 2-(benzoyloxy)-4-phenyl-	1591-17-9	NI26441
263	77	103	160	104	76	245	116	**263**	100	72	62	44	37	36	32	29		17	13	2	1	–	–	–	–	–	–	–	–	–	1,3-Isoindoledione, N-(2-phenylallyl)-		NI26442
90	173	116	63	263	129			**263**	100	40	25	20	8	7				17	13	2	1	–	–	–	–	–	–	–	–	–	4,7-Methanoisoindole-1,3-dione, 1,3,3a,4,7,7a-hexahydro-2-phenyl-8-vinylidene-		LI05497
90	173	116	263	63	129			**263**	100	35	13	11	10	4				17	13	2	1	–	–	–	–	–	–	–	–	–	4,7-Methanoisoindole-1,3-dione, 4-ethynyl-1,3,3a,4,7,7a-hexahydro-2-phenyl-		LI05498
119	91	263	65	120	89	264	90	**263**	100	31	19	12	10	5	4	4		17	13	2	1	–	–	–	–	–	–	–	–	–	2-Oxazolin-5-one, 4-benzylidene-2-p-tolyl-	16352-73-1	NI26443
119	91	263	65	89	90	116	130	**263**	100	31	19	12	5	4	2	1		17	13	2	1	–	–	–	–	–	–	–	–	–	2-Oxazolin-5-one, 4-benzylidene-2-p-tolyl-		LI05499
203	204	88	218	176				**263**	100	18	11	10	9				**0.00**	17	13	2	1	–	–	–	–	–	–	–	–	–	9-Phenanthrenecarboxaldehyde, O-acetyloxime, (E)-	51873-99-5	NI26444
141	107	140	79	139	115	263	52	**263**	100	50	46	32	23	21	18	16		17	13	2	1	–	–	–	–	–	–	–	–	–	3-Pyridinecarboxylic acid, 1-naphthalenemethyl ester	86328-55-4	NI26445
248	263	249	220	234	58	57	42	**263**	100	49	33	25	23	17	17	16		17	29	1	1	–	–	–	–	–	–	–	–	–	Anilinium, 3,5-di-tert-butyl-2-hydroxy-N,N,N-trimethyl-, hydroxide, inner salt	53442-15-2	NI26446
248	263	249	57					**263**	100	43	20	9						17	29	1	1	–	–	–	–	–	–	–	–	–	Anilinium, 3,5-di-tert-butyl-4-hydroxy-N,N,N-trimethyl-, hydroxide, inner salt	61219-67-8	NI26447
123	108	263	136	124	81	120	80	**263**	100	70	48	39	19	15	11	11		17	29	1	1	–	–	–	–	–	–	–	–	–	1H-Pyrrole, 5-methyl-2-(1-oxododecyl)-		NI26448
165	263	235	103	77	104	166	105	**263**	100	90	72	64	44	43	34	30		18	17	1	1	–	–	–	–	–	–	–	–	–	Oxazole, 4,5-diphenyl-2-propyl-	20662-95-7	NI26449
165	235	103	263	77	104	166	105	**263**	100	72	65	63	44	42	34	30		18	17	1	1	–	–	–	–	–	–	–	–	–	Oxazole, 4,5-diphenyl-2-propyl-	20662-95-7	LI05500
96	55	41	67	83	178	248	206	**263**	100	78	39	34	27	19	7	6	**4.50**	18	33	–	1	–	–	–	–	–	–	–	–	–	2H-Pyrrole, 3,4-dihydro-5-methyl-4-[(5Z)-tridecenyl]-		NI26450
70	44	191	31	192	189	18	59	**263**	100	76	69	24	15	15	12	9	**3.04**	19	21	–	1	–	–	–	–	–	–	–	–	–	5H-Dibenzo[a,d]cycloheptene-5-propanamine, N-methyl-	438-60-8	NI26451
263	248	194	221	165	69	206	180	**263**	100	99	99	73	71	69	67	66		19	21	–	1	–	–	–	–	–	–	–	–	–	19-Norretinal nitrile, 9,10,11,12-tetradehydro-, all-trans-		NI26452
203	202	263	45	220	218	215	219	**263**	100	88	63	56	50	40	40	38		19	21	–	1	–	–	–	–	–	–	–	–	–	1-Propylamine, 3-(10,11-dihydro-5H-dibenzo[a,d]cyclohepten-5-ylidene)-N-methyl-	72-69-5	MS05841
44	57	43	71	41	55	45	56	**263**	100	13	12	7	6	5	5	4	**2.40**	19	21	–	1	–	–	–	–	–	–	–	–	–	1-Propylamine, 3-(10,11-dihydro-5H-dibenzo[a,d]cyclohepten-5-ylidene)-N-methyl-	72-69-5	NI26453
202	44	203	215	220	204	189	219	**263**	100	99	78	51	49	44	39	34	**13.90**	19	21	–	1	–	–	–	–	–	–	–	–	–	1-Propylamine, 3-(10,11-dihydro-5H-dibenzo[a,d]cyclohepten-5-ylidene)-N-methyl-	72-69-5	NI26454
97	36	99	132	134	28	38	35	**264**	100	82	64	51	48	30	27	25	**0.00**	–	–	–	–	–	–	6	–	–	–	–	–	2	Aluminium, di-μ-chlorotetrachlorodi-	13845-12-0	NI26455
27	187	106	108	185	189	107	105	**264**	100	72	69	63	33	31	23	23	**0.71**	2	3	–	–	–	–	–	3	–	–	–	–	–	Ethane, 1,1,2-tribromo-	78-74-0	NI26456
187	185	105	107	189	186	106	184	**264**	100	55	54	51	48	38	23	21	**2.71**	2	3	–	–	–	–	–	3	–	–	–	–	–	Ethane, 1,1,2-tribromo-	78-74-0	IC03439
264	64	100	82	69	81	63	131	**264**	100	94	64	56	48	40	40	36		4	–	–	–	2	8	–	–	–	–	–	–	–	1,2-Dithiane, perfluoro-		MS05842
97	225	195	151	51	50	29	79	**264**	100	95	93	81	55	34	31	28	**0.00**	5	1	2	–	–	9	–	–	–	–	–	–	–	Acetic acid, trifluoro-, 2,2,2-trifluoro-1-(trifluoromethyl)ethyl ester	42031-15-2	NI26457
185	45	139	93	63	109	46	61	**264**	100	93	82	75	64	58	50	34	**2.00**	5	12	4	–	4	–	–	–	–	–	–	–	–	6-Oxa-2,4,8,9-tetrathiadecane 2,9,9-trioxide		MS05843
185	45	139	93	63	109	46	61	**264**	100	95	78	75	63	59	49	33	**2.00**	5	12	4	–	4	–	–	–	–	–	–	–	–	6-Oxa-2,4,8,9-tetrathiadecane 2,9,9-trioxide		LI05501
266	268	264	165	36	167	270	230	**264**	100	65	63	29	29	28	23	18		6	1	1	–	–	–	5	–	–	–	–	–	–	Phenol, pentachloro-	87-86-5	IC03440
266	268	264	165	167	95	130	60	**264**	100	70	68	54	52	44	30	28		6	1	1	–	–	–	5	–	–	–	–	–	–	Phenol, pentachloro-	87-86-5	MS05844
266	264	268	165	167	95	130	202	**264**	100	66	63	59	52	44	33	31		6	1	1	–	–	–	5	–	–	–	–	–	–	Phenol, pentachloro-	87-86-5	NI26458
264	65	69	127	129	180	249	217	**264**	100	67	66	61	32	30	26	25		7	3	–	–	1	6	–	–	–	–	1	–	–	Phosphinothioic fluoride, methyl(pentafluorophenyl)-	53327-23-4	NI26459

MASS TO CHARGE RATIOS								M.W.	INTENSITIES								Parent	C	H	O	N	S	F	Cl	Br	I	Si	P	B	X	COMPOUND NAME	CAS Reg No	No
246	264	63	218	53	91	62	127	**264**	100	78	42	34	12	10	10	9		7	5	3	–	–	–	–	–	–	–	–	–	1	**Benzoic acid, 2-hydroxy-5-iodo-**	119-30-2	**LI05502**
246	264	63	218	91	53	62	109	**264**	100	79	46	34	12	12	11	9		7	5	3	–	–	–	–	–	–	–	–	–	1	**Benzoic acid, 2-hydroxy-5-iodo-**	119-30-2	**MS05845**
246	264	63	218	53	90	62	248	**264**	100	80	40	33	12	10	10	9		7	5	3	–	–	–	–	–	–	–	–	–	1	**Benzoic acid, 2-hydroxy-5-iodo-**	119-30-2	**NI26460**
63	266	251	75	264	223	268	62	**264**	100	89	52	49	44	44	40	40		7	6	1	–	–	–	–	2	–	–	–	–	–	**Benzene, 1,3-dibromo-2-methoxy-**	38603-09-7	**NI26461**
63	266	223	251	264	62	268	75	**264**	100	58	41	37	30	30	29	29		7	6	1	–	–	–	–	2	–	–	–	–	–	**Benzene, 2,4-dibromo-1-methoxy-**	21702-84-1	**NI26463**
266	251	268	264	223	253	249	221	**264**	100	78	59	55	55	38	34	25		7	6	1	–	–	–	–	2	–	–	–	–	–	**Benzene, 2,4-dibromo-1-methoxy-**	21702-84-1	**NI26462**
63	266	251	264	62	268	223	75	**264**	100	85	53	40	40	39	35	32		7	6	1	–	–	–	–	2	–	–	–	–	–	**Benzene, 2,4-dibromo-1-methoxy-**	21702-84-1	**NI26464**
77	51	185	78	187	50	266	53	**264**	100	98	58	55	53	51	48	45	**26.73**	7	6	1	–	–	–	–	2	–	–	–	–	–	**Phenol, 2,6-dibromo-4-methyl-**	2432-14-6	**NI26465**
266	264	268	109	231	229	133	124	**264**	100	80	49	15	12	12	12	12		8	–	–	2	–	–	4	–	–	–	–	–	–	**1,2-Benzenedicarbonitrile, 3,4,5,6-tetrachloro-**	1953-99-7	**NI26466**
266	264	268	109	124	231	229	98	**264**	100	78	47	23	19	12	12	10		8	–	–	2	–	–	4	–	–	–	–	–	–	**1,3-Benzenedicarbonitrile, 2,4,5,6-tetrachloro-**	1897-45-6	**NI26467**
266	264	268	109	124	133	270	229	**264**	100	80	48	19	15	11	10	9		8	–	–	2	–	–	4	–	–	–	–	–	–	**1,3-Benzenedicarbonitrile, 2,4,5,6-tetrachloro-**	1897-45-6	**NI26468**
266	264	268	109	124	231	229	133	**264**	100	85	52	30	23	15	15	15		8	–	–	2	–	–	4	–	–	–	–	–	–	**1,3-Benzenedicarbonitrile, 2,4,5,6-tetrachloro-**	1897-45-6	**NI26469**
106	91	105	77	79	170	159	157	**264**	100	81	55	32	26	12	11	11	**0.20**	8	10	–	–	–	–	–	2	–	–	–	–	–	**Cyclopropane, 2-cyclopropyl-1-(dibromomethylene)-2-methyl-**		**MS05846**
106	91	105	77	79	170	157	159	**264**	100	73	59	30	27	10	8	7	**0.20**	8	10	–	–	–	–	–	2	–	–	–	–	–	**Cyclopropane, 2-cyclopropyl-3,3-dibromo-2-methyl-1-methylene-**		**MS05847**
100	139	80	264	65	109	165	63	**264**	100	57	42	41	36	29	25	25		8	10	5	2	–	1	–	–	–	–	1	–	–	**Phosphoramidic acid, 2-fluoroethyl p-nitrophenyl diester**		**MS05848**
264	63	154	184	266	265	81	77	**264**	100	18	16	12	11	11	11	11		8	12	4	2	2	–	–	–	–	–	–	–	–	**Hydrazine, [2,4-bis(methylsulphonyl)phenyl]-**	57396-91-5	**NI26470**
142	127	126	48	47	46	141	112	**264**	100	50	23	20	20	19	13	13	**0.00**	8	13	–	2	–	–	–	–	–	–	–	–	1	**Pyridine, 3-trimethylammonium, iodide**		**MS05849**
127	142	78	57	55	60	93	73	**264**	100	63	54	52	52	45	42	38	**0.00**	8	13	–	2	–	–	–	–	–	–	–	–	1	**Pyridine, 4-trimethylammonium, iodide**		**MS05850**
202	155	96	71	69	229	70	85	**264**	100	70	67	56	36	34	25	16	**6.99**	8	13	–	4	2	–	1	–	–	–	–	–	–	**1,3,5-Triazin-2-amine, N,N-dimethyl-4-(2-chloroethylthio)-6-methylthio-**	79923-83-4	**NI26471**
214	181	155	213	99	114	96	154	**264**	100	46	22	19	18	16	16	14	**2.99**	8	13	–	4	2	–	1	–	–	–	–	–	–	**1,3,5-Triazine-2(1H)-thione, 1-(2-chloroethyl)-4-dimethylamino-6-methylthio-**	79923-85-6	**NI26472**
73	189	131	191	45	116	264	43	**264**	100	46	33	22	22	22	19	19		8	24	–	–	1	–	–	–	–	4	–	–	–	**Sulphacyclopentasilane, 2,2,3,3,4,4,5,5-octamethyl-**	83272-65-5	**NI26473**
164	163	69	145	104	264	114	103	**264**	100	61	38	32	31	30	22	20		9	4	–	–	–	8	–	–	–	–	–	–	–	**Bicyclo[2.2.2]oct-2-ene, 5-methylene-1,2,3,4,7,7,8,8-octafluoro-**		**NI26474**
164	163	264	133	249	165	121	149	**264**	100	62	31	19	13	13	13	12		9	7	1	–	–	7	–	–	–	–	–	–	–	**Bicyclo[2.2.2]oct-2-ene, 2-methoxy-1,3,4,5,5,6,6-heptafluoro-**		**NI26475**
264	262	233	249	76	247	231	260	**264**	100	98	86	81	75	71	67	56		9	10	1	–	–	–	–	–	–	–	–	–	1	**Acetophenone, 4′-(methyltelluryl)-**	32294-61-4	**NI26476**
184	183	169	185	80	155	82	151	**264**	100	90	24	20	18	17	17	16	**0.00**	9	13	–	–	2	–	–	1	–	–	–	–	–	**1,2-Benzodithiol-1-ium, 3-ethyl-4,5,6,7-tetrahydro-, bromide**	55836-84-5	**NI26477**
184	155	183	169	80	156	82	59	**264**	100	92	89	66	53	52	52	38	**0.00**	9	13	–	–	2	–	–	1	–	–	–	–	–	**4H-Cyclohepta-1,2-dithiol-1-ium, 5,6,7,8-tetrahydro-3-methyl-, bromide**	55836-81-2	**NI26478**
87	57	41	31	69	180	178	181	**264**	100	69	34	25	21	11	11	9	**0.00**	9	13	4	–	–	–	–	1	–	–	–	–	–	**2H,4H-1,3-Dioxin-4-one, 5-bromo-2-tert-butyl-6-(hydroxymethyl)-, (2R)-**		**DD00795**
41	55	36	104	69	29	91	67	**264**	100	68	63	54	54	51	46	46	**0.00**	9	16	–	–	–	–	4	–	–	–	–	–	–	**Nonane, 1,1,1,9-tetrachloro-**		**MS05851**
61	39	76	59	147	149	146	41	**264**	100	76	64	64	62	53	45	40	**0.80**	9	19	–	–	–	–	–	–	–	–	1	–	1	**Palladium, bis(η^3-allyl)(trimethylphosphine)-**	74811-11-3	**NI26479**
43	58	209	179	207	181	149	235	**264**	100	58	25	25	24	24	20	19	**4.00**	9	20	1	–	–	–	–	–	–	–	–	–	1	**Stannane, acetyl(methylene)triethyl-**		**LI05503**
266	264	268	194	196	98	97	229	**264**	100	80	47	47	29	19	17	15		10	4	–	–	–	–	4	–	–	–	–	–	–	**Naphthalene, 1,2,3,4-tetrachloro-**	20020-02-4	**NI26480**
266	264	268	194	97	196	98	133	**264**	100	80	47	45	30	28	27	20		10	4	–	–	–	–	4	–	–	–	–	–	–	**Naphthalene, 1,2,3,4-tetrachloro-**	20020-02-4	**NI26481**
266	264	268	194	196	267	270	229	**264**	100	78	48	28	18	11	10	9		10	4	–	–	–	–	4	–	–	–	–	–	–	**Naphthalene, 1,3,5,7-tetrachloro-**	53555-64-9	**NI26482**
266	264	268	194	196	267	270	229	**264**	100	77	48	33	21	11	10	9		10	4	–	–	–	–	4	–	–	–	–	–	–	**Naphthalene, 1,4,6,7-tetrachloro-**	55720-43-9	**NI26483**
104	264	76	50	102	262	158	75	**264**	100	77	54	38	37	36	27	23		10	4	2	2	–	–	–	–	–	–	–	–	1	**Naphtho[2,3-c][1,2,5]selenadiazol-4,9-dione**	83274-08-2	**NI26484**
206	167	208	236	141	114	102	178	**264**	100	55	46	33	18	13	10	8	**0.00**	10	10	2	–	–	–	–	–	–	–	–	–	1	**Ruthenium, carbonylcyclopentadienyl(σ-allyl)-**		**NI26485**
191	193	264	266	141	73	53	97	**264**	100	69	40	30	23	22	14	13		10	10	4	–	–	–	2	–	–	–	–	–	–	**Acetic acid, (2,5-dichloro-4-methoxyphenoxy)-, methyl ester**	75627-00-8	**NI26486**
178	196	179	149	135	77	51	230	**264**	100	99	68	22	21	18	18	16	**0.00**	10	10	4	–	–	–	2	–	–	–	–	–	–	**Benzoic acid, 3,5-dichloro-4,6-dimethoxy-2-methyl-**	5859-29-0	**NI26487**
233	235	137	234	149	109	237	113	**264**	100	66	17	13	13	13	12	12	**3.50**	10	10	4	–	–	–	2	–	–	–	–	–	–	**Benzoic acid, 2,5-dichloro-3,6-dimethoxy-, methyl ester**	62059-39-6	**NI26488**
233	264	235	249	266	251	87	53	**264**	100	69	60	56	43	35	33	31		10	10	4	–	–	–	2	–	–	–	–	–	–	**Benzoic acid, 2,5-dichloro-3,6-dimethoxy-, methyl ester**	62059-39-6	**NI26489**
189	220	45	97	15	264	233	133	**264**	100	54	20	8	8	6	6	6		10	10	4	–	–	–	2	–	–	–	–	–	–	**Benzoic acid, 3,5-dichloro-4-hydroxyethoxy-, methyl ester**		**IC03441**
45	31	29	18	28	14	44	15	**264**	100	81	64	61	60	57	53	45	**0.00**	10	12	3	1	–	–	2	–	–	–	–	–	–	**Acetic acid, (2,4-dichlorophenoxy)-, compd. with N-methylmethylamine (1:1)**	2008-39-1	**NI26490**
60	264	57	174	55	135	68	175	**264**	100	33	20	18	16	13	13	12		10	12	3	6	–	–	–	–	–	–	–	–	–	**Adenosine, 8,2′-aminoanhydro-**		**NI26491**
80	79	93	41	39	67	77	91	**264**	100	49	34	24	24	19	18	16	**0.18**	10	16	4	–	2	–	–	–	–	–	–	–	–	**Cyclobuta[1,2-b:4,3-b′]bisthiopyran, decahydro-, 1,1,8,8-tetraoxide, (4α,4bβ,8aβ,8bα)-**	74630-21-0	**NI26492**
179	151	235	149	121	207	177	120	**264**	100	63	52	37	31	21	11	10	**0.00**	10	24	–	–	–	–	–	–	–	–	–	–	1	**Tin, n-butyltriethyl-**	17582-53-5	**NI26493**
179	235	151	121	207	149	120	177	**264**	100	56	40	27	18	11	8	5	**0.30**	10	24	–	–	–	–	–	–	–	–	–	–	1	**Tin, n-butyltriethyl-**	17582-53-5	**LI05504**
43	179	177	58	235	149	175	151	**264**	100	69	56	40	37	36	34	32	**0.55**	10	24	–	–	–	–	–	–	–	–	–	–	1	**Tin, n-butyltriethyl-**	17582-53-5	**MS05852**
179	177	149	151	207	147	175	205	**264**	100	78	71	58	54	50	49	41	**3.64**	10	24	–	–	–	–	–	–	–	–	–	–	1	**Tin, sec-butyltriethyl-**		**MS05853**
178	176	149	151	207	147	174	205	**264**	100	80	72	59	55	51	49	41	**3.10**	10	24	–	–	–	–	–	–	–	–	–	–	1	**Tin, sec-butyltriethyl-**		**MS05854**
57	179	207	177	149	205	151	147	**264**	100	66	58	51	50	43	38	34	**2.40**	10	24	–	–	–	–	–	–	–	–	–	–	1	**Tin, tert-butyltriethyl-**		**MS05855**
57	179	207	209	177	149	205	151	**264**	100	71	62	60	56	54	46	41	**0.00**	10	24	–	–	–	–	–	–	–	–	–	–	1	**Tin, tert-butyltriethyl-**		**MS05856**
135	179	133	221	177	137	219	27	**264**	100	93	74	68	65	59	51	51	**2.55**	10	24	–	–	–	–	–	–	–	–	–	–	1	**Tin, methyltriisopropyl-**		**MS05857**
208	57	29	209	178	56	28	27	**264**	100	95	76	52	25	22	16	16	**12.00**	11	12	4	4	–	–	–	–	–	–	–	–	–	**1H-1,2,4-Triazole, 5-ethyl-3-(5-nitro-2-furanyl)-1-propanoyl-**	35732-74-2	**NI26494**

MASS TO CHARGE RATIOS	M.W.	INTENSITIES	Parent	C	H	O	N	S	F	Cl	Br	I	Si	P	B	X	COMPOUND NAME	CAS Reg No	No
152 139 126 55 112 69 109 264	264	100 35 32 31 28 24 23 4		11	12	4	4	–	–	–	–	–	–	–	–	–	Uracil, 1,1′-trimethylenebis-		LI05505
147 148 69 49 63 149 77 62	264	100 95 68 43 25 15 12 12	0.00	11	13	–	–	1	4	–	–	–	–	–	1	–	Benzo[b]thiophenium, 1-ethyl-2-methyl-, tetrafluoroborate		LI05506
162 147 161 163 128 150 115 164	264	100 70 40 23 12 10 8 7	0.00	11	13	–	–	1	4	–	–	–	–	–	1	–	Benzo[b]thiophenium, 1,2,3-trimethyl-, tetrafluoroborate		LI05507
132 264 84 77 190 78 160 83	264	100 86 53 36 35 33 32 32		11	16	2	6	–	–	–	–	–	–	–	–	–	4-Pyrazino[2,3-g]pteridinecarboxylic acid, 5,5a,6,7,8,9,9a,10-octahydro-, ethyl ester		LI05508
88 101 75 45 59 57 73 55	264	100 66 49 43 27 27 26 18	0.00	11	20	7	–	–	–	–	–	–	–	–	–	–	Methyl (methyl 2,3,4-tri-O-methyl-D-galactopyranoside)uronate		LI05509
101 88 75 73 45 85 89 102	264	100 60 25 14 14 10 7 6	0.00	11	20	7	–	–	–	–	–	–	–	–	–	–	Methyl (methyl 2,3,4-tri-O-methyl-α-D-glucopyranoside)uronate	52729-97-2	NI26495
101 88 75 45 73 163 87 85	264	100 69 49 37 22 16 14 13	0.00	11	20	7	–	–	–	–	–	–	–	–	–	–	Methyl (methyl 2,3,4-tri-O-methyl-D-glucopyranoside)uronate		LI05510
101 88 75 45 73 85 89 102	264	100 64 26 17 16 12 10 5	0.00	11	20	7	–	–	–	–	–	–	–	–	–	–	Methyl (methyl 2,3,4-tri-O-methyl-α-D-mannopyranoside)uronate	40147-18-0	NI26496
101 88 75 43 117 103 102 73	264	100 58 32 18 8 6 6 6	0.00	11	20	7	–	–	–	–	–	–	–	–	–	–	α-D-Xyloside, methyl 2,3,4-tri-O-methyl-5-acetoxy-		LI05511
194 97 191 88 55 120 69 41	264	100 89 79 71 60 50 47 40	9.00	11	21	1	–	2	–	–	–	–	–	1	–	–	Phosphetane-1-thioacetic acid, 2,2,3,4,4-pentamethyl-, methyl ester, 1-oxide		NI26497
74 49 79 87 55 41 61 43	264	100 65 55 43 28 27 22 20	1.04	11	21	2	–	–	–	–	1	–	–	–	–	–	Decanoic acid, 10-bromo-, methyl ester	26825-94-5	NI26498
88 133 101 73 45 131 75 115	264	100 71 56 37 23 17 15 14	0.00	11	24	5	–	–	–	–	–	–	1	–	–	–	Lyxopyranose, 2,3,4-tri-O-methyl-1-O-(trimethylsilyl)-	20561-82-4	NI26499
88 133 101 73 115 45 89 75	264	100 74 58 34 25 21 20 13	0.00	11	24	5	–	–	–	–	–	–	1	–	–	–	Xylopyranose, 2,3,4-tri-O-methyl-1-O-(trimethylsilyl)-	55101-95-6	NI26500
264 249 263 250 245 242 223 236	264	100 50 27 8 7 6 6 5		12	6	–	–	–	6	–	–	–	–	–	–	–	Naphthalene, 2,3-dimethyl-1,4,5,6,7,8-hexafluoro-		NI26501
264 221 43 265 69 236 194 218	264	100 74 39 23 22 18 15 12		12	8	7	–	–	–	–	–	–	–	–	–	–	1,4-Naphthalenedione, 2-acetyl-3,5,6,8-tetrahydroxy-	3718-80-7	NI26502
264 221 43 265 69 236 194 218	264	100 74 40 24 22 20 16 12		12	8	7	–	–	–	–	–	–	–	–	–	–	1,4-Naphthalenedione, 3-acetyl-2,5,7,8-tetrahydroxy-		LI05512
264 221 236 249 218 265 60	264	100 44 31 21 18 13 12		12	8	7	–	–	–	–	–	–	–	–	–	–	1,4-Naphthalenedione, 6-acetyl-2,3,5,7-tetrahydroxy-		LI05513
105 77 51 106 73 264 50 91	264	100 37 14 8 7 6 5 4		12	9	1	2	1	–	1	–	–	–	–	–	–	Acetophenone, 2-[(6-chloro-3-pyridazinyl)thio]-	18592-51-3	NI26503
184 264 92 139 185 265 163 186	264	100 86 25 14 13 9 8 6		12	9	3	–	1	–	–	–	–	–	1	–	–	Thiophosphoric acid, O,O′-(2,2′-biphenylylene)-	66300-70-7	NI26504
154 51 152 77 151 153 50 155	264	100 33 20 20 19 12 12 11	5.00	12	10	–	–	–	–	1	–	–	–	–	–	1	Arsine, chlorodiphenyl-	712-48-1	NI26505
177 264 89 88 28 81 178 161	264	100 46 36 16 15 13 12 10		12	12	1	2	2	–	–	–	–	–	–	–	–	4-Thiazolidinone, 5-[[4-(dimethylamino)phenyl]methylene]-2-thioxo-	536-17-4	NI26506
125 109 93 264 64 80 66 81	264	100 96 94 51 50 49 44 43		12	12	3	2	1	–	–	–	–	–	–	–	–	Benzene, 4′-hydroxy-1,1′-sulphonylbis[3-amino-	14571-19-8	NI26507
264 135 77 246 204 43 205 51	264	100 78 66 40 40 39 25 23		12	12	3	2	1	–	–	–	–	–	–	–	–	5-Imidazolidinecarboxylic acid, 4-oxo-3-phenyl-2-thioxo-, ethyl ester		LI05514
117 102 148 76 77 147 104 75	264	100 69 68 66 56 55 51 37	2.00	12	12	3	2	1	–	–	–	–	–	–	–	–	Phthalimide, N-[(2-acetamidoethyl)thio]-		LI05515
59 114 264 77 113 119 118 43	264	100 80 65 62 57 48 45 45		12	12	3	2	1	–	–	–	–	–	–	–	–	1,3-Thiazine-6-carboxylic acid, 4-oxo-2-phenyliminoperhydro-, methyl ester		MS05858
91 106 93 77 69 117 41 95	264	100 85 66 54 49 48 47 40	1.50	12	15	3	–	–	3	–	–	–	–	–	–	–	Spiro[bicyclo[3.2.1]octane-6,1′-cyclopropan]-exo-3-ol, endo-2-(trifluoroacetoxy)-		MS05859
150 108 82 28 95 41 77 69	264	100 57 56 38 30 29 28 26	0.03	12	15	3	–	–	3	–	–	–	–	–	–	–	Tricyclo[4.3.1.0^{3,7}]decan-exo-8-ol, 3-(trifluoroacetoxy)-		MS05860
57 135 41 108 191 73 56 45	264	100 56 38 25 23 20 20 20	0.00	12	21	4	–	–	–	1	–	–	–	–	–	–	Butanedioic acid, chloro-, bis(1-methylpropyl) ester	57983-51-4	NI26508
55 115 116 60 73 199 264 45	264	100 89 83 65 55 54 47 31		12	24	–	–	3	–	–	–	–	–	–	–	–	1,2,4-Trithiolane, 3,5-dipentyl-	73861-41-3	NI26509
45 43 89 44 87 59 58 88	264	100 46 38 32 23 20 20 13	0.00	12	24	6	–	–	–	–	–	–	–	–	–	–	1,4,7,10,13,16-Hexaoxacyclooctadecane	17455-13-9	NI26510
45 43 89 28 44 73 87 59	264	100 46 42 31 28 22 19 18	0.09	12	24	6	–	–	–	–	–	–	–	–	–	–	1,4,7,10,13,16-Hexaoxacyclooctadecane	17455-13-9	NI26511
89 151 96 83 54 67 70 57	264	100 56 53 53 53 42 39 39	0.00	12	24	6	–	–	–	–	–	–	–	–	–	–	3-Bis(hydroxymethyl)-1,5,8,11-tetraoxacyclotridecane		MS05861
101 75 88 45 144 131 73 114	264	100 34 26 21 19 9 9 8	4.00	12	24	6	–	–	–	–	–	–	–	–	–	–	myo-Inositol, 1,2,3,4,5,6-hexa-O-methyl-	2075-22-1	NI26512
249 165 207 121 79 163 250 123	264	100 91 41 35 32 20 18 17	0.50	12	28	4	–	–	–	–	–	–	1	–	–	–	Silicic acid, tetrapropyl ester	682-01-9	NI26513
36 200 228 171 229 111	264	100 42 36 36 21 14	0.00	13	9	2	–	1	–	1	–	–	–	–	–	–	Benzopyrilium chloride, 7-hydroxy-2-(2-thienyl)-		LI05516
228 165 264 266 230 229 265 263	264	100 71 50 33 33 18 17 13		13	10	–	–	–	–	2	–	–	1	–	–	–	9-Silaanthracene, 9,9-dichloro-9,10-dihydro-	32306-76-6	NI26514
264 165 166 266 228 265 230 229	264	100 97 74 67 49 20 17 14		13	10	–	–	–	–	2	–	–	1	–	–	–	9-Silaphenanthrene, 9,9-dichloro-9,10-dihydro-	76470-26-3	NI26515
264 248 229 62 266 263 51 250	264	100 78 65 62 58 53 52 50		13	10	–	2	–	–	2	–	–	–	–	–	–	5H-Dibenzo[c,f][1,2]diazepine, 3,8-dichloro-6,11-dihydro-	955-66-8	NI26516
264 199 265 263 246 200 247 107	264	100 17 14 10 5 5 4 3		13	10	3	–	–	1	–	–	–	–	1	–	–	Phenoxaphosphinic acid, 2-fluoro-8-methyl-		NI26517
123 232 109 139 125 124 218 122	264	100 42 34 14 14 13 12 8	0.00	13	12	2	–	2	–	–	–	–	–	–	–	–	Benzenesulphinothioic acid, 4-methoxy-, O-phenyl ester	56196-21-5	NI26518
139 109 96 95 140 248 124 218	264	100 23 16 13 12 11 11 10	0.85	13	12	2	–	2	–	–	–	–	–	–	–	–	Benzenesulphinothioic acid, 4-methoxy-, S-phenyl ester	26974-29-8	NI26519
139 109 96 95 248 218 124 155	264	100 24 18 14 13 12 12 7	1.00	13	12	2	–	2	–	–	–	–	–	–	–	–	Benzenesulphinothioic acid, 4-methoxy-, S-phenyl ester	26974-29-8	LI05517
134 128 135 43 39 89 63 45	264	100 15 10 10 10 6 6 6	2.00	13	12	4	–	1	–	–	–	–	–	–	–	–	Benzo[b]cyclobuta[d]thiophene-1,2-dicarboxylic acid, 1,2,2a,7b-tetrahydro-1-methyl-	34244-75-2	NI26520
134 128 135 44 40 90 89 46	264	100 15 11 10 10 7 7 6	2.00	13	12	4	–	1	–	–	–	–	–	–	–	–	Benzo[b]cyclobuta[d]thiophene-1,2-dicarboxylic acid, 1,2,2a,7b-tetrahydro-1-methyl-	34244-75-2	LI05518
60 180 222 136 181 57 69 55	264	100 56 12 6 5 5 4 4	2.55	13	12	6	–	–	–	–	–	–	–	–	–	–	2-Propenoic acid, 3-[3,4-bis(acetyloxy)phenyl]-	13788-48-2	MS05862
71 131 69 103 159 153 42 72	264	100 62 36 31 28 26 26 24	23.00	13	13	2	2	–	–	1	–	–	–	–	–	–	4H-1,3-Oxazine-5-carboxaldehyde, 4-(p-chlorophenyl)-2-dimethylamino-	104409-68-9	NI26521
143 44 144 115 186 142 43 42	264	100 20 17 17 13 10 10 8	0.00	13	13	2	2	–	–	1	–	–	–	–	–	–	3H-Pyrido[3,4-b]indolium, 2-(carboxymethyl)-4,9-dihydro-, chloride	37986-34-8	MS05863
167 98 168 182 149 221 264 108	264	100 45 21 17 9 8 5 5		13	16	–	2	2	–	–	–	–	–	–	–	–	Benzothiazole, 2-[(cyclohexylamino)thio]-		IC03442
115 158 43 145 143 88 116 77	264	100 99 90 85 49 46 32 22	0.80	13	16	2	2	1	–	–	–	–	–	–	–	–	Ethanol, 2-(2-acetimidothiazolid-3-yl)-1-phenyl-		IC03443
264 231 236 249 204 163 203 189	264	100 52 50 37 24 19 17 14		13	16	2	2	1	–	–	–	–	–	–	–	–	Quinazoline, 6,7-dimethoxy-4-ethylthio-2-methyl-	86749-93-1	NI26522

MASS TO CHARGE RATIOS								M.W.	INTENSITIES								Parent	C	H	O	N	S	F	Cl	Br	I	Si	P	B	X	COMPOUND NAME	CAS Reg No	No
105	220	207	77	164	51	117	115	**264**	100	48	48	43	27	10	8	8	**0.25**	13	16	2	2	1	–	–	–	–	–	–	–	–	Thiazolidine, 3-acetonyl-2-benzimido-		IC03444
44	174	86	43	247	160	175	42	**264**	100	26	25	23	13	13	10	10	**1.70**	13	16	4	2	–	–	–	–	–	–	–	–	–	1,3-Benzodioxole, 5-(hydroxyiminomethyl)-6-[2-(N-acetyl-N-methylamino)ethyl]-		NI26523
91	118	84	92	155	65	102	90	**264**	100	74	51	37	30	21	20	13	**0.21**	13	16	4	2	–	–	–	–	–	–	–	–	–	L-Glutamine, N^2-(phenylacetyl)-	28047-15-6	NI26524
30	104	105	77	189	106	78	56	**264**	100	72	68	68	32	32	32	28	**0.00**	13	16	4	2	–	–	–	–	–	–	–	–	–	Isooxazolizidine, 1,1-dimethyl-8-nitro-6-phenyl-		LI05519
30	104	105	77	78	189	41	56	**264**	100	72	67	67	35	33	30	29	**0.00**	13	16	4	2	–	–	–	–	–	–	–	–	–	2H-Isoxazolo[2,3-b]isoxazole, 4,4-dimethyl-2-phenyl-3a-nitroperhydro-		LI05520
91	65	92	77	190	118	43	93	**264**	100	78	54	33	26	19	17	13	**6.00**	13	16	4	2	–	–	–	–	–	–	–	–	–	Propanedioic acid, 2-phenylhydrazono-, diethyl ester		NI26525
264	119	147	219	190	163	101		**264**	100	95	72	52	50	31	14			13	16	4	2	–	–	–	–	–	–	–	–	–	1H-Pyrrole-2-carboxylic acid, 3-cyano-1-(2-ethoxycarbonylethyl)-, ethyl ester		LI05521
191	233	161	234	43	105			**264**	100	77	39	37	24	23			**1.39**	13	16	4	2	–	–	–	–	–	–	–	–	–	5-Pyrrolidone-2-carboxylic acid, 2-Hydroxymethyl-3-(3-pyridyl)-, ethyl ester		MS05864
163	136	264	137	164	162	151	179	**264**	100	86	76	48	32	26	26	24		13	17	5	–	–	–	–	–	–	–	–	1	–	Benzeneacetic acid, α,4-dihydroxy-3-methoxy-, butyl boronate		MS05865
120	105	191	222	79	194	149	41	**264**	100	60	44	41	41	32	32	29	**0.00**	13	18	3	3	–	–	–	–	–	–	–	–	–	Bicyclo[3.3.0]oct-1-en-8-acetic acid, 2-ethyl-3-oxo-, methyl ester, trans-		LI05522
83	41	55	43	69	81	95	138	**264**	100	72	58	48	46	44	43	34	**0.88**	13	26	1	–	–	–	1	–	–	–	1	–	–	Phosphonochloridous acid, isopropyl-, 5-methyl-2-isopropylcyclohexyl ester	74630-89-0	NI26526
59	73	117	45	41	31	103	32	**264**	100	50	27	18	16	11	8	8	**0.00**	13	28	5	–	–	–	–	–	–	–	–	–	–	Tetrapropylene glycol methyl ether		DO01056
83	69	55	97	73	103	57	75	**264**	100	61	57	41	35	25	21	20	**0.00**	13	29	1	–	–	–	1	–	–	1	–	–	–	Decanol, 10-chloro-, trimethylsilyl ether	34714-01-7	NI26527
97	125	181	111	165	137	265		**264**	100	14	6	3	2	2	1		**0.00**	13	29	3	–	–	–	–	–	–	–	1	–	–	Phosphonic acid, methyl-, 3,3-dimethylbutyl ester		NI26528
123	97	124	83	165	80	43	69	**264**	100	97	65	44	16	15	14	12	**0.00**	13	29	3	–	–	–	–	–	–	–	1	–	–	Phosphonic acid, methyl-, 3,3-dimethylbutyl ester		MS05866
119	264	65	266	173	91	28	145	**264**	100	77	56	50	35	26	25	24		14	10	1	–	–	–	2	–	–	–	–	–	–	Benzophenone, methyldichloro-		IC03445
227	195	166	228	262	264			**264**	100	99	53	36	8	4				14	15	2	1	–	–	1	–	–	–	–	–	–	Benz[b]indolizine-10-carboxylic acid, 9-chloro-6,7,8,9-tetrahydro-, methyl ester		LI05523
264	91	108	125	249	139	140	238	**264**	100	97	60	59	53	52	49	37		14	16	3	–	1	–	–	–	–	–	–	–	–	Furan, 2-isopropyl-3-[(4-methylphenyl)sulphonyl]-	118356-42-6	DD00796
91	79	81	65	109	77	53	39	**264**	100	95	75	74	56	50	50	50	**0.00**	14	16	3	–	1	–	–	–	–	–	–	–	–	2,5-Hexadienal, 2-methyl-3-[(4-methylphenyl)sulphonyl]-, (E)-	118356-37-9	DD00797
39	41	79	65	91	81	53	109	**264**	100	74	66	58	47	44	37	29	**0.00**	14	16	3	–	1	–	–	–	–	–	–	–	–	2,5-Hexadienal, 5-methyl-3-[(4-methylphenyl)sulphonyl]-, (E)-	118356-36-8	DD00798
222	43	179	131	180	264	163	119	**264**	100	45	35	26	23	17	12	11		14	16	5	–	–	–	–	–	–	–	–	–	–	Benzeneprop-2-enol, 4′-hydroxy-3′-methoxy-, diacetate, (E)-		NI26529
175	176	59	264	163	69	71	177	**264**	100	84	63	31	25	24	23	22		14	16	5	–	–	–	–	–	–	–	–	–	–	2H-1-Benzopyran-2-one, 6-(2,3-dihydroxy-3-methylbutyl)-7-hydroxy-, (S)-	20126-72-1	NI26530
43	208	166	194	264	41	222	57	**264**	100	25	25	13	8	6	5	5		14	16	5	–	–	–	–	–	–	–	–	–	–	4,7-Ethanoisobenzofuran-1,3-dione, 5-(acetyloxy)-3a,4,7,7a-tetrahydro-8,8-dimethyl-	52918-82-8	NI26531
105	45	77	218	46	44	106	146	**264**	100	66	53	24	23	20	17	14	**0.47**	14	16	5	–	–	–	–	–	–	–	–	–	–	Propanedioic acid, benzoyl-, diethyl ester	1087-97-4	NI26532
167	97	160	28	58	135	161	107	**264**	100	51	24	23	18	18	16	13	**4.34**	14	16	5	–	–	–	–	–	–	–	–	–	–	2H,6H-1,4,5-Trioxadicyclopent[a,hi]inden-2-one, 3-ethylidene,3,3a,4a,7,7a,9b-hexahydro-6-methoxy-	93376-10-4	NI26533
185	91	184	129	155	115	143	128	**264**	100	79	74	43	42	41	33	29	**28.30**	14	17	–	–	–	–	–	1	–	–	–	–	–	Bicyclo[2.2.2]octane, 1-bromo-4-phenyl-	714-68-1	NI26534
41	116	123	42	264	148	55	39	**264**	100	72	53	52	49	46	44	44		14	20	1	2	1	–	–	–	–	–	–	–	–	4H-1,3-Thiazine, 5,6-dihydro-2-[p-methoxyphenylamino(imino)]-4,4,6-trimethyl-	53004-45-8	NI26535
264	203	75	73	77	219	265	165	**264**	100	75	57	54	35	30	25	25		14	20	3	–	–	–	–	–	–	1	–	–	–	Cinnamic acid, 3-(trimethylsilyl)oxy-, ethyl ester		MS05867
221	249	75	73	118	264	90	45	**264**	100	91	84	74	54	19	19	18		14	20	3	–	–	–	–	–	–	1	–	–	–	Cinnamic acid, 3-(trimethylsilyl)oxy-, ethyl ester		MS05868
70	44	177	43	145	178	137	89	**264**	100	46	45	29	20	14	13	12	**5.00**	14	20	3	2	–	–	–	–	–	–	–	–	–	1,4-Butanediamine, N-feruloyl-, (E)-	501-13-3	NI26536
249	81	250	183	79	185	264	184	**264**	100	36	21	21	16	12	10	10		14	20	3	2	–	–	–	–	–	–	–	–	–	2,4,6(1H,3H,5H)-Pyrimidinetrione, 5-(1-cyclohexen-1-yl)-1-ethyl-3,5-dimethyl-	55044-42-3	NI26537
235	169	236	178	112	79	121	77	**264**	100	42	16	7	5	5	4	3	**0.00**	14	20	3	2	–	–	–	–	–	–	–	–	–	2,4,6(1H,3H,5H)-Pyrimidinetrione, 5-(1-cyclohexen-1-yl)-5-ethyl-1,3-dimethyl-	891-90-7	NI26538
235	169	236	79	81	178	41	77	**264**	100	29	15	10	9	8	8	7	**0.00**	14	20	3	2	–	–	–	–	–	–	–	–	–	2,4,6(1H,3H,5H)-Pyrimidinetrione, 5-(1-cyclohexen-1-yl)-5-ethyl-1,3-dimethyl-	891-90-7	LI05524
109	42	80	56	27	135	98	93	**264**	100	35	32	32	19	18	16	16	**0.50**	14	20	3	2	–	–	–	–	–	–	–	–	–	Urea, N-(4-hydroxy-2-methylcyclohexyl)-N′-(4-hydroxyphenyl)-	25546-04-7	LI05525
109	43	80	56	57	53	27	41	**264**	100	34	32	31	21	19	19	18	**0.53**	14	20	3	2	–	–	–	–	–	–	–	–	–	Urea, N-(4-hydroxy-2-methylcyclohexyl)-N′-(4-hydroxyphenyl)-	25546-04-7	NI26539
220	73	221	264	147	45	74	58	**264**	100	61	18	17	9	7	5	5		14	21	–	2	–	1	–	–	–	1	–	–	–	1H-Indole-3-methylamine, 5-fluoro-N,N-dimethyl-1-(trimethylsilyl)-		NI26540
236	208	264	237	179	180	265	104	**264**	100	26	25	25	11	9	9	9		15	8	3	2	–	–	–	–	–	–	–	–	–	Dibenzo[de,h][1,6]naphthyridin-6-one, 5,10-dihydroxy-	92631-70-4	NI26541
104	76	264	50	265	132	105	75	**264**	100	61	49	21	9	8	7	6		15	8	3	2	–	–	–	–	–	–	–	–	–	Indolazo[1,2-b]phthalazine, 5,7,13-trioxo-		IC03446
264	187	77	263	188	91	65	92	**264**	100	68	36	23	12	12	12	10		15	12	1	4	–	–	–	–	–	–	–	–	–	1H-Pyrazole-4,5-dione, 1-phenyl-, 4-(phenylhydrazone)	16335-50-5	NI26542
192	264	165	178	193	234	235	166	**264**	100	78	77	63	44	33	31	21		15	12	1	4	–	–	–	–	–	–	–	–	–	1,2,4-Triazine, 3-amino-5,6-diphenyl-, 1-oxide	4512-01-0	NI26543
192	264	165	178	265	193	234	235	**264**	100	78	77	63	57	44	33	31		15	12	1	4	–	–	–	–	–	–	–	–	–	1,2,4-Triazine, 3-amino-5,6-diphenyl-, 1-oxide	4512-01-0	LI05526
247	165	264	178	248	193	166	235	**264**	100	80	71	23	21	20	20	17		15	12	1	4	–	–	–	–	–	–	–	–	–	1,2,4-Triazine, 3-amino-5,6-diphenyl-, 2-oxide	13162-95-3	NI26544
247	165	264	178	248	166	193	176	**264**	100	79	70	23	21	20	19	16		15	12	1	4	–	–	–	–	–	–	–	–	–	1,2,4-Triazine, 3-amino-5,6-diphenyl-, 2-oxide	13162-95-3	LI05527
105	77	264	51	106	103	104	265	**264**	100	60	43	14	13	9	8	7		15	12	1	4	–	–	–	–	–	–	–	–	–	1,2,4-Triazole, 3-benzamido-5-phenyl-		MS05869

MASS TO CHARGE RATIOS	M.W.	INTENSITIES	Parent	C	H	O	N	S	F	Cl	Br	I	Si	P	B	X	COMPOUND NAME	CAS Reg No	No
264 266 194 179 249 214 251 229	**264**	100 65 50 45 33 26 22 22		15	14	–	–	–	–	2	–	–	–	–	–	–	**1,1'-Biphenyl, 2',4'-dichloro-2,4,6-trimethyl-**	75601-31-9	**NI26545**
95 43 67 110 94 79 85 55	**264**	100 99 54 46 38 38 36 34	**11.00**	15	20	2	–	1	–	–	–	–	–	–	–	–	**4-Hexen-1-ol, 4-methyl-6-(phenylthio)-, acetate, (E)-**	67884-56-4	**NI26546**
249 43 193 209 41 264 69 18	**264**	100 94 64 60 33 32 26 20		15	20	4	–	–	–	–	–	–	–	–	–	–	**Acetophenone, 2,6-dimethoxy-4-hydroxy-3-(3-methylbut-2-enyl)-**		**MS05870**
57 208 163 41 79 209 107 95	**264**	100 44 22 22 11 10 10 7	**3.40**	15	20	4	–	–	–	–	–	–	–	–	–	–	**2-Adamantanecarboxylic acid, 4,8-dioxo-, tert-butyl ester**	19305-81-8	**MS05871**
43 201 142 115 91 185 173 57	**264**	100 63 37 35 35 32 32 30	**1.70**	15	20	4	–	–	–	–	–	–	–	–	–	–	**Azuleno[5,6-c]furan-1(3H)-one, 4,4a,5,6,7,9-hexahydro-3,4-dihydroxy-6,6,8-trimethyl-**	62824-37-7	**NI26547**
91 185 228 218 213 191 246 203	**264**	100 29 26 25 20 19 19 19	**4.00**	15	20	4	–	–	–	–	–	–	–	–	–	–	**Azuleno[4,5-b]furan-2(3H)-one, decahydro-6,8-dihydroxy-6-methyl-3,9-bis(methylene)-**	101221-74-3	**NI26548**
166 105 77 107 81 106 99 55	**264**	100 94 58 42 37 36 34 33	**0.00**	15	20	4	–	–	–	–	–	–	–	–	–	–	**Benzeneacetic acid, α-hydroxy-α-(1-hydroxy-1-cyclohexyl)-, methyl ester**		**DD00799**
43 189 186 161 92 264 204 77	**264**	100 96 90 72 38 22 14 14		15	20	4	–	–	–	–	–	–	–	–	–	–	**Benzene, 4-hydroxymethyl-5-(2-hydroxybut-3-en-2-yl)-1-methoxy-2-methyl-, acetate**		**MS05872**
99 83 43 71 166 41 105 121	**264**	100 45 43 39 20 19 16 13	**9.00**	15	20	4	–	–	–	–	–	–	–	–	–	–	**1,3-Cyclohexadiene, 1-formyl-4,6,6-trimethyl-5-[(E)-(hydroxymethyl)-2-butenoyloxy]-**		**LI05528**
99 71 81 249 264 232 204	**264**	100 50 25 7 1 1 1		15	20	4	–	–	–	–	–	–	–	–	–	–	**1,4-Cyclohexadiene, 4-formyl-1,3,3-trimethyl-6-[(E)-3-(hydroxymethyl)-2-butenoyloxy]-**		**LI05529**
99 166 71 43 121 105 151 41	**264**	100 44 36 35 18 17 12 12	**2.00**	15	20	4	–	–	–	–	–	–	–	–	–	–	**1,4-Cyclohexadiene, 4-formyl-1,3,3-trimethyl-6-[(E)-3-(hydroxymethyl)-2-butenoyloxy]-**		**LI05530**
248 233 191 95 249 234 79 41	**264**	100 81 73 24 16 15 15 15	**0.00**	15	20	4	–	–	–	–	–	–	–	–	–	–	**3-Cyclopentene-1,2-dione, 4-hydroxy-5-(1-hydroxy-3-methylbutylidene)-3-(3-methyl-2-butenylidene)-**	38574-25-3	**NI26549**
41 264 43 57 39 55 69 152	**264**	100 50 47 42 42 40 36 30		15	20	4	–	–	–	–	–	–	–	–	–	–	**3,5-Cyclopentenedione, 1-hydroxy-2-isopentanoyl-4-isopentenyl-**		**MS05873**
41 264 43 57 55 69 152 53	**264**	100 52 47 44 39 36 30 21		15	20	4	–	–	–	–	–	–	–	–	–	–	**3,5-Cyclopentenedione, 1-hydroxy-2-isopentanoyl-4-isopentenyl-**		**NI26550**
43 41 264 55 193 39 91 124	**264**	100 34 31 23 19 19 18 15		15	20	4	–	–	–	–	–	–	–	–	–	–	**Eudesma-1,4-dien-12-oic acid, 6a-hydroxy-3-oxo-, γ-lactone, (11S)-**		**MS05874**
55 57 83 246 95 119 213 202	**264**	100 88 84 66 32 24 18 17	**2.00**	15	20	4	–	–	–	–	–	–	–	–	–	–	**Eudesma-3,11(13)-dien-12-oic acid, 1α,6a,9β-trihydroxy-, γ-lactone**		**MS05875**
84 86 59 74 228 246 119 264	**264**	100 66 18 10 9 7 7 6		15	20	4	–	–	–	–	–	–	–	–	–	–	**Eudesma-3,11(13)-dien-12-oic acid, 1β,6a,9α-trihydroxy-, γ-lactone**		**MS05876**
249 203 231 175 264 250 185 221	**264**	100 32 18 18 16 16 12 4		15	20	4	–	–	–	–	–	–	–	–	–	–	**4aH-Eudesman-12-oic acid, 6a-hydroxy-1-oxo-, γ-lactone**		**MS05877**
151 95 69 55 41 173 67 53	**264**	100 40 39 38 34 33 33 31	**25.00**	15	20	4	–	–	–	–	–	–	–	–	–	–	**Guaia-1(10),4-dien-12-oic acid, 8-hydroxy-3-oxo-, γ-lactone**		**MS05878**
43 77 105 79 39 51 41 107	**264**	100 48 47 32 20 19 15 14	**0.00**	15	20	4	–	–	–	–	–	–	–	–	–	–	**6-Heptenoic acid, 2,3-dihydroxy-3-methyl-2-phenyl-, methyl ester**		**DD00800**
132 105 91 119 246 77 218 264	**264**	100 41 40 27 25 20 19 5		15	20	4	–	–	–	–	–	–	–	–	–	–	**3-Marasmenone, 15-hydroxy-**		**MS05879**
220 91 190 164 119 118 264 105	**264**	100 99 91 75 74 74 72 63		15	20	4	–	–	–	–	–	–	–	–	–	–	**15-Marasmenone, 1-α-hydroxy-**		**MS05880**
208 132 105 119 91 69 104 264	**264**	100 61 59 57 53 44 43 40		15	20	4	–	–	–	–	–	–	–	–	–	–	**15-Marasmenone, 3-β-hydroxy-**		**MS05881**
113 41 135 43 80 79 95 91	**264**	100 26 24 20 18 18 15 12	**4.00**	15	20	4	–	–	–	–	–	–	–	–	–	–	**1,4-Methano-3-benzoxepin-5,7(2H,4H)dione, 1,8,9,9a-tetrahydro-10-hydroxy-2,2,9,9a-tetramethyl-**	55924-03-3	**NI26551**
91 44 105 147 131 41 55 119	**264**	100 93 86 73 57 57 55 52	**3.00**	15	20	4	–	–	–	–	–	–	–	–	–	–	**7-Oxabicyclo[4.3.0]nonan-8-one, 5-(2-carboxyethylidene)-4-isopropylidene-9-methyl-**		**LI05531**
190 162 134 111 191 208 91 41	**264**	100 28 23 17 15 13 13 13	**3.94**	15	20	4	–	–	–	–	–	–	–	–	–	–	**2,4-Pentadienoic acid, 5-(1-hydroxy-2,6,6-trimethyl-4-oxo-2-cyclohexen-1-yl)-3-methyl-, [S-(Z,E)]-**	21293-29-8	**MS05882**
160 133 104 91 88 39 115 32	**264**	100 57 42 29 26 25 24 24	**6.00**	15	20	4	–	–	–	–	–	–	–	–	–	–	**Propanedioic acid, (2-phenylethyl)-, diethyl ester**	6628-68-8	**NI26552**
135 208 191 163 133 164 151 264	**264**	100 21 21 17 15 6 6 4		15	20	4	–	–	–	–	–	–	–	–	–	–	**Propanoic acid, 2,3-epoxy-2-methyl-2-(4-methoxyphenyl)-, tert-butyl ester**		**NI26553**
264 163 136 236 180 179 207 208	**264**	100 49 37 36 34 32 30 29		15	20	4	–	–	–	–	–	–	–	–	–	–	**2-Propenoic acid, 3-(3,4-diethoxyphenyl)-, ethyl ester**	75332-49-9	**NI26554**
41 91 39 55 105 43 69 79	**264**	100 94 67 61 60 58 53 45	**1.00**	15	20	4	–	–	–	–	–	–	–	–	–	–	**Schkuhridin b**		**MS05883**
122 138 79 123 77 55 107 53	**264**	100 86 69 58 46 45 45 45	**0.00**	15	20	4	–	–	–	–	–	–	–	–	–	–	**Trichothec-9-en-8-one, 12,13-epoxy-3α-hydroxy-**		**NI26555**
41 79 43 77 122 39 91 53	**264**	100 63 57 51 50 50 46 37	**0.00**	15	20	4	–	–	–	–	–	–	–	–	–	–	**Trichothec-9-en-8-one, 12,13-epoxy-4β-hydroxy-**		**NI26556**
264 93 86 58 57 41 43 77	**264**	100 77 55 43 39 31 30 24		15	24	–	2	1	–	–	–	–	–	–	–	–	**Thiourea, N,N-bis(2-methylpropyl)-N'-phenyl-**	15093-48-8	**NI26557**
75 249 117 73 130 91 129 132	**264**	100 90 87 69 69 66 42 35	**27.00**	15	24	2	–	–	–	–	–	–	1	–	–	–	**Benzenehexanoic acid, trimethylsilyl ester**		**NI26558**
176 265 263 220	**264**	100 32 14 5	**0.00**	15	24	2	2	–	–	–	–	–	–	–	–	–	**Benzoic acid, 4-(butylamino)-, 2-(dimethylamino)ethyl ester**	94-24-6	**NI26559**
136 218 204 137 264 219 122 265	**264**	100 70 28 27 16 15 13 3		15	24	2	2	–	–	–	–	–	–	–	–	–	**Cyclohexane, 1-aci-nitro-2-(cyclohexylimino)-3-(2-propenyl)-**		**MS05884**
152 264 165 166 56 112 113 247	**264**	100 77 58 33 30 27 26 21		15	24	2	2	–	–	–	–	–	–	–	–	–	**7,14-Methano-4H,6H-dipyrido[1,2-a:1',2'-e][1,5]diazocin-4-one, dodecahydro-9-hydroxy-**		**LI05532**
152 153 84 264 236 55 123 154	**264**	100 74 51 35 34 31 26 25		15	24	2	2	–	–	–	–	–	–	–	–	–	**7,14-Methano-2H,6H-dipyrido[1,2-a:1',2'-e][1,5]diazocin-6-one, dodecahydro-9-hydroxy-, [7R-(7α,7aβ,9β,14α,14aα)]-**	2636-61-5	**NI26560**
264 136 134 263 150 98 149 55	**264**	100 59 35 33 32 26 24 21		15	24	2	2	–	–	–	–	–	–	–	–	–	**7,14-Methano-4H,6H-dipyrido[1,2-a:1',2'-e][1,5]diazocin-4-one, dodecahydro-2-hydroxy-, [2S-(2α,7α,7aβ,14α,14aα)]-**	23360-87-4	**NI26561**
138 264 83 97 55 152 110 111	**264**	100 63 63 58 51 50 50 44		15	24	2	2	–	–	–	–	–	–	–	–	–	**2,6-Piperidinedione, 1-[(octahydro-2H-quinolizin-1-yl)methyl]-, (1R-cis)-**	18688-40-9	**NI26562**
138 152 83 264 110 97 111 55	**264**	100 60 54 50 45 44 37 35		15	24	2	2	–	–	–	–	–	–	–	–	–	**2,6-Piperidinedione, 1-[(octahydro-2H-quinolizin-1-yl)methyl]-, (1R-cis)-**	18688-40-9	**LI05533**

MASS TO CHARGE RATIOS	M.W.	INTENSITIES	Parent	C	H	O	N	S	F	Cl	Br	I	Si	P	B	X	COMPOUND NAME	CAS Reg No	No
138 97 83 110 152 264 263 222	264	100 57 57 49 48 35 19 17		15	24	2	2	–	–	–	–	–	–	–	–	–	2,6-Piperidinedione, 1-[(octahydro-2H-quinolizin-1-yl)methyl]-, (1S-trans)-	80374-24-9	NI26563
152 112 55 41 150 245 42 134	264	100 84 84 80 62 57 55 54	30.06	15	24	2	2	–	–	–	–	–	–	–	–	–	Sparteine, 13-epi-hydroxy-2-oxo-		IC03447
265 247 263 264 266 248 152 114	264	100 60 31 29 17 10 6 6		15	24	2	2	–	–	–	–	–	–	–	–	–	Sparteine, 13-epi-hydroxy-2-oxo-		NI26564
98 97 264 126 136 55 41 236	264	100 89 69 35 31 30 30 27		15	24	2	2	–	–	–	–	–	–	–	–	–	Sparteine, 10-hydroxy-13-oxo-		LI05534
247 99 98 264 97 248 55 69	264	100 92 77 39 27 19 19 15		15	24	2	2	–	–	–	–	–	–	–	–	–	Sparteine, 12-hydroxy-13-oxo-		LI05535
265 247 263 152 264 266 114 246	264	100 33 25 25 24 18 18 14		15	24	2	2	–	–	–	–	–	–	–	–	–	Sparteine, 13α-hydroxy-2-oxo-	15358-48-2	NI26566
152 264 246 55 165 112 166 134	264	100 60 52 47 44 34 30 29		15	24	2	2	–	–	–	–	–	–	–	–	–	Sparteine, 13α-hydroxy-2-oxo-	15358-48-2	NI26565
98 97 264 126 136 41 55 236	264	100 89 69 35 32 31 30 28		15	24	2	2	–	–	–	–	–	–	–	–	–	Sparteine, 14α-hydroxy-17-oxo-	18108-62-8	NI26567
264 76 180 208 152 265 104 50	264	100 42 32 28 28 19 18 16		16	8	4	–	–	–	–	–	–	–	–	–	–	[1,1′-Biisobenzofuran]-3,3′-(1H,1′H)-dione	4281-21-4	MS05885
264 234 265 152 150 208 151 132	264	100 82 51 51 47 36 33 31		16	8	4	–	–	–	–	–	–	–	–	–	–	Cycloocta[1,2-b:4,3-b′:5,6-b″:8,7-b‴]tetrafuran		NI26568
217 264 216 218 215 140 51 64	264	100 98 75 29 26 26 26 25		16	12	2	2	–	–	–	–	–	–	–	–	–	Cyclopropanecarbonitrile, 1-(p-nitrophenyl)-2-phenyl-		LI05536
217 264 216 77 220 51 221 115	264	100 98 44 38 38 32 30 24		16	12	2	2	–	–	–	–	–	–	–	–	–	1H-Indole, 3-(2-nitroethenyl)-1-phenyl-	63050-02-2	NI26569
132 264 133 90 105 89 118 104	264	100 86 42 37 19 19 18 16		16	12	2	2	–	–	–	–	–	–	–	–	–	Phthalazino[2,3-b]phthalazine-5,12(7H,14H)-dione	21626-89-1	MS05886
132 264 133 90 89 105 118 77	264	100 63 47 44 27 20 18 17		16	12	2	2	–	–	–	–	–	–	–	–	–	Phthalazino[2,3-b]phthalazine-5,12(7H,14H)-dione	21626-89-1	NI26570
264 130 131 102 77 195 196 51	264	100 51 44 42 41 28 26 18		16	12	2	2	–	–	–	–	–	–	–	–	–	3-Pyrazolin-5-one, 4-benzylidene-3-hydroxy-1-phenyl-	15083-26-8	NI26571
264 234 206 263 221 205 220 235	264	100 57 54 33 31 26 24 24		16	16	–	4	–	–	–	–	–	–	–	–	–	3H-1,3,4-Benzotriazepine, 2-methylamino-3-methyl-5-phenyl-		MS05887
133 41 132 67 39 27 77 79	264	100 96 62 55 30 19 15 12	0.00	16	16	–	4	–	–	–	–	–	–	–	–	–	Dispiro[4.1.4.1]dodecane-6,6,12,12-tetracarbonitrile	74764-30-0	NI26572
264 249 118 77 103 265 76 51	264	100 62 60 53 25 19 14 14		16	16	–	4	–	–	–	–	–	–	–	–	–	Ethylene, 1,2-bis(methylazo)-1,2-diphenyl-		LI05537
235 263 264 39 27 41 77 51	264	100 76 40 30 30 30 27 26		16	16	–	4	–	–	–	–	–	–	–	–	–	Pyrido[1,2-a]diazepin-10-amine, 9-cyano-8-phenyl-2,3,4,5-tetrahydro-		NI26573
264 249 106 263 79 265 77 250	264	100 98 28 27 25 24 21 18		16	16	–	4	–	–	–	–	–	–	–	–	–	Pyrido[1″,2″:1′,2′]imidazo[4′,5′:4,5]imidazo[1,2-a]pyridine, 1,3,8,10-tetramethyl-		MS05888
264 145 131 144 119 265 132 263	264	100 41 32 26 26 19 19 16		16	16	–	4	–	–	–	–	–	–	–	–	–	Quinoxalino[2,3-b]quinoxaline, 5,6-dimethylene-5,5a,6,11,11a,12-hexahydro-		IC03448
118 77 264 119 43 103 51 76	264	100 18 14 9 9 8 6 4		16	16	–	4	–	–	–	–	–	–	–	–	–	s-Tetrazine, 1,2-dihydro-1,2-dimethyl-3,6-diphenyl-		MS05889
264 118 77 265 103 119 132 91	264	100 72 23 18 8 7 5 4		16	16	–	4	–	–	–	–	–	–	–	–	–	s-Tetrazine, 1,4-dihydro-1,4-dimethyl-3,6-diphenyl-		MS05890
264 118 91 265 65 119 117 90	264	100 88 28 20 14 9 8 6		16	16	–	4	–	–	–	–	–	–	–	–	–	s-Tetrazine, 1,2-dihydro-3,6-di-m-tolyl-		MS05891
264 118 91 265 117 65 119 146	264	100 82 22 20 11 11 8 7		16	16	–	4	–	–	–	–	–	–	–	–	–	s-Tetrazine, 1,2-dihydro-3,6-di-m-tolyl-		MS05892
44 45 42 43 91 263 40 30	264	100 88 51 46 36 34 29 27	12.70	16	22	–	2	–	–	–	–	–	–	–	2	–	1,2-Diborane(4)diamine, N,N,N′,N′-tetramethyl-1,2-diphenyl-	19127-90-3	NI26574
147 189 163 264 176 43 57 161	264	100 66 51 36 28 25 20 19		16	24	3	–	–	–	–	–	–	–	–	–	–	Acetic acid, 4-(3-tert-butyl-4-hydroxyphenyl)butyl-		NI26575
249 264 250 103 177 263 233 117	264	100 24 17 13 12 11 11 11		16	24	3	–	–	–	–	–	–	–	–	–	–	Benzoic acid, 3,5-di-tert-butyl-4-hydroxy-, methyl ester	2511-22-0	NI26576
249 221 250 43 219 29 203 91	264	100 18 17 11 8 7 6 5	0.50	16	24	3	–	–	–	–	–	–	–	–	–	–	1,4-Benzopyran-3-carboxylic acid, 5,6,7,8-tetrahydro-2,4,4,7-tetramethyl-, ethyl ester		LI05538
249 221 219 203	264	100 19 9 6	0.00	16	24	3	–	–	–	–	–	–	–	–	–	–	1,4-Benzopyran-3-carboxylic acid, 5,6,7,8-tetrahydro-2,4,4,7-tetramethyl-, ethyl ester		LI05539
83 134 55 166 98 125 79 28	264	100 27 26 19 18 17 13 12	6.00	16	24	3	–	–	–	–	–	–	–	–	–	–	1-Cyclohexene-1-carboxylic acid, 4-(1,5-dimethyl-3-oxohex-4-enyl)-, methyl ester		LI05540
83 55 134 166 98 125 79 29	264	100 28 27 19 18 17 14 13	6.70	16	24	3	–	–	–	–	–	–	–	–	–	–	1-Cyclohexene-1-carboxylic acid, 4-(1,5-dimethyl-3-oxohex-4-enyl)-, methyl ester		MS05893
43 79 55 93 39 29 121 264	264	100 24 21 19 19 17 8 1		16	24	3	–	–	–	–	–	–	–	–	–	–	1,1-Dimethyl-3-acetoxy-4,4-diallylcyclohexane-5-one		LI05541
68 43 264 41 83 249 55 95	264	100 88 70 61 38 37 35 34		16	24	3	–	–	–	–	–	–	–	–	–	–	Naphtho[2,1-b]furan-2,8-dione, decahydro-3a,6,6,9a-tetramethyl-, (3aR,5aS,9aS,9bR)-	29064-99-1	NI26577
264 249 150 135 43 83 95 41	264	100 67 56 54 48 47 43 37		16	24	3	–	–	–	–	–	–	–	–	–	–	Naphtho[2,1-b]furan-2,8-dione, decahydro-3aβ,6,6,9aβ-tetramethyl-, (3aR,5aS,9aS,9bR)-		LI05542
264 249 150 135 43 83 95 41	264	100 71 58 56 50 48 45 38		16	24	3	–	–	–	–	–	–	–	–	–	–	Naphtho[2,1-b]furan-2,8-dione, decahydro-3a,6,6,9a-tetramethyl-, (3aS,5aS,9aS,9bR)-	29065-00-7	NI26578
264 249 150 135 83 95 107 137	264	100 69 58 51 48 43 36 35		16	24	3	–	–	–	–	–	–	–	–	–	–	Naphtho[2,1-b]furan-2,8-dione, decahydro-3a,6,6,9a-tetramethyl-, (3aS,5aS,9aS,9bR)-	29065-00-7	NI26579
68 43 264 41 83 55 95 67	264	100 85 70 60 39 35 33 33		16	24	3	–	–	–	–	–	–	–	–	–	–	Naphtho[2,1-b]furan-2,8-dione, decahydro-3aβ,6,6,9aβ-tetramethyl-, (3aS,5aS,9aS,9bR)-		LI05543
91 107 92 87 74 55 158 41	264	100 28 24 20 15 12 10 10	4.00	16	24	3	–	–	–	–	–	–	–	–	–	–	Octanoic acid, 8-(phenylmethoxy)-, methyl ester	31662-21-2	NI26580
91 127 107 158 67 79 81 65	264	100 22 18 12 10 8 7 7	0.00	16	24	3	–	–	–	–	–	–	–	–	–	–	1-Oxacycloheptane, 7-[2-(benzyloxy)ethyl]-2-(hydroxymethyl)-, (2S,7R)-		DD00801
91 127 107 233 158 67 55 65	264	100 11 10 8 8 8 7 6	0.00	16	24	3	–	–	–	–	–	–	–	–	–	–	1-Oxacycloheptane, 7-[2-(benzyloxy)ethyl]-2-(hydroxymethyl)-, (2S,7S)-		DD00802
96 56 264 168 153 112 152	264	100 37 14 13 13 13 8		16	24	3	–	–	–	–	–	–	–	–	–	–	1-Oxaspiro[5.5]undec-9-ene-4,8-dione, 2,3,5,7,9,11-hexamethyl-		LI05544
123 95 93 79 232 189 179 174	264	100 38 38 38 25 25 25 23	0.00	16	24	3	–	–	–	–	–	–	–	–	–	–	2,4-Pentadienoic acid, 3-methyl-5-(1,2-epoxy-2,6,6-trimethylcyclohexyl)-, methyl ester, (Z,E)-		LI05545

MASS TO CHARGE RATIOS								M.W.	INTENSITIES								Parent	C	H	O	N	S	F	Cl	Br	I	Si	P	B	X	COMPOUND NAME	CAS Reg No	No
73	75	117	41	56	127	135	57	264	100	53	51	45	44	32	30	26	0.00	16	28	1	–	–	–	–	–	–	1	–	–	–	2-Propen-1-ol, (E)-1-(7,7-dimethyl-cis-bicyclo[4.2.0]oct-2-en-3-yl)-3-(trimethylsilyl)-		DD00803
72	58	73	44	71	42	86	70	264	100	12	5	4	3	3	2	2	0.00	16	28	1	2	–	–	–	–	–	–	–	–	–	Benzeneethanamine, 4-[3-(dimethylamino)propoxy]-α,N,N-trimethyl-		MS05894
249	264	221	222	250	41	57	193	264	100	64	52	44	30	27	26	22		16	28	1	2	–	–	–	–	–	–	–	–	–	2(1H)-Pyrazinone, 3,5,6-tri-tert-butyl-	39950-99-7	NI26581
264	266	236	200	100	265	201	118	264	100	33	30	26	19	18	17	12		17	9	1	–	–	–	1	–	–	–	–	–	–	Chlorobenzanthrone		IC03449
264	235	234	202	263	221	203	187	264	100	29	23	14	12	12	11	11		17	12	1	–	1	–	–	–	–	–	–	–	–	5H-Naphtho[1,8-be]thiophen-5-one, 3,4-dihydro-2-phenyl-		LI05546
115	77	264	144	220	51	219	127	264	100	86	74	63	51	47	45	38		17	12	3	–	–	–	–	–	–	–	–	–	–	Carbonic acid, 1-naphthyl phenyl ester	17145-98-1	NI26582
264	235	77	265	67	121	105	236	264	100	87	49	45	39	22	22	20		17	12	3	–	–	–	–	–	–	–	–	–	–	Chromone, 3-benzoyl-2-methyl-		MS05895
171	115	264	114	172	65	39	142	264	100	58	13	11	10	8	8	8		17	12	3	–	–	–	–	–	–	–	–	–	–	2-Naphthalenecarboxylic acid, 3-hydroxy-, phenyl ester	7260-11-9	NI26583
144	115	85	120	121	92	145	43	264	100	32	27	24	13	13	11	10	2.29	17	12	3	–	–	–	–	–	–	–	–	–	–	Salicylic acid, 2-naphthyl ester		DO01057
263	236	194	235	221	208	264	77	264	100	85	66	65	62	54	40	36		17	16	1	2	–	–	–	–	–	–	–	–	–	1,5-Benzodiazocin-2(1H)-one, 3,4-dihydro-1-methyl-6-phenyl-	42717-80-6	NI26584
77	180	264	263	104	181	172	78	264	100	58	24	20	19	12	12	9		17	16	1	2	–	–	–	–	–	–	–	–	–	1H-Imidazole, 4,5-dihydro-5-hydroxy-5-methyl-4-methylene-1,2-diphenyl-		LI05547
157	78	158	264	77	107	79	51	264	100	22	18	17	16	13	12	7		17	16	1	2	–	–	–	–	–	–	–	–	–	1H-Imidazole-1-ethanol, α,2-diphenyl-	58275-49-3	NI26585
158	28	77	157	107	44	79	103	264	100	42	24	20	14	14	12	11	0.10	17	16	1	2	–	–	–	–	–	–	–	–	–	1H-Imidazole-1-ethanol, α,β-diphenyl-, (R*,R*)-	34572-27-5	NI26586
158	157	28	77	43	79	107	105	264	100	30	30	24	20	18	14	14	0.20	17	16	1	2	–	–	–	–	–	–	–	–	–	1H-Imidazole-1-ethanol, α,β-diphenyl-, (R*,S*)-	34600-46-9	NI26587
264	263	77	247	180	144	262	265	264	100	95	78	25	21	20	18	15		17	16	1	2	–	–	–	–	–	–	–	–	–	1H-Imidazole, 4-(hydroxymethyl)-5-methyl-1,2-diphenyl-		LI05548
56	264	165	82	235	166	83	265	264	100	31	13	13	10	9	9	7		17	16	1	2	–	–	–	–	–	–	–	–	–	1H-Imidazol-5(4H)-one, 1,2-dimethyl-4,4-diphenyl-	24133-92-4	NI26588
264	249	165	77	118	263	208	265	264	100	66	51	47	42	38	23	21		17	16	1	2	–	–	–	–	–	–	–	–	–	1H-Imidazol-5(4H)-one, 1,2-dimethyl-5,5-diphenyl-	37927-94-9	NI26589
145	264	117	247	103	77	130	132	264	100	36	36	25	17	14	10	6		17	16	1	2	–	–	–	–	–	–	–	–	–	4H-Isoquino[2,1-e]-1,2,5-oxadiazine, 6,7-dihydro-2-phenyl-		NI26590
161	264	119	132	178	130	133	91	264	100	71	51	43	32	31	29	27		17	16	1	2	–	–	–	–	–	–	–	–	–	1,2,4-Oxadiazole, 5-(2,4-dimethylphenyl)-3-(2-methylphenyl)-		NI26591
158	157	107	130	79	77	159	103	264	100	76	18	15	13	12	12	6	0.00	17	16	1	2	–	–	–	–	–	–	–	–	–	1,3-Oxazolidine, 4-methyl-cis-5-phenyl-2-(4-cyanophenyl)-		MS05896
158	157	130	159	107	77	79	116	264	100	75	14	12	9	9	7	6	0.00	17	16	1	2	–	–	–	–	–	–	–	–	–	1,3-Oxazolidine, 4-methyl-trans-5-phenyl-2-(4-cyanophenyl)-		MS05897
222	119	264	235	223	206	205	90	264	100	60	40	25	20	20	14	11		17	16	1	2	–	–	–	–	–	–	–	–	–	Propyl 2-phenyl-4-quinazolinyl ether		LI05549
144	187	77	264	145	91	104	103	264	100	77	48	19	19	12	11	11		17	16	1	2	–	–	–	–	–	–	–	–	–	2(1H)-Pyrimidinone, 3,4-dihydro-3,4-diphenyl-6-methyl-		LI05550
263	118	264	91	128	130	119	261	264	100	81	74	73	71	64	43	38		17	16	1	2	–	–	–	–	–	–	–	–	–	2(1H)-Pyrimidinone, 3,4-dihydro-3-phenyl-6-tolyl-		LI05551
264	247	221	263	77	248	76	232	264	100	89	73	67	55	39	37	36		17	16	1	2	–	–	–	–	–	–	–	–	–	Quinoxaline, 2-isopropyl-3-phenyl-, 4-oxide	16007-79-7	NI26592
43	189	107	93	161	105	204	135	264	100	86	86	75	71	71	58	53	0.00	17	28	2	–	–	–	–	–	–	–	–	–	–	5-Azulenemethanol, 1,2,3,4,5,6,7,8-octahydro-α,α,3,8-tetramethyl-, acetate, [3S-(3α,5α,8α)]-	134-28-1	NI26593
43	161	93	81	121	41	107	105	264	100	62	42	41	37	35	30	30	2.00	17	28	2	–	–	–	–	–	–	–	–	–	–	Cyclohexanol, 2-isopropenyl-6-isopropyl-3-methyl-3-vinyl-, acetate, [1R-(1α,2α,3β,6α)]-	69350-62-5	NI26594
43	93	161	81	41	123	204	69	264	100	58	57	49	36	33	32	32	1.00	17	28	2	–	–	–	–	–	–	–	–	–	–	Cyclohexanol, 2-isopropenyl-6-isopropyl-3-methyl-3-vinyl-, acetate, [1R-(1α,2β,3α,6α)]-	69350-63-6	NI26595
43	204	81	41	80	95	123	148	264	100	88	74	73	67	66	62	61	1.01	17	28	2	–	–	–	–	–	–	–	–	–	–	1H-Cyclopenta[a]pentalen-7-ol, decahydro-3,3,4,7a-tetramethyl-, acetate	57984-03-9	NI26596
69	41	81	128	121	141	82	43	264	100	61	47	42	22	21	21	18	5.40	17	28	2	–	–	–	–	–	–	–	–	–	–	2,6,10-Dodecatrienoic acid, 3,7,11-trimethyl-, ethyl ester, (E,E)-	19954-66-6	NI26597
69	81	128	41	121	82	43	55	264	100	47	42	30	23	22	18	14	5.50	17	28	2	–	–	–	–	–	–	–	–	–	–	2,6,10-Dodecatrienoic acid, 3,7,11-trimethyl-, ethyl ester, (E,E)-	19954-66-6	LI05552
69	81	43	41	128	121	67	55	264	100	47	43	43	28	27	22	22	5.00	17	28	2	–	–	–	–	–	–	–	–	–	–	2,6,10-Dodecatrienoic acid, 3,7,11-trimethyl-, ethyl ester, (E,E)-	19954-66-6	LI05553
69	81	41	43	128	121	67	54	264	100	47	43	42	28	27	22	22	5.20	17	28	2	–	–	–	–	–	–	–	–	–	–	2,6,10-Dodecatrienoic acid, 3,7,11-trimethyl-, ethyl ester, (Z,Z)-	20085-73-8	NI26598
69	43	41	93	81	68	107	80	264	100	43	39	31	25	24	14	12	2.70	17	28	2	–	–	–	–	–	–	–	–	–	–	2,6,10-Dodecatrien-1-ol, 3,7,11-trimethyl-, acetate, (E,E)-	4128-17-0	WI01318
41	69	93	28	43	55	79	81	264	100	91	72	67	47	40	32	26	0.00	17	28	2	–	–	–	–	–	–	–	–	–	–	2,6,10-Dodecatrien-1-ol, 3,7,11-trimethyl-, acetate, (E,E)-	4128-17-0	LI05554
43	41	204	161	105	107	91	189	264	100	69	67	67	64	58	53	52	1.30	17	28	2	–	–	–	–	–	–	–	–	–	–	Guai-1(5)-en-11-ol, acetate		IC03450
67	79	41	80	81	93	55	91	264	100	91	70	65	62	58	50	48	9.04	17	28	2	–	–	–	–	–	–	–	–	–	–	4,7,10-Hexadecatrienoic acid, methyl ester	17364-31-7	NI26599
79	41	67	93	55	95	80	91	264	100	72	68	51	50	40	36	34	22.04	17	28	2	–	–	–	–	–	–	–	–	–	–	7,10,13-Hexadecatrienoic acid, methyl ester	56554-30-4	NI26600
67	79	41	80	93	55	81	91	264	100	92	88	58	57	56	46	37	11.14	17	28	2	–	–	–	–	–	–	–	–	–	–	Hexadecatrienoic acid, methyl ester	37822-81-4	NI26601
43	161	109	204	189	107	93	41	264	100	77	77	64	47	43	37	24	1.00	17	28	2	–	–	–	–	–	–	–	–	–	–	1-Naphthalenol, decahydro-4a-methyl-8-methylene-2-isopropyl-, acetate, [1S-(1α,2β,4aα,8aα)]-	69297-57-0	NI26602
85	126	43	95	67	55	107	189	264	100	91	87	47	23	23	19	13	0.00	17	28	2	–	–	–	–	–	–	–	–	–	–	1-Pentenol, 3-methyl-5-(2,6,6-trimethyl-1-cyclohexen-1-yl)-, acetate, (E)-		DD00804
123	43	84	95	67	55	107	148	264	100	97	92	57	32	30	27	23	0.00	17	28	2	–	–	–	–	–	–	–	–	–	–	1-Pentenol, 3-methyl-5-(2,6,6-trimethyl-1-cyclohexen-1-yl)-, acetate, (Z)-		DD00805
249	41	119	55	69	153	264	43	264	100	81	73	61	57	56	52	48		17	28	2	–	–	–	–	–	–	–	–	–	–	Propanoic acid, 2,5,5-trimethyldecalin-1-enyl-		LI05555
249	191	55	264	69	95	41	119	264	100	79	57	56	53	50	50	47		17	28	2	–	–	–	–	–	–	–	–	–	–	Propanoic acid, 5,5,9-trimethyldecalin-1-enyl-		LI05556
69	81	121	41	93	68	137	85	264	100	38	36	34	14	14	13	11	5.79	17	28	2	–	–	–	–	–	–	–	–	–	–	3,7,11-Tridecatrienoic acid, 4,8,12-trimethyl-, methyl ester, (E,E)-	36237-69-1	NI26603
69	81	121	68	67	93	39	55	264	100	40	29	18	17	15	15	14	9.52	17	28	2	–	–	–	–	–	–	–	–	–	–	3,7,11-Tridecatrienoic acid, 4,8,12-trimethyl-, methyl ester, (Z,E)-	36237-70-4	NI26604
96	111	41	264	110	126	55	69	264	100	79	68	53	47	43	42	40		17	28	2	–	–	–	–	–	–	–	–	–	–	14,15,16-Trinor-8βH-labdan-13-oic acid, 8-hydroxy-, δ-lactone	2466-24-2	NI26605
96	111	41	264	110	126	55	191	264	100	77	66	52	46	42	42	40		17	28	2	–	–	–	–	–	–	–	–	–	–	14,15,16-Trinor-8βH-labdan-13-oic acid, 8-hydroxy-, δ-lactone	2466-24-2	LI05557
96	111	110	264	126	191	221	207	264	100	86	56	54	42	40	38	22		17	28	2	–	–	–	–	–	–	–	–	–	–	14,15,16-Trinor-8βH-labdan-13-oic acid, 8-hydroxy-, δ-lactone	2466-24-2	NI26606
192	43	136	41	191	82	67	249	264	100	63	60	57	53	53	48	47	18.02	17	28	2	–	–	–	–	–	–	–	–	–	–	14,15,16-Trinorlabdan-13-oic acid, 8-hydroxy-, δ-lactone	468-84-8	NI26607
192	43	136	41	82	191	67	69	264	100	63	62	57	55	53	48	47	20.00	17	28	2	–	–	–	–	–	–	–	–	–	–	14,15,16-Trinorlabdan-13-oic acid, 8-hydroxy-, δ-lactone	468-84-8	LI05558

MASS TO CHARGE RATIOS	M.W.	INTENSITIES	Parent	C	H	O	N	S	F	Cl	Br	I	Si	P	B	X	COMPOUND NAME	CAS Reg No	No
249 57 264 41 193 29 250 265	264	100 86 70 34 32 19 17 13		17	29	–	–	–	–	–	–	–	–	1	–	–	Phosphorin, 2,4,6-tri-tert-butyl-	17420-29-0	NI26608
264 266 228 265 226 152 227 202	264	100 33 24 20 16 10 9 8		18	13	–	–	–	–	1	–	–	–	–	–	–	1,1′:4′,1′′-Terphenyl, 4-chloro-	1762-83-0	NI26609
264 249 234 235 265 248 117 247	264	100 66 50 28 21 20 20 18		18	16	–	–	1	–	–	–	–	–	–	–	–	7,8-Dimethyl-9,10-dihydronaphtho[2,3-b]benzothiophene		AI01871
264 249 234 265 117 248 247 250	264	100 72 38 20 17 15 15 14		18	16	–	–	1	–	–	–	–	–	–	–	–	9,10-Dimethyl-6,7-dihydro-11-thiabenzo[b]fluorene		AI01872
249 221 264 250 193 165 152 76	264	100 27 21 19 12 8 8 8		18	16	2	–	–	–	–	–	–	–	–	–	–	9,10-Anthracenedione, 2-tert-butyl-	84-47-9	NI26610
264 263 132 237 121 221 165 247	264	100 17 6 5 5 4 3 2		18	16	2	–	–	–	–	–	–	–	–	–	–	Benzofuran, 2-(4-hydroxyphenyl)-3-methyl-5-propenyl-, (E)-		MS05898
105 264 249 205 77 203 202 189	264	100 75 72 45 30 27 26 20		18	16	2	–	–	–	–	–	–	–	–	–	–	2,3-Butadienoic acid, 2-methyl-4,4-diphenyl-, methyl ester		DD00806
264 205 129 219 204 191 165 91	264	100 90 37 36 28 28 27 27		18	16	2	–	–	–	–	–	–	–	–	–	–	2(3H)-Furanone, dihydro-3-(2,2-diphenylvinyl)-		DD00807
205 191 91 203 204 264 202 219	264	100 52 47 46 43 32 32 21		18	16	2	–	–	–	–	–	–	–	–	–	–	Propanoic acid, 3-(1,2:4,5-dibenzo-1,4-cycloheptadien-11-ylidene)-		MS05899
191 208 264 189 190 179 57 29	264	100 76 48 48 31 30 25 19		18	16	2	–	–	–	–	–	–	–	–	–	–	Propanoic acid, 9-phenanthrenemethyl ester	92174-33-9	NI26611
107 106 132 149 91 43 65 264	264	100 83 80 46 39 29 17 13		18	20	–	2	–	–	–	–	–	–	–	–	–	2,3-Butanedionylidenebis(p-toluidine)	19215-50-0	NI26612
44 45 264 57 43 41 208 222	264	100 21 19 15 15 10 9 8		18	20	–	2	–	–	–	–	–	–	–	–	–	2,5-Ethano-2H-azocino[4,3-b]indole, 4-ethylidene-1,3,4,5,6,7-hexahydro-6-methylene-, [R-(E)]-	3463-93-2	NI26613
264 208 222 221 44 235 41 57	264	100 61 50 39 36 33 31 29		18	20	–	2	–	–	–	–	–	–	–	–	–	2,5-Ethano-2H-azocino[4,3-b]indole, 4-ethylidene-1,3,4,5,6,7-hexahydro-6-methylene-, [R-(E)]-	3463-93-2	NI26614
249 54 137 177 192 193 250 149	264	100 55 43 30 27 18 16 16	0.00	18	32	1	–	–	–	–	–	–	–	–	–	–	1H-Naphtho[2,1-b]pyran, dodecahydro-3,4a,7,7,10a-pentamethyl-	6252-26-2	NI26615
41 67 29 55 43 81 27 54	264	100 70 56 53 39 38 38 37	4.00	18	32	1	–	–	–	–	–	–	–	–	–	–	9,12-Octadecadienal	26537-70-2	NI26616
41 43 68 82 96 57 110 264	264	100 39 24 18 12 11 7 4		18	32	1	–	–	–	–	–	–	–	–	–	–	9,12-Octadecadienal	26537-70-2	LI05559
67 81 41 55 95 54 68 82	264	100 72 65 58 46 46 42 37	7.74	18	32	1	–	–	–	–	–	–	–	–	–	–	9,17-Octadecadienal, (Z)-	56554-35-9	NI26617
43 80 67 79 41 45 55 81	264	100 84 82 75 74 67 49 46	9.64	18	32	1	–	–	–	–	–	–	–	–	–	–	6,9,12-Octadecatrien-1-ol	56630-94-5	NI26618
79 67 41 80 55 108 95 93	264	100 76 63 61 58 53 51 51	9.51	18	32	1	–	–	–	–	–	–	–	–	–	–	9,12,15-Octadecatrien-1-ol	2774-90-5	WI01319
80 79 94 81 55 41 67 93	264	100 56 37 32 31 27 26 24	0.14	18	32	1	–	–	–	–	–	–	–	–	–	–	17-Octadecen-14-yn-1-ol	18202-28-3	NI26619
43 69 41 55 221 53 80 67	264	100 91 84 40 30 20 19 18	7.01	18	32	1	–	–	–	–	–	–	–	–	–	–	9,13-Pentadecadien-2-one, 6,10,14-trimethyl-, (E)-	31138-32-6	NI26620
117 264 115 119 91 145 43 65	264	100 79 78 64 54 51 27 17		19	20	1	–	–	–	–	–	–	–	–	–	–	1-Buten-3-one, 2-(2-tolyl)-4-(2,6-xylyl)-		LI05560
193 206 180 165 115 167 264 178	264	100 99 83 49 48 43 36 35		19	20	1	–	–	–	–	–	–	–	–	–	–	Cyclohexanone, 6-methyl-2,2-diphenyl-	50592-52-4	NI26621
264 173 133 120 144 265 105 174	264	100 53 43 35 32 23 18 11		19	20	1	–	–	–	–	–	–	–	–	–	–	6-Heptenophenone, 7-phenyl-	32363-57-8	NI26622
105 133 264 120 77 117 144 129	264	100 75 67 67 45 35 30 27		19	20	1	–	–	–	–	–	–	–	–	–	–	6-Heptenophenone, 7-phenyl-	32363-57-8	NI26623
81 95 67 55 43 109 123 69	264	100 94 92 91 77 71 66 53	5.01	19	36	–	–	–	–	–	–	–	–	–	–	–	Bicyclo[4.3.0]nonane, 3-butyl-4-hexyl-		AI01873
95 41 81 55 67 29 109 179	264	100 82 81 74 65 53 51 48	20.89	19	36	–	–	–	–	–	–	–	–	–	–	–	Bicyclo[4.3.0]nonane, 8-butyl-3-hexyl-		AI01874
41 95 81 55 43 67 109 123	264	100 99 81 77 68 63 50 46	19.73	19	36	–	–	–	–	–	–	–	–	–	–	–	Bicyclo[4.3.0]nonane, 8-butyl-3-hexyl-		AI01875
96 180 55 41 81 67 95 97	264	100 94 86 84 69 65 52 51	4.71	19	36	–	–	–	–	–	–	–	–	–	–	–	Bicyclo[4.3.0]nonane, δ-butyl-7-hexyl-		AI01876
96 180 67 81 95 97 69 82	264	100 91 74 71 57 44 43 36	3.95	19	36	–	–	–	–	–	–	–	–	–	–	–	Bicyclo[4.3.0]nonane, δ-butyl-7-hexyl-		AI01877
123 81 41 67 43 55 95 122	264	100 83 55 50 38 35 29 24	9.01	19	36	–	–	–	–	–	–	–	–	–	–	–	Bicyclo[4.3.0]nonane, 3-decyl-		AI01878
123 81 67 41 55 122 95 43	264	100 79 46 40 27 26 26 26	10.30	19	36	–	–	–	–	–	–	–	–	–	–	–	Bicyclo[4.3.0]nonane, 3-decyl-		AI01879
81 96 123 41 67 55 43 97	264	100 82 80 80 72 59 52 36	16.75	19	36	–	–	–	–	–	–	–	–	–	–	–	Bicyclo[4.3.0]nonane, 8-decyl-		AI01880
81 43 41 95 68 55 67 82	264	100 65 61 59 55 53 45 40	7.00	19	36	–	–	–	–	–	–	–	–	–	–	–	1,2-Dioctylcyclopropene	1089-40-3	NI26624
55 83 41 180 69 82 96 67	264	100 96 83 45 44 42 36 35	0.48	19	36	–	–	–	–	–	–	–	–	–	–	–	Heptane, 1,1-dicyclohexyl-		AI01881
264 207 193 265 191 168 208 178	264	100 39 23 21 20 19 18 17		20	24	–	–	–	–	–	–	–	–	–	–	–	Chrysene, 1,2,3,4,4a,11,12,12a-octahydro-7,12a-dimethyl-	75758-27-9	NI26625
249 264 234 117 132 28 144 91	264	100 53 41 35 28 27 22 18		20	24	–	–	–	–	–	–	–	–	–	–	–	Dibenzo[a,e]cyclooctatetraene, 5,6,11,12-tetrahydro-2,3,8,9-tetramethyl-	27742-96-7	DD00808
249 157 264 251 117 105 142 115	264	100 35 24 23 19 18 15 11		20	24	–	–	–	–	–	–	–	–	–	–	–	1H-Indene, 2,3-dihydro-1,1,3,5-tetramethyl-3-(4-tolyl)-		AI01882
145 159 264 104 160 91 117 105	264	100 76 56 55 35 27 26 23		20	24	–	–	–	–	–	–	–	–	–	–	–	[2.2]Paracyclophane, butyl-		LI05561
264 117 134 119 105 131 132 265	264	100 49 33 29 29 27 26 24		20	24	–	–	–	–	–	–	–	–	–	–	–	[3.3]Paracyclophane, 5-ethyl-		LI05562
249 234 264 219 221 203 206 191	264	100 69 48 19 13 9 8 6		20	24	–	–	–	–	–	–	–	–	–	–	–	Phenanthrene, 9,10-dihydro-3,6,9,9,10,10-hexamethyl-		LI05563
69 92 47 180 85 135 114 31	265	100 48 33 28 23 17 15 13	0.00	4	–	1	1	–	8	1	–	–	–	–	–	–	Propylamine, 4-chloro-2-formyloctafluoro-		LI05564
166 113 78 85 27 28 93 112	265	100 83 70 48 26 22 18 7	0.00	4	4	–	1	–	9	–	–	–	1	–	–	–	Ethylamine, N,N-bis(trifluoromethyl)-2-(trifluorosilyl)-	28871-59-2	NI26626
64 256 128 102 160 57 96 43	265	100 66 61 57 48 38 28 25	0.00	4	6	–	2	4	–	–	–	–	–	–	–	1	Manganese, [[1,2-ethanediylbis[carbamodithioato]](2-)]-	12427-38-2	NI26627
250 252 230 265 267 232 254 180	265	100 63 47 25 15 15 13 10		4	12	–	3	–	–	2	–	–	–	3	–	–	1,1,3,3-Tetramethyl-5,5-dichlorocyclotriphosphazene	86748-46-1	NI26628
204 251 267 202 94 141 143 253	265	100 94 90 86 76 72 68 64	56.00	5	–	1	1	–	–	5	–	–	–	–	–	–	Pyridine, pentachloro-, N-oxide		MS05900
85 265 113 180 208 246 153	265	100 93 42 37 12 8 3		8	6	2	3	1	3	–	–	–	–	–	–	–	4,6-Dimethyl-2-perfluoromethyl-4,5,6,8-tetrahydro-5,7-dioxothiazolo[4,5-d]pyrimidine		LI05565
183 185 76 50 40 75 155 157	265	100 95 45 44 40 39 35 31	0.00	10	8	1	3	–	–	–	1	–	–	–	–	–	1-(4-Bromophenacyl)-1,2,4-triazole		NI26629
183 219 156 184 157 41 221 164	265	100 82 79 61 53 50 24 21	1.00	10	8	2	5	–	–	1	–	–	–	–	–	–	1-Cyclopentene-1,5,5-trinitrile, 2-amino-3-chloro-4-ethyl-3-nitro-		MS05901
207 67 81 265 150 72 56 58	265	100 72 44 39 31 27 23 20		10	11	2	5	1	–	–	–	–	–	–	–	–	7-Methyl-8-isothiocyanatomethyltheophylline		LI05566

MASS TO CHARGE RATIOS								M.W.	INTENSITIES								Parent	C	H	O	N	S	F	Cl	Br	I	Si	P	B	X	COMPOUND NAME	CAS Reg No	No
151	265	176	152	164	124	43	69	**265**	100	81	20	20	16	16	16	13		10	11	4	5	–	–	–	–	–	–	–	–	–	Furo[2′,3′:4,5]oxazolo[3,2-e]purine-8-methanol, 4-amino-6a,7,8,9a-tetrahydro-7-hydroxy-, [6aS-(6aα,7α,8β,9aα)]-	13089-44-6	NI26630
105	43	166	85	183	100	250	167	**265**	100	63	47	29	28	25	24	11	**0.00**	10	14	5	–	–	–	–	–	–	–	–	–	1	Vanadium, bis(1-methyl-2-acetoxyvinyl)-		IC03451
265	166	250	183	43	167	67	223	**265**	100	96	32	32	32	22	22	11		10	14	5	–	–	–	–	–	–	–	–	–	1	Vanadium, oxobis(2,4-pentanedionato-O,O′)-	3153-26-2	NI26631
265	248	244	216	206	123	221	245	**265**	100	88	71	55	54	39	38	37		11	8	1	1	–	5	–	–	–	–	–	–	–	2-Propenal, 2-(pentafluorophenyl)-3-(dimethylamino)-		NI26632
186	131	265	63	187	51	77	158	**265**	100	41	24	14	12	12	10	8		11	11	3	3	1	–	–	–	–	–	–	–	–	1-(5-Methylsulphonyl-2-furyl)-2-(6-amino-3-pyridazyl)ethylene		MS05902
136	135	162	108	248	81	137	119	**265**	100	49	23	18	10	10	9	7	**4.00**	11	15	3	5	–	–	–	–	–	–	–	–	–	Adenine, 9-[β-(2α,3α-dihydroxy-4β-hydroxymethylcyclopentyl)]-		LI05567
81	84	82	110	181	136	83	95	**265**	100	99	98	96	94	92	92	91	**7.00**	11	15	3	5	–	–	–	–	–	–	–	–	–	Histidylamide, [(2-oxopyrrolidin-5-yl)oxy]-		NI26633
73	128	115	75	43	86	88	58	**265**	100	75	71	27	23	19	17	12	**0.20**	11	20	5	1	–	1	–	–	–	–	–	–	–	α-D-Glucopyranoside, methyl 2-(acetylamino)-2,6-dideoxy-6-fluoro-3,4-di-O-methyl-	41341-97-3	NI26634
73	115	128	43	75	86	58	88	**265**	100	55	48	26	20	16	14	12	**0.09**	11	20	5	1	–	1	–	–	–	–	–	–	–	α-D-Glucopyranoside, methyl 2-(acetylamino)-2,6-dideoxy-6-fluoro-3,4-di-O-methyl-	41341-97-3	NI26635
133	69	232	265	165	233	68	67	**265**	100	56	44	36	11	6	5	3		11	24	2	1	1	–	–	–	–	–	1	–	–	2-(Hexylamino)-5,5-dimethyl-1,3,2-dioxaphosphorinane 2-sulphide	884-92-4	NI26636
39	221	220	84	265	223	222	82	**265**	100	91	80	70	62	46	38	31		12	8	2	1	1	–	1	–	–	–	–	–	–	2-Propenoic acid, 3-[2-(4-chlorophenyl)-4-thiazolyl]-	21278-79-5	NI26637
166	205	234	59	265	203	147	174	**265**	100	85	75	72	63	42	42	40		12	11	6	1	–	–	–	–	–	–	–	–	–	Propanedioic acid, [(4-nitrophenyl)methylene]-, dimethyl ester	38323-22-7	NI26638
43	162	204	29	77	27	118	93	**265**	100	70	54	36	33	30	24	22	**0.93**	12	15	2	3	1	–	–	–	–	–	–	–	–	Carbamothioic acid, [(acetylamino)(phenylamino)methylene]-, S-ethyl ester	74793-51-4	NI26639
186	68	264	53	119	265	79	92	**265**	100	76	55	51	43	41	25	17		12	15	2	3	1	–	–	–	–	–	–	–	–	5H-Cyclopenta[d]pyridazine, 1-[(2,3-butadienyl)amino]-6,7-dihydro-4-(methylsulphonyl)-		MS05903
265	186	213	119	226	65	79	92	**265**	100	95	70	60	51	29	17	14		12	15	2	3	1	–	–	–	–	–	–	–	–	5H-Cyclopenta[d]pyridazine, 1-[(3-butynyl)amino]-6,7-dihydro-4-(methylsulphonyl)-		MS05904
234	265	221	190	235	266	222	218	**265**	100	79	62	26	17	11	9	7		12	15	4	3	–	–	–	–	–	–	–	–	–	1(2H)-Phthalazinone, 4-[(2-hydroxyethyl)amino]-6,7-dimethoxy-	71932-68-8	NI26640
191	265	220	219	202	193	237	53	**265**	100	78	52	47	38	35	32	31		12	15	4	3	–	–	–	–	–	–	–	–	–	4H-Pyrido[1,2-a]pyrimidine-3-carboxylic acid, 6,7,8,9-tetrahydro-9-(hydroxyimino)-6-methyl-4-oxo-, ethyl ester	64399-30-0	NI26641
125	111	75	127	90	249	113	63	**265**	100	86	57	44	35	25	25	25	**13.00**	13	9	1	1	–	–	2	–	–	–	–	–	–	Aniline, 4-chloro-N-(4-chlorobenzylidene)-, N-oxide		LI05568
125	111	75	127	90	139	249	113	**265**	100	86	57	44	35	26	25	25	**13.00**	13	9	1	1	–	–	2	–	–	–	–	–	–	Aniline, 4-chloro-N-(4-chlorobenzylidene)-, N-oxide		MS05905
139	111	141	75	265	267	113	50	**265**	100	43	35	25	21	15	12	11		13	9	1	1	–	–	2	–	–	–	–	–	–	Benzamide, 4-chloro-N-(4-chlorophenyl)-		MS05906
230	265	267	232	111	139	75	154	**265**	100	52	34	34	33	32	27	26		13	9	1	1	–	–	2	–	–	–	–	–	–	Benzophenone, 2-amino-2′,5′-dichloro-	2958-36-3	NI26642
265	77	235	188	28	44	266	250	**265**	100	83	52	47	29	15	10	8		13	11	–	7	–	–	–	–	–	–	–	–	–	1H-Pyrazolo[4,3-e][1,2,4]triazolo[4,3-a][1,2,4]triazine, 3,8-dimethyl-1-phenyl-	95571-09-8	NI26643
43	265	118	250	70	216	230	44	**265**	100	35	33	24	22	19	16	16		13	12	1	1	1	–	1	–	–	–	–	–	–	Thiazole, 5-acetyl-4-(chloromethyl)-2-(4-methylphenyl)-	41991-34-8	NI26644
222	224	265	267	144	185	186	187	**265**	100	97	62	58	54	30	24	20		13	16	–	1	–	–	–	1	–	–	–	–	–	Carbazole, 1,2,3,4,4a,9a-hexahydro-6-bromo-9-methyl-, trans-		LI05569
82	111	236	156	128	129	219	77	**265**	100	88	86	74	73	60	58	57	**56.16**	13	16	3	1	–	–	–	–	–	–	1	–	–	3-Phenylpropenenitrile, 2-(diethoxyphosphinyl)-	18896-73-6	NI26645
73	91	116	75	250	74	89	45	**265**	100	17	16	11	10	10	8	8	**6.80**	13	19	3	1	–	–	–	–	–	1	–	–	–	Benzenepropanoic acid, α-(methoxyimino)-, trimethylsilyl ester	55494-13-8	NI26646
73	116	91	89	45	75	74	65	**265**	100	99	99	56	51	47	40	33	**19.00**	13	19	3	1	–	–	–	–	–	1	–	–	–	Benzenepropanoic acid, α-(methoxyimino)-, trimethylsilyl ester	55494-13-8	NI26647
73	116	91	89	45	75	74	65	**265**	100	25	25	14	13	12	10	8	**4.75**	13	19	3	1	–	–	–	–	–	1	–	–	–	Benzenepropanoic acid, α-(methoxyimino)-, trimethylsilyl ester	55494-13-8	NI26648
73	91	75	221	250	30	118	105	**265**	100	64	26	19	19	16	14	14	**13.21**	13	19	3	1	–	–	–	–	–	1	–	–	–	Glycine, N-phenylacetyl-, trimethylsilyl ester		NI26649
188	88	132	32	56	42	44	74	**265**	100	64	24	22	14	13	11	10	**0.50**	13	19	3	1	–	–	–	–	–	1	–	–	–	2,9,10-Trioxa-6-aza-1-silatricyclo[4.3.3.$0^{1,6}$]dodecane, 1-phenyl-		MS05907
91	42	172	265	40	133	65	132	**265**	100	63	57	29	26	26	13	12		14	19	–	1	2	–	–	–	–	–	–	–	–	1-Benzyl-4-(1,3-dithiolan-2-yl)piperidine		NI26650
55	83	118	41	91	200	77	159	**265**	100	91	65	65	63	45	43	37	**0.00**	14	19	2	1	1	–	–	–	–	–	–	–	–	1,2,3-Oxathiazolidine, 3-cyclohexyl-5-phenyl-, 2-oxide	38584-00-8	MS05908
41	57	44	252	267	43	56	55	**265**	100	85	81	65	33	32	23	23	**0.00**	14	19	4	1	–	–	–	–	–	–	–	–	–	1,2-Benzoquinone, 3,5-di-tert-butyl-6-nitro-		MS05909
43	58	146	44	56	265	42	77	**265**	100	27	24	21	20	14	14	8		14	19	4	1	–	–	–	–	–	–	–	–	–	1,3-Cyclopentadiene-1,3-dicarboxylic acid, 5-[1-(dimethylamino)ethylidene]-2-methyl-, dimethyl ester	14469-80-8	NI26651
91	108	42	107	79	57	40	28	**265**	100	71	46	41	40	24	23	23	**4.28**	14	19	4	1	–	–	–	–	–	–	–	–	–	DL-Leucine, N-[(phenylmethoxy)carbonyl]-	3588-60-1	NI26652
105	77	163	122	28	233	106	58	**265**	100	24	19	17	14	9	7	7	**0.00**	14	19	4	1	–	–	–	–	–	–	–	–	–	Methyl N-benzoyl-α-daunosaminide		NI26653
91	57	148	59	40	92	29	120	**265**	100	96	55	46	23	21	14	12	**0.00**	14	19	4	1	–	–	–	–	–	–	–	–	–	L-Phenylalanine, N-(tert-butoxycarbonyl)-	13734-34-4	NI26654
220	208	265	192	236	164	191	221	**265**	100	47	44	33	32	22	21	17		14	19	4	1	–	–	–	–	–	–	–	–	–	3,5-Pyridinedicarboxylic acid, 2,4,6-trimethyl-, diethyl ester		NI26655
116	70	176	131	75	117	144	42	**265**	100	56	31	23	23	19	17	15	**11.92**	14	23	2	1	–	–	–	–	–	1	–	–	–	Morpholine, 3-methyl-2-phenyl-N-(trimethylsiloxy)-		NI26656
174	73	75	192	77	45	175	74	**265**	100	67	40	34	21	15	14	12	**0.00**	14	23	2	1	–	–	–	–	–	1	–	–	–	Phenylalanine, N-(trimethylsilyl)-, ethyl ester		MS05910
98	140	127	126	112	109	84	70	**265**	100	75	65	60	60	50	40	40	**9.00**	14	23	2	3	–	–	–	–	–	–	–	–	–	Acetamide, N-(2-cyanoethyl)-N-[3-(hexahydro-2-oxo-1H-azepin-1-yl)propyl]-	67370-64-3	NI26657
174	73	91	86	58	250	118	59	**265**	100	39	22	13	10	9	8	6	**0.20**	14	27	–	1	–	–	–	–	–	2	–	–	–	Silanamine, 1,1,1-trimethyl-N-(2-phenylethyl)-N-(trimethylsilyl)-	58367-45-6	NI26659
174	73	91	86	59	175	45	176	**265**	100	68	50	26	20	16	14	13	**0.00**	14	27	–	1	–	–	–	–	–	2	–	–	–	Silanamine, 1,1,1-trimethyl-N-(2-phenylethyl)-N-(trimethylsilyl)-	58367-45-6	MS05911
174	73	91	175	86	58	250	118	**265**	100	39	22	20	13	10	9	8	**0.20**	14	27	–	1	–	–	–	–	–	2	–	–	–	Silanamine, 1,1,1-trimethyl-N-(2-phenylethyl)-N-(trimethylsilyl)-	58367-45-6	NI26658
131	265	264	77	136	129	108	63	**265**	100	88	61	53	35	34	32	30		15	11	–	3	1	–	–	–	–	–	–	–	–	Benzimidazole, 2-(benzothiazol-2-yl)-1-methyl-		IC03452
178	265	76	179	152	103	266	51	**265**	100	37	22	20	7	7	6	5		15	11	–	3	1	–	–	–	–	–	–	–	–	1,2,4-Triazine-3(2H)-thione, 5,6-diphenyl-	37469-24-2	NI26660

MASS TO CHARGE RATIOS								M.W.	INTENSITIES								Parent	C	H	O	N	S	F	Cl	Br	I	Si	P	B	X	COMPOUND NAME	CAS Reg No	No
178	265	76	179	152	103	151	51	**265**	100	37	22	20	7	7	5	5		15	11	–	3	1	–	–	–	–	–	–	–	–	1,2,4-Triazine-3(2H)-thione, 5,6-diphenyl-	37469-24-2	LI05570
129	51	102	50	76	131	77	103	**265**	100	66	55	45	44	35	34	28	**14.00**	15	11	2	3	–	–	–	–	–	–	–	–	–	Isoquinolinium, 2-(2-nitroanilino)-, hydroxide, inner salt	31383-01-4	NI26661
129	102	65	63	50	92	75	74	**265**	100	50	39	38	28	27	27	27	**16.00**	15	11	2	3	–	–	–	–	–	–	–	–	–	Isoquinolinium, 2-(3-nitroanilino)-, hydroxide, inner salt	31383-03-6	NI26662
129	63	102	265	128	235	130	75	**265**	100	42	39	24	24	23	23	22		15	11	2	3	–	–	–	–	–	–	–	–	–	Isoquinolinium, 2-(4-nitroanilino)-, hydroxide, inner salt	31383-05-8	NI26663
91	77	103	93	65	265	51	64	**265**	100	82	66	34	33	27	24	23		15	11	2	3	–	–	–	–	–	–	–	–	–	4,5-Isoxazoledione, 3-phenyl-, 4-(phenylhydrazone)	4865-99-0	MS05912
77	51	172	105	94	103	76	50	**265**	100	35	29	21	15	14	14	14	**6.10**	15	11	2	3	–	–	–	–	–	–	–	–	–	as-Triazine, 5,6-diphenoxy-		IC03453
165	178	248	221	249	265	193	166	**265**	100	77	53	42	37	24	21	20		15	11	2	3	–	–	–	–	–	–	–	–	–	as-Triazin-3(4H)-one, 5,6-diphenyl-, 2-oxide	13162-99-7	NI26664
55	82	41	83	68	56	91	28	**265**	100	91	62	56	51	30	28	23	**0.00**	15	20	1	1	–	–	1	–	–	–	–	–	–	2H-1,3-Benzoxazine, 6-chloro-3-cyclohexyl-3,4-dihydro-8-methyl-	6640-34-2	NI26665
137	237	55	191	121	93	100	41	**265**	100	97	75	68	62	57	52	48	**31.00**	15	23	3	1	–	–	–	–	–	–	–	–	–	2-Cyclohexene-6-carboxylic acid, 1-(1,3,4,5,6,7-hexahydro-2H-azepin-1-yl)-, ethyl ester		MS05913
93	137	100	91	121	77	55	41	**265**	100	71	71	67	63	60	54	52	**8.00**	15	23	3	1	–	–	–	–	–	–	–	–	–	2-Cyclohexene-6-carboxylic acid, 3-methyl-1-(2-oxo-piperidinyl)-, ethyl ester		MS05914
150	265	162	222	151	233	149	83	**265**	100	20	19	13	13	7	6	6		15	23	3	1	–	–	–	–	–	–	–	–	–	Methyl α-(2-piperidinocyclopentenyl)acetoacetate		MS05915
150	84	222	265	162	190	151	70	**265**	100	29	23	22	18	16	16	11		15	23	3	1	–	–	–	–	–	–	–	–	–	Methyl α-(2-pyrrolidinocyclohexenyl)acetoacetate		MS05916
166	222	265	178	137	43	165	164	**265**	100	91	38	29	24	20	16	14		15	23	3	1	–	–	–	–	–	–	–	–	–	2,4-Pentanedione, 3-(2-morpholinocyclohexenyl)-		MS05917
72	30	32	41	56	43	57	73	**265**	100	79	25	20	13	9	6	6	**0.00**	15	23	3	1	–	–	–	–	–	–	–	–	–	2-Propanol, 1-[(1-methylethyl)amino]-3-[2-(2-propenyloxy)phenoxy]-	6452-71-7	NI26666
164	85	181	123	136	180	150	265	**265**	100	74	47	39	33	25	8	5		15	23	3	1	–	–	–	–	–	–	–	–	–	Pyridine, 2-methoxy-6-[4-[(tetrahydropyran-2-yl)oxy]butyl]-		NI26667
91	106	97	195	112	55	152	132	**265**	100	79	63	47	27	18	16	15	**4.00**	15	24	1	1	–	–	–	–	–	–	1	–	–	Phosphetane, 1-benzylamino-2,2,3,4,4-pentamethyl-, 1-oxide		NI26668
176	44	58	220	133	177			**265**	100	53	48	36	25	22			**6.82**	15	27	1	3	–	–	–	–	–	–	–	–	–	Phenol, 2,4,6-tris[(dimethylamino)methyl]-	90-72-2	IC03454
176	58	44	133	42	177	105	220	**265**	100	68	62	33	22	21	19	18	**5.81**	15	27	1	3	–	–	–	–	–	–	–	–	–	Phenol, 2,4,6-tris[(dimethylamino)methyl]-	90-72-2	NI26670
176	58	44	220	265	177	133	105	**265**	100	41	35	24	23	21	21	13		15	27	1	3	–	–	–	–	–	–	–	–	–	Phenol, 2,4,6-tris[(dimethylamino)methyl]-	90-72-2	NI26669
217	249	265	236	237	248	218	216	**265**	100	77	53	45	25	20	20	20		16	11	1	1	1	–	–	–	–	–	–	–	–	6,7-Benzophenothiazine-5-oxide		NI26671
121	77	265	51	102	123	76	205	**265**	100	23	13	10	9	6	5	1		16	11	1	1	1	–	–	–	–	–	–	–	–	5(4H)-Thiazolone, 4-benzylidene-2-phenyl-	16446-30-3	LI05571
121	77	265	51	102	122	123	76	**265**	100	23	13	10	9	8	6	5		16	11	1	1	1	–	–	–	–	–	–	–	–	5(4H)-Thiazolone, 4-benzylidene-2-phenyl-	16446-30-3	NI26672
264	265	236	28	152	76	151	208	**265**	100	93	40	38	23	18	17	17		16	11	3	1	–	–	–	–	–	–	–	–	–	4H-Anthra[1,2-d][1,3]oxazine-7,12-dione, 1,2-dihydro-		NI26673
135	265	136	102	146				**265**	100	9	8	6	2					16	11	3	1	–	–	–	–	–	–	–	–	–	Benzoic acid, 4-cyano-, 4-acetylphenyl ester	53327-07-4	NI26674
265	250	77	266	132	222	76	251	**265**	100	61	19	18	18	15	11	10		16	11	3	1	–	–	–	–	–	–	–	–	–	Furacridone	62541-22-4	NI26675
145	221	205	206	77	222	43	191	**265**	100	83	74	43	42	35	35	28	**20.00**	16	15	1	3	–	–	–	–	–	–	–	–	–	Hydrazinecarboxamide, 2-(1,3-diphenyl-2-propenylidene)-	16983-74-7	NI26676
116	119	115	129	130	93	117	91	**265**	100	47	44	43	32	31	28	28	**2.00**	16	15	1	3	–	–	–	–	–	–	–	–	–	3H-Indeno[2,1-c]-1,2-diazete-1-carboxanilide, 1,2,2a,7b-tetrahydro-		LI05572
91	77	159	92	65	158	51	248	**265**	100	46	46	40	37	33	28	24	**4.09**	16	15	1	3	–	–	–	–	–	–	–	–	–	4H-1,2,4-Triazole, 3-benzyl-4-benzyloxy-		NI26677
232	223	181	250	191	41	265	27	**265**	100	53	51	47	44	32	21	19		16	27	–	1	1	–	–	–	–	–	–	–	–	Pyridine, 2,4-dibutyl-6-isopropylthio-	74630-76-5	NI26678
232	222	149	265	223	190	236	181	**265**	100	68	49	34	33	32	29	26		16	27	–	1	1	–	–	–	–	–	–	–	–	Pyridine, 2,4-dibutyl-6-isopropylthio-	74630-76-5	NI26679
41	112	55	67	43	98	94	42	**265**	100	89	82	44	44	40	38	30	**0.00**	16	27	2	1	–	–	–	–	–	–	–	–	–	1-Cyclododecanecarbonitrile, 1-(3-hydroxypropyl)-2-oxo-		DD00809
98	41	55	81	84	97	112	96	**265**	100	89	76	57	49	44	41	40	**0.00**	16	27	2	1	–	–	–	–	–	–	–	–	–	5-Oxacyclohexadecane-1-carbonitrile, 6-oxo-		DD00810
152	164	236	123	248	135	265	122	**265**	100	66	26	24	20	20	18	15		16	27	2	1	–	–	–	–	–	–	–	–	–	Serratinane-8,13-diol, (8α,13S)-	5635-95-0	LI05573
152	164	236	124	137	248	265	122	**265**	100	67	26	23	21	20	18	15		16	27	2	1	–	–	–	–	–	–	–	–	–	Serratinane-8,13-diol, (8α,13S)-	5635-95-0	NI26680
230	265	264	101	202	267	266	231	**265**	100	79	37	29	25	24	24	18		17	12	–	1	–	–	1	–	–	–	–	–	–	Quinoline, 4-(4-chlorostyryl)-	4594-89-2	NI26681
230	265	264	101	267	266	202	231	**265**	100	80	47	30	25	25	25	19		17	12	–	1	–	–	1	–	–	–	–	–	–	Quinoline, 4-(4-chlorostyryl)-, trans-		LI05574
266	224	252	206	225	223	107	267	**265**	100	32	12	11	5	4	4	3	**0.29**	17	15	2	1	–	–	–	–	–	–	–	–	–	Acetamide, N-acetyl-N-9H-fluoren-2-yl-	642-65-9	NI26683
181	223	180	265	43	182	165	152	**265**	100	58	29	20	17	16	13	12		17	15	2	1	–	–	–	–	–	–	–	–	–	Acetamide, N-acetyl-N-9H-fluoren-2-yl-	642-65-9	NI26682
181	43	223	180	152	165	15	182	**265**	100	57	44	38	23	20	18	17	**13.46**	17	15	2	1	–	–	–	–	–	–	–	–	–	Acetamide, N-acetyl-N-9H-fluoren-2-yl-	642-65-9	NI26684
265	208	250	181	152	153	247	266	**265**	100	67	61	50	45	34	33	19		17	15	2	1	–	–	–	–	–	–	–	–	–	17-Azaestra-1,3,5(10),6,8-pentene-11,16-dione		NI26685
209	237	210	180	208	265	77	104	**265**	100	15	15	15	12	10	10	7		17	15	2	1	–	–	–	–	–	–	–	–	–	3(2H)-Cyclopent[b]indolone, 1,3a,4,8b-tetrahydro-8b-hydroxy-3a-phenyl-	90732-23-3	NI26686
91	222	194	43	193	77	119	89	**265**	100	59	42	17	15	13	12	12	**11.00**	17	15	2	1	–	–	–	–	–	–	–	–	–	2-Isoxazoline, 5-acetyl-3,4-diphenyl-		NI26687
146	222	43	77	223	105	162	206	**265**	100	74	54	52	50	42	40	30	**6.00**	17	15	2	1	–	–	–	–	–	–	–	–	–	2-Isozazoline, 4-acetyl-3,5-diphenyl-		NI26688
105	77	104	91	51	28	103	174	**265**	100	57	56	50	34	31	22	19	**2.99**	17	15	2	1	–	–	–	–	–	–	–	–	–	4H-1,3-Oxazin-5-one, 4-benzyl-5,6-dihydro-2-phenyl-		NI26689
160	132	244	130	105	77	103	85	**265**	100	97	41	40	37	27	24	20	**13.47**	17	15	2	1	–	–	–	–	–	–	–	–	–	3-Pyrrolidinone, 5-benzoyl-1-phenyl-	31397-02-1	NI26690
265	264	235	236	250	222			**265**	100	84	23	13	9	5				17	15	2	1	–	–	–	–	–	–	–	–	–	4(1H)-Quinolinone, 6-methoxy-1-methyl-2-phenyl-	6878-08-6	LI05575
265	264	235	148	120	236	65	163	**265**	100	71	24	22	21	16	15	15		17	15	2	1	–	–	–	–	–	–	–	–	–	4(1H)-Quinolinone, 6-methoxy-1-methyl-2-phenyl-	6878-08-6	NI26691
45	265	220	190	165	116	219	192	**265**	100	61	39	17	10	9	8	8		17	15	2	1	–	–	–	–	–	–	–	–	–	Stilbene, 2-(methoxymethoxy)-4′-cyano-, (E)-		MS05918
45	265	220	190	192	165	116	235	**265**	100	42	27	13	7	7	6	4		17	15	2	1	–	–	–	–	–	–	–	–	–	Stilbene, 2-(methoxymethoxy)-4′-cyano-, (Z)-		MS05919
265	73	250	266	45	193	251	234	**265**	100	98	34	25	19	12	7	7		17	19	–	1	–	–	–	–	–	1	–	–	–	1H-Indole, 2-phenyl-1-(trimethylsilyl)-		NI26692
266	196	264	224					**265**	100	51	8	8					**0.00**	17	19	–	3	–	–	–	–	–	–	–	–	–	2-Imidazoline, 2-[(N-benzylanilino)methyl]-	91-75-8	NI26693
84	91	182	85	77	55	83	65	**265**	100	15	6	6	6	6	4	3	**1.40**	17	19	–	3	–	–	–	–	–	–	–	–	–	2-Imidazoline, 2-[(N-benzylanilino)methyl]-	91-75-8	NI26694
84	58	71	43	164	41	59	83	**265**	100	99	95	92	83	81	77	75	**69.06**	17	31	1	1	–	–	–	–	–	–	–	–	–	2-Naphthalenemethanol, 1-(dimethylamino)-1,2,3,4,4a,5,6,8a-octahydro-α,α,4a,8a-tetramethyl-	56701-22-5	NI26695

MASS TO CHARGE RATIOS								M.W.	INTENSITIES								Parent	C	H	O	N	S	F	Cl	Br	I	Si	P	B	X	COMPOUND NAME	CAS Reg No	No
265	222	250	237	194	179	266	208	**265**	100	62	43	38	29	24	20	20		18	19	1	1	–	–	–	–	–	–	–	–	–	2-Azabenzo[8,9]tricyclo[5.2.2.$0^{1,5}$]undeca-5,8,10-trien-3-one, 2-ethyl-4,4-dimethyl-, (1RS,7SR)-		DD00811
105	265	77	167	222	118	188	160	**265**	100	54	48	25	23	22	18	18		18	19	1	1	–	–	–	–	–	–	–	–	–	Aziridine, 2-benzoyl-1-isopropyl-3-phenyl-	23846-11-9	NI26696
208	209	105	77	180	57	165	182	**265**	100	67	43	43	35	35	31	18	**14.01**	18	19	1	1	–	–	–	–	–	–	–	–	–	Aziridinone, 1-tert-butyl-3,3-diphenyl-	18150-78-2	NI26697
198	122	265	177	142	154	67	266	**265**	100	63	37	28	19	17	14	7		18	19	1	1	–	–	–	–	–	–	–	–	–	3-Butynamide, N-ethyl-2,2-dimethyl-N-(1-naphthalenyl)-		DD00812
144	265	145	250	159	158	77	143	**265**	100	99	85	83	46	21	19	16		18	19	1	1	–	–	–	–	–	–	–	–	–	3H-Indole, 2-(2-hydroxyphenethyl)-3,3-dimethyl-	78744-82-8	NI26698
145	265	144	107	250	77	158	159	**265**	100	52	48	38	31	8	7	4		18	19	1	1	–	–	–	–	–	–	–	–	–	3H-Indole, 2-(4-hydroxyphenethyl)-3,3-dimethyl-	78744-90-8	NI26699
131	103	265	77	135	120	132	51	**265**	100	37	30	30	24	14	11	10		18	19	1	1	–	–	–	–	–	–	–	–	–	N-Isopropyl-3-phenylpropenanilide		TR00290
234	91	265	264	130	146	115	132	**265**	100	34	33	29	25	19	16	15		18	19	1	1	–	–	–	–	–	–	–	–	–	Isoquinoline, 3,4-dihydro-1-(2-methoxyphenethyl)-	78764-69-9	NI26700
162	104	147	119	91	77	105	103	**265**	100	99	85	72	64	46	42	34	**9.00**	18	19	1	1	–	–	–	–	–	–	–	–	–	Oxazole, 2,5-dihydro-2,2-dimethyl-5-(4-methylphenyl)-4-phenyl-	51304-31-5	NI26701
66	199	134	133	79	106	93	78	**265**	100	39	35	26	22	15	10	9	**3.50**	18	19	1	1	–	–	–	–	–	–	–	–	–	2-Propenenitrile, 3-[3-(bicyclo[2.2.1]hept-5-en-2-ylcarbonyl)bicyclo[2.2.1]hept-5-en-2-yl]-, [1α,2α(Z),3α(1R*,2S*,4R),4α]-	67401-00-7	NI26702
98	167	55	165	265	152	99	166	**265**	100	21	19	18	14	8	6	5		18	19	1	1	–	–	–	–	–	–	–	–	–	Pyrrolidine, diphenylacetyl-	60678-46-8	NI26703
57	43	97	110	41	55	96	124	**265**	100	87	81	65	61	52	51	46	**0.50**	18	35	–	1	–	–	–	–	–	–	–	–	–	Octadecanenitrile	638-65-3	NI26704
43	41	57	29	55	97	70	27	**265**	100	90	74	57	53	45	27	27	**1.70**	18	35	–	1	–	–	–	–	–	–	–	–	–	Octadecanenitrile	638-65-3	WI01320
96	55	83	41	43	69	70	111	**265**	100	80	60	36	35	28	18	9	**6.50**	18	35	–	1	–	–	–	–	–	–	–	–	–	2H-Pyrrole, 3,4-dihydro-5-methyl-4-tridecyl-		NI26705
42	91	118	174	105	70	77	132	**265**	100	84	83	56	45	30	27	18	**7.00**	19	23	–	1	–	–	–	–	–	–	–	–	–	Azetidine, 3,3-dimethyl-1-phenylethyl-2-phenyl-	22606-99-1	LI05576
42	118	91	174	105	70	77	117	**265**	100	84	83	56	45	30	27	18	**7.01**	19	23	–	1	–	–	–	–	–	–	–	–	–	Azetidine, 3,3-dimethyl-1-phenylethyl-2-phenyl-	22606-99-1	NI26706
181	77	180	182	104	265	91	78	**265**	100	37	32	19	17	13	10	10		19	23	–	1	–	–	–	–	–	–	–	–	–	Azetidine, 3-tert-butyl-1,2-diphenyl-, cis-		LI05577
77	181	41	145	105	104	91	51	**265**	100	95	69	60	43	39	36	36	**30.03**	19	23	–	1	–	–	–	–	–	–	–	–	–	Azetidine, 3-tert-butyl-1,2-diphenyl-, cis-2,3-	22628-29-1	NI26707
77	181	145	41	51	180	105	104	**265**	100	95	69	68	45	43	43	40	**32.00**	19	23	–	1	–	–	–	–	–	–	–	–	–	Azetidine, 3-tert-butyl-1,2-diphenyl-, trans-		LI05578
181	77	180	182	104	265	91	78	**265**	100	38	33	18	17	12	10	10		19	23	–	1	–	–	–	–	–	–	–	–	–	Azetidine, 3-tert-butyl-1,2-diphenyl-, trans-2,3-	22628-30-4	NI26708
250	196	41	146	265	131	91	181	**265**	100	94	65	63	58	58	58	56		19	23	–	1	–	–	–	–	–	–	–	–	–	9,10-Didehydro-19-norretinal nitrile		NI26709
82	266	69	165	133	64	76	63	**266**	100	84	39	34	31	29	20	18		2	–	–	–	4	6	–	–	–	–	–	–	–	Bis(trifluoromethyl)tetrasulphane		NI26710
117	85	132	66	247	31	98	63	**266**	100	42	39	22	17	17	16	16	**0.00**	2	–	2	–	2	4	2	–	–	–	–	–	–	2,4-Dichloro-1,1,2,4-tetrafluoro-3,3-dioxo-1,3-dithietane		MS05920
29	27	28	26	202	237	200	230	**266**	100	84	26	15	14	12	12	11	**7.70**	2	5	–	–	–	–	1	–	–	–	–	–	1	Mercury, chloroethyl-	107-27-7	NI26711
54	232	230	268	234	116	266	233	**266**	100	51	48	19	18	16	13	12		4	6	1	–	–	–	4	–	–	2	–	–	–	1,1,3,3-Tetrachloro-1,3-disila-2-oxacyclo-5-heptene		NI26712
268	270	266	272	233	231	199	201	**266**	100	64	61	21	19	16	14	12		6	–	–	–	–	1	5	–	–	–	–	–	–	Benzene, pentachlorofluoro-	319-87-9	NI26713
28	98	182	154	126	96	95	92	**266**	100	14	13	11	11	10	9	9	**6.37**	6	–	6	–	–	–	–	–	–	–	–	–	1	Molybdenum hexacarbonyl	13939-06-5	NI26714
98	182	126	154	266	110	77	63	**266**	100	89	82	63	32	25	18	17		6	–	6	–	–	–	–	–	–	–	–	–	1	Molybdenum hexacarbonyl	13939-06-5	MS05921
30	75	74	63	231	79	62	38	**266**	100	92	90	51	34	33	27	26	**0.00**	6	3	6	2	1	–	1	–	–	–	–	–	–	Benzenesulphonyl chloride, 2,4-dinitro-	1656-44-6	NI26715
231	169	155	154	232	203	185	267	**266**	100	20	13	11	8	8	7	6	**0.00**	6	3	6	2	1	–	1	–	–	–	–	–	–	Benzenesulphonyl chloride, 2,4-dinitro-	1656-44-6	NI26716
266	231	236	57	73	55	174	127	**266**	100	34	31	26	17	16	15	14		7	5	4	4	–	3	–	–	–	–	–	–	–	1,3-Benzenediamine, 2,4-dinitro-6-(trifluoromethyl)-	38949-19-8	NI26717
268	266	214	212	270	241	216	142	**266**	100	80	63	55	52	30	30	26		8	2	–	2	–	–	4	–	–	–	–	–	–	Quinoxaline, 5,6,7,8-tetrachloro-	3495-42-9	NI26718
251	249	105	76	247	77	266	264	**266**	100	97	73	71	61	49	48	46		8	8	2	–	–	–	–	–	–	–	–	–	1	Benzoic acid, 2-(methyltelluryl)-	22261-92-3	NI26719
179	177	175	149	178	176	137	147	**266**	100	39	22	21	16	15	13	12	**0.00**	8	18	2	–	–	–	–	–	–	–	–	–	1	Stannane, (acetyloxy)triethyl-	1907-13-7	NI26720
237	235	179	177	233	236	175	149	**266**	100	76	61	24	22	15	14	13	**0.00**	8	18	2	–	–	–	–	–	–	–	–	–	1	Stannane, (acetyloxy)triethyl-	1907-13-7	NI26721
125	127	268	266	126	270	128	117	**266**	100	34	5	3	3	2	2	2		9	9	–	–	–	–	2	1	–	–	–	–	–	3-Bromo-1,2-dichloro-1-phenylpropane		NI26722
59	131	43	126	133	191	127	117	**266**	100	92	88	45	38	30	26	26	**16.77**	9	14	1	–	4	–	–	–	–	–	–	–	–	2-Oxa-4,6,8,9-tetrathiaadamantane, 1,3,5,7-tetramethyl-, (±)-	17497-22-2	NI26723
73	69	75	132	117	139	137	97	**266**	100	78	73	36	32	29	29	27	**0.40**	9	19	2	–	–	–	–	1	–	1	–	–	–	Hexanoic acid, 6-bromo-, trimethylsilyl ester	34176-76-6	NI26724
165	251	151	135	167	233	150	137	**266**	100	55	40	25	20	17	10	5	**0.00**	9	22	1	–	–	–	–	–	–	–	–	–	1	1-Hexanol, 6-(trimethylstannyl)-	53044-14-7	NI26725
197	266	69	155	133	85	75	101	**266**	100	50	45	42	37	32	25	20		10	6	1	–	1	3	1	–	–	–	–	–	–	3-Buten-2-one, 4-(4-chlorophenyl)-1,1,1-trifluoro-4-mercapto-	53765-00-7	NI26726
45	177	176	250	47	46	44	70	**266**	100	99	99	84	74	74	68	61	**6.72**	10	6	3	2	2	–	–	–	–	–	–	–	–	Dioxyluciferin		NI26727
235	266	149	236	150	205	177	59	**266**	100	18	13	11	6	5	5	3		10	6	4	–	–	4	–	–	–	–	–	–	–	1,2-Benzenedicarboxylic acid, 3,4,5,6-tetrafluoro-, dimethyl ester	1024-59-5	NI26729
235	266	236	149	205	177	98	59	**266**	100	18	11	8	6	4	4	3		10	6	4	–	–	4	–	–	–	–	–	–	–	1,2-Benzenedicarboxylic acid, 3,4,5,6-tetrafluoro-, dimethyl ester	1024-59-5	NI26728
235	266	236	148	205	149	117	98	**266**	100	19	11	10	6	6	5	5		10	6	4	–	–	4	–	–	–	–	–	–	–	1,2-Benzenedicarboxylic acid, 3,4,5,6-tetrafluoro-, dimethyl ester	1024-59-5	LI05579
266	77	162	267	78	106	122	134	**266**	100	28	16	14	13	12	10	7		10	6	7	2	–	–	–	–	–	–	–	–	–	Chromone, 6-hydroxy-2-methyl-5,7-dinitro-	30253-11-3	NI26730
69	153	125	152	41	97	266	169	**266**	100	50	49	45	30	27	20	20		10	9	5	–	–	3	–	–	–	–	–	–	–	Penicillic acid, trifluoroacetate		NI26731
169	181	145	63	159	93	57	251	**266**	100	34	32	32	29	29	29	27	**2.00**	10	13	–	–	–	–	3	–	–	1	–	–	–	Silane, trimethyl(α,α,α-trichloro-2-tolyl)-	30610-55-0	NI26732
136	135	164	194	178	108	148	72	**266**	100	38	32	15	14	13	11	10	**2.00**	10	14	3	6	–	–	–	–	–	–	–	–	–	Adenosine, 3'-amino-3'-deoxy-		LI05580
136	135	164	194	178	108	72	114	**266**	100	38	32	15	15	13	10	9	**1.00**	10	14	3	6	–	–	–	–	–	–	–	–	–	Adenosine, 3'-amino-3'-deoxy-		MS05922
136	178	135	79	164	194	227	108	**266**	100	58	39	27	25	24	15	11	**0.30**	10	14	3	6	–	–	–	–	–	–	–	–	–	Adenosine, 5'-amino-5'-deoxy-		MS05923
231	41	39	55	217	233	49	168	**266**	100	51	50	41	38	36	32	27	**0.00**	10	16	–	2	1	–	2	–	–	–	–	–	–	2,5-Bis(2-chloro-1,1-dimethylethyl)-1,3,4-thiadiazole	96594-29-5	NI26733

MASS TO CHARGE RATIOS	M.W.	INTENSITIES	Parent	C	H	O	N	S	F	Cl	Br	I	Si	P	B	X	COMPOUND NAME	CAS Reg No	No
67 41 68 39 30 27 134 53	266	100 45 22 22 22 12 10 10	1.00	10	16	2	2	–	–	2	–	–	–	–	–	–	Cyclopentane, 1-chloro-2-nitroso-, dimer	6866-01-9	NI26734
43 91 266 42 202 65 139 118	266	100 93 74 37 32 28 20 18		11	14	2	4	1	–	–	–	–	–	–	–	–	Δ^3-1,2,3-Triazoline, 1,2-dimethyl-5-(4-tolylsulphonylimino)-		MS05924
43 91 121 266 111 65 82 139	266	100 90 79 75 34 32 31 26		11	14	2	4	1	–	–	–	–	–	–	–	–	1,2,3-Triazole, 1-methyl-5-[methyl(4-tolylsulphonyl)amino]-		MS05925
164 266 149 206 76 117 177 77	266	100 55 50 47 36 33 29 29		11	14	4	4	–	–	–	–	–	–	–	–	–	Butanal, 3-methyl-, (2,4-dinitrophenyl)hydrazone	2256-01-1	NI26736
266 55 79 75 57 152 122 78	266	100 73 69 69 68 62 58 58		11	14	4	4	–	–	–	–	–	–	–	–	–	Butanal, 3-methyl-, (2,4-dinitrophenyl)hydrazone	2256-01-1	NI26737
43 41 55 266 69 77 39 79	266	100 74 33 32 32 30 28 23		11	14	4	4	–	–	–	–	–	–	–	–	–	Butanal, 3-methyl-, (2,4-dinitrophenyl)hydrazone	2256-01-1	NI26735
42 43 266 69 87 39 55 63	266	100 61 36 25 23 18 16 15		11	14	4	4	–	–	–	–	–	–	–	–	–	2-Butanone, 3-methyl-, (2,4-dinitrophenyl)hydrazone	3077-97-2	NI26738
194 266 209 132 110 137 165 181	266	100 96 46 40 40 30 28 27		11	14	4	4	–	–	–	–	–	–	–	–	–	1,4-Dioxino[2,3-g]pteridine-4-carboxylic acid, 5,5a,7,8,9a,10-hexahydro-, ethyl ester		LI05581
41 29 27 206 28 30 69 55	266	100 62 57 55 49 48 42 37	35.03	11	14	4	4	–	–	–	–	–	–	–	–	–	Pentanal, (2,4-dinitrophenyl)hydrazone	2057-84-3	NI26739
40 41 231 69 43 55 44 266	266	100 99 46 43 39 29 25 14		11	14	4	4	–	–	–	–	–	–	–	–	–	Pentanal, (2,4-dinitrophenyl)hydrazone	2057-84-3	NI26740
41 42 43 266 87 39 69 181	266	100 89 58 51 24 24 20 18		11	14	4	4	–	–	–	–	–	–	–	–	–	2-Pentanone, (2,4-dinitrophenyl)hydrazone	1636-82-4	NI26741
56 41 266 57 87 117 69 181	266	100 71 40 34 29 26 26 22		11	14	4	4	–	–	–	–	–	–	–	–	–	3-Pentanone, (2,4-dinitrophenyl)hydrazone	1636-83-5	NI26742
162 221 237 135 163 161 136 107	266	100 73 67 61 50 50 49 49	36.00	11	14	4	4	–	–	–	–	–	–	–	–	–	Purin-6-ol, 9-(2-ethoxy-1,3-dioxan-5-yl)-, cis-	91487-79-5	NI26743
162 136 163 221 161 135 107 57	266	100 70 66 65 59 52 42 30	22.00	11	14	4	4	–	–	–	–	–	–	–	–	–	Purin-6-ol, 9-(2-ethoxy-1,3-dioxan-5-yl)-, trans-	91487-80-8	NI26744
249 41 55 168 209 223 185 67	266	100 71 52 51 46 39 38 35	28.10	11	14	4	4	–	–	–	–	–	–	–	–	–	2-Pyridinamine, N-cyclohexyl-3,5-dinitro-	26820-41-7	NI26745
207 193 55 179 72 125 112 69	266	100 88 57 52 47 37 35 34	1.99	11	18	2	6	–	–	–	–	–	–	–	–	–	1,3,5-Triazine-2,4-diamine, 6-cyanomethoxy-N,N-diethyl-N′-methoxymethyl-		NI26746
75 88 43 73 89 87 45 174	266	100 99 87 63 52 52 52 44	0.00	11	19	6	–	–	1	–	–	–	–	–	–	–	α-D-Glucopyranoside, methyl 4-deoxy-4-fluoro-2,3-di-O-methyl-, acetate	30694-48-5	NI26747
137 183 165 109 99 179 81 207	266	100 88 53 27 17 14 14 13	0.00	11	23	5	–	–	–	–	–	–	–	1	–	–	Ethyl 2-(diisopropoxyphosphinyl)propanoate		DD00813
208 266 28 209 168 141 103 64	266	100 58 45 23 23 17 15 15		12	6	–	6	1	–	–	–	–	–	–	–	–	4-Cyano-2-thiocyanomethyl-1,3,6-triazacyclo[3.3.3]azine		NI26748
168 266 268 169 267 231 270 269	266	100 65 45 13 12 9 8 8		12	8	1	–	–	–	2	–	–	1	–	–	–	9-Oxa-10-silaanthracene, 10,10-dichloro-9,10-dihydro-		NI26749
266 268 231 267 270 269 233 152	266	100 67 27 18 13 12 9 8		12	8	1	–	–	–	2	–	–	1	–	–	–	9-Oxa-10-silaphenanthrene, 10,10-dichloro-9,10-dihydro-	20470-98-8	NI26750
111 75 231 90 50 63 113 233	266	100 80 75 36 35 32 31 24	0.10	12	8	1	2	–	–	2	–	–	–	–	–	–	Diazene, bis(2-chlorophenyl)-, 1-oxide	13556-84-8	NI26751
111 75 113 63 50 90 99 125	266	100 62 32 29 28 27 21 18	8.60	12	8	1	2	–	–	2	–	–	–	–	–	–	Diazene, bis(3-chlorophenyl)-, 1-oxide	139-24-2	NI26752
111 75 125 113 90 63 127 99	266	100 44 43 31 31 19 18 15	14.72	12	8	1	2	–	–	2	–	–	–	–	–	–	Diazene, bis(4-chlorophenyl)-, 1-oxide	614-26-6	NI26753
111 250 266 252 139 113 268 75	266	100 57 40 37 33 32 26 23		12	8	1	2	–	–	2	–	–	–	–	–	–	Diazene, bis(4-chlorophenyl)-, 1-oxide	614-26-6	IC03455
266 268 168 84 267 139 269 128	266	100 31 17 17 13 9 4 3		12	8	3	–	–	–	1	–	–	–	1	–	–	2,2′-Biphenylylenephosphoric acid chloride	52258-06-7	NI26754
266 268 251 253 223 225 113 101	266	100 99 67 66 53 52 23 23		12	11	2	–	–	–	–	1	–	–	–	–	–	Naphthalene, 1-bromo-2,6-dimethoxy-	25315-10-0	NI26755
193 195 266 112 194 55 268 76	266	100 33 26 19 13 10 8 7		12	11	3	2	–	–	1	–	–	–	–	–	–	4H-Pyrido[1,2-a]pyrimidine-3-acetic acid, 7-chloro-4-oxo-, ethyl ester	50609-68-2	NI26756
220 165 193 166 266 194 192 222	266	100 53 46 46 41 40 34 33		12	11	3	2	–	–	1	–	–	–	–	–	–	2(1H)-Quinoxalinone, 3-(ethoxycarbonylmethylene)-6-chloro-		MS05926
266 205 77 192 61 136 135 51	266	100 83 79 47 41 32 32 28		12	14	1	2	2	–	–	–	–	–	–	–	–	Imidazoline, 5-[2-(methylthio)ethyl]-4-oxo-3-phenyl-2-thioxo-		LI05582
119 91 205 161 29 65 62 43	266	100 46 31 17 15 14 13 13	0.13	12	14	3	2	1	–	–	–	–	–	–	–	–	Carbamothioic acid, [[(3-methylbenzoyl)amino]carbonyl]-, S-ethyl ester	74793-43-4	NI26757
69 187 119 91 77 150 266 213	266	100 35 26 10 7 4 3 3		12	14	3	2	1	–	–	–	–	–	–	–	–	5H-Cyclopenta[d]pyridazine, 1-(2,3-butadienyloxy)-6,7-dihydro-4-(methylsulphonyl)-		MS05927
83 214 66 95 149 135 199 187	266	100 8 6 5 4 3 2 2	1.00	12	14	3	2	1	–	–	–	–	–	–	–	–	5H-Cyclopenta[d]pyridazine, 1-(3-butynyloxy)-6,7-dihydro-4-(methylsulphonyl)-		MS05928
139 96 167 140 95 124 42 141	266	100 12 11 9 9 8 8 5	2.60	12	14	3	2	1	–	–	–	–	–	–	–	–	Sydnone, 3-[2-(4-methoxyphenylthio)ethyl]-4-methyl-	53492-58-3	NI26758
91 107 108 65 79 77 92 51	266	100 33 30 20 19 10 10 7	0.30	12	14	5	2	–	–	–	–	–	–	–	–	–	L-Asparagine, N^2-[(phenylmethoxy)carbonyl]-	2304-96-3	NI26759
41 55 39 77 67 53 63 51	266	100 47 42 35 35 35 27 26	20.13	12	14	5	2	–	–	–	–	–	–	–	–	–	Phenol, 2-cyclohexyl-4,6-dinitro-	131-89-5	NI26760
41 266 231 55 28 39 193 67	266	100 63 61 51 49 48 47 45		12	14	5	2	–	–	–	–	–	–	–	–	–	Phenol, 2-cyclohexyl-4,6-dinitro-	131-89-5	NI26761
251 194 220 209 266 195 165 81	266	100 80 60 53 43 35 28 25		12	18	3	4	–	–	–	–	–	–	–	–	–	[4,5′-Bipyrimidine]-2,2′,4′(1H,1′H,3′H)-trione, 3,4,5,6-tetrahydro-1,1′,3,3′-tetramethyl-	25064-49-7	NI26762
68 75 234 79 73 147 67 144	266	100 39 37 36 33 30 16 11	0.00	12	22	1	2	–	–	–	–	–	2	–	–	–	3-Pyridinecarboxamide, N,N-bis(trimethylsilyl)-	69688-12-6	NI26763
45 89 145 101 59 75 133 71	266	100 82 81 78 70 60 48 41	0.00	12	26	6	–	–	–	–	–	–	–	–	–	–	Galactitol, hexa-O-methyl-		LI05583
45 145 101 89 75 59 133 177	266	100 87 87 81 67 57 50 48	0.00	12	26	6	–	–	–	–	–	–	–	–	–	–	D-Glucitol, hexa-O-methyl-		LI05584
101 89 75 59 45 145 133 88	266	100 96 91 71 65 56 56 51	0.00	12	26	6	–	–	–	–	–	–	–	–	–	–	D-Mannitol, 1,2,3,4,5,6-hexa-O-methyl-	20746-37-6	LI05585
188 247 89 158 232	266	100 80 47 27 20	0.00	12	26	6	–	–	–	–	–	–	–	–	–	–	D-Mannitol, 1,2,3,4,5,6-hexa-O-methyl-	20746-37-6	NI26764
59 58 103 45 31 87 29 89	266	100 29 27 12 10 9 8 7	0.00	12	26	6	–	–	–	–	–	–	–	–	–	–	2,5,8,11,14,17-Octadecanehexone	1191-87-3	NI26765
45 73 89 72 59 87 44 43	266	100 51 40 37 28 17 16 14	0.00	12	26	6	–	–	–	–	–	–	–	–	–	–	3,6,9,12,15,18-Octadecanehexone		IC03456
101 115 71 43 57 55 127 75	266	100 90 82 82 73 73 72 72	0.00	12	26	6	–	–	–	–	–	–	–	–	–	–	1,3,5,7,9,11-Undecanehexol, 2-methyl-	85888-01-3	NI26766
99 155 57 211 41 125 137 55	266	100 17 11 9 9 7 6 3	0.00	12	27	4	–	–	–	–	–	–	–	1	–	–	Phosphoric acid, tributyl ester	126-73-8	NI26767
99 155 41 57 29 125 55 56	266	100 27 19 16 14 9 7 6	0.00	12	27	4	–	–	–	–	–	–	–	1	–	–	Phosphoric acid, tributyl ester	126-73-8	IC03457
173 43 231 136 209 174 101 208	266	100 35 25 18 12 10 10 7	0.00	13	11	4	–	–	–	1	–	–	–	–	–	–	1,3-Dioxane-4,6-dione, 5-(2-chlorobenzylidene)-2,2-dimethyl-	15851-88-4	LI05586
173 208 43 136 164 180 209 210	266	100 52 45 43 33 31 21 18	13.21	13	11	4	–	–	–	1	–	–	–	–	–	–	1,3-Dioxane-4,6-dione, 5-(4-chlorobenzylidene)-2,2-dimethyl-	15851-87-3	NI26768

MASS TO CHARGE RATIOS								M.W.	INTENSITIES								Parent	C	H	O	N	S	F	Cl	Br	I	Si	P	B	X	COMPOUND NAME	CAS Reg No	No
180	266	76	224	123	207	151	182	**266**	100	80	77	76	72	63	40	35		13	11	4	–	–	–	1	–	–	–	–	–	–	1H-Indene-1-carboxylic acid, 1-chloro-2,3-dihydro-2,3-dioxo, 1-methylethyl ester	71369-18-1	NI26769
180	76	123	224	207	266	182	151	**266**	100	79	70	64	58	50	35	32		13	11	4	–	–	–	1	–	–	–	–	–	–	1H-Indene-2-carboxylic acid, 2-chloro-2,3-dihydro-1,3-dioxo-, 1-methylethyl ester	71369-20-5	NI26770
266	231	268	233	140	77	195	104	**266**	100	98	67	33	30	26	23	19		13	12	–	2	–	–	2	–	–	–	–	–	–	Aniline, 4,4′-methylenebis[2-chloro-	101-14-4	IC03458
231	266	268	140	195	233	77	229	**266**	100	58	37	36	32	31	27	24		13	12	–	2	–	–	2	–	–	–	–	–	–	Aniline, 4,4′-methylenebis[2-chloro-	101-14-4	NI26771
192	224	193	134	235	162	133	225	**266**	100	84	74	64	18	18	13	11	**11.00**	13	14	6	–	–	–	–	–	–	–	–	–	–	1,2-Benzenedicarboxylic acid, 3(or 5)-(acetyloxy)-5(or 3)-methyl-, dimethyl ester	56728-05-3	MS05929
163	76	104	50	77	74	29	92	**266**	100	67	49	40	38	22	21	20	**3.00**	13	14	6	–	–	–	–	–	–	–	–	–	–	1,2-Benzenedicarboxylic acid, 2-ethoxy-2-oxoethyl methyl ester	85-71-2	NI26772
137	135	136	102	138	130	109	77	**266**	100	22	15	12	10	7	6	6	**5.65**	13	14	6	–	–	–	–	–	–	–	–	–	–	3-Hydroxy-3-methoxycarbonyl-4-(4-methoxyphenyl)-4-butanolide		MS05930
182	123	43	17	224	183	266	124	**266**	100	61	53	34	16	16	7	7		13	14	6	–	–	–	–	–	–	–	–	–	–	Phenylacetic acid, 3,4-diacetoxy-, methyl ester		MS05931
43	123	165	182	15	224	17	121	**266**	100	61	28	26	25	19	18	17	**4.02**	13	14	6	–	–	–	–	–	–	–	–	–	–	Phenylacetic acid, α,3-diacetoxy-, methyl ester		MS05932
123	43	17	165	121	192	15	224	**266**	100	83	57	26	23	21	12	10	**4.27**	13	14	6	–	–	–	–	–	–	–	–	–	–	Phenylacetic acid, α,4-diacetoxy-, methyl ester		MS05933
235	266	236	175	234	31	176	15	**266**	100	17	13	9	6	6	5	5		13	14	6	–	–	–	–	–	–	–	–	–	–	2,4,5-Toluenetricarboxylic acid, trimethyl ester		MS05934
114	251	113	99	59	253	266	41	**266**	100	78	74	62	30	27	19	18		13	15	–	2	1	–	1	–	–	–	–	–	–	4H-1,3-Thiazine, 4,4,6-trimethyl-2-[3-chlorophenylamino(imino)]-	85298-48-2	NI26773
91	92	79	81	77	65	177	175	**266**	100	47	32	16	15	13	12	12	**0.00**	13	15	1	–	–	–	–	1	–	–	–	–	–	2-Cyclohexene, 1-(benzyloxy)-2-bromo-		DD00814
82	109	137	69	97	89	81	57	**266**	100	82	66	58	22	19	18	18	**1.00**	13	15	1	2	1	1	–	–	–	–	–	–	–	Pyrrolo[1,2-b][1,2,4]oxadiazole-2(1H)-thione, 1-(4-fluorophenyl)tetrahydro-5,5-dimethyl-	50455-86-2	NI26774
235	139	96	236	111	217	141	238	**266**	100	28	20	17	14	10	10	6	**4.00**	13	15	2	2	–	–	1	–	–	–	–	–	–	3-Isoxazolamine, 5-(4-chlorophenyl)-N-(1,1-dimethyl-2-hydroxyethyl)-		MS05935
58	178	235	180	237	266	150	111	**266**	100	95	94	36	33	32	27	19		13	15	2	2	–	–	1	–	–	–	–	–	–	5-Isoxazolamine, 3-(4-chlorophenyl)-N-(1,1-dimethyl-2-hydroxyethyl)-		MS05936
160	173	145	161	174	266	117	158	**266**	100	60	17	14	13	11	11	10		13	15	2	2	–	–	1	–	–	–	–	–	–	Tryptamine, N-(1-chloroacetyl)-5-methoxy-		MS05937
75	251	223	73	103	149	76	221	**266**	100	32	31	27	17	15	15	11	**3.22**	13	18	4	–	–	–	–	–	–	1	–	–	–	1,2-Benzenedicarboxylic acid, ethyl trimethylsilyl ester	55530-57-9	NI26775
73	117	147	75	45	221	88	118	**266**	100	83	63	23	17	12	12	11	**0.00**	13	18	4	–	–	–	–	–	–	1	–	–	–	1,2-Benzenedicarboxylic acid, ethyl trimethylsilyl ester	55530-57-9	NI26776
103	31	75	47	45	104	188	89	**266**	100	98	80	78	60	41	40	40	**0.00**	13	18	4	2	–	–	–	–	–	–	–	–	–	4,4-Diethoxy-1-(2-nitrophenyl)-2-azabut-1-ene		LI05587
45	237	157	83	141	67	111	156	**266**	100	64	48	48	33	33	32	28	**9.00**	13	18	4	2	–	–	–	–	–	–	–	–	–	2,4,6(1H,3H,5H)-Pyrimidinetrione, 5-(1-cyclohepten-1-yl)-5-ethyl-1″,2′-epoxy-		NI26777
97	170	171	251	41	55	152	69	**266**	100	95	85	54	42	39	28	26	**19.00**	13	18	4	2	–	–	–	–	–	–	–	–	–	2,4,6(1H,3H,5H)-Pyrimidinetrione, 1,3,5-trimethyl-5-(7-oxabicyclo[4.1.0]hept-1-yl)-	57304-98-0	NI26778
108	79	77	107	97	159	96	90	**266**	100	93	89	83	75	72	56	46	**4.00**	13	19	2	2	–	–	–	–	–	–	1	–	–	1H-1,3,2-Benzodiazaphosphole, octahydro-2-(3-methylphenoxy)-, 2-oxide	84957-79-9	NI26779
158	43	166	79	266	112	160	67	**266**	100	67	65	47	43	43	39	35		13	20	2	–	–	–	–	–	–	–	–	–	1	Nickel, [(1,4,5-η)-4-cycloocten-1-yl](2,4-pentanedionato-O,O′)-	12086-04-3	NI26780
176	125	136	124	123	150	26	51	**266**	100	72	51	37	36	32	22	17	**7.00**	13	22	2	4	–	–	–	–	–	–	–	–	–	L-Ornithine, N^5-(4,6-dimethyl-2-pyrimidinyl)-, ethyl ester	21175-12-2	LI05588
176	125	136	124	123	150	67	42	**266**	100	74	51	38	38	33	21	20	**7.22**	13	22	2	4	–	–	–	–	–	–	–	–	–	L-Ornithine, N^5-(4,6-dimethyl-2-pyrimidinyl)-, ethyl ester	21175-12-2	NI26781
100	166	72	29	266	167	42	152	**266**	100	80	70	35	30	20	20	18		13	22	2	4	–	–	–	–	–	–	–	–	–	Pyrimidine, 2-(dimethylamino)-4-(diethylaminocarbonyloxy)-5,6-dimethyl-		IC03459
266	265	210	238	175	267	183	76	**266**	100	78	69	68	36	16	14	13		14	6	2	2	1	–	–	–	–	–	–	–	–	Anthra[1,2-c][1,2,5]thiadiazole, 6,11-dihydro-6,11-dioxo-		IC03460
94	267	266	137	43	95	183	65	**266**	100	66	44	43	31	14	13	11		14	10	2	4	–	–	–	–	–	–	–	–	–	2-Oxabicyclo[2.2.1]heptane-4,7,7-tricarbonitrile, 5-(2-furyl)-3-imino-1-methyl-		MS05938
266	209	127	154	267	181	208	128	**266**	100	36	33	26	18	11	8	8		14	10	2	4	–	–	–	–	–	–	–	–	–	Pyrrolo[1,2-a]-1,3,5-triazine-8-carbonitrile, 1,2,3,4-tetrahydro-3-methyl-2,4-dioxo-7-phenyl-	54450-42-9	NI26782
165	235	237	199	31	164	163	63	**266**	100	80	60	23	22	18	18	16	**4.00**	14	12	1	–	–	–	2	–	–	–	–	–	–	Benzeneethanol, 4-chloro-β-(4-chlorophenyl)-	2642-82-2	NI26783
251	139	43	253	111	141	178	75	**266**	100	88	86	62	32	30	27	26	**12.30**	14	12	1	–	–	–	2	–	–	–	–	–	–	Benzenemethanol, 4-chloro-2-(4-chlorophenyl)-2-methyl-		MS05939
251	139	43	253	111	141	178	75	**266**	100	88	86	62	32	30	27	25	**12.11**	14	12	1	–	–	–	2	–	–	–	–	–	–	Benzenemethanol, 4-chloro-α-(4-chlorophenyl)-α-methyl-	80-06-8	NI26784
139	43	251	253	178	111	141	248	**266**	100	85	75	45	38	35	32	25	**10.85**	14	12	1	–	–	–	2	–	–	–	–	–	–	Benzenemethanol, 4-chloro-α-(4-chlorophenyl)-α-methyl-	80-06-8	NI26785
139	165	155	75	111	141	91	157	**266**	100	86	79	31	30	29	29	26	**22.10**	14	12	1	–	–	–	2	–	–	–	–	–	–	Benzene, 1,1′-(methoxymethylene)bis[4-chloro-	55702-41-5	NI26786
72	44	192	43	220	131	103	77	**266**	100	98	75	61	49	38	31	29	**7.40**	14	15	3	–	–	–	1	–	–	–	–	–	–	2H-Pyran-3-carboxaldehyde, 4-(4-chlorophenyl)-6-ethoxy-5,6-dihydro-	84735-94-4	NI26787
91	65	39	223	41	131	92	43	**266**	100	43	35	32	31	29	27	25	**0.00**	14	18	3	–	1	–	–	–	–	–	–	–	–	Furan, 2,5-dihydro-2-isopropyl-3-[(4-methylphenyl)sulphonyl]-	118356-38-0	DD00815
91	77	65	92	139	79	39	93	**266**	100	53	46	43	40	34	34	33	**0.00**	14	18	3	–	1	–	–	–	–	–	–	–	–	2,5-Hexadien-1-ol, 2-methyl-3-[(4-methylphenyl)sulphonyl]-, (E)-	118356-30-2	DD00816
91	39	41	65	92	77	43	55	**266**	100	94	83	68	53	52	40	38	**0.00**	14	18	3	–	1	–	–	–	–	–	–	–	–	2,5-Hexadien-1-ol, 5-methyl-3-[(4-methylphenyl)sulphonyl]-, (E)-	118356-21-1	DD00817
43	123	115	45	150	110	109	39	**266**	100	15	15	11	10	10	10	6	**4.00**	14	18	3	–	1	–	–	–	–	–	–	–	–	2-Propanone, 1-acetoxy-1-isopropyl-3-(phenylthio)-		MS05940
43	110	85	57	139	181	109	41	**266**	100	16	16	14	11	8	8	8	**2.50**	14	18	3	–	1	–	–	–	–	–	–	–	–	2-Propanone, 1-acetoxy-3-isopropyl-1-(phenylthio)-		MS05941
165	43	55	123	109	110	87	39	**266**	100	71	55	43	16	13	12	11	**9.00**	14	18	3	–	1	–	–	–	–	–	–	–	–	2-Propanone, 3-acetoxy-1-isopropyl-1-(phenylthio)-		MS05942
206	119	266	91	193	207	178	234	**266**	100	78	67	62	58	57	57	49		14	18	5	–	–	–	–	–	–	–	–	–	–	1,3-Adamantanedicarboxylic acid, 4-oxo-, dimethyl ester	19930-87-1	NI26788
179	105	117	99	266	107	265	77	**266**	100	84	76	67	64	32	26	24		14	18	5	–	–	–	–	–	–	–	–	–	–	α-D-Arabinohexopyranoside, methyl 4,6-O-benzylidene-2-deoxy-		NI26789
223	205	266	69	43	224	251	41	**266**	100	50	35	14	14	13	8	7		14	18	5	–	–	–	–	–	–	–	–	–	–	1,2,5-Benzenetriol, 2,6-dibutanoyl-		MS05943
193	43	207	266	208	195	59	179	**266**	100	96	93	90	85	68	65	45		14	18	5	–	–	–	–	–	–	–	–	–	–	Benzofuran, 7-acetyl-2,3-dihydro-6-hydroxy-2-isopropanol-4-methoxy-	52117-68-7	NI26790

MASS TO CHARGE RATIOS								M.W.	INTENSITIES								Parent	C	H	O	N	S	F	Cl	Br	I	Si	P	B	X	COMPOUND NAME	CAS Reg No	No
205	45	234	192	206	178	55	179	266	100	81	71	69	62	56	55	49	26.00	14	18	5	–	–	–	–	–	–	–	–	–	–	1,3-Cyclohexanedione, 2-(2-hydroxy-6-oxo-1-cyclohexen-1-yl)-2-(methoxymethyl)-	29817-42-3	NI26791
217	200	266	202	201	175	248	173	266	100	86	45	26	25	23	22	20		14	18	5	–	–	–	–	–	–	–	–	–	–	1H-Inden-1-one, 2,3-dihydro-2,3-dihydroxy-6-(2-hydroxyethyl)-5=(hydroxymethyl)-2,7-dimethyl-, (2S-trans)-	84299-76-3	NI26792
251	105	223	69	91	79	66	68	266	100	67	51	43	41	39	35	30	0.00	14	18	5	–	–	–	–	–	–	–	–	–	–	4,10-Methano-1,3-dioxolo[4,5-h][3]benzoxepin-9(8H)-one, octahydro-5,8-epoxy-2,2-dimethyl-		NI26793
200	163	57	73	248	91	55	84	266	100	72	60	49	48	48	44	39	7.00	14	18	5	–	–	–	–	–	–	–	–	–	–	Pterosin T, 11-hydroxy-		NI26794
56	69	41	84	55	42	43	39	266	100	59	50	32	28	22	21	21	19.09	14	19	1	2	–	–	1	–	–	–	–	–	–	4H-1,3-Oxazine, 2-[(3-chlorophenyl)imino]tetrahydro-3,4,4,6-tetramethyl-	53004-31-2	NI26795
84	141	168	251	50	41	140	56	266	100	80	79	51	49	46	43	34	20.29	14	19	1	2	–	–	1	–	–	–	–	–	–	4H-1,3-Oxazine, 5,6-dihydro-4,4,6-trimethyl-2-[(N-methyl-3-chlorophenyl)amino]-	53004-38-9	NI26796
205	266	192	73	75	193	206	179	266	100	67	30	30	25	22	18	18		14	22	3	–	–	–	–	–	–	1	–	–	–	Benzenepropanoic acid, 3-(trimethylsilyloxy)-, ethyl ester		MS05944
179	266	73	192	180	177	75	103	266	100	41	40	21	15	14	13	11		14	22	3	–	–	–	–	–	–	1	–	–	–	Benzenepropanoic acid, 4-(trimethylsilyloxy)-, ethyl ester		MS05945
91	117	73	85	75	132	160	92	266	100	34	31	20	20	16	10	9	0.00	14	22	3	–	–	–	–	–	–	1	–	–	–	Butanoic acid, 4-(benzyloxy)-, trimethylsilyl ester	21273-17-6	NI26797
134	117	75	251	91	266	73	119	266	100	57	47	33	30	28	24	21		14	22	3	–	–	–	–	–	–	1	–	–	–	Butanoic acid, 4-(2-methoxyphenyl)-, trimethylsilyl ester		NI26798
134	75	121	251	266	135	77	73	266	100	28	21	17	14	12	11	11		14	22	3	–	–	–	–	–	–	1	–	–	–	Butanoic acid, 4-(4-methoxyphenyl)-, trimethylsilyl ester		MS05946
209	266	236	179	193	251	210	223	266	100	80	70	46	37	35	20	12		14	22	3	–	–	–	–	–	–	1	–	–	–	2-Butanone, 4-[3-methoxy-4-[(trimethylsilyl)oxy]phenyl]-	56700-87-9	MS05947
173	73	75	157	55	151	83	174	266	100	66	36	33	33	32	15	14	2.00	14	22	3	–	–	–	–	–	–	1	–	–	–	Pentanoic acid, 5-phenoxy-, trimethylsilyl ester	21273-11-0	NI26799
173	73	157	75	266	251	151	55	266	100	67	57	57	48	33	33	33		14	22	3	–	–	–	–	–	–	1	–	–	–	Pentanoic acid, 5-phenoxy-, trimethylsilyl ester	21273-11-0	NI26800
72	30	151	43	109	56	73	28	266	100	39	25	25	23	15	13	12	1.70	14	22	3	2	–	–	–	–	–	–	–	–	–	Acetanilide, 4'-[2-hydroxy-3-(isopropylamino)propoxy]-	6673-35-4	NI26801
72	151	43	109	56	73	222	108	266	100	18	18	10	7	6	5	5	0.60	14	22	3	2	–	–	–	–	–	–	–	–	–	Acetanilide, 4'-[2-hydroxy-3-(isopropylamino)propoxy]-	6673-35-4	MS05948
72	151	43	109	56	57	108	73	266	100	17	15	11	8	6	5	5	0.70	14	22	3	2	–	–	–	–	–	–	–	–	–	Acetanilide, 4'-[2-hydroxy-3-(isopropylamino)propoxy]-	6673-35-4	IC03461
169	195	57	209	41	251	112	55	266	100	80	65	64	60	21	20	18	4.00	14	22	3	2	–	–	–	–	–	–	–	–	–	Barbituric acid, 5-allyl-1,3-dimethyl-5-neopentyl-		LI05589
72	30	32	43	56	41	57	107	266	100	57	18	10	9	7	6	6	0.00	14	22	3	2	–	–	–	–	–	–	–	–	–	Benzeneacetamide, 4-[2-hydroxy-3-[(1-methylethyl)amino]propoxy]-	29122-68-7	NI26802
197	124	266	151	98	56	29	69	266	100	97	72	59	52	52	35	28		14	22	3	2	–	–	–	–	–	–	–	–	–	4H-Pyrido[1,2-a]pyrimidine-3-carboxylic acid, 1-ethyl-1,6,7,8,9,9a-hexahydro-6-methyl-4-oxo-, ethyl ester	30238-41-6	NI26803
196	195	181	138	111	197	110	41	266	100	75	47	20	14	12	11	9	1.25	14	22	3	2	–	–	–	–	–	–	–	–	–	2,4,6(1H,3H,5H)-Pyrimidinetrione, 1,3-dimethyl-5-(1-methylbutyl)-5-(2-propenyl)-	28239-49-8	NI26804
196	195	41	181	43	111	138	52	266	100	66	40	25	21	20	19	16	4.00	14	22	3	2	–	–	–	–	–	–	–	–	–	2,4,6(1H,3H,5H)-Pyrimidinetrione, 1,3-dimethyl-5-(1-methylbutyl)-5-(2-propenyl)-	28239-49-8	NI26805
196	195	181	41	111	138	43	197	266	100	62	26	26	20	18	17	13	3.00	14	22	3	2	–	–	–	–	–	–	–	–	–	2,4,6(1H,3H,5H)-Pyrimidinetrione, 1,3-dimethyl-5-(1-methylbutyl)-5-(2-propenyl)-	28239-49-8	NI26806
238	266	237	239	267	77	182	210	266	100	89	27	17	15	11	10	8		15	10	3	2	–	–	–	–	–	–	–	–	–	2-Anthraquinonecarboxaldehyde, 1,4-diamino-		NI26807
77	262	121	105	189	161	65	92	266	100	71	49	43	31	25	18	14		15	10	3	2	–	–	–	–	–	–	–	–	–	4H-1-Benzopyran-4-one, 2-hydroxy-3-phenylazo-		IC03462
266	165	238	164	163	192	191	267	266	100	78	65	57	50	45	35	22		15	10	3	2	–	–	–	–	–	–	–	–	–	Formamide, N-[(2-nitrofluoren-9-ylidene)methyl]-	27398-54-5	NI26808
119	266	91	64	120	63	51	41	266	100	44	18	10	8	3	3	2		15	10	3	2	–	–	–	–	–	–	–	–	–	2,4,5-Imidazolinetrione, 1,3-diphenyl-		MS05949
266	167	140	265	192	220	237	180	266	100	67	67	64	62	60	46	30		15	10	3	2	–	–	–	–	–	–	–	–	–	6H-Indolo[3,2,1-de][1,5]naphthyridin-6-one, 5-hydroxy-6-methoxy-	18110-86-6	NI26809
266	223	250	179	267	207	152	151	266	100	29	25	24	19	18	18	13		15	10	3	2	–	–	–	–	–	–	–	–	–	6H-Indolo[3,2,1-de][1,5]naphthyridin-6-one, 1-methoxy-, 3-oxide	77370-00-4	NI26810
105	77	266	119	89	145	76	267	266	100	81	61	42	36	21	12	11		15	10	3	2	–	–	–	–	–	–	–	–	–	Isoxazole, 4-nitro-3,5-diphenyl-	53215-16-0	NI26811
105	266	77	144	150	106	89	104	266	100	79	62	14	13	8	8	7		15	10	3	2	–	–	–	–	–	–	–	–	–	Isoxazole, 3-(4-nitrophenyl)-5-phenyl-	31108-56-2	LI05590
105	266	77	144	150	267	106	89	266	100	79	62	14	13	12	8	8		15	10	3	2	–	–	–	–	–	–	–	–	–	Isoxazole, 3-(4-nitrophenyl)-5-phenyl-	31108-56-2	NI26812
266	120	42	105	77	119	267	104	266	100	44	32	27	25	20	19	19		15	14	1	4	–	–	–	–	–	–	–	–	–	Benzo-1,2,3-triazin-4-one, 3-(4-dimethylaminophenyl)-		MS05950
191	41	123	55	107	95	43	53	266	100	91	78	78	71	64	58	46	8.92	15	22	4	–	–	–	–	–	–	–	–	–	–	Azuleno[4,5-b]furan-2(3H)-one, decahydro-8,9-dihydroxy-6,9a-dimethyl-3-methylene-, [3aS-(3aα,6β,6aα,8α,9α,9aβ,9bα)]-	5090-67-5	NI26813
123	43	41	79	95	77	55	53	266	100	90	12	6	5	5	5	5	0.00	15	22	4	–	–	–	–	–	–	–	–	–	–	3-Buten-2-one, 4-[4-(acetyloxy)-2,2,6-trimethyl-7-oxabicyclo[4.1.0]hept-1-yl]-	38274-02-1	NI26814
100	123	107	121	124	122	151	149	266	100	79	72	45	39	35	28	21	4.40	15	22	4	–	–	–	–	–	–	–	–	–	–	Chamissonin, 11βH-dihydro-		NI26815
43	41	95	55	69	248	59	191	266	100	48	28	27	20	19	18	17	0.00	15	22	4	–	–	–	–	–	–	–	–	–	–	5-Desoxylactarolide B		NI26816
31	96	41	109	69	68	81	137	266	100	72	16	14	13	13	12	10	0.00	15	22	4	–	–	–	–	–	–	–	–	–	–	Furo[3,4-b]furan-2,6(3H,4H)-dione, dihydro-3-methylene-4-octyl-		NI26817
31	96	41	43	68	109	81	58	266	100	59	27	18	16	14	14	14	0.00	15	22	4	–	–	–	–	–	–	–	–	–	–	Furo[3,4-b]furan-2,6(3H,4H)-dione, dihydro-3-methylene-4-octyl-, [3aR-(3aα,4α,6aα)]-	20223-76-1	NI26818
166	43	105	107	77	79	58	134	266	100	49	43	42	24	20	20	10	0.00	15	22	4	–	–	–	–	–	–	–	–	–	–	Heptanoic acid, 2-phenyl-2,3-dihydroxy-3-methyl-, methyl ester		DD00818
166	105	43	106	77	107	101	59	266	100	51	23	22	22	15	11	11	0.00	15	22	4	–	–	–	–	–	–	–	–	–	–	Hexanoic acid, 2-phenyl-2,3-dihydroxy-3,5-dimethyl-, methyl ester		DD00819
69	57	42	197	85	124	55	71	266	100	80	56	49	38	31	30	29	15.00	15	22	4	–	–	–	–	–	–	–	–	–	–	Humulinic acid	520-40-1	NI26819
41	69	57	43	197	85	124	55	266	100	62	47	35	30	22	20	18	10.00	15	22	4	–	–	–	–	–	–	–	–	–	–	Humulinic acid	520-40-1	LI05591
41	69	57	43	197	85	124	55	266	100	60	49	33	29	22	19	18	9.00	15	22	4	–	–	–	–	–	–	–	–	–	–	Humulinic acid	520-40-1	NI26820

MASS TO CHARGE RATIOS								M.W.	INTENSITIES								Parent	C	H	O	N	S	F	Cl	Br	I	Si	P	B	X	COMPOUND NAME	CAS Reg No	No
85	153	182	84	166	136	183	86	**266**	100	21	16	13	12	10	6	6	**0.20**	15	22	4	–	–	–	–	–	–	–	–	–	–	Indeno[3a,4-b]oxiren-2(1aH)-one, hexahydro-4a-methyl-5-[(tetrahydro-2H-pyran-2-yl)oxy]-, (1aα,4aβ,5β,7aR*)-	57279-15-9	NI26821
109	55	139	81	123	140	95	107	**266**	100	67	56	50	47	44	43	40	**0.30**	15	22	4	–	–	–	–	–	–	–	–	–	–	Isotrichodermol, 7-hydroxy-		NI26822
123	109	122	55	81	95	53	69	**266**	100	95	81	80	72	70	69	67	**0.40**	15	22	4	–	–	–	–	–	–	–	–	–	–	Isotrichodermol, 8-hydroxy-		NI26823
173	91	55	118	81	69	248	57	**266**	100	99	81	71	65	64	57	57	**3.00**	15	22	4	–	–	–	–	–	–	–	–	–	–	Marasmene, 1α,15-dihydroxy-		MS05951
173	105	91	172	69	220	55	79	**266**	100	80	71	63	58	57	49	46	**6.00**	15	22	4	–	–	–	–	–	–	–	–	–	–	Marasmene, 3α,15-dihydroxy-		MS05952
113	43	41	95	91	80	67	55	**266**	100	24	20	18	9	9	9	9	**3.00**	15	22	4	–	–	–	–	–	–	–	–	–	–	1,4-Methano-3-benzoxepin-5-(4H)-one, 1,2,7,8,9,9a-hexahydro-7α,10-dihydroxy-2,2,9,9a-tetramethyl-		NI26824
113	31	59	74	95	109	123	91	**266**	100	68	34	20	16	8	5	4	**4.00**	15	22	4	–	–	–	–	–	–	–	–	–	–	1,4-Methano-3-benzoxepin-5-(4H)-one, 1,2,7,8,9,9a-hexahydro-7β,10-dihydroxy-2,2,9,9a-tetramethyl-		NI26825
59	148	43	151	123	109	41	135	**266**	100	98	68	60	52	50	48	43	**10.00**	15	22	4	–	–	–	–	–	–	–	–	–	–	1(2H)-Naphthalenone, 2,3;8,8a-bisepoxy-3,4,4a,5,6,7,8,8a-octahydro-4-(1-hydroxy-1-methylethyl)-4a,5-dimethyl-,[2S-(2α,3α,4α,4aβ,5β,8α,8aα)]-		NI26826
97	41	238	69	83	109	95	81	**266**	100	60	45	39	35	31	30	28	**17.00**	15	22	4	–	–	–	–	–	–	–	–	–	–	1(2H)-Naphthalenone, 2,3-epoxy-8a,4-epoxymethano-3,4,4a,5,6,7,8,8a-octahydro-8-hydroxy-4a,5,10,10-tetramethyl-, [2S-(2α,3α,4α,4aβ,5β,8aα)]-		NI26827
43	83	151	95	148	59			**266**	100	90	63	46	40	35			**0.00**	15	22	4	–	–	–	–	–	–	–	–	–	–	2-Nonen-4-one, 9-(3-furanyl)-6,9-dihydroxy-2,6-dimethyl-	36203-88-0	NI26828
137	43	123	206	124	81	41	266	**266**	100	86	59	57	38	32	24	21		15	22	4	–	–	–	–	–	–	–	–	–	–	Octanoic acid, 8-(5-acetyl-2-furyl)-, methyl ester		NI26829
178	43	133	79	77	71	195	163	**266**	100	95	51	34	33	27	17	16	**1.80**	15	22	4	–	–	–	–	–	–	–	–	–	–	Pernetic acid D		NI26830
178	43	133	79	77	71	195	163	**266**	100	91	44	35	30	26	14	13	**1.70**	15	22	4	–	–	–	–	–	–	–	–	–	–	Pernetic acid E		NI26831
125	69	126	41	43	59	55	39	**266**	100	31	21	14	10	9	7	6	**3.20**	15	22	4	–	–	–	–	–	–	–	–	–	–	2H-Pyran-2-one, 4-methoxy-6-nonanoyl-		MS05953
55	173	83	248	266	233	219	230	**266**	100	66	50	26	8	5	5	3		15	22	4	–	–	–	–	–	–	–	–	–	–	Santamarin, 11β,13-dihydro-9α-hydroxy-		MS05954
124	82	43	41	125	83	55	109	**266**	100	75	37	30	22	21	20	17	**12.90**	15	22	4	–	–	–	–	–	–	–	–	–	–	Trichothec-9-ene-3,13-diol, 11,12-epoxy-	90044-33-0	NI26832
41	43	91	77	105	79	93	55	**266**	100	77	71	59	57	50	49	49	**10.18**	15	22	4	–	–	–	–	–	–	–	–	–	–	Trichothec-9-ene-4β,15-diol, 12,13-epoxy-		NI26833
91	175	69	160	97	55	266	57	**266**	100	75	32	27	25	22	16	16		15	23	2	–	–	–	–	–	–	–	1	–	–	Phosphetane, 1-benzyloxy-2,2,3,4,4-pentamethyl-, 1-oxide		NI26834
55	83	73	94	266	151	103	75	**266**	100	99	92	68	56	48	45	37		15	26	2	–	–	–	–	–	–	1	–	–	–	Silane, trimethyl[(6-phenoxyhexyl)oxy]-	16654-55-0	NI26835
55	83	73	94	266	151	75	173	**266**	100	99	88	66	56	47	38	32		15	26	2	–	–	–	–	–	–	1	–	–	–	Silane, trimethyl[(6-phenoxyhexyl)oxy]-	16654-55-0	NI26836
178	139	266	220	185	97	236	224	**266**	100	74	46	37	35	31	26	25		15	26	2	2	–	–	–	–	–	–	–	–	–	Cyclohexanimine, N-cyclohexyl-1-isopropyl-6-aci-nitro-		MS05955
266	264	265	267	131	268	220	131.5	**266**	100	33	27	21	20	10	10	8		16	10	–	–	2	–	–	–	–	–	–	–	–	Benzo[1,2-b:4,3-b']dithiophene, 1-phenyl-		AI01883
266	204	165	221	82	249	150	110	**266**	100	40	27	14	9	7	6	6		16	10	4	–	–	–	–	–	–	–	–	–	–	1,8-Anthracenedicarboxylic acid		NI26837
265	267	266	229	115	231	102	268	**266**	100	35	29	20	19	15	11	9		16	11	–	2	–	–	1	–	–	–	–	–	–	Phthalazine, 1-chloro-4-(2-phenylethenyl)-, (E)-		MS05956
77	102	231	265	266	103	267	128	**266**	100	60	59	46	38	30	22	20		16	11	–	2	–	–	1	–	–	–	–	–	–	Quinoxaline, 2-chloro-3-(2-phenylethenyl)-, (E)-		MS05957
266	265	239	267	77	223	91	238	**266**	100	38	25	24	11	10	10	9		16	14	–	2	1	–	–	–	–	–	–	–	–	1H-1,4-Benzodiazepine-2-thione, 2,3-dihydro-7-methyl-5-phenyl-	2888-60-0	NI26838
165	166	207	266	234	82	233	83	**266**	100	69	68	31	27	21	19	16		16	14	–	2	1	–	–	–	–	–	–	–	–	2-Imidazoline-4-thione, 2-methyl-5,5-diphenyl-	79220-95-4	NI26839
266	249	234	77	76	220	267	152	**266**	100	36	18	17	15	15	14	13		16	14	2	2	–	–	–	–	–	–	–	–	–	9,10-Anthracenedione, 1,4-bis(methylamino)-	2475-44-7	NI26840
266	249	267	265	234	44	238	220	**266**	100	27	19	16	9	9	8	8		16	14	2	2	–	–	–	–	–	–	–	–	–	9,10-Anthracenedione, 1,4-bis(methylamino)-	2475-44-7	NI26841
120	266	91	59	118	119	58	55	**266**	100	45	41	41	36	32	30	25		16	14	2	2	–	–	–	–	–	–	–	–	–	1H-1,4-Benzodiazepine-2,5-dione, 3,4-dihydro-4-(2-methylphenyl)-		NI26842
149	133	224	266	39	147	134	53	**266**	100	99	98	62	45	39	38	33		16	14	2	2	–	–	–	–	–	–	–	–	–	12,14-Dioxabicyclo[8.2.2]tetradeca-1(13),5,10-triene-5,6-dicarbonitrile, pseudo-gem-(E)-		NI26843
133	266	149	39	134	53	147	91	**266**	100	34	27	25	21	18	14	12		16	14	2	2	–	–	–	–	–	–	–	–	–	12,14-Dioxabicyclo[8.2.2]tetradeca-1(13),5,10-triene-5,6-dicarbonitrile, pseudo-ortho-(E)-		NI26844
223	118	194	165	266	195	77	189	**266**	100	88	72	30	27	27	27	18		16	14	2	2	–	–	–	–	–	–	–	–	–	2,4-Imidazolidinedione, 1-methyl-5,5-diphenyl-	6859-11-6	LI05592
223	118	194	165	266	77	195	222	**266**	100	90	72	30	28	28	27	20		16	14	2	2	–	–	–	–	–	–	–	–	–	2,4-Imidazolidinedione, 1-methyl-5,5-diphenyl-	6859-11-6	NI26845
180	266	104	237	209	77	189	181	**266**	100	88	49	47	43	32	29	24		16	14	2	2	–	–	–	–	–	–	–	–	–	2,4-Imidazolidinedione, 3-methyl-5,5-diphenyl-	4224-00-4	NI26846
180	104	266	77	237	57	209	71	**266**	100	60	48	40	39	38	26	23		16	14	2	2	–	–	–	–	–	–	–	–	–	2,4-Imidazolidinedione, 3-methyl-5,5-diphenyl-	4224-00-4	NI26847
180	266	104	237	209	77	189	181	**266**	100	88	52	49	44	32	30	25		16	14	2	2	–	–	–	–	–	–	–	–	–	2,4-Imidazolidinedione, 3-methyl-5,5-diphenyl-	4224-00-4	NI26848
266	237	180	194	118	208	104	165	**266**	100	80	75	60	58	55	38	25		16	14	2	2	–	–	–	–	–	–	–	–	–	2,4-Imidazolidinedione, 5-(4-methylphenyl)-5-phenyl-	51169-17-6	NI26849
180	237	194	223	118	266	77	208	**266**	100	76	71	68	62	61	46	45		16	14	2	2	–	–	–	–	–	–	–	–	–	2,4-Imidazolidinedione, 5-(4-methylphenyl)-5-phenyl-	51169-17-6	NI26850
105	91	43	78	175	266	65	224	**266**	100	43	35	33	17	15	12	10		16	14	2	2	–	–	–	–	–	–	–	–	–	Imidazo[1,2-a]pyridinium, 2-acetyl-1-benzyl-2,3-dihydro-3-oxo, hydroxide, inner salt	11066-85-6	LI05593
266	133	69	267	132	105	81	77	**266**	100	78	26	16	15	15	13	11		16	14	2	2	–	–	–	–	–	–	–	–	–	4-(4-Methoxybenzyl)-3-cinnolinol		AI01884
121	266	122	267	89	77	78	63	**266**	100	33	7	5	5	5	3	2		16	14	2	2	–	–	–	–	–	–	–	–	–	2-(4-Methoxybenzyl)-3(2H)-cinnolinone		AI01885
133	266	132	105	134	104	77	267	**266**	100	73	22	21	14	13	13	12		16	14	2	2	–	–	–	–	–	–	–	–	–	1-(4-Methoxybenzylideneamino)-2-indolinone		AI01886
123	266	144	89	116	115	122	143	**266**	100	42	24	21	8	6	3	2		16	14	2	2	–	–	–	–	–	–	–	–	–	4'-Methoxyindole-2-carboxyanilide		LI05594
123	144	266	116	89	143	122	115	**266**	100	55	25	17	5	4	4	4		16	14	2	2	–	–	–	–	–	–	–	–	–	4'-Methoxyindole-3-carboxyanilide		LI05595
77	183	266	105	51	91	28	64	**266**	100	50	41	19	19	12	11	9		16	14	2	2	–	–	–	–	–	–	–	–	–	3,5-Pyrazolidinedione, 4-methyl-1,2-diphenyl-		MS05958

MASS TO CHARGE RATIOS								M.W.	INTENSITIES								Parent	C	H	O	N	S	F	Cl	Br	I	Si	P	B	X	COMPOUND NAME	CAS Reg No	No
251	266	249	91	252	65	132	267	266	100	66	46	30	17	16	15	12		16	14	2	2	–	–	–	–	–	–	–	–	–	4(3H)-Quinazolinone, 6-hydroxy-2-methyl-3-(2-methylphenyl)-	5060-51-5	NI26851
251	266	91	249	132	92	252	63	266	100	82	51	44	20	17	15	15		16	14	2	2	–	–	–	–	–	–	–	–	–	4(3H)-Quinazolinone, 6-hydroxy-2-methyl-3-(2-methylphenyl)-	5060-51-5	NI26852
224	266	251	76	249	77	267	143	266	100	78	60	22	20	20	13	12		16	14	2	2	–	–	–	–	–	–	–	–	–	4(3H)-Quinazolinone, 3-(2-hydroxy-5-methylphenyl)-2-methyl-		NI26853
251	224	266	77	59	76	73	50	266	100	59	36	36	36	27	27	23		16	14	2	2	–	–	–	–	–	–	–	–	–	4(3H)-Quinazolinone, 3-(2-hydroxy-6-methylphenyl)-2-methyl-		NI26854
251	266	249	77	252	148	83	76	266	100	44	35	20	17	12	11	10		16	14	2	2	–	–	–	–	–	–	–	–	–	4(3H)-Quinazolinone, 3-(3-hydroxy-2-methylphenyl)-2-methyl-	5060-63-9	NI26855
251	266	249	77	143	252	76	267	266	100	50	40	25	23	18	14	9		16	14	2	2	–	–	–	–	–	–	–	–	–	4(3H)-Quinazolinone, 3-(4-hydroxy-2-methylphenyl)-2-methyl-	5060-52-6	NI26856
266	160	77	235	251	146	76	247	266	100	88	60	50	39	38	33	26		16	14	2	2	–	–	–	–	–	–	–	–	–	4(3H)-Quinazolinone, 3-[2-(hydroxymethyl)phenyl]-2-methyl-	5060-50-4	LI05596
160	266	235	77	251	146	247	76	266	100	57	51	45	35	29	28	23		16	14	2	2	–	–	–	–	–	–	–	–	–	4(3H)-Quinazolinone, 3-[2-(hydroxymethyl)phenyl]-2-methyl-	5060-50-4	NI26857
235	266	265	76	249	77	102	65	266	100	96	61	48	39	30	21	20		16	14	2	2	–	–	–	–	–	–	–	–	–	4(3H)-Quinazolinone, 3-(2-methyl-6-methoxyphenyl)-		NI26858
233	234	77	231	76	250	232	50	266	100	45	32	31	30	29	29	22	2.00	16	14	2	2	–	–	–	–	–	–	–	–	–	Quinoxaline, 2-ethyl-3-phenyl-, 1,4-dioxide	15034-22-7	NI26859
146	121	131	145	132	266	160	105	266	100	77	66	48	35	31	15	13		16	18	–	4	–	–	–	–	–	–	–	–	–	2-Benzimidazolamine, N,1-dimethyl-N-[2-(methylamino)phenyl]-	93767-91-0	NI26860
133	266	120	106	147	119	118	132	266	100	98	65	52	48	40	35	34		16	18	–	4	–	–	–	–	–	–	–	–	–	Ethylenediamine, N,N′-bis(2-aminobenzylidene)-		LI05597
266	265	267	145	252	134	238	160	266	100	51	20	12	11	8	7	4		16	18	–	4	–	–	–	–	–	–	–	–	–	Imidazo[1,2-a]pyridin-2-amine, 5-ethyl-N-(6-ethyl-2-pyridinyl)-		MS05959
134	57	41	107	79	167	166	43	266	100	48	46	37	37	27	26	26	0.00	16	26	3	–	–	–	–	–	–	–	–	–	–	1-Cyclohexene-1-carboxylic acid, 4-(1,5-dimethyl-3-oxohexyl)-, methyl ester		MS05960
134	57	41	107	79	85	43	167	266	100	64	47	41	37	30	25	20	0.00	16	26	3	–	–	–	–	–	–	–	–	–	–	1-Cyclohexene-1-carboxylic acid, 4-(1,5-dimethyl-3-oxohexyl)-, methyl ester, [S-(R*,R*)]-	26462-72-6	NI26861
206	177	43	149	137	121	191	164	266	100	57	30	26	23	22	21	15	1.00	16	26	3	–	–	–	–	–	–	–	–	–	–	2-Cyclohexen-1-one, 3-[3-(acetyloxy)butyl]-2,4,4,5-tetramethyl-	68931-39-5	NI26862
81	43	41	69	71	114	82	55	266	100	96	53	52	51	43	41	37	2.00	16	26	3	–	–	–	–	–	–	–	–	–	–	2,6-Dodecadienoic acid, 10,11-epoxy-3,7,11-trimethyl-, methyl ester, (E,E)-	5299-11-6	NI26863
41	55	43	69	57	83	97	67	266	100	97	90	82	56	53	39	37	1.50	16	26	3	–	–	–	–	–	–	–	–	–	–	2,5-Furandione, 3-(dodecenyl)dihydro-	25377-73-5	NI26864
141	130	128	95	55	69	137	81	266	100	46	34	34	34	32	31	31	0.50	16	26	3	–	–	–	–	–	–	–	–	–	–	Naphthalene-2(1H)-spiro-2′-oxirane-1-carboxylic acid, octahydro-5,5,8a-trimethyl-, methyl ester, (1β,4aα,8aβ,2(1′)α)-	104758-39-6	NI26865
43	121	123	122	222	139	135	233	266	100	57	50	39	38	38	27	26	9.00	16	26	3	–	–	–	–	–	–	–	–	–	–	Naphtho[2,1-b]furan-2(1H)-one, decahydro-8-hydroxy-3a,6,6,9a-tetramethyl-, (3aR,5aS,8R,9aS,9bR)-	29065-33-6	NI26866
43	41	123	69	55	67	222	139	266	100	81	50	50	45	40	38	38	9.01	16	26	3	–	–	–	–	–	–	–	–	–	–	Naphtho[2,1-b]furan-2(1H)-one, decahydro-8-hydroxy-3a,6,6,9a-tetramethyl-, (3aR,5aS,8R,9aS,9bR)-	29065-33-6	NI26867
134	266	59	119	44	139	135	121	266	100	35	35	32	23	22	20	20		16	26	3	–	–	–	–	–	–	–	–	–	–	Naphtho[2,1-b]furan-2(1H)-one, decahydro-8-hydroxy-3a,6,6,9a-tetramethyl-, (3aR,5aS,8R,9aS,9bR)-	29065-33-6	LI05598
134	59	266	119	45	139	135	121	266	100	37	34	32	21	20	19	19		16	26	3	–	–	–	–	–	–	–	–	–	–	Naphtho[2,1-b]furan-2(1H)-one, decahydro-8-hydroxy-3a,6,6,9a-tetramethyl-, (3aS,5aS,8R,9aS,9bR)-	29080-98-6	NI26869
43	41	69	123	55	67	222	139	266	100	80	50	49	45	40	38	38	9.00	16	26	3	–	–	–	–	–	–	–	–	–	–	Naphtho[2,1-b]furan-2(1H)-one, decahydro-8-hydroxy-3a,6,6,9a-tetramethyl-, (3aS,5aS,8R,9aS,9bR)-	29080-98-6	LI05599
134	266	119	135	139	109	121	107	266	100	38	30	26	25	24	23	19		16	26	3	–	–	–	–	–	–	–	–	–	–	Naphtho[2,1-b]furan-2(1H)-one, decahydro-8-hydroxy-3a,6,6,9a-tetramethyl-, (3aS,5aS,8R,9aS,9bR)-	29080-98-6	NI26868
123	105	43	77	207	41	122	51	266	100	94	63	48	47	34	22	20	0.00	16	26	3	–	–	–	–	–	–	–	–	–	–	Orthobenzoic acid, tripropyl ester	6946-85-6	NI26870
107	191	234	139	206	43	266	248	266	100	54	47	47	39	29	4	2		16	26	3	–	–	–	–	–	–	–	–	–	–	Pernetic acid A, methyl ester		NI26871
77	57	103	121	91	266	135	192	266	100	96	66	65	46	24	20	8		16	26	3	–	–	–	–	–	–	–	–	–	–	1,2-Propanediol, 3-[(1,5-tridecadien-3-ynyl)oxy]-, (–)-[(S)-(Z,Z)]-		DD00820
81	43	85	71	41	95	135	82	266	100	98	89	80	63	55	54	53	2.00	16	26	3	–	–	–	–	–	–	–	–	–	–	2,6-Tridecadienoic acid, 10,11-epoxy-7-ethyl-3,11-dimethyl-, methyl ester		LI05600
43	83	113	114	181	196	55	156	266	100	60	38	35	34	25	25	18	10.00	16	30	1	2	–	–	–	–	–	–	–	–	–	Heptanamide, N-(2-cyanoethyl)-N-hexyl-	67138-89-0	NI26872
266	237	137	165	265	115	152	128	266	100	10	7	6	5	4	3	3		17	14	3	–	–	–	–	–	–	–	–	–	–	Benzofuran, 2-(2,4-dihydroxyphenyl)-5-propenyl-, (E)-		MS05961
266	151	251	249	265	115	116	152	266	100	93	40	40	34	18	16	13		17	14	3	–	–	–	–	–	–	–	–	–	–	3(2H)-Benzofuranone, 6-methoxy-2-(2-methylbenzylidene)-	77764-85-3	NI26873
265	266	251	252	150	132	122	115	266	100	94	75	13	9	8	7	7		17	14	3	–	–	–	–	–	–	–	–	–	–	3(2H)-Benzofuranone, 6-methoxy-2-(3-methylbenzylidene)-	77764-86-4	NI26874
265	266	251	150	252	250	147	223	266	100	91	51	10	9	7	6	5		17	14	3	–	–	–	–	–	–	–	–	–	–	3(2H)-Benzofuranone, 6-methoxy-2-(4-methylbenzylidene)-	77764-87-5	NI26875
251	266	265	252	223	267	237	165	266	100	56	28	22	19	15	13	13		17	14	3	–	–	–	–	–	–	–	–	–	–	7H-1-Benzopyran-1-one, 5-methoxy-6-methyl-2-phenyl-		MS05962
134	270	135	149	121	148	119	91	266	100	29	15	10	9	7	6	5	0.00	17	14	3	–	–	–	–	–	–	–	–	–	–	4H-1-Benzopyran-4-one, 3-(4-methoxyphenyl)-7-methyl-	54889-75-7	NI26876
105	77	266	121	265	178	51	206	266	100	50	33	30	20	10	10	9		17	14	3	–	–	–	–	–	–	–	–	–	–	Cinnamic acid, α-benzoyl-, methyl		LI05601
191	31	192	189	165	266	190	179	266	100	38	24	24	16	15	12	11		17	14	3	–	–	–	–	–	–	–	–	–	–	Glycolic acid, 9-phenanthrenemethyl ester	92174-35-1	NI26877
105	77	106	69	161	51	147	120	266	100	34	20	18	11	10	6	5	2.80	17	14	3	–	–	–	–	–	–	–	–	–	–	1,3,5-Pentanetrione, 1,5-diphenyl-	1467-40-9	NI26878
106	118	77	266	91	224	107	196	266	100	70	61	42	40	39	39	30		17	18	1	2	–	–	–	–	–	–	–	–	–	Acetic acid, 2-methyl-2-phenyl-1-(2-phenylethenyl)hydrazide	67134-55-8	NI26879
249	146	130	117	42	103	104	77	266	100	63	29	17	17	14	13	13	0.80	17	18	1	2	–	–	–	–	–	–	–	–	–	Acetophenone, α-(1,2,3,4-tetrahydro-2-isoquinolyl)-, (E)-oxime		NI26880
235	266	251	221	182	206	193	195	266	100	74	60	22	21	14	14	10		17	18	1	2	–	–	–	–	–	–	–	–	–	Canthine, 6-ethyl-4,5-dihydro-6-methoxy-		LI05602
77	180	248	131	247	104	249	174	266	100	73	58	32	29	25	23	21	8.00	17	18	1	2	–	–	–	–	–	–	–	–	–	5-Hydroxy-4,5-dimethyl-1,2-diphenyl-4,5-dihydroimidazole		LI05603
148	266	133	132	107	91	267	106	266	100	77	22	19	19	16	15	14		17	18	1	2	–	–	–	–	–	–	–	–	–	3-Pyrazolidinone, 2,4-diphenyl-5,5-dimethyl-		NI26881
265	266	210	209	238	167	224	223	266	100	80	60	52	25	24	20	18		17	18	1	2	–	–	–	–	–	–	–	–	–	Pyrrolo[2,1-a]pyrido[1,3-l,m]-β-carboline, 1,2,3,5,6,12,13,14-octahydro-12-oxo-		MS05963

MASS TO CHARGE RATIOS								M.W.	INTENSITIES								Parent	C	H	O	N	S	F	Cl	Br	I	Si	P	B	X	COMPOUND NAME	CAS Reg No	No
181	195	266	97	123	111	110	182	266	100	90	32	23	16	16	15	14		17	30	–	–	1	–	–	–	–	–	–	–	–	Thiophene, 2-heptyl-5-hexyl-		AI01887
67	41	82	55	74	87	81	66	266	100	31	30	20	15	13	13	12	5.64	17	30	2	–	–	–	–	–	–	–	–	–	–	2-Cyclopentene-1-undecanoic acid, methyl ester	24828-56-6	NI26882
67	41	55	82	39	74	59	43	266	100	32	18	15	12	11	10	10	0.30	17	30	2	–	–	–	–	–	–	–	–	–	–	2-Cyclopentene-1-undecanoic acid, methyl ester	24828-56-6	WI01321
69	81	43	123	95	41	223	109	266	100	70	55	42	40	36	25	22	0.00	17	30	2	–	–	–	–	–	–	–	–	–	–	2,6,10-Dodecatrien-1-ol, dihydro-3,7,11-trimethyl-, acetate		LI05604
67	41	81	55	54	95	79	68	266	100	83	74	71	47	45	41	36	16.04	17	30	2	–	–	–	–	–	–	–	–	–	–	7,10-Hexadecadienoic acid, methyl ester	16106-03-9	NI26883
67	81	41	55	54	82	95	68	266	100	73	68	62	45	44	41	38	13.74	17	30	2	–	–	–	–	–	–	–	–	–	–	9,12-Hexadecadienoic acid, methyl ester	2462-80-8	NI26884
67	41	55	81	43	28	69	95	266	100	95	86	81	74	74	54	49	13.64	17	30	2	–	–	–	–	–	–	–	–	–	–	Hexadecadienoic acid, methyl ester	29961-54-4	NI26885
96	81	67	55	54	95	110	41	266	100	93	63	48	39	38	36	31	2.20	17	30	2	–	–	–	–	–	–	–	–	–	–	11-Hexadecynoic acid, methyl ester	55000-41-4	WI01322
43	41	87	109	55	69	251	127	266	100	95	94	84	74	71	69	54	2.18	17	30	2	–	–	–	–	–	–	–	–	–	–	6α-Hydroxy-8,13-oxido-14,15,16-trisnorlabdane		MS05964
251	43	41	127	55	99	69	44	266	100	84	73	65	54	44	42	38	3.00	17	30	2	–	–	–	–	–	–	–	–	–	–	6β-Hydroxy-8,13-oxido-14,15,16-trisnorlabdane		MS05965
221	222	266	137	177	97	109	195	266	100	78	73	69	67	58	56	55		17	30	2	–	–	–	–	–	–	–	–	–	–	Naphtho[2,1-d][1,3]dioxepin, dodecahydro-5a,8,8,11a-tetramethyl-, [5aR-(5aα,7aβ,11aα,11bβ)]-	38419-77-1	NI26886
221	222	266	137	177	95	195	123	266	100	77	74	69	66	59	55	54		17	30	2	–	–	–	–	–	–	–	–	–	–	Naphtho[2,1-d][1,3]dioxepin, dodecahydro-5a,8,8,11a-tetramethyl-, [5aR-(5aα,7aβ,11aα,11bβ)]-	38419-77-1	NI26887
123	69	125	122	67	82	81	83	266	100	85	69	60	60	50	41	39	0.15	17	30	2	–	–	–	–	–	–	–	–	–	–	Propane, 2-cyclohexyl-2-(4-acetoxycyclohexyl)-		IC03463
123	43	69	41	122	67	81	55	266	100	66	60	51	50	48	44	44	0.49	17	30	2	–	–	–	–	–	–	–	–	–	–	Propane, 2-cyclohexyl-2-(4-acetoxycyclohexyl)-		IC03464
123	69	43	67	55	41	81	122	266	100	66	61	47	46	46	44	42	0.51	17	30	2	–	–	–	–	–	–	–	–	–	–	Propane, 2-cyclohexyl-2-(4-acetoxycyclohexyl)-		IC03465
266	251	119	107	265	223	121	131	266	100	12	9	8	7	7	6	5		18	18	2	–	–	–	–	–	–	–	–	–	–	Benzofuran, 2,3-dihydro-2-(4-hydroxyphenyl)-3-methyl-5-propenyl-, (2R,3R)-(E)-		MS05966
105	266	77	250	267	191	51	115	266	100	92	40	20	17	14	10	10		18	18	2	–	–	–	–	–	–	–	–	–	–	1,3-Butadiene, 1,4-dimethoxy-1,4-diphenyl-	54848-75-8	NI26888
222	91	165	221	167	179	77	266	266	100	64	53	48	48	41	36	31		18	18	2	–	–	–	–	–	–	–	–	–	–	4H-Cyclohepta[cd]benzofuran-2-one, 5,7,8,9,9a,9b-hexahydro-4-phenyl-, (±)-(4α,9aα,9bα)-		DD00821
197	43	181	223	165	266	152	240	266	100	95	69	43	43	25	23	12		18	18	2	–	–	–	–	–	–	–	–	–	–	1,4:5,8-Diepoxy-1,5,9,10-tetramethyl-1,4,5,8-tetrahydroanthracene		DD00822
43	197	181	165	85	223	266	152	266	100	46	37	27	26	23	19	15		18	18	2	–	–	–	–	–	–	–	–	–	–	1,4:5,8-Diepoxy-1,8,9,10-tetramethyl-1,4,5,8-tetrahydroanthracene		DD00823
266	223	210	197	209	267	195	238	266	100	32	32	29	25	22	15	14		18	18	2	–	–	–	–	–	–	–	–	–	–	Estra-1,3,5,7,9-pentaen-17-one, 3-hydroxy-	517-09-9	NI26889
266	223	210	197	209	195	181	165	266	100	33	29	29	26	14	12	11		18	18	2	–	–	–	–	–	–	–	–	–	–	Estra-1,3,5,7,9-pentaen-17-one, 3-hydroxy-	517-09-9	NI26890
266	223	197	210	267	209	195	165	266	100	28	28	26	22	21	11	11		18	18	2	–	–	–	–	–	–	–	–	–	–	Estra-1,3,5,7,9-pentaen-17-one, 3-hydroxy-	517-09-9	NI26891
105	77	133	106	51	122	131	50	266	100	88	86	77	74	65	56	55	7.92	18	18	2	–	–	–	–	–	–	–	–	–	–	1,5-Hexadiene-3,4-diol, 1,6-diphenyl-	4403-20-7	NI26892
43	105	77	146	51	120	45	61	266	100	58	44	17	15	12	12	8	1.20	18	18	2	–	–	–	–	–	–	–	–	–	–	1,6-Hexanedione, 1,6-diphenyl-	3375-38-0	NI26893
266	251	267	223	133	165	180	252	266	100	35	20	17	17	9	8	7		18	18	2	–	–	–	–	–	–	–	–	–	–	Phenanthrene, 3,6-dimethoxy-9,10-dimethyl-		MS05967
221	266	28	233	222	248	178	202	266	100	69	60	39	38	36	24	18		18	18	2	–	–	–	–	–	–	–	–	–	–	9-Phenanthreneethanol, 10-hydroxy-α,β-dimethyl-	62108-20-7	NI26894
57	51	41	39	50	65	53	63	266	100	74	27	27	21	19	19	15	8.00	18	19	–	–	–	–	–	–	–	–	1	–	–	Phosphine, (3,3-dimethyl-1-butynyl)diphenyl-	33730-51-7	NI26895
184	266	183	41	267	43	223	167	266	100	87	53	37	33	31	26	24		18	22	–	2	–	–	–	–	–	–	–	–	–	Benzylamine, 4-(cyclohexylamino)-α-phenyl-		IC03466
223	222	208	207	206	224	204	205	266	100	38	36	27	24	18	15	11	7.01	18	22	–	2	–	–	–	–	–	–	–	–	–	9H-Carbazole-2-ethylamine, 3-ethyl-N,1-dimethyl-	55044-78-5	NI26896
136	266	130	123	144	122	138	124	266	100	31	22	20	16	13	9	8		18	22	–	2	–	–	–	–	–	–	–	–	–	Condyfolan, 14,19-didehydro-, (14E)-	3890-05-9	NI26898
136	42	266	123	41	130	144	122	266	100	36	34	30	28	25	22	22		18	22	–	2	–	–	–	–	–	–	–	–	–	Condyfolan, 14,19-didehydro-, (14E)-	3890-05-9	NI26897
194	209	223	180	266	208	237	195	266	100	79	73	71	70	66	56	40		18	22	–	2	–	–	–	–	–	–	–	–	–	Dasycarpidan, 1-methylene-	517-81-7	NI26899
209	194	266	180	44	237	208	195	266	100	89	82	80	76	60	45	40		18	22	–	2	–	–	–	–	–	–	–	–	–	Dasycarpidan, 1-methylene-	517-81-7	NI26900
235	195	234	208	44	193	266	71	266	100	97	94	90	71	65	55	52		18	22	–	2	–	–	–	–	–	–	–	–	–	5H-Dibenz[b,f]azepine-5-propanamine, 10,11-dihydro-N-methyl-	50-47-5	NI26901
195	235	234	193	44	208	194	18	266	100	69	69	64	63	53	52	37	28.42	18	22	–	2	–	–	–	–	–	–	–	–	–	5H-Dibenz[b,f]azepine-5-propanamine, 10,11-dihydro-N-methyl-	50-47-5	NI26902
195	235	234	208	193	266	194	84	266	100	94	78	70	70	64	58	57		18	22	–	2	–	–	–	–	–	–	–	–	–	5H-Dibenz[b,f]azepine-5-propanamine, 10,11-dihydro-N-methyl-	50-47-5	NI26903
223	77	224	160	106	104	91	51	266	100	19	15	7	6	6	5	5	3.63	18	22	–	2	–	–	–	–	–	–	–	–	–	Imidazolidine, 1,3-diphenyl-2-propyl-	55320-82-6	NI26904
266	223	91	212	116	169	181	225	266	100	60	60	45	45	35	34	27		18	22	–	2	–	–	–	–	–	–	–	–	–	2,4-Pentadienenitrile, 2-(N-allyl-N-butylamino)-5-phenyl-, (2E,4E)-	117679-48-8	DD00824
266	175	91	223	115	212	181	169	266	100	98	98	58	51	45	34	34		18	22	–	2	–	–	–	–	–	–	–	–	–	2,4-Pentadienenitrile, 2-(N-allyl-N-butylamino)-5-phenyl-, (2Z,4E)-	117679-38-6	DD00825
146	91	119	118	266	65	120	105	266	100	71	68	55	49	26	22	16		18	22	–	2	–	–	–	–	–	–	–	–	–	Piperazine, 1,4-bis(3-tolyl)-		IC03467
167	195	189	267					266	100	12	11	5					0.00	18	22	–	2	–	–	–	–	–	–	–	–	–	Piperazine, 1-(diphenylmethyl)-4-methyl-	82-92-8	NI26906
99	56	167	266	194	165	207	195	266	100	66	60	43	43	38	36	36		18	22	–	2	–	–	–	–	–	–	–	–	–	Piperazine, 1-(diphenylmethyl)-4-methyl-	82-92-8	NI26905
98	112	55	84	69	238	124	266	266	100	36	35	30	20	5	5	4		18	34	1	–	–	–	–	–	–	–	–	–	–	Cyclobutanone, 2-tetradecyl-	35493-47-1	NI26907
41	67	55	95	81	82	68	43	266	100	76	76	62	62	56	55	54	0.00	18	34	1	–	–	–	–	–	–	–	–	–	–	1-Cyclopropeneheptanol, 2-octyl-		NI26908
124	95	96	55	81	69	97	41	266	100	83	68	67	64	53	36	29	0.00	18	34	1	–	–	–	–	–	–	–	–	–	–	Bis[2-(4-methylcyclohexyl)ethyl] ether		DO01058
67	81	55	41	82	95	54	68	266	100	82	80	68	59	57	49	47	16.61	18	34	1	–	–	–	–	–	–	–	–	–	–	9,12-Octadecadien-1-ol	1577-52-2	WI01323
95	83	97	109	123	81	68	111	266	100	88	83	82	63	63	58	55	13.60	18	34	1	–	–	–	–	–	–	–	–	–	–	9,12-Octadecadien-1-ol, (Z,Z)-	506-43-4	NI26909
43	57	83	69	41	55	82	97	266	100	85	75	62	62	59	53	52	0.00	18	34	1	–	–	–	–	–	–	–	–	–	–	2-Octadecenal	56554-96-2	NI26910
43	57	83	84	69	41	55	28	266	100	89	63	61	58	58	54	54	0.00	18	34	1	–	–	–	–	–	–	–	–	–	–	3-Octadecenal	56554-99-5	NI26911
57	43	83	69	84	41	55	97	266	100	97	70	59	56	55	52	50	0.00	18	34	1	–	–	–	–	–	–	–	–	–	–	4-Octadecenal	56554-98-4	NI26912
57	43	83	69	84	97	55	41	266	100	92	78	65	61	56	52	52	0.34	18	34	1	–	–	–	–	–	–	–	–	–	–	5-Octadecenal	56554-88-2	NI26913

MASS TO CHARGE RATIOS								M.W.	INTENSITIES								Parent	C	H	O	N	S	F	Cl	Br	I	Si	P	B	X	COMPOUND NAME	CAS Reg No	No
57	43	83	41	82	68	55	97	266	100	96	77	56	54	51	50	49	0.00	18	34	1	–	–	–	–	–	–	–	–	–	–	6-Octadecenal	56554-97-3	NI26914
57	43	82	83	41	68	55	44	266	100	96	66	61	58	52	51	51	0.00	18	34	1	–	–	–	–	–	–	–	–	–	–	7-Octadecenal	56555-00-1	NI26915
57	43	82	41	83	68	55	44	266	100	97	71	63	62	58	56	49	0.20	18	34	1	–	–	–	–	–	–	–	–	–	–	8-Octadecenal	56554-94-0	NI26916
41	55	43	29	69	67	57	54	266	100	78	58	47	36	34	27	27	2.20	18	34	1	–	–	–	–	–	–	–	–	–	–	9-Octadecenal	5090-41-5	NI26917
41	43	57	82	68	96	110	248	266	100	58	27	18	18	12	4	3	2.20	18	34	1	–	–	–	–	–	–	–	–	–	–	9-Octadecenal, (Z)-	2423-10-1	LI05605
55	41	69	43	67	83	57	81	266	100	88	66	64	55	53	49	46	3.34	18	34	1	–	–	–	–	–	–	–	–	–	–	9-Octadecenal, (Z)-	2423-10-1	NI26918
57	43	82	41	55	68	44	83	266	100	97	84	69	66	61	60	59	0.30	18	34	1	–	–	–	–	–	–	–	–	–	–	10-Octadecenal	56554-92-8	NI26919
43	57	82	41	55	44	68	83	266	100	94	77	74	71	56	55	52	0.74	18	34	1	–	–	–	–	–	–	–	–	–	–	11-Octadecenal	56554-95-1	NI26920
82	57	43	68	55	41	96	44	266	100	94	82	64	64	61	60	60	0.00	18	34	1	–	–	–	–	–	–	–	–	–	–	12-Octadecenal	56554-91-7	NI26921
82	57	43	55	68	96	44	41	266	100	90	74	66	61	60	59	58	0.20	18	34	1	–	–	–	–	–	–	–	–	–	–	13-Octadecenal	56554-90-6	NI26922
82	57	43	55	96	68	41	44	266	100	88	71	66	62	60	58	57	0.24	18	34	1	–	–	–	–	–	–	–	–	–	–	14-Octadecenal	56554-89-3	NI26923
82	57	43	55	44	68	96	41	266	100	91	79	67	62	61	60	60	0.00	18	34	1	–	–	–	–	–	–	–	–	–	–	15-Octadecenal	56554-93-9	NI26924
82	57	43	55	96	68	41	83	266	100	84	74	66	63	59	56	54	0.20	18	34	1	–	–	–	–	–	–	–	–	–	–	16-Octadecenal	56554-87-1	NI26925
82	57	43	55	96	68	41	83	266	100	85	74	69	61	59	58	54	0.20	18	34	1	–	–	–	–	–	–	–	–	–	–	17-Octadecenal	56554-86-0	NI26926
43	55	41	57	83	67	266	153	266	100	99	90	75	69	68	62	62		18	34	1	–	–	–	–	–	–	–	–	–	–	2-Undecenal, 2-heptyl-		IC03468
84	69	183	41	105	77	43	208	266	100	57	31	20	19	18	11	9	2.00	19	22	1	–	–	–	–	–	–	–	–	–	–	Oxetane, 2,2,3,3-tetramethyl-4,4-diphenyl-	42245-06-7	NI26927
115	266	136	148	117	267	91	121	266	100	99	51	29	24	22	19	16		19	22	1	–	–	–	–	–	–	–	–	–	–	[3.3]-Paracyclophane, 6-methoxy-	24770-02-3	NI26928
251	197	69	165	55	43	153	83	266	100	76	67	61	58	56	54	47	25.00	19	22	1	–	–	–	–	–	–	–	–	–	–	9,10,11,12-Tetradehydro-19-norretinal (all-trans)		NI26929
41	43	57	83	55	69	97	29	266	100	97	84	76	73	71	60	59	0.45	19	38	–	–	–	–	–	–	–	–	–	–	–	1-Cyclopentyl-2-hexyloctane		AI01888
41	43	57	83	55	69	97	71	266	100	90	83	73	72	66	61	43	0.81	19	38	–	–	–	–	–	–	–	–	–	–	–	1-Cyclopentyl-2-hexyloctane		AI01889
43	41	55	57	29	83	69	97	266	100	99	91	72	57	54	54	53	8.51	19	38	–	–	–	–	–	–	–	–	–	–	–	1-Nonadecene	18435-45-5	NI26930
83	57	55	82	69	43	41	71	266	100	98	88	81	66	64	57	48	0.00	19	38	–	–	–	–	–	–	–	–	–	–	–	Tridecane, 3-cyclohexyl-	13151-88-7	NI26931
57	83	55	82	43	41	71	97	266	100	95	85	69	64	52	51	41	0.00	19	38	–	–	–	–	–	–	–	–	–	–	–	Tridecane, 4-cyclohexyl-	13151-89-8	NI26932
57	83	55	43	82	71	97	41	266	100	79	75	63	60	57	51	47	0.00	19	38	–	–	–	–	–	–	–	–	–	–	–	Tridecane, 5-cyclohexyl-	13151-90-1	NI26933
57	83	55	43	82	71	41	97	266	100	78	78	60	58	56	52	44	0.00	19	38	–	–	–	–	–	–	–	–	–	–	–	Tridecane, 6-cyclohexyl-	13151-91-2	NI26934
57	55	83	82	43	41	71	69	266	100	81	76	63	62	61	55	52	0.00	19	38	–	–	–	–	–	–	–	–	–	–	–	Tridecane, 7-cyclohexyl-	13151-92-3	NI26935
55	41	43	83	57	82	29	71	266	100	97	94	88	88	71	60	54	0.86	19	38	–	–	–	–	–	–	–	–	–	–	–	Tridecane, 7-cyclohexyl-	13151-92-3	AI01890
55	41	43	83	57	82	71	182	266	100	97	86	79	77	65	44	40	2.80	19	38	–	–	–	–	–	–	–	–	–	–	–	Tridecane, 7-cyclohexyl-	13151-92-3	AI01891
133	105	134	104	132	41	93	117	266	100	15	10	10	8	6	5	4	1.36	20	26	–	–	–	–	–	–	–	–	–	–	–	Benzene, 1,1′-(1,1,2,2-tetramethyl-1,2-ethanediyl)bis[4-methyl-	734-17-8	NI26936
266	223	238	237	197	224	195	181	266	100	93	53	37	31	30	25	24		20	26	–	–	–	–	–	–	–	–	–	–	–	Benzo[b]fluoranthene, 1,2,3,3a,3b,4,5,6,7,7a,9,10,11,12-tetradecahydro-		NI26937
266	238	267	265	209	237	223	165	266	100	66	19	19	16	14	14	14		20	26	–	–	–	–	–	–	–	–	–	–	–	Benzo[a]pyrene, 1,2,3,3a,4,5,7,8,9,10,11,12,12a-tetradecahydro-	95676-40-7	NI26938
266	251	236	221	237	252	235	222	266	100	83	42	33	18	17	16	12		20	26	–	–	–	–	–	–	–	–	–	–	–	1,1′-Biphenyl, 2,2′,3,3′,5,5′,6,6′-octamethyl-		LI05606
197	198	266	169	41	195	167	153	266	100	17	9	9	6	5	5	5		20	26	–	–	–	–	–	–	–	–	–	–	–	1H-Cyclopent[e]-as-indacene, 1-cyclopentyl-2,3,4,5,6,7,8,9-octahydro-		IC03469
266	197	171	184	169	267	198	170	266	100	98	51	32	24	23	21	19		20	26	–	–	–	–	–	–	–	–	–	–	–	1H-Cyclopent[e]-as-indacene, 2-cyclopentyl-2,3,4,5,6,7,8,9-octahydro-		IC03470
266	238	223	210	237	29	267	28	266	100	92	34	32	29	28	21	21		20	26	–	–	–	–	–	–	–	–	–	–	–	Perylene, 1,2,3,3a,4,5,6,7,8,9,9a,10,11,12-tetradecahydro-	7594-86-7	AI01892
266	238	223	237	210	267	265	239	266	100	73	27	24	24	21	15	14		20	26	–	–	–	–	–	–	–	–	–	–	–	Perylene, 1,2,3,3a,4,5,6,7,8,9,9a,10,11,12-tetradecahydro-	7594-86-7	AI01893
266	265	268	267	263	133	252	264	266	100	39	28	28	18	17	10	9		21	14	–	–	–	–	–	–	–	–	–	–	–	Benz[j]aceanthrylene, 3-methyl-	3343-10-0	WI01324
232	200	267	85	63	130	50	45	267	100	58	53	37	35	22	21	17		–	–	–	1	1	1	4	–	–	–	2	–	–	Phosphorimidic dichloride fluoride, (dichlorophosphinothioyl)-	27352-09-6	MS05968
35	69	36	136	64	117	101	32	267	100	99	98	66	60	54	38	38	0.00	1	–	1	1	1	3	3	–	–	–	1	–	–	Phosphorimidic trichloride, [(trifluoromethyl)sulphinyl]-	51735-87-6	NI26939
150	135	122	267	269	65	92	52	267	100	28	28	17	16	12	11	9		9	8	2	1	–	–	3	–	–	–	–	–	–	Acetamide, 2,2,2-trichloro-N-(2-methoxyphenyl)-	4257-82-3	NI26940
150	107	122	77	267	269	151	92	267	100	37	25	25	12	11	9	9		9	8	2	1	–	–	3	–	–	–	–	–	–	Acetamide, 2,2,2-trichloro-N-(3-methoxyphenyl)-	4306-33-6	NI26941
122	150	267	269	149	123	52	40	267	100	47	20	19	9	9	9	8		9	8	2	1	–	–	3	–	–	–	–	–	–	Acetamide, 2,2,2-trichloro-N-(4-methoxyphenyl)-	4257-81-2	NI26942
58	71	42	44	72	56	99	55	267	100	75	75	70	62	61	41	35	2.91	9	10	5	1	–	–	–	–	–	–	–	–	1	Manganese, π-(dimethylamino)propionyltetracarbonyl-		MS05969
267	269	241	239	77	76	92	120	267	100	99	80	80	80	60	54	44		10	6	3	1	–	–	–	1	–	–	–	–	–	1H-1-Benzazepine-2,5-dione, 4-bromo-3-hydroxy-	52280-66-7	NI26943
91	269	267	65	119	270	268		267	100	22	22	13	11	3	3			10	10	1	3	–	–	–	1	–	–	–	–	–	Anhydro-5-bromo-4-hydroxy-1-methyl-3-(4-tolyl)-1,2,3-triazolium hydroxide		LI05607
44	82	204	57	156	124	110	168	267	100	85	80	60	55	40	30	25	0.00	10	12	1	1	–	–	3	–	–	–	–	–	–	4-Piperidineethanol, α-(dichloromethyl)-3-vinyl-, hydrochloride, [3R-[3α,4α(S*)]]-	35189-38-9	NI26944
178	164	52	109	119	67	43	149	267	100	61	56	42	27	23	23	22	0.40	10	13	4	5	–	–	–	–	–	–	–	–	–	Adenine, 2-β-D-ribofuranosyl-		NI26945
164	178	267	176	43	165	149	54	267	100	96	25	20	17	14	14	13		10	13	4	5	–	–	–	–	–	–	–	–	–	Adenine, 8-β-D-ribofuranosyl-		NI26946
135	136	178	164	108	148	57	119	267	100	45	43	38	28	9	9	7	4.20	10	13	4	5	–	–	–	–	–	–	–	–	–	Adenine, 9α-D-xylofuranosyl-	2946-52-3	LI05608
135	136	178	164	108	148	43	57	267	100	45	42	38	27	8	8	7	4.00	10	13	4	5	–	–	–	–	–	–	–	–	–	Adenine, 9α-D-xylofuranosyl-	2946-52-3	NI26947
135	136	164	178	108	148	43	57	267	100	94	65	38	32	13	12	10	5.20	10	13	4	5	–	–	–	–	–	–	–	–	–	Adenine, 9β-D-xylofuranosyl-		LI05609
135	136	164	108	43	178	148	54	267	100	79	70	49	37	36	23	23	3.40	10	13	4	5	–	–	–	–	–	–	–	–	–	Adenosine	58-61-7	NI26950

MASS TO CHARGE RATIOS								M.W.	INTENSITIES								Parent	C	H	O	N	S	F	Cl	Br	I	Si	P	B	X	COMPOUND NAME	CAS Reg No	No
135	164	136	108	178	237	148	44	267	100	87	75	36	34	10	9	8	1.57	10	13	4	5	–	–	–	–	–	–	–	–	–	Adenosine	58-61-7	NI26949
135	164	136	28	108	178	27	29	267	100	57	56	34	26	21	16	15	2.34	10	13	4	5	–	–	–	–	–	–	–	–	–	Adenosine	58-61-7	NI26948
164	44	28	45	17	267	149	54	267	100	84	83	81	76	50	44	44		10	13	4	5	–	–	–	–	–	–	–	–	–	Formycin	6742-12-7	LI05610
81	44	99	43	73	57	98	42	267	100	83	80	67	64	62	61	61	0.00	10	13	4	5	–	–	–	–	–	–	–	–	–	Guanosine, 2′-deoxy-	961-07-9	NI26951
135	148	179	178	136	149	108	67	267	100	82	62	60	55	42	34	18	12.00	10	13	4	5	–	–	–	–	–	–	–	–	–	Lentysine methyl ester		MS05970
79	78	106	51	52	29	49	27	267	100	53	30	23	17	15	14	14	0.00	10	14	2	1	–	4	–	–	–	–	–	1	–	2-(Ethoxycarbonyl)-N-ethylpyridine, boron tetrafluoride		NI26952
147	104	76	46	194	50	45	103	267	100	78	68	52	37	35	22	20	10.00	11	9	5	1	1	–	–	–	–	–	–	–	–	Acetic acid, [(1,3-dihydro-1,3-dioxo-2H-isoindol-2-yl)sulphinyl]-, methyl ester	42300-59-4	NI26953
249	44	223	143	115	159	267	114	267	100	83	82	54	46	45	28	27		11	9	5	1	1	–	–	–	–	–	–	–	–	2-Naphthalenecarboxylic acid, 4-(aminosulphonyl)-1-hydroxy-	64415-15-2	NI26954
105	267	252	79	104	107	77	91	267	100	93	85	34	31	29	21	17		11	10	1	1	–	5	–	–	–	–	–	–	–	4-Methylbenzylamine pentafluoropropionate		MS05971
104	91	105	65	119	28	176	92	267	100	52	17	9	8	8	6	5	2.20	11	10	1	1	–	5	–	–	–	–	–	–	–	Propanamide, 2,2,3,3,3-pentafluoro-N-(2-phenylethyl)-		MS05972
104	91	105	65	78	92	77	51	267	100	62	15	13	8	6	6	5	0.00	11	10	1	1	–	5	–	–	–	–	–	–	–	Propanamide, 2,2,3,3,3-pentafluoro-N-(2-phenylethyl)-		MS05973
71	209	267	208	45	236	211	210	267	100	90	87	85	56	54	39	39		11	10	1	3	1	–	1	–	–	–	–	–	–	4-Thiazoleacetic acid, 2-(4-chlorophenyl)-, hydrazide	33313-15-4	NI26955
160	188	76	161	104	77	50	41	267	100	51	30	27	21	20	20	14	12.86	11	10	2	1	–	–	–	1	–	–	–	–	–	Phthalimide, N-(3-bromopropyl)-	5460-29-7	IC03471
160	188	76	161	104	50	77	41	267	100	44	39	22	21	21	18	13	5.58	11	10	2	1	–	–	–	1	–	–	–	–	–	Phthalimide, N-(3-bromopropyl)-	5460-29-7	NI26956
150	43	118	96	123	86	60	151	267	100	33	22	22	21	18	18	13	2.00	11	13	1	3	2	–	–	–	–	–	–	–	–	N-(2-Acetamidoethylthio)-2-aminobenzothiazole		LI05611
113	94	268	126	141	72	111	95	267	100	24	20	20	17	17	16	15	0.00	11	13	3	3	1	–	–	–	–	–	–	–	–	Benzenesulphonamide, 4-amino-N-(4,5-dimethyl-2-oxazolyl)-	729-99-7	NI26957
115	88	116	56	60	43	77	103	267	100	66	50	28	16	14	11	9	2.40	11	13	3	3	1	–	–	–	–	–	–	–	–	3-[2-Hydroxy-2-(4-nitrophenyl)ethyl]-2-iminothiazolidine		IC03472
92	156	108	65	267	157	93	66	267	100	81	63	49	11	10	9	8		11	13	3	3	1	–	–	–	–	–	–	–	–	Sulphaisoxazole		LI05612
166	69	41	29	27	96	167	39	267	100	26	26	16	12	11	7	7	2.20	11	16	3	1	–	3	–	–	–	–	–	–	–	L-Proline, 1-(trifluoroacetyl)-, butyl ester	2563-29-3	NI26958
166	41	167	57	69	56	194	96	267	100	65	63	43	32	20	17	16	8.15	11	16	3	1	–	3	–	–	–	–	–	–	–	L-Proline, 1-(trifluoroacetyl)-, sec-butyl ester	58072-45-0	NI26959
166	167	57	41	69	96	194	53	267	100	14	14	14	11	6	4	3	2.00	11	16	3	1	–	3	–	–	–	–	–	–	–	L-Proline, 1-(trifluoroacetyl)-, sec-butyl ester, (R)-	55056-63-8	NI26960
114	72	127	79	70	115	84	139	267	100	22	11	11	10	8	6	5	0.00	11	26	2	1	1	–	–	–	–	–	1	–	–	O-Ethyl O-diisopropylaminoethyl methylphosphonothionate		MS05974
268	128	127	141	252	114	266		267	100	40	25	20	16	14	12		0.00	11	26	2	1	1	–	–	–	–	–	1	–	–	O-Ethyl S-diisopropylaminoethyl methylphosphonothiolate		NI26962
114	72	127	70	85	84	83	167	267	100	25	23	16	13	13	13	8	0.00	11	26	2	1	1	–	–	–	–	–	1	–	–	O-Ethyl S-diisopropylaminoethyl methylphosphonothiolate		NI26961
114	72	70	85	84	83	167	127	267	100	25	16	13	13	13	8	3	0.00	11	26	2	1	1	–	–	–	–	–	1	–	–	O-Ethyl S-diisopropylaminoethyl methylphosphonothiolate		MS05975
222	224	267	45	85	223	269	59	267	100	39	36	34	27	18	16	16		12	10	2	1	1	–	1	–	–	–	–	–	–	2-Thiazoleacetic acid, 2-(4-chlorophenyl)-4-methyl-	17969-41-4	NI26963
71	45	208	267	69	59	138	75	267	100	89	56	44	22	22	21	21		12	10	2	1	1	–	1	–	–	–	–	–	–	4-Thiazoleacetic acid, 2-(4-chlorophenyl)-, methyl ester	17969-38-9	NI26964
222	224	267	45	223	71	269	85	267	100	88	71	67	51	39	28	27		12	10	2	1	1	–	1	–	–	–	–	–	–	4-Thiazolepropanoic acid, 2-(4-chlorophenyl)-	23856-05-5	NI26965
139	111	149	128	267	113	140	57	267	100	46	28	20	15	10	9	8		12	10	4	1	–	–	1	–	–	–	–	–	–	Benzamide, 3-chloro-N-[dihydro-4-(hydroxymethyl)-2-oxo-3-furanyl]-	72361-14-9	LI05613
139	111	141	128	267	113	140	75	267	100	49	28	20	15	11	9	8		12	10	4	1	–	–	1	–	–	–	–	–	–	Benzamide, 3-chloro-N-[dihydro-4-(hydroxymethyl)-2-oxo-3-furanyl]-	72361-14-9	NI26966
267	148	269	176	191	113	178	99	267	100	36	34	29	26	26	22	18		12	10	4	1	–	–	1	–	–	–	–	–	–	Naphthalene, 1-chloro-2,6-dimethoxy-5-nitro-	25314-98-1	NI26967
267	148	269	176	191	113	178	174	267	100	35	33	29	26	26	21	17		12	10	4	1	–	–	1	–	–	–	–	–	–	Naphthalene, 1-chloro-2,6-dimethoxy-5-nitro-	25314-98-1	LI05614
105	267	121	122	223	166	222	224	267	100	52	42	32	28	20	18	8		12	13	2	1	2	–	–	–	–	–	–	–	–	4-Thiazolidinone, 5-ethoxy-3-methyl-5-phenyl-2-thioxo-	116927-96-9	DD00826
43	175	93	267	147	65	39	77	267	100	91	45	42	29	17	17	12		12	13	4	1	1	–	–	–	–	–	–	–	–	1,4-Oxathiin-3-carboxamide, 5,6-dihydro-2-methyl-N-phenyl-, 4,4-dioxide	5259-88-1	NI26968
42	119	43	183	56	84	91	239	267	100	25	23	14	10	9	7	7	1.76	12	13	4	1	1	–	–	–	–	–	–	–	–	3-Pyrrolidinone, 1-(4-acetylbenzenesulphonyl)-		NI26969
91	108	107	43	79	77	42	51	267	100	85	77	76	67	35	23	17	2.87	12	13	6	1	–	–	–	–	–	–	–	–	–	L-Aspartic acid, N-[benzyloxycarbonyl]-	1152-61-0	NI26970
57	89	41	224	226	210	212	185	267	100	15	15	14	13	13	12	9	0.50	12	14	1	1	–	–	–	1	–	–	–	–	–	2-Aziridinone, 3-(4-bromophenyl)-1-tert-butyl-	27147-97-3	NI26971
160	159	223	267	145	90	91	63	267	100	99	58	36	33	24	19	18		12	17	2	3	1	–	–	–	–	–	–	–	–	5-Benzimidazolesulphonamide, N,1-diethyl-2-methyl-	4979-75-3	NI26972
194	267	223	45	221	268	238	208	267	100	90	35	33	26	16	16	16		12	21	2	5	–	–	–	–	–	–	–	–	–	1,3,5-Triazine-2-carboxylic acid, 4,6-bis(diethylamino)-		NI26973
267	69	247	235	45	63	75	50	267	100	72	39	32	28	24	19	19		13	8	–	1	1	3	–	–	–	–	–	–	–	10H-Phenothiazine, 2-(trifluoromethyl)-	92-30-8	NI26974
267	266	235	268	69	247	234	248	267	100	34	22	17	14	12	7	7		13	8	–	1	1	3	–	–	–	–	–	–	–	10H-Phenothiazine, 2-(trifluoromethyl)-	92-30-8	NI26975
102	130	51	52	103	76	75	50	267	100	69	18	12	11	8	8	8	1.50	13	9	2	5	–	–	–	–	–	–	–	–	–	1-(2-Cyanophenyl)-3-(2-nitrophenyl)triazene		LI05615
173	175	267	145	269	147	109	177	267	100	64	25	18	16	12	11	10		13	11	1	1	–	–	2	–	–	–	–	–	–	Pyrrole, 1-(2,6-dichlorobenzoyl)-2,5-dimethyl-	95883-12-8	NI26976
209	211	59	210	71	138	45	72	267	100	30	21	12	12	10	9	7	3.30	13	14	1	1	1	–	1	–	–	–	–	–	–	4-Thiazoleethanol, 2-(4-chlorophenyl)-α,α-dimethyl-	33313-14-3	NI26977
138	179	165	41	130	137	42	140	267	100	71	59	51	48	45	42	33	27.00	13	16	2	2	–	–	1	–	–	–	–	–	–	3-Imidazoline, 2,2,5,5-tetramethyl-4-(4-chlorophenyl)-, 3-oxide 1-oxile		NI26978
91	208	176	267	88	117	86	134	267	100	31	29	12	12	10	10	9		13	17	3	1	1	–	–	–	–	–	–	–	–	L-Cysteine, N-acetyl-S-benzyl-, methyl ester	77549-14-5	NI26979
176	191	132	160	159	117	190	130	267	100	59	46	38	31	29	28	26	0.00	13	17	5	1	–	–	–	–	–	–	–	–	–	Dimethyl hydroxywilfordic acid		NI26980
91	161	93	65	41	133	267	29	267	100	63	34	29	23	20	18	14		13	18	3	1	–	–	–	–	–	–	1	–	–	1,2-Azaphosphinane, 6-benzyloxy-2-ethoxy-2,3,4,5-tetrahydro-, 2-oxide	93939-03-8	NI26981
91	136	80	108	65	41	29	267	267	100	86	50	49	42	42	38	37		13	18	3	1	–	–	–	–	–	–	1	–	–	1,2-Azaphosphinan-6-one, 1-benzyl-2-ethoxy-, 2-oxide	93939-02-7	NI26982
79	147	91	52	73	176	51	50	267	100	85	73	73	34	32	32	16	10.40	13	25	1	1	–	–	–	–	–	2	–	–	–	Benzyloxyamine, N,N-bis(trimethylsilyl)-		MS05976
267	164	77	105	134	88	103	51	267	100	82	45	34	32	31	26	24		14	9	3	3	–	–	–	–	–	–	–	–	–	5-Phenyl-3-(4-nitrophenyl)-1,2,4-oxadiazole		NI26983
44	51	45	220	50	221	42	27	267	100	69	52	31	27	26	22	20	0.00	14	16	–	1	–	2	–	–	–	–	1	–	–	Phosphorane, difluorodiphenyl(dimethylamino)-		LI05616
224	155	91	267	42	198	212	112	267	100	84	84	31	31	27	23	18		14	21	2	1	1	–	–	–	–	–	–	–	–	Benzenesulphonamide, N-cyclohexyl-N,4-dimethyl-	57186-74-0	NI26984
152	164	267	235	226	126	153	122	267	100	39	31	24	20	14	13	13		14	21	4	1	–	–	–	–	–	–	–	–	–	Methyl α-(2-morpholinocyclopentenyl)acetoacetate		MS05977
225	43	29	41	79	152	107	55	267	100	77	41	38	19	18	18	18	14.00	14	21	4	1	–	–	–	–	–	–	–	–	–	3-Oxa-2-aza[3.3.3]propellane-4-carboxylic acid, 2-acetyl-, ethyl ester		NI26985

MASS TO CHARGE RATIOS	M.W.	INTENSITIES	Parent	C	H	O	N	S	F	Cl	Br	I	Si	P	B	X	COMPOUND NAME	CAS Reg No	No
252 206 251 178 196 253 224 150	**267**	100 50 28 20 18 15 14 13	**0.00**	14	21	4	1	–	–	–	–	–	–	–	–	–	3,5-Pyridinedicarboxylic acid, 1,4-dihydro-4-isopropyl-, diethyl ester	54833-75-9	NI26986
252 196 222 253 224 267 223 206	**267**	100 22 19 14 14 3 3 3		14	21	4	1	–	–	–	–	–	–	–	–	–	3,5-Pyridinedicarboxylic acid, 1,4-dihydro-2,4,6-trimethyl-, diethyl ester	632-93-9	LI05617
252 196 222 253 224 29 42 179	**267**	100 21 19 14 14 7 4 3	**1.01**	14	21	4	1	–	–	–	–	–	–	–	–	–	3,5-Pyridinedicarboxylic acid, 1,4-dihydro-2,4,6-trimethyl-, diethyl ester	632-93-9	NI26987
252 196 222 253 224 179 178 150	**267**	100 22 20 15 15 4 4 4	**3.91**	14	21	4	1	–	–	–	–	–	–	–	–	–	3,5-Pyridinedicarboxylic acid, 1,4-dihydro-2,4,6-trimethyl-, diethyl ester	632-93-9	NI26988
44 59 41 55 69 57 81 67	**267**	100 64 62 57 37 34 30 25	**5.00**	14	25	2	3	–	–	–	–	–	–	–	–	–	Cyclohexane, 2-(dimethylhydrazono)-3-[4-hexenyl]-1-aci-nitro-, (E,E)-		MS05978
267 223 249 44 69 57 55 43	**267**	100 58 50 29 24 21 21 21		15	9	4	1	–	–	–	–	–	–	–	–	–	9,10-Anthracenedione, 1-amino-2-carboxy-		IC03473
267 165 221 222 250 209 164 163	**267**	100 63 58 31 30 25 25 24		15	9	4	1	–	–	–	–	–	–	–	–	–	9,10-Anthracenedione, 2-methyl-1-nitro-	129-15-7	NI26989
267 164 89 239 237 193 104	**267**	100 30 12 8 8 5 3		15	9	4	1	–	–	–	–	–	–	–	–	–	Isocoumarin, 3-(4-nitrophenyl)-		LI05618
77 51 105 99 50 267 135 74	**267**	100 45 28 25 20 15 13 11		15	10	–	3	–	–	1	–	–	–	–	–	–	Quinoline, 7-chloro-4-(phenylazo)-	90929-82-1	NI26990
178 267 75 89 179 269 77 51	**267**	100 50 25 24 20 16 11 11		15	10	–	3	–	–	1	–	–	–	–	–	–	1,2,4-Triazine, 3-chloro-5,6-diphenyl-	34177-11-2	NI26991
267 234 237 77 239 205 223 268	**267**	100 51 33 32 29 22 20 19		15	13	–	3	1	–	–	–	–	–	–	–	–	3H-1,3,4-Benzotriazepine-2-thione, 1,2-dihydro-3-methyl-5-phenyl-		MS05979
118 77 267 28 266 51 268	**267**	100 60 58 42 18 16 10		15	13	–	3	1	–	–	–	–	–	–	–	–	1,4-Diphenyl-5-methyl-1,2,4-thiazole-3-thione		LI05619
267 91 194 268 266 158 234 269	**267**	100 80 30 20 11 11 10 7		15	13	–	3	1	–	–	–	–	–	–	–	–	1,2,4-Triazole, 1,3-diphenyl-5-methylthio-	51384-17-9	NI26992
267 91 164 268 266 77 180 269	**267**	100 95 20 18 14 12 10 6		15	13	–	3	1	–	–	–	–	–	–	–	–	1,2,4-Triazole, 1,5-diphenyl-3-methylthio-	72234-23-2	NI26993
120 69 28 68 30 43 119 121	**267**	100 53 34 29 17 16 15 12	**0.00**	15	13	–	3	1	–	–	–	–	–	–	–	–	1H-1,2,3-Triazolium, 4-mercapto-5-methyl-1,3-diphenyl-, hydroxide, inner salt	56701-45-2	NI26994
225 267 182 155 252 197 154 195	**267**	100 45 25 20 10 8 8 5		15	13	2	3	–	–	–	–	–	–	–	–	–	Acetamide, N-(7-methoxy-1-phenazinyl)-	18450-07-2	LI05620
225 154 267 182 155 226 252 197	**267**	100 75 45 25 20 14 10 8		15	13	2	3	–	–	–	–	–	–	–	–	–	Acetamide, N-(7-methoxy-1-phenazinyl)-	18450-07-2	NI26995
267 239 238 210 98 268 211 196	**267**	100 78 53 41 20 19 18 15		15	13	2	3	–	–	–	–	–	–	–	–	–	Carbazole, 3,6-bis(N-formamido)-9-methyl-	98785-98-9	NI26996
133 108 227 54 267 174 104 77	**267**	100 94 87 71 47 27 16 16		15	17	–	5	–	–	–	–	–	–	–	–	–	N,N,N′-Tris(2-cyanoethyl)-1,3-diaminobenzene		IC03474
139 107 167 79 86 81 58 77	**267**	100 61 54 50 21 14 14 12	**0.00**	15	25	3	1	–	–	–	–	–	–	–	–	–	1-Cyclopentenecarboxamide, N,N-diisopropyl-4-(methoxycarbonyl)-4-methyl-	110426-83-0	DD00827
125 137 136 164 151 150 149 135	**267**	100 91 91 64 64 55 55 55	**0.00**	15	25	3	1	–	–	–	–	–	–	–	–	–	Ethaneperoxoic acid, 1-cyano-1,4,8-trimethyl-7-nonenyl ester	58422-65-4	NI26997
110 166 85 182 183 167 111 164	**267**	100 52 50 40 28 18 15 15	**2.30**	15	25	3	1	–	–	–	–	–	–	–	–	–	Pyridine, 2,5-dihydro-6-methoxy-2-[4-[(tetrahydro-2H-pyran-2-yl)oxy]butyl]-		NI26998
91 267 146 92 116 89 103 102	**267**	100 20 13 13 12 9 7 5		16	13	1	1	1	–	–	–	–	–	–	–	–	Isoxazole, 5-(benzylthio)-3-phenyl-	28884-17-5	NI26999
265 207 264 208 77 209 266 132	**267**	100 80 73 73 73 70 53 47	**36.00**	16	13	1	1	1	–	–	–	–	–	–	–	–	1,3-Thiazin-2-one, 3,6-dihydro-4,6-diphenyl-		NI27000
117 151 267 250 144 118 104 107	**267**	100 38 24 14 11 11 11 9		16	13	3	1	–	–	–	–	–	–	–	–	–	2′-Amino-6-methoxyaurone	77764-94-4	NI27001
267 266 251 117 252 224 151 223	**267**	100 93 44 15 11 6 5 5		16	13	3	1	–	–	–	–	–	–	–	–	–	3′-Amino-6-methoxyaurone	77764-95-5	NI27002
267 266 117 106 224 251 207 151	**267**	100 35 19 8 5 4 3 3		16	13	3	1	–	–	–	–	–	–	–	–	–	4′-Amino-6-methoxyaurone	77764-96-6	NI27003
145 146 105 267 77 135 51 103	**267**	100 89 53 43 41 35 24 14		16	13	3	1	–	–	–	–	–	–	–	–	–	Benzeneacetonitrile, α-(benzoyloxy)-4-methoxy-	6948-58-9	NI27004
145 146 267 77 135 105 51 103	**267**	100 90 85 82 70 50 46 30		16	13	3	1	–	–	–	–	–	–	–	–	–	Benzeneacetonitrile, α-(benzoyloxy)-4-methoxy-	6948-58-9	LI05621
105 192 193 182 165 181 183 166	**267**	100 99 56 53 34 9 8 8	**1.00**	16	13	3	1	–	–	–	–	–	–	–	–	–	Ethaneperoxoic acid, cyanodiphenylmethyl ester	58422-81-4	NI27005
267 148 268 252 120 144 115 89	**267**	100 55 45 32 27 25 23 22		16	13	3	1	–	–	–	–	–	–	–	–	–	2(1H)-Quinolinone, 4-(4-methoxyphenoxy)-	68903-83-3	NI27006
120 107 160 77 121 76 104 267	**267**	100 42 25 15 9 7 6 4		16	13	3	1	–	–	–	–	–	–	–	–	–	Tyramine, N-phthaloyl-	64985-42-8	NI27007
30 118 77 43 267 181 57 223	**267**	100 92 79 79 77 73 72 70		16	17	1	3	–	–	–	–	–	–	–	–	–	11H-Dibenzo[b,e][1,4]diazepin-11-one, 5-(3-aminopropyl)-5,10-dihydro-	13450-73-2	NI27008
120 267 42 77 105 148 268 79	**267**	100 65 28 19 18 15 14 14		16	17	1	3	–	–	–	–	–	–	–	–	–	N,N-Dimethyl-4-(4-acetylphenylazo)aniline		LI05622
146 267 105 77 162 145 147 118	**267**	100 72 39 35 30 23 18 15		16	17	1	3	–	–	–	–	–	–	–	–	–	4-(Dimethylamino)benzaldehyde benzoyl hydrazone		LI05623
267 250 266 173 226 252 188 225	**267**	100 62 55 29 22 20 19 15		16	17	1	3	–	–	–	–	–	–	–	–	–	Quinoline, 2-(3,5-dimethylpyrazol-1-yl)-4-methyl-6-methoxy-		MS05980
124 82 83 57 41 94 42 43	**267**	100 32 25 23 21 19 17 15	**7.10**	16	29	2	1	–	–	–	–	–	–	–	–	–	Pentanoic acid, 2-propyl-, 8-methyl-8-azabicyclo[3.2.1]oct-3-yl ester, endo-	25333-49-7	NI27009
84 128 59 55 142 252 268 113	**267**	100 13 10 6 4 3 2 2	**0.00**	16	33	–	1	–	–	–	–	–	1	–	–	–	2,2-Dimethyl-1-(1-pyrrolidinyl)-2-sila-3-undecene, (E)-		DD00828
84 142 59 73 55 268 128 113	**267**	100 76 14 12 7 3 3 3	**0.00**	16	33	–	1	–	–	–	–	–	1	–	–	–	2,2-Dimethyl-1-(1-pyrrolidinyl)-2-sila-3-vinylnonane		DD00829
98 43 41 71 57 55 99 56	**267**	100 99 34 28 24 21 10 10	**0.00**	16	33	–	3	–	–	–	–	–	–	–	–	–	Undecane, 6-azido-6-pentyl-	51677-32-8	NI27010
267 163 116 115 161 150 91 137	**267**	100 54 50 41 21 16 16 14		17	15	–	–	–	–	–	–	–	–	–	–	1	Titanium, (η^8-cyclooctatetraene)(η^5-indenyl)-	54538-57-7	NI27011
132 77 27 117 51 181 39 91	**267**	100 99 93 91 62 46 35 33	**17.92**	17	17	–	1	1	–	–	–	–	–	–	–	–	Azetidin-3-thione, 3,3-dimethyl-1,4-diphenyl-	100679-09-2	NI27012
225 238 226 267 152 139 224 184	**267**	100 17 17 14 7 7 6 6		17	17	–	1	1	–	–	–	–	–	–	–	–	2-Butyl-6-phenylbenzothiazole		AI01894
267 266 105 77 162 268 190 252	**267**	100 86 59 38 35 21 21 16		17	17	2	1	–	–	–	–	–	–	–	–	–	Aniline, N-(1-methyl-2-benzoylethylidene)-4-methoxy-		NI27013
43 227 267 139 211 152 140 168	**267**	100 51 15 6 5 5 4 3		17	17	2	1	–	–	–	–	–	–	–	–	–	Butanenitrile, 3-(4′-methoxy-1,1-biphenyl-4-yl)-3-hydroxy-	87588-67-8	NI27014
267 252 236 221 126 133 209 224	**267**	100 26 12 8 7 6 5 4		17	17	2	1	–	–	–	–	–	–	–	–	–	9H-Carbazole, 1-allyl-2,8-dimethoxy-		MS05981
120 131 148 103 77 107 121 267	**267**	100 82 72 26 18 14 12 9		17	17	2	1	–	–	–	–	–	–	–	–	–	Cinnamamide, N-(4-hydroxyphenethyl)-	20384-14-9	LI05624
120 131 148 103 77 107 121 132	**267**	100 81 52 26 17 14 13 11	**8.47**	17	17	2	1	–	–	–	–	–	–	–	–	–	Cinnamamide, N-(4-hydroxyphenethyl)-	20384-14-9	NI27015
105 77 267 57 51 106 43 71	**267**	100 41 16 12 10 8 8 7		17	17	2	1	–	–	–	–	–	–	–	–	–	Cinnamic acid, N-benzyl-2-amino-, methyl ester		NI27016
239 162 250 238 267 240 210 268	**267**	100 91 26 26 24 17 7 6		17	17	2	1	–	–	–	–	–	–	–	–	–	Cyclobuta[c]quinolin-2a,8b-diol, 1,2,3,4-tetrahydro-3-phenyl-, (2aR*,3R*,8bS*)-	90732-18-6	NI27017
176 91 31 148 120 65 267 77	**267**	100 60 35 33 10 9 8 7		17	17	2	1	–	–	–	–	–	–	–	–	–	4(3H)-Isoquinolinone, 2-benzyl-1,2-dihydro-7-methoxy-		MS05982

MASS TO CHARGE RATIOS								M.W.	INTENSITIES								Parent	C	H	O	N	S	F	Cl	Br	I	Si	P	B	X	COMPOUND NAME	CAS Reg No	No
104	91	105	63	65	77	79	103	**267**	100	66	62	53	38	31	16	14	**7.67**	17	17	2	1	–	–	–	–	–	–	–	–	–	**6,8-Nonadiynamide, 2,3-epoxy-N-(2-phenylethyl)-, cis-**		**NI27018**
147	267	134	146	266	268	121	148	**267**	100	70	51	32	20	16	13	11		17	17	2	1	–	–	–	–	–	–	–	–	–	**2-Propen-1-one, 3-[4-(dimethylamino)phenyl]-1-(2-hydroxyphenyl)-**	6342-97-8	**NI27019**
267	252	42	268	251	236	208	193	**267**	100	41	22	20	15	13	13	13		17	21	–	3	–	–	–	–	–	–	–	–	–	**9H-Carbazole-3,6-diamine, N,N-dimethyl-9-methyl-**		**NI27020**
121	120	106	267	147	122	77	146	**267**	100	19	17	12	10	10	10	7		17	21	–	3	–	–	–	–	–	–	–	–	–	**N,N′-Dixylylguanidine**		**IC03475**
55	45	41	57	69	43	70	97	**267**	100	85	85	77	62	62	54	46	**0.00**	17	33	1	1	–	–	–	–	–	–	–	–	–	**2-Hydroxy-16-cyanohexadecane**		**IC03476**
267	146	208	224	225	209	268	43	**267**	100	78	47	36	26	24	21	15		18	21	1	1	–	–	–	–	–	–	–	–	–	**Acetamide, N,N-bis(2,4-dimethylphenyl)-**	52812-80-3	**NI27021**
224	267	41	27	180	225	43	90	**267**	100	29	23	23	18	17	16	14		18	21	1	1	–	–	–	–	–	–	–	–	–	**Aniline, 4-butyl-N-[(4-methoxyphenyl)methylene]-**	26227-73-6	**NI27022**
91	105	77	106	267	176	104	51	**267**	100	79	33	30	22	20	20	19		18	21	1	1	–	–	–	–	–	–	–	–	–	**Benzenepropanamide, 2-methyl-N-[(S)-1-phenylethyl]-**		**DD00830**
93	105	104	77	66	92	91	31	**267**	100	63	54	54	26	25	24	21	**0.60**	18	21	1	1	–	–	–	–	–	–	–	–	–	**2H-1,3-Benzoxazine, 6-tert-butyl-3,4-dihydro-3-phenyl-**	55955-97-0	**NI27023**
210	224	267	196	57	211	209	43	**267**	100	86	49	49	38	28	22	18		18	21	1	1	–	–	–	–	–	–	–	–	–	**3-Methyl-2-(2-pentylidenecyclopropyl)-4-quinolinone**		**MS05983**
162	42	239	163	267	252	266	164	**267**	100	34	28	28	24	18	17	17		18	21	1	1	–	–	–	–	–	–	–	–	–	**Morphinan, 5,6,7,8-tetradehydro-3-methoxy-17-methyl-**	55000-50-5	**NI27024**
208	44	42	210	165	209	267	266	**267**	100	61	57	50	50	38	31	22		18	21	1	1	–	–	–	–	–	–	–	–	–	**Morphinan, 6,7,8,14-tetradehydro-3-methoxy-17-methyl-**	55000-51-6	**NI27025**
66	67	136	91	80	77	267	137	**267**	100	11	10	10	9	9	8	8		18	21	1	1	–	–	–	–	–	–	–	–	–	**3-(5-Norbornen-2-yl)-4-cyano-3-hydroxytricyclo[5.2.1.0^{2,6}]dodec-8-ene**		**NI27026**
250	268	190	266					**267**	100	21	17	7					**0.00**	18	21	1	1	–	–	–	–	–	–	–	–	–	**2-Piperidinemethanol, α,α-diphenyl-**	467-60-7	**NI27027**
85	84	183	107	77	56	55	184	**267**	100	75	71	44	23	19	19	12	**5.91**	18	21	1	1	–	–	–	–	–	–	–	–	–	**4-Piperidinemethanol, α,α-diphenyl-**	115-46-8	**NI27028**
250	190	268	266	183				**267**	100	61	49	15	5				**0.00**	18	21	1	1	–	–	–	–	–	–	–	–	–	**4-Piperidinemethanol, α,α-diphenyl-**	115-46-8	**NI27029**
91	120	105	77	40	41	65	92	**267**	100	49	20	19	15	13	11	9	**0.00**	18	21	1	1	–	–	–	–	–	–	–	–	–	**1-Propanone, 2,2-dimethyl-3-(N-benzylamino)-1-phenyl-**	110871-26-6	**DD00831**
30	44	55	41	43	45	56	154	**267**	100	15	11	11	9	8	7	6	**4.21**	18	37	–	1	–	–	–	–	–	–	–	–	–	**1-Cyanoheptadecane**		**IC03477**
55	56	154	67	69	45	81	54	**267**	100	64	58	45	41	38	32	28	**12.91**	18	37	–	1	–	–	–	–	–	–	–	–	–	**9-Octadecen-1-amine**	1838-19-3	**NI27030**
176	91	58	119	85	65	77	70	**267**	100	95	43	19	12	10	6	5	**0.00**	19	25	–	1	–	–	–	–	–	–	–	–	–	**N,α,α′-trimethyldiphenethylamine**		**NI27031**
267	266	268	133	132	265	264	118	**267**	100	24	21	16	13	10	7	3		20	13	–	1	–	–	–	–	–	–	–	–	–	**1H-Dibenzo[a,i]carbazole**	42226-53-9	**NI27032**
267	265	266	268	264	149	43	41	**267**	100	32	22	21	18	16	10	9		20	13	–	1	–	–	–	–	–	–	–	–	–	**7H-Dibenzo[c,g]carbazole**	194-59-2	**NI27033**
267	268	266	133.5	132.5	265	133	264	**267**	100	21	16	15	7	6	5	4		20	13	–	1	–	–	–	–	–	–	–	–	–	**7H-Dibenzo[c,g]carbazole**	194-59-2	**AI01895**
267	133	265	266	268	118	264	132	**267**	100	41	40	31	21	19	17	6		20	13	–	1	–	–	–	–	–	–	–	–	–	**7H-Dibenzo[c,g]carbazole**	194-59-2	**NI27034**
267	133	266	268	118	265	119	264	**267**	100	72	25	21	12	11	8	8		20	13	–	1	–	–	–	–	–	–	–	–	–	**13H-Dibenzo[a,i]carbazole**	239-64-5	**NI27035**
267	133	266	268	119	265	132	264	**267**	100	63	26	21	18	13	13	9		20	13	–	1	–	–	–	–	–	–	–	–	–	**13H-Dibenzo[a,i]carbazole**	239-64-5	**NI27036**
267	266	268	133	265	264	132	239	**267**	100	23	23	11	10	7	7	3		20	13	–	1	–	–	–	–	–	–	–	–	–	**13H-Dibenzo[a,i]carbazole**	239-64-5	**NI27037**
191	189	193	81	79	93	172	91	**268**	100	51	49	7	7	5	4	4	**0.45**	1	–	–	–	–	1	–	3	–	–	–	–	–	**Methane, tribromofluoro-**	353-54-8	**DO01059**
191	189	192	193	190	194	195	196	**268**	100	78	50	44	43	18	12	3	**0.00**	1	–	–	–	–	1	–	3	–	–	–	–	–	**Methane, tribromofluoro-**	353-54-8	**NI27038**
268	141	127	254	140	139	134	63.5	**268**	100	68	39	12	7	6	2	1		1	2	–	–	–	–	–	–	2	–	–	–	–	**Diiodomethane**		**TR00291**
235	233	117	119	69	85	237	121	**268**	100	78	77	74	69	56	48	24	**0.00**	3	–	–	–	–	3	5	–	–	–	–	–	–	**1,1,1,2,2-Pentachloro-3,3,3-trifluoropropane**		**DO01060**
253	223	208	239	238	222	252	209	**268**	100	53	28	22	21	4	3	2	**0.66**	4	12	–	–	–	–	–	–	–	–	–	–	1	**Tetramethyllead**		**MS05984**
253	223	208	251	221	252	222	206	**268**	100	85	66	52	50	41	40	33	**0.21**	4	12	–	–	–	–	–	–	–	–	–	–	1	**Tetramethyllead**		**AI01896**
69	268	199	155	148	180	269	101	**268**	100	77	21	11	10	9	7	7		6	–	–	–	1	7	–	–	–	–	1	–	–	**Phosphonothioic difluoride, (pentafluorophenyl)-**	53327-21-2	**NI27039**
270	233	235	76	268	85	111	73	**268**	100	87	83	82	77	66	57	54		6	–	–	4	–	–	4	–	–	–	–	–	–	**Pyrimido[5,4-d]pyrimidine, tetrachloro-**	32980-71-5	**MS05985**
270	272	74	268	191	75	189	110	**268**	100	69	55	44	42	41	32	20		6	3	–	–	–	–	1	2	–	–	–	–	–	**1-Chloro-3,5-dibromobenzene**		**AI01897**
155	157	76	75	221	219	50	74	**268**	100	97	94	94	89	89	86	44	**22.00**	7	6	2	–	1	–	1	1	–	–	–	–	–	**Benzene, 1-bromo-4-[(chloromethyl)sulphonyl]-**	54091-06-4	**NI27040**
199	268	51	200	249	69	99	31	**268**	100	21	9	8	7	7	5	5		8	1	–	–	–	9	–	–	–	–	–	–	–	**1,1,1,2-Tetrafluoro-2-pentafluorophenylethane**		**IC03478**
150	76	78	92	104	222	75	268	**268**	100	70	42	37	35	30	28	17		8	4	3	4	2	–	–	–	–	–	–	–	–	**Benzamide, N-5H-1,3,2,4-dithia(3-SIV)diazol-5-ylidene-4-nitro-**	57726-56-4	**NI27041**
206	75	189	63	15	253	268	205	**268**	100	73	72	71	62	56	53	46		8	9	4	–	2	–	1	–	–	–	–	–	–	**Benzene, 1-chloro-2,4-bis(methylsulphonyl)-**	17481-98-0	**NI27042**
268	224	44	176	160	270	226	162	**268**	100	95	81	80	79	69	66	53		8	10	2	2	1	–	2	–	–	–	–	–	–	**4-(N,N-Dimethylsulphonamido)-2,6-dichloroaniline**		**IC03479**
226	152	169	209	173	228	154	190	**268**	100	88	79	68	68	65	61	58	**5.03**	8	10	4	2	–	–	2	–	–	–	–	–	–	**1,3-Dimethyl-6-acetoxy-5,5-dichloro-5,6-dihydrouracil**		**NI27043**
233	235	239	241	41	203	58	42	**268**	100	65	53	51	46	37	36	35	**5.00**	8	11	–	4	–	–	3	–	–	–	–	–	–	**2-Methyl-4-trichloromethyl-6-propylamino-s-triazine**		**LI05625**
59	117	268	145	204	119	45	58	**268**	100	90	39	31	26	24	24	23		8	12	–	–	5	–	–	–	–	–	–	–	–	**2,4,6,8,9-Pentathiaadamantane, 1,3,5-trimethyl-**	17443-91-3	**NI27044**
59	268	131	204	145	144	99	127	**268**	100	48	45	41	27	15	15	14		8	12	–	–	5	–	–	–	–	–	–	–	–	**2,4,6,8,9-Pentathiaadamantane, 1,3,7-trimethyl-**	57274-30-3	**NI27045**
59	117	268	45	131	204	119	145	**268**	100	55	35	25	23	20	20	14		8	12	–	–	5	–	–	–	–	–	–	–	–	**2,4,6,8,9-Pentathiaadamantane, 1,5,10-trimethyl-**	17443-95-7	**NI27046**
59	268	117	45	85	103	203	171	**268**	100	50	44	31	14	13	12	12		8	12	–	–	5	–	–	–	–	–	–	–	–	**2,4,6,8-Tetrathiaadamantane-1-thiol, 3,5-dimethyl-, (±)-**	57289-10-8	**NI27047**
233	73	235	209	268	211	270	145	**268**	100	73	60	43	42	40	29	24		9	7	3	–	–	–	3	–	–	–	–	–	–	**Acetic acid, (2,4,5-trichlorophenoxy)-, methyl ester**	1928-37-6	**NI27048**
45	73	59	233	235	211	209	97	**268**	100	97	51	50	37	35	33	32	**27.40**	9	7	3	–	–	–	3	–	–	–	–	–	–	**Acetic acid, (2,4,5-trichlorophenoxy)-, methyl ester**	1928-37-6	**NI27049**
45	73	109	233	146	145	74	148	**268**	100	96	37	35	34	32	23	21	**6.00**	9	7	3	–	–	–	3	–	–	–	–	–	–	**Acetic acid, (2,4,5-trichlorophenoxy)-, methyl ester**	1928-37-6	**NI27050**
198	196	45	97	200	55	62	43	**268**	100	96	70	48	36	31	30	29	**19.45**	9	7	3	–	–	–	3	–	–	–	–	–	–	**Propanoic acid, 2-(2,4,5-trichlorophenoxy)-**	93-72-1	**NI27051**
196	198	97	200	45	270	169	268	**268**	100	95	40	29	21	19	19	18		9	7	3	–	–	–	3	–	–	–	–	–	–	**Propanoic acid, 2-(2,4,5-trichlorophenoxy)-**	93-72-1	**NI27052**
196	198	200	270	268	97	45	27	**268**	100	89	25	18	17	17	15	14		9	7	3	–	–	–	3	–	–	–	–	–	–	**Propanoic acid, 2-(2,4,5-trichlorophenoxy)-**	93-72-1	**NI27053**
57	43	84	41	29	213	87	126	**268**	100	78	59	58	58	50	45	32	**1.00**	9	14	3	2	–	–	2	–	–	–	–	–	–	**Hydrouracil, 3-tert-butyl-5,5-dichloro-6-hydroxy-6-methyl-**	24577-78-4	**NI27054**

MASS TO CHARGE RATIOS								M.W.	INTENSITIES								Parent	C	H	O	N	S	F	Cl	Br	I	Si	P	B	X	COMPOUND NAME	CAS Reg No	No
225	223	163	161	221	241	224	159	268	100	71	59	43	40	33	27	27	1.36	9	21	–	–	–	1	–	–	–	–	–	–	1	Tin, fluorotripropyl-		MS05986
143	151	41	43	227	209	109	185	268	100	72	51	50	47	44	43	33	0.00	9	21	4	–	–	–	–	–	–	–	–	–	1	Arsenic acid, tripropyl ester	15606-96-9	NI27055
270	268	189	52	50	161	77	105	268	100	98	85	44	35	33	28	23		10	5	4	–	–	–	–	1	–	–	–	–	–	1,4-Naphthalenedione, 6-bromo-5,8-dihydroxy-	78226-78-5	NI27056
43	197	51	162	39	75	199	63	268	100	62	57	54	43	42	41	40	26.04	10	8	–	–	–	–	4	–	–	–	–	–	–	1,6-Methano-1H-indene, 2,3,4,5-tetrachloro-3a,6,7,7a-tetrahydro-	57160-06-2	NI27057
162	197	128	127	199	63	164	126	268	100	80	63	56	51	33	32	28	2.27	10	8	–	–	–	–	4	–	–	–	–	–	–	Naphthalene, 1,2,3,4-tetrachlorotetrahydro-	28801-89-0	NI27058
137	136	165	179	238	29	57	73	268	100	98	88	70	50	50	47	41	1.00	10	12	5	4	–	–	–	–	–	–	–	–	–	Allopurinol riboside	16220-07-8	NI27059
71	43	41	202	55	63	77	200	268	100	27	25	13	12	11	10	9	6.75	10	12	5	4	–	–	–	–	–	–	–	–	–	Butanal, 4-hydroxy-, (2,4-dinitrophenyl)hydrazone	1708-33-4	NI27060
43	181	266	63	182	42	78	62	268	100	28	18	9	6	6	4	4	0.00	10	12	5	4	–	–	–	–	–	–	–	–	–	2-Butanone, 3-hydroxy-, (2,4-dinitrophenyl)hydrazone	35015-94-2	NI27061
43	42	181	266	30	182	63	75	268	100	26	22	18	14	8	8	7	0.00	10	12	5	4	–	–	–	–	–	–	–	–	–	2-Butanone, 3-hydroxy-, (2,4-dinitrophenyl)hydrazone	35015-94-2	NI27062
28	27	136	54	53	43	29	44	268	100	45	44	22	10	9	9	8	0.00	10	12	5	4	–	–	–	–	–	–	–	–	–	Inosine	58-63-9	NI27064
136	54	57	53	137	81	74	55	268	100	35	13	7	5	5	4	4	0.00	10	12	5	4	–	–	–	–	–	–	–	–	–	Inosine	58-63-9	NI27063
136	137	135	52	44	53	45	56	268	100	58	58	36	32	18	18	14	0.12	10	12	5	4	–	–	–	–	–	–	–	–	–	4H-Pyrazolo[3,4-d]pyrimidin-4-one, 1,5-dihydro-1-β-D-ribofuranosyl-	16220-07-8	NI27065
165	44	31	179	29	28	43	54	268	100	71	66	59	59	56	48	36	9.40	10	12	5	4	–	–	–	–	–	–	–	–	–	7H-Pyrazolo[4,3-d]pyrimidin-7-one, 1,4-dihydro-3-β-D-ribofuranosyl-	13877-76-4	NI27066
106	133	107	41	268	45	61	74	268	100	33	32	29	26	26	21	20		10	20	–	–	4	–	–	–	–	–	–	–	–	1,4,8,11-Tetrathiacyclotetradecane	24194-61-4	NI27067
69	41	135	42	70	71	55	93	268	100	30	22	21	20	13	13	9	0.35	10	20	4	–	2	–	–	–	–	–	–	–	–	2,5-Dipropyl-1,4-dithiane-1,1,4,4-tetraoxide		IC03480
43	57	71	85	41	55	29	141	268	100	75	58	55	37	23	17	13	2.00	10	21	–	–	–	–	–	–	1	–	–	–	–	Decane, 1-iodo-	2050-77-3	LI05626
57	43	41	71	85	55	29	39	268	100	97	68	56	55	36	20	17	0.50	10	21	–	–	–	–	–	–	1	–	–	–	–	Decane, 1-iodo-	2050-77-3	NI27068
43	57	71	41	85	55	29	56	268	100	73	51	43	34	24	21	17	0.10	10	21	–	–	–	–	–	–	1	–	–	–	–	Nonane, 1-iodo-2-methyl-		NI27069
102	114	198	106	103	100	165	110	268	100	30	13	4	4	2	1	1	0.00	10	21	6	–	–	–	–	–	–	–	1	–	–	Butyl (3-methoxycarbonyl)propyl methyl phosphate		NI27070
171	115	172	268	270	189	116	128	268	100	64	39	33	32	24	24	19		11	9	1	–	1	–	–	1	–	–	–	–	–	3H-Naphtho[1,8-bc]thiophene-5-ol, 2-bromo-4,5-dihydro-		LI05627
105	104	77	103	51	119	78	268	268	100	98	63	58	56	54	54	49		11	9	2	–	–	5	–	–	–	–	–	–	–	1-Phenylethyl pentafluoropropionate	55702-47-1	NI27071
93	204	66	79	105	80	268	107	268	100	86	76	74	68	65	59	25		11	9	2	2	1	–	1	–	–	–	–	–	–	Pyridinium, 1-[[(4-chlorophenyl)sulphonyl]amino]-, hydroxide, inner salt	55000-35-6	LI05628
93	204	66	79	80	125	268	75	268	100	86	76	74	69	68	59	46		11	9	2	2	1	–	1	–	–	–	–	–	–	Pyridinium, 1-[[(4-chlorophenyl)sulphonyl]amino]-, hydroxide, inner salt	55000-35-6	NI27072
161	268	270	89	162	189	51	63	268	100	28	26	15	13	13	12	12		11	9	3	–	–	–	–	1	–	–	–	–	–	2H-1-Benzopyran-2-one, 4-(bromomethyl)-7-methoxy-	35231-44-8	NI27073
42	43	41	65	92	106	93	134	268	100	86	63	54	22	19	18	15	0.00	11	12	4	2	1	–	–	–	–	–	–	–	–	3-Azetidinone, N-[(4-acetylamino)benzenesulphonyl]-	80076-95-5	NI27074
167	268	77	81	137	152	66	53	268	100	40	31	30	27	25	25	18		11	12	4	2	1	–	–	–	–	–	–	–	–	1,2,3-Thiadiazolium, 4-hydroxy-3-(3,4,5-trimethoxyphenyl)-, hydroxide, inner salt	41175-64-8	NI27075
141	113	95	41	67	59	43	127	268	100	92	53	39	38	37	33	32	0.00	11	15	3	–	–	2	1	–	–	–	–	–	–	Cyclopropanecarboxylic acid, 3-(2-chloro-3,3-difluoro-1-hydroxy-2-propenyl)-2,2-dimethyl-, ethyl ester	106908-94-5	DD00832
184	185	268	43	221	142	186	41	268	100	35	31	22	19	18	15	15		11	16	–	4	2	–	–	–	–	–	–	–	–	Thiazolo[5,4-d]pyrimidine, 5-amino-2-(hexylthio)-	19844-42-9	NI27076
160	133	91	132	53	188	232	42	268	100	42	27	22	22	20	14	12	1.00	11	16	4	4	–	–	–	–	–	–	–	–	–	4-Pteridinecarboxylic acid, 5,6,7,8-tetrahydro-6,7-dihydroxy-6,7-dimethyl-, ethyl ester	17445-70-4	NI27077
160	133	91	132	53	188	232	161	268	100	42	27	22	22	20	14	12	1.00	11	16	4	4	–	–	–	–	–	–	–	–	–	4-Pteridinecarboxylic acid, 5,6,7,8-tetrahydro-6,7-dihydroxy-6,7-dimethyl-, ethyl ester	17445-70-4	LI05629
232	188	233	234	268	182	160	161	268	100	81	72	37	36	36	36	17		11	16	4	4	–	–	–	–	–	–	–	–	–	4-Pteridinecarboxylic acid, 5,6,7,8-tetrahydro-6,7-dihydroxy-7,8-dimethyl-, ethyl ester	69796-09-4	NI27078
81	95	183	82	185	138	41	123	268	100	89	74	51	50	50	50	45	0.11	11	22	1	–	–	–	2	–	–	1	–	–	–	Silane, dichloro(4-menth-3-yloxy)methyl-	24799-65-3	NI27079
268	213	78	114	77	157	241	91	268	100	43	30	27	25	21	20	20		12	4	4	4	–	–	–	–	–	–	–	–	–	2,6-Diamino-4,8-dihydro-4,8-dioxobenzo[1,2-b:4,5-b′]difuran-3,7-dicarbonitrile		MS05987
77	51	141	99	73	63	50	268	268	100	37	32	25	14	14	9	8		12	9	3	–	1	–	1	–	–	–	–	–	–	Benzenesulphonic acid, 4-chlorophenyl ester	80-38-6	NI27081
77	141	268	51	57	270	56	99	268	100	65	22	16	12	9	9	8		12	9	3	–	1	–	1	–	–	–	–	–	–	Benzenesulphonic acid, 4-chlorophenyl ester	80-38-6	NI27080
141	268	109	65	44	160	75	39	268	100	70	60	50	50	40	38	38		12	9	3	–	1	–	1	–	–	–	–	–	–	Phenol, 4-[(4-chlorophenyl)sulphonyl]-	7402-62-2	IC03481
91	92	268	141	65	109	93	160	268	100	92	91	72	56	49	49	47		12	9	3	–	1	–	1	–	–	–	–	–	–	Phenol, 4-[(4-chlorophenyl)sulphonyl]-	7402-62-2	NI27082
152	28	268	127	44	99	193	154	268	100	91	76	54	44	37	35	35		12	13	3	2	–	–	1	–	–	–	–	–	–	Butyric acid, α-(2-dichlorophenylhydrazino)-β-oxo-, ethyl ester		IC03482
120	43	151	92	65	45	72	121	268	100	99	61	31	25	20	15	9	4.00	12	16	3	2	1	–	–	–	–	–	–	–	–	N-(2-Acetamidoethylthio)-4-carbomethoxyaniline		LI05630
41	61	56	45	53	47	79	93	268	100	64	38	33	28	25	23	20	13.00	12	16	3	2	1	–	–	–	–	–	–	–	–	2,4,6-Pyrimidinetrione, 5,5-diallyl-1-(methylthiomethyl)-		MS05988
253	91	77	115	41	268	43	39	268	100	63	44	42	40	33	32	32		12	16	5	2	–	–	–	–	–	–	–	–	–	Benzene, 5-tert-butyl-1,3-dinitro-4-methoxy-2-methyl-		NI27083
39	41	40	42	143	67	185	52	268	100	71	28	25	17	14	13	13	1.46	12	20	–	–	–	–	–	–	–	–	–	–	2	Chromium, tetra-2-propenyldi-	12295-17-9	NI27084
185	143	52	144	226	93	186	39	268	100	96	91	68	56	54	53	40	19.26	12	20	–	–	–	–	–	–	–	–	–	–	2	Chromium, tetra-2-propenyldi-	12295-17-9	NI27085
157	213	41	155	43	268	39	95	268	100	80	28	22	21	19	16	15		12	21	–	–	–	–	–	–	–	–	–	–	1	Rhodium, tris[(1,2,3-η)-2-methyl-2-propenyl]-	74811-17-9	NI27086
183	185	141	143	41	43	113	69	268	100	70	60	39	38	36	29	26	3.90	12	26	–	–	–	–	2	–	–	1	–	–	–	Silane, dichlorodihexyl-	18204-93-8	NI27087
125	127	268	126	89	270	99	63	268	100	44	13	12	10	9	6	6		13	10	–	–	1	–	2	–	–	–	–	–	–	Benzene, 1-chloro-4-[[(4-chlorophenyl)methyl]thio]-	103-17-3	NI27088
125	145	127	157	173	147	269	141	268	100	47	41	33	23	18	15	15	8.30	13	10	–	–	1	–	2	–	–	–	–	–	–	Benzene, 1-chloro-4-[[(4-chlorophenyl)methyl]thio]-	103-17-3	NI27089
125	127	89	63	108	99	268	39	268	100	34	19	15	9	9	8	8		13	10	–	–	1	–	2	–	–	–	–	–	–	Benzene, 1-chloro-4-[[(4-chlorophenyl)methyl]thio]-	103-17-3	NI27090
190	192	268	270	194	65	63	51	268	100	60	47	31	11	11	10	10		13	10	2	–	–	–	2	–	–	–	–	–	–	1,3-Benzenediol, 4,6-dichloro-2-benzyl-	52956-22-6	NI27091
141	268	270	140	115	139	156	127	268	100	47	29	14	12	9	7	7		13	10	2	–	–	–	2	–	–	–	–	–	–	1-Naphthalenemethyl dichloroacetate	86328-50-9	NI27092

MASS TO CHARGE RATIOS								M.W.	INTENSITIES								Parent	C	H	O	N	S	F	Cl	Br	I	Si	P	B	X	COMPOUND NAME	CAS Reg No	No
128	141	268	152	77	215	63	75	**268**	100	62	32	32	31	30	29	28		13	10	2	–	–	–	2	–	–	–	–	–	–	**Phenol, 2,2′-methylenebis[4-chloro-**	97-23-4	**NI27093**
165	142	166	77	127	51	161	96	**268**	100	72	56	50	31	30	18	12	**0.00**	13	12	–	–	–	2	–	–	–	–	2	–	–	**Phosphane, μ-methylenedifluorodiphenyldi-**		**MS05989**
135	106	160	107	133	108	104	53	**268**	100	93	86	84	38	30	29	26	**12.75**	13	12	1	6	–	–	–	–	–	–	–	–	–	**Urea, N-(2-methylphenyl)-N′-1H-purin-6-yl-**	32725-50-1	**NI27094**
96	91	155	68	55	54	41	43	**268**	100	79	48	26	23	23	21	19	**3.60**	13	16	4	–	1	–	–	–	–	–	–	–	–	**Cyclohexanone, 4-[[(4-methylphenyl)sulphonyl]oxy]-**	23511-04-8	**NI27095**
145	236	205	193	268	167	237	177	**268**	100	81	65	53	44	42	38	35		13	16	6	–	–	–	–	–	–	–	–	–	–	**Dimethyl (1α,2α,5α)-3-methoxy-7-oxobicyclo[3.3.0]oct-3-ene -2,4-dicarboxylate**		**MS05990**
116	100	41	268	127	42	60	61	**268**	100	98	91	58	54	48	45	40		13	17	–	2	1	–	1	–	–	–	–	–	–	**4H-1,3-Thiazine, 4,4,6-trimethyl-2-[3-chlorophenylamino(imino)]-5,6-dihydro-**	53004-50-5	**NI27096**
73	194	209	268	75	179	253	91	**268**	100	50	47	44	28	26	19	12		13	20	4	–	–	–	–	–	–	1	–	–	–	**Benzeneacetic acid, 2,3-dimethoxy-, trimethylsilyl ester**		**MS05991**
151	209	73	268	121	75	77	253	**268**	100	61	61	36	24	23	21	20		13	20	4	–	–	–	–	–	–	1	–	–	–	**Benzeneacetic acid, 2,4-dimethoxy-, trimethylsilyl ester**		**MS05992**
174	73	175	75	96	59	147	86	**268**	100	63	18	12	11	11	10	10	**0.00**	13	20	4	–	–	–	–	–	–	1	–	–	–	**Benzeneacetic acid, 3,4-dimethoxy-, trimethylsilyl ester**	27750-60-3	**NI27097**
209	268	210	269	151	211	253	215	**268**	100	67	55	34	26	16	14	14		13	20	4	–	–	–	–	–	–	1	–	–	–	**Benzeneacetic acid, 3,4-dimethoxy-, trimethylsilyl ester**	27750-60-3	**NI27098**
73	209	268	151	75	253	210	74	**268**	100	66	35	35	22	16	11	9		13	20	4	–	–	–	–	–	–	1	–	–	–	**Benzeneacetic acid, 3,4-dimethoxy-, trimethylsilyl ester**	27750-60-3	**NI27099**
73	75	268	209	44	253	43	89	**268**	100	39	21	16	14	13	12	11		13	20	4	–	–	–	–	–	–	1	–	–	–	**Benzeneacetic acid, 3,5-dimethoxy-, trimethylsilyl ester**		**MS05993**
73	209	45	89	59	210	75	135	**268**	100	88	30	27	19	17	17	16	**0.00**	13	20	4	–	–	–	–	–	–	1	–	–	–	**Benzeneacetic acid, 2-methoxy-α-[(trimethylsilyl)oxy]-, methyl ester**	55590-94-8	**NI27100**
268	238	209	253	269	179	239	210	**268**	100	52	42	37	21	16	10	7		13	20	4	–	–	–	–	–	–	1	–	–	–	**Benzeneacetic acid, 3-methoxy-4-[(trimethylsilyl)oxy]-, methyl ester**	15964-84-8	**NI27101**
73	179	238	209	268	253	45	59	**268**	100	89	82	62	44	37	30	21		13	20	4	–	–	–	–	–	–	1	–	–	–	**Benzeneacetic acid, 3-methoxy-4-[(trimethylsilyl)oxy]-, methyl ester**	15964-84-8	**MS05994**
179	238	73	209	268	253	45	149	**268**	100	92	84	73	47	35	24	22		13	20	4	–	–	–	–	–	–	1	–	–	–	**Benzeneacetic acid, 3-methoxy-4-[(trimethylsilyl)oxy]-, methyl ester**	15964-84-8	**NI27102**
73	209	89	45	59	75	135	43	**268**	100	26	26	22	20	10	8	8	**1.05**	13	20	4	–	–	–	–	–	–	1	–	–	–	**Benzeneacetic acid, 3-methoxy-α-[(trimethylsilyl)oxy]-, methyl ester**	55590-93-7	**NI27103**
179	238	209	268	73	253	149	180	**268**	100	87	50	39	36	33	19	15		13	20	4	–	–	–	–	–	–	1	–	–	–	**Benzeneacetic acid, 4-methoxy-3-[(trimethylsilyl)oxy]-, methyl ester**	15964-85-9	**MS05995**
238	179	268	209	253	73	239	149	**268**	100	94	83	76	38	37	18	17		13	20	4	–	–	–	–	–	–	1	–	–	–	**Benzeneacetic acid, 4-methoxy-3-[(trimethylsilyl)oxy]-, methyl ester**	15964-85-9	**NI27104**
209	73	135	210	89	208	45	77	**268**	100	65	36	24	20	17	13	12	**1.00**	13	20	4	–	–	–	–	–	–	1	–	–	–	**Benzeneacetic acid, 4-methoxy-α-[(trimethylsilyl)oxy]-, methyl ester**	55000-34-5	**NI27105**
209	210	73	253	211	89	225	135	**268**	100	18	17	5	5	4	3	3	**0.50**	13	20	4	–	–	–	–	–	–	1	–	–	–	**Benzeneacetic acid, 4-methoxy-α-[(trimethylsilyl)oxy]-, methyl ester**	55000-34-5	**NI27106**
84	126	43	127	96	85	56	41	**268**	100	51	14	5	5	5	5	4	**0.00**	13	20	4	2	–	–	–	–	–	–	–	–	–	**Acetamide, N-formyl-N-[4-(1-formyl-2-piperidinyl)-4-oxobutyl]-**	54966-13-1	**NI27107**
253	268	73	254	119	45	269	133	**268**	100	73	49	25	24	21	18	13		13	24	2	–	–	–	–	–	–	2	–	–	–	**Benzene, 1-methyl-3,5-bis[(trimethylsilyl)oxy]-**	89267-67-4	**NI27108**
75	251	223	73	103	149	76	221	**268**	100	32	31	27	19	15	15	11	**0.00**	13	24	2	–	–	–	–	–	–	2	–	–	–	**Benzene, 1-[(trimethylsilyl)oxy]-2-[[(trimethylsilyl)oxy]methyl]-**		**NI27109**
73	179	268	147	253	267	45	180	**268**	100	92	42	32	31	27	17	15		13	24	2	–	–	–	–	–	–	2	–	–	–	**Benzene, 1-[(trimethylsilyl)oxy]-4-[[(trimethylsilyl)oxy]methyl]-**	18401-58-6	**NI27110**
73	179	268	253	147	267	45	180	**268**	100	89	48	32	29	28	22	14		13	24	2	–	–	–	–	–	–	2	–	–	–	**Benzyl alcohol, α,4-bis[(trimethylsilyl)oxy]-**	18401-58-6	**NI27111**
138	146	139	147	102	265			**268**	100	42	8	3	2	1			**0.00**	14	8	4	2	–	–	–	–	–	–	–	–	–	**Benzoic acid, 4-cyano-, 3-nitrophenyl ester**	53327-06-3	**NI27112**
102	138	103	139	268	146			**268**	100	61	10	5	1	1				14	8	4	2	–	–	–	–	–	–	–	–	–	**Benzoic acid, 4-cyano-, 4-nitrophenyl ester**	32792-60-2	**NI27113**
28	268	54	240	26	44	269	82	**268**	100	99	53	47	26	25	19	19		14	8	4	2	–	–	–	–	–	–	–	–	–	**2(5H)-Furanone, 5,5′-(1,4-phenylenedinitrilo)bis-**	55000-36-7	**NI27114**
268	54	269	28	240	26	55	198	**268**	100	21	17	16	11	11	9	7		14	8	4	2	–	–	–	–	–	–	–	–	–	**1H-Pyrrole-2,5-dione, 1,1′-(1,3-phenylene)bis-**	3006-93-7	**MS05996**
268	54	269	28	240	26	55	82	**268**	100	21	17	16	11	11	9	7		14	8	4	2	–	–	–	–	–	–	–	–	–	**1H-Pyrrole-2,5-dione, 1,1′-(1,3-phenylene)bis-**	3006-93-7	**NI27115**
134	104	119	76	267	91	64	77	**268**	100	38	31	21	18	18	15	12	**3.00**	14	10	3	3	–	–	–	–	–	–	–	–	–	**5-(2-Nitrophenyl)-3-phenyl-1,2,4-oxadiazole**		**NI27116**
119	267	91	76	64	268	63	50	**268**	100	68	26	18	18	13	10	10		14	10	3	3	–	–	–	–	–	–	–	–	–	**5-(3-Nitrophenyl)-3-phenyl-1,2,4-oxadiazole**		**NI27117**
119	267	91	64	76	50	268	103	**268**	100	70	27	19	17	13	12	11		14	10	3	3	–	–	–	–	–	–	–	–	–	**5-(4-Nitrophenyl)-3-phenyl-1,2,4-oxadiazole**		**NI27118**
212	121	56	91	65	240	213	208	**268**	100	95	70	49	26	20	16	13	**0.39**	14	12	2	–	–	–	–	–	–	–	–	–	1	**Iron, benzyldicarbonyl-π-cyclopentadienyl-**	12093-91-3	**NI27119**
212	121	56	208	91	240	210	156	**268**	100	54	49	43	40	39	29	27	**8.99**	14	12	2	–	–	–	–	–	–	–	–	–	1	**Iron, 4-tolyldicarbonyl-π-cyclopentadienyl-**	33308-66-6	**NI27120**
268	196	129	169	224	223	142	52	**268**	100	94	88	59	35	29	21	15		14	12	2	4	–	–	–	–	–	–	–	–	–	**Pyrazolo[5,1-c]-as-triazine-3-carboxylic acid, 4-phenyl-, ethyl ester**	6757-48-8	**NI27121**
268	196	129	69	224	223	142	103	**268**	100	94	88	59	35	29	21	15		14	12	2	4	–	–	–	–	–	–	–	–	–	**Pyrazolo[5,1-c]-as-triazine-3-carboxylic acid, 4-phenyl-, ethyl ester**	6757-48-8	**LI05631**
108	160	52	78	65	268	93	145	**268**	100	12	10	9	8	7	3	2		14	16	2	–	–	–	–	–	–	–	–	–	1	**Chromium, bis(methoxyphenyl)-**	57820-92-5	**NI27122**
208	180	209	79	91	131	149	59	**268**	100	57	49	44	43	36	35	34	**13.00**	14	20	5	–	–	–	–	–	–	–	–	–	–	**1,3-Adamantanedicarboxylic acid, 4-hydroxy-, dimethyl ester**	19930-86-0	**NI27123**
136	268	121	45	43	80	269	108	**268**	100	69	39	31	25	23	15	15		14	20	5	–	–	–	–	–	–	–	–	–	–	**1,4,7,10,13-Benzopentaoxacyclopentadecin, 2,3,5,6,8,9,11,12-octahydro-**	14098-44-3	**NI27124**
57	43	138	41	123	152	170	105	**268**	100	83	74	56	53	48	40	38	**2.00**	14	20	5	–	–	–	–	–	–	–	–	–	–	**1,3-Cyclohexadiene-1-carboxylic acid, 5-(acetyloxy)-6-methoxy-, tert-butyl ester, trans-**	55723-83-6	**NI27125**
148	166	149	81	167	80	209	151	**268**	100	71	45	21	18	18	15	15	**3.00**	14	20	5	–	–	–	–	–	–	–	–	–	–	**Cyclopenta[c]pyran-4-methanol, 1-(acetyloxy)-1,4a,5,6,7,7a-hexahydro-7-methyl-, acetate, [1S-(1α,4aα,7α,7aα)]-**	89289-90-7	**NI27126**
236	153	111	95	121	94	107	237	**268**	100	60	37	34	29	29	28	23	**12.00**	14	20	5	–	–	–	–	–	–	–	–	–	–	**Hexanoic acid, 6-[5-(methoxycarbonyl)methyl-2-furyl]-, methyl ester**		**NI27127**
150	73	87	163	57	105	121	117	**268**	100	51	48	40	34	31	29	24	**5.00**	14	20	5	–	–	–	–	–	–	–	–	–	–	**α-L-Rhamnopyranoside, methyl 2-O-benzyl-**	75336-81-1	**NI27128**
128	150	116	151	100	88	73	121	**268**	100	88	65	52	42	39	39	34	**3.43**	14	20	5	–	–	–	–	–	–	–	–	–	–	**α-L-Rhamnopyranoside, methyl 3-O-benzyl-**	75336-82-2	**NI27129**
87	148	74	150	73	45	163	75	**268**	100	85	74	58	58	38	33	31	**28.19**	14	20	5	–	–	–	–	–	–	–	–	–	–	**α-L-Rhamnopyranoside, methyl 4-O-benzyl-**	25019-69-6	**NI27130**
253	255	57	268	270	254	256	29	**268**	100	98	85	22	21	14	13	9		14	21	–	–	–	–	–	1	–	–	–	–	–	**Benzene, 1-bromo-3,5-di-tert-butyl-**	22385-77-9	**NI27131**
151	97	55	268	69	111	41	47	**268**	100	71	36	28	25	22	22	13		14	21	1	–	1	–	–	–	–	–	1	–	–	**1-Phenylthio-2,2,3,4,4-pentamethylphosphetan 1-oxide**		**NI27132**
184	169	170	185	112	183	55	41	**268**	100	98	43	40	37	34	26	21	**0.00**	14	24	3	2	–	–	–	–	–	–	–	–	–	**2,4,6(1H,3H,5H)-Pyrimidinetrione, 5-ethyl-5-hexyl-1,3-dimethyl-**	55000-37-8	**NI27133**
73	225	147	75	195	45	155	226	**268**	100	69	47	38	27	24	22	17	**12.71**	14	28	1	–	–	–	–	–	–	2	–	–	–	**Silane, trimethyl[[1-[(trimethylsilyl)ethynyl]cyclohexyl]oxy]-**	74810-49-4	**NI27134**
268	251	269	195	223	139	150	75	**268**	100	28	18	14	13	13	11	10		15	8	5	–	–	–	–	–	–	–	–	–	–	**Fluorenone-2,7-dicarboxylic acid**		**IC03483**

MASS TO CHARGE RATIOS								M.W.	INTENSITIES								Parent	C	H	O	N	S	F	Cl	Br	I	Si	P	B	X	COMPOUND NAME	CAS Reg No	No
136	268	91	137	108	106	269	152	**268**	100	50	22	14	14	12	10	7		15	12	1	2	1	–	–	–	–	–	–	–	–	**4H-1,3-Benzothiazin-4-one, 2-benzylamino-2,3-dihydro-**		**NI27135**
268	180	104	165	77	239	51	225	**268**	100	84	75	56	55	43	31	26		15	12	1	2	1	–	–	–	–	–	–	–	–	**4-Imidazolidinone, 5,5-diphenyl-2-thioxo-**	21083-47-6	**NI27136**
269	237	212	182	104	91	270	130	**268**	100	73	43	24	21	21	18	18	**0.00**	15	12	1	2	1	–	–	–	–	–	–	–	–	**4-Imidazolidinone, 5,5-diphenyl-2-thioxo-**	21083-47-6	**NI27137**
91	268	121	250	90	134	235	102	**268**	100	95	48	45	22	13	9	9		15	12	1	2	1	–	–	–	–	–	–	–	–	**2-Phenyl-3-hydroxythiazolidino[3,2-a]benzimidazole**		**LI05632**
268	250	269	222	249	251	221	28	**268**	100	71	18	18	15	15	14	14		15	12	3	2	–	–	–	–	–	–	–	–	–	**Anthraquinone, 1,4-diamino-2-(hydroxymethyl)-**		**NI27138**
177	119	91	65	28	120	196	239	**268**	100	81	70	38	38	29	28	25	**17.49**	15	12	3	2	–	–	–	–	–	–	–	–	–	**2,4-Imidazolidinedione, 5-(4-hydroxyphenyl)-5-phenyl-**	2784-27-2	**NI27139**
239	268	196	197	77	104	180	39	**268**	100	99	87	26	26	22	17	17		15	12	3	2	–	–	–	–	–	–	–	–	–	**2,4-Imidazolidinedione, 5-(4-hydroxyphenyl)-5-phenyl-**	2784-27-2	**NI27140**
196	239	268	120	77	225	65	104	**268**	100	87	73	70	51	44	44	41		15	12	3	2	–	–	–	–	–	–	–	–	–	**2,4-Imidazolidinedione, 5-(4-hydroxyphenyl)-5-phenyl-**	2784-27-2	**LI05633**
196	239	268	120	65	77	180	104	**268**	100	75	65	44	21	20	16	15		15	12	3	2	–	–	–	–	–	–	–	–	–	**2,4-Imidazolidinedione, 5-(4-hydroxyphenyl)-5-phenyl-, (±)-**	57464-82-1	**NI27141**
131	103	77	268	132	51	102	63	**268**	100	32	18	11	9	6	5	3		15	12	3	2	–	–	–	–	–	–	–	–	–	**3′-Nitro-3-phenylpropenanilide**		**TR00292**
120	212	268	171	224	202	89	228	**268**	100	91	59	51	23	16	13	12		15	12	3	2	–	–	–	–	–	–	–	–	–	**Pyrazolo[1,5-a]pyridine-4,7-dione, 5-ethoxy-3-phenyl-**		**NI27142**
268	128	168	223	196	69	195	129	**268**	100	86	63	35	35	30	23	23		15	12	3	2	–	–	–	–	–	–	–	–	–	**1H-Pyrimido[1,2-a]quinoline-2-carboxylic acid, 1-oxo-, ethyl ester**	37611-65-7	**NI27143**
226	268	76	227	253	144	77	50	**268**	100	40	19	17	12	8	8	7		15	12	3	2	–	–	–	–	–	–	–	–	–	**4(3H)-Quinazolinone, 3-(2,6-dihydroxyphenyl)-2-methyl-**		**NI27144**
208	209	268	194	224	77	207	251	**268**	100	65	56	53	35	31	30	28		15	16	1	4	–	–	–	–	–	–	–	–	–	**5-Methyl-2-aminobenzophenone semicarbazone**		**MS05997**
268	253	43	121	91	270	109	107	**268**	100	75	41	40	28	23	23	20		15	21	2	–	–	–	1	–	–	–	–	–	–	**Laurencial**	86330-91-8	**NI27145**
71	43	138	111	180	197	179	41	**268**	100	98	76	58	52	36	27	27	**0.00**	15	24	4	–	–	–	–	–	–	–	–	–	–	**Cyclohexanepropanoic acid, β,4-dimethyl-1-(2-methyl-1-oxopropyl)-3-oxo-**	55700-39-5	**NI27146**
268	253	198	57	41	124	43	197	**268**	100	71	66	59	59	52	47	46		15	24	4	–	–	–	–	–	–	–	–	–	–	**2-Cyclopenten-1-one, 3,4-dihydroxy-5-(3-methylbutyl)-2-(3-methyl-1-oxobutyl)-, (4S-cis)-**	13855-87-3	**NI27147**
85	166	140	84	183	167	86	150	**268**	100	53	33	21	13	12	11	10	**0.00**	15	24	4	–	–	–	–	–	–	–	–	–	–	**7,7a-Dihydro-5β-hydroxy-4,7bα-oxa-1β-tetrahydropyranyloxy-6H-indan**		**NI27148**
43	87	123	73	109	137	95	41	**268**	100	37	36	24	23	15	13	10	**0.00**	15	24	4	–	–	–	–	–	–	–	–	–	–	**1,4-Dioxane-2-methanol, α-[[3,3-dimethyl-2-isopropyloxiranyl]vinyl]-α-methyl-**	63922-55-4	**NI27149**
184	84	183	129	71	43	185	41	**268**	100	26	22	16	16	14	12	10	**1.00**	15	24	4	–	–	–	–	–	–	–	–	–	–	**2(5H)-Furanone, 3-butyryl-5-hexyl-4-hydroxy-5-methyl-**	21494-12-2	**NI27150**
184	84	183	129	71	43	69	41	**268**	100	26	22	16	16	14	10	10	**1.00**	15	24	4	–	–	–	–	–	–	–	–	–	–	**2(5H)-Furanone, 3-butyryl-5-hexyl-4-hydroxy-5-methyl-**	21494-12-2	**LI05634**
85	140	166	86	84	107	183	167	**268**	100	7	6	5	5	4	3	3	**0.00**	15	24	4	–	–	–	–	–	–	–	–	–	–	**Indeno[3a,4-b]oxiren-2-ol, octahydro-4a-methyl-5-[(tetrahydro-2H-pyran-2-yl)oxy]-, (1aα,2β,4aβ,5β,7aS*)-**	67920-65-4	**NI27151**
111	59	210	110	95	55	167	109	**268**	100	80	50	50	49	40	31	31	**4.00**	15	24	4	–	–	–	–	–	–	–	–	–	–	**1-Naphthalenol, 2,3:8,8a-bisepoxydecahydro-4-(2-hydroxy-2-isopropyl)-4a,5-dimethyl-, [2R-(1ζ,2α,3α,4α,4aβ,5β,8β,8aβ)]-**		**NI27152**
43	113	165	85	55	97	250	183	**268**	100	87	70	62	55	45	38	38	**3.00**	15	24	4	–	–	–	–	–	–	–	–	–	–	**2,6,10,14-Pentadecanetetrone**	36452-83-2	**NI27153**
43	161	210	69	192	57	175	55	**268**	100	53	36	30	29	27	24	24	**2.00**	15	24	4	–	–	–	–	–	–	–	–	–	–	**Spiro[4,5]dec-6-en-8-one, 2-(1,2-dihydroxy-1-methylethyl)-6,10-dimethyl-9-hydroxy-**	84321-96-0	**NI27154**
129	268	100	196	182	56	154	168	**268**	100	55	50	44	42	40	34	25		15	28	2	2	–	–	–	–	–	–	–	–	–	**Acetamide, N-[3-(2-oxoazacycloundec-1-yl)propyl]-**	67370-81-4	**NI27155**
225	56	70	112	100	98	84	126	**268**	100	16	8	5	5	4	4	3	**3.00**	15	28	2	2	–	–	–	–	–	–	–	–	–	**1,5-Diazacyclopentadecan-6-one, 1-acetyl-**	67370-83-6	**NI27156**
72	196	154	198	41	126	44	109	**268**	100	37	20	15	9	8	8	7	**0.20**	15	28	2	2	–	–	–	–	–	–	–	–	–	**Propanediamide, N,N,N′,N′-tetramethyl-2-(1-methylbutyl)-2-(2-propenyl)-**	42948-62-9	**NI27157**
72	196	154	198	126	224	109	152	**268**	100	37	20	15	8	7	7	5	**0.30**	15	28	2	2	–	–	–	–	–	–	–	–	–	**Propanediamide, N,N,N′,N′-tetramethyl-2-(1-methylbutyl)-2-(2-propenyl)-**	42948-62-9	**LI05635**
233	268	176	234	76	270	205	129	**268**	100	73	41	34	29	26	24	18		16	9	2	–	–	–	1	–	–	–	–	–	–	**2-Chloro-3-phenyl-1,4-napthoquinone**		**LI05636**
268	121	234	269	77	267	191	235	**268**	100	50	24	21	21	20	18	17		16	12	–	–	2	–	–	–	–	–	–	–	–	**1,4-Dithiin, 2,5-diphenyl-**	4159-03-9	**NI27158**
204	268	102	129	205	128	64	235	**268**	100	34	30	18	16	7	7	6		16	12	2	–	1	–	–	–	–	–	–	–	–	**Benzo[b]naphtho[1,2-d]thiophene, 7a,11b-dihydro-, 7,7-dioxide**	19076-30-3	**NI27159**
163	91	77	135	105	134	164	268	**268**	100	40	30	23	16	12	11	7		16	12	2	–	1	–	–	–	–	–	–	–	–	**2-Benzoyl-5-phenyl-1,3-oxathiole**		**NI27160**
191	268	36	221	223	235	145	178	**268**	100	75	21	18	12	10	10	8		16	12	2	–	1	–	–	–	–	–	–	–	–	**4-Phenyl-3-thia-3,4-dihydro-1-naphthoic acid**		**LI05637**
102	204	105	268	202	203	220	101	**268**	100	32	30	23	20	15	10	10		16	12	2	–	1	–	–	–	–	–	–	–	–	**2H-Thiete, 2-benzylidene-4-phenyl-, 1,1-dioxide**		**LI05638**
102	204	105	44	268	202	77	76	**268**	100	28	28	20	19	18	18	18		16	12	2	–	1	–	–	–	–	–	–	–	–	**2H-Thiete, 2-benzylidene-4-phenyl-, 1,1-dioxide**		**NI27161**
268	269	102	253	77	165	225	51	**268**	100	14	7	7	7	6	6	4		16	12	4	–	–	–	–	–	–	–	–	–	–	**9,10-Anthracenedione, 1,4-dihydroxy-2,3-dimethyl-**	25060-18-8	**NI27162**
268	237	152	139	239	209	269	251	**268**	100	34	27	26	22	22	18	15		16	12	4	–	–	–	–	–	–	–	–	–	–	**9,10-Anthracenedione, 1,5-dimethoxy-**	6448-90-4	**NI27163**
268	237	152	139	209	239	251	180	**268**	100	35	26	25	23	22	14	14		16	12	4	–	–	–	–	–	–	–	–	–	–	**9,10-Anthracenedione, 1,5-dimethoxy-**	6448-90-4	**LI05639**
253	268	251	152	139	254	209	180	**268**	100	48	25	18	18	17	12	10		16	12	4	–	–	–	–	–	–	–	–	–	–	**9,10-Anthracenedione, 1,8-dimethoxy-**	6407-55-2	**LI05640**
253	268	251	159	254	139	236	209	**268**	100	42	22	19	18	15	10	10		16	12	4	–	–	–	–	–	–	–	–	–	–	**9,10-Anthracenedione, 1,8-dimethoxy-**	6407-55-2	**IC03484**
253	268	251	139	152	254	151	180	**268**	100	47	26	19	18	17	13	10		16	12	4	–	–	–	–	–	–	–	–	–	–	**9,10-Anthracenedione, 1,8-dimethoxy-**	6407-55-2	**NI27164**
225	240	268	270	197	57	54	226	**268**	100	73	58	52	34	31	25	24		16	12	4	–	–	–	–	–	–	–	–	–	–	**1,2-Anthracenedione, 7-hydroxy-8-methoxy-5-methyl-**		**LI05641**
268	222	250	239	269	197	194	249	**268**	100	29	26	17	15	11	10	9		16	12	4	–	–	–	–	–	–	–	–	–	–	**1,4-Anthracenedione, 10-hydroxy-5-methoxy-2-methyl-**	84542-45-0	**NI27165**
268	269	57	55	69	239	71	210	**268**	100	36	35	30	26	25	25	22		16	12	4	–	–	–	–	–	–	–	–	–	–	**9,10-Anthracenedione, 1-hydroxy-3-methoxy-6-methyl-**		**NI27166**
268	267	225	252	251	106	122	118	**268**	100	98	12	10	8	8	7	7		16	12	4	–	–	–	–	–	–	–	–	–	–	**2(3H)-Benzofuranone, 4-(hydroxybenzylidene)-6-methoxy-**	77764-93-3	**NI27167**
150	268	118	251	151	134	267	122	**268**	100	84	47	44	29	20	16	14		16	12	4	–	–	–	–	–	–	–	–	–	–	**3(2H)-Benzofuranone, 2-(hydroxybenzylidene)-6-methoxy-**	77764-91-1	**NI27168**
267	268	251	106	252	225	239	122	**268**	100	78	31	19	16	16	14	14		16	12	4	–	–	–	–	–	–	–	–	–	–	**3(2H)-Benzofuranone, 3-(hydroxybenzylidene)-6-methoxy-**	77764-92-2	**NI27169**
268	148	121	269	120	93	65	211	**268**	100	79	72	22	19	14	13	5		16	12	4	–	–	–	–	–	–	–	–	–	–	**2H-1-Benzopyran-2-one, 4-hydroxy-3-(4-methoxyphenyl)-**	26151-95-1	**LI05642**
268	148	121	269	120	93	65	76	**268**	100	79	72	19	19	14	11	9		16	12	4	–	–	–	–	–	–	–	–	–	–	**2H-1-Benzopyran-2-one, 4-hydroxy-3-(4-methoxyphenyl)-**	26151-95-1	**NI27170**

MASS TO CHARGE RATIOS	M.W.	INTENSITIES	Parent	C	H	O	N	S	F	Cl	Br	I	Si	P	B	X	COMPOUND NAME	CAS Reg No	No
151 268 118 269 152 150 134 63	**268**	100 56 33 12 11 9 7 5		16	12	4	-	-	-	-	-	-	-	-	-	-	**2H-1-Benzopyran-2-one, 4-hydroxy-7-methoxy-3-phenyl-**	2555-24-0	**NI27171**
151 268 118 269 152 150 134 108	**268**	100 57 34 13 12 9 8 5		16	12	4	-	-	-	-	-	-	-	-	-	-	**2H-1-Benzopyran-2-one, 4-hydroxy-7-methoxy-3-phenyl-**	2555-24-0	**LI05643**
268 239 269 267 225 138 95 238	**268**	100 29 19 12 12 9 9 8		16	12	4	-	-	-	-	-	-	-	-	-	-	**4H-1-Benzopyran-4-one, 5-hydroxy-7-methoxy-2-phenyl-**	520-28-5	**NI27172**
268 132 267 269 117 240 225 43	**268**	100 63 25 20 15 12 9 9		16	12	4	-	-	-	-	-	-	-	-	-	-	**4H-1-Benzopyran-4-one, 7-hydroxy-2-(4-methoxyphenyl)-**	487-24-1	**NI27173**
268 151 267 118 122 150 269 107	**268**	100 68 46 32 23 21 18 14		16	12	4	-	-	-	-	-	-	-	-	-	-	**4H-1-Benzopyran-4-one, 3-(4-hydroxyphenyl)-7-methoxy-**	486-63-5	**NI27174**
226 195 227 167 268 139 138 43	**268**	100 22 16 14 13 8 6 5		16	12	4	-	-	-	-	-	-	-	-	-	-	**6-Biphenylenecarboxylic acid, 2-acetoxy-, methyl ester**	93103-43-6	**NI27175**
226 43 197 268 141	**268**	100 17 10 8 6		16	12	4	-	-	-	-	-	-	-	-	-	-	**6H-Dibenzo[b,d]pyran-6-one, 3-acetyloxy-1-methyl-**		**LI05644**
167 43 209 168 268 184 183 166	**268**	100 45 31 30 21 21 17 13		16	16	2	2	-	-	-	-	-	-	-	-	-	**Acetamide, N,N′-[1,1′-biphenyl]-2,2′-diylbis-**	29325-49-3	**NI27176**
253 267 134 268 120 77 92 135	**268**	100 99 99 92 89 81 68 61		16	16	2	2	-	-	-	-	-	-	-	-	-	**Aniline, N,N′-1,2-ethanediylidenebis[4-methoxy-**	24764-91-8	**NI27177**
268 161 267 134 269 77 135 241	**268**	100 60 52 25 18 17 13 12		16	16	2	2	-	-	-	-	-	-	-	-	-	**Benzaldehyde, 4-methoxy-, [(4-methoxyphenyl)methylene]hydrazone**	2299-73-2	**NI27178**
161 268 267 134 77 269 92 241	**268**	100 99 45 30 25 18 15 14		16	16	2	2	-	-	-	-	-	-	-	-	-	**Benzaldehyde, 4-methoxy-, [(4-methoxyphenyl)methylene]hydrazone**	2299-73-2	**NI27179**
268 161 267 134 77 92 241 240	**268**	100 99 45 30 25 15 14 12		16	16	2	2	-	-	-	-	-	-	-	-	-	**Benzaldehyde, 4-methoxy-, [(4-methoxyphenyl)methylene]hydrazone**	2299-73-2	**LI05645**
119 237 91 77 150 131 104 134	**268**	100 61 59 33 28 23 21 20	**12.00**	16	16	2	2	-	-	-	-	-	-	-	-	-	**Benzamide, 2-methoxy-, [(2-methoxyphenyl)methylene]hydrazone**	17745-81-2	**LI05646**
119 237 91 77 150 131 104 92	**268**	100 61 59 33 28 23 21 20	**12.01**	16	16	2	2	-	-	-	-	-	-	-	-	-	**Benzamide, 2-methoxy-, [(2-methoxyphenyl)methylene]hydrazone**	17745-81-2	**NI27180**
91 268 163 92 269 164	**268**	100 50 25 10 8 4		16	16	2	2	-	-	-	-	-	-	-	-	-	**2H-Benzimidazolin-2-one, 1-benzyl-5-ethoxy-**	28643-55-2	**NI27181**
91 268 161 65 92 269 177 239	**268**	100 38 15 15 10 7 7 4		16	16	2	2	-	-	-	-	-	-	-	-	-	**1H-Benzimidazol-4-ol, 5-ethoxy-1-benzyl-**	55000-39-0	**NI27182**
105 77 51 268 106 163 43 42	**268**	100 40 10 9 8 7 6 5		16	16	2	2	-	-	-	-	-	-	-	-	-	**Benzoic acid, 1-benzoyl-2,2-dimethylhydrazide**	30859-86-0	**NI27183**
105 77 51 268 163 43 42 269	**268**	100 40 10 9 7 6 5 3		16	16	2	2	-	-	-	-	-	-	-	-	-	**Benzoic acid, 1-benzoyl-2,2-dimethylhydrazide**	30859-86-0	**LI05647**
105 77 106 136 163 51 135 134	**268**	100 31 9 7 6 5 4 2	**0.50**	16	16	2	2	-	-	-	-	-	-	-	-	-	**Benzoic acid, 2-benzoyl-1,2-dimethylhydrazide**	1226-43-3	**MS05998**
268 224 154 180 192 223 221 207	**268**	100 71 55 51 49 43 41 41		16	16	2	2	-	-	-	-	-	-	-	-	-	**Ergoline-8-carboxylic acid, 9,10-didehydro-6-methyl-, (8β)-**		**MS05999**
91 268 161 65 177 239	**268**	100 38 15 15 7 4		16	16	2	2	-	-	-	-	-	-	-	-	-	**1H-Indazol-3-ol, 1-benzyl-5-ethoxy-**		**LI05648**
121 130 122 104 120 76 105 79	**268**	100 82 78 77 72 70 65 65	**49.00**	16	16	2	2	-	-	-	-	-	-	-	-	-	**1H-Isoindole-1,3(2H)-dione, 2-(9-azabicyclo[6.1.0]non-4-en-9-yl)-, (1α,4Z,8α)-**	66387-75-5	**NI27184**
105 106 162 41 251 227 40 268	**268**	100 55 41 30 18 14 11 10		16	16	2	2	-	-	-	-	-	-	-	-	-	**3-Methyl-4-(4-methoxyphenyl)-5-phenyl-1,2-4-oxadiazole**		**LI05649**
107 193 121 134 122 148 120 268	**268**	100 82 79 71 71 57 56 52		16	16	2	2	-	-	-	-	-	-	-	-	-	**Phenol, 2,2′-[1,2-ethanediylbis(nitrilomethylidene)]bis-**	94-93-9	**IC03485**
268 107 147 122 134 121 148 120	**268**	100 67 59 55 53 47 45 35		16	16	2	2	-	-	-	-	-	-	-	-	-	**Phenol, 2,2′-[1,2-ethanediylbis(nitrilomethylidene)]bis-**	94-93-9	**LI05650**
268 107 147 122 134 148 121 149	**268**	100 65 60 58 55 45 45 40		16	16	2	2	-	-	-	-	-	-	-	-	-	**Phenol, 2,2′-[1,2-ethanediylbis(nitrilomethylidyne)]bis-**	94-93-9	**NI27185**
146 77 253 235 120 65 51 160	**268**	100 94 88 61 48 43 41 40	**12.50**	16	16	2	2	-	-	-	-	-	-	-	-	-	**4(3H)-Quinazolinone, 1,2-dihydro-3-[2-(hydroxymethyl)phenyl]-2-methyl-**		**NI27186**
253 254 268 255 195 269 239 237	**268**	100 27 10 9 7 3 3 2		16	20	-	-	-	-	-	-	-	2	-	-	-	**9,9,10,10-Tetramethyl-9,10-disila-9,10-dihydroanthracene**	33022-24-1	**NI27187**
268 53 269 253 134 93 212 267	**268**	100 16 12 11 8 8 7 5		16	20	-	4	-	-	-	-	-	-	-	-	-	**5,8,13,16-Tetramethyl[2.2](2,5)pyrazinophane, (±)-pseudo-geminal-**	123673-27-8	**DD00833**
268 53 269 253 134 93 135 267	**268**	100 19 12 10 10 8 6 5		16	20	-	4	-	-	-	-	-	-	-	-	-	**5,8,13,16-Tetramethyl[2.2](2,5)pyrazinophane, (±)-pseudo-ortho-**	123624-88-4	**DD00834**
57 108 85 41 67 109 127 81	**268**	100 96 59 53 52 51 47 45	**0.00**	16	28	3	-	-	-	-	-	-	-	-	-	-	**Cyclohexanecarboxylic acid, 4-(1,5-dimethyl-3-oxohexyl)-, methyl ester, cis-**	17904-29-9	**NI27188**
43 41 55 31 166 110 27 69	**268**	100 50 30 19 17 17 17 14	**0.36**	16	28	3	-	-	-	-	-	-	-	-	-	-	**5-Cyclohexanone, 1,1-dimethyl-3-acetoxy-4,4-dipropyl-**		**LI05651**
125 69 59 109 81 107 55 80	**268**	100 41 41 33 33 27 27 22	**0.00**	16	28	3	-	-	-	-	-	-	-	-	-	-	**2,4-Dodecadienoic acid, 11-hydroxy-3,7,11-trimethyl-, methyl ester, (E,E)-**	69687-75-8	**NI27189**
110 95 69 41 113 111 141 98	**268**	100 68 67 28 23 18 16 14	**0.00**	16	28	3	-	-	-	-	-	-	-	-	-	-	**2-Pentenoic acid, 5-[(1,5-dimethyl-4-hexenyl)oxy]-3-methyl-, ethyl ester, (E)-**	53311-39-0	**NI27190**
41 69 81 95 83 209 43 129	**268**	100 95 89 87 83 76 76 75	**0.00**	16	28	3	-	-	-	-	-	-	-	-	-	-	**2,4,4,7-Tetramethyloctahydrobenzo[b]pyran-3-carboxylic acid, ethyl ester**		**LI05652**
89 227 61 81 137 90 101 85	**268**	100 31 23 14 13 8 6 6	**0.44**	16	32	1	-	-	-	-	-	-	1	-	-	-	**Silane, ethylmethyl[[5-methyl-2-isopropyl-cyclohexy]loxy]-2-propenyl-, (1α,2β,5α)-**	74841-60-4	**NI27191**
58 57 71 97 83 111 126 224	**268**	100 95 90 42 42 24 20 17	**17.00**	16	32	1	2	-	-	-	-	-	-	-	-	-	**Azacyclotridecan-2-one, 1-[3-(methylamino)propyl]-**	67370-86-9	**NI27192**
58 70 98 83 126 112 114 155	**268**	100 75 42 40 35 32 30 24	**23.00**	16	32	1	2	-	-	-	-	-	-	-	-	-	**1,5-Diazacycloheptadecan-6-one, 5-methyl-**	67370-87-0	**NI27193**
191 105 268 77 208 209 131 269	**268**	100 84 79 47 36 22 20 15		17	16	3	-	-	-	-	-	-	-	-	-	-	**Benzenepropanoic acid, 4-benzoyl-, methyl ester**	54411-77-7	**NI27194**
148 268 120 267 135 147 121 239	**268**	100 97 83 57 57 24 23 22		17	16	3	-	-	-	-	-	-	-	-	-	-	**4H-1-Benzopyran-4-one, 2-(4-ethoxyphenyl)-2,3-dihydro-**	69687-98-5	**NI27195**
268 253 135 267 225 237 161 160	**268**	100 40 37 32 26 23 22 21		17	16	3	-	-	-	-	-	-	-	-	-	-	**4,4′-Dimethoxychalcone**		**MS06000**
268 165 225 152	**268**	100 16 10 8		17	16	3	-	-	-	-	-	-	-	-	-	-	**2,4-Dimethoxy-4′-formyl-stilbene**		**LI05653**
225 268 267 184 43 197 169 226	**268**	100 88 65 55 31 27 22 20		17	20	1	2	-	-	-	-	-	-	-	-	-	**2-Acetyloctahydroindolo[2,3-a]quinolizine**		**LI05654**
225 43 268 184 267 197 130 223	**268**	100 86 82 68 58 40 34 22		17	20	1	2	-	-	-	-	-	-	-	-	-	**2-Acetyloctahydroindolo[2,3-a]quinolizine**		**LI05655**
134 77 251 148 240 104 51 85	**268**	100 81 52 48 38 30 29 27	**20.66**	17	20	1	2	-	-	-	-	-	-	-	-	-	**Benzophenone, 2,4′-bis(dimethylamino)-**		**IC03486**
148 30 29 28 134 132 77 252	**268**	100 88 76 76 37 33 27 24	**1.44**	17	20	1	2	-	-	-	-	-	-	-	-	-	**Benzophenone, 4,4′-bis(dimethylamino)-**		**MS06001**
268 148 267 269 224 251 120 133	**268**	100 73 22 19 16 11 11 9		17	20	1	2	-	-	-	-	-	-	-	-	-	**Benzophenone, 4,4′-bis(dimethylamino)-**		**IC03487**
211 268 198 183 196 78 168 225	**268**	100 74 63 58 57 46 44 43		17	20	1	2	-	-	-	-	-	-	-	-	-	**Dasycarpidan-1-one**	2825-08-3	**NI27196**
44 211 268 42 198 196 183 41	**268**	100 91 67 61 53 50 49 39		17	20	1	2	-	-	-	-	-	-	-	-	-	**Dasycarpidan-1-one**	2825-08-3	**NI27197**
268 211 239 225 269 198 196 267	**268**	100 43 31 27 23 20 17 16		17	20	1	2	-	-	-	-	-	-	-	-	-	**Dasycarpidan-1-one**	2825-08-3	**NI27198**

MASS TO CHARGE RATIOS								M.W.	INTENSITIES								Parent	C	H	O	N	S	F	Cl	Br	I	Si	P	B	X	COMPOUND NAME	CAS Reg No	No
58	139	210	140	128	127	141	130	**268**	100	35	34	31	31	26	25	20	**5.13**	17	20	1	2	–	–	–	–	–	–	–	–	–	Dibenz[b,e][1,4]oxazepine-5(11H)-ethanamine, N,N-dimethyl-	16882-89-6	NI27199
124	268	144	225	239	196	182	210	**268**	100	66	17	16	12	12	10	6		17	20	1	2	–	–	–	–	–	–	–	–	–	3-Ethyl-1,2,5,6-tetrahydro-1-methyl-4-pyridyl indol-2-yl ketone		LI05656
268	95	67	41	201	174	103	253	**268**	100	72	30	26	24	22	19	18		17	20	1	2	–	–	–	–	–	–	–	–	–	N^2-(Hexa-2E,4-dienoyl)-N^1,N^1-dimethyl-2-phenylacrylamidine		NI27200
69	60	267	268	97	169	129	184	**268**	100	86	79	60	42	40	33	20		17	20	1	2	–	–	–	–	–	–	–	–	–	Indolo[2,3-a]quinolizin-2(1H)-one, 3-ethyl-3,4,6,7,12,12b-hexahydro-, trans-		DD00835
268	99	155	170	70	253	91	77	**268**	100	80	61	39	30	29	19	15		17	20	1	2	–	–	–	–	–	–	–	–	–	1H-Isoindol-1-one, 3-(dimethylamino)-3a,4,5,7a-tetrahydro-5-methyl-3a-phenyl-	75378-99-3	NI27201
268	171	168	157	158	170	169	197	**268**	100	63	51	45	43	42	39	37		17	20	1	2	–	–	–	–	–	–	–	–	–	1H-Pyrido[3,2-c]carbazole, 3,4-epoxy-4a-ethyl-2,3,4,4a,5,6-hexahydro-	80249-75-8	NI27202
41	74	55	87	69	43	67	39	**268**	100	99	80	52	52	32	23	22	**1.00**	17	32	2	–	–	–	–	–	–	–	–	–	–	Cyclopentaneundecanoic acid, methyl ester	25779-85-5	WI01325
96	97	55	81	67	41	155	71	**268**	100	49	41	40	20	15	12	11	**0.00**	17	32	2	–	–	–	–	–	–	–	–	–	–	Decanoic acid, cyclohexanemethyl ester	2890-71-3	NI27203
100	43	85	41	113	55	29	57	**268**	100	77	70	30	19	19	16	15	**1.75**	17	32	2	–	–	–	–	–	–	–	–	–	–	2,4-Heptadecanedione	64042-18-8	NI27204
73	60	43	57	270	55	41	69	**268**	100	86	86	68	63	60	56	38	**0.00**	17	32	2	–	–	–	–	–	–	–	–	–	–	Heptadecenoic acid	26265-99-6	NI27205
41	43	87	55	81	57	113	69	**268**	100	94	77	76	42	39	35	34	**0.00**	17	32	2	–	–	–	–	–	–	–	–	–	–	2-Hexadecenoic acid, methyl ester, (E)-	2825-81-2	NI27206
116	73	221	117	28	44	220	45	**268**	100	42	10	10	9	7	4	4	**0.01**	17	32	2	–	–	–	–	–	–	–	–	–	–	2-Hexadecenoic acid, methyl ester, (Z)-	2825-80-1	NI27207
55	41	43	74	69	67	83	87	**268**	100	90	71	67	52	51	42	41	**0.57**	17	32	2	–	–	–	–	–	–	–	–	–	–	7-Hexadecenoic acid, methyl ester, (Z)-	56875-67-3	NI27208
55	69	74	87	97	83	96	84	**268**	100	88	81	75	74	74	67	65	**1.94**	17	32	2	–	–	–	–	–	–	–	–	–	–	9-Hexadecenoic acid, methyl ester, (Z)-	1120-25-8	NI27211
40	54	42	68	73	28	66	26	**268**	100	88	51	48	46	44	28	26	**3.84**	17	32	2	–	–	–	–	–	–	–	–	–	–	9-Hexadecenoic acid, methyl ester, (Z)-	1120-25-8	NI27209
55	41	43	69	74	83	84	67	**268**	100	96	61	50	44	38	35	35	**0.00**	17	32	2	–	–	–	–	–	–	–	–	–	–	9-Hexadecenoic acid, methyl ester, (Z)-	1120-25-8	NI27210
55	74	69	41	84	83	97	87	**268**	100	66	60	60	51	49	48	47	**17.52**	17	32	2	–	–	–	–	–	–	–	–	–	–	11-Hexadecenoic acid, methyl ester	55000-42-5	WI01326
219	268	204	203	202	205	91	220	**268**	100	41	27	23	23	22	21	20		18	17	–	–	–	–	1	–	–	–	–	–	–	5H-Dibenzo[a,d]cycloheptene, 5-(3-chloropropylidene)-10,11-dihydro-	14330-49-5	NI27212
268	235	220	269	221	236	222	207	**268**	100	57	36	20	16	12	10	10		18	20	–	–	1	–	–	–	–	–	–	–	–	Dibenzo[c,e]thiepin, 5,7-dihydro-1,3,9,11-tetramethyl-	31066-96-3	NI27213
253	268	41	128	39	91	143	117	**268**	100	78	68	51	38	33	29	23		18	20	–	–	1	–	–	–	–	–	–	–	–	3,3-Dimethyl-2-phenyl-1-butenyl phenyl sulphide		DD00836
128	159	116	77	91	237	266	51	**268**	100	50	50	50	40	27	25	20	**0.00**	18	20	–	–	1	–	–	–	–	–	–	–	–	3-Phenyl-4-(phenylthio)-3-hexene		DD00837
107	163	77	79	57	105	41	51	**268**	100	40	32	27	25	21	13	10	**0.00**	18	20	2	–	–	–	–	–	–	–	–	–	–	Acetophenone, 2-isobutoxy-2-phenyl-	22499-12-3	NI27214
268	239	145	107	240	253	115	221	**268**	100	53	24	13	9	7	7	5		18	20	2	–	–	–	–	–	–	–	–	–	–	Benzofuran, 2,3-dihydro-2-(4-hydroxyphenyl)-3-methyl-5-propyl-, (2R,3R)-		MS06002
119	118	133	77	120	91	105	79	**268**	100	60	25	13	10	10	9	7	**0.60**	18	20	2	–	–	–	–	–	–	–	–	–	–	Benzoic acid, 2,4-dimethyl-, (2,4-dimethylphenyl)methyl ester	55000-43-6	NI27215
118	119	133	77	91	105	120	79	**268**	100	72	34	14	11	10	7	7	**0.30**	18	20	2	–	–	–	–	–	–	–	–	–	–	Benzoic acid, 2,4-dimethyl-, (2,5-dimethylphenyl)methyl ester	55000-44-7	NI27216
133	119	268	77	105	91	134	79	**268**	100	74	17	16	12	12	11	8		18	20	2	–	–	–	–	–	–	–	–	–	–	Benzoic acid, 2,4-dimethyl-, (3,5-dimethylphenyl)methyl ester	55000-45-8	NI27217
119	118	133	120	77	105	91	103	**268**	100	46	20	10	9	8	8	6	**2.70**	18	20	2	–	–	–	–	–	–	–	–	–	–	Benzoic acid, 2,5-dimethyl-, (2,4-dimethylphenyl)methyl ester	55000-48-1	NI27218
118	119	133	77	91	105	120	79	**268**	100	89	29	15	12	11	9	7	**0.70**	18	20	2	–	–	–	–	–	–	–	–	–	–	Benzoic acid, 2,5-dimethyl-, (2,5-dimethylphenyl)methyl ester	55000-49-2	NI27219
118	119	133	77	91	105	120	103	**268**	100	89	29	15	12	11	9	7	**0.67**	18	20	2	–	–	–	–	–	–	–	–	–	–	Benzoic acid, 2,5-dimethyl-, (2,5-dimethylphenyl)methyl ester	55000-49-2	AM00137
118	119	133	105	91	77	117	120	**268**	100	74	30	10	10	10	8	7	**0.40**	18	20	2	–	–	–	–	–	–	–	–	–	–	Benzoic acid, 3,5-dimethyl-, (2,4-dimethylphenyl)methyl ester	55000-46-9	NI27220
133	119	268	134	105	77	91	79	**268**	100	32	18	13	13	12	10	7		18	20	2	–	–	–	–	–	–	–	–	–	–	Benzoic acid, 3,5-dimethyl-, (3,5-dimethylphenyl)methyl ester	55000-47-0	NI27221
91	55	268	92	269	253	177	83	**268**	100	27	17	10	5	5	5	3		18	20	2	–	–	–	–	–	–	–	–	–	–	Bicyclo[3,3,1]nonane-8,9-dione, 4-benzylidene-2,2-dimethyl-		IC03488
268	223	195	180	165	179	178	269	**268**	100	77	50	41	40	20	18	16		18	20	2	–	–	–	–	–	–	–	–	–	–	[1,1′-Biphenyl]-4-carboxylic acid, 2′,4′,6′-trimethyl-, ethyl ester	66818-60-8	NI27222
268	253	145	269	121	147	165	137	**268**	100	26	26	20	16	11	10	9		18	20	2	–	–	–	–	–	–	–	–	–	–	1-Butene, 1,2-bis(4-methoxyphenyl)-		MS06003
135	133	268	134	147	77	136	269	**268**	100	81	48	14	12	12	11	10		18	20	2	–	–	–	–	–	–	–	–	–	–	1-Butene, 2,3-bis(4-methoxyphenyl)-		MS06004
268	269	145	253	238	126	121	119	**268**	100	19	14	13	13	8	8	8		18	20	2	–	–	–	–	–	–	–	–	–	–	2-Butene, 2,3-bis(4-methoxyphenyl)-		LI05657
268	269	145	253	238	137	135	126	**268**	100	19	14	13	13	10	10	8		18	20	2	–	–	–	–	–	–	–	–	–	–	2-Butene, 2,3-bis(4-methoxyphenyl)-		MS06005
163	105	117	135	118	43	91	77	**268**	100	95	83	49	45	22	18	17	**0.01**	18	20	2	–	–	–	–	–	–	–	–	–	–	1,3-Dioxolane, 2-(1-phenylethyl)-4-methyl-4-phenyl-		IC03489
209	268	235	224	197	211	210	195	**268**	100	83	48	44	39	37	36	31		18	20	2	–	–	–	–	–	–	–	–	–	–	Estra-1,3,5,7,9-pentaene-3,17-diol, (17β)-	1423-97-8	NI27223
209	268	235	224	197	211	210	165	**268**	100	83	49	45	38	37	36	31		18	20	2	–	–	–	–	–	–	–	–	–	–	Estra-1,3,5,7,9-pentaen-17-one, 3-hydroxy-		LI05658
268	170	157	144	269	183	211	158	**268**	100	32	27	23	21	20	19	16		18	20	2	–	–	–	–	–	–	–	–	–	–	Estra-1,3,5(10),6-tetraen-17-one, 3-hydroxy-	2208-12-0	NI27224
268	157	144	170	41	158	55	183	**268**	100	67	55	40	40	31	28	26		18	20	2	–	–	–	–	–	–	–	–	–	–	Estra-1,3,5(10),6-tetraen-17-one, 3-hydroxy-	2208-12-0	NI27226
268	157	144	170	158	211	183	145	**268**	100	41	30	29	22	18	17	13		18	20	2	–	–	–	–	–	–	–	–	–	–	Estra-1,3,5(10),6-tetraen-17-one, 3-hydroxy-	2208-12-0	NI27225
268	144	157	211	41	183	115	158	**268**	100	67	49	32	29	28	28	26		18	20	2	–	–	–	–	–	–	–	–	–	–	Estra-1,3,5(10),7-tetraen-17-one, 3-hydroxy-	474-86-2	NI27227
268	144	157	211	158	170	183	225	**268**	100	36	32	24	20	18	17	13		18	20	2	–	–	–	–	–	–	–	–	–	–	Estra-1,3,5(10),7-tetraen-17-one, 3-hydroxy-	474-86-2	NI27228
268	269	211	144	183	157	170	158	**268**	100	22	20	19	15	15	14	12		18	20	2	–	–	–	–	–	–	–	–	–	–	Estra-1,3,5(10),7-tetraen-17-one, 3-hydroxy-	474-86-2	NI27229
268	253	210	211	269	197	251	43	**268**	100	38	30	22	21	15	13	11		18	20	2	–	–	–	–	–	–	–	–	–	–	Estra-1,3,5(10),9(11)-tetraen-17-one, 3-hydroxy-	1089-80-1	NI27230
268	253	210	211	269	197	251	235	**268**	100	42	38	28	21	18	16	15		18	20	2	–	–	–	–	–	–	–	–	–	–	Estra-1,3,5(10),9(11)-tetraen-17-one, 3-hydroxy-	1089-80-1	NI27231
268	240	211	198	172	269	61	160	**268**	100	30	29	29	23	18	16	15		18	20	2	–	–	–	–	–	–	–	–	–	–	Estra-1,3,5(10),15-tetraen-17-one, 3-hydroxy-, 14α-		NI27232
172	268	173	160	145	171	269	157	**268**	100	72	51	23	23	17	15	15		18	20	2	–	–	–	–	–	–	–	–	–	–	Estra-1,3,5(10),15-tetraen-17-one, 3-hydroxy-, 14β-		NI27233
164	268	197	211	151	152	180	114	**268**	100	90	85	75	75	71	65	61		18	20	2	–	–	–	–	–	–	–	–	–	–	Ethylene, 1,1-bis(4-ethoxyphenyl)-	5031-92-5	NI27234
154	153	152	155	185	76	77	227	**268**	100	35	26	23	22	17	12	10	**0.00**	18	20	2	–	–	–	–	–	–	–	–	–	–	5-Hexenoic acid, 3-(1-naphthyl)-, ethyl ester		DD00838
205	268	199	131	269	206	267	197	**268**	100	94	29	22	19	18	11	9		18	20	2	–	–	–	–	–	–	–	–	–	–	1,1-Bis(4-hydroxyphenyl)cyclohexane		IC03490

MASS TO CHARGE RATIOS								M.W.	INTENSITIES								Parent	C	H	O	N	S	F	Cl	Br	I	Si	P	B	X	COMPOUND NAME	CAS Reg No	No
181	268	195	165	194	179	182	180	**268**	100	30	27	23	18	17	16	13		18	20	2	–	–	–	–	–	–	–	–	–	–	Methyl 3-(1,2,3,4-tetrahydro-9-anthryl)propanoate		AI01898
172	155	127	55	41	81	268	67	**268**	100	33	27	13	8	8	7	6		18	20	2	–	–	–	–	–	–	–	–	–	–	2-Naphthalenecarboxylic acid, cyclohexanemethyl ester		NI27235
195	196	178	43	77	105	91	51	**268**	100	86	16	16	15	13	13	13	**10.40**	18	20	2	–	–	–	–	–	–	–	–	–	–	1,2-Naphthalenediol, 1,2,3,4-tetrahydro-3,3-dimethyl-1-phenyl-	51086-37-4	NI27236
195	196	77	268	105	91	179	178	**268**	100	85	15	13	13	13	9	8		18	20	2	–	–	–	–	–	–	–	–	–	–	1,2-Naphthalenediol, 1,2,3,4-tetrahydro-3,3-dimethyl-1-phenyl-, cis-	56588-42-2	NI27237
197	212	103	91	237	198	77	119	**268**	100	21	20	17	15	15	15	12	**0.00**	18	20	2	–	–	–	–	–	–	–	–	–	–	Oxirane, [[4-(1-methyl-1-phenylethyl)phenoxy]methyl]-	61578-04-9	NI27238
240	211	134	44	212	141	241	183	**268**	100	52	48	48	41	27	21	21	**3.00**	18	20	2	–	–	–	–	–	–	–	–	–	–	Pentacyclo[9.6.1.0^{2,10}.0^{5,10}.0^{12,17}]octadeca-4,12(17)-diene-3,18-dione, endo-		NI27239
107	145	268	239	77	121	159	39	**268**	100	69	54	34	28	27	21	19		18	20	2	–	–	–	–	–	–	–	–	–	–	Phenol, 4,4′-(1,2-diethyl-1,2-ethenediyl)bis-	6898-97-1	NI27240
91	160	120	30	70	106	177	92	**268**	100	54	27	25	20	18	15	10	**9.00**	18	24	–	2	–	–	–	–	–	–	–	–	–	N,N′-Dibenzylputrescine		LI05659
134	106	135	268	77	269	104	91	**268**	100	12	10	9	6	2	2	2		18	24	–	2	–	–	–	–	–	–	–	–	–	N,N′-Diethyl-N,N′-diphenylethylenediamine		IC03491
253	211	268	167	210	168	169	41	**268**	100	94	58	54	40	34	33	21		18	24	–	2	–	–	–	–	–	–	–	–	–	4-(Diisopropylamino)diphenylamine		IC03492
211	43	167	268	183	224	41	40	**268**	100	38	31	29	28	27	22	19		18	24	–	2	–	–	–	–	–	–	–	–	–	4-(1-Methylpentylamino)diphenylamine		IC03493
134	135	91	42	118	77	18	179	**268**	100	10	8	8	7	4	4	3	**0.00**	18	24	–	2	–	–	–	–	–	–	–	–	–	N,N,N′,N′-Tetramethyl-1,2-diphenylethylenediamine, (±)-		DD00839
134	135	42	118	91	77	179	178	**268**	100	10	7	6	6	3	2	2	**0.00**	18	24	–	2	–	–	–	–	–	–	–	–	–	N,N,N′,N′-Tetramethyl-1,2-diphenylethylenediamine, meso-		DD00840
182	211	44	42	194	180	41	167	**268**	100	91	77	74	62	53	52	49	**30.69**	18	24	–	2	–	–	–	–	–	–	–	–	–	Uleine, dihydro-	19046-20-9	NI27241
43	82	41	55	96	68	69	83	**268**	100	47	40	34	20	17	16	15	**0.00**	18	36	1	–	–	–	–	–	–	–	–	–	–	Cyclododecane, hexyloxy-		DD00841
55	43	41	82	96	83	69	68	**268**	100	94	93	86	27	20	20	20	**0.00**	18	36	1	–	–	–	–	–	–	–	–	–	–	Cyclopentadecane, propoxy-		DD00842
251	43	141	69	97	55	81	123	**268**	100	82	33	30	24	24	23	22	**0.00**	18	36	1	–	–	–	–	–	–	–	–	–	–	3-Dodecanol, 3-cyclopentyl-2-methyl-		DD00843
43	41	57	82	68	96	110	250	**268**	100	84	54	29	21	15	4	3	**0.30**	18	36	1	–	–	–	–	–	–	–	–	–	–	Octadecanal	638-66-4	LI05660
43	41	57	55	29	82	44	69	**268**	100	84	54	49	47	29	27	23	**0.34**	18	36	1	–	–	–	–	–	–	–	–	–	–	Octadecanal	638-66-4	NI27242
43	41	57	55	82	69	29	83	**268**	100	86	80	74	63	43	42	40	**0.44**	18	36	1	–	–	–	–	–	–	–	–	–	–	Octadecanal	638-66-4	NI27243
72	57	43	29	73	41	239	55	**268**	100	98	62	61	49	46	34	34	**2.70**	18	36	1	–	–	–	–	–	–	–	–	–	–	3-Octadecanone	18261-92-2	WI01327
82	55	41	69	67	96	110	124	**268**	100	98	86	83	83	76	31	18	**1.10**	18	36	1	–	–	–	–	–	–	–	–	–	–	9-Octadecen-1-ol, (Z)-	143-28-2	IC03494
41	43	55	57	69	82	29	83	**268**	100	84	83	52	47	38	38	37	**0.07**	18	36	1	–	–	–	–	–	–	–	–	–	–	9-Octadecen-1-ol, (Z)-	143-28-2	AI01899
55	82	41	96	69	67	81	43	**268**	100	85	69	66	64	61	58	52	**0.40**	18	36	1	–	–	–	–	–	–	–	–	–	–	9-Octadecen-1-ol, (Z)-	143-28-2	WI01328
43	58	71	57	59	55	41	69	**268**	100	96	54	46	42	33	27	26	**1.00**	18	36	1	–	–	–	–	–	–	–	–	–	–	2-Pentadecanone, 6,10,14-trimethyl-	502-69-2	NI27244
43	58	71	57	59	41	55	69	**268**	100	90	45	43	41	37	35	25	**0.90**	18	36	1	–	–	–	–	–	–	–	–	–	–	2-Pentadecanone, 6,10,14-trimethyl-	502-69-2	NI27245
225	171	115	268	91	145	129	55	**268**	100	87	69	61	61	57	57	56		19	24	1	–	–	–	–	–	–	–	–	–	–	9,10-Didehydro-19-norretinal (all-trans)		NI27246
109	145	240	115	105	131	129	67	**268**	100	99	84	78	75	67	67	63	**25.00**	19	24	1	–	–	–	–	–	–	–	–	–	–	1(2H)-Naphthalenone, octahydro-8aβ-methyl-2-(4-methylbenzylidene)-	17429-46-8	NI27247
57	71	43	85	112	183	41	56	**268**	100	78	50	27	24	21	20	18	**1.45**	19	40	–	–	–	–	–	–	–	–	–	–	–	Hexadecane, 2,6,10-trimethyl-		MS06006
57	43	71	85	41	55	99	29	**268**	100	70	66	38	38	21	16	16	**2.00**	19	40	–	–	–	–	–	–	–	–	–	–	–	Nonadecane	629-92-5	NI27248
57	71	43	85	113	99	41	55	**268**	100	65	58	47	29	16	15	13	**1.15**	19	40	–	–	–	–	–	–	–	–	–	–	–	Nonadecane	629-92-5	IC03495
57	43	71	85	41	29	55	56	**268**	100	72	51	30	28	20	19	12	**0.40**	19	40	–	–	–	–	–	–	–	–	–	–	–	Nonadecane	629-92-5	NI27249
43	57	41	71	55	29	85	56	**268**	100	75	43	34	24	24	20	17	**2.84**	19	40	–	–	–	–	–	–	–	–	–	–	–	Octadecane, 2-methyl-	1560-88-9	NI27250
57	43	71	85	41	55	56	29	**268**	100	87	59	40	29	21	19	15	**1.73**	19	40	–	–	–	–	–	–	–	–	–	–	–	Octadecane, 2-methyl-	1560-88-9	MS06007
43	71	57	70	85	41	55	225	**268**	100	99	85	51	40	31	26	19	**0.66**	19	40	–	–	–	–	–	–	–	–	–	–	–	Octadecane, 4-methyl-		LI05661
43	57	85	71	84	41	55	56	**268**	100	84	74	48	39	28	21	18	**0.53**	19	40	–	–	–	–	–	–	–	–	–	–	–	Octadecane, 5-methyl-		LI05662
57	71	43	85	41	113	55	56	**268**	100	82	50	34	28	20	20	18	**0.20**	19	40	–	–	–	–	–	–	–	–	–	–	–	Pentadecane, 2,6,10,13-tetramethyl-		MS06008
71	57	43	85	41	113	56	55	**268**	100	84	65	38	30	24	24	24	**0.34**	19	40	–	–	–	–	–	–	–	–	–	–	–	Pentadecane, 2,6,10,14-tetramethyl-	1921-70-6	NI27251
57	43	71	41	55	85	56	113	**268**	100	89	68	45	27	24	23	15	**0.24**	19	40	–	–	–	–	–	–	–	–	–	–	–	Pentadecane, 2,6,10,14-tetramethyl-	1921-70-6	AI01900
71	85	113	112	99	183	127	70	**268**	100	45	34	27	24	15	12	10	**1.10**	19	40	–	–	–	–	–	–	–	–	–	–	–	Pentadecane, 2,6,10,14-tetramethyl-	1921-70-6	NI27252
57	43	29	71	41	85	55	182	**268**	100	99	66	64	49	44	33	24	**0.07**	19	40	–	–	–	–	–	–	–	–	–	–	–	Tridecane, 7-hexyl-		AI01901
268	239	78	252	69	28	269	18	**268**	100	42	32	29	26	23	22	22		20	12	1	–	–	–	–	–	–	–	–	–	–	4,5-Dihydro-4,5-epoxybenzopyrene		NI27253
268	145	119	105	91	159	121	93	**268**	100	84	76	61	52	38	32	32		20	28	–	–	–	–	–	–	–	–	–	–	–	Anhydroretinol		LI05663
155	268	156	141	28	183	269	198	**268**	100	51	23	22	18	14	11	11		20	28	–	–	–	–	–	–	–	–	–	–	–	Naphthalene, 2-butyl-3-hexyl-		AI01902
155	268	141	156	29	27	165	41	**268**	100	55	18	17	17	12	11	11		20	28	–	–	–	–	–	–	–	–	–	–	–	Naphthalene, 2-butyl-3-hexyl-		AI01904
155	268	141	156	41	269	183	43	**268**	100	47	19	18	12	10	10	10		20	28	–	–	–	–	–	–	–	–	–	–	–	Naphthalene, 2-butyl-3-hexyl-		AI01903
197	268	154	155	153	141	41	269	**268**	100	85	33	29	22	21	21	19		20	28	–	–	–	–	–	–	–	–	–	–	–	Naphthalene, 2-butyl-8-hexyl-		AI01906
197	268	154	155	153	141	198	269	**268**	100	76	31	29	20	20	19	17		20	28	–	–	–	–	–	–	–	–	–	–	–	Naphthalene, 2-butyl-8-hexyl-		AI01905
197	268	154	155	153	29	269	141	**268**	100	93	34	27	23	21	19	19		20	28	–	–	–	–	–	–	–	–	–	–	–	Naphthalene, 7-butyl-1-hexyl-		AI01907
197	268	155	154	141	198	153	41	**268**	100	65	35	33	25	23	23	19		20	28	–	–	–	–	–	–	–	–	–	–	–	Naphthalene, 7-butyl-1-hexyl-		AI01908
197	268	155	154	141	198	28	269	**268**	100	73	32	27	25	21	20	16		20	28	–	–	–	–	–	–	–	–	–	–	–	Naphthalene, 7-butyl-1-hexyl-		AI01909
169	268	167	170	29	269	153	152	**268**	100	74	20	19	18	15	13	13		20	28	–	–	–	–	–	–	–	–	–	–	–	Naphthalene, 1,4-dimethyl-5-octyl-		AI01910
169	268	170	167	29	155	153	41	**268**	100	50	23	20	15	13	12	12		20	28	–	–	–	–	–	–	–	–	–	–	–	Naphthalene, 1,4-dimethyl-5-octyl-		AI01911
169	268	167	170	41	269	153	152	**268**	100	61	21	20	15	13	13	12		20	28	–	–	–	–	–	–	–	–	–	–	–	Naphthalene, 1,4-dimethyl-5-octyl-		AI01912
169	170	268	269	155	171	153	44	**268**	100	76	61	14	12	10	10	9		20	28	–	–	–	–	–	–	–	–	–	–	–	Naphthalene, 2,6-dimethyl-3-octyl-		AI01913
169	170	268	41	29	155	269	153	**268**	100	75	49	14	14	13	11	11		20	28	–	–	–	–	–	–	–	–	–	–	–	Naphthalene, 2,6-dimethyl-3-octyl-		AI01914

MASS TO CHARGE RATIOS								M.W.	INTENSITIES								Parent	C	H	O	N	S	F	Cl	Br	I	Si	P	B	X	COMPOUND NAME	CAS Reg No	No
169	268	170	269	155	153	29	152	**268**	100	64	57	14	12	12	10	9		20	28	–	–	–	–	–	–	–	–	–	–	–	Naphthalene, 2,6-dimethyl-3-octyl-		MS06009
268	253	252	267	269	126	265	134	**268**	100	40	39	24	21	21	12	9		21	16	–	–	–	–	–	–	–	–	–	–	–	Benz[j]aceanthrylene, 1,2-dihydro-3-methyl-	56-49-5	MS06010
268	126	252	253	267	269	134	133	**268**	100	40	38	36	24	22	20	19		21	16	–	–	–	–	–	–	–	–	–	–	–	Benz[j]aceanthrylene, 1,2-dihydro-3-methyl-	56-49-5	NI27254
268	267	191	265	189	252	126	165	**268**	100	42	24	18	18	14	14	10		21	16	–	–	–	–	–	–	–	–	–	–	–	1,1-Diphenylindene		LI05664
268	269	267	265	191	189	165		**268**	100	25	25	15	15	13	10			21	16	–	–	–	–	–	–	–	–	–	–	–	2,3-Diphenylindene		LI05665
269	268	168	187	131	225	149	130	**269**	100	96	83	41	32	10	7	6		2	6	–	1	–	5	–	–	–	–	–	–	1	(Dimethylamino)tellurium pentafluoride		LI05666
134	69	70	42	250	185	234	200	**269**	100	50	24	23	3	3	1	1	**0.00**	3	–	–	1	1	8	1	–	–	–	–	–	–	N-[(1,1,2,3,3,3-Hexafluoro-2-chloropropyl)thio]difluorimide		LI05667
106	108	46	53	65	118	93	91	**269**	100	97	72	49	36	26	24	24	**2.60**	3	1	2	3	–	–	–	2	–	–	–	–	–	2,4-Dibromo-5-nitroimidazole		IC03496
43	98	85	128	127	42	84	70	**269**	100	15	12	11	9	7	6	6	**0.70**	6	8	3	1	1	5	–	–	–	–	–	–	–	(2-Acetylacetoacetamido)sulphur pentafluoride	90598-19-9	NI27255
61	209	56	254	208	250			**269**	100	67	30	13	7	4			**0.00**	8	10	1	1	–	7	–	–	–	–	–	–	–	N-(2-Fluoro-2-methylpropyl)-2-(trifluoromethyl)trifluoropropionamide		LI05668
140	74	106	196	81	168	46	227	**269**	100	85	83	71	50	50	39	33	**5.27**	8	16	3	1	2	–	–	–	–	–	1	–	–	Phosphoramidic acid, (4-methyl-1,3-dithiolan-2-ylidene)-, diethyl ester	950-10-7	NI27256
140	106	196	74	41	168	81	227	**269**	100	75	68	66	53	52	43	32	**4.80**	8	16	3	1	2	–	–	–	–	–	1	–	–	Phosphoramidic acid, (4-methyl-1,3-dithiolan-2-ylidene)-, diethyl ester	950-10-7	NI27257
44	211	73	226	195	135	45	75	**269**	100	54	27	19	16	15	13	12	**5.00**	8	24	3	1	–	–	–	–	–	2	1	–	–	Phosphonic acid, (1-aminoethyl)-, bis(trimethylsilyl) ester	53121-99-6	NI27258
44	211	73	226	195	135	75	133	**269**	100	54	27	19	16	15	12	9	**5.00**	8	24	3	1	–	–	–	–	–	2	1	–	–	Phosphonic acid, (1-aminoethyl)-, bis(trimethylsilyl) ester	53121-99-6	NI27259
60	188	189	59	103	58	44	28	**269**	100	72	64	38	31	21	21	19	**0.00**	9	8	–	3	1	–	–	1	–	–	–	–	–	Pyrido[2,3-d]thiazolo[3,2-b]pyridazin-4-ium, 2,3-dihydro-, bromide	18592-68-2	NI27260
60	188	189	59	103	58	44	45	**269**	100	73	64	38	30	21	21	19	**0.00**	9	8	–	3	1	–	–	1	–	–	–	–	–	Pyrido[2,3-d]thiazolo[3,2-b]pyridazin-4-ium, 2,3-dihydro-, bromide	18592-68-2	NI27261
188	189	60	190	82	80	161	103	**269**	100	90	63	32	29	28	27	25	**0.00**	9	8	–	3	1	–	–	1	–	–	–	–	–	Pyrido[3,2-d]thiazolo[3,2-b]pyridazin-4-ium, 2,3-dihydro-, bromide	18592-67-1	NI27262
220	92	174	269	77	221	118	91	**269**	100	28	27	25	11	10	10	10		10	8	4	3	–	–	1	–	–	–	–	–	–	Imidazo[1,2-a]pyrimidinium, 3-(chloroacetyl)-2-hydroxy-1-methyl-6-nitro-, hydroxide, inner salt	11070-38-5	NI27263
220	92	174	269	77	221	118	51	**269**	100	27	26	24	11	10	10	10		10	8	4	3	–	–	1	–	–	–	–	–	–	Imidazo[1,2-a]pyrimidinium, 3-(chloroacetyl)-2-hydroxy-1-methyl-6-nitro-,hydroxide, inner salt	11070-38-5	LI05669
77	269	106	42	145	44	254	81	**269**	100	99	87	87	80	79	70	50		10	12	–	1	–	5	–	–	–	1	–	–	–	Silanamine, N,N,1,1-tetramethyl-1-(pentafluorophenyl)-	23761-75-3	NI27264
165	18	28	44	31	179	29	54	**269**	100	99	85	71	66	59	59	36	**9.35**	10	13	5	4	–	–	–	–	–	–	–	–	–	Formycin B		LI05670
43	86	40	44	42	41	70	234	**269**	100	60	46	39	34	33	30	27	**0.00**	10	17	1	1	1	–	2	–	–	–	–	–	–	Carbamothioic acid, diisopropyl-, S-(2,3-dichloro-2-propenyl) ester	2303-16-4	NI27265
43	86	41	27	42	128	39	234	**269**	100	80	48	37	32	26	22	20	**0.00**	10	17	1	1	1	–	2	–	–	–	–	–	–	Carbamothioic acid, diisopropyl-, S-(2,3-dichloro-2-propenyl) ester	2303-16-4	NI27266
225	71	45	269	227	154	72	271	**269**	100	77	57	43	37	36	22	17		11	8	3	1	1	–	1	–	–	–	–	–	–	4-Thiazoleacetic acid, 2-(3-chloro-4-hydroxyphenyl)-	20403-74-1	IC03497
71	45	269	227	154	225	72	271	**269**	100	75	59	53	49	43	29	23		11	8	3	1	1	–	1	–	–	–	–	–	–	4-Thiazoleacetic acid, 2-(3-chloro-4-hydroxyphenyl)-	20403-74-1	NI27267
269	240	267	224	270	238	222	116	**269**	100	50	48	40	35	26	19	19		11	11	2	1	–	–	–	–	–	–	–	–	1	Selenolo[2,3-b]pyridine-2-carboxylic acid, 3-methyl-, ethyl ester	55108-59-3	NI27268
104	76	147	41	179	105	50	130	**269**	100	96	80	64	56	49	44	38	**2.00**	11	11	3	1	2	–	–	–	–	–	–	–	–	2H-Isoindole-2-sulphinothioic acid, 1,3-dihydro-1,3-dioxo-, S-isopropyl ester	42300-60-7	NI27269
195	227	43	269	137	93	53	59	**269**	100	88	85	40	25	20	20	15		11	11	7	1	–	–	–	–	–	–	–	–	–	Methyl 2-acetoxy-3-nitro-5-methoxybenzoate		LI05671
55	168	43	84	56	41	69	42	**269**	100	89	71	26	24	23	22	20	**0.00**	11	18	3	1	–	3	–	–	–	–	–	–	–	β-Alanine, N-(trifluoroacetyl)-, 1-methylpentyl ester	57983-13-8	NI27270
43	140	141	85	41	42	69	57	**269**	100	52	31	23	17	11	10	10	**0.00**	11	18	3	1	–	3	–	–	–	–	–	–	–	L-Alanine, N-(trifluoroacetyl)-, 1-methylpentyl ester	57983-75-2	NI27271
43	140	141	85	41	42	69	168	**269**	100	68	50	28	24	21	15	13	**0.00**	11	18	3	1	–	3	–	–	–	–	–	–	–	Glycine, N-methyl-N-(trifluoroacetyl)-, 1-methylpentyl ester	57983-78-5	NI27272
168	55	57	169	184	166	114	100	**269**	100	26	15	13	12	10	7	6	**2.00**	11	18	3	1	–	3	–	–	–	–	–	–	–	D-Isovaline, N-(trifluoroacetyl)-, sec-butyl ester	54889-84-8	NI27273
168	55	57	41	42	184	169	69	**269**	100	82	48	36	27	17	17	16	**0.00**	11	18	3	1	–	3	–	–	–	–	–	–	–	L-Isovaline, N-(trifluoroacetyl)-, sec-butyl ester	57983-15-0	NI27274
55	168	29	41	126	57	27	169	**269**	100	95	63	49	41	35	27	21	**0.00**	11	18	3	1	–	3	–	–	–	–	–	–	–	L-Norvaline, N-(trifluoroacetyl)-, butyl ester	54889-85-9	NI27275
168	55	57	169	126	41	140	114	**269**	100	97	90	68	42	33	24	16	**0.00**	11	18	3	1	–	3	–	–	–	–	–	–	–	D-Norvaline, N-(trifluoroacetyl)-, sec-butyl ester	54889-86-0	NI27276
168	55	57	169	126	41	140	114	**269**	100	98	83	61	49	49	28	18	**0.00**	11	18	3	1	–	3	–	–	–	–	–	–	–	L-Norvaline, N-(trifluoroacetyl)-, sec-butyl ester	57983-22-9	NI27277
196	74	140	55	57	41	56	83	**269**	100	48	45	44	42	39	31	23	**0.00**	11	18	3	1	–	3	–	–	–	–	–	–	–	Pentanoic acid, 4-[(trifluoroacetyl)amino]-, 1-methylpropyl ester	57983-38-7	NI27278
55	168	41	29	57	69	153	27	**269**	100	95	53	47	40	25	23	20	**0.00**	11	18	3	1	–	3	–	–	–	–	–	–	–	L-Valine, N-(trifluoroacetyl)-, butyl ester	1478-96-2	NI27279
168	55	57	41	169	56	171	153	**269**	100	89	85	80	51	40	19	14	**0.00**	11	18	3	1	–	3	–	–	–	–	–	–	–	L-Valine, N-(trifluoroacetyl)-, sec-butyl ester	16974-92-8	NI27280
254	73	269	255	147	112	256	45	**269**	100	28	27	21	12	12	9	8		11	23	1	3	–	–	–	–	–	2	–	–	–	4-Pyrimidinamine, 5-methyl-N-(trimethylsilyl)-2-[(trimethylsilyl)oxy]-	32865-88-6	NI27281
170	168	269	169	127	154	83	155	**269**	100	99	89	71	61	59	40	40		12	12	2	1	1	–	1	–	–	–	–	–	–	1-Naphthalenesulphonyl chloride, 5-(dimethylamino)-	605-65-2	NI27282
155	154	44	77	50	127	156	51	**269**	100	67	35	21	20	18	15	14	**0.00**	12	12	4	1	–	–	1	–	–	–	–	–	–	Pyridinium, 1-methyl-2-phenyl-, perchlorate	52806-04-9	NI27283
181	166	135	107	369	120	79	182	**269**	100	36	29	28	23	23	22	17	**0.00**	12	15	6	1	–	–	–	–	–	–	–	–	–	14-Nitro-2,3-benzo-1,4,7,10-tetraoxacyclododeca-2-ene		MS06011
84	42	41	55	85	56	63	75	**269**	100	13	6	6	4	3	2	1	**0.67**	12	16	1	1	–	–	–	1	–	–	–	–	–	Pyrrolidine, 1-[2-(4-bromophenoxy)ethyl]-	1081-73-8	NI27284
213	121	186	56	93	212	39	66	**269**	100	98	96	60	53	51	50	47	**1.56**	13	11	2	1	–	–	–	–	–	–	–	–	1	Iron, π-cyclopentadienyl-π-(2-methylpyridyl)dicarbonyl-		MS06012
269	122	210	56	174	152	121	204	**269**	100	89	64	57	43	42	35	17		13	11	2	1	–	–	–	–	–	–	–	–	1	Methyl 2-cyano-1-ferrocenecarboxylate		NI27285
269	122	210	56	121	174	152	204	**269**	100	73	65	63	50	46	32	26		13	11	2	1	–	–	–	–	–	–	–	–	1	Methyl 3-cyano-1-ferrocenecarboxylate		NI27286
79	204	52	203	202	51	49	101	**269**	100	95	69	65	48	38	36	28	**0.00**	13	12	–	1	–	4	–	–	–	–	–	1	–	Pyridinium, 1-styryl-, tetrafluoroborate(1-)	579-54-4	NI27287
119	91	118	269	120	225	115	105	**269**	100	20	13	12	12	8	6	5		13	19	3	1	1	–	–	–	–	–	–	–	–	4H-1,3,2,6-Dioxathiazocine, 5,6,7,8-tetrahydro-4,8-dimethyl-6-(4-tolyl)-, 2-oxide	70291-66-6	NI27288

MASS TO CHARGE RATIOS								M.W.	INTENSITIES								Parent	C	H	O	N	S	F	Cl	Br	I	Si	P	B	X	COMPOUND NAME	CAS Reg No	No
152	94	210	43	168	80	269	138	**269**	100	69	54	33	23	22	18	15		13	19	5	1	–	–	–	–	–	–	–	–	–	2-Oxa-7-azatricyclo[4.4.0.0^{3,8}]decane-7-carboxylic acid, 4-(acetyloxy)-, ethyl ester	49656-68-0	NI27289
168	43	153	152	269	154	170	124	**269**	100	49	45	38	36	26	25	23		13	19	5	1	–	–	–	–	–	–	–	–	–	3-Oxa-10-azatricyclo[4.3.1.0^{2,4}]decane-10-carboxylic acid, 7-(acetyloxy)-, ethyl ester, (1α,2β,4β,6α,7β)-	49656-56-6	NI27290
83	82	178	39	55	53	150	210	**269**	100	99	94	81	80	67	60	57	**1.30**	13	19	5	1	–	–	–	–	–	–	–	–	–	Maleic acid, [(1-ethoxy-3-methyl-2-butenyl)imino]-, dimethyl ester		NI27291
178	210	41	150	53	68	67	39	**269**	100	67	47	20	18	13	13	13	**10.00**	13	19	5	1	–	–	–	–	–	–	–	–	–	Maleic acid, [(1-methoxy-2,3-dimethyl-2-butenyl)imino]-, dimethyl ester		NI27292
152	80	43	108	269	81	82	94	**269**	100	16	13	8	7	7	5	4		13	19	5	1	–	–	–	–	–	–	–	–	–	2,5-Methanofuro[3,2-b]pyridine-4(2H)-carboxylic acid, 8-(acetyloxy)hexahydro-, ethyl ester, (2α,3aβ,5α,7aβ,8S*)-	49656-72-6	NI27293
152	80	43	269	108	81	124	82	**269**	100	15	14	8	8	7	5	5		13	19	5	1	–	–	–	–	–	–	–	–	–	2,5-Methanofuro[3,2-b]pyridine-4(2H)-carboxylic acid, 8-(acetyloxy)hexahydro-, ethyl ester, (2α,3aβ,5α,7aβ,8R*)-	49656-67-9	NI27294
194	166	237	210	59	178	150	154	**269**	100	99	63	34	29	24	24	22	**0.00**	13	19	5	1	–	–	–	–	–	–	–	–	–	Spiro[5-oxopyrrolidine-2,1′-cyclohexane]-3,4-dicarboxylic acid, dimethyl ester, trans-		DD00844
141	77	197	156	140	184	169	104	**269**	100	88	72	70	56	51	42	35	**27.00**	13	20	3	1	–	–	–	–	–	–	1	–	–	4-(P-Ethoxy-P-phenylphosphinoyl)-N-methylbutanamide	93939-07-2	NI27295
212	75	138	41	68	213	73	67	**269**	100	66	30	29	19	17	12	11	**0.00**	13	23	3	1	–	–	–	–	–	1	–	–	–	3-Isoxazolepropanoic acid, 5-methyl-, [dimethyl(1,1-dimethylethyl)silyl] ester		NI27296
154	254	212	129	73	226	269	169	**269**	100	78	73	41	35	33	32	30		13	32	–	1	–	–	–	–	–	2	–	1	–	[Bis(trimethylsilyl)methylene](diisopropylamino)borane		NI27297
91	269	92	65	270	66			**269**	100	25	10	10	4	1				14	11	3	3	–	–	–	–	–	–	–	–	–	1H-Benzimidazol-4-ol, 5-nitro-1-benzyl-	54889-67-7	NI27298
91	269	92	223	105	104	97	129	**269**	100	33	18	10	9	9	9	8		14	11	3	3	–	–	–	–	–	–	–	–	–	2H-Benzimidazol-2-one, 1,3-dihydro-5-nitro-1-benzyl-	2215-55-6	NI27299
269	43	170	127	77	155	103	51	**269**	100	97	60	56	55	45	35	35		14	11	3	3	–	–	–	–	–	–	–	–	–	Imidazo[1,5-b]pyridazine-5,7(3H,6H)-dione, 4-acetyl-2-phenyl-		NI27300
269	105	77	226	270	51	225	141	**269**	100	80	51	50	18	17	14	9		14	11	3	3	–	–	–	–	–	–	–	–	–	Imidazo[1,5-b]pyridazine-5,7(3H,6H)-dione, 4-benzoyl-2-methyl-		NI27301
269	198	28	158	240	252	130	125	**269**	100	39	34	19	15	11	11	10		14	11	3	3	–	–	–	–	–	–	–	–	–	1-Methyloxindolidene-3-(2,4-dihydroxy-6-pyrimidinyl)methine		LI05672
105	77	121	76	269	252			**269**	100	35	30	3	2	2				14	11	3	3	–	–	–	–	–	–	–	–	–	2-Nitrobenzaldehyde benzoyl hydrazone		LI05673
269	270	176	224	138	121	204	251	**269**	100	75	50	36	29	29	24	22		14	15	1	1	–	–	–	–	–	–	–	–	1	anti-1,2-(α-Ketotetramethylene)ferrocene oxime		MS06013
269	270	176	224	138	204	121	253	**269**	100	81	50	42	36	34	34	26		14	15	1	1	–	–	–	–	–	–	–	–	1	syn-1,2-(α-Ketotetramethylene)ferrocene oxime		MS06014
45	77	44	160	188	91	117	49	**269**	100	24	24	23	16	15	14	12	**1.37**	14	20	2	1	–	–	1	–	–	–	–	–	–	Acetamide, 2-chloro-N-(2,6-diethylphenyl)-N-(methoxymethyl)-	15972-60-8	NI27302
45	160	188	146	132	91	77	118	**269**	100	29	25	12	10	10	10	9	**1.84**	14	20	2	1	–	–	1	–	–	–	–	–	–	Acetamide, 2-chloro-N-(2,6-diethylphenyl)-N-(methoxymethyl)-	15972-60-8	NI27303
45	160	188	237	162	224	161	146	**269**	100	45	43	14	12	11	11	11	**9.03**	14	20	2	1	–	–	1	–	–	–	–	–	–	Acetamide, 2-chloro-N-(2,6-diethylphenyl)-N-(methoxymethyl)-	15972-60-8	NI27304
84	85	56	91	28	55	36	30	**269**	100	9	6	5	5	4	4	3	**0.00**	14	20	2	1	–	–	1	–	–	–	–	–	–	2-Piperidineacetic acid, α-phenyl-, methyl ester, hydrochloride	298-59-9	NI27305
41	59	119	159	87	43	55	93	**269**	100	89	82	74	74	72	60	58	**42.20**	14	23	2	1	1	–	–	–	–	–	–	–	–	Methyl (+)-thujan-3-ylspiro-2′-thiazolidine-4-carboxylate	73154-86-6	NI27306
181	182	196	45	151	44	88	167	**269**	100	40	31	13	11	9	8	7	**3.84**	14	23	4	1	–	–	–	–	–	–	–	–	–	Benzenemethanamine, 3,4,5-trimethoxy-N-(3-methoxypropyl)-	34274-04-9	NI27307
154	226	269	86	238	227	125	155	**269**	100	89	22	18	14	14	14	13		14	23	4	1	–	–	–	–	–	–	–	–	–	Methyl α-(1-methyl-2-morpholinobutenyl)acetoacetate		MS06015
57	58	98	142	101	157	113	89	**269**	100	42	40	34	29	19	12	12	**0.00**	14	27	2	3	–	–	–	–	–	–	–	–	–	1,4,2-Dioxazolidine, 2-tert-butyl-3,5-bis(tert-butylimino)-		MS06016
177	269	77	51	192	45	121	39	**269**	100	93	60	53	33	25	22	22		15	11	–	1	2	–	–	–	–	–	–	–	–	Aniline, (2-thio-3-benzo[b]thienylidene)-	55136-30-6	NI27308
269	192	77	51	121	159	165	268	**269**	100	85	80	69	44	41	33	29		15	11	–	1	2	–	–	–	–	–	–	–	–	Aniline, (3-thio-2-benzo[b]thienylidene)-	55136-26-0	NI27309
269	118	268	134.5	270	236	95.5	111.5	**269**	100	49	40	33	25	25	19	11		15	11	–	1	2	–	–	–	–	–	–	–	–	Thieno[3′,4′:5,6]thiopyrano[4,3-b]indole, 1,3-dimethyl-	17820-13-2	LI05674
269	118	268	270	236	117	235	223	**269**	100	49	40	25	25	10	9	8		15	11	–	1	2	–	–	–	–	–	–	–	–	Thieno[3′,4′:5,6]thiopyrano[4,3-b]indole, 1,3-dimethyl-	17820-13-2	NI27310
269	270	183	268	212	240	211	44	**269**	100	81	36	34	34	32	31	18		15	11	4	1	–	–	–	–	–	–	–	–	–	1,3-Dioxolo[4,5-b]acridin-10(5H)-one, 11-methoxy-5-methyl-		MS06017
269	183	212	154	182	77	127	240	**269**	100	67	60	50	42	37	35	34		15	11	4	1	–	–	–	–	–	–	–	–	–	10-Methyl-2,3-methylenedioxy-1-hydroxyacridone		MS06018
147	269	268	121	120	65	270	93	**269**	100	85	52	48	34	26	14	13		15	11	4	1	–	–	–	–	–	–	–	–	–	2-Propen-1-one, 1-(2-hydroxyphenyl)-3-(4-nitrophenyl)-	38270-09-6	NI27311
104	105	61	77	51	78	164	194	**269**	100	56	37	36	28	24	23	18	**4.00**	15	15	–	3	1	–	–	–	–	–	–	–	–	2,6-Diphenyl-4-thiohexahydro-sym-triazine		IC03498
120	42	269	77	148	105	79	65	**269**	100	55	43	23	21	19	18	14		15	15	2	3	–	–	–	–	–	–	–	–	–	Benzoic acid, 4-[4-(dimethylamino)phenylazo]-		LI05675
208	269	180	222	207	152	252	209	**269**	100	65	53	46	38	33	31	29		15	15	2	3	–	–	–	–	–	–	–	–	–	3-Carbazolamine, N,N,9-trimethyl-4-nitro-	94127-25-0	NI27312
269	268	270	183	170	143	171	115	**269**	100	54	17	12	12	11	10	10		15	15	2	3	–	–	–	–	–	–	–	–	–	Uracil, 1,3-dimethyl-5-(3-methyl-1H-indol-2-yl)-		MS06019
158	56	85	268	213	112	74	57	**269**	100	45	36	33	33	28	24	20	**12.20**	15	27	3	1	–	–	–	–	–	–	–	–	–	Tris(3,3-dimethyl-2-oxetanyl)amine		MS06020
104	105	78	269	193	77	103	115	**269**	100	68	27	26	26	22	19	14		16	15	1	1	1	–	–	–	–	–	–	–	–	1,3-Oxazine-2-thione, tetrahydro-4,6-diphenyl-	86071-96-7	NI27313
198	252	180	269	186	253	199	221	**269**	100	85	46	28	24	20	20	17		16	15	1	1	1	–	–	–	–	–	–	–	–	Phenothiazine, 10-(2-methyl-2-propenyl)-, 5-oxide	31037-04-4	NI27314
145	150	254	269	109	149	144	91	**269**	100	93	54	36	34	29	28	27		16	15	1	1	1	–	–	–	–	–	–	–	–	11-Oxa-2-thia-9-azatetracyclo[8.7.1.0^{3,8}.0^{12,17}]octadeca-3,5,7,12,14,16-hexaene, 10-methyl-		MS06021
91	105	227	77	43	92	210	50	**269**	100	66	30	22	19	9	8	7	**4.00**	16	15	3	1	–	–	–	–	–	–	–	–	–	N-(α-Acetoxybenzylidene)-O-benzylhydroxylamine		LI05676
123	105	77	269	120	149	146	91	**269**	100	77	57	37	24	23	12	9		16	15	3	1	–	–	–	–	–	–	–	–	–	N-(3-Anisyl)-2-benzoylacetamide		LI05677
123	149	105	77	269	120	146	91	**269**	100	82	82	68	53	12	10	9		16	15	3	1	–	–	–	–	–	–	–	–	–	N-(4-Anisyl)-2-benzoylacetamide		LI05678
269	181	214	186	268	240	185	224	**269**	100	40	37	32	24	21	15	14		16	15	3	1	–	–	–	–	–	–	–	–	–	Apohemanthamine	6900-81-8	NI27315
93	94	105	91	225	44	224	77	**269**	100	7	4	4	3	3	1	1	**0.88**	16	15	3	1	–	–	–	–	–	–	–	–	–	3-Carboxy-N,3-diphenylpropionamide		IC03499
168	167	166	165	169	194	153	154	**269**	100	99	99	99	97	96	91	90	**39.00**	16	15	3	1	–	–	–	–	–	–	–	–	–	Diphenylacetylglycine	65707-74-6	NI27316
269	254	226	196	270	211	183	255	**269**	100	56	24	23	19	18	14	8		16	15	3	1	–	–	–	–	–	–	–	–	–	10-Methyl-1,3-dimethoxyacridone		MS06022
241	212	240	211	181	269	226		**269**	100	64	31	20	19	11	11			16	15	3	1	–	–	–	–	–	–	–	–	–	Oxodemethoxyhaemanthamine		LI05679

MASS TO CHARGE RATIOS								M.W.	INTENSITIES								Parent	C	H	O	N	S	F	Cl	Br	I	Si	P	B	X	COMPOUND NAME	CAS Reg No	No
269	268	252	160	228	176	161	227	**269**	100	41	26	26	20	18	16	7		16	16	–	3	–	1	–	–	–	–	–	–	–	2-(3,4,5-Trimethylpyrazol-1-yl)-4-methyl-6-fluoroquinoline		MS06023
73	88	90	44	43	55	41	30	**269**	100	95	37	32	32	28	27	27	**0.00**	16	31	2	1	–	–	–	–	–	–	–	–	–	N-Allyl-4-oxatetradecanamide		MS06024
41	55	43	56	57	67	69	84	**269**	100	97	58	42	33	31	22	21	**0.00**	16	31	2	1	–	–	–	–	–	–	–	–	–	Cyclododecanone, 6-(aminomethyl)-(3-hydroxypropyl)-		DD00845
55	41	82	69	44	56	67	68	**269**	100	99	71	61	59	55	54	53	**0.00**	16	31	2	1	–	–	–	–	–	–	–	–	–	2-Oxacyclohexadecanone, 6-(aminomethyl)-		DD00846
57	41	43	69	142	85	56	100	**269**	100	65	64	42	39	30	28	24	**0.00**	16	31	2	1	–	–	–	–	–	–	–	–	–	Propanamide, N-(5-oxoundecyl)-2,2-dimethyl-	116437-32-2	DD00847
71	125	55	170	212	126	111	86	**269**	100	88	70	46	33	30	30	30	**23.00**	16	35	–	3	–	–	–	–	–	–	–	–	–	N[1],N[2],N[3]-Triisopentylguanidine		LI05680
267	209	178	30	269	240	91	191	**269**	100	64	57	55	53	53	52	48		17	19	–	1	1	–	–	–	–	–	–	–	–	1H-2-Benzazepine, 6-(methylthio)-5-phenyl-2,3,4,5-tetrahydro-		MS06025
269	178	31	192	225	91	205	115	**269**	100	64	52	34	33	31	30	23		17	19	–	1	1	–	–	–	–	–	–	–	–	1H-2-Benzazepine, 8-(methylthio)-5-phenyl-2,3,4,5-tetrahydro-		MS06026
132	178	91	117	105	269	77	45	**269**	100	71	69	46	35	21	15	13		17	19	–	1	1	–	–	–	–	–	–	–	–	Cinnamylamine, N-[(3-methylthio)benzyl]-		MS06027
117	104	209	91	210	118	115	237	**269**	100	62	60	56	44	27	27	24	**16.00**	17	19	2	1	–	–	–	–	–	–	–	–	–	trans-3-Acetoxy-6-cyano-trans-[8]paracycloph-4-ene		NI27317
269	226	165	211	193	268	254	240	**269**	100	87	59	42	35	22	16	8		17	19	2	1	–	–	–	–	–	–	–	–	–	Apogalantamine methyl ester		LI05681
239	91	72	240	165	167	166	105	**269**	100	50	26	21	21	16	13	11	**0.00**	17	19	2	1	–	–	–	–	–	–	–	–	–	Benzeneacetamide, α-(hydroxymethyl)-N,N-dimethyl-	72572-09-9	NI27318
105	122	164	77	106	79	104	120	**269**	100	30	26	21	17	15	14	9	**7.00**	17	19	2	1	–	–	–	–	–	–	–	–	–	Carbamic acid, (α-methylbenzyl)-, α-methylbenzyl ester	32692-88-9	LI05682
105	122	164	77	106	79	104	51	**269**	100	30	25	21	16	14	13	7	**7.00**	17	19	2	1	–	–	–	–	–	–	–	–	–	Carbamic acid, (α-methylbenzyl)-, α-methylbenzyl ester	32692-88-9	NI27319
105	122	164	77	106	79	104	51	**269**	100	30	25	20	16	14	8	7	**6.00**	17	19	2	1	–	–	–	–	–	–	–	–	–	Carbamic acid, (α-methylbenzyl)-, α-methylbenzyl ester	32692-88-9	NI27320
143	269	99	144	171	81	115	270	**269**	100	28	20	14	13	13	12	6		17	19	2	1	–	–	–	–	–	–	–	–	–	Cyclohexanecarboxamide, 1-hydroxy-N-(1-naphthyl)-		MS06028
192	226	269	268	193	227	150	149	**269**	100	22	20	16	13	4	4	4		17	19	2	1	–	–	–	–	–	–	–	–	–	3,5-Diacetyl-4-phenyl-2,6-dimethyl-1,4-dihydropyridine	20970-67-6	LI05683
192	226	269	268	43	193	254	42	**269**	100	23	21	16	14	13	5	4		17	19	2	1	–	–	–	–	–	–	–	–	–	3,5-Diacetyl-4-phenyl-2,6-dimethyl-1,4-dihydropyridine	20970-67-6	NI27321
192	43	268	226	67	269	193	42	**269**	100	34	17	17	16	14	14	9		17	19	2	1	–	–	–	–	–	–	–	–	–	3,5-Diacetyl-4-phenyl-2,6-dimethyl-1,4-dihydropyridine	20970-67-6	NI27322
47	130	48	49	129	70	41	118	**269**	100	54	41	33	31	27	20	18	**9.00**	17	19	2	1	–	–	–	–	–	–	–	–	–	4,4a-Dihydro-1-isopropylidene-4,4-dimethyl-1,3-oxazino[3,4-a]quinolin-3-one		LI05684
58	42	77	105	30	72	56	43	**269**	100	52	38	25	22	20	19	19	**4.40**	17	19	2	1	–	–	–	–	–	–	–	–	–	4-(2-Dimethylaminoethoxy)benzophenone		IC03500
198	199	269	197	141	255	170	184	**269**	100	54	32	24	22	21	16	12		17	19	2	1	–	–	–	–	–	–	–	–	–	Hasubanan, N-methyl-7-oxo-, DL-		NI27323
209	91	267	57	269	120	71	165	**269**	100	97	87	69	51	48	40	37		17	19	2	1	–	–	–	–	–	–	–	–	–	Isoquinoline, 5,6-dimethoxy-4-phenyl-1,2,3,4-tetrahydro-		MS06029
240	209	269	165	225	36	153	178	**269**	100	90	46	25	14	12	9	8		17	19	2	1	–	–	–	–	–	–	–	–	–	Isoquinoline, 6,7-dimethoxy-4-phenyl-1,2,3,4-tetrahydro-		MS06030
162	163	135	105	77	132	79	91	**269**	100	29	25	9	9	5	5	4	**0.00**	17	19	2	1	–	–	–	–	–	–	–	–	–	1,3-Oxazolidine, 4-methyl-5-cis-phenyl-2-(4-methoxyphenyl)-		MS06031
162	163	135	77	105	91	79	148	**269**	100	39	27	10	10	7	5	5	**0.00**	17	19	2	1	–	–	–	–	–	–	–	–	–	1,3-Oxazolidine, 4-methyl-5-trans-phenyl-2-(4-methoxyphenyl)-		MS06032
226	269	145	106	187	146	240	268	**269**	100	48	39	37	35	27	26	23		17	23	–	3	–	–	–	–	–	–	–	–	–	1-Cyclohexyl-7,8,9,10-tetrahydro-1H-benzo[1,2,e][2,1-c]-as-triazine		LI05685
254	269	255	270	268	83	224	238	**269**	100	75	20	16	15	15	14	10		17	23	–	3	–	–	–	–	–	–	–	–	–	2,3,5,6-Tetraethyl-1,7-dihydrodipyrrolo[2,3-b:3',2'-e]pyridine		NI27324
170	142	128	184	60	55	43	69	**269**	100	93	93	88	59	57	54	53	**0.40**	17	35	1	1	–	–	–	–	–	–	–	–	–	7-Acetamido-7-methyltetradecane		MS06033
184	43	55	142	156	198	84	69	**269**	100	59	58	53	52	49	45	45	**0.10**	17	35	1	1	–	–	–	–	–	–	–	–	–	6-Acetamido-6-propyldodecane		MS06034
170	72	171	73	74	69	184	142	**269**	100	65	13	11	10	9	6	6	**2.10**	17	35	1	1	–	–	–	–	–	–	–	–	–	Formamide, N,N-dioctyl-	6280-57-5	NI27325
269	270	140	271	268	242	112	111	**269**	100	13	12	6	5	5	5	5		18	11	–	3	–	–	–	–	–	–	–	–	–	5H-Benz[g]indolo[2,3-b]quinoxaline	249-06-9	NI27326
269	270	140	268	267	126	121	114	**269**	100	10	9	5	4	4	4	4		18	11	–	3	–	–	–	–	–	–	–	–	–	5H-Benz[g]indolo[2,3-b]quinoxaline	249-06-9	NI27327
148	178	105	135	91	43	121	44	**269**	100	69	67	35	34	30	25	23	**1.00**	18	23	1	1	–	–	–	–	–	–	–	–	–	Benzeneethanamine, 3-methoxy-N-methyl-N-(2-phenylethyl)-	52059-58-2	NI27328
148	105	121	135	42	77	44	78	**269**	100	56	35	27	25	22	18	16	**1.00**	18	23	1	1	–	–	–	–	–	–	–	–	–	Benzeneethanamine, 4-methoxy-N-methyl-N-(2-phenylethyl)-	52059-57-1	NI27329
213	269	252	226	184	159	198	240	**269**	100	96	77	25	25	24	16	10		18	23	1	1	–	–	–	–	–	–	–	–	–	7H-Cyclohepta[c]isoquinolin-5(6H)-one, 6-butyl-8,9,10,11-tetrahydro-		DD00848
81	41	104	189	79	53	91	43	**269**	100	32	31	30	23	18	18	17	**5.32**	18	23	1	1	–	–	–	–	–	–	–	–	–	2,6,8-Decatrienamide, N-(2-phenylethyl)-, (E,Z,E)-		NI27330
105	269	104	79	91	93	55	41	**269**	100	67	52	42	38	30	27	27		18	23	1	1	–	–	–	–	–	–	–	–	–	3,6,8-Decatrienamide, N-(2-phenylethyl)-, (E,Z,E)-		NI27331
58	73	45	59	166	165	46	181	**269**	100	38	12	10	7	7	7	5	**0.00**	18	23	1	1	–	–	–	–	–	–	–	–	–	Ethanamine, N,N-dimethyl-2-[(2-methylphenyl)phenylmethoxy]-	83-98-7	NI27332
181	209	270	268	221				**269**	100	17	7	6	3				**0.00**	18	23	1	1	–	–	–	–	–	–	–	–	–	Ethanamine, N,N-dimethyl-2-[(2-methylphenyl)phenylmethoxy]-	83-98-7	NI27333
226	188	187	269	198	200	227	212	**269**	100	81	80	62	60	22	19	14		18	23	1	1	–	–	–	–	–	–	–	–	–	Isoindolin-3-one, 1-cyclohexylidene-4,5,7-trimethyl-		IC03501
254	196	212	269	91	43	108	119	**269**	100	83	69	56	45	29	20	17		18	23	1	1	–	–	–	–	–	–	–	–	–	2-Isoxazoline, 5,5-dimethyl-4-(2-methyl-1-propenylidene)-3-(2,4,6-trimethylphenyl)-		DD00849
269	59	226	148	268	42	270	44	**269**	100	50	36	33	32	28	21	20		18	23	1	1	–	–	–	–	–	–	–	–	–	Morphinan, 5,6-didehydro-3-methoxy-17-methyl-, (9α,13α,14α)-	2640-17-7	NI27334
59	269	148	203	268	270	254	60	**269**	100	81	34	22	21	18	16	15		18	23	1	1	–	–	–	–	–	–	–	–	–	Morphinan, 6,7-didehydro-3-methoxy-17-methyl-, (14α)-	55028-44-9	NI27335
170	198	226	55	43	41	57	224	**269**	100	90	73	22	21	20	17	16	**0.20**	18	39	–	1	–	–	–	–	–	–	–	–	–	N-Ethyl-6-propyl-6-tetradecanamine		MS06035
198	199	128	43	170	226	44	58	**269**	100	15	15	13	11	8	5	4	**2.00**	18	39	–	1	–	–	–	–	–	–	–	–	–	1-Hexanamine, N,N-dihexyl-	102-86-3	NI27337
198	199	128	43	226	44	41	58	**269**	100	15	15	13	8	5	5	4	**2.00**	18	39	–	1	–	–	–	–	–	–	–	–	–	1-Hexanamine, N,N-dihexyl-	102-86-3	NI27336
198	128	199	43	44	114	58	269	**269**	100	18	15	8	7	6	5	4		18	39	–	1	–	–	–	–	–	–	–	–	–	1-Hexanamine, N,N-dihexyl-	102-86-3	NI27338
156	170	84	43	41	171	157	55	**269**	100	94	17	17	17	13	12	9	**3.00**	18	39	–	1	–	–	–	–	–	–	–	–	–	Nonylamine, 1-heptyl-N,N-dimethyl-	24539-84-2	NI27340
156	170	84	43	41	44	42	157	**269**	100	94	17	17	16	13	13	12	**2.00**	18	39	–	1	–	–	–	–	–	–	–	–	–	Nonylamine, 1-heptyl-N,N-dimethyl-	24539-84-2	NI27339
156	170	58	43	44	41	42	57	**269**	100	94	61	60	48	42	25	22	**4.00**	18	39	–	1	–	–	–	–	–	–	–	–	–	Nonylamine, N-methyl-N-octyl-	24552-02-1	NI27341
30	44	57	43	58	28	71	41	**269**	100	48	28	27	22	21	17	16	**0.00**	18	39	–	1	–	–	–	–	–	–	–	–	–	1-Octadecanamine	124-30-1	NI27342
44	30	57	156	41	212	269	43	**269**	100	92	31	26	20	14	13	13		18	39	–	1	–	–	–	–	–	–	–	–	–	Bis(3,5,5-trimethylhexyl)amine		IC03502
269	149	270	240	241	239	167	213	**269**	100	26	23	12	10	8	7	5		19	11	1	1	–	–	–	–	–	–	–	–	–	Indolo[3,2,1-de]phenanthridin-13-one		LI05686

MASS TO CHARGE RATIOS								M.W.	INTENSITIES								Parent	C	H	O	N	S	F	Cl	Br	I	Si	P	B	X	COMPOUND NAME	CAS Reg No	No
165	269	166	51	270	77	192	164	**269**	100	90	25	16	15	13	12	10		20	15	–	1	–	–	–	–	–	–	–	–	–	**Aniline, N-(diphenylvinylidene)-**	14181-84-1	**NI27343**
147	87	116	69	201	149	203	85	**270**	100	57	43	41	34	32	22	17	**0.56**	4	–	–	–	–	8	2	–	–	–	–	–	–	**Dichlorooctafluorobutane**		**DO01061**
95	130	36	142	132	60	165	141	**270**	100	78	70	60	48	46	45	42	**11.21**	5	–	–	–	–	–	6	–	–	–	–	–	–	**1,3-Cyclopentadiene, 1,2,3,4,5,5-hexachloro-**	77-47-4	**IC03503**
36	237	239	235	95	38	130	35	**270**	100	87	57	54	33	30	27	27	**9.90**	5	–	–	–	–	–	6	–	–	–	–	–	–	**1,3-Cyclopentadiene, 1,2,3,4,5,5-hexachloro-**	77-47-4	**IC03504**
237	239	235	95	130	66	60	117	**270**	100	64	63	41	28	28	28	27	**6.00**	5	–	–	–	–	–	6	–	–	–	–	–	–	**1,3-Cyclopentadiene, 1,2,3,4,5,5-hexachloro-**	77-47-4	**NI27344**
270	194	114	76	70	108	271	176	**270**	100	73	33	28	25	20	14	11		5	–	–	–	3	6	–	–	–	–	–	–	–	**1,2-Bis(trifluoromethyl)vinylene trithiocarbonate**		**LI05687**
191	193	59	272	83	122	120	153	**270**	100	99	35	23	15	13	13	12	**12.00**	5	4	3	–	–	–	–	2	–	–	–	–	–	**3,5-Dibromo-4-methoxy-2,5-dihydrofuran-2-one**		**NI27345**
205	203	270	39	268	65	271	37	**270**	100	42	32	15	14	9	2	2		5	5	–	–	–	–	–	–	–	–	–	–	1	**Thallium, (η^5-2,4-cyclopentadien-1-yl)-**	34822-90-7	**NI27346**
55	27	39	85	121	119	191	193	**270**	100	20	11	11	8	8	7	7	**0.00**	6	8	2	–	–	–	–	2	–	–	–	–	–	**2-Propenoic acid, 2,3-dibromopropyl ester**	19660-16-3	**NI27347**
143	270	30	74	272	178	159	69	**270**	100	88	87	40	30	30	29	24		7	2	4	2	–	3	1	–	–	–	–	–	–	**Benzene, 2-chloro-1,3-dinitro-5-(trifluoromethyl)-**	393-75-9	**NI27348**
71	43	243	245	42	236	41	223	**270**	100	50	36	25	17	15	15	14	**1.00**	7	8	2	–	–	4	2	–	–	–	–	–	–	**Butanoic acid, 1,3-dichloro-1,1,3,3-tetrafluoropropyl ester**		**LI05688**
69	76	52	270	75	272	18	112	**270**	100	69	52	51	40	38	37	35		8	3	1	2	–	3	2	–	–	–	–	–	–	**1H-Benzimidazol-4-ol, 5,6-dichloro-2-(trifluoromethyl)-**	19690-31-4	**NI27349**
172	199	237	235	174	201	272	137	**270**	100	73	68	68	68	49	40	35	**31.00**	8	6	–	–	–	–	4	–	–	1	–	–	–	**1,1,3,3-Tetrachloro-1-silaindan**		**NI27350**
77	129	131	108	51	64	116	69	**270**	100	75	27	27	22	17	9	8	**0.03**	8	6	–	–	–	7	1	–	–	–	–	–	–	**Cyclobutane, 2-chloro-1,1,2-trifluoro-3-(1,1,2,2-tetrafluoro-3-butenyl)-, cis-**	35207-96-6	**MS06036**
77	129	108	131	51	64	116	69	**270**	100	81	30	28	22	17	8	7	**0.03**	8	6	–	–	–	7	1	–	–	–	–	–	–	**Cyclobutane, 2-chloro-1,1,2-trifluoro-3-(1,1,2,2-tetrafluoro-3-butenyl)-, trans-**	35207-95-5	**MS06037**
102	47	48	55	45	30	242	158	**270**	100	94	79	53	41	32	31	19	**0.30**	8	7	5	–	1	–	–	–	–	–	–	–	1	**Manganese, tetracarbonyl-π-[(3-methylthio)propyl]-**		**MS06038**
59	151	146	118	60	150	58	178	**270**	100	44	35	18	8	7	7	6	**3.62**	8	14	–	–	5	–	–	–	–	–	–	–	–	**2,4,6,8,9-Pentathiabicyclo[3.3.1]nonane, 1,3,5,7-tetramethyl-**	57274-33-6	**NI27351**
28	36	235	41	27	57	43	26	**270**	100	80	75	70	70	60	60	60	**0.00**	8	14	1	2	–	–	3	–	–	–	–	1	–	**2-Chloro-N-(2-chloro-1-iminoethyl)acetic acid imide, butylchloroboryl ester**		**NI27352**
43	41	69	55	57	39	111	70	**270**	100	97	91	79	53	45	29	24	**0.17**	8	16	–	–	–	–	–	2	–	–	–	–	–	**Octane, 1,2-dibromo-**		**IC03505**
69	41	55	57	43	29	111	39	**270**	100	90	70	62	62	53	52	40	**0.00**	8	16	–	–	–	–	–	2	–	–	–	–	–	**Octane, 1,2-dibromo-**		**MS06039**
69	55	41	111	57	43	135	137	**270**	100	82	68	55	47	35	27	25	**0.23**	8	16	–	–	–	–	–	2	–	–	–	–	–	**Octane, 1,8-dibromo-**	4549-32-0	**NI27353**
55	69	41	111	39	135	137	57	**270**	100	99	99	37	36	34	33	33	**0.00**	8	16	–	–	–	–	–	2	–	–	–	–	–	**Octane, 1,8-dibromo-**	4549-32-0	**NI27354**
73	255	147	45	66	75	74	59	**270**	100	37	18	13	12	11	10	8	**0.00**	8	22	4	–	1	–	–	–	–	2	–	–	–	**2-[(Trimethylsilyl)oxy]ethyl trimethylsilyl sulphonate**		**NI27355**
168	170	89	133	69	63	62	49	**270**	100	40	32	30	18	12	12	12	**0.00**	9	8	–	–	1	4	1	–	–	–	–	1	–	**5-Chloro-1-methylbenzo[b]thiophenium tetrafluoroborate**		**LI05689**
270	92	156	65	108	206	109	107	**270**	100	89	60	53	44	23	22	21		9	10	2	4	2	–	–	–	–	–	–	–	–	**Sulphamethazole**	144-82-1	**LI05690**
74	93	115	59	66	28	60	64	**270**	100	58	52	34	27	24	20	18	**0.00**	9	10	2	4	2	–	–	–	–	–	–	–	–	**Sulphamethazole**	144-82-1	**NI27356**
120	242	214	149	75	39			**270**	100	36	30	29	25	21			**17.86**	10	5	2	–	–	3	–	–	–	–	–	–	1	**Iron, dicarbonyl(η^5-2,4-cyclopentadien-1-yl)(3,3,3-trifluoro-1-propynyl)-**	12307-00-5	**NI27357**
144	115	116	43	58	89	63	45	**270**	100	82	41	15	15	13	13	12	**1.01**	10	7	1	–	–	–	–	–	1	–	–	–	–	**2-Iodo-1-naphthol**	113855-57-5	**NI27358**
77	141	206	51	78	142	270	143	**270**	100	81	33	28	19	15	13	13		10	7	4	2	1	1	–	–	–	–	–	–	–	**1-Benzenesulphonyl-5-fluorouracil**		**NI27359**
96	42	242	270	41	54	69	43	**270**	100	63	54	36	24	16	13	13		10	9	–	6	–	3	–	–	–	–	–	–	–	**Pyrimido[5,4-e]-1,2,4-triazine, 7-(1-pyrrolidinyl)-5-(trifluoromethyl)-**	32709-25-4	**NI27360**
96	42	242	270	41	54	69	215	**270**	100	64	55	36	24	17	14	10		10	9	–	6	–	3	–	–	–	–	–	–	–	**Pyrimido[5,4-e]-1,2,4-triazine, 7-(1-pyrrolidinyl)-5-(trifluoromethyl)-**	32709-25-4	**LI05691**
42	122	50	186	76	75	41	43	**270**	100	13	12	10	9	8	7	7	**0.39**	10	10	5	2	1	–	–	–	–	–	–	–	–	**3-Pyrrolidinone, 1-(4-nitrobenzenesulphonyl)-**	83133-04-4	**NI27361**
270	272	184	186	199	197	185	183	**270**	100	99	85	60	38	36	35	31		10	11	2	2	–	–	–	1	–	–	–	–	–	**1-(4-Bromophenyl)-3-methoxyimidazolidinone**		**NI27362**
205	234	191	69	95	206	94	219	**270**	100	58	58	50	30	28	25	18	**11.99**	10	15	1	6	–	–	1	–	–	–	–	–	–	**1,3,5-Triazine-2,4-diamine, 6-(2-chloroethoxy)-N-cyano-N′,N′-diethyl-**	79788-25-3	**NI27363**
69	89	61	147	129	43	270	45	**270**	100	99	67	44	21	21	20	17		10	22	4	–	2	–	–	–	–	–	–	–	–	**D-Arabino-hexose, 2-deoxy-, diethyl mercaptal**	3650-68-8	**NI27364**
57	123	122	41	149	105	55	197	**270**	100	45	37	25	22	9	9	7	**0.10**	10	22	4	–	2	–	–	–	–	–	–	–	–	**1,2-Bis(butylsulphonyl)ethane**	16823-95-3	**LI05692**
57	123	122	41	241	149	213	55	**270**	100	45	36	25	24	22	11	10	**1.50**	10	22	4	–	2	–	–	–	–	–	–	–	–	**1,2-Bis(butylsulphonyl)ethane**	16823-95-3	**NI27365**
135	105	45	75	136	107	104	43	**270**	100	58	28	26	25	24	16	16	**15.22**	10	22	4	–	2	–	–	–	–	–	–	–	–	**L-Mannose, 6-deoxy-, diethyl mercaptal**	6748-70-5	**NI27366**
57	213	270	157	135	215	214	79	**270**	100	92	29	29	19	8	8	7		10	24	–	–	2	–	–	–	–	–	2	–	–	**1,2-Dimethyl-1,2-dibutyldiphosphane disulphide**		**MS06040**
251	271	253	171	189	207	252	139	**270**	100	67	24	20	15	14	13	10	**6.74**	11	7	5	–	1	1	–	–	–	–	–	–	–	**2-Naphthoic acid, 4-(fluorosulphonyl)-1-hydroxy-**	839-78-1	**NI27367**
224	222	270	168	226	220	221	225	**270**	100	51	30	29	20	20	19	16		11	10	3	–	–	–	–	–	–	–	–	–	1	**Benzo[b]selenophene-2-carboxylic acid, 3-hydroxy-, ethyl ester**	39857-43-7	**NI27368**
105	80	82	77	81	79	51	45	**270**	100	89	86	45	41	40	26	22	**0.00**	11	11	3	–	–	–	–	1	–	–	–	–	–	**Benzenepentanoic acid, γ-bromo-δ-oxo-**	39826-14-7	**NI27369**
57	185	183	75	76	50	157	155	**270**	100	89	84	29	25	24	16	16	**1.70**	11	11	3	–	–	–	–	1	–	–	–	–	–	**Propanoic acid, 4-bromophenacyl ester**		**NI27370**
240	184	239	42	270	56	77	90	**270**	100	96	70	65	52	44	30	27		11	14	4	2	1	–	–	–	–	–	–	–	–	**1-(4-Mesylphenyl)-3-methoxyimidazolidinone**		**NI27371**
105	160	161	104	93	225	77	89	**270**	100	90	79	55	41	40	33	27	**1.00**	11	15	4	2	–	–	–	–	–	–	1	–	–	**Phosphonic acid, [3-(aminocarbonyl)-2-phenyl-2-aziridinyl]-, dimethyl ester**	33866-51-2	**NI27372**
255	253	105	251	225	254			**270**	100	76	53	46	39	34			**2.40**	11	18	–	–	–	–	–	–	–	–	–	–	1	**2,6-Dimethylphenyltrimethyltin**		**MS06041**
165	105	163	90	161	164	162	120	**270**	100	90	76	75	46	29	26	26	**11.80**	11	18	–	–	–	–	–	–	–	–	–	–	1	**2-Methylbenzyltrimethyltin**		**MS06042**
165	163	91	119	161	120	164	162	**270**	100	72	51	49	45	41	29	25	**12.10**	11	18	–	–	–	–	–	–	–	–	–	–	1	**3-Methylbenzyltrimethyltin**		**MS06043**
165	163	270	161	119	164	162	91	**270**	100	77	72	46	32	28	25	23		11	18	–	–	–	–	–	–	–	–	–	–	1	**4-Methylbenzyltrimethyltin**		**MS06044**
111	45	110	225	125	99	57	69	**270**	100	48	25	24	24	18	16	15	**0.49**	11	20	6	–	–	–	–	–	–	–	–	2	–	**α-D-Glucofuranose, 6-O-methyl-, cyclic 1,2:3,5-bis(ethylboronate)**	74792-97-5	**NI27373**

MASS TO CHARGE RATIOS								M.W.	INTENSITIES								Parent	C	H	O	N	S	F	Cl	Br	I	Si	P	B	X	COMPOUND NAME	CAS Reg No	No
111	45	29	43	99	27	110	57	**270**	100	86	71	63	54	42	35	35	**0.00**	11	20	6	–	–	–	–	–	–	–	–	2	–	α-D-Mannofuranoside, methyl, cyclic 2,3:5,6-bis(ethylboronate)	74810-68-7	NI27374
112	111	98	61	84	143	28	70	**270**	100	57	35	30	21	20	15	12	**0.03**	11	20	6	–	–	–	–	–	–	–	–	2	–	α-D-Mannopyranoside, methyl, cyclic 2,3:4,6-bis(ethyl boronate)	61553-44-4	NI27375
73	226	270	255	211	153	147	227	**270**	100	42	24	22	18	15	13	12		11	22	2	2	–	–	–	–	–	2	–	–	–	1H-Imidazole-4-acetic acid, 1-(trimethylsilyl)-, trimethylsilyl ester	56114-70-6	NI27376
143	270	184	127	222	159	108	254	**270**	100	87	82	79	75	72	71	70		12	8	1	–	1	–	2	–	–	–	–	–	–	Benzene, 2-chloro-1-[(4-chlorophenyl)sulphinyl]-	74941-79-0	NI27377
143	222	75	270	224	145	127	108	**270**	100	55	45	40	35	35	35	30		12	8	1	–	1	–	2	–	–	–	–	–	–	Benzene, 1,1′-sulphinylbis[4-chloro-	3085-42-5	IC03506
143	75	222	224	145	108	270	127	**270**	100	70	65	42	41	38	37	37		12	8	1	–	1	–	2	–	–	–	–	–	–	Benzene, 1,1′-sulphinylbis[4-chloro-	3085-42-5	NI27378
143	222	75	145	127	270	224	50	**270**	100	60	56	41	40	39	38	31		12	8	1	–	1	–	2	–	–	–	–	–	–	Benzene, 1,1′-sulphinylbis[4-chloro-	3085-42-5	NI27379
127	109	97	69	81	110	98	111	**270**	100	84	45	39	35	34	26	25	**10.00**	12	14	7	–	–	–	–	–	–	–	–	–	–	1,6-Anhydro-3,4-O-[5-(hydroxymethyl)-2-furfurylidene]-β-D-galactopyranose	95881-40-6	NI27380
191	81	65	91	135	109	158	270	**270**	100	99	85	75	62	62	60	27		12	15	–	–	1	–	–	1	–	–	–	–	–	Cyclopropane, 1-(1-bromopropyl)-1-(phenylthio)-		DD00850
169	171	85	90	89	41	101	29	**270**	100	97	65	32	27	21	19	19	**1.40**	12	15	2	–	–	–	–	1	–	–	–	–	–	4-Bromobenzyl pyranyl ether	17100-68-4	NI27381
43	99	172	174	71	41	272	270	**270**	100	99	83	79	52	17	14	14		12	15	2	–	–	–	–	1	–	–	–	–	–	Hexanoic acid, 4-bromophenyl ester	56052-27-8	NI27382
270	144	83	69	43	85	116	58	**270**	100	43	43	39	38	35	21	14		12	18	3	2	1	–	–	–	–	–	–	–	–	2-Isopropylimino-3-isopropyl-6-methoxycarbonyl-2,3-dihydro-1,3-thiazin-4-one		MS06045
41	255	61	123	270	95	56	43	**270**	100	98	92	84	79	70	55	50		12	18	3	2	1	–	–	–	–	–	–	–	–	2,4,6(1H,3H,5H)-Pyrimidinetrione, 5-allyl-1-(methylthiomethyl)-5-isopropyl-		MS06046
91	108	30	155	206	65	107	99	**270**	100	91	83	47	42	30	24	20	**13.00**	12	18	3	2	1	–	–	–	–	–	–	–	–	Urea, N-butyl-N′-(4-tolylsulphonyl)-		LI05693
150	270	196	138	240	224	178	149	**270**	100	83	32	17	17	14	5	4		13	6	5	2	–	–	–	–	–	–	–	–	–	9H-Fluoren-9-one, 2,6-dinitro-		NI27383
270	150	196	224	271	30	178	138	**270**	100	52	25	22	16	13	12	11		13	6	5	2	–	–	–	–	–	–	–	–	–	9H-Fluoren-9-one, 2,7-dinitro-	31551-45-8	NI27384
238	270	164	239	161	166	220		**270**	100	78	62	44	22	12	10			13	18	4	–	1	–	–	–	–	–	–	–	–	Methyl 4-(benzylthio)-α-D-xylopyranoside		LI05694
150	149	182	178	160	210	192	270	**270**	100	65	45	35	35	33	20	10		13	18	6	–	–	–	–	–	–	–	–	–	–	Cyclopenta[c]pyran-4-carboxylic acid, 6-(acetyloxy)-1,4a,5,6,7,7a-hexahydro-1-hydroxy-7-methyl-, methyl ester, (1α,4aα,5α,7α,7aα)-(±)-	29748-09-2	MS06047
43	57	56	55	70	69	71	83	**270**	100	57	52	38	36	34	24	20	**0.00**	13	22	–	–	–	3	1	–	–	–	–	–	–	2-Tridecene, 1-chloro-1,1,2-trifluoro-		DD00851
43	56	57	41	70	69	55	71	**270**	100	97	77	66	64	56	55	32	**0.00**	13	22	–	–	–	3	1	–	–	–	–	–	–	2-Tridecene, 2-chloro-1,1,1-trifluoro-, (Z)-	108400-10-8	DD00852
169	184	69	41	45	185	224	55	**270**	100	89	85	48	46	44	30	29	**0.00**	13	22	4	2	–	–	–	–	–	–	–	–	–	N,N′-Dimethyl-3′-hydroxypentobarbitone		LI05695
70	85	128	112	57	99	139	84	**270**	100	33	29	22	21	19	18	15	**15.00**	13	26	2	4	–	–	–	–	–	–	–	–	–	N^2,N^6-Bis(dimethylaminomethylene)lysine methyl ester		MS06048
69	55	73	151	182	70	59	147	**270**	100	20	19	17	14	11	10	9	**0.00**	13	30	–	–	–	–	–	–	–	3	–	–	–	Silane, (3-methyl-1,2-propadien-1-yl-3-ylidene)tris[trimethyl-	24295-79-2	NI27385
148	73	64	149	56	111	150	83	**270**	100	73	18	15	14	11	9	8	**0.00**	13	30	–	–	–	–	–	–	–	3	–	–	–	Silane, (3-methyl-1-propyn-1-yl-3-ylidene)tris[trimethyl-	24295-78-1	NI27386
121	77	270	51	180	109	122	271	**270**	100	44	38	26	14	12	10	9		14	10	–	2	2	–	–	–	–	–	–	–	–	4,5-Diphenyl-2-thio-1,3,4-thiadiazole		LI05696
177	77	51	199	121	178	76	230	**270**	100	70	35	21	20	18	16	13	**0.00**	14	10	–	2	2	–	–	–	–	–	–	–	–	1,2,3-Thiadiazolium, 5-mercapto-3,4-diphenyl-, hydroxide, inner salt	56909-20-7	NI27387
121	119	117	270	77	122	82	47	**270**	100	33	31	26	18	10	9	9		14	10	–	2	2	–	–	–	–	–	–	–	–	2-(Thiobenzoylamino)benzothiazole		MS06049
212	109	270	77	51	213	110	65	**270**	100	59	48	36	19	16	16	12		14	10	2	2	1	–	–	–	–	–	–	–	–	Sydnone, 3-phenyl-4-(phenylthio)-	69978-09-2	NI27388
270	271	135	241	272	269	196	135.5	**270**	100	16	10	3	2	2	2	1		14	10	4	2	–	–	–	–	–	–	–	–	–	1,8-Diamino-4,5-dihydroxyanthraquinone		IC03507
227	228	135	105	91	134	183	151	**270**	100	20	16	15	12	11	9	9	**0.35**	14	10	4	2	–	–	–	–	–	–	–	–	–	Diazenedicarboxylic acid, diphenyl ester	2449-14-1	NI27389
158	270	186	131	77	103	104	242	**270**	100	61	27	27	26	24	17	16		14	10	4	2	–	–	–	–	–	–	–	–	–	7,8-Dimethyl-2,3-dihydroxyphenazine-1,4-dione		MS06050
44	18	158	270	156	184	141	131	**270**	100	82	27	15	12	9	9	8		14	10	4	2	–	–	–	–	–	–	–	–	–	7,8-Dimethylphenazine-1,2,3,4-tetrone		MS06051
105	150	77	106	270	164	51	88	**270**	100	55	38	30	28	21	17	11		14	10	4	2	–	–	–	–	–	–	–	–	–	1,4,2-Dioxazole, 3-(4-nitrophenyl)-5-phenyl-	2289-99-8	NI27390
119	91	77	270	64	51	105	120	**270**	100	25	23	21	16	16	15	10		14	10	4	2	–	–	–	–	–	–	–	–	–	1,4,2-Dioxazole, 5-(4-nitrophenyl)-3-phenyl-	55076-23-8	NI27391
91	90	105	104	77	63	165	164	**270**	100	68	65	60	52	46	45	45	**8.01**	14	10	4	2	–	–	–	–	–	–	–	–	–	Stilbene, 2,4-dinitro-	2486-13-7	NI27393
91	90	105	104	77	165	164	63	**270**	100	70	66	62	53	51	48	48	**14.11**	14	10	4	2	–	–	–	–	–	–	–	–	–	Stilbene, 2,4-dinitro-	2486-13-7	NI27392
270	178	165	176	177	166	271	76	**270**	100	31	29	26	20	19	17	15		14	10	4	2	–	–	–	–	–	–	–	–	–	Stilbene, 4,4′-dinitro-	2501-02-2	NI27394
270	241	207	131	111	206	214	272	**270**	100	67	61	56	56	42	36	33		14	11	–	4	–	–	1	–	–	–	–	–	–	3-Methyl-4-(2-chlorophenylimino)-3,4-dihydro-1,2,3-benzotriazine		LI05697
270	56	199	71	163	227	121	120	**270**	100	22	20	18	14	10	10	9		14	14	2	–	–	–	–	–	–	–	–	–	1	Ferrocene, 1,1′-diacetyl-	1273-94-5	NI27395
270	205	187	271	252	206	103	138	**270**	100	88	26	17	15	11	11	10		14	14	2	–	–	–	–	–	–	–	–	–	1	Iron, π-cyclopentadienyl(4,5,6,7-tetrahydro-4-hydroxy-7-oxo-π-indenyl)-, endo-	12126-89-5	NI27396
270	205	175	121	56	122	272	207	**270**	100	73	58	50	42	33	25	25		14	14	2	–	–	–	–	–	–	–	–	–	1	Methyl cis-3-ferrocenylpropenoate		MS06052
270	205	175	121	56	122	207	89	**270**	100	72	58	50	42	33	25	17		14	14	2	–	–	–	–	–	–	–	–	–	1	Methyl cis-3-ferrocenylpropenoate		LI05698
270	205	175	121	56	122	147	90	**270**	100	75	38	29	21	18	7	7		14	14	2	–	–	–	–	–	–	–	–	–	1	Methyl trans-3-ferrocenylpropenoate		LI05699
270	205	175	121	56	271	122	206	**270**	100	75	38	29	21	18	18	11		14	14	2	–	–	–	–	–	–	–	–	–	1	Methyl trans-3-ferrocenylpropenoate		MS06053
270	158	271	185	157	241	159	43	**270**	100	50	48	47	36	22	18	18		14	14	2	4	–	–	–	–	–	–	–	–	–	Benzo[g]pteridine-2,4(1H,3H)-dione, 1,3,7,8-tetramethyl-	14684-48-1	NI27397
213	270	214	185	184	170	171	158	**270**	100	90	55	44	38	36	34	34		14	14	2	4	–	–	–	–	–	–	–	–	–	Benzo[g]pteridine-2,4(3H,10H)-dione, 3,7,8,10-tetramethyl-	18636-32-3	NI27398
214	156	216	41	28	155	78	56	**270**	100	65	38	32	28	25	18	18	**2.15**	14	16	–	2	–	–	–	–	–	–	–	–	1	Nickel, (2,2′-bipyridine-N,N′)-1,4-butanediyl-	62320-23-4	NI27399
55	165	164	178	207	182	210	152	**270**	100	60	58	53	44	41	39	39	**15.00**	14	22	5	–	–	–	–	–	–	–	–	–	–	1,3-Cyclohexanedipropanoic acid, 2-oxo-, dimethyl ester		MS06054
45	77	43	120	31	89	65	39	**270**	100	20	12	11	10	9	9	8	**0.50**	14	22	5	–	–	–	–	–	–	–	–	–	–	Tetraethyleneglycol monophenyl ether		IC03508
115	142	76	128	51	78	77	39	**270**	100	78	33	31	22	19	19	19	**0.09**	15	6	–	6	–	–	–	–	–	–	–	–	–	Tricyclo[3.2.2.0^{2,4}]non-8-ene-3,3,6,6,7,7-hexacarbonitrile	62199-55-7	NI27400

MASS TO CHARGE RATIOS								M.W.	INTENSITIES								Parent	C	H	O	N	S	F	Cl	Br	I	Si	P	B	X	COMPOUND NAME	CAS Reg No	No
270	165	105	149	57	77	43	121	270	100	78	60	33	29	28	26	23		15	10	1	–	2	–	–	–	–	–	–	–	–	1,3-Oxathiole-2-thione, 4,5-diphenyl-	54199-62-1	NI27401
270	252	224	196	28	241	139	271	270	100	71	35	32	23	21	19	18		15	10	5	–	–	–	–	–	–	–	–	–	–	9,10-Anthracenedione, 1,4-dihydroxy-2-(hydroxymethyl)-	22296-59-9	NI27402
224	270	252	115	168	171	199	225	270	100	65	33	25	22	21	20	19		15	10	5	–	–	–	–	–	–	–	–	–	–	9,10-Anthracenedione, 2,5-dihydroxy-1-methoxy-		LI05700
270	271	139	69	213	242	51	77	270	100	15	11	10	9	8	7	7		15	10	5	–	–	–	–	–	–	–	–	–	–	9,10-Anthracenedione, 1,3,8-trihydroxy-6-methyl-	518-82-1	NI27403
241	270	139	213	168	224	196	155	270	100	70	22	13	10	9	8	7		15	10	5	–	–	–	–	–	–	–	–	–	–	9,10-Anthracenedione, 1,5,8-trihydroxy-3-methyl-		NI27404
242	213	270	243	197	241	214	139	270	100	28	21	16	16	15	15	13		15	10	5	–	–	–	–	–	–	–	–	–	–	2,3-Benzofurandione, 6-hydroxy-4-(4-hydroxybenzyl)-	14309-90-1	NI27405
149	119	94	65	240	121	270	76	270	100	60	32	26	25	24	22	20		15	10	5	–	–	–	–	–	–	–	–	–	–	Benzophenone-4,4′-dicarboxylic acid	964-68-1	NI27406
270	153	121	152	118	124	269	242	270	100	23	21	18	15	12	11	10		15	10	5	–	–	–	–	–	–	–	–	–	–	4H-1-Benzopyran-4-one, 5,7-dihydroxy-2-(4-hydroxyphenyl)-	520-36-5	MS06055
270	153	242	124	121	152	118	269	270	100	22	19	18	16	16	14	13		15	10	5	–	–	–	–	–	–	–	–	–	–	4H-1-Benzopyran-4-one, 5,7-dihydroxy-2-(4-hydroxyphenyl)-	520-36-5	NI27407
118	270	153	152	220	165	105	124	270	100	65	45	40	16	15	13	11		15	10	5	–	–	–	–	–	–	–	–	–	–	2H-1-Benzopyran-2-one, 4,5,7-trihydroxy-3-phenyl-	4222-02-0	NI27408
270	77	105	269	69	242	213	241	270	100	80	50	42	40	34	30	20		15	10	5	–	–	–	–	–	–	–	–	–	–	4H-1-Benzopyran-4-one, 3,5,7-trihydroxy-2-phenyl-		MS06056
242	227	270	243	199	171			270	100	28	21	16	16	11				15	10	5	–	–	–	–	–	–	–	–	–	–	2-Hydroxy-3-methoxy-1-formylxanthone		LI05701
270	227	225	224	269	228	239	199	270	100	54	11	10	9	8	6	6		15	10	5	–	–	–	–	–	–	–	–	–	–	3-Hydroxy-2-methoxy-4-formylxanthone		LI05702
270	196	77	241	252	126	269	253	270	100	22	22	18	14	14	12	12		15	10	5	–	–	–	–	–	–	–	–	–	–	1-Methoxy-2,3-(methylenedioxy)xanthone		NI27409
270	227	210	241	224	271	199	115	270	100	40	32	20	18	17	14	13		15	10	5	–	–	–	–	–	–	–	–	–	–	Purpurin 1-methyl ether		LI05703
235	269	241	270	243	242	207	76	270	100	83	71	69	53	53	45	45		15	11	1	2	–	–	1	–	–	–	–	–	–	2H-1,4-Benzodiazepin-2-one, 6-chloro-1,3-dihydro-5-phenyl-		MS06057
242	270	269	241	243	271	244	77	270	100	89	80	77	48	40	34	32		15	11	1	2	–	–	1	–	–	–	–	–	–	2H-1,4-Benzodiazepin-2-one, 7-chloro-1,3-dihydro-5-phenyl-	1088-11-5	MS06058
270	242	269	241	271	243	272	244	270	100	97	80	78	49	46	36	34		15	11	1	2	–	–	1	–	–	–	–	–	–	2H-1,4-Benzodiazepin-2-one, 7-chloro-1,3-dihydro-5-phenyl-	1088-11-5	NI27410
242	270	269	241	243	271	244	103	270	100	88	78	67	48	39	34	30		15	11	1	2	–	–	1	–	–	–	–	–	–	2H-1,4-Benzodiazepin-2-one, 7-chloro-1,3-dihydro-5-phenyl-	1088-11-5	NI27411
242	241	269	243	270	244	77	235	270	100	77	64	50	43	34	32	29		15	11	1	2	–	–	1	–	–	–	–	–	–	2H-1,4-Benzodiazepin-2-one, 8-chloro-1,3-dihydro-5-phenyl-		MS06059
241	269	242	243	270	235	207	103	270	100	99	83	71	68	49	47	43		15	11	1	2	–	–	1	–	–	–	–	–	–	2H-1,4-Benzodiazepin-2-one, 9-chloro-1,3-dihydro-5-phenyl-		MS06060
172	270	235	241	269	242	243	271	270	100	99	99	84	63	50	42	37		15	11	1	2	–	–	1	–	–	–	–	–	–	2H-1,4-Benzodiazepin-2-one, 5-(2-chlorophenyl)-1,3-dihydro-		MS06061
144	143	89	270	127	116	115	129	270	100	48	36	29	17	12	12	5		15	11	1	2	–	–	1	–	–	–	–	–	–	3′-Chloroindole-2-carboxanilide		LI05704
144	127	89	116	270	143	129	115	270	100	15	15	14	11	5	5	5		15	11	1	2	–	–	1	–	–	–	–	–	–	3′-Chloroindole-3-carboxanilide		LI05705
144	89	270	143	127	116	115	272	270	100	38	32	15	14	14	13	9		15	11	1	2	–	–	1	–	–	–	–	–	–	4′-Chloroindole-2-carboxanilide		LI05706
144	127	116	89	270	129	115	272	270	100	29	17	17	14	9	5	4		15	11	1	2	–	–	1	–	–	–	–	–	–	4′-Chloroindole-3-carboxanilide		LI05707
235	75	76	111	50	236	90	77	270	100	33	32	31	27	25	17	14	6.71	15	11	1	2	–	–	1	–	–	–	–	–	–	4(3H)-Quinazolinone, 3-(2-chlorophenyl)-2-methyl-	340-57-8	NI27412
119	91	135	65	93	270	77	51	270	100	62	36	29	24	19	15	11		15	14	1	2	1	–	–	–	–	–	–	–	–	Benzamide, 3-methyl-N-[(phenylamino)thioxomethyl]-	56437-99-1	NI27413
160	115	270	146	161	102	147	145	270	100	20	16	16	15	13	11	11		15	14	1	2	1	–	–	–	–	–	–	–	–	1H-[1]Benzothieno[3,2-d]azonine-3-carbonitrile, 2,3,4,5,6,7-hexahydro-7-oxo-	99659-22-0	NI27414
270	269	127	160	69	115	271	41	270	100	82	49	36	34	32	23	19		15	14	1	2	1	–	–	–	–	–	–	–	–	4(1H)-Pyrimidinone, tetrahydro-1-(1-naphthyl)-6-methyl-6-thio-	87973-73-7	NI27415
123	78	106	149	147	270	121	92	270	100	37	31	23	21	17	8	8		15	14	3	2	–	–	–	–	–	–	–	–	–	N-(3-Anisolyl)-2-isonicotinylacetamide		LI05708
123	149	78	106	270	147	148	92	270	100	37	35	29	24	19	7	6		15	14	3	2	–	–	–	–	–	–	–	–	–	N-(4-Anisolyl)-2-isonicotinylacetamide		LI05709
123	106	149	92	147	270	148	121	270	100	23	9	9	7	4	4	3		15	14	3	2	–	–	–	–	–	–	–	–	–	N-(3-Anisolyl)-2-nicotinylacetamide		LI05710
77	65	270	149	91	105	51	29	270	100	52	40	35	29	27	25	24		15	14	3	2	–	–	–	–	–	–	–	–	–	Benzoic acid, 4-(phenylazoxy)-, ethyl ester	31389-05-6	NI27416
270	184	156	128	271	213	115	219	270	100	24	24	20	18	17	11	8		15	14	3	2	–	–	–	–	–	–	–	–	–	2,4,6(1H,3H,5H)-Pyrimidinetrione, 5-[(E)-cinnamylidene]-1,3-dimethyl-		MS06062
198	130	270	225	197	199	77	103	270	100	52	51	46	17	14	14	13		15	14	3	2	–	–	–	–	–	–	–	–	–	1H-Pyrimido[1,2-a]quinoline-2-carboxylic acid, 5,6-dihydro-1-oxo-, ethyl ester	63455-50-5	NI27417
43	41	57	144	71	55	129	85	270	100	67	60	41	33	30	21	15	10.42	15	26	–	–	2	–	–	–	–	–	–	–	–	2-Ethyl-5-(nonylthio)thiophene		AI01915
41	55	67	43	81	45	69	95	270	100	92	51	40	37	32	31	29	0.00	15	26	4	–	–	–	–	–	–	–	–	–	–	Carbonic acid, ethyl [1-methyl-3-(2-oxocyclooct-1-yl)propyl] ester		NI27418
155	41	55	127	109	81	43	67	270	100	61	58	56	53	48	45	40	2.00	15	26	4	–	–	–	–	–	–	–	–	–	–	Cyclooctane-1-carboxylic acid, 1-(3-hydroxybutyl)-2-oxo-, ethyl ester		NI27419
41	55	43	155	81	142	67	127	270	100	87	60	55	50	48	43	42	3.00	15	26	4	–	–	–	–	–	–	–	–	–	–	11-Dodecanolide, 8-(ethoxycarbonyl)-		NI27420
97	143	69	41	115	55	159	113	270	100	93	62	46	39	29	18	12	3.00	15	26	4	–	–	–	–	–	–	–	–	–	–	Hexanoic acid, 6-[(2,2-dimethyl-5-oxocyclopentyl)oxy]-, ethyl ester		NI27421
160	68	115	55	41	161	133	99	270	100	70	63	45	45	39	35	27	0.00	15	26	4	–	–	–	–	–	–	–	–	–	–	Propanedioic acid, (1,1-dimethyl-5-hexenyl)-, diethyl ester	54752-00-0	NI27422
83	81	95	138	57	69	55	139	270	100	50	48	46	36	30	30	22	0.10	15	26	4	–	–	–	–	–	–	–	–	–	–	Propanoic acid, 2-(acetyloxy)-, 5-methyl-2-isopropylcyclohexyl ester	64870-66-2	NI27423
206	85	91	205	101	128	191	151	270	100	64	62	54	44	36	33	28	0.00	16	14	2	–	1	–	–	–	–	–	–	–	–	5,5-Diphenyl-2-thiabicyclo[2.1.0]pentane-2,2-dioxide		LI05711
165	226	225	270	121	211	166	198	270	100	73	70	62	55	36	34	18		16	14	2	–	1	–	–	–	–	–	–	–	–	1,3-Oxathiolan-5-one, 2-methyl-4,4-diphenyl-	18648-76-5	NI27424
170	198	114	91	77	270	126	63	270	100	81	42	38	34	28	28	24		16	14	4	–	–	–	–	–	–	–	–	–	–	1,2-Acenaphthylenedione, cyclic 1,1:2,2-bis(1,2-ethanediyl acetal)		LI05712
154	270	126	182	242	170	91	198	270	100	67	51	37	28	24	20	19		16	14	4	–	–	–	–	–	–	–	–	–	–	1,2-Acenaphthylenedione, cyclic 1,2:1,2-bis(1,2-ethanediyl acetal)		LI05713
270	239	77	269	163	51	91	240	270	100	86	78	76	59	37	35	19		16	14	4	–	–	–	–	–	–	–	–	–	–	1,3-Benzodioxole, 6-methyl-5-(3-methoxybenzoyl)-	52806-40-3	NI27425
270	239	269	143	135	240	271	255	270	100	86	76	60	35	19	17	15		16	14	4	–	–	–	–	–	–	–	–	–	–	1,3-Benzodioxole, 6-methyl-5-(3-methoxybenzoyl)-	52806-40-3	NI27426
269	270	239	135	77	163	253	240	270	100	93	87	40	38	21	20	19		16	14	4	–	–	–	–	–	–	–	–	–	–	1,3-Benzodioxole, 6-methyl-5-(4-methoxybenzoyl)-	52806-35-6	NI27427
270	269	255	271	148	105	147	161	270	100	42	20	17	12	12	9	8		16	14	4	–	–	–	–	–	–	–	–	–	–	6H-Benzofuro[3,2-c][1]benzopyran-3-ol, 6a,11a-dihydro-9-methoxy-, (6aR-cis)-	32383-76-9	MS06063
270	269	255	148	147	149	161	134	270	100	40	33	31	23	19	16	12		16	14	4	–	–	–	–	–	–	–	–	–	–	6H-Benzofuro[3,2-c][1]benzopyran-3-ol, 6a,11a-dihydro-9-methoxy-, (6aR-cis)-	32383-76-9	NI27428

MASS TO CHARGE RATIOS								M.W.	INTENSITIES								Parent	C	H	O	N	S	F	Cl	Br	I	Si	P	B	X	COMPOUND NAME	CAS Reg No	No
105	270	77	149	106	239	89	90	270	100	15	15	13	8	5	5	4		16	14	4	–	–	–	–	–	–	–	–	–	–	Benzoic acid, 4-(methoxycarbonyl)benzyl ester		IC03509
105	77	149	270	121	239	165	225	270	100	41	14	12	12	5	3	1		16	14	4	–	–	–	–	–	–	–	–	–	–	Benzoic acid, 4-(methoxycarbonyl)benzyl ester		NI27429
135	105	77	270	136	65	92	51	270	100	55	16	9	9	9	8	7		16	14	4	–	–	–	–	–	–	–	–	–	–	Benzoic acid, 2-methoxy-, 2-oxo-2-phenylethyl ester	55153-19-0	NI27430
105	135	77	270	107	106	92	29	270	100	40	30	24	12	10	9	7		16	14	4	–	–	–	–	–	–	–	–	–	–	Benzoic acid, 3-methoxy-, 2-oxo-2-phenylethyl ester	55153-18-9	NI27431
135	105	77	270	136	91	51	64	270	100	78	32	22	11	11	7	6		16	14	4	–	–	–	–	–	–	–	–	–	–	Benzoic acid, 4-methoxy-, 2-oxo-2-phenylethyl ester	55153-14-5	NI27432
239	152	270	151	76	59	165	153	270	100	93	52	35	26	18	16	15		16	14	4	–	–	–	–	–	–	–	–	–	–	4,4′-Bis(carboxymethyl)biphenyl		IC03510
239	270	152	240	104	76	151	271	270	100	71	40	17	17	15	14	13		16	14	4	–	–	–	–	–	–	–	–	–	–	4,4′-Bis(carboxymethyl)biphenyl		IC03511
255	183	77	43	270	106	105	256	270	100	37	34	31	28	28	28	18		16	14	4	–	–	–	–	–	–	–	–	–	–	1-Dibenzofurancarboxylic acid, 3,9b-dihydro-4,9b-dimethyl-3-oxo-, methyl ester	40801-45-4	NI27433
270	241	44	43	28	128	197	69	270	100	59	35	27	26	21	20	18		16	14	4	–	–	–	–	–	–	–	–	–	–	6H-Dibenzo[b,d]pyran-6-one, 3,7-dimethoxy-1-methyl-		LI05714
170	198	114	91	77	270	126	63	270	100	80	42	39	34	28	28	24		16	14	4	–	–	–	–	–	–	–	–	–	–	Dispiro[1,3-dioxolane-2,1′(2′H)-acenaphthylene-2′,2″-[1,3]dioxolane]	30339-98-1	NI27434
154	270	126	182	243	170	91	155	270	100	67	50	37	27	24	20	18		16	14	4	–	–	–	–	–	–	–	–	–	–	Dispiro[1,3-dioxolane-2,1′(2′H)-acenaphthylene-2′,2″-[1,3]dioxolane]	30339-98-1	NI27435
163	270	152	76	211	50	151	59	270	100	97	77	53	35	34	33	30		16	14	4	–	–	–	–	–	–	–	–	–	–	2,5-Etheno[4.2.2]propella-3,7,9-triene-3,4-dicarboxylic acid, dimethyl ester		NI27436
105	77	106	51	227	78	148	50	270	100	21	7	6	2	2	1	1	0.01	16	14	4	–	–	–	–	–	–	–	–	–	–	Ethylene glycol dibenzoate	94-49-5	IC03512
105	122	77	51	28	106	50	129	270	100	35	34	11	10	8	5	4	0.00	16	14	4	–	–	–	–	–	–	–	–	–	–	Ethylene glycol dibenzoate	94-49-5	DO01062
105	77	51	106	226	50	78	148	270	100	25	8	8	5	2	2	2	0.20	16	14	4	–	–	–	–	–	–	–	–	–	–	Ethylene glycol dibenzoate	94-49-5	NI27437
270	255	202	271	227	215	69	78	270	100	26	25	17	11	11	10	8		16	14	4	–	–	–	–	–	–	–	–	–	–	7H-Furo[3,2-g][1]benzopyran-7-one, 9-hydroxy-4-(3-methyl-2-butenyl)-	642-05-7	NI27438
270	255	202	271	227	215	69	199	270	100	27	25	16	10	10	10	8		16	14	4	–	–	–	–	–	–	–	–	–	–	7H-Furo[3,2-g][1]benzopyran-7-one, 9-hydroxy-4-(3-methyl-2-butenyl)-	642-05-7	LI05715
69	41	202	270	255	174	67	53	270	100	79	77	25	24	23	20	19		16	14	4	–	–	–	–	–	–	–	–	–	–	7H-Furo[3,2-g][1]benzopyran-7-one, 4-[(3-methyl-2-butenyl)oxy]-	482-45-1	NI27439
239	270	211	196	152	271	240	76	270	100	80	51	19	14	13	13	10		16	14	4	–	–	–	–	–	–	–	–	–	–	Isophthalic acid, 5-phenyl-, dimethyl ester		IC03513
270	255	227	271	88	128	199	144	270	100	47	37	17	17	13	12	12		16	14	4	–	–	–	–	–	–	–	–	–	–	1H-Naphtho[2,1-b]pyran-1-one, 7,8-dimethoxy-2-methyl-	54889-69-9	NI27440
255	227	270	187	104	76	256	44	270	100	70	48	28	25	20	19	13		16	14	4	–	–	–	–	–	–	–	–	–	–	4H-Naphtho[1,2-b]pyran-4-one, 5,6-dimethoxy-2-methyl-	32454-45-8	NI27441
255	227	270	187	256	228	271	185	270	100	71	47	36	18	10	6	6		16	14	4	–	–	–	–	–	–	–	–	–	–	4H-Naphtho[1,2-b]pyran-4-one, 5,6-dimethoxy-2-methyl-	32454-45-8	NI27442
255	227	270	187	256	241	228		270	100	70	48	34	18	10	10			16	14	4	–	–	–	–	–	–	–	–	–	–	4H-Naphtho[1,4-b]pyran-4-one, 5,6-dimethoxy-2-methyl-		LI05716
180	91	179	181	107	92	270	108	270	100	79	34	14	13	11	5	1		16	14	4	–	–	–	–	–	–	–	–	–	–	Oxalic acid, benzyl ester		MS06064
180	179	178	252	165	181	226	270	270	100	64	34	25	24	19	14	10		16	14	4	–	–	–	–	–	–	–	–	–	–	1,2-Phenanthrenedicarboxylic acid, 1,2,3,4-tetrahydro-	74764-58-2	NI27443
104	76	77	50	91	79	107	122	270	100	40	24	22	21	21	18	17	0.00	16	14	4	–	–	–	–	–	–	–	–	–	–	Phthalic acid, 2-methylbenzyl hydrogen ester		DO01063
166	270	193	69	269	77	103	167	270	100	75	57	33	27	25	24	22		16	14	4	–	–	–	–	–	–	–	–	–	–	2-Propen-1-one, 1-(2,4-dihydroxy-5-methoxyphenyl)-3-phenyl-		MS06065
193	270	166	138	269	95	167	51	270	100	93	78	49	48	35	26	19		16	14	4	–	–	–	–	–	–	–	–	–	–	2-Propen-1-one, 1-(2,6-dihydroxy-4-methoxyphenyl)-3-phenyl-, (E)-	18956-15-5	MS06066
163	91	270	239	77	135	252	225	270	100	59	19	9	8	8	3	2		16	14	4	–	–	–	–	–	–	–	–	–	–	Terephthalic acid, benzyl methyl ester	24318-74-9	NI27444
242	207	244	208	243	165	270	269	270	100	91	25	19	17	17	15	13		16	15	–	2	–	–	1	–	–	–	–	–	–	1H-1,4-Benzodiazepine, 7-chloro-2,3-dihydro-1-methyl-5-phenyl-	2898-12-6	NI27445
270	272	271	105	91	269	65	104	270	100	33	30	28	27	25	11	10		16	15	–	2	–	–	1	–	–	–	–	–	–	1-(4-Tolyl)-3-(4-chlorophenyl)-2-pyrazoline		MS06067
270	93	237	91	271	44	118	45	270	100	51	47	43	36	36	32	27		16	18	–	2	1	–	–	–	–	–	–	–	–	Thiourea, N-phenyl-N′-(3-phenylpropyl)-	15093-43-3	NI27446
255	197	198	121	77	105	147	256	270	100	56	48	39	33	27	26	23	4.00	16	18	2	–	–	–	–	–	–	1	–	–	–	Benzophenone, 2-(trimethylsiloxy)-	33342-95-9	NI27447
191	270	105	77	255	73	147	271	270	100	93	88	51	45	45	30	23		16	18	2	–	–	–	–	–	–	1	–	–	–	Benzophenone, 4-(trimethylsiloxy)-	33663-72-8	NI27448
270	225	226	197	181	271	198	182	270	100	53	38	33	24	23	14	9		16	18	2	–	–	–	–	–	–	1	–	–	–	9,9-Diethoxy-9-silafluorene	52897-51-5	NI27449
270	271	144	154	155	167	197	127	270	100	20	20	19	11	10	8	8		16	18	2	2	–	–	–	–	–	–	–	–	–	Dihydroxylysergic acid		NI27450
270	154	144	167	155	197	127	168	270	100	27	22	15	14	11	10	9		16	18	2	2	–	–	–	–	–	–	–	–	–	Dihydroxylysergic acid		MS06068
154	44	183	184	42	127	170	270	270	100	64	50	48	43	36	30	25		16	18	2	2	–	–	–	–	–	–	–	–	–	6,8-Dimethyl-8,9-epoxy-10-hydroxyergoline		LI05717
154	184	42	170	270	130	115	28	270	100	98	71	64	42	42	35	28		16	18	2	2	–	–	–	–	–	–	–	–	–	6,8-Dimethyl-8,9-epoxy-10-hydroxyergoline		LI05718
76	104	148	95	162	201	105	80	270	100	90	70	65	56	33	30	28	5.00	16	18	2	2	–	–	–	–	–	–	–	–	–	1H-Isoindole-1,3(2H)-dione, 2-(9-azabicyclo[6.1.0]non-9-yl)-, cis-	66387-74-4	NI27451
131	70	130	102	172	270	132	76	270	100	65	52	33	19	18	13	13		16	18	2	2	–	–	–	–	–	–	–	–	–	4-Isopropylidene-1,1-dimethyl-4H-1,3,4-oxadiazino[4,3-a]cinnol-2(1H)-one		LI05719
118	91	106	136	119	148	135	65	270	100	58	35	34	33	28	28	20	5.00	16	22	–	4	–	–	–	–	–	–	–	–	–	N,N′-Bis[2-(aminomethyl)phenyl]ethylenediamine		LI05720
112	110	103	70	104	77	173	41	270	100	91	60	55	47	40	35	34	0.00	16	22	–	4	–	–	–	–	–	–	–	–	–	1H-1,2,3-Triazole, 4-ethyl-1-(1-phenylvinyl)-5-(1-pyrrolidinyl)-4,5-dihydro-	77726-83-1	NI27452
69	110	55	97	115	41	95	143	270	100	89	66	62	56	52	46	45	0.00	16	30	3	–	–	–	–	–	–	–	–	–	–	Ethyl 7-oxa-7-(2-ethyl-2-methylcyclopentyl)heptanoate		NI27453
127	57	55	43	60	29	73	98	270	100	66	40	40	24	24	16	13	0.00	16	30	3	–	–	–	–	–	–	–	–	–	–	Octanoic anhydride		LI05721
116	43	117	57	41	129	55	53	270	100	32	26	25	25	23	20	17	3.20	16	30	3	–	–	–	–	–	–	–	–	–	–	Pentadecanoic acid, 3-oxo-, methyl ester	54889-72-4	WI01329
126	158	111	71	155	143	101	98	270	100	87	53	41	34	30	30	27	7.99	16	30	3	–	–	–	–	–	–	–	–	–	–	Pentadecanoic acid, 6-oxo-, methyl ester	13328-28-4	NI27454
43	58	55	41	98	74	213	69	270	100	57	57	42	35	35	34	31	6.29	16	30	3	–	–	–	–	–	–	–	–	–	–	Pentadecanoic acid, 14-oxo-, methyl ester	54889-71-3	NI27455
55	115	57	71	41	56	101	70	270	100	27	17	16	14	13	12	12	0.10	16	30	3	–	–	–	–	–	–	–	–	–	–	2,5,5,11-Tetramethyl-8-isopropyl-4,7,9-trioxa-1,11-dodecadiene		MS06069
113	195	43	255	72	85	41	81	270	100	62	40	36	33	23	21	18	0.00	16	30	3	–	–	–	–	–	–	–	–	–	–	7-Tridecanone, 1,2-isopropylidenedioxy-, (S)-		DD00853
57	43	71	155	44	41	253	55	270	100	95	55	45	45	45	40	32	1.25	16	34	1	2	–	–	–	–	–	–	–	–	–	Dioctylnitrosamine		MS06070
157	128	114	142	270	170	214	87	270	100	85	85	50	40	40	30	25		16	34	1	2	–	–	–	–	–	–	–	–	–	Heptanamide, N-(3-aminopropyl)-N-hexyl-	67138-90-3	NI27456

MASS TO CHARGE RATIOS	M.W.	INTENSITIES	Parent	C	H	O	N	S	F	Cl	Br	I	Si	P	B	X	COMPOUND NAME	CAS Reg No	No
199 170 114 55 100 70 87 128	**270**	100 52 48 44 30 20 18 14	**4.00**	16	34	1	2	–	–	–	–	–	–	–	–	–	**Heptanamide, N-[3-(hexylamino)propyl]-**	67139-02-0	**NI27457**
270 255 227 242 152 171 165 271	**270**	100 80 80 60 22 21 21 18		17	18	3	–	–	–	–	–	–	–	–	–	–	**1,4-Anthracenedione, 5,6,7,8-tetrahydro-2-methoxy-5,5-dimethyl-**	54889-74-6	**NI27458**
91 270 92 65 32 271 39 179	**270**	100 14 8 6 4 3 3 1		17	18	3	–	–	–	–	–	–	–	–	–	–	**Benzaldehyde, 4-benzyloxy-2-methoxy-3,6-dimethyl-**		**MS06071**
93 92 120 65 148 51 77 63	**270**	100 15 14 13 11 11 10 8	**0.00**	17	18	3	–	–	–	–	–	–	–	–	–	–	**Benzenepropanoic acid, 2-(phenylmethoxy)-, methyl ester**	56052-50-7	**NI27459**
214 270 197 198 168 43 77 141	**270**	100 98 98 65 55 53 48 43		17	18	3	–	–	–	–	–	–	–	–	–	–	**Benzoic acid, 3-phenoxy-, butyl ester**		**IC03514**
148 120 270 121 135 252 149 251	**270**	100 64 23 16 15 13 13 7		17	18	3	–	–	–	–	–	–	–	–	–	–	**2H-1-Benzopyran-4-ol, 2-(4-ethoxyphenyl)-3,4-dihydro-**	72594-24-2	**NI27460**
197 198 211 154 152 153 77 59	**270**	100 31 30 26 22 19 18 17	**16.00**	17	18	3	–	–	–	–	–	–	–	–	–	–	**[1,1′-Biphenyl]-4-acetic acid, α-hydroxy-α-methyl-, ethyl ester**		**NI27461**
183 105 208 77 88 182 44 45	**270**	100 42 28 14 8 5 5 4	**3.00**	17	18	3	–	–	–	–	–	–	–	–	–	–	**1,4-Dioxane-2-methanol, α,α-diphenyl-**	24455-14-9	**NI27462**
134 270 135 149 121 148 119 91	**270**	100 30 16 11 9 6 5 4		17	18	3	–	–	–	–	–	–	–	–	–	–	**7-Methoxy-3-(4-methoxyphenyl)-3,4-dihydro-2H-1-benzopyran**		**LI05722**
270 166 117 138 91 239 179 43	**270**	100 88 30 29 29 23 22 22		17	18	3	–	–	–	–	–	–	–	–	–	–	**5-Methoxy-6-methylflavan-7-ol**		**MS06072**
155 237 252 127 156 195 210 270	**270**	100 67 49 46 29 21 15 9		17	18	3	–	–	–	–	–	–	–	–	–	–	**1-Naphthalenepentanoic acid, β,β-dimethyl-δ-oxo-**	64184-12-9	**NI27463**
155 170 127 270 237 252	**270**	100 56 46 23 19 10		17	18	3	–	–	–	–	–	–	–	–	–	–	**2-Naphthalenepentanoic acid, β,β-dimethyl-δ-oxo-**	64184-11-8	**NI27464**
270 255 227 242 152 171 165 128	**270**	100 80 80 61 22 21 21 20		17	18	3	–	–	–	–	–	–	–	–	–	–	**1,4-Phenanthrenedione, 5,6,7,8-tetrahydro-3-methoxy-8,8-dimethyl-**	30684-14-1	**NI27465**
270 45 197 225 165 164 121 238	**270**	100 87 71 68 36 31 28 18		17	18	3	–	–	–	–	–	–	–	–	–	–	**Stilbene, 2-(methoxymethoxy)-4′-methoxy-, (E)-**		**MS06073**
270 45 197 225 165 164 121 238	**270**	100 77 71 67 42 34 27 17		17	18	3	–	–	–	–	–	–	–	–	–	–	**Stilbene, 2-(methoxymethoxy)-4′-methoxy-, (Z)-**		**MS06074**
148 133 174 109 270 91 96 77	**270**	100 81 63 57 47 45 43 38		17	18	3	–	–	–	–	–	–	–	–	–	–	**Tetracyclo[7.6.0.0^{2,7}.0^{10,14}]pentadeca-6,12-diene-4,8,11-trione, 9,12-dimethyl-, (1R,2R,9S,10S,14R)-**		**DD00854**
134 270 108 105 119 91 162 79	**270**	100 68 50 44 40 40 30 22		17	18	3	–	–	–	–	–	–	–	–	–	–	**Tetracyclo[7.6.0.0^{2,7}.0^{10,14}]pentadeca-5,12-diene-4,8,11-trione, 7,10-trimethyl-, (1R,2R,7S,9S,10S,14R)-**		**DD00855**
97 98 144 117 130 270 143 115	**270**	100 95 20 15 13 10 10 10		17	22	1	2	–	–	–	–	–	–	–	–	–	**cis-N-Acetyl-indolin-2-one-3,7′-spiro[1′-azabicyclo[4.3.0]nonane]**		**LI05723**
97 98 144 270 130 117 143 269	**270**	100 95 23 18 15 15 10 4		17	22	1	2	–	–	–	–	–	–	–	–	–	**trans-N-Acetyl-indolin-2-one-3,7′-spiro[1′-azabicyclo[4.3.0]nonane]**		**LI05724**
270 167 144 170 122 166 107 184	**270**	100 64 55 40 28 25 20 15		17	22	1	2	–	–	–	–	–	–	–	–	–	**2-Acetyl-1,2-seco-1,2-butano-1,2,3,4-tetrahydro-β-carboline**		**LI05725**
253 270 148 254 122 42 237 269	**270**	100 40 39 34 27 16 15 11		17	22	1	2	–	–	–	–	–	–	–	–	–	**Benzenemethanol, α-phenyl-4,4′-bis(dimethylamino)-**		**IC03515**
253 270 254 122 126 269 271 237	**270**	100 69 44 30 16 14 13 12		17	22	1	2	–	–	–	–	–	–	–	–	–	**Benzenemethanol, α-phenyl-4,4′-bis(dimethylamino)-**		**IC03516**
84 157 186 158 85 131 91 185	**270**	100 63 48 34 31 28 28 27	**22.00**	17	22	1	2	–	–	–	–	–	–	–	–	–	**2-Cyclopentenone, 2-(4-methylanilino)-4-piperidino-**		**MS06075**
118 120 195 117 119 180 187 158	**270**	100 92 87 61 58 37 33 31	**19.00**	17	22	1	2	–	–	–	–	–	–	–	–	–	**Dasycarpidol**	2744-46-9	**NI27466**
58 180 71 182 44 183 167 181	**270**	100 80 71 50 42 40 24 22	**0.00**	17	22	1	2	–	–	–	–	–	–	–	–	–	**Ethanamine, N,N-dimethyl-2-[1-phenyl-1-(2-pyridinyl)ethoxy]-**	469-21-6	**NI27468**
58 71 180 167 73 182 72 181	**270**	100 50 13 10 10 6 6 5	**0.00**	17	22	1	2	–	–	–	–	–	–	–	–	–	**Ethanamine, N,N-dimethyl-2-[1-phenyl-1-(2-pyridinyl)ethoxy]-**	469-21-6	**NI27467**
58 71 42 44 43 51 30 45	**270**	100 60 51 26 26 22 21 16	**0.00**	17	22	1	2	–	–	–	–	–	–	–	–	–	**Ethanamine, N,N-dimethyl-2-[1-phenyl-1-(2-pyridinyl)ethoxy]-**	469-21-6	**NI27469**
150 270 120 271 151 121 272	**270**	100 62 54 11 10 6 1		17	22	1	2	–	–	–	–	–	–	–	–	–	**Ethylenediamine, N-(3-methoxyphenyl)-N,N′-dimethyl-N′-phenyl-**	32857-39-9	**NI27470**
150 120 151 77 121 135 118 42	**270**	100 61 11 9 7 6 6 5	**5.00**	17	22	1	2	–	–	–	–	–	–	–	–	–	**Ethylenediamine, N-(3-methoxyphenyl)-N,N′-dimethyl-N′-phenyl-**	32857-39-9	**NI27471**
150 120 151 77 121 135 118 270	**270**	100 61 11 9 7 6 6 5		17	22	1	2	–	–	–	–	–	–	–	–	–	**Ethylenediamine, N-(3-methoxyphenyl)-N,N′-dimethyl-N′-phenyl-**	32857-39-9	**LI05726**
150 120 270 151 77 135 121 42	**270**	100 39 17 11 8 7 5 5		17	22	1	2	–	–	–	–	–	–	–	–	–	**Ethylenediamine, N-(4-methoxyphenyl)-N,N′-dimethyl-N′-phenyl-**	32869-57-1	**NI27472**
175 144 82 96 158 270 55 160	**270**	100 69 63 60 44 36 30 27		17	22	1	2	–	–	–	–	–	–	–	–	–	**9-Methyl-2,3,4,5,6,7,8,9-octahydro-2,6-methano-1H-azecino[5,4-b]indol-8-ol**		**LI05727**
105 165 77 136 270 123 269 253	**270**	100 94 74 67 65 48 32 29		17	22	1	2	–	–	–	–	–	–	–	–	–	**Pyridine, 1-benzoyl-1,2,3,4-tetrahydro-5-(2-piperidinyl)-**	47077-87-2	**NI27473**
81 67 41 43 55 95 82 157	**270**	100 73 73 64 57 56 39 38	**0.34**	17	31	–	–	–	–	1	–	–	–	–	–	–	**4-Heptadecyne, 1-chloro-**	56554-73-5	**NI27474**
81 67 54 95 41 55 82 43	**270**	100 93 71 70 64 62 48 44	**0.84**	17	31	–	–	–	–	1	–	–	–	–	–	–	**6-Heptadecyne, 1-chloro-**	56599-59-8	**NI27475**
81 67 95 54 41 55 82 69	**270**	100 90 74 67 64 56 50 42	**0.70**	17	31	–	–	–	–	1	–	–	–	–	–	–	**7-Heptadecyne, 1-chloro-**	56554-74-6	**NI27476**
81 67 95 55 41 54 82 109	**270**	100 97 83 70 70 65 53 49	**1.04**	17	31	–	–	–	–	1	–	–	–	–	–	–	**7-Heptadecyne, 17-chloro-**	56554-75-7	**NI27477**
158 87 186 55 143 115 43 41	**270**	100 99 68 34 30 28 27 27	**16.00**	17	34	2	–	–	–	–	–	–	–	–	–	–	**Decanoic acid, 2-hexyl-, methyl ester**		**MS06076**
43 73 41 60 57 55 29 270	**270**	100 84 77 75 69 63 46 38		17	34	2	–	–	–	–	–	–	–	–	–	–	**Heptadecanoic acid**	506-12-7	**NI27479**
73 60 43 57 270 55 41 69	**270**	100 86 86 68 63 60 56 38		17	34	2	–	–	–	–	–	–	–	–	–	–	**Heptadecanoic acid**	506-12-7	**NI27478**
74 43 75 41 55 87 57 270	**270**	100 26 20 20 19 15 14 13		17	34	2	–	–	–	–	–	–	–	–	–	–	**Hexadecanoic acid, methyl ester**	112-39-0	**IC03517**
74 87 270 143 75 43 55 41	**270**	100 85 22 19 19 18 17 14		17	34	2	–	–	–	–	–	–	–	–	–	–	**Hexadecanoic acid, methyl ester**	112-39-0	**WI01330**
74 87 43 41 75 55 270 57	**270**	100 66 40 32 28 28 22 18		17	34	2	–	–	–	–	–	–	–	–	–	–	**Hexadecanoic acid, methyl ester**	112-39-0	**IC03518**
57 70 71 97 56 55 125 41	**270**	100 31 28 26 21 20 18 18	**0.00**	17	34	2	–	–	–	–	–	–	–	–	–	–	**Hexanoic acid, 3,5,5-trimethyl-, 2,4,4-trimethylpentyl ester**		**IC03519**
57 41 70 55 56 87 71 43	**270**	100 42 36 34 32 29 29 29	**0.19**	17	34	2	–	–	–	–	–	–	–	–	–	–	**Hexanoic acid, 3,5,5-trimethyl-, 2,4,4-trimethylpentyl ester**		**IC03520**
41 29 43 57 55 69 31 58	**270**	100 78 77 74 73 40 31 30	**0.00**	17	34	2	–	–	–	–	–	–	–	–	–	–	**Oxirane, [(tetradecyloxy)methyl]-**	38954-75-5	**NI27480**
88 101 43 41 55 73 70 57	**270**	100 46 40 34 28 21 21 20	**0.14**	17	34	2	–	–	–	–	–	–	–	–	–	–	**Pentadecanoic acid, ethyl ester**	41114-00-5	**NI27481**
74 43 87 29 57 41 55 44	**270**	100 66 61 53 51 42 41 33	**0.00**	17	34	2	–	–	–	–	–	–	–	–	–	–	**Pentadecanoic acid, 10-methyl-, methyl ester**		**LI05728**
74 87 270 55 43 41 57 143	**270**	100 52 49 35 35 34 30 22		17	34	2	–	–	–	–	–	–	–	–	–	–	**Pentadecanoic acid, 13-methyl-, methyl ester**	5487-50-3	**WI01331**
74 87 43 55 57 41 75 69	**270**	100 66 30 23 20 20 19 16	**9.94**	17	34	2	–	–	–	–	–	–	–	–	–	–	**Pentadecanoic acid, 14-methyl-, methyl ester**	5129-60-2	**NI27482**
74 43 41 87 55 29 57 69	**270**	100 89 75 71 58 33 28 27	**10.95**	17	34	2	–	–	–	–	–	–	–	–	–	–	**Pentadecanoic acid, 14-methyl-, methyl ester**	5129-60-2	**NI27484**
74 87 270 43 55 41 143 69	**270**	100 73 45 39 37 33 23 22		17	34	2	–	–	–	–	–	–	–	–	–	–	**Pentadecanoic acid, 14-methyl-, methyl ester**	5129-60-2	**NI27483**
43 83 61 69 97 57 70 55	**270**	100 87 86 66 65 58 52 50	**0.10**	17	34	2	–	–	–	–	–	–	–	–	–	–	**Pentadecyl acetate**		**IC03521**

MASS TO CHARGE RATIOS								M.W.	INTENSITIES								Parent	C	H	O	N	S	F	Cl	Br	I	Si	P	B	X	COMPOUND NAME	CAS Reg No	No
88	101	57	43	41	55	69	71	270	100	56	20	18	18	15	15	10	0.70	17	34	2	–	–	–	–	–	–	–	–	–	–	Tetradecanoic acid, 2,4-dimethyl-, methyl ester		NI27485
43	41	102	60	57	55	228	73	270	100	61	58	56	45	42	35	26	2.05	17	34	2	–	–	–	–	–	–	–	–	–	–	Tetradecanoic acid, isopropyl ester		IC03522
73	43	45	41	55	57	74	42	270	100	15	11	9	7	6	5	2	0.00	17	34	2	–	–	–	–	–	–	–	–	–	–	2-Tetradecyl-1,3-dioxolane		IC03523
73	43	45	41	55	57	74	56	270	100	11	9	9	7	6	4	3	0.00	17	34	2	–	–	–	–	–	–	–	–	–	–	2-Tetradecyl-1,3-dioxolane		IC03524
87	74	55	43	41	57	270	88	270	100	57	51	48	46	43	41	41		17	34	2	–	–	–	–	–	–	–	–	–	–	Tridecanoic acid, 2,6,10-trimethyl-, methyl ester		LI05729
87	74	57	71	157	84	213	69	270	100	47	17	15	12	11	9	7	1.82	17	34	2	–	–	–	–	–	–	–	–	–	–	Tridecanoic acid, 4,8,12-trimethyl-, methyl ester	10339-74-9	NI27486
100	58	44	42	41	57	43	101	270	100	82	76	33	24	19	15	7	0.00	17	38	–	2	–	–	–	–	–	–	–	–	–	Methanediamine, N,N,N′,N′-tetrabutyl-	20280-10-8	NI27487
270	255	272	271	105	257	191	205	270	100	51	33	20	20	15	15	9		18	19	–	–	–	–	1	–	–	–	–	–	–	Benzene, 1,1′-(chloroethenylidene)bis[4-ethyl-	14720-90-2	NI27488
120	119	44	270	151	41	57	55	270	100	73	43	39	30	24	21	21		18	22	–	–	1	–	–	–	–	–	–	–	–	Bis(3,5-dimethylbenzyl) sulphide	23566-27-0	NI27489
135	110	91	57	41	105	213	104	270	100	53	51	44	32	31	29	23	0.00	18	22	–	–	1	–	–	–	–	–	–	–	–	3,3-Dimethyl-2-phenylbutyl phenyl sulphide		DD00856
211	270	44	212	209	195	40	224	270	100	43	32	29	26	14	14	13		18	22	2	–	–	–	–	–	–	–	–	–	–	1-Azulenepropanoic acid, 3,8-dimethyl-5-isopropyl-	35336-11-9	MS06077
91	107	73	179	71	79	77	108	270	100	34	20	9	7	6	4	3	0.00	18	22	2	–	–	–	–	–	–	–	–	–	–	Benzene, 1,1′-[1,4-butanediylbis(oxymethylene)]bis-	6282-65-1	NI27490
91	107	73	92	179	65	55	71	270	100	35	20	14	9	9	8	7	0.00	18	22	2	–	–	–	–	–	–	–	–	–	–	Benzene, 1,1′-[1,4-butanediylbis(oxymethylene)]bis-	6282-65-1	NI27492
91	107	73	92	55	71	65	77	270	100	34	20	17	11	10	7	6	0.00	18	22	2	–	–	–	–	–	–	–	–	–	–	Benzene, 1,1′-[1,4-butanediylbis(oxymethylene)]bis-	6282-65-1	NI27491
135	77	115	105	91	41	121	57	270	100	13	12	12	12	12	10	10	0.00	18	22	2	–	–	–	–	–	–	–	–	–	–	Benzenemethanol, α-tert-butyl-α-phenyl-2-methoxy-		MS06078
121	104	77	91	79	105	122	129	270	100	33	27	21	14	13	12	11	0.00	18	22	2	–	–	–	–	–	–	–	–	–	–	Benzenepentanol, α-phenyl-ϵ-methoxy-		MS06079
91	162	107	147	92	65	270	43	270	100	33	20	13	9	8	6	5		18	22	2	–	–	–	–	–	–	–	–	–	–	Benzenepropanol, α,α-dimethyl-2-(phenylmethoxy)-	56052-47-2	NI27493
91	105	77	106	92	107	28	51	270	100	45	36	30	29	20	19	14	0.00	18	22	2	–	–	–	–	–	–	–	–	–	–	1,2-Bis(benzyloxy)butane		DO01064
91	28	105	77	107	106	92	51	270	100	70	58	55	47	47	28	21	0.00	18	22	2	–	–	–	–	–	–	–	–	–	–	1,3-Bis(benzyloxy)butane		DO01065
91	105	77	106	92	28	107	51	270	100	42	34	28	28	23	21	13	0.00	18	22	2	–	–	–	–	–	–	–	–	–	–	2,3-Bis(benzyloxy)butane		DO01066
181	209	157	224	268	270	158	211	270	100	24	21	19	17	16	15	13		18	22	2	–	–	–	–	–	–	–	–	–	–	1H-Cyclopent[a]anthracene-3,8-diol, 2,3β,3a,4,5,6,11,11bβ-octahydro-3aα-methyl-, (±)-	17292-24-9	NI27494
135	256	241	43	148	270	255	91	270	100	72	67	49	48	35	30	27		18	22	2	–	–	–	–	–	–	–	–	–	–	3,5-Dimethyl-4-hydroxybenzyl 2,4,6-trimethylphenyl ether		NI27495
209	151	224	268	270	158	211	181	270	100	85	80	68	66	60	55	41		18	22	2	–	–	–	–	–	–	–	–	–	–	Estra-1,3,5(10),7(8)-tetraen-3,17-diol		LI05730
270	157	158	144	145	170	211	159	270	100	70	65	50	34	31	20	20		18	22	2	–	–	–	–	–	–	–	–	–	–	Estra-1,3,5(10),6-tetraene-3,17-diol, (17β)-	7291-41-0	NI27496
270	211	160	147	158	157	145	197	270	100	36	25	15	14	13	13	12		18	22	2	–	–	–	–	–	–	–	–	–	–	Estra-1,3,5(10),9(11)-tetraene-3,17-diol, (17β)-	791-69-5	NI27497
270	252	146	160	172	158	159	145	270	100	67	53	48	35	32	29	19		18	22	2	–	–	–	–	–	–	–	–	–	–	Estra-1,3,5(10)-trien-3-ol, 16,17-epoxy-, (16α,17α)-	472-56-0	NI27498
270	160	172	227	146	159	145	133	270	100	56	41	35	30	22	20	20		18	22	2	–	–	–	–	–	–	–	–	–	–	Estra-1,3,5(10)-trien-3-ol, 16,17-epoxy-, (16β,17β)-	472-57-1	NI27499
172	270	271	173	159	146	157	145	270	100	59	25	21	20	14	13	13		18	22	2	–	–	–	–	–	–	–	–	–	–	Estra-1,3,5(10)-trien-16-one, 3-hydroxy-		NI27500
270	146	185	172	213	159	145	271	270	100	41	36	31	22	22	22	21		18	22	2	–	–	–	–	–	–	–	–	–	–	Estra-1,3,5(10)-trien-17-one, 3-hydroxy-	53-16-7	NI27502
270	146	185	172	271	213	159	145	270	100	32	30	24	20	17	17	16		18	22	2	–	–	–	–	–	–	–	–	–	–	Estra-1,3,5(10)-trien-17-one, 3-hydroxy-	53-16-7	NI27501
270	271	146	213	185	172	160	159	270	100	18	18	14	14	13	12	12		18	22	2	–	–	–	–	–	–	–	–	–	–	Estra-1,3,5(10)-trien-17-one, 3-hydroxy-	53-16-7	TR00293
270	199	160	159	271	115	186	41	270	100	25	23	22	19	13	12	11		18	22	2	–	–	–	–	–	–	–	–	–	–	Gona-1,3,5(10)-trien-17-one, 3-methoxy-	4147-10-8	NI27503
270	83	199	91	55	115	41	77	270	100	37	34	30	28	27	24	22		18	22	2	–	–	–	–	–	–	–	–	–	–	Gona-1,3,5(10)-trien-17-one, 3-methoxy-, (13α)-	4248-04-8	NI27504
255	270	256	271	128	153	141	115	270	100	59	20	12	10	6	6	6		18	22	2	–	–	–	–	–	–	–	–	–	–	s-Indacene-1,7-dione, 2,3,5,6-tetrahydro-3,3,4,5,5,8-hexamethyl-	55591-16-7	NI27505
115	141	270	155	129	183	197	196	270	100	95	80	70	70	14	4	4		18	22	2	–	–	–	–	–	–	–	–	–	–	Methyl heptadeca-7E,9E,15E-trien-11,13-diyn-1-oate		LI05731
255	270	148	135	256	271	149	91	270	100	61	43	22	18	14	11	10		18	22	2	–	–	–	–	–	–	–	–	–	–	2,3,3′,5,5′-Pentamethyl-4,4′-dihydroxydiphenylmethane	29366-00-5	MS06080
135	107	134	136	41	91	77	108	270	100	55	15	11	9	7	6	5	2.54	18	22	2	–	–	–	–	–	–	–	–	–	–	Phenol, 4,4′-(1,2-diethyl-1,2-ethanediyl)bis-, (R*,S*)-	84-16-2	NI27506
135	107	134	136	77	91	108	120	270	100	72	21	12	10	8	5	5	0.42	18	22	2	–	–	–	–	–	–	–	–	–	–	Phenol, 4,4′-(1,2-diethyl-1,2-ethanediyl)bis-, (R*,S*)-	84-16-2	NI27507
241	270	213	242	128	171	115	271	270	100	77	32	23	21	19	18	17		18	22	2	–	–	–	–	–	–	–	–	–	–	Spiro[2-cyclohexene-1,1′-[1H]indene]-4,6′(2′H)-dione, 2′,3′-diethyl-4′,5′-dihydro-	54889-77-9	NI27508
160	270	161	145	187	91	227	131	270	100	62	19	18	16	16	13	12		18	22	2	–	–	–	–	–	–	–	–	–	–	1,4-Triphenylenedione, 4a,4b,5,6,7,8,9,10,11,12,12a,12b-dodecahydro-	33283-44-2	LI05732
160	270	161	145	187	91	271	227	270	100	63	18	18	16	16	13	13		18	22	2	–	–	–	–	–	–	–	–	–	–	1,4-Triphenylenedione, 4a,4b,5,6,7,8,9,10,11,12,12a,12b-dodecahydro-	33283-44-2	NI27509
149	105	270	121	135	28			270	100	98	80	41	19	15				18	22	2	–	–	–	–	–	–	–	–	–	–	1,2-Bis(2,6-xylyloxy)ethane		LI05733
197	199	126	128	127	184	115	198	270	100	99	91	82	71	68	51	48	8.00	18	26	–	2	–	–	–	–	–	–	–	–	–	Tricyclo[5.2.1.0^{4,10}]deca-1(10),2,5,8-tetraene-4,7-diamine, N,N′-dimethyl-		NI27510
70	57	55	43	83	69	41	71	270	100	95	83	71	67	67	64	60	0.00	18	38	1	–	–	–	–	–	–	–	–	–	–	1-Heptadecanol, 15-methyl-		LI05734
56	69	55	57	129	41	43	83	270	100	96	96	80	72	70	68	52	0.00	18	38	1	–	–	–	–	–	–	–	–	–	–	7-Heptadecanol, 7-methyl-	55723-93-8	NI27511
143	171	69	43	55	41	57	56	270	100	99	99	99	98	94	93	81	0.00	18	38	1	–	–	–	–	–	–	–	–	–	–	8-Heptadecanol, 8-methyl-	55723-92-7	NI27512
45	43	84	55	97	83	69	71	270	100	97	78	76	68	66	64	61	0.00	18	38	1	–	–	–	–	–	–	–	–	–	–	Hexadecane, 1-methoxy-13-methyl-	51166-34-8	NI27513
43	41	57	55	82	69	83	29	270	100	85	71	67	42	40	39	35	0.01	18	38	1	–	–	–	–	–	–	–	–	–	–	1-Octadecanol		AI01916
43	55	57	41	83	69	56	97	270	100	90	75	75	70	70	50	45	0.00	18	38	1	–	–	–	–	–	–	–	–	–	–	1-Octadecanol		IC03525
57	43	82	55	83	69	96	97	270	100	66	62	52	48	46	43	39	0.00	18	38	1	–	–	–	–	–	–	–	–	–	–	1-Octadecanol		IC03526
57	71	69	41	83	43	55	98	270	100	38	28	18	16	14	13	12	0.00	18	38	1	–	–	–	–	–	–	–	–	–	–	Bis(3,5,5-trimethylhexyl) ether		IC03527
57	70	71	41	69	43	55	83	270	100	58	31	25	20	18	14	12	0.10	18	38	1	–	–	–	–	–	–	–	–	–	–	Bis(3,5,5-trimethylhexyl) ether		IC03528
57	83	69	41	71	43	56	55	270	100	13	12	11	10	10	9	9	0.08	18	38	1	–	–	–	–	–	–	–	–	–	–	5,7,7-Trimethyl-2-(1,3,3-trimethylbutyl)-1-octanol		DO01067
269	270	271	77	268	51	135	139	270	100	89	17	15	12	11	10	4		19	14	–	2	–	–	–	–	–	–	–	–	–	1H-Benzimidazole, 1,2-diphenyl-	2622-67-5	NI27514

MASS TO CHARGE RATIOS								M.W.	INTENSITIES								Parent	C	H	O	N	S	F	Cl	Br	I	Si	P	B	X	COMPOUND NAME	CAS Reg No	No
269	270	77	268	51	135	166	139	**270**	100	89	15	12	11	10	4	4		19	14	–	2	–	–	–	–	–	–	–	–	–	**1H-Benzimidazole, 1,2-diphenyl-**	2622-67-5	**LI05735**
270	269	167	166	271	139	91	134	**270**	100	36	29	27	21	16	10	9		19	14	–	2	–	–	–	–	–	–	–	–	–	**Carbazole, 3-(benzylideneamino)-**	14384-36-2	**NI27515**
172	171	77	165	105	157	145	135	**270**	100	86	83	80	80	72	72	70	**0.00**	19	26	1	–	–	–	–	–	–	–	–	–	–	**Acetophenone, α-(2,2,4,6,6-pentamethylcyclohexylidene)-**	72101-50-9	**NI27516**
41	270	91	159	79	161	55	187	**270**	100	92	82	79	56	53	52	45		19	26	1	–	–	–	–	–	–	–	–	–	–	**Androsta-3,5-dien-7-one**	32222-21-2	**NI27517**
91	79	146	147	93	77	41	105	**270**	100	98	88	80	75	75	67	65	**53.30**	19	26	1	–	–	–	–	–	–	–	–	–	–	**Androsta-4,16-dien-3-one**	4075-07-4	**MS06082**
147	146	255	91	79	93	77	124	**270**	100	92	88	77	71	54	54	50	**0.00**	19	26	1	–	–	–	–	–	–	–	–	–	–	**Androsta-4,16-dien-3-one**	4075-07-4	**MS06081**
255	213	159	171	184	270	199		**270**	100	66	53	44	40	35	20			19	26	1	–	–	–	–	–	–	–	–	–	–	**4b,8-Dimethyl-2-isopropyl-4b,5,6,7,8,8a,9,10-octahydrophenanthren-7-one**		**LI05736**
270	271	173	199	160	95	174	159	**270**	100	21	21	9	9	9	7	7		19	26	1	–	–	–	–	–	–	–	–	–	–	**Estra-1,3,5(10)-triene, 3-methoxy-**	14550-57-3	**NI27518**
185	199	171	44	157	186	270		**270**	100	33	27	26	24	15	14			19	26	1	–	–	–	–	–	–	–	–	–	–	**Naphthalene, 2-nonyloxy-**		**IC03529**
270	214	107	213	271	239	215	241	**270**	100	73	53	45	20	16	15	11		20	14	1	–	–	–	–	–	–	–	–	–	–	**Benzo[a]pyren-7(8H)-one, 9,10-dihydro-**	3331-46-2	**NI27519**
270	241	239	269	240	119	242	271	**270**	100	78	58	42	25	24	23	20		20	14	1	–	–	–	–	–	–	–	–	–	–	**Di-trans-1,2:3,4-dibenzo[7,8-c]furo[10]annulene**		**MS06083**
270	241	239	269	240	242	255		**270**	100	78	58	42	25	23	12			20	14	1	–	–	–	–	–	–	–	–	–	–	**Di-trans-1,2:3,4-dibenzo[7,8-c]furo[10]annulene**		**LI05737**
239	270	241	226	242	255	240	269	**270**	100	97	71	59	58	52	42	37		20	14	1	–	–	–	–	–	–	–	–	–	–	**7-Formyl-12-methylbenz[a]anthracene**		**MS06084**
270	241	271	239	77	165	135	51	**270**	100	23	19	19	19	18	15	9		20	14	1	–	–	–	–	–	–	–	–	–	–	**Isobenzofuran, 1,3-diphenyl-**	5471-63-6	**NI27520**
269	270	226	239	241	237	242	215	**270**	100	68	43	40	29	14	13	10		20	14	1	–	–	–	–	–	–	–	–	–	–	**7-Methyl-12-formylbenz[a]anthracene**		**MS06085**
269	270	267	239	119.5	241	226	266	**270**	100	69	62	32	25	17	17	16		20	14	1	–	–	–	–	–	–	–	–	–	–	**7-Methyl-12-formylbenz[a]anthracene**		**MS06086**
270	241	271	127	44	115	242	28	**270**	100	27	22	22	22	17	15	13		20	14	1	–	–	–	–	–	–	–	–	–	–	**Bis(1-naphthyl) ether**	607-52-3	**IC03530**
158	115	143	159	116	63	89	127	**270**	100	87	43	13	9	9	8	7	**0.00**	20	14	1	–	–	–	–	–	–	–	–	–	–	**Bis(1-naphthyl) ether**	607-52-3	**NI27521**
270	269	115	127	271	241	126	77	**270**	100	30	29	26	23	17	13	13		20	14	1	–	–	–	–	–	–	–	–	–	–	**Bis(1-naphthyl) ether**	607-52-3	**IC03531**
270	269	268	134	239	271	39	51	**270**	100	98	63	39	30	19	12	9		20	14	1	–	–	–	–	–	–	–	–	–	–	**9H-Xanthene, 9-benzylidene-**		**DD00857**
91	270	94	79	41	28	29	129	**270**	100	62	56	54	52	52	49	46		20	30	–	–	–	–	–	–	–	–	–	–	–	**Bi-1,4-cyclohexadien-1-yl, 2,2′-dibutyl-**	61142-53-8	**NI27522**
255	41	69	43	159	173	185	270	**270**	100	78	63	63	54	46	33	32		20	30	–	–	–	–	–	–	–	–	–	–	–	**Phenanthrene, 1,2,3,4,4a,9,10,10a-octahydro-1,1,4a-trimethyl-7-isopropyl-, (4aS-trans)-**	19407-28-4	**LI05738**
255	159	173	270	69	43	185	41	**270**	100	70	39	38	36	35	32	32		20	30	–	–	–	–	–	–	–	–	–	–	–	**Phenanthrene, 1,2,3,4,4a,9,10,10a-octahydro-1,1,4a-trimethyl-7-isopropyl-, (4aS-trans)-**	19407-28-4	**NI27524**
255	270	173	159	43	256	185	69	**270**	100	34	32	32	24	20	16	14		20	30	–	–	–	–	–	–	–	–	–	–	–	**Phenanthrene, 1,2,3,4,4a,9,10,10a-octahydro-1,1,4a-trimethyl-7-isopropyl-, (4aS-trans)-**	19407-28-4	**NI27523**
241	270	239	242	271	240	226	121	**270**	100	82	26	23	19	8	8	8		21	18	–	–	–	–	–	–	–	–	–	–	–	**Benz[a]anthracene, 8-propyl-**	54889-82-6	**WI01332**
241	270	239	242	271	120.5	107.5	226	**270**	100	82	26	23	19	10	9	8		21	18	–	–	–	–	–	–	–	–	–	–	–	**Benz[a]anthracene, 8-propyl-**	54889-82-6	**AM00138**
178	179	176	152	177	89	180	151	**270**	100	23	19	13	12	11	8	8	**2.00**	21	18	–	–	–	–	–	–	–	–	–	–	–	**Bicyclo[4.2.1]nona-2,4,7-triene, 7,8-diphenyl-**	54049-09-1	**NI27525**
179	178	270	192	91	191	180	271	**270**	100	75	60	46	24	17	15	14		21	18	–	–	–	–	–	–	–	–	–	–	–	**2,3-Diphenylindan**		**MS06087**
179	270	192	91	191	180	271	165	**270**	100	60	46	24	17	15	14	14		21	18	–	–	–	–	–	–	–	–	–	–	–	**2,3-Diphenylindan**		**IC03532**
270	239	242	215	241	229	226	112	**270**	100	20	19	12	11	11	11	9		21	18	–	–	–	–	–	–	–	–	–	–	–	**8,9,10,11-Tetrahydro-7H-cyclohepta[a]pyrene**		**LI05739**
271	231	230	69	270	216	197	114	**271**	100	75	45	30	25	23	18	14		3	5	–	3	–	5	–	–	–	–	3	–	–	**2-(1-Propenyl)pentafluorocyclotriphosphazene**		**MS06088**
231	271	230	69	212	216	171	114	**271**	100	69	18	14	12	7	7	7		3	5	–	3	–	5	–	–	–	–	3	–	–	**2-(2-Propenyl)pentafluorocyclotriphosphazene**		**MS06089**
192	69	91	92	37	273	271	81	**271**	100	97	48	42	24	22	22	21		5	4	–	1	–	6	–	1	–	–	–	–	–	**1-Bromo-1-perfluorodimethylamino-prop-1-ene**		**LI05740**
69	273	271	39	104	96	192	119	**271**	100	74	74	34	29	19	16	10		5	4	–	1	–	6	–	1	–	–	–	–	–	**cis-1-[Bis(trifluoromethyl)amino]-2-bromo-prop-1-ene**		**LI05741**
69	273	271	39	104	192	96	121	**271**	100	91	91	36	31	22	15	13		5	4	–	1	–	6	–	1	–	–	–	–	–	**trans-1-[Bis(trifluoromethyl)amino]-2-bromo-prop-1-ene**		**LI05742**
236	117	119	30	238	50	69	79	**271**	100	96	94	87	70	52	51	39	**17.00**	7	4	2	1	1	–	3	–	–	–	–	–	–	**Trichloromethyl 4-nitrophenyl sulphide**	713-66-6	**NI27526**
201	203	238	78	166	105	273	140	**271**	100	66	33	20	16	13	12	12	**0.00**	7	5	–	3	–	–	4	–	–	–	–	–	–	**3,5,6,8-Tetrachloro-7-methyl-5,6-dihydro-s-triazolo[4,3-a]pyridine**		**LI05743**
271	91	77	116	51	89	64	272	**271**	100	32	30	25	16	12	11	10		8	6	–	3	–	–	–	–	1	–	–	–	–	**1,2,3-Triazole, 2-phenyl-4-iodo-**		**NI27527**
61	211	59	256	210	252			**271**	100	11	8	5	3	1			**0.00**	8	9	–	1	–	8	–	–	–	–	–	–	–	**1-(2-Fluoro-2-methylpropylimino)-1-fluoro-2,2-bis(trifluoromethyl)ethane**		**LI05744**
270	77	116	89	51	143	63	103	**271**	100	54	41	34	30	16	15	13	**12.00**	9	6	1	1	–	–	–	–	1	–	–	–	–	**4-Iodo-3-phenylisoxazole**		**MS06090**
226	134	241	166	271	179			**271**	100	58	48	42	41	39				9	9	7	3	–	–	–	–	–	–	–	–	–	**2,4-Dinitrophenylserine**		**LI05745**
210	208	255	269	182	227	253	183	**271**	100	52	42	33	29	26	21	21	**5.85**	10	9	3	1	–	–	–	–	–	–	–	–	1	**Selenolo[2,3-b]pyridine-2-carboxylic acid, ethyl ester, 7-oxide**	55108-61-7	**NI27528**
225	271	169	227	222	221	269	199	**271**	100	36	32	20	18	18	17	17		10	9	3	1	–	–	–	–	–	–	–	–	1	**Selenolo[2,3-b]pyridine-2-carboxylic acid, 3-hydroxy-, ethyl ester**	41323-15-3	**NI27529**
129	142	102	128	130	127	51	75	**271**	100	46	25	16	15	15	14	10	**0.00**	10	10	–	1	–	–	–	–	1	–	–	–	–	**Isoquinolinium, 2-methyl-, iodide**	3947-77-1	**NI27530**
129	142	102	127	128	130	51	50	**271**	100	46	23	20	17	12	11	11	**0.00**	10	10	–	1	–	–	–	–	1	–	–	–	–	**Quinolinium, 1-methyl-, iodide**	3947-76-0	**NI27531**
61	28	179	75	41	196	105	78	**271**	100	78	77	76	70	53	50	49	**19.00**	10	13	4	3	1	–	–	–	–	–	–	–	–	**N-[1-(Methylthio)propyl]-2,4-dinitroaniline**		**MS06091**
57	41	29	153	69	152	56	27	**271**	100	74	73	53	37	30	17	16	**0.00**	10	16	4	1	–	3	–	–	–	–	–	–	–	**L-Threonine, N-(trifluoroacetyl)-, butyl ester**	23403-41-0	**NI27532**
241	154	45	96	71	44	169	226	**271**	100	96	95	80	80	72	64	60	**39.99**	10	17	4	5	–	–	–	–	–	–	–	–	–	**Acetic acid, [4-(dimethylamino)-6-(methoxyamino)-1,3,5-triazinyloxy]-, ethyl ester**	78927-38-5	**NI27533**
271	106	134	149	78	256	87	51	**271**	100	90	83	79	69	61	54	44		11	7	3	1	–	–	2	–	–	–	–	–	–	**N-(Anisol-4-yl)-2,3-dichloromaleimide**		**IC03533**
271	93	59	77	195	65	95	53	**271**	100	57	40	29	25	21	19	18		11	13	7	1	–	–	–	–	–	–	–	–	–	**Benzoic acid, 3,4,5-trimethoxy-2-nitro-, methyl ester**	5081-42-5	**NI27534**

MASS TO CHARGE RATIOS								M.W.	INTENSITIES								Parent	C	H	O	N	S	F	Cl	Br	I	Si	P	B	X	COMPOUND NAME	CAS Reg No	No
256	58	45	213	43	271	226	212	**271**	100	47	38	35	34	32	30	30		11	21	1	5	1	–	–	–	–	–	–	–	–	1,3,5-Triazine-2,4-diamine, N-(3-methoxypropyl)-N′-isopropyl-6-(methylthio)-	841-06-5	NI27535
187	55	159	81	146	188	78	56	**271**	100	57	19	15	12	10	10	8	**0.69**	12	10	3	1	–	–	–	–	–	–	–	–	1	(3-Cyanopropyl)cymantrene	64848-76-6	NI27536
172	174	187	129	144	189	131	78	**271**	100	33	32	30	26	11	10	9	**0.00**	12	14	4	1	–	–	1	–	–	–	–	–	–	Butanamide, N-(4-chloro-2,5-dimethoxyphenyl)-3-oxo-	4433-79-8	NI27537
45	143	43	169	145	229	78	187	**271**	100	93	78	63	33	30	28	27	**5.00**	12	14	4	1	–	–	1	–	–	–	–	–	–	Carbamic acid, [2-(acetyloxy)-5-chlorophenyl]-, isopropyl ester	34061-88-6	NI27538
155	91	108	197	207	107	171	151	**271**	100	95	55	41	19	13	10	6	**0.00**	12	17	4	1	1	–	–	–	–	–	–	–	–	Carbamic acid, (4-tolylsulphonyl)-, isobutyl ester	32363-27-2	LI05746
155	91	43	108	197	58	42	41	**271**	100	95	77	55	41	28	24	23	**0.00**	12	17	4	1	1	–	–	–	–	–	–	–	–	Carbamic acid, (4-tolylsulphonyl)-, isobutyl ester	32363-27-2	NI27539
155	91	108	171	197	107	151	207	**271**	100	84	54	43	39	15	13	8	**0.00**	12	17	4	1	1	–	–	–	–	–	–	–	–	Carbamic acid, (4-tolylsulphonyl)-, sec-butyl ester	32363-29-4	LI05747
155	91	108	171	197	45	65	107	**271**	100	84	54	43	39	35	24	15	**0.00**	12	17	4	1	1	–	–	–	–	–	–	–	–	Carbamic acid, (4-tolylsulphonyl)-, sec-butyl ester	32363-29-4	NI27540
91	155	212	108	171	197	151	107	**271**	100	56	47	15	13	12	7	5	**0.00**	12	17	4	1	1	–	–	–	–	–	–	–	–	Carbamic acid, (4-tolylsulphonyl)-, tert-butyl ester	18303-04-3	LI05748
91	41	155	212	57	59	44	108	**271**	100	79	56	47	38	29	19	14	**0.00**	12	17	4	1	1	–	–	–	–	–	–	–	–	Carbamic acid, (4-tolylsulphonyl)-, tert-butyl ester	18303-04-3	NI27541
184	142	87	109	71	57	55	83	**271**	100	59	39	35	20	14	14	12	**0.00**	12	17	6	1	–	–	–	–	–	–	–	–	–	D-Gluconitrile, 3,6-dideoxy-2,4,5-tri-O-acetyl-		NI27542
93	178	109	94	180	91	271	212	**271**	100	72	70	53	24	23	17	16		12	18	2	1	1	–	–	–	–	–	1	–	–	N-(Dimethylthiophosphinyl)phenylalanine methyl ester		MS06092
271	77	270	104	51	272	78	252	**271**	100	92	37	34	27	15	10	8		13	6	–	1	–	5	–	–	–	–	–	–	–	N-(Pentafluorophenylmethylene)aniline		MS06093
228	150	78	44	254	104	271	182	**271**	100	34	33	33	28	23	12	12		13	9	4	3	–	–	–	–	–	–	–	–	–	2-Pyridinecarboxamide, 5-(3-nitrobenzoyl)-	82465-63-2	NI27543
107	271	42	134	78	52	79	39	**271**	100	70	31	27	24	23	20	19		13	9	4	3	–	–	–	–	–	–	–	–	–	2-Pyridinone, N-(3-nitrophenyl)-3-cyano-4-methyl-6-hydroxy-		IC03534
271	107	134	243	78	272	42	79	**271**	100	88	36	30	20	16	16	15		13	9	4	3	–	–	–	–	–	–	–	–	–	2-Pyridinone, N-(4-nitrophenyl)-3-cyano-4-methyl-6-hydroxy-		IC03535
91	86	65	58	30	92	101	148	**271**	100	54	14	12	10	10	9	8	**0.00**	13	22	–	1	–	–	–	1	–	–	–	–	–	Triethylbenzylammonium bromide	5197-95-5	NI27544
100	113	56	55	41	28	42	30	**271**	100	89	29	27	21	18	16	14	**2.57**	13	25	1	3	1	–	–	–	–	–	–	–	–	Thiourea, N-cyclohexyl-N′-[2-(4-morpholinyl)ethyl]-	21545-54-0	NI27545
272	114	131	238	100	185	88	173	**271**	100	49	40	34	24	21	18	17	**1.80**	13	25	1	3	1	–	–	–	–	–	–	–	–	Thiourea, N-cyclohexyl-N′-[2-(4-morpholinyl)ethyl]-	21545-54-0	NI27546
147	76	104	125	77	51	50	130	**271**	100	60	58	46	42	36	34	26	**14.00**	14	9	3	1	1	–	–	–	–	–	–	–	–	1H-Isoindole-1,3(2H)-dione, 2-(phenylsulphinyl)-	40167-15-5	NI27547
137	271	135	134	77	155	136	52	**271**	100	72	52	39	31	27	19	17		14	13	3	3	–	–	–	–	–	–	–	–	–	Benzaldehyde, 4-methoxy-, (2-nitrophenyl)hydrazone		LI05749
137	271	135	134	77	155	136	107	**271**	100	72	53	38	31	27	20	15		14	13	3	3	–	–	–	–	–	–	–	–	–	Benzaldehyde, 4-methoxy-, (2-nitrophenyl)hydrazone		LI05750
271	134	77	108	107	92	136	64	**271**	100	46	45	42	28	27	25	25		14	13	3	3	–	–	–	–	–	–	–	–	–	Benzaldehyde, 4-methoxy-, (4-nitrophenyl)hydrazone	5880-63-7	NI27548
228	271	229	226	198	182	56	215	**271**	100	18	14	13	13	12	10	8		14	25	4	1	–	–	–	–	–	–	–	–	–	Ethyl 2-butyl-3-[[(ethoxycarbonyl)methyl]amino]crotonate		MS06094
270	271	272	128	101	77	75	273	**271**	100	60	38	36	27	15	15	13		15	10	–	1	1	–	1	–	–	–	–	–	–	2-(4-Chlorophenylthio)quinoline		MS06095
105	150	77	271	51	149	106	273	**271**	100	38	38	30	25	16	15	11		15	10	2	1	–	–	1	–	–	–	–	–	–	Benzeneacetonitrile, α-(benzoyloxy)-4-chloro-	19788-56-8	LI05751
105	150	77	271	51	149	106	152	**271**	100	20	20	16	12	8	7	6		15	10	2	1	–	–	1	–	–	–	–	–	–	Benzeneacetonitrile, α-(benzoyloxy)-4-chloro-	19788-56-8	NI27549
271	179	272	163	107	136	46		**271**	100	18	16	9	7	6	4			15	13	4	1	–	–	–	–	–	–	–	–	–	Benzeneacetic acid, 4-nitro-, 3-methylphenyl ester	53218-12-5	NI27550
271	163	107	46	272	136	179	164	**271**	100	72	19	17	16	12	6	5		15	13	4	1	–	–	–	–	–	–	–	–	–	Benzeneacetic acid, 4-nitro-, 4-methylphenyl ester	53274-19-4	NI27551
166	271	167	122	46				**271**	100	32	8	8	2					15	13	4	1	–	–	–	–	–	–	–	–	–	Benzenemethanol, 3-methyl-, 4-nitrobenzoate	53218-08-9	NI27552
166	167	122	271	46				**271**	100	8	5	3	2					15	13	4	1	–	–	–	–	–	–	–	–	–	Benzenemethanol, 4-methyl-, 4-nitrobenzoate	53218-09-0	NI27553
121	109	151	65	93	43	39	271	**271**	100	84	60	30	21	20	15	13		15	13	4	1	–	–	–	–	–	–	–	–	–	Benzoic acid, 2-hydroxy-, 4-(acetylamino)phenyl ester	118-57-0	NI27554
228	256	185	129	271	128	102	157	**271**	100	78	65	39	34	22	22	20		15	13	4	1	–	–	–	–	–	–	–	–	–	2,3-Dimethoxy-1-hydroxyacridone		MS06096
172	271	215	171	44	41	146	89	**271**	100	71	68	44	27	25	22	15		15	17	2	3	–	–	–	–	–	–	–	–	–	2-Pyrimidinecarboxamide, 5-(4-butoxyphenyl)-		NI27555
165	122	94	271	178	131	137	253	**271**	100	97	80	65	45	34	23	21		15	17	2	3	–	–	–	–	–	–	–	–	–	2-Pyrrolidinone, 3-(hydroxyphenylmethyl)-4-(1-methyl-1H-imidazol-4-yl)-, [3R-[3α(S*),4β]]-		NI27556
202	41	271	119	203	185	160	28	**271**	100	38	22	17	16	13	13	13		15	21	–	5	–	–	–	–	–	–	–	–	–	Adenine, N,N-bis(3-methyl-2-butenyl)-	5122-43-0	NI27557
202	41	134	135	28	160	119	203	**271**	100	28	26	23	17	16	14	12	**10.23**	15	21	–	5	–	–	–	–	–	–	–	–	–	1H-Purin-6-amine, N-(3,7-dimethyl-2,6-octadienyl)-, (E)-	14714-89-7	NI27558
57	170	128	155	41	43	86	198	**271**	100	94	53	51	51	46	41	33	**0.00**	15	29	3	1	–	–	–	–	–	–	–	–	–	Carbamic acid, 4-oxodecyl-, tert-butyl ester	116437-37-7	DD00858
229	271	108	230	228	69	272	109	**271**	100	72	43	34	19	16	15	11		16	14	–	1	1	1	–	–	–	–	–	–	–	1,5-Benzothiazepine, 2,3-dihydro-2-methyl-4-(4-fluorophenyl)-		NI27559
105	150	77	271	51	91	83	137	**271**	100	71	58	17	16	10	7	6		16	17	1	1	1	–	–	–	–	–	–	–	–	Benzamide, N-[2-[3-(methylthio)phenyl]ethyl]-		MS06097
109	227	185	152	271	184	199	200	**271**	100	43	37	27	26	22	20	15		16	17	1	1	1	–	–	–	–	–	–	–	–	Benzo[d]thieno[2,3-g]azecin-13-one, 4,5,6,7,8,13-hexahydro-6-methyl-		NI27560
123	81	124	105	77	166	43	80	**271**	100	67	38	33	23	17	17	13	**0.00**	16	17	3	1	–	–	–	–	–	–	–	–	–	Bicyclo[2.2.1]hept-2-ene, 1-(acetylamino)-6-benzoyloxy-		LI05752
271	243	214	187	170	115	144	228	**271**	100	50	50	18	17	17	10	8		16	17	3	1	–	–	–	–	–	–	–	–	–	1,4-Butanedione, 1-(7-allyl-6-methoxyindol-2-yl)-		MS06098
271	199	187	228	270	216	254	215	**271**	100	51	40	14	12	12	7	7		16	17	3	1	–	–	–	–	–	–	–	–	–	Crinan-3-ol, 1,2-didehydro-, (3α)-	510-67-8	LI05753
271	199	187	56	57	228	272	115	**271**	100	46	41	35	27	21	18	18		16	17	3	1	–	–	–	–	–	–	–	–	–	Crinan-3-ol, 1,2-didehydro-, (3α)-	510-67-8	NI27561
227	271	213	228	270	242	212	200	**271**	100	81	39	30	29	20	16	16		16	17	3	1	–	–	–	–	–	–	–	–	–	Crinan-11-ol, 1,2-didehydro-, (5α,11R,13β,19α)-	561-19-3	NI27562
202	203	96	41	53	201	67	271	**271**	100	88	86	74	47	44	42	38		16	17	3	1	–	–	–	–	–	–	–	–	–	4,7-Epoxy-1H-isoindole-1,3(2H)-dione, hexahydro-3a,7a-dimethyl-2-phenyl-, [3aR-(3aα,4β,7β,7aα)]-	54355-51-0	NI27563
117	253	105	77	118	148	134	135	**271**	100	70	47	35	20	17	12	10	**0.00**	16	17	3	1	–	–	–	–	–	–	–	–	–	Ethylamine, N-benzoyl-2-hydroxy-2-(4-methoxyphenyl)-, (±)-		LI05754
226	271	227	252	272	253	225	77	**271**	100	91	60	59	17	13	13	13		16	17	3	1	–	–	–	–	–	–	–	–	–	Galanthan-1-ol, 3,12-didehydro-9,10-[methylenebis(oxy)]-, (1α)-	477-12-3	NI27564
242	241	271	225	211	137	210	243	**271**	100	79	63	57	37	27	23	16		16	17	3	1	–	–	–	–	–	–	–	–	–	7-Isoquinolinol, 1,2,3,4-tetrahydro-4-(4-hydroxyphenyl)-6-methoxy-, (S)-	56847-08-6	NI27565
254	272	271	270					**271**	100	41	16	5						16	17	3	1	–	–	–	–	–	–	–	–	–	Morphinan-3,6-diol, 7,8-didehydro-4,5-epoxy-, (5α,6α)-	466-97-7	NI27566
271	115	201	203	143	202	116	173	**271**	100	95	93	26	26	24	17	15		16	17	3	1	–	–	–	–	–	–	–	–	–	Pyrrolidine, 1-piperoyl-, (E,E)-		MS06099
201	271	115	173	143	55	43	70	**271**	100	84	52	37	22	20	20	15		16	17	3	1	–	–	–	–	–	–	–	–	–	Pyrrolidine, 1-piperoyl-, (E,E)-		LI05755

MASS TO CHARGE RATIOS								M.W.	INTENSITIES								Parent	C	H	O	N	S	F	Cl	Br	I	Si	P	B	X	COMPOUND NAME	CAS Reg No	No
215	187	271	170	198	216	189	272	**271**	100	83	50	28	27	17	13	9		16	17	3	1	–	–	–	–	–	–	–	–	–	**1H-Pyrrolo[1,2-a]indole-9-carboxylic acid, 2,3-dihydro-3-oxo-, tert-butyl ester**		**NI27567**
223	81	80	79	120	91	95	226	**271**	100	90	87	86	80	77	66	60	**25.00**	16	17	3	1	–	–	–	–	–	–	–	–	–	**Tricyclo[3.3.1.1^{3,7}]decanone, 1-(nitrophenyl)-**	40621-19-0	**NI27568**
151	227	228	212	211	115	91	41	**271**	100	67	50	19	19	19	19	17	**9.00**	16	21	1	3	–	–	–	–	–	–	–	–	–	**Cyclooctanone, 2-benzylidene-, semicarbazone**	16983-73-6	**NI27569**
116	41	43	117	55	42	29	56	**271**	100	25	22	18	17	17	17	16	**8.00**	16	33	–	1	1	–	–	–	–	–	–	–	–	**Thiomorpholine, 4-dodecyl-**		**IC03536**
215	271	243	256	216	131	103	172	**271**	100	52	24	18	18	16	10	8		17	21	2	1	–	–	–	–	–	–	–	–	–	**1,8-Acridinedione, 3,3,6,6-tetramethyl-1,2,3,4,5,6,7,8-octahydro-**	27361-25-7	**NI27570**
57	197	171	215	128	59	271	170	**271**	100	65	62	48	39	32	18	17		17	21	2	1	–	–	–	–	–	–	–	–	–	**Carbamic acid, [1-(3-methyl-1H-inden-1-ylidene)ethyl]-, tert-butyl ester**	50585-35-8	**NI27571**
132	57	271	79	105	104	99	177	**271**	100	54	33	18	13	12	8	7		17	21	2	1	–	–	–	–	–	–	–	–	–	**2-Indoleacetic acid, cyclohexanemethyl ester**		**NI27572**
132	271	131	41	130	70	69	133	**271**	100	40	35	34	32	30	25	24		17	21	2	1	–	–	–	–	–	–	–	–	–	**1,3-Oxazino[3,4-a]quinolin-3-one, 4,4a,5,6-tetrahydro-1-isopropylidene-4,4-dimethyl-**		**LI05756**
72	128	100	29	271	44	127	115	**271**	100	63	40	27	26	24	19	17		17	21	2	1	–	–	–	–	–	–	–	–	–	**Propanamide, N,N-diethyl-2-(1-naphthalenyloxy)-**	15299-99-7	**NI27573**
105	70	77	202	41	42	168	81	**271**	100	70	60	43	27	20	17	10	**0.99**	17	21	2	1	–	–	–	–	–	–	–	–	–	**Spirocyclohexane-1,4′-[1,3-oxazin-6-one], 4′,5′-dihydro-5′,5′-dimethyl-2′-phenyl-**		**NI27574**
44	42	96	122	115	94	41	271	**271**	100	83	80	53	33	32	32	29		18	25	1	1	–	–	–	–	–	–	–	–	–	**Aporphine, hexahydro-3-methoxy-N-methyl-, (4bS,7aS,7bR,10aS)-**		**NI27575**
42	44	150	271	122	41	96	115	**271**	100	95	59	51	44	42	41	39		18	25	1	1	–	–	–	–	–	–	–	–	–	**Aporphine, hexahydro-3-methoxy-N-methyl-, (4bS,7aS,7bS,10aS)-**		**NI27576**
95	67	123	148	176	204	79	96	**271**	100	30	23	22	16	15	13	9	**5.00**	18	25	1	1	–	–	–	–	–	–	–	–	–	**(2-Cyanoethyl)norbornan-2-yl norbornan-2-yl ketone**		**NI27577**
45	44	43	226	150	271	41	55	**271**	100	90	78	49	34	33	32	30		18	25	1	1	–	–	–	–	–	–	–	–	–	**Morphinan, 3-methoxy-6-methyl-, (6α)-**	54889-81-5	**NI27578**
59	42	150	44	128	115	171	41	**271**	100	43	23	23	16	16	14	14	**12.40**	18	25	1	1	–	–	–	–	–	–	–	–	–	**Morphinan, 3-methoxy-17-methyl-, (9α,13α,14α)-**	125-71-3	**NI27579**
59	212	213	74	82	145	80	70	**271**	100	54	22	17	10	8	8	7	**0.00**	18	25	1	1	–	–	–	–	–	–	–	–	–	**Morphinan, 3-methoxy-17-methyl-, (9α,13α,14α)-**	125-71-3	**NI27580**
59	42	271	150	44	171	128	115	**271**	100	40	32	32	24	20	17	17		18	25	1	1	–	–	–	–	–	–	–	–	–	**Morphinan, 3-methoxy-17-methyl-, (9α,13α,14α)-**	125-71-3	**NI27581**
200	228	159	271	172	118	254	214	**271**	100	98	93	88	41	36	26	25		18	25	1	1	–	–	–	–	–	–	–	–	–	**Spiro[cyclohexane-1,3′(2′H)-isoquinolin]-1′-one, 1′,4′-dihydro-2′-butyl-**	110773-49-4	**DD00859**
271	51	106	242	108	122	121	94	**271**	100	46	42	37	33	32	29	28		19	13	1	1	–	–	–	–	–	–	–	–	–	**Benzonitrile, 2-hydroxy-4,6-diphenyl-**	55249-87-1	**DD00860**
226	271	255	254	227	224	113	225	**271**	100	96	85	68	42	29	26	22		19	13	1	1	–	–	–	–	–	–	–	–	–	**Benzo[c]phenanthrene-1-carboxamide**	56909-18-3	**NI27582**
127	155	271	77	139	270	272	140	**271**	100	94	85	33	24	21	18	17		19	13	1	1	–	–	–	–	–	–	–	–	–	**Isoxazole, 5-(1-naphthyl)-3-phenyl-**	1035-93-4	**NI27583**
127	155	271	77	139	270	140	128	**271**	100	94	85	33	24	21	17	12		19	13	1	1	–	–	–	–	–	–	–	–	–	**Isoxazole, 5-(1-naphthyl)-3-phenyl-**	1035-93-4	**LI05757**
28	256	271	157	41	172	152	173	**271**	100	71	42	34	34	30	23	22		19	29	–	1	–	–	–	–	–	–	–	–	–	**1H-Azepine, hexahydro-1-(1,2,3,4-tetrahydro-4,5,8-trimethyl-2-naphthyl)-**	23853-67-0	**NI27584**
269	267	254	41	270	157	141	268	**271**	100	49	46	29	25	23	19	18	**8.51**	19	29	–	1	–	–	–	–	–	–	–	–	–	**1H-Azepine, hexahydro-1-(1,2,3,4-tetrahydro-4,5,8-trimethyl-2-naphthyl)-**	23853-67-0	**WI01333**
69	165	127	196	146	203	184	50	**272**	100	70	47	34	21	16	13	13	**0.00**	1	–	–	–	–	7	–	–	1	–	–	–	–	**Tetrafluoro(trifluoromethyl)iodine**		**MS06100**
117	85	63	51	66	136	70	31	**272**	100	65	35	20	17	16	16	14	**0.00**	2	–	–	–	2	6	2	–	–	–	–	–	–	**1,3-Dithietane, 2,4-dichloro-1,1,2,3,3,4-hexafluoro-**		**DD00861**
143	220	64	131	271	51			**272**	100	27	15	13	9	5			**0.00**	3	2	–	–	–	4	–	2	–	–	–	–	–	**Propane, 1,2-dibromo-1,1,3,3-tetrafluoro-**		**LI05758**
195	193	69	113	111	131	129	31	**272**	100	99	32	27	27	10	10	10	**3.00**	3	2	–	–	–	4	–	2	–	–	–	–	–	**Propane, 1,2-dibromo-1,3,3,3-tetrafluoro-, erythro-**		**LI05759**
195	193	69	51	113	111	131	129	**272**	100	99	74	59	58	58	20	20	**16.00**	3	2	–	–	–	4	–	2	–	–	–	–	–	**Propane, 1,2-dibromo-1,3,3,3-tetrafluoro-, threo-**		**LI05760**
44	149	151	45	131	133	53	106	**272**	100	79	72	62	41	40	36	25	**2.90**	4	2	4	–	–	–	–	2	–	–	–	–	–	**2-Butenedioic acid, 2,3-dibromo-, (Z)-**	608-37-7	**NI27585**
229	43	227	231	28	135	228	133	**272**	100	83	72	67	58	49	31	29	**9.40**	5	13	–	–	–	–	–	1	–	–	–	–	1	**Dimethylisopropyltin bromide**		**MS06101**
59	136	272	45	207	131	274	150	**272**	100	23	22	10	9	8	6	6		6	8	–	–	6	–	–	–	–	–	–	–	–	**2,4,6,8,9,10-Hexathiaadamantane, 1,3-dimethyl-**	13787-68-3	**NI27586**
73	45	207	150	122	131	272	103	**272**	100	69	28	25	20	17	16	12		6	8	–	–	6	–	–	–	–	–	–	–	–	**2,4,6,8,9,10-Hexathiaadamantane, 1-ethyl-**	57274-64-3	**NI27587**
29	193	195	41	69	121	122	120	**272**	100	94	92	80	66	38	37	37	**0.71**	6	10	2	–	–	–	–	2	–	–	–	–	–	**Butanoic acid, 2,3-dibromo-, ethyl ester**	609-11-0	**NI27588**
59	167	135	105	91	165	137	121	**272**	100	61	58	28	28	25	23	22	**0.63**	6	20	6	–	–	–	–	–	–	3	–	–	–	**Trisilane, 1,1,1,3,3,3-hexamethoxy-**	53518-41-5	**NI27589**
165	167	118	130	274	230	272	202	**272**	100	95	58	56	52	45	43	43		7	–	3	–	–	–	4	–	–	–	–	–	–	**Carbonic acid, cyclic 3,4,5,6-tetrachloro-O-phenylene ester**	711-62-6	**NI27590**
272	145	127	273	125	95	75	253	**272**	100	33	11	8	8	8	8	7		7	4	–	–	–	3	–	–	1	–	–	–	–	**Benzene, 1-iodo-2-(trifluoromethyl)-**	444-29-1	**NI27591**
57	58	101	56	100	40	30	28	**272**	100	25	23	22	21	18	14	13	**0.00**	7	13	3	–	–	–	–	–	1	–	–	–	–	**2H-Pyran-4-ol, tetrahydro-2-(iodomethyl)-6-methoxy-**	56701-56-5	**NI27592**
77	51	145	69	75	95	78	57	**272**	100	16	5	5	4	3	3	3	**0.05**	8	5	–	–	–	9	–	–	–	–	–	–	–	**1,5-Octadiene, 3,3,4,4,7,7,8,8,8-nonafluoro-**	40723-75-9	**MS06102**
77	51	69	233					**272**	100	16	5	1					**0.00**	8	5	–	–	–	9	–	–	–	–	–	–	–	**1,5-Octadiene, 3,3,4,4,7,7,8,8,8-nonafluoro-**	40723-75-9	**LI05761**
257	272	181	89	255	137	105	109	**272**	100	94	72	52	47	42	42	36		8	6	–	–	–	5	–	–	–	–	–	–	1	**Arsine, dimethyl(pentafluorophenyl)-**	60575-47-5	**NI27593**
79	129	64	51	131	77	110	59	**272**	100	45	19	15	14	14	13	9	**0.02**	8	8	–	–	–	7	1	–	–	–	–	–	–	**Cyclobutane, 2-chloro-1,1,2-trifluoro-3-(1,1,2,2-tetrafluorobutyl)-, cis-**	35207-99-9	**MS06103**
79	129	64	131	110	77	51	116	**272**	100	52	16	15	15	14	12	8	**0.02**	8	8	–	–	–	7	1	–	–	–	–	–	–	**Cyclobutane, 2-chloro-1,1,2-trifluoro-3-(1,1,2,2-tetrafluorobutyl)-, trans-**	35207-98-8	**MS06104**
51	238	236	274	272	87	52	240	**272**	100	90	85	73	55	50	48	40		8	8	–	2	–	–	4	–	–	–	–	–	–	**Pyrazine, 2,3,5,6-tetrakis(chloromethyl)-**	123624-91-9	**DD00862**
97	79	41	81	80	69	43	55	**272**	100	91	47	42	31	31	24	21	**0.00**	8	16	6	–	2	–	–	–	–	–	–	–	–	**1,2-Cyclohexanediol, dimethanesulphonate**	55724-11-3	**NI27594**
43	186	97	65	228	153	142	129	**272**	100	31	22	20	19	14	13	11	**0.00**	8	17	4	–	2	–	–	–	–	–	1	–	–	**Phosphorodithioic acid, O,O-diethyl, S-(1-acetoxyethyl)-**		**MS06105**
43	186	97	65	228	153	142		**272**	100	31	22	20	19	14	13		**0.00**	8	17	4	–	2	–	–	–	–	–	1	–	–	**Phosphorodithioic acid, O,O-diethyl, S-(2-acetoxyethyl)-**		**MS06106**
57	139	41	137	196	29	194	135	**272**	100	62	48	48	33	32	31	30	**1.10**	8	18	–	–	–	2	–	–	–	–	–	–	1	**Stannane, dibutyldifluoro-**	563-25-7	**NI27595**
70	93	162	179	30	94	65	272	**272**	100	57	26	17	16	15	14	11		8	22	–	2	2	–	–	–	–	–	2	–	–	**1,4-Butanediamine, N,N′-bis(dimethylthiophosphinyl)-**		**MS06107**
225	223	160	159	63	125	161	162	**272**	100	92	58	58	50	48	40	38	**32.79**	9	8	1	–	–	–	4	–	–	–	–	–	–	**Benzene, 1-(chloromethyl)-3-methoxy-5-methyl-trichloro-**	74313-16-9	**NI27596**
147	73	200	77	272	199	148	45	**272**	100	97	52	43	18	18	18	17		9	9	–	–	1	5	–	–	–	1	–	–	–	**Silane, trimethyl[(pentafluorophenyl)thio]-**	22577-87-3	**NI27597**

MASS TO CHARGE RATIOS								M.W.	INTENSITIES								Parent	C	H	O	N	S	F	Cl	Br	I	Si	P	B	X	COMPOUND NAME	CAS Reg No	No
191	192	53	80	82	52	42	193	272	100	97	77	66	64	50	40	32	0.00	9	9	1	2	1	–	–	1	–	–	–	–	–	Thiazolo[3,2-a]pyridinium, 3-cyano-2,3-dihydro-8-hydroxy-5-methyl-, bromide	22989-59-9	NI27598
55	132	188	105	272	80	216	77	272	100	29	28	23	20	19	15	11		10	5	4	2	–	–	–	–	–	–	–	–	1	Manganese, tricarbonyl(diazoacetylcyclopentadienyl)-		LI05762
140	26	215	55	121	75	122	56	272	100	35	25	22	20	20	18	18	12.00	10	7	2	–	–	3	–	–	–	–	–	–	1	Iron, dicarbonyl(η^5-2,4-cyclopentadien-1-yl)(3,3,3-trifluoro-1-propenyl)-	56271-18-2	NI27599
192	82	80	128	127	81	79	77	272	100	74	74	49	34	28	28	16	0.00	10	9	–	–	2	–	–	1	–	–	–	–	–	1,2-Dithiol-1-ium, 3-methyl-5-phenyl-, bromide	55836-87-8	NI27600
257	255	253	227	108	254	225	256	272	100	77	73	38	30	27	24	21	7.10	10	16	1	–	–	–	–	–	–	–	–	–	1	Tin, (3-methoxyphenyl)trimethyl-		MS06108
257	255	108	253	256	254	225	78	272	100	76	52	46	34	28	24	23	6.30	10	16	1	–	–	–	–	–	–	–	–	–	1	Tin, (4-methoxyphenyl)trimethyl-		MS06109
272	274	120	148	244	246	226	224	272	100	97	61	50	31	31	30	27		11	10	2	–	–	1	–	1	–	–	–	–	–	2-Propenoic acid, 2-fluoro-3-(4-bromophenyl)-, ethyl ester, (E)-		NI27601
43	55	41	84	105	77	99	169	272	100	84	58	54	51	50	48	45	3.00	11	12	3	–	–	–	–	–	–	–	–	–	1	2(3H)-Furanone, 4,5-dihydro-5-(hydroxymethyl)-3-(phenylseleno)-		NI27602
169	140	127	129	126	139	236	170	272	100	98	71	35	33	31	27	27	3.60	11	16	6	2	–	–	–	–	–	–	–	–	–	D-Mannitol, 3,6-anhydro-1-deoxy-1-(3,4-dihydro-5-methyl-1(2H)-pyrimidinyl)-	35823-83-7	NI27603
43	111	124	44	28	110	42	29	272	100	61	21	19	16	15	8	8	0.00	11	17	7	–	–	–	–	–	–	–	–	1	–	β-D-Ribopyranose, cyclic 2,4-(ethylboronate) 1,3-diacetate	74779-74-1	NI27604
202	272	236	274	238	173	204	237	272	100	50	49	48	31	31	26	17		12	7	1	–	–	–	3	–	–	–	–	–	–	[1,1′-Biphenyl]-2-ol, 2′,5,5′-trichloro-	59852-81-2	NI27605
272	274	173	87	86	276	137	138	272	100	95	77	34	32	32	30	26		12	7	1	–	–	–	3	–	–	–	–	–	–	[1,1′-Biphenyl]-4-ol, 2,2′,5′-trichloro-	53905-33-2	NI27606
272	274	276	202	173	139	273	275	272	100	98	30	30	18	18	11	10		12	7	1	–	–	–	3	–	–	–	–	–	–	[1,1′-Biphenyl]-4-ol, 2′,4′,6′-trichloro-	14962-28-8	NI27607
272	274	173	276	273	275	201	175	272	100	97	34	32	14	13	11	11		12	7	1	–	–	–	3	–	–	–	–	–	–	[1,1′-Biphenyl]-4-ol, 3,4′,5-trichloro-	4400-06-0	NI27608
103	77	51	104	188	78	132	50	272	100	90	88	68	62	58	52	51	8.00	12	8	4	–	–	–	–	–	–	–	–	–	1	Iron, tricarbonyl(1-formyl-1,3,5,7-cyclooctatetraene)-	11064-31-6	MS06110
150	39	64	51	92	50	63	272	272	100	34	31	24	23	20	19	18		12	8	4	4	–	–	–	–	–	–	–	–	–	Azobenzene, 2,2-dinitro-	4171-33-9	NI27609
122	76	150	75	272	50	92	64	272	100	67	60	58	48	36	27	19		12	8	4	4	–	–	–	–	–	–	–	–	–	Azobenzene, 4,4′-dinitro-	3646-57-9	MS06111
122	150	76	75	272	92	50	64	272	100	50	46	45	37	27	22	11		12	8	4	4	–	–	–	–	–	–	–	–	–	Azobenzene, 4,4′-dinitro-	3646-57-9	NI27610
270	105	166	77	102	76	177	127	272	100	70	54	45	42	40	31	30	1.00	12	8	4	4	–	–	–	–	–	–	–	–	–	Phenazine, 5,10-dihydro-1,3-dinitro-	18450-20-9	NI27611
28	52	160	106	54	67	39	80	272	100	80	54	25	19	17	13	11	5.92	12	12	4	–	–	–	–	–	–	–	–	–	1	Chromium, tetracarbonyl[(1,2,5,6-η)-1,5-cyclooctadiene]-	12301-34-7	NI27612
139	197	111	85	141	199	75	230	272	100	84	55	48	40	33	26	25	11.00	12	13	1	–	2	–	1	–	–	–	–	–	–	2-Propene(dithioic) acid, 3-(4-chlorophenyl)-3-hydroxy-, propyl ester	56298-83-0	NI27613
91	59	65	39	92	101	55	41	272	100	61	23	17	16	14	14	14	12.00	12	16	2	–	–	–	–	–	–	–	–	–	1	Butanoic acid, 3-(benzylselenyl)-3-methyl-	1599-29-7	LI05763
91	59	65	39	92	100	55	41	272	100	30	11	9	8	7	7	7	5.67	12	16	2	–	–	–	–	–	–	–	–	–	1	Butanoic acid, 3-(benzylselenyl)-3-methyl-	1599-29-7	NI27614
110	97	139	81	152	73	60	86	272	100	99	90	40	36	33	32	27	0.00	12	16	7	–	–	–	–	–	–	–	–	–	–	D-Arabino-hex-1-enitol, 1,5-anhydro-2-deoxy-, triacetate	2873-29-2	MS06112
153	213	111	95	154	214	83	97	272	100	54	26	21	10	6	6	5	0.00	12	16	7	–	–	–	–	–	–	–	–	–	–	D-Arabino-hex-1-enitol, 1,5-anhydro-2-deoxy-, triacetate	2873-29-2	NI27615
43	180	153	222	212	42	121	211	272	100	32	30	19	16	11	9	8	2.30	12	16	7	–	–	–	–	–	–	–	–	–	–	Furo[3,2-b]furan-3,6-dicarboxylic acid, 3a,5,6,6a-tetrahydro-5-hydroxy-2,5-dimethyl-, dimethyl ester		MS06113
67	94	82	41	65	44	66	55	272	100	96	81	77	50	46	42	40	26.92	12	17	3	–	1	–	–	–	–	–	1	–	–	1,3,2-Dioxaphosphorinane, 4,4,6-trimethyl-2-phenoxy-, 2-sulphide	33719-81-2	MS06114
58	272	43	69	59	244	87	126	272	100	98	89	53	40	20	17	16		12	20	3	2	1	–	–	–	–	–	–	–	–	4H-1,3-Thiazine-4-one, tetrahydro-2-isopropylimino-3-isopropyl-6-(methoxycarbonyl)-		MS06115
57	72	115	58	113	56	114	127	272	100	89	82	78	77	67	47	32	2.06	12	24	3	4	–	–	–	–	–	–	–	–	–	1-Oxa-4,6,9,11-tetraazacyclotridecane-5,10-dione, 4,6,9,11-tetramethyl-		NI27616
105	77	51	272	117	50	167	106	272	100	61	24	14	11	9	8	6		13	5	1	–	–	5	–	–	–	–	–	–	–	Benzophenone, 2,3,4,5,6-pentafluoro-	1536-23-8	NI27617
105	77	93	51	179	75	149	103	272	100	92	37	18	16	13	10	7	2.60	13	8	5	2	–	–	–	–	–	–	–	–	–	Benzophenone, 2,4-dinitro-		NI27618
272	274	193	166	63	273	194	90	272	100	99	29	22	16	13	13	13		13	9	–	2	–	–	–	1	–	–	–	–	–	1H-Pyrrolo[2,3-b]pyridine, 3-bromo-2-phenyl-	23616-58-2	NI27619
272	81	82	153	77	192	96	135	272	100	68	44	31	29	20	18	14		13	12	1	4	1	–	–	–	–	–	–	–	–	2-Imidazolidinethione, 5-(4-imidazolylmethyl)-4-oxo-3-phenyl-		LI05764
272	122	152	56	177	121	120	150	272	100	92	54	53	44	22	22	14		13	12	3	–	–	–	–	–	–	–	–	–	1	Ferrocenecarboxylic acid, 2-formyl-, methyl ester		NI27620
272	122	56	121	152	177	120	207	272	100	79	59	52	43	38	35	21		13	12	3	–	–	–	–	–	–	–	–	–	1	Ferrocenecarboxylic acid, 3-formyl-, methyl ester		NI27621
272	145	143	273	157	146	255	186	272	100	65	17	16	16	11	10	8		13	12	3	4	–	–	–	–	–	–	–	–	–	Benzimidazolium, 1,2-dimethyl-3-(1,2,3,4-tetrahydro-6-hydroxy-2,4-dioxo-5-pyrimidinyl)-, hydroxide, inner salt	27355-90-4	NI27622
43	57	56	258	69	244	40	215	272	100	97	80	57	54	38	34	31	11.77	13	12	3	4	–	–	–	–	–	–	–	–	–	Benzo[g]pteridine-2,4(3H,10H)-dione, 10-ethyl-8-hydroxy-7-methyl-	56196-96-4	NI27623
93	91	89	87	135	41	117	65	272	100	79	52	52	47	42	22	22	0.00	13	20	4	–	1	–	–	–	–	–	–	–	–	Cyclohexanesulphonic acid, 2,2-diallyl-3-oxo-		LI05765
189	95	161	59	55	41	129	190	272	100	33	13	13	12	12	9	8	0.00	13	20	6	–	–	–	–	–	–	–	–	–	–	1,3-Dioxolane-4,5-dicarboxylic acid, 2-cyclohexyl-, dimethyl ester, (4R,5R)-		DD00863
142	272	181	88	139	195	241	141	272	100	97	95	91	86	82	46	37		13	20	6	–	–	–	–	–	–	–	–	–	–	3-Oxabicyclo[4.3.0]nonane-2-carboxylic acid, 8,8-dimethoxy-2-methyl-4-oxo-, methyl ester, (1α,2β,6α)-		MS06116
43	94	136	95	81	152	79	170	272	100	47	23	21	19	16	13	12	0.00	13	20	6	–	–	–	–	–	–	–	–	–	–	Propanoic acid, 3-(3,4-diacetoxycyclohexyl)-		NI27624
72	114	44	115	230	31	30	56	272	100	79	20	18	13	12	12	11	0.15	13	24	4	2	–	–	–	–	–	–	–	–	–	Glycine, N-(N-acetylvalyl)-, butyl ester		AI01917
43	112	70	130	30	229	143	84	272	100	90	88	71	47	41	33	26	5.00	13	24	4	2	–	–	–	–	–	–	–	–	–	Putrescine, N,N′-diacetyl-N-(α-acetoxypropyl)-		LI05766
72	73	30	171	100	43	41	29	272	100	24	21	13	13	11	6	6	0.79	13	24	4	2	–	–	–	–	–	–	–	–	–	L-Valine, N-(N-acetylglycyl)-, butyl ester	55712-40-8	NI27625
73	75	185	147	272	45	81	55	272	100	42	30	27	24	18	14	11		13	28	2	–	–	–	–	–	–	2	–	–	–	Cyclohexane, bis(trimethylsilyloxy)methylene-		MS06117
136	108	272	82	228	64	76	244	272	100	93	56	28	26	18	16	14		14	8	2	–	2	–	–	–	–	–	–	–	–	Thiosalicylic acid, anhydride, bis-		MS06118
272	273	108	136	170	77	244	113	272	100	13	10	8	6	6	5	5		14	8	6	–	–	–	–	–	–	–	–	–	–	9,10-Anthracenedione, 1,2,5,8-tetrahydroxy-	81-61-8	NI27626
272	273	136	274	271	243	142	108	272	100	15	10	3	2	2	2	2		14	8	6	–	–	–	–	–	–	–	–	–	–	9,10-Anthracenedione, 1,4,5,8-tetrahydroxy-		IC03537
272	274	244	216	203	188	273	245	272	100	99	95	68	60	53	50	50		14	8	6	–	–	–	–	–	–	–	–	–	–	4H-Naphtho[2,3-b]pyran-4,6,9-trione, 5,8-dihydroxy-2-methyl-	62681-86-1	NI27627

MASS TO CHARGE RATIOS								M.W.	INTENSITIES								Parent	C	H	O	N	S	F	Cl	Br	I	Si	P	B	X	COMPOUND NAME	CAS Reg No	No
272	274	273	153	154	181	126	152	**272**	100	35	20	20	19	16	16	15		14	9	2	2	–	–	1	–	–	–	–	–	–	9,10-Anthracenedione, 1,4-diaminochloro-		IC03538
139	111	75	140	113	51	77	50	**272**	100	57	23	20	20	20	19	14	**11.00**	14	9	2	2	–	–	1	–	–	–	–	–	–	1,3,4-Oxadiazolium, 3-(4-chlorophenyl)-5-hydroxy-2-phenyl-, hydroxide, inner salt	24660-42-2	NI27628
77	107	272	151	94	108	78	165	**272**	100	72	51	43	33	20	19	2		14	12	4	2	–	–	–	–	–	–	–	–	–	Acetamide, N-(3-nitrophenyl)-2-phenoxy-		MS06119
77	107	272	151	94	108	78	165	**272**	100	83	53	40	26	22	18	3		14	12	4	2	–	–	–	–	–	–	–	–	–	Acetamide, N-(4-nitrophenyl)-2-phenoxy-		MS06120
150	122	242	77	151	65	107	44	**272**	100	22	20	12	10	9	7	7	**0.00**	14	12	4	2	–	–	–	–	–	–	–	–	–	Benzamide, 4-methoxy-3-nitro-N-phenyl-	97-32-5	NI27629
153	102	77	44	103	197	228	28	**272**	100	90	66	65	62	60	53	36	**17.00**	14	12	4	2	–	–	–	–	–	–	–	–	–	1H-Imidazole-4,5-dicarboxylic acid, 1-styryl-, 4-methyl ester, (E)-	56382-59-3	NI27630
77	45	272	108	80	51	50	81	**272**	100	76	74	71	71	71	56	50		14	12	4	2	–	–	–	–	–	–	–	–	–	8H-Imidazo[5,1-c][1,4]oxazine-1-carboxylic acid, 5,6-dihydro-8-oxo-6-phenyl-, methyl ester	56382-68-4	NI27631
272	244	273	159	170	158	76	103	**272**	100	32	19	17	16	16	14	11		14	12	4	2	–	–	–	–	–	–	–	–	–	Phenazine, 1,2,3,4-tetrahydroxy-7,8-dimethyl-		MS06121
272	273	134	136	135	56	166	270	**272**	100	18	18	16	8	8	7	6		14	16	2	–	–	–	–	–	–	–	–	–	1	Ferrocene, α-methoxycarbonyl-1,1-dimethyl-		MS06122
272	273	134	136	135	166	56	270	**272**	100	18	18	16	9	7	7	5		14	16	2	–	–	–	–	–	–	–	–	–	1	Ferrocene, β-methoxycarbonyl-1,1-dimethyl-		MS06123
117	272	236	115	116	138	273	118	**272**	100	72	34	24	18	17	12	10		14	16	2	–	–	–	–	–	–	–	–	–	1	Iron, π-cyclopentadienyl(4,5,6,7-tetrahydro-4,7-dihydroxy-5-phenyl-π-indenyl)-, endo,endo-	12127-08-1	NI27632
117	272	236	115	116	138	254	273	**272**	100	95	60	36	31	27	23	16		14	16	2	–	–	–	–	–	–	–	–	–	1	Iron, π-cyclopentadienyl(4,5,6,7-tetrahydro-4,7-dihydroxy-5-phenyl-π-indenyl)-, endo,exo-	12127-09-2	NI27633
272	91	119	273	120	92			**272**	100	80	70	16	8	6				14	16	2	4	–	–	–	–	–	–	–	–	–	Piperidine, 1-[(4,5-dihydro-5-oxo-1-phenyl-1H-1,2,4-triazol-3-yl)carbonyl]-	32589-64-3	NI27634
216	201	218	272	203	274	181	166	**272**	100	83	43	30	24	11	6	4		14	22	1	–	–	–	1	–	–	–	1	–	–	4-tert-Butylphenyl-tert-butylphosphinyl chloride		LI05767
121	43	60	120	48	47	91	164	**272**	100	88	78	44	42	31	28	5	**2.00**	14	24	1	–	2	–	–	–	–	–	–	–	–	3-Cyclohexen-1-ol, 2,2-dipropyl-, xanthate		LI05768
143	111	116	83	101	199	167	223	**272**	100	69	33	32	13	8	4	3	**0.00**	14	24	5	–	–	–	–	–	–	–	–	–	–	Dodecanedioic acid, 3,6-epoxy-, dimethyl ester		NI27635
257	109	72	151	197	43	127	169	**272**	100	53	41	32	28	26	23	22	**0.00**	14	24	5	–	–	–	–	–	–	–	–	–	–	Nonanoic acid, 8,9-di-O-(isopropylidenedioxy)-3-oxo-, ethyl ester, (S)-		DD00864
141	75	157	73	156	91	99	142	**272**	100	41	28	27	20	20	18	13	**1.00**	14	28	3	–	–	–	–	–	–	1	–	–	–	1-Butene, 3-(tert-butylacetoxy)-3-methyl-2-(trimethylsiloxy)-		NI27636
73	191	59	117	103	257	89	227	**272**	100	34	21	14	14	12	5	4	**0.00**	14	28	3	–	–	–	–	–	–	1	–	–	–	1,5-Hexadiene, 1-[dimethyl[2-(2-ethoxyethoxy)ethoxy]silyl]-, (E)-		DD00865
244	103	242	105	101	240	243	41	**272**	100	96	79	79	78	58	31	31	**0.00**	14	30	–	–	–	–	–	–	–	–	–	–	1	Germacycloundecane, 1,1-diethyl-	56438-29-0	NI27637
272	241	271	165	135	107	77	92	**272**	100	79	50	45	32	16	14	13		15	12	5	–	–	–	–	–	–	–	–	–	–	Benzoic acid, 4-hydroxy-3-(3-methoxybenzoyl)-		NI27638
149	123	77	79	105	151	227	65	**272**	100	88	72	64	56	44	36	29	**0.00**	15	12	5	–	–	–	–	–	–	–	–	–	–	Benzoic acid, 4,4′-(hydroxymethylene)[bis-		NI27639
77	197	65	121	141	94	51	63	**272**	100	70	49	44	39	38	36	30	**11.00**	15	12	5	–	–	–	–	–	–	–	–	–	–	Benzoic acid, 3-[(phenoxycarbonyl)oxy]-, methyl ester	54644-59-6	NI27640
240	137	213	239	198	272	212	77	**272**	100	99	70	64	60	52	41	40		15	12	5	–	–	–	–	–	–	–	–	–	–	Benzoic acid, β-resorcyloyl-, methyl ester	21147-36-4	NI27641
239	149	163	65	195	240	194	165	**272**	100	24	21	20	19	16	11	11	**0.00**	15	12	5	–	–	–	–	–	–	–	–	–	–	Benzophenone, 3,4′-dicarboxy-6-methyl-		NI27642
197	77	272	65	51	141	198	121	**272**	100	60	38	22	22	20	16	16		15	12	5	–	–	–	–	–	–	–	–	–	–	Carbonic acid, phenyl ester, ester with methyl 4-hydroxybenzoate	17175-12-1	NI27643
272	257	173	201	44	127	115	145	**272**	100	29	29	28	21	19	17	13		15	12	5	–	–	–	–	–	–	–	–	–	–	6H-Dibenzo[b,d]pyran-6-one, 3,7-dihydroxy-4-methoxy-1-methyl-		LI05769
272	243	115	69	201	199	77	63	**272**	100	31	26	26	21	18	17	16		15	12	5	–	–	–	–	–	–	–	–	–	–	6H-Dibenzo[b,d]pyran-6-one, 3,7-dihydroxy-9-methoxy-1-methyl-	23452-05-3	NI27644
272	69	273	55	57	201	149	77	**272**	100	19	18	16	15	14	12	11		15	12	5	–	–	–	–	–	–	–	–	–	–	6H-Dibenzo[b,d]pyran-6-one, 3,7-dihydroxy-9-methoxy-1-methyl-	23452-05-3	NI27645
272	28	273	187	229	274	243	115	**272**	100	41	18	11	4	3	2	2		15	12	5	–	–	–	–	–	–	–	–	–	–	6H-Dibenzo[b,d]pyran-6-one, 7,9-dihydroxy-3-methoxy-1-methyl-	56771-85-8	MS06124
153	120	272	271	179	166	107	152	**272**	100	71	63	29	26	25	21	20		15	12	5	–	–	–	–	–	–	–	–	–	–	Flavanone, 4′,5,7-trihydroxy-		MS06125
272	230	273	202	201	115	187	159	**272**	100	53	15	6	5	5	4	4		15	12	5	–	–	–	–	–	–	–	–	–	–	1H-Naphtho[2,3-c]pyran-1-one, 3,4-dihydro-9,10-dihydroxy-7-methoxy-3-methylene	80503-54-4	NI27646
272	243	273	242	229	244	136	43	**272**	100	31	17	12	12	4	4	4		15	12	5	–	–	–	–	–	–	–	–	–	–	4H-Naphtho[2,3-b]pyran-4-one, 5,6-dihydroxy-8-methoxy-2-methyl-		MS06126
272	243	273	229	242	189	271	244	**272**	100	32	16	13	12	5	4	4		15	12	5	–	–	–	–	–	–	–	–	–	–	4H-Naphtho[2,3-b]pyran-4-one, 5,6-dihydroxy-8-methoxy-2-methyl-		LI05770
43	76	126	97	75	63	272	77	**272**	100	46	42	40	29	19	18	15		15	12	5	–	–	–	–	–	–	–	–	–	–	Spiro[3-acetoxy-2,3-dihydrofuran-2,2′-(4′-hexa-2″,4″-diynylidene-3′,6′-dioxa-bicyclo[3.1.0]hexane)]		LI05771
272	273	229	243	124	271	69	77	**272**	100	19	13	12	10	7	7	4		15	12	5	–	–	–	–	–	–	–	–	–	–	9H-Xanthen-9-one, 1,3-dihydroxy-6-methoxy-8-methyl-	22938-77-8	LI05772
272	273	229	243	124	271	69	77	**272**	100	16	13	11	9	8	7	4		15	12	5	–	–	–	–	–	–	–	–	–	–	9H-Xanthen-9-one, 1,3-dihydroxy-6-methoxy-8-methyl-	22938-77-8	NI27647
272	243	273	242	271	229	69	77	**272**	100	57	18	15	11	11	9	8		15	12	5	–	–	–	–	–	–	–	–	–	–	9H-Xanthen-9-one, 1,6-dihydroxy-3-methoxy-8-methyl-	3569-83-3	LI05773
272	243	273	242	271	229	244	69	**272**	100	57	17	15	11	10	9	9		15	12	5	–	–	–	–	–	–	–	–	–	–	9H-Xanthen-9-one, 1,6-dihydroxy-3-methoxy-8-methyl-	3569-83-3	NI27648
257	272	229	186	243	141	121	77	**272**	100	98	75	51	48	43	33	31		15	12	5	–	–	–	–	–	–	–	–	–	–	9H-Xanthen-9-one, 1-hydroxy-2,3-dimethoxy-		NI27649
242	227	272	199	171	257	213		**272**	100	63	56	25	22	16	16			15	12	5	–	–	–	–	–	–	–	–	–	–	9H-Xanthen-9-one, 3-hydroxy-2,4-dimethoxy-		LI05774
139	132	131	140	41	231	40	270	**272**	100	61	43	39	20	14	11	4	**2.30**	15	13	1	2	–	–	1	–	–	–	–	–	–	1,2,4-Oxadiazole, 3-methyl-4-phenyl-5-(4-chlorophenyl)-		LI05775
118	243	104	77	129	231	130	128	**272**	100	99	53	44	39	36	35	33	**11.21**	15	16	3	2	–	–	–	–	–	–	–	–	–	Barbituric acid, 5-allyl-1,3-dimethyl-5-phenyl-	28239-53-4	NI27650
272	243	118	104	77	129	231	130	**272**	100	89	89	47	40	35	33	31		15	16	3	2	–	–	–	–	–	–	–	–	–	Barbituric acid, 5-allyl-1,3-dimethyl-5-phenyl-	28239-53-4	NI27651
131	272	130	132	226	273	199	77	**272**	100	60	58	21	11	10	9	7		15	16	3	2	–	–	–	–	–	–	–	–	–	1H-Pyrimido[1,2-a]quinoline-2-carboxylic acid, 4,4a,5,6-tetrahydro-1-oxo-, ethyl ester	63658-84-4	NI27652
225	182	258	226	154	167	257	183	**272**	100	39	33	31	22	21	20	20	**0.00**	15	16	3	2	–	–	–	–	–	–	–	–	–	Urea, N-methoxy-N′-(3′-methoxy[1,1′-biphenyl]-2-yl)-	56666-85-4	NI27653
225	257	154	210	226	182	31	127	**272**	100	34	31	27	25	13	13	12	**0.00**	15	16	3	2	–	–	–	–	–	–	–	–	–	Urea, N-methoxy-N′-(4′-methoxy[1,1′-biphenyl]-2-yl)-	56667-14-2	NI27654
138	272	56	121	41	186	73	65	**272**	100	94	90	67	42	42	34	27		15	20	1	–	–	–	–	–	–	–	–	–	1	(1-Hydroxypentyl)ferrocene		NI27655

MASS TO CHARGE RATIOS								M.W.	INTENSITIES								Parent	C	H	O	N	S	F	Cl	Br	I	Si	P	B	X	COMPOUND NAME	CAS Reg No	No
81	124	69	93	55	163	109	41	272	100	57	46	41	39	37	37	26	0.00	15	25	2	–	–	–	1	–	–	–	–	–	–	Acetic acid, chloro-, (±)-1,2α,3,4,4aα,5,6,7,8,8aβ-decahydro-5,5,8aβ-trimethyl-2β-naphthalenyl ester		DD00866
187	87	155	99	143	188	156	55	272	100	97	95	40	34	18	17	17	0.00	15	28	4	–	–	–	–	–	–	–	–	–	–	1,3-Dioxolane, 2-hexyl-2′-methyl-2,2′-ethylenebis-	960-23-6	NI27656
125	143	101	199	41	55	56	69	272	100	99	79	75	68	62	59	43	0.00	15	28	4	–	–	–	–	–	–	–	–	–	–	Heptanedioic acid, dibutyl ester	51238-94-9	NI27657
143	41	57	55	69	73	101	56	272	100	66	56	54	40	35	33	33	0.00	15	28	4	–	–	–	–	–	–	–	–	–	–	Heptanedioic acid, di-sec-butyl ester	57983-34-3	NI27658
199	153	69	227	195	83	96	111	272	100	47	37	28	28	28	24	23	0.00	15	28	4	–	–	–	–	–	–	–	–	–	–	Hexanedioic acid, 3-hexyl-, ethyl methyl ester		NI27659
199	143	125	101	157	41	170	69	272	100	51	42	23	20	17	16	16	0.00	15	28	4	–	–	–	–	–	–	–	–	–	–	Hexanedioic acid, 3-methyl-, dibutyl ester	56051-58-2	NI27660
143	57	41	101	125	69	199	55	272	100	31	30	26	25	25	23	20	0.00	15	28	4	–	–	–	–	–	–	–	–	–	–	Hexanedioic acid, 3-methyl-, di-sec-butyl ester	57983-09-2	NI27661
143	41	57	69	125	101	55	73	272	100	34	30	27	24	23	22	21	0.00	15	28	4	–	–	–	–	–	–	–	–	–	–	Hexanedioic acid, 3-methyl-, di-sec-butyl ester	57983-09-2	NI27662
173	216	227	199	157	127	272		272	100	74	32	32	21	15	3			15	28	4	–	–	–	–	–	–	–	–	–	–	Malonic acid, dibutyl-, diethyl ester	596-75-8	LI05776
173	227	157	127	142	169	160	125	272	100	33	20	15	10	7	7	7	3.01	15	28	4	–	–	–	–	–	–	–	–	–	–	Malonic acid, dibutyl-, diethyl ester	596-75-8	NI27663
133	185	43	55	29	41	132	42	272	100	76	75	59	58	49	32	29	0.16	15	28	4	–	–	–	–	–	–	–	–	–	–	Malonic acid, pentyl-, ethyl 2-pentyl ester		IC03539
115	199	97	143	171	144	41	69	272	100	76	34	29	20	20	18	17	0.00	15	28	4	–	–	–	–	–	–	–	–	–	–	Pentanedioic acid, 2,2-dimethyl-, dibutyl ester	56051-61-7	NI27664
143	199	101	170	157	41	56	114	272	100	99	90	78	78	50	43	42	0.77	15	28	4	–	–	–	–	–	–	–	–	–	–	Pentanedioic acid, 3,3-dimethyl-, dibutyl ester	56051-59-3	NI27665
115	143	57	97	41	69	56	116	272	100	90	49	43	38	28	23	18	0.00	15	28	4	–	–	–	–	–	–	–	–	–	–	Pentanedioic acid, 2,2-dimethyl-, di-sec-butyl ester	57983-30-9	NI27666
143	57	41	56	115	74	69	142	272	100	76	67	52	40	35	31	30	0.00	15	28	4	–	–	–	–	–	–	–	–	–	–	Pentanedioic acid, 2,4-dimethyl-, di-sec-butyl ester	57983-50-3	NI27667
143	101	142	57	41	83	56	43	272	100	43	40	38	34	30	29	23	0.40	15	28	4	–	–	–	–	–	–	–	–	–	–	Pentanedioic acid, 3,3-dimethyl-, di-sec-butyl ester	57983-32-1	NI27668
98	74	55	84	41	69	199	241	272	100	67	61	48	46	41	38	36	0.00	15	28	4	–	–	–	–	–	–	–	–	–	–	Tridecanedioic acid, dimethyl ester	1472-87-3	LI05777
98	74	55	84	41	43	69	83	272	100	89	72	64	54	46	42	39	0.00	15	28	4	–	–	–	–	–	–	–	–	–	–	Tridecanedioic acid, dimethyl ester	1472-87-3	NI27669
98	199	84	241	112	74	167	126	272	100	48	41	32	32	32	21	16	0.60	15	28	4	–	–	–	–	–	–	–	–	–	–	Tridecanedioic acid, dimethyl ester	1472-87-3	NI27670
88	69	55	41	126	97	185	59	272	100	48	44	35	27	27	25	25	0.00	15	28	4	–	–	–	–	–	–	–	–	–	–	Undecanedioic acid, 2,4-dimethyl-, dimethyl ester, (2S,4S)-	85547-48-4	NI27671
104	75	207	73	222	91	105	45	272	100	88	53	46	27	25	12	10	0.00	15	32	2	–	–	–	–	–	–	1	–	–	–	Dodecanoic acid, trimethylsilyl ester	55520-95-1	NI27674
73	75	117	257	43	132	41	129	272	100	83	52	51	42	39	35	25	4.20	15	32	2	–	–	–	–	–	–	1	–	–	–	Dodecanoic acid, trimethylsilyl ester	55520-95-1	NI27673
73	75	117	257	132	55	129	41	272	100	78	63	39	30	22	21	21	2.43	15	32	2	–	–	–	–	–	–	1	–	–	–	Dodecanoic acid, trimethylsilyl ester	55520-95-1	NI27672
126	178	185	170	72	128	127	272	272	100	71	51	48	48	44	43	39		15	32	2	2	–	–	–	–	–	–	–	–	–	L-Lysine, N[2],N[6]-diisopropyl-, isopropyl ester	31552-12-2	MS06127
127	87	115	59	99	243	128	71	272	100	75	74	47	32	19	15	11	0.00	15	36	–	–	–	–	–	–	–	2	–	–	–	Propane, 1,1-bis(triethylsilyl)-		MS06128
272	271	270	135	128	273	269	135.5	272	100	87	44	19	19	18	15	15		16	14	1	2	–	–	–	–	–	–	–	2	–	2-Bora-1-azanaphthalene, 2,2′-bis[oxy-		LI05778
180	167	193	207	129	181	272	165	272	100	49	42	25	17	16	10	10		16	16	2	–	1	–	–	–	–	–	–	–	–	Thiophene, tetrahydro-3,3-diphenyl-, 1,1-dioxide		LI05779
272	257	77	115	258	239	79	243	272	100	51	44	30	29	28	28	27		16	16	4	–	–	–	–	–	–	–	–	–	–	9,10-Anthracenedione, 1,2,3,4-tetrahydro-5,8-dimethoxy-		NI27675
107	272	108	134	240	166	165	147	272	100	63	52	50	34	22	20	9		16	16	4	–	–	–	–	–	–	–	–	–	–	Benzoic acid, 2,3-dihydroxy-6-(2-phenylethyl)-, methyl ester	75112-82-2	NI27676
91	272	240	65	92	241	121	211	272	100	97	61	34	32	18	12	11		16	16	4	–	–	–	–	–	–	–	–	–	–	Benzoic acid, 2-hydroxy-6-methyl-4-benzyloxy-, methyl ester	22375-05-9	NI27677
135	92	77	107	64	136	63	270	272	100	21	20	14	11	9	7	6	0.00	16	16	4	–	–	–	–	–	–	–	–	–	–	Benzoin, 4,4′-dimethoxy-	119-52-8	NI27678
273	134	135	181	274	271	77	180	272	100	35	26	21	17	14	12	11	0.00	16	16	4	–	–	–	–	–	–	–	–	–	–	Benzophenone, 2,4′-dihydroxy-4,6-dimethoxy-2′-methyl-		LI05780
150	272	137	138	135	273	151	123	272	100	66	40	19	15	12	12	11		16	16	4	–	–	–	–	–	–	–	–	–	–	2H-1-Benzopyran-7-ol, 3,4-dihydro-3-(2-hydroxy-4-methoxyphenyl)-	56701-24-7	MS06129
150	272	137	149	107	138	121	135	272	100	33	31	22	20	18	18	16		16	16	4	–	–	–	–	–	–	–	–	–	–	2H-1-Benzopyran-7-ol, 3,4-dihydro-3-(4-hydroxy-2-methoxyphenyl)-	75172-32-6	NI27679
105	82	77	43	125	124	167	106	272	100	59	52	48	44	44	8	8	0.20	16	16	4	–	–	–	–	–	–	–	–	–	–	Bicyclo[2.2.1]hept-2-ene, 1-acetoxy-6-benzoyloxy-		LI05781
241	272	43	157	215	226	242	185	272	100	65	36	26	25	19	18	17		16	16	4	–	–	–	–	–	–	–	–	–	–	1,3-Butanedione, 1-(3,4-dimethoxy-2-naphthalenyl)-	55255-49-7	NI27680
212	240	169	43	213	115	211	197	272	100	26	25	23	21	19	17	16	15.00	16	16	4	–	–	–	–	–	–	–	–	–	–	1,3-Cyclohexadiene-1-carboxylic acid, 6-(2-hydroxyphenyl)-4,6-dimethyl-5-oxo-, methyl ester	40801-43-2	NI27681
159	77	43	103	202	229	272	257	272	100	70	70	50	40	30	1	1		16	16	4	–	–	–	–	–	–	–	–	–	–	1,4-Naphthoquinone, 2-methyl-3-(3-hydroperoxy-3-methyl-1-butenyl)-, trans-		LI05782
213	272	119	127	126	155	117	111	272	100	44	26	20	18	17	17	16		16	16	4	–	–	–	–	–	–	–	–	–	–	2-Propenoic acid, 3-(2,7-dimethoxy-1-naphthyl)-, methyl ester	89229-15-2	NI27682
93	118	119	65	136	38	106	63	272	100	92	61	27	22	15	10	9	0.00	16	16	4	–	–	–	–	–	–	–	–	–	–	4-Tolualdehyde, diperoxide		LI05783
132	272	257	254	117	220	130	255	272	100	69	60	28	27	27	24	22		16	17	–	2	–	–	1	–	–	–	–	–	–	Isoquinoline, 1-[5-chloro-2-(methylamino)phenyl]-1,2,3,4-tetrahydro-, (±)-	23495-28-5	NI27683
91	272	229	65	44	41	39	106	272	100	41	22	16	16	14	13	11		16	20	–	2	1	–	–	–	–	–	–	–	–	Cyclohexanespiro-4′-(5′-methylene-2′-benzylimino-1′,3′-thiazolidine)		NI27684
194	195	139	151	229	272	77	167	272	100	99	69	44	43	41	38	34		16	20	2	–	–	–	–	–	–	1	–	–	–	Silane, diethoxydiphenyl-	2553-19-7	NI27685
186	157	158	187	131	185	143	271	272	100	75	69	43	36	28	21	16	15.00	16	20	2	2	–	–	–	–	–	–	–	–	–	2-Cyclopentenone, 2-(4-methylanilino)-4-morpholino-		MS06130
94	137	43	136	28	118	105	120	272	100	92	87	62	43	38	34	32	6.00	16	20	2	2	–	–	–	–	–	–	–	–	–	2,6-Methano-1(1H)-benzazonine, 3-acetyl-4-amino-7-hydroxy-2,5,6,7-tetrahydro-7-methyl-		LI05784
201	216	272	229	77	202	92	91	272	100	66	64	33	24	13	13	12		16	20	2	2	–	–	–	–	–	–	–	–	–	2-Pyrazolin-5-one, 4-hexanoyl-3-methyl-1-phenyl-	18667-23-7	MS06131
198	130	199	143	156	168	182	115	272	100	20	19	17	11	10	8	8	3.00	16	20	2	2	–	–	–	–	–	–	–	–	–	Rufomycin B1		LI05785
130	41	198	131	77	69	129	103	272	100	22	20	12	7	7	6	5	2.28	16	20	2	2	–	–	–	–	–	–	–	–	–	L-Tryptophan, 1-(1,1-dimethylallyl)-	20880-68-6	NI27686
198	156	154	155	199	169	168	129	272	100	51	32	21	16	16	14	14	6.30	16	20	2	2	–	–	–	–	–	–	–	–	–	L-Tryptophan, 4-(3,3-dimethylallyl)-	5017-33-4	NI27687
201	216	272	229	77	202	92	91	272	100	68	64	31	22	12	11	10		16	20	2	2	–	–	–	–	–	–	–	–	–	L-Tryptophan, 4-hexanoyl-3-methyl-1-phenyl-		LI05786

MASS TO CHARGE RATIOS								M.W.	INTENSITIES								Parent	C	H	O	N	S	F	Cl	Br	I	Si	P	B	X	COMPOUND NAME	CAS Reg No	No
162	242	272	164	137	138	103	270	**272**	100	81	74	35	32	23	23	22		16	21	–	–	–	–	–	–	–	–	–	–	1	**Cobalt, [(1,2,5,6-η)-1,5-cyclooctadiene][(1,2,3,3a,6a-η)-1,4,5,6-tetrahydro-1-pentalenyl]-**	38959-18-1	**NI27689**
242	162	272	163	164	137	138	103	**272**	100	80	79	67	62	54	35	32		16	21	–	–	–	–	–	–	–	–	–	–	1	**Cobalt, [(1,2,5,6-η)-1,5-cyclooctadiene][(1,2,3,3a,6a-η)-1,4,5,6-tetrahydro-1-pentalenyl]-**	38959-18-1	**NI27688**
79	80	95	93	77	81	94	91	**272**	100	73	69	35	31	21	19	15	**1.00**	16	24	–	4	–	–	–	–	–	–	–	–	–	**9-Azabicyclo[6.1.0]non-4-ene, 9,9′-azobis-, [1α,4Z,8α,9[E(1′R*,4′Z,8′S*)]]-**	66387-82-4	**NI27690**
260	259	261	131	103	133	129	102	**272**	100	29	19	16	7	5	5	3		16	24	–	4	–	–	–	–	–	–	–	–	–	**6H,13H-Pyrazino[1,2-a:4,5-a′]bisimidazole**		**IC03540**
99	71	43	131	55	41	57	100	**272**	100	45	33	12	11	9	8	7	**0.96**	16	32	1	–	1	–	–	–	–	–	–	–	–	**Hexanethioic acid, S-decyl ester**	2432-81-7	**NI27691**
43	57	41	87	71	55	69	56	**272**	100	57	35	31	28	28	22	19	**0.10**	16	32	3	–	–	–	–	–	–	–	–	–	–	**Ethane, 1-acetoxy-2-(dodecyloxy)-**		**MS06132**
43	104	55	41	98	57	99	29	**272**	100	96	79	73	58	57	55	46	**7.44**	16	32	3	–	–	–	–	–	–	–	–	–	–	**Tetradecanoic acid, 2-hydroxyethyl ester**	22122-18-5	**NI27692**
103	43	74	71	41	57	55	69	**272**	100	36	25	24	20	18	18	14	**0.10**	16	32	3	–	–	–	–	–	–	–	–	–	–	**Tetradecanoic acid, 3-hydroxy-13-methyl-, methyl ester**		**LI05787**
257	103	75	73	258	69	55	97	**272**	100	98	83	61	21	14	14	13	**0.70**	16	36	1	–	–	–	–	–	–	1	–	–	–	**Silane, trimethyl(2-ethylundecyloxy)-**		**IC03541**
257	103	75	73	258	55	69	89	**272**	100	72	70	49	22	10	9	8	**0.70**	16	36	1	–	–	–	–	–	–	1	–	–	–	**Silane, trimethyl(2-methyldodecyloxy)-**		**IC03542**
117	73	75	43	41	29	55	27	**272**	100	68	67	39	39	27	19	14	**0.00**	16	36	1	–	–	–	–	–	–	1	–	–	–	**Silane, trimethyl(tridecyloxy)-**	56554-32-6	**NI27693**
257	75	258	73	103	83	89	69	**272**	100	38	22	20	17	9	8	8	**0.40**	16	36	1	–	–	–	–	–	–	1	–	–	–	**Silane, trimethyl(tridecyloxy)-**	56554-32-6	**IC03543**
272	178	214	165	68	243	76	205	**272**	100	27	10	10	10	8	7	5		17	12	–	4	–	–	–	–	–	–	–	–	–	**Pyrazolo[3,2-c]-as-triazine, 3,4-diphenyl-**		**LI05788**
272	178	273	214	165	68	243	76	**272**	100	27	18	10	10	10	8	7		17	12	–	4	–	–	–	–	–	–	–	–	–	**Pyrazolo[5,1-c]-s-triazine, 3,4-diphenyl-**	6726-61-0	**NI27694**
272	271	165	273	245	77	51	190	**272**	100	87	35	20	19	19	13	11		17	12	–	4	–	–	–	–	–	–	–	–	–	**s-Triazolo[4,3-a]pyrazine, 5,6-diphenyl-**	6969-76-2	**NI27695**
272	271	165	245	77	51	190	115	**272**	100	87	35	19	19	13	11	10		17	12	–	4	–	–	–	–	–	–	–	–	–	**s-Triazolo[4,3-a]pyrazine, 5,6-diphenyl-**	6969-76-2	**LI05789**
257	43	229	227	211	272	199		**272**	100	33	23	8	5	3	3			17	20	3	–	–	–	–	–	–	–	–	–	–	**4H-Pyran-3-carboxylic acid, 4,4,6-trimethyl-2-phenyl-, ethyl ester**		**LI05790**
272	110	82	138	243	215	107	55	**272**	100	65	60	30	20	20	18	15		17	24	1	2	–	–	–	–	–	–	–	–	–	**1-Azabicyclo[2.2.2]octane-2-carboxamide, 5-ethyl-N-(2-methylphenyl)-, (2α,4α,5α)-(±)-**	75364-49-7	**NI27696**
123	272	108	121	257	149	150	122	**272**	100	86	80	67	65	61	58	32		17	24	1	2	–	–	–	–	–	–	–	–	–	**Bis(3,5-dimethyl-4-ethyl-2-pyrrolyl)ketone**	13228-22-3	**NI27697**
271	228	273	202					**272**	100	80	40	20					**0.00**	17	24	1	2	–	–	–	–	–	–	–	–	–	**Isoquinoline, 3-butyl-1-[2-(dimethylamino)ethoxy]-**	86-80-6	**NI27698**
55	41	28	69	43	83	57	97	**272**	100	96	74	73	73	63	54	46	**9.65**	17	33	–	–	–	–	1	–	–	–	–	–	–	**7-Heptadecene, 1-chloro-**	56554-78-0	**NI27699**
55	41	69	83	43	57	56	97	**272**	100	88	75	63	63	55	47	45	**8.44**	17	33	–	–	–	–	1	–	–	–	–	–	–	**7-Heptadecene, 1-chloro-**	56554-78-0	**NI27700**
55	41	69	28	43	83	57	56	**272**	100	80	79	69	64	60	41	38	**10.54**	17	33	–	–	–	–	1	–	–	–	–	–	–	**7-Heptadecene, 17-chloro-**	56554-79-1	**NI27701**
55	43	41	69	83	57	42	28	**272**	100	96	93	70	59	50	46	45	**9.64**	17	33	–	–	–	–	1	–	–	–	–	–	–	**8-Heptadecene, 1-chloro-**	56554-80-4	**NI27702**
75	71	241	41	55	43	82	57	**272**	100	13	9	7	6	6	5	5	**0.00**	17	36	2	–	–	–	–	–	–	–	–	–	–	**Pentadecane, 1,1-dimethoxy-**		**IC03544**
57	43	71	55	41	83	69	63	**272**	100	76	58	56	50	46	45	38	**0.00**	17	36	2	–	–	–	–	–	–	–	–	–	–	**Pentadecane, 1-(2-hydroxyethoxy)-**		**IC03545**
75	71	83	241	82	97	96	41	**272**	100	34	20	17	16	13	12	9	**0.00**	17	36	2	–	–	–	–	–	–	–	–	–	–	**Tetradecane, 1,1-dimethoxy-2-methyl-**		**IC03546**
57	43	56	71	55	41	69	45	**272**	100	73	68	65	53	49	47	40	**0.00**	17	36	2	–	–	–	–	–	–	–	–	–	–	**Tetradecane, 1-(2-hydroxyethoxy)-2-methyl-**		**IC03547**
272	200	228	100	199	273	174	201	**272**	100	92	52	31	27	20	19	16		18	8	3	–	–	–	–	–	–	–	–	–	–	**Dibenzo[c,e]oxepin-5,7-dione, 6,6′-diethynyl-**		**LI05791**
160	57	127	272	39	129	69	28	**272**	100	99	85	84	81	75	72	66		18	24	–	–	1	–	–	–	–	–	–	–	–	**1-Naphthyl octyl sulphide**		**AI01918**
160	272	39	116	69	71	162	129	**272**	100	86	80	63	63	57	56	55		18	24	–	–	1	–	–	–	–	–	–	–	–	**2-Naphthyl octyl sulphide**		**AI01919**
160	121	108	272	161	83	55	41	**272**	100	87	46	31	22	20	20	20		18	24	2	–	–	–	–	–	–	–	–	–	–	**Bicyclo[2.2.1]heptan-2-ol, 3-[(4-methoxyphenyl)methylene]-1,7,7-trimethyl-, exo-**	38108-50-8	**MS06133**
185	198	272	158	199	240	186	241	**272**	100	83	41	26	23	22	22	17		18	24	2	–	–	–	–	–	–	–	–	–	–	**Cyclohepta[1,2-a]naphthalene-1-carboxylic acid, octahydro-, methyl ester**		**LI05792**
254	74	225	173	239	186	226	91	**272**	100	98	51	45	40	36	30	30	**0.00**	18	24	2	–	–	–	–	–	–	–	–	–	–	**Cyclopropa[3,4]cyclohepta[1,2-a]naphthalen-10-ol, 1,1a,1b,2,3,7b,8,9,10,10a-decahydro-5-methoxy-10-methyl-**	57983-82-1	**NI27703**
272	213	172	160	146	273	186	159	**272**	100	29	25	25	22	20	14	14		18	24	2	–	–	–	–	–	–	–	–	–	–	**Estra-1,3,5(10)-triene-3,17-diol, (17α)-**	57-91-0	**NI27704**
272	213	160	273	172	159	146	133	**272**	100	26	23	18	18	18	16	15		18	24	2	–	–	–	–	–	–	–	–	–	–	**Estra-1,3,5(10)-triene-3,17-diol, (17β)-**	50-28-2	**TR00294**
272	213	172	146	160	270	186	159	**272**	100	48	43	41	40	29	22	22		18	24	2	–	–	–	–	–	–	–	–	–	–	**Estra-1,3,5(10)-triene-3,17-diol, (17β)-**	50-28-2	**NI27705**
272	213	160	146	273	172	159	133	**272**	100	32	32	22	21	20	19	17		18	24	2	–	–	–	–	–	–	–	–	–	–	**Estra-1,3,5(10)-triene-3,17-diol, (17β)-**	50-28-2	**NI27706**
272	110	273	91	186	107	228	105	**272**	100	66	41	41	32	28	26	26		18	24	2	–	–	–	–	–	–	–	–	–	–	**Estr-4-ene-3,7-dione, (9β)-**	55622-62-3	**NI27707**
272	110	91	150	160	123	105	273	**272**	100	98	49	31	27	25	24	23		18	24	2	–	–	–	–	–	–	–	–	–	–	**Estr-4-ene-3,17-dione**	734-32-7	**NI27708**
272	110	186	228	244	273	97	122	**272**	100	46	32	31	24	23	20	17		18	24	2	–	–	–	–	–	–	–	–	–	–	**Estr-4-ene-3,17-dione**	734-32-7	**NI27709**
272	110	91	41	79	186	228	55	**272**	100	55	34	32	30	29	26	24		18	24	2	–	–	–	–	–	–	–	–	–	–	**Estr-4-ene-3,17-dione**	734-32-7	**NI27710**
272	215	244	97	57	273	270	110	**272**	100	44	34	34	26	24	21	20		18	24	2	–	–	–	–	–	–	–	–	–	–	**Estr-5(10)-ene-3,17-dione**	3962-66-1	**NI27711**
129	128	141	41	155	115	55	91	**272**	100	74	72	67	63	61	61	60	**2.24**	18	24	2	–	–	–	–	–	–	–	–	–	–	**5,8,11-Heptadecatriynoic acid, methyl ester**	56554-57-5	**NI27712**
272	57	257	273	41	29	258	201	**272**	100	43	33	23	13	9	7	4		18	24	2	–	–	–	–	–	–	–	–	–	–	**Naphthalene, 2,6-di-tert-butyl-1,5-dihydroxy-**		**IC03548**
272	147	173	229	161	121	273	115	**272**	100	91	37	29	24	23	16	15		18	24	2	–	–	–	–	–	–	–	–	–	–	**1-Phenanthrenecarboxaldehyde, 1,2,3,4,4a,9,10,10a-octahydro-6-methoxy-1,4a-dimethyl-, [1S-(1α,4aα,10aβ)]-**	16826-83-8	**NI27713**
272	257	187	189	175	215	258	188	**272**	100	93	40	39	28	21	15	12		18	24	2	–	–	–	–	–	–	–	–	–	–	**9(1H)-Phenanthrenone, 2,3,4,4a,10,10a-hexahydro-6-hydroxy-1,1,4a,7-tetramethyl-, (4aS-trans)-**	561-95-5	**NI27714**

MASS TO CHARGE RATIOS								M.W.	INTENSITIES								Parent	C	H	O	N	S	F	Cl	Br	I	Si	P	B	X	COMPOUND NAME	CAS Reg No	No
257	215	57	272	41	71	43	55	272	100	46	32	29	19	17	17	15		18	24	2	–	–	–	–	–	–	–	–	–	–	2(1H)-Phenanthrenone, 3,4,4a,9,10,10a-hexahydro-7-methoxy-1,1,4a-trimethyl-	55255-51-1	NI27715
197	257	198	131	272	141	129	128	272	100	19	17	16	14	14	14	13		18	24	2	–	–	–	–	–	–	–	–	–	–	Podocarpa-8,11,13-trien-15-oic acid, methyl ester	3650-04-2	NI27716
197	240	212	141	198	143	257	129	272	100	59	42	20	17	17	13	12	4.60	18	24	2	–	–	–	–	–	–	–	–	–	–	Podocarpa-8,11,13-trien-15-oic acid, methyl ester, (5β)-	24035-56-1	NI27717
197	257	198	141	129	128	117	272	272	100	40	27	22	17	16	16	15		18	24	2	–	–	–	–	–	–	–	–	–	–	Podocarpa-8,11,13-trien-16-oic acid, methyl ester	2651-34-5	NI27718
197	198	129	143	128	141	131	257	272	100	20	20	19	19	17	15	14	8.40	18	24	2	–	–	–	–	–	–	–	–	–	–	Podocarpa-8,11,13-trien-16-oic acid, methyl ester, (5β)-	3745-36-6	NI27719
272	112	162	187	145	161	91	56	272	100	68	55	39	38	34	33	32		18	24	2	–	–	–	–	–	–	–	–	–	–	1,4-Triphenylenedione, 2,3,4a,4b,5,6,7,8,9,10,11,12,12a,12b-tetradecahydro-	5981-14-6	NI27720
272	57	273	107	41	82	229	55	272	100	26	19	15	14	13	12	12		18	28	–	2	–	–	–	–	–	–	–	–	–	Benzene, 1,4-bis(cyclohexylamino)-		IC03549
186	187							272	100	82							0.00	18	28	–	2	–	–	–	–	–	–	–	–	–	Piperidine, 1,1′-[1,2-phenylenebis(methylene)]bis-	55076-37-4	NI27721
57	43	71	55	41	85	69	56	272	100	92	57	47	47	35	30	20	3.80	18	37	–	–	–	1	–	–	–	–	–	–	–	Octadecyl fluoride		IC03550
114	170	272	113	102	273	244	88	272	100	97	94	26	26	20	20	18		19	12	2	–	–	–	–	–	–	–	–	–	–	4H-Naphtho[1,2-b]pyran-4-one, 2-phenyl-	604-59-1	NI27722
169	271	104	77	168	167	51	66	272	100	51	37	33	31	22	17	16	15.37	19	16	–	2	–	–	–	–	–	–	–	–	–	Benzenecarboximidamide, N,N-diphenyl-	33345-17-4	NI27723
272	271	273	228	136	114	257	229	272	100	28	21	10	9	6	5	5		19	16	–	2	–	–	–	–	–	–	–	–	–	Benz[cd]indole, 2-[4-(dimethylamino)phenyl]-		IC03551
272	77	180	273	51	271	181	165	272	100	76	75	22	20	17	14	11		19	16	–	2	–	–	–	–	–	–	–	–	–	Benzophenone, phenylhydrazone	574-61-8	NI27724
272	180	45	77	46	43	273	271	272	100	85	70	49	30	24	22	20		19	16	–	2	–	–	–	–	–	–	–	–	–	Benzophenone, phenylhydrazone	574-61-8	LI05793
272	167	273	168	136	135	77	32	272	100	26	22	10	10	8	7	6		19	16	–	2	–	–	–	–	–	–	–	–	–	Diphenylamine, 4-benzylideneamino-		IC03552
218	178	272	219	191	273	179	176	272	100	56	45	19	19	10	8	7		19	16	–	2	–	–	–	–	–	–	–	–	–	Fluorene, 9,9-bis(2-cyanoethyl)-		IC03553
122	79	67	81	95	109	91	230	272	100	96	88	74	70	68	65	61	26.26	19	28	1	–	–	–	–	–	–	–	–	–	–	Androst-1-en-3-one, (5α)-	3248-10-0	NI27725
108	41	122	109	55	68	79	95	272	100	76	66	64	44	40	39	37	20.00	19	28	1	–	–	–	–	–	–	–	–	–	–	Androst-2-en-1-one, (5α)-	25824-61-7	NI27726
41	177	79	81	91	67	55	93	272	100	99	75	72	71	70	70	55	55.00	19	28	1	–	–	–	–	–	–	–	–	–	–	Androst-2-en-11-one, (5α)-	54548-20-8	NI27727
41	218	79	91	272	67	55	93	272	100	79	72	64	57	53	52	50		19	28	1	–	–	–	–	–	–	–	–	–	–	Androst-2-en-17-one, (5α)-	963-75-7	NI27728
218	272	190	91	161	79	273	93	272	100	98	35	27	25	23	22	20		19	28	1	–	–	–	–	–	–	–	–	–	–	Androst-2-en-17-one, (5α)-	963-75-7	NI27730
218	272	190	161	273	219	162	147	272	100	99	35	25	22	18	13	12		19	28	1	–	–	–	–	–	–	–	–	–	–	Androst-2-en-17-one, (5α)-	963-75-7	NI27729
124	79	67	149	91	93	81	77	272	100	69	66	64	59	48	47	47	33.03	19	28	1	–	–	–	–	–	–	–	–	–	–	Androst-4-en-3-one	2872-90-4	NI27731
124	149	231	272	41	79	95	93	272	100	63	49	47	30	28	27	27		19	28	1	–	–	–	–	–	–	–	–	–	–	Androst-4-en-3-one	2872-90-4	NI27732
272	55	105	149	271	95	79	91	272	100	57	48	46	45	44	43	41		19	28	1	–	–	–	–	–	–	–	–	–	–	Androst-5-en-4-one	13583-72-7	NI27733
55	272	149	95	79	91	67	257	272	100	79	79	73	73	70	68	65		19	28	1	–	–	–	–	–	–	–	–	–	–	Androst-5-en-4-one	13583-72-7	NI27734
41	257	272	55	91	161	177	67	272	100	63	59	51	49	44	42	40		19	28	1	–	–	–	–	–	–	–	–	–	–	Androst-8-en-11-one, (5α)-	54498-82-7	NI27735
147	257	41	40	91	272	67	55	272	100	41	32	31	23	21	18	18		19	28	1	–	–	–	–	–	–	–	–	–	–	Androst-9(11)-en-12-one, (5α)-	4354-37-4	NI27736
272	257	94	79	149	147	107	93	272	100	80	45	40	36	35	30	30		19	28	1	–	–	–	–	–	–	–	–	–	–	Androst-16-en-3-one, (5α)-		MS06135
272	149	257	94	147	107	124	105	272	100	94	93	89	87	86	80	80		19	28	1	–	–	–	–	–	–	–	–	–	–	Androst-16-en-3-one, (5α)-		MS06134
272	257	94	147	149	95	93	107	272	100	91	73	61	55	55	49	47		19	28	1	–	–	–	–	–	–	–	–	–	–	Androst-16-en-3-one, (5α)-		LI05794
177	272	95	96	109	178	190	133	272	100	94	90	88	87	69	58	36		19	28	1	–	–	–	–	–	–	–	–	–	–	Cyclopentanone, 2,5-bis[(E)-cyclohexylmethylene]-		MS06136
272	257	115	77	243	273	217	229	272	100	30	30	26	22	21	19	18		20	16	1	–	–	–	–	–	–	–	–	–	–	4H-Cyclopenta[b]furan, 6-methyl-4,5-diphenyl-	86738-92-3	NI27737
272	115	77	257	243	217	229	215	272	100	38	34	25	25	22	21	21		20	16	1	–	–	–	–	–	–	–	–	–	–	6H-Cyclopenta[b]furan, 6-methyl-4,5-diphenyl-	86738-91-2	NI27738
272	228	243	273	165	202	195	152	272	100	33	31	22	17	11	11	11		20	16	1	–	–	–	–	–	–	–	–	–	–	Cyclopropene, 3-(2-furyl)-3-methyl-1,2-diphenyl-	86738-90-1	NI27739
272	229	114	228	257	113	78	101	272	100	53	47	41	38	31	29	28		20	16	1	–	–	–	–	–	–	–	–	–	–	5,6-Dihydro-5,6-epoxy-7,12-dimethylbenz[a]anthracene		NI27740
257	272	228	226	258	113	259	114	272	100	68	44	33	23	23	21	20		20	16	1	–	–	–	–	–	–	–	–	–	–	7,12-Dihydro-7-methylene-12-hydroxy-12-methylbenz[a]anthracene		NI27741
272	257	239	242	226	255	228	202	272	100	54	44	40	39	31	26	18		20	16	1	–	–	–	–	–	–	–	–	–	–	1-Hydroxy-7,12-dimethylbenz[a]anthracene		MS06137
272	257	239	226	228	255	215	202	272	100	38	26	21	19	15	12	12		20	16	1	–	–	–	–	–	–	–	–	–	–	2-Hydroxy-7,12-dimethylbenz[a]anthracene		MS06138
272	257	239	228	226	255	215	202	272	100	43	18	15	15	14	7	7		20	16	1	–	–	–	–	–	–	–	–	–	–	3-Hydroxy-7,12-dimethylbenz[a]anthracene		MS06139
272	257	239	226	228	229	202	215	272	100	33	25	21	20	19	16	15		20	16	1	–	–	–	–	–	–	–	–	–	–	4-Hydroxy-7,12-dimethylbenz[a]anthracene		MS06140
228	272	243	239	226	255	215	202	272	100	63	46	40	21	20	11	10		20	16	1	–	–	–	–	–	–	–	–	–	–	7-Hydroxymethyl-12-methylbenz[a]anthracene		MS06141
228	243	272	239	226	255	257	215	272	100	65	60	51	23	10	7	6		20	16	1	–	–	–	–	–	–	–	–	–	–	7-Methyl-12-hydroxymethylbenz[a]anthracene		MS06142
272	165	271	195	152	270	243	239	272	100	12	11	7	7	5	5	5		20	16	1	–	–	–	–	–	–	–	–	–	–	Stilbene, 2-hydroxy-4′-phenyl-, (E)-		MS06143
81	106	80	91	137	244	119	69	272	100	97	64	58	53	52	50	49	25.00	20	32	–	–	–	–	–	–	–	–	–	–	–	Atis-15-ene, (5β,8α,9β,10α,12α)-		MS06144
81	105	104	80	91	137	234	120	272	100	96	66	62	56	51	50	48	24.00	20	32	–	–	–	–	–	–	–	–	–	–	–	Atis-15-ene, (5β,8α,9β,10α,12α)-		LI05795
257	69	91	81	105	55	79	93	272	100	58	52	50	45	45	38	35	25.00	20	32	–	–	–	–	–	–	–	–	–	–	–	Atis-16-ene, (5β,8α,9β,10α,12α)-		MS06145
272	161	133	159	273	147	145	173	272	100	55	31	27	24	24	19	18		20	32	–	–	–	–	–	–	–	–	–	–	–	Benzocyclododecene, 2,3-diethyl-4a,5,6,7,8,9,10,11,12,13-decahydro-	61141-65-9	NI27742
68	93	81	57	121	107	272	257	272	100	73	60	52	50	50	31	29		20	32	–	–	–	–	–	–	–	–	–	–	–	Cembrene A, (R,all-E)-		DD00867
81	55	41	44	107	149	67	43	272	100	67	59	58	40	39	39	36	10.00	20	32	–	–	–	–	–	–	–	–	–	–	–	Cyclohexene, 1-methyl-4-(1,5,9-trimethyl-1,5,9-decatrienyl)-	56248-11-4	NI27743
93	40	91	81	105	119	42	107	272	100	98	93	88	87	83	81	75	50.00	20	32	–	–	–	–	–	–	–	–	–	–	–	1,3,6,10-Cyclotetradecatetraene, 14-isopropyl-3,7,11-trimethyl-, [S-(E,Z,E,E)]-		LI05796
68	93	81	121	107	41	67	272	272	100	73	64	55	53	43	41	38		20	32	–	–	–	–	–	–	–	–	–	–	–	1,5,9-Cyclotetradecatriene, 1,5,9-trimethyl-12-isopropylene-	38748-84-4	NI27744
131	132	43	41	29	91	272	129	272	100	11	8	8	8	7	6	5		20	32	–	–	–	–	–	–	–	–	–	–	–	Decane, 1-(3-methyl-1-indanyl)-	33425-50-2	WI01334

MASS TO CHARGE RATIOS								M.W.	INTENSITIES								Parent	C	H	O	N	S	F	Cl	Br	I	Si	P	B	X	COMPOUND NAME	CAS Reg No	No
272	41	120	106	91	104	105	55	**272**	100	92	88	88	88	72	68	62		20	32	–	–	–	–	–	–	–	–	–	–	–	Fillocladine		LI05797
69	41	93	81	107	55	79	91	**272**	100	58	50	50	28	28	22	18	**1.80**	20	32	–	–	–	–	–	–	–	–	–	–	–	Hexadeca-1,3,6,10,14-pentaene, 3,7,11,15-tetramethyl-, (E,E,E)-		MS06146
69	41	93	81	55	67	79	95	**272**	100	54	39	38	20	19	16	15	**2.60**	20	32	–	–	–	–	–	–	–	–	–	–	–	Hexadeca-1,6,10,14-tetraene, 3-methylene-7,11,15-trimethyl-, (E,E)-		MS06147
94	106	119	163	257	272	229	187	**272**	100	67	29	23	22	19	10	10		20	32	–	–	–	–	–	–	–	–	–	–	–	ent-Isokaurene		MS06148
94	106	105	91	119	93	69	54	**272**	100	64	36	33	28	28	26	23	**20.00**	20	32	–	–	–	–	–	–	–	–	–	–	–	Isopanrene		LI05798
257	69	41	91	105	81	55	125	**272**	100	88	86	75	69	68	65	64	**48.00**	20	32	–	–	–	–	–	–	–	–	–	–	–	ent-Kaurene		MS06149
120	272	106	91	107	119	105	41	**272**	100	80	75	49	48	44	43	38		20	32	–	–	–	–	–	–	–	–	–	–	–	Kaur-15-ene		LI05799
120	106	272	41	107	119	91	105	**272**	100	87	71	64	54	51	51	50		20	32	–	–	–	–	–	–	–	–	–	–	–	Kaur-15-ene		MS06150
94	106	105	91	119	93	69	55	**272**	100	65	36	33	28	28	26	24	**21.00**	20	32	–	–	–	–	–	–	–	–	–	–	–	Kaur-15-ene		MS06151
120	106	91	107	105	272	119	41	**272**	100	85	59	58	58	55	49	38		20	32	–	–	–	–	–	–	–	–	–	–	–	Kaur-15-ene, (5α,9α,10β)-	511-85-3	NI27745
257	272	229	125	123	91	69	105	**272**	100	77	54	47	47	45	45	44		20	32	–	–	–	–	–	–	–	–	–	–	–	Kaur-16-ene	562-28-7	NI27746
41	69	91	55	81	229	79	272	**272**	100	78	76	74	66	59	56	55		20	32	–	–	–	–	–	–	–	–	–	–	–	Kaur-16-ene	562-28-7	MS06152
69	257	91	81	105	55	123	125	**272**	100	95	83	80	70	70	65	63	**59.14**	20	32	–	–	–	–	–	–	–	–	–	–	–	Kaur-16-ene	562-28-7	MS06153
272	229	257	91	41	69	81	123	**272**	100	59	53	51	43	39	38	36		20	32	–	–	–	–	–	–	–	–	–	–	–	Kaur-16-ene, (8β,13β)-	20070-61-5	NI27748
272	257	94	91	106	105	229	41	**272**	100	86	57	54	53	48	47	41		20	32	–	–	–	–	–	–	–	–	–	–	–	Kaur-16-ene, (8β,13β)-	20070-61-5	NI27747
257	41	81	55	69	95	93	105	**272**	100	77	62	57	56	54	50	46	**31.83**	20	32	–	–	–	–	–	–	–	–	–	–	–	Labda-8(20),12,14-triene	5957-33-5	NI27749
257	41	272	81	258	55	105	95	**272**	100	27	26	23	22	22	21	20		20	32	–	–	–	–	–	–	–	–	–	–	–	Labda-8(20),13(16),14-triene	511-02-4	NI27750
69	272	229	109	203	121	161	147	**272**	100	54	31	26	24	22	17	17		20	32	–	–	–	–	–	–	–	–	–	–	–	Myrcene dimer		LI05800
131	117	187	129	29	215	41	43	**272**	100	96	67	51	45	42	32	31	**18.32**	20	32	–	–	–	–	–	–	–	–	–	–	–	Naphthalene, 1-butyl-4-hexyl-1,2,3,4-tetrahydro-	33509-81-8	WI01335
159	272	145	160	41	43	129	55	**272**	100	28	18	16	16	15	10	8		20	32	–	–	–	–	–	–	–	–	–	–	–	Naphthalene, 2-butyl-3-hexyl-1,2,3,4-tetrahydro-		AI01920
187	41	188	131	43	272	129	57	**272**	100	17	16	15	13	11	10	8		20	32	–	–	–	–	–	–	–	–	–	–	–	Naphthalene, 2-butyl-8-hexyl-1,2,3,4-tetrahydro-		AI01921
187	131	188	41	272	57	43	129	**272**	100	17	16	14	10	9	9	8		20	32	–	–	–	–	–	–	–	–	–	–	–	Naphthalene, 2-butyl-8-hexyl-1,2,3,4-tetrahydro-		AI01922
159	272	43	41	145	160	129	273	**272**	100	36	17	17	15	14	9	8		20	32	–	–	–	–	–	–	–	–	–	–	–	Naphthalene, 6-butyl-7-hexyl-1,2,3,4-tetrahydro-		AI01923
159	272	29	145	160	41	43	27	**272**	100	41	17	14	13	13	11	10		20	32	–	–	–	–	–	–	–	–	–	–	–	Naphthalene, 6-butyl-7-hexyl-1,2,3,4-tetrahydro-		AI01924
187	29	188	131	41	272	57	129	**272**	100	20	18	16	16	12	12	11		20	32	–	–	–	–	–	–	–	–	–	–	–	Naphthalene, 7-butyl-1-hexyl-1,2,3,4-tetrahydro-		AI01925
131	132	91	41	129	29	141	43	**272**	100	12	9	8	7	7	6	6	**5.30**	20	32	–	–	–	–	–	–	–	–	–	–	–	Naphthalene, 1-decyl-1,2,3,4-tetrahydro-	55255-57-7	NI27751
145	146	272	131	41	43	117	91	**272**	100	33	28	25	20	19	11	9		20	32	–	–	–	–	–	–	–	–	–	–	–	Naphthalene, 6-decyl-1,2,3,4-tetrahydro-		AI01926
145	272	146	131	41	29	43	117	**272**	100	45	29	22	18	18	15	12		20	32	–	–	–	–	–	–	–	–	–	–	–	Naphthalene, 6-decyl-1,2,3,4-tetrahydro-		AI01927
41	29	43	131	27	91	143	129	**272**	100	96	85	70	56	55	46	45	**0.00**	20	32	–	–	–	–	–	–	–	–	–	–	–	Naphthalene, 1,2,3,4-tetrahydro-2,6-dimethyl-7-octyl-		AI01928
159	160	29	272	144	41	129	43	**272**	100	13	7	6	6	6	5	5		20	32	–	–	–	–	–	–	–	–	–	–	–	Naphthalene, 1,2,3,4-tetrahydro-5,8-dimethyl-1-octyl-		AI01929
159	160	41	144	43	272	157	129	**272**	100	13	7	6	6	5	5	5		20	32	–	–	–	–	–	–	–	–	–	–	–	Naphthalene, 1,2,3,4-tetrahydro-5,8-dimethyl-1-octyl-		AI01931
159	160	144	41	272	129	43	143	**272**	100	13	6	6	5	5	5	4		20	32	–	–	–	–	–	–	–	–	–	–	–	Naphthalene, 1,2,3,4-tetrahydro-5,8-dimethyl-1-octyl-		AI01930
134	135	106	105	93	91	122	69	**272**	100	81	68	68	68	55	53	53	**43.00**	20	32	–	–	–	–	–	–	–	–	–	–	–	17-Norkaur-15-ene, 13-methyl-, (8β,13β)-		MS06154
134	135	93	106	105	272	122	148	**272**	100	87	67	62	61	60	52	39		20	32	–	–	–	–	–	–	–	–	–	–	–	ent-17-Norkaur-15-ene, 13-methyl-, (8β,13β)-		MS06155
272	135	41	67	81	79	229	136	**272**	100	77	50	45	35	34	29	26		20	32	–	–	–	–	–	–	–	–	–	–	–	Perylene, perhydro-		AI01932
272	135	41	67	79	81	229	273	**272**	100	54	47	36	32	31	24	22		20	32	–	–	–	–	–	–	–	–	–	–	–	Perylene, perhydro-		AI01933
257	41	105	161	68	55	91	81	**272**	100	84	63	56	56	55	54	46	**27.66**	20	32	–	–	–	–	–	–	–	–	–	–	–	Phenanthrene, 1,2,3,4,4a,5,6,7,8,9,10,10a-dodecahydro-1,1,4a-trimethyl-7-vinyl-	55255-56-6	NI27752
257	41	80	81	55	93	79	91	**272**	100	46	40	33	33	28	27	26	**21.50**	20	32	–	–	–	–	–	–	–	–	–	–	–	Phenanthrene, 1,2,3,4,4a,4b,5,6,7,8,8a,9-dodecahydro-1,1,4b,7-tetramethyl-7-vinyl-, 4aS-(4aα,4bβ,7β,10aβ)-	1686-67-5	MS06156
257	272	258	80	136	121	81	93	**272**	100	30	22	18	15	14	13	12		20	32	–	–	–	–	–	–	–	–	–	–	–	Phenanthrene, 1,2,3,4,4a,4b,5,6,7,8,8a,9-dodecahydro-1,1,4b,7-tetramethyl-7-vinyl-, 4aS-(4aα,4bβ,7β,10aβ)-	1686-67-5	NI27754
257	80	136	81	258	272	93	55	**272**	100	36	29	29	27	26	26	25		20	32	–	–	–	–	–	–	–	–	–	–	–	Phenanthrene, 1,2,3,4,4a,4b,5,6,7,8,8a,9-dodecahydro-1,1,4b,7-tetramethyl-7-vinyl-, 4aS-(4aα,4bβ,7β,10aβ)-	1686-67-5	NI27753
176	161	104	133	149	91	257	93	**272**	100	80	46	42	39	38	33	31	**26.82**	20	32	–	–	–	–	–	–	–	–	–	–	–	Phenanthrene, 1,2,3,4,4b,5,8,8a,9,10-decahydro-7-isopropyl-1,1,4b-trimethyl-	31197-60-1	NI27755
109	148	105	41	119	257	133	124	**272**	100	97	59	56	52	50	43	42	**31.00**	20	32	–	–	–	–	–	–	–	–	–	–	–	ent-Pimara-7,15-diene		MS06157
137	136	81	257	95	123	91	272	**272**	100	40	34	33	28	25	23	22		20	32	–	–	–	–	–	–	–	–	–	–	–	ent-Pimara-8(14),15-diene		MS06158
257	161	41	105	69	272	187	119	**272**	100	49	42	35	35	30	27	23		20	32	–	–	–	–	–	–	–	–	–	–	–	ent-Pimara-8,15-diene		MS06159
272	106	257	105	91	94	120	93	**272**	100	75	55	54	52	46	45	42		20	32	–	–	–	–	–	–	–	–	–	–	–	Podocarpa-6,13-diene, 13-isopropyl-	5939-62-8	NI27756
109	41	105	55	80	119	257	91	**272**	100	82	62	60	54	53	50	50	**34.22**	20	32	–	–	–	–	–	–	–	–	–	–	–	Podocarp-7-ene, 13β-methyl-13-vinyl-	1686-66-4	NI27757
272	257	109	148	105	119	243	133	**272**	100	97	63	46	41	39	30	30		20	32	–	–	–	–	–	–	–	–	–	–	–	Podocarp-8(14)-ene, 13β-methyl-13-vinyl-	1686-56-2	NI27758
137	41	81	95	136	54	91	79	**272**	100	65	63	55	50	46	45	37	**16.66**	20	32	–	–	–	–	–	–	–	–	–	–	–	Podocarp-8(14)-ene, 13β-methyl-13-vinyl-	1686-56-2	NI27759
41	216	105	93	119	91	69	272	**272**	100	90	90	77	76	72	72	68		20	32	–	–	–	–	–	–	–	–	–	–	–	Trachylobane		MS06160
243	165	244	166	167	91	115	245	**272**	100	37	29	8	7	6	5	4	**0.20**	21	20	–	–	–	–	–	–	–	–	–	–	–	Benzene, 1,1′,1″-propylidynetris-	54889-83-7	NI27760
272	257	242	273	129	128	111	144	**272**	100	45	40	27	19	17	13	12		21	20	–	–	–	–	–	–	–	–	–	–	–	Naphthalene, 1-(2,4,6-trimethylstyryl)-		LI05801

MASS TO CHARGE RATIOS								M.W.	INTENSITIES								Parent	C	H	O	N	S	F	Cl	Br	I	Si	P	B	X	COMPOUND NAME	CAS Reg No	No
272	257	242	273	144	129	128	241	272	100	44	32	24	24	24	17	11		21	20	–	–	–	–	–	–	–	–	–	–	–	Naphthalene, 2-(2,4,6-trimethylstyryl)-		LI05802
167	165	168	272	91	152	166	105	272	100	21	19	17	13	11	9	9		21	20	–	–	–	–	–	–	–	–	–	–	–	Phenylpropane, 1,3-diphenyl-		AI01934
275	277	273	133	279	120	98	240	273	100	69	59	35	23	21	21	20		7	–	–	1	–	–	5	–	–	–	–	–	–	Benzonitrile, pentachloro-	20925-85-3	NI27761
275	273	277	133	240	279	238	203	273	100	64	64	23	20	18	16	14		7	–	–	1	–	–	5	–	–	–	–	–	–	Benzonitrile, pentachloro-	20925-85-3	NI27762
104	71	43	147	36	42	44	58	273	100	64	54	42	30	25	21	19	0.00	7	16	2	3	2	–	1	–	–	–	–	–	–	Carbamothioic acid, S,S'-[2-(dimethylamino)-1,3-propanediyl] ester, monohydrochloride	15263-52-2	NI27763
104	71	147	42	44	70	56	102	273	100	74	47	44	36	35	31	24	0.00	7	16	2	3	2	–	1	–	–	–	–	–	–	Carbamothioic acid, S,S'-[2-(dimethylamino)-1,3-propanediyl] ester, monohydrochloride	15263-52-2	NI27764
195	238	118	152	84	197	196	150	273	100	23	16	15	15	9	6	5	0.00	7	16	2	3	2	–	1	–	–	–	–	–	–	Carbamothioic acid, S,S'-[2-(dimethylamino)-1,3-propanediyl] ester, monohydrochloride	15263-52-2	NI27765
125	93	131	59	45	143	30	28	273	100	85	76	68	61	38	38	38	21.61	7	16	4	1	2	–	–	–	–	–	1	–	–	Phosphorodithioic acid, S-[2-[(2-methoxyethyl)amino]-2-oxoethyl] O,O-dimethyl ester	919-76-6	NI27766
275	115	277	273	196	194	137	136	273	100	50	40	40	16	16	13	6		8	5	–	1	–	–	–	2	–	–	–	–	–	1H-Indole, 5,7-dibromo-		NI27767
275	115	196	194	273	277	114	116	273	100	96	71	63	51	50	34	28		8	5	–	1	–	–	–	2	–	–	–	–	–	1H-Indole, dibromo-	65216-78-6	NI27768
273	77	30	53	75	243	110	109	273	100	73	73	52	51	30	29	20		8	7	8	3	–	–	–	–	–	–	–	–	–	Benzene, 1,3-dimethoxy-2,4,6-trinitro-		IC03554
36	221	44	38	175	28	89	35	273	100	78	77	32	28	27	21	11	0.00	9	8	3	3	1	–	1	–	–	–	–	–	–	2(3H)-Thiazolimine, 3-hydroxy-4-(3-nitrophenyl)-, monohydrochloride	58275-70-0	NI27769
221	36	89	175	191	38	222	148	273	100	44	41	31	19	16	12	12	0.00	9	8	3	3	1	–	1	–	–	–	–	–	–	2(3H)-Thiazolimine, 3-hydroxy-4-(4-nitrophenyl)-, monohydrochloride	58275-69-7	NI27770
186	238	96	69	142	171	273	83	273	100	60	34	26	22	19	18	16		9	12	1	5	1	–	1	–	–	–	–	–	–	1,3,5-Triazin-2-amine, 4-(2-chloro-2-cyanoethylthio)-6-methoxy-N,N-dimethyl-		NI27771
186	238	113	69	158	87	142	85	273	100	72	26	23	20	16	14	14	11.99	9	12	1	5	1	–	1	–	–	–	–	–	–	1,3,5-Triazin-2-amine, 4-(2-chloro-2-cyanoethylthio)-N-ethyl-6-methoxy-		NI27772
186	56	121	49	39	187	82	95	273	100	41	39	32	14	11	10	9	0.00	10	10	–	–	–	4	–	–	–	–	–	1	1	Ferrocenium tetrafluoroborate	1282-37-7	NI27773
43	87	186	170	214	76	75	121	273	100	65	51	21	17	11	11	8	1.51	10	11	6	1	1	–	–	–	–	–	–	–	–	Ethanol, 2-[(4-nitrophenyl)sulphonyl]-, acetate	38476-91-4	NI27774
43	87	186	170	214	76	75	121	273	100	66	52	22	18	13	13	8	1.00	10	11	6	1	1	–	–	–	–	–	–	–	–	Ethanol, 2-[(4-nitrophenyl)sulphonyl]-, acetate		LI05803
273	96	166	150	275	165	167	258	273	100	74	38	37	32	32	28	22		10	16	2	5	–	–	1	–	–	–	–	–	–	1,3,5-Triazine-2-carboxylic acid, 4,6-bis(dimethylamino)-, 2-chloroethyl ester		NI27775
43	65	41	45	57	55	63	47	273	100	74	60	47	46	46	42	36	9.10	10	19	2	5	1	–	–	–	–	–	–	–	–	1,3,5-Triazine-2,4-diamine, N,N'-diisopropyl-6-(methylsulphonyl)-	55723-98-3	NI27776
186	41	201	188	43	273	152	203	273	100	91	83	73	64	60	60	52		11	9	3	1	–	–	2	–	–	–	–	–	–	2,4-Oxazolidinedione, 3-(3,5-dichlorophenyl)-5,5-dimethyl-	24201-58-9	NI27777
41	186	43	188	201	42	187	203	273	100	90	80	57	54	50	40	33	22.78	11	9	3	1	–	–	2	–	–	–	–	–	–	2,4-Oxazolidinedione, 3-(3,5-dichlorophenyl)-5,5-dimethyl-	24201-58-9	NI27778
106	204	161	78	273	150	51	162	273	100	88	48	48	39	24	24	16		11	10	2	3	–	3	–	–	–	–	–	–	–	1,1,1-Trifluoro-4-oxopent-2-ylidene-isonicotinohydrazide		LI05804
154	274	156	121					273	100	40	34	12					0.00	11	12	3	1	1	–	1	–	–	–	–	–	–	4H-1,3-Thiazin-4-one, 2-(4-chlorophenyl)tetrahydro-3-methyl-, 1,1-dioxide	80-77-3	NI27780
152	42	98	154	153	28	174	69	273	100	62	58	42	32	31	28	24	0.00	11	12	3	1	1	–	1	–	–	–	–	–	–	4H-1,3-Thiazin-4-one, 2-(4-chlorophenyl)tetrahydro-3-methyl-, 1,1-dioxide	80-77-3	NI27779
239	241	270	78	272	105	251	192	273	100	65	40	27	26	19	16	16	3.66	11	13	1	3	–	–	2	–	–	–	–	–	–	1,3,5-Triazine, 4,6-dichloro-2-(2-methoxyanilino)-		MS06161
224	200	182	159	273	255			273	100	22	15	11	3	2				11	15	7	1	–	–	–	–	–	–	–	–	–	Isoxazole, 2-methyl-3-(ethoxycarbonyl)-2,3-bis(methoxycarbonyl)- 2,3-dihydro-		LI05805
257	259	274	230	232				273	100	98	25	19	18				0.00	11	16	2	1	–	–	–	1	–	–	–	–	–	Benzeneethylamine, 4-bromo-2,5-dimethoxy-α-methyl-	32156-26-6	NI27781
73	156	45	75	43	147	157	74	273	100	77	32	20	18	14	10	10	0.00	11	23	3	1	–	–	–	–	–	2	–	–	–	L-Proline, 5-oxo-1-(trimethylsilyl)-, trimethylsilyl ester	30274-77-2	MS06162
156	73	45	147	157	75	74	258	273	100	83	18	15	14	10	8	7	0.00	11	23	3	1	–	–	–	–	–	2	–	–	–	L-Proline, 5-oxo-1-(trimethylsilyl)-, trimethylsilyl ester	30274-77-2	NI27782
156	73	147	157	45	75	230	74	273	100	78	19	15	13	10	8	7	0.10	11	23	3	1	–	–	–	–	–	2	–	–	–	L-Proline, 5-oxo-1-(trimethylsilyl)-, trimethylsilyl ester	30274-77-2	NI27783
155	127	77	273	126	78	75	196	273	100	96	26	16	16	15	15	12		12	7	1	3	2	–	–	–	–	–	–	–	–	2-Naphthalenecarboxamide, N-5H-1,3,2,4-dithia(3-SIV)diazol-5-ylidene-	57726-57-5	NI27784
273	226	227	180	75	79	181	51	273	100	52	47	47	45	34	30	28		12	7	5	3	–	–	–	–	–	–	–	–	–	10H-Phenoxazine, 1,8-dinitro-	39522-54-8	MS06163
139	31	111	141	28	45	245	273	273	100	62	52	40	37	35	24	23		12	8	1	5	–	–	1	–	–	–	–	–	–	Benzamide, 4-chloro-N-1H-purin-6-yl-	36383-87-6	NI27785
124	273	83	82	94	42	140	96	273	100	43	41	23	17	15	12	12		12	19	2	1	2	–	–	–	–	–	–	–	–	1aH,5aH-Tropan-3-ol, 1,2-dithiolane-3-carboxylate		LI05806
43	176	144	41	273	134	67	98	273	100	62	50	49	42	36	28	24		12	19	4	1	1	–	–	–	–	–	–	–	–	L-Cysteine, N-acetyl-S-(2-oxocyclohexyl)-, methyl ester	77549-11-2	NI27786
97	69	214	177	43	88	86	41	273	100	30	25	21	18	17	17	14	1.00	12	19	4	1	1	–	–	–	–	–	–	–	–	L-Cysteine, N-acetyl-S-(3-oxocyclohexyl)-, methyl ester	77549-17-8	NI27787
97	214	69	88	43	181	86	180	273	100	98	57	47	45	40	36	31	0.00	12	19	4	1	1	–	–	–	–	–	–	–	–	L-Cysteine, N-acetyl-S-(4-oxocyclohexyl)-, methyl ester	82971-47-9	NI27788
210	178	134	122	135	136	211	94	273	100	84	78	23	15	14	13	12	0.00	12	19	4	1	1	–	–	–	–	–	–	–	–	1H-Pyrrolizine-7a(5H)-propanoic acid, tetrahydro-3-(methylsulphonyl)-5-oxo-, methyl ester		DD00868
195	194	180	43	178	41	55	56	273	100	95	45	37	34	31	20	12	11.00	12	19	4	1	1	–	–	–	–	–	–	–	–	8-Oxa-3-thiatricyclo[5.2.0.0^{7,9}]nonane-3,3-dioxide, 7-methyl-1-(4-morpholino)-		NI27789
44	94	43	60	61	81	85	153	273	100	99	99	91	78	76	61	56	10.00	12	19	6	1	–	–	–	–	–	–	–	–	–	2-Acetamido-4,6-di-O-acetyl-1,5-anhydro-2,3-dideoxy-D-arabino(or D-ribo)-hexitol		NI27790
43	125	107	55	185	97	83	167	273	100	85	25	20	13	13	13	11	0.00	12	19	6	1	–	–	–	–	–	–	–	–	–	1,3-Cyclohexanediol, 2,5-dimethyl-2-nitro-, (1R,2R,3S,5R)-, diacetate		DD00869

MASS TO CHARGE RATIOS	M.W.	INTENSITIES	Parent	C	H	O	N	S	F	Cl	Br	I	Si	P	B	X	COMPOUND NAME	CAS Reg No	No
43 125 107 55 185 167 126 95	273	100 81 45 12 9 8 8 8	0.00	12	19	6	1	–	–	–	–	–	–	–	–	–	1,3-Cyclohexanediol, 4,6-dimethyl-2-nitro-, (1R,2R,3S,4R,6S)-, diacetate		DD00870
43 72 44 60 86 114 101 102	273	100 33 30 24 20 15 15 12	0.00	12	19	6	1	–	–	–	–	–	–	–	–	–	β-Daunosamine, triacetyl-		NI27791
143 101 117 88 127 112 44 45	273	100 87 49 28 23 23 23 21	0.00	12	19	6	1	–	–	–	–	–	–	–	–	–	D-Galactonitrile, 4,5-di-O-acetyl-2,3-di-O-methyl-6-deoxy-		NI27792
127 96 112 101 131 186 126 154	273	100 99 97 97 86 62 61 52	0.00	12	19	6	1	–	–	–	–	–	–	–	–	–	D-Gluconitrile, 3,5-di-O-acetyl-2,4-di-O-methyl-6-deoxy-		NI27793
130 102 68 184 59 43 86 57	273	100 78 70 69 67 60 55 37	0.10	12	19	6	1	–	–	–	–	–	–	–	–	–	2-Pyrrolidinone, 3-hydroxy-1-[2,3-O-isopropylidene-β-D-ribofuranosyl]-, (S)-	64018-70-8	NI27794
58 134 82 80 91 119 42 79	273	100 88 41 41 30 26 23 15	0.00	12	20	1	1	–	–	–	1	–	–	–	–	–	Benzeneethanaminium, 4-methoxy-N,N,N-trimethyl-, bromide	6948-08-9	NI27795
156 73 157 45 147 75 84 74	273	100 58 16 11 8 7 6 6	0.00	12	27	2	1	–	–	–	–	–	2	–	–	–	Cyclopentanecarboxylic acid, 1-amino-, bis(trimethylsilyl)-	56273-05-3	NI27796
73 45 156 59 43 18 75 29	273	100 42 18 16 14 13 12 11	0.00	12	27	2	1	–	–	–	–	–	2	–	–	–	2-Piperidinecarboxylic acid, 1-(trimethylsilyl)-, trimethylsilyl ester	55255-44-2	NI27798
156 73 157 45 84 59 158 75	273	100 49 15 11 6 6 5 5	0.00	12	27	2	1	–	–	–	–	–	2	–	–	–	2-Piperidinecarboxylic acid, 1-(trimethylsilyl)-, trimethylsilyl ester	55255-44-2	NI27797
156 73 45 157 59 43 75 74	273	100 75 23 14 11 10 8 7	0.00	12	27	2	1	–	–	–	–	–	2	–	–	–	2-Piperidinecarboxylic acid, 1-(trimethylsilyl)-, trimethylsilyl ester	55255-44-2	MS06164
273 275 237 202 119 274 239 238	273	100 67 38 25 19 17 13 13		13	5	–	3	–	–	2	–	–	–	–	–	–	Pyridine, 2,6-dichloro-3,5-dicyano-4-phenyl-		MS06165
273 229 40 68 28 146 119 274	273	100 47 39 32 30 19 19 17		13	11	2	3	1	–	–	–	–	–	–	–	–	7H-1,3,4-Thiadiazolo[3,2-a]pyrimidin-7-one, 2-(2-ethoxyphenyl)-		NI27799
273 180 182 181 65 63 152 91	273	100 86 27 27 21 12 11 11		13	11	4	3	–	–	–	–	–	–	–	–	–	Aniline, N-(3-methylphenyl)-2,4-dinitro-	964-79-4	NI27800
91 105 273	273	100 80 51		13	11	4	3	–	–	–	–	–	–	–	–	–	Benzylamine, 2,4-dinitrophenylhydrazone		LI05807
173 175 230 232 145 147 109 177	273	100 67 37 24 22 14 13 12	0.00	13	17	1	1	–	–	2	–	–	–	–	–	–	Benzamide, N,N-diisopropyl-, dichloro-	74367-07-0	NI27802
184 274 276 149 185 275 173	273	100 94 63 19 14 13 13	0.00	13	17	1	1	–	–	2	–	–	–	–	–	–	Benzamide, N,N-diisopropyl-, dichloro-	74367-07-0	NI27801
110 188 129 97 59 42 41 43	273	100 84 70 42 30 29 28 27	0.00	13	23	5	1	–	–	–	–	–	–	–	–	–	4-Oxazolidinecarboxylic acid, 2-tert-butyl-3-formyl-4-(1-hydroxyisopropyl)-, methyl ester, (2R,4R)-		DD00871
145 18 130 75 73 41 69 55	273	100 68 56 53 51 27 26 25	2.56	13	27	3	1	–	–	–	–	–	1	–	–	–	Octanoic acid, trimethylsilyl-7-ketoxime, ethyl ester		NI27803
158 73 189 258 57 30 145 172	273	100 84 65 65 47 47 39 31	4.12	13	27	3	1	–	–	–	–	–	1	–	–	–	Octanoylglycine, trimethylsilyl-		NI27804
139 150 76 51 123 43 45 77	273	100 69 41 41 31 31 28 27	0.10	14	11	3	1	1	–	–	–	–	–	–	–	–	Benzenecarbothioic acid, 2-nitro-, S-(4-methylphenyl) ester	52909-88-3	NI27805
150 104 76 273 123 77 45 43	273	100 40 35 26 12 10 10 9		14	11	3	1	1	–	–	–	–	–	–	–	–	Benzenecarbothioic acid, 3-nitro-, S-(4-methylphenyl) ester	52909-89-4	NI27806
150 104 76 43 91 273 45 123	273	100 38 29 20 19 11 11 9		14	11	3	1	1	–	–	–	–	–	–	–	–	Benzenecarbothioic acid, 4-nitro-, S-(4-methylphenyl) ester	28122-84-1	NI27807
231 43 232 167 186 273 233 198	273	100 25 16 12 10 6 6 6		14	11	3	1	1	–	–	–	–	–	–	–	–	10H-Phenothiazine, 10-acetyl-, 5,5-dioxide	1220-99-1	LI05808
231 43 232 167 186 233 183 182	273	100 25 15 12 8 6 5 5	5.00	14	11	3	1	1	–	–	–	–	–	–	–	–	10H-Phenothiazine, 10-acetyl-, 5,5-dioxide	1220-99-1	NI27808
231 232 167 184 233 198 273 183	273	100 15 12 9 6 6 5 5		14	11	3	1	1	–	–	–	–	–	–	–	–	10H-Phenothiazine, 10-acetyl-, 5,5-dioxide	1220-99-1	MS06166
150 104 243 273 120	273	100 32 31 13 13		14	11	5	1	–	–	–	–	–	–	–	–	–	Benzoic acid, 4-nitro-, phenoxymethyl ester		LI05809
273 272 228 244 243 230 77 245	273	100 55 49 33 31 27 27 26		14	11	5	1	–	–	–	–	–	–	–	–	–	1,3-Dioxolo[4,5-g]furo[2,3-b]quinoline, 4,9-dimethoxy-	522-06-5	NI27809
239 205 254 240 273	273	100 99 90 64 15		14	12	1	3	–	–	1	–	–	–	–	–	–	3-Amino-6-chloro-3,4-dihydro-4-hydroxy-4-phenylquinazoline		LI05810
217 41 273 121 57 119 144 218	273	100 39 33 31 23 18 16 14		14	15	3	3	–	–	–	–	–	–	–	–	–	Pyrimidine, 5-nitro-2-(4-butoxyphenyl)-		NI27810
151 86 237 36 152 73 220 56	273	100 70 37 28 27 20 16 15	0.00	14	24	2	1	–	–	1	–	–	–	–	–	–	Hexanamine, (3,4-dimethoxyphenyl)-, hydrochloride		NI27811
214 45 59 29 174 102 72 140	273	100 91 66 59 58 57 50 39	0.00	14	31	2	1	–	–	–	–	–	1	–	–	–	Bis(2-ethoxyethyl)amine, N-[(allyldimethylsilyl)methyl]-		DD00872
219 246 77 117 191 105 164 257	273	100 89 62 56 55 41 40 36	0.00	15	7	1	5	–	–	–	–	–	–	–	–	–	4H-Pyran-3,4,4,5-tetracarbonitrile, 2-amino-6-phenyl-	120446-35-7	DD00873
127 105 77 129 273 120 91 275	273	100 72 50 31 21 21 8 7		15	12	2	1	–	–	1	–	–	–	–	–	–	Acetamide, 2-benzoyl-N-(3-chlorophenyl)-		LI05811
127 105 129 120 273 77 275 146	273	100 96 32 32 22 10 8 8		15	12	2	1	–	–	1	–	–	–	–	–	–	Acetamide, 2-benzoyl-N-(4-chlorophenyl)-		LI05812
196 119 198 273 105 120 197 77	273	100 78 70 64 49 32 31 28		15	12	2	1	–	–	1	–	–	–	–	–	–	4-Chloroacetamidobenzophenone	13788-59-5	NI27812
152 180 109 77 106 78 65 51	273	100 98 25 22 18 18 18 18	0.00	15	15	2	1	1	–	–	–	–	–	–	–	–	Benzene, 4-[ethoxy(phenylthio)methyl]-1-nitro-		DD00874
194 273 48 195 165 36 274 192	273	100 33 18 17 8 7 6 6		15	15	2	1	1	–	–	–	–	–	–	–	–	Methanesulphonamide, N-(9,10-dihydrophenanthren-2-yl)-		MS06167
118 91 119 210 77 125 182 94	273	100 32 30 26 25 24 23 23	14.00	15	15	2	1	1	–	–	–	–	–	–	–	–	Sulphoximine, S-methyl-N-(2-methylbenzoyl)-S-phenyl-	54090-91-4	NI27814
118 91 119 210 77 125 156 94	273	100 32 29 25 25 24 22 22	15.14	15	15	2	1	1	–	–	–	–	–	–	–	–	Sulphoximine, S-methyl-N-(2-methylbenzoyl)-S-phenyl-	54090-91-4	NI27813
272 200 131 273 242 201 169 143	273	100 99 94 87 59 43 39 39		15	15	4	1	–	–	–	–	–	–	–	–	–	1H-1-Benzazepine-3,4-dicarboxylic acid, 1-methyl-, dimethyl ester	34132-46-2	NI27815
155 273 214 213 274 241 181 128	273	100 86 28 20 15 15 15 15		15	15	4	1	–	–	–	–	–	–	–	–	–	Maleic acid, 2-(3-methylindol-2-yl)-, dimethyl ester		NI27816
186 231 185 214 213 104 43 76	273	100 99 48 40 36 36 36 32	0.44	15	15	4	1	–	–	–	–	–	–	–	–	–	4-Phthalimido-2-butenoic acid, isopropyl ester		NI27817
77 118 199 67 69 273 104 51	273	100 97 81 65 60 50 50 50		15	15	4	1	–	–	–	–	–	–	–	–	–	3-Pyridinecarboxylic acid, 1,6-dihydro-4-hydroxy-2-methyl-6-oxo-1-phenyl-, ethyl ester	1153-83-9	NI27818
163 52 78 79 77 69 51 50	273	100 84 45 42 39 34 28 28	0.49	15	15	4	1	–	–	–	–	–	–	–	–	–	Tricyclo[4.2.2.0^{2,5}]deca-7,9-diene-7,8-dicarboxylic acid, 3-cyano-, dimethyl ester	20185-30-2	NI27819
163 53 78 79 77 69 51 91	273	100 84 46 42 39 34 28 27	1.00	15	15	4	1	–	–	–	–	–	–	–	–	–	Tricyclo[4.2.2.0^{2,5}]deca-7,9-diene-9,10-dicarboxylic acid, 3-cyano-, dimethyl ester		LI05813
78 113 59 131 163 77 144 79	273	100 34 22 19 18 18 16 16	5.00	15	15	4	1	–	–	–	–	–	–	–	–	–	Tricyclo[4.4.4.0^{2,5}]deca-7,9-diene-9,10-dicarboxylic acid, 3-cyano-, dimethyl ester		LI05814
78 150 91 79 106 104 44 39	273	100 44 21 19 18 11 11 11	0.00	15	15	4	1	–	–	–	–	–	–	–	–	–	Tricyclo[4.2.0.0^{2,4}]octan-5-ol, 4-nitrobenzoate, (1α,2β,4β,5α,6α)-	53600-58-1	NI27820
78 28 113 163 131 79 77 145	273	100 32 17 9 9 8 8 7	1.70	15	15	4	1	–	–	–	–	–	–	–	–	–	Tricyclo[4.2.2.0^{5,8}]oct-9-ene-2,3-dicarboxylic acid, 7-cyano-, dimethyl ester		IC03555
228 273 237 229 165 274 238 230	273	100 47 38 30 16 10 10 6		15	16	–	1	–	–	1	–	–	1	–	–	–	9-Silaanthracene, 9-(dimethylamino)-9-chloro-9,10-dihydro-		NI27821

MASS TO CHARGE RATIOS								M.W.	INTENSITIES								Parent	C	H	O	N	S	F	Cl	Br	I	Si	P	B	X	COMPOUND NAME	CAS Reg No	No
258	273	180	260	195	275	259	274	**273**	100	70	52	34	31	24	21	15		15	16	–	1	–	–	1	–	–	1	–	–	–	9-Aza-10-silaanthracene, 9-ethyl-10-chloro-10-methyl-9,10-dihydro-		NI27822
88	73	44	98	75	143	142	101	**273**	100	86	80	70	68	32	20	17	**6.33**	15	19	2	3	–	–	–	–	–	–	–	–	–	L-Tryptophan, N-[(dimethylamino)methylene]-, methyl ester	59824-39-4	NI27823
136	230	273	231	109	150	149	274	**273**	100	76	72	54	49	43	32	14		16	16	–	1	1	1	–	–	–	–	–	–	–	1,5-Benzothiazepine, 2-methyl-4-(4-fluorophenyl)-2,3,4,5-tetrahydro-	78031-26-2	NI27824
139	273	105	141	111	258	275	274	**273**	100	49	41	33	23	19	16	12		16	16	1	1	–	–	1	–	–	–	–	–	–	Benzamide, 4-chloro-N-methyl-N-(2-methylbenzyl)-		LI05815
139	273	105	141	275	274	111	258	**273**	100	48	40	33	25	22	22	18		16	16	1	1	–	–	1	–	–	–	–	–	–	Benzamide, 4-chloro-N-methyl-N-(α-methylbenzyl)-	24456-01-7	NI27825
166	167	168	169	139	103	44	77	**273**	100	61	39	20	20	14	14	11	**0.00**	16	16	1	1	–	–	1	–	–	–	–	–	–	1,3-Oxazolidine, 4-methyl-cis-5-phenyl-2-(4-chlorophenyl)-		MS06168
166	167	168	139	77	103	89	132	**273**	100	81	40	23	16	16	15	14	**0.00**	16	16	1	1	–	–	1	–	–	–	–	–	–	1,3-Oxazolidine, 4-methyl-trans-5-phenyl-2-(4-chlorophenyl)-		MS06169
58	273	240	176	74	274	184	154	**273**	100	42	36	23	20	16	15	14		16	19	1	1	1	–	–	–	–	–	–	–	–	Benzo[d]thieno[2,3-g]azecin-13-ol, 4,5,6,7,8,13-hexahydro-6-methyl-		NI27826
273	98	57	70	97	58	201	230	**273**	100	57	47	44	39	33	30	27		16	19	1	1	1	–	–	–	–	–	–	–	–	[1]Benzothieno[3,2-d]azecin-8-one, 1,2,3,4,5,6,7,8-octahydro-3-methyl-	99659-33-3	NI27827
229	255	273	214	240	198	186	170	**273**	100	98	95	43	40	23	19	18		16	19	3	1	–	–	–	–	–	–	–	–	–	1-Butanone, 1-(7-allyl-6-methoxyindol-2-yl)-4-hydroxy-		MS06170
43	216	188	273	186	258	174	77	**273**	100	93	71	47	32	28	23	18		16	19	3	1	–	–	–	–	–	–	–	–	–	Carbazole, 6,7-methylenedioxy-9-acetyl-cis-4a-methyl-1,2,3,4,4a,9a-hexahydro-		MS06171
273	245	228	202	272	226	256	200	**273**	100	15	12	12	11	11	9	8		16	19	3	1	–	–	–	–	–	–	–	–	–	Crinan-1-ol, (1α)-	41928-89-6	NI27828
273	201	256	272	200	185	174	202	**273**	100	31	18	17	16	15	13	11		16	19	3	1	–	–	–	–	–	–	–	–	–	Crinan-3-ol, (3α)-	10438-97-8	NI27829
175	273	174	148	173	256	149	230	**273**	100	44	39	36	20	13	13	12		16	19	3	1	–	–	–	–	–	–	–	–	–	Desoxydihydropancracine		LI05816
242	273	258	240	202	215	194		**273**	100	83	33	26	5	4	4			16	19	3	1	–	–	–	–	–	–	–	–	–	α-Erythroidine		LI05817
242	273	258	240	215	202	194		**273**	100	84	32	18	7	6	3			16	19	3	1	–	–	–	–	–	–	–	–	–	β-Erythroidine		LI05818
105	131	168	133	199	198	132	200	**273**	100	15	9	9	7	5	4	1	**0.00**	16	19	3	1	–	–	–	–	–	–	–	–	–	Ethaneperoxoic acid, cyclohexylbenzyl ester	58422-78-9	NI27830
273	186	226	218	213	198	258	156	**273**	100	39	34	24	17	14	10	9		16	19	3	1	–	–	–	–	–	–	–	–	–	1H-Indole-2-carboxylic acid, 6-methoxy-7-(3-methylbut-2-enyl)-, methyl ester		MS06172
273	228	230	274	218	200	184	156	**273**	100	42	36	15	15	13	11	11		16	19	3	1	–	–	–	–	–	–	–	–	–	1H-Indole-3-carboxylic acid, 5-methoxy-1-methyl-2-[(E)-1-pentenyl]-		DD00875
202	273	55	77	43	111	53	105	**273**	100	18	17	16	16	13	12	9		16	19	3	1	–	–	–	–	–	–	–	–	–	1H-Isoindole-1,3(2H)-dione, hexahydro-4-hydroxy-3a,7a-dimethyl-2-phenyl-, [3aS-(3aα,4β,7aα)]-	54345-99-2	NI27831
44	43	42	55	41	115	56	77	**273**	100	71	47	41	41	40	39	38	**25.02**	16	19	3	1	–	–	–	–	–	–	–	–	–	Morphinan-3,14-diol, 4,5-epoxy-, (5α)-	55255-50-0	NI27832
104	192	105	55	77	41	83	78	**273**	100	25	19	17	12	12	10	10	**5.00**	16	19	3	1	–	–	–	–	–	–	–	–	–	2,3-Morpholinedione, 4-cyclohexyl-6-phenyl-	37154-90-8	MS06173
115	201	173	273	143	96	172	135	**273**	100	60	50	30	27	25	20	12		16	19	3	1	–	–	–	–	–	–	–	–	–	2,4-Pentadienamide, 5-(1,3-benzodioxol-5-yl)-N-(2-methylpropyl)-, (Z,Z)-	79097-19-1	NI27833
115	201	173	273	143	172	96	216	**273**	100	50	46	34	27	25	16	10		16	19	3	1	–	–	–	–	–	–	–	–	–	2,4-Pentenediamide, 5-(1,3-benzodioxol-5-yl)-N-(2-methylpropyl)-, (E,E)-	5950-12-9	NI27834
201	173	115	273	174	172	143	96	**273**	100	82	79	75	35	34	32	28		16	19	3	1	–	–	–	–	–	–	–	–	–	Piperamide, N-isobutyl-		LI05819
273	201	173	202	274	130	186	230	**273**	100	99	47	42	31	30	24	21		16	19	3	1	–	–	–	–	–	–	–	–	–	2H-Pyrano[4,3-b]indol-1-one, 1,4-dihydro-8-methoxy-5-methyl-3-propyl-		DD00876
105	188	77	148	120	160	245	86	**273**	100	25	25	22	21	17	10	10	**2.00**	16	19	3	1	–	–	–	–	–	–	–	–	–	2-Pyrrolidone, 1-(2-methoxy-4-methyl-3-oxo-2-phenylcyclobutyl)-		NI27835
152	121	77	153	83	96	91	124	**273**	100	21	15	12	10	8	8	6	**1.00**	16	19	3	1	–	–	–	–	–	–	–	–	–	2-Pyrrolidone, 1-(4-methoxy-2-methyl-4-phenyl-3-oxo-1-butenyl)-		NI27836
79	77	256	120	91	168	130	115	**273**	100	85	72	72	72	70	66	55	**46.00**	16	19	3	1	–	–	–	–	–	–	–	–	–	Tricyclo[3.3.1.1^{3,7}]decan-2-ol, 1-(nitrophenyl)-	40621-20-3	NI27837
145	131	104	91	41	77	103	144	**273**	100	99	71	35	32	31	26	24	**20.60**	16	21	2	2	–	–	–	–	–	–	–	–	–	5,5-Dimethyl-4-phenylspiro[1,2]hexane-3′-imidazoline-3′-oxide-1′-oxile		NI27838
140	111	91	273	112	43	106	69	**273**	100	95	37	27	22	14	12	12		16	23	1	3	–	–	–	–	–	–	–	–	–	2-Pyrazoline-1-carboxamide, N-benzyl-4-ethyl-3-propyl-	24769-36-6	NI27839
245	217	273	216	126	114	246	77	**273**	100	55	50	27	26	26	20	19		17	11	1	3	–	–	–	–	–	–	–	–	–	Naphtho[2,3-d]-1,2,3-triazin-4(3H)-one, 3-phenyl-	19275-09-3	NI27840
245	217	273	216	114	126	77	246	**273**	100	55	48	27	27	26	21	20		17	11	1	3	–	–	–	–	–	–	–	–	–	Naphtho[2,3-d]-1,2,3-triazin-4(3H)-one, 3-phenyl-	19275-09-3	LI05820
272	129	273	83	189	245	117	85	**273**	100	90	35	24	20	18	16	16		17	11	1	3	–	–	–	–	–	–	–	–	–	1,2,3-Triazole, 5-phenyl-4-(3-phenyl-2-propynoyl)-		NI27841
182	148	105	103	42	139	184	77	**273**	100	88	67	37	36	34	33	29	**1.00**	17	20	–	1	–	–	1	–	–	–	–	–	–	Benzeneethylamine, 3-chloro-N-methyl-N-(2-phenylethyl)-	52059-56-0	NI27842
148	182	105	139	103	44	42	184	**273**	100	58	56	29	23	21	19	18	**1.00**	17	20	–	1	–	–	1	–	–	–	–	–	–	Benzeneethylamine, 4-chloro-N-methyl-N-(2-phenylethyl)-	52059-55-9	NI27843
44	182	125	184	36	127	43	30	**273**	100	23	16	8	7	5	5	5	**0.00**	17	20	–	1	–	–	1	–	–	–	–	–	–	Benzenepropylamine, 4-chloro-N-(2-phenylethyl)-	55255-78-2	NI27844
273	257	150	152	153	114	104	215	**273**	100	29	8	7	6	6	6	5		17	21	–	–	–	–	–	–	–	–	–	–	1	Titanium, η-cyclooctatetraene-η-(1,1-dimethylethyl)cyclopentadienyl-	55672-84-9	NI27845
214	91	82	83	42	215	43	94	**273**	100	43	42	31	29	24	23	21	**18.91**	17	23	2	1	–	–	–	–	–	–	–	–	–	8-Azabicyclo[3.2.1]octan-3-ol, 8-methyl-2-benzyl-, acetate (ester)	55925-27-4	NI27846
91	104	150	105	107	77	65	149	**273**	100	52	40	26	18	15	9	7	**0.10**	17	23	2	1	–	–	–	–	–	–	–	–	–	Diisopropylcyanomethyl 3-phenylpropionate		LI05821
82	97	103	42	77	72	91	115	**273**	100	99	37	33	26	24	16	13	**4.20**	17	23	2	1	–	–	–	–	–	–	–	–	–	Ethyl cis-2-(dimethylamino)-1-phenyl-3-cyclohexene-1-carboxylate		MS06174
82	97	103	77	72	42	91	115	**273**	100	99	49	26	26	21	18	17	**10.70**	17	23	2	1	–	–	–	–	–	–	–	–	–	Ethyl trans-2-(dimethylamino)-1-phenyl-3-cyclohexene-1-carboxylate		MS06175
173	152	139	96	121	273	158	100	**273**	100	82	63	55	44	41	30	27		17	23	2	1	–	–	–	–	–	–	–	–	–	2,4-Hexadienamide, N-isobutyl-6-(4-methoxyphenyl)-, (2E,4E)-	25090-18-0	NI27847
173	152	174	139	96	273	121	158	**273**	100	81	67	53	44	36	34	28		17	23	2	1	–	–	–	–	–	–	–	–	–	2,4-Hexadienamide, N-isobutyl-6-(4-methoxyphenyl)-, (2E,4E)-	25090-18-0	NI27848
173	273	174	57	159	158	41	115	**273**	100	52	45	36	34	26	19	16		17	23	2	1	–	–	–	–	–	–	–	–	–	3,5-Hexadienamide, N-isobutyl-6-(4-methoxyphenyl)-, (3E,5E)-	83029-40-7	NI27849
81	137	95	93	119	67	69	91	**273**	100	79	62	57	29	26	20	20	**7.60**	17	23	2	1	–	–	–	–	–	–	–	–	–	Isoborneolphenyl urethane		MS06176
72	45	144	115	43	73	46	44	**273**	100	13	11	11	7	6	6	6	**0.90**	17	23	2	1	–	–	–	–	–	–	–	–	–	1-Isopropylamino-2-methyl-3-(1-naphthyloxy)propan-2-ol		MS06177
72	158	115	56	43	129	73	41	**273**	100	16	14	10	10	6	6	6	**4.60**	17	23	2	1	–	–	–	–	–	–	–	–	–	1-Isopropylamino-3-(4-methyl-1-naphthyloxy)propan-2-ol		MS06178
72	141	58	43	158	115	142	73	**273**	100	39	11	10	9	9	8	6	**1.40**	17	23	2	1	–	–	–	–	–	–	–	–	–	1-Isopropylamino-3-(1-naphthylmethyloxy)propan-2-ol		IC03556
72	141	58	43	158	115	142	41	**273**	100	38	11	9	8	8	7	7	**1.40**	17	23	2	1	–	–	–	–	–	–	–	–	–	1-Isopropylamino-3-(1-naphthylmethyloxy)propan-2-ol		MS06179
72	102	144	115	43	73	130	127	**273**	100	10	9	9	7	6	5	5	**3.50**	17	23	2	1	–	–	–	–	–	–	–	–	–	1-Isopropylamino-3-(1-naphthyloxy)butan-2-ol		MS06180
72	130	115	70	43	144	73	41	**273**	100	34	9	7	7	6	5	4	**0.50**	17	23	2	1	–	–	–	–	–	–	–	–	–	1-Isopropylamino-4-(1-naphthyloxy)butan-2-ol		MS06181

MASS TO CHARGE RATIOS								M.W.	INTENSITIES								Parent	C	H	O	N	S	F	Cl	Br	I	Si	P	B	X	COMPOUND NAME	CAS Reg No	No
166	273	122	228	44	202	244	274	**273**	100	66	46	27	20	18	15	12		17	23	2	1	–	–	–	–	–	–	–	–	–	**6-Isoquinolinol, 2-methyl-4aα-(3-methoxyphenyl)-2,3,4,4a,5,6,7,8-octahydro-, (6α)-**	118724-84-8	**DD00877**
79	44	57	273	115	96	202	165	**273**	100	67	36	21	9	8	6	6		17	23	2	1	–	–	–	–	–	–	–	–	–	**6-Isoquinolinone, 2-methyl-4aα-(3-methoxyphenyl)-1,2,3,4,4a,5,6,7,8,8aα-decahydro-**	61528-05-0	**DD00878**
273	71	57	44	150	165	274	202	**273**	100	72	65	40	32	28	19	13		17	23	2	1	–	–	–	–	–	–	–	–	–	**6-Isoquinolinone, 2-methyl-4aα-(3-methoxyphenyl)-1,2,3,4,4a,5,6,7,8,8aβ-decahydro-**	61528-04-9	**DD00879**
93	43	41	80	69	55	71	121	**273**	100	94	50	41	34	26	24	23	**0.00**	17	23	2	1	–	–	–	–	–	–	–	–	–	**1,6-Octadien-3-ol, 3,7-dimethyl-, 2-aminobenzoate**	7149-26-0	**NI27850**
226	273	113	215	227	112	224	100	**273**	100	93	75	65	49	39	25	22		18	11	2	1	–	–	–	–	–	–	–	–	–	**Chrysene, 6-nitro-**	7496-02-8	**NI27851**
226	215	273	227	113	224	228	225	**273**	100	86	80	56	30	26	20	20		18	11	2	1	–	–	–	–	–	–	–	–	–	**Chrysene, nitro-**		**IC03557**
273	143	115	102	217	245			**273**	100	78	69	69	12	9				18	11	2	1	–	–	–	–	–	–	–	–	–	**Furan-2-one, 2,5-dihydro-5-(α-cyanobenzylidene-3-phenyl)-**		**LI05822**
273	217	216	77	63	204	272	76	**273**	100	43	39	34	30	29	27	24		18	11	2	1	–	–	–	–	–	–	–	–	–	**1H-Indene-1,3(2H)-dione, 2-(2-quinolinyl)-**	83-08-9	**NI27852**
273	76	228	229	50	140	104	274	**273**	100	84	50	40	26	21	21	19		18	11	2	1	–	–	–	–	–	–	–	–	–	**1H-Isoindole-1,3(2H)-dione, 2-(1-naphthalenyl)-**	5333-99-3	**NI27853**
272	273	180	228	126	274	229	152	**273**	100	93	37	21	18	16	9	9		18	11	2	1	–	–	–	–	–	–	–	–	–	**1,8-Naphthalimide, N-phenyl-**		**MS06182**
272	273	271	257	256	135.5	129	274	**273**	100	70	33	27	19	17	14	13		18	15	–	3	–	–	–	–	–	–	–	–	–	**Benz[a]imidazo[1,2-c]carbazole, 2,3-dihydro-8-methyl-**		**MS06183**
168	167	273	169	77	274	78	51	**273**	100	58	57	35	15	12	7	4		18	15	–	3	–	–	–	–	–	–	–	–	–	**Diphenylamine, 4-phenylazo-**	101-75-7	**NI27854**
168	167	77	273	51	169	166	165	**273**	100	60	46	35	25	14	11	11		18	15	–	3	–	–	–	–	–	–	–	–	–	**Diphenylamine, 4-phenylazo-**	101-75-7	**LI05823**
167	168	77	273	51	166	169	115	**273**	100	97	42	37	15	14	12	8		18	15	–	3	–	–	–	–	–	–	–	–	–	**Diphenylamine, 4-phenylazo-**	101-75-7	**NI27855**
135	41	161	43	39	79	83	77	**273**	100	25	22	22	20	18	16	15	**0.80**	18	27	1	1	–	–	–	–	–	–	–	–	–	**2-Aziridinone, 1-(1-adamantyl)-3-(1-methylcyclopentyl)-**	26905-18-0	**NI27856**
55	82	41	83	68	56	67	28	**273**	100	95	62	57	52	30	19	19	**0.00**	18	27	1	1	–	–	–	–	–	–	–	–	–	**2H-1,3-Benzoxazine, 6-tert-butyl-3-cyclohexyl-3,4-dihydro-**	5451-32-1	**NI27857**
122	105	77	57	174	273	258	216	**273**	100	40	15	12	10	4	4	4		18	27	1	1	–	–	–	–	–	–	–	–	–	**Cyclohexylamine, N-benzoyl-4-tert-butyl-1-methyl-, cis-**		**LI05824**
122	105	77	174	57	273	258	216	**273**	100	53	17	15	12	5	5	5		18	27	1	1	–	–	–	–	–	–	–	–	–	**Cyclohexylamine, N-benzoyl-4-tert-butyl-1-methyl-, trans-**		**LI05825**
180	77	181	51	45	60	105	44	**273**	100	44	17	9	9	8	5	4	**3.96**	19	15	1	1	–	–	–	–	–	–	–	–	–	**Benzenecarboximidic acid, N-phenyl-, phenyl ester**	15940-86-0	**NI27858**
131	143	103	273	77	115	132	144	**273**	100	79	49	39	27	24	10	9		19	15	1	1	–	–	–	–	–	–	–	–	–	**Phenylpropenamide, N-(1-naphthyl)-**		**TR00295**
143	131	273	103	77	115	144	274	**273**	100	99	51	46	25	22	12	10		19	15	1	1	–	–	–	–	–	–	–	–	–	**Phenylpropenamide, N-(2-naphthyl)-**		**TR00296**
273	91	182	274	181	77	180	104	**273**	100	85	25	23	20	15	10	10		20	19	–	1	–	–	–	–	–	–	–	–	–	**Aniline, N,N-dibenzyl-**		**IC03558**
197	195	198	278	147	145	31	93	**274**	100	71	26	19	16	12	11	8	**1.13**	2	–	–	–	–	3	1	2	–	–	–	–	–	**Ethane, 1,1-dibromo-1-chloro-2,2,2-trifluoro-**		**IC03559**
191	195	199	207	147	209	145	176	**274**	100	77	25	10	9	7	7	5	**0.00**	2	–	–	–	–	3	1	2	–	–	–	–	–	**Ethane, 1,1-dibromo-1-chloro-2,2,2-trifluoro-**		**IC03560**
77	274	127	51	27	69	65	141	**274**	100	56	50	49	48	38	32	18		4	4	–	–	–	5	–	–	1	–	–	–	–	**Butane, 1,1,1,2,2-pentafluoro-4-iodo-**	40723-80-6	**NI27859**
151	153	45	133	135	80	82	71	**274**	100	85	63	51	48	41	39	36	**0.00**	4	4	4	–	–	–	–	2	–	–	–	–	–	**Butanedioic acid, 2,3-dibromo-**	526-78-3	**NI27860**
205	274	59	155	61	45	57	136	**274**	100	54	34	33	30	26	18	15		4	6	–	–	–	6	–	–	–	–	1	–	1	**Arsine, dimethylphosphinobis(trifluoromethyl)-**		**LI05826**
105	274	103	109	131	89	124	69	**274**	100	34	33	20	17	17	10	10		4	6	–	–	–	6	–	–	–	–	1	–	1	**Arsine, dimethylphosphinobis(trifluoromethyl)-**		**LI05827**
144	72	76	102	34	30	32	60	**274**	100	97	95	94	47	39	26	23	**0.00**	4	6	–	2	4	–	–	–	–	–	–	–	1	**Zinc, [[1,2-ethanediylbis[carbamodithioato]](2-)]-**	12122-67-7	**NI27861**
103	131	143	104	105	71	132	77	**274**	100	27	9	6	5	3	2	2	**0.00**	4	6	–	2	4	–	–	–	–	–	–	–	1	**Zinc, [[1,2-ethanediylbis[carbamodithioato]](2-)]-**	12122-67-7	**NI27862**
227	107	105	242	103	89	259	123	**274**	100	90	90	67	39	35	28	28	**18.00**	4	12	–	–	2	–	–	–	–	–	–	–	2	**Arsinodithioic acid, dimethyl-, anhydrosulphide with dimethylarsinothious acid**	7208-00-6	**MS06184**
39	67	41	232	234	27	274	276	**274**	100	61	60	58	56	51	41	40		5	8	–	–	–	–	–	1	1	–	–	–	–	**1-Pentene, 1-bromo-1-iodo-**	55683-02-8	**MS06185**
31	121	36	93	38	32	65	39	**274**	100	22	10	6	6	6	5	2	**0.00**	5	16	1	8	–	–	2	–	–	–	–	–	–	**Hydrazinecarboximidamide, 2,2′-(1-methyl-1,2-ethanediylidene)bis-, dihydrochloride, monohydrate**	24968-67-0	**NI27863**
41	39	69	89	81	67	274	147	**274**	100	30	4	3	2	2	1	1		6	5	1	–	1	6	1	–	–	–	–	–	–	**1-Propene, 3-[[1-chloro-2,2,2-trifluoro-1-(trifluoromethyl)ethyl]sulphinyl]-**		**DD00880**
69	179	121	247	246	168	126	110	**274**	100	63	44	38	38	31	31	25	**2.50**	6	6	2	2	–	6	–	–	–	–	–	2	–	**2,4-Dimethyl-3-(trifluoroacetyl)-6-(trifluoromethyl)-1-oxa-3,5-diaza-2,4-dibora-5-cyclohexene**		**NI27864**
205	255	69	93	117	155	136	274	**274**	100	34	21	19	15	12	8	7		7	–	–	–	–	10	–	–	–	–	–	–	–	**Cyclobutene, 1,2-dimethyl-3-methylene-perfluoro-**		**LI05828**
63	64	29	31	30	95	131	69	**274**	100	19	17	12	11	10	9	9	**0.10**	7	6	2	–	–	8	–	–	–	–	–	–	–	**1,3-Dioxonane, 5,5,6,6,7,7,8,8-octafluoro-**	36301-47-0	**NI27865**
132	223	221	239	276	241	193	195	**274**	100	23	21	18	18	17	15	15	**12.98**	8	6	2	–	–	–	4	–	–	–	–	–	–	**1,4-Benzenedimethanol, 2,3,5,6-tetrachloro-**	7154-26-9	**NI27866**
276	261	259	274	36	218	278	263	**274**	100	94	77	75	54	51	47	44		8	6	2	–	–	–	4	–	–	–	–	–	–	**Benzene, 1,2,3,4-tetrachloro-5,6-dimethoxy-**	944-61-6	**MS06186**
261	276	259	274	263	209	87	211	**274**	100	80	80	62	50	42	41	40		8	6	2	–	–	–	4	–	–	–	–	–	–	**Benzene, 1,2,4,5-tetrachloro-3,6-dimethoxy-**	944-78-5	**NI27867**
77	51	177	197	127	113	69	95	**274**	100	47	17	9	7	6	6	4	**2.90**	8	7	–	–	–	9	–	–	–	–	–	–	–	**1-Octene, 3,3,4,4,7,7,8,8,8-nonafluoro-**	35208-07-2	**MS06187**
77	51	177	197	127	113	69	274	**274**	100	47	17	9	7	6	6	3		8	7	–	–	–	9	–	–	–	–	–	–	–	**1-Octene, 3,3,4,4,7,7,8,8,8-nonafluoro-**	35208-07-2	**LI05829**
154	194	51	80	65	79	52	165	**274**	100	91	48	45	40	36	33	30	**8.21**	8	7	4	2	–	–	–	1	–	–	–	–	–	**Acetamide, 2-bromo-N-(2-hydroxy-5-nitrophenyl)-**	3947-58-8	**NI27868**
57	29	41	55	27	274	272	218	**274**	100	56	44	19	19	15	13	13		8	18	–	–	–	–	–	–	–	–	–	–	2	**Diselenide, dibutyl**	20333-40-8	**NI27869**
54	67	82	55	79	41	81	42	**274**	100	98	66	66	61	61	36	24	**0.00**	8	18	6	–	2	–	–	–	–	–	–	–	–	**1,6-Hexanediol dimethanesulphonate**	4239-24-1	**NI27870**
83	56	79	123	55	43	41	15	**274**	100	70	67	56	54	34	30	20	**0.00**	8	18	6	–	2	–	–	–	–	–	–	–	–	**2,5-Hexanediol, dimethanesulphonate**	55-93-6	**NI27871**
88	89	60	61	97	274	142	186	**274**	100	37	24	23	19	13	13	11		8	19	2	–	3	–	–	–	–	–	1	–	–	**Phosphorodithioic acid, O,O-diethyl S-[2-(ethylthio)ethyl] ester**	298-04-4	**NI27872**
185	159	155	175	172	221	242	177	**274**	100	60	35	26	25	12	11	11	**0.00**	8	19	2	–	3	–	–	–	–	–	1	–	–	**Phosphorodithioic acid, O,O-diethyl S-[2-(ethylthio)ethyl] ester**	298-04-4	**NI27873**

MASS TO CHARGE RATIOS								M.W.	INTENSITIES								Parent	C	H	O	N	S	F	Cl	Br	I	Si	P	B	X	COMPOUND NAME	CAS Reg No	No
88	89	60	97	142	186	153	61	**274**	100	49	34	31	28	21	21	21	**14.80**	8	19	2	–	3	–	–	–	–	–	1	–	–	Phosphorodithioic acid, O,O-diethyl S-[2-(ethylthio)ethyl] ester	298-04-4	NI27874
29	45	27	141	88	97	60	91	**274**	100	58	52	50	48	35	32	31	**0.40**	8	19	4	–	2	–	–	–	–	–	1	–	–	Phosphorothioic acid, O,O-diethyl O-[2-(ethylsulphinyl)ethyl] ester	5286-73-7	NI27875
29	109	27	141	45	81	197	61	**274**	100	83	73	70	65	63	49	41	**0.00**	8	19	4	–	2	–	–	–	–	–	1	–	–	Phosphorothioic acid, O,O-diethyl S-[2-(ethylsulphinyl)ethyl] ester	2496-92-6	NI27876
65	81	99	127	145	155	162	161	**274**	100	79	74	64	59	49	48	42	**2.90**	8	20	6	–	–	–	–	–	–	–	2	–	–	Diphosphoric(III,V) acid, tetraethyl ester	682-24-6	NI27877
65	99	81	127	145	162	155	161	**274**	100	86	86	77	72	60	58	52	**2.70**	8	20	6	–	–	–	–	–	–	–	2	–	–	Hypophosphoric acid, tetraethyl ester	679-37-8	NI27878
275	277	239	278	241	276	178	154	**274**	100	91	35	29	27	20	18	15	**10.80**	9	5	–	4	–	–	3	–	–	–	–	–	–	1,3,5-Triazin-2-amine, 4,6-dichloro-N-(2-chlorophenyl)-	101-05-3	NI27879
239	241	178	143	90	180	111	87	**274**	100	67	49	47	16	15	13	12	**6.00**	9	5	–	4	–	–	3	–	–	–	–	–	–	1,3,5-Triazin-2-amine, 4,6-dichloro-N-(2-chlorophenyl)-	101-05-3	NI27881
239	241	178	143	75	87	18	99	**274**	100	62	47	46	31	25	25	23	**13.26**	9	5	–	4	–	–	3	–	–	–	–	–	–	1,3,5-Triazin-2-amine, 4,6-dichloro-N-(2-chlorophenyl)-	101-05-3	NI27880
183	185	155	75	157	163	76	87	**274**	100	99	71	71	69	68	68	40	**2.10**	9	7	1	–	2	–	–	1	–	–	–	–	–	2-Propenedithioic acid, 3-(3-bromophenyl)-3-hydroxy-	55256-28-5	NI27882
185	85	76	75	155	162	50	241	**274**	100	94	90	88	72	71	63	51	**23.17**	9	7	1	–	2	–	–	1	–	–	–	–	–	2-Propenedithioic acid, 3-(4-bromophenyl)-3-hydroxy-	55256-29-6	NI27883
116	115	195	197	58	63	89	117	**274**	100	84	53	51	19	13	11	10	**0.95**	9	8	–	–	–	–	–	2	–	–	–	–	–	1H-Indene, 1,2-dibromo-2,3-dihydro-	20357-79-3	NI27884
42	186	122	215	58	76	75	50	**274**	100	81	76	51	37	35	26	20	**5.00**	9	10	6	2	1	–	–	–	–	–	–	–	–	1,3,5-Dioxazine, 5-(3-nitrophenylsulphonyl)dihydro-		NI27885
176	134	82	80	18	89	90	177	**274**	100	85	29	28	26	23	21	19	**0.00**	9	11	1	2	1	–	–	1	–	–	–	–	–	2-Thiazolamine, 4-phenyl-, monohydrobromide, monohydrate	52253-69-7	NI27886
177	205	178	217	179	176	162	76	**274**	100	14	11	4	4	3	2	2	**0.00**	9	11	1	2	1	–	–	1	–	–	–	–	–	2-Thiazolamine, 4-phenyl-, monohydrobromide, monohydrate	52253-69-7	NI27887
68	137	39	106	69	187	75	156	**274**	100	99	64	50	37	28	28	23	**0.00**	10	5	1	–	–	7	–	–	–	–	–	–	–	11-Oxatetracyclo[6.2.1.0[2,7].0[3,6]]undec-9-ene, 3,4,4,5,5,6,7-heptafluoro-		NI27888
134	52	91	162	190	218	274	246	**274**	100	99	82	36	19	13	12	12		10	6	6	–	–	–	–	–	–	–	–	–	1	Chromium, (methoxy-1-propynylmethylene)pentacarbonyl-		NI27889
143	57	73	133	125	55	43	82	**274**	100	75	70	44	44	40	40	29	**2.41**	10	14	7	2	–	–	–	–	–	–	–	–	–	Uridine, 5-(hydroxymethyl)-	30414-00-7	NI27890
274	182	229	181	103	275	183	245	**274**	100	96	27	24	24	15	14	12		11	6	5	4	–	–	–	–	–	–	–	–	–	1H-Pyrido[3,2-b][1,4]benzoxazine, 7,9-dinitro-	39522-51-5	MS06188
172	190	55	144	72	274	218	89	**274**	100	92	80	49	38	35	26	16		11	7	5	–	–	–	–	–	–	–	–	–	1	Manganese, tricarbonyl-π-cyclopentadienylacrylic acid		LI05830
246	174	274	248	56	187	230	173	**274**	100	83	63	63	63	58	54	46		11	12	2	2	–	–	2	–	–	–	–	–	–	1-(3,4-Dichlorophenyl)-3-ethoxyimidazolidinone		NI27891
43	111	82	215	153	144	71	170	**274**	100	42	40	20	17	11	11	9	**0.00**	11	14	8	–	–	–	–	–	–	–	–	–	–	β-D-Xylofuranose, 3-C-methyl-, cyclic 3,5-carbonate 1,2-diacetate	70723-16-9	NI27892
91	28	92	41	105	65	55	39	**274**	100	28	12	10	8	8	8	8	**1.00**	11	15	–	–	–	–	–	–	1	–	–	–	–	Benzene, (3-iodo-2,2-dimethylpropyl)-	40548-64-9	NI27893
73	147	75	45	259	148	83	74	**274**	100	94	26	20	16	15	12	9	**0.30**	11	22	4	–	–	–	–	–	–	2	–	–	–	Butanedioic acid, methylene-, bis(trimethylsilyl) ester	55494-04-7	NI27894
147	73	75	45	148	259	149	74	**274**	100	90	18	18	16	14	10	9	**0.00**	11	22	4	–	–	–	–	–	–	2	–	–	–	Butanedioic acid, methylene-, bis(trimethylsilyl) ester	55494-04-7	NI27895
147	73	75	259	148	45	149	74	**274**	100	79	19	15	14	13	9	7	**0.00**	11	22	4	–	–	–	–	–	–	2	–	–	–	Butanedioic acid, methylene-, bis(trimethylsilyl) ether	55494-04-7	MS06189
147	73	45	75	148	259	28	74	**274**	100	100	27	22	16	12	12	9	**0.00**	11	22	4	–	–	–	–	–	–	2	–	–	–	2-Butenedioic acid, methyl-, bis(trimethylsilyl) ester, (Z)-	77220-12-3	NI27896
73	147	259	217	75	68	230	45	**274**	100	74	55	44	41	31	26	18	**0.00**	11	22	4	–	–	–	–	–	–	2	–	–	–	2-Pentenedioic acid, bis(trimethylsilyl) ester	55255-43-1	NI27897
73	147	68	217	259	75	45	230	**274**	100	47	37	23	22	19	14	11	**0.00**	11	22	4	–	–	–	–	–	–	2	–	–	–	2-Pentenedioic acid, bis(trimethylsilyl) ester	55255-43-1	NI27898
73	147	68	75	45	217	259	74	**274**	100	36	34	26	14	12	11	9	**0.00**	11	22	4	–	–	–	–	–	–	2	–	–	–	2-Pentenedioic acid, bis(trimethylsilyl) ester	55255-43-1	NI27899
73	147	157	259	45	75	43	148	**274**	100	71	38	31	24	13	12	11	**0.00**	11	22	4	–	–	–	–	–	–	2	–	–	–	2,4-Pentenoic acid, 2,4-bis[(trimethylsilyl)oxy]-		NI27900
73	147	259	157	45	75	43	74	**274**	100	55	34	32	26	17	14	10	**0.00**	11	22	4	–	–	–	–	–	–	2	–	–	–	2,4-Pentenoic acid, 2,4-bis[(trimethylsilyl)oxy]-		NI27901
73	130	75	157	115	44	131	89	**274**	100	96	66	43	40	36	34	32	**0.90**	11	23	5	–	–	–	–	–	–	1	–	1	–	α-L-Galactopyranoside, methyl 6-deoxy-2-O-(trimethylsilyl)-, cyclic methylboronate	54400-92-9	NI27902
106	154	159	102	131	174	172	138	**274**	100	89	75	66	64	54	52	40	**22.00**	11	30	–	4	–	–	–	–	–	2	–	–	–	Cyclodisilazane-1,3-diamine, N,N,N′,N′,2,2,4-heptamethyl-4-isopropyl-	66436-32-6	NI27903
190	151	117	115	91	116	103	218	**274**	100	85	71	62	55	48	46	45	**0.00**	12	10	4	–	–	–	–	–	–	–	–	–	1	Iron, tricarbonyl(1,3,5,7-cyclooctatetraene-1-methanol)-	11064-43-0	MS06190
190	232	149	148	191	122	66	189	**274**	100	67	36	21	19	19	18	16	**2.00**	12	14	2	6	–	–	–	–	–	–	–	–	–	Pteridine, 2,4-bis(acetylamino)-6,7-dimethyl-		LI05831
256	274	238	221	258	276	240	203	**274**	100	94	84	80	34	30	28	22		12	15	3	–	1	–	1	–	–	–	–	–	–	α-D-Xylopyranose, 4-S-benzyl-3-chloro-3-deoxy-4-thio-	32848-94-5	NI27904
88	275	145	277	116	86	173	276	**274**	100	40	31	25	11	10	8	7	**0.59**	12	16	1	2	–	–	2	–	–	–	–	–	–	Urea, N-butyl-N′-(3,4-dichlorophenyl)-N-methyl-	555-37-3	NI27907
57	44	32	114	187	58	29	41	**274**	100	72	67	47	23	19	19	17	**10.50**	12	16	1	2	–	–	2	–	–	–	–	–	–	Urea, N-butyl-N′-(3,4-dichlorophenyl)-N-methyl-	555-37-3	NI27905
57	44	114	41	58	42	43	125	**274**	100	85	58	55	43	25	14	13	**6.71**	12	16	1	2	–	–	2	–	–	–	–	–	–	Urea, N-butyl-N′-(3,4-dichlorophenyl)-N-methyl-	555-37-3	NI27906
258	113	112	214	196	181	183	259	**274**	100	96	92	60	44	35	27	24	**0.00**	12	18	5	–	1	–	–	–	–	–	–	–	–	Benzene, 2-(2-hydroxyethylthio)-1,4-bis(2-hydroxyethoxy)-	93885-75-7	NI27908
43	259	59	113	201	100	71	85	**274**	100	30	18	10	8	6	6	4	**0.00**	12	18	7	–	–	–	–	–	–	–	–	–	–	α-D-Galactopyranuronic acid, 1,2:3,4-di-O-isopropylidene-	25253-46-7	LI05832
43	259	59	113	41	201	100	71	**274**	100	29	19	12	10	9	8	7	**0.00**	12	18	7	–	–	–	–	–	–	–	–	–	–	α-D-Galactopyranuronic acid, 1,2:3,4-di-O-isopropylidene-	25253-46-7	NI27909
43	259	44	59	113	41	29	100	**274**	100	28	22	19	12	10	9	8	**0.00**	12	18	7	–	–	–	–	–	–	–	–	–	–	α-D-Galactopyranuronic acid, 1,2:3,4-di-O-isopropylidene-	25253-46-7	MS06191
43	127	187	85	83	215	113	112	**274**	100	99	80	73	72	43	36	34	**0.00**	12	18	7	–	–	–	–	–	–	–	–	–	–	D-Glucofuranoside, acetyl 2,5-di-O-acetyl-3,6-dideoxy-		NI27910
43	83	84	112	189	57	55	103	**274**	100	99	30	26	25	23	20	16	**0.00**	12	18	7	–	–	–	–	–	–	–	–	–	–	D-Glucopyranoside, acetyl 2,4-di-O-acetyl-3,6-dideoxy-		NI27911
45	111	158	127	157	125	171	215	**274**	100	45	26	25	20	19	13	7	**0.00**	12	18	7	–	–	–	–	–	–	–	–	–	–	7-Oxabicyclo[2.2.1]heptane, 2-endo,5-exo-diacetyl-2-exo,6-endo-dimethoxy-, (1RS,2RS,4RS,5RS,6SR)-		DD00881
215	155	127	141	216	243	143	218	**274**	100	48	26	22	18	15	12	11	**0.00**	12	18	7	–	–	–	–	–	–	–	–	–	–	2-Oxabicyclo[2.2.1]heptane, 3-exo,7-syn-diacetyl-6,6-dimethoxy-, (1RS,3SR,4RS,7RS)-		DD00882
43	172	155	215	59	143	116	83	**274**	100	95	85	84	49	33	30	29	**3.00**	12	18	7	–	–	–	–	–	–	–	–	–	–	2,3-Pyrandicarboxylic acid, tetrahydro-3-hydroxy-6-oxo-2,4,5-trimethyl-, dimethyl ester	54370-19-3	NI27912
57	139	128	138	79	77	203	201	**274**	100	58	54	26	14	14	12	12	**0.00**	12	19	2	–	–	–	–	1	–	–	–	–	–	Bicyclo[2.1.1]hexane-1-carboxylic acid, 4-(bromomethyl)-, tert-butyl ester		DD00883
171	42	56	70	55	97	184	44	**274**	100	41	39	31	22	20	16	16	**2.00**	12	22	5	2	–	–	–	–	–	–	–	–	–	Ethyl 2-[1-(4-ethoxycarbonyl)piperazinyl]ethyl carbonate		MS06192
73	147	259	75	45	141	43	169	**274**	100	43	36	32	23	16	12	10	**1.83**	12	26	3	–	–	–	–	–	–	2	–	–	–	Acetoacetic acid, 2-ethylbis(trimethylsilyl)-		NI27913

MASS TO CHARGE RATIOS	M.W.	INTENSITIES	Parent	C	H	O	N	S	F	Cl	Br	I	Si	P	B	X	COMPOUND NAME	CAS Reg No	No
73 147 259 45 75 141 43 74	**274**	100 36 35 30 25 16 13 10	**1.75**	12	26	3	–	–	–	–	–	–	2	–	–	–	Acetoacetic acid, 2-ethylbis(trimethylsilyl)-		NI27914
147 73 259 55 148 75 45 77	**274**	100 62 25 20 16 16 16 11	**5.30**	12	26	3	–	–	–	–	–	–	2	–	–	–	2-Hexenoic acid, 2-[(trimethylsilyl)oxy]-, trimethylsilyl ester	69796-61-8	NI27915
147 73 45 259 148 75 69 149	**274**	100 83 23 20 18 16 16 11	**6.98**	12	26	3	–	–	–	–	–	–	2	–	–	–	Pentanoic acid, 4-methyl-2-oxo-, bis(trimethylsilyl)-	72088-30-3	NI27916
73 147 259 75 45 245 83 185	**274**	100 75 40 15 15 12 12 10	**2.00**	12	26	3	–	–	–	–	–	–	2	–	–	–	Pentanoic acid, 2-methyl-3-[(trimethylsilyl)oxy]-, trimethylsilyl ester		NI27917
73 75 57 148 85 77 259 147	**274**	100 23 23 14 14 13 12 11	**1.60**	12	26	3	–	–	–	–	–	–	2	–	–	–	2-Pentenoic acid, 3-methyl-2-[(trimethylsilyl)oxy]-, trimethylsilyl ester	55320-09-7	NI27918
73 147 45 75 148 259 69 156	**274**	100 98 30 22 16 15 14 10	**0.00**	12	26	3	–	–	–	–	–	–	2	–	–	–	2-Pentenoic acid, 3-methyl-2-[(trimethylsilyl)oxy]-, trimethylsilyl ester	55320-09-7	MS06193
73 259 147 83 45 260 184 185	**274**	100 74 59 21 18 18 18 14	**9.65**	12	26	3	–	–	–	–	–	–	2	–	–	–	2-Pentenoic acid, 3-methyl-2-[(trimethylsilyl)oxy]-, trimethylsilyl ester	55320-09-7	NI27919
95 138 110 136 137 274 81 67	**274**	100 54 35 15 10 8 5 5		13	10	5	2	–	–	–	–	–	–	–	–	–	Acetamide, N-(3-nitrophenyl)-2-(2-furoyl)-		LI05833
95 138 110 274 136 137 67 81	**274**	100 42 21 12 9 7 4 3		13	10	5	2	–	–	–	–	–	–	–	–	–	Acetamide, N-(3-nitrophenyl)-2-(3-furoyl)-		LI05834
95 138 110 136 274 137 81 164	**274**	100 45 36 14 10 10 4 3		13	10	5	2	–	–	–	–	–	–	–	–	–	Acetamide, N-(4-nitrophenyl)-2-(2-furoyl)-		LI05835
95 138 110 274 136 67 137 81	**274**	100 39 21 12 11 6 5 4		13	10	5	2	–	–	–	–	–	–	–	–	–	Acetamide, N-(4-nitrophenyl)-2-(3-furoyl)-		LI05836
107 77 65 79 274 91 181 108	**274**	100 20 20 19 13 12 11 9		13	10	5	2	–	–	–	–	–	–	–	–	–	Benzene, 1-(2-methylphenoxy)-2,4-dinitro-	2363-26-0	NI27920
195 274 276 92 65 93 66 275	**274**	100 95 92 88 62 47 27 21		13	11	–	2	–	–	–	1	–	–	–	–	–	Benzaldehyde, 2-bromo-, phenylhydrazone	10407-11-1	LI05837
195 274 276 92 65 93 66 275	**274**	100 95 91 88 60 44 26 19		13	11	–	2	–	–	–	1	–	–	–	–	–	Benzaldehyde, 2-bromo-, phenylhydrazone	10407-11-1	NI27921
55 190 134 188 79 56 77 78	**274**	100 41 26 21 20 18 9 8	**6.29**	13	15	3	–	–	–	–	–	–	–	–	–	1	Cymantrene, pentyl-	12126-78-2	NI27922
18 41 27 43 127 105 91 45	**274**	100 56 45 38 27 26 22 22	**0.00**	13	22	2	–	2	–	–	–	–	–	–	–	–	O-[Isopropoxy(cyclohexylidene)methyl] S-ethyl dithiocarbonate		DD00884
43 97 101 59 85 274 74 201	**274**	100 82 61 37 29 28 23 14		13	22	4	–	1	–	–	–	–	–	–	–	–	L-Arabino-1-hexene-3,4,5,6-tetrol, 1-(methylthio)-3,4:5,6-di-O-isopropylidene-, (cis + trans)		LI05838
101 259 141 115 126 113 173 243	**274**	100 50 20 16 13 13 11 2	**0.00**	13	22	6	–	–	–	–	–	–	–	–	–	–	β-D-Allofuranoside, methyl 2,3:5,6-di-O-isopropylidene-		LI05839
43 73 45 155 29 41 27 55	**274**	100 68 61 33 29 20 18 16	**0.00**	13	22	6	–	–	–	–	–	–	–	–	–	–	Hexanoic acid, 2-(acetyloxy)-1-[(acetyloxy)methyl]ethyl ester	56554-37-1	NI27923
73 45 43 41 29 55 155 57	**274**	100 97 67 28 24 18 13 13	**0.00**	13	22	6	–	–	–	–	–	–	–	–	–	–	Hexanoic acid, 2,3-bis(acetyloxy)propyl ester	56554-36-0	NI27924
205 73 74 157 72 233 117 61	**274**	100 66 47 29 17 15 15 15	**0.00**	13	26	2	2	1	–	–	–	–	–	–	–	–	Methionine, N-(dimethylaminomethylene)-, carboxyneopentyl ester		NI27925
41 43 56 55 39 84 58 71	**274**	100 92 76 51 32 28 27 25	**0.00**	13	26	4	2	–	–	–	–	–	–	–	–	–	Carbamic acid, butyl-, 2-[[(aminocarbonyl)oxy]methyl]-2-methylpentyl ester	4268-36-4	NI27926
55 72 97 118 158 56 41 57	**274**	100 81 67 61 58 40 36 35	**8.61**	13	26	4	2	–	–	–	–	–	–	–	–	–	Carbamic acid, butyl-, 2-[[(aminocarbonyl)oxy]methyl]-2-methylpentyl ester	4268-36-4	NI27927
95 73 147 184 217	**274**	100 32 28 21 15	**0.00**	13	30	2	–	–	–	–	–	–	2	–	–	–	Silane, [(5-methyl-1,3-cyclohexanediyl)bis(oxy)]bis[trimethyl-, (1α,3α,5α)-	58800-82-1	NI27928
95 73 147 217 184	**274**	100 33 26 15 13	**0.00**	13	30	2	–	–	–	–	–	–	2	–	–	–	Silane, [(5-methyl-1,3-cyclohexanediyl)bis(oxy)]bis[trimethyl-, (1α,3α,5β)-	58800-80-9	NI27929
184 73 95 217 147	**274**	100 79 15 13 11	**0.00**	13	30	2	–	–	–	–	–	–	2	–	–	–	Silane, [(5-methyl-1,3-cyclohexanediyl)bis(oxy)]bis[trimethyl-, (1α,3β,5α)-	58751-75-0	NI27930
105 77 51 50 198 106 78 226	**274**	100 44 28 16 13 8 8 6	**0.00**	14	10	2	–	2	–	–	–	–	–	–	–	–	Dibenzoyl disulphide	644-32-6	NI27931
106 105 77 51 50 198 78 226	**274**	100 99 44 28 16 13 8 6	**0.00**	14	10	2	–	2	–	–	–	–	–	–	–	–	Dibenzoyl disulphide	644-32-6	MS06194
136 152 108 154 96 105 69 122	**274**	100 72 43 24 24 18 18 17	**3.20**	14	10	4	–	1	–	–	–	–	–	–	–	–	Benzoic acid, 2,2′-disulpho-		IC03561
274 231 259 232 246 245 203 137	**274**	100 12 10 9 7 7 6 6		14	10	6	–	–	–	–	–	–	–	–	–	–	Anhydrofusarubin, O-demethyl-		LI05840
274 245 256 43 228 275 244 200	**274**	100 35 31 22 19 16 16 15		14	10	6	–	–	–	–	–	–	–	–	–	–	Naphtho[2,3-b]furan-5,8-dione, 4,9-dihydroxy-7-methoxy-2-methyl-		NI27932
274 276 165 167 195 194 166 152	**274**	100 99 75 43 35 33 32 27		14	11	1	–	–	–	–	1	–	–	–	–	–	Stilbene, 2-hydroxy-4′-bromo-, 2-hydroxy-		MS06195
119 139 274 111 93 141 91 43	**274**	100 35 33 22 17 11 11 11		14	11	2	2	–	–	1	–	–	–	–	–	–	Benzamide, 3-chloro-N-[(phenylamino)carbonyl]-	56438-00-7	NI27933
274 108 276 259 245 231 215 77	**274**	100 35 33 20 18 18 18 12		14	11	2	2	–	–	1	–	–	–	–	–	–	3H-Indazol-3-one, 5-chloro-1,2-dihydro-2-(4-methoxyphenyl)-	28561-72-0	LI05841
274 108 276 259 245 231 215 275	**274**	100 35 33 20 18 18 18 15		14	11	2	2	–	–	1	–	–	–	–	–	–	3H-Indazol-3-one, 5-chloro-1,2-dihydro-2-(4-methoxyphenyl)-	28561-72-0	NI27934
127 78 129 106 147 153 92 274	**274**	100 33 32 30 19 11 11 8		14	11	2	2	–	–	1	–	–	–	–	–	–	2-Isonicotinylacetamide, N-(4-chlorophenyl)-		LI05842
127 106 78 129 147 92 274 121	**274**	100 36 36 31 13 13 12 11		14	11	2	2	–	–	1	–	–	–	–	–	–	2-Isonicotinylacetamide, N-(4-chlorophenyl)-		LI05843
127 106 78 129 147 121 92 148	**274**	100 46 44 31 15 14 12 10	**9.00**	14	11	2	2	–	–	1	–	–	–	–	–	–	2-Nicotinylacetamide, N-(3-chlorophenyl)-		LI05844
127 106 78 129 121 274 92 147	**274**	100 49 47 36 23 15 12 11		14	11	2	2	–	–	1	–	–	–	–	–	–	2-Nicotinylacetamide, N-(4-chlorophenyl)-		LI05845
274 108 123 273 259 153 107 243	**274**	100 95 92 90 77 50 46 44		14	14	2	2	1	–	–	–	–	–	–	–	–	Benzene, 4,4′-dimethoxy-azothiobis-		LI05846
77 213 51 197 137 109 67 132	**274**	100 79 74 71 53 50 48 45	**28.00**	14	14	2	2	1	–	–	–	–	–	–	–	–	2-(Phenylsulphinyl)phenylacetamide oxime		MS06196
227 274 200 171 145 143 245 183	**274**	100 98 94 87 84 80 78 78		14	14	4	2	–	–	–	–	–	–	–	–	–	4H-Pyrazolo[5,1-c][1,4]benzoxazine-2-carboxylic acid, 3-(hydroxymethyl)-, ethyl ester		MS06197
232 172 144 171 143 173 274 43	**274**	100 92 62 46 30 21 18 17		14	14	4	2	–	–	–	–	–	–	–	–	–	2,3-Quinoxalinedimethanol, diacetate	20128-12-5	NI27935
105 231 152 77 82 106 169 151	**274**	100 78 64 39 36 23 21 20	**8.00**	14	14	4	2	–	–	–	–	–	–	–	–	–	Uracil, 3-(2-benzoxyethyl)-1-methyl-		NI27936
92 31 29 27 91 65 39 51	**274**	100 56 46 46 37 24 24 20	**1.09**	14	16	–	–	–	–	–	–	–	–	–	–	1	Zirconium, bis(methylbenzene)-		NI27937
164 92 274 91 256 147 121 165	**274**	100 77 69 36 23 14 13 12		14	18	2	–	–	–	–	–	–	–	–	–	1	Ferrocene, 1,1′-bis(1-hydroxyethyl)-	12289-46-2	NI27938
189 161 42 204 41 103 274 215	**274**	100 83 63 50 47 43 40 38		14	18	2	4	–	–	–	–	–	–	–	–	–	2-Pyridinecarboxylic acid, 3-[5-(dimethylamino)-4,4-dimethyl-4H-imidazol-2-yl]-, methyl ester		DD00885

MASS TO CHARGE RATIOS								M.W.	INTENSITIES								Parent	C	H	O	N	S	F	Cl	Br	I	Si	P	B	X	COMPOUND NAME	CAS Reg No	No
274	70	161	189	131	103	163	148	274	100	71	43	36	28	25	21	21		14	18	2	4	–	–	–	–	–	–	–	–	–	3-Pyridinecarboxylic acid, 2-[5-(dimethylamino)-4,4-dimethyl-4H-imidazol-2-yl]-, methyl ester		DD00886
45	89	73	43	81	99	75	87	274	100	65	38	32	28	27	27	26	1.39	14	26	5	–	–	–	–	–	–	–	–	–	–	Cyclohexyl-15-crown-5	17454-48-7	NI27939
28	31	72	59	43	29	45	44	274	100	89	81	60	50	48	45	28	0.00	14	26	5	–	–	–	–	–	–	–	–	–	–	Decanedioic acid, 2-ethoxyethyl monoester		DO01068
55	69	83	41	97	43	57	56	274	100	78	62	57	49	44	34	33	1.95	14	27	–	–	–	–	–	1	–	–	–	–	–	1-Tetradecene, 14-bromo-	74646-31-4	NI27940
131	73	75	95	132	259	69	43	274	100	93	44	24	20	19	16	16	0.00	14	30	3	–	–	–	–	–	–	1	–	–	–	Octanoic acid, 7-(trimethylsiloxy)-7-methyl-, ethyl ester		NI27941
231	201	89	57	43	41	58	42	274	100	65	59	49	35	32	25	18	0.99	14	30	3	–	–	–	–	–	–	1	–	–	–	Silane, vinyltributoxy-	18402-29-4	NI27942
218	274	246	205	219	232	206	231	274	100	92	44	36	30	26	26	18		15	10	–	6	–	–	–	–	–	–	–	–	–	Quinazoline, 2-phenyl-4-(1H-tetrazol-5-yl)-	67824-29-7	NI27943
105	139	77	118	106	111	141	75	274	100	19	17	16	9	8	6	5	2.00	15	11	3	–	–	–	1	–	–	–	–	–	–	Benzoic acid, 2-chloro-, 2-oxo-2-phenylethyl ester	55153-25-8	NI27944
105	77	139	111	118	106	51	75	274	100	23	10	10	9	9	9	7	2.00	15	11	3	–	–	–	1	–	–	–	–	–	–	Benzoic acid, 3-chloro-, 2-oxo-2-phenylethyl ester	55184-84-4	NI27945
105	77	139	118	111	51	106	75	274	100	23	16	11	11	10	9	8	3.00	15	11	3	–	–	–	1	–	–	–	–	–	–	Benzoic acid, 4-chloro-, 2-oxo-2-phenylethyl ester	54797-45-4	NI27946
215	217	152	274	216	151	276	214	274	100	35	30	21	18	9	7	6		15	11	3	–	–	–	1	–	–	–	–	–	–	Fluorene-9-carboxylic acid, 2-chloro-9-hydroxy-, methyl ester		MS06198
43	109	123	231	110	121	218	122	274	100	35	28	26	23	16	13	7	4.00	15	14	1	–	2	–	–	–	–	–	–	–	–	2-Propanone, 1,1-bis(phenylthio)-	69753-44-2	NI27947
151	256	257	122	272	124	213	274	274	100	72	45	38	29	27	20	15		15	14	5	–	–	–	–	–	–	–	–	–	–	Benzophenone, 2,2′-dihydroxy-4,4′-dimethoxy-	131-54-4	IC03562
151	274	257	124	152	108	150	95	274	100	21	20	14	9	8	6	6		15	14	5	–	–	–	–	–	–	–	–	–	–	Benzophenone, 2,2′-dihydroxy-4,4′-dimethoxy-	131-54-4	NI27948
151	270	138	271	165	150	152	149	274	100	67	32	31	30	25	23	20	0.00	15	14	5	–	–	–	–	–	–	–	–	–	–	Benzophenone, 2′,6,6′-trihydroxy-2,4′-dimethyl-		LI05847
203	274	160	89	175	76	186	69	274	100	19	11	8	7	6	5	5		15	14	5	–	–	–	–	–	–	–	–	–	–	2H-1-Benzopyran-2-one, 6-[(3,3-dimethyloxiranyl)carbonyl]-7-methoxy-, (–)-	63975-56-4	NI27949
162	134	274	242	243	113	81	53	274	100	37	16	15	13	11	11	11		15	14	5	–	–	–	–	–	–	–	–	–	–	2-Butenoic acid, 2-methyl-4-[(2-oxo-2H-1-benzopyran-7-yl)oxy]-, methyl ester	38971-87-8	MS06199
77	138	79	274	91	93	51	122	274	100	60	40	38	38	31	31	26		15	14	5	–	–	–	–	–	–	–	–	–	–	Carbonic acid, di-O-methoxyphenyl ester		LI05848
259	274	216	201	173	89	213	145	274	100	89	39	37	26	26	22	20		15	14	5	–	–	–	–	–	–	–	–	–	–	Dibenzofuran-2-ol, 1,2,3-trimethoxy-		NI27950
259	274	216	201	89	173	88	137	274	100	86	44	42	36	31	27	21		15	14	5	–	–	–	–	–	–	–	–	–	–	Dibenzofuran-2-ol, 1,3,4-trimethoxy-	88256-05-7	NI27951
259	274	216	201	89	173	88	213	274	100	97	54	54	52	43	36	29		15	14	5	–	–	–	–	–	–	–	–	–	–	Dibenzofuran-3-ol, 1,2,4-trimethoxy-	88256-04-6	NI27952
274	259	201	216	173	89	213	275	274	100	90	39	35	26	25	24	20		15	14	5	–	–	–	–	–	–	–	–	–	–	Dibenzofuran-3-ol, 1,2,4-trimethoxy-	88256-04-6	NI27953
274	259	201	216	173	89	275	145	274	100	94	38	37	29	28	18	17		15	14	5	–	–	–	–	–	–	–	–	–	–	Dibenzofuran-4-ol, 1,2,3-trimethoxy-	88256-06-8	NI27954
259	274	43	256	244	260	241	216	274	100	40	29	20	17	16	13	13		15	14	5	–	–	–	–	–	–	–	–	–	–	Furo[3,2-g][1]benzopyran-2-propan-2-ol, 4-methoxy-7-oxo-		NI27955
259	274	233	231	245	203			274	100	46	19	11	8	7				15	14	5	–	–	–	–	–	–	–	–	–	–	Karenin		MS06200
257	258	255	43	256	274	259	276	274	100	86	57	55	54	33	18	16		15	15	1	2	–	–	1	–	–	–	–	–	–	Diphenylamine, 5-chloro-2-(N-methylacetamido)-	94483-61-1	NI27956
256	99	255	257	82	258	162	67	274	100	55	54	39	36	34	32	30	27.60	15	15	1	2	–	–	1	–	–	–	–	–	–	1,2-Oxazine, 4,5-dimethyl-3,6-dihydro-N-(7-chloro-4-quinolinyl)-	90929-81-0	NI27957
259	274	105	260	275	120	118	91	274	100	73	20	15	11	11	10	10		15	15	3	–	–	–	–	–	–	–	1	–	–	10H-Phenoxaphosphine, 2-ethyl-10-hydroxy-8-methyl-, 10-oxide	36360-91-5	NI27958
274	102	275	91	104	105	209	195	274	100	16	15	15	11	9	8	8		15	15	3	–	–	–	–	–	–	–	1	–	–	10H-Phenoxaphosphine, 2,6,7-trimethyl-, 10-oxide		NI27959
217	218	132	146	117	246	103	118	274	100	26	25	24	23	19	19	13	1.00	15	18	3	2	–	–	–	–	–	–	–	–	–	2,4(3H,5H)-Pyrimidinedione, 6-ethoxy-5-ethyl-3-methyl-5-phenyl-	55255-47-5	NI27960
246	146	117	118	247	175	103	77	274	100	46	33	28	21	20	15	14	3.50	15	18	3	2	–	–	–	–	–	–	–	–	–	2,4,6(1H,3H,5H)-Pyrimidinetrione, 1,5-diethyl-3-methyl-5-phenyl-	55255-46-4	NI27961
246	146	117	118	218	247	175	91	274	100	39	21	16	13	12	11	9	4.51	15	18	3	2	–	–	–	–	–	–	–	–	–	2,4,6(1H,3H,5H)-Pyrimidinetrione, 1,5-diethyl-3-methyl-5-phenyl-	55255-46-4	NI27962
123	108	274	42	78	124	97	55	274	100	50	28	17	16	15	10	10		15	18	3	2	–	–	–	–	–	–	–	–	–	2-Quinuclidinecarboxamide, N-(4-methoxyphenyl)-3-oxo-	34291-66-2	LI05849
123	108	274	42	124	78	44	41	274	100	50	27	15	14	14	12	10		15	18	3	2	–	–	–	–	–	–	–	–	–	2-Quinuclidinecarboxamide, N-(4-methoxyphenyl)-3-oxo-	34291-66-2	NI27963
273	183	135	272	73	274	45	121	274	100	86	78	67	53	39	20	16		15	19	1	–	–	–	–	–	–	1	1	–	–	Phosphinous acid, diphenyl-, trimethysilyl ester		IC03563
201	43	55	41	69	57	83	29	274	100	87	83	74	65	52	41	39	10.43	15	30	2	–	1	–	–	–	–	–	–	–	–	4-Thiahexadecanoic acid		IC03564
43	71	57	113	41	55	85	97	274	100	78	61	35	32	22	18	7	0.00	15	30	4	–	–	–	–	–	–	–	–	–	–	1,2-Butanediol, 2-(2,2-diisopropyl-1,3-dioxolan-4-yl)-3,3-dimethyl-		DD00887
43	183	98	57	41	55	74	29	274	100	93	82	81	80	72	70	49	1.00	15	30	4	–	–	–	–	–	–	–	–	–	–	Dodecanoic acid, 2,3-dihydroxypropyl ester	142-18-7	WI01337
183	134	97	73	243	201	83	200	274	100	78	58	53	44	44	32	25	1.00	15	30	4	–	–	–	–	–	–	–	–	–	–	Dodecanoic acid, 2,3-dihydroxypropyl ester	142-18-7	WI01336
134	183	243	74	201	200	98	214	274	100	99	94	55	47	46	45	40	2.00	15	30	4	–	–	–	–	–	–	–	–	–	–	Dodecanoic acid, 2-hydroxy-1-(hydroxymethyl)ethyl ester	1678-45-1	WI01338
43	183	57	41	55	74	27	29	274	100	98	79	74	66	63	55	54	0.94	15	30	4	–	–	–	–	–	–	–	–	–	–	Dodecanoic acid, 2-hydroxy-1-(hydroxymethyl)ethyl ester	1678-45-1	NI27964
143	139	96	95	121	179	161	109	274	100	37	18	18	12	10	10	8	0.00	15	30	4	–	–	–	–	–	–	–	–	–	–	2,5,6-Pentanetriol, 2-(2,3-dimethyl-3-hydroxycyclopentyl)-6-methyl-		LI05850
57	136	137	41	55	95	80	81	274	100	72	37	12	11	9	9	8	4.61	15	31	–	–	1	–	–	–	–	–	1	–	–	Phosphine sulphide, tert-butylmethyl(methylisopropylcyclohexyl)-	74710-03-5	NI27965
211	133	191	210	165	209	109	219	274	100	93	69	68	67	49	25	21	6.81	16	12	1	–	–	1	1	–	–	–	–	–	–	Cyclopropanecarbonyl chloride, 1-fluoro-2,2-diphenyl-	56701-17-8	NI27966
91	77	51	105	65	106	39	78	274	100	48	30	28	24	23	21	19	0.40	16	15	2	–	–	–	1	–	–	–	–	–	–	2-Butanone, 3-chloro-4-hydroxy-1,4-diphenyl-	34737-56-9	NI27967
45	229	274	165	166	231	201	276	274	100	46	42	37	20	15	15	14		16	15	2	–	–	–	1	–	–	–	–	–	–	Stilbene, 2-(methoxymethoxy)-4′-chloro-, (E)-		MS06201
45	274	229	165	166	276	231	125	274	100	39	39	34	17	13	13	11		16	15	2	–	–	–	1	–	–	–	–	–	–	Stilbene, 2-(methoxymethoxy)-4′-chloro-, (Z)-		MS06202
105	104	91	106	103	78	77	79	274	100	64	50	39	28	28	27	18	4.73	16	18	–	–	–	–	–	–	–	–	–	–	1	Zinc, bis(2-phenylethyl)-	74685-32-8	NI27968
122	274	153	121	91	107	123	275	274	100	38	16	15	11	10	9	7		16	18	2	–	1	–	–	–	–	–	–	–	–	Bis(2-hydroxy-3,5-dimethylphenyl) sulphide		IC03565
274	152	275	153	151	122	121	154	274	100	31	19	10	9	8	8	7		16	18	2	–	1	–	–	–	–	–	–	–	–	Bis(4-hydroxy-2,5-dimethylphenyl) sulphide		IC03566
208	274	193	195	256	91	151	105	274	100	99	86	83	82	70	66	66		16	18	2	–	1	–	–	–	–	–	–	–	–	2,2,5,5-Tetramethyldiphenyl sulphone	6632-44-6	NI27969
208	193	256	195	151	209	105	180	274	100	92	91	83	77	70	69	68	68.03	16	18	2	–	1	–	–	–	–	–	–	–	–	2,2′,4,4′-Tetramethyldiphenyl sulphone	5184-75-8	NI27970
256	195	274	208	193	209	165	194	274	100	87	82	82	72	72	68	68		16	18	2	–	1	–	–	–	–	–	–	–	–	2,3,4,6-Tetramethyldiphenyl sulphone	21128-93-8	NI27971

MASS TO CHARGE RATIOS	M.W.	INTENSITIES	Parent	C	H	O	N	S	F	Cl	Br	I	Si	P	B	X	COMPOUND NAME	CAS Reg No	No
256 208 193 165 179 91 194 239	**274**	100 89 81 68 67 66 64 64	**25.64**	16	18	2	–	1	–	–	–	–	–	–	–	–	2,4,4,6-Tetramethyldiphenyl sulphone	5184-64-5	NI27972
213 119 91 214 121 137 43	**274**	100 21 20 15 14 13 10	**0.01**	16	18	4	–	–	–	–	–	–	–	–	–	–	2,3-Butanediol, 2,3-bis(hydroxyphenyl)-		LI05851
259 274 231 219 260 275	**274**	100 63 35 29 17 12		16	18	4	–	–	–	–	–	–	–	–	–	–	Coumarin, 5,7-dimethoxy-8-(1,1-dimethylallyl)-	17245-25-9	LI05853
259 274 231 206 219	**274**	100 66 25 21 20		16	18	4	–	–	–	–	–	–	–	–	–	–	Coumarin, 5,7-dimethoxy-8-(1,1-dimethylallyl)-	17245-25-9	LI05852
259 274 231 219 28 217 229 161	**274**	100 67 55 48 40 28 24 22		16	18	4	–	–	–	–	–	–	–	–	–	–	Coumarin, 5,7-dimethoxy-8-(1,1-dimethylallyl)-	17245-25-9	NI27973
201 274 200 69 118 229 187 160	**274**	100 85 52 36 33 18 18 16		16	18	4	–	–	–	–	–	–	–	–	–	–	3-Cyclohexenecarboxylic acid, 4-hydroxy-6-methyl-2-oxo-3-phenyl-, ethyl ester		LI05854
104 155 128 77 115 91 183 185	**274**	100 87 80 74 70 54 44 44	**15.10**	16	18	4	–	–	–	–	–	–	–	–	–	–	2H-1,4:4a,7-Dimethanonaphthalene-5,6-dicarboxylic acid, 1,3,4,7-tetrahydro-, dimethyl ester, (1α,4α,4aβ,7β)-		NI27974
227 242 155 156 214 243 128 184	**274**	100 86 83 51 35 32 32 28	**3.60**	16	18	4	–	–	–	–	–	–	–	–	–	–	Dispiro[2.2.4.2]dodeca-4,7,11-triene, trans-7,8-diacetoxy-		NI27975
55 130 156 90 104 117 143 173	**274**	100 87 75 66 64 56 55 50	**0.80**	16	18	4	–	–	–	–	–	–	–	–	–	–	1,3a-Etheno-4,7-methano-3aH-indene-2,3-dicarboxylic acid, 1,4,5,6,7,7a-hexahydro-, dimethyl ester, (1α,3aα,4β,7β,7aβ)-		NI27976
259 274 241 260 244 275 243 57	**274**	100 53 23 19 15 11 11 10		16	18	4	–	–	–	–	–	–	–	–	–	–	Isohemigossypol 1-methyl ether		NI27977
128 65 155 214 78 187 104 116	**274**	100 95 94 93 92 88 75 73	**4.90**	16	18	4	–	–	–	–	–	–	–	–	–	–	4,7-Methano-1,3,3a-metheno-3aH-cycloprop[c]indene-1,9(1aH)-dicarboxylic acid, hexahydro-, dimethyl ester		NI27978
118 155 129 64 143 85 214 78	**274**	100 74 64 64 62 58 55 53	**2.40**	16	18	4	–	–	–	–	–	–	–	–	–	–	1H-2,5-Methano-1,5a,6-methenocycloprop[a]indene-6,6a-dicarboxylic acid, hexahydro-, dimethyl ester (1α,1aβ,1bβ,2β,5β,5aα,6α,6aβ,7R*)-		NI27979
55 89 130 77 117 155 185 104	**274**	100 79 77 76 76 74 73 64	**7.30**	16	18	4	–	–	–	–	–	–	–	–	–	–	2,5-Methano-1,1b,5a-methenocycloprop[a]indene-1,7(1aH)-dicarboxylic acid, hexahydro-, dimethyl ester, (1α,1aα,1bβ,2β,5β,5aβ,6aα,7S*)-		NI27980
155 128 215 183 115 59 187 233	**274**	100 87 78 65 56 46 40 37	**4.30**	16	18	4	–	–	–	–	–	–	–	–	–	–	1,4-Methano-1H-benzocycloheptene-5,6-dicarboxylic acid, 2,3,4,4a-tetrahydro-, dimethyl ester, (1α,4α,4aβ)-(±)-		NI27981
274 241 256 115 259 141 242 128	**274**	100 95 43 26 24 17 17 15		16	18	4	–	–	–	–	–	–	–	–	–	–	1-Naphthalenecarboxaldehyde, 2,8-dihydroxy-3-methoxy-6-methyl-4-isopropyl-	50399-95-6	NI27982
135 274 131 77 161 148 51 192	**274**	100 41 32 26 20 16 14 11		16	18	4	–	–	–	–	–	–	–	–	–	–	2,4-Nonadienoic acid, 9-(3,4-methylenedioxyphenyl)-, (2E,4E)-	96386-57-1	NI27983
231 219 274 259 189 205 206 177	**274**	100 88 41 24 16 6 5 5		16	18	4	–	–	–	–	–	–	–	–	–	–	Peucenin-7-methyl ether		MS06203
211 256 78 169 77 75 152 195	**274**	100 30 21 16 13 13 12 11	**2.00**	16	18	4	–	–	–	–	–	–	–	–	–	–	Propanoic acid, 2-hydroxy-2-[6-(1-hydroxy-1-methylethyl)naphthalen-2-yl]-		NI27984
229 155 91 59 77 183 143 105	**274**	100 72 68 64 61 60 56 54	**35.00**	16	18	4	–	–	–	–	–	–	–	–	–	–	1,1′-Bis(tricyclo[3.1.0.0^{2,6}]hexane)-6,6′-dicarboxylic acid, dimethyl ester		MS06204
120 154 77 121 42 105 104 274	**274**	100 13 12 10 7 6 6 5		16	19	–	2	–	–	1	–	–	–	–	–	–	Ethylenediamine, N-(3-chlorophenyl)-N,N′-dimethyl-N′-phenyl-	32857-45-7	NI27985
120 154 274 121 77 156 105 42	**274**	100 19 10 10 10 6 5 5		16	19	–	2	–	–	1	–	–	–	–	–	–	Ethylenediamine, N-(4-chlorophenyl)-N,N′-dimethyl-N′-phenyl-	32857-42-4	NI27986
203 58 205 72 204 167 202 168	**274**	100 53 34 26 16 13 11 8	**0.40**	16	19	–	2	–	–	1	–	–	–	–	–	–	2-Pyridinepropanamine, γ-(4-chlorophenyl)-N,N-dimethyl-	132-22-9	NI27987
203 58 205 204 72 167 202 168	**274**	100 50 32 19 17 15 13 10	**0.00**	16	19	–	2	–	–	1	–	–	–	–	–	–	2-Pyridinepropanamine, γ-(4-chlorophenyl)-N,N-dimethyl-	132-22-9	NI27988
203 58 205 72 204 42 167 202	**274**	100 92 33 26 15 13 13 12	**0.00**	16	19	–	2	–	–	1	–	–	–	–	–	–	2-Pyridinepropanamine, γ-(4-chlorophenyl)-N,N-dimethyl-	132-22-9	NI27989
203 58 43 57 205 71 72 32	**274**	100 56 48 40 33 27 23 22	**1.40**	16	19	–	2	–	–	1	–	–	–	–	–	–	2-Pyridinepropanamine, γ-(4-chlorophenyl)-N,N-dimethyl-, (S)-	25523-97-1	NI27990
43 168 72 27 99 15 57 42	**274**	100 21 21 9 8 8 6 4	**0.00**	16	19	2	–	–	–	–	–	–	–	1	–	–	2-Butanol, 2-(diphenylphosphenoxy)-		IC03567
256 213 214 215 185 174 274 144	**274**	100 85 60 54 14 12 9 9		16	22	2	2	–	–	–	–	–	–	–	–	–	Azepino[2,3-b]indol-5a-(1H)-ol, 1-acetyl-10-ethyl-2,3,4,5,10,10a-hexahydro-		LI05855
245 274 203 120 146 112 185 259	**274**	100 74 44 26 25 24 18 13		16	22	2	2	–	–	–	–	–	–	–	–	–	Azepino[2,3-b]indol-5a-(1H)-ol, 10-acetyl-1-ethyl-2,3,4,5,10,10a-hexahydro-		LI05856
133 71 132 203 274 131 43 134	**274**	100 43 38 35 27 13 13 10		16	22	2	2	–	–	–	–	–	–	–	–	–	Quinoxaline, 1,2,3,4-tetrahydro-2,3-bis(tetrahydro-2-furyl)-	18503-40-7	NI27991
274 164 137 138 124 67 166 79	**274**	100 50 49 48 48 42 41 37		16	23	–	–	–	–	–	–	–	–	–	–	1	Cobalt, [(1,2,5,6-η)-1,5-cyclooctadiene][(1,2,3,5,6-η)-2,5-cyclooctadien-1-yl]-	55836-27-6	NI27992
54 67 80 39 79 41 66 27	**274**	100 84 49 43 38 23 17 13	**1.96**	16	24	–	–	–	–	–	–	–	–	–	–	1	Nickel, bis[(1,2,5,6-η)-1,5-cyclooctadiene]-	1295-35-8	NI27993
231 73 274 58 74 142 275 59	**274**	100 94 17 10 8 6 4 4		16	26	–	2	–	–	–	–	–	1	–	–	–	1H-Indole-3-methylamine, 5-ethyl-N,N-dimethyl-1-(trimethylsilyl)-		NI27994
137 54 42 67 41 94 95 68	**274**	100 66 65 56 55 53 45 45	**24.77**	16	32	–	2	–	–	–	–	–	–	–	2	–	9-Borabicyclo[3.3.1]nonan-9-amine, dimer	65930-25-8	NI27995
78 41 64 189 55 29 176 92	**274**	100 31 29 21 21 21 17 16	**10.10**	16	35	1	–	–	–	–	–	–	–	1	–	–	Phosphine oxide, dioctyl-		IC03568
181 78 197 196 274 273 79 182	**274**	100 78 57 48 43 16 16 15		17	14	–	4	–	–	–	–	–	–	–	–	–	1,2,4-Triazolo[4,3-a]pyridine, 2,3-dihydro-3-phenyl-2-(2-pyridinyl)-	16744-83-5	NI27996
201 274 171 128 256 115 241 225	**274**	100 61 33 22 17 17 11 11		17	22	3	–	–	–	–	–	–	–	–	–	–	2-Butanol, 4-(2,7-dimethoxy-1-naphthyl)-2-methyl-		NI27997
91 139 111 168 179 107 79 85	**274**	100 52 30 20 18 16 16 15	**0.00**	17	22	3	–	–	–	–	–	–	–	–	–	–	8,10-Dioxabicyclo[5.2.1]-3-decene, 7-[2-(benzyloxy)ethyl]-, (1S,7S)-		DD00888
213 274 157 259 133 159 214 107	**274**	100 69 45 31 21 19 18 15		17	22	3	–	–	–	–	–	–	–	–	–	–	1-Phenanthrenecarboxylic acid, 1,2,3,4,4a,9,10,10a-octahydro-6-hydroxy-1,4a-dimethyl-, [1S-(1α,4aα,10aβ)]-	5947-49-9	NI27998
195 267 223 194 43 94 193 268	**274**	100 98 52 30 28 27 24 21	**0.00**	18	10	3	–	–	–	–	–	–	–	–	–	–	Benzoic acid, 2-ethynyl-, anhydride	56772-05-5	NI27999
274 246 189 94 76 273 218 190	**274**	100 97 97 47 35 31 22 10		18	10	3	–	–	–	–	–	–	–	–	–	–	1H-Indene-1,3(2H)-dione, 2-(2,3-dihydro-3-oxo-1H-inden-1-ylidene)-	1707-95-5	NI28000
105 143 51 106 50 274 115 77	**274**	100 10 10 8 8 7 4 4		18	14	1	2	–	–	–	–	–	–	–	–	–	1-Isoquinolinecarbonitrile, 2-benzoyl-1,2-dihydro-1-methyl-	16576-32-2	NI28001

MASS TO CHARGE RATIOS	M.W.	INTENSITIES	Parent	C	H	O	N	S	F	Cl	Br	I	Si	P	B	X	COMPOUND NAME	CAS Reg No	No
274 28 44 273 245 137 102 117	274	100 99 52 27 14 14 9 7		18	14	1	2	–	–	–	–	–	–	–	–	–	1-Methyloxindolidene-3-(indol-3-yl)methine		LI05857
274 41 79 67 55 95 197 109	274	100 92 64 52 51 47 34 33		18	26	–	–	1	–	–	–	–	–	–	–	–	3-(Phenylthio)-1,3-dodecadiene, (E)-	123289-39-4	DD00889
41 67 95 109 175 81 55 274	274	100 69 67 64 62 54 47 24		18	26	–	–	1	–	–	–	–	–	–	–	–	3-(Phenylthio)-1,3-dodecadiene, (Z)-	123289-32-7	DD00890
259 203 274 241	274	100 27 26 24		18	26	2	–	–	–	–	–	–	–	–	–	–	Cyclohepta[a]naphthalen-7β-ol, 5,6,6aβ,11a-tetrahydro-3-methoxy-8,11aβ-dimethyl-		LI05858
122 121 134 135 274 246	274	100 64 61 50 27 21		18	26	2	–	–	–	–	–	–	–	–	–	–	Cyclohexanecarboxaldehyde, 1-trans-3-dimethyl-2-(3-methoxyphenylethyl)-		LI05859
122 121 134 274 135	274	100 57 31 22 21		18	26	2	–	–	–	–	–	–	–	–	–	–	Cyclohexanecarboxaldehyde, 1-trans-3-dimethyl-2-(3-methoxyphenylethyl)-		LI05860
110 274 215 256 147 275 231 107	274	100 67 41 40 36 26 26 26		18	26	2	–	–	–	–	–	–	–	–	–	–	Estr-4-en-3-one, 17β-hydroxy-	434-22-0	NI28003
110 90 165 147 160 124 105 92	274	100 20 18 18 13 12 12 11	0.00	18	26	2	–	–	–	–	–	–	–	–	–	–	Estr-4-en-3-one, 17β-hydroxy-	434-22-0	NI28002
110 274 91 41 79 55 105 107	274	100 94 46 39 37 36 33 32		18	26	2	–	–	–	–	–	–	–	–	–	–	Estr-4-en-3-one, 17β-hydroxy-	434-22-0	NI28004
274 90 215 110 160 122 105 92	274	100 37 31 29 28 26 22 22		18	26	2	–	–	–	–	–	–	–	–	–	–	Estr-4-en-3-one, 17β-hydroxy-, (8α)-	3218-21-1	NI28005
110 274 90 105 275 215 147 92	274	100 80 40 33 29 29 26 25		18	26	2	–	–	–	–	–	–	–	–	–	–	Estr-4-en-3-one, 17β-hydroxy-, (9β)-	4374-03-2	NI28006
259 15 143 57 260 274 41 215	274	100 25 22 22 19 16 11 10		18	26	2	–	–	–	–	–	–	–	–	–	–	1H-Indene-4-acetic acid, 6-tert-butyl-2,3-dihydro-1,1-dimethyl-, methyl ester	55591-15-6	NI28007
121 122 134 202 274	274	100 75 68 58 8		18	26	2	–	–	–	–	–	–	–	–	–	–	6-Nonene, 1,2-epoxy-2,6-dimethyl-9-(3-methoxyphenyl)-, cis-		LI05861
153 231 274 121 134 122	274	100 91 65 52 46 32		18	26	2	–	–	–	–	–	–	–	–	–	–	6-Oxabicyclo[3.2.1]octane, 1,5-dimethyl-syn-8-(3-methoxyphenylethyl)-		LI05862
91 71 79 107 161 81 41 67	274	100 16 15 14 9 8 7 6	0.00	18	26	2	–	–	–	–	–	–	–	–	–	–	1-Oxacyclooct-5-ene, 2-[2-(benzyloxy)ethyl]-8-ethyl-, (2R,8R)-		DD00891
274 147 161 241 173 259	274	100 89 85 53 46 38		18	26	2	–	–	–	–	–	–	–	–	–	–	1-Phenanthrenemethanol, 1,2,3,4a,9,10,10aα-octahydro-5-methoxy-1,4aβ-dimethyl-		LI05863
274 259 241 147 161 173	274	100 88 84 26 24 18		18	26	2	–	–	–	–	–	–	–	–	–	–	1-Phenanthrenemethanol, 1,2,3,4a,9,10,10aα-octahydro-7-methoxy-1,4aβ-dimethyl-		LI05864
259 274 147 241 161 173	274	100 56 40 33 30 26		18	26	2	–	–	–	–	–	–	–	–	–	–	1-Phenanthrenemethanol, 1,2,3,4a,9,10,10aα-octahydro-7-methoxy-1,4aβ-dimethyl-		LI05865
274 161 147 121 173 41 241 55	274	100 97 95 57 53 48 45 37		18	26	2	–	–	–	–	–	–	–	–	–	–	1-Phenanthrenemethanol, 1,2,3,4,4a,9,10,10a-octahydro-6-methoxy-1,4a-dimethyl-, [1S-(1α,4aα,10aβ)]-	16826-86-1	NI28008
241 259 173 274 147 171 189 260	274	100 96 49 42 39 30 27 22		18	26	2	–	–	–	–	–	–	–	–	–	–	2-Phenanthrenol, 1,2,3,4,4a,9,10,10a-octahydro-7-methoxy-1,1,4a-trimethyl-	55255-52-2	NI28009
245 219 261 91 276 18 105 41	274	100 74 67 42 41 40 39 39	0.50	18	26	2	–	–	–	–	–	–	–	–	–	–	3(2H)-Phenanthrenone, 1,4,4a,4b,5,6,7,8,10,10a-decahydro-1-(hydroxymethyl)-4a-methyl-, [1R-(1α,4aβ,4bα,7α,10aα)]-	57289-37-9	NI28010
170 274 171 275 142 104 197 103	274	100 62 15 13 12 12 10 7		19	14	2	–	–	–	–	–	–	–	–	–	–	5,6-Benzoflavanone		IC03569
229 231 202 274 259 215	274	100 71 23 15 8 7		19	14	2	–	–	–	–	–	–	–	–	–	–	7,12-Epoxy-7,12-dihydro-12-methylpleiaden-7-ol		LI05866
259 226 274 260 201 202 97 275	274	100 92 34 27 27 22 9 7		19	14	2	–	–	–	–	–	–	–	–	–	–	7-Hydroxy-7-methylbenz[a]anthrone		MS06205
259 274 260 202 101 258 127 275	274	100 22 19 19 9 7 6 5		19	14	2	–	–	–	–	–	–	–	–	–	–	12-Hydroxy-12-methylbenz[a]anthrone		MS06206
274 231 246 202 215 203 232 51	274	100 95 44 44 30 26 25 23		19	14	2	–	–	–	–	–	–	–	–	–	–	1-Hydroxy-3-phenyl-4-methylnaphthyl-2-acetolactone		NI28011
181 152 153 182 154 151 127 76	274	100 28 21 13 3 3 3 3	1.60	19	14	2	–	–	–	–	–	–	–	–	–	–	Phenyl diphenyl-4-carboxylate		IC03570
197 198 77 115 51 257 139 141	274	100 13 12 11 7 6 5 5	4.79	19	14	2	–	–	–	–	–	–	–	–	–	–	9H-Xanthen-9-ol, 9-phenyl-	596-38-3	NI28012
274 196 165 197 154 152 192 166	274	100 97 88 46 31 31 27 25		19	15	–	–	–	–	–	–	–	–	1	–	–	9,10-Dihydro-9-phenyl-9-phosphaanthracene		MS06207
259 274 260 215 275 243 244 261	274	100 56 27 22 15 9 6 5		19	18	–	–	–	–	–	–	–	1	–	–	–	7,7-Dimethyl-7-sila-7,12-dihydropleiadene		NI28013
259 260 197 105 274 181 53 261	274	100 25 18 18 13 11 8 7		19	18	–	–	–	–	–	–	–	1	–	–	–	Silane, methyltriphenyl-	791-29-7	NI28014
274 259 273 230 275 231 202 101	274	100 36 31 22 21 21 11 9		19	18	–	2	–	–	–	–	–	–	–	–	–	Quinoline, 4-[4-(dimethylamino)styryl]-	897-55-2	NI28015
274 259 273 275 230 231 202 101	274	100 35 32 24 24 23 12 9		19	18	–	2	–	–	–	–	–	–	–	–	–	Quinoline, 4-[4-(dimethylamino)styryl]-, trans-		LI05867
274 255 273 41 43 55 128 256	274	100 62 56 53 50 35 33 31		19	18	–	2	–	–	–	–	–	–	–	–	–	Quinoline, 2-(1-methyl-1,2,3,4-tetrahydroquinolyl)-		LI05868
41 55 203 95 67 81 79 256	274	100 64 51 49 48 43 38 34	34.00	19	30	1	–	–	–	–	–	–	–	–	–	–	Androstan-1-one, (5α)-	1755-29-9	NI28016
41 203 95 67 81 79 259 256	274	100 53 50 48 45 38 34 34	34.03	19	30	1	–	–	–	–	–	–	–	–	–	–	Androstan-1-one, (5α)-	1755-29-9	NI28017
216 55 67 95 231 79 81 274	274	100 91 60 54 44 42 39 35		19	30	1	–	–	–	–	–	–	–	–	–	–	Androstan-2-one, (5α)-	1225-48-5	NI28018
202 274 203 95 80 78 93 121	274	100 76 71 54 47 34 27 26		19	30	1	–	–	–	–	–	–	–	–	–	–	Androstan-3-one, (5α)-	1224-95-9	NI28020
202 274 41 203 55 95 81 67	274	100 76 72 70 67 54 47 45		19	30	1	–	–	–	–	–	–	–	–	–	–	Androstan-3-one, (5α)-	1224-95-9	NI28019
202 203 274 95 231 187 204 107	274	100 66 41 21 18 18 16 16		19	30	1	–	–	–	–	–	–	–	–	–	–	Androstan-3-one, (5α)-	1224-95-9	NI28021
98 274 95 93 41 81 91 55	274	100 82 64 61 57 53 49 47		19	30	1	–	–	–	–	–	–	–	–	–	–	Androstan-4-one, (5α)-	13583-70-5	NI28022
98 274 95 81 93 91 41 79	274	100 83 64 63 62 59 57 39		19	30	1	–	–	–	–	–	–	–	–	–	–	Androstan-4-one, (5α)-	13583-70-5	LI05869
111 43 55 274 259 67 81 95	274	100 27 23 20 20 20 18 17		19	30	1	–	–	–	–	–	–	–	–	–	–	Androstan-4-one, (5β)-	13583-71-6	LI05870
274 259 55 111 41 81 67 79	274	100 94 80 79 52 46 46 44		19	30	1	–	–	–	–	–	–	–	–	–	–	Androstan-4-one, (5β)-	13583-71-6	NI28023
111 43 55 67 274 259 81 79	274	100 28 23 21 20 20 19 17		19	30	1	–	–	–	–	–	–	–	–	–	–	Androstan-4-one, (5β)-	13583-71-6	NI28024
259 219 95 67 55 274 79 135	274	100 86 75 75 72 60 50 48		19	30	1	–	–	–	–	–	–	–	–	–	–	Androstan-6-one, (5α)-	3676-06-0	NI28025

MASS TO CHARGE RATIOS	M.W.	INTENSITIES	Parent	C	H	O	N	S	F	Cl	Br	I	Si	P	B	X	COMPOUND NAME	CAS Reg No	No
259 219 95 274 135 123 93 91	**274**	100 86 74 60 48 41 41 27		19	30	1	–	–	–	–	–	–	–	–	–	–	**Androstan-6-one, (5α)-**	3676-06-0	**NI28027**
259 274 55 81 219 67 95 41	**274**	100 77 54 53 50 50 49 48		19	30	1	–	–	–	–	–	–	–	–	–	–	**Androstan-6-one, (5α)-**	3676-06-0	**NI28026**
40 178 90 41 55 95 81 67	**274**	100 99 99 79 58 51 50 47	**21.00**	19	30	1	–	–	–	–	–	–	–	–	–	–	**Androstan-7-one, (5α)-**	2232-18-0	**NI28028**
203 81 79 256 274 259 108 125	**274**	100 83 72 66 65 63 61 59		19	30	1	–	–	–	–	–	–	–	–	–	–	**Androstan-7-one, (5α)-**	2232-18-0	**NI28029**
90 178 41 55 95 81 67 135	**274**	100 99 80 59 52 52 48 45	**22.02**	19	30	1	–	–	–	–	–	–	–	–	–	–	**Androstan-7-one, (5α)-**	2232-18-0	**NI28030**
164 41 274 177 55 149 67 81	**274**	100 81 55 53 53 49 48 47		19	30	1	–	–	–	–	–	–	–	–	–	–	**Androstan-11-one, (5α)-**	1755-32-4	**NI28032**
164 177 41 274 81 55 67 149	**274**	100 68 51 48 39 38 37 36		19	30	1	–	–	–	–	–	–	–	–	–	–	**Androstan-11-one, (5α)-**	1755-32-4	**NI28031**
164 177 151 274 41 81 55 68	**274**	100 39 34 31 20 15 15 14		19	30	1	–	–	–	–	–	–	–	–	–	–	**Androstan-11-one, (5β)-**	894-64-4	**NI28033**
41 67 55 81 79 95 93 43	**274**	100 70 63 54 35 30 30 29	**22.53**	19	30	1	–	–	–	–	–	–	–	–	–	–	**Androstan-12-one, (5α)-**	3676-09-3	**NI28034**
274 55 259 41 275 97 79 67	**274**	100 25 23 21 20 20 20 20		19	30	1	–	–	–	–	–	–	–	–	–	–	**Androstan-15-one, (5α)-**	734-68-9	**NI28036**
97 274 81 95 79 109 93 245	**274**	100 74 65 52 49 47 42 41		19	30	1	–	–	–	–	–	–	–	–	–	–	**Androstan-15-one, (5α)-**	734-68-9	**NI28035**
97 274 67 81 95 109 79 93	**274**	100 73 50 47 39 38 38 32		19	30	1	–	–	–	–	–	–	–	–	–	–	**Androstan-15-one, (5α)-**	734-68-9	**LI05871**
97 259 81 274 79 93 98 95	**274**	100 16 15 12 12 9 8 8		19	30	1	–	–	–	–	–	–	–	–	–	–	**Androstan-15-one, (5α,14β)-**	734-69-0	**NI28037**
257 272 145 201 105 258 41 91	**274**	100 44 23 22 22 21 20 18	**2.00**	19	30	1	–	–	–	–	–	–	–	–	–	–	**Androstan-15-one, (5α,14β)-**	734-69-0	**NI28038**
274 259 275 256 260 55 97 149	**274**	100 99 46 43 39 33 32 28		19	30	1	–	–	–	–	–	–	–	–	–	–	**Androstan-15-one, (5α,14β)-**	734-69-0	**NI28039**
274 217 55 81 66 95 259 109	**274**	100 57 41 38 38 32 30 28		19	30	1	–	–	–	–	–	–	–	–	–	–	**Androstan-16-one, (5α)-**	1032-16-2	**NI28040**
274 217 259 109 216 275 218 110	**274**	100 58 30 27 25 20 20 17		19	30	1	–	–	–	–	–	–	–	–	–	–	**Androstan-16-one, (5α)-**	1032-16-2	**NI28041**
274 230 109 67 81 55 241 148	**274**	100 51 51 47 40 38 32 32		19	30	1	–	–	–	–	–	–	–	–	–	–	**Androstan-17-one**		**MS06208**
274 109 66 54 230 218 81 108	**274**	100 41 39 37 34 29 25 24		19	30	1	–	–	–	–	–	–	–	–	–	–	**Androstan-17-one, (5α)-**	963-74-6	**NI28042**
274 109 67 230 41 55 108 81	**274**	100 49 42 35 35 34 31 31		19	30	1	–	–	–	–	–	–	–	–	–	–	**Androstan-17-one, (5α)-**	963-74-6	**NI28044**
274 109 67 55 108 81 230 95	**274**	100 88 74 62 60 58 53 50		19	30	1	–	–	–	–	–	–	–	–	–	–	**Androstan-17-one, (5α)-**	963-74-6	**NI28043**
97 274 67 109 41 55 110 81	**274**	100 60 30 27 27 26 24 24		19	30	1	–	–	–	–	–	–	–	–	–	–	**Androstan-17-one, (5α,14β)-**	22430-34-8	**NI28045**
274 55 67 81 108 109 79 95	**274**	100 90 88 58 57 52 46 41		19	30	1	–	–	–	–	–	–	–	–	–	–	**Androstan-17-one, (5β)-**	1912-61-4	**NI28046**
274 220 79 91 41 93 176 55	**274**	100 94 43 42 40 38 34 32		19	30	1	–	–	–	–	–	–	–	–	–	–	**Androst-2-en-17-ol, (5α,17β)-**	2639-53-4	**NI28047**
274 94 148 259 79 93 241 149	**274**	100 88 80 72 66 65 60 48		19	30	1	–	–	–	–	–	–	–	–	–	–	**Androst-16-en-3-ol, (3α,5α)-**	1153-51-1	**NI28048**
274 94 79 148 93 259 91 241	**274**	100 99 94 92 91 77 76 73		19	30	1	–	–	–	–	–	–	–	–	–	–	**Androst-16-en-3-ol, (3α,5α)-**	1153-51-1	**MS06209**
94 107 148 149 105 274 259 147	**274**	100 87 79 54 50 46 45 40		19	30	1	–	–	–	–	–	–	–	–	–	–	**Androst-16-en-3-ol, (3β,5α)-**		**MS06211**
274 79 94 93 41 148 91 259	**274**	100 96 87 83 83 79 79 73		19	30	1	–	–	–	–	–	–	–	–	–	–	**Androst-16-en-3-ol, (3β,5α)-**		**MS06210**
91 65 104 183 118 81 133 56	**274**	100 13 10 8 3 3 2 2	**0.00**	19	30	1	–	–	–	–	–	–	–	–	–	–	**Cyclododecane, (benzyloxy)-**		**DD00892**
274 97 110 67 81 55 149 41	**274**	100 64 36 36 33 33 26 26		19	30	1	–	–	–	–	–	–	–	–	–	–	**D-Homo-18-norandrostan-17a-one, (5α,13ξ)-**	54482-47-2	**NI28049**
261 109 105 41 121 55 91 69	**274**	100 57 54 49 44 42 37 35	**0.00**	19	30	1	–	–	–	–	–	–	–	–	–	–	**1-Phenanthrenemethanol, 7-vinyl-1,2,3,4,4a,4b,5,6,7,8,10,10a-dodecahydro-4a,7-dimethyl-, [1R-(1α,4aβ,4bα,7α,10aα)]-**	57119-13-8	**NI28050**
152 153 276 173 243 258 137 56	**274**	100 21 19 15 11 8 8 8	**0.00**	19	30	1	–	–	–	–	–	–	–	–	–	–	**9-Phenanthrenol, 7-vinyl-1,2,3,4,4a,5,6,7,8,9,10,10a-dodecahydro-1,4a,7-trimethyl-, [1R-(1α,4aβ,7β,9β,10aα)]-**	57289-36-8	**NI28051**
259 274 217 136 93 91 41 121	**274**	100 55 29 25 21 19 19 18		19	30	1	–	–	–	–	–	–	–	–	–	–	**2(1H)-Phenanthrenone, 3,4,4a,4b,5,6,7,8,10,10a-decahydro-1,1,4a,7,7-pentamethyl-, [4aR-(4aα,4bβ,10aβ)]-**	7715-44-8	**NI28052**
152 261 276 153 109 135 41 258	**274**	100 92 83 75 69 58 56 53	**0.00**	19	30	1	–	–	–	–	–	–	–	–	–	–	**2H-Phenanthro[8a,9-b]oxirene, 3-vinyldodecahydro-3,7,10b-trimethyl-3-vinyl-, 3R-(3α,4aR*,5aβ,6aβ,7β,10aα,10bβ)-**	57289-35-7	**NI28053**
274 111 41 259 98 67 42 95	**274**	100 88 58 52 45 40 40 38		19	30	1	–	–	–	–	–	–	–	–	–	–	**5α-Pregnan-3-one, oxime**		**LI05872**
129 43 29 274 146 128 118 41	**274**	100 97 90 77 75 73 73 72		20	18	1	–	–	–	–	–	–	–	–	–	–	**[1,2′-Binaphthalen]-1′(2′H)-one, 3,3′,4,4′-tetrahydro-**	23804-16-2	**NI28054**
273 274 115 217 275 128 218 91	**274**	100 90 27 20 19 14 12 11		20	18	1	–	–	–	–	–	–	–	–	–	–	**Cyclohexanone, 2,6-dibenzylidene-**		**MS06212**
129 146 274 128 131 115 130 145	**274**	100 89 31 24 20 20 17 11		20	18	1	–	–	–	–	–	–	–	–	–	–	**5,13:6,12-Dimethanodibenzo[a,f]cyclodecen-7-one, 5,6,7,12,13,14-hexahydro-**		**AI01935**
244 165 243 167 197 274 166 105	**274**	100 63 62 45 42 33 29 18		20	18	1	–	–	–	–	–	–	–	–	–	–	**Methyl trityl ether**	596-31-6	**NI28055**
244 165 243 167 197 274 166 245	**274**	100 63 62 45 42 33 29 18		20	18	1	–	–	–	–	–	–	–	–	–	–	**Methyl trityl ether**	596-31-6	**LI05873**
183 105 77 260 165 154 182 155	**274**	100 76 57 51 24 23 21 21	**0.00**	20	18	1	–	–	–	–	–	–	–	–	–	–	**Triphenylmethanol**		**LI05874**
95 191 41 109 55 67 81 190	**274**	100 78 75 63 58 51 42 37	**8.93**	20	34	–	–	–	–	–	–	–	–	–	–	–	**Anthracene, 9-cyclohexyltetrahydro-**		**AI01936**
191 109 67 81 190 82 83 79	**274**	100 77 75 57 49 31 29 29	**11.60**	20	34	–	–	–	–	–	–	–	–	–	–	–	**Anthracene, 9-cyclohexyltetrahydro-**		**AI01937**
161 217 274 57 218 162 41 28	**274**	100 87 25 25 20 17 15 11		20	34	–	–	–	–	–	–	–	–	–	–	–	**Benzene, 1,2,3,4-tetramethyl-5,6-dineopentyl-**	6668-20-8	**NI28056**
161 217 57 41 29 274 218 162	**274**	100 77 30 29 28 19 17 17		20	34	–	–	–	–	–	–	–	–	–	–	–	**Benzene, 1,2,3,4-tetramethyl-5,6-dineopentyl-**	6668-20-8	**AI01938**
217 161 218 57 41 29 274 160	**274**	100 22 19 18 15 15 11 6		20	34	–	–	–	–	–	–	–	–	–	–	–	**Benzene, 1,2,3,5-tetramethyl-4,6-dineopentyl-**		**AI01939**
161 217 28 57 162 41 273 218	**274**	100 40 34 26 17 14 11 9	**3.00**	20	34	–	–	–	–	–	–	–	–	–	–	–	**Benzene, 1,2,4,5-tetramethyl-3,6-dineopentyl-**	33770-83-1	**NI28057**
161 217 57 41 29 162 274 218	**274**	100 50 35 21 21 16 14 11		20	34	–	–	–	–	–	–	–	–	–	–	–	**Benzene, 1,2,4,5-tetramethyl-3,6-dineopentyl-**	33770-83-1	**AI01940**
81 137 123 69 95 55 80 122	**274**	100 63 41 34 27 25 25 22	**1.41**	20	34	–	–	–	–	–	–	–	–	–	–	–	**Bicyclo[2.2.1]heptane, 2,2′-bis[1,3,3-trimethyl-**		**NI28058**
274 217 259 81 95 109 67 55	**274**	100 60 54 44 41 34 32 32		20	34	–	–	–	–	–	–	–	–	–	–	–	**D-Homoandrostane, (5α)-**	35575-26-9	**NI28059**
95 55 69 259 109 81 41 67	**274**	100 92 88 82 81 76 76 75	**40.00**	20	34	–	–	–	–	–	–	–	–	–	–	–	**D-Homoandrostane, (5α,13α)-**	54482-31-4	**NI28060**

MASS TO CHARGE RATIOS								M.W.	INTENSITIES								Parent	C	H	O	N	S	F	Cl	Br	I	Si	P	B	X	COMPOUND NAME	CAS Reg No	No
91	92	274	43	41	105	29	57	274	100	98	29	19	15	12	11	10		20	34	–	–	–	–	–	–	–	–	–	–	–	Tetradecane, 1-phenyl-	1459-10-5	AI01941
92	91	274	43	57	41	133	93	274	100	58	16	15	11	10	8	8		20	34	–	–	–	–	–	–	–	–	–	–	–	Tetradecane, 1-phenyl-	1459-10-5	WI01339
105	91	106	43	29	41	31	274	274	100	23	15	14	13	12	11	10		20	34	–	–	–	–	–	–	–	–	–	–	–	Tetradecane, 2-phenyl-		IC03571
105	106	91	274	41	104	43	29	274	100	13	11	7	5	4	4	4		20	34	–	–	–	–	–	–	–	–	–	–	–	Tetradecane, 2-phenyl-		AI01942
91	132	119	117	104	105	92	41	274	100	49	49	36	22	21	17	17	4.19	20	34	–	–	–	–	–	–	–	–	–	–	–	Tetradecane, 3-phenyl-		MS06213
91	119	105	41	43	29	245	274	274	100	52	20	20	19	19	14	11		20	34	–	–	–	–	–	–	–	–	–	–	–	Tetradecane, 3-phenyl-		IC03572
91	133	105	43	41	29	231	274	274	100	27	21	14	14	13	10	9		20	34	–	–	–	–	–	–	–	–	–	–	–	Tetradecane, 4-phenyl-		IC03573
91	105	147	41	29	43	161	274	274	100	23	14	12	11	10	8	7		20	34	–	–	–	–	–	–	–	–	–	–	–	Tetradecane, 5-phenyl-		IC03574
91	175	189	105	104	92	274	43	274	100	18	15	14	10	8	7	7		20	34	–	–	–	–	–	–	–	–	–	–	–	Tetradecane, 7-phenyl-		AI01943
119	91	41	120	43	29	105	27	274	100	18	16	10	9	8	7	4	1.70	20	34	–	–	–	–	–	–	–	–	–	–	–	Tridecane, 2-methyl-2-phenyl-	27854-41-7	WI01340
69	83	97	54	31	104	64	48	275	100	16	9	5	5	2	2	2	0.00	4	–	3	1	1	7	–	–	–	–	–	–	–	Propane, 1,1,1,3,3,3-hexafluoro-2-cyano-2-(O-fluorosulphonyl)-		LI05875
69	196	108	277	275	125	123	111	275	100	70	23	6	6	3	3	1		4	1	–	1	–	7	–	1	–	–	–	–	–	Vinylamine, 1-bromo-2-fluoro-N,N-bis(trifluoromethyl)-, cis-		LI05876
69	196	58	277	275	108	176	174	275	100	66	58	24	24	23	5	5		4	1	–	1	–	7	–	1	–	–	–	–	–	Vinylamine, 1-bromo-2-fluoro-N,N-bis(trifluoromethyl)-, trans-		LI05877
127	70	89	150	105	31	30	29	275	100	86	45	31	30	17	17	16	8.70	5	10	4	1	1	5	–	–	–	–	–	–	–	(Trimethoxyacetamido)sulphur pentafluoride	90598-20-2	NI28061
277	88	275	279	62	61	53	37	275	100	93	51	48	36	34	26	26		7	3	1	1	–	–	–	2	–	–	–	–	–	Benzonitrile, 3,5-dibromo-4-hydroxy-	1689-84-5	NI28063
88	62	61	53	37	277	63	18	275	100	49	45	35	32	29	26	22	13.72	7	3	1	1	–	–	–	2	–	–	–	–	–	Benzonitrile, 3,5-dibromo-4-hydroxy-	1689-84-5	NI28062
126	218	216	67	69	41	28	180	275	100	99	99	99	98	93	81	57	4.00	9	13	3	1	–	3	1	–	–	–	–	–	–	Leucine, β-chloro-N-(trifluoroacetyl)-, methyl ester		LI05878
153	277	275	87	196	99	198	122	275	100	97	97	88	82	53	51	40		10	4	2	1	–	–	3	–	–	–	–	–	–	N-(4-Chlorophenyl)-2,3-dichloromaleicisoimide		MS06214
109	81	99	65	63	149	91	139	275	100	88	41	35	31	27	27	24	12.00	10	14	6	1	–	–	–	–	–	–	1	–	–	Phosphoric acid, diethyl 4-nitrophenyl ester	311-45-5	NI28064
109	29	149	81	275	139	65	99	275	100	57	46	46	40	37	32	30		10	14	6	1	–	–	–	–	–	–	1	–	–	Phosphoric acid, diethyl 4-nitrophenyl ester	311-45-5	NI28066
109	81	149	99	139	65	127	91	275	100	93	64	60	52	41	35	33	31.00	10	14	6	1	–	–	–	–	–	–	1	–	–	Phosphoric acid, diethyl 4-nitrophenyl ester	311-45-5	NI28065
211	240	69	96	239	71	181	210	275	100	86	60	54	48	48	46	44	17.99	10	18	2	5	–	–	1	–	–	–	–	–	–	1,3,5-Triazine-2,4-diamine, 6-(2-chloroethoxy)-N-methoxy-N′,N′-dimethyl-	80197-65-5	NI28067
126	43	80	96	228	125	79	168	275	100	84	52	42	38	27	25	23	4.00	11	17	3	1	2	–	–	–	–	–	–	–	–	Pyridine, 1-acetyl-2-acetoxy-1,2,3,6-tetrahydro-3,6-bis(methylthio)-		MS06215
177	41	176	156	43	55	57	69	275	100	64	60	57	54	43	42	31	1.20	11	18	1	1	–	5	–	–	–	–	–	–	–	1-Octylamine, N-pentafluoropropanoyl ester		MS06216
73	147	75	145	160	129	70	260	275	100	61	54	46	45	42	38	28	2.37	11	25	3	1	–	–	–	–	–	2	–	–	–	Butanoic acid, α-(aminocarbonyl)-, bis(trimethylsilyl)		NI28068
73	147	158	45	75	59	260	232	275	100	37	22	18	16	14	14	13	7.15	11	25	3	1	–	–	–	–	–	2	–	–	–	Butanoic acid, 3-methyl-2-[(trimethylsiloxy)imino]-, trimethylsilyl ester		NI28069
73	102	158	147	45	75	260	57	275	100	73	52	26	24	22	18	17	5.56	11	25	3	1	–	–	–	–	–	2	–	–	–	Glycine, N-(1-oxopropyl)-N-(trimethylsilyl)-, trimethylsilyl ester	55557-16-9	NI28070
158	68	73	75	159	45	41	69	275	100	89	43	30	15	12	12	11	0.00	11	25	3	1	–	–	–	–	–	2	–	–	–	Proline, (trimethylsiloxy)(trimethylsilyl)-	70125-41-6	NI28071
275	182	154	228	276	183	63	127	275	100	18	16	15	15	14	8	8		12	9	5	3	–	–	–	–	–	–	–	–	–	Phenol, 4-[(2,4-dinitrophenyl)amino]-	119-15-3	NI28072
112	154	77	127	114	163	275	51	275	100	63	40	36	33	30	13	9		12	10	–	–	–	–	2	–	–	–	–	–	1	Vanadium, bis(chlorobenzene)-	53966-10-2	NI28073
107	275	51	77	39	50	256	91	275	100	43	14	11	7	5	3	3		12	10	–	1	–	4	–	–	–	–	1	–	–	Phosphorane, tetrafluoro(diphenylamino)-		LI05879
241	166	242	167	140	151	120.5	91	275	100	53	22	19	17	14	12	8	2.50	12	10	2	1	–	–	–	–	–	–	–	–	1	Phenarsazine, 5,10-dihydro-10-hydroxy-, 10-oxide	4733-19-1	LI05880
241	166	242	167	140	151	139	240	275	100	53	20	20	17	12	12	11	2.00	12	10	2	1	–	–	–	–	–	–	–	–	1	Phenarsazine, 5,10-dihydro-10-hydroxy-, 10-oxide	4733-19-1	NI28074
162	65	91	69	77	131	103	216	275	100	62	52	47	38	33	25	14	1.00	12	12	3	1	–	3	–	–	–	–	–	–	–	L-Phenylalanine, N-(trifluoroacetyl)-, methyl ester	23635-30-5	LI05881
91	162	65	69	77	131	103	119	275	100	96	59	44	36	31	24	12	1.52	12	12	3	1	–	3	–	–	–	–	–	–	–	L-Phenylalanine, N-(trifluoroacetyl)-, methyl ester	23635-30-5	NI28075
148	191	55	146	93	70	91	219	275	100	78	57	46	21	20	19	13	0.00	12	14	3	1	–	–	–	–	–	–	–	–	1	Cymantrene, [1-(dimethylamino)ethyl]-	68520-45-6	NI28076
161	163	275	277	217	165	219	162	275	100	65	25	15	12	12	9	8		12	15	2	1	–	–	2	–	–	–	–	–	–	Pentanamide, N-(3,4-dichlorophenyl)-3-hydroxy-2-methyl-	37764-61-7	NI28077
161	163	59	61	275	277	47	71	275	100	65	30	27	25	16	14	12		12	15	2	1	–	–	2	–	–	–	–	–	–	Pentanamide, N-(3,4-dichlorophenyl)-3-hydroxy-2-methyl-	37764-61-7	NI28078
43	72	28	155	88	128	85	111	275	100	85	73	63	55	48	48	44	0.00	12	21	4	1	1	–	–	–	–	–	–	–	–	D-Arabino-hexitol, 2,6-anhydro-1-deoxy-4,5-O-isopropylidene-, dimethylcarbamothioate	69688-58-0	NI28079
43	198	81	257	98	86	88	216	275	100	80	75	35	20	20	18	16	0.00	12	21	4	1	1	–	–	–	–	–	–	–	–	L-Cysteine, trans-S-(2-hydroxycyclohexyl)-N-acetyl-, methyl ester	78414-55-8	NI28080
45	129	161	145	101	88	87	131	275	100	91	71	52	51	46	32	26	0.00	12	21	6	1	–	–	–	–	–	–	–	–	–	D-Galactonitrile, 5-O-acetyl-2,3,4,6-tetra-O-methyl-	84659-28-9	NI28081
202	59	42	130	116	203	275	56	275	100	78	52	39	20	19	12	11		12	21	6	1	–	–	–	–	–	–	–	–	–	Glycine, N,N-bis(2-ethoxy-2-oxoethyl)-, ethyl ester	16669-54-8	MS06217
129	43	45	161	101	88	87	71	275	100	81	71	67	38	35	35	12	0.00	12	21	6	1	–	–	–	–	–	–	–	–	–	D-Mannonitrile, 5-O-acetyl-2,3,4,6-tetra-O-methyl-		MS06218
101	59	43	260	73	175	160	142	275	100	70	70	34	17	16	14	13	0.00	12	21	6	1	–	–	–	–	–	–	–	–	–	D-Mannose, 2,3:5,6-di-O-isopropylidene-, oxime	35023-80-4	NI28082
41	219	56	151	136	123	191	177	275	100	48	47	28	20	20	17	17	3.00	12	22	4	1	–	–	–	–	–	–	1	–	–	Butanamide, 4-[P-ethoxy-P-(2-propenyloxy)phosphinoyl]-N-(2-propenyl)-		NI28083
73	158	232	45	114	75	43	159	275	100	86	23	21	12	11	11	10	0.00	12	29	2	1	–	–	–	–	–	2	–	–	–	Butanoic acid, 2-amino-2,3-dimethyl-, N-(trimethylsilyl)-, trimethylsilyl ester		MS06219
73	158	218	45	41	75	29	100	275	100	64	31	20	12	11	11	10	0.00	12	29	2	1	–	–	–	–	–	2	–	–	–	Butanoic acid, 2-amino-3,3-dimethyl-, N-(trimethylsilyl)-, trimethylsilyl ester		MS06220
102	73	75	147	170	103	260	30	275	100	49	24	13	11	11	9	9	3.06	12	29	2	1	–	–	–	–	–	2	–	–	–	Hexanoic acid, 6-amino-, bis(trimethylsilyl)-	56272-91-4	NI28084
102	73	147	260	170	45	96	59	275	100	50	13	11	11	8	6	5	3.00	12	29	2	1	–	–	–	–	–	2	–	–	–	Hexanoic acid, 6-amino-, bis(trimethylsilyl)-	56272-91-4	MS06221
102	73	75	147	170	260	218	188	275	100	50	24	16	13	11	3	3	2.50	12	29	2	1	–	–	–	–	–	2	–	–	–	Hexanoic acid, 6-amino-, bis(trimethylsilyl)-	56272-91-4	LI05882

MASS TO CHARGE RATIOS								M.W.	INTENSITIES								Parent	C	H	O	N	S	F	Cl	Br	I	Si	P	B	X	COMPOUND NAME	CAS Reg No	No
158	73	218	159	147	45	75	74	**275**	100	50	16	15	7	7	6	5	**0.00**	12	29	2	1	–	–	–	–	–	2	–	–	–	L-Isoleucine, N-(trimethylsilyl)-, trimethylsilyl ester	7483-92-3	NI28085
158	73	45	218	159	29	41	147	**275**	100	90	20	16	14	14	11	10	**0.00**	12	29	2	1	–	–	–	–	–	2	–	–	–	L-Isoleucine, N-(trimethylsilyl)-, trimethylsilyl ester	7483-92-3	MS06222
158	73	218	86	147	232	100	75	**275**	100	30	22	10	8	6	5	5	**2.00**	12	29	2	1	–	–	–	–	–	2	–	–	–	L-Isoleucine, N-(trimethylsilyl)-, trimethylsilyl ester	7483-92-3	MS06223
158	73	159	147	160	86	75	45	**275**	100	34	16	12	5	5	5	4	**0.13**	12	29	2	1	–	–	–	–	–	2	–	–	–	L-Leucine, N-(trimethylsilyl)-, trimethylsilyl ester	7364-46-7	NI28087
158	73	45	159	59	147	43	41	**275**	100	73	17	14	11	8	7	7	**0.00**	12	29	2	1	–	–	–	–	–	2	–	–	–	L-Leucine, N-(trimethylsilyl)-, trimethylsilyl ester	7364-46-7	MS06224
158	73	159	147	75	45	77	43	**275**	100	38	14	6	5	5	4	4	**0.00**	12	29	2	1	–	–	–	–	–	2	–	–	–	L-Leucine, N-(trimethylsilyl)-, trimethylsilyl ester	7364-46-7	NI28086
158	73	45	159	59	29	147	74	**275**	100	72	15	14	10	8	7	7	**0.00**	12	29	2	1	–	–	–	–	–	2	–	–	–	L-Norleucine, N-(trimethylsilyl)-, trimethylsilyl ester	55320-11-1	MS06225
158	73	159	147	75	45	86	59	**275**	100	37	14	8	8	7	5	5	**2.00**	12	29	2	1	–	–	–	–	–	2	–	–	–	L-Norleucine, N-(trimethylsilyl)-, trimethylsilyl ester	55320-11-1	MS06226
73	158	45	77	59	47	75	43	**275**	100	86	23	17	15	14	13	12	**0.00**	12	29	2	1	–	–	–	–	–	2	–	–	–	L-Norleucine, N-(trimethylsilyl)-, trimethylsilyl ester	55320-11-1	NI28088
275	247	276	177	219	146	248	230	**275**	100	56	16	14	10	10	7	7		13	9	6	1	–	–	–	–	–	–	–	–	–	Pyrano[2,3-e]benzoxazole-5-carboxylic acid, 2,7-dioxo-, ethyl ester		LI05883
275	247	276	146	219	177	72	230	**275**	100	26	15	14	10	7	6	5		13	9	6	1	–	–	–	–	–	–	–	–	–	Pyrano[3,2-f]benzoxazole-7-carboxylic acid, 2,5-dioxo-, ethyl ester		LI05884
217	275	139	63	183	107	185	152	**275**	100	66	65	19	16	16	5	5		13	10	–	1	2	–	–	–	–	–	1	–	–	Diphenylphosphinothioic isothiocyanate		MS06227
277	275	276	274	77	105	98	195	**275**	100	97	89	83	83	58	39	31		13	10	1	1	–	–	–	1	–	–	–	–	–	5-Bromo-2-aminobenzophenone		MS06228
275	144	274	276	172	159	215	145	**275**	100	20	19	14	14	9	7	7		13	13	2	3	1	–	–	–	–	–	–	–	–	4,6(1H,5H)-Pyrimidinedione, 5-[4-(dimethylamino)benzylidene]-2,3-dihydro-2-thioxo-		NI28089
214	231	39	96	95	94	174	136	**275**	100	94	74	69	55	55	50	50	**27.00**	13	13	4	3	–	–	–	–	–	–	–	–	–	β-(Furfurylideneimino)furfurylmalonamide		LI05885
122	154	41	77	107	59	275	91	**275**	100	70	38	17	15	14	13	13	**13.00**	13	16	2	1	–	3	–	–	–	–	–	–	–	Acetamide, N-[2-(2,6-dimethylphenoxy)-1-isopropyl]-2,2,2-trifluoro-		NI28090
122	154	41	28	105	121	107	59	**275**	100	97	34	21	16	12	11	11	**4.59**	13	16	2	1	–	3	–	–	–	–	–	–	–	Acetamide, N-[2-(2,6-dimethylphenoxy)-1-isopropyl]-2,2,2-trifluoro-		MS06229
260	275	219	41	234	68	43	261	**275**	100	84	45	30	26	24	24	22		13	21	–	7	–	–	–	–	–	–	–	–	–	1,3,5-Triazine-2,4,6-triamine, N-allyl-N-cyano-N',N''-bis(isopropyl)-	67704-83-0	NI28091
77	247	51	129	246	39	275	50	**275**	100	41	40	34	14	10	10	10		14	9	–	7	–	–	–	–	–	–	–	–	–	Pyrazolo[4,5-e]pyrrolo[1,2-a]tetrazolo[4,5-c]pyrazine, N-phenyl-	94714-50-8	NI28092
111	123	126	275	149	83	152	97	**275**	100	83	77	48	13	10	7	5		14	13	3	1	1	–	–	–	–	–	–	–	–	Acetamide, N-(3-anisolyl)-2-(2-thienoyl)-		LI05886
111	149	123	275	83	126	97	152	**275**	100	94	80	44	33	30	30	7		14	13	3	1	1	–	–	–	–	–	–	–	–	Acetamide, N-(4-anisolyl)-2-(3-thienoyl)-		LI05887
91	155	108	211	275	247	182	168	**275**	100	33	18	8	5	5	5	4		14	13	3	1	1	–	–	–	–	–	–	–	–	Formamide, N-(4-tolylsulphonyl)-N-phenyl-		MS06230
243	215	275	156	244	183	129	112	**275**	100	42	33	32	31	19	19	14		14	13	5	1	–	–	–	–	–	–	–	–	–	Oxindole, 3-[1,2-bis(methoxycarbonyl)ethylidene]-		LI05888
87	43	145	89	101	88	42	128	**275**	100	69	6	6	5	5	5	4	**1.60**	14	13	5	1	–	–	–	–	–	–	–	–	–	2-Quinolinecarboxylic acid, 4-(2-acetoxyethoxy)-		IC03575
211	274	128	101	157	183	187	242	**275**	100	60	39	35	31	29	25	22	**12.00**	14	13	5	1	–	–	–	–	–	–	–	–	–	4-Quinolinecarboxylic acid, 3-hydroxy-2-(methoxycarbonyl)methyl-, methyl ester		LI05889
235	163	121	55	43	216	72	93	**275**	100	99	99	99	98	64	64	48	**46.00**	14	17	3	3	–	–	–	–	–	–	–	–	–	3-Anilineacetamide, N-(carboxyethyl)-N-(cyanoethyl)-		IC03576
253	254	252	175	255	180	209		**275**	100	18	4	3	2	2	1		**0.00**	15	12	2	2	–	–	–	–	–	–	–	–	1	2,4-Imidazolidinedione, 5,5-diphenyl-, monosodium salt	630-93-3	NI28093
58	165	42	164	132	44	275	150	**275**	100	59	50	49	42	42	35	33		15	17	2	1	1	–	–	–	–	–	–	–	–	1,3-Benzenediol, 4-[2-(dimethylaminomethyl)phenylthio]-		MS06231
120	155	91	275	77	211	104	65	**275**	100	79	65	43	35	29	19	18		15	17	2	1	1	–	–	–	–	–	–	–	–	Benzenesulphonamide, N-ethyl-4-methyl-N-phenyl-	1821-40-5	NI28094
91	81	65	169	141	140	112	79	**275**	100	82	45	40	31	30	21	21	**0.00**	15	17	2	1	1	–	–	–	–	–	–	–	–	Cyclohexanespiro-2'-[4-(benzylthio)-1,3-oxazol-5(2H)-one]		DD00893
216	157	273	217	184	115	158	156	**275**	100	62	45	19	16	14	13	12	**0.90**	15	17	4	1	–	–	–	–	–	–	–	–	–	1H-1-Benzazepine-3,4-dicarboxylic acid, 4,5-dihydro-1-methyl-, dimethyl ester	36132-32-8	NI28095
150	67	218	151	80	109	108	104	**275**	100	50	40	39	39	34	33	31	**0.00**	15	17	4	1	–	–	–	–	–	–	–	–	–	Bicyclo[2.2.1]heptan-2-exo-ol, endo-2-methyl-, 4-nitrobenzoate		MS06232
200	201	275	202	244	185	226	186	**275**	100	55	33	23	21	13	12	12		15	17	4	1	–	–	–	–	–	–	–	–	–	2-(2,3-Dihydro-4-methoxyfuro[2,3-b]quinolin-2-yl)-1,2-propanediol		LI05890
226	188	275	189	244	176	134	214	**275**	100	64	58	58	17	17	16	15		15	17	4	1	–	–	–	–	–	–	–	–	–	2-(2,3-Dihydro-4-methoxyfuro[2,3-b]quinolin-2-yl)-1,2-propanediol		LI05891
275	158	174	130	186	230	202	276	**275**	100	63	34	33	26	19	18	17		15	17	4	1	–	–	–	–	–	–	–	–	–	Indole-1,3-dicarboxylic acid, 2-methyl-, diethyl ester		DD00894
149	275	164	77	188	91	39	216	**275**	100	38	23	18	16	16	13	12		15	17	4	1	–	–	–	–	–	–	–	–	–	2,6-Piperidinedione, 4-[2-(2-hydroxy-3,5-dimethylphenyl)-2-oxoethyl]-	526-02-3	NI28096
215	178	275	97	77	216	93	69	**275**	100	62	50	30	19	17	16	13		15	21	–	3	1	–	–	–	–	–	–	–	–	1H-Pyrazole-1-carbothioamide, 4-ethyl-4,5-dihydro-N-phenyl-5-propyl-	55255-48-6	NI28097
111	275	77	135	136	93	173	69	**275**	100	67	43	32	23	19	16	16		15	21	–	3	1	–	–	–	–	–	–	–	–	2-Pyrazoline-1-carboxanilide, 4-ethyl-3-propylthio-	24769-34-4	NI28098
202	73	275	203	45	75	204	43	**275**	100	97	38	22	14	13	9	9		15	21	2	1	–	–	–	–	–	1	–	–	–	1H-Indole-3-acetic acid, 1-(trimethylsilyl)-, ethyl ester		NI28099
202	73	275	203	75	45	204	130	**275**	100	45	26	17	7	7	5	5		15	21	2	1	–	–	–	–	–	1	–	–	–	1H-Indoleacetic acid, 1-(trimethylsilyl)-, ethyl ester		MS06233
202	73	201	275	203	200	45	273	**275**	100	64	30	28	24	22	11	9		15	21	2	1	–	–	–	–	–	1	–	–	–	1H-Indole-3-propanoic acid, 1-(trimethylsilyl)-, methyl ester	56196-72-6	NI28100
235	206	149	275	151	236	205	163	**275**	100	21	18	17	17	15	11	9		15	21	2	3	–	–	–	–	–	–	–	–	–	Acetamide, N-[3-[(2-cyanoethyl)ethylamino]-4-ethoxyphenyl]-	67905-64-0	NI28101
219	133	57	77	92	175	104	105	**275**	100	82	76	72	58	37	35	27	**0.00**	15	21	2	3	–	–	–	–	–	–	–	–	–	1,2,4-Oxadiazolidin-5-one, 2-tert-butyl-3-(isopropylamino)-4-phenyl-		MS06234
276	219	218	277	275	274	220	162	**275**	100	23	22	17	17	10	5	3		15	21	2	3	–	–	–	–	–	–	–	–	–	Physostigmine	57-47-6	NI28103
218	174	160	161	275	219	276	217	**275**	100	45	41	37	34	23	18	17		15	21	2	3	–	–	–	–	–	–	–	–	–	Physostigmine	57-47-6	NI28102
218	275	174	160	161	219	276	217	**275**	100	62	45	41	37	23	18	17		15	21	2	3	–	–	–	–	–	–	–	–	–	Physostigmine	57-47-6	NI28104
275	229	228	202	201	176	175	151	**275**	100	42	28	27	23	17	13	12		16	9	2	3	–	–	–	–	–	–	–	–	–	2-Quinolinecarbonitrile, 4-(4-nitrophenyl)-		LI05892
274	245	77	275	246	51	218	220	**275**	100	80	29	21	16	16	9	8		16	11	1	4	–	–	–	–	–	–	–	–	–	Pyrazolo[3,4-b]quinoxaline, 3-formyl-1-phenyl-		MS06235
275	102	90	218	192	219	103	91	**275**	100	78	59	43	32	24	24	22		16	13	–	5	–	–	–	–	–	–	–	–	–	Anhydro-2-ethyl-4-(2-cyanoanilino)-1,2,3-benzotriazinium hydroxide		LI05893
102	130	103	131	75	76	51	77	**275**	100	60	16	13	9	8	7	6	**3.00**	16	13	–	5	–	–	–	–	–	–	–	–	–	3-Ethyl-1,3-bis(2-cyanophenyl)triazene		LI05894
58	98	100	147	242	115	275	128	**275**	100	55	27	24	22	20	19	16		16	21	1	1	1	–	–	–	–	–	–	–	–	[1]Benzothieno[3,2-d]azecin-8-ol, 1,2,3,4,5,6,7,8-octahydro-3-methyl-	99659-32-2	NI28105
72	115	73	160	43	159	173	41	**275**	100	8	5	4	4	3	2	2	**0.80**	16	21	1	1	1	–	–	–	–	–	–	–	–	2-Propanol, 1-(isopropylamino)-3-(1-naphthylthio)-		MS06236
41	170	172	55	43	258	44	42	**275**	100	99	66	54	54	51	45	43	**17.86**	16	21	3	1	–	–	–	–	–	–	–	–	–	Annotine	5096-59-3	NI28106

MASS TO CHARGE RATIOS								M.W.	INTENSITIES								Parent	C	H	O	N	S	F	Cl	Br	I	Si	P	B	X	COMPOUND NAME	CAS Reg No	No
124	276	125	277	126	258	135	110	**275**	100	11	11	2	2	1	1	1	**0.79**	16	21	3	1	–	–	–	–	–	–	–	–	–	Benzeneacetic acid, α-hydroxy-, 8-methyl-8-azabicyclo[3.2.1]oct-3-yl ester, endo-(±)-	87-00-3	NI28107
124	105	135	258	276				**275**	100	14	6	3	2				**0.00**	16	21	3	1	–	–	–	–	–	–	–	–	–	Benzeneacetic acid, α-hydroxy-, 8-methyl-8-azabicyclo[3.2.1]oct-3-yl ester, endo-(±)-	87-00-3	NI28108
105	133	131	120	190	134	201	200	**275**	100	20	7	7	6	5	2	2	**0.00**	16	21	3	1	–	–	–	–	–	–	–	–	–	Ethaneperoxoic acid, 1-cyano-4,4-dimethyl-1-phenylpentyl ester	58422-75-6	NI28109
135	123	70	97	122	150	77	83	**275**	100	49	40	39	35	34	31	25	**24.00**	16	21	3	1	–	–	–	–	–	–	–	–	–	Isoelaeocarpicine, (+)-		LI05895
135	275	136	77	174	175	105	51	**275**	100	12	10	10	5	4	4	4		16	21	3	1	–	–	–	–	–	–	–	–	–	Δ-Piperlonguminine, β,γ-dihydro-		LI05896
57	135	275	117	174	161	122	115	**275**	100	87	71	36	31	24	23	23		16	21	3	1	–	–	–	–	–	–	–	–	–	Δ-Piperlonguminine, β,γ-dihydro-		LI05897
205	275	190	204	276	206	177	55	**275**	100	71	51	25	14	14	11	11		16	21	3	1	–	–	–	–	–	–	–	–	–	2(1H)-Quinolone, 3-ethyl-4-hydroxy-6-(isopentyloxy)-	22048-16-4	NI28110
190	275	247	205	260	276	204	177	**275**	100	80	41	31	28	13	12	12		16	21	3	1	–	–	–	–	–	–	–	–	–	2(1H)-Quinolone, 3-ethyl-4-hydroxy-7-(isopentyloxy)-	22048-17-5	NI28111
205	190	275	204	206	55	69	71	**275**	100	86	57	40	21	19	13	11		16	21	3	1	–	–	–	–	–	–	–	–	–	2(1H)-Quinolone, 3-ethyl-4-hydroxy-8-(isopentyloxy)-	22048-18-6	NI28112
205	275	190	204	276	206	177	55	**275**	100	71	51	25	14	14	11	11		16	21	3	1	–	–	–	–	–	–	–	–	–	2(1H)-Quinolone, 3-ethyl-4-hydroxy-6-(pentyloxy)-		LI05898
190	275	205	260	247	276	204	177	**275**	100	80	31	28	14	13	12	12		16	21	3	1	–	–	–	–	–	–	–	–	–	2(1H)-Quinolone, 3-ethyl-4-hydroxy-7-(pentyloxy)-		LI05899
205	190	275	204	206	55	69	218	**275**	100	86	57	40	21	19	13	11		16	21	3	1	–	–	–	–	–	–	–	–	–	2(1H)-Quinolone, 3-ethyl-4-hydroxy-8-(pentyloxy)-		LI05900
275	276	188	247	217	189	162	81	**275**	100	20	17	14	11	11	11	9		17	9	3	1	–	–	–	–	–	–	–	–	–	8H-Benzo[g]-1,3-benzodioxolo[6,5,4-de]quinolin-8-one	475-75-2	NI28113
104	172	76	276	247	77	214	190	**275**	100	72	48	32	25	24	19	11	**4.60**	17	9	3	1	–	–	–	–	–	–	–	–	–	Naphth[2,1-d]isoxazole-4,5-dione, 3-phenyl-	23767-19-3	NI28114
233	275	205	234	232	276	177	151	**275**	100	45	25	18	16	9	6	6		17	13	1	3	–	–	–	–	–	–	–	–	–	Acetamide, N-1H-phenanthro[9,10-d]imidazol-2-yl-	37052-15-6	NI28115
275	274	103	131	77	246	78	91	**275**	100	86	71	61	53	34	29	27		17	13	1	3	–	–	–	–	–	–	–	–	–	1,2,3-Triazole, 4-cinnamoyl-5-phenyl-		NI28116
105	164	69	57	128	106	55	41	**275**	100	49	45	39	33	31	31	28	**2.00**	17	25	2	1	–	–	–	–	–	–	–	–	–	Carbamic acid, (α-methylbenzyl)-, 1-pentylallyl ester	32589-38-1	NI28117
105	164	69	57	165	128	55	106	**275**	100	50	45	39	35	33	33	32	**3.00**	17	25	2	1	–	–	–	–	–	–	–	–	–	Carbamic acid, (α-methylbenzyl)-, 1-pentylallyl ester	32589-38-1	NI28118
44	71	204	58	275	121	187	91	**275**	100	97	58	47	37	19	14	13		17	25	2	1	–	–	–	–	–	–	–	–	–	6-Isoquinolinol, 2-methyl-4a-(3-methoxyphenyl)-1,2,3,4,4a,5,6,7,8,8a-decahydro-, (4aα,6α,8aα)-	61528-24-3	DD00895
71	57	275	204	91	121	167	260	**275**	100	79	34	33	20	19	16	7		17	25	2	1	–	–	–	–	–	–	–	–	–	6-Isoquinolinol, 2-methyl-4a-(3-methoxyphenyl)-1,2,3,4,4a,5,6,7,8,8a-decahydro-, (4aα,6α,8aβ)-	61528-20-9	DD00896
71	91	57	275	258	167	121	187	**275**	100	92	28	20	16	14	12	10		17	25	2	1	–	–	–	–	–	–	–	–	–	6-Isoquinolinol, 2-methyl-4a-(3-methoxyphenyl)-1,2,3,4,4a,5,6,7,8,8a-decahydro-, (4aα,6β,8aα)-	61528-23-2	DD00897
71	275	258	57	167	121	276	187	**275**	100	78	38	37	31	16	14	12		17	25	2	1	–	–	–	–	–	–	–	–	–	6-Isoquinolinol, 2-methyl-4a-(3-methoxyphenyl)-1,2,3,4,4a,5,6,7,8,8a-decahydro-, (4aα,6β,8aβ)-	61528-21-0	DD00898
164	66	111	112	110	98	67	91	**275**	100	59	40	25	23	21	19	17	**16.65**	17	25	2	1	–	–	–	–	–	–	–	–	–	4,7-Methano-1H-isoindole-1,3(2H)-dione, 2-(2-ethylhexyl)-3a,4,7,7a-tetrahydro-	113-48-4	AI01944
66	164	111	67	110	98	91	41	**275**	100	72	55	52	35	33	25	25	**5.08**	17	25	2	1	–	–	–	–	–	–	–	–	–	4,7-Methano-1H-isoindole-1,3(2H)-dione, 2-(2-ethylhexyl)-3a,4,7,7a-tetrahydro-	113-48-4	NI28119
43	178	275	232	176	152	151	205	**275**	100	46	14	13	7	4	3	3		18	13	2	1	–	–	–	–	–	–	–	–	–	2-Acetoxy-2-cyanospiro[cyclopropane-1,9′-fluorene]	87656-19-7	NI28120
244	275	215	216	217	189			**275**	100	99	24	23	21	12				18	13	2	1	–	–	–	–	–	–	–	–	–	Pyrrolo[1,2-f]phenanthridine-1-carboxylic acid, methyl ester		LI05901
275	227	228	276	229	230	114	113	**275**	100	38	20	17	16	12	9	9		18	13	2	1	–	–	–	–	–	–	–	–	–	1,1′:4′,1″-Terphenyl, 4-nitro-	10355-53-0	NI28121
275	228	276	226	229	202	152	227	**275**	100	29	19	18	16	14	13	12		18	13	2	1	–	–	–	–	–	–	–	–	–	1,1′:4′,1″-Terphenyl, 4-nitro-	10355-53-0	NI28122
275	274	276	260	183	233	273	92	**275**	100	27	18	5	5	3	2	2		18	17	–	3	–	–	–	–	–	–	–	–	–	2,2′:6′,2″-Terpyridine, 4,4′,4″-trimethyl-	33354-75-5	NI28123
275	274	276	260	183	137	233	92	**275**	100	27	18	5	5	4	3	2		18	17	–	3	–	–	–	–	–	–	–	–	–	2,2′:6′,2″-Terpyridine, 4,4′,4″-trimethyl-	33354-75-5	NI28124
275	202	260	203					**275**	100	57	24	22						18	29	1	1	–	–	–	–	–	–	–	–	–	2-Azaandrostan-3-one		LI05902
275	260	232	247	233				**275**	100	65	46	34	22					18	29	1	1	–	–	–	–	–	–	–	–	–	3-Azaandrostan-2-one		LI05903
275	218	260	230					**275**	100	26	25	16						18	29	1	1	–	–	–	–	–	–	–	–	–	16-Azaandrostan-17-one, (5α)-		LI05904
275	136	274	137	67	55	123	276	**275**	100	90	45	27	27	27	25	21		18	29	1	1	–	–	–	–	–	–	–	–	–	17a-Aza-18-nor-D-homo-androstan-17-one, (5α)-		LI05905
275	189	136	67	56	55	260	216	**275**	100	42	40	25	25	24	20	19		18	29	1	1	–	–	–	–	–	–	–	–	–	17a-Aza-18-nor-D-homo-androstan-17-one, (5α,13α)-		LI05906
275	136	274	137	67	55	123	81	**275**	100	90	45	27	27	27	26	21		18	29	1	1	–	–	–	–	–	–	–	–	–	Naphtho[2,1-f]quinolin-2(1H)-one, 3,4,4aα,4bβ,5,6,6aα,7,8,9,10,10a,10bα,11,12,12aβ-hexadecahydro-10aβ-methyl-	21171-75-5	NI28125
275	189	136	67	56	55	276	260	**275**	100	42	39	25	25	24	20	20		18	29	1	1	–	–	–	–	–	–	–	–	–	Naphtho[2,1-f]quinolin-2(1H)-one, 3,4,4aα,4bβ,5,6,6aα,7,8,9,10,10a,10bα,11,12,12aβ-hexadecahydro-10aβ-methyl-	21171-75-5	NI28126
164	246	41	55	136	163	245	67	**275**	100	90	64	54	39	25	22	19	**13.39**	18	34	–	1	–	–	–	–	–	–	–	1	–	Boranamine, N-cyclohexyl-N-1-cycloocten-1-yl-1,1-diethyl-	74810-32-5	NI28127
246	247	232	217	248	233	218	216	**275**	100	99	55	18	17	10	8	8	**0.00**	19	17	1	1	–	–	–	–	–	–	–	–	–	12-Aza-5-oxa-benz[a]anthracene, 5,6-dihydro-6,6,7-trimethyl-		NI28128
256	275	180	257	276	167	244	77	**275**	100	81	29	23	20	19	18	18		19	17	1	1	–	–	–	–	–	–	–	–	–	Benzyl alcohol, 2-(diphenylamino)-	53044-25-0	NI28129
275	258	276	273	274	259	77	167	**275**	100	33	21	19	18	17	10	9		19	17	1	1	–	–	–	–	–	–	–	–	–	Benzyl alcohol, 4-(diphenylamino)-	25069-40-3	NI28130
275	247	246	274	129	248	128	167	**275**	100	66	39	37	19	14	13	11		19	17	1	1	–	–	–	–	–	–	–	–	–	Pyridine, 2,3-dimethyl-4-phenyl-6-phenoxy-		DD00899
150	55	151	68	79	67	122	82	**275**	100	12	11	9	8	7	6	5	**2.00**	19	33	–	1	–	–	–	–	–	–	–	–	–	Pyrrolizidine, 3-(8-nonenyl)-5-[(E)-1-propenyl]-		NI28131
145	144	275	184	91	130	260	115	**275**	100	69	63	33	31	24	22	21		20	21	–	1	–	–	–	–	–	–	–	–	–	Carbazolenine, 1-benzyl-1,2,3,4-tetrahydro-11-methyl-		MS06237

MASS TO CHARGE RATIOS								M.W.	INTENSITIES								Parent	C	H	O	N	S	F	Cl	Br	I	Si	P	B	X	COMPOUND NAME	CAS Reg No	No
217	260	189	58	218	216	204	273	**275**	100	84	67	47	44	32	31	30	**0.00**	20	21	–	1	–	–	–	–	–	–	–	–	–	1H-Dibenz[e,g]isoindole, 2,3-dihydro-2-tert-butyl-	110028-15-4	DD00900
217	215	214	216	213	202	200	261	**276**	100	85	65	53	40	39	33	26	**2.38**	3	6	2	–	–	–	–	–	–	–	–	–	1	Mercury, (acetato-O)methyl-	108-07-6	NI28132
69	278	276	209	93	207	147	197	**276**	100	58	46	46	37	35	25	21		4	–	–	–	–	6	1	1	–	–	–	–	–	Hexafluoro-2-bromo-3-chloro-2-butene, cis-		IC03577
53	135	133	54	41	82	80	39	**276**	100	83	81	22	19	15	15	15	**0.00**	4	6	2	–	1	–	–	2	–	–	–	–	–	trans-3,4-Dibromosulpholene		LI05907
69	98	92	50	202	124	114	133	**276**	100	89	4	4	3	3	3	1	**0.00**	5	1	1	2	–	9	–	–	–	–	–	–	–	1,1,6,6,6-Hexafluoro-5-(trifluoromethyl)-2,5-diazahex-2-en-4-one		LI05908
44	58	29	86	114	70	142	69	**276**	100	89	49	31	26	18	17	15	**0.00**	6	12	12	–	–	–	–	–	–	–	–	–	–	Dodecahydroxycyclohexane		MS06238
73	146	276	274	45	74	272	186	**276**	100	20	13	12	11	8	7	4		6	18	–	–	–	–	–	–	–	2	–	–	1	Hexamethyldisilatellurane		MS06239
167	276	73	75	93	261	152	137	**276**	100	99	45	34	26	18	18	11		6	18	2	–	2	–	–	–	–	1	2	–	–	Dimethylsilylenebis(dimethylthiophosphinate)		NI28133
97	29	125	27	153	199	65	45	**276**	100	99	68	64	54	36	33	29	**0.00**	7	17	3	–	3	–	–	–	–	–	1	–	–	Phosphorodithioic acid, O,O-diethyl S-[(ethylsulphinyl)methyl] ester	2588-03-6	NI28134
29	27	109	81	45	75	46	183	**276**	100	88	63	61	55	46	42	38	**0.00**	7	17	5	–	2	–	–	–	–	–	1	–	–	Phosphorothioic acid, O,O-diethyl S-[(ethylsulphonyl)methyl] ester	2588-06-9	NI28135
109	183	45	46	81	139	75	47	**276**	100	64	59	48	43	34	33	30	**0.00**	7	17	5	–	2	–	–	–	–	–	1	–	–	Phosphorothioic acid, O,O-diethyl S-[(ethylsulphonyl)methyl] ester	2588-06-9	NI28136
183	185	157	155	50	76	75	200	**276**	100	99	28	28	26	24	24	10	**5.00**	8	6	1	–	–	–	–	2	–	–	–	–	–	Phenacetyl bromide, 4-bromo-	99-73-0	IC03578
185	183	75	76	50	157	155	90	**276**	100	96	37	34	30	29	28	27	**1.60**	8	6	1	–	–	–	–	2	–	–	–	–	–	Phenacetyl bromide, 4-bromo-	99-73-0	NI28137
79	51	77	59	197	177	69	29	**276**	100	13	9	8	7	6	4	4	**0.05**	8	9	–	–	–	9	–	–	–	–	–	–	–	Octane, 1,1,1,2,2,5,5,6,6-nonafluoro-	35262-54-5	MS06240
79	51	77	59	197	177	238	237	**276**	100	13	9	8	7	6	1	1	**0.05**	8	9	–	–	–	9	–	–	–	–	–	–	–	Octane, 1,1,1,2,2,5,5,6,6-nonafluoro-	35262-54-5	NI28138
125	241	97	197	81	43	53	141	**276**	100	92	86	64	63	51	48	47	**0.00**	9	9	2	–	–	3	2	–	–	–	–	–	–	4-(1,1-Dichloro-2,2,2-trifluoroethyl)-6,6-dimethyl-3-oxabicyclo[3.1.0]hexan-2-one	107686-54-4	DD00901
278	197	199	280	276	117	115	51	**276**	100	60	59	49	45	37	33	26		9	10	–	–	–	–	–	2	–	–	–	–	–	Benzene, 2,4-dibromo-1,3,5-trimethyl-	6942-99-0	MS06241
146	276	102	148	278	277	83	147	**276**	100	57	21	17	15	15	14	10		9	12	–	–	4	–	–	–	–	1	–	–	–	Tetrathiafulvalene, 2-(trimethylsilyl)-		MS06242
139	156	111	75	50	212	117	63	**276**	100	40	32	16	9	8	7	6	**3.00**	9	12	4	2	2	–	–	–	–	–	–	–	–	3,6-Bis(methylsulphonyl)-4,5-cyclopenteno-1,2-diazine		MS06243
27	168	60	29	110	108	28	169	**276**	100	99	99	88	70	70	65	57	**0.00**	9	13	1	2	1	–	–	1	–	–	–	–	–	Thiazolo[3,2-c]pyrimidin-4-ium, 8-ethoxy-2,3-dihydro-5-methyl-, bromide	24611-16-3	LI05909
27	168	60	29	110	108	28	167	**276**	100	99	99	88	70	70	65	57	**0.00**	9	13	1	2	1	–	–	1	–	–	–	–	–	Thiazolo[3,2-c]pyrimidin-4-ium, 8-ethoxy-2,3-dihydro-5-methyl-, bromide	24611-16-3	NI28139
221	223	41	29	278	276	40	38	**276**	100	99	62	38	30	30	23	19		9	13	1	2	1	–	–	1	–	–	–	–	–	Uracil, 3-bromo-3-sec-butyl-6-methyl-4-thio-		LI05910
221	223	42	29	54	39	278	276	**276**	100	99	43	36	32	29	28	27		9	13	1	2	1	–	–	1	–	–	–	–	–	Uracil, 5-bromo-3-sec-butyl-6-methyl-2-thio-	24561-91-9	LI05911
223	221	42	29	55	278	39	276	**276**	100	97	41	34	29	27	27	26		9	13	1	2	1	–	–	1	–	–	–	–	–	Uracil, 5-bromo-3-sec-butyl-6-methyl-2-thio-	24561-91-9	NI28140
223	221	41	29	278	276	40	38	**276**	100	97	60	37	30	29	22	19		9	13	1	2	1	–	–	1	–	–	–	–	–	Uracil, 5-bromo-3-sec-butyl-6-methyl-4-thio-	24561-92-0	NI28141
207	276	149	208	192	209	108	103	**276**	100	59	23	12	12	10	10	9		10	7	–	2	2	3	–	–	–	–	–	–	–	Benzothiazoline, 3-methyl-2-[(trifluorothioacetyl)imino]-		MS06244
150	104	76	92	50	75	120	69	**276**	100	35	29	18	16	15	12	10	**4.00**	10	7	4	2	–	3	–	–	–	–	–	–	–	Acetamide, 2,2,2-trifluoro-N-(4-nitrophenacyl)-	29205-55-8	MS06245
276	176	95	175	81	233	133	69	**276**	100	76	71	57	52	43	24	24		10	10	2	–	–	6	–	–	–	–	–	–	–	2,3-Dimethoxy-1,4,5,5,6,6-hexafluorobicyclo[2.2.2]oct-2-ene		NI28142
276	135	195	63	107	216	133	111	**276**	100	43	42	31	28	24	23	22		10	13	–	–	1	–	–	–	–	–	1	–	1	4-Phenyl-4-seleno-1,4-thiaphosphorinane		MS06246
149	121	79	67	93	150	41	77	**276**	100	80	51	26	21	13	12	11	**1.60**	10	13	1	–	–	–	–	–	1	–	–	–	–	Tricyclo[3.3.1.1^{3,7}]decanone, 4-iodo-, ($1\alpha,3\beta,4\alpha,5\alpha,7\beta$)-	56781-85-2	MS06247
149	121	79	67	93	77	41	150	**276**	100	78	60	30	24	15	14	13	**0.50**	10	13	1	–	–	–	–	–	1	–	–	–	–	Tricyclo[3.3.1.1^{3,7}]decanone, 4-iodo-, ($1\alpha,3\beta,4\beta,5\alpha,7\beta$)-	56781-86-3	MS06248
111	75	191	175	113	50	18	128	**276**	100	50	45	40	31	30	29	27	**0.00**	10	13	3	2	1	–	1	–	–	–	–	–	–	Benzenesulphonamide, 4-chloro-N-[(propylamino)carbonyl]-	94-20-2	NI28143
204	276	261	72	229	206	131	58	**276**	100	46	28	20	15	15	15	13		10	16	1	2	3	–	–	–	–	–	–	–	–	4-Isothiazolecarboxamide, N,N-diethyl-3,5-bis(methylthio)-	37572-36-4	NI28144
86	57	118	87	85	72	56	117	**276**	100	81	63	63	59	56	46	45	**21.65**	10	16	5	2	1	–	–	–	–	–	–	–	–	1,7-Dioxa-10-thia-4,13-diazacyclopentadecane-5,9,12-trione	74229-40-6	NI28145
241	212	276	225	229	247	198	213	**276**	100	99	67	50	44	43	38	37		10	17	1	4	1	–	1	–	–	–	–	–	–	1,3,5-Triazin-2-amine, 4-(2-chloroethoxy)-N,N-diethyl-6-methylthio-	65248-20-6	NI28146
229	276	247	261	231	249	278	167	**276**	100	71	70	51	42	38	36	33		10	17	1	4	1	–	1	–	–	–	–	–	–	1,3,5-Triazin-2(1H)-one, 1-(2-chloroethyl)-4-diethylamino-6-methylthio-	65248-22-8	NI28147
57	41	29	103	45	203	58	205	**276**	100	14	13	11	6	5	5	4	**0.00**	10	19	2	–	–	–	3	–	–	–	–	–	–	Chloral di-sec-butyl acetal		DO01069
67	165	66	135	261	232	151	185	**276**	100	64	47	21	13	12	9	6	**1.00**	10	20	1	–	–	–	–	–	–	–	–	–	1	Norbornan-2-endo-ol, anti-7-trimethylstannyl-		NI28148
165	66	151	67	135	261	150	183	**276**	100	77	70	68	50	47	16	10	**8.00**	10	20	1	–	–	–	–	–	–	–	–	–	1	Norbornan-2-endo-ol, syn-7-trimethylstannyl-		NI28149
67	165	66	135	261	232	151	185	**276**	100	81	26	25	17	15	9	6	**2.00**	10	20	1	–	–	–	–	–	–	–	–	–	1	Norbornan-2-exo-ol, anti-7-trimethylstannyl-		NI28150
167	165	66	261	135	67	137	150	**276**	100	33	31	29	19	19	13	5	**0.00**	10	20	1	–	–	–	–	–	–	–	–	–	1	Norbornan-2-exo-ol, syn-7-trimethylstannyl-		NI28151
102	73	70	43	100	60	89	75	**276**	100	28	19	17	14	13	12	12	**0.00**	10	20	5	2	–	–	–	–	–	1	–	–	–	Butanedioic acid, 2-[[[(trimethylsilyl)amino]carbonyl]amino]-, dimethyl ester	56051-84-4	NI28152
276	97	39	52	79	95	51	94	**276**	100	58	35	34	33	29	23	22		11	8	5	4	–	–	–	–	–	–	–	–	–	2-Furancarboxaldehyde, (2,4-dinitrophenyl)hydrazone	2074-02-4	NI28153
174	192	72	55	150	83	60	79	**276**	100	63	57	50	34	30	21	16	**5.00**	11	9	5	–	–	–	–	–	–	–	–	–	1	Manganese, tricarbonyl(2-carboxyethyl)cyclopentadienyl-		LI05912
100	142	74	114	59	215	86	101	**276**	100	49	43	39	37	35	35	32	**0.00**	11	20	6	2	–	–	–	–	–	–	–	–	–	α-D-Glucopyranoside, methyl 2,3-diacetamido-2,3-dideoxy-	21893-19-6	NI28154
100	59	142	139	84	101	60	114	**276**	100	52	41	37	36	33	33	32	**0.00**	11	20	6	2	–	–	–	–	–	–	–	–	–	β-D-Glucopyranoside, methyl 2,3-diacetamido-2,3-dideoxy-	7688-01-9	NI28155
149	151	233	147	231	150	148	135	**276**	100	86	64	61	49	41	36	34	**3.10**	11	24	–	–	–	–	–	–	–	–	–	–	1	Tin, cyclohexylisopropyldimethyl-		MS06249
233	191	147	231	189	41	145	43	**276**	100	97	81	76	71	64	61	52	**5.10**	11	24	–	–	–	–	–	–	–	–	–	–	1	Tin, triisopropylvinyl-		MS06250
165	135	261	276	120	231	150	121	**276**	100	18	13	12	7	6	6	3		11	24	–	–	–	–	–	–	–	–	–	–	1	Tin, 1,3-xylyltrimethyl-		LI05913
165	135	276	261	121	150	120	231	**276**	100	14	13	11	7	6	5	4		11	24	–	–	–	–	–	–	–	–	–	–	1	Tin, 1,4-xylyltrimethyl-		LI05914
147	73	75	55	45	148	261	69	**276**	100	88	26	20	17	16	15	13	**0.00**	11	24	4	–	–	–	–	–	–	2	–	–	–	Butanedioic acid, methyl-, bis(trimethylsilyl) ester	55557-26-1	NI28156
246	147	143	103	73	247	261	190	**276**	100	29	25	24	24	23	18	8	**0.00**	11	24	4	–	–	–	–	–	–	2	–	–	–	2,4(3H,5H)-Furandione, 3-(hydroxymethyl)-2,4-bis-O-(trimethylsilyl)-		LI05915

MASS TO CHARGE RATIOS								M.W.	INTENSITIES								Parent	C	H	O	N	S	F	Cl	Br	I	Si	P	B	X	COMPOUND NAME	CAS Reg No	No
246	147	73	143	103	247	261	248	276	100	29	25	24	24	21	16	9	0.00	11	24	4	–	–	–	–	–	–	2	–	–	–	2(3H)-Furanone, dihydro-3-[(trimethylsilyl)oxy]-3-[[(trimethylsilyl)oxy]methyl]-	38166-06-2	NI28157
146	261	74	157	203	147	262	98	276	100	62	40	33	17	15	12	11	0.41	11	24	4	–	–	–	–	–	–	2	–	–	–	Pentanedioic acid, bis(trimethylsilyl) ester	55494-07-0	NI28160
73	147	55	75	261	45	158	148	276	100	95	59	49	28	21	19	16	0.00	11	24	4	–	–	–	–	–	–	2	–	–	–	Pentanedioic acid, bis(trimethylsilyl) ester	55494-07-0	NI28158
73	147	75	261	55	45	158	129	276	100	80	73	46	45	25	21	19	0.80	11	24	4	–	–	–	–	–	–	2	–	–	–	Pentanedioic acid, bis(trimethylsilyl) ester	55494-07-0	NI28159
73	75	103	45	97	47	28	147	276	100	88	29	29	22	19	16	12	0.00	11	24	4	–	–	–	–	–	–	2	–	–	–	Pentonic acid, 2-deoxy-3,5-bis-O-(trimethylsilyl)-, γ-lactone	74742-34-0	NI28161
73	75	147	45	130	115	97	59	276	100	35	32	28	18	10	10	10	1.32	11	24	4	–	–	–	–	–	–	2	–	–	–	Pentonic acid, 5-deoxy-2,3-bis-O-(trimethylsilyl)-, γ-lactone	74742-33-9	NI28162
147	73	55	75	45	217	148	43	276	100	66	28	26	18	15	14	12	0.00	11	24	4	–	–	–	–	–	–	2	–	–	–	Propanedioic acid, dimethyl-, bis(trimethylsilyl) ester	55557-23-8	MS06251
147	73	55	75	148	45	217	149	276	100	75	34	18	16	16	12	10	0.00	11	24	4	–	–	–	–	–	–	2	–	–	–	Propanedioic acid, dimethyl-, bis(trimethylsilyl) ester	55557-23-8	NI28163
73	147	70	75	45	55	41	69	276	100	63	29	24	21	19	15	13	0.00	11	24	4	–	–	–	–	–	–	2	–	–	–	Propanedioic acid, dimethyl-, bis(trimethylsilyl) ester	55557-23-8	NI28164
147	73	55	75	148	45	217	149	276	100	75	34	18	16	16	12	10	0.00	11	24	4	–	–	–	–	–	–	2	–	–	–	Propanedioic acid, ethyl-, bis(trimethylsilyl) ester	55557-24-9	NI28165
73	147	70	75	55	45	148	41	276	100	85	31	22	22	21	14	14	0.00	11	24	4	–	–	–	–	–	–	2	–	–	–	Propanedioic acid, ethyl-, bis(trimethylsilyl) ester	55557-24-9	NI28167
147	73	55	75	148	217	45	149	276	100	72	25	19	16	15	15	10	0.00	11	24	4	–	–	–	–	–	–	2	–	–	–	Propanedioic acid, ethyl-, bis(trimethylsilyl) ester	55557-24-9	NI28166
73	75	147	45	130	115	97	59	276	100	35	32	28	18	10	10	10	1.33	11	24	4	–	–	–	–	–	–	2	–	–	–	D-Ribonic acid, 5-deoxy-2,3-bis-O-(trimethylsilyl)-, γ-lactone	62338-19-6	NI28168
92	166	139	69	78	51	108	182	276	100	98	84	81	74	74	62	61	57.31	12	8	4	2	1	–	–	–	–	–	–	–	–	2,2′-Dinitrodiphenyl sulphide	1223-31-0	NI28169
276	184	246	139	125	183	68	152	276	100	83	78	76	68	66	60	59		12	8	4	2	1	–	–	–	–	–	–	–	–	4,4′-Dinitrodiphenyl sulphide	22100-66-9	NI28170
276	126	260	155	30	77	184	230	276	100	21	14	12	11	9	8	7		12	8	6	2	–	–	–	–	–	–	–	–	–	4,2′-Dihydroxy-3,3′-dinitrobiphenyl	97851-13-3	NI28171
276	126	63	53	77	155	230	184	276	100	19	18	14	13	12	10	10		12	8	6	2	–	–	–	–	–	–	–	–	–	4,4′-Dihydroxy-3,3′-dinitrobiphenyl	66041-61-0	NI28172
243	168	276	183	184	244	139	169	276	100	70	48	25	20	18	15	14		12	9	3	–	–	–	–	–	–	–	–	–	1	10H-Phenoxarsine, 10-hydroxy-, 10-oxide	4846-20-2	NI28173
276	138	243	108	139	169	277	278	276	100	59	48	28	22	14	10	7		12	12	–	4	2	–	–	–	–	–	–	–	–	2,11-Dithia[3.3](2,6)pyrazinophane, (±)-	123624-96-4	DD00902
139	107	276	137	138	170	140	106	276	100	28	16	14	13	11	10	8		12	12	–	4	2	–	–	–	–	–	–	–	–	2,11-Dithia[3.3](2,5)pyrazinophane, pseudo-geminal-	123673-28-9	DD00903
139	107	276	231	170	137	106	79	276	100	16	11	11	10	10	9	9		12	12	–	4	2	–	–	–	–	–	–	–	–	2,11-Dithia[3.3](2,5)pyrazinophane, pseudo-ortho-	123624-97-5	DD00904
164	110	192	220	46	248	79	165	276	100	73	44	37	29	22	13	12	10.00	12	12	4	–	–	–	–	–	–	–	–	–	1	Iron, tetracarbonyl[(1,2-η)-1,5-cyclooctadiene]-	65582-74-3	NI28174
190	218	91	160	134	246	92	274	276	100	50	50	36	32	30	18	16	0.00	12	12	4	–	–	–	–	–	–	–	–	–	1	Iron, tricarbonyl[bicyclo[3.2.2]nona-6,8-dien-3-one]-		NI28175
162	56	220	192	91	190	176	276	276	100	75	53	35	33	25	18	7		12	12	4	–	–	–	–	–	–	–	–	–	1	Iron, tricarbonyl[(1,3,4,5-η)-1,4-dimethyl-2-oxo-4-cycloheptene-1,4-diyl]-		DD00905
261	80	41	39	28	168	276	214	276	100	68	46	44	42	39	38	35		12	12	4	4	–	–	–	–	–	–	–	–	–	2,4-Hexadienal, 2,4-dinitrophenylhydrazone		MS06252
57	29	41	28	43	27	185	42	276	100	92	90	46	38	28	25	24	6.60	12	14	3	–	–	–	2	–	–	–	–	–	–	Acetic acid, (2,4-dichlorophenoxy)-, butyl ester	94-80-4	NI28176
57	276	278	29	185	175	162	177	276	100	90	60	50	24	24	22	17		12	14	3	–	–	–	2	–	–	–	–	–	–	Acetic acid, (2,4-dichlorophenoxy)-, butyl ester	94-80-4	NI28177
57	41	29	175	162	276	177	164	276	100	33	28	16	13	12	10	8		12	14	3	–	–	–	2	–	–	–	–	–	–	Acetic acid, (2,4-dichlorophenoxy)-, isobutyl ester	1713-15-1	AI01945
57	41	43	42	39	175	162	111	276	100	78	29	26	16	15	15	14	6.80	12	14	3	–	–	–	2	–	–	–	–	–	–	Acetic acid, (2,4-dichlorophenoxy)-, isobutyl ester	1713-15-1	NI28178
91	134	107	92	105	97	169	136	276	100	22	21	10	9	9	7	7	0.67	12	14	3	–	–	–	2	–	–	–	–	–	–	2-Benzyloxy-2,2-dichloroethyl propionate		DO01070
276	261	247	164	233	71	277	28	276	100	80	79	34	22	22	18	17		12	16	2	6	–	–	–	–	–	–	–	–	–	2,2′-Bis(dimethylamino)-1,3′,4′,6-tetrahydro-4′,6-dioxo-4,5′-bispyrimidine		NI28179
43	74	116	129	85	115	114	87	276	100	97	94	59	59	48	40	40	0.00	12	20	7	–	–	–	–	–	–	–	–	–	–	Fucofuranoside, methyl 2,5-di-O-acetyl-3-O-methyl-		LI05916
43	87	129	74	114	28	156	15	276	100	66	46	17	16	13	8	7	0.00	12	20	7	–	–	–	–	–	–	–	–	–	–	Fucofuranoside, methyl 3,4-di-O-acetyl-2-O-methyl-		NI28180
43	87	129	74	114	113	156	85	276	100	64	49	20	19	9	8	8	0.00	12	20	7	–	–	–	–	–	–	–	–	–	–	Fucopyranoside, methyl 3,4-di-O-acetyl-2-O-methyl-		LI05917
74	116	85	129	115	87	114	113	276	100	94	85	55	48	39	37	36	0.00	12	20	7	–	–	–	–	–	–	–	–	–	–	L-Galactofuranoside, methyl 6-deoxy-3-O-methyl-, diacetate	55821-19-7	NI28181
87	43	129	114	83	115	113	85	276	100	99	78	55	50	47	43	42	0.00	12	20	7	–	–	–	–	–	–	–	–	–	–	α-L-Galactofuranoside, methyl 3,5-di-O-acetyl-2-O-methyl-6-deoxy-		NI28182
87	71	130	74	88	117	129	111	276	100	99	76	54	46	38	35	33	0.00	12	20	7	–	–	–	–	–	–	–	–	–	–	D-Glucitol, 1,5-anhydro-2,6-di-O-acetyl-3,4-di-O-methyl-	102850-66-8	NI28183
74	75	101	87	156	71	125	117	276	100	94	88	51	43	43	41	38	0.00	12	20	7	–	–	–	–	–	–	–	–	–	–	D-Glucitol, 1,5-anhydro-3,6-di-O-acetyl-2,4-di-O-methyl-	97275-53-1	NI28184
85	71	74	156	75	103	97	143	276	100	81	78	77	76	72	69	68	0.00	12	20	7	–	–	–	–	–	–	–	–	–	–	D-Glucitol, 1,5-anhydro-4,6-di-O-acetyl-2,3-di-O-methyl-	102803-54-3	NI28185
43	74	116	88	87	129	96	113	276	100	99	58	41	41	35	25	22	0.00	12	20	7	–	–	–	–	–	–	–	–	–	–	Mannofuranoside, methyl 6-deoxy-2-O-methyl-, diacetate	29839-03-0	NI28186
43	129	87	28	114	115	113	83	276	100	48	42	35	30	20	20	20	0.00	12	20	7	–	–	–	–	–	–	–	–	–	–	Mannofuranoside, methyl 6-deoxy-3-O-methyl-, diacetate	29836-38-2	NI28187
43	116	75	74	85	45	59	101	276	100	99	99	99	55	30	29	26	0.00	12	20	7	–	–	–	–	–	–	–	–	–	–	α-L-Mannopyranoside, methyl 6-deoxy-3-O-methyl-, diacetate	56083-41-1	NI28188
74	43	72	129	87	59	116	71	276	100	99	88	87	70	27	26	26	0.00	12	20	7	–	–	–	–	–	–	–	–	–	–	α-D-Mannopyranoside, methyl 2,3-di-O-acetyl-4-O-methyl-6-deoxy-		NI28189
74	43	115	75	85	45	101	59	276	100	99	44	44	25	15	14	11	0.00	12	20	7	–	–	–	–	–	–	–	–	–	–	α-D-Mannopyranoside, methyl 2,4-di-O-acetyl-3-O-methyl-6-deoxy-	56083-41-1	NI28190
74	43	116	88	87	129	113	83	276	100	99	49	40	40	29	17	15	0.00	12	20	7	–	–	–	–	–	–	–	–	–	–	α-Mannopyranoside, methyl 3,4-di-O-acetyl-2-O-methyl-6-deoxy-	63527-42-4	NI28191
145	99	101	189	89	45	87	70	276	100	75	68	38	30	30	28	20	0.10	12	20	7	–	–	–	–	–	–	–	–	–	–	1,4,7,10,13-Pentaoxacycloheptadecane-14,17-dione		MS06253
157	115	29	42	111	87	27	129	276	100	44	38	13	12	12	12	11	0.00	12	20	7	–	–	–	–	–	–	–	–	–	–	1,2,3-Propanetricarboxylic acid, 2-hydroxy-, triethyl ester	77-93-0	NI28192
157	115	203	29	43	28	158	111	276	100	34	27	21	16	9	8	8	0.00	12	20	7	–	–	–	–	–	–	–	–	–	–	1,2,3-Propanetricarboxylic acid, 2-hydroxy-, triethyl ester	77-93-0	NI28193
43	129	87	114	101	74	113	83	276	100	45	40	37	23	23	22	22	0.00	12	20	7	–	–	–	–	–	–	–	–	–	–	Rhamnofuranoside, methyl 2,5-di-O-acetyl-3-O-methyl-		LI05918
74	43	116	88	87	129	96	157	276	100	99	65	45	45	40	29	25	0.00	12	20	7	–	–	–	–	–	–	–	–	–	–	Rhamnofuranoside, methyl 3,5-di-O-acetyl-2-O-methyl-		LI05919
75	73	201	219	93	99	161	203	276	100	26	22	16	14	10	9	7	0.00	12	21	3	–	–	–	1	–	–	1	–	–	–	7-Oxabicyclo[4.1.0]hept-2-en-5-ol, 2-chloro-4-[[(1,1-dimethylethyl)dimethylsilyl]oxy]-, (1β,4β,5α,6β)-		DD00906
74	87	179	123	248	143	151	111	276	100	80	77	29	28	27	23	16	0.00	12	21	5	–	–	–	–	–	–	–	1	–	–	3-(Diethoxyphosphinyl)octahydro-2-benzofuranone		DD00907
79	87	43	109	189	107	151	122	276	100	84	71	56	50	44	43	39	20.00	12	21	5	–	–	–	–	–	–	–	1	–	–	Dimethyl (2Z-E,4E)-6,6-ethylenedioxy-2-methyl-2,4-heptadienylphosphonate	80283-35-8	NI28194

MASS TO CHARGE RATIOS	M.W.	INTENSITIES	Parent	C	H	O	N	S	F	Cl	Br	I	Si	P	B	X	COMPOUND NAME	CAS Reg No	No
73 117 147 75 45 232 118 74	276	100 87 30 22 14 10 10 9	0.00	12	28	3	–	–	–	–	–	–	2	–	–	–	Butanoic acid, 2,2-dimethyl-3-[(trimethylsilyl)oxy]-, trimethylsilyl ester	55530-56-8	NI28195
73 147 146 75 130 261 103 131	276	100 98 34 31 22 21 20 20	0.00	12	28	3	–	–	–	–	–	–	2	–	–	–	Butanoic acid, 2,3-dimethyl-4-[(trimethylsilyl)oxy]-, trimethylsilyl ester		NI28196
73 159 147 233 247 157 217 171	276	100 84 64 18 14 12 10 9	0.00	12	28	3	–	–	–	–	–	–	2	–	–	–	Butanoic acid, 2-ethyl-2-[(trimethylsilyl)oxy]-, trimethylsilyl ester		MS06254
73 117 147 75 45 74 59 41	276	100 65 39 39 21 10 10 9	0.00	12	28	3	–	–	–	–	–	–	2	–	–	–	Butanoic acid, 2-ethyl-3-[(trimethylsilyl)oxy]-, trimethylsilyl ester		NI28197
73 147 159 75 45 59 133 74	276	100 49 48 26 23 15 13 9	0.00	12	28	3	–	–	–	–	–	–	2	–	–	–	Hexanoic acid, 2-[(trimethylsilyl)oxy]-, trimethylsilyl ester	54890-07-2	NI28198
73 159 147 45 75 103 69 160	276	100 66 38 16 15 15 15 9	0.00	12	28	3	–	–	–	–	–	–	2	–	–	–	Hexanoic acid, 2-[(trimethylsilyl)oxy]-, trimethylsilyl ester	54890-07-2	NI28199
73 147 145 75 233 219 45 74	276	100 51 41 18 14 11 10 9	0.00	12	28	3	–	–	–	–	–	–	2	–	–	–	Hexanoic acid, 3-[(trimethylsilyl)oxy]-, trimethylsilyl ester		NI28200
117 73 75 171 147 69 204 261	276	100 90 43 35 31 28 22 19	0.00	12	28	3	–	–	–	–	–	–	2	–	–	–	Hexanoic acid, 5-[(trimethylsilyl)oxy]-, trimethylsilyl ester	79628-63-0	NI28201
117 73 75 171 69 204 129 261	276	100 90 40 33 29 20 20 18	0.00	12	28	3	–	–	–	–	–	–	2	–	–	–	Hexanoic acid, 5-[(trimethylsilyl)oxy]-, trimethylsilyl ester	79628-63-0	NI28202
147 73 261 75 69 171 149 55	276	100 85 57 53 53 34 22 19	2.19	12	28	3	–	–	–	–	–	–	2	–	–	–	Hexanoic acid, 6-[(trimethylsilyl)oxy]-, trimethylsilyl ester		NI28203
131 73 147 218 75 132 261 247	276	100 97 58 21 18 12 11 10	0.00	12	28	3	–	–	–	–	–	–	2	–	–	–	Pentanoic acid, 2-methyl-3-[(trimethylsilyl)oxy]-, trimethylsilyl ester		NI28204
73 131 147 75 218 45 132 74	276	100 79 42 16 15 13 10 10	0.00	12	28	3	–	–	–	–	–	–	2	–	–	–	Pentanoic acid, 2-methyl-3-[(trimethylsilyl)oxy]-, trimethylsilyl ester		NI28205
73 131 147 75 45 218 74 115	276	100 68 34 15 13 11 10 9	0.00	12	28	3	–	–	–	–	–	–	2	–	–	–	Pentanoic acid, 2-methyl-3-[(trimethylsilyl)oxy]-, trimethylsilyl ester		NI28206
73 147 159 75 45 59 133 41	276	100 34 33 28 24 13 12 11	0.00	12	28	3	–	–	–	–	–	–	2	–	–	–	Pentanoic acid, 3-methyl-2-[(trimethylsilyl)oxy]-, trimethylsilyl ester	54890-09-4	NI28207
73 159 147 103 69 75 45 28	276	100 91 43 19 17 16 15 14	0.00	12	28	3	–	–	–	–	–	–	2	–	–	–	Pentanoic acid, 3-methyl-2-[(trimethylsilyl)oxy]-, trimethylsilyl ester	54890-09-4	NI28208
73 145 147 247 75 115 74 45	276	100 60 43 29 25 10 9 9	0.00	12	28	3	–	–	–	–	–	–	2	–	–	–	Pentanoic acid, 3-methyl-3-[(trimethylsilyl)oxy]-, trimethylsilyl ester		NI28209
73 103 159 147 75 45 69 43	276	100 76 72 44 18 18 17 11	0.00	12	28	3	–	–	–	–	–	–	2	–	–	–	Pentanoic acid, 4-methyl-2-[(trimethylsilyl)oxy]-, trimethylsilyl ester	54890-08-3	NI28210
73 159 103 147 75 45 160 69	276	100 87 71 53 18 15 13 12	0.00	12	28	3	–	–	–	–	–	–	2	–	–	–	Pentanoic acid, 4-methyl-2-[(trimethylsilyl)oxy]-, trimethylsilyl ester	54890-08-3	NI28211
73 147 233 145 75 189 261 74	276	100 33 33 32 15 12 10 9	0.00	12	28	3	–	–	–	–	–	–	2	–	–	–	Pentanoic acid, 4-methyl-3-[(trimethylsilyl)oxy]-, trimethylsilyl ester		NI28212
147 73 103 261 75 148 45 177	276	100 82 22 21 21 16 14 13	0.10	12	28	3	–	–	–	–	–	–	2	–	–	–	2-Propenoic acid, 2-propyl-3-[(trimethylsilyl)oxy]-, trimethylsilyl ester		NI28213
150 276 220 75 185 278 109 222	276	100 99 91 91 66 65 59 57		13	6	1	2	–	–	2	–	–	–	–	–	–	11H-Dibenzo[c,f][1,2]diazepin-11-one, 3,8-dichloro-	958-00-9	NI28214
150 276 75 220 278 185 110 277	276	100 98 91 88 66 66 58 54		13	6	1	2	–	–	2	–	–	–	–	–	–	11H-Dibenzo[c,f][1,2]diazepin-11-one, 3,8-dichloro-	958-00-9	NI28215
150 75 220 276 74 222 185 110	276	100 99 86 72 72 57 57 43		13	6	1	2	–	–	2	–	–	–	–	–	–	11H-Dibenzo[c,f][1,2]diazepin-11-one, 3,8-dichloro-	958-00-9	NI28216
125 127 276 90 111 128 75 151	276	100 43 18 18 14 13 12 9		13	9	3	2	–	–	1	–	–	–	–	–	–	Aniline, 4-chloro-N-(4-nitrobenzylidene)-, N-oxide		MS06255
51 77 194 119 66 44 169 91	276	100 85 84 58 43 32 29 28	10.09	13	12	–	2	–	–	–	–	–	–	–	–	1	Selenourea, N,N-diphenyl-	21347-28-4	NI28217
93 194 103 77 94 91 119 76	276	100 24 16 10 8 7 5 4	3.04	13	12	–	2	–	–	–	–	–	–	–	–	1	Selenourea, N,N′-diphenyl-	16519-43-0	NI28218
93 195 77 194 92 51 276 66	276	100 88 77 46 42 35 33 28		13	12	–	2	–	–	–	–	–	–	–	–	1	Selenourea, N,N′-diphenyl-	16519-43-0	MS06256
192 234 120 163 148 191 121 193	276	100 84 57 39 37 26 21 15	10.51	13	12	5	2	–	–	–	–	–	–	–	–	–	2,4-Imidazolidinedione, 1,3-diacetyl-5-(4-hydroxyphenyl)-, (±)-	56818-00-9	MS06257
119 190 93 91 248 134 174 148	276	100 99 42 36 34 34 33 25	0.00	13	16	3	–	–	–	–	–	–	–	–	–	1	Iron, tricarbonyl[(1,2,3,4-η)-1-methyl-4-isopropyl-1,3-cyclohexadiene]-	51509-55-8	NI28219
190 174 148 248 119 188 56 218	276	100 48 46 36 34 30 24 20	0.00	13	16	3	–	–	–	–	–	–	–	–	–	1	Iron, tricarbonyl[(1,2,3,4-η)-2-methyl-5-isopropyl-1,3-cyclohexadiene]-	33519-41-4	NI28220
192 190 124 220 174 258 188 218	276	100 85 72 58 50 36 27 15	0.00	13	16	3	–	–	–	–	–	–	–	–	–	1	Iron, tricarbonyl(η^4-7-methyl-3-methylene-1,6-octadiene)-	33434-20-7	NI28221
58 218 104 103 59 30 115 77	276	100 64 48 44 40 25 19 18	4.00	13	16	3	4	–	–	–	–	–	–	–	–	–	4-Isoxazolecarboxamide, 4,5-dihydro-N-methyl-5-[[(methylamino)carbonyl]amino]-3-phenyl-, cis-	35053-71-5	NI28222
276 189 233 161 277 232 147 43	276	100 40 29 27 18 18 18 15		13	16	3	4	–	–	–	–	–	–	–	–	–	2,4(3H,5H)-Pyrimidinedione, 6-amino-3-(dimethylamino)-5-(4-methoxyphenyl)-	39265-99-1	NI28223
168 276 246 273 247 182 208 205	276	100 97 44 24 22 21 17 14		13	17	–	–	–	–	–	–	–	–	–	–	1	Rhodium, (1,5-cyclooctadiene)-π-cyclopentadienyl-	32610-45-0	NI28224
276 168 246 273 208 247 275 182	276	100 96 39 30 23 22 21 20		13	17	–	–	–	–	–	–	–	–	–	–	1	Rhodium, (1,5-cyclooctadiene)-π-cyclopentadienyl-	32610-45-0	MS06258
96 138 67 41 55 81 39 109	276	100 55 55 53 42 37 28 16	0.00	13	24	4	–	1	–	–	–	–	–	–	–	–	1-Methosulphato-2,2-dipropylcyclohexan-3-one		LI05920
73 187 179 146 105 89 276 229	276	100 34 28 27 9 9 5 3		13	28	–	–	2	–	–	–	–	1	–	–	–	1-Hexene, 2-[3-(methylthio)propylthio]-1-(trimethylsilyl)-	80403-00-5	NI28225
73 114 99 147 156 105 276 261	276	100 66 25 24 23 23 11 2		13	28	–	–	2	–	–	–	–	1	–	–	–	1-Hexene, 2-[3-(methylthio)-3-(trimethylsilyl)propylthio]-		NI28226
73 187 179 146 276 188 89 121	276	100 32 30 30 17 14 12 10		13	28	–	–	2	–	–	–	–	1	–	–	–	1-Pentene, 4-methyl-2-[3-(methylthio)propylthio]-1-(trimethylsilyl)-	80403-04-9	NI28227
73 114 99 156 276 105 147 155	276	100 41 26 24 21 21 18 13		13	28	–	–	2	–	–	–	–	1	–	–	–	1-Pentene, 4-methyl-2-[3-(methylthio)-3-(trimethylsilyl)propylthio]-		NI28228
55 147 97 73 75 103 149 69	276	100 67 45 39 24 22 19 12	0.10	13	32	2	–	–	–	–	–	–	2	–	–	–	3,11-Dioxa-2,12-disilatridecane, 2,2,12,12-tetramethyl-	54494-07-4	NI28229
55 147 73 75 97 149 103 45	276	100 71 52 45 39 22 22 16	0.10	13	32	2	–	–	–	–	–	–	2	–	–	–	3,11-Dioxa-2,12-disilatridecane, 2,2,12,12-tetramethyl-	54494-07-4	NI28230
220 222 185 213 150 276 221 187	276	100 63 37 26 21 17 17 13		14	6	2	–	–	–	2	–	–	–	–	–	–	9,10-Anthracenedione, 1,4-dichloro-		IC03579
150 276 75 248 28 278 220 74	276	100 93 82 68 67 60 59 55		14	6	2	–	–	–	2	–	–	–	–	–	–	9,10-Anthracenedione, 1,5-dichloro-	82-46-2	NI28231
150 276 75 220 278 74 222 248	276	100 93 89 74 60 59 48 46		14	6	2	–	–	–	2	–	–	–	–	–	–	9,10-Anthracenedione, 1,8-dichloro-	82-43-9	NI28232
241 206 137 243 102 75 205 139	276	100 60 39 33 28 22 20 20	15.01	14	10	–	2	–	–	2	–	–	–	–	–	–	Benzaldehyde, 2-chloro-, [(2-chlorophenyl)methylene]hydrazone	5328-80-3	NI28233
165 276 111 278 137 275 167 139	276	100 58 40 38 35 33 33 33		14	10	–	2	–	–	2	–	–	–	–	–	–	Benzaldehyde, 3-chloro-, [(3-chlorophenyl)methylene]hydrazone	6971-97-7	NI28234
165 276 111 278 137 139 275 167	276	100 58 39 37 34 33 32 32		14	10	–	2	–	–	2	–	–	–	–	–	–	Benzaldehyde, 4-chloro-, [(4-chlorophenyl)methylene]hydrazone		LI05921
43 190 45 69 42 216 171 114	276	100 85 69 38 38 35 35 35	0.00	14	12	2	–	2	–	–	–	–	–	–	–	–	Ethanethioic acid, S,S′-1,8-naphthalenediyl ester	54964-87-3	NI28235
139 107 91 65 276 169 77 79	276	100 96 60 51 39 27 27 26		14	12	4	–	1	–	–	–	–	–	–	–	–	Benzoic acid, 4-(4-tolylsulphonyl)-		IC03580
91 178 106 150 105 276 43 65	276	100 73 59 56 32 28 28 25		14	12	6	–	–	–	–	–	–	–	–	–	–	4,8-Etheno-1H,3H-benzo[1,2-c:4,5-c′]difuran-1,3,5,7-tetrone, 3a,4,4a,7a,8,8a-hexahydro-4,9-dimethyl-	32251-35-7	MS06259
276 261 205 233 247 69 28 137	276	100 52 46 38 26 26 21 18		14	12	6	–	–	–	–	–	–	–	–	–	–	3H,5H-Furo[3,2-g][1]benzopyran-2,5-dione, 4.9-dimethoxy-7-methyl-		NI28236
261 276 232 247 44 233 205 203	276	100 89 41 35 20 19 18 15		14	12	6	–	–	–	–	–	–	–	–	–	–	5H-Furo[3,2-g][1]benzopyran-5-one, 7-(hydroxymethyl)-4,9-dimethoxy-		MS06260

MASS TO CHARGE RATIOS								M.W.	INTENSITIES								Parent	C	H	O	N	S	F	Cl	Br	I	Si	P	B	X	COMPOUND NAME	CAS Reg No	No
261	276	232	247	205	233	262	203	276	100	89	44	34	20	19	14	14		14	12	6	–	–	–	–	–	–	–	–	–	–	5H-Furo[3,2-g][1]benzopyran-5-one, 7-(hydroxymethyl)-4,9-dimethoxy-		LI05922
276	233	190	277	261	147	175	218	276	100	44	19	18	15	10	8	5		14	12	6	–	–	–	–	–	–	–	–	–	–	7H-Furo[3,2-g][1]benzopyran-7-one, 4,5,6-trimethoxy-	18646-71-4	LI05923
276	233	190	277	261	147	175	234	276	100	44	19	17	14	9	7	6		14	12	6	–	–	–	–	–	–	–	–	–	–	7H-Furo[3,2-g][1]benzopyran-7-one, 4,5,6-trimethoxy-	18646-71-4	NI28237
276	233	277	190	148	261	162	63	276	100	44	18	14	12	10	6	5		14	12	6	–	–	–	–	–	–	–	–	–	–	7H-Furo[3,2-g][1]benzopyran-7-one, 5,6,9-trimethoxy-	18646-72-5	NI28238
276	233	277	190	147	261	162	161	276	100	44	19	15	13	10	7	5		14	12	6	–	–	–	–	–	–	–	–	–	–	7H-Furo[3,2-g][1]benzopyran-7-one, 5,6,9-trimethoxy-	18646-72-5	LI05924
219	43	234	276	216	153	67	69	276	100	87	57	46	46	32	29	14		14	12	6	–	–	–	–	–	–	–	–	–	–	2-Methyl-3-acetyl-5-hydroxy-7-acetoxychromone		MS06261
219	234	276	216	153	67	69	220	276	100	57	47	46	31	27	13	12		14	12	6	–	–	–	–	–	–	–	–	–	–	2-Methyl-3-acetyl-5-hydroxy-7-acetoxychromone		LI05925
247	276	249	248	278	277	250	215	276	100	40	37	17	16	8	6	6		14	13	–	2	1	–	1	–	–	–	–	–	–	7-Phenothiazinamine, 3-chloro-10-ethyl-		NI28239
276	212	211	277	243	197	91	121	276	100	28	26	21	16	13	12	9		14	13	2	–	1	–	–	–	–	–	1	–	–	10H-Phenothiaphosphine, 10-hydroxy-2,7-dimethyl-, 10-oxide	54964-91-9	NI28240
127	123	108	276	129	149	80	65	276	100	64	52	37	33	18	18	15		14	13	2	2	–	–	1	–	–	–	–	–	–	N-(2-Chlorophenyl)-N′-(2-methoxyphenyl)urea		MS06262
123	127	108	276	153	149	129	80	276	100	92	72	38	38	29	29	26		14	13	2	2	–	–	1	–	–	–	–	–	–	N-(3-Chlorophenyl)-N′-(2-methoxyphenyl)urea		MS06263
150	44	135	91	45	149	123	39	276	100	95	68	53	39	36	22	20	0.00	14	16	2	2	1	–	–	–	–	–	–	–	–	4-Pyrazolecarboxylic acid, 5-methyl-1-[2-(4-methylphenylthio)ethyl]-		NI28241
121	92	212	91	107	94	65	53	276	100	70	33	27	25	24	23	18	7.00	14	16	2	2	1	–	–	–	–	–	–	–	–	Pyridinium, 2,6-dimethyl-1-[[(4-methylphenyl)sulphonyl]amino]-, hydroxide, inner salt	55124-89-5	NI28242
121	92	212	107	94	196	106	211	276	100	70	33	25	24	17	15	7	7.00	14	16	2	2	1	–	–	–	–	–	–	–	–	Pyridinium, 2,6-dimethyl-1-[[(4-methylphenyl)sulphonyl]amino]-, hydroxide, inner salt	55124-89-5	LI05926
118	90	40	77	219	74	51	63	276	100	79	61	46	38	38	37	35	0.00	14	16	2	2	1	–	–	–	–	–	–	–	–	Pyrimidinium-4-olate, 2-(ethylthio)-3,6-dihydro-1,3-dimethyl-6-oxo-5-phenyl-		DD00908
110	276	28	111	195	82	81	80	276	100	53	28	25	23	21	20	19		14	16	4	2	–	–	–	–	–	–	–	–	–	1H-Pyrrole-2,5-dione, 1,1′-(1,6-hexanediyl)bis-	4856-87-5	NI28243
110	82	276	111	80	81	195	98	276	100	26	25	25	25	21	20	14		14	16	4	2	–	–	–	–	–	–	–	–	–	1H-Pyrrole-2,5-dione, 1,1′-(1,6-hexanediyl)bis-	4856-87-5	NI28244
43	159	105	160	104	261	91	59	276	100	54	48	45	39	32	23	19	1.23	14	17	5	–	–	–	–	–	–	–	–	1	–	1,2-O-Isopropylidene-α-D-xylofuranose 3,5-benzeneboronate	53829-60-0	NI28245
103	31	75	59	55	47	29	28	276	100	47	39	36	36	36	30	28	0.00	14	28	5	–	–	–	–	–	–	–	–	–	–	3-Hexanone, 1,5,6,6-tetraethoxy-	56667-13-1	NI28246
57	70	55	69	43	41	56	83	276	100	53	50	43	42	34	26	23	0.00	14	29	–	–	–	–	–	1	–	–	–	–	–	1-Bromo-1-isobutyl-4-ethyloctane		LI05927
57	43	135	71	55	41	69	85	276	100	88	82	62	48	47	40	34	0.50	14	29	–	–	–	–	–	1	–	–	–	–	–	1-Bromotetradecane		LI05928
57	43	135	137	71	55	41	69	276	100	91	82	81	55	50	50	40	0.53	14	29	–	–	–	–	–	1	–	–	–	–	–	1-Bromotetradecane		MS06264
73	117	75	55	69	83	132	77	276	100	78	53	50	45	38	36	32	1.20	14	29	2	–	–	1	–	–	–	1	–	–	–	Undecanoic acid, 11-fluoro-, trimethylsilyl ester	26305-97-5	NI28247
247	57	203	41	233	91	42	43	276	100	57	32	31	31	28	28	23	0.00	14	32	3	–	–	–	–	–	–	1	–	–	–	Ethyltributoxysilane	17957-38-9	NI28248
86	190	116	58	87	191	72	56	276	100	82	26	24	18	11	8	8	0.00	14	32	3	2	–	–	–	–	–	–	–	–	–	1,7-Bis(diethylamino)-2,6-dihydroxy-4-oxaheptane		MS06265
77	135	169	261	276	259	107		276	100	63	62	48	41	29	27			15	13	3	–	–	–	1	–	–	–	–	–	–	2,2′-Dihydroxy-3′-chloro-5,5′-dimethylbenzophenone		LI05929
208	69	209	193	207	180	190	123	276	100	90	33	29	16	14	11	11	1.40	15	16	5	–	–	–	–	–	–	–	–	–	–	2H-1-Benzopyran-2-one, 8-hydroxy-6-methoxy-7-[(3-methyl-2-butenyl)oxy]-		NI28249
105	77	91	79	67	81	126	123	276	100	59	21	20	16	14	13	12	0.00	15	16	5	–	–	–	–	–	–	–	–	–	–	1H-Cyclopenta[c]furan-1-one, 4-[(benzoyloxy)methyl]hexahydro-5-hydroxy-, [3aR-(3aα,4α,5β,6aα)]-		NI28250
205	189	276	175					276	100	35	28	25						15	16	5	–	–	–	–	–	–	–	–	–	–	(+)-cis-Khellactone methyl ether		LI05930
205	189	175	276					276	100	36	23	18						15	16	5	–	–	–	–	–	–	–	–	–	–	(–)-trans-Khellactone methyl ether		LI05931
276	261	233	246	229				276	100	66	40	14	12					15	16	5	–	–	–	–	–	–	–	–	–	–	2-Naphthalenecarboxaldehyde, 4,5,6,8-tetramethoxy-		LI05932
276	261	218	233	147	174	201	88	276	100	59	36	32	32	23	22	20		15	16	5	–	–	–	–	–	–	–	–	–	–	2-Naphthalenecarboxylic acid, 5,6,7-trimethoxy-, methyl ester	23673-53-2	NI28251
276	261	218	233	147	277	201	88	276	100	58	35	31	30	20	20	19		15	16	5	–	–	–	–	–	–	–	–	–	–	2-Naphthalenecarboxylic acid, 5,6,7-trimethoxy-, methyl ester	23673-53-2	MS06266
151	41	69	43	53	126	111	109	276	100	93	84	67	53	31	26	16	14.00	15	16	5	–	–	–	–	–	–	–	–	–	–	3H-Oxireno[8,8a]azuleno[4,5-b]furan-3,8(4aH)-dione, 5,6,9a,9b-tetrahydro-6-hydroxy-1,4a,7-trimethyl-, [4aS-(3aS*,4aα,6α,9aα,9bβ)]-	86948-40-5	NI28252
200	258	218	174	115	144	116	276	276	100	55	53	53	49	45	45	39		15	16	5	–	–	–	–	–	–	–	–	–	–	Piperonylic acid, 6-(4-methoxy-1-cyclohexen-1-yl)-	16831-74-6	NI28253
175	217	134	91	276	43	176	121	276	100	95	63	54	53	52	48	48		15	16	5	–	–	–	–	–	–	–	–	–	–	Porelladiolide, 3α,4α-epoxy-		NI28254
69	135	77	110	119	83	91	126	276	100	78	71	66	58	56	34	24	1.00	15	20	1	2	1	–	–	–	–	–	–	–	–	Pyrrolo[1,2-b][1,2,4]oxadiazole-2(1H)-thione, tetrahydro-5,5,7,7-tetramethyl-1-phenyl-	50455-65-7	NI28255
114	134	261	162	39	113	99	51	276	100	62	52	38	37	35	31	30	21.39	15	20	1	2	1	–	–	–	–	–	–	–	–	4H-1,3-Thiazine, 4,4,6-trimethyl-2-[(4-ethoxyphenyl)amino(imino)]-	37591-11-0	NI28256
261	162	147	113	99	59	262	276	276	100	36	31	21	18	18	17	13		15	20	1	2	1	–	–	–	–	–	–	–	–	4H-1,3-Thiazine, 4,4,6-trimethyl-2-(N-methyl-4-methoxyphenylamino)-	37489-66-0	NI28257
58	18	43	42	15	30			276	100	15	11	7	7	5			0.00	15	20	3	2	–	–	–	–	–	–	–	–	–	1H-Indole-3-ethanamine, 1-(acetyloxy)-5-methoxy-N,N-dimethyl-	55030-13-2	NI28258
58	276	59	218	277	278	261		276	100	10	4	3	2	1	1			15	24	1	2	–	–	–	–	–	1	–	–	–	Indole, 3-[2-(dimethylamino)ethyl]-4-(trimethylsiloxy)-	17943-16-7	NI28259
83	55	41	69	57	43	81	95	276	100	53	51	41	41	41	27	26	13.07	15	30	–	–	–	–	1	–	–	–	1	–	–	Phosphinous chloride, (2,2-dimethylpropyl)[methylisopropylcyclohexyl]-	74645-94-6	NI28260
104	206	207	233	77	193	277	103	276	100	42	40	36	34	28	24	24	20.00	16	12	1	4	–	–	–	–	–	–	–	–	–	2-Oxabicyclo[2.2.1]heptane-4,7,7-tricarbonitrile, 3-imino-1-methyl-5-phenyl-		MS06267
107	109	151	123	126	276	260		276	100	63	50	23	22	2	2			16	20	2	–	1	–	–	–	–	–	–	–	–	Bicyclo[2.2.1]heptan-2-one, 5,5-dimethyl-6-[(phenylsulphinyl)methyl]-		MS06268
228	149	107	260	109	259	108	231	276	100	72	65	63	49	40	40	32	6.00	16	20	2	–	1	–	–	–	–	–	–	–	–	Bicyclo[3.2.1]oct-2-en-1-ol, 4,4-dimethyl-2-(phenylsulphinyl)-, (1RS,5RS,S_S,R_S)-		MS06269
182	95	123	40	79	41	121	39	276	100	93	88	72	70	66	65	60	26.00	16	20	2	–	1	–	–	–	–	–	–	–	–	Cyclohexenepropanoic acid, α-(phenylthio)-, methyl ester	74367-00-3	NI28261

MASS TO CHARGE RATIOS	M.W.	INTENSITIES	Parent	C	H	O	N	S	F	Cl	Br	I	Si	P	B	X	COMPOUND NAME	CAS Reg No	No
135 122 245 123 136 43 258 149	**276**	100 16 14 14 13 13 12 12	**0.00**	16	20	2	–	1	–	–	–	–	–	–	–	–	**7-Oxabicyclo[2.2.1]heptane-2-exo-methanol, (±)-3-exo-vinyl-4-methyl-5-endo-(phenylthio)-**	117626-59-2	**DD00909**
105 77 234 106 43 154 78 51	**276**	100 20 18 9 9 3 3 3	**2.00**	16	20	4	–	–	–	–	–	–	–	–	–	–	**3-Acetyl-3-(benzoyloxy)heptan-2-one**		**MS06270**
28 43 105 42 77 41 67 57	**276**	100 58 47 40 37 26 25 24	**0.00**	16	20	4	–	–	–	–	–	–	–	–	–	–	**5,10-Benzocyclooctenediol, 5,6,7,8,9,10-hexahydro-, diacetate, (5R-trans)-**	56728-03-1	**NI28262**
231 185 91 232 202 93 79 77	**276**	100 36 18 16 12 12 10 10	**8.00**	16	20	4	–	–	–	–	–	–	–	–	–	–	**3,5,1,7-[1,2,3,4]Butanetetraylnaphthalene-1,6(2H)-dicarboxylic acid, octahydro-**	75314-17-9	**NI28263**
129 276 29 157 130 128 185 115	**276**	100 31 30 27 21 19 16 13		16	20	4	–	–	–	–	–	–	–	–	–	–	**3-Butene-1,1-dicarboxylic acid, 3-phenyl-, diethyl ester**		**NI28264**
104 129 202 29 184 203 128 105	**276**	100 29 12 12 11 10 10 10	**6.00**	16	20	4	–	–	–	–	–	–	–	–	–	–	**1,1-Cyclobutanedicarboxylic acid, 3-phenyl-, diethyl ester**	1769-72-8	**NI28265**
175 276 217 233 45 91 162 41	**276**	100 92 83 43 43 42 39 35		16	20	4	–	–	–	–	–	–	–	–	–	–	**Cyclodeca[b]furan-6-carboxylic acid, 4,5,6,7,8,11-hexahydro-3,10-dimethyl-4-oxo-, methyl ester, [R-(E)]-**	56051-41-3	**NI28266**
43 91 41 53 77 93 39 121	**276**	100 86 72 59 53 49 49 48	**11.20**	16	20	4	–	–	–	–	–	–	–	–	–	–	**Deoxysericealactone**		**MS06271**
157 129 117 156 91 216 115 128	**276**	100 52 43 37 28 26 22 20	**1.00**	16	20	4	–	–	–	–	–	–	–	–	–	–	**trans-7,8-Diacetoxydispiro[2.2.4.2]dodeca-4,11-diene**		**NI28267**
219 276 220 189 149 121 43 41	**276**	100 22 13 8 4 4 4 4		16	20	4	–	–	–	–	–	–	–	–	–	–	**Dihydropeucenin-7-methyl ester**		**MS06272**
220 219 189 276 232 41 43 221	**276**	100 86 22 21 15 15 14 13		16	20	4	–	–	–	–	–	–	–	–	–	–	**Dihydropeucenin-7-methyl ether**		**MS06273**
175 276 217 233 45 91 162 44	**276**	100 94 84 44 44 42 39 35		16	20	4	–	–	–	–	–	–	–	–	–	–	**3,11-Dimethyl-7-methoxycarbonyl-13-oxabicyclo[8.3.0]trideca-1(10),3,11-trien-9-one**		**LI05933**
124 69 93 276 77 41 66 39	**276**	100 64 50 38 24 23 19 17		16	20	4	–	–	–	–	–	–	–	–	–	–	**2(5H)-Furanone, 5-(1,3-hexadienyl)-4-methoxy-5-methyl-3-(1-oxo-2-butenyl)-, (E,E,E)-(+)-**		**LI05934**
135 276 188 202 161 203 131 148	**276**	100 57 33 18 14 12 12 11		16	20	4	–	–	–	–	–	–	–	–	–	–	**2-Heptenoic acid, 7-(3,4-methylenedioxyphenyl)-, ethyl ester, (2E)-**	83029-44-1	**NI28268**
175 276 217 45 233 91 162 199	**276**	100 92 83 44 43 42 39 26		16	20	4	–	–	–	–	–	–	–	–	–	–	**Linderalactone keto ester**		**MS06274**
79 152 93 80 124 203 106 230	**276**	100 82 62 60 46 45 45 26	**0.00**	16	20	4	–	–	–	–	–	–	–	–	–	–	**4,7-Methano-1H-indene-1,4-dicarboxylic acid, 3a,4,7,7a-tetrahydro-1,8-dimethyl-, 1-ethyl hydrogen ester**		**DD00910**
91 108 78 92 107 76 129 51	**276**	100 25 25 23 20 16 11 9	**0.00**	16	20	4	–	–	–	–	–	–	–	–	–	–	**α-D-Ribo-hex-5-enofuranose, 3-O-benzyl-5,6-dideoxy-1,2-O-isopropylidene-**	23537-43-1	**NI28269**
217 129 157 91 185 77 115 128	**276**	100 99 94 69 62 59 58 55	**16.00**	16	20	4	–	–	–	–	–	–	–	–	–	–	**Spiro[2.5]oct-4-ene-5,8-dicarboxylic acid, 8-(1,3-butadienyl)-, dimethyl ester**		**MS06275**
164 218 152 219 165 276 121 108	**276**	100 84 84 68 63 45 45 42		16	24	2	2	–	–	–	–	–	–	–	–	–	**Lycocernuine, 14,15-didehydro-**	36101-39-0	**NI28270**
276 56 134 150 148 135 41 163	**276**	100 95 82 75 74 61 59 50		16	24	2	2	–	–	–	–	–	–	–	–	–	**4H-1,3-Oxazine, 5,6-dihydro-3,4,4,6-tetramethyl-2-(4-ethoxyphenylimino)-**	53004-26-5	**NI28271**
122 261 276 275 150 151 84 178	**276**	100 81 80 79 66 66 37 34		16	24	2	2	–	–	–	–	–	–	–	–	–	**4H-1,3-Oxazine, 5,6-dihydro-4,4,6-trimethyl-2-(N-methyl-4-ethoxyphenylamino)-**	53004-33-4	**NI28272**
41 67 55 54 82 81 68 69	**276**	100 79 79 77 71 45 42 22	**1.00**	16	28	–	4	–	–	–	–	–	–	–	–	–	**9-Azabicyclo[6.1.0]nonane, 9,9′-azobis-, [1α,8α,9[E(1′R*,8′S*)]]-**	66387-81-3	**NI28273**
276 230 229 202 228 154 217 275	**276**	100 71 52 44 39 24 23 20		17	12	2	2	–	–	–	–	–	–	–	–	–	**Quinoline, 4-(4-nitrostyryl)-**	4594-97-2	**NI28274**
145 146 276 103 261 158 133 131	**276**	100 31 25 23 21 21 21 16		17	16	–	4	–	–	–	–	–	–	–	–	–	**1H-Pyrrolo[2,3-b]pyridine, 3,3′-methylenebis[2-methyl-**	23709-49-1	**NI28275**
245 246 261 43 276 161 41 189	**276**	100 19 17 17 9 7 7 5		17	24	3	–	–	–	–	–	–	–	–	–	–	**3-Buten-2-one, 4-(2,6-dimethoxy-4-pentylphenyl)-**	3411-09-4	**NI28276**
91 145 131 173 190 146 119 57	**276**	100 92 83 81 68 66 17 16	**0.00**	17	24	3	–	–	–	–	–	–	–	–	–	–	**1,3-Dioxan-4-one, 5-benzyl-2-tert-butyl-5,6-dimethyl-, (2R,5S,6R)-**		**DD00911**
127 185 91 160 98 107 129 170	**276**	100 63 62 49 46 32 26 25	**4.00**	17	24	3	–	–	–	–	–	–	–	–	–	–	**DL-Gluco-3-heptulose, 1,5-anhydro-2,4,6-trideoxy-2,4,6-trimethyl-7-O-benzyl-**		**NI28277**
91 69 109 83 82 92 87 85	**276**	100 42 15 13 13 11 11 11	**2.00**	17	24	3	–	–	–	–	–	–	–	–	–	–	**DL-Mannohept-1-enitol, 1,5-anhydro-2,4,6-trideoxy-2,4,6-trimethyl-7-O-benzyl-**		**NI28278**
91 79 71 107 65 41 67 55	**276**	100 16 13 12 8 8 7 7	**0.00**	17	24	3	–	–	–	–	–	–	–	–	–	–	**1-Oxacyclooct-4-ene, 8-[2-(benzyloxy)ethyl]-2-(hydroxymethyl)-, (2S,8R)-**		**DD00912**
91 79 71 107 245 159 81 142	**276**	100 18 17 15 9 9 9 6	**0.00**	17	24	3	–	–	–	–	–	–	–	–	–	–	**1-Oxacyclooct-4-ene, 8-[2-(benzyloxy)ethyl]-2-(hydroxymethyl)-, (2S,8S)-**		**DD00913**
58 98 85 219 84 41 42 55	**276**	100 51 45 13 12 12 11 10	**0.00**	17	28	1	2	–	–	–	–	–	–	–	–	–	**2-Propanamine, N-methyl-1-[4-[2-(1-piperidyl)ethoxy]phenyl]-**		**MS06276**
72 84 42 73 56 98 44 205	**276**	100 6 6 5 4 3 3 2	**0.00**	17	28	1	2	–	–	–	–	–	–	–	–	–	**2-Propanamine, N,N-dimethyl-1-[4-[2-(1-pyrrolidinyl)ethoxy]phenyl]-**		**MS06277**
178 179 89 176 276 76 88 177	**276**	100 18 14 13 10 10 8 7		18	12	3	–	–	–	–	–	–	–	–	–	–	**Furano[9,10:3′,4′]anthracene-12,14-dione, 12,14-dihydro-**		**IC03581**
234 276 231 202 247 203 220 248	**276**	100 73 43 42 33 24 20 19		18	12	3	–	–	–	–	–	–	–	–	–	–	**Indan-3-one, 1-hydroxy-1-(inden-3-on-1-yl)-**		**LI05935**
248 276 191 189 176 165 134 192	**276**	100 30 30 25 16 15 15 12		18	12	3	–	–	–	–	–	–	–	–	–	–	**Phenanthro[1,2-b]furan-10,11-dione, 1,6-dimethyl-**	568-73-0	**NI28279**
248 249 276 191 277 190 176 165	**276**	100 35 31 31 23 22 15 15		18	12	3	–	–	–	–	–	–	–	–	–	–	**Phenanthro[1,2-b]furan-10,11-dione, 1,6-dimethyl-**	568-73-0	**NI28280**
248 191 189 276 192 176 165 177	**276**	100 43 39 29 18 16 10 9		18	12	3	–	–	–	–	–	–	–	–	–	–	**Phenanthro[1,2-b]furan-10,11-dione, 1,6-dimethyl-**	568-73-0	**NI28281**
276 261 189 220 248 219 191 139	**276**	100 70 26 23 21 20 20 17		18	12	3	–	–	–	–	–	–	–	–	–	–	**Phenanthro[3,2-b]furan-7,11-dione, 4,8-dimethyl-**	20958-17-2	**NI28282**
276 261 189 248 220 262 219 191	**276**	100 70 27 24 23 20 20 20		18	12	3	–	–	–	–	–	–	–	–	–	–	**Phenanthro[3,2-b]furan-7,11-dione, 4,8-dimethyl-**	20958-17-2	**NI28283**
276 277 138 219 165 191 189 139	**276**	100 22 16 13 13 10 8 8		18	12	3	–	–	–	–	–	–	–	–	–	–	**19,20,21-Trioxatetracyclo[14.2.1.1^{4,7},1^{10,13}]heneicos-2,4,6,8,10,12,14,16,18-nonaene**		**LI05936**
199 276 122 77 200 45 197 198	**276**	100 31 31 25 19 17 10 9		18	16	1	–	–	–	–	–	–	1	–	–	–	**Silanol, triphenyl-**	791-31-1	**NI28284**
105 143 115 276 77 79 120 247	**276**	100 83 69 55 45 23 16 14		18	16	1	2	–	–	–	–	–	–	–	–	–	**2-Naphthalenol, 1-[(2,4-dimethylphenyl)azo]-**	3118-97-6	**NI28285**

MASS TO CHARGE RATIOS								M.W.	INTENSITIES								Parent	C	H	O	N	S	F	Cl	Br	I	Si	P	B	X	COMPOUND NAME	CAS Reg No	No
104	105	71	57	43	69	55	85	**276**	100	31	8	8	7	6	6	5	**0.00**	18	28	2	–	–	–	–	–	–	–	–	–	–	**Decanoic acid, 2-phenylethyl ester**		**LI05937**
41	55	67	81	79	68	217	43	**276**	100	99	79	68	63	56	53	47	**38.62**	18	28	2	–	–	–	–	–	–	–	–	–	–	**Estran-3-one, 17-hydroxy-, (5α,17β)-**	1434-85-1	**NI28286**
276	258	277	202	214	201	199	187	**276**	100	24	20	18	15	10	10	9		18	28	2	–	–	–	–	–	–	–	–	–	–	**Estran-17-one, 3-hydroxy-, (3α,5α)-**	1225-01-0	**NI28287**
205	206	163	121	276	261	55	42	**276**	100	16	12	12	6	6	5	4		18	28	2	–	–	–	–	–	–	–	–	–	–	**5,9-Methanobenzocycloocten-4(1H)-one, 2,3,5,6,7,8,9,10-octahydro-5-hydroxy-2,2,7,7,9-pentamethyl-**	6244-16-2	**NI28288**
261	109	108	95	55	218	43	93	**276**	100	42	37	36	36	35	34	29	**13.00**	18	28	2	–	–	–	–	–	–	–	–	–	–	**17-Oxaandrostan-16-one, (5α)-**	54482-41-6	**NI28289**
91	141	107	127	185	123	159	170	**276**	100	47	45	28	23	20	16	13	**0.00**	18	28	2	–	–	–	–	–	–	–	–	–	–	**1-Oxacycloheptane, 2-[2-(benzyloxy)ethyl]-7-propyl-, (2R,7R)-**		**DD00914**
91	108	43	57	92	32	41	55	**276**	100	74	37	30	28	26	21	17	**0.70**	18	28	2	–	–	–	–	–	–	–	–	–	–	**Undecanoic acid, benzyl ester**		**NI28290**
91	117	173	105	161	131	205	145	**276**	100	16	14	14	13	9	7	5	**10.00**	18	28	2	–	–	–	–	–	–	–	–	–	–	**Undecanoic acid, 6-phenyl-, methyl ester**		**LI05938**
91	147	187	105	117	276	169	219	**276**	100	22	19	13	12	11	9	7		18	28	2	–	–	–	–	–	–	–	–	–	–	**Undecanoic acid, 7-phenyl-, methyl ester**		**LI05939**
91	133	183	105	117	276	201	233	**276**	100	26	18	11	9	7	7	3		18	28	2	–	–	–	–	–	–	–	–	–	–	**Undecanoic acid, 8-phenyl-, methyl ester**		**LI05940**
91	191	197	215	271	117	105	131	**276**	100	50	13	11	10	10	9	8	**0.00**	18	28	2	–	–	–	–	–	–	–	–	–	–	**Undecanoic acid, 9-phenyl-, methyl ester**		**LI05941**
105	91	276	244	119	74	216	229	**276**	100	14	12	12	6	5	3	2		18	28	2	–	–	–	–	–	–	–	–	–	–	**Undecanoic acid, 10-phenyl-, methyl ester**		**LI05942**
43	219	58	41	276	57	28	85	**276**	100	51	31	25	20	20	18	15		18	32	–	2	–	–	–	–	–	–	–	–	–	**1,4-Benzenediamine, N,N'-bis(1,3-dimethylbutyl)-**		**IC03582**
215	262	216	244	231	202	258	245	**276**	100	55	48	38	29	27	18	17	**10.95**	19	16	2	–	–	–	–	–	–	–	–	–	–	**Benz[a]anthracene-5,6-diol, 5,6-dihydro-7-methyl-**		**NI28291**
276	258	229	215	245	230	228	202	**276**	100	86	81	81	71	44	42	42		19	16	2	–	–	–	–	–	–	–	–	–	–	**Benz[a]anthracene-5,6-diol, 5,6-dihydro-7-methyl-**		**NI28292**
230	276	215	229	202	258	231	204	**276**	100	66	44	43	39	30	28	28		19	16	2	–	–	–	–	–	–	–	–	–	–	**Benz[a]anthracene-3,4-diol, 3,4-dihydro-11-methyl-, trans-**	82735-47-5	**NI28293**
215	276	229	202	230	258	245	216	**276**	100	85	85	66	47	38	38	30		19	16	2	–	–	–	–	–	–	–	–	–	–	**Benz[a]anthracene-5,6-diol, 5,6-dihydro-11-methyl-, trans-**	82735-48-6	**NI28294**
230	215	229	202	228	258	276	101	**276**	100	84	60	45	39	38	34	30		19	16	2	–	–	–	–	–	–	–	–	–	–	**Benz[a]anthracene-8,9-diol, 8,9-dihydro-11-methyl-, trans-**	82735-49-7	**NI28295**
230	233	202	215	276	232	231	229	**276**	100	91	82	74	62	61	46	42		19	16	2	–	–	–	–	–	–	–	–	–	–	**Benz[a]anthracene-10,11-diol, 10,11-dihydro-11-methyl-, trans-**	82735-50-0	**NI28296**
141	91	142	115	276	139	140	157	**276**	100	14	13	11	9	5	5	4		19	16	2	–	–	–	–	–	–	–	–	–	–	**Benzeneacetic acid, 1-naphthalenemethyl ester**	36707-29-6	**NI28297**
69	191	41	189	276	208	190	179	**276**	100	67	64	38	27	22	22	20		19	16	2	–	–	–	–	–	–	–	–	–	–	**2-Butenoic acid, 9-phenanthrenemethyl ester**	92174-36-2	**NI28298**
191	276	189	208	190	165	41	179	**276**	100	28	26	23	14	14	14	10		19	16	2	–	–	–	–	–	–	–	–	–	–	**3-Butenoic acid, 9-phenanthrenemethyl ester**	92174-34-0	**NI28299**
276	40	277	261	215	44	233	203	**276**	100	27	21	19	19	14	13	12		19	16	2	–	–	–	–	–	–	–	–	–	–	**2,5-Cyclohexadien-1-one, 2-methoxy-4,4-diphenyl-**	96283-91-9	**NI28300**
248	143	51	105	276	249	128	115	**276**	100	59	48	42	41	25	20	14		19	16	2	–	–	–	–	–	–	–	–	–	–	**2H-Pyran-2-one, 4,5-dimethyl-3,6-diphenyl-**	33731-56-5	**NI28301**
185	183	276	91	275	277	186	184	**276**	100	76	72	28	16	15	14	11		19	17	–	–	–	–	–	–	–	–	1	–	–	**Phosphine, benzyldiphenyl-**	7650-91-1	**NI28302**
262	275	183	276	36	263	108	185	**276**	100	63	55	21	21	19	13	12		19	17	–	–	–	–	–	–	–	–	1	–	–	**Phosphorane, methylenetriphenyl-**	3487-44-3	**LI05943**
262	275	183	276	36	263	261	185	**276**	100	65	56	24	21	20	20	14		19	17	–	–	–	–	–	–	–	–	1	–	–	**Phosphorane, methylenetriphenyl-**	3487-44-3	**NI28303**
276	275	277	134	261	117	186	130	**276**	100	27	25	21	12	6	5	5		19	20	–	2	–	–	–	–	–	–	–	–	–	**Cyclopropanecarbonitrile, 2-[4-(dimethylamino)phenyl]-1-4-tolyl-**	32589-51-8	**LI05944**
276	275	277	134	261	185	118	77	**276**	100	27	25	20	12	6	6	5		19	20	–	2	–	–	–	–	–	–	–	–	–	**Cyclopropanecarbonitrile, 2-[4-(dimethylamino)phenyl]-1-4-tolyl-**	32589-51-8	**NI28304**
243	95	258	162	108	81	41	67	**276**	100	97	83	83	75	75	74	68	**2.50**	19	32	1	–	–	–	–	–	–	–	–	–	–	**Androstan-3-ol, (3α,5β)-**	15360-53-9	**NI28305**
41	243	95	55	81	276	67	93	**276**	100	99	92	91	77	74	72	58		19	32	1	–	–	–	–	–	–	–	–	–	–	**Androstan-3-ol, (3β,5α)-**	1224-92-6	**NI28306**
81	243	95	41	55	67	276	107	**276**	100	99	99	94	90	85	80	72		19	32	1	–	–	–	–	–	–	–	–	–	–	**Androstan-3-ol, (3β,5α)-**	1224-92-6	**NI28307**
258	108	243	95	81	43	41	67	**276**	100	97	96	95	77	75	70	65	**3.00**	19	32	1	–	–	–	–	–	–	–	–	–	–	**Androstan-3-ol, (3β,5β)-**	15360-52-8	**NI28308**
276	217	149	55	261	95	81	41	**276**	100	69	45	35	34	32	32	32		19	32	1	–	–	–	–	–	–	–	–	–	–	**Androstan-17-ol, (5α,17β)-**	1225-43-0	**NI28309**
276	217	149	55	261	109	81	95	**276**	100	68	44	36	35	33	33	32		19	32	1	–	–	–	–	–	–	–	–	–	–	**Androstan-17-ol, (5α,17β)-**	1225-43-0	**LI05945**
217	276	149	55	109	81	67	95	**276**	100	81	78	58	57	56	56	52		19	32	1	–	–	–	–	–	–	–	–	–	–	**Androstan-17-ol, (5α,17β)-**	1225-43-0	**NI28310**
276	217	261	55	151	95	109	81	**276**	100	57	34	26	25	25	24	24		19	32	1	–	–	–	–	–	–	–	–	–	–	**Androstan-17-ol, (5β,17β)-**	10328-72-0	**NI28311**
276	217	261	81	55	151	95	67	**276**	100	57	33	26	26	25	25	23		19	32	1	–	–	–	–	–	–	–	–	–	–	**Androstan-17-ol, (5β,17β)-**	10328-72-0	**LI05946**
205	121	206	41	135	276	71	43	**276**	100	49	16	11	10	6	6	6		19	32	1	–	–	–	–	–	–	–	–	–	–	**5,9-Methanobenzocycloocten-5(1H)-ol, 2,3,4,6,7,8,9,10-octahydro-2,2,4,7,7,9-hexamethyl-**	16004-84-5	**NI28312**
218	109	148	108	43	95	175	81	**276**	100	98	89	75	70	69	67	67	**5.00**	19	32	1	–	–	–	–	–	–	–	–	–	–	**D-Norandrostane-16-methanol, (5α,16β)-**	54411-60-8	**NI28313**
191	192	123	95	109	137	135	81	**276**	100	50	47	33	32	31	20	20	**6.00**	19	32	1	–	–	–	–	–	–	–	–	–	–	**15-Norlabd-13-ene, 8,12-epoxy-, (12RS)-**		**MS06278**
71	41	79	67	93	55	95	80	**276**	100	70	53	37	27	27	24	23	**17.04**	19	32	1	–	–	–	–	–	–	–	–	–	–	**1,9,12,15-Octadecatetraene, 1-methoxy-**	56847-00-8	**NI28314**
247	248	261	276	57	262	231	80	**276**	100	19	15	11	4	3	3	2		19	32	1	–	–	–	–	–	–	–	–	–	–	**Phenol, 2,6-di-tert-butyl-4-(1,1-dimethylpropyl)-**		**IC03583**
136	243	93	107	121	41	96	91	**276**	100	56	34	29	27	26	22	19	**13.60**	19	32	1	–	–	–	–	–	–	–	–	–	–	**Podocarp-7-en-3α-ol, 13,13-dimethyl-**	4807-60-7	**NI28315**
136	93	243	107	121	276	69	41	**276**	100	61	49	47	43	41	38	38		19	32	1	–	–	–	–	–	–	–	–	–	–	**Podocarp-7-en-3β-ol, 13,13-dimethyl-**	5366-68-7	**NI28316**
136	243	276	93	121	261	107	137	**276**	100	52	40	31	28	27	25	19		19	32	1	–	–	–	–	–	–	–	–	–	–	**Podocarp-7-en-3β-ol, 13,13-dimethyl-**	5366-68-7	**NI28317**
172	276	171	104	169	115	128	119	**276**	100	62	54	46	34	31	18	18		20	20	1	–	–	–	–	–	–	–	–	–	–	**Benzo[d][2,2]paracyclophan-4(3H)-dione, 1,2-dihydro-**		**LI05947**
276	202	277	261	220	205	101	189	**276**	100	19	18	18	17	17	17	15		20	20	1	–	–	–	–	–	–	–	–	–	–	**Benzo[e]pyren-9(2H)-one, 1,3,6,7,8,10,11,12-octahydro-**	68151-08-6	**NI28318**
55	83	41	97	193	192	67	69	**276**	100	74	64	62	37	36	30	26	**0.48**	20	36	–	–	–	–	–	–	–	–	–	–	–	**1,1,2-Tricyclohexylethane**		**AI01946**
276	91	117	77	92	93	129	115	**276**	100	58	33	33	27	26	25	25		21	24	–	–	–	–	–	–	–	–	–	–	–	**Bicyclo[2.2.1]heptadiene, trimer**	54965-16-1	**NI28319**
276	91	277	138	92	274	137	275	**276**	100	30	25	25	22	20	18	10		22	12	–	–	–	–	–	–	–	–	–	–	–	**Anthanthrene**	191-26-4	**MS06279**
276	138	137	136	277	274	124	275	**276**	100	99	85	39	25	23	15	12		22	12	–	–	–	–	–	–	–	–	–	–	–	**Anthanthrene**	191-26-4	**NI28320**
276	138	137	277	274	136	275	272	**276**	100	34	27	24	20	12	11	6		22	12	–	–	–	–	–	–	–	–	–	–	–	**Benzo[ghi]perylene**		**NI28321**
276	138	137	277	274	136	275	272	**276**	100	37	28	25	22	12	10	6		22	12	–	–	–	–	–	–	–	–	–	–	–	**1,12-Benzperylene**	191-24-2	**NI28322**

MASS TO CHARGE RATIOS								M.W.	INTENSITIES								Parent	C	H	O	N	S	F	Cl	Br	I	Si	P	B	X	COMPOUND NAME	CAS Reg No	No
276	138	274	137	277	275	272	137.5	276	100	32	27	23	22	15	7	7		22	12	–	–	–	–	–	–	–	–	–	–	–	1,12-Benzperylene	191-24-2	MS06280
276	138	277	274	137	275	136	138.5	276	100	28	27	19	18	8	8	7		22	12	–	–	–	–	–	–	–	–	–	–	–	Indeno[1,2,3-cd]pyrene	193-39-5	MS06281
276	138	277	274	137	275	136	124	276	100	28	27	19	18	8	8	6		22	12	–	–	–	–	–	–	–	–	–	–	–	Indeno[1,2,3-cd]pyrene	193-39-5	NI28323
30	28	44	46	16	127	153	14	277	100	66	62	12	4	3	2	2	0.00	1	–	6	3	–	–	–	–	1	–	–	–	–	Methane, iodotrinitro-	630-70-6	NI28324
31	64	69	109	45	52	50	83	277	100	76	30	18	18	17	16	13	0.00	3	–	–	5	–	9	–	–	–	–	–	–	–	N-[2,3-Bis(difluoroamino)-2,3-difluoro-1-aziridyl]difluoroaminofluoromethylenimine		LI05948
69	85	31	54	104	50	97	47	277	100	9	5	4	3	2	1	1	0.30	4	–	2	1	–	8	–	–	–	–	1	–	–	2-Cyano-2-(μ-oxodifluorophosphoryl)hexafluoropropane		LI05949
166	69	127	125	198	78	110	46	277	100	68	19	19	18	18	10	10	0.35	4	3	–	1	–	7	–	1	–	–	–	–	–	1-Bromo-1-fluoro-2-[bis(trifluoromethyl)amino]ethane		LI05950
69	177	91	125	277	258	110	81	277	100	44	24	23	19	7	7	7		6	4	1	1	–	9	–	–	–	–	–	–	–	1-Propenamine, N,N-bis(trifluoromethyl)-1-methoxy-3,3,3-trifluoro-, (E)-		LI05951
69	177	81	277	96	125	89	91	277	100	24	12	11	6	5	5	4		6	4	1	1	–	9	–	–	–	–	–	–	–	1-Propenamine, N,N-bis(trifluoromethyl)-2-methoxy-3,3,3-trifluoro-		LI05952
277	196	198	279	42	157	155	275	277	100	88	86	79	78	74	63	47		8	8	–	1	–	–	–	1	–	–	–	–	1	N-(4-Bromophenyl)selenoacetamide		MS06282
42	41	277	279	281	195	193	158	277	100	23	20	19	6	6	6	6		9	6	1	3	–	–	3	–	–	–	–	–	–	1-(2,4,6-Trichlorophenyl)-3-amino-5-pyrazolone		IC03584
55	218	43	189	176	236	235	98	277	100	69	50	23	16	16	16	15	0.00	9	12	3	1	–	5	–	–	–	–	–	–	–	β-Alanine, N-pentafluoropropionyl-		MS06283
190	43	191	41	45	119	140	27	277	100	56	43	15	13	12	10	7	0.00	9	12	3	1	–	5	–	–	–	–	–	–	–	D-Alanine, N-pentafluoropropionyl-, isopropyl ester		MS06284
190	43	191	41	119	45	218	27	277	100	50	43	12	12	11	7	7	0.00	9	12	3	1	–	5	–	–	–	–	–	–	–	L-Alanine, N-pentafluoropropionyl-, isopropyl ester		MS06285
43	190	191	41	45	119	27	42	277	100	86	35	22	19	13	13	9	0.00	9	12	3	1	–	5	–	–	–	–	–	–	–	L-Alanine, N-pentafluoropropionyl-, isopropyl ester		NI28325
190	43	191	119	42	235	41	218	277	100	60	31	17	16	14	13	11	4.90	9	12	3	1	–	5	–	–	–	–	–	–	–	Sarcosine, N-pentafluoropropionyl-, isopropyl ester		MS06286
278	279	218	101	280	246	115	94	277	100	9	3	3	1	1	1	1	0.00	9	12	3	1	–	5	–	–	–	–	–	–	–	Valine, methyl ester, N-pentafluoropropionate		NI28326
79	125	47	109	63	93	51	62	277	100	95	87	86	73	40	33	31	0.00	9	12	5	1	1	–	–	–	–	–	1	–	–	Phosphorothioic acid, O,O-dimethyl O-(3-methyl-4-nitrophenyl) ester	122-14-5	NI28327
125	109	277	260	79	93	47	63	277	100	88	86	43	29	24	23	18		9	12	5	1	1	–	–	–	–	–	1	–	–	Phosphorothioic acid, O,O-dimethyl O-(3-methyl-4-nitrophenyl) ester	122-14-5	IC03585
109	125	277	260	79	47	93	63	277	100	98	90	44	38	37	31	19		9	12	5	1	1	–	–	–	–	–	1	–	–	Phosphorothioic acid, O,O-dimethyl O-(3-methyl-4-nitrophenyl) ester	122-14-5	NI28328
79	80	52	82	96	83	81	55	277	100	71	60	51	40	37	35	35	0.00	9	15	5	3	1	–	–	–	–	–	–	–	–	2H-Pyran-2-methanol, 3-azido-6-ethoxy-3,6-dihydro-, methanesulphonate (ester)	56248-60-3	NI28329
242	162	74	244	164	88	42	224	277	100	52	38	36	33	33	30	25	9.00	10	9	4	1	–	–	2	–	–	–	–	–	–	Glycine, N-[(2,4-dichlorophenoxy)acetyl]-	17212-10-1	NI28330
43	111	171	129	101	99	55	112	277	100	95	51	39	26	19	13	11	0.00	10	15	6	1	1	–	–	–	–	–	–	–	–	3,5-Thianediol, 4-methyl-4-nitro-, diacetate, (3R,4S,5S)-		DD00915
73	160	45	75	43	59	74	147	277	100	67	24	11	10	9	8	7	0.00	10	23	2	1	1	–	–	–	–	2	–	–	–	Thiazolidine-4-carboxylic acid, N-(trimethylsilyl)-, trimethylsilyl ester		MS06287
277	279	28	36	98	242	44	27	277	100	65	63	55	33	27	26	21		11	5	–	5	–	–	2	–	–	–	–	–	–	7,9-Dichloro-4-cyano-2-methyl-1,3,6-triazacyclo[3.3.3]azine		NI28331
261	218	246	262	234	219	220	247	277	100	70	65	12	12	12	8	7	0.00	11	10	2	1	1	3	–	–	–	–	–	–	–	2H-1,4-Benzothiazin-3(4H)-one, 4-hydroxy-2,2-dimethyl-6-(trifluoromethyl)-	4875-09-6	NI28332
234	277	232	218	261	246	235	217	277	100	70	29	23	21	15	12	12		11	10	2	1	1	3	–	–	–	–	–	–	–	2H-1,4-Benzothiazin-3(4H)-one, 4-hydroxy-2,2-dimethyl-6-(trifluoromethyl)-	4875-09-6	NI28333
93	43	277	77	65	88	39	66	277	100	38	13	12	10	10	8	7		11	10	2	1	1	3	–	–	–	–	–	–	–	Butanamide, 3-oxo-N-phenyl-α-[(trifluoromethyl)thio]-		MS06288
93	43	277	77	88	65	39	66	277	100	38	13	12	10	10	8	7		11	10	2	1	1	3	–	–	–	–	–	–	–	Butanamide, 3-oxo-N-phenyl-α-[(trifluoromethyl)thio]-		DD00916
193	235	192	176	105	165	146	147	277	100	58	52	28	25	18	18	17	8.00	11	11	4	5	–	–	–	–	–	–	–	–	–	2-Acetylamino-7-acetoxymethyl-3,4-dihydropteridin-4-one		LI05953
218	59	117	45	186	69	185	147	277	100	50	37	37	29	26	25	21	16.00	11	19	5	1	1	–	–	–	–	–	–	–	–	L-Cysteine, N-acetyl-S-(1-methyl-3-carboxypropyl)-, methyl ester		NI28334
218	45	126	186	43	88	117	147	277	100	80	47	42	29	25	24	19	16.00	11	19	5	1	1	–	–	–	–	–	–	–	–	L-Cysteine, N-acetyl-S-(2-methyl-3-carboxypropyl)-, methyl ester		NI28335
174	73	147	100	175	86	45	59	277	100	70	20	19	18	17	14	12	0.00	11	31	1	1	–	–	–	–	–	3	–	–	–	Silanamine, 1,1,1-trimethyl-N-(trimethylsilyl)-N-[2-[(trimethylsilyl)oxy]ethyl]-	5630-81-9	NI28336
233	198	235	199	201	154	277	279	277	100	62	42	38	34	32	31	30		12	8	–	1	1	–	–	1	–	–	–	–	–	Phenothiazine, 2-bromo-	66820-95-9	NI28337
152	277	279	76	153	151	102	168	277	100	53	52	21	15	15	12	9		12	8	2	1	–	–	–	1	–	–	–	–	–	1,1′-Biphenyl, 4-bromo-3-nitro-	27701-66-2	NI28338
241	242	240	167	243	140	139	214	277	100	38	11	11	6	6	6	4	0.00	12	9	–	1	–	–	1	–	–	–	–	–	1	Phenarsazine, 10-chloro-5,10-dihydro-	578-94-9	NI28339
242	167	277	166	279	168	138.5	121	277	100	67	38	24	15	10	10	8		12	9	–	1	–	–	1	–	–	–	–	–	1	Phenarsazine, 10-chloro-5,10-dihydro-	578-94-9	LI05954
242	167	277	166	279	168	139	140	277	100	66	37	23	13	10	10	9		12	9	–	1	–	–	1	–	–	–	–	–	1	Phenarsazine, 10-chloro-5,10-dihydro-	578-94-9	NI28340
91	43	65	58	92	124	40	89	277	100	30	11	11	10	7	7	6	0.00	12	14	–	1	–	–	3	–	–	–	–	–	–	Benzylamine, N-(3,3,3-trichloro-2,2-dimethyl-1-propylidene)-		MS06289
135	136	178	164	108	55	41	83	277	100	48	40	28	23	15	12	8	0.70	12	15	3	5	–	–	–	–	–	–	–	–	–	9-(4-C-Cyclopropyl-α-D-xylo-tetrofuranosyl)adenine		LI05955
135	136	178	164	108	55	41	206	277	100	80	41	30	28	20	15	11	3.00	12	15	3	5	–	–	–	–	–	–	–	–	–	9-(4-C-Cyclopropyl-β-D-xylo-tetrofuranosyl)adenine		LI05956
151	125	81	138	95	111	112	68	277	100	78	71	44	43	40	39	39	15.00	12	15	3	5	–	–	–	–	–	–	–	–	–	1-(3-(Thymin-1-yl)propyl)cytosine		LI05957
73	115	75	45	43	128	71	101	277	100	80	38	38	30	26	20	16	0.00	12	23	6	1	–	–	–	–	–	–	–	–	–	α-D-Galactopyranoside, methyl 2-acetamido-2-deoxy-3,4,6-tri-O-methyl-	10049-88-4	NI28341
115	73	75	128	45	43	117	101	277	100	63	41	32	18	15	14	13	0.00	12	23	6	1	–	–	–	–	–	–	–	–	–	α-D-Galactopyranoside, methyl 2-acetamido-2-deoxy-3,4,6-tri-O-methyl-	10049-88-4	NI28342
73	115	75	71	45	128	43	101	277	100	74	42	35	35	26	25	20	0.00	12	23	6	1	–	–	–	–	–	–	–	–	–	β-D-Galactopyranoside, methyl 2-acetamido-2-deoxy-3,4,6-tri-O-methyl-	7437-12-9	NI28343
115	73	75	128	88	45	86	43	277	100	95	73	35	26	24	18	18	0.00	12	23	6	1	–	–	–	–	–	–	–	–	–	α-D-Glucopyranoside, methyl 2-acetamido-2-deoxy-3,4,6-tri-O-methyl-		NI28344
73	115	75	45	128	88	86	71	277	100	88	56	42	32	23	17	17	0.00	12	23	6	1	–	–	–	–	–	–	–	–	–	α-D-Glucopyranoside, methyl 2-acetamido-2-deoxy-3,4,6-tri-O-methyl-		NI28345
73	115	75	71	45	43	88	128	277	100	67	58	43	42	32	27	23	0.00	12	23	6	1	–	–	–	–	–	–	–	–	–	β-D-Glucopyranoside, methyl 2-acetamido-2-deoxy-3,4,6-tri-O-methyl-	6195-86-4	NI28346
89	128	73	115	86	45	43	101	277	100	70	63	58	46	46	46	23	0.00	12	23	6	1	–	–	–	–	–	–	–	–	–	D-Mannofuranoside, methyl 2-acetamido-2-deoxy-3,5,6-tri-O-methyl-	54890-06-1	NI28347

MASS TO CHARGE RATIOS								M.W.	INTENSITIES								Parent	C	H	O	N	S	F	Cl	Br	I	Si	P	B	X	COMPOUND NAME	CAS Reg No	No
73	115	75	71	45	128	43	88	277	100	71	56	35	34	26	26	23	0.00	12	23	6	1	–	–	–	–	–	–	–	–	–	β-D-Mannopyranoside, methyl 2-acetamido-2-deoxy-3,4,6-tri-O-methyl-	7384-32-9	NI28348
277	107	91	139	65	79	155	77	277	100	84	74	73	45	45	44	39		13	11	4	1	1	–	–	–	–	–	–	–	–	4-Nitro-4′-methyldiphenyl sulphone	4094-37-5	NI28349
43	202	219	101	129	59	175	75	277	100	97	86	60	48	40	36	23	1.46	13	11	6	1	–	–	–	–	–	–	–	–	–	1,3-Dioxane-4,6-dione, 2,2-dimethyl-5-[(3-nitrophenyl)methylene]-	16286-25-2	NI28350
218	63	77	185	140	108	217	277	277	100	77	63	53	53	47	30	23		13	12	–	1	2	–	–	–	–	–	1	–	–	Thiocarbamoyl diphenylphosphino sulphide		LI05958
93	277	279	94	106	78	79	39	277	100	9	7	7	6	6	4	3		13	12	1	1	–	–	–	1	–	–	–	–	–	2′-Hydroxy-5′-bromodihydro-2-stilbazole		NI28351
127	111	99	277	201	39	80	126	277	100	61	56	47	47	45	43	40		13	12	2	3	–	–	1	–	–	–	–	–	–	4-[2-(2-Chlorophenyl)hydrazono]-1-methylethylidene-3-methylisoxazol-5-one		IC03586
84	44	42	219	55	110	109	277	277	100	50	50	43	43	25	22	17		13	15	2	3	1	–	–	–	–	–	–	–	–	Sydnone, 4-(phenylthio)-3-(1-piperidinyl)-	68486-73-7	NI28352
114	262	113	99	59	39	115	41	277	100	71	55	50	38	27	27	25	13.69	13	15	2	3	1	–	–	–	–	–	–	–	–	4H-1,3-Thiazine, 4,4,6-trimethyl-2-[4-nitrophenylamino(imino)]-	85298-50-6	NI28353
73	232	233	262	76	75	160	77	277	100	89	47	40	30	27	26	24	0.00	13	15	4	1	–	–	–	–	–	1	–	–	–	Glycine, phthalyl-, trimethylsilyl ester		NI28354
174	277	175	158	91	60	77	32	277	100	33	16	13	13	13	10	9		13	15	4	3	–	–	–	–	–	–	–	–	–	4-α-D-Lyxofuranosyl-2-phenyl-1,2,3-osotriazole		MS06290
174	277	91	175	60	77	158	92	277	100	28	24	18	18	17	16	10		13	15	4	3	–	–	–	–	–	–	–	–	–	4-β-D-Lyxofuranosyl-2-phenyl-1,2,3-osotriazole		MS06291
277	234	279	236	164	40	44	278	277	100	96	64	62	37	18	15	14		14	9	1	1	–	–	2	–	–	–	–	–	–	Acridine, 6,9-dichloro-2-methoxy-	86-38-4	NI28355
100	67	91	126	65	277	74	155	277	100	80	67	60	30	19	18	15		14	15	1	1	2	–	–	–	–	–	–	–	–	4-(Benzylthio)-1,3-thiazol-5(2H)-one-2-spirocyclopentane		DD00917
76	60	84	85	112	78	128	277	277	100	64	44	18	14	10	6	4		14	15	1	1	2	–	–	–	–	–	–	–	–	1,3-Dithiol-1-ium, 4-hydroxy-5-phenyl-2-(1-piperidinyl)-, hydroxide, inner salt	75365-80-9	NI28356
80	77	221	277	220	53	51	41	277	100	52	31	21	19	19	19	18		14	15	3	1	1	–	–	–	–	–	–	–	–	3-Oxa-10-azatricyclo[4.3.1.0^{2,4}]dec-7-ene, 10-(phenylsulphonyl)-, (1α,2α,4α,6α)-	50267-09-9	NI28357
91	277	141	92	233	169	77	115	277	100	89	72	72	69	69	47	42		14	15	3	1	1	–	–	–	–	–	–	–	–	4-Phenoxy-N,N-dimethylbenzenesulphonamide		IC03587
233	277	203	176	174	162	188	231	277	100	42	41	40	30	27	25	22		14	15	5	1	–	–	–	–	–	–	–	–	–	Benzeneacetic acid, 2-(2,5-dioxo-1-pyrrolidinyl)-5-methoxy-4-methyl-		NI28358
43	57	105	55	41	45	187	72	277	100	92	91	86	83	73	67	55	0.00	14	15	5	1	–	–	–	–	–	–	–	–	–	2-Naphthalenepropanal, 6-methoxy-2-nitro-1-oxo-1,2,3,4-tetrahydro-		DD00918
107	186	91	79	65	180	108	44	277	100	55	52	16	12	10	10	8	0.00	14	16	3	1	–	–	–	–	–	–	1	–	–	Phosphoramidic acid, bis(4-tolyl) ester	56830-89-8	MS06292
107	186	91	79	65	180	108	77	277	100	55	51	16	10	8	7	6	0.00	14	16	3	1	–	–	–	–	–	–	1	–	–	Phosphoramidic acid, dibenzyl ester		LI05959
277	218	278	219	79	279	220	93	277	100	90	22	19	11	7	5	5		14	19	3	1	–	–	–	–	–	1	–	–	–	1H-Indole-3-acetic acid, 5-[(trimethylsilyl)oxy]-, methyl ester	72101-35-0	NI28359
143	233	174	218	144	115	89	234	277	100	52	52	22	21	20	19	10	0.00	14	23	1	1	–	–	–	–	–	2	–	–	–	1H-Indole, 1-(trimethylsilyl)-5-[(trimethylsilyl)oxy]-	55712-57-7	NI28361
73	277	262	45	278	263	279	74	277	100	99	53	35	27	15	10	10		14	23	1	1	–	–	–	–	–	2	–	–	–	1H-Indole, 1-(trimethylsilyl)-5-[(trimethylsilyl)oxy]-	55712-57-7	NI28360
277	262	278	279	263	130	123	131	277	100	73	35	27	18	18	12	10		14	23	1	1	–	–	–	–	–	2	–	–	–	1H-Indole, 1-(trimethylsilyl)-5-[(trimethylsilyl)oxy]-	55712-57-7	NI28362
277	180	152	165	153	164	76	69	277	100	58	41	26	26	16	16	15		15	10	1	1	–	3	–	–	–	–	–	–	–	Acetamide, N-9H-fluoren-2-yl-2,2,2-trifluoro-	363-17-7	NI28363
128	103	277	149	111	250	109	129	277	100	99	23	20	20	15	15	14		15	11	1	5	–	–	–	–	–	–	–	–	–	1H-Indol-4(5H)-one, 6,7-dihydro-5-(1,1,2,2-tetracyanoethyl)-3-methyl-		NI28364
277	276	102	181	155	93	77	221	277	100	43	18	11	11	11	10	7		15	11	1	5	–	–	–	–	–	–	–	–	–	Pyrimido[1,2-a]purin-10(9H)-one, 7-phenyl-1-methyl-		NI28365
212	277	129	213	64	130	211	210	277	100	42	21	19	16	15	12	11		15	17	2	–	–	–	–	–	–	–	–	–	1	Titanium, bis(η^5-2,4-cyclopentadien-1-yl)(2,4-pentanedionato-O,O′)-	74811-12-4	NI28366
41	99	71	55	67	81	137	168	277	100	93	70	53	26	22	14	8	0.00	15	19	2	1	1	–	–	–	–	–	–	–	–	2H-Pyran-2-one, tetrahydro-6-(2-pentenyl)-3-(2-pyridinylthio)-	75314-22-6	NI28367
156	95	94	77	217	277	126	96	277	100	82	82	73	70	58	45	8		15	19	4	1	–	–	–	–	–	–	–	–	–	(±)-6β,7β-Dihydroxytropan-3α-yl benzoate		MS06293
43	41	55	56	39	54	77	91	277	100	70	65	63	59	50	46	41	2.00	15	19	4	1	–	–	–	–	–	–	–	–	–	1,3-Isobenzofurandione, 3a,4,7,7a-tetrahydro-4-(1,3,4,5,6,7-hexahydro-2-oxo-2H-azepin-1-yl)-6-methyl-		MS06294
105	160	77	43	51	219	57	187	277	100	88	88	50	24	22	20	19	0.00	15	19	4	1	–	–	–	–	–	–	–	–	–	4-Oxazolecarboxylic acid, 4,5-dihydro-4-(1-hydroxyisopropyl)-5-methyl-2-phenyl-, methyl ester, (4R,5S)-		DD00919
99	135	77	142	141	277	136	41	277	100	75	60	55	38	29	23	23		15	23	–	3	1	–	–	–	–	–	–	–	–	1-Pyrazolidinecarboxanilide, 4-ethyl-3-propylthio-	24769-38-8	NI28368
57	73	42	277	103	187	56	43	277	100	28	28	18	16	15	15	14		15	23	2	1	–	–	–	–	–	1	–	–	–	4-Piperidinecarboxylic acid, 4-phenyl-, trimethylsilyl ester	54964-92-0	NI28369
41	193	57	235	277	153	84	55	277	100	94	59	47	45	43	36	35		15	23	2	3	–	–	–	–	–	–	–	–	–	5-Nonanone, (4-nitrophenyl)hydrazone	14090-63-2	NI28371
193	41	277	57	63	153	235	84	277	100	94	78	77	64	50	45	40		15	23	2	3	–	–	–	–	–	–	–	–	–	5-Nonanone, (4-nitrophenyl)hydrazone	14090-63-2	NI28370
193	41	277	57	235	153	84	55	277	100	93	76	76	50	48	39	38		15	23	2	3	–	–	–	–	–	–	–	–	–	5-Nonanone, (4-nitrophenyl)hydrazone	14090-63-2	LI05960
174	277	129	77	278				277	100	58	50	40	12					16	11	–	3	1	–	–	–	–	–	–	–	–	3,4-Diphenylthiazolo[2,3-c]-s-triazole		LI05961
41	277	152	55	123	43	162	42	277	100	79	66	65	63	54	51	51		16	23	3	1	–	–	–	–	–	–	–	–	–	Annotine, 2,3-dihydro-	5630-43-3	NI28372
105	57	160	176	77	120	56	41	277	100	72	43	42	41	30	23	16	0.00	16	23	3	1	–	–	–	–	–	–	–	–	–	Carbamic acid, (5-phenyl-5-oxopentyl)-, tert-butyl ester	116437-42-4	DD00920
277	177	220	164	176	57	262	192	277	100	70	58	48	38	38	33	30		16	23	3	1	–	–	–	–	–	–	–	–	–	Isoquinoline, 2-tert-butoxy-1,2,3,4-tetrahydro-6,7-dimethoxy-		DD00921
122	277	262	97	82	109	260	276	277	100	48	48	37	35	22	17	15		16	23	3	1	–	–	–	–	–	–	–	–	–	Rudrakine		NI28373
123	150	97	165	249	122	166	125	277	100	96	85	71	31	28	16	14	4.00	16	23	3	1	–	–	–	–	–	–	–	–	–	Serratinine, 13-dehydro-		LI05962
123	150	97	165	249	122	166	193	277	100	96	84	70	30	27	16	14	2.98	16	23	3	1	–	–	–	–	–	–	–	–	–	Serratinine, 14-deoxy-14-oxo-	18331-13-0	NI28374
142	41	140	100	44	143	98	57	277	100	16	15	14	11	10	10	9	0.00	16	36	–	1	–	–	1	–	–	–	–	–	–	Tetra-N-butylammonium chloride	1112-67-0	NI28375
277	249	105	77	200	51	206	129	277	100	96	75	62	35	22	16	10		17	11	3	1	–	–	–	–	–	–	–	–	–	Homophthalimide, 4-phenacylidene-		LI05963
130	102	105	77	103	174	76	119	277	100	84	64	35	34	31	25	14	3.60	17	11	3	1	–	–	–	–	–	–	–	–	–	Spiro[1,4,2-dioxazole-5,1′(2′H)-naphthalen]-2′-one, 3-phenyl-	23767-21-7	NI28376
105	130	102	77	103	76	51	131	277	100	85	85	29	24	20	12	9	6.80	17	11	3	1	–	–	–	–	–	–	–	–	–	Spiro[1,4,2-dioxazole-5,2′(1′H)-naphthalen]-1′-one, 3-phenyl-	23767-20-6	NI28377
276	91	275	92	77	44	51	116	277	100	58	57	55	55	51	33	27	4.54	17	15	1	3	–	–	–	–	–	–	–	–	–	2-Hydroxy-4,6-di-3-tolyl-s-triazine		MS06295
118	117	40	116	91	29	119	90	277	100	99	78	66	63	48	43	38	25.00	17	15	1	3	–	–	–	–	–	–	–	–	–	2-Hydroxy-4,6-di-4-tolyl-s-triazine		MS06296
91	157	277	182	158	65	278	167	277	100	88	83	37	22	21	19	12		17	15	1	3	–	–	–	–	–	–	–	–	–	Imidazole-2-carboxanilide, 1-benzyl-	13189-14-5	NI28378

MASS TO CHARGE RATIOS								M.W.	INTENSITIES								Parent	C	H	O	N	S	F	Cl	Br	I	Si	P	B	X	COMPOUND NAME	CAS Reg No	No
277	276	278	118	89	130	91	77	**277**	100	59	20	17	15	14	14	13		17	15	1	3	–	–	–	–	–	–	–	–	–	**Phenol, 4-[5-amino-6-benzylpyrazinyl]-**	37156-84-6	**NI28379**
164	146	277	148	147	165	278	260	**277**	100	74	55	20	14	12	10	10		17	27	2	1	–	–	–	–	–	–	–	–	–	**Acetophenone, 2-hydroxy-5-nonyl-, (E)-oxime**		**DD00922**
203	277	178	262	220	204	41	57	**277**	100	49	38	37	20	19	16	13		17	27	2	1	–	–	–	–	–	–	–	–	–	**2,6-Di-tert-butyl-4-(acetamidomethyl)phenol**		**IC03588**
220	57	41	221	206	58	15	29	**277**	100	46	30	25	23	23	19	17	**9.00**	17	27	2	1	–	–	–	–	–	–	–	–	–	**Carbamic acid, 2,6-di-tert-butyl-4-methylphenyl ester**	1918-11-2	**NI28380**
205	57	220	41	58	55	206	105	**277**	100	53	49	32	26	17	15	12	**0.00**	17	27	2	1	–	–	–	–	–	–	–	–	–	**Carbamic acid, 2,6-di-tert-butyl-4-methylphenyl ester**	1918-11-2	**NI28381**
105	150	132	165	83	106	164	59	**277**	100	82	79	52	52	46	45	44	**5.00**	17	27	2	1	–	–	–	–	–	–	–	–	–	**Carbamic acid, (α-methylbenzyl)-, 1-ethylhexyl ester**	32589-45-0	**NI28383**
105	150	132	165	83	164	106	59	**277**	100	80	80	51	50	45	45	43	**7.01**	17	27	2	1	–	–	–	–	–	–	–	–	–	**Carbamic acid, (α-methylbenzyl)-, 1-ethylhexyl ester**	32589-45-0	**NI28384**
132	165	147	164	150	83	277	101	**277**	100	90	70	60	58	39	17	17		17	27	2	1	–	–	–	–	–	–	–	–	–	**Carbamic acid, (α-methylbenzyl)-, 1-ethylhexyl ester**	32589-45-0	**NI28382**
105	57	71	106	150	70	41	55	**277**	100	57	52	45	40	40	36	35	**0.07**	17	27	2	1	–	–	–	–	–	–	–	–	–	**Carbamic acid, (α-methylbenzyl)-, 1-ethyl-1-methylpentyl ester**	32589-47-2	**NI28385**
105	57	71	106	55	150	70	41	**277**	100	58	52	45	45	41	40	37	**0.10**	17	27	2	1	–	–	–	–	–	–	–	–	–	**Carbamic acid, (α-methylbenzyl)-, 1-ethyl-1-methylpentyl ester**	32589-47-2	**NI28386**
105	57	71	106	150	70	166	41	**277**	100	58	53	46	42	41	36	36	**0.00**	17	27	2	1	–	–	–	–	–	–	–	–	–	**Carbamic acid, (α-methylbenzyl)-, 1-ethyl-1-methylpentyl ester**	32589-47-2	**LI05964**
150	105	164	106	165	43	120	71	**277**	100	88	86	57	45	28	22	21	**5.94**	17	27	2	1	–	–	–	–	–	–	–	–	–	**Carbamic acid, (α-methylbenzyl)-, 1-methylheptyl ester**	32589-44-9	**NI28387**
133	277	165	134	178	147	160	260	**277**	100	65	55	45	35	35	31	21		17	27	2	1	–	–	–	–	–	–	–	–	–	**1-Decanone, 1-(2-hydroxy-5-methylphenyl)-, (E)-oxime**		**DD00923**
58	277	44	57	43	42	276	204	**277**	100	90	50	42	39	38	30	29		17	27	2	1	–	–	–	–	–	–	–	–	–	**Paniculatine**		**MS06297**
248	249	247	250	216	217	204	102	**277**	100	93	27	20	13	5	5	5	**0.00**	18	15	–	1	1	–	–	–	–	–	–	–	–	**6,6-Dimethyl-5,6-dihydro-12-aza-5-thiabenz[a]anthracene**		**NI28388**
277	276	230	244	278	243	242	262	**277**	100	51	31	25	21	20	15	13		18	15	–	1	1	–	–	–	–	–	–	–	–	**Pyridine, 2-(methylthio)-3,6-diphenyl-**	74663-72-2	**NI28389**
43	180	165	217	234	179	178	77	**277**	100	96	94	86	77	60	38	22	**0.70**	18	15	2	1	–	–	–	–	–	–	–	–	–	**1-Cyclopropanecarbonitrile, 1-acetoxy-2,2-diphenyl-**	87656-16-4	**NI28390**
103	131	205	149	102	277	104	128	**277**	100	88	84	80	65	58	50	46		18	15	2	1	–	–	–	–	–	–	–	–	–	**Dicinnamamide, (E)-**	26091-02-1	**NI28391**
179	128	205	277	178	77	129	51	**277**	100	92	87	66	65	52	49	48		18	15	2	1	–	–	–	–	–	–	–	–	–	**Naphthalene-2,3-dicarboximide, 1,2,3,4-tetrahydro-1-phenyl-**		**MS06298**
277	135	107	249	142	138.5			**277**	100	50	7	5	2	1				18	15	2	1	–	–	–	–	–	–	–	–	–	**3-(3-Phenylethyl)-5-(4-methoxyphenyl)isoxazole**		**LI05965**
277	165	249	278	203	204	248	276	**277**	100	26	21	20	19	16	15	13		18	15	2	1	–	–	–	–	–	–	–	–	–	**2-Propenoic acid, 2-cyano-3-(3-phenylphenyl)-, ethyl ester**		**NI28392**
276	277	197	169	224	278	43	198	**277**	100	84	72	22	18	17	17	13		18	19	–	3	–	–	–	–	–	–	–	–	–	**2′-Cyanospiro(1,2,3,4,6,7,12,12b-octahydroindolo[2,3-a]quinolizine-1,1′-cyclopropane)**		**NI28393**
277	276	262	220	221	278	234	164	**277**	100	24	22	18	15	13	13	8		18	31	1	1	–	–	–	–	–	–	–	–	–	**6-Azaandrostan-17-ol, (5β,17β)-**	55056-58-1	**NI28394**
277	276	262	220	221	278	234	124	**277**	100	25	23	19	15	14	14	8		18	31	1	1	–	–	–	–	–	–	–	–	–	**6-Azaandrostan-17-ol, (5β,17β)-**	55056-58-1	**LI05966**
44	192	206	147	45	58	220	135	**277**	100	96	61	55	42	40	34	33	**28.70**	18	31	1	1	–	–	–	–	–	–	–	–	–	**4-Nonyl-2-(dimethylaminomethyl)phenol**		**IC03589**
113	55	41	43	70	85	56	72	**277**	100	30	29	26	22	21	12	11	**7.04**	18	31	1	1	–	–	–	–	–	–	–	–	–	**Pyrrolidine, 1-(1-oxo-5,8-tetradecadienyl)-**	56666-40-1	**NI28395**
91	65	92	170	39	186	51	121	**277**	100	21	17	13	13	11	9	7	**2.00**	19	19	1	1	–	–	–	–	–	–	–	–	–	**1-Benzyl-2-(4-methoxybenzyl)pyrrole**		**LI05967**
128	105	57	129	43	104	157	55	**277**	100	90	78	73	72	72	70	66	**24.95**	19	19	1	1	–	–	–	–	–	–	–	–	–	**2,4-Undecadiene-8,10-diynamide, N-(2-phenylethyl)-, (Z,E)-**		**NI28396**
105	91	43	129	30	57	128	55	**277**	100	29	28	27	25	25	25	24	**1.76**	19	19	1	1	–	–	–	–	–	–	–	–	–	**3,5-Undecadiene-8,10-diynamide, N-(2-phenylethyl)-, (E,E)-**		**NI28397**
91	277	130	129	131	82	186	115	**277**	100	99	94	94	92	84	78	68		19	24	–	1	–	–	–	–	–	–	–	1	–	**1-Benzyl-4-boryl-3-methyl-4-phenylpiperidine**		**MS06299**
91	130	186	116	131	115	54	40	**277**	100	32	26	20	16	16	16	16	**11.90**	19	24	–	1	–	–	–	–	–	–	–	1	–	**1-Benzyl-4-boryl-3-methyl-4-phenylpiperidine**		**MS06300**
44	30	55	41	180	85	43	56	**277**	100	70	55	55	50	28	24	20	**6.00**	19	35	–	1	–	–	–	–	–	–	–	–	–	**Piperidine, 3,5-bis(cyclohexylmethyl)-**		**IC03590**
152	194	67	56	68	82	69	195	**277**	100	81	30	19	18	16	16	13	**3.00**	19	35	–	1	–	–	–	–	–	–	–	–	–	**Pyrrolidine, 2-(5-hexenyl)-5-(8-nonenyl)-, trans-**		**NI28398**
152	234	41	153	55	67	235	70	**277**	100	43	16	14	10	9	7	6	**3.00**	19	35	–	1	–	–	–	–	–	–	–	–	–	**Pyrrolizidine, 3-(8-nonenyl)-5-propyl-**		**NI28399**
91	277	262	144	65	234	130	235	**277**	100	26	19	15	11	10	10	7		20	23	–	1	–	–	–	–	–	–	–	–	–	**Carbazole, cis-9-benzyl-4a-methyl-1,2,3,4,4a,9a-hexahydro-**		**MS06301**
58	59	202	218	217	215	203	91	**277**	100	5	3	2	2	2	2	2	**0.00**	20	23	–	1	–	–	–	–	–	–	–	–	–	**1-Propanamine, 3-(10,11-dihydro-5H-dibenzo[a,d]cyclohepten-5-ylidene)-N,N-dimethyl-**	50-48-6	**NI28400**
58	59	30	57	275	202	42	28	**277**	100	5	4	3	2	2	2	2	**0.30**	20	23	–	1	–	–	–	–	–	–	–	–	–	**1-Propanamine, 3-(10,11-dihydro-5H-dibenzo[a,d]cyclohepten-5-ylidene)-N,N-dimethyl-**	50-48-6	**NI28401**
58	42	59	202	203	30	215	189	**277**	100	5	5	5	3	3	3	2	**0.13**	20	23	–	1	–	–	–	–	–	–	–	–	–	**1-Propanamine, 3-(10,11-dihydro-5H-dibenzo[a,d]cyclohepten-5-ylidene)-N,N-dimethyl-**	50-48-6	**NI28402**
276	277	262	105	220	278	200	205	**277**	100	85	48	27	23	22	18	14		20	23	–	1	–	–	–	–	–	–	–	–	–	**Pyrrolidine, 1,5-dimethyl-3,3-diphenyl-2-ethylidene-**	30223-73-5	**NI28403**
277	278	138	276	275	124	249	279	**277**	100	22	17	14	12	11	5	5		21	11	–	1	–	–	–	–	–	–	–	–	–	**Azabenzo[ghi]perylene**		**NI28404**
166	164	201	168	129	131	203	199	**278**	100	75	58	49	47	45	38	37	**0.06**	2	–	–	–	–	–	5	1	–	–	–	–	–	**Pentachlorobromoethane**		**IC03591**
278	127	69	113	146	259	203	190	**278**	100	16	12	8	7	3	3	1		3	1	–	–	–	6	–	–	1	–	–	–	–	**Propane, 1,1,1,3,3,3-hexafluoro-2-iodo-**	4141-91-7	**NI28405**
151	153	127	280	278	221	219	59	**278**	100	99	62	55	55	50	50	42		3	4	2	–	–	–	–	1	1	–	–	–	–	**Methyl bromoiodoacetate**		**NI28406**
119	201	199	53	93	52	120	200	**278**	100	78	77	73	52	52	27	15	**0.00**	4	6	–	–	–	2	–	2	–	1	–	–	–	**1-Silacyclopentane, 1,1-difluoro-3,4-dibromo-**	75292-55-6	**NI28407**
93	94	42	91	160	278	92	31	**278**	100	83	64	59	54	53	51	49		4	6	4	–	–	–	–	–	–	–	–	–	2	**Acetic acid, 2,2′-diselenobis-**	16066-50-5	**NI28408**
69	42	44	46	76	43	96	194	**278**	100	84	82	79	78	78	77	51	**6.60**	5	6	–	2	1	8	–	–	–	–	–	–	–	**(Dimethylamino)sulphur(IV) fluoro(heptafluoroisopropylimide)**		**LI05968**
250	95	222	123	205	141	81	109	**278**	100	77	60	55	51	46	44	43	**19.57**	5	12	3	–	–	–	–	–	1	–	1	–	–	**Phosphonic acid, (iodomethyl)-, diethyl ester**	10419-77-9	**NI28409**
243	245	179	181	74	109	143	108	**278**	100	96	96	91	66	58	53	45	**24.44**	6	2	2	–	1	–	4	–	–	–	–	–	–	**Benzenesulphonyl chloride, 2,4,5-trichloro-**	15945-07-0	**NI28410**
145	43	181	165	69	77	209	115	**278**	100	72	55	46	44	34	23	20	**1.30**	6	3	2	–	–	9	–	–	–	–	–	–	–	**Acetic acid, trifluoro-, 2,2,2-trifluoro-1-methyl-1-(trifluoromethyl)ethyl ester**	42031-16-3	**NI28411**

MASS TO CHARGE RATIOS								M.W.	INTENSITIES								Parent	C	H	O	N	S	F	Cl	Br	I	Si	P	B	X	COMPOUND NAME	CAS Reg No	No
31	41	160	278	276	158	29	57	278	100	30	26	25	23	23	23	18		6	14	2	–	–	–	–	–	–	–	–	–	2	1-Propanol, 3,3'-diselenobis-	23243-47-2	NI28412
245	243	280	247	282	217	44	278	278	100	92	79	58	57	57	57	55		7	3	1	–	–	–	5	–	–	–	–	–	–	Benzenemethanol, 2,3,4,5,6-pentachloro-	16022-69-8	NI28413
281	279	283	280	282	285	278	266	278	100	66	59	27	21	18	14	10		7	3	1	–	–	–	5	–	–	–	–	–	–	Benzene, pentachloromethoxy-	1825-21-4	NI28414
237	265	280	278	267	263	235	282	278	100	96	74	62	62	62	57	52		7	3	1	–	–	–	5	–	–	–	–	–	–	Benzene, pentachloromethoxy-	1825-21-4	NI28415
265	280	237	267	263	282	278	95	278	100	97	86	64	63	62	60	60		7	3	1	–	–	–	5	–	–	–	–	–	–	Benzene, pentachloromethoxy-	1825-21-4	NI28416
197	199	278	280	198	171	173	92	278	100	98	65	51	48	44	37	36		7	7	–	2	–	–	–	1	–	–	–	–	1	4-Bromophenylselenourea		MS06302
117	65	244	39	137	109	89	264	278	100	90	48	35	21	19	18	17	**0.10**	7	7	–	2	2	5	–	–	–	–	–	–	–	(3-Phenylthioureido)sulphur pentafluoride	90598-07-5	NI28417
261	259	263	257	119	117	262	265	278	100	97	71	47	30	28	27	23	**10.41**	7	18	–	2	–	–	–	–	–	–	–	–	2	Bis(trimethylgermyl)carbodiimide		MS06303
217	278	250	69	129	180	117	168	278	100	71	62	56	35	21	14	12		8	5	–	–	1	6	–	–	–	–	1	–	–	Phosphinothioic fluoride, ethyl(pentafluorophenyl)-	53327-24-5	NI28418
278	277	127	79	51	29	53	50	278	100	53	36	35	29	28	25	24		8	7	3	–	–	–	–	–	1	–	–	–	–	Benzaldehyde, 4-hydroxy-3-iodo-5-methoxy-	5438-36-8	NI28419
246	278	63	218	247	53	92	91	278	100	99	31	29	17	13	11	11		8	7	3	–	–	–	–	–	1	–	–	–	–	Benzoic acid, 2-hydroxy-5-iodo-, methyl ester	4068-75-1	MS06304
246	278	218	63	247	53	90	62	278	100	99	31	31	17	12	10	10		8	7	3	–	–	–	–	–	1	–	–	–	–	Benzoic acid, 2-hydroxy-5-iodo-, methyl ester	4068-75-1	NI28420
246	278	218	63	247	53	90	91	278	100	99	30	30	16	11	10	9		8	7	3	–	–	–	–	–	1	–	–	–	–	Benzoic acid, 2-hydroxy-5-iodo-, methyl ester	4068-75-1	LI05969
107	109	63	280	64	65	172	174	278	100	97	32	21	17	15	15	14	**11.17**	8	8	1	–	–	–	–	2	–	–	–	–	–	Benzene, 1-bromo-4-(2-bromoethoxy)-	18800-30-1	NI28421
265	280	267	263	282	278	91	120	278	100	63	57	57	35	35	31	25		8	8	1	–	–	–	–	2	–	–	–	–	–	2,6-Dibromo-4-ethylphenol		MS06305
265	280	263	267	278	282	91	63	278	100	89	53	51	48	45	32	32		8	8	1	–	–	–	–	2	–	–	–	–	–	Phenol, 4,6-dibromo-2-ethyl-		LI05970
151	111	83	233	165	123	277	141	278	100	49	37	19	8	8	7	6	**0.00**	8	14	5	–	–	3	–	–	–	–	1	–	–	2,2,2-Trifluoroethyl 2-[(ethoxyphosphinyl)oxy]-2-methylpropanoate		DD00924
274	259	212	214	137	105	181	180	278	100	36	8	6	5	5	4	4	**0.00**	8	28	–	4	–	–	–	–	–	–	–	–	1	Methylamine, N-methyl-, molybdenum(4+) salt	33851-46-6	NI28422
231	246	247	262	203	261	263	278	278	100	77	50	46	16	15	10	4		9	11	2	–	–	–	–	–	1	–	–	–	–	1,3-Dihydro-1-hydroxy-3,3-dimethyl-1,2-benziodoxole		MS06306
135	136	142	58	56	127	121	65	278	100	90	83	39	39	27	27	27	**0.00**	9	15	–	2	–	–	–	–	1	–	–	–	–	2-Methyl-4-(dimethylamino)pyridine methiodide		MS06307
142	136	135	121	65	92	127	105	278	100	63	48	37	30	25	24	14	**0.00**	9	15	–	2	–	–	–	–	1	–	–	–	–	3-Methyl-4-(dimethylamino)pyridine methiodide		MS06308
41	43	69	44	39	55	53	70	278	100	92	47	44	41	39	28	26	**0.00**	9	15	3	2	–	–	–	1	–	–	–	–	–	Butanamide, N-[(acetylamino)carbonyl]-2-bromo-2-ethyl-	77-66-7	NI28423
221	73	263	222	223	264	262	147	278	100	77	40	23	12	10	10	7	**0.00**	9	27	3	–	–	–	–	–	–	3	–	1	–	Tris(trimethylsilyl)borate	4325-85-3	NI28424
73	221	263	222	223	74	45	133	278	100	76	21	18	10	9	9	8	**0.00**	9	27	3	–	–	–	–	–	–	3	–	1	–	Tris(trimethylsilyl)borate	4325-85-3	NI28425
73	221	263	222	133	223	175	74	278	100	88	27	21	12	12	9	9	**0.00**	9	27	3	–	–	–	–	–	–	3	–	1	–	Tris(trimethylsilyl)borate	4325-85-3	NI28426
278	115	145	165	116	178	177	169	278	100	91	82	64	64	27	27	27		10	6	–	–	–	8	–	–	–	–	–	–	–	1,2,3,3,4,4,5,6-Octafluorotricyclo[4.4.0.0^{2,5}]dec-8-ene		NI28427
76	104	50	75	102	264	74	51	278	100	94	85	50	45	32	30	28	**0.00**	10	6	1	4	–	–	–	–	–	–	–	–	1	Naphtho[2,3-c][1,2,5]selenadiazole-4,9-dione, monohydrazone	83274-12-8	NI28428
51	77	30	50	278	39	123	142	278	100	86	76	65	56	53	46	44		10	6	6	4	–	–	–	–	–	–	–	–	–	1,1'-Dioxo-4,4'-dinitro-2,2'-bipyridyl		IC03592
86	87	118	90	147	103	119	74	278	100	79	75	39	36	30	26	26	**7.96**	10	14	7	–	1	–	–	–	–	–	–	–	–	1,4,7,10-Tetraoxa-13-thiacyclopentadecane-3,11,14-trione	79687-35-7	NI28429
278	125	109	169	93	279	153	79	278	100	32	27	19	15	13	11	10		10	15	3	–	2	–	–	–	–	–	1	–	–	Fenthion	55-38-9	NI28430
278	125	109	168	153	169	93	79	278	100	42	41	34	34	30	25	19		10	15	3	–	2	–	–	–	–	–	1	–	–	Phosphorothioic acid, O,O-dimethyl O-[3-methyl-4-(methylthio)phenyl] ester	55-38-9	NI28431
109	40	125	278	93	79	47	63	278	100	97	93	71	54	50	44	32		10	15	3	–	2	–	–	–	–	–	1	–	–	Phosphorothioic acid, O,O-dimethyl O-[3-methyl-4-(methylthio)phenyl] ester	55-38-9	NI28432
15	109	79	263	127	51	45	39	278	100	91	27	25	20	18	18	18	**6.20**	10	15	5	–	1	–	–	–	–	–	1	–	–	Phosphoric acid, dimethyl 3-methyl-4-(methylsulphinyl)phenyl ester	6552-13-2	NI28433
109	263	262	79	278	127	44	77	278	100	89	24	10	8	5	2	2		10	15	5	–	1	–	–	–	–	–	1	–	–	Phosphoric acid, dimethyl 3-methyl-4-(methylsulphinyl)phenyl ester	6552-13-2	NI28434
114	86	42	56	44	146	171	88	278	100	62	46	43	40	37	36	33	**0.00**	10	18	7	2	–	–	–	–	–	–	–	–	–	Glycine, N-[2-[bis(carboxymethyl)amino]ethyl]-N-(2-hydroxyethyl)-	150-39-0	NI28435
249	247	191	245	149	250	189	147	278	100	76	60	40	40	38	34	30	**10.00**	10	22	1	–	–	–	–	–	–	–	–	–	1	Triethyl-ethylene-acetyl-stannane		LI05971
73	147	75	43	45	204	117	177	278	100	51	29	19	18	13	13	12	**0.00**	10	22	5	–	–	–	–	–	–	2	–	–	–	Acetic acid, 2,2'-oxybis-, bis(trimethylsilyl) ester	55557-25-0	NI28436
73	130	103	116	75	117	101	147	278	100	69	66	62	48	29	21	10	**0.00**	10	26	3	2	–	–	–	–	–	2	–	–	–	Diethanolamine, N-nitroso-O,O'-bis(trimethylsilyl)-	69981-29-9	NI28437
278	280	50	75	199	102	279	59	278	100	99	24	22	20	20	18	18		11	7	–	2	1	–	–	1	–	–	–	–	–	4-Isothiazolecarbonitrile, 3-(4-bromophenyl)-5-methyl-	19762-79-9	LI05972
278	280	50	75	199	102	76	59	278	100	99	24	22	20	20	18	18		11	7	–	2	1	–	–	1	–	–	–	–	–	4-Isothiazolecarbonitrile, 3-(4-bromophenyl)-5-methyl-	19762-79-9	NI28438
44	125	167	43	46	278	127	42	278	100	76	50	50	43	40	23	23		11	11	1	6	–	–	1	–	–	–	–	–	–	1H-Pyrazole-1-carboximidamide, 4-[(3-chlorophenyl)hydrazono]-4,5-dihydro-3-methyl-5-oxo-	42541-27-5	LI05973
42	176	131	117	88	219	103	145	278	100	19	8	7	7	5	5	4	**0.00**	11	15	7	–	–	1	–	–	–	–	–	–	–	α-D-Arabinopyranose, 2-deoxy-2-fluoro-, triacetate	55000-57-2	NI28439
43	128	115	170	157	97	69	86	278	100	13	6	5	4	4	4	3	**0.00**	11	15	7	–	–	1	–	–	–	–	–	–	–	β-D-Xylopyranose, 3-deoxy-3-fluoro-, triacetate	20409-33-0	NI28440
59	278	131	127	126	280	245	159	278	100	98	60	38	24	19	19	17		11	18	–	–	4	–	–	–	–	–	–	–	–	2,4,6,8-Tetrathiatricyclo[3.3.1.1^{3,7}]decane, 1,3,5,7,9-pentamethyl-	17749-59-6	NI28441
146	86	164	69	60	278	116	101	278	100	83	81	63	54	49	48	44		11	22	4	2	1	–	–	–	–	–	–	–	–	1,4,7,10-Tetraoxa-13,15-diazacycloheptadecane-14-thione	82718-54-5	NI28442
146	86	164	69	60	278	101	116	278	100	83	81	62	53	49	44	40		11	22	4	2	1	–	–	–	–	–	–	–	–	1,4,7,10-Tetraoxa-13,15-diazacycloheptadecane-14-thione	82718-54-5	NI28443
164	276	137	192	100	166	75	136	278	100	52	52	46	40	32	29	21	**20.00**	12	7	4	2	–	–	1	–	–	–	–	–	–	7-Chloro-2,3-dihydroxyphenazine-1,4-quinone		MS06309
278	250	280	279	176	252	204	165	278	100	36	32	15	13	12	11	11		12	7	4	2	–	–	1	–	–	–	–	–	–	7-Chloro-1,2,3,4-tetrahydroxyphenazine		MS06310
278	280	237	202	239	136	150	101	278	100	57	36	27	25	21	20	18		12	8	–	4	–	–	2	–	–	–	–	–	–	Imidazo[1,2-b]-1,2,4-triazine, 2-methyl-3,7-dichloro-6-phenyl-		NI28444
168	243	278	139	169	280	244	113	278	100	38	23	19	15	7	6	4		12	8	1	–	–	–	1	–	–	–	–	–	1	10H-Phenoxarsine, 10-chloro-	2865-70-5	NI28445
141	137	277	261	278	142	138	279	278	100	52	47	47	21	9	8	5		12	10	4	2	1	–	–	–	–	–	–	–	–	Benzenesulphanilide, 2'-nitro-		LI05974
141	277	278	137	142	143	279	138	278	100	95	20	12	10	4	2	2		12	10	4	2	1	–	–	–	–	–	–	–	–	Benzenesulphanilide, 3'-nitro-		LI05975
137	277	141	278	142	138	279	143	278	100	97	78	40	10	8	6	3		12	10	4	2	1	–	–	–	–	–	–	–	–	Benzenesulphanilide, 4'-nitro-		LI05976

MASS TO CHARGE RATIOS								M.W.	INTENSITIES								Parent	C	H	O	N	S	F	Cl	Br	I	Si	P	B	X	COMPOUND NAME	CAS Reg No	No
278	108	92	156	65	140	80	93	278	100	77	65	54	54	45	33	24		12	10	4	2	1	–	–	–	–	–	–	–	–	4-Nitro-4′-aminodiphenyl sulphone	1948-92-1	NI28446
278	113	114	115	126	101	100	144	278	100	35	25	22	18	16	15	14		12	10	6	2	–	–	–	–	–	–	–	–	–	Naphthalene, 2,6-dimethoxy-1,5-dinitro-	25314-97-0	NI28447
278	112	113	114	126	101	100	129	278	100	35	25	23	18	16	15	13		12	10	6	2	–	–	–	–	–	–	–	–	–	Naphthalene, 2,6-dimethoxy-1,5-dinitro-	25314-97-0	NI28448
186	117	28	119	79	69	81	91	278	100	99	99	96	87	64	63	58	0.50	12	13	4	–	–	3	–	–	–	–	–	–	–	endo-8′-(Trifluoroacetoxy)spiro[cyclopropane-1,6′-exo-[3]oxatricyclo[3.3.1.0^{2,4}]nonan]-exo-7′-ol		MS06311
120	118	251	249	197	147	116	119	278	100	75	69	53	46	42	42	36	0.00	12	14	–	–	–	–	–	–	–	–	–	–	1	Phenyltrivinyltin		MS06312
214	279	156	213	278	124	215	92	278	100	79	45	42	34	29	16	16		12	14	2	4	1	–	–	–	–	–	–	–	–	Benzenesulphonamide, 4-amino-N-(4,6-dimethyl-2-pyrimidinyl)-	57-68-1	NI28449
214	213	92	65	215	108	123	93	278	100	64	27	17	14	11	8	8	0.00	12	14	2	4	1	–	–	–	–	–	–	–	–	Benzenesulphonamide, 4-amino-N-(4,6-dimethyl-2-pyrimidinyl)-	57-68-1	NI28450
137	110	278	138	83	44	51	42	278	100	24	17	10	7	7	6	6		12	14	2	4	1	–	–	–	–	–	–	–	–	Benzenesulphonamide, N-[2-(dimethylamino)-5-pyrimidinyl]-	14233-47-7	NI28451
108	80	140	107	81	278	109	64	278	100	66	50	48	30	29	24	23		12	14	2	4	1	–	–	–	–	–	–	–	–	3,3′,4,4′-Tetraaminodiphenyl sulphone	13224-79-8	NI28452
278	193	42	180	178	68	279	221	278	100	65	57	46	30	24	19	14		12	14	4	4	–	–	–	–	–	–	–	–	–	5,5′-Biuracil, 1,1′,3,3′-tetramethyl-	7033-42-3	LI05977
278	193	42	180	178	66	279	65	278	100	65	56	46	30	23	19	15		12	14	4	4	–	–	–	–	–	–	–	–	–	5,5′-Biuracil, 1,1′,3,3′-tetramethyl-	7033-42-3	NI28453
278	277	166	42	82	279	193	81	278	100	51	34	29	19	18	18	18		12	14	4	4	–	–	–	–	–	–	–	–	–	5,6′-Biuracil, 1,1′,3,3′-tetramethyl-		NI28454
41	61	39	55	75	63	81	67	278	100	82	67	54	53	51	50	50	26.66	12	14	4	4	–	–	–	–	–	–	–	–	–	Cyclohexanone, (2,4-dinitrophenyl)hydrazone	1589-62-4	NI28455
81	278	99	79	67	55	69	96	278	100	79	62	57	55	48	38	28		12	14	4	4	–	–	–	–	–	–	–	–	–	Cyclohexanone, (2,4-dinitrophenyl)hydrazone	1589-62-4	IC03593
278	81	55	99	41	67	79	69	278	100	49	49	38	38	32	31	30		12	14	4	4	–	–	–	–	–	–	–	–	–	Cyclohexanone, (2,4-dinitrophenyl)hydrazone	1589-62-4	NI28456
263	95	278	81	41	39	55	94	278	100	58	54	52	52	51	49	43		12	14	4	4	–	–	–	–	–	–	–	–	–	Mesityl oxide, (2,4-dinitrophenyl)hydrazone		IC03594
162	147	161	47	49	128	50	48	278	100	98	93	62	48	47	26	19	0.00	12	15	–	–	1	4	–	–	–	–	–	1	–	1-Ethyl-2,3-dimethylbenzo[b]thiophenium tetrafluoroborate		LI05978
176	161	175	49	177	115	162	128	278	100	90	44	18	17	15	13	10	0.00	12	15	–	–	1	4	–	–	–	–	–	1	–	1,2,3,5-Tetramethylbenzo[b]thiophenium tetrafluoroborate		LI05979
118	43	129	45	101	161	102	87	278	100	99	62	48	34	30	28	25	0.00	12	22	7	–	–	–	–	–	–	–	–	–	–	Arabinitol, 1,4-di-O-acetyl-2,3,5-tri-O-methyl-	14537-14-5	NI28457
117	43	129	101	161	87	45	71	278	100	74	44	37	21	20	18	15	0.00	12	22	7	–	–	–	–	–	–	–	–	–	–	Arabinitol, 1,4-di-O-acetyl-2,3,5-tri-O-methyl-	14537-14-5	NI28458
88	75	43	45	101	87	85	71	278	100	84	35	21	14	14	13	10	0.00	12	22	7	–	–	–	–	–	–	–	–	–	–	α-D-Galactopyranoside, methyl 4-O-acetyl-2,3,6-tri-O-methyl-	67310-40-1	NI28459
75	43	74	71	88	101	45	102	278	100	75	51	51	46	31	27	20	0.00	12	22	7	–	–	–	–	–	–	–	–	–	–	α-D-Glucopyranoside, methyl 2-O-acetyl-3,4,6-tri-O-methyl-	24904-97-0	NI28460
74	101	71	116	43	75	45	88	278	100	99	85	70	66	59	44	31	0.00	12	22	7	–	–	–	–	–	–	–	–	–	–	α-D-Glucopyranoside, methyl 3-O-acetyl-2,4,6-tri-O-methyl-	24904-95-8	NI28461
75	74	43	71	101	45	88	102	278	100	75	67	63	34	31	27	22	0.00	12	22	7	–	–	–	–	–	–	–	–	–	–	β-D-Glucopyranoside, methyl 2-O-acetyl-3,4,6-tri-O-methyl-	24904-98-1	NI28462
75	88	69	101	161	45	59	73	278	100	94	81	69	32	28	16	16	0.00	12	22	7	–	–	–	–	–	–	–	–	–	–	α-D-Mannofuranoside, methyl 5-O-acetyl-2,3,6-tri-O-methyl-	64244-05-9	NI28463
75	88	74	71	43	101	45	89	278	100	60	50	45	40	34	31	18	0.00	12	22	7	–	–	–	–	–	–	–	–	–	–	α-D-Mannopyranoside, methyl 2-O-acetyl-3,4,6-tri-O-methyl-	72922-28-2	NI28464
88	101	75	43	73	45	89	87	278	100	53	41	18	12	10	7	7	0.00	12	22	7	–	–	–	–	–	–	–	–	–	–	α-D-Mannopyranoside, methyl 6-O-acetyl-2,3,4-tri-O-methyl-	72922-16-8	NI28465
117	129	101	58	87	43	57	55	278	100	58	49	41	25	22	20	20	0.00	12	22	7	–	–	–	–	–	–	–	–	–	–	Ribitol, 1,4-di-O-acetyl-2,3,5-tri-O-methyl-	84925-40-6	NI28466
85	55	56	97	41	84	57	69	278	100	37	23	20	18	11	10	9	0.00	12	23	2	–	–	–	–	1	–	–	–	–	–	2H-Pyran, 2-[(7-bromoheptyl)oxy]tetrahydro-	10160-25-5	NI28467
85	55	56	41	96	57	43	67	278	100	31	25	20	14	11	8	6	0.00	12	23	2	–	–	–	–	1	–	–	–	–	–	2H-Pyran, 2-[(7-bromoheptyl)oxy]tetrahydro-	10160-25-5	NI28468
74	87	55	41	43	69	59	29	278	100	51	36	35	21	14	14	12	0.54	12	23	2	–	–	–	–	1	–	–	–	–	–	Undecanoic acid, 11-bromo-, methyl ester	6287-90-7	NI28469
74	87	55	43	41	69	199	59	278	100	46	25	22	22	13	10	9	0.47	12	23	2	–	–	–	–	1	–	–	–	–	–	Undecanoic acid, 11-bromo-, methyl ester	6287-90-7	MS06313
43	111	99	40	87	45	127	81	278	100	55	55	55	50	50	45	45	4.55	12	23	5	–	–	–	–	–	–	–	1	–	–	Phosphonic acid, [[2-(1-methyl-1-propenyl)-1,3-dioxolan-2-yl]methyl]-, diethyl ester, (E)-	54543-05-4	NI28470
137	57	249	138	179	149	139	77	278	100	96	69	50	39	37	33	27	0.00	12	30	3	–	–	–	–	–	–	2	–	–	–	Disiloxane, 1,3-diethyl-1,3-dibutoxy-	73420-24-3	NI28471
278	280	279	213	75	208	178	177	278	100	59	40	36	28	27	26	25		13	8	1	2	–	–	2	–	–	–	–	–	–	11H-Dibenzo[c,f][1,2]diazepine, 3,8-dichloro-, 5-oxide	958-71-4	NI28472
278	280	213	177	75	163	63	178	278	100	65	43	43	41	39	39	29		13	8	1	2	–	–	2	–	–	–	–	–	–	11H-Dibenzo[c,f][1,2]diazepine, 3,8-dichloro-, 5-oxide	958-71-4	LI05980
152	45	278	215	151	280	75	187	278	100	51	50	47	31	29	29	27		13	8	1	2	–	–	2	–	–	–	–	–	–	11H-Dibenzo[c,f][1,2]diazepin-11-ol, 3,8-dichloro-	23469-56-9	NI28473
278	280	243	63	279	164	75	111	278	100	65	64	60	54	50	50	33		13	8	1	2	–	–	2	–	–	–	–	–	–	11H-Dibenzo[c,f][1,2]diazepin-11-one, 3,8-dichloro-5,6-dihydro-	23469-53-6	NI28474
278	183	167	155	279	185	96	123	278	100	62	62	24	21	18	17	16		13	8	1	2	1	2	–	–	–	–	–	–	–	Thiocyanic acid, 4-amino-2-fluoro-5-(4-fluorophenoxy)phenyl ester	19239-05-5	MS06314
278	235	43	69	208	53	279	217	278	100	64	63	28	18	18	15	10		13	10	7	–	–	–	–	–	–	–	–	–	–	3-Acetyl-2-hydroxy-7-methoxynaphthazarin		LI05981
236	43	69	53	165	137	237	95	278	100	92	40	20	16	16	14	14	8.00	13	10	7	–	–	–	–	–	–	–	–	–	–	1,4-Naphthalenedione, 2-(acetyloxy)-5,8-dihydroxy-7-methoxy-	54725-00-7	NI28475
278	235	43	69	208	53	279	57	278	100	62	61	28	18	18	15	10		13	10	7	–	–	–	–	–	–	–	–	–	–	1,4-Naphthoquinone, 2-acetyl-3,5,8-trihydroxy-6-methoxy-	15254-72-5	NI28476
141	157	129	140	115	128	127	139	278	100	78	34	27	24	23	22	19	13.00	13	11	2	–	–	–	–	1	–	–	–	–	–	1-Naphthalenemethyl monobromoacetate	59956-74-0	NI28477
184	278	92	185	139	168	279	186	278	100	40	16	14	11	9	7	5		13	11	3	–	1	–	–	–	–	–	1	–	–	O,O′-(2,2′-Biphenylene)thiophosphoric acid O-methyl ester	76045-10-8	NI28478
278	232	168	139	279	233	280	199	278	100	100	23	17	13	12	9	7		13	11	3	–	1	–	–	–	–	–	1	–	–	O,O′-(2,2′-Biphenylene)thiophosphoric acid S-methyl ester	77671-17-1	NI28479
218	159	145	91	187	204	119	278	278	100	80	31	24	22	19	17	1		13	14	5	2	–	–	–	–	–	–	–	–	–	2-Quinazolineacetic acid, 1,2,3,4-tetrahydro-2-(methoxycarbonyl)-4-oxo-, methyl ester	17244-35-8	LI05982
219	159	145	187	205	160	130	89	278	100	96	31	29	25	24	15	14	0.00	13	14	5	2	–	–	–	–	–	–	–	–	–	2-Quinazolineacetic acid, 1,2,3,4-tetrahydro-2-(methoxycarbonyl)-4-oxo-, methyl ester	17244-35-8	NI28480
278	92	260	93	279	77	187	65	278	100	74	43	31	22	20	18	12		13	18	3	4	–	–	–	–	–	–	–	–	–	D-Arabino-hexos-2-ulose, 3,6-anhydro-, 2-(methylhydrazone) 1-(phenylhydrazone)	50611-48-8	NI28481
43	28	125	83	39	54	27	56	278	100	62	58	52	39	38	38	29	2.35	13	22	1	2	–	–	–	–	–	–	–	–	1	Iron, (η^4-1,3-butadiene)carbonyl[N,N′-1,2-ethanediylidenebis[2-propanamine]-N,N′]-	72142-40-6	NI28482
249	263	165	207	179	121	79	163	278	100	93	92	46	35	32	27	25	0.70	13	30	4	–	–	–	–	–	–	1	–	–	–	Silicic acid, butyl tripropyl ester	55683-12-0	NI28483

MASS TO CHARGE RATIOS	M.W.	INTENSITIES	Parent	C	H	O	N	S	F	Cl	Br	I	Si	P	B	X	COMPOUND NAME	CAS Reg No	No
43 194 149 57 196 41 263 236	278	100 89 50 35 30 20 18 16	9.00	14	11	4	–	–	–	1	–	–	–	–	–	–	1,2-Naphthalenediol, 4-chloro-, diacetate	57396-88-0	NI28484
278 91 279 213 170 95 97 199	278	100 17 15 12 11 11 10 9		14	12	3	–	–	1	–	–	–	–	1	–	–	2,4-Dimethyl-8-fluorophenoxaphosphinic acid		NI28485
91 213 45 65 123 278 92 77	278	100 21 19 14 6 5 5 5		14	14	–	–	3	–	–	–	–	–	–	–	–	Trisulphide, dibenzyl	6493-73-8	NI28486
182 171 107 91 169 167 65 168	278	100 97 96 72 56 46 35 30	16.39	14	14	1	–	–	–	–	–	–	–	–	–	1	4,4′-Dimethyldiphenyl selenoxide	25862-12-8	NI28487
182 171 107 91 169 167 262 65	278	100 97 96 72 56 46 41 35	17.00	14	14	1	–	–	–	–	–	–	–	–	–	1	4,4′-Dimethyldiphenyl selenoxide	25862-12-8	LI05983
278 140 125 109 45 65 111 139	278	100 71 39 32 29 27 26 22		14	14	2	–	2	–	–	–	–	–	–	–	–	Bis(2-methoxyphenyl) disulphide	13920-94-0	NI28488
139 278 140 124 279 141 280 125	278	100 39 11 10 5 5 4 2		14	14	2	–	2	–	–	–	–	–	–	–	–	Bis(4-methoxyphenyl) disulphide	5335-87-5	NI28489
139 278 140 124 279 141 280 214	278	100 38 10 10 6 6 4 1		14	14	2	–	2	–	–	–	–	–	–	–	–	Bis(4-methoxyphenyl) disulphide	5335-87-5	LI05984
91 139 40 278 124 64 92 105	278	100 59 59 24 24 24 23 22		14	14	2	–	2	–	–	–	–	–	–	–	–	S-(4-Methylphenyl) 4-methylbenzenethiosulphonate		DD00925
139 91 123 278 155 214 198 184	278	100 47 40 33 31 8 5 3		14	14	2	–	2	–	–	–	–	–	–	–	–	Bis(4-methylphenyl) thiosulphonate	2943-42-2	LI05985
139 91 123 278 155 140 121 214	278	100 46 39 33 30 9 9 7		14	14	2	–	2	–	–	–	–	–	–	–	–	Bis(4-phenylphenyl) thiosulphonate	2943-42-2	NI28490
107 155 77 278 79 249 109 28	278	100 93 59 53 51 47 36 30		14	14	4	–	1	–	–	–	–	–	–	–	–	Benzenemethanol, 4,4′-sulphonylbis-	52123-62-3	NI28491
134 135 39 44 136 90 89 63	278	100 11 8 7 6 6 6 5	0.00	14	14	4	–	1	–	–	–	–	–	–	–	–	Benzo[b]cyclobuta[d]thiophene-1,2-dicarboxylic acid, 1,2,2a,7b-tetrahydro-1,2-dimethyl-	57156-87-3	NI28492
155 123 278 27 44 64 69 279	278	100 92 68 22 21 18 17 14		14	14	4	–	1	–	–	–	–	–	–	–	–	4,4′-Dimethoxydiphenyl sulphone		IC03595
233 278 215 234 43 276 219 279	278	100 24 19 18 18 6 4 3		14	14	6	–	–	–	–	–	–	–	–	–	–	2-Benzofuranacetic acid, 7-acetyl-4,6-dihydroxy-3,5-dimethyl-	21987-08-6	LI05986
233 278 215 43 234 276 219 263	278	100 24 18 18 17 5 4 3		14	14	6	–	–	–	–	–	–	–	–	–	–	2-Benzofuranacetic acid, 7-acetyl-4,6-dihydroxy-3,5-dimethyl-	21987-08-6	NI28493
44 40 278 247 219 279 190 64	278	100 85 70 25 21 14 14 12		14	14	6	–	–	–	–	–	–	–	–	–	–	Dimethyl 4,4′-dimethyl-2,2′-bifuran-5,5′-dicarboxylate		MS06315
134 43 15 106 236 194 105 137	278	100 53 27 23 21 21 16 13	2.39	14	14	6	–	–	–	–	–	–	–	–	–	–	Methyl 2,4-diacetoxycinnamate		NI28494
217 278 139 183 218 279 199 61	278	100 50 25 21 15 13 13 13		14	15	–	–	2	–	–	–	–	–	1	–	–	Phosphine sulphide, [(methylthio)methyl]diphenyl-	10428-60-1	NI28495
157 129 238 236 128 64 209 211	278	100 69 56 56 43 42 38 37	36.63	14	15	1	–	–	–	–	1	–	–	–	–	–	2-Cyclohexen-1-one, 4-(4-bromophenyl)-4,5-dimethyl-	17429-37-7	NI28496
199 200 115 143 144 183 167 145	278	100 79 62 53 43 23 20 17	16.00	14	15	1	–	–	–	–	1	–	–	–	–	–	Cyclopent[a]inden-8(1H)-one, 8a-bromo-2,3,3a,8a-tetrahydro-2,2-dimethyl-	64129-23-3	NI28497
43 41 219 53 39 79 80 124	278	100 47 34 29 24 22 21 14	3.20	14	18	4	2	–	–	–	–	–	–	–	–	–	Barbituric acid, α-DL-1-methyl-5-allyl-5-(1-methyl-4-hydroxypentyn-2-yl)-	80832-89-9	NI28498
105 106 79 204 91 278 132 107	278	100 68 60 48 40 37 24 12		14	18	4	2	–	–	–	–	–	–	–	–	–	2-(2-Methylphenyl)hydrazonopropanedioic acid, diethyl ester		NI28499
135 87 192 136 278 77 86 70	278	100 38 11 10 7 7 6 5		14	18	4	2	–	–	–	–	–	–	–	–	–	Morpholine, 4-[N-(4-methoxybenzamido)acetyl]-		NI28500
139 112 98 41 55 54 140 69	278	100 19 19 18 17 10 9 9	3.13	14	22	2	4	–	–	–	–	–	–	–	–	–	Ethanedioic acid, bis(cyclohexylidenehydrazide)	370-81-0	NI28501
43 45 29 85 89 41 55 82	278	100 98 38 33 31 29 21 19	0.00	14	30	5	–	–	–	–	–	–	–	–	–	–	Tetraethylene glycol monohexyl ether		IC03596
83 97 69 55 73 103 75 41	278	100 99 89 84 51 35 34 27	0.00	14	31	1	–	–	–	1	–	–	1	–	–	–	11-Chloroundecyl trimethylsilyl ether	26305-82-8	NI28502
120 55 102 222 121 76 66 223	278	100 71 63 48 10 10 10 8	5.99	15	11	2	–	–	–	–	–	–	–	–	–	1	Manganese, dicarbonylcyclopentadienylstyrylidene-	60039-69-2	NI28503
187 261 277 278 262 91 78 188	278	100 44 35 30 15 12 11 10		15	13	–	2	–	3	–	–	–	–	–	–	–	Dibenzo[b,f][1,4]diazocine, 5,6,11,12-tetrahydro-2-(trifluoromethyl)-	27188-36-9	NI28504
41 28 55 77 69 91 198 278	278	100 99 70 66 45 40 33 30		15	18	3	–	1	–	–	–	–	–	–	–	–	2-Cyclohexen-1-one, 5-(hydroxymethyl)-4,4-dimethyl-2-(phenylsulphinyl)-, (5RS,R_S,S_S)-		MS06316
91 139 123 43 65 278 155 41	278	100 66 60 51 47 42 40 38		15	18	3	–	1	–	–	–	–	–	–	–	–	Furan, 2-isopropyl-4-methyl-3-[(4-methylphenyl)sulphonyl]-	118356-45-9	DD00926
139 43 122 91 278 263 65 140	278	100 97 73 68 64 52 47 28		15	18	3	–	1	–	–	–	–	–	–	–	–	Furan, 2-isopropyl-5-methyl-3-[(4-methylphenyl)sulphonyl]-	118356-44-8	DD00927
83 85 47 87 84 48 45 43	278	100 71 15 12 10 10 7 7	6.60	15	18	5	–	–	–	–	–	–	–	–	–	–	1H-Azuleno[1,8-bc:4,5-b′]difuran-1,8(2H)-dione, decahydro-9c-hydroxy-9b-methyl-7-methylene-, [2aR-(2aα,4aα,6aα,9aα,9bβ,9cα)]-	28625-28-7	NI28505
189 177 220 190 159 59 131 278	278	100 64 30 30 20 20 19 18		15	18	5	–	–	–	–	–	–	–	–	–	–	2H-1-Benzopyran-2-one, 6-(2,3-dihydroxy-3-methylbutyl)-7-methoxy-	28095-18-3	NI28506
189 177 190 59 220 278 191 219	278	100 59 48 42 38 33 24 18		15	18	5	–	–	–	–	–	–	–	–	–	–	2H-1-Benzopyran-2-one, 6-(2,3-dihydroxy-3-methylbutyl)-7-methoxy-, (±)-	36149-96-9	NI28507
84 55 177 53 105 91 121 260	278	100 51 37 37 27 24 20 17	5.00	15	18	5	–	–	–	–	–	–	–	–	–	–	Cyclodeca[b]furan-10-carboxylic acid, 7α-hydroxy-2,3,3a,4,7,8,11,11a-octahydro-6-methyl-3-methylene-2-oxo-, [3aR-(3aR*,5E,9Z,11aS*)]-		NI28508
43 123 138 108 236 180 164 208	278	100 72 43 25 22 21 15 13	7.64	15	18	5	–	–	–	–	–	–	–	–	–	–	4,7-Ethanoisobenzofuran-1,3-dione, 6-(acetyloxy)-3a,4,7,7a-tetrahydro-4,8,8-dimethyl-	29339-48-8	NI28509
132 29 278 233 160 204 161 27	278	100 93 87 63 48 37 33 30		15	18	5	–	–	–	–	–	–	–	–	–	–	α-(4-Methoxybenzylidene)malonic acid, diethyl ester		IC03597
135 107 43 29 278	278	100 51 38 32 4		15	18	5	–	–	–	–	–	–	–	–	–	–	Oxiraneacetic acid, 2-(ethoxycarbonyl)-3-phenyl-, ethyl ester		LI05987
137 278 138 163 131 127 279 128	278	100 67 39 36 26 24 14 14		15	18	5	–	–	–	–	–	–	–	–	–	–	2H-Pyran-2-one, 5,6-dihydro-6-[2-(3-hydroxy-4-methoxyphenyl)ethyl]-4-methoxy-, (S)-	38146-59-7	NI28510
222 121 135 165 193 109 122 117	278	100 71 69 60 31 29 26 24	0.00	15	22	1	–	1	–	–	–	–	1	–	–	–	Silane, (Z,E)-[1-ethylidene-3-methoxy-2-(phenylthio)-2-propenyl]trimethyl-		NI28511
97 41 39 67 136 42 64 43	278	100 69 41 34 24 23 23 23	20.99	15	22	1	2	1	–	–	–	–	–	–	–	–	4H-1,3-Thiazine, 4,4,6-trimethyl-2-(N-methyl-4-methoxyphenylamino)-5,6-dihydro-	53004-51-6	NI28512
84 83 121 122 96 260 204 229	278	100 42 36 31 28 27 27 25	1.00	15	22	3	2	–	–	–	–	–	–	–	–	–	2(1H)-Pyridinone, 6-[octahydro-1-(hydroxymethyl)-2H-quinolizin-3-yl]-, N-oxide, (–)-		NI28513

MASS TO CHARGE RATIOS								M.W.	INTENSITIES								Parent	C	H	O	N	S	F	Cl	Br	I	Si	P	B	X	COMPOUND NAME	CAS Reg No	No
169	249	250	183	112	81	79	41	278	100	94	37	18	15	10	9	9	0.00	15	22	3	2	–	–	–	–	–	–	–	–	–	2,4,6(1H,3H,5H)-Pyrimidinetrione, 5-(1-cyclohepten-1-yl)-5-ethyl-1,3-dimethyl-	54965-32-1	NI28514
234	278	75	206	178	235	44	74	278	100	94	56	38	26	19	19	18		16	6	5	–	–	–	–	–	–	–	–	–	–	Anthraquinone-2,3-dicarboxylic acid anhydride		IC03598
105	278	77	51	131	106	279		278	100	54	54	14	10	8	7			16	10	3	2	–	–	–	–	–	–	–	–	–	Furazan, dibenzoyl-	10349-12-9	NI28515
91	77	278	65	119	105	51	89	278	100	17	14	13	13	11	11	11		16	14	1	4	–	–	–	–	–	–	–	–	–	1-Benzyl-4-benzoylamino-1,2,4-triazolium ylide	25857-87-8	NI28516
91	278	119	159	158	105	132	118	278	100	14	12	11	11	11	10	6		16	14	1	4	–	–	–	–	–	–	–	–	–	1-Benzyl-4-benzoylimino-1,2,4-triazolium ylide		MS06317
278	77	201	93	279	65	91	67	278	100	65	42	30	20	18	17	15		16	14	1	4	–	–	–	–	–	–	–	–	–	1-Phenyl-3-methyl-4-phenylazo-pyrazol-5-one		IC03599
91	278	132	77	92	65	118	279	278	100	85	57	24	19	18	17	15		16	14	1	4	–	–	–	–	–	–	–	–	–	3H-Pyrazol-3-one, 4-[(4-aminophenyl)imino]-2,4-dihydro-5-methyl-2-phenyl-	13617-67-9	NI28517
278	77	201	93	279	65	91	67	278	100	99	92	82	77	63	51	49		16	14	1	4	–	–	–	–	–	–	–	–	–	5-Pyrazolone, 1-phenyl-3-methyl-4-(phenylazo)-		NI28518
104	91	117	129	125	207	105	103	278	100	57	31	25	24	13	12	12	0.50	16	16	–	–	–	–	2	–	–	–	–	–	–	Butane, 1,4-dichloro-1,4-diphenyl-	3617-23-0	NI28519
166	278	121	165	167	168	45	134	278	100	46	21	17	13	10	10	9		16	22	–	–	2	–	–	–	–	–	–	–	–	2-(Octylthio)benzo[b]thiophene		AI01947
166	278	121	165	167	45	168	134	278	100	33	20	17	14	12	10	10		16	22	–	–	2	–	–	–	–	–	–	–	–	3-(Octylthio)benzo[b]thiophene		AI01948
166	107	105	79	77	55	113	68	278	100	82	58	55	48	26	25	25	0.00	16	22	4	–	–	–	–	–	–	–	–	–	–	Benzeneacetic acid, α-hydroxy-α-(1-hydroxy-1-cycloheptyl)-, methyl ester		DD00928
149	150	223	41	104	76	57	29	278	100	9	6	6	5	5	5	3	0.80	16	22	4	–	–	–	–	–	–	–	–	–	–	1,2-Benzenedicarboxylic acid, butyl isobutyl ester	17851-53-5	NI28520
149	29	41	150	57	223	56	205	278	100	11	10	9	7	6	6	4	1.18	16	22	4	–	–	–	–	–	–	–	–	–	–	1,2-Benzenedicarboxylic acid, dibutyl ester	84-74-2	IC03600
149	150	223	205	57	41	29	56	278	100	9	6	5	5	5	5	4	0.00	16	22	4	–	–	–	–	–	–	–	–	–	–	1,2-Benzenedicarboxylic acid, dibutyl ester	84-74-2	WI01341
149	41	29	28	57	27	104	56	278	100	28	26	22	11	10	9	9	0.70	16	22	4	–	–	–	–	–	–	–	–	–	–	1,2-Benzenedicarboxylic acid, dibutyl ester	84-74-2	WI01342
56	41	205	167	223	149	55	48	278	100	56	48	45	39	30	24	24	1.00	16	22	4	–	–	–	–	–	–	–	–	–	–	1,4-Benzenedicarboxylic acid, dibutyl ester		LI05988
149	57	150	41	29	223	104	56	278	100	23	10	8	8	7	5	5	0.00	16	22	4	–	–	–	–	–	–	–	–	–	–	1,2-Benzenedicarboxylic acid, diisobutyl ester	84-69-5	WI01343
149	57	41	223	29	150	104	76	278	100	25	15	11	11	10	9	7	0.19	16	22	4	–	–	–	–	–	–	–	–	–	–	1,2-Benzenedicarboxylic acid, diisobutyl ester	84-69-5	IC03601
149	57	28	56	41	150	29	104	278	100	66	36	32	30	25	23	10	0.00	16	22	4	–	–	–	–	–	–	–	–	–	–	1,2-Benzenedicarboxylic acid, diisobutyl ester	84-69-5	NI28521
149	57	150	223	41	29	163	104	278	100	23	9	8	8	8	6	5	0.07	16	22	4	–	–	–	–	–	–	–	–	–	–	1,3-Benzenedicarboxylic acid, diisobutyl ester		DO01071
56	43	205	57	278	279	222	149	278	100	41	34	23	18	17	12	7		16	22	4	–	–	–	–	–	–	–	–	–	–	1,4-Benzenedicarboxylic acid, diisobutyl ester		LI05989
149	70	167	57	41	112	83	55	278	100	38	26	25	18	17	16	16	0.00	16	22	4	–	–	–	–	–	–	–	–	–	–	1,2-Benzenedicarboxylic acid, mono(2-ethylhexyl) ester	4376-20-9	NI28522
117	204	130	190	278	118	191	207	278	100	68	51	34	28	24	23	20		16	22	4	–	–	–	–	–	–	–	–	–	–	1,4-Benzenepropanoic acid, diethyl ester		IC03602
205	149	223	167	56	57	104	121	278	100	48	44	44	39	29	14	12	0.20	16	22	4	–	–	–	–	–	–	–	–	–	–	1,4-Dicarboxylic acid, diisobutyl ester		LI05990
205	149	223	167	56	57	206	104	278	100	48	44	44	39	29	14	14	0.00	16	22	4	–	–	–	–	–	–	–	–	–	–	1,4-Dicarboxylic acid, diisobutyl ester		DO01072
231	159	187	278	91	263	149	105	278	100	48	44	39	39	37	37	35		16	22	4	–	–	–	–	–	–	–	–	–	–	1,2-Dimethoxycarbonyl-3,4-dimethyl-6-isobutenyl-1,4-cyclohexadiene		LI05991
83	263	233	264	278	203	31	91	278	100	99	48	19	17	14	13	12		16	22	4	–	–	–	–	–	–	–	–	–	–	3-(2,6-Dimethoxy-6,6′-dimethylpyrano[2′,3′:3,4]phenyl)-1-propanol		NI28523
124	43	28	161	55	137	119	67	278	100	81	76	72	51	33	32	30	5.00	16	22	4	–	–	–	–	–	–	–	–	–	–	2-(8-Hydroxy-3-hydroxymethyl-7-methylocta-2,6-dienyl)hydroquinone		NI28524
159	218	246	113	91	145	133	158	278	100	77	47	43	38	36	32	28	23.02	16	22	4	–	–	–	–	–	–	–	–	–	–	as-Indacene-4,5-dicarboxylic acid, 1,2,3,3aα,4α,5α,5aα,6,7,8-decahydro-, dimethyl ester	29365-45-5	NI28525
218	159	246	219	131	247	91	117	278	100	83	28	20	20	15	13	12	12.01	16	22	4	–	–	–	–	–	–	–	–	–	–	as-Indacene-4,5-dicarboxylic acid, 1,2,3,3aα,4α,5β,5aα,6,7,8-decahydro-, dimethyl ester	29365-47-7	NI28526
159	218	131	41	160	219	91	77	278	100	72	18	17	14	12	12	11	11.01	16	22	4	–	–	–	–	–	–	–	–	–	–	as-Indacene-4,5-dicarboxylic acid, 1,2,3,3aα,4β,5β,5aα,6,7,8-decahydro-, dimethyl ester	29365-46-6	NI28527
208	43	71	176	105	278	165	247	278	100	60	39	10	8	6	3	1		16	22	4	–	–	–	–	–	–	–	–	–	–	1H-Indene-5-carboxylic acid, 2,3,3a,6,7,7a-hexahydro-3-hydroxy-1-methyl-3a-(2-methyl-1-oxopropyl)-, methyl ester, [1R-(1α,3α,3aα,7aα)]-		NI28528
278	43	108	177	121	109	161	91	278	100	83	63	60	60	55	43	37		16	22	4	–	–	–	–	–	–	–	–	–	–	Methyl neolinderanate		MS06318
139	140	95	97	43	57	110	41	278	100	60	12	11	9	7	6	5	1.00	16	22	4	–	–	–	–	–	–	–	–	–	–	4,5-Octanediol, 4,5-di-2-furanyl-	56335-75-2	NI28529
28	190	134	125	162	91	112	41	278	100	99	29	29	27	14	13	13	0.00	16	22	4	–	–	–	–	–	–	–	–	–	–	2,4-Pentadienoic acid, 5-(1-hydroxy-2,6,6-trimethyl-4-oxo-2-cyclohexen-1-yl)-3-methyl-, methyl ester, (R)-(Z,E)-	7200-31-9	NI28531
29	43	41	190	55	69	39	27	278	100	47	37	24	22	18	18	17	0.50	16	22	4	–	–	–	–	–	–	–	–	–	–	2,4-Pentadienoic acid, 5-(1-hydroxy-2,6,6-trimethyl-4-oxo-2-cyclohexen-1-yl)-3-methyl-, methyl ester, (R)-(Z,E)-	7200-31-9	NI28530
153	57	55	43	41	69	154	125	278	100	17	16	16	15	14	12	11	3.50	16	22	4	–	–	–	–	–	–	–	–	–	–	2H-Pyran-3-carboxaldehyde, 4-methoxy-6-nonyl-2-oxo-		MS06319
73	160	75	117	161	74	43	45	278	100	23	15	13	10	9	9	8	2.90	16	26	2	–	–	–	–	–	–	1	–	–	–	Benzeneacetic acid, α-methyl-4-isobutyl-, trimethylsilyl ester	74810-89-2	NI28532
91	75	73	129	92	81	172	131	278	100	32	29	16	13	13	10	10	0.00	16	26	2	–	–	–	–	–	–	1	–	–	–	Cyclohexane, 1-benzyloxy-4-trimethylsiloxy-	31404-18-9	NI28534
91	75	73	129	92	81	172	179	278	100	33	29	17	14	14	10	9		16	26	2	–	–	–	–	–	–	1	–	–	–	Cyclohexane, 1-benzyloxy-4-trimethylsiloxy-	31404-18-9	LI05992
91	172	179	92	188	170	129	82	278	100	71	59	31	27	26	25	23	1.00	16	26	2	–	–	–	–	–	–	1	–	–	–	Cyclohexane, 1-benzyloxy-4-trimethylsiloxy-	31404-18-9	NI28533
172	91	179	188	170	92	187	97	278	100	95	88	74	60	59	46	31	1.00	16	26	2	–	–	–	–	–	–	1	–	–	–	Cyclohexane, 1-benzyloxy-4-trimethylsiloxy-, cis-	31447-55-9	NI28535
91	73	92	129	75	81	172	131	278	100	23	16	13	12	10	9	9	0.00	16	26	2	–	–	–	–	–	–	1	–	–	–	Cyclohexane, 1-benzyloxy-4-trimethylsiloxy-, cis-	31447-55-9	NI28536
91	73	92	129	75	188	170	131	278	100	24	19	12	12	11	10	9	0.00	16	26	2	–	–	–	–	–	–	1	–	–	–	Cyclohexane, 1-benzyloxy-4-trimethylsiloxy-, trans-	31447-56-0	NI28537
188	170	187	91	92	97	171	131	278	100	66	55	53	51	34	15	15	1.00	16	26	2	–	–	–	–	–	–	1	–	–	–	Cyclohexane, 1-benzyloxy-4-trimethylsiloxy-, trans-	31447-56-0	NI28538

MASS TO CHARGE RATIOS								M.W.	INTENSITIES								Parent	C	H	O	N	S	F	Cl	Br	I	Si	P	B	X	COMPOUND NAME	CAS Reg No	No
135	137	245	75	59	143	109	202	278	100	99	97	73	70	61	58	52	0.00	16	26	2	–	–	–	–	–	–	1	–	–	–	3-Hexene-2,5-diol, 2,5-dimethyl-3-(dimethylphenylsilyl)-, cis-		NI28539
58	160	245	189	278	205	45	260	278	100	41	36	34	28	24	24	21		16	26	2	2	–	–	–	–	–	–	–	–	–	Eserethole methine		MS06320
219	220	278	249	221	165	261	152	278	100	59	42	22	22	19	16	15		16	26	2	2	–	–	–	–	–	–	–	–	–	Lycocernuine	6871-55-2	NI28540
105	77	278	250	129	173	234	51	278	100	46	19	13	13	12	11	11		17	10	4	–	–	–	–	–	–	–	–	–	–	Phencyclidenehomophthalic anhydride		LI05993
278	280	139	279	244	242	243	107	278	100	32	25	23	13	13	10	10		17	11	–	2	–	–	1	–	–	–	–	–	–	Dibenzo[b,h][1,6]naphthyridine, 2-chloro-6-methyl-	4240-91-9	NI28541
278	280	139	279	121.5	244	242	107	278	100	32	24	23	17	13	12	10		17	11	–	2	–	–	1	–	–	–	–	–	–	Dibenzo[b,h][1,6]naphthyridine, 2-chloro-6-methyl-	4240-91-9	LI05994
135	247	77	92	248	219	136	63	278	100	71	37	19	14	13	9	6	2.42	17	14	2	2	–	–	–	–	–	–	–	–	–	Benzamide, N-1-isoquinolinyl-2-methoxy-	40339-89-7	NI28542
171	278	135	277	249	107	77	128	278	100	84	84	63	46	43	42	38		17	14	2	2	–	–	–	–	–	–	–	–	–	Benzamide, N-1-isoquinolinyl-3-methoxy-	40339-90-0	NI28543
135	278	171	277	77	92	128	136	278	100	46	32	24	23	18	17	14		17	14	2	2	–	–	–	–	–	–	–	–	–	Benzamide, N-1-isoquinolinyl-4-methoxy-	40339-91-1	NI28544
278	116	221	131	83	248	85	220	278	100	87	80	67	49	42	33	27		17	14	2	2	–	–	–	–	–	–	–	–	–	1H-1,4-Benzodiazepine-2,5-dione, 3,4-dihydro-4-methyl-3-benzylene-	31965-37-4	LI05995
278	116	221	131	83	248	85	90	278	100	87	80	67	50	41	32	30		17	14	2	2	–	–	–	–	–	–	–	–	–	1H-1,4-Benzodiazepine-2,5-dione, 3,4-dihydro-4-methyl-3-benzylene-	31965-37-4	NI28545
236	278	237	207	178	279	235	131	278	100	48	16	13	10	9	9	6		17	14	2	2	–	–	–	–	–	–	–	–	–	N-(3-Benzylidene-2-oxo-1-indolinyl)acetamide		AI01949
263	278	132	264	146	279	133	104	278	100	95	82	32	31	29	22	22		17	14	2	2	–	–	–	–	–	–	–	–	–	7-Methyldiftalone		MS06321
277	278	279	77	234	104	263	139	278	100	82	15	15	12	10	8	8		17	14	2	2	–	–	–	–	–	–	–	–	–	4-Phenyl-6-(4-methoxyphenyl)-2-pyrimidol		LI05996
233	278	77	234	51	279	116	103	278	100	59	36	19	16	10	9	6		17	14	2	2	–	–	–	–	–	–	–	–	–	1H-Pyrazole-4-acetic acid, 1,5-diphenyl-	32702-05-9	NI28546
118	77	278	105	119	51	279	78	278	100	39	28	11	10	9	5	4		17	14	2	2	–	–	–	–	–	–	–	–	–	Pyrazole, 1-phenyl-3-benzoyl-4-hydroxy-5-methyl-	95884-24-5	NI28547
278	77	130	102	105	209	51	121	278	100	78	48	32	24	18	17	16		17	14	2	2	–	–	–	–	–	–	–	–	–	3,5-Pyrazolidinedione, 4-benzylidene-2-methyl-1-phenyl-	106584-28-5	NI28548
278	105	277	77	200	91	67	279	278	100	57	50	49	38	23	21	20		17	14	2	2	–	–	–	–	–	–	–	–	–	3H-Pyrazol-3-one, 4-benzoyl-2,4-dihydro-5-methyl-2-phenyl-	4551-69-3	MS06322
277	278	104	77	103	279	235	76	278	100	97	74	39	20	19	16	16		17	14	2	2	–	–	–	–	–	–	–	–	–	2(1H)-Pyrimidinone, 5-methoxy-4,6-diphenyl-	28567-84-2	NI28549
224	263	209	225	264	180	167	210	278	100	81	30	23	20	15	15	12	0.50	17	18	–	2	–	–	–	–	–	1	–	–	–	10-Sila-2-azaanthracene, 10,10-dimethyl-9-(β-cyanoethyl)-9,10-dihydro-	84841-67-8	NI28550
78	51	50	52	91	93	42	77	278	100	31	24	23	22	18	16	14	13.00	17	18	–	4	–	–	–	–	–	–	–	–	–	4-Benzyl-2-imino-6-phenyliminopiperazine		MS06323
172	278	173	77	106	103	91	279	278	100	54	52	32	20	15	10	10		17	18	–	4	–	–	–	–	–	–	–	–	–	1,2,3-Triazole, 2-phenyl-4-[methyl(phenyl)aminomethyl]-5-methyl-		NI28551
152	278	135	153	107	77	136	92	278	100	63	35	24	11	9	6	6		17	26	3	–	–	–	–	–	–	–	–	–	–	Benzoic acid, 3-methoxy-, nonyl ester	69833-39-2	NI28552
152	135	153	107	278	77	136	92	278	100	35	24	11	10	9	6	6		17	26	3	–	–	–	–	–	–	–	–	–	–	Benzoic acid, 4-methoxy-, nonyl ester	55469-25-5	NI28553
57	263	41	278	147	91	55	264	278	100	99	42	24	24	18	17	16		17	26	3	–	–	–	–	–	–	–	–	–	–	3-(3,5-Di-tert-butyl-4-hydroxyphenyl)propionic acid		IC03603
137	278	179	151	119	122	124	107	278	100	50	10	10	8	7	6	5		17	26	3	–	–	–	–	–	–	–	–	–	–	3-Decanone, 1-(4-hydroxy-3-methoxyphenyl)-	27113-22-0	MS06324
109	278	205	55	81	67	95	53	278	100	58	55	37	35	31	28	28		17	26	3	–	–	–	–	–	–	–	–	–	–	8α-Ethoxy-10αH-eremophilenolide		NI28554
201	219	123	137	278	260	149		278	100	89	63	61	59	52	30			17	26	3	–	–	–	–	–	–	–	–	–	–	Perhydro-6,9a-dimethyl-6-methoxycarbonyl-3-oxo-cyclopenta[a]naphthalene		LI05997
94	54	69	83	98	95	73	84	278	100	30	24	15	13	13	13	12	5.00	17	26	3	–	–	–	–	–	–	–	–	–	–	Undecanoic acid, 11-phenoxy-	7170-44-7	NI28555
72	44	86	73	40	71	100	58	278	100	9	8	5	5	4	3	3	0.00	17	30	1	2	–	–	–	–	–	–	–	–	–	2-Propanamine, N,N-dimethyl-1-[4-[2-(diethylamino)ethoxy]phenyl]-		MS06325
235	278	250	248	179	139	178	263	278	100	53	48	30	28	28	21	18		18	14	3	–	–	–	–	–	–	–	–	–	–	Anthra[2,1-b]furan-4,5-dione, 1,2-dihydro-1,10-dimethyl-	54964-96-4	NI28556
278	263	235	179	139	250	178	279	278	100	78	39	33	33	28	24	19		18	14	3	–	–	–	–	–	–	–	–	–	–	Anthra[2,3-b]furan-4,11-dione, 2,3-dihydro-3,6-dimethyl-	54964-97-5	NI28557
278	277	127	151	201	139	263	115	278	100	65	36	26	17	11	10	10		18	14	3	–	–	–	–	–	–	–	–	–	–	3(2H)-Benzofuranone, 2-cinnamylidene-6-methoxy-	77765-15-2	NI28558
278	263	235	179	139	250	178	207	278	100	78	39	33	33	28	24	15		18	14	3	–	–	–	–	–	–	–	–	–	–	4,8-Dimethyl-10-oxa-7,9,10,11-tetrahydro-15H-cyclopenta[a]phenanthrene-11,12-dione		LI05998
154	250	278	260	263	82			278	100	73	32	15	5	4				18	14	3	–	–	–	–	–	–	–	–	–	–	(1,3-Dioxo-2,3-dihydrophenalene)-2-spiro-2′-(4′-methyl-3′,6′-dihydro-2′H-pyran)		LI05999
171	277	278	276	199	279	115	172	278	100	56	37	27	18	15	15	12		18	14	3	–	–	–	–	–	–	–	–	–	–	1,5-Naphthalenediol, 4-(4-hydroxystyryl)-	27667-40-9	NI28559
235	250	280	179	178	207	169	152	278	100	49	37	33	31	29	21	19	16.00	18	14	3	–	–	–	–	–	–	–	–	–	–	Phenanthreno[1,2-b]furan-10,11-dione, 1,2-dihydro-1,6-dimethyl-, (–)-	87205-99-0	NI28560
235	278	250	248	115	179	139	178	278	100	53	48	30	30	28	28	21		18	14	3	–	–	–	–	–	–	–	–	–	–	Phenanthro[2,1-b]furan-4,5-dione, 2,3-dihydro-3,9-dimethyl-	56804-88-7	NI28561
278	263	235	179	139	250	178	140	278	100	78	39	33	33	28	24	22		18	14	3	–	–	–	–	–	–	–	–	–	–	Phenanthro[3,2-b]furan-7,11-dione, 8,9-dihydro-4,8-dimethyl-	20958-18-3	NI28562
278	248	235	263	178	189	191	179	278	100	53	37	34	34	29	26	26		18	14	3	–	–	–	–	–	–	–	–	–	–	Phenanthro[1,2-b]furan-10,11-dione, 6,7,8,9-tetrahydro-1-methyl-6-methylene-	67656-29-5	NI28563
175	160	161	103	131	77	263	104	278	100	99	98	98	83	83	66	17	0.00	18	14	3	–	–	–	–	–	–	–	–	–	–	2-Propen-1-one, 1-(4-methoxy-5-benzofuranyl)-3-phenyl-	64280-20-2	NI28564
277	183	278	185	201	77	51	107	278	100	72	56	44	18	16	13	11		18	15	1	–	–	–	–	–	–	–	1	–	–	Phosphine, diphenylphenoxy-		IC03604
277	278	77	201	183	199	279	51	278	100	50	30	19	13	12	10	10		18	15	1	–	–	–	–	–	–	–	1	–	–	Phosphine oxide, triphenyl-	791-28-6	NI28565
277	278	77	201	183	51	199	279	278	100	57	36	26	22	22	19	14		18	15	1	–	–	–	–	–	–	–	1	–	–	Phosphine oxide, triphenyl-	791-28-6	IC03605
277	278	77	201	199	51	183	185	278	100	45	24	21	19	15	13	9		18	15	1	–	–	–	–	–	–	–	1	–	–	Phosphine oxide, triphenyl-	791-28-6	IC03606
277	250	249	208	235	278	222	223	278	100	92	76	76	76	69	59	45		18	18	1	2	–	–	–	–	–	–	–	–	–	1,5-Benzodiazocin-2-one, 1,2,3,4-tetrahydro-1,8-dimethyl-6-phenyl-	63563-54-2	NI28566
249	278	208	165	77	250	279	146	278	100	56	20	20	19	18	11	10		18	18	1	2	–	–	–	–	–	–	–	–	–	2-Imidazolin-4-one, 1-ethyl-2-methyl-5,5-diphenyl-	94514-22-4	NI28567
278	70	42	165	166	207	249	279	278	100	99	50	40	37	34	24	23		18	18	1	2	–	–	–	–	–	–	–	–	–	2-Imidazolin-5-one, 1-ethyl-2-methyl-4,4-diphenyl-	94514-26-8	NI28568
278	167	263	165	277	279	208	118	278	100	88	64	36	28	24	24	22		18	18	1	2	–	–	–	–	–	–	–	–	–	2-Imidazolin-4-one, 1-methyl-2-ethyl-5,5-diphenyl-	94514-23-5	NI28569
70	278	165	42	249	166	279	82	278	100	50	20	20	15	15	10	9		18	18	1	2	–	–	–	–	–	–	–	–	–	2-Imidazolin-5-one, 1-methyl-2-ethyl-4,4-diphenyl-	94514-24-6	NI28570
219	220	236	206	261	77	76	231	278	100	45	43	38	31	25	25	24	10.01	18	18	1	2	–	–	–	–	–	–	–	–	–	Quinoxaline, 2-butyl-3-phenyl-, 1-oxide	18992-52-4	NI28571
278	165	222	277	279	221	77	166	278	100	39	31	28	21	17	12	8		18	19	2	–	–	–	–	–	–	–	–	1	–	Benzoin butyl boronate		MS06326

MASS TO CHARGE RATIOS	M.W.	INTENSITIES	Parent	C	H	O	N	S	F	Cl	Br	I	Si	P	B	X	COMPOUND NAME	CAS Reg No	No
219 159 160 145 220 119 45 201	**278**	100 28 27 20 18 17 16 15	**2.73**	18	30	2	–	–	–	–	–	–	–	–	–	–	**Benzene, 1,4-bis(1,1-dimethyl-3-hydroxybutyl)-**		**MS06327**
139 140 96 83 234 205 278 260	**278**	100 88 78 76 4 4 4 3		18	30	2	–	–	–	–	–	–	–	–	–	–	**Dodecane, 6,7-bis(3-oxo-1-propenyl)-**	95452-48-5	**NI28572**
218 278 123 236 203 95 221 99	**278**	100 83 78 74 69 64 62 62		18	30	2	–	–	–	–	–	–	–	–	–	–	**6H-3,10b-Epoxy-1H-naphtho[2,1-b]pyran, decahydro-3,4a,7,7,10a-pentamethyl-, [3S-(3α,4aβ,6aα,10aβ,10bα)]-**	38419-69-1	**NI28573**
278 109 123 236 203 221 151 99	**278**	100 81 73 70 65 59 59 59		18	30	2	–	–	–	–	–	–	–	–	–	–	**6H-3,10b-Epoxy-1H-naphtho[2,1-b]pyran, decahydro-3,4a,7,7,10a-pentamethyl-, [3S-(3α,4aβ,6aα,10aβ,10bα)]-**	38419-69-1	**NI28574**
278 263 235 248 150 125 111 123	**278**	100 84 80 75 75 65 63 55		18	30	2	–	–	–	–	–	–	–	–	–	–	**2H-3,5a-Epoxynaphth[2,1-b]oxepin, dodecahydro-3,8,8,11a-tetramethyl-, [3R-(3α,5aα,7aβ,11aα,11bβ)]-**	38419-74-8	**NI28575**
278 263 235 248 125 123 97 95	**278**	100 84 80 75 65 55 53 50		18	30	2	–	–	–	–	–	–	–	–	–	–	**2H-3,5a-Epoxynaphth[2,1-b]oxepin, dodecahydro-3,8,8,11a-tetramethyl-, [3R-(3α,5aα,7aβ,11aα,11bβ)]-**	38419-74-8	**NI28576**
218 190 175 137 109 203 189 121	**278**	100 95 90 86 79 75 75 74	**27.00**	18	30	2	–	–	–	–	–	–	–	–	–	–	**5H-3,5a-Epoxynaphth[2,1-c]oxepin, dodecahydro-3,8,8,11a-tetramethyl-, [3R-(3α,5aα,7aβ,11aβ,11bα)]-**	1153-35-1	**NI28577**
218 190 137 203 189 121 95 81	**278**	100 94 86 74 74 73 66 65	**26.74**	18	30	2	–	–	–	–	–	–	–	–	–	–	**5H-3,5a-Epoxynaphth[2,1-c]oxepin, dodecahydro-3,8,8,11a-tetramethyl-, [3R-(3α,5aα,7aβ,11aβ,11bα)]-**	1153-35-1	**NI28578**
190 218 43 109 83 174 69 95	**278**	100 95 61 57 56 53 53 52	**11.00**	18	30	2	–	–	–	–	–	–	–	–	–	–	**5H-3,5a-Epoxynaphth[2,1-c]oxepin, dodecahydro-3,8,8,11a-tetramethyl-, [3S-(3α,5aα,7aα,11aβ,11bα)]-**	1153-34-0	**NI28579**
190 218 83 121 189 123 81 41	**278**	100 74 43 37 31 28 28 27	**8.52**	18	30	2	–	–	–	–	–	–	–	–	–	–	**5H-3,5a-Epoxynaphth[2,1-c]oxepin, dodecahydro-3,8,8,11a-tetramethyl-, [3S-(3α,5aα,7aα,11aβ,11bα)]-**	1153-34-0	**NI28580**
278 201 260 203 216 219 279 234	**278**	100 38 33 28 26 24 22 12		18	30	2	–	–	–	–	–	–	–	–	–	–	**Estrane-3,17-diol, (3α,5α,17β)-**	10002-95-6	**NI28581**
67 55 57 41 169 81 125 109	**278**	100 66 60 54 44 39 34 25	**0.00**	18	30	2	–	–	–	–	–	–	–	–	–	–	**Ethane, 1,2-bis(2,7-octadienyloxy)-**		**IC03607**
67 79 93 80 41 81 91 55	**278**	100 88 74 73 65 59 51 48	**17.04**	18	30	2	–	–	–	–	–	–	–	–	–	–	**5,8,11-Heptadecatrienoic acid, methyl ester**	22117-08-4	**NI28582**
43 247 121 135 189 153 260 278	**278**	100 29 28 18 17 15 14 13		18	30	2	–	–	–	–	–	–	–	–	–	–	**19-Hydroxy-15,16-dinorlabol-8(17)-en-13-one**		**NI28583**
107 79 28 77 108 116 18 51	**278**	100 55 43 35 28 24 21 15	**0.00**	18	30	2	–	–	–	–	–	–	–	–	–	–	**Linolenic acid**		**DO01073**
110 43 41 44 55 111 57 29	**278**	100 25 19 16 15 12 11 9	**5.45**	18	30	2	–	–	–	–	–	–	–	–	–	–	**Phenol, 3-dodecyloxy-**		**IC03608**
278 263 248 249 279 247 124 234	**278**	100 58 45 29 22 22 16 15		19	18	–	–	1	–	–	–	–	–	–	–	–	**6,7,8-Trimethyl-9,10-dihydronaphtho[2,3-b]benzothiophene**		**AI01950**
204 219 40 278 205 236 178 173	**278**	100 86 76 73 54 42 42 42		19	18	2	–	–	–	–	–	–	–	–	–	–	**2-Cyclohexen-1-one, 2-methoxy-4,4-diphenyl-**	96250-82-7	**NI28584**
105 183 165 115 193 149 278 111	**278**	100 49 33 22 21 17 16 14		19	18	2	–	–	–	–	–	–	–	–	–	–	**2-Cyclohexen-1-one, 2-methoxy-6,6-diphenyl-**		**DD00929**
278 250 235 279 251 125 191 165	**278**	100 79 27 24 19 14 13 11		19	18	2	–	–	–	–	–	–	–	–	–	–	**17H-Cyclopenta[a]phenanthren-17-one, 11,12,13,16-tetrahydro-3-methoxy-13-methyl-, (S)-**	56614-59-6	**NI28585**
278 279 263 235 209 165 277 250	**278**	100 25 19 11 11 11 9 9		19	18	2	–	–	–	–	–	–	–	–	–	–	**Estra-1,3,5,7,9,15-hexaen-17-one, 3-methoxy-**	56588-53-5	**NI28586**
278 41 156 170 115 169 77 249	**278**	100 70 52 50 42 41 35 34		19	22	–	2	–	–	–	–	–	–	–	–	–	**Aspidospermidine, 1,2,6,7-tetradehydro-, (5α,12β,19α)-**	32975-46-5	**NI28587**
43 45 278 60 148 42 41 58	**278**	100 84 78 67 59 23 19 17		19	22	–	2	–	–	–	–	–	–	–	–	–	**Condyfolan, 14,19-didehydro-16-methylene-, (14E)-**	56293-10-8	**NI28588**
278 148 122 130 144 77 43 263	**278**	100 44 41 39 38 35 30 28		19	22	–	2	–	–	–	–	–	–	–	–	–	**Curan, 16,17,19,20-tetradehydro-**	56053-17-9	**NI28589**
208 249 41 29 43 42 27 278	**278**	100 97 54 53 46 44 39 37		19	22	–	2	–	–	–	–	–	–	–	–	–	**Eburnamenine**	517-30-6	**MS06328**
208 249 278 193 250 206 209 248	**278**	100 97 31 24 21 21 17 16		19	22	–	2	–	–	–	–	–	–	–	–	–	**Eburnamenine**	517-30-6	**NI28590**
209 208 278 207 193 194 84 200	**278**	100 91 59 37 22 19 18 17		19	22	–	2	–	–	–	–	–	–	–	–	–	**Pyridine, 2-[1-(4-methylphenyl)-3-(1-pyrrolidinyl)-1-propenyl]-, (E)-**	486-12-4	**NI28591**
119 69 91 107 109 121 177 207	**278**	100 98 88 70 63 57 48 46	**4.00**	19	34	1	–	–	–	–	–	–	–	–	–	–	**Friedo-15-norlabd-9-en-8-ol, (10S)-**		**MS06329**
207 147 141 139 123 109 151 149	**278**	100 99 97 96 92 88 82 75	**9.00**	19	34	1	–	–	–	–	–	–	–	–	–	–	**15-Norlabdane, 5,8-epoxy-**		**MS06330**
43 42 191 235 192 95 137 109	**278**	100 96 72 40 30 30 23 22	**2.00**	19	34	1	–	–	–	–	–	–	–	–	–	–	**15-Norlabdane, 8,12-epoxy-, (12RS)-**		**MS06331**
71 41 278 67 79 80 55 81	**278**	100 56 51 39 30 28 26 25		19	34	1	–	–	–	–	–	–	–	–	–	–	**1,9,12-Octadecatriene, 1-methoxy-**	56554-59-7	**NI28592**
245 260 69 41 136 123 55 95	**278**	100 85 59 56 55 55 52 48	**6.74**	19	34	1	–	–	–	–	–	–	–	–	–	–	**8β-Podocarpan-7α-ol, 13,13-dimethyl-**	1224-29-9	**NI28594**
245 260 69 136 123 55 95 175	**278**	100 63 60 57 56 53 49 42	**5.00**	19	34	1	–	–	–	–	–	–	–	–	–	–	**8β-Podocarpan-7α-ol, 13,13-dimethyl-**	1224-29-9	**NI28593**
139 123 140 69 41 55 95 109	**278**	100 79 76 74 69 67 66 55	**53.48**	19	34	1	–	–	–	–	–	–	–	–	–	–	**8β-Podocarpan-7β-ol, 13,13-dimethyl-**	1224-28-8	**NI28595**
139 175 123 140 69 55 95 109	**278**	100 95 81 76 75 68 67 55	**53.00**	19	34	1	–	–	–	–	–	–	–	–	–	–	**8β-Podocarpan-7β-ol, 13,13-dimethyl-**	1224-28-8	**NI28596**
91 278 57 105 129 115 221 161	**278**	100 62 59 32 25 25 24 23		20	22	1	–	–	–	–	–	–	–	–	–	–	**Asperenone**	18810-05-4	**NI28597**
129 92 128 157 91 77 51 115	**278**	100 88 79 77 58 47 42 41	**0.00**	20	22	1	–	–	–	–	–	–	–	–	–	–	**Asperyellone**		**MS06332**
156 173 104 278 141 174 115 81	**278**	100 50 47 45 39 31 30 24		20	22	1	–	–	–	–	–	–	–	–	–	–	**Benzo[d][2,2]paracyclophane, 1,2,3,4-tetrahydro-4-hydroxy-**		**LI06000**
180 206 193 165 167 115 179 178	**278**	100 90 59 44 38 32 31 30	**23.20**	20	22	1	–	–	–	–	–	–	–	–	–	–	**Cyclohexanone, 2,2-dimethyl-6,6-diphenyl-**	50592-53-5	**NI28598**
105 91 260 129 77 278 231 279	**278**	100 98 78 76 68 34 33 7		20	22	1	–	–	–	–	–	–	–	–	–	–	**4-Cycloocten-1-ol, 1,5-diphenyl-**	119770-23-9	**DD00930**
278 117 105 91 160 159 131 118	**278**	100 63 54 51 43 39 34 31		20	22	1	–	–	–	–	–	–	–	–	–	–	**Tricyclo[10.2.2.2^{5,8}]octadeca-5,7,12,14,15,17-hexaene, 6-acetyl-**	24777-35-3	**NI28599**
41 95 55 81 67 43 193 69	**278**	100 94 92 74 63 63 59 56	**6.89**	20	38	–	–	–	–	–	–	–	–	–	–	–	**2-Butyl-3-hexylbicyclo[4.4.0]decane**		**AI01951**
41 95 55 81 67 43 193 69	**278**	100 92 85 81 68 66 60 54	**34.93**	20	38	–	–	–	–	–	–	–	–	–	–	–	**2-Butyl-8-hexylbicyclo[4.4.0]decane**		**AI01952**
57 83 41 69 97 43 55 111	**278**	100 97 78 73 69 57 55 47	**3.60**	20	38	–	–	–	–	–	–	–	–	–	–	–	**Decane, 5,6-bis(2,2-dimethylpropylidene)-, (E,Z)-**	55712-56-6	**MS06333**
137 81 95 41 55 43 67 136	**278**	100 75 65 56 40 39 37 32	**11.20**	20	38	–	–	–	–	–	–	–	–	–	–	–	**2-Decyldecahydronaphthalene**		**AI01953**
137 81 95 41 43 55 67 136	**278**	100 69 63 47 39 36 34 30	**10.31**	20	38	–	–	–	–	–	–	–	–	–	–	–	**6-Decyldecahydronaphthalene**		**AI01954**
109 55 41 165 95 81 43 67	**278**	100 73 72 69 63 58 52 38	**26.18**	20	38	–	–	–	–	–	–	–	–	–	–	–	**2,5-Dimethyl-7-octylbicyclo[4.4.0]decane**		**AI01955**

MASS TO CHARGE RATIOS	M.W.	INTENSITIES	Parent	C	H	O	N	S	F	Cl	Br	I	Si	P	B	X	COMPOUND NAME	CAS Reg No	No
109 55 165 95 41 81 43 67	278	100 63 60 58 58 54 39 37	25.10	20	38	–	–	–	–	–	–	–	–	–	–	–	2,5-Dimethyl-7-octylbicyclo[4.4.0]decane		AI01956
109 55 165 41 81 95 29 67	278	100 84 74 74 67 59 48 44	43.48	20	38	–	–	–	–	–	–	–	–	–	–	–	2,5-Dimethyl-7-octylbicyclo[4.4.0]decane		AI01957
165 109 95 55 41 81 43 83	278	100 88 74 57 53 41 36 34	4.56	20	38	–	–	–	–	–	–	–	–	–	–	–	3,8-Dimethyl-4-octylbicyclo[4.4.0]decane		AI01958
43 82 96 81 55 41 57 67	278	100 99 86 78 76 73 60 54	2.00	20	38	–	–	–	–	–	–	–	–	–	–	–	1-Eicosyne	765-27-5	LI06001
82 43 96 81 55 41 57 67	278	100 98 83 80 72 70 58 53	2.00	20	38	–	–	–	–	–	–	–	–	–	–	–	1-Eicosyne	765-27-5	NI28600
67 43 41 82 81 68 55 95	278	100 68 67 63 60 57 55 49	0.10	20	38	–	–	–	–	–	–	–	–	–	–	–	3-Eicosyne	61886-66-6	NI28601
67 109 82 81 95 68 110 43	278	100 92 84 78 77 56 55 50	3.20	20	38	–	–	–	–	–	–	–	–	–	–	–	3-Eicosyne	61886-66-6	NI28602
81 67 54 55 95 43 41 82	278	100 79 67 62 59 54 54 52	0.38	20	38	–	–	–	–	–	–	–	–	–	–	–	5-Eicosyne	74685-31-7	NI28603
81 67 55 43 95 82 41 57	278	100 96 86 81 72 72 71 56	0.37	20	38	–	–	–	–	–	–	–	–	–	–	–	9-Eicosyne	71899-38-2	NI28604
83 138 139 97 57 41 69 43	278	100 75 60 53 51 30 28 18	1.00	20	38	–	–	–	–	–	–	–	–	–	–	–	Hexane, 3,4-dicyclohexyl-2,5-dimethyl-, meso-	62678-52-8	NI28605
83 97 69 41 57 55 43 111	278	100 74 73 65 64 52 51 50	2.91	20	38	–	–	–	–	–	–	–	–	–	–	–	3,5-Octadiene, 4,5-dibutyl-2,2,7,7-tetramethyl-, (Z,Z)-		MS06334
91 69 131 117 41 104 105 208	278	100 43 28 23 23 16 15 14	1.60	21	26	–	–	–	–	–	–	–	–	–	–	–	Benzene, 1,1′-[1-(2,2-dimethyl-3-butenyl)-1,3-propanediyl]bis-	61142-62-9	NI28606
278 276 279 138.5 137.5 277 274 136.5	278	100 82 22 21 18 15 11 6		22	14	–	–	–	–	–	–	–	–	–	–	–	2,3-Benzchrysene		MS06335
278 276 279 139 138 277 274 125	278	100 32 22 21 18 15 11 7		22	14	–	–	–	–	–	–	–	–	–	–	–	2,3-Benzchrysene		IC03609
278 279 139 276 138 274 277 137	278	100 24 20 15 9 5 4 4		22	14	–	–	–	–	–	–	–	–	–	–	–	Benzo[a]naphthacene		NI28607
278 279 276 139 138 137 125 274	278	100 20 20 18 16 6 3 3		22	14	–	–	–	–	–	–	–	–	–	–	–	Benzo[b]triphenylene	215-58-7	NI28608
278 276 279 139 138 137 277 274	278	100 23 22 22 21 11 8 7		22	14	–	–	–	–	–	–	–	–	–	–	–	Benzo[b]triphenylene	215-58-7	NI28609
278 279 139 276 138 140 137 125	278	100 24 24 16 13 6 6 6		22	14	–	–	–	–	–	–	–	–	–	–	–	Dibenz[a,h]anthracene	53-70-3	AI01959
278 279 139 276 138 137 125 274	278	100 24 24 16 13 7 7 5		22	14	–	–	–	–	–	–	–	–	–	–	–	Dibenz[a,h]anthracene	53-70-3	NI28610
278 139 279 276 138 32 137 125	278	100 27 26 25 19 17 13 12		22	14	–	–	–	–	–	–	–	–	–	–	–	Dibenz[a,h]anthracene	53-70-3	NI28611
278 279 276 139 138 274 277 137	278	100 24 16 15 8 5 5 4		22	14	–	–	–	–	–	–	–	–	–	–	–	Dibenz[a,j]anthracene		NI28612
277 278 276 274 279 272 273 280	278	100 87 80 25 17 5 4 3		22	14	–	–	–	–	–	–	–	–	–	–	–	Indeno[1,2,3-cd]pyrene, 1,2-dihydro-		NI28613
278 279 276 274 139 138 137 125	278	100 2 2 1 1 1 1 1		22	14	–	–	–	–	–	–	–	–	–	–	–	Pentacene	135-48-8	NI28614
278 279 276 139 277 280 274 275	278	100 25 18 13 6 4 3 2		22	14	–	–	–	–	–	–	–	–	–	–	–	Pentacene	135-48-8	NI28615
85 69 104 79 50 244 66 200	279	100 39 37 32 16 14 12 10	7.80	–	–	–	1	1	3	1	1	–	–	2	–	–	Phosphorimidic bromide difluoride, (chlorofluorophosphinothioyl)-	27352-11-0	MS06336
260 258 279 187 241 277 225 203	279	100 53 16 12 10 9 8 7		–	–	1	–	–	4	–	–	–	–	–	–	1	Rhenium oxide tetrafluoride		NI28616
69 63 146 64 101 115 46 279	279	100 29 22 19 12 10 10 8		3	–	–	1	3	7	–	–	–	–	–	–	–	N-(Trifluoromethylthio)(trifluoromethyldithio)fluoromethimine		NI28617
75 74 281 235 223 154 156 283	279	100 98 42 36 29 26 25 22	20.46	6	3	2	1	–	–	–	2	–	–	–	–	–	Benzene, 1,4-dibromo-2-nitro-	3460-18-2	NI28618
110 109 127 66 65 39 47 84	279	100 41 23 21 21 12 11 11	2.40	7	6	1	1	2	5	–	–	–	–	–	–	–	[(Phenylthio)formamido]sulphur pentafluoride	89590-17-0	NI28619
261 279 205 78 233 262 51 79	279	100 75 38 27 14 13 9 7		7	6	3	1	–	–	–	–	1	–	–	–	–	Salicylic acid, 3-amino-5-iodo-	22738-81-4	NI28620
110 109 81 111 127 53 82 54	279	100 21 21 15 14 12 12 12	1.20	7	6	3	1	1	5	–	–	–	–	–	–	–	[(4-Hydroxyphenoxy)formamido]sulphur pentafluoride	90597-98-1	NI28621
279 59 117 214 281 246 45 28	279	100 54 48 29 20 20 17 16		9	13	1	1	4	–	–	–	–	–	–	–	–	2,4,6,8-Tetrathiatricyclo[3.3.1.1^{3,7}]decane-1-carboxamide, 3,5-dimethyl-, (±)-	57289-09-5	NI28622
137 142 136 42 122 127 94 65	279	100 89 57 46 43 36 31 19	0.00	9	14	1	1	–	–	–	–	1	–	–	–	–	Anilinium, 2-hydroxy-N,N,N-trimethyl-, iodide	21405-07-2	NI28623
136 137 142 121 94 65 42 39	279	100 76 15 11 7 7 7 7	0.00	9	14	1	1	–	–	–	–	1	–	–	–	–	Anilinium, 3-hydroxy-N,N,N-trimethyl-, iodide	2498-27-3	NI28624
136 137 142 121 127 120 65 39	279	100 78 69 25 21 12 12 10	0.00	9	14	1	1	–	–	–	–	1	–	–	–	–	Anilinium, 4-hydroxy-N,N,N-trimethyl-, iodide	6545-97-7	NI28625
222 44 86 236 43 28 279 264	279	100 63 37 32 25 16 14 10		10	15	1	1	–	6	–	–	–	–	–	–	–	Propanamide, 3,3,3-trifluoro-N,N-diisopropyl-2-(trifluoromethyl)-	50837-72-4	NI28626
60 73 43 44 61 41 57 69	279	100 93 91 78 61 58 55 54	0.00	10	17	6	1	1	–	–	–	–	–	–	–	–	Desulphosinigrin		MS06337
128 96 95 137 94 127 142 41	279	100 97 96 93 92 91 82 58	1.70	10	18	–	1	–	–	–	–	1	–	–	–	–	Norazadamantane, N-methylammonium iodide		LI06002
73 60 57 71 43 155 55 41	279	100 92 59 58 56 55 52 44	0.30	10	20	2	–	–	–	–	–	–	–	–	–	1	Decanoic acid, silver(1+) salt	13126-67-5	NI28627
167 152 168 69 60 151 43 41	279	100 21 15 15 15 14 12 12	0.00	11	12	4	1	–	3	–	–	–	–	–	–	–	Acetamide, 2,2,2-trifluoro-N-[2-hydroxy-2-(4-hydroxy-3-methoxyphenyl)ethyl]-	56805-13-1	MS06338
207 81 67 279 150 72 82 208	279	100 43 40 39 31 30 16 15		11	13	2	5	1	–	–	–	–	–	–	–	–	1H-Purine-2,6-dione, 7-ethyl-3,7-dihydro-8-(isothiocyanatomethyl)-1,3-dimethyl-	27042-75-7	NI28628
264 73 279 265 45 192 84 74	279	100 66 28 25 15 10 10 10		11	21	–	5	–	–	–	–	–	2	–	–	–	Adenine, 6,9-bis(trimethylsilyl)-		MS06339
264 73 279 265 75 192 147 266	279	100 43 28 24 22 10 10 9		11	21	–	5	–	–	–	–	–	2	–	–	–	Adenine, N,7-bis(trimethylsilyl)-	17963-53-0	NI28629
93 264 73 75 95 279 77 265	279	100 49 47 43 36 20 18 12		11	21	–	5	–	–	–	–	–	2	–	–	–	Adenine, N,7-bis(trimethylsilyl)-	17963-53-0	NI28630
279 221 107 249 280 234 75 91	279	100 19 18 16 14 10 9 7		12	9	7	1	–	–	–	–	–	–	–	–	–	4H-1-Benzopyran-2-carboxylic acid, 6-hydroxy-5-nitro-4-oxo-, ethyl ester	30095-84-2	NI28631
279 251 234 280 233 221 205 206	279	100 28 16 14 13 12 12 11		12	9	7	1	–	–	–	–	–	–	–	–	–	4H-1-Benzopyran-2-carboxylic acid, 7-hydroxy-6-nitro-4-oxo-, ethyl ester	5823-37-0	NI28632
279 251 233 234 76 280 221 249	279	100 82 44 21 15 14 14 11		12	9	7	1	–	–	–	–	–	–	–	–	–	4H-1-Benzopyran-2-carboxylic acid, 7-hydroxy-8-nitro-4-oxo-, ethyl ester	30192-14-4	NI28633
279 281 127 251 253 234 236 29	279	100 95 77 57 57 51 47 46		12	10	2	1	–	–	–	1	–	–	–	–	–	2-Propenoic acid, 2-cyano-3-(3-bromophenyl)-, ethyl ester	24398-79-6	NI28634

MASS TO CHARGE RATIOS								M.W.	INTENSITIES								Parent	C	H	O	N	S	F	Cl	Br	I	Si	P	B	X	COMPOUND NAME	CAS Reg No	No
279	281	127	234	236	251	253	155	279	100	98	76	54	51	51	51	43		12	10	2	1	–	–	–	1	–	–	–	–	–	2-Propenoic acid, 2-cyano-3-(4-bromophenyl)-, ethyl ester	18861-58-0	NI28635
64	44	197	48	198	196	199	107	279	100	44	36	27	27	24	22	17	0.00	12	13	3	3	1	–	–	–	–	–	–	–	–	Benzenesulphonic acid, 5-amino-2-[(4-aminophenyl)amino]-	119-70-0	NI28636
56	86	279	110	221	109	77	55	279	100	72	33	32	22	19	15	13		12	13	3	3	1	–	–	–	–	–	–	–	–	Sydnone, 3-(4-morpholinyl)-4-(phenylthio)-	68486-71-5	NI28637
111	127	126	129	279	83	152	281	279	100	96	57	30	22	12	9	8		13	10	2	1	1	–	1	–	–	–	–	–	–	N-(3-Chlorophenyl)-2-(thien-2-oyl)acetamide		LI06003
111	127	126	129	97	83	279	153	279	100	99	46	31	28	28	25	11		13	10	2	1	1	–	1	–	–	–	–	–	–	N-(3-Chlorophenyl)-2-(thien-3-oyl)acetamide		LI06004
111	127	126	279	129	83	281	97	279	100	73	67	32	24	13	11	7		13	10	2	1	1	–	1	–	–	–	–	–	–	N-(4-Chlorophenyl)-2-(thien-2-oyl)acetamide		LI06005
111	127	83	126	279	129	281	97	279	100	88	80	54	28	28	10	8		13	10	2	1	1	–	1	–	–	–	–	–	–	N-(4-Chlorophenyl)-2-(thien-3-oyl)acetamide		LI06006
127	128	88	42	279	191	58	126	279	100	67	42	41	20	18	13	11		13	13	4	1	1	–	–	–	–	–	–	–	–	1,3,5-Dioxazine, 5-(2-naphthylsulphonyl)dihydro-	56221-13-7	NI28638
205	187	132	104	279	172	130	200	279	100	95	40	18	10	10	10	5		13	13	4	1	1	–	–	–	–	–	–	–	–	2H-Isoindole-2-acetic acid, 1,3-dihydro-α-[2-(methylthio)ethyl]-1,3-dioxo-, (±)-		LI06007
187	205	132	61	104	76	130	186	279	100	93	67	61	46	38	23	22	6.64	13	13	4	1	1	–	–	–	–	–	–	–	–	2H-Isoindole-2-acetic acid, 1,3-dihydro-α-[2-(methylthio)ethyl]-1,3-dioxo-, (±)-	52881-96-6	NI28639
43	208	55	71	70	128	181	210	279	100	67	56	54	30	26	25	24	5.00	13	14	2	3	–	–	1	–	–	–	–	–	–	1-(4-Chlorophenoxy)-3-methyl-1-(1,2,4-triazol-1-yl)butanone		MS06340
155	96	139	91	140	89	118	144	279	100	96	95	95	26	20	18	12	0.00	13	17	2	3	1	–	–	–	–	–	–	–	–	1,2,3-Triazole, 5-butyl-1-(4-methylphenylsulphonyl)-		DD00931
279	77	132	188	29	250	46	278	279	100	86	36	32	23	16	16	12		14	13	–	7	–	–	–	–	–	–	–	–	–	1H-Pyrazolo[4,3-e][1,2,4]triazolo[4,3-a][1,2,4]triazine, 8-ethyl-3-methyl-1-phenyl-		NI28640
223	279	232	164	149	237	250	121	279	100	78	61	48	43	30	22	17		14	17	1	1	2	–	–	–	–	–	–	–	–	Thiazole, 2-(butylthio)-4-(4-methoxyphenyl)-	69426-52-4	NI28641
77	220	279	117	119	94	141	41	279	100	99	89	73	72	70	56	56		14	17	3	1	1	–	–	–	–	–	–	–	–	2-Oxa-6-azatricyclo[3.3.1.1^{3,7}]decane, 6-(phenylsulphonyl)-	50267-22-6	NI28642
237	174	158	279	128	91	77	105	279	100	66	65	30	19	10	10	8		14	17	3	1	1	–	–	–	–	–	–	–	–	6,7-Cyclopentenoindole, N-acetyl-2,3-dihydro-4-(methylsulphonyl)-		MS06341
43	15	135	30	77	51	42	39	279	100	28	14	9	8	7	7	6	0.00	14	17	5	1	–	–	–	–	–	–	–	–	–	Acetamide, N-[2-(acetyloxy)-2-[4-(acetyloxy)phenyl]ethyl]-	55044-38-7	NI28643
149	220	43	88	135	178	150	77	279	100	99	64	25	21	19	19	18	2.11	14	17	5	1	–	–	–	–	–	–	–	–	–	D-Alanine, N-acetyl-3-(3-formyl-4-methoxyphenyl)-, methyl ester		NI28644
234	148	215	221	71	159	189	231	279	100	70	49	41	40	31	30	28	0.00	14	17	5	1	–	–	–	–	–	–	–	–	–	1(2H)-Naphthalenone, 3,4-dihydro-2-(3-hydroxypropyl)-6-methoxy-2-nitro-		DD00932
162	91	88	120	131	59	163	101	279	100	93	86	66	48	30	28	20	3.00	14	17	5	1	–	–	–	–	–	–	–	–	–	Phenylalanine, N-malonyl-N-methyl-, methyl ester		LI06008
105	77	207	206	106	161	134	28	279	100	23	15	10	10	8	7	7	4.15	14	17	5	1	–	–	–	–	–	–	–	–	–	Propanedioic acid, benzamido-, diethyl ester		NI28645
89	107	130	149	113	173	43	160	279	100	51	50	43	38	25	22	21	5.30	14	17	5	1	–	–	–	–	–	–	–	–	–	Serine, O,N-diacetyl-β-phenyl-, methyl ester		NI28646
170	171	42	199	128	115	77	57	279	100	23	13	11	7	6	5	5	0.00	14	18	–	1	–	–	–	1	–	–	–	–	–	3-Phenyltropidine hydrobromide		MS06342
170	171	42	199	115	128	85	91	279	100	30	19	14	9	8	7	6	0.00	14	18	–	1	–	–	–	1	–	–	–	–	–	3α-Phenyltropidine hydrobromide		MS06343
170	171	199	42	128	115	142	94	279	100	39	16	16	12	10	7	7	0.00	14	18	–	1	–	–	–	1	–	–	–	–	–	3β-Phenyltropidine hydrobromide		MS06344
91	73	77	45	92	75	65	74	279	100	21	15	13	8	8	8	7	0.00	14	21	3	1	–	–	–	–	–	1	–	–	–	Butanoic acid, 2-[(phenylmethoxy)imino]-, trimethylsilyl ester	55520-91-7	NI28647
73	75	104	91	105	148	131	189	279	100	68	58	37	25	24	20	18	10.00	14	21	3	1	–	–	–	–	–	1	–	–	–	Glycine, phenylpropionyl-, trimethylsilyl-	85538-32-5	NI28648
72	180	84	135	44	57	45	41	279	100	55	24	21	14	13	13	12	2.16	14	21	3	3	–	–	–	–	–	–	–	–	–	Carbamic acid, tert-butyl-, 3-[[(dimethylamino)carbonyl]amino]phenyl ester	4849-32-5	NI28650
72	180	135	45	44	84	57	41	279	100	72	38	22	21	20	18	11	1.58	14	21	3	3	–	–	–	–	–	–	–	–	–	Carbamic acid, tert-butyl-, 3-[[(dimethylamino)carbonyl]amino]phenyl ester	4849-32-5	NI28649
125	154	127	156	89	36			279	100	55	38	19	10	10			0.00	15	15	–	1	–	–	2	–	–	–	–	–	–	Benzeneethanamine, 4-chloro-N-(4-chlorophenyl)-		LI06009
179	135	207	163	136	134	189	161	279	100	49	42	39	37	35	27	27	1.90	15	21	4	1	–	–	–	–	–	–	–	–	–	2-(1-Adamantyl)-2-ethoxyoxazolidine-4,5-dione		LI06010
208	206	209	280	204	207	165	178	279	100	39	13	10	8	6	4	3	0.00	15	25	2	1	–	–	–	–	–	1	–	–	–	Salsolidine, (trimethylsilyl)-		NI28651
188	73	100	59	189	91	45	190	279	100	74	44	24	19	14	14	13	0.00	15	29	–	1	–	–	–	–	–	2	–	–	–	Silanamine, 1,1,1-trimethyl-N-(1-methyl-2-phenylethyl)-N-(trimethylsilyl)-, (±)-	57346-67-5	NI28652
209	279	219	220	237	236	43	65	279	100	77	74	58	52	52	39	27		16	13	2	3	–	–	–	–	–	–	–	–	–	1-Acetyl-2-oxoindoline-3-spiro-2′-(1′,2′-dihydrobenzimidazole)		LI06011
279	148	147	278	280	119	133	105	279	100	57	40	22	20	19	18	18		16	13	2	3	–	–	–	–	–	–	–	–	–	2-Aminodiftalone		MS06345
279	118	36	147	105	104	134	90	279	100	70	70	56	48	38	30	30		16	13	2	3	–	–	–	–	–	–	–	–	–	3-Aminodiftalone		MS06346
91	194	92	279	119	195	280		279	100	20	8	5	4	3	1			16	13	2	3	–	–	–	–	–	–	–	–	–	Δ^2-1,2,4-Triazolin-5-one, 4-acetyl-1,3-diphenyl-	32589-67-6	NI28653
91	117	279	90	118	78	119	103	279	100	80	77	49	47	38	35	35		16	13	2	3	–	–	–	–	–	–	–	–	–	4-Phenyl-2,4,6-triaza[5.4.2.0^{2,6}]-tricyclotrideca-8,10,12-triene-3,5-dione		LI06012
165	262	279	178	190	166	191	179	279	100	70	45	31	27	25	18	18		16	13	2	3	–	–	–	–	–	–	–	–	–	as-Triazine, 3-methoxy-5,6-diphenyl-, 2-oxide	13332-65-5	LI06013
165	262	279	178	190	166	179	176	279	100	70	45	31	27	25	18	18		16	13	2	3	–	–	–	–	–	–	–	–	–	as-Triazine, 3-methoxy-5,6-diphenyl-, 2-oxide	13332-65-5	NI28654
41	262	43	174	192	55	42	279	279	100	77	65	58	57	54	52	50		16	25	3	1	–	–	–	–	–	–	–	–	–	Annotinol, dihydro-	5096-60-6	NI28655
41	93	55	91	77	39	96	79	279	100	96	91	72	58	52	44	44	2.00	16	25	3	1	–	–	–	–	–	–	–	–	–	1-Cyclohexene-4-carboxylic acid, 1-methyl-3-(1,3,4,5,6,7-hexahydro-2H-azepin-1-yl)-, ethyl ester		MS06347
235	112	208	44	43	194	279	220	279	100	46	44	28	20	18	14	14		16	25	3	1	–	–	–	–	–	–	–	–	–	Dendroban-12-one, 10-hydroxy-	29414-86-6	LI06014
236	112	208	44	43	194	220	279	279	100	47	45	28	22	21	17	16		16	25	3	1	–	–	–	–	–	–	–	–	–	Dendroban-12-one, 10-hydroxy-	29414-86-6	NI28657
236	208	112	28	194	279	44	237	279	100	42	40	25	18	17	17	15		16	25	3	1	–	–	–	–	–	–	–	–	–	Dendroban-12-one, 10-hydroxy-	29414-86-6	NI28656
192	174	175	193	208	146	190	176	279	100	30	24	20	16	15	10	8	6.45	16	25	3	1	–	–	–	–	–	–	–	–	–	Lycopodan-8-one, 5,12-dihydroxy-15-methyl-, (5β,15S)-	3175-92-6	NI28658
164	236	176	165	237	279	247	204	279	100	71	17	13	12	11	11	10		16	25	3	1	–	–	–	–	–	–	–	–	–	Methyl α-(2-piperidinocyclohexenyl)acetoacetate		MS06348
123	85	178	195	136	194	179	279	279	100	99	89	70	58	48	22	13		16	25	3	1	–	–	–	–	–	–	–	–	–	Pyridine, 2-methoxy-6-[5-[(tetrahydro-2h-pyran-2-yl)oxy]pentyl]-		NI28659

MASS TO CHARGE RATIOS								M.W.	INTENSITIES								Parent	C	H	O	N	S	F	Cl	Br	I	Si	P	B	X	COMPOUND NAME	CAS Reg No	No
152	251	153	150	234	123	99	122	**279**	100	25	13	13	12	8	7	6	**4.00**	16	25	3	1	–	–	–	–	–	–	–	–	–	**Serratinan-5-one, 8,13-dihydroxy-, (8α,13S)-**	5545-99-3	**LI06015**
152	251	153	150	234	123	99	110	**279**	100	24	13	13	12	8	7	6	**1.49**	16	25	3	1	–	–	–	–	–	–	–	–	–	**Serratinan-5-one, 8,13-dihydroxy-, (8α,13S)-**	5545-99-3	**NI28660**
279	32	251	77	192	252	265	264	**279**	100	73	64	15	12	12	12	9		17	13	3	1	–	–	–	–	–	–	–	–	–	**4(1H)-Quinolinone, 2-(1,3-benzodioxol-5-yl)-1-methyl-**	485-61-0	**NI28661**
83	159	144	77	93	115	51	91	**279**	100	67	48	36	29	18	16	15	**7.00**	17	17	1	3	–	–	–	–	–	–	–	–	–	**3-Buten-2-one, 4-phenyl-, 4-phenylsemicarbazone**	.17014-34-5	**NI28662**
136	279	135	89	144	120	116	115	**279**	100	33	24	12	7	4	4	3		17	17	1	3	–	–	–	–	–	–	–	–	–	**4′-(Dimethylamino)indole-2-carboxanilide**		**LI06016**
200	91	243	84	242	186	166	201	**279**	100	48	26	25	23	21	18	17	**0.00**	17	26	–	1	–	–	1	–	–	–	–	–	–	**Piperidine, 1-(1-phenylcyclohexyl)-, hydrochloride**	956-90-1	**NI28663**
161	91	162	90	118	89	65	63	**279**	100	35	22	18	10	7	6	3	**0.00**	18	17	2	1	–	–	–	–	–	–	–	–	–	**Benzoic acid, 4-butyl-, 4-cyanophenyl ester**	38690-77-6	**NI28664**
109	279	81	143	169	79	251	201	**279**	100	64	40	20	11	11	5	5		18	17	2	1	–	–	–	–	–	–	–	–	–	**Cyclohexanecarboxamide, 1-hydroxy-N-(1-naphthyl)-**		**LI06017**
109	279	81	143	280	115	169	79	**279**	100	65	40	20	14	13	11	11		18	17	2	1	–	–	–	–	–	–	–	–	–	**2-Cyclohexenylglyoxamide, N-(1-naphthyl)-**		**MS06349**
105	182	77	183	43	97	220	160	**279**	100	64	32	29	16	15	12	11	**0.00**	18	17	2	1	–	–	–	–	–	–	–	–	–	**3-Aza-4,6-dioxabicyclo[3.2.0]hept-2-ene, 2,5-dimethyl-7,7-diphenyl-**		**NI28665**
91	222	194	57	29	193	279	77	**279**	100	71	57	33	18	16	15	15		18	17	2	1	–	–	–	–	–	–	–	–	–	**3,4-Diphenyl-5-propionyl-2-isoxazoline**		**NI28666**
57	146	223	29	222	77	105	206	**279**	100	86	81	68	58	41	25	24	**5.00**	18	17	2	1	–	–	–	–	–	–	–	–	–	**3,5-Diphenyl-4-propionyl-2-isoxazoline**		**NI28667**
239	279	278	240	197	280	41	45	**279**	100	50	36	29	23	15	8	6		18	21	–	3	–	–	–	–	–	–	–	–	–	**Indolo[2,3-a]quinolizine, 1-(2-cyanoethyl)-1,2,3,4,6,7,12,12b-octahydro-**		**NI28668**
239	279	278	240	169	197	170	41	**279**	100	43	34	19	16	15	13	12		18	21	–	3	–	–	–	–	–	–	–	–	–	**Indolo[2,3-a]quinolizine, 1-(2-cyanoethyl)-1,2,3,4,6,7,12,12b-octahydro-**		**NI28669**
85	113	126	55	70	279	41	43	**279**	100	94	57	40	28	25	21	17		18	33	1	1	–	–	–	–	–	–	–	–	–	**Pyrrolidine, 1-(1-oxo-9-tetradecenyl)-**	56600-04-5	**NI28670**
132	91	117	118	77	105	133	115	**279**	100	68	48	26	13	12	11	11	**1.00**	19	21	1	1	–	–	–	–	–	–	–	–	–	**2-Azetidinone, 3,3-dimethyl-4-phenyl-1-(2-phenylethyl)-**	54965-33-2	**NI28671**
181	180	279	145	77	182	280	160	**279**	100	32	27	25	20	17	10	10		19	21	1	1	–	–	–	–	–	–	–	–	–	**2-Azetidinone, 3-tert-butyl-1,4-diphenyl-, cis-**	17324-22-0	**NI28672**
145	160	181	279	180	77	146	91	**279**	100	60	45	41	31	31	17	17		19	21	1	1	–	–	–	–	–	–	–	–	–	**2-Azetidinone, 3-tert-butyl-1,4-diphenyl-, trans-**	20903-59-7	**NI28673**
105	77	222	206	279	223	118	91	**279**	100	44	40	38	35	28	28	21		19	21	1	1	–	–	–	–	–	–	–	–	–	**Aziridine, 1-benzoyl-1-tert-butyl-3-phenyl-**	20847-27-2	**NI28674**
91	121	277	49	170	65	186	77	**279**	100	27	23	23	18	18	16	12	**0.00**	19	21	1	1	–	–	–	–	–	–	–	–	–	**1-Benzyl-2-(4-methoxybenzyl)-2,5-dihydropyrrole**		**LI06018**
105	77	28	106	51	91	196	146	**279**	100	26	14	8	6	5	4	3	**0.00**	19	21	1	1	–	–	–	–	–	–	–	–	–	**1,3-Diphenyl-2-morpholinopropene**		**MS06350**
194	279	133	118	195	146	280	150	**279**	100	87	38	30	28	23	22	20		19	21	1	1	–	–	–	–	–	–	–	–	–	**2,6-Diphenyl-4-piperidone**		**IC03610**
279	264	224	248	180	222	208	204	**279**	100	40	35	25	25	20	20	20		19	21	1	1	–	–	–	–	–	–	–	–	–	**1-Methoxy-2-(3-methylbut-2-enyl)-3-methylcarbazole**		**NI28675**
118	105	77	117	162	103	78	51	**279**	100	83	60	21	16	13	13	12	**4.94**	19	21	1	1	–	–	–	–	–	–	–	–	–	**4H-1,3-Oxazine, 4,4,6-trimethyl-2,6-diphenyl-5,6-dihydro-**	91875-66-0	**NI28676**
58	59	42	30	165	178	43	57	**279**	100	6	5	3	2	2	2	1	**0.00**	19	21	1	1	–	–	–	–	–	–	–	–	–	**1-Propanamine, 3-dibenz[b,e]oxepin-11(6H)-ylidene-N,N-dimethyl-**	1668-19-5	**NI28677**
66	148	213	147	79	77	120	91	**279**	100	51	30	24	20	12	11	10	**3.00**	19	21	1	1	–	–	–	–	–	–	–	–	–	**2-Propenenitrile, 3-[3-(bicyclo[2.2.1]hept-5-en-2-ylcarbonyl)bicyclo[2.2.1]hept-5-en-2-yl]-2-methyl-, [1α,2α(Z),3α(1R*,2S*,4R),4α]-**	67401-03-0	**NI28678**
105	77	146	96	182	97	264	279	**279**	100	58	42	37	34	34	6	2		19	21	1	1	–	–	–	–	–	–	–	–	–	**1-Pyrazoline, 2-(2-hydroxy-1-methyl-2,2-diphenylethyl)-**		**NI28679**
237	251	279	280	184	156	208	223	**279**	100	94	76	48	47	43	32	29		19	23	–	2	–	–	–	–	–	–	–	–	–	**3,4-Dehydroquebrachamine**		**LI06019**
168	180	96	236	181	208	279	110	**279**	100	83	80	71	40	39	28	23		19	37	–	1	–	–	–	–	–	–	–	–	–	**1-Azaspiro[5.5]undecane, 7-butyl-2-pentyl-, [6R-[6α(S*),7β]]-**	63983-65-3	**NI28680**
43	41	57	29	55	97	70	71	**279**	100	84	78	54	51	41	28	24	**2.20**	19	37	–	1	–	–	–	–	–	–	–	–	–	**Nonadecanenitrile**	28623-46-3	**WI01344**
152	236	237	153	67	69	82	70	**279**	100	68	12	12	12	8	6	6	**4.00**	19	37	–	1	–	–	–	–	–	–	–	–	–	**Pyrrolizidine, 3-nonyl-5-propyl-**		**NI28681**
129	91	148	44	128	220	106	104	**279**	100	83	82	48	43	42	34	32	**1.40**	20	25	–	1	–	–	–	–	–	–	–	–	–	**2-Naphthalenamine, 1,2,3,4-tetrahydro-N-isopropyl-N-benzyl-**	54965-39-8	**WI01345**
279	280	139.5	277	278	139	125	140	**279**	100	23	14	9	8	6	4	3		21	13	–	1	–	–	–	–	–	–	–	–	–	**Dibenz[a,j]acridine**	224-42-0	**AI01960**
279	280	277	278	139	125	57	43	**279**	100	22	13	11	6	5	5	5		21	13	–	1	–	–	–	–	–	–	–	–	–	**Dibenz[a,j]acridine**	224-42-0	**NI28682**
279	280	277	278	43	139	125	41	**279**	100	23	13	11	7	6	5	5		21	13	–	1	–	–	–	–	–	–	–	–	–	**Dibenz[a,j]acridine**	224-42-0	**NI28683**
203	121	93	123	95	201	205	31	**280**	100	77	74	56	54	53	50	33	**0.00**	2	3	1	–	–	–	–	3	–	–	–	–	–	**Ethanol, 2,2,2-tribromo-**	75-80-9	**NI28684**
69	119	147	117	64	100	97	31	**280**	100	41	16	14	10	6	6	6	**0.00**	4	–	3	–	1	8	–	–	–	–	–	–	–	**Trifluoromethylsulphinyl pentafluoropropionate**		**LI06020**
55	140	167	56	280	83	84	75	**280**	100	65	30	21	19	19	15	11		5	–	5	–	–	3	–	–	–	1	–	–	1	**Manganese, (trifluorosilyl)pentacarbonyl-**		**LI06021**
282	247	284	245	280	249	79	246	**280**	100	82	67	61	60	42	39	35		6	1	–	–	1	–	5	–	–	–	–	–	–	**Benzenethiol, pentachloro-**	133-49-3	**NI28686**
282	284	280	247	245	212	286	249	**280**	100	73	59	48	34	29	28	27		6	1	–	–	1	–	5	–	–	–	–	–	–	**Benzenethiol, pentachloro-**	133-49-3	**NI28685**
69	166	53	39	167	41	27	153	**280**	100	51	44	29	27	25	25	24	**0.00**	8	6	4	–	–	6	–	–	–	–	–	–	–	**1,4-Bis(trifluoroacetoxy)but-1-ene**		**MS06351**
77	249	280	156	51	50	91	117	**280**	100	92	51	30	24	24	18	16		8	8	1	–	–	–	–	–	–	–	–	–	2	**2-Methoxy-1,3-benzodiselenole**		**DD00933**
80	63	139	245	142	218	280	109	**280**	100	96	86	74	72	67	45	40		8	10	5	2	–	–	1	–	–	–	1	–	–	**2-Chloroethyl (4-nitrophenyl)phosphoramidate**		**MS06352**
29	43	55	41	83	56	27	39	**280**	100	97	92	92	83	81	75	46	**0.00**	8	13	1	–	–	4	–	1	–	–	–	–	–	**3-Octanol, 1-bromo-1,1,2,2-tetrafluoro-**	74810-73-4	**NI28687**
41	69	43	29	42	27	87	95	**280**	100	86	73	72	59	52	46	40	**0.00**	8	13	1	–	–	4	–	1	–	–	–	–	–	**4-Octanol, 1-bromo-1,1,2,2-tetrafluoro-**	74825-24-4	**NI28688**
140	43	28	139	60	125	110	30	**280**	100	69	69	54	44	38	38	37	**8.00**	8	28	–	8	–	–	–	–	–	–	–	4	–	**Octamethyl hydrazinoborane tetramer**		**NI28689**
206	194	250	278	128	222	66	39	**280**	100	41	36	34	28	20	20	13		9	10	2	–	1	–	–	–	–	–	–	–	1	**π-Cyclopentadienyl-π-2-thiaprop-1-enyldicarbonyl molybdenum**		**MS06353**
280	59	117	215	247	282	45	103	**280**	100	58	37	35	20	18	17	14		9	12	2	–	4	–	–	–	–	–	–	–	–	**2,4,6,8-Tetrathiatricyclo[3.3.1.1^{3,7}]decane-1-carboxylic acid, 3,5-dimethyl-**	55955-60-7	**NI28690**
138	110	171	109	156	143	139	127	**280**	100	24	22	16	15	15	13	13	**8.33**	9	13	6	–	1	–	–	–	–	–	1	–	–	**Phosphorothioic acid, S-[(5-methoxy-4-oxo-4H-pyran-2-yl)methyl] O,O-dimethyl ester**	2778-04-3	**NI28691**
170	245	155	247	280	171	141	282	**280**	100	94	56	35	32	30	24	19		9	14	2	4	–	–	2	–	–	–	–	–	–	**1,3,5-Triazin-2-amine, 4-(2,3-dichloropropoxy)-6-methoxy-N,N-dimethyl-**	74468-17-0	**NI28692**
149	207	205	151	147	251	179	177	**280**	100	92	70	70	70	66	65	48	**0.00**	9	20	2	–	–	–	–	–	–	–	–	–	1	**Triethyl-ethylene(methoxycarbonyl)stannane**		**LI06022**

MASS TO CHARGE RATIOS	M.W.	INTENSITIES	Parent	C	H	O	N	S	F	Cl	Br	I	Si	P	B	X	COMPOUND NAME	CAS Reg No	No
43 251 209 249 207 149 151 74	**280**	100 80 70 60 60 57 56 56	**0.00**	9	20	2	–	–	–	–	–	–	–	–	–	1	**Triethyl-methylene-acetoxy-stannane**		**LI06023**
195 251 223 119 105 133 209 89	**280**	100 96 90 54 49 46 35 31	**19.42**	9	24	–	–	–	–	–	–	–	–	–	–	2	**Digermane, 1,1,1-triethyl-2,2,2-trimethyl-**	7788-37-6	**NI28693**
76 140 75 64 39	**280**	100 43 33 28 26	**0.00**	10	8	–	–	–	8	–	–	–	–	–	–	–	**2,2,3,3,9,9,10,10-Octafluorodispiro[3.2.3.0]decane**		**LI06024**
249 183 198 218 203 143 248 153	**280**	100 62 55 51 46 46 43 38	**0.00**	10	8	4	4	1	–	–	–	–	–	–	–	–	**2H-1,2,3-Thiadiazine, 2-(2,4-dinitrophenyl)-6-methyl-**	57954-47-9	**NI28694**
135 28 44 100 108 56 54 29	**280**	100 67 54 29 28 24 20 18	**0.00**	10	12	4	6	–	–	–	–	–	–	–	–	–	**L-Threonine, N-[(1H-purin-6-ylamino)carbonyl]-**	33422-66-1	**NI28695**
121 77 55 141 73 47 41 91	**280**	100 91 91 74 45 38 27 26	**0.00**	10	14	2	–	–	6	–	–	–	–	–	–	–	**5-Nonanol, 1,1,1,9,9,9-hexafluoro-, formate**	54699-53-5	**NI28696**
91 147 51 77 119 65 131 55	**280**	100 66 41 37 24 24 22 22	**5.00**	11	11	3	–	1	3	–	–	–	–	–	–	–	**1,3-Cyclodecadien-5-yn-1-yl triflate**		**MS06354**
147 91 280 131 119 115 107 69	**280**	100 71 28 27 20 19 16 16		11	11	3	–	1	3	–	–	–	–	–	–	–	**1,8-Cyclodecadien-6-yn-1-yl triflate**		**MS06355**
147 91 280 131 119 107 69 65	**280**	100 66 40 29 24 18 16 13		11	11	3	–	1	3	–	–	–	–	–	–	–	**5,6,7,8-Tetrahydro-1-naphthyl triflate**		**MS06356**
281 216 156 188 126	**280**	100 3 3 2 2	**0.00**	11	12	3	4	1	–	–	–	–	–	–	–	–	**Benzenesulphonamide, 4-amino-N-(6-methoxy-3-pyridazinyl)-**	80-35-3	**NI28697**
81 111 41 79 39 140 80 53	**280**	100 81 50 47 43 34 34 34	**1.58**	11	12	5	–	–	–	–	–	–	–	–	–	1	**Iron, tricarbonyl[(2,3,4,5-η)-methyl 2,4-heptadienoate]-**	74779-86-5	**NI28698**
28 43 83 125 196 41 194 56	**280**	100 43 28 26 23 15 14 14	**2.60**	11	16	3	2	–	–	–	–	–	–	–	–	1	**Iron, tricarbonyl[N,N′-1,2-ethanediylidenebis[2-propanamine]-N,N′]-**	54446-62-7	**NI28699**
117 59 173 280 100 131 132 41	**280**	100 85 83 56 50 35 22 16		11	20	–	--	4	–	–	–	–	–	–	–	–	**2,4,6,8-Tetrathiabicyclo[3.3.1]nonane, 1,3,3,5,7,7-hexamethyl-**	57289-15-3	**NI28700**
265 73 280 266 45 281 193 267	**280**	100 70 41 24 22 12 11 10		11	20	1	4	–	–	–	–	–	2	–	–	–	**9H-Purine, 9-(trimethylsilyl)-6-[(trimethylsilyl)oxy]-**	17962-89-9	**NI28701**
93 73 75 95 265 280 79 77	**280**	100 42 40 35 29 14 10 9		11	20	1	4	–	–	–	–	–	2	–	–	–	**6H-Purin-6-one, 1,7-dihydro-1,7-bis(trimethylsilyl)-**	55622-58-7	**NI28702**
280 152 153 151 76 281 63 51	**280**	100 75 62 25 21 15 12 12		12	9	–	–	–	–	–	–	1	–	–	–	–	**1,1′-Biphenyl, 2-iodo-**		**NI28703**
280 152 281 153 154 151 282 150	**280**	100 24 14 10 5 5 1 1		12	9	–	–	–	–	–	–	1	–	–	–	–	**1,1′-Biphenyl, 4-iodo-**	1591-31-7	**NI28704**
184 280 168 185 92 139 281 186	**280**	100 47 22 15 15 13 7 6		12	9	2	–	2	–	–	–	–	–	1	–	–	**O,O′-(2,2′-Biphenylylene)dithiophosphoric acid**	68560-86-1	**NI28705**
153 280 89 282 152 154 63 56	**280**	100 52 34 30 17 13 13 9		12	10	–	–	–	–	2	–	–	–	–	–	1	**Dichlorovinyl ferrocene**		**LI06025**
141 53 97 27 112 80 52 69	**280**	100 36 22 22 20 20 20 16	**5.03**	12	12	2	2	2	–	–	–	–	–	–	–	–	**3-Pyridinol, 2,2′-dithiobis[6-methyl-**	23003-31-8	**NI28706**
156 44 155 28 157 159 145 158	**280**	100 83 81 80 73 68 68 61	**13.99**	12	13	2	4	–	–	1	–	–	–	–	–	–	**4-Pyrazolecarboxylic acid, 5(3)-amino-3(5)-(4-chlorophenylamino)-, ethyl ester**	112606-35-6	**NI28707**
83 55 41 206 43 280 77 159	**280**	100 73 48 33 33 26 24 17		12	16	4	4	–	–	–	–	–	–	–	–	–	**Hexanal, (2,4-dinitrophenyl)hydrazone**	1527-97-5	**NI28708**
83 55 41 206 43 77 280 149	**280**	100 71 60 40 33 31 27 22		12	16	4	4	–	–	–	–	–	–	–	–	–	**Hexanal, (2,4-dinitrophenyl)hydrazone**	1527-97-5	**IC03611**
43 41 42 55 29 28 178 83	**280**	100 66 50 42 40 35 30 22	**18.02**	12	16	4	4	–	–	–	–	–	–	–	–	–	**2-Hexanone, (2,4-dinitrophenyl)hydrazone**	2348-17-6	**NI28709**
41 203 55 43 117 220 39 280	**280**	100 97 63 59 57 48 36 34		12	16	4	4	–	–	–	–	–	–	–	–	–	**Pentenal, 2-methyl-, (2,4-dinitrophenyl)hydrazone**		**IC03612**
162 235 135 234 163 280 107 192	**280**	100 34 28 23 23 21 21 19		12	16	4	4	–	–	–	–	–	–	–	–	–	**Purine, 6-hydroxy-9-(2-ethoxy-2-methyl-1,3-dioxan-5-yl)-**	91487-81-9	**NI28710**
151 197 123 179 43 99 69 221	**280**	100 62 54 47 21 15 13 10	**0.00**	12	25	5	–	–	–	–	–	–	–	1	–	–	**Butanoic acid, 2-(diisopropoxyphosphinyl)-, ethyl ester**		**DD00934**
207 179 123 29 138 41 151 81	**280**	100 78 55 45 40 38 35 31	**4.00**	12	25	5	–	–	–	–	–	–	–	1	–	–	**Diethyl (1-butoxycarbonylpropyl)phosphonate**		**LI06026**
193 152 125 165 179 235 151 68	**280**	100 72 70 63 60 58 50 48	**3.49**	12	25	5	–	–	–	–	–	–	–	1	–	–	**Hexanoic acid, 6-(diethoxyphosphinyl)-, ethyl ester**	54965-29-6	**NI28711**
127 28 129 153 90 155 280 65	**280**	100 37 32 30 11 10 8 8		13	10	1	2	–	–	2	–	–	–	–	–	–	**N,N′-Bis(2-chlorophenyl)urea**		**MS06357**
128 129 127 130 131 153 111 75	**280**	100 75 45 33 18 16 13 11	**3.75**	13	10	1	2	–	–	2	–	–	–	–	–	–	**N,N′-Bis(3-chlorophenyl)urea**	13208-31-6	**NI28712**
262 264 152 75 280 76 74 126	**280**	100 75 59 55 40 39 37 35		13	10	1	2	–	–	2	–	–	–	–	–	–	**5H-Dibenzo[c,f][1,2]diazepin-11-ol, 3,8-dichloro-6,11-dihydro-**	23469-60-5	**NI28713**
143 116 203 30 144 131 171 130	**280**	100 61 31 31 24 23 18 18	**0.00**	13	13	2	–	–	–	–	1	–	–	–	–	–	**1H-Cyclopenta[3,4]cyclobuta[1,2]benzene-1-carboxylic acid, 3-bromo-2,3,3a,7b-tetrahydro-, methyl ester**	56771-45-0	**NI28714**
280 123 77 265 187 281 279 170	**280**	100 57 36 30 27 17 14 11		13	13	3	–	1	–	–	–	–	–	1	–	–	**Phosphorothioic acid, O-methyl O,O-diphenyl ester**	5138-73-8	**LI06027**
280 77 109 94 187 171 279 155	**280**	100 52 39 30 26 23 17 16		13	13	3	–	1	–	–	–	–	–	1	–	–	**Phosphorothioic acid, O-methyl O,O-diphenyl ester**	5138-73-8	**NI28715**
211 108 280 91 147 175 213 119	**280**	100 96 80 78 73 70 69 61		13	13	3	2	–	–	1	–	–	–	–	–	–	**1-Chloro-1-[[2-[(4-hydroxy-2-butynyl)oxy]phenyl]hydrazono]acetone**		**MS06358**
91 108 79 44 107 77 84 65	**280**	100 55 49 35 32 28 22 22	**0.00**	13	16	5	2	–	–	–	–	–	–	–	–	–	**L-Glutamine, N²-[(benzyloxy)carbonyl]-**	2650-64-8	**NI28716**
91 108 173 92 280 107 104 90	**280**	100 28 26 15 10 10 9 9		13	16	5	2	–	–	–	–	–	–	–	–	–	**Glycine, (benzyloxycarbonyl)glycyl-, methyl ester**		**LI06028**
43 57 41 55 69 149 71 97	**280**	100 83 72 69 61 44 35 33	**0.00**	13	16	5	2	–	–	–	–	–	–	–	–	–	**(N-Salicoyl-β-alanyl)glycine methyl ester**		**DD00935**
73 280 77 45 282 265 74 281	**280**	100 99 63 42 40 37 33 26		13	17	1	2	–	–	1	–	–	1	–	–	–	**3H-Pyrazol-3-one, 4-chloro-1,2-dihydro-5-methyl-2-phenyl-1-(trimethylsilyl)-**	57396-96-0	**NI28717**
41 56 78 45 191 57 101 55	**280**	100 64 54 49 36 34 29 29	**1.40**	13	28	–	–	3	–	–	–	–	–	–	–	–	**Butane, 1,1′,1′′-[methylidynetris(thio)]tris-**	16754-60-2	**NI28718**
41 56 78 45 191 57 101 55	**280**	100 65 54 48 37 35 29 26	**0.00**	13	28	–	–	3	–	–	–	–	–	–	–	–	**Butane, 1,1′,1′′-[methylidynetris(thio)]tris-**	16754-60-2	**NI28719**
235 237 165 82 280 199	**280**	100 62 62 26 19 18		14	10	2	–	–	–	2	–	–	–	–	–	–	**Benzeneacetic acid, 4-chloro-α-(4-chlorophenyl)-**	83-05-6	**NI28720**
125 169 127 142 171 201 126 217	**280**	100 36 33 17 14 12 9 8	**0.00**	14	10	2	–	–	–	2	–	–	–	–	–	–	**Benzeneacetic acid, 4-chloro-α-(4-chlorophenyl)-**	83-05-6	**NI28721**
139 111 141 75 250 252 113 249	**280**	100 41 33 33 22 15 15 13	**7.40**	14	10	2	–	–	–	2	–	–	–	–	–	–	**[1,1′-Biphenyl]-2-carboxylic acid, 4,4′-dichloro-, methyl ester**	55702-42-6	**NI28722**
210 280 282 181 152 211 245 281	**280**	100 76 51 30 28 15 14 13		14	10	2	–	–	–	2	–	–	–	–	–	–	**1,1-Dichloro-2,2-bis(4-hydroxyphenyl)ethylene**		**NI28723**
112 140 141 280 111 180 97 59	**280**	100 79 77 34 27 24 24 16		14	16	2	–	2	–	–	–	–	–	–	–	–	**4H-Thiopyran-4-one, 2,6-dimethyl-, dimer**	55649-32-6	**NI28724**
149 29 177 76 104 235 50 105	**280**	100 44 36 32 29 21 13 11	**4.00**	14	16	6	–	–	–	–	–	–	–	–	–	–	**1,2-Benzenedicarboxylic acid, 2-ethoxy-2-oxoethyl ethyl ester**	84-72-0	**NI28725**
161 189 193 248 220 216 77 188	**280**	100 51 39 23 23 23 21 20	**16.34**	14	16	6	–	–	–	–	–	–	–	–	–	–	**2-Carbomethoxy-3-hydroxy-4-(3-carbomethoxypropoxy)toluene**		**NI28726**
121 138 166 44 207 29 41 87	**280**	100 66 59 46 44 35 26 25	**17.03**	14	16	6	–	–	–	–	–	–	–	–	–	–	**Malonic acid, 2,2-dimethyl-, carboxyphenyl ethyl ester**		**IC03613**
178 43 107 147 179 15 220 77	**280**	100 90 57 24 16 12 11 6	**0.56**	14	16	6	–	–	–	–	–	–	–	–	–	–	**Methyl 3-(4-acetoxyphenyl)-2-acetoxypropionate**		**MS06359**
178 107 43 147 179 220 77 15	**280**	100 56 54 23 16 13 8 8	**1.00**	14	16	6	–	–	–	–	–	–	–	–	–	–	**Methyl 3-(4-acetoxyphenyl)-2-acetoxypropionate**		**NI28727**
196 43 123 136 238 197 86 29	**280**	100 46 35 26 14 13 12 12	**7.66**	14	16	6	–	–	–	–	–	–	–	–	–	–	**Methyl 3-(3,4-diacetoxyphenyl)propionate**		**MS06360**

MASS TO CHARGE RATIOS	M.W.	INTENSITIES	Parent	C	H	O	N	S	F	Cl	Br	I	Si	P	B	X	COMPOUND NAME	CAS Reg No	No
166 113 99 128 280 168 114 115	280	100 92 82 56 46 35 32 32		14	17	–	2	1	–	1	–	–	–	–	–	–	4H-1,3-Thiazine, 3,4,4,6-tetramethyl-2-(3-chlorophenylimino)-2,3-dihydro-	85298-44-8	NI28728
91 92 81 79 77 95 65 93	280	100 45 39 17 15 13 12 10	0.00	14	17	1	–	–	–	–	1	–	–	–	–	–	Cyclohexene, 1-(benzyloxymethyl)-2-bromo-		DD00936
99 280 139 85 65 140 67 97	280	100 66 61 47 43 36 36 33		14	20	–	2	2	–	–	–	–	–	–	–	–	8-Thiabicyclo[3.2.1]octan-3-one, 8-thiabicyclo[3.2.1]oct-3-ylidenehydrazone	40697-29-8	NI28729
280 191 265 73 221 281 75 151	280	100 92 53 49 30 23 18 13		14	20	4	–	–	–	–	–	–	1	–	–	–	Cinnamic acid, 3,4-dimethoxy-, trimethylsilyl ester	27750-71-6	NI28730
250 280 73 251 281 219 249 117	280	100 85 45 22 20 18 17 15		14	20	4	–	–	–	–	–	–	1	–	–	–	Cinnamic acid, 3-methoxy-4-(trimethylsiloxy)-, methyl ester	27798-70-5	NI28731
280 250 281 265 251 282 249 73	280	100 67 21 21 13 6 5 4		14	20	4	–	–	–	–	–	–	1	–	–	–	Cinnamic acid, 3-methoxy-4-(trimethylsiloxy)-, methyl ester	27798-70-5	NI28732
250 280 73 219 251 265 249 281	280	100 57 35 21 19 14 12 12		14	20	4	–	–	–	–	–	–	1	–	–	–	Cinnamic acid, 3-methoxy-4-(trimethylsiloxy)-, methyl ester	27798-70-5	MS06361
250 280 73 251 281 219 249 59	280	100 90 55 23 20 20 17 17		14	20	4	–	–	–	–	–	–	1	–	–	–	Cinnamic acid, 4-methoxy-3-(trimethylsiloxy)-, methyl ester	32342-05-5	NI28733
137 280 250 195 164 163 136 177	280	100 57 40 39 38 34 28 17		14	20	4	2	–	–	–	–	–	–	–	–	–	1,2-Bis(morpholin-2-yl)-4,5-dihydroxybenzene		IC03614
98 280 252 193	280	100 32 26 11		14	20	4	2	–	–	–	–	–	–	–	–	–	Pyrazolo[1,2-a]pyrazol-1,3,5,7-tetraone, 2,2,6,6-tetraethyl-		LI06029
141 125 42 58 140 84 57 82	280	100 99 61 42 33 27 18 15	6.00	14	24	2	4	–	–	–	–	–	–	–	–	–	4,4-Dimethyl-6-(4,4,6-trimethyl-2-oxo-hexahydropyrimid-6-ylmethylene)-2-oxo-1,2,3,4-tetrahydropyrimidine		LI06030
69 99 100 42 41 140 141 39	280	100 82 71 54 46 36 29 25	11.00	14	24	2	4	–	–	–	–	–	–	–	–	–	1,3,3,3a,6,8,8,8a-Octamethyl-3,3a,8,8a-tetrahydro-diimidazo[1,5-b:1′,5′-e][1.4.2.5]dioxadiazine		MS06362
178 73 179 102 45 180 74 59	280	100 70 23 20 7 6 6 5	3.00	14	28	–	2	–	–	–	–	–	2	–	–	–	3,6-Diaza-2,7-disilaoctane, 2,2,7,7-tetramethyl-3-phenyl-	17814-47-0	NI28734
174 73 175 86 125 176 265 59	280	100 39 19 13 9 8 7 7	4.00	14	28	–	2	–	–	–	–	–	2	–	–	–	Ethylenediamine, N′-phenyl-N,N-bis(trimethylsilyl)-	17814-46-9	NI28735
280 122 180 143 136 165 223 176	280	100 57 51 51 50 36 17 16		14	28	–	4	–	–	–	–	–	1	–	–	–	Hydrazine, 1,1′-[tert-butylphenylsilylene]bis[2,2-dimethyl-	66436-28-0	NI28736
69 43 111 41 85 84 127 27	280	100 78 51 20 16 13 11 11	4.70	14	28	–	6	–	–	–	–	–	–	–	–	–	3,6-Bis(diisopropylamino)-1,2,4,5-tetrazine		MS06363
43 97 55 127 41 70 27 30	280	100 86 77 37 37 29 22 17	8.80	14	28	–	6	–	–	–	–	–	–	–	–	–	3,6-Bis(dipropylamino)-1,2,4,5-tetrazine		MS06364
280 281 77 93 65 91 78 92	280	100 41 36 30 28 24 24 22		15	12	2	4	–	–	–	–	–	–	–	–	–	Pyrazolidinetrione, phenyl-, 4-(phenylhydrazone)	21272-26-4	NI28737
159 233 263 232 42 160 234 158	280	100 68 50 25 23 15 13 13	6.00	15	12	2	4	–	–	–	–	–	–	–	–	–	3,5-Pyridinedicarbonitrile, 4-(2-nitrophenyl)-2,6-dimethyl-1,4-dihydro-	53055-16-6	NI28738
280 209 127 154 281 181 155 128	280	100 47 40 25 19 9 9 9		15	12	2	4	–	–	–	–	–	–	–	–	–	Pyrrolo[1,2-a]-1,3,5-triazine-8-carbonitrile, 3-ethyl-1,2,3,4-tetrahydro-2,4-dioxo-7-phenyl-	54450-43-0	NI28739
280 91 119 281 92 120	280	100 52 30 16 4 3		15	12	2	4	–	–	–	–	–	–	–	–	–	1H-1,2,4-Triazole-3-carboxamide, 4,5-dihydro-5-oxo-N,1-diphenyl-	32589-62-1	NI28740
265 267 231 280 103 77 282 51	280	100 52 52 40 39 24 22 15		15	14	1	–	–	–	2	–	–	–	–	–	–	2,6-Dichloro-4-cumylphenol		IC03615
280 279 281 173 140 239 198 97	280	100 46 18 12 6 5 5 4		15	16	–	6	–	–	–	–	–	–	–	–	–	2-Methyl-4-amino-6-(2-amino-6-methylpyrimid-4-ylamino)quinoline		IC03616
105 185 221 77 223 79 167 157	280	100 94 85 81 35 26 24 22	0.00	15	17	3	–	–	–	1	–	–	–	–	–	–	Benzeneacetic acid, α-(1-chloro-2-cyclohexen-1-yl)-α-hydroxy-, methyl ester		DD00937
280 189 245 217 207 191 190 282	280	100 75 69 69 48 46 46 33		15	17	3	–	–	–	1	–	–	–	–	–	–	Naphtho[1,2-b]furan-2,8(3H)-dione, 7-chloro-3a,4,5,9b-tetrahydro-3,5a,9-trimethyl-, [3S-(3α,3aα,5aβ,9bβ)]-	28136-94-9	NI28741
245 280 207 201 265 210 189 199	280	100 81 80 52 44 31 29 23		15	17	3	–	–	–	1	–	–	–	–	–	–	Naphtho[1,2-b]furan-2,8(3H,4H)-dione, 9-(chloromethyl)-3a,5,5a,9b-tetrahydro-3,5a-dimethyl-, [3S-(3α,3aα,5aβ,9bβ)]-	28624-59-1	NI28742
189 133 190 131 188 161 160 145	280	100 27 14 5 3 3 3 3	1.00	15	20	3	–	1	–	–	–	–	–	–	–	–	Acetic acid, [(2,4,6-triethylbenzoyl)thio]-	67902-78-7	NI28743
43 57 29 41 143 110 18 15	280	100 10 7 6 4 4 4 4	0.01	15	20	3	–	1	–	–	–	–	–	–	–	–	2-Acetoxy-3-tert-butyl-5-phenyl-propanethioate		MS06365
43 123 87 150 129 69 41 110	280	100 40 36 20 19 19 18 17	11.00	15	20	3	–	1	–	–	–	–	–	–	–	–	1-Acetoxy-1-tert-butyl-3-phenylthio-2-propanone		MS06366
43 57 99 139 110 41 71 69	280	100 10 4 2 2 2 1 1	0.20	15	20	3	–	1	–	–	–	–	–	–	–	–	1-Phenylthio-1-acetoxy-3-tert-butyl-2-propanone		MS06367
43 69 179 41 57 182 29 123	280	100 74 59 43 29 19 19 18	9.00	15	20	3	–	1	–	–	–	–	–	–	–	–	1-Phenylthio-1-tert-butyl-3-acetoxy-2-propanone		MS06368
53 55 69 84 67 83 97 220	280	100 98 90 80 73 67 64 60	0.00	15	20	5	–	–	–	–	–	–	–	–	–	–	1β,10α:4α,5β-Diepoxy-8β-hydroxy-glechoma-8α,12-olide		NI28744
122 265 175 121 205 249 135 123	280	100 72 72 72 58 56 56 53	13.07	15	20	5	–	–	–	–	–	–	–	–	–	–	Azuleno[6,5-b]furan-2,5-dione, 3,3a,4,4a,7a,8,9,9a-octahydro-3-hydroxy-3-(hydroxymethyl)-4a,8-dimethyl-	51292-63-8	NI28745
205 178 234 192 206 179 150 55	280	100 72 63 63 57 43 43 42	4.00	15	20	5	–	–	–	–	–	–	–	–	–	–	1,3-Cyclohexanedione, 2-(ethoxymethyl)-2-(2-hydroxy-6-oxo-1-cyclohexen-1-yl)-	29817-43-4	NI28746
91 115 207 262 189 280	280	100 58 56 51 43 1		15	20	5	–	–	–	–	–	–	–	–	–	–	Diethyl 2-benzylmalate		LI06031
165 280 135 235 91	280	100 20 11 2 2		15	20	5	–	–	–	–	–	–	–	–	–	–	Ethyl γ-(2,4-dimethoxy-6-methylphenyl)acetoacetate		LI06032
97 188 161 111 206 122 218 244	280	100 92 83 78 75 48 41 20	2.00	15	20	5	–	–	–	–	–	–	–	–	–	–	Nigellic acid	91897-25-5	NI28747
43 121 125 41 122 55 139 69	280	100 45 41 41 39 36 31 31	5.69	15	20	5	–	–	–	–	–	–	–	–	–	–	2,4-Pentadienoic acid, 5-(8-hydroxy-1,5-dimethyl-3-oxo-6-oxabicyclo[3.2.1]oct-8-yl)-3-methyl-, [1R-[1α,5α,8S*(2Z,4E)]]-	24394-14-7	NI28748
161 133 280 207 162 105 59 208	280	100 78 53 48 47 26 17 13		15	20	5	–	–	–	–	–	–	–	–	–	–	Propanedioic acid, α-(2-methoxyphenyl)-α-methyl-, diethyl ester	118657-12-8	DD00938
176 91 105 265 120 133 252 148	280	100 99 76 65 39 30 25 25	0.00	15	20	5	–	–	–	–	–	–	–	–	–	–	Propanedioic acid, α-(2-phenylethyl)-α-hydroxy-, diethyl ester	118598-70-2	DD00939
252 280 251 180 207 175 163 177	280	100 97 63 62 31 29 29 28		15	20	5	–	–	–	–	–	–	–	–	–	–	2-Propenoic acid, 3-(4-ethoxy-3,5-dimethoxyphenyl)-, ethyl ester	75332-50-2	NI28749
91 131 85 59 43 57 92 68	280	100 35 29 23 18 17 12 10	0.00	15	20	5	–	–	–	–	–	–	–	–	–	–	β-D-Ribofuranoside, benzyl 2,3-O-isopropylidene-	23276-32-6	NI28750
125 112 95 123 219 111 138 93	280	100 95 95 67 51 47 44 42	6.71	15	20	5	–	–	–	–	–	–	–	–	–	–	Spiro[7H-cyclohepta[b]furan-7,2′(5′H)-furan]-2,5′(3H)-dione, octahydro-8-hydroxy-6,8-dimethyl-3-methylene-	29135-44-2	NI28751

MASS TO CHARGE RATIOS								M.W.	INTENSITIES								Parent	C	H	O	N	S	F	Cl	Br	I	Si	P	B	X	COMPOUND NAME	CAS Reg No	No
55	125	57	138	95	69	123	97	**280**	100	79	63	54	54	45	44	42	**10.01**	15	20	5	–	–	–	–	–	–	–	–	–	–	Spiro[7H-cyclohepta[b]furan-7,2′(5′H)-furan]-2,5′(3H)-dione, octahydro-8-hydroxy-6,8-dimethyl-3-methylene-, [3aS-(3aα,6β,7α,8α,8aα)]-	3533-47-9	NI28752
125	112	138	95	81	97	93	79	**280**	100	55	49	42	38	33	33	31	**12.73**	15	20	5	–	–	–	–	–	–	–	–	–	–	Spiro[7H-cyclohepta[b]furan-7,2′(5′H)-furan]-2,5′(3H)-dione, octahydro-8-hydroxy-6,8-dimethyl-3-methylene-, [3aS-(3aα,6β,7α,8α,8aα)]-	3533-47-9	NI28753
43	55	125	45	41	95	123	57	**280**	100	97	70	62	62	55	47	45	**7.00**	15	20	5	–	–	–	–	–	–	–	–	–	–	Spiro[7H-cyclohepta[b]furan-7,2′(5′H)-furan]-2,5′(3H)-dione, octahydro-8-hydroxy-6,8-dimethyl-3-methylene-, [3aS-(3aα,6β,7α,8β,8aα)]-	27696-09-9	NI28755
219	138	165	177	139	191	207	176	**280**	100	74	54	51	50	40	37	37	**13.98**	15	20	5	–	–	–	–	–	–	–	–	–	–	Spiro[7H-cyclohepta[b]furan-7,2′(5′H)-furan]-2,5′(3H)-dione, octahydro-8-hydroxy-6,8-dimethyl-3-methylene-, [3aS-(3aα,6β,7α,8β,8aα)]-	27696-09-9	NI28754
187	73	69	171	75	94	97	151	**280**	100	95	75	73	59	38	33	25	**3.00**	15	24	3	–	–	–	–	–	–	1	–	–	–	Hexanoic acid, 6-phenoxy-, trimethylsilyl ester	21273-12-1	NI28756
75	121	190	91	73	265	134	117	**280**	100	97	93	76	73	65	54	53	**49.30**	15	24	3	–	–	–	–	–	–	1	–	–	–	Pentanoic acid, 5-(2-methoxyphenyl)-, trimethylsilyl ester		NI28757
75	121	134	73	147	190	265	280	**280**	100	99	80	68	44	41	40	34		15	24	3	–	–	–	–	–	–	1	–	–	–	Pentanoic acid, 5-(3-methoxyphenyl)-, trimethylsilyl ester		NI28758
121	75	134	190	147	265	73	77	**280**	100	30	28	23	21	18	17	16	**16.40**	15	24	3	–	–	–	–	–	–	1	–	–	–	Pentanoic acid, 5-(4-methoxyphenyl)-, trimethylsilyl ester		MS06369
72	165	109	57	56	43	108	41	**280**	100	22	19	13	9	8	6	6	**0.40**	15	24	3	2	–	–	–	–	–	–	–	–	–	1-Isopropylamino-3-(4-propionamidophenoxy)propan-2-ol		MS06370
263	154	150	136	138	98	280	264	**280**	100	73	50	40	38	32	28	28		15	24	3	2	–	–	–	–	–	–	–	–	–	2,6-Piperidinedione, 1-[(octahydro-2H-quinolizin-1-yl)methyl]-, N-oxide, (1S-trans)-	80324-07-8	NI28759
223	280	86	58	224	167	139	195	**280**	100	29	25	22	18	18	17	14		16	12	3	2	–	–	–	–	–	–	–	–	–	2-(Amino)-N-(9,10-dioxaanthracen-2-yl)-ethanamide		MS06371
280	119	91	133	120	65	132	92	**280**	100	93	93	50	20	13	12	9		16	12	3	2	–	–	–	–	–	–	–	–	–	1-Benzyl-3-phenyl-2,4,5-trioxoimidazolidine		MS06372
280	237	281	263	43	164	77	238	**280**	100	25	18	15	9	8	7	6		16	12	3	2	–	–	–	–	–	–	–	–	–	1,4-Diamino-2-acetylanthraquinone		IC03617
280	132	148	149	134	281	133	90	**280**	100	60	48	34	28	20	20	20		16	12	3	2	–	–	–	–	–	–	–	–	–	2-Hydroxydiftalone		MS06373
280	118	148	106	90	132	105	133	**280**	100	55	52	49	45	36	32	27		16	12	3	2	–	–	–	–	–	–	–	–	–	3-Hydroxydiftalone		MS06374
105	280	116	133	77	89	91	281	**280**	100	98	93	83	73	60	35	18		16	12	3	2	–	–	–	–	–	–	–	–	–	Isoxazole, 3-(4′-tolyl)-5-phenyl-4-nitro-	66692-94-2	NI28760
119	91	280	77	103	120	135	281	**280**	100	87	87	31	26	18	18	17		16	12	3	2	–	–	–	–	–	–	–	–	–	Isoxazole, 5-(4′-tolyl)-3-phenyl-4-nitro-	66692-96-4	NI28761
280	115	143	204	205	234	206	178	**280**	100	18	17	12	9	8	4	4		16	12	3	2	–	–	–	–	–	–	–	–	–	1-(4-Nitroanilinomethylene)-2-indanone		MS06375
81	137	95	41	69	67	93	265	**280**	100	61	28	27	22	16	15	12	**7.42**	16	22	–	–	–	–	1	–	–	–	1	–	–	Phosphinous chloride, phenyl(1,7,7-trimethylbicyclo[2.2.1]hept-2-yl)-	74630-27-6	NI28762
139	41	79	97	39	91	77	107	**280**	100	51	47	40	39	33	33	29	**22.30**	16	24	–	–	2	–	–	–	–	–	–	–	–	1,2,3,4,5,7,8,9,10,11,12,14-Dodecahydrodibenzo[c,h][1,6]dithiacyclodecin		MS06376
41	119	67	55	81	79	84	43	**280**	100	88	88	85	80	80	77	71	**2.90**	16	24	4	–	–	–	–	–	–	–	–	–	–	Brefeldin A	20350-15-6	MS06377
119	79	67	117	81	91	93	84	**280**	100	81	77	73	73	70	67	65	**10.55**	16	24	4	–	–	–	–	–	–	–	–	–	–	Brefeldin A	20350-15-6	NI28763
41	79	67	119	81	55	84	93	**280**	100	82	82	80	68	68	65	59	**3.10**	16	24	4	–	–	–	–	–	–	–	–	–	–	Brefeldin A	20350-15-6	NI28764
125	189	248	91	140	161	221	79	**280**	100	70	66	58	56	53	50	47	**30.00**	16	24	4	–	–	–	–	–	–	–	–	–	–	1-Cyclohexene-1,4-dicarboxylic acid, 4-(1-butenyl)-3-ethyl-, dimethyl ester		MS06378
29	206	41	252	105	91	280	119	**280**	100	48	40	37	31	29	27	27		16	24	4	–	–	–	–	–	–	–	–	–	–	Cyclooctane, 1,5-bis(ethoxycarbonylmethylene)-		NI28765
43	41	119	91	67	160	79	31	**280**	100	20	13	12	11	6	6	5	**2.00**	16	24	4	–	–	–	–	–	–	–	–	–	–	2,2-Diallylcyclohexane-1,3-diol diacetate		LI06033
196	124	164	93	69	41	43	66	**280**	100	69	65	43	43	36	26	22	**3.00**	16	24	4	–	–	–	–	–	–	–	–	–	–	2(5H)-Furanone, 3-crotonoyl-5-hexyl-4-methoxy-5-methyl-, (E)-	22628-12-2	NI28766
99	165	155	59	163	91	151	139	**280**	100	79	50	46	43	29	24	22	**5.00**	16	24	4	–	–	–	–	–	–	–	–	–	–	5-Heptanone, 6-hydroxy-2-(2,6-dimethoxyphenyl)-6-methyl-		LI06034
192	43	133	177	71	209	280	249	**280**	100	67	61	30	30	15	3	1		16	24	4	–	–	–	–	–	–	–	–	–	–	1H-Indene-5-carboxylic acid, 2,3,3a,6,7,7a-hexahydro-3-hydroxy-1-methyl-3a-(2-methyl-1-oxopropyl)-, methyl ester [1R-(1α,3α,3aα,7aα)]-		NI28767
166	161	91	133	81	115	206	240	**280**	100	45	43	39	39	29	27	23	**13.00**	16	24	4	–	–	–	–	–	–	–	–	–	–	Malonic acid, 2-[2-(1-cyclohexenyl)-2-propenyl]-, diethyl ester		MS06379
113	43	136	95	91	41	107	79	**280**	100	16	12	12	12	12	10	10	**4.00**	16	24	4	–	–	–	–	–	–	–	–	–	–	7α-Methoxynardofuran		NI28768
252	223	153	251	123	140	280	167	**280**	100	99	92	78	54	48	46	44		16	24	4	–	–	–	–	–	–	–	–	–	–	4-Methoxy-6-nonyl-2-oxo-2H-pyran-3-carboxaldehyde		MS06380
73	75	41	67	57	55	95	81	**280**	100	61	48	43	41	41	39	37	**0.00**	16	28	2	–	–	–	–	–	–	1	–	–	–	2-Propen-1-one, 1-[4-(3-hydroxypropyl)-4-methyl-1-cyclohexenyl]-3-(trimethylsilyl)-, (E)-		DD00940
55	73	97	94	280	75	151	103	**280**	100	49	48	47	33	25	24	24		16	28	2	–	–	–	–	–	–	1	–	–	–	Silane, trimethyl[(7-phenoxyheptyl)oxy]-	16654-57-2	NI28769
178	224	153	234	280	97	152	179	**280**	100	27	26	24	22	22	15	14		16	28	2	2	–	–	–	–	–	–	–	–	–	2-(Cyclohexylimino)-3-butyl-1-aci-nitrocyclohexane		MS06381
43	251	280	77	67	137	105	281	**280**	100	87	76	59	55	38	34	32		17	12	4	–	–	–	–	–	–	–	–	–	–	4H-1-Benzopyran-4-one, 3-benzoyl-7-hydroxy-2-methyl-		MS06382
207	165	166	208	82	280	167	83	**280**	100	85	76	17	17	11	11	11		17	16	–	2	1	–	–	–	–	–	–	–	–	Imidazole, 2-methyl-5,5-diphenyl-4-(methylthio)-	24133-96-8	NI28770
222	165	280	207	166	223	56	224	**280**	100	53	42	28	22	20	18	15		17	16	–	2	1	–	–	–	–	–	–	–	–	2-Imidazoline-4-thione, 1,2-dimethyl-5,5-diphenyl-	37927-97-2	NI28771
165	166	207	280	56	82	281	83	**280**	100	88	63	62	32	18	14	14		17	16	–	2	1	–	–	–	–	–	–	–	–	2-Imidazoline-5-thione, 1,2-dimethyl-4,4-diphenyl-	24134-06-3	NI28772
91	238	280	147	236	119	92	239	**280**	100	71	51	37	11	10	10	9		17	16	2	2	–	–	–	–	–	–	–	–	–	Acetamide, N-(3-benzyl-2-oxo-1-indolinyl)-		AI01961
196	280	238	43	167	168	239	197	**280**	100	59	39	21	12	10	9	7		17	16	2	2	–	–	–	–	–	–	–	–	–	Acetamide, N,N′-9H-fluorene-2,7-diylbis-	304-28-9	NI28773
194	237	195	238	236	193	209	210	**280**	100	83	24	14	5	5	2	1	**0.62**	17	16	2	2	–	–	–	–	–	–	–	–	–	1-Acetyl-5,5-diphenylimidazolid-4-one		MS06383
251	280	72	134	77	265	208	175	**280**	100	62	61	47	37	29	22	20		17	16	2	2	–	–	–	–	–	–	–	–	–	2,4-Imidazolidinedione, 1,3-dimethyl-5,5-diphenyl-	6456-01-5	NI28774
118	194	77	203	280	165	195	51	**280**	100	53	41	33	21	20	17	16		17	16	2	2	–	–	–	–	–	–	–	–	–	2,4-Imidazolidinedione, 1,3-dimethyl-5,5-diphenyl-	6456-01-5	NI28775
203	118	280	194	223	195	77	165	**280**	100	82	76	76	31	29	26	25		17	16	2	2	–	–	–	–	–	–	–	–	–	2,4-Imidazolidinedione, 1,3-dimethyl-5,5-diphenyl-	6456-01-5	NI28776
180	209	280	181	77	104	251	165	**280**	100	59	36	33	33	28	24	22		17	16	2	2	–	–	–	–	–	–	–	–	–	2,4-Imidazolidinedione, 3-ethyl-5,5-diphenyl-	39588-47-1	NI28777
263	160	42	118	280	146	103	43	**280**	100	95	59	53	48	48	38	38		17	16	2	2	–	–	–	–	–	–	–	–	–	1(2H)-Isoquinolinone, 3,4-dihydro-2-(2-hydroximino-2-phenylethyl)-, (Z)-		NI28778

MASS TO CHARGE RATIOS	M.W.	INTENSITIES	Parent	C	H	O	N	S	F	Cl	Br	I	Si	P	B	X	COMPOUND NAME	CAS Reg No	No
252 237 165 206 221 115 178 191	**280**	100 37 31 16 15 15 14 13	**5.00**	17	16	2	2	–	–	–	–	–	–	–	–	–	**1-Phenyl-7,8-dimethoxy-1H-2,3-benzodiazepine**		**NI28779**
280 77 252 253 165 238 237 51	**280**	100 21 19 16 15 14 12 12		17	16	2	2	–	–	–	–	–	–	–	–	–	**1-Phenyl-7,8-dimethoxy-5H-2,3-benzodiazepine**		**NI28780**
77 280 105 182 51 118 132 41	**280**	100 78 69 23 15 14 12 12		17	16	2	2	–	–	–	–	–	–	–	–	–	**3,5-Pyrazolidinedione, 4,4-dimethyl-1,2-diphenyl-**	54719-43-6	**NI28781**
77 280 105 41 51 182 118 281	**280**	100 86 55 16 15 14 14 13		17	16	2	2	–	–	–	–	–	–	–	–	–	**3,5-Pyrazolidinedione, 4,4-dimethyl-1,2-diphenyl-**	54719-43-6	**MS06384**
104 77 203 280 249 132 105 194	**280**	100 83 68 53 45 44 28 24		17	16	2	2	–	–	–	–	–	–	–	–	–	**2(1H)-Pyrimidinone, 3,4-dihydro-4,6-diphenyl-5-methoxy-2-oxo-**		**LI06035**
280 279 224 223 196 167 195 281	**280**	100 93 47 41 29 29 23 19		17	16	2	2	–	–	–	–	–	–	–	–	–	**Pyrrolo[2,1-a]pyrido[1,3-l,m]-beta-carboline-3,12-dione, 1,2,3,5,6,12,13,14-octahydro-**		**MS06385**
251 57 55 45 43 41 280 224	**280**	100 60 60 60 60 60 40 40		17	16	2	2	–	–	–	–	–	–	–	–	–	**4(3H)-Quinazolinone, 2-ethyl-3-(2-hydroxy-6-methylphenyl)-**		**NI28782**
209 225 229 41 231 57 211 227	**280**	100 82 65 50 41 40 33 28	**0.68**	17	25	1	–	–	–	1	–	–	–	–	–	–	**Benzene, 1-chloro-3,5-di-tert-butyl-2-(2-propenyloxy)-**	55955-96-9	**NI28783**
165 121 43 71 41 93 81 119	**280**	100 81 73 32 29 26 26 23	**14.00**	17	28	3	–	–	–	–	–	–	–	–	–	–	**3-Cyclohexene-1-propanoic acid, β,4-dimethyl-1-(2-methyl-1-oxopropyl)-, ethyl ester**	37829-31-5	**NI28784**
85 43 127 91 41 79 55 238	**280**	100 63 23 23 23 18 16 15	**3.00**	17	28	3	–	–	–	–	–	–	–	–	–	–	**1,6,10-Dodecatrien-3-ol, 9-acetoxy-3,7,11-trimethyl-**		**MS06386**
134 121 132 93 119 125 105 41	**280**	100 70 64 54 53 49 23 21	**0.74**	17	28	3	–	–	–	–	–	–	–	–	–	–	**2-Furanmethanol, tetrahydro-α,α,5-trimethyl-5-(4-methyl-3-cyclohexen-1-yl)-, acetate**	55712-54-4	**NI28785**
57 81 99 55 95 43 41 85	**280**	100 92 83 64 59 57 55 52	**3.00**	17	28	3	–	–	–	–	–	–	–	–	–	–	**Methyl 10,11-epoxy-3E,7E,11Z-trimethyltrideca-2,6-dienoate**		**LI06036**
57 77 93 105 121 280 131 206	**280**	100 56 16 15 12 11 3 1		17	28	3	–	–	–	–	–	–	–	–	–	–	**1,2-Propanediol, 3-[(12-methyl-1,5-tridecadien-3-ynyl)oxy]-, (Z,Z)-**		**DD00941**
107 280 77 91 121 135 105 103	**280**	100 77 57 51 34 24 23 23		17	28	3	–	–	–	–	–	–	–	–	–	–	**1,2-Propanediol, 3-[(1,5-tetradecadien-3-ynyl)oxy]-, (Z,E)-**		**MS06387**
91 121 77 280 135 206 149 178	**280**	100 48 45 22 13 7 4 3		17	28	3	–	–	–	–	–	–	–	–	–	–	**1,2-Propanediol, 3-[(1,5-tetradecadien-3-ynyl)oxy]-, (Z,Z)-(–)-**		**DD00942**
280 281 88 279 75 140 228 87	**280**	100 40 22 20 12 11 10 10		18	8	–	4	–	–	–	–	–	–	–	–	–	**2,3-Dicyanodibenzo[f,h]quinoxaline**		**MS06388**
170 129 171 184 115 142 141 169	**280**	100 76 68 65 58 47 38 30	**0.00**	18	16	3	–	–	–	–	–	–	–	–	–	–	**2-Acetyl-1,4,4a,5,8,8a-hexahydro-3-phenyl-1,4:5,8-trans-diepoxynaphthalene**		**NI28786**
280 265 140 251 165 151 253 237	**280**	100 35 7 4 4 4 1 1		18	16	3	–	–	–	–	–	–	–	–	–	–	**Benzofuran, 2-(2-hydroxy-4-methoxyphenyl)-5-propenyl-, (E)-**		**MS06389**
280 165 115 221 121 279 194 253	**280**	100 12 12 10 9 8 5 4		18	16	3	–	–	–	–	–	–	–	–	–	–	**Benzofuran, 2-(4-hydroxyphenyl)-7-methoxy-5-propenyl-, (E)-**		**MS06390**
105 280 77 279 206 281 178 207	**280**	100 92 57 41 29 19 15 14		18	16	3	–	–	–	–	–	–	–	–	–	–	**Ethyl benzylidinebenzoylacetate**		**LI06037**
161 265 160 105 77 159 145 120	**280**	100 95 90 65 64 54 45 37	**34.00**	18	20	1	2	–	–	–	–	–	–	–	–	–	**1,2-Benzenediamine, 3-(benzoylacetonylidene)-4,5-dimethyl-**		**NI28787**
281 282 236 279 264 159 283 237	**280**	100 23 19 3 3 3 1 1	**0.00**	18	20	1	2	–	–	–	–	–	–	–	–	–	**1H-Indole-3-methanamine, N,N-dimethyl-5-benzyloxy-**	1453-97-0	**NI28788**
280 42 167 168 196 115 239 122	**280**	100 63 60 57 53 47 40 40	**0.00**	18	20	1	2	–	–	–	–	–	–	–	–	–	**1,5-Methano-1H-azonino[4,3-b]indole, 4-ethylidene-2-methyl-7-oxo-2,3,4,5,6,7-hexahydro-, (Z)-**		**MS06391**
171 169 170 172 144 143 142 168	**280**	100 44 28 13 7 7 7 4	**4.00**	18	20	1	2	–	–	–	–	–	–	–	–	–	**3-(1,2,3,4-Tetrahydro-β-carbolin-1-ylmethyl)-2-cyclohexen-1-one**		**NI28789**
90 91 279 118 89 107 202 117	**280**	100 80 60 44 40 31 29 28	**12.00**	18	21	2	–	–	–	–	–	–	–	–	1	–	**meso-Hydrobenzoin butylboronate**		**MS06392**
195 57 280 41 181 29 209 153	**280**	100 77 49 33 26 26 20 20		18	29	–	–	–	–	1	–	–	–	–	–	–	**Chloro-tri-sec-butylbenzene**		**LI06038**
195 280 196 97 123 111 110 197	**280**	100 21 16 12 10 9 8 6		18	32	–	–	1	–	–	–	–	–	–	–	–	**Thiophene, 2,5-diheptyl-**		**AI01962**
81 96 41 55 67 95 43 73	**280**	100 59 56 52 46 31 30 29	**3.04**	18	32	2	–	–	–	–	–	–	–	–	–	–	**Bicyclo[3.1.0]hexane-2-undecanoic acid, methyl ester**	10152-73-5	**NI28790**
67 41 55 39 43 82 81 68	**280**	100 34 19 14 10 9 9 9	**1.00**	18	32	2	–	–	–	–	–	–	–	–	–	–	**2-Cyclopentene-1-undecanoic acid, ethyl ester**	3552-12-3	**WI01346**
41 55 67 81 29 54 43 95	**280**	100 94 93 74 73 57 57 54	**13.00**	18	32	2	–	–	–	–	–	–	–	–	–	–	**4-Hydroxyoctadec-9-enolide**		**LI06039**
67 55 81 28 41 69 95 68	**280**	100 95 79 79 71 56 55 53	**7.70**	18	32	2	–	–	–	–	–	–	–	–	–	–	**9,12-Octadecadienoic acid, (Z,Z)-**	60-33-3	**NI28792**
81 95 67 96 82 109 100 68	**280**	100 94 74 67 65 52 44 43	**35.00**	18	32	2	–	–	–	–	–	–	–	–	–	–	**9,12-Octadecadienoic acid, (Z,Z)-**	60-33-3	**NI28791**
67 55 41 81 54 68 95 82	**280**	100 77 76 73 55 51 45 45	**4.70**	18	32	2	–	–	–	–	–	–	–	–	–	–	**9,12-Octadecadienoic acid, (Z,Z)-**	60-33-3	**NI28793**
69 55 83 108 80 43 81 41	**280**	100 58 56 54 48 31 30 30	**5.00**	18	32	2	–	–	–	–	–	–	–	–	–	–	**5,9-Tridecadien-1-ol, 10-propyl-, acetate**	18654-86-9	**LI06040**
69 55 83 108 80 43 41 67	**280**	100 59 57 54 48 31 31 30	**5.46**	18	32	2	–	–	–	–	–	–	–	–	–	–	**5,9-Tridecadien-1-ol, 10-propyl-, acetate**	18654-86-9	**NI28794**
69 83 80 108 55 93 95 81	**280**	100 58 43 42 37 33 28 28	**9.00**	18	32	2	–	–	–	–	–	–	–	–	–	–	**5,9-Tridecadien-1-ol, 10-propyl-, acetate, (E)-**	10297-61-7	**LI06042**
69 83 55 43 41 125 67 57	**280**	100 65 43 28 24 17 12 11	**1.60**	18	32	2	–	–	–	–	–	–	–	–	–	–	**5,9-Tridecadien-1-ol, 10-propyl-, acetate, (E)-**	10297-61-7	**LI06041**
69 41 55 67 83 80 79 81	**280**	100 96 87 59 53 53 42 41	**0.00**	18	32	2	–	–	–	–	–	–	–	–	–	–	**5,9-Tridecadien-1-ol, 10-propyl-, acetate, (E)-**	10297-61-7	**NI28795**
117 198 116 83 199 115 55 280	**280**	100 99 37 34 32 28 27 25		18	33	–	–	–	–	–	–	–	–	1	–	–	**Phosphine, tricyclohexyl-**	2622-14-2	**NI28796**
252 280 145 121 237 91 105 253	**280**	100 58 38 36 32 20 19 17		19	20	2	–	–	–	–	–	–	–	–	–	–	**Cyclohexanone, 6-furfurylidene-2,3-dimethyl-2-phenyl-**	17429-57-1	**NI28797**
280 281 252 210 251 165 237 209	**280**	100 22 13 10 9 9 8 8		19	20	2	–	–	–	–	–	–	–	–	–	–	**11H-Cyclopenta[a]phenanthren-17-ol, 12,13,16,17-tetrahydro-3-methoxy-13-methyl-, (13S-cis)-**	33760-56-4	**NI28798**
280 224 165 209 281 223 166 178	**280**	100 42 24 20 19 18 12 11		19	20	2	–	–	–	–	–	–	–	–	–	–	**Estra-1,3,5,7,9-pentaen-17-one, 3-methoxy-**	3907-67-3	**NI28799**
280 237 224 211 165 44 41 223	**280**	100 40 37 34 34 34 32 27		19	20	2	–	–	–	–	–	–	–	–	–	–	**Estra-1,3,5,7,9-pentaen-17-one, 3-methoxy-**	3907-67-3	**NI28800**
280 281 237 224 211 223 278 165	**280**	100 26 26 21 21 17 13 13		19	20	2	–	–	–	–	–	–	–	–	–	–	**Estra-1,3,5,7,9-pentaen-17-one, 3-methoxy-**	3907-67-3	**NI28801**
280 265 281 266 132.5 222 140 249	**280**	100 79 22 16 14 12 12 8		19	20	2	–	–	–	–	–	–	–	–	–	–	**9-Ethyl-3,6-dimethoxy-10-methylphenanthrene**		**MS06393**
280 251 224 209 195 223 44 165	**280**	100 41 39 28 28 26 24 23		19	20	2	–	–	–	–	–	–	–	–	–	–	**Gona-1,3,5,7,9-pentaen-17-one, 13-ethyl-3-hydroxy-**	30788-51-3	**NI28802**
280 251 224 195 209 223 165 194	**280**	100 46 34 29 24 23 16 15		19	20	2	–	–	–	–	–	–	–	–	–	–	**Gona-1,3,5,7,9-pentaen-17-one, 13-ethyl-3-hydroxy-, (13α)-**	69853-73-2	**NI28803**
280 251 281 224 223 209 195 194	**280**	100 33 23 22 17 16 16 10		19	20	2	–	–	–	–	–	–	–	–	–	–	**Gona-1,3,5,7,9-pentaen-17-one, 13-ethyl-3-hydroxy-, (14β)-**	51262-24-9	**NI28804**
91 105 77 103 134 43 131 151	**280**	100 88 82 77 70 61 58 56	**10.00**	19	20	2	–	–	–	–	–	–	–	–	–	–	**1-Hepten-3-one, 5-hydroxy-1,7-diphenyl-**		**LI06043**
225 197 165 105 91 166 128 115	**280**	100 64 42 41 31 18 10 10	**0.00**	19	20	2	–	–	–	–	–	–	–	–	–	–	**4-Hexenoic acid, 2,2-diphenyl-, methyl ester, (E)-**		**DD00943**

MASS TO CHARGE RATIOS	M.W.	INTENSITIES	Parent	C	H	O	N	S	F	Cl	Br	I	Si	P	B	X	COMPOUND NAME	CAS Reg No	No
225 165 197 105 91 166 226 129	**280**	100 70 54 35 33 20 18 12	**0.00**	19	20	2	–	–	–	–	–	–	–	–	–	–	**4-Hexenoic acid, 2,2-diphenyl-, methyl ester, (Z)-**		**DD00944**
129 207 91 206 128 191 205 178	**280**	100 55 54 51 25 19 17 16	**0.00**	19	20	2	–	–	–	–	–	–	–	–	–	–	**4-Hexenoic acid, 3,3-diphenyl-, methyl ester, (E)-**		**DD00945**
129 91 207 206 128 191 205 178	**280**	100 57 52 52 26 19 15 15	**0.00**	19	20	2	–	–	–	–	–	–	–	–	–	–	**4-Hexenoic acid, 3,3-diphenyl-, methyl ester, (Z)-**		**DD00946**
91 143 115 280 193 129 128 105	**280**	100 49 49 46 44 37 29 18		19	20	2	–	–	–	–	–	–	–	–	–	–	**3-Pentenoic acid, 4,5-diphenyl-2-methyl-, methyl ester, (E)-**		**DD00947**
206 178 59 177 205 192 176 41	**280**	100 70 65 45 39 33 33 30	**0.00**	19	20	2	–	–	–	–	–	–	–	–	–	–	**4-Phenanthrenylmethyl-tert-butyl peroxide**	56909-17-2	**NI28805**
118 117 119 174 91 77 115 105	**280**	100 48 36 23 16 11 10 9	**4.00**	19	20	2	–	–	–	–	–	–	–	–	–	–	**4H-Pyran-4-one, dihydro-3,5-dimethyl-2,6-diphenyl-**		**IC03618**
41 42 43 130 115 55 128 44	**280**	100 47 34 25 25 18 15 14	**1.00**	19	24	–	2	–	–	–	–	–	–	–	–	–	**Aspidofractinine**	6871-25-6	**NI28806**
135 121 144 280 122 107 130 251	**280**	100 75 52 50 48 30 27 24		19	24	–	2	–	–	–	–	–	–	–	–	–	**Aspidospermidine, 6,7-dehydro-**		**LI06044**
280 210 41 43 251 58 42 208	**280**	100 87 83 67 52 51 43 42		19	24	–	2	–	–	–	–	–	–	–	–	–	**Aspidospermidine, 1,2-didehydro-, (5α,12β,19α)-**	56245-51-3	**NI28807**
136 144 280 158 123 137 157 122	**280**	100 27 26 22 22 11 10 10		19	24	–	2	–	–	–	–	–	–	–	–	–	**Condyfolan, 14,19-didehydro-1-methyl-, (14E)-**	69796-62-9	**NI28808**
150 280 144 122 69 130 151 152	**280**	100 32 25 20 15 14 13 12		19	24	–	2	–	–	–	–	–	–	–	–	–	**Condyfolan, 16-methylene-**	2670-56-6	**NI28809**
55 280 43 41 144 182 57 196	**280**	100 78 71 70 51 49 42 40		19	24	–	2	–	–	–	–	–	–	–	–	–	**Curan, 16,17-didehydro-**	56053-18-0	**NI28810**
189 146 94 96 280	**280**	100 96 52 20 16		19	24	–	2	–	–	–	–	–	–	–	–	–	**Despropionyl-fentanyl**		**LI06045**
189 146 94 96 280	**280**	100 92 52 24 16		19	24	–	2	–	–	–	–	–	–	–	–	–	**Despropionyl-fentanyl**		**LI06046**
58 234 85 235 195 193 42 194	**280**	100 42 39 34 19 19 19 17	**10.30**	19	24	–	2	–	–	–	–	–	–	–	–	–	**5H-Dibenz[b,f]azepine-5-propanamine, 10,11-dihydro-N,N-dimethyl-**	50-49-7	**NI28812**
58 235 85 234 195 280 193 208	**280**	100 89 79 70 38 31 27 25		19	24	–	2	–	–	–	–	–	–	–	–	–	**5H-Dibenz[b,f]azepine-5-propanamine, 10,11-dihydro-N,N-dimethyl-**	50-49-7	**NI28811**
280 210 279 251 41 140 281 252	**280**	100 92 69 65 26 25 19 19		19	24	–	2	–	–	–	–	–	–	–	–	–	**Eburnamenine, 14,15-dihydro-**	47122-74-7	**NI28813**
136 135 122 149 195 156 41 265	**280**	100 49 37 33 28 27 26 21	**16.00**	19	24	–	2	–	–	–	–	–	–	–	–	–	**Ibogamine**	481-87-8	**NI28814**
280 124 156 41 281 195 168 169	**280**	100 76 60 32 22 21 20 19		19	24	–	2	–	–	–	–	–	–	–	–	–	**Ibogamine, (2α,5β,6α,18β)-**	1673-99-0	**NI28815**
124 122 280 125 71 60 82 69	**280**	100 18 14 12 10 9 8 7		19	24	–	2	–	–	–	–	–	–	–	–	–	**Indole, 2-[1-(3-ethylidene-1-methyl-4-piperidyl)vinyl]-3-methyl-**	1850-33-5	**NI28816**
146 189 44 42 118 105 57 43	**280**	100 55 42 31 17 17 15 13	**4.60**	19	24	–	2	–	–	–	–	–	–	–	–	–	**4-Piperidinamine, N-phenyl-1-(2-phenylethyl)-**	21409-26-7	**NI28817**
209 280 194 180 251 208 181 265	**280**	100 93 88 80 64 58 43 41		19	24	–	2	–	–	–	–	–	–	–	–	–	**Uleine, N-demethyl-N-ethyl-**	3620-78-8	**NI28818**
69 83 97 98 81 95 280 67	**280**	100 98 91 88 82 81 79 75		19	36	1	–	–	–	–	–	–	–	–	–	–	**Cyclopropaneoctanal, 2-octyl-**	56196-06-6	**NI28819**
67 55 81 96 109 121 107 135	**280**	100 92 81 38 21 9 8 5	**0.00**	19	36	1	–	–	–	–	–	–	–	–	–	–	**12-Nonadecyn-1-ol**		**DD00948**
71 41 107 55 98 43 67 82	**280**	100 58 43 30 26 25 24 20	**3.74**	19	36	1	–	–	–	–	–	–	–	–	–	–	**1,9-Octadecadiene, 1-methoxy-**	56554-63-3	**NI28820**
55 57 41 43 29 127 97 69	**280**	100 77 72 51 45 38 38 33	**1.50**	19	36	1	–	–	–	–	–	–	–	–	–	–	**1-Octanone, 1-(2-octylcyclopropyl)-**	54965-36-5	**NI28821**
280 140 279 281 139 126 176 150	**280**	100 25 24 21 10 5 5 4		20	12	–	2	–	–	–	–	–	–	–	–	–	**Dibenzo[a,c]phenazine**	215-64-5	**NI28822**
280 140 279 281 139 126 50 176	**280**	100 28 27 22 21 7 4 4		20	12	–	2	–	–	–	–	–	–	–	–	–	**Dibenzo[a,c]phenazine**	215-64-5	**NI28823**
81 92 93 91 104 172 55 80	**280**	100 85 84 77 47 44 31 24	**0.00**	20	24	1	–	–	–	–	–	–	–	–	–	–	**2,4-Dibenzylcyclohexanol**		**DO01074**
43 55 69 41 57 97 83 81	**280**	100 98 89 82 81 80 72 53	**2.00**	20	40	–	–	–	–	–	–	–	–	–	–	–	**Cyclohexane, 1-(1,5-dimethylhexyl)-4-(4-methylpentyl)-**	56009-20-2	**NI28824**
55 69 57 236 97 83 41 43	**280**	100 89 81 80 79 71 70 67	**5.00**	20	40	–	–	–	–	–	–	–	–	–	–	–	**Cyclotetradecane, 1,7,11-trimethyl-4-isopropyl-**	1786-12-5	**NI28825**
55 57 69 43 41 97 83 71	**280**	100 90 85 77 75 66 63 48	**0.90**	20	40	–	–	–	–	–	–	–	–	–	–	–	**Cyclotetradecane, 1,7,11-trimethyl-4-isopropyl-**	1786-12-5	**MS06394**
43 41 55 57 97 83 69 29	**280**	100 90 82 73 61 54 50 49	**7.11**	20	40	–	–	–	–	–	–	–	–	–	–	–	**1-Eicosene**	3452-07-1	**WI01347**
57 69 55 43 41 83 97 56	**280**	100 94 93 91 89 74 63 53	**2.34**	20	40	–	–	–	–	–	–	–	–	–	–	–	**3-Eicosene, (E)-**	74685-33-9	**NI28826**
55 57 43 69 83 41 97 56	**280**	100 77 70 64 62 54 51 45	**1.95**	20	40	–	–	–	–	–	–	–	–	–	–	–	**5-Eicosene, (E)-**	74685-30-6	**NI28827**
57 55 43 83 69 41 97 56	**280**	100 99 92 85 83 67 64 57	**2.54**	20	40	–	–	–	–	–	–	–	–	–	–	–	**9-Eicosene, (E)-**	74685-29-3	**NI28828**
43 55 57 41 56 70 69 71	**280**	100 89 81 72 71 51 50 45	**5.24**	20	40	–	–	–	–	–	–	–	–	–	–	–	**2-Hexadecene, 2,6,10,14-tetramethyl-**	56554-34-8	**NI28829**
70 69 57 55 43 41 71 83	**280**	100 65 56 55 51 50 46 36	**9.10**	20	40	–	–	–	–	–	–	–	–	–	–	–	**trans-Phytene**		**NI28830**
280 281 126 125 140 113 112 250	**280**	100 22 22 20 17 17 14 13		21	12	1	–	–	–	–	–	–	–	–	–	–	**7H-Indeno[2,1-a]anthracen-7-one**	27582-45-2	**NI28831**
280 165 279 278 277 276 281 265	**280**	100 67 46 41 31 31 22 18		22	16	–	–	–	–	–	–	–	–	–	–	–	**1-(4-Azulenyl)-2-(6-azulenyl)ethylene**		**NI28832**
280 279 165 277 276 278 152 252	**280**	100 90 55 41 39 33 28 26		22	16	–	–	–	–	–	–	–	–	–	–	–	**1,2-Bis(4-azulenyl)ethylene**	94154-55-9	**NI28833**
280 279 202 203 165 281 138 278	**280**	100 73 46 43 24 20 16 15		22	16	–	–	–	–	–	–	–	–	–	–	–	**Cinnamal fluorene**		**NI28834**
280 279 278 281 277 265 263 274	**280**	100 72 56 22 22 18 9 8		22	16	–	–	–	–	–	–	–	–	–	–	–	**Dibenz[a,h]anthracene, 5,6-dihydro-**	153-34-4	**NI28835**
280 278 279 138 277 276 265 252	**280**	100 74 59 58 49 44 38 15		22	16	–	–	–	–	–	–	–	–	–	–	–	**Dibenzo[g]chrysene, 13,14-dihydro-**		**DD00949**
280 279 281 203 202 278 276 178	**280**	100 33 23 18 18 15 14 12		22	16	–	–	–	–	–	–	–	–	–	–	–	**α,α-Diphenylbenzofulvene**		**LI06047**
280 279 281 203 202 278 276 138	**280**	100 33 23 17 17 15 13 12		22	16	–	–	–	–	–	–	–	–	–	–	–	**1H-Indene, 1-(diphenylmethylene)-**	13245-90-4	**NI28836**
279 280 276 277 278 252 138 274	**280**	100 81 76 49 26 23 22 19		22	16	–	–	–	–	–	–	–	–	–	–	–	**Indeno[1,2,3-cd]pyrene, 1,2,6,6a-tetrahydro-**		**NI28837**
278 28 277 279 202 201 276 44	**280**	100 24 23 22 14 14 10 10	**3.00**	22	16	–	–	–	–	–	–	–	–	–	–	–	**Naphthalene, 1,7-diphenyl-**	970-06-9	**NI28838**
280 279 278 281 276 140 139 277	**280**	100 60 41 26 17 17 17 16		22	16	–	–	–	–	–	–	–	–	–	–	–	**1,2-Bis(2-naphthyl)ethylene**		**LI06048**
69 85 135 196 108 47 87 31	**281**	100 37 30 29 26 24 12 11	**0.00**	4	–	2	1	–	8	1	–	–	–	–	–	–	**Perfluoro[O-formyl-N-methyl-N-(2-chloroethyl)hydroxylamine]**		**LI06049**
124 281 251 59 186 98 65 189	**281**	100 66 58 23 21 13 4 3		5	5	1	1	–	–	–	–	1	–	–	–	1	**Cyclopentadienylnitrosyliodocobalt**		**LI06050**
156 43 125 93 281 157 18 15	**281**	100 25 24 20 19 12 12 12		6	12	2	5	2	–	–	–	–	–	1	–	–	**Phosphorodithioic acid, S-[(4,6-diamino-1,3,5-triazin-2-yl)methyl] O,O-dimethyl ester**	78-57-9	**NI28839**

MASS TO CHARGE RATIOS								M.W.	INTENSITIES								Parent	C	H	O	N	S	F	Cl	Br	I	Si	P	B	X	COMPOUND NAME	CAS Reg No	No
156	93	281	125	43	157	63	55	**281**	100	45	37	32	26	24	11	11		6	12	2	5	2	–	–	–	–	–	1	–	–	Phosphorodithioic acid, S-[(4,6-diamino-1,3,5-triazin-2-yl)methyl] O,O-dimethyl ester	78-57-9	NI28840
127	65	77	119	154	43	93	188	**281**	100	87	60	54	49	43	41	32	**13.50**	7	5	1	1	1	5	1	–	–	–	–	–	–	(Chlorophenoxymethylenimino)sulphur pentafluoride	81439-21-6	NI28841
146	164	90	147	65	283	281	44	**281**	100	28	25	11	7	6	6	5		9	6	3	1	–	–	3	–	–	–	–	–	–	Benzoic acid, 2-[(trichloroacetyl)amino]-	4257-77-6	NI28842
164	65	136	44	121	165	63	64	**281**	100	23	21	17	11	9	6	5	**5.00**	9	6	3	1	–	–	3	–	–	–	–	–	–	Benzoic acid, 3-[(trichloroacetyl)amino]-	56177-39-0	NI28843
164	136	65	44	165	121	283	281	**281**	100	7	6	5	4	4	3	3		9	6	3	1	–	–	3	–	–	–	–	–	–	Benzoic acid, 4-[(trichloroacetyl)amino]-	56177-40-3	NI28844
60	80	82	281	57	81	79	55	**281**	100	71	68	41	30	29	28	26		10	11	3	5	1	–	–	–	–	–	–	–	–	8,2′-Thioanhydroadenosine		NI28845
281	86	252	58	117	181	114	98	**281**	100	47	46	39	33	25	23	22		11	8	2	1	–	5	–	–	–	–	–	–	–	2-Propenal, 2-(pentafluorophenoxy)-3-(dimethylamino)-		NI28846
236	190	189	144	281	143			**281**	100	17	15	15	13	11				11	11	6	3	–	–	–	–	–	–	–	–	–	L-Proline, 2,4-dinitrophenyl-		LI06051
83	128	44	223	221	159	129	102	**281**	100	31	20	15	15	14	14	14	**5.00**	11	12	1	3	–	–	–	1	–	–	–	–	–	3-Buten-2-one, 4-(2-bromophenyl)-, semicarbazone	17014-27-6	NI28847
149	178	150	120	121	43	93	57	**281**	100	41	37	37	26	24	22	22	**6.80**	11	15	4	5	–	–	–	–	–	–	–	–	–	Adenosine, N,6-didehydro-1,6-dihydro-1-methyl-	15763-06-1	NI28848
149	93	120	121	28	30	148	36	**281**	100	46	45	35	29	28	24	23	**2.96**	11	15	4	5	–	–	–	–	–	–	–	–	–	Adenosine, N-methyl-	1867-73-8	NI28849
135	45	136	192	29	164	146	43	**281**	100	84	81	54	47	38	38	38	**3.48**	11	15	4	5	–	–	–	–	–	–	–	–	–	Adenosine, 2′-O-methyl-	2140-79-6	NI28850
135	136	192	164	146	251	108	45	**281**	100	89	74	47	47	27	24	19	**6.01**	11	15	4	5	–	–	–	–	–	–	–	–	–	Adenosine, 2′-O-methyl-	2140-79-6	NI28851
136	135	146	192	164	251	45	108	**281**	100	99	79	78	52	27	25	24	**5.60**	11	15	4	5	–	–	–	–	–	–	–	–	–	Adenosine, 2′-O-methyl-	2140-79-6	NI28852
135	136	164	178	220	108	44	43	**281**	100	98	96	30	24	21	16	16	**5.00**	11	15	4	5	–	–	–	–	–	–	–	–	–	Adenosine, 3′-O-methyl-	10300-22-8	NI28853
135	136	164	178	220	108	71	266	**281**	100	98	96	30	26	24	18	17	**7.00**	11	15	4	5	–	–	–	–	–	–	–	–	–	Adenosine, 3′-O-methyl-	10300-22-8	LI06052
136	137	178	164	108	194	281	88	**281**	100	77	31	29	11	9	8	7		11	15	4	5	–	–	–	–	–	–	–	–	–	9H-Purin-6-amine, 9-(3-C-methyl-β-D-xylofuranosyl)-	70723-11-4	NI28854
177	74	135	86	44	43	146	60	**281**	100	84	81	76	68	58	51	43	**0.45**	11	23	3	1	2	–	–	–	–	–	–	–	–	D-Xylose, 4-acetamido-4,5-dideoxy-, diethyl mercaptal	3509-44-2	NI28855
281	127	159	143	233	99	108	75	**281**	100	89	78	77	62	54	49	47		12	8	3	1	1	–	1	–	–	–	–	–	–	4-Nitro-4′-chlorodiphenyl sulphoxide	24535-53-3	NI28856
283	281	126	127	98	202	174	282	**281**	100	99	87	84	76	60	48	46		12	9	1	1	–	1	–	1	–	–	–	–	–	Aniline, 5-bromo-2-(4-fluorophenoxy)-	31081-30-8	MS06395
190	280	119	281	253	45	140	69	**281**	100	60	25	12	10	10	9	8		12	12	1	1	–	5	–	–	–	–	–	–	–	Amphetamine pentafluoropropionate		MS06396
160	202	76	104	77	50	130	201	**281**	100	29	18	15	14	13	12	9	**0.98**	12	12	2	1	–	–	–	1	–	–	–	–	–	1H-Isoindole-1,3(2H)-dione, 2-(4-bromobutyl)-	5394-18-3	NI28857
224	239	43	281	266	178	225	132	**281**	100	32	29	23	13	13	11	11		12	15	5	3	–	–	–	–	–	–	–	–	–	Acetanilide, 4′-tert-butyl-2′,6′-dinitro-	22503-15-7	NI28858
180	55	67	69	181	82	57	54	**281**	100	36	22	15	8	8	7	6	**1.40**	12	18	3	1	–	3	–	–	–	–	–	–	–	2-Piperidinecarboxylic acid, 1-(trifluoroacetyl)-, butyl ester	54965-28-5	NI28859
180	55	67	181	41	208	69	82	**281**	100	71	60	46	36	16	15	14	**7.30**	12	18	3	1	–	3	–	–	–	–	–	–	–	2-Piperidinecarboxylic acid, 1-(trifluoroacetyl)-, sec-butyl ester	54965-27-4	NI28862
180	208	41	55	224	57	225	67	**281**	100	57	49	47	45	40	35	25	**7.91**	12	18	3	1	–	3	–	–	–	–	–	–	–	2-Piperidinecarboxylic acid, 1-(trifluoroacetyl)-, sec-butyl ester	54965-27-4	NI28860
180	55	41	67	181	57	69	126	**281**	100	17	16	13	11	8	7	4	**2.00**	12	18	3	1	–	3	–	–	–	–	–	–	–	2-Piperidinecarboxylic acid, 1-(trifluoroacetyl)-, sec-butyl ester	54965-27-4	NI28861
180	55	178	224	58	208	41	128	**281**	100	61	60	56	51	50	47	41	**25.02**	12	18	3	1	–	3	–	–	–	–	–	–	–	4-Piperidinecarboxylic acid, 1-(trifluoroacetyl)-, sec butyl ester, (S)-	55044-14-9	NI28863
180	41	181	57	67	55	69	56	**281**	100	20	14	14	13	9	8	4	**1.37**	12	18	3	1	–	3	–	–	–	–	–	–	–	Proline, 3-methyl-1-(trifluoroacetyl)-, sec-butyl ester, cis-	57983-47-8	NI28864
180	41	67	57	181	55	69	56	**281**	100	18	14	14	12	11	8	4	**1.81**	12	18	3	1	–	3	–	–	–	–	–	–	–	Proline, 3-methyl-1-(trifluoroacetyl)-, sec-butyl ester, trans-	57983-48-9	NI28865
166	43	41	167	69	71	55	42	**281**	100	41	20	19	19	15	13	9	**0.00**	12	18	3	1	–	3	–	–	–	–	–	–	–	L-Proline, 1-(trifluoroacetyl)-, 1-methylbutyl ester	57983-46-7	NI28866
281	205	76	50	102	282	75	218	**281**	100	62	36	27	23	18	18	17		13	7	3	5	–	–	–	–	–	–	–	–	–	2,1,3-Oxadiazole, 4-(1-benzimidazolyl)-7-nitro-	91485-32-4	NI28867
221	223	84	43	222	45	71	220	**281**	100	33	30	30	29	25	24	11	**7.80**	13	12	2	1	1	–	1	–	–	–	–	–	–	4-Thiazoleethanol, 2-(4-chlorophenyl)-, acetate	27551-11-7	NI28868
91	126	54	65	155	39	80	82	**281**	100	72	63	29	28	16	13	12	**0.00**	13	15	4	1	1	–	–	–	–	–	–	–	–	2(3H)-Furanone, dihydro-2-(tosylamino)-3-vinyl-, (2RS,3RS)-		DD00950
91	54	126	155	65	55	39	82	**281**	100	73	72	29	26	14	14	13	**0.00**	13	15	4	1	1	–	–	–	–	–	–	–	–	2(3H)-Furanone, dihydro-2-(tosylamino)-3-vinyl-, (2RS,3SR)-		DD00951
165	222	43	133	77	105	121	36	**281**	100	73	59	49	26	15	15	15	**0.70**	13	15	6	1	–	–	–	–	–	–	–	–	–	D-Alanine, N-acetyl-3-(3-carboxy-4-methoxyphenyl)-		NI28869
91	108	79	84	107	65	44	77	**281**	100	47	30	27	24	21	18	18	**0.60**	13	15	6	1	–	–	–	–	–	–	–	–	–	L-Glutamic acid, N-[(benzyloxy)carbonyl]-	1155-62-0	NI28870
154	70	128	43	85	112	41	130	**281**	100	78	59	35	26	23	22	20	**0.10**	13	16	2	3	–	–	1	–	–	–	–	–	–	1-(4-Chlorophenoxy)-3-methyl-1-(1,2,4-triazol-1-yl)butan-2-ol		MS06397
109	136	137	41	281	28	167	43	**281**	100	69	59	42	40	39	37	27		13	17	5	2	–	–	–	–	–	–	–	–	–	1H-Pyrrol-1-yloxy, 3-[[(2,5-dioxo-1-pyrrolidinyl)oxy]carbonyl]-2,5-dihydro-2,2,5,5-tetramethyl-	37558-29-5	NI28871
193	73	281	194	89	195	135	75	**281**	100	25	22	19	11	6	6	6		13	19	4	1	–	–	–	–	–	1	–	–	–	Glycine, N-[4-[(trimethylsilyl)oxy]benzoyl]-, methyl ester	55638-48-7	NI28872
29	43	252	41	27	18	57	77	**281**	100	66	62	52	46	43	39	29	**5.60**	13	19	4	3	–	–	–	–	–	–	–	–	–	Aniline, N-(1-ethylpropyl)-3,4-dimethyl-2,6-dinitro-	40487-42-1	NI28873
266	224	41	132	263	43	264	178	**281**	100	90	73	59	49	44	43	39	**27.20**	13	19	4	3	–	–	–	–	–	–	–	–	–	Aniline, 4-tert-butyl-N-isopropyl-2,6-dinitro-	33629-45-7	NI28874
266	224	263	132	264	178	248	281	**281**	100	90	49	49	43	39	37	27		13	19	4	3	–	–	–	–	–	–	–	–	–	Aniline, 4-tert-butyl-N-isopropyl-2,6-dinitro-	33629-45-7	NI28875
75	73	281	266	45	192	76	282	**281**	100	45	25	14	13	12	8	6		13	23	2	1	–	–	–	–	–	2	–	–	–	Benzaldehyde, 4-[(trimethylsilyl)oxy]-, (trimethylsilyl)oxime		NI28876
73	266	250	176	147	45	75	267	**281**	100	79	53	39	35	33	30	19	**3.41**	13	23	2	1	–	–	–	–	–	2	–	–	–	Benzamide, N-(trimethylsilyl)-2-[(trimethylsilyl)oxy]-	55887-58-6	NI28877
266	73	267	268	75	45	281	147	**281**	100	41	25	10	10	10	8	6		13	23	2	1	–	–	–	–	–	2	–	–	–	Benzoic acid, 2-amino-, bis(trimethylsilyl)-	56272-87-8	NI28878
266	73	267	45	268	134	75	147	**281**	100	66	25	18	10	10	10	8	**6.72**	13	23	2	1	–	–	–	–	–	2	–	–	–	Benzoic acid, 2-[(trimethylsilyl)amino]-, trimethylsilyl ester	18406-07-0	NI28879
266	267	73	268	281	192	147	149	**281**	100	23	15	10	7	3	3	2		13	23	2	1	–	–	–	–	–	2	–	–	–	Benzoic acid, 2-[(trimethylsilyl)amino]-, trimethylsilyl ester	18406-07-0	NI28880
73	147	75	43	45	117	204	177	**281**	100	51	29	19	18	13	13	12	**0.00**	13	23	2	1	–	–	–	–	–	2	–	–	–	Benzoic acid, 2-[(trimethylsilyl)amino]-, trimethylsilyl ester	18406-07-0	NI28881
266	192	281	222	45	267	73	282	**281**	100	86	83	42	36	24	22	20		13	23	2	1	–	–	–	–	–	2	–	–	–	Benzoic acid, 4-[(trimethylsilyl)amino]-, trimethylsilyl ester	18406-05-8	NI28882
266	281	192	73	222	282	267	193	**281**	100	93	53	49	37	23	22	9		13	23	2	1	–	–	–	–	–	2	–	–	–	Benzoic acid, 4-[(trimethylsilyl)amino]-, trimethylsilyl ester	18406-05-8	NI28883
120	105	186	246	106	53	228	81	**281**	100	91	60	55	45	36	33	31	**6.50**	14	16	3	1	–	–	1	–	–	–	–	–	–	Cyclobutanecarboxylic acid, 3-chloro-1-[(1-phenylethyl)carbamoyl]-	85318-11-2	NI28884
101	43	252	148	58	266	59	194	**281**	100	79	31	30	19	15	13	9	**6.50**	14	19	5	1	–	–	–	–	–	–	–	–	–	3-C-Cyanomethyl-3-desoxy-1,2,5,6-di-O-isopropylidene-α-D-erythro-hex-3-enofuranose		NI28885

MASS TO CHARGE RATIOS								M.W.	INTENSITIES								Parent	C	H	O	N	S	F	Cl	Br	I	Si	P	B	X	COMPOUND NAME	CAS Reg No	No
195	281	196	66	81	152	77	53	**281**	100	17	17	9	8	7	7	6		14	19	5	1	–	–	–	–	–	–	–	–	–	Morpholine, 4-(3,4,5-trimethoxybenzoyl)-	635-41-6	NI28886
212	43	41	55	127	69	57	126	**281**	100	86	78	73	62	56	42	38	**1.60**	14	26	1	1	–	3	–	–	–	–	–	–	–	Dodecane, 1-[(trifluoroacetyl)amino]-		MS06398
102	73	103	147	45	179	75	74	**281**	100	58	12	8	8	7	6	5	**0.00**	14	27	1	1	–	–	–	–	–	2	–	–	–	Phenylethanolamine, bis(trimethylsilyl)-		MS06399
73	281	266	209	251	45	194	75	**281**	100	76	58	57	39	39	36	25		14	27	1	1	–	–	–	–	–	2	–	–	–	Silanamine, N-[2,6-dimethyl-4-[(trimethylsilyl)oxy]phenyl]-1,1,1-trimethyl-	72088-09-6	NI28887
101	266	282	193	194	210	267	221	**281**	100	39	33	32	13	12	11	8	**0.00**	14	27	1	1	–	–	–	–	–	2	–	–	–	Tyramine, bis(trimethylsilyl)-		NI28888
102	73	266	103	40	75	45	59	**281**	100	52	11	11	9	8	8	6	**0.00**	14	27	1	1	–	–	–	–	–	2	–	–	–	Tyramine, bis(trimethylsilyl)-		MS06400
248	249	77	172	105	51	250	89	**281**	100	47	19	16	13	8	7	6	**1.00**	15	11	1	3	1	–	–	–	–	–	–	–	–	Acetophenone, 2-(pyrido[3,4-d]pyridazin-5-ylthio)-	18599-27-4	NI28889
248	105	249	77	172	176	51	89	**281**	100	54	40	33	18	13	12	8	**6.80**	15	11	1	3	1	–	–	–	–	–	–	–	–	Acetophenone, 2-(pyrido[3,4-d]pyridazin-8-ylthio)-	18599-28-5	NI28890
253	280	281	206	252	234	254	205	**281**	100	99	88	68	62	50	42	41		15	11	3	3	–	–	–	–	–	–	–	–	–	2H-1,4-Benzodiazepin-2-one, 1,3-dihydro-7-nitro-5-phenyl-	146-22-5	NI28891
251	205	281	252	234	76	77	235	**281**	100	62	61	41	41	30	28	25		15	11	3	3	–	–	–	–	–	–	–	–	–	2H-1-Benzodiazepin-2-one, 1,3-dihydro-5-(2-nitrophenyl)-		MS06401
253	281	280	234	206	252	254	205	**281**	100	72	59	44	41	38	36	28		15	11	3	3	–	–	–	–	–	–	–	–	–	2H-1-Benzodiazepin-2-one, 1,3-dihydro-5-(3-nitrophenyl)-		MS06402
280	253	281	206	252	234	251	254	**281**	100	95	83	62	60	58	49	48		15	11	3	3	–	–	–	–	–	–	–	–	–	2H-1-Benzodiazepin-2-one, 1,3-dihydro-7-nitro-5-phenyl-	146-22-5	NI28892
234	280	253	206	264	281	254	205	**281**	100	99	95	85	65	60	55	48		15	11	3	3	–	–	–	–	–	–	–	–	–	2H-1-Benzodiazepin-2-one, 1,3-dihydro-7-nitro-5-phenyl-	146-22-5	MS06403
144	89	281	143	116	115	117	138	**281**	100	40	31	18	17	14	5	4		15	11	3	3	–	–	–	–	–	–	–	–	–	3′-Nitroindole-2-carboxanilide		LI06053
144	116	89	281	143	115	117	137	**281**	100	14	11	9	4	4	3	1		15	11	3	3	–	–	–	–	–	–	–	–	–	3′-Nitroindole-3-carboxanilide		LI06054
144	89	281	115	143	116	117	122	**281**	100	40	31	18	17	17	7	3		15	11	3	3	–	–	–	–	–	–	–	–	–	4′-Nitroindole-2-carboxanilide		LI06055
144	116	89	281	143	115	117	138	**281**	100	20	13	10	5	5	4	2		15	11	3	3	–	–	–	–	–	–	–	–	–	4′-Nitroindole-3-carboxanilide		LI06056
91	120	92	36	65	121	103	77	**281**	100	58	10	8	7	6	3	3	**0.00**	15	17	–	1	–	–	2	–	–	–	–	–	–	Phenethylamine, N-benzyl-4-chloro-, hydrochloride	13622-43-0	NI28893
114	57	224	58	88	59	70	281	**281**	100	90	34	33	31	31	27	26		15	23	4	1	–	–	–	–	–	–	–	–	–	2,3-Benzo-10-methyl-1,4,7,13-tetraoxa-10-azacyclopentadecane		MS06404
86	266	57	117.5	41	70	87	71	**281**	100	35	29	16	12	11	8	8	**0.10**	15	23	4	1	–	–	–	–	–	–	–	–	–	1-(4-Methoxycarbonylphenoxy)-3-tert-butylaminopropan-2-ol		MS06405
166	238	281	167	178	239	165	136	**281**	100	65	34	13	12	10	9	9		15	23	4	1	–	–	–	–	–	–	–	–	–	Methyl α-(2-morpholinocyclohexenyl)acetoacetate		MS06406
168	238	183	151	152	281			**281**	100	64	52	51	23	6				15	23	4	1	–	–	–	–	–	–	–	–	–	Methyl N-(1-oxohexyl)-α-acetyl-Δ²,α-pyrrolidineacetate		LI06057
84	55	41	69	126	43	112	263	**281**	100	65	56	52	47	35	34	33	**1.05**	15	23	4	1	–	–	–	–	–	–	–	–	–	2,6-Piperidinedione, 4-[2-(3,5-dimethyl-2-oxocyclohexyl)-2-hydroxyethyl]-, [1S-[1α(S*),3α,5β]]-	66-81-9	NI28894
84	41	55	69	126	43	82	83	**281**	100	94	76	62	47	43	42	38	**0.00**	15	23	4	1	–	–	–	–	–	–	–	–	–	2,6-Piperidinedione, 4-[2-(3,5-dimethyl-2-oxocyclohexyl)-2-hydroxyethyl]-, [1S-[1α(S*),3α,5β]]-	66-81-9	IC03619
266	238	210	236	162	164	208	192	**281**	100	49	45	27	20	12	11	7	**5.00**	15	23	4	1	–	–	–	–	–	–	–	–	–	3,5-Pyridinedicarboxylic acid, 1,4-dihydro-1,2,4,6-tetramethyl-, diethyl ester	14258-08-3	NI28895
156	111	113	83	126	70	168	55	**281**	100	37	15	15	10	10	9	9	**1.00**	15	27	2	3	–	–	–	–	–	–	–	–	–	1H-1,4-Diazepine-5,7(2H,6H)-dione, 3-(dimethylamino)-2,2-dimethyl-6,6-isopropyl-	69315-97-5	NI28896
124	194	238	97	95	266	225	111	**281**	100	85	33	33	30	28	28	24	**24.00**	15	27	2	3	–	–	–	–	–	–	–	–	–	L-Histidine, N,1-diisopropyl-, isopropyl ester	56771-69-8	MS06407
29	237	42	253	41	138	113	56	**281**	100	56	36	34	34	15	15	15	**1.00**	15	27	2	3	–	–	–	–	–	–	–	–	–	4H-Imidazole-2-acetic acid, 5-(dimethylamino)-α,α-diethyl-4,4-dimethyl-, ethyl ester	69315-94-2	NI28897
131	281	145	144	163	160	77	146	**281**	100	98	98	98	94	85	74	70		16	15	–	3	1	–	–	–	–	–	–	–	–	Thiazolo[4,5-b]quinoxaline, 9-methyl-2-phenyl-3a,4,9,9a-tetrahydr0-	99538-35-9	NI28898
266	130	102	267	222	204	203	268	**281**	100	99	97	95	91	73	37	32	**12.10**	16	15	2	1	–	–	–	–	–	1	–	–	–	Phthalimide, N-(dimethylphenylsilyl)-	31634-67-0	NI28899
281	250	237	130	77	221	282	251	**281**	100	75	75	36	22	19	18	16		16	15	2	3	–	–	–	–	–	–	–	–	–	1(2H)-Phthalazinone, 4-(2-hydroxyethylamino)-2-phenyl-		NI28900
84	141	127	81	43	40	69	95	**281**	100	89	86	68	65	50	45	41	**0.50**	16	25	4	–	–	–	–	–	–	–	–	–	–	1-Menthyl 2-propoxy-5,6-dihydropyran-6-carboxylate		LI06058
192	174	193	137	208	190	151	146	**281**	100	19	16	10	8	7	7	7	**2.42**	16	27	3	1	–	–	–	–	–	–	–	–	–	Lycopodane-5,8,12-triol, 15-methyl-, (5β,8R,15S)-	3175-91-5	NI28901
110	85	196	180	197	166	281	252	**281**	100	61	47	35	21	9	1	1		16	27	3	1	–	–	–	–	–	–	–	–	–	Pyridine, 2,5-dihydro-6-methoxy-2-[5-[(tetrahydro-2H-pyran-2-yl)oxy]pentyl]-		NI28902
152	180	134	140	123	220	162	153	**281**	100	27	21	15	14	13	12	12	**10.45**	16	27	3	1	–	–	–	–	–	–	–	–	–	Serratinane-5,8,13-triol, (5α,8α,13S)-	5532-22-9	NI28903
204	219	43	160	176	65	132	220	**281**	100	96	46	39	19	19	18	14	**8.00**	17	15	3	1	–	–	–	–	–	–	–	–	–	Benzoic acid, 4-(1-methyl-2-benzoylethylideneamino)-		NI28904
266	251	281	43	57	267	41	175	**281**	100	25	24	24	20	19	18	14		17	15	3	1	–	–	–	–	–	–	–	–	–	3H-Furo[2,3-b]pyrano[2,3-h]quinoline, 7-methoxy-3,3-dimethyl-	80151-78-6	NI28905
266	251	281	267	252	195	222	194	**281**	100	28	19	18	6	6	5	5		17	15	3	1	–	–	–	–	–	–	–	–	–	Furo[2,3-b]quinoline, 7-(1,1-dimethylpropynyloxy)-4-methoxy-	81781-88-6	NI28906
237	282	283	238	254	210	204	266	**281**	100	96	16	15	9	6	5	4	**0.00**	17	15	3	1	–	–	–	–	–	–	–	–	–	1H-Indole-3-acetic acid, 5-(phenylmethoxy)-	4382-53-0	NI28907
281	238	44	71	249	210	118	223	**281**	100	95	88	61	60	59	59	46		17	19	1	3	–	–	–	–	–	–	–	–	–	11H-Dibenzo[b,e][1,4]diazepin-11-one, 5,10-dihydro-5-[3-(methylamino)propyl]-	13450-70-9	NI28908
281	264	239	173	188	266	172	240	**281**	100	22	21	18	13	10	8	6		17	19	1	3	–	–	–	–	–	–	–	–	–	Quinoline, 2-(3,4,5-trimethylpyrazol-1-yl)-4-methyl-6-methoxy-		MS06408
91	120	162	43	132	282	106	105	**281**	100	58	27	24	15	12	10	10	**1.12**	18	19	2	1	–	–	–	–	–	–	–	–	–	Acetamide, N-[(3-formyl-2,4,6-cycloheptatrien-1-yl)methyl]-N-benzyl-	52895-47-3	MS06409
281	225	224	210	226	282	266	252	**281**	100	54	33	30	22	20	16	16		18	19	2	1	–	–	–	–	–	–	–	–	–	6-Azaestra-1,3,5(10),6,8-pentaen-17-one, 3-methoxy-	5144-20-7	NI28909
281	238	225	212	224	210	282	252	**281**	100	65	60	50	34	29	20	20		18	19	2	1	–	–	–	–	–	–	–	–	–	6-Azaestra-1,3,5(10),6,8-pentaen-17-one, 3-methoxy-	5144-20-7	NI28910
281	152	253	282	209	153	181	266	**281**	100	32	31	20	20	19	16	15		18	19	2	1	–	–	–	–	–	–	–	–	–	9,10-Dehydro-6-deoxyindolinocodeine		LI06059
280	281	250	276	221	237	252	249	**281**	100	69	44	23	21	14	13	11		18	19	2	1	–	–	–	–	–	–	–	–	–	4H-Dibenzo[de,g]quinoline, 5,6,6a,7-tetrahydro-1,2-dimethoxy-, (R)-	4846-19-9	NI28911
280	281	152	165	139	250	140	115	**281**	100	96	35	34	29	27	24	22		18	19	2	1	–	–	–	–	–	–	–	–	–	4H-Dibenzo[de,g]quinoline, 5,6,6a,7-tetrahydro-1,2-dimethoxy-, (R)-	4846-19-9	NI28912

MASS TO CHARGE RATIOS								M.W.	INTENSITIES								Parent	C	H	O	N	S	F	Cl	Br	I	Si	P	B	X	COMPOUND NAME	CAS Reg No	No
281	253	165	209	153	266	181	238	**281**	100	62	42	39	38	33	33	27		18	19	2	1	–	–	–	–	–	–	–	–	–	7a,9c-(Iminoethano)phenanthro[4,5-bcd]furan, 4aα,5-dihydro-3-methoxy-12-methyl-	24695-70-3	NI28914
281	253	152	209	153	266	181	238	**281**	100	31	31	19	18	15	15	13		18	19	2	1	–	–	–	–	–	–	–	–	–	7a,9c-(Iminoethano)phenanthro[4,5-bcd]furan, 4aα,5-dihydro-3-methoxy-12-methyl-	24695-70-3	NI28913
145	144	137	130	77	281	103	91	**281**	100	68	50	50	44	15	13	13		18	19	2	1	–	–	–	–	–	–	–	–	–	3H-Indole, 2,3-dimethyl-3-(3-methoxy-4-hydroxybenzyl)-		LI06060
105	134	121	120	280	132	253	118	**281**	100	63	25	22	20	20	17	15	**5.00**	18	19	2	1	–	–	–	–	–	–	–	–	–	Oxiranecarboxamide, 3-methyl-3-phenyl-N-[(S)-1-phenylethyl]-		NI28915
117	281	91	264	115	121	130	105	**281**	100	99	95	56	51	48	45	39		18	19	2	1	–	–	–	–	–	–	–	–	–	Tricyclo[10.2.2.$2^{5,8}$]octadeca-5,7,12,14,15,17-hexaene, 5-nitro-		LI06061
117	281	91	264	115	121	130	105	**281**	100	99	95	56	51	48	45	39		18	19	2	1	–	–	–	–	–	–	–	–	–	Tricyclo[10.2.2.$2^{5,8}$]octadeca-5,7,12,14,15,17-hexaene, 6-nitro-	24777-32-0	NI28916
281	266	282	250	141	267	126	237	**281**	100	35	20	12	9	9	9	9		18	23	–	3	–	–	–	–	–	–	–	–	–	3,6-Carbazolediamine, N,N′-dimethyl-9-ethyl-	57103-04-5	NI28917
154	55	41	141	42	281	155	70	**281**	100	34	24	19	16	14	14	13		18	35	1	1	–	–	–	–	–	–	–	–	–	Morpholine, 4-cyclododecyl-2,6-dimethyl-	1593-77-7	NI28918
41	55	59	83	43	85	72	69	**281**	100	75	71	63	59	43	43	39	**4.54**	18	35	1	1	–	–	–	–	–	–	–	–	–	9-Octadecenamide	3322-62-1	NI28919
43	59	41	55	28	72	69	57	**281**	100	99	57	56	55	48	31	30	**2.40**	18	35	1	1	–	–	–	–	–	–	–	–	–	9-Octadecenamide, (Z)-	301-02-0	NI28920
59	72	55	41	43	69	57	83	**281**	100	57	31	30	25	18	16	11	**11.00**	18	35	1	1	–	–	–	–	–	–	–	–	–	9-Octadecenamide, (Z)-	301-02-0	IC03620
59	72	41	55	43	29	44	57	**281**	100	51	49	42	41	24	21	20	**2.00**	18	35	1	1	–	–	–	–	–	–	–	–	–	9-Octadecenamide, (Z)-	301-02-0	NI28921
281	280	227	113.5	100	282	226	228	**281**	100	42	32	24	23	22	17	11		19	11	–	3	–	–	–	–	–	–	–	–	–	1-(1,3,5-Triazin-2-yl)pyrene		IC03621
239	222	266	146	281	43	223	224	**281**	100	99	91	68	62	62	43	33		19	23	1	1	–	–	–	–	–	–	–	–	–	Acetamide, N-(2,6-dimethylphenyl)-N-(2,4,6-trimethylphenyl)-	68014-54-0	NI28922
121	281	225	251	122	92	282	160	**281**	100	33	16	13	13	10	8	8		19	23	1	1	–	–	–	–	–	–	–	–	–	Azetidine, 1-(4-methoxybenzyl)-3,3-dimethyl-2-phenyl-	22606-98-0	NI28923
281	238	266	69	41	225	201	169	**281**	100	93	78	67	57	35	33	31		19	23	1	1	–	–	–	–	–	–	–	–	–	3-Formyl-2-(1,1-dimethylprop-2-enyl)-6-(3,3-dimethylprop-2-enyl)indole		MS06410
281	160	44	42	282	280	147	226	**281**	100	73	28	25	21	18	17	15		19	23	1	1	–	–	–	–	–	–	–	–	–	14α-Morphinan, 7,8-didehydro-3-methoxy-17-methyl-6-methylene-	17939-34-3	NI28924
281	160	44	282	280	42	226	147	**281**	100	26	25	22	21	20	14	14		19	23	1	1	–	–	–	–	–	–	–	–	–	Morphinan, 7,8-didehydro-3-methoxy-17-methyl-6-methylene-, (–)-	1816-06-4	NI28925
99	114	98	167	70	165	96	168	**281**	100	41	32	27	16	13	10	9	**1.40**	19	23	1	1	–	–	–	–	–	–	–	–	–	Piperidine, 4-(diphenylmethoxy)-1-methyl-	147-20-6	NI28926
165	282	109	280					**281**	100	15	13	12					**0.00**	19	23	1	1	–	–	–	–	–	–	–	–	–	Piperidine, 4-(diphenylmethoxy)-1-methyl-	147-20-6	NI28927
91	105	120	119	281	77	162	103	**281**	100	86	43	31	28	25	18	14		19	23	1	1	–	–	–	–	–	–	–	–	–	Propanamide, N,2-dimethyl-3-phenyl-N-[(S)-1-phenylethyl]-		DD00952
112	113	55	110	43	41	42	44	**281**	100	8	6	5	5	5	3	2	**1.30**	19	39	–	1	–	–	–	–	–	–	–	–	–	Cyclohexanamine, N-(2-methyldodec-1-yl)-		IC03622
112	113	110	41	30	55	43	42	**281**	100	12	9	9	9	7	7	5	**3.50**	19	39	–	1	–	–	–	–	–	–	–	–	–	Cyclohexanamine, N-(tridec-1-yl)-		IC03623
154	196	69	55	68	197	82	56	**281**	100	88	22	19	16	14	12	12	**1.00**	19	39	–	1	–	–	–	–	–	–	–	–	–	Pyrrolidine, 2-hexyl-5-nonyl-		NI28928
94	84	95	110	55	108	57	82	**281**	100	70	33	12	12	10	10	7	**0.00**	19	39	–	1	–	–	–	–	–	–	–	–	–	Pyrrolidine, 1-methyl-2-tetradecyl-	3447-09-4	NI28929
44	100	167	266	58	280	281	57	**281**	100	62	29	20	14	13	11	11		20	27	–	1	–	–	–	–	–	–	–	–	–	Benzenepropylamine, N-tert-butyl-α-methyl-γ-phenyl-	15793-40-5	NI28930
281	91	282	280	253	252	203	204	**281**	100	29	23	22	21	15	14	12		21	15	–	1	–	–	–	–	–	–	–	–	–	Benzene, 1-phenyl-4-(2-cyano-2-phenylethenyl)	27869-56-3	NI28931
178	105	77	51	281	152	114	103	**281**	100	21	21	16	10	6	6	5		21	15	–	1	–	–	–	–	–	–	–	–	–	Cyclopropenone, N-phenyl-2,3-diphenyl-, imine		MS06411
155	127	254	282	128	156	141	129	**282**	100	23	14	5	4	2	2	1		2	4	–	–	–	–	–	–	2	–	–	–	–	Ethane, 1,2-diiodo-	624-73-7	NI28932
69	113	163	94	144	194	175	75	**282**	100	50	32	31	19	16	12	12	**2.58**	3	–	–	–	–	9	–	–	–	–	–	–	1	Tris(trifluoromethyl)arsine		MS06412
185	155	165	205	120	170	150	135	**282**	100	53	46	29	22	19	18	12	**0.00**	4	9	–	–	–	–	3	–	–	–	–	–	1	Stannane, trimethyl(trichloromethyl)-	13340-12-0	NI28933
100	131	282	63	113	69	82	263	**282**	100	47	43	39	20	20	15	14		5	–	–	–	1	10	–	–	–	–	–	–	–	Perfluorothiane		MS06413
223	253	208	221	209	267	252	251	**282**	100	86	84	54	50	48	47	47	**0.00**	5	14	–	–	–	–	–	–	–	–	–	–	1	Plumbane, ethyltrimethyl-		AI01963
284	286	282	288	142	249	250	36	**282**	100	79	52	34	21	20	15	15		6	–	–	–	–	–	6	–	–	–	–	–	–	Benzene, hexachloro-	118-74-1	IC03624
284	286	282	288	142	249	107	144	**282**	100	79	51	34	30	24	20	16		6	–	–	–	–	–	6	–	–	–	–	–	–	Benzene, hexachloro-	118-74-1	WI01348
285	287	283	284	286	289	282	288	**282**	100	89	58	45	39	34	21	17		6	–	–	–	–	–	6	–	–	–	–	–	–	Benzene, hexachloro-	118-74-1	NI28934
284	286	282	142	249	288	107	144	**282**	100	80	60	40	37	32	29	25		6	–	–	–	–	–	6	–	–	–	–	–	–	1,3-Cyclopentadiene, 1,2,3,4-tetrachloro-5-(dichloromethylene)-	6317-25-5	NI28935
284	249	286	251	247	282	214	142	**282**	100	94	84	58	57	55	53	53		6	–	–	–	–	–	6	–	–	–	–	–	–	1,2-Dichloro-3,4-bis(dichloromethylene)cyclobutene		LI06062
142	249	247	251	284	286	282	253	**282**	100	69	42	41	38	30	20	14		6	–	–	–	–	–	6	–	–	–	–	–	–	Bis(trichlorovinyl)acetylene		LI06063
249	118	247	120	251	83	59	155	**282**	100	93	80	59	51	48	41	40	**0.00**	6	3	2	–	–	–	5	–	–	–	–	–	–	2,4-Pentadienoic acid, 2,3,4,5,5-pentachloro-, methyl ester	57661-00-4	NI28936
282	284	157	155	76	75	50	74	**282**	100	98	39	39	27	24	23	11		6	4	–	–	–	–	–	1	1	–	–	–	–	1-Bromo-2-iodobenzene		MS06414
282	284	155	157	50	76	75	74	**282**	100	99	53	52	38	33	33	19		6	4	–	–	–	–	–	1	1	–	–	–	–	1-Bromo-3-iodobenzene		MS06415
282	284	155	157	75	76	50	74	**282**	100	97	45	43	28	27	27	17		6	4	–	–	–	–	–	1	1	–	–	–	–	1-Bromo-4-iodobenzene		MS06416
141	71	77	113	107	154	247		**282**	100	70	50	22	22	12	10		**0.00**	6	6	4	–	–	–	4	–	–	–	–	–	–	4,4′,5,5′-Tetrachlorobidioxolan-2-yl		LI06064
77	282	51	27	155	47	135	91	**282**	100	17	17	13	12	10	7	7		6	7	–	–	–	4	–	–	1	–	–	–	–	1-Hexene, 3,3,4,4-tetrafluoro-6-iodo-	40723-84-0	NI28937
233	235	77	133	103	153	27	39	**282**	100	96	82	59	57	51	43	36	**0.00**	6	7	–	–	–	4	–	–	1	–	–	–	–	1-Hexene, 5,5,6,6-tetrafluoro-3-iodo-	74793-39-8	NI28938
36	161	196	66	135	39	65	224	**282**	100	65	51	38	36	29	25	19	**3.07**	8	5	3	–	–	–	1	–	–	–	–	–	1	π-Cyclopentadienyltricarbonylmolybdenum monochloride		MS06417
165	121	91	90	284	282	63	64	**282**	100	30	21	15	11	11	9	8		8	5	3	2	–	–	3	–	–	–	–	–	–	Acetamide, 2,2,2-trichloro-N-(2-nitrophenyl)-	23326-83-2	NI28939
165	137	166	76	64	91	75	63	**282**	100	14	10	10	7	6	6	6	**6.00**	8	5	3	2	–	–	3	–	–	–	–	–	–	Acetamide, 2,2,2-trichloro-N-(3-nitrophenyl)-	56177-38-9	NI28940
165	166	284	282	76	64	63	75	**282**	100	13	6	6	6	6	6	5		8	5	3	2	–	–	3	–	–	–	–	–	–	Acetamide, 2,2,2-trichloro-N-(4-nitrophenyl)-	4306-32-5	NI28941
133	42	44	226	85	282	132	90	**282**	100	55	51	40	40	36	33	19		8	20	3	4	–	–	–	–	–	–	2	–	–	Bis-2-(2-oxo-1,3-dimethyl-1,3,2-diazaphospholidinyl) ether		IC03625
207	73	267	251	193	191	133	208	**282**	100	55	23	21	21	21	21	18	**0.00**	8	26	3	–	–	–	–	–	–	4	–	–	–	Tetrasiloxane, 1,1,1,3,5,7,7,7-octamethyl-		NI28942

MASS TO CHARGE RATIOS								M.W.	INTENSITIES								Parent	C	H	O	N	S	F	Cl	Br	I	Si	P	B	X	COMPOUND NAME	CAS Reg No	No
182	163	132	106	164	107	213	183	**282**	100	50	30	17	14	14	10	10	**3.00**	9	3	–	–	–	9	–	–	–	–	–	–	–	3-Trifluoromethyl-1,6,7,7,8,8-hexafluorobicyclo[2.2.2]octa-2,5-diene		NI28943
168	254	226	282	196	142	115	103	**282**	100	88	86	58	58	31	18	16		9	7	4	–	–	–	–	–	–	–	–	–	1	Rhodium(I), η^5-(methoxycarbonyl)cyclopentadienylbis(η-carbonyl)-	73249-58-8	NI28944
198	55	119	93	150	118	70	226	**282**	100	59	41	25	22	20	14	12	**9.99**	9	7	5	–	1	–	–	–	–	–	–	–	1	Cymantrene, (methylsulphonyl)-	12108-15-5	NI28945
91	92	105	104	65	282	154	152	**282**	100	9	8	5	5	1	1	1		9	11	4	–	2	–	1	–	–	–	–	–	–	Benzene, [[[[(chloromethyl)sulphonyl]methyl]sulphonyl]methyl]-	58751-72-7	NI28946
59	131	218	126	282	159	191	99	**282**	100	67	65	42	41	40	35	32		9	14	–	–	5	–	–	–	–	–	–	–	–	2,4,6,8,9-Pentathiaadamantane, 1,3,5,7-tetramethyl-	17443-92-4	NI28947
59	282	131	218	223	191	284	101	**282**	100	53	49	31	22	22	21	20		9	14	–	–	5	–	–	–	–	–	–	–	–	2,4,6,8,9-Pentathiaadamantane, 1,5,10,10-tetramethyl-	17443-96-8	NI28948
34	59	33	31	39	32	45	282	**282**	100	85	47	29	22	21	19	18	**17.97**	9	14	–	–	5	–	–	–	–	–	–	–	–	2,4,6,8-Tetrathiaadamantane-1-thiol, 3,5,7-trimethyl-	57274-35-8	NI28949
87	45	196	198	43	41	27	39	**282**	100	71	67	66	49	47	35	23	**16.81**	10	9	3	–	–	–	3	–	–	–	–	–	–	Butanoic acid, 4-(2,4,5-trichlorophenoxy)-	93-80-1	NI28950
45	43	41	87	198	196	39	42	**282**	100	96	82	71	47	47	31	26	**1.20**	10	9	3	–	–	–	3	–	–	–	–	–	–	Butanoic acid, 4-(2,4,5-trichlorophenoxy)-	93-80-1	NI28951
59	196	198	87	55	97	200	62	**282**	100	75	72	45	45	40	23	22	**10.00**	10	9	3	–	–	–	3	–	–	–	–	–	–	Propanoic acid, 2-(2,4,5-trichlorophenoxy)-, methyl ester	4841-20-7	NI28953
198	196	59	55	87	223	225	200	**282**	100	97	72	59	53	36	35	30	**20.51**	10	9	3	–	–	–	3	–	–	–	–	–	–	Propanoic acid, 2-(2,4,5-trichlorophenoxy)-, methyl ester	4841-20-7	NI28952
59	55	198	196	87	43	97	62	**282**	100	47	39	38	33	23	20	14	**9.37**	10	9	3	–	–	–	3	–	–	–	–	–	–	Propanoic acid, 2-(2,4,5-trichlorophenoxy)-, methyl ester	4841-20-7	NI28954
196	198	59	55	223	225	87	200	**282**	100	90	64	28	27	27	27	26	**21.73**	10	9	3	–	–	–	3	–	–	–	–	–	–	Silvex, methyl ester		NI28955
101	55	83	59	43	29	45	27	**282**	100	95	45	27	26	26	20	19	**11.78**	10	18	4	–	–	–	–	–	–	–	–	–	1	Selenobisvaleric acid	20846-68-8	NI28956
165	137	109	102	118	92	83	209	**282**	100	76	44	7	6	6	6	4	**0.00**	10	19	7	–	–	–	–	–	–	–	1	–	–	Diethyl 2,2′-[phosphinylidenebis(oxy)]bispropanoate		DD00953
203	284	282	175	91	53	147	204	**282**	100	85	81	44	25	20	20	17		11	7	4	–	–	–	–	1	–	–	–	–	–	1,4-Naphthalenedione, 2-bromo-5,8-dihydroxy-3-methyl-	104506-30-1	NI28957
169	171	284	282	240	238	172	170	**282**	100	99	40	40	13	13	8	8		11	11	2	2	–	–	–	1	–	–	–	–	–	Δ^2-1,3,4-Oxadiazolin-5-one, 4-(p-bromophenyl)-2-isopropyl-	28740-57-0	NI28958
282	148	176	132	193	210	222	105	**282**	100	92	72	62	47	43	41	38		11	14	3	4	1	–	–	–	–	–	–	–	–	Ethyl 5,5a,7,8,9a,10-hexahydro-1,4-thioxino[2,3-g]pteridine-4-carboxylate		LI06065
263	41	42	39	55	53	95	81	**282**	100	70	51	51	47	41	37	36	**0.00**	11	14	5	4	–	–	–	–	–	–	–	–	–	Butanal, 3-hydroxy-3-methyl-, (2,4-dinitrophenyl)hydrazone	56335-69-4	NI28959
263	41	42	39	53	81	95	55	**282**	100	63	44	41	40	38	37	31	**0.00**	11	14	5	4	–	–	–	–	–	–	–	–	–	Butanal, 3-hydroxy-3-methyl-, (2,4-dinitrophenyl)hydrazone	56335-69-4	NI28960
251	41	217	55	282	167	164	252	**282**	100	58	54	35	27	21	19	15		11	14	5	4	–	–	–	–	–	–	–	–	–	Hydroxypivaldehyde (2,4-dinitrophenyl)hydrazone		IC03626
29	151	28	43	30	42	58	44	**282**	100	47	47	31	18	16	15	15	**0.90**	11	14	5	4	–	–	–	–	–	–	–	–	–	Inosine, 1-methyl-	2140-73-0	NI28961
150	42	136	151	43	44	122	54	**282**	100	27	26	18	18	15	12	11	**0.95**	11	14	5	4	–	–	–	–	–	–	–	–	–	Inosine, 1-methyl-	2140-73-0	NI28962
85	71	41	55	31	43	67	27	**282**	100	60	30	25	21	20	19	18	**17.00**	11	14	5	4	–	–	–	–	–	–	–	–	–	Pentanal, 5-hydroxy-, (2,4-dinitrophenyl)hydrazone	3638-33-3	NI28963
85	71	41	67	55	57	43	77	**282**	100	61	38	23	23	18	18	17	**15.00**	11	14	5	4	–	–	–	–	–	–	–	–	–	Pentanal, 5-hydroxy-, (2,4-dinitrophenyl)hydrazone	3638-33-3	NI28964
226	168	254	196	282	142	103	115	**282**	100	86	62	34	30	27	19	13		11	15	2	–	–	–	–	–	–	–	–	–	1	Rhodium(I), η^5-(methoxycarbonyl)cyclopentadienylbis(η^2-ethene)-	36005-12-6	NI28965
69	233	125	107	79	111	83	121	**282**	100	73	58	58	53	48	33	28	**5.70**	11	19	4	–	1	–	1	–	–	–	–	–	–	2,9-Decanedione, 1-chloro-10-(methylsulphonyl)-	76078-77-8	NI28966
57	43	71	85	41	55	155	29	**282**	100	77	50	36	31	21	12	12	**0.80**	11	23	–	–	–	–	–	–	1	–	–	–	–	Undecyl iodide		LI06066
184	282	92	284	139	185	168	283	**282**	100	65	28	25	15	14	13	10		12	8	2	–	1	–	1	–	–	–	1	–	–	O,O′-(2,2′-Biphenylylene)thiophosphoric acid chloride	61335-18-0	NI28967
282	218	284	139	220	109	283	138	**282**	100	58	37	27	19	19	17	15		12	8	2	–	1	–	1	–	–	–	1	–	–	Phenothiaphosphine, 2-chloro-10-hydroxy-, 10-oxide	31121-36-5	NI28968
201	282	266	284	166	268	203	236	**282**	100	75	63	49	43	42	33	31		12	8	2	2	–	–	2	–	–	–	–	–	–	N-Hydroxy-2,3′-dichloro-4-nitrosodiphenylamine		IC03627
141	125	97	152	234	153	109	282	**282**	100	26	9	3	2	2	2	1		12	10	4	–	2	–	–	–	–	–	–	–	–	Diphenyl disulphone		LI06067
282	211	267	239	283	53	140	69	**282**	100	30	27	18	14	10	9	9		12	10	8	–	–	–	–	–	–	–	–	–	–	1,4-Naphthalenedione, 2,3,5,8-tetrahydroxy-6,7-dimethoxy-	54725-02-9	NI28969
211	140	282	239	168	267	155	154	**282**	100	38	32	29	28	21	21	20		12	10	8	–	–	–	–	–	–	–	–	–	–	1,4-Naphthalenedione, 2,5,7,8-tetrahydroxy-3,6-dimethoxy-	14090-99-4	NI28970
211	282	140	239	168	267	155	154	**282**	100	50	37	29	27	20	20	20		12	10	8	–	–	–	–	–	–	–	–	–	–	1,4-Naphthalenedione, 2,5,7,8-tetrahydroxy-3,6-dimethoxy-	14090-99-4	NI28971
56	43	65	42	58	92	55	93	**282**	100	96	48	24	21	17	16	14	**1.19**	12	14	4	2	1	–	–	–	–	–	–	–	–	3-Azetidinone, 1-[(4-acetylamino)benzenesulphonyl]-2-methyl-		NI28972
42	43	198	65	134	56	92	41	**282**	100	87	30	30	18	16	15	13	**1.93**	12	14	4	2	1	–	–	–	–	–	–	–	–	3-Pyrrolidinone, 1-[(4-acetylamino)benzenesulphonyl]-	80076-99-9	NI28973
81	83	82	193	124	282	150	69	**282**	100	17	8	7	6	4	4	3		12	14	6	2	–	–	–	–	–	–	–	–	–	3,6-Epoxy-2H,8H-pyrimido[6,1-b][1,3]oxazocine-8,10(9H)-dione, 4-(acetyloxy)-3,4,5,6-tetrahydro-11-methyl-, [3R-(3α,4β,6α)]-	41547-71-1	NI28974
253	77	197	105	49	47	119	93	**282**	100	96	85	46	36	32	21	20	**7.10**	12	15	–	–	–	5	–	–	–	1	–	–	–	Silane, triethyl(pentafluorophenyl)-	35369-98-3	NI28975
109	77	125	81	51	129	91	137	**282**	100	63	53	51	36	16	16	15	**6.92**	12	15	4	2	–	–	–	–	–	–	1	–	–	3,5-Dioxa-6-aza-4-phosphaoct-6-ene-8-nitrile, 4-ethoxy-7-phenyl-, 4-oxide	14816-17-2	NI28976
282	249	253	254	267	240	41	125	**282**	100	68	52	49	48	33	31	23		12	18	–	4	2	–	–	–	–	–	–	–	–	Thiazolo[5,4-d]pyrimidine, 2-(butylamino)-5-(propylthio)-	19844-52-1	NI28977
99	127	139	111	98	126	57	83	**282**	100	85	38	27	26	22	19	12	**0.00**	12	20	6	–	–	–	–	–	–	–	–	2	–	α-D-Glucofuranuronic acid, γ-lactone, cyclic 1,2-(ethylboronate) 5-(diethylborinate)	74779-81-0	NI28978
55	43	181	41	69	97	113	196	**282**	100	86	70	60	58	54	51	50	**0.50**	12	20	6	–	–	–	–	–	–	–	–	2	–	Tartaric acid butyl boronate		MS06418
99	127	155	125	153	183	113	111	**282**	100	12	10	10	6	2	1	1	**0.00**	12	27	5	–	–	–	–	–	–	–	1	–	–	Dibutyl 3-hydroxybutyl phosphate	89197-69-3	NI28979
282	236	237	281	263	145	44	93	**282**	100	94	76	74	70	28	22	16		13	9	2	2	–	3	–	–	–	–	–	–	–	3-Pyridinecarboxylic acid, 2-[[3-(trifluoromethyl)phenyl]amino]-	4394-00-7	NI28980
55	227	197	142	141	255	283	282	**282**	100	57	57	53	44	7	1	1		13	9	3	1	–	–	–	–	–	–	–	–	1	Manganese(1+), dicarbonylnitrosyl[(1,2,3,4,5-η)-1-phenyl-2,4-cyclopentadien-1-yl]-	56784-33-9	MS06419
139	281	282	119	124	66	190	214	**282**	100	90	72	67	55	42	40	36		13	10	–	–	–	3	–	–	–	–	–	–	1	Cobalt, (η^5-2,4-cyclopentadien-1-yl)[(1,2,3,4-η)-5-(3,3,3-trifluoro-1-propynyl)-1,3-cyclopentadiene]-	55991-82-7	NI28981
93	77	282	284	110	175	161	65	**282**	100	87	70	69	69	64	56	54		13	15	–	–	1	–	–	1	–	–	–	–	–	1,3-Heptadiene, 7-bromo-3-(phenylthio)-, (E)-	123289-42-9	DD00954
175	91	77	284	282	65	110	147	**282**	100	89	81	76	76	76	55	47		13	15	–	–	1	–	–	1	–	–	–	–	–	1,3-Heptadiene, 7-bromo-3-(phenylthio)-, (Z)-	123289-35-0	DD00955
253	43	41	251	27	184	105	211	**282**	100	69	54	50	48	34	32	27	**26.25**	13	18	–	2	–	–	–	–	–	–	–	–	1	Selenocyanic acid, 4-(dipropylamino)phenyl ester	22037-09-8	NI28982
253	43	41	251	27	184	105	282	**282**	100	69	53	49	48	33	32	26		13	18	–	2	–	–	–	–	–	–	–	–	1	Selenocyanic acid, 4-(dipropylamino)phenyl ester	22037-09-8	LI06068
41	43	103	39	104	56	119	120	**282**	100	92	58	52	46	39	37	32	**0.43**	13	18	1	2	2	–	–	–	–	–	–	–	–	N-Acetyl-N-tert-butyldisulphido-N′-benzylidenehydrazine		IC03628

MASS TO CHARGE RATIOS								M.W.	INTENSITIES								Parent	C	H	O	N	S	F	Cl	Br	I	Si	P	B	X	COMPOUND NAME	CAS Reg No	No
165	166	73	267	122	75	77	107	282	100	10	9	6	4	3	3	3	0.00	13	18	5	–	–	–	–	–	–	1	–	–	–	Formic acid, 3,4-dimethoxybenzoyl-, trimethylsilyl ester		NI28983
267	73	268	269	126	282	195	75	282	100	28	23	10	6	4	4	3		13	22	3	–	–	–	–	–	–	2	–	–	–	Benzaldehyde, 2,4-bis(trimethylsiloxy)-	33617-38-8	NI28984
267	268	73	269	282	126	249	117	282	100	25	18	10	9	4	3	3		13	22	3	–	–	–	–	–	–	2	–	–	–	Benzaldehyde, 2,5-bis[(trimethylsilyl)oxy]-	56114-69-3	NI28985
267	268	73	269	45	282	117	126	282	100	32	28	13	12	10	6	5		13	22	3	–	–	–	–	–	–	2	–	–	–	Benzaldehyde, 2,5-bis[(trimethylsilyl)oxy]-	56114-69-3	MS06420
73	267	45	135	43	74	268	75	282	100	33	25	10	10	9	8	8	0.00	13	22	3	–	–	–	–	–	–	2	–	–	–	Benzoic acid, 2-[(trimethylsilyl)oxy]-, trimethylsilyl ester	3789-85-3	NI28986
267	73	268	269	45	193	135	75	282	100	99	25	13	12	9	9	8	0.00	13	22	3	–	–	–	–	–	–	2	–	–	–	Benzoic acid, 2-[(trimethylsilyl)oxy]-, trimethylsilyl ester	3789-85-3	NI28987
73	267	45	268	135	74	75	43	282	100	51	19	12	9	9	8	8	0.00	13	22	3	–	–	–	–	–	–	2	–	–	–	Benzoic acid, 2-[(trimethylsilyl)oxy]-, trimethylsilyl ester	3789-85-3	NI28988
267	73	193	282	223	268	45	75	282	100	58	43	37	32	22	13	10		13	22	3	–	–	–	–	–	–	2	–	–	–	Benzoic acid, 3-[(trimethylsilyl)oxy]-, trimethylsilyl ester	3782-84-1	MS06421
267	73	193	282	223	268	45	126	282	100	81	48	40	40	24	22	14		13	22	3	–	–	–	–	–	–	2	–	–	–	Benzoic acid, 3-[(trimethylsilyl)oxy]-, trimethylsilyl ester	3782-84-1	NI28989
267	73	282	193	223	268	45	186	282	100	80	42	38	31	23	20	19		13	22	3	–	–	–	–	–	–	2	–	–	–	Benzoic acid, 3-[(trimethylsilyl)oxy]-, trimethylsilyl ester	3782-84-1	NI28990
267	223	193	73	282	268	77	75	282	100	61	48	45	22	21	14	14		13	22	3	–	–	–	–	–	–	2	–	–	–	Benzoic acid, 4-[(trimethylsilyl)oxy]-, trimethylsilyl ester	2078-13-9	MS06422
73	267	193	223	282	268	45	126	282	100	90	60	59	26	21	21	17		13	22	3	–	–	–	–	–	–	2	–	–	–	Benzoic acid, 4-[(trimethylsilyl)oxy]-, trimethylsilyl ester	2078-13-9	NI28991
267	223	193	73	75	77	93	282	282	100	56	52	51	46	30	29	26		13	22	3	–	–	–	–	–	–	2	–	–	–	Benzoic acid, 4-[(trimethylsilyl)oxy]-, trimethylsilyl ester	2078-13-9	NI28992
126	70	152	139	125	127	44	41	282	100	64	44	42	30	26	24	22	1.05	13	24	5	–	–	–	–	–	–	–	–	2	–	β-L-Arabinopyranose 1,2:3,4-bis(butaneboronate)	82284-59-1	NI28993
139	127	126	83	41	43	55	97	282	100	82	57	48	48	46	42	40	1.90	13	24	5	–	–	–	–	–	–	–	–	2	–	α-D-Xylofuranose, cyclic 1,2:3,5-bis(butylboronate)	52572-01-7	NI28994
139	127	126	125	138	97	152	113	282	100	71	55	33	32	32	31	29	1.49	13	24	5	–	–	–	–	–	–	–	–	2	–	α-D-Xylofuranose, cyclic 1,2:3,5-bis(butylboronate)	52572-01-7	NI28995
212	176	214	284	282	213	247	177	282	100	50	33	25	25	18	13	12		14	9	–	–	–	–	3	–	–	–	–	–	–	Benzene, 1-chloro-2-[2-chloro-1-(4-chlorophenyl)vinyl]-	14835-94-0	NI28996
212	88	176	75	214	105	87	106	282	100	74	66	51	33	29	28	26	26.05	14	9	–	–	–	–	3	–	–	–	–	–	–	Benzene, 1-chloro-2-[2-chloro-1-(4-chlorophenyl)vinyl]-	14835-94-0	NI28997
171	173	285	283	247	249	137	139	282	100	63	42	41	35	22	21	15	4.60	14	9	–	–	–	–	3	–	–	–	–	–	–	Ethylene, 2-chloro-1,1-bis(4-chlorophenyl)-	1022-22-6	NI28998
212	176	282	284	214	247	213	75	282	100	54	48	47	33	18	17	16		14	9	–	–	–	–	3	–	–	–	–	–	–	Ethylene, 2-chloro-1,1-bis(4-chlorophenyl)-	1022-22-6	NI28999
212	282	284	176	214	88	75	286	282	100	68	67	49	42	41	36	24		14	9	–	–	–	–	3	–	–	–	–	–	–	Ethylene, 2-chloro-1,1-bis(4-chlorophenyl)-	1022-22-6	NI29000
282	281	283	42	134	43	77	265	282	100	38	18	18	13	10	8	8		14	10	1	4	1	–	–	–	–	–	–	–	–	Pyrazolo[3,4-d]thiazolo[3,2-a]pyrimidin-4(2H)-one, 2-methyl-6-phenyl-		NI29001
198	142	141	226	121	56	254	115	282	100	84	53	51	44	42	24	23	0.08	14	10	3	–	–	–	–	–	–	–	–	–	1	Iron, π-cyclopentadienylbenzoyldicarbonyl-		MS06423
237	192	282	136	165	239	164	101	282	100	70	64	52	40	34	33	26		14	15	4	–	–	–	1	–	–	–	–	–	–	Malonic acid, (2-chlorobenzylidene)-, diethyl ester	6768-21-4	NI29003
173	282	247	284	237	219	174	101	282	100	64	57	22	21	12	11	10		14	15	4	–	–	–	1	–	–	–	–	–	–	Malonic acid, (2-chlorobenzylidene)-, diethyl ester	6768-20-3	NI29002
81	91	65	155	55	127	99	282	282	100	87	44	27	25	24	19	11		14	18	4	–	1	–	–	–	–	–	–	–	–	Cyclohexanone, 2-methyl-6-[(4-toluenesulphonyl)oxy]-		DD00956
105	204	131	264	237	282			282	100	40	35	15	10	5				14	18	4	–	1	–	–	–	–	–	–	–	–	Ethyl 3-benzoyl-3-(methylsulphinyl)butyrate		LI06069
91	41	39	65	79	77	55	53	282	100	75	71	69	63	49	49	44	0.00	14	18	4	–	1	–	–	–	–	–	–	–	–	2,5-Heptadiene-1,4-diol, 3-[(4-methylphenyl)sulphonyl]-, (E,E)-	118356-22-2	DD00957
149	150	41	104	76	205	105	56	282	100	10	9	5	5	3	3	3	0.05	14	18	6	–	–	–	–	–	–	–	–	–	–	1,2-Benzenedicarboxylic acid, bis(2-methoxyethyl) ester	117-82-8	NI29004
208	209	193	240	235	194	43	149	282	100	80	64	63	53	41	38	36	4.00	14	18	6	–	–	–	–	–	–	–	–	–	–	Bicyclo[2.2.2]oct-2-ene-2,3-dicarboxylic acid, 1-hydroxy-8,8-dimethyl-5-oxo-, dimethyl ester	25864-65-7	MS06424
181	282	223	250	191	251	239	149	282	100	51	46	29	24	21	18	16		14	18	6	–	–	–	–	–	–	–	–	–	–	Bicyclo[3.3.0]oct-3-ene-2,4-dicarboxylic acid, 3-methoxy-2-methyl-7-oxo, dimethyl ester, (1α,2β,5α)-		MS06425
43	194	236	240	165	211	237	212	282	100	78	53	43	41	40	22	20	9.00	14	18	6	–	–	–	–	–	–	–	–	–	–	3-Furancarboxylic acid, 2-methyl-5-[1-(ethoxycarbonyl)-2-oxopropyl]-, ethyl ester		MS06426
43	194	236	165	29	237	166	121	282	100	62	36	34	23	13	11	10	3.00	14	18	6	–	–	–	–	–	–	–	–	–	–	Furo[3,2-b]furan-3,6-dicarboxylic acid, 3a,6a-dihydro-2,5-dimethyl-, diethyl ester, cis-		MS06427
100	41	97	267	141	140	281	60	282	100	82	77	58	53	42	40	39	32.99	14	19	–	2	1	–	1	–	–	–	–	–	–	4H-1,3-Thiazine, 4,4,6-trimethyl-2-(N-methyl-3-chlorophenylamino)-5,6-dihydro-	53004-56-1	NI29005
209	73	75	210	211	179	77	45	282	100	64	20	15	11	10	9	9	0.00	14	22	4	–	–	–	–	–	–	1	–	–	–	Benzeneacetic acid, 2-methoxy-α-[(trimethylsilyl)oxy]-, ethyl ester		MS06428
105	209	282	252	179	77	73	134	282	100	54	39	37	37	34	31	24		14	22	4	–	–	–	–	–	–	1	–	–	–	Benzeneacetic acid, 3-methoxy-4-[(trimethylsilyl)oxy]-, ethyl ester	55590-92-6	NI29006
209	282	252	207	73	179	267	75	282	100	87	69	68	65	62	38	32		14	22	4	–	–	–	–	–	–	1	–	–	–	Benzeneacetic acid, 3-methoxy-4-[(trimethylsilyl)oxy]-, ethyl ester	55590-92-6	MS06429
209	73	210	75	211	45	135	131	282	100	53	15	9	7	7	6	6	0.00	14	22	4	–	–	–	–	–	–	1	–	–	–	Benzeneacetic acid, 4-methoxy-α-[(trimethylsilyl)oxy]-, ethyl ester		MS06430
282	164	73	151	121	75	267	283	282	100	83	56	56	26	25	24	21		14	22	4	–	–	–	–	–	–	1	–	–	–	Benzenepropanoic acid, 2,5-dimethoxy-, trimethylsilyl ester		NI29007
151	164	282	73	75	165	152	149	282	100	60	39	16	10	9	9	9		14	22	4	–	–	–	–	–	–	1	–	–	–	Benzenepropanoic acid, 3,5-dimethoxy-, trimethylsilyl ester		MS06431
86	267	96	72	116	282	136	238	282	100	80	60	53	45	40	40	30		14	22	4	2	–	–	–	–	–	–	–	–	–	1-(tert-Butylamino)-3-(3-methyl-2-nitrophenoxy)-2-propanol		MS06432
169	212	156	183	41	43	112	55	282	100	85	62	52	31	26	20	20	0.00	14	22	4	2	–	–	–	–	–	–	–	–	–	2,4,6(1H,3H,5H)-Pyrimidinetrione, 1,3-dimethyl-5-(1-methylbutyl)-5-(oxiranylmethyl)-	57397-45-2	NI29008
128	55	83	235	41	69	237	42	282	100	58	50	50	49	41	29	27	18.30	14	24	3	3	–	–	–	–	–	–	–	–	–	3-Imidazoline, 4-(cyclohexylcarbamoyl)-2,2,5,5-tetramethyl-, 3-oxide 1-oxile		NI29009
73	103	179	147	45	193	75	74	282	100	36	20	18	16	15	11	9	8.83	14	26	2	–	–	–	–	–	–	2	–	–	–	Benzene, 2-[(trimethylsilyl)oxy]-1-[2-[(trimethylsilyl)oxy]ethyl]-		NI29010
179	73	282	180	193	267	103	75	282	100	54	18	15	12	11	10	7		14	26	2	–	–	–	–	–	–	2	–	–	–	Benzene, 4-[(trimethylsilyl)oxy]-1-[2-[(trimethylsilyl)oxy]ethyl]-		MS06433
179	73	282	180	267	193	103	75	282	100	43	18	15	12	12	8	6		14	26	2	–	–	–	–	–	–	2	–	–	–	Benzene, 4-[(trimethylsilyl)oxy]-1-[2-[(trimethylsilyl)oxy]ethyl]-		MS06434
73	282	147	45	283	75	74	92	282	100	43	23	15	11	9	9	7		14	26	2	–	–	–	–	–	–	2	–	–	–	Bicyclo[4.2.0]octa-3,7-diene, 7,8-bis(trimethylsiloxy)-	36461-33-3	NI29011
267	140	126	209	57	27	84	281	282	100	21	19	18	8	8	6	1	0.30	14	32	–	2	–	–	–	–	–	–	–	–	2	Dialuminium, tetramethyldi-μ-1-piperidinyl-	36523-33-8	MS06435
226	121	56	105	222	254	227	91	282	100	79	58	40	27	20	18	17	0.49	15	14	2	–	–	–	–	–	–	–	–	–	1	Iron, (2-methylbenzyl)dicarbonyl-π-cyclopentadienyl-	74102-45-7	NI29012

MASS TO CHARGE RATIOS								M.W.	INTENSITIES								Parent	C	H	O	N	S	F	Cl	Br	I	Si	P	B	X	COMPOUND NAME	CAS Reg No	No
121	104	105	56	78	103	77	91	**282**	100	91	82	72	43	41	37	30	**0.00**	15	14	2	–	–	–	–	–	–	–	–	–	1	Iron, (1-phenylethyl)dicarbonyl-π-cyclopentadienyl-	25153-31-5	NI29013
226	56	225	160	121	198	227	104	**282**	100	63	57	46	31	22	15	11	**5.13**	15	14	2	–	–	–	–	–	–	–	–	–	1	Iron, (2-phenylethyl)dicarbonyl-π-cyclopentadienyl-	32760-31-9	NI29014
197	282	198	153	254	226	170	125	**282**	100	88	73	66	58	51	50	30		15	22	5	–	–	–	–	–	–	–	–	–	–	Benzoic acid, 3,4,5-triethoxy-, ethyl ester	31374-71-7	NI29015
69	41	85	140	264	214	44	57	**282**	100	37	35	28	27	24	22	21	**4.24**	15	22	5	–	–	–	–	–	–	–	–	–	–	2-Cyclopenten-1-one, 3,4,5-trihydroxy-5-(3-methyl-2-butenyl)-2-(3-methyl-1-oxobutyl)-	469-30-7	NI29016
69	41	43	57	42	214	140	56	**282**	100	82	68	51	36	31	25	21	**4.00**	15	22	5	–	–	–	–	–	–	–	–	–	–	2-Cyclopenten-1-one, 3,4,5-trihydroxy-5-(3-methyl-2-butenyl)-2-(3-methyl-1-oxobutyl)-	469-30-7	NI29017
101	43	267	179	73	137	133	91	**282**	100	34	28	18	13	9	8	8	**0.00**	15	22	5	–	–	–	–	–	–	–	–	–	–	6-Decyne-1-carboxylic acid, 9,10-di-O-isopropylidene-9,10-dihydroxy-3-oxo-, ethyl ester, (S)-		DD00958
250	153	95	121	111	107	94	59	**282**	100	58	47	45	43	37	36	31	**4.00**	15	22	5	–	–	–	–	–	–	–	–	–	–	Heptanoic acid, 7-[5-(methoxycarbonyl)methyl-2-furyl]-, methyl ester		NI29018
91	90	160	97	282	41	69	55	**282**	100	36	28	26	25	25	20	20		15	23	1	–	1	–	–	–	–	–	1	–	–	1-Benzylthio-2,2,3,4,4-pentamethylphosphetan 1-oxide		NI29019
57	156	86	29	41	30	282	100	**282**	100	44	34	23	18	16	15	14		15	23	1	2	–	–	1	–	–	–	–	–	–	N-(3-Chlorophenyl)-N′,N′-dibutylurea		MS06436
98	99	70	42	36	96	55	44	**282**	100	7	5	4	3	1	1	1	**0.00**	15	23	1	2	–	–	1	–	–	–	–	–	–	2-Piperidinecarboxamide, N-(2,6-dimethylphenyl)-1-methyl-, monohydrochloride	1722-62-9	NI29020
98	99	70	42	36	120	96	55	**282**	100	7	5	3	3	1	1	1	**0.00**	15	23	1	2	–	–	1	–	–	–	–	–	–	2-Piperidinecarboxamide, N-(2,6-dimethylphenyl)-1-methyl-, monohydrochloride	1722-62-9	NI29021
282	230	178	151	231	141	177	89	**282**	100	69	37	22	12	10	6	6		16	6	–	6	–	–	–	–	–	–	–	–	–	2,3-Dicyanodipyrido[2,3-f:3′,2′-h]quinoxaline		MS06437
281	282	283	284	77	104	247	76	**282**	100	75	45	27	26	21	17	8		16	11	1	2	–	–	1	–	–	–	–	–	–	2(1H)-Pyrimidinone, 5-chloro-4,6-diphenyl-	28567-83-1	NI29022
110	118	282	90	227	184	124	213	**282**	100	57	32	21	18	18	16	13		16	14	1	2	1	–	–	–	–	–	–	–	–	Benzo[d]thieno[2,3-g]azecine-6-carbonitrile, 4,5,6,7,8,13-hexahydro-13-oxo-		NI29023
77	118	177	51	178	104	91	150	**282**	100	82	68	68	64	40	32	29	**0.00**	16	14	1	2	1	–	–	–	–	–	–	–	–	2-Imidazolidinethione, 1-benzoyl-3-phenyl-	55044-49-0	NI29024
91	282	77	136	135	92	65	191	**282**	100	58	39	21	17	17	10	8		16	14	1	2	1	–	–	–	–	–	–	–	–	2-Imidazolidinethione, 5-benzyl-4-oxo-3-phenyl-		LI06070
282	223	240	205	165	283	194	193	**282**	100	80	50	36	24	18	12	12		16	14	3	2	–	–	–	–	–	–	–	–	–	Acetamide, N-(9,10-dihydro-1-nitro-2-phenanthryl)-	18264-80-7	NI29025
240	47	282	48	49	194	241	165	**282**	100	61	52	31	23	20	18	18		16	14	3	2	–	–	–	–	–	–	–	–	–	Acetamide, N-(9,10-dihydro-3-nitro-2-phenanthryl)-	18264-87-4	NI29026
93	120	43	92	147	121	162	106	**282**	100	58	46	38	33	27	25	25	**25.00**	16	14	3	2	–	–	–	–	–	–	–	–	–	Acetamide, N-[2-[2-(2-nitrophenyl)vinyl]phenyl]-	69395-25-1	NI29027
240	282	194	47	165	55	210	252	**282**	100	55	45	35	26	24	21	20		16	14	3	2	–	–	–	–	–	–	–	–	–	7-Acetylamino-2-nitro-9,10-dihydrophenanthrene		MS06438
176	130	148	175	163	77	104	177	**282**	100	70	27	25	18	16	12	11	**3.00**	16	14	3	2	–	–	–	–	–	–	–	–	–	2-(2-Hydroxy-2-phenylethyl)-4-hydroxy-1-phthalazone		LI06071
209	282	128	210	181	283	237	101	**282**	100	37	27	17	7	6	5	4		16	14	3	2	–	–	–	–	–	–	–	–	–	1H-Pyrimido[1,2-a]quinoline-2-acetic acid, 1-oxo-, ethyl ester	50609-54-6	NI29028
282	196	128	212	223	168	195	283	**282**	100	79	67	50	37	37	19	17		16	14	3	2	–	–	–	–	–	–	–	–	–	1H-Pyrimido[1,2-a]quinoline-2-carboxylic acid, 1-oxo-, isopropyl ester	63455-52-7	NI29029
128	282	223	196	168	69	212	129	**282**	100	75	58	57	53	40	28	24		16	14	3	2	–	–	–	–	–	–	–	–	–	1H-Pyrimido[1,2-a]quinoline-2-carboxylic acid, 1-oxo-, propyl ester	63455-51-6	NI29030
226	282	160	253	227	76	130	77	**282**	100	39	37	32	17	17	15	11		16	14	3	2	–	–	–	–	–	–	–	–	–	4(3H)-Quinazolinone, 3-(2,6-dihydroxyphenyl)-2-ethyl-		NI29031
251	282	136	146	76	252	236	221	**282**	100	68	68	28	23	20	18	18		16	14	3	2	–	–	–	–	–	–	–	–	–	4(3H)-Quinazolinone, 3-(2,6-dimethoxyphenyl)-		NI29032
208	282	223	209	221	91	120	193	**282**	100	40	35	23	21	17	15	13		16	18	1	4	–	–	–	–	–	–	–	–	–	5-Methyl-2-N-methylaminobenzophenone semicarbazone		MS06439
107	133	121	119	120	78	282	79	**282**	100	62	55	52	43	40	37	35		16	18	1	4	–	–	–	–	–	–	–	–	–	1-Pyrid-2-yl-4-anilinocarbonylpiperazine		IC03629
198	166	95	199	167	180	251	41	**282**	100	80	28	21	21	19	18	16	**7.01**	16	26	4	–	–	–	–	–	–	–	–	–	–	2(5H)-Furanone, 3-butyryl-5-hexyl-4-methoxy-5-methyl-	21494-11-1	NI29033
119	55	83	101	67	74	165	156	**282**	100	52	35	34	14	10	4	4	**0.00**	16	26	4	–	–	–	–	–	–	–	–	–	–	Malonic acid, methyl-, dicyclohexyl ester	73742-26-4	NI29034
222	43	180	135	122	121	81	80	**282**	100	81	72	72	64	48	48	48	**19.77**	16	26	4	–	–	–	–	–	–	–	–	–	–	Oxacyclotetradec-10-en-2-one, 5-acetoxy-14-methyl-		NI29035
101	83	57	119	99	67	82	56	**282**	100	85	67	43	37	31	27	23	**0.00**	16	26	4	–	–	–	–	–	–	–	–	–	–	Succinic acid, dicyclohexyl ester		IC03630
101	119	83	55	41	82	99	67	**282**	100	98	60	46	28	23	22	19	**0.00**	16	26	4	–	–	–	–	–	–	–	–	–	–	Succinic acid, dicyclohexyl ester		IC03631
58	28	74	136	114	239	57	140	**282**	100	95	73	58	50	40	29	27	**16.10**	16	30	2	2	–	–	–	–	–	–	–	–	–	Propanamide, N-[1-methyl-1-[4-methyl-4-[(1-oxopropyl)amino]cyclohexyl]ethyl]-	35046-17-4	NI29036
58	224	73	116	102	140	225	95	**282**	100	59	31	26	24	9	8	7	**0.00**	16	34	–	2	–	–	–	–	–	1	–	–	–	3-[Dimethyl[[[2-(N,N-dimethylamino)ethyl]-N-methylamino]methyl]silyl]-6-methyl-1,5-heptadiene		DD00959
119	91	163	120	65	135	251	282	**282**	100	26	13	12	8	3	2	1		17	14	4	–	–	–	–	–	–	–	–	–	–	4-Acetoxy-4′-methylbenzil		IC03632
251	282	252	236	151	141	131	106	**282**	100	30	18	8	6	5	4	4		17	14	4	–	–	–	–	–	–	–	–	–	–	Aurone, 2′,6-dimethoxy-	77764-83-1	NI29037
281	282	251	267	239	266	252	141	**282**	100	85	33	8	8	6	6	6		17	14	4	–	–	–	–	–	–	–	–	–	–	Aurone, 3′,6-dimethoxy-	77764-84-2	NI29038
282	281	132	239	251	267	266	141	**282**	100	72	24	14	12	11	6	6		17	14	4	–	–	–	–	–	–	–	–	–	–	Aurone, 4′,6-dimethoxy-	36685-48-0	NI29039
223	221	250	222	282	165	224	178	**282**	100	80	77	44	37	36	18	16		17	14	4	–	–	–	–	–	–	–	–	–	–	2-Benzofuranacetic acid, α-(2-hydroxyphenyl)-, methyl ester	40800-99-5	NI29040
221	250	222	282	223	165	251	44	**282**	100	93	78	59	37	34	24	15		17	14	4	–	–	–	–	–	–	–	–	–	–	3-Benzofurancarboxylic acid, 2-[(2-hydroxyphenyl)methyl]-, methyl ester	40800-98-4	NI29041
282	281	267	283	141	268	266	139	**282**	100	48	48	20	16	9	8	5		17	14	4	–	–	–	–	–	–	–	–	–	–	6H-Benzofuro[3,2-c][1]benzopyran, 3,9-dimethoxy-	1433-08-5	NI29042
207	282	208	151	179	209	43	239	**282**	100	37	31	18	17	12	11	10		17	14	4	–	–	–	–	–	–	–	–	–	–	3-Biphenylenecarboxylic acid, 2-(acetoxymethyl)-, methyl ester	93103-70-9	NI29043
282	223	240	283	164	163	43	224	**282**	100	70	26	19	19	17	14	12		17	14	4	–	–	–	–	–	–	–	–	–	–	6-Biphenylenecarboxylic acid, 2-(acetoxymethyl)-, methyl ester	93103-71-0	NI29044
181	180	282	179	43	151	152	153	**282**	100	78	64	48	40	37	31	25		17	14	4	–	–	–	–	–	–	–	–	–	–	Biphenylene, 2-(diacetoxymethyl)-		NI29045
131	103	77	104	282	136	223	91	**282**	100	62	49	45	39	37	33	20		17	14	4	–	–	–	–	–	–	–	–	–	–	Cryptocaryone		NI29046
282	132	281	239	89	117	283	63	**282**	100	68	32	28	26	21	18	17		17	14	4	–	–	–	–	–	–	–	–	–	–	Flavone, 4′,7-dimethoxy-	20979-50-4	NI29047
282	281	236	253	69	149	77	209	**282**	100	60	45	43	40	30	30	27		17	14	4	–	–	–	–	–	–	–	–	–	–	Flavone, 5,7-dimethoxy-	21392-57-4	NI29048

MASS TO CHARGE RATIOS								M.W.	INTENSITIES								Parent	C	H	O	N	S	F	Cl	Br	I	Si	P	B	X	COMPOUND NAME	CAS Reg No	No
282	283	150	132	69	122	81	83	**282**	100	19	19	18	17	11	10	9		17	14	4	–	–	–	–	–	–	–	–	–	–	Flavone, 3-hydroxy-4′-methoxy-7-methyl-		MS06440
282	283	150	132	69	122	81	77	**282**	100	19	19	18	17	11	10	9		17	14	4	–	–	–	–	–	–	–	–	–	–	Flavone, 5-hydroxy-4′-methoxy-7-methyl-	33500-23-1	NI29049
132	104	282	133	103	131	283	149	**282**	100	26	25	10	7	5	4	4		17	14	4	–	–	–	–	–	–	–	–	–	–	Isocoumarin, trans-3-(3,4-methylenedioxyphenyl)-4-methyl-3,4-dihydro-		LI06072
282	281	180	141	267	165	251	236	**282**	100	70	18	16	14	10	7	6		17	14	4	–	–	–	–	–	–	–	–	–	–	Isoflavone, 6,7-dimethoxy-	24195-17-3	NI29050
282	267	201	264	281	67	283	43	**282**	100	74	53	34	28	28	19	17		17	14	4	–	–	–	–	–	–	–	–	–	–	1H-Naphtho[2,1-b]pyran-1-one, 3-acetyl-7-methoxy-2-methyl-	55044-51-4	NI29051
282	267	201	67	281	264	283	44	**282**	100	68	48	35	30	22	19	19		17	14	4	–	–	–	–	–	–	–	–	–	–	1H-Naphtho[2,1-b]pyran-1-one, 3-acetyl-8-methoxy-2-methyl-	55044-50-3	NI29052
240	282	239	241	264	102	128	77	**282**	100	30	16	15	13	12	11	9		17	14	4	–	–	–	–	–	–	–	–	–	–	2-Propylquinizarin		MS06441
180	264	43	222	181	197	282	265	**282**	100	61	38	35	34	18	16	11		17	18	2	2	–	–	–	–	–	–	–	–	–	Acetamide, N,N′-(methylenedi-2,1-phenylene)bis-	52812-76-7	NI29053
282	240	197	198	43	106	182	239	**282**	100	89	84	79	79	74	28	22		17	18	2	2	–	–	–	–	–	–	–	–	–	Bis(4-acetamidophenyl)methane		IC03633
184	156	282	239	170	169	281	253	**282**	100	40	25	25	25	25	20	15		17	18	2	2	–	–	–	–	–	–	–	–	–	2-Acetyl-1,2-epoxyoctahydroindolo[2,3-a]quinolizine		LI06073
282	93	94	135	79	179	225	91	**282**	100	62	38	33	33	30	24	24		17	18	2	2	–	–	–	–	–	–	–	–	–	1-(1-Adamantyl)-2-cyano-5-nitrobenzene		NI29054
282	94	226	225	93	135	79	95	**282**	100	63	32	32	28	24	16	12		17	18	2	2	–	–	–	–	–	–	–	–	–	1-(1-Adamantyl)-3-cyano-5-nitrobenzene		NI29055
282	94	226	225	93	79	135	67	**282**	100	75	30	30	28	26	25	14		17	18	2	2	–	–	–	–	–	–	–	–	–	1-(1-Adamantyl)-4-cyano-3-nitrobenzene		NI29056
186	187	282	144	145	143	96	115	**282**	100	94	92	92	72	48	48	42		17	18	2	2	–	–	–	–	–	–	–	–	–	9-Methyl-2,3,4,5,6,7,8,9-octahydro-2,6-methano-1H-azecino[5,4-b]indole-4,8-dione		LI06074
267	282	149	44	78	161	121	226	**282**	100	33	22	14	13	12	11	7		17	18	2	2	–	–	–	–	–	–	–	–	–	2-Pyridinecarboxamide, 5-(4-tert-butylbenzoyl)-	82465-59-6	NI29057
267	282	268	283	251	269	195	73	**282**	100	54	29	16	13	11	11	9		17	22	–	–	–	–	–	–	–	2	–	–	–	9-Silafluorene, 4-trimethylsilyl-9,9-dimethyl-	58263-56-2	NI29058
237	282	238	194	195	165	283	239	**282**	100	71	47	24	23	21	17	13		17	22	–	2	–	–	–	–	–	1	–	–	–	9-Silaanthracen-9-amine, N,N-dimethyl-9,10-dihydro-		NI29059
224	225	282	238	209	265	226	210	**282**	100	44	27	27	27	21	19	15		17	22	–	2	–	–	–	–	–	1	–	–	–	10-Sila-2-azaanthracene-9-propanamine, 10,10-dimethyl-9,10-dihydro-	84841-69-0	NI29060
135	164	136	121	60	163	68	119	**282**	100	95	61	25	24	21	21	15	**0.00**	17	22	–	4	–	–	–	–	–	–	–	–	–	N,N′-Bis[4-(N,N-dimethyl)anilino]formamidine	2611-98-5	NI29061
73	125	69	81	55	109	107	79	**282**	100	34	28	22	21	17	17	16	**0.30**	17	30	3	–	–	–	–	–	–	–	–	–	–	2,4-Dodecadienoic acid, 11-methoxy-3,7,11-trimethyl-, methyl ester, (E,E)-	40596-68-7	NI29062
85	55	41	56	29	43	97	84	**282**	100	69	59	30	30	29	28	27	**0.00**	17	30	3	–	–	–	–	–	–	–	–	–	–	3-Dodecyn-1-ol, 12-[(tetrahydro-2H-pyran-2-yl)oxy]-	64031-52-3	NI29063
101	113	43	267	79	224	165	72	**282**	100	20	16	13	8	6	6	6	**0.00**	17	30	3	–	–	–	–	–	–	–	–	–	–	4-Tetradecen-8-one, 1,2-di-O-isopropylidene-1,2-dihydroxy-, (S)-(Z)-		DD00960
281	282	103	76	127	254	228	114	**282**	100	99	98	29	13	10	8	8		18	10	–	4	–	–	–	–	–	–	–	–	–	2,3-Dicyano-5,6-diphenylpyrazine		MS06442
78	77	52	51	50	39	79	76	**282**	100	13	13	12	10	8	7	5	**0.00**	18	16	1	–	–	–	–	–	–	–	–	1	1	Borate(1-), hydroxytriphenyl-, sodium, (T-4)-	12113-07-4	NI29064
220	191	192	103	28	219	179	282	**282**	100	76	56	40	40	36	36	28		18	18	1	–	1	–	–	–	–	–	–	–	–	Dimethylsulphonium 2-oxo-3,4-diphenyl-cis-3-butenylide		LI06075
282	57	91	173	149	56	172	110	**282**	100	96	89	83	81	65	51	42		18	18	1	–	1	–	–	–	–	–	–	–	–	5-Phenyl-2-(phenylthio)cyclobutanone		NI29065
282	238	237	165	223	195	119	115	**282**	100	73	51	10	8	4	4	4		18	18	3	–	–	–	–	–	–	–	–	–	–	Benzofuran, 2-(4-hydroxyphenyl)-5-(3-hydroxypropyl)-3-methyl-		MS06443
223	250	77	191	105	224	103	91	**282**	100	49	46	39	34	17	13	13	**9.60**	18	18	3	–	–	–	–	–	–	–	–	–	–	Cyclopropanecarboxylic acid, 2-methoxy-2,3-diphenyl-, methyl ester, (1α,2α,3β)-	55700-34-0	NI29066
223	250	77	191	105	224	145	121	**282**	100	52	42	37	29	17	12	12	**3.50**	18	18	3	–	–	–	–	–	–	–	–	–	–	Cyclopropanecarboxylic acid, 2-methoxy-2,3-diphenyl-, methyl ester, (1α,2β,3β)-	37730-15-7	NI29067
181	152	28	182	153	29	165	41	**282**	100	28	22	21	15	15	11	10	**6.58**	18	18	3	–	–	–	–	–	–	–	–	–	–	9H-Fluorene-9-carboxylic acid, 9-hydroxy-, butyl ester	2314-09-2	NI29068
152	282	182	181	153	151	71	167	**282**	100	99	99	99	99	81	61	44		18	18	3	–	–	–	–	–	–	–	–	–	–	Isobutyric acid 4-phenylphenacyl ester		NI29069
254	282	209	210	208	255	211	237	**282**	100	85	39	26	20	19	16	10		18	18	3	–	–	–	–	–	–	–	–	–	–	1′,2′,3′,7′,8′,9′,10′,10′a-Octahydrocyclohepta[de]naphthalen-3′-ylidenesuccinic anhydride		LI06076
57	91	135	226	148	282	253	165	**282**	100	50	28	17	12	10	8	6		18	18	3	–	–	–	–	–	–	–	–	–	–	1,3-Pentanedione, 1-(3-methoxyphenyl)-2-phenyl-		DD00961
105	165	223	115	193	180	205	178	**282**	100	62	48	33	27	27	21	19	**0.00**	18	18	3	–	–	–	–	–	–	–	–	–	–	Pentanoic acid, 5-oxo-2,2-diphenyl-, methyl ester		DD00962
172	130	110	282	111	173	158	108	**282**	100	46	44	43	36	26	20	14		18	22	1	2	–	–	–	–	–	–	–	–	–	Apparicine, 16(S)-hydroxy-16,22-dihydro-	94061-32-2	NI29070
91	105	106	77	120	174	30	92	**282**	100	77	51	32	28	19	13	10	**2.00**	18	22	1	2	–	–	–	–	–	–	–	–	–	N-Benzoyl-N′-benzylputrescine		LI06077
96	282	110	109	94	70	97	144	**282**	100	92	68	65	51	40	38	35		18	22	1	2	–	–	–	–	–	–	–	–	–	3,9-Dimethyl-2,3,4,5,6,7,8,9-octahydro-2,6-methano-1H-azecino[5,4-b]-indol-8-one		LI06078
175	176	77	79	148	160	42	131	**282**	100	18	15	11	10	10	7	7	**3.10**	18	22	1	2	–	–	–	–	–	–	–	–	–	1,3-Oxazolidine, 4-methyl-cis-5-phenyl-2-[4-(dimethylamino)phenyl]-		MS06444
175	176	148	160	131	132	133	161	**282**	100	23	14	11	7	6	5	5	**4.40**	18	22	1	2	–	–	–	–	–	–	–	–	–	1,3-Oxazolidine, 4-methyl-trans-5-phenyl-2-[4-(dimethylamino)phenyl]-		MS06445
55	41	57	69	74	43	83	59	**282**	100	89	71	70	63	48	40	36	**6.14**	18	34	2	–	–	–	–	–	–	–	–	–	–	Cyclopropaneoctanoic acid, 2-hexyl-, methyl ester	10152-61-1	NI29071
85	43	55	57	83	69	264	97	**282**	100	61	56	49	46	46	45	45	**10.71**	18	34	2	–	–	–	–	–	–	–	–	–	–	2(3H)-Furanone, dihydro-5-tetradecyl-	502-26-1	NI29072
85	55	43	41	69	57	97	83	**282**	100	45	45	39	35	31	28	28	**2.30**	18	34	2	–	–	–	–	–	–	–	–	–	–	2(3H)-Furanone, dihydro-5-tetradecyl-	502-26-1	WI01349
43	41	55	85	29	57	69	83	**282**	100	68	65	64	57	54	35	33	**1.56**	18	34	2	–	–	–	–	–	–	–	–	–	–	2(3H)-Furanone, dihydro-5-tetradecyl-	502-26-1	IC03634
59	15	132	129	101	69	55	41	**282**	100	79	62	37	37	32	30	25	**0.00**	18	34	2	–	–	–	–	–	–	–	–	–	–	7-Hexadecenoic acid, 14-methyl-, methyl ester, [R-(Z)]-	55101-07-0	NI29073
55	69	56	74	41	250	43	83	**282**	100	88	73	66	64	54	54	52	**24.02**	18	34	2	–	–	–	–	–	–	–	–	–	–	11-Hexadecenoic acid, 15-methyl-, methyl ester	55044-54-7	WI01350
222	43	282	96	82	57	124	83	**282**	100	46	35	35	35	34	30	30		18	34	2	–	–	–	–	–	–	–	–	–	–	1-Hexadecen-1-ol, acetate	1787-09-3	NI29074
43	28	45	60	78	91	56	41	**282**	100	74	42	30	28	24	21	21	**0.00**	18	34	2	–	–	–	–	–	–	–	–	–	–	1-Hexadecen-1-ol, acetate	1787-09-3	NI29075
55	83	98	211	264	182	111	71	**282**	100	43	39	28	24	23	19	19	**3.00**	18	34	2	–	–	–	–	–	–	–	–	–	–	13-Hydroxyoctadecanoic acid lactone		LI06079
283	284	251	285	115	250	249	97	**282**	100	24	6	3	3	1	1	1	**0.00**	18	34	2	–	–	–	–	–	–	–	–	–	–	Methyl cis-9,10-methylenehexadecanoate		NI29076
55	74	87	41	43	83	69	75	**282**	100	99	72	71	50	40	23	17	**3.50**	18	34	2	–	–	–	–	–	–	–	–	–	–	Methyl cyclohexyldecanoate		LI06080

MASS TO CHARGE RATIOS								M.W.	INTENSITIES								Parent	C	H	O	N	S	F	Cl	Br	I	Si	P	B	X	COMPOUND NAME	CAS Reg No	No
283	284	251	285	250	281	249	96	**282**	100	19	4	2	2	1	1	1	**0.00**	18	34	2	–	–	–	–	–	–	–	–	–	–	Methyl heptadec-10-enoate		NI29077
55	69	43	41	83	57	73	60	**282**	100	73	59	58	53	48	47	39	**0.50**	18	34	2	–	–	–	–	–	–	–	–	–	–	9-Octadecenoic acid (Z)-	112-80-1	NI29078
41	55	43	69	83	67	57	54	**282**	100	95	65	51	34	33	33	27	**0.50**	18	34	2	–	–	–	–	–	–	–	–	–	–	9-Octadecenoic acid (Z)-	112-80-1	WI01351
99	55	43	70	41	71	83	69	**282**	100	58	53	43	43	39	33	33	**5.00**	18	34	2	–	–	–	–	–	–	–	–	–	–	2H-Pyran-2-one, tetrahydro-6-tridecyl-	1227-51-6	WI01352
282	133	210	209	165	191	211	166	**282**	100	78	72	61	58	49	37	37		19	19	1	–	–	1	–	–	–	–	–	–	–	Furan, 2-(2-fluoro-3,3-diphenyl-2-propenyl)tetrahydro-	56701-23-6	NI29079
237	128	159	280	91	77	204	189	**282**	100	95	90	60	40	30	10	9	**0.00**	19	22	–	–	1	–	–	–	–	–	–	–	–	3-Hexene, 2-methyl-6-phenyl-4-(phenylthio)-		DD00963
134	282	149	283	171				**282**	100	99	84	47	23					19	22	2	–	–	–	–	–	–	–	–	–	–	Androsta-1,4,6-triene-3,17-dione	633-35-2	NI29080
282	211	212	239	283	115	42	165	**282**	100	31	29	25	21	20	20	19		19	22	2	–	–	–	–	–	–	–	–	–	–	1(2H)-Chrysenone, 3,4,4a,5,6,11,12,12a-octahydro-8-methoxy-, (4aS-cis)-	58072-55-2	NI29081
141	55	81	41	115	97	158	140	**282**	100	43	25	19	15	12	11	8	**5.50**	19	22	2	–	–	–	–	–	–	–	–	–	–	Cyclohexaneacetic acid, 1-naphthalenylmethyl ester	86328-54-3	NI29082
105	77	91	264	133	183	173	120	**282**	100	38	33	23	20	16	13	11	**0.00**	19	22	2	–	–	–	–	–	–	–	–	–	–	Cyclohexanol, 2-[(hydroxy)diphenylmethyl]-, (1S,2R)-		DD00964
183	83	105	85	149	77	115	193	**282**	100	61	58	41	25	22	14	13	**0.00**	19	22	2	–	–	–	–	–	–	–	–	–	–	Cyclohexanol, 2-[(hydroxy)diphenylmethyl]-, (1S,2S)-		DD00965
282	160	254	225	283	186	187	159	**282**	100	44	25	23	21	19	16	13		19	22	2	–	–	–	–	–	–	–	–	–	–	14,15-Dehydro-8α-isoestrone methyl ether		LI06081
282	160	254	283	225	239	161	171	**282**	100	78	54	41	28	18	14	13		19	22	2	–	–	–	–	–	–	–	–	–	–	14,15-Dehydro-8α-isoestrone methyl ether		LI06082
267	268	238	231	41	43	282	39	**282**	100	20	19	12	10	9	8	7		19	22	2	–	–	–	–	–	–	–	–	–	–	6H-Dibenzo[b,d]pyran-1-ol, 6,6,9-trimethyl-3-propyl-	33745-21-0	NI29083
282	184	158	197	171	283	185	159	**282**	100	45	42	34	24	22	11	11		19	22	2	–	–	–	–	–	–	–	–	–	–	Estra-1,3,5(10),6-tetraen-17-one, 3-hydroxy-1-methyl-	3129-08-6	NI29084
282	171	184	157	158	283	197	267	**282**	100	33	31	31	27	22	22	15		19	22	2	–	–	–	–	–	–	–	–	–	–	Estra-1,3,5(10),6-tetraen-17-one, 3-hydroxy-6-methyl-	10506-91-9	NI29085
282	160	254	283	225	239	173	161	**282**	100	39	27	20	14	9	7	7		19	22	2	–	–	–	–	–	–	–	–	–	–	8α-Estra-1,3,5(10),14-tetraen-17-one, 3-methoxy-	10003-02-8	NI29086
282	225	148	283	239	226	171	165	**282**	100	25	24	22	20	13	11	11		19	22	2	–	–	–	–	–	–	–	–	–	–	Estra-1,3,5(10),8-tetraen-17-one, 3-methoxy-	6885-44-5	NI29087
282	148	283	225	239	226	254	171	**282**	100	24	21	20	17	9	8	8		19	22	2	–	–	–	–	–	–	–	–	–	–	Estra-1,3,5(10),8-tetraen-17-one, 3-methoxy-	6885-44-5	NI29088
282	267	283	224	225	211	284	265	**282**	100	37	22	22	19	15	13	13		19	22	2	–	–	–	–	–	–	–	–	–	–	Estra-1,3,5(10),9(11)-tetraen-17-one, 3-methoxy-	1670-49-1	NI29089
282	225	283	174	148	160	147	173	**282**	100	23	22	21	20	16	11	10		19	22	2	–	–	–	–	–	–	–	–	–	–	Estra-1,3,5(10),15-tetraen-17-one, 3-methoxy-	17748-68-4	NI29090
282	41	80	91	115	77	79	174	**282**	100	99	76	74	69	69	63	52		19	22	2	–	–	–	–	–	–	–	–	–	–	Estra-1,3,5(10),15-tetraen-17-one, 3-methoxy-	17748-68-4	NI29091
91	105	133	264	104	92	282	117	**282**	100	60	46	41	39	20	16	16		19	22	2	–	–	–	–	–	–	–	–	–	–	3-Heptanone, 5-hydroxy-1,7-diphenyl-	55836-43-6	NI29092
282	159	267	283	121	135	147	126	**282**	100	37	32	21	20	12	11	9		19	22	2	–	–	–	–	–	–	–	–	–	–	2,3-Bis(4-methoxyphenyl)pent-2-ene		MS06446
141	55	186	282	97	41	139	115	**282**	100	92	45	26	26	24	17	17		19	22	2	–	–	–	–	–	–	–	–	–	–	2-Naphthaleneacetic acid, cyclohexanemethyl ester	86328-82-7	NI29093
158	124	143	122	159	108	44	123	**282**	100	40	6	6	4	4	4	3	**2.38**	19	26	–	2	–	–	–	–	–	–	–	–	–	Apparicine, Nb-methyltetrahydro-	3484-71-7	NI29094
124	282	254	130	125	152	144	143	**282**	100	17	12	10	10	6	5	5		19	26	–	2	–	–	–	–	–	–	–	–	–	Aspidospermidine	2912-09-6	NI29095
124	282	254	125	130	144	69	67	**282**	100	13	12	11	6	4	4	4		19	26	–	2	–	–	–	–	–	–	–	–	–	Aspidospermidine	2912-09-6	NI29096
124	282	190	281	130	138	254	152	**282**	100	32	22	21	20	9	8	7		19	26	–	2	–	–	–	–	–	–	–	–	–	Aspidospermidine, 7β-ethyl-5-desethyl-		LI06083
148	105	45	77	135	42	149	44	**282**	100	58	34	19	15	14	13	13	**2.00**	19	26	–	2	–	–	–	–	–	–	–	–	–	Benzeneethanamine, 4-(dimethylamino)-N-methyl-N-(2-phenylethyl)-	52059-43-5	NI29097
73	182	209	194	180	167	208	181	**282**	100	63	29	28	24	19	15	13	**1.92**	19	26	–	2	–	–	–	–	–	–	–	–	–	1H-Carbazole-2-ethanamine, 3-ethyl-2,3,4,9-tetrahydro-N,N-dimethyl-1-methylene-	55320-35-9	NI29098
152	69	144	153	67	154	84	136	**282**	100	26	17	15	13	12	12	11	**9.00**	19	26	–	2	–	–	–	–	–	–	–	–	–	Condyfolan, 16-methyl-	2772-71-6	NI29099
211	282	41	55	40	105.5	29	212	**282**	100	59	58	47	27	22	21	20		19	26	–	2	–	–	–	–	–	–	–	–	–	4-(1,3-Dimethylpentylamino)diphenylamine		IC03635
41	211	40	55	282	105.5	39	43	**282**	100	86	84	22	21	15	14	12		19	26	–	2	–	–	–	–	–	–	–	–	–	4-(1,4-Dimethylpentylamino)diphenylamine		IC03636
282	110	125	124	157	138	126	143	**282**	100	92	80	80	51	48	48	42		19	26	–	2	–	–	–	–	–	–	–	–	–	2H-3,7-Methanoazacycloundecino[5,4-b]indole, 7-ethyl-1,4,5,6,7,8,9,10-octahydro-, (R)-	4850-21-9	NI29100
282	125	110	124	42	157	41	58	**282**	100	96	94	78	71	59	57	48		19	26	–	2	–	–	–	–	–	–	–	–	–	2H-3,7-Methanoazacycloundecino[5,4-b]indole, 7-ethyl-1,4,5,6,7,8,9,10-octahydro-, (R)-	4850-21-9	NI29101
71	85	69	83	82	96	97	282	**282**	100	28	27	19	17	15	14	13		19	38	1	–	–	–	–	–	–	–	–	–	–	2-Nonadecanone	629-66-3	WI01353
58	43	59	71	41	57	55	85	**282**	100	85	72	46	27	26	26	15	**10.00**	19	38	1	–	–	–	–	–	–	–	–	–	–	2-Nonadecanone	629-66-3	LI06084
55	96	69	83	110	124	138	264	**282**	100	49	49	30	18	12	7	4	**0.00**	19	38	1	–	–	–	–	–	–	–	–	–	–	12-Nonadecen-1-ol, (Z)-		DD00966
71	82	41	43	68	96	55	67	**282**	100	37	30	21	19	18	14	13	**2.04**	19	38	1	–	–	–	–	–	–	–	–	–	–	1-Octadecene, 1-methoxy-	26537-06-4	NI29102
71	41	55	43	67	82	29	81	**282**	100	55	29	27	23	18	18	17	**0.00**	19	38	1	–	–	–	–	–	–	–	–	–	–	9-Octadecene, 1-methoxy-, (E)-	56847-01-9	NI29103
86	69	71	43	57	55	41	87	**282**	100	89	88	82	81	81	67	59	**10.30**	19	38	1	–	–	–	–	–	–	–	–	–	–	3-Pentadecanone, 2,6,10,14-tetramethyl-		NI29104
128	72	71	43	81	57	41	55	**282**	100	94	92	91	90	88	79	77	**1.40**	19	38	1	–	–	–	–	–	–	–	–	–	–	5-Pentadecanone, 2,6,10,14-tetramethyl-		NI29105
72	71	57	43	69	55	85	41	**282**	100	100	100	100	99	82	75	75	**2.90**	19	38	1	–	–	–	–	–	–	–	–	–	–	7-Pentadecanone, 2,6,10,14-tetramethyl-		NI29106
282	283	254	226	113	224	112	225	**282**	100	24	18	18	16	15	10	8		20	10	2	–	–	–	–	–	–	–	–	–	–	Benzo[3,4]cyclobuta[1,2-b]anthraquinone		IC03637
282	141	283	112	253	224	16		**282**	100	29	15	14	7	7	6			20	10	2	–	–	–	–	–	–	–	–	–	–	Perixanthenoxanthene		IC03638
282	281	283	280	284				**282**	100	64	20	10	3					20	14	–	2	–	–	–	–	–	–	–	–	–	15H-Dibenzo[c,e]benzimidazo[1,2-a]azepine		LI06085
282	281	141	280	140	254	178	151	**282**	100	23	20	13	12	9	8	7		20	14	–	2	–	–	–	–	–	–	–	–	–	Pyrrolo[2,3-c]carbazole, 2-phenyl-		NI29107
282	281	179	283	76	178	50	77	**282**	100	57	27	20	18	13	11	10		20	14	–	2	–	–	–	–	–	–	–	–	–	Quinoxaline, 2,3-diphenyl-	1684-14-6	NI29108
282	281	76	179	50	77	178	283	**282**	100	60	44	41	35	23	22	19		20	14	–	2	–	–	–	–	–	–	–	–	–	Quinoxaline, 2,3-diphenyl-	1684-14-6	LI06086
122	161	267	282	105	91	147	109	**282**	100	93	52	45	37	30	28	26		20	26	1	–	–	–	–	–	–	–	–	–	–	Androsta-1,4,13-trien-3-one, 17,17-dimethyl-		NI29109
267	282	268	98	126	57	41	112	**282**	100	22	21	13	7	7	6	5		20	26	1	–	–	–	–	–	–	–	–	–	–	Bis(4-tert-butylphenyl) ether		DO01075
133	134	105	149	135	132	117	91	**282**	100	49	38	19	9	7	7	7	**0.00**	20	26	1	–	–	–	–	–	–	–	–	–	–	Bis[2-(3-ethylphenyl)ethyl] ether		DO01076

MASS TO CHARGE RATIOS								M.W.	INTENSITIES								Parent	C	H	O	N	S	F	Cl	Br	I	Si	P	B	X	COMPOUND NAME	CAS Reg No	No
267	282	57	41	283	268	211	226	282	100	99	42	23	22	20	17	14		20	26	1	–	–	–	–	–	–	–	–	–	–	1-Hydroxy-2,6-di-tert-butylbiphenyl		IC03639
43	91	225	55	57	105	282	267	282	100	69	60	59	57	55	50	46		20	26	1	–	–	–	–	–	–	–	–	–	–	Spiro[2.4]heptan-4-one, 5-[(E)-2,2-dimethylpropylidene]-1,1-dimethyl-2-phenyl-		MS06447
71	85	69	126	225	70	127	83	282	100	67	24	22	21	14	13	13	0.05	20	42	–	–	–	–	–	–	–	–	–	–	–	5-Butylhexadecane		AI01964
57	43	71	85	41	29	55	56	282	100	91	67	48	42	28	27	15	5.10	20	42	–	–	–	–	–	–	–	–	–	–	–	Eicosane		AI01966
43	57	71	41	85	55	56	42	282	100	97	57	46	38	28	15	13	1.38	20	42	–	–	–	–	–	–	–	–	–	–	–	Eicosane		AI01965
57	43	71	85	41	55	99	29	282	100	66	63	42	21	17	12	12	4.80	20	42	–	–	–	–	–	–	–	–	–	–	–	Eicosane		IC03640
57	71	43	85	41	55	113	56	282	100	80	58	40	29	24	19	19	0.34	20	42	–	–	–	–	–	–	–	–	–	–	–	Hexadecane, 2,6,10,14-tetramethyl-	638-36-8	NI29111
57	43	71	41	55	85	56	29	282	100	83	58	44	30	24	23	23	1.44	20	42	–	–	–	–	–	–	–	–	–	–	–	Hexadecane, 2,6,10,14-tetramethyl-	638-36-8	NI29110
57	72	44	86	42	55	56	70	282	100	70	63	33	28	21	17	11	1.00	20	42	–	–	–	–	–	–	–	–	–	–	–	Hexadecane, 2,6,10,14-tetramethyl-	638-36-8	NI29112
57	43	71	44	41	55	56	85	282	100	99	93	49	47	33	30	24	0.06	20	42	–	–	–	–	–	–	–	–	–	–	–	Hexadecane, 2,6,11,15-tetramethyl-		AI01967
71	43	57	70	85	41	55	239	282	100	96	87	52	43	30	27	21	0.60	20	42	–	–	–	–	–	–	–	–	–	–	–	4-Methylnonadecane		LI06087
57	43	41	71	29	55	85	98	282	100	73	38	37	30	29	26	19	0.04	20	42	–	–	–	–	–	–	–	–	–	–	–	4-Propylheptadecane		AI01968
282	205	177	176	77	281	283	151	282	100	77	61	58	36	34	21	19		21	14	1	–	–	–	–	–	–	–	–	–	–	Anthracene, 9-benzoyl-	1564-53-0	NI29113
282	254	283	252	126	253	239	113	282	100	24	21	21	21	18	12	12		21	14	1	–	–	–	–	–	–	–	–	–	–	Benz[j]aceanthrylen-1(2H)-one, 3-methyl-	3343-07-5	MS06448
282	281	252	253	126	265	250	125	282	100	62	40	27	23	13	11	9		21	14	1	–	–	–	–	–	–	–	–	–	–	2,3-Diphenylinden-1-one		MS06449
282	281	252	253	126	265	250	113	282	100	63	41	27	23	14	12	8		21	14	1	–	–	–	–	–	–	–	–	–	–	2,3-Diphenylinden-1-one		IC03641
282	252	253	281	254	126	250	283	282	100	65	56	34	32	27	25	23		21	14	1	–	–	–	–	–	–	–	–	–	–	10-Phenyl-9-anthraldehyde		MS06450
267	268	282	211	126	29	181	165	282	100	22	12	10	7	7	6	5		21	30	–	–	–	–	–	–	–	–	–	–	–	1,1,4,4,7,7-Hexamethyltrindan		AI01969
267	268	211	282	126	181	195	165	282	100	76	24	20	12	9	8	7		21	30	–	–	–	–	–	–	–	–	–	–	–	1,1,4,4,9,9-Hexamethyltrindan		AI01970
171	227	226	41	29	129	55	115	282	100	94	44	40	39	31	29	27	15.84	21	30	–	–	–	–	–	–	–	–	–	–	–	1,3,5-Tris(3-methyl-3-butenyl)benzene		AI01971
141	282	142	41	43	283	115	128	282	100	52	31	16	15	12	10	5		21	30	–	–	–	–	–	–	–	–	–	–	–	1-Undecylnaphthalene		AI01972
141	282	142	29	41	43	115	27	282	100	36	36	15	13	11	8	8		21	30	–	–	–	–	–	–	–	–	–	–	–	1-Undecylnaphthalene		AI01973
141	282	142	41	43	283	115	128	282	100	39	33	16	14	9	9	5		21	30	–	–	–	–	–	–	–	–	–	–	–	1-Undecylnaphthalene		AI01974
282	254	252	278	239	253	283	241	282	100	60	35	25	24	23	22	17		22	18	–	–	–	–	–	–	–	–	–	–	–	Benzo[b]triphenylene, 10,11,12,13-tetrahydro-	25486-89-9	NI29114
282	263	283	265	267	252	266	126	282	100	77	25	25	22	20	15	11		22	18	–	–	–	–	–	–	–	–	–	–	–	1,1′-Binaphthalene, 2,2′-dimethyl-	32834-84-7	NI29115
282	283	265	267	252	266	263	126	282	100	25	25	22	20	15	11	11		22	18	–	–	–	–	–	–	–	–	–	–	–	1,1′-Binaphthalene, 2,2′-dimethyl-	32834-84-7	NI29116
282	266	267	281	252	283	265	126	282	100	47	40	27	24	23	20	19		22	18	–	–	–	–	–	–	–	–	–	–	–	1,1′-Binaphthalene, 3,3′-dimethyl-	34042-82-5	NI29117
282	252	267	265	253	283	126	266	282	100	58	46	27	26	25	24	18		22	18	–	–	–	–	–	–	–	–	–	–	–	1,1′-Binaphthalene, 8,8′-dimethyl-	32693-05-3	NI29118
282	254	252	283	253	279	281	265	282	100	41	29	21	21	19	18	18		22	18	–	–	–	–	–	–	–	–	–	–	–	Dibenz[a,h]anthracene, 1,2,3,4-tetrahydro-	153-39-9	NI29119
282	265	281	278	283	266	267	279	282	100	36	32	27	24	22	19	15		22	18	–	–	–	–	–	–	–	–	–	–	–	Dibenz[a,h]anthracene, 5,6,12,13-tetrahydro-	153-31-1	NI29120
267	282	252	265	266	281	253	263	282	100	69	43	25	20	13	10	9		22	18	–	–	–	–	–	–	–	–	–	–	–	1,4-Dimethylanthracene benzyne A end ring adduct		LI06088
282	267	266	265	252	281	253	256	282	100	87	28	28	26	22	10	8		22	18	–	–	–	–	–	–	–	–	–	–	–	1,4-Dimethylanthracene benzyne C end ring adduct		LI06089
267	282	252	265	266	263	281	253	282	100	94	40	30	24	14	12	12		22	18	–	–	–	–	–	–	–	–	–	–	–	1,4-Dimethylanthracene benzyne centre ring adduct		LI06090
282	267	252	266	281	265	263		282	100	96	32	25	23	23	9			22	18	–	–	–	–	–	–	–	–	–	–	–	2,6-Dimethylanthracene benzyne centre ring adduct		LI06091
267	252	282	253	265	266	263		282	100	70	67	22	15	9	6			22	18	–	–	–	–	–	–	–	–	–	–	–	9,10-Dimethylanthracene benzyne centre ring adduct		LI06092
282	267	252	266	281	265	263		282	100	97	33	26	24	23	11			22	18	–	–	–	–	–	–	–	–	–	–	–	2,6-Dimethylanthracene benzyne end ring adduct		LI06093
267	282	252	253	265	266	263		282	100	68	68	45	16	11	6			22	18	–	–	–	–	–	–	–	–	–	–	–	9,10-Dimethylanthracene benzyne end ring adduct		LI06094
141	282	115	139					282	100	23	14	5						22	18	–	–	–	–	–	–	–	–	–	–	–	Ethane, 1,2-di-1-naphthyl-	15374-45-5	NI29121
141	282	115	142	283	139	140	116	282	100	25	17	12	6	4	1	1		22	18	–	–	–	–	–	–	–	–	–	–	–	Ethane, 1,2-di-2-naphthyl-	21969-45-9	NI29122
239	282	276	226	240	253	283	227	282	100	76	36	33	30	23	20	19		22	18	–	–	–	–	–	–	–	–	–	–	–	Indeno[1,2,3-cd]pyrene, 6b,7,8,9,10,10a-hexahydro-	95676-43-0	NI29123
282	254	239	253	252	281	276	283	282	100	93	86	71	62	53	41	24		22	18	–	–	–	–	–	–	–	–	–	–	–	Indeno[1,2,3-cd]pyrene, 1,2,7,8,9,10-hexahydro-		NI29124
282	253	252	254	281	283	239	277	282	100	49	36	34	26	24	24	18		22	18	–	–	–	–	–	–	–	–	–	–	–	Indeno[1,2,3-cd]pyrene, hexahydro-		NI29125
282	252	126	267	239				282	100	17	17	15	12					22	18	–	–	–	–	–	–	–	–	–	–	–	7-Methyl-10,11-dihydro-9H-cyclohepta[a]pyrene		LI06095
178	104	282						282	100	3	1							22	18	–	–	–	–	–	–	–	–	–	–	–	11-Phenyl-9,10-ethano-9,10-dihydroanthracene		LI06096
268	187	282	168	267				283	100	32	22	5	4				0.00	3	8	–	1	–	5	–	–	–	–	–	–	1	(Methylethylamino)tellurium pentafluoride		LI06097
69	131	46	64	100	63	32	47	283	100	92	90	88	75	64	64	61	0.00	4	2	1	1	1	9	–	–	–	–	–	–	–	1-Butanesulphinamide, 1,1,2,2,3,3,4,4,4-nonafluoro-	51735-84-3	NI29126
114	264	119	145	100	176	164	214	283	100	20	20	18	16	14	14	8	0.23	5	–	–	1	–	11	–	–	–	–	–	–	–	Perfluorobutylmethyleneimine		TR00297
69	114	31	50	264	119	100	214	283	100	68	20	13	10	10	10	6	0.00	5	–	–	1	–	11	–	–	–	–	–	–	–	Perfluorobutylmethyleneimine		LI06098
283	75	110	74	237	225	126	253	283	100	91	83	71	38	27	22	19		6	3	2	1	–	–	1	–	1	–	–	–	–	Benzene, 1-chloro-2-iodo-5-nitro-	41252-96-4	NI29127
285	283	214	107	142	241	287	212	283	100	76	57	55	53	50	49	46		8	1	2	1	–	–	4	–	–	–	–	–	–	Tetrachlorophthalimide		IC03642
36	38	80	82	39	68	73	81	283	100	32	32	31	19	14	12	11	0.00	8	13	2	2	–	–	1	1	–	–	–	–	–	3-Imidazoline, 4-(bromochloromethyl)-2,2,5,5-tetramethyl-, 3-oxide 1-oxile		NI29128
156	81	88	76	59	114	15	283	283	100	70	61	48	17	15	10	5		8	14	2	1	–	–	–	–	1	–	–	–	–	Methyl N-trans-2-iodocyclohexylcarbamate		LI06099

MASS TO CHARGE RATIOS								M.W.	INTENSITIES								Parent	C	H	O	N	S	F	Cl	Br	I	Si	P	B	X	COMPOUND NAME	CAS Reg No	No
270	268	285	272	283	255	253	31	**283**	100	80	50	46	39	35	25	19		9	5	1	1	–	–	4	–	–	–	–	–	–	**1H-Isoindole, 4,5,6,7-tetrachloro-3-methoxy-**	110568-85-9	**DD00967**
88	248	72	250	124	283			**283**	100	68	62	47	9	5				9	8	1	1	1	–	3	–	–	–	–	–	–	**O-2,3,6-Trichlorophenyl dimethylthiocarbamate**		**LI06100**
226	227	43	41	42	169	71	70	**283**	100	98	79	55	42	41	34	34	**0.00**	9	12	1	1	–	7	–	–	–	–	–	–	–	**1-Aminopentane, N-(heptafluorobutyryl)-**		**MS06451**
58	211	73	135	195	45	75	133	**283**	100	37	29	15	13	12	11	9	**2.19**	9	26	3	1	–	–	–	–	–	2	1	–	–	**Phosphonic acid, (1-aminopropyl)-, bis(trimethylsilyl) ester**	53044-30-7	**NI29129**
58	211	73	135	195	75	226	133	**283**	100	37	29	15	13	11	9	9	**2.20**	9	26	3	1	–	–	–	–	–	2	1	–	–	**Phosphonic acid, (1-aminopropyl)-, bis(trimethylsilyl) ester**	53044-30-7	**NI29130**
58	211	73	195	45	75	226	133	**283**	100	37	29	13	12	11	9	9	**2.20**	9	26	3	1	–	–	–	–	–	2	1	–	–	**Phosphonic acid, (1-aminopropyl)-, bis(trimethylsilyl) ester**	53044-30-7	**NI29131**
44	28	43	151	57	45	29	108	**283**	100	66	45	37	18	16	14	12	**1.53**	10	13	5	5	–	–	–	–	–	–	–	–	–	**Adenosine, 1,2-dihydro-2-oxo-**	1818-71-9	**NI29132**
73	151	43	44	57	60	71	42	**283**	100	95	83	78	77	58	52	50	**0.00**	10	13	5	5	–	–	–	–	–	–	–	–	–	**Guanosine**	118-00-3	**NI29133**
151	43	57	109	73	54	110	44	**283**	100	41	31	17	15	15	14	14	**0.00**	10	13	5	5	–	–	–	–	–	–	–	–	–	**Guanosine**	118-00-3	**NI29134**
166	268	239	73	170	147	269	240	**283**	100	96	68	64	28	20	15	12	**1.90**	10	16	3	1	–	3	–	–	–	1	–	–	–	**L-Proline, 1-(trifluoroacetyl)-, trimethylsilyl ester**	52558-87-9	**NI29135**
136	91	121	77	65	283	107	78	**283**	100	77	48	28	25	25	23	22		11	10	2	1	–	5	–	–	–	–	–	–	–	**Benzylamine, 2-methoxy-N-(pentafluoropropionyl)-**		**NI29136**
121	283	136	77	282	252	119	78	**283**	100	81	19	15	14	13	13	12		11	10	2	1	–	5	–	–	–	–	–	–	–	**Benzylamine, 4-methoxy-N-(pentafluoropropionyl)-**		**NI29137**
121	283	233	136	77	78	282	91	**283**	100	61	20	15	13	12	9	8		11	10	2	1	–	5	–	–	–	–	–	–	–	**Benzylamine, 4-methoxy-N-(pentafluoropropionyl)-**		**MS06452**
238	43	134	55	283	166	194	146	**283**	100	45	23	17	16	13	8	7		11	13	6	3	–	–	–	–	–	–	–	–	–	**Valine, 2,4-dinitrophenyl-**		**LI06101**
182	41	164	29	40	69	27	56	**283**	100	49	47	44	38	31	25	17	**0.00**	11	16	4	1	–	3	–	–	–	–	–	–	–	**L-Proline, 4-hydroxy-1-(trifluoroacetyl)-, butyl ester, trans-**	23403-45-4	**NI29138**
182	41	164	29	40	69	27	44	**283**	100	48	47	43	38	31	25	17	**0.00**	11	16	4	1	–	3	–	–	–	–	–	–	–	**L-Proline, 4-hydroxy-1-(trifluoroacetyl)-, butyl ester, trans-**	23403-45-4	**NI29139**
283	285	202	139	50	75	63	76	**283**	100	75	65	41	27	25	25	23		12	7	3	1	–	–	2	–	–	–	–	–	–	**Benzene, 2,4-dichloro-1-(4-nitrophenoxy)-**	1836-75-5	**NI29141**
283	285	202	139	50	63	204	64	**283**	100	64	53	27	23	18	17	15		12	7	3	1	–	–	2	–	–	–	–	–	–	**Benzene, 2,4-dichloro-1-(4-nitrophenoxy)-**	1836-75-5	**NI29142**
283	50	285	63	202	75	64	74	**283**	100	81	73	70	52	48	42	40		12	7	3	1	–	–	2	–	–	–	–	–	–	**Benzene, 2,4-dichloro-1-(4-nitrophenoxy)-**	1836-75-5	**NI29140**
130	144	143	283	115	131	77	281	**283**	100	32	22	14	12	11	7	6		12	13	2	1	–	–	–	–	–	–	–	–	1	**Acetic acid, [[2-(1H-indol-3-yl)ethyl]seleno]-**	1140-74-5	**LI06102**
130	283	115	131	281	77	117	103	**283**	100	13	11	10	6	6	5	4		12	13	2	1	–	–	–	–	–	–	–	–	1	**Acetic acid, [[2-(1H-indol-3-yl)ethyl]seleno]-**	1140-74-5	**NI29143**
192	193	95	109	206	111	194	110	**283**	100	63	16	15	14	14	12	12	**0.92**	12	14	4	3	–	1	–	–	–	–	–	–	–	**1,2,3,4-Butanetetrol, 1-[2-(2-fluorophenyl)-2H-1,2,3-triazol-4-yl]-, [1R-(1R*,2S*,3R*)]-**	51306-42-4	**NI29144**
43	126	158	98	127	70	68	55	**283**	100	76	42	32	29	27	23	23	**1.00**	12	17	5	3	–	–	–	–	–	–	–	–	–	**5′-Amino-5′-deoxy-3′-O-acetylthymidine**		**MS06453**
43	140	182	55	199	42	153	102	**283**	100	92	76	75	46	41	37	31	**0.63**	12	20	3	1	–	3	–	–	–	–	–	–	–	**L-Alanine, N-methyl-N-(trifluoroacetyl)-, 1-methylpentyl ester**	57983-57-0	**NI29145**
154	41	57	155	69	182	56	140	**283**	100	90	76	43	14	10	10	8	**0.00**	12	20	3	1	–	3	–	–	–	–	–	–	–	**Butanoic acid, 2-[(trifluoroacetyl)amino]-, 1-methylpentyl ester**	57983-58-1	**NI29146**
182	69	43	140	41	42	84	200	**283**	100	74	52	48	27	22	18	16	**0.00**	12	20	3	1	–	3	–	–	–	–	–	–	–	**Butanoic acid, 3-[(trifluoroacetyl)amino]-, 1-methylpentyl ester**	57983-79-6	**NI29147**
210	69	41	57	44	56	126	55	**283**	100	74	53	38	30	29	28	27	**0.00**	12	20	3	1	–	3	–	–	–	–	–	–	–	**Hexanoic acid, 6-[(trifluoroacetyl)amino]-, 1-methylpropyl ester**	57983-24-1	**NI29148**
69	41	182	57	153	171	56	39	**283**	100	60	47	47	21	15	7	7	**0.00**	12	20	3	1	–	3	–	–	–	–	–	–	–	**L-Isoleucine, N-(trifluoroacetyl)-, butyl ester**	2505-28-4	**NI29150**
284	285	182	254	210	286			**283**	100	13	8	3	3	1			**0.00**	12	20	3	1	–	3	–	–	–	–	–	–	–	**L-Isoleucine, N-(trifluoroacetyl)-, butyl ester**	2505-28-4	**NI29149**
69	41	182	57	153	171	227	114	**283**	100	61	48	47	21	15	8	8	**0.00**	12	20	3	1	–	3	–	–	–	–	–	–	–	**L-Isoleucine, N-(trifluoroacetyl)-, butyl ester**	2505-28-4	**NI29151**
69	41	57	182	153	171	56	154	**283**	100	72	66	54	24	21	21	12	**0.00**	12	20	3	1	–	3	–	–	–	–	–	–	–	**L-Isoleucine, N-(trifluoroacetyl)-, sec-butyl ester**	16974-97-3	**NI29152**
69	57	41	182	183	153	171	126	**283**	100	60	55	45	22	19	11	11	**0.00**	12	20	3	1	–	3	–	–	–	–	–	–	–	**L-Isoleucine, N-(trifluoroacetyl)-, sec-butyl ester**	16974-97-3	**NI29153**
168	55	43	42	41	71	169	69	**283**	100	99	64	34	25	20	15	12	**0.00**	12	20	3	1	–	3	–	–	–	–	–	–	–	**L-Isovaline, N-(trifluoroacetyl)-, 1-methylbutyl ester**	57983-39-8	**NI29154**
69	41	43	182	29	140	57	27	**283**	100	47	46	36	36	32	26	17	**0.00**	12	20	3	1	–	3	–	–	–	–	–	–	–	**L-Leucine, N-(trifluoroacetyl)-, butyl ester**	2796-38-5	**NI29155**
284	285	254	182	240	210	286		**283**	100	15	8	5	3	3	1		**0.00**	12	20	3	1	–	3	–	–	–	–	–	–	–	**L-Leucine, N-(trifluoroacetyl)-, butyl ester**	2796-38-5	**NI29156**
69	57	140	41	182	43	183	56	**283**	100	76	64	62	50	40	39	21	**0.00**	12	20	3	1	–	3	–	–	–	–	–	–	–	**L-Leucine, N-(trifluoroacetyl)-, sec-butyl ester**	16974-96-2	**NI29157**
69	41	29	182	27	57	43	114	**283**	100	74	72	31	30	28	21	13	**0.00**	12	20	3	1	–	3	–	–	–	–	–	–	–	**L-Norleucine, N-(trifluoroacetyl)-, butyl ester**	55044-47-8	**NI29158**
69	182	57	41	183	114	126	43	**283**	100	82	76	63	38	30	29	21	**0.00**	12	20	3	1	–	3	–	–	–	–	–	–	–	**L-Norleucine, N-(trifluoroacetyl)-, sec-butyl ester**	57983-76-3	**NI29159**
142	71	240	43	241	125	252	41	**283**	100	26	25	25	13	10	5	5	**2.50**	12	20	3	1	–	3	–	–	–	–	–	–	–	**4-Oxazolidinone, 2-methoxy-5,5-diisopropyl-3-(2,2,2-trifluoroethyl)-**	56440-39-2	**NI29160**
43	55	168	71	169	41	153	69	**283**	100	91	73	41	34	27	14	12	**0.00**	12	20	3	1	–	3	–	–	–	–	–	–	–	**L-Valine, N-(trifluoroacetyl)-, 1-methylbutyl ester**	57983-40-1	**NI29161**
268	73	269	283	282	224	270	75	**283**	100	30	23	21	13	12	9	6		12	21	3	1	–	–	–	–	–	2	–	–	–	**3-Pyridinecarboxylic acid, 6-[(trimethylsilyl)oxy]-, trimethylsilyl ester**	36972-82-4	**NI29162**
227	236	168	283	229	241	41	89	**283**	100	63	47	42	42	26	26	21		13	14	–	1	2	–	1	–	–	–	–	–	–	**Thiazole, 2-(butylthio)-4-(4-chlorophenyl)-**	69390-12-1	**NI29163**
41	141	181	143	51	283	183	142	**283**	100	67	31	23	20	16	11	11		13	14	4	1	–	–	1	–	–	–	–	–	–	**Glycine, N-[[3-chloro-4-(2-propenyloxy)phenyl]acetyl]-**	54139-62-7	**NI29164**
91	108	79	107	61	77	75	131	**283**	100	70	70	53	47	42	39	27	**5.13**	13	17	4	1	1	–	–	–	–	–	–	–	–	**DL-Methionine, N-[(phenylmethoxy)carbonyl]-**	4434-61-1	**NI29165**
284	104	266	267	240	285	176	91	**283**	100	82	73	14	12	10	10	8	**0.00**	13	17	4	1	1	–	–	–	–	–	–	–	–	**DL-Methionine, N-[(phenylmethoxy)carbonyl]-**	4434-61-1	**NI29166**
283	237	182	209	127	154	284	153	**283**	100	43	12	11	8	6	5	4		14	9	4	3	–	–	–	–	–	–	–	–	–	**9,10-Anthracenedione, 1,4-diamino-5-nitro-**	82-33-7	**NI29167**
225	179	76	283	226	77	167	103	**283**	100	52	43	14	13	9	4	4		14	9	4	3	–	–	–	–	–	–	–	–	–	**Sydnone, 3-(4-nitrophenyl)-4-phenyl-**	69978-04-7	**NI29168**
73	44	42	72	43	74	58	41	**283**	100	75	23	7	7	5	5	3	**1.70**	14	13	2	5	–	–	–	–	–	–	–	–	–	**7-Cyano-5,8-dimethoxy-2-ethyl-1,3,4,9-tetraazaphenalene**		**MS06454**
236	28	56	238	29	129	139	44	**283**	100	48	37	37	35	32	32	23	**3.30**	14	18	3	1	–	–	1	–	–	–	–	–	–	**4-Piperidin-1-ol, 4-(4-chlorophenyl)-N-(ethoxycarbonyl)-**		**NI29169**
69	108	134	175	191	119	204	150	**283**	100	8	6	5	4	3	2	2	**0.00**	14	21	3	1	1	–	–	–	–	–	–	–	–	**N-(2-Methyl-2-propenoyl)bornane-10,2-sultam**		**DD00968**
142	73	59	170	111	283	82	268	**283**	100	66	54	53	48	40	39	35		14	25	3	1	–	–	–	–	–	1	–	–	–	**2-Pyrrolidinone, 1-[(Z)-4-(trimethylsilyl)-3-butenyl]-3,4-(isopropylidenedioxy)-, (3R,4R)-**	119795-72-1	**DD00969**
204	203	285	283	88	177	176	102	**283**	100	63	62	62	49	39	29	24		15	10	–	1	–	–	–	1	–	–	–	–	–	**Benzene, 1-bromo-3-(2-cyano-2-phenylethenyl)**	7379-63-7	**NI29170**
204	285	283	203	88	177	176	102	**283**	100	64	62	61	49	40	29	26		15	10	–	1	–	–	–	1	–	–	–	–	–	**Benzene, 1-bromo-4-(2-cyano-2-phenylethenyl)**	66998-58-1	**NI29171**
283	161	147	131	133	132	236	77	**283**	100	67	58	44	40	34	33	32		15	13	3	3	–	–	–	–	–	–	–	–	–	**1H-Benzimidazole-2-methanol, 1-methyl-α-(3-nitrophenyl)-**	6692-36-0	**NI29172**

MASS TO CHARGE RATIOS								M.W.	INTENSITIES								Parent	C	H	O	N	S	F	Cl	Br	I	Si	P	B	X	COMPOUND NAME	CAS Reg No	No
105	147	162	136	121	106	118	91	**283**	100	19	14	12	12	12	11	4	**0.00**	15	13	3	3	–	–	–	–	–	–	–	–	–	**Benzoic acid, 2-[(benzoylamino)carbonyl]hydrazide**	16956-44-8	**NI29173**
283	43	170	77	127	155	284	103	**283**	100	51	34	28	26	24	20	20		15	13	3	3	–	–	–	–	–	–	–	–	–	**3H-Imidazo[1,5-b]pyridazine-5,7-dione, 4-acetyl-6-methyl-2-phenyl-**		**NI29174**
283	105	77	226	177	284	51	225	**283**	100	68	43	25	20	18	14	12		15	13	3	3	–	–	–	–	–	–	–	–	–	**3H-Imidazo[1,5-b]pyridazine-5,7-dione, 4-benzoyl-2,6-dimethyl-**		**NI29175**
85	55	41	57	43	84	67	76	**283**	100	32	29	24	23	22	20	16	**0.16**	15	13	3	3	–	–	–	–	–	–	–	–	–	**1H-Naphtho[2,3-d]triazole-4,9-dione, 4,9-dihydro-1-(2-pyranyl)-**		**NI29176**
161	132	283	131	150	104	226	282	**283**	100	55	46	42	39	29	25	22		15	13	3	3	–	–	–	–	–	–	–	–	–	**1,2,4-Oxadiazole, 3-methyl-4-phenyl-5-(3-nitrophenyl)-**		**LI06103**
41	57	29	43	27	132	148	77	**283**	100	89	72	67	52	48	47	44	**1.50**	15	22	2	1	–	–	1	–	–	–	–	–	–	**Acetamide, 2-chloro-N-(2,6-dimethylphenyl)-N-[(2-methylpropoxy)methyl]-**	24353-58-0	**NI29177**
162	45	41	18	238	29	15	77	**283**	100	76	33	27	26	26	25	20	**0.20**	15	22	2	1	–	–	1	–	–	–	–	–	–	**Acetamide, 2-chloro-N-(2-ethyl-6-methylphenyl)-N-(2-methoxyisopropyl)-**	51218-45-2	**NI29178**
75	73	268	117	224	55	269	169	**283**	100	77	68	48	25	21	16	16	**0.00**	15	29	2	1	–	–	–	–	–	1	–	–	–	**Undecanoic acid, 11-cyano-, trimethylsilyl ester**	22396-24-3	**NI29179**
75	73	268	117	224	55	269	129	**283**	100	78	70	48	24	21	15	15	**0.00**	15	29	2	1	–	–	–	–	–	1	–	–	–	**Undecanoic acid, 11-cyano-, trimethylsilyl ester**	22396-24-3	**LI06104**
177	283	147	89	146	65	192	102	**283**	100	65	52	26	25	25	22	21		16	13	–	1	2	–	–	–	–	–	–	–	–	**Aniline, N-(2-thio-3-benzo[b]thienylidene)-4-methyl-**	92886-26-5	**NI29180**
283	268	154	77	182	127	76	240	**283**	100	68	43	35	26	26	24	23		16	13	4	1	–	–	–	–	–	–	–	–	–	**1,3-Dioxolo[4,5-c]acridin-6(11H)-one, 4-methoxy-11-methyl-**	14482-04-3	**NI29181**
283	209	237	77	238	224	212	167	**283**	100	94	82	41	37	35	34	34		16	13	4	1	–	–	–	–	–	–	–	–	–	**Evoxanthine**	477-82-7	**MS06455**
209	255	237	283	212	210	77	238	**283**	100	94	71	66	46	45	43	39		16	13	4	1	–	–	–	–	–	–	–	–	–	**Evoxanthine**	477-82-7	**NI29182**
121	210	91	77	76	122	255	211	**283**	100	17	16	12	12	11	8	8	**2.00**	16	13	4	1	–	–	–	–	–	–	–	–	–	**3-(4-Methoxybenzylamino)phthalic anhydride**		**LI06105**
211	238	283	240	239	184	128	186	**283**	100	31	26	19	18	18	17	16		16	13	4	1	–	–	–	–	–	–	–	–	–	**2-Naphthalenebutanoic acid, γ-cyano-1,4-dihydro-3-methyl-1,4-dioxo-**	56051-45-7	**NI29183**
121	136	149	135	283	104	134	77	**283**	100	65	60	58	35	33	25	23		16	13	4	1	–	–	–	–	–	–	–	–	–	**2-Propen-1-one, 1-(4-methoxyphenyl)-3-(2-nitrophenyl)-**	22396-06-1	**NI29184**
105	77	135	104	42	43	51	44	**283**	100	93	77	45	45	35	34	18	**10.00**	16	17	–	3	1	–	–	–	–	–	–	–	–	**1,3,5-Triazine-2-thione, hexahydro-5-methyl-1,3-diphenyl-**		**NI29185**
238	77	283	180	254	135	224	178	**283**	100	92	34	32	23	14	9	9		16	17	2	1	–	–	–	–	–	1	–	–	–	**1-Oxa-3-aza-2-silacyclopentan-5-one, 2,2-dimethyl-3,4-diphenyl-**	57954-44-6	**NI29186**
268	283	105	197	104	254	91	213	**283**	100	96	59	49	39	38	36	26		16	17	2	1	–	–	–	–	–	1	–	–	–	**1-Oxa-3-aza-2-silacyclopentan-5-one, 2,4-dimethyl-2,3-diphenyl-**	21648-70-4	**NI29187**
283	282	284	130	141.5	268	168	115	**283**	100	67	19	17	9	8	6	5		16	17	2	3	–	–	–	–	–	–	–	–	–	**Uracil, 1,3,6-trimethyl-5-(3-methyl-1H-indol-2-yl)-**		**MS06456**
113	126	72	71	55	43	41	224	**283**	100	36	27	27	21	19	12	10	**0.00**	16	29	3	1	–	–	–	–	–	–	–	–	–	**1-Aziridinedecanoic acid, 2,2-dimethyl-ι-oxo-, ethyl ester**	56817-99-3	**MS06457**
283	285	284	240	140	121	64	77	**283**	100	40	27	25	25	23	21	18		17	14	1	1	–	–	1	–	–	–	–	–	–	**Cyclopropanecarbonitrile, 1-(4-chlorophenyl)-2-(p-methoxyphenyl)-**	32589-54-1	**NI29188**
283	285	284	140	240	121	77	248	**283**	100	39	27	25	24	23	18	17		17	14	1	1	–	–	1	–	–	–	–	–	–	**Cyclopropanecarbonitrile, 1-(4-chlorophenyl)-2-(p-methoxyphenyl)-**	32589-54-1	**LI06106**
241	283	226	198	242	284	197	227	**283**	100	69	43	31	25	13	9	8		17	17	1	1	1	–	–	–	–	–	–	–	–	**1,5-Benzothiazepine, 2-methyl-4-(4-methoxyphenyl)-2,3-dihydro-**	74148-63-3	**NI29189**
283	239	268	240	266	196	168	226	**283**	100	85	83	79	48	46	39	34		17	17	3	1	–	–	–	–	–	–	–	–	–	**9-Acridanone, 1-ethoxy-3-methoxy-10-methyl-**	17014-42-5	**NI29190**
166	167	284	168	165	91	90	153	**283**	100	99	97	97	97	96	93	92	**57.00**	17	17	3	1	–	–	–	–	–	–	–	–	–	**Alanine, diphenylacetyl-**		**NI29191**
283	148	105	120	77	284	94	41	**283**	100	82	82	35	32	26	23	13		17	17	3	1	–	–	–	–	–	–	–	–	–	**Benzoic acid, 3-(dimethylamino)-, 2-oxo-2-phenylethyl ester**	55153-17-8	**NI29192**
148	283	105	149	77	284	42	51	**283**	100	41	17	13	13	8	7	5		17	17	3	1	–	–	–	–	–	–	–	–	–	**Benzoic acid, 4-(dimethylamino)-, 2-oxo-2-phenylethyl ester**	55153-13-4	**NI29193**
283	28	282	254	284	239	255	240	**283**	100	57	40	34	21	10	7	6		17	17	3	1	–	–	–	–	–	–	–	–	–	**Crotonosine**	2241-43-2	**MS06458**
283	254	282	284	239	255	240	28	**283**	100	91	40	21	19	12	12	10		17	17	3	1	–	–	–	–	–	–	–	–	–	**Crotonosine**	2241-43-2	**NI29194**
283	282	254	284	239	255	240	211	**283**	100	40	35	22	10	6	6	5		17	17	3	1	–	–	–	–	–	–	–	–	–	**Crotonosine**	2241-43-2	**LI06107**
215	69	41	283	216	187	186	159	**283**	100	30	27	20	14	12	8	7		17	17	3	1	–	–	–	–	–	–	–	–	–	**Furo[2,3-b]quinoline, 7-(3-methyl-2-butenoxy)-4-methoxy-**	85802-28-4	**NI29195**
215	41	200	216	69	172	283	156	**283**	100	19	17	14	13	8	6	4		17	17	3	1	–	–	–	–	–	–	–	–	–	**Furo[2,3-b]quinoline, 7-(3-methyl-2-butenoxy)-4-methoxy-**	85802-28-4	**NI29196**
215	283	216	284	172	186	158	106	**283**	100	22	13	4	4	3	3	2		17	17	3	1	–	–	–	–	–	–	–	–	–	**Furo[2,3-b]quinolin-4(9H)-one, 7-methoxy-9-(3-methyl-2-butenyl)-**	18904-40-0	**NI29197**
91	119	43	65	39	77	51	92	**283**	100	65	43	15	15	14	11	10	**0.00**	17	17	3	1	–	–	–	–	–	–	–	–	–	**Hydroxylamine, N-(α-acetoxy-4-methylbenzylidene)-O-benzyl-**		**LI06108**
161	91	77	106	105	176	133	51	**283**	100	46	39	30	29	17	16	16	**15.00**	17	17	3	1	–	–	–	–	–	–	–	–	–	**Hydroxylamine, O-Benzyl-N-(4-methoxycinnamoyl)-**		**LI06109**
283	268	266	212	240	284	269	222	**283**	100	80	51	33	32	20	19	19		17	17	3	1	–	–	–	–	–	–	–	–	–	**6-Aza-2β-hydroxy-8-methoxy-2a-methyl-5-oxo-2a,3,4,5-tetrahydroaceanthrene**		**LI06110**
147	120	164	119	107	91	148	283	**283**	100	64	60	18	16	14	12	10		17	17	3	1	–	–	–	–	–	–	–	–	–	**N-(4-Hydroxyphenyl)-β-ethyl-4-hydroxycinnamamide**		**LI06111**
176	177	107	174	118	91	175	117	**283**	100	12	4	3	3	3	2	2	**0.13**	17	17	3	1	–	–	–	–	–	–	–	–	–	**Norcinnamolaurine**		**MS06459**
91	105	133	77	146	92	119	120	**283**	100	80	44	40	34	31	29	23	**21.00**	17	17	3	1	–	–	–	–	–	–	–	–	–	**2-Oxazolidinone, 4-hydroxy-3-phenyl-4-(2-phenylethyl)-**	52512-35-3	**NI29198**
58	57	59	69	55	283	81	71	**283**	100	32	22	20	15	12	12	12		17	17	3	1	–	–	–	–	–	–	–	–	–	**9-Oximo-2,7-diethoxyfluorene**		**MS06460**
43	283	123	108	149	95	240	134	**283**	100	49	41	33	30	26	24	21		17	17	3	1	–	–	–	–	–	–	–	–	–	**4-Pentynamide, 2-acetyl-N-(2-methoxyphenyl)-2-(2-propynyl)-**	55760-06-0	**NI29199**
93	121	283	104	77	251	163	103	**283**	100	51	25	21	20	13	12	12		17	17	3	1	–	–	–	–	–	–	–	–	–	**Propanoic acid, 3-phenyl-2-(N-phenylamido)-, methyl ester**		**IC03643**
147	120	164	119	107	91	65	148	**283**	100	64	59	18	16	16	12	11	**10.17**	17	17	3	1	–	–	–	–	–	–	–	–	–	**2-Propenamide, 3-(4-hydroxyphenyl)-N-[2-(4-hydroxyphenyl)ethyl]-**	20375-37-5	**NI29200**
283	239	128	127	255	227	226	238	**283**	100	86	81	42	40	37	33	30		17	21	1	3	–	–	–	–	–	–	–	–	–	**1-Methyl-3-amino-7-(dimethylamino)phenoxazine**		**IC03644**
192	193	176	150	149	134	282	283	**283**	100	13	7	7	6	6	5	4		18	21	2	1	–	–	–	–	–	–	–	–	–	**3,5-Diacetyl-4-benzyl-2,6-dimethyl-1,4-dihydropyridine**		**LI06112**
144	143	70	41	128	115	42	213	**283**	100	92	79	46	21	18	18	16	**9.00**	18	21	2	1	–	–	–	–	–	–	–	–	–	**4,4a-Dihydro-1-isopropylidene-4,4,4a-trimethyl-1,3-oxazino[3,4-a]quinolin-3-one**		**LI06113**
143	144	70	213	41	115	283		**283**	100	88	34	21	21	18	16			18	21	2	1	–	–	–	–	–	–	–	–	–	**1,11b-Dihydro-4-isopropylidene-1,1,11b-trimethyl-1,3-oxazino[4,3-a]isoquinolin-2-one**		**LI06114**
144	143	70	213	41	283	115	42	**283**	100	92	29	25	20	17	15	11		18	21	2	1	–	–	–	–	–	–	–	–	–	**1,11b-Dihydro-4-isopropylidene-1,1,6-trimethyl-1,3-oxazino[4,3-a]isoquinolin-2-one**		**LI06115**
192	283	282	121	134	240	191	190	**283**	100	73	65	54	45	31	25	17		18	21	2	1	–	–	–	–	–	–	–	–	–	**Isoquinoline, 1,2,3,4-tetrahydro-6-methoxy-2-methyl-7-benzyloxy-**	15778-79-7	**NI29201**
191	192	190	120	162	177	163	121	**283**	100	30	19	18	11	10	10	10	**4.02**	18	21	2	1	–	–	–	–	–	–	–	–	–	**Isoquinoline, 1,2,3,4-tetrahydro-7-methoxy-2-methyl-8-benzyloxy-**	36646-87-4	**NI29202**

MASS TO CHARGE RATIOS								M.W.	INTENSITIES								Parent	C	H	O	N	S	F	Cl	Br	I	Si	P	B	X	COMPOUND NAME	CAS Reg No	No
283	238	210	45	165	177	194	134	**283**	100	77	50	23	20	17	16	12		18	21	2	1	–	–	–	–	–	–	–	–	–	**Stilbene, 2-(methoxymethoxy)-4′-(dimethylamino)-, (E)-**		**MS06461**
283	238	210	165	45	177	194	134	**283**	100	67	40	17	15	13	12	11		18	21	2	1	–	–	–	–	–	–	–	–	–	**Stilbene, 2-(methoxymethoxy)-4′-(dimethylamino)-, (Z)-**		**MS06462**
241	40	283	92	159	147	215	38	**283**	100	71	57	57	44	43	38	37		18	25	–	3	–	–	–	–	–	–	–	–	–	**1-Cyclohexyl-12-methyl-7,8,9,10-tetrahydro-1H-benzo[1,2-e]pyrido[2,1-c]-as-triazine**		**LI06116**
163	120	77	147	148	105	104	283	**283**	100	52	25	19	15	15	15	12		18	25	–	3	–	–	–	–	–	–	–	–	–	**1,3-Phenylenediamine, N,N,N′-trimethyl-N′-[2-(N-methylanilino)ethyl]-**	32869-59-3	**NI29203**
163	283	148	164	120	147	77	42	**283**	100	23	18	13	12	9	8	6		18	25	–	3	–	–	–	–	–	–	–	–	–	**1,4-Phenylenediamine, N,N,N′-trimethyl-N′-[2-(N-methylanilino)ethyl]-**	32869-56-0	**NI29204**
184	170	142	198	212	240	60	55	**283**	100	96	96	86	82	60	45	45	**0.30**	18	37	1	1	–	–	–	–	–	–	–	–	–	**5-Acetamido-5-propyltridecane**		**MS06463**
142	184	43	44	268	170	143	30	**283**	100	52	29	26	17	12	11	11	**3.00**	18	37	1	1	–	–	–	–	–	–	–	–	–	**N,N-Dioctylacetamide**		**MS06464**
59	72	28	43	57	55	41	44	**283**	100	44	34	27	18	16	16	15	**4.00**	18	37	1	1	–	–	–	–	–	–	–	–	–	**Stearamide**		**MS06465**
59	72	43	60	57	41	55	73	**283**	100	41	22	15	13	12	11	10	**6.96**	18	37	1	1	–	–	–	–	–	–	–	–	–	**Stearamide**		**IC03645**
203	73	204	147	217	216	205	74	**283**	100	92	22	17	11	10	10	8	**0.00**	18	37	1	1	–	–	–	–	–	–	–	–	–	**Tetradecanamide, N-(2-methylpropyl)-**	74420-88-5	**NI29205**
282	105	77	283	208	51	280	206	**283**	100	76	67	55	41	23	22	14		19	13	–	3	–	–	–	–	–	–	–	–	–	**3,5-Pyridinedicarbonitrile, 1,4-dihydro-2,6-diphenyl-**	15520-99-7	**NI29206**
185	67	105	104	43	77	41	42	**283**	100	73	48	41	40	32	30	30	**18.09**	19	25	1	1	–	–	–	–	–	–	–	–	–	**Cyclohexanespiro-4′-(2′-phenyl-2′-oxazoline)-5′-spirocyclohexane**		**NI29207**
283	266	227	184	213	198	284	240	**283**	100	84	52	29	26	23	21	20		19	25	1	1	–	–	–	–	–	–	–	–	–	**Cycloocta[c]isoquinolin-5(6H)-one, 6-butyl-7,8,9,10,11,12-hexahydro-**	120637-38-9	**DD00970**
283	59	226	162	42	284	282	43	**283**	100	51	47	43	34	24	24	21		19	25	1	1	–	–	–	–	–	–	–	–	–	**Morphinan, 5,6-didehydro-3-methoxy-6,17-dimethyl-**	3964-06-5	**NI29208**
59	283	162	203	268	60	282	284	**283**	100	61	31	20	17	17	16	13		19	25	1	1	–	–	–	–	–	–	–	–	–	**Morphinan, 6,7-didehydro-3-methoxy-6,17-dimethyl-**	3894-27-7	**NI29209**
59	283	162	282	171	284	203	226	**283**	100	65	44	20	15	14	14	11		19	25	1	1	–	–	–	–	–	–	–	–	–	**Morphinan, 3-methoxy-17-methyl-6-methylene-**	47109-56-8	**NI29210**
283	282	256	176	157	43	41	57	**283**	100	62	54	44	43	42	40	37		19	25	1	1	–	–	–	–	–	–	–	–	–	**Morphinan-3-ol, 17-(2-propenyl)-**	152-02-3	**NI29212**
200	176	157	57	85	56	84	198	**283**	100	95	93	64	43	41	38	37	**0.00**	19	25	1	1	–	–	–	–	–	–	–	–	–	**Morphinan-3-ol, 17-(2-propenyl)-**	152-02-3	**NI29211**
284	282	285	283	286	256			**283**	100	31	20	19	3	1				19	25	1	1	–	–	–	–	–	–	–	–	–	**Morphinan-3-ol, 17-(2-propenyl)-**	152-02-3	**NI29213**
44	58	43	41	57	55	42	29	**283**	100	29	14	11	10	7	5	4	**1.20**	19	41	–	1	–	–	–	–	–	–	–	–	–	**1-Octadecanamine, N-methyl-**	2439-55-6	**NI29214**
282	283	120	121	92	254	65	55	**283**	100	65	65	53	35	32	31	30		20	13	1	1	–	–	–	–	–	–	–	–	–	**Isoindolo[2,1-f]phenanthridin-10(14bH)-one**		**NI29215**
268	283	269	55	41	81	67	104	**283**	100	40	19	14	13	12	12	11		20	29	–	1	–	–	–	–	–	–	–	–	–	**5α-Androst-16-ene-17-carbonitrile**	7704-88-3	**NI29216**
44	268	255	283	282	269	241	240	**283**	100	86	60	44	22	18	13	13		20	29	–	1	–	–	–	–	–	–	–	–	–	**Quinoline, 3-pentyl-6,8-dipropyl-**	10372-06-2	**NI29217**
268	255	283	212	282	199	198	211	**283**	100	51	30	21	19	16	14	13		20	29	–	1	–	–	–	–	–	–	–	–	–	**Quinoline, 6-pentyl-3,8-dipropyl-**	7661-54-3	**NI29218**
227	254	240	241	198	255	228	211	**283**	100	70	62	24	19	16	14	13	**8.61**	20	29	–	1	–	–	–	–	–	–	–	–	–	**Quinoline, 8-pentyl-3,6-dipropyl-**	7634-75-5	**NI29219**
283	268	267	133	284	282	134	266	**283**	100	48	32	27	25	22	17	15		21	17	–	1	–	–	–	–	–	–	–	–	–	**7H-Dibenzo[b,g]carbazole, 7a,8-dihydro-7a-methyl-**	56114-47-7	**NI29220**
283	284	267	134	282	266			**283**	100	25	19	9	8	6				21	17	–	1	–	–	–	–	–	–	–	–	–	**1H-Indole, 1-methyl-2,3-diphenyl-**	6121-45-5	**MS06466**
47	85	114	28	265	198	113	67	**284**	100	83	16	16	12	11	5	5	**0.01**	–	1	–	–	–	9	–	–	–	4	–	–	–	**1,1,1,3,3,3-Hexafluoro-2-(trifluorosilyl)trisilane**		**MS06467**
85	117	119	251	87	253	249	121	**284**	100	55	53	34	32	22	21	17	**0.00**	3	–	–	–	–	2	6	–	–	–	–	–	–	**1,1-Difluorohexachloropropane**		**DO01077**
286	288	269	284	81	37	45	160	**284**	100	53	52	50	48	37	36	33		5	2	2	–	1	–	–	2	–	–	–	–	–	**2-Thiophenecarboxylic acid, 3,5-dibromo-**	7311-68-4	**NI29221**
286	288	269	284	81	37	45	162	**284**	100	53	52	50	48	37	36	33		5	2	2	–	1	–	–	2	–	–	–	–	–	**2-Thiophenecarboxylic acid, 3,5-dibromo-**	7311-68-4	**MS06468**
70	56	84	160	120	122	158	162	**284**	100	37	33	21	19	18	11	10	**4.00**	5	6	2	2	–	–	–	2	–	–	–	–	–	**2,4-Imidazolidinedione, 1,3-dibromo-5,5-dimethyl-**	77-48-5	**NI29222**
284	249	286	251	247	282	214	142	**284**	100	94	84	58	57	55	53	53		6	2	–	–	–	–	6	–	–	–	–	–	–	**Cyclobutane, 1,2-dichloro-3,4-bis(dichloromethylene)-**	55044-46-7	**NI29223**
63	28	49	62	27	65	124	107	**284**	100	67	34	28	28	26	24	21	**0.00**	6	12	4	–	–	–	3	–	–	–	1	–	–	**Tris(2-chloroethyl) phosphate**		**IC03646**
249	63	251	143	205	65	223	99	**284**	100	96	66	41	39	34	32	26	**1.53**	6	12	4	–	–	–	3	–	–	–	1	–	–	**Tris(2-chloroethyl) phosphate**		**IC03647**
63	170	172	157	62	75	50	155	**284**	100	32	31	29	28	27	24	23	**15.83**	7	6	3	–	1	–	1	1	–	–	–	–	–	**2-Bromo-5-methoxy-benzenesulphonyl chloride**		**IC03648**
69	41	39	119	121	87	205	207	**284**	100	60	37	13	13	12	10	10	**0.00**	7	10	2	–	–	–	–	2	–	–	–	–	–	**2-Propenoic acid, 2-methyl-, 2,3-dibromopropyl ester**	3066-70-4	**NI29224**
44	107	115	242	142	150	240	71	**284**	100	67	64	62	51	50	48	48	**21.62**	8	–	3	–	–	–	4	–	–	–	–	–	–	**1,3-Isobenzofurandione, 4,5,6,7-tetrachloro-**	117-08-8	**WI01354**
55	41	69	135	57	137	83	39	**284**	100	84	65	34	34	33	31	28	**0.00**	9	18	–	–	–	–	–	2	–	–	–	–	–	**Nonane, 1,9-dibromo-**	4549-33-1	**NI29225**
73	67	79	97	93	65	55	69	**284**	100	65	50	46	35	25	23	18	**0.00**	9	18	1	–	–	–	1	1	–	1	–	–	–	**2H-Pyran, 3-[bromochloro(trimethylsilyl)methyl]-3,4,5,6-tetrahydro-**		**DD00971**
241	163	155	199	43	197	161	153	**284**	100	47	47	45	42	36	35	33	**6.10**	9	21	–	–	–	–	1	–	–	–	–	–	1	**Tripropyltin chloride**		**MS06469**
126	286	63	284	288	74	50	75	**284**	100	68	39	36	34	29	26	19		10	6	–	–	–	–	–	2	–	–	–	–	–	**Naphthalene, 2,3-dibromo-**	13214-70-5	**NI29226**
147	238	148	91	43	77	149	58	**284**	100	25	11	9	9	7	6	6	**0.00**	10	10	–	–	1	4	1	–	–	–	–	1	–	**5-Chloro-1-ethylbenzo[b]thiophenium tetrafluoroborate**		**LI06117**
168	284	248	196	142	286	103	194	**284**	100	78	76	33	30	25	18	17		10	10	1	–	–	–	1	–	–	–	–	–	1	**Rhodium(I), (η^5-formylcyclopentadienyl)(η^4-2-chlorobutadiene)-**	101076-32-8	**NI29227**
61	284	87	63	237	74	41	75	**284**	100	84	79	71	61	52	52	50		10	12	4	4	1	–	–	–	–	–	–	–	–	**3-(Methylthio)propionaldehyde dinitrophenylhydrazone**		**MS06470**
57	29	43	73	31	44	60	71	**284**	100	58	55	54	44	39	38	29	**0.00**	10	12	6	4	–	–	–	–	–	–	–	–	–	**Xanthosine**	146-80-5	**NI29229**
152	109	57	73	54	43	60	55	**284**	100	73	52	45	42	28	27	22	**0.00**	10	12	6	4	–	–	–	–	–	–	–	–	–	**Xanthosine**	146-80-5	**NI29228**
57	152	73	54	109	43	60	71	**284**	100	69	63	46	42	42	34	32	**0.08**	10	12	6	4	–	–	–	–	–	–	–	–	–	**Xanthosine**	146-80-5	**NI29230**
72	225	88	126	55	69	130	28	**284**	100	64	25	23	21	17	15	15	**2.50**	10	15	4	2	–	3	–	–	–	–	–	–	–	**L-Valine, N-[N-(trifluoroacetyl)glycyl]-, methyl ester**	5680-78-4	**NI29231**
91	155	65	220	156	284	157	89	**284**	100	93	30	26	14	10	10	10		11	9	4	2	1	1	–	–	–	–	–	–	–	**Uracil, 1-(4-toluenesulphonyl)-5-fluoro-**		**NI29232**
86	56	136	57	134	135	28	85	**284**	100	47	41	38	36	34	26	23	**0.51**	11	12	3	2	2	–	–	–	–	–	–	–	–	**2-Benzthiazyl sulphonmorpholide**		**IC03649**
43	163	76	104	162	51	103	105	**284**	100	93	86	73	67	30	21	19	**7.27**	11	12	3	2	2	–	–	–	–	–	–	–	–	**δ^2-1,3,4-Thidiazolidine, 4-acetyl-2-methylsulphonyl-5-phenyl-**	86738-34-3	**NI29233**
284	185	242	243	143	144	183	157	**284**	100	89	81	42	37	36	35	35		11	17	2	–	–	–	–	–	–	–	–	–	1	**Rhodium, (2,4-pentanedionato-O,O′)bis(η^3-2-propenyl)-**	12307-52-7	**NI29234**

MASS TO CHARGE RATIOS								M.W.	INTENSITIES								Parent	C	H	O	N	S	F	Cl	Br	I	Si	P	B	X	COMPOUND NAME	CAS Reg No	No
73	269	284	147	45	154	99	285	**284**	100	94	92	52	36	27	27	23		11	20	3	2	–	–	–	–	–	2	–	–	–	4-Pyrimidinecarboxaldehyde, 2,6-bis[(trimethylsilyl)oxy]-	56272-56-1	NI29235
73	269	256	127	45	241	255	100	**284**	100	64	40	39	32	24	22	21	**8.79**	11	20	3	2	–	–	–	–	–	2	–	–	–	5-Pyrimidinecarboxaldehyde, 2,4-bis[(trimethylsilyl)oxy]-	56272-57-2	NI29236
163	43	119	121	133	284	89	164	**284**	100	27	20	19	17	15	15	13		11	24	4	–	2	–	–	–	–	–	–	–	–	D-Arabinose, dipropyl mercaptal	3767-33-7	NI29237
135	119	136	118	191	161	284	177	**284**	100	24	15	12	10	10	8	8		11	24	4	–	2	–	–	–	–	–	–	–	–	Galactose, 6-deoxy-2-O-methyl-, diethyl mercaptal	55836-49-2	LI06118
135	119	136	118	137	191	161	117	**284**	100	36	25	21	16	13	13	13	**9.13**	11	24	4	–	2	–	–	–	–	–	–	–	–	Galactose, 6-deoxy-2-O-methyl-, diethyl mercaptal	55836-49-2	NI29238
284	220	96	250	95	285	110	160	**284**	100	71	20	17	17	15	15	12		12	7	2	–	1	2	–	–	–	–	1	–	–	10H-Phenothiaphosphine, 2,7-difluoro-10-hydroxy-, 10-oxide	55255-69-1	NI29239
284	286	220	285	110	157	267	222	**284**	100	33	15	12	9	6	5	5		12	7	3	–	–	1	1	–	–	–	1	–	–	2-Chloro-8-fluorophenoxaphosphinic acid		NI29240
218	284	152	219	285	286	220	283	**284**	100	77	28	18	16	6	4	3		12	8	–	–	–	4	–	–	–	2	–	–	–	9,10-Disilaanthracene, 9,9,10,10-tetrafluoro-9,10-dihydro-		NI29241
77	108	143	125	141	159	284	220	**284**	100	80	68	66	62	46	40	39		12	9	2	–	2	–	1	–	–	–	–	–	–	Benzenesulphonothioic acid, 4-chloro-, S-phenyl ester	1142-97-8	NI29242
77	108	143	125	141	159	284	220	**284**	100	82	68	67	62	47	40	40		12	9	2	–	2	–	1	–	–	–	–	–	–	Benzenesulphonothioic acid, 4-chloro-, S-phenyl ester	1142-97-8	LI06119
125	75	284	157	65	109	50	159	**284**	100	51	47	47	47	38	37	35		12	9	2	–	2	–	1	–	–	–	–	–	–	4-Chloro-4'-mercaptodiphenyl sulphone		IC03650
117	133	105	218	77	91	79	55	**284**	100	62	53	45	26	18	18	14	**0.00**	12	10	1	–	–	6	–	–	–	–	–	–	–	Homocubane, 4-(hexafluoroisopropoxy)-	84785-05-7	NI29243
284	286	77	109	185	285	65	50	**284**	100	38	30	23	20	18	10	8		12	10	2	–	1	–	1	–	–	–	1	–	–	Diphenyl phosphorochloridothioate		LI06120
211	284	209	212	282	183	208	102	**284**	100	70	48	42	36	36	24	23		12	12	3	–	–	–	–	–	–	–	–	–	1	Benzo[b]selenophene-2-carboxylic acid, 3-methoxy-, ethyl ester	39857-44-8	NI29244
43	185	183	71	41	75	76	39	**284**	100	76	72	45	30	19	16	15	**0.50**	12	13	3	–	–	–	–	1	–	–	–	–	–	Butanoic acid, 4-bromophenacyl ester		NI29245
45	183	185	43	75	50	76	157	**284**	100	61	61	23	22	19	18	12	**0.60**	12	13	3	–	–	–	–	1	–	–	–	–	–	Propanoic acid, 2-methyl-, 4-bromophenacyl ester		NI29246
145	44	139	56	153	121	45	154	**284**	100	99	83	82	75	72	50	34	**9.20**	12	16	4	2	1	–	–	–	–	–	–	–	–	Alanine, N-[2-(4-methoxyphenylthio)ethyl]-N-nitroso-		NI29247
71	69	284	101	111	110	99	125	**284**	100	88	71	52	50	40	33	29		12	16	6	2	–	–	–	–	–	–	–	–	–	3,6-Epoxy-2H,8H-pyrimido[6,1-b][1,3]oxazocine-8,10(9H)-dione, 3,4,5,6-tetrahydro-4,5-dimethoxy-9-methyl-, [3R-(3α,4β,5β,6α)]-	41547-69-7	NI29248
173	68	59	113	43	69	71	269	**284**	100	76	70	67	56	52	48	47	**14.00**	12	16	6	2	–	–	–	–	–	–	–	–	–	Uridine, 2',3'-O-isopropylidine-	362-43-6	LI06121
173	68	59	113	43	69	71	269	**284**	100	75	68	65	55	50	45	43	**14.01**	12	16	6	2	–	–	–	–	–	–	–	–	–	Uridine, 2',3'-O-isopropylidine-	362-43-6	NI29250
209	174	58	141	59	57	54	53	**284**	100	95	73	63	60	60	55	50	**11.01**	12	16	6	2	–	–	–	–	–	–	–	–	–	Uridine, 2',3'-O-isopropylidine-	362-43-6	NI29249
58	204	59	159	146	205	160	57	**284**	100	15	4	3	3	2	1	1	**0.10**	12	17	4	2	–	–	–	–	–	–	1	–	–	1H-Indol-4-ol, 3-[2-(dimethylamino)ethyl]-, dihydrogen phosphate (ester)	520-52-5	NI29251
119	175	174	118	239	93	77	44	**284**	100	60	55	51	39	30	21	21	**3.00**	12	17	4	2	–	–	–	–	–	–	1	–	–	Phosphonic acid, [3-(aminocarbonyl)-2-(4-methylphenyl)-2-aziridinyl]-, dimethyl ester	33866-50-1	NI29252
119	175	174	118	239	93	91	71	**284**	100	60	55	51	39	30	21	21	**3.00**	12	17	4	2	–	–	–	–	–	–	1	–	–	Phosphonic acid, [3-(aminocarbonyl)-2-(4-methylphenyl)-2-aziridinyl]-, dimethyl ester	33866-50-1	LI06122
161	159	157	243	241	160	158	239	**284**	100	79	47	46	33	33	29	23	**1.40**	12	20	–	–	–	–	–	–	–	–	–	–	1	Stannane, tetra-2-propenyl-	7393-43-3	NI29253
43	111	124	110	113	57	29	123	**284**	100	98	78	22	20	20	19	18	**0.00**	12	22	6	–	–	–	–	–	–	–	–	2	–	L-Mannitol, 1-deoxy-, cyclic 3,4:5,6-bis(ethylboronate) 2-acetate	74779-69-4	NI29254
269	284	73	113	270	147	285	127	**284**	100	54	35	26	24	20	13	11		12	24	2	2	–	–	–	–	–	2	–	–	–	Pyrimidine, 4,5-dimethyl-2,6-bis[(trimethylsilyl)oxy]-	31111-32-7	NI29255
269	284	73	127	285	283	270	45	**284**	100	56	52	35	24	23	22	14		12	24	2	2	–	–	–	–	–	2	–	–	–	Pyrimidine, 5-ethyl-2,4-bis[(trimethylsilyl)oxy]-	31167-05-2	NI29256
269	73	270	197	45	271	211	170	**284**	100	78	27	27	21	13	12	11	**10.28**	12	28	–	2	–	–	–	–	–	3	–	–	–	1H-Pyrazole, 1,3,4-tris(trimethylsilyl)-	52805-97-7	NI29257
242	73	269	284	243	270	244	45	**284**	100	67	59	36	29	17	14	14		12	28	–	2	–	–	–	–	–	3	–	–	–	1H-Pyrazole, 1,4,5-tris(trimethylsilyl)-	52805-98-8	NI29258
139	173	141	175	111	284	286	75	**284**	100	44	34	28	25	24	23	7		13	7	1	–	–	–	3	–	–	–	–	–	–	Benzophenone, 2,4,4'-trichloro-	33146-57-5	LI06123
139	173	141	175	111	284	286	145	**284**	100	46	34	29	26	25	24	9		13	7	1	–	–	–	3	–	–	–	–	–	–	Benzophenone, 2,4,4'-trichloro-	33146-57-5	NI29260
139	173	141	175	111	284	286	145	**284**	100	43	33	28	25	24	23	9		13	7	1	–	–	–	3	–	–	–	–	–	–	Benzophenone, 2,4,4'-trichloro-	33146-57-5	NI29259
139	111	173	141	75	284	286	175	**284**	100	38	35	33	33	28	27	24		13	7	1	–	–	–	3	–	–	–	–	–	–	Benzophenone, 3,4,4'-trichloro-	33093-42-4	NI29261
90	284	63	76	50	39	254	208	**284**	100	98	48	45	36	33	32	26		13	8	4	4	–	–	–	–	–	–	–	–	–	Aniline, N,N'-methanetetraylbis[4-nitro-	738-66-9	NI29262
202	284	109	154	77	51	249	186	**284**	100	72	72	64	46	45	31	24		13	10	1	–	1	–	2	–	–	–	–	–	–	Diphenylsulphoxonium dichloromethylide		MS06471
151	108	152	284	184	136	153	41	**284**	100	11	9	8	8	7	5	3		13	16	3	–	2	–	–	–	–	–	–	–	–	Butanoic acid, 4-[[(4-methoxyphenyl)thioxomethyl]thio]-, methyl ester	36122-41-5	MS06472
137	211	29	109	27	165	81	55	**284**	100	65	22	16	16	13	11	10	**3.71**	13	16	7	–	–	–	–	–	–	–	–	–	–	1,3-Cyclopentanediglyoxylic acid, 2-oxo-, diethyl ester	22358-21-0	NI29263
105	77	73	122	123				**284**	100	31	30	14	10				**0.00**	13	16	7	–	–	–	–	–	–	–	–	–	–	β-D-Glucopyranose 1-benzoate	21056-52-0	NI29264
95	205	77	69	109	149	284	161	**284**	100	87	83	72	57	54	43	41		13	17	–	–	1	–	–	1	–	–	–	–	–	Cyclopropane, 1-(1-bromo-2-methylpropyl)-1-(phenylthio)-		DD00972
284	282	204	283	203	177	281	285	**284**	100	48	41	40	35	33	30	26		14	8	–	2	–	–	–	–	–	–	–	–	1	Phenanthro[1,2-c][1.2.5]selenadiazole	219-40-9	NI29265
284	204	177	58	75	150	257	142	**284**	100	81	58	37	34	33	31	13		14	8	–	2	–	–	–	–	–	–	–	–	1	Phenanthro[9,10-c][1.2.5]selenadiazole		DD00973
284	238	285	194	239	162	286	210	**284**	100	33	16	8	5	3	2	2		14	8	5	2	–	–	–	–	–	–	–	–	–	2-(3-Hydroxy-1,2-dihydropyrid-2-ylidene)-5-nitro-1,3-indandione		IC03651
284	226	127	63	154	254	285	126	**284**	100	50	35	26	25	22	20	20		14	8	5	2	–	–	–	–	–	–	–	–	–	1-Hydroxy-4-nitro-5-aminoanthraquinone		IC03652
44	103	105	284	240	28	77	30	**284**	100	51	46	41	34	32	31	23		14	8	5	2	–	–	–	–	–	–	–	–	–	6-Hydroxy-4-nitro-2-phenylisatogen		LI06124
77	284	286	285	205	283	258	260	**284**	100	68	66	31	26	20	19	18		14	9	–	2	–	–	–	1	–	–	–	–	–	3-Bromo-N-phenylbenzimidoyl cyanide		MS06473
201	166	165	203	178	202	179	76	**284**	100	97	97	75	52	31	29	25	**2.50**	14	11	–	–	–	–	3	–	–	–	–	–	–	Benzene, 1-chloro-4-(2,2-dichloro-1-phenylethyl)-	6952-08-5	NI29266
235	165	237	82	75	101	199	51	**284**	100	77	54	29	27	18	17	17	**6.00**	14	11	–	–	–	–	3	–	–	–	–	–	–	Benzene, 1,1'-(2-chloroethylidene)bis[4-chloro-	2642-80-0	NI29267
173	175	139	177	174	249	176	125	**284**	100	62	12	10	9	5	5	5	**0.00**	14	11	–	–	–	–	3	–	–	–	–	–	–	Benzene, 1,1'-(2-chloroethylidene)bis[4-chloro-	2642-80-0	NI29268
105	284	215	77	197	129	141	128	**284**	100	85	48	42	21	16	15	14		14	11	3	–	–	3	–	–	–	–	–	–	–	Cyclopentanone, 2-benzoyl-5-(trifluoroacetyl)-	62185-60-8	NI29269
239	284	254	256	211	226	180	283	**284**	100	83	60	58	50	48	35	27		14	12	3	2	–	–	–	–	–	1	–	–	–	10-Sila-2-azaanthrone, 4-nitro-10,10-dimethyl-9,10-dihydro-	99506-31-7	NI29270
256	254	210	180	224	269	238	239	**284**	100	92	82	65	62	60	47	43	**32.00**	14	12	3	2	–	–	–	–	–	1	–	–	–	10-Sila-2-azaanthrone, 5-nitro-10,10-dimethyl-9,10-dihydro-	99506-33-9	NI29271
269	284	223	180	211	153	154	152	**284**	100	42	20	9	5	5	4	3		14	12	3	2	–	–	–	–	–	1	–	–	–	10-Sila-2-azaanthrone, 6-nitro-10,10-dimethyl-9,10-dihydro-		NI29272

MASS TO CHARGE RATIOS	M.W.	INTENSITIES	Parent	C	H	O	N	S	F	Cl	Br	I	Si	P	B	X	COMPOUND NAME	CAS Reg No	No
269 284 223 180 239 224 211 254	**284**	100 51 22 8 5 5 4 3		14	12	3	2	–	–	–	–	–	1	–	–	–	10-Sila-2-azaanthrone, 7-nitro-10,10-dimethyl-9,10-dihydro-	99506-34-0	NI29273
284 254 239 269 180 210 166 224	**284**	100 22 22 20 20 12 12 10		14	12	3	2	–	–	–	–	–	1	–	–	–	10-Sila-2-azaanthrone, 8-nitro-10,10-dimethyl-9,10-dihydro-	99506-35-1	NI29274
134 284 92 43 65 162 285 91	**284**	100 76 55 42 31 21 13 13		14	12	3	4	–	–	–	–	–	–	–	–	–	4-(N-Acetamido)-4-nitroazobenzene		IC03653
43 256 213 171 242 77 156 157	**284**	100 99 89 78 76 68 60 53	**8.57**	14	12	3	4	–	–	–	–	–	–	–	–	–	Benzo[g]pteridine-10(2H)-acetaldehyde, 3,4-dihydro-7,8-dimethyl-2,4-dioxo-	4250-90-2	NI29275
284 181 98 70 103 180 105 104	**284**	100 88 88 81 78 77 76 70		14	16	1	6	–	–	–	–	–	–	–	–	–	Imidazo[1,2-a][1,3,5]triazin-4(8H)-one, 8-(benzylideneamino)-2-(dimethylamino)-6,7-dihydro-		NI29276
193 235 206 242 194 201 200 248	**284**	100 60 26 20 20 18 15 14	**4.16**	14	17	4	–	–	–	1	–	–	–	–	–	–	2-Chloromethyl-2-[2-(4-acetoxyphenyl)ethyl]-1,3-dioxolane		MS06474
227 85 57 284 41 228 29 229	**284**	100 45 36 21 19 18 13 12		14	20	2	–	2	–	–	–	–	–	–	–	–	2-Butanone, 1,1′-(1,3-dithietane-2,4-diylidene)bis[3,3-dimethyl-	19018-14-5	NI29277
43 91 41 65 39 139 157 92	**284**	100 95 86 58 58 52 50 35	**0.00**	14	20	4	–	1	–	–	–	–	–	–	–	–	2-Hexene-1,4-diol, 5-methyl-3-[(4-methylphenyl)sulphonyl]-, (E)-	118356-23-3	DD00974
125 140 113 169 145 253 43 83	**284**	100 87 69 60 45 44 41 40	**37.00**	14	20	6	–	–	–	–	–	–	–	–	–	–	Bicyclo[2.2.2]octane-2,3-dicarboxylic acid, 1-hydroxy-8,8-dimethyl-5-oxo-, dimethyl ester	25864-66-8	MS06475
193 175 284 207 115 234 253 252	**284**	100 46 29 23 21 15 13 11		14	20	6	–	–	–	–	–	–	–	–	–	–	Bicyclo[3.3.0]oct-3-ene-2,4-dicarboxylic acid, 7-hydroxy-3-methoxy-2-methyl-, dimethyl ester, (1α,2β,5α,7β)-		MS06476
73 86 146 100 74 144 121 175	**284**	100 99 55 48 46 45 40 39	**11.13**	14	20	6	–	–	–	–	–	–	–	–	–	–	α-D-Glucopyranoside, methyl 3-O-benzyl-	40010-11-5	NI29278
73 87 75 100 163 45 82 61	**284**	100 86 85 80 77 56 46 44	**29.72**	14	20	6	–	–	–	–	–	–	–	–	–	–	α-D-Glucopyranoside, methyl 4-O-benzyl-	23392-31-6	NI29279
163 73 150 133 164 121 115 97	**284**	100 28 24 12 11 11 10 10	**1.42**	14	20	6	–	–	–	–	–	–	–	–	–	–	D-Glucopyranoside, methyl 2-O-benzyl-	53958-32-0	NI29280
252 87 73 60 133 162 71 74	**284**	100 96 39 38 33 31 30 25	**10.69**	14	20	6	–	–	–	–	–	–	–	–	–	–	D-Glucopyranoside, methyl 6-O-benzyl-	23392-32-7	NI29281
224 206 79 119 91 55 192 225	**284**	100 68 60 56 52 47 46 43	**3.33**	14	20	6	–	–	–	–	–	–	–	–	–	–	Tricyclo[3.3.1.1^{3,7}]decane-1,3-dicarboxylic acid, 4,8-dihydroxy-, dimethyl ester	55724-17-9	NI29282
269 271 57 284 286 41 272 270	**284**	100 98 49 45 44 20 18 18		14	21	1	–	–	–	–	1	–	–	–	–	–	4-Bromo-2,6-di-tert-butylphenol		IC03654
240 56 141 55 105 77 159 187	**284**	100 83 45 43 42 42 37 30	**4.19**	14	21	4	–	–	–	–	–	–	–	1	–	–	1,4,6,9-Tetraoxa-5-phospha(5-PV)spiro[4.4]nonane, 2,3,7,8-tetramethyl-5-phenyl-	74810-56-3	NI29283
96 142 155 71 143 82 42 29	**284**	100 58 39 18 17 16 13 12	**11.50**	14	24	4	2	–	–	–	–	–	–	–	–	–	1,2-Ethanediamine, N,N′-bis[1-(ethoxycarbonyl)prop-1-en-2-yl]-		IC03655
96 142 155 143 71 82 97 42	**284**	100 58 37 20 19 16 15 12	**8.00**	14	24	4	2	–	–	–	–	–	–	–	–	–	1,2-Ethanediamine, N,N′-bis[1-(ethoxycarbonyl)prop-1-en-2-yl]-		NI29284
99 142 43 69 113 141 42 57	**284**	100 53 45 43 35 34 34 27	**2.50**	14	28	2	4	–	–	–	–	–	–	–	–	–	1,2,4,5-Tetrazine, 1,4-diacetyl-2,5-dibutylhexahydro-	35028-99-0	NI29285
142 141 143 284 199 86 99 84	**284**	100 74 37 31 30 19 16 13		14	28	2	4	–	–	–	–	–	–	–	–	–	1,2,4,5-Tetrazine, 1,4-diacetyl-2,5-dibutylhexahydro-	35028-99-0	NI29286
196 284 181 122 155 197 285 123	**284**	100 66 40 40 29 24 22 14		14	32	–	–	–	–	–	–	–	3	–	–	–	Silane, (1-ethyl-1,2-propadien-1-yl-3-ylidene)tris[trimethyl-	61227-93-8	NI29287
121 118 284 77 122 285 163 123	**284**	100 59 34 27 11 7 7 6		15	12	–	2	2	–	–	–	–	–	–	–	–	Benzothiazole, 2-(thiobenzoylmethylamino)-		MS06477
284 121 251 181 207 77 285 149	**284**	100 60 59 53 50 30 19 15		15	12	–	2	2	–	–	–	–	–	–	–	–	Benzothiazoline, 3-methyl-2-(thiobenzoylimino)-		MS06478
284 225 165 224 252 285 193 104	**284**	100 60 54 42 35 20 19 14		15	12	–	2	2	–	–	–	–	–	–	–	–	2,4-Imidazolidinedithione, 5,5-diphenyl-	2032-13-5	NI29288
194 77 135 51 220 219 195 91	**284**	100 75 64 40 32 17 17 17	**16.00**	15	12	–	2	2	–	–	–	–	–	–	–	–	1,3,4-Thiadiazolium, 5-mercapto-2-(4-methylphenyl)-3-phenyl-, hydroxide, inner salt	26229-05-0	NI29289
148 284 137 136 108 105 133 120	**284**	100 83 43 40 31 23 17 17		15	12	2	2	1	–	–	–	–	–	–	–	–	4H-1,3-Benzothiazin-4-one, 2,3-dihydro-2-[(2-methoxyphenyl)amino]-		NI29290
148 284 133 137 136 108 285 149	**284**	100 65 34 32 15 13 12 10		15	12	2	2	1	–	–	–	–	–	–	–	–	4H-1,3-Benzothiazin-4-one, 2,3-dihydro-2-[(4-methoxyphenyl)amino]-		NI29291
198 226 77 115 102 199 121 120	**284**	100 26 22 21 14 14 11 10	**3.20**	15	12	2	2	1	–	–	–	–	–	–	–	–	2,4-Pentadienoic acid, 2-cyano-5-phenyl-5-thiocyanato-, ethyl ester		MS06479
168 128 284 64 253 140 101 77	**284**	100 52 42 35 33 20 19 14		15	12	4	2	–	–	–	–	–	–	–	–	–	Imidazo[2,1-a]isoquinoline-2,3-dicarboxylic acid, dimethyl ester	58275-55-1	NI29292
105 138 77 146 120 284 91 147	**284**	100 74 60 14 13 10 10 7		15	12	4	2	–	–	–	–	–	–	–	–	–	N-(3-Nitrophenyl)-2-benzoylacetamide		LI06125
105 77 138 146 91 284 120 147	**284**	100 62 50 17 13 11 11 7		15	12	4	2	–	–	–	–	–	–	–	–	–	N-(4-Nitrophenyl)-2-benzoylacetamide		LI06126
284 283 268 285 286 220 77 219	**284**	100 58 41 36 34 30 23 15		15	13	–	4	–	–	1	–	–	–	–	–	–	3H-1,3,4-Benzotriazepine, 7-chloro-2-(methylamino)-5-phenyl-		MS06480
284 192 221 286 90 102 285 193	**284**	100 67 50 35 30 25 20 19		15	13	–	4	–	–	1	–	–	–	–	–	–	2-Ethyl-1,2,3-benzotriazinium 4-(2-chloroanilide)		LI06127
119 93 91 64 136 120 77 66	**284**	100 94 50 31 25 25 25 25	**0.00**	15	16	2	4	–	–	–	–	–	–	–	–	–	Methylenebis(3-phenylurea)		LI06128
184 93 65 108 151 191 284 176	**284**	100 44 44 34 33 32 29 5		15	16	2	4	–	–	–	–	–	–	–	–	–	ω-Phenylsemicarbazidoacetanilide		LI06129
108 93 65 176 284 151 191	**284**	100 78 78 53 6 6 5		15	16	2	4	–	–	–	–	–	–	–	–	–	ω-Phenylsemicarbazidoacetanilide		LI06130
55 128 142 69 169 101 155 115	**284**	100 99 96 95 91 65 58 38	**18.70**	15	16	2	4	–	–	–	–	–	–	–	–	–	1H-1,2,3-Triazolo[4,5-c]quinoline-1-hexanoic acid	52799-13-0	NI29293
199 201 179 214 216 171 136 249	**284**	100 76 67 59 40 35 32 28	**11.11**	15	18	1	–	–	–	2	–	–	–	–	–	–	1-Nonen-3-one, 1-(2,4-dichlorophenyl)-	36383-93-4	NI29294
249 199 179 201 251 171 135 136	**284**	100 61 41 40 32 29 26 25	**5.85**	15	18	1	–	–	–	2	–	–	–	–	–	–	1-Nonen-3-one, 1-(2,6-dichlorophenyl)-	36383-94-5	NI29295
199 201 179 214 136 216 171 41	**284**	100 91 79 73 70 49 41 37	**23.96**	15	18	1	–	–	–	2	–	–	–	–	–	–	1-Nonen-3-one, 1-(3,4-dichlorophenyl)-	36383-95-6	NI29296
227 195 266 135 167 163 168 284	**284**	100 95 85 77 68 68 52 41		15	24	3	–	1	–	–	–	–	–	–	–	–	Bicyclo[3.3.0]octane-1-carboxylic acid, 2-(butylthio)-4-methyl-8-oxo-, methyl ester, (1RS,2RS,4RS,5SR)-		MS06481
227 195 93 135 167 107 163 284	**284**	100 90 90 86 83 60 46 45		15	24	3	–	1	–	–	–	–	–	–	–	–	Bicyclo[3.3.0]octane-1-carboxylic acid, 2-(butylthio)-4-methyl-8-oxo-, methyl ester, (1RS,2RS,4SR,5SR)-		MS06482
141 135 195 227 284 163 167 168	**284**	100 68 66 65 62 55 50 43		15	24	3	–	1	–	–	–	–	–	–	–	–	Bicyclo[3.3.0]octane-1-carboxylic acid, 2-(butylthio)-4-methyl-8-oxo-, methyl ester, (1RS,2SR,4RS,5SR)-		MS06483
143 111 195 141 284 227 162 163	**284**	100 80 35 30 15 13 12 10		15	24	3	–	1	–	–	–	–	–	–	–	–	1-Cyclopentanecarboxylic acid, 2-[3-(butylthio)-1-methylprop-2-enyl]-5-oxo-, methyl ester, (1RS,1′SR,2SR,2′E)-		MS06484

MASS TO CHARGE RATIOS								M.W.	INTENSITIES								Parent	C	H	O	N	S	F	Cl	Br	I	Si	P	B	X	COMPOUND NAME	CAS Reg No	No
57	87	141	109	113	111	210	197	**284**	100	65	40	35	19	17	10	9	**7.00**	15	24	3	–	1	–	–	–	–	–	–	–	–	1-Cyclopentanecarboxylic acid, 2-[3-(tert-butylthio)-1-methylprop-2-enyl]-5-oxo-, methyl ester, (1RS,1'RS,2RS,2'E)-		MS06485
101	269	73	181	121	43	79	163	**284**	100	17	13	11	9	8	6	5	**0.00**	15	24	5	–	–	–	–	–	–	–	–	–	–	6-Decenoic acid, 9,10-di-O-isopropylidene-9,10-dihydroxy-3-oxo-, ethyl ester, (S)-(Z)-		DD00975
182	69	43	110	111	156	71	44	**284**	100	87	63	35	33	26	22	22	**0.00**	15	24	5	–	–	–	–	–	–	–	–	–	–	2,4(3H,5H)-Furandione, 5-hexyl-3-(2-hydroxy-1-oxobutyl)-5-methyl-	55255-65-7	NI29297
284	249	77	221	286	135	123	193	**284**	100	67	64	42	40	30	30	28		16	9	3	–	–	–	1	–	–	–	–	–	–	1,4-Naphthoquinone, 2-chloro-3-phenoxy-	71369-17-0	NI29298
284	185	115	255	128	139	213	256	**284**	100	65	64	52	49	43	40	35		16	12	5	–	–	–	–	–	–	–	–	–	–	1,4-Anthracenedione, 5,10-dihydroxy-2-methoxy-7-methyl-	74815-58-0	NI29299
44	43	57	284	83	69	41	55	**284**	100	29	17	12	11	11	11	10		16	12	5	–	–	–	–	–	–	–	–	–	–	1,4-Anthracenedione, 7,9-dihydroxy-5-methoxy-2-methyl-	84542-43-8	NI29300
284	128	51	139	241	185	63	77	**284**	100	46	46	44	40	35	35	34		16	12	5	–	–	–	–	–	–	–	–	–	–	9,10-Anthracenedione, 1,8-dihydroxy-3-methoxy-6-methyl-	521-61-9	NI29301
284	255	241	139	267	256	213	152	**284**	100	71	18	15	13	11	9	7		16	12	5	–	–	–	–	–	–	–	–	–	–	9,10-Anthracenedione, 4,8-dihydroxy-1-methoxy-2-methyl-		NI29302
256	284	213	185	285	257	255	157	**284**	100	91	87	23	17	17	15	14		16	12	5	–	–	–	–	–	–	–	–	–	–	2,3-Benzofurandione, 4-(4-hydroxybenzyl)-6-methoxy-	14309-91-2	NI29303
284	283	238	285	137	118	255	142	**284**	100	31	15	9	9	5	5	3		16	12	5	–	–	–	–	–	–	–	–	–	–	4H-1-Benzopyran-4-one, 7-hydroxy-3-(4-hydroxyphenyl)-5-methoxy-		NI29304
284	255	285	283	241	138	166	95	**284**	100	30	19	13	13	10	9	9		16	12	5	–	–	–	–	–	–	–	–	–	–	Flavone, 4',5-dihydroxy-7-methoxy-	437-64-9	NI29305
284	256	138	241	95	167	166	118	**284**	100	24	12	10	10	9	9	7		16	12	5	–	–	–	–	–	–	–	–	–	–	Flavone, 4',5-dihydroxy-7-methoxy-	437-64-9	NI29306
284	153	285	254	131	124	77	69	**284**	100	50	21	19	16	16	13	13		16	12	5	–	–	–	–	–	–	–	–	–	–	Flavone, 5,7-dihydroxy-2'-methoxy-	10458-35-2	NI29307
283	132	284	241	282	152	133	128	**284**	100	21	19	10	9	8	6	6		16	12	5	–	–	–	–	–	–	–	–	–	–	Flavone, 5,7-dihydroxy-4'-methoxy-	480-44-4	NI29308
284	132	152	283	241	124	117	135	**284**	100	29	12	9	9	9	9	6		16	12	5	–	–	–	–	–	–	–	–	–	–	Flavone, 5,7-dihydroxy-4'-methoxy-	480-44-4	MS06486
284	69	269	241	266	139	103	77	**284**	100	89	68	55	47	27	15	15		16	12	5	–	–	–	–	–	–	–	–	–	–	Flavone, 5,7-dihydroxy-6-methoxy-	480-11-5	NI29309
191	286	284	205	203	202	204	176	**284**	100	43	43	24	20	19	18	10		16	13	–	–	–	–	–	1	–	–	–	–	–	2-(9-Anthryl)ethyl bromide		MS06487
283	284	285	91	286	249	268	248	**284**	100	73	43	31	24	17	9	8		16	13	1	2	–	–	1	–	–	–	–	–	–	3H-1,4-Benzodiazepine, 7-chloro-2-methoxy-5-phenyl-		MS06488
256	283	284	255	257	258	285	222	**284**	100	87	69	42	41	35	23	18		16	13	1	2	–	–	1	–	–	–	–	–	–	1H-1,4-Benzodiazepin-2-one, 7-chloro-2,3-dihydro-1-methyl-5-phenyl-		MS06489
284	256	283	285	255	257	258	286	**284**	100	98	78	45	41	40	35	33		16	13	1	2	–	–	1	–	–	–	–	–	–	1H-1,4-Benzodiazepin-2-one, 7-chloro-2,3-dihydro-1-methyl-5-phenyl-		LI06131
241	243	283	242	284	285	244	76	**284**	100	87	73	62	30	25	25	19		16	13	1	2	–	–	1	–	–	–	–	–	–	1H-1,4-Benzodiazepin-2-one, 7-chloro-2,3-dihydro-3-methyl-5-phenyl-	4699-82-5	NI29310
256	283	284	257	285	258	255	221	**284**	100	85	69	47	40	40	39	34		16	13	1	2	–	–	1	–	–	–	–	–	–	2H-1,4-Benzodiazepin-2-one, 7-chloro-1,3-dihydro-1-methyl-5-phenyl-	439-14-5	NI29311
256	110	89	51	63	165	77	91	**284**	100	90	83	82	75	68	67	59	**43.82**	16	13	1	2	–	–	1	–	–	–	–	–	–	2H-1,4-Benzodiazepin-2-one, 7-chloro-1,3-dihydro-1-methyl-5-phenyl-	439-14-5	NI29312
256	283	284	255	285	257	258	286	**284**	100	87	82	43	42	41	37	27		16	13	1	2	–	–	1	–	–	–	–	–	–	2H-1,4-Benzodiazepin-2-one, 7-chloro-1,3-dihydro-1-methyl-5-phenyl-	439-14-5	NI29313
284	241	69	268	39	102	142	129	**284**	100	46	37	36	32	24	21	19		16	16	3	2	–	–	–	–	–	–	–	–	–	Benzonitrile, 2-[(2,4,6-trimethoxyphenyl)amino]-	34913-35-4	NI29314
57	83	43	55	129	71	69	118	**284**	100	84	79	71	64	63	55	40	**29.00**	16	16	3	2	–	–	–	–	–	–	–	–	–	Benzonitrile, 3-[(2,4,6-trimethoxyphenyl)amino]-	34913-36-5	NI29315
178	161	177	131	77	132	107	105	**284**	100	51	45	44	29	23	23	19	**0.00**	16	16	3	2	–	–	–	–	–	–	–	–	–	1,3-Oxazolidine, 4-methyl-cis-5-phenyl-2-(4-nitrophenyl)-		MS06490
178	131	161	177	77	132	130	179	**284**	100	39	35	32	13	12	11	11	**0.00**	16	16	3	2	–	–	–	–	–	–	–	–	–	1,3-Oxazolidine, 4-methyl-trans-5-phenyl-2-(4-nitrophenyl)-		MS06491
207	284	179	255	43	161	211	202	**284**	100	42	30	27	27	20	17	14		16	16	3	2	–	–	–	–	–	–	–	–	–	4H-Pyran-5-carboxylic acid, 2-amino-3-cyano-4-phenyl-6-methyl-, ethyl ester	72568-47-9	NI29316
151	149	147	74	150	228	78	72	**284**	100	72	52	48	40	37	35	34	**12.00**	16	18	–	–	–	–	–	–	–	–	–	–	1	Germacyclopentane, 1,1-diphenyl-	4514-06-1	NI29317
93	212	92	65	77	66	94	91	**284**	100	14	10	9	9	7	9	4	**0.00**	16	20	1	2	–	–	–	–	–	1	–	–	–	Urea, N,N'-diphenyl-N-(trimethylsilyl)-	1154-84-3	NI29318
43	41	57	55	59	39	69	56	**284**	100	94	62	54	45	28	26	24	**4.36**	16	28	–	–	2	–	–	–	–	–	–	–	–	2-(Decylthio)-5-ethylthiophene		AI01975
43	55	95	41	138	121	79	164	**284**	100	23	22	18	13	8	8	7	**0.00**	16	28	4	–	–	–	–	–	–	–	–	–	–	2,2-Dipropylcyclohexane-trans-1,3-diol diacetate		LI06132
131	101	73	269	139	189	107	157	**284**	100	51	51	15	9	6	5	4	**0.20**	16	28	4	–	–	–	–	–	–	–	–	–	–	1-O-(2-Methoxy-4-nonynyl)-2,3-O-isopropylideneglycerol		NI29319
97	143	69	55	41	125	159	239	**284**	100	71	30	20	17	13	11	2	**1.00**	16	28	4	–	–	–	–	–	–	–	–	–	–	7-Oxaheptanoic acid, 7-(2-ethyl-2-methyl-5-oxocyclopentyl)-, ethyl ester		NI29320
30	43	72	73	41	60	55	86	**284**	100	52	25	24	20	15	15	14	**6.00**	16	32	2	2	–	–	–	–	–	–	–	–	–	N,N'-Diacetyl-1,12-diaminododecane		LI06133
249	284	178	286	248	221	219	250	**284**	100	73	48	27	23	20	20	19		17	13	2	–	–	–	1	–	–	–	–	–	–	5,10-Methano-5H-dibenzo[a,d]cyclohepten-11-one, 12-chloro-10,11-dihydro-7-methoxy-, anti-		NI29321
249	284	178	250	286	248	221	176	**284**	100	79	56	49	30	30	28	23		17	13	2	–	–	–	1	–	–	–	–	–	–	5,11-Methano-5H-dibenzo[a,d]cyclohepten-10-one, 12-chloro-10,11-dihydro-7-methoxy-, anti-		NI29322
249	248	284	178	205	286	250	189	**284**	100	48	46	31	21	17	15	13		17	13	2	–	–	–	1	–	–	–	–	–	–	5,10-Methano-5H-dibenzo[a,d]cyclohepten-11-one, 12-chloro-10,11-dihydro-2-methoxy-, syn-		NI29323
284	249	219	178	248	286	189	250	**284**	100	63	50	45	44	31	25	21		17	13	2	–	–	–	1	–	–	–	–	–	–	5,10-Methano-5H-dibenzo[a,d]cyclohepten-11-one, 12-chloro-10,11-dihydro-7-methoxy-, syn-		NI29324
249	248	284	250	178	205	286	206	**284**	100	71	46	30	26	21	17	14		17	13	2	–	–	–	1	–	–	–	–	–	–	5,11-Methano-5H-dibenzo[a,d]cyclohepten-10-one, 12-chloro-10,11-dihydro-7-methoxy-, syn-		NI29325
191	189	284	82	190	179	165	208	**284**	100	29	26	22	21	15	15	14		17	13	2	–	–	–	1	–	–	–	–	–	–	9-Phenanthrenemethyl monochloroacetate	92174-37-3	NI29326
165	198	284	121	166	240	225	239	**284**	100	90	65	50	32	24	24	20		17	16	2	–	1	–	–	–	–	–	–	–	–	1,3-Oxathiolan-5-one, 2,2-dimethyl-4,4-diphenyl-	31061-71-9	NI29327
165	211	121	240	239	284	166	198	**284**	100	66	66	64	53	40	30	22		17	16	2	–	1	–	–	–	–	–	–	–	–	1,3-Oxathiolan-5-one, 2-ethyl-4,4-diphenyl-	18648-77-6	NI29328
163	105	284	77	121	135	91	253	**284**	100	84	30	21	10	10	7	7		17	16	4	–	–	–	–	–	–	–	–	–	–	1,4-Benzenedicarboxylic acid, methyl 4-methylbenzyl ester	67801-55-2	NI29329
251	266	252	269	178	284	267	179	**284**	100	23	19	12	6	5	5	5		17	16	4	–	–	–	–	–	–	–	–	–	–	1,2-Benzenedicarboxylic acid, 4-[2-(2-phenylpropyl)]-		IC03656
284	283	269	148	285	161	133	253	**284**	100	34	24	22	20	18	5	4		17	16	4	–	–	–	–	–	–	–	–	–	–	6H-Benzofuro[3,2-c][1]benzopyran, 6a,11a-dihydro-3,9-dimethoxy-, (6aR-cis)-	606-91-7	NI29330

MASS TO CHARGE RATIOS								M.W.	INTENSITIES								Parent	C	H	O	N	S	F	Cl	Br	I	Si	P	B	X	COMPOUND NAME	CAS Reg No	No
119	149	91	77	284	253	266	105	284	100	88	59	17	16	12	11	7		17	16	4	–	–	–	–	–	–	–	–	–	–	Benzoic acid, 2-methyl-, [4-(methoxycarbonyl)phenyl]methyl ester		NI29331
119	44	91	149	43	120	90	89	284	100	32	25	15	13	11	11	11	9.14	17	16	4	–	–	–	–	–	–	–	–	–	–	Benzoic acid, 4-methyl-, [4-(methoxycarbonyl)phenyl]methyl ester	55044-52-5	MS06492
119	284	91	149	120	89	59	90	284	100	13	13	12	10	5	5	4		17	16	4	–	–	–	–	–	–	–	–	–	–	Benzoic acid, 4-methyl-, [4-(methoxycarbonyl)phenyl]methyl ester	55044-52-5	IC03657
119	91	149	121	284	77	105	253	284	100	30	14	14	10	8	6	4		17	16	4	–	–	–	–	–	–	–	–	–	–	Benzoic acid, 4-methyl-, [4-(methoxycarbonyl)phenyl]methyl ester	55044-52-5	NI29332
284	152	180	104	283	207	124	241	284	100	55	50	40	30	30	30	5		17	16	4	–	–	–	–	–	–	–	–	–	–	Flavanone, 5,7-dihydroxy-6,8-dimethyl-		NI29333
180	284	152	137	207	181	285	283	284	100	51	21	20	14	14	10	9		17	16	4	–	–	–	–	–	–	–	–	–	–	Flavanone, 5,7-dimethoxy-	1036-72-2	NI29334
269	284	229	241	270	69	41	285	284	100	75	33	20	19	17	16	16		17	16	4	–	–	–	–	–	–	–	–	–	–	7H-Furo[3,2-g][1]benzopyran-7-one, 9-methoxy-4-(1,1-dimethylallyl)-	34155-80-1	NI29335
284	216	269	285	241	229	282	115	284	100	22	21	18	12	12	9	7		17	16	4	–	–	–	–	–	–	–	–	–	–	7H-Furo[3,2-g][1]benzopyran-7-one, 9-methoxy-4-(3-methyl-2-butenyl)-	10523-54-3	NI29336
284	216	269	285	241	229	282	68	284	100	22	20	18	13	12	9	9		17	16	4	–	–	–	–	–	–	–	–	–	–	7H-Furo[3,2-g][1]benzopyran-7-one, 9-methoxy-4-(3-methyl-2-butenyl)-	10523-54-3	NI29337
266	137	265	284	91	119	77	238	284	100	82	71	66	60	54	52	44		17	16	4	–	–	–	–	–	–	–	–	–	–	3(3′)-Hydroxy-7-methoxyisochromano[4,3:2,3]chroman		MS06493
237	252	284	269	238	204	119	69	284	100	42	35	22	18	18	18	18		17	16	4	–	–	–	–	–	–	–	–	–	–	2H-Naphtho[1,2-b]pyran-5-carboxylic acid, 6-hydroxy-2,2-dimethyl-		NI29338
105	77	51	106	227	162	179	104	284	100	26	8	8	6	5	4	4	0.40	17	16	4	–	–	–	–	–	–	–	–	–	–	1,3-Propanediol dibenzoate	2451-86-7	NI29339
284	251	164	285	240	267	283	148	284	100	63	25	21	21	19	17	15		17	20	–	2	1	–	–	–	–	–	–	–	–	Methanethione, bis[4-(dimethylamino)phenyl]-	1226-46-6	NI29340
72	73	42	198	213	180	56	43	284	100	6	4	3	2	2	2	2	1.78	17	20	–	2	1	–	–	–	–	–	–	–	–	10H-Phenothiazine-10-ethanamine, N,N,α-trimethyl-	60-87-7	NI29341
72	73	42	213	284	198	56	180	284	100	77	59	42	40	39	36	31		17	20	–	2	1	–	–	–	–	–	–	–	–	10H-Phenothiazine-10-ethanamine, N,N,α-trimethyl-	60-87-7	NI29342
285	284	286	287	257	283	255	200	284	100	22	21	7	3	2	2	2		17	20	–	2	1	–	–	–	–	–	–	–	–	10H-Phenothiazine-10-ethanamine, N,N,α-trimethyl-	60-87-7	NI29343
58	42	86	284	198	44	238	85	284	100	94	93	74	67	65	48	45		17	20	–	2	1	–	–	–	–	–	–	–	–	10H-Phenothiazine-10-propanamine, N,N-dimethyl-	58-40-2	NI29346
58	284	85	238	198	239	84	199	284	100	18	18	13	9	8	8	7		17	20	–	2	1	–	–	–	–	–	–	–	–	10H-Phenothiazine-10-propanamine, N,N-dimethyl-	58-40-2	NI29344
284	58	86	238	199	85	285	198	284	100	41	38	35	34	25	24	21		17	20	–	2	1	–	–	–	–	–	–	–	–	10H-Phenothiazine-10-propanamine, N,N-dimethyl-	58-40-2	NI29345
269	91	283	193	284	134.5	89		284	100	11	6	4	3	1	1			17	20	2	–	–	–	–	–	–	1	–	–	–	Acetophenone, 2-phenyl-2′-(trimethylsiloxy)-	33342-96-0	LI06134
269	255	270	73	256	91	147	127	284	100	55	25	13	12	11	9	9	3.00	17	20	2	–	–	–	–	–	–	1	–	–	–	Acetophenone, 2-phenyl-2′-(trimethylsiloxy)-	33342-96-0	NI29347
154	284	182	169	225	167	283	183	284	100	95	83	76	64	63	58	58		17	20	2	2	–	–	–	–	–	–	–	–	–	Azecino[4,5,6-cd]indole-11-carboxylic acid, 2,6,7,10,11,12-hexahydro-8-methyl-, methyl ester	55702-34-6	NI29348
193	105	148	42	91	104	194	103	284	100	33	32	16	15	14	13	13	1.00	17	20	2	2	–	–	–	–	–	–	–	–	–	Benzeneethanamine, N-methyl-3-nitro-N-(2-phenylethyl)-	52059-40-2	NI29349
193	105	148	42	91	77	78	104	284	100	58	49	32	27	25	21	20	1.00	17	20	2	2	–	–	–	–	–	–	–	–	–	Benzeneethanamine, N-methyl-4-nitro-N-(2-phenylethyl)-	52118-15-7	NI29350
121	91	284	133	107	57	28	43	284	100	81	74	61	59	45	43	43		17	20	2	2	–	–	–	–	–	–	–	–	–	4-Butyl-4′-methoxy-NNO-azoxybenzene	31401-35-1	NI29351
107	77	135	57	91	284	43	92	284	100	47	33	26	21	20	19	19		17	20	2	2	–	–	–	–	–	–	–	–	–	4-Butyl-4′-methoxy-ONN-azoxybenzene	31401-36-2	NI29352
154	284	182	169	225	167	283	183	284	100	95	83	76	64	63	58	58		17	20	2	2	–	–	–	–	–	–	–	–	–	Clavicipitic acid methyl ester		NI29353
134	120	284	147	285	148	135	106	284	100	18	16	15	8	8	8	6		17	20	2	2	–	–	–	–	–	–	–	–	–	2-(N-Ethylanilino)ethyl 4-aminobenzoate		IC03658
284	227	240	269	239	225	285	213	284	100	48	42	36	30	29	28	23		17	20	2	2	–	–	–	–	–	–	–	–	–	17β-Hydroxy-6,7-diazaequilenin-3-methyl ether		LI06135
145	70	115	144	186	146	284	214	284	100	67	44	41	17	11	10	6		17	20	2	2	–	–	–	–	–	–	–	–	–	4-Isopropylidene-1,1,7-trimethyl-4H-[1,3,4]oxadiazino[4,3-a]cinnolin-2(1H)-one		LI06136
283	284	225	169	223	170	197	157	284	100	90	32	32	30	30	28	26		17	20	2	2	–	–	–	–	–	–	–	–	–	2-(Methoxycarbonyl)octahydroindolo[2,3-a]quinolizine		LI06138
283	284	197	225	170	156	169		284	100	90	42	33	17	15	12			17	20	2	2	–	–	–	–	–	–	–	–	–	2-(Methoxycarbonyl)octahydroindolo[2,3-a]quinolizine		LI06137
130	154	161	144	284	143			284	100	49	20	4	3	3				17	20	2	2	–	–	–	–	–	–	–	–	–	1,2,3,6-Tetrahydro-1-[2-(indol-3-yl)ethyl]-4-(methoxycarbonyl)pyridine		LI06139
268	198	212	251	82	144	143	199	284	100	50	45	34	20	18	18	15	0.00	17	20	2	2	–	–	–	–	–	–	–	–	–	17,18,19-Trinorsarpagan-3,10-diol, 1-methyl-	55030-10-9	NI29354
55	41	85	67	81	29	54	68	284	100	73	62	51	38	38	35	34	0.00	17	32	3	–	–	–	–	–	–	–	–	–	–	3-Dodecen-1-ol, 12-[(tetrahydro-2H-pyran-2-yl)oxy]-, (Z)-	69494-16-2	NI29355
57	43	41	71	55	85	69	95	284	100	79	49	48	42	26	20	19	0.00	17	32	3	–	–	–	–	–	–	–	–	–	–	Hexadecanoic acid, 2-oxo-, methyl ester	55836-30-1	NI29356
69	124	95	55	97	115	41	143	284	100	81	78	63	60	58	57	42	0.00	17	32	3	–	–	–	–	–	–	–	–	–	–	7-Oxaheptanoic acid, 7-(2-propyl-2-methylcyclopentyl)-, ethyl ester		NI29357
144	123	43	141	81	55	67	40	284	100	93	89	65	48	37	31	24	0.73	17	32	3	–	–	–	–	–	–	–	–	–	–	2-Oxocanemethanol, 8-ethyl-α-pentyl-, acetate	55101-94-5	NI29358
29	144	95	57	43	41	55	102	284	100	92	92	89	89	81	77	73	0.80	17	32	3	–	–	–	–	–	–	–	–	–	–	Tridecanoic acid, 2,6-dimethyl-3-oxo-, ethyl ester		NI29359
213	184	128	114	170	101	70	100	284	100	44	35	35	20	20	19	18	8.00	17	36	1	2	–	–	–	–	–	–	–	–	–	Heptanamide, N-[3-(hexylamino)propyl]-N-methyl-	67139-04-2	NI29360
58	114	128	70	156	184	142	170	284	100	56	50	40	36	31	31	25	25.00	17	36	1	2	–	–	–	–	–	–	–	–	–	Heptanamide, N-hexyl-N-[3-(methylamino)propyl]-	67178-87-4	NI29361
106	57	156	41	108	148	42	56	284	100	60	44	44	32	29	16	15	0.00	17	36	1	2	–	–	–	–	–	–	–	–	–	Urea, tetrabutyl-	4559-86-8	NI29362
156	141	115	157	128	155	139	116	284	100	22	19	15	10	8	8	5	5.00	18	12	–	4	–	–	–	–	–	–	–	–	–	1,4-Methanonaphthalene-2,2,3,3-tetracarbonitrile, 1,4-dihydro-9-isopropylidene-	55723-84-7	NI29363
284	226	155	238	240	239	195	181	284	100	51	33	26	24	19	18	18		18	20	3	–	–	–	–	–	–	–	–	–	–	Allogibberic acid		MS06494
284	239	145	107	115	131	133	91	284	100	35	22	22	13	11	10	10		18	20	3	–	–	–	–	–	–	–	–	–	–	Benzofuran, 2,3-dihydro-2-(4-hydroxyphenyl)-5-(3-hydroxypropyl)-3-methyl-, (2R,3R)-		MS06495
284	269	165	152	155	153	238	115	284	100	29	27	22	21	21	19	17		18	20	3	–	–	–	–	–	–	–	–	–	–	5H-Dibenzo[a,c]cycloheptene, 6,7-dihydro-1,2,3-trimethoxy-		MS06496
284	227	173	240	228	186	285	174	284	100	32	27	23	23	21	20	20		18	20	3	–	–	–	–	–	–	–	–	–	–	Estra-1,3,5(10)-triene-6,17-dione, 3-hydroxy-	1476-34-2	NI29364
284	161	285	132	160	237	209	121	284	100	27	20	19	18	17	17	13		18	20	3	–	–	–	–	–	–	–	–	–	–	Estra-1,3,5(10)-triene-7,17-dione, 3-hydroxy-	2464-15-5	NI29365
146	240	284	145	172	241	131	144	284	100	79	52	42	24	20	18	17		18	20	3	–	–	–	–	–	–	–	–	–	–	Estra-1,3,5(10)-triene-11,17-dione, 3-hydroxy-, (9β)-	17391-44-5	NI29366
141	153	115	128	284	198	152	238	284	100	48	42	32	30	23	20	18		18	20	3	–	–	–	–	–	–	–	–	–	–	2-Furanpropanoic acid, tetrahydro-α-(1-naphthalenylmethyl)-	25379-26-4	NI29367
162	147	163	107	133	121	148	134	284	100	71	11	10	9	9	7	7	5.00	18	20	3	–	–	–	–	–	–	–	–	–	–	Furan, 2,3-trans-3,4-cis-4,5-cis-2,5-bis(4-hydroxyphenyl)-3,4-dimethyltetrahydro-		MS06497

MASS TO CHARGE RATIOS	M.W.	INTENSITIES	Parent	C	H	O	N	S	F	Cl	Br	I	Si	P	B	X	COMPOUND NAME	CAS Reg No	No
284 200 201 107 199 57 44 41	284	100 78 5 5 4 3 2 2		18	20	3	–	–	–	–	–	–	–	–	–	–	Pentanoic acid, 4-[(4-hydroxyphenyl)methyl]phenyl ester	55724-85-1	NI29368
134 123 133 284 162 161 115 91	284	100 38 27 13 13 13 13 9		18	20	3	–	–	–	–	–	–	–	–	–	–	1-Propanol, 1-(4-hydroxyphenyl)-2-[4-(E)-propenylphenoxy]-, erythro-		MS06498
91 73 179 65 77 180 92 284	284	100 99 68 20 16 12 12 8		18	24	1	–	–	–	–	–	–	1	–	–	–	1-[(Trimethylsilyl)oxy]-4-(3-phenylpropyl)benzene		DD00976
96 284 283 285 142 141 124 41	284	100 45 39 9 7 7 5 4		18	24	1	2	–	–	–	–	–	–	–	–	–	(5ξ,19ξ)-20,21-Dinoraspidospermidine, 17-methoxy-	36459-11-7	NI29369
266 194 267 180 265 223 143 71	284	100 31 22 16 15 13 12 10	1.98	18	24	1	2	–	–	–	–	–	–	–	–	–	Uleine, 1,13-dihydro-13-hydroxy-	4532-71-2	NI29370
122 123 108 107 42 120 124 41	284	100 59 32 6 6 5 4 4	0.99	18	29	–	2	–	–	–	–	–	–	–	1	–	Borane, ethylbis(2,3,4,5-tetramethylpyrrol-1-yl)-	28916-15-6	NI29371
59 97 227 111 169 127 125 213	284	100 45 25 25 14 13 12 11	0.00	18	36	2	–	–	–	–	–	–	–	–	–	–	1,3-Dioxolane, 2,2-dimethyl-4-pentyl-5-octyl-		LI06140
96 97 83 69 124 68 85 82	284	100 88 70 64 54 52 45 44	2.58	18	36	2	–	–	–	–	–	–	–	–	–	–	1,3-Dioxolane, 4-hexyl-2-methyl-5-octyl-, (2α,4α,5α)-	56051-69-5	NI29372
73 45 43 74 41 57 55 283	284	100 5 5 4 4 3 3 2	0.44	18	36	2	–	–	–	–	–	–	–	–	–	–	1,3-Dioxolane, 2-pentadecyl-	4360-57-0	NI29373
55 41 69 43 67 57 81 82	284	100 88 70 61 56 52 47 39	0.00	18	36	2	–	–	–	–	–	–	–	–	–	–	9,10-Epoxyoctadecan-1-ol, cis-	13980-12-6	NI29374
74 87 43 41 55 75 57 143	284	100 66 42 32 27 25 17 12	5.76	18	36	2	–	–	–	–	–	–	–	–	–	–	Heptadecanoic acid, methyl ester	1731-92-6	NI29375
74 87 43 55 143 57 41 75	284	100 67 27 23 20 18 18 16	12.40	18	36	2	–	–	–	–	–	–	–	–	–	–	Heptadecanoic acid, methyl ester	1731-92-6	MS06499
74 87 43 55 41 75 57 69	284	100 57 46 30 29 18 16 14	1.00	18	36	2	–	–	–	–	–	–	–	–	–	–	Heptadecanoic acid, methyl ester	1731-92-6	NI29376
88 101 43 41 57 55 69 59	284	100 37 25 22 19 19 11 7	3.00	18	36	2	–	–	–	–	–	–	–	–	–	–	Hexadecanoic acid, 2-methyl-, methyl ester	2490-53-1	NI29377
88 43 55 57 41 69 284 101	284	100 17 13 12 12 11 10 4		18	36	2	–	–	–	–	–	–	–	–	–	–	Hexadecanoic acid, 2-methyl-, methyl ester	2490-53-1	MS06500
88 101 43 41 55 57 29 284	284	100 38 31 27 22 15 13 12		18	36	2	–	–	–	–	–	–	–	–	–	–	Hexadecanoic acid, 2-methyl-, methyl ester	2490-53-1	NI29378
74 101 75 43 41 55 284 57	284	100 96 30 30 30 27 18 17		18	36	2	–	–	–	–	–	–	–	–	–	–	Hexadecanoic acid, 3-methyl-, methyl ester		NI29379
74 43 41 87 55 69 57 75	284	100 35 30 29 27 25 21 17	9.90	18	36	2	–	–	–	–	–	–	–	–	–	–	Hexadecanoic acid, 5-methyl-, methyl ester		NI29380
74 43 87 55 41 57 157 129	284	100 86 77 57 56 43 39 24	8.00	18	36	2	–	–	–	–	–	–	–	–	–	–	Hexadecanoic acid, 7-methyl-, methyl ester		NI29381
74 87 43 55 57 75 41 69	284	100 67 28 27 25 22 22 19	15.64	18	36	2	–	–	–	–	–	–	–	–	–	–	Hexadecanoic acid, 14-methyl-, methyl ester	2490-49-5	NI29383
74 87 43 55 41 57 69 29	284	100 68 50 46 41 37 21 20	20.00	18	36	2	–	–	–	–	–	–	–	–	–	–	Hexadecanoic acid, 14-methyl-, methyl ester	2490-49-5	NI29382
74 87 43 75 55 41 284 57	284	100 81 32 21 21 21 19 18		18	36	2	–	–	–	–	–	–	–	–	–	–	Hexadecanoic acid, 15-methyl-, methyl ester	6929-04-0	NI29384
43 55 41 57 83 69 61 97	284	100 43 37 29 28 28 27 19	0.00	18	36	2	–	–	–	–	–	–	–	–	–	–	1-Hexadecanol, acetate	629-70-9	WI01355
43 83 55 97 57 69 61 41	284	100 51 48 44 43 42 39 39	0.00	18	36	2	–	–	–	–	–	–	–	–	–	–	1-Hexadecanol, acetate	629-70-9	NI29385
43 55 83 69 57 61 41 97	284	100 59 51 50 48 44 40 37	0.48	18	36	2	–	–	–	–	–	–	–	–	–	–	1-Hexadecanol, acetate	629-70-9	IC03659
159 126 43 57 55 41 71 56	284	100 85 84 55 50 43 42 37	6.40	18	36	2	–	–	–	–	–	–	–	–	–	–	Nonanoic acid, nonyl ester		IC03660
73 57 55 87 101 217 85 70	284	100 86 64 63 48 41 41 40	8.22	18	36	2	–	–	–	–	–	–	–	–	–	–	3-Octadecene-1,2-diol	69502-95-0	NI29386
88 101 43 55 41 57 73 70	284	100 55 52 34 33 26 23 22	2.00	18	36	2	–	–	–	–	–	–	–	–	–	–	Palmitic acid, ethyl ester	628-97-7	NI29387
88 101 69 73 89 83 71 70	284	100 55 17 15 14 10 9 9	4.60	18	36	2	–	–	–	–	–	–	–	–	–	–	Palmitic acid, ethyl ester	628-97-7	WI01356
88 101 43 55 73 57 41 157	284	100 93 50 40 37 35 33 32	23.20	18	36	2	–	–	–	–	–	–	–	–	–	–	Palmitic acid, ethyl ester	628-97-7	MS06501
43 41 102 55 60 57 242 69	284	100 65 47 45 44 44 27 23	2.34	18	36	2	–	–	–	–	–	–	–	–	–	–	Pentadecanoic acid, isopropyl ester		IC03661
43 73 28 60 57 55 41 71	284	100 99 99 82 81 63 52 42	13.20	18	36	2	–	–	–	–	–	–	–	–	–	–	Stearic acid	57-11-4	NI29388
44 73 43 57 60 55 69 71	284	100 24 24 22 20 16 14 13	7.00	18	36	2	–	–	–	–	–	–	–	–	–	–	Stearic acid	57-11-4	NI29389
57 43 73 60 55 71 18 41	284	100 88 87 83 59 56 49 43	7.58	18	36	2	–	–	–	–	–	–	–	–	–	–	Stearic acid	57-11-4	IC03662
56 57 41 43 55 211 73 29	284	100 83 64 57 38 37 37 37	10.64	18	36	2	–	–	–	–	–	–	–	–	–	–	Tetradecanoic acid, butyl ester		IC03663
74 284 43 55 41 69 57 87	284	100 49 47 44 42 38 37 36		18	36	2	–	–	–	–	–	–	–	–	–	–	Tetradecanoic acid, 2,6,10-trimethyl-, methyl ester		LI06141
74 43 57 87 69 55 71 41	284	100 55 53 50 44 39 32 32	3.82	18	36	2	–	–	–	–	–	–	–	–	–	–	Tetradecanoic acid, 5,9,13-trimethyl-, methyl ester	56196-55-5	NI29390
91 284 55 41 240 121 67 53	284	100 79 75 75 66 59 59 54		19	24	2	–	–	–	–	–	–	–	–	–	–	Androsta-1,4-diene-3,11-dione	54498-90-7	NI29391
122 123 121 91 28 41 159 284	284	100 31 17 17 15 14 13 12		19	24	2	–	–	–	–	–	–	–	–	–	–	Androsta-1,4-diene-3,17-dione	897-06-3	NI29392
136 284 149 91 41 132 108 56	284	100 57 42 32 24 22 22 20		19	24	2	–	–	–	–	–	–	–	–	–	–	Androsta-4,6-diene-3,17-dione	633-34-1	NI29393
136 284 149 107 91 41 132 240	284	100 57 42 35 33 24 22 20		19	24	2	–	–	–	–	–	–	–	–	–	–	Androsta-4,6-diene-3,17-dione	633-34-1	LI06142
285 134 284 133 151 121 171	284	100 61 60 47 43 36 26		19	24	2	–	–	–	–	–	–	–	–	–	–	Androsta-1,4,6-trien-3-one, 17-hydroxy-, (17β)-	4075-12-1	NI29394
43 242 91 167 83 130 225 284	284	100 82 60 52 49 48 44 42		19	24	2	–	–	–	–	–	–	–	–	–	–	3-Benzylideneisobornyl acetate	55122-51-5	MS06502
91 193 92 69 107 79 65 108	284	100 40 14 10 8 8 8 7	1.00	19	24	2	–	–	–	–	–	–	–	–	–	–	Butane, 1-(benzyloxy)-2-[(benzyloxy)methyl]-	33498-90-7	NI29395
91 193 92 69 87 43 29 284	284	100 40 14 10 3 2 2 1		19	24	2	–	–	–	–	–	–	–	–	–	–	Butane, 1-(benzyloxy)-2-[(benzyloxy)methyl]-	33498-90-7	LI06143
284 199 41 285 173 160 55 115	284	100 43 22 20 18 17 15 14		19	24	2	–	–	–	–	–	–	–	–	–	–	1(2H)-Chrysenone, 3,4,4a,4b,5,6,10b,11,12,12a-decahydro-8-methoxy-, [4aS-(4aα,4bβ,10bα,12aβ)]-	58165-59-6	NI29396
284 227 91 105 41 159 55 117	284	100 64 44 30 28 26 26 24		19	24	2	–	–	–	–	–	–	–	–	–	–	5,19-Cyclo-5β-androst-6-ene-3,17-dione	33585-88-5	NI29397
284 242 91 93 41 117 227 79	284	100 88 62 51 43 42 40 39		19	24	2	–	–	–	–	–	–	–	–	–	–	6β,19-Cycloandrost-4-ene-3,17-dione	2352-95-6	NI29398
269 284 147 270 127 285 41 91	284	100 24 23 21 7 6 6 5		19	24	2	–	–	–	–	–	–	–	–	–	–	2,2-Bis(3,5-dimethyl-4-hydroxyphenyl)propane		IC03664
284 158 171 184 159 285 172 185	284	100 38 37 36 26 22 21 16		19	24	2	–	–	–	–	–	–	–	–	–	–	Estra-1,3,5(10),6-tetraene-3,17β-diol, 1-methyl-	10506-68-0	NI29399
284 199 160 186 285 200 159 134	284	100 63 34 27 26 16 10 10		19	24	2	–	–	–	–	–	–	–	–	–	–	Estra-1,3,5(10),15-tetraen-17-ol, 3-methoxy-, (17β)-	17980-88-0	NI29400
284 199 160 186 285 159 145 173	284	100 63 51 30 22 18 17 15		19	24	2	–	–	–	–	–	–	–	–	–	–	Estra-1,3,5(10)-trien-17-one, 3-hydroxy-1-methyl-	4011-48-7	NI29401
284 199 160 186 285 200 159 173	284	100 83 41 39 22 19 18 16		19	24	2	–	–	–	–	–	–	–	–	–	–	Estra-1,3,5(10)-trien-17-one, 3-hydroxy-1-methyl-	4011-48-7	NI29402
284 199 120 285 242 186 160 145	284	100 30 28 22 21 19 19 16		19	24	2	–	–	–	–	–	–	–	–	–	–	Estra-1,3,5(10)-trien-17-one, 3-hydroxy-6-methyl-	30627-17-9	NI29403
284 160 199 285 186 159 174 173	284	100 38 32 22 19 15 14 13		19	24	2	–	–	–	–	–	–	–	–	–	–	Estra-1,3,5(10)-trien-17-one, 3-methoxy-	1624-62-0	NI29404

MASS TO CHARGE RATIOS								M.W.	INTENSITIES								Parent	C	H	O	N	S	F	Cl	Br	I	Si	P	B	X	COMPOUND NAME	CAS Reg No	No
284	160	41	199	186	115	91	55	**284**	100	50	43	40	26	24	22	21		19	24	2	–	–	–	–	–	–	–	–	–	–	Estra-1,3,5(10)-trien-17-one, 3-methoxy-	1624-62-0	NI29405
284	160	41	199	186	115	91	55	**284**	100	50	43	37	23	23	21	20		19	24	2	–	–	–	–	–	–	–	–	–	–	Estra-1,3,5(10)-trien-17-one, 3-methoxy-	1624-62-0	NI29406
284	160	199	186	159	285	115	173	**284**	100	83	61	31	30	20	20	16		19	24	2	–	–	–	–	–	–	–	–	–	–	Estra-1,3,5(10)-trien-17-one, 3-methoxy-, (8α)-	13865-88-8	NI29407
284	160	199	41	285	227	186	174	**284**	100	36	29	23	21	20	19	19		19	24	2	–	–	–	–	–	–	–	–	–	–	Estra-1,3,5(10)-trien-17-one, 3-methoxy-, (8α,9β)-	19592-58-6	NI29408
284	186	41	199	160	97	159	115	**284**	100	73	55	54	42	37	32	32		19	24	2	–	–	–	–	–	–	–	–	–	–	Estra-1,3,5(10)-trien-17-one, 3-methoxy-, (8α,13α)-	58072-53-0	NI29409
284	160	199	186	285	97	115	41	**284**	100	35	29	22	21	20	17	14		19	24	2	–	–	–	–	–	–	–	–	–	–	Estra-1,3,5(10)-trien-17-one, 3-methoxy-, (9β,13α)-	58072-52-9	NI29410
283	41	97	199	55	91	115	186	**284**	100	77	49	47	44	35	34	28	**21.21**	19	24	2	–	–	–	–	–	–	–	–	–	–	Estra-1,3,5(10)-trien-17-one, 3-methoxy-, (13α)-	17554-55-1	NI29411
284	186	199	187	285	173	161	160	**284**	100	57	37	27	23	15	15	14		19	24	2	–	–	–	–	–	–	–	–	–	–	Estra-1,3,5(10)-trien-17-one, 3-methoxy-, (14β)-	17748-69-5	NI29412
148	284	269	149	285	270	105	91	**284**	100	48	48	22	10	8	5	5		19	24	2	–	–	–	–	–	–	–	–	–	–	2,2′,3,3′,5,5′-Hexamethyl-4,4′-dihydroxydiphenylmethane	29366-02-7	NI29413
148	284	269	149	285	270	91	41	**284**	100	48	47	26	10	10	7	7		19	24	2	–	–	–	–	–	–	–	–	–	–	2,2′,3,3′,5,5′-Hexamethyl-4,4′-dihydroxydiphenylmethane	29366-02-7	MS06503
162	284	269	135	163	285	270	150	**284**	100	99	90	42	25	22	19	19		19	24	2	–	–	–	–	–	–	–	–	–	–	2,3,3′,5,5′,6-Hexamethyl-4,4′-dihydroxydiphenylmethane		MS06504
57	28	41	43	105	163	42	29	**284**	100	42	27	26	14	13	13	11	**0.00**	19	24	2	–	–	–	–	–	–	–	–	–	–	2a,3,4,4a,4b,5,8,8a-Octahydro-4,5-diacetyl-4,8-dimethylcyclobuta[1′,5′]cyclopenta[1′,2′:3,4]cyclobuta[1,2]benzene	56909-21-8	NI29414
91	108	87	92	69	193	85	65	**284**	100	21	20	13	12	10	6	6	**0.00**	19	24	2	–	–	–	–	–	–	–	–	–	–	Pentane, bis(benzyloxy)-	53150-24-6	NI29415
227	284	242	269	147	241	228	173	**284**	100	54	38	29	25	18	18	14		19	24	2	–	–	–	–	–	–	–	–	–	–	Spiro[cyclobutane-1,2′(1′H)-phenanthren]-2-one, 3′,4′,4′a,9′,10,10′a-hexahydro-7′-methoxy-4′a-methyl-, (2′α,4′aα,10′aβ)-	74231-51-9	NI29416
284	57	213	269	41	241	285	29	**284**	100	73	40	36	26	22	21	16		19	25	–	–	–	–	–	–	–	–	1	–	–	Phosphorin, 2,6-di-tert-butyl-4-phenyl-	17420-27-8	NI29417
182	73	284	183	211	157	167	156	**284**	100	92	73	46	44	36	29	29		19	28	–	2	–	–	–	–	–	–	–	–	–	1H-Carbazole-2-ethanamine, 3-ethyl-2,3,4,9-tetrahydro-N,N,1-trimethyl-	55373-99-4	NI29418
152	110	82	284	132	130	139	55	**284**	100	96	67	65	64	58	22	18		19	28	–	2	–	–	–	–	–	–	–	–	–	α-1,2′,3′,4′,10,11-Hexahydro-9-deoxycinchonine		LI06144
152	110	284	132	82	130	139	55	**284**	100	85	77	58	57	56	20	17		19	28	–	2	–	–	–	–	–	–	–	–	–	β-1′,2′,3′,4′,10,11-Hexahydro-9-deoxycinchonine		LI06145
43	41	57	55	29	69	83	71	**284**	100	80	76	70	51	42	39	32	**0.00**	19	40	1	–	–	–	–	–	–	–	–	–	–	Nonadecanol		MS06505
59	69	55	43	57	71	41	58	**284**	100	90	82	82	81	73	57	51	**0.00**	19	40	1	–	–	–	–	–	–	–	–	–	–	2-Pentadecanol, 2,6,10,14-tetramethyl-		NI29419
129	69	43	111	57	55	71	83	**284**	100	91	89	75	74	73	62	53	**0.00**	19	40	1	–	–	–	–	–	–	–	–	–	–	6-Pentadecanol, 2,6,10,14-tetramethyl-		NI29420
284	285	142	282	141	286	283	239	**284**	100	23	16	15	10	7	5	4		20	12	–	–	1	–	–	–	–	–	–	–	–	Dinaphtho[1,2-b:1′,2′-d]thiophene		AI01976
284	285	282	28	286	283	142	141	**284**	100	22	19	10	7	7	7	6		20	12	–	–	1	–	–	–	–	–	–	–	–	Dinaphtho[1,2-b:1′,2′-d]thiophene		MS06506
284	285	282	142	28	286	141	252	**284**	100	23	14	13	13	6	6	5		20	12	–	–	1	–	–	–	–	–	–	–	–	Dinaphtho[1,2-b:2′,1′-d]thiophene		MS06507
284	283	282	141	285	141.5	140	142	**284**	100	91	85	42	22	13	13	12		20	12	–	–	1	–	–	–	–	–	–	–	–	Dinaphtho[2,1-b:1′,2′-d]thiophene		MS06508
284	282	283	141	285	142	141.5	140	**284**	100	84	80	62	26	22	21	18		20	12	–	–	1	–	–	–	–	–	–	–	–	Dinaphtho[2,1-b:1′,2′-d]thiophene		AI01977
284	285	283	180	152	142	141	181	**284**	100	22	21	17	8	6	6	6		20	16	–	2	–	–	–	–	–	–	–	–	–	Carbazole, 3-(benzylideneamino)-9-methyl-	35473-72-4	NI29421
269	284	173	241	159	185	270	43	**284**	100	76	37	36	26	23	20	19		20	28	1	–	–	–	–	–	–	–	–	–	–	Dehydroabietinal	13601-88-2	NI29422
159	241	185	143	141	129	43	117	**284**	100	35	33	21	19	18	17	15	**5.00**	20	28	1	–	–	–	–	–	–	–	–	–	–	4-Epidehydroabietal		LI06146
91	95	105	69	119	55	109	81	**284**	100	86	73	73	69	67	57	55	**9.00**	20	28	1	–	–	–	–	–	–	–	–	–	–	Retinal	116-31-4	NI29423
284	91	119	105	95	41	43	173	**284**	100	69	67	61	59	57	56	50		20	28	1	–	–	–	–	–	–	–	–	–	–	Retinal	116-31-4	LI06147
284	95	119	173	105	91	41	43	**284**	100	58	55	53	50	48	48	41		20	28	1	–	–	–	–	–	–	–	–	–	–	Retinal, 9-cis-	514-85-2	NI29424
284	95	119	173	105	91	41	69	**284**	100	57	55	52	50	48	48	41		20	28	1	–	–	–	–	–	–	–	–	–	–	Retinal, 9-cis-	514-85-2	LI06148
266	284	267	265	252	263	268	255	**284**	100	67	61	47	26	21	15	15		21	16	1	–	–	–	–	–	–	–	–	–	–	Benz[j]aceanthrylen-1-ol, 1,2-dihydro-3-methyl-	3342-98-1	MS06509
284	266	265	267	241	252	269	239	**284**	100	87	41	37	30	28	23	22		21	16	1	–	–	–	–	–	–	–	–	–	–	Benz[j]aceanthrylen-2-ol, 1,2-dihydro-3-methyl-	3308-64-3	MS06510
284	207	178	105	179	77	285	208	**284**	100	72	53	39	28	27	22	11		21	16	1	–	–	–	–	–	–	–	–	–	–	Benzophenone, 4-styryl-	74685-66-8	NI29425
207	284	178	208	283	285	165	179	**284**	100	60	24	18	15	13	8	7		21	16	1	–	–	–	–	–	–	–	–	–	–	2,2-Diphenylchromene		MS06511
283	284	165	285	90	128	141	89	**284**	100	65	18	14	13	11	9	6		21	16	1	–	–	–	–	–	–	–	–	–	–	1(2H)-Naphthalenone, 3,4-dihydro-2-(1-naphthalenylmethylene)-	55760-09-3	NI29426
283	284	141	128	165	285	90	89	**284**	100	67	15	15	14	13	13	7		21	16	1	–	–	–	–	–	–	–	–	–	–	1(2H)-Naphthalenone, 3,4-dihydro-2-(2-naphthalenylmethylene)-	55723-88-1	NI29427
108	79	93	80	119	91	95	67	**284**	100	95	53	40	39	37	30	27	**0.02**	21	32	–	–	–	–	–	–	–	–	–	–	–	cis-3,6,9,12,15,18-Heneicosahexaene		LI06149
266	267	282	281	268	283	265	284	**284**	100	27	6	6	4	2	2	1		22	20	–	–	–	–	–	–	–	–	–	–	–	Benzo[c]phenanthrene, 1,5,8,12-tetramethyl-	21297-24-5	NI29428
241	284	278	283	282	242	254	255	**284**	100	91	59	42	40	35	30	29		22	20	–	–	–	–	–	–	–	–	–	–	–	Dibenz[a,h]anthracene, 1,2,3,4,4a,14b-hexahydro-	72390-47-7	NI29429
284	256	285	278	241	255	283	253	**284**	100	34	26	23	23	19	17	17		22	20	–	–	–	–	–	–	–	–	–	–	–	Dibenz[a,h]anthracene, 1,2,3,4,12,13-hexahydro-	153-32-2	NI29430
116	284	141	115	117	142	129	169	**284**	100	59	45	41	37	36	32	31		22	20	–	–	–	–	–	–	–	–	–	–	–	5,12:6,11-Dimethanodibenzo[b,h]biphenylene, 5,5a,5b,6,11,11a,11b,12-octahydro-	55176-83-5	NI29431
284	241	239	226	242	227	285	255	**284**	100	81	37	36	33	26	22	21		22	20	–	–	–	–	–	–	–	–	–	–	–	Indeno[1,2,3-cd]pyrene, 1,2,6b,7,8,9,10,10a-octahydro-		NI29432
243	165	244	228	166	239	241	215	**284**	100	31	23	5	5	4	3	3	**0.40**	22	20	–	–	–	–	–	–	–	–	–	–	–	4,4,4-Triphenyl-1-butene	16876-20-3	NI29433
180	179	181	178	166	142	115	141	**284**	100	60	40	32	8	6	6	4	**1.50**	22	20	–	–	–	–	–	–	–	–	–	–	–	1,2,3-Triphenylcyclobutane		LI06150
69	112	46	67	31	164	146	250	**285**	100	24	22	13	8	6	4	3	**2.50**	3	–	–	1	1	7	2	–	–	–	–	–	–	N-(Heptafluoroisopropyl)sulphur dichloroimide		LI06151
69	46	285	96	184	115	78	82	**285**	100	96	62	58	51	44	32	29		3	–	–	1	2	9	–	–	–	–	–	–	–	Bis(trifluoromethylthio)trifluoromethylamine		NI29434
270	272	250	252	274	285	287	289	**285**	100	95	56	35	30	27	26	8		3	9	–	3	–	–	3	–	–	–	3	–	–	1,1,3-Trimethyl-3,5,5-trichlorocyclotriphosphazene	107651-59-2	NI29435
69	121	75	190	285	266	50	31	**285**	100	67	50	38	37	37	35	23		6	–	–	3	–	9	–	–	–	–	–	–	–	1,3,5-Triazine, 2,4,6-tris(trifluoromethyl)-	368-66-1	NI29436

MASS TO CHARGE RATIOS								M.W.	INTENSITIES								Parent	C	H	O	N	S	F	Cl	Br	I	Si	P	B	X	COMPOUND NAME	CAS Reg No	No
69	121	285	266	76	190	31	50	285	100	42	28	28	19	16	8	7		6	–	–	3	–	9	–	–	–	–	–	–	–	1,3,5-Triazine, 2,4,6-tris(trifluoromethyl)-	368-66-1	LI06152
93	65	127	39	285	89	46	94	285	100	84	23	20	18	10	9	7		6	5	1	1	2	6	–	–	–	–	–	–	–	S-Fluoro-N-pentafluorosulphanyl-S-phenoxysulphilimine		NI29437
285	287	289	87	209	211	270	175	285	100	97	35	35	31	29	24	18		8	6	4	1	–	–	3	–	–	–	–	–	–	Benzene, 1,2,4-trichloro-3,6-dimethoxy-5-nitro-	35282-83-8	NI29438
286	256	113	127	254	101	288	287	285	100	31	29	28	17	11	10	9	0.00	9	7	4	3	2	–	–	–	–	–	–	–	–	2-Thiazolamine, 5-[(4-nitrophenyl)sulphonyl]-	39565-05-4	NI29440
99	285	57	55	28	72	50	71	285	100	75	43	43	39	22	21	18		9	7	4	3	2	–	–	–	–	–	–	–	–	2-Thiazolamine, 5-[(4-nitrophenyl)sulphonyl]-	39565-05-4	NI29439
89	130	216	285					285	100	99	54	7						9	8	–	3	–	–	–	–	1	–	–	–	–	2-Azido-1-iodo-1-phenylpropene		LI06153
115	97	73	65	43	121	125	93	285	100	70	38	37	37	36	32	27	21.07	9	20	3	1	2	–	–	–	–	–	1	–	–	Phosphorodithioic acid, O,O-diethyl S-(2-isopropylamino) ester	2275-18-5	NI29441
105	285	77	89	130	51	63	103	285	100	53	36	29	19	12	11	10		10	8	1	1	–	–	–	–	1	–	–	–	–	4-Iodo-3-methyl-5-phenylisoxazole		LI06154
105	285	77	89	130	51	63	106	285	100	53	36	29	19	12	11	10		10	8	1	1	–	–	–	–	1	–	–	–	–	4-Iodo-3-methyl-5-phenylisoxazole		MS06512
73	226	270	241	186	113	56	271	285	100	71	68	28	21	14	14	12	0.00	10	18	3	1	–	3	–	–	–	1	–	–	–	L-Valine, N-(trifluoroacetyl)-, trimethylsilyl ester	52558-81-3	NI29442
129	142	156	102	128	51	127	130	285	100	34	33	25	21	19	13	11	0.00	11	12	–	1	–	–	–	–	1	–	–	–	–	Isoquinoline ethiodide		MS06513
143	115	142	128	144	127	116	89	285	100	28	23	21	15	10	7	6	0.00	11	12	–	1	–	–	–	–	1	–	–	–	–	Lepidine methiodide		MS06514
143	142	127	115	141	144	58.5	71.5	285	100	87	19	18	16	15	11	9		11	12	–	1	–	–	–	–	1	–	–	–	–	6-Methylquinoline methiodide		IC03665
143	142	127	115	144	141	58.5	158	285	100	66	22	18	15	11	10	9	0.00	11	12	–	1	–	–	–	–	1	–	–	–	–	7-Methylquinoline methiodide		IC03666
144	143	128	116	158	142	58.5	159	285	100	66	22	17	14	11	10	9	0.00	11	12	–	1	–	–	–	–	1	–	–	–	–	7-Methylquinoline methiodide		MS06515
129	156	102	51	130	76	127	50	285	100	53	21	16	12	8	7	7	0.00	11	12	–	1	–	–	–	–	1	–	–	–	–	Quinoline ethiodide		IC03667
129	156	102	128	51	130	76	127	285	100	53	20	18	15	11	8	7	0.00	11	12	–	1	–	–	–	–	1	–	–	–	–	Quinoline ethiodide		MS06516
157	143	115	128	156	142	127	116	285	100	46	42	32	29	28	21	19	0.00	11	12	–	1	–	–	–	–	1	–	–	–	–	Quinolinium, 1,2-dimethyl-, iodide	876-87-9	NI29443
60	44	43	45	29	42	31	28	285	100	68	65	39	16	15	8	7	0.00	11	15	6	3	–	–	–	–	–	–	–	–	–	Cytidine, N-acetyl-	3768-18-1	NI29444
201	164	136	44	42	43	202	165	285	100	99	98	98	85	67	59	55	33.00	11	15	6	3	–	–	–	–	–	–	–	–	–	3,4-Dimethoxy-6-nitro-N-nitrosoadrenalin		LI06155
286	287	288	256					285	100	14	6	2					0.00	13	19	4	1	1	–	–	–	–	–	–	–	–	Benzoic acid, 4-[(dipropylamino)sulphonyl]-	57-66-9	NI29445
256	121	185	214	257	65	43	41	285	100	53	51	16	14	12	11	8	2.00	13	19	4	1	1	–	–	–	–	–	–	–	–	Benzoic acid, 4-[(dipropylamino)sulphonyl]-	57-66-9	NI29446
286	256	268	284	185				285	100	23	4	3	3				0.00	13	19	4	1	1	–	–	–	–	–	–	–	–	Benzoic acid, 4-[(dipropylamino)sulphonyl]-	57-66-9	NI29447
165	155	91	57	108	43	41	65	285	100	98	74	64	42	38	28	23	0.00	13	19	4	1	1	–	–	–	–	–	–	–	–	Carbamic acid, methyl[(4-methylphenyl)sulphonyl]-, tert-butyl ester	56805-36-8	NI29448
91	155	108	197	221	107	151		285	100	94	44	43	23	9	7		0.00	13	19	4	1	1	–	–	–	–	–	–	–	–	Carbamic acid, [(4-methylphenyl)sulphonyl]-, 2,2-dimethylpropyl ester	32363-28-3	LI06156
91	155	57	108	197	43	65	41	285	100	94	52	44	43	38	31	31	0.00	13	19	4	1	1	–	–	–	–	–	–	–	–	Carbamic acid, [(4-methylphenyl)sulphonyl]-, 2,2-dimethylpropyl ester	32363-28-3	NI29449
70	55	59	73	171	155	42	44	285	100	88	74	72	59	57	48	43	0.00	13	19	4	1	1	–	–	–	–	–	–	–	–	Carbamic acid, (4-tolylsulphonyl)-, tert-pentyl ester	32363-31-8	NI29450
43	270	101	131	212	271	152	55	285	100	83	56	33	14	13	12	12	0.00	13	19	6	1	–	–	–	–	–	–	–	–	–	3-C-Cyano-1,2,5,6-di-O-isopropylidene-α-D-allofuranose		NI29451
170	138	197	229	165	198	285	214	285	100	78	75	55	55	45	40	20		13	19	6	1	–	–	–	–	–	–	–	–	–	[Bis(methoxycarbonyl)methyl]methoxycarbonyl-ketene-tert-butylimine		LI06157
285	286	239	211	183	127	126	155	285	100	16	15	12	10	6	6	4		14	7	6	1	–	–	–	–	–	–	–	–	–	1,4-Dihydroxy-5-nitroanthraquinone		IC03668
138	106	92	78	148	147	121	285	285	100	99	97	92	47	43	15	9		14	11	4	3	–	–	–	–	–	–	–	–	–	Acetamide, N-(4-nitrophenyl)-2-isonicotinyl-		LI06158
106	138	92	78	148	147	121	164	285	100	92	83	81	38	31	14	12	5.00	14	11	4	3	–	–	–	–	–	–	–	–	–	Acetamide, N-(3-nitrophenyl)-2-nicotinyl-		LI06159
138	106	78	92	147	148	121	164	285	100	95	87	49	34	23	17	12	4.00	14	11	4	3	–	–	–	–	–	–	–	–	–	Acetamide, N-(4-nitrophenyl)-2-nicotinyl-		LI06160
268	285	165	192	203	205	191	204	285	100	93	51	37	33	21	21	20		14	11	4	3	–	–	–	–	–	–	–	–	–	2-Amino-1,3-dinitro-9,10-dihydrophenanthrene		MS06517
268	285	165	204	177	192	222	191	285	100	57	37	35	21	19	18	18		14	11	4	3	–	–	–	–	–	–	–	–	–	2-Amino-1,7-dinitro-9,10-dihydrophenanthrene		MS06518
285	165	193	255	239	286	192	209	285	100	26	23	21	17	15	13	12		14	11	4	3	–	–	–	–	–	–	–	–	–	2-Amino-3,7-dinitro-9,10-dihydrophenanthrene		MS06519
225	285	154	153	226	77	155	105	285	100	67	31	20	18	17	15	15		14	11	4	3	–	–	–	–	–	–	–	–	–	3H-Imidazo[1,5-b]pyridazine-5,7(6H)-dione, 4-(methoxycarbonyl)-2-phenyl-		NI29452
150	43	119	242	45	104	79	41	285	100	84	68	46	40	34	34	32	5.00	14	11	4	3	–	–	–	–	–	–	–	–	–	Urea, 1-(4-nitrobenzoyl)-3-phenyl-	26972-00-9	NI29453
84	56	107	77	166	188	207	221	285	100	18	7	5	2	1	0	0	0.00	14	20	3	1	–	–	1	–	–	–	–	–	–	Methyl threo-DL-2-(4-hydroxyphenyl)-2-(2-piperidyl)acetate, hydrochloride		NI29454
192	206	190	207	41	84	43	164	285	100	90	55	45	41	33	29	23	7.00	14	23	3	1	1	–	–	–	–	–	–	–	–	8-Oxa-3-thiatricyclo[5.2.0.0^{7,9}]nonane, 5,7-dimethyl-1-(1-piperidino)-, 3,3-dioxide	104195-55-3	NI29455
192	206	207	41	70	43	179	193	285	100	54	30	24	22	22	20	19	4.00	14	23	3	1	1	–	–	–	–	–	–	–	–	8-Oxa-3-thiatricyclo[5.2.0.0^{7,9}]nonane, 5,5,7-trimethyl-1-(1-pyrrolidinyl)-, 3,3-dioxide	104195-58-6	NI29456
226	154	285	227	254	166	155	86	285	100	49	19	15	6	5	5	3		14	23	5	1	–	–	–	–	–	–	–	–	–	Dimethyl α-(1-methyl-2-morpholinobutenyl)malonate		MS06520
203	204	176	201	202	177	175	82	285	100	20	14	13	12	12	10	9	6.00	15	11	–	1	–	–	–	–	–	–	–	–	1	9-Phenanthrenoselenocarboxamide		MS06521
50	77	110	103	65	83	66	51	285	100	62	55	55	55	45	45	45	10.34	15	11	3	1	1	–	–	–	–	–	–	–	–	1H-Isoindole-1,3(2H)-dione, 2-(benzylsulphinyl)-	40167-14-4	NI29457
147	40	104	76	91	139	50	43	285	100	62	57	52	34	29	25	25	14.00	15	11	3	1	1	–	–	–	–	–	–	–	–	1H-Isoindole-1,3(2H)-dione, 2-[(4-methylphenyl)sulphinyl]-	42300-58-3	NI29458
121	243	77	285	214	43	39	122	285	100	76	26	24	21	16	12	10		15	11	3	1	1	–	–	–	–	–	–	–	–	Thiazolo[3,2-a]pyridinium, 8-(acetyloxy)-3-hydroxy-2-phenyl-, hydroxide, inner salt	32002-92-9	NI29459
105	77	106	51	76	118	151	150	285	100	18	8	7	6	5	4	4	0.00	15	11	5	1	–	–	–	–	–	–	–	–	–	Acetophenone, 2-[(2-nitrobenzoyl)oxy]-	55153-33-8	NI29460
105	76	120	51	39	255	65	92	285	100	34	21	17	12	11	11	10	0.00	15	11	5	1	–	–	–	–	–	–	–	–	–	Acetophenone, 2-[(3-nitrobenzoyl)oxy]-	55153-32-7	NI29461
105	77	255	106	120	76	51	50	285	100	21	12	8	6	6	6	5	0.00	15	11	5	1	–	–	–	–	–	–	–	–	–	Acetophenone, 2-[(4-nitrobenzoyl)oxy]-	7254-22-0	NI29462
105	77	106	255	120	135	76	104	285	100	40	15	14	11	10	10	9	5.00	15	11	5	1	–	–	–	–	–	–	–	–	–	Acetophenone, 4′-nitro-2-(benzoyloxy)-	55153-34-9	NI29463
267	150	164	104	89	120	239	193	285	100	70	25	25	12	9	8	8	0.10	15	11	5	1	–	–	–	–	–	–	–	–	–	2-Carboxybenzyl 4-nitrophenyl ketone		LI06161

MASS TO CHARGE RATIOS								M.W.	INTENSITIES								Parent	C	H	O	N	S	F	Cl	Br	I	Si	P	B	X	COMPOUND NAME	CAS Reg No	No
270	285	240	144	271	286	241	89	**285**	100	55	47	28	17	11	8	7		15	11	5	1	–	–	–	–	–	–	–	–	–	1,3-Dioxolo[4,5-b]acridin-10(5H)-one, 11-hydroxy-4-methoxy-	17014-49-2	NI29464
268	269	285	270	77	233	284	91	**285**	100	71	50	42	34	30	29	26		15	12	1	3	–	–	1	–	–	–	–	–	–	3H-1,4-Benzodiazepin-2-amine, 7-chloro-5-phenyl-, 4-oxide	7722-15-8	MS06522
268	269	285	270	77	233	284	91	**285**	100	83	59	50	40	35	34	31		15	12	1	3	–	–	1	–	–	–	–	–	–	3H-1,4-Benzodiazepin-2-amine, 7-chloro-5-phenyl-, 4-oxide	7722-15-8	NI29465
135	285	137	287	155	136	286	168	**285**	100	40	33	13	13	8	6	5		15	12	1	3	–	–	1	–	–	–	–	–	–	Δ-1,2,4-Triazolin-5-one, 1-(4-chlorophenyl)-3-methyl-4-phenyl-	32589-65-4	NI29466
150	285	122	151	134.5	65	107	92	**285**	100	19	13	9	5	5	4	4		15	15	3	3	–	–	–	–	–	–	–	–	–	4-Amino-2-methoxy-4-carbamoylacetanilide		IC03669
226	285	210	227	211	44	213	198	**285**	100	58	36	24	16	13	12	12		15	15	3	3	–	–	–	–	–	–	–	–	–	2-Methoxy-4-(2,2-dicyanovinyl)-5-methyl-N-(2-carboxyethyl)aniline		IC03670
137	191	203	168	159	234	219	257	**285**	100	19	18	9	9	7	5	4	**1.00**	15	15	3	3	–	–	–	–	–	–	–	–	–	Propenoic acid, 2-azido-3-[4-(1,1-dimethylprop-2-ynyloxy)phenyl]-, methyl ester		MS06523
59	146	67	240	114	89	204	100	**285**	100	81	62	56	30	26	19	19	**0.00**	15	31	2	1	–	–	–	–	–	1	–	–	–	1,5-Hexadiene, 3-[[[bis(2-methoxyethyl)amino]methyl]dimethylsilyl]-		DD00977
139	141	130	131	111	285	170	77	**285**	100	35	31	21	18	15	10	9		16	12	2	1	–	–	1	–	–	–	–	–	–	1-(4-Chlorobenzoyloxy)-2-methylindole		LI06162
135	285	287	286	69	77	136	142.5	**285**	100	85	29	15	11	10	9	6		16	12	2	1	–	–	1	–	–	–	–	–	–	3-(4-Chlorophenyl)-5-(4-methoxyphenyl)isoxazole		LI06163
285	139	69	287	146	286	81	41	**285**	100	58	48	33	31	19	19	18		16	12	2	1	–	–	1	–	–	–	–	–	–	3-(4-Methoxyphenyl)-5-(4-chlorophenyl)isoxazole		LI06164
130	129	285	284	164	155	220		**285**	100	46	28	2	2	2	1			16	15	2	1	1	–	–	–	–	–	–	–	–	Quinoline, 1,2-dihydro-1-(4-tolylsulphonyl)-	13268-54-7	NI29467
91	92	285	65	253	242	164	162	**285**	100	9	3	3	1	1	1	1		16	15	4	1	–	–	–	–	–	–	–	–	–	Benzene, 2-methoxy-1-(2-nitroethenyl)-3-(phenylmethoxy)-	74810-83-6	NI29468
285	237	255	211	139	238	209	286	**285**	100	88	34	26	22	20	18	16		16	15	4	1	–	–	–	–	–	–	–	–	–	[1,1′-Biphenyl]-4-carboxylic acid, 4′-methoxy-2′-[(methylamino)carbonyl]-	53366-26-0	NI29469
270	285	242	199	284	271	170	171	**285**	100	74	59	39	23	23	18	17		16	15	4	1	–	–	–	–	–	–	–	–	–	10-Methyl-2,3-dimethoxy-1-hydroxyacridone		MS06524
45	224	285	240	165	194	225	166	**285**	100	29	25	20	18	7	6	6		16	15	4	1	–	–	–	–	–	–	–	–	–	Stilbene, 2-(methoxymethoxy)-4′-nitro-, (E)-		MS06525
45	224	285	240	165	194	166	225	**285**	100	44	34	32	25	12	9	8		16	15	4	1	–	–	–	–	–	–	–	–	–	Stilbene, 2-(methoxymethoxy)-4′-nitro-, (Z)-		MS06526
242	229	243	285	134	270	256	150	**285**	100	75	60	50	32	25	25	16		16	19	–	3	1	–	–	–	–	–	–	–	–	Benzothiazole, 2-(4-butyl-3,5-dimethylpyrazol-1-yl)-		DD00978
241	286	200	284	214				**285**	100	37	34	21	17				**0.00**	16	19	–	3	1	–	–	–	–	–	–	–	–	10H-Pyrido[3,2-b][1,4]benzothiazine-10-ethanamine, N,N,α-trimethyl-	482-15-5	NI29470
58	285	214	200	86	227	85	213	**285**	100	33	28	27	20	15	15	12		16	19	–	3	1	–	–	–	–	–	–	–	–	10H-Pyrido[3,2-b][1,4]benzothiazine-10-propanamine, N,N-dimethyl-	303-69-5	NI29471
286	241	285	284	269				**285**	100	66	65	7	3					16	19	–	3	1	–	–	–	–	–	–	–	–	10H-Pyrido[3,2-b][1,4]benzothiazine-10-propanamine, N,N-dimethyl-	303-69-5	NI29472
120	121	77	285	119	105	104	91	**285**	100	10	7	6	6	4	4	3		16	19	2	3	–	–	–	–	–	–	–	–	–	Ethylenediamine, N,N′-dimethyl-N-(4-nitrophenyl)-N′-phenyl-	32857-47-9	NI29473
57	43	41	56	184	113	115	128	**285**	100	43	33	30	29	26	23	20	**0.00**	16	31	3	1	–	–	–	–	–	–	–	–	–	Carbamic acid, (5-oxoundecyl)-, tert-butyl ester	116437-38-8	DD00979
242	161	285	136	243	109	150	91	**285**	100	82	70	68	66	17	15	14		17	19	1	1	1	–	–	–	–	–	–	–	–	1,5-Benzothiazepine, 2-methyl-4-(4-methoxyphenyl)-2,3,4,5-tetrahydro-	78031-27-3	NI29474
285	215	115	230	286	253	201	187	**285**	100	46	35	24	20	20	19	18		17	19	3	1	–	–	–	–	–	–	–	–	–	Crinan, 1,2-didehydro-3-methoxy-, (3α)-	468-22-4	NI29475
286	254	285	284	287	270	255	214	**285**	100	76	73	43	20	20	20	6		17	19	3	1	–	–	–	–	–	–	–	–	–	Crinan, 1,2-didehydro-3-methoxy-, (3α)-	468-22-4	NI29476
216	96	217	285	43	41	215	55	**285**	100	64	56	50	50	40	38	30		17	19	3	1	–	–	–	–	–	–	–	–	–	4,7-Epoxy-1H-isoindole-1,3(2H)-dione, hexahydro-3a,7a-dimethyl-2-(4-methylphenyl)-	55637-46-2	NI29477
254	270	285	252	206	214	227		**285**	100	46	44	32	31	22	12			17	19	3	1	–	–	–	–	–	–	–	–	–	Erysopine		LI06165
210	242	211	285	225	191	181	241	**285**	100	92	83	80	76	74	71	61		17	19	3	1	–	–	–	–	–	–	–	–	–	5-Isoquinolinol, 1,2,3,4-tetrahydro-4-(4-hydroxyphenyl)-6-methoxy-2-methyl-, (S)-	93915-33-4	NI29478
178	163	179	177	176	120	118	162	**285**	100	17	11	2	2	2	2	1	**0.35**	17	19	3	1	–	–	–	–	–	–	–	–	–	7-Isoquinolinol, 1,2,3,4-tetrahydro-6-methoxy-1-salicyl-	24656-61-9	NI29479
178	163	179	135	134	120	118	107	**285**	100	16	12	3	3	2	2	2	**0.20**	17	19	3	1	–	–	–	–	–	–	–	–	–	7-Isoquinolinol, 1,2,3,4-tetrahydro-6-methoxy-1-salicyl-	24656-61-9	NI29480
242	241	225	211	210	181	285	227	**285**	100	87	65	55	37	32	29	25		17	19	3	1	–	–	–	–	–	–	–	–	–	7-Isoquinolinol, 1,2,3,4-tetrahydro-2-methyl-4-(4-hydroxyphenyl)-6-methoxy-		MS06527
285	44	228	229	42	214	36	286	**285**	100	33	30	26	23	21	21	18		17	19	3	1	–	–	–	–	–	–	–	–	–	Morphinan-6-one, 4,5-epoxy-3-hydroxy-17-methyl-, (5α)-	466-99-9	NI29481
285	162	215	42	286	124	284	174	**285**	100	38	31	21	20	19	18	16		17	19	3	1	–	–	–	–	–	–	–	–	–	Morphine, (–)-	57-27-2	NI29482
268	286	269	285	287	284	270	267	**285**	100	27	19	9	5	5	3	3		17	19	3	1	–	–	–	–	–	–	–	–	–	Morphine, (–)-	57-27-2	NI29483
285	215	162	124	115	70	77	286	**285**	100	34	32	27	26	23	22	20		17	19	3	1	–	–	–	–	–	–	–	–	–	Morphine, (–)-	57-27-2	WI01357
285	201	115	173	202	174	200	286	**285**	100	95	44	39	29	24	21	20		17	19	3	1	–	–	–	–	–	–	–	–	–	Piperidine, 1-[5-(1,3-benzodioxol-5-yl)-1-oxo-2,4-pentadienyl]-, (E,E)-	94-62-2	NI29484
201	115	285	173	143	202	171	174	**285**	100	92	65	42	32	29	26	25		17	19	3	1	–	–	–	–	–	–	–	–	–	Piperidine, 1-[5-(1,3-benzodioxol-5-yl)-1-oxo-2,4-pentadienyl]-, (E,E)-	94-62-2	NI29485
130	70	171	216	170	129	115	77	**285**	100	80	68	54	40	38	15	15	**9.00**	17	19	3	1	–	–	–	–	–	–	–	–	–	1,2,3,4-Tetrahydro-1,1,3,3-tetramethyl-11bH-isoquinolino[2,1-b][1,2]oxazepine-2,4-dione		LI06166
121	117	286	241					**285**	100	30	24	22					**0.00**	17	23	1	3	–	–	–	–	–	–	–	–	–	1,2-Ethanediamine, N-[(4-methoxyphenyl)methyl]-N′,N′-dimethyl-N-(2-pyridinyl)-	91-84-9	NI29486
121	58	72	79	71	28	78	42	**285**	100	89	25	17	17	15	14	12	**0.00**	17	23	1	3	–	–	–	–	–	–	–	–	–	1,2-Ethanediamine, N-[(4-methoxyphenyl)methyl]-N′,N′-dimethyl-N-(2-pyridinyl)-	91-84-9	NI29487
121	58	72	71	215	214	122	78	**285**	100	55	16	11	10	10	9	8	**3.20**	17	23	1	3	–	–	–	–	–	–	–	–	–	1,2-Ethanediamine, N-[(4-methoxyphenyl)methyl]-N′,N′-dimethyl-N-(2-pyridinyl)-	91-84-9	NI29488
116	74	56	43	41	30	70	45	**285**	100	13	12	7	7	7	6	5	**0.00**	17	35	2	1	–	–	–	–	–	–	–	–	–	2-Dodecyl-3-(2-hydroxyethyl)oxazolidine		IC03671
43	40	91	58	105	56	77	107	**285**	100	84	40	31	28	14	12	10	**0.00**	18	20	–	1	–	–	1	–	–	–	–	–	–	Benzylamine, N-(3-chloro-2,2-dimethyl-1-phenylpropylidene)-		MS06528
220	219	162	104	221	91	204	66	**285**	100	54	28	22	20	20	18	18	**0.00**	18	20	–	1	–	–	1	–	–	–	–	–	–	1-Propylamine, 3-(10,11-dihydro-5H-dibenzo[a,d]cyclohepten-5-ylidene)-, hydrochloride	25887-71-2	NI29489
161	285	284	57	41	55	149	71	**285**	100	84	83	53	51	44	40	29		18	23	2	1	–	–	–	–	–	–	–	–	–	8-Azaestra-1,3,5(10)-trien-17-one, 3-methoxy-, (14β)-	58072-51-8	NI29490

MASS TO CHARGE RATIOS								M.W.	INTENSITIES								Parent	C	H	O	N	S	F	Cl	Br	I	Si	P	B	X	COMPOUND NAME	CAS Reg No	No
285	168	197	169	212	167	228	202	**285**	100	98	70	70	63	47	46	44		18	23	2	1	–	–	–	–	–	–	–	–	–	**3-Indolecarboxylic acid, 2-(2-methyl-1-hexenyl)-, ethyl ester**		**DD00980**
42	285	44	115	128	77	165	164	**285**	100	55	41	39	29	29	27	26		18	23	2	1	–	–	–	–	–	–	–	–	–	**Morphinan-6α-ol, 7,8-didehydro-3-methoxy-N-methyl-**	3205-45-6	**NI29491**
285	59	286	164	44	228	69	55	**285**	100	28	21	13	10	9	7	7		18	23	2	1	–	–	–	–	–	–	–	–	–	**Morphinan-6-one, 3-methoxy-17-methyl-**	2246-20-0	**NI29492**
213	212	214	229	228	285	215	43	**285**	100	99	95	78	62	42	30	23		18	23	2	1	–	–	–	–	–	–	–	–	–	**Morphinan-6-one, 3-methoxy-17-methyl-, (14α)-**	55298-15-2	**NI29493**
142	184	43	44	268	143	30	41	**285**	100	52	29	26	17	11	11	10	**3.00**	18	39	1	1	–	–	–	–	–	–	–	–	–	**N,N-Dioctylacetamide**		**LI06167**
285	286	284	283	142.5	255	142	150	**285**	100	22	10	10	10	7	5	4		19	11	–	1	1	–	–	–	–	–	–	–	–	**Benzo[e][1]benzothiopyrano[4,3-b]indole**	846-35-5	**LI06168**
285	286	284	283	255	142	150	145	**285**	100	22	10	10	7	5	4	2		19	11	–	1	1	–	–	–	–	–	–	–	–	**Benzo[e][1]benzothiopyrano[4,3-b]indole**	846-35-5	**NI29494**
285	286	284	142	255	145	283	240	**285**	100	22	16	12	7	6	4	4		19	11	–	1	1	–	–	–	–	–	–	–	–	**Benzo[g][1]benzothiopyrano[4,3-b]indole**	10023-23-1	**NI29495**
285	142.5	286	284	142	120.5	255	145	**285**	100	29	22	16	12	7	6	6		19	11	–	1	1	–	–	–	–	–	–	–	–	**Benzo[g][1]benzothiopyrano[4,3-b]indole**	10023-23-1	**LI06169**
285	193	90	194	77	286	284	28	**285**	100	91	40	32	31	22	18	17		19	15	–	3	–	–	–	–	–	–	–	–	–	**1-Anilino-2-phenylbenzimidazole**		**IC03672**
285	286	284	142	283	128	266	135	**285**	100	27	11	7	7	5	4	4		19	15	–	3	–	–	–	–	–	–	–	–	–	**1H-Pyrrolo[2,3-b]pyridin-6-amine, 2,3-diphenyl-**		**NI29496**
96	176	218	285	122	144	162	190	**285**	100	50	46	36	36	33	31	27		19	27	1	1	–	–	–	–	–	–	–	–	–	**1-Azaspiro[5.5]undecan-8-ol, 7-(1-buten-3-ynyl)-2-(4-pentynyl)-, [6R-[6α(R*),7β(Z),8α]]-**	63983-63-1	**NI29497**
95	67	162	190	123	218	79	163	**285**	100	41	25	21	19	18	18	14	**8.00**	19	27	1	1	–	–	–	–	–	–	–	–	–	**[(2-Cyanopropyl)norbornan-2-yl](norbornan-2-yl) ketone**		**NI29498**
42	59	164	115	285	128	44	171	**285**	100	82	48	42	40	40	35	32		19	27	1	1	–	–	–	–	–	–	–	–	–	**Morphinan, 3-methoxy-N,6α-dimethyl-, (–)-**	2246-03-9	**NI29499**
41	45	217	70	69	110	202	285	**285**	100	84	63	56	51	34	28	23		19	27	1	1	–	–	–	–	–	–	–	–	–	**Pentazocine**	359-83-1	**NI29500**
217	45	70	110	270	202	69	285	**285**	100	58	36	33	27	27	24	22		19	27	1	1	–	–	–	–	–	–	–	–	–	**Pentazocine**	359-83-1	**MS06529**
200	285	268	159	228	242	172	214	**285**	100	93	75	70	50	34	29	23		19	27	1	1	–	–	–	–	–	–	–	–	–	**Spiro[cycloheptane-1,3′(2′H)-isoquinolin]-1′-one, 1′,4′-dihydro-2′-butyl-**	120637-19-6	**DD00981**
281	266	41	285	57	124	282	283	**285**	100	42	40	31	28	26	24	21		20	31	–	1	–	–	–	–	–	–	–	–	–	**1H-Azepine, 1-[tert-butyl-1,2,3,4-tetrahydro-2-naphthalenyl]hexahydro-**	72101-32-7	**WI01358**
281	270	41	112	285	43	283	266	**285**	100	89	85	71	63	55	54	49		20	31	–	1	–	–	–	–	–	–	–	–	–	**1H-Azepine, 1-[1,2,3,4-tetrahydro-4-methyl-6-isopropyl-2-naphthalenyl]hexahydro-**	55256-26-3	**WI01359**
285	284	270	241	134.5	121.5	257	133.5	**285**	100	79	18	15	13	10	9	9		21	19	–	1	–	–	–	–	–	–	–	–	–	**8,10-Diethylbenzo[a]acridine**		**MS06530**
285	284	270	257	122	254	133.5	133	**285**	100	70	17	13	11	10	10	10		21	19	–	1	–	–	–	–	–	–	–	–	–	**9,11-Diethylbenzo[c]acridine**		**MS06531**
285	284	270	208	286	77	210	191	**285**	100	42	39	36	26	15	12	12		21	19	–	1	–	–	–	–	–	–	–	–	–	**1H-Pyrrole, 1-methyl-2-(1-methyl-2,3-diphenylcyclopropyl)-**	86738-93-4	**NI29501**
127	105	89	67	86	70	32	48	**286**	100	92	27	20	19	12	8	4	**0.00**	–	–	2	–	2	10	–	–	–	–	–	–	–	**Sulphur pentafluoride peroxide**		**LI06170**
69	153	85	135	31	87	101	93	**286**	100	61	56	36	21	18	13	12	**0.00**	4	–	–	–	–	7	3	–	–	–	–	–	–	**2,2,3-Trichloroheptafluorobutane**		**IC03673**
85	87	151	69	153	135	101	166	**286**	100	32	31	28	20	16	8	7	**0.00**	4	–	–	–	–	7	3	–	–	–	–	–	–	**Trichloroheptafluorobutane**		**DO01078**
46	200	202	217	198	28	44	215	**286**	100	74	48	39	37	30	20	18	**0.50**	4	4	3	2	–	–	–	2	–	–	–	–	–	**2,4(1H,3H)-Pyrimidinedione, 5,5-dibromodihydro-6-hydroxy-**	1124-83-0	**NI29502**
46	200	202	217	198	28	44	215	**286**	100	75	48	40	39	33	24	21	**0.50**	4	4	3	2	–	–	–	2	–	–	–	–	–	**2,4(1H,3H)-Pyrimidinedione, 5,5-dibromodihydro-6-hydroxy-**	1124-83-0	**MS06532**
46	30	29	28	43	27	55	44	**286**	100	41	19	12	9	8	5	5	**0.00**	4	6	11	4	–	–	–	–	–	–	–	–	–	**2-Nitro-2-nitrooxymethyl-1,3-propane-diol dinitrate**		**MS06533**
286	101	86	271	214	77	30	73	**286**	100	89	87	60	39	33	31	29		4	12	–	2	1	4	–	–	–	1	2	–	–	**Phosphoramidothioic difluoride, [difluoro[methyl(trimethylsilyl)amino]phosphoranylidene]-**	27351-99-1	**MS06534**
82	286	284	117	242	45	282	240	**286**	100	85	75	63	50	46	45	45		5	2	–	–	2	–	–	–	–	–	–	–	2	**Thieno[2,3-d]-1,3-diseleno-2-thione**	95928-39-5	**NI29503**
82	45	117	57	81	69	80	112	**286**	100	95	63	53	48	38	35	32	**26.00**	5	2	–	–	2	–	–	–	–	–	–	–	2	**Thieno[3,4-d]-1,3-diseleno-2-thione**	87207-40-7	**NI29504**
257	255	259	199	198	149	197	201	**286**	100	67	64	49	43	42	33	32	**6.58**	6	15	–	–	–	–	–	1	–	–	–	–	1	**Triethyltin bromide**		**MS06535**
288	286	253	251	290	188	190	152	**286**	100	75	56	55	53	48	33	27		7	6	–	–	–	–	4	–	–	2	–	–	–	**1,3-Disilaindan, 1,1,3,3-tetrachloro-**	16538-74-2	**NI29505**
59	150	286	131	136	118	58	18	**286**	100	10	8	7	6	5	5	5		7	10	–	–	6	–	–	–	–	–	–	–	–	**2,4,6,8,9,10-Hexathiaadamantane, 1,3,5-trimethyl-**	14870-38-3	**NI29506**
143	89	53	26	224	180	106	160	**286**	100	66	33	28	18	18	16	15	**11.25**	7	10	–	–	6	–	–	–	–	–	–	–	–	**2,4,6,8,9,10-Hexathiatricyclo[3.3.1.1^{3,7}]decane, 1-propyl-**	57274-29-0	**NI29507**
217	286	31	69	117	218	167	267	**286**	100	20	13	10	9	8	7	6		8	–	–	–	–	10	–	–	–	–	–	–	–	**Decafluoroethylbenzene**		**IC03674**
288	207	209	102	290	286	128	75	**286**	100	96	93	89	53	52	46	34		8	4	–	2	–	–	–	2	–	–	–	–	–	**Quinoxaline, 2,3-dibromo-**	23719-78-0	**NI29508**
122	174	146	202	286	116	131	159	**286**	100	40	27	18	11	8	7	5		8	10	4	–	2	–	–	–	–	–	–	–	1	**Chromium tetracarbonyl(2,5-dithiahexane)**		**LI06171**
111	112	97	140	125	107	77	59	**286**	100	62	40	23	20	13	11	10	**0.00**	8	13	–	–	1	6	–	–	–	–	1	–	–	**1-Ethyl-2,5-dimethylthiophenium hexafluorophosphate**		**LI06172**
153	92	135	44	199	286	243	18	**286**	100	93	91	90	82	78	74	71		8	24	3	4	–	–	–	–	–	–	2	–	–	**Diphosphoramide, octamethyl-**	152-16-9	**NI29510**
44	135	92	42	46	153	199	18	**286**	100	62	53	52	47	43	26	26	**8.40**	8	24	3	4	–	–	–	–	–	–	2	–	–	**Diphosphoramide, octamethyl-**	152-16-9	**NI29509**
44	135	199	153	243	42	286	92	**286**	100	70	46	38	35	35	30	24		8	24	3	4	–	–	–	–	–	–	2	–	–	**Diphosphoramide, octamethyl-**	152-16-9	**IC03675**
147	228	226	151	148	245	247	82	**286**	100	99	99	99	93	82	79	70	**30.03**	9	11	2	4	–	–	–	1	–	–	–	–	–	**Pyrazolo[5,1-c]-as-triazin-4-ol, 6-acetyl-4-(bromomethyl)-4,6-dihydro-3-methyl-**	6730-55-8	**NI29511**
286	288	256	87	177	106	240	124	**286**	100	68	52	50	33	27	27	25		10	4	4	2	–	–	2	–	–	–	–	–	–	**N-(4-Nitrophenyl)-2,3-dichloromaleicisomide**		**MS06536**
286	284	282	288	283	285	287	280	**286**	100	88	47	39	32	23	19	18		10	6	–	–	–	–	–	–	–	–	–	–	2	**Naphtho[1,8-cd]-1,2-diselenole**	36579-71-2	**NI29512**
286	284	126	206	282	283	288	204	**286**	100	90	55	54	51	33	30	26		10	6	–	–	–	–	–	–	–	–	–	–	2	**Selenolo[2,3-b][1]benzoselenophene**	40197-97-5	**NI29513**
286	284	282	206	126	283	288	204	**286**	100	95	54	48	43	35	31	26		10	6	–	–	–	–	–	–	–	–	–	–	2	**Selenolo[3,2-b][1]benzoselenophene**	40198-00-3	**NI29514**
158	78	157	119	55	79	60	99	**286**	100	31	27	16	15	10	10	9	**0.00**	10	11	–	2	–	–	–	–	1	–	–	–	–	**Pyrido[1,2-a]pyrimidin-5-ium, 2,4-dimethyl-, iodide**	50993-75-4	**NI29515**
207	165	208	124	41	167	166	122	**286**	100	16	14	14	11	7	6	5	**0.00**	10	11	3	2	–	–	–	1	–	–	–	–	–	**Barbituric acid, 5-allyl-5-(2-bromoallyl)-**	561-86-4	**NI29516**
79	147	78	77	51	59	127		**286**	100	18	16	11	10	8	7		**0.00**	10	14	–	–	–	8	–	–	–	–	–	–	–	**Decane, 3,3,4,4,7,7,8,8-octafluoro-**	40723-69-1	**LI06173**

MASS TO CHARGE RATIOS								M.W.	INTENSITIES								Parent	C	H	O	N	S	F	Cl	Br	I	Si	P	B	X	COMPOUND NAME	CAS Reg No	No
79	147	78	77	51	59	127	65	286	100	18	16	11	10	8	7	5	0.05	10	14	–	–	–	8	–	–	–	–	–	–	–	Decane, 3,3,4,4,7,7,8,8-octafluoro-	40723-69-1	MS06537
199	251	286	155	96	171	169	71	286	100	78	60	52	43	40	38	38		10	15	–	6	1	–	1	–	–	–	–	–	–	1,3,5-Triazine-2,4-diamine, 6-(2-chloro-2-cyanoethylthio)-N,N′-diethyl-	84055-76-5	NI29517
199	96	251	155	286	71	288	170	286	100	86	85	74	71	28	22	20		10	15	–	6	1	–	1	–	–	–	–	–	–	1,3,5-Triazine-2,4-diamine, 6-(2-chloro-2-cyanoethylthio)-N,N,N′,N′-tetramethyl-	84055-75-4	NI29518
271	256	151	105	286	103	101	77	286	100	42	40	26	25	19	19	17		10	16	–	–	–	–	–	–	–	–	–	–	2	Arsine, 1,2-phenylenebis[dimethyl-	13246-32-7	LI06174
271	256	286	167	227	182	241	89	286	100	57	16	14	8	8	7	7		10	16	–	–	–	–	–	–	–	–	–	–	2	Arsine, 1,2-phenylenebis[dimethyl-	13246-32-7	MS06538
135	105	136	107	286	75	61	45	286	100	38	34	25	17	16	15	11		10	22	5	–	2	–	–	–	–	–	–	–	–	D-Glucose, diethyl mercaptal	1941-52-2	NI29519
43	42	198	286	134	93	156	135	286	100	85	78	71	58	57	54	46		11	14	5	2	1	–	–	–	–	–	–	–	–	1,3,5-Dioxazine, 5-[4-(acetylamino)phenylsulphonyl]dihydro-	56221-11-5	NI29520
286	288	223	290	160	225	287	143	286	100	98	38	32	30	25	14	13		12	5	2	–	–	–	3	–	–	–	–	–	–	Dibenzo[b,e][1,4]dioxin, 1,2,4-trichloro-	39227-58-2	NI29521
286	288	290	223	287	160	225	143	286	100	93	29	20	16	15	14	14		12	5	2	–	–	–	3	–	–	–	–	–	–	Dibenzo[b,e][1,4]dioxin, 1,2,4-trichloro-	39227-58-2	AI01978
143	286	288	108	145	144	290	99	286	100	69	50	43	37	13	12	11		12	8	–	–	2	–	2	–	–	–	–	–	–	Bis(4-chlorophenyl) disulphide	1142-19-4	DO01079
143	286	108	145	288	144	99	290	286	100	49	46	36	35	11	11	8		12	8	–	–	2	–	2	–	–	–	–	–	–	Bis(4-chlorophenyl) disulphide	1142-19-4	NI29522
159	286	111	288	75	131	152	127	286	100	87	69	62	60	59	46	41		12	8	2	–	1	–	2	–	–	–	–	–	–	(2-Chlorophenyl) (4-chlorophenyl) sulphone	38980-51-7	NI29523
159	111	75	161	286	99	288	50	286	100	38	36	34	32	23	22	13		12	8	2	–	1	–	2	–	–	–	–	–	–	Bis(3-chlorophenyl) sulphone		IC03676
159	286	161	288	111	75	160	287	286	100	81	75	67	43	30	23	15		12	8	2	–	1	–	2	–	–	–	–	–	–	Bis(4-chlorophenyl) sulphone	80-07-9	NI29524
159	161	111	286	113	288	131	50	286	100	37	33	30	11	10	10	8		12	8	2	–	1	–	2	–	–	–	–	–	–	Bis(4-chlorophenyl) sulphone	80-07-9	IC03677
159	286	111	288	75	131	125	50	286	100	97	71	64	62	53	42	40		12	8	2	–	1	–	2	–	–	–	–	–	–	Bis(4-chlorophenyl) sulphone	80-07-9	NI29525
65	77	78	184	158	129	104	286	286	100	77	58	55	52	48	42	40		12	11	1	–	–	–	1	–	–	–	–	–	1	2-Cyclohexen-1-one, 3-chloro-6-(phenylseleno)-		DD00982
194	286	121	181	166	195	288	287	286	100	66	59	29	15	14	13	11		12	14	–	–	4	–	–	–	–	–	–	–	–	Phenylglyoxal bis-ethylenethioacetal		MS06539
138	148	175	151	108	149	135	162	286	100	88	85	54	54	50	45	32	4.00	12	14	1	8	–	–	–	–	–	–	–	–	–	9-[(3-Cytos-1-yl)propyl]adenine		LI06175
194	286	240	95	69	166	138		286	100	33	33	22	19	15	11			12	14	4	–	2	–	–	–	–	–	–	–	–	Diethyl 2,5-dimercaptoterephthalate		LI06176
172	174	44	41	205	29	213	215	286	100	96	92	26	20	19	18	17	10.73	12	15	3	–	–	–	–	1	–	–	–	–	–	Ethyl 2-(4-bromophenoxy)-2-methylpropionate		IC03678
117	102	191	213	110	134	132	174	286	100	43	41	36	35	29	26	19	0.00	12	18	4	2	1	–	–	–	–	–	–	–	–	Butanoic acid, 2-sulphamoyl-4-phenyl-, ethyl ester		DD00983
169	73	147	170	243	171	75	148	286	100	35	27	13	12	5	5	4	1.90	12	22	4	–	–	–	–	–	–	2	–	–	–	2-Furanacetic acid, α-[(trimethylsilyl)oxy]-, trimethylsilyl ester	55517-39-0	NI29526
169	73	147	170	243	45	75	74	286	100	58	30	14	12	9	7	5	2.70	12	22	4	–	–	–	–	–	–	2	–	–	–	2-Furanacetic acid, α-[(trimethylsilyl)oxy]-, trimethylsilyl ester	55517-39-0	NI29527
73	147	271	123	169	75	28	45	286	100	69	42	33	28	24	21	20	2.94	12	22	4	–	–	–	–	–	–	2	–	–	–	2-Furancarboxylic acid, 5-[[(trimethylsilyl)oxy]methyl]-, trimethylsilyl ester	55517-40-3	NI29530
75	147	73	77	271	169	123	197	286	100	80	61	59	54	34	34	19	3.46	12	22	4	–	–	–	–	–	–	2	–	–	–	2-Furancarboxylic acid, 5-[[(trimethylsilyl)oxy]methyl]-, trimethylsilyl ester	55517-40-3	NI29529
271	147	73	169	272	197	123	75	286	100	70	60	32	23	23	18	16	8.60	12	22	4	–	–	–	–	–	–	2	–	–	–	2-Furancarboxylic acid, 5-[[(trimethylsilyl)oxy]methyl]-, trimethylsilyl ester	55517-40-3	NI29528
73	147	271	75	45	79	272	148	286	100	86	71	47	23	19	17	14	1.60	12	22	4	–	–	–	–	–	–	2	–	–	–	2,4-Hexadienedioic acid, bis(trimethylsilyl) ester, (E,E)-	55517-43-6	NI29532
271	73	147	75	272	169	45	79	286	100	60	54	34	24	13	13	12	3.60	12	22	4	–	–	–	–	–	–	2	–	–	–	2,4-Hexadienedioic acid, bis(trimethylsilyl) ester, (E,E)-	55517-43-6	NI29531
73	271	147	45	272	75	128	74	286	100	81	26	21	19	14	9	9	0.00	12	22	4	–	–	–	–	–	–	2	–	–	–	4H-Pyran-4-one, 5-[(trimethylsilyl)oxy]-2-[[(trimethylsilyl)oxy]methyl]-	55557-21-6	NI29533
271	73	272	45	199	75	43	273	286	100	39	20	18	15	11	10	9	3.00	12	22	4	–	–	–	–	–	–	2	–	–	–	Bis(trimethylsilyl)-5-hydroxymaltol		LI06177
169	73	75	89	45	138	137	72	286	100	87	39	34	26	19	17	13	6.40	12	22	4	2	–	–	–	–	–	1	–	–	–	2-Heptenoic acid, 4,6-dioxo-, dimethoxime, trimethylsilyl		NI29534
243	211	187	131	43	31	41	181	286	100	26	26	13	11	11	10	7	0.09	12	27	4	–	–	–	–	–	–	–	–	–	1	Tris(isobutoxo)oxovanadium	19120-62-8	NI29535
44	172	200	174	155	127	202	75	286	100	79	71	49	47	36	21	21	16.00	13	6	6	2	–	–	–	–	–	–	–	–	–	7-Carboxyphenazine-1,2,3-4-tetrone		MS06540
271	273	287	173	57	288	245	243	286	100	73	57	51	42	39	36	36	0.00	13	9	1	–	–	–	3	–	–	–	–	–	–	1,1′-Biphenyl, 2,3′,4-trichloro-4′-methoxy-	74298-92-3	NI29536
271	273	288	173	286	243	245	275	286	100	91	72	64	62	49	39	29		13	9	1	–	–	–	3	–	–	–	–	–	–	1,1′-Biphenyl, 3,4′,5-trichloro-4-methoxy-	74298-91-2	NI29537
202	148	230	121	56	258	134	172	286	100	81	69	62	61	47	33	24	0.59	13	10	4	–	–	–	–	–	–	–	–	–	1	Iron, (5-methyl-2-furanoyl)dicarbonyl-π-cyclopentadienyl-	59308-01-9	NI29538
286	77	107	79	51	78	104	165	286	100	77	67	52	41	31	29	20		13	10	4	4	–	–	–	–	–	–	–	–	–	Benzaldehyde 2,4-dinitrophenylhydrazone		IC03679
286	107	79	51	78	105	63	165	286	100	57	47	47	34	33	30	25		13	10	4	4	–	–	–	–	–	–	–	–	–	Benzaldehyde 2,4-dinitrophenylhydrazone		LI06178
103	161	77	51	129	67	125	187	286	100	56	42	37	33	32	28	20	3.00	13	10	4	4	–	–	–	–	–	–	–	–	–	5(4H)-Isoxazolone, 4-[(4,5-dihydro-3-methyl-5-oxo-4-isoxazolyl)azo]-3-phenyl-	56666-65-0	MS06541
184	92	67	79	185	139	152	69	286	100	67	26	19	15	15	13	10	0.00	13	11	–	–	1	4	–	–	–	–	–	1	–	5-Methyldibenzo[b,d]thiophenium tetrafluoroborate		LI06179
134	28	79	202	230	67	55	39	286	100	79	57	37	31	30	26	21	9.81	13	15	2	2	–	–	–	–	–	–	–	–	1	Manganese, dicarbonyl(2,3-diazabicyclo[2.2.1]hept-2-ene-N^2,N^3)[(1,2,3,4,5-η)-1-methyl-2,4-cyclopentadien-1-yl]-	74764-12-8	NI29539
70	113	42	43	98	56	58	44	286	100	78	32	14	8	8	6	6	0.10	13	16	1	2	–	–	2	–	–	–	–	–	–	Piperazine, 1-[(2,4-dichlorobenzoyl)methyl]-4-methyl-		MS06542
91	115	65	92	41	39	59	286	286	100	41	19	17	17	12	10	8		13	18	2	–	–	–	–	–	–	–	–	–	1	Butanoic acid, 2,2-dimethyl-4-(benzylseleno)-	3709-09-9	LI06180
91	115	65	92	41	39	59	286	286	100	20	10	8	8	6	5	4		13	18	2	–	–	–	–	–	–	–	–	–	1	Butanoic acid, 2,2-dimethyl-4-(benzylseleno)-	3709-09-9	NI29540
91	65	286	41	39	69	55	115	286	100	17	14	12	10	7	7	6		13	18	2	–	–	–	–	–	–	–	–	–	1	Butanoic acid, 3,3-dimethyl-4-(benzylseleno)-	3709-11-3	LI06181
91	65	286	41	39	284	69	55	286	100	8	7	6	5	3	3	3		13	18	2	–	–	–	–	–	–	–	–	–	1	Butanoic acid, 3,3-dimethyl-4-(benzylseleno)-	3709-11-3	NI29541
207	161	179	137	28	286	18	133	286	100	75	38	38	34	28	27	26		13	18	5	–	1	–	–	–	–	–	–	–	–	5-Benzofuranol, 2-ethoxy-2,3-dihydro-3,3-dimethyl-, methanesulphonate, (±)-	26225-79-6	NI29542

MASS TO CHARGE RATIOS								M.W.	INTENSITIES								Parent	C	H	O	N	S	F	Cl	Br	I	Si	P	B	X	COMPOUND NAME	CAS Reg No	No
43	15	29	161	207	137	79	27	**286**	100	96	91	85	83	57	53	47	**13.31**	13	18	5	–	1	–	–	–	–	–	–	–	–	5-Benzofuranol, 2-ethoxy-2,3-dihydro-3,3-dimethyl-, methanesulphonate, (±)-	26225-79-6	NI29543
43	28	145	44	103	113	71	42	**286**	100	23	20	11	10	8	7	6	**0.00**	13	18	7	–	–	–	–	–	–	–	–	–	–	4-Cyclopropyl-3,2,1-triacetyllyxose		NI29544
186	152	153	214	179	260	195	242	**286**	100	88	73	43	40	20	20	14	**0.00**	13	18	7	–	–	–	–	–	–	–	–	–	–	Malonic acid, (4-oxo-6,8-dioxabicyclo[3.2.1]oct-2-yl)-, diethyl ester		NI29545
128	69	143	73	169	75	159	41	**286**	100	98	87	69	52	44	39	35	**14.01**	13	19	3	–	–	–	1	–	–	1	–	–	–	Propanoic acid, 2-(4-chlorophenoxy)-2-methyl-, trimethylsilyl ester	55255-67-9	NI29546
43	56	57	41	70	69	55	84	**286**	100	72	66	64	55	50	49	28	**0.00**	13	22	–	–	–	2	2	–	–	–	–	–	–	1,2-Dichloro-1,1-difluoro-2-tridecene		DD00984
43	174	187	41	55	57	155	281	**286**	100	89	72	35	33	31	28	24	**15.30**	13	22	2	–	–	4	–	–	–	–	–	–	–	2,2,3,3-Tetrafluoropropyl decanoate		MS06543
43	174	187	55	57	155	243	71	**286**	100	90	73	33	30	29	27	20	**15.00**	13	22	2	–	–	4	–	–	–	–	–	–	–	2,2,3,3-Tetrafluoropropyl decanoate		LI06182
73	271	147	169	75	197	45	286	**286**	100	52	34	25	22	18	15	12		13	26	3	–	–	–	–	–	–	2	–	–	–	3-Hexenoic acid, 3-methyl-5-[(trimethylsilyl)oxy]-, trimethylsilyl ester		NI29547
286	288	151	75	76	150	260	258	**286**	100	99	86	63	61	46	35	35		14	7	2	–	–	–	–	1	–	–	–	–	–	1-Bromoanthraquinone		IC03680
103	183	286	76	51	287	143	156	**286**	100	95	76	33	15	12	8	4		14	10	–	2	–	–	–	–	–	–	–	–	1	3,4-Diphenyl-1,2,5-selenadiazole	19768-00-4	DD00985
183	103	286	181	284	76	185	179	**286**	100	97	63	50	33	31	20	20		14	10	–	2	–	–	–	–	–	–	–	–	1	3,4-Diphenyl-1,2,5-selenadiazole	19768-00-4	NI29548
183	103	181	76	179	185	180	286	**286**	100	70	50	23	19	18	17	16		14	10	–	2	–	–	–	–	–	–	–	–	1	3,5-Diphenyl-1,2,4-selenadiazole	19768-00-4	MS06544
105	77	103	51	286	76	62	50	**286**	100	75	47	31	19	18	16	15		14	10	3	2	1	–	–	–	–	–	–	–	–	4,4-Diphenyl-1,2,3,5-oxathiadiazine-2,2-dioxide		MS06545
239	165	163	164	193	181	139	240	**286**	100	76	32	22	21	19	19	14	**0.00**	14	10	5	2	–	–	–	–	–	–	–	–	–	Phenol, 2-(2,4-dinitrostyryl)-		NI29549
121	165	122	107	164	120	90	163	**286**	100	95	93	45	33	32	27	27	**11.96**	14	10	5	2	–	–	–	–	–	–	–	–	–	Phenol, 3-(2,4-dinitrostyryl)-	81308-27-2	NI29550
286	285	119	168	193	180	179	167	**286**	100	20	17	9	1	1	1	1		14	14	3	4	–	–	–	–	–	–	–	–	–	α-[(2-Amino-1,2-dicyanovinyl)imino]-3,4,5-trimethoxytoluene		NI29551
286	172	132	129	287	102	143	173	**286**	100	28	28	24	17	16	15	13		14	14	3	4	–	–	–	–	–	–	–	–	–	Benzimidazolium, 1-methyl-3-(1,2,3,4-tetrahydro-6-hydroxy-1,3-dimethyl-2,4-dioxo-5-pyrimidinyl)-, hydroxide, inner salt	27355-89-1	NI29552
43	144	115	41	241	85	143	170	**286**	100	37	19	17	13	12	11	9	**0.00**	14	14	3	4	–	–	–	–	–	–	–	–	–	2-Oxabicyclo[2.2.1]heptane-6-carboxylic acid, 1,5-dimethyl-3-imino-4,7,7-tricyano-, ethyl ester		MS06546
69	29	31	43	195	123	98	149	**286**	100	78	71	69	64	53	50	38	**0.35**	14	22	6	–	–	–	–	–	–	–	–	–	–	Diethyl α,α′-diacetyl-β-methylglutarate		LI06183
200	69	199	169	113	141	55	41	**286**	100	95	82	82	67	63	52	45	**0.00**	14	22	6	–	–	–	–	–	–	–	–	–	–	6,10-Dioxaspiro[5.4]decane-2,3-dicarboxylic acid, 8,8-dimethyl-, dimethyl ester, cis-		DD00986
97	153	213	70	41	43	69	181	**286**	100	86	71	45	44	43	35	34	**1.00**	14	22	6	–	–	–	–	–	–	–	–	–	–	2-Propoxy-6,6-bis(ethoxycarbonyl)-5,6-dihydro-2H-pyran		LI06184
113	69	41	112	86	45	114	39	**286**	100	90	42	23	14	10	9	9	**0.17**	14	22	6	–	–	–	–	–	–	–	–	–	–	Triethylene glycol dimethacrylate		IC03681
72	114	115	44	43	142	29	55	**286**	100	92	43	29	16	13	11	10	**0.27**	14	26	4	2	–	–	–	–	–	–	–	–	–	L-Alanine, N-(N-acetyl-L-valyl)-, butyl ester	55712-45-3	NI29553
86	128	43	44	41	30	29	27	**286**	100	61	32	28	24	24	21	14	**0.00**	14	26	4	2	–	–	–	–	–	–	–	–	–	Glycine, N-(N-acetyl-L-leucyl)-, butyl ester	55712-44-2	NI29554
86	30	73	100	185	43	72	29	**286**	100	27	24	13	12	12	11	9	**0.57**	14	26	4	2	–	–	–	–	–	–	–	–	–	L-Isoleucine, N-(N-acetylglycyl)-, butyl ester	55712-42-0	NI29555
86	73	30	185	43	100	87	72	**286**	100	21	18	12	9	7	6	6	**0.82**	14	26	4	2	–	–	–	–	–	–	–	–	–	L-Leucine, N-(N-acetylglycyl)-, butyl ester	62167-66-2	NI29556
72	44	87	86	43	41	114	29	**286**	100	94	88	51	18	15	13	13	**1.50**	14	26	4	2	–	–	–	–	–	–	–	–	–	L-Valine, N-(N-acetyl-L-alanyl)-, butyl ester	55712-41-9	NI29557
257	255	253	103	256	243	101	241	**286**	100	75	57	33	32	29	27	24	**0.10**	14	28	1	–	–	–	–	–	–	–	–	–	1	Germacycloundecan-6-one, 1,1-diethyl-	17973-67-0	NI29558
286	153	134	152	258	229	154	163	**286**	100	93	59	32	27	18	18	14		15	10	6	–	–	–	–	–	–	–	–	–	–	4H-1-Benzopyran-4-one, 5,7-dihydroxy-2-(3,4-dihydroxyphenyl)-	491-70-3	NI29559
286	121	69	258	229	285	93	77	**286**	100	48	45	26	26	24	22	22		15	10	6	–	–	–	–	–	–	–	–	–	–	4H-1-Benzopyran-4-one, 3,5,7-trihydroxy-2-(4-hydroxyphenyl)-	520-18-3	NI29560
286	121	69	93	257	229	258	108	**286**	100	29	13	10	8	8	7	6		15	10	6	–	–	–	–	–	–	–	–	–	–	4H-1-Benzopyran-4-one, 3,5,7-trihydroxy-2-(4-hydroxyphenyl)-	520-18-3	NI29561
286	285	121	258	257	229	153	93	**286**	100	17	14	8	7	6	5	5		15	10	6	–	–	–	–	–	–	–	–	–	–	4H-1-Benzopyran-4-one, 3,5,7-trihydroxy-2-(4-hydroxyphenyl)-	520-18-3	MS06547
44	149	286	201	166	120	45	132	**286**	100	96	68	32	32	27	27	26		15	10	6	–	–	–	–	–	–	–	–	–	–	Biphenyl-2,4′,5-tricarboxylic acid		NI29562
286	153	119	131	169	181	270	134	**286**	100	71	67	55	40	36	28	28		15	10	6	–	–	–	–	–	–	–	–	–	–	7-[O-β-D-Glucopyranosyl-(1-2)-β-D-glucopyranosyloxy]-3′,4′,5-isoflavone		NI29563
286	285	287	288	223	259	77	252	**286**	100	40	35	33	25	25	11	10		15	11	–	2	1	–	1	–	–	–	–	–	–	1H-1,4-Benzodiazepine-2-thione, 7-chloro-2,3-dihydro-5-phenyl-	4547-02-8	NI29564
287	285	286	288	207	131	103	77	**286**	100	93	79	77	73	72	66	59		15	11	1	–	–	–	–	1	–	–	–	–	–	2-Propen-1-one, 1-(4-bromophenyl)-3-phenyl-	2403-27-2	NI29565
77	288	286	207	105	102	287	51	**286**	100	99	99	99	90	90	60	45		15	11	1	–	–	–	–	1	–	–	–	–	–	2-Propen-1-one, 3-(4-bromophenyl)-1-phenyl-	1774-66-9	NI29566
268	239	77	267	205	233	269	240	**286**	100	64	64	62	58	53	40	36	**0.00**	15	11	2	2	–	–	1	–	–	–	–	–	–	2H-1,4-Benzodiazepin-2-one, 7-chloro-1,3-dihydro-3-hydroxy-5-phenyl-	604-75-1	NI29567
77	205	233	239	267	51	268	75	**286**	100	96	82	70	59	57	48	37	**2.50**	15	11	2	2	–	–	1	–	–	–	–	–	–	2H-1,4-Benzodiazepin-2-one, 7-chloro-1,3-dihydro-3-hydroxy-5-phenyl-	604-75-1	NI29568
269	287	271	257					**286**	100	40	35	5					**0.00**	15	11	2	2	–	–	1	–	–	–	–	–	–	2H-1,4-Benzodiazepin-2-one, 7-chloro-1,3-dihydro-3-hydroxy-5-phenyl-	604-75-1	NI29569
242	269	241	270	243	271	244	272	**286**	100	92	84	69	54	40	34	23	**13.01**	15	11	2	2	–	–	1	–	–	–	–	–	–	2H-1,4-Benzodiazepin-2-one, 7-chloro-1,3-dihydro-5-phenyl-, 4-oxide	963-39-3	NI29570
258	257	286	285	259	260	287	288	**286**	100	69	63	54	40	36	28	20		15	11	2	2	–	–	1	–	–	–	–	–	–	2H-1,4-Benzodiazepin-2-one, 7-chloro-1,3-dihydro-5-phenyl-, 4-oxide	963-39-3	MS06548
237	77	51	286	181	238	288	78	**286**	100	68	49	38	25	18	13	12		15	11	2	2	–	–	1	–	–	–	–	–	–	Imidazo[1,2-a]pyridinium, 3-(chloroacetyl)-2,3-dihydro-2-oxo-1-phenyl-, hydroxide, inner salt	11066-37-8	NI29571
251	252	286	76	288	143	271	77	**286**	100	20	10	8	4	4	3	3		15	11	2	2	–	–	1	–	–	–	–	–	–	4(3H)-Quinazolinone, 3-(2-chloro-3-hydroxyphenyl)-2-methyl-	51837-88-8	NI29572
251	252	286	76	143	77	288	271	**286**	100	17	9	8	7	4	3	3		15	11	2	2	–	–	1	–	–	–	–	–	–	4(3H)-Quinazolinone, 3-(2-chloro-4-hydroxyphenyl)-2-methyl-	29909-23-7	NI29573
251	252	76	286	271	143	77	288	**286**	100	18	5	4	3	3	2	2		15	11	2	2	–	–	1	–	–	–	–	–	–	4(3H)-Quinazolinone, 3-(2-chloro-5-hydroxyphenyl)-2-methyl-	51837-91-3	NI29574
251	286	288	271	76	252	143	117	**286**	100	64	21	19	17	16	11	7		15	11	2	2	–	–	1	–	–	–	–	–	–	4(3H)-Quinazolinone, 3-(2-chloro-6-hydroxyphenyl)-2-methyl-	51837-92-4	NI29575
251	286	252	111	288	271	152	113	**286**	100	45	16	16	15	12	8	6		15	11	2	2	–	–	1	–	–	–	–	–	–	4(3H)-Quinazolinone, 3-(2-chlorophenyl)-5-hydroxy-2-methyl-	51837-93-5	NI29576
251	286	252	288	111	154	271	76	**286**	100	44	18	15	14	9	8	8		15	11	2	2	–	–	1	–	–	–	–	–	–	4(3H)-Quinazolinone, 3-(2-chlorophenyl)-6-hydroxy-2-methyl-	29909-22-6	NI29577
251	286	252	78	288	111	152	271	**286**	100	20	18	11	7	6	5	4		15	11	2	2	–	–	1	–	–	–	–	–	–	4(3H)-Quinazolinone, 3-(2-chlorophenyl)-7-hydroxy-2-methyl-	51837-90-2	NI29578
251	286	152	252	288	111	154	160	**286**	100	57	30	28	19	12	10	9		15	11	2	2	–	–	1	–	–	–	–	–	–	4(3H)-Quinazolinone, 3-(2-chlorophenyl)-8-hydroxy-2-methyl-	51837-89-9	NI29579

MASS TO CHARGE RATIOS								M.W.	INTENSITIES								Parent	C	H	O	N	S	F	Cl	Br	I	Si	P	B	X	COMPOUND NAME	CAS Reg No	No
131	130	91	77	103	286	44	32	**286**	100	67	67	41	38	33	30	26		15	14	2	2	1	–	–	–	–	–	–	–	–	N-(3,4-Dihydrophthalazinyl)-4-tolyl sulphone		MS06549
136	163	286	104	76	120	150	92	**286**	100	60	42	30	30	28	25	20		15	14	4	2	–	–	–	–	–	–	–	–	–	1,4-Benzanisidide, N-methyl-4-nitro-	1092-52-0	MS06550
242	50	211	77	157	43	103	115	**286**	100	83	72	61	52	44	42	39	**3.00**	15	14	4	2	–	–	–	–	–	–	–	–	–	1H-Imidazole-4,5-dicarboxylic acid, 2-methyl-1-(2-phenylvinyl)-, 4-methyl ester, (E)-	56382-57-1	NI29580
134	195	102	165	77	286	103	223	**286**	100	50	50	45	45	40	34	23		15	14	4	2	–	–	–	–	–	–	–	–	–	1H-Imidazole-4,5-dicarboxylic acid, 1-(2-phenylvinyl)-, dimethyl ester, (E)-	56382-61-7	NI29581
170	286	104	103	198	255	256	227	**286**	100	52	30	26	24	15	14	14		15	14	4	2	–	–	–	–	–	–	–	–	–	4,5-Pyrimidinedicarboxylic acid, 6-methyl-2-phenyl-, dimethyl ester	61416-99-7	NI29582
169	69	57	101	41	55	83	29	**286**	100	84	51	43	42	35	24	21	**0.00**	15	26	3	–	1	–	–	–	–	–	–	–	–	1,3-Dioxolane-4-carbothioic acid, 2-tert-butyl-4-(2-propenyl)-, S-tert-butyl ester, (2R,4S)-		DD00987
143	111	116	55	83	43	69	59	**286**	100	49	29	29	20	18	16	14	**1.00**	15	26	5	–	–	–	–	–	–	–	–	–	–	Tridecanedioic acid, 3,6-epoxy-, dimethyl ester		NI29583
129	102	258	257	130	156	18	75	**286**	100	71	56	40	25	21	21	21	**0.00**	16	10	–	6	–	–	–	–	–	–	–	–	–	s-Tetrazino[1,6-a:4,3-a']diquinoxaline	93764-49-9	NI29584
285	286	287	288	251	270	150	122	**286**	100	78	52	27	22	10	9	8		16	11	3	–	–	–	1	–	–	–	–	–	–	Aurone, 3'-chloro-6-methoxy-	77764-89-7	NI29585
285	286	287	288	251	122	106	107	**286**	100	92	49	30	22	10	8	6		16	11	3	–	–	–	1	–	–	–	–	–	–	Aurone, 4'-chloro-6-methoxy-	77764-90-0	NI29586
204	205	286	77	102	91	128	129	**286**	100	84	73	61	60	60	59	54		16	14	–	–	–	–	–	–	–	–	–	–	1	Bis-β-styryl selenide, (E,E)-		DD00988
205	206	286	77	102	128	204	129	**286**	100	65	62	53	52	25	23	22		16	14	–	–	–	–	–	–	–	–	–	–	1	Bis-β-styryl selenide, (Z,E)-		DD00989
205	206	77	286	102	103	128	129	**286**	100	64	64	57	51	32	29	25		16	14	–	–	–	–	–	–	–	–	–	–	1	Bis-β-styryl selenide, (Z,Z)-		DD00990
144	145	77	103	131	51	18	115	**286**	100	80	80	75	68	35	33	27	**24.00**	16	14	3	–	1	–	–	–	–	–	–	–	–	trans-4-Phenyl-1-phenylsulphonyl-3-buten-2-one		LI06185
286	257	243	229	57	69	43	55	**286**	100	40	32	28	26	21	20	18		16	14	5	–	–	–	–	–	–	–	–	–	–	Aflatoxin D_1		MS06551
286	270	243	285	227	257	143	255	**286**	100	30	12	8	8	7	7	7		16	14	5	–	–	–	–	–	–	–	–	–	–	9(10H)-Anthracenone, 1,4,8-trihydroxy-6-methoxy-3-methyl-	84542-56-3	NI29587
43	159	160	202	244	184	156	174	**286**	100	86	76	51	25	24	23	22	**0.00**	16	14	5	–	–	–	–	–	–	–	–	–	–	7(8H)-Benzocyclooctenone, 5,10-diacetoxy-	93103-68-5	NI29588
286	258	241	227	268	271	267	229	**286**	100	50	35	30	26	21	15	12		16	14	5	–	–	–	–	–	–	–	–	–	–	6H-Benzofuro[3,2-c][1]benzopyran-3,6a(11aH)-diol, 9-methoxy-, cis-	75362-93-5	NI29589
267	268	269	271	258	286	253	252	**286**	100	92	17	12	10	8	8	7		16	14	5	–	–	–	–	–	–	–	–	–	–	6H-Benzofuro[3,2-c][1]benzopyran-6a,9(11aH)-diol, 3-methoxy-, cis-	75330-97-1	NI29590
255	286	285	179	135	147	92	77	**286**	100	79	56	45	25	18	16	13		16	14	5	–	–	–	–	–	–	–	–	–	–	Benzoic acid, 4-hydroxy-3-(3-methoxybenzoyl)-, methyl ester		NI29591
135	136	105	152	151	121	120	106	**286**	100	5	3	1	1	1	1	1	**0.30**	16	14	5	–	–	–	–	–	–	–	–	–	–	Benzoic acid, 4-(2-methoxybenzoyloxy)-, methyl ester		NI29592
255	286	285	189	227	135	147	256	**286**	100	89	63	45	25	25	20	13		16	14	5	–	–	–	–	–	–	–	–	–	–	Benzoic acid, 4-(3-methoxybenzoyloxy)-, methyl ester		NI29593
91	136	119	149	150	105	92	77	**286**	100	90	73	61	49	20	16	16	**0.00**	16	14	5	–	–	–	–	–	–	–	–	–	–	Benzoic acid, 4,4'-[oxybis(methylene)]bis-	55255-64-6	NI29594
215	216	243	184	185	244	156	128	**286**	100	77	66	63	61	47	44	38	**34.00**	16	14	5	–	–	–	–	–	–	–	–	–	–	1aH-Benz[b]oxireno[h]benzofuran-4-carboxylic acid, 2,4a-dihydro-1a,4a-dimethyl-2-oxo-, methyl ester	40801-46-5	NI29595
137	150	135	55	77	69	81	107	**286**	100	55	44	40	30	29	24	23	**13.00**	16	14	5	–	–	–	–	–	–	–	–	–	–	Chalcone, 2',3,4'-trihydroxy-4-methoxy-	13323-67-6	NI29596
227	183	159	181	286	209	152	168	**286**	100	77	60	58	54	34	31	30		16	14	5	–	–	–	–	–	–	–	–	–	–	1,2-Dibenzofurandicarboxylic acid, 1,9b-dihydro-4,9b-dimethyl-	56247-78-0	NI29597
167	286	120	285	166	180	193	138	**286**	100	86	58	44	38	35	34	33		16	14	5	–	–	–	–	–	–	–	–	–	–	Flavanone, 4',5-dihydroxy-7-methoxy-		MS06552
167	91	257	286	120	166	179	138	**286**	100	22	21	14	14	13	10	9		16	14	5	–	–	–	–	–	–	–	–	–	–	Flavanone, 5-methoxy-3,7-dihydroxy-	87620-04-0	NI29598
105	77	167	286	269	166	268	138	**286**	100	40	30	25	15	13	12	8		16	14	5	–	–	–	–	–	–	–	–	–	–	Flavanone, 7-methoxy-2,5-dihydroxy-	35486-66-9	NI29599
286	167	138	166	285	120	287	193	**286**	100	67	47	40	32	25	17	14		16	14	5	–	–	–	–	–	–	–	–	–	–	Flavanone, 5,7,4'-trihydroxy-8-methyl-, (2R)-		NI29600
202	174	203	89	90	63	146	145	**286**	100	25	24	10	6	6	4	4	**0.00**	16	14	5	–	–	–	–	–	–	–	–	–	–	7H-Furo[3,2-g][1]benzopyran-7-one, 9-[(4-hydroxy-3-methyl-2-butenyl)oxy]-, (E)-	65853-14-7	NI29601
286	187	215	71	201	216	43	202	**286**	100	44	40	30	27	25	25	16		16	14	5	–	–	–	–	–	–	–	–	–	–	7H-Furo[3,2-g][1]benzopyran-7-one, 4-(3-methyl-2-oxobutoxy)-	5058-15-1	NI29602
255	286	112	254	168	287	84	70	**286**	100	54	22	15	11	9	8	8		16	14	5	–	–	–	–	–	–	–	–	–	–	4,4'-Bis(methoxycarbonyl)diphenyl ether		IC03682
285	209	286	105	77	45	287	51	**286**	100	65	57	32	27	10	9	7		16	14	5	–	–	–	–	–	–	–	–	–	–	Methyl α-(3-hydroxy-4-benzoylphenoxy)acetate		IC03683
167	286	285	193	180	166	120	129	**286**	100	74	67	43	28	22	18	15		16	14	5	–	–	–	–	–	–	–	–	–	–	2-Propen-1-one, 1-(2,4-dihydroxy-6-methoxyphenyl)-3-(2-hydroxyphenyl)-	69707-17-1	NI29603
286	136	44	271	39	243	77	215	**286**	100	80	28	27	24	21	21	19		16	14	5	–	–	–	–	–	–	–	–	–	–	Spiro[benzofuran-2(3H),1'-[2,5]cyclohexadiene]-3,4'-dione, 2'-hydroxy-6-methoxy-4,6'-dimethyl-	26891-80-5	NI29604
286	287	258	285	243	257	129	144	**286**	100	21	20	18	6	5	5	3		16	14	5	–	–	–	–	–	–	–	–	–	–	9H-Xanthen-9-one, 1-hydroxy-3,6-dimethoxy-8-methyl-	15222-53-4	NI29605
286	287	258	285	243	130	257	143	**286**	100	21	21	9	7	6	5	4		16	14	5	–	–	–	–	–	–	–	–	–	–	9H-Xanthen-9-one, 1-hydroxy-3,6-dimethoxy-8-methyl-	15222-53-4	LI06186
286	257	287	243	256	285	200	199	**286**	100	38	18	13	11	9	9	9		16	14	5	–	–	–	–	–	–	–	–	–	–	9H-Xanthen-9-one, 1-hydroxy-3,6-dimethoxy-8-methyl-	15222-53-4	NI29606
271	286	121	272	243	228	213	170	**286**	100	27	26	15	14	13	11	7		16	14	5	–	–	–	–	–	–	–	–	–	–	9H-Xanthen-9-one, 1,2,3-trimethoxy-		LI06187
286	271	243	186	77	135	269	214	**286**	100	88	28	24	18	17	12	12		16	14	5	–	–	–	–	–	–	–	–	–	–	9H-Xanthen-9-one, 1,2,3-trimethoxy-		NI29607
104	286	288	287	289	184	182	105	**286**	100	70	63	13	11	11	11	11		16	15	–	–	–	–	–	1	–	–	–	–	–	4-Bromo[2.2]paracyclophane		LI06188
104	286	288	287	289	183	181	105	**286**	100	70	64	13	12	12	11	11		16	15	–	–	–	–	–	1	–	–	–	–	–	Tricyclo[8.2.2^{4,7}]hexadeca-4,6,10,12,13,15-hexaene, 2-bromo-	38274-86-1	NI29608
91	121	106	136	286	165	195	77	**286**	100	92	92	91	55	12	7	6		16	18	1	2	1	–	–	–	–	–	–	–	–	N-Benzyl-N'-(4-methoxybenzyl)thiourea		NI29609
165	209	105	43	78	77	123	44	**286**	100	85	37	36	32	29	22	19	**0.20**	16	18	3	–	–	–	–	–	–	1	–	–	–	2,2-Diphenyl-1,3,6-trioxa-2-silacyclooctane		MS06553
286	213	241	287	242	197	214	288	**286**	100	32	27	24	13	10	7	5		16	18	3	–	–	–	–	–	–	1	–	–	–	9-Oxa-10-silaphenanthrene, 10,10-diethoxy-9,10-dihydro-		NI29610
244	201	286	185	184	157	227	226	**286**	100	63	51	50	32	24	2	2		16	18	3	2	–	–	–	–	–	–	–	–	–	Azepino[2,3-b]indol-5a(1H)-ol, 1-acetyl-2,3,4,5-tetrahydro-, acetate		LI06189
240	77	241	286	212	69	55	118	**286**	100	34	25	24	17	17	14	11		16	18	3	2	–	–	–	–	–	–	–	–	–	5-Pyrazolone, 4-(1-Ethoxycarbonylisopropylidene)-3-methyl-1-phenyl-		IC03684
271	286	241	255	45	272	242	226	**286**	100	71	57	30	24	23	23	21		16	18	3	2	–	–	–	–	–	–	–	–	–	9H-Pyrido[3,4-b]indole, 1-(methoxyethyl)-4,8-dimethoxy-	82652-19-5	NI29611

MASS TO CHARGE RATIOS								M.W.	INTENSITIES								Parent	C	H	O	N	S	F	Cl	Br	I	Si	P	B	X	COMPOUND NAME	CAS Reg No	No
213	286	241	257	105	197	104	185	286	100	56	22	21	10	9	8	6		16	18	3	2	–	–	–	–	–	–	–	–	–	Pyrrolo[2,1-b]quinazolin-10-one, 1,2,3,10-tetrahydro-4-ethoxycarbonyl-5-methyl-		LI06190
72	241	286	44	92	90	77	226	286	100	38	23	11	10	10	10	7		16	18	3	2	–	–	–	–	–	–	–	–	–	Urea, N′-[4-(4-methoxyphenoxy)phenyl]-N,N-dimethyl-	14214-32-5	NI29612
121	58	72	78	71	215	122	77	286	100	68	14	13	13	11	9	9	8.91	16	22	1	4	–	–	–	–	–	–	–	–	–	1,2-Ethanediamine, N-[(4-methoxyphenyl)methyl]-N′,N′-dimethyl-N-2-pyrimidinyl-	91-85-0	NI29613
121	287	285	242	216				286	100	29	14	14	8				0.00	16	22	1	4	–	–	–	–	–	–	–	–	–	1,2-Ethanediamine, N-[(4-methoxyphenyl)methyl]-N′,N′-dimethyl-N-2-pyrimidinyl-	91-85-0	NI29614
81	137	41	95	55	67	69	93	286	100	62	42	41	28	27	22	21	8.00	16	28	–	–	–	–	1	–	–	–	1	–	–	Phosphinous chloride, cyclohexyl(1,7,7-trimethylbicyclo[2.2.1]hept-2-yl)-	74630-28-7	NI29615
69	101	213	59	111	55	181	83	286	100	85	78	56	50	46	44	44	0.00	16	30	4	–	–	–	–	–	–	–	–	–	–	Decanedioic acid, 3,4,7,8-tetramethyl-, dimethyl ester	57289-32-4	NI29616
271	101	43	41	116	57	155	71	286	100	60	58	31	23	23	21	21	0.00	16	30	4	–	–	–	–	–	–	–	–	–	–	Decanoic acid, (2,2-dimethyl-1,3-dioxolan-4-yl)methyl ester	20620-23-9	NI29617
171	227	213	153	74	195	83	69	286	100	51	46	44	40	35	27	24	0.00	16	30	4	–	–	–	–	–	–	–	–	–	–	Decanoic acid, 4-(methoxycarbonylmethyl)-, isopropyl ester		NI29618
29	55	98	41	43	241	69	88	286	100	89	83	72	56	55	55	53	1.58	16	30	4	–	–	–	–	–	–	–	–	–	–	Diethyl dodecan-1,12-dioate		IC03685
98	241	199	69	55	88	153	84	286	100	63	59	49	48	47	43	41	0.00	16	30	4	–	–	–	–	–	–	–	–	–	–	Diethyl dodecan-1,12-dioate		IC03686
43	57	41	55	42	44	82	56	286	100	65	56	44	37	33	32	27	0.00	16	30	4	–	–	–	–	–	–	–	–	–	–	1,1-Dodecanediol, diacetate	56438-07-4	NI29619
98	84	55	41	43	60	83	97	286	100	65	53	36	33	22	20	15	0.00	16	30	4	–	–	–	–	–	–	–	–	–	–	Hexadecanedioic acid	505-54-4	NI29620
157	41	115	57	55	56	139	83	286	100	56	49	46	39	32	31	30	0.00	16	30	4	–	–	–	–	–	–	–	–	–	–	Hexanedioic acid, 3-ethyl-, diisopropyl ester	57983-54-7	NI29621
131	73	101	109	141	271	254	157	286	100	45	23	13	10	8	8	4	0.10	16	30	4	–	–	–	–	–	–	–	–	–	–	1-O-(2-Methoxy-4-nonenyl)-2,3-O-isopropylideneglycerol		NI29622
157	213	115	138	139	111	56	55	286	100	86	45	35	28	27	27	25	0.00	16	30	4	–	–	–	–	–	–	–	–	–	–	Octanedioic acid, dibutyl ester	16090-77-0	NI29623
157	57	41	55	56	83	138	69	286	100	56	53	51	48	28	26	25	0.00	16	30	4	–	–	–	–	–	–	–	–	–	–	Octanedioic acid, di-sec-butyl ester	57983-35-4	NI29624
157	115	41	127	57	73	55	69	286	100	61	38	33	31	22	20	19	0.00	16	30	4	–	–	–	–	–	–	–	–	–	–	Pentanedioic acid, 3-ethyl-3-methyl-, di-sec-butyl ester	57983-53-6	NI29625
43	71	55	41	56	42	57	69	286	100	59	33	23	13	9	7	6	0.00	16	30	4	–	–	–	–	–	–	–	–	–	–	Propanoic acid, 2-methyl-, 2-ethyl-1-propyl-1,3-propanediyl ester	74367-30-9	NI29626
71	43	72	56	70	69	41	55	286	100	16	6	6	5	5	5	4	0.00	16	30	4	–	–	–	–	–	–	–	–	–	–	Propanoic acid, 2-methyl-, 1-tert-butyl-2-methyl-1,3-propanediyl ester	74381-40-1	NI29627
129	169	95	81	41	183	104	55	286	100	49	40	37	29	27	27	24	0.50	16	30	4	–	–	–	–	–	–	–	–	–	–	12-Tridecenoic acid, 2,4-dimethoxy-, methyl ester	55255-62-4	WI01360
43	41	42	56	44	55	70	57	286	100	71	65	63	41	38	30	30	0.00	16	34	2	2	–	–	–	–	–	–	–	–	–	Dioctyl nitramine		MS06554
58	85	84	86	98	100	42	71	286	100	52	28	20	17	15	12	10	3.30	16	38	–	4	–	–	–	–	–	–	–	–	–	Permethylspermine		MS06555
158	143	128	115	76	51	118	103	286	100	20	13	13	10	7	6	6	3.00	17	10	1	4	–	–	–	–	–	–	–	–	–	1,1,2,2-Tetracyano-4-methyl-1,2,3,11-tetrahydrodibenzofuran		LI06191
91	195	65	107	123	92	45	39	286	100	40	18	15	14	13	11	9	2.00	17	18	2	–	1	–	–	–	–	–	–	–	–	Benzyl 2-(benzylthio)propionate		MS06556
105	168	91	77	117	237	219	286	286	100	74	35	22	20	9	3	2		17	18	2	–	1	–	–	–	–	–	–	–	–	3,5-Diphenyl-3-hydroxytetrahydrothiopyran-1-oxide		LI06192
171	51	43	244	45	115	172	128	286	100	70	65	31	18	17	13	12	3.50	17	18	4	–	–	–	–	–	–	–	–	–	–	2-Acetoxy-8-(2-methoxycarbonylethyl)-7-methylnaphthalene		LI06193
121	122	286	107	91	148	135	165	286	100	37	35	9	9	5	4	3		17	18	4	–	–	–	–	–	–	–	–	–	–	Benzoic acid, 2-methoxy-6-[2-(4-methoxyphenyl)ethyl]-	36640-14-9	NI29628
164	151	286	121	149	152	107	165	286	100	74	52	44	36	33	16	14		17	18	4	–	–	–	–	–	–	–	–	–	–	2H-1-Benzopyran-7-ol, 3-(2,4-dimethoxyphenyl)-3,4-dihydro-	56701-26-9	NI29629
105	108	77	136	106	258	109	41	286	100	34	17	12	11	7	7	7	3.30	17	18	4	–	–	–	–	–	–	–	–	–	–	exo-5-Benzoyloxy-1,7,7-trimethylbicyclo[2.2.1]heptane-2,3-dione		MS06557
225	107	226	131	18	120	121	77	286	100	40	36	33	21	18	16	13	6.00	17	18	4	–	–	–	–	–	–	–	–	–	–	1,3-Bis(4-hydroxyphenyl)pent-1-ene-4,5-diol		LI06194
285	136	123	163	286	147	201	240	286	100	41	40	23	18	18	17	16		17	18	4	–	–	–	–	–	–	–	–	–	–	6-Oxaestra-1,3,5(10)-triene-7,17-dione, 3-hydroxy-, (±)-		MS06558
225	107	131	121	120	77	149	136	286	100	41	32	17	17	14	8	8	6.96	17	18	4	–	–	–	–	–	–	–	–	–	–	4-Pentene-1,2-diol, 3,5-bis(4-hydroxyphenyl)-, [S-[R*,R*-(E)]]-	7288-11-1	NI29630
150	137	286	149	138	148	151	121	286	100	84	41	25	18	15	12	12		17	18	4	–	–	–	–	–	–	–	–	–	–	Phenol, 2-(3,4-dihydro-7-methoxy-2H-1-benzopyran-3-yl)-5-methoxy-	3722-59-6	NI29631
120	192	121	226	107	123	286	149	286	100	63	60	54	53	45	40	37		17	18	4	–	–	–	–	–	–	–	–	–	–	2H-Pyran-3-ol, tetrahydro-4,6-bis(4-hydroxyphenyl)-, [3S-(3α,4β,6β)]-	2141-07-3	NI29632
120	192	121	107	123	286	149	226	286	100	63	59	53	44	40	38	25		17	18	4	–	–	–	–	–	–	–	–	–	–	2H-Pyran-3-ol, tetrahydro-4,6-bis(4-hydroxyphenyl)-, [3S-(3α,4β,6β)]-	2141-07-3	LI06195
159	131	97	158	104	133	129	226	286	100	82	49	48	45	42	35	34	15.70	17	18	4	–	–	–	–	–	–	–	–	–	–	2H-Pyran-2-one, 6-[2-(acetyloxy)-4-phenyl-3-butenyl]-5,6-dihydro-	34427-30-0	NI29633
43	131	97	158	104	133	130	129	286	100	96	58	56	52	51	47	41	18.36	17	18	4	–	–	–	–	–	–	–	–	–	–	2H-Pyran-2-one, 6-[2-(acetyloxy)-4-phenyl-3-butenyl]-5,6-dihydro-	34427-30-0	MS06559
56	85	201	165	241	166	58	203	286	100	92	59	44	27	25	20	19	3.53	17	19	–	2	–	–	1	–	–	–	–	–	–	Piperazine, 1-[(4-chlorophenyl)phenylmethyl]-	303-26-4	NI29634
56	201	85	165	241	166	203	42	286	100	84	76	75	39	39	26	20	5.14	17	19	–	2	–	–	1	–	–	–	–	–	–	Piperazine, 1-[(4-chlorophenyl)phenylmethyl]-	303-26-4	NI29635
49	143	144	84	286	241	130	86	286	100	91	79	74	71	70	65	63		17	22	2	2	–	–	–	–	–	–	–	–	–	4-Piperidone, trans-5-ethyl-2-(2-indolyl)-, ethylene acetal		DD00991
286	287	143	124	221	161	147	59	286	100	18	11	11	10	10	8	8		17	23	–	–	–	–	–	–	–	–	–	–	1	Cobalt, (η^5-2,4-cyclopentadien-1-yl)[(1,2,3,4,5,6-η)-hexamethylbenzene]-	74811-00-0	NI29636
239	71	85	240	83	95	97	286	286	100	43	26	18	14	13	9	8		17	34	1	–	1	–	–	–	–	–	–	–	–	2-Heptadecanone, 1-mercapto-	76078-79-0	NI29637
43	41	55	57	69	83	97	90	286	100	90	68	64	44	34	23	23	5.00	17	34	3	–	–	–	–	–	–	–	–	–	–	Hexadecanoic acid, 2-hydroxy-, methyl ester	16742-51-1	NI29638
55	57	69	43	83	41	97	29	286	100	96	88	87	74	68	59	59	1.00	17	34	3	–	–	–	–	–	–	–	–	–	–	Hexadecanoic acid, 2-hydroxy-, methyl ester	16742-51-1	NI29639
103	43	41	74	71	55	57	61	286	100	57	29	27	27	22	20	16	0.00	17	34	3	–	–	–	–	–	–	–	–	–	–	Hexadecanoic acid, 3-hydroxy-, methyl ester	51883-36-4	NI29640
169	201	172	55	87	74	173	69	286	100	79	69	60	57	46	26	23	1.22	17	34	3	–	–	–	–	–	–	–	–	–	–	Hexadecanoic acid, 10-hydroxy-, methyl ester	56247-30-4	NI29641
169	201	172	55	87	74	173	97	286	100	79	69	60	58	45	23	21	0.00	17	34	3	–	–	–	–	–	–	–	–	–	–	Hexadecanoic acid, 10-hydroxy-, methyl ester	56247-30-4	LI06196
183	55	87	186	41	215	83	74	286	100	71	62	51	50	46	42	40	0.40	17	34	3	–	–	–	–	–	–	–	–	–	–	Hexadecanoic acid, 11-hydroxy-, methyl ester	60368-18-5	NI29642
87	74	242	143	97	45	98	96	286	100	86	57	44	39	39	33	29	0.00	17	34	3	–	–	–	–	–	–	–	–	–	–	Hexadecanoic acid, 15-hydroxy-, methyl ester	55823-13-7	LI06197
242	74	87	45	69	43	236	83	286	100	82	79	56	27	26	22	20	0.40	17	34	3	–	–	–	–	–	–	–	–	–	–	Hexadecanoic acid, 15-hydroxy-, methyl ester	55823-13-7	NI29643
74	98	84	87	75	97	83	69	286	100	77	48	45	45	34	34	34	0.00	17	34	3	–	–	–	–	–	–	–	–	–	–	Hexadecanoic acid, 16-hydroxy-, methyl ester		LI06198
115	116	286	89	63	230	57.5	39	286	100	65	27	19	15	13	13	9		18	14	–	–	–	–	–	–	–	–	–	–	1	Di-π-indenyliron		MS06560
77	181	105	286	51	76	78	50	286	100	43	40	36	21	18	8	8		18	14	–	4	–	–	–	–	–	–	–	–	–	4-Benzeneazoazobenzene		MS06561

MASS TO CHARGE RATIOS								M.W.	INTENSITIES								Parent	C	H	O	N	S	F	Cl	Br	I	Si	P	B	X	COMPOUND NAME	CAS Reg No	No
286	285	244	165	287	245	142	77	**286**	100	42	39	25	22	18	13	13		18	14	–	4	–	–	–	–	–	–	–	–	–	**s-Triazolo[4,3-a]pyrazine, 3-methyl-5,6-diphenyl-**	33709-78-3	**LI06199**
286	285	244	165	287	245	142	77	**286**	100	40	39	25	22	18	12	12		18	14	–	4	–	–	–	–	–	–	–	–	–	**s-Triazolo[4,3-a]pyrazine, 3-methyl-5,6-diphenyl-**	33709-78-3	**NI29644**
286	227	174	173	287	147	160	186	**286**	100	51	48	44	21	16	12	11		18	22	3	–	–	–	–	–	–	–	–	–	–	**Estra-1,3,5(10)-trien-6-one, 3,17-dihydroxy-, (17β)-**	571-92-6	**NI29645**
268	148	211	286	269	225	158	212	**286**	100	48	29	22	20	16	15	13		18	22	3	–	–	–	–	–	–	–	–	–	–	**Estra-1,3,5(10)-trien-16-one, 3,17-dihydroxy-, (17α)-**	6199-65-1	**NI29646**
213	286	214	172	133	287	173	159	**286**	100	68	43	20	15	14	12	12		18	22	3	–	–	–	–	–	–	–	–	–	–	**Estra-1,3,5(10)-trien-16-one, 3,17-dihydroxy-, (17β)-**	566-75-6	**NI29647**
213	286	214	133	159	172	173	145	**286**	100	60	37	23	17	16	15	14		18	22	3	–	–	–	–	–	–	–	–	–	–	**Estra-1,3,5(10)-trien-16-one, 3,17-dihydroxy-, (17β)-**	566-75-6	**NI29648**
286	123	162	287	175	188	229	161	**286**	100	22	20	19	12	11	10	10		18	22	3	–	–	–	–	–	–	–	–	–	–	**Estra-1,3,5(10)-trien-17-one, 2,3-dihydroxy-**	362-06-1	**NI29649**
268	148	211	286	269	225	158	212	**286**	100	51	31	21	21	20	16	14		18	22	3	–	–	–	–	–	–	–	–	–	–	**Estra-1,3,5(10)-trien-17-one, 3,6-dihydroxy-, (6β)-**	1229-25-0	**NI29650**
286	146	186	170	172	97	185	145	**286**	100	67	28	28	27	27	24	22		18	22	3	–	–	–	–	–	–	–	–	–	–	**Estra-1,3,5(10)-trien-17-one, 3,11-dihydroxy-, (11β)-**	6803-21-0	**NI29651**
286	61	70	284	146	145	160	97	**286**	100	40	36	35	33	27	23	22		18	22	3	–	–	–	–	–	–	–	–	–	–	**Estra-1,3,5(10)-trien-17-one, 3,12-dihydroxy-, (12β)-**	17736-14-0	**NI29652**
270	158	157	144	145	170	211	271	**286**	100	61	38	32	28	25	23	19	**3.80**	18	22	3	–	–	–	–	–	–	–	–	–	–	**Estra-1,3,5(10)-trien-17-one, 3,14-dihydroxy-**	5949-46-2	**NI29653**
286	159	137	187	186	268	146	215	**286**	100	79	69	35	35	31	31	25		18	22	3	–	–	–	–	–	–	–	–	–	–	**Estra-1,3,5(10)-trien-17-one, 3,15-dihydroxy-, (15α)-**	2208-13-1	**NI29654**
286	159	146	186	187	145	287	133	**286**	100	75	46	44	29	26	21	21		18	22	3	–	–	–	–	–	–	–	–	–	–	**Estra-1,3,5(10)-trien-17-one, 3,15-dihydroxy-, (15α)-**	2208-13-1	**NI29655**
286	214	172	159	213	146	145	133	**286**	100	98	51	46	42	27	22	22		18	22	3	–	–	–	–	–	–	–	–	–	–	**Estra-1,3,5(10)-trien-17-one, 3,16-dihydroxy-**	18186-49-7	**NI29656**
213	286	214	172	133	287	146	159	**286**	100	61	38	18	16	12	12	11		18	22	3	–	–	–	–	–	–	–	–	–	–	**Estra-1,3,5(10)-trien-17-one, 3,16-dihydroxy-, (16α)-**	566-76-7	**NI29657**
115	141	129	155	286	212	213	226	**286**	100	95	78	70	32	24	23	21		18	22	3	–	–	–	–	–	–	–	–	–	–	**Methyl (7E,9E,15E)-3-hydroxyhepta-7,9,15-trien-11,13-diyn-1-oate**		**LI06200**
165	43	121	42	149	166	134	124	**286**	100	76	69	60	54	45	42	38	**1.70**	18	26	1	2	–	–	–	–	–	–	–	–	–	**1,2,3,4,5,6,7,8-Octahydro-1-(4-methoxybenzyl)-2,3-dimethylphthalazine**		**MS06562**
57	55	43	41	69	71	83	97	**286**	100	99	92	73	67	63	49	39	**3.80**	18	38	–	–	1	–	–	–	–	–	–	–	–	**1-Octadecanethiol**	2885-00-9	**NI29658**
57	71	157	43	41	113	55	56	**286**	100	82	77	53	37	31	25	20	**0.00**	18	38	2	–	–	–	–	–	–	–	–	–	–	**1,1-Dioctyloxyethane**		**IC03687**
57	71	43	63	85	55	45	83	**286**	100	68	67	51	47	47	40	37	**0.14**	18	38	2	–	–	–	–	–	–	–	–	–	–	**Ethanol, 2-(hexadecyloxy)-**	2136-71-2	**NI29659**
57	43	71	85	83	45	69	55	**286**	100	72	71	49	45	45	44	44	**0.00**	18	38	2	–	–	–	–	–	–	–	–	–	–	**Ethanol, 2-(hexadecyloxy)-**	2136-71-2	**DO01080**
57	43	71	55	63	85	45	41	**286**	100	80	70	51	48	47	46	44	**0.14**	18	38	2	–	–	–	–	–	–	–	–	–	–	**Ethanol, 2-(hexadecyloxy)-**	2136-71-2	**NI29660**
71	28	41	43	82	29	55	68	**286**	100	55	40	23	22	17	15	13	**0.00**	18	38	2	–	–	–	–	–	–	–	–	–	–	**Hexadecane, 1,1-dimethoxy-**	2791-29-9	**NI29661**
71	41	29	82	31	43	32	68	**286**	100	50	42	29	28	26	20	17	**0.00**	18	38	2	–	–	–	–	–	–	–	–	–	–	**Hexadecane, 1,1-dimethoxy-**	2791-29-9	**NI29662**
71	82	96	75	110	254	222	124	**286**	100	22	9	4	3	2	1	1	**0.00**	18	38	2	–	–	–	–	–	–	–	–	–	–	**Hexadecane, 1,1-dimethoxy-**	2791-29-9	**LI06201**
57	43	55	83	97	69	41	56	**286**	100	95	77	75	62	62	58	50	**0.00**	18	38	2	–	–	–	–	–	–	–	–	–	–	**1,2-Octadecanediol**		**DO01081**
105	286	77	181	154	51	287	127	**286**	100	68	60	56	25	18	16	13		19	14	1	2	–	–	–	–	–	–	–	–	–	**Carbazole, 3-(benzamido)-**	109693-64-3	**NI29663**
142	286	243	141	115	155	128	129	**286**	100	90	90	90	90	80	80	70		19	26	2	–	–	–	–	–	–	–	–	–	–	**1-Acetoxyheptadeca-8,10-diene-4,6-diyne**		**LI06202**
122	69	286	91	147	93	54	123	**286**	100	22	18	10	9	7	7	6		19	26	2	–	–	–	–	–	–	–	–	–	–	**Androsta-1,4-diene-3-one, 17-hydroxy-, (17β)-**		**LI06203**
41	286	55	91	67	79	81	77	**286**	100	57	54	46	43	40	37	37		19	26	2	–	–	–	–	–	–	–	–	–	–	**Androst-1-ene-3,11-dione, (5α)-**	54498-86-1	**NI29664**
28	122	244	286	109	107	121	41	**286**	100	99	55	46	37	31	30	29		19	26	2	–	–	–	–	–	–	–	–	–	–	**Androst-1-ene-3,17-dione**	21507-41-5	**MS06563**
122	41	79	55	244	109	67	91	**286**	100	90	53	53	52	46	45	39	**36.00**	19	26	2	–	–	–	–	–	–	–	–	–	–	**Androst-1-ene-3,17-dione, (5α)-**	571-40-4	**NI29665**
135	80	137	109	258	136	79	41	**286**	100	59	55	47	44	44	44	41	**19.41**	19	26	2	–	–	–	–	–	–	–	–	–	–	**Androst-4-ene-3,6-dione**	604-25-1	**NI29666**
135	136	80	137	109	79	258	41	**286**	100	75	60	57	47	44	43	41	**20.20**	19	26	2	–	–	–	–	–	–	–	–	–	–	**Androst-4-ene-3,6-dione**	604-25-1	**NI29667**
122	91	165	151	286	242	105	95	**286**	100	58	56	49	38	31	31	28		19	26	2	–	–	–	–	–	–	–	–	–	–	**Androst-4-ene-3,11-dione**	570-29-6	**NI29668**
286	124	244	41	91	79	148	107	**286**	100	65	60	41	40	40	39	35		19	26	2	–	–	–	–	–	–	–	–	–	–	**Androst-4-ene-3,17-dione**	63-05-8	**NI29669**
91	92	286	124	107	244	93	105	**286**	100	51	40	34	32	31	27	26		19	26	2	–	–	–	–	–	–	–	–	–	–	**Androst-4-ene-3,17-dione**	63-05-8	**WI01361**
124	79	91	148	286	109	107	105	**286**	100	90	73	68	67	60	59	47		19	26	2	–	–	–	–	–	–	–	–	–	–	**Androst-4-ene-3,17-dione**	63-05-8	**NI29670**
286	177	287	91	79	104	97	93	**286**	100	33	27	27	24	21	21	20		19	26	2	–	–	–	–	–	–	–	–	–	–	**Androst-5(6)-ene-3,17-dione**		**LI06204**
203	41	43	39	91	77	53	67	**286**	100	47	31	26	24	20	20	19	**4.87**	19	26	2	–	–	–	–	–	–	–	–	–	–	**1,3-Benzenediol, 2-[3-methyl-6-isopropenyl-2-cyclohexen-1-yl]-5-propyl-, (1R-trans)-**	24274-48-4	**NI29671**
81	287	165	135	109	121	107	285	**286**	100	58	46	38	20	15	11	10	**9.00**	19	26	2	–	–	–	–	–	–	–	–	–	–	**1,3-Benzenediol, 2-[3-methyl-6-isopropenyl-2-cyclohexen-1-yl]-5-propyl-, (1R-trans)-**	24274-48-4	**NI29672**
150	242	228	79	91	55	41	286	**286**	100	90	74	66	64	53	53	49		19	26	2	–	–	–	–	–	–	–	–	–	–	**6β,19-Cyclo-5α-androstane-3,17-dione**	5295-63-6	**NI29673**
110	41	55	91	286	79	124	163	**286**	100	72	70	65	62	58	51	48		19	26	2	–	–	–	–	–	–	–	–	–	–	**14,18-Cyclo-5α,14β-androstane-3,17-dione**	100024-29-1	**NI29674**
41	43	271	203	243	39	286	81	**286**	100	50	37	36	33	31	25	23		19	26	2	–	–	–	–	–	–	–	–	–	–	**6H-Dibenzo[b,d]pyran-1-ol, 6a,7,8,10a-tetrahydro-6,6,9-trimethyl-3-propyl-, (6aR-trans)-**	31262-37-0	**NI29675**
203	286	243	163	218	194	271		**286**	100	95	57	38	36	33	19			19	26	2	–	–	–	–	–	–	–	–	–	–	**6H-Dibenzo[b,d]pyran-1-ol, 6a,7,8,10a-tetrahydro-6,6,9-trimethyl-3-propyl-, (6aR-trans)-**	31262-37-0	**LI06205**
286	160	147	173	227	287	174	159	**286**	100	40	32	25	23	21	20	20		19	26	2	–	–	–	–	–	–	–	–	–	–	**Estra-1,3,5(10)-triene-3,17-diol, 1-methyl-, (17β)-**	3597-38-4	**NI29676**
213	286	160	133	159	228	145	107	**286**	100	93	93	67	63	37	31	31		19	26	2	–	–	–	–	–	–	–	–	–	–	**Estra-1,3,5(10)-triene-3,17-diol, 17-methyl-, (17β)-**	302-76-1	**NI29677**
286	160	186	200	199	173	159	287	**286**	100	52	46	28	27	26	23	20		19	26	2	–	–	–	–	–	–	–	–	–	–	**Estra-1,3,5(10)-trien-17-ol, 3-methoxy-, (8α,17β)-**	6570-46-3	**NI29678**
286	186	160	284	287	173	174	147	**286**	100	25	23	22	20	20	19	16		19	26	2	–	–	–	–	–	–	–	–	–	–	**Estra-1,3,5(10)-trien-17-ol, 3-methoxy-, (17β)-**	1035-77-4	**NI29680**
286	186	287	160	227	200	199	173	**286**	100	28	22	19	15	15	14	14		19	26	2	–	–	–	–	–	–	–	–	–	–	**Estra-1,3,5(10)-trien-17-ol, 3-methoxy-, (17β)-**	1035-77-4	**NI29679**
286	287	186	173	160	147	199	174	**286**	100	19	19	16	16	11	10	9		19	26	2	–	–	–	–	–	–	–	–	–	–	**Estra-1,3,5(10)-trien-17-ol, 3-methoxy-, (17β)-**	1035-77-4	**NI29681**
287	286	285	81	245	231	288	165	**286**	100	33	33	33	28	28	26	15		19	26	2	–	–	–	–	–	–	–	–	–	–	**3-Propyl-Δ^9-tetrahydrocannabinol**		**NI29682**

MASS TO CHARGE RATIOS	M.W.	INTENSITIES	Parent	C	H	O	N	S	F	Cl	Br	I	Si	P	B	X	COMPOUND NAME	CAS Reg No	No
286 287 252 143 126 253 165 121	**286**	100 22 10 5 5 4 3 3		20	14	–	–	1	–	–	–	–	–	–	–	–	**Benzo[c]thiophene, 1,3-diphenyl-**	16587-39-6	**NI29683**
286 287 252 288 284 143 126 253	**286**	100 23 11 7 7 6 5 4		20	14	–	–	1	–	–	–	–	–	–	–	–	**Benzo[c]thiophene, 1,3-diphenyl-**	16587-39-6	**AI01979**
286 285 284 282 287 283 141 252	**286**	100 50 39 24 23 20 19 16		20	14	–	–	1	–	–	–	–	–	–	–	–	**5,6-Dihydrodinaphtho[2,1;1′,2′]thiophene**		**AI01980**
77 209 105 286 152 51 210 76	**286**	100 87 68 40 36 32 13 12		20	14	2	–	–	–	–	–	–	–	–	–	–	**Benzene, 1,2-dibenzoyl-**	1159-86-0	**NI29684**
105 209 286 77 287 181 210 152	**286**	100 65 62 38 12 12 9 8		20	14	2	–	–	–	–	–	–	–	–	–	–	**Benzene, 1,4-dibenzoyl-**	3016-97-5	**NI29685**
286 178 181 287 257 152 176 177	**286**	100 50 34 22 19 18 16 13		20	14	2	–	–	–	–	–	–	–	–	–	–	**Benzo[b]benzo[3,4]cyclobuta[1,2-e][1,4]dioxin, 4b,10a-dihydro-4b-phenyl-**	53486-88-7	**NI29686**
286 105 209 181 77 287 257 178	**286**	100 46 45 26 22 22 19 18		20	14	2	–	–	–	–	–	–	–	–	–	–	**1,4-Benzodioxin, 2,3-diphenyl-**	75694-46-1	**NI29687**
286 239 257 287 268 120 115 267	**286**	100 37 25 23 20 20 17 16		20	14	2	–	–	–	–	–	–	–	–	–	–	**[1,1′-Binaphthalene]-2,2′-diol**	602-09-5	**NI29688**
286 287 119.5 115 257 192 239 144	**286**	100 22 22 19 16 15 14 14		20	14	2	–	–	–	–	–	–	–	–	–	–	**[1,1′-Binaphthalene]-2,2′-diol**	602-09-5	**IC03688**
268 239 269 286 78 252 240 237	**286**	100 61 33 30 28 16 15 15		20	14	2	–	–	–	–	–	–	–	–	–	–	**4,5-Dihydro-4,5-dihydroxybenzopyrene**		**NI29689**
229 228 286 257 226 239 224 237	**286**	100 64 56 40 40 38 12 9		20	14	2	–	–	–	–	–	–	–	–	–	–	**7-Formyl-12-hydroxymethylbenz[a]anthracene**		**MS06564**
286 239 228 226 229 285 255 257	**286**	100 98 95 93 86 67 38 34		20	14	2	–	–	–	–	–	–	–	–	–	–	**7-Hydroxymethyl-12-formylbenz[a]anthracene**		**MS06565**
286 255 226 113 287 224 256 28	**286**	100 91 45 26 18 18 17 16		20	14	2	–	–	–	–	–	–	–	–	–	–	**1-Triphenylenecarboxylic acid, methyl ester**	56909-16-1	**NI29690**
178 286 77 51 39 78 107 209	**286**	100 42 41 41 32 31 30 28		20	15	–	–	–	–	–	–	–	–	1	–	–	**Phosphine, diphenyl(phenylethynyl)-**	7608-17-5	**NI29691**
208 286 285 271 209 79 207 105	**286**	100 73 49 40 38 36 31 30		20	18	–	2	–	–	–	–	–	–	–	–	–	**Aniline, 2,5-dimethyl-N-(phenyl-2-pyridinylmethylene)-**	74432-14-7	**NI29692**
285 286 284 143 144 142	**286**	100 60 60 50 40 30		20	18	–	2	–	–	–	–	–	–	–	–	–	**8-Methyl-4-(3-quinolinyl)-8-methyl-1,4-dihydroquinoline**		**LI06206**
271 286 255 91 42 79 159 105	**286**	100 94 58 50 50 46 40 30		20	30	1	–	–	–	–	–	–	–	–	–	–	**4b,8-Dimethyl-8-hydroxymethyl-12-methylidene-2,3,4,4a,4b,5,6,7,8,10-decahydro-1H-2,10a-ethanophenanthrene**		**LI06207**
286 271 189 69 175 201 187 55	**286**	100 97 55 55 51 45 32 27		20	30	1	–	–	–	–	–	–	–	–	–	–	**Ferruginol**		**LI06209**
286 271 69 189 175 201 187 43	**286**	100 97 57 55 51 45 32 27		20	30	1	–	–	–	–	–	–	–	–	–	–	**Ferruginol**		**LI06208**
257 271 123 91 286 121 241 93	**286**	100 51 51 48 43 42 37 37		20	30	1	–	–	–	–	–	–	–	–	–	–	**1-Phenanthrenecarboxaldehyde, 7-vinyl-1,2,3,4,4a,4b,5,6,7,9,10,10a-dodecahydro-1,4a,7-trimethyl-, [1R-(1α,4aβ,4bα,7β,10aα)]-**	472-39-9	**NI29693**
117 271 286 173 159 272 185 254	**286**	100 91 40 38 28 19 18 16		20	30	1	–	–	–	–	–	–	–	–	–	–	**1-Phenanthrenemethanol, 1,2,3,4,4a,9,10,10a-octahydro-1,4a-dimethyl-7-isopropyl-, [1R-(1α,4aβ,10aα)]-**	3772-55-2	**NI29694**
271 159 253 286 173 211 272 185	**286**	100 53 52 36 33 27 22 21		20	30	1	–	–	–	–	–	–	–	–	–	–	**1-Phenanthrenemethanol, 1,2,3,4,4a,9,10,10a-octahydro-1,4a-dimethyl-7-isopropyl-, [1S-(1α,4aα,10aβ)]-**	24035-43-6	**NI29695**
271 175 201 286 272 189 69 41	**286**	100 65 41 39 26 25 18 18		20	30	1	–	–	–	–	–	–	–	–	–	–	**2-Phenanthrenol, 4b,5,6,7,8,8a,9,10-octahydro-4b,8,8-trimethyl-1-isopropyl-, (4bS-trans)-**	511-15-9	**NI29696**
271 175 286 201 272 189 69 41	**286**	100 66 38 38 24 22 16 16		20	30	1	–	–	–	–	–	–	–	–	–	–	**2-Phenanthrenol, 4b,5,6,7,8,8a,9,10-octahydro-4b,8,8-trimethyl-1-isopropyl-, (4bS-trans)-**	511-15-9	**NI29697**
286 105 119 41 91 55 81 79	**286**	100 82 74 73 68 66 61 51		20	30	1	–	–	–	–	–	–	–	–	–	–	**Podocarp-7-en-3-one, 13β-methyl-13-vinyl-**	7715-48-2	**NI29698**
69 55 41 105 91 255 119 107	**286**	100 92 91 76 69 68 68 59	**56.96**	20	30	1	–	–	–	–	–	–	–	–	–	–	**Retinol**	68-26-8	**NI29699**
91 105 119 77 145 268 79 93	**286**	100 75 73 68 59 56 50 48	**2.80**	20	30	1	–	–	–	–	–	–	–	–	–	–	**Retinol**	68-26-8	**WI01362**
119 91 69 41 105 55 255 43	**286**	100 97 95 93 89 77 76 71	**67.00**	20	30	1	–	–	–	–	–	–	–	–	–	–	**Retinol**	68-26-8	**NI29700**
147 55 203 81 41 91 134 162	**286**	100 90 89 81 80 79 77 73	**64.51**	20	30	1	–	–	–	–	–	–	–	–	–	–	**Sandaracopimara-7,15-dien-6-one**		**NI29701**
165 286 179 192 178 180 208 195	**286**	100 86 86 83 77 73 45 45		21	18	1	–	–	–	–	–	–	–	–	–	–	**2H-1-Benzopyran, 3,4-dihydro-2,2-diphenyl-**	10419-28-0	**MS06566**
286 192 179 165 195 180 178 208	**286**	100 50 42 42 40 32 32 30		21	18	1	–	–	–	–	–	–	–	–	–	–	**2H-1-Benzopyran, 3,4-dihydro-2,2-diphenyl-**	10419-28-0	**NI29702**
286 165 192 179 178 180 195 208	**286**	100 76 73 67 64 47 46 41		21	18	1	–	–	–	–	–	–	–	–	–	–	**2H-1-Benzopyran, 3,4-dihydro-2,2-diphenyl-**	10419-28-0	**NI29703**
181 286 195 192 182 208 152 91	**286**	100 42 27 25 22 19 18 15		21	18	1	–	–	–	–	–	–	–	–	–	–	**2H-1-Benzopyran, 3,4-dihydro-2,4-diphenyl-**		**MS06567**
215 229 202 241 257 286 258	**286**	100 99 93 35 28 25 24		21	18	1	–	–	–	–	–	–	–	–	–	–	**1,5-Bis(2-ethynylcyclohexenyl)-1,4-pentadiyn-3-one**		**LI06210**
141 286 115 90 142 128 287 89	**286**	100 44 26 20 16 14 11 11		21	18	1	–	–	–	–	–	–	–	–	–	–	**1(2H)-Naphthalenone, 3,4-dihydro-2-(1-naphthalenylmethyl)-**	55723-90-5	**NI29704**
142 129 141 270 128 115 286 271	**286**	100 85 76 68 44 38 19 17		21	18	1	–	–	–	–	–	–	–	–	–	–	**1(2H)-Naphthalenone, 3,4-dihydro-2-(2-naphthalenylmethyl)-**	55723-89-2	**NI29705**
105 167 77 103 165 286 51 152	**286**	100 51 34 23 21 13 13 9		21	18	1	–	–	–	–	–	–	–	–	–	–	**1-Propanone, 1,3,3-triphenyl-**	606-86-0	**NI29706**
131 132 286 41 43 117 91 115	**286**	100 34 27 23 22 17 9 8		21	34	–	–	–	–	–	–	–	–	–	–	–	**1H-Indene, 5-dodecyl-2,3-dihydro-**	55256-23-0	**WI01363**
41 136 135 67 95 81 55 286	**286**	100 84 83 82 58 57 52 51		21	34	–	–	–	–	–	–	–	–	–	–	–	**Perhydrodibenzo[a,i]fluorene**		**AI01982**
41 135 136 67 95 81 55 79	**286**	100 82 80 76 54 53 51 48	**49.20**	21	34	–	–	–	–	–	–	–	–	–	–	–	**Perhydrodibenzo[a,i]fluorene**		**AI01981**
271 286 176 105 161 272 91 106	**286**	100 66 50 35 24 22 22 21		21	34	–	–	–	–	–	–	–	–	–	–	–	**Pregn-11-ene, (5β)-**	6673-73-0	**NI29707**
257 41 55 271 93 109 81 67	**286**	100 44 32 24 23 21 21 21	**19.19**	21	34	–	–	–	–	–	–	–	–	–	–	–	**Pregn-14-ene, (5β)-**	54411-80-2	**NI29708**
122 271 286 81 107 95 55 121	**286**	100 89 66 57 53 47 42 41		21	34	–	–	–	–	–	–	–	–	–	–	–	**Pregn-17(20)-ene, (5α,17Z)-**	14964-36-4	**NI29709**
286 258 287 215 278 245 229 228	**286**	100 38 24 24 20 19 18 15		22	22	–	–	–	–	–	–	–	–	–	–	–	**Benzo[b]triphenylene, 1,2,3,4,10,11,12,13-octahydro-**	35281-24-4	**NI29710**
286 258 245 215 287 229 243 228	**286**	100 27 27 25 24 23 20 17		22	22	–	–	–	–	–	–	–	–	–	–	–	**Chrysene, 11-but-3-enyl-1,2,3,4-tetrahydro-**		**LI06211**
195 286 91 141 167 165 115 229	**286**	100 89 68 67 52 45 44 39		22	22	–	–	–	–	–	–	–	–	–	–	–	**Cycloheptane, 2-methylene-1,3-dibenzylidene-**		**MS06568**
177 179 178 152 151 203 202 286	**286**	100 92 92 25 18 14 14 8		22	22	–	–	–	–	–	–	–	–	–	–	–	**11-Cyclohex-3-enyl-9,10-ethanoanthracene**		**IC03689**
243 286 229 228 215 244 287 278	**286**	100 99 70 55 37 29 25 24		22	22	–	–	–	–	–	–	–	–	–	–	–	**Dibenz[a,h]anthracene, 1,2,3,4,4a,5,6,14b-octahydro-**	95138-20-8	**NI29711**
286 271 256 165 287 128 115 272	**286**	100 34 34 27 26 18 18 17		22	22	–	–	–	–	–	–	–	–	–	–	–	**Naphthalene, 2-methyl-1-(2,4,6-trimethylstyryl)-**		**LI06212**
286 91 256 131 115 229 271 165	**286**	100 24 23 14 14 13 11 11		22	22	–	–	–	–	–	–	–	–	–	–	–	**2,6-Octadien-4-yne, 2,7-dimethyl-3,6-diphenyl-**		**NI29712**

MASS TO CHARGE RATIOS								M.W.	INTENSITIES								Parent	C	H	O	N	S	F	Cl	Br	I	Si	P	B	X	COMPOUND NAME	CAS Reg No	No
165	167	166	152	168	243	51	115	**286**	100	97	51	22	14	13	13	10	**0.00**	22	22	–	–	–	–	–	–	–	–	–	–	–	Propane, 2-methyl-1,1,1-triphenyl-	29379-48-4	WI01364
57	127	89	104	232	230	282		**287**	100	38	32	17	4	2	0		**0.00**	6	10	2	1	2	5	–	–	–	–	–	–	–	S,S-Bis(allyloxy)-N-pentafluorosulphanyl sulphilimine		NI29713
241	181	91	66	30	224	90	74	**287**	100	33	31	31	29	28	28	28	**4.00**	7	5	8	5	–	–	–	–	–	–	–	–	–	Aniline, N-methyl-N,2,4,6-tetranitro-	479-45-8	LI06213
77	194	51	242	92	62	104	46	**287**	100	88	79	68	55	48	33	30	**0.00**	7	5	8	5	–	–	–	–	–	–	–	–	–	Aniline, N-methyl-N,2,4,6-tetranitro-	479-45-8	NI29714
87	109	145	79	58	142	112	88	**287**	100	18	16	15	15	14	11	11	**1.23**	8	18	4	1	2	–	–	–	–	–	1	–	–	Phosphorothioic acid, O,O-dimethyl S-[2-[[1-methyl-2-(methylamino)-2-oxoethyl]thio]ethyl] ester	2275-23-2	NI29715
289	287	129	291	208	128	274	127	**287**	100	46	46	31	24	23	20	13		9	7	–	1	–	–	–	2	–	–	–	–	–	1H-Indole, dibromomethyl-	65216-80-0	NI29716
64	48	77	41	204	51	135	203	**287**	100	56	50	32	30	26	25	17	**0.00**	10	11	4	2	2	–	–	–	–	–	–	–	–	2-Imidazolidinethione, 4-oxo-3-phenyl-5-sulphomethyl-		LI06214
208	287	289	168	141	64	209	103	**287**	100	63	62	25	22	20	17	17		11	6	–	5	–	–	–	1	–	–	–	–	–	1,3,6,9b-Tetraazaphenalene-4-carbonitrile, 2-(bromomethyl)-		NI29717
287	289	64	167	248	246	104	28	**287**	100	96	37	29	28	28	19	19		11	6	–	5	–	–	–	1	–	–	–	–	–	1,3,6,9b-Tetraazaphenalene-4-carbonitrile, 7-bromo-2-methyl-	37160-09-1	NI29718
287	289	167	64	104	288	290	208	**287**	100	99	36	22	16	14	13	13		11	6	–	5	–	–	–	1	–	–	–	–	–	1,3,6,9b-Tetraazaphenalene-4-carbonitrile, 9-bromo-2-methyl-	37160-10-4	NI29719
93	147	66	92	55	78	65	39	**287**	100	37	37	21	17	15	14	14	**2.03**	11	6	5	1	–	–	–	–	–	–	–	–	1	π-2-(2-Pyridyl)acetyltetracarbonylmanganese		MS06569
91	155	132	43	65	92	139	90	**287**	100	87	70	66	30	18	13	11	**9.75**	11	13	4	1	2	–	–	–	–	–	–	–	–	3H-1,2,3-Oxathiazole, 4,5-dimethyl-3-[(4-methylphenyl)sulphonyl]-, 2-oxide	52938-18-8	NI29720
135	42	199	28	76	58	104	29	**287**	100	85	62	41	33	33	31	27	**2.00**	11	13	6	1	1	–	–	–	–	–	–	–	–	1,3,5-Dioxazine, 5-[(4-methoxycarbonyl)phenylsulphonyl]dihydro-	56221-12-6	NI29721
68	220	69	287	272	67	232	202	**287**	100	51	44	31	20	16	10	3		11	14	6	1	–	–	–	–	–	–	1	–	–	1,3,2-Dioxaphosphorinane, 5,5-dimethyl-2-(4-nitrophenoxy)-, 2-oxide		NI29722
135	79	93	136	134	107	67	44	**287**	100	26	23	18	12	11	11	11	**0.00**	11	16	2	–	–	–	–	–	–	–	–	–	1	Tricyclo[3.3.1.1^{3,7}]decane-1-carboxylic acid, silver(1+) salt	75112-78-6	NI29723
59	55	28	126	54	27	72	42	**287**	100	96	68	52	35	32	25	25	**0.00**	11	17	6	3	–	–	–	–	–	–	–	–	–	2,4(1H,3H)-Pyrimidinedione, 1-(2-amino-2-deoxy-D-galactofuranosyl)-5-methyl-	56710-84-0	NI29724
175	287	203	102	75	129	259	101	**287**	100	77	55	52	42	40	25	25		12	5	6	3	–	–	–	–	–	–	–	–	–	7-Nitro-2,3-dihydroxyphenazine-1,4-quinone		MS06570
122	75	76	150	50	92	64	63	**287**	100	71	59	49	49	35	34	27	**0.30**	12	9	4	5	–	–	–	–	–	–	–	–	–	4-Nitro-N-(4-nitrophenylazo)aniline		IC03690
149	148	135	162	82	54	108	287	**287**	100	79	26	14	14	13	12	10		12	13	2	7	–	–	–	–	–	–	–	–	–	9-[3-(Uracil-1-yl)propyl]adenine		LI06215
125	135	211	134	131	91	27	208	**287**	100	66	60	38	37	33	33	31	**0.77**	12	14	3	1	1	–	1	–	–	–	–	–	–	Benzenepropanoic acid, α-[(aminocarbonyl)thio]-β-chloro-, ethyl ester	37073-56-6	NI29725
135	134	91	162	77	208	107	103	**287**	100	59	53	37	27	15	15	12	**0.00**	12	14	3	1	1	–	1	–	–	–	–	–	–	1,3-Oxathiolane-4-carboxylic acid, 2-imino-5-phenyl-, ethyl ester, hydrochloride	37073-50-0	NI29726
171	228	107	44	229	187	41	77	**287**	100	99	43	32	20	20	20	18	**0.20**	12	17	5	1	1	–	–	–	–	–	–	–	–	tert-Butyl 4-anisylsulphonylcarbamate	55125-01-4	NI29727
228	171	107	44	187	243	77	92	**287**	100	99	35	34	22	19	18	15	**0.30**	12	17	5	1	1	–	–	–	–	–	–	–	–	tert-Butyl 4-anisylsulphonylcarbamate	55125-01-4	LI06216
244	101	138	72	59	102	96	60	**287**	100	61	54	46	34	28	26	25	**12.99**	12	17	7	1	–	–	–	–	–	–	–	–	–	D-Altrosane, 1,6-anhydro-2,4-di-O-acetyl-3-acetamido-3-deoxy-	80970-56-5	NI29728
101	59	98	114	139	60	140	96	**287**	100	56	42	29	22	20	19	19	**0.29**	12	17	7	1	–	–	–	–	–	–	–	–	–	D-Altrosane, 1,6-anhydro-3,4-di-O-acetyl-2-acetamido-2-deoxy-	93250-09-0	NI29729
73	170	69	41	45	147	75	55	**287**	100	35	27	26	22	18	17	10	**3.97**	12	25	3	1	–	–	–	–	–	2	–	–	–	Glycine, N-(2-methyl-1-oxo-2-propenyl)-N-(trimethylsilyl)-, trimethylsilyl ester	55557-27-2	NI29730
73	69	102	170	75	45	272	41	**287**	100	82	58	51	27	26	25	20	**1.73**	12	25	3	1	–	–	–	–	–	2	–	–	–	Glycine, N-(1-oxo-2-butenyl)-N-(trimethylsilyl)-, trimethylsilyl ester	55557-28-3	NI29731
111	55	83	177	43	86	269	88	**287**	100	53	34	32	30	28	22	18	**4.00**	13	21	4	1	1	–	–	–	–	–	–	–	–	L-Cysteine, N-acetyl-S-(3-oxocycloheptyl)-, methyl ester		NI29732
208	194	209	43	41	181	55	67	**287**	100	98	81	63	57	38	32	29	**23.00**	13	21	4	1	1	–	–	–	–	–	–	–	–	8-Oxa-3-thiatricyclo[5.2.0.0^{7,9}]nonane, 5,7-dimethyl-1-(4-morpholino)-, 3,3-dioxide		NI29733
196	69	41	140	55	224	172	255	**287**	100	50	50	45	42	34	30	24	**20.00**	13	21	6	1	–	–	–	–	–	–	–	–	–	Maleic acid, [[5-(methoxycarbonyl)pentyl]amino]-, dimethyl ester		NI29734
57	43	81	97	85	139	98	232	**287**	100	34	23	20	11	3	3	1	**0.00**	13	21	6	1	–	–	–	–	–	–	–	–	–	Propanoic acid, 2,2-dimethyl-, 2-nitro-2-cyclohexen-1-yl ester, (R)-		DD00992
287	286	289	288	223	77	229	126	**287**	100	53	38	35	34	29	29	18		14	10	–	3	1	–	1	–	–	–	–	–	–	3H-1,3,4-Benzothiazepine-2-thione, 7-chloro-1,2-dihydro-5-phenyl-		MS06571
196	91	77	287	92	104	65	150	**287**	100	44	16	14	13	12	12	10		14	13	4	3	–	–	–	–	–	–	–	–	–	N-(2-Phenylethyl)-2,4-dinitroaniline		MS06572
57	187	214	41	186	194	90	58	**287**	100	25	22	22	16	14	12	10	**0.00**	14	16	2	1	–	3	–	–	–	–	–	–	–	1-Aziridinecarboxylic acid, 3-phenyl-2-(trifluoromethyl)-, tert-butyl ester, cis-		DD00993
57	187	59	41	231	58	188	117	**287**	100	40	24	12	8	5	4	4	**0.00**	14	16	2	1	–	3	–	–	–	–	–	–	–	1-Propene, 2-[(tert-butoxycarbonyl)amino]-1-phenyl-3,3,3-trifluoro-, (Z)-		DD00994
145	245	244	213	144	116	157	132	**287**	100	99	96	94	93	93	72	57	**28.00**	14	25	5	1	–	–	–	–	–	–	–	–	–	Dibutyl N-acetyliminodiacetate		NI29735
214	78	216	287	104	101	86	121	**287**	100	38	35	29	23	23	23	13		15	10	1	1	1	–	1	–	–	–	–	–	–	Thiazolium, 3-(4-chlorophenyl)-4-hydroxy-2-phenyl-, hydroxide, inner salt	26245-44-3	NI29736
287	269	289	252	271	288	234	151	**287**	100	50	33	21	20	20	18	18		15	10	3	1	–	–	1	–	–	–	–	–	–	Anthraquinone, 1-amino-4-chloro-2-(hydroxymethyl)-	56594-21-9	NI29737
146	91	65	51	77	287	147	89	**287**	100	40	35	28	22	19	13	9		15	13	3	1	1	–	–	–	–	–	–	–	–	3H-Indol-3-one, 1-(benzenesulphonyl)-1,2-dihydro-5-methyl-	83372-78-5	NI29738
132	77	91	51	65	287	50	76	**287**	100	60	50	35	21	18	16	14		15	13	3	1	1	–	–	–	–	–	–	–	–	3H-Indol-3-one, 1,2-dihydro-1-[(4-methylphenyl)sulphonyl]-	22394-27-0	NI29739
287	163	288	46	136	179	164	123	**287**	100	40	17	14	7	6	2	1		15	13	5	1	–	–	–	–	–	–	–	–	–	Benzeneacetic acid, 4-nitro-, 4-methoxyphenyl ester	53218-13-6	NI29740
166	167	287	122	46				**287**	100	8	6	2	1					15	13	5	1	–	–	–	–	–	–	–	–	–	Benzenemethanol, 3-methoxy-, 4-nitrobenzoate	53218-07-8	NI29741
166	167	122	287	46				**287**	100	8	5	3	1					15	13	5	1	–	–	–	–	–	–	–	–	–	Benzenemethanol, 4-methoxy-, 4-nitrobenzoate	53218-10-3	NI29742
219	253	256	254	287	268			**287**	100	80	57	46	7	7				15	14	1	3	–	–	1	–	–	–	–	–	–	3-Amino-6-chloro-3,4-dihydro-4-hydroxy-2-methyl-4-phenylquinazoline		LI06217
132	69	191	159	259	242	172	147	**287**	100	89	81	66	23	22	12	12	**4.00**	15	17	3	3	–	–	–	–	–	–	–	–	–	Propenoic acid, 2-azido-3-[4-(1,1-dimethylallyloxy)phenyl]-, methyl ester		MS06573
114	69	55	130	100	228	174	86	**287**	100	64	46	43	35	14	11	11	**0.00**	15	33	2	1	–	–	–	–	–	1	–	–	–	1-Butene, 1-[[[bis(2-ethoxyethyl)amino]methyl]dimethylsilyl]-, (Z)-		DD00995

MASS TO CHARGE RATIOS								M.W.	INTENSITIES								Parent	C	H	O	N	S	F	Cl	Br	I	Si	P	B	X	COMPOUND NAME	CAS Reg No	No
69	55	114	100	242	130	288	232	287	100	72	51	36	27	22	20	20	0.00	15	33	2	1	–	–	–	–	–	1	–	–	–	1-Butene, 3-[[[bis(2-ethoxyethyl)amino]methyl]dimethylsilyl]-		DD00996
245	287	247	108	246	289	69	288	287	100	58	45	35	28	21	14	11		16	14	–	1	1	–	1	–	–	–	–	–	–	1,5-Benzothiazepine, 2-methyl-4-(4-chlorophenyl)-2,3-dihydro-	78031-24-0	NI29743
252	77	105	140	107	142	253	125	287	100	65	63	62	41	27	26	26	5.00	16	14	2	1	–	–	1	–	–	–	–	–	–	Acetamide, 2-benzoyl-N-(2-chlorobenzyl)-		NI29744
104	91	118	155	105	287	132	78	287	100	42	22	20	20	8	8	8		16	17	2	1	1	–	–	–	–	–	–	–	–	Azetidine, 1-[(4-methylphenyl)sulphonyl]-2-phenyl-	38455-36-6	LI06218
104	91	118	155	105	132	287	78	287	100	40	24	20	20	12	11	10		16	17	2	1	1	–	–	–	–	–	–	–	–	Azetidine, 1-[(4-methylphenyl)sulphonyl]-2-phenyl-	38455-36-6	NI29745
194	152	184	229	213	287	168	77	287	100	92	66	59	58	54	41	36		16	17	2	1	1	–	–	–	–	–	–	–	–	Benzamide, N-isopropyl-2-phenylsulphinyl-	65838-75-7	NI29746
146	287	91	77	288	147	131	119	287	100	63	15	12	11	11	10	9		16	17	2	1	1	–	–	–	–	–	–	–	–	3H-3-Benzazepine, 6,7,8,9-tetrahydro-3-(phenylsulphonyl)-	20646-45-1	NI29747
146	287	91	77					287	100	58	19	19						16	17	2	1	1	–	–	–	–	–	–	–	–	4,5-Butylene-N-benzenesulphonylazepine		LI06219
167	91	168	123	122	80	288	76	287	100	11	10	3	3	3	2	2	0.00	16	17	2	1	1	–	–	–	–	–	–	–	–	L-Cysteine, S-(diphenylmethyl)-	5191-80-0	NI29748
167	165	152	166	168	44	42	198	287	100	97	37	30	29	19	18	15	1.14	16	17	2	1	1	–	–	–	–	–	–	–	–	L-Cysteine, S-(diphenylmethyl)-	5191-80-0	NI29749
132	131	287	286	155	222	223	139	287	100	33	10	9	4	3	1	1		16	17	2	1	1	–	–	–	–	–	–	–	–	Isoquinoline, 1,2,3,4-tetrahydro-2-[(4-methylphenyl)sulphonyl]-	20335-69-7	NI29750
132	131	286	287	155	222	223		287	100	33	8	7	4	2	1			16	17	2	1	1	–	–	–	–	–	–	–	–	Quinoline, 1,2,3,4-tetrahydro-1-[(4-methylphenyl)sulphonyl]-	24310-24-5	NI29751
145	214	144	287	18	130	240	168	287	100	90	56	54	36	35	21	21		16	17	4	1	–	–	–	–	–	–	–	–	–	3,4-Benzo-2-azabicyclo[3.2.0]hepta-3,6-diene-6,7-dicarboxylic acid, 1,5-dimethyl-, dimethyl ester		NI29752
287	214	243	203	269	185	215	244	287	100	52	46	41	33	27	24	19		16	17	4	1	–	–	–	–	–	–	–	–	–	Crinine, 6α-hydroxy-	80665-67-4	NI29753
270	288	49	271	286	51	287	289	287	100	85	28	24	18	18	13	12		16	17	4	1	–	–	–	–	–	–	–	–	–	Galanthan-1,2-diol, 3,12-didehydro-9,10-[methylenebis(oxy)]-, (1α,2β)-	476-28-8	NI29754
226	227	287	286	268	228	288	256	287	100	99	70	31	26	16	11	11		16	17	4	1	–	–	–	–	–	–	–	–	–	Galanthan-1,2-diol, 3,12-didehydro-9,10-[methylenebis(oxy)]-, (1α,2β)-	476-28-8	NI29755
287	184	214	185	213	168	243	169	287	100	97	75	67	43	24	23	20		16	17	4	1	–	–	–	–	–	–	–	–	–	Indole[2,3-b]furan-2-one, 3a,8-dihydro-3a,8,8a-trimethyl-3-exo-(methoxycarbonylmethylidene)-		NI29756
287	205	204	288	256	202	228	191	287	100	76	18	14	14	14	12	6		16	17	4	1	–	–	–	–	–	–	–	–	–	1H-Indole, 1-[7-methoxycarbonyl-5a-methyl-2-oxo-3,3a,4,5,5a,10-hexahydro-2H-furo[2,3;2,3]cyclopent-2-yl]-		IC03691
228	168	287	129	128	155	144	115	287	100	99	63	34	33	26	25	25		16	17	4	1	–	–	–	–	–	–	–	–	–	Maleic acid, 2-(2,3-dimethyl-3H-indol-3-yl)-, dimethyl ester	80265-22-1	NI29757
287	57	243	55	199	185	286	270	287	100	24	22	21	20	20	19	18		16	17	4	1	–	–	–	–	–	–	–	–	–	Pancracine	21416-14-8	LI06220
287	57	243	185	55	286	199	288	287	100	25	22	21	21	20	20	18		16	17	4	1	–	–	–	–	–	–	–	–	–	Pancracine	21416-14-8	NI29758
287	199	185	288	214	270	243	226	287	100	24	23	20	20	20	19	12		16	17	4	1	–	–	–	–	–	–	–	–	–	Pancracine, (3α)-	1354-81-0	NI29759
104	131	96	91	227	105	43	132	287	100	73	53	52	38	35	27	17	5.00	16	17	4	1	–	–	–	–	–	–	–	–	–	Pipermethystine		MS06574
128	86	41	43	42	29	27	85	287	100	90	85	80	45	43	42	41	6.51	16	33	3	1	–	–	–	–	–	–	–	–	–	N-[2-(Hexyloxy)ethyloxy]acetyl-N,N-dipropylamine		NI29760
74	43	256	104	57	41	56	55	287	100	15	12	11	11	10	8	8	0.81	16	33	3	1	–	–	–	–	–	–	–	–	–	Lauric diethanolamide		IC03692
174	97	112	176	128	244	175	159	287	100	89	50	40	38	30	13	13	5.00	16	34	1	1	–	–	–	–	–	–	1	–	–	1-(Dibutylamino)-2,2,3,4,4-pentamethylphosphetan 1-oxide		NI29761
244	174	97	159	188	176	141	112	287	100	25	25	15	9	9	8	6	4.00	16	34	1	1	–	–	–	–	–	–	1	–	–	1-(Dibutylamino)-2,2,3,4,4-pentamethylphosphetan 1-oxide		NI29762
287	286	178	77	165	230	202	89	287	100	45	28	22	14	11	11	11		17	13	–	5	–	–	–	–	–	–	–	–	–	2-Amino-6,7-diphenylpyrazolo[5,1-c][1,2,4]-triazine		NI29763
287	288	286	102	77	231	104	230	287	100	40	40	23	17	14	13	12		17	13	–	5	–	–	–	–	–	–	–	–	–	4-Phenyl-5-(5-phenylpyrazol-3-yl)-1,2,3-triazole		MS06575
94	122	121	83	82	42	95	138	287	100	82	71	60	46	37	35	31	27.64	17	21	3	1	–	–	–	–	–	–	–	–	–	Benzeneacetic acid, α-(hydroxymethyl)-, 8-methyl-8-azabicyclo[3.2.1]oct-6-en-3-yl ester	54725-48-3	NI29764
94	122	121	81	80	42	95	138	287	100	81	71	59	46	38	36	31	27.00	17	21	3	1	–	–	–	–	–	–	–	–	–	Benzeneacetic acid, α-(hydroxymethyl)-, 8-methyl-8-azabicyclo[3.2.1]oct-6-en-3-yl ester	54725-48-3	NI29765
105	77	287	51	42	106	41	182	287	100	44	12	11	11	9	8	7		17	21	3	1	–	–	–	–	–	–	–	–	–	3-(1-Benzoyl-3-allylpiperid-4-yl)propionic acid		NI29766
130	43	200	131	129	71	41		287	100	35	13	11	9	9	7		0.00	17	21	3	1	–	–	–	–	–	–	–	–	–	2-(1-Carboxyisopropyl)-1,2-dihydro-1-isobutyrylquinoline		LI06221
287	256	228	226	272	227	244	202	287	100	25	25	25	19	18	15	13		17	21	3	1	–	–	–	–	–	–	–	–	–	Crinan, 1-methoxy-, (1α)-	41928-92-1	NI29767
287	201	256	202	200	185	226	286	287	100	36	27	17	16	14	12	10		17	21	3	1	–	–	–	–	–	–	–	–	–	Crinan, 3-methoxy-, (3α)-	21051-68-3	NI29768
287	258	259	288	257	232	231	286	287	100	54	30	19	18	18	14	12		17	21	3	1	–	–	–	–	–	–	–	–	–	Crinan, 7-methoxy-	3660-64-8	NI29769
217	139	148	105	287	202	77	120	287	100	76	55	55	49	48	48	45		17	21	3	1	–	–	–	–	–	–	–	–	–	Cyclobutanone, 2-methoxy-4,4-dimethyl-3-(2-oxopyrrolidinyl)-2-phenyl-		NI29770
149	105	150	210	205	161	176	151	287	100	40	10	4	2	2	1	1	0.00	17	21	3	1	–	–	–	–	–	–	–	–	–	1,3-Dioxolane-2-heptanenitrile, α-methyl-δ-oxo-2-phenyl-	58422-90-5	NI29771
218	287	133	230	41	70	258	134	287	100	98	89	60	60	54	49	38		17	21	3	1	–	–	–	–	–	–	–	–	–	5-Indolizidinone, 6-(4-methoxyphenyl)-7-ethoxy-6,7-dehydro-		MS06576
287	44	42	45	65	43	70	39	287	100	69	58	42	35	31	27	27		17	21	3	1	–	–	–	–	–	–	–	–	–	Morphinan-3,6-diol, 4,5-epoxy-17-methyl-, (5α,6β)-	26626-12-0	NI29772
44	287	57	115	55	43	77	41	287	100	38	34	32	29	27	25	24		17	21	3	1	–	–	–	–	–	–	–	–	–	Morphinan-3,14-diol, 4,5-epoxy-17-methyl-, (5α)-	6801-26-9	NI29773
115	286	55	77	42	41	56	43	287	100	94	89	88	82	78	77	74	19.02	17	21	3	1	–	–	–	–	–	–	–	–	–	Morphinan-14-ol, 4,5-epoxy-3-methoxy-, (5α)-	55256-27-4	NI29774
287	242	243	286	268	288	266	269	287	100	84	70	52	50	20	18	16		17	21	3	1	–	–	–	–	–	–	–	–	–	Pluvine		LI06222
159	172	242	287	130	186	214	200	287	100	50	18	13	10	9	4	4		17	21	3	1	–	–	–	–	–	–	–	–	–	2-Quinolinehexanoic acid, 1,4-dihydro-4-oxo-	98752-01-3	NI29775
285	269	286	140	271	114	287	268	287	100	68	20	17	14	10	6	6		18	13	1	3	–	–	–	–	–	–	–	–	–	2-Oxo-3,3-bis-(1,2-diaminonaphthyl)indole		LI06223
285	269	286	140	270	114	268	115	287	100	53	19	15	13	8	5	5	2.00	18	13	1	3	–	–	–	–	–	–	–	–	–	Spiro[indoline-3,2′-[2H]naphth[1,2-d]imidazol]-2-one, 1′,3′-dihydro-	21943-54-4	NI29776
93	66	65	78	39	92	94	46.5	287	100	40	20	12	12	10	7	6	3.20	18	18	–	3	–	–	–	–	–	–	–	1	–	Trianilinoborane		IC03693
140	153	154	96	230	105	138	244	287	100	35	28	27	21	18	17	15	13.00	18	25	2	1	–	–	–	–	–	–	–	–	–	Acetone, 1-(octahydro-6-hydroxy-7-methyl-6-phenyl-5-indolizinyl)-, (5α,6β,7β,8aα)-(±)-	50906-93-9	NI29777

MASS TO CHARGE RATIOS								M.W.	INTENSITIES								Parent	C	H	O	N	S	F	Cl	Br	I	Si	P	B	X	COMPOUND NAME	CAS Reg No	No
287	231	189	91	160	117	105	79	**287**	100	71	68	63	49	35	35	31		18	25	2	1	–	–	–	–	–	–	–	–	–	**1H-Dicyclohept[e,g]isoindole-1,3(2H)-dione, 3a,3b,4,5,6,7,8,9,10,11,12,13,13a,13b-tetradecahydro-, (3aα,3bα,13aα,13bα)-**	51624-46-5	**NI29778**
140	212	153	287	96	244	70	154	**287**	100	57	31	28	28	26	25	24		18	25	2	1	–	–	–	–	–	–	–	–	–	**Furo[3,2-e]indolizin-2-ol, decahydro-2,4-dimethyl-3a-phenyl-, (2α,3aβ,4α,5aβ,9aβ)-(±)-**	50906-92-8	**NI29779**
72	144	115	102	43	41	73	127	**287**	100	27	13	10	8	8	7	6	**0.30**	18	25	2	1	–	–	–	–	–	–	–	–	–	**1-Isopropylamino-3-methyl-3-(1-naphthyloxy)butan-2-ol**		**MS06577**
287	231	189	91	159	39	118	105	**287**	100	70	67	63	50	43	37	37		18	25	2	1	–	–	–	–	–	–	–	–	–	**Maleimide-1,1′-bicycloheptenyl endo adduct**		**LI06224**
59	42	287	166	44	115	128	41	**287**	100	57	39	34	30	27	23	18		18	25	2	1	–	–	–	–	–	–	–	–	–	**14α-Morphinan-6α-ol, 3-methoxy-17-methyl-**	18067-41-9	**NI29780**
59	42	287	166	44	115	128	43	**287**	100	73	41	40	39	38	31	29		18	25	2	1	–	–	–	–	–	–	–	–	–	**14α-Morphinan-6β-ol, 3-methoxy-17-methyl-**	17948-43-5	**NI29781**
150	59	287	42	286	44	151	148	**287**	100	55	40	26	21	19	14	14		18	25	2	1	–	–	–	–	–	–	–	–	–	**Morphinan-4-ol, 3-methoxy-17-methyl-**	3327-79-5	**NI29782**
44	42	115	57	41	43	128	55	**287**	100	71	23	21	21	18	17	17	**15.01**	18	25	2	1	–	–	–	–	–	–	–	–	–	**Morphinan-14-ol, 3-methoxy-17-methyl-**	1639-74-3	**NI29784**
230	270	44	212	213	187	42	115	**287**	100	84	49	37	32	30	29	18	**0.00**	18	25	2	1	–	–	–	–	–	–	–	–	–	**Morphinan-14-ol, 3-methoxy-17-methyl-**	1639-74-3	**NI29783**
196	287	91	197	65	288	168	93	**287**	100	49	27	16	12	11	10	10		19	17	–	3	–	–	–	–	–	–	–	–	–	**3,6-Carbazolediamine, 9-benzyl-**	84892-75-1	**NI29785**
96	134	162	106	176	120	202	287	**287**	100	70	49	40	38	38	36	29		19	29	1	1	–	–	–	–	–	–	–	–	–	**1-Azaspiro[5.5]undecan-8-ol, 7-(1,3-butadienyl)-2-(4-pentynyl)-, [6R-[6α(R*),7β(Z),8α]]-**	63983-64-2	**NI29786**
98	99	218	85	219	131	84	69	**287**	100	8	7	5	3	3	3	2	**0.90**	19	29	1	1	–	–	–	–	–	–	–	–	–	**1-Piperidinepropanol, α-cyclopentyl-α-phenyl-**	77-39-4	**NI29787**
288	286	204	210	270				**287**	100	50	36	8	3				**0.00**	19	29	1	1	–	–	–	–	–	–	–	–	–	**1-Pyrrolidinepropanol, α-cyclohexyl-α-phenyl-**	77-37-2	**NI29788**
84	55	42	204	85	202	205	77	**287**	100	12	11	8	8	7	6	6	**0.39**	19	29	1	1	–	–	–	–	–	–	–	–	–	**1-Pyrrolidinepropanol, α-cyclohexyl-α-phenyl-**	77-37-2	**NI29789**
242	222	83	85	243	122	287	286	**287**	100	50	40	27	21	21	12	11		19	29	1	1	–	–	–	–	–	–	–	–	–	**Pyrrolo[1,2-a]quinoline-1-ethanol, dodecahydro-6-(2-penten-4-ynyl)-, [1S-[1α,3aβ,5aα,6α(Z),9aα]]-**	55893-12-4	**NI29790**
77	105	91	122	185	65	79	74	**287**	100	95	62	57	27	24	18	17	**0.00**	20	17	1	1	–	–	–	–	–	–	–	–	–	**Acetophenone, 2-[1-benzyl-2(1H)-pyridylidene]-**	14759-36-5	**NI29791**
167	165	168	152	77	166	169	120	**287**	100	76	69	47	47	44	39	31	**13.21**	20	17	1	1	–	–	–	–	–	–	–	–	–	**Benzeneacetamide, N,α-diphenyl-**	4695-14-1	**NI29792**
168	169	167	51	91	77	170	65	**287**	100	72	63	63	55	46	39	34	**11.71**	20	17	1	1	–	–	–	–	–	–	–	–	–	**Benzeneacetamide, N,N-diphenyl-**	33675-70-6	**NI29793**
169	91	168	167	287	51	65	77	**287**	100	48	16	16	15	15	13	12		20	17	1	1	–	–	–	–	–	–	–	–	–	**Benzeneacetamide, N,N-diphenyl-**	33675-70-6	**MS06578**
244	243	210	167	245	258	242	241	**287**	100	72	53	48	20	15	15	12	**3.40**	21	21	–	1	–	–	–	–	–	–	–	–	–	**Ethylamine, N-trityl-**	7370-34-5	**NI29794**
244	243	165	210	167	166	245	242	**287**	100	72	66	53	48	29	20	15	**3.40**	21	21	–	1	–	–	–	–	–	–	–	–	–	**Ethylamine, N-trityl-**	7370-34-5	**MS06579**
287	96	215	286	288	243	229	230	**287**	100	62	48	44	30	26	23	20		21	21	–	1	–	–	–	–	–	–	–	–	–	**Piperidine, 4-(5H-dibenzo[a,d]cyclohepten-5-ylidene)-1-methyl-**	129-03-3	**NI29795**
285	194	207	270	202	286	165	200	**287**	100	68	48	36	35	31	24	23	**7.30**	21	21	–	1	–	–	–	–	–	–	–	–	–	**Pyridine, 2,5-dimethyl-3,4-dibenzyl-**	50610-06-5	**NI29796**
287	286	288	272	284	208	195	194	**287**	100	94	33	26	19	16	15	13		21	21	–	1	–	–	–	–	–	–	–	–	–	**Pyridine, 2,5-dimethyl-4,6-dibenzyl-**		**NI29797**
210	287	196	91	92	65	36	211	**287**	100	80	71	66	58	46	30	21		21	21	–	1	–	–	–	–	–	–	–	–	–	**Tribenzylamine**	620-40-6	**IC03695**
91	210	196	287	65	92	211	181	**287**	100	24	20	18	18	12	4	4		21	21	–	1	–	–	–	–	–	–	–	–	–	**Tribenzylamine**	620-40-6	**NI29798**
91	210	287	196	92	65	288	211	**287**	100	30	21	20	13	9	5	5		21	21	–	1	–	–	–	–	–	–	–	–	–	**Tribenzylamine**	620-40-6	**IC03694**
69	253	36	96	35	252	237	65	**288**	100	99	80	60	50	40	40	30	**0.00**	4	1	1	2	–	6	2	–	–	–	–	1	–	**2,2-Dichloro-4,6-bis(trifluoromethyl)-1-oxa-3-azonia-5-aza-2-borata-3,5-cyclohexadiene**		**NI29799**
63	161	163	65	64	83	130	48	**288**	100	89	59	29	27	13	10	10	**0.00**	4	7	4	–	2	–	3	–	–	–	–	–	–	**3,3,5-Trichloro-2,4-dithiahexane 2,2,4,4-tetroxide**		**MS06580**
69	119	131	31	181	100	93	50	**288**	100	16	8	7	6	4	3	2	**0.00**	5	–	–	–	–	12	–	–	–	–	–	–	–	**Dodecafluoroisopentane**		**AI01983**
69	119	169	31	131	100	181	50	**288**	100	30	10	9	7	7	3	3	**0.03**	5	–	–	–	–	12	–	–	–	–	–	–	–	**Dodecafluoropentane**		**AI01984**
109	183	181	219	217	111	51	288	**288**	100	96	96	93	80	79	46	3		6	6	–	–	–	–	6	–	–	–	–	–	–	**Cyclohexane, 1,2,3,4,5,6-hexachloro-**		**LI06225**
181	183	109	217	215	111	219	252	**288**	100	99	80	78	63	59	51	30	**3.00**	6	6	–	–	–	–	6	–	–	–	–	–	–	**Cyclohexane, 1,2,3,4,5,6-hexachloro-**		**LI06226**
219	181	183	217	109	111	221	187	**288**	100	91	90	81	78	46	45	35	**0.75**	6	6	–	–	–	–	6	–	–	–	–	–	–	**Cyclohexane, 1,2,3,4,5,6-hexachloro-, (1α,2α,3α,4β,5α,6β)-**	319-86-8	**AI01985**
181	183	219	111	109	217	51	218	**288**	100	93	70	66	59	55	48	47	**0.00**	6	6	–	–	–	–	6	–	–	–	–	–	–	**Cyclohexane, 1,2,3,4,5,6-hexachloro-, (1α,2α,3α,4β,5α,6β)-**	319-86-8	**NI29801**
109	111	181	183	219	51	217	85	**288**	100	84	79	77	67	58	53	47	**0.00**	6	6	–	–	–	–	6	–	–	–	–	–	–	**Cyclohexane, 1,2,3,4,5,6-hexachloro-, (1α,2α,3α,4β,5α,6β)-**	319-86-8	**NI29800**
219	181	183	109	217	111	221	36	**288**	100	97	90	85	78	68	47	44		6	6	–	–	–	–	6	–	–	–	–	–	–	**Cyclohexane, 1,2,3,4,5,6-hexachloro-, (1α,2α,3α,4β,5β,6β)-**	6108-10-7	**NI29802**
181	183	219	109	217	111	221	185	**288**	100	93	77	63	62	58	35	29	**1.90**	6	6	–	–	–	–	6	–	–	–	–	–	–	**Cyclohexane, 1,2,3,4,5,6-hexachloro-, (1α,2α,3β,4α,5α,6β)-**	58-89-9	**NI29803**
219	181	183	109	217	111	221	185	**288**	100	99	94	87	82	66	45	29	**4.07**	6	6	–	–	–	–	6	–	–	–	–	–	–	**Cyclohexane, 1,2,3,4,5,6-hexachloro-, (1α,2α,3β,4α,5α,6β)-**	58-89-9	**AI01986**
181	183	109	51	111	219	50	217	**288**	100	98	98	91	88	72	67	56	**0.00**	6	6	–	–	–	–	6	–	–	–	–	–	–	**Cyclohexane, 1,2,3,4,5,6-hexachloro-, (1α,2α,3β,4α,5α,6β)-**	58-89-9	**NI29804**
181	183	219	217	43	111	109	221	**288**	100	88	84	69	49	45	44	41	**0.00**	6	6	–	–	–	–	6	–	–	–	–	–	–	**Cyclohexane, 1,2,3,4,5,6-hexachloro-, (1α,2α,3β,4α,5β,6β)-**	319-84-6	**MS06581**
181	183	219	109	217	111	51	221	**288**	100	96	89	76	71	51	45	44	**0.58**	6	6	–	–	–	–	6	–	–	–	–	–	–	**Cyclohexane, 1,2,3,4,5,6-hexachloro-, (1α,2α,3β,4α,5β,6β)-**	319-84-6	**AI01987**
183	181	109	111	219	51	217	50	**288**	100	91	86	84	75	71	61	51	**0.00**	6	6	–	–	–	–	6	–	–	–	–	–	–	**Cyclohexane, 1,2,3,4,5,6-hexachloro-, (1α,2α,3β,4α,5β,6β)-**	319-84-6	**NI29805**
219	217	183	181	221	147	218	185	**288**	100	80	70	70	44	37	29	23	**0.00**	6	6	–	–	–	–	6	–	–	–	–	–	–	**Cyclohexane, 1,2,3,4,5,6-hexachloro-, (1α,2β,3α,4β,5α,6β)-**	319-85-7	**NI29806**
109	219	181	183	217	111	221	185	**288**	100	95	88	85	78	67	46	31	**2.60**	6	6	–	–	–	–	6	–	–	–	–	–	–	**Cyclohexane, 1,2,3,4,5,6-hexachloro-, (1α,2β,3α,4β,5α,6β)-**	319-85-7	**AI01988**
109	183	181	111	51	219	85	217	**288**	100	88	88	82	78	66	57	49	**0.00**	6	6	–	–	–	–	6	–	–	–	–	–	–	**Cyclohexane, 1,2,3,4,5,6-hexachloro-, (1α,2β,3α,4β,5α,6β)-**	319-85-7	**NI29807**
288	290	287	289	286	89	285	180	**288**	100	80	78	60	54	34	30	25		9	5	1	–	–	–	–	1	–	–	–	–	1	**Benzo[b]selenophene-2-carboxaldehyde, 3-bromo-**	26581-53-3	**NI29808**
102	211	209	75	101	76	290	130	**288**	100	88	87	32	31	29	26	25	**12.54**	9	6	1	–	–	–	–	2	–	–	–	–	–	**1H-Inden-1-one, 2,2-dibromo-2,3-dihydro-**	7749-02-2	**NI29809**

MASS TO CHARGE RATIOS								M.W.	INTENSITIES								Parent	C	H	O	N	S	F	Cl	Br	I	Si	P	B	X	COMPOUND NAME	CAS Reg No	No
160	132	78	131	51	79	52	159	288	100	80	70	57	40	22	16	8	0.00	9	9	1	2	–	–	–	–	1	–	–	–	–	4H-Pyrido[1,2-a]pyrimidin-4-one, 2-methyl-, monohydriodide	50993-69-6	NI29810
57	29	41	103	97	27	65	231	288	100	57	44	35	31	25	23	21	1.80	9	21	2	–	3	–	–	–	–	–	1	–	–	Phosphorodithioic acid, S-[(tert-butylthio)methyl] O,O-diethyl ester	13071-79-9	NI29811
57	231	41	103	97	153	65	125	288	100	43	22	18	17	16	13	12	5.90	9	21	2	–	3	–	–	–	–	–	1	–	–	Phosphorodithioic acid, S-[(tert-butylthio)methyl] O,O-diethyl ester	13071-79-9	NI29812
152	159	125	261	177	233	124	97	288	100	85	52	42	32	31	25	25	21.00	9	22	6	–	–	–	–	–	–	–	2	–	–	Tetraethyl methylenediphosphonate		MS06582
106	91	288	77	64	44	51	290	288	100	56	45	35	15	15	13	12		10	6	2	–	–	5	1	–	–	–	–	–	–	1,1,1,3,3-Pentafluoro-3-chloro-2-propyl benzoate	10315-84-1	LI06227
106	91	288	77	122	158	136	78	288	100	55	44	34	9	6	6	4		10	6	2	–	–	5	1	–	–	–	–	–	–	1,1,1,3,3-Pentafluoro-3-chloro-2-propyl benzoate	10315-84-1	NI29813
288	290	253	71	45	255	100	207	288	100	98	98	92	73	57	47	36		10	6	2	2	1	–	2	–	–	–	–	–	–	Thiazole, 4-(chloromethyl)-2-(4-chloro-3-nitrophenyl)-	33451-02-4	NI29814
290	127	126	288	162	97	91	56	288	100	99	97	85	70	42	38	36		10	7	–	–	–	–	3	–	–	–	–	–	1	Ferrocene, 1,2,3-trichloro-	33306-52-4	LI06228
127	290	126	288	162	97	91	56	288	100	98	95	82	65	40	35	35		10	7	–	–	–	–	3	–	–	–	–	–	1	Ferrocene, 1,2,3-trichloro-	33306-52-4	NI29816
127	126	290	288	162	97	91	56	288	100	98	96	83	66	38	35	35		10	7	–	–	–	–	3	–	–	–	–	–	1	Ferrocene, 1,2,3-trichloro-	33306-52-4	NI29815
225	290	255	199	190	164	160		288	100	53	19	14	11	8	6		0.00	10	8	–	–	–	–	2	–	–	–	–	–	1	Zirconium bis(π-cyclopentadienyl)dichloride		LI06229
288	115	131	208	103	75	126	224	288	100	39	35	21	17	17	16	13		10	8	6	–	2	–	–	–	–	–	–	–	–	1,5-Disulphonaphthalene		IC03696
76	183	185	155	85	241	75	157	288	100	99	98	97	97	96	96	95	52.99	10	9	1	–	2	–	–	1	–	–	–	–	–	2-Propene(dithioic) acid, 3-(4-bromophenyl)-3-hydroxy-, methyl ester	56247-41-7	NI29817
130	288	129	257	102	41	287	289	288	100	89	60	52	34	13	12	11		10	9	2	–	–	–	–	–	1	–	–	–	–	Propenoic acid, (4-iodophenyl)-, methyl ester	101364-82-3	NI29818
129	116	115	51	290	128	44	77	288	100	93	80	54	51	49	34	33	28.00	10	10	–	–	–	–	–	2	–	–	–	–	–	Benzene, [bromo(1-bromocyclopropyl)methyl]-	41893-70-3	MS06583
44	91	155	65	43	270			288	100	91	55	18	18	6			0.00	10	12	6	2	1	–	–	–	–	–	–	–	–	5-(4-Toluenesulphonoxy)hydantoic acid		LI06230
41	67	103	54	82	56	69	55	288	100	60	54	43	38	38	35	35	0.21	10	13	–	–	–	4	–	1	–	–	–	–	–	1,7-Decadiene, 10-bromo-9,9,10,10-tetrafluoro-	74792-99-7	NI29819
241	195	70	125	167	97	225	226	288	100	66	66	54	49	49	44	41	8.00	10	16	6	4	–	–	–	–	–	–	–	–	–	2,5-Piperazinediacetamide, N,N′-dihydroxy-N,N′-dimethyl-3,6-dioxo-	69833-33-6	NI29820
290	288	218	292	254	256	220	289	288	100	98	67	37	36	24	24	16		12	7	–	–	1	–	3	–	–	–	–	–	–	Benzene, 2,4-dichloro-1-[(4-chlorophenyl)thio]-	55759-88-1	NI29821
288	242	28	18	51	214	44	142	288	100	69	39	30	24	23	22	17		12	8	5	4	–	–	–	–	–	–	–	–	–	1-(2-Furyl)-1-(2-amino-1,3,4-oxadiazol-5-yl)-2-(5-nitro-2-furyl)ethylene		MS06584
186	204	55	72	144	158	288	103	288	100	89	86	46	39	35	31	21		12	9	5	–	–	–	–	–	–	–	–	–	1	1-(Carboxyisopropenyl)cyclopentadienylmanganese tricarbonyl		LI06231
184	158	77	182	288	156	78	183	288	100	92	73	50	48	45	43	34		12	13	1	–	–	–	1	–	–	–	–	–	1	1-Cyclohexen-3-ol, 1-chloro-4-(phenylseleno)-, (3β,4β)-		DD00997
69	75	28	144	29	84	85	31	288	100	66	52	44	44	36	22	19	0.10	12	16	8	–	–	–	–	–	–	–	–	–	–	Acetic acid, 2,2′-(3,6-dimethoxy-1,4-dioxane-2,5-diylidene)bis-, dimethyl ester	61141-93-3	NI29822
116	141	187	128	127	230	229	179	288	100	52	41	30	28	20	18	13	0.50	12	16	8	–	–	–	–	–	–	–	–	–	–	β-D-Glucopyranose, 2,3,4-tri-O-acetyl-1,6-anhydro-		NI29823
43	81	98	115	102	70	69	112	288	100	10	8	6	4	4	4	3	0.00	12	16	8	–	–	–	–	–	–	–	–	–	–	D-Glucopyranose, 2,3,4-tri-O-acetyl-1,6-anhydro-		NI29824
81	98	115	71	70	69	157	102	288	100	81	58	51	47	47	45	35	0.00	12	16	8	–	–	–	–	–	–	–	–	–	–	β-D-Talopyranose, 1,6-anhydro-, triacetate	14661-16-6	NI29825
15	224	169	257	196	59	113	256	288	100	74	72	71	43	41	38	35	0.18	12	16	8	–	–	–	–	–	–	–	–	–	–	1,2,3,4-Tetramethoxycarbonylcyclobutane		MS06585
75	73	215	129	273	131	141	103	288	100	87	72	42	37	19	15	14	1.00	12	20	6	–	–	–	–	–	–	1	–	–	–	D-Glucofuranurono-6,3-lactone, 1,2-O-isopropylidene-5-O-(trimethylsilyl)-		NI29826
57	116	115	99	58	87	86	114	288	100	74	67	67	66	57	56	44	3.50	12	20	6	2	–	–	–	–	–	–	–	–	–	7,13-Dimethyl-1,4,10-trioxa-7,13-diazacyclopentadecane-2,6,14-trione		MS06586
57	116	115	99	58	86	114	245	288	100	74	67	67	66	56	44	39	3.53	12	20	6	2	–	–	–	–	–	–	–	–	–	1,4,10-Trioxa-7,13-diazacyclopentadecane-2,6,14-trione, 7,13-dimethyl-	87989-14-8	NI29827
137	273	152	42	84	288	151	41	288	100	77	56	51	48	41	39	34		12	21	4	2	–	–	–	–	–	–	1	–	–	Phosphoric acid, diethyl 6-methyl-2-isopropyl-4-pyrimidinyl ester	962-58-3	NI29828
137	273	288	134	152	151	260	29	288	100	95	63	34	32	32	27	22		12	21	4	2	–	–	–	–	–	–	1	–	–	Phosphoric acid, diethyl 6-methyl-2-isopropyl-4-pyrimidinyl ester	962-58-3	NI29829
175	231	147	149	121	120	205	261	288	100	90	32	24	21	12	7	2	0.10	12	24	–	–	–	–	–	–	–	–	–	–	1	Dibutyldivinyltin		LI06232
73	75	147	45	82	54	74	53	288	100	37	33	25	16	16	9	8	0.00	12	24	4	–	–	–	–	–	–	2	–	–	–	2-Hexenedioic acid, bis(trimethylsilyl) ester, (E)-	55494-10-5	NI29830
73	75	147	45	273	82	54	74	288	100	46	41	20	14	14	13	11	0.00	12	24	4	–	–	–	–	–	–	2	–	–	–	2-Hexenedioic acid, bis(trimethylsilyl) ester, (E)-	55494-10-5	NI29831
73	147	75	82	45	54	74	273	288	100	27	24	14	14	12	9	6	0.38	12	24	4	–	–	–	–	–	–	2	–	–	–	3-Hexenedioic acid, bis(trimethylsilyl) ester, (E)-		NI29832
73	147	75	82	45	273	244	231	288	100	38	28	24	15	14	13	11	0.75	12	24	4	–	–	–	–	–	–	2	–	–	–	2-Pentenedioic acid, 2-methyl-, bis(trimethylsilyl) ester		NI29833
73	147	82	75	273	244	231	74	288	100	99	77	69	31	26	24	23	0.00	12	24	4	–	–	–	–	–	–	2	–	–	–	2-Pentenedioic acid, 2-methyl-, bis(trimethylsilyl) ester, (E)-		NI29834
147	73	82	75	109	244	183	273	288	100	99	61	47	27	21	21	18	0.00	12	24	4	–	–	–	–	–	–	2	–	–	–	2-Pentenedioic acid, 2-methyl-, bis(trimethylsilyl) ester, (Z)-		NI29835
73	147	75	198	82	231	273	109	288	100	69	39	29	29	24	20	16	0.00	12	24	4	–	–	–	–	–	–	2	–	–	–	2-Pentenedioic acid, 3-methyl-, bis(trimethylsilyl) ester	55887-63-3	MS06587
73	147	75	45	82	198	74	109	288	100	48	25	18	16	10	10	8	0.00	12	24	4	–	–	–	–	–	–	2	–	–	–	2-Pentenedioic acid, 3-methyl-, bis(trimethylsilyl) ester	55887-63-3	NI29836
73	147	75	82	45	198	231	109	288	100	49	47	32	26	18	14	13	0.79	12	24	4	–	–	–	–	–	–	2	–	–	–	2-Pentenedioic acid, 3-methyl-, bis(trimethylsilyl) ester	55887-63-3	NI29837
147	73	273	198	75	82	183	109	288	100	97	42	39	37	36	27	26	2.00	12	24	4	–	–	–	–	–	–	2	–	–	–	2-Pentenedioic acid, 3-methyl-, bis(trimethylsilyl) ester, (E)-	55125-08-1	NI29838
73	147	75	198	82	231	273	109	288	100	69	39	29	29	24	20	16	0.00	12	24	4	–	–	–	–	–	–	2	–	–	–	2-Pentenedioic acid, 3-methyl-, bis(trimethylsilyl) ester, (Z)-	55125-09-2	NI29839
73	75	89	109	45	139	55	181	288	100	49	47	43	43	35	28	25	16.10	12	24	4	2	–	–	–	–	–	1	–	–	–	Heptanoic acid, 4,6-dioxo-, dimethoxime, trimethylsilyl ester	85863-67-8	NI29840
273	185	73	129	59	187	113	45	288	100	99	88	18	17	14	9	8	0.00	12	32	–	–	–	–	–	–	–	4	–	–	–	1,1,3,3,5,5,7,7-Octamethyl-1,3,5,7-tetrasilacyclooctane		LI06233
273	185	97	73	274	186	275	129	288	100	99	99	88	32	23	19	18	0.00	12	32	–	–	–	–	–	–	–	4	–	–	–	1,1,3,3,5,5,7,7-Octamethyl-1,3,5,7-tetrasilacyclooctane		MS06588
175	286	202	75	258	90	288		288	100	61	41	24	17	17	15			13	8	6	2	–	–	–	–	–	–	–	–	–	7-Carboxy-2,3-dihydroxyphenazine-1,4-quinone		MS06589
288	260	186	75	130	289	175	174	288	100	77	45	29	27	26	26	25		13	8	6	2	–	–	–	–	–	–	–	–	–	7-Carboxy-1,2,3,4-tetrahydroxyphenazine		MS06590
232	141	260	234	56	196	115	91	288	100	58	40	33	29	27	21	16	11.99	13	9	2	–	–	–	1	–	–	–	–	–	1	Iron, (4-chlorophenyl)dicarbonyl-π-cyclopentadienyl-	12282-67-6	NI29841
147	288	77	53	51	148	52	224	288	100	19	19	12	12	10	10	9		13	12	2	4	1	–	–	–	–	–	–	–	–	Pyrazolo[5,1-c]-as-triazine, 4,6-dihydro-3-methyl-4-methylene-6-(phenylsulphonyl)-	6849-75-8	NI29842
56	174	232	176	105	91	204	148	288	100	99	71	59	30	29	28	24	0.00	13	12	4	–	–	–	–	–	–	–	–	–	1	Iron, tricarbonyl-C,5,6,C-η-(2,3-dimethyl-5,6-dimethylidene-7-oxabicyclo[2.2.1]hept-2-ene)-, (1R,4S,5R,6S)-		DD00998

MASS TO CHARGE RATIOS								M.W.	INTENSITIES								Parent	C	H	O	N	S	F	Cl	Br	I	Si	P	B	X	COMPOUND NAME	CAS Reg No	No
204	202	232	148	174	189	176	108	**288**	100	96	78	75	62	57	30	29	**0.00**	13	12	4	–	–	–	–	–	–	–	–	–	1	Iron, tricarbonyl-C,5,6,C-η-(2-endo-methyl-3,5,6-trimethylidene-7-oxabicyclo[2.2.1]heptane)-, (1RS,2SR,4SR,5RS,6SR)-		DD00999
186	134	131	129	103	232	151	204	**288**	100	60	55	36	36	30	30	24	**3.00**	13	12	4	–	–	–	–	–	–	–	–	–	1	Iron, tricarbonyl(α-methyl-1,3,5,7-cyclooctatetraene-1-methanol)-	11065-38-6	MS06591
194	168	259	91	142	103	288	260	**288**	100	99	42	32	25	24	22	21		13	13	1	–	–	–	–	–	–	–	–	–	1	Rhodium, (η^4-2-formylnorbornadiene)(η^5-cyclopentadienyl)-	77906-37-7	NI29843
194	168	259	288	103	142	91	192	**288**	100	87	80	79	35	30	29	22		13	13	1	–	–	–	–	–	–	–	–	–	1	Rhodium, (η^4-norbornadiene)(η^5-formylcyclopentadienyl)-	82512-29-6	NI29844
204	55	93	148	56	205	119	84	**288**	100	72	28	22	15	10	10	10	**7.99**	13	13	4	–	–	–	–	–	–	–	–	–	1	Cymantrene, (1-oxopentyl)-	67606-67-1	NI29845
125	77	153	288	51	138	68	98	**288**	100	45	37	22	22	18	18	14		13	16	–	6	1	–	–	–	–	–	–	–	–	1,3,5-Triazin-2-amine, 4-[(3-phenyl)thioureido]-6-propyl-		NI29846
46	108	43	288	136	246	167	42	**288**	100	74	36	29	24	20	19	16		13	16	2	6	–	–	–	–	–	–	–	–	–	1H-Pyrazole-1-carboximidamide, 4-[(4-ethoxyphenyl)hydrazono]-4,5-dihydro-3-methyl-5-oxo-	42541-28-6	NI29847
46	108	43	288	136	246	167	81	**288**	100	74	36	29	24	20	19	16		13	16	2	6	–	–	–	–	–	–	–	–	–	1H-Pyrazole-1-carboximidamide, 4-[(4-ethoxyphenyl)hydrazono]-4,5-dihydro-3-methyl-5-oxo-	42541-28-6	LI06234
43	128	72	86	188	30	187	253	**288**	100	59	50	28	25	25	20	18	**5.50**	13	18	1	2	–	–	2	–	–	–	–	–	–	N-(2,5-Dichlorophenyl)-N′,N′-dipropylurea		MS06592
43	127	28	30	86	187	41	72	**288**	100	55	38	29	26	19	19	15	**7.30**	13	18	1	2	–	–	2	–	–	–	–	–	–	N-(3,4-Dichlorophenyl)-N′,N′-dipropylurea		MS06593
131	276	59	43	147	113	169	133	**288**	100	56	55	49	35	29	28	27	**0.00**	13	20	1	–	3	–	–	–	–	–	–	–	–	2,4,6-Trithiatricyclo[3.3.1.1^{3,7}]decan-8-one, 1,7,9,9,10,10-hexamethyl-, (±)-	57289-40-4	NI29848
186	246	170	169	109	84	228	288	**288**	100	68	32	30	30	30	20	4		13	20	7	–	–	–	–	–	–	–	–	–	–	Cyclohexanecarboxylic acid, 2,4-bis(acetyloxy)-1-hydroxy-3-methyl-, methyl ester	75364-50-0	NI29849
43	273	59	113	85	127	100	71	**288**	100	89	39	29	29	28	27	26	**0.60**	13	20	7	–	–	–	–	–	–	–	–	–	–	α-D-Galactopyranuronic acid, 1,2:3,4-bis-O-isopropylidene-, methyl ester	18524-41-9	MS06594
43	273	59	85	113	127	100	71	**288**	100	93	39	30	29	27	26	26	**2.11**	13	20	7	–	–	–	–	–	–	–	–	–	–	α-D-Galactopyranuronic acid, 1,2:3,4-bis-O-isopropylidene-, methyl ester	18524-41-9	NI29850
43	273	59	85	127	113	100	141	**288**	100	95	38	30	25	25	24	20	**0.00**	13	20	7	–	–	–	–	–	–	–	–	–	–	α-D-Galactopyranuronic acid, 1,2:3,4-bis-O-isopropylidene-, methyl ester	18524-41-9	LI06235
57	153	107	152	189	187	79	234	**288**	100	93	61	49	25	25	20	10	**0.00**	13	21	2	–	–	–	–	1	–	–	–	–	–	Bicyclo[3.1.1]heptane-1-carboxylic acid, 5-(bromomethyl)-, tert-butyl ester		DD01000
57	125	111	234	232	107	235	233	**288**	100	55	34	32	32	26	10	10	**0.00**	13	21	2	–	–	–	–	1	–	–	–	–	–	Cyclobutanecarboxylic acid, 1-(3-bromopropyl)-3-methylene-, tert-butyl ester		DD01001
91	43	45	41	204	65	123	39	**288**	100	73	64	60	56	47	35	34	**3.30**	13	21	3	–	1	–	–	–	–	–	1	–	–	Phosphorothioic acid, O,O-diisopropyl S-(phenylmethyl) ester	26087-47-8	NI29851
73	183	273	75	147	45	155	43	**288**	100	37	37	35	25	21	17	12	**5.02**	13	28	3	–	–	–	–	–	–	2	–	–	–	2-Butenoic acid, 2-isopropyl-3-[(trimethylsilyl)oxy]-, trimethylsilyl ester		NI29852
171	73	147	245	172	75	45	81	**288**	100	60	35	20	15	15	10	9	**0.00**	13	28	3	–	–	–	–	–	–	2	–	–	–	1-Cyclohexanecarboxylic acid, 1-[(trimethylsilyl)oxy]-, trimethylsilyl ester		MS06595
147	73	75	81	129	148	273	142	**288**	100	66	24	22	20	18	16	8	**1.20**	13	28	3	–	–	–	–	–	–	2	–	–	–	1-Cyclohexanecarboxylic acid, 2-[(trimethylsilyl)oxy]-, trimethylsilyl ester		MS06596
73	75	129	81	273	170	183	116	**288**	100	77	60	49	43	29	28	25	**7.02**	13	28	3	–	–	–	–	–	–	2	–	–	–	1-Cyclohexanecarboxylic acid, 4-[(trimethylsilyl)oxy]-, trimethylsilyl ester, cis-		NI29853
73	81	273	129	75	147	183	170	**288**	100	83	64	63	60	54	30	29	**4.78**	13	28	3	–	–	–	–	–	–	2	–	–	–	1-Cyclohexanecarboxylic acid, 4-[(trimethylsilyl)oxy]-, trimethylsilyl ester, trans-		NI29854
73	45	97	75	74	69	29	57	**288**	100	18	13	10	8	6	5	5	**0.00**	13	28	3	–	–	–	–	–	–	2	–	–	–	2-Pentenoic acid, 2-ethyl-3-[(trimethylsilyl)oxy]-, trimethylsilyl ester		NI29855
208	180	224	151	152	76	150	77	**288**	100	82	39	35	29	25	20	19	**7.00**	14	8	5	–	1	–	–	–	–	–	–	–	–	1-Sulphoanthraquinone		IC03697
227	121	77	180	226	181	91	79	**288**	100	55	46	44	37	29	23	22	**20.69**	14	12	5	2	–	–	–	–	–	–	–	–	–	Benzene, 1-(2-ethylphenoxy)-2,4-dinitro-	53044-48-7	NI29856
288	290	196	198	92	157	155	93	**288**	100	91	44	42	42	33	33	27		14	13	–	2	–	–	–	1	–	–	–	–	–	4-Bromoacetophenone phenylhydrazone		LI06236
59	114	55	41	15	146	95	43	**288**	100	81	78	68	63	59	35	32	**0.00**	14	24	6	–	–	–	–	–	–	–	–	–	–	1,2,8-Octanetricarboxylic acid, trimethyl ester	52323-07-6	NI29857
73	97	75	215	147	217	55	45	**288**	100	55	40	34	24	22	18	17	**0.14**	14	28	4	–	–	–	–	–	–	1	–	–	–	Propanedioic acid, dipropyl-, ethyl trimethylsilyl ester		NI29858
58	45	145	36	44	42	29	30	**288**	100	70	18	12	9	9	7	5	**0.30**	14	28	4	2	–	–	–	–	–	–	–	–	–	2,3-Butanedicarboxylic acid, 1,4-bis(dimethylamino)-, diethyl ester		NI29859
73	144	88	59	231	41	57	115	**288**	100	94	30	24	23	17	15	14	**0.00**	14	36	–	2	–	–	–	–	–	2	–	–	–	Ethylenediamine, N,N′-bis(dimethyl-tert-butylsilyl)-	21934-57-6	NI29860
105	107	77	79	212	106	51	78	**288**	100	72	16	12	9	8	7	5	**0.00**	15	12	2	–	2	–	–	–	–	–	–	–	–	1,3-Oxathiolane-2-thione, 4-hydroxy-4,5-diphenyl-	54699-18-2	NI29861
153	107	259	134	136	288	165	77	**288**	100	42	41	38	33	30	18	11		15	12	6	–	–	–	–	–	–	–	–	–	–	4H-1-Benzopyran-4-one, 2,3-dihydro-3,5,7-trihydroxy-2-(4-hydroxyphenyl)-	5150-32-3	NI29862
107	153	134	136	259	288	69	165	**288**	100	89	74	67	64	53	47	33		15	12	6	–	–	–	–	–	–	–	–	–	–	2H-1-Benzopyran-4(3H)-one, 2-(4-hydroxyphenyl)-3,5,7-trihydroxy-, (2R,3R)-	480-20-6	NI29863
203	205	175	218	189	288	229	261	**288**	100	80	70	43	36	34	21	8		15	12	6	–	–	–	–	–	–	–	–	–	–	3,6-Dioxabicyclo[3.1.0]hexan-2-one, 4-(7-methoxy-2-oxo-2H-1-benzopyran-6-yl)-1-methyl-, (1R*,4S*,5R*)-		NI29864
119	91	125	93	288	152	54	127	**288**	100	81	72	51	40	37	25	24		15	13	2	2	–	–	1	–	–	–	–	–	–	N-(4-Chlorobenzoyl)-N′-phenylurea		LI06237
227	288	242	211	257	183	199	273	**288**	100	87	63	61	40	24	19	13		15	16	4	2	–	–	–	–	–	–	–	–	–	4H-Pyrazolo[5,1-c][1,4]benzoxazine-2-carboxylic acid, 3-(methoxymethyl)-, ethyl ester		MS06597
232	175	103	118	245	91	144	115	**288**	100	30	29	26	22	21	20	18	**6.00**	15	16	4	2	–	–	–	–	–	–	–	–	–	2,4,6(1H,3H,5H)-Pyrimidinetrione, 1,3-dimethyl-5-(oxiranylmethyl)-5-phenyl-	57305-00-7	NI29865
227	259	192	169	117	152	229	91	**288**	100	98	91	90	88	63	34	29	**0.00**	15	17	1	–	–	–	–	–	–	–	–	–	1	Propyldiphenylarsane oxide		DD01002
157	288	198	185	73	273	169	103	**288**	100	80	68	48	45	44	27	20		15	20	2	2	–	–	–	–	–	1	–	–	–	1H-Pyrano[3,4-b]quinoxaline, 3,4-dihydro-3-[(trimethylsilyl)oxymethyl]-		NI29866

MASS TO CHARGE RATIOS								M.W.	INTENSITIES								Parent	C	H	O	N	S	F	Cl	Br	I	Si	P	B	X	COMPOUND NAME	CAS Reg No	No
273	288	177	161	44	133	176	274	288	100	44	38	26	25	21	19	17		15	20	2	4	–	–	–	–	–	–	–	–	–	4-(4,5-Dihydro-5-oxo-3-methyloxazol-4-ylidenehydrazino)-3-methyl-N,N-diethylaniline		IC03698
119	63	163	107	79	253	135	190	288	100	79	78	34	31	20	19	10	0.00	15	32	3	–	–	–	–	–	–	1	–	–	–	1-Nonene, 1-(triethoxysilyl)-, (E)-		DD01003
43	57	55	29	28	27	41	105	288	100	88	87	83	80	61	39	37	6.00	15	33	–	–	–	–	–	–	–	–	–	–	1	Arsine, tripentyl-	5852-59-5	NI29867
86	260	72	147	84	176	112	100	288	100	42	36	27	21	20	20	18	2.00	15	33	1	2	–	–	–	–	–	–	1	–	–	1-[1-Amino-3-(diethylamino)propane]-2,2,3,4,4-pentamethylphosphetan 1-oxide		NI29868
121	91	115	258	288				288	100	40	18	13	10					16	16	5	–	–	–	–	–	–	–	–	–	–	Benzoic acid, 2-hydroxy-6-methoxy-, (2-methoxybenzyl) ester		LI06238
273	134	135	181	274	271	77	180	288	100	35	26	21	17	14	12	11	2.00	16	16	5	–	–	–	–	–	–	–	–	–	–	Benzophenone, 2,4′-dihydroxy-4,6-dimethoxy-2′-methyl-	4650-75-3	NI29869
151	165	138	273	164	124	150	39	288	100	99	38	30	27	23	20	15	11.01	16	16	5	–	–	–	–	–	–	–	–	–	–	Benzophenone, 2,2′,4-trihydroxy-4′-methoxy-6,6′-dimethyl-	21147-34-2	NI29870
151	165	138	273	164	124	150	152	288	100	99	38	30	27	23	20	15	11.00	16	16	5	–	–	–	–	–	–	–	–	–	–	Benzophenone, 2,2′,4-trihydroxy-4′-methoxy-6,6′-dimethyl-	21147-34-2	LI06239
151	270	138	271	165	150	152	149	288	100	67	32	31	30	25	23	20	13.01	16	16	5	–	–	–	–	–	–	–	–	–	–	Benzophenone, 2,2′,6-trihydroxy-4′-methoxy-4,6′-dimethyl-	21147-33-1	NI29871
213	228	187	176	214	288	175	231	288	100	29	27	21	15	14	13	12		16	16	5	–	–	–	–	–	–	–	–	–	–	Columbianetin acetate		LI06240
243	212	169	211	183	215	115	152	288	100	93	88	80	76	65	52	38	8.00	16	16	5	–	–	–	–	–	–	–	–	–	–	1-Dibenzofurancarboxylic acid, 4a,9b-dihydro-4a-hydroperoxy-4,9b-dimethyl-, methyl ester	40801-42-1	NI29872
288	273	88	215	230	187	289	243	288	100	88	46	42	36	32	19	17		16	16	5	–	–	–	–	–	–	–	–	–	–	Dibenzofuran, 1,2,3,4-tetramethoxy-	88256-07-9	NI29873
213	176	175	228	214	288	91	77	288	100	47	32	21	15	14	10	10		16	16	5	–	–	–	–	–	–	–	–	–	–	Lomatin acetate		LI06241
137	151	152	81	109	138	123	122	288	100	63	62	15	13	12	10	10	0.40	16	16	5	–	–	–	–	–	–	–	–	–	–	3-Methoxy-4-hydroxybenzyl 2-methoxy-4-formylphenyl ether		NI29874
288	273	115	77	245	213	128	39	288	100	47	32	27	25	24	24	23		16	16	5	–	–	–	–	–	–	–	–	–	–	1,4-Naphthoquinone, 8-formyl-7-hydroxy-5-isopropyl-2-methoxy-3-methyl-		NI29875
220	219	69	41	221	270	218	191	288	100	60	50	27	13	7	7	7	4.00	16	16	5	–	–	–	–	–	–	–	–	–	–	Shikonin		LI06242
209	179	194	290	288	193	178	197	288	100	68	63	39	39	37	33	20		16	17	–	–	–	–	–	1	–	–	–	–	–	Naphthalene, 1-(2-bromopropenyl)-2,3,4-trimethyl-	26193-57-7	MS06598
125	127	288	77	107	106	290	126	288	100	34	25	14	11	10	9	9		16	17	1	2	–	–	1	–	–	–	–	–	–	Urea, 1-(4-chlorobenzyl)-1,3-dimethyl-3-phenyl-		NI29876
273	288	129	274	289	258	287	165	288	100	73	17	15	12	8	7	7		16	17	3	–	–	–	–	–	–	–	1	–	–	10H-Phenoxaphosphine, 2,8-diethyl-10-hydroxy-, 10-oxide	36360-92-6	NI29877
288	91	103	105	104	289	111	112	288	100	44	21	20	18	17	17	16		16	17	3	–	–	–	–	–	–	–	1	–	–	2,4,6,8-Tetramethylphenoxaphosphinic acid		NI29878
107	126	163	57	215	98	288		288	100	76	54	40	22	18	1			16	20	3	2	–	–	–	–	–	–	–	–	–	1-(2-tert-Butyloxy-2-phenylethyl)-3-hydroxy-6-pyridazone		LI06243
288	229	273	289	199	259	213		288	100	47	22	17	14	11	10			16	20	3	2	–	–	–	–	–	–	–	–	–	(3-Methyl-4-ethoxycarbonyl-2-pyrrolyl)(3-ethyl-4-methyl-5-oxo-2-pyrrolylidene)methane		LI06244
146	260	118	117	103	261	91	232	288	100	99	38	34	26	24	17	16	3.50	16	20	3	2	–	–	–	–	–	–	–	–	–	2,4,6(1H,3H,5H)-Pyrimidinetrione, 1,3,5-triethyl-5-phenyl-	38024-60-1	NI29879
144	215	145	143	288	216	184	42	288	100	61	14	10	9	9	7	5		16	20	3	2	–	–	–	–	–	–	–	–	–	L-Tryptophan, N-acetyl-N,1-dimethyl-, methyl ester	55836-36-7	NI29880
144	215	145	143	288	216	184	102	288	100	60	14	10	9	9	7	5		16	20	3	2	–	–	–	–	–	–	–	–	–	L-Tryptophan, N-acetyl-N,1-dimethyl-, methyl ester	55836-36-7	NI29881
41	55	43	145	29	69	61	27	288	100	96	73	68	60	57	42	36	30.04	16	32	2	–	1	–	–	–	–	–	–	–	–	Octanoic acid, 8-(octylthio)-	56909-05-8	NI29882
91	132	288	77	289	64	102	128	288	100	95	78	18	17	12	10	9		17	12	1	4	–	–	–	–	–	–	–	–	–	Benzonitrile, 4-[(1,5-dihydro-3-methyl-5-oxo-1-phenyl-4H-pyrazol-4-ylidene)amino]-	27807-96-1	NI29883
91	132	288	77	289	64	102	128	288	100	94	77	18	15	12	10	9		17	12	1	4	–	–	–	–	–	–	–	–	–	1-Phenyl-3-methyl-4-(4-cyanophenylimino)-2-pyrazolin-5-one		MS06599
288	287	77	289	172	144	185	89	288	100	78	46	35	31	28	25	25		17	12	1	4	–	–	–	–	–	–	–	–	–	4-Phenyl-5-(3-phenylisoxazol-5-yl)-1,2,3-triazole		NI29884
288	287	77	289	172	144	117	89	288	100	78	46	35	31	28	27	25		17	12	1	4	–	–	–	–	–	–	–	–	–	4-Phenyl-5-(3-phenylisoxazol-5-yl)-1,2,3-triazole		MS06600
153	152	288	217	154	290	65	15	288	100	92	44	34	30	29	23	22		17	14	–	–	–	–	2	–	–	–	–	–	–	cis-2,3-Dichloro-endo-5,6-(1,8-naphthylene)norbornane		NI29885
152	153	57	288	55	43	41	69	288	100	91	70	49	49	49	42	40		17	14	–	–	–	–	2	–	–	–	–	–	–	cis-2,3-Dichloro-exo-5,6-(1,8-naphthylene)norbornane		NI29886
152	153	288	290	187	217	102	66	288	100	93	84	54	30	27	21	18		17	14	–	–	–	–	2	–	–	–	–	–	–	trans-2,3-Dichloro-exo-5,6-(1,8-naphthylene)norbornane		NI29887
270	222	209	223	194	165	208	193	288	100	88	84	72	67	66	65	64	49.32	17	20	2	–	1	–	–	–	–	–	–	–	–	2,2′,4,4′,6-Pentamethyldiphenyl sulphone	18103-61-2	NI29888
253	288	120	209	105	222	193	104	288	100	92	88	85	79	78	78	76		17	20	2	–	1	–	–	–	–	–	–	–	–	2,4,6-Trimethyl-2′-ethyldiphenyl sulphone		NI29889
270	195	222	209	193	165	180	223	288	100	85	79	75	72	71	66	65	49.04	17	20	2	–	1	–	–	–	–	–	–	–	–	2,4,6-Trimethyl-3′-ethyldiphenyl sulphone		NI29890
252	145	270	209	237	253	112	77	288	100	61	35	33	32	29	27	20	11.80	17	20	4	–	–	–	–	–	–	–	–	–	–	2,2′-Dihydroxy-3,3′-dihydroxymethyl-5,5′-dimethyldiphenylmethane		MS06601
123	138	231	82	228	43	246	180	288	100	58	32	32	25	25	24	19	14.00	17	20	4	–	–	–	–	–	–	–	–	–	–	1,4-Ethanonaphthalene-5,8-dione, 1-acetoxy-3,10,10-trimethyl-1,4,4a,8a-tetrahydro-		LI06245
43	82	180	123	108	109	54	190	288	100	86	48	42	39	31	28	20	9.00	17	20	4	–	–	–	–	–	–	–	–	–	–	1,4-Ethanonaphthalene-5,8-dione, 3-acetoxy-1,10,10-trimethyl-1,4,4a,8a-tetrahydro-		LI06246
288	273	258	257	115	289	91	274	288	100	83	41	38	20	18	16	15		17	20	4	–	–	–	–	–	–	–	–	–	–	Isohemogossypol-1,2-dimethyl ether		NI29891
228	213	43	137	246	217	176	91	288	100	44	35	22	16	14	14	9	5.12	17	20	4	–	–	–	–	–	–	–	–	–	–	Naphtho[2,3-b]furan-9(4H)-one, 4-(acetyloxy)-4a,5,6,7-tetrahydro-3,4a,5-trimethyl-, [4S-(4α,4aα,5α)]-	24405-79-6	NI29892
135	131	77	288	148	161	51	206	288	100	27	26	23	16	15	11	10		17	20	4	–	–	–	–	–	–	–	–	–	–	2,4-Nonadienoic acid, 9-(3,4-methylenedioxyphenyl)-, methyl ester, (2E,4E)-	83029-47-4	NI29893
246	245	32	105	199	129	228	247	288	100	75	75	52	48	48	47	17	0.00	17	20	4	–	–	–	–	–	–	–	–	–	–	Zaluzanin D	16838-85-0	DD01004
160	245	142	98	44	147	42	86	288	100	72	49	49	38	35	31	28	4.27	17	24	2	2	–	–	–	–	–	–	–	–	–	N-[(N,N-Diisobutylamino)methyl]phthalimide		IC03699
43	57	55	41	69	271	71	83	288	100	99	71	61	51	49	48	44	0.60	17	36	1	–	1	–	–	–	–	–	–	–	–	Sulphoxide, hexadecyl methyl	3079-32-1	WI01365
57	43	71	55	41	85	69	83	288	100	73	69	47	45	44	32	29	0.00	17	36	3	–	–	–	–	–	–	–	–	–	–	1,2-Propanediol, 3-(tetradecyloxy)-	1561-06-4	NI29894

MASS TO CHARGE RATIOS	M.W.	INTENSITIES	Parent	C	H	O	N	S	F	Cl	Br	I	Si	P	B	X	COMPOUND NAME	CAS Reg No	No
288 287 289 144 255 77 244 227	**288**	100 60 26 10 9 9 8 8		18	12	–	2	1	–	–	–	–	–	–	–	–	2-Pyridinethione, 3-cyano-4,6-diphenyl-		NI29895
288 94 195 156 287 77 194 289	**288**	100 72 35 35 29 24 22 19		18	12	2	2	–	–	–	–	–	–	–	–	–	Phenol, 4,4′-(2,3-diisocyano-1,3-butadiene-1,4-diyl)bis-	580-74-5	NI29896
288 94 287 39 195 156 77 194	**288**	100 63 29 25 24 24 24 21		18	12	2	2	–	–	–	–	–	–	–	–	–	Phenol, 4,4′-(2,3-diisocyano-1,3-butadiene-1,4-diyl)bis-	580-74-5	NI29897
126 288 107 150 149 105 93 79	**288**	100 72 67 51 50 31 31 31		18	21	2	–	–	1	–	–	–	–	–	–	–	Estra-1,4-diene-3,17-dione, 10-fluoro-	794-13-8	NI29898
126 288 150 149 163 125 138 145	**288**	100 72 51 50 29 24 23 22		18	21	2	–	–	1	–	–	–	–	–	–	–	Estra-1,4-diene-3,17-dione, 10-fluoro-	794-13-8	LI06247
273 217 161 41 55 77 288	**288**	100 8 7 7 6 5 4		18	24	3	–	–	–	–	–	–	–	–	–	–	Acetaldehyde anhydrodimedone		LI06248
171 227 255 128 213 119 207 185	**288**	100 81 80 46 40 36 35 31	**6.00**	18	24	3	–	–	–	–	–	–	–	–	–	–	Acetophenone, 3-[hydroxyisopent-2(E)-enyl]-5-(isopent-2-enyl)-4-hydroxy-	77370-28-6	NI29899
288 123 289 175 229 176 162 188	**288**	100 22 20 16 15 15 14 11		18	24	3	–	–	–	–	–	–	–	–	–	–	Estra-1,3,5(10)-triene-2,3,17-triol, (17β)-	362-05-0	NI29900
270 158 157 144 211 170 271 159	**288**	100 60 36 27 24 22 21 18	**6.29**	18	24	3	–	–	–	–	–	–	–	–	–	–	Estra-1,3,5(10)-triene-3,6,17-triol, (6α,17β)-	1229-24-9	NI29901
146 288 172 145 186 175 147 173	**288**	100 93 66 33 31 30 26 25		18	24	3	–	–	–	–	–	–	–	–	–	–	Estra-1,3,5(10)-triene-3,11,17-triol, (11α,17β)-	1464-61-5	NI29902
288 83 85 160 159 185 158 147	**288**	100 79 54 29 28 27 25 23		18	24	3	–	–	–	–	–	–	–	–	–	–	Estra-1,3,5(10)-triene-3,15,17-triol, (15α,17β)-	570-30-9	NI29903
288 160 146 213 289 172 159 202	**288**	100 33 30 28 20 20 20 18		18	24	3	–	–	–	–	–	–	–	–	–	–	Estra-1,3,5(10)-triene-3,16,17-triol, (16α,17β)-	50-27-1	NI29904
288 160 146 159 289 133 213 145	**288**	100 28 26 23 20 20 19 18		18	24	3	–	–	–	–	–	–	–	–	–	–	Estra-1,3,5(10)-triene-3,16,17-triol, (16α,17β)-	50-27-1	NI29905
273 77 41 217 55 161 43 91	**288**	100 21 21 17 17 12 11 10	**4.00**	18	24	3	–	–	–	–	–	–	–	–	–	–	3,4,5,6,7,9-Hexahydro-3,3,6,6,9-pentamethyl-1H-xanthene-1,8(2H)-dione		LI06249
109 134 202 145 174 119 117 220	**288**	100 51 36 31 29 29 25 18	**2.52**	18	24	3	–	–	–	–	–	–	–	–	–	–	6-Octynoic acid, 8-[3-oxo-2-(pent-2-en-1-yl)cyclopent-4-enyl]-		NI29906
213 288 43 157 83 214 18 159	**288**	100 47 45 28 25 22 17 16		18	24	3	–	–	–	–	–	–	–	–	–	–	1-Phenanthrenecarboxylic acid, 1,2,3,4,4a,9,10,10a-octahydro-6-hydroxy-1,4a-dimethyl-, methyl ester, [1S-(1α,4aα,10aβ)]-	4614-56-6	NI29907
213 288 214 157 159 147 133 289	**288**	100 54 22 21 14 14 14 11		18	24	3	–	–	–	–	–	–	–	–	–	–	1-Phenanthrenecarboxylic acid, 1,2,3,4,4a,9,10,10a-octahydro-6-hydroxy-1,4a-dimethyl-, methyl ester, [1S-(1α,4aα,10aβ)]-	4614-56-6	NI29908
43 57 41 55 71 29 69 83	**288**	100 90 75 65 47 44 38 27	**0.50**	18	37	–	–	–	–	1	–	–	–	–	–	–	Octadecane, 1-chloro-	3386-33-2	WI01366
57 71 43 85 55 91 69 41	**288**	100 62 60 41 37 32 32 29	**2.00**	18	37	–	–	–	–	1	–	–	–	–	–	–	Octadecane, 1-chloro-	3386-33-2	LI06250
288 101 202 100 201 94 259 189	**288**	100 91 68 60 38 38 37 36		19	12	3	–	–	–	–	–	–	–	–	–	–	8-Methoxybenz[a]anthracene-7,12-dione		MS06602
288 189 94 100 95 187 63 188	**288**	100 91 37 35 34 31 27 26		19	12	3	–	–	–	–	–	–	–	–	–	–	9-Methoxybenz[a]anthracene-7,12-dione		MS06603
288 189 40 94 289 260 287 100	**288**	100 53 23 22 20 20 19 18		19	12	3	–	–	–	–	–	–	–	–	–	–	10-Methoxybenz[a]anthracene-7,12-dione		MS06604
288 101 202 100 95 94 189 201	**288**	100 73 59 54 48 46 41 34		19	12	3	–	–	–	–	–	–	–	–	–	–	11-Methoxybenz[a]anthracene-7,12-dione		MS06605
171 287 288 286 259 199 115 128	**288**	100 56 39 29 21 18 15 12		19	12	3	–	–	–	–	–	–	–	–	–	–	2H-Naphth[1,8-bc]oxepin-2-one, 7-hydroxy-3-phenyl-	27785-55-3	LI06251
171 287 288 286 261 199 289 118	**288**	100 56 37 28 21 18 15 15		19	12	3	–	–	–	–	–	–	–	–	–	–	2H-Naphth[1,8-bc]oxepin-2-one, 7-hydroxy-3-phenyl-	27785-55-3	NI29909
288 228 183 211 28 289 107 108	**288**	100 28 27 25 23 22 20 18		19	13	1	–	–	–	–	–	–	–	1	–	–	9,10-Dihydro-9-phenyl-9-phosphaanthracen-10-one		MS06606
194 288 165 94 167 78 152 115	**288**	100 78 52 45 35 25 15 9		19	16	1	2	–	–	–	–	–	–	–	–	–	N-(α-Pyridyl)-4-biphenylacetamide		MS06607
169 77 51 167 168 39 65 64	**288**	100 64 63 54 52 38 30 18	**14.25**	19	16	1	2	–	–	–	–	–	–	–	–	–	Urea, triphenyl-	5663-04-7	NI29910
179 83 137 164 161 180 147 149	**288**	100 77 39 26 18 14 12 11	**6.90**	19	25	1	–	–	1	–	–	–	–	–	–	–	2,4-Cyclohexadien-1-ol, 6-(3-fluorobenzyl)-1,2,3,4,5,6-hexamethyl-	84615-09-8	NI29911
288 124 41 270 244 55 111 79	**288**	100 43 40 31 27 27 22 18		19	28	2	–	–	–	–	–	–	–	–	–	–	Androstane-1,17-dione, (5α)-	10455-05-7	LI06252
288 124 41 270 244 55 111 289	**288**	100 42 39 29 28 27 22 20		19	28	2	–	–	–	–	–	–	–	–	–	–	Androstane-1,17-dione, (5α)-	10455-05-7	NI29912
55 41 135 67 81 79 95 93	**288**	100 85 81 64 60 52 39 31	**10.83**	19	28	2	–	–	–	–	–	–	–	–	–	–	Androstane-2,11-dione, (5α)-	1449-57-6	NI29913
41 55 124 67 81 95 288 79	**288**	100 86 71 60 59 48 43 42		19	28	2	–	–	–	–	–	–	–	–	–	–	Androstane-3,11-dione, (5α)-	3510-00-7	NI29914
55 41 67 79 81 91 93 122	**288**	100 82 81 63 54 47 46 36	**17.30**	19	28	2	–	–	–	–	–	–	–	–	–	–	Androstane-3,17-dione	5982-99-0	NI29915
288 244 217 124 255 229 270 97	**288**	100 53 42 34 27 26 23 23		19	28	2	–	–	–	–	–	–	–	–	–	–	Androstane-3,17-dione, (5α)-	846-46-8	NI29917
288 41 217 67 55 244 81 79	**288**	100 45 40 40 40 36 30 29		19	28	2	–	–	–	–	–	–	–	–	–	–	Androstane-3,17-dione, (5α)-	846-46-8	NI29916
288 244 217 55 289 124 41 97	**288**	100 31 23 22 20 19 19 16		19	28	2	–	–	–	–	–	–	–	–	–	–	Androstane-3,17-dione, (5α)-	846-46-8	NI29918
288 41 55 67 79 81 244 93	**288**	100 61 56 51 41 39 36 34		19	28	2	–	–	–	–	–	–	–	–	–	–	Androstane-3,17-dione, (5β)-	1229-12-5	NI29919
288 244 55 41 67 289 81 79	**288**	100 25 25 22 20 18 17 14		19	28	2	–	–	–	–	–	–	–	–	–	–	Androstane-3,17-dione, (5β)-	1229-12-5	NI29920
288 244 174 122 199 162 255 217	**288**	100 49 33 31 30 27 26 23		19	28	2	–	–	–	–	–	–	–	–	–	–	Androstane-3,17-dione, (5β)-	1229-12-5	NI29921
288 246 122 289 121 109 123 107	**288**	100 55 53 21 19 19 18 18		19	28	2	–	–	–	–	–	–	–	–	–	–	Androst-1-en-3-one, 17-hydroxy-, (5α,17β)-	65-06-5	NI29922
122 288 270 109 108 133 123 121	**288**	100 65 49 35 22 21 18 18		19	28	2	–	–	–	–	–	–	–	–	–	–	Androst-1-en-3-one, 17-hydroxy-, (5β,17β)-	10529-96-1	NI29923
91 105 288 41 255 79 55 107	**288**	100 89 86 86 85 81 72 67		19	28	2	–	–	–	–	–	–	–	–	–	–	Androst-2-en-17-one, 3-hydroxy-, (5β)-	57289-70-0	NI29924
124 288 91 109 147 123 105 93	**288**	100 32 25 24 23 22 20 20		19	28	2	–	–	–	–	–	–	–	–	–	–	Androst-4-en-3-one, 17-hydroxy-, (10α,17β)-	604-39-7	NI29925
124 150 96 98 55 109 94 272	**288**	100 60 37 33 31 30 24 22	**0.00**	19	28	2	–	–	–	–	–	–	–	–	–	–	Androst-4-en-3-one, 17-hydroxy-, (17α)-	481-30-1	NI29926
124 288 147 91 79 41 228 55	**288**	100 94 91 57 48 48 46 40		19	28	2	–	–	–	–	–	–	–	–	–	–	Androst-4-en-3-one, 17-hydroxy-, (17α)-	481-30-1	NI29927
124 288 246 41 55 147 91 79	**288**	100 89 43 36 32 30 30 30		19	28	2	–	–	–	–	–	–	–	–	–	–	Androst-4-en-3-one, 17-hydroxy-, (17β)-	58-22-0	NI29929
124 91 288 246 105 93 147 107	**288**	100 84 76 59 56 49 46 39		19	28	2	–	–	–	–	–	–	–	–	–	–	Androst-4-en-3-one, 17-hydroxy-, (17β)-	58-22-0	NI29928
124 288 246 147 41 91 79 28	**288**	100 53 38 32 28 27 26 26		19	28	2	–	–	–	–	–	–	–	–	–	–	Androst-4-en-3-one, 17-hydroxy-, (17β)-	58-22-0	NI29930
91 105 79 270 255 107 77 67	**288**	100 77 74 72 64 61 58 56	**52.85**	19	28	2	–	–	–	–	–	–	–	–	–	–	Androst-5-en-17-one, 3-hydroxy-, (3β)-	53-43-0	WI01367
288 255 270 107 203 105 145 159	**288**	100 65 59 59 56 46 40 35		19	28	2	–	–	–	–	–	–	–	–	–	–	Androst-5-en-17-one, 3-hydroxy-, (3β)-	53-43-0	NI29931
288 255 270 107 91 41 105 79	**288**	100 76 64 59 55 55 53 51		19	28	2	–	–	–	–	–	–	–	–	–	–	Androst-5-en-17-one, 3-hydroxy-, (3β)-	53-43-0	NI29932
288 71 55 231 110 91 79 67	**288**	100 71 62 60 55 53 48 44		19	28	2	–	–	–	–	–	–	–	–	–	–	Estr-4-en-3-one, 17-hydroxy-17-methyl-, (17β)-	514-61-4	NI29933

MASS TO CHARGE RATIOS	M.W.	INTENSITIES	Parent	C	H	O	N	S	F	Cl	Br	I	Si	P	B	X	COMPOUND NAME	CAS Reg No	No
288 255 231 270 91 110 256 79	**288**	100 67 45 36 31 30 23 23		19	28	2	–	–	–	–	–	–	–	–	–	–	**Estr-4-en-3-one, 17α-hydroxy-17-methyl-**	10582-09-9	**NI29934**
110 288 161 90 270 289 215 105	**288**	100 89 31 22 19 18 18 18		19	28	2	–	–	–	–	–	–	–	–	–	–	**D-Homoestr-4-en-3-one, 17-hydroxy-, (8α,17β)-**	55659-69-3	**NI29935**
110 288 161 90 246 270 107 289	**288**	100 67 22 22 19 15 15 13		19	28	2	–	–	–	–	–	–	–	–	–	–	**D-Homoestr-4-en-3-one, 17-hydroxy-, (9β,17β)-**	55659-70-6	**NI29936**
110 179 161 122 288 90 94 105	**288**	100 23 22 20 19 16 14 11		19	28	2	–	–	–	–	–	–	–	–	–	–	**D-Homoestr-4-en-3-one, 17-hydroxy-, (17β)-**	13421-98-2	**NI29937**
110 288 161 91 246 107 270 160	**288**	100 69 23 21 20 15 13 13		19	28	2	–	–	–	–	–	–	–	–	–	–	**19-Nor-homotestosterone**		**LI06253**
288 55 79 124 105 91 93 107	**288**	100 53 49 48 48 48 43 42		19	28	2	–	–	–	–	–	–	–	–	–	–	**17β-Hydroxyandrost-5(6)-en-3-one**		**LI06254**
288 255 231 270 91 110 79 71	**288**	100 67 45 36 31 29 22 22		19	28	2	–	–	–	–	–	–	–	–	–	–	**17β-Methyl-19-nortestosterone**		**LI06255**
121 123 275 136 107 91 81 109	**288**	100 65 40 40 40 40 40 38	**3.00**	19	28	2	–	–	–	–	–	–	–	–	–	–	**1-Phenanthrenecarboxylic acid, 7-ethenyl-1,2,3,4,4a,4b,5,6,7,9,10,10a-dodecahydro-4a,7-dimethyl-, [1R-(1α,4aβ,4bα,7β,10aα)]-**	57289-55-1	**NI29938**
56 44 233 273 43 288 57 41	**288**	100 60 40 30 30 20 18 18		19	32	–	2	–	–	–	–	–	–	–	–	–	**Androst-5-ene-3β,17β-diamine**	6748-15-8	**NI29939**
288 258 273 289 211 290 239 197	**288**	100 79 77 25 15 13 11 11		20	16	–	–	1	–	–	–	–	–	–	–	–	**Thiophene, 2-(3-methyl-1,2-diphenylcycloprop-1-en-3-yl)-**	86738-97-8	**NI29940**
288 273 289 274 271 287 211 197	**288**	100 75 23 17 17 15 15 15		20	16	–	–	1	–	–	–	–	–	–	–	–	**Thiophene, 3-(3-methyl-1,2-diphenylcycloprop-1-en-3-yl)-**	86738-99-0	**NI29941**
241 228 229 239 226 288 215 202	**288**	100 41 35 25 25 23 13 13		20	16	2	–	–	–	–	–	–	–	–	–	–	**Benz[a]anthracene, 7,12-bis(hydroxymethyl)-**		**MS06608**
244 288 259 271 215 226 239 202	**288**	100 92 43 29 24 23 16 13		20	16	2	–	–	–	–	–	–	–	–	–	–	**Benz[a]anthracene, 2-hydroxy-7-(hydroxymethyl)-12-methyl-**		**MS06609**
244 288 259 215 271 226 239 202	**288**	100 94 43 32 26 26 20 20		20	16	2	–	–	–	–	–	–	–	–	–	–	**Benz[a]anthracene, 3-hydroxy-7-(hydroxymethyl)-12-methyl-**		**MS06610**
244 288 259 215 226 239 271 202	**288**	100 66 58 58 36 28 27 23		20	16	2	–	–	–	–	–	–	–	–	–	–	**Benz[a]anthracene, 4-hydroxy-7-(hydroxymethyl)-12-methyl-**		**MS06611**
269 270 253 255 239 288 244 226	**288**	100 84 84 69 61 58 53 53		20	16	2	–	–	–	–	–	–	–	–	–	–	**Benz[a]anthracene, 2-hydroxy-7-methyl-12-(hydroxymethyl)-**		**MS06612**
288 244 259 271 215 226 239 202	**288**	100 84 50 27 25 20 16 11		20	16	2	–	–	–	–	–	–	–	–	–	–	**Benz[a]anthracene, 3-hydroxy-7-methyl-12-(hydroxymethyl)-**		**MS06613**
288 244 259 215 226 202 271 239	**288**	100 80 57 43 31 26 23 22		20	16	2	–	–	–	–	–	–	–	–	–	–	**Benz[a]anthracene, 4-hydroxy-7-methyl-12-(hydroxymethyl)-**		**MS06614**
181 182 77 106 105 51 183 76	**288**	100 53 19 15 14 10 9 5	**0.00**	20	16	2	–	–	–	–	–	–	–	–	–	–	**4H-1,3-Benzodioxin, 2,4-diphenyl-**	56182-53-7	**NI29942**
91 288 92 271 184 287 183 181	**288**	100 92 74 61 51 47 47 29		20	16	2	–	–	–	–	–	–	–	–	–	–	**[1]Benzopyrano[2,3-b]xanthene, 5a,6,12a,13-tetrahydro-**		**IC03700**
141 131 103 77 288 115 140 242	**288**	100 65 48 48 43 32 24 22		20	16	2	–	–	–	–	–	–	–	–	–	–	**Cinnamic acid, 1-naphthalenemethyl ester**	86328-60-1	**NI29943**
256 241 202 239 245 215 226 288	**288**	100 40 29 26 22 20 14 4		20	16	2	–	–	–	–	–	–	–	–	–	–	**7,12-Dimethylbenz[a]anthracene-7,12-endoperoxide**		**MS06615**
95 178 288 108 152 206 165 215	**288**	100 50 30 25 23 15 15 14		20	16	2	–	–	–	–	–	–	–	–	–	–	**2-[2-(4-Methyl-3-furyl)vinyl]-2′-formylbiphenyl**		**MS06616**
95 178 288 108 150 206 165 215	**288**	100 50 30 25 23 15 15 14		20	16	2	–	–	–	–	–	–	–	–	–	–	**2-[2-(4-Methyl-3-furyl)vinyl]-2′-formylbiphenyl**		**LI06256**
288 232 88 176 260 246 202 101	**288**	100 24 8 6 5 5 5 5		20	16	2	–	–	–	–	–	–	–	–	–	–	**3,4,7,8-Tetrahydrodibenzo[b,h]biphenylene-1(2H),10(9H)-dione**		**LI06257**
181 288 182 152 287 77 105 243	**288**	100 23 16 15 11 9 8 6		20	16	2	–	–	–	–	–	–	–	–	–	–	**Xanthene, 9-(4-methoxyphenyl)-**	19234-05-0	**NI29944**
232 260 217 245 202 215 261 233	**288**	100 68 56 35 33 31 30 24	**0.10**	20	20	–	2	–	–	–	–	–	–	–	–	–	**Benzo[c]cyclopenta[f]-1,2-diazepine, 1,2,3,3a-tetrahydro-7-methyl-10-(4-methylphenyl)-**		**NI29945**
287 288 115 129 289 128 273 130	**288**	100 66 27 15 12 9 7 4		20	20	–	2	–	–	–	–	–	–	–	–	–	**Pyrazine, 2,5-diethyl-3,6-diphenyl-**	21798-29-8	**NI29946**
111 32 273 288 41 55 95 81	**288**	100 51 29 17 16 12 11 11		20	32	1	–	–	–	–	–	–	–	–	–	–	**5α,14β-Androstan-15-one, 17α-methyl-**	13864-69-2	**NI29947**
111 41 110 55 67 43 81 95	**288**	100 24 22 18 14 12 11 10	**6.06**	20	32	1	–	–	–	–	–	–	–	–	–	–	**5α,14β-Androstan-15-one, 17β-methyl-**	14012-11-4	**NI29948**
111 55 81 67 288 109 95 79	**288**	100 85 74 74 72 64 59 50		20	32	1	–	–	–	–	–	–	–	–	–	–	**5α-Androstan-15-one, 17β-methyl-**	14012-10-3	**NI29949**
216 217 55 41 67 81 288 95	**288**	100 74 60 59 50 42 36 36		20	32	1	–	–	–	–	–	–	–	–	–	–	**5α-Androstan-16-one, 17β-methyl-**	25780-31-8	**NI29950**
216 217 288 121 218 201 273 231	**288**	100 74 37 25 11 11 9 9		20	32	1	–	–	–	–	–	–	–	–	–	–	**Androstan-16-one, 17-methyl-, (5α,17α)-**	56051-42-4	**NI29951**
81 43 55 109 41 67 28 95	**288**	100 80 68 51 51 46 30 29	**14.14**	20	32	1	–	–	–	–	–	–	–	–	–	–	**5α-Androst-15-en-17β-ol, 17-methyl-**	13864-65-8	**NI29952**
105 257 288 123 135 134 119 229	**288**	100 98 87 83 80 76 76 74		20	32	1	–	–	–	–	–	–	–	–	–	–	**ent-Beyer-15-en-18-ol**	20107-90-8	**NI29953**
41 69 203 163 95 55 43 81	**288**	100 93 44 40 37 29 20 17	**11.00**	20	32	1	–	–	–	–	–	–	–	–	–	–	**Geranylgeranial**		**LI06258**
288 81 109 95 273 55 217 111	**288**	100 36 35 34 32 31 30 29		20	32	1	–	–	–	–	–	–	–	–	–	–	**D-Homoandrostan-17a-one, (5α)-**	10147-56-5	**NI29954**
288 217 95 81 55 67 111 109	**288**	100 39 38 38 37 36 35 33		20	32	1	–	–	–	–	–	–	–	–	–	–	**D-Homoandrostan-17a-one, (5α)-**	10147-56-5	**NI29955**
288 230 81 231 67 95 55 41	**288**	100 87 55 48 47 45 45 41		20	32	1	–	–	–	–	–	–	–	–	–	–	**D-Homoandrostan-17-one, (5α)-**	19897-22-4	**NI29956**
82 41 81 55 69 68 67 53	**288**	100 99 98 97 68 68 67 49	**21.00**	20	32	1	–	–	–	–	–	–	–	–	–	–	**2-Isopropyl-5,9,13-trimethylcyclotetradeca-5,9,13-trienone**		**NI29957**
41 81 55 91 123 79 93 67	**288**	100 93 90 79 73 68 63 61	**12.00**	20	32	1	–	–	–	–	–	–	–	–	–	–	**Kaurenol**		**LI06259**
105 81 91 93 41 106 55 79	**288**	100 85 80 79 78 72 67 58	**0.00**	20	32	1	–	–	–	–	–	–	–	–	–	–	**Monoginol**		**LI06260**
41 81 137 95 55 69 273 123	**288**	100 82 78 71 70 69 66 64	**41.00**	20	32	1	–	–	–	–	–	–	–	–	–	–	**ent-D-Norbeyerane-15-carboxaldehyde, exo-**		**MS06617**
230 217 218 203 148 109 83 175	**288**	100 48 47 34 33 33 30 28	**16.00**	20	32	1	–	–	–	–	–	–	–	–	–	–	**D(15)-Norpregnan-20-one, (5α)-**	32318-97-1	**NI29958**
255 288 105 119 148 133 107 91	**288**	100 98 60 57 44 44 41 41		20	32	1	–	–	–	–	–	–	–	–	–	–	**Podocarp-7-en-3β-ol, 13β-methyl-13-vinyl-**	4752-56-1	**NI29959**
135 41 92 106 90 43 55 78	**288**	100 34 28 27 27 26 25 21	**5.00**	20	32	1	–	–	–	–	–	–	–	–	–	–	**Podocarp-8(14)-en-3β-ol, 13β-methyl-13-vinyl-**	4728-30-7	**NI29960**
55 81 41 257 91 123 93 79	**288**	100 98 96 92 87 81 77 77	**76.92**	20	32	1	–	–	–	–	–	–	–	–	–	–	**4,25-Secoobscurinervan-4-ol, 22-ethyl-15,16-dimethoxy-, 25-acetate, (4β,22α)-**	72101-12-3	**NI29961**
91 57 72 105 43 41 55 92	**288**	100 66 33 25 21 20 19 18	**3.81**	20	32	1	–	–	–	–	–	–	–	–	–	–	**3-Tridecanone, 7-methyl-13-phenyl-**	16964-45-7	**NI29962**
91 57 104 72 214 105 92 288	**288**	100 59 54 38 25 24 24 23		20	32	1	–	–	–	–	–	–	–	–	–	–	**3-Tridecanone, 8-methyl-13-phenyl-**	56630-77-4	**NI29963**
288 287 115 289 91 217 128 141	**288**	100 79 37 25 25 21 19 16		21	20	1	–	–	–	–	–	–	–	–	–	–	**2,7-Dibenzylidenecycloheptanone**		**MS06618**
287 288 115 217 91 128 289 218	**288**	100 87 56 28 27 25 18 18		21	20	1	–	–	–	–	–	–	–	–	–	–	**2,6-Dibenzylidene-4-methylcyclohexanone**		**MS06619**
231 259 243 287 271	**288**	100 52 25 11 6	**0.00**	21	20	1	–	–	–	–	–	–	–	–	–	–	**1,5-Bis(2-ethynylcyclohexenyl)-1,4-pentadiyn-3-ol**		**LI06261**
244 165 243 167 166 105 211 245	**288**	100 69 60 55 31 25 24 20	**19.02**	21	20	1	–	–	–	–	–	–	–	–	–	–	**Trityl ethyl ether**	968-39-8	**NI29964**

MASS TO CHARGE RATIOS								M.W.	INTENSITIES								Parent	C	H	O	N	S	F	Cl	Br	I	Si	P	B	X	COMPOUND NAME	CAS Reg No	No
105	211	165	243	77	183	244	288	**288**	100	69	56	54	53	33	24	22		21	20	1	–	–	–	–	–	–	–	–	–	–	Trityl ethyl ether	968-39-8	IC03701
105	91	118	106	131	104	43	41	**288**	100	12	11	9	8	7	6	6	**4.80**	21	36	–	–	–	–	–	–	–	–	–	–	–	Benzene, pentadecyl-	2131-18-2	WI01368
119	105	91	120	71	133	118	106	**288**	100	47	34	11	10	9	8	7	**1.30**	21	36	–	–	–	–	–	–	–	–	–	–	–	Benzene, pentadecyl-	2131-18-2	WI01369
137	81	41	95	136	67	55	69	**288**	100	75	68	66	52	49	45	28	**16.79**	21	36	–	–	–	–	–	–	–	–	–	–	–	Bis(decahydro-1-naphthyl)methane		AI01989
217	55	288	95	81	109	67	41	**288**	100	85	82	75	75	65	63	55		21	36	–	–	–	–	–	–	–	–	–	–	–	D-Dihomoandrostane, (5α)-	35575-54-3	NI29965
105	91	118	131	104	43	41	288	**288**	100	12	11	8	7	6	6	5		21	36	–	–	–	–	–	–	–	–	–	–	–	2-Phenylpentadecane	4534-66-1	AM00139
105	106	91	288	104	79	77	103	**288**	100	16	13	6	6	4	4	2		21	36	–	–	–	–	–	–	–	–	–	–	–	2-Phenylpentadecane	4534-66-1	WI01370
91	119	259	41	43	105	288	120	**288**	100	65	15	13	11	10	9	9		21	36	–	–	–	–	–	–	–	–	–	–	–	3-Phenylpentadecane		AI01990
91	133	92	104	245	117	105	41	**288**	100	30	13	12	9	9	8	8	**7.86**	21	36	–	–	–	–	–	–	–	–	–	–	–	4-Phenylpentadecane		AI01991
91	161	105	217	104	92	288	41	**288**	100	25	13	12	9	9	7	7		21	36	–	–	–	–	–	–	–	–	–	–	–	6-Phenylpentadecane		AI01992
91	189	105	104	92	43	41	29	**288**	100	31	14	9	8	7	7	7	**6.90**	21	36	–	–	–	–	–	–	–	–	–	–	–	8-Phenylpentadecane		AI01993
288	218	217	149	109	95	81	55	**288**	100	71	56	48	34	32	32	31		21	36	–	–	–	–	–	–	–	–	–	–	–	Pregnane, (5α)-	641-85-0	NI29968
288	218	217	149	109	95	81	55	**288**	100	69	54	46	32	31	31	30		21	36	–	–	–	–	–	–	–	–	–	–	–	Pregnane, (5α)-	641-85-0	NI29966
218	217	288	149	273	81	109	95	**288**	100	99	65	50	44	37	34	33		21	36	–	–	–	–	–	–	–	–	–	–	–	Pregnane, (5α)-	641-85-0	NI29967
55	41	81	67	217	95	40	43	**288**	100	93	79	75	64	60	58	57	**30.00**	21	36	–	–	–	–	–	–	–	–	–	–	–	Pregnane, (5β)-	481-26-5	NI29970
55	81	67	217	95	109	218	79	**288**	100	78	74	65	59	50	46	41	**31.00**	21	36	–	–	–	–	–	–	–	–	–	–	–	Pregnane, (5β)-	481-26-5	NI29969
217	218	81	273	95	288	109	151	**288**	100	99	58	52	50	48	42	41		21	36	–	–	–	–	–	–	–	–	–	–	–	Pregnane, (5β)-	481-26-5	NI29971
57	120	176	232	288	231	273	39	**288**	100	89	67	44	35	34	27	27		21	36	–	–	–	–	–	–	–	–	–	–	–	1,3,5-Trineopentylbenzene		LI06262
91	43	77	41	39	51	165	65	**288**	100	52	32	28	28	23	22	22	**14.30**	22	24	–	–	–	–	–	–	–	–	–	–	–	Azulene, 1,4-dimethyl-7-isopropyl-2-(phenylmethyl)-	39665-57-1	MS06620
288	245	115	91	289	215	197	117	**288**	100	38	38	32	26	26	22	22		22	24	–	–	–	–	–	–	–	–	–	–	–	Benzene, 1,1′-[1,4-diisopropyl-1,2,3-butatriene-1,4-diyl]bis-	56805-10-8	MS06621
288	260	259	245	276	289	101	202	**288**	100	35	31	30	24	23	17	16		22	24	–	–	–	–	–	–	–	–	–	–	–	Benzo[ghi]perylene, dodecahydro-		NI29972
288	273	260	57	245	43	259	91	**288**	100	78	70	50	41	39	36	36		22	24	–	–	–	–	–	–	–	–	–	–	–	Chrysene, 11-butyl-1,2,3,4-tetrahydro-		LI06263
288	245	278	287	259	289	246	286	**288**	100	57	32	27	22	21	20	19		22	24	–	–	–	–	–	–	–	–	–	–	–	Dibenz[a,h]anthracene, 1,2,3,4,4a,5,6,12,13,14b-decahydro-		NI29973
202	203	200	204	201	245	214	189	**288**	100	65	15	14	14	11	7	7	**3.00**	22	24	–	–	–	–	–	–	–	–	–	–	–	Pyrene, 10b,10c-dihydro-10b,10c-dipropyl-, trans-	28816-94-6	NI29974
230	291	167	165	289	228	215	231	**289**	100	97	90	82	79	73	67	63		7	3	3	1	–	–	4	–	–	–	–	–	–	Benzene, 1,2,3,4-tetrachloro-5-methoxy-6-nitro-	69576-80-3	NI29975
165	291	261	167	259	87	211	209	**289**	100	93	88	82	73	70	69	67	**66.90**	7	3	3	1	–	–	4	–	–	–	–	–	–	Benzene, 1,2,4,5-tetrachloro-3-methoxy-6-nitro-	2438-88-2	NI29976
44	51	42	246	244	53	52	45	**289**	100	39	39	27	27	18	14	14	**0.00**	9	8	3	1	1	–	–	1	–	–	–	–	–	Thiazolo[3,2-a]pyridinium, 7-bromo-3-carboxy-2,3-dihydro-8-hydroxy-5-methyl-, hydroxide, inner salt	22989-29-3	NI29977
202	96	69	155	204	71	81	97	**289**	100	80	68	60	58	52	34	30	**15.99**	9	12	–	5	2	–	1	–	–	–	–	–	–	1,3,5-Triazin-2-amine, 4-(2-chloro-2-cyanoethylthio)-N,N-dimethyl-6-methylthio-	84055-79-8	NI29978
57	149	147	41	29	28	43	205	**289**	100	98	98	67	47	44	35	28	**0.00**	9	12	3	3	–	–	–	1	–	–	–	–	–	2H,4H-1,3-Dioxin-4-one, 6-(azidomethyl)-5-bromo-2-tert-butyl-, (2R)-		DD01005
230	231	176	119	202	289	180	229	**289**	100	8	7	7	3	3	3	2		10	12	3	1	–	5	–	–	–	–	–	–	–	2-Piperidinecarboxylic acid, N-(pentafluoropropionyl)-, methyl ester		NI29979
230	142	289	229	258	271	119	110	**289**	100	37	27	24	18	17	15	13		10	12	3	1	–	5	–	–	–	–	–	–	–	3-Piperidinecarboxylic acid, N-(pentafluoropropionyl)-, methyl ester		NI29980
142	230	289	229	100	119	228	110	**289**	100	67	47	38	37	30	19	18		10	12	3	1	–	5	–	–	–	–	–	–	–	4-Piperidinecarboxylic acid, N-(pentafluoropropionyl)-, methyl ester		NI29981
226	246	289	183	41	55	228	59	**289**	100	92	70	60	60	54	45	45		10	16	3	5	–	–	1	–	–	–	–	–	–	1,3,5-Triazine-2,4-diamine, 6-chloro-N-butyl-N′-methoxy-N′-(methoxycarbonyl)-		NI29982
145	289	88	288	291	90	173	290	**289**	100	49	40	18	12	10	8	8		11	7	1	3	–	3	1	–	–	–	–	–	–	3(2H)-Pyridazinone, 5-amino-4-chloro-2-[3-(trifluoromethyl)phenyl]-	23576-24-1	NI29983
254	87	123	256	218	125	108	51	**289**	100	53	50	43	37	33	29	26	**17.45**	11	9	–	1	2	–	2	–	–	–	–	–	–	Benzothiazole, 2-[(3,3-dichloro-2-methyl-2-propenyl)thio]-	55976-01-7	NI29984
125	46	247	44	43	42	138	97	**289**	100	77	67	43	43	43	30	27	**23.00**	11	11	3	7	–	–	–	–	–	–	–	–	–	1H-Pyrazole-1-carboximidamide, 4,5-dihydro-3-methyl-4-[(4-nitrophenyl)hydrazono]-5-oxo-	42541-26-4	NI29985
125	46	247	44	43	42	138	167	**289**	100	77	67	43	43	43	30	27	**23.00**	11	11	3	7	–	–	–	–	–	–	–	–	–	1H-Pyrazole-1-carboximidamide, 4,5-dihydro-3-methyl-4-[(4-nitrophenyl)hydrazono]-5-oxo-		LI06264
194	223	69	180	195	83	229	181	**289**	100	70	50	45	35	35	34	30	**10.99**	11	20	2	5	–	–	1	–	–	–	–	–	–	1,3,5-Triazine-2,4-diamine, 6-(2-chloroethoxy)-N,N-diethyl-N′-methoxy-N′-methyl-	101132-51-8	NI29986
229	258	231	260	223	194	230	289	**289**	100	60	33	26	22	22	20	19		11	20	2	5	–	–	1	–	–	–	–	–	–	1,3,5-Triazin-2(1H)-one, 1-(2-chloroethyl)-4-(diethylamino)-6-(N-methoxy-N-methylamino)-		NI29987
257	145	75	287	259	129	203	175	**289**	100	98	95	94	92	91	90	89	**77.00**	12	7	6	3	–	–	–	–	–	–	–	–	–	7-Nitro-1,2,3,4-tetrahydroxyphenazine		MS06622
91	205	113	55	92	146	65	70	**289**	100	77	73	57	49	46	29	23	**0.00**	12	12	4	1	–	–	–	–	–	–	–	–	1	Cymantrene, [1-(acetylamino)ethyl]-	12193-73-6	NI29988
162	55	205	289	107	56	79	233	**289**	100	42	35	23	22	14	12	8		12	12	4	1	–	–	–	–	–	–	–	–	1	Cymantrene, [1-(dimethylaminomethyl)-2-formyl]-	70614-50-5	NI29989
78	43	67	95	55	185	167	126	**289**	100	92	52	45	12	9	8	8	**0.00**	12	19	7	1	–	–	–	–	–	–	–	–	–	1,3-Cyclohexanediol, 5-ethoxy-2-nitro-, diacetate, (1R,2R,3S,5S)-		DD01006
177	43	170	41	176	55	69	56	**289**	100	64	62	56	55	54	32	29	**0.00**	12	20	1	1	–	5	–	–	–	–	–	–	–	1-Aminononane, N-(pentafluoropropionyl)-		MS06623
73	102	172	43	75	45	274	71	**289**	100	60	40	38	21	20	19	18	**9.12**	12	27	3	1	–	–	–	–	–	2	–	–	–	Glycine, N-(2-methyl-1-oxopropyl)-N-(trimethylsilyl)-, trimethylsilyl ester	55530-55-7	NI29990

MASS TO CHARGE RATIOS	M.W.	INTENSITIES	Parent	C	H	O	N	S	F	Cl	Br	I	Si	P	B	X	COMPOUND NAME	CAS Reg No	No
73 102 172 75 43 147 274 45	289	100 78 61 49 46 38 30 23	15.20	12	27	3	1	–	–	–	–	–	2	–	–	–	Glycine, N-(2-methyl-1-oxopropyl)-N-(trimethylsilyl)-, trimethylsilyl ester	55530-55-7	MS06624
73 102 172 274 43 75 147 176	289	100 69 48 39 32 31 30 29	7.20	12	27	3	1	–	–	–	–	–	2	–	–	–	Glycine, N-(1-oxobutyl)-N-(trimethylsilyl)-, trimethylsilyl ester	55494-12-7	NI29991
73 102 172 45 75 43 147 274	289	100 71 30 24 23 23 17 15	3.00	12	27	3	1	–	–	–	–	–	2	–	–	–	Glycine, N-(1-oxobutyl)-N-(trimethylsilyl)-, trimethylsilyl ester	55494-12-7	NI29992
73 147 75 56 45 200 172 59	289	100 51 38 18 18 17 16 15	2.46	12	27	3	1	–	–	–	–	–	2	–	–	–	Pentanoic acid, 3-methyl-2-[(trimethylsilyl)oxyimino]-, trimethylsilyl ester	74173-63-0	NI29993
73 147 45 75 56 59 74 172	289	100 24 20 20 15 14 11 10	1.32	12	27	3	1	–	–	–	–	–	2	–	–	–	Pentanoic acid, 3-methyl-2-[(trimethylsilyl)oxyimino]-, trimethylsilyl ester	74173-63-0	NI29994
73 147 75 56 200 45 172 274	289	100 30 27 17 16 16 14 13	2.25	12	27	3	1	–	–	–	–	–	2	–	–	–	Pentanoic acid, 3-methyl-2-[(trimethylsilyl)oxyimino]-, trimethylsilyl ester	74173-63-0	NI29995
73 147 45 75 200 59 74 43	289	100 25 22 21 17 15 11 10	0.36	12	27	3	1	–	–	–	–	–	2	–	–	–	Pentanoic acid, 4-methyl-2-[(trimethylsilyl)oxyimino]-, trimethylsilyl ester		NI29996
115 289 142 57 55 71 127 69	289	100 79 58 55 37 34 32 32		13	8	1	1	–	5	–	–	–	–	–	–	–	Propanamide, 2,2,3,3,3-pentafluoro-N-1-naphthalenyl-	21970-67-2	NI29997
289 115 57 142 127 71 55 69	289	100 84 72 64 52 44 40 36		13	8	1	1	–	5	–	–	–	–	–	–	–	Propanamide, 2,2,3,3,3-pentafluoro-N-2-naphthalenyl-	21970-68-3	NI29998
164 163 210 82 289 291 152 243	289	100 98 78 44 36 36 35 18		13	8	2	1	–	–	–	1	–	–	–	–	–	Fluorene, 7-bromo-2-nitro-		NI29999
210 164 163 152 211 165 82 162	289	100 70 56 16 16 13 7 7	0.00	13	8	2	1	–	–	–	1	–	–	–	–	–	Fluorene, 9-bromo-2-nitro-	53172-79-5	NI30000
289 288 242 60 28 59 290 44	289	100 86 31 31 27 19 18 15		13	11	3	3	1	–	–	–	–	–	–	–	–	Thiazolo[3,2-c]pyrimidin-4-ium, 2,3-dihydro-8-hydroxy-5-methyl-7-(p-nitrophenyl)-, hydroxide, inner salt	24644-39-1	NI30001
105 44 140 77 51 106 141 78	289	100 87 81 68 17 15 10 10	0.00	13	11	5	3	–	–	–	–	–	–	–	–	–	8-Benzamido-2,4,7-trioxo-6-oxa-1,3-diazaspiro[4.4]nonane		LI06265
289 77 291 290 288 194 209 210	289	100 94 93 80 80 80 63 60		13	12	–	3	–	–	–	1	–	–	–	–	–	5-Bromo-2-aminobenzophenonehydrazone		MS06625
43 57 44 30 127 202 41 203	289	100 81 71 65 41 39 39 27	0.00	13	23	4	1	1	–	–	–	–	–	–	–	–	Malonic acid, thio-, S-ester with N-(2-mercaptoethyl)octanamide	21044-08-6	NI30002
115 72 73 98 96 142 216 56	289	100 92 64 32 21 20 17 17	0.00	13	23	6	1	–	–	–	–	–	–	–	–	–	D-Galactopyranoside, methyl 3-O-acetyl-2,6-dideoxy-2-(N-methylacetamido)-4-O-methyl-		NI30003
73 172 29 232 45 41 114 173	289	100 75 24 21 21 16 14 11	0.00	13	31	2	1	–	–	–	–	–	2	–	–	–	N-Trimethylsilyl-α-methylisoleucine, trimethylsilyl ester		MS06626
172 73 45 173 147 100 75 74	289	100 75 18 15 9 9 9 8	0.00	13	31	2	1	–	–	–	–	–	2	–	–	–	N-Trimethylsilyl-α-methylleucine, trimethylsilyl ester		MS06627
289 291 290 144.5 292 226 127 260	289	100 33 15 6 5 3 3 2		14	8	4	1	–	–	1	–	–	–	–	–	–	1,4-Dihydroxy-6-chloro-8-aminoanthraquinone		IC03702
91 102 289 77 103 119 76 139	289	100 99 91 33 26 14 10 7		14	11	4	1	1	–	–	–	–	–	–	–	–	Benzene, 1-nitro-4-[(2-phenylvinyl)sulphonyl]-	38367-00-9	NI30004
289 232 290 261 260 233 160 77	289	100 32 17 16 14 6 5 5		14	11	6	1	–	–	–	–	–	–	–	–	–	Ethyl 3,4-dihydro-3,6-dioxopyrano[3,2-g][1,4]benzoxazine-8-carboxylate		LI06266
289 232 261 290 233 246 220 94	289	100 40 20 17 8 5 5 5		14	11	6	1	–	–	–	–	–	–	–	–	–	Ethyl 3,4-dihydro-3,8-dioxopyrano[2,3-f][1,4]benzoxazine-6-carboxylate		LI06267
217 139 289 183 63 185 107 152	289	100 56 38 35 33 27 19 7		14	12	–	1	2	–	–	–	–	–	1	–	–	Cyanomethyl diphenylphosphinodithioate		MS06628
289 291 193 77 290 105 288 209	289	100 99 93 88 82 82 72 36		14	12	1	1	–	–	–	1	–	–	–	–	–	5-Bromo-2-N-methylaminobenzophenone		MS06629
56 84 121 186 55 112 111 96	289	100 94 90 89 66 52 51 50	0.54	14	19	2	1	–	–	–	–	–	–	–	–	1	π-Cyclopentadienyl-π-2-ethylpiperidyl(dicarbonyl)iron		MS06630
289 155 290 274 209 291 210 182	289	100 26 17 10 8 6 6 4		15	15	3	1	1	–	–	–	–	–	–	–	–	Benzenesulphonamide, 4-methyl-N-(2-oxo-2-phenylethyl)-	30057-92-2	NI30005
289 155 274 209 210 182 180	289	100 26 10 8 6 5 3		15	15	3	1	1	–	–	–	–	–	–	–	–	N-Tosyl-2-aminoacetophenone		LI06268
260 289 261 258 143 115 202 89	289	100 32 22 13 11 9 8 7		15	15	5	1	–	–	–	–	–	–	–	–	–	Dimethyl 1-propionylindolizine-2,3-dicarboxylate		NI30006
289 274 246 290 275 228 200 144	289	100 39 39 17 7 7 7 7		15	15	5	1	–	–	–	–	–	–	–	–	–	Ethyl 3,4-dihydro-3-methyl-8-oxopyrano[2,3-f][1,4]benzoxazine-6-carboxylate		LI06269
274 289 230 258 244 259 245 275	289	100 83 37 35 31 24 18 17		15	15	5	1	–	–	–	–	–	–	–	–	–	Isoacronycidine	476-27-7	NI30007
274 289 230 244 259 245 275 231	289	100 83 37 31 24 18 17 15		15	15	5	1	–	–	–	–	–	–	–	–	–	Isoacronycidine	476-27-7	LI06270
146 205 43 247 289 145 147 117	289	100 52 42 39 14 13 11 10		15	15	5	1	–	–	–	–	–	–	–	–	–	Methyl 5-acetoxy-N-acetylindole-3-acetate		MS06631
289 161 148 254 291 162 147 107	289	100 91 66 62 37 37 37 28		15	16	1	3	–	–	1	–	–	–	–	–	–	4-Pyridinecarboxaldehyde, 5-(2-chloroethyl)-3-hydroxy-2-methyl-, phenylhydrazone	72088-08-5	NI30008
249 216 289 43 175 250 54 45	289	100 55 37 22 16 15 13 13		15	19	3	3	–	–	–	–	–	–	–	–	–	N-Cyanoethyl-N-methoxycarbonylethyl-3-acetamidoaniline		IC03703
289 162 287 59 137 124 78 285	289	100 96 66 62 47 47 45 25		15	20	1	1	–	–	–	–	–	–	–	–	1	Cyclopentadienyl(5-exo-morpholinocyclohexa-1,3-diene)cobalt		LI06271
136 289 246 247 109 248 291 165	289	100 56 53 42 25 24 21 19		16	16	–	1	1	–	1	–	–	–	–	–	–	1,5-Benzothiazepine, 2-methyl-4-(4-chlorophenyl)-2,3,4,5-tetrahydro-	78031-28-4	NI30009
105 77 85 217 140 167 57 51	289	100 81 79 76 70 68 56 52	15.00	16	19	4	1	–	–	–	–	–	–	–	–	–	Acetic acid, benzoyloxy-sec-butylcyano-, ethyl ester		LI06272
105 122 77 84 123 44 151 179	289	100 40 30 14 12 12 10 9	0.00	16	19	4	1	–	–	–	–	–	–	–	–	–	Acetic acid, benzoyloxy-sec-butylcyano-, ethyl ester		LI06273
105 122 77 123 44 85 43 41	289	100 38 33 16 16 15 15 14	0.00	16	19	4	1	–	–	–	–	–	–	–	–	–	Acetic acid, isobutyl(benzoyloxy)cyano-, ethyl ester		LI06274
290 168 124 196 272	289	100 81 11 8 3	0.00	16	19	4	1	–	–	–	–	–	–	–	–	–	8-Azabicyclo[3.2.1]octane-2-carboxylic acid, 3-(benzoyloxy)-8-methyl-, [1R-(exo,exo)]-	519-09-5	NI30010
39 44 42 51 41 77 81 40	289	100 84 75 48 46 39 36 30	0.01	16	19	4	1	–	–	–	–	–	–	–	–	–	8-Azabicyclo[3.2.1]octane-2-carboxylic acid, 3-(benzoyloxy)-8-methyl-, [1R-(exo,exo)]-	519-09-5	NI30011
289 202 272 201 288 224 217 173	289	100 20 17 10 9 9 9 7		16	19	4	1	–	–	–	–	–	–	–	–	–	Crinane, dihydroxy-		LI06275
105 77 84 44 51 41 106 57	289	100 29 14 14 10 10 8 8	0.00	16	19	4	1	–	–	–	–	–	–	–	–	–	Hexanoic acid, 2-cyano-2-hydroxy-, ethyl ester, benzoate (ester)	19788-60-4	NI30012
288 289 269 226 254 227 271 270	289	100 45 16 16 11 11 10 8		16	19	4	1	–	–	–	–	–	–	–	–	–	Licorine, dihydro-		LI06276
135 217 136 163 160 218 162 134	289	100 49 45 10 10 9 8 8	1.96	16	19	4	1	–	–	–	–	–	–	–	–	–	4,5-Oxazolidinedione, 2-(1-adamantyl)-2-(2-propynoxy)-		LI06277

MASS TO CHARGE RATIOS								M.W.	INTENSITIES								Parent	C	H	O	N	S	F	Cl	Br	I	Si	P	B	X	COMPOUND NAME	CAS Reg No	No
175	289	174	148	149	115	272	230	**289**	100	90	38	25	20	20	17	7		16	19	4	1	–	–	–	–	–	–	–	–	–	Pancracine, dihydro-		LI06278
175	289	174	176	290	173	148	256	**289**	100	67	31	20	13	12	11	9		16	19	4	1	–	–	–	–	–	–	–	–	–	Pancracine, dihydro-, (3α)-		NI30013
201	115	289	230	59	173	172	143	**289**	100	83	60	43	43	42	32	24		16	19	4	1	–	–	–	–	–	–	–	–	–	2,4-Pentadienamide, 5-(1,3-benzodioxol-5-yl)-N-(2-hydroxyisobutyl)-, (E,E)-		NI30014
105	77	85	217	140	167	57	51	**289**	100	81	80	76	71	69	58	53	**16.27**	16	19	4	1	–	–	–	–	–	–	–	–	–	Valeric acid, 2-cyano-2-hydroxy-3-methyl-, ethyl ester, benzoate (ester)	19788-62-6	NI30015
105	77	44	85	43	41	54	51	**289**	100	32	15	14	14	14	12	12	**0.00**	16	19	4	1	–	–	–	–	–	–	–	–	–	Valeric acid, 2-cyano-2-hydroxy-4-methyl-, ethyl ester, benzoate (ester)	19788-61-5	NI30016
72	289	139	185	183	217	63	107	**289**	100	48	20	16	15	11	10	7	**2.00**	16	20	–	1	1	–	–	–	–	–	1	–	–	N-Butyl-P,P-diphenylphosphinothioic amide		MS06632
289	215	153	113	183	181	78	150	**289**	100	14	14	14	12	11	11	10		16	21	–	–	–	–	–	–	–	1	–	–	1	Titanium, η^8-cyclooctatetraene-η^5-[(trimethylsilyl)cyclopentadienyl]-	55960-19-5	NI30017
202	73	289	203	200	45	290	201	**289**	100	59	29	21	8	8	7	7		16	23	2	1	–	–	–	–	–	1	–	–	–	1H-Indole-3-butanoic acid, 1-(trimethylsilyl)-, methyl ester	55590-99-3	NI30018
266	281	192	222	44	193	267	116	**289**	100	98	72	43	32	22	21	21	**0.00**	16	23	2	1	–	–	–	–	–	1	–	–	–	1H-Indole-2-carboxylic acid, 5-ethyl-1-(trimethylsilyl)-, ethyl ester	74367-43-4	NI30019
246	274	171	156	289	216	73	247	**289**	100	76	62	41	37	25	20	19		16	23	2	1	–	–	–	–	–	1	–	–	–	1H-Indole-2-carboxylic acid, 5-ethyl-1-(trimethylsilyl)-, ethyl ester	74367-43-4	NI30020
192	274	83	55	234	246	135	206	**289**	100	93	81	41	36	33	24	21	**9.00**	16	23	2	3	–	–	–	–	–	–	–	–	–	Alchornidine		LI06279
57	233	177	119	84	83	58	91	**289**	100	15	15	15	15	15	13	11	**0.70**	16	23	2	3	–	–	–	–	–	–	–	–	–	2-tert-Butyl-3-tert-butylimino-5-phenylimino-1,4-dioxazolidine		MS06633
57	133	177	83	77	93	91	233	**289**	100	80	41	26	26	22	20	14	**0.00**	16	23	2	3	–	–	–	–	–	–	–	–	–	2-tert-Butyl-3-tert-butylimino-4-phenyl-1,2,4-oxadiazolidin-5-one		MS06634
91	146	216	289	65	258	55	198	**289**	100	35	30	11	10	9	6	5		17	23	3	1	–	–	–	–	–	–	–	–	–	4H-Azepin-4-one, 1-benzyl-5-[2-(methoxycarbonyl)ethyl]-1,2,3,5,6,7-hexahydro-		DD01007
124	290	288	140					**289**	100	2	1	1					**0.00**	17	23	3	1	–	–	–	–	–	–	–	–	–	Benzeneacetic acid, α-(hydroxymethyl)-, 8-methyl-8-azabicyclo[3.2.1]oct-3-yl ester endo-(±)-	51-55-8	NI30021
124	82	28	42	83	94	96	67	**289**	100	43	36	30	24	22	18	18	**7.81**	17	23	3	1	–	–	–	–	–	–	–	–	–	Benzeneacetic acid, α-(hydroxymethyl)-, 8-methyl-8-azabicyclo[3.2.1]oct-3-yl ester endo-(±)-	51-55-8	NI30022
124	82	94	83	289	125	140	96	**289**	100	21	19	17	14	10	9	9		17	23	3	1	–	–	–	–	–	–	–	–	–	Benzeneacetic acid, α-(hydroxymethyl)-, 8-methyl-8-azabicyclo[3.2.1]oct-3-yl ester endo-(±)-	51-55-8	NI30023
124	83	82	94	289	42	96	125	**289**	100	24	23	20	17	13	12	11		17	23	3	1	–	–	–	–	–	–	–	–	–	Benzeneacetic acid, α-(hydroxymethyl)-, 8-methyl-8-azabicyclo[3.2.1]oct-3-yl ester, [3(S)-endo]-	101-31-5	NI30024
124	125	290	291	110	96	121	103	**289**	100	10	8	2	2	2	1	1	**0.62**	17	23	3	1	–	–	–	–	–	–	–	–	–	Benzeneacetic acid, α-(hydroxymethyl)-, 8-methyl-8-azabicyclo[3.2.1]oct-3-yl ester, [3(S)-endo]-	101-31-5	NI30025
120	246	105	28	95	44	43	29	**289**	100	80	73	36	29	26	26	22	**13.24**	17	23	3	1	–	–	–	–	–	–	–	–	–	Cyclobutanecarboxylic acid, 3-methyl-1-[[(1-phenylethyl)amino]carbonyl]-, ethyl ester	55649-63-3	NI30026
120	246	105	28	95	44	43	29	**289**	100	79	71	37	29	25	25	22	**11.11**	17	23	3	1	–	–	–	–	–	–	–	–	–	Cyclobutanecarboxylic acid, 3-methyl-1-[[(1-phenylethyl)amino]carbonyl]-, ethyl ester	55649-63-3	NI30027
120	105	246	113	77	67	104	79	**289**	100	82	63	26	22	21	20	17	**5.50**	17	23	3	1	–	–	–	–	–	–	–	–	–	Cyclobutanecarboxylic acid, 3-methyl-1-[[(R)-(1-phenylethyl)amino]carbonyl]-, ethyl ester	85317-95-9	NI30028
120	105	43	81	172	104	79	289	**289**	100	84	48	45	31	25	24	22		17	23	3	1	–	–	–	–	–	–	–	–	–	1-Cyclobutanol, 3-methyl-1-[[(1-phenylethyl)amino]carbonyl]-, acetate	85318-02-1	NI30029
105	120	204	133	131	121	214	215	**289**	100	72	17	11	11	8	7	5	**0.00**	17	23	3	1	–	–	–	–	–	–	–	–	–	Ethaneperoxoic acid, 1-cyano-1-phenyloctyl ester	58422-91-6	NI30030
230	43	105	77	109	104	231	69	**289**	100	54	51	36	21	21	17	15	**0.00**	17	23	3	1	–	–	–	–	–	–	–	–	–	4-Oxazolecarboxylic acid, 4,5-dihydro-5-methyl-4-(3-methylbutyl)-2-phenyl-, methyl ester, (4R,5R)-		DD01008
233	204	218	219	289	176	162	205	**289**	100	90	78	72	65	32	28	20		17	23	3	1	–	–	–	–	–	–	–	–	–	2(1H)-Quinolinone, 3-ethyl-4-hydroxy-7-methoxy-1-(3-methylbutyl)-	18904-41-1	NI30031
289	204	274	219	290	218	261	205	**289**	100	75	34	34	18	11	9	9		17	23	3	1	–	–	–	–	–	–	–	–	–	2(1H)-Quinolinone, 3-ethyl-4-hydroxy-1-methyl-7-(4-methylbutoxy)-		LI06280
219	289	218	204	246	290	220	203	**289**	100	81	79	53	18	14	14	14		17	23	3	1	–	–	–	–	–	–	–	–	–	2(1H)-Quinolinone, 6-methoxy-3-ethyl-4-isopentyloxy-	22048-14-2	NI30032
204	219	218	289	191	246	220	205	**289**	100	74	49	45	33	18	17	12		17	23	3	1	–	–	–	–	–	–	–	–	–	2(1H)-Quinolinone, 7-methoxy-3-ethyl-4-isopentyloxy-	22048-15-3	NI30033
204	219	218	289	191	246	220	290	**289**	100	74	49	45	33	18	17	12		17	23	3	1	–	–	–	–	–	–	–	–	–	2(1H)-Quinolinone, 7-methoxy-3-ethyl-4-isopentyloxy-	22048-15-3	LI06281
127	289	140	205	135	154	148	84	**289**	100	75	39	34	30	25	25	25		17	23	3	1	–	–	–	–	–	–	–	–	–	Tetrahydropiperine		LI06282
289	144	118	89	261				**289**	100	85	80	80	4					18	11	3	1	–	–	–	–	–	–	–	–	–	α-Phenyl-β-hydroxy-γ-cyanobenzylidene-Δ(α,β)-butenolide		LI06283
289	259	232	290	260	231	102	165	**289**	100	99	26	24	22	22	22	20		18	15	1	3	–	–	–	–	–	–	–	–	–	Pyrimidine, 2-(hydroxyiminoethyl)-4,6-diphenyl-	79691-73-9	NI30034
274	288	246	232	164	136	81	91	**289**	100	75	66	66	45	43	42	38	**16.99**	18	27	2	1	–	–	–	–	–	–	–	–	–	6-Azaandrost-4-en-7-one, 17β-hydroxy-	16373-59-4	NI30035
289	203	274	55	41	178	57	232	**289**	100	93	49	47	41	34	28	23		18	27	2	1	–	–	–	–	–	–	–	–	–	N-(3,5-Di-tert-butyl-4-hydroxybenzyl)acrylamide		IC03704
86	99	91	144	58	56	41	87	**289**	100	16	14	11	9	9	7	6	**0.00**	18	27	2	1	–	–	–	–	–	–	–	–	–	Cyclopentanecarboxylic acid, 1-phenyl-, 2-(diethylamino)ethyl ester	77-22-5	NI30036
57	217	216	132	160	117	41	85	**289**	100	76	54	54	27	22	19	17	**0.00**	18	27	2	1	–	–	–	–	–	–	–	–	–	1-Isoquinolinemethanol, u-2-tert-butoxy-α-ethyl-1,2,3,4-tetrahydro-α-methyl-		DD01009
91	133	84	117	105	103	115	132	**289**	100	98	41	31	22	19	17	16	**1.40**	18	27	2	1	–	–	–	–	–	–	–	–	–	Pentanoic acid, 3-methyl-2-[(1-methyl-1-phenylpropyl)imino]-, ethyl ester	49707-59-7	NI30037
121	204	148	84	29	28	57	85	**289**	100	41	28	18	14	12	11	10	**0.00**	18	27	2	1	–	–	–	–	–	–	–	–	–	4-(3-Piperidin-1-ylpropionyl)-1-butoxybenzene		NI30038
98	290	288	205	234				**289**	100	15	11	3	2				**0.00**	18	27	2	1	–	–	–	–	–	–	–	–	–	1-Propanone, 1-(4-butoxyphenyl)-3-(1-piperidinyl)-	586-60-7	NI30039
120	271	145	204	219	91	57	117	**289**	100	95	95	84	45	40	40	30	**10.00**	18	27	2	1	–	–	–	–	–	–	–	–	–	Valeramide, N-(α-valerylphenethyl)-	27820-20-8	NI30040
180	121	77	289	122	288			**289**	100	48	29	5	4	3				19	15	–	1	1	–	–	–	–	–	–	–	–	Benzenecarbothioamide, N,N-diphenyl-	17435-12-0	NI30041
196	168	77	43	289	57	55	29	**289**	100	58	45	40	39	35	35	32		19	15	2	1	–	–	–	–	–	–	–	–	–	Carbamic acid, diphenyl-, phenyl ester	5416-45-5	NI30042
196	289	168	77	167	51	197	290	**289**	100	37	35	31	14	14	13	9		19	15	2	1	–	–	–	–	–	–	–	–	–	Carbamic acid, diphenyl-, phenyl ester	5416-45-5	NI30043

MASS TO CHARGE RATIOS								M.W.	INTENSITIES								Parent	C	H	O	N	S	F	Cl	Br	I	Si	P	B	X	COMPOUND NAME	CAS Reg No	No
91	65	51	77	108	78	115	205	**289**	100	35	32	18	15	13	12	11	**0.00**	19	15	2	1	–	–	–	–	–	–	–	–	–	2-Furancarbonitrile, 2,5-dihydro-3,4-dibenzyl-5-oxo-	122949-78-4	DD01010
274	275	150	276	110	55	81	53	**289**	100	20	12	4	4	4	3	3	**2.00**	19	31	1	1	–	–	–	–	–	–	–	–	–	17a-Aza-5α-androstan-17-one		LI06284
289	41	124	95	55	67	272	79	**289**	100	69	66	59	54	53	48	45		19	31	1	1	–	–	–	–	–	–	–	–	–	Androstan-2-one, oxime, (5α)-	14614-11-0	LI06285
289	41	124	95	55	67	272	79	**289**	100	69	66	57	53	51	49	44		19	31	1	1	–	–	–	–	–	–	–	–	–	Androstan-2-one, oxime, (5α)-	14614-11-0	NI30044
41	112	55	67	79	81	91	53	**289**	100	99	92	90	77	64	52	52	**12.00**	19	31	1	1	–	–	–	–	–	–	–	–	–	Androstan-3-one, oxime, (5α)-	14546-37-3	NI30045
289	41	124	95	55	67	272	79	**289**	100	70	67	58	54	52	47	44		19	31	1	1	–	–	–	–	–	–	–	–	–	Androstan-3-one, oxime, (5α)-	14546-37-3	NI30046
272	289	41	95	56	55	81	67	**289**	100	68	40	37	35	31	30	30		19	31	1	1	–	–	–	–	–	–	–	–	–	Androstan-4-one, oxime, (5α)-	54156-18-2	NI30047
41	55	79	67	81	53	95	29	**289**	100	72	63	59	46	42	41	38	**29.00**	19	31	1	1	–	–	–	–	–	–	–	–	–	Androstan-6-one, oxime, (5α)-	54156-19-3	NI30048
41	55	95	67	289	81	79	272	**289**	100	81	78	77	73	65	57	50		19	31	1	1	–	–	–	–	–	–	–	–	–	Androstan-7-one, oxime, (5α)-	54156-20-6	NI30049
41	55	166	67	81	79	53	29	**289**	100	99	94	91	72	59	49	45	**19.00**	19	31	1	1	–	–	–	–	–	–	–	–	–	Androstan-11-one, oxime, (5α)-	14475-43-5	NI30050
41	55	166	67	81	79	53	95	**289**	100	99	94	91	71	59	48	44	**18.82**	19	31	1	1	–	–	–	–	–	–	–	–	–	Androstan-11-one, oxime, (5α)-	14475-43-5	NI30051
217	274	258	55	67	41	289	81	**289**	100	47	36	36	34	33	32	32		19	31	1	1	–	–	–	–	–	–	–	–	–	Androstan-16-one, oxime, (5α)-	54156-21-7	NI30052
200	272	28	41	55	67	81	109	**289**	100	99	50	43	35	31	27	26	**6.00**	19	31	1	1	–	–	–	–	–	–	–	–	–	Androstan-17-one, oxime, (5α)-	1035-62-7	NI30053
57	41	121	42	69	56	55	43	**289**	100	60	32	27	21	18	13	11	**0.20**	19	31	1	1	–	–	–	–	–	–	–	–	–	Aziridinone, 1-tert-butyl-3-(3,5,7-trimethyltricyclo[3.3.1.1^{3,7}]dec-1-yl)-	18606-19-4	NI30054
105	77	135	134	43	41	148	106	**289**	100	31	26	22	12	12	10	8	**2.97**	19	31	1	1	–	–	–	–	–	–	–	–	–	Benzamide, N-dodecyl-	33140-65-7	NI30055
105	228	77	289	106	288	148	43	**289**	100	15	15	8	7	5	4	4		19	31	1	1	–	–	–	–	–	–	–	–	–	Benzamide, N,N-dihexyl-	53044-17-0	NI30056
86	85	56	57	289	41	95	55	**289**	100	82	46	33	24	22	20	19		19	31	1	1	–	–	–	–	–	–	–	–	–	Deoxysamandarine		LI06287
86	85	56	57	289	41	95	55	**289**	100	84	46	34	25	23	22	20		19	31	1	1	–	–	–	–	–	–	–	–	–	Deoxysamandarine		LI06286
274	275	150	276	110	55	67	53	**289**	100	21	12	4	4	4	3	3	**2.14**	19	31	1	1	–	–	–	–	–	–	–	–	–	Naphtho[2,1-f]quinolin-2(1H)-one, 3,4,4aα,4bβ,5,6,6aα,7,8,9,10,10a,10bα,11,12,12a-hexadecahydro-10aβ,12aβ-dimethyl-	17556-10-4	NI30057
244	222	245	122	202	204	148	176	**289**	100	49	21	20	10	9	7	4	**4.00**	19	31	1	1	–	–	–	–	–	–	–	–	–	Pyrrolo[1,2-a]quinoline-1-ethanol, dodecahydro-6-(2,4-pentadienyl)-, [1Z-[1α,3aβ,5aα,6α(S),9aα]]-	63983-58-4	NI30058
44	57	128	261	230	43	96	58	**289**	100	72	39	34	33	32	26	26	**12.00**	19	31	1	1	–	–	–	–	–	–	–	–	–	Samanone		LI06288
258	77	259	169	180	104	168	51	**289**	100	24	21	13	9	9	8	8	**6.62**	20	19	1	1	–	–	–	–	–	–	–	–	–	[1,1′-Biphenyl]ethanol, ar′-amino-β-phenyl-	61142-84-5	NI30059
289	288	261	150	260	151	121	115	**289**	100	75	42	37	32	24	17	15		20	19	1	1	–	–	–	–	–	–	–	–	–	Pyridine, 2,4,5-trimethyl-3-phenyl-6-phenoxy-		DD01011
288	289	77	115	260	51	261	105	**289**	100	98	72	36	31	28	26	20		20	19	1	1	–	–	–	–	–	–	–	–	–	2(1H)-Pyridinone, 1,4,5-trimethyl-3,6-diphenyl-	52148-67-1	MS06635
288	289	115	290	202	260	273	261	**289**	100	80	20	14	11	7	6	6		20	19	1	1	–	–	–	–	–	–	–	–	–	4(1H)-Pyridinone, 1,2,6-trimethyl-3,5-diphenyl-	42215-29-2	MS06636
70	81	206	41	55	29	95	67	**289**	100	69	51	48	31	28	27	18	**1.09**	20	35	–	1	–	–	–	–	–	–	–	–	–	2,7-Octadien-1-amine, 2,7-dimethyl-N-(2,7-dimethyl-1,7-octadien-3-yl)-		NI30060
81	41	30	55	70	29	95	96	**289**	100	73	57	46	45	40	39	32	**2.76**	20	35	–	1	–	–	–	–	–	–	–	–	–	2,7-Octadien-1-amine, 2,7-dimethyl-N-(2,7-dimethyl-2,7-octadien-1-yl)-		NI30061
81	41	70	30	55	29	96	121	**289**	100	73	52	49	43	38	33	33	**4.82**	20	35	–	1	–	–	–	–	–	–	–	–	–	2,7-Octadien-1-amine, 2,7-dimethyl-N-(2,7-dimethyl-2,7-octadien-1-yl)-, (E,Z)-	77984-60-2	NI30062
213	211	215	132	147	145	134	178	**290**	100	62	45	37	33	25	24	10	**1.08**	2	–	–	–	–	2	2	2	–	–	–	–	–	1,2-Difluoro-1,2-dichloro-1,2-dibromoethane		DO01082
171	173	76	175	36	256	112	78	**290**	100	99	57	32	28	25	23	18	**0.00**	4	4	–	2	–	–	6	–	–	–	–	–	–	Ethanimidamide, 2,2-dichloro-N-(trichloroethenyl)-, monohydrochloride	40645-79-2	NI30063
131	43	45	100	58	119	69	169	**290**	100	59	31	25	24	13	13	9	**0.00**	6	2	4	–	–	8	–	–	–	–	–	–	–	Hexanedioic acid, octafluoro-	336-08-3	NI30064
129	255	47	257	162	131	63	50	**290**	100	92	75	63	42	37	33	29	**20.34**	7	6	2	–	1	–	3	–	–	–	1	–	–	Phosphorochloridothioic acid, O-(2,4-dichlorophenyl) O-methyl ester	946-04-3	NI30065
56	87	234	135	127	166	206	97	**290**	100	84	60	60	60	52	50	50	**2.00**	7	7	4	–	–	–	–	1	–	–	–	–	1	(2-Methoxy-π-allyl)iron tricarbonyl bromide		MS06637
28	178	74	57	55	150	56	31	**290**	100	99	99	99	99	80	66	50	**25.00**	8	2	5	–	–	3	–	–	–	–	–	–	1	Manganese, tetracarbonyl(3,3,3-trifluoro-2-formyl-1-propenyl-C,O)-, (OC-6-23)-	56271-17-1	NI30066
292	294	290	75	185	183	74	29	**290**	100	54	54	42	21	21	20	12		8	4	2	–	–	–	–	2	–	–	–	–	–	3,3′-Dibromo-2,2′-bifuran		NI30067
211	213	292	76	185	183	38	290	**290**	100	99	69	56	47	47	46	38		8	4	2	–	–	–	–	2	–	–	–	–	–	5,5′-Dibromo-2,2′-bifuran		MS06638
29	97	27	125	61	153	65	213	**290**	100	70	69	51	37	32	31	23	**0.10**	8	19	3	–	3	–	–	–	–	–	1	–	–	Phosphorodithioic acid, O,O-diethyl S-[2-(ethylsulphinyl)ethyl] ester	2497-07-6	NI30068
45	153	197	141	97	125	121	65	**290**	100	79	77	64	58	42	31	26	**0.00**	8	19	5	–	2	–	–	–	–	–	1	–	–	Phosphorothioic acid, O,O-diethyl O-[2-(ethylsulfonyl)ethyl] ester	4891-54-7	NI30069
29	27	65	97	45	28	93	121	**290**	100	68	47	45	39	32	28	25	**0.00**	8	19	5	–	2	–	–	–	–	–	1	–	–	Phosphorothioic acid, O,O-diethyl O-[2-(ethylsulphonyl)ethyl] ester	4891-54-7	NI30070
29	197	109	141	27	81	45	61	**290**	100	78	78	69	69	58	52	35	**0.00**	8	19	5	–	2	–	–	–	–	–	1	–	–	Phosphorothioic acid, O,O-diethyl S-[2-(ethylsulphonyl)ethyl] ester	2496-91-5	NI30071
63	65	121	93	97	217	170	81	**290**	100	99	68	67	60	48	32	31	**16.70**	8	20	5	–	1	–	–	–	–	–	2	–	–	Thiohypophosphoric acid, tetraethyl ester	5935-34-2	NI30072
99	155	43	127	81	109	82	45	**290**	100	96	68	57	48	37	32	20	**0.00**	8	20	7	–	–	–	–	–	–	–	2	–	–	Diphosphoric acid, tetraethyl ester	107-49-3	NI30073
99	127	81	82	29	111	27	45	**290**	100	68	58	45	44	38	36	23	**0.00**	8	20	7	–	–	–	–	–	–	–	2	–	–	Diphosphoric acid, tetraethyl ester	107-49-3	NI30074
161	27	263	99	29	81	179	26	**290**	100	86	73	73	73	68	64	64	**5.45**	8	20	7	–	–	–	–	–	–	–	2	–	–	Diphosphoric acid, tetraethyl ester	107-49-3	IC03705
140	186	56	121	234	177	65	75	**290**	100	49	46	39	38	21	16	15	**1.93**	9	5	4	–	–	3	–	–	–	–	–	–	1	Iron, (trifluoroacetoxy)dicarbonyl-π-cyclopentadienyl-	37048-23-0	NI30075
57	56	55	58	275	117	290	164	**290**	100	93	57	54	41	37	34	21		10	9	1	–	–	6	–	–	–	–	1	–	–	Phosphine oxide, tert-butylfluoro(pentafluorophenyl)-	53381-02-5	NI30076
262	160	290	76	88	116	173	264	**290**	100	68	62	37	30	29	27	19		10	10	2	–	4	–	–	–	–	–	–	–	–	2-Methyl-6(7)-ethoxycarbonyltetrathiafulvalene		MS06639
163	79	107	164	91	41	55	77	**290**	100	43	23	15	15	12	11	7	**0.50**	10	11	2	–	–	–	–	–	1	–	–	–	–	2,6-Adamantanedione, 4-iodo-, (1R)-	19305-95-4	MS06640
220	71	43	290	41	93	64	27	**290**	100	87	77	37	14	13	12	12		10	11	2	–	–	–	–	–	1	–	–	–	–	Butyric acid, 3-iodophenyl ester	29052-07-1	NI30077

MASS TO CHARGE RATIOS								M.W.	INTENSITIES								Parent	C	H	O	N	S	F	Cl	Br	I	Si	P	B	X	COMPOUND NAME	CAS Reg No	No
43	71	220	64	41	93	63	65	**290**	100	57	38	19	19	16	16	15	**3.70**	10	11	2	–	–	–	–	–	1	–	–	–	–	Butyric acid, 4-iodophenyl ester	29052-08-2	NI30078
74	87	230	290	84	69	217	96	**290**	100	63	54	41	27	27	25	24		10	11	2	–	–	–	–	–	1	–	–	–	–	Propanoic acid, (4-iodophenyl)-, methyl ester	33994-44-4	NI30079
99	28	71	98	72	39	59	45	**290**	100	40	31	25	24	13	11	10	**0.00**	10	11	4	2	1	–	1	–	–	–	–	–	–	2H-1,2,3-Thiadiazine, 6-methyl-2-phenyl-, monoperchlorate	57954-50-4	NI30080
90	91	64	38	289	114	50	62	**290**	100	10	9	7	6	4	4	3	**0.00**	10	12	–	–	–	–	–	2	–	–	–	–	–	Benzene, [3-bromo-2-(bromomethyl)propyl]-	35694-75-8	NI30081
117	91	131	104	77	78	116	132	**290**	100	97	56	44	44	33	31	29	**0.00**	10	12	–	–	–	–	–	2	–	–	–	–	–	Cyclopropane, 3,3-dibromo-2,2-dicyclopropyl-1-methylene-		MS06641
117	91	131	104	77	132	79	78	**290**	100	99	57	48	48	43	43	33	**0.00**	10	12	–	–	–	–	–	2	–	–	–	–	–	Cyclopropane, 1-(dibromomethylene)-2,2-dicyclopropyl-		MS06642
230	232	174	176	56	231	229	290	**290**	100	65	57	35	26	21	17	14		11	12	1	2	1	–	2	–	–	–	–	–	–	1-(3,4-Dichlorophenyl)-3-(ethylthio)imidazolidinone		NI30082
207	179	29	133	77	290	208	104	**290**	100	60	40	15	14	12	12	10		11	12	3	2	–	–	2	–	–	–	–	–	–	N-Dichloroacetyl-N-ethyl-2-methyl-5-nitroaniline		IC03706
91	148	147	92	94	76	63	77	**290**	100	30	16	12	10	9	9	8	**0.00**	11	12	3	2	–	–	2	–	–	–	–	–	–	DL-Phenylalanine, N-glycyl-, dichloro-	56247-33-7	NI30083
89	73	275	187	159	59	231	215	**290**	100	64	50	46	46	42	33	32	**0.00**	11	18	7	–	–	–	–	–	–	1	–	–	–	2,3-Furandicarboxylic acid, tetrahydro-5-oxo-3-[(trimethylsilyl)oxy]-, dimethyl ester	56051-88-8	NI30084
275	73	187	189	203	191	59	219	**290**	100	51	29	23	15	15	15	14	**0.10**	11	30	1	–	–	–	–	–	–	4	–	–	–	1-Oxa-2,4,6,8-tetrasilacyclooctane, 2,2,4,4,6,6,8,8-octamethyl-	24590-70-3	LI06289
275	73	187	189	59	191	203	219	**290**	100	51	29	23	19	17	15	14	**0.10**	11	30	1	–	–	–	–	–	–	4	–	–	–	1-Oxa-2,4,6,8-tetrasilacyclooctane, 2,2,4,4,6,6,8,8-octamethyl-	24590-70-3	NI30085
292	290	294	257	255	220	222	296	**290**	100	76	51	14	14	12	9	6		12	6	–	–	–	–	4	–	–	–	–	–	–	1,1′-Biphenyl, 2,2′,3,3′-tetrachloro-	38444-93-8	NI30087
220	292	222	255	257	290	110	150	**290**	100	70	63	61	59	57	46	39		12	6	–	–	–	–	4	–	–	–	–	–	–	1,1′-Biphenyl, 2,2′,3,3′-tetrachloro-	38444-93-8	NI30086
220	292	110	290	222	255	257	150	**290**	100	83	72	66	64	51	49	49		12	6	–	–	–	–	4	–	–	–	–	–	–	1,1′-Biphenyl, 2,2′,3,3′-tetrachloro-	38444-93-8	NI30088
292	220	290	222	294	110	150	111	**290**	100	81	79	52	47	41	32	24		12	6	–	–	–	–	4	–	–	–	–	–	–	1,1′-Biphenyl, 2,2′,3,4-tetrachloro-	52663-59-9	NI30089
220	292	222	290	110	150	255	92	**290**	100	76	66	62	56	45	44	44		12	6	–	–	–	–	4	–	–	–	–	–	–	1,1′-Biphenyl, 2,2′,3,4′-tetrachloro-	36559-22-5	NI30090
220	292	290	110	222	92	111	150	**290**	100	96	74	72	63	52	52	50		12	6	–	–	–	–	4	–	–	–	–	–	–	1,1′-Biphenyl, 2,2′,3,5-tetrachloro-	70362-46-8	NI30091
220	292	222	290	255	257	110	150	**290**	100	73	68	60	50	48	48	43		12	6	–	–	–	–	4	–	–	–	–	–	–	1,1′-Biphenyl, 2,2′,3,5′-tetrachloro-	41464-39-5	NI30092
292	290	220	257	255	222	296	224	**290**	100	77	12	10	10	9	7	4		12	6	–	–	–	–	4	–	–	–	–	–	–	1,1′-Biphenyl, 2,2′,3,5′-tetrachloro-	41464-39-5	NI30093
220	222	150	292	290	255	257	127	**290**	100	63	57	40	33	29	27	24		12	6	–	–	–	–	4	–	–	–	–	–	–	1,1′-Biphenyl, 2,2′,3,5′-tetrachloro-	41464-39-5	NI30094
220	292	222	110	290	255	257	92	**290**	100	73	67	64	56	54	53	48		12	6	–	–	–	–	4	–	–	–	–	–	–	1,1′-Biphenyl, 2,2′,3,6′-tetrachloro-	41464-47-5	NI30095
220	150	222	292	290	149	184	123	**290**	100	69	65	38	35	30	28	26		12	6	–	–	–	–	4	–	–	–	–	–	–	1,1′-Biphenyl, 2,2′,4,4′-tetrachloro-	2437-79-8	NI30096
292	290	294	220	110	222	92	111	**290**	100	79	48	47	36	30	25	22		12	6	–	–	–	–	4	–	–	–	–	–	–	1,1′-Biphenyl, 2,2′,4,4′-tetrachloro-	2437-79-8	NI30097
292	290	220	150	294	127	222	128	**290**	100	78	56	52	42	36	34	31		12	6	–	–	–	–	4	–	–	–	–	–	–	1,1′-Biphenyl, 2,2′,4,5′-tetrachloro-	41464-40-8	NI30099
220	150	222	149	128	292	127	123	**290**	100	86	62	43	34	32	31	31	**25.81**	12	6	–	–	–	–	4	–	–	–	–	–	–	1,1′-Biphenyl, 2,2′,4,5′-tetrachloro-	41464-40-8	NI30098
292	220	290	222	149	294	150	110	**290**	100	93	77	62	57	48	20	15		12	6	–	–	–	–	4	–	–	–	–	–	–	1,1′-Biphenyl, 2,2′,4,5′-tetrachloro-	41464-40-8	NI30100
220	222	292	290	150	110	294	255	**290**	100	75	74	61	47	46	38	35		12	6	–	–	–	–	4	–	–	–	–	–	–	1,1′-Biphenyl, 2,2′,4,6-tetrachloro-	62796-65-0	NI30101
292	220	290	222	149	294	150	110	**290**	100	98	79	66	66	24	24	20		12	6	–	–	–	–	4	–	–	–	–	–	–	1,1′-Biphenyl, 2,2′,5,5′-tetrachloro-	35693-99-3	NI30104
220	150	222	149	292	184	146	127	**290**	100	76	63	35	33	28	28	28		12	6	–	–	–	–	4	–	–	–	–	–	–	1,1′-Biphenyl, 2,2′,5,5′-tetrachloro-		NI30103
292	290	230	294	232	293	291	184	**290**	100	77	70	49	44	16	13	13		12	6	–	–	–	–	4	–	–	–	–	–	–	1,1′-Biphenyl, 2,2′,5,5′-tetrachloro-	35693-99-3	NI30102
292	220	290	222	110	255	294	257	**290**	100	96	80	59	50	45	44	44		12	6	–	–	–	–	4	–	–	–	–	–	–	1,1′-Biphenyl, 2,2′,5,6′-tetrachloro-	41464-41-9	NI30105
292	220	290	222	110	294	150	92	**290**	100	97	77	64	60	46	44	41		12	6	–	–	–	–	4	–	–	–	–	–	–	1,1′-Biphenyl, 2,2′,5,6′-tetrachloro-	41464-41-9	NI30106
220	150	222	292	290	149	127	123	**290**	100	72	65	53	42	29	27	27		12	6	–	–	–	–	4	–	–	–	–	–	–	1,1′-Biphenyl, 2,2′,6,6′-tetrachloro-	15968-05-5	NI30108
292	290	220	294	222	110	150	111	**290**	100	77	64	24	22	13	12	8		12	6	–	–	–	–	4	–	–	–	–	–	–	1,1′-Biphenyl, 2,2′,6,6′-tetrachloro-	15968-05-5	NI30107
292	290	220	294	110	222	92	150	**290**	100	81	70	49	45	43	38	34		12	6	–	–	–	–	4	–	–	–	–	–	–	1,1′-Biphenyl, 2,2′,6,6′-tetrachloro-	15968-05-5	NI30109
292	220	290	222	110	294	92	150	**290**	100	83	80	54	50	49	38	33		12	6	–	–	–	–	4	–	–	–	–	–	–	1,1′-Biphenyl, 2,3,3′,4′-tetrachloro-	41464-43-1	NI30110
292	220	290	222	110	294	150	92	**290**	100	95	81	57	55	48	43	43		12	6	–	–	–	–	4	–	–	–	–	–	–	1,1′-Biphenyl, 2,3,3′,5′-tetrachloro-	41464-49-7	NI30111
220	150	222	292	290	184	128	127	**290**	100	73	70	67	54	33	31	31		12	6	–	–	–	–	4	–	–	–	–	–	–	1,1′-Biphenyl, 2,3,4,5-tetrachloro-	33284-53-6	NI30112
292	290	220	110	294	222	150	111	**290**	100	81	65	53	49	41	36	33		12	6	–	–	–	–	4	–	–	–	–	–	–	1,1′-Biphenyl, 2,3,4,5-tetrachloro-	33284-53-6	NI30113
292	220	290	294	222	150	185	110	**290**	100	80	79	48	48	26	14	14		12	6	–	–	–	–	4	–	–	–	–	–	–	1,1′-Biphenyl, 2,3,4′,6-tetrachloro-	52663-58-8	NI30114
150	222	292	127	184	128	290	149	**290**	100	96	56	54	50	49	47	42		12	6	–	–	–	–	4	–	–	–	–	–	–	1,1′-Biphenyl, 2,3,5,6-tetrachloro-	33284-54-7	NI30115
292	290	220	294	222	110	150	92	**290**	100	80	63	49	40	40	30	25		12	6	–	–	–	–	4	–	–	–	–	–	–	1,1′-Biphenyl, 2,3,5,6-tetrachloro-	33284-54-7	NI30116
292	290	294	293	220	296	222	288	**290**	100	79	53	28	7	6	5	2		12	6	–	–	–	–	4	–	–	–	–	–	–	1,1′-Biphenyl, 2,3′,4,4′-tetrachloro-	32598-10-0	NI30117
292	290	294	293	296	295			**290**	100	80	48	14	11	7				12	6	–	–	–	–	4	–	–	–	–	–	–	1,1′-Biphenyl, 2,3′,4,4′-tetrachloro-	32598-10-0	NI30118
292	290	220	294	110	222	92	111	**290**	100	78	56	49	40	35	28	25		12	6	–	–	–	–	4	–	–	–	–	–	–	1,1′-Biphenyl, 2,3′,4,4′-tetrachloro-	32598-10-0	NI30119
292	290	220	294	110	222	92	150	**290**	100	79	56	49	35	34	24	24		12	6	–	–	–	–	4	–	–	–	–	–	–	1,1′-Biphenyl, 2,3′,4,6-tetrachloro-	60233-24-1	NI30120
292	290	220	110	294	222	92	111	**290**	100	78	61	52	48	40	37	34		12	6	–	–	–	–	4	–	–	–	–	–	–	1,1′-Biphenyl, 2,3′,4′,5-tetrachloro-	32598-11-1	NI30121
292	290	220	294	222	110	150	111	**290**	100	78	73	48	48	38	26	26		12	6	–	–	–	–	4	–	–	–	–	–	–	1,1′-Biphenyl, 2,3′,4′,5-tetrachloro-	32598-11-1	NI30122
220	150	222	149	292	127	128	184	**290**	100	88	64	45	41	40	38	37	**33.54**	12	6	–	–	–	–	4	–	–	–	–	–	–	1,1′-Biphenyl, 2,3′,4′,5-tetrachloro-	32598-11-1	NI30123
292	220	290	222	110	294	92	150	**290**	100	88	75	55	48	46	38	32		12	6	–	–	–	–	4	–	–	–	–	–	–	1,1′-Biphenyl, 2,3′,4′,6-tetrachloro-	41464-46-4	NI30124
292	290	220	294	222	110	150	92	**290**	100	76	49	43	32	23	20	15		12	6	–	–	–	–	4	–	–	–	–	–	–	1,1′-Biphenyl, 2,3′,5,5′-tetrachloro-	41464-42-0	NI30125
292	290	220	294	222	110	150	74	**290**	100	71	60	50	40	32	31	25		12	6	–	–	–	–	4	–	–	–	–	–	–	1,1′-Biphenyl, 2,3′,5,5′-tetrachloro-	41464-42-0	NI30126
292	290	220	294	110	222	92	150	**290**	100	79	57	48	41	37	28	27		12	6	–	–	–	–	4	–	–	–	–	–	–	1,1′-Biphenyl, 2,4,4′,6-tetrachloro-	32598-12-2	NI30127
292	290	220	294	222	110	150	74	**290**	100	78	53	45	34	23	21	15		12	6	–	–	–	–	4	–	–	–	–	–	–	1,1′-Biphenyl, 2,4,4′,6-tetrachloro-	32598-12-2	NI30128

MASS TO CHARGE RATIOS								M.W.	INTENSITIES								Parent	C	H	O	N	S	F	Cl	Br	I	Si	P	B	X	COMPOUND NAME	CAS Reg No	No
220	150	292	146	184	128	290	127	**290**	100	74	42	38	36	36	34	34		12	6	–	–	–	–	4	–	–	–	–	–	–	**1,1′-Biphenyl, 3,3′,4,4′-tetrachloro-**	32598-13-3	**NI30129**
292	290	220	222	294	110	92	111	**290**	100	84	69	48	47	47	34	28		12	6	–	–	–	–	4	–	–	–	–	–	–	**1,1′-Biphenyl, 3,3′,4,4′-tetrachloro-**	32598-13-3	**NI30130**
292	290	220	294	222	110	111	92	**290**	100	75	56	45	37	34	25	24		12	6	–	–	–	–	4	–	–	–	–	–	–	**1,1′-Biphenyl, 3,3′,4,4′-tetrachloro-**	32598-13-3	**NI30131**
292	290	220	294	110	222	92	74	**290**	100	84	77	48	48	46	36	31		12	6	–	–	–	–	4	–	–	–	–	–	–	**1,1′-Biphenyl, 3,3′,4,5′-tetrachloro-**	41464-48-6	**NI30132**
292	290	220	294	222	110	150	92	**290**	100	83	76	48	47	39	25	25		12	6	–	–	–	–	4	–	–	–	–	–	–	**1,1′-Biphenyl, 3,3′,5,5′-tetrachloro-**	33284-52-5	**NI30133**
220	73	222	150	110	91	185	111	**290**	100	68	66	66	66	44	38	38	**28.57**	12	6	–	–	–	–	4	–	–	–	–	–	–	**1,1′-Biphenyl, tetrachloro-**	26914-33-0	**NI30134**
292	290	220	294	222	110	92	75	**290**	100	78	56	47	38	21	17	17		12	6	–	–	–	–	4	–	–	–	–	–	–	**1,1′-Biphenyl, tetrachloro-**	26914-33-0	**NI30136**
292	290	220	294	222	110	92	74	**290**	100	79	49	47	33	26	18	18		12	6	–	–	–	–	4	–	–	–	–	–	–	**1,1′-Biphenyl, tetrachloro-**	26914-33-0	**NI30135**
262	264	263	200	261	63	137	75	**290**	100	63	31	27	26	26	23	19	**15.00**	12	9	–	4	–	–	2	–	–	–	–	1	–	**Δ^2-Tetrazaboroline, 1,4-bis(4-chlorophenyl)-**	28149-53-3	**NI30137**
91	148	206	66	65	119	122	56	**290**	100	50	47	40	36	30	29	28	**1.69**	12	10	5	–	–	–	–	–	–	–	–	–	1	**Iron, [(2-methoxycarbonyl)norbornadiene]tricarbonyl-**		**NI30138**
55	146	91	72	206	164	65	188	**290**	100	69	45	44	21	21	20	17	**0.00**	12	11	5	–	–	–	–	–	–	–	–	–	1	**2-(Carboxyisopropyl)cyclopentadienylmanganese tricarbonyl**		**LI06290**
115	128	85	87	69	86	170	157	**290**	100	68	68	38	37	35	32	31	**0.00**	12	18	8	–	–	–	–	–	–	–	–	–	–	**L-Arabinofuranoside, methyl 2,3,5-tri-O-acetyl-**	90244-44-3	**NI30139**
43	115	103	145	128	86	73	44	**290**	100	10	9	8	3	3	2	2	**0.00**	12	18	8	–	–	–	–	–	–	–	–	–	–	**1,2,3,4-Butanetetrol, tetraacetate, (R*,S*)-**	7208-40-4	**NI30140**
43	115	145	103	217	128	86	44	**290**	100	14	11	11	9	5	4	3	**0.00**	12	18	8	–	–	–	–	–	–	–	–	–	–	**1,2,3,4-Butanetetrol, tetraacetate, [S-(R*,R*)]-**	49560-29-4	**NI30141**
73	45	99	127	91	126	43	71	**290**	100	88	75	57	41	36	36	33	**3.00**	12	18	8	–	–	–	–	–	–	–	–	–	–	**2,3-Bis(ethoxycarbonyl)-1,4,5,8-tetraoxadecalin**		**LI06291**
128	69	170	115	157	68	86	103	**290**	100	62	55	43	38	34	32	27	**0.00**	12	18	8	–	–	–	–	–	–	–	–	–	–	**α-D-Xylopyranoside, methyl 2,3,4-tri-O-acetyl-**	20880-54-0	**NI30142**
128	69	170	115	157	86	68	103	**290**	100	91	50	43	34	34	34	29	**0.00**	12	18	8	–	–	–	–	–	–	–	–	–	–	**β-D-Xylopyranoside, methyl 2,3,4-tri-O-acetyl-**	13007-37-9	**NI30143**
128	69	115	170	157	86	68	103	**290**	100	90	49	48	37	32	31	29	**0.00**	12	18	8	–	–	–	–	–	–	–	–	–	–	**β-D-Xylopyranoside, methyl 2,3,4-tri-O-acetyl-**	13007-37-9	**NI30144**
142	141	171	184	98	220	241	128	**290**	100	81	42	20	20	17	14	11	**0.00**	12	19	4	2	–	–	1	–	–	–	–	–	–	**5-(3-Chloro-2-hydroxypropyl)-5-(1-methylbutyl)barbituric acid**		**LI06292**
249	113	177	63	103	145	290	85	250	**290**	100	51	37	29	18	17	13	11		12	20	–	–	2	–	–	–	–	–	2	–	–	**Tetraallyldiphosphane disulphide**		**MS06643**
179	177	233	43	149	261	151	175	**290**	100	75	63	55	51	49	46	45	**3.10**	12	26	–	–	–	–	–	–	–	–	–	–	1	**Cyclohexyltriethyltin**		**MS06644**
247	161	245	163	205	159	243	203	**290**	100	94	91	70	63	63	57	50	**1.70**	12	26	–	–	–	–	–	–	–	–	–	–	1	**Cyclopropyltriisopropyltin**		**MS06645**
165	247	163	135	161	245	137	133	**290**	100	84	78	49	48	45	35	32	**2.00**	12	26	–	–	–	–	–	–	–	–	–	–	1	**Methylethylisopropylcyclohexyltin**		**MS06646**
73	147	75	69	231	45	148	275	**290**	100	91	29	27	19	18	15	12	**0.25**	12	26	4	–	–	–	–	–	–	2	–	–	–	**Butanedioic acid, 2,2-dimethyl-, bis(trimethylsilyl) ester**	55530-54-6	**NI30145**
147	73	75	69	275	148	45	231	**290**	100	97	29	29	19	16	16	13	**0.00**	12	26	4	–	–	–	–	–	–	2	–	–	–	**Butanedioic acid, 2,3-dimethyl-, bis(trimethylsilyl) ester**	55557-29-4	**NI30146**
73	75	111	147	275	55	141	172	**290**	100	82	74	56	44	32	27	20	**0.00**	12	26	4	–	–	–	–	–	–	2	–	–	–	**Hexanedioic acid, bis(trimethylsilyl) ester**	18105-31-2	**MS06647**
111	275	73	147	172	141	55	159	**290**	100	59	56	53	39	34	26	25	**0.53**	12	26	4	–	–	–	–	–	–	2	–	–	–	**Hexanedioic acid, bis(trimethylsilyl) ester**	18105-31-2	**NI30147**
73	75	111	77	55	147	141	45	**290**	100	67	61	40	39	24	20	18	**0.20**	12	26	4	–	–	–	–	–	–	2	–	–	–	**Hexanedioic acid, bis(trimethylsilyl) ester**	18105-31-2	**NI30148**
73	147	75	55	275	204	172	83	**290**	100	65	43	39	24	19	19	18	**0.00**	12	26	4	–	–	–	–	–	–	2	–	–	–	**Pentanedioic acid, 2-methyl-, bis(trimethylsilyl) ester**	55530-53-5	**NI30149**
73	69	147	75	275	172	45	159	**290**	100	95	74	46	29	18	18	12	**0.00**	12	26	4	–	–	–	–	–	–	2	–	–	–	**Pentanedioic acid, 3-methyl-, bis(trimethylsilyl) ester**	55517-41-4	**NI30150**
147	73	69	231	75	45	275	246	**290**	100	97	43	37	22	19	17	16	**0.00**	12	26	4	–	–	–	–	–	–	2	–	–	–	**Propanedioic acid, ethylmethyl-, bis(trimethylsilyl) ester**		**NI30151**
147	73	55	217	275	148	248	45	**290**	100	58	38	29	19	15	11	11	**0.00**	12	26	4	–	–	–	–	–	–	2	–	–	–	**Propanedioic acid, propyl-, bis(trimethylsilyl) ester**		**NI30152**
147	73	55	217	148	75	275	45	**290**	100	77	35	27	19	16	14	12	**0.00**	12	26	4	–	–	–	–	–	–	2	–	–	–	**Propanedioic acid, propyl-, bis(trimethylsilyl) ester**		**NI30153**
73	217	131	129	135	117	290	177	**290**	100	62	47	42	37	20	17	17		12	34	–	–	–	–	–	–	–	4	–	–	–	**Bis(hexamethyldisilane)**		**MS06648**
77	290	149	51	225	105	226	108	**290**	100	90	60	36	24	20	16	13		13	10	2	2	2	–	–	–	–	–	–	–	–	**Benzenesulphonamide, N-2-benzothiazolyl-**	13068-60-5	**NI30154**
93	64	44	66	226	65	48	39	**290**	100	62	45	39	25	24	19	15	**0.00**	13	10	2	2	2	–	–	–	–	–	–	–	–	**Benzothiazole sulphanilide**		**IC03707**
111	138	126	290	152	83	153	97	**290**	100	44	30	17	12	9	4	4		13	10	4	2	1	–	–	–	–	–	–	–	–	**N-(3-Nitrophenyl)-2-thien-2-oylacetamide**		**LI06293**
111	138	126	83	97	290	164	152	**290**	100	67	25	20	13	10	9	9		13	10	4	2	1	–	–	–	–	–	–	–	–	**N-(3-Nitrophenyl)-2-thien-3-oylacetamide**		**LI06294**
111	138	126	152	83	290	153	97	**290**	100	47	22	13	11	7	5	5		13	10	4	2	1	–	–	–	–	–	–	–	–	**N-(4-Nitrophenyl)-2-thien-2-oylacetamide**		**LI06295**
111	138	126	83	97	152	290	164	**290**	100	51	21	19	13	11	10	9		13	10	4	2	1	–	–	–	–	–	–	–	–	**N-(4-Nitrophenyl)-2-thien-3-oylacetamide**		**LI06296**
290	273	243	274	255	196	291	51	**290**	100	97	27	19	17	16	15	14		13	10	6	2	–	–	–	–	–	–	–	–	–	**Bis(2-hydroxy-5-nitrophenyl)methane**		**IC03708**
77	106	51	63	105	170	157	172	**290**	100	42	36	35	33	32	32	30	**29.13**	13	11	1	2	–	–	–	1	–	–	–	–	–	**Azobenzene, 4-bromo-2-methoxy-**	18277-96-8	**NI30155**
77	92	157	155	135	76	51	75	**290**	100	66	62	61	49	42	40	32	**24.62**	13	11	1	2	–	–	–	1	–	–	–	–	–	**Azobenzene, 4′-bromo-2-methoxy-**	18277-97-9	**NI30156**
147	283	73	45	148	284	151	149	**290**	100	54	49	16	15	14	13	10	**0.00**	13	14	2	–	1	–	–	–	–	2	–	–	–	**Thiosalicylic acid, S-(trimethylsilyl)-, trimethylsilyl ester**		**NI30157**
168	259	286	285	283	142	103	233	**290**	100	43	30	26	25	22	21	18	**0.00**	13	15	1	–	–	–	–	–	–	–	–	–	1	**Rhodium, [η^4-2-(hydroxymethyl)norbornadiene](η^5-cyclopentadienyl)-**	67420-86-4	**NI30158**
105	91	134	77	79	188	119	78	**290**	100	96	86	47	46	45	44	37	**1.69**	13	15	4	–	–	–	–	–	–	–	–	–	1	**Cymantrene, (1-ethyl-1-hydroxypropyl)-**	69364-35-8	**NI30159**
133	161	29	207	91	165	27	105	**290**	100	97	86	79	63	55	50	47	**21.05**	13	16	3	–	–	–	2	–	–	–	–	–	–	**1,4-Cyclohexadiene-1-propanoic acid, 3-(dichloromethyl)-3-methyl-6-oxo-, ethyl ester**	56700-86-8	**MS06649**
43	59	117	113	142	84	86	275	**290**	100	96	74	53	50	49	47	46	**0.00**	13	22	7	–	–	–	–	–	–	–	–	–	–	**1,2:3,4-Di-O-isopropylidene-D-manno-heptulofuranose**		**NI30160**
86	159	115	42	45	70	87	203	**290**	100	75	52	50	49	40	37	30	**0.01**	13	22	7	–	–	–	–	–	–	–	–	–	–	**1,4,7,10,13-Pentaoxacyclooctadecane-14,18-dione**		**MS06650**
96	55	81	69	97	67	177	179	**290**	100	54	44	35	34	32	13	12	**0.00**	13	23	2	–	–	–	–	1	–	–	–	–	–	**Hexanoic acid, 6-bromo-, cyclohexanemethyl ester**	86328-78-1	**NI30161**
102	58	101	73	57	88	59	115	**290**	100	64	49	40	38	35	32	29	**4.25**	13	26	5	2	–	–	–	–	–	–	–	–	–	**1,4,7,10-Tetraoxa-13,15-diazacycloheptadecan-14-one, 13,15-dimethyl-**	89913-93-9	**NI30162**
117	73	147	75	69	231	275	118	**290**	100	91	51	22	17	17	12	12	**0.00**	13	30	3	–	–	–	–	–	–	2	–	–	–	**Butanoic acid, 2-isopropyl-3-[(trimethylsilyl)oxy]-, trimethylsilyl ester**		**NI30163**
173	73	147	174	103	275	75	247	**290**	100	63	46	16	12	11	11	9	**0.12**	13	30	3	–	–	–	–	–	–	2	–	–	–	**Heptanoic acid, 2-[(trimethylsilyl)oxy]-, trimethylsilyl ester**		**NI30164**
131	73	129	75	275	69	55	132	**290**	100	56	29	28	22	14	12	12	**0.00**	13	30	3	–	–	–	–	–	–	2	–	–	–	**Hexanoic acid, 5-methyl-5-[(trimethylsilyl)oxy]-, trimethylsilyl ester**		**NI30165**
131	73	75	69	129	55	143	132	**290**	100	94	49	30	29	21	18	17	**0.00**	13	30	3	–	–	–	–	–	–	2	–	–	–	**Hexanoic acid, 5-methyl-5-[(trimethylsilyl)oxy]-, trimethylsilyl ester**		**NI30166**

MASS TO CHARGE RATIOS								M.W.	INTENSITIES								Parent	C	H	O	N	S	F	Cl	Br	I	Si	P	B	X	COMPOUND NAME	CAS Reg No	No
73	145	147	75	115	45	247	74	290	100	38	20	17	11	10	9	9	0.00	13	30	3	–	–	–	–	–	–	2	–	–	–	Pentanoic acid, 2,4-dimethyl-3-[(trimethylsilyl)oxy]-, trimethylsilyl ester		NI30167
73	131	147	75	232	45	132	217	290	100	84	30	18	14	12	11	10	0.00	13	30	3	–	–	–	–	–	–	2	–	–	–	Pentanoic acid, 2-ethyl-3-[(trimethylsilyl)oxy]-, trimethylsilyl ester		NI30168
73	147	113	129	45	217	59	75	290	100	45	39	16	12	10	10	9	2.42	13	34	1	–	–	–	–	–	–	3	–	–	–	Silane, [1-[(trimethylsilyl)oxy]butylidene]bis[trimethyl-	56919-94-9	NI30169
173	175	145	109	147	177	290	174	290	100	64	17	12	11	11	10	8		14	8	1	2	–	–	2	–	–	–	–	–	–	2-Propenenitrile, 3-[1-(2,6-dichlorobenzoyl)pyrrol-2-yl]-		NI30170
139	119	111	76	135	290	141	289	290	100	82	48	43	40	34	31	30		14	11	1	2	1	–	1	–	–	–	–	–	–	Benzamide, 3-chloro-N-[(phenylamino)thioxomethyl]-	56437-96-8	NI30171
43	248	108	290	140	65	58	92	290	100	75	70	63	50	27	27	23		14	14	3	2	1	–	–	–	–	–	–	–	–	Acetamide, N-[4-[(4-aminophenyl)sulfonyl]phenyl]-	565-20-8	NI30172
206	248	205	290	120	148	249	207	290	100	83	32	28	23	16	11	11		14	14	5	2	–	–	–	–	–	–	–	–	–	2,4-Imidazolidinedione, 1-acetyl-5-[4-(acetyloxy)phenyl]-3-methyl-, (±)-	55822-93-0	NI30173
191	159	40	105	42	132	146	56	290	100	99	46	44	41	26	25	20	0.00	14	18	3	4	–	–	–	–	–	–	–	–	–	Carbamic acid, [1-[(butylamino)carbonyl]-1H-benzimidazol-2-yl]-, methyl ester	17804-35-2	NI30174
290	259	275	243	291	123	200	81	290	100	43	31	22	17	16	14	11		14	18	3	4	–	–	–	–	–	–	–	–	–	2,4-Pyrimidinediamine, 5-[(3,4,5-trimethoxyphenyl)methyl]-	738-70-5	NI30175
75	43	73	101	99	275	59	189	290	100	71	46	32	29	28	26	20	0.00	14	26	6	–	–	–	–	–	–	–	–	–	–	2-Deoxy-3,4:5,6-di-O-isopropylidene-L-arabino-hexose dimethyl acetal		LI06297
72	73	45	111	59	55	155	44	290	100	68	53	46	35	30	27	22	0.00	14	26	6	–	–	–	–	–	–	–	–	–	–	Hexanedioic acid, bis(2-ethoxyethyl) ester	109-44-4	NI30176
87	115	260	189	59	88	217	75	290	100	75	63	57	54	28	26	26	0.00	14	34	2	–	–	–	–	–	–	2	–	–	–	4,7-Dioxa-3,8-disiladecane, 3,3,8,8-tetraethyl-	13175-68-3	NI30177
69	147	73	75	55	103	149	40	290	100	71	55	48	47	23	22	15	0.00	14	34	2	–	–	–	–	–	–	2	–	–	–	3,12-Dioxa-2,13-disilatetradecane, 2,2,13,13-tetramethyl-	16654-42-5	NI30178
275	69	276	259	245	289	147	55	290	100	28	22	22	22	17	15	11	0.00	14	34	2	–	–	–	–	–	–	2	–	–	–	3,12-Dioxa-2,13-disilatetradecane, 2,2,13,13-tetramethyl-	16654-42-5	NI30179
69	147	55	73	75	149	111	103	290	100	53	37	25	15	14	13	11	0.22	14	34	2	–	–	–	–	–	–	2	–	–	–	3,12-Dioxa-2,13-disilatetradecane, 2,2,13,13-tetramethyl-	16654-42-5	NI30180
69	147	73	55	75	103	149	44	290	100	72	56	50	48	24	22	16	1.00	14	34	2	–	–	–	–	–	–	2	–	–	–	1,8-Bis[(trimethylsilyl)oxy]octane		LI06298
73	117	74	45	75	147	115	59	290	100	9	9	9	8	7	7	7	0.00	14	34	2	–	–	–	–	–	–	2	–	–	–	1,2-Bis(γ-trimethylsilylpropoxy)ethane		IC03709
248	233	43	290	249	201	133	234	290	100	45	21	17	13	10	9	7		15	14	6	–	–	–	–	–	–	–	–	–	–	6-Acetoxy-5,7-dimethoxy-2-naphthoic acid		MS06651
248	136	63	203	164	108	249	202	290	100	72	38	33	23	19	16	15	8.70	15	14	6	–	–	–	–	–	–	–	–	–	–	6-Acetoxy-3-hydroxymethyl-7-methoxy-4-oxo-1,2,3,4-tetrahydro-2-naphthoic acid lactone		MS06652
139	123	152	290	149	110	77	124	290	100	72	55	33	33	29	23	22		15	14	6	–	–	–	–	–	–	–	–	–	–	2H-1-Benzopyran-3,5,7-triol, 2-(3,4-dihydroxyphenyl)-3,4-dihydro-, (2R-cis)-	490-46-0	NI30181
139	123	152	77	290	124	149	127	290	100	92	55	28	27	21	16	15		15	14	6	–	–	–	–	–	–	–	–	–	–	2H-1-Benzopyran-3,5,7-triol, 2-(3,4-dihydroxyphenyl)-3,4-dihydro-, (2R-trans)-	154-23-4	NI30182
177	114	135	290	147	148	131	89	290	100	58	56	41	33	19	14	12		15	14	6	–	–	–	–	–	–	–	–	–	–	2H-Pyran-2-one, 6-[2-(1,3-benzodioxol-5-yl)vinyl]-5,6-dihydro-5-hydroxy-4-methoxy-, [5R-[5α,6α(E)]]-	52525-97-0	NI30183
177	114	147	135	272	131	103	89	290	100	73	48	42	39	39	24	24	22.00	15	14	6	–	–	–	–	–	–	–	–	–	–	2H-Pyran-2-one, 6-[2-(1,3-benzodioxol-5-yl)vinyl]-5,6-dihydro-5-hydroxy-4-methoxy-, [5S-[5α,6β(E)]]-	52525-98-1	NI30184
28	230	229	193	139	201	160	258	290	100	94	85	83	80	74	70	62	59.22	15	14	6	–	–	–	–	–	–	–	–	–	–	2H,4aH-1,4,5-Trioxadicyclopent[a,hi]indene-7-carboxylic acid, 3-ethylidene-3,3a,7a,9b-tetrahydro-2-oxo-, methyl ester, [3aS-(3E,3aα,4aβ,7aβ,9aR*,9bβ)]-	77-16-7	NI30185
201	230	160	193	290	139	258	261	290	100	97	89	81	77	69	58	53		15	14	6	–	–	–	–	–	–	–	–	–	–	2H,4aH-1,4,5-Trioxadicyclopent[a,hi]indene-7-carboxylic acid, 3-ethylidene-3,3a-7a,9b-tetrahydro-2-oxo-, methyl ester , [3aS-(3Z,3aα,4aβ,7aβ,9aR*,9bβ)]-	31298-76-7	NI30186
290	211	212	227	213	197	191	259	290	100	46	40	33	20	20	20	19		15	15	2	–	1	–	–	–	–	–	1	–	–	10H-Phenothiaphosphine, 10-methoxy-2,7-dimethyl-, 10-oxide	54833-25-9	NI30187
291	74	246	293	163	172	292	115	290	100	51	34	32	31	23	20	20	9.65	15	15	2	2	–	–	1	–	–	–	–	–	–	Urea, N′-[4-(4-chlorophenoxy)phenyl]-N,N-dimethyl-	1982-47-4	NI30189
72	245	44	290	45	40	247	75	290	100	20	13	11	10	8	7	7		15	15	2	2	–	–	1	–	–	–	–	–	–	Urea, N′-[4-(4-chlorophenoxy)phenyl]-N,N-dimethyl-	1982-47-4	NI30188
72	44	245	290	75	71	45	42	290	100	13	11	9	9	7	6	6		15	15	2	2	–	–	1	–	–	–	–	–	–	Urea, N′-[4-(4-chlorophenoxy)phenyl]-N,N-dimethyl-	1982-47-4	NI30190
44	86	173	43	246	160	174	161	290	100	63	44	37	30	18	15	15	0.00	15	18	4	2	–	–	–	–	–	–	–	–	–	1,3-Benzodioxole, 5-(acetyliminomethyl)-6-[2-(N-acetyl-N-methylamino)ethyl]-		NI30191
69	115	130	41	91	116	77	78	290	100	78	73	73	50	43	40	40	0.00	15	18	4	2	–	–	–	–	–	–	–	–	–	Butanoic acid, 4-cyano-2-ethyl-2-(4-nitrophenyl)-, ethyl ester		MS06653
116	290	117	115	291	149	171	130	290	100	23	17	16	4	4	3	3		15	18	4	2	–	–	–	–	–	–	–	–	–	4a,9b-Dihydro-2-ethoxy-4H-indeno[2,1-e]-1,3,4-oxadiazine-4-carboxylic acid, ethyl ester		LI06299
110	82	54	41	111	55	94	95	290	100	69	48	45	40	39	33	26	24.02	15	18	4	2	–	–	–	–	–	–	–	–	–	Maleimide, N,N′-heptamethylenedi-	28537-72-6	NI30192
118	189	102	248	206	133	134	147	290	100	78	64	58	53	50	39	25	17.00	15	18	4	2	–	–	–	–	–	–	–	–	–	2-Propenoic acid, 2-(acetyloxy)-3-phenyl-, 1-acetyl-2,2-dimethylhydrazide	66190-57-6	NI30193
261	290	233	148	262	291	133	176	290	100	80	55	44	26	15	11	9		15	18	4	2	–	–	–	–	–	–	–	–	–	2,4,6(1H,3H,5H)-Pyrimidinetrione, 5-ethyl-5-(4-methoxyphenyl)-1,3-dimethyl-	55125-17-2	NI30194
261	290	148	233	262	176	108	44	290	100	79	59	58	50	21	18	17		15	18	4	2	–	–	–	–	–	–	–	–	–	2,4,6(1H,3H,5H)-Pyrimidinetrione, 5-ethyl-5-(4-methoxyphenyl)-1,3-dimethyl-	55125-17-2	NI30195
261	73	135	262	275	190	176	45	290	100	58	22	21	13	10	9	7	1.37	15	22	2	2	–	–	–	–	–	1	–	–	–	2,4-Imidazolidinedione, 5-ethyl-3-methyl-5-phenyl-1-(trimethylsilyl)-	57396-66-4	NI30196
43	41	89	44	131	42	47	117	290	100	42	26	19	12	5	4	2	0.00	15	30	5	–	–	–	–	–	–	–	–	–	–	Acetone, 1,1,3,3-tetrapropoxy-		NI30197
151	290	139	292	141	111			290	100	26	22	12	7	6				16	15	3	–	–	–	1	–	–	–	–	–	–	2-Chlorophenyl 3,4-dimethoxybenzyl ketone		LI06300
226	211	290	195	225	210	196	224	290	100	96	83	76	72	71	67	63		16	18	3	–	1	–	–	–	–	–	–	–	–	2,4,6-Trimethyl-3′-methoxydiphenyl sulphone		NI30198
258	272	290	224	165	195	209	211	290	100	97	90	89	81	81	79	77		16	18	3	–	1	–	–	–	–	–	–	–	–	2,4,6-Trimethyl-4′-methoxydiphenyl sulphone	84113-59-7	NI30199
153	138	124	93	137	77	125	109	290	100	60	45	45	42	29	28	24	14.00	16	18	5	–	–	–	–	–	–	–	–	–	–	Benzenemethanol, 4-hydroxy-3-methoxy-α-[(2-methoxyphenoxy)methyl]-		NI30200

MASS TO CHARGE RATIOS								M.W.	INTENSITIES								Parent	C	H	O	N	S	F	Cl	Br	I	Si	P	B	X	COMPOUND NAME	CAS Reg No	No
222	69	78	223	207	194	176	79	**290**	100	96	63	61	54	27	22	18	**3.00**	16	18	5	–	–	–	–	–	–	–	–	–	–	2H-1-Benzopyran-2-one, 6,8-dimethoxy-7-[(3-methyl-2-butenyl)oxy]-	57419-60-0	NI30201
43	108	106	107	220	28	109	290	**290**	100	77	47	38	27	26	25	13		16	18	5	–	–	–	–	–	–	–	–	–	–	4,7-Ethanoisobenzofuran-1,3-dione, 6-(acetyloxy)-3a,4,7,7a-tetrahydro-5-methyl-8-isopropenyl-	52918-81-7	NI30202
94	95	137	136	158	111	66	172	**290**	100	30	25	22	18	8	7	5	**0.10**	16	18	5	–	–	–	–	–	–	–	–	–	–	4H-Furo[3,2-c]pyran, 6,7-dihydro-6-methyl-6-[[2-(2-oxopropyl)furan-3-yl]methoxy]-		NI30203
154	137	93	65	122	125	94	123	**290**	100	65	43	37	32	30	22	19	**0.20**	16	18	5	–	–	–	–	–	–	–	–	–	–	3-Methoxy-4-hydroxybenzyl 2-methoxy-4-(hydroxymethyl)phenyl ether		NI30204
69	41	136	290	107	137	106	105	**290**	100	65	22	10	7	4	3	2		16	18	5	–	–	–	–	–	–	–	–	–	–	3-Methoxy-4-methacrylyloxybenzyl methacrylate		NI30205
290	275	77	232	189	291	247	69	**290**	100	75	40	26	25	19	17	16		16	18	5	–	–	–	–	–	–	–	–	–	–	2,3,4,5-Tetramethoxydiphenyl ether	88037-82-5	NI30206
84	290	206	177	85	137	178	208	**290**	100	33	32	26	21	20	13	12		16	19	1	2	–	–	1	–	–	–	–	–	–	2-Cyclopentenone, 2-(4-chloroanilino)-4-piperidino-		MS06654
202	291	204	289	255				**290**	100	73	35	18	17				**0.00**	16	19	1	2	–	–	1	–	–	–	–	–	–	Ethanamine, 2-[(4-chlorophenyl)-2-pyridinylmethoxy]-N,N-dimethyl-	486-16-8	NI30207
58	71	167	72	42	202	59	45	**290**	100	62	8	7	5	4	4	3	**0.00**	16	19	1	2	–	–	1	–	–	–	–	–	–	Ethanamine, 2-[(4-chlorophenyl)-2-pyridinylmethoxy]-N,N-dimethyl-	486-16-8	NI30208
58	71	167	72	42	202	203	45	**290**	100	74	11	10	8	6	5	5	**0.00**	16	19	1	2	–	–	1	–	–	–	–	–	–	Ethanamine, 2-[(4-chlorophenyl)-2-pyridinylmethoxy]-N,N-dimethyl-, (–)-	5560-77-0	NI30209
275	148	133	176	41	113	39	65	**290**	100	49	36	30	29	27	25	23	**15.39**	16	22	1	2	1	–	–	–	–	–	–	–	–	4H-1,3-Thiazin-2-amine, 4,4,6-trimethyl-N-methyl-N-(4-ethoxyphenyl)-	85298-53-9	NI30210
176	147	148	290	128	119	177	113	**290**	100	82	52	29	23	21	18	15		16	22	1	2	1	–	–	–	–	–	–	–	–	4H-1,3-Thiazine, 3,4,4,6-tetramethyl-2-(4-ethoxyphenylimino)-2,3-dihydro-	37489-65-9	NI30211
247	97	137	109	229	205			**290**	100	32	21	16	6	5			**0.00**	16	22	3	2	–	–	–	–	–	–	–	–	–	6,9a-Dimethyl-2,9-dioxospiroperhydroazuleno[4,5-b]furan-3,3′-(1)-pyrazoline		LI06301
247	97	137	109	262	229	205		**290**	100	34	24	20	6	6	5		**0.00**	16	22	3	2	–	–	–	–	–	–	–	–	–	6,9a-Dimethyl-2,9-dioxospiroperhydroazuleno[4,5-b]furan-3,3′-(1)-pyrazoline		LI06302
105	233	77	42	234	106	51	41	**290**	100	76	35	13	11	11	5	5	**0.00**	16	22	3	2	–	–	–	–	–	–	–	–	–	4-Imidazolidinone, 1-benzoyl-2-tert-butyl-5-methoxy-3-methyl-, (2S,5R)-		DD01012
57	71	43	41	55	290	178	29	**290**	100	80	74	49	36	28	27	27		16	34	–	–	2	–	–	–	–	–	–	–	–	Dioctyl disulphide	822-27-5	NI30212
73	103	113	45	75	86	85	129	**290**	100	87	41	38	30	24	19	16	**0.00**	16	34	4	–	–	–	–	–	–	–	–	–	–	1,1,3,5-Tetraethoxy-2,4-dimethylhexane		LI06303
126	154	182	250	76	74	210	63	**290**	100	41	30	26	25	24	22	19	**9.30**	17	10	3	2	–	–	–	–	–	–	–	–	–	2,7-Bis(2-cyanomethoxy)-fluoren-9-one		MS06655
290	289	291	271	76	190			**290**	100	24	19	13	8	7				17	10	3	2	–	–	–	–	–	–	–	–	–	1,3-Dioxo-4-(3-oxo-2,3-dihydroisoindol-1-ylidene)-1,2,3,4-tetrahydroisoquinoline		IC03710
290	56	77	129	121	105	115	133	**290**	100	64	55	29	29	29	25	23		17	14	1	–	–	–	–	–	–	–	–	–	1	Ferrocene, benzoyl-	1272-44-2	NI30213
57	104	77	78	105	103	91	246	**290**	100	40	18	15	11	11	6	5	**3.00**	17	14	1	4	–	–	–	–	–	–	–	–	–	2-Oxabicyclo[2.2.1]heptane-4,7,7-tricarbonitrile, 1-ethyl-3-imino-5-phenyl-		MS06656
230	55	57	91	53	202	119	201	**290**	100	99	95	76	74	58	58	56	**26.00**	17	22	4	–	–	–	–	–	–	–	–	–	–	Eudesma-4(15),7(11)-dien-8,12-olide, 1-acetoxy-, (1β,8α)-		NI30214
91	43	65	59	105	92	77	55	**290**	100	32	12	10	8	8	8	8	**0.00**	17	22	4	–	–	–	–	–	–	–	–	–	–	Furo[2,3-d]-1,3-dioxole, 6-(benzyloxy)tetrahydro-2,2-dimethyl-5-(2-propenyl)-, [3aR-(3aα,5α,6α,6aα)]-	120417-97-2	DD01013
91	43	191	92	65	59	125	77	**290**	100	20	10	8	7	6	4	4	**0.00**	17	22	4	–	–	–	–	–	–	–	–	–	–	Furo[2,3-d]-1,3-dioxole, 6-(benzyloxy)tetrahydro-2,2-dimethyl-5-(2-propenyl)-, [3aR-(3aα,5β,6α,6aα)]-	120417-96-1	DD01014
91	43	249	92	65	129	149	162	**290**	100	51	28	22	20	11	8	5	**0.00**	17	22	4	–	–	–	–	–	–	–	–	–	–	Furo[2,3-d]-1,3-dioxole, 6-(benzyloxy)tetrahydro-2,2-dimethyl-5-(2-propenyl)-, [3aR-(3aα,5β,6β,6aα)]-	89755-56-6	DD01015
230	133	60	231	91	119	145	157	**290**	100	37	27	18	18	16	14	13		17	22	4	–	–	–	–	–	–	–	–	–	–	4ξ,5ξ,7ξ-Guaia-1(10),11(13)-dien-12-oic acid, 4,8-dihydroxy-, γ-lactone, acetate	1461-36-5	NI30215
230	119	231	157	118	91	215	173	**290**	100	21	18	15	13	13	11	11	**0.00**	17	22	4	–	–	–	–	–	–	–	–	–	–	Guaia-1(10),4(15)-dieno-12,6-lactone, 3-acetoxy-, (3α,6α,11S)-	82206-91-5	DD01016
230	43	201	188	44	125	41	124	**290**	100	89	74	62	48	47	45	41	**23.16**	17	22	4	–	–	–	–	–	–	–	–	–	–	3H-Oxireno[8,8a]naphtho[2,3-b]furan-5-ol, 1a,2,4,4a,5,9-hexahydro-4,4a,6-trimethyl-, acetate	56246-43-6	NI30216
290	178	83	206	262	291	207	219	**290**	100	73	70	46	22	16	15	14		17	22	4	–	–	–	–	–	–	–	–	–	–	Spiro[benzofuran-2(4H),1′-cyclohexane]-2′,4,6′-trione, 3,5,6,7-tetrahydro-4′,4′,6,6-tetramethyl-	1984-51-6	NI30217
221	73	63	222	130	175	145	129	**290**	100	65	16	13	10	5	5	3	**0.00**	17	26	2	2	–	–	–	–	–	–	–	–	–	N-(Dimethylaminomethylene)phenylalanine, neopentyl ester		NI30218
221	73	129	61	249	222	261	175	**290**	100	22	21	19	15	12	5	4	**0.00**	17	26	2	2	–	–	–	–	–	–	–	–	–	N-(Dimethylaminomethylene)phenylalanine, neopentyl ester		NI30219
58	84	97	43	133	42	98	83	**290**	100	62	38	17	16	16	15	14	**4.00**	17	26	2	2	–	–	–	–	–	–	–	–	–	N,N,N′-Trimethyl-N′-(4-methoxy-cis-cinnamoyl)putrescine		LI06304
290	291	145	292	288	258	245	44	**290**	100	21	12	10	9	7	7	6		18	10	–	–	2	–	–	–	–	–	–	–	–	Benzo[1,2-b:3,4-b′]bis[1]benzothiophene		MS06657
290	291	145	292	258	245	288	247	**290**	100	21	11	10	8	7	5	3		18	10	–	–	2	–	–	–	–	–	–	–	–	Benzo[2,1-b:3,4-b′]bis[1]benzothiophene		MS06658
290	291	145	292	288	245	258	144	**290**	100	21	14	10	8	5	4	4		18	10	–	–	2	–	–	–	–	–	–	–	–	Benzo[1,2-b:3,4-b′]bisthianaphthene		AI01994
290	288	291	144	289	292	145	258	**290**	100	30	20	16	11	10	9	6		18	10	–	–	2	–	–	–	–	–	–	–	–	5,8-Dithiaindeno[2,1-c]fluorene		AI01995
290	291	145	292	258	245	288	146	**290**	100	21	18	10	5	5	4	3		18	10	–	–	2	–	–	–	–	–	–	–	–	6,12-Dithiaindeno[1,2-b]fluorene		AI01996
290	291	145	292	245	288	258	69	**290**	100	21	17	11	5	4	4	3		18	10	–	–	2	–	–	–	–	–	–	–	–	11,12-Dithiaindeno[2,1-a]fluorene		AI01997
290	104	55	29	27	69			**290**	100	93	57	51	51	50				18	10	4	–	–	–	–	–	–	–	–	–	–	2,2′-Bis(1,3-dioxoindanyl)		IC03711
290	145	89	291	117	234	178	146	**290**	100	78	36	19	16	10	8	7		18	10	4	–	–	–	–	–	–	–	–	–	–	Furo[3,2-b]furan-2,5-dione, 3,6-diphenyl-	6273-79-6	LI06305
145	89	290	117	63	291	234	146	**290**	100	91	86	28	24	15	11	10		18	10	4	–	–	–	–	–	–	–	–	–	–	Furo[3,2-b]furan-2,5-dione, 3,6-diphenyl-	6273-79-6	NI30220
262	111	178	233	112	290	176	207	**290**	100	53	43	39	34	22	22	21		18	10	4	–	–	–	–	–	–	–	–	–	–	4-Phenyl-5-hydroxy-5,7-benzocoumaran-2,3-dione		MS06659
290	77	197	289	291	196	94	51	**290**	100	96	48	24	19	16	10	7		18	15	3	–	–	–	–	–	–	–	–	1	–	Boric acid, triphenyl ester	1095-03-0	NI30221

MASS TO CHARGE RATIOS	M.W.	INTENSITIES	Parent	C	H	O	N	S	F	Cl	Br	I	Si	P	B	X	COMPOUND NAME	CAS Reg No	No
145 290 159 77 146 158 291 92	**290**	100 31 16 15 11 10 7 7		18	18	–	4	–	–	–	–	–	–	–	–	–	1,2-Bis(2-methylbenzimidazol-1-yl)ethane		IC03712
290 246 219 55 41 67 272 79	**290**	100 59 41 40 33 32 29 27		18	26	3	–	–	–	–	–	–	–	–	–	–	4-Oxa-5α-androstane-3,17-dione		MS06660
178 161 133 134 41 179 43 55	**290**	100 56 15 14 13 13 8 8	**3.20**	18	26	3	–	–	–	–	–	–	–	–	–	–	2-Propenoic acid, 3-(4-methoxyphenyl)-, 2-ethylhexyl ester	5466-77-3	NI30222
112 152 91 97 167 109 258 111	**290**	100 33 30 29 25 21 20 20	**0.30**	18	26	3	–	–	–	–	–	–	–	–	–	–	2H-Pyran, 3,6-dihydro-6-methoxy-3,5-dimethyl-2-[1-methyl-2-(phenylmethoxy)ethyl]-, (±)-[2α(S*),3α,6β]-		NI30223
178 107 162 84 135 234 161 150	**290**	100 65 59 43 40 37 23 23	**20.00**	18	30	1	2	–	–	–	–	–	–	–	–	–	1-tert-Butyl-3-methyl-4,5,6,7-tetrahydroindole-2-carboxylic acid tert-butyl amide		MS06661
58 98 233 41 42 55 127 99	**290**	100 48 8 8 7 6 4 4	**0.00**	18	30	1	2	–	–	–	–	–	–	–	–	–	2-Propanamine, N-methyl-1-[4-[3-(1-piperidyl)propoxy]phenyl]-		MS06662
262 290 247 131 176 263 248 219	**290**	100 94 84 25 23 21 17 16		19	14	3	–	–	–	–	–	–	–	–	–	–	1,4-Pentadiyn-3-one, 1,5-bis(4-methoxyphenyl)-	34793-63-0	MS06663
290 262 247 263 291 176 219 131	**290**	100 82 56 21 18 12 11 11		19	14	3	–	–	–	–	–	–	–	–	–	–	2,4-Pentadiyn-1-one, 1,5-bis(4-methoxyphenyl)-	29372-67-6	MS06664
213 136 290 183 214 181 105 59	**290**	100 65 58 52 20 20 19 17		19	18	1	–	–	–	–	–	–	1	–	–	–	Silane, methoxytriphenyl-	1829-41-0	NI30224
147 290 289 148 145 291 144 136	**290**	100 35 25 11 10 8 7 7		19	18	1	2	–	–	–	–	–	–	–	–	–	1,2-Benzenediamine, (2-hydroxynaphthylidene)-4,5-dimethyl-		NI30225
43 232 107 28 200 55 290 105	**290**	100 98 61 50 32 23 22 22		19	30	2	–	–	–	–	–	–	–	–	–	–	8-Acetonyl-2-methoxy-4a-methyl-1,2,3,4,4a,4b,5,6,8a,9,10,10a-dodecahydrophenanthrene		LI06306
91 93 81 79 105 55 67 107	**290**	100 89 89 88 83 81 76 72	**48.04**	19	30	2	–	–	–	–	–	–	–	–	–	–	Androstan-17-ol, 2,3-epoxy-, (2α,3α,5α,17β)-	965-66-2	NI30226
290 231 291 233 232 272 246 94	**290**	100 29 21 12 12 9 9 5		19	30	2	–	–	–	–	–	–	–	–	–	–	Androstan-3-one, 17-hydroxy-, (5α,17β)-	521-18-6	NI30227
290 231 55 41 81 67 95 79	**290**	100 91 44 42 36 35 29 28		19	30	2	–	–	–	–	–	–	–	–	–	–	Androstan-3-one, 17-hydroxy-, (5α,17β)-	521-18-6	NI30229
55 41 67 79 81 231 93 91	**290**	100 89 70 66 63 54 54 53	**30.50**	19	30	2	–	–	–	–	–	–	–	–	–	–	Androstan-3-one, 17-hydroxy-, (5α,17β)-	521-18-6	NI30228
231 290 124 163 123 73 232 121	**290**	100 52 30 29 29 24 23 21		19	30	2	–	–	–	–	–	–	–	–	–	–	Androstan-3-one, 17-hydroxy-, (5β,17β)-	571-22-2	NI30230
290 55 41 220 81 43 247 67	**290**	100 50 46 42 34 33 32 32		19	30	2	–	–	–	–	–	–	–	–	–	–	Androstan-3-one, 17-hydroxy-, (5β,17β)-	571-22-2	NI30231
290 220 247 291 233 229 231 272	**290**	100 39 28 23 16 13 12 11		19	30	2	–	–	–	–	–	–	–	–	–	–	Androstan-3-one, 17-hydroxy-, (5β,17β)-	571-22-2	NI30232
41 55 177 81 67 164 79 147	**290**	100 80 74 70 65 55 55 51	**50.00**	19	30	2	–	–	–	–	–	–	–	–	–	–	Androstan-11-one, 3-hydroxy-, (3β,5α)-	570-27-4	NI30233
67 81 79 151 91 95 93 177	**290**	100 73 59 53 52 49 43 39	**3.00**	19	30	2	–	–	–	–	–	–	–	–	–	–	Androstan-11-one, 19-hydroxy-, (5α)-	564-29-4	NI30234
273 255 291 274 271 289 290 257	**290**	100 35 32 27 26 13 10 9		19	30	2	–	–	–	–	–	–	–	–	–	–	Androstan-17-one, 3-hydroxy-, (3α,5α)-	53-41-8	NI30235
67 79 91 107 93 81 108 105	**290**	100 92 88 85 82 81 66 63	**44.14**	19	30	2	–	–	–	–	–	–	–	–	–	–	Androstan-17-one, 3-hydroxy-, (3α,5α)-	53-41-8	WI01371
290 41 107 108 67 79 55 93	**290**	100 73 72 70 63 58 58 53		19	30	2	–	–	–	–	–	–	–	–	–	–	Androstan-17-one, 3-hydroxy-, (3α,5α)-	53-41-8	NI30236
41 290 67 55 79 93 81 107	**290**	100 94 86 80 73 67 62 61		19	30	2	–	–	–	–	–	–	–	–	–	–	Androstan-17-one, 3-hydroxy-, (3α,5β)-	53-42-9	NI30237
272 218 109 161 257 147 121 119	**290**	100 57 55 53 52 49 49 44	**18.92**	19	30	2	–	–	–	–	–	–	–	–	–	–	Androstan-17-one, 3-hydroxy-, (3α,5β)-	53-42-9	WI01372
290 41 55 67 81 79 93 107	**290**	100 72 65 52 41 41 40 38		19	30	2	–	–	–	–	–	–	–	–	–	–	Androstan-17-one, 3-hydroxy-, (3α,5β)-	53-42-9	NI30238
290 291 246 272 257 165 275 216	**290**	100 20 18 8 8 8 7 6		19	30	2	–	–	–	–	–	–	–	–	–	–	Androstan-17-one, 3-hydroxy-, (3β,5α)-	481-29-8	NI30240
290 107 108 79 81 93 91 95	**290**	100 99 74 74 70 68 64 46		19	30	2	–	–	–	–	–	–	–	–	–	–	Androstan-17-one, 3-hydroxy-, (3β,5α)-	481-29-8	WI01373
290 107 108 41 67 55 246 79	**290**	100 37 33 32 31 28 25 25		19	30	2	–	–	–	–	–	–	–	–	–	–	Androstan-17-one, 3-hydroxy-, (3β,5α)-	481-29-8	NI30239
217 108 107 93 79 81 290 91	**290**	100 75 75 66 66 63 58 57		19	30	2	–	–	–	–	–	–	–	–	–	–	Androstan-17-one, 3-hydroxy-, (3β,5β)-	571-31-3	WI01374
216 41 290 55 97 109 67 43	**290**	100 95 88 86 75 66 61 57		19	30	2	–	–	–	–	–	–	–	–	–	–	Androstan-17-one, 12-hydroxy-, (5α,12β)-	32222-25-6	NI30241
221 290 203 220 202 176 161 222	**290**	100 69 58 56 24 22 19 14		19	30	2	–	–	–	–	–	–	–	–	–	–	Androst-2-ene-1,17-diol, (1α,5α,17β)-	1229-06-7	NI30242
290 221 70 203 43 41 220 55	**290**	100 98 78 74 73 51 47 47		19	30	2	–	–	–	–	–	–	–	–	–	–	Androst-2-ene-1β,17β-diol, (5α)-	20112-26-9	NI30243
290 272 257 179 239 161 205 187	**290**	100 67 39 36 29 27 22 11		19	30	2	–	–	–	–	–	–	–	–	–	–	Androst-5-ene-3,17-diol		LI06307
272 290 239 257 161 179 187 205	**290**	100 65 62 49 40 27 14 13		19	30	2	–	–	–	–	–	–	–	–	–	–	Androst-5-ene-3,17-diol		LI06308
239 161 272 290 257 107 179 187	**290**	100 88 78 75 70 59 50 44		19	30	2	–	–	–	–	–	–	–	–	–	–	Androst-5-ene-3,17-diol, (3β,17α)-	1963-03-7	NI30244
290 272 205 107 179 291 257 105	**290**	100 38 24 23 21 20 20 20		19	30	2	–	–	–	–	–	–	–	–	–	–	Androst-5-ene-3,17-diol, (3β,17β)-	521-17-5	NI30245
57 91 105 79 55 290 107 145	**290**	100 94 93 80 78 70 70 64		19	30	2	–	–	–	–	–	–	–	–	–	–	Androst-5-ene-3,17-diol, (3β,17β)-	521-17-5	LI06309
290 272 91 107 105 41 79 205	**290**	100 62 54 51 50 46 45 42		19	30	2	–	–	–	–	–	–	–	–	–	–	Androst-5-ene-3,17-diol, (3β,17β)-	521-17-5	NI30246
290 246 43 107 55 41 272 57	**290**	100 40 28 27 27 27 24 21		19	30	2	–	–	–	–	–	–	–	–	–	–	Androst-8(14)-ene-3β,17β-diol, (5α)-	20112-27-0	NI30247
91 290 105 28 41 29 55 15	**290**	100 99 99 96 66 56 54 52		19	30	2	–	–	–	–	–	–	–	–	–	–	Benzeneoctanoic acid, 2-butyl-, methyl ester	13397-99-4	NI30248
275 276 290 57 257 41	**290**	100 25 23 17 13 6		19	30	2	–	–	–	–	–	–	–	–	–	–	Benzoic acid, 2,4,6-tris-tert-butyl-	66415-27-8	NI30249
191 290 275 151 165 138 57	**290**	100 95 75 71 24 24 24		19	30	2	–	–	–	–	–	–	–	–	–	–	cis-4-tert-Butyl-1-(2,4-dimethoxyphenyl)-1-methylcyclohexane		LI06310
191 290 151 275 138 165 57 109	**290**	100 88 62 39 31 23 19 15		19	30	2	–	–	–	–	–	–	–	–	–	–	trans-4-tert-Butyl-1-(2,4-dimethoxyphenyl)-1-methylcyclohexane		LI06311
91 108 32 43 57 92 41 55	**290**	100 78 43 40 30 29 27 20	**1.50**	19	30	2	–	–	–	–	–	–	–	–	–	–	Dodecanoic acid, phenylmethyl ester	140-25-0	NI30250
91 210 225 65 92 148 118 226	**290**	100 22 20 17 9 7 7 3	**0.00**	19	30	2	–	–	–	–	–	–	–	–	–	–	Dodecanoic acid, phenylmethyl ester	140-25-0	NI30251
41 55 67 290 81 231 79 43	**290**	100 66 54 51 51 48 44 40		19	30	2	–	–	–	–	–	–	–	–	–	–	5α-Estran-3-one, 17β-hydroxy-2α-methyl-	4267-75-8	NI30252
218 275 55 219 290 109 67 81	**290**	100 41 20 18 17 17 17 14		19	30	2	–	–	–	–	–	–	–	–	–	–	D-Homo-17a-oxaandrostan-17-one, (5α)-	2466-25-3	NI30253
105 290 91 183 133 258 201 147	**290**	100 80 73 37 27 20 20 20		19	30	2	–	–	–	–	–	–	–	–	–	–	Methyl 8-(2-butylphenyl)octanoate		LI06312
105 132 104 131 290 91 230 117	**290**	100 57 43 36 32 32 29 25		19	30	2	–	–	–	–	–	–	–	–	–	–	Methyl 1-(2-decylphenyl)acetate		LI06313
105 258 91 131 290 173 117 259	**290**	100 57 57 46 43 36 32 25		19	30	2	–	–	–	–	–	–	–	–	–	–	Methyl 5-(2-heptylphenyl)pentanoate		LI06314
105 91 290 161 173 131 187 117	**290**	100 47 40 18 17 16 15 15		19	30	2	–	–	–	–	–	–	–	–	–	–	Methyl 6-(2-hexylphenyl)hexanoate		LI06315
146 91 105 119 117 290 131 118	**290**	100 90 67 53 53 40 23 23		19	30	2	–	–	–	–	–	–	–	–	–	–	Methyl 3-(2-nonylphenyl)propionate		LI06316

MASS TO CHARGE RATIOS	M.W.	INTENSITIES	Parent	C	H	O	N	S	F	Cl	Br	I	Si	P	B	X	COMPOUND NAME	CAS Reg No	No
288 110 71 231 91 79 55 67	**290**	100 87 83 68 64 54 50 43	**3.00**	19	30	2	–	–	–	–	–	–	–	–	–	–	**17α-Methyl-19-nortestosterone**		**NI30254**
105 290 91 187 147 258 106 201	**290**	100 50 33 23 17 13 13 10		19	30	2	–	–	–	–	–	–	–	–	–	–	**Methyl 7-(2-pentylphenyl)heptanoate**	13397-97-2	**LI06317**
41 105 290 91 32 40	**290**	100 99 85 81 75 74		19	30	2	–	–	–	–	–	–	–	–	–	–	**Methyl 7-(2-pentylphenyl)heptanoate**	13397-97-2	**NI30255**
105 91 133 290 119 41 134 117	**290**	100 76 64 52 42 31 29 29		19	30	2	–	–	–	–	–	–	–	–	–	–	**Methyl 9-(2-propylphenyl)nonanoate**	17670-86-9	**NI30256**
131 150 259 118 91 290 163 149	**290**	100 93 71 46 46 43 39 36		19	30	2	–	–	–	–	–	–	–	–	–	–	**Methyl 2-undecylbenzoate**		**LI06318**
218 148 109 175 203 108 95 55	**290**	100 65 60 58 55 53 42 42	**16.00**	19	30	2	–	–	–	–	–	–	–	–	–	–	**D-Norandrostane-16-carboxylic acid, (5α,16β)-**	32319-09-8	**NI30257**
91 105 117 41 43 79 77 55	**290**	100 71 61 55 50 47 45 43	**0.00**	19	30	2	–	–	–	–	–	–	–	–	–	–	**2,5-Octadecadiynoic acid, methyl ester**	57156-91-9	**NI30258**
43 41 27 42 29 39 32 31	**290**	100 89 84 66 60 59 23 23	**0.00**	19	30	2	–	–	–	–	–	–	–	–	–	–	**3,6-Octadecadiynoic acid, methyl ester**	56554-43-9	**NI30259**
105 41 43 91 55 117 131 67	**290**	100 82 80 78 62 56 55 47	**12.04**	19	30	2	–	–	–	–	–	–	–	–	–	–	**4,7-Octadecadiynoic acid, methyl ester**	18202-20-5	**NI30260**
249 105 91 117 41 43 119 55	**290**	100 33 33 30 29 28 23 23	**3.33**	19	30	2	–	–	–	–	–	–	–	–	–	–	**5,8-Octadecadiynoic acid, methyl ester**	18202-21-6	**NI30261**
91 41 131 105 117 55 79 77	**290**	100 78 68 68 66 65 60 60	**1.04**	19	30	2	–	–	–	–	–	–	–	–	–	–	**6,9-Octadecadiynoic acid, methyl ester**	56847-03-1	**NI30262**
43 78 41 91 42 105 55 79	**290**	100 99 83 65 65 52 52 44	**1.04**	19	30	2	–	–	–	–	–	–	–	–	–	–	**8,11-Octadecadiynoic acid, methyl ester**	18202-23-8	**NI30263**
91 105 119 92 106 134 55 41	**290**	100 96 64 62 54 53 49 49	**0.60**	19	30	2	–	–	–	–	–	–	–	–	–	–	**10,13-Octadecadiynoic acid, methyl ester**	18202-24-9	**NI30264**
43 120 105 41 91 205 57 42	**290**	100 94 82 80 77 73 58 55	**0.14**	19	30	2	–	–	–	–	–	–	–	–	–	–	**11,14-Octadecadiynoic acid, methyl ester**	56554-58-6	**NI30265**
43 28 91 106 29 41 55 44	**290**	100 99 97 83 72 65 50 46	**0.00**	19	30	2	–	–	–	–	–	–	–	–	–	–	**12,15-Octadecadiynoic acid, methyl ester**	57156-95-3	**NI30266**
74 55 41 81 87 67 43 69	**290**	100 63 60 47 43 37 33 28	**0.00**	19	30	2	–	–	–	–	–	–	–	–	–	–	**13,16-Octadecadiynoic acid, methyl ester**	56846-98-1	**NI30267**
257 290 122 121 275 107 261 96	**290**	100 77 65 65 59 56 49 48		19	30	2	–	–	–	–	–	–	–	–	–	–	**2-Phenanthrenecarboxaldehyde, 1,2,3,4,4a,4b,5,6,7,8,8a,9-dodecahydro-7-hydroxy-2,4b,8,8-tetramethyl-**	57289-56-2	**NI30268**
255 290 253 254 252 126 240 239	**290**	100 72 50 42 38 36 28 26		20	15	–	–	–	–	1	–	–	–	–	–	–	**2-(2-Chlorophenyl)-1,1-diphenylethylene**		**LI06319**
255 290 253 254 252 126 270 239	**290**	100 71 48 42 37 36 28 26		20	15	–	–	–	–	1	–	–	–	–	–	–	**2-(2-Chlorophenyl)-1,1-diphenylethylene**		**MS06665**
290 255 254 253 292 126 291 178	**290**	100 48 38 38 34 26 24 24		20	15	–	–	–	–	1	–	–	–	–	–	–	**2-(4-Chlorophenyl)-1,1-diphenylethylene**		**LI06320**
290 255 254 253 292 252 126 178	**290**	100 48 36 35 33 27 27 23		20	15	–	–	–	–	1	–	–	–	–	–	–	**2-(4-Chlorophenyl)-1,1-diphenylethylene**		**MS06666**
248 157 249 290 43 91 115 55	**290**	100 31 19 16 16 15 13 11		20	18	2	–	–	–	–	–	–	–	–	–	–	**1,5-Cyclohexadien-1-ol, 4,4-diphenyl-, acetate**	33795-14-1	**NI30269**
272 229 228 257 273 226 239 114	**290**	100 46 36 35 22 16 13 12	**1.31**	20	18	2	–	–	–	–	–	–	–	–	–	–	**5,6-Dihydro-5,6-dihydroxy-7,12-dimethylbenz[a]anthracene**		**NI30270**
229 290 243 257 215 202 272 239	**290**	100 78 59 45 38 26 9 9		20	18	2	–	–	–	–	–	–	–	–	–	–	**5,6-Dihydro-7,12-dimethylbenz[a]anthracene-cis-5,6-diol**		**MS06667**
290 202 244 229 215 257 272 275	**290**	100 88 84 73 65 25 22 13		20	18	2	–	–	–	–	–	–	–	–	–	–	**3,4-Dihydro-7,12-dimethylbenz[a]anthracene-trans-3,4-diol**		**MS06668**
229 290 243 257 215 202 272 239	**290**	100 60 58 38 38 29 20 10		20	18	2	–	–	–	–	–	–	–	–	–	–	**5,6-Dihydro-7,12-dimethylbenz[a]anthracene-trans-5,6-diol**		**MS06669**
229 244 290 203 202 215 272 257	**290**	100 74 52 52 51 34 28 24		20	18	2	–	–	–	–	–	–	–	–	–	–	**8,9-Dihydro-7,12-dimethylbenz[a]anthracene-trans-8,9-diol**		**MS06670**
229 290 244 202 203 272 215 257	**290**	100 81 77 76 73 50 46 44		20	18	2	–	–	–	–	–	–	–	–	–	–	**10,11-Dihydro-7,12-dimethylbenz[a]anthracene-trans-10,11-diol**		**MS06671**
229 245 290 247 43 230 228 202	**290**	100 72 46 40 38 35 35 26		20	18	2	–	–	–	–	–	–	–	–	–	–	**5,6-Dihydro-5,6-dimethyl-5,6-dihydroxychrysene**		**NI30271**
275 260 257 130 289 276 115.5 113	**290**	100 57 44 40 34 22 13 13	**5.70**	20	18	2	–	–	–	–	–	–	–	–	–	–	**cis-7,12-Dihydroxy-7,12-dimethylbenz[a]anthracene**		**MS06672**
275 257 289 276 130 258 261 259	**290**	100 57 21 21 17 14 11 11	**5.56**	20	18	2	–	–	–	–	–	–	–	–	–	–	**trans-7,12-Dihydroxy-7,12-dimethylbenz[a]anthracene**		**MS06673**
217 105 290 41 43 29 229 167	**290**	100 96 69 59 58 52 46 44		20	18	2	–	–	–	–	–	–	–	–	–	–	**Hydrocinnamic acid, O-(2-naphthylmethyl)-**	23804-23-1	**NI30272**
141 290 133 105 115 79 77 43	**290**	100 36 28 12 11 9 7 6		20	18	2	–	–	–	–	–	–	–	–	–	–	**1-Naphthalenemethyl 2,6-dimethylbenzoate**	86328-56-5	**NI30273**
217 105 290 41 43	**290**	100 96 69 59 58		20	18	2	–	–	–	–	–	–	–	–	–	–	**1-(Naphth-2-ylmethyl)-2-(2-ethoxycarbonyl)benzene**		**LI06321**
290 291 220 201 219 272 246 231	**290**	100 22 20 16 15 14 12 8		20	34	1	–	–	–	–	–	–	–	–	–	–	**5α-Androstan-17β-ol, 1β-methyl-**	5846-88-8	**NI30274**
108 95 55 41 149 81 94 67	**290**	100 98 73 66 61 61 58 57	**11.00**	20	34	1	–	–	–	–	–	–	–	–	–	–	**Androstan-3-ol, 9-methyl-, (3β,5α)-**	54550-11-7	**NI30275**
122 121 43 41 135 91 123 29	**290**	100 37 17 17 12 11 10 10	**9.05**	20	34	1	–	–	–	–	–	–	–	–	–	–	**Anisole, 3-tridecyl-**	22161-93-9	**NI30276**
55 257 95 275 137 192 177 93	**290**	100 96 86 76 62 47 41 33	**0.00**	20	34	1	–	–	–	–	–	–	–	–	–	–	**5β,8βH,9βH,10α-Labd-14-ene, 8,13-epoxy-**	19123-29-6	**NI30277**
272 123 69 81 94 93 109 92	**290**	100 97 66 59 57 52 42 38	**9.01**	20	34	1	–	–	–	–	–	–	–	–	–	–	**16βH-Kauran-16-ol**	5354-44-9	**NI30278**
93 76 59 121 107 272 95 136	**290**	100 94 85 75 72 63 58 52	**0.00**	20	34	1	–	–	–	–	–	–	–	–	–	–	**3,7,11-Cyclotetradecatriene-1-methanol, α,α,4,8,12-pentamethyl-, [R-(E,E,E)]-**	53915-41-6	**DD01017**
93 133 69 91 105 272 107 121	**290**	100 85 83 69 64 59 59 56	**20.00**	20	34	1	–	–	–	–	–	–	–	–	–	–	**2,6,10-Cyclotetradecatrien-1-ol, 3,7,11-trimethyl-14-isopropyl-**	39012-00-5	**NI30279**
69 41 81 123 95 93 68 67	**290**	100 63 47 46 37 26 25 17	**1.50**	20	34	1	–	–	–	–	–	–	–	–	–	–	**Digeranyl ether**		**LI06322**
69 41 55 109 67 95 82 81	**290**	100 96 69 47 40 39 37 36	**14.00**	20	34	1	–	–	–	–	–	–	–	–	–	–	**Geranylcitronellal**		**LI06323**
137 271 123 273 135 149 121 109	**290**	100 72 56 50 48 43 40 30	**0.00**	20	34	1	–	–	–	–	–	–	–	–	–	–	**Geranylgeraniol**		**NI30280**
166 109 41 55 43 69 91 83	**290**	100 88 31 25 25 19 16 16	**1.40**	20	34	1	–	–	–	–	–	–	–	–	–	–	**Grindelene**		**NI30281**
91 272 120 55 105 93 79 119	**290**	100 99 79 78 75 63 62 59	**6.00**	20	34	1	–	–	–	–	–	–	–	–	–	–	**12-Hydroxy-5,8,8,8a,12-pentamethylperhydro-1H-2,10a-ethenophenanthrene**		**LI06324**
120 272 106 44 91 105 107 119	**290**	100 89 72 72 70 59 57 53		20	34	1	–	–	–	–	–	–	–	–	–	–	**12-Hydroxy-5,8,8,8a,12-pentamethylperhydro-1H-2,10a-ethenophenanthrene**		**LI06325**
247 123 290 248 137 110 109 95	**290**	100 38 26 21 18 9 7 7		20	34	1	–	–	–	–	–	–	–	–	–	–	**13-Hydroxystevane**		**MS06674**
272 229 161 81 41 93 44 91	**290**	100 74 65 56 55 50 49 44	**4.00**	20	34	1	–	–	–	–	–	–	–	–	–	–	**1-Isopropyl-4,8,12-trimethyl-2,7,11-cyclotetradecatriene-4-ol**		**LI06326**
272 232 123 134 257 137 105 290	**290**	100 77 75 42 40 40 33 21		20	34	1	–	–	–	–	–	–	–	–	–	–	**Kauran-16-ol**	5524-17-4	**LI06327**
272 94 123 69 232 81 257 91	**290**	100 79 70 59 58 51 50 49	**17.02**	20	34	1	–	–	–	–	–	–	–	–	–	–	**Kauran-16-ol**	5524-17-4	**NI30282**
69 41 81 93 95 71 68 83	**290**	100 76 75 58 57 57 54 41	**2.00**	20	34	1	–	–	–	–	–	–	–	–	–	–	**Linalyl geranyl ether**		**LI06328**

MASS TO CHARGE RATIOS								M.W.	INTENSITIES								Parent	C	H	O	N	S	F	Cl	Br	I	Si	P	B	X	COMPOUND NAME	CAS Reg No	No
137	81	95	69	41	93	55	43	**290**	100	68	59	48	48	43	42	42	**0.00**	20	34	1	–	–	–	–	–	–	–	–	–	–	**1-Naphthalenepropanol, α-vinyldecahydro-α,5,5,8a-tetramethyl-2-methylene-, [1S-[1α(R*),4aβ,8aα]]-**	1438-62-6	**NI30283**
137	95	81	71	257	93	109	69	**290**	100	78	76	72	60	53	49	45	**1.35**	20	34	1	–	–	–	–	–	–	–	–	–	–	**1-Naphthalenepropanol, α-vinyldecahydro-α,5,5,8a-tetramethyl-2-methylene-, [1S-[1α(S*),4aβ,8aα]]-**	596-85-0	**NI30284**
275	256	81	192	56	137	177	95	**290**	100	80	52	49	47	43	39	39	**1.31**	20	34	1	–	–	–	–	–	–	–	–	–	–	**1H-Naphtho[2,1-b]pyran, 3-vinyldodecahydro-3,4a,7,7,10a-pentamethyl-, [3R-(3α,4aβ,6aα,10aβ,10bα)]-**	596-84-9	**NI30285**
275	257	81	192	55	137	177	95	**290**	100	74	51	49	47	43	39	39	**1.24**	20	34	1	–	–	–	–	–	–	–	–	–	–	**1H-Naphtho[2,1-b]pyran, 3-vinyldodecahydro-3,4a,7,7,10a-pentamethyl-, [3R-(3α,4aβ,6aα,10aβ,10bα)]-**	596-84-9	**NI30286**
275	257	192	81	55	43	177	137	**290**	100	77	29	27	24	24	21	20	**0.00**	20	34	1	–	–	–	–	–	–	–	–	–	–	**1H-Naphtho[2,1-b]pyran, 3-vinyldodecahydro-3,4a,7,7,10a-pentamethyl-, [3S-(3α,4aα,6aβ,10aα,10bβ)]-**	1227-93-6	**NI30287**
259	137	275	109	95	81	257	149	**290**	100	62	52	44	43	40	36	34	**17.00**	20	34	1	–	–	–	–	–	–	–	–	–	–	**ent-D-Norbeyerane-endo-15-methanol**		**MS06675**
137	109	275	95	272	257	232	290	**290**	100	61	60	57	43	42	40	28		20	34	1	–	–	–	–	–	–	–	–	–	–	**ent-D-Norbeyerane-exo-15-methanol**		**MS06676**
257	121	133	93	119	135	107	41	**290**	100	97	81	79	75	73	71	71	**0.00**	20	34	1	–	–	–	–	–	–	–	–	–	–	**Verticellol**		**NI30288**
141	142	115	290	291	139	135	91	**290**	100	13	9	7	2	2	2	1		21	22	1	–	–	–	–	–	–	–	–	–	–	**Naphthalene, [(4-tert-butylphenoxy)methyl]-**	74646-44-9	**NI30289**
193	290	248	179	194	291	192	233	**290**	100	66	61	32	28	19	18	15		21	22	1	–	–	–	–	–	–	–	–	–	–	**Spiro[cyclobutane-1,16′-estra[1,3,5,7,9]pentaen]-2-one, (13ξ,14ξ)-**	74310-23-9	**NI30290**
191	69	95	81	55	123	109	137	**290**	100	49	44	40	38	32	32	29	**21.78**	21	38	–	–	–	–	–	–	–	–	–	–	–	**15-Methylisocopalane, (13α)-**		**NI30291**
55	41	83	206	97	69	111	67	**290**	100	79	52	38	37	28	25	25	**0.47**	21	38	–	–	–	–	–	–	–	–	–	–	–	**1,1,3-Tricyclohexylpropane**		**AI01998**
290	247	291	233	248	191	205	203	**290**	100	85	23	20	19	19	15	13		22	26	–	–	–	–	–	–	–	–	–	–	–	**Dibenz[a,h]anthracene, 1,2,3,4,7b,8,9,10,11,11a,12,13-dodecahydro-**		**NI30292**
91	143	199	69	95	129	105	41	**290**	100	24	21	19	17	14	14	12	**3.50**	22	26	–	–	–	–	–	–	–	–	–	–	–	**3,3-Dimethyl-6,6-dibenzylbicyclo[3.1.0]hexane**		**LI06329**
41	43	39	29	27	55	91	69	**290**	100	57	52	50	41	38	36	26	**2.00**	22	26	–	–	–	–	–	–	–	–	–	–	–	**1,1-Dimethyl-4,4-dibenzylcyclohex-2-ene**		**LI06330**
145	40	146	91	144	117	115	130	**290**	100	28	19	17	10	10	7	6	**3.00**	22	26	–	–	–	–	–	–	–	–	–	–	–	**trans,trans-1,6-Diphenyl-3,4-diethyl-1,5-hexadiene**		**MS06677**
275	290	276	102	291	202	130	116	**290**	100	47	24	21	12	9	9	8		22	26	–	–	–	–	–	–	–	–	–	–	–	**Phenanthrene, 3,9-di-tert-butyl-**	55125-03-6	**WI01375**
256	258	64	63	46	48	260	36	**291**	100	81	64	62	50	49	23	16	**1.30**	–	–	3	3	3	–	3	–	–	–	–	–	–	**α-Sulphanuric chloride**		**LI06331**
278	276	292	293	290	280	291	256	**291**	100	77	63	61	50	50	48	45		1	4	–	3	–	–	4	–	–	–	3	–	–	**1-Methyl-1-hydridotetrachlorocyclotriphosphazene**	68351-74-6	**NI30293**
293	295	291	133	214	212	106	62	**291**	100	52	52	28	19	19	19	11		7	3	–	1	1	–	–	2	–	–	–	–	–	**Thieno[2,3-c]pyridine, 2,3-dibromo-**	28783-19-9	**NI30294**
293	295	291	133	106	294	214	212	**291**	100	55	50	21	14	10	9	9		7	3	–	1	1	–	–	2	–	–	–	–	–	**Thieno[3,2-c]pyridine, 2,3-dibromo-**	28783-20-2	**NI30295**
291	274	119	118	117	91	147	39	**291**	100	85	54	53	38	35	31	19		9	10	2	1	–	–	–	–	1	–	–	–	–	**Benzene, 2-iodo-1,3,5-trimethyl-4-nitro-**	56701-30-5	**MS06678**
118	103	96	94	77	76	212	214	**291**	100	73	52	42	37	29	20	19	**0.00**	9	11	–	1	–	–	–	2	–	–	–	–	–	**N,N-Dimethyl-α-bromobenziminium bromide**	41182-91-6	**NI30296**
232	43	41	204	176	190	249	102	**291**	100	65	48	39	33	28	27	24	**0.00**	10	14	3	1	–	5	–	–	–	–	–	–	–	**Butanoic acid, 4-(pentafluoropropionylamino)-, isopropyl ester**		**MS06679**
204	41	43	205	176	190	119	27	**291**	100	50	47	36	9	7	7	6	**0.00**	10	14	3	1	–	5	–	–	–	–	–	–	–	**Butanoic acid, 2-(pentafluoropropionylamino)-, isopropyl ester, D-**		**MS06680**
204	41	43	205	176	190	119	27	**291**	100	49	44	38	9	8	6	5	**0.00**	10	14	3	1	–	5	–	–	–	–	–	–	–	**Butanoic acid, 2-(pentafluoropropionylamino)-, isopropyl ester, L-**		**MS06681**
109	97	65	291	139	137	125	63	**291**	100	99	49	44	37	37	35	32		10	14	5	1	1	–	–	–	–	–	1	–	–	**Phosphorothioic acid, O,O-diethyl O-(4-nitrophenyl) ester**	56-38-2	**NI30297**
109	291	29	97	139	137	40	125	**291**	100	85	85	71	63	60	50	45		10	14	5	1	1	–	–	–	–	–	1	–	–	**Phosphorothioic acid, O,O-diethyl O-(4-nitrophenyl) ester**	56-38-2	**NI30298**
97	109	29	65	139	137	125	63	**291**	100	90	73	43	36	30	30	30	**29.00**	10	14	5	1	1	–	–	–	–	–	1	–	–	**Phosphorothioic acid, O,O-diethyl O-(4-nitrophenyl) ester**	56-38-2	**NI30299**
73	147	45	84	172	89	59	43	**291**	100	49	37	32	27	21	20	16	**0.00**	10	21	5	1	–	–	–	–	–	2	–	–	–	**Propanedioic acid, (methoxyimino)-, bis(trimethylsilyl) ester**	55494-03-6	**NI30300**
44	256	56	72	88	43	70	42	**291**	100	59	59	48	37	37	36	35	**9.00**	11	11	4	1	–	–	2	–	–	–	–	–	–	**L-Alanine, N-[(2,4-dichlorophenoxy)acetyl]-**	50648-96-9	**NI30301**
291	56	29	220	42	43	248	152	**291**	100	71	70	65	53	45	44	38		11	22	4	3	–	–	–	–	–	–	1	–	–	**1,3-Diethyl-2-diethylphosphorylimidohydantoin**		**LI06332**
147	73	218	276	148	86	174	45	**291**	100	82	53	43	16	15	15	12	**0.00**	11	25	4	1	–	–	–	–	–	2	–	–	–	**Propanedioic acid, amino-, dimethyl ester, bis(trimethylsilyl)-**		**NI30302**
73	174	45	59	147	175	43	86	**291**	100	54	28	19	16	13	13	12	**0.00**	11	29	2	1	–	–	–	–	–	3	–	–	–	**Glycine, N,N-bis(trimethylsilyl)-, trimethylsilyl ester**	5630-82-0	**MS06682**
174	73	147	175	248	86	176	45	**291**	100	48	19	18	15	10	8	8	**0.00**	11	29	2	1	–	–	–	–	–	3	–	–	–	**Glycine, N,N-bis(trimethylsilyl)-, trimethylsilyl ester**	5630-82-0	**NI30303**
174	73	147	175	248	86	176	276	**291**	100	51	21	18	16	13	10	8	**1.43**	11	29	2	1	–	–	–	–	–	3	–	–	–	**Glycine, N,N-bis(trimethylsilyl)-, trimethylsilyl ester**	5630-82-0	**NI30304**
91	148	44	291	45	60	43	57	**291**	100	75	36	29	21	14	14	11		12	9	2	3	2	–	–	–	–	–	–	–	–	**2-(Benzylthio)-7-hydroxy-5H-1,3,4-thiadiazolo[3,2-a]pyrimidin-5-one**		**MS06683**
255	276	169	182	108	184	291	278	**291**	100	65	60	59	45	23	22	22		12	19	3	1	–	–	1	–	–	–	1	–	–	**Phosphoramidic acid, methyl-, 2-chloro-4-tert-butylphenyl methyl ester**	299-86-5	**NI30305**
256	108	182	276	169	41	184	77	**291**	100	85	77	46	42	42	25	16	**16.00**	12	19	3	1	–	–	1	–	–	–	1	–	–	**Phosphoramidic acid, methyl-, 2-chloro-4-tert-butylphenyl methyl ester**	299-86-5	**NI30306**
291	56	276	123	42	292	18	98	**291**	100	62	50	33	19	17	15	13		12	21	–	9	–	–	–	–	–	–	–	–	–	**2,4,6-Trisisopropylidenehydrazine-s-triazine**		**IC03713**
73	174	45	86	175	59	133	114	**291**	100	88	19	17	15	15	10	9	**0.00**	12	33	1	1	–	–	–	–	–	3	–	–	–	**1-[N,N-Bis(trimethylsilyl)amino]-2-trimethylsiloxypropane**		**MS06684**
291	263	292	235	162	293	259	275	**291**	100	28	14	9	9	7	7	6		13	9	5	1	1	–	–	–	–	–	–	–	–	**Ethyl 5-oxo-2-thioxopyrano[3,2-f]benzoxazole-7-carboxylate**		**LI06333**
291	263	292	193	235	162	293	264	**291**	100	56	16	12	9	9	8	8		13	9	5	1	1	–	–	–	–	–	–	–	–	**Ethyl 7-oxo-2-thioxopyrano[2,3-e]benzoxazole-5-carboxylate**		**LI06334**
138	32	41	17	154	137	139	59	**291**	100	41	29	27	22	15	13	12	**11.50**	13	16	3	1	–	3	–	–	–	–	–	–	–	**1-Methyl-N-trifluoroacetyl-2-(4-hydroxy-2,6-xylyloxy)ethylamine**		**NI30307**
120	154	138	41	171	77	69	291	**291**	100	80	50	46	42	34	34	30	**3.00**	13	16	3	1	–	3	–	–	–	–	–	–	–	**1-Methyl-N-trifluoroacetyl-2-(α-hydroxy-2,6-xylyloxy)ethylamine**		**NI30308**
93	119	155	66	91	127	41	65	**291**	100	49	40	38	30	24	24	23	**2.00**	13	17	1	5	1	–	–	–	–	–	–	–	–	**5-(3-Guanidinopropyl)-4-oxo-3-phenyl-2-thioxoimidazolidine**		**LI06335**
87	129	142	45	72	43	42	88	**291**	100	56	45	45	32	32	20	18	**0.40**	13	25	6	1	–	–	–	–	–	–	–	–	–	**α-D-Galactopyranoside, methyl 2-(acetylmethylamino)-2-deoxy-3,4,6-tri-O-methyl-**	36668-35-6	**NI30309**

MASS TO CHARGE RATIOS								M.W.	INTENSITIES								Parent	C	H	O	N	S	F	Cl	Br	I	Si	P	B	X	COMPOUND NAME	CAS Reg No	No
87	142	129	45	88	72	43	98	**291**	100	56	50	44	38	32	32	30	**0.20**	13	25	6	1	–	–	–	–	–	–	–	–	–	α-D-Glucopyranoside, methyl 2-(acetylmethylamino)-2-deoxy-3,4,6-tri-O-methyl-	36757-11-6	NI30310
84	207	151	41	123	179	154	85	**291**	100	49	45	32	27	19	14	12	**4.00**	13	26	4	1	–	–	–	–	–	–	1	–	–	Butanamide, 4-(diethoxyphosphinoyl)-N-ethyl-N-(2-propenyl)-	93939-10-7	NI30311
169	94	291	293	228	141	140	212	**291**	100	93	74	49	28	27	26	18		14	7	2	1	–	–	2	–	–	–	–	–	–	N-(Naphthyl)-2,3-dichloromaleicisomide		MS06685
163	121	179	291	129	46	293	136	**291**	100	14	6	5	5	3	2	1		14	10	4	1	–	–	1	–	–	–	–	–	–	Benzeneacetic acid, 4-nitro-, 4-chlorophenyl ester	53218-11-4	NI30312
166	167	122	46	293	291			**291**	100	8	6	6	1	1				14	10	4	1	–	–	1	–	–	–	–	–	–	Benzenemethanol, 3-chloro-, 4-nitrobenzoate	53218-05-6	NI30313
276	113	99	230	41	277	59	39	**291**	100	34	31	28	19	18	17	13	**0.99**	14	17	2	3	1	–	–	–	–	–	–	–	–	4H-1,3-Thiazin-2-amine, 4,4,6-trimethyl-N-methyl-N-(4-nitrophenyl)-		NI30314
113	99	128	291	114	276	90	115	**291**	100	91	66	31	30	28	25	25		14	17	2	3	1	–	–	–	–	–	–	–	–	4H-1,3-Thiazine, 3,4,4,6-tetramethyl-2-(4-nitrophenylimino)-2,3-dihydro-	85298-46-0	NI30315
107	84	40	77	168	163	39	55	**291**	100	41	20	18	10	10	8	7	**0.00**	14	17	4	3	–	–	–	–	–	–	–	–	–	L-Tyrosinamide, 5-oxo-L-prolyl-	34368-03-1	NI30316
81	140	91	291	65	79	114	72	**291**	100	80	48	28	27	22	20	15		15	17	1	1	2	–	–	–	–	–	–	–	–	Cyclohexanespiro-2'-[4-(benzylthio)-1,3-thiazol-5(2H)-one]		DD01018
122	136	277	291					**291**	100	69	10	8						15	17	3	1	1	–	–	–	–	–	–	–	–	Benzenesulphonamide, N-(4-methoxyphenyl)-N,4-dimethyl-	21969-24-4	NI30317
91	149	105	84	290	291	77	142	**291**	100	77	61	50	42	36	30	23		15	17	5	1	–	–	–	–	–	–	–	–	–	1,3-Dioxino[4',5':5,6]pyrano[3,2-d]oxazol-9-ol, 3a,4a,5,8a,9,9a-hexahydro-2-methyl-7-phenyl-, stereoisomer	25539-42-8	NI30318
91	149	84	105	290	77	107	291	**291**	100	95	92	68	36	26	24	22		15	17	5	1	–	–	–	–	–	–	–	–	–	1,3-Dioxino[4',5':5,6]pyrano[3,2-d]oxazol-9-ol, 3a,4a,5,8a,9,9a-hexahydro-2-methyl-7-phenyl-, stereoisomer	23626-71-3	NI30319
149	273	188	91	77	164	150	53	**291**	100	22	22	21	21	15	15	13	**6.74**	15	17	5	1	–	–	–	–	–	–	–	–	–	2,6-Piperidinedione, 3-hydroxy-4-[2-(2-hydroxy-3,5-dimethylphenyl)-2-oxoethyl]-	13091-95-7	NI30320
203	231	259	176	291	177	232	162	**291**	100	73	59	52	49	38	36	24		15	17	5	1	–	–	–	–	–	–	–	–	–	1H-Pyrrolo[1,2-a]indole-9-carboxylic acid, 2,3,9α,9aα-tetrahydro-9a-hydroxy-7-methoxy-6-methyl-3-oxo-, methyl ester, (±)-		NI30321
174	291	73	175	159	131	75	158	**291**	100	17	14	13	8	6	6	5		15	21	3	1	–	–	–	–	–	1	–	–	–	1H-Indole-3-acetic acid, 5-methoxy-2-methyl-, trimethylsilyl ester	54833-72-6	NI30322
232	291	73	233	230	292	231	234	**291**	100	46	42	21	12	11	9	6		15	21	3	1	–	–	–	–	–	1	–	–	–	1H-Indole-3-acetic acid, 5-methoxy-1-(trimethylsilyl)-, methyl ester	55591-01-0	NI30323
130	131	291	292	232	147	79	73	**291**	100	12	9	2	2	2	2	2		15	21	3	1	–	–	–	–	–	1	–	–	–	1H-Indole-1-propanoic acid, α-[(trimethylsilyl)oxy]-, methyl ester	54833-73-7	NI30324
174	175	160	161	291	162	176	234	**291**	100	92	56	40	23	15	14	13		15	21	3	3	–	–	–	–	–	–	–	–	–	Geneserine		LI06336
202	73	233	89	129	59	203	45	**291**	100	82	39	38	32	23	22	15	**0.00**	15	25	1	1	–	–	–	–	–	2	–	–	–	1-(Trimethylsilyl)-3-[(trimethylsilyloxy)methyl]indole		NI30325
291	165	194	292	179	177	152	44	**291**	100	23	19	18	15	12	10	9		16	12	1	1	–	3	–	–	–	–	–	–	–	Acetamide, N-(9,10-dihydro-2-phenanthryl)-2,2,2-trifluoro-	18264-89-6	NI30326
105	77	106	186	51	109	160	90	**291**	100	29	10	9	9	7	6	6	**0.00**	16	12	1	1	–	3	–	–	–	–	–	–	–	Aziridine, 1-benzoyl-3-phenyl-2-(trifluoromethyl)-, cis-		DD01019
77	105	89	90	52	78	76	117	**291**	100	43	13	12	11	8	7	6	**0.00**	16	12	1	1	–	3	–	–	–	–	–	–	–	Benzamide, N-(3,3,3-trifluoro-1-phenyl-1-propen-2-yl)-, (Z)-		DD01020
185	105	147	77	40	291	119	91	**291**	100	36	31	30	30	29	27	24		16	12	1	1	–	3	–	–	–	–	–	–	–	2-Oxazoline, 2,5-diphenyl-4-(trifluoromethyl)-, cis-		DD01021
185	291	77	105	103	186	51	166	**291**	100	42	32	27	24	21	14	13		16	12	1	1	–	3	–	–	–	–	–	–	–	2-Oxazoline, 2,5-diphenyl-4-(trifluoromethyl)-, trans-		DD01022
104	103	77	291	129	130	51	76	**291**	100	73	61	47	47	19	18	16		16	13	1	5	–	–	–	–	–	–	–	–	–	s-Triazine-2-carboxylic acid, 4,6-diphenyl-, hydrazide	29366-69-6	NI30327
118	42	119	86	71	77	96	135	**291**	100	79	63	57	49	43	38	36	**0.00**	16	21	2	1	1	–	–	–	–	–	–	–	–	5-Oxa-8-thia-6-azaspiro[3.4]octan-2-one, 1,1,3,3,6-pentamethyl-7-phenyl-	50455-59-9	NI30328
128	121	70	291	135	142	148	41	**291**	100	38	24	21	14	14	13	12		16	21	4	1	–	–	–	–	–	–	–	–	–	2-Pyrrolidineacetic acid, 1-[2-(4-methoxyphenyl)acetyl]-, methyl ester		MS06686
165	70	84	248	291	68	126	112	**291**	100	91	53	50	33	16	15	11		16	21	4	1	–	–	–	–	–	–	–	–	–	Pyrrolidine, 1-acetyl-2-[2-(3,4-dimethoxyphenyl)-2-oxoethyl]-, (±)-	67257-60-7	NI30329
71	70	42	103	73	291	57	44	**291**	100	17	16	14	14	12	11	10		16	25	2	1	–	–	–	–	–	1	–	–	–	4-Piperidinecarboxylic acid, 1-methyl-4-phenyl-, trimethylsilyl ester	54833-74-8	NI30330
256	291	292	102	57	293	75	229	**291**	100	51	22	20	20	17	15	14		17	10	–	3	–	–	1	–	–	–	–	–	–	Quinoxaline, 2-chloro-3-[2-(4-cyanophenyl)ethenyl]-, (E)-		MS06687
276	291	277	248	292	274	77	164	**291**	100	76	19	18	15	15	13	12		17	13	2	3	–	–	–	–	–	–	–	–	–	4-Amino-2-cyano-1-ethylaminoanthraquinone		IC03714
291	256	127	244	128	52	292	243	**291**	100	48	42	24	23	22	19	19		17	13	2	3	–	–	–	–	–	–	–	–	–	1-Naphthaldehyde, (2-nitrophenyl)hydrazone	19263-85-5	NI30331
119	129	91	64	177	291	120	77	**291**	100	43	42	29	27	23	18	17		17	13	2	3	–	–	–	–	–	–	–	–	–	1,2,2a,7b-Tetrahydro-3H-indeno[2,1-c]1,2-diazete-1,2-dicarboxylic acid, N-phenylimide		LI06337
58	55	28	90	89	118	42	71	**291**	100	28	27	24	24	22	21	20	**0.00**	17	25	3	1	–	–	–	–	–	–	–	–	–	Benzeneacetic acid, α-(1-hydroxycyclopentyl)-, 2-(dimethylamino)ethyl ester	512-15-2	NI30332
91	57	41	65	56	39	85	156	**291**	100	20	20	14	12	11	10	8	**0.00**	17	25	3	1	–	–	–	–	–	–	–	–	–	Carbamic acid, (5-oxononyl)-, benzyl ester	116437-36-6	DD01023
291	44	234	292	180	233	151	203	**291**	100	76	53	23	23	22	16	14		17	25	3	1	–	–	–	–	–	–	–	–	–	4-(N-Methylaminoethyl)-4-(3,4-dimethoxyphenyl)cyclohex-2-en-1-ol		NI30333
143	76	115	104	50	18	116	38	**291**	100	68	61	61	50	33	27	19	**2.20**	18	13	3	1	–	–	–	–	–	–	–	–	–	Benzoic acid, 2-[(1-naphthalenylamino)carbonyl]-	132-66-1	NI30334
291	276	262	219	290	218	176	151	**291**	100	93	58	58	35	27	27	26		18	13	3	1	–	–	–	–	–	–	–	–	–	2-O,N-Dimethylliriodendronine		NI30335
57	276	41	43	277	291	91	55	**291**	100	93	52	39	17	13	9	9		18	29	2	1	–	–	–	–	–	–	–	–	–	Benzene, 2,4,6-tri-tert-butylnitro-	4074-25-3	NI30336
105	55	70	106	71	43	73	41	**291**	100	78	77	62	54	54	47	45	**0.00**	18	29	2	1	–	–	–	–	–	–	–	–	–	Carbamic acid, (α-methylbenzyl)-, 1-ethyl-1-methylhexyl ester	32589-48-3	NI30337
106	70	166	165	126	73	105	71	**291**	100	85	81	64	58	48	45	33	**0.00**	18	29	2	1	–	–	–	–	–	–	–	–	–	Carbamic acid, (α-methylbenzyl)-, 1-ethyl-1-methylhexyl ester	32589-48-3	NI30338
105	55	70	106	43	71	73	41	**291**	100	79	76	62	56	55	45	45	**0.00**	18	29	2	1	–	–	–	–	–	–	–	–	–	Carbamic acid, (1-phenylethyl)-, 3-methyloctyl ester	54833-47-5	NI30339
105	55	70	106	43	71	73	41	**291**	100	78	77	62	55	54	45	43	**0.00**	18	29	2	1	–	–	–	–	–	–	–	–	–	Carbamic acid, (1-phenylethyl)-, 3-methyloctyl ester	54833-47-5	LI06338
93	99	135	66	41	71	55	43	**291**	100	36	18	14	14	12	12	12	**3.80**	18	29	2	1	–	–	–	–	–	–	–	–	–	Dodecanamide, 5-hydroxy-N-phenyl-, (S)-	64527-03-3	NI30340
248	291	262	249	292	247	263	293	**291**	100	84	36	23	20	9	8	7		19	17	–	1	1	–	–	–	–	–	–	–	–	6,7-Benzophenothiazine, 10-propyl-	24025-35-2	NI30341
291	258	276	230	231	292	263	259	**291**	100	81	44	37	28	23	18	17		19	17	–	1	1	–	–	–	–	–	–	–	–	Pyridine, 2-[(1,1-diphenylethyl)thio]-	62238-36-2	NI30342
291	258	230	262	231	276	292	259	**291**	100	78	54	42	31	24	22	18		19	17	–	1	1	–	–	–	–	–	–	–	–	Pyridine, 2-(ethylthio)-3,6-diphenyl-	74685-35-1	NI30343
291	258	276	230	231	292	263	259	**291**	100	71	42	36	29	22	18	16		19	17	–	1	1	–	–	–	–	–	–	–	–	Pyridine, 2-(ethylthio)-4,6-diphenyl-	74663-73-3	NI30344
263	248	262	264	230	249	247	260	**291**	100	70	65	23	17	15	14	11	**0.00**	19	17	–	1	1	–	–	–	–	–	–	–	–	6,6,7-Trimethyl-5,6-dihydro-12-aza-5-thiabenz[a]anthracene		NI30345

MASS TO CHARGE RATIOS								M.W.	INTENSITIES								Parent	C	H	O	N	S	F	Cl	Br	I	Si	P	B	X	COMPOUND NAME	CAS Reg No	No
91	135	156	105	65	291	44	41	**291**	100	33	27	27	20	17	17	16		19	17	2	1	–	–	–	–	–	–	–	–	–	1-Benzyl-2-(4-anisoyl)pyrrole		LI06339
291	292	204	205	260	233	290	248	**291**	100	21	14	11	10	10	9	9		19	17	2	1	–	–	–	–	–	–	–	–	–	3,4-Bis(4-methoxyphenyl)pyridine		MS06688
121	115	171	170	291	142	106	273	**291**	100	99	91	61	36	32	24	16		19	17	2	1	–	–	–	–	–	–	–	–	–	2-Naphthalenecarboxamide, N-(2-ethylphenyl)-3-hydroxy-	68911-98-8	NI30346
91	200	65	291	92	172	41	292	**291**	100	13	12	12	3	0	0	0		19	17	2	1	–	–	–	–	–	–	–	–	–	2(1H)-Pyridinone, 3-(phenylmethoxy)-1-(phenylmethyl)-	24016-11-3	NI30347
291	276	185	144	292	259	130	171	**291**	100	71	42	26	24	22	22	19		19	21	–	3	–	–	–	–	–	–	–	–	–	Aniline, N,N-dimethyl-2-(1,2,3,9-tetrahydropyrrolo[2,1-b]quinazolin-3-yl)-	33903-13-8	NI30348
291	56	44	276	57	43	58	96	**291**	100	92	92	73	72	54	47	39		19	33	1	1	–	–	–	–	–	–	–	–	–	3-Aza-A-homoandrostan-16-ol, (5β,16β)-	22614-24-0	NI30349
44	262	291	96	72	55	95	41	**291**	100	27	16	6	6	6	5	5		19	33	1	1	–	–	–	–	–	–	–	–	–	Samanol		LI06340
181	172	291	180	81	182	104	77	**291**	100	39	36	25	24	22	17	16		20	21	1	1	–	–	–	–	–	–	–	–	–	2-Azaspiro[3.5]nonane, 1,3-diphenyl-		NI30350
57	43	83	56	47	85	219	69	**291**	100	80	63	52	51	46	26	18	**0.00**	20	21	1	1	–	–	–	–	–	–	–	–	–	1H-Dibenz[e,g]isoindole, 2-tert-butyl-2,3-dihydro-, N-oxide	110028-17-6	DD01024
235	103	165	248	77	291	104	105	**291**	100	77	73	69	58	43	41	29		20	21	1	1	–	–	–	–	–	–	–	–	–	Oxazole, 2-pentyl-4,5-diphenyl-	20662-96-8	NI30351
235	103	165	248	77	291	104	249	**291**	100	77	73	69	58	43	41	29		20	21	1	1	–	–	–	–	–	–	–	–	–	Oxazole, 2-pentyl-4,5-diphenyl-	20662-96-8	MS06689
158	130	91	105	144	129	115	77	**291**	100	79	69	55	41	38	36	36	**30.60**	20	26	–	1	–	–	–	–	–	–	–	1	–	4-Boryl-3-methyl-1-phenethyl-4-phenylpiperidine		MS06690
130	158	186	291	105	144	292	157	**291**	100	99	65	60	60	58	47	47		20	26	–	1	–	–	–	–	–	–	–	1	–	4-Boryl-3-methyl-1-phenethyl-4-phenylpiperidine		MS06691
96	55	83	41	67	206	69	234	**291**	100	64	40	38	25	20	20	8	**4.90**	20	37	–	1	–	–	–	–	–	–	–	–	–	2H-Pyrrole, 3,4-dihydro-5-methyl-4-[(7Z)-pentadecenyl]-		NI30352
59	87	147	175	28	115	88	69	**292**	100	72	39	34	32	31	17	17	**0.00**	2	1	2	–	–	6	–	–	–	–	2	–	1	Bis(trifluorophosphine)carbonylcobalt hydride		MS06692
43	42	47	175	61	101	214		**292**	100	98	88	53	44	42	6		**1.10**	2	6	2	2	–	–	4	–	–	–	2	–	–	N,N-Dimethyl-N′,N′-bis(dichlorophosphoryl)hydrazine		LI06341
145	143	187	129	177	179	127	147	**292**	100	91	81	80	68	63	61	37	**0.00**	3	2	–	–	–	–	5	1	–	–	–	–	–	1-Bromo-2,3,4,4,4-pentachlorobutane		LI06342
107	105	215	187	213	185	217	189	**292**	100	99	49	37	28	23	22	19	**0.60**	3	3	1	–	–	–	–	3	–	–	–	–	–	1,1,3-Tribromo-1,2-epoxypropane		NI30353
134	136	99	73	133	63	135	101	**292**	100	63	59	41	40	40	31	18	**0.00**	5	4	–	–	–	–	2	2	–	–	–	–	–	1,3-Dibromo-2,2-dichlorobicyclo[1.1.1]pentane	119327-49-0	DD01025
292	277	185	230	170	138	184	215	**292**	100	81	10	7	6	5	4	3		5	15	–	–	–	–	–	–	–	–	7	–	–	Pentamethyl-heptaphosphane(5)	79180-89-5	NI30354
59	67	131	69	116	135	147	85	**292**	100	15	14	12	7	5	4	4	**0.00**	6	4	2	–	–	6	2	–	–	–	–	–	–	Pentanoic acid, 2,2,3,4,4,5-hexafluoro-3,5-dichloro-, methyl ester		LI06343
266	268	264	165	167	95	130	270	**292**	100	63	63	36	33	24	22	21	**0.00**	7	1	2	–	–	–	5	–	–	–	–	–	–	Benzoic acid, pentachloro-	1012-84-6	NI30355
294	277	292	296	279	275	36	298	**292**	100	95	64	63	59	59	21	20		7	1	2	–	–	–	5	–	–	–	–	–	–	Benzoic acid, pentachloro-	1012-84-6	NI30356
180	28	208	165	93	135	264	292	**292**	100	85	76	49	41	40	20	18		7	9	7	–	–	–	–	–	–	–	1	–	1	Iron, tetracarbonyl(trimethylphosphite-P)-	14878-71-8	NI30357
41	69	64	134	87	45	101	102	**292**	100	93	87	78	76	73	58	53	**0.00**	7	16	4	–	4	–	–	–	–	–	–	–	–	Methanesulphonothioic acid, 1,5-pentanediyl ester		MS06693
29	97	125	153	27	65	199	18	**292**	100	81	57	50	45	30	29	21	**1.00**	7	17	4	–	3	–	–	–	–	–	1	–	–	Phosphorodithioic acid, O,O-diethyl S-[(ethylsulphonyl)methyl] ester	2588-04-7	NI30358
266	268	264	294	270	29	296	292	**292**	100	63	63	15	15	12	9	9		8	5	1	–	–	–	5	–	–	–	–	–	–	Benzene, (pentachloroethoxy)-	56818-03-2	MS06694
266	268	264	270	27	294	167	165	**292**	100	64	63	21	20	18	17	17	**13.00**	8	5	1	–	–	–	5	–	–	–	–	–	–	Benzene, pentachloroethoxy-	10463-10-2	IC03715
266	268	264	165	167	130	270	237	**292**	100	72	67	28	27	21	20	17	**10.05**	8	5	1	–	–	–	5	–	–	–	–	–	–	Benzene, pentachloroethoxy-	10463-10-2	NI30359
75	74	263	261	265	235	294	154	**292**	100	93	82	37	35	24	23	21	**14.72**	8	6	2	–	–	–	–	2	–	–	–	–	–	Benzoic acid, 2,5-dibromo-, methyl ester	57381-43-8	NI30360
279	281	277	294	296	292	251	170	**292**	100	50	50	33	17	17	16	10		8	6	2	–	–	–	–	2	–	–	–	–	–	3,5-Dibromo-4-oxyacetophenone		MS06695
277	278	279	260	261	131	244	205	**292**	100	32	18	17	13	10	9	7	**0.00**	8	28	–	4	–	–	–	–	–	4	–	–	–	Cyclotetrasilazane, 2,2,4,4,6,6,8,8-octamethyl-	1020-84-4	LI06344
277	278	279	260	261	131	244	205	**292**	100	86	48	46	35	27	24	18	**1.10**	8	28	–	4	–	–	–	–	–	4	–	–	–	Cyclotetrasilazane, 2,2,4,4,6,6,8,8-octamethyl-	1020-84-4	NI30361
155	157	221	219	76	75	228	230	**292**	100	98	67	64	54	54	28	27	**22.00**	9	9	4	–	1	–	–	1	–	–	–	–	–	Acetic acid, [(4-bromophenyl)sulphonyl]-, methyl ester	50397-65-4	NI30362
61	46	294	60	62	45	292	63	**292**	100	51	14	13	12	12	11	10		9	10	2	2	–	–	1	1	–	–	–	–	–	Urea, N′-(4-bromo-3-chlorophenyl)-N-methoxy-N-methyl-	13360-45-7	NI30364
61	46	294	292	206	60	45	63	**292**	100	21	9	6	6	6	5	4		9	10	2	2	–	–	1	1	–	–	–	–	–	Urea, N′-(4-bromo-3-chlorophenyl)-N-methoxy-N-methyl-	13360-45-7	NI30363
68	155	39	124	224	69	93	174	**292**	100	32	29	19	16	16	13	10	**0.00**	10	4	1	–	–	8	–	–	–	–	–	–	–	11-Oxatetracyclo[6.2.1.0^{2,7}.0^{3,6}]undec-9-ene, 2,3,4,4,5,5,6,7-octafluoro-		NI30365
28	52	208	56	82	124	112	84	**292**	100	30	25	25	21	15	14	12	**0.00**	10	4	7	–	–	–	–	–	–	–	–	–	1	Iron, tetracarbonyl[(6,7-η)-3-oxabicyclo[3.2.0]hept-6-ene-2,4-dione]-	72339-50-5	NI30366
114	72	55	70	146	292	41	115	**292**	100	54	34	28	23	17	16	13		10	16	–	2	4	–	–	–	–	–	–	–	–	Pyrrolidine, 1,1′-(dithiodicarbonothioyl)bis-	496-08-2	NI30367
230	229	257	263	183	292	201	277	**292**	100	52	47	46	36	26	26	24		10	17	–	4	2	–	1	–	–	–	–	–	–	1,3,5-Triazin-2-amine, 4-(2-chloroethylthio)-N,N-diethyl-6-(methylthio)-	79923-84-5	NI30368
292	93	167	151	166	137	107	65	**292**	100	96	79	68	60	45	29	29		10	18	–	2	2	–	–	–	–	–	2	–	–	N,N′-Bis(dimethylthiophosphinyl)-1,3-phenylenediamine		MS06696
292	93	107	199	200	65	167	108	**292**	100	76	67	55	17	9	7	5		10	18	–	2	2	–	–	–	–	–	2	–	–	N,N′-Bis(dimethylthiophosphinyl)-1,4-phenylenediamine		MS06697
103	86	102	128	60	292	69	248	**292**	100	99	71	69	59	52	51	35		10	20	2	4	2	–	–	–	–	–	–	–	–	1,4-Dioxa-7,9,12,14-tetraazacyclohexadecane-8,13-dithione		NI30369
147	86	69	60	103	130	62	85	**292**	100	79	37	37	36	35	35	28	**18.00**	10	20	2	4	2	–	–	–	–	–	–	–	–	1,9-Dioxa-4,6,12,14-tetraazacyclohexadecane-5,13-dithione	74804-39-0	NI30370
146	73	75	189	190	147	148	131	**292**	100	50	30	29	24	15	8	8	**2.19**	10	20	6	–	–	–	–	–	–	2	–	–	–	1,4-Dioxane-2,5-dione, 3,6-bis[(trimethylsilyl)oxy]-	55638-47-6	NI30371
73	75	130	103	45	59	42	116	**292**	100	34	26	25	21	20	20	19	**0.00**	10	24	4	2	–	–	–	–	–	2	–	–	–	Glycine, N-nitroso-N-[2-(trimethylsiloxy)ethyl]-, trimethylsilyl ester		NI30372
277	73	189	261	43	131	41	205	**292**	100	25	17	8	7	6	6	5	**0.01**	10	28	2	–	–	–	–	–	–	4	–	–	–	1,5-Dioxa-2,4,6,8-tetrasilacyclooctane, 2,2,4,4,6,6,8,8-octamethyl-	15261-06-0	LI06345
277	73	189	279	278	261	43	131	**292**	100	25	17	14	9	7	7	6	**0.01**	10	28	2	–	–	–	–	–	–	4	–	–	–	1,5-Dioxa-2,4,6,8-tetrasilacyclooctane, 2,2,4,4,6,6,8,8-octamethyl-	15261-06-0	NI30373
192	68	67	163	86	150	292	41	**292**	100	44	36	26	26	21	19	16		11	8	–	–	–	8	–	–	–	–	–	–	–	Tricyclo[5.2.2.0^{2,6}]undec-8-ene, 1,7,8,9,10,10,11,11-octafluoro-		NI30374
43	77	85	71	292	207	91	130	**292**	100	45	40	39	37	17	11	6		11	14	1	–	–	–	–	–	–	–	–	–	1	Furan, tetrahydro-2-[(phenyltelluro)methyl]-		DD01026
251	193	83	264	69	81	77	139	**292**	100	51	44	40	39	26	26	24	**1.00**	11	14	2	–	–	6	–	–	–	–	–	–	–	Cyclopropane, 1,1′-[bis(2,2,2-trifluoroethoxy)methylene]bis-	70106-36-4	NI30375
73	75	74	45	93	233	292	235	**292**	100	22	10	8	8	6	5	4		11	14	3	–	–	–	2	–	–	1	–	–	–	Acetic acid, 2,4-dichlorophenoxy-, trimethylsilyl ester	34113-76-3	NI30376
61	76	59	161	39	163	27	160	**292**	100	64	62	55	53	50	46	44	**1.12**	11	23	–	–	–	–	–	–	–	–	1	–	1	Palladium, bis[(1,2,3-η)-2-methyl-2-propenyl](trimethylphosphine)-	74811-10-2	NI30377

MASS TO CHARGE RATIOS								M.W.	INTENSITIES								Parent	C	H	O	N	S	F	Cl	Br	I	Si	P	B	X	COMPOUND NAME	CAS Reg No	No
73	55	245	147	45	75	18	157	**292**	100	65	37	34	21	16	10	10	**8.65**	11	24	3	–	1	–	–	–	–	2	–	–	–	2-Butenoic acid, 4-(methylthio)-2-[(trimethylsilyl)oxy]-, trimethylsilyl ester		NI30378
77	51	65	125	182	211	93	152	**292**	100	47	43	43	38	30	27	24	**3.01**	12	8	5	2	1	–	–	–	–	–	–	–	–	2,4-Dinitrodiphenyl sulphoxide	19093-39-1	NI30379
105	77	93	106	292	45	43	277	**292**	100	14	9	7	6	4	4	2		12	11	3	–	1	3	–	–	–	–	–	–	–	Acetic acid, α-[(trifluoromethyl)thio]benzoyl-, ethyl ester		DD01027
55	153	80	124	97	111	67	177	**292**	100	85	84	61	49	47	41	9	**0.00**	12	13	5	–	–	–	–	–	–	–	–	–	1	Manganese, (OC-6-23)-tetracarbonyl[1-methyl-1-oxo-2-heptenyl]-	122967-42-4	DD01028
192	194	274	157	276	248	196	250	**292**	100	67	43	42	29	29	16	14	**14.00**	12	14	4	–	–	–	2	–	–	–	–	–	–	Benzoic acid, 3,5-dichloro-2,4-dihydroxy-6-pentyl-	64185-27-9	NI30380
105	216	77	236	155	81	292	101	**292**	100	71	56	32	30	23	22	21		12	15	4	–	–	2	–	–	–	–	1	–	–	Diethyl 2,2-difluoro-1-phenylethenyl phosphate		DD01029
59	292	205	141	131	147	113	41	**292**	100	70	62	54	44	32	32	24		12	20	–	–	4	–	–	–	–	–	–	–	–	2,4,6,8-Tetrathiatricyclo[3.3.1.1^{3,7}]decane, 1,3,9,9,10,10-hexamethyl-, (±)-	57289-41-5	NI30381
86	45	161	87	42	205	89	70	**292**	100	48	45	43	42	19	17	12	**0.80**	12	20	8	–	–	–	–	–	–	–	–	–	–	1,4,7,10,13,16-Hexaoxacyclooctadecane-14,18-dione		MS06698
75	74	88	43	87	129	116	101	**292**	100	48	45	43	33	30	30	30	**0.09**	12	20	8	–	–	–	–	–	–	–	–	–	–	Methyl (methyl 2-O-acetyl-3,4-di-O-methyl-α-D-mannopyranoside)uronate	87910-35-8	NI30382
74	101	43	75	116	88	117	85	**292**	100	92	88	75	47	33	30	30	**0.00**	12	20	8	–	–	–	–	–	–	–	–	–	–	Methyl (methyl 3-O-acetyl-2,4-di-O-methyl-α-D-mannopyranoside)uronate	87910-34-7	NI30383
75	88	43	129	87	101	73	45	**292**	100	80	41	35	22	16	15	15	**0.09**	12	20	8	–	–	–	–	–	–	–	–	–	–	Methyl (methyl 4-O-acetyl-2,3-di-O-methyl-α-D-mannopyranoside)uronate	87910-33-6	NI30384
43	88	201	101	83	73	71	81	**292**	100	64	37	34	22	17	15	13	**0.00**	12	20	8	–	–	–	–	–	–	–	–	–	–	D-Xylopyranose, 5-C-(acetyloxy)-2,3,4-tri-O-methyl-, acetate	69502-92-7	NI30385
200	94	147	80	79	199	65	77	**292**	100	49	33	27	22	16	16	15	**0.00**	12	22	4	–	–	–	–	–	–	–	2	–	–	4,7-Phosphinidenephosphindole, 1,8-diethoxyoctahydro-, 1,8-dioxide	16182-90-4	MS06699
179	181	178	292	177	180	294	151	**292**	100	60	31	29	29	24	20	9		12	27	–	–	–	–	–	–	–	–	–	–	1	Stibine, tri-tert-butyl-	13787-35-4	NI30386
179	181	178	292	180	294	235	236	**292**	100	62	32	30	26	22	9	6		12	27	–	–	–	–	–	–	–	–	–	–	1	Stibine, tri-tert-butyl-	13787-35-4	NI30387
151	179	207	121	263	149	235	120	**292**	100	91	86	83	54	51	36	32	**0.00**	12	28	–	–	–	–	–	–	–	–	–	–	1	Dibutyldiethyltin	20525-62-6	NI30388
135	193	137	133	191	29	27	249	**292**	100	84	79	66	65	56	53	43	**0.00**	12	28	–	–	–	–	–	–	–	–	–	–	1	Dibutylmethylisopropyltin		MS06700
207	249	165	121	163	120	135	292	**292**	100	98	93	51	50	25	15	5		12	28	–	–	–	–	–	–	–	–	–	–	1	Tetraisopropyltin		LI06346
207	163	165	249	205	208	247	161	**292**	100	92	82	80	73	72	62	62	**3.30**	12	28	–	–	–	–	–	–	–	–	–	–	1	Tetraisopropyltin		MS06701
207	165	163	205	247	161	203	121	**292**	100	84	76	75	51	50	44	34	**0.30**	12	28	–	–	–	–	–	–	–	–	–	–	1	Tetrapropyltin		MS06702
207	165	249	121	163	120	123	135	**292**	100	93	66	56	26	16	13	8	**0.10**	12	28	–	–	–	–	–	–	–	–	–	–	1	Tetrapropyltin		LI06347
207	165	249	121	163	123	120	135	**292**	100	82	68	30	20	11	11	7	**0.30**	12	28	–	–	–	–	–	–	–	–	–	–	1	Tetrapropyltin		LI06348
292	75	150	294	110	234	293	164	**292**	100	70	64	63	60	53	52	50		13	6	2	2	–	–	2	–	–	–	–	–	–	11H-Dibenzo[c,f][1,2]diazepin-11-one, 3,8-dichloro-, 5-oxide	960-86-1	NI30389
292	294	246	277	293	231	196	248	**292**	100	36	30	24	18	18	16	15		13	9	2	2	1	–	1	–	–	–	–	–	–	10H-Phenothiazine, 3-chloro-10-methyl-7-nitro-	35076-84-7	NI30390
292	294	246	277	293	231	248	196	**292**	100	40	34	27	22	19	17	17		13	9	2	2	1	–	1	–	–	–	–	–	–	10H-Phenothiazine, 3-chloro-10-methyl-7-nitro-	35076-84-7	LI06349
77	141	294	292	247	63	51	65	**292**	100	54	30	30	28	28	28	27		13	9	3	–	–	–	–	1	–	–	–	–	–	Carbonic acid, 4-bromophenyl phenyl ester	17175-13-2	NI30391
77	93	141	51	110	119	66	65	**292**	100	60	52	27	19	18	16	15	**1.00**	13	12	2	2	2	–	–	–	–	–	–	–	–	Benzenesulphonamide, N-[(phenylamino)thioxomethyl]-	24539-87-5	LI06350
77	93	141	51	119	110	66	65	**292**	100	59	52	27	18	17	17	15	**2.00**	13	12	2	2	2	–	–	–	–	–	–	–	–	Benzenesulphonamide, N-[(phenylamino)thioxomethyl]-	24539-87-5	NI30392
292	148	261	105	133	85	120	147	**292**	100	73	58	32	31	30	28	19		13	12	4	2	1	–	–	–	–	–	–	–	–	1,3-Thiazin-4-one, 2-(2-methoxyphenylimino)-6-(methoxycarbonyl)-2,3-dihydro-		MS06703
217	139	292	183	63	107	185	152	**292**	100	51	48	44	22	12	11	5		13	13	–	2	2	–	–	–	–	–	1	–	–	1-Diphenylphosphinothioylthiourea		MS06704
277	275	43	257	273	255	276	128	**292**	100	78	55	51	47	38	37	31	**12.60**	13	16	–	–	–	–	–	–	–	–	–	–	1	1-Naphthyltrimethyltin		MS06705
277	275	28	247	273	276	245	274	**292**	100	65	51	43	42	33	31	24	**9.70**	13	16	–	–	–	–	–	–	–	–	–	–	1	2-Naphthyltrimethyltin		MS06706
70	57	41	42	44	83	98	43	**292**	100	81	57	38	37	33	32	31	**2.00**	13	16	4	4	–	–	–	–	–	–	–	–	–	Cyclohexanone, 3-methyl-, (2,4-dinitrophenyl)hydrazone	4864-46-4	NI30393
292	41	55	69	95	113	67	152	**292**	100	85	82	72	68	60	48	22		13	16	4	4	–	–	–	–	–	–	–	–	–	Cyclohexanone, 4-methyl-, (2,4-dinitrophenyl)hydrazone	5138-32-9	LI06351
292	55	41	69	95	113	81	67	**292**	100	84	84	73	68	60	57	48		13	16	4	4	–	–	–	–	–	–	–	–	–	Cyclohexanone, 4-methyl-, (2,4-dinitrophenyl)hydrazone	5138-32-9	NI30394
292	55	41	69	95	113	81	67	**292**	100	86	82	74	68	60	58	50		13	16	4	4	–	–	–	–	–	–	–	–	–	Cyclohexanone, 4-methyl-, (2,4-dinitrophenyl)hydrazone	5138-32-9	MS06707
166	110	96	153	55	140	167	54	**292**	100	83	77	65	65	60	51	48	**18.02**	13	16	4	4	–	–	–	–	–	–	–	–	–	2,4(1H,3H)-Pyrimidinedione, 1,1′-(1,3-propanediyl)bis[5-methyl-	22917-75-5	NI30395
161	176	162	177	147	115	160	141	**292**	100	98	20	14	14	14	10	9	**0.00**	13	17	–	–	1	4	–	–	–	–	–	1	–	1-Ethyl-2,3,5-trimethylbenzo[b]thiophenium tetrafluoroborate		LI06352
111	77	139	105	83	45	47	263	**292**	100	99	79	68	41	35	31	6	**0.00**	13	19	2	–	–	3	–	–	–	1	–	–	–	2,5-Cyclohexadien-1-one, 4-(triethylsilyloxy)-4-(trifluoromethyl)-	120120-28-7	DD01030
101	131	117	89	36	43	115	72	**292**	100	78	69	60	58	52	49	27	**0.00**	13	24	7	–	–	–	–	–	–	–	–	–	–	Hexitol, 1-deoxy-3,4,5-tri-O-methyl-, diacetate	75330-98-2	NI30396
292	294	29	291	263	43	69	258	**292**	100	50	19	11	11	11	10	8		14	9	5	–	–	–	1	–	–	–	–	–	–	9H-Xanthen-9-one, 2-chloro-1,3,6-trihydroxy-8-methyl-		MS06708
292	294	293	291	263	29	69	77	**292**	100	33	18	9	9	9	6	5		14	9	5	–	–	–	1	–	–	–	–	–	–	9H-Xanthen-9-one, 4-chloro-1,3,6-trihydroxy-8-methyl-		MS06709
292	249	43	221	246	294	293	264	**292**	100	52	31	29	24	20	17	17		14	12	7	–	–	–	–	–	–	–	–	–	–	1,4-Naphthoquinone, 2-acetyl-3,5-dihydroxy-6,8-dimethoxy-	14090-52-9	NI30397
292	43	231	69	222	293	264	53	**292**	100	59	28	24	21	17	15	14		14	12	7	–	–	–	–	–	–	–	–	–	–	1,4-Naphthoquinone, 2-acetyl-3,8-dihydroxy-5,6-dimethoxy-	14090-53-0	NI30398
292	43	249	231	69	222	293	264	**292**	100	59	47	27	24	21	17	15		14	12	7	–	–	–	–	–	–	–	–	–	–	1,4-Naphthoquinone, 2-acetyl-3,8-dihydroxy-5,6-dimethoxy-	14090-53-0	LI06353
292	249	43	221	246	231	293	264	**292**	100	51	31	29	25	17	16	16		14	12	7	–	–	–	–	–	–	–	–	–	–	1,4-Naphthoquinone, 2-acetyl-3,8-dihydroxy-5,6-dimethoxy-	14090-53-0	NI30399
105	77	146	106	51	65	149	292	**292**	100	29	8	8	8	7	5	3		14	13	2	–	–	–	–	1	–	–	–	–	–	Bicyclo[2.2.1]hept-2-ene, 1-bromo-6-benzoyloxy-		LI06354
292	245	117	145	91	174	61	146	**292**	100	84	80	61	60	59	51	36		14	16	3	2	1	–	–	–	–	–	–	–	–	2,4,6-Pyrimidinetrione, 5-ethyl-1-(methylthiomethyl)-5-phenyl-		MS06710
91	157	141	65	58	201	292	107	**292**	100	83	68	42	22	18	13	13		14	16	3	2	1	–	–	–	–	–	–	–	–	4-Thiazolecarboxamide, 2-[2-(benzyloxy)ethoxymethyl]-		MS06711

MASS TO CHARGE RATIOS								M.W.	INTENSITIES								Parent	C	H	O	N	S	F	Cl	Br	I	Si	P	B	X	COMPOUND NAME	CAS Reg No	No
152	157	126	158	292	125	91	109	**292**	100	78	77	67	49	44	30	29		14	16	3	2	1	–	–	–	–	–	–	–	–	Urea, N-(exo-bicyclo[4.1.0]hept-3-en-7-yl)-N′-[2-(methoxycarbonyl)thiophen-3-yl]-	85156-53-2	NI30400
59	233	116	89	232	63	115	77	**292**	100	84	56	50	32	31	28	26	**0.00**	14	16	5	2	–	–	–	–	–	–	–	–	–	4a,8a-Methanophthalazine-1,4-dicarboxylic acid, 1,2-dihydro-1-methoxy-, dimethyl ester		DD01031
43	159	172	73	158	171	105	160	**292**	100	23	16	13	6	4	3	2	**0.00**	14	17	6	–	–	–	–	–	–	–	–	1	–	1,3,2-Dioxaborolane-4,5-dimethanol, 2-phenyl-, diacetate, trans-	74793-34-3	NI30401
60	73	130	129	45	227	292	162	**292**	100	70	54	33	28	26	12	7		14	28	–	–	3	–	–	–	–	–	–	–	–	1,2,4-Trithiolane, 3,5-dihexyl-	73861-43-5	NI30402
73	117	132	75	55	69	83	145	**292**	100	82	61	56	38	36	35	24	**2.60**	14	29	2	–	–	–	1	–	–	1	–	–	–	Undecanoic acid, 11-chloro-, trimethylsilyl ester	26305-98-6	NI30403
263	165	179	277	221	121	79	207	**292**	100	50	46	41	30	21	20	19	**0.30**	14	32	4	–	–	–	–	–	–	1	–	–	–	Silicic acid (H4SiO4), dibutyl dipropyl ester	55683-16-4	NI30404
208	153	152	236	292	151	209	264	**292**	100	79	53	24	22	16	13	10		15	9	3	–	–	–	–	–	–	–	–	–	1	Manganese, tricarbonyl(1-hydro-s-indacene)-		MS06712
208	209	152	77	44	153	151	105	**292**	100	18	17	8	8	7	5	4	**0.00**	15	10	2	–	–	–	2	–	–	–	–	–	–	10-(Dichloromethyl)-10-hydroxyanthrone		IC03716
136	139	292	91	65	137	51	125	**292**	100	49	37	37	23	19	15	13		15	13	2	–	1	–	1	–	–	–	–	–	–	(1-Chloro-2-phenylethenyl) (4-methylphenyl) sulphone		DD01032
245	260	261	144	246	115	187	174	**292**	100	61	35	18	13	13	12	11	**9.00**	15	16	6	–	–	–	–	–	–	–	–	–	–	Benzo[b]furan-5,6-dicarboxylic acid, 4-(methoxymethyl)-2-methyl-, dimethyl ester		NI30405
177	43	32	206	220	39	77	51	**292**	100	89	74	48	45	43	39	39	**28.90**	15	16	6	–	–	–	–	–	–	–	–	–	–	6H-Dibenzo[b,d]pyran-6-one, 2,3,4,4a-tetrahydro-2,3,7-trihydroxy-9-methoxy-4a-methyl-, $(2\alpha,3\beta,4a\beta)$-(±)-	29752-43-0	NI30406
177	292	219	206	220	248	204	175	**292**	100	84	84	82	81	64	32	28		15	16	6	–	–	–	–	–	–	–	–	–	–	6H-Dibenzo[b,d]pyran-6-one, 2,3,4,4a-tetrahydro-2,3,7-trihydroxy-9-methoxy-4a-methyl-, $(2\alpha,3\beta,4a\beta)$-(±)-	29752-43-0	NI30407
43	77	39	105	133	115	116	81	**292**	100	12	10	10	9	9	9	9	**2.10**	15	16	6	–	–	–	–	–	–	–	–	–	–	2,4,6,8-Nonatetraene, 1,1-diacetoxy-9,9-diformyl-	104970-92-5	NI30408
206	178	177	208	207	143	292	205	**292**	100	53	44	40	33	26	23	20		15	17	2	2	–	–	1	–	–	–	–	–	–	2-Cyclopenten-1-one, 2-(4-chloroanilino)-4-morpholino-		MS06713
116	117	176	104	130	177	115	103	**292**	100	25	16	14	13	9	8	6	**0.40**	15	20	4	2	–	–	–	–	–	–	–	–	–	1,2-Hydrazinedicarboxylic acid, indan-2-yl-, diethyl ester		LI06355
81	69	163	95	55	41	107	123	**292**	100	98	79	55	55	43	40	27	**0.00**	15	23	2	–	–	3	–	–	–	–	–	–	–	Naphthalen-2β-ol, 1,2α,3,4,4aα,5,6,7,8,8aβ-decahydro-5,5,8aβ-trimethyl-, trifluoroacetate, (±)-		DD01033
292	190	114	218	264	76	186	104	**292**	100	18	12	10	9	9	8	5		16	8	4	2	–	–	–	–	–	–	–	–	–	4-Cyano-3-(4-nitrophenyl)isocoumarin		LI06356
264	139	266	111	141	251	152	253	**292**	100	99	59	35	33	28	20	19	**3.03**	16	14	1	–	–	–	2	–	–	–	–	–	–	Benzenemethanol, 4-chloro-α-(4-chlorophenyl)-α-cyclopropyl-	14088-71-2	NI30409
132	91	41	65	292	74	58	99	**292**	100	42	24	11	7	7	7	6		16	20	1	–	2	–	–	–	–	–	–	–	–	6,10-Dithiaspiro[4.5]decan-1-one, 2-benzyl-2-methyl-	59579-82-7	NI30410
277	247	292	278	131	217	77	52	**292**	100	22	18	16	5	4	3	3		16	20	5	–	–	–	–	–	–	–	–	–	–	2H-1-Benzopyran, 6-acetyl-5,7,8-trimethoxy-2,2-dimethyl-	482-07-5	NI30411
99	81	55	43	151	292	179	150	**292**	100	39	15	13	12	11	10	9		16	20	5	–	–	–	–	–	–	–	–	–	–	Cladosporin		MS06714
145	218	219	77	105	159	146	131	**292**	100	39	28	23	22	17	16	14	**3.80**	16	20	5	–	–	–	–	–	–	–	–	–	–	1,2-Cyclopropanedicarboxylic acid, 3-methoxy-3-phenyl-, diethyl ester	1472-10-2	NI30412
145	218	77	146	247	219	105	159	**292**	100	46	20	17	15	15	15	14	**0.20**	16	20	5	–	–	–	–	–	–	–	–	–	–	1,2-Cyclopropanedicarboxylic acid, 3-methoxy-3-phenyl-, diethyl ester, $(1\alpha,2\beta,3\alpha)$-	42756-22-9	NI30413
152	194	43	153	222	150	137	135	**292**	100	68	67	40	8	8	8	6	**2.93**	16	20	5	–	–	–	–	–	–	–	–	–	–	4,7-Ethanoisobenzofuran-1,3-dione, 5-(acetyloxy)-3a,4,7,7a-tetrahydro-4-methyl-7-isopropyl-	52918-77-1	NI30414
152	43	137	194	82	153	41	135	**292**	100	93	35	15	12	10	10	9	**5.34**	16	20	5	–	–	–	–	–	–	–	–	–	–	4,7-Ethanoisobenzofuran-1,3-dione, 5-(acetyloxy)-3a,4,7,7a-tetrahydro-7-methyl-4-isopropyl-	52918-78-2	NI30415
151	292	177	152	293	127	77	67	**292**	100	81	30	28	15	14	9	9		16	20	5	–	–	–	–	–	–	–	–	–	–	2H-Pyran-2-one, 6-[2-(3,4-dimethoxyphenyl)ethyl]-5,6-dihydro-4-methoxy-, (S)-	38146-60-0	NI30416
53	91	41	67	65	77	39	51	**292**	100	97	95	94	94	91	89	85	**0.00**	16	20	5	–	–	–	–	–	–	–	–	–	–	Sericealactone		MS06715
53	91	41	67	65	77	43	39	**292**	100	96	94	93	93	90	90	89	**0.00**	16	20	5	–	–	–	–	–	–	–	–	–	–	Sericealactone		LI06357
105	169	83	174	274	256	69	218	**292**	100	46	38	26	15	15	11	5	**0.00**	16	20	5	–	–	–	–	–	–	–	–	–	–	2,3,8-Trioxabicyclo[4.4.0]decan-7-one, 1-hydroxy-6,9,9-trimethyl-4-phenyl-		DD01034
262	263	261	291	292	260	290	65	**292**	100	72	57	25	19	18	15	14		16	34	–	–	–	–	–	–	–	–	–	6	–	2,4,6,8,9,10-Hexaboratricyclo[3.3.1.1^{3,7}]decane, 2,4,6,8,9,10-hexaethyl-	57387-91-4	NI30417
292	142	92	107	170	115	293	146	**292**	100	86	49	37	31	29	26	21		17	16	1	4	–	–	–	–	–	–	–	–	–	1-(4-Aminonaphthylazo)-2-methoxy-4-aminobenzene		IC03717
292	91	132	223	77	159	131	293	**292**	100	88	48	36	34	23	21	20		17	16	1	4	–	–	–	–	–	–	–	–	–	2-Pyrazolin-5-one, 1-phenyl-3-methyl-4-[4-(methylamino)phenylimino]-		MS06716
41	43	152	83	93	163	91	123	**292**	100	70	35	31	26	25	23	21	**10.00**	17	24	4	–	–	–	–	–	–	–	–	–	–	Anhydroportentol		LI06358
292	165	83	41	43	124	39	293	**292**	100	73	45	43	32	27	22	20		17	24	4	–	–	–	–	–	–	–	–	–	–	1,3-Cyclohexanedione, 2,2′-methylenebis[5,5-dimethyl-	2181-22-8	LI06359
292	165	293	83	18	277	125	124	**292**	100	45	32	22	22	14	10	10		17	24	4	–	–	–	–	–	–	–	–	–	–	1,3-Cyclohexanedione, 2,2′-methylenebis[5,5-dimethyl-	2181-22-8	NI30418
83	56	55	292	152	165	140	41	**292**	100	49	42	36	28	26	25	25		17	24	4	–	–	–	–	–	–	–	–	–	–	1,3-Cyclohexanedione, 2,2′-methylenebis[5,5-dimethyl-	2181-22-8	IC03718
157	232	175	190	189	292			**292**	100	68	60	59	16	2				17	24	4	–	–	–	–	–	–	–	–	–	–	1-Cyclohexene, 4-acetoxy-1-(3-acetoxybut-1-ynyl)-2,6,6-trimethyl-		LI06360
43	190	159	121	71	41	134	59	**292**	100	72	60	54	52	52	38	34	**0.00**	17	24	4	–	–	–	–	–	–	–	–	–	–	2-Naphthalenepropanoic acid, 1,2,3,4-tetrahydro-β-hydroxy-6-methoxy-α,α-dimethyl-, methyl ester	2473-18-9	NI30419
43	190	159	121	71	41	134	59	**292**	100	70	58	52	50	50	38	34	**0.00**	17	24	4	–	–	–	–	–	–	–	–	–	–	2-Naphthalenepropionic acid, 1,2,3,4-tetrahydro-β-hydroxy-6-methoxy-α,α-dimethyl-, methyl ester	2473-18-9	LI06361
163	149	70	77	55	57	181	76	**292**	100	56	51	33	27	20	18	15	**0.00**	17	24	4	–	–	–	–	–	–	–	–	–	–	Phthalic acid, methyl 2-ethylhexyl ester		NI30420
43	124	93	108	41	107	91	109	**292**	100	41	40	31	29	25	24	20	**7.80**	17	24	4	–	–	–	–	–	–	–	–	–	–	Trichothec-9-en-3-ol, 12,13-epoxy-, acetate, (3α)-	91423-90-4	NI30421
292	93	41	200	71	136	57	42	**292**	100	44	35	34	28	25	24	24		17	28	–	2	1	–	–	–	–	–	–	–	–	Thiourea, N,N-bis(3-methylbutyl)-N′-phenyl-	56438-21-2	NI30422
277	75	117	91	73	129	145	132	**292**	100	95	79	72	71	46	40	38	**33.80**	17	28	2	–	–	–	–	–	–	1	–	–	–	Benzeneoctanoic acid, trimethylsilyl ester		NI30423

MASS TO CHARGE RATIOS	M.W.	INTENSITIES	Parent	C	H	O	N	S	F	Cl	Br	I	Si	P	B	X	COMPOUND NAME	CAS Reg No	No
292 202 189 146 203 293 184 190	**292**	100 44 29 25 22 20 16 9		18	12	2	–	1	–	–	–	–	–	–	–	–	19,20-Dioxa-21-thiatetracyclo[14.2.1.1^{4,7}.1^{10,13}]heneicos-2,4,6,8,10,12,14,16,18-nonaene		LI06362
291 128 292 275 127 119 102 63	**292**	100 95 92 58 53 35 35 31		18	12	4	–	–	–	–	–	–	–	–	–	–	Chromone-6-carboxylic acid, 2-styryl-, trans-		MS06717
118 292 294 89 63 293 291 275	**292**	100 58 22 22 15 14 14 14		18	12	4	–	–	–	–	–	–	–	–	–	–	2,5-Cyclohexadiene-1,4-dione, 2,5-bis(4-hydroxyphenyl)-	16688-88-3	NI30424
292 293 89 118 129 90 77 63	**292**	100 19 17 11 10 7 6 6		18	12	4	–	–	–	–	–	–	–	–	–	–	2,5-Cyclohexadiene-1,4-dione, 2,5-dihydroxy-3,6-diphenyl-	548-59-4	NI30425
264 292 178 179 165 189 235 265	**292**	100 68 26 25 25 21 19 18		18	12	4	–	–	–	–	–	–	–	–	–	–	Phenanthro[1,2-b]furan-10,11-dione, 1-(hydroxymethyl)-6-methyl-	76829-01-1	NI30426
292 215 294 293 217 228 216 295	**292**	100 66 36 26 22 15 12 9		18	13	–	–	–	–	1	–	–	1	–	–	–	9-Chloro-9-phenyl-9-silafluorene	18766-52-4	NI30427
142 259 291 143 292 115 277 169	**292**	100 71 56 48 47 37 33 30		18	16	–	2	1	–	–	–	–	–	–	–	–	Benzothiazole, 6-methyl-2-(6-methylquinol-2-yl)-2,6-dihydro-		IC03719
146 292 91 147 293 118 117 77	**292**	100 99 35 25 23 19 15 8		18	16	2	2	–	–	–	–	–	–	–	–	–	3,3′-Bioxindole, N,N′-dimethyl-		LI06363
105 182 183 68 187 43 77 181	**292**	100 60 38 34 29 28 28 11	**1.40**	18	16	2	2	–	–	–	–	–	–	–	–	–	6-Oxa-2,4-diazabicyclo[3.2.0]hept-2-ene, 4-acetyl-7,7-diphenyl-		NI30428
291 292 264 262 222 263 249 265	**292**	100 86 77 40 40 24 18 16		18	16	2	2	–	–	–	–	–	–	–	–	–	Dragabine		NI30429
159 105 292 77 293 160 143 51	**292**	100 61 60 24 12 11 9 8		18	16	2	2	–	–	–	–	–	–	–	–	–	5(4H)-Oxazolonone, 4-[4-(dimethylamino)benzylidene]-2-phenyl-	1564-29-0	NI30430
159 105 292 77 143 134 51 146	**292**	100 61 60 24 9 8 8 5		18	16	2	2	–	–	–	–	–	–	–	–	–	5(4H)-Oxazolonone, 4-[4-(dimethylamino)benzylidene]-2-phenyl-	1564-29-0	LI06364
247 292 77 248 51 293 204 123	**292**	100 54 35 19 12 10 9 9		18	16	2	2	–	–	–	–	–	–	–	–	–	1H-Pyrazole-4-acetic acid, 3-methyl-1,5-diphenyl-	32701-85-2	NI30431
77 292 93 200 119 51 172 144	**292**	100 51 40 27 18 18 16 15		18	16	2	2	–	–	–	–	–	–	–	–	–	3,5-Pyrazolidinedione, 1,2-diphenyl-4-propylidene-	2810-68-6	NI30432
261 292 77 115 185 262 53 81	**292**	100 60 28 27 19 17 16 14		18	16	2	2	–	–	–	–	–	–	–	–	–	5-Pyrazolone, 1-phenyl-3-methyl-4-(2-methoxybenzylidene)-		NI30433
150 215 187 292 291 185 293 149	**292**	100 76 71 70 50 18 16 14		18	16	2	2	–	–	–	–	–	–	–	–	–	2-Pyrimidinone, 5-benzoyl-4-methyl-6-phenyl-1,2,3,4-tetrahydro-		LI06365
104 132 292 263 103 77 293 105	**292**	100 54 45 36 17 12 10 9		18	20	–	4	–	–	–	–	–	–	–	–	–	s-Tetrazine, 1,2-diethyl-1,2-dihydro-3,6-diphenyl-		MS06718
292 127 293 147 179 233 274 128	**292**	100 37 20 20 18 12 7 4		18	25	2	–	–	1	–	–	–	–	–	–	–	Estra-5(10)-en-6-one, 3β-fluoro-17β-hydroxy-	33547-53-4	NI30434
292 127 233 179 150 274 259 272	**292**	100 53 28 27 26 15 6 3		18	25	2	–	–	1	–	–	–	–	–	–	–	Estra-5(10)-en-6-one, 3β-fluoro-17β-hydroxy-	33547-53-4	NI30435
292 81 80 41 193 43 175 55	**292**	100 94 40 34 32 32 28 28		18	28	3	–	–	–	–	–	–	–	–	–	–	C-Nor-5β,13α-androstan-11-one, 3α,17β-dihydroxy-	13976-73-3	NI30436
152 135 153 292 107 77 136 109	**292**	100 37 24 10 6 6 4 4		18	28	3	–	–	–	–	–	–	–	–	–	–	Benzoic acid, 3-methoxy-, decyl ester	69833-38-1	NI30437
277 191 292 57 28 219 147 278	**292**	100 72 66 37 24 23 20 19		18	28	3	–	–	–	–	–	–	–	–	–	–	Butanoic acid, 1-(3,5-di-tert-butyl-4-hydroxyphenyl)-		IC03720
221 41 137 179 43 55 292 222	**292**	100 41 33 29 29 18 16 14		18	28	3	–	–	–	–	–	–	–	–	–	–	5,9-Methanobenzocycloocten-1(2H)-one, 3,4,5,6,7,8,9,10-octahydro-3,5-dihydroxy-3,3,7,7,9-pentamethyl-	54833-46-4	NI30438
221 39 203 222 175 121 41 55	**292**	100 39 24 16 16 16 14 10	**7.35**	18	28	3	–	–	–	–	–	–	–	–	–	–	5,9-Methanobenzocycloocten-1(2H)-one, 3,4,5,6,7,8,9,10-octahydro-5,10-dihydroxy-3,3,7,7,9-pentamethyl-	54833-45-3	NI30439
153 197 113 292 233 277 163 93	**292**	100 69 67 33 24 19 17 16		18	28	3	–	–	–	–	–	–	–	–	–	–	Methyl 13-deisopropyl-1,11-epoxyabietanoate		NI30440
149 177 135 122 133 292 164 217	**292**	100 99 50 44 33 31 31 27		18	28	3	–	–	–	–	–	–	–	–	–	–	Octanoic acid, 8-[3-oxo-2-(2-penten-1-yl)cyclopent-1-enyl]-		NI30441
95 107 163 109 224 149 121 178	**292**	100 48 37 36 24 24 22 17	**10.70**	18	28	3	–	–	–	–	–	–	–	–	–	–	Octanoic acid, 8-[3-oxo-2-(2-penten-1-yl)cyclopent-4-enyl]-		NI30442
232 43 108 134 178 41 55 93	**292**	100 68 50 45 39 32 28 25	**24.09**	18	28	3	–	–	–	–	–	–	–	–	–	–	1(2H)-Phenanthrenone, 7-(acetyloxy)dodecahydro-2,4b-dimethyl-, [2S-(2α,4aα,4bβ,7β,8aα,10aβ)]-	41853-35-4	NI30443
57 44 277 41 147 29 43 55	**292**	100 61 60 56 43 31 30 29	**20.00**	18	28	3	–	–	–	–	–	–	–	–	–	–	Propionic acid, 3-(3,5-di-tert-butyl-4-hydroxyphenyl)-, methyl ester		IC03721
73 277 57 292 45 41 75 278	**292**	100 46 25 23 13 12 12 11		18	32	1	–	–	–	–	–	–	1	–	–	–	3,5-Di-tert-butyl-4-trimethylsiloxytoluene		NI30444
250 42 292 55 41 192 235 221	**292**	100 58 55 51 39 37 36 26		18	32	1	2	–	–	–	–	–	–	–	–	–	1,3-Diazaadamantan-6-one, 5,7-dipentyl-		MS06719
292 247 263 215 187 234 202 115	**292**	100 47 38 32 32 21 11 11		19	16	1	–	1	–	–	–	–	–	–	–	–	2H-Naphtho[1,8-bc]thiophene, 5-ethoxy-2-phenyl-	10245-70-2	LI06366
292 187 247 215 263 234 115 202	**292**	100 95 94 94 75 63 33 32		19	16	1	–	1	–	–	–	–	–	–	–	–	2H-Naphtho[1,8-bc]thiophene, 5-ethoxy-2-phenyl-	10245-70-2	NI30445
141 107 115 77 79 105 292 139	**292**	100 46 19 18 17 14 12 8		19	16	3	–	–	–	–	–	–	–	–	–	–	Benzeneacetic acid, α-hydroxy-, 1-naphthalenemethyl ester	86328-57-6	NI30446
141 135 139 140 77 115 107 292	**292**	100 55 14 12 12 11 9 6		19	16	3	–	–	–	–	–	–	–	–	–	–	Benzoic acid, 3-methoxy-, 1-naphthalenemethyl ester	86328-59-8	NI30447
105 77 292 187 146 264 274 215	**292**	100 32 15 6 4 3 1 1		19	16	3	–	–	–	–	–	–	–	–	–	–	1-Cyclopentanone, 2,5-dibenzoyl-		LI06367
154 274 264 96 292 277	**292**	100 26 17 14 6 6		19	16	3	–	–	–	–	–	–	–	–	–	–	Spiro[phenalene-2,2′(2′H)-pyran], 1,3-dioxo-4′,5′-dimethyl-2,3,3′,6′-tetrahydro-		LI06368
201 77 291 51 202 91 65 47	**292**	100 36 20 20 18 12 10 10	**10.00**	19	17	1	–	–	–	–	–	–	–	1	–	–	Benzyldiphenylphosphine oxide		LI06369
201 77 291 51 202 91 292 199	**292**	100 36 20 20 18 12 10 10		19	17	1	–	–	–	–	–	–	–	1	–	–	Benzyldiphenylphosphine oxide		MS06720
117 145 84 249 292 77 107 106	**292**	100 99 88 77 61 55 50 44		19	20	1	2	–	–	–	–	–	–	–	–	–	Acetic acid, 1-(3,4-dihydro-2-naphthalenyl)-2-methyl-2-phenylhydrazide	67134-52-5	NI30448
249 222 264 221 236 263 77 55	**292**	100 73 68 67 40 39 39 38	**0.00**	19	20	1	2	–	–	–	–	–	–	–	–	–	1,5-Benzodiazocin-2-one, 1-ethyl-1,2,3,4-tetrahydro-8-methyl-6-phenyl-	63563-55-3	NI30449
292 134 291 210 77 133 120	**292**	100 72 56 32 32 28 28		19	20	1	2	–	–	–	–	–	–	–	–	–	Meloscin		LI06370
292 291 182 143 169 183 184 156	**292**	100 96 67 53 49 40 36 23		19	20	1	2	–	–	–	–	–	–	–	–	–	1,5-Methanoazocino[1′,2′:1,2]pyrido[3,4-b]indol-2(1H)-one, 5,6,8,9,14,14b-hexahydro-4-methyl-, (1α,5α,14bα)-	51598-51-7	NI30450
292 249 145 186 106 107 115 185	**292**	100 58 54 41 41 39 36 35		19	20	1	2	–	–	–	–	–	–	–	–	–	Naphthalene, 1-acetyl-2-(2-methyl-2-phenylhydrazino)-3,4-dihydro-	67134-58-1	NI30451
105 264 144 292 77 159 69 158	**292**	100 52 51 42 28 25 24 23		19	20	1	2	–	–	–	–	–	–	–	–	–	Pyrrolo[2,3-b]indole, 1-benzoyl-1,2,3,3a,8,8a-hexahydro-3a,8-dimethyl-, (3aS-cis)-	54833-65-7	NI30452
169 53 42 41 156 292 79 43	**292**	100 99 99 84 82 75 57 54		19	20	1	2	–	–	–	–	–	–	–	–	–	Yohimban-17-one, 18,19-didehydro-	54725-64-3	NI30453
292 233 248 274 259 277 163 131	**292**	100 53 45 15 15 10 10 10		19	29	1	–	–	1	–	–	–	–	–	–	–	Estr-5(10)-en-17-ol, 3-fluoro-6-methyl-, (3α,6β,17β)-	24467-83-2	LI06371
292 233 248 293 274 259 145 131	**292**	100 53 45 20 15 15 10 10		19	29	1	–	–	1	–	–	–	–	–	–	–	Estr-5(10)-en-17-ol, 3-fluoro-6-methyl-, (3α,6β,17β)-	24467-83-2	NI30454
292 248 233 293 274 259 277 131	**292**	100 48 48 20 16 16 13 10		19	29	1	–	–	1	–	–	–	–	–	–	–	Estr-5(10)-en-17-ol, 3-fluoro-6-methyl-, (3β,6β,17β)-	24467-84-3	NI30455

MASS TO CHARGE RATIOS								M.W.	INTENSITIES								Parent	C	H	O	N	S	F	Cl	Br	I	Si	P	B	X	COMPOUND NAME	CAS Reg No	No
292	248	233	274	259	277	163	145	**292**	100	48	48	16	16	13	10	10		19	29	1	–	–	1	–	–	–	–	–	–	–	Estr-5(10)-en-17-ol, 3-fluoro-6-methyl-, (3β,6β,17β)-	24467-84-3	LI06372
292	233	248	293	274	259	277	91	**292**	100	50	47	19	16	16	13	12		19	29	1	–	–	1	–	–	–	–	–	–	–	Estr-5(10)-en-17-ol, 3-fluoro-6-methyl-, (3β,6β,17β)-	24467-84-3	NI30456
274	292	215	275	201	256	230	259	**292**	100	43	25	22	18	16	15	11		19	32	2	–	–	–	–	–	–	–	–	–	–	Androstane-1,17-diol, (1α,5α,17β)-	7417-23-4	NI30457
256	259	95	81	55	67	107	41	**292**	100	75	73	70	70	63	60	60	**5.00**	19	32	2	–	–	–	–	–	–	–	–	–	–	Androstane-3,11-diol, (3β,5α,11β)-	54200-09-8	NI30458
233	274	259	41	55	81	93	107	**292**	100	98	96	95	94	80	78	75	**37.00**	19	32	2	–	–	–	–	–	–	–	–	–	–	Androstane-3,16-diol, (3β,5α,16α)-	22630-49-5	NI30459
41	55	67	292	233	215	81	107	**292**	100	99	84	83	78	77	77	74		19	32	2	–	–	–	–	–	–	–	–	–	–	Androstane-3,17-diol	25126-76-5	NI30460
292	233	215	217	293	277	274	165	**292**	100	28	26	24	22	22	18	18		19	32	2	–	–	–	–	–	–	–	–	–	–	Androstane-3,17-diol, (3α,5α,17β)-	1852-53-5	NI30461
292	41	233	215	107	93	55	67	**292**	100	60	57	57	56	56	56	55		19	32	2	–	–	–	–	–	–	–	–	–	–	Androstane-3,17-diol, (3α,5α,17β)-	1852-53-5	NI30462
274	215	41	55	67	93	81	79	**292**	100	96	87	81	73	72	70	67	**18.98**	19	32	2	–	–	–	–	–	–	–	–	–	–	Androstane-3,17-diol, (3α,5β,17β)-	1851-23-6	NI30463
215	274	256	217	230	241	107	147	**292**	100	85	52	40	36	34	27	26	**7.11**	19	32	2	–	–	–	–	–	–	–	–	–	–	Androstane-3,17-diol, (3α,5β,17β)-	1851-23-6	NI30464
292	233	293	215	217	165	277	166	**292**	100	26	20	20	19	18	14	14		19	32	2	–	–	–	–	–	–	–	–	–	–	Androstane-3,17-diol, (3β,5α,17β)-	571-20-0	NI30465
215	233	292	165	166	217	107	121	**292**	100	99	93	59	56	53	50	46		19	32	2	–	–	–	–	–	–	–	–	–	–	Androstane-3,17-diol, (3β,5α,17β)-	571-20-0	NI30466
177	246	231	123	136	81	277	137	**292**	100	60	59	58	50	49	42	42	**32.46**	19	32	2	–	–	–	–	–	–	–	–	–	–	1,4-Epoxy-1H,3H-naphth[8a,1-c]oxocin, dodecahydro-4,7,10,10-tetramethyl-	55570-95-1	NI30467
177	246	231	123	146	95	109	81	**292**	100	78	76	76	65	65	63	63	**42.00**	19	32	2	–	–	–	–	–	–	–	–	–	–	3,6-Epoxy-2H-naphth[2,1-c]oxocin, tetradecahydro-3,9,9,12a-tetramethyl-, [3R-(3α,6α,6aβ,8aα,12aβ,12bα)]-	43125-55-9	NI30468
107	43	71	41	81	67	55	91	**292**	100	94	84	75	72	66	59	45	**31.00**	19	32	2	–	–	–	–	–	–	–	–	–	–	C-Homo-D-dinorandrostane, 3β-methoxy-13α-hydroxy-		LI06373
81	292	28	79	227	91	107	133	**292**	100	93	78	77	73	73	69	65		19	32	2	–	–	–	–	–	–	–	–	–	–	C-Homo-D-dinorandrostane, 3β-methoxy-13β-hydroxy-		LI06374
81	93	71	121	161	107	135	119	**292**	100	93	75	63	26	26	22	22	**0.04**	19	32	2	–	–	–	–	–	–	–	–	–	–	1-Naphthalenepropanol, α-vinyldecahydro-8a-(hydroxymethyl)-α,5-dimethyl-2-methylene-	55649-42-8	NI30469
81	107	93	71	121	135	119	177	**292**	100	70	70	59	56	29	29	28	**10.00**	19	32	2	–	–	–	–	–	–	–	–	–	–	1-Naphthalenol, decahydro-5-(5-hydroxy-3-methyl-3-pentenyl)-1,4a-dimethyl-6-methylene-, [1R-[1α,4aβ,5β(E),8aα]]-	51513-06-5	NI30470
43	91	79	41	55	87	121	74	**292**	100	91	87	86	75	57	43	41	**14.00**	19	32	2	–	–	–	–	–	–	–	–	–	–	Nonanoic acid, 9-(2-propylcyclohexa-3,5-dienyl)-, methyl ester		LI06375
81	80	95	79	67	55	109	94	**292**	100	99	98	97	96	92	88	66	**0.00**	19	32	2	–	–	–	–	–	–	–	–	–	–	5,9,12-Octadecatrienoic acid, methyl ester		NI30471
41	67	79	55	39	91	81	80	**292**	100	54	52	47	33	30	30	30	**4.84**	19	32	2	–	–	–	–	–	–	–	–	–	–	6,9,12-Octadecatrienoic acid, methyl ester	2676-41-7	NI30472
41	67	79	80	55	81	30	29	**292**	100	95	78	75	60	59	52	47	**46.39**	19	32	2	–	–	–	–	–	–	–	–	–	–	6,9,12-Octadecatrienoic acid, methyl ester	2676-41-7	NI30473
67	79	41	80	81	55	93	43	**292**	100	85	82	73	66	66	55	42	**23.04**	19	32	2	–	–	–	–	–	–	–	–	–	–	6,9,12-Octadecatrienoic acid, methyl ester	2676-41-7	NI30474
41	55	39	79	67	43	59	81	**292**	100	53	49	48	44	39	25	24	**4.74**	19	32	2	–	–	–	–	–	–	–	–	–	–	9,12,15-Octadecatrienoic acid, methyl ester	7361-80-0	NI30475
79	95	108	292	93	67	81	80	**292**	100	81	79	77	72	54	47	45		19	32	2	–	–	–	–	–	–	–	–	–	–	9,12,15-Octadecatrienoic acid, methyl ester	7361-80-0	NI30476
79	95	292	108	93	67	107	121	**292**	100	90	81	80	78	61	58	52		19	32	2	–	–	–	–	–	–	–	–	–	–	9,12,15-Octadecatrienoic acid, methyl ester	7361-80-0	NI30477
79	67	41	55	93	81	80	95	**292**	100	78	61	57	47	43	42	40	**1.00**	19	32	2	–	–	–	–	–	–	–	–	–	–	9,12,15-Octadecatrienoic acid, methyl ester, (Z,Z,Z)-	301-00-8	NI30479
293	163	137	151	123	291	149	109	**292**	100	50	48	43	43	35	35	33	**13.00**	19	32	2	–	–	–	–	–	–	–	–	–	–	9,12,15-Octadecatrienoic acid, methyl ester, (Z,Z,Z)-	301-00-8	NI30478
43	42	41	57	55	71	39	93	**292**	100	64	53	31	20	19	17	14	**1.14**	19	32	2	–	–	–	–	–	–	–	–	–	–	9-Octadecen-12-ynoic acid, methyl ester	56847-05-3	NI30480
150	79	80	93	81	55	67	41	**292**	100	94	92	49	41	39	38	33	**8.00**	19	32	2	–	–	–	–	–	–	–	–	–	–	11-Octadecen-9-ynoic acid, methyl ester, (E)-		NI30481
80	79	94	81	55	93	41	67	**292**	100	51	39	29	26	22	22	20	**0.24**	19	32	2	–	–	–	–	–	–	–	–	–	–	17-Octadecen-14-ynoic acid, methyl ester	18202-19-2	NI30482
106	89	78	44	292	80	40	91	**292**	100	67	63	60	57	57	57	53		19	32	2	–	–	–	–	–	–	–	–	–	–	Phenanthrene, 8-(2-hydroxypropyl)-2-methoxy-4a-methyl-1,2,3,4,4a,4b,5,6,8a,9,10,10a-dodecahydro-		LI06376
67	80	41	93	81	55	121	95	**292**	100	46	33	22	22	19	15	15	**2.00**	19	32	2	–	–	–	–	–	–	–	–	–	–	6-Tridecenoic acid, 13-(2-cyclopenten-1-yl)-, methyl ester	24828-60-2	WI01376
117	146	292	91	103	264	77	277	**292**	100	94	56	54	36	28	26	23		20	20	2	–	–	–	–	–	–	–	–	–	–	[2.2](4,7)-1H,3H-Benzo[c]furanophane		NI30483
160	292	134	133	132	91	105	77	**292**	100	98	98	58	57	52	32	30		20	20	2	–	–	–	–	–	–	–	–	–	–	Dicyclopent[a,c]anthracene, 1,2,3,4,5,6,6a,6b,7,12,12a,12b-dodecahydro-		NI30484
167	292	168	165	166	293	233	205	**292**	100	21	18	11	5	4	1	1		20	20	2	–	–	–	–	–	–	–	–	–	–	3-Oxabicyclo[3.2.1]octan-2-one, 8-(diphenylmethyl)-, syn-		DD01035
246	167	292	217	205	218	233	129	**292**	100	60	59	45	41	39	36	36		20	20	2	–	–	–	–	–	–	–	–	–	–	3-Oxatricyclo[4.3.0.0^{4,9}]nonan-5-ol, 2,2-diphenyl-		DD01036
191	57	292	189	208	165	190	179	**292**	100	44	24	17	12	9	8	6		20	20	2	–	–	–	–	–	–	–	–	–	–	Propanoic acid, 2,2-dimethyl-, 9-phenanthrenemethyl ester	81558-09-0	NI30485
186	167	91	97	157	105	107	175	**292**	100	52	43	27	23	20	17	14	**0.00**	20	20	2	–	–	–	–	–	–	–	–	–	–	Spiro[bicyclo[3.1.0]hexane-2,2′-[1,3]dioxolane], 4′,5′-diphenyl-		DD01037
146	277	105	79	77	147	28	131	**292**	100	39	29	20	17	13	11	9	**1.43**	20	24	–	2	–	–	–	–	–	–	–	–	–	Aniline, N,N′-(1,2-dimethyl-1,2-ethanediylidene)bis[2,6-dimethyl-	49673-40-7	NI30487
146	277	105	79	77	147	278	103	**292**	100	37	29	15	15	13	8	8	**1.05**	20	24	–	2	–	–	–	–	–	–	–	–	–	Aniline, N,N′-(1,2-dimethyl-1,2-ethanediylidene)bis[2,6-dimethyl-	49673-40-7	NI30486
292	41	291	43	57	42	129	55	**292**	100	75	46	38	35	35	33	32		20	24	–	2	–	–	–	–	–	–	–	–	–	Aspidofractinine, 3-methylene-, (2α,5α)-	55103-49-6	NI30488
134	172	135	156	185	171	158	169	**292**	100	78	71	66	57	57	47	38	**0.00**	20	24	–	2	–	–	–	–	–	–	–	–	–	2,20-Cycloaspidospermidine, 6,7-didehydro-3-methyl-, (2α,3β,5α,12β,19α,20R)-	56053-28-2	NI30489
145	146	147	131	144	130	77	162	**292**	100	75	64	59	55	42	40	30	**16.70**	20	24	–	2	–	–	–	–	–	–	–	–	–	2,5-Dimethylacetophenone azine		LI06377
250	194	251	292	136	70	249	222	**292**	100	96	88	61	48	45	41	37		20	24	–	2	–	–	–	–	–	–	–	–	–	1H-Indolizino[8,1-cd]carbazole, 3a,5a,10a-(1,2,3-propanetriyl)-2,3,3a,4,5,5a,10a,11,12,13a-decahydro-		MS06721
69	41	81	95	55	136	67	68	**292**	100	54	36	22	21	20	16	15	**5.38**	20	36	1	–	–	–	–	–	–	–	–	–	–	6,10,14-Hexadecatrien-1-ol, 3,7,11,15-tetramethyl-, [R-(E,E)]-	36237-66-8	NI30490
137	95	81	259	69	55	121	41	**292**	100	56	50	36	36	34	32	32	**1.40**	20	36	1	–	–	–	–	–	–	–	–	–	–	1-Naphthalenepropanol, α-vinyldecahydro-α,5,5,8a-tetramethyl-2-methylene-, dihydro deriv., [1S-[1α(S*),4aβ,8aα]]-	72360-94-2	NI30491

MASS TO CHARGE RATIOS								M.W.	INTENSITIES								Parent	C	H	O	N	S	F	Cl	Br	I	Si	P	B	X	COMPOUND NAME	CAS Reg No	No
245	137	263	95	43	81	69	109	**292**	100	45	43	28	27	26	25	23	**0.00**	20	36	1	–	–	–	–	–	–	–	–	–	–	1H-Naphtho[2,1-b]pyran, 3-ethyldodecahydro-3,4a,7,7,10a-pentamethyl-, [3S-(3α,4aβ,6aα,10aβ,10bα)]-	57129-66-5	NI30492
137	81	41	95	43	55	67	136	**292**	100	57	47	46	35	34	32	22	**16.40**	21	40	–	–	–	–	–	–	–	–	–	–	–	Bicyclo[4.4.0]decane, 2-undecyl-		AI01999
137	81	95	41	55	67	43	69	**292**	100	59	48	46	36	33	32	23	**7.96**	21	40	–	–	–	–	–	–	–	–	–	–	–	Bicyclo[4.4.0]decane, 2-undecyl-		AI02000
28	29	71	55	43	41	83	44	**292**	100	36	12	12	12	12	11	10	**0.00**	21	40	–	–	–	–	–	–	–	–	–	–	–	Cyclohexane, 1,1'-[1-(2,2-dimethylbutyl)-1,3-propanediyl]bis-	61142-63-0	NI30493
105	236	118	144	106	131	81	221	**292**	100	29	25	17	13	12	12	11	**2.60**	22	28	–	–	–	–	–	–	–	–	–	–	–	Benzene, 1,1'-[1-(3-methyl-3-butenyl)-1,3-propanediyl]bis[4-methyl-	74685-54-4	NI30494
249	292	250	264	263	293	291	221	**292**	100	92	65	52	42	21	19	19		22	28	–	–	–	–	–	–	–	–	–	–	–	Benzo[ghi]perylene, hexadecahydro-		NI30495
91	105	97	92	55	292	41	119	**292**	100	29	29	24	17	12	12	10		22	28	–	–	–	–	–	–	–	–	–	–	–	Cyclohexane, 1,1-dimethyl-4,4-dibenzyl-		LI06378
292	291	91	105	141	129	79	55	**292**	100	30	13	7	6	6	6	5		22	28	–	–	–	–	–	–	–	–	–	–	–	1,5,7,11-Ethanediylidene-3,5:9,11-dimethanochrysene, hexadecahydro-	27745-90-0	NI30496
262	247	277	157	217	232	202	292	**292**	100	64	51	50	24	22	19	17		22	28	–	–	–	–	–	–	–	–	–	–	–	Bis(3,4,4,5-tetramethyl-2,5-cyclohexadien-1-ylidene)ethylene		NI30497
113	132	78	151	75	45	32	46	**293**	100	37	25	20	14	14	14	13	**0.00**	2	2	–	1	2	6	–	–	–	–	–	–	1	1,3,2-Dithiazolium hexafluoroarsenate(V)		NI30498
69	28	85	104	31	50	47	147	**293**	100	39	15	10	7	6	5	3	**0.00**	3	–	2	3	–	8	–	–	–	–	1	–	–	2-Azido-2-(μ-oxodifluorophosphoryl)hexafluoropropane		LI06379
69	101	117	293	31	205	128	32	**293**	100	25	18	14	8	6	6	6		4	–	1	1	1	8	–	–	–	–	1	–	–	2-Cyano-2-(μ-oxodifluorothiophosphoryl)hexafluoropropane		LI06380
69	28	166	78	99	97	115	113	**293**	100	62	32	32	17	17	10	10	**0.00**	5	7	–	1	–	6	2	–	–	1	–	–	–	Ethanamine, 2-(dichloromethylsilyl)-N,N-bis(trifluoromethyl)-	28871-57-0	NI30499
69	28	166	78	99	117	149		**293**	100	62	32	32	17	10	3		**0.00**	5	7	–	1	–	6	2	–	–	1	–	–	–	Ethanamine, 2-(dichloromethylsilyl)-N,N-bis(trifluoromethyl)-	28871-57-0	LI06381
142	237	107	214	144	36	212	235	**293**	100	88	82	65	65	65	63	57	**22.65**	6	–	2	1	–	–	5	–	–	–	–	–	–	Benzene, pentachloronitro-	82-68-8	NI30500
295	249	214	237	142	212	247	30	**293**	100	99	97	95	89	81	74	72	**71.50**	6	–	2	1	–	–	5	–	–	–	–	–	–	Benzene, pentachloronitro-	82-68-8	NI30501
295	249	237	297	293	214	247	251	**293**	100	82	72	64	64	55	50	48		6	–	2	1	–	–	5	–	–	–	–	–	–	Benzene, pentachloronitro-	82-68-8	NI30502
211	293	253	278	252	197	212	251	**293**	100	70	57	46	44	41	33	24		6	10	–	3	–	4	–	–	–	–	3	–	–	2,2-Di(1-propenyl)tetrafluorocyclotriphosphazene		MS06722
42	28	293	43	17	41	59	60	**293**	100	62	54	48	46	40	38	36		6	10	–	6	2	–	–	–	–	–	–	–	1	Copper, [[2,2'-(1,2-dimethyl-1,2-ethanediylidene)bis[hydrazinecarbothioamidato]](2-)-N^2,N$^{2''}$,S,S']-	24457-73-6	NI30503
293	124	192	187	164	234	224	147	**293**	100	16	9	7	7	6	6	4		6	10	–	6	2	–	–	–	–	–	–	–	1	Copper, [[2,2'-(1,2-dimethyl-1,2-ethanediylidene)bis[hydrazinecarbothioamidato]](2-)-N^2,N$^{2''}$,S,S']-	24457-73-6	LI06382
261	78	51	293	205	52	79	50	**293**	100	91	50	47	47	37	29	26		8	8	3	1	–	–	–	–	1	–	–	–	–	Salicylic acid, 3-amino-5-iodo-, methyl ester	22621-47-2	NI30504
261	78	51	293	205	52	79	50	**293**	100	93	51	48	46	37	28	26		8	8	3	1	–	–	–	–	1	–	–	–	–	Salicylic acid, 3-amino-5-iodo-, methyl ester	22621-47-2	LI06383
261	78	51	293	205	52	79	50	**293**	100	96	53	47	47	36	28	27		8	8	3	1	–	–	–	–	1	–	–	–	–	Salicylic acid, 3-amino-5-iodo-, methyl ester	22621-47-2	MS06723
293	221	86	29	220	69	237	193	**293**	100	58	58	30	25	18	17	16		9	6	3	3	1	3	–	–	–	–	–	–	–	7H-1,3,4-Thiadiazolo[3,2-a]pyrimidine-5-carboxylic acid, 7-oxo-2-(trifluoromethyl)-, ethyl ester		NI30505
134	142	44	107	165	127	78	106	**293**	100	99	62	50	49	25	22	15	**0.00**	9	12	2	1	–	–	–	–	1	–	–	–	–	Pyridinium, 2-(2-methoxy-2-oxoethyl)-1-methyl-, iodide	39998-18-0	NI30506
142	166	44	108	107	167	152	92	**293**	100	99	99	75	70	63	50	34	**0.00**	9	12	2	1	–	–	–	–	1	–	–	–	–	Pyridinium, 3-(2-methoxy-2-oxoethyl)-1-methyl-, iodide	14996-89-5	NI30507
134	142	44	107	165	127	106	77	**293**	100	99	71	51	43	28	11	11	**0.00**	9	12	2	1	–	–	–	–	1	–	–	–	–	Pyridinium, 4-(2-methoxy-2-oxoethyl)-1-methyl-, iodide	39998-19-1	NI30508
214	91	200	154	228	198	184	92	**293**	100	56	29	22	16	15	15	14	**0.00**	9	12	3	1	1	–	–	1	–	–	–	–	–	[(2-Hydroxy-5-nitrophenyl)methyl]dimethylsulphonium bromide	28611-73-6	NI30509
73	278	263	191	60	46	293		**293**	100	25	14	6	5	5	3			9	27	–	3	1	–	–	–	–	3	–	–	–	Tris(N-trimethylsilylimido)sulphur		LI06384
293	59	131	117	156	228	185	125	**293**	100	94	38	32	28	23	23	22		10	15	1	1	4	–	–	–	–	–	–	–	–	2,4,6,8-Tetrathiatricyclo[3.3.1.1^{3,7}]decane-1-carboxamide, 3,5,7-trimethyl-	57289-08-4	NI30510
140	219	81	112	78	124	79	41	**293**	100	75	53	34	34	31	22	17	**0.00**	10	15	5	1	2	–	–	–	–	–	–	–	–	Thiocyanic acid, 2-ethoxy-3,6-dihydro-6-[[(methylsulphonyl)oxy]methyl]-2H-pyran-3-yl ester	56248-61-4	NI30511
260	258	295	293	65	297	262	223	**293**	100	74	64	56	39	31	31	25		11	7	–	1	–	–	4	–	–	–	–	–	–	1H-Pyrrole, 1-(2,3,4,6-tetrachloro-5-methylphenyl)-	54833-29-3	NI30512
177	81	219	151	82	80	109	123	**293**	100	79	58	47	41	41	29	21	**0.00**	11	19	6	1	1	–	–	–	–	–	–	–	–	Acetamide, N-[2-ethoxy-3,6-dihydro-6-[[(methylsulphonyl)oxy]methyl]-2H-pyran-3-yl]-	56248-05-6	NI30513
176	73	77	61	128	104	75	147	**293**	100	44	31	29	26	19	16	12	**11.00**	11	27	2	1	1	–	–	–	–	2	–	–	–	Methionine, N-(trimethylsilyl)-, trimethylsilyl ester		MS06724
73	61	176	45	47	128	77	75	**293**	100	32	31	31	28	27	22	22	**1.00**	11	27	2	1	1	–	–	–	–	2	–	–	–	Methionine, N-(trimethylsilyl)-, trimethylsilyl ester		MS06725
60	102	73	144	85	72	112	278	**293**	100	68	14	13	13	8	4	2	**0.00**	12	23	7	1	–	–	–	–	–	–	–	–	–	D-Glucitol, 1-acetamido-2,3-O-isopropylidene-1-O-methyl-	83937-97-7	NI30514
91	155	65	63	92	293	156	64	**293**	100	65	19	10	8	6	6	6		13	11	5	1	1	–	–	–	–	–	–	–	–	Toluenesulphonic acid, 4-nitrophenyl ester		LI06385
88	77	293	73	89	90	42	51	**293**	100	9	5	5	5	4	4	3		13	15	1	3	2	–	–	–	–	–	–	–	–	Carbamodithioic acid, dimethyl-, 3-methyl-5-oxo-1-phenyl-2-pyrazolin-4-yl ester		NI30515
105	77	106	51	174	78	76	50	**293**	100	89	76	76	72	63	61	60	**9.78**	13	15	5	3	–	–	–	–	–	–	–	–	–	N-Benzoylaspartylurea methyl ester		LI06386
87	70	114	150	120	86	104	57	**293**	100	64	63	56	29	28	25	20	**2.90**	13	15	5	3	–	–	–	–	–	–	–	–	–	Morpholine, 4-[N-(4-nitrobenzamido)acetyl]-		NI30516
43	118	115	131	101	175	162	206	**293**	100	23	18	15	10	8	6	1	**0.00**	13	25	7	–	–	–	–	–	–	–	–	–	–	Fucitol, 2,3,4-tri-O-methyl-1,5-di-O-acetyl-		NI30517
44	179	209	178	180	152	76	36	**293**	100	52	37	19	15	12	12	12	**0.00**	14	12	4	1	–	–	1	–	–	–	–	–	–	Phenanthridinium, 5-methyl-, perchlorate	52806-10-7	NI30518
96	77	68	41	293	80	117	119	**293**	100	62	36	32	26	22	21	19		14	15	4	1	1	–	–	–	–	–	–	–	–	3,8-Dioxa-11-azatetracyclo[4.4.1.0^{2,4}.0^{7,9}]undecane, 11-(phenylsulphonyl)-, (1α,2α,4α,6α,7β,9β)-	50267-12-4	NI30519
57	208	85	210	181	128	110	130	**293**	100	44	25	15	13	13	13	5	**5.00**	14	16	2	3	–	–	1	–	–	–	–	–	–	2-Butanone, 1-(4-chlorophenoxy)-3,3-dimethyl-1-(1H-1,2,4-triazol-1-yl)-	43121-43-3	NI30520
57	41	85	208	70	128	110	69	**293**	100	24	20	19	8	8	7	7	**1.62**	14	16	2	3	–	–	1	–	–	–	–	–	–	2-Butanone, 1-(4-chlorophenoxy)-3,3-dimethyl-1-(1H-1,2,4-triazol-1-yl)-	43121-43-3	NI30521

MASS TO CHARGE RATIOS								M.W.	INTENSITIES								Parent	C	H	O	N	S	F	Cl	Br	I	Si	P	B	X	COMPOUND NAME	CAS Reg No	No
43	45	15	42	29	182	31	60	293	100	84	39	15	13	8	7	5	1.40	14	16	2	3	–	–	1	–	–	–	–	–	–	5-Pyrimidinecarboxylic acid, 2-amino-6-(4-chlorophenyl)-5,6-dihydro-4-methyl-, ethyl ester		NI30522
293	77	28	216	55	188	43	279	293	100	78	45	34	23	21	19	18		15	15	–	7	–	–	–	–	–	–	–	–	–	1H-Pyrazolo[4,3-e][1,2,4]triazolo[4,3-a][1,2,4]triazine, 3-methyl-1-phenyl-8-propyl-		NI30523
293	292	133	161	134	294	65	92	293	100	49	36	26	23	18	18	15		15	15	–	7	–	–	–	–	–	–	–	–	–	1,3,5-Triazine-2,4-diamine, N,N′-bis(3-aminophenyl)-	33933-63-0	NI30524
220	219	139	111	275	154	222	108	293	100	58	58	49	42	35	33	33	8.14	15	16	3	1	–	–	1	–	–	–	–	–	–	3-Butenoic acid, 2-(2-aminoethylidene)-4-(4-chlorobenzoyl)-, ethyl ester		IC03722
97	41	81	71	69	55	79	64	293	100	53	27	24	23	21	18	12	0.00	15	19	3	1	1	–	–	–	–	–	–	–	–	2H-Pyran-2-one, tetrahydro-6-(2-pentenyl)-3-(2-pyridinylsulphinyl)-	66971-98-0	NI30525
149	43	191	56	233	148	42	107	293	100	59	54	48	42	23	21	20	0.00	15	19	5	1	–	–	–	–	–	–	–	–	–	Acetamide, N-(β,4-dihydroxyphenethyl)-N-methyl-, diacetate (ester)	14383-57-4	NI30526
174	293	29	219	220	28	147	31	293	100	85	43	40	35	35	30	30		15	19	5	1	–	–	–	–	–	–	–	–	–	Diethyl 2-(4-methoxyanilino)fumarate		MS06726
101	85	55	163	41	147	119	91	293	100	90	90	89	87	80	80	71	0.00	15	19	5	1	–	–	–	–	–	–	–	–	–	3-Oxabicyclo[8.4.0]tetradeca-1(10),11,13-trien-2-one, 12-methoxy-4-methyl-7-nitro-		DD01038
91	73	92	45	77	75	40	65	293	100	18	8	8	7	7	5	4	0.00	15	23	3	1	–	–	–	–	–	1	–	–	–	Butanoic acid, 3-methyl-2-[(phenylmethoxy)imino]-, trimethylsilyl ester	55520-96-2	NI30527
91	73	92	77	75	45	65	40	293	100	16	8	7	7	7	4	3	0.00	15	23	3	1	–	–	–	–	–	1	–	–	–	Pentanoic acid, 2-[(phenylmethoxy)imino]-, trimethylsilyl ester	55520-94-0	NI30528
166	208	209	96	194	167	70	42	293	100	60	50	26	20	16	16	10	7.00	15	23	3	3	–	–	–	–	–	–	–	–	–	Pyrimidine, 4-acetoxy-5-butyl-2-(N-ethylacetamido)-6-methyl-		IC03723
143	115	293	171	75	76	63	128	293	100	81	31	14	13	11	10	9		16	11	3	3	–	–	–	–	–	–	–	–	–	2-Naphthalenol, 1-[(4-nitrophenyl)azo]-	6410-10-2	NI30529
104	77	293	81	89	103	109	105	293	100	69	53	36	18	17	15	15		16	11	3	3	–	–	–	–	–	–	–	–	–	2(1H)-Pyrimidinone, 5-nitro-4,6-diphenyl-	37673-86-2	NI30530
293	292	246	294	247	104	77	262	293	100	67	33	20	17	11	11	8		16	11	3	3	–	–	–	–	–	–	–	–	–	2(1H)-Pyrimidinone, 4-(4-nitrophenyl)-6-phenyl-	22114-24-5	NI30531
293	292	246	294	247	104	77	262	293	100	68	55	22	18	11	11	9		16	11	3	3	–	–	–	–	–	–	–	–	–	2(1H)-Pyrimidinone, 4-phenyl-6-(4-nitrophenyl)-		LI06387
195	293	90	196	76	54	82	50	293	100	68	55	53	50	50	40	40		16	11	3	3	–	–	–	–	–	–	–	–	–	4(3H)-Quinazolinone, 2-(2-furyl)-3-(3-methyl-5-isoxazolyl)-	90059-44-2	NI30532
168	139	103	170	141	77	36	169	293	100	90	38	35	32	16	15	12	0.00	16	17	–	1	–	–	2	–	–	–	–	–	–	Diphenethylamine, 4,4′-dichloro-	13026-02-3	NI30533
168	139	103	166	141	77	36	30	293	100	91	39	34	32	16	14	11	0.00	16	17	–	1	–	–	2	–	–	–	–	–	–	Diphenethylamine, 4,4′-dichloro-	13026-02-3	LI06388
105	293	77	51	294	106	130	78	293	100	47	45	16	9	8	4	4		17	11	4	1	–	–	–	–	–	–	–	–	–	2(3H)-Furanone, 3-(4-nitrobenzylidene)-5-phenyl-	54833-77-1	NI30534
105	293	77	51	130	263	247	76	293	100	47	45	16	4	3	3	3		17	11	4	1	–	–	–	–	–	–	–	–	–	2(3H)-Furanone, 3-(4-nitrobenzylidene)-5-phenyl-	54833-77-1	LI06389
293	265	292	264	222	294	223	69	293	100	63	30	23	12	10	9	8		17	15	2	3	–	–	–	–	–	–	–	–	–	1H-1,4-Benzodiazepin-2-one, 7-(acetylamino)-2,3-dihydro-5-phenyl-	4928-03-4	NI30535
91	132	293	77	64	92	133	294	293	100	73	63	34	18	17	13	12		17	15	2	3	–	–	–	–	–	–	–	–	–	2-Pyrazolin-5-one, 1-phenyl-3-methyl-4-(4-methoxyphenylimino)-		MS06727
145	144	132	105	146	104	117	161	293	100	73	70	36	35	26	20	12	0.00	17	15	2	3	–	–	–	–	–	–	–	–	–	2,3-Pyridinedicarboximide, N-[(1,2,3,4-tetrahydroisoquinolin-2-yl)methyl]-		NI30536
145	132	144	77	50	104	105	78	293	100	99	71	57	55	51	45	44	0.00	17	15	2	3	–	–	–	–	–	–	–	–	–	3,4-Pyridinedicarboximide, N-[(1,2,3,4-tetrahydroisoquinolin-2-yl)methyl]-		NI30537
116	117	178	115	177	91	78	149	293	100	41	36	18	14	11	9	5	3.00	17	15	2	3	–	–	–	–	–	–	–	–	–	1,2,4-Triazolidine-3,5-dione, 1-(2-indanyl)-4-phenyl-		LI06390
41	81	86	79	57	39	150	69	293	100	72	67	48	34	33	28	24	0.20	17	27	3	1	–	–	–	–	–	–	–	–	–	1-Cyclohexene-4-carboxylic acid, 6-butyl-3-(5-oxo-3,4-dihydro-2H-pyrrolyl)-, ethyl ester		MS06728
230	245	102	246	231	293	248	202	293	100	98	43	25	20	19	19	15		18	15	1	1	1	–	–	–	–	–	–	–	–	10-Ethyl-6,7-benzophenothiazine-5-oxide		NI30538
77	105	171	170	51	122	104	39	293	100	75	68	35	35	28	24	21	5.00	18	15	3	1	–	–	–	–	–	–	–	–	–	2,6-Dioxa-3-azabicyclo[3.3.0]-7-octene, 4-benzoyl-3-phenyl-		NI30539
105	146	120	77	160	174	147	133	293	100	89	68	39	19	17	11	10	2.00	18	15	3	1	–	–	–	–	–	–	–	–	–	1H-Isoindole-1,3(2H)-dione, 2-(4-oxo-4-phenylbutyl)-	7347-68-4	NI30540
91	118	92	44	131	90	77	89	293	100	58	30	30	26	26	21	20	2.00	18	15	3	1	–	–	–	–	–	–	–	–	–	6H-1,3-Oxazin-6-one, 2,5-dibenzyl-4-hydroxy-		NI30541
248	293	45	44	43	165	249	294	293	100	73	39	39	26	24	20	11		18	15	3	1	–	–	–	–	–	–	–	–	–	2-Oxazolepropionic acid, 4,5-diphenyl-		MS06729
105	77	51	106	188	144	78	147	293	100	58	18	8	8	7	5	5	3.70	18	15	3	1	–	–	–	–	–	–	–	–	–	2-Pyrrolidinone, N,4-dibenzoyl-		NI30542
201	293	77	92	65	51	202	39	293	100	48	26	22	17	14	13	10		18	16	1	1	–	–	–	–	–	–	1	–	–	1H-Azepine, 1-(diphenylphosphino)-, 1-oxide	56909-19-4	NI30543
169	168	201	77	170	105	202	51	293	100	63	48	15	13	9	6	5	3.38	18	16	1	1	–	–	–	–	–	–	1	–	–	Diphenyl(pyridin-2-ylmethyl)phosphine oxide		IC03724
293	91	118	237	264				293	100	60	25	15	10					18	19	1	3	–	–	–	–	–	–	–	–	–	3H-Indolo[3,2-g]indolizin-3-one, 1-pyrrolidino-5,6,11,11b-tetrahydro-		LI06391
207	208	205	206	103	76	57	45	293	100	34	22	20	14	9	6	6	6.00	18	19	1	3	–	–	–	–	–	–	–	–	–	1H-Pyrrolo[2,3-b]pyridine, 3-(morpholinomethyl)-2-phenyl-	27257-23-4	NI30544
105	224	293	77	225	70	106	190	293	100	70	33	22	20	19	11	8		19	19	2	1	–	–	–	–	–	–	–	–	–	2-Azetidinone, 1-benzoyl-3,3,4-trimethyl-4-phenyl-	77064-42-7	NI30545
209	180	238	293	210	278	250	220	293	100	70	60	45	40	15	12	12		19	19	2	1	–	–	–	–	–	–	–	–	–	Carbazole, 2-(γ,γ-dimethylallyl)-N-formyl-1-hydroxy-3-methyl-		NI30546
293	250	222	236	278	188	208	260	293	100	76	67	65	48	44	40	14		19	19	2	1	–	–	–	–	–	–	–	–	–	Indizoline		NI30547
293	160	91	161	104	77	105	294	293	100	89	73	46	38	28	26	23		19	19	2	1	–	–	–	–	–	–	–	–	–	1H-Isoindole-1,3(2H)-dione, 2-(5-phenylpentyl)-	53429-14-4	NI30548
209	238	293	210	180	265	278	250	293	100	87	52	45	35	10	8	5		19	19	2	1	–	–	–	–	–	–	–	–	–	Pyrano[2,3-a]carbazole, 11-formyl-2,2,5-trimethyl-3,4-dihydro-		NI30549
144	267	292	43	224	115	294	279	293	100	74	55	35	34	33	26	26	14.65	19	23	–	3	–	–	–	–	–	–	–	–	–	11H-Indolo[3,2-g]indolizine, 1-ethyl-1-(2-cyanoethyl)-1,2,3,5,6,11b-hexahydro-		NI30550
144	293	224	115	295	291	254	221	293	100	28	17	17	9	9	8	6		19	23	–	3	–	–	–	–	–	–	–	–	–	11H-Indolo[3,2-g]indolizine, 1-ethyl-1-(2-cyanoethyl)-1,2,3,5,6,11b-hexahydro-		NI30551
162	121	132	147	293	120	77	42	293	100	98	77	59	51	33	29	29		19	23	–	3	–	–	–	–	–	–	–	–	–	1,3,5-Triazapenta-1,4-diene, 1,5-bis(3,5-dimethylphenyl)-3-methyl-		NI30552
248	222	250	249	122	293	148	176	293	100	32	16	15	9	5	4	3		19	35	1	1	–	–	–	–	–	–	–	–	–	Pyrrolo[1,2-a]quinoline-1-ethanol, dodecahydro-6-pentyl-, [1S-(1α,3aβ,5aα,6α,9aα)]-	63983-59-5	NI30553
232	217	215	202	231	216	117	203	293	100	72	46	35	33	30	22	21	0.00	20	23	1	1	–	–	–	–	–	–	–	–	–	Amitriptyline N-oxide		MS06730
117	105	236	77	70	278	57	104	293	100	66	40	34	26	24	18	13	9.52	20	23	1	1	–	–	–	–	–	–	–	–	–	Azetidine, 1-tert-butyl-2-phenyl-3-benzoyl-	55530-46-6	NI30554

MASS TO CHARGE RATIOS								M.W.	INTENSITIES								Parent	C	H	O	N	S	F	Cl	Br	I	Si	P	B	X	COMPOUND NAME	CAS Reg No	No
236	117	105	77	278	146	235	70	**293**	100	79	67	37	23	21	19	19	**15.00**	20	23	1	1	–	–	–	–	–	–	–	–	–	Azetidine, 1-tert-butyl-2-phenyl-3-benzoyl-, cis-	10231-03-5	LI06392
236	117	105	77	278	146	237	70	**293**	100	79	67	37	23	21	19	19	**15.37**	20	23	1	1	–	–	–	–	–	–	–	–	–	Azetidine, 1-tert-butyl-2-phenyl-3-benzoyl-, cis-	10231-03-5	NI30555
117	105	236	77	70	238	57	104	**293**	100	66	40	34	25	24	17	16	**7.99**	20	23	1	1	–	–	–	–	–	–	–	–	–	Azetidine, 1-tert-butyl-2-phenyl-3-benzoyl-, trans-	10235-75-3	NI30556
55	82	41	83	68	58	28	67	**293**	100	90	59	55	50	29	29	18	**0.00**	20	23	1	1	–	–	–	–	–	–	–	–	–	2H-1,3-Benzoxazine, 3-cyclohexyl-3,4-dihydro-6-phenyl-	35183-40-5	NI30557
293	163	131	175	133	121	132	117	**293**	100	92	78	74	69	58	49	43		20	23	1	1	–	–	–	–	–	–	–	–	–	[3,3]Paracyclophane, 5-acetamido-		LI06393
293	163	131	175	133	121	132	117	**293**	100	92	78	74	69	58	49	43		20	23	1	1	–	–	–	–	–	–	–	–	–	[3,3]Paracyclophane, 6-acetamido-	24777-34-2	NI30558
174	105	111	182	77	293	96	216	**293**	100	93	53	43	43	32	16	9		20	23	1	1	–	–	–	–	–	–	–	–	–	1-Pyrazoline, 2-(2-hydroxy-2,2-diphenylethyl)-4,4-dimethyl-		NI30559
43	41	57	55	29	97	70	69	**293**	100	79	75	54	53	32	26	23	**0.20**	20	39	–	1	–	–	–	–	–	–	–	–	–	Eicosanenitrile	4616-73-3	WI01377
127	89	64	275	172	129	294	44	**294**	100	22	9	6	4	4	3	3		1	–	–	2	2	10	–	–	–	–	–	–	–	Methanediylidenebis(nitrilo)bis(sulphur pentafluoride)	58776-14-0	NI30560
66	105	69	47	28	127	44	86	**294**	100	70	66	45	25	19	17	13	**0.00**	2	–	3	–	1	10	–	–	–	–	–	–	–	cis-Tetrafluoro(trifluoromethoxy)(trifluoromethylhydroperoxidato) sulphur		MS06731
133	135	63	126	98	163	161	128	**294**	100	99	65	55	43	38	38	37	**2.20**	2	4	–	–	–	–	6	–	–	2	–	–	–	Silane, 1,2-ethanediylbis[trichloro-	2504-64-5	NI30561
117	119	129	121	127	132	145	189	**294**	100	93	48	35	34	31	23	17	**0.00**	3	1	–	–	–	2	4	1	–	–	–	–	–	1-Bromo-3,3-difluoro-2,4,4,4-tetrachlorobutane		LI06394
296	294	69	169	167	113	127	87	**294**	100	99	75	67	67	59	46	41		3	1	–	–	–	5	1	–	1	–	–	–	–	2-Chloro-1,1,1,3,3-pentafluoro-3-iodopropane		LI06395
69	296	294	169	167	127	87	85	**294**	100	67	67	50	50	17	13	13		3	1	–	–	–	5	1	–	1	–	–	–	–	3-Chloro-1,1,1,2,2-pentafluoro-3-iodopropane		LI06396
107	109	27	28	136	138	29	15	**294**	100	97	54	39	15	14	8	8	**0.05**	4	8	3	–	1	–	–	2	–	–	–	–	–	Bis(2-bromoethyl) sulphite		IC03725
57	65	41	237	202	55	294	56	**294**	100	60	57	18	15	13	12	12		4	9	–	–	–	–	1	–	–	–	–	–	1	Butylmercuric chloride		LI06397
294	225	144	129	69	89	279	239	**294**	100	68	67	38	34	22	21	19		5	9	–	–	–	–	–	–	–	–	–	–	3	1,2,6-Triarsatricyclo[2.2.1.0^{2,6}]heptane, 4-methyl-		MS06732
117	294	167	31	127	98	79	93	**294**	100	85	54	34	33	30	24	23		6	–	–	–	–	5	–	–	1	–	–	–	–	Benzene, pentafluoroiodo-	827-15-6	NI30562
117	294	167	31	127	98	80	148	**294**	100	83	52	33	32	30	21	14		6	–	–	–	–	5	–	–	1	–	–	–	–	Benzene, pentafluoroiodo-	827-15-6	NI30564
294	117	167	93	98	31	148	295	**294**	100	51	47	13	11	10	9	8		6	–	–	–	–	5	–	–	1	–	–	–	–	Benzene, pentafluoroiodo-	827-15-6	NI30563
296	62	63	90	168	250	170	294	**294**	100	75	74	69	65	58	57	52		6	4	2	2	–	–	–	2	–	–	–	–	–	Aniline, 2,4-dibromo-6-nitro-	827-23-6	NI30565
266	296	170	264	168	90	73	268	**294**	100	54	53	50	49	48	44	42	**27.00**	6	4	2	2	–	–	–	2	–	–	–	–	–	Aniline, 2,6-dibromo-4-nitro-	827-94-1	NI30566
266	90	296	63	62	168	170	52	**294**	100	98	90	90	87	69	68	65	**43.47**	6	4	2	2	–	–	–	2	–	–	–	–	–	Aniline, 2,6-dibromo-4-nitro-	827-94-1	NI30567
296	68	215	294	298	40	217	120	**294**	100	83	64	56	53	29	21	16		6	4	2	2	–	–	–	2	–	–	–	–	–	2,5-Cyclohexadiene-1,4-dione, 2,5-diamino-3,6-dibromo-	27344-26-9	NI30568
127	167	124	275	128	117	119	89	**294**	100	89	85	25	24	21	20	20	**0.50**	6	10	–	2	1	8	–	–	–	–	–	–	–	[1-(Diethylamino)-2,2,2-trifluoroethylidenimino]sulphur pentafluoride	81439-17-0	NI30569
296	45	246	294	298	47	244	263	**294**	100	79	70	60	59	58	53	39		7	3	–	–	1	–	5	–	–	–	–	–	–	Benzene, pentachloro(methylthio)-	1825-19-0	NI30570
296	298	294	246	244	248	263	281	**294**	100	86	72	57	42	37	36	30		7	3	–	–	1	–	5	–	–	–	–	–	–	Benzene, pentachloro(methylthio)-	1825-19-0	NI30571
278	62	63	276	61	280	143	141	**294**	100	81	55	54	47	45	39	37	**16.55**	7	4	3	–	–	–	–	2	–	–	–	–	–	Salicylic acid, 3,5-dibromo-		LI06398
296	294	298	215	89	217	179	297	**294**	100	60	47	45	33	30	27	10		8	5	1	–	–	–	2	1	–	1	–	–	–	1-Oxa-2-silanaphthalene, 2,2-dichloro-3-bromo-1,2-dihydro-	73060-36-3	NI30572
279	73	191	55	43	263	41	57	**294**	100	22	19	10	10	8	8	6	**0.02**	9	26	3	–	–	–	–	–	–	4	–	–	–	1,3,5-Trioxa-2,4,6,8-tetrasilacyclooctane, 2,2,4,4,6,6,8,8-octamethyl-	19939-05-0	LI06399
279	73	191	281	55	43	263	41	**294**	100	22	19	14	10	10	8	8	**0.02**	9	26	3	–	–	–	–	–	–	4	–	–	–	1,3,5-Trioxa-2,4,6,8-tetrasilacyclooctane, 2,2,4,4,6,6,8,8-octamethyl-	19939-05-0	NI30573
294	167	140	114	102	76			**294**	100	76	54	39	34	32				10	6	2	4	–	–	–	–	–	–	–	–	1	Naphtho[2,3-c][1,2,5]selenadiazole-4,9-dione, dioxime		NI30574
59	294	28	117	131	229	185	45	**294**	100	68	42	33	27	23	19	16		10	14	2	–	4	–	–	–	–	–	–	–	–	2,4,6,8-Tetrathiaadamantane-1-carboxylic acid, 3,5,7-trimethyl-	21404-62-6	NI30575
109	215	104	294	79	63	89	51	**294**	100	37	36	33	25	23	20	16		10	15	6	–	1	–	–	–	–	–	1	–	–	Phosphoric acid, dimethyl 3-methyl-4-(methylsulphonyl)phenyl ester	14086-35-2	NI30576
67	81	135	53	93	79	296	294	**294**	100	63	54	40	34	30	0	0		10	16	–	–	–	–	–	2	–	–	–	–	–	2,8-Decadiene, 1,10-dibromo-, (Z,Z)-		MS06733
111	138	81	109	82	93	97	29	**294**	100	99	71	66	57	48	47	47	**0.00**	10	19	4	2	1	–	–	–	–	–	1	–	–	Phosphorothioic acid, S-[2-[(1-cyanoisopropyl)amino]-2-oxoethyl] O,O-diethyl ester	3734-95-0	NI30577
221	147	73	191	222	148	117	223	**294**	100	99	58	25	24	16	15	13	**0.20**	10	30	2	–	–	–	–	–	–	4	–	–	–	3,6-Dioxa-2,4,5,7-tetrasilaoctane, 2,2,4,4,5,5,7,7-octamethyl-	4342-25-0	NI30578
245	247	217	219	147	73	249	182	**294**	100	90	80	76	75	40	37	32	**26.82**	11	6	1	–	–	–	4	–	–	–	–	–	–	Naphthalene, 2,4,5,8-tetrachloro-1-methoxy-	74298-69-4	NI30579
69	111	55	225	57	87	59	41	**294**	100	31	24	18	11	10	8	8	**0.00**	11	16	2	–	–	6	–	–	–	–	–	–	–	1,3-Dioxolane, 4-butyl-5-ethyl-2,2-bis(trifluoromethyl)-, cis-	38363-90-5	MS06734
69	111	55	225	87	59	57	41	**294**	100	29	26	18	14	12	12	9	**0.00**	11	16	2	–	–	6	–	–	–	–	–	–	–	1,3-Dioxolane, 4-butyl-5-ethyl-2,2-bis(trifluoromethyl)-, trans-	38363-91-6	NI30580
69	111	55	73	225	56	207	57	**294**	100	37	35	23	20	14	10	10	**0.00**	11	16	2	–	–	6	–	–	–	–	–	–	–	1,3-Dioxolane, 4,5-dipropyl-2,2-bis(trifluoromethyl)-, cis-	38363-92-7	MS06735
69	55	111	73	225	56	207	57	**294**	100	43	39	38	24	19	14	10	**0.00**	11	16	2	–	–	6	–	–	–	–	–	–	–	1,3-Dioxolane, 4,5-dipropyl-2,2-bis(trifluoromethyl)-, trans-	38363-93-8	MS06736
69	55	111	225	70	43	81	57	**294**	100	34	25	21	21	16	15	14	**0.00**	11	16	2	–	–	6	–	–	–	–	–	–	–	1,3-Dioxolane, 4-hexyl-2,2-bis(trifluoromethyl)-	38274-66-7	MS06737
69	111	55	225	83	57	56	68	**294**	100	29	29	20	18	14	10	8	**0.00**	11	16	2	–	–	6	–	–	–	–	–	–	–	1,3-Dioxolane, 4-methyl-5-pentyl-2,2-bis(trifluoromethyl)-, cis-	38424-78-1	MS06738
69	55	225	111	83	57	56	101	**294**	100	35	32	30	25	12	12	11	**0.00**	11	16	2	–	–	6	–	–	–	–	–	–	–	1,3-Dioxolane, 4-methyl-5-pentyl-2,2-bis(trifluoromethyl)-, trans-	38363-89-2	MS06739
39	65	94	201	91	66	63	52	**294**	100	79	61	58	35	33	33	33	**6.00**	12	10	5	2	1	–	–	–	–	–	–	–	–	Phenyl 4-amino-3-nitrophenylsulphonate		IC03726
145	147	295	117	105	239	119	161	**294**	100	32	24	16	12	9	8	7	**0.00**	12	13	3	–	1	3	–	–	–	–	–	–	–	1,3-Cycloundecadien-5-yn-1-yl triflate		MS06740
194	294	168	169	195	192	103	91	**294**	100	46	23	20	19	11	8	8		12	15	2	–	–	–	–	–	–	–	–	–	1	Rhodium, (η^4-norbornadiene)acetylacetonato-O,O′-	32354-50-0	NI30581
164	134	192	163	148	206	294	222	**294**	100	74	52	28	16	14	11	11		12	18	3	6	–	–	–	–	–	–	–	–	–	Puromycin nucleoside		MS06741
160	135	44	294	104	92	161	77	**294**	100	42	41	30	15	14	13	8		12	19	1	6	–	–	–	–	–	–	1	–	–	Phosphonic diamide, P-(5-amino-3-phenyl-1H-1,2,4-triazol-1-yl)-N,N,N′,N′-tetramethyl-	1031-47-6	NI30582

MASS TO CHARGE RATIOS								M.W.	INTENSITIES								Parent	C	H	O	N	S	F	Cl	Br	I	Si	P	B	X	COMPOUND NAME	CAS Reg No	No
124	208	151	85	95	109	57	193	**294**	100	86	23	16	14	10	8	5	**0.00**	12	23	6	–	–	–	–	–	–	–	1	–	–	**Phosphonic acid, [3-(isopropylacetoxy)-3-methyl-2-oxobutyl]-, dimethyl ester**		**NI30583**
45	121	58	76	41	253	131	219	**294**	100	18	14	12	9	8	8	3	**0.00**	12	26	6	–	–	–	–	–	–	1	–	–	–	**Allyltris(2-methoxyethoxy)silane**		**DD01039**
153	151	279	135	237	43	179	133	**294**	100	60	55	53	39	37	36	31	**0.00**	12	30	4	–	–	–	–	–	–	2	–	–	–	**Disiloxane, 1,1,1-triisopropoxy-3,3,3-trimethyl-**		**MS06742**
73	117	116	103	75	74	101	45	**294**	100	42	20	13	10	9	7	7	**0.00**	12	30	4	–	–	–	–	–	–	2	–	–	–	**3,6,9,12-Tetraoxa-2,13-disilatetradecane, 2,2,13,13-tetramethyl-**	62185-58-4	**NI30584**
75	152	278	150	215	74	294	265	**294**	100	99	58	50	48	42	40	36		13	8	2	2	–	–	2	–	–	–	–	–	–	**1H-Dibenzo[c,f][1,2]diazepin-11-ol, 3,8-dichloro-, 5-oxide**	23469-59-2	**NI30585**
184	294	215	168	185	139	295	186	**294**	100	45	36	24	14	12	8	7		13	11	2	–	2	–	–	–	–	–	1	–	–	**O,O′-(2,2′-Biphenylylene)dithiophosphoric acid methyl ester**	76045-04-0	**NI30586**
108	107	80	153	142	105	294	81	**294**	100	54	46	30	26	17	17	16		13	12	2	2	–	–	1	–	–	–	1	–	–	**1H-1,3,2-Benzodiazaphosphole, 2-(4-chloro-3-methylphenoxy)-2,3-dihydro , 2-oxide**		**NI30587**
294	43	41	70	181	71	165	231	**294**	100	64	62	53	27	22	20	19		13	18	4	4	–	–	–	–	–	–	–	–	–	**4-Heptanone, (2,4-dinitrophenyl)hydrazone**	1655-41-0	**NI30588**
294	43	55	41	70	181	71	165	**294**	100	65	64	62	54	27	23	20		13	18	4	4	–	–	–	–	–	–	–	–	–	**4-Heptanone, (2,4-dinitrophenyl)hydrazone**	1655-41-0	**NI30589**
206	55	43	149	97	41	50	57	**294**	100	81	72	65	64	58	56	53	**39.60**	13	18	4	4	–	–	–	–	–	–	–	–	–	**Hexanal, 4-methyl-, (2,4-dinitrophenyl)hydrazone**	14093-70-0	**NI30590**
43	71	42	55	294	69	57	44	**294**	100	96	89	57	52	47	46	42		13	18	4	4	–	–	–	–	–	–	–	–	–	**2-Pentanone, 3,4-dimethyl-, (2,4-dinitrophenyl)hydrazone**	55103-79-2	**NI30591**
43	45	89	101	59	205	71	113	**294**	100	90	82	65	52	25	25	22	**0.00**	13	26	7	–	–	–	–	–	–	–	–	–	–	**Glucitol, 1,2,3,5,6-penta-O-methyl-4-O-acetyl-**		**NI30592**
88	75	101	45	193	89	73	59	**294**	100	80	70	25	19	17	12	8	**0.00**	13	26	7	–	–	–	–	–	–	–	–	–	–	**β-Glycero-D-glucoheptopyranoside, methyl 2,3,4,6,7-penta-O-methyl-**		**NI30593**
88	101	45	89	193	73	59	145	**294**	100	91	45	32	13	12	10	9	**0.00**	13	26	7	–	–	–	–	–	–	–	–	–	–	**α-L-Glycero-D-mannoheptopyranoside, methyl 2,3,4,6,7-penta-O-methyl-**	103488-98-8	**NI30594**
251	145	173	78	294	44	121	225	**294**	100	89	74	66	55	37	18	14		14	9	2	2	–	3	–	–	–	–	–	–	–	**2-Pyridinecarboxamide, 5-(3-trifluoromethylbenzoyl)-**	82465-61-0	**NI30595**
251	294	145	173	44	78	225	275	**294**	100	69	59	54	25	24	16	8		14	9	2	2	–	3	–	–	–	–	–	–	–	**2-Pyridinecarboxamide, 5-(4-trifluoromethylbenzoyl)-**	82465-60-9	**NI30596**
294	296	295	91	165	97	229	149	**294**	100	32	16	16	14	13	9	9		14	12	3	–	–	–	1	–	–	–	1	–	–	**10H-Phenoxaphosphine, 8-chloro-2,4-dimethyl-10-hydroxy-, 10-oxide**		**NI30597**
279	294	281	296	295	280	293	215	**294**	100	83	34	28	16	15	7	6		14	12	3	–	–	–	1	–	–	–	1	–	–	**10H-Phenoxaphosphine, 2-chloro-8-ethyl-10-hydroxy-, 10-oxide**	36360-90-4	**NI30598**
214	199	171	294	128	292	107	187	**294**	100	86	28	22	15	11	11	9		14	14	2	–	–	–	–	–	–	–	–	–	1	**Bis(4-anisyl) selenide**	22216-66-6	**NI30599**
261	233	294	205	276	235	265	220	**294**	100	92	88	84	79	50	44	39		14	14	7	–	–	–	–	–	–	–	–	–	–	**4H-1-Benzopyran-6-carboxylic acid, 5,8-dimethoxy-7-hydroxy-2-methyl-4 oxo**		**NI30600**
265	263	207	205	276	261	220	247	**294**	100	42	25	23	22	21	21	20	**9.00**	14	14	7	–	–	–	–	–	–	–	–	–	–	**5H-Furo[3,2-g]benzopyran-5-one, 2,3-dihydroxy-2,3-dihydro-4,9-dimethoxy-7-methyl-, cis-**		**NI30601**
94	294	77	173	110	109	295	65	**294**	100	96	36	25	21	21	16	13		14	15	3	–	1	–	–	–	–	–	1	–	–	**Diphenyl phosphoroethoxythioate**		**LI06400**
210	294	238	266	211	68	181	114	**294**	100	87	86	63	62	61	55	33		14	15	3	2	–	–	1	–	–	–	–	–	–	**1,3-Benzenedicarbonitrile, 5-chloro-2,4,6-triethoxy-**		**DD01040**
211	108	91	294	213	175	119	103	**294**	100	97	81	71	67	67	59	59		14	15	3	2	–	–	1	–	–	–	–	–	–	**1-Chloro-1-[[2-[(4-methoxy-2-butynyl)oxy]phenyl]hydrazono]acetone**		**MS06743**
145	104	238	87	53	77	144	117	**294**	100	99	77	71	34	32	23	22	**1.00**	14	18	1	2	2	–	–	–	–	–	–	–	–	**4-Thia-1-azabicyclo[3.2.0]heptan-7-one, 6-exo-amino-2,2-dimethyl-3-endo-(methylthio)-5-phenyl-**		**NI30602**
54	53	52	107	44	43	77	42	**294**	100	16	11	5	2	2	1	1	**0.11**	14	18	5	2	–	–	–	–	–	–	–	–	–	**L-Alanine, N-(N-acetyl-L-tyrosyl)-**	56272-47-0	**NI30603**
121	122	95	294	148	220	107	43	**294**	100	65	30	27	26	20	17	16		14	18	5	2	–	–	–	–	–	–	–	–	–	**2-(4-Methoxyphenyl)hydrazonopropanedioic acid, diethyl ester**		**NI30604**
43	174	150	122	234	132	133	149	**294**	100	75	45	23	18	13	12	11	**2.89**	14	18	5	2	–	–	–	–	–	–	–	–	–	**4,5-Pyridinedimethanol, 2-methyl-3-amino-, triacetate**		**NI30605**
188	88	144	132	32	42	294	56	**294**	100	14	10	8	5	5	5	4		14	22	3	2	–	–	–	–	–	1	–	–	–	**2,9,10-Trioxa-6-aza-1-silatricyclo[4.3.3.$0^{1,6}$]dodecane, 1-(phenylaminomethyl)-**		**MS06744**
45	57	89	29	59	44	87	41	**294**	100	34	31	16	15	15	13	13	**0.00**	14	30	6	–	–	–	–	–	–	–	–	–	–	**3,6,9,12,15-Pentaoxanonadecan-1-ol**	1786-94-3	**NI30606**
238	210	89	173	266	294	208	56	**294**	100	48	39	37	36	35	30	26		15	10	3	–	–	–	–	–	–	–	–	–	1	**Iron, (2-benzofuranyl)dicarbonyl-π-cyclopentadienyl-**	59308-00-8	**NI30607**
15	235	165	237	59	199	18	163	**294**	100	67	58	42	22	19	16	14	**10.21**	15	12	2	–	–	–	2	–	–	–	–	–	–	**Benzeneacetic acid, 4-chloro-α-(4-chlorophenyl)-, methyl ester**	5359-38-6	**NI30608**
276	193	165	91	179	294	161	111	**294**	100	86	63	61	53	49	45	43		15	15	2	–	1	–	1	–	–	–	–	–	–	**2,4,6-Trimethyl-4′-chlorodiphenyl sulphone**		**NI30609**
252	91	43	41	69	39	65	92	**294**	100	58	46	46	45	39	36	29	**0.00**	15	18	4	–	1	–	–	–	–	–	–	–	–	**2(5H)-Furanone, 4-isopropyl-2-methyl-3-[(4-methylphenyl)sulphonyl]-**	118356-51-7	**DD01041**
192	252	203	204	149	210	161	234	**294**	100	88	33	21	21	18	18	17	**12.00**	15	18	6	–	–	–	–	–	–	–	–	–	–	**3-Acetoxy-1-(4-acetoxy-3,5-dimethoxyphenyl)propene**		**LI06401**
207	221	239	263	56	208	162	161	**294**	100	39	27	25	15	11	10	10	**2.00**	15	18	6	–	–	–	–	–	–	–	–	–	–	**1,2,4-Benzenetricarboxylic acid, 4-butyl 1,2-dimethyl ester**	43049-07-6	**NI30610**
221	249	193	29	222	250	148	27	**294**	100	46	27	21	15	10	9	9	**8.01**	15	18	6	–	–	–	–	–	–	–	–	–	–	**1,2,4-Benzenetricarboxylic acid, triethyl ester**	14230-18-3	**NI30611**
249	221	73	193	266	210	29	250	**294**	100	36	27	19	18	16	16	15	**13.78**	15	18	6	–	–	–	–	–	–	–	–	–	–	**1,3,5-Benzenetricarboxylic acid, triethyl ester**	4105-92-4	**NI30612**
232	203	263	235	262	261	294	115	**294**	100	86	64	39	20	17	16	16		15	18	6	–	–	–	–	–	–	–	–	–	–	**1H-2-Benzopyran-5,6-dicarboxylic acid, 3,4-dihydro-1-methoxy-3-methyl-, dimethyl ester**		**NI30613**
263	231	262	115	264	232	205	116	**294**	100	81	58	21	15	14	13	13	**1.00**	15	18	6	–	–	–	–	–	–	–	–	–	–	**1H-2-Benzopyran-7,8-dicarboxylic acid, 3,4-dihydro-1-methoxy-3-methyl-, dimethyl ester**		**NI30614**
43	42	55	82	175	85	124	294	**294**	100	34	31	18	16	14	13	12		15	18	6	–	–	–	–	–	–	–	–	–	–	**1-(2-Hydroxy-5-oxocyclohexenylmethyl)-2-hydroxy-3-acetoxycyclohexen-5-one**		**LI06402**
43	44	139	71	55	91	79	77	**294**	100	61	60	36	33	26	23	21	**6.00**	15	18	6	–	–	–	–	–	–	–	–	–	–	**3,6-Methano-8H-1,5,7-trioxacyclopenta[ij]cycloprop[a]azulene-4,8(3H)-dione, hexahydro-9-hydroxy-8b-methyl-9-isopropyl-, [1aR-(1aα,2aβ,3β,6β,6aβ,8aS*8bβ,9R*)]-**	62655-15-6	**NI30615**
119	91	140	137	158	263	235	264	**294**	100	31	30	14	11	10	6	2	**0.00**	15	18	6	–	–	–	–	–	–	–	–	–	–	**α-D-erythro-Pentofuranose, 1-(methoxycarbonyl)-2-deoxy-3-O-(4-toluoyl)**		**NI30616**
119	91	140	137	263	158	235	264	**294**	100	49	36	17	13	10	7	4	**0.00**	15	18	6	–	–	–	–	–	–	–	–	–	–	**α-L-erythro-Pentofuranose, 1-(methoxycarbonyl)-2-deoxy-3-O-(4-toluoyl)**		**NI30617**
119	91	140	263	136	158	264	235	**294**	100	21	6	6	4	2	1	1	**0.00**	15	18	6	–	–	–	–	–	–	–	–	–	–	**β-D-erythro-Pentofuranose, 1-(methoxycarbonyl)-2-deoxy-3-O-(4-toluoyl)**		**NI30618**

MASS TO CHARGE RATIOS								M.W.	INTENSITIES								Parent	C	H	O	N	S	F	Cl	Br	I	Si	P	B	X	COMPOUND NAME	CAS Reg No	No
119	91	81	263	137	136	264	158	**294**	100	30	10	8	6	3	2	2	**0.00**	15	18	6	–	–	–	–	–	–	–	–	–	–	β-L-erythro-Pentofuranose, 1-(methoxycarbonyl)-2-deoxy-3-O-(4-toluoyl)		NI30619
91	123	279	129	65	107	161	219	**294**	100	11	10	10	10	7	6	5	**0.00**	15	18	6	–	–	–	–	–	–	–	–	–	–	α-D-Ribofuranuronic acid, 1,2-O-isopropylidene-3-O-benzyl-	120417-90-5	DD01042
201	216	296	294	281	279	172	115	**294**	100	54	42	42	38	38	30	30		15	19	1	–	–	–	–	1	–	–	–	–	–	Benzene, 1-bromo-4-methoxy-2-methyl-5-(1-methyl-2-methylenecyclopentyl)-	55956-26-8	NI30620
279	59	226	74	201	174	237	200	**294**	100	82	75	74	72	67	65	65	**0.00**	15	19	1	–	–	–	–	1	–	–	–	–	–	Laurinterol		LI06403
129	115	145	155	141	159	169	197	**294**	100	60	54	40	39	21	18	16	**0.10**	15	19	1	–	–	–	–	1	–	–	–	–	–	3,6,9-Pentadecatrien-1-yne, 12-bromo-5,13-epoxy-, (3Z,6Z,9Z)-		MS06745
226	228	174	215	159	173	172	77	**294**	100	86	69	52	41	28	19	19	**12.00**	15	19	1	–	–	–	–	1	–	–	–	–	–	Phenol, 4-bromo-5-(1,2-dimethylbicyclo[3.1.0]hex-2-yl)-2-methyl-, [1S-(1α,2β,5α)]-	103439-83-4	NI30621
173	200	215	185	159	296	294	158	**294**	100	41	22	22	20	17	16	12		15	19	1	–	–	–	–	1	–	–	–	–	–	Phenol, 4-bromo-2-methyl-5-(1,2,2-trimethyl-3-cyclopenten-1-yl)-, (R)-	103439-84-5	NI30622
127	83	140	199	126	99	129	98	**294**	100	69	58	42	39	35	33	30	**3.00**	15	19	2	2	–	–	1	–	–	–	–	–	–	Pyrrolo[1,2-b][1,2,4]oxadiazol-2(1H)-one, 1-(4-chlorophenyl)tetrahydro-5,5,7,7-tetramethyl-	39931-33-4	NI30623
73	294	205	45	293	189	115	75	**294**	100	42	33	18	18	16	14	12		15	26	2	–	–	–	–	–	–	2	–	–	–	2-Propene, 3-[4-(trimethylsiloxy)phenyl]-1-(trimethylsiloxy)-		NI30624
279	73	280	132	281	294	45	263	**294**	100	54	32	15	14	13	12	10		15	30	–	–	–	–	–	–	–	3	–	–	–	Silane, 1,3,5-benzenetriyltris[trimethyl-	5624-60-2	NI30625
206	73	294	207	97	191	295	208	**294**	100	95	33	23	15	13	10	10		15	30	–	–	–	–	–	–	–	3	–	–	–	Silane, (1-methyl-1,2-pentadien-4-yne-1,3,5-triyl)tris[trimethyl-	61227-81-4	NI30626
73	129	100	205	178	114	102	59	**294**	100	58	55	28	28	20	19	17	**5.00**	15	30	–	2	–	–	–	–	–	2	–	–	–	1,3-Propanediamine, N-phenyl-N,N′-bis(trimethylsilyl)-	54550-15-1	NI30627
133	73	174	132	106	294	147	59	**294**	100	76	63	57	33	29	25	23		15	30	–	2	–	–	–	–	–	2	–	–	–	1,3-Propanediamine, N′-phenyl-N,N-bis(trimethylsilyl)-	54550-16-2	NI30628
147	294	105	120	104	148	36	295	**294**	100	90	51	36	33	25	20	19		16	14	2	4	–	–	–	–	–	–	–	–	–	3,10-Diaminodiftalone		MS06746
294	93	295	172	77	94	92	217	**294**	100	45	33	28	16	14	13	11		16	14	2	4	–	–	–	–	–	–	–	–	–	1H-Pyrazole-4,5-dione, 3-(hydroxymethyl)-1-phenyl-, 4-(phenylhydrazone)	35426-87-0	NI30629
133	134	103	90	118	85	294	104	**294**	100	72	64	62	50	40	39	34		16	14	2	4	–	–	–	–	–	–	–	–	–	s-Tetrazine, 3,6-bis(4-methoxyphenyl)-	14141-66-3	NI30630
147	146	294	159	160	105	91	104	**294**	100	62	57	55	49	36	36	33		16	16	4	–	–	–	–	–	–	–	–	2	–	[1,3,2]Dioxaborino[5,4-d]-1,3,2-dioxaborin, tetrahydro-2,6-diphenyl-, cis-	74841-65-9	NI30631
294	147	160	159	105	293	91	104	**294**	100	94	69	54	49	47	45	43		16	16	4	–	–	–	–	–	–	–	–	2	–	[1,3,2]Dioxaborino[5,4-d]-1,3,2-dioxaborin, tetrahydro-2,6-diphenyl-, trans-	38572-02-0	NI30632
147	146	91	294	105	104	293	148	**294**	100	57	37	26	14	14	12	10		16	16	4	–	–	–	–	–	–	–	–	2	–	4,4′-Bi-1,3,2-dioxaborolane, 2,2′-diphenyl-, (R*,S*)-	71166-92-2	NI30633
145	91	97	106	294	41	132	119	**294**	100	61	45	43	39	36	25	25		16	22	1	–	2	–	–	–	–	–	–	–	–	2-Benzyl-2-methyl-6,10-dithiaspiro[4.5]decan-1-ol	92640-86-3	NI30634
174	148	191	159	118	105	234	294	**294**	100	96	45	35	22	12	11	10		16	22	5	–	–	–	–	–	–	–	–	–	–	Benzeneethanol, 2-[(acetyloxy)methyl]-3-methoxy-α,6-dimethyl-, acetate, (S)-	67549-66-0	NI30635
43	93	45	176	69	175	124	91	**294**	100	68	49	38	38	37	32	32	**19.02**	16	22	5	–	–	–	–	–	–	–	–	–	–	2(5H)-Furanone, 5-(1,3-hexadienyl)-3-(2-hydroxy-1-oxobutyl)-4-methoxy-5-methyl-, (E,E)-	54833-27-1	NI30636
43	125	122	121	41	94	69	55	**294**	100	85	64	56	49	47	42	39	**16.09**	16	22	5	–	–	–	–	–	–	–	–	–	–	2,4-Pentadienoic acid, 5-(8-hydroxy-1,5-dimethyl-3-oxo-6-oxabicyclo[3.2.1]oct-8-yl)-3-methyl-, methyl ester, [1R-[1α,5α,8S*(2Z,4E)]]-	41670-48-8	NI30637
125	122	294	139	121	167	154	163	**294**	100	73	60	55	50	40	38	35		16	22	5	–	–	–	–	–	–	–	–	–	–	2,4-Pentadienoic acid, 5-(8-hydroxy-1,5-dimethyl-3-oxo-6-oxabicyclo[3.2.1]oct-8-yl)-3-methyl-, methyl ester, [1R-[1α,5α,8S*(2Z,4E)]]-	41670-48-8	NI30638
91	45	99	145	100	71	43	129	**294**	100	38	29	18	15	15	12	10	**0.00**	16	22	5	–	–	–	–	–	–	–	–	–	–	β-D-Ribofuranoside, benzyl 5-O-methyl-2,3-O-isopropylidene-	64018-52-6	NI30639
237	265	219	294	43	18	210	223	**294**	100	58	56	36	32	30	28	25		16	22	5	–	–	–	–	–	–	–	–	–	–	2,6-Valerylphloroglucinol		MS06747
237	163	73	238	294	164	221	165	**294**	100	73	31	24	21	14	13	12		16	26	3	–	–	–	–	–	–	1	–	–	–	Benzeneacetic acid, 4-(tert-butyldimethylsilyloxy)-, ethyl ester		MS06748
73	94	75	185	83	201	55	294	**294**	100	81	72	67	53	49	33	28		16	26	3	–	–	–	–	–	–	1	–	–	–	Heptanoic acid, 7-phenoxy-, trimethylsilyl ester	21273-13-2	NI30641
73	94	75	185	85	294	55	201	**294**	100	81	72	69	54	43	33	31		16	26	3	–	–	–	–	–	–	1	–	–	–	Heptanoic acid, 7-phenoxy-, trimethylsilyl ester	21273-13-2	NI30640
121	91	75	73	279	294	117	204	**294**	100	58	54	44	43	37	36	26		16	26	3	–	–	–	–	–	–	1	–	–	–	Hexanoic acid, (2-methoxyphenyl)-, trimethylsilyl ester		NI30642
121	294	279	75	122	73	176	117	**294**	100	19	17	16	10	9	5	4		16	26	3	–	–	–	–	–	–	1	–	–	–	Hexanoic acid, (3-methoxyphenyl)-, trimethylsilyl ester		NI30643
72	179	116	137	136	43	56	180	**294**	100	45	25	21	15	15	13	8	**1.50**	16	26	3	2	–	–	–	–	–	–	–	–	–	2-Propanol, 1-(4-acetamido-2,6-dimethylphenoxy)-3-(isopropylamino)-		MS06749
72	179	116	137	136	43	56	73	**294**	100	46	26	21	15	15	14	8	**1.50**	16	26	3	2	–	–	–	–	–	–	–	–	–	2-Propanol, 1-(4-acetamido-2,6-dimethylphenoxy)-3-(isopropylamino)-		IC03727
224	223	41	225	125	209	109	43	**294**	100	80	18	16	16	15	14	14	**5.00**	16	26	3	2	–	–	–	–	–	–	–	–	–	2,4,6(1H,3H,5H)-Pyrimidinetrione, 1,3-diethyl-5-(1-methylbutyl)-5-(2-propenyl)-	54833-76-0	NI30644
294	195	295	223	251	264	296	151	**294**	100	34	26	24	18	10	9	9		17	10	3	–	1	–	–	–	–	–	–	–	–	Benzo[b]naphtho[2,3-d]thiophene-6,11-dione, 8-methoxy-	31247-67-3	NI30645
294	195	295	223	251	296	208	151	**294**	100	35	26	25	19	10	9	9		17	10	3	–	1	–	–	–	–	–	–	–	–	Benzo[b]naphtho[2,3-d]thiophene-6,11-dione, 8-methoxy-	31247-67-3	LI06404
45	46	44	124	57	91	55	123	**294**	100	78	25	15	15	12	11	10	**9.00**	17	10	5	–	–	–	–	–	–	–	–	–	–	2-Indenone, 3-phenyl-6,7-dicarboxy-	92241-98-0	NI30646
294	136	117	108	295	116	137	130	**294**	100	87	64	23	20	15	11	8		17	14	1	2	1	–	–	–	–	–	–	–	–	4H-5,6-Benzothiazin-4-one, 2-(1,2,3,4-tetrahydroisoquinolinyl)-		NI30647
91	139	141	111	162	92	43	75	**294**	100	38	19	18	13	10	10	9	**0.00**	17	14	3	2	–	–	–	–	–	–	–	–	–	N-(α-Acetoxy-4-cyanobenzylidene)-O-benzylhydroxylamine		LI06405
223	55	277	224	251	276	195	91	**294**	100	50	35	22	20	14	14	13	**1.10**	17	14	3	2	–	–	–	–	–	–	–	–	–	2-(Amino)-N-(9,10-dioxaanthracen-1-yl)-propanamide		MS06750
133	102	91	104	119	43	120	132	**294**	100	69	63	61	60	53	49	45	**14.80**	17	14	3	2	–	–	–	–	–	–	–	–	–	2-Hydroxy-1,3-bis(2-cyanophenoxy)propane		IC03728
252	294	253	251	295	165	223	83	**294**	100	61	17	12	11	9	8	5		17	14	3	2	–	–	–	–	–	–	–	–	–	Indolin-2-one, N-acetamido-3-(4-hydroxybenzylidene)-		AI02001
294	162	163	132	18	295	293	134	**294**	100	58	54	23	21	20	20	20		17	14	3	2	–	–	–	–	–	–	–	–	–	2-Methoxydiftalone		MS06751
294	162	118	133	90	120	134	18	**294**	100	66	44	31	28	27	25	21		17	14	3	2	–	–	–	–	–	–	–	–	–	3-Methoxydiftalone		MS06752

MASS TO CHARGE RATIOS								M.W.	INTENSITIES								Parent	C	H	O	N	S	F	Cl	Br	I	Si	P	B	X	COMPOUND NAME	CAS Reg No	No
91	199	62	118	160	248	105	61	**294**	100	99	94	84	81	40	37	37	**0.00**	17	14	3	2	–	–	–	–	–	–	–	–	–	3-Pentanone, 2-diazo-1,5-diphenyl-4,5-epoxy-1-hydroxy-		MS06753
132	133	104	262	294	235	130	77	**294**	100	33	17	16	14	14	13	13		17	14	3	2	–	–	–	–	–	–	–	–	–	1(2H)-Phthalazinone, 2-(2-(methoxycarbonyl)benzyl)-	54648-85-0	NI30648
183	77	294	43	51	184	93	105	**294**	100	86	67	35	18	15	11	7		17	14	3	2	–	–	–	–	–	–	–	–	–	3,5-Pyrazolidinedione, 4-acetyl-1,2-diphenyl-	6139-79-3	NI30649
294	160	161	226	225	117	77	145	**294**	100	49	34	30	22	22	22	21		17	14	3	2	–	–	–	–	–	–	–	–	–	3,5-Pyrazolidinedione, 4-(4-methoxybenzylidene)-1-phenyl-	87059-94-7	NI30650
208	294	165	77	166	104	180	222	**294**	100	85	60	45	31	26	21	19		17	14	3	2	–	–	–	–	–	–	–	–	–	2,4,6(1H,3H,5H)-Pyrimidinetrione, 5-methyl-1,3-diphenyl-	739-50-4	NI30651
235	294	219	236	278	218	43	217	**294**	100	50	33	25	19	16	16	15		17	14	3	2	–	–	–	–	–	–	–	–	–	2(1H)-Quinolone, 3-(α-acetamidophenyl)-1-hydroxy-	54833-78-2	NI30652
235	294	219	236	278	218	43	252	**294**	100	50	33	25	19	16	16	15		17	14	3	2	–	–	–	–	–	–	–	–	–	2(1H)-Quinolone, 3-(α-acetamidophenyl)-1-hydroxy-	54833-78-2	MS06754
119	161	42	294	90	91	92	237	**294**	100	76	74	68	61	55	50	38		17	14	3	2	–	–	–	–	–	–	–	–	–	Spiro[3H-1,4-benzodiazepine-3,2′-oxirane]-2,5(1H,4H)-dione, 4-methyl-3′ phenyl-, cis-(–)-	20007-87-8	LI06406
119	161	42	294	90	92	91	118	**294**	100	75	73	68	60	50	45	38		17	14	3	2	–	–	–	–	–	–	–	–	–	Spiro[3H-1,4-benzodiazepine-3,2′-oxirane]-2,5(1H,4H)-dione, 4-methyl-3′ phenyl-, cis-(–)-	20007-87-8	NI30653
279	217	132	294	280	90	295	205	**294**	100	59	49	48	22	22	13	13		17	18	1	2	–	–	–	–	–	1	–	–	–	Quinoxaline, 2-phenyl-3-[(trimethylsilyl)oxy]-	80782-67-8	NI30654
43	121	125	109	123	95	294	107	**294**	100	24	16	15	15	14	14	13		17	26	4	–	–	–	–	–	–	–	–	–	–	3α-Acetylcuauhtemone		NI30655
159	43	131	91	174	41	144	105	**294**	100	36	34	29	25	22	21	21	**0.00**	17	26	4	–	–	–	–	–	–	–	–	–	–	3-Cyclohexen-1-ol, 4-[3-(acetyloxy)-1-butenyl]-3,5,5-trimethyl-, acetate	68573-21-7	NI30656
137	150	294	151	138	194	122	119	**294**	100	55	30	20	20	15	11	11		17	26	4	–	–	–	–	–	–	–	–	–	–	3-Decanone, 5-hydroxy-1-(4-hydroxy-3-methoxyphenyl)-, (S)-	23513-14-6	MS06755
94	95	123	128	109	191	163	234	**294**	100	98	75	58	58	38	34	21	**0.00**	17	26	4	–	–	–	–	–	–	–	–	–	–	Hanalpinol acetate		DD01043
43	248	41	122	109	55	95	81	**294**	100	41	36	23	22	22	20	16	**7.00**	17	26	4	–	–	–	–	–	–	–	–	–	–	Lactarorufin A, 3-ethyl-		NI30657
294	184	185	295	51	77	152	186	**294**	100	60	44	20	19	11	10	8		18	14	–	–	2	–	–	–	–	–	–	–	–	1,2-Bis(phenylthio)benzene		DD01044
294	263	88	235	102	220	116	203	**294**	100	52	27	25	20	19	19	18		18	14	4	–	–	–	–	–	–	–	–	–	–	1,8-Anthracenedicarboxylic acid, dimethyl ester		NI30658
148	129	130	276	260				**294**	100	27	21	2	2				**0.00**	18	14	4	–	–	–	–	–	–	–	–	–	–	1,1′-Dihydroxy-3,3′-dioxo-1,1′-biindanyl		LI06407
261	259	278	232	230	260	247	244	**294**	100	29	22	14	14	10	10	10	**6.50**	18	14	4	–	–	–	–	–	–	–	–	–	–	8,9-Epoxy-2,15ξ-dihydroxy-11-methyl-8,9-secogona-1,3,5,7,9,11,13-hepten 17-one		NI30659
223	238	152	50	165	224	76	151	**294**	100	45	31	27	17	16	15	14	**0.00**	18	14	4	–	–	–	–	–	–	–	–	–	–	9,10-Ethanoanthracene-11,12-dione, 9,10-dihydro-9,10-dimethoxy-	122949-71-7	DD01045
105	77	189	191	294	53	51	217	**294**	100	31	20	16	12	12	12	9		18	14	4	–	–	–	–	–	–	–	–	–	–	3-Furancarboxylic acid, 2,5-dihydro-5-oxo-2,2-diphenyl-, methyl ester	33545-32-3	MS06756
147	69	77	105	189	91	51	148	**294**	100	99	49	29	28	20	13	11	**3.10**	18	14	4	–	–	–	–	–	–	–	–	–	–	1,3,4,6-Hexanetetrone, 1,6-diphenyl-	53454-78-7	NI30660
294	76	279	104	251	180	208	236	**294**	100	67	60	56	38	25	19	18		18	14	4	–	–	–	–	–	–	–	–	–	–	1,4-Naphthalenedione, 2-(2,5-dimethoxyphenyl)-		MS06757
294	276	265	165	266	248	251	191	**294**	100	69	58	58	57	44	35	33		18	14	4	–	–	–	–	–	–	–	–	–	–	Phenanthro[1,2-b]furan-10,11-dione, hydroxy-6,7,8,9-tetrahydro-1-methyl 6-methylene		NI30661
154	217	63	294	181	219	51	155	**294**	100	83	48	37	29	28	23	23		18	15	–	–	–	–	1	–	–	1	–	–	–	Silane, chlorotriphenyl-	76-86-8	NI30662
294	293	185	183	295	139	262	184	**294**	100	46	42	39	20	10	9	6		18	15	–	–	1	–	–	–	–	–	1	–	–	Phosphine sulphide, triphenyl-	3878-45-3	NI30664
294	183	185	293	262	295	43	152	**294**	100	55	47	44	35	20	17	14		18	15	–	–	1	–	–	–	–	–	1	–	–	Phosphine sulphide, triphenyl-	3878-45-3	NI30663
201	293	94	294	77	170	141	91	**294**	100	77	70	41	37	25	16	15		18	15	2	–	–	–	–	–	–	–	1	–	–	Phenyl diphenylphosphinate		MS06758
120	119	194	209	92	65	150	75	**294**	100	93	86	63	53	40	39	23	**0.00**	18	18	2	–	–	–	–	–	–	1	–	–	–	Anthranilic acid, trimethylsilyl ester		NI30665
194	209	237	195	210	236	238	294	**294**	100	41	33	33	14	11	4	1		18	18	2	2	–	–	–	–	–	–	–	–	–	1-Acetyl-2-methyl-5,5-diphenylimidazolid-4-one		MS06759
140	294	154	155	196	127			**294**	100	81	45	38	19	9				18	18	2	2	–	–	–	–	–	–	–	–	–	Deacetylcyclopiazonic acid		LI06408
105	294	252	106	295	147	253	103	**294**	100	19	9	9	3	3	2	2		18	18	2	2	–	–	–	–	–	–	–	–	–	N-[3-(4-Methoxybenzylidene)-2-oxo-1-indolinyl]acetamide		AI02002
91	294	175	203	120	65	103	77	**294**	100	42	40	22	20	14	11	10		18	18	2	2	–	–	–	–	–	–	–	–	–	Phenylalanine anhydride		IC03729
91	175	294	203	120	65	28	92	**294**	100	29	26	15	14	10	9	7		18	18	2	2	–	–	–	–	–	–	–	–	–	2,5-Piperazinedione, 3,6-dibenzyl-		LI06409
183	294	77	184	105	91	51	69	**294**	100	67	67	16	12	8	8	7		18	18	2	2	–	–	–	–	–	–	–	–	–	3,5-Pyrazolidinedione, 4-isopropyl-1,2-diphenyl-	1093-68-1	NI30666
265	263	91	77	294	45	50	160	**294**	100	64	50	46	36	36	32	27		18	18	2	2	–	–	–	–	–	–	–	–	–	4(3H)-Quinazolinone, 2-ethyl-3-(2-methyl-6-methoxyphenyl)-		NI30667
294	293	147	295	266	174	145	148	**294**	100	27	25	21	17	15	9	8		18	22	–	4	–	–	–	–	–	–	–	–	–	4-(Dimethylamino)benzylidene azine		LI06410
294	266	279	293	32	280	295	133	**294**	100	88	64	44	35	23	22	22		18	22	–	4	–	–	–	–	–	–	–	–	–	Imidazo[1,2-a]pyridin-2-amine, 5-isopropyl-N-(6-isopropyl-2-pyridinyl)-		MS06760
43	41	73	55	81	139	96	56	**294**	100	90	77	69	67	62	41	41	**0.00**	18	30	3	–	–	–	–	–	–	–	–	–	–	Non-2-enoic anhydride		IC03730
223	135	224	45	57	107	91	77	**294**	100	56	14	14	13	11	10	6	**2.38**	18	30	3	–	–	–	–	–	–	–	–	–	–	Ethanol, 2-[2-[4-(1,1,3,3-tetramethylbutyl)phenoxy]ethoxy]-	2315-61-9	NI30668
41	57	55	43	95	82	85	114	**294**	100	72	62	60	57	55	49	47	**10.00**	18	30	3	–	–	–	–	–	–	–	–	–	–	Methyl 12,14-dihomojuvenate		LI06411
57	95	41	55	114	83	67	81	**294**	100	99	79	78	52	50	49	48	**1.20**	18	30	3	–	–	–	–	–	–	–	–	–	–	2,6-Nonadienoic acid, 7-ethyl-9-(3-ethyl-3-methyloxiranyl)-3-methyl-, methyl ester, [2R-[2α(2E,6E),3α]]-	13804-51-8	MS06761
79	95	103	121	294	220	163	276	**294**	100	41	32	26	7	4	3	2		18	30	3	–	–	–	–	–	–	–	–	–	–	1,2-Propanediol, 3-[(13-methyl-1,5-tetradecadien-3-ynyl)oxy]-, (Z,Z)-		DD01046
43	91	121	77	131	149	135	294	**294**	100	84	24	16	11	8	4	3		18	30	3	–	–	–	–	–	–	–	–	–	–	1,2-Propanediol, 3-[(1,5-pentadecadien-3-ynyl)oxy]-, (Z,Z)-		DD01047
69	73	93	75	41	81	80	107	**294**	100	99	65	62	58	45	35	27	**3.00**	18	34	1	–	–	–	–	–	–	1	–	–	–	2,6,10-Dodecatriene, 3,7,11-trimethyl-1-(trimethylsilyloxy)-	57305-02-9	NI30670
69	73	75	41	81	93	107	135	**294**	100	99	92	75	56	53	37	32	**3.50**	18	34	1	–	–	–	–	–	–	1	–	–	–	2,6,10-Dodecatriene, 3,7,11-trimethyl-1-(trimethylsilyloxy)-	57305-02-9	NI30669
69	73	41	75	81	93	135	68	**294**	100	95	55	49	28	26	16	14	**1.07**	18	34	1	–	–	–	–	–	–	1	–	–	–	2,6,10-Dodecatriene, 3,7,11-trimethyl-1-(trimethylsilyloxy)-, (E,E)-		NI30671
73	75	131	41	204	161	43	45	**294**	100	68	41	37	36	35	33	27	**4.43**	18	34	1	–	–	–	–	–	–	1	–	–	–	4H-Eremophil-(10)-ene, 11-(trimethylsilyloxy)-, (4β,5α)-	28272-09-5	MS06762
69	73	75	41	93	81	135	103	**294**	100	97	56	45	38	38	21	19	**1.50**	18	34	1	–	–	–	–	–	–	1	–	–	–	Silane, trimethyl[(3,7,11-trimethyl-2,6,10-dodecatrienyl)oxy]-	57305-02-9	NI30672
73	131	204	161	189	279	294		**294**	100	57	36	35	12	5	4			18	34	1	–	–	–	–	–	–	1	–	–	–	Vanenanol trimethylsilyl ether		LI06412
294	293	147	267	121	251	235	165	**294**	100	10	6	4	4	3	3	2		19	18	3	–	–	–	–	–	–	–	–	–	–	Benzofuran, 2-(4-hydroxyphenyl)-7-methoxy-3-methyl-5(E)-propenyl-		MS06763

MASS TO CHARGE RATIOS								M.W.	INTENSITIES								Parent	C	H	O	N	S	F	Cl	Br	I	Si	P	B	X	COMPOUND NAME	CAS Reg No	No
191	208	45	189	179	192	294	165	**294**	100	44	38	36	28	26	22	20		19	18	3	–	–	–	–	–	–	–	–	–	–	9-Phenanthrenemethyl 4-hydroxybutanoate	92174-39-5	NI30673
261	294	251	165	279	178	116	233	**294**	100	74	69	52	47	43	29	28		19	18	3	–	–	–	–	–	–	–	–	–	–	Phenanthro[1,2-b]furan-10,11-dione, 6,7,8,9-tetrahydro-1,6,6-trimethyl-	568-72-9	LI06413
261	294	251	165	279	178	116	152	**294**	100	73	68	52	45	43	30	28		19	18	3	–	–	–	–	–	–	–	–	–	–	Phenanthro[1,2-b]furan-10,11-dione, 6,7,8,9-tetrahydro-1,6,6-trimethyl-	568-72-9	NI30674
261	149	279	233	294	165	251	189	**294**	100	70	67	67	59	59	50	50		19	18	3	–	–	–	–	–	–	–	–	–	–	Phenanthro[1,2-b]furan-10,11-dione, 6,7,8,9-tetrahydro-1,6,6-trimethyl-	568-72-9	NI30675
279	294	251	265	132	165	280	128	**294**	100	92	53	31	23	22	20	20		19	18	3	–	–	–	–	–	–	–	–	–	–	Phenanthro[3,2-b]furan-7,11-dione, 1,2,3,4-tetrahydro-4,4,8-trimethyl-	20958-15-0	NI30676
279	294	251	265	132	165	280	128	**294**	100	93	43	32	24	22	21	21		19	18	3	–	–	–	–	–	–	–	–	–	–	Phenanthro[3,2-b]furan-7,11-dione, 1,2,3,4-tetrahydro-4,4,8-trimethyl-	20958-15-0	LI06414
294	169	293	168	130	143	170	117	**294**	100	78	60	42	28	26	18	16		19	22	1	2	–	–	–	–	–	–	–	–	–	Ajmalan-17-ol, 19,20-didehydro-1-demethyl-, (17R,19E)-	68160-76-9	NI30677
294	238	237	209	96	265	156	143	**294**	100	27	27	27	18	13	13	13		19	22	1	2	–	–	–	–	–	–	–	–	–	Aspidofractinine, 3-oxo-	19634-37-8	LI06415
110	41	96	43	294	42	55	109	**294**	100	91	62	56	55	51	44	42		19	22	1	2	–	–	–	–	–	–	–	–	–	Aspidofractinine, 3-oxo-	19634-37-8	NI30678
123	266	294	158	144	138	156	157	**294**	100	51	47	37	35	30	23	20		19	22	1	2	–	–	–	–	–	–	–	–	–	Aspidofractinine, 6-oxo-	19634-36-7	NI30679
294	238	123	251	95	266	144	109	**294**	100	75	67	50	43	33	31	28		19	22	1	2	–	–	–	–	–	–	–	–	–	Aspidofractinine, 17-oxo-, (±)-	120385-51-5	DD01048
109	294	251	280	115	156	194	143	**294**	100	97	94	62	46	41	40	36		19	22	1	2	–	–	–	–	–	–	–	–	–	Aspidofractinine, 20-oxo-		LI06416
41	109	149	43	42	294	57	55	**294**	100	66	58	49	46	41	30	30		19	22	1	2	–	–	–	–	–	–	–	–	–	Aspidospermidin-20-one, 1,2-didehydro-	56053-19-1	NI30680
142	88	100	130	41	44	42	57	**294**	100	76	72	51	48	43	40	39	**0.00**	19	22	1	2	–	–	–	–	–	–	–	–	–	Cinchonan-9-ol, (8α,9R)-	485-71-2	NI30681
136	128	81	130	101	129	143	77	**294**	100	43	39	34	33	29	26	26	**0.00**	19	22	1	2	–	–	–	–	–	–	–	–	–	Cinchonan-9-ol, (9S)-	118-10-5	WI01378
136	294	159	295	143	137	253	95	**294**	100	71	19	16	10	10	9	5		19	22	1	2	–	–	–	–	–	–	–	–	–	Cinchonan-9-ol, (9S)-	118-10-5	NI30682
41	225	123	168	84	167	70	42	**294**	100	99	93	78	73	62	58	54	**43.00**	19	22	1	2	–	–	–	–	–	–	–	–	–	Curan-17-al, 2,16-didehydro-	6872-34-0	NI30683
293	144	130	77	41	279	225	143	**294**	100	45	42	38	31	24	24	23	**0.00**	19	22	1	2	–	–	–	–	–	–	–	–	–	Curan-14-ol, 16,17,19,20-tetradehydro-, (19E)-	56053-21-5	NI30684
58	294	85	221	220	249	223	72	**294**	100	19	16	11	10	9	9	7		19	22	1	2	–	–	–	–	–	–	–	–	–	10H-Dibenz[b,f]azepin-10-one, 5-[3-(dimethylamino)propyl]-5,11-dihydro-	796-29-2	NI30685
294	293	265	224	237	166	295	167	**294**	100	67	34	34	31	22	21	17		19	22	1	2	–	–	–	–	–	–	–	–	–	Eburnamenin-14(15H)-one	474-00-0	NI30687
45	294	69	59	293	44	43	41	**294**	100	37	30	30	28	22	22	22		19	22	1	2	–	–	–	–	–	–	–	–	–	Eburnamenin-14(15H)-one	474-00-0	NI30686
294	293	265	237	224	295	42	166	**294**	100	71	35	33	27	22	18	16		19	22	1	2	–	–	–	–	–	–	–	–	–	Eburnamenin-14(15H)-one, (+)-		MS06764
294	293	265	237	224	42	295	166	**294**	100	70	35	34	29	22	21	20		19	22	1	2	–	–	–	–	–	–	–	–	–	Eburnamenin-14(15H)-one, (–)-		MS06765
294	293	265	237	224	295	166	42	**294**	100	72	35	33	28	21	19	18		19	22	1	2	–	–	–	–	–	–	–	–	–	Eburnamenin-14(15H)-one, (±)-		MS06766
48	91	49	294	145	117	65	77	**294**	100	5	5	3	2	2	2	1		19	22	1	2	–	–	–	–	–	–	–	–	–	1H-Indole-3-ethanamine, N,N-dimethyl-5-(phenylmethoxy)-	25390-67-4	NI30688
143	144	136	130	164	294	151	152	**294**	100	18	15	11	7	6	5	3		19	22	1	2	–	–	–	–	–	–	–	–	–	1H-Quinolino[8a,1-a]-β-carboline-7-one, 2,3,4,4a,5,6,7,9,10,15-decahydro		MS06767
294	293	169	168	263	295	41	156	**294**	100	86	81	58	36	22	19	12		19	22	1	2	–	–	–	–	–	–	–	–	–	Sarpagan-17-ol	604-99-9	NI30689
294	293	169	168	263	295	170	182	**294**	100	88	76	56	35	22	15	14		19	22	1	2	–	–	–	–	–	–	–	–	–	Sarpagan-17-ol	604-99-9	NI30690
294	293	43	295	61	44	169	57	**294**	100	97	26	23	21	19	16	13		19	22	1	2	–	–	–	–	–	–	–	–	–	Yohimban-17-one	523-14-8	NI30691
67	82	41	74	81	55	87	80	**294**	100	46	24	19	18	18	17	17	**4.00**	19	34	2	–	–	–	–	–	–	–	–	–	–	2-Cyclopentene-1-tridecanoic acid, methyl ester, (S)-	24828-59-9	WI01379
81	41	95	55	67	79	43	93	**294**	100	60	57	53	43	34	34	30	**18.00**	19	34	2	–	–	–	–	–	–	–	–	–	–	1-Cyclopropene-1-heptanoic acid, 2-octyl-, methyl ester	5026-66-4	NI30692
261	262	276	55	41	191	109	95	**294**	100	24	16	8	8	6	6	6	**0.00**	19	34	2	–	–	–	–	–	–	–	–	–	–	Norketodihydromanool		NI30693
41	43	55	79	111	67	29	81	**294**	100	99	76	75	57	56	55	54	**31.04**	19	34	2	–	–	–	–	–	–	–	–	–	–	2,5-Octadecadienoic acid, methyl ester	56846-97-0	NI30694
80	79	81	67	41	93	55	43	**294**	100	83	64	53	49	44	40	39	**0.00**	19	34	2	–	–	–	–	–	–	–	–	–	–	3,6-Octadecadienoic acid, methyl ester	57156-92-0	NI30695
80	79	220	67	41	93	81	43	**294**	100	70	57	56	55	54	52	49	**28.34**	19	34	2	–	–	–	–	–	–	–	–	–	–	4,7-Octadecadienoic acid, methyl ester	56630-29-6	NI30696
41	43	29	67	79	55	27	81	**294**	100	92	92	80	69	62	58	49	**9.05**	19	34	2	–	–	–	–	–	–	–	–	–	–	5,8-Octadecadienoic acid, methyl ester	56630-74-1	NI30698
67	80	41	81	79	43	55	94	**294**	100	88	87	80	78	72	71	66	**41.04**	19	34	2	–	–	–	–	–	–	–	–	–	–	5,8-Octadecadienoic acid, methyl ester	56630-74-1	NI30697
67	81	41	55	95	79	82	43	**294**	100	82	74	64	56	55	48	46	**33.04**	19	34	2	–	–	–	–	–	–	–	–	–	–	6,9-Octadecadienoic acid, methyl ester	56599-55-4	NI30699
67	81	41	55	95	79	82	80	**294**	100	79	75	63	54	53	46	44	**22.94**	19	34	2	–	–	–	–	–	–	–	–	–	–	6,9-Octadecadienoic acid, methyl ester	56599-55-4	NI30700
67	81	41	55	95	43	82	79	**294**	100	90	72	70	67	63	54	48	**31.04**	19	34	2	–	–	–	–	–	–	–	–	–	–	7,10-Octadecadienoic acid, methyl ester	56554-24-6	NI30701
67	81	55	95	41	82	68	96	**294**	100	85	69	66	64	57	51	47	**27.00**	19	34	2	–	–	–	–	–	–	–	–	–	–	8,11-Octadecadienoic acid, methyl ester	56599-58-7	NI30702
67	81	41	55	95	68	82	69	**294**	100	79	65	63	60	46	45	42	**22.04**	19	34	2	–	–	–	–	–	–	–	–	–	–	8,11-Octadecadienoic acid, methyl ester	56599-58-7	NI30703
67	41	81	55	43	95	79	82	**294**	100	67	61	50	45	43	35	34	**22.04**	19	34	2	–	–	–	–	–	–	–	–	–	–	9,11-Octadecadienoic acid, methyl ester, (E,E)-	13038-47-6	NI30704
67	81	55	41	95	82	54	79	**294**	100	86	66	58	57	49	48	38	**27.52**	19	34	2	–	–	–	–	–	–	–	–	–	–	9,12-Octadecadienoic acid (Z,Z)-, methyl ester	112-63-0	NI30705
67	81	294	95	82	96	55	68	**294**	100	98	85	73	62	57	56	52		19	34	2	–	–	–	–	–	–	–	–	–	–	9,12-Octadecadienoic acid (Z,Z)-, methyl ester	112-63-0	NI30706
67	81	55	41	95	82	68	54	**294**	100	82	64	61	53	49	48	44	**21.43**	19	34	2	–	–	–	–	–	–	–	–	–	–	9,12-Octadecadienoic acid (Z,Z)-, methyl ester	112-63-0	NI30707
41	67	55	81	29	79	43	54	**294**	100	82	67	49	45	35	29	27	**7.00**	19	34	2	–	–	–	–	–	–	–	–	–	–	9,12-Octadecadienoic acid, methyl ester, (E,E)-	2566-97-4	NI30708
67	41	55	81	95	82	68	54	**294**	100	91	72	67	45	45	40	38	**2.00**	19	34	2	–	–	–	–	–	–	–	–	–	–	9,12-Octadecadienoic acid, methyl ester, (E,E)-	2566-97-4	NI30709
41	55	67	81	69	82	39	29	**294**	100	65	47	30	26	24	24	23	**4.54**	19	34	2	–	–	–	–	–	–	–	–	–	–	9,15-Octadecadienoic acid, methyl ester	56630-73-0	NI30710
41	55	67	82	81	69			**294**	100	83	77	62	59	38			**15.19**	19	34	2	–	–	–	–	–	–	–	–	–	–	9,15-Octadecadienoic acid, methyl ester, (Z,Z)-	17309-05-6	NI30711
67	55	41	81	82	95	54	96	**294**	100	98	93	83	56	55	53	52	**24.50**	19	34	2	–	–	–	–	–	–	–	–	–	–	10,13-Octadecadienoic acid, methyl ester	56554-62-2	LI06417
67	81	55	41	95	82	54	68	**294**	100	97	80	68	59	55	51	49	**27.44**	19	34	2	–	–	–	–	–	–	–	–	–	–	10,13-Octadecadienoic acid, methyl ester	56554-62-2	NI30713
67	55	41	81	82	95	54	68	**294**	100	97	92	85	53	51	49	48	**23.04**	19	34	2	–	–	–	–	–	–	–	–	–	–	10,13-Octadecadienoic acid, methyl ester	56554-62-2	NI30712
67	81	55	41	82	95	96	54	**294**	100	85	67	64	63	52	50	44	**26.04**	19	34	2	–	–	–	–	–	–	–	–	–	–	11,14-Octadecadienoic acid, methyl ester	56554-61-1	NI30714
67	41	81	55	28	82	95	68	**294**	100	72	70	70	68	65	56	55	**7.04**	19	34	2	–	–	–	–	–	–	–	–	–	–	12,15-Octadecadienoic acid, methyl ester	57156-97-5	NI30715
55	41	74	68	67	81	69	87	**294**	100	91	74	57	49	47	47	42	**5.04**	19	34	2	–	–	–	–	–	–	–	–	–	–	13,16-Octadecadienoic acid, methyl ester	56846-99-2	NI30716

MASS TO CHARGE RATIOS								M.W.	INTENSITIES								Parent	C	H	O	N	S	F	Cl	Br	I	Si	P	B	X	COMPOUND NAME	CAS Reg No	No
68	55	81	82	41	67	95	69	**294**	100	84	82	68	67	65	50	50	**17.04**	19	34	2	–	–	–	–	–	–	–	–	–	–	14,17-Octadecadienoic acid, methyl ester	56554-60-0	NI30717
80	94	154	81	41	67	79	43	**294**	100	67	57	54	48	46	45	45	**1.84**	19	34	2	–	–	–	–	–	–	–	–	–	–	6-Octadecynoic acid, methyl ester	2777-64-2	NI30718
41	94	55	67	43	81	29	79	**294**	100	99	68	67	64	52	49	45	**3.84**	19	34	2	–	–	–	–	–	–	–	–	–	–	7-Octadecynoic acid, methyl ester	18545-06-7	NI30719
41	67	55	81	43	68	29	95	**294**	100	94	86	83	66	58	58	54	**1.94**	19	34	2	–	–	–	–	–	–	–	–	–	–	8-Octadecynoic acid, methyl ester	18545-05-6	NI30721
28	41	81	67	150	55	95	43	**294**	100	99	96	95	94	85	73	70	**8.74**	19	34	2	–	–	–	–	–	–	–	–	–	–	8-Octadecynoic acid, methyl ester	18545-05-6	NI30720
41	55	43	67	81	29	68	54	**294**	100	75	73	70	67	48	45	37	**1.24**	19	34	2	–	–	–	–	–	–	–	–	–	–	9-Octadecynoic acid, methyl ester	1120-32-7	NI30723
81	67	55	68	41	95	82	54	**294**	100	95	86	74	71	70	67	60	**0.54**	19	34	2	–	–	–	–	–	–	–	–	–	–	9-Octadecynoic acid, methyl ester	1120-32-7	NI30722
81	67	55	95	41	68	54	82	**294**	100	96	87	74	72	67	62	58	**0.44**	19	34	2	–	–	–	–	–	–	–	–	–	–	10-Octadecynoic acid, methyl ester	26543-36-2	NI30724
41	67	55	81	124	95	82	54	**294**	100	92	91	85	63	63	56	56	**1.64**	19	34	2	–	–	–	–	–	–	–	–	–	–	11-Octadecynoic acid, methyl ester	26543-37-3	NI30725
165	194	196	166	207	167	238	27	**294**	100	88	78	73	46	45	39	22	**3.81**	20	22	2	–	–	–	–	–	–	–	–	–	–	Cyclobutanone, 2-(2-hydroxy-sec-butyl)-3,3-diphenyl-	62338-10-7	NI30726
294	121	160	147	295	159	91	89	**294**	100	44	42	34	23	18	18	18		20	22	2	–	–	–	–	–	–	–	–	–	–	Cyclohexane, 2,3-bis(4-methoxyphenyl)-		MS06768
294	121	160	147	295	186	159	91	**294**	100	44	42	34	23	18	18	18		20	22	2	–	–	–	–	–	–	–	–	–	–	Cyclohexane, 2,3-bis(4-methoxyphenyl)-		LI06418
294	121	295	235	147	173	186	159	**294**	100	27	23	13	13	9	8	7		20	22	2	–	–	–	–	–	–	–	–	–	–	Cyclohexene, 1,2-bis(4-methoxyphenyl)-		MS06769
294	237	265	223	238	295	165	209	**294**	100	72	36	24	23	22	22	13		20	22	2	–	–	–	–	–	–	–	–	–	–	Gona-1,3,5,7,9-pentaen-17-one, 13-ethyl-3-methoxy-	29017-48-9	NI30727
294	237	265	295	238	223	165	225	**294**	100	61	30	24	17	17	15	9		20	22	2	–	–	–	–	–	–	–	–	–	–	Gona-1,3,5,7,9-pentaen-17-one, 13-ethyl-3-methoxy-	29017-48-9	NI30728
294	238	265	223	165	237	209	295	**294**	100	38	35	28	28	23	22	21		20	22	2	–	–	–	–	–	–	–	–	–	–	Gona-1,3,5,7,9-pentaen-17-one, 13-ethyl-3-methoxy-	29017-48-9	NI30729
294	265	238	223	295	237	209	165	**294**	100	36	27	23	22	18	17	16		20	22	2	–	–	–	–	–	–	–	–	–	–	Gona-1,3,5,7,9-pentaen-17-one, 13-ethyl-3-methoxy-, (13α)-	69853-74-3	NI30730
294	265	295	238	223	237	165	209	**294**	100	23	22	20	15	12	11	10		20	22	2	–	–	–	–	–	–	–	–	–	–	Gona-1,3,5,7,9-pentaen-17-one, 13-ethyl-3-methoxy-, (14β)-	51262-25-0	NI30731
131	130	115	41	91	43	29	141	**294**	100	28	28	28	27	24	22	21	**8.45**	20	22	2	–	–	–	–	–	–	–	–	–	–	Hydrocinnamic acid, O-[(1,2,3,4-tetrahydro-2-naphthyl)methyl]-	23804-21-9	NI30732
131	130	115	41	91	294			**294**	100	28	28	28	27	1				20	22	2	–	–	–	–	–	–	–	–	–	–	Hydrocinnamic acid, O-[(1,2,3,4-tetrahydro-2-naphthyl)methyl]-		LI06419
252	122	294	134	117	91	253	133	**294**	100	45	34	34	33	33	20	19		20	22	2	–	–	–	–	–	–	–	–	–	–	[3.3]Paracyclophane, 5-acetoxy-		LI06420
252	122	294	134	117	91	295	253	**294**	100	45	34	34	33	33	23	20		20	22	2	–	–	–	–	–	–	–	–	–	–	[3.3]Paracyclophane, 6-acetoxy-	24770-05-6	NI30733
294	117	105	131	91	118	279	115	**294**	100	74	53	51	50	48	38	31		20	22	2	–	–	–	–	–	–	–	–	–	–	[3.3]Paracyclophane-5-carboxylic acid, methyl ester		LI06421
294	117	105	131	91	118	279	119	**294**	100	74	53	51	50	48	38	32		20	22	2	–	–	–	–	–	–	–	–	–	–	[3.3]Paracyclophane-6-carboxylic acid, methyl ester	24777-37-5	NI30734
294	279	295	280	264	265	132	147	**294**	100	47	22	14	14	13	12	7		20	22	2	–	–	–	–	–	–	–	–	–	–	Phenanthrene, 9,10-diethyl-3,6-dimethoxy-		MS06770
135	158	121	107	122	144	294	265	**294**	100	58	46	44	39	31	30	10		20	26	–	2	–	–	–	–	–	–	–	–	–	Aspidospermidine, Nα-methyl-6,7-dehydro-		LI06422
58	249	193	208	234	194	99	84	**294**	100	31	17	16	13	11	11	10	**6.55**	20	26	–	2	–	–	–	–	–	–	–	–	–	5H-Dibenz[b,f]azepine-5-propanamine, 10,11-dihydro-N,N,β-trimethyl-	739-71-9	NI30736
58	249	208	234	99	193	194	248	**294**	100	41	23	17	17	15	12	11	**7.77**	20	26	–	2	–	–	–	–	–	–	–	–	–	5H-Dibenz[b,f]azepine-5-propanamine, 10,11-dihydro-N,N,β-trimethyl-	739-71-9	NI30737
58	249	208	99	234	248	193	294	**294**	100	59	29	19	17	16	16	13		20	26	–	2	–	–	–	–	–	–	–	–	–	5H-Dibenz[b,f]azepine-5-propanamine, 10,11-dihydro-N,N,β-trimethyl-	739-71-9	NI30735
136	252	251	294	253	164	130	144	**294**	100	38	28	17	12	11	10	7		20	26	–	2	–	–	–	–	–	–	–	–	–	1H-Indolizino[8,1-cd]carbazole, 3a-(2-propenyl)-2,3,3a,4,5,5a,6,11,12,13a-decahydro-		MS06771
209	211	84	126	210	128	127	115	**294**	100	70	60	54	29	23	22	12	**1.00**	20	26	–	2	–	–	–	–	–	–	–	–	–	Tricyclo[5.2.1.0[4,10]]deca-1(10),2,5,8-tetraene, 4,7-dipiperidino-		NI30738
130	164	144	162	93	44	131	135	**294**	100	91	91	78	65	52	43	39	**35.10**	20	26	–	2	–	–	–	–	–	–	–	–	–	Tryptamine, N-(+)-thujylidene-		NI30739
294	293	147	146	292	264	277	118	**294**	100	42	17	15	14	10	9	7		21	14	–	2	–	–	–	–	–	–	–	–	–	Indeno[2,1-a]fluorene-6-carbonitrile, 5-amino-		MS06772
294	295	190	163	164	293	191	89	**294**	100	22	21	10	9	7	4	2		21	14	–	2	–	–	–	–	–	–	–	–	–	1H-Phenanthro[9,10-d]imidazole, 2-phenyl-	6931-31-3	NI30740
294	190	163	164	293	191	292	189	**294**	100	21	10	9	7	4	2	2		21	14	–	2	–	–	–	–	–	–	–	–	–	1H-Phenanthro[9,10-d]imidazole, 9-phenyl-		LI06423
91	71	92	105	43	119	57	117	**294**	100	50	26	21	17	14	14	12	**7.00**	22	30	–	–	–	–	–	–	–	–	–	–	–	Hexane, 2,2,5-trimethyl-5-benzyl-6-phenyl-		LI06424
178	116	294						**294**	100	13	1							23	18	–	–	–	–	–	–	–	–	–	–	–	9,10-(1,2-Indano)-9,10-dihydroanthracene		LI06425
82	63	44	69	31	50	127	133	**295**	100	24	18	9	8	4	2	1	**0.00**	3	–	–	1	3	6	1	–	–	–	–	–	–	N-(Trifluoromethylthio)(trifluoromethyldithio)chloromethimine		NI30741
61	79	69	226	76	32	149	117	**295**	100	99	84	62	58	58	48	48	**0.00**	3	3	1	1	2	3	3	–	–	–	–	–	–	Methanesulphinamide, 1,1,1-trifluoro-N-methyl-N-[(trichloromethyl)thio]	51735-85-4	NI30742
91	64	205	203	295	293	279	277	**295**	100	94	13	5	2	1	1	0		6	4	–	1	–	–	–	–	–	–	–	–	1	Thallium, eta[5]-(cyanocyclopentadienyl)-	63544-93-4	NI30743
296	279	295	281					**295**	100	5	4	2						7	6	4	3	2	–	1	–	–	–	–	–	–	2H-1,2,4-Benzothiadiazine-7-sulphonamide, 6-chloro-, 1,1-dioxide	58-94-6	NI30744
295	268	297	97	57	270	62	64	**295**	100	70	44	43	35	30	27	25		7	6	4	3	2	–	1	–	–	–	–	–	–	2H-1,2,4-Benzothiadiazine-7-sulphonamide, 6-chloro-, 1,1-dioxide	58-94-6	NI30745
295	268	297	97	57	270	62	124	**295**	100	70	44	43	35	30	27	25		7	6	4	3	2	–	1	–	–	–	–	–	–	2H-1,2,4-Benzothiadiazine-7-sulphonamide, 6-chloro-, 1,1-dioxide	58-94-6	MS06773
104	117	260	119	76	50	130	262	**295**	100	50	48	48	45	37	28	25	**12.50**	9	4	2	1	1	–	3	–	–	–	–	–	–	1H-Isoindole-1,3(2H)-dione, 2-[(trichloromethyl)thio]-	133-07-3	NI30747
76	104	50	130	117	119	260	79	**295**	100	92	65	55	43	41	36	27	**7.16**	9	4	2	1	1	–	3	–	–	–	–	–	–	1H-Isoindole-1,3(2H)-dione, 2-[(trichloromethyl)thio]-	133-07-3	NI30746
148	147	176	260	262	188	149	175	**295**	100	35	26	23	17	15	11	8	**4.25**	9	4	2	1	1	–	3	–	–	–	–	–	–	1H-Isoindole-1,3(2H)-dione, 2-[(trichloromethyl)thio]-	133-07-3	NI30748
70	73	295	280	135	121	42	41	**295**	100	25	9	9	9	7	7	7		10	26	3	1	–	–	–	–	–	2	1	–	–	Phosphonic acid, [[isopropylideneamino]methyl]-, bis(trimethylsilyl) ester	55108-63-9	NI30749
77	141	231	184	78	108	125	295	**295**	100	64	18	14	10	9	8	7		12	9	4	1	2	–	–	–	–	–	–	–	–	Benzenesulphonothioic acid, 4-nitro-, S-phenyl ester	3660-41-1	NI30750
77	141	231	184	108	78	295	125	**295**	100	65	20	15	12	12	8	8		12	9	4	1	2	–	–	–	–	–	–	–	–	Benzenethiosulphonic acid, S-(4-nitrophenyl) ester		LI06426
295	167	296	168	83.5	84	169	77	**295**	100	26	14	9	8	7	6	3		12	10	–	1	–	–	–	–	1	–	–	–	–	Iododiphenylamine		IC03731
43	71	280	282	45	297	295	72	**295**	100	54	33	32	31	28	28	18		12	10	1	1	1	–	–	1	–	–	–	–	–	Thiazole, 5-acetyl-2-(4-bromophenyl)-4-methyl-	71338-62-0	NI30751
105	77	190	119	51	106	42	69	**295**	100	57	15	14	14	11	10	5	**0.00**	12	10	2	1	–	5	–	–	–	–	–	–	–	Ethylamine, 2-oxo-2-phenyl-N-methyl-N-(pentafluoropropionyl)-		MS06774

MASS TO CHARGE RATIOS								M.W.	INTENSITIES								Parent	C	H	O	N	S	F	Cl	Br	I	Si	P	B	X	COMPOUND NAME	CAS Reg No	No
223	221	114	29	297	295	88	27	**295**	100	99	80	75	70	70	50	45		12	10	3	1	–	–	–	1	–	–	–	–	–	4(1H)-Quinolone, 6-bromo-2-ethoxycarbonyl-		MS06775
134	192	163	164	148	206	295	44	**295**	100	56	53	49	28	27	20	11		12	17	4	5	–	–	–	–	–	–	–	–	–	Adenosine, N,N-dimethyl-	2620-62-4	NI30752
134	192	163	164	148	206	295	120	**295**	100	57	55	49	28	27	22	13		12	17	4	5	–	–	–	–	–	–	–	–	–	Adenosine, N,N-dimethyl-	2620-62-4	LI06427
44	134	36	28	163	148	45	42	**295**	100	67	67	50	45	39	39	26	**6.30**	12	17	4	5	–	–	–	–	–	–	–	–	–	Adenosine, N,N-dimethyl-	2620-62-4	NI30753
58	43	210	195	295	280	221	179	**295**	100	60	46	33	20	20	20	20		12	21	2	7	–	–	–	–	–	–	–	–	–	1,3,5-Triazine-2,4-diamine, 6-(3-acetylureido)-N,N′-diisopropyl-		NI30754
91	104	176	132	92	65	41	119	**295**	100	21	15	14	10	8	8	7	**5.60**	13	14	1	1	–	5	–	–	–	–	–	–	–	Butylamine, 4-phenyl-N-(pentafluoropropionyl)-		MS06776
204	160	42	118	91	154	119	205	**295**	100	30	26	21	13	11	10	7	**0.00**	13	14	1	1	–	5	–	–	–	–	–	–	–	Methamphetamine, N-(pentafluoropropionyl)-		MS06777
166	43	167	41	69	85	55	42	**295**	100	36	22	18	14	9	7	6	**0.23**	13	20	3	1	–	3	–	–	–	–	–	–	–	L-Proline, 1-(trifluoroacetyl)-, 1-methylpentyl ester	57983-56-9	NI30755
296	278	298	280	297	295	89	61	**295**	100	67	58	44	31	27	25	17		14	11	2	1	–	–	2	–	–	–	–	–	–	Benzoic acid, 2-[(2,6-dichloro-3-methylphenyl)amino]-	644-62-2	NI30756
295	77	264	220	43	41	57	55	**295**	100	53	50	50	38	37	29	29		14	17	4	1	1	–	–	–	–	–	–	–	–	2-Oxa-6-azatricyclo[3.3.1.1^{3,7}]decan-4-ol, 6-(phenylsulphonyl)-, (1α,3β,4β,5α,7β)-	50267-32-8	NI30757
117	116	69	73	44	57	60	43	**295**	100	56	23	21	19	15	13	12	**0.00**	14	17	6	1	–	–	–	–	–	–	–	–	–	Benzeneacetonitrile, α-(β-D-glucopyranosyloxy)-, (R)-	99-18-3	NI30758
43	129	128	91	117	115	130	189	**295**	100	96	47	36	23	14	13	11	**0.00**	14	17	6	1	–	–	–	–	–	–	–	–	–	1,3-Propanediol, 2-benzyl-2-nitro-, diacetate		DD01049
43	151	193	123	106	253	152	235	**295**	100	56	30	26	18	16	11	8	**0.63**	14	17	6	1	–	–	–	–	–	–	–	–	–	Vitamin B6, triacetate	10030-93-0	NI30759
113	129	251	296					**295**	100	23	17	10					**0.00**	14	18	–	3	1	–	1	–	–	–	–	–	–	1,2-Ethanediamine, N-[(5-chloro-2-thienyl)methyl]-N′,N′-dimethyl-N-2-pyridinyl-	148-65-2	NI30761
58	131	72	71	79	42	28	30	**295**	100	18	16	15	12	11	11	8	**1.30**	14	18	–	3	1	–	1	–	–	–	–	–	–	1,2-Ethanediamine, N-[(5-chloro-2-thienyl)methyl]-N′,N′-dimethyl-N-2-pyridinyl-	148-65-2	NI30760
55	82	41	83	68	56	28	67	**295**	100	93	62	56	51	48	23	19	**0.00**	14	18	1	1	–	–	–	1	–	–	–	–	–	2H-1,3-Benzoxazine, 6-bromo-3-cyclohexyl-3,4-dihydro-	6638-88-6	NI30762
112	168	57	128	70	41	43	130	**295**	100	57	57	51	33	22	19	18	**0.00**	14	18	2	3	–	–	1	–	–	–	–	–	–	1H-1,2,4-Triazole-1-ethanol, β-(4-chlorophenoxy)-α-tert-butyl-		NI30763
295	111	82	186	158	249	138	266	**295**	100	99	97	72	51	46	46	43		14	18	4	1	–	–	–	–	–	–	1	–	–	2-Propenenitrile, 3-(3-methoxyphenyl)-2-(diethoxyphosphinyl)-		NI30764
266	224	267	220	250	132	57	178	**295**	100	19	16	15	12	12	12	10	**8.50**	14	21	4	3	–	–	–	–	–	–	–	–	–	Aniline, N-sec-butyl-4-tert-butyl-2,6-dinitro-	33629-47-9	NI30765
266	267	224	220	295	250	277	43	**295**	100	14	13	10	9	9	8	8		14	21	4	3	–	–	–	–	–	–	–	–	–	Aniline, N-sec-butyl-4-tert-butyl-2,6-dinitro-	33629-47-9	NI30766
73	45	206	75	43	116	280	74	**295**	100	33	32	21	20	19	11	10	**7.20**	14	25	2	1	–	–	–	–	–	2	–	–	–	Acetamide, N-(trimethylsilyl)-N-[4-[(trimethylsilyl)oxy]phenyl]-	55530-61-5	NI30767
206	73	280	295	116	75	207	45	**295**	100	96	75	55	20	20	18	17		14	25	2	1	–	–	–	–	–	2	–	–	–	Acetamide, N-(trimethylsilyl)-N-[4-[(trimethylsilyl)oxy]phenyl]-	55530-61-5	MS06778
73	206	280	295	75	116	281	296	**295**	100	53	45	43	16	15	13	12		14	25	2	1	–	–	–	–	–	2	–	–	–	Acetamide, N-(trimethylsilyl)-N-[4-[(trimethylsilyl)oxy]phenyl]-	55530-61-5	NI30768
178	73	179	45	147	75	74	106	**295**	100	98	17	16	11	9	9	7	**5.26**	14	25	2	1	–	–	–	–	–	2	–	–	–	Glycine, N-phenyl-N-(trimethylsilyl)-, trimethylsilyl ester	55887-86-0	NI30769
73	178	45	43	179	75	74	59	**295**	100	80	32	16	13	11	11	11	**0.00**	14	25	2	1	–	–	–	–	–	2	–	–	–	Glycine, N-phenyl-N-(trimethylsilyl)-, trimethylsilyl ester	55887-86-0	MS06779
280	266	223	252	253	77	222	151	**295**	100	89	60	56	53	40	26	25	**0.00**	15	22	2	1	–	1	–	–	–	1	–	–	–	Benzamide, N,N-diethyl-3-fluoro-6-formyl-2-(trimethylsilyl)-	121425-03-4	DD01050
239	141	197	77	211	41	156	140	**295**	100	83	62	62	59	52	44	32	**15.00**	15	22	3	1	–	–	–	–	–	–	1	–	–	Butanamide, 4-(P-ethoxy-P-phenylphosphinoyl)-N-(prop-2′-enyl)-	93939-04-9	NI30770
116	44	73	117	45	75	180	42	**295**	100	68	53	10	10	7	6	4	**0.00**	15	29	1	1	–	–	–	–	–	2	–	–	–	Ethylamine, 2-[2-(trimethylsilyloxy)phenyl]-N-(trimethylsilyl)-N-methyl-		MS06780
119	120	46	59	148	61	295	41	**295**	100	19	17	17	17	12	11	8		16	13	3	3	–	–	–	–	–	–	–	–	–	2,4-Imidazolidinedione, 1-(phenylcarbamoyl)-3-phenyl-		NI30771
211	295	253	210	155	184	183	182	**295**	100	60	30	20	12	9	7	6		16	13	3	3	–	–	–	–	–	–	–	–	–	Phenazine, 1-acetamido-7-acetoxy-	18450-10-7	LI06428
211	295	253	210	212	155	184	296	**295**	100	60	30	20	12	12	9	8		16	13	3	3	–	–	–	–	–	–	–	–	–	Phenazine, 1-acetamido-7-acetoxy-	18450-10-7	NI30772
94	95	122	55	42	83	138	57	**295**	100	95	39	26	23	20	19	16	**6.00**	16	25	4	1	–	–	–	–	–	–	–	–	–	6β-Propanoyloxy-3α-tigloyloxytropane		MS06781
252	206	251	178	196	253	224	150	**295**	100	49	28	20	19	15	14	13	**0.00**	16	25	4	1	–	–	–	–	–	–	–	–	–	3,5-Pyridinedicarboxylic acid, 1,4-dihydro-2,6-dimethyl-4-isopropyl-, diethyl ester	1539-32-8	NI30773
252	206	251	178	196	253	224	205	**295**	100	50	29	20	19	15	14	14	**0.00**	16	25	4	1	–	–	–	–	–	–	–	–	–	3,5-Pyridinedicarboxylic acid, 1,4-dihydro-2,6-dimethyl-4-isopropyl-, diethyl ester	1539-32-8	LI06429
115	235	230	234	101	91	203	236	**295**	100	76	45	29	29	28	26	24	**6.87**	17	13	2	1	1	–	–	–	–	–	–	–	–	6H-6,12-Methanodibenzo[b,f]thiocin-12-carbonitrile, 11,12-dihydro-, 5,5-dioxide		MS06782
180	77	295	181	51				**295**	100	46	16	16	16					17	13	2	1	1	–	–	–	–	–	–	–	–	1,3-Thiazol-4-one, 5-acetyl-2,3-diphenyl-		LI06430
105	295	77	265	106	296	51	174	**295**	100	69	43	20	14	13	13	4		17	13	4	1	–	–	–	–	–	–	–	–	–	D,L-Glyceronontrile, 2,3-di-O-benzoyl-	84348-15-2	NI30774
125	248	96	230	254	244	153	184	**295**	100	97	94	91	88	88	88	86	**0.00**	17	17	2	3	–	–	–	–	–	–	–	–	–	α-(Aminomethylene)-α-(phenylacetamido)acetanilide		MS06783
85	57	42	77	252	183	44	91	**295**	100	78	56	54	18	17	16	13	**9.00**	17	17	2	3	–	–	–	–	–	–	–	–	–	3,5-Pyrazolidinedione, 4-(dimethylamino)-1,2-diphenyl-	57488-07-0	NI30775
113	105	77	115	114	182	165	183	**295**	100	17	11	10	7	3	3	3	**0.00**	18	17	1	1	1	–	–	–	–	–	–	–	–	6-Oxa-4-thia-2-azabicyclo[3.2.0]hept-2-ene, 1,3-dimethyl-7,7-diphenyl-		NI30776
131	91	103	253	77	132	43	51	**295**	100	95	40	34	34	20	20	11	**6.00**	18	17	3	1	–	–	–	–	–	–	–	–	–	N-(α-Acetoxycinnamylidene)-O-benzylhydroxylamine		LI06431
239	210	267	236	167	295	266	240	**295**	100	84	59	43	30	24	19	15		18	17	3	1	–	–	–	–	–	–	–	–	–	1-Benzazocine-3,6-dione, 8-methoxy-2-phenyl-1,2,4,5-tetrahydro-	90732-25-5	NI30777
59	152	153	163	151	142	165	141	**295**	100	77	67	56	27	19	15	10	**0.00**	18	17	3	1	–	–	–	–	–	–	–	–	–	Ethaneperoxoic acid, 1-cyano-2-phenyl-1-benzylethyl ester	58422-68-7	NI30778
295	294	278	280	264	279	238	237	**295**	100	42	24	19	11	10	8	7		18	17	3	1	–	–	–	–	–	–	–	–	–	Isoquinoline, 3-methoxy-1-(2,4-dimethoxyphenyl)-		LI06432
176	252	105	295	43	77	236	253	**295**	100	67	50	45	45	39	30	29		18	17	3	1	–	–	–	–	–	–	–	–	–	2-Isoxazoline, 5-acetyl-3-(4-methoxyphenyl)-4-phenyl-		NI30779
91	252	224	295	223	43	92	90	**295**	100	39	28	25	14	14	10	10		18	17	3	1	–	–	–	–	–	–	–	–	–	2-Isoxazoline, 5-acetyl-3-(4-methoxyphenyl)-4-phenyl-		NI30780
295	238	211	280	296	277	32	183	**295**	100	46	28	21	20	14	14	12		18	17	3	1	–	–	–	–	–	–	–	–	–	17-Aza-3-methoxyestra-1,3,5(10),6,8-pentene-11,16-dione		NI30781
295	32	264	294	276	63	252	277	**295**	100	90	70	64	64	31	28	22		18	17	3	1	–	–	–	–	–	–	–	–	–	4(1H)-Quinolinone, 3,6-dimethoxy-1-methyl-2-phenyl-	30426-61-0	NI30782
295	179	147.5	116	265	250	190	151	**295**	100	4	4	4	3	3	3	3		18	17	3	1	–	–	–	–	–	–	–	–	–	Stilbene, 2,4,6-trimethoxy-4′-cyano-		LI06433
57	190	295	277	252	91	165	105	**295**	100	46	29	22	19	18	15	14		18	17	3	1	–	–	–	–	–	–	–	–	–	(±)-Ushinsunine		DD01051

MASS TO CHARGE RATIOS								M.W.	INTENSITIES								Parent	C	H	O	N	S	F	Cl	Br	I	Si	P	B	X	COMPOUND NAME	CAS Reg No	No
58	224	71	209	225	72	195	59	**295**	100	34	9	7	6	6	4	4	**0.37**	18	21	1	3	–	–	–	–	–	–	–	–	–	11H-Dibenzo[b,e][1,4]diazepin-11-one, 10-[2-(dimethylamino)ethyl]-5,10-dihydro-5-methyl-	4498-32-2	NI30783
58	295	118	249	85	296	221	57	**295**	100	98	49	40	25	17	15	13		18	21	1	3	–	–	–	–	–	–	–	–	–	11H-Dibenzo[b,e][1,4]diazepin-11-one, 5-[3-(dimethylamino)propyl]-5,10-dihydro-	13450-72-1	NI30784
97	226	225	295	124	169	171		**295**	100	57	54	50	50	37	24			18	21	1	3	–	–	–	–	–	–	–	–	–	1H-Indolo[3,2-g]indolizin-3-one, 2,3,5,6,11,11b-hexahydro-		LI06434
186	201	145	161	217	295	57	78	**295**	100	56	36	35	23	18	15	14		18	21	1	3	–	–	–	–	–	–	–	–	–	Isoxazolo[2,3-a]indole-2,3-dicarbonitrile, 7-tert-butyl-2,3,3a,4-tetrahydro-4,4-dimethyl-, rel-(2S,3S,3aR)-		DD01052
295	221	181	196	223	180	222	207	**295**	100	51	39	36	32	28	21	21		18	21	1	3	–	–	–	–	–	–	–	–	–	Lysergic acid ethylamide		MS06784
280	295	254	253	172	129	173	188	**295**	100	56	23	10	9	6	5	4		18	21	1	3	–	–	–	–	–	–	–	–	–	Quinoline, 2-(4-ethyl-3,5-dimethylpyrazol-1-yl)-4-methyl-6-methoxy-		MS06785
58	59	42	204	202	203	221	295	**295**	100	4	1	1	1	1	1	1		19	21	–	1	1	–	–	–	–	–	–	–	–	1-Propanamine, 3-dibenzo[b,e]thiepin-11(6H)-ylidene-N,N-dimethyl-	113-53-1	NI30785
132	121	117	133	91	131	77	122	**295**	100	44	32	11	9	6	6	4	**1.00**	19	21	2	1	–	–	–	–	–	–	–	–	–	2-Azetidinone, 1-[(4-methoxyphenyl)methyl]-3,3-dimethyl-4-phenyl-	54833-66-8	NI30786
143	169	109	67	115	80	141	79	**295**	100	70	56	45	36	30	26	26	**0.00**	19	21	2	1	–	–	–	–	–	–	–	–	–	Carbamic acid, (N-endo-2-methyl-exo-2-norbornyl)-N-naphthyl-, ethyl ester		MS06786
169	109	143	67	141	140	60	114	**295**	100	75	61	57	35	28	22	20	**7.40**	19	21	2	1	–	–	–	–	–	–	–	–	–	Carbamic acid, N-(exo-2-methyl-endo-2-norbornyl)-N-naphthyl-, ethyl ester		MS06787
294	295	43	221	264	44	278	237	**295**	100	67	42	34	25	22	21	21		19	21	2	1	–	–	–	–	–	–	–	–	–	4H-Dibenzo[de,g]quinoline, 5,6,6a,7-tetrahydro-10,11-dimethoxy-6-methyl-, (R)-	38291-33-7	NI30787
294	295	292	190	278	174	276	296	**295**	100	36	10	9	6	6	5	4		19	21	2	1	–	–	–	–	–	–	–	–	–	(±)-2,3-Dimethoxyberbine		MS06788
295	207	264	296	157	208	280	209	**295**	100	49	28	23	19	18	16	13		19	21	2	1	–	–	–	–	–	–	–	–	–	Estra-1,3,5,7,9-pentaen-17-one, 3-hydroxy-, O-methyloxime	3896-64-8	NI30788
295	280	296	199	198	278	235	222	**295**	100	30	20	18	17	11	9	8		19	21	2	1	–	–	–	–	–	–	–	–	–	1'H-Estra-1,3,5(10),9(11)-tetraeno[1,10,9,11-bcd]pyrrol-17-one, 3-methoxy-	13211-82-0	NI30789
180	179	43	152	151	250	71	196	**295**	100	76	48	23	19	15	11	6	**5.00**	19	21	2	1	–	–	–	–	–	–	–	–	–	Phenanthridine, 6-ethoxy-5-isobutyryl-5,6-dihydro-		LI06435
91	118	204	120	105	278	208	104	**295**	100	78	75	75	62	59	37	33	**11.00**	19	21	2	1	–	–	–	–	–	–	–	–	–	4-Piperidone, 1-benzyl-3(ax)-hydroxy-3-methyl-6-phenyl-	56642-31-0	NI30790
91	118	204	120	105	278	208	104	**295**	100	77	55	36	34	31	19	17	**11.00**	19	21	2	1	–	–	–	–	–	–	–	–	–	4-Piperidone, 1-benzyl-3(eq)-hydroxy-3-methyl-6-phenyl-	56642-34-3	NI30791
113	55	43	41	70	126	57	71	**295**	100	29	28	28	20	19	17	14	**2.24**	19	37	1	1	–	–	–	–	–	–	–	–	–	Pyrrolidine, 1-(12-methyl-1-oxotetradecyl)-	56630-61-6	NI30792
113	43	55	41	126	70	71	98	**295**	100	40	25	25	19	18	14	10	**2.34**	19	37	1	1	–	–	–	–	–	–	–	–	–	Pyrrolidine, 1-(13-methyl-1-oxotetradecyl)-	56630-52-5	NI30793
113	43	55	41	125	70	71	98	**295**	100	44	24	24	19	18	13	10	**2.34**	19	37	1	1	–	–	–	–	–	–	–	–	–	Pyrrolidine, 1-(13-methyl-1-oxotetradecyl)-	56630-52-5	NI30794
113	43	41	55	70	126	71	98	**295**	100	39	27	25	21	19	15	10	**2.20**	19	37	1	1	–	–	–	–	–	–	–	–	–	Pyrrolidine, 1-(1-oxopentadecyl)-	56630-55-8	NI30795
99	98	86	112	43	70	41	55	**295**	100	81	30	22	18	17	17	15	**3.80**	19	37	1	1	–	–	–	–	–	–	–	–	–	2-Pyrrolidinone, 1-(3,7,11-trimethyldodecyl)-	63913-37-1	NI30796
295	296	266	267	133	294	265	264	**295**	100	27	11	9	6	5	5	5		21	13	1	1	–	–	–	–	–	–	–	–	–	Dibenz[c,h]acridin-7(14H)-one		NI30797
295	294	147.5	239	264	134	278	237	**295**	100	43	42	36	24	22	18	16		21	13	1	1	–	–	–	–	–	–	–	–	–	Indeno[2,1-a]fluorene-6-carbonitrile, 5-hydroxy-		MS06789
295	164	163	165	294	266	190	162	**295**	100	28	21	4	2	2	2	2		21	13	1	1	–	–	–	–	–	–	–	–	–	Phenanthro[9,10-d]oxazole, 2-phenyl-	4410-14-4	LI06436
295	164	296	163	165	162	63	51	**295**	100	28	22	21	4	2	2	2		21	13	1	1	–	–	–	–	–	–	–	–	–	Phenanthro[9,10-d]oxazole, 2-phenyl-	4410-14-4	NI30798
58	204	91	59	36	35	115	114	**295**	100	10	6	5	4	4	3	3	**1.00**	21	29	–	1	–	–	–	–	–	–	–	–	–	1-Hexanamine, 4,N,N-trimethyl-5,6-diphenyl-		MS06790
178	72	44	114	59	295	152	51	**295**	100	56	35	24	23	14	11	10		22	17	–	1	–	–	–	–	–	–	–	–	–	Cyclopropenone imine, N-(4-methylphenyl)-2,3-diphenyl-		MS06791
296	202	217	298	215	294	281	200	**296**	100	75	73	67	67	65	64	60		1	3	–	–	–	–	–	1	–	–	–	–	1	Mercury, bromomethyl-	506-83-2	NI30799
85	117	119	217	219	131	129	263	**296**	100	84	82	75	73	49	49	39	**0.00**	3	–	–	–	–	4	3	1	–	–	–	–	–	1,1,1-Trichloro-2,2,3,3-tetrafluoro-3-bromopropane		IC03732
296	69	169	127	277	227	177	150	**296**	100	28	11	9	4	4	4	4		3	–	–	–	–	7	–	–	1	–	–	–	–	Propane, 1,1,1,2,3,3,3-heptafluoro-2-iodo-	677-69-0	NI30800
128	41	169	127	296	254			**296**	100	76	69	53	35	8				3	6	–	–	–	–	–	–	2	–	–	–	–	Propane, 1,3-diiodo-		LI06437
265	263	267	59	264	45	299	262	**296**	100	66	61	58	55	49	43	40	**20.05**	3	7	1	–	–	–	1	–	–	–	–	–	1	Mercury, chloro(2-methoxyethyl)-	123-88-6	NI30801
46	30	44	43	42	149	75	57	**296**	100	90	75	73	42	30	25	20	**0.00**	4	8	8	8	–	–	–	–	–	–	–	–	–	1,3,5,7-Tetrazocine, octahydro-1,3,5,7-tetranitro-	2691-41-0	NI30802
46	30	44	42	148	75	128	43	**296**	100	65	27	25	12	10	9	7	**0.00**	4	8	8	8	–	–	–	–	–	–	–	–	–	1,3,5,7-Tetrazocine, octahydro-1,3,5,7-tetranitro-	2691-41-0	NI30803
46	30	44	42	148	128	103	120	**296**	100	65	28	25	12	9	8	7	**0.00**	4	8	8	8	–	–	–	–	–	–	–	–	–	1,3,5,7-Tetrazocine, octahydro-1,3,5,7-tetranitro-	2691-41-0	MS06792
209	239	267	296	238	210	268	237	**296**	100	39	38	27	17	5	4	2		6	15	–	–	–	–	–	–	–	–	–	–	1	Bismuthine, triethyl-	617-77-6	NI30805
209	238	27	29	267	296	26	28	**296**	100	47	43	39	35	18	12	9		6	15	–	–	–	–	–	–	–	–	–	–	1	Bismuthine, triethyl-	617-77-6	NI30804
267	223	208	237	209	269	252	224	**296**	100	95	78	31	29	18	15	4	**2.00**	6	16	–	–	–	–	–	–	–	–	–	–	1	Plumbane, diethyldimethyl-	1762-27-2	NI30806
267	223	208	265	221	207	266	222	**296**	100	94	89	50	50	45	44	43	**1.10**	6	16	–	–	–	–	–	–	–	–	–	–	1	Plumbane, diethyldimethyl-	1762-27-2	AI02003
263	261	265	85	84	191	107	96	**296**	100	66	62	26	25	24	24	23	**0.00**	7	2	–	–	–	–	6	–	–	–	–	–	–	Bicyclo[2.2.1]hepta-2,5-diene, 1,2,3,4,7,7-hexachloro-	3389-71-7	NI30807
263	261	265	85	191	107	96	84	**296**	100	7	6	3	2	2	2	2	**0.15**	7	2	–	–	–	–	6	–	–	–	–	–	–	Bicyclo[2.2.1]hepta-2,5-diene, 1,2,3,4,7,7-hexachloro-	3389-71-7	NI30808
165	75	73	223	167	74	180	166	**296**	100	99	99	87	70	67	67	64	**4.58**	8	24	–	–	2	–	–	–	–	4	–	–	–	1,4-Dithiacyclohexasilane, 2,2,3,3,5,5,6,6-octamethyl-	66907-27-5	NI30809
281	282	283	265	73	207	284	249	**296**	100	28	18	6	6	5	4	4	**0.15**	8	24	4	–	–	–	–	–	–	4	–	–	–	Cyclotetrasiloxane, octamethyl-	556-67-2	IC03733
281	283	73	193	282	265	207	191	**296**	100	14	10	9	7	7	7	6	**0.10**	8	24	4	–	–	–	–	–	–	4	–	–	–	Cyclotetrasiloxane, octamethyl-	556-67-2	NI30810
281	282	283	265	73	207	193	191	**296**	100	28	17	13	9	6	6	6	**0.00**	8	24	4	–	–	–	–	–	–	4	–	–	–	Cyclotetrasiloxane, octamethyl-	556-67-2	NI30811
267	268	211	183	269	137	239	119	**296**	100	26	26	18	16	16	15	12	**0.89**	8	24	4	–	–	–	–	–	–	4	–	–	–	Cyclotetrasiloxane, 1,3,5,7-tetraethyl-	16066-10-7	NI30812

MASS TO CHARGE RATIOS								M.W.	INTENSITIES								Parent	C	H	O	N	S	F	Cl	Br	I	Si	P	B	X	COMPOUND NAME	CAS Reg No	No
281	231	246	261	283	203	139	296	**296**	100	70	51	32	31	27	10	1		9	10	1	–	–	–	1	–	1	–	–	–	–	1,2-Benziodoxole, 1-chloro-1,3-dihydro-3,3-dimethyl-		MS06793
69	41	68	55	67	29	27	42	**296**	100	59	55	22	17	17	15	12	**0.00**	9	10	4	–	–	6	–	–	–	–	–	–	–	Pentane, 1,5-bis(trifluoroacetoxy)-		MS06794
55	69	41	169	29	39	27	56	**296**	100	57	19	13	13	8	8	6	**0.00**	9	10	4	–	–	6	–	–	–	–	–	–	–	Propane, 1,3-bis(trifluoroacetoxy)-2,2-dimethyl-		MS06795
196	127	177	128	84	277	145	125	**296**	100	93	21	16	13	11	11	9	**0.00**	10	5	–	–	–	9	–	–	–	–	–	–	–	Bicyclo[2.2.2]octa-2,5-diene, 2-methyl-3-(trifluoromethyl)-1,4,7,7,8,8-hexafluoro-		NI30813
121	254	43	296	94	93	18	15	**296**	100	50	46	20	16	11	11	10		10	11	3	2	1	3	–	–	–	–	–	–	–	Acetamide, N-[4-methyl-3-[[(trifluoromethyl)sulphonyl]amino]phenyl]-	47000-92-0	NI30814
43	75	117	41	69	296	76	61	**296**	100	46	41	28	14	12	8	8		10	15	1	–	–	6	–	–	–	–	1	–	–	Phosphine, diisopropyl[3,3,3-trifluoro-1-oxo-2-(trifluoromethyl)propyl]-	38079-85-5	NI30815
43	75	117	42	69	296	76	61	**296**	100	47	41	30	15	14	9	9		10	15	1	–	–	6	–	–	–	–	1	–	–	Phosphine, diisopropyl[3,3,3-trifluoro-1-oxo-2-(trifluoromethyl)propyl]-	38079-85-5	LI06438
59	145	113	205	58	141	131	61	**296**	100	34	25	22	13	12	12	10	**4.40**	10	16	–	–	5	–	–	–	–	–	–	–	–	2,4,6,8,9-Pentathiaadamantane, 1,3,5,7,10-pentamethyl-	17443-93-5	NI30816
116	88	44	148	60	56	72	117	**296**	100	49	43	40	36	29	17	16	**11.81**	10	20	–	2	4	–	–	–	–	–	–	–	–	Thioperoxydicarbonic diamide, tetraethyl-	97-77-8	NI30817
60	116	88	296	44	148	77	72	**296**	100	75	38	8	8	5	5	5		10	20	–	2	4	–	–	–	–	–	–	–	–	Thioperoxydicarbonic diamide, tetraethyl-	97-77-8	IC03734
116	73	180	147	101	296	45	203	**296**	100	91	68	40	25	23	22	20		10	24	2	–	2	–	–	–	–	2	–	–	–	Dithioerythritol (oxidized), bis(trimethylsilyl)-		NI30818
116	73	180	147	101	296	75	59	**296**	100	95	64	34	24	18	18	16		10	24	2	–	2	–	–	–	–	2	–	–	–	Dithiothreitol (oxidized), bis(trimethylsilyl)-		NI30819
207	193	281	237	209	208	191	133	**296**	100	37	28	26	26	23	11	6	**0.00**	10	28	4	–	–	–	–	–	–	3	–	–	–	3,3-Diethoxy-1,1,1,5,5,5-hexamethyltrisiloxane		MS06796
112	158	76	296	298	160	114	295	**296**	100	82	82	50	38	32	32	25		11	6	–	4	1	–	2	–	–	–	–	–	–	3H-1,2,4-Thiadiazolo[4,3-a]pyridine, 6-chloro-3-(5-chloropyrid-2-ylimino)		LI06439
148	28	56	212	240	91	77	39	**296**	100	77	51	45	44	31	19	16	**4.14**	11	9	4	–	–	–	1	–	–	–	–	–	1	Iron, tricarbonyl[(2,3,4,5-η)-2-chloro-4,5-dimethyl-2,4-cyclohexadien-1-one]-	59688-23-2	NI30820
43	41	27	42	15	29	209	39	**296**	100	53	51	29	27	20	19	18	**9.20**	11	11	3	–	–	–	3	–	–	–	–	–	–	Acetic acid, (2,4,5-trichlorophenoxy)-, isopropyl ester	93-78-7	NI30821
209	211	196	198	74	145	109	181	**296**	100	95	66	63	41	40	40	39	**35.52**	11	11	3	–	–	–	3	–	–	–	–	–	–	Acetic acid, (2,4,5-trichlorophenoxy)-, isopropyl ester	93-78-7	NI30822
59	15	101	41	29	69	42	39	**296**	100	95	94	47	27	25	22	17	**0.40**	11	11	3	–	–	–	3	–	–	–	–	–	–	Butanoic acid, 4-(2,4,5-trichlorophenoxy)-, methyl ester	25333-21-5	NI30823
59	101	41	69	42	39	97	36	**296**	100	73	29	18	11	9	5	5	**0.41**	11	11	3	–	–	–	3	–	–	–	–	–	–	Butanoic acid, 4-(2,4,5-trichlorophenoxy)-, methyl ester	25333-21-5	NI30824
59	101	41	42	40	198	196	97	**296**	100	74	34	12	9	8	8	6	**0.00**	11	11	3	–	–	–	3	–	–	–	–	–	–	Butanoic acid, 4-(2,4,5-trichlorophenoxy)-, methyl ester	25333-21-5	NI30825
296	246	227	265	277	196	163	31	**296**	100	21	17	11	9	8	5	5		12	–	–	–	–	8	–	–	–	–	–	–	–	Octafluorobiphenylene		LI06440
296	298	270	268	299	297	167	169	**296**	100	99	25	25	16	16	14	13		12	9	4	–	–	–	–	1	–	–	–	–	–	4H-1-Benzopyran-2-carboxylic acid, 6-bromo-4-oxo-, ethyl ester	33513-40-5	NI30826
270	268	49	84	298	296	86	51	**296**	100	99	68	50	50	45	36	34		12	9	4	–	–	–	–	1	–	–	–	–	–	1,2-Naphthalenedione, 7-bromo-8-hydroxy-4-methoxy-6-methyl-		NI30827
47	204	77	46	41	203	135	92	**296**	100	60	59	47	42	30	29	26	**2.00**	12	12	3	2	2	–	–	–	–	–	–	–	–	Imidazolidin-4-one, 5-(carboxymethylthiomethyl)-3-phenyl-2-thioxo-		LI06441
43	44	28	181	206	31	45	29	**296**	100	27	26	17	15	12	10	10	**0.63**	12	12	7	2	–	–	–	–	–	–	–	–	–	Butyric acid, 3-oxo-2-(2,4-dinitrophenyl)-, ethyl ester		MS06797
169	171	298	296	254	252	172	170	**296**	100	99	30	30	10	10	8	8		12	13	2	2	–	–	–	1	–	–	–	–	–	Δ²-1,3,4-Oxadiazolin-5-one, 4-(4-bromophenyl)-2-butyl-	28740-55-8	NI30828
169	171	298	296	271	254	252	170	**296**	100	99	33	33	33	11	11	8		12	13	2	2	–	–	–	1	–	–	–	–	–	Δ²-1,3,4-Oxadiazolin-5-one, 4-(4-bromophenyl)-2-isobutyl-	28740-58-1	NI30829
180	73	281	296	75	181	74	84	**296**	100	71	26	20	18	10	10	7		12	20	3	4	–	–	–	–	–	1	–	–	–	Xanthine, 7-[2-(trimethylsilyloxy)ethyl]-1,3-dimethyl-		NI30830
73	117	75	45	252	281	44	74	**296**	100	50	20	12	10	10	9	9	**2.82**	12	20	3	4	–	–	–	–	–	1	–	–	–	Xanthine, 7-[2-(trimethylsilyloxy)propyl]-1-methyl-		NI30831
99	98	111	197	112	196	110	28	**296**	100	48	41	19	17	12	9	9	**0.12**	12	23	6	–	–	–	–	–	–	–	–	3	–	Galactitol, cyclic 1,6:2,3:4,5-tris(ethylboronate)	58881-48-4	NI30832
99	98	111	197	196	124	97	43	**296**	100	43	13	11	8	5	5	5	**0.03**	12	23	6	–	–	–	–	–	–	–	–	3	–	D-Glucitol, cyclic 1,3:2,4:5,6-tris(ethylboronate)	74779-72-9	NI30833
99	98	197	196	111	43	97	57	**296**	100	41	10	8	6	6	5	5	**0.00**	12	23	6	–	–	–	–	–	–	–	–	3	–	D-Mannitol, cyclic 1,2:3,4:5,6-tris(ethylboronate)	59184-36-0	NI30834
57	43	71	85	41	55	69	29	**296**	100	77	56	34	34	28	16	15	**0.50**	12	25	–	–	–	–	–	–	1	–	–	–	–	Dodecane, iodo-		LI06442
57	43	71	41	85	55	56	79	**296**	100	86	51	49	31	31	26	23	**0.06**	12	25	–	–	–	–	–	–	1	–	–	–	–	Undecane, 1-iodo-2-methyl-	73105-67-6	NI30835
75	169	65	159	111	50	76	51	**296**	100	84	75	71	65	62	50	42	**20.83**	13	9	4	–	1	–	1	–	–	–	–	–	–	Benzoic acid, 4-[(4-chlorophenyl)sulphonyl]-	37940-65-1	NI30836
121	92	94	232	107	106	234	122	**296**	100	71	24	23	21	14	8	7	**5.00**	13	13	2	2	1	–	1	–	–	–	–	–	–	Pyridinium, 1-[[(4-chlorophenyl)sulphonyl]amino]-2,6-dimethyl-, hydroxide, inner salt	55044-56-9	NI30837
121	92	94	232	107	65	91	106	**296**	100	71	24	23	21	18	16	14	**5.00**	13	13	2	2	1	–	1	–	–	–	–	–	–	Pyridinium, 1-[[(4-chlorophenyl)sulphonyl]amino]-2,6-dimethyl-, hydroxide, inner salt	55044-56-9	NI30838
296	298	283	281	148	149	77	297	**296**	100	99	96	96	18	17	15	15		13	13	3	–	–	–	–	1	–	–	–	–	–	1-Naphthalenol, 2-bromo-5,8-dimethoxy-3-methyl-	101459-25-0	NI30839
91	108	155	171	65	56	232	39	**296**	100	55	35	30	25	20	15	15	**0.10**	13	16	4	2	1	–	–	–	–	–	–	–	–	Azetidinone, N-[(4-methylphenyl)sulphonylaminocarbonyl]-4,4-dimethyl-		NI30840
254	239	44	41	91	77	191	296	**296**	100	85	40	33	25	25	21	20		13	16	6	2	–	–	–	–	–	–	–	–	–	Phenol, 6-tert-butyl-3-methyl-2,4-dinitro-, acetate (ester)	2487-01-6	NI30841
42	84	41	56	171	173	43	199	**296**	100	83	45	40	31	30	28	26	**18.99**	13	17	1	2	–	–	–	1	–	–	–	–	–	4H-1,3-Oxazine, 4,4,6-trimethyl-2-[4-bromophenylamino(imino)]-5,6-dihydro-	37723-83-4	NI30842
184	185	296	41	43	249	142	29	**296**	100	71	28	27	25	19	17	15		13	20	–	4	2	–	–	–	–	–	–	–	–	Thiazolo[5,4-d]pyrimidine, 5-amino-2-(octylthio)-	19844-43-0	NI30843
254	296	267	249	41	240	263	184	**296**	100	87	55	51	45	38	35	30		13	20	–	4	2	–	–	–	–	–	–	–	–	Thiazolo[5,4-d]pyrimidine, 2-(butylamino)-5-(butylthio)-	19844-53-2	NI30844
99	69	113	98	267	57	125	41	**296**	100	57	36	26	21	21	17	16	**0.04**	13	27	5	–	–	–	–	–	–	–	–	3	–	Xylitol, cyclic 1,2:3,5-bis(ethylboronate) 4-(diethylborinate)	58066-76-5	NI30845
296	76	240	104	50	295	268	297	**296**	100	64	47	47	35	34	29	22		14	4	2	2	2	–	–	–	–	–	–	–	–	Naphtho[2,3-b]-1,4-dithiin-2,3-dicarbonitrile, 5,10-dihydro-5,10-dioxo-	3347-22-6	NI30846
79	36	52	38	51	50	261	78	**296**	100	75	58	27	23	15	13	12	**0.00**	14	9	–	6	–	–	1	–	–	–	–	–	–	N-[6-Amino-3,5-dicyano-4-(cyanomethyl)-2-pyridyl]pyridinium chloride		IC03735
215	173	81	217	175	54	159	75	**296**	100	85	67	64	56	44	27	20	**4.39**	14	14	1	2	–	–	2	–	–	–	–	–	–	1H-Imidazole, 1-[2-(2,4-dichlorophenyl)-2-(2-propenyloxy)ethyl]-	35554-44-0	NI30847
194	59	176	43	195	151	150	44	**296**	100	88	66	52	50	45	42	42	**29.00**	14	16	7	–	–	–	–	–	–	–	–	–	–	5H-1,4-Benzodioxepin-6-carboxylic acid, 7-hydroxy-3-(1-hydroxyisopropyl)-9-methyl-5-oxo-2,3-dihydro-		LI06443
153	43	151	254	152	195	93	222	**296**	100	41	30	15	12	11	10	9	**2.19**	14	16	7	–	–	–	–	–	–	–	–	–	–	Methyl (3-methoxy-5-acetoxyphenyl)-α-acetoxyacetate		MS06798
133	105	148	55	75	191	298	296	**296**	100	53	35	22	17	6	2	2		14	16	7	–	–	–	–	–	–	–	–	–	–	5,8-Naphthoquinone, 1,4-epoxy-2,2,3,3-tetramethoxy-1,2,3,4-tetrahydro-		LI06444

MASS TO CHARGE RATIOS								M.W.	INTENSITIES								Parent	C	H	O	N	S	F	Cl	Br	I	Si	P	B	X	COMPOUND NAME	CAS Reg No	No
79	296	298	91	110	107	93	191	**296**	100	78	77	65	58	56	50	31		14	17	–	–	1	–	–	1	–	–	–	–	–	8-Bromo-3-(phenylthio)-1,3-octadiene, (E)-	123289-41-8	DD01053
175	161	298	296	79	91	110	147	**296**	100	63	62	60	52	45	39	37		14	17	–	–	1	–	–	1	–	–	–	–	–	8-Bromo-3-(phenylthio)-1,3-octadiene, (Z)-	123289-34-9	DD01054
73	147	75	45	77	253	164	149	**296**	100	34	33	22	21	16	14	13	**10.40**	14	24	3	–	–	–	–	–	–	2	–	–	–	Benzeneacetic acid, 2-[(trimethylsilyl)oxy]-, trimethylsilyl ester	27750-52-3	MS06799
73	147	164	296	253	74	149	75	**296**	100	29	15	13	11	8	6	6		14	24	3	–	–	–	–	–	–	2	–	–	–	Benzeneacetic acid, 2-[(trimethylsilyl)oxy]-, trimethylsilyl ester	27750-52-3	NI30848
73	147	253	296	164	281	149	297	**296**	100	39	34	25	25	15	9	7		14	24	3	–	–	–	–	–	–	2	–	–	–	Benzeneacetic acid, 2-[(trimethylsilyl)oxy]-, trimethylsilyl ester	27750-52-3	NI30849
73	75	164	45	147	74	281	296	**296**	100	22	14	12	11	7	6	5		14	24	3	–	–	–	–	–	–	2	–	–	–	Benzeneacetic acid, 3-[(trimethylsilyl)oxy]-, trimethylsilyl ester	27750-55-6	MS06801
73	75	77	164	147	296	281	45	**296**	100	30	18	16	13	11	11	10		14	24	3	–	–	–	–	–	–	2	–	–	–	Benzeneacetic acid, 3-[(trimethylsilyl)oxy]-, trimethylsilyl ester	27750-55-6	MS06800
73	75	164	147	45	74	296	281	**296**	100	22	17	12	10	9	9	8		14	24	3	–	–	–	–	–	–	2	–	–	–	Benzeneacetic acid, 3-[(trimethylsilyl)oxy]-, trimethylsilyl ester	27750-55-6	NI30850
73	179	75	296	164	281	252	74	**296**	100	26	18	14	13	12	12	9		14	24	3	–	–	–	–	–	–	2	–	–	–	Benzeneacetic acid, 4-[(trimethylsilyl)oxy]-, trimethylsilyl ester	27750-57-8	NI30851
73	79	75	93	179	77	52	164	**296**	100	27	27	25	24	18	14	11	**8.79**	14	24	3	–	–	–	–	–	–	2	–	–	–	Benzeneacetic acid, 4-[(trimethylsilyl)oxy]-, trimethylsilyl ester	27750-57-8	NI30852
73	179	252	75	164	296	281	45	**296**	100	26	16	16	14	12	12	9		14	24	3	–	–	–	–	–	–	2	–	–	–	Benzeneacetic acid, 4-[(trimethylsilyl)oxy]-, trimethylsilyl ester	27750-57-8	MS06802
179	73	147	180	253	45	149	74	**296**	100	60	28	17	11	8	6	5	**0.00**	14	24	3	–	–	–	–	–	–	2	–	–	–	Benzeneacetic acid, α-[(trimethylsilyl)oxy]-, trimethylsilyl ester	2078-19-5	NI30854
179	73	147	180	45	75	40	74	**296**	100	58	23	14	9	8	7	5	**0.16**	14	24	3	–	–	–	–	–	–	2	–	–	–	Benzeneacetic acid, α-[(trimethylsilyl)oxy]-, trimethylsilyl ester	2078-19-5	MS06803
179	73	147	180	45	253	75	74	**296**	100	92	27	17	14	9	8	8	**0.00**	14	24	3	–	–	–	–	–	–	2	–	–	–	Benzeneacetic acid, α-[(trimethylsilyl)oxy]-, trimethylsilyl ester	2078-19-5	NI30853
126	70	125	113	43	170	69	127	**296**	100	29	26	15	9	8	8	8	**0.64**	14	26	5	–	–	–	–	–	–	–	–	2	–	α-L-Fucopyranose, 1,2:3,4-bis(butaneboronate)	82284-60-4	NI30855
126	70	125	75	43	114	44	69	**296**	100	60	41	37	29	28	27	23	**0.00**	14	26	5	–	–	–	–	–	–	–	–	2	–	α-L-Galactopyranose, 6-deoxy-, cyclic 1,2:3,4-bis(butylboronate)	53829-55-3	NI30856
268	296	134	162	135	90	163	69	**296**	100	75	69	63	14	14	9	9		15	8	1	2	2	–	–	–	–	–	–	–	–	Bis(2-benzothiazolyl) ketone	4464-60-2	LI06445
268	296	134	162	135	90	108	69	**296**	100	75	69	63	14	14	9	9		15	8	1	2	2	–	–	–	–	–	–	–	–	Bis(2-benzothiazolyl) ketone	4464-60-2	NI30857
104	177	296	74	93	77	135	51	**296**	100	80	52	43	35	32	20	13		15	12	1	4	1	–	–	–	–	–	–	–	–	Urea, N-phenyl-N′-(5-phenyl-1,2,4-thiadiazol-3-yl)-	42053-81-6	NI30858
155	142	296	154	298	90	108	157	**296**	100	97	85	58	57	48	43	35		15	14	2	–	–	–	2	–	–	–	–	–	–	Bis(3-chloro-2-hydroxy-5-methylphenyl)methane		LI06447
155	142	296	154	91	107			**296**	100	98	86	58	49	41				15	14	2	–	–	–	2	–	–	–	–	–	–	Bis(3-chloro-2-hydroxy-5-methylphenyl)methane		LI06446
281	283	296	153	298	282	284	28	**296**	100	66	33	22	21	15	10	9		15	14	2	–	–	–	2	–	–	–	–	–	–	2,2-Bis(3-chloro-4-hydroxyphenyl)propane	79-98-1	NI30859
189	296	281	260	244	243	240	239	**296**	100	99	99	99	99	99	99	99		15	17	4	–	–	–	1	–	–	–	–	–	–	2H-1-Benzopyran-2-one, 7-methoxy-8-(2-chloro-3-hydroxy-3-methylbutyl)		NI30860
91	100	122	43	59	238	164	45	**296**	100	99	79	79	77	71	50	45	**2.00**	15	20	4	–	1	–	–	–	–	–	–	–	–	α-D-Xylopyranose, 4-S-benzyl-1,2-O-isopropylidene-4-thio-	32848-89-8	NI30861
152	41	136	43	125	153	296	137	**296**	100	75	66	63	46	38	34	30		15	20	6	–	–	–	–	–	–	–	–	–	–	Isodeoxynivalenol		NI30862
98	91	72	79	77	278	186	181	**296**	100	78	65	64	57	50	50	50	**0.00**	15	20	6	–	–	–	–	–	–	–	–	–	–	Marasmal		MS06804
43	147	41	191	97	55	119	93	**296**	100	82	60	42	39	29	28	28	**0.50**	15	20	6	–	–	–	–	–	–	–	–	–	–	5,8-Methano-1,7-dioxacyclopent[cd]azulene-2,6-dione, octahydro-2a,9-dihydroxy-8b-methyl-9-isopropyl-, [2aR-(2aα,4aα,5α,8α,8aα,8bα,9R*)]-	70219-70-4	NI30863
41	44	98	79	248	91	107	135	**296**	100	99	80	80	78	69	67	62	**6.39**	15	20	6	–	–	–	–	–	–	–	–	–	–	Trichothec-9-en-8-one, 12,13-epoxy-3,7,15-trihydroxy-, (3α,7α)-	51481-10-8	NI30864
248	135	91	107	79	77	98	137	**296**	100	62	60	60	60	59	52	49	**1.00**	15	20	6	–	–	–	–	–	–	–	–	–	–	Trichothec-9-en-8-one, 12,13-epoxy-3,7,15-trihydroxy-, (3α,7α)-	51481-10-8	NI30865
135	201	121	148	149	115	133	159	**296**	100	58	43	35	23	20	17	15	**0.00**	15	21	1	–	–	–	–	1	–	–	–	–	–	Phenol, 2-(3-bromo-2,2,5-trimethylcyclopentyl)-5-methyl-, (1R*,3R*)-		MS06805
164	151	296	75	281	165	149	73	**296**	100	26	20	19	15	11	10	10		15	24	4	–	–	–	–	–	–	1	–	–	–	Butanoic acid, (3,4-dimethoxyphenyl)-, trimethylsilyl ester		NI30866
72	237	85	135	238	167	91	79	**296**	100	48	12	9	8	8	7	7	**0.80**	15	24	4	2	–	–	–	–	–	–	–	–	–	1-Cyclohexene-1-acetic acid, α-[[[(dimethylamino)carbonyl]methylamino]carbonyl]-α-methyl-, methyl ester	42948-64-1	NI30867
72	237	135	167	107	180	252	192	**296**	100	48	9	8	7	3	2	2	**1.00**	15	24	4	2	–	–	–	–	–	–	–	–	–	1-Cyclohexene-1-acetic acid, α-[[[(dimethylamino)carbonyl]methylamino]carbonyl]-α-methyl-, methyl ester	42948-64-1	LI06448
73	193	147	75	194	45	74	43	**296**	100	98	26	18	17	14	9	6	**0.00**	15	28	2	–	–	–	–	–	–	2	–	–	–	Benzene, [1,3-bis(trimethylsiloxy)prop-2-yl]-		NI30868
206	73	191	89	207	179	59	75	**296**	100	63	61	24	23	18	17	16	**9.01**	15	28	2	–	–	–	–	–	–	2	–	–	–	Benzene, 1-(trimethylsiloxy)-4-[3-(trimethylsiloxy)propyl]-		NI30869
56	43	55	83	61	54	99	82	**296**	100	80	67	40	40	40	27	27	**0.00**	15	28	2	4	–	–	–	–	–	–	–	–	–	Urea, 1-cyclohexyl-3-[[[(cyclohexylamino)carbonyl]amino]methyl]-		LI06449
296	120	76	297	240	298	104	192	**296**	100	24	22	20	13	12	10	8		16	8	2	–	2	–	–	–	–	–	–	–	–	Benzo[b]thiophen-3(2H)-one, 2-(3-oxobenzo[b]thien-2(3H)-ylidene)-, (E)-	3844-31-3	MS06806
296	120	76	297	240	298	104	50	**296**	100	24	22	21	14	13	12	9		16	8	2	–	2	–	–	–	–	–	–	–	–	Benzo[b]thiophen-3(2H)-one, 2-(3-oxobenzo[b]thien-2(3H)-ylidene)-, (E)-	3844-31-3	NI30870
296	120	240	297	298	295	232	195	**296**	100	38	25	20	11	11	11	10		16	8	2	–	2	–	–	–	–	–	–	–	–	Benzo[c]thiophen-1(3H)-one, 3-(3-oxobenzo[c]thien-1(3H)-ylidene)-, (E)-	23667-32-5	NI30871
180	208	252	75	44	150	152	151	**296**	100	81	65	60	60	57	52	49	**3.30**	16	8	6	–	–	–	–	–	–	–	–	–	–	1,6-Anthraquinonedicarboxylic acid		IC03736
234	252	150	75	180	139	208	279	**296**	100	71	47	39	32	27	16	7	**0.50**	16	8	6	–	–	–	–	–	–	–	–	–	–	1,8-Anthraquinonedicarboxylic acid		NI30872
105	77	51	106	278	129	122	131	**296**	100	55	18	9	6	2	2	1	**0.00**	16	12	4	2	–	–	–	–	–	–	–	–	–	Butanetetrone, diphenyl-, 2,3-dioxime	55232-36-5	NI30873
148	149	93	65	77	51	92	39	**296**	100	81	63	59	52	44	33	29	**22.00**	16	12	4	2	–	–	–	–	–	–	–	–	–	2,2′-Diformyloxanilide		LI06450
121	56	117	91	225	240	148	118	**296**	100	65	62	61	57	55	46	46	**0.18**	16	16	2	–	–	–	–	–	–	–	–	–	1	Iron, (3-phenylpropyl)dicarbonyl-π-cyclopentadienyl-	60003-79-4	NI30874
177	119	91	77	106	162	203	131	**296**	100	71	34	30	27	17	8	4	**0.00**	16	16	2	4	–	–	–	–	–	–	–	–	–	2-Imidazolidinone, 3-(N-phenylcarbamoylamino)-1-phenyl-		LI06451
134	296	133	297	119	91	135	77	**296**	100	85	25	16	14	12	11	10		16	16	2	4	–	–	–	–	–	–	–	–	–	1,2,4,5-Tetrazine, 3,6-bis(4-methoxyphenyl)-1,2-dihydro-		MS06807
179	236	223	239	207	296	55	265	**296**	100	71	34	30	19	18	17	16		16	24	5	–	–	–	–	–	–	–	–	–	–	Furan-2-propionic acid, 3-(methoxycarbonyl)-4-methyl-5-pentyl-, methyl ester		MS06808
73	81	72	126	53	206	85	55	**296**	100	93	47	38	35	31	30	30	**0.50**	16	24	5	–	–	–	–	–	–	–	–	–	–	Glechomanolide, 1β,10α-epoxy-4-methoxy-8-hydroxy-		NI30875
264	153	121	95	111	94	107	265	**296**	100	48	47	45	39	31	28	25	**3.00**	16	24	5	–	–	–	–	–	–	–	–	–	–	Octanoic acid, 8-[5-(methoxycarbonyl)methyl-2-furyl]-, methyl ester		NI30876

MASS TO CHARGE RATIOS								M.W.	INTENSITIES								Parent	C	H	O	N	S	F	Cl	Br	I	Si	P	B	X	COMPOUND NAME	CAS Reg No	No
296	295	267	279	297	234	202	121	**296**	100	38	32	26	25	21	21	21		17	12	1	–	2	–	–	–	–	–	–	–	–	Acetaldehyde, (4,5-diphenyl-3H-1,2-dithiol-3-ylidene)-	38443-47-9	NI30877
296	219	77	105	297	121	263	51	**296**	100	66	60	36	22	21	18	18		17	12	1	–	2	–	–	–	–	–	–	–	–	Acetophenone, 2-(4-phenyl-3H-1,2-dithiol-3-ylidene)-	1033-62-1	NI30879
105	296	77	219	279	297	191	51	**296**	100	63	45	27	22	15	15	10		17	12	1	–	2	–	–	–	–	–	–	–	–	Acetophenone, 2-(4-phenyl-3H-1,2-dithiol-3-ylidene)-	5368-02-5	NI30878
296	297	121	268	234	202	279	267	**296**	100	22	22	21	16	15	13	13		17	12	1	–	2	–	–	–	–	–	–	–	–	Benzeneacetaldehyde, α-(5-phenyl-3H-1,2-dithiol-3-ylidene)-	20365-46-2	NI30880
265	296	150	266	209	297	75	117	**296**	100	84	24	17	15	14	13	12		17	12	5	–	–	–	–	–	–	–	–	–	–	Fluorenone, 3,4-bis(methoxycarbonyl)-		NI30881
265	296	297	150	209	266	237	117	**296**	100	94	19	16	15	14	14	11		17	12	5	–	–	–	–	–	–	–	–	–	–	Fluorenone, 3,6-bis(methoxycarbonyl)-		IC03737
28	45	65	77	91	148	219	104	**296**	100	49	48	25	20	15	12	11	**10.00**	17	12	5	–	–	–	–	–	–	–	–	–	–	1-Indanone, 3-phenyl-6,7-dicarboxy-	92241-99-1	NI30882
230	232	151	150	152	231	66	39	**296**	100	90	76	24	16	14	14	13	**4.00**	17	13	–	–	–	–	–	1	–	–	–	–	–	7H-Cyclopenta[3,4]cyclobut[1,2-a]acenapthylene, 3-bromo-6b,6c,9a,9b-tetrahydro-, (6bα,6cβ,9aβ,9bα)-	50682-98-9	NI30883
261	77	296	262	28	185	51	128	**296**	100	90	76	75	40	33	33	30		17	13	1	2	–	–	1	–	–	–	–	–	–	5-Pyrazolone, 1-phenyl-3-methyl-4-(2-chlorobenzylidene)-	93259-79-1	NI30884
295	296	89	297	32	298	63	231	**296**	100	68	53	45	27	24	15	9		17	13	1	2	–	–	1	–	–	–	–	–	–	Pyrimidine, 2-chloro-5-methoxy-4,6-diphenyl-	94496-24-9	NI30885
104	77	51	91	205	119	65	135	**296**	100	66	44	40	32	25	23	23	**15.00**	17	16	1	2	1	–	–	–	–	–	–	–	–	4H-1,3-Thiazin-5-one, 4-benzyl-5,6-dihydro-2-phenylamino-	86387-93-1	NI30886
296	180	267	210	104	77	134	297	**296**	100	52	42	34	32	23	21	20		17	16	3	2	–	–	–	–	–	–	–	–	–	2,4-Imidazolidinedione, 5-(3-methoxyphenyl)-3-methyl-5-phenyl-	54833-60-2	NI30887
296	267	219	210	134	180	77	297	**296**	100	65	60	60	39	31	25	21		17	16	3	2	–	–	–	–	–	–	–	–	–	2,4-Imidazolidinedione, 5-(4-methoxyphenyl)-3-methyl-5-phenyl-	54833-61-3	NI30888
128	223	296	196	168	240	212	129	**296**	100	95	60	60	47	46	34	24		17	16	3	2	–	–	–	–	–	–	–	–	–	1H-Pyrimido[1,2-a]quinoline-2-carboxylic acid, 1-oxo-, butyl ester	63455-53-8	NI30889
296	265	90	160	76	77	281	136	**296**	100	83	83	67	67	58	50	50		17	16	3	2	–	–	–	–	–	–	–	–	–	4(3H)-Quinazolinone, 3-(2,6-dimethoxyphenyl)-2-methyl-		NI30890
188	159	108	127	296	189	173	79	**296**	100	66	52	49	47	42	38	33		17	19	1	–	–	3	–	–	–	–	–	–	–	2,5-Methano-1H-inden-7-ol, octahydro-7-[3-(trifluoromethyl)phenyl]-	40571-17-3	NI30891
108	107	67	93	80	237	224	160	**296**	100	31	24	21	17	11	11	9	**8.00**	17	19	1	–	–	3	–	–	–	–	–	–	–	Tricyclo[3.3.1.1^{3,7}]decan-2-ol, 1-[3-(trifluoromethyl)phenyl]-	40571-18-4	NI30892
179	71	138	43	208	111	225	151	**296**	100	83	74	69	39	35	31	30	**0.00**	17	28	4	–	–	–	–	–	–	–	–	–	–	Cyclohexanepropanoic acid, β,4-dimethyl-1-(2-methyl-1-oxopropyl)-3-oxo, ethyl ester	37730-50-0	NI30893
71	43	138	208	111	225	151	179	**296**	100	95	94	47	45	44	44	43	**0.00**	17	28	4	–	–	–	–	–	–	–	–	–	–	Cyclohexanepropanoic acid, β,4-dimethyl-1-(2-methyl-1-oxopropyl)-3-oxo, ethyl ester, [1α(S*),4α]-(±)-	37730-49-7	NI30894
55	41	112	111	69	101	43	98	**296**	100	95	87	80	51	44	42	33	**0.00**	17	28	4	–	–	–	–	–	–	–	–	–	–	Cyclotetradecanecarboxylic acid, 1-methyl-2,5-dioxo-, methyl ester		DD01055
29	67	41	160	55	79	27	121	**296**	100	64	46	41	36	33	31	26	**0.00**	17	28	4	–	–	–	–	–	–	–	–	–	–	Diethyl (2,6-dimethyl-2,7-octadienyl)malonate		MS06809
29	41	27	67	39	55	53	43	**296**	100	83	80	45	38	37	33	23	**0.00**	17	28	4	–	–	–	–	–	–	–	–	–	–	Diethyl (1,5-dimethyl-1-vinyl-5-hexenyl)malonate		MS06810
115	55	83	87	67	41	133	82	**296**	100	29	17	16	15	15	12	10	**0.02**	17	28	4	–	–	–	–	–	–	–	–	–	–	Glutaric acid, dicyclohexyl ester		IC03738
115	133	55	41	83	67	42	99	**296**	100	42	24	17	15	10	10	9	**0.00**	17	28	4	–	–	–	–	–	–	–	–	–	–	Glutaric acid, dicyclohexyl ester		IC03739
55	43	71	113	141	97	193	278	**296**	100	96	75	67	56	48	44	38	**1.00**	17	28	4	–	–	–	–	–	–	–	–	–	–	2,6,10,14-Heptadecanetetrone	55110-16-2	NI30895
97	145	69	115	41	209	55	159	**296**	100	35	25	21	17	14	11	5	**0.00**	17	28	4	–	–	–	–	–	–	–	–	–	–	Hexanoic acid, 6-[(2-allyl-2-methyl-5-oxocyclopentyl)oxy]-, ethyl ester		NI30896
83	81	113	55	41	69	95	57	**296**	100	82	78	49	37	35	33	31	**0.50**	17	28	4	–	–	–	–	–	–	–	–	–	–	Pyran-6-carboxylic acid, 2-methoxy-5,6-dihydro-, 1-menthyl ester		MS06811
83	81	113	55	41	95	139	127	**296**	100	82	78	49	37	33	30	30	**0.50**	17	28	4	–	–	–	–	–	–	–	–	–	–	Pyran-6-carboxylic acid, 2-methoxy-5,6-dihydro-, 1-menthyl ester		LI06452
129	268	224	182	100	296	155	210	**296**	100	90	85	80	75	55	50	45		17	32	2	2	–	–	–	–	–	–	–	–	–	Acetamide, N-[3-(2-oxoazacyclotridec-1-yl)propyl]-	67171-81-7	NI30897
253	70	100	112	210	196	86	84	**296**	100	10	8	6	5	5	5	5	**4.00**	17	32	2	2	–	–	–	–	–	–	–	–	–	1,5-Diazacycloheptadecan-6-one, 1-acetyl-	67171-86-2	NI30898
73	155	99	239	166	151	125	74	**296**	100	22	16	14	11	11	10	8	**4.00**	17	36	–	–	–	–	–	–	–	2	–	–	–	Silane, (3-methyl-1,2-butadienylidene)bis-tert-butyldimethyl-	61228-15-7	NI30899
240	239	238	268	227	241	201	251	**296**	100	54	37	36	29	17	11	10	**0.10**	18	14	–	2	–	2	–	–	–	–	–	–	–	Benzo[c]cyclopenta[f]-1,2-diazepine, 1,2,3,3a-tetrahydro-7-fluoro-10-(4-fluorophenyl)-	56005-89-1	NI30900
296	135	297	115	248	205	91	298	**296**	100	28	23	21	19	14	14	11		18	16	–	–	2	–	–	–	–	–	–	–	–	1,4-Dithiin, 2,5-bis(4-methylphenyl)-	37989-48-3	NI30901
221	296	249	235	234	222	250	236	**296**	100	93	93	85	50	37	32	27		18	16	–	–	2	–	–	–	–	–	–	–	–	Thioxanthene, 9-(tetrahydrothiopyranyliden-4-yl)-		MS06812
267	249	268	239	224	223	240	211	**296**	100	37	28	20	13	13	10	7	**3.00**	18	16	4	–	–	–	–	–	–	–	–	–	–	9,10-Anthracenedione, 1,8-diethoxy-	16294-26-1	LI06453
267	249	268	239	224	223	240	250	**296**	100	37	27	19	12	12	10	9	**2.92**	18	16	4	–	–	–	–	–	–	–	–	–	–	9,10-Anthracenedione, 1,8-diethoxy-	16294-26-1	NI30902
296	224	281	268	253	223	225	267	**296**	100	36	34	33	28	27	23	22		18	16	4	–	–	–	–	–	–	–	–	–	–	9,10-Anthrancenedione, 1,5-diethoxy-		LI06454
296	237	137	295	253	152	115	267	**296**	100	7	6	5	5	4	4	3		18	16	4	–	–	–	–	–	–	–	–	–	–	Benzofuran, 2-(2,4-dihydroxyphenyl)-7-methoxy-5(E)-propenyl-		MS06813
194	165	43	195	296	193	152	166	**296**	100	26	22	19	17	14	9	7		18	16	4	–	–	–	–	–	–	–	–	–	–	Biphenylene, 2,3-bis(acetoxymethyl)-	93103-44-7	NI30903
296	237	165	43	297	178	238	163	**296**	100	77	28	23	20	19	14	12		18	16	4	–	–	–	–	–	–	–	–	–	–	Biphenylene, 2,6-bis(acetoxymethyl)-	93103-45-8	NI30904
148	131	147	103	91	250	77	149	**296**	100	54	51	24	17	14	12	10	**0.00**	18	16	4	–	–	–	–	–	–	–	–	–	–	1,2-Cyclobutanedicarboxylic acid, 3,4-diphenyl-	4482-52-4	NI30905
148	147	103	91	131	149	77	102	**296**	100	55	16	14	12	11	9	8	**0.12**	18	16	4	–	–	–	–	–	–	–	–	–	–	1,3-Cyclobutanedicarboxylic acid, 2,4-diphenyl-	4462-95-7	NI30906
91	79	108	107	77	28	51	92	**296**	100	97	84	75	67	61	35	32	**0.00**	18	16	4	–	–	–	–	–	–	–	–	–	–	Maleic ester, dibenzyl ester		DO01083
296	252	165	253	152	278	235	265	**296**	100	64	62	54	50	49	42	39		18	16	4	–	–	–	–	–	–	–	–	–	–	Phenanthro[1,2-b]furan-10,11-dione, 6,7,8,9-tetrahydro-1,6-dimethyl-hydroxy		NI30907
254	296	217	256	267	218	232	255	**296**	100	58	35	33	32	32	25	20		18	17	–	2	–	–	1	–	–	–	–	–	–	Azepinoindole, 5-(4-chlorophenyl)hexahydro-		NI30908
296	298	297	295	134	77	146	118	**296**	100	37	30	27	20	10	9	8		18	17	–	2	–	–	1	–	–	–	–	–	–	Cyclopropanecarbonitrile, 1-(3-chlorophenyl)-2-[(dimethylamino)phenyl]-	54833-14-6	NI30909
296	298	297	295	134	78	146	140	**296**	100	37	30	27	20	11	10	8		18	17	–	2	–	–	1	–	–	–	–	–	–	Cyclopropanecarbonitrile, 1-(3-chlorophenyl)-2-[4-(dimethylamino)phenyl]-		LI06455
296	298	297	295	134	146	77	118	**296**	100	35	30	27	18	9	9	7		18	17	–	2	–	–	1	–	–	–	–	–	–	Cyclopropanecarbonitrile, 1-(4-chlorophenyl)-2-[(dimethylamino)phenyl]-	54833-15-7	NI30910
296	298	297	295	134	78	146	140	**296**	100	37	30	27	19	10	9	7		18	17	–	2	–	–	1	–	–	–	–	–	–	Cyclopropanecarbonitrile, 1-(4-chlorophenyl)-2-[4-(dimethylamino)phenyl]-		LI06456

MASS TO CHARGE RATIOS								M.W.	INTENSITIES								Parent	C	H	O	N	S	F	Cl	Br	I	Si	P	B	X	COMPOUND NAME	CAS Reg No	No
97	296	199	98	212	55	198	96	**296**	100	91	84	69	55	55	47	44		18	20	–	2	1	–	–	–	–	–	–	–	–	10H-Phenothiazine, 10-[(1-methyl-3-pyrrolidinyl)methyl]-	1982-37-2	NI30911
97	98	55	82	199	198	296	180	**296**	100	95	94	45	37	34	32	32		18	20	–	2	1	–	–	–	–	–	–	–	–	10H-Phenothiazine, 10-[(1-methyl-3-pyrrolidinyl)methyl]-	1982-37-2	NI30912
97	296	199	98	212	55	198	180	**296**	100	91	84	69	55	55	47	44		18	20	–	2	1	–	–	–	–	–	–	–	–	10H-Phenothiazine, 10-[(1-methyl-3-pyrrolidinyl)methyl]-	1982-37-2	LI06457
84	296	85	42	180	212	55	41	**296**	100	7	6	6	4	3	3	2		18	20	–	2	1	–	–	–	–	–	–	–	–	10H-Phenothiazine, 10-[2-(1-pyrrolidinyl)ethyl]-	84-08-2	NI30913
267	77	132	103	109	91	104	196	**296**	100	43	42	27	14	14	10	9	**0.00**	18	20	–	2	1	–	–	–	–	–	–	–	–	1,3,4-Thiadiazole, 2,2-diethyl-2,3-dihydro-3,5-diphenyl-	53067-48-4	NI30914
267	77	132	103	268	91	109	76	**296**	100	44	43	27	23	15	14	11	**9.14**	18	20	–	2	1	–	–	–	–	–	–	–	–	1,3,4-Thiadiazole, 2,2-diethyl-2,3-dihydro-3,5-diphenyl-	53067-48-4	NI30915
148	77	92	296	149	281	107	64	**296**	100	24	16	15	10	9	9	9		18	20	2	2	–	–	–	–	–	–	–	–	–	Aniline, N,N′-(1,2-dimethyl-1,2-ethanediylidene)bis[4-methoxy-	19215-52-2	NI30916
91	296	240	92	297	241			**296**	100	27	20	10	5	4				18	20	2	2	–	–	–	–	–	–	–	–	–	2-Benzimidazolinone, 1-benzyl-5-butoxy-	28643-56-3	NI30917
91	296	240	92	65	57	55	297	**296**	100	28	23	19	12	10	7	6		18	20	2	2	–	–	–	–	–	–	–	–	–	1H-Benzimidazol-4-ol, 5-butoxy-1-benzyl-	54833-24-8	NI30918
120	296	295	121	174	297	149	252	**296**	100	88	30	28	20	19	19	15		18	20	2	2	–	–	–	–	–	–	–	–	–	20,21-Dinoraspidospermidin-10-one, 5,19-didehydro-17-methoxy-	36528-89-9	NI30919
91	296	240	92	65	57	55	256	**296**	100	28	23	19	12	10	7	5		18	20	2	2	–	–	–	–	–	–	–	–	–	Indazole, 1-benzyl-5-butoxy-3-hydroxy-		LI06458
45	296	251	104	103	77	118	297	**296**	100	69	56	36	26	21	16	13		18	20	2	2	–	–	–	–	–	–	–	–	–	2-Methoxyacetophenone azine		TR00298
162	296	121	175	145	122	77	42	**296**	100	66	65	45	42	29	25	25		18	20	2	2	–	–	–	–	–	–	–	–	–	Phenol, 2,2′-[(1,1-dimethyl-1,2-ethanediyl)bis(nitrilomethylidyne)]bis-	30180-37-1	NI30920
214	104	43	76	296	215	55	41	**296**	100	76	46	37	18	18	18	15		18	20	2	2	–	–	–	–	–	–	–	–	–	Pyridine, 1,1′-(1,2-phenylenedicarbonyl)bis[1,2,3,4-tetrahydro-	52881-76-2	NI30921
214	104	76	296	55	215	43	105	**296**	100	85	35	24	20	16	14	11		18	20	2	2	–	–	–	–	–	–	–	–	–	Pyridine, 1,1′-(1,4-phenylenedicarbonyl)bis[1,2,3,4-tetrahydro-	52881-77-3	NI30922
162	296	121	175	145	122	77	107	**296**	100	66	65	46	41	28	22	17		18	20	2	2	–	–	–	–	–	–	–	–	–	N,N′-Bis(salicylidene)-2,2-dimethylethylenediamine		LI06459
83	278	85	236	279	264	235	263	**296**	100	71	56	54	37	22	22	15	**1.20**	18	20	2	2	–	–	–	–	–	–	–	–	–	Tetrahydrodragabine		NI30923
157	172	142	66	141	230	128	91	**296**	100	61	48	42	40	39	25	25	**21.27**	18	22	–	–	–	–	–	–	–	–	–	–	1	(π-Cyclopentadienyl)[-π-(4-2H-cyclopentadienyl-1,2,3,4-tetramethylcyclobut-2-enyl)]nickel		MS06814
218	163	162	78	55	296	79	217	**296**	100	25	25	25	18	16	6	5		18	25	–	–	–	–	–	–	–	–	–	–	1	Cyclohexadienyl(hexamethylbenzene)manganese		MS06815
169	135	155	141	121	107	41	183	**296**	100	94	48	44	42	42	37	35	**2.27**	18	29	1	–	–	–	1	–	–	–	–	–	–	Phenol, 2-chloro-4-dodecyl-		MS06816
43	41	55	253	69	81	95	67	**296**	100	42	41	17	17	15	14	14	**1.56**	18	32	3	–	–	–	–	–	–	–	–	–	–	Cyclohexadecanone, 2-acetoxy-		IC03740
59	91	176	94	179	152	111	100	**296**	100	49	35	30	29	29	28	26	**0.00**	18	32	3	–	–	–	–	–	–	–	–	–	–	2,4-Dodecadienoic acid, 11-hydroxy-3,7,11-trimethyl-, isopropyl ester, (E,E)-	40596-70-1	NI30924
104	97	103	95	265	237	121	107	**296**	100	61	47	38	34	34	34	33	**3.49**	18	32	3	–	–	–	–	–	–	–	–	–	–	4-Hexadecynoic acid, 2-methoxy-, methyl ester	40924-17-2	NI30925
98	55	41	43	67	83	69	95	**296**	100	86	79	50	37	27	26	25	**3.00**	18	32	3	–	–	–	–	–	–	–	–	–	–	Oxacyclotetradec-10-en-2-one, 13-hydroxy-14-pentyl-, [13S-(10Z,13R*,14R*)]-	63872-94-6	NI30926
214	41	132	133	55	113	215	83	**296**	100	52	47	43	38	24	20	19	**11.15**	18	33	1	–	–	–	–	–	–	–	1	–	–	Phosphine oxide, tricyclohexyl-	13689-19-5	NI30927
296	281	137	253	150	171	151	131	**296**	100	15	10	5	5	4	4	4		19	20	3	–	–	–	–	–	–	–	–	–	–	Benzofuran, 2,3-dihydro-2-(4-hydroxy-3-methoxyphenyl)-3-methyl-5(E)-propenyl-, (2R,3R)-		MS06817
296	107	121	281	202	119	189	267	**296**	100	8	7	5	5	5	2	1		19	20	3	–	–	–	–	–	–	–	–	–	–	Benzofuran, 2,3-dihydro-2-(4-hydroxyphenyl)-7-methoxy-3-methyl-5(E)-propenyl-, (2R,3R)-		MS06818
148	105	116	43	115	77	28	133	**296**	100	43	25	19	19	18	16	12	**9.72**	19	20	3	–	–	–	–	–	–	–	–	–	–	1,4-Dioxin, 2,3-dihydro-2,5-dimethyl-3,6-diphenyl-2-methoxy-		NI30928
183	209	224	141	198	165	296	153	**296**	100	98	74	71	69	44	34	34		19	20	3	–	–	–	–	–	–	–	–	–	–	2,5-Furandione, 3-[3,8-dimethyl-5-isopropyl-1-azulenyl]dihydro-	33658-55-8	MS06819
194	167	165	195	166	240	196	105	**296**	100	59	31	30	30	28	21	19	**0.11**	19	20	3	–	–	–	–	–	–	–	–	–	–	Oxiranecarboxylic acid, 2,2-diphenyl-, tert-butyl ester		NI30929
253	296	251	268	254	171	297	165	**296**	100	67	35	30	20	15	14	12		19	20	3	–	–	–	–	–	–	–	–	–	–	Phenanthro[1,2-b]furan-10,11-dione, 1,2,6,7,8,9-hexahydro-1,6,6-trimethyl-, (–)-	35825-57-1	NI30930
253	296	251	268	254	171	297	169	**296**	100	67	38	31	20	15	14	12		19	20	3	–	–	–	–	–	–	–	–	–	–	Phenanthro[1,2-b]furan-10,11-dione, 1,2,6,7,8,9-hexahydro-1,6,6-trimethyl-, (–)-	35825-57-1	LI06460
296	281	253	267	297	235	263	165	**296**	100	45	43	37	23	19	17	17		19	20	3	–	–	–	–	–	–	–	–	–	–	Phenanthro[3,2-b]furan-7,11-dione, 1,2,3,4,8,9-hexahydro-4,4,8-trimethyl-, (+)-	22550-15-8	NI30931
296	281	253	267	297	235	263	165	**296**	100	45	45	37	23	19	16	16		19	20	3	–	–	–	–	–	–	–	–	–	–	Phenanthro[3,2-b]furan-7,11-dione, 1,2,3,4,8,9-hexahydro-4,4,8-trimethyl-, (+)-	22550-15-8	LI06461
148	105	205	43	120	42	149	57	**296**	100	42	25	20	16	13	12	11	**1.00**	19	24	1	2	–	–	–	–	–	–	–	–	–	Acetamide, N-[4-[2-[methyl(2-phenylethyl)amino]ethyl]phenyl]-	52059-42-4	NI30932
223	225	296	295	169	184	156	197	**296**	100	94	77	75	24	19	19	15		19	24	1	2	–	–	–	–	–	–	–	–	–	Antirhine		MS06820
109	268	296	124	174	41	108	269	**296**	100	77	50	41	23	20	16	15		19	24	1	2	–	–	–	–	–	–	–	–	–	Aspidofractinin-17-ol	55724-56-6	NI30933
138	268	252	160	130	296	269	110	**296**	100	31	26	16	16	14	12	12		19	24	1	2	–	–	–	–	–	–	–	–	–	Aspidospermidine, 19,21-epoxy-	36459-35-5	NI30934
136	296	123	174	122	160	137	173	**296**	100	37	23	16	16	14	13	9		19	24	1	2	–	–	–	–	–	–	–	–	–	Condyfolan, 14,19-didehydro-12-methoxy-, (14E)-	54833-53-3	NI30935
140	143	156	296	144	122	124	130	**296**	100	58	56	49	42	39	23	22		19	24	1	2	–	–	–	–	–	–	–	–	–	Conoflorine		LI06462
170	296	295	156	169	69	223	71	**296**	100	67	64	64	56	33	32	26		19	24	1	2	–	–	–	–	–	–	–	–	–	Corynan-17-ol, 18,19-didehydro-	23443-70-1	NI30936
166	144	296	130	143	136	167	122	**296**	100	47	43	34	30	21	14	13		19	24	1	2	–	–	–	–	–	–	–	–	–	Curan-17-ol, 19,20-didehydro-, (19E)-	13941-27-0	NI30937
138	296	55	159	126	267	82	110	**296**	100	56	38	25	23	18	18	17		19	24	1	2	–	–	–	–	–	–	–	–	–	10,11-Dihydrocinchonine		LI06463
249	208	296	193	278	206	295	250	**296**	100	97	44	33	29	28	27	18		19	24	1	2	–	–	–	–	–	–	–	–	–	Eburnamenin-14-ol, 14,15-dihydro-, (14α)-	473-99-4	MS06821
208	249	278	193	206	250	209	248	**296**	100	94	30	24	22	19	18	15	**0.00**	19	24	1	2	–	–	–	–	–	–	–	–	–	Eburnamenin-14-ol, 14,15-dihydro-, (14α)-	473-99-4	NI30938
82	215	296	162	297	200	174	148	**296**	100	69	64	58	25	18	16	15		19	24	1	2	–	–	–	–	–	–	–	–	–	Indole, 5-methoxy-1-methyl-2-[(8-methyl-8-azabicyclo[3.2.1]octan-3-ylidene)methyl]-		DD01056

MASS TO CHARGE RATIOS								M.W.	INTENSITIES								Parent	C	H	O	N	S	F	Cl	Br	I	Si	P	B	X	COMPOUND NAME	CAS Reg No	No
279	144	149	114	91	130	113	177	296	100	98	95	83	67	60	50	45	0.00	19	24	1	2	–	–	–	–	–	–	–	–	–	3H-Indolizino[8,7-b]indole-2-propanol, 2-ethyl-2,5,6,11-tetrahydro-	55670-05-8	NI30939
249	208	278	193	206	296	250	295	296	100	96	35	28	26	23	20	18		19	24	1	2	–	–	–	–	–	–	–	–	–	Isoeburnamine		MS06822
227	171	228	296	226	78	128	41	296	100	55	35	18	16	15	13	12		19	24	1	2	–	–	–	–	–	–	–	–	–	Morphinan-17-carbonitrile, 3-methoxy-6-methyl-, (6α)-	54833-52-2	NI30940
131	165	296	103	137	136	123	110	296	100	80	68	52	41	39	33	28		19	24	1	2	–	–	–	–	–	–	–	–	–	Pyridine, 1,2,3,4-tetrahydro-1-(1-oxo-3-phenyl-2-propenyl)-5-(2-piperidinyl)-, (E)-	6793-63-1	NI30941
267	138	296	295	268				296	100	60	40	24	24					19	24	1	2	–	–	–	–	–	–	–	–	–	Tetrahydromeloscin		LI06464
55	41	28	69	83	43	97	296	296	100	83	83	73	72	71	61	58		19	36	2	–	–	–	–	–	–	–	–	–	–	Cyclohexaneheptanoic acid, 2-pentyl-, methyl ester	16847-10-2	NI30942
83	55	43	42	69	74	41	30	296	100	96	96	96	91	71	62	56	47.61	19	36	2	–	–	–	–	–	–	–	–	–	–	Cyclohexaneoctanoic acid, 2-butyl-, methyl ester	16847-05-5	NI30943
74	87	41	55	43	69	83	57	296	100	68	52	50	47	32	18	18	8.01	19	36	2	–	–	–	–	–	–	–	–	–	–	Cyclopentanetridecanoic acid, methyl ester	24828-61-3	WI01380
74	87	41	55	43	69	75	57	296	100	70	53	52	48	33	21	19	8.14	19	36	2	–	–	–	–	–	–	–	–	–	–	Cyclopentanetridecanoic acid, methyl ester	24828-61-3	NI30944
43	41	87	55	113	57	81	69	296	100	91	87	69	52	45	40	40	7.72	19	36	2	–	–	–	–	–	–	–	–	–	–	2-Hexadecenoic acid, 2,3-dimethyl-, methyl ester, (E)-	57289-45-9	NI30945
113	43	41	55	81	100	87	57	296	100	47	41	35	32	30	28	24	22.48	19	36	2	–	–	–	–	–	–	–	–	–	–	2-Hexadecenoic acid, 2,3-dimethyl-, methyl ester, (Z)-	57289-46-0	NI30946
100	43	85	41	55	113	57	29	296	100	71	62	29	21	18	16	15	1.34	19	36	2	–	–	–	–	–	–	–	–	–	–	2,4-Nonadecanedione	16577-69-8	NI30947
113	43	41	55	87	74	100	57	296	100	60	53	46	38	35	33	31	11.04	19	36	2	–	–	–	–	–	–	–	–	–	–	2-Octadecenoic acid, methyl ester	14435-34-8	NI30949
43	41	113	55	29	57	81	27	296	100	89	58	48	47	35	26	20	7.94	19	36	2	–	–	–	–	–	–	–	–	–	–	2-Octadecenoic acid, methyl ester	14435-34-8	NI30948
74	96	87	55	97	83	84	43	296	100	55	55	49	48	46	45	45	4.44	19	36	2	–	–	–	–	–	–	–	–	–	–	3-Octadecenoic acid, methyl ester	56599-33-8	NI30950
74	43	41	55	96	67	69	57	296	100	87	84	77	54	51	44	42	2.64	19	36	2	–	–	–	–	–	–	–	–	–	–	4-Octadecenoic acid, methyl ester	56555-10-3	NI30951
74	41	43	55	96	67	81	84	296	100	73	71	66	49	46	42	41	3.14	19	36	2	–	–	–	–	–	–	–	–	–	–	5-Octadecenoic acid, methyl ester	56554-45-1	NI30952
43	41	55	74	29	67	81	57	296	100	95	54	51	46	40	33	32	2.34	19	36	2	–	–	–	–	–	–	–	–	–	–	5-Octadecenoic acid, methyl ester	56554-45-1	NI30953
74	41	55	43	84	69	87	81	296	100	98	97	85	55	54	53	47	3.04	19	36	2	–	–	–	–	–	–	–	–	–	–	6-Octadecenoic acid, methyl ester	52355-31-4	NI30955
55	74	84	69	96	81	67	83	296	100	91	66	57	54	53	51	49	5.34	19	36	2	–	–	–	–	–	–	–	–	–	–	6-Octadecenoic acid, methyl ester	52355-31-4	NI30954
41	55	43	74	84	69	81	67	296	100	91	83	81	52	49	46	46	3.84	19	36	2	–	–	–	–	–	–	–	–	–	–	6-Octadecenoic acid, methyl ester, (Z)-	2777-58-4	NI30957
41	55	43	74	67	81	69	84	296	100	94	76	56	52	43	41	39	3.11	19	36	2	–	–	–	–	–	–	–	–	–	–	6-Octadecenoic acid, methyl ester, (Z)-	2777-58-4	NI30956
41	55	43	74	264	84	96	69	296	100	90	85	79	67	65	55	51	19.04	19	36	2	–	–	–	–	–	–	–	–	–	–	6-Octadecenoic acid, methyl ester, (Z)-	2777-58-4	NI30958
55	41	43	74	69	84	83	87	296	100	89	76	74	58	51	49	46	5.80	19	36	2	–	–	–	–	–	–	–	–	–	–	7-Octadecenoic acid, methyl ester	57396-98-2	NI30959
55	41	74	43	69	83	87	84	296	100	88	71	71	70	50	48	48	6.04	19	36	2	–	–	–	–	–	–	–	–	–	–	8-Octadecenoic acid, methyl ester	2345-29-1	NI30960
55	69	43	41	264	83	97	87	296	100	71	68	63	58	53	49	49	15.00	19	36	2	–	–	–	–	–	–	–	–	–	–	8-Octadecenoic acid, methyl ester, (E)-	26528-50-7	NI30961
55	39	69	41	74	83	43	73	296	100	84	75	72	70	61	58	53	8.61	19	36	2	–	–	–	–	–	–	–	–	–	–	9-Octadecenoic acid (Z)-, methyl ester	112-62-9	NI30962
55	69	74	83	96	97	41	84	296	100	79	69	67	64	62	59	58	1.21	19	36	2	–	–	–	–	–	–	–	–	–	–	9-Octadecenoic acid (Z)-, methyl ester	112-62-9	NI30964
55	264	41	69	74	43	83	87	296	100	88	85	65	64	56	53	44	27.68	19	36	2	–	–	–	–	–	–	–	–	–	–	9-Octadecenoic acid (Z)-, methyl ester	112-62-9	NI30963
55	41	69	74	43	83	84	87	296	100	75	59	55	52	48	41	40	10.00	19	36	2	–	–	–	–	–	–	–	–	–	–	9-Octadecenoic acid, methyl ester	2462-84-2	IC03741
55	41	69	74	43	83	87	84	296	100	85	69	68	63	51	45	42	5.94	19	36	2	–	–	–	–	–	–	–	–	–	–	9-Octadecenoic acid, methyl ester	2462-84-2	NI30965
41	55	43	69	74	83	84	67	296	100	96	72	43	40	35	30	30	0.00	19	36	2	–	–	–	–	–	–	–	–	–	–	9-Octadecenoic acid, methyl ester, (E)-	1937-62-8	NI30966
55	69	74	83	41	43	84	87	296	100	99	98	93	87	86	82	77	4.50	19	36	2	–	–	–	–	–	–	–	–	–	–	9-Octadecenoic acid, methyl ester, (E)-	1937-62-8	NI30967
55	41	69	43	74	83	87	97	296	100	72	60	59	55	45	37	34	3.24	19	36	2	–	–	–	–	–	–	–	–	–	–	10-Octadecenoic acid, methyl ester	13481-95-3	NI30968
55	41	43	69	74	83	87	264	296	100	78	72	66	61	47	37	35	10.30	19	36	2	–	–	–	–	–	–	–	–	–	–	10-Octadecenoic acid, methyl ester, (E)-	13038-45-4	NI30969
55	41	43	69	74	29	67	87	296	100	86	59	53	40	33	28	24	2.24	19	36	2	–	–	–	–	–	–	–	–	–	–	11-Octadecenoic acid, methyl ester	52380-33-3	NI30970
55	41	69	74	43	83	87	97	296	100	75	70	58	54	46	40	37	5.24	19	36	2	–	–	–	–	–	–	–	–	–	–	11-Octadecenoic acid, methyl ester, (Z)-	1937-63-9	NI30971
55	41	69	74	83	43	87	97	296	100	77	72	54	43	43	37	35	6.64	19	36	2	–	–	–	–	–	–	–	–	–	–	12-Octadecenoic acid, methyl ester	56554-46-2	NI30972
55	41	69	74	43	83	87	97	296	100	63	54	51	39	36	35	30	3.54	19	36	2	–	–	–	–	–	–	–	–	–	–	13-Octadecenoic acid, methyl ester	56554-47-3	NI30973
264	74	96	84	97	98	87	83	296	100	77	68	65	63	62	57	53	19.70	19	36	2	–	–	–	–	–	–	–	–	–	–	13-Octadecenoic acid, methyl ester, (Z)-	13058-55-4	NI30974
55	41	69	74	83	43	87	97	296	100	73	56	54	39	38	36	32	3.44	19	36	2	–	–	–	–	–	–	–	–	–	–	14-Octadecenoic acid, methyl ester	56554-48-4	NI30975
55	41	69	74	87	43	83	97	296	100	94	73	67	45	45	43	36	3.24	19	36	2	–	–	–	–	–	–	–	–	–	–	15-Octadecenoic acid, methyl ester	4764-72-1	NI30976
55	41	74	69	43	87	83	97	296	100	62	61	57	43	40	36	31	2.44	19	36	2	–	–	–	–	–	–	–	–	–	–	16-Octadecenoic acid, methyl ester	56554-49-5	NI30977
74	55	41	87	69	43	83	97	296	100	99	81	61	58	57	39	31	1.00	19	36	2	–	–	–	–	–	–	–	–	–	–	17-Octadecenoic acid, methyl ester	18654-84-7	NI30978
74	55	41	87	69	43	83	97	296	100	97	74	64	59	54	42	34	1.14	19	36	2	–	–	–	–	–	–	–	–	–	–	17-Octadecenoic acid, methyl ester	18654-84-7	NI30979
101	74	55	44	69	57	42	75	296	100	82	52	49	37	35	35	26	0.00	19	36	2	–	–	–	–	–	–	–	–	–	–	Pseudoisoprenoid acid		NI30980
88	69	101	55	41	83	97	111	296	100	93	80	72	49	46	41	39	0.00	19	36	2	–	–	–	–	–	–	–	–	–	–	13-Tetradecenoic acid, 2,4,6,8-tetramethyl-, methyl ester		LI06465
88	69	101	55	41	83	97	111	296	100	93	79	70	49	45	40	39	0.16	19	36	2	–	–	–	–	–	–	–	–	–	–	13-Tetradecenoic acid, 2,4,6,8-tetramethyl-, methyl ester, (all-R)-(–)-	27829-59-0	NI30981
88	101	69	55	83	41	111	97	296	100	72	66	49	39	38	28	23	2.70	19	36	2	–	–	–	–	–	–	–	–	–	–	13-Tetradecenoic acid, 2,4,6,8-tetramethyl-, methyl ester, (all-R)-(–)-	27829-59-0	WI01381
88	101	69	55	83	41	111	43	296	100	85	83	58	53	48	29	28	2.50	19	36	2	–	–	–	–	–	–	–	–	–	–	13-Tetradecenoic acid, 2,4,6,8-tetramethyl-, methyl ester, (2S,4R,6R,8R)-(+)-	27829-58-9	WI01382
41	55	43	57	29	69	71	83	296	100	90	84	66	52	42	41	36	0.20	19	36	2	–	–	–	–	–	–	–	–	–	–	10-Undecenoic acid, octyl ester	28080-85-5	WI01383
296	91	119	147	145	95	77	41	296	100	36	32	24	20	20	20	18		20	24	2	–	–	–	–	–	–	–	–	–	–	Crocetindial		MS06823
84	43	190	69	41	57	197	179	296	100	99	84	60	39	36	35	30	0.20	20	24	2	–	–	–	–	–	–	–	–	–	–	1,3-Dioxolane, 2-tert-butyl-2-methyl-4,5-diphenyl-	74793-55-8	NI30982
122	270	107	278	213	228	158	263	296	100	53	38	35	35	33	31	30	25.14	20	24	2	–	–	–	–	–	–	–	–	–	–	4,6-Estradien-3-one, 17α-ethynyl-17β-hydroxy-		NI30983

MASS TO CHARGE RATIOS								M.W.	INTENSITIES								Parent	C	H	O	N	S	F	Cl	Br	I	Si	P	B	X	COMPOUND NAME	CAS Reg No	No
296	198	172	211	297	185	173	199	**296**	100	45	42	38	22	21	14	13		20	24	2	–	–	–	–	–	–	–	–	–	–	**Estra-1,3,5(10),6-tetraen-17-one, 3-hydroxy-1,2-dimethyl-**	1818-09-3	**NI30984**
296	297	281	238	239	225	223	279	**296**	100	22	22	16	13	11	9	8		20	24	2	–	–	–	–	–	–	–	–	–	–	**Estra-1,3,5(10),9(11)-tetraen-17-one, 3-methoxy-1-methyl-**	69796-63-0	**NI30985**
91	296	105	262	133	117	159	191	**296**	100	53	50	32	28	20	17	6		20	24	2	–	–	–	–	–	–	–	–	–	–	**3-Heptanone, 5-methoxy-1,7-diphenyl-**		**DD01057**
296	281	104	187	160	297	225	211	**296**	100	49	39	30	26	25	25	24		20	24	2	–	–	–	–	–	–	–	–	–	–	**D-Homoestra-1,3,5(10),14-tetraen-17a-one, 3-methoxy-**		**LI06466**
296	187	186	297	212	281	172	159	**296**	100	45	38	23	18	15	15	15		20	24	2	–	–	–	–	–	–	–	–	–	–	**D-Homoestra-1,3,5(10),14-tetraen-17a-one, 3-methoxy-**		**LI06467**
296	187	186	297	278	212	263	159	**296**	100	44	37	23	18	18	16	15		20	24	2	–	–	–	–	–	–	–	–	–	–	**D-Homoestra-1,3,5(10),15-tetraen-17a-one, 3-methoxy-**	54844-27-8	**NI30986**
296	225	281	226	297	173	172	115	**296**	100	50	39	21	20	20	18	18		20	24	2	–	–	–	–	–	–	–	–	–	–	**D-Homoestra-1,3,5(10),8(14)-tetraen-17-one, 3-methoxy-, (9β)-**	54844-26-7	**NI30987**
160	296	254	225	172	159	171	161	**296**	100	86	45	21	20	18	17	16		20	24	2	–	–	–	–	–	–	–	–	–	–	**D-Homoestra-1,3,5(10),14-tetraen-17-one, 3-methoxy-, (8α)-**	54869-07-7	**NI30988**
160	296	225	161	281	254	173	115	**296**	100	84	26	24	22	20	20	20		20	24	2	–	–	–	–	–	–	–	–	–	–	**D-Homoestra-1,3,5(10),14-tetraen-17-one, 3-methoxy-, (8α,9α)-**		**LI06468**
296	281	225	160	297	211	187	159	**296**	100	58	27	23	22	18	18	18		20	24	2	–	–	–	–	–	–	–	–	–	–	**D-Homoestra-1,3,5(10),14-tetraen-17-one, 3-methoxy-, (8α,9β)-**	54833-57-7	**NI30989**
296	281	225	160	297	212	187	159	**296**	100	58	28	26	25	19	19	19		20	24	2	–	–	–	–	–	–	–	–	–	–	**D-Homoestra-1,3,5(10),14-tetraen-17-one, 3-methoxy-, (8α,9β)-**	54833-57-7	**LI06469**
160	296	161	225	294	281	172	159	**296**	100	77	28	23	22	22	22	18		20	24	2	–	–	–	–	–	–	–	–	–	–	**D-Homoestra-1,3,5(10),14-tetraen-17-one, 3-methoxy-, (9β)-**	54869-08-8	**NI30990**
281	173	296	282	149	297	189	121	**296**	100	53	43	24	10	9	9	9		20	24	2	–	–	–	–	–	–	–	–	–	–	**4-Indanol, 3-(3-hydroxy-4-methylphenyl)-1,1,3,5-tetramethyl-**		**NI30991**
281	296	282	173	297	189	165	133	**296**	100	28	21	13	7	6	6	5		20	24	2	–	–	–	–	–	–	–	–	–	–	**5-Indanol, 1-(3-hydroxy-4-methylphenyl)-1,3,3,6-tetramethyl-**		**NI30992**
296	121	267	159	297	281	173	135	**296**	100	38	31	31	23	13	12	11		20	24	2	–	–	–	–	–	–	–	–	–	–	**3,4-Bis(4-methoxyphenyl)hex-3-ene**		**MS06824**
213	160	296	159	133	228	145	146	**296**	100	40	38	23	23	16	14	12		20	24	2	–	–	–	–	–	–	–	–	–	–	**19-Norpregna-1,3,5(10)-trien-20-yne-3,17-diol, (17α)-**	57-63-6	**NI30994**
213	296	160	270	214	159	133	172	**296**	100	44	40	30	22	18	18	17		20	24	2	–	–	–	–	–	–	–	–	–	–	**19-Norpregna-1,3,5(10)-trien-20-yne-3,17-diol, (17α)-**	57-63-6	**NI30993**
241	91	205	296	150	105	281	137	**296**	100	95	81	74	37	34	26	26		20	24	2	–	–	–	–	–	–	–	–	–	–	**Phenol, 3-methoxy-2-(3-methylbut-2-enyl)-5-(2-phenylethyl)-**	70610-10-5	**NI30995**
296	121	159	267	297	173	281	135	**296**	100	61	55	51	24	23	21	19		20	24	2	–	–	–	–	–	–	–	–	–	–	**Stilbene, α,β-diethyl-4,4′-dimethoxy-**	130-79-0	**NI30996**
296	251	45	223	281	208	190	165	**296**	100	68	43	13	10	10	10	10		20	24	2	–	–	–	–	–	–	–	–	–	–	**Stilbene, 2-(methoxymethoxy)-4′-tert-butyl-, (E)-**		**MS06825**
296	251	45	223	165	208	190	207	**296**	100	98	63	25	19	18	18	16		20	24	2	–	–	–	–	–	–	–	–	–	–	**Stilbene, 2-(methoxymethoxy)-4′-tert-butyl-, (Z)-**		**MS06826**
183	140	175	218	105	184	112	181	**296**	100	30	29	25	22	21	16	15	**0.10**	20	28	–	–	–	–	–	–	–	1	–	–	–	**Diphenyloctylsilane**		**IC03742**
41	79	77	91	67	93	296	55	**296**	100	95	81	72	72	66	64	57		20	28	–	2	–	–	–	–	–	–	–	–	–	**Adamantanone azine**		**MS06827**
124	296	125	268	41	69	55	43	**296**	100	41	24	23	23	21	21	21		20	28	–	2	–	–	–	–	–	–	–	–	–	**Aspidospermidine, 1-methyl-**	2671-47-8	**NI30997**
124	296	125	144	152	268	158	148	**296**	100	16	11	9	7	5	5	4		20	28	–	2	–	–	–	–	–	–	–	–	–	**Aspidospermidine, 1-methyl-**	2671-47-8	**NI30998**
124	107	295	296	150	135	125	77	**296**	100	56	54	44	32	12	12	11		20	28	–	2	–	–	–	–	–	–	–	–	–	**Aspidospermidine, 1-methyl-, (19α)-**	2816-23-1	**NI30999**
157	295	158	171	41	199	138	170	**296**	100	40	40	31	20	19	18	17	**0.00**	20	28	–	2	–	–	–	–	–	–	–	–	–	**2,20-Cyclo-8,9-secoaspidospermidine, 3-methyl-, (2α,3β,5α,12β,19α,20R)**	56053-29-3	**NI31000**
211	41	296	43	212	105.5	167	40	**296**	100	86	49	29	21	20	19	19		20	28	–	2	–	–	–	–	–	–	–	–	–	**4-(1,5-Dimethylhexylamino)diphenylamine**		**IC03743**
134	162	163	91	135	118	28	79	**296**	100	99	60	35	21	19	19	16	**0.00**	20	28	–	2	–	–	–	–	–	–	–	–	–	**1,2-Ethanediamine, l-N,N-diethyl-N′,N′-dimethyl-1,2-diphenyl-**		**DD01058**
148	149	132	42	105	91	146	77	**296**	100	26	13	9	7	7	4	3	**0.00**	20	28	–	2	–	–	–	–	–	–	–	–	–	**1,2-Ethanediamine, meso-N,N,N′,N′-tetramethyl-1,2-bis(4-methylphenyl)-**		**DD01059**
148	149	132	91	105	42	178	119	**296**	100	30	12	10	9	7	6	5	**0.00**	20	28	–	2	–	–	–	–	–	–	–	–	–	**1,2-Ethanediamine, (±)-N,N,N′,N′-tetramethyl-1,2-bis(4-methylphenyl)-**		**DD01060**
162	134	91	42	28	118	29	77	**296**	100	95	53	41	31	28	25	24	**0.00**	20	28	–	2	–	–	–	–	–	–	–	–	–	**1,2-Ethanediamine, u-N,N-diethyl-N′,N′-dimethyl-1,2-diphenyl-**		**DD01061**
41	57	29	55	267	42	70	225	**296**	100	56	48	34	28	25	24	21	**17.42**	20	28	–	2	–	–	–	–	–	–	–	–	–	**4-(1-Ethyl-3-methylpentylamino)diphenylamine**		**IC03744**
171	124	42	296	110	170	41	158	**296**	100	79	50	47	47	45	38	32		20	28	–	2	–	–	–	–	–	–	–	–	–	**2H-3,7-Methanoazacycloundecino[5,4-b]indole, 7-ethyl-1,4,5,6,7,8,9,10-octahydro-10-methyl-, (R)-**	2671-48-9	**NI31001**
124	41	144	158	296	44	43	55	**296**	100	99	87	66	61	60	59	58		20	28	–	2	–	–	–	–	–	–	–	–	–	**Vallesamidine**	19637-77-5	**NI31002**
71	43	55	41	69	81	97	83	**296**	100	72	63	54	53	32	30	30	**0.00**	20	40	1	–	–	–	–	–	–	–	–	–	–	**Cyclotetradecanol, 1,7,11-trimethyl-4-isopropyl-, (–)-**	20489-83-2	**NI31003**
56	72	42	73	28	40	85	54	**296**	100	99	64	56	51	39	31	31	**1.90**	20	40	1	–	–	–	–	–	–	–	–	–	–	**3-Eicosanone**	2955-56-8	**NI31004**
125	111	139	123	151	137	153	113	**296**	100	98	50	45	37	33	27	23	**0.00**	20	40	1	–	–	–	–	–	–	–	–	–	–	**2-Hexadecen-1-ol, 3,7,11,15-tetramethyl-**		**NI31005**
57	71	123	81	140	196	97	95	**296**	100	95	69	68	48	35	35	35	**12.00**	20	40	1	–	–	–	–	–	–	–	–	–	–	**2-Hexadecen-1-ol, 3,7,11,15-tetramethyl-**	102608-53-7	**NI31006**
71	43	57	41	55	123	69	81	**296**	100	39	35	27	26	24	24	22	**3.20**	20	40	1	–	–	–	–	–	–	–	–	–	–	**2-Hexadecen-1-ol, 3,7,11,15-tetramethyl-, [R-[R*,R*,(E)]]-**	150-86-7	**WI01384**
71	43	57	123	81	95	111	196	**296**	100	38	35	25	22	15	8	6	**5.00**	20	40	1	–	–	–	–	–	–	–	–	–	–	**2-Hexadecen-1-ol, 3,7,11,15-tetramethyl-, [R-[R*,R*,(E)]]-**	150-86-7	**LI06470**
43	57	41	55	29	69	83	71	**296**	100	86	76	66	56	42	38	37	**0.20**	20	40	1	–	–	–	–	–	–	–	–	–	–	**Octadecane, 1-(ethenyloxy)-**	930-02-9	**WI01385**
239	115	296	268	144	237	63	240	**296**	100	82	69	63	44	37	33	25		21	12	2	–	–	–	–	–	–	–	–	–	–	**13H-Dibenzo[b,i]xanthen-13-one**		**IC03745**
119	120	106	239	134	296	237	163	**296**	100	50	47	43	27	20	20	13		21	12	2	–	–	–	–	–	–	–	–	–	–	**13H-Indeno[1,2-b]anthracene-6,11-quinone**		**LI06471**
296	239	240	268	267	163	148	134	**296**	100	80	20	15	12	6	5	3		21	12	2	–	–	–	–	–	–	–	–	–	–	**13H-Indeno[2,1-a]anthracene-7,12-quinone**		**LI06472**
296	165	295	89	297	148	166	63	**296**	100	81	42	34	23	14	13	10		21	16	–	2	–	–	–	–	–	–	–	–	–	**1H-Imidazole, 2,4,5-triphenyl-**	484-47-9	**NI31007**
41	91	146	160	115	194	234	296	**296**	100	51	41	35	30	14	13	12		21	16	–	2	–	–	–	–	–	–	–	–	–	**Propanedinitrile, (5,6-diphenylbicyclo[3.1.0]hex-2-ylidene)-, (1α,5α,6α)-**	70238-85-6	**NI31008**
296	91	219	205	191	297	165	115	**296**	100	59	42	36	29	25	21	20		21	16	–	2	–	–	–	–	–	–	–	–	–	**Propanedinitrile, (5,6-diphenylbicyclo[3.1.0]hex-2-ylidene)-, (1α,5α,6β)-**	70286-10-1	**NI31009**
43	57	71	41	85	55	56	42	**296**	100	92	55	50	38	29	15	13	**1.61**	21	44	–	–	–	–	–	–	–	–	–	–	–	**Heneicosane**	629-94-7	**AI02004**
57	71	85	43	296	99	113	70	**296**	100	78	57	43	28	19	13	11		21	44	–	–	–	–	–	–	–	–	–	–	–	**Heneicosane**	629-94-7	**NI31010**
85	71	56	54	55	70	69	113	**296**	100	99	99	95	72	51	49	35	**18.05**	21	44	–	–	–	–	–	–	–	–	–	–	–	**Heptadecane, 2,6,10,14-tetramethyl-**	18344-37-1	**NI31011**
71	85	57	97	83	111	70	99	**296**	100	99	99	94	88	82	56	52	**0.00**	21	44	–	–	–	–	–	–	–	–	–	–	–	**Heptadecane, 2,6,10,14-tetramethyl-**	18344-37-1	**LI06473**
57	43	71	41	55	85	56	70	**296**	100	89	63	35	27	23	21	12	**0.05**	21	44	–	–	–	–	–	–	–	–	–	–	–	**Heptadecane, 2,6,10,14-tetramethyl-**	18344-37-1	**MS06828**
57	71	43	85	55	41	56	69	**296**	100	74	58	43	29	22	19	14	**0.80**	21	44	–	–	–	–	–	–	–	–	–	–	–	**Heptadecane, 2,6,10,15-tetramethyl-**		**MS06829**
43	57	71	41	85	55	69	56	**296**	100	99	64	50	45	34	17	17	**0.04**	21	44	–	–	–	–	–	–	–	–	–	–	–	**8-Hexylpentadecane**		**AI02005**

MASS TO CHARGE RATIOS								M.W.	INTENSITIES								Parent	C	H	O	N	S	F	Cl	Br	I	Si	P	B	X	COMPOUND NAME	CAS Reg No	No
57	43	71	85	41	55	56	69	**296**	100	85	60	40	31	25	21	16	**1.55**	21	44	–	–	–	–	–	–	–	–	–	–	–	2-Methyleicosane		MS06830
57	43	41	71	56	55	85	267	**296**	100	74	49	41	37	30	27	15	**0.28**	21	44	–	–	–	–	–	–	–	–	–	–	–	3-Methyleicosane		AI02006
57	43	41	71	56	29	55	85	**296**	100	71	44	43	35	32	27	19	**0.24**	21	44	–	–	–	–	–	–	–	–	–	–	–	3-Methyleicosane		AI02007
57	43	41	71	56	55	85	69	**296**	100	77	49	39	37	29	25	19	**0.20**	21	44	–	–	–	–	–	–	–	–	–	–	–	3-Methyleicosane		AI02008
71	57	70	97	85	41	55	253	**296**	100	87	52	43	43	30	28	21	**0.47**	21	44	–	–	–	–	–	–	–	–	–	–	–	4-Methyleicosane		LI06474
43	57	85	71	84	41	55	56	**296**	100	86	79	48	43	27	24	18	**0.00**	21	44	–	–	–	–	–	–	–	–	–	–	–	5-Methyleicosane		LI06475
43	57	71	41	85	55	56	69	**296**	100	96	54	50	38	31	17	13	**0.16**	21	44	–	–	–	–	–	–	–	–	–	–	–	10-Methyleicosane		AI02009
57	43	71	85	41	55	99	56	**296**	100	82	58	43	38	26	15	15	**0.09**	21	44	–	–	–	–	–	–	–	–	–	–	–	10-Methyleicosane		AI02010
141	142	211	55	41	225	29	43	**296**	100	17	8	8	8	7	7	5	**4.87**	22	32	–	–	–	–	–	–	–	–	–	–	–	Naphthalene, 2,3-dihexyl-	74646-30-3	NI31012
142	296	141	154	155	115	153	127	**296**	100	83	76	64	56	27	25	9		23	20	–	–	–	–	–	–	–	–	–	–	–	Naphthalene, 1,1′-(1,3-propanediyl)bis-	14564-86-4	NI31013
100	127	70	109	119	150	81	69	**297**	100	19	12	8	7	6	4	4	**0.30**	4	–	2	1	1	9	–	–	–	–	–	–	–	(Tetrafluorosuccinimido)sulphur pentafluoride	80409-49-0	NI31014
77	297	282	250	73	47	147	69	**297**	100	90	77	58	34	25	20	17		4	9	–	3	1	4	–	–	–	1	2	–	–	Phosphoramidothioic difluoride, [difluoro[[(trimethylsilyl)imidocarbonyl]amino]phosphoranylidene]-	27351-98-0	MS06831
69	44	119	77	48	219	131	169	**297**	100	94	79	75	61	56	50	46	**0.00**	5	4	1	1	1	9	–	–	–	–	–	–	–	1-Butanesulphinamide, 1,1,2,2,3,3,4,4,4-nonafluoro-N-methyl-	51735-83-2	NI31015
299	301	53	265	267	266	297	36	**297**	100	81	81	75	55	50	46	34		6	2	1	1	–	–	1	2	–	–	–	–	–	2,6-Dibromo-1,4-benzoquinone chlorimine		IC03746
257	43	255	259	52	299	69	176	**297**	100	67	48	45	16	15	14	11	**8.00**	6	5	1	1	1	–	–	2	–	–	–	–	–	2,4-Dibromo-5-acetamidothiophene		IC03747
170	109	135	127	172	65	188	262	**297**	100	96	88	67	33	23	21	17	**10.80**	7	5	–	1	2	5	1	–	–	–	–	–	–	[Chloro(phenylthio)methylenimino]sulphur pentafluoride	89590-15-8	NI31016
269	64	205	297	271	43	44	31	**297**	100	92	49	45	42	40	38	36		7	8	4	3	2	–	1	–	–	–	–	–	–	2H-1,2,4-Benzothiadiazine-7-sulphonamide, 6-chloro-3,4-dihydro-, 1,1-dioxide	58-93-5	NI31017
298	269	300	271	281	299	283	133	**297**	100	65	42	27	25	15	12	12	**11.53**	7	8	4	3	2	–	1	–	–	–	–	–	–	2H-1,2,4-Benzothiadiazine-7-sulphonamide, 6-chloro-3,4-dihydro-, 1,1-dioxide	58-93-5	NI31018
262	125	79	63	47	109	93	62	**297**	100	69	42	42	41	25	23	20	**0.00**	8	9	5	1	1	–	1	–	–	–	1	–	–	Phosphorothioic acid, O-(2-chloro-4-nitrophenyl) O,O-dimethyl ester	2463-84-5	NI31019
262	125	79	47	109	93	63	216	**297**	100	64	35	23	21	21	18	13	**0.00**	8	9	5	1	1	–	1	–	–	–	1	–	–	Phosphorothioic acid, O-(2-chloro-4-nitrophenyl) O,O-dimethyl ester	2463-84-5	NI31020
109	125	79	297	128	47	63	93	**297**	100	97	43	40	34	27	23	20		8	9	5	1	1	–	1	–	–	–	1	–	–	Phosphorothioic acid, O-(3-chloro-4-nitrophenyl) O,O-dimethyl ester	500-28-7	NI31021
127	125	109	79	47	18	63	15	**297**	100	79	74	47	40	40	37	37	**32.03**	8	12	5	1	2	–	–	–	–	–	1	–	–	Phosphorothioic acid, O-[4-(aminosulphonyl)phenyl] O,O-dimethyl ester	115-93-5	NI31022
72	225	99	170	169	71	197	122	**297**	100	60	58	26	23	19	16	16	**0.00**	9	20	–	3	1	5	–	–	–	–	–	–	–	[Bis(diethylamino)methylenimino]sulphur pentafluoride	81439-14-7	NI31023
211	195	226	282	297	73	135	133	**297**	100	85	60	57	47	46	34	25		9	24	4	1	–	–	–	–	–	2	1	–	–	Phosphonic acid, [(acetylamino)methyl]-, bis(trimethylsilyl) ester	55108-80-0	NI31024
269	271	243	241	270	236	234	273	**297**	100	89	84	64	55	54	48	44	**0.00**	10	7	1	1	–	–	4	–	–	–	–	–	–	1H-Isoindole, 4,5,6,7-tetrachloro-3-ethoxy-	110568-88-2	DD01062
108	91	107	79	82	90	77	84	**297**	100	80	55	42	30	24	22	19	**0.70**	10	10	3	1	–	–	3	–	–	–	–	–	–	Urethane, (1-hydroxy-2,2,2-trichloroethyl)benzyl-	6798-34-1	NI31025
43	42	175	122	297	76	75	50	**297**	100	36	23	22	17	15	11	10		10	11	4	5	1	–	–	–	–	–	–	–	–	1,2-Dimethyl-5-[4-(nitrophenyl)sulphonylimino]-Δ^3-1,2,4-triazoline		MS06832
43	82	42	111	175	54	76	232	**297**	100	54	40	37	32	27	24	22	**10.00**	10	11	4	5	1	–	–	–	–	–	–	–	–	1-Methyl-5-[methyl[4-(nitrophenyl)sulphonyl]amino]-1,2,3-triazole		MS06833
43	227	226	41	56	169	55	69	**297**	100	99	67	40	39	35	34	33	**0.00**	10	14	1	1	–	7	–	–	–	–	–	–	–	1-Hexanamine, N-heptafluorobutanoyl-		MS06834
72	73	211	75	135	195	147	133	**297**	100	31	29	14	12	10	9	9	**1.30**	10	28	3	1	–	–	–	–	–	2	1	–	–	Phosphonic acid, (1-aminobutyl)-, bis(trimethylsilyl) ester	53044-31-8	NI31026
72	73	211	75	135	45	195	43	**297**	100	31	29	14	12	12	10	9	**1.30**	10	28	3	1	–	–	–	–	–	2	1	–	–	Phosphonic acid, (1-aminobutyl)-, bis(trimethylsilyl) ester	53044-31-8	NI31027
72	73	211	75	195	147	135	226	**297**	100	26	25	10	7	7	7	6	**1.20**	10	28	3	1	–	–	–	–	–	2	1	–	–	Phosphonic acid, (1-amino-2-methylpropyl)-, bis(trimethylsilyl) ester	53044-33-0	NI31028
72	73	55	75	71	45	195	147	**297**	100	26	12	10	9	8	7	7	**1.20**	10	28	3	1	–	–	–	–	–	2	1	–	–	Phosphonic acid, (1-amino-2-methylpropyl)-, bis(trimethylsilyl) ester	53044-33-0	NI31029
72	73	211	55	75	71	45	147	**297**	100	26	25	12	10	9	8	7	**1.20**	10	28	3	1	–	–	–	–	–	2	1	–	–	Phosphonic acid, (1-amino-2-methylpropyl)-, bis(trimethylsilyl) ester	53044-33-0	NI31030
297	299	268	266	165	163	84	218	**297**	100	99	94	94	68	68	54	48		11	8	2	1	1	–	–	1	–	–	–	–	–	4-Isothiazolecarboxylic acid, 5-bromo-3-phenyl-, methyl ester		LI06476
297	135	77	298	296	65	51	67	**297**	100	82	16	12	12	10	10	9		11	8	3	1	–	5	–	–	–	–	–	–	–	Benzylamine, 3,4-methylenedioxy-N-(pentafluoropropionyl)-		MS06835
135	164	136	108	60	178	73	58	**297**	100	90	58	40	32	23	19	12	**1.00**	11	15	3	5	1	–	–	–	–	–	–	–	–	5′-S-Methyl-5′-thioadenosine		LI06477
135	164	136	108	62	178	73	60	**297**	100	91	58	40	32	24	20	13	**1.00**	11	15	3	5	1	–	–	–	–	–	–	–	–	5′-S-Methyl-5′-thioadenosine		MS06836
27	165	28	194	169	166	43	135	**297**	100	97	52	40	38	35	29	28	**7.28**	11	15	5	5	–	–	–	–	–	–	–	–	–	Adenosine, 2-methoxy-	24723-77-1	NI31031
128	142	127	44	27	43	28	41	**297**	100	80	70	59	27	25	19	17	**0.00**	11	15	5	5	–	–	–	–	–	–	–	–	–	Guanosine, 1-methyl-	2140-65-0	NI31032
28	165	27	136	110	44	43	57	**297**	100	77	67	28	25	22	22	21	**0.00**	11	15	5	5	–	–	–	–	–	–	–	–	–	Guanosine, N-methyl-	2140-77-4	NI31033
44	29	151	43	31	45	41	27	**297**	100	27	22	19	19	16	16	12	**2.31**	11	15	5	5	–	–	–	–	–	–	–	–	–	Guanosine, 2′-O-methyl-	2140-71-8	NI31035
71	87	151	88	59	43			**297**	100	95	75	56	51	51			**9.47**	11	15	5	5	–	–	–	–	–	–	–	–	–	Guanosine, 2′-O-methyl-	2140-71-8	NI31034
136	135	164	178	108	81	266	148	**297**	100	73	47	21	17	17	7	6	**3.00**	11	15	5	5	–	–	–	–	–	–	–	–	–	9H-Purin-6-amine, 9-[3-C-(hydroxymethyl)-β-D-xylofuranosyl]-	70723-08-9	NI31036
135	164	108	60	136	43	57	73	**297**	100	89	29	27	21	20	19	17	**5.10**	11	15	5	5	–	–	–	–	–	–	–	–	–	9H-Purin-6-amine, 9-β-D-allopyranosyl-		LI06478
135	136	108	57	43	73	54	81	**297**	100	47	26	24	18	14	13	10	**0.80**	11	15	5	5	–	–	–	–	–	–	–	–	–	9H-Purin-6-amine, 9-β-D-psicofuranosyl-	1874-54-0	LI06479
135	136	108	57	43	73	54	53	**297**	100	47	26	24	17	14	13	10	**1.00**	11	15	5	5	–	–	–	–	–	–	–	–	–	9H-Purin-6-amine, 9-β-D-psicofuranosyl-	1874-54-0	NI31037
135	102	60	162	43	72	174	136	**297**	100	55	44	42	31	23	18	14	**0.85**	11	23	4	1	2	–	–	–	–	–	–	–	–	L-Arabinose, 5-(acetylamino)-5-deoxy-, diethyl mercaptal	55955-63-0	NI31038
135	43	218	60	90	72	126	158	**297**	100	94	91	56	52	38	32	26	**0.00**	11	23	4	1	2	–	–	–	–	–	–	–	–	D-Ribose, 3-acetamido-3-deoxy-, diethyl mercaptal	3509-32-8	NI31039
296	297	77	269	174	147	129	102	**297**	100	80	33	23	18	18	18	16		12	6	1	3	1	3	–	–	–	–	–	–	–	5H-1,3,4-Thiadiazolo[3,2-a]pyrimidin-5-one, 7-phenyl-2-(trifluoromethyl)-	94605-00-2	NI31040
102	297	269	77	298	296	103	116	**297**	100	85	22	17	13	13	12	10		12	6	1	3	1	3	–	–	–	–	–	–	–	7H-1,3,4-Thiadiazolo[3,2-a]pyrimidin-7-one, 5-phenyl-2-(trifluoromethyl)-	82077-81-4	NI31041

MASS TO CHARGE RATIOS								M.W.	INTENSITIES								Parent	C	H	O	N	S	F	Cl	Br	I	Si	P	B	X	COMPOUND NAME	CAS Reg No	No
297	127	159	111	75	170	175	76	**297**	100	88	88	76	64	57	53	52		12	8	4	1	1	–	1	–	–	–	–	–	–	4-Nitro-4′-chlorodiphenyl sulphone	39055-84-0	NI31042
159	126	173	142	298	140	95	199	**297**	100	71	44	38	32	28	28	27	**0.00**	12	11	4	1	2	–	–	–	–	–	–	–	–	Benzenesulphonamide, N-(phenylsulphonyl)-	2618-96-4	NI31043
77	140	51	141	233	94	92	78	**297**	100	68	65	59	21	19	18	18	**5.39**	12	11	4	1	2	–	–	–	–	–	–	–	–	Benzenesulphonamide, N-(phenylsulphonyl)-	2618-96-4	NI31044
155	154	142	77	156	127	51	128	**297**	100	67	24	14	13	12	11	9	**0.00**	12	12	–	1	–	–	–	–	1	–	–	–	–	Pyridinium, 1-methyl-2-phenyl-, iodide	52806-02-7	NI31045
134	91	121	119	135	297	65	77	**297**	100	98	55	44	18	17	12	9		12	12	2	1	–	5	–	–	–	–	–	–	–	Ethanamine, (2-methoxyphenyl)-N-(pentafluoropropionyl)-		MS06837
134	121	91	135	297	78	119	77	**297**	100	58	31	23	20	16	12	12		12	12	2	1	–	5	–	–	–	–	–	–	–	Ethanamine, (2-methoxyphenyl)-N-(pentafluoropropionyl)-		MS06838
121	134	135	122	119	91	78	77	**297**	100	65	12	9	8	8	8	8	**7.20**	12	12	2	1	–	5	–	–	–	–	–	–	–	Ethanamine, (3-methoxyphenyl)-N-(pentafluoropropionyl)-		MS06839
43	282	77	41	30	128	115	91	**297**	100	54	26	25	25	24	23	23	**5.00**	12	15	6	3	–	–	–	–	–	–	–	–	–	Benzene, 1-tert-butyl-3,5-dimethyl-2,4,6-trinitro-		NI31046
252	57	69	134	297	196			**297**	100	41	23	17	13	11				12	15	6	3	–	–	–	–	–	–	–	–	–	Isoleucine, 2,4-dinitrophenyl-		LI06480
252	43	196	134	297	210			**297**	100	96	28	13	12	10				12	15	6	3	–	–	–	–	–	–	–	–	–	Leucine, 2,4-dinitrophenyl-		LI06481
252	43	41	134	164	297	196		**297**	100	40	35	19	14	9	8			12	15	6	3	–	–	–	–	–	–	–	–	–	Norleucine, 2,4-dinitrophenyl-		LI06482
238	239	254	166	297	146	119	130	**297**	100	15	13	12	11	7	7	6		12	15	6	3	–	–	–	–	–	–	–	–	–	L-Valine, N-(2,4-dinitrophenyl)-, methyl ester	10420-77-6	NI31047
297	205	251	298	267	206	102	221	**297**	100	64	19	16	10	10	10	8		13	7	4	5	–	–	–	–	–	–	–	–	–	5H-Benzomidazo[1,2-a]benzimidazole, 7,9-dinitro-		IC03748
130	144	143	115	131	297	77	117	**297**	100	38	22	11	10	9	6	5		13	15	2	1	–	–	–	–	–	–	–	–	1	Propionic acid, 3-[(2-indol-3-ylethyl)selenyl]-	1144-33-8	NI31048
72	297	266	225	135	63	42	210	**297**	100	31	18	13	12	12	11	8		13	15	5	1	1	–	–	–	–	–	–	–	–	Thiocarbamic acid, N,N-dimethyl-, S-2,5-di(methyoxycarbonyl)phenyl ester		LI06483
196	137	59	57	191	190	71	69	**297**	100	89	65	62	56	38	36	35	**17.00**	13	15	7	1	–	–	–	–	–	–	–	–	–	Benzeneacetic acid, 5-hydroxy-4-methoxy-2-(3-nitro-1-oxopropyl)-, methyl ester	40992-01-6	NI31049
29	43	41	27	57	268	39	30	**297**	100	74	48	42	35	30	30	25	**2.00**	13	19	5	3	–	–	–	–	–	–	–	–	–	Benzenemethanol, 4-[(1-ethylpropyl)amino]-2-methyl-3,5-dinitro-	56750-76-6	NI31050
266	267	57	220	174	132	191	190	**297**	100	10	8	7	5	5	4	4	**2.50**	13	19	5	3	–	–	–	–	–	–	–	–	–	1-Propanol, 2-[[4-tert-butyl-2,6-dinitrophenyl]amino]-	55702-43-7	NI31051
69	43	41	182	71	183	153	154	**297**	100	77	42	40	35	21	15	14	**0.00**	13	22	3	1	–	3	–	–	–	–	–	–	–	L-Isoleucine, N-(trifluoroacetyl)-, 1-methylbutyl ester	57983-41-2	NI31052
168	55	43	41	42	169	85	69	**297**	100	55	51	19	18	15	11	10	**0.00**	13	22	3	1	–	3	–	–	–	–	–	–	–	L-Isovaline, N-(trifluoroacetyl)-, 1-methylpentyl ester	57983-77-4	NI31053
43	69	140	71	41	182	183	70	**297**	100	69	43	31	27	25	16	11	**0.00**	13	22	3	1	–	3	–	–	–	–	–	–	–	L-Leucine, N-(trifluoroacetyl)-, 1-methylbutyl ester	57983-42-3	NI31054
208	238	67	41	69	240	55	141	**297**	100	80	53	45	43	30	26	25	**2.00**	13	22	3	1	–	3	–	–	–	–	–	–	–	Methyl 2,2-dibutyl-2-trifluoroacetamidoacetate		LI06484
43	55	168	169	85	41	57	56	**297**	100	72	54	33	29	28	15	11	**0.00**	13	22	3	1	–	3	–	–	–	–	–	–	–	L-Valine, N-(trifluoroacetyl)-, 1-methylpentyl ester	57983-80-9	NI31055
297	254	299	256	298	255	232	102	**297**	100	52	37	18	17	7	6	6		14	7	1	1	–	3	1	–	–	–	–	–	–	Acridine, 1,2,4-trifluoro-3-methoxy-9-chloro-		NI31056
296	298	297	299	300	91	89	116	**297**	100	99	87	37	34	16	12	11		14	10	–	1	–	–	3	–	–	–	–	–	–	2-(2,4,6-Trichlorophenyl)-2-azaindane		MS06840
91	57	82	155	67	44	108	197	**297**	100	98	75	67	45	42	27	24	**0.00**	14	19	4	1	1	–	–	–	–	–	–	–	–	Carbamic acid, (4-tolylsulphonyl)-, cyclohexyl ester	18303-08-7	NI31057
91	155	197	171	151	107	108	233	**297**	100	67	24	12	9	8	5	3	**0.00**	14	19	4	1	1	–	–	–	–	–	–	–	–	Carbamic acid, (4-tolylsulphonyl)-, cyclohexyl ester	18303-08-7	LI06485
58	97	191	190	261	203	189	42	**297**	100	87	81	55	18	17	16	10	**0.00**	14	20	–	3	1	–	1	–	–	–	–	–	–	1,2-Ethanediamine, N,N-dimethyl-N′-2-pyridinyl-N′-(2-thienylmethyl)-, monohydrochloride	135-23-9	NI31058
73	208	192	297	207	282	45	75	**297**	100	54	37	32	32	31	21	15		14	27	2	1	–	–	–	–	–	2	–	–	–	Pyridine, 2,4-dimethyl-3-(trimethylsiloxy)-5-[(trimethlysiloxy)methyl]-		NI31059
297	296	270	250	298	77	91	223	**297**	100	36	16	16	15	10	9	8		15	11	2	3	1	–	–	–	–	–	–	–	–	1H-1,4-Benzodiazepine-2-thione, 2,3-dihydro-7-nitro-5-phenyl-	35628-44-5	NI31060
77	128	156	155	297	51	127	101	**297**	100	90	68	45	36	35	33	31		15	11	2	3	1	–	–	–	–	–	–	–	–	1-Phthalazinecarbonitrile, 1,2-dihydro-2-(phenylsulphonyl)-	21452-56-2	NI31061
232	262	77	204	205	75	177	104	**297**	100	81	81	60	55	45	35	30	**0.00**	15	11	4	3	–	–	–	–	–	–	–	–	–	2H-1,4-Benzodiazepin-2-one, 1,3-dihydro-3-hydroxy-7-nitro-5-phenyl-	16442-74-3	NI31062
77	145	119	281	183	105	51	91	**297**	100	38	31	27	23	23	23	22	**20.00**	15	11	4	3	–	–	–	–	–	–	–	–	–	3,5-Pyrazolidinedione, 4-nitro-1,2-diphenyl-	6148-09-0	NI31063
84	55	297	218	205	134	233	108	**297**	100	40	4	4	3	3	2	2		15	23	3	1	1	–	–	–	–	–	–	–	–	Bornane-10,2-sultam, N-[(E)-2-methyl-2-butenoyl]-		DD01063
166	238	297	163	165	167	164	295	**297**	100	33	32	15	14	13	9	8		15	23	5	1	–	–	–	–	–	–	–	–	–	Dimethyl (2-morpholinocyclohexenyl)malonate		MS06841
83	182	240	269	225	297			**297**	100	50	30	17	11	9				15	27	3	3	–	–	–	–	–	–	–	–	–	1,5,9-Triazacyclododecane-4,8,12-trione, 2,2,6,6,10,10-hexamethyl-	42492-21-7	NI31064
150	119	122	151	107	152	208	251	**297**	100	35	30	22	9	8	4	3	**0.00**	16	11	5	1	–	–	–	–	–	–	–	–	–	Aurone, 6-methoxy-2′-nitro-	77764-80-8	NI31065
297	296	250	251	208	152	106	236	**297**	100	76	62	28	27	15	12	11		16	11	5	1	–	–	–	–	–	–	–	–	–	Aurone, 6-methoxy-3′-nitro-	77764-81-9	NI31066
297	250	296	251	280	208	152	106	**297**	100	86	76	33	19	16	13	9		16	11	5	1	–	–	–	–	–	–	–	–	–	Aurone, 6-methoxy-4′-nitro-	77764-82-0	NI31067
297	149	266	167	105	57	217	279	**297**	100	65	31	29	24	24	21	18		16	11	5	1	–	–	–	–	–	–	–	–	–	4-Benzofurancarboxylic acid, 2-(4-nitrophenyl)-, methyl ester	41019-65-2	NI31068
297	233	147	267	295	298	175	232	**297**	100	31	28	21	20	19	19	18		16	15	3	3	–	–	–	–	–	–	–	–	–	Benzimidazole, 2-(α-hydroxy-3-nitrobenzyl)-5,6-dimethyl-		LI06486
150	132	131	104	282	165	175	297	**297**	100	54	33	28	19	19	5	2		16	15	3	3	–	–	–	–	–	–	–	–	–	1,2,4-Oxadiazole, 3,5-dimethyl-5-(3-nitrophenyl)-4-phenyl-		LI06487
77	205	268	51	121	204	165	190	**297**	100	99	87	60	48	43	40	38	**2.00**	17	15	–	1	2	–	–	–	–	–	–	–	–	Aniline, N-(3-ethylthio-2-benzo[b]thienylidene)-	92886-22-1	NI31069
91	297	220	180	104	179	117	77	**297**	100	32	18	18	16	15	15	15		17	15	–	1	2	–	–	–	–	–	–	–	–	1,3-Thiazole-5(4H)-thione, 4-benzyl-4-methyl-2-phenyl-		DD01064
164	255	131	136	91	297	206	109	**297**	100	46	33	29	20	14	8	6		17	15	2	1	1	–	–	–	–	–	–	–	–	2H-1,4-Benzothiazin-3(4H)-one, 4-acetyl-2-benzyl-		MS06842
238	170	297	265	178	180	194		**297**	100	59	28	22	17	13	7			17	15	4	1	–	–	–	–	–	–	–	–	–	Benzo[g]indolizine-1-acetic acid, 2-(methoxycarbonyl)-, methyl ester		LI06488
91	107	105	206	118	191	130	79	**297**	100	29	28	25	9	5	5	5	**0.00**	17	15	4	1	–	–	–	–	–	–	–	–	–	3-Oxazolidinecarboxylic acid, 5-oxo-2-phenyl-, benzyl ester, (±)-		DD01065
118	194	193	104	119	91	77	105	**297**	100	87	58	53	47	35	33	27	**6.00**	17	19	–	3	1	–	–	–	–	–	–	–	–	s-Triazine-6-thione, 1,5-dimethyl-2,4-diphenylhexahydro-		IC03749
297	296	298	268	106	130	118	282	**297**	100	43	19	13	13	12	11	9		17	19	2	3	–	–	–	–	–	–	–	–	–	Uracil, 1,3-dimethyl-6-ethyl-5-(3-methyl-1H-indol-2-yl)-		MS06843
297	253	282	208	167	210	211	224	**297**	100	67	66	60	56	55	49	40		18	19	3	1	–	–	–	–	–	–	–	–	–	Benzo[k,l]acridin-6-one, 3β-hydroxy-3a-methyl-9-methoxy-1,2,3,3a,4,5,6-heptahydro-		LI06489
91	120	77	190	108	147	79	51	**297**	100	59	58	35	25	19	17	17	**0.00**	18	19	3	1	–	–	–	–	–	–	–	–	–	Carbamic acid, (4-phenyl-4-oxobutyl)-, benzyl ester	69352-30-3	DD01066
190	191	149	175	132	131	107	81	**297**	100	18	4	2	2	2	2	2	**0.10**	18	19	3	1	–	–	–	–	–	–	–	–	–	Cinnamolaurine		MS06844

MASS TO CHARGE RATIOS								M.W.	INTENSITIES								Parent	C	H	O	N	S	F	Cl	Br	I	Si	P	B	X	COMPOUND NAME	CAS Reg No	No
297	296	268	254	298	42	239	28	**297**	100	46	41	25	24	12	10	10		18	19	3	1	–	–	–	–	–	–	–	–	–	Crotonosine, N-methyl-		MS06845
192	223	202	279	269	236	297	268	**297**	100	48	48	41	32	24	20	20		18	19	3	1	–	–	–	–	–	–	–	–	–	Cyclobut[c]quinoline, 2a,8b-dihydroxy-7-methoxy-3-phenyl-1,2,3,4-tetrahydro-, (2aR*,3R*,8bS*)-	90820-24-9	NI31070
266	297	282	264	218	226	239		**297**	100	72	45	21	20	12	8			18	19	3	1	–	–	–	–	–	–	–	–	–	Erythraline		LI06490
297	42	298	296	58	282	241	240	**297**	100	53	20	19	18	17	15	15		18	19	3	1	–	–	–	–	–	–	–	–	–	Morphinan-3-ol, 6,7,8,14-tetradehydro-4,5-epoxy-6-methoxy-17-methyl-, (5α)-	467-04-9	NI31071
297	42	94	298	115	229	59	44	**297**	100	23	22	18	17	16	13	11		18	19	3	1	–	–	–	–	–	–	–	–	–	Morphinan-6-one, 7,8-didehydro-4,5-epoxy-3-methoxy-17-methyl-, (5α)-	467-13-0	LI06491
155	127	156	297	240	128	126	41	**297**	100	30	12	5	4	4	4	3		18	19	3	1	–	–	–	–	–	–	–	–	–	5-Oxazolidinone, 2-tert-butyl-3-(1-naphthoyl)-, (±)-		DD01067
155	127	57	156	41	43	240	56	**297**	100	31	14	12	9	7	6	5	**0.00**	18	19	3	1	–	–	–	–	–	–	–	–	–	5-Oxazolidinone, 2-tert-butyl-3-(2-naphthoyl)-, (±)-		DD01068
44	254	297	251	238	221	42	266	**297**	100	44	38	21	18	12	11	10		18	19	3	1	–	–	–	–	–	–	–	–	–	1,5-Phenanthrenediol, 6-methoxy-4-[2-(methylamino)ethyl]-	476-68-6	NI31072
282	297	184	105	254	210	264	226	**297**	100	29	22	12	6	5	4	4		18	19	3	1	–	–	–	–	–	–	–	–	–	1H,7H-Pyrano[2,3-g]indole, 7,7-dimethyl-2-(3-methylsuccinyl)-		MS06846
268	297	296	298	269	253	225	254	**297**	100	79	42	21	18	13	13	11		18	19	3	1	–	–	–	–	–	–	–	–	–	Spiro[2,5-cyclohexadiene-1,7′(1′H)-cyclopent[ij]isoquinolin]-4-one, 2′,3′,8′,8′a-tetrahydro-5′,6′-dimethoxy-, (R)-	2810-21-1	NI31073
297	268	296	298	269	253	254	225	**297**	100	44	36	18	9	6	5	5		18	19	3	1	–	–	–	–	–	–	–	–	–	Spiro[2,5-cyclohexadiene-1,7′(1′H)-cyclopent[ij]isoquinolin]-4-one, 2′,3′,8′,8′a-tetrahydro-5′,6′-dimethoxy-, (R)-	2810-21-1	LI06492
297	268	296	298	269	253	254	225	**297**	100	46	37	20	9	7	6	6		18	19	3	1	–	–	–	–	–	–	–	–	–	Spiro[2,5-cyclohexadiene-1,7′(1′H)-cyclopent[ij]isoquinolin]-4-one, 2′,3′,8′,8′a-tetrahydro-5′,6′-dimethoxy-, (R)-	2810-21-1	MS06847
297	296	268	298	254	42	239	165	**297**	100	45	43	25	25	13	10	8		18	19	3	1	–	–	–	–	–	–	–	–	–	Spiro[2,5-cyclohexadiene-1,7′(1′H)-cyclopent[ij]isoquinolin]-4-one, 2′,3′,8′,8′a-tetrahydro-5′-hydroxy-6′-methoxy-1′-methyl-, (R)-	10214-76-3	NI31074
297	296	268	254	298	42	269	239	**297**	100	45	40	24	23	12	9	9		18	19	3	1	–	–	–	–	–	–	–	–	–	Spiro[2,5-cyclohexadiene-1,7′(1′H)-cyclopent[ij]isoquinolin]-4-one, 2′,3′,8′,8′a-tetrahydro-5′-hydroyx-6′-methoxy-1′-methyl-, (R)-	10214-76-3	LI06493
86	57	84	85	212	128	127	51	**297**	100	34	29	14	12	12	12	12	**0.05**	18	19	3	1	–	–	–	–	–	–	–	–	–	9-Xanthenone, 4-[4-(methylamino)butoxy]-		MS06848
201	297	96	77	202	51	296	47	**297**	100	70	56	46	32	18	17	17		18	20	1	1	–	–	–	–	–	–	1	–	–	N-Cyclohexylidenediphenylphosphinamide		MS06849
239	297	296	197	170	298	240	41	**297**	100	96	58	46	29	22	20	19		18	23	1	3	–	–	–	–	–	–	–	–	–	Indolo[2,3-a]quinolizine, 1-(2-carbamoylethyl)-1,2,3,4,6,7,12,12b-octahydro		NI31075
114	43	240	96	44	115	58	283	**297**	100	20	14	11	11	8	6	4	**2.00**	18	35	2	1	–	–	–	–	–	–	–	–	–	2-Dodecanone, 12-(5-hydroxy-6-methyl-2-piperidinyl)-, [2S-(2α,5α,6α)]-	5227-24-7	LI06494
114	96	240	115	70	84	69	67	**297**	100	18	10	8	8	6	6	5	**1.00**	18	35	2	1	–	–	–	–	–	–	–	–	–	2-Dodecanone, 12-(5-hydroxy-6-methyl-2-piperidinyl)-, [2S-(2α,5α,6α)]-	5227-24-7	NI31077
298	114	296	299	280	297	96	115	**297**	100	90	60	25	17	15	10	8		18	35	2	1	–	–	–	–	–	–	–	–	–	2-Dodecanone, 12-(5-hydroxy-6-methyl-2-piperidinyl)-, [2S-(2α,5α,6α)]-	5227-24-7	NI31076
297	77	298	269	268	148.5	296	270	**297**	100	27	23	20	13	8	6	5		19	11	1	3	–	–	–	–	–	–	–	–	–	1a-Phenyl-1a,2-dihydroisoxazolo[4,3]phenazine		MS06850
297	296	282	268	298	174	269	200	**297**	100	63	62	51	23	17	15	15		19	23	2	1	–	–	–	–	–	–	–	–	–	6-Azaestra-1,3,5(10),8(14)-tetraen-17-one, 3-methoxy-6-methyl-	54869-09-9	LI06495
297	296	282	269	298	174	283	200	**297**	100	63	62	50	23	17	14	14		19	23	2	1	–	–	–	–	–	–	–	–	–	6-Azaestra-1,3,5(10),8(14)-tetraen-17-one, 3-methoxy-6-methyl-	54869-09-9	NI31078
134	91	163	106	135	44	105	77	**297**	100	99	85	44	43	30	25	19	**9.00**	19	23	2	1	–	–	–	–	–	–	–	–	–	2H-1,3-Benzoxazine, 3,4-dihydro-3-(2-hydroxy-3,5-dimethylbenzyl)-6,8-dimethyl-		NI31079
228	186	41	174	229	105	69	67	**297**	100	71	47	42	37	35	33	25	**24.39**	19	23	2	1	–	–	–	–	–	–	–	–	–	Carbostyril, 3-(3-methyl-2-butenyl)-4-[(3-methyl-2-butenyl)oxy]-	18118-29-1	NI31080
157	158	41	70	156	227	297		**297**	100	67	34	27	20	17	12			19	23	2	1	–	–	–	–	–	–	–	–	–	4,4a-Dihydro-1-isopropylidene-4,4,4a,8-tetramethyl-1,3-oxazino[3,4-a]quinolin-3-one		LI06496
297	296	186	160	298	187	200	184	**297**	100	81	35	22	20	17	11	9		19	23	2	1	–	–	–	–	–	–	–	–	–	1′H-Estra-1,3,5(10),9(11)-tetraeno[1,10,9,11-bcd]pyrrol-17-one, 9α,11β-dihydro-3-methoxy-	13871-40-4	NI31081
297	157	96	91	92	266	144	126	**297**	100	52	51	47	39	35	28	27		19	23	2	1	–	–	–	–	–	–	–	–	–	Estra-1,3,5(10),6-tetraen-17-one, 3-hydroxy-, O-methyloxime	69833-89-2	NI31082
297	139	266	126	298	157	144	282	**297**	100	32	27	23	22	20	18	15		19	23	2	1	–	–	–	–	–	–	–	–	–	Estra-1,3,5(10),7-tetraen-17-one, 3-hydroxy-, O-methyloxime	69834-04-4	NI31083
297	296	266	298	176	42	165	163	**297**	100	64	24	22	21	19	16	16		19	23	2	1	–	–	–	–	–	–	–	–	–	Morphinan, 5,6,8,14-tetradehydro-3,6-dimethoxy-17-methyl-	1092-95-1	NI31084
148	206	121	43	57	42	149	77	**297**	100	40	30	22	20	16	12	11	**1.00**	19	23	2	1	–	–	–	–	–	–	–	–	–	Phenol, 4-[2-[methyl(2-phenylethyl)amino]ethyl]-, acetate (ester)	52059-48-0	NI31085
128	129	115	202	70	43	41	55	**297**	100	9	8	6	6	6	6	5	**1.47**	19	39	1	1	–	–	–	–	–	–	–	–	–	Morpholine, 2,6-dimethyl-4-tridecyl-	24602-86-6	NI31086
128	43	41	42	57	55	44	18	**297**	100	44	29	21	20	19	17	16	**0.60**	19	39	1	1	–	–	–	–	–	–	–	–	–	Morpholine, 2,6-dimethyl-4-tridecyl-	24602-86-6	NI31087
131	73	75	132	147	144	28	204	**297**	100	53	13	11	9	9	9	8	**0.00**	19	39	1	1	–	–	–	–	–	–	–	–	–	Tetradecanamide, N-pentyl-	74420-90-9	NI31088
125	250	297	124	267	239	251	112	**297**	100	56	52	48	45	43	40	26		20	11	2	1	–	–	–	–	–	–	–	–	–	Benzo[a]pyrene, 6-nitro-	63041-90-7	NI31089
250	125	297	239	267	124	251	119	**297**	100	98	77	71	67	55	43	32		20	11	2	1	–	–	–	–	–	–	–	–	–	Benzo[e]pyrene, 3-nitro-	81340-58-1	NI31090
297	240	114	63	298	269	241	75	**297**	100	41	32	26	25	25	23	20		20	11	2	1	–	–	–	–	–	–	–	–	–	14H-Naphtho[1′,2′:5,6]pyrano[2,3-b]quinoline, 14-oxo-		MS06851
282	283	77	51	127	280	152	141	**297**	100	23	14	8	8	6	6	5	**2.90**	20	15	–	3	–	–	–	–	–	–	–	–	–	3,5-Pyridinedicarbonitrile, 1,4-dihydro-2,6-diphenyl-4-methyl-	15521-04-7	NI31091
165	269	166	51	77	270	192	164	**297**	100	92	25	22	20	17	10	10	**5.00**	20	15	–	3	–	–	–	–	–	–	–	–	–	1H-1,2,3-Triazole, 1,4,5-triphenyl-	33471-63-5	NI31092
269	165	270	166	28	77	51	190	**297**	100	97	21	20	17	11	11	6	**0.00**	20	15	–	3	–	–	–	–	–	–	–	–	–	1H-1,2,3-Triazole, 1,4,5-triphenyl-	33471-63-5	NI31094
165	269	166	270	190	268	164	77	**297**	100	99	34	28	9	8	8	8	**8.00**	20	15	–	3	–	–	–	–	–	–	–	–	–	1H-1,2,3-Triazole, 1,4,5-triphenyl-	33471-63-5	NI31093
297	296	77	180	91	298	194	51	**297**	100	70	57	38	28	26	26	22		20	15	–	3	–	–	–	–	–	–	–	–	–	4H-1,2,4-Triazole, 3,4,5-triphenyl-	4073-72-7	NI31095
118	160	43	136	77	41	282	196	**297**	100	42	42	20	17	14	12	11	**0.00**	20	27	1	1	–	–	–	–	–	–	–	–	–	Cyclopentane, 1-[2-(hydroxyimino)-2-phenylethylidene]-2,2,3,3,4,4-hexamethyl-5-methylene-		DD01069

MASS TO CHARGE RATIOS								M.W.	INTENSITIES								Parent	C	H	O	N	S	F	Cl	Br	I	Si	P	B	X	COMPOUND NAME	CAS Reg No	No
149	110	41	84	67	297	213	212	**297**	100	62	52	49	45	44	42	42		20	27	1	1	–	–	–	–	–	–	–	–	–	**1,2-Oxazaspiro[4.4]non-2-ene, 6,6,7,7,8,8-hexamethyl-9-methylene-3-phenyl-**		**DD01070**
142	44	30	43	57	143	41	71	**297**	100	44	18	14	12	10	9	8	**0.00**	20	43	–	1	–	–	–	–	–	–	–	–	–	**1-Hexanamine, N-(2,2-diethylhexyl)-2,2-diethyl-**	56667-18-6	**NI31096**
58	44	43	28	59	57	41	30	**297**	100	7	6	5	4	4	4	4	**0.00**	20	43	–	1	–	–	–	–	–	–	–	–	–	**1-Octadecanamine, N,N-dimethyl-**	124-28-7	**NI31097**
58	84	43	59	57	55	41	71	**297**	100	6	6	4	4	3	3	2	**0.30**	20	43	–	1	–	–	–	–	–	–	–	–	–	**1-Octadecanamine, N,N-dimethyl-**	124-28-7	**IC03750**
286	267	209	49	268	265	285	287	**297**	100	92	44	42	35	27	26	23		21	15	–	–	–	1	–	–	–	–	–	1	–	**1,2,3-Triphenylcyclopropenylfluoroborate**		**LI06497**
105	77	297	180	165	89	51	106	**297**	100	53	40	22	20	18	10	8		21	15	1	1	–	–	–	–	–	–	–	–	–	**Isoxazole, 3,4,5-triphenyl-**	22020-72-0	**NI31098**
297	105	180	77	298	165	89	106	**297**	100	92	33	33	24	15	9	7		21	15	1	1	–	–	–	–	–	–	–	–	–	**Isoxazole, 3,4,5-triphenyl-**	22020-72-0	**LI06498**
297	165	241	298	77	149	242	89	**297**	100	26	26	17	16	10	9	9		21	15	1	1	–	–	–	–	–	–	–	–	–	**Oxazole, 2-[1,1′-biphenyl]-4-yl-5-phenyl-**	852-37-9	**NI31099**
297	165	166	298	269	89	77	51	**297**	100	80	47	25	23	21	16	7		21	15	1	1	–	–	–	–	–	–	–	–	–	**Oxazole, triphenyl-**	573-34-2	**LI06499**
297	165	166	298	269	89	77	105	**297**	100	80	47	25	23	21	16	8		21	15	1	1	–	–	–	–	–	–	–	–	–	**Oxazole, triphenyl-**	573-34-2	**NI31100**
297	165	166	298	269	89	77	108	**297**	100	80	47	25	23	21	16	8		21	15	1	1	–	–	–	–	–	–	–	–	–	**Oxazole, triphenyl-**	573-34-2	**MS06852**
143	142	156	297	41	43	157	144	**297**	100	93	59	21	18	17	14	13		21	31	–	1	–	–	–	–	–	–	–	–	–	**Quinoline, 6-dodecyl-**	56259-08-6	**NI31101**
279	281	227	278	260	262	258	241	**298**	100	93	86	46	14	13	12	12	**0.00**	–	–	–	–	–	6	–	–	–	–	–	–	1	**Tungsten hexafluoride**	7783-82-6	**AM00140**
279	281	277	278	260	262	241	222	**298**	100	93	86	46	14	13	13	12	**0.00**	–	–	–	–	–	6	–	–	–	–	–	–	1	**Tungsten hexafluoride**	7783-82-6	**NI31102**
117	119	121	79	151	149	265	267	**298**	100	95	30	22	18	18	12	8	**1.00**	2	–	–	–	2	–	6	–	–	–	–	–	–	**Dichloro[(trichloromethyl)thio]methanesulphenyl chloride**	91631-93-5	**NI31103**
117	79	149	114	44	263	82	47	**298**	100	58	17	17	17	15	14	11	**4.10**	2	–	–	–	2	–	6	–	–	–	–	–	–	**Bis(trichloromethyl) disulphide**		**LI06500**
205	207	203	155	45	69	153	201	**298**	100	72	70	53	46	42	39	36	**0.00**	4	6	2	–	–	3	1	–	–	–	–	–	1	**Stannane, chlorodimethyl[(trifluoroacetyl)oxy]-**	38186-20-8	**NI31104**
165	135	133	64	150	69	48	32	**298**	100	55	40	29	25	24	24	24	**0.00**	4	9	2	–	1	3	–	–	–	–	–	–	1	**Stannane, trimethyl[[(trifluoromethyl)sulphinyl]oxy]-**	51735-76-3	**NI31105**
147	83	69	215	113	33	67	64	**298**	100	58	30	23	21	12	11	11	**0.00**	5	3	2	–	1	9	–	–	–	–	–	–	–	**2,2,2-Trifluoroethyl (1,1,1,3,3,3-hexafluoroisopropyl) sulphone**		**DD01071**
120	298	238	60	121	180	150	91	**298**	100	99	99	87	84	82	80	72		6	18	–	6	–	–	–	–	–	–	4	–	–	**2,4,6,8,9,10-Hexaaza-1,3,5,7-tetraphosphatricyclo[3.3.1.1^{3,7}]decane, 2,4,6,8,9,10-hexamethyl-**	10369-17-2	**NI31106**
171	143	155	75	50	127	298	31	**298**	100	62	47	43	30	20	15	15		7	7	3	–	–	–	–	1	–	–	–	–	1	**Benzeneselenonic acid, 4-bromo-, methyl ester**	25633-05-0	**LI06501**
171	173	143	145	155	157	75	76	**298**	100	98	62	61	46	45	42	35	**14.84**	7	7	3	–	–	–	–	1	–	–	–	–	1	**Benzeneselenonic acid, 4-bromo-, methyl ester**	25633-05-0	**NI31107**
55	64	48	39	94	43	40	41	**298**	100	65	46	32	31	21	19	16	**0.00**	8	3	7	–	1	–	–	–	–	–	–	–	1	**Manganese, pentacarbonyl(2-propynylsulphonyl)-**	72403-09-9	**NI31108**
31	298	248	229	112	179	279	217	**298**	100	38	34	23	21	18	17	15		9	–	–	–	–	10	–	–	–	–	–	–	–	**Indene, decafluoro-3a,7a-dihydro-**	13385-25-6	**NI31109**
45	109	15	125	157	155	108	79	**298**	100	88	77	65	46	41	30	28	**8.10**	9	12	3	–	2	–	1	–	–	–	1	–	–	**Phosphorothioic acid, S-[[(4-chlorophenyl)thio]methyl] O,O-dimethyl ester**	7332-32-3	**NI31110**
298	147	207	211	73	299	283	148	**298**	100	97	56	54	54	36	25	15		9	27	3	–	–	–	–	–	–	3	1	–	–	**Silanol, trimethyl-, phosphite (3:1)**	1795-31-9	**NI31111**
109	189	187	298	45	191	185	58	**298**	100	91	45	30	24	21	19	18		10	6	–	2	2	–	–	–	–	–	–	–	1	**3,5-Bis(2-thienyl)-1,2,4-selenadiazole**		**MS06853**
69	41	67	39	230	228	68	70	**298**	100	46	25	11	9	8	7	6	**6.00**	10	18	–	–	–	–	–	–	–	–	–	–	2	**Dicyclopentyl diselenide**		**MS06854**
43	41	85	69	55	29	238	39	**298**	100	37	32	23	19	16	12	10	**0.86**	10	19	2	–	–	–	–	–	1	–	–	–	–	**1-Iodo-2-octyl acetate**		**IC03751**
55	41	69	83	57	137	135	71	**298**	100	72	62	48	45	39	39	38	**0.00**	10	20	–	–	–	–	–	2	–	–	–	–	–	**Decane, 1,10-dibromo-**	4101-68-2	**NI31112**
55	83	69	41	57	97	43	29	**298**	100	85	64	61	31	30	29	23	**0.00**	10	20	–	–	–	–	–	2	–	–	–	–	–	**Decane, 4,5-dibromo-, (R*,R*)-**	61141-70-6	**NI31113**
69	41	133	213	145	68	56	55	**298**	100	55	42	29	29	24	23	23	**5.20**	10	20	6	–	–	–	–	–	–	–	2	–	–	**2,2′-Bi-1,3,2-dioxaphosphorinane, 5,5,5′,5′-tetramethyl-, 2,2′-dioxide**	16368-06-2	**NI31114**
129	137	171	75	73	169	283	219	**298**	100	74	70	64	64	48	38	28	**0.00**	10	22	4	–	2	–	–	–	–	1	–	–	–	**Hexanoic acid, 6-[(methylsulphonyl)thio]-, trimethylsilyl ester**	76078-73-4	**NI31115**
140	116	120	238	162	158	182	154	**298**	100	91	74	72	66	66	65	58	**11.00**	10	28	–	4	–	2	–	–	–	2	–	–	–	**Hydrazine, 1-[(tert-butyl)difluorosilyl]-1-[(2,2-dimethylhydrazino)dimethylsilyl]-2,2-dimethyl-**	66436-31-5	**NI31116**
206	132	160	105	133	298	104	94	**298**	100	98	85	48	28	25	20	12		11	14	2	4	2	–	–	–	–	–	–	–	–	**Ethyl 5,5a,7,8,9a,10-hexahydro-1,4-dithiino[2,3-g]pteridine-4-carboxylate**		**LI06502**
73	77	45	74	81	47	43	49	**298**	100	69	32	25	18	18	18	15	**5.00**	11	15	–	–	–	5	–	–	–	2	–	–	–	**Disilane, pentamethyl(pentafluorophenyl)-**	33558-55-3	**NI31117**
27	26	25	68	43	41	42	28	**298**	100	71	17	13	13	11	9	8	**0.00**	11	16	5	5	–	–	–	–	–	–	–	–	–	**1H-Purinium, 2-amino-6,9-dihydro-7-methyl-6-oxo-9-β-D-ribofuranosyl-**	20244-86-4	**NI31118**
171	139	172	73	140	83	55	41	**298**	100	18	11	6	4	4	4	3	**0.00**	11	23	1	–	–	–	–	–	1	–	–	–	–	**Decane, 3-methoxy-8-iodo-**	20599-96-6	**NI31119**
73	83	55	41	69	171	97	45	**298**	100	57	46	35	32	30	22	21	**0.00**	11	23	1	–	–	–	–	–	1	–	–	–	–	**Decane, 3-methoxy-8-iodo-**	20599-96-6	**NI31120**
168	298	296	139	294	169	63	293	**298**	100	40	37	33	23	18	12	9		12	8	1	–	–	–	–	–	–	–	–	–	1	**Phenoxatellurin**		**LI06503**
77	97	129	157	125	31	103	51	**298**	100	74	62	48	44	29	28	28	**10.32**	12	15	3	2	1	–	–	–	–	–	1	–	–	**3,5-Dioxa-6-aza-4-phosphaoct-6-ene-8-nitrile, 4-ethoxy-7-phenyl-, 4-sulphide**	14816-18-3	**NI31121**
146	298	157	156	129	118	90	97	**298**	100	57	46	44	39	36	30	23		12	15	3	2	1	–	–	–	–	–	1	–	–	**Phosphorothioic acid, O,O-diethyl O-2-quinoxalinyl ester**	13593-03-8	**NI31122**
57	146	40	41	118	176	97	160	**298**	100	57	53	44	29	28	28	26	**3.80**	12	15	3	2	1	–	–	–	–	–	1	–	–	**Phosphorothioic acid, O,O-diethyl O-2-quinoxalinyl ester**	13593-03-8	**NI31123**
283	165	105	227	135	77	297	253	**298**	100	63	36	34	32	31	20	16	**1.00**	12	18	1	–	–	–	–	–	–	–	–	–	1	**(2-Benzoylethyl)trimethyltin**		**LI06504**
73	81	147	169	170	69	129	155	**298**	100	68	26	25	23	23	18	15	**14.00**	12	18	5	2	–	–	–	–	–	1	–	–	–	**3,6-Epoxy-2H,8H-pyrimido[6,1-b][1,3]oxazocine-8,10(9H)-dione, 3,4,5,6-tetrahydro-4-[(trimethylsilyl)oxy]-, [3R-(3α,4β,6α)]-**	41547-75-5	**NI31124**
43	97	55	41	77	69	70	44	**298**	100	90	86	69	68	53	48	46	**0.02**	12	20	4	–	–	–	2	–	–	–	–	–	–	**1,2-Bis(chloroacetoxy)octane**		**IC03752**
43	225	111	124	125	224	169	99	**298**	100	75	72	46	45	37	29	28	**0.20**	12	20	7	–	–	–	–	–	–	–	–	2	–	**β-D-Fructopyranose, cyclic 2,3:4,5-bis(ethylboronate) 1-acetate**	74779-80-9	**NI31125**
43	111	101	124	99	73	110	125	**298**	100	70	53	25	21	21	20	18	**0.88**	12	20	7	–	–	–	–	–	–	–	–	2	–	**α-L-Sorbofuranose, cyclic 2,3:4,6-bis(ethylboronate) 1-acetate**	74779-77-4	**NI31126**

MASS TO CHARGE RATIOS								M.W.	INTENSITIES								Parent	C	H	O	N	S	F	Cl	Br	I	Si	P	B	X	COMPOUND NAME	CAS Reg No	No
99	155	127	144	109	298	166	272	**298**	100	81	72	15	14	8	4	3		12	21	4	–	–	2	–	–	–	–	1	–	–	Diethyl 1-cyclohexyl-2,2-difluoroethenyl phosphate		DD01072
57	41	153	55	209	97	63	39	**298**	100	83	38	29	27	25	20	19	**3.50**	12	27	–	–	3	–	–	–	–	–	1	–	–	Phosphorotrithious acid, tributyl ester	150-50-5	NI31127
57	41	209	153	56	55	97	122	**298**	100	82	59	47	39	35	34	29	**28.00**	12	27	–	–	3	–	–	–	–	–	1	–	–	Phosphorotrithious acid, tributyl ester	150-50-5	NI31128
57	209	41	298	88	153	56	55	**298**	100	89	77	46	31	30	28	25		12	27	–	–	3	–	–	–	–	–	1	–	–	Phosphorotrithious acid, tributyl ester	150-50-5	NI31129
126	124	196	223	151	153	169	194	**298**	100	65	40	21	19	17	12	11	**0.00**	12	27	6	–	–	–	–	–	–	–	1	–	–	Butyl bis(3-hydroxybutyl) phosphate	89197-71-7	NI31130
149	255	107	298	73	213	214	65	**298**	100	74	70	55	48	37	29	29		12	28	–	–	2	–	–	–	–	–	2	–	–	Tetrapropyldiphosphane disulphide		MS06855
298	232	204	253	279	252	233	280	**298**	100	34	24	22	18	18	16	12		13	9	3	2	–	3	–	–	–	–	–	–	–	3-Pyridinecarboxylic acid, 2-[[4-hydroxy-3-(trifluoromethyl)phenyl]amino]-	42240-82-4	NI31131
298	44	252	253	297	279	40	280	**298**	100	97	77	58	51	44	40	39		13	9	3	2	–	3	–	–	–	–	–	–	–	3-Pyridinecarboxylic acid, 5-hydroxy-2-[[3-(trifluoromethyl)phenyl]amino]-	42240-83-5	NI31132
185	183	57	41	85	39	75	76	**298**	100	99	90	61	52	29	24	22	**0.00**	13	15	3	–	–	–	–	1	–	–	–	–	–	Pentanoic acid, 4-bromophenacyl ester		NI31133
135	224	99	298	152	126	127	111	**298**	100	57	53	39	36	35	13	13		13	15	4	2	–	–	1	–	–	–	–	–	–	2-(2-Chlorophenyl)hydrazonopropanedioic acid, diethyl ester		NI31134
103	73	75	101	104	74	41	129	**298**	100	86	18	10	8	8	8	4	**2.00**	13	19	1	–	–	–	–	1	–	1	–	–	–	Silane, [[3-(3-bromophenyl)-3-butenyl]oxy]trimethyl-	55724-00-0	NI31135
105	161	160	104	77	197	89	65	**298**	100	97	93	67	40	35	33	33	**2.00**	13	19	4	2	–	–	–	–	–	–	1	–	–	Phosphonic acid, [3-(aminocarbonyl)-2-phenyl-2-aziridinyl]-, diethyl ester	33866-52-3	NI31136
207	179	149	211	205	177	209	147	**298**	100	79	78	75	64	62	57	56	**10.10**	13	22	–	–	–	–	–	–	–	–	–	–	1	Benzyltriethyltin		NI31137
181	73	147	182	45	75	298	283	**298**	100	81	28	16	16	14	10	9		13	26	2	2	–	–	–	–	–	2	–	–	–	5-Pyrazolepropanoic acid, 3-methyl-N-(trimethylsilyl)-, trimethylsilyl ester		NI31138
298	138	150	75	222	166	194	240	**298**	100	83	47	42	37	30	26	21		14	6	6	2	–	–	–	–	–	–	–	–	–	1,5-Dinitroanthraquinone		IC03753
298	138	150	75	166	194	281	30	**298**	100	73	59	48	35	27	24	23		14	6	6	2	–	–	–	–	–	–	–	–	–	1,8-Dinitroanthraquinone		IC03754
77	91	79	135	51	64	52	63	**298**	100	93	80	72	66	46	45	32	**0.00**	14	10	4	4	–	–	–	–	–	–	–	–	–	Benzaldehyde, 2-nitro-, [(2-nitrophenyl)methylene]hydrazone	1929-19-7	NI31139
77	91	135	79	51	52	64	105	**298**	100	82	78	75	75	45	40	20	**5.00**	14	10	4	4	–	–	–	–	–	–	–	–	–	Benzaldehyde, 2-nitro-, [(2-nitrophenyl)methylene]hydrazone	1929-19-7	NI31140
176	298	76	130	103	89	205	50	**298**	100	54	42	34	28	26	24	20		14	10	4	4	–	–	–	–	–	–	–	–	–	Benzaldehyde, 3-nitro-, [(3-nitrophenyl)methylene]hydrazone	1567-91-5	NI31141
176	298	76	251	130	103	297	177	**298**	100	50	33	26	26	25	22	12		14	10	4	4	–	–	–	–	–	–	–	–	–	Benzaldehyde, 4-nitro-, [(4-nitrophenyl)methylene]hydrazone	2143-99-9	NI31142
176	298	76	251	130	103	297	205	**298**	100	50	33	26	26	25	22	9		14	10	4	4	–	–	–	–	–	–	–	–	–	Benzaldehyde, 4-nitro-, [(4-nitrophenyl)methylene]hydrazone	2143-99-9	LI06506
176	298	76	130	103	89	205	50	**298**	100	54	42	34	28	26	24	20		14	10	4	4	–	–	–	–	–	–	–	–	–	Benzaldehyde, 4-nitro-, [(4-nitrophenyl)methylene]hydrazone	2143-99-9	LI06505
91	135	132	167	119	107	150	151	**298**	100	42	22	22	18	18	14	13	**0.70**	14	18	7	–	–	–	–	–	–	–	–	–	–	1,6-Cycloheptadiene-1-carboxylic acid, 3-[bis(methoxycarbonyl)methoxy], methyl ester		MS06856
81	225	112	87	109	111	141	113	**298**	100	96	94	80	79	71	69	53	**10.00**	14	18	7	–	–	–	–	–	–	–	–	–	–	β-D-Galactopyranose, 1,6-anhydro-3,4-O-[5-(methoxymethyl)-2-furfurylidene]-2-O-methyl-	95881-42-8	NI31143
121	136	93	85	145	127	97	163	**298**	100	89	58	46	28	25	25	10	**0.00**	14	18	7	–	–	–	–	–	–	–	–	–	–	β-D-Glucopyranoside, 4-acetylphenyl-		MS06857
136	43	121	73	60	57	85	93	**298**	100	99	83	66	61	60	53	46	**0.30**	14	18	7	–	–	–	–	–	–	–	–	–	–	β-D-Glucopyranoside, 4-acetylphenyl-		MS06858
43	206	81	126	158	207	179	178	**298**	100	90	82	71	44	43	43	38	**1.00**	14	18	7	–	–	–	–	–	–	–	–	–	–	2-Propenedioic acid, 2-(2-acetoxy-4-oxocyclopentyl)-, dimethyl ester, (E)		MS06859
55	27	81	82	126	56	127	108	**298**	100	8	5	5	4	3	3	3	**0.00**	14	18	7	–	–	–	–	–	–	–	–	–	–	2-Propenoic acid, 2-(hydroxymethyl)-2-[[(1-oxo-2-propenyl)oxy]methyl]-1,3-propanediyl ester	3524-68-3	NI31144
285	57	287	41	219	189	91	302	**298**	100	99	96	72	59	33	28	26	**0.00**	14	19	2	–	–	–	–	1	–	–	–	–	–	1,2-Benzoquinone, 3,5-di-tert-butyl-6-bromo-		MS06860
239	240	73	298	241	283	165	89	**298**	100	19	13	7	6	5	3	3		14	22	5	–	–	–	–	–	–	1	–	–	–	Benzeneacetic acid, 3,4-dimethoxy-α-[(trimethylsilyl)oxy]-, methyl ester	2911-70-8	NI31145
73	239	298	181	166	209	283	75	**298**	100	93	67	67	32	26	23	23		14	22	5	–	–	–	–	–	–	1	–	–	–	Benzeneacetic acid, 2,3,4-trimethoxy-, trimethylsilyl ester		MS06861
239	73	298	181	283	240	299	75	**298**	100	91	67	36	22	18	14	14		14	22	5	–	–	–	–	–	–	1	–	–	–	Benzeneacetic acid, 3,4,5-trimethoxy-, trimethylsilyl ester		MS06862
57	85	74	111	149	267	67	125	**298**	100	43	40	35	30	25	21	20	**4.00**	14	22	5	2	–	–	–	–	–	–	–	–	–	3-(tert-Butylamino)-1-(4-hydroxy-3-methyl-2-nitrophenoxy)-2-propanol		MS06863
196	228	169	43	41	197	112	267	**298**	100	14	10	8	7	5	5	2	**0.30**	14	22	5	2	–	–	–	–	–	–	–	–	–	N,N′-Dimethyl-5-(1-methylbutyl)-5-methoxycarbonylmethylbarbituric acid		LI06507
73	298	209	268	283	267	179	45	**298**	100	64	61	36	26	25	25	18		14	26	3	–	–	–	–	–	–	2	–	–	–	Benzene, 3-methoxy-2-(trimethylsiloxy)-1-[(trimethoxysiloxy)methyl]-		NI31146
298	43	199	121	56	41	81	233	**298**	100	80	49	48	43	17	15	14		15	14	3	–	–	–	–	–	–	–	–	–	1	Ferrocene, [(dihydro-2,5-dioxofuranyl)methyl]-		NI31147
121	91	56	242	212	122	65	186	**298**	100	63	51	38	27	25	25	17	**0.00**	15	14	3	–	–	–	–	–	–	–	–	–	1	Iron, (2-methoxybenzyl)dicarbonyl-π-cyclopentadienyl-	64494-50-4	NI31148
207	131	91	298	92	117	104	77	**298**	100	36	26	9	8	6	6	6		15	14	3	4	–	–	–	–	–	–	–	–	–	7-[N-(1-Phenyl-2-propyl)amino]-4-nitrobenzo-2-oxa-1,3-diazole		LI06508
93	254	108	92	77	280	65	298	**298**	100	96	96	96	79	47	37	34		15	14	3	4	–	–	–	–	–	–	–	–	–	Propanedioic acid, (phenylhydrazono)-, mono(2-phenylhydrazide)	51471-61-5	NI31149
52	214	215	28	53	298	147	242	**298**	100	68	17	17	12	8	7	6		15	18	3	–	–	–	–	–	–	–	–	–	1	Chromium, tricarbonyl[(1,2,3,4,5,6-η)-hexamethylbenzene]-	12088-11-8	NI31150
157	139	91	255	43	155	92	83	**298**	100	66	61	36	27	20	20	20	**0.00**	15	22	4	–	1	–	–	–	–	–	–	–	–	2-Hexene-1,4-diol, 2,5-dimethyl-3-[(4-methylphenyl)sulphonyl]-, (E)-	118356-31-3	DD01073
45	138	72	137	73	59	53	52	**298**	100	37	31	27	25	20	16	16	**0.00**	15	22	6	–	–	–	–	–	–	–	–	–	–	1,3-Benzodioxole, 5-[1-[2-(2-ethoxyethoxy)ethoxy]ethoxy]-	51-14-9	NI31151
41	97	43	125	143	109	166	191	**298**	100	96	96	79	77	49	43	27	**2.00**	15	22	6	–	–	–	–	–	–	–	–	–	–	3,6-Methano-5H-oxireno[3,4]cyclopent[1,2-d]oxepin-5-one, octahydro-2,6a-dihydroxy-1a-(hydroxymethyl)-1b-methyl-8-isopropyl-	70219-71-5	NI31152
109	81	95	125	55	79	108	91	**298**	100	75	67	59	56	52	46	39	**2.40**	15	22	6	–	–	–	–	–	–	–	–	–	–	Trichothecan-8-one, 12,13-epoxy-3,7,15-trihydroxy-, (3α,7α)-		NI31153
173	109	43	191	91	41	145	221	**298**	100	99	98	87	87	85	80	75	**3.75**	15	22	6	–	–	–	–	–	–	–	–	–	–	Trichothec-9-ene-3,4,8,15-tetrol, 12,13-epoxy-, (3α,4β,8α)-	34114-99-3	NI31154
161	203	218	175	163	189	176	219	**298**	100	72	63	47	33	24	20	18	**0.00**	15	23	1	–	–	–	–	1	–	–	–	–	–	Phenol, 2,6-di-tert-butyl-4-bromomethyl-		IC03755
298	97	85	83	95	111	81	253	**298**	100	83	76	60	58	56	47	40		16	10	–	–	3	–	–	–	–	–	–	–	–	4,7-Di-2-thienylbenzo[b]thiophene		LI06509
176	283	285	88	300	298	177	175	**298**	100	32	32	22	20	20	15	13		16	11	1	–	–	–	–	1	–	–	–	–	–	Phenanthrene, 3-acetyl-9-bromo-		NI31155

MASS TO CHARGE RATIOS								M.W.	INTENSITIES								Parent	C	H	O	N	S	F	Cl	Br	I	Si	P	B	X	COMPOUND NAME	CAS Reg No	No
298	132	133	89	90	300	118	299	**298**	100	84	46	42	40	35	29	27		16	11	2	2	–	–	1	–	–	–	–	–	–	3-Chlorodiftalone		MS06864
283	298	148	149	268	284	285	150	**298**	100	34	31	28	27	21	12	12		16	14	–	2	2	–	–	–	–	–	–	–	–	Benzothiazole, 2,3-dihydro-3-methyl-2-(3-methyl-2(3H)-benzothiazolylidene)-	2786-70-1	NI31156
225	298	165	193	224	102	192	166	**298**	100	96	78	70	44	30	21	20		16	14	–	2	2	–	–	–	–	–	–	–	–	2H-Imidazole-2-thione, 1,5-dihydro-4-(methylthio)-5,5-diphenyl-	3718-53-4	NI31157
224	239	165	298	225	121	166	240	**298**	100	46	35	29	18	17	10	8		16	14	–	2	2	–	–	–	–	–	–	–	–	4H-Imidazole-4-thione, 1,5-dihydro-2-(methylthio)-5,5-diphenyl-	4110-12-7	NI31158
298	210	165	299	300	239	121	118	**298**	100	57	40	22	11	10	10	10		16	14	–	2	2	–	–	–	–	–	–	–	–	2,4-Imidazolidinedithione, 1-methyl-5,5-diphenyl-	6826-03-5	NI31159
298	225	224	165	299	121	166	226	**298**	100	64	47	45	21	14	13	12		16	14	–	2	2	–	–	–	–	–	–	–	–	2,4-Imidazolidinedithione, 3-methyl-5,5-diphenyl-	16116-38-4	NI31160
77	105	119	51	147	166	145	118	**298**	100	71	53	40	23	19	18	17	0.00	16	14	2	2	1	–	–	–	–	–	–	–	–	Carbamimidothioic acid, N-(oxophenylacetyl)-N′-phenyl-, methyl ester	74810-30-3	NI31161
298	225	269	265	238	270	299	300	**298**	100	44	42	34	29	23	20	8		16	14	2	2	1	–	–	–	–	–	–	–	–	1,4-Diamino-2-(ethylthio)anthraquinone		IC03756
206	64	207	191	91	205	128	74	**298**	100	25	19	18	17	15	9	5	0.00	16	14	2	2	1	–	–	–	–	–	–	–	–	4,4-Diphenyl-6-thia-2,3-diazabicyclo[3.2.0]hept-2-ene-6,6-dioxide		LI06510
107	192	77	298	136	108	193	51	**298**	100	53	22	11	7	6	5	5		16	14	2	2	1	–	–	–	–	–	–	–	–	2-Imidazolidinethione, 5-(4-hydroxybenzyl)-4-oxo-3-phenyl-		LI06511
298	252	282	280	251	250	264	296	**298**	100	68	47	39	33	29	29	21		16	14	4	2	–	–	–	–	–	–	–	–	–	9,10-Anthracenedione, 1,8-diamino-2,7-bis(hydroxymethyl)-		NI31162
91	77	51	176	105	50	106	92	**298**	100	34	25	15	15	15	14	10	4.00	16	14	4	2	–	–	–	–	–	–	–	–	–	O-Benzyl-N-(4-nitrocinnamoyl)hydroxylamine		LI06512
149	77	119	103	91	51	150	102	**298**	100	53	43	32	28	11	10	10	0.00	16	14	4	2	–	–	–	–	–	–	–	–	–	cis-1,2-Bis(4-nitrophenyl)cyclobutane		LI06513
149	77	119	103	91	150	102	133	**298**	100	27	24	22	15	11	6	5	0.00	16	14	4	2	–	–	–	–	–	–	–	–	–	trans-1,2-Bis(4-nitrophenyl)cyclobutane		LI06514
234	228	298	282	269	241	283	235	**298**	100	46	42	35	31	22	19	18		16	15	–	4	–	–	1	–	–	–	–	–	–	1H-1,3,4-Benzotriazepin-2-amine, 7-chloro-N,1-dimethyl-5-phenyl-		MS06865
298	282	300	284	283	77	299	297	**298**	100	81	33	29	28	28	27	24		16	15	–	4	–	–	1	–	–	–	–	–	–	3H-1,3,4-Benzotriazepin-2-amine, 7-chloro-N-ethyl-5-phenyl-		MS06866
282	298	284	77	300	283	44	299	**298**	100	71	36	28	24	22	17	15		16	15	–	4	–	–	1	–	–	–	–	–	–	3H-1,3,4-Benzotriazepin-2-amine, 7-chloro-N,N-dimethyl-5-phenyl-		MS06867
298	238	121	56	172	225	173	171	**298**	100	99	63	43	41	35	26	26		16	18	2	–	–	–	–	–	–	–	–	–	1	Ferrocene, (4-acetyloxybut-1-enyl)-		MS06868
298	121	56	269	213	185	129	65	**298**	100	35	30	20	20	12	12	11		16	18	2	–	–	–	–	–	–	–	–	–	1	Ferrocene, 1,1′-bis(1-oxopropyl)-	1274-01-7	NI31163
239	251	283	298	210	240	149	209	**298**	100	78	65	62	47	33	27	25		16	18	2	2	–	–	–	–	–	1	–	–	–	10-Sila-2-azaanthracene, 9,10-dihydro-9-(methoxycarbonylamino)-10,10-dimethyl-		NI31164
283	298	148	133	299	284	105	119	**298**	100	98	74	39	18	18	14	13		16	18	2	4	–	–	–	–	–	–	–	–	–	N,N-Diethyl-4-(4-nitrophenylazo)aniline		IC03757
91	106	28	44	78	51	177	79	**298**	100	82	70	56	47	39	27	25	2.70	16	18	2	4	–	–	–	–	–	–	–	–	–	4-Pyridinecarboxylic acid, 2-[3-oxo-3-[benzylamino]propyl]hydrazide	51-12-7	NI31165
299	192	91	138					**298**	100	30	29	18					0.00	16	18	2	4	–	–	–	–	–	–	–	–	–	4-Pyridinecarboxylic acid, 2-[3-oxo-3-[benzylamino]propyl]hydrazide	51-12-7	NI31166
135	298	96	190	108	176	163	299	**298**	100	97	58	20	17	16	13	12		16	18	2	4	–	–	–	–	–	–	–	–	–	Pyrimidine, 2-[4-(1,3-benzodioxol-5-ylmethyl)-1-piperazinyl]-	3605-01-4	NI31167
135	298	108	56	190	164	177	77	**298**	100	39	36	35	15	13	12	12		16	18	2	4	–	–	–	–	–	–	–	–	–	Pyrimidine, 2-[4-(1,3-benzodioxol-5-ylmethyl)-1-piperazinyl]-	3605-01-4	NI31168
170	138	43	41	45	214	69	67	**298**	100	75	56	30	25	23	23	22	0.00	16	26	5	–	–	–	–	–	–	–	–	–	–	2,4(3H,5H)-Furandione, 3-hexyl-5-(2-methoxy-1-oxobutyl)-3-methyl-	54934-68-8	NI31169
270	272	298	200	300	271	99	135	**298**	100	67	33	29	22	18	18	16		17	8	1	–	–	–	2	–	–	–	–	–	–	1,4-Pentadiyn-3-one, 1,5-bis(4-chlorophenyl)-	34793-65-2	MS06869
270	272	298	40	300	187	44	200	**298**	100	63	52	47	33	33	26	21		17	8	1	–	–	–	2	–	–	–	–	–	–	2,4-Pentadiyn-1-one, 1,5-bis(4-chlorophenyl)-	29372-69-8	MS06870
298	269	237	223	184	115	280	252	**298**	100	40	28	26	23	20	19	19		17	14	5	–	–	–	–	–	–	–	–	–	–	1,4-Anthracenedione, 10-hydroxy-2,5-dimethoxy-7-methyl-	84542-53-0	NI31170
298	252	280	237	269	135	281	299	**298**	100	70	49	27	26	24	22	21		17	14	5	–	–	–	–	–	–	–	–	–	–	9,10-Anthracenedione, 1-hydroxy-3,8-dimethoxy-6-methyl-	23610-20-0	NI31171
298	270	299	241	28	283	269	271	**298**	100	46	24	20	17	10	10	8		17	14	5	–	–	–	–	–	–	–	–	–	–	9,10-Anthracenedione, 1,6,8-trihydroxy-3-propyl-		MS06871
179	298	120	297	205	147	107	91	**298**	100	85	74	36	28	23	18	18		17	14	5	–	–	–	–	–	–	–	–	–	–	Benzaldehyde, 4-hydroxy-5-[3-(4-hydroxyphenyl)-1-oxo-2-propenyl]-2-methoxy-, (E)-	65621-10-5	NI31172
164	137	300	163	165	151	133	150	**298**	100	57	41	38	12	12	12	10	0.00	17	14	5	–	–	–	–	–	–	–	–	–	–	1,3-Benzodioxol-5-ol, 6-(7-methoxy-4H-1-benzopyran-3-yl)-	54934-76-8	NI31173
108	298	135	77	44	299	152	92	**298**	100	84	76	21	18	16	15	14		17	14	5	–	–	–	–	–	–	–	–	–	–	3(2H)-Benzofuranone, 6-methoxy-2-(4-methoxybenzoyl)-	56588-15-9	NI31174
163	298	267	118	135	76	164	105	**298**	100	31	20	14	12	11	9	8		17	14	5	–	–	–	–	–	–	–	–	–	–	Benzophenone-4,4′-dicarboxylic acid, dimethyl ester	1233-73-4	IC03758
163	298	267	135	76	239	118	104	**298**	100	32	23	15	12	10	10	8		17	14	5	–	–	–	–	–	–	–	–	–	–	Benzophenone-4,4′-dicarboxylic acid, dimethyl ester	1233-73-4	NI31175
166	138	298	297	221	104	193	270	**298**	100	52	45	35	25	25	22	20		17	14	5	–	–	–	–	–	–	–	–	–	–	4H-1-Benzopyran-4-one, 2,3-dihydro-5,7-dihydroxy-6-formyl-8-methyl-2-phenyl-		NI31176
137	253	69	298	135	81	71	83	**298**	100	91	70	68	42	37	31	27		17	14	5	–	–	–	–	–	–	–	–	–	–	4H-1-Benzopyran-4-one, 2-(3,4-dimethoxyphenyl)-7-hydroxy-	33513-36-9	NI31177
137	253	69	298	135	81	71	162	**298**	100	91	70	68	42	37	31	27		17	14	5	–	–	–	–	–	–	–	–	–	–	4H-1-Benzopyran-4-one, 2-(3,4-dimethoxyphenyl)-7-hydroxy-	33513-36-9	MS06872
255	298	43	121	187	256	241	227	**298**	100	66	43	40	38	34	30	27		17	14	5	–	–	–	–	–	–	–	–	–	–	2H-1-Benzopyran-2-one, 3-[1-(2-furanyl)-3-oxobutyl]-4-hydroxy-	117-52-2	NI31178
255	43	121	298	65	92	39	187	**298**	100	71	64	38	27	26	24	23		17	14	5	–	–	–	–	–	–	–	–	–	–	2H-1-Benzopyran-2-one, 3-[1-(2-furanyl)-3-oxobutyl]-4-hydroxy-	117-52-2	NI31179
166	138	298	193	270	104	221	183	**298**	100	48	45	24	22	20	19	18		17	14	5	–	–	–	–	–	–	–	–	–	–	4H-1-Benzopyran-4-one, 2,3-hydro-5,7-dihydroxy-8-formyl-6-methyl-5-phenyl-		NI31180
181	298	180	299	182	149	136	105	**298**	100	52	14	10	10	8	5	4		17	14	5	–	–	–	–	–	–	–	–	–	–	2H-1-Benzopyran-2-one, 4-hydroxy-5,7-dimethoxy-3-phenyl-	4222-00-8	NI31181
298	283	153	181	297	269	255	103	**298**	100	69	35	17	13	13	11	9		17	14	5	–	–	–	–	–	–	–	–	–	–	4H-1-Benzopyran-4-one, 5-hydroxy-6,7-dimethoxy-2-phenyl-	740-33-0	NI31182
69	255	283	298	105	153	77	281	**298**	100	40	26	24	23	22	21	20		17	14	5	–	–	–	–	–	–	–	–	–	–	4H-1-Benzopyran-4-one, 7-hydroxy-5,6-dimethoxy-2-phenyl-		NI31183
282	283	239	89	133	94	66	127	**298**	100	49	14	11	10	9	9	8	0.00	17	14	5	–	–	–	–	–	–	–	–	–	–	4H-1-Benzopyran-4-one, 5-hydroxy-7-methoxy-2-(3-methoxyphenyl)-	57396-77-7	NI31184
298	135	269	132	297	166	255	138	**298**	100	19	15	14	11	11	9	9		17	14	5	–	–	–	–	–	–	–	–	–	–	4H-1-Benzopyran-4-one, 5-hydroxy-7-methoxy-2-(4-methoxyphenyl)-	5128-44-9	MS06873
298	135	132	269	95	166	255	138	**298**	100	17	17	16	14	12	10	10		17	14	5	–	–	–	–	–	–	–	–	–	–	4H-1-Benzopyran-4-one, 5-hydroxy-7-methoxy-2-(4-methoxyphenyl)-	5128-44-9	NI31185
298	135	299	132	297	269	255	166	**298**	100	18	16	16	8	8	8	6		17	14	5	–	–	–	–	–	–	–	–	–	–	4H-1-Benzopyran-4-one, 5-hydroxy-7-methoxy-2-(4-methoxyphenyl)-	5128-44-9	NI31186
267	298	268	181	299	180	208	152	**298**	100	52	16	12	9	9	8	8		17	14	5	–	–	–	–	–	–	–	–	–	–	1,2-Dibenzofurandicarboxylic acid, 4-methyl-, dimethyl ester		LI06515

MASS TO CHARGE RATIOS								M.W.	INTENSITIES								Parent	C	H	O	N	S	F	Cl	Br	I	Si	P	B	X	COMPOUND NAME	CAS Reg No	No
298	148	299	297	162	283	161	175	**298**	100	20	16	16	15	10	8	6		17	14	5	–	–	–	–	–	–	–	–	–	–	**6H-[1,3]Dioxolo[5,6]benzofuro[3,2-c][1]benzopyran, 6a,12a-dihydro-3-methoxy-, (6aR-cis)-**	524-97-0	**NI31187**
298	148	297	162	281	161	175	133	**298**	100	19	16	15	9	9	7	5		17	14	5	–	–	–	–	–	–	–	–	–	–	**6H-[1,3]Dioxolo[5,6]benzofuro[3,2-c][1]benzopyran, 6a,12a-dihydro-3-methoxy-, (6aR-cis)-**	524-97-0	**NI31188**
298	90	180	181	89	137	269	255	**298**	100	47	41	36	32	30	29	29		17	14	5	–	–	–	–	–	–	–	–	–	–	**Spiro[benzofuran-2(3H),2′-oxiran]-3-one, 4,6-dimethoxy-3′-phenyl-**	35405-27-7	**NI31189**
298	152	90	299	241	89	180	270	**298**	100	24	24	20	20	17	14	12		17	14	5	–	–	–	–	–	–	–	–	–	–	**Spiro[benzofuran-2(3H),2′-oxiran]-3-one, 6,7-dimethoxy-3′-phenyl-**	10173-80-5	**LI06516**
298	90	152	299	241	89	269	180	**298**	100	25	24	21	19	16	12	12		17	14	5	–	–	–	–	–	–	–	–	–	–	**Spiro[benzofuran-2(3H),2′-oxiran]-3-one, 6,7-dimethoxy-3′-phenyl-**	10173-80-5	**NI31190**
241	243	242	297	298	244	77	299	**298**	100	74	58	36	24	23	17	13		17	15	1	2	–	–	1	–	–	–	–	–	–	**1H-1,4-Benzodiazepin-2-one, 7-chloro-3-ethyl-2,3-dihydro-5-phenyl-**	28611-27-0	**NI31191**
297	270	255	298	228	243	242	78	**298**	100	97	92	81	72	66	62	39		17	15	1	2	–	–	1	–	–	–	–	–	–	**1,5-Benzodiazocin-2-one, 8-chloro-1,2,3,4-tetrahydro-1-methyl-6-phenyl-**	17954-18-6	**NI31192**
226	198	194	199	227	298	193	154	**298**	100	38	29	14	14	10	6	5		17	18	1	2	1	–	–	–	–	–	–	–	–	**10H-Phenothiazine-10-acetamide, N,N,β-trimethyl-**		**NI31193**
169	185	141	140	82	69	114	55	**298**	100	59	50	47	43	33	28	27	**0.10**	17	18	1	2	1	–	–	–	–	–	–	–	–	**Pyrrolo[1,2-b][1,2,4]oxadiazole-2(1H)-thione, tetrahydro-5,5-dimethyl-1-(1-naphthalenyl)-**	50455-84-0	**NI31194**
91	77	104	79	254	51	146	108	**298**	100	21	18	12	11	11	10	9	**0.02**	17	18	3	2	–	–	–	–	–	–	–	–	–	**N-Benzyloxycarbonyl-α-phenylalanine amide**		**IC03759**
91	65	147	92	79	77	39	51	**298**	100	19	14	14	11	11	10	9	**0.16**	17	18	3	2	–	–	–	–	–	–	–	–	–	**N-Benzyloxycarbonyl-β-phenylalanine amide**		**IC03760**
225	226	298	224	98	167	168	223	**298**	100	18	7	7	6	5	4	3		17	18	3	2	–	–	–	–	–	–	–	–	–	**Propanoic acid, 3-(3-oxo-1,2,3,5,6,11b-hexahydro-11H-pyrrolo[2,1-a]-β-carbolin-11b-yl)-**		**MS06874**
100	113	186	210	36	211	198	87	**298**	100	54	50	19	14	13	12	12	**0.00**	17	22	1	4	–	–	–	–	–	–	–	–	–	**3-Pyridazinamine, N-(2-morpholinoethyl)-4-methyl-6-phenyl-**		**NI31195**
132	298	43	205	133	134	55	105	**298**	100	34	18	15	12	11	11	9		17	30	–	–	2	–	–	–	–	–	–	–	–	**Eremophilan-8-one thioketal**		**LI06517**
75	61	60	41	59	57	47	150	**298**	100	66	63	61	60	59	57	50	**3.00**	17	30	–	–	2	–	–	–	–	–	–	–	–	**Spiro[1,3-dithiolan-2,1′-(4-isopropyl-4a,5-dimethyldecahydronaphthalene)], (4′S,4a′R,5′R,8a′S)-**		**NI31196**
139	225	252	207	226	97	280	157	**298**	100	98	71	71	53	52	46	46	**33.84**	17	30	4	–	–	–	–	–	–	–	–	–	–	**3-Furanacetic acid, 5-decyltetrahydro-α-methyl-2-oxo-**	54934-67-7	**NI31197**
139	97	157	112	132	114	110	225	**298**	100	51	45	45	25	25	14	9	**3.00**	17	30	4	–	–	–	–	–	–	–	–	–	–	**3-Furanacetic acid, 5-decyltetrahydro-α-methyl-2-oxo-**	54934-67-7	**LI06518**
225	97	114	143	226	109	154	252	**298**	100	94	63	59	58	46	43	39	**10.94**	17	30	4	–	–	–	–	–	–	–	–	–	–	**3-Furancarboxylic acid, tetrahydro-4-methyl-5-oxo-2-undecyl-**	480-71-7	**NI31198**
225	97	114	143	109	226	154	252	**298**	100	93	62	59	45	40	40	37	**9.00**	17	30	4	–	–	–	–	–	–	–	–	–	–	**3-Furancarboxylic acid, tetrahydro-4-methyl-5-oxo-2-undecyl-**	480-71-7	**LI06519**
97	143	69	115	55	41	159	125	**298**	100	86	40	34	28	22	9	5	**1.00**	17	30	4	–	–	–	–	–	–	–	–	–	–	**Hexanoic acid, 6-[(2-propyl-2-methyl-5-oxocyclopentyl)oxy]-, ethyl ester**		**NI31199**
73	75	55	117	41	129	283	67	**298**	100	99	60	55	46	39	35	20	**4.45**	17	34	2	–	–	–	–	–	–	1	–	–	–	**9-Tetradecenoic acid, trimethylsilyl ester**		**NI31200**
58	240	72	102	116	241	140	173	**298**	100	83	32	18	13	11	9	7	**0.00**	17	38	–	2	–	–	–	–	–	1	–	–	–	**3-[Dimethyl[[[2-(N,N-dimethylamino)ethyl]-N-methylamino]methyl]silyl]-1-nonene**		**DD01074**
298	228	300	113	263	226	299	227	**298**	100	78	65	60	57	55	31	22		18	12	–	–	–	–	2	–	–	–	–	–	–	**1-(2,6-Dichlorostyryl)naphthalene**		**LI06520**
228	226	263	113	227	229	262	265	**298**	100	45	40	37	26	18	13	13	**9.79**	18	12	–	–	–	–	2	–	–	–	–	–	–	**1,1′:2′,1″-Terphenyl, 2,5-dichloro-**	61577-02-4	**NI31201**
298	300	226	228	113	227	299	150	**298**	100	64	48	48	29	19	19	13		18	12	–	–	–	–	2	–	–	–	–	–	–	**1,1′:2′,1″-Terphenyl, 2,5-dichloro-**	61577-02-4	**NI31202**
298	300	226	228	113	299	227	150	**298**	100	64	41	39	32	19	17	14		18	12	–	–	–	–	2	–	–	–	–	–	–	**1,1′:4′;1″-Terphenyl, 2,5-dichloro-**	61576-83-8	**NI31203**
270	269	28	298	179	268	271	78	**298**	100	67	52	50	50	28	20	19		18	15	–	4	–	–	–	–	–	–	–	1	–	**1,2,5-Triphenylcyclotetrazenoborane**		**MS06875**
298	270	299	242	297	300	41	271	**298**	100	34	22	13	12	11	8	7		18	18	–	–	2	–	–	–	–	–	–	–	–	**1,2,3,4,8,9,10,11-Octahydrobenzo[1,2-b:3,4-b′]bisthianaphthene**		**AI02011**
165	254	211	121	253	166	298	44	**298**	100	60	57	57	45	37	28	22		18	18	2	–	1	–	–	–	–	–	–	–	–	**1,3-Oxathiolan-5-one, 4,4-diphenyl-2-propyl-**	31061-70-8	**NI31204**
163	298	251	164	299	126	77	252	**298**	100	48	33	12	10	10	9	7		18	18	4	–	–	–	–	–	–	–	–	–	–	**Benzoic acid, 2-(4-hydroxy-3-isopropylbenzoyl)-, methyl ester**	83938-71-0	**NI31205**
163	298	251	77	91	92	40	164	**298**	100	49	43	28	24	13	13	12		18	18	4	–	–	–	–	–	–	–	–	–	–	**Benzoic acid, 2-(4-hydroxy-3-isopropylbenzoyl)-, methyl ester**	83938-71-0	**NI31206**
162	131	119	298	91	163	134	225	**298**	100	36	22	21	15	14	10	9		18	18	4	–	–	–	–	–	–	–	–	–	–	**[2]Benzopyrano[4,3-b][1]benzopyran, 5,6a,7,12a-tetrahydro-6a,10-dimethoxy-**	14991-63-0	**NI31207**
121	149	267	179	178	239	77	91	**298**	100	85	42	24	21	19	17	14	**6.00**	18	18	4	–	–	–	–	–	–	–	–	–	–	**Butanedioic acid, 2,3-diphenyl-, dimethyl ester**	19020-59-8	**NI31208**
239	197	103	240	165	180	179	178	**298**	100	47	32	15	15	12	12	12	**9.90**	18	18	4	–	–	–	–	–	–	–	–	–	–	**Butanedioic acid, α,α-diphenyl-, dimethyl ester**		**IC03761**
121	28	149	266	105	18	77	91	**298**	100	93	91	64	55	52	41	31	**3.80**	18	18	4	–	–	–	–	–	–	–	–	–	–	**Butanedioic acid, diphenyl-, dimethyl ester, DL-**		**IC03762**
28	105	149	121	77	91	266	106	**298**	100	74	58	58	52	47	41	39	**3.80**	18	18	4	–	–	–	–	–	–	–	–	–	–	**Butanedioic acid, diphenyl-, dimethyl ester, meso-**		**IC03763**
105	54	77	71	176	193	106	55	**298**	100	39	24	21	10	9	8	3	**0.88**	18	18	4	–	–	–	–	–	–	–	–	–	–	**1,4-Butanediol, dibenzoate**	19224-27-2	**NI31209**
105	54	77	71	31	176	106	193	**298**	100	41	28	23	9	7	7	6	**0.06**	18	18	4	–	–	–	–	–	–	–	–	–	–	**1,4-Butanediol, dibenzoate**	19224-27-2	**IC03764**
298	166	123	194	297	221	104	151	**298**	100	75	65	60	25	25	25	15		18	18	4	–	–	–	–	–	–	–	–	–	–	**Desmethoxymatteucinol-7-methyl ether**		**NI31210**
105	149	297	91	298	106	191	162	**298**	100	92	48	42	39	25	20	20		18	18	4	–	–	–	–	–	–	–	–	–	–	**Erythritol, 1,3:2,4-di-O-benzylidene-**	4148-59-8	**NI31211**
149	90	105	150	106	77	107	91	**298**	100	50	43	13	11	9	6	5	**2.20**	18	18	4	–	–	–	–	–	–	–	–	–	–	**Erythritol, 1,3:2,4-di-O-benzylidene-**	4148-59-8	**NI31212**
105	150	149	106	77	92	91	191	**298**	100	60	55	48	35	24	24	20	**10.00**	18	18	4	–	–	–	–	–	–	–	–	–	–	**Erythritol, 1,3:2,4-di-O-benzylidene-**	4148-59-8	**LI06521**
141	153	198	115	298	142	168	152	**298**	100	53	30	26	19	16	14	14		18	18	4	–	–	–	–	–	–	–	–	–	–	**2-Furanpropanoic acid, tetrahydro-α-(1-naphthalenylmethyl)-5-oxo-**	55760-28-6	**NI31213**
149	118	121	298	90	150	267	89	**298**	100	22	13	11	10	9	8	7		18	18	4	–	–	–	–	–	–	–	–	–	–	**Bis(4,4′-methoxycarbonyl)dibenzyl**		**IC03765**
45	211	298	256	183	165	150	107	**298**	100	72	66	52	45	24	22	20		18	18	4	–	–	–	–	–	–	–	–	–	–	**Stilbene, 2-(methoxymethoxy)-4′-acetoxy-, (Z)-**		**MS06876**
45	211	298	256	183	165	150	107	**298**	100	46	33	31	31	17	17	17		18	18	4	–	–	–	–	–	–	–	–	–	–	**Stilbene, 2-(methoxymethoxy)-4′acetoxy-, (E)-**		**MS06877**
45	298	253	165	194	149	192	166	**298**	100	84	75	35	32	31	17	17		18	18	4	–	–	–	–	–	–	–	–	–	–	**Stilbene, 2-(methoxymethoxy)-4′-(carboxymethyl)-, (E)-**		**MS06878**
45	253	298	165	194	166	149	192	**298**	100	61	58	36	25	16	16	14		18	18	4	–	–	–	–	–	–	–	–	–	–	**Stilbene, 2-(methoxymethoxy)-4′-(carboxymethyl)-, (Z)-**		**MS06879**

MASS TO CHARGE RATIOS								M.W.	INTENSITIES								Parent	C	H	O	N	S	F	Cl	Br	I	Si	P	B	X	COMPOUND NAME	CAS Reg No	No
162	131	119	298	91	163	134	77	**298**	100	36	22	21	15	14	9	9		18	18	4	–	–	–	–	–	–	–	–	–	–	5,6a,7,12a-Tetrahydro-3,6a-dimethoxy[2]benzopyrano[4,3-b][1]benzopyran		MS06880
105	297	149	91	298	106	191	162	**298**	100	96	92	85	78	51	40	40		18	18	4	–	–	–	–	–	–	–	–	–	–	L-Threitol 1,3:2,4-di-O-benzylidene-		LI06522
86	58	180	87	298	198	42	212	**298**	100	62	56	52	43	34	30	24		18	22	–	2	1	–	–	–	–	–	–	–	–	10H-Phenothiazine-10-ethanamine, N,N-diethyl-	60-91-3	NI31214
86	298	87	83	58	30	180	85	**298**	100	10	6	6	6	5	4	4		18	22	–	2	1	–	–	–	–	–	–	–	–	10H-Phenothiazine-10-ethanamine, N,N-diethyl-	60-91-3	NI31215
58	298	198	180	100	59	42	30	**298**	100	12	8	6	5	4	4	4		18	22	–	2	1	–	–	–	–	–	–	–	–	10H-Phenothiazine-10-propanamine, N,N,β-trimethyl-	84-96-8	NI31216
58	45	43	298	42	41	57	46	**298**	100	36	20	12	11	10	9	9		18	22	–	2	1	–	–	–	–	–	–	–	–	10H-Phenothiazine-10-propanamine, N,N,β-trimethyl-	84-96-8	NI31217
58	72	298	198	180	100	59	42	**298**	100	22	15	9	7	7	7	7		18	22	–	2	1	–	–	–	–	–	–	–	–	10H-Phenothiazine-10-propanamine, N,N,β-trimethyl-	84-96-8	NI31218
239	154	182	298	197	167	196	183	**298**	100	60	46	39	35	31	28	28		18	22	2	2	–	–	–	–	–	–	–	–	–	Azecino[4,5,6-cd]indole-11-carboxylic acid, 2,6,7,10,11,12-hexahydro-6,8-dimethyl-, methyl ester	55702-37-9	NI31219
182	167	253	298	180	183	196	194	**298**	100	37	31	26	25	23	19	16		18	22	2	2	–	–	–	–	–	–	–	–	–	Brafouedine	104021-43-4	NI31220
239	154	182	298	197	167	196	183	**298**	100	60	46	39	35	31	28	28		18	22	2	2	–	–	–	–	–	–	–	–	–	Clavicipitic acid, 6-methyl-, methyl ester		NI31221
180	298	167	194	182	168	183	254	**298**	100	95	77	62	59	55	43	41		18	22	2	2	–	–	–	–	–	–	–	–	–	Compactinervine, de(methoxycarbonyl)-	2111-94-6	NI31222
297	43	298	169	168	41	55	170	**298**	100	99	76	65	45	43	39	37		18	22	2	2	–	–	–	–	–	–	–	–	–	2-(Hydroxymethyl)octahydroindolo[2,3-a]quinolizine		LI06523
43	169	170	156	168	237	297	154	**298**	100	89	49	49	45	41	39	37	**31.00**	18	22	2	2	–	–	–	–	–	–	–	–	–	2-(Hydroxymethyl)octahydroindolo[2,3-a]quinolizine		LI06524
298	223	130	172	195	194	182	167	**298**	100	54	43	35	33	32	31	31		18	22	2	2	–	–	–	–	–	–	–	–	–	Isobrafouedine		NI31223
131	187	130	117	111	90	132	116	**298**	100	42	42	26	20	15	13	11	**0.00**	18	22	2	2	–	–	–	–	–	–	–	–	–	2,3-Naphthalenedicarboximide, endo-1,2,3,4-tetrahydro-1,4-imino-N-methyl-9-neopentyl-	123542-28-9	DD01075
211	212	105	210	209	167	91	168	**298**	100	18	11	6	6	4	4	3	**2.00**	18	22	2	2	–	–	–	–	–	–	–	–	–	Propanoic acid, 3-(1,2,3,5,6,11b-hexahydro-11H-pyrrolo[2,1-a]-β-carbolin 11b-yl)-, methyl ester		MS06881
227	84	298	168	167	239	214	195	**298**	100	99	35	18	17	16	15	11		18	22	2	2	–	–	–	–	–	–	–	–	–	1H-Pyrrolo[2,3-d]carbazole-6-carboxylic acid, 3-ethyl-2,3,3a,4,5,7-hexahydro-, methyl ester	91085-35-7	NI31224
128	211	126	213	214	127	299	87	**298**	100	92	90	86	72	65	47	38	**14.00**	18	22	2	2	–	–	–	–	–	–	–	–	–	Tricyclo[5.2.1.0^{4,10}]deca-1(10),2,5,8-tetraene, 4,7-dimorpholino-		NI31225
55	67	41	81	69	54	79	43	**298**	100	83	75	55	43	39	36	36	**0.70**	18	31	1	–	–	–	1	–	–	–	–	–	–	9,12-Octadecadienoyl chloride, (Z,Z)-	7459-33-8	NI31226
116	117	43	129	57	41	55	74	**298**	100	31	28	22	22	20	19	12	**3.30**	18	34	3	–	–	–	–	–	–	–	–	–	–	Heptadecanoic acid, 3-oxo-, methyl ester	54934-65-5	WI01386
43	55	41	143	57	170	200	83	**298**	100	94	70	63	58	57	53	50	**0.00**	18	34	3	–	–	–	–	–	–	–	–	–	–	Heptadecanoic acid, 8-oxo-, methyl ester	54575-44-9	NI31227
112	114	153	185	152	227	111	242	**298**	100	91	89	71	58	51	47	46	**6.00**	18	34	3	–	–	–	–	–	–	–	–	–	–	Heptadecanoic acid, 12-oxo-, methyl ester		LI06525
104	111	266	95	239	83	81	71	**298**	100	23	11	11	10	10	10	8	**0.95**	18	34	3	–	–	–	–	–	–	–	–	–	–	4-Hexadecenoic acid, 2-methoxy-, methyl ester	40924-23-0	NI31228
55	67	41	81	95	69	43	82	**298**	100	73	73	55	41	41	41	36	**0.00**	18	34	3	–	–	–	–	–	–	–	–	–	–	9-Octadecenoic acid, 12-hydroxy-, (R,Z)-		IC03766
55	41	43	69	67	57	155	80	**298**	100	74	70	69	63	63	61	50	**0.54**	18	34	3	–	–	–	–	–	–	–	–	–	–	Oxiraneoctanoic acid, 3-octyl-, cis-	24560-98-3	NI31229
87	130	143	55	57	101	43	41	**298**	100	97	72	53	51	47	46	44	**6.00**	18	34	3	–	–	–	–	–	–	–	–	–	–	Pentadecanoic acid, 6-methyl-3-oxo-, methyl ester		NI31230
57	55	41	67	81	69	43	95	**298**	100	42	37	28	27	27	23	20	**0.20**	18	34	3	–	–	–	–	–	–	–	–	–	–	Propanoic acid, 2-hydroxy-, 9(Z)-pentadecenyl ester		MS06882
43	58	72	114	85	128	298	156	**298**	100	70	60	20	18	15	12	11		18	38	1	2	–	–	–	–	–	–	–	–	–	Heptanamide, N-[3-(ethylamino)propyl]-N-hexyl-	67138-97-0	NI31231
227	128	58	114	72	298	198	85	**298**	100	43	40	24	23	18	18	15		18	38	1	2	–	–	–	–	–	–	–	–	–	Heptanamide, N-ethyl-N-[3-(hexylamino)propyl]-	67139-06-4	NI31232
69	41	162	81	93	163	134	67	**298**	100	46	32	31	25	21	14	13	**0.00**	19	22	3	–	–	–	–	–	–	–	–	–	–	2H-1-Benzopyran-2-one, 7-[(3,7-dimethyl-2,6-octadienyl)oxy]-, (E)-	495-02-3	NI31233
224	267	298	280	297	97	223	165	**298**	100	94	81	81	61	61	59	41		19	22	3	–	–	–	–	–	–	–	–	–	–	Estra-1,3,5(10),9(11)-tetraen-17-one, 12-hydroxy-3-methoxy-, (12α)-	29082-67-5	NI31234
186	298	241	171	240	159	299	187	**298**	100	96	39	35	34	31	24	20		19	22	3	–	–	–	–	–	–	–	–	–	–	Estra-1,3,5(10),9(11)-tetraen-17-one, 14-hydroxy-3-methoxy-		LI06526
186	298	241	240	171	159	299	187	**298**	100	97	40	36	36	32	24	21		19	22	3	–	–	–	–	–	–	–	–	–	–	Estra-1,3,5(10),9(11)-tetraen-17-one, 14-hydroxy-3-methoxy-, (8α,14β)-	17908-45-1	NI31235
238	91	155	41	298	239	128	77	**298**	100	81	75	74	70	59	56	56		19	22	3	–	–	–	–	–	–	–	–	–	–	Methyl allo-gibberellate		IC03767
297	225	197	184	169	156	223	144	**298**	100	69	43	25	25	24	20	14	**0.00**	19	26	1	2	–	–	–	–	–	–	–	–	–	Antirhine, dihydro-		MS06883
280	297	251	168	298	169	170	124	**298**	100	90	88	82	74	72	63	60		19	26	1	2	–	–	–	–	–	–	–	–	–	Corynan-17-ol	2270-72-6	NI31236
170	297	298	225	55	197	169	184	**298**	100	77	63	47	44	32	24	21		19	26	1	2	–	–	–	–	–	–	–	–	–	10-Desoxy-18,19-dihydro-15-epi-hunterburnine		MS06884
112	298	283	140	299	158	122	171	**298**	100	83	29	29	20	14	14	11		19	26	1	2	–	–	–	–	–	–	–	–	–	20,21-Dinoraspidospermidin-5-ol, 1-ethyl-	55162-78-2	NI31237
184	156	185	169	168	130	183	144	**298**	100	36	24	19	15	14	13	12	**12.00**	19	26	1	2	–	–	–	–	–	–	–	–	–	1- or 2-[2-(Hydroxymethyl)butyl]-2,3,5,6,11,11b-hexahydro-1H-indolizino [8,7-b]indole		LI06527
184	156	297	298	223	221	225	169	**298**	100	23	19	18	18	13	12	12		19	26	1	2	–	–	–	–	–	–	–	–	–	1-[2-(Hydroxymethyl)butyl]-2,3,5,6,11,11b-hexahydro-1H-indolizino[8,7-b]indole		LI06528
205	111	297	110	298	198	149	170	**298**	100	84	78	75	67	64	62	39		19	26	1	2	–	–	–	–	–	–	–	–	–	2-[2-(Hydroxymethyl)butyl]-2,3,5,6,11,11b-hexahydro-1H-indolizino[8,7-b]indole		LI06529
123	298	156	268	110	280	144	169	**298**	100	30	30	25	25	23	20	13		19	26	1	2	–	–	–	–	–	–	–	–	–	1H-Indole-3-ethanol, 2-(5-ethyl-1-azabicyclo[2.2.2]oct-2-yl)-, [1S-(1α,2α,4α,5β)]-	10283-68-8	NI31238
124	59	45	42	43	41	123	74	**298**	100	36	33	31	20	20	19	19	**7.27**	19	26	1	2	–	–	–	–	–	–	–	–	–	1H-Indole-2-ethanol, β-(3-ethylidene-1-methyl-4-piperidinyl)-3-methyl-	1850-29-9	NI31240
124	44	174	125	122	123	57	43	**298**	100	11	9	9	8	7	5	5	**2.42**	19	26	1	2	–	–	–	–	–	–	–	–	–	1H-Indole-2-ethanol, β-(3-ethylidene-1-methyl-4-piperidinyl)-3-methyl-	1850-29-9	NI31239
149	184	133	167	170	156	129	140	**298**	100	78	47	35	29	21	20	16	**8.01**	19	26	1	2	–	–	–	–	–	–	–	–	–	1H-Indolizino[8,7-b]indole-1-propanol, 1-ethyl-2,3,5,6,11,11b-hexahydro-	55670-07-0	NI31241
184	156	223	297	298	221	169	225	**298**	100	22	20	17	16	13	13	11		19	26	1	2	–	–	–	–	–	–	–	–	–	1H-Indolizino[8,7-b]indole-1-propanol, β-ethyl-2,3,5,6,11,11b-hexahydro-	55670-04-7	NI31242
205	112	297	111	298	199	149	97	**298**	100	83	80	75	70	63	61	50		19	26	1	2	–	–	–	–	–	–	–	–	–	1H-Indolizino[8,7-b]indole-2-propanol, β-ethyl-2,3,5,6,11,11b-hexahydro-	14058-65-2	NI31243

MASS TO CHARGE RATIOS								M.W.	INTENSITIES								Parent	C	H	O	N	S	F	Cl	Br	I	Si	P	B	X	COMPOUND NAME	CAS Reg No	No
149	170	184	167	130	120	297	156	**298**	100	55	37	35	35	34	27	19	**18.02**	19	26	1	2	–	–	–	–	–	–	–	–	–	1H-Indolizino[8,7-b]indole-2-propanol, 2-ethyl-2,3,5,6,11,11b-hexahydro-, cis-	19621-67-1	NI31244
170	171	184	297	140	154	298	142	**298**	100	75	40	33	30	25	23	20		19	26	1	2	–	–	–	–	–	–	–	–	–	1H-Indolizino[8,7-b]indole-2-propanol, 2-ethyl-2,3,5,6,11,11b-hexahydro-, trans-	19621-66-0	NI31245
87	43	41	59	57	55	88	58	**298**	100	10	8	6	6	6	5	5	**0.94**	19	38	2	–	–	–	–	–	–	–	–	–	–	1,3-Dioxane, 2-pentadecyl-	17352-27-1	NI31246
59	283	83	97	43	69	57	241	**298**	100	73	34	32	25	23	23	22	**0.00**	19	38	2	–	–	–	–	–	–	–	–	–	–	1,3-Dioxolane, 4-hexyl-2,2-dimethyl-5-octyl-	54934-58-6	NI31247
87	59	41	43	28	88	57	55	**298**	100	12	9	7	5	4	4	4	**0.30**	19	38	2	–	–	–	–	–	–	–	–	–	–	1,3-Dioxolane, 4-methyl-2-pentadecyl-	54950-56-0	NI31248
283	72	43	83	97	57	69	284	**298**	100	48	45	43	39	29	28	22	**0.00**	19	38	2	–	–	–	–	–	–	–	–	–	–	1,3-Dioxolane, 2,2,4-trimethyl-5-tridecyl-	55712-73-7	NI31249
88	101	298	43	89	55	57	41	**298**	100	55	35	26	23	22	18	18		19	38	2	–	–	–	–	–	–	–	–	–	–	Heptadecanoic acid, ethyl ester	14010-23-2	NI31250
74	101	75	43	28	55	41	57	**298**	100	91	30	24	21	20	19	17	**10.80**	19	38	2	–	–	–	–	–	–	–	–	–	–	Heptadecanoic acid, 3-methyl-, methyl ester		MS06885
74	40	87	44	43	57	55	69	**298**	100	80	74	61	53	44	41	35	**28.70**	19	38	2	–	–	–	–	–	–	–	–	–	–	Heptadecanoic acid, 9-methyl-, methyl ester		MS06886
74	298	143	199	255	185	157	153	**298**	100	69	55	45	23	20	17	15		19	38	2	–	–	–	–	–	–	–	–	–	–	Heptadecanoic acid, 9-methyl-, methyl ester		LI06530
74	87	43	57	55	41	69	71	**298**	100	66	55	42	41	34	28	23	**16.70**	19	38	2	–	–	–	–	–	–	–	–	–	–	Heptadecanoic acid, 10-methyl-, methyl ester		MS06887
74	298	87	43	55	255	143	41	**298**	100	98	71	60	44	38	38	31		19	38	2	–	–	–	–	–	–	–	–	–	–	Heptadecanoic acid, 14-methyl-, methyl ester		MS06888
74	298	87	43	55	255	143	41	**298**	100	98	71	60	44	38	38	31		19	38	2	–	–	–	–	–	–	–	–	–	–	Heptadecanoic acid, 14-methyl-, methyl ester, (±)-	57274-45-0	NI31251
74	87	298	57	55	43	41	143	**298**	100	72	45	39	36	35	30	26		19	38	2	–	–	–	–	–	–	–	–	–	–	Heptadecanoic acid, 15-methyl-, methyl ester		MS06889
74	87	28	43	75	298	55	41	**298**	100	71	32	30	25	24	23	21		19	38	2	–	–	–	–	–	–	–	–	–	–	Heptadecanoic acid, 16-methyl-, methyl ester	5129-61-3	MS06890
74	43	87	55	41	75	57	69	**298**	100	60	53	31	29	20	19	17	**1.00**	19	38	2	–	–	–	–	–	–	–	–	–	–	Heptadecanoic acid, 16-methyl-, methyl ester	5129-61-3	NI31252
43	83	55	57	69	61	97	41	**298**	100	57	50	48	46	45	43	38	**0.24**	19	38	2	–	–	–	–	–	–	–	–	–	–	1-Heptadecanol, acetate	822-20-8	NI31253
43	87	55	41	57	97	83	69	**298**	100	40	36	32	31	30	29	28	**0.00**	19	38	2	–	–	–	–	–	–	–	–	–	–	2-Heptadecanol, acetate	56599-51-0	NI31254
102	87	43	115	41	55	57	29	**298**	100	35	34	31	29	26	17	15	**5.10**	19	38	2	–	–	–	–	–	–	–	–	–	–	Hexadecanoic acid, 2-ethyl-, methyl ester		MS06891
102	43	41	87	55	115	57	29	**298**	100	39	34	32	29	28	16	13	**12.81**	19	38	2	–	–	–	–	–	–	–	–	–	–	Hexadecanoic acid, 2-ethyl-, methyl ester		NI31255
43	41	102	57	60	55	256	69	**298**	100	62	48	45	43	43	35	23	**2.97**	19	38	2	–	–	–	–	–	–	–	–	–	–	Hexadecanoic acid, isopropyl ester	142-91-6	IC03768
102	43	41	60	73	57	129	55	**298**	100	83	51	49	40	38	37	36	**0.00**	19	38	2	–	–	–	–	–	–	–	–	–	–	Hexadecanoic acid, isopropyl ester	142-91-6	NI31256
43	73	57	41	60	55	298	29	**298**	100	75	72	70	69	61	41	39		19	38	2	–	–	–	–	–	–	–	–	–	–	Nonadecanoic acid	646-30-0	NI31257
43	73	57	60	41	55	29	69	**298**	100	77	76	75	67	63	37	36	**15.94**	19	38	2	–	–	–	–	–	–	–	–	–	–	Nonadecanoic acid	646-30-0	NI31258
87	74	143	43	55	75	57	41	**298**	100	80	45	33	27	24	18	16	**3.94**	19	38	2	–	–	–	–	–	–	–	–	–	–	Octadecanoic acid, methyl ester	112-61-8	NI31259
74	87	43	298	55	41	57	75	**298**	100	67	55	41	41	39	28	25		19	38	2	–	–	–	–	–	–	–	–	–	–	Octadecanoic acid, methyl ester	112-61-8	IC03769
74	87	43	41	75	57	55	143	**298**	100	65	27	22	19	17	17	16	**6.97**	19	38	2	–	–	–	–	–	–	–	–	–	–	Octadecanoic acid, methyl ester	112-61-8	IC03770
70	43	71	41	55	211	57	229	**298**	100	53	35	23	20	14	13	10	**6.00**	19	38	2	–	–	–	–	–	–	–	–	–	–	Tetradecanoic acid, 3-methylbutyl ester	62488-24-8	NI31260
298	299	265	149	296	264	297	266	**298**	100	31	16	15	10	10	9	5		20	12	–	1	1	–	–	–	–	–	–	–	–	Dibenzo[c,h]phenothiazinyl		MS06892
221	297	298	269	222	193	192	90	**298**	100	90	82	50	50	38	25	18		20	14	1	2	–	–	–	–	–	–	–	–	–	3-Benzoyl-2-phenylpyrazolo[1,5-a]pyridine	17408-39-8	NI31261
127	138	298	137	115	114	77	128	**298**	100	60	41	35	21	21	17	16		20	14	1	2	–	–	–	–	–	–	–	–	–	Diazene, 1-naphthalenyl-2-naphthalenyl-, 1-oxide	54833-56-6	NI31262
181	298	105	152	77	241	153	165	**298**	100	86	68	67	61	50	30	20		20	14	1	2	–	–	–	–	–	–	–	–	–	1,3,4-Oxadiazole, 2-[1,1′-biphenyl]-4-yl-5-phenyl-	852-38-0	NI31263
297	298	281	282	299	77	78	252	**298**	100	80	59	35	16	16	13	9		20	14	1	2	–	–	–	–	–	–	–	–	–	Phthalazine, 1,4-diphenyl-, 2-oxide	19415-35-1	NI31264
298	165	297	299	77	166	164	163	**298**	100	68	27	21	13	12	7	6		20	15	2	–	–	–	–	–	–	–	–	1	–	1,3,2-Dioxaborole, 2,4,5-triphenyl-	4844-17-1	NI31265
64	213	270	160	41	298	42	149	**298**	100	99	78	78	57	54	51	47		20	26	2	–	–	–	–	–	–	–	–	–	–	D-Dihomoestra-1,3,5(10)-trien-17b-one, 3-hydroxy-	34365-33-8	NI31266
298	198	185	172	173	299	201	186	**298**	100	34	34	31	23	22	18	16		20	26	2	–	–	–	–	–	–	–	–	–	–	Estra-1,3,5(10),6-tetraene-3,17-diol, 1,2-dimethyl-, (17β)-	1818-13-9	NI31267
298	213	174	200	299	159	173	214	**298**	100	60	48	25	22	16	15	12		20	26	2	–	–	–	–	–	–	–	–	–	–	Estra-1,3,5(10)-trien-17-one, 3-hydroxy-1,2-dimethyl-	1094-07-1	NI31268
298	174	213	41	299	187	173	200	**298**	100	45	29	28	23	23	20	18		20	26	2	–	–	–	–	–	–	–	–	–	–	Estra-1,3,5(10)-trien-17-one, 1-methoxy-4-methyl-	57983-89-8	NI31269
298	174	213	200	159	173	299	41	**298**	100	88	73	30	26	25	23	20		20	26	2	–	–	–	–	–	–	–	–	–	–	Estra-1,3,5(10)-trien-17-one, 3-methoxy-1-methyl-	2684-40-4	NI31270
298	213	174	200	299	173	214	187	**298**	100	67	54	27	23	17	13	13		20	26	2	–	–	–	–	–	–	–	–	–	–	Estra-1,3,5(10)-trien-17-one, 3-methoxy-1-methyl-	2684-40-4	NI31271
298	160	241	299	172	159	146	41	**298**	100	53	40	36	34	34	34	29		20	26	2	–	–	–	–	–	–	–	–	–	–	Gona-1,3,5(10)-trien-17-one, 3-hydroxy-13-propyl-	39667-85-1	NI31272
298	160	241	299	146	172	159	133	**298**	100	31	25	22	21	19	18	17		20	26	2	–	–	–	–	–	–	–	–	–	–	Gona-1,3,5(10)-trien-17-one, 3-hydroxy-13-propyl-	39667-85-1	NI31273
162	148	298	283	163	299	284	119	**298**	100	81	61	37	19	13	8	8		20	26	2	–	–	–	–	–	–	–	–	–	–	2,2′,3,3′,5,5′,6-Heptamethyl-4,4′-dihydroxydiphenylmethane	29366-03-8	MS06893
162	148	298	283	149	163	299	284	**298**	100	80	60	40	35	16	12	10		20	26	2	–	–	–	–	–	–	–	–	–	–	2,2′,3,3′,5,5′,6-Heptamethyl-4,4′-dihydroxydiphenylmethane	29366-03-8	NI31274
283	298	241	284	255	41	299	187	**298**	100	68	36	22	22	19	17	17		20	26	2	–	–	–	–	–	–	–	–	–	–	Hinokion-1-ene		LI06531
255	256	298	237	239	91	299	135	**298**	100	19	14	6	5	5	4	4		20	26	2	–	–	–	–	–	–	–	–	–	–	1,1-Bis(2-hydroxy-3,5-dimethylphenyl)-2-methylpropane		IC03771
91	110	79	231	272	298	41	77	**298**	100	90	80	79	78	77	72	64		20	26	2	–	–	–	–	–	–	–	–	–	–	19-Norpregn-4-en-20-yn-3-one, 17-hydroxy-, (17α)-	68-22-4	NI31275
231	110	298	162	272	215	160	121	**298**	100	96	88	58	57	54	48	43		20	26	2	–	–	–	–	–	–	–	–	–	–	19-Norpregn-4-en-20-yn-3-one, 17-hydroxy-, (17α)-	68-22-4	NI31276
283	298	241	284	255	41	299	159	**298**	100	68	36	22	22	19	18	17		20	26	2	–	–	–	–	–	–	–	–	–	–	Podocarpa-1,8,11,13-tetraen-3-one, 12-hydroxy-13-isopropyl-	18326-19-7	NI31277
298	283	215	213	229	187	195	227	**298**	100	74	62	59	54	32	29	26		20	26	2	–	–	–	–	–	–	–	–	–	–	Podocarpa-5,8,11,13-tetraen-7-one, 13-hydroxy-14-isopropyl-	3810-52-4	NI31278
43	57	55	56	41	70	83	71	**298**	100	94	74	64	63	62	60	60	**0.00**	20	42	1	–	–	–	–	–	–	–	–	–	–	Didecyl ether	2456-28-2	DO01084
57	43	41	71	85	55	29	83	**298**	100	95	67	63	54	53	45	42	**0.00**	20	42	1	–	–	–	–	–	–	–	–	–	–	Didecyl ether	2456-28-2	WI01387
43	57	55	56	41	70	83	69	**298**	100	95	73	64	63	62	60	59	**0.00**	20	42	1	–	–	–	–	–	–	–	–	–	–	Didecyl ether	2456-28-2	NI31279
41	55	57	97	83	69	43	29	**298**	100	98	84	79	77	71	63	50	**0.00**	20	42	1	–	–	–	–	–	–	–	–	–	–	1-Eicosanol	629-96-9	WI01388

MASS TO CHARGE RATIOS								M.W.	INTENSITIES								Parent	C	H	O	N	S	F	Cl	Br	I	Si	P	B	X	COMPOUND NAME	CAS Reg No	No
43	57	41	55	29	69	83	71	**298**	100	77	77	69	48	41	39	33	**0.00**	20	42	1	–	–	–	–	–	–	–	–	–	–	1-Eicosanol	629-96-9	MS06894
45	57	43	55	97	83	69	41	**298**	100	34	34	24	19	18	18	18	**0.23**	20	42	1	–	–	–	–	–	–	–	–	–	–	2-Eicosanol	4340-76-5	NI31280
45	57	43	55	97	71	69	83	**298**	100	82	76	57	55	51	49	47	**0.55**	20	42	1	–	–	–	–	–	–	–	–	–	–	2-Eicosanol, (±)-	34019-45-9	NI31281
57	43	69	55	71	56	70	83	**298**	100	78	73	67	61	47	46	41	**0.60**	20	42	1	–	–	–	–	–	–	–	–	–	–	1-Hexadecanol, 3,7,11,15-tetramethyl-	645-72-7	WI01389
57	43	55	71	41	69	85	83	**298**	100	78	63	60	53	47	39	39	**0.00**	20	42	1	–	–	–	–	–	–	–	–	–	–	2-Octyldodecan-1-ol		IC03772
194	69	89	57	195	116	77	55	**298**	100	21	20	19	18	17	17	16	**0.00**	21	18	–	2	–	–	–	–	–	–	–	–	–	Methanediamine, 1-phenyl-N,N'-bis(phenylmethylene)-	92-29-5	NI31282
181	298	44	73	296	208	153	42	**298**	100	71	69	68	42	40	39	31		21	18	–	2	–	–	–	–	–	–	–	–	–	1-Phenyl-3,3-diphenyl-2-pyrazoline		MS06895
77	298	91	115	105	103	221	65	**298**	100	84	61	40	39	37	36	35		21	18	–	2	–	–	–	–	–	–	–	–	–	2-Propen-1-one, 1,3-diphenyl-, phenylhydrazone	37799-62-5	NI31283
255	227	242	256	298	43	241	41	**298**	100	53	50	21	20	16	13	13		21	30	1	–	–	–	–	–	–	–	–	–	–	s-Indacen-1(2H)-one, 3,5,6,7-tetrahydro-3,3,5,5-tetramethyl-8-(3-methylbutyl)-	55712-64-6	NI31284
298	241	283	213	171	173	215	199	**298**	100	75	65	42	27	25	19	17		21	30	1	–	–	–	–	–	–	–	–	–	–	Phenanthrene, 1,2,3,4,4a,10a-hexahydro-7-methoxy-1,1,4a-trimethyl-8-isopropyl-	54833-49-7	NI31285
91	43	105	107	79	121	298	81	**298**	100	94	85	69	69	59	57	52		21	30	1	–	–	–	–	–	–	–	–	–	–	Pregna-3,5-dien-20-one	1093-87-4	NI31286
283	284	298	43	69	111	225	41	**298**	100	24	21	10	7	5	4	4		22	34	–	–	–	–	–	–	–	–	–	–	–	1,1,4,4,5,5,8,8-Octamethyl-1,2,3,4,5,6,7,8-octahydroanthracene		NI31287
119	214	69	230	164	114	142	92	**299**	100	97	71	63	32	26	10	6	**0.00**	5	–	1	1	–	11	–	–	–	–	–	–	–	N,N-Bis(pentafluoroethyl)carbamoyl fluoride		MS06896
69	100	280	31	133	114	112	50	**299**	100	34	9	7	6	5	5	5	**1.00**	5	–	1	1	–	11	–	–	–	–	–	–	–	Perfluoro(tetrahydro-2-methyl-2H-1,2-oxazine)		LI06532
69	92	47	187	169	114	192	119	**299**	100	30	24	14	12	6	4	4	**0.00**	5	–	1	1	–	11	–	–	–	–	–	–	–	N-(Trifluoromethyl)-N-(heptafluoropropyl)carbamoyl fluoride		DD01076
152	115	117	151	264	180	153	266	**299**	100	39	28	14	13	11	11	9	**0.00**	9	8	2	1	1	–	3	–	–	–	–	–	–	1H-Isoindole-1,3(2H)-dione, 3a,4,7,7a-tetrahydro-2-[(trichloromethyl)thio]-	133-06-2	NI31288
79	77	80	149	78	39	107	117	**299**	100	25	21	19	18	18	12	11	**1.00**	9	8	2	1	1	–	3	–	–	–	–	–	–	1H-Isoindole-1,3(2H)-dione, 3a,4,7,7a-tetrahydro-2-[(trichloromethyl)thio]-	133-06-2	NI31289
79	149	80	78	117	119	77	107	**299**	100	30	26	25	24	19	17	13	**0.00**	9	8	2	1	1	–	3	–	–	–	–	–	–	1H-Isoindole-1,3(2H)-dione, 3a,4,7,7a-tetrahydro-2-[(trichloromethyl)thio]-	133-06-2	NI31290
30	184	113	126	127	173	56	211	**299**	100	81	23	22	18	15	15	13	**2.00**	9	12	5	3	–	3	–	–	–	–	–	–	–	Glycine, N-[N-[N-(trifluoroacetyl)glycyl]glycyl]-, methyl ester	651-18-3	LI06533
30	184	126	113	127	173	56	211	**299**	100	69	23	23	19	16	16	14	**2.01**	9	12	5	3	–	3	–	–	–	–	–	–	–	Glycine, N-[N-[N-(trifluoroacetyl)glycyl]glycyl]-, methyl ester	651-18-3	NI31291
44	179	63	43	78	90	105	89	**299**	100	26	17	16	14	13	11	11	**1.20**	10	9	8	3	–	–	–	–	–	–	–	–	–	L-Aspartic acid, N-(2,4-dinitrophenyl)-	7683-81-0	MS06897
172	77	103	91	173	64	51	40	**299**	100	18	14	13	13	8	8	4	**1.80**	10	10	–	3	–	–	–	–	1	–	–	–	–	1,2,3-Triazole, 4-(iodomethyl)-5-methyl-2-phenyl-		NI31292
127	72	264	138	109	28	67	193	**299**	100	61	46	35	20	20	18	15	**0.00**	10	19	5	1	–	–	1	–	–	–	1	–	–	Phosphoric acid, 2-chloro-3-(diethylamino)-1-methyl-3-oxo-1-propenyl dimethyl ester	13171-21-6	NI31294
127	72	264	138	109	67	15	44	**299**	100	84	42	29	21	21	20	18	**0.00**	10	19	5	1	–	–	1	–	–	–	1	–	–	Phosphoric acid, 2-chloro-3-(diethylamino)-1-methyl-3-oxo-1-propenyl dimethyl ester	13171-21-6	NI31293
300	73	127	302	72	141	125	140	**299**	100	59	58	33	28	23	20	19	**0.00**	10	19	5	1	–	–	1	–	–	–	1	–	–	Phosphoric acid, 2-chloro-3-(diethylamino)-1-methyl-3-oxo-1-propenyl dimethyl ester	13171-21-6	NI31295
103	299	179	196	73	117	75	105	**299**	100	86	72	71	50	43	33	20		11	17	5	3	–	–	–	–	–	1	–	–	–	2,4-Dinitrophenylethanolamine, trimethylsilyl-		NI31296
226	73	284	255	227	91	69	240	**299**	100	83	60	19	16	15	15	11	**0.00**	11	20	3	1	–	3	–	–	–	1	–	–	–	L-Isoleucine, N-(trifluoroacetyl)-, trimethylsilyl ester	52558-83-5	NI31297
212	73	284	213	69	285	255	74	**299**	100	81	44	13	9	7	7	7	**0.00**	11	20	3	1	–	3	–	–	–	1	–	–	–	L-Leucine, N-(trifluoroacetyl)-, trimethylsilyl ester	52558-82-4	NI31298
209	45	71	211	69	138	102	75	**299**	100	95	88	44	28	26	26	25	**5.80**	12	10	2	1	2	–	1	–	–	–	–	–	–	Acetic acid, [[[2-(4-chlorophenyl)-4-thiazolyl]methyl]thio]-	33451-03-5	NI31299
130	148	243	299	209	194	102	76	**299**	100	45	30	26	21	20	20	16		12	14	4	1	1	–	–	–	–	–	1	–	–	Phosphonothioic acid, (1,3-dihydro-1,3-dioxo-2H-isoindol-2-yl)-, O,O-diethyl ester	5131-24-8	NI31300
91	43	55	44	69	231	57	232	**299**	100	50	29	29	25	21	21	17	**8.00**	12	17	2	3	2	–	–	–	–	–	–	–	–	2-(Butylthio)-7-hydroxy-6-isopropyl-5H-1,3,4-thiadiazolo[3,2-a]pyrimidin-5-one		MS06898
82	73	110	154	182	75	83	45	**299**	100	54	49	47	38	34	18	15	**0.00**	12	25	2	3	–	–	–	–	–	2	–	–	–	Histidine, bis(trimethylsilyl)-	71241-10-6	NI31301
73	99	45	75	98	72	143	58	**299**	100	90	58	39	22	15	11	11	**6.76**	13	25	3	5	–	–	–	–	–	–	–	–	–	L-Ornithine, N^2-[(dimethylamino)methylene]-N^5-[[[(dimethylamino)methylene]amino]carbonyl]-, methyl ester	59824-38-3	NI31302
299	149	107	43	80	177	108	300	**299**	100	81	47	32	25	18	17	10		14	13	3	5	–	–	–	–	–	–	–	–	–	4-(4-Nitrophenylazo)-3-acetamidoaniline		IC03773
149	148	44	176	135	122	254	299	**299**	100	60	43	38	25	15	12	6		14	17	1	7	–	–	–	–	–	–	–	–	–	1-[3-(Aden-9-yl)propyl]-3-carbamoyl-1,4-dihydropyridine		LI06534
119	118	41	57	59	136	117	91	**299**	100	54	50	49	43	33	33	27	**0.00**	14	21	4	1	1	–	–	–	–	–	–	–	–	Carbamic acid, [(2,4,6-trimethylphenyl)sulphonyl]-, tert-butyl ester	56830-76-3	NI31303
270	135	199	49	271	99	98	48	**299**	100	77	48	22	17	16	14	13	**1.20**	14	21	4	1	1	–	–	–	–	–	–	–	–	N,N-Dipropyl-4-(methoxycarbonyl)benzenesulphonamide		LI06535
155	144	91	112	184	96	126	268	**299**	100	91	80	44	37	24	8	4	**0.00**	14	21	4	1	1	–	–	–	–	–	–	–	–	Hexanoic acid, 6-[[(4-methylphenyl)sulphonyl]amino]-, methyl ester	67370-67-6	NI31304
270	135	199	44	271	104	103	43	**299**	100	75	46	20	15	14	13	13	**0.00**	14	21	4	1	1	–	–	–	–	–	–	–	–	Methyl 4-[(dipropylamino)sulphonyl]benzoate		LI06536
86	99	58	184	87	90	56	30	**299**	100	21	8	7	7	3	3	3	**0.12**	14	22	2	3	–	–	1	–	–	–	–	–	–	Benzamide, 4-amino-5-chloro-N-[2-(diethylamino)ethyl]-2-methoxy-	364-62-5	NI31305
208	181	91	209	299	77	65	41	**299**	100	17	12	8	4	4	4	3		15	10	–	1	–	5	–	–	–	–	–	–	–	Phenylethylamine, N-[(pentafluorophenyl)methylene]-		MS06900
208	91	181	209	65	299	77	168	**299**	100	28	18	10	8	7	7	6		15	10	–	1	–	5	–	–	–	–	–	–	–	Phenylethylamine, N-[(pentafluorophenyl)methylene]-		MS06899
208	91	181	209	65	77	168	51	**299**	100	28	18	10	9	7	6	5	**6.50**	15	10	–	1	–	5	–	–	–	–	–	–	–	Phenylethylamine, N-[(pentafluorophenyl)methylene]-		LI06537

MASS TO CHARGE RATIOS	M.W.	INTENSITIES	Parent	C	H	O	N	S	F	Cl	Br	I	Si	P	B	X	COMPOUND NAME	CAS Reg No	No
271 273 301 299 165 191 89 90	299	100 99 72 72 41 33 30 22		15	10	1	1	–	–	–	1	–	–	–	–	–	Benz[d][1,3]oxazepine, 2-(4-bromophenyl)-	19062-91-0	NI31306
273 271 299 301 165 192 191 89	299	100 99 94 92 44 40 34 31		15	10	1	1	–	–	–	1	–	–	–	–	–	Benz[d][1,3]oxazepine, 2-phenyl-7-bromo-	19062-90-9	NI31307
105 77 301 299 220 51 180 106	299	100 50 21 21 15 15 13 8		15	10	1	1	–	–	–	1	–	–	–	–	–	3,5-Diphenyl-4-bromoisoxazole		LI06538
299 301 219 300 298 191 271 273	299	100 98 97 79 66 39 34 32		15	10	1	1	–	–	–	1	–	–	–	–	–	Quinoline, 6-bromo-2-phenyl-, 1-oxide	19051-07-1	NI31308
225 299 154 105 227 77 44 155	299	100 87 35 35 27 27 25 21		15	13	4	3	–	–	–	–	–	–	–	–	–	3H-Imidazo[1,5-b]pyridazine-5,7(6H)-dione, 4-(ethoxycarbonyl)-2-phenyl-	72481-73-3	NI31309
239 299 154 77 240 103 300 105	299	100 70 42 35 20 19 14 14		15	13	4	3	–	–	–	–	–	–	–	–	–	3H-Imidazo[1,5-b]pyridazine-5,7(6H)-dione, 4-(methoxycarbonyl)-6-methyl-2-phenyl-		NI31310
299 210 300 254 271 211 239 227	299	100 26 20 18 16 16 12 8		15	13	4	3	–	–	–	–	–	–	–	–	–	Pyrrolo[1,2-a]-1,3,5-triazine-7-carboxylic acid, 1,2,3,4-tetrahydro-2,4-dioxo-8-phenyl-, ethyl ester	54449-26-2	NI31311
220 206 221 204 41 43 55 193	299	100 64 50 32 23 20 17 12	11.00	15	25	3	1	1	–	–	–	–	–	–	–	–	8-Oxa-3-thiatricyclo[5.2.0.0^{7,9}]nonane, 1-(1-hexahydroazepino)-5,7-dimethyl-, 3,3-dioxide	104195-56-4	NI31312
206 220 221 193 41 43 207 55	299	100 67 35 25 24 20 17 16	5.00	15	25	3	1	1	–	–	–	–	–	–	–	–	8-Oxa-3-thiatricyclo[5.2.0.0^{7,9}]nonane, 5,5,7-trimethyl-1-(1-piperidino)-, 3,3-dioxide	104195-59-7	NI31313
127 46 59 43 41 95 140 55	299	100 46 37 31 27 25 15 13	5.00	15	25	5	1	–	–	–	–	–	–	–	–	–	1,2-Benzisoxazole, octahydro-2-[2,3-O-isopropylidene-β-D-ribofuranosyl], (3aR-cis)-	64018-62-8	NI31314
127 59 56 43 41 95 71 55	299	100 70 53 49 31 26 24 22	6.00	15	25	5	1	–	–	–	–	–	–	–	–	–	1,2-Benzisoxazole, octahydro-2-[2,3-O-isopropylidene-β-D-ribofuranosyl], (3aS-cis)-	64044-17-3	NI31315
138 93 43 44 139 41 94 80	299	100 70 38 32 30 28 27 17	2.00	15	25	5	1	–	–	–	–	–	–	–	–	–	Echinatine	480-83-1	NI31316
138 93 43 139 44 94 80 120	299	100 70 38 30 30 25 17 11	2.00	15	25	5	1	–	–	–	–	–	–	–	–	–	Echinatine	480-83-1	LI06539
198 156 82 124 43 36 199 157	299	100 84 40 20 17 12 8 8	1.00	15	25	5	1	–	–	–	–	–	–	–	–	–	2-Piperidineacetic acid, 1-acetyl-6-[1-(acetyloxy)propyl]-, methyl ester	54984-46-2	NI31317
299 192 121 77 159 92 64 108	299	100 50 48 42 40 40 40 36		16	13	1	1	2	–	–	–	–	–	–	–	–	Aniline, N-(3-thio-2-benzo[t]thienylidene)-4-methoxy-	71643-96-4	NI31318
150 267 104 299 165 120 118 239	299	100 60 20 15 7 7 7 2		16	13	5	1	–	–	–	–	–	–	–	–	–	Acetophenone, 4′-nitro-2-[2-(methoxycarbonyl)phenyl]-		LI06540
135 179 299 136 163 46	299	100 85 12 10 2 1		16	13	5	1	–	–	–	–	–	–	–	–	–	Benzeneacetic acid, 4-nitro-, 4-acetylphenyl ester	53327-13-2	NI31319
256 284 299 238 254 115 127 210	299	100 96 90 78 53 50 45 44		16	13	5	1	–	–	–	–	–	–	–	–	–	1,4-Dimethoxy-2,3-methylenedioxyacridone		MS06901
299 300 242 134.5 149.5 132 77 31	299	100 83 74 63 61 59 59 59		16	13	5	1	–	–	–	–	–	–	–	–	–	1,3-Dioxolo[4,5-b]acridin-10(5H)-one, 6,11-dimethoxy-5-methyl-		MS06902
284 299 254 158 285 300 170 77	299	100 83 38 28 17 14 12 10		16	13	5	1	–	–	–	–	–	–	–	–	–	1,3-Dioxolo[4,5-c]acridin-6(11H)-one, 5-hydroxy-4-methoxy-11-methyl-	517-76-0	NI31320
299 241 183 268 267	299	100 74 52 50 49		16	13	5	1	–	–	–	–	–	–	–	–	–	2-Methyl-3,9-bis(methoxycarbonyl)furo[3,2-b]quinoline		LI06541
284 299 254 158 285 300 115 77	299	100 59 34 24 20 13 12 10		16	13	5	1	–	–	–	–	–	–	–	–	–	10-Methyl-4-methoxy-2,3-methylenedioxy-1-hydroxyacridone		MS06903
192 165 193 299 301 190 27 166	299	100 42 17 15 14 10 8 7		16	14	–	1	–	–	–	1	–	–	–	–	–	Benzeneacetonitrile, α-(2-bromoethyl)-α-phenyl-	39186-58-8	NI31321
220 254 239 205 282 297 240	299	100 70 55 44 36 35 33	0.00	16	14	1	3	–	–	1	–	–	–	–	–	–	3-Acetamido-6-chloro-3,4-dihydro-4-hydroxy-4-phenylquinazoline		LI06542
283 282 284 285 220 77 241 205	299	100 87 51 32 20 17 14 14	6.62	16	14	1	3	–	–	1	–	–	–	–	–	–	3H-1,4-Benzodiazepin-2-amine, 7-chloro-N-methyl-5-phenyl-, 4-oxide	58-25-3	NI31322
282 283 284 299 285 77 56 41	299	100 72 44 35 24 23 20 19		16	14	1	3	–	–	1	–	–	–	–	–	–	3H-1,4-Benzodiazepin-2-amine, 7-chloro-N-methyl-5-phenyl-, 4-oxide	58-25-3	NI31323
282 283 284 285 299 77 241 91	299	100 72 40 24 22 18 15 12		16	14	1	3	–	–	1	–	–	–	–	–	–	3H-1,4-Benzodiazepin-2-amine, 7-chloro-N-methyl-5-phenyl-, 4-oxide	58-25-3	NI31324
77 63 103 51 104 299 76 105	299	100 69 58 53 40 35 27 24		16	14	1	3	–	–	1	–	–	–	–	–	–	1,2,4-Triazine-4-carbonyl chloride, 1,4,5,6-tetrahydro-1,3-diphenyl-		NI31325
28 125 127 205 283 111 89 139	299	100 78 43 33 22 22 22 17	6.99	16	14	1	3	–	–	1	–	–	–	–	–	–	1,2,4-Triazole, 1-(4-chlorobenzyl)-5-methyl-3-phenyl-, 4-oxide		NI31326
210 178 211 226 209 179 165 299	299	100 50 24 12 12 12 12 5		17	17	2	1	1	–	–	–	–	–	–	–	–	Propanoic acid, 2-(6,11-dihydrodibenzo[b,e]thiepin-11-ylamino)-		MS06904
44 72 228 171 299 229 43 57	299	100 95 33 14 11 11 8 8		17	17	2	1	1	–	–	–	–	–	–	–	–	Thioxanthen-9-one, 4-[(3-methylamino)propoxy]-		MS06905
271 181 211 270 299 240 153 272	299	100 47 23 20 10 10 10 9		17	17	4	1	–	–	–	–	–	–	–	–	–	Crinan-11-one, 1,2-didehydro-3-methoxy-, (3α,5α,13β,19α)-	1797-92-8	NI31327
271 181 270 211 240 238 256 299	299	100 43 31 20 10 10 5 4		17	17	4	1	–	–	–	–	–	–	–	–	–	Crinan-11-one, 1,2-didehydro-3-methoxy-, (3α,5α,13β,19α)-	1797-92-8	LI06543
271 181 211 270 272 153 240 152	299	100 36 27 18 14 13 10 9	3.41	17	17	4	1	–	–	–	–	–	–	–	–	–	Crinan-11-one, 1,2-didehydro-3-methoxy-, (3β,5α,13β,19α)-	1472-75-9	NI31328
271 181 270 211 240 238 299 256	299	100 48 31 21 8 7 6 3		17	17	4	1	–	–	–	–	–	–	–	–	–	Crinan-11-one, 1,2-didehydro-3-methoxy-, (3β,5α,13β,19α)-	1472-75-9	LI06544
256 299 228 282 115 257 229 270	299	100 99 98 32 30 28 28 24		17	17	4	1	–	–	–	–	–	–	–	–	–	Demethylcephalotaxinone	51020-45-2	NI31329
176 299 298 255 256 44 42 177	299	100 75 73 73 64 49 44 28		17	17	4	1	–	–	–	–	–	–	–	–	–	5H-Dibenzo[a,d]cycloheptene-2,3,7,8-tetrol, 5-[(methylamino)methyl]-	481-90-3	NI31330
215 299 59 85 200 216 228 300	299	100 59 37 28 24 23 16 12		17	17	4	1	–	–	–	–	–	–	–	–	–	Furo[2,3-b]quinoline, 7-(2,3-epoxy-3-methylbutoxy)-4-methoxy-	85802-29-5	NI31331
109 108 110 111 82 42 94 81	299	100 20 8 4 3 3 2 1	0.08	17	17	4	1	–	–	–	–	–	–	–	–	–	Lycorenan-7-one, 1-methyl-9,10-[methylenebis(oxy)]-	568-40-1	NI31332
167 168 152 165 166 212 153 169	299	100 88 79 78 71 69 40 38	0.00	17	17	4	1	–	–	–	–	–	–	–	–	–	Serine, diphenylacetyl-	65707-86-0	NI31333
256 299 257 284 300 164 270 229	299	100 91 24 20 18 10 8 8		17	21	–	3	1	–	–	–	–	–	–	–	–	Benzothiazole, 2-(4-butyl-3,5-dimethylpyrazol-1-yl)-6-methyl-		DD01077
222 78 44 178 77 42 41 91	299	100 41 28 18 18 15 14 13	0.25	17	21	2	1	–	–	–	–	–	1	–	–	–	2,2-Diphenyl-6-methyl-1,3-dioxa-6-aza-2-silacyclooctane		MS06906
105 77 51 222 104 91 44 78	299	100 54 53 51 51 43 35 28	19.00	17	21	2	1	–	–	–	–	–	1	–	–	–	2-Methyl-2,6-diphenyl-1,3-dioxa-6-aza-2-silacyclooctane		MS06907
167 267 268 168 223 154 184 221	299	100 32 24 21 17 17 16 13	10.00	17	21	2	3	–	–	–	–	–	–	–	–	–	10α-Methoxydihydroisolysergamide		MS06908
167 184 154 267 223 268 299 186	299	100 45 29 27 25 24 23 23		17	21	2	3	–	–	–	–	–	–	–	–	–	10β-Methoxydihydroisolysergamide		MS06909
167 184 154 267 223 268 221 168	299	100 40 37 36 35 25 21 21	16.00	17	21	2	3	–	–	–	–	–	–	–	–	–	10α-Methoxydihydrolysergamide		MS06910
167 267 154 184 223 168 224 207	299	100 32 28 23 22 22 19 13	9.00	17	21	2	3	–	–	–	–	–	–	–	–	–	10β-Methoxydihydrolysergamide		MS06911
113 57 198 129 86 41 226 142	299	100 87 84 48 43 41 36 28	0.00	17	33	3	1	–	–	–	–	–	–	–	–	–	Carbamic acid, (6-oxododecyl)-, tert-butyl ester	116437-39-9	DD01078
102 144 87 256 113 137 145 184	299	100 67 58 36 27 25 18 16	4.00	17	33	3	1	–	–	–	–	–	–	–	–	–	Tridecanoic acid, 3-acetylamino-12-methyl-, methyl ester		NI31334
91 65 299 90 92 77 63 39	299	100 13 12 6 4 3 3 3		18	13	–	5	–	–	–	–	–	–	–	–	–	Pyrimido[1,2-a]benzimidazole-2-carbonitrile, 3-(benzylamino)-		MS06912

MASS TO CHARGE RATIOS								M.W.	INTENSITIES								Parent	C	H	O	N	S	F	Cl	Br	I	Si	P	B	X	COMPOUND NAME	CAS Reg No	No
264	299	179	178	86	214	301	77	**299**	100	72	72	63	37	29	25	14		18	18	1	1	–	–	1	–	–	–	–	–	–	1-Chloro-2-(4-morpholino)-1,2-diphenylethane		LI06545
239	183	165	152	115	77	270	141	**299**	100	45	45	45	45	45	35	35	**30.00**	18	21	3	1	–	–	–	–	–	–	–	–	–	4αH-Cherylline, 2-demethyl-O,O-dimethyl-	23330-75-8	MS06913
299	298	162	300	124	42	229	59	**299**	100	88	30	21	21	20	18	16		18	21	3	1	–	–	–	–	–	–	–	–	–	Codeine	76-57-3	NI31335
162	229	124	59	81	188	214	70	**299**	100	93	93	71	53	49	47	41	**0.00**	18	21	3	1	–	–	–	–	–	–	–	–	–	Codeine	76-57-3	NI31336
299	162	229	124	300	298	214	42	**299**	100	38	28	22	20	15	15	14		18	21	3	1	–	–	–	–	–	–	–	–	–	Codeine	76-57-3	NI31337
268	299	284	220	266	228	241		**299**	100	58	50	18	15	11	10			18	21	3	1	–	–	–	–	–	–	–	–	–	Erysodine	7290-03-1	LI06546
268	299	284	269	215	266	300	228	**299**	100	58	52	28	17	14	12	10		18	21	3	1	–	–	–	–	–	–	–	–	–	Erysodine	7290-03-1	NI31338
268	284	299	266	220	228	241		**299**	100	49	46	15	12	8	6			18	21	3	1	–	–	–	–	–	–	–	–	–	Erysodine	7290-03-1	LI06547
299	282	300	132	284	134	298	133	**299**	100	33	20	17	16	16	13	13		18	21	3	1	–	–	–	–	–	–	–	–	–	Hasubanan-9-ol, 7,8-didehydro-4,5-epoxy-3-methoxy-17-methyl-, (5α,9α,13β,14β)-	24695-69-0	NI31339
299	300	242	270	282	256	284	115	**299**	100	23	20	18	10	10	9	9		18	21	3	1	–	–	–	–	–	–	–	–	–	Indolinocodeine	738-91-0	NI31340
299	298	256	42	257	241	228	216	**299**	100	91	69	16	13	10	10	10		18	21	3	1	–	–	–	–	–	–	–	–	–	Linearsine		MS06914
299	96	242	243	282	108	300	228	**299**	100	78	70	65	33	26	23	19		18	21	3	1	–	–	–	–	–	–	–	–	–	Metathebainone	510-66-7	NI31341
299	254	300	255	284	298	243	150	**299**	100	42	18	13	11	8	7	6		18	21	3	1	–	–	–	–	–	–	–	–	–	Morphinan-6-ol, 8,14-didehydro-4,5-epoxy-3-methoxy-17-methyl-, (5α,6α)-	467-14-1	LI06548
299	254	255	284	197	58	300	243	**299**	100	88	27	20	20	20	18	18		18	21	3	1	–	–	–	–	–	–	–	–	–	Morphinan-6-ol, 8,14-didehydro-4,5-epoxy-3-methoxy-17-methyl-, (5α,6α)-	467-14-1	NI31342
222	194	299	223	256	43	270	226	**299**	100	35	15	14	13	10	8	7		18	21	3	1	–	–	–	–	–	–	–	–	–	Nicotinic acid, 5-acetyl-1,4-dihydro-2,6-dimethyl-4-phenyl-, ethyl ester	20970-69-8	NI31343
201	299	173	202	130	300	186	158	**299**	100	63	49	36	26	24	22	19		18	21	3	1	–	–	–	–	–	–	–	–	–	Spiro[cyclohexane-1,3′-(8-methoxy-5-methyl-1,4-dihydro-1-oxo-2H-pyrano[4,3-b]indole)]		DD01079
98	201	256	77	216	142	202	299	**299**	100	82	35	29	25	22	20	17		18	22	1	1	–	–	–	–	–	–	1	–	–	N-Cyclohexyldiphenylphosphinamide		MS06915
104	105	77	51	91	78	103	65	**299**	100	57	49	13	10	8	6	6	**0.00**	18	37	2	1	–	–	–	–	–	–	–	–	–	Hexadecanamide, N-(2-hydroxyethyl)-	544-31-0	NI31344
59	55	72	43	41	44	69	214	**299**	100	80	75	60	55	35	25	21	**0.00**	18	37	2	1	–	–	–	–	–	–	–	–	–	12-Hydroxystearamide		IC03774
300	301	302	282	298	283			**299**	100	26	15	6	2	1			**0.00**	18	37	2	1	–	–	–	–	–	–	–	–	–	4-Octadecene-1,3-diol, 2-amino-, [R-[R*,S*,(E)]]-	123-78-4	NI31345
60	57	43	55	70	61	268	41	**299**	100	39	39	34	33	32	30	22	**1.12**	18	37	2	1	–	–	–	–	–	–	–	–	–	4-Octadecene-1,3-diol, 2-amino-, [R*,S*,(E)]-(±)-	2733-29-1	NI31346
161	299	162	271	270	298	160	146	**299**	100	18	16	14	11	9	9	6		19	25	2	1	–	–	–	–	–	–	–	–	–	8-Azaestra-1,3,5(10)-trien-17-one, 18-methyl-3-methoxy-		IC03775
105	137	164	299	77	136	243	135	**299**	100	53	50	33	33	20	20	20		19	25	2	1	–	–	–	–	–	–	–	–	–	1-Azaspiro[3.5]nonan-2-one, 1-benzoyl-6,6,8,8-tetramethyl-	77064-45-0	NI31347
91	30	208	190	92	102	43	65	**299**	100	24	15	10	10	8	7	6	**0.00**	19	25	2	1	–	–	–	–	–	–	–	–	–	1-Butanamine, 4-benzyloxy-3-[benzyloxymethyl]-	33498-87-2	NI31348
91	30	208	190	92	102	100	84	**299**	100	24	15	10	10	8	3	3	**0.00**	19	25	2	1	–	–	–	–	–	–	–	–	–	1-Butanamine, 4-benzyloxy-3-[benzyloxymethyl]-	33498-87-2	LI06549
299	213	172	268	300	211	159	133	**299**	100	40	35	21	20	20	16	16		19	25	2	1	–	–	–	–	–	–	–	–	–	Estra-1,3,5(10)-trien-16-one, 3-hydroxy-, O-methyloxime		NI31349
268	299	91	269	92	211	159	84	**299**	100	42	26	24	21	20	16	14		19	25	2	1	–	–	–	–	–	–	–	–	–	Estra-1,3,5(10)-trien-17-one, 3-hydroxy-, O-methyloxime	3342-64-1	NI31350
44	42	43	150	59	41	115	141	**299**	100	58	29	28	23	21	20	18	**14.91**	19	25	2	1	–	–	–	–	–	–	–	–	–	8H-5,12b-(Iminoethano)-1H-phenanthro[3,2-d][1,3]dioxin, 2,3,4,4a,5,6-hexahydro-15-methyl-, [4aR-(4aα,5α,12bα)]-	54984-45-1	NI31351
284	299	59	285	300	178	162	42	**299**	100	64	24	23	15	15	14	14		19	25	2	1	–	–	–	–	–	–	–	–	–	Morphinan, 5,6-didehydro-3,6-dimethoxy-17-methyl-	1038-07-9	NI31352
299	298	300	267	265	133	266	301	**299**	100	30	25	24	19	17	12	8		20	13	–	1	1	–	–	–	–	–	–	–	–	7H-Dibenzo[c,h]phenothiazine		MS06916
252	299	126	250	253	125	113	251	**299**	100	59	35	34	32	23	20	14		20	13	2	1	–	–	–	–	–	–	–	–	–	Fluorene, 9-(3-nitrobenzylidene)-	4421-51-6	NI31353
252	299	253	250	126	125	113	300	**299**	100	64	40	30	18	17	16	13		20	13	2	1	–	–	–	–	–	–	–	–	–	Fluorene, 9-(4-nitrobenzylidene)-	6954-71-8	NI31354
299	252	253	250	126	282	251	124	**299**	100	99	60	27	24	18	17	16		20	13	2	1	–	–	–	–	–	–	–	–	–	2-Nitrotriptycene		LI06550
254	299	282	76	270	104	255	300	**299**	100	90	51	50	32	26	22	19		20	13	2	1	–	–	–	–	–	–	–	–	–	2-Phthalimidobiphenyl		LI06551
254	299	282	76	270	104	255	50	**299**	100	90	51	50	32	26	22	14		20	13	2	1	–	–	–	–	–	–	–	–	–	2-Phthalimidobiphenyl		MS06917
299	152	300	255	76	254	104	102	**299**	100	35	22	21	20	13	8	8		20	13	2	1	–	–	–	–	–	–	–	–	–	4-Phthalimidobiphenyl		LI06552
299	152	28	300	255	76	44	254	**299**	100	35	24	22	21	20	18	13		20	13	2	1	–	–	–	–	–	–	–	–	–	4-Phthalimidobiphenyl		MS06918
299	93	125	91	79	153	137	105	**299**	100	90	71	63	61	60	55	45		20	29	1	1	–	–	–	–	–	–	–	–	–	Androsta-4,16-dien-3-one, O-methyloxime	56196-48-6	NI31355
45	41	231	70	69	110	216	186	**299**	100	73	63	40	34	31	29	29	**20.40**	20	29	1	1	–	–	–	–	–	–	–	–	–	2,6-Methano-3-benzazocine, 1,2,3,4,5,6-hexahydro-8-methoxy-6,11-dimethyl-3-(3-methyl-2-butenyl)-, (2α,6α,11R*)-	74810-91-6	NI31356
110	282	41	105	299	94	119	91	**299**	100	68	61	57	48	45	44	43		20	29	1	1	–	–	–	–	–	–	–	–	–	Retinal, oxime, all-trans-, syn-		MS06919
110	299	162	201	163	176	300	255	**299**	100	84	50	38	37	23	19	15		20	29	1	1	–	–	–	–	–	–	–	–	–	Retinamide	20638-84-0	NI31357
29	104	45	193	106	77	78	178	**299**	100	94	44	31	25	24	22	20	**0.00**	21	17	1	1	–	–	–	–	–	–	–	–	–	Isoxazole, 4,5-dihydro-3,4,5-triphenyl-, cis-	17669-25-9	NI31358
193	28	178	180	179	194	77	103	**299**	100	24	23	20	20	13	10	7	**3.50**	21	17	1	1	–	–	–	–	–	–	–	–	–	Isoxazole, 4,5-dihydro-3,4,5-triphenyl-, trans-	4894-25-1	NI31359
178	105	77	297	51	180	179	89	**299**	100	77	56	25	19	17	15	13	**1.00**	21	17	1	1	–	–	–	–	–	–	–	–	–	2-Propen-1-one, 3-(N-phenylamino)-1,2-diphenyl-		MS06920
222	118	299	90	223	270	119	203	**299**	100	81	34	27	17	14	13	10		21	17	1	1	–	–	–	–	–	–	–	–	–	2(1H)-Quinolinone, 3,4-dihydro-3,3-diphenyl-		MS06921
77	51	207	222	129	91	206	191	**299**	100	76	68	51	51	32	27	20	**17.30**	22	21	–	1	–	–	–	–	–	–	–	–	–	Aniline, N-[[2-isopropylphenyl]benzylene]-	19103-10-7	NI31360
77	207	222	129	51	91	192	284	**299**	100	99	82	68	42	34	32	31	**31.00**	22	21	–	1	–	–	–	–	–	–	–	–	–	Aniline, N-[[2-isopropylphenyl]benzylene]-	19103-10-7	NI31361
77	299	180	256	51	222	298	207	**299**	100	69	48	43	32	27	20	18		22	21	–	1	–	–	–	–	–	–	–	–	–	Aniline, N-[[3-isopropylphenyl]benzylene]-	32388-75-3	NI31362
77	299	222	180	256	207	51	298	**299**	100	93	72	66	52	36	36	25		22	21	–	1	–	–	–	–	–	–	–	–	–	Aniline, N-[[4-isopropylphenyl]benzylene]-	18864-77-2	NI31363
299	298	271	126.5	127	141.5	134.5	133.5	**299**	100	76	20	14	13	12	11	11		22	21	–	1	–	–	–	–	–	–	–	–	–	8,10-Diethyl-12-methylbenzo[a]acridine		MS06922

MASS TO CHARGE RATIOS								M.W.	INTENSITIES								Parent	C	H	O	N	S	F	Cl	Br	I	Si	P	B	X	COMPOUND NAME	CAS Reg No	No
299	298	271	283	269	296	284	254	299	100	82	35	22	20	19	14	14		22	21	–	1	–	–	–	–	–	–	–	–	–	9,11-Diethyl-7-methylbenzo[c]acridine		MS06923
69	89	175	173	302	300	127	75	300	100	52	49	49	30	30	30	30		3	1	–	–	1	8	–	1	–	–	–	–	–	(2-Bromo-3,3,3-trifluoro-1-propenyl)pentafluorosulphur		MS06924
43	217	215	84	57	214	42	216	300	100	36	34	28	28	24	23	21	7.56	3	6	–	4	–	–	–	–	–	–	–	–	1	Mercury, (cyanoguanidinato-N')methyl-	502-39-6	NI31364
84	43	42	202	200	68	199	85	300	100	87	57	55	44	34	31	31	5.10	3	6	–	4	–	–	–	–	–	–	–	–	1	Mercury, (cyanoguanidinato-N')methyl-	502-39-6	NI31365
69	124	143	93	105	231	193	81	300	100	56	46	30	25	16	16	10	0.00	4	–	–	–	–	10	–	–	–	–	2	–	–	2-Butene, 1,1,1,4,4,4-hexafluoro-2,3-bis(difluorophosphino)-		DD01080
131	69	100	31	231	181	93	119	300	100	70	29	22	21	21	15	6	0.70	6	–	–	–	–	12	–	–	–	–	–	–	–	Dodecafluorocyclohexane		AI02012
101	300	76	103	69	63	302	265	300	100	78	78	70	62	56	54	46		6	–	–	–	1	5	2	–	–	–	1	–	–	Phosphonothioic dichloride, (pentafluorophenyl)-	53327-26-7	NI31366
302	266	264	300	304	268	267	166	300	100	92	88	73	53	36	33	33		8	8	–	–	–	–	4	–	–	2	–	–	–	1,3-Disilaindan, 1,1,3,3-tetrachloro-5-methyl-	54113-94-9	NI31367
59	300	150	118	209	302	117	151	300	100	47	41	28	25	12	8	7		8	12	–	–	6	–	–	–	–	–	–	–	–	2,4,6,8,9,10-Hexathiatricyclo[3.3.1.1^{3,7}]decane, 1,3,5,7-tetramethyl-	6327-74-8	NI31368
59	300	150	118	209	145	302	117	300	100	46	41	29	26	19	13	8		8	12	–	–	6	–	–	–	–	–	–	–	–	2,4,6,8,9,10-Hexathiatricyclo[3.3.1.1^{3,7}]decane, 1,3,5,7-tetramethyl-	6327-74-8	NI31369
59	131	235	191	236	99	300	76	300	100	40	20	14	13	13	12	11		8	12	–	–	6	–	–	–	–	–	–	–	–	2,4,6,8,9-Pentathiatricyclo[3.3.1.1^{3,7}]decane-1-thiol, 3,5,7-trimethyl-	57274-31-4	NI31370
59	117	58	149	144	45	176	85	300	100	17	14	11	6	5	4	4	3.35	8	12	–	–	6	–	–	–	–	–	–	–	–	2,4,6,8,9-Pentathiatricyclo[3.3.1.1^{3,7}]decane-1-thiol, 3,5,7-trimethyl-	57274-31-4	NI31371
83	55	73	285	103	75	185	215	300	100	78	42	25	25	25	21	19	0.10	9	21	1	–	–	–	–	–	1	1	–	–	–	Hexanol, 6-iodo-, trimethylsilyl ether	26306-01-4	NI31372
166	55	74	216	244	300	138	92	300	100	47	33	21	17	13	13	11		10	4	4	–	–	3	–	–	–	–	–	–	1	Cymantrene, (trifluoroacetyl)-	12128-79-9	NI31373
302	113	300	304	63	114	87	62	300	100	89	43	39	31	27	20	20		10	6	1	–	–	–	–	2	–	–	–	–	–	2-Naphthalenol, 1,6-dibromo-	16239-18-2	NI31374
302	113	304	300	63	114	87	62	300	100	88	54	48	40	38	27	25		10	6	1	–	–	–	–	2	–	–	–	–	–	2-Naphthalenol, 1,6-dibromo-	16239-18-2	NI31375
113	302	300	114	304	63	56	194	300	100	90	43	38	38	36	21	20		10	6	1	–	–	–	–	2	–	–	–	–	–	2-Naphthalenol, 1,6-dibromo-	16239-18-2	NI31376
300	204	108	69	133	81	52	231	300	100	67	58	33	21	12	12	11		10	6	2	2	–	6	–	–	–	–	–	–	–	Bis(2,2,2-trifluoroacetamide), N,N'-4-phenylene-	404-28-4	NI31377
255	165	137	109	143	45	121	43	300	100	52	43	43	36	32	26	25	0.00	10	25	5	–	–	–	–	–	–	–	–	–	1	Arsenic, pentaethoxy-	5954-41-6	NI31378
142	141	115	116	104	221	63	91	300	100	70	57	40	18	11	7	7	0.00	11	10	–	–	–	–	–	2	–	–	–	–	–	Spiro[2.2]pentane, 4,4-dibromo-1-phenyl-		MS06925
88	116	115	77	56	103	60	78	300	100	84	73	49	47	29	29	24	0.80	11	13	1	2	1	–	–	1	–	–	–	–	–	3-(4-Bromophenyl)-β-hydroxyethyl-2-iminothiazolidine		IC03776
117	75	73	84	97	103	44	129	300	100	83	82	43	41	39	39	32	0.00	11	22	6	–	–	–	–	–	–	1	–	2	–	α-D-Galactopyranose, 6-O-(trimethylsilyl)-, cyclic 1,2:3,4-bis(methylboronate)	54400-98-5	NI31379
73	117	103	75	97	129	171	43	300	100	99	80	74	58	57	39	35	0.00	11	22	6	–	–	–	–	–	–	1	–	2	–	Glucopyranose, 6-O-(trimethylsilyl)-, cyclic 1,2:3,5-bis(methylboronate)		NI31380
97	129	75	171	73	103	85	96	300	100	97	72	67	60	34	32	31	0.00	11	22	6	–	–	–	–	–	–	1	–	2	–	α-D-Mannopyranose, 1-O-(trimethylsilyl)-, cyclic 2,3:4,6-bis(methylboronate)	56246-46-9	NI31381
135	119	136	63	61	75	118	87	300	100	28	26	22	16	14	13	13	5.61	11	24	5	–	2	–	–	–	–	–	–	–	–	D-Glucose, 2-O-methyl-, diethyl mercaptal	3767-34-8	NI31382
135	75	136	45	107	103	105	87	300	100	57	48	35	26	23	20	19	8.82	11	24	5	–	2	–	–	–	–	–	–	–	–	D-Glucose, 3-O-methyl-, diethyl mercaptal	3554-76-5	NI31383
300	302	236	266	96	267	97	301	300	100	40	25	22	17	15	15	13		12	7	2	–	1	1	1	–	–	–	1	–	–	10H-Phenothiaphosphine, 2-chloro-8-fluoro-10-hydroxy-, 10-oxide		LI06553
300	302	236	266	96	267	97	95	300	100	40	25	22	17	15	15	13		12	7	2	–	1	1	1	–	–	–	1	–	–	10H-Phenothiaphosphine, 7-chloro-2-fluoro-10-hydroxy-, 10-oxide	54934-84-8	NI31384
285	147	204	213	161	205	300	89	300	100	99	71	66	65	61	55	51		12	12	2	–	1	–	–	–	–	–	–	–	1	Benzo[b]thiophene, 3-(methylseleno)-2-(1,3-dioxolan-2-yl)-	71740-02-8	NI31385
184	128	43	127	141	77	185	169	300	100	47	35	26	21	19	16	13	0.00	12	13	1	–	–	–	–	–	1	–	–	–	–	2-Phenyl-4,6-dimethylpyrilium iodide		LI06554
91	127	176	178	141	105	139	149	300	100	39	24	15	9	9	8	7	0.00	12	16	–	–	–	–	4	–	–	–	–	–	–	Bicyclo[2.2.2]oct-2-ene, 5,6,7,8-tetrakis(chloromethyl)-, (1R,4S,5R,6R,7S,8S)-		DD01081
112	113	36	143	122	85	45	38	300	100	34	34	30	29	25	22	14	0.00	12	17	1	4	1	–	1	–	–	–	–	–	–	Thiamine	59-43-8	NI31386
112	113	85	143	114	264	233	207	300	100	34	34	28	9	1	1	1	0.00	12	17	1	4	1	–	1	–	–	–	–	–	–	Thiamine	59-43-8	LI06555
285	147	73	286	75	287	148	300	300	100	57	39	21	10	9	9	8		12	20	5	–	–	–	–	–	–	2	–	–	–	2,5-Furandicarboxylic acid, bis(trimethylsilyl) ester	55494-09-2	NI31387
147	73	285	45	75	286	148	71	300	100	87	69	20	17	16	16	16	4.47	12	20	5	–	–	–	–	–	–	2	–	–	–	2,5-Furandicarboxylic acid, bis(trimethylsilyl) ester	55494-09-2	NI31388
285	147	93	73	95	286	75	148	300	100	74	74	46	26	23	16	11	6.32	12	20	5	–	–	–	–	–	–	2	–	–	–	2,5-Furandicarboxylic acid, bis(trimethylsilyl) ester	55494-09-2	NI31389
155	99	127	173	217	244	285	300	300	100	97	75	25	12	8	3	2		12	23	4	–	–	2	–	–	–	–	1	–	–	Diethyl 2,2-difluoro-1-hexylethenyl phosphate		DD01082
150	300	104	76	301	18	120	151	300	100	90	30	17	13	13	9	8		13	8	3	4	1	–	–	–	–	–	–	–	–	3-(4-Nitrobenzoylimino)-1,3α-diaza-2-thiaindene		MS06926
241	166	59	167	242	140	151	120.5	300	100	62	33	32	31	16	11	11	2.50	13	9	–	2	1	–	–	–	–	–	–	–	1	Thiocyanic acid, 5,10-dihydro-10-phenarsazinyl ester	4808-20-2	LI06556
241	166	59	167	242	140	139	240	300	100	62	33	22	20	15	13	12	2.00	13	9	–	2	1	–	–	–	–	–	–	–	1	Thiocyanic acid, 5,10-dihydro-10-phenarsazinyl ester	4808-20-2	NI31390
216	138	55	161	217	112	218	128	300	100	37	17	16	13	9	5	5	1.99	13	9	3	–	1	–	–	–	–	–	–	–	1	Cymantrene, (2-thienylmethyl)-	82057-19-0	NI31391
253	47	139	147	255	111	141	113	300	100	99	76	69	40	30	20	4	3.00	13	13	2	–	2	–	1	–	–	–	–	–	–	5-(4-Chlorophenyl)-1,1-bis(methylthio)-3-hydroxy-1,3-pentadien-5-one		MS06927
253	139	147	255	111	141	300	113	300	100	72	63	40	30	12	3	3		13	13	2	–	2	–	1	–	–	–	–	–	–	1,3-Pentadien-5-one, 5-(2-chlorophenyl)-1,1-bis(methylthio)-3-hydroxy-		MS06928
156	216	158	258	43	198	157	218	300	100	85	80	73	71	65	59	46	2.80	13	13	6	–	–	–	1	–	–	–	–	–	–	3-Chloro-2,5-dihydroxybenzyl alcohol triacetate		MS06929
158	113	55	125	159	114	101	100	300	100	45	30	26	22	21	20	18	5.32	13	20	6	2	–	–	–	–	–	–	–	–	–	1,5-Dioxa-10,13-diazacycloheptadecane-6,9,14,17-tetrone	79688-09-8	NI31392
101	44	32	169	26	55	43	142	300	100	40	32	27	27	26	25	23	5.00	13	20	6	2	–	–	–	–	–	–	–	–	–	Uridine, 3-methyl-2',3',5'-tri-O-methyl-	53657-37-7	NI31393
183	73	285	257	75	184	147	45	300	100	58	22	22	21	16	16	13	3.93	13	24	4	–	–	–	–	–	–	2	–	–	–	Fumarylacetone, bis(trimethylsilyl)-		NI31394
183	73	75	257	147	184	45	285	300	100	76	20	18	16	15	14	13	2.50	13	24	4	–	–	–	–	–	–	2	–	–	–	2-Heptenoic acid, 4,6-dioxo-, bis(trimethylsilyl)-, (E)-		NI31395
183	73	184	45	300	257	178	66	300	100	65	16	11	10	8	8	8		13	24	4	–	–	–	–	–	–	2	–	–	–	2-Heptenoic acid, 4,6-dioxo-, bis(trimethylsilyl)-, (Z)-		NI31396
285	73	43	117	169	184	75	41	300	100	75	64	60	58	50	50	44	0.68	13	24	4	2	–	–	–	–	–	1	–	–	–	2,4,6(1H,3H,5H)-Pyrimidinetrione, 5-ethyl-1,3-dimethyl-5-[2-[(trimethylsilyl)oxy]ethyl]-	57346-58-4	NI31397

MASS TO CHARGE RATIOS								M.W.	INTENSITIES								Parent	C	H	O	N	S	F	Cl	Br	I	Si	P	B	X	COMPOUND NAME	CAS Reg No	No
300	242	108	268	301	302	224	167	300	100	44	25	24	18	14	11	10		14	8	–	2	3	–	–	–	–	–	–	–	–	2,2′-Dibenzthiazyl sulphide		IC03777
77	272	51	300	63	273	245	196	300	100	61	60	26	10	4	4	4		14	9	1	2	–	–	–	1	–	–	–	–	–	N-(5-Bromo-2-oxo-3-indolinylidine)aniline		LI06557
272	75	300	155	118	90	63	64	300	100	45	44	22	16	14	12	10		14	9	1	2	–	–	–	1	–	–	–	–	–	N-(2-Oxo-3-indolinylidine)-3-bromoaniline		LI06558
272	300	75	155	90	63	118	64	300	100	62	50	20	16	14	13	12		14	9	1	2	–	–	–	1	–	–	–	–	–	N-(2-Oxo-3-indolinylidine)-4-bromoaniline		LI06559
300	118	77	103	301	51	121	78	300	100	99	99	98	93	93	86	76		14	12	4	4	–	–	–	–	–	–	–	–	–	Acetophenone, (2,4-dinitrophenyl)hydrazone	1677-87-8	IC03778
77	300	103	121	165	118	78	265	300	100	55	47	21	20	18	18	13		14	12	4	4	–	–	–	–	–	–	–	–	–	Acetophenone, (2,4-dinitrophenyl)hydrazone	1677-87-8	NI31398
300	77	103	118	301	121	51	78	300	100	60	51	22	17	14	14	13		14	12	4	4	–	–	–	–	–	–	–	–	–	Acetophenone, (2,4-dinitrophenyl)hydrazone	1677-87-8	IC03779
53	164	78	65	64	89	90	283	300	100	88	78	60	52	49	37	23	17.50	14	12	4	4	–	–	–	–	–	–	–	–	–	Azobenzene, 2,2′-dimethyl-6,6′-dinitro-		NI31399
300	117	91	118	301	121	90	65	300	100	52	38	30	20	20	15	15		14	12	4	4	–	–	–	–	–	–	–	–	–	Benzaldehyde, 2-methyl-, (2,4-dinitrophenyl)hydrazone	1773-44-0	NI31400
300	121	91	93	79	78	65	178	300	100	38	38	22	22	20	20	18		14	12	4	4	–	–	–	–	–	–	–	–	–	Benzaldehyde, 3-methyl-, (2,4-dinitrophenyl)hydrazone	2880-05-9	NI31401
121	300	91	65	78	93	79	119	300	100	70	60	39	35	31	30	26		14	12	4	4	–	–	–	–	–	–	–	–	–	Benzaldehyde, 4-methyl-, (2,4-dinitrophenyl)hydrazone	2571-00-8	LI06560
121	300	91	65	78	93	79	63	300	100	70	60	38	35	32	30	25		14	12	4	4	–	–	–	–	–	–	–	–	–	Benzaldehyde, 4-methyl-, (2,4-dinitrophenyl)hydrazone	2571-00-8	NI31402
194	177	77	79	107	117	105	106	300	100	87	86	84	66	61	38	29	15.00	14	12	4	4	–	–	–	–	–	–	–	–	–	7-[N-(1-Phenylethyl)amino]-4-nitrobenzo-2-oxa-1,3-diazole		LI06561
184	47	92	49	79	185	139	91	300	100	57	48	38	23	15	15	13	0.00	14	13	–	–	1	4	–	–	–	–	–	1	–	5-Ethyldibenzo[b,d]thiophenium tetrafluoroborate		LI06562
300	105	180	301	285	98	152	299	300	100	84	25	24	20	20	16	14		14	16	2	6	–	–	–	–	–	–	–	–	–	Imidazo[1,2-a][1,3,5]triazin-4(8H)-one, 8-benzamido-2-(dimethylamino)-6,7-dihydro-		NI31403
105	300	77	121	88	181	167	138	300	100	40	40	24	22	18	17	16		14	16	2	6	–	–	–	–	–	–	–	–	–	Oxazolo[3,2-a][1,3,5]triazin-2-amine, 4-(2-benzoylhydrazono)-N,N-dimethyl-6,7-dihydro-		NI31404
91	92	65	41	300	59	39	69	300	100	8	8	7	5	5	5	4		14	20	2	–	–	–	–	–	–	–	–	–	1	Pentanoic acid, 3,3-dimethyl-5-(benzylseleno)-	3709-13-5	NI31405
91	92	65	41	59	300	39	69	300	100	17	17	15	11	10	10	9		14	20	2	–	–	–	–	–	–	–	–	–	1	Pentanoic acid, 3,3-dimethyl-5-(benzylseleno)-	3709-13-5	LI06563
43	167	29	139	211	240	194	165	300	100	27	26	19	13	12	9	9	2.00	14	20	7	–	–	–	–	–	–	–	–	–	–	Furo[3,2-b]furan-3,6-dicarboxylic acid, 3a,5,6,6a-tetrahydro-5-hydroxy-2,5-dimethyl-, diethyl ester		MS06930
118	91	79	109	196	172	199	254	300	100	78	27	25	22	15	10	6	0.00	14	21	5	–	–	–	–	–	–	–	1	–	–	Benzeneacetic acid, α-(diethoxyphosphinyl)-, ethyl ester		DD01083
108	123	109	139	124	121	125	122	300	100	62	53	30	30	18	17	16	0.00	14	24	5	2	–	–	–	–	–	–	–	–	–	3,6,9,12,15-Pentaoxa-19,20-diazabicyclo[15.2.1]icosa-1(20),17-diene, 19-methyl-	118713-61-4	DD01084
71	236	45	99	59	157	84	211	300	100	84	78	34	33	30	28	25	0.81	14	24	5	2	–	–	–	–	–	–	–	–	–	Thymidine, 5,6-dihydro-3,5-dimethyl-3′,5′-di-O-methyl-	51844-39-4	NI31406
142	70	143	98	300	114	68	170	300	100	44	31	31	21	19	11	9		14	24	5	2	–	–	–	–	–	–	–	–	–	L-Valine, N-(1-carboxy-L-prolyl)-, ethyl methyl ester	21026-88-0	NI31407
103	101	131	105	44	159	99	129	300	100	86	65	65	64	55	53	48	0.00	14	26	2	–	–	–	–	–	–	–	–	–	1	Germacycloundecane-6,7-dione, 1,1-diethyl-	18100-65-7	NI31408
300	302	105	272	270	165	77	193	300	100	99	45	40	40	38	38	32		15	9	2	–	–	–	–	1	–	–	–	–	–	4-Bromo-3-phenylisocoumarin		LI06564
139	119	300	111	141	77	75	123	300	100	39	37	37	33	28	22	17		15	9	3	2	–	–	1	–	–	–	–	–	–	Isoxazole, 5-(4′-chlorophenyl)-3-phenyl-4-nitro-	66692-98-6	NI31409
210	151	77	300	236	51	135	211	300	100	58	48	22	21	19	18	17		15	12	1	2	2	–	–	–	–	–	–	–	–	1,3,4-Thiadiazolium, 5-mercapto-2-(4-methoxyphenyl)-3-phenyl-, hydroxide, inner salt	34009-55-7	NI31410
210	151	77	300	236	51	135	235	300	100	58	48	22	21	19	18	17		15	12	1	2	2	–	–	–	–	–	–	–	–	1,3,4-Thiadiazolium, 5-mercapto-2-(4-methoxyphenyl)-3-phenyl-, hydroxide, inner salt	34009-55-7	LI06565
135	136	121	300	165	134	164	63	300	100	75	42	20	20	18	13	13		15	12	5	2	–	–	–	–	–	–	–	–	–	Anisole, 4-(2,4-dinitrostyryl)-	22396-03-8	NI31411
105	77	162	139	161	51	65	39	300	100	47	39	21	19	17	16	13	0.00	15	12	5	2	–	–	–	–	–	–	–	–	–	Glycine, N-benzoyl-, 3-nitrophenyl ester	2979-53-5	NI31412
105	77	162	51	161	139	134	65	300	100	43	43	14	14	14	14	10	0.00	15	12	5	2	–	–	–	–	–	–	–	–	–	Hippuric acid p-nitrophenyl ester	3101-11-9	NI31413
210	56	216	214	134	242	212	272	300	100	80	74	73	63	47	41	38	34.40	15	16	3	–	–	–	–	–	–	–	–	–	1	Iron, tricarbonyl[(5,6,7,8-η)-1,2,3,4,4a,4b,8a,8b-octahydrobiphenylene]-	41447-58-9	MS06931
300	145	146	143	102	301	283	157	300	100	99	28	22	19	18	13	13		15	16	3	4	–	–	–	–	–	–	–	–	–	Benzimidazolium, 1,2-dimethyl-, 3-(hexahydro-1,3-dimethyl-2,4,6-trioxo-5 pyrimidinylide)	27355-95-9	NI31414
272	300	187	215	273	214	243	160	300	100	55	36	35	24	22	18	15		15	16	3	4	–	–	–	–	–	–	–	–	–	Benzo[g]pteridine-2,4(3H,10H)-dione, 10-ethyl-8-methoxy-3,7-dimethyl-	56210-78-7	NI31415
43	156	158	143	56	157	129	184	300	100	38	35	23	19	18	17	12	0.00	15	16	3	4	–	–	–	–	–	–	–	–	–	6-Ethoxycarbonyl-5-ethyl-3-imino-1-methyl-4,7,7-tricyano-2-oxabicyclo[2.2.1]heptane		MS06932
100	67	101	200	71	300	41	135	300	100	28	20	15	13	12	11	8		15	24	–	–	3	–	–	–	–	–	–	–	–	6,12,18-Trithiatrispiro[4.1.4.1.4.1]octadecane	177-61-7	NI31416
109	43	55	95	82	67	152	178	300	100	79	56	53	46	44	36	28	0.00	15	24	4	–	1	–	–	–	–	–	–	–	–	3,6-Dioxo-1,1,10,10a-tetramethyl-perhydronaphtho[2,1-d][1,3,2]dioxathin		LI06566
103	109	164	143	141	195	217	283	300	100	55	40	40	40	8	6	4	0.00	15	24	4	–	1	–	–	–	–	–	–	–	–	Methyl (1RS,1′RS,2RS,2′E,R(S)S(S))-2-[3′-(butylsulphinyl)-1′-methylprop-2′-enyl]-5-oxocyclopentane-1-carboxylate		MS06933
103	109	164	143	141	283	268	251	300	100	50	40	40	40	6	3	3	0.00	15	24	4	–	1	–	–	–	–	–	–	–	–	Methyl (1RS,1′RS,2RS,2′E,R(S)S(S))-2-[3′-(butylsulphinyl)-1′-methylprop-2′-enyl]-5-oxocyclopentane-1-carboxylate		MS06934
43	29	55	103	283	86	109	269	300	100	99	69	50	48	39	37	36	8.00	15	24	4	–	1	–	–	–	–	–	–	–	–	Methyl (1RS,1′RS,2RS,2′E,R(S)S(S))-2-[3′-(tert-butylsulphinyl)-1′-methylprop-2′-enyl]-5-oxocyclopentane-1-carboxylate		MS06935
188	57	41	56	43	160	58	42	300	100	96	91	27	27	16	16	13	0.00	15	24	4	–	1	–	–	–	–	–	–	–	–	2-Thiophenecarboxylic acid, 4,5-di-tert-butoxy-, ethyl ester	5556-17-2	NI31417
97	227	41	70	153	181	57	125	300	100	89	52	51	49	38	34	22	1.00	15	24	6	–	–	–	–	–	–	–	–	–	–	2-Butoxy-6,6-bis-(ethoxycarbonyl)-5,6-dihydro-2H-pyran		LI06567
177	191	253	221	251	208	193	175	300	100	87	82	74	71	67	48	48	5.00	15	24	6	–	–	–	–	–	–	–	–	–	–	Methyl (1α,2β,5α)-4-hydroxymethyl-3,7,7-trimethoxy-2-methyl bicyclo[3.3.0.]oct-3-ene-2-carboxylate		MS06936
72	111	171	141	139	132	230	199	300	100	44	33	14	14	14	9	9	0.00	15	28	4	2	–	–	–	–	–	–	–	–	–	L-Alanine, N-(N-acetyl-L-leucyl)-, butyl ester	62167-77-5	LI06568

MASS TO CHARGE RATIOS								M.W.	INTENSITIES								Parent	C	H	O	N	S	F	Cl	Br	I	Si	P	B	X	COMPOUND NAME	CAS Reg No	No
86	128	44	129	28	43	156	41	300	100	88	55	48	37	17	15	14	0.47	15	28	4	2	–	–	–	–	–	–	–	–	–	L-Alanine, N-(N-acetyl-L-leucyl)-, butyl ester	62167-77-5	NI31418
111	171	97	102	103	141	139	129	300	100	75	48	41	34	32	32	32	0.00	15	28	4	2	–	–	–	–	–	–	–	–	–	Hexanoic acid, 2-[[[(dimethylamino)carbonyl]methylamino]carbonyl]-2-ethyl-3-methyl-, methyl ester	42948-65-2	NI31419
173	69	149	55	133	174	175	150	300	100	96	73	39	29	19	10	10	0.00	15	32	2	–	–	–	–	–	–	2	–	–	–	Pentane-1,5-diol, bis(allyldimethylsilyl) ether		NI31420
202	300	302	203	301	303	265	304	300	100	63	43	20	15	10	9	7		16	10	–	–	–	–	2	–	–	1	–	–	–	1,1-Dichloro-1-sila-2,3-benzophenalane	49789-80-2	NI31421
299	301	300	202	302	51	28	200	300	100	77	57	44	37	32	23	21		16	10	–	2	–	–	2	–	–	–	–	–	–	3,4-Dichloro-5,6-diphenylpyridazine	77892-06-9	NI31422
134	121	300	242	78	135	77	65	300	100	34	29	12	10	10	7	7		16	12	6	–	–	–	–	–	–	–	–	–	–	[1,4]Benzodioxino[2,3-b]-syn-[1,4]benzodioxino[2,3-e]-5a,6a,12a,13a-tetrahydro[1,4]dioxin	91201-71-7	NI31423
285	301	257	120	229	300	299	286	300	100	66	30	27	25	23	20	20		16	12	6	–	–	–	–	–	–	–	–	–	–	4H-1-Benzopyran-4-one, 3,7-dihydroxy-2-(3-hydroxy-4-methoxyphenyl)-	57396-72-2	NI31424
137	57	43	149	300	55	60	31	300	100	99	96	83	76	70	56	56		16	12	6	–	–	–	–	–	–	–	–	–	–	4H-1-Benzopyran-4-one, 5,7-dihydroxy-2-(4-hydroxyphenyl)-3-methoxy-	1592-70-7	NI31425
257	69	300	282	285	81	95	119	300	100	99	98	63	62	49	45	32		16	12	6	–	–	–	–	–	–	–	–	–	–	4H-1-Benzopyran-4-one, 5,7-dihydroxy-2-(4-hydroxyphenyl)-3-methoxy-	1592-70-7	MS06937
69	257	300	282	285	81	95	119	300	100	99	98	63	62	49	45	32		16	12	6	–	–	–	–	–	–	–	–	–	–	4H-1-Benzopyran-4-one, 5,7-dihydroxy-2-(4-hydroxyphenyl)-6-methoxy-	1447-88-7	NI31426
300	257	285	282	167	139	299	271	300	100	71	68	61	17	16	14	10		16	12	6	–	–	–	–	–	–	–	–	–	–	4H-1-Benzopyran-4-one, 5,7-dihydroxy-2-(4-hydroxyphenyl)-6-methoxy-	1447-88-7	NI31427
148	120	300	149	153	91	105	89	300	100	22	15	11	8	7	3	3		16	12	6	–	–	–	–	–	–	–	–	–	–	2H-1-Benzopyran-2-one, 4,5,7-trihydroxy-3-(4-methoxyphenyl)-	4376-81-2	NI31428
300	285	299	257	136	135	271	229	300	100	13	9	7	7	7	6	6		16	12	6	–	–	–	–	–	–	–	–	–	–	4H-1-Benzopyran-4-one, 3,5,7-trihydroxy-2-(4-methoxyphenyl)-		MS06938
300	285	271	258	229	201	301	171	300	100	98	98	88	88	88	85	85		16	12	6	–	–	–	–	–	–	–	–	–	–	4H-Naphtho[2,3-b]pyran-4,6,9-trione, 5,8-dimethoxy-2-methyl-	19206-58-7	NI31429
300	237	273	227	302	77	301	275	300	100	50	50	40	34	14	13	12		16	13	–	2	1	–	1	–	–	–	–	–	–	7-Chloro-2,3-dihydro-1-methyl-5-phenyl-1,4-benzodiazepine-2-thione	2898-13-7	NI31430
299	300	301	91	302	252	117	77	300	100	41	37	19	10	8	3	3		16	13	–	2	1	–	1	–	–	–	–	–	–	7-Chloro-2-methylmercapto-5-phenyl-3H-1,4-benzodiazepine		NI31431
300	77	258	51	255	259	256	283	300	100	85	67	58	52	47	38	38		16	13	2	2	–	–	1	–	–	–	–	–	–	1H-1,5-Benzodiazepine-2,4(3H,5H)-dione, 7-chloro-1-methyl-5-phenyl-	22316-47-8	NI31432
271	273	300	272	256	255	257	104	300	100	26	20	19	19	16	15	11		16	13	2	2	–	–	1	–	–	–	–	–	–	2H-1,4-Benzodiazepin-2-one, 7-chloro-1,3-dihydro-3-hydroxy-1-methyl-5-phenyl-	846-50-4	NI31433
271	273	272	256	77	300	255	257	300	100	34	19	19	19	18	13	12		16	13	2	2	–	–	1	–	–	–	–	–	–	2H-1,4-Benzodiazepin-2-one, 7-chloro-1,3-dihydro-3-hydroxy-1-methyl-5-phenyl-	846-50-4	NI31434
300	299	301	77	165	283	152	255	300	100	92	50	38	30	25	21	19		16	13	2	2	–	–	1	–	–	–	–	–	–	Diazepam N-oxide	2888-64-4	LI06570
300	299	301	77	302	165	283	152	300	100	94	51	37	33	28	25	20		16	13	2	2	–	–	1	–	–	–	–	–	–	Diazepam N-oxide	2888-64-4	LI06569
77	165	51	151	105	152	76	110	300	100	74	58	54	49	46	45	41	36.17	16	13	2	2	–	–	1	–	–	–	–	–	–	Diazepam N-oxide	2888-64-4	NI31435
91	105	78	28	18	300	65	209	300	100	93	47	26	19	17	17	15		16	13	2	2	–	–	1	–	–	–	–	–	–	1H-Imidazo[1,2-a]pyridin-4-ium, 2-(chloroacetyl)-3-hydroxy-1-benzyl-, hydroxide, inner salt	31757-85-4	NI31436
91	105	78	65	300	209	39	181	300	100	85	43	16	15	13	10	9		16	13	2	2	–	–	1	–	–	–	–	–	–	1H-Imidazo[1,2-a]pyridin-4-ium, 2-(chloroacetyl)-3-hydroxy-1-benzyl-, hydroxide, inner salt	31757-85-4	LI06571
91	251	300	302	65	252	301	118	300	100	68	49	16	16	10	9	9		16	13	2	2	–	–	1	–	–	–	–	–	–	1H-Imidazo[1,2-a]pyridin-4-ium, 3-(chloroacetyl)-2-hydroxy-1-benzyl-, hydroxide, inner salt	31757-80-9	NI31437
91	251	300	65	92	78	51	302	300	100	22	16	16	7	6	6	5		16	13	2	2	–	–	1	–	–	–	–	–	–	2H-Imidazo[1,2-a]pyridin-4-ium, 3-(chloroacetyl)-2-hydroxy-1-benzyl-, hydroxide, inner salt		LI06572
144	145	91	115	116	65	92	77	300	100	65	65	50	33	33	28	28	2.50	16	16	2	2	1	–	–	–	–	–	–	–	–	Benzenesulphonic acid, 4-methyl-, (3-phenyl-2-propenylidene)hydrazide	7318-33-4	NI31438
145	146	300	91	103	77	159	144	300	100	12	8	8	6	6	5	5		16	16	2	2	1	–	–	–	–	–	–	–	–	Indole-3-sulphonamide, 2-methyl-N-(p-tolyl)-		IC03780
133	167	135	107	104	78	39	122	300	100	83	51	49	48	44	43	40	0.00	16	16	4	2	–	–	–	–	–	–	–	–	–	Carbamic acid, (3-methylphenyl)-, 3-[(methoxycarbonyl)amino]phenyl ester	13684-63-4	NI31439
168	301	241	167	169	196	134	243	300	100	80	74	14	10	8	8	5	0.00	16	16	4	2	–	–	–	–	–	–	–	–	–	Carbamic acid, (3-methylphenyl)-, 3-[(methoxycarbonyl)amino]phenyl ester	13684-63-4	NI31440
167	133	135	104	132	122	105	78	300	100	96	64	53	36	35	24	24	2.04	16	16	4	2	–	–	–	–	–	–	–	–	–	Carbamic acid, (3-methylphenyl)-, 3-[(methoxycarbonyl)amino]phenyl ester	13684-63-4	NI31441
119	91	109	181	122	64	29	18	300	100	66	60	47	38	38	35	28	0.20	16	16	4	2	–	–	–	–	–	–	–	–	–	Carbamic acid, [3-[[(phenylamino)carbonyl]oxy]phenyl]-, ethyl ester	13684-56-5	NI31442
300	301	252	269	182	181	284	267	300	100	19	19	16	16	11	10	10		16	16	4	2	–	–	–	–	–	–	–	–	–	Dimethyl 2,2′-diaminobiphenyl-4,4′-dicarboxylate		LI06573
300	237	219	301	238	236	269	209	300	100	90	45	21	21	19	17	17		16	16	4	2	–	–	–	–	–	–	–	–	–	Dimethyl 2,2′-diaminobiphenyl-6,6′-dicarboxylate		LI06574
270	271	144	154	155	167	197	127	300	100	20	20	19	11	10	8	8	0.00	16	16	4	2	–	–	–	–	–	–	–	–	–	Ergoline-8-carboxylic acid, 9,10-didehydro-6-methyl-, dihydroxy deriv., (8β)-	72088-21-2	NI31443
105	77	195	45	139	51	46	121	300	100	71	61	36	30	26	24	18	0.10	16	16	4	2	–	–	–	–	–	–	–	–	–	β-(2-Nitroanilino)-β-ethoxy-acetophenone		MS06939
229	227	154	209	152	155	228	117	300	100	56	54	36	32	30	24	24	0.00	16	17	1	–	–	–	–	–	–	–	–	–	1	(3-Butenyl)diphenylarsane oxide		DD01085
285	300	226	286	211	269	301	287	300	100	32	28	27	21	14	9	8		16	20	2	–	–	–	–	–	–	2	–	–	–	2,2,6,6-Tetramethyl-1,5-dioxa-2,6-disila-3,4:7,8-dibenzocyclooctane		NI31444
269	300	92	270	149	105	65	45	300	100	48	38	19	13	11	10	10		16	20	2	4	–	–	–	–	–	–	–	–	–	Azobenzene, 4,4′-diamino-, N,N-bis(2-hydroxyethyl)-		IC03781
83	85	47	61	36	119	49	48	300	100	99	99	86	74	66	47	46	0.00	16	28	1	–	2	–	–	–	–	–	–	–	–	Cyclohexanone, 2-[2-(1,3-dithiolan-2-yl)propyl]-6-methyl-3-isopropyl-	56772-19-1	NI31445
143	111	116	83	55	41	43	101	300	100	65	42	33	33	21	17	14	0.00	16	28	5	–	–	–	–	–	–	–	–	–	–	Dimethyl 3,6-epoxytetradecanedioate		NI31446
133	69	119	73	157	107	59	149	300	100	77	28	20	19	12	10	9	0.00	16	32	3	–	–	–	–	–	–	1	–	–	–	(E)-1-[Dimethyl[2-(2-ethoxyethoxy)ethoxy]silyl]-6-methyl-1,5-heptadiene		DD01086
133	69	73	157	107	149	241	225	300	100	22	16	15	11	8	7	7	0.00	16	32	3	–	–	–	–	–	–	1	–	–	–	3-[Dimethyl[2-(2-ethoxyethoxy)ethoxy]silyl]-6-methyl-1,5-heptadiene		DD01087
43	135	103	77	45	107	110	51	300	100	47	47	39	38	34	21	21	1.50	17	16	3	–	1	–	–	–	–	–	–	–	–	1-Acetoxy-1-phenyl-3-phenylthio-2-propanone		MS06940

MASS TO CHARGE RATIOS								M.W.	INTENSITIES								Parent	C	H	O	N	S	F	Cl	Br	I	Si	P	B	X	COMPOUND NAME	CAS Reg No	No
159	141	131	145	158	129	116	117	**300**	100	98	98	95	90	90	90	88	**25.00**	17	16	3	–	1	–	–	–	–	–	–	–	–	3-Methyl-4-phenyl-1-phenylsulphonyl-3-buten-2-one		LI06575
115	91	65	116	145	144	77	39	**300**	100	95	59	48	45	38	37	34	**0.00**	17	16	3	–	1	–	–	–	–	–	–	–	–	3-[(4-Methylphenyl)sulphonyl]-2-phenyl-2,5-dihydrofuran	118356-39-1	DD01088
43	135	103	110	77	45	91	240	**300**	100	22	21	15	15	14	13	11	**1.50**	17	16	3	–	1	–	–	–	–	–	–	–	–	1-Phenylthio-1-acetoxy-3-phenyl-2-propanone		MS06941
199	43	200	77	105	121	51	45	**300**	100	80	15	13	12	9	9	8	**4.00**	17	16	3	–	1	–	–	–	–	–	–	–	–	1-Phenylthio-1-phenyl-3-acetoxy-2-propanone		MS06942
300	149	285	272	151	282	241	255	**300**	100	74	57	41	40	29	28	25		17	16	5	–	–	–	–	–	–	–	–	–	–	6H-Benzofuro[3,2-c][1]benzopyran-6a(11aH)-ol, 3,9-dimethoxy-, (6aR-cis)-	3187-52-8	NI31447
198	215	183	197	216	199	182	201	**300**	100	35	32	30	27	20	20	16	**0.00**	17	16	5	–	–	–	–	–	–	–	–	–	–	2trans,3cis-Diacetoxy-2-methoxy-2,3-dihydrobiphenylene	93103-67-4	NI31448
232	217	69	231	189	300	160	203	**300**	100	44	22	10	5	3	3	2		17	16	5	–	–	–	–	–	–	–	–	–	–	5-[(γ,γ-Dimethylalloxy)methoxy]psoralen		LI06576
134	121	300	299	166	193	119	167	**300**	100	72	71	33	20	17	16	13		17	16	5	–	–	–	–	–	–	–	–	–	–	5-Hydroxy-4′,7-dimethoxyflavanone	69097-96-7	MS06943
134	121	300	91	166	95	299	119	**300**	100	48	40	25	24	23	21	21		17	16	5	–	–	–	–	–	–	–	–	–	–	5-Hydroxy-4′,7-dimethoxyflavanone	69097-96-7	NI31449
300	257	186	128	93	285	214	115	**300**	100	40	14	13	13	12	11	11		17	16	5	–	–	–	–	–	–	–	–	–	–	Di-O-methyl-autumnariniol		LI06577
67	68	53	217	232	39	27	41	**300**	100	71	63	51	48	42	41	39	**0.00**	17	16	5	–	–	–	–	–	–	–	–	–	–	Phellopterin	2543-94-4	LI06578
232	217	189	231	160	69	188	203	**300**	100	98	24	15	15	15	10	8	**3.00**	17	16	5	–	–	–	–	–	–	–	–	–	–	Phellopterin	2543-94-4	LI06579
300	285	299	257	271	65	229	241	**300**	100	40	31	30	20	16	13	12		17	16	5	–	–	–	–	–	–	–	–	–	–	2,3,6-Trimethoxy-10,11-dihydrodibenz[b,f]oxepin-10-one		LI06580
282	300	283	254	270	239	271	299	**300**	100	86	34	29	25	25	24	20		17	16	5	–	–	–	–	–	–	–	–	–	–	Xanthen-9-one, 1,3,6-trimethoxy-8-methyl-	15222-54-5	NI31450
282	300	283	254	271	270	299	239	**300**	100	90	30	22	20	18	17	17		17	16	5	–	–	–	–	–	–	–	–	–	–	Xanthen-9-one, 1,3,6-trimethoxy-8-methyl-	15222-54-5	NI31451
58	212	84	213	199	180	284	238	**300**	100	79	16	15	14	14	11	11	**10.44**	17	20	1	2	1	–	–	–	–	–	–	–	–	10H-Phenothiazine-10-propanamine, N,N-dimethyl-, 5-oxide	146-21-4	NI31452
58	300	18	86	254	215	214	17	**300**	100	55	54	36	16	15	14	11		17	20	1	2	1	–	–	–	–	–	–	–	–	10H-Phenothiazin-3-ol, 10-[3-(dimethylamino)propyl]-	316-85-8	NI31453
172	300	171	41	173	228	57	146	**300**	100	41	21	17	13	13	11	9		17	20	3	2	–	–	–	–	–	–	–	–	–	Pyrimidine-2-carboxylic acid, 5-butoxyphenyl-, ethyl ester	106307-73-7	NI31454
227	153	171	195	83	69	96	57	**300**	100	42	38	35	20	19	18	16	**0.00**	17	32	4	–	–	–	–	–	–	–	–	–	–	Butyl 4-(methoxycarbonylmethyl)decanoate		NI31455
173	29	55	230	41	69	43	127	**300**	100	38	36	34	27	21	21	13	**0.00**	17	32	4	–	–	–	–	–	–	–	–	–	–	Diethyl 2,2-dipentylmalonate		IC03782
98	55	88	112	41	74	69	43	**300**	100	95	91	85	71	68	68	49	**6.00**	17	32	4	–	–	–	–	–	–	–	–	–	–	Dimethyl 2-methyltetradecane-1,14-dioate		LI06581
98	74	55	44	42	84	69	87	**300**	100	72	70	53	53	52	45	34	**1.00**	17	32	4	–	–	–	–	–	–	–	–	–	–	Dimethyl pentadeca-1,15-dioate		LI06582
227	171	152	55	41	56	129	57	**300**	100	94	56	52	49	40	37	37	**0.00**	17	32	4	–	–	–	–	–	–	–	–	–	–	Nonanedioic acid, dibutyl ester	2917-73-9	NI31456
171	227	152	55	56	41	129	125	**300**	100	77	68	52	41	37	34	32	**0.00**	17	32	4	–	–	–	–	–	–	–	–	–	–	Nonanedioic acid, dibutyl ester	2917-73-9	NI31457
41	56	55	57	29	27	227	39	**300**	100	59	56	47	45	44	37	32	**0.09**	17	32	4	–	–	–	–	–	–	–	–	–	–	Nonanedioic acid, dibutyl ester	2917-73-9	AI02013
171	57	227	152	157	56	41	129	**300**	100	81	50	36	35	34	32	31	**0.00**	17	32	4	–	–	–	–	–	–	–	–	–	–	Nonanedioic acid, di-isobutyl ester		DO01085
171	41	57	55	43	56	73	83	**300**	100	62	58	47	32	29	25	17	**0.00**	17	32	4	–	–	–	–	–	–	–	–	–	–	Nonanedioic acid, di-sec-butyl ester	57983-36-5	NI31458
97	114	111	132	156	264	170	282	**300**	100	93	78	68	65	57	56	53	**9.00**	17	32	4	–	–	–	–	–	–	–	–	–	–	Roccellic acid	29838-46-8	LI06583
98	55	29	255	41	88	69	43	**300**	100	85	70	60	55	54	50	50	**0.70**	17	32	4	–	–	–	–	–	–	–	–	–	–	Tridecanedioic acid, diethyl ester	15423-05-9	NI31459
43	82	55	96	68	69	67	83	**300**	100	51	41	34	34	29	25	24	**0.00**	17	32	4	–	–	–	–	–	–	–	–	–	–	1,13-Tridecanediol, diacetate	42236-70-4	NI31460
285	117	132	73	145	75	286	129	**300**	100	92	70	64	41	25	22	13	**8.70**	17	36	2	–	–	–	–	–	–	1	–	–	–	Tetradecanoic acid, trimethylsilyl ester	18603-17-3	NI31461
73	285	75	117	132	129	43	41	**300**	100	94	82	64	53	44	44	31	**12.50**	17	36	2	–	–	–	–	–	–	1	–	–	–	Tetradecanoic acid, trimethylsilyl ester	18603-17-3	NI31462
73	75	117	285	132	43	129	41	**300**	100	77	68	46	39	30	28	27	**3.81**	17	36	2	–	–	–	–	–	–	1	–	–	–	Tetradecanoic acid, trimethylsilyl ester	18603-17-3	NI31463
163	77	29	51	50	41	39	27	**300**	100	76	56	55	47	44	40	26	**0.00**	18	15	–	–	–	2	–	–	–	–	1	–	–	Triphenyldifluorophosphorane		LI06584
57	159	102	103	77	119	117	158	**300**	100	53	38	27	23	21	20	14	**0.00**	18	20	2	–	1	–	–	–	–	–	–	–	–	3,3-Dimethyl-2-phenyl-1-butenyl phenyl sulphone		DD01089
121	300	122	268	269	161	179		**300**	100	68	50	19	18	14	7			18	20	4	–	–	–	–	–	–	–	–	–	–	Benzoic acid, 2-methoxy-6-[2-(4-methoxyphenyl)ethyl]-, methyl ester	63898-02-2	NI31464
164	149	300	151	121	152	165	301	**300**	100	60	52	48	44	24	14	13		18	20	4	–	–	–	–	–	–	–	–	–	–	2H-1-Benzopyran, 3-(2,4-dimethoxyphenyl)-3,4-dihydro-6-methoxy-	56701-27-0	MS06944
164	149	151	121	300	152	165	137	**300**	100	68	45	45	41	27	14	14		18	20	4	–	–	–	–	–	–	–	–	–	–	2H-1-Benzopyran, 3-(2,4-dimethoxyphenyl)-3,4-dihydro-7-methoxy-	58822-02-9	NI31465
43	227	300	76	228	221	169	168	**300**	100	63	13	11	10	10	7	3		18	20	4	–	–	–	–	–	–	–	–	–	–	Ethyl 2-(4′-methoxy-1,1′-biphenyl-4-yl)-2-hydroxypropanoate		NI31466
300	190	241	177	243	229	242	110	**300**	100	69	66	37	35	24	19	17		18	20	4	–	–	–	–	–	–	–	–	–	–	(±)-17β-Hydroxy-3-methoxy-6-oxaestra-1,3,5(10),8(9)-tetraen-7-one		MS06945
45	223	300	224	165	149	181	131	**300**	100	65	35	25	12	12	10	10		18	20	4	–	–	–	–	–	–	–	–	–	–	(E)-2,2′-Bis(methoxymethoxy)stilbene		MS06946
45	300	255	197	165	194	223	225	**300**	100	51	22	12	9	8	7	6		18	20	4	–	–	–	–	–	–	–	–	–	–	(E)-2,4′-Bis(methoxymethoxy)stilbene		MS06947
300	224	45	223	149	165	131	181	**300**	100	76	70	53	46	8	8	7		18	20	4	–	–	–	–	–	–	–	–	–	–	(Z)-2,2′-Bis(methoxymethoxy)stilbene		MS06948
300	45	255	197	223	165	194	225	**300**	100	99	37	22	15	15	12	11		18	20	4	–	–	–	–	–	–	–	–	–	–	(Z)-2,4′-Bis(methoxymethoxy)stilbene		MS06949
300	150	137	177	148	41	215	124	**300**	100	40	35	20	20	18	16	16		18	20	4	–	–	–	–	–	–	–	–	–	–	(±)-3-Methoxy-6-oxaestra-1,3,5(10)-trien-7,17-dione		MS06950
91	300	92	65	149	268	176	77	**300**	100	13	10	9	7	6	5	5		18	20	4	–	–	–	–	–	–	–	–	–	–	Methyl 4-benzyloxy-2-hydroxy-3,5,6-trimethylbenzoate		MS06951
43	107	91	166	79	77	105	92	**300**	100	45	35	32	25	25	22	12	**0.00**	18	20	4	–	–	–	–	–	–	–	–	–	–	Methyl 2-phenyl-2,3-dihydroxy-3-benzylbutanoate		DD01090
254	282	255	283	238	237	300	236	**300**	100	84	19	17	12	10	9	9		18	20	4	–	–	–	–	–	–	–	–	–	–	1′,2′,3′,7′,8′,9′,10′,10′a-Octahydrocyclohepta[de]naphthalen-3′-ylidenesuccinic acid		LI06585
91	115	131	118	121	300	103	152	**300**	100	81	54	46	38	35	34	33		18	20	4	–	–	–	–	–	–	–	–	–	–	Phenol, 2,3-dimethoxy-4-[3-(2-methoxyphenyl)-2-propenyl]-, (Z)-	20362-18-9	NI31467
264	193	72	71	220	165	192	265	**300**	100	99	71	57	35	29	25	24	**0.00**	18	21	–	2	–	–	1	–	–	–	–	–	–	Dibenzo[c,f]pyrazino[1,2-a]azepine, 1,2,3,4,10,14b-hexahydro-2-methyl-, monohydrochloride	21535-47-7	NI31468
201	265	301	299	203				**300**	100	56	51	36	35				**0.00**	18	21	–	2	–	–	1	–	–	–	–	–	–	Piperazine, 1-[(4-chlorophenyl)benzyl]-4-methyl-	82-93-9	NI31469
99	56	72	165	300	228	229	242	**300**	100	52	39	31	28	24	23	22		18	21	–	2	–	–	1	–	–	–	–	–	–	Piperazine, 1-[(4-chlorophenyl)benzyl]-4-methyl-	82-93-9	NI31470
300	112	301	140	299	178	160	122	**300**	100	60	21	16	12	11	7	7		18	24	2	2	–	–	–	–	–	–	–	–	–	20,21-Dinoraspidospermidin-5-ol, 17-methoxy-	55162-77-1	NI31471

MASS TO CHARGE RATIOS								M.W.	INTENSITIES								Parent	C	H	O	N	S	F	Cl	Br	I	Si	P	B	X	COMPOUND NAME	CAS Reg No	No
227	77	29	300	41	55	65	91	300	100	40	35	33	21	12	10	9		18	24	2	2	–	–	–	–	–	–	–	–	–	4-Ethoxycarbonyl-3-phenyl-2,3-diaza[3.3.3]propellane		NI31472
143	130	115	171	99	300	42	70	300	100	54	48	45	35	26	22	16		18	24	2	2	–	–	–	–	–	–	–	–	–	trans-5-Ethyl-2-(2-indolyl)-1-methyl-4-piperidone ethylene acetal		DD01091
107	44	79	29	77	41	195	55	300	100	57	54	52	49	42	36	20	2.00	18	24	2	2	–	–	–	–	–	–	–	–	–	1-Phenylazo-5-(ethoxycarbonylmethyl)bicyclo[3.3.0]octane		NI31473
84	41	160	105	216	215	188	77	300	100	45	37	35	34	34	22	21	14.00	18	24	2	2	–	–	–	–	–	–	–	–	–	Piperidine, 1,1′-(1,2-phenylenedicarbonyl)bis-	38256-33-6	NI31474
299	216	104	300	105	41	84	42	300	100	80	73	62	47	44	37	35		18	24	2	2	–	–	–	–	–	–	–	–	–	Piperidine, 1,1′-(1,4-phenylenedicarbonyl)bis-	15088-30-9	NI31475
43	104	55	41	98	57	117	28	300	100	99	69	65	60	56	46	39	11.04	18	36	3	–	–	–	–	–	–	–	–	–	–	Hexadecanoic acid, 2-hydroxyethyl ester	4219-49-2	NI31476
241	97	83	111	69	57	242	71	300	100	40	36	23	21	20	19	17	1.94	18	36	3	–	–	–	–	–	–	–	–	–	–	Hexadecanoic acid, 2-methoxy-, methyl ester	16725-36-3	NI31477
89	282	71	83	264	95	222	149	300	100	85	80	57	55	54	42	38	0.00	18	36	3	–	–	–	–	–	–	–	–	–	–	3-Hydroxy-16-methylheptadecanoic acid		NI31478
57	43	88	41	55	69	58	83	300	100	43	33	30	28	20	18	15	0.00	18	36	3	–	–	–	–	–	–	–	–	–	–	2-Hydroxypentadecyl propanoate		MS06952
55	197	283	69	43	41	57	83	300	100	86	84	70	70	52	46	44	0.00	18	36	3	–	–	–	–	–	–	–	–	–	–	Stearic acid, 12-hydroxy-		IC03783
300	285	31	257	301	152	135	150	300	100	69	58	28	22	18	13	12		19	12	2	2	–	–	–	–	–	–	–	–	–	5-Methoxy-7-oxo-7H-benzimidazo[2,1-a]benz[de]isoquinoline		IC03784
300	244	299	165	245	76	142	51	300	100	41	38	37	27	21	19	12		19	16	–	4	–	–	–	–	–	–	–	–	–	s-Triazolo[4,3-a]pyrazine, 3-ethyl-5,6-diphenyl-	33590-25-9	LI06586
300	244	299	165	245	76	301	142	300	100	41	38	37	27	21	20	19		19	16	–	4	–	–	–	–	–	–	–	–	–	s-Triazolo[4,3-a]pyrazine, 3-ethyl-5,6-diphenyl-	33590-25-9	NI31479
271	300	272	253	269	301	270	254	300	100	78	50	40	20	18	15	10		19	24	3	–	–	–	–	–	–	–	–	–	–	Androst-4-en-19-al, 3,17-dioxo-	968-49-0	NI31480
122	300	41	121	123	91	55	79	300	100	44	19	17	15	14	13	11		19	24	3	–	–	–	–	–	–	–	–	–	–	Androst-4-en-3,11,17-trione	382-45-6	NI31481
91	103	101	85	193	92	209	194	300	100	45	33	33	20	20	19	15	0.70	19	24	3	–	–	–	–	–	–	–	–	–	–	1-Butanol, 4-(benzyloxy)-3-[(benzyloxy)methyl]-		LI06587
91	107	103	79	108	43	65	85	300	100	46	45	41	36	36	35	33	0.00	19	24	3	–	–	–	–	–	–	–	–	–	–	1-Butanol, 4-(benzyloxy)-3-[(benzyloxy)methyl]-	33498-88-3	NI31482
107	133	120	282	300	175	161	160	300	100	80	68	25	13	7	2	2		19	24	3	–	–	–	–	–	–	–	–	–	–	(–)-Centrolobol		LI06588
227	300	228	186	301	160	173	147	300	100	99	51	40	21	21	19	15		19	24	3	–	–	–	–	–	–	–	–	–	–	Estra-1,3,5(10)-trien-16-one, 17-hydroxy-3-methoxy-, (17β)-	24721-15-1	NI31483
300	301	215	176	137	298	150	97	300	100	21	10	10	9	7	7	7		19	24	3	–	–	–	–	–	–	–	–	–	–	Estra-1,3,5(10)-trien-17-one, 2-hydroxy-3-methoxy-	5976-63-6	NI31484
300	301	215	176	97	202	150	137	300	100	22	14	13	7	6	6	6		19	24	3	–	–	–	–	–	–	–	–	–	–	Estra-1,3,5(10)-trien-17-one, 3-hydroxy-2-methoxy-	362-08-3	NI31485
300	301	176	137	215	150	189	97	300	100	21	18	18	9	9	8	8		19	24	3	–	–	–	–	–	–	–	–	–	–	Estra-1,3,5(10)-trien-17-one, 3-hydroxy-2-methoxy-	362-08-3	NI31486
268	269	170	157	211	158	148	183	300	100	21	20	15	14	11	11	10	4.01	19	24	3	–	–	–	–	–	–	–	–	–	–	Estra-1,3,5(10)-trien-17-one, 3-hydroxy-6-methoxy-, (6α)-	74299-17-5	NI31487
300	268	269	170	157	301	270	150	300	100	79	28	26	26	21	20	20		19	24	3	–	–	–	–	–	–	–	–	–	–	Estra-1,3,5(10)-trien-17-one, 3-hydroxy-6-methoxy-, (6β)-	74299-14-2	NI31488
300	186	187	161	227	301	160	174	300	100	43	31	23	22	21	20	17		19	24	3	–	–	–	–	–	–	–	–	–	–	Estra-1,3,5(10)-trien-17-one, 3-hydroxy-14-methoxy-, (14β)-		LI06589
300	214	146	301	172	268	159	158	300	100	27	22	21	17	14	14	14		19	24	3	–	–	–	–	–	–	–	–	–	–	Estra-1,3,5(10)-trien-17-one, 3-hydroxy-15-methoxy-, (15β)-	74299-15-3	NI31489
300	301	176	215	202	97	243	190	300	100	22	11	10	7	7	6	5		19	24	3	–	–	–	–	–	–	–	–	–	–	Estra-1,3,5(10)-trien-17-one, 4-hydroxy-3-methoxy-	5976-62-5	NI31490
160	300	41	199	159	55	161	97	300	100	69	56	35	33	28	25	25		19	24	3	–	–	–	–	–	–	–	–	–	–	Estra-1,3,5(10)-trien-17-one, 11-hydroxy-3-methoxy-, (11α)-	5210-12-8	NI31491
300	160	199	301	186	159	200	184	300	100	65	34	22	22	21	19	17		19	24	3	–	–	–	–	–	–	–	–	–	–	Estra-1,3,5(10)-trien-17-one, 11-hydroxy-3-methoxy-, (11α)-	5210-12-8	NI31492
282	267	300	224	225	160	43	159	300	100	89	61	51	49	37	34	33		19	24	3	–	–	–	–	–	–	–	–	–	–	Estra-1,3,5(10)-trien-17-one, 12-hydroxy-3-methoxy-, (12α)-	29461-94-7	NI31493
282	300	283	225	148	239	254	115	300	100	46	18	17	17	15	12	9		19	24	3	–	–	–	–	–	–	–	–	–	–	Estra-1,3,5(10)-trien-17-one, 14-hydroxy-3-methoxy-, (8α,9β,14β)-	15909-05-4	LI06590
282	300	283	225	148	239	254	147	300	100	46	18	17	17	15	11	9		19	24	3	–	–	–	–	–	–	–	–	–	–	Estra-1,3,5(10)-trien-17-one, 14-hydroxy-3-methoxy-, (8α,9β,14β)-	15909-05-4	NI31494
160	300	282	159	161	184	115	225	300	100	82	62	53	42	39	36	31		19	24	3	–	–	–	–	–	–	–	–	–	–	Estra-1,3,5(10)-trien-17-one, 14-hydroxy-3-methoxy-, (8α,14β)-	15909-06-5	NI31495
199	300	200	171	158	184	141	185	300	100	22	19	15	10	9	6	5		19	24	3	–	–	–	–	–	–	–	–	–	–	2-Naphthalenepropanoic acid, β-ethyl-3-methoxy-α,α-dimethyl-, methyl ester	57289-67-5	NI31496
199	171	300	200	184	158	141	185	300	100	20	19	16	9	9	6	5		19	24	3	–	–	–	–	–	–	–	–	–	–	2-Naphthalenepropanoic acid, β-ethyl-5-methoxy-α,α-dimethyl-, methyl ester	57274-56-3	NI31497
199	200	171	300	184	141	158	128	300	100	15	12	11	6	6	5	4		19	24	3	–	–	–	–	–	–	–	–	–	–	2-Naphthalenepropanoic acid, β-ethyl-6-methoxy-α,α-dimethyl-, methyl ester	22290-97-7	NI31498
199	171	300	200	158	184	141	185	300	100	20	18	15	10	7	7	6		19	24	3	–	–	–	–	–	–	–	–	–	–	2-Naphthalenepropanoic acid, β-ethyl-7-methoxy-α,α-dimethyl-, methyl ester	57289-68-6	NI31499
199	171	200	300	184	158	141	185	300	100	25	15	12	12	8	8	6		19	24	3	–	–	–	–	–	–	–	–	–	–	2-Naphthalenepropanoic acid, β-ethyl-8-methoxy-α,α-dimethyl-, methyl ester	57289-69-7	NI31500
300	138	271	163	55	243	110	82	300	100	63	52	42	38	31	28	22		19	28	1	2	–	–	–	–	–	–	–	–	–	5′,6′,7′,8′,10,11-Hexahydrocinchonine		LI06591
168	110	82	132	130	300	139	55	300	100	80	73	58	38	34	25	22		19	28	1	2	–	–	–	–	–	–	–	–	–	α-1′,2′,3′,4′,10,11-Hexahydrocinchonine		LI06592
168	110	132	82	300	130	139	55	300	100	55	52	44	40	38	18	16		19	28	1	2	–	–	–	–	–	–	–	–	–	β-1′,2′,3′,4′,10,11-Hexahydrocinchonine		LI06593
78	170	171	79	52	77	70	168	300	100	73	26	25	24	18	18	17	0.58	19	28	1	2	–	–	–	–	–	–	–	–	–	N-[β-(3′-Indolyl)ethyl]-4-(1-hydroxymethyl-propyl)piperidine		MS06953
143	140	156	171	145	142	131	124	300	100	87	83	65	65	65	60	49	2.00	19	28	1	2	–	–	–	–	–	–	–	–	–	2-Pyrrolidinepropanol, 3-ethyl-1-[2-(1H-indol-3-yl)ethyl]-	55670-06-9	NI31501
43	41	55	69	57	173	83	159	300	100	79	69	48	45	35	35	33	5.20	19	40	–	–	1	–	–	–	–	–	–	–	–	Decyl nonyl sulphide		MS06954
43	55	41	57	29	69	131	56	300	100	97	88	72	61	47	44	41	6.77	19	40	–	–	1	–	–	–	–	–	–	–	–	Dodecyl heptyl sulphide		MS06955
43	41	55	57	61	69	83	97	300	100	83	79	66	63	53	47	31	9.58	19	40	–	–	1	–	–	–	–	–	–	–	–	Methyl octadecyl sulphide		MS06956
57	71	28	43	45	85	83	55	300	100	64	64	63	52	50	48	40	0.00	19	40	2	–	–	–	–	–	–	–	–	–	–	1-Hexadecyloxy-2-propanol		DO01086
56	43	54	83	96	68	41	70	300	100	98	87	80	79	77	77	58	1.28	19	40	2	–	–	–	–	–	–	–	–	–	–	1,2-Nonadecanediol	39516-65-9	NI31502
97	83	69	55	71	43	41	255	300	100	97	94	90	73	70	60	56	0.61	19	40	2	–	–	–	–	–	–	–	–	–	–	2,3-Nonadecanediol	54934-55-3	NI31503
255	256	224	223	225	300	226	196	300	100	64	42	28	20	17	16	16		20	12	3	–	–	–	–	–	–	–	–	–	–	Fluoran	596-24-7	LI06594
255	256	224	223	225	300	257	196	300	100	65	42	29	20	19	17	17		20	12	3	–	–	–	–	–	–	–	–	–	–	Fluoran	596-24-7	NI31504

MASS TO CHARGE RATIOS	M.W.	INTENSITIES	Parent	C	H	O	N	S	F	Cl	Br	I	Si	P	B	X	COMPOUND NAME	CAS Reg No	No
154 223 300 155 301 224 105 153	**300**	100 54 49 14 13 11 9 5		20	16	1	–	–	–	–	–	–	1	–	–	–	2,2-Diphenyl-1-oxa-2-sila-1,2-dihydronaphthalene	63503-13-9	NI31505
223 300 224 301 130 239 165 77	**300**	100 67 17 16 11 8 7 7		20	16	1	2	–	–	–	–	–	–	–	–	–	2-Amino-3,3-diphenylphthalimidine		MS06957
91 300 194 104 105 223 180 120	**300**	100 75 48 19 13 9 7 5		20	16	1	2	–	–	–	–	–	–	–	–	–	1,3,4-Oxadiazole, 4,5-dihydro-2,4,5-triphenyl-	20561-17-5	NI31506
91 300 194 301 104 105 195 223	**300**	100 75 48 19 19 13 10 9		20	16	1	2	–	–	–	–	–	–	–	–	–	1,3,4-Oxadiazole, 4,5-dihydro-2,4,5-triphenyl-	20561-17-5	NI31507
300 160 43 41 115 29 301 55	**300**	100 92 38 36 31 28 23 20		20	28	–	–	1	–	–	–	–	–	–	–	–	Decyl 1-naphthyl sulphide	5092-82-0	AI02014
160 115 43 300 41 54 161 128	**300**	100 26 22 21 19 14 13 12		20	28	–	–	1	–	–	–	–	–	–	–	–	Decyl 1-naphthyl sulphide	5092-82-0	NI31508
160 300 129 39 116 302 162 69	**300**	100 83 78 78 64 57 55 45		20	28	–	–	1	–	–	–	–	–	–	–	–	Decyl 2-naphthyl sulphide		AI02015
122 121 43 91 41 55 107 105	**300**	100 58 58 38 31 26 22 22	**1.34**	20	28	2	–	–	–	–	–	–	–	–	–	–	Androsta-1,4-dien-3-one, 17-hydroxy-17-methyl-, (17α)-	33526-40-8	NI31509
122 121 161 147 108 107 134 123	**300**	100 44 27 22 18 16 14 13	**2.95**	20	28	2	–	–	–	–	–	–	–	–	–	–	Androsta-1,4-dien-3-one, 17-hydroxy-17-methyl-, (17β)-	72-63-9	NI31510
241 200 187 141 127 225 82 267	**300**	100 53 47 32 20 12 11 10	**9.00**	20	28	2	–	–	–	–	–	–	–	–	–	–	6,7-Dehydrosalviol		MS06958
300 119 117 121 231 165 41 244	**300**	100 98 98 57 45 42 42 41		20	28	2	–	–	–	–	–	–	–	–	–	–	1H-Dibenzo[a,d]cycloheptene-6,7-diol, 2,3,5,10,11,11a-hexahydro-11-dimethyl-8-(1-methylethyl)-, (S)-	88515-76-8	NI31511
199 217 61 214 256 282 70 300	**300**	100 90 69 62 53 52 47 45		20	28	2	–	–	–	–	–	–	–	–	–	–	17α-Ethynyl-5(10)-estrene-3α,17β-diol		NI31512
199 217 282 256 147 159 131 160	**300**	100 89 88 60 59 54 54 48	**45.60**	20	28	2	–	–	–	–	–	–	–	–	–	–	17α-Ethynyl-5(10)-estrene-3β,17β-diol		NI31513
274 230 215 300 267 256 217 162	**300**	100 62 46 40 39 28 28 28		20	28	2	–	–	–	–	–	–	–	–	–	–	17α-Ethynyl-17β-hydroxy-5α-estran-3-one		NI31514
274 215 233 230 232 124 300 160	**300**	100 74 59 56 52 39 36 31		20	28	2	–	–	–	–	–	–	–	–	–	–	17α-Ethynyl-17β-hydroxy-5β-estran-3-one		NI31515
285 300 239 91 143 42 131 117	**300**	100 72 48 35 29 28 26 25		20	28	2	–	–	–	–	–	–	–	–	–	–	Kaura-9(11),16-dien-18-oic acid, (4α)-		LI06595
285 300 239 91 143 41 131 79	**300**	100 76 48 35 29 28 26 25		20	28	2	–	–	–	–	–	–	–	–	–	–	Kaura-9(11),16-dien-18-oic acid, (4α)-	22338-67-6	NI31516
91 285 41 105 79 77 55 131	**300**	100 76 66 64 53 46 46 45	**26.00**	20	28	2	–	–	–	–	–	–	–	–	–	–	Kaura-9(11),16-dien-19-oic acid		LI06596
300 282 227 173 174 147 301 160	**300**	100 45 34 32 24 23 21 20		20	28	2	–	–	–	–	–	–	–	–	–	–	3-Methoxy-17β-methyl-1,3,5(10)-estratriene-17α-ol		LI06597
43 110 91 41 300 79 55 84	**300**	100 40 31 28 26 26 25 24		20	28	2	–	–	–	–	–	–	–	–	–	–	19-Norpregn-4-ene-3,20-dione	472-54-8	NI31518
43 110 91 41 300 79 55 84	**300**	100 99 78 72 66 66 63 60		20	28	2	–	–	–	–	–	–	–	–	–	–	19-Norpregn-4-ene-3,20-dione	472-54-8	NI31517
285 239 197 300 286 240 141 155	**300**	100 87 29 26 21 18 16 12		20	28	2	–	–	–	–	–	–	–	–	–	–	1-Phenanthrenecarboxylic acid, 1,2,3,4,4a,9,10,10a-octahydro-1,4a-dimethyl-7-isopropyl-, [1R-(1α,4aβ,10aα)]-	1740-19-8	NI31519
285 239 300 197 286 240 141 155	**300**	100 71 34 27 23 15 15 9		20	28	2	–	–	–	–	–	–	–	–	–	–	1-Phenanthrenecarboxylic acid, 1,2,3,4,4a,9,10,10a-octahydro-1,4a-dimethyl-7-isopropyl-, [1S-(1α,4aα,10aβ)]-	5155-70-4	NI31520
300 285 243 199 43 201 301 187	**300**	100 69 66 28 25 24 22 21		20	28	2	–	–	–	–	–	–	–	–	–	–	2(1H)-Phenanthrenone, 3,4,4a,9,10,10a-hexahydro-6-hydroxy-1,1,4a-trimethyl-7-isopropyl-, (4aS-trans)-	472-37-7	NI31521
300 285 243 301 199 201 125 18	**300**	100 53 39 16 13 11 11 11		20	28	2	–	–	–	–	–	–	–	–	–	–	2(1H)-Phenanthrenone, 3,4,4a,9,10,10a-hexahydro-6-hydroxy-1,1,4a-trimethyl-7-isopropyl-, (4aS-trans)-	472-37-7	NI31523
300 285 243 301 199 201 125 43	**300**	100 53 42 17 14 12 11 10		20	28	2	–	–	–	–	–	–	–	–	–	–	2(1H)-Phenanthrenone, 3,4,4a,9,10,10a-hexahydro-6-hydroxy-1,1,4a-trimethyl-7-isopropyl-, (4aS-trans)-	472-37-7	NI31522
285 300 243 286 201 301 175 159	**300**	100 42 32 21 13 10 10 9		20	28	2	–	–	–	–	–	–	–	–	–	–	2(1H)-Phenanthrenone, 3,4,4a,9,10,10a-hexahydro-7-hydroxy-1,1,4a-trimethyl-8-isopropyl-, (4aS-trans)-	6755-93-7	NI31524
300 267 285 215 197 203 301 173	**300**	100 56 36 30 27 23 21 12		20	28	2	–	–	–	–	–	–	–	–	–	–	2(1H)-Phenanthrenone, 3,4,4a,9,10,10a-hexahydro-7-hydroxy-1,1,4a-trimethyl-8-isopropyl-, (4aS-trans)-	6755-93-7	NI31525
285 283 300 243 286 43 201 41	**300**	100 51 41 36 22 18 16 16		20	28	2	–	–	–	–	–	–	–	–	–	–	2(1H)-Phenanthrenone, 3,4,4a,9,10,10a-hexahydro-7-hydroxy-1,1,4a-trimethyl-8-isopropyl-, (4aS-trans)-	6755-93-7	NI31526
285 300 217 215 203 286 243 41	**300**	100 74 31 26 25 22 20 18		20	28	2	–	–	–	–	–	–	–	–	–	–	9(1H)-Phenanthrenone, 2,3,4,4a,10,10a-hexahydro-6-hydroxy-1,1,4a-trimethyl-7-isopropyl-, (4aS-trans)-	511-05-7	NI31527
285 43 41 69 300 55 217 215	**300**	100 79 70 65 56 50 46 46		20	28	2	–	–	–	–	–	–	–	–	–	–	9(1H)-Phenanthrenone, 2,3,4,4a,10,10a-hexahydro-6-hydroxy-1,1,4a-trimethyl-7-isopropyl-, (4aS-trans)-	511-05-7	NI31528
258 243 300 259 161 147 173 244	**300**	100 56 31 20 18 16 14 11		20	28	2	–	–	–	–	–	–	–	–	–	–	Podocarpa-8,11,13-trien-12-ol, 13-methyl-, acetate	15340-78-0	NI31529
300 105 91 145 119 107 41 159	**300**	100 46 40 37 36 36 35 33		20	28	2	–	–	–	–	–	–	–	–	–	–	Retinoic acid	302-79-4	LI06598
300 105 91 145 119 107 41 69	**300**	100 47 41 39 38 38 37 35		20	28	2	–	–	–	–	–	–	–	–	–	–	Retinoic acid	302-79-4	NI31530
285 300 243 286 201 301 175 159	**300**	100 41 31 20 12 9 9 9		20	28	2	–	–	–	–	–	–	–	–	–	–	Totarol-3-one		LI06599
300 267 285 215 197 203 301 173	**300**	100 56 36 30 26 23 20 12		20	28	2	–	–	–	–	–	–	–	–	–	–	Totarol-7-one		LI06600
300 282 269 239 126 254 228 113	**300**	100 48 34 32 32 29 28 23		21	16	2	–	–	–	–	–	–	–	–	–	–	Benz[j]aceanthrylene-1,2-diol, 1,2-dihydro-3-methyl-	15544-91-9	MS06959
91 300 181 301 166 92 65 103	**300**	100 56 22 14 11 10 10 9		21	20	–	2	–	–	–	–	–	–	–	–	–	Benzaldehyde, dibenzylhydrazone	21136-32-3	NI31531
209 300 91 257 210 301 193 166	**300**	100 71 34 19 18 17 10 8		21	20	–	2	–	–	–	–	–	–	–	–	–	3-(N,N-Dimethylamino)-9-benzylcarbazole	94127-16-9	NI31532
41 95 55 67 81 96 80 79	**300**	100 77 70 62 58 56 55 54	**25.38**	21	32	1	–	–	–	–	–	–	–	–	–	–	Androst-1-en-3-one, 4,4-dimethyl-, (5α)-	54550-04-8	NI31533
300 285 257 245 82 95 93 91	**300**	100 66 52 50 45 40 40 39		21	32	1	–	–	–	–	–	–	–	–	–	–	Androst-2-en-17-one, 4,4-dimethyl-, (5α)-	53286-38-7	NI31534
124 41 55 95 67 105 91 43	**300**	100 75 51 40 36 35 34 34	**29.00**	21	32	1	–	–	–	–	–	–	–	–	–	–	Androst-5-en-3-one, 4,4-dimethyl-	5062-43-1	NI31535
300 285 41 189 43 215 203 55	**300**	100 87 36 30 25 23 23 19		21	32	1	–	–	–	–	–	–	–	–	–	–	Phenanthrene, 1,2,3,4,4a,9,10,10a-octahydro-6-methoxy-1,1,4a-trimethyl-7-isopropyl-, (4aS-trans)-	10064-26-3	NI31536

MASS TO CHARGE RATIOS								M.W.	INTENSITIES								Parent	C	H	O	N	S	F	Cl	Br	I	Si	P	B	X	COMPOUND NAME	CAS Reg No	No
285	300	189	286	215	301	203	201	**300**	100	54	36	21	21	11	11	6		21	32	1	–	–	–	–	–	–	–	–	–	–	Phenanthrene, 1,2,3,4,4a,9,10,10a-octahydro-7-methoxy-1,1,4a-trimethyl-8-isopropyl-, (4aS-trans)-	15340-83-7	NI31537
285	300	189	286	215	301	203	201	**300**	100	54	36	20	20	10	10	8		21	32	1	–	–	–	–	–	–	–	–	–	–	Phenanthrene, 1,2,3,4,4a,9,10,10a-octahydro-7-methoxy-1,1,4a-trimethyl-8-isopropyl-, (4aS-trans)-	15340-83-7	NI31538
91	105	79	107	267	41	300	93	**300**	100	85	79	77	73	72	71	70		21	32	1	–	–	–	–	–	–	–	–	–	–	Pregna-5,17(20)-dien-3-ol, (3β,17E)-	1159-25-7	NI31539
121	41	43	55	44	57	231	67	**300**	100	26	23	21	19	16	15	11	**6.66**	21	32	1	–	–	–	–	–	–	–	–	–	–	5α-Pregn-9(11)-en-12-one	4354-35-2	NI31540
43	257	41	55	67	300	93	81	**300**	100	43	33	27	23	22	22	21		21	32	1	–	–	–	–	–	–	–	–	–	–	5α-Pregn-16-en-20-one	3752-04-3	NI31541
271	300	41	93	55	91	107	81	**300**	100	53	51	48	46	39	35	35		21	32	1	–	–	–	–	–	–	–	–	–	–	Pregn-14-en-3-one, (5β)-	54411-79-9	NI31542
202	109	135	121	136	105	107	300	**300**	100	76	69	62	59	56	52	51		21	32	1	–	–	–	–	–	–	–	–	–	–	Pregn-17(20)-en-16-one, (5α,17E)-	54548-13-9	NI31543
285	108	286	135	121	300	104	106	**300**	100	29	24	24	24	22	22	21		21	32	1	–	–	–	–	–	–	–	–	–	–	Pregn-17(20)-en-16-one, (5α,17Z)-	54548-12-8	NI31544
191	269	282	91	165	192	77	105	**300**	100	44	33	32	26	19	17	12	**0.60**	22	20	1	–	–	–	–	–	–	–	–	–	–	1-(Hydroxymethyl)-2,3,3-triphenylcyclopropane		IC03785
244	165	243	167	166	105	245	242	**300**	100	58	47	46	28	24	19	12	**3.80**	22	20	1	–	–	–	–	–	–	–	–	–	–	Trityl allyl ether	1235-22-9	NI31545
244	165	243	167	166	105	242	152	**300**	100	58	47	46	28	24	12	11	**4.00**	22	20	1	–	–	–	–	–	–	–	–	–	–	Trityl allyl ether	1235-22-9	LI06601
300	230	285	271	95	109	121	93	**300**	100	56	50	42	42	40	38	36		22	36	–	–	–	–	–	–	–	–	–	–	–	D-Homopregn-17a(20)-ene, (5α,17aE)-	54482-44-9	NI31546
229	57	230	300	173	41	243	29	**300**	100	22	19	5	5	5	4	4		22	36	–	–	–	–	–	–	–	–	–	–	–	1,1,3,4-Tetramethyl-6-tert-butyl-3-(2,2-dimethylpropyl)-2,2-dihydroindene		AI02016
91	141	300	167	229	115	209	217	**300**	100	93	77	66	57	57	52	49		23	24	–	–	–	–	–	–	–	–	–	–	–	2-Methylene-1,3-bis(phenylmethylene)cyclooctane		MS06960
243	165	244	241	239	228	166	242	**300**	100	27	23	5	4	4	4	3	**0.00**	23	24	–	–	–	–	–	–	–	–	–	–	–	Propane, 2,2-dimethyl-1,1,1-triphenyl-	24523-60-2	NI31547
300	150	149	301	148	149.5	298	150.5	**300**	100	64	46	25	24	21	16	16		24	12	–	–	–	–	–	–	–	–	–	–	–	Coronene	191-07-1	MS06961
300	150	149	301	148	298	299	296	**300**	100	64	46	25	24	16	10	6		24	12	–	–	–	–	–	–	–	–	–	–	–	Coronene	191-07-1	NI31548
69	196	266	46	67	231	32	212	**301**	100	30	27	26	25	16	15	11	**0.05**	3	–	–	1	1	6	3	–	–	–	–	–	–	N-(α-Chlorohexafluoroisopropyl)sulphur dichlorodimide		LI06602
183	155	76	75	139	255	78	119	**301**	100	46	43	36	12	10	10	9	**8.20**	8	4	1	3	2	–	–	1	–	–	–	–	–	Benzamide, 4-bromo-N-(5H-1,3,2,4-dithiadiazol-5-ylidene)-	57726-54-2	NI31549
115	57	303	62	114	88	87	194	**301**	100	66	64	59	53	53	51	43	**35.93**	9	5	1	1	–	–	–	2	–	–	–	–	–	8-Quinolinol, 5,7-dibromo-	521-74-4	NI31550
72	201	71	44	45	39	130	36	**301**	100	95	84	39	38	22	21	19	**0.00**	10	8	4	3	1	–	1	–	–	–	–	–	–	Pyrido[2,3-d]thiazolo[3,2-b]pyridazin-4-ium, 3-methyl-, perchlorate	18592-56-8	NI31551
160	132	77	104	76	105	50	51	**301**	100	75	72	28	20	17	15	13	**5.00**	10	12	4	3	1	–	–	–	–	–	1	–	–	O,O-Dimethyl (4-oxo-1,2,3-benzotriazin-3(4H)-yl-methyl)-phosphorothiolate	961-22-8	NI31552
160	15	76	77	104	18	50	109	**301**	100	39	37	36	27	23	22	14	**3.50**	11	12	5	1	1	–	–	–	–	–	1	–	–	Phosphorothioic acid, S-[(1,3-dihydro-1,3-dioxo-2H-isoindol-2-yl)methyl] O,O-dimethyl ester	3735-33-9	NI31553
58	72	43	42	301	71	44	59	**301**	100	99	84	73	62	46	45	40		11	15	1	3	3	–	–	–	–	–	–	–	–	5-Acetyl-4-amino-3-(2-dimethylaminoethylthio)thieno[2,3-c]isothiazole	97090-68-1	NI31554
58	71	44	43	72	56	45	59	**301**	100	95	80	75	70	35	35	30	**6.00**	11	15	1	3	3	–	–	–	–	–	–	–	–	5-Acetyl-4-amino-3-(2-dimethylaminoethylthio)thieno[3,2-d]isothiazole	97106-67-7	NI31555
195	301	180	137	125	150	178	149	**301**	100	48	45	26	25	18	12	12		11	15	5	3	1	–	–	–	–	–	–	–	–	3-Imino-4-(3,4,5-trimethoxyphenyl)-1,2,5-thiadiazolidine 1,1-dioxide		DD01092
272	301	286	213	185	214	69	142	**301**	100	86	84	64	60	50	50	44		11	16	1	5	1	–	1	–	–	–	–	–	–	4-(2-Chloro-2-cyanoethylthio)-N,N-diethyl-6-methoxy-1,3,5-triazin-2-amine	84055-80-1	NI31556
61	29	41	57	27	75	153	69	**301**	100	68	56	42	31	24	19	17	**3.30**	11	18	3	1	1	3	–	–	–	–	–	–	–	L-Methionine, N-(trifluoroacetyl)-, butyl ester	2340-90-1	NI31558
302	228	303	304	229	230			**301**	100	45	15	5	4	1			**0.00**	11	18	3	1	1	3	–	–	–	–	–	–	–	L-Methionine, N-(trifluoroacetyl)-, butyl ester	2340-90-1	NI31557
129	85	59	72	102	130	58	36	**301**	100	24	20	19	17	8	7	3	**0.00**	11	19	3	5	1	–	–	–	–	–	–	–	–	1,1,4-Trimethyl-4-(aminomethylene)-2-tetrazenium-4-toluenesulphonate		LI06603
91	210	105	30	28	132	92	41	**301**	100	33	17	12	12	9	9	9	**3.80**	12	10	–	1	–	7	–	–	–	–	–	–	–	2-Phenyl-N-(heptafluoropropylmethylene)ethylamine		MS06962
45	286	55	73	256	210	270	243	**301**	100	66	25	21	20	19	18	16	**5.86**	12	23	2	5	1	–	–	–	–	–	–	–	–	1,3,5-Triazine-2,4-diamine, N,N'-bis(3-methoxypropyl)-6-(methylthio)-	845-52-3	NI31559
155	157	76	75	90	50	63	31	**301**	100	97	91	74	62	53	39	38	**24.10**	13	8	1	3	–	–	–	1	–	–	–	–	–	3-(4-Bromophenyl)benzo-1,2,3-triazin-4-one		MS06963
150	123	92	105	301	273	180	167	**301**	100	30	22	4	2	2	2	1		13	11	4	5	–	–	–	–	–	–	–	–	–	Benzene, 1,1'-(3-methyl-1-triazene-1,3-diyl)bis[2-nitro-	64229-79-4	NI31560
213	244	76	301	214	243	227	272	**301**	100	82	74	46	46	40	32	26		13	14	2	3	–	3	–	–	–	–	–	–	–	1H-Benzimidazole, 2-ethyl-7-nitro-1-propyl-5-(trifluoromethyl)-	55702-44-8	NI31561
149	148	162	135	301	108	176	96	**301**	100	81	13	12	10	8	6	6		13	15	2	7	–	–	–	–	–	–	–	–	–	9-[3-(Thym-1-yl)propyl]adenine		LI06604
181	124	201	171	182	57	137	107	**301**	100	52	28	16	14	14	10	7	**7.00**	13	19	5	1	1	–	–	–	–	–	–	–	–	tert-Butyl N-methyl-N-(4-anisolylsulphonyl)-carbamate	54934-83-7	LI06605
181	124	201	182	171	57	137	107	**301**	100	39	29	15	15	13	8	6	**6.01**	13	19	5	1	1	–	–	–	–	–	–	–	–	tert-Butyl N-methyl-N-(4-anisolylsulphonyl)-carbamate	54934-83-7	NI31562
155	113	131	112	89	85	95	214	**301**	100	88	76	68	59	42	36	33	**0.00**	13	19	7	1	–	–	–	–	–	–	–	–	–	2,3,5-Tri-O-acetyl-4-O-methyl-6-deoxy-D-gluconitrile		NI31563
143	101	117	83	142	84	159	203	**301**	100	75	71	68	52	48	39	28	**0.00**	13	19	7	1	–	–	–	–	–	–	–	–	–	2,4,5-Tri-O-acetyl-3-O-methyl-6-deoxy-D-gluconitrile		NI31564
129	145	103	87	69	85	130	113	**301**	100	85	56	36	33	31	28	20	**0.00**	13	19	7	1	–	–	–	–	–	–	–	–	–	3,4,5-Tri-O-acetyl-2-O-methyl-6-deoxy-D-gluconitrile		NI31565
73	69	83	55	75	286	184	147	**301**	100	55	51	41	30	29	22	13	**5.60**	13	27	3	1	–	–	–	–	–	2	–	–	–	Glycine, N-(2-methyl-1-oxo-2-butenyl)-N-(trimethylsilyl)-, trimethylsilyl ester, (E)-	55517-45-8	NI31566
73	102	183	83	285	45	156	75	**301**	100	79	77	63	45	30	29	27	**5.81**	13	27	3	1	–	–	–	–	–	2	–	–	–	Glycine, N-(3-methyl-1-oxo-2-butenyl)-N-(trimethylsilyl)-, trimethylsilyl ester	55649-86-0	NI31567
301	227	115	75	199	69	114	101	**301**	100	40	40	34	31	30	21	20		14	7	7	1	–	–	–	–	–	–	–	–	–	1,2,4-Trihydroxy-5-nitroanthraquinone		IC03786
301	303	57	43	55	69	71	139	**301**	100	99	47	37	32	28	27	25		14	8	2	1	–	–	–	1	–	–	–	–	–	1-Amino-4-bromoanthraquinone		IC03787
303	301	304	302	102	173	171	146	**301**	100	98	16	15	9	3	3	1		14	8	2	1	–	–	–	1	–	–	–	–	–	Benzoic acid, 4-cyano-, 4-bromophenyl ester	53327-10-9	NI31568
301	303	183	185	257	146	76	259	**301**	100	95	43	41	26	26	26	25		14	8	2	1	–	–	–	1	–	–	–	–	–	2-(4'-Bromophenyl)-4H-3,1-benzoxazin-4-one		MS06964

MASS TO CHARGE RATIOS	M.W.	INTENSITIES	Parent	C	H	O	N	S	F	Cl	Br	I	Si	P	B	X	COMPOUND NAME	CAS Reg No	No
301 303 286 284 76 166 183 155	**301**	100 99 55 55 50 23 17 14		14	8	2	1	–	–	–	1	–	–	–	–	–	2-(4′-Bromophenyl)isatogen		MS06965
150 135 77 93 104 76 50 51	**301**	100 69 61 55 45 45 36 35	9.00	14	11	3	3	1	–	–	–	–	–	–	–	–	Benzamide, 4-nitro-N-[(phenylamino)thioxomethyl]-	56437-98-0	NI31569
301 86 229 103 302 68 104 29	**301**	100 29 23 20 17 15 14 14		14	11	3	3	1	–	–	–	–	–	–	–	–	Ethyl 7-oxo-2-phenyl-7H-1,3,4-thiadiazolo[3,2-A]pyrimidine-5-carboxylate	68787-54-2	NI31570
150 104 301 76 151 120 92 50	**301**	100 28 20 17 8 8 8 4		14	11	5	3	–	–	–	–	–	–	–	–	–	Benzamide, N-methyl-4-nitro-N-(4-nitrophenyl)-	33675-69-3	MS06966
92 91 45 18 65 93 43 39	**301**	100 60 33 13 10 8 7 6	0.00	14	14	1	1	1	3	–	–	–	–	–	–	–	Benzo[b]cyclohepta[e][1,4]thiazine, 6,7,8,9,10,11-hexahydro-2-(trifluoromethyl)-, 5-oxide	54969-25-4	NI31571
208 222 43 41 223 195 209 55	**301**	100 79 55 47 43 36 25 23	6.00	14	23	4	1	1	–	–	–	–	–	–	–	–	5,5,7-Trimethyl-1-(4-morpholino)-8-oxa-3-thiatricyclo[5.2.0.0(7,9)]nonane-3,3-dioxide	104195-57-5	NI31572
102 73 75 103 59 45 30 74	**301**	100 47 14 10 7 7 6 5	0.00	14	31	2	1	–	–	–	–	–	2	–	–	–	trans-4-(Aminomethyl)-cyclohexanecarboxylic acid di-TMS		MS06967
73 110 272 41 240 142 101 80	**301**	100 77 72 70 44 44 40 38	17.41	14	33	–	1	2	–	–	–	–	–	–	2	–	Di-boron, [μ-(ethanethiolato)](ethanethiolato)triethyl[μ-[1-(methylimino)propyl-C:N]]-	36487-35-1	NI31573
210 181 32 211 91 182 65 40	**301**	100 99 50 13 11 8 4 4	0.33	15	12	–	1	–	5	–	–	–	–	–	–	–	N-(Pentafluorophenylmethyl)-2-phenylethylamine		MS06968
301 268 273 77 303 271 223 302	**301**	100 55 42 40 39 37 32 21		15	12	–	3	1	–	1	–	–	–	–	–	–	7-Chloro-3-methyl-5-phenyl-2-thioxo-1,2-dihydro-3H-1,3,4-benzothiazepine		MS06969
300 301 302 303 285 77 254 51	**301**	100 83 50 31 26 17 15 10		15	12	–	3	1	–	1	–	–	–	–	–	–	7-Chloro-2-methylthio-5-phenyl-3H-1,3,4-benzothiazepine		MS06970
131 103 77 303 301 132 171 173	**301**	100 26 15 14 14 9 8 7		15	12	1	1	–	–	–	1	–	–	–	–	–	4′-Bromo-3-phenylpropenanilide		TR00299
158 73 189 30 145 286 99 172	**301**	100 93 83 60 56 49 29 28	3.46	15	31	3	1	–	–	–	–	–	1	–	–	–	Decanoylglycine, trimethylsilyl-		NI31574
91 129 130 102 65 51 76 39	**301**	100 57 44 27 17 12 10 9	0.78	16	15	–	1	–	–	–	–	–	–	–	–	1	Indole, 3-[(benzylselenyl)methyl]-	21903-67-3	NI31575
146 91 65 301 147 41 89 63	**301**	100 69 48 19 17 13 13 12		16	15	3	1	1	–	–	–	–	–	–	–	–	2,3-Dihydro-5-methyl-1-(4-methylbenzenesulphonyl)-3-oxoindole		NI31576
146 301 155 180 194 237 236 209	**301**	100 57 40 7 6 4 4 3		16	15	3	1	1	–	–	–	–	–	–	–	–	4(1H)-Quinolinone, 2,3-dihydro-1-[(4-methylphenyl)sulphonyl]-		LI06606
160 43 202 161 159 131 301 77	**301**	100 47 17 11 4 4 3 3		16	15	5	1	–	–	–	–	–	–	–	–	–	1-(N-Acetyl-N-methyl)carbamoyl-4-acetoxynaphthalene		MS06971
160 202 132 161 301 104 244 259	**301**	100 34 19 18 6 4 3 2		16	15	5	1	–	–	–	–	–	–	–	–	–	1-(N-Acetyl-N-methyl)carbamoyl-5-acetoxynaphthalene		LI06607
160 43 202 131 161 159 301 102	**301**	100 47 34 20 19 10 7 6		16	15	5	1	–	–	–	–	–	–	–	–	–	1-(N-Acetyl-N-methyl)carbamoyl-5-acetoxynaphthalene		MS06972
269 77 301 270 183 154 51 211	**301**	100 97 68 51 45 38 35 34		16	15	5	1	–	–	–	–	–	–	–	–	–	Dimethyl 1,2-dihydro-3-methyl-2-oxo-1-phenyl-4,5-pyridinedicarboxylate		DD01093
286 301 243 159 158 287 258 131	**301**	100 70 38 31 31 25 22 20		16	15	5	1	–	–	–	–	–	–	–	–	–	10-Methyl-2,3-dimethoxy-1,4-dihydroxyacridone		MS06973
59 146 69 256 89 114 204 100	**301**	100 83 54 50 33 30 23 15	0.00	16	35	2	1	–	–	–	–	–	1	–	–	–	(E)-1-[[[Bis(2-methoxyethyl)amino]methyl]dimethylsilyl]-1-heptene		DD01094
146 59 69 256 204 114 89 100	**301**	100 91 65 56 47 36 23 15	0.00	16	35	2	1	–	–	–	–	–	1	–	–	–	3-[[[Bis(2-methoxyethyl)amino]methyl]dimethylsilyl]-1-heptene		DD01095
140 142 105 125 77 283 127 161	**301**	100 37 36 34 28 11 10 9	7.00	17	16	2	1	–	–	1	–	–	–	–	–	–	3-Benzoyl-N-(4-chlorobenzyl)propionamide		NI31577
125 135 105 301 131 107 103 177	**301**	100 81 49 41 28 17 15 8		17	19	2	1	1	–	–	–	–	–	–	–	–	Ethyl 3-(2-aminophenylthio)-3-phenylpropionate		MS06974
146 91 224 155 301	**301**	100 67 30 20 10		17	19	2	1	1	–	–	–	–	–	–	–	–	2-Phenylpyrrolidine 4-toluenesulphonamide		LI06608
301 231 300 284 272 286 273	**301**	100 18 14 10 9 4 4		17	19	4	1	–	–	–	–	–	–	–	–	–	Buphanamine		LI06609
269 181 268 240 224 211 225 209	**301**	100 32 28 26 15 12 11 7	4.00	17	19	4	1	–	–	–	–	–	–	–	–	–	Crinan-11-ol, 1,2-didehydro-3-methoxy-, (3α,5α,11R,13β,19α)-	639-41-8	NI31578
269 240 270 268 211 225 224 153	**301**	100 28 21 21 18 13 11 8	1.53	17	19	4	1	–	–	–	–	–	–	–	–	–	Crinan-11-ol, 1,2-didehydro-3-methoxy-, (3α,5α,11R,13β,19α)-	639-41-8	NI31579
269 181 227 301 268 240 224 211	**301**	100 41 40 36 34 28 20 18		17	19	4	1	–	–	–	–	–	–	–	–	–	Crinan-11-ol, 1,2-didehydro-3-methoxy-, (3α,5α,11S,13β,19α)-	41855-34-9	NI31580
301 227 225 181 269 257 224 211	**301**	100 61 45 36 33 33 25 21		17	19	4	1	–	–	–	–	–	–	–	–	–	Crinan-11-ol, 1,2-didehydro-3-methoxy-, (3β,5α,11R,13β,19α)-	466-75-1	NI31581
301 227 181 225 257 269 224 211	**301**	100 80 50 49 46 41 41 26		17	19	4	1	–	–	–	–	–	–	–	–	–	Crinan-11-ol, 1,2-didehydro-3-methoxy-, (3β,5α,11R,13β,19α)-	466-75-1	LI06610
301 227 181 257 269 224 211 225	**301**	100 80 50 46 41 41 26 25		17	19	4	1	–	–	–	–	–	–	–	–	–	Crinan-11-ol, 1,2-didehydro-3-methoxy-, (3β,5α,11R,13β,19α)-	466-75-1	NI31582
301 227 257 181 224 269 225 272	**301**	100 78 53 53 49 44 35 33		17	19	4	1	–	–	–	–	–	–	–	–	–	Crinan-11-ol, 1,2-didehydro-3-methoxy-, (3β,5α,11S,13β,19α)-	1472-76-0	NI31583
224 225 192 270 242 164 77 42	**301**	100 13 11 5 5 5 5 4	3.90	17	19	4	1	–	–	–	–	–	–	–	–	–	Dimethyl 4-phenyl-2,6-dimethyl-1,4-dihydropyridine-3,5-dicarboxylate	70677-78-0	NI31584
225 226 192 242 270 42 77 164	**301**	100 13 10 6 5 5 5 5	4.00	17	19	4	1	–	–	–	–	–	–	–	–	–	Dimethyl 4-phenyl-2,6-dimethyl-1,4-dihydropyridine-3,5-dicarboxylate	70677-78-0	NI31585
246 231 257 187 270 252 115 258	**301**	100 41 20 17 14 11 11 10	0.00	17	19	4	1	–	–	–	–	–	–	–	–	–	6α-Hydroxybuphaninisine		NI31586
43 301 188 231 94 189 79 77	**301**	100 56 40 28 28 24 21 21		17	19	4	1	–	–	–	–	–	–	–	–	–	1H-Isoindole-1,3(2H)-dione, 4-(acetyloxy)hexahydro-3a-methyl-2-phenyl-, [3aR-(3aα,4β,7aα)]-	54346-15-5	NI31587
56 130 159 158 301 145 218 84	**301**	100 47 33 16 12 6 5 5		17	19	4	1	–	–	–	–	–	–	–	–	–	6,7-Bis(methoxycarbonyl)-1,2,5-trimethyl-3,4-benzo-2-azabicyclo[3.2.0]hepta-3,6-diene		NI31588
226 118 241 301 242 200 72 134	**301**	100 72 54 20 20 16 12 10		17	19	4	1	–	–	–	–	–	–	–	–	–	1-Methyl-12-acetoxy-khaplofoline		LI06611
105 77 150 70 179 301 51 151	**301**	100 64 58 52 46 37 31 27		17	19	4	1	–	–	–	–	–	–	–	–	–	Morphinan-3,6,14-triol, 7,8-didehydro-4,5-epoxy-17-methyl-, (5α,6α)-	3371-56-0	NI31589
301 270 257 43 44 42 302 57	**301**	100 68 29 25 24 22 20 18		17	19	4	1	–	–	–	–	–	–	–	–	–	Pancracine, O^2-methyl-	642-52-4	NI31590
301 270 42 302 223 300 286 271	**301**	100 62 21 20 16 14 14 14		17	19	4	1	–	–	–	–	–	–	–	–	–	Pancracine, O^2-methyl-, (2β)-	485-57-4	NI31591
301 243 214 199 302 185 57 55	**301**	100 65 27 24 21 20 17 17		17	19	4	1	–	–	–	–	–	–	–	–	–	Pancracine, O^3-methyl-	574-05-0	NI31592
301 243 214 199 302 185 57 55	**301**	100 64 26 25 22 20 15 15		17	19	4	1	–	–	–	–	–	–	–	–	–	Pancracine, O^3-methyl-	574-05-0	LI06612
301 199 270 300 185 243 241 214	**301**	100 25 23 18 17 16 13 13		17	19	4	1	–	–	–	–	–	–	–	–	–	Pancracine, O^3-methyl-, (2β,3α)-	485-58-5	NI31593
301 243 269 214 185 270 302 199	**301**	100 61 41 28 22 20 18 17		17	19	4	1	–	–	–	–	–	–	–	–	–	Pancracine, O^3-methyl-, (3α)-	485-59-6	NI31594
240 286 172 69 77 159 301 105	**301**	100 97 75 75 72 65 62 62		17	19	4	1	–	–	–	–	–	–	–	–	–	3-Pyridinecarboxylic acid, 1-(2,4-dimethylphenyl)-1,6-dihydro-4-hydroxy-2-methyl-6-oxo-, ethyl ester	10230-60-1	NI31595

MASS TO CHARGE RATIOS								M.W.	INTENSITIES								Parent	C	H	O	N	S	F	Cl	Br	I	Si	P	B	X	COMPOUND NAME	CAS Reg No	No
84	91	77	36	85	182	55	65	301	100	18	9	9	8	7	6	4	0.00	17	20	–	3	–	–	1	–	–	–	–	–	–	1H-Imidazole-2-methanamine, 4,5-dihydro-N-phenyl-N-benzyl-, monohydrochloride	2508-72-7	NI31596
71	146	156	117	103	44	203	42	301	100	93	68	37	29	25	23	22	8.00	17	23	2	3	–	–	–	–	–	–	–	–	–	1H-1,4-Diazepine-5,7(2H,6H)-dione, 3-(dimethylamino)-6-ethyl-2,2-dimethyl-6-phenyl-	69315-95-3	NI31597
301	218	242	176	259	160	258	198	301	100	59	46	43	39	28	26	24		18	23	3	1	–	–	–	–	–	–	–	–	–	O-Acetylalolycopine		LI06613
67	81	55	95	41	82	43	68	301	100	99	81	73	68	63	61	57	0.00	18	23	3	1	–	–	–	–	–	–	–	–	–	Erythrinan-16-ol, 1,6-didehydro-3,15-dimethoxy-, (3β)-	7236-36-4	NI31598
242	241	182	228	210	212	200	243	301	100	39	33	30	22	19	16	15	12.69	18	23	3	1	–	–	–	–	–	–	–	–	–	Erythrinan-16-ol, 1,6-didehydro-3,15-dimethoxy-, (3β)-	7236-36-4	NI31599
301	244	229	245	300	302	286	272	301	100	40	27	26	25	20	20	12		18	23	3	1	–	–	–	–	–	–	–	–	–	Hasubanan-6-ol, 4,5-epoxy-3-methoxy-17-methyl-, (5α,6α,13β,14β)-	4892-49-3	LI06614
301	44	42	36	244	245	229	55	301	100	64	46	42	41	35	31	30		18	23	3	1	–	–	–	–	–	–	–	–	–	Hasubanan-6-ol, 4,5-epoxy-3-methoxy-17-methyl-, (5α,6α,13β,14β)-	4892-49-3	NI31600
109	108	110	192	82	191	55	42	301	100	12	8	2	2	1	1	1	0.16	18	23	3	1	–	–	–	–	–	–	–	–	–	Lycorenan, 9,10-dimethoxy-1-methyl-	13255-14-6	NI31601
130	43	129	200	41	131	71		301	100	32	22	19	12	11	9		0.20	18	23	3	1	–	–	–	–	–	–	–	–	–	2-(1-Methoxycarbonyl-1-methylethyl)-1,2-dihydro-1-isobutylquinoline		LI06615
105	302	151	133	123	89	301	143	301	100	18	14	11	6	5	4	4		18	23	3	1	–	–	–	–	–	–	–	–	–	Morphinan-6-ol, 4,5-epoxy-3-methoxy-17-methyl-, (5α,6α)-	125-28-0	NI31602
301	44	42	59	70	164	31	302	301	100	40	38	37	31	26	24	21		18	23	3	1	–	–	–	–	–	–	–	–	–	Morphinan-6-ol, 4,5-epoxy-3-methoxy-17-methyl-, (5α,6α)-	125-28-0	NI31603
214	301	147	88	100	199	215	188	301	100	66	59	28	26	25	20	20		18	23	3	1	–	–	–	–	–	–	–	–	–	2-Morpholyl-3-cyclohexenyl-5-hydroxy-coumaran		IC03788
214	301	152	215	230	184	241	186	301	100	29	24	17	15	12	9	9		18	23	3	1	–	–	–	–	–	–	–	–	–	Serratinine, 6,7,8,16-tetradehydro-8-deoxy-, acetate	5532-12-7	NI31604
214	301	152	215	230	184	242	186	301	100	24	24	17	15	12	10	10		18	23	3	1	–	–	–	–	–	–	–	–	–	Serratinine, 6,7,8,16-tetradehydro-8-deoxy-, acetate		LI06616
72	73	229	230	75	153	45	43	301	100	24	11	5	5	4	4	4	0.10	18	27	1	1	–	–	–	–	–	1	–	–	–	2-Naphthaleneethanamine, N-isopropyl-β-[(trimethylsilyl)oxy]-	54934-74-6	NI31605
180	104	77	78	226	301	123	165	301	100	46	33	20	15	11	10	8		19	13	–	1	–	–	1	–	–	–	–	1	–	Diphenylmethyleneamino-phenyl-chloroborane		LI06617
301	18	168	302	28	300	44	77	301	100	75	26	22	12	7	6	5		19	15	1	3	–	–	–	–	–	–	–	–	–	4-(1,2-Dihydroxyquinol-2-ylidene)-3-methyl-1-phenyl-5-pyrazolone		IC03789
105	77	41	55	196	30	27	39	301	100	63	36	33	21	20	18	17	11.00	19	27	2	1	–	–	–	–	–	–	–	–	–	2-Benzoyl-12-dodecanelactam	102222-15-1	NI31606
59	301	180	300	242	162	60	58	301	100	39	22	16	10	10	10	9		19	27	2	1	–	–	–	–	–	–	–	–	–	Morphinan-6-ol, 3-methoxy-6,17-dimethyl-, (6α)-	2246-06-2	NI31607
209	77	180	301	255	254	165	152	301	100	25	19	12	12	12	9	9		20	15	2	1	–	–	–	–	–	–	–	–	–	3-Anilino-3-phenylphthalide		LI06618
105	77	51	78	39	79	301	69	301	100	58	32	24	23	21	17	17		20	15	2	1	–	–	–	–	–	–	–	–	–	N,N-Dibenzoyl-aniline		IC03790
105	77	197	51	198	39	65	106	301	100	63	48	23	10	10	9	8	1.00	20	15	2	1	–	–	–	–	–	–	–	–	–	N,N-Dibenzoyl-aniline		MS06975
209	210	152	77	301	105	284		301	100	16	11	11	5	5	2			20	15	2	1	–	–	–	–	–	–	–	–	–	3-Hydroxy-2,3-diphenyl-2,3-dihydro-1-oxoisoindole		LI06619
209	301	142	170	141	114	152	115	301	100	52	26	25	22	19	14	13		20	15	2	1	–	–	–	–	–	–	–	–	–	3-Isopropenyloxy-2-naphthanilide		NI31608
141	130	115	301	127	89	154	139	301	100	13	12	8	7	6	4	4		20	15	2	1	–	–	–	–	–	–	–	–	–	1-Naphthalenemethyl indole-2-carboxylate	86328-61-2	NI31609
41	232	130	89	301	77	233	117	301	100	48	34	32	28	26	24	23		20	19	–	3	–	–	–	–	–	–	–	–	–	2-(3-Indolyl)-1-(2-(1-pyrrolinyl))indoline		MS06976
91	196	301	77	105	104	302	65	301	100	96	83	50	33	23	18	18		20	19	–	3	–	–	–	–	–	–	–	–	–	4-(N-Methyl-benzylamino)-azobenzene		MS06977
41	130	89	144	117	301	158	232	301	100	35	25	23	20	14	12	11		20	19	–	3	–	–	–	–	–	–	–	–	–	2,3,5,6-Tetrahydro-5-(3-indolyl)-1H-pyrrolo[2,1-b][1,3]benzodiazepine		MS06978
79	93	41	270	91	77	81	105	301	100	84	65	62	62	46	42	38	24.30	20	31	1	1	–	–	–	–	–	–	–	–	–	Androst-16-en-3-one, O-methyloxime, (5β)-	56196-49-7	NI31610
274	55	202	259	203	301	258	275	301	100	81	78	77	74	67	61	60		20	31	1	1	–	–	–	–	–	–	–	–	–	3-Cyanoandrostan-3-ol		LI06620
301	180	259	258	210	182	43	165	301	100	98	85	71	42	40	29	27		21	19	1	1	–	–	–	–	–	–	–	–	–	Acetamide, N-phenyl-N-[2-benzylphenyl]-	52812-81-4	NI31611
169	170	168	167	301	196	302	132	301	100	54	52	39	27	16	10	7		21	19	1	1	–	–	–	–	–	–	–	–	–	Benzeneacetamide, 4-methyl-N,N-diphenyl-	55955-87-8	NI31612
131	193	208	91	209	77	132	115	301	100	63	46	34	31	14	13	12	6.50	22	23	–	1	–	–	–	–	–	–	–	–	–	Benzenemethanamine, 2-isopropyl-N,α-diphenyl-	19103-11-8	NI31613
131	193	208	91	209	178	179	77	301	100	63	46	34	31	19	15	14	7.00	22	23	–	1	–	–	–	–	–	–	–	–	–	Benzenemethanamine, 2-isopropyl-N,α-diphenyl-	19103-11-8	NI31614
209	167	165	77	210	301	178	51	301	100	25	25	25	20	17	15	12		22	23	–	1	–	–	–	–	–	–	–	–	–	Benzenemethanamine, 3-isopropyl-N,α-diphenyl-	32388-78-6	NI31615
209	167	165	77	210	301	178	166	301	100	25	25	25	20	17	15	12		22	23	–	1	–	–	–	–	–	–	–	–	–	Benzenemethanamine, 3-isopropyl-N,α-diphenyl-	32388-78-6	NI31616
209	77	167	165	210	65	194	51	301	100	24	18	17	15	15	13	11	6.80	22	23	–	1	–	–	–	–	–	–	–	–	–	Benzenemethanamine, 4-isopropyl-N,α-diphenyl	23431-27-8	NI31617
219	301	299	159	220	302	55	41	301	100	86	45	45	22	21	19	19		22	23	–	1	–	–	–	–	–	–	–	–	–	1-Cyclohexyl-2,4-diphenylpyrrole		MS06979
182	56	212	242	272	147	116	91	302	100	92	50	42	3	3	3	2	1.00	–	–	4	4	–	–	2	–	–	–	–	–	2	Di-iron, di-μ-chlorotetranitroso-		MS06980
306	304	144	302	146	308	225	198	302	100	99	33	32	32	30	30	23		3	1	–	2	–	–	–	3	–	–	–	–	–	1H-Pyrazole, 3,4,5-tribromo-	17635-44-8	NI31618
101	85	103	31	87	135	151	69	302	100	84	64	34	27	26	23	18	0.00	4	–	–	–	–	6	4	–	–	–	–	–	–	1,1,2,3,3,4-Hexafluoro-1,2,4,4-tetrachlorobutane		TR00300
85	151	87	153	135	101	147	69	302	100	35	34	22	19	13	12	11	0.00	4	–	–	–	–	6	4	–	–	–	–	–	–	1,2,3,4-Tetrachloro-1,1,2,3,4,4-hexafluorobutane		DO01087
151	153	69	85	101	155	103	87	302	100	64	44	22	12	10	8	7	0.00	4	–	–	–	–	6	4	–	–	–	–	–	–	2,2,3,3-Tetrachlorohexafluorobutane		DO01088
119	69	302	64	183	63	31	82	302	100	54	47	36	31	20	13	12		4	–	–	–	2	10	–	–	–	–	–	–	–	Bis(pentafluoroethyl) disulphide	679-77-6	NI31619
219	69	131	119	267	100	63	48	302	100	78	73	65	59	56	46	43	6.50	4	–	1	–	1	9	1	–	–	–	–	–	–	1-Butanesulphinyl chloride, 1,1,2,2,3,3,4,4,4-nonafluoro-	41006-41-1	NI31620
301	202	76	266	303	112	110	203	302	100	98	57	56	43	26	21	15	8.30	6	4	2	–	1	–	1	–	1	–	–	–	–	Benzenesulphonyl chloride, 2-iodo-	63059-29-0	NI31621
145	85	93	125	58	47	63	146	302	100	74	27	25	18	18	14	12	5.20	6	11	4	2	3	–	–	–	–	–	1	–	–	Phosphorodithioic acid, S-[(5-methoxy-2-oxo-1,3,4-thiadiazol-3(2H)-yl)methyl] O,O-dimethyl ester	950-37-8	NI31622
15	85	145	18	93	58	47	125	302	100	79	70	46	31	31	25	21	1.50	6	11	4	2	3	–	–	–	–	–	1	–	–	Phosphorodithioic acid, S-[(5-methoxy-2-oxo-1,3,4-thiadiazol-3(2H)-yl)methyl] O,O-dimethyl ester	950-37-8	NI31623
145	85	18	28	93	125	146	15	302	100	54	17	15	11	10	7	7	5.11	6	11	4	2	3	–	–	–	–	–	1	–	–	Phosphorodithioic acid, S-[(5-methoxy-2-oxo-1,3,4-thiadiazol-3(2H)-yl)methyl] O,O-dimethyl ester	950-37-8	NI31624

MASS TO CHARGE RATIOS								M.W.	INTENSITIES								Parent	C	H	O	N	S	F	Cl	Br	I	Si	P	B	X	COMPOUND NAME	CAS Reg No	No
45	107	142	18	108	36	71	73	**302**	100	22	18	18	16	16	14	13	**4.30**	8	2	4	–	–	–	4	–	–	–	–	–	–	**1,4-Benzenedicarboxylic acid, 2,3,5,6-tetrachloro-**	2136-79-0	**NI31625**
216	214	218	74	179	181	108	109	**302**	100	78	47	18	17	16	16	12	**0.00**	8	2	4	–	–	–	4	–	–	–	–	–	–	**1,4-Benzenedicarboxylic acid, 2,3,5,6-tetrachloro-**	2136-79-0	**NI31626**
247	245	214	107	212	286	142	249	**302**	100	78	67	54	53	51	50	47	**0.00**	8	2	4	–	–	–	4	–	–	–	–	–	–	**3,4,5,6-Tetrachlorobenzene-1,2-dicarboxylic acid**		**NI31627**
55	170	90	119	93	172	92	302	**302**	100	80	49	47	38	28	18	14		8	4	5	–	1	–	1	–	–	–	–	–	1	**(Chlorosulphonyl)cymantrene**	12247-52-8	**NI31628**
119	117	115	121	118	123	89	184	**302**	100	78	53	31	25	21	19	16	**9.00**	9	9	1	–	–	5	–	–	–	–	–	–	1	**Germane, trimethyl(pentafluorophenoxy)-**	22530-02-5	**NI31629**
69	149	95	302	92	177	52	83	**302**	100	36	33	28	27	25	21	20		10	4	4	–	–	6	–	–	–	–	–	–	–	**2-Hydroxyphenol, bis(trifluroacetyl)-**	23529-06-8	**NI31630**
69	302	92	246	189	111	149	281	**302**	100	34	22	18	12	12	9	8		10	4	4	–	–	6	–	–	–	–	–	–	–	**3-Hydroxyphenol, bis(trifluroacetyl)-**	34065-72-0	**NI31631**
302	69	177	149	205	189	246	303	**302**	100	99	78	36	21	20	12	11		10	4	4	–	–	6	–	–	–	–	–	–	–	**4-Hydroxyphenol, bis(trifluroacetyl)-**	34065-73-1	**NI31632**
302	300	274	271	272	273	298	270	**302**	100	91	62	61	58	56	54	46		10	6	1	–	–	–	–	–	–	–	–	–	2	**Selenolo[3,4-b][1]benzoselenophen-3(1H)-one**	39827-01-5	**NI31633**
196	36	269	267	198	233	39	231	**302**	100	99	96	77	68	53	53	50	**18.43**	10	7	–	–	–	–	5	–	–	–	–	–	–	**1,4-Ethenopentalene, 1,2,3,7,8-pentachloro-1,3a,4,5,6,6a-hexahydro-**	57206-19-6	**NI31634**
196	198	231	233	269	51	98	99	**302**	100	69	58	56	44	41	39	38	**3.66**	10	7	–	–	–	–	5	–	–	–	–	–	–	**1,4-Ethenopentalene, 2,4,5,6,8-pentachloro-1,2,3,3a,4,6a-hexahydro-**	69743-75-5	**NI31635**
196	269	198	267	36	231	233	99	**302**	100	75	66	59	57	48	46	39	**19.57**	10	7	–	–	–	–	5	–	–	–	–	–	–	**1,4-Ethenopentalene, 2,5,6,7,8-pentachloro-1,2,3,3a,4,6a-hexahydro-**	69743-76-6	**NI31636**
196	198	304	51	36	233	231	75	**302**	100	67	65	62	58	57	57	53	**41.15**	10	7	–	–	–	–	5	–	–	–	–	–	–	**1,6-Methano-1H-indene, 2,3,4,5,7-pentachloro-3a,6,7,7a-tetrahydro-**	69743-79-9	**NI31637**
196	36	198	231	233	75	51	50	**302**	100	74	69	68	66	64	63	49	**22.56**	10	7	–	–	–	–	5	–	–	–	–	–	–	**1,6-Methano-1H-indene, 2,3,4,5,8-pentachloro-3a,6,7,7a-tetrahydro-**	69743-72-2	**NI31638**
196	198	39	65	75	51	231	36	**302**	100	67	44	35	34	34	32	32	**0.32**	10	7	–	–	–	–	5	–	–	–	–	–	–	**1,4,5-Metheno-1H-cyclopropa[a]pentalene, 1,3,5,5a,6-pentachlorooctahydro-**	69743-78-8	**NI31639**
186	193	302	121	272	56	151	112	**302**	100	95	67	65	53	19	11	5		10	10	2	2	–	–	–	–	–	–	–	–	2	**Di-iron, π-di(cyclopentadienyl)-μ-dinitrosyl-**		**LI06621**
106	73	107	61	133	134	135	302	**302**	100	60	47	44	42	31	30	21		10	19	–	–	4	–	1	–	–	–	–	–	–	**6-Chloro-1,4,8,11-tetrathiacyclotetradecane**	91325-67-6	**NI31640**
116	73	101	147	43	93	45	59	**302**	100	98	51	38	24	23	19	13	**0.00**	10	24	2	–	–	–	2	–	–	2	–	–	–	**Butane, 1,4-dichloro-2,3-bis(trimethylsiloxy)-**	67177-73-5	**NI31641**
165	173	166	109	257	137	138	81	**302**	100	32	30	29	19	18	15	14	**3.00**	10	24	6	–	–	–	–	–	–	–	2	–	–	**Tetraethyl ethane-1,2-diphosphonate**		**MS06981**
91	155	65	56	172	107	88	302	**302**	100	53	34	16	11	11	11	5		11	14	6	2	1	–	–	–	–	–	–	–	–	**Methyl 5-(4-toluenesulphonoxy)hydantoate**		**LI06622**
175	127	104	72	230	283	103	77	**302**	100	84	65	59	52	32	32	32	**0.90**	11	15	–	2	1	5	–	–	–	–	–	–	–	**[α-(Diethylamino)benzylidenimino]sulphur pentafluoride**		**NI31642**
167	247	249	303	223				**302**	100	93	92	68	59				**0.00**	11	15	3	2	–	–	–	1	–	–	–	–	–	**2,4,6(1H,3H,5H)-Pyrimidinetrione, 5-(2-bromo-2-propenyl)-5-sec-butyl-**	1142-70-7	**NI31643**
175	111	177	302	113	75	28	304	**302**	100	90	39	27	27	25	24	19		12	8	3	–	1	–	2	–	–	–	–	–	–	**Benzenesulphonic acid, 4-chloro-, 4-chlorophenyl ester**	80-33-1	**NI31644**
111	175	75	99	177	113	63	73	**302**	100	87	56	37	31	31	24	23	**14.20**	12	8	3	–	1	–	2	–	–	–	–	–	–	**Benzenesulphonic acid, 4-chloro-, 4-chlorophenyl ester**	80-33-1	**NI31645**
111	175	75	99	113	177	73	63	**302**	100	75	53	51	32	28	23	23	**22.35**	12	8	3	–	1	–	2	–	–	–	–	–	–	**Benzenesulphonic acid, 4-chloro-, 4-chlorophenyl ester**	80-33-1	**NI31646**
77	141	51	63	50	133	18	73	**302**	100	67	61	27	23	21	17	14	**5.50**	12	8	3	–	1	–	2	–	–	–	–	–	–	**Phenol, 2,4-dichloro-, benzenesulphonate**	97-16-5	**NI31647**
188	123	246	190	248	28	125	250	**302**	100	59	47	39	36	25	22	13	**7.66**	12	10	2	–	–	–	–	–	–	–	–	–	2	**Di-nickel, di-μ-carbonylbis(η^5-2,4-cyclopentadienyl)-**	12170-92-2	**NI31648**
246	123	186	244	302	97	58	274	**302**	100	57	36	33	11	11	8	6		12	10	2	–	–	–	–	–	–	–	–	–	2	**Di-nickel, π-di-cyclopentadienyl-dicarbonyl-**		**MS06982**
125	108	127	110	85	79	114	136	**302**	100	78	64	53	52	51	47	42	**20.00**	12	12	3	2	–	–	2	–	–	–	–	–	–	**4-Hexenoic acid, 3,5-dichloro-6-hydroxy-2-(phenylhydrazono)-**	56771-75-6	**MS06983**
41	175	57	29	177	112	39	27	**302**	100	82	71	69	50	38	33	28	**9.00**	12	12	3	2	–	–	2	–	–	–	–	–	–	**1,3,4-Oxadiazol-2(3H)-one, 3-(2,4-dichloro-5-hydroxyphenyl)-5-tert-butyl-**	39807-19-7	**NI31649**
111	151	150	55	69	122	135	191	**302**	100	97	77	53	52	50	36	34	**0.00**	12	14	2	8	–	–	–	–	–	–	–	–	–	**9-[3-(Cytos-1-yl)propyl]guanine**		**LI06623**
211	201	243	258	199	173	141	115	**302**	100	49	44	21	18	16	13	13	**0.60**	12	14	9	–	–	–	–	–	–	–	–	–	–	**2-Methoxy-3,4-bis(methoxycarbonyl)-4-methoxycarbonylmethyl-1,4-but-2 enolide**		**LI06624**
130	131	103	184	183	104	77	145	**302**	100	85	45	37	35	35	35	32	**0.00**	12	15	5	2	–	–	1	–	–	–	–	–	–	**(E)-2-(2-Hydroximinopropyl)-3,4-dihydroisoquinolinium perchlorate**		**NI31650**
184	183	130	131	103	77	104	115	**302**	100	72	47	33	22	22	21	13	**0.00**	12	15	5	2	–	–	1	–	–	–	–	–	–	**(Z)-2-(2-Hydroximinopropyl)-3,4-dihydroisoquinolinium perchlorate**		**NI31651**
155	272	132	287	170	171	302	211	**302**	100	15	12	5	5	3	2	2		12	18	7	2	–	–	–	–	–	–	–	–	–	**Uridine, 5-(1-methoxyethyl)-**	68972-46-3	**NI31652**
302	135	287	274	85	221	164	166	**302**	100	88	67	35	33	29	14	10		12	19	–	–	–	–	–	–	–	1	1	–	1	**1-Phenyl-4,4-dimethyl-1-seleno-1,4-phosphasilacyclohexane**		**MS06984**
302	135	287	300	274	85	285	221	**302**	100	86	69	46	35	33	33	30		12	19	–	–	–	–	–	–	–	1	1	–	1	**1-Phenyl-4,4-dimethyl-1-seleno-1,4-phosphasilacyclohexane**		**MS06985**
73	215	147	258	45	169	216	259	**302**	100	46	38	31	18	13	9	9	**0.00**	12	22	5	–	–	–	–	–	–	2	–	–	–	**2-Methyl-4-ketoglutaconate, bis(trimethylsilyl)-**		**NI31653**
73	215	147	258	45	169	75	43	**302**	100	44	38	25	22	13	13	10	**0.00**	12	22	5	–	–	–	–	–	–	2	–	–	–	**2-Methyl-4-ketoglutaconate, bis(trimethylsilyl)-**		**NI31654**
287	288	302	211	73	135	195	147	**302**	100	25	20	15	10	9	8	8		12	23	3	–	–	–	–	–	–	2	1	–	–	**Phosphonic acid, phenyl-, bis(trimethylsilyl) ester**	42449-24-1	**NI31656**
287	302	211	73	135	195	147	133	**302**	100	20	15	10	9	8	8	6		12	23	3	–	–	–	–	–	–	2	1	–	–	**Phosphonic acid, phenyl-, bis(trimethylsilyl) ester**	42449-24-1	**NI31655**
57	125	43	71	177	85	41	129	**302**	100	97	72	62	48	47	47	46	**0.30**	12	25	–	–	–	–	3	–	–	1	–	–	–	**Silane, trichlorododecyl-**	4484-72-4	**NI31657**
164	134	192	163	148	206	294	222	**302**	100	74	51	28	16	14	11	11	**0.00**	12	26	3	6	–	–	–	–	–	–	–	–	–	**2-Furanmethanol, 3-amino-5-[6-(dimethylamino)octahydro-9H-purin-9-yl]tetrahydro-4-hydroxy-**	54833-68-0	**NI31658**
302	123	77	91	303	152	78	65	**302**	100	52	20	19	17	17	15	15		13	10	5	4	–	–	–	–	–	–	–	–	–	**Benzaldehyde, 2-hydroxy-, (2,4-dinitrophenyl)hydrazone**	1160-76-5	**NI31659**
302	65	79	78	39	93	77	63	**302**	100	54	45	45	44	40	38	38		13	10	5	4	–	–	–	–	–	–	–	–	–	**Benzaldehyde, 3-hydroxy-, (2,4-dinitrophenyl)hydrazone**	1160-77-6	**NI31660**
302	83	123	85	65	31	78	77	**302**	100	54	44	43	32	27	24	23		13	10	5	4	–	–	–	–	–	–	–	–	–	**Benzaldehyde, 4-hydroxy-, (2,4-dinitrophenyl)hydrazone**	1160-78-7	**NI31661**
302	123	152	120	303	121	65	78	**302**	100	38	17	17	15	15	15	12		13	10	5	4	–	–	–	–	–	–	–	–	–	**Benzaldehyde, 4-hydroxy-, (2,4-dinitrophenyl)hydrazone**	1160-78-7	**NI31662**
247	189	249	243	191	245	55	227	**302**	100	62	61	47	42	30	28	12	**9.03**	13	12	4	–	–	–	2	–	–	–	–	–	–	**Acetic acid, [2,3-dichloro-4-(2-methylene-1-oxobutyl)phenoxy]-**	58-54-8	**MS06986**
157	73	169	75	56	287	158	55	**302**	100	74	30	26	18	15	14	12	**6.87**	13	26	4	–	–	–	–	–	–	2	–	–	–	**heptanoic acid, 4,6-dioxo-, bis(trimethylsilyl)-**	101329-13-9	**NI31663**
157	73	169	75	287	158	45	43	**302**	100	78	24	21	14	13	12	8	**5.40**	13	26	4	–	–	–	–	–	–	2	–	–	–	**heptanoic acid, 4,6-dioxo-, bis(trimethylsilyl)-**	101329-13-9	**NI31664**
105	181	77	302	257	51	258	106	**302**	100	94	67	63	32	27	21	16		14	7	2	–	–	5	–	–	–	–	–	–	–	**Benzoic acid, pentafluorobenzyl ester**		**MS06987**
179	138	163	180	136	139	302	164	**302**	100	73	32	10	8	6	4	2		14	10	6	2	–	–	–	–	–	–	–	–	–	**Benzeneacetic acid, 4-nitro-, 3-nitrophenyl ester**	21997-26-2	**NI31665**
138	46	139	179	136	302	163		**302**	100	55	8	6	4	2	1			14	10	6	2	–	–	–	–	–	–	–	–	–	**Benzeneacetic acid, 4-nitro-, 4-nitrophenyl ester**	35665-94-2	**NI31666**

MASS TO CHARGE RATIOS								M.W.	INTENSITIES								Parent	C	H	O	N	S	F	Cl	Br	I	Si	P	B	X	COMPOUND NAME	CAS Reg No	No
166	122	167	302	46				**302**	100	26	8	1	1					14	10	6	2	–	–	–	–	–	–	–	–	–	**Benzenemethanol, 4-nitro-, 4-nitrobenzoate**	3481-11-6	**NI31667**
205	302	304	119	118	185	77	147	**302**	100	86	71	58	53	43	36	35		14	11	1	2	–	–	–	1	–	–	–	–	–	**2-Benzimidazolemethanol, α-(4-bromophenyl)-**	7104-86-1	**NI31668**
260	216	302	215	242	43	243	261	**302**	100	99	67	59	46	32	18	14		14	14	2	4	1	–	–	–	–	–	–	–	–	**Carbamic acid, [2-(4-thiazolyl)-1H-benzimidazol-5-yl]-, isopropyl ester**	26097-80-3	**NI31669**
171	99	45	215	87	127	70	89	**302**	100	80	80	30	22	20	20	10	**0.05**	14	22	7	–	–	–	–	–	–	–	–	–	–	**trans-Cyclotetra[O]-1,4,7,10,13-pentaoxacycloheptadecane-14,17-dione**		**MS06988**
99	55	27	45	100	43	87	70	**302**	100	77	9	8	6	6	5	4	**0.00**	14	22	7	–	–	–	–	–	–	–	–	–	–	**2-Propenoic acid, oxybis(2,1-ethanediyloxy-2,1-ethanediyl) ester**	17831-71-9	**NI31670**
103	101	41	44	55	43	105	131	**302**	100	80	76	68	60	59	56	55	**0.00**	14	28	2	–	–	–	–	–	–	–	–	–	1	**Germacycloundecan-6-one, 1,1-diethyl-7-hydroxy-**	17973-65-8	**NI31671**
73	287	147	45	75	288	74	197	**302**	100	34	26	19	19	10	9	8	**2.77**	14	30	3	–	–	–	–	–	–	2	–	–	–	**2-Ethyl-3-oxohexanoic acid, bis(trimethylsilyl)-**		**NI31672**
73	147	55	287	185	148	75	149	**302**	100	99	22	17	17	16	15	9	**3.80**	14	30	3	–	–	–	–	–	–	2	–	–	–	**Octanoic acid, 2-oxo-, bis(trimethylsilyl)-**	56272-73-2	**NI31673**
77	75	287	44	73	47	57	43	**302**	100	55	40	35	35	23	21	20	**3.60**	14	30	3	–	–	–	–	–	–	2	–	–	–	**(E)-2-Propyl-3-oxopentanoic acid, bis(trimethylsilyl)-**		**NI31674**
73	77	287	147	75	183	47	45	**302**	100	99	72	67	64	31	28	28	**5.80**	14	30	3	–	–	–	–	–	–	2	–	–	–	**(Z)-2-Propyl-3-oxopentanoic acid, bis(trimethylsilyl)-**		**NI31675**
73	287	183	147	155	273	75	288	**302**	100	60	27	24	21	20	17	16	**7.18**	14	30	3	–	–	–	–	–	–	2	–	–	–	**2-Propyl-3-oxopentanoic acid, bis(trimethylsilyl)-**		**NI31676**
73	170	75	81	147	129	171	95	**302**	100	96	79	54	47	41	38	35	**0.71**	14	30	3	–	–	–	–	–	–	2	–	–	–	**4-Trimethylsiloxycyclohexylacetate, trimethylsilyl-**		**NI31677**
73	75	170	129	287	81	197	117	**302**	100	88	83	35	32	31	29	25	**2.45**	14	30	3	–	–	–	–	–	–	2	–	–	–	**4-Trimethylsiloxycyclohexylacetate, trimethylsilyl-**		**NI31678**
302	301	137	153	273	274	245	109	**302**	100	16	12	8	7	6	4	4		15	10	7	–	–	–	–	–	–	–	–	–	–	**3,3′,4′,5,7-Pentahydroxyflavone**		**MS06989**
302	303	301	137	69	153	128	286	**302**	100	34	34	28	27	20	17	12		15	10	7	–	–	–	–	–	–	–	–	–	–	**3,3′,4′,5,7-Pentahydroxyflavone**		**LI06625**
302	168	140	112	134	274			**302**	100	70	23	21	11	6				15	10	7	–	–	–	–	–	–	–	–	–	–	**5,7,8,3′,4′-Pentahydroxyflavone**		**LI06626**
147	120	302	121	304	303	102	301	**302**	100	78	61	59	57	54	53	47		15	11	2	–	–	–	–	1	–	–	–	–	–	**2-Propen-1-one, 3-(4-bromophenyl)-1-(2-hydroxyphenyl)-**	16635-18-0	**NI31679**
107	267	226	227	180	120	302	91	**302**	100	99	99	50	18	18	17	17		15	14	5	2	–	–	–	–	–	–	–	–	–	**Benzene, 1-[(2-isopropyl)phenoxy]-2,4-dinitro-**	32101-55-6	**NI31680**
44	42	302	304	147	224	193	226	**302**	100	50	42	41	32	30	29	25		15	15	–	2	–	–	–	1	–	–	–	–	–	**N-Methyl-3-(3-pyridyl)-3-(4-bromophenyl)prop-2-enamine**		**NI31681**
214	194	246	210	274	215	195	211	**302**	100	87	33	33	16	12	10	4	**0.00**	15	18	3	–	–	–	–	–	–	–	–	–	1	**Iron, (di-1-cyclohexen-1-yl)tricarbonyl-**	12216-15-8	**NI31682**
43	267	41	187	91	128	117	115	**302**	100	51	50	37	37	22	22	22	**1.32**	15	23	2	–	1	–	1	–	–	–	–	–	–	**Benzenesulphonyl chloride, 2,4,6-triisopropyl-**	6553-96-4	**NI31683**
71	43	41	201	27	72	28	143	**302**	100	36	9	6	6	5	4	3	**0.00**	15	26	6	–	–	–	–	–	–	–	–	–	–	**Butanoic acid, 1,2,3-propanetriyl ester**	60-01-5	**NI31684**
71	43	41	27	201	42	39	29	**302**	100	56	21	12	11	9	7	5		15	26	6	–	–	–	–	–	–	–	–	–	–	**Butanoic acid, 1,2,3-propanetriyl ester**	60-01-5	**NI31685**
71	43	201	143	215	72	131	144	**302**	100	84	78	37	32	28	25	23	**0.14**	15	26	6	–	–	–	–	–	–	–	–	–	–	**Butanoic acid, 1,2,3-propanetriyl ester**	60-01-5	**NI31686**
143	101	169	58	287	85	111	83	**302**	100	94	61	60	37	35	33	17	**0.00**	15	26	6	–	–	–	–	–	–	–	–	–	–	**D-Glucitol, 1,2:3,4:5,6-tris-O-isopropylidene-**	4239-88-7	**NI31687**
101	59	85	83	111	143	287	115	**302**	100	99	67	67	50	40	25	25	**0.00**	15	26	6	–	–	–	–	–	–	–	–	–	–	**D-Glucitol, 1,3:2,4:5,6-tris-O-isopropylidene-**	817389-92-7	**NI31688**
71	43	41	27	72	201	143	39	**302**	100	76	12	8	5	5	3	3	**0.00**	15	26	6	–	–	–	–	–	–	–	–	–	–	**Tri-isobutyrin**		**NI31689**
101	43	169	111	85	288	57	287	**302**	100	93	84	65	58	43	40	33	**0.00**	15	26	6	–	–	–	–	–	–	–	–	–	–	**D-Mannitol, 1,2:3,4:5,6-tris-O-isopropylidene-**	3969-59-3	**LI06627**
287	143	102	101	43	169	111	85	**302**	100	91	83	59	27	25	19	18	**0.00**	15	26	6	–	–	–	–	–	–	–	–	–	–	**D-Mannitol, 1,2:3,4:5,6-tris-O-isopropylidene-**	3969-59-3	**NI31690**
143	59	101	85	111	287	169	58	**302**	100	79	77	38	26	20	20	15	**0.00**	15	26	6	–	–	–	–	–	–	–	–	–	–	**D-Mannitol, 1,2:3,4:5,6-tris-O-isopropylidene-**	3969-59-3	**NI31691**
59	157	101	115	143	85	111	72	**302**	100	61	45	30	25	21	20	15	**0.00**	15	26	6	–	–	–	–	–	–	–	–	–	–	**D-Mannitol, 1,2:3,6:4,5-tris-O-isopropylidene-**	91667-55-9	**NI31692**
43	173	113	131	71	128	122	81	**302**	100	40	30	29	27	21	21	18	**0.00**	15	26	6	–	–	–	–	–	–	–	–	–	–	**Nonane-1,6,8-triacetate**		**NI31693**
73	99	182	153	105	147	114	302	**302**	100	34	27	20	20	19	19	11		15	30	–	–	2	–	–	–	–	1	–	–	–	**1-[3-Methylthio-3-(trimethylsilyl)propylthio]-1-cyclohexylethene**	80403-07-2	**NI31694**
130	116	160	55	158	84	86	131	**302**	100	36	32	31	27	24	20	19	**1.97**	15	30	4	2	–	–	–	–	–	–	–	–	–	**L-Valine, N,N′-trimethylenedi-, dimethyl ester**	23179-85-3	**NI31695**
302	159	303	237	115	143	131	144	**302**	100	62	20	20	20	13	11	8		16	11	3	–	1	1	–	–	–	–	–	–	–	**4-(4-Fluorophenylsulphonyl)-1-naphthol**		**IC03791**
274	301	302	273	275	69	235	77	**302**	100	65	63	56	14	13	11	9		16	12	2	2	–	2	–	–	–	–	–	–	–	**7-Difluoromethoxy-2,3-dihydro-5-phenyl-1H-1,4-benzodiazepin-2-one**	55384-53-7	**NI31696**
192	302	218	193	191	97	274	194	**302**	100	61	19	13	13	11	10	10		16	14	–	–	3	–	–	–	–	–	–	–	–	**4,7-Di-2-thienyl-4,5,6,7-tetrahydrobenzo[b]thiophene**		**LI06628**
165	121	166	65	93	164	39	122	**302**	100	56	11	11	10	7	7	5	**1.80**	16	14	6	–	–	–	–	–	–	–	–	–	–	**Benzoic acid, 4-hydroxy-, 2-(4-carboxyphenoxy)-, ethyl ester**		**IC03792**
191	218	302	54	77	60	69	190	**302**	100	91	67	63	50	45	40	37		16	14	6	–	–	–	–	–	–	–	–	–	–	**3(9bH)-Dibenzofuranone, 2-acetyl-1,7,9-trihydroxy-8,9b-dimethyl-**	55682-76-3	**NI31697**
260	232	217	302	261	55	69	77	**302**	100	85	82	80	75	58	54	52		16	14	6	–	–	–	–	–	–	–	–	–	–	**3(9bH)-Dibenzofuranone, 6-acetyl-1,7,9-trihydroxy-8,9b-dimethyl-**	51827-48-6	**NI31698**
211	243	15	127	155	59	28	18	**302**	100	90	85	66	64	55	32	29	**5.50**	16	14	6	–	–	–	–	–	–	–	–	–	–	**Dimethyl 3-oxo-1,3,4,5-tetrahydro-naphtho[1,2-c]furan-4,5-dicarboxylate**		**MS06990**
137	150	302	153	135	179	301	124	**302**	100	89	75	65	60	28	26	21		16	14	6	–	–	–	–	–	–	–	–	–	–	**Flavanone, 3,5,7-trihydroxy-4′-methoxy-**	3570-69-2	**NI31699**
302	31	256	115	228	227	29	303	**302**	100	36	25	22	20	20	19	16		16	14	6	–	–	–	–	–	–	–	–	–	–	**2H-Furo[3′,2′:4,5]furo[2,3-h]-1-benzopyran-2-one, 7a,10a-dihydro-4-(2-hydroxyethyl)-5-methoxy-**	23315-33-5	**NI31700**
65	121	122	39	93	152	76	150	**302**	100	98	95	74	72	66	58	45	**0.40**	16	14	6	–	–	–	–	–	–	–	–	–	–	**1,4-Bis(2-hydroxyphenyl)buta-2,3-diol-1,4-dione**		**IC03793**
302	303	259	216	273	244	137	287	**302**	100	17	16	9	6	6	6	5		16	14	6	–	–	–	–	–	–	–	–	–	–	**1-Hydroxy-3,5,6-trimethoxyxanthone**		**MS06991**
302	287	284	273	258	269	259	256	**302**	100	40	40	31	22	9	9	7		16	14	6	–	–	–	–	–	–	–	–	–	–	**1-Hydroxy-3,5,8-trimethoxyxanthone**		**NI31701**
302	44	301	273	229	64	256	258	**302**	100	99	58	42	35	30	29	21		16	14	6	–	–	–	–	–	–	–	–	–	–	**3-Hydroxy-1,5,6-trimethoxyxanthone**		**LI06629**
302	89	175	145	69	117	228	259	**302**	100	76	73	59	47	41	31	30		16	14	6	–	–	–	–	–	–	–	–	–	–	**2H-Pyran-2-one, 6-[2-(1,3-benzodioxol-5-yl)ethenyl]-4,5-dimethoxy-, (E)-**	52525-96-9	**NI31702**
176	43	218	175	177	28	149	105	**302**	100	47	29	15	13	13	12	12	**5.00**	16	14	6	–	–	–	–	–	–	–	–	–	–	**1,2,4-Triacetoxynaphthalene**		**MS06992**
269	302	136	63	165	270	71	133	**302**	100	60	45	35	33	21	16	16		16	15	–	–	2	–	–	–	–	–	1	–	–	**Bis(4-vinylphenyl)dithiophosphinic acid**	85417-01-2	**NI31703**
285	284	165	79	93	67	81	175	**302**	100	67	67	61	38	32	31	30	**19.00**	16	18	4	2	–	–	–	–	–	–	–	–	–	**1-(1-Adamantyl)-2,4-dinitrobenzene**		**NI31704**
302	94	93	245	246	79	135	67	**302**	100	61	25	24	22	18	16	10		16	18	4	2	–	–	–	–	–	–	–	–	–	**1-(1-Adamantyl)-3,4-dinitrobenzene**		**NI31705**
302	94	93	246	79	245	135	81	**302**	100	75	38	32	30	29	26	19		16	18	4	2	–	–	–	–	–	–	–	–	–	**1-(1-Adamantyl)-3,5-dinitrobenzene**		**NI31706**
224	192	225	164	149	211	160	271	**302**	100	14	13	5	4	4	3	3	**1.60**	16	18	4	2	–	–	–	–	–	–	–	–	–	**Dimethyl 1,4-dihydro-2,6-dimethyl-4-(2-pyridyl)pyridine-3,5-dicarboxylate**		**NI31707**

MASS TO CHARGE RATIOS								M.W.	INTENSITIES								Parent	C	H	O	N	S	F	Cl	Br	I	Si	P	B	X	COMPOUND NAME	CAS Reg No	No
259	246	229	91	260	273	77	302	**302**	100	46	39	34	18	16	16	6		16	19	1	–	–	–	–	–	–	–	–	–	1	Butyldiphenylarsane oxide		DD01096
80	56	54	79	53	55	41	259	**302**	100	86	80	79	65	63	63	57	**0.00**	16	22	2	4	–	–	–	–	–	–	–	–	–	Methyl 2-[5-(dimethylamino)-4-isopropyl-4-methyl-4H-imidazol-2-yl]pyridine-3-carboxylate		DD01097
259	56	260	228	42	103	41	227	**302**	100	87	60	45	45	43	42	27	**0.00**	16	22	2	4	–	–	–	–	–	–	–	–	–	Methyl 3-[5-(dimethylamino)-4-isopropyl-4-methyl-4H-imidazol-2-yl]pyridine-2-carboxylate		DD01098
73	302	199	303	184	45	287	200	**302**	100	54	51	16	13	12	11	11		16	26	–	2	–	–	–	–	–	2	–	–	–	1,8-Naphthalenediamine, N,N′-bis(trimethylsilyl)-	33285-90-4	NI31708
302	199	73	303	184	45	287	200	**302**	100	95	47	29	25	23	20	19		16	26	–	2	–	–	–	–	–	2	–	–	–	1,8-Naphthalenediamine, N,N′-bis(trimethylsilyl)-		LI06630
189	133	245	55	29	89	91	103	**302**	100	61	47	13	11	10	9	7	**0.00**	16	36	–	–	–	–	–	–	–	–	–	–	1	Tetrabutylgermane		MS06993
302	259	241	217	199	284	245	213	**302**	100	92	50	25	22	18	10	9		17	18	5	–	–	–	–	–	–	–	–	–	–	7-Acetyl-2-(diacetylmethyl)-6-hydroxy-3,5-dimethylbenzofuran		NI31709
302	259	241	269	260	218	284	242	**302**	100	58	30	22	21	14	12	11		17	18	5	–	–	–	–	–	–	–	–	–	–	7-Acetyl-2-(diacetylmethyl)-6-hydroxy-3,5-dimethylbenzofuran		MS06994
302	260	218	259	241	284	242	176	**302**	100	54	19	15	13	12	10	7		17	18	5	–	–	–	–	–	–	–	–	–	–	7-Acetyl-3-(diacetylmethyl)-6-hydroxy-2,5-dimethylbenzofuran		MS06995
190	189	131	242	175	77	103	161	**302**	100	33	29	26	17	17	13	12	**2.13**	17	18	5	–	–	–	–	–	–	–	–	–	–	Auraptenol acetate	68148-01-6	NI31710
165	302	303	271					**302**	100	91	24	24						17	18	5	–	–	–	–	–	–	–	–	–	–	Benzophenone, 3,3′,4,4′-tetramethoxy-	4131-03-7	NI31711
180	302	167	168	135	123	133	107	**302**	100	56	33	28	14	13	12	12		17	18	5	–	–	–	–	–	–	–	–	–	–	2H-1-Benzopyran-7-ol, 3,4-dihydro-3-(2-hydroxy-3,4-dimethoxyphenyl)-	73353-84-1	NI31712
180	168	302	167	133	165	135	107	**302**	100	59	46	42	42	40	33	24		17	18	5	–	–	–	–	–	–	–	–	–	–	2H-1-Benzopyran-7-ol, 3,4-dihydro-3-(3-hydroxy-2,4-dimethoxyphenyl)-	27213-18-9	NI31713
150	121	151	153	135	256	302	137	**302**	100	26	18	17	12	10	7	5		17	18	5	–	–	–	–	–	–	–	–	–	–	4′,7-Dimethoxy-2,3-cis-flavan-3,4-cis-diol		LI06631
150	121	151	153	256	135	302	137	**302**	100	26	18	14	13	11	7	6		17	18	5	–	–	–	–	–	–	–	–	–	–	4′,7-Dimethoxy-2,3-trans-flavan-3,4-cis-diol		LI06632
150	121	151	153	256	135	302	137	**302**	100	37	20	19	13	10	9	5		17	18	5	–	–	–	–	–	–	–	–	–	–	4′,7-Dimethoxy-2,3-trans-flavan-3,4-trans-diol		LI06633
185	73	243	141	75	302	115	45	**302**	100	82	51	27	23	22	17	14		17	22	3	–	–	–	–	–	–	1	–	–	–	2-Naphthaleneacetic acid, 6-methoxy-α-methyl-, trimethylsilyl ester, (+)	74793-83-2	NI31714
185	73	243	302	287	75	141	244	**302**	100	69	66	47	21	16	15	15		17	22	3	–	–	–	–	–	–	1	–	–	–	2-Naphthaleneacetic acid, 6-methoxy-α-methyl-, trimethylsilyl ester, (+)	74793-83-2	NI31715
134	211	98	228	271	229	242	74	**302**	100	96	79	75	66	49	37	34	**2.00**	17	34	4	–	–	–	–	–	–	–	–	–	–	Tetradecanoic acid, 2,3-dihydroxypropyl ester	589-68-4	WI01390
43	211	57	41	98	55	74	84	**302**	100	82	80	76	75	70	66	51	**0.00**	17	34	4	–	–	–	–	–	–	–	–	–	–	Tetradecanoic acid, 2,3-dihydroxypropyl ester	589-68-4	WI01391
211	134	98	271	228	229	74	84	**302**	100	87	64	52	51	44	44	28	**2.00**	17	34	4	–	–	–	–	–	–	–	–	–	–	Tetradecanoic acid, 2-hydroxy-1-(hydroxymethyl)ethyl ester	3443-83-2	WI01393
43	211	57	98	41	55	74	84	**302**	100	96	87	77	75	73	71	52	**1.00**	17	34	4	–	–	–	–	–	–	–	–	–	–	Tetradecanoic acid, 2-hydroxy-1-(hydroxymethyl)ethyl ester	3443-83-2	WI01392
77	76	51	103	50	130	302	102	**302**	100	74	50	36	31	25	23	18		18	14	1	4	–	–	–	–	–	–	–	–	–	3-(3-Methyl-1-phenylpyrazol-5-yl)-quinazolin-4(3H)-one		MS06996
223	302	165	119	221	287	224	208	**302**	100	77	65	65	63	62	56	55		18	22	2	–	1	–	–	–	–	–	–	–	–	Diphenylsulphone, 2,2′,4,4′,6,6′-hexamethyl-	3112-79-6	NI31716
123	302	124	43	45	303	137	77	**302**	100	29	27	17	7	6	6	6		18	22	4	–	–	–	–	–	–	–	–	–	–	1,2-Benzenediol, 4,4′-(2,3-dimethyl-1,4-butanediyl)bis-	500-38-9	NI31717
104	117	183	242	91	115	143	78	**302**	100	93	61	53	52	32	31	31	**4.00**	18	22	4	–	–	–	–	–	–	–	–	–	–	cis-3,6-Diacetoxy-trans-[8]paracycloph-4-ene		NI31718
104	117	242	183	91	182	243	270	**302**	100	99	95	86	51	36	35	33	**7.00**	18	22	4	–	–	–	–	–	–	–	–	–	–	trans-3,6-Diacetoxy-trans-[8]paracycloph-4-ene		NI31719
227	133	228	151	107				**302**	100	20	19	19	18				**0.01**	18	22	4	–	–	–	–	–	–	–	–	–	–	2,3-Dihydroxy-2,3-di(4′-hydroxyphenyl)hexane		LI06634
135	105	151	77	43	133	121	152	**302**	100	97	82	76	69	35	35	28	**0.01**	18	22	4	–	–	–	–	–	–	–	–	–	–	2,3-Dihydroxy-2,3-di(2′-methoxyphenyl)butane		LI06635
75	270	227	255	119	271	135	113	**302**	100	60	60	25	17	14	12	12	**0.00**	18	22	4	–	–	–	–	–	–	–	–	–	–	1,1-Dimethoxy-2,2-bis(4-methoxyphenyl)ethane		MS06997
151	75	135	270	227	152	136	121	**302**	100	20	17	12	12	12	8	5	**0.00**	18	22	4	–	–	–	–	–	–	–	–	–	–	DL-1,2-Dimethoxy-1,2-bis(4-methoxyphenyl)ethane		MS06998
302	55	41	67	79	91	81	256	**302**	100	88	54	44	43	37	37	36		18	22	4	–	–	–	–	–	–	–	–	–	–	1β,2β-Epoxy-10β-hydroxy-4-estrene-3, 17-dione	109613-05-0	MS06999
302	55	41	67	91	79	71	110	**302**	100	67	67	50	49	49	49	41		18	22	4	–	–	–	–	–	–	–	–	–	–	4β,5β-Epoxy-10β-hydroxy-1-oestrene-3,17-dione		MS07000
151	121	57	152	93	65	122	150	**302**	100	60	20	14	11	7	6	5	**0.10**	18	22	4	–	–	–	–	–	–	–	–	–	–	3,4-Hexanediol, 3,4-bis(4-hydroxyphenyl)-	7507-01-9	NI31720
43	123	55	119	124	147	41	138	**302**	100	78	68	62	39	38	36	32	**20.00**	18	22	4	–	–	–	–	–	–	–	–	–	–	10β-Hydroxy-17a-oxa-D-homo-1,4-estradiene- 3,17-dione	109613-04-9	MS07001
302	177	137	150	124	256	215	202	**302**	100	64	33	30	27	14	14	14		18	22	4	–	–	–	–	–	–	–	–	–	–	(±)-17β-Hydroxy-3-methoxy-6-oxa-8α-estra-1,3,5(10)-trien-7-one		MS07002
151	135	75	152	136	227	270	121	**302**	100	22	13	12	10	9	8	7	**0.00**	18	22	4	–	–	–	–	–	–	–	–	–	–	meso-1,2-Dimethoxy-1,2-bis(4-methoxyphenyl)ethane		MS07003
151	78	152	302	86	91	123	303	**302**	100	96	80	60	17	16	13	12		18	22	4	–	–	–	–	–	–	–	–	–	–	3,3′,5,5′-Tetramethoxydibenzyl		LI06636
193	32	137	194	43	41	82	54	**302**	100	53	26	15	14	12	10	9	**0.07**	18	26	–	2	1	–	–	–	–	–	–	–	–	2-Diethylamino-4-phenylthiooct-2-ene nitrile		NI31721
142	174	259	57	41	29	42	30	**302**	100	69	58	39	36	31	22	22	**1.25**	18	26	2	2	–	–	–	–	–	–	–	–	–	N-[2-(N,N-Di-isobutylamino)ethyl]phthalimide		IC03794
302	160	228	186	254	173	159	171	**302**	100	52	28	27	26	24	19	15		19	23	2	–	–	1	–	–	–	–	–	–	–	Estra-1,3,5(10)-trien-17-one, 16-fluoro-3-methoxy-, (16β)-	2383-29-1	NI31722
302	160	228	186	254	173	159	171	**302**	100	26	14	14	13	12	9	8		19	23	2	–	–	1	–	–	–	–	–	–	–	Estra-1,3,5(10)-trien-17-one, 16-fluoro-3-methoxy-, (16β)-	2383-29-1	NI31723
157	175	302	217					**302**	100	40	25	14						19	26	3	–	–	–	–	–	–	–	–	–	–	cis-3-Acetoxyheptadeca-1,9-dione-4,6-diyn-8-ol		LI06637
94	119	135	79	41	81	122	43	**302**	100	31	24	24	24	20	18	17	**9.00**	19	26	3	–	–	–	–	–	–	–	–	–	–	α-DL-trans-Allethrin		MS07004
302	273	55	41	67	79	93	303	**302**	100	50	50	42	30	27	22	20		19	26	3	–	–	–	–	–	–	–	–	–	–	Androstane-3,6,17-trione, (5α)-	2243-05-2	NI31724
302	55	273	41	67	79	93	108	**302**	100	52	51	43	30	28	23	20		19	26	3	–	–	–	–	–	–	–	–	–	–	Androstane-3,6,17-trione, (5α)-	2243-05-2	LI06638
302	273	123	137	108	207	93	109	**302**	100	90	87	66	58	54	44	43		19	26	3	–	–	–	–	–	–	–	–	–	–	Androstane-3,6,17-trione, (5α)-	2243-05-2	MS07005
302	123	55	41	67	79	206	93	**302**	100	56	38	34	28	25	24	23		19	26	3	–	–	–	–	–	–	–	–	–	–	Androstane-3,6,17-trione, (5β)-	26991-58-2	NI31725
302	109	122	191	41	287	124	121	**302**	100	50	49	40	40	39	35	28		19	26	3	–	–	–	–	–	–	–	–	–	–	Androstane-3,6,17-trione, (5β)-	26991-58-2	NI31726
255	302	256	41	55	79	81	67	**302**	100	90	38	30	28	17	13	13		19	26	3	–	–	–	–	–	–	–	–	–	–	Androstane-3,7,17-trione, (5α)-	4147-15-3	LI06639
255	137	256	302	124	91	131	133	**302**	100	46	39	26	21	16	15	14		19	26	3	–	–	–	–	–	–	–	–	–	–	Androstane-3,7,17-trione, (5α)-	4147-15-3	MS07006
255	302	256	41	55	303	79	43	**302**	100	90	37	29	27	19	17	13		19	26	3	–	–	–	–	–	–	–	–	–	–	Androstane-3,7,17-trione, (5α)-	4147-15-3	NI31727
302	124	41	55	43	109	79	67	**302**	100	70	40	34	33	13	13	13		19	26	3	–	–	–	–	–	–	–	–	–	–	Androstane-3,11,17-trione, (5α)-	1482-70-8	LI06640
302	124	41	55	43	109	79	67	**302**	100	69	39	34	32	22	22	22		19	26	3	–	–	–	–	–	–	–	–	–	–	Androstane-3,11,17-trione, (5α)-	1482-70-8	NI31728

MASS TO CHARGE RATIOS								M.W.	INTENSITIES								Parent	C	H	O	N	S	F	Cl	Br	I	Si	P	B	X	COMPOUND NAME	CAS Reg No	No
302	109	122	287	191	41	124	165	**302**	100	51	50	40	40	40	35	27		19	26	3	–	–	–	–	–	–	–	–	–	–	Androstane-3,11,17-trione, (5β)-		LI06641
302	110	55	41	67	123	79	81	**302**	100	84	65	63	49	37	36	32		19	26	3	–	–	–	–	–	–	–	–	–	–	Androstane-3,12,17-trione, (5α)-	4171-01-1	NI31729
302	110	55	41	97	67	123	81	**302**	100	56	46	42	34	30	24	24		19	26	3	–	–	–	–	–	–	–	–	–	–	Androstane-3,12,17-trione, (5β)-	53604-37-8	NI31730
302	163	123	41	55	124	91	189	**302**	100	81	48	38	33	32	27	26		19	26	3	–	–	–	–	–	–	–	–	–	–	Androst-4-ene-3,17-dione, 11-hydroxy-, (11β)-	382-44-5	NI31731
302	177	124	123	109	97	284	260	**302**	100	84	70	32	31	26	25	25		19	26	3	–	–	–	–	–	–	–	–	–	–	Androst-4-ene-3,17-dione, 12-hydroxy-, (12β)-	53604-49-2	NI31732
302	260	109	124	79	91	55	107	**302**	100	88	49	45	43	40	32	31		19	26	3	–	–	–	–	–	–	–	–	–	–	Androst-4-ene-3,17-dione, 15-hydroxy-, (15α)-	566-08-5	NI31733
302	91	107	124	105	229	230	215	**302**	100	42	35	32	32	26	22	20		19	26	3	–	–	–	–	–	–	–	–	–	–	Androst-4-ene-3,17-dione, 18-hydroxy-	7121-60-0	MS07007
272	302	273	81	79	303	253	77	**302**	100	50	19	8	8	6	6	6		19	26	3	–	–	–	–	–	–	–	–	–	–	Androst-4-ene-3,17-dione, 19-hydroxy-	510-64-5	NI31734
123	41	107	136	79	43	81	91	**302**	100	37	31	28	25	22	18	17	**7.70**	19	26	3	–	–	–	–	–	–	–	–	–	–	Cyclopropanecarboxylic acid, 2,2-dimethyl-3-(2-methyl-1-propenyl)-, 2-methyl-4-oxo-3-(2-propenyl)-2-cyclopenten-1-yl ester	584-79-2	NI31735
123	136	107	168	153	151	302		**302**	100	32	32	14	10	7	4			19	26	3	–	–	–	–	–	–	–	–	–	–	Cyclopropanecarboxylic acid, 2,2-dimethyl-3-(2-methyl-1-propenyl)-, 2-methyl-4-oxo-3-(2-propenyl)-2-cyclopenten-1-yl ester	584-79-2	LI06642
123	91	40	43	41	79	81	107	**302**	100	79	73	55	48	42	36	34	**0.80**	19	26	3	–	–	–	–	–	–	–	–	–	–	Cyclopropanecarboxylic acid, 2,2-dimethyl-3-(2-methyl-1-propenyl)-, 2-methyl-4-oxo-3-(2-propenyl)-2-cyclopenten-1-yl ester	584-79-2	NI31736
123	41	79	43	107	136	81	91	**302**	100	55	49	43	42	41	39	34	**1.88**	19	26	3	–	–	–	–	–	–	–	–	–	–	Cyclopropanecarboxylic acid, 2,2-dimethyl-3-(2-methyl-1-propenyl)-, 2-methyl-4-oxo-3-(2-propenyl)-2-cyclopenten-1-yl ester	28434-00-6	NI31737
302	303	202	176	137	175	215	189	**302**	100	30	14	12	12	9	8	8		19	26	3	–	–	–	–	–	–	–	–	–	–	Estra-1,3,5(10)-triene-3,17-diol, 2-methoxy-, (17β)-	362-07-2	NI31738
302	303	137	189	176	202	175	190	**302**	100	22	22	11	11	9	9	8		19	26	3	–	–	–	–	–	–	–	–	–	–	Estra-1,3,5(10)-triene-3,17-diol, 2-methoxy-, (17β)-	362-07-2	NI31739
270	158	157	271	211	145	170	159	**302**	100	40	24	21	15	15	14	14	**5.28**	19	26	3	–	–	–	–	–	–	–	–	–	–	Estra-1,3,5(10)-triene-3,17-diol, 6-methoxy-, (6α,17β)-	74312-88-2	NI31740
302	303	243	190	202	176	300	189	**302**	100	22	13	10	9	9	6	6		19	26	3	–	–	–	–	–	–	–	–	–	–	Estra-1,3,5(10)-triene-4,17-diol, 3-methoxy-, (17β)-	5976-66-9	NI31741
284	172	171	159	285	184	158	225	**302**	100	39	33	25	22	22	22	17	**1.24**	19	26	3	–	–	–	–	–	–	–	–	–	–	Estra-1,3,5(10)-triene-6,17-diol, 3-methoxy-, (6α,17β)-	7291-47-6	NI31742
302	160	186	200	189	159	173	161	**302**	100	66	57	33	29	23	21	21		19	26	3	–	–	–	–	–	–	–	–	–	–	Estra-1,3,5(10)-triene-11,17-diol, 3-methoxy-, (11α,17β)-	10516-35-5	NI31743
302	173	199	174	303	160	159	161	**302**	100	32	28	21	20	20	20	16		19	26	3	–	–	–	–	–	–	–	–	–	–	Estra-1,3,5(10)-triene-15,17-diol, 3-methoxy-, (15α,17β)-	28715-36-8	NI31744
302	160	173	303	215	186	174	216	**302**	100	29	23	21	20	20	20	18		19	26	3	–	–	–	–	–	–	–	–	–	–	Estra-1,3,5(10)-triene-16,17-diol, 3-methoxy-, (16α,17β)-	1474-53-9	NI31745
302	227	303	228	171	173	161	44	**302**	100	96	18	18	15	11	10	10		19	26	3	–	–	–	–	–	–	–	–	–	–	Methyl 12-ketomethylpodocarpate		NI31746
145	149	133	91	187	159	201	129	**302**	100	98	79	69	69	55	52	47	**22.80**	19	26	3	–	–	–	–	–	–	–	–	–	–	Methyl 8-[3-oxo-2-(pent-2-en-1-yl)cyclopent-1-enyl]oct-6-ynoate		NI31747
91	159	131	117	116	171	105	132	**302**	100	58	48	44	34	26	26	25	**0.50**	19	26	3	–	–	–	–	–	–	–	–	–	–	17-Octadecene-9,11-diynoic acid, 8-oxo-, methyl ester	75112-83-3	NI31748
159	289	205	160	206	105	204	158	**302**	100	23	22	20	19	18	16	14	**2.40**	19	26	3	–	–	–	–	–	–	–	–	–	–	1-Phenanthrenecarboxylic acid, 7-vinyl-1,2,3,4,4a,4b,5,6,7,8,10,10a-dodecahydro-4a,7-dimethyl-3-oxo-, [1R-(1α,4aβ,4bα,7α,10aα)]-	57362-20-6	NI31749
259	150	109	243	91	55	93	81	**302**	100	99	99	95	90	74	73	71	**44.00**	19	26	3	–	–	–	–	–	–	–	–	–	–	Zoapatline ketone		LI06643
121	137	95	108	136	273	163	302	**302**	100	98	60	47	43	42	34	31		19	30	1	–	–	–	–	–	–	1	–	–	–	1,7,7-Trimethylbicyclo[2.2.1]heptane, [(2-ethylmethylphenylsilyl)oxy]-	74764-31-1	NI31750
302	301	152	151	285	139	153	202	**302**	100	83	72	15	13	12	8	7		20	14	3	–	–	–	–	–	–	–	–	–	–	6-Methoxy-2-(1-naphthylmethylene)-3(2H)-benzofuranone	77764-97-7	NI31751
302	301	152	139	286	259	202	150	**302**	100	95	21	8	7	6	6	6		20	14	3	–	–	–	–	–	–	–	–	–	–	6-Methoxy-2-(2-naphthylmethylene)-3(2H)-benzofuranone	77764-98-8	NI31752
191	95	189	302	190	165	179	178	**302**	100	59	27	26	17	12	10	6		20	14	3	–	–	–	–	–	–	–	–	–	–	9-Phenanthrenemethyl furan-2-carboxylate	92174-41-9	NI31753
91	93	77	79	65	78	75	92	**302**	100	99	82	70	63	55	50	45	**0.00**	20	18	1	2	–	–	–	–	–	–	–	–	–	Acetamide, N-phenyl-2-[1-benzyl-2(1H)-pyridinylidene]-	54934-72-4	NI31754
77	104	93	209	51	105	66	65	**302**	100	99	96	58	54	50	42	29	**0.00**	20	18	1	2	–	–	–	–	–	–	–	–	–	2,2-Dianilinoacetophenone	66749-88-0	NI31755
180	77	51	181	105	78	53	50	**302**	100	50	17	16	4	4	4	3	**0.53**	20	18	1	2	–	–	–	–	–	–	–	–	–	O-3-(2,6-Dimethylpyridyl)-N-phenylbenzimidate		NI31756
302	287	303	69	217	123	119	145	**302**	100	32	21	18	18	17	17	16		20	27	1	–	–	1	–	–	–	–	–	–	–	Retinoyl fluoride (all-trans)	83802-77-1	NI31757
287	302	257	41	55	288	109	67	**302**	100	86	60	37	35	33	31	30		20	30	2	–	–	–	–	–	–	–	–	–	–	Androst-16-ene-17-carboxylic acid, (5α)-	54411-93-7	NI31758
123	136	302	124	137	41	135	79	**302**	100	99	91	58	53	47	40	38		20	30	2	–	–	–	–	–	–	–	–	–	–	Androst-1-en-3-one, 17-hydroxy-1-methyl-, (5α,17β)-	153-00-4	LI06644
41	57	55	43	44	53	123	67	**302**	100	99	67	62	58	45	33	33	**2.66**	20	30	2	–	–	–	–	–	–	–	–	–	–	Androst-1-en-3-one, 17-hydroxy-1-methyl-, (5α,17β)-	153-00-4	NI31759
302	229	245	124	303	161	284	202	**302**	100	31	27	27	22	17	16	16		20	30	2	–	–	–	–	–	–	–	–	–	–	Androst-4-en-3-one, 17-hydroxy-17-methyl-, (17β)-	58-18-4	NI31761
302	43	124	41	55	303	71	79	**302**	100	76	58	38	37	34	31	29		20	30	2	–	–	–	–	–	–	–	–	–	–	Androst-4-en-3-one, 17-hydroxy-17-methyl-, (17β)-	58-18-4	NI31760
136	302	246	41	28	231	81	108	**302**	100	69	60	59	52	51	51	43		20	30	2	–	–	–	–	–	–	–	–	–	–	Ent-15,16-beyeranedione		MS07008
95	123	81	41	302	94	55	96	**302**	100	71	71	71	50	50	50	43		20	30	2	–	–	–	–	–	–	–	–	–	–	Cascarillone	59742-40-4	NI31762
151	97	55	109	81	67	95	41	**302**	100	77	47	36	25	23	21	20	**20.54**	20	30	2	–	–	–	–	–	–	–	–	–	–	α-Dicarvelone	18457-34-6	MS07009
98	96	84	110	191	136	176	150	**302**	100	91	56	41	33	29	23	20	**9.00**	20	30	2	–	–	–	–	–	–	–	–	–	–	(12R)-15,16-Epoxy-12-hydroxy-ent-cleroda-4(18),13(16),14-triene		NI31763
191	136	96	93	122	110	108	150	**302**	100	27	27	22	21	20	20	18	**5.00**	20	30	2	–	–	–	–	–	–	–	–	–	–	16-Hydroxy-ent-cleroda-5(10),13-dien-15-oic acid γ-lactone		NI31764
91	79	93	105	81	133	107	131	**302**	100	85	78	75	64	60	59	57	**39.83**	20	30	2	–	–	–	–	–	–	–	–	–	–	Kaur-16-en-18-oic acid, (4β)-	20316-84-1	NI31765
302	121	133	105	162	148	149	120	**302**	100	97	80	74	70	67	65	64		20	30	2	–	–	–	–	–	–	–	–	–	–	Linifoliol		MS07010
87	43	91	41	105	79	77	145	**302**	100	63	57	37	35	33	33	27	**7.00**	20	30	2	–	–	–	–	–	–	–	–	–	–	2-Methyl-2-[(1-E,3Z-E,5E)-4-methyl-6-(2,6,6-trimethyl-1-cyclohexenyl)-1,3,5-hexatrienyl]-1,3-dioxolane	80283-34-7	NI31766
302	231	215	110	85	303	255	273	**302**	100	75	37	33	33	25	25	22		20	30	2	–	–	–	–	–	–	–	–	–	–	19-Norpregn-4-en-3-one, 17-hydroxy-, (17α)-	52-78-8	NI31767
91	105	302	77	79	121	93	67	**302**	100	78	62	56	55	49	49	45		20	30	2	–	–	–	–	–	–	–	–	–	–	1-Phenanthrenecarboxylic acid, 1,2,3,4,4a,4b,5,6,10,10a-decahydro-1,4a-dimethyl-7-isopropyl-, [1R-(1α,4aβ,4bα,10aα)]-	514-10-3	NI31768

MASS TO CHARGE RATIOS								M.W.	INTENSITIES								Parent	C	H	O	N	S	F	Cl	Br	I	Si	P	B	X	COMPOUND NAME	CAS Reg No	No
135	136	302	148	41	91	55	93	**302**	100	27	26	20	17	16	15	13		20	30	2	–	–	–	–	–	–	–	–	–	–	**1-Phenanthrenecarboxylic acid, 1,2,3,4,4a,4b,5,6,7,9,10,10a-dodecahydro-1,4a-dimethyl-7-isopropylidene-, [1R-(1α,4aβ,4bα,10aα)]-**	471-77-2	**NI31769**
287	302	121	167	257	139	105	93	**302**	100	92	68	33	31	25	24	24		20	30	2	–	–	–	–	–	–	–	–	–	–	**1-Phenanthrenecarboxylic acid, 7-vinyl-1,2,3,4,4a,4b,5,6,7,9,10,10a-dodecahydro-1,4a,7-trimethyl-, [1R-(1α,4aβ,4bα,7β,10aα)]-**	127-27-5	**NI31770**
269	302	199	227	201	270	43	287	**302**	100	76	27	26	21	20	20	18		20	30	2	–	–	–	–	–	–	–	–	–	–	**2,6-Phenanthrenediol, 1,2,3,4,4a,9,10,10a-octahydro-1,1,4a-trimethyl-7-isopropyl-, [2S-(2α,4aα,10aβ)]-**	564-73-8	**NI31771**
78	175	287	43	302	201	271	41	**302**	100	81	77	31	30	27	24	24		20	30	2	–	–	–	–	–	–	–	–	–	–	**2,7-Phenanthrenediol, 1,2,3,4,4a,9,10,10a-octahydro-1,1,4a-trimethyl-8-isopropyl-, [2S-(2α,4aα,10aβ)]-**	3772-56-3	**NI31772**
287	302	269	175	288	201	303	227	**302**	100	87	65	38	21	20	19	17		20	30	2	–	–	–	–	–	–	–	–	–	–	**2,7-Phenanthrenediol, 1,2,3,4,4a,9,10,10a-octahydro-1,1,4a-trimethyl-8-isopropyl-, [2S-(2α,4aα,10aβ)]-**	3772-56-3	**NI31773**
269	270	287	302	175	43	159	201	**302**	100	27	20	19	17	16	14	11		20	30	2	–	–	–	–	–	–	–	–	–	–	**2,7-Phenanthrenediol, 1,2,3,4,4a,9,10,10a-octahydro-1,1,4a-trimethyl-8-isopropyl-, [2S-(2α,4aα,10aβ)]-**	3772-56-3	**LI06645**
284	269	227	199	157	285	185	213	**302**	100	80	64	49	28	25	25	23	**12.10**	20	30	2	–	–	–	–	–	–	–	–	–	–	**2,10-Phenanthrenediol, 4b,5,6,7,8,8a,9,10-octahydro-4b,8,8-trimethyl-1-isopropyl-, [4bS-(4bα,8aβ,10β)]-**	6811-52-5	**NI31774**
302	108	271	109	303	131	81	91	**302**	100	78	43	21	19	16	15	11		20	30	2	–	–	–	–	–	–	–	–	–	–	**Phenanthro[3,2-b]furan-4-methanol, 1,2,3,4,4a,5,6,6a,7,11,11a,11b-dodecahydro-4,7,11-trimethyl-**	472-33-3	**NI31776**
302	108	271	303	109	131	81	133	**302**	100	72	51	23	23	16	16	12		20	30	2	–	–	–	–	–	–	–	–	–	–	**Phenanthro[3,2-b]furan-4-methanol, 1,2,3,4,4a,5,6,6a,7,11,11a,11b-dodecahydro-4,7,11-trimethyl-**	472-33-3	**NI31775**
269	270	287	302	175	43	159	201	**302**	100	27	21	19	17	16	14	11		20	30	2	–	–	–	–	–	–	–	–	–	–	**Podocarpa-8,11,13-triene-3α,13-diol, 14-isopropyl-**	18325-87-6	**NI31777**
269	270	287	302	175	159	201	27	**302**	100	28	21	19	18	14	12	11		20	30	2	–	–	–	–	–	–	–	–	–	–	**Podocarpa-8,11,13-triene-3α,13-diol, 14-isopropyl-**	18325-87-6	**NI31778**
269	199	284	227	157	185	159	213	**302**	100	89	85	70	62	55	45	35	**12.90**	20	30	2	–	–	–	–	–	–	–	–	–	–	**Podocarpa-8,11,13-triene-7β,13-diol, 14-isopropyl-**	24338-19-0	**NI31779**
96	108	191	124	178	136	150	288	**302**	100	94	61	50	41	34	23	19	**15.00**	20	30	2	–	–	–	–	–	–	–	–	–	–	**Solidagolactone**	41943-73-1	**NI31780**
201	276	215	202	216	258	203	61	**302**	100	96	87	85	71	66	57	57	**41.30**	20	30	2	–	–	–	–	–	–	–	–	–	–	**17α-Vinyl-5α-estrane-3α,17β-diol**		**NI31781**
201	276	215	258	214	202	216	61	**302**	100	92	77	70	66	66	56	56	**42.85**	20	30	2	–	–	–	–	–	–	–	–	–	–	**17α-Vinyl-5α-estrane-3β,17β-diol**		**NI31782**
201	215	214	202	232	216	258	203	**302**	100	70	61	56	54	52	49	48	**19.67**	20	30	2	–	–	–	–	–	–	–	–	–	–	**17α-Vinyl-5β-estrane-3α,17β-diol**		**NI31783**
201	215	202	147	258	124	146	216	**302**	100	51	43	42	40	40	37	35	**20.22**	20	30	2	–	–	–	–	–	–	–	–	–	–	**17α-Vinyl-5β-estrane-3β,17β-diol**		**NI31784**
70	233	302	71	56	57	45	41	**302**	100	18	7	7	7	4	3	3		20	34	–	2	–	–	–	–	–	–	–	–	–	**Androst-5-ene-3,17-diamine, N[3]-methyl-, (3β,17β)-**	2842-69-5	**NI31785**
267	268	265	263	132	302	266	189	**302**	100	25	14	7	7	6	6	6		21	15	–	–	–	–	1	–	–	–	–	–	–	**Cyclopropene, 3-chloro-1,2,3-triphenyl-**		**LI06646**
284	302	255	239	241	256	240	269	**302**	100	79	76	62	43	41	41	35		21	18	2	–	–	–	–	–	–	–	–	–	–	**Benz[j]aceanthrylene-11,12-diol, 1,2,11,12-tetrahydro-3-methyl-**	3343-01-9	**MS07011**
259	243	260	165	244	105	77	166	**302**	100	85	60	59	39	19	18	16	**11.01**	21	18	2	–	–	–	–	–	–	–	–	–	–	**Benzenemethanol, α,α-diphenyl-, acetate**	971-85-7	**NI31786**
105	77	302	197	106	303	91	51	**302**	100	24	19	11	8	4	3	3		21	18	2	–	–	–	–	–	–	–	–	–	–	**2-Benzyl-4-methylphenyl benzoate**		**MS07012**
158	159	144	115	127	128	116	145	**302**	100	53	47	10	8	7	7	6	**0.20**	21	18	2	–	–	–	–	–	–	–	–	–	–	**1,8:2,7-Dimethanodibenzo[a,e]cyclobuta[c]cycloocten-13-one, 1,2,2a,7,8,12b-hexahydro-1-methoxy-**	55649-64-4	**NI31787**
105	107	93	92	77	79	28	180	**302**	100	81	76	50	47	41	25	22	**0.00**	21	18	2	–	–	–	–	–	–	–	–	–	–	**1,2-Diphenylethyl benzoate**		**DO01089**
207	284	107	77	208	283	178	183	**302**	100	34	29	25	22	19	15	10	**5.00**	21	18	2	–	–	–	–	–	–	–	–	–	–	**1,3-Diphenyl-3-(2-hydroxyphenyl)-propan-1-one**		**MS07013**
274	245	231	259	246	215	216	275	**302**	100	64	54	41	32	32	23	22	**5.00**	21	22	–	2	–	–	–	–	–	–	–	–	–	**4,5,6,7-Tetrahydro-3,3-di-4-tolyl-3H-indazole**	56005-80-2	**NI31788**
175	41	55	95	67	81	43	302	**302**	100	95	70	61	53	52	51	43		21	34	1	–	–	–	–	–	–	–	–	–	–	**Androstan-3-one, 4,4-dimethyl-, (5α)-**	898-96-4	**NI31789**
219	302	151	95	109	81	204	41	**302**	100	94	93	87	70	67	61	59		21	34	1	–	–	–	–	–	–	–	–	–	–	**Androstan-6-one, 4,4-dimethyl-, (5α)-**	14295-33-1	**NI31790**
219	302	151	95	109	81	204	55	**302**	100	94	94	85	69	66	59	59		21	34	1	–	–	–	–	–	–	–	–	–	–	**Androstan-6-one, 4,4-dimethyl-, (5α)-**	14295-33-1	**LI06647**
219	220	259	109	55	41	135	94	**302**	100	19	17	12	10	10	9	9	**4.06**	21	34	1	–	–	–	–	–	–	–	–	–	–	**Androstan-6-one, 4,4-dimethyl-, (5β)-**	14295-32-0	**NI31791**
219	220	259	109	134	54	41	95	**302**	100	17	15	12	9	9	9	8	**4.00**	21	34	1	–	–	–	–	–	–	–	–	–	–	**Androstan-6-one, 4,4-dimethyl-, (5β)-**	14295-32-0	**LI06648**
218	109	67	108	95	81	79	93	**302**	100	97	90	74	72	71	57	49	**28.00**	21	34	1	–	–	–	–	–	–	–	–	–	–	**Androstan-17-one, 16,16-dimethyl-, (5α)-**	56052-96-1	**NI31792**
41	43	95	55	269	67	91	81	**302**	100	74	65	65	60	49	44	44	**20.00**	21	34	1	–	–	–	–	–	–	–	–	–	–	**Androst-5-en-3-ol, 4,4-dimethyl-, (3β)-**	7673-17-8	**NI31793**
43	41	55	67	80	81	79	91	**302**	100	97	65	59	58	54	50	45	**4.00**	21	34	1	–	–	–	–	–	–	–	–	–	–	**13,21-Cyclo-18-norpregnan-20-ol, 20-methyl-, (5α,20S)-**	56324-74-4	**NI31794**
217	274	302	287	96	79	56	134	**302**	100	65	44	35	25	25	25	24		21	34	1	–	–	–	–	–	–	–	–	–	–	**D-Dihomoandrostan-17a-one, (5α)-**	32319-05-4	**NI31795**
217	302	95	287	109	81	149	55	**302**	100	69	55	35	35	35	34	29		21	34	1	–	–	–	–	–	–	–	–	–	–	**D-Dihomoandrostan-17b-one, (5α)-**	32319-07-6	**NI31796**
260	67	95	81	287	79	69	83	**302**	100	97	93	77	72	69	69	65	**57.57**	21	34	1	–	–	–	–	–	–	–	–	–	–	**5α-Estran-3-one, 2α-isopropyl-**	1975-31-1	**NI31797**
260	95	67	81	79	83	69	91	**302**	100	80	60	49	49	40	40	39	**17.00**	21	34	1	–	–	–	–	–	–	–	–	–	–	**5α-Estran-3-one, 2β-isopropyl-**	4507-79-3	**NI31798**
55	231	232	81	67	302	95	79	**302**	100	88	72	63	54	42	42	38		21	34	1	–	–	–	–	–	–	–	–	–	–	**Pregnan-3-one, (5α)-**	14778-11-1	**NI31799**
192	302	41	55	149	109	67	81	**302**	100	99	75	58	50	47	46	44		21	34	1	–	–	–	–	–	–	–	–	–	–	**Pregnan-11-one, (5α)-**	26039-99-6	**NI31800**
192	204	302	179	109	56	193	81	**302**	100	39	38	26	19	18	15	14		21	34	1	–	–	–	–	–	–	–	–	–	–	**Pregnan-11-one, (5β)-**	26729-46-4	**NI31801**
192	205	302	179	41	109	55	193	**302**	100	39	38	26	21	19	18	16		21	34	1	–	–	–	–	–	–	–	–	–	–	**Pregnan-11-one, (5β)-**	26729-46-4	**NI31802**
233	287	302	273	234	81	67	109	**302**	100	32	31	21	19	19	15	14		21	34	1	–	–	–	–	–	–	–	–	–	–	**Pregnan-12-one, (5α)-**	6022-46-4	**NI31803**
233	44	81	41	55	67	109	40	**302**	100	70	54	54	51	45	40	37	**24.48**	21	34	1	–	–	–	–	–	–	–	–	–	–	**Pregnan-12-one, (5α,17α)-**	5618-24-6	**NI31804**
233	302	41	273	287	81	67	55	**302**	100	67	38	36	35	29	24	22		21	34	1	–	–	–	–	–	–	–	–	–	–	**Pregnan-12-one, (5β)-**	17996-90-6	**NI31805**
41	55	44	125	43	81	67	40	**302**	100	83	83	77	67	54	53	42	**40.83**	21	34	1	–	–	–	–	–	–	–	–	–	–	**Pregnan-15-one, (5α)-**	14012-13-6	**NI31806**

MASS TO CHARGE RATIOS								M.W.	INTENSITIES								Parent	C	H	O	N	S	F	Cl	Br	I	Si	P	B	X	COMPOUND NAME	CAS Reg No	No
125	55	41	67	44	124	81	109	302	100	33	32	20	20	18	17	16	12.00	21	34	1	–	–	–	–	–	–	–	–	–	–	Pregnan-15-one, (5α,14β)-	14304-60-0	NI31807
216	217	121	302	135	201	123	122	302	100	74	35	30	26	15	13	13		21	34	1	–	–	–	–	–	–	–	–	–	–	Pregnan-16-one, (5α)-	42921-26-6	NI31808
302	43	84	217	81	55	95	67	302	100	97	91	84	52	52	51	48		21	34	1	–	–	–	–	–	–	–	–	–	–	Pregnan-20-one, (5α)-	848-62-4	NI31810
302	217	287	95	43	109	55	81	302	100	39	29	29	28	26	24	22		21	34	1	–	–	–	–	–	–	–	–	–	–	Pregnan-20-one, (5α)-	848-62-4	NI31809
217	43	84	302	95	109	81	67	302	100	95	82	68	63	62	61	52		21	34	1	–	–	–	–	–	–	–	–	–	–	Pregnan-20-one, (5α)-	848-62-4	NI31811
55	41	44	43	67	71	217	81	302	100	98	84	81	73	73	66	63	44.40	21	34	1	–	–	–	–	–	–	–	–	–	–	Pregnan-20-one, (5α,17α)-	7704-90-7	NI31812
217	302	68	72	81	110	96	287	302	100	68	62	59	58	51	51	46		21	34	1	–	–	–	–	–	–	–	–	–	–	Pregnan-20-one, (5α,17α)-	7704-90-7	NI31813
43	55	41	44	67	71	217	81	302	100	62	61	51	48	45	41	39	28.12	21	34	1	–	–	–	–	–	–	–	–	–	–	Pregnan-20-one, (5α,17α)-	7704-90-7	NI31814
273	230	302	93	91	176	95	57	302	100	99	86	64	62	61	52	52		21	34	1	–	–	–	–	–	–	–	–	–	–	Pregn-2-en-17-ol, (5α,17ξ)-	53286-39-8	NI31815
302	284	303	287	269	217	191	81	302	100	25	22	20	20	20	19	6		21	34	1	–	–	–	–	–	–	–	–	–	–	Pregn-5-en-3-ol, (3β)-	2862-58-0	NI31816
284	257	93	41	81	55	91	67	302	100	73	72	70	69	67	63	62	12.24	21	34	1	–	–	–	–	–	–	–	–	–	–	Pregn-14-en-20-ol, (5α,20S)-	54411-99-3	NI31817
67	217	81	109	79	95	93	91	302	100	99	89	78	77	73	63	61	39.00	21	34	1	–	–	–	–	–	–	–	–	–	–	13,17-seco-5α-Pregn-13(18)-en-20-one	31239-27-7	NI31818
302	77	105	287	303	165	197	51	302	100	84	76	69	24	24	19	19		22	22	1	–	–	–	–	–	–	–	–	–	–	Azulene, 1,4-dimethyl-7-isopropyl-, 2-benzoyl-	39665-56-0	MS07014
105	287	302	51	165	77	288	259	302	100	59	56	50	43	38	30	28		22	22	1	–	–	–	–	–	–	–	–	–	–	Azulene, 2,4-dimethyl-7-isopropyl-, benzoyl-	72361-25-2	MS07015
285	77	287	51	105	286	141	128	302	100	90	47	47	26	23	19	17	1.70	22	22	1	–	–	–	–	–	–	–	–	–	–	Azulene, 3,8-dimethyl-5-isopropyl-, 1-benzoyl-	22190-65-4	MS07016
284	105	178	91	191	179	192	77	302	100	96	75	59	45	41	39	36	0.00	22	22	1	–	–	–	–	–	–	–	–	–	–	Benzenepropanol, 4-methyl-β,β-diphenyl-	54934-50-8	NI31819
302	274	287	259	303	165	229	186	302	100	47	42	32	23	20	19	12		22	22	1	–	–	–	–	–	–	–	–	–	–	2,4-Cyclohexadien-1-one, 2,3,4,5-tetramethyl-6,6-diphenyl-	74764-49-1	NI31820
302	274	91	287	115	259	261	129	302	100	20	20	12	12	11	9	9		22	22	1	–	–	–	–	–	–	–	–	–	–	1,1-Dimethyl-2-phenyl-5-[(E)-phenylmethylene]spiro[2.4]heptan-4-one		MS07017
287	183	302	198	105	288	209	197	302	100	73	67	36	33	23	22	21		22	22	1	–	–	–	–	–	–	–	–	–	–	2,4-Di(1-phenylethyl)phenol		IC03795
243	105	244	165	183	225	302	77	302	100	69	48	48	45	38	33	25		22	22	1	–	–	–	–	–	–	–	–	–	–	Propyl trityl ether	13594-77-9	NI31821
91	92	43	117	41	29	131	57	302	100	81	51	45	43	35	30	25	23.01	22	38	–	–	–	–	–	–	–	–	–	–	–	Benzene, hexadecyl-	1459-09-2	MS07018
92	91	302	43	41	105	57	104	302	100	99	32	22	15	12	12	10		22	38	–	–	–	–	–	–	–	–	–	–	–	Benzene, hexadecyl-	1459-09-2	AI02017
92	91	43	104	302	57	41	105	302	100	55	19	18	14	13	12	9		22	38	–	–	–	–	–	–	–	–	–	–	–	Benzene, hexadecyl-	1459-09-2	WI01394
245	57	246	190	134	302	189	41	302	100	78	59	56	52	36	30	23		22	38	–	–	–	–	–	–	–	–	–	–	–	Benzene, 1,3,5-tris(2,2-dimethylpropyl)-2-methyl-	40572-12-1	MS07019
137	81	95	136	41	55	67	69	302	100	94	87	80	73	66	51	38	1.92	22	38	–	–	–	–	–	–	–	–	–	–	–	1,1-Di-(1′-decahydronaphthyl)-ethane		AI02018
137	81	95	136	67	55	69	135	302	100	67	59	47	47	45	26	23	15.20	22	38	–	–	–	–	–	–	–	–	–	–	–	1,2-Di-(1′-decahydronaphthyl)-ethane		AI02019
41	82	67	55	83	81	95	69	302	100	83	65	60	59	58	54	47	6.63	22	38	–	–	–	–	–	–	–	–	–	–	–	1,5-Dicyclopentyl-3-(2-cyclopentylethyl)-2-pentene		AI02020
217	302	55	95	81	218	67	40	302	100	87	77	68	67	65	65	60		22	38	–	–	–	–	–	–	–	–	–	–	–	D-Homopregnane, (5α)-	35575-28-1	NI31822
119	91	41	120	43	105	29	55	302	100	16	15	10	10	7	7	4	1.30	22	38	–	–	–	–	–	–	–	–	–	–	–	Pentadecane, 2-methyl-2-phenyl-	29138-94-1	WI01395
105	106	91	302	43	41	104	29	302	100	13	10	7	5	5	4	3		22	38	–	–	–	–	–	–	–	–	–	–	–	2-Phenylhexadecane		AI02021
91	189	203	105	104	92	43	41	302	100	18	15	15	10	9	8	7	7.41	22	38	–	–	–	–	–	–	–	–	–	–	–	8-Phenylhexadecane		AI02022
302	303	300	298	151	150	149	137	302	100	3	3	1	1	1	1	1		24	14	–	–	–	–	–	–	–	–	–	–	–	Dibenzo[def,p]chrysene	191-30-0	NI31823
302	303	300	151	298	150	149	137	302	100	3	2	2	1	1	1	1		24	14	–	–	–	–	–	–	–	–	–	–	–	Dibenzo[fg,op]naphthacene	192-51-8	NI31824
302	303	300	298	151	150	149	137	302	100	3	2	1	1	1	1	1		24	14	–	–	–	–	–	–	–	–	–	–	–	Naphtho[1,2,3,4-def]chrysene	192-65-4	NI31825
302	150	151	303	300	149	137	298	302	100	36	33	25	24	19	9	7		24	14	–	–	–	–	–	–	–	–	–	–	–	Naphtho[1,2,3,4-def]chrysene	192-65-4	NI31826
43	28	186	188	47	82	36	62	303	100	45	40	39	37	32	25	22	0.00	4	3	–	3	–	–	6	–	–	–	–	–	–	N,N-Bis(1-imino-2,2,2-trichloroethyl)amine		NI31827
117	119	50	121	76	75	63	70	303	100	95	59	46	44	28	17	17	0.00	7	4	4	1	1	–	3	–	–	–	–	–	–	Trichloromethyl 4-nitrophenyl sulphone	87721-12-8	NI31828
158	184	119	190	69	126	57	91	303	100	45	42	37	29	26	26	25	18.80	9	12	5	1	1	3	–	–	–	–	–	–	–	L-Cysteine, S-(2-methoxy-2-oxoethyl)-N-(trifluoroacetyl)-, methyl ester	40027-86-9	MS07020
66	161	65	39	189	217	135	245	303	100	95	84	79	67	40	34	27	0.00	10	7	4	1	–	–	–	–	–	–	–	–	1	π-Cyclopentadienyl-π-methlisocyano-tricarbonyl molybdenum		MS07021
242	240	258	256	257	255	75	133	303	100	98	97	95	64	52	39	38	2.40	10	10	1	1	2	–	–	1	–	–	–	–	–	3H-Indol-3-one, 6-bromo-1,2-dihydro-2,2-bis(methylthio)-	50630-71-2	NI31829
42	50	75	76	56	221	155	219	303	100	10	9	9	8	7	7	6	0.99	10	10	3	1	1	–	–	1	–	–	–	–	–	1-(4-Bromobenzenesulphonyl)pyrrolidin-3-one	83133-01-1	NI31830
86	268	128	270	143	145	70	58	303	100	22	18	15	13	12	8	6	0.00	10	16	1	1	1	–	3	–	–	–	–	–	–	Carbamic acid, diisopropylthio-, S-(2,3,3-trichloro-2-propenyl) ester	2303-17-5	NI31831
43	86	41	27	42	128	70	58	303	100	92	61	42	36	35	21	17	0.00	10	16	1	1	1	–	3	–	–	–	–	–	–	Carbamothioic acid, diisopropyl-, S-(2,3,3-trichloro-2-propenyl) ester	2303-17-5	NI31832
231	303	76	203	288	248	49	40	303	100	48	47	27	25	25	18	14		11	14	1	1	–	–	–	–	1	–	–	–	–	Benzamide, N-tert-butyl-3-iodo-	42498-37-3	NI31833
231	303	76	288	248	203	247	104	303	100	40	30	21	18	18	15	13		11	14	1	1	–	–	–	–	1	–	–	–	–	Benzamide, N-tert-butyl-4-iodo-	42498-36-2	NI31834
216	217	43	41	119	71	303	166	303	100	11	10	10	8	7	5	4		11	14	3	1	–	5	–	–	–	–	–	–	–	N-Pentafluoropropionyl-D-proline isopropyl ester		MS07022
216	166	217	43	41	119	71	69	303	100	13	11	11	11	9	7	6	4.40	11	14	3	1	–	5	–	–	–	–	–	–	–	N-Pentafluoropropionyl-L-proline isopropyl ester		MS07023
69	303	68	149	236	67	165	133	303	100	53	35	23	14	11	8	5		11	14	5	1	1	–	–	–	–	–	1	–	–	5,5-Dimethyl-2-(4-nitrophenoxy)-1,3,2-dioxaphosphorinane 2-sulphide	892-04-6	NI31835
188	84	303	239	43	274			303	100	34	28	25	24	1				11	17	1	3	3	–	–	–	–	–	–	–	–	Spiro[3,3a-dimethylperhydrofuro[2,3-d]thiazole-2,4′-(1′,3′-dimethylimidazolidine-2′,5′-dithione)]		LI06649
145	102	18	303	42	28	15	173	303	100	84	60	39	34	28	28	26		12	9	1	3	–	3	1	–	–	–	–	–	–	3(2H)-Pyridazinone, 4-chloro-5-(methylamino)-2-[3-(trifluoromethyl)phenyl]-	27314-13-2	NI31836
43	200	27	68	28	58	41	42	303	100	97	95	75	69	60	58	56	0.00	12	10	–	5	–	–	–	1	–	–	–	–	–	1H-Purin-6-amine, [(2-bromophenyl)methyl]-	74421-49-1	NI31837
28	27	41	43	57	144	18	29	303	100	55	51	47	42	39	35	34	0.00	12	10	–	5	–	–	–	1	–	–	–	–	–	1H-Purin-6-amine, [(4-bromophenyl)methyl]-	74421-50-4	NI31838

MASS TO CHARGE RATIOS								M.W.	INTENSITIES								Parent	C	H	O	N	S	F	Cl	Br	I	Si	P	B	X	COMPOUND NAME	CAS Reg No	No
303	305	184	186	90	120	89	119	303	100	96	69	61	56	52	50	42		12	10	–	5	–	–	–	1	–	–	–	–	–	1H-Purin-6-amine, N-[(4-bromophenyl)methyl]-	67023-55-6	NI31839
43	72	60	42	83	129	45	102	303	100	20	18	17	16	15	11	9	0.00	12	21	6	3	–	–	–	–	–	–	–	–	–	2,4,6-Tris(acetoxymethyl)hexahydro-1,3,5-triazine		NI31840
187	171	100	86	188	129	84	143	303	100	15	15	12	10	8	8	7	0.00	12	36	–	3	–	–	–	–	–	–	–	–	3	Tri-aluminium, tri[μ-(diethylaminato)]hexahydro-	32106-39-1	NI31841
285	83	235	43	81	202	56	175	303	100	51	48	26	23	22	18	17	0.00	13	17	2	7	–	–	–	–	–	–	–	–	–	1H-Pyrrole-2-carboxamide, 4-[[(5-amino-3,4-dihydro-2H-pyrrol-2-yl)carbonyl]amino]-N-(3-amino-3-imino-1-propenyl)-	37913-77-2	NI31842
129	161	45	43	87	101	126	71	303	100	64	55	38	33	31	23	13	0.00	13	21	7	1	–	–	–	–	–	–	–	–	–	2,5-Di-O-acetyl-3,4,6-tri-O-methyl-D-gluconitrile	35439-79-3	NI31843
43	45	129	101	161	71	87	112	303	100	86	58	56	28	27	27	26	0.00	13	21	7	1	–	–	–	–	–	–	–	–	–	3,5-Di-O-acetyl-2,4,6-tri-O-methyl-D-gluconitrile	35439-35-1	NI31844
129	87	99	45	88	189	73	74	303	100	61	57	52	43	23	17	16	0.00	13	21	7	1	–	–	–	–	–	–	–	–	–	5,6-Di-O-acetyl-2,3,4-tri-O-methyl-D-gluconitrile	35439-77-1	NI31845
129	43	161	45	101	87	126	145	303	100	98	65	50	32	31	20	8	0.00	13	21	7	1	–	–	–	–	–	–	–	–	–	2,5-Di-O-acetyl-3,4,6-tri-O-methyl-D-mannonitrile		MS07024
43	101	129	112	45	186	161	154	303	100	67	50	35	35	32	24	21	0.00	13	21	7	1	–	–	–	–	–	–	–	–	–	3,5-Di-O-acetyl-2,4,6-tri-O-methyl-D-mannonitrile		MS07025
43	87	113	45	99	129	233	131	303	100	48	44	42	35	33	18	18	0.03	13	21	7	1	–	–	–	–	–	–	–	–	–	4,5-Di-O-acetyl-2,3,6-tri-O-methyl-D-mannonitrile		MS07026
43	113	87	45	99	129	233	131	303	100	50	42	34	29	28	18	16	0.00	13	21	7	1	–	–	–	–	–	–	–	–	–	4,5-Di-O-acetyl-2,3,6-tri-O-methyl-D-mannonitrile		NI31846
129	43	87	189	99	88	45	71	303	100	98	53	39	38	29	19	10	0.00	13	21	7	1	–	–	–	–	–	–	–	–	–	5,6-Di-O-acetyl-2,3,4-tri-O-methyl-D-mannonitrile		MS07027
171	113	184	288	216	145	199	231	303	100	82	64	56	35	34	31	29	27.00	13	21	7	1	–	–	–	–	–	–	–	–	–	N-tert-Butyl-2,3,3-trimethoxycarbonylpropanamide		LI06650
131	59	43	69	41	71	72	31	303	100	98	67	33	32	28	24	23	4.00	13	21	7	1	–	–	–	–	–	–	–	–	–	5-Isoxazolidinecarboxylic acid, 2-[2,3-O-isopropylidene-β-D-ribofuranosyl]-, methyl ester, (R)-	64018-69-5	NI31847
43	101	184	114	74	157	129	87	303	100	65	43	41	36	32	32	27	0.71	13	21	7	1	–	–	–	–	–	–	–	–	–	α-D-Mannopyranoside, methyl 4-acetamido-4,6-dideoxy-, 2,3-diacetate	20197-44-8	NI31848
43	101	184	114	74	129	157	97	303	100	75	43	41	36	33	32	26	1.00	13	21	7	1	–	–	–	–	–	–	–	–	–	α-D-Mannopyranoside, methyl 4-acetamido-4,6-dideoxy-, 2,3-diacetate	20197-44-8	LI06651
101	200	102	60	123	87	140	98	303	100	49	47	46	38	35	31	31	0.00	13	21	7	1	–	–	–	–	–	–	–	–	–	Methyl 2,4-di-O-acetyl-3-acetamido-3,6-dideoxy-α-D-allopyranoside	28079-83-6	NI31849
114	82	124	98	141	156	72	184	303	100	53	44	43	42	36	36	35	0.00	13	21	7	1	–	–	–	–	–	–	–	–	–	Methyl 3,5-di-O-acetyl-2-acetamido-2,6-dideoxy-D-galactofuranoside	93250-08-9	NI31850
101	143	100	130	102	83	244	184	303	100	20	12	10	10	10	9	9	0.00	13	21	7	1	–	–	–	–	–	–	–	–	–	Methyl 3,4-di-O-acetyl-2-acetamido-2,6-dideoxy-α-D-galactopyranoside	93250-07-8	NI31851
101	59	143	83	114	60	99	130	303	100	34	30	18	16	15	14	13	0.00	13	21	7	1	–	–	–	–	–	–	–	–	–	Methyl 3,4-di-O-acetyl-2-acetamido-2,6-dideoxy-α-D-glucopyranoside	39686-91-4	NI31852
43	101	74	102	129	57	243	59	303	100	48	30	23	20	20	19	19	1.00	13	21	7	1	–	–	–	–	–	–	–	–	–	α-D-Talopyranoside, methyl 1-acetamido-4,6-dideoxy-, 2,3-diacetate		LI06652
43	101	74	102	129	99	57	59	303	100	49	29	22	20	20	20	19	0.72	13	21	7	1	–	–	–	–	–	–	–	–	–	α-L-Talopyranoside, methyl 4-acetamido-4,6-dideoxy-, 2,3-diacetate	17495-07-7	NI31853
44	154	80	43	29	18	122	45	303	100	83	74	63	56	55	52	46	28.92	13	22	3	1	1	–	–	–	–	–	1	–	–	Phosphoramidic acid, isopropyl-, ethyl 3-methyl-4-(methylthio)phenyl ester	22224-92-6	NI31854
73	57	102	75	45	186	41	29	303	100	50	48	24	22	18	16	16	2.98	13	29	3	1	–	–	–	–	–	2	–	–	–	Glycine, N-(2-methyl-1-oxobutyl)-N-(trimethylsilyl)-, trimethylsilyl ester	55530-60-4	NI31855
73	102	288	75	147	57	45	130	303	100	50	23	22	21	20	19	19	3.70	13	29	3	1	–	–	–	–	–	2	–	–	–	Glycine, N-(3-methyl-1-oxobutyl)-N-(trimethylsilyl)-, trimethylsilyl ester	55520-90-6	NI31856
73	102	186	75	57	45	176	261	303	100	73	28	25	20	20	18	17	2.14	13	29	3	1	–	–	–	–	–	2	–	–	–	Glycine, N-(1-oxopentyl)-N-(trimethylsilyl)-, trimethylsilyl ester	55530-59-1	NI31857
303	305	178	176	304	306	177	179	303	100	96	60	22	16	15	15	11		14	10	2	1	–	–	–	1	–	–	–	–	–	2-Bromo-3-nitro-9,10-dihydrophenanthrene		MS07028
303	272	305	274	165	164	244	246	303	100	99	98	98	24	20	18	17		14	10	2	1	–	–	–	1	–	–	–	–	–	Methyl 7-bromobenzo[e]inolizine-3-carboxylate		LI06653
105	198	106	77	51	244	78	302	303	100	66	44	44	40	29	23	15	4.88	14	13	5	3	–	–	–	–	–	–	–	–	–	5-(2-Benzamido-2-methoxycarbonyl-ethylidene)-hydantoin		LI06654
254	118	256	63	255	132	117	91	303	100	44	34	15	15	13	13	13	5.10	14	19	2	1	–	–	2	–	–	–	–	–	–	Benzenebutanoic acid, 4-[bis(2-chloroethyl)amino]-	305-03-3	NI31858
254	256	118	255	303	44	305	132	303	100	27	17	15	10	7	6	6		14	19	2	1	–	–	2	–	–	–	–	–	–	Benzenebutanoic acid, 4-[bis(2-chloroethyl)amino]-	305-03-3	NI31859
111	29	41	42	18	175	27	75	303	100	98	84	76	63	61	55	46	1.60	14	22	2	1	1	–	1	–	–	–	–	–	–	Benzenesulphonamide, N,N-dibutyl-4-chloro-	127-59-3	NI31860
186	73	187	45	59	43	75	147	303	100	63	16	11	8	8	7	6	0.00	14	33	2	1	–	–	–	–	–	2	–	–	–	2-Amino-caprylic acid di-TMS		MS07029
303	275	304	44	302	191	64	286	303	100	34	20	20	16	16	16	12		15	10	3	1	1	1	–	–	–	–	–	–	–	6-Quinolinesulphonyl fluoride, 4-hydroxy-2-phenyl-	31241-71-1	NI31861
77	171	132	51	107	50	303	76	303	100	70	64	44	43	26	23	21		15	13	4	1	1	–	–	–	–	–	–	–	–	2,3-Dihydro-1-(4-methoxybenzenesulphonyl)-3-oxoindole	83372-77-4	NI31862
162	77	51	63	163	303	64	92	303	100	38	23	15	13	13	13	12		15	13	4	1	1	–	–	–	–	–	–	–	–	2,3-Dihydro-5-methoxy-1-phenylsulphonyl-3-oxoindole	83372-79-6	NI31863
139	91	303	65	140	102	239	148	303	100	19	10	10	8	7	4	4		15	13	4	1	1	–	–	–	–	–	–	–	–	1-Methyl-4-[[2-(4-nitrophenyl)ethenyl]sulphonyl]benzene		DD01099
91	28	119	92	106	65	118	107	303	100	18	17	15	13	12	11	9	0.00	15	14	1	1	–	–	–	1	–	–	–	–	–	2H-1,3-Benzoxazine, 6-bromo-3,4-dihydro-3-benzyl-	55955-91-4	NI31864
157	63	77	29	169	141	47	51	303	100	77	70	65	56	55	53	48	7.20	15	14	2	1	1	–	–	–	–	–	1	–	–	Phosphonothioic acid, phenyl-, O-(4-cyanophenyl) O-ethyl ester	13067-93-1	NI31865
88	196	128	101	114	71	127	197	303	100	59	59	40	37	28	26	25	1.00	15	29	5	1	–	–	–	–	–	–	–	–	–	N-Cyclopentyl-2,3,4,6-tetra-O-methyl-D-glucosylamine		NI31866
229	303	69	51	77	67	41	42	303	100	77	52	48	40	40	35	31		16	17	5	1	–	–	–	–	–	–	–	–	–	Nicotinic acid, 1,6-dihydro-4-hydroxy-1-(4-methoxyphenyl)-2-methyl-6-oxo-, ethyl ester	1159-49-5	NI31867
115	174	144	116	244	91	117	145	303	100	86	56	53	37	33	32	26	0.00	16	17	5	1	–	–	–	–	–	–	–	–	–	Sureothinic acid, methyl ester		NI31868
288	73	290	77	289	75	105	207	303	100	45	38	29	23	21	19	17	11.42	16	18	1	1	–	–	1	–	–	1	–	–	–	2-Amino-5-chlorobenzophenone, trimethylsilyl-		NI31869
303	73	288	304	200	45	305	289	303	100	66	40	30	11	11	10	10		16	25	1	1	–	–	–	–	–	2	–	–	–	Silylamine, 1,1,1-trimethyl-N-[5-(trimethylsiloxy)-1-naphthyl]-	33285-82-4	NI31870
303	288	73	304	289	305	45	290	303	100	49	34	29	15	11	6	5		16	25	1	1	–	–	–	–	–	2	–	–	–	Silylamine, 1,1,1-trimethyl-N-[6-(trimethylsiloxy)-2-naphthyl]-	33285-85-7	NI31871
165	164	150	132	58	44	45	166	303	100	54	52	45	42	28	21	16	7.00	17	21	2	1	1	–	–	–	–	–	–	–	–	N-[2-(2,4-Dimethoxyphenylthio)benzyl]-N,N-dimethylamine		MS07030
182	82	83	105	94	77	96	303	303	100	95	41	35	33	31	27	25		17	21	4	1	–	–	–	–	–	–	–	–	–	8-Azabicyclo[3.2.1]octane-2-carboxylic acid, 3-(benzoyloxy)-8-methyl-, methyl ester, [1R-(exo,exo)]-	50-36-2	NI31872
82	105	81	78	77	51	94	80	303	100	53	46	44	42	32	31	29	0.08	17	21	4	1	–	–	–	–	–	–	–	–	–	8-Azabicyclo[3.2.1]octane-2-carboxylic acid, 3-(benzoyloxy)-8-methyl-, methyl ester, [1R-(exo,exo)]-	50-36-2	NI31873
82	28	182	83	77	42	94	105	303	100	45	45	32	31	30	26	25	7.80	17	21	4	1	–	–	–	–	–	–	–	–	–	8-Azabicyclo[3.2.1]octane-2-carboxylic acid, 3-(benzoyloxy)-8-methyl-, methyl ester, [1R-(exo,exo)]-	50-36-2	NI31874

MASS TO CHARGE RATIOS								M.W.	INTENSITIES								Parent	C	H	O	N	S	F	Cl	Br	I	Si	P	B	X	COMPOUND NAME	CAS Reg No	No
138	304	139	305	156	303	154	306	**303**	100	40	9	8	5	4	2	1		17	21	4	1	–	–	–	–	–	–	–	–	–	Benzeneacetic acid, α-(hydroxymethyl)-, 9-methyl-3-oxa-9-azatricyclo[3.3.1.0(2,4)]non-7-yl ester, [7(S)-(1α,2β,4β,5α,7β)]-	51-34-3	NI31875
94	138	108	42	154	136	303	97	**303**	100	73	50	44	43	43	40	25		17	21	4	1	–	–	–	–	–	–	–	–	–	Benzeneacetic acid, α-(hydroxymethyl)-, 9-methyl-3-oxa-9-azatricyclo[3.3.1.02,4]non-7-yl ester, [7(S)-(1α,2β,4β,5α,7β)]-	51-34-3	NI31876
303	175	174	272	149	229	148	185	**303**	100	53	38	26	23	21	19	10		17	21	4	1	–	–	–	–	–	–	–	–	–	Dihydrococcinine		LI06655
175	303	272	62	57	55	56	174	**303**	100	90	85	70	70	50	49	41		17	21	4	1	–	–	–	–	–	–	–	–	–	Dihydromontanine	21446-29-7	LI06656
272	62	57	175	55	56	303	174	**303**	100	84	84	59	59	56	52	47		17	21	4	1	–	–	–	–	–	–	–	–	–	Dihydromontanine	21446-29-7	NI31877
57	247	203	158	303	174	248	304	**303**	100	89	74	63	54	45	41	27		17	21	4	1	–	–	–	–	–	–	–	–	–	Ethyl 1-(tert-butoxycarbonyl)-2-methylindole-3-carboxylate		DD01100
136	95	150	121	93	110	104	108	**303**	100	92	61	55	48	34	23	23	**5.40**	17	21	4	1	–	–	–	–	–	–	–	–	–	Isoborneol 4-nitrobenzoate		MS07031
192	174	56	70	55	41	69	43	**303**	100	62	41	40	33	30	29	29	**1.00**	17	21	4	1	–	–	–	–	–	–	–	–	–	1H-Isoindole-5-carboxylic acid, 2,3-dihydro-1,3-dioxo-, octyl ester	33975-33-6	NI31878
303	70	58	42	57	216	304	43	**303**	100	71	39	34	27	23	20	16		17	21	4	1	–	–	–	–	–	–	–	–	–	Morphinan-3,6,14-triol, 4,5-epoxy-17-methyl-, (5α,6β)-	54934-75-7	NI31879
124	218	202	219	176	261	203	70	**303**	100	67	34	18	17	15	12	12	**9.00**	17	21	4	1	–	–	–	–	–	–	–	–	–	Pyrrolidine, 1-acetyl-2-[2-[4-(acetyloxy)-3-methoxyphenyl]vinyl]-, (E)-(±)-	67257-63-0	NI31880
173	121	259	159	174	153	303	43	**303**	100	56	51	51	49	48	35	34		17	25	2	3	–	–	–	–	–	–	–	–	–	1-Nonen-3-one, 1-(4-methoxyphenyl)-, semicarbazone	18281-97-5	NI31881
205	303	262	76	179	206	77	90	**303**	100	55	41	25	23	21	20	18		18	13	2	3	–	–	–	–	–	–	–	–	–	2-Phenyl-3-(3-methyl-5-isoxazolyl)-4(3H)-quinazolinone		NI31882
135	154	57	77	303	41	168	203	**303**	100	33	22	20	15	15	14	12		18	25	3	1	–	–	–	–	–	–	–	–	–	(2E)-N-Isobutyl-7-(3,4-methylenedioxyphenyl)hept-2-enamide	83029-38-3	NI31883
116	72	188	56	173	187	43	117	**303**	100	66	42	14	12	11	11	10	**7.70**	18	25	3	1	–	–	–	–	–	–	–	–	–	1-(4-Methoxy-2-methylnaphthyl-1-oxy)-3-isopropylamino-propan-2-ol		MS07032
181	152	303	153	76	182	151	50	**303**	100	44	36	29	15	12	12	8		19	13	3	1	–	–	–	–	–	–	–	–	–	4-Nitro-4′-phenylbenzophenone		IC03796
180	104	77	78	226	303	123	154	**303**	100	46	33	20	15	11	10	8		19	15	–	1	–	–	1	–	–	–	–	1	–	Borane, chloro[(diphenylmethylene)amino]phenyl-	17814-63-0	NI31884
130	131	129	69	132	77	172	76	**303**	100	51	36	24	16	16	15	15	**4.00**	19	17	1	3	–	–	–	–	–	–	–	–	–	1′-Acetyl-2′-(1,2-dihydroquinoxalin-2-yl)-1′,2′-dihydroquinoline	95234-58-5	NI31885
130	131	129	172	132	76	102	103	**303**	100	59	44	35	20	20	19	18	**2.00**	19	17	1	3	–	–	–	–	–	–	–	–	–	2′-Acetyl-1′-(1,2-dihydroquinoxalin-2-yl)-1′,2′-dihydroquinoline		NI31886
174	83	82	303	173	109	185		**303**	100	99	99	92	74	41	11			19	17	1	3	–	–	–	–	–	–	–	–	–	1,2,4-Triphenylsemicarbazide		LI06657
303	246	110	208	95	81	67	55	**303**	100	55	40	39	37	35	31	31		19	29	2	1	–	–	–	–	–	–	–	–	–	5α-Androstane-15,17-dione, 16-aza-D-homo-	123239-85-0	DD01101
244	109	95	286	55	81	67	41	**303**	100	61	48	37	36	29	29	29	**0.00**	19	29	2	1	–	–	–	–	–	–	–	–	–	5α-Androstane-15,16-dione 15-oxime	123239-84-9	DD01102
258	109	95	67	81	55	41	218	**303**	100	75	54	51	48	43	39	29	**0.00**	19	29	2	1	–	–	–	–	–	–	–	–	–	5α-Androstane-16,17-dione 16-oxime	1158-62-9	DD01103
139	303	137	123	124	112	286	111	**303**	100	92	70	67	65	64	61	52		19	29	2	1	–	–	–	–	–	–	–	–	–	Androst-4-en-3-one, 17-hydroxy-, oxime, (17β)-	6911-95-1	NI31887
164	43	55	105	106	40	288	120	**303**	100	62	58	53	52	44	30	29	**10.01**	19	29	2	1	–	–	–	–	–	–	–	–	–	Carbamic acid, (α-methylbenzyl)-, 9-decenyl ester	32589-36-9	NI31888
132	105	95	81	71	55	150	41	**303**	100	88	82	82	74	73	71	67	**6.00**	19	29	2	1	–	–	–	–	–	–	–	–	–	Carbamic acid, (α-methylbenzyl)-, 4-menth-3-yl ester	33027-13-3	NI31889
132	95	81	105	83	71	55	41	**303**	100	83	83	80	77	75	75	68	**8.01**	19	29	2	1	–	–	–	–	–	–	–	–	–	Carbamic acid, (1-phenylethyl)-, 2-methyl-5-isopropylcyclohexyl ester	54934-73-5	NI31890
137	138	272	139	303	302	255	108	**303**	100	52	42	42	31	28	20	18		19	29	2	1	–	–	–	–	–	–	–	–	–	Estr-4-en-3-one, 17β-hydroxy-, methyloxime		NI31891
44	57	43	275	244	96	41	55	**303**	100	52	32	24	18	17	17	16	**8.00**	19	29	2	1	–	–	–	–	–	–	–	–	–	Samandione		LI06658
196	168	303	77	167	51	65	93	**303**	100	52	35	35	26	16	9	7		20	17	2	1	–	–	–	–	–	–	–	–	–	Carbamic acid, diphenyl-, 4-tolyl ester		LI06659
126	303	272	113	288	100	87	273	**303**	100	47	23	19	15	12	12	11		20	33	1	1	–	–	–	–	–	–	–	–	–	Androstan-3-one, O-methyloxime, (5α)-	50512-85-1	NI31892
58	275	71	43	260	44	57	110	**303**	100	33	29	20	18	18	13	9	**8.00**	20	33	1	1	–	–	–	–	–	–	–	–	–	Samanone, N-methyl-		LI06660
93	135	94	43	41	303	57	55	**303**	100	16	12	11	7	6	6	5		20	33	1	1	–	–	–	–	–	–	–	–	–	Tetradecanamide, N-phenyl-	622-56-0	NI31893
274	41	55	136	192	81	165	273	**303**	100	82	78	46	40	30	29	24	**9.58**	20	38	–	1	–	–	–	–	–	–	–	1	–	Boranamine, N-1-cyclodecen-1-yl-N-cyclohexyl-1,1-diethyl-	74810-34-7	NI31894
226	303	105	77	286	227	304	148	**303**	100	60	55	29	25	17	14	11		21	21	1	1	–	–	–	–	–	–	–	–	–	Benzenemethanol, 4-(dimethylamino)-α,α-diphenyl-	1719-05-7	NI31895
182	77	93	104	105	183	79	303	**303**	100	24	23	22	18	15	15	12		21	21	1	1	–	–	–	–	–	–	–	–	–	Benzenepropanol, α-phenyl-γ-(phenylamino)-	4566-58-9	NI31896
274	303	302	275	202	203	246	115	**303**	100	99	71	50	17	14	12	12		21	21	1	1	–	–	–	–	–	–	–	–	–	2(1H)-Pyridinone, 1-ethyl-4,6-dimethyl-3,5-diphenyl-	52148-69-3	MS07033
288	303	244	273	215				**303**	100	98	26	14	8					21	21	1	1	–	–	–	–	–	–	–	–	–	[1,1′:3′,1″-Terphenyl]-5′-aminium, 2′-hydroxy-N,N,N-trimethyl-, hydroxide, inner salt	61219-71-4	NI31897
84	110	41	56	55	71	67	85	**303**	100	47	18	14	12	11	9	8	**6.47**	21	37	–	1	–	–	–	–	–	–	–	–	–	Androstan-3-amine, N-ethyl-, (3β,5α)-	55320-46-2	NI31898
84	43	260	303	71	41	85	70	**303**	100	88	86	85	77	75	67	57		21	37	–	1	–	–	–	–	–	–	–	–	–	Androstan-1-amine, N,N-dimethyl-, (1β,5α)-	54498-43-0	NI31899
124	111	58	98	71	55	41	303	**303**	100	67	53	50	48	38	37	35		21	37	–	1	–	–	–	–	–	–	–	–	–	Androstan-11-amine, N,N-dimethyl-, (5α,11α)-	54498-46-3	NI31900
84	71	58	44	124	41	43	55	**303**	100	99	86	79	67	61	50	47	**40.00**	21	37	–	1	–	–	–	–	–	–	–	–	–	Androstan-16-amine, N,N-dimethyl-, (5α)-	54498-45-2	NI31901
41	55	84	67	300	275	109	81	**303**	100	78	76	73	72	72	58	54	**2.89**	21	37	–	1	–	–	–	–	–	–	–	–	–	Androstan-17-amine, N,N-dimethyl-, (5α,17β)-	54498-44-1	NI31902
286	82	56	69	57	40	55	43	**303**	100	99	91	66	51	39	36	33	**27.00**	21	37	–	1	–	–	–	–	–	–	–	–	–	Pregnan-3-amine, (3α)-	54984-47-3	LI06661
286	82	56	69	57	41	55	44	**303**	100	98	89	65	50	38	36	34	**27.02**	21	37	–	1	–	–	–	–	–	–	–	–	–	Pregnan-3-amine, (3α)-	54984-47-3	NI31903
303	302	151	304	150	301	300	138	**303**	100	29	24	24	15	9	8	6		23	13	–	1	–	–	–	–	–	–	–	–	–	Azadibenzopyrene	104219-69-4	NI31904
306	142	212	304	214	107	144	216	**304**	100	99	91	87	85	74	72	52		6	–	4	2	–	–	4	–	–	–	–	–	–	2,3,5,6-Tetrachloro-1,4-dinitrobenzene		IC03797
146	71	120	65	101	199	197	80	**304**	100	86	71	48	43	41	41	39	**0.40**	6	11	2	–	–	–	–	2	–	–	1	–	–	Phospholane, 3,4-dibromo-1-ethoxy-, 1-oxide	16182-88-0	MS07034
88	44	58	43	120	304	72	306	**304**	100	28	25	22	20	19	17	16		6	12	–	2	4	–	–	–	–	–	–	–	1	Zinc, bis(dimethylcarbamodithioato-S,S′)-	137-30-4	LI06662
88	44	58	43	120	304	72	42	**304**	100	28	25	22	20	19	17	16		6	12	–	2	4	–	–	–	–	–	–	–	1	Zinc, bis(dimethylcarbamodithioato-S,S′)-	137-30-4	NI31905
121	56	183	248	276	304	186	66	**304**	100	43	30	29	24	20	16	15		7	5	2	–	–	–	–	–	1	–	–	–	1	Cyclopentadienylirondicarbonyl iodide		NI31906

MASS TO CHARGE RATIOS								M.W.	INTENSITIES								Parent	C	H	O	N	S	F	Cl	Br	I	Si	P	B	X	COMPOUND NAME	CAS Reg No	No
269	76	64	63	304	221	205	271	**304**	100	44	44	43	26	22	19	7		7	6	5	–	2	–	2	–	–	–	–	–	–	**Anisole-2,4-disulphonyl chloride**		**IC03798**
149	260	245	97	179	125	81	47	**304**	100	99	82	57	55	52	51	45	**14.60**	8	6	–	–	–	5	–	1	–	1	–	–	–	**Silane, bromodimethyl(pentafluorophenyl)-**	23761-73-1	**NI31907**
109	269	271	79	97	47	62	93	**304**	100	50	33	28	15	14	10	8	**3.38**	8	8	4	–	–	–	3	–	–	–	1	–	–	**Fenchlorophos oxygen analog**		**NI31908**
28	192	71	42	74	51	55	56	**304**	100	99	99	90	65	52	45	20	**10.00**	9	4	5	–	–	3	–	–	–	–	–	–	1	**Manganese, tetracarbonyl[3,3,3-trifluoro-2-formyl-1-methyl-1-propenyl-C,O]-**	56211-38-2	**NI31909**
41	70	42	183	125	127	213	143	**304**	100	69	69	64	54	49	47	44	**0.00**	9	9	2	2	–	3	2	–	–	–	–	–	–	**1-(1,1-Dichloro-2,2,2-trifluoroethyl)-3-methyl-2-butenyl diazoacetate**	107686-53-3	**DD01104**
73	200	199	168	97	72	43	169	**304**	100	96	69	31	31	30	28	27	**0.00**	9	24	8	–	–	–	–	–	–	–	–	4	–	**Boronic acid, methanetetrayltetrakis-, octamethyl ester**	17936-80-0	**NI31910**
304	302	289	287	101	261	89	259	**304**	100	88	83	76	76	55	55	50		10	8	1	–	–	–	–	–	–	–	–	–	2	**Benzo[b]selenophene-3-carboxaldehyde, 2-(methylseleno)-**	39857-05-1	**NI31911**
55	269	271	116	41	273	91	233	**304**	100	98	94	48	33	30	30	27	**0.20**	10	16	–	2	–	–	4	–	–	–	–	–	–	**1,1′-Biazetidine, 2,2,2′,2′-tetrachloro-3,3,3′,3′-tetramethyl-**	74779-83-2	**NI31912**
55	269	271	116	41	273	91	233	**304**	100	96	95	50	35	32	30	29	**0.20**	10	16	–	2	–	–	4	–	–	–	–	–	–	**1,2-Bis(1,3-dichloro-2,2-dimethylpropylidene)hydrazine**		**NI31913**
248	165	180	135	93	276	56	304	**304**	100	67	61	43	28	14	13	12		10	17	5	–	–	–	–	–	–	–	1	–	1	**Iron, [dicarbonyl(1,2,3,4-η)-1,3-pentadiene](trimethyl phosphite-P)-**	74810-97-2	**NI31914**
162	127	191	304	120	204	106	111	**304**	100	71	50	49	46	39	28	21		10	24	–	2	–	4	–	–	–	2	–	–	–	**Hydrazine, 1,1-bis[difluoro(1-methylpropyl)silyl]-2,2-dimethyl-**	66436-30-4	**NI31915**
160	276	304	116	217	76	71	45	**304**	100	79	74	55	44	32	31	27		11	12	2	–	4	–	–	–	–	–	–	–	–	**2,6(7)-Dimethyl-7(6)ethoxycarbonyltetrathiafulvalene**		**MS07035**
91	92	172	170	307	65	39	146	**304**	100	14	10	10	6	6	5	4	**0.00**	11	14	–	–	–	–	–	2	–	–	–	–	–	**Benzene, [3-bromo-2-(bromomethyl)-isobutyl]-**	40548-55-8	**NI31916**
170	118	168	116	166	114	169	117	**304**	100	84	77	66	64	52	34	22	**9.00**	11	24	–	–	–	–	–	–	–	–	–	–	2	**Germane, (3,3-dimethyl-1-propyne-1,3-diyl)bis[trimethyl-**	61228-12-4	**NI31917**
306	304	171	308	153	243	121	241	**304**	100	75	51	48	41	34	33	32		12	4	1	–	–	–	4	–	–	–	–	–	–	**Dibenzofuran, 2,3,7,8-tetrachloro-**	51207-31-9	**NI31918**
285	287	289	300	302	286	304	288	**304**	100	95	35	17	16	14	13	13		12	4	1	–	–	–	4	–	–	–	–	–	–	**Dibenzofuran, 2,4,6,8-tetrachloro-**	58802-19-0	**NI31919**
202	174	55	220	72	304	203	93	**304**	100	74	62	46	30	18	12	12		12	9	6	–	–	–	–	–	–	–	–	–	1	**3-Carboxypropionyl-cyclopentadienylmanganese tricarbonyl**		**LI06663**
61	256	304	56	102	257	69	98	**304**	100	59	34	26	19	18	13	13		12	20	3	2	2	–	–	–	–	–	–	–	–	**5,5-Diethyl-1,3-bis(methylthiomethyl)pyrimidinetrione**		**MS07036**
101	83	102	160	100	70	85	72	**304**	100	89	74	72	62	58	43	42	**1.22**	12	20	7	2	–	–	–	–	–	–	–	–	–	**1,4,10,13-Tetraoxa-7,16-diazacyclooctadecane-2,6,17-trione**	74229-38-2	**NI31920**
101	102	160	100	70	85	72	71	**304**	100	74	72	62	59	43	43	41	**0.00**	12	20	7	2	–	–	–	–	–	–	–	–	–	**1,4,10,13-Tetraoxa-7,16-diazacyclooctadecane-2,6,17-trione**		**MS07037**
137	179	304	93	199	153	135	43	**304**	100	94	64	48	44	42	35	34		12	21	3	2	1	–	–	–	–	–	1	–	–	**Phosphorothioic acid, O,O-diethyl O-[6-methyl-2-isopropyl-4-pyrimidinyl] ester**	333-41-5	**NI31921**
179	137	304	152	29	27	43	41	**304**	100	74	71	62	51	47	40	37		12	21	3	2	1	–	–	–	–	–	1	–	–	**Phosphorothioic acid, O,O-diethyl O-[6-methyl-2-isopropyl-4-pyrimidinyl] ester**	333-41-5	**NI31922**
137	179	152	93	199	97	153	66	**304**	100	73	54	37	34	33	30	26	**9.20**	12	21	3	2	1	–	–	–	–	–	1	–	–	**Phosphorothioic acid, O,O-diethyl O-[6-methyl-2-isopropyl-4-pyrimidinyl] ester**	333-41-5	**NI31923**
83	85	115	102	58	101	57	72	**304**	100	74	33	31	21	19	18	17	**0.57**	12	24	5	4	–	–	–	–	–	–	–	–	–	**1,4,7-Trioxa-10,12,15,17-tetraazacyclononadecane-11,16-dione**		**NI31924**
122	76	30	63	64	75	92	50	**304**	100	55	53	50	49	48	41	40	**29.76**	13	8	7	2	–	–	–	–	–	–	–	–	–	**Carbonic acid, bis(4-nitrophenyl) ester**	5070-13-3	**NI31925**
146	178	220	55	188	302	148	91	**304**	100	76	64	55	47	37	35	30	**1.00**	13	13	5	–	–	–	–	–	–	–	–	–	1	**1-Methoxycarbonylisopropyl-cyclopentadienylmanganese tricarbonyl**		**LI06664**
157	115	74	87	71	83	102	100	**304**	100	72	69	62	54	53	48	41	**0.00**	13	20	8	–	–	–	–	–	–	–	–	–	–	**Methyl 2,3,4-tri-O-acetyl-6-deoxy-α-D-glucopyranoside**	19940-17-1	**NI31926**
115	217	85	83	87	157	71	113	**304**	100	35	31	26	25	23	22	19	**0.00**	13	20	8	–	–	–	–	–	–	–	–	–	–	**Methyl 2,3,5-tri-O-acetyl-6-deoxy-L-galactofuranoside**	24333-02-6	**NI31927**
43	72	114	112	103	28	44	18	**304**	100	7	6	6	6	6	5	4	**0.00**	13	20	8	–	–	–	–	–	–	–	–	–	–	**Peracetyl pentaerythritol**		**IC03799**
59	149	70	58	184	116	87	29	**304**	100	65	47	24	22	17	17	16	**0.03**	13	24	2	2	2	–	–	–	–	–	–	–	–	**1-(2,2-Bis(ethylthio)propionyl)-N-methyl(2S)-pyrrolidine-2-carboxamide**		**NI31928**
30	100	230	104	61	156	43	56	**304**	100	89	84	76	71	60	57	55	**15.98**	13	24	4	2	1	–	–	–	–	–	–	–	–	**N-(N-Acetylglycyl)methionine butyl ester**		**AI02023**
88	43	101	73	45	75	169	115	**304**	100	58	30	25	25	22	19	19	**0.00**	13	24	6	2	–	–	–	–	–	–	–	–	–	**Galactopyranoside, methyl 4,6-diacetamido-4,6-dideoxy-2,3-di-O-methyl-**	6490-75-1	**NI31929**
73	147	83	75	245	28	45	289	**304**	100	83	34	26	25	18	16	15	**0.00**	13	28	4	–	–	–	–	–	–	2	–	–	–	**Diethylmalonic acid, bis(trimethylsilyl) ester**		**NI31930**
73	75	55	125	147	155	69	289	**304**	100	75	34	32	30	29	26	19	**0.00**	13	28	4	–	–	–	–	–	–	2	–	–	–	**Heptanedioic acid, bis(trimethylsilyl) ester**	55530-58-0	**NI31931**
73	75	125	289	69	147	55	45	**304**	100	87	35	34	26	21	19	19	**0.50**	13	28	4	–	–	–	–	–	–	2	–	–	–	**Hexanedioic acid, 3-methyl-, bis(trimethylsilyl) ester**	55520-93-9	**NI31932**
73	75	125	69	55	45	147	97	**304**	100	82	35	33	24	21	15	13	**0.00**	13	28	4	–	–	–	–	–	–	2	–	–	–	**Hexanedioic acid, 3-methyl-, bis(trimethylsilyl) ester**	55520-93-9	**NI31933**
73	83	147	75	289	117	186	45	**304**	100	80	51	43	20	17	16	16	**0.00**	13	28	4	–	–	–	–	–	–	2	–	–	–	**Pentanedioic acid, 3,3-dimethyl-, bis(trimethylsilyl) ester**	55887-57-5	**NI31934**
73	102	84	156	200	128	304	160	**304**	100	78	77	41	37	25	22	20		13	32	2	2	–	–	–	–	–	2	–	–	–	**L-Lysine, bis(trimethylsilyl)-, methyl ester**	25737-21-7	**MS07038**
73	102	84	156	200	128	304	160	**304**	100	75	75	40	35	25	20	18		13	32	2	2	–	–	–	–	–	2	–	–	–	**L-Lysine, bis(trimethylsilyl)-, methyl ester**	25737-21-7	**LI06665**
73	102	84	156	200	128	74	304	**304**	100	77	77	41	35	25	25	20		13	32	2	2	–	–	–	–	–	2	–	–	–	**L-Lysine, bis(trimethylsilyl)-, methyl ester**	25737-21-7	**NI31935**
201	73	145	129	202	203	146	217	**304**	100	84	74	31	24	13	13	10	**0.00**	13	36	–	–	–	–	–	–	–	4	–	–	–	**2,2,4,4,6,6,8,8-Octamethyl-1,3,5,7-tetrasilanonane**		**MS07039**
235	237	165	304	306	199	82	236	**304**	100	65	61	44	30	24	19	17		14	9	–	–	–	3	2	–	–	–	–	–	–	**Ethane, 2,2-bis(4-chlorophenyl)-1,1,1-trifluoro-**	361-07-9	**NI31936**
151	136	153	108	96	152	137	109	**304**	100	56	18	16	14	9	6	6	**0.10**	14	12	2	2	2	–	–	–	–	–	–	–	–	**Bis(2-carbamoylphenyl) disulphide**		**IC03800**
304	166	305	165	69	120	92	91	**304**	100	43	17	17	14	12	10	10		14	12	6	2	–	–	–	–	–	–	–	–	–	**1,2-Bis(4-nitrophenoxy)ethane**		**IC03801**
229	228	230	231	77	194	51	304	**304**	100	71	41	33	24	20	13	11		14	13	–	4	1	–	1	–	–	–	–	–	–	**5-Chloro-2-aminobenzophenone thiosemicarbazone**		**MS07040**
220	56	248	204	304	249	221	276	**304**	100	99	98	25	14	14	13	3		14	16	4	–	–	–	–	–	–	–	–	–	1	**Tricarbonyl[(1,3,4,5-η)-6-isopropyl-3-methyl-2-oxo-4-cycloheptene-1,3-diyl]iron**		**DD01105**
304	105	66	168	91	85	77	79	**304**	100	99	99	95	92	80	75	73		14	17	1	–	–	–	–	–	–	–	–	–	1	**[η^4-2-(Methoxymethyl)norbornadiene](η^5-cyclopentadienyl)rhodium**	101006-49-9	**NI31937**
87	41	43	29	18	69	143	42	**304**	100	60	49	49	37	28	24	21	**0.40**	14	18	3	–	–	–	2	–	–	–	–	–	–	**Butanoic acid, 4-(2,4-dichlorophenoxy)-, butyl ester**	6753-24-8	**NI31938**
87	41	57	43	69	42	40	45	**304**	100	65	61	58	44	27	26	19	**0.20**	14	18	3	–	–	–	2	–	–	–	–	–	–	**Butanoic acid, 4-(2,4-dichlorophenoxy)-, sec-butyl ester**	51550-64-2	**NI31939**
244	286	245	122	43	108	193	82	**304**	100	24	14	10	9	7	6	6	**0.00**	14	19	2	2	–	3	–	–	–	–	–	–	–	**Pyridine, 1-acetyl-1,2,3,4-tetrahydro-5-[1-(trifluoroacetyl)-2-piperidinyl]-**	54966-10-8	**NI31940**
304	289	44	275	261	64	305	290	**304**	100	76	60	40	34	21	18	15		14	20	2	6	–	–	–	–	–	–	–	–	–	**2,2′-Bis(dimethylamino)-4′,6-dimethoxy-4,5′-bipyrimidine**		**NI31941**

MASS TO CHARGE RATIOS	M.W.	INTENSITIES	Parent	C	H	O	N	S	F	Cl	Br	I	Si	P	B	X	COMPOUND NAME	CAS Reg No	No
304 289 246 275 305 303 290 43	**304**	100 76 42 19 18 18 15 10		14	20	2	6	–	–	–	–	–	–	–	–	–	**6,6′-Bis(dimethylamino)-2,2′-dimethoxy-4,4′-bipyrimidine**		**NI31942**
45 99 173 89 87 42 217 129	**304**	100 28 15 14 9 6 6 5	**0.09**	14	24	7	–	–	–	–	–	–	–	–	–	–	**1,4,7,10,13-Pentaoxacyclononadecane-14,19-dione**		**MS07041**
43 71 145 73 75 41 103 129	**304**	100 84 69 35 22 16 14 13	**0.00**	14	28	5	–	–	–	–	–	–	1	–	–	–	**1,3-Diisobutyrin, trimethylsilyl-**		**NI31943**
100 70 56 58 88 57 118 102	**304**	100 94 88 88 82 82 76 76	**29.40**	14	28	5	2	–	–	–	–	–	–	–	–	–	**3,7-Dihydroxy-5,12,17-trioxa-1,9-diazabicyclo[7.5.5]nonadecane**		**MS07042**
73 145 147 75 275 146 45 16	**304**	100 95 33 28 16 13 10 9	**0.00**	14	32	3	–	–	–	–	–	–	2	–	–	–	**2-Ethyl-3-trimethylsiloxy-3-methylpentanoic acid, trimethylsilyl ester**		**NI31944**
73 145 75 261 147 129 217 74	**304**	100 58 20 19 19 12 11 9	**0.00**	14	32	3	–	–	–	–	–	–	2	–	–	–	**2-Ethyl-3-trimethylsiloxy-4-methylpentanoic acid, trimethylsilyl ester**		**NI31945**
73 147 173 233 289 75 247 189	**304**	100 56 38 38 21 19 15 10	**0.00**	14	32	3	–	–	–	–	–	–	2	–	–	–	**3-Trimethylsiloxyoctanoic acid, trimethylsilyl ester**		**NI31946**
117 73 75 55 147 217 199 118	**304**	100 82 40 23 15 14 12 12	**0.00**	14	32	3	–	–	–	–	–	–	2	–	–	–	**7-Trimethylsiloxyoctanoic acid, trimethylsilyl ester**		**NI31947**
117 73 75 18 55 131 147 217	**304**	100 84 49 31 26 20 16 11	**0.40**	14	32	3	–	–	–	–	–	–	2	–	–	–	**7-Trimethylsiloxyoctanoic acid, trimethylsilyl ester**		**NI31948**
73 75 55 289 147 199 97 149	**304**	100 87 67 53 50 26 18 18	**0.00**	14	32	3	–	–	–	–	–	–	2	–	–	–	**8-Trimethylsiloxyoctanoic acid, trimethylsilyl ester**		**NI31949**
73 77 131 147 75 47 55 49	**304**	100 86 79 36 29 19 12 12	**0.00**	14	32	3	–	–	–	–	–	–	2	–	–	–	**3-Trimethylsiloxy-2-propylpentanoic acid, trimethylsilyl ester**		**NI31950**
131 73 147 217 75 246 132 275	**304**	100 86 33 19 17 15 12 12	**0.00**	14	32	3	–	–	–	–	–	–	2	–	–	–	**3-Trimethylsiloxy-2-propylpentanoic acid, trimethylsilyl ester**		**NI31951**
73 131 147 75 41 55 217 45	**304**	100 72 37 24 15 12 11 11	**0.00**	14	32	3	–	–	–	–	–	–	2	–	–	–	**3-Trimethylsiloxy-2-propylpentanoic acid, trimethylsilyl ester**		**MS07043**
73 147 55 75 289 41 185 45	**304**	100 70 41 32 26 25 20 13	**0.00**	14	32	3	–	–	–	–	–	–	2	–	–	–	**5-Trimethylsiloxy-2-propylpentanoic acid, trimethylsilyl ester**		**MS07044**
77 73 147 55 75 289 185 47	**304**	100 88 62 56 35 30 22 19	**0.00**	14	32	3	–	–	–	–	–	–	2	–	–	–	**5-Trimethylsiloxy-2-propylpentanoic acid, trimethylsilyl ester**		**NI31952**
269 271 189 225 306 304 190 270	**304**	100 99 91 56 28 22 21 19		15	10	–	–	–	–	1	1	–	–	–	–	–	**9-Bromo-10-chloromethylanthracene**		**LI06666**
225 189 227 226 306 228 304 271	**304**	100 37 35 22 7 7 6 3		15	10	–	–	–	–	1	1	–	–	–	–	–	**9-Bromomethyl-10-chloroanthracene**		**LI06667**
127 128 304 77	**304**	100 8 7 6		15	10	–	–	–	6	–	–	–	–	–	–	–	**1,3-Diphenylhexafluoropropane**		**LI06668**
153 123 275 152 304 150 69 124	**304**	100 53 34 31 28 25 20 13		15	12	7	–	–	–	–	–	–	–	–	–	–	**2H-1-Benzopyran-4-one, 2-(3,4-dihydroxyphenyl)-3,4-dihydro-3,5,7-trihydroxy-**		**LI06669**
153 123 275 152 149 304 150 165	**304**	100 46 37 36 33 30 20 18		15	12	7	–	–	–	–	–	–	–	–	–	–	**(+)-Dihydroquercetin**	480-18-2	**NI31953**
153 123 275 152 304 150 69 165	**304**	100 52 34 31 28 25 20 14		15	12	7	–	–	–	–	–	–	–	–	–	–	**(+)-Dihydroquercetin**	480-18-2	**NI31954**
153 123 275 152 304 150 165 124	**304**	100 42 28 25 23 21 12 11		15	12	7	–	–	–	–	–	–	–	–	–	–	**Dihydroquercetin**	5117-01-1	**NI31955**
106 107 91 223 222 65 304 77	**304**	100 77 71 63 48 27 22 19		15	16	–	2	–	–	–	–	–	–	–	–	1	**N,N′-Bis(4-tolyl)selenourea**		**MS07045**
257 125 140 77 289 258 65 51	**304**	100 56 46 23 16 16 12 12	**10.00**	15	16	1	2	2	–	–	–	–	–	–	–	–	**Sulphoximine, S-methyl-N-[(methylthio)(phenylimino)methyl]-S-phenyl-**	55759-87-0	**NI31956**
257 125 140 77 258 289 110 65	**304**	100 55 46 22 16 15 13 11	**10.07**	15	16	1	2	2	–	–	–	–	–	–	–	–	**Sulphoximine, S-methyl-N-[(methylthio)(phenylimino)methyl]-S-phenyl-**	55759-87-0	**NI31957**
77 149 121 105 91 51 65 92	**304**	100 97 44 44 36 36 31 22	**16.00**	15	16	3	2	1	–	–	–	–	–	–	–	–	**N′-Methyl-N′-phenyl-N-tosylformohydrazide**		**MS07046**
289 303 290 168 103 105 121 304	**304**	100 19 17 10 10 9 8 7		15	21	–	–	–	–	–	–	–	–	–	–	1	**Rhodium, π-cyclopentadienyl(1,2,3,4,5-pentamethyl-1,3-cyclopentadiene)-, endo-**	32698-19-4	**NI31958**
303 168 103 302 289 142 304 152	**304**	100 45 30 27 24 23 20 14		15	21	–	–	–	–	–	–	–	–	–	–	1	**Rhodium, π-cyclopentadienyl(1,2,3,4,5-pentamethyl-1,3-cyclopentadiene)-, exo-**	33503-63-8	**NI31959**
168 304 302 275 273 300 103 169	**304**	100 70 70 41 31 29 25 22		15	21	–	–	–	–	–	–	–	–	–	–	1	**Rhodium, (η^5-2,4-cyclopentandien-1-yl)(η^4-1,2-diethenylcyclohexane)-, stereoisomer**	74810-93-8	**NI31960**
289 290 73 291 304 137 75 74	**304**	100 27 17 11 8 6 3 3		15	24	1	2	–	–	–	–	–	2	–	–	–	**8-Quinolinamine, N-(trimethylsilyl)-6-[(trimethylsilyl)oxy]-**	36972-87-9	**NI31961**
29 105 41 143 141 55 304 275	**304**	100 61 57 52 50 47 44 43		15	28	2	–	2	–	–	–	–	–	–	–	–	**O-(1-Butoxy-2-butyl-2-ethylvinyl) S-ethyl dithiocarbonate**		**DD01106**
173 71 261 43 145 115 74 117	**304**	100 90 58 56 36 18 14 10	**0.00**	15	28	6	–	–	–	–	–	–	–	–	–	–	**Ethyl 2-hydroxy-2-(2,2-diisopropyl-5-methoxy-1,3-dioxolan-4-yl)propanoate**		**DD01107**
69 83 147 55 73 103 149 57	**304**	100 58 50 48 44 31 19 13	**0.40**	15	36	2	–	–	–	–	–	–	2	–	–	–	**1,9-Bis(trimethylsiloxy)nonane**		**NI31962**
165 220 166 248 55 304 163 164	**304**	100 34 10 9 8 6 5 4		16	9	3	–	–	–	–	–	–	–	–	–	1	**Manganese, tricarbonyl[(4a,4b,8a,9,9b-η)-9H-fluoren-9-yl]-**	31760-87-9	**NI31963**
304 306 269 241 305 176 276 178	**304**	100 66 25 25 23 21 20 16		16	10	2	–	–	–	2	–	–	–	–	–	–	**1,4-Dichloro-2,3-dimethylanthraquinone**		**IC03802**
269 233 271 205 227 135 176 163	**304**	100 40 36 27 25 22 20 20	**10.00**	16	10	2	–	–	–	2	–	–	–	–	–	–	**Naphthalene, 2,3-dichloro-1,2-dihydro-4-hydroxy-1-oxo-2-phenyl-**	71531-36-7	**NI31964**
105 139 77 111 106 51 75 140	**304**	100 50 19 10 6 6 5 4	**0.06**	16	13	4	–	–	–	1	–	–	–	–	–	–	**Ethyl glycol-p-monochloro-dibenzoate**		**IC03803**
166 139 137 304 138 167 151 69	**304**	100 82 50 33 24 17 10 10		16	16	6	–	–	–	–	–	–	–	–	–	–	**2H-1-Benzopyran-3,5,7-triol, 3,4-dihydro-2-(4-hydroxy-3-methoxyphenyl), cis-(±)-**	39689-34-4	**MS07047**
304 258 257 240 305 273 243 241	**304**	100 36 25 23 17 14 13 13		16	16	6	–	–	–	–	–	–	–	–	–	–	**Cassialactone**	80489-64-1	**NI31965**
43 112 220 154 262 82 192 162	**304**	100 88 72 57 55 28 25 22	**1.60**	16	16	6	–	–	–	–	–	–	–	–	–	–	**1,3-Diacetoxy-1,4,4a,8a-tetrahydro-1,4-ethanonaphthalene-5,8-dione**		**LI06670**
28 119 91 145 118 136 43 39	**304**	100 66 64 63 60 50 50 25	**10.00**	16	16	6	–	–	–	–	–	–	–	–	–	–	**Dimethyl 1-acetyl-1,4-dihydro-1-hydroxy-2,3-naphthalenedicarboxylate**		**MS07048**
128 84 113 103 143 77 135 115	**304**	100 96 77 34 32 29 23 23	**11.00**	16	16	6	–	–	–	–	–	–	–	–	–	–	**5,6-cis-5-Methoxymethysticin**	52525-99-2	**NI31966**
128 113 84 85 143 115 103 135	**304**	100 69 32 20 19 14 14 12	**8.00**	16	16	6	–	–	–	–	–	–	–	–	–	–	**5,6-trans-5-Methoxymethysticin**	52526-00-8	**NI31967**
105 77 182 79 122 106 78 304	**304**	100 66 32 20 17 16 14 1		16	16	6	–	–	–	–	–	–	–	–	–	–	**Methyl [3aS-(3aα,4α,5β,6aα)]-hexahydro-5-(benzoyloxy)-1-oxo-1H-cyclopenta[c]furan-4-carboxylate**		**NI31968**
105 77 199 304 182 150 138 167	**304**	100 28 10 7 4 4 3 2		16	16	6	–	–	–	–	–	–	–	–	–	–	**Methyl [3aS-(3aα,4β,5β,6aα)]-hexahydro-5-(benzoyloxy)-1-oxo-1H-cyclopenta[c]furan-4-carboxylate**		**NI31969**
141 168 110 196 142 169 213 245	**304**	100 95 89 85 73 71 70 69	**0.00**	16	16	6	–	–	–	–	–	–	–	–	–	–	**2-Oxospiro[tetrahydrofuran-3,3′-tricyclo[3.2.2.0^{2,4}]nona-6′,8′-dien]6′,7′-dicarboxylic acid dimethyl ester**	81003-59-0	**NI31970**
198 225 170 91 55 197 141 304	**304**	100 58 32 31 30 25 23 17		16	17	1	–	–	–	–	1	–	–	–	–	–	**2(3H)-Naphthalenone, 1-bromo-4,4a,5,6,7,8-hexahydro-4a-phenyl-**	55162-75-9	**NI31971**

MASS TO CHARGE RATIOS	M.W.	INTENSITIES	Parent	C	H	O	N	S	F	Cl	Br	I	Si	P	B	X	COMPOUND NAME	CAS Reg No	No
105 77 167 289 304	**304**	100 48 25 11 1		16	20	2	2	1	–	–	–	–	–	–	–	–	S,S-(-)-N,N′-α-Phenylethylsulphuric acid diamide		LI06671
113 81 146 55 90 82 163 120	**304**	100 60 44 27 26 20 16 16	**5.00**	16	20	4	2	–	–	–	–	–	–	–	–	–	Benzoic acid, 4-(tetrahydro-5,5-dimethyl-2-oxopyrrolo[1,2-b][1,2,4]oxadiazol-1(2H)-yl)-, ethyl ester	50455-85-1	NI31972
141 200 129 184 142 115 145 157	**304**	100 60 55 50 45 39 35 24	**0.00**	16	20	4	2	–	–	–	–	–	–	–	–	–	9a,10a-Dihydroxy-1,2,3,4,4a,9,10,10a-(trans-4a,10a)-octahydrophenanthrene dicarbamate	53447-06-6	NI31973
200 145 141 129 115 185 157 184	**304**	100 62 61 58 37 32 28 25	**0.00**	16	20	4	2	–	–	–	–	–	–	–	–	–	9a,10E-Dihydroxy-1,2,3,4,4a,9,10,10a-(trans-4a,10a)-octahydrophenanthrene dicarbamate	53447-04-4	NI31974
141 200 129 184 182 128 157 115	**304**	100 78 49 44 40 40 37 37	**0.00**	16	20	4	2	–	–	–	–	–	–	–	–	–	9E,10a-Dihydroxy-1,2,3,4,4a,9,10,10a-(trans-4a,10a)-octahydrophenanthrene dicarbamate	53447-07-7	NI31975
200 157 141 182 129 158 115 91	**304**	100 88 56 55 55 52 42 36	**0.00**	16	20	4	2	–	–	–	–	–	–	–	–	–	9E,10E-Dihydroxy-1,2,3,4,4a,9,10,10a-(trans-4a,10a)-octahydrophenanthrene dicarbamate	53447-05-5	NI31976
110 82 304 41 111 55 54 108	**304**	100 35 28 26 24 22 22 15		16	20	4	2	–	–	–	–	–	–	–	–	–	N,N′-Octamethylenebismaleimide	28537-73-7	NI31977
304 73 305 289 306 45 201 74	**304**	100 53 28 15 11 7 6 5		16	24	2	–	–	–	–	–	–	2	–	–	–	Silane, [1,5-naphthalenediylbis(oxy)]bis[trimethyl-	1032-28-6	NI31978
303 289 73 305 290 137 306 217	**304**	100 60 38 30 17 15 12 10	**0.00**	16	24	2	–	–	–	–	–	–	2	–	–	–	Silane, (2,7-naphthylenedioxy)bis[trimethyl-	1226-72-8	NI31979
304 289 305 290 137 306 231 218	**304**	100 60 29 16 14 11 8 8		16	24	2	–	–	–	–	–	–	2	–	–	–	Silane, (2,7-naphthylenedioxy)bis[trimethyl-	1226-72-8	NI31980
102 73 203 202 103 204 45 74	**304**	100 78 56 31 11 10 10 6	**3.40**	16	28	–	2	–	–	–	–	–	2	–	–	–	Aminoethylindole, 1,3-bis(trimethylsilyl)-		NI31981
174 73 175 86 176 59 130 289	**304**	100 65 26 21 13 13 11 10	**0.80**	16	28	–	2	–	–	–	–	–	2	–	–	–	Aminoethylindole, bis(trimethylsilyl)-		NI31982
55 43 41 69 83 70 84 57	**304**	100 95 87 86 77 74 73 68	**6.48**	16	32	3	–	1	–	–	–	–	–	–	–	–	1,4-Hexadecanesultone		MS07049
55 67 82 96 69 79 110 208	**304**	100 82 73 57 46 39 23 15	**0.00**	16	32	3	–	1	–	–	–	–	–	–	–	–	Methyl 8-pentadecene-1-sulphonate		DD01108
155 95 131 55 57 69 67 109	**304**	100 44 35 23 21 19 16 13	**0.00**	16	32	5	–	–	–	–	–	–	–	–	–	–	Hexadecanoic acid, 9,10,16-trihydroxy-, (R*,S*)-(±)-	533-87-9	NI31983
43 41 57 55 71 69 135 137	**304**	100 91 79 63 38 35 28 27	**0.02**	16	33	–	–	–	–	–	1	–	–	–	–	–	1-Bromohexadecane		AI02024
57 43 71 55 41 135 69 85	**304**	100 93 53 52 52 50 37 31	**1.27**	16	33	–	–	–	–	–	1	–	–	–	–	–	1-Bromohexadecane		IC03804
57 43 71 69 55 41 135 85	**304**	100 87 52 38 35 35 28 28	**0.00**	16	33	–	–	–	–	–	1	–	–	–	–	–	1-Bromohexadecane		IC03805
304 305 138 277 221 193 164 76	**304**	100 20 7 5 5 5 5 4		17	8	4	2	–	–	–	–	–	–	–	–	–	2-Amino-3-cyano-5-hydroxy-6,11-dioxo-anthra[1,2-b]furan		IC03806
77 104 150 51 121 103 116 89	**304**	100 64 55 41 39 39 35 33	**0.00**	17	14	1	–	–	–	2	–	–	–	–	–	–	1-(2-Chloro-2-methoxyethenyl)-4-(1-chloro-2-phenylethenyl)benzene		DD01109
208 164 209 189 304 306 234 233	**304**	100 67 45 24 19 13 12 12		17	14	1	–	–	–	2	–	–	–	–	–	–	anti-11-syn-12-Dichloro-2-methoxy-9,10-dihydro-9,10-ethanoanthracene		NI31984
208 164 209 189 304 104 234 233	**304**	100 70 48 19 14 14 13 13		17	14	1	–	–	–	2	–	–	–	–	–	–	anti-anti-11,12-Dichloro-2-methoxy-9,10-dihydro-9,10-ethanoanthracene		NI31985
208 164 209 189 304 104 165 234	**304**	100 71 65 32 22 19 17 16		17	14	1	–	–	–	2	–	–	–	–	–	–	syn-11-anti-12-Dichloro-2-methoxy-9,10-dihydro-9,10-ethanoanthracene		NI31986
208 164 209 189 304 234 233 104	**304**	100 75 50 20 15 15 12 12		17	14	1	–	–	–	2	–	–	–	–	–	–	syn-syn-11,12-Dichloro-2-methoxy-9,10-dihydro-9,10-ethanoanthracene		NI31987
162 133 232 204 176 244 304 287	**304**	100 70 53 45 42 11 3 2		17	20	5	–	–	–	–	–	–	–	–	–	–	10-Acetoxy-8,9-epoxythymol 3-angelate		NI31988
43 215 260 93 91 176 189 79	**304**	100 69 56 30 28 26 25 24	**0.28**	17	20	5	–	–	–	–	–	–	–	–	–	–	Acetyloxyparthenin		NI31989
43 41 39 91 80 27 45 77	**304**	100 45 37 27 24 20 18 17	**0.00**	17	20	5	–	–	–	–	–	–	–	–	–	–	Cyclodeca[b]furan-2,9(3H,4H)-dione, 4-(acetyloxy)-3a,7,8,10,11,11a-hexahydro-6-methyl-3,10-bis(methylene)-, [3aR-(3aR*,4R*,5E,11aS*)]-	34226-88-5	NI31990
193 84 112 43 83 79 56 69	**304**	100 82 76 74 71 68 65 64	**0.00**	17	20	5	–	–	–	–	–	–	–	–	–	–	5-Isobenzofurancarboxylic acid, 1,3-dihydro-1,3-dioxo-, octyl ester	22485-49-0	NI31991
44 108 91 57 201 115 109 131	**304**	100 86 23 23 21 15 15 14	**1.00**	17	20	5	–	–	–	–	–	–	–	–	–	–	5-Methoxycarbonyl-9-(3-oxobutenyl)-oxatricycloundecen-2-one		MS07050
185 105 83 77 43 85 244 157	**304**	100 93 27 26 25 18 17 12	**0.00**	17	20	5	–	–	–	–	–	–	–	–	–	–	Methyl α-[1-(acetyloxy)-2-cyclohexen-1-yl]-α-hydroxybenzeneacetate		DD01110
192 206 193 218 204 176 98 41	**304**	100 31 22 17 9 9 9 8	**7.00**	17	24	3	2	–	–	–	–	–	–	–	–	–	2-Pyrrolidinone, 1-[2-(1,2,3,4-tetrahydro-6,7-dimethoxy-1-isoquinolinyl)ethyl]-	47216-65-9	MS07051
168 304 137 169 136 305 108 140	**304**	100 57 26 14 14 13 13 11		18	12	1	2	1	–	–	–	–	–	–	–	–	2-α-Naphthylimino-2,3-dihydro-4H-1,3-benzothiazin-4-one	75096-84-3	NI31992
168 304 137 108 169 141 136 305	**304**	100 55 30 16 14 14 13 12		18	12	1	2	1	–	–	–	–	–	–	–	–	2-β-Naphthylimino-2,3-dihydro-4H-1,3-benzothiazin-4-one	56580-25-7	NI31993
304 305 73 232 260 152 219 120	**304**	100 21 21 17 14 9 8 7		18	12	3	2	–	–	–	–	–	–	–	–	–	3-Hydroxyquino-(4-aminophthalone)		IC03807
262 304 263 144 104 92 233 91	**304**	100 29 20 15 14 13 12 12		18	16	1	4	–	–	–	–	–	–	–	–	–	Acetic acid, phenyl[(1-phenyl-1H-pyrazol-3-yl)methylene]hydrazide	55649-75-7	NI31994
131 43 132 77 51 76 133 103	**304**	100 88 67 59 50 48 34 30	**2.00**	18	16	1	4	–	–	–	–	–	–	–	–	–	1′-Acetyl-2′-(1,2-dihydroquinoxalin-2-yl)-1′,2′-dihydroquinoxaline	95234-57-4	NI31995
289 259 304 94 290 243 57 56	**304**	100 56 26 23 20 17 16 15		18	24	4	–	–	–	–	–	–	–	–	–	–	2,6-Di-tert-butyl-4-carboxyacryloyl-phenol		IC03808
202 187 159 240 225 131 183 91	**304**	100 85 46 39 35 30 25 22	**4.00**	18	24	4	–	–	–	–	–	–	–	–	–	–	3-[4-Hydroxyisopent-2(Z)-enyl]-5-[4-hydroxyisopent-2-(E)-enyl]-4-hydroxyacetophenone		NI31996
179 161 56 55 43 41 70 69	**304**	100 26 26 22 22 20 17 16	**9.00**	18	24	4	–	–	–	–	–	–	–	–	–	–	5-Isobenzofurancarboxylic acid, 1,3-dihydro-1-oxo-, nonyl ester	54699-45-5	NI31997
179 161 56 55 180 41 69 43	**304**	100 24 13 12 11 11 10 10	**10.00**	18	24	4	–	–	–	–	–	–	–	–	–	–	5-Isobenzofurancarboxylic acid, 1,3-dihydro-3-oxo-, nonyl ester	54699-44-4	NI31998
131 103 41 162 77 174 37 51	**304**	100 61 57 56 40 37 32 28	**0.15**	18	24	4	–	–	–	–	–	–	–	–	–	–	Malonic acid, benzylidene-, di-tert-butyl ester	32046-34-7	NI31999
152 153 124 96 81 108 28 29	**304**	100 42 10 10 10 9 9 8	**0.00**	18	28	2	2	–	–	–	–	–	–	–	–	–	meso-N,N,N′,N′-Tetraethyl-1,2-bis(2-furyl)ethylenediamine		DD01111
152 153 160 124 108 96 81 28	**304**	100 55 26 22 19 19 18 15	**0.00**	18	28	2	2	–	–	–	–	–	–	–	–	–	(±)-N,N,N′,N′-Tetraethyl-1,2-bis(2-furyl)ethylenediamine		DD01112
304 229 303 227 28 152 211 228	**304**	100 84 81 49 40 38 37 35		19	13	2	–	–	–	–	–	–	–	1	–	–	9,10-Dihydro-9-phenyl-9-phosphaanthracen-10-one 9-oxide		MS07052
199 304 105 227 184 289 305 303	**304**	100 87 83 30 24 19 18 17		19	16	2	2	–	–	–	–	–	–	–	–	–	1,2-Dihydro-3-benzoyl-4-methyl-9H-pyrrolo[2,1-b]quinazoline-9-one		LI06672
277 278 77 262 183 201 51 199	**304**	100 66 25 18 18 17 15 12	**1.00**	19	17	–	2	–	–	–	–	–	–	1	–	–	N-Methylidene-N[1]-triphenylphosphoranylidene azine		MS07053
277 278 77 262 183 201 199 185	**304**	100 66 25 18 18 17 12 10	**1.00**	19	17	–	2	–	–	–	–	–	–	1	–	–	N-Methylidene-N[1]-triphenylphosphoranylidene azine		LI06673

MASS TO CHARGE RATIOS	M.W.	INTENSITIES	Parent	C	H	O	N	S	F	Cl	Br	I	Si	P	B	X	COMPOUND NAME	CAS Reg No	No
147 83 162 161 179 125 77 85	**304**	100 89 51 49 43 40 37 36	**1.70**	19	25	1	–	–	–	1	–	–	–	–	–	–	1-Hydroxy-6-(4′-chlorobenzyl)-1,2,3,4,5,6-hexamethylcyclohexa-2,4-diene	84615-08-7	NI32000
304 137 275 245 305 83 276 289	**304**	100 45 39 34 22 22 18 17		19	28	3	–	–	–	–	–	–	–	–	–	–	Androstane-3,6-dione, 17-hydroxy-, (5α,17β)-	899-39-8	NI32001
304 55 43 41 57 67 275 79	**304**	100 53 48 44 29 26 23 23		19	28	3	–	–	–	–	–	–	–	–	–	–	Androstane-3,6-dione, 17-hydroxy-, (5α,17β)-	899-39-8	NI32002
304 151 55 164 41 81 67 305	**304**	100 35 35 30 26 22 22 21		19	28	3	–	–	–	–	–	–	–	–	–	–	Androstane-3,6-dione, 17-hydroxy-, (5β,17β)-	53512-50-8	NI32003
304 55 41 124 43 180 81 305	**304**	100 34 34 28 28 21 21 20		19	28	3	–	–	–	–	–	–	–	–	–	–	Androstane-3,11-dione, 17-hydroxy-, (5α,17β)-	32694-37-4	NI32004
304 41 55 124 43 180 305 81	**304**	100 34 32 27 27 21 20 20		19	28	3	–	–	–	–	–	–	–	–	–	–	Androstane-3,11-dione, 17-hydroxy-, (5α,17β)-	32694-37-4	LI06674
304 193 122 124 289 109 55 41	**304**	100 40 40 34 33 33 33 33		19	28	3	–	–	–	–	–	–	–	–	–	–	Androstane-3,11-dione, 17-hydroxy-, (5β,17β)-	1420-71-9	NI32005
304 193 122 289 124 109 55 41	**304**	100 40 40 32 32 32 32 32		19	28	3	–	–	–	–	–	–	–	–	–	–	Androstane-3,11-dione, 17-hydroxy-, (5β,17β)-	1420-71-9	LI06675
304 55 41 123 67 305 93 43	**304**	100 42 33 26 22 20 19 18		19	28	3	–	–	–	–	–	–	–	–	–	–	Androstane-3,17-dione, 6-hydroxy-, (5α,6α)-	49643-95-0	NI32006
304 55 41 43 67 230 124 79	**304**	100 44 40 38 24 21 21 20		19	28	3	–	–	–	–	–	–	–	–	–	–	Androstane-3,17-dione, 6-hydroxy-, (5α,6β)-	7801-26-5	NI32007
304 55 41 67 286 93 81 79	**304**	100 97 65 49 43 40 39 39		19	28	3	–	–	–	–	–	–	–	–	–	–	Androstane-3,17-dione, 6-hydroxy-, (5β,6α)-	5225-43-4	NI32008
304 55 41 43 67 233 286 197	**304**	100 50 45 30 25 24 22 22		19	28	3	–	–	–	–	–	–	–	–	–	–	Androstane-3,17-dione, 6-hydroxy-, (5β,6β)-	51467-38-0	NI32009
304 55 97 41 305 95 81 79	**304**	100 32 27 27 20 17 17 17		19	28	3	–	–	–	–	–	–	–	–	–	–	Androstane-3,17-dione, 11-hydroxy-, (5α,11α)-		LI06676
304 55 41 286 271 97 43 305	**304**	100 38 37 31 27 26 22 21		19	28	3	–	–	–	–	–	–	–	–	–	–	Androstane-3,17-dione, 11-hydroxy-, (5α,11β)-		LI06677
286 304 122 55 41 147 123 109	**304**	100 89 82 67 59 37 37 36		19	28	3	–	–	–	–	–	–	–	–	–	–	Androstane-3,17-dione, 11-hydroxy-, (5β,11α)-		LI06678
304 55 41 43 260 97 67 305	**304**	100 42 40 25 23 21 21 20		19	28	3	–	–	–	–	–	–	–	–	–	–	Androstane-3,17-dione, 11-hydroxy-, (5β,11β)-		LI06679
304 97 260 55 41 43 230 81	**304**	100 52 48 45 30 26 24 22		19	28	3	–	–	–	–	–	–	–	–	–	–	Androstane-3,17-dione, 12-hydroxy-, (5α,12α)-	53604-44-7	NI32010
304 97 230 55 41 43 305 81	**304**	100 41 35 33 26 22 21 16		19	28	3	–	–	–	–	–	–	–	–	–	–	Androstane-3,17-dione, 12-hydroxy-, (5α,12β)-	53608-28-9	NI32011
286 55 41 81 43 289 109 67	**304**	100 47 39 31 31 29 24 23	**3.03**	19	28	3	–	–	–	–	–	–	–	–	–	–	Androstane-3,17-dione, 12-hydroxy-, (5β,12α)-	53604-45-8	NI32012
304 55 41 43 97 81 67 79	**304**	100 76 45 40 33 29 25 20		19	28	3	–	–	–	–	–	–	–	–	–	–	Androstane-3,17-dione, 12-hydroxy-, (5β,12β)-	53604-46-9	NI32013
304 55 95 67 93 41 79 108	**304**	100 38 35 32 30 27 25 22		19	28	3	–	–	–	–	–	–	–	–	–	–	Androstane-6,17-dione, 3-hydroxy-, (3α,5α)-	53512-53-1	NI32014
139 95 55 67 93 41 79 43	**304**	100 70 60 54 52 52 45 38	**0.00**	19	28	3	–	–	–	–	–	–	–	–	–	–	Androstane-6,17-dione, 3-hydroxy-, (3β,5α)-	4601-47-2	NI32015
233 286 304 55 234 41 136 67	**304**	100 37 21 18 16 16 15 14		19	28	3	–	–	–	–	–	–	–	–	–	–	Androstane-6,17-dione, 3-hydroxy-, (3β,5β)-	53512-52-0	NI32016
304 147 41 107 55 109 108 79	**304**	100 38 35 30 29 28 26 26		19	28	3	–	–	–	–	–	–	–	–	–	–	Androstane-11,17-dione, 3-hydroxy-, (3α,5α)-	1231-82-9	NI32017
304 147 41 107 109 55 108 79	**304**	100 37 35 30 27 27 25 25		19	28	3	–	–	–	–	–	–	–	–	–	–	Androstane-11,17-dione, 3-hydroxy-, (3α,5α)-	1231-82-9	LI06680
232 191 286 150 271 121 109 163	**304**	100 72 53 50 48 38 34 33	**25.14**	19	28	3	–	–	–	–	–	–	–	–	–	–	Androstane-11,17-dione, 3-hydroxy-, (3α,5β)-	739-27-5	NI32018
232 304 191 286 41 271 150 55	**304**	100 99 80 70 56 50 50 45		19	28	3	–	–	–	–	–	–	–	–	–	–	Androstane-11,17-dione, 3-hydroxy-, (3α,5β)-	739-27-5	NI32019
304 41 191 147 305 55 107 79	**304**	100 25 22 22 20 20 17 17		19	28	3	–	–	–	–	–	–	–	–	–	–	Androstane-11,17-dione, 3-hydroxy-, (3β,5α)-	7090-90-6	LI06681
304 41 191 147 305 55 79 107	**304**	100 25 23 23 20 20 19 18		19	28	3	–	–	–	–	–	–	–	–	–	–	Androstane-11,17-dione, 3-hydroxy-, (3β,5α)-	7090-90-6	NI32020
304 191 150 109 108 107 41 163	**304**	100 99 87 52 51 47 43 40		19	28	3	–	–	–	–	–	–	–	–	–	–	Androstane-11,17-dione, 3-hydroxy-, (3β,5β)-	28336-28-9	LI06682
191 304 150 109 108 107 41 163	**304**	100 99 87 52 50 44 44 41		19	28	3	–	–	–	–	–	–	–	–	–	–	Androstane-11,17-dione, 3-hydroxy-, (3β,5β)-	28336-28-9	NI32021
110 304 55 97 286 67 41 81	**304**	100 90 73 70 68 57 53 51		19	28	3	–	–	–	–	–	–	–	–	–	–	Androstane-12,17-dione, 3-hydroxy-, (3α,5β)-	53604-39-0	NI32022
110 304 41 97 55 67 123 79	**304**	100 44 27 25 23 22 21 18		19	28	3	–	–	–	–	–	–	–	–	–	–	Androstane-12,17-dione, 3-hydroxy-, (3β,5α)-	53604-38-9	NI32023
217 199 105 91 79 81 41 218	**304**	100 73 64 38 31 27 27 26	**17.00**	19	28	3	–	–	–	–	–	–	–	–	–	–	17-Oxo-4-nor-3,5-seco-5-androsten-3-oic acid methyl ester		MS07054
124 304 262 109 163 162 145 123	**304**	100 49 47 28 24 22 22 22		19	28	3	–	–	–	–	–	–	–	–	–	–	Androst-4-en-3-one, 16,17-dihydroxy-, (16α,17β)-	63-01-4	NI32024
304 213 286 271 231 232 214 199	**304**	100 58 53 41 39 36 35 27		19	28	3	–	–	–	–	–	–	–	–	–	–	Androst-5-en-17-one, 3,16-dihydroxy-, (3β,16α)-	1232-73-1	NI32025
256 255 286 257 91 97 145 143	**304**	100 46 33 20 16 15 13 11	**2.98**	19	28	3	–	–	–	–	–	–	–	–	–	–	Androst-5-en-17-one, 3,19-dihydroxy-, (3β)-	2857-45-6	NI32026
161 91 105 133 147 119 176 162	**304**	100 68 54 45 37 28 20 20	**2.00**	19	28	3	–	–	–	–	–	–	–	–	–	–	9,11-Octadecadiynoic acid, 8-oxo-, methyl ester	75125-35-8	NI32027
55 91 87 105 161 67 117 131	**304**	100 92 91 59 49 48 47 45	**0.00**	19	28	3	–	–	–	–	–	–	–	–	–	–	17-Octadecene-9,11-diynoic acid, 8-hydroxy-, methyl ester	6084-79-3	NI32028
137 275 121 226 138 81 276 41	**304**	100 40 29 13 13 10 9 9	**0.15**	19	32	1	–	–	–	–	–	–	1	–	–	–	Silane, ethylmethyl[[5-methyl-2-isopropylcyclohexyl]oxy]phenyl-	74646-00-7	NI32029
137 275 121 138 219 226 276 149	**304**	100 36 22 13 10 9 8 8	**0.24**	19	32	1	–	–	–	–	–	–	1	–	–	–	Silane, ethylmethyl[[5-methyl-2-isopropylcyclohexyl]oxy]phenyl-, (1α,2α,5β)-	74841-58-0	NI32030
72 98 44 73 42 41 233 126	**304**	100 8 6 5 4 3 2 2	**0.00**	19	32	1	2	–	–	–	–	–	–	–	–	–	N,N-Dimethyl-1-Æ4-[3-(1-piperidyl)propoxy]phenylŒ-2-propylamine		MS07055
304 99 77 200 149 91 87 71	**304**	100 19 11 10 10 10 10 10		20	8	–	4	–	–	–	–	–	–	–	–	–	10,11-Dicyanophenanthro[4,5-fgh]quinoxaline		MS07056
304 196 181 305 275 194 170 197	**304**	100 57 37 23 22 17 16 10		20	13	2	–	–	1	–	–	–	–	–	–	–	Benzo[b]benzo[3,4]cyclobuta[1,2-e][1,4]dioxin, 4b-(4-fluorophenyl)-4b,10a dihydro-	58330-09-9	NI32031
117 215 119 149 216 245 244 121	**304**	100 99 97 70 57 41 38 34	**32.43**	20	16	3	–	–	–	–	–	–	–	–	–	–	cis-5,6-Dihydro-5,6-dihydroxychrysene monomethyl orthoformate		NI32032
257 304 226 215 244 269 239 202	**304**	100 50 21 17 16 11 11 11		20	16	3	–	–	–	–	–	–	–	–	–	–	2-Hydroxy-7,12-bis(dihydroxymethyl)benz[a]anthracene		MS07057
257 304 215 245 226 244 202 239	**304**	100 46 25 23 23 18 17 13		20	16	3	–	–	–	–	–	–	–	–	–	–	3-Hydroxy-7,12-bis(dihydroxymethyl)benz[a]anthracene		MS07058
257 226 215 304 202 239 245 269	**304**	100 46 46 30 27 23 20 16		20	16	3	–	–	–	–	–	–	–	–	–	–	4-Hydroxy-7,12-bis(dihydroxymethyl)benz[a]anthracene		MS07059
272 304 215 244 288 73 273 256	**304**	100 56 48 35 34 30 25 17		20	16	3	–	–	–	–	–	–	–	–	–	–	Methyl 3,5-diphenyl-2-hydroxybenzoate		MS07060
105 77 199 304 286 106 51 152	**304**	100 31 14 9 9 8 6 5		20	16	3	–	–	–	–	–	–	–	–	–	–	2,2′-Methylenebisphenol benzoate		NI32033
105 77 182 106 51 78 28 50	**304**	100 79 38 20 20 10 10 9	**0.00**	20	16	3	–	–	–	–	–	–	–	–	–	–	1,2,4-Trioxolane, 3,3,5-triphenyl-	23246-12-0	NI32034
227 304 305 228 152 165 77 195	**304**	100 84 17 16 10 9 8 7		20	17	1	–	–	–	–	–	–	–	1	–	–	2,8-Dimethyl-10-phenylphenoxaphosphine		LI06683
304 227 305 228 152 195 165 196	**304**	100 92 23 16 9 5 4 3		20	17	1	–	–	–	–	–	–	–	1	–	–	2,8-Dimethyl-10-phenylphenoxaphosphine		LI06684
183 150 227 77 181 226 199 105	**304**	100 59 55 47 43 38 37 32	**17.30**	20	20	1	–	–	–	–	–	–	1	–	–	–	Silane, ethoxytriphenyl-	1516-80-9	NI32035

MASS TO CHARGE RATIOS								M.W.	INTENSITIES								Parent	C	H	O	N	S	F	Cl	Br	I	Si	P	B	X	COMPOUND NAME	CAS Reg No	No
103	261	262	91	131	304	42	77	**304**	100	90	63	56	47	40	39	38		20	20	1	2	–	–	–	–	–	–	–	–	–	5,7-Diphenyl-6-oxo-1,3-diazaadamantane		MS07061
144	145	304	149	159	158	130	146	**304**	100	60	48	39	36	26	17	14		20	20	1	2	–	–	–	–	–	–	–	–	–	Pyrrolo[1,2-a:5,4-b']diindol-12(5aH)-one, 6,10b,11,11a-tetrahydro-5a,10b,11a-trimethyl-	24628-65-7	NI32036
304	184	185	156	77	170	130	275	**304**	100	81	75	50	38	37	36	35		20	20	1	2	–	–	–	–	–	–	–	–	–	Spiro[cyclohexane-1,3'-[3H]indole]-2'-carboxanilide	15308-85-7	NI32037
304	43	41	55	93	81	105	95	**304**	100	80	60	56	54	54	46	46		20	32	2	–	–	–	–	–	–	–	–	–	–	5α-Androstan-17β-ol, 2α,3α-epoxy-3-methyl-	26737-12-2	NI32038
245	81	70	93	95	105	91	79	**304**	100	79	78	77	74	73	72	71	**0.00**	20	32	2	–	–	–	–	–	–	–	–	–	–	5α-Androstan-17β-ol, 2β,3β-epoxy-2-methyl-	16427-03-5	NI32039
304	43	41	81	55	93	95	67	**304**	100	88	71	56	56	55	45	45		20	32	2	–	–	–	–	–	–	–	–	–	–	5α-Androstan-17β-ol, 2β,3β-epoxy-3-methyl-	16394-67-5	NI32040
41	55	218	304	67	81	43	69	**304**	100	62	51	42	40	36	36	33		20	32	2	–	–	–	–	–	–	–	–	–	–	Androstan-3-one, 17-hydroxy-1-methyl-, (1α,5α,17β)-	1424-00-6	NI32041
235	236	41	218	55	43	219	67	**304**	100	50	38	21	21	16	15	15	**10.47**	20	32	2	–	–	–	–	–	–	–	–	–	–	Androstan-3-one, 17-hydroxy-1-methyl-, (1β,5α,17β)-	1232-57-1	NI32042
41	55	43	67	81	79	95	93	**304**	100	64	50	42	39	31	29	29	**15.68**	20	32	2	–	–	–	–	–	–	–	–	–	–	Androstan-3-one, 17-hydroxy-2-methyl-, (2β,5β,17β)-	54630-68-1	NI32043
220	304	176	202	221	161	187	289	**304**	100	34	31	26	17	16	14	12		20	32	2	–	–	–	–	–	–	–	–	–	–	Androst-1-en-3,17-diol, 1-methyl-, (3β,5α,17β)-	56051-67-3	NI32044
287	269	288	285	303	270	286	283	**304**	100	86	27	26	23	19	11	10	**9.80**	20	32	2	–	–	–	–	–	–	–	–	–	–	Androst-5-en-3,17-diol, 17-methyl-, (3β,17α,17β)-		NI32045
286	201	287	268	271	253	225		**304**	100	63	21	20	10	9	7		**0.00**	20	32	2	–	–	–	–	–	–	–	–	–	–	1,8-Azulenediol, 7-(1,5-dimethyl-4-hexenyl)-1,3a,4,5,6,7,8,8a-octahydro-1-methyl-4-methylene-, [1R-[1α,3aα,7β(R*,8β,8aβ)]]-	63908-63-4	NI32046
95	107	121	123	109	191	122	136	**304**	100	54	35	33	32	27	21	20	**1.00**	20	32	2	–	–	–	–	–	–	–	–	–	–	rel-(5S,8R,9S,10R)-ent-Cleroda-3,13-dien-15-oic acid		MS07062
304	41	55	43	67	245	81	79	**304**	100	52	39	34	31	28	26	23		20	32	2	–	–	–	–	–	–	–	–	–	–	Estran-3-one, 17-hydroxy-4,4-dimethyl-, (5α,17β)-	2059-40-7	NI32047
133	273	121	135	107	105	109	117	**304**	100	98	77	38	37	33	29	27	**0.80**	20	32	2	–	–	–	–	–	–	–	–	–	–	Isopimara-8(14),15-diene-17,19-diol		MS07063
123	81	286	55	109	257	94	79	**304**	100	98	83	77	73	70	68	63	**17.02**	20	32	2	–	–	–	–	–	–	–	–	–	–	Kauran-18-al, 16-hydroxy-, (4β)-	55255-87-3	NI32048
81	95	107	93	109	121	135	43	**304**	100	91	79	73	71	63	58	57	**0.40**	20	32	2	–	–	–	–	–	–	–	–	–	–	Labda-8(20),13-dien-15-oic acid, (E)-(+)-	24470-48-2	NI32049
289	271	206	43	55	81	151	122	**304**	100	53	35	32	29	28	24	24	**1.82**	20	32	2	–	–	–	–	–	–	–	–	–	–	Labd-14-en-2-one, 8,13-epoxy-, (13R)-	1231-34-1	NI32050
81	55	43	67	41	95	79	93	**304**	100	69	54	52	37	35	31	30	**0.20**	20	32	2	–	–	–	–	–	–	–	–	–	–	Labd-14-en-3-one, 8,13-epoxy-	26729-54-4	NI32051
81	289	123	114	95	304	230	191	**304**	100	92	92	92	70	45	44	41		20	32	2	–	–	–	–	–	–	–	–	–	–	Methyl 19-norlabda-8(20),13-dien-15-oate		MS07064
218	203	175	148	109	108	43	55	**304**	100	63	54	49	49	44	42	40	**21.00**	20	32	2	–	–	–	–	–	–	–	–	–	–	D-Norandrostane-16-carboxylic acid, methyl ester, (5α,16β)-	54411-59-5	NI32052
218	109	148	108	43	175	203	55	**304**	100	57	53	49	43	38	36	30	**1.00**	20	32	2	–	–	–	–	–	–	–	–	–	–	D-Norandrostan-16-ol, acetate, (5α,16β)-	54411-62-0	NI32053
81	55	93	67	41	123	109	95	**304**	100	69	62	59	59	58	57	55	**24.00**	20	32	2	–	–	–	–	–	–	–	–	–	–	17-Norkaur-18-oic acid, 13-methyl-, (4α,8β,13β)-		LI06685
289	275	304	121	243	109	107	105	**304**	100	56	41	39	31	28	27	27		20	32	2	–	–	–	–	–	–	–	–	–	–	1-Phenanthrenecarboxylic acid, 7-ethyl-1,2,3,4,4a,4b,5,6,7,9,10,10a-dodecahydro-1,4a,7-trimethyl-, [1R-(1α,4aβ,4bα,7α,10aα)]-	4807-69-6	NI32054
121	93	133	109	255	148	119	286	**304**	100	57	56	49	46	41	41	40	**27.00**	20	32	2	–	–	–	–	–	–	–	–	–	–	8(14),15-Pimaradiene-3β,18-diol		NI32055
57	205	85	41	220	44	58	55	**304**	100	34	30	26	19	11	7	5	**4.00**	20	32	2	–	–	–	–	–	–	–	–	–	–	Propanoic acid, 2,2-dimethyl-, 2,6-di-tert-butyl-4-methylphenyl ester	54644-43-8	NI32056
233	304	82	234	305	289	107	161	**304**	100	58	21	17	13	10	10	7		20	36	–	2	–	–	–	–	–	–	–	–	–	1,4-Benzenediamine, N,N'-bis(1,4-dimethylpentyl)-	3081-14-9	IC03809
233	304	81	234	45	305	289	43	**304**	100	53	22	17	17	13	10	9		20	36	–	2	–	–	–	–	–	–	–	–	–	1,4-Benzenediamine, N,N'-bis(1,4-dimethylpentyl)-	3081-14-9	IC03810
233	81	304	107	234	43	120	135	**304**	100	30	22	19	17	11	10	10		20	36	–	2	–	–	–	–	–	–	–	–	–	1,4-Benzenediamine, N,N'-bis(1,4-dimethylpentyl)-	3081-14-9	NI32057
158	203	159	128	144	149	115	160	**304**	100	98	98	35	32	25	22	19	**4.00**	21	20	2	–	–	–	–	–	–	–	–	–	–	1,8:2,7-Dimethanodibenzo[a,e]cyclobuta[c]cycloocten-13-ol, 1,2,2a,7,8,12b hexahydro-1-methoxy-	55649-65-5	NI32058
304	195	194	91	149	115	167	129	**304**	100	68	64	62	57	41	36	35		21	20	2	–	–	–	–	–	–	–	–	–	–	(1R*,2S*,7S*,8S*,11S*,13S*)-13-Methyl-11-phenylhexacyclo[6.5.1.0^{1,10}.0^{2,7}.0^{4,10}.0^{5,9}]tetradeca-3,6-dione	123542-68-7	DD01113
69	191	41	208	189	97	304	190	**304**	100	98	57	43	24	22	19	13		21	20	2	–	–	–	–	–	–	–	–	–	–	9-Phenanthrenemethyl cyclopentylcarboxylate	92174-40-8	NI32059
91	304	213	135	65	92	305	303	**304**	100	30	18	14	12	8	7	7		21	21	–	–	–	–	–	–	–	–	1	–	–	Phosphine, tribenzyl-	7650-89-7	NI32060
289	304	290	197	303	165	305	196	**304**	100	60	22	18	17	14	13	13		21	21	–	–	–	–	–	–	–	–	1	–	–	Phosphine, tris(2-methylphenyl)-	6163-58-2	NI32061
304	211	122	303	78	197	305	165	**304**	100	36	27	26	26	24	23	22		21	21	–	–	–	–	–	–	–	–	1	–	–	Phosphine, tris(3-methylphenyl)-	6224-63-1	NI32062
304	305	211	122	303	78	197	183	**304**	100	22	20	18	15	15	12	8		21	21	–	–	–	–	–	–	–	–	1	–	–	Phosphine, tris(3-methylphenyl)-	6224-63-1	NI32063
289	304	305	290	248	91	115	133	**304**	100	99	23	22	17	11	8	7		21	21	–	–	–	–	–	–	–	–	1	–	–	Phosphorin, 4-tert-butyl-2,6-diphenyl-	17420-26-7	NI32064
41	43	55	95	81	109	67	69	**304**	100	81	68	67	58	53	52	43	**14.00**	21	36	1	–	–	–	–	–	–	–	–	–	–	5α-Androstan-3β-ol, 4,4-dimethyl-	7673-19-0	NI32065
218	286	148	217	203	109	204	175	**304**	100	86	77	75	71	70	65	65	**0.02**	21	36	1	–	–	–	–	–	–	–	–	–	–	D(15)-Norpregnan-20-ol, 20-methyl-, (5α)-	54411-61-9	NI32066
108	107	304	121	43	109	41	305	**304**	100	27	24	9	9	8	8	6		21	36	1	–	–	–	–	–	–	–	–	–	–	3-Pentadecyl phenol		IC03811
234	55	81	233	215	304	41	95	**304**	100	80	79	73	71	70	69	68		21	36	1	–	–	–	–	–	–	–	–	–	–	Pregnan-3β-ol	13164-34-6	NI32067
286	233	55	82	257	271	215	304	**304**	100	46	46	45	32	24	12	8		21	36	1	–	–	–	–	–	–	–	–	–	–	Pregnan-12-ol		LI06686
257	286	82	55	271	233	287	215	**304**	100	41	35	35	27	10	9	9	**0.70**	21	36	1	–	–	–	–	–	–	–	–	–	–	Pregnan-12-ol		LI06687
273	67	81	109	95	79	163	69	**304**	100	81	75	68	68	45	37	36	**0.66**	21	36	1	–	–	–	–	–	–	–	–	–	–	Pregnan-18-ol, (5α)-	56053-09-9	NI32068
69	81	136	123	95	204	107	149	**304**	100	73	45	44	33	29	29	24	**0.00**	21	36	1	–	–	–	–	–	–	–	–	–	–	(3E,7E,11E)-4,8,12,16-Tetramethylheptadeca-3,7,11,15-tetraen-1-ol	118495-31-1	DD01114
304	277	278	303	302				**304**	100	63	49	43	13					22	12	–	2	–	–	–	–	–	–	–	–	–	9,10-Dicyano-anthracene benzyne centre ring adduct		LI06688
304	303	302	277	278				**304**	100	50	21	18	14					22	12	–	2	–	–	–	–	–	–	–	–	–	9,10-Dicyano-anthracene benzyne end ring adduct		LI06689
91	65	41	171	39	77	43	304	**304**	100	17	14	12	12	7	7	5		22	24	1	–	–	–	–	–	–	–	–	–	–	1,1-Dimethyl-4,4-dibenzyl-cyclohex-2-ene-5-one		LI06690
91	129	83	65	127	115	304	77	**304**	100	49	19	17	14	14	10	10		22	24	1	–	–	–	–	–	–	–	–	–	–	1,1-Dimethyl-3-(1,1-dibenzylmethylene)-cyclopentan-5-one		LI06691
41	83	69	55	97	111	67	125	**304**	100	90	90	88	69	55	44	27	**0.20**	22	40	–	–	–	–	–	–	–	–	–	–	–	1,5-Dicyclopentyl-3-(2-cyclopentylethyl)-pentane		AI02025
191	95	123	55	81	69	109	192	**304**	100	36	34	33	32	31	23	21	**20.35**	22	40	–	–	–	–	–	–	–	–	–	–	–	15-Ethyl-(13αH)-isocopalane		NI32069

MASS TO CHARGE RATIOS	M.W.	INTENSITIES	Parent	C	H	O	N	S	F	Cl	Br	I	Si	P	B	X	COMPOUND NAME	CAS Reg No	No
191 192 304 55 165 189 57 190	**304**	100 74 34 22 17 16 13 12		23	28	–	–	–	–	–	–	–	–	–	–	–	9-Nonylphenanthrene		MS07065
91 219 304	**304**	100 75 48		23	28	–	–	–	–	–	–	–	–	–	–	–	1-Phenylheptadeca-2,9-diene-4,6-diyne		LI06692
302 150 303 300 152 151 304 301	**304**	100 46 42 34 31 30 20 13		24	16	–	–	–	–	–	–	–	–	–	–	–	Acenaphthylidene, 1,1′-bis-	2435-82-7	NI32070
304 303 302 305 300 301 289 151	**304**	100 20 18 9 9 5 5 5		24	16	–	–	–	–	–	–	–	–	–	–	–	Diphenylacenaphthalene		IC03812
292 290 294 307 305 270 272 309	**305**	100 76 49 33 26 24 23 17		2	6	–	3	–	–	4	–	–	–	3	–	–	1,1-Dimethyltetrachlorocyclotriphosphazene	6204-32-6	NI32071
270 272 292 307 274 290 305 294	**305**	100 97 35 32 32 29 25 18		2	6	–	3	–	–	4	–	–	–	3	–	–	cis-1,3-Dimethyl-1,3,5,5-tetrachlorocyclotriphosphazene	94235-14-0	NI32072
270 272 274 307 292 305 290 60	**305**	100 96 32 26 23 21 18 15		2	6	–	3	–	–	4	–	–	–	3	–	–	trans-1,3-Dimethyl-1,3,5,5-tetrachlorocyclotriphosphazene	94235-15-1	NI32073
278 279 277 276 280 281 242 270	**305**	100 89 78 75 45 42 35 25	**13.64**	2	6	–	3	–	–	4	–	–	–	3	–	–	1-Ethyl-1-hydridotetrachlorocyclotriphosphazene	71982-84-8	NI32074
256 258 140 139 138 260 141 270	**305**	100 95 69 64 35 28 24 23	**8.00**	8	8	1	1	–	–	4	–	–	–	1	–	–	N-(2-Chloroethyl)-N-dichlorophosphinyl-2-chloroaniline		IC03813
248 249 39 304 27 305 41 250	**305**	100 88 88 87 80 75 68 59		8	14	–	–	–	–	–	–	–	–	–	–	1	Platinum, bis[(1,2,3-η)-2-butenyl]-	63817-36-7	NI32075
272 275 66 57 41 274 55 69	**305**	100 94 94 94 86 79 79 56	**29.41**	8	14	–	–	–	–	–	–	–	–	–	–	1	Platinum, π-cyclopentadienyltrimethyl-		MS07066
150 127 305 115 114 87 123 88	**305**	100 60 54 51 48 43 40 32		9	5	1	1	–	–	1	–	1	–	–	–	–	8-Quinolinol, 5-chloro-7-iodo-	130-26-7	NI32076
91 225 227 146 104 103	**305**	100 48 47 33 20 13	**0.00**	9	9	1	1	–	–	–	2	–	–	–	–	–	N-Benzyl-2,2-dibromoacetamide		LI06693
246 93 108 121 94 79 152 78	**305**	100 29 23 18 18 18 16 15	**0.50**	9	15	5	1	–	3	–	–	–	–	1	–	–	Butanoic acid, 4-[(methoxymethyl)phosphinyl]-2-[(trifluoroacetyl)amino]-, methyl ester, (S)-	56772-24-8	MS07067
134 142 163 135 127 160 65 39	**305**	100 88 80 37 37 17 17 15	**0.00**	11	16	1	1	–	–	–	–	1	–	–	–	–	5,5-Dimethyl-2,3,4,5-tetrahydro-1,5-benzoxazepinium iodide		LI06694
55 218 43 176 219 41 164 190	**305**	100 100 65 50 42 19 18 18	**0.00**	11	16	3	1	–	5	–	–	–	–	–	–	–	N(O,S)-Pentafluoropropionyl-D-norvaline isopropyl ester		MS07068
218 55 43 219 203 41 164 204	**305**	100 86 54 35 25 17 13 9	**0.00**	11	16	3	1	–	5	–	–	–	–	–	–	–	N(O,S)-Pentafluoropropionyl-D-valine isopropyl ester		MS07069
218 55 43 176 219 164 190 41	**305**	100 96 63 50 43 19 18 17	**0.00**	11	16	3	1	–	5	–	–	–	–	–	–	–	N(O,S)-Pentafluoropropionyl-L-norualine isopropyl ester		MS07070
218 55 43 219 203 41 164 204	**305**	100 74 46 35 25 15 13 9	**0.00**	11	16	3	1	–	5	–	–	–	–	–	–	–	N(O,S)-Pentafluoropropionyl-L-valine isopropyl ester		MS07071
177 179 43 30 72 178 180 219	**305**	100 98 80 80 73 23 22 18	**0.00**	11	16	4	1	–	–	–	1	–	–	–	–	–	N-[[(2R)-5-Bromo-2-tert-butyl-4-oxo-2H,4H-1,3-dioxin-6-yl]methyl]acetamide		DD01115
290 305 180 125 262 72 42 233	**305**	100 66 42 27 24 21 20 19		11	20	3	3	1	–	–	–	–	–	1	–	–	2-Diethylamino-3-[methoxy(methylthio)phosphinyloxy]-6-methyl-pyrimidine		IC03814
125 42 93 72 67 79 276 109	**305**	100 84 72 58 57 52 50 49	**39.13**	11	20	3	3	1	–	–	–	–	–	1	–	–	Phosphorothioic acid, O-[2-(diethylamino)-6-methyl-4-pyrimidinyl] O,O-dimethyl ester	29232-93-7	NI32077
109 137 28 18 246 290 110 305	**305**	100 63 53 40 39 26 26 24		11	20	3	3	1	–	–	–	–	–	1	–	–	Phosphorothioic acid, O-[2-(diethylamino)-6-methyl-4-pyrimidinyl] O,O-dimethyl ester	29232-93-7	NI32078
101 73 116 97 75 157 144 117	**305**	100 88 84 76 69 61 61 51	**0.00**	12	27	4	1	–	–	–	–	–	2	–	–	–	α-D-Glucopyranose, 2-amino-3,6-anhydro-2-deoxy-1,4-bis-O-(trimethylsilyl)-	56196-81-7	NI32079
73 174 248 133 147 59 45 75	**305**	100 43 35 24 21 21 20 16	**0.00**	12	31	2	1	–	–	–	–	–	3	–	–	–	β-Alanine, tris(trimethylsilyl)-		MS07072
73 174 248 45 147 75 59 86	**305**	100 48 32 28 23 18 18 14	**0.00**	12	31	2	1	–	–	–	–	–	3	–	–	–	β-Alanine, tris(trimethylsilyl)-		MS07073
174 248 73 290 147 86 100 232	**305**	100 75 54 52 29 11 9 8	**5.00**	12	31	2	1	–	–	–	–	–	3	–	–	–	β-Alanine, tris(trimethylsilyl)-		MS07074
219 277 161 247 220 133 56 134	**305**	100 42 22 18 13 13 9 7	**0.00**	13	15	4	1	–	–	–	–	–	–	–	–	1	Iron, tricarbonyl[4-[(2,3,4,5-η)-2,4-cyclohexadien-1-yl]morpholine]-	12307-93-6	NI32080
83 125 43 81 56 124 44 109	**305**	100 25 22 16 16 15 15 14	**0.00**	13	19	2	7	–	–	–	–	–	–	–	–	–	1H-Pyrrole-2-carboxamide, 5-[[(5-amino-3,4-dihydro-2H-pyrrol-2-yl)carbonyl]amino]-N-(3-amino-3-iminopropyl)-	61233-56-5	NI32081
115 73 75 43 45 74 116 58	**305**	100 98 87 63 29 16 13 12	**0.00**	13	23	7	1	–	–	–	–	–	–	–	–	–	α-D-Galactopyranoside, methyl 2-acetamido-2-deoxy-3,6-di-O-methyl-, 4-acetate	17296-11-6	NI32082
115 73 75 43 45 74 58 84	**305**	100 98 87 63 28 15 11 10	**0.00**	13	23	7	1	–	–	–	–	–	–	–	–	–	α-D-Galactopyranoside, methyl 2-acetamido-2-deoxy-3,6-di-O-methyl-, 4-acetate	17296-11-6	MS07075
101 43 71 143 45 102 59 246	**305**	100 62 31 30 30 25 21 19	**0.00**	13	23	7	1	–	–	–	–	–	–	–	–	–	α-D-Galactopyranoside, methyl 2-acetamido-2-deoxy-4,6-di-O-methyl-, 3-acetate	17296-09-2	NI32083
101 71 128 143 59 98 102 86	**305**	100 45 35 30 24 18 16 12	**0.00**	13	23	7	1	–	–	–	–	–	–	–	–	–	Methyl 3-O-acetyl-2-acetamido-2-deoxy-4,6-di-O-methyl-α-D-glucopyranoside	80831-96-5	NI32084
73 115 75 130 116 74 84 98	**305**	100 96 40 34 19 11 11 10	**0.00**	13	23	7	1	–	–	–	–	–	–	–	–	–	Methyl 4-O-acetyl-2-acetamido-2-deoxy-3,6-di-O-methyl-α-D-glucopyranoside		NI32085
73 115 75 43 128 88 87 86	**305**	100 81 47 38 30 21 14 13	**0.00**	13	23	7	1	–	–	–	–	–	–	–	–	–	Methyl 6-O-acetyl-2-acetamido-2-deoxy-3,4-di-O-methyl-α-D-glucopyranoside	55976-34-6	NI32086
149 65 75 184 138 110 150 93	**305**	100 21 20 20 18 16 15 14	**5.00**	14	8	5	1	–	–	1	–	–	–	–	–	–	Benzoic acid, 2-(4-chloro-3-nitrobenzoyl)-	85-54-1	NI32087
91 305 65 106 182 104 131 77	**305**	100 99 49 48 27 27 23 23		14	10	–	5	–	3	–	–	–	–	–	–	–	2-Pteridinamine, N-benzyl-4-(trifluoromethyl)-	32706-27-7	NI32088
77 287 286 271 272 230 231 200	**305**	100 100 73 58 52 51 48 47	**32.40**	14	12	1	3	1	–	1	–	–	–	–	–	–	3-Amino-6-chloro-4-hydroxy-4-phenyl-2-thioxo-1,2,3,4-tetrahydroquinazoline		MS07076
105 77 51 106 43 78 44 42	**305**	100 95 88 80 71 66 65 61	**7.30**	14	15	5	3	–	–	–	–	–	–	–	–	–	2-Oxo-4-benzoyl-6a-hydroxyperhydropyrrolo[2,3-d]imidazole-5-carboxylic acid methyl ester		LI06695

MASS TO CHARGE RATIOS	M.W.	INTENSITIES	Parent	C	H	O	N	S	F	Cl	Br	I	Si	P	B	X	COMPOUND NAME	CAS Reg No	No
43 112 56 188 41 170 60 146	**305**	100 69 57 55 44 43 43 40	**0.00**	14	27	4	1	1	–	–	–	–	–	–	–	–	6-Acetylamino-4,6-dideoxy-1-hexylthio-α-DL-lyxo-hexopyranoside		NI32089
270 305 242 272 307 271 152 236	**305**	100 54 42 36 35 27 21 17		15	9	2	1	–	–	2	–	–	–	–	–	–	1-Amino-4-chloro-2-(chloromethyl)anthraquinone	56594-25-3	NI32090
231 122 148 305 260 190	**305**	100 17 15 14 13 10		15	12	4	1	–	–	1	–	–	–	–	–	–	6-Chloro-3-(3-carbethoxyethyl)-4-cyanoisocoumarin		LI06696
107 76 77 106 246 122 108 123	**305**	100 44 36 27 25 24 24 18	**2.10**	15	15	2	1	2	–	–	–	–	–	–	–	–	Carbamodithioic acid, [(4-hydroxyphenyl)methyl]-, (4-hydroxyphenyl)methyl ester	14191-94-7	NI32091
77 119 305 136 141 91 80 92	**305**	100 47 33 26 25 18 18 15		15	15	4	1	1	–	–	–	–	–	–	–	–	Ethyl 3-[1-(benzenesulphonyl)pyrrol-2-yl]prop-2-enoate	95883-03-7	NI32092
83 98 90 164 69 126 141 85	**305**	100 60 52 49 45 42 31 25	**5.00**	15	19	4	3	–	–	–	–	–	–	–	–	–	Pyrrolo[1,2-b][1,2,4]oxadiazol-2(1H)-one, tetrahydro-5,5,7,7-tetramethyl-1-(4-nitrophenyl)-	39931-31-2	NI32093
290 73 45 291 147 305 216 75	**305**	100 98 26 24 18 17 12 12		15	23	2	1	–	–	–	–	–	2	–	–	–	1H-Indole-2-carboxylic acid, 1-(trimethylsilyl)-, trimethylsilyl ester	55255-80-6	NI32094
73 45 216 143 246 305 115 217	**305**	100 37 29 27 26 24 19 17		15	23	2	1	–	–	–	–	–	2	–	–	–	1H-Indole-5-carboxylic acid, 1-(trimethylsilyl)-, trimethylsilyl ester	55255-79-3	NI32095
305 202 159 103 104 185 118 200	**305**	100 88 35 27 25 24 23 20		16	15	–	7	–	–	–	–	–	–	–	–	–	Benzaldehyde, [2-(4,6-diamino-s-triazin-2-yl)phenyl]hydrazone	29366-80-1	NI32096
233 290 169 141 115 77 185 305	**305**	100 50 38 29 21 21 20 15		16	19	3	1	1	–	–	–	–	–	–	–	–	4-Phenoxy-N,N-diethyl-benzensulphonamide		IC03815
131 273 200 143 274 128 241 115	**305**	100 92 72 30 20 20 19 19	**5.70**	16	19	5	1	–	–	–	–	–	–	–	–	–	1H-1-Benzazepine-3,4-dicarboxylic acid, 2,3-dihydro-2-methoxy-1-methyl , dimethyl ester	36132-34-0	NI32097
145 146 105 101 77 205 104 91	**305**	100 80 68 58 30 26 22 18	**0.00**	16	19	5	1	–	–	–	–	–	–	–	–	–	4,5-(2-Deoxy-5,6-O-isopropylidene-α-D-glucofurano)-2-phenyl-2-oxazoline		LI06697
145 146 105 101 77 104 91 205	**305**	100 80 66 57 26 23 12 11	**1.58**	16	19	5	1	–	–	–	–	–	–	–	–	–	Furo[3,2-d]oxazol-6-ol, 5-(2,2-dimethyl-1,3-dioxolan-4-yl)-3a,5,6,6a-tetrahydro-2-phenyl-, [3aS-[3aα,5α(S*),6α,6aα]]-	23661-45-2	NI32098
181 84 69 105 261 91 77 71	**305**	100 84 69 39 32 31 31 30	**15.91**	16	19	5	1	–	–	–	–	–	–	–	–	–	1H-Pyrrole-2-carboxylic acid, 1-[2-(3,4,5-trimethoxyphenyl)ethyl]-	19717-25-0	NI32099
73 202 45 203 74 305 75 43	**305**	100 73 19 14 9 8 7 6		16	27	1	1	–	–	–	–	–	2	–	–	–	1H-Indole, 1-(trimethylsilyl)-3-[2-[(trimethylsilyl)oxy]ethyl]-	55334-85-5	NI32100
202 73 203 305 45 130 75 40	**305**	100 53 16 13 7 5 5 5		16	27	1	1	–	–	–	–	–	2	–	–	–	1H-Indole, 1-(trimethylsilyl)-3-[2-[(trimethylsilyl)oxy]ethyl]-		NI32101
305 103 77 102 116 277 203 144	**305**	100 43 37 35 32 28 28 28		17	11	1	3	1	–	–	–	–	–	–	–	–	2,5-Diphenyl-7H-1,3,4-thiadiazolo[3,2-a]pyrimidin-7-one		NI32102
305 304 174 306 103 121 77 277	**305**	100 66 28 23 19 15 15 10		17	11	1	3	1	–	–	–	–	–	–	–	–	2,7-Diphenyl-5H-1,3,4-thiadiazolo[3,2-a]pyrimidin-5-one	82090-63-9	NI32103
105 199 77 200 198 103 51 106	**305**	100 45 32 22 10 9 9 7	**0.00**	17	14	1	1	–	3	–	–	–	–	–	–	–	1-Benzoyl-2-methyl-3-phenyl-2-(trifluoromethyl)aziridine		DD01116
305 240 167 185 289 211 137 121	**305**	100 64 44 35 34 28 26 21		17	15	1	1	–	–	–	–	–	–	–	–	1	anti-Benzoylferrocene oxime		MS07077
305 240 167 137 185 77 56 129	**305**	100 61 52 30 29 20 18 16		17	15	1	1	–	–	–	–	–	–	–	–	1	syn-Benzoylferrocene oxime		MS07078
305 240 167 137 185 77 56 289	**305**	100 61 52 30 29 20 18 15		17	15	1	1	–	–	–	–	–	–	–	–	1	syn-Benzoylferrocene oxime		LI06698
142 43 121 184 57 84 68 77	**305**	100 41 32 19 18 17 17 10	**2.00**	17	23	4	1	–	–	–	–	–	–	–	–	–	1-Acetyl-3,4-isopropylidenedioxy-2-(4-methoxybenzyl)-pyrrolidine		LI06699
145 115 146 130 133 147 103 104	**305**	100 90 83 60 35 29 27 25	**2.00**	17	23	4	1	–	–	–	–	–	–	–	–	–	2-(1,1-Diethoxycarbonylmethyl)-3,3-dimethylindoline		MS07079
247 175 218 201 59 305 248 131	**305**	100 66 63 60 58 52 40 35		17	23	4	1	–	–	–	–	–	–	–	–	–	Ethyl 2-(2-hydroxy-2-methylpropyl)-5-methoxyindole-3-carboxylate		DD01117
220 248 91 160 221 115 57 41	**305**	100 60 47 15 14 9 9 9	**0.00**	17	23	4	1	–	–	–	–	–	–	–	–	–	Methyl (2R,4S)-4-benzyl-2-tert-butyl-3-formyloxazolidine-4-carboxylate		DD01118
70 113 112 263 91 137 68 77	**305**	100 46 19 15 13 12 12 10	**6.00**	17	23	4	1	–	–	–	–	–	–	–	–	–	Pyrrolidine, 1-acetyl-2-[2-[4-(acetyloxy)-3-methoxyphenyl]ethyl]-, (±)-	67257-65-2	NI32104
73 129 305 128 304 276 103 232	**305**	100 99 54 52 46 35 34 32		17	27	2	1	–	–	–	–	–	1	–	–	–	4-Piperidinecarboxylic acid, 4-phenyl-1-(trimethylsilyl)-, ethyl ester	55255-40-8	NI32105
73 129 305 128 304 276 103 232	**305**	100 67 36 34 31 23 23 21		17	27	2	1	–	–	–	–	–	1	–	–	–	4-Piperidinecarboxylic acid, 4-phenyl-1-(trimethylsilyl)-, ethyl ester	55255-40-8	NI32106
305 277 163 306 138 278 81 276	**305**	100 90 23 21 19 17 16 15		18	11	4	1	–	–	–	–	–	–	–	–	–	5H-Benzo[g]-1,3-benzodioxolo[6,5,4-de]quinoline-5,6(7H)-dione, 7-methyl	55610-01-0	NI32107
305 290 306 262 260 176 234 291	**305**	100 36 23 20 17 13 11 9		18	11	4	1	–	–	–	–	–	–	–	–	–	8H-Benzo[g]-1,3-benzodioxolo[6,5,4-de]quinolin-8-one, 4-methoxy-	3912-57-0	NI32108
28 140 218 191 277 105 165 115	**305**	100 90 83 61 60 35 32 28	**1.31**	18	15	2	3	–	–	–	–	–	–	–	–	–	Methyl 2-azido-5,5-diphenylpenta-2,4-dienoate		NI32109
305 304 306 218 303 217 245 143	**305**	100 85 20 18 15 14 12 12		18	15	2	3	–	–	–	–	–	–	–	–	–	3-Pyridinecarboxylic acid, 5-(4,9-dihydro-3H-pyrido[3,4-b]indol-1-yl)-, methyl ester	26238-84-6	NI32110
77 130 233 129 103 104 234 76	**305**	100 97 62 62 62 53 38 33	**28.03**	18	15	2	3	–	–	–	–	–	–	–	–	–	s-Triazine-2-carboxylic acid, 4,6-diphenyl-, ethyl ester	29366-67-4	NI32111
245 246 290 219 262 204 305 178	**305**	100 60 58 51 50 46 41 40		18	27	3	1	–	–	–	–	–	–	–	–	–	Acetyldebenzoylalopecurine		LI06700
270 244 288 216 287 230 173 271	**305**	100 38 36 26 22 21 21 19	**6.46**	18	27	3	1	–	–	–	–	–	–	–	–	–	Cyclohexanol, 1,1′-(4-methyl-2,6-pyridinediyl)bis-, N-oxide	17117-09-8	MS07080
196 168 305 167 197 77 149 57	**305**	100 54 43 34 20 18 17 15		19	15	1	1	1	–	–	–	–	–	–	–	–	Carbamic acid, diphenylthio-, S-phenyl ester	13509-40-5	NI32112
57 41 55 106 43 107 69 71	**305**	100 96 89 83 82 74 64 58	**3.00**	19	15	3	1	–	–	–	–	–	–	–	–	–	N-Methyl-N-phenyl-2-methyl-1,4-naphthoquinone-3-carboxamide		MS07081
201 77 202 51 47 305 180 78	**305**	100 93 92 55 34 27 15 14		19	16	1	1	–	–	–	–	–	–	1	–	–	N-Benzylidenediphenylphosphinamide		MS07082
305 144 290 273 274 185 306 77	**305**	100 82 64 38 31 30 25 25		19	19	1	3	–	–	–	–	–	–	–	–	–	Pyrrolo[2,1-b]quinazolin-9(1H)-one, 3-[2-(dimethylamino)phenyl]-2,3-dihydro-	33903-15-0	NI32113
248 287 249 305 217 147 288 242	**305**	100 29 17 13 12 10 8 8		19	31	2	1	–	–	–	–	–	–	–	–	–	Androstan-18-amide, 17-hydroxy-, (17β)-	55298-13-0	NI32114
305 273 91 234 55 306 79 67	**305**	100 37 28 26 23 22 22 21		19	31	2	1	–	–	–	–	–	–	–	–	–	16-Azaandrostan-17-one, 3-methoxy-, (3β)-	55399-19-4	NI32115
305 273 91 234 55 306 79 93	**305**	100 37 28 26 23 22 21 20		19	31	2	1	–	–	–	–	–	–	–	–	–	16-Azaandrostan-17-one, 3-methoxy-, (3β)-	55399-19-4	LI06701
305 290 203 178 306 43 41 169	**305**	100 65 37 31 23 22 19 17		19	31	2	1	–	–	–	–	–	–	–	–	–	N-(3,5-Di-tert-butyl-4-hydroxybenzyl)propionamide		IC03816
85 86 56 305 57 41 55 81	**305**	100 93 49 36 36 17 16 15		19	31	2	1	–	–	–	–	–	–	–	–	–	3-Aza-A-homoandrostan-16-ol, 1,4-epoxy-, (1α,4α,5β,16β)-		LI06702
85 86 56 305 57 41 44 81	**305**	100 94 49 37 37 17 16 15		19	31	2	1	–	–	–	–	–	–	–	–	–	3-Aza-A-homoandrostan-16-ol, 1,4-epoxy-, (1α,4α,5β,16β)-		LI06703
98 55 60 43 41 141 56 83	**305**	100 70 69 56 55 47 40 29	**16.72**	19	35	–	3	–	–	–	–	–	–	–	–	–	N,N′,N″-Tricyclohexylguanidine		IC03817
228 57 229 305 77 137 115 304	**305**	100 32 27 19 15 13 13 12		20	19	–	1	1	–	–	–	–	–	–	–	–	11b-Phenyl-1,2,3,5,6,11b-hexahydro[1]benzothieno[3,2-g]indolizine	99659-37-7	NI32116
262 231 305 230 272 263 232 306	**305**	100 96 67 53 46 43 19 16		20	19	–	1	1	–	–	–	–	–	–	–	–	Pyridine, 2-isopropylthio-4,6-diphenyl-	74663-74-4	NI32117

MASS TO CHARGE RATIOS								M.W.	INTENSITIES								Parent	C	H	O	N	S	F	Cl	Br	I	Si	P	B	X	COMPOUND NAME	CAS Reg No	No
272	305	263	231	290	230	273	219	**305**	100	50	46	45	44	32	22	15		20	19	–	1	1	–	–	–	–	–	–	–	–	**Pyridine, 2-isopropylthio-4,6-diphenyl-**	74663-74-4	**NI32118**
276	275	277	104	105	77	115	91	**305**	100	24	20	15	13	13	7	7	**1.20**	20	24	1	1	–	–	–	–	–	–	–	1	–	**Boron, (2-benzimidoylpropiophenonato)diethyl-**	21205-45-8	**NI32119**
113	55	126	41	70	43	72	67	**305**	100	69	57	55	46	41	34	29	**17.04**	20	35	1	1	–	–	–	–	–	–	–	–	–	**Pyrrolidine, 1-(1-oxo-7,10-hexadecadienyl)-**	56666-41-2	**NI32120**
105	184	77	169	305	18	43	28	**305**	100	70	54	51	25	25	23	22		21	23	1	1	–	–	–	–	–	–	–	–	–	**6-Methyl-4,5-tetramethylene-2,6-diphenyl-5,6-dihydro-4H-1,3-oxazine**		**NI32121**
148	146	222	224	150	220	226	144	**306**	100	96	75	74	73	62	55	50	**2.78**	–	10	–	–	–	–	–	–	–	–	–	–	4	**Tetragermane**	14691-47-5	**MS07083**
179	181	31	100	158	127	177	208	**306**	100	98	46	44	37	36	28	25	**0.00**	2	–	–	–	–	4	–	1	1	–	–	–	–	**Ethane, 1-bromo-1,1,2,2-tetrafluoro-2-iodo-**	421-70-5	**NI32122**
179	181	69	31	100	127	129	50	**306**	100	99	94	83	62	35	25	25	**18.99**	2	–	–	–	–	4	–	1	1	–	–	–	–	**Ethane, 1-bromo-1,1,2,2-tetrafluoro-2-iodo-**	421-70-5	**NI32123**
229	227	85	129	131	231	69	87	**306**	100	78	58	49	47	24	20	19	**0.15**	3	1	–	–	–	4	1	2	–	–	–	–	–	**1,2-Dibromo-3-chloro-1,1,3,3-tetrafluoropropane**		**DO01090**
147	149	229	67	41	53	133	39	**306**	100	99	92	92	73	72	58	58	**6.15**	5	9	–	–	–	–	–	3	–	–	–	–	–	**1,1,1-Tris(bromomethyl)-ethane**		**IC03818**
308	148	98	79	306	310	31	117	**306**	100	97	60	60	50	48	41	30		6	–	–	–	–	4	–	2	–	–	–	–	–	**Benzene, 1,2-dibromo-3,4,5,6-tetrafluoro-**	827-08-7	**LI06704**
308	148	79	98	310	306	31	117	**306**	100	98	58	55	50	50	42	32		6	–	–	–	–	4	–	2	–	–	–	–	–	**Benzene, 1,2-dibromo-3,4,5,6-tetrafluoro-**	827-08-7	**NI32124**
308	148	306	310	229	227	79	118	**306**	100	82	52	50	25	25	19	17		6	–	–	–	–	4	–	2	–	–	–	–	–	**Benzene, 1,3-dibromo-2,4,5,6-tetrafluoro-**	1559-87-1	**NI32125**
308	148	306	310	229	227	118	79	**306**	100	83	51	50	24	24	19	19		6	–	–	–	–	4	–	2	–	–	–	–	–	**Benzene, 1,3-dibromo-2,4,5,6-tetrafluoro-**	1559-87-1	**LI06705**
308	148	31	306	310	98	79	227	**306**	100	76	70	57	55	35	30	26		6	–	–	–	–	4	–	2	–	–	–	–	–	**Benzene, 1,4-dibromo-2,3,5,6-tetrafluoro-**	344-03-6	**LI06706**
308	148	31	306	310	98	79	117	**306**	100	75	70	58	55	32	30	25		6	–	–	–	–	4	–	2	–	–	–	–	–	**Benzene, 1,4-dibromo-2,3,5,6-tetrafluoro-**	344-03-6	**NI32126**
73	55	45	27	29	108	28	107	**306**	100	72	47	46	30	14	13	12	**3.15**	6	10	4	–	–	–	–	–	–	–	–	–	2	**3,3′-Diselenodipropionic acid**	7370-58-3	**NI32127**
57	183	185	167	165	123	125	166	**306**	100	40	39	16	16	16	14	8	**0.00**	6	12	4	–	–	–	–	2	–	–	–	–	–	**D-Mannitol, 1,6-dibromo-1,6-dideoxy-**	488-41-5	**NI32128**
69	96	56	29	104	46	42	51	**306**	100	74	70	68	66	65	47	22	**0.50**	7	10	–	2	1	8	–	–	–	–	–	–	–	**Imidosulphurous amide fluoride, N,N-diethyl- N′-(heptafluoroisopropyl)-**		**LI06707**
55	45	43	306	304	216	214	29	**306**	100	62	27	18	16	13	12	12		8	18	2	–	–	–	–	–	–	–	–	–	2	**1-Butanol, 3,3′-diselenobis-**	23243-49-4	**NI32129**
55	73	43	31	41	71	40	27	**306**	100	51	29	28	15	11	11	11	**4.37**	8	18	2	–	–	–	–	–	–	–	–	–	2	**1-Butanol, 4,4′-diselenobis-**	23243-48-3	**NI32130**
55	31	43	41	306	304	73	160	**306**	100	47	29	26	24	21	16	15		8	18	2	–	–	–	–	–	–	–	–	–	2	**1-Propanol, 3,3′-diselenobis[2-methyl-**	23243-50-7	**NI32131**
41	55	45	148	65	147	115	87	**306**	100	88	82	81	71	49	45	41	**0.00**	8	18	4	–	4	–	–	–	–	–	–	–	–	**Methanesulphonothioic acid 1,6-hexanediyl ester**		**MS07084**
29	97	27	61	153	125	65	213	**306**	100	69	69	58	55	48	42	41	**0.20**	8	19	4	–	3	–	–	–	–	–	1	–	–	**Phosphorodithioic acid, O,O-diethyl S-[2-(ethylsulphonyl)ethyl] ester**	2497-06-5	**NI32132**
121	93	65	97	306	125	154	122	**306**	100	56	47	41	36	17	6	5		8	20	4	–	2	–	–	–	–	–	2	–	–	**Thiohypophosphoric acid, tetraethyl ester**	5935-39-7	**NI32133**
159	28	109	160	64	119	48	36	**306**	100	23	13	10	6	5	5	5	**0.00**	9	7	2	–	1	3	2	–	–	–	–	–	–	**Sulphone, dichloromethyl 4-(trifluoromethyl)benzyl**	15894-29-8	**NI32134**
264	246	63	43	218	62	306	247	**306**	100	99	33	23	19	15	13	13		9	7	4	–	–	–	–	–	1	–	–	–	–	**Benzoic acid, 2-(acetyloxy)-5-iodo-**	1503-54-4	**MS07085**
246	264	63	43	218	62	247	53	**306**	100	99	32	27	19	13	12	11	**9.20**	9	7	4	–	–	–	–	–	1	–	–	–	–	**Benzoic acid, 2-(acetyloxy)-5-iodo-**	1503-54-4	**NI32135**
306	69	208	85	86	179	265	245	**306**	100	68	59	45	41	36	25	25		9	8	3	2	–	6	–	–	–	–	–	–	–	**2-Imidazolidinone, 1-acetyl-3-[3,3,3-trifluoro-1-oxo-2-(trifluoromethyl)propyl]-**	61709-50-0	**NI32136**
57	56	69	306	217	58	76	55	**306**	100	11	7	3	3	3	2	2		10	9	–	–	1	6	–	–	–	–	1	–	–	**Phosphinothioic fluoride, tert-butyl(pentafluorophenyl)-**	53327-25-6	**NI32137**
100	71	218	88	60	188	87	70	**306**	100	70	61	44	25	21	21	12	**1.00**	10	19	3	1	–	–	3	–	–	–	–	–	–	**2-Ethoxy-3-hydroxy-4-dimethylamino-6-trichloromethyl-1-oxacyclohexane**		**LI06708**
218	216	203	201	55	65	63	94	**306**	100	95	30	30	28	17	17	16	**4.00**	11	12	3	–	–	–	1	1	–	–	–	–	–	**Propanoic acid, 3-chloro-, 4-bromo-2-methoxy-5-methylphenyl ester**	41194-86-9	**NI32138**
91	67	79	105	147	81	77	119	**306**	100	67	58	39	38	38	32	28	**4.00**	11	16	–	–	–	–	–	2	–	–	–	–	–	**3,3-Dibromotricyclo[8.1.0.0^{2,4}]undecane**		**MS07086**
308	306	310	236	238	173	207	312	**306**	100	67	28	17	10	8	7	4		12	6	1	–	–	–	4	–	–	–	–	–	–	**Phenol, 3-phenyl-2′,3′,4′,5′-tetrachloro-**	67651-37-0	**NI32139**
308	306	310	236	238	207	173	309	**306**	100	71	32	17	9	7	6	4		12	6	1	–	–	–	4	–	–	–	–	–	–	**Phenol, 4-phenyl-2′,3′,4′,5′-tetrachloro-**	67651-34-7	**NI32140**
67	81	41	82	39	163	69	54	**306**	100	78	67	49	45	39	36	34	**29.40**	12	10	–	–	–	8	–	–	–	–	–	–	–	**1,8,9,10,11,11,12,12-Octafluorotricyclo[6.2.2.0^{2,7}]dodec-9-ene**		**NI32141**
306	109	127	124	110	184	125	111	**306**	100	54	19	17	11	10	7	7		12	10	4	4	1	–	–	–	–	–	–	–	–	**2-Acetylthiophene 2,4,-dinitrophenylhydrazone**		**LI06709**
307	273	277	289	274	290	259	308	**306**	100	22	8	6	4	3	3	2	**0.00**	12	10	6	4	–	–	–	–	–	–	–	–	–	**1H-Pyrrole-2,5-dione, 1-[4-(dimethylamino)-3,5-dinitrophenyl]-**	3475-74-9	**NI32143**
54	289	42	82	213	26	66	15	**306**	100	85	78	76	57	54	43	38	**33.23**	12	10	6	4	–	–	–	–	–	–	–	–	–	**1H-Pyrrole-2,5-dione, 1-[4-(dimethylamino)-3,5-dinitrophenyl]-**	3475-74-9	**NI32142**
43	58	44	144	55	117	185	143	**306**	100	9	8	7	7	6	5	5	**4.00**	12	11	1	4	–	–	–	1	–	–	–	–	–	**3-(N-Bromoimino)-4,7,7-tricyano-1,5,5-trimethyl-2-oxabicyclo[2.2.1]heptane**		**MS07087**
43	171	129	85	144	102	84	59	**306**	100	99	62	56	49	46	35	33	**0.00**	12	18	7	–	1	–	–	–	–	–	–	–	–	**Xylopyranoside, methyl 4-thio-, triacetate, α-D-**	32848-88-7	**NI32144**
215	73	89	101	143	59	153	43	**306**	100	75	74	62	61	61	53	47	**0.00**	12	22	7	–	–	–	–	–	–	1	–	–	–	**1,2,3-Propanetricarboxylic acid, 2-[(trimethylsilyl)oxy]-, trimethyl ester**	55590-87-9	**NI32145**
117	147	88	219	218	191	129	291	**306**	100	47	24	21	17	15	14	11	**0.00**	12	26	5	–	–	–	–	–	–	2	–	–	–	**Lactic acid dimer, bis(trimethylsilyl)-**		**NI32146**
73	117	147	75	28	45	55	88	**306**	100	83	19	18	13	12	12	11	**0.00**	12	26	5	–	–	–	–	–	–	2	–	–	–	**Lactic acid dimer, bis(trimethylsilyl)-**		**NI32147**
196	227	225	107	195	79	184	197	**306**	100	83	81	60	35	35	35	33	**13.56**	13	10	5	2	1	–	–	–	–	–	–	–	–	**2,4-Dinitro-4′-methyldiphenylsulphoxide**	966-71-2	**NI32148**
96	181	153	213	95	306	154	193	**306**	100	85	79	58	55	53	29	16		13	11	3	4	–	–	1	–	–	–	–	–	–	**trans-6-Chloro-9-(2-furyl-1,3-dioxan-5-yl)purine**		**NI32149**
125	153	68	111	95	138	306	69	**306**	100	39	35	23	23	19	15	15		13	15	–	6	1	1	–	–	–	–	–	–	–	**2-Amino-4-[3-(4-fluorophenyl)thioureido]-6-propyl-1,3,5-triazine**		**NI32150**
306	181	280	103	196	104	206	78	**306**	100	64	62	55	31	28	27	27		13	15	2	–	–	–	–	–	–	–	–	–	1	**(η^4-Cyclooctatetraene)acetylacetonatorhodium**	12307-95-8	**NI32151**
155	99	127	151	152	306	230	258	**306**	100	48	44	39	35	20	4	2		13	17	4	–	–	2	–	–	–	–	1	–	–	**Diethyl 1-benzyl-2,2-difluoroethenyl phosphate**		**DD01119**
129	101	87	43	161	189	71	99	**306**	100	39	38	35	32	16	15	13	**0.00**	13	22	8	–	–	–	–	–	–	–	–	–	–	**1,2,4-Tri-O-acetyl-3,5-di-O-methylarabinitol**	77983-81-4	**NI32152**
117	43	129	87	101	99	173	71	**306**	100	57	44	27	18	18	13	13	**0.00**	13	22	8	–	–	–	–	–	–	–	–	–	–	**1,4,5-Tri-O-acetyl-2,3-di-O-methylarabinitol**	19318-21-9	**NI32153**
43	129	87	161	101	189	71	57	**306**	100	96	40	23	23	18	15	15	**0.00**	13	22	8	–	–	–	–	–	–	–	–	–	–	**1,2,4-Tri-O-acetyl-3,5-di-O-methylribitol**	84925-33-7	**NI32154**

MASS TO CHARGE RATIOS								M.W.	INTENSITIES								Parent	C	H	O	N	S	F	Cl	Br	I	Si	P	B	X	COMPOUND NAME	CAS Reg No	No
117	43	58	87	113	99	45	129	**306**	100	46	23	23	18	13	11	10	**0.00**	13	22	8	–	–	–	–	–	–	–	–	–	–	1,3,4-Tri-O-acetyl-2,5-di-O-methylribitol	84925-31-5	NI32155
43	71	45	87	102	88	129	101	**306**	100	99	99	86	48	45	43	40	**0.00**	13	22	8	–	–	–	–	–	–	–	–	–	–	β-D-Mannopyranose, 2,4,6-tri-O-methyl-, diacetate	55255-81-7	NI32156
75	43	74	45	116	87	103	141	**306**	100	99	94	40	38	21	18	16	**0.00**	13	22	8	–	–	–	–	–	–	–	–	–	–	Methyl 2,4-di-O-acetyl-3,6-di-O-methyl-α-D-glucopyranoside	24905-05-3	NI32157
75	43	74	116	45	87	103	115	**306**	100	99	97	30	24	21	14	14	**0.00**	13	22	8	–	–	–	–	–	–	–	–	–	–	Methyl 2,4-di-O-acetyl-3,6-di-O-methyl-β-D-glucopyranoside	24905-08-6	NI32158
88	75	43	101	45	161	85	71	**306**	100	69	50	33	20	14	13	13	**0.00**	13	22	8	–	–	–	–	–	–	–	–	–	–	Methyl 5,6-di-O-acetyl-2,3-di-O-methyl-α-D-mannofuranoside	64244-12-8	NI32159
71	43	74	87	129	102	45	101	**306**	100	99	93	92	91	66	50	40	**0.00**	13	22	8	–	–	–	–	–	–	–	–	–	–	Methyl 2,3-di-O-acetyl-4,6-di-O-methyl-α-D-mannopyranoside	53919-56-5	NI32160
75	88	43	74	87	71	101	116	**306**	100	79	66	62	46	42	26	23	**0.00**	13	22	8	–	–	–	–	–	–	–	–	–	–	Methyl 2,6-di-O-acetyl-3,4-di-O-methyl-α-D-mannopyranoside	72922-29-3	NI32161
74	43	116	87	113	129	85	112	**306**	100	99	45	44	41	40	30	29	**0.00**	13	22	8	–	–	–	–	–	–	–	–	–	–	Methyl 3,4-di-O-acetyl-2,6-di-O-methyl-α-D-mannopyranoside	23089-68-1	NI32162
74	43	116	101	88	71	75	87	**306**	100	99	85	85	67	62	60	35	**0.00**	13	22	8	–	–	–	–	–	–	–	–	–	–	Methyl 3,6-di-O-acetyl-2,4-di-O-methyl-α-D-mannopyranoside	72922-21-5	NI32163
88	75	43	73	45	74	101	84	**306**	100	58	44	12	10	9	9	9	**0.00**	13	22	8	–	–	–	–	–	–	–	–	–	–	Methyl 4,6-di-O-acetyl-2,3-di-O-methyl-α-D-mannopyranoside	72962-69-7	NI32164
43	118	129	87	102	189	162	233	**306**	100	85	46	31	18	15	5	1	**0.00**	13	22	8	–	–	–	–	–	–	–	–	–	–	2,3-Di-O-methyl-1,4,5-tri-O-acetyl-arabinitol		NI32165
59	117	89	73	101	129	175	233	**306**	100	71	42	27	20	10	7	6	**1.00**	13	22	8	–	–	–	–	–	–	–	–	–	–	Triethyl (S)-2-(carboxymethoxymethyl)oxydiacetate		NI32166
193	29	191	135	137	249	133	189	**306**	100	80	78	71	63	56	49	46	**0.00**	13	30	–	–	–	–	–	–	–	–	–	–	1	Methyltributyltin		MS07088
71	43	203	73	147	41	75	205	**306**	100	70	43	41	22	12	11	8	**0.00**	13	30	4	–	–	–	–	–	–	2	–	–	–	1-Mono-isobutyrin, bis(trimethylsilyl)-		NI32167
73	43	71	129	147	145	75	103	**306**	100	70	52	51	48	38	31	30	**0.00**	13	30	4	–	–	–	–	–	–	2	–	–	–	2-Mono-isobutyrin, bis(trimethylsilyl)-		NI32168
73	189	147	103	143	75	85	105	**306**	100	78	46	38	36	33	30	25	**0.00**	13	30	4	–	–	–	–	–	–	2	–	–	–	Pentanoic acid, 3-methyl-3,5-bis[(trimethylsilyl)oxy]-, methyl ester	56051-93-5	NI32169
306	308	44	307	28	310	36	309	**306**	100	63	46	20	17	11	11	10		14	8	2	2	–	–	2	–	–	–	–	–	–	1,4-Diamino-2,3-dichloro-anthraquinone		IC03819
105	238	77	222	103	51	49	119	**306**	100	79	65	45	41	29	27	20	**0.00**	14	10	2	2	–	3	–	–	–	–	–	1	–	3,5-Diphenyl-4-trifluoroboroxy-1,2,4-oxadiazole	22250-22-2	NI32170
105	77	103	51	49	119	238	50	**306**	100	65	40	29	27	20	16	14	**0.00**	14	10	2	2	–	3	–	–	–	–	–	1	–	3-Aza-A-homoandrostan-16-ol, (5β,16β)-		LI06710
136	152	108	154	96	69	153	137	**306**	100	70	40	28	22	17	14	10	**0.00**	14	10	4	–	2	–	–	–	–	–	–	–	–	Dithiosalycylic acid		IC03821
136	152	108	69	154	96	50	39	**306**	100	54	48	25	24	24	14	13	**2.06**	14	10	4	–	2	–	–	–	–	–	–	–	–	Dithiosalycylic acid		IC03820
169	65	76	50	39	51	121	75	**306**	100	46	15	15	14	13	12	10	**7.69**	14	10	6	–	1	–	–	–	–	–	–	–	–	Diphenyl sulphone-4,4′-dicarboxylic acid		IC03822
83	306	222	55	138	167	307	194	**306**	100	72	72	34	16	15	12	11		14	10	8	–	–	–	–	–	–	–	–	–	–	2,5-Dihydroxy-3-(2,5-dihydroxy-3,6-dioxo-4-methyl-cyclohex-1,4-dienyl)-6-methyl-1,4-benzoquinone		IC03823
43	306	263	245	236	69	307	53	**306**	100	94	40	20	20	16	15	11		14	10	8	–	–	–	–	–	–	–	–	–	–	1,4-Naphthoquinone, 2,7-diacetyl-3,5,6,8-tetrahydroxy-	3404-89-5	NI32171
258	260	91	77	79	103	102	89	**306**	100	99	65	57	42	27	20	15	**5.00**	14	11	1	–	1	–	–	1	–	–	–	–	–	4-Bromophenyl styryl sulphoxide	38366-98-2	LI06711
258	91	77	179	103	102	89	119	**306**	100	65	57	42	27	20	15	12	**5.00**	14	11	1	–	1	–	–	1	–	–	–	–	–	4-Bromophenyl styryl sulphoxide	38366-98-2	NI32172
91	155	93	65	124	92	77	66	**306**	100	57	50	31	21	17	16	16	**0.10**	14	14	2	2	2	–	–	–	–	–	–	–	–	Benzenesulphonamide, 4-methyl-N-[(phenylamino)thioxomethyl]-	6361-95-1	NI32173
91	155	93	65	124	135	92	66	**306**	100	56	49	31	22	17	17	16	**0.06**	14	14	2	2	2	–	–	–	–	–	–	–	–	Benzenesulphonamide, 4-methyl-N-[(phenylamino)thioxomethyl]-	6361-95-1	NI32174
306	67	55	109	95	81	79	165	**306**	100	91	78	38	29	26	18	14		14	18	4	4	–	–	–	–	–	–	–	–	–	Cyclooctanone (2,4-dinitrophenyl)hydrazone		IC03824
290	306	259	275	123	289	243		**306**	100	50	33	27	22	19	17			14	18	4	4	–	–	–	–	–	–	–	–	–	2,4-Diamino-5-(3,4,5-trimethoxybenzyl)-pyrimidine-1-oxide		LI06712
289	290	306	259	275	243	273	257	**306**	100	45	13	12	10	8	7	7		14	18	4	4	–	–	–	–	–	–	–	–	–	2,4-Diamino-5-(3,4,5-trimethoxybenzyl)-pyrimidine-3-oxide		LI06713
168	278	246	306	276	39	208	142	**306**	100	48	35	26	24	19	18	17		14	19	1	–	–	–	–	–	–	–	–	–	1	Rhodium, carbonyl[(1,2-η)-cyclooctene](η^5-2,4-cyclopentadien-1-yl)-	74810-96-1	NI32175
29	117	89	45	57	44	27	41	**306**	100	80	62	45	40	34	31	28	**0.00**	14	26	7	–	–	–	–	–	–	–	–	–	–	Di-sec-butyl diglycolcarbonate		NI32176
55	69	227	41	43	83	57	97	**306**	100	79	76	76	63	50	48	44	**5.04**	14	27	2	–	–	–	–	1	–	–	–	–	–	13-Bromotetradecanoic acid		NI32177
232	56	100	114	70	58	57	144	**306**	100	94	93	59	51	51	50	44	**0.00**	14	30	5	2	–	–	–	–	–	–	–	–	–	13-(5-Amino-3-oxapentyl)-1,4,7,10-tetraoxa-13-azacyclopentadecane		MS07089
306	308	307	242	28	277	263	276	**306**	100	35	19	16	15	13	12	9		15	11	5	–	–	–	1	–	–	–	–	–	–	4-Chloro-1,6-dihydroxy-3-methoxy-8-methyl-9H-xanthen-9-one		MS07090
306	304	307	308	277	278	309	279	**306**	100	94	41	39	37	16	14	14		15	11	5	–	–	–	1	–	–	–	–	–	–	7-Chloro-1,6-dihydroxy-3-methoxy-8-methyl-9H-xanthen-9-one		MS07091
275	165	277	169	142	306	126	111	**306**	100	13	11	10	10	5	3	3		15	11	5	–	–	–	1	–	–	–	–	–	–	3-(5-Chloro-2-methoxybenzoyl)-4-hydroxybenzoic acid		NI32178
91	171	135	65	77	133	106	39	**306**	100	60	48	29	27	22	20	17	**0.00**	15	14	2	–	–	–	–	–	–	–	–	–	1	2-Toluic acid, α-(benzylselenyl)-	15855-31-9	NI32179
91	135	65	77	133	306	92	39	**306**	100	48	29	27	21	20	17	17		15	14	2	–	–	–	–	–	–	–	–	–	1	2-Toluic acid, α-(benzylselenyl)-	15855-31-9	LI06714
91	155	65	306	63	39	156	92	**306**	100	70	14	12	6	6	5	5		15	14	5	–	1	–	–	–	–	–	–	–	–	(4-Methoxycarboxyphenyl) 4-toluenesulphonate		IC03825
306	232	149	249	179	121	277	235	**306**	100	50	41	27	27	27	23	23		15	14	7	–	–	–	–	–	–	–	–	–	–	7-Ethoxy-2-hydroxy-8-methoxy-3-methyl-5,6-methylenedioxy-1,4-naphthoquinone	75629-09-3	NI32180
306	235	263	43	278	207	69	245	**306**	100	76	69	54	34	23	23	22		15	14	7	–	–	–	–	–	–	–	–	–	–	1,4-Naphthoquinone, 2-acetyl-3-hydroxy-5,6,8-trimethoxy-	14090-54-1	NI32181
306	291	227	307	292	258	169	121	**306**	100	91	25	18	15	15	15	15		15	15	3	–	1	–	–	–	–	–	1	–	–	Phenothiaphosphine, 2,8,10-trimethyl-, 5,5,10-trioxide	24059-97-0	NI32182
174	91	117	259	306	61	146	56	**306**	100	72	69	65	56	38	38	34		15	18	3	2	1	–	–	–	–	–	–	–	–	5-Ethyl-3-methyl-1-methylthiomethyl-5-phenyl-2,4,6-pyrimidinetrione		MS07092
247	219	187	116	59	248	159	89	**306**	100	24	23	19	18	13	12	12	**0.00**	15	18	5	2	–	–	–	–	–	–	–	–	–	Dimethyl 1-ethoxy-1,2-dihydro-4a,8a-methanophthalazine-1,4-dicarboxylate		DD01120
43	159	59	105	147	41	104	291	**306**	100	93	54	51	35	35	25	25	**0.23**	15	19	6	–	–	–	–	–	–	–	–	1	–	1,2-O-Isopropylidene-α-D-glucofuranose 3,5-benzeneboronate	14927-09-4	NI32183
147	291	277	306	115	146	74	133	**306**	100	62	58	56	37	31	27	24		15	22	3	–	–	–	–	–	–	2	–	–	–	4-[Bis(trimethylsilyl)methyl]phthalic anhydride	85370-82-7	DD01121
81	69	163	93	124	55	107	41	**306**	100	57	52	47	46	40	38	23	**0.00**	15	24	2	–	–	–	2	–	–	–	–	–	–	($\pm$)-1,2α,3,4,4aα,5,6,7,8,8aβ-Decahydro-5,5,8aβ-trimethyl-2β-naphthalenyl dichloroacetate		DD01122
43	196	28	125	41	83	67	82	**306**	100	69	62	60	60	59	54	43	**4.13**	15	26	1	2	–	–	–	–	–	–	–	–	1	Iron, carbonyl[(1,2,3,4-η)-2,3-dimethyl-1,3-butadiene][N,N′-1,2-ethanediylidenebis[2-propanamine]-N,N′]-	72142-41-7	NI32184
59	103	87	42	41	102	101	43	**306**	100	32	16	16	13	10	9	9	**0.00**	15	30	6	–	–	–	–	–	–	–	–	–	–	γ-Methoxybutyraldehyde trimer		NI32185

MASS TO CHARGE RATIOS								M.W.	INTENSITIES								Parent	C	H	O	N	S	F	Cl	Br	I	Si	P	B	X	COMPOUND NAME	CAS Reg No	No
43	155	117	172	116	101	85	111	306	100	73	42	34	33	32	28	22	0.00	15	30	6	–	–	–	–	–	–	–	–	–	–	2H-Pyran-4-ol, 6-(dimethoxymethyl)-2-(2,3-dimethoxypropyl)tetrahydro-3,3-dimethyl-		NI32186
277	179	221	165	79	291	278	121	306	100	79	42	32	26	22	20	18	0.40	15	34	4	–	–	–	–	–	–	1	–	–	–	Silicic acid, tributyl propyl ester	55683-20-0	NI32187
306	104	200	134	77	107	40	186	306	100	89	78	78	78	57	43	33		16	13	1	2	–	3	–	–	–	–	–	–	–	cis-2-(Phenylamino)-5-phenyl-4-(trifluoromethyl)-2-oxazoline		DD01123
306	134	200	77	104	237	107	91	306	100	85	53	50	48	37	31	30		16	13	1	2	–	3	–	–	–	–	–	–	–	trans-2-(Phenylamino)-5-phenyl-4-(trifluoromethyl)-2-oxazoline		DD01124
78	77	306	186	134	51	52	200	306	100	37	36	35	26	25	24	20		16	13	1	2	–	3	–	–	–	–	–	–	–	cis-1-(Phenylcarbamoyl)-3-phenyl-2-(trifluoromethyl)aziridine		DD01125
306	109	141	110	169	137	65	142	306	100	14	13	10	8	6	6	5		16	18	4	–	1	–	–	–	–	–	–	–	–	4,4'-Diethoxydiphenyl sulphone		IC03826
135	148	113	77	84	306	85	51	306	100	50	46	40	35	31	25	23		16	18	6	–	–	–	–	–	–	–	–	–	–	2H-Pyran-2-one, 6-[2-(1,3-benzodioxol-5-yl)ethyl]-5,6-dihydro-4,5-dimethoxy-, (5R-cis)-	52526-01-9	NI32188
135	139	136	113	148	306	77	85	306	100	49	42	40	34	28	21	17		16	18	6	–	–	–	–	–	–	–	–	–	–	2H-Pyran-2-one, 6-[2-(1,3-benzodioxol-5-yl)ethyl]-5,6-dihydro-4,5-dimethoxy-, (5S-trans)-	52526-02-0	NI32189
91	171	227	251	253	129	228	115	306	100	65	61	52	51	48	39	38	25.00	16	19	1	–	–	–	–	1	–	–	–	–	–	2(1H)-Naphthalenone, 1α-bromo-3,4,4a,5,6,7,8,8aα-octahydro-4aβ-phenyl	22844-38-8	NI32190
306	274	153	137	247	244			306	100	54	33	20	5	5				16	22	–	2	2	–	–	–	–	–	–	–	–	13H-1,4-Benzodithieno[2,3-b]pyrido[3',2'-g]indole		LI06715
43	73	45	101	70	72	177	222	306	100	64	62	61	47	41	38	36	0.00	16	22	4	2	–	–	–	–	–	–	–	–	–	trans-N-Acetyl-N'-feruloylputrescine		NI32191
72	121	117	59	146	91	73	44	306	100	12	10	8	5	5	4	4	1.00	16	22	4	2	–	–	–	–	–	–	–	–	–	Benzeneacetic acid, α-[[[(dimethylamino)carbonyl]methylamino]carbonyl] α-ethyl-, methyl ester	42948-66-3	NI32192
72	121	117	146	91	249	177	306	306	100	10	7	4	4	3	3	1		16	22	4	2	–	–	–	–	–	–	–	–	–	Benzeneacetic acid, α-[[[(dimethylamino)carbonyl]methylamino]carbonyl] α-ethyl-, methyl ester		LI06717
72	121	117	146	249	177	129	306	306	100	12	10	5	4	3	3	1		16	22	4	2	–	–	–	–	–	–	–	–	–	Benzeneacetic acid, α-[[[(dimethylamino)carbonyl]methylamino]carbonyl] α-ethyl-, methyl ester		LI06716
103	306	77	233	129	104	259	193	306	100	35	31	27	15	15	13	10		17	14	–	4	1	–	–	–	–	–	–	–	–	1,6-Diphenyl-2-(methylthio)-1H-imidazo[1,2-b][1,2,4]triazole		MS07093
306	241	223	139	307	242	195	140	306	100	47	34	31	18	8	8	8		17	14	2	–	–	–	–	–	–	–	–	–	1	2-Ferrocenylbenzoic acid		LI06718
306	241	223	139	307	56	242	195	306	100	53	39	35	20	11	9	9		17	14	2	–	–	–	–	–	–	–	–	–	1	2-Ferrocenylbenzoic acid		MS07094
306	121	56	129	241	95	128	214	306	100	31	25	20	19	17	15	10		17	14	2	–	–	–	–	–	–	–	–	–	1	1-(β-Fur-2-ylacroyl)ferrocene		LI06719
275	276	220	247	245	65	51	77	306	100	39	29	18	18	12	12	9	0.00	17	14	2	4	–	–	–	–	–	–	–	–	–	1,2-Ethanediol, 1-(1-phenyl-1H-pyrazolo[3,4-b]quinoxalin-3-yl)-, (S)-	31504-88-8	NI32193
217	132	91	246	105	218	175	119	306	100	65	45	43	36	28	28	24	0.00	17	22	5	–	–	–	–	–	–	–	–	–	–	3-Oxo-11-acetoxydeoxyoreadone		MS07095
57	149	162	85	133	191	204	105	306	100	59	52	37	17	11	8	8	1.20	17	22	5	–	–	–	–	–	–	–	–	–	–	10-Acetoxy-8,9-epoxythymol 3-isovalerate		NI32194
145	218	173	91	105	53	119	190	306	100	74	64	49	44	42	41	38	9.00	17	22	5	–	–	–	–	–	–	–	–	–	–	1β-Acetoxy-8β-hydroxyeudesma-4(15),7(11)-dien-8α,12-olide		NI32195
172	262	132	202	119	105	157	173	306	100	76	48	41	39	39	36	35	0.30	17	22	5	–	–	–	–	–	–	–	–	–	–	3-Acetoxy-marasmen-15-one		MS07096
123	43	246	95	55	53	79	141	306	100	69	35	32	26	22	21	20	0.00	17	22	5	–	–	–	–	–	–	–	–	–	–	Azuleno[4,5-b]furan-2,9-dione, 9a-[(acetyloxy)methyl]decahydro-6-methyl 3-methylene-, [3aS-(3aα,6β,6aα,9aβ,9bα)]-	31299-06-6	NI32196
207	179	41	53	109	79	69	43	306	100	74	46	43	37	37	37	37	9.14	17	22	5	–	–	–	–	–	–	–	–	–	–	Benz[e]azulene-3,8-dione, 3a,4,6a,7,9,10,10a,10b-octahydro-3a,10a-dihydroxy-5-(hydroxymethyl)-2,10-dimethyl-, (3aα,6aα,10β,10aβ,10bβ)-(+)-	25578-89-6	NI32197
108	91	121	93	97	107	149	150	306	100	83	80	77	68	65	51	49	5.63	17	22	5	–	–	–	–	–	–	–	–	–	–	Cyclodeca[b]furan-2(3H)-one, 9-(acetyloxy)-3a,4,5,8,9,11a-hexahydro-4-hydroxy-6,10-dimethyl-3-methylene-	69745-81-9	NI32198
43	188	204	176	189	91	79	123	306	100	27	22	21	17	17	16	15	0.00	17	22	5	–	–	–	–	–	–	–	–	–	–	2H-Cyclohepta[b]furan-2-one, 6-[1-(acetyloxy)-3-oxobutyl]-3,3a,4,7,8,8a-hexahydro-7-methyl-3-methylene-	580-49-4	NI32199
43	188	204	176	189	135	79	91	306	100	28	21	21	17	16	15	13	0.08	17	22	5	–	–	–	–	–	–	–	–	–	–	2H-Cyclohepta[b]furan-2-one, 6-[1-(acetyloxy)-3-oxobutyl]-3,3a,4,7,8,8a-hexahydro-7-methyl-3-methylene-	580-49-4	NI32200
137	196	306	121	105	288	191	195	306	100	72	62	40	34	33	33	28		17	22	5	–	–	–	–	–	–	–	–	–	–	3β,20-Epoxy-3α-hydroxy-14-oxo-9β-podocarpan-19,6β-olide		NI32201
99	82	121	231	306	291	249	246	306	100	47	31	19	3	1	1	1		17	22	5	–	–	–	–	–	–	–	–	–	–	4-Formyl-1,3,3-trimethyl-6-(3-acetoxymethyl-crotonoyloxy)-cyclohexa-1,4 diene		LI06720
43	82	141	81	306	231	246	291	306	100	75	55	40	11	3	2	1		17	22	5	–	–	–	–	–	–	–	–	–	–	4-Formyl-1,5,5-trimethyl-6-(2-acetoxymethyl-crotonoyloxy)-cyclohexa-1,3 diene		LI06721
163	126	57	164	76	81	58	181	306	100	83	45	29	16	15	13	12	0.00	17	22	5	–	–	–	–	–	–	–	–	–	–	Methyl 2-ethyl-4-ketohexylphthalate		NI32202
163	149	126	83	77	108	164	55	306	100	32	23	20	17	17	14	14	0.00	17	22	5	–	–	–	–	–	–	–	–	–	–	Methyl 2-ethyl-5-ketohexylphthalate		NI32203
132	228	60	91	246	117	118	90	306	100	83	71	61	55	55	51	37	0.00	17	22	5	–	–	–	–	–	–	–	–	–	–	Oxireno[2,3]azuleno[6,5-b]furan-5(1aH)-one, decahydro-7a-hydroxy-1a,7-dimethyl-3-methylene-, acetate	4955-25-3	NI32204
91	43	92	142	141	65	105	99	306	100	41	15	9	9	8	6	6	0.00	17	22	5	–	–	–	–	–	–	–	–	–	–	[3aR-(3aα,5α,6α,6aα)]-Tetrahydro-2,2-dimethyl-5-(2-oxopropyl)-6-(benzyloxy)furo[2,3-d]-1,3-dioxole	120418-01-1	DD01126
91	43	141	92	65	245	169	100	306	100	33	10	8	7	4	3	3	0.00	17	22	5	–	–	–	–	–	–	–	–	–	–	[3aR-(3aα,5β,6β,6aα)]-Tetrahydro-2,2-dimethyl-5-(2-oxopropyl)-6-(benzyloxy)furo[2,3-d]-1,3-dioxole	120417-93-8	DD01127
209	179	223	210	304	277	291	251	306	100	33	30	26	14	10	6	3	0.00	17	26	3	–	–	–	–	–	–	1	–	–	–	4-Hepten-3-one, 1-[3-methoxy-4-[(trimethylsilyl)oxy]phenyl]-	56700-89-1	MS07097
213	70	43	44	112	30	143	100	306	100	32	32	31	22	21	18	15	2.00	17	26	3	2	–	–	–	–	–	–	–	–	–	N,N'-Diacetyl-N-(γ-phenoxypropyl)putrescine		LI06722
234	72	206	160	135	306	235	305	306	100	98	54	38	29	26	24	21		17	26	3	2	–	–	–	–	–	–	–	–	–	N,N-Diethyl-2-(diethylcarbamoyl)-4-methoxybenzamide	85370-64-5	DD01128

MASS TO CHARGE RATIOS								M.W.	INTENSITIES								Parent	C	H	O	N	S	F	Cl	Br	I	Si	P	B	X	COMPOUND NAME	CAS Reg No	No
137	121	278	161	246	175	260	222	**306**	100	77	57	55	39	39	36	34	**0.00**	17	26	3	2	–	–	–	–	–	–	–	–	–	4,8a-Dimethyl-7-(3-methoxycarbonyl-1-pyrazolin-3-yl)-1-oxo-perhydroazulene		LI06723
291	81	292	79	136	221	156	306	**306**	100	26	19	10	9	8	7	6		17	26	3	2	–	–	–	–	–	–	–	–	–	1-Pentyl-hexobarbital		NI32205
306	161	118	89	307	77	76	105	**306**	100	91	60	37	22	21	19	18		18	10	5	–	–	–	–	–	–	–	–	–	–	2(3H)-Benzofuranone, 3-(3-hydroxy-5-oxo-4-phenyl-2(5H)-furanylidene)-	10091-92-6	NI32206
277	278	306	297	165	139	249	163	**306**	100	67	47	47	28	16	14	14		18	10	5	–	–	–	–	–	–	–	–	–	–	6,8-Dideoxyversicolorin	119998-28-6	DD01129
306	307	305	308	272	262	245	140	**306**	100	26	17	9	7	7	7	7		18	11	–	2	1	1	–	–	–	–	–	–	–	6-(4-Fluorophenyl)-4-phenyl-3-cyanopyridine-2(1H)-thione		NI32207
105	306	201	173	94	149	133	289	**306**	100	97	84	72	57	48	38	18		18	14	3	2	–	–	–	–	–	–	–	–	–	2-Benzoyl-3-hydroxymethyl-4h-pyrazolo[5,1-c][1,4]benzoxazine		MS07098
152	306	227	151	153	51	229	154	**306**	100	29	27	14	10	9	7	7		18	15	–	–	–	–	–	–	–	–	–	–	1	Arsine, triphenyl-	603-32-7	NI32208
152	306	227	154	151	153	229	307	**306**	100	31	30	16	12	9	8	5		18	15	–	–	–	–	–	–	–	–	–	–	1	Arsine, triphenyl-	603-32-7	NI32209
152	51	151	77	227	306	153	50	**306**	100	33	30	28	24	20	17	12		18	15	–	–	–	–	–	–	–	–	–	–	1	Arsine, triphenyl-	603-32-7	NI32210
306	240	66	307	91	121	238	56	**306**	100	46	28	21	21	18	15	14		18	18	1	–	–	–	–	–	–	–	–	–	1	Ferrocenyl(2-norbornadienyl)carbinole		NI32211
91	172	134	306	42	107	77	65	**306**	100	67	59	57	48	40	40	35		18	18	1	4	–	–	–	–	–	–	–	–	–	3-Amino-4-(4-dimethylaminobenzylidene)-1-phenyl-2-pyrazolin-5-one		NI32212
306	92	237	307	173	145	77	132	**306**	100	59	43	23	23	22	21	16		18	18	1	4	–	–	–	–	–	–	–	–	–	3H-Pyrazol-3-one, 4-[[4-(dimethylamino)phenyl]imino]-2,4-dihydro-5-methyl-2-phenyl-	1456-89-9	NI32213
223	224	167	117	179	178	165	193	**306**	100	23	13	12	11	11	11	8	**3.27**	18	20	–	–	–	–	2	–	–	–	–	–	–	Ethane, 1,1-dichloro-2,2-bis(4-ethylphenyl)-	72-56-0	NI32215
223	224	165	179	167	178	193	115	**306**	100	55	22	19	19	18	17	16	**4.60**	18	20	–	–	–	–	2	–	–	–	–	–	–	Ethane, 1,1-dichloro-2,2-bis(4-ethylphenyl)-	72-56-0	NI32214
223	224	165	115	179	178	193	77	**306**	100	19	9	8	7	7	6	6	**0.00**	18	20	–	–	–	–	2	–	–	–	–	–	–	Ethane, 1,1-dichloro-2,2-bis(4-ethylphenyl)-	72-56-0	NI32216
306	107	108	93	114	79	121	163	**306**	100	77	68	61	52	37	33	30		18	20	2	–	–	2	–	–	–	–	–	–	–	Estra-1,4-diene-3,17-dione, 4,10-difluoro-	32307-51-0	NI32217
306	149	121	163	119	150	145	143	**306**	100	73	33	29	29	26	25	25		18	20	2	–	–	2	–	–	–	–	–	–	–	Estra-1,4-diene-3,17-dione, 4,10β-difluoro-		LI06724
126	306	288	125	181	153	167	231	**306**	100	61	18	18	12	10	9	6		18	20	2	–	–	2	–	–	–	–	–	–	–	Estra-1,4-diene-3,17-dione, 10,16α-difluoro-	32307-46-3	NI32218
126	107	306	93	105	91	125	79	**306**	100	87	61	34	19	19	18	16		18	20	2	–	–	2	–	–	–	–	–	–	–	Estra-1,4-diene-3,17-dione, 10,16α-difluoro-	32307-46-3	NI32219
233	291	273	306	234	57	46	147	**306**	100	31	28	22	18	18	15	10		18	26	4	–	–	–	–	–	–	–	–	–	–	3-(3,5-Di-tert-butyl-4-hydroxy)benzoylpropionic acid		IC03827
154	98	167	83	141	55	306	69	**306**	100	89	54	51	48	35	26	23		18	26	4	–	–	–	–	–	–	–	–	–	–	2-Cyclohexen-1-one, 2,2'-ethylenebis[3-hydroxy-5,5-dimethyl-	22381-61-9	NI32220
177	260	306	245	178	179	217	139	**306**	100	20	19	12	12	11	7	7		18	26	4	–	–	–	–	–	–	–	–	–	–	9,11-Dihydroxy-6-(1'-ethoxy-1'-isopropyl)-2-oxatricyclo[6.4.0.1^{3,7}]trideca-1(8),9,11-triene		LI06725
291	262	247	306					**306**	100	43	27	12						18	26	4	–	–	–	–	–	–	–	–	–	–	1,4a-Dimethyl-1-ethoxycarbonyl-7-hydroxy-8-oxo-1,2,3,4,4a,5,6,7,8,9,10,10a-dodecahydro-phenanthrene		LI06726
121	126	246	306					**306**	100	56	25	19						18	26	4	–	–	–	–	–	–	–	–	–	–	1,4a-Dimethyl-1-ethoxycarbonyl-8-hydroxy-7-oxo-1,2,3,4,4a,4b,5,6,7,9,10,10a-dodecahydro-phenanthrene		LI06727
83	82	56	55	166	140	110	54	**306**	100	95	64	64	59	24	24	24	**6.70**	18	26	4	–	–	–	–	–	–	–	–	–	–	1,1-Bis(2,6-dioxo-4,4-dimethylcyclohexyl)ethane		IC03828
149	150	43	237	71	55	104	70	**306**	100	38	26	12	11	10	9	9	**0.00**	18	26	4	–	–	–	–	–	–	–	–	–	–	Dipentyl phthalate	131-18-0	NI32221
149	42	41	43	55	70	150	29	**306**	100	16	15	14	13	11	10	7	**0.00**	18	26	4	–	–	–	–	–	–	–	–	–	–	Dipentyl phthalate	131-18-0	WI01396
149	43	41	150	55	42	237	104	**306**	100	19	14	9	7	7	6	6	**0.00**	18	26	4	–	–	–	–	–	–	–	–	–	–	Dipentyl phthalate	131-18-0	MS07099
241	270	213	242	128	171	115	271	**306**	100	77	32	22	21	19	18	17	**0.00**	18	26	4	–	–	–	–	–	–	–	–	–	–	3,4-Hexanediol, 3,4-bis(4-oxocyclohexen-1-yl)-		LI06728
288	245	233	204	263	306			**306**	100	92	91	88	59	50				18	26	4	–	–	–	–	–	–	–	–	–	–	2-Hydroxy-4-methoxy-3-(3-methylbut-2-enyl)-6-(1-pentyl)benzoic acid		NI32222
306	215	55	159	41	233	164	120	**306**	100	49	40	39	35	30	30	28		18	26	4	–	–	–	–	–	–	–	–	–	–	6β-Hydroxy-4-oxa-5β-androstane-3,17-dione		MS07100
246	187	145	247	188	306	276	185	**306**	100	86	27	18	15	13	10	10		18	26	4	–	–	–	–	–	–	–	–	–	–	9,10-Phenanthrenedicarboxylic acid, 1,2,3,4,5,6,7,8,8a,9,10,10a-dodecahydro-, dimethyl ester	55255-39-5	NI32223
187	246	274	145	186	306	113	275	**306**	100	80	65	38	32	31	27	22		18	26	4	–	–	–	–	–	–	–	–	–	–	9,10-Phenanthrenedicarboxylic acid, 1,2,3,4,5,6,7,8,8a,9,10,10a-dodecahydro-, dimethyl ester	55255-39-5	NI32224
246	187	274	214	145	247	275	41	**306**	100	84	82	37	36	28	20	18	**11.01**	18	26	4	–	–	–	–	–	–	–	–	–	–	9,10-Phenanthrenedicarboxylic acid, 1,2,3,4,5,6,7,8,8a,9,10,10a-dodecahydro-, dimethyl ester	55255-39-5	NI32225
307	263	120	142	170				**306**	100	70	54	47	29				**0.00**	18	30	2	2	–	–	–	–	–	–	–	–	–	1-Propanol, 3-(dibutylamino)-, 4-aminobenzoate	149-16-6	NI32226
306	305	141	289	291	115	169	307	**306**	100	84	72	58	41	36	26	20		19	14	4	–	–	–	–	–	–	–	–	–	–	trans-6-Methoxy-2-(4-methylstyryl)chromone		MS07101
119	67	68	91	187	77	120	93	**306**	100	73	68	59	53	40	33	33	**0.00**	19	18	2	2	–	–	–	–	–	–	–	–	–	N-(2,2-Dimethyl-3-butynoyl)-N,N'-diphenylurea		DD01130
219	277	218	248	289	221	206	205	**306**	100	64	59	57	53	53	50	49	**43.00**	19	18	2	2	–	–	–	–	–	–	–	–	–	Mappicine	54318-59-1	NI32227
238	239	149	306	210	155	154	237	**306**	100	70	60	53	53	46	46	40		19	18	2	2	–	–	–	–	–	–	–	–	–	Pyridine, 3-[5-[4-[(3-methyl-2-butenyl)oxy]phenyl]-2-oxazolyl]-	17190-80-6	NI32228
140	306	142	262	93	41	91	120	**306**	100	96	78	70	63	61	54	52		19	27	2	–	–	1	–	–	–	–	–	–	–	Androstan-17-one, 5,6-epoxy-3-fluoro-, (3α,5α,6α)-	28344-36-7	NI32229
306	142	262	93	41	91	123	120	**306**	100	82	73	66	64	57	55	55		19	27	2	–	–	1	–	–	–	–	–	–	–	Androstan-17-one, 5,6-epoxy-3-fluoro-, (3α,5α,6α)-	28344-36-7	MS07102
306	93	140	91	41	143	108	79	**306**	100	39	32	32	31	28	28	25		19	27	2	–	–	1	–	–	–	–	–	–	–	Androstan-17-one, 5,6-epoxy-3-fluoro-, (3α,5β,6β)-	28344-37-8	MS07103
306	142	93	91	79	123	55	143	**306**	100	88	52	52	52	44	44	42		19	27	2	–	–	1	–	–	–	–	–	–	–	Androstan-17-one, 5,6-epoxy-3-fluoro-, (3β,5α,6α)-	40242-88-4	MS07104
306	142	140	93	91	79	123	55	**306**	100	88	87	52	52	52	44	44		19	27	2	–	–	1	–	–	–	–	–	–	–	Androstan-17-one, 5,6-epoxy-3-fluoro-, (3β,5α,6α,17β)-	40242-88-4	NI32230
272	91	81	93	105	55	67	107	**306**	100	74	73	72	68	63	56	55	**0.00**	19	30	1	–	1	–	–	–	–	–	–	–	–	Androstan-17-ol, 2,3-epithio-, (2α,3α,5α,17β)-	2363-58-8	NI32231
91	93	41	79	105	55	81	67	**306**	100	97	97	95	82	77	76	69	**0.00**	19	30	1	–	1	–	–	–	–	–	–	–	–	Androstan-17-ol, 2,3-epithio-, (2β,3β,5α,17β)-	2136-25-6	NI32232
306	272	108	55	66	107	79	81	**306**	100	54	48	44	38	37	37	35		19	30	1	–	1	–	–	–	–	–	–	–	–	Androstan-17-one, 3-mercapto-		LI06729
288	55	41	306	43	81	244	95	**306**	100	99	89	72	67	62	55	54		19	30	3	–	–	–	–	–	–	–	–	–	–	Androstan-2-one, 11,17-dihydroxy-, (5β,11β,17β)-	55267-53-3	NI32233

MASS TO CHARGE RATIOS	M.W.	INTENSITIES	Parent	C	H	O	N	S	F	Cl	Br	I	Si	P	B	X	COMPOUND NAME	CAS Reg No	No
286 306 55 41 229 81 43 93	306	100 81 63 46 38 36 34 33		19	30	3	–	–	–	–	–	–	–	–	–	–	Androstan-3-one, 6,17-dihydroxy-, (5α,6α,17β)-	49644-02-2	NI32234
288 55 43 41 306 57 93 81	306	100 92 75 73 62 44 42 42		19	30	3	–	–	–	–	–	–	–	–	–	–	Androstan-3-one, 6,17-dihydroxy-, (5α,6β,17β)-	18529-66-3	NI32235
288 55 306 41 81 93 67 43	306	100 54 50 48 34 32 32 32		19	30	3	–	–	–	–	–	–	–	–	–	–	Androstan-3-one, 6,17-dihydroxy-, (5α,6β,17β)-	18529-66-3	NI32236
55 288 306 41 93 81 43 95	306	100 90 83 64 52 51 47 46		19	30	3	–	–	–	–	–	–	–	–	–	–	Androstan-3-one, 6,17-dihydroxy-, (5β,6α,17β)-	51467-36-8	NI32237
288 306 55 41 43 93 81 95	306	100 73 54 43 34 31 31 27		19	30	3	–	–	–	–	–	–	–	–	–	–	Androstan-3-one, 7,17-dihydroxy-, (5β,7α,17β)-	55365-27-0	NI32238
288 306 41 81 95 55 93 108	306	100 67 65 52 50 50 43 42		19	30	3	–	–	–	–	–	–	–	–	–	–	Androstan-3-one, 11,17-dihydroxy-, (5α,11α,17β)-	25788-56-1	NI32239
288 229 244 41 55 306 43 81	306	100 52 50 45 42 34 28 25		19	30	3	–	–	–	–	–	–	–	–	–	–	Androstan-3-one, 11,17-dihydroxy-, (5α,11β,17β)-	7801-30-1	NI32240
122 288 55 41 43 81 109 123	306	100 85 76 65 52 44 43 42	**30.75**	19	30	3	–	–	–	–	–	–	–	–	–	–	Androstan-3-one, 11,17-dihydroxy-, (5β,11α,17β)-	32693-31-5	NI32241
122 288 55 41 43 81 109 123	306	100 86 74 64 52 43 42 41	**30.00**	19	30	3	–	–	–	–	–	–	–	–	–	–	Androstan-3-one, 11,17-dihydroxy-, (5β,11α,17β)-	32693-31-5	LI06730
288 55 41 306 43 81 244 122	306	100 99 88 71 64 62 54 53		19	30	3	–	–	–	–	–	–	–	–	–	–	Androstan-3-one, 11,17-dihydroxy-, (5β,11β,17β)-		LI06731
273 288 55 41 81 43 244 111	306	100 85 44 33 30 28 24 24	**0.91**	19	30	3	–	–	–	–	–	–	–	–	–	–	Androstan-3-one, 12,17-dihydroxy-, (5α,12α,17β)-	53604-42-5	NI32242
288 55 81 41 247 43 289 111	306	100 37 29 28 26 24 21 19	**13.03**	19	30	3	–	–	–	–	–	–	–	–	–	–	Androstan-3-one, 12,17-dihydroxy-, (5α,12β,17β)-	53608-27-8	NI32243
273 288 55 41 81 43 244 95	306	100 74 45 37 33 31 24 22	**15.76**	19	30	3	–	–	–	–	–	–	–	–	–	–	Androstan-3-one, 12,17-dihydroxy-, (5β,12α,17β)-	53608-25-6	NI32244
306 43 55 95 41 93 307 67	306	100 32 29 26 25 20 19 19		19	30	3	–	–	–	–	–	–	–	–	–	–	Androstan-6-one, 3,17-dihydroxy-, (3α,5α,17β)-	49644-08-8	NI32245
306 55 41 139 43 95 67 93	306	100 30 28 27 27 26 22 21		19	30	3	–	–	–	–	–	–	–	–	–	–	Androstan-6-one, 3,17-dihydroxy-, (3β,5α,17β)-	17328-37-9	NI32246
235 288 55 41 306 136 95 43	306	100 43 24 22 21 21 21 18		19	30	3	–	–	–	–	–	–	–	–	–	–	Androstan-6-one, 3,17-dihydroxy-, (3β,5β,17β)-	53512-51-9	NI32247
306 193 180 147 41 55 107 108	306	100 43 36 33 28 27 25 24		19	30	3	–	–	–	–	–	–	–	–	–	–	Androstan-11-one, 3,17-dihydroxy-, (3α,5α,17β)-	32693-28-0	NI32248
306 193 180 147 41 107 55 108	306	100 42 35 32 27 25 25 22		19	30	3	–	–	–	–	–	–	–	–	–	–	Androstan-11-one, 3,17-dihydroxy-, (3α,5α,17β)-	32693-28-0	LI06732
193 234 306 288 41 93 55 108	306	100 99 85 70 62 57 55 52		19	30	3	–	–	–	–	–	–	–	–	–	–	Androstan-11-one, 3,17-dihydroxy-, (3α,5β,17β)-	1158-94-7	NI32249
306 193 41 55 147 107 93 79	306	100 50 45 40 38 35 33 32		19	30	3	–	–	–	–	–	–	–	–	–	–	Androstan-11-one, 3,17-dihydroxy-, (3β,5α,17β)-	32693-29-1	NI32250
306 193 41 55 147 107 93 79	306	100 50 44 40 37 32 32 31		19	30	3	–	–	–	–	–	–	–	–	–	–	Androstan-11-one, 3,17-dihydroxy-, (3β,5α,17β)-	32693-29-1	LI06733
193 180 108 55 41 306 107 93	306	100 50 49 40 40 36 35 34		19	30	3	–	–	–	–	–	–	–	–	–	–	Androstan-11-one, 3,17-dihydroxy-, (3β,5β,17β)-	32810-92-7	NI32251
306 41 55 107 67 124 93 79	306	100 56 54 45 39 38 38 36		19	30	3	–	–	–	–	–	–	–	–	–	–	Androstan-17-one, 3,11-dihydroxy-, (3α,5α,11α)-	7801-12-9	NI32252
306 41 55 107 93 67 81 79	306	100 54 53 44 38 37 36 36		19	30	3	–	–	–	–	–	–	–	–	–	–	Androstan-17-one, 3,11-dihydroxy-, (3α,5α,11α)-	7801-12-9	LI06734
306 288 273 107 41 106 270 43	306	100 34 31 29 28 25 20 20		19	30	3	–	–	–	–	–	–	–	–	–	–	Androstan-17-one, 3,11-dihydroxy-, (3α,5α,11β)-		LI06735
124 288 55 106 41 306 107 95	306	100 99 72 70 69 67 63 60		19	30	3	–	–	–	–	–	–	–	–	–	–	Androstan-17-one, 3,11-dihydroxy-, (3α,5β,11α)-	2551-74-8	NI32253
271 289 253 272 287 290 269 305	306	100 61 43 38 18 16 12 9	**5.90**	19	30	3	–	–	–	–	–	–	–	–	–	–	Androstan-17-one, 3,11-dihydroxy-, (3α,5β,11β)-	2551-74-8	NI32254
306 41 55 67 97 93 81 107	306	100 50 48 34 33 31 31 30		19	30	3	–	–	–	–	–	–	–	–	–	–	Androstan-17-one, 3,11-dihydroxy-, (3α,5β,11β)-	2551-74-8	LI06736
306 41 55 107 97 277 67 43	306	100 32 29 24 23 21 20 20		19	30	3	–	–	–	–	–	–	–	–	–	–	Androstan-17-one, 3,11-dihydroxy-, (3β,5α,11α)-	25848-75-3	NI32255
306 41 55 107 97 277 43 307	306	100 30 27 23 21 20 20 18		19	30	3	–	–	–	–	–	–	–	–	–	–	Androstan-17-one, 3,11-dihydroxy-, (3β,5α,11α)-	25848-75-3	LI06737
306 55 41 273 107 288 106 97	306	100 36 36 31 31 30 27 25		19	30	3	–	–	–	–	–	–	–	–	–	–	Androstan-17-one, 3,11-dihydroxy-, (3β,5α,11β)-		LI06738
306 124 41 55 108 43 107 67	306	100 65 62 61 54 47 43 41		19	30	3	–	–	–	–	–	–	–	–	–	–	Androstan-17-one, 3,11-dihydroxy-, (3β,5β,11α)-	32212-55-8	LI06739
306 124 41 55 108 43 107 67	306	100 66 63 62 55 49 44 42		19	30	3	–	–	–	–	–	–	–	–	–	–	Androstan-17-one, 3,11-dihydroxy-, (3β,5β,11α)-	32212-55-8	NI32256
306 55 41 107 288 97 273 106	306	100 29 29 27 26 26 25 23		19	30	3	–	–	–	–	–	–	–	–	–	–	Androstan-17-one, 3,11-dihydroxy-, (3β,5β,11β)-		LI06740
270 55 41 43 81 97 288 260	306	100 52 49 42 39 36 35 32	**9.70**	19	30	3	–	–	–	–	–	–	–	–	–	–	Androstan-17-one, 3,12-dihydroxy-, (3α,5β,12α)-	10448-51-8	NI32257
306 97 232 41 55 43 107 81	306	100 45 39 38 35 33 26 25		19	30	3	–	–	–	–	–	–	–	–	–	–	Androstan-17-one, 3,12-dihydroxy-, (3β,5α,12β)-	848-63-5	NI32258
306 97 262 55 41 43 307 81	306	100 33 30 25 25 23 21 20		19	30	3	–	–	–	–	–	–	–	–	–	–	Androstan-17-one, 3,12-dihydroxy-, (3β,5α,12β)-	848-63-5	NI32259
288 255 145 273 237 270 105 143	306	100 51 32 28 28 27 26 25	**19.25**	19	30	3	–	–	–	–	–	–	–	–	–	–	Androst-5-ene-3,11,17-triol, (3β,11β,17β)-	3924-22-9	NI32260
138 57 139 43 306 71 55 69	306	100 68 62 59 46 34 31 25		19	30	3	–	–	–	–	–	–	–	–	–	–	Benzoic acid, 4-(dodecyloxy)-	2312-15-4	NI32261
152 135 153 107 77 92 136 109	306	100 44 27 16 15 12 7 7	**0.00**	19	30	3	–	–	–	–	–	–	–	–	–	–	Benzoic acid, 3-methoxy-, undecyl ester	69833-35-8	NI32262
152 135 153 77 136 154 107 92	306	100 46 27 8 5 4 4 4	**0.00**	19	30	3	–	–	–	–	–	–	–	–	–	–	Benzoic acid, 4-methoxy-, undecyl ester	69833-36-9	NI32263
196 154 125 69 178 181 127 97	306	100 97 90 79 49 44 35 32	**11.00**	19	30	3	–	–	–	–	–	–	–	–	–	–	Nonadeca-2,18-diene-4,7,10-trione		NI32264
79 55 121 43 41 120 93 67	306	100 88 80 77 69 45 45 40	**36.24**	19	30	3	–	–	–	–	–	–	–	–	–	–	8-Nonenoic acid, 9-(1,3,6-nonatrienyloxy)-, methyl ester, (E,E,Z,Z)-	52077-22-2	MS07105
87 91 101 107 105 144 95 93	306	100 59 44 43 41 39 39 37	**0.30**	19	30	3	–	–	–	–	–	–	–	–	–	–	9,11-Octadecadiynoic acid, 8-hydroxy-, methyl ester	6084-80-6	NI32265
141 173 95 91 275 306 291 247	306	100 53 49 20 2 1 1 1		19	30	3	–	–	–	–	–	–	–	–	–	–	10,12-Octadecadiynoic acid, 8-hydroxy-, methyl ester	75112-84-4	NI32266
43 121 105 107 41 91 122 93	306	100 61 58 57 57 55 42 42	**7.70**	19	30	3	–	–	–	–	–	–	–	–	–	–	2-Phenanthrenecarboxaldehyde, 1,2,3,4,4a,4b,5,6,7,8,8a,9-dodecahydro-7-hydroxy-2,4b,8,8-tetramethyl-, monohydroxy-	72088-19-8	NI32267
169 168 125 126 55 47 79 43	306	100 88 48 31 15 14 12 11	**1.67**	19	31	1	–	–	–	–	–	–	–	1	–	–	Phosphinous acid, isopropylphenyl-, 2-isopropyl-5-methylcyclohexyl ester, (1α,2β,5α)-	62376-13-0	NI32268
215 202 216 306 231 244 260 288	306	100 89 83 67 59 43 38 26		20	18	3	–	–	–	–	–	–	–	–	–	–	3,4-Dihydro-7-hydroxymethyl-12-methylbenz[a]anthracene-trans-3,4-diol		MS07106
245 215 229 202 258 259 270 288	306	100 87 75 42 35 34 13 8	**7.00**	20	18	3	–	–	–	–	–	–	–	–	–	–	5,6-Dihydro-7-hydroxymethyl-12-methylbenz[a]anthracene-trans-5,6-diol		MS07107
215 259 242 202 245 270 228 226	306	100 94 64 61 58 56 44 41	**24.00**	20	18	3	–	–	–	–	–	–	–	–	–	–	8,9-Dihydro-7-hydroxymethyl-12-methylbenz[a]anthracene-trans-8,9-diol		MS07108
215 202 259 229 306 242 288 275	306	100 90 61 61 48 43 32 29		20	18	3	–	–	–	–	–	–	–	–	–	–	10,11-Dihydro-7-hydroxymethyl-12-methylbenz[a]anthracene-trans-10,11-diol		MS07109
215 202 244 259 306 229 226 242	306	100 70 65 60 41 31 30 26		20	18	3	–	–	–	–	–	–	–	–	–	–	3,4-Dihydro-7-methyl-12-hydroxymethylbenz[a]anthracene-trans-3,4-diol		MS07110
215 245 202 228 242 270 306 288	306	100 60 55 47 40 37 23 11		20	18	3	–	–	–	–	–	–	–	–	–	–	5,6-Dihydro-7-methyl-12-hydroxymethylbenz[a]anthracene-trans-5,6-diol		MS07111

MASS TO CHARGE RATIOS								M.W.	INTENSITIES								Parent	C	H	O	N	S	F	Cl	Br	I	Si	P	B	X	COMPOUND NAME	CAS Reg No	No
215	202	259	244	227	288	306	272	**306**	100	92	79	72	60	47	41	23		20	18	3	–	–	–	–	–	–	–	–	–	–	8,9-Dihydro-7-methyl-12-hydroxymethylbenz[a]anthracene-trans-8,9-diol		MS07112
259	215	202	227	288	245	270	306	**306**	100	58	47	45	40	40	31	26		20	18	3	–	–	–	–	–	–	–	–	–	–	10,11-Dihydro-7-methyl-12-hydroxymethylbenz[a]anthracene-trans-10,11-diol		MS07113
186	130	187	306	77	131	118	93	**306**	100	26	16	6	5	4	4	4		20	22	1	2	–	–	–	–	–	–	–	–	–	Spiro[cyclohexane-1,3′-indoline]-2′-carboxanilide	15317-65-4	LI06741
186	130	187	306	77	144	131	83	**306**	100	26	16	6	5	4	4	4		20	22	1	2	–	–	–	–	–	–	–	–	–	Spiro[cyclohexane-1,3′-indoline]-2′-carboxanilide	15317-65-4	NI32269
186	130	187	306					**306**	100	26	16	6						20	22	1	2	–	–	–	–	–	–	–	–	–	Spiro[cyclohexane-1,3′-indoline]-2′-carboxanilide	15317-65-4	MS07114
191	189	190	109	192	163	120	263	**306**	100	92	82	67	57	57	45	39	**15.00**	20	34	2	–	–	–	–	–	–	–	–	–	–	(5R,8R,9S,10R)-Ent-3-cleroden-15-oic acid		MS07115
81	43	41	107	95	55	69	93	**306**	100	94	63	57	54	51	44	43	**0.00**	20	34	2	–	–	–	–	–	–	–	–	–	–	4,8,13-Cyclotetradecatriene-1,3-diol, 12-isopropyl-1,5,9-trimethyl-	7220-78-2	NI32270
41	67	55	43	79	80	81	93	**306**	100	96	81	81	78	75	63	47	**3.74**	20	34	2	–	–	–	–	–	–	–	–	–	–	8,11,14-Eicosatrienoic acid, (Z,Z,Z)-	1783-84-2	NI32271
291	273	205	109	177	135	121	254	**306**	100	72	71	60	57	48	48	40	**17.00**	20	34	2	–	–	–	–	–	–	–	–	–	–	19-Hydroxy-13-epimanoyl oxide	77096-88-9	NI32272
95	79	109	191	175	235	250	288	**306**	100	40	23	11	11	3	2	1	**0.20**	20	34	2	–	–	–	–	–	–	–	–	–	–	Isolinaridiol		NI32273
123	257	109	81	95	93	65	55	**306**	100	99	80	74	72	57	57	57	**3.00**	20	34	2	–	–	–	–	–	–	–	–	–	–	Kaurane-16,18-diol, (4β)-	24532-72-7	NI32274
69	109	95	153	81	93	71	55	**306**	100	93	67	64	62	58	57	50	**0.00**	20	34	2	–	–	–	–	–	–	–	–	–	–	Labda-8(20),14-dien-6,13-diol, (6α,13S)-	1438-66-0	NI32275
69	109	95	153	81	93	71	55	**306**	100	76	56	53	51	48	47	42	**0.00**	20	34	2	–	–	–	–	–	–	–	–	–	–	Labda-8(20),14-dien-6,13-diol, (6α,13S)-	1438-66-0	NI32276
42	123	95	69	55	81	43	109	**306**	100	89	89	80	77	72	70	51	**2.00**	20	34	2	–	–	–	–	–	–	–	–	–	–	Labda-8(20),14-dien-7,13-diol, (7α,13S)-		LI06742
81	95	93	55	41	79	107	67	**306**	100	76	75	73	66	54	52	52	**0.00**	20	34	2	–	–	–	–	–	–	–	–	–	–	Labda-8(20),13-diene-15,19-diol, (E)-	1857-24-5	NI32277
41	123	69	55	81	43	109	95	**306**	100	89	80	78	72	70	51	49	**2.28**	20	34	2	–	–	–	–	–	–	–	–	–	–	Labda-8(20),14-diene-7,13-diol	69782-87-2	NI32278
257	121	109	107	288	189	135	123	**306**	100	62	53	53	49	49	49	47	**0.00**	20	34	2	–	–	–	–	–	–	–	–	–	–	Labda-8(20),14-diene-7,13-diol, (13S)-	72347-66-1	NI32279
81	95	93	71	107	109	121	105	**306**	100	58	56	52	40	34	30	26	**0.06**	20	34	2	–	–	–	–	–	–	–	–	–	–	Labda-8(20),14-diene-13,18-diol, (13S)-	4549-12-6	NI32280
257	189	109	121	107	95	153	93	**306**	100	44	39	38	37	37	36	36	**1.92**	20	34	2	–	–	–	–	–	–	–	–	–	–	Labda-8(20),14-diene-13,19-diol	1908-44-7	NI32281
81	95	93	71	107	121	109	135	**306**	100	85	72	63	54	49	49	31	**0.20**	20	34	2	–	–	–	–	–	–	–	–	–	–	Labda-8(20),14-diene-13,19-diol, (13S)-	3650-30-4	NI32282
182	109	122	124	95	191	123	81	**306**	100	73	58	33	33	31	30	21	**18.37**	20	34	2	–	–	–	–	–	–	–	–	–	–	Labd-7-en-15-oic acid	468-82-6	NI32283
306	288	235	69	291	109	123	81	**306**	100	76	73	63	56	56	53	49		20	34	2	–	–	–	–	–	–	–	–	–	–	Labd-14-en-6-ol, 8,13-epoxy-, (6α,13β)-		NI32284
79	67	41	95	81	55	93	108	**306**	100	83	74	71	61	60	56	51	**20.02**	20	34	2	–	–	–	–	–	–	–	–	–	–	Linolenic acid, ethyl ester	1191-41-9	WI01397
257	81	95	93	41	107	153	109	**306**	100	74	66	63	63	55	50	49	**2.50**	20	34	2	–	–	–	–	–	–	–	–	–	–	1-Naphthalenepentanol, decahydro-5-(hydroxymethyl)-5,8a-dimethyl-γ,2-bis(methylene)-, (1α,4aβ,5α,8aα)-	57289-28-8	NI32285
43	277	55	41	67	81	69	259	**306**	100	98	61	50	35	29	29	28	**0.00**	20	34	2	–	–	–	–	–	–	–	–	–	–	8H-Naphtho[1,2-b]pyran-8-one, 3-ethyldodecahydro-3,4a,7,7,10a-pentamethyl-, [3R-(3α,4aβ,6aα,10aβ,10bα)]-	55836-76-5	NI32286
259	277	43	203	260	151	135	55	**306**	100	83	34	26	20	20	20	18	**0.00**	20	34	2	–	–	–	–	–	–	–	–	–	–	10H-Naphtho[2,1-b]pyran-10-one, 3-ethyldodecahydro-3,4a,7,7,10a-pentamethyl-, 3S-(3α,4aβ,6aα,10aβ,10bα)-	1231-33-0	NI32287
55	81	69	43	67	109	95	41	**306**	100	96	88	88	81	77	73	73	**2.31**	20	34	2	–	–	–	–	–	–	–	–	–	–	Sandaracopimar-15-ene-6,8-diol, (6β,8β)-		NI32288
55	81	69	216	275	67	258	79	**306**	100	86	85	79	68	68	63	62	**4.11**	20	34	2	–	–	–	–	–	–	–	–	–	–	Sandaracopimar-15-ene-8,18-diol, (8β)-		NI32289
288	69	163	41	55	150	95	81	**306**	100	85	76	68	62	60	58	56	**6.40**	20	34	2	–	–	–	–	–	–	–	–	–	–	Sandaracopimar-15-ene-8-11-diol (8β,11α)-		NI32290
91	167	306	246	247	275			**306**	100	95	90	90	70	12				21	22	2	–	–	–	–	–	–	–	–	–	–	1-Cyclohexanecarboxylic acid, 2-benzhydrylidene-, methyl ester		LI06743
200	91	167	105	180	179	165	110	**306**	100	58	47	34	25	25	19	16	**0.00**	21	22	2	–	–	–	–	–	–	–	–	–	–	2,3-Diphenyl-1,4-dioxaspiro[4.6]undec-6-ene	117583-53-6	DD01131
129	128	91	115	117	306			**306**	100	67	31	28	27	1				21	22	2	–	–	–	–	–	–	–	–	–	–	Hydrocinnamic acid, 2-[(3,4-dihydro-2-naphthyl)methyl]-, methyl ester	23796-81-8	LI06744
129	91	115	117	141	15	142	118	**306**	100	31	28	27	24	24	23	18	**17.41**	21	22	2	–	–	–	–	–	–	–	–	–	–	Hydrocinnamic acid, 2-[(3,4-dihydro-2-naphthyl)methyl]-, methyl ester	23796-81-8	NI32291
184	47	306	186	49	170	135	200	**306**	100	97	91	87	75	70	65	64		21	26	–	2	–	–	–	–	–	–	–	–	–	2,20-Cycloaspidospermidine, 6,7-didehydro-1,3-dimethyl-, (2α,3β,5α,12β,19α,20R)-	56053-37-3	NI32292
306	307	278	250	248	249	279	251	**306**	100	25	18	18	10	8	4	4		22	10	2	–	–	–	–	–	–	–	–	–	–	Dibenzo[def,mno]chrysene-6,12-dione	641-13-4	NI32293
306	305	152	307	153	135	303	151	**306**	100	90	28	21	20	11	8	7		22	14	–	2	–	–	–	–	–	–	–	–	–	14H-[4.5]Benzoisoquino[2,1-a]perimidine		LI06745
175	103	306	145	176	131	77	119	**306**	100	34	21	20	15	11	9	8		22	26	1	–	–	–	–	–	–	–	–	–	–	3-Buten-2-one, 1-(4-tert-butyl-2,6-xylyl)-3-phenyl-	29906-96-5	NI32294
57	97	41	55	83	153	69	67	**306**	100	47	43	40	29	17	11	11	**0.00**	22	42	–	–	–	–	–	–	–	–	–	–	–	Cyclohexane, 1,1′-(1,2-diisopropyl-1,2-ethanediyl)bis-, (R*,R*)-(±)-	65149-85-1	NI32295
97	57	83	153	55	71	41	70	**306**	100	99	42	33	29	27	20	18	**1.00**	22	42	–	–	–	–	–	–	–	–	–	–	–	Cyclohexane, 1,1′-(1,2-diisopropyl-1,2-ethanediyl)bis-, (R*,S*)-	65149-86-2	NI32296
152	97	83	55	153	96	41	57	**306**	100	74	58	57	45	42	29	26	**6.00**	22	42	–	–	–	–	–	–	–	–	–	–	–	Cyclohexane, 1,1′-(1,2-diisopropyl-1,2-ethanediyl)bis-, (R*,S*)-	65149-86-2	NI32297
306	138	276	307	277	305	137	130	**306**	100	36	35	25	21	15	10	10		23	14	1	–	–	–	–	–	–	–	–	–	–	7H-Benzo[c]fluorenone, 5-phenyl-		LI06746
178	141	306	289	305	165	152	151	**306**	100	62	43	25	25	22	21	19		24	18	–	–	–	–	–	–	–	–	–	–	–	1-(1-Azulenyl)-4-(4-azulenyl)buta-1,3-diene	94154-64-0	NI32298
178	128	202	306	204	152	151	203	**306**	100	56	35	30	30	24	23	22		24	18	–	–	–	–	–	–	–	–	–	–	–	6,7-Benzotetracyclo[3.3.0^{1,5}.0^{2,4}.0^{3,8}]octane, 1,2-diphenyl-		LI06747
306	229	228	305	178	291	289	307	**306**	100	94	67	64	39	27	27	26		24	18	–	–	–	–	–	–	–	–	–	–	–	1,4-Methanonaphthalene, 9-(diphenylmethylene)-1,4-dihydro-	28591-78-8	NI32299
306	307	153	289	308	302	303	151	**306**	100	27	15	6	4	4	3	3		24	18	–	–	–	–	–	–	–	–	–	–	–	m,p-Quaterphenyl	1166-19-4	NI32300
306	307	153	289	153.5	152	228	226	**306**	100	26	19	4	4	4	3	3		24	18	–	–	–	–	–	–	–	–	–	–	–	m,p-Quaterphenyl	1166-19-4	MS07116
306	307	153	289	228	153.5	152	226	**306**	100	26	18	5	4	4	4	3		24	18	–	–	–	–	–	–	–	–	–	–	–	m-Quaterphenyl	1166-18-3	MS07117
306	307	153	289	152	51	39	77	**306**	100	25	14	4	4	4	4	3		24	18	–	–	–	–	–	–	–	–	–	–	–	m-Quaterphenyl	1166-18-3	WI01398
306	153	289	302	228	226	276	202	**306**	100	15	6	5	5	5	3	3		24	18	–	–	–	–	–	–	–	–	–	–	–	m-Quaterphenyl	1166-18-3	MS07118
306	307	305	215	289	229	291	228	**306**	100	26	19	18	17	13	12	12		24	18	–	–	–	–	–	–	–	–	–	–	–	o,m-Quaterphenyl		MS07119
306	307	229	228	305	145	289	215	**306**	100	25	23	20	17	17	16	14		24	18	–	–	–	–	–	–	–	–	–	–	–	o,o-Quaterphenyl		MS07120

MASS TO CHARGE RATIOS								M.W.	INTENSITIES								Parent	C	H	O	N	S	F	Cl	Br	I	Si	P	B	X	COMPOUND NAME	CAS Reg No	No
306	307	215	305	289	229	228	291	**306**	100	26	23	16	16	11	11	10		24	18	–	–	–	–	–	–	–	–	–	–	–	o,p-Quaterphenyl		MS07121
306	307	153	153.5	152	308	289	151	**306**	100	26	26	6	4	3	3	3		24	18	–	–	–	–	–	–	–	–	–	–	–	p,p-Quaterphenyl		MS07122
306	204	203	202	307	51	39	289	**306**	100	44	31	30	28	21	18	14		24	18	–	–	–	–	–	–	–	–	–	–	–	m-Terphenyl, 4′-phenyl-	1165-53-3	NI32301
306	288	307	305	291	289	290	302	**306**	100	75	25	17	16	16	11	9		24	18	–	–	–	–	–	–	–	–	–	–	–	m-Terphenyl, 4′-phenyl-	1165-53-3	MS07123
306	307	228	289	226	153	77	302	**306**	100	26	8	7	7	6	6	4		24	18	–	–	–	–	–	–	–	–	–	–	–	m-Terphenyl, 5′-phenyl-	612-71-5	NI32302
306	307	289	302	303	290	228	153	**306**	100	26	8	5	4	4	4	4		24	18	–	–	–	–	–	–	–	–	–	–	–	m-Terphenyl, 5′-phenyl-	612-71-5	NI32303
306	305	307	289	290	291	265	145	**306**	100	30	25	25	18	17	15	15		24	18	–	–	–	–	–	–	–	–	–	–	–	o-Terphenyl, 3′-phenyl-		MS07124
306	276	305	277	307	303	138	302	**306**	100	45	38	35	31	23	21	19		24	18	–	–	–	–	–	–	–	–	–	–	–	Tetracyclo[14.2.2.2^{4,7}.2^{10,13}]tetracosa-2,4,6,8,10,12,14,16,18,19,21,23-dodecaene	16337-16-9	NI32304
113	132	59	78	103	118	151	64	**307**	100	63	39	19	15	13	12	9	**0.00**	3	4	–	1	2	6	–	–	–	–	–	–	1	4-Methyl-1,3,2-dithiazolium hexafluoroarsenate(V)		NI32305
180	69	155	28	27	78	307	166	**307**	100	57	49	21	19	17	15	14		4	4	–	1	–	6	–	–	1	–	–	–	–	Ethylamine, N,N-bis(trifluoromethyl)-2-iodo-		LI06748
69	176	174	157	155	78	166	307	**307**	100	54	54	37	37	34	19	13		5	2	–	1	–	8	–	1	–	–	–	–	–	1-Propene, 2-bromo-1,1-difluoro-3,3-bis(trifluoromethyl)-		LI06749
308	309	276	310	248	160	147	146	**307**	100	10	2	1	1	1	1	1	**0.00**	9	10	5	1	–	5	–	–	–	–	–	–	–	Aspartic acid, N-pentafluoropropionate, dimethyl ester		NI32306
214	228	91	198	230	232	166	274	**307**	100	99	86	76	39	31	31	30	**0.00**	10	14	3	1	1	–	–	1	–	–	–	–	–	Sulphonium, [(2-methoxy-5-nitrophenyl)methyl]dimethyl-, bromide	64415-08-3	NI32307
84	41	56	85	198	196	97	55	**307**	100	16	9	8	7	7	6	3		11	8	3	1	–	–	3	–	–	–	–	–	–	5-Pyrrolidinone, 2-(2,4,5-trichlorophenoxycarbonyl)-		MS07125
162	164	254	42	175	272	145	100	**307**	100	72	60	42	39	32	31	31	**3.00**	11	11	5	1	–	–	2	–	–	–	–	–	–	L-Serine, N-[(2,4-dichlorophenoxy)acetyl]-	50648-97-0	NI32308
305	307	261	232	216	306	215	56	**307**	100	34	25	16	16	13	13	12		11	12	4	3	–	3	–	–	–	–	–	–	–	Aniline, N,N-diethyl-2,6-dinitro-4-(trifluoromethyl)-	5254-27-3	NI32309
307	260	309	261	214	290	262	75	**307**	100	71	35	35	29	26	26	18		12	6	5	3	–	–	1	–	–	–	–	–	–	10H-Phenoxazine, 3-chloro-1,8-dinitro-	39522-56-0	MS07126
93	64	168	92	259	138	65	44	**307**	100	65	62	59	56	47	46	38	**8.09**	12	9	5	3	1	–	–	–	–	–	–	–	–	2,4-Dinitro-4′-aminodiphenylsulphoxide	75317-02-1	NI32310
81	206	82	29	40	41	54	69	**307**	100	81	32	29	27	24	18	16	**9.30**	12	16	3	3	–	3	–	–	–	–	–	–	–	L-Histidine, N-(trifluoroacetyl)-, butyl ester	55255-83-9	NI32311
86	73	93	75	58	44	42	95	**307**	100	82	65	55	37	26	25	24	**14.74**	12	21	1	5	–	–	–	–	–	2	–	–	–	4(1H)-Pteridinone, 2-amino-, bis(trimethylsilyl)-	56272-92-5	NI32312
73	60	43	57	55	41	183	71	**307**	100	86	83	70	69	68	55	39	**0.16**	12	24	2	–	–	–	–	–	–	–	–	–	1	Dodecanoic acid, silver(1+) salt	18268-45-6	NI32313
73	190	75	128	45	29	47	59	**307**	100	66	56	34	30	22	13	12	**2.30**	12	29	2	1	1	–	–	–	–	2	–	–	–	Butanoic acid, 4-(ethylthio)-2-[(trimethylsilyl)amino]-, trimethylsilyl ester	55255-82-8	MS07127
190	73	75	128	191	147	45	219	**307**	100	67	49	26	15	12	12	11	**8.60**	12	29	2	1	1	–	–	–	–	2	–	–	–	Butanoic acid, 4-(ethylthio)-2-[(trimethylsilyl)amino]-, trimethylsilyl ester	55255-82-8	NI32314
73	45	61	75	190	142	74	43	**307**	100	30	19	17	16	14	12	12		12	29	2	1	1	–	–	–	–	2	–	–	–	Butanoic acid, 2-methyl-4-(methylthio)-2-[(trimethylsilyl)amino]-, trimethylsilyl ester		MS07128
186	43	131	307	187	51	28	158	**307**	100	34	18	12	12	7	7	6		13	13	4	3	1	–	–	–	–	–	–	–	–	1-(5-Methylsulphonyl-2-furyl)-2-(6-acetamido-3-pyridazyl)ethylene		MS07129
135	59	60	164	136	41	218	108	**307**	100	45	39	35	33	28	27	17	**3.92**	13	17	4	5	–	–	–	–	–	–	–	–	–	Adenosine, 2′,3′-O-isopropylidene-	362-75-4	NI32315
135	164	59	218	136	108	219	204	**307**	100	34	30	29	27	15	6	6	**0.90**	13	17	4	5	–	–	–	–	–	–	–	–	–	Adenosine, 2′,3′-O-isopropylidene-	362-75-4	NI32316
164	163	43	149	307	218	59	165	**307**	100	96	75	33	29	26	23	22		13	17	4	5	–	–	–	–	–	–	–	–	–	8-(2,3-O-Isopropylidene-β-D-ribofuranosyl)adenine		NI32317
118	91	146	90	159	75	132	119	**307**	100	61	56	22	15	11	10	10	**1.00**	14	9	2	7	–	–	–	–	–	–	–	–	–	2,3-Pyrazinedicarbonitrile, 5-amino-6-[(2,3-dihydro-3-hydroxy-2-oxo-1H-indol-3-yl)amino]-	52197-20-3	NI32318
271	272	43	60	289	45	228	58	**307**	100	69	63	32	31	24	15	14	**4.00**	14	13	7	1	–	–	–	–	–	–	–	–	–	Narciclasine		MS07130
217	139	243	183	185	63	107	152	**307**	100	54	46	26	23	16	10	6		14	14	1	1	2	–	–	–	–	–	1	–	–	Phosphinodithioic acid, diphenyl-, S-carbamoylmethyl-		MS07131
264	186	53	226	68	119	79	307	**307**	100	91	36	28	21	17	14	10		14	17	3	3	1	–	–	–	–	–	–	–	–	N-Acetyl-6-[(2,3-butadienyl)amino]-4,5-cyclopenteno-3-(methylsulphonyl)-1,2-diazine		MS07132
226	186	53	240	77	65	119	57	**307**	100	52	33	26	24	24	20	19	**3.00**	14	17	3	3	1	–	–	–	–	–	–	–	–	N-Acetyl-6-[(3-butynyl)amino]-4,5-cyclopenteno-3-(methylsulphonyl)-1,2-diazine		MS07133
128	216	156	188	307	247	248	202	**307**	100	82	71	50	41	12	11	10		15	17	6	1	–	–	–	–	–	–	–	–	–	Aniline-2-acetic acid, N-1,2-bis(methoxycarbonyl)vinyl-, methyl ester		LI06750
216	188	202	130	234	307			**307**	100	67	40	40	27	25				15	17	6	1	–	–	–	–	–	–	–	–	–	Indole-2-acetic acid, 2,3-dihydro-2,3-bis(methoxycarbonyl)-, methyl ester		LI06751
72	165	235	237	111	199	163	164	**307**	100	46	42	27	22	17	12	11	**9.50**	16	15	1	1	–	–	2	–	–	–	–	–	–	Acetamide, N,N-dimethyl-di(4-chlorophenyl)-		NI32319
190	161	292	278	174	249	307	274	**307**	100	67	42	31	27	21	9	9		16	21	5	1	–	–	–	–	–	–	–	–	–	5-Benzylamino-1,2-O-isopropylidene-5-deoxy-L-idohexodialdo-1,4β-3α,6β-difuranose		LI06752
292	262	72	278	235	234	293	221	**307**	100	58	56	52	46	25	22	20	**0.00**	16	25	3	1	–	–	–	–	–	1	–	–	–	N,N-Diethyl-6-formyl-3-methoxy-2-(trimethylsilyl)benzamide	85370-76-9	DD01132
143	115	307	128	171	77	114	101	**307**	100	74	20	20	18	18	17	17		17	13	3	3	–	–	–	–	–	–	–	–	–	2-Naphthalenol, 1-[(4-methyl-2-nitrophenyl)azo]-	2425-85-6	NI32320
292	142	293	217	307	216	130	146	**307**	100	25	19	19	13	11	10	9		17	14	–	1	–	–	–	–	–	–	–	–	1	Benzo[c]phenarsazine, 7,12-dihydro-7-methyl-	13493-34-0	NI32321
110	216	180	306	307	217	181	91	**307**	100	99	37	13	12	11	8	8		17	16	1	1	–	3	–	–	–	–	–	–	–	Acetamide, N-(1,2-diphenylethyl)-2,2,2-trifluoro-N-methyl-	24456-03-9	LI06753
110	216	180	217	181	118	91	76	**307**	100	98	36	10	7	7	7	5	**0.10**	17	16	1	1	–	3	–	–	–	–	–	–	–	Acetamide, N-(1,2-diphenylethyl)-2,2,2-trifluoro-N-methyl-	24456-03-9	NI32322
44	182	125	184	127				**307**	100	22	17	9	4					17	19	–	1	–	–	2	–	–	–	–	–	–	Phenylpropylamine, 4′-chloro-N-[(4-chlorophenyl)ethyl]-		LI06754
307	308	309	274	153.5	241	228	214	**307**	100	32	11	11	11	10	6	6		18	13	–	1	2	–	–	–	–	–	–	–	–	19,20-Dithia-21-azatetracyclo[14.2.1.1^{4,7}.1^{10,13}]heneicos-2,4,6,8,10,12,14,16,18-nonaene		LI06755
307	236	292	221	264	208	250	278	**307**	100	19	15	14	13	8	7	5		18	13	4	1	–	–	–	–	–	–	–	–	–	Hallacridone	109897-77-0	NI32323

MASS TO CHARGE RATIOS								M.W.	INTENSITIES								Parent	C	H	O	N	S	F	Cl	Br	I	Si	P	B	X	COMPOUND NAME	CAS Reg No	No
290	145	89	291	117	234	178	146	**307**	100	75	33	19	15	10	8	8	**6.00**	18	13	4	1	–	–	–	–	–	–	–	–	–	**Pulvinamide**		**LI06756**
307	308	134	306	261	146	118	77	**307**	100	23	14	12	7	6	6	5		18	17	2	3	–	–	–	–	–	–	–	–	–	**Cyclopropanecarbonitrile, 2-[4-(dimethylamino)phenyl]-1-(4-nitrophenyl)-**	28752-34-3	**NI32324**
307	308	121	134	306	261	147	145	**307**	100	24	21	14	13	8	5	5		18	17	2	3	–	–	–	–	–	–	–	–	–	**Cyclopropanecarbonitrile, 2-[4-(dimethylamino)phenyl]-1-(4-nitrophenyl)-**	28752-34-3	**LI06757**
199	200	144	307	129	169	143	115	**307**	100	20	10	9	9	8	8	6		18	17	2	3	–	–	–	–	–	–	–	–	–	**11H-6,10-Metheno[1,6]diazacyclotridecino[10,9-b]indol-11-one, 1,2,3,4,5,12-hexahydro-5-hydroxy-**	38940-74-8	**NI32325**
77	215	42	307	43	105	96	110	**307**	100	85	84	75	32	25	20	19		18	17	2	3	–	–	–	–	–	–	–	–	–	**3,5-Pyrazolidinedione, 4-[(dimethylamino)methylene]-1,2-diphenyl-**	6139-74-8	**NI32326**
249	104	77	106	250	91	188	105	**307**	100	57	40	29	20	19	17	17	**10.00**	18	17	2	3	–	–	–	–	–	–	–	–	–	**Pyrimidin-2-one, 1,2,3,4-tetrahydro-5-acetamido-4,6-diphenyl-**		**LI06758**
307	279	264	145	171	306	130	77	**307**	100	80	39	39	35	27	24	19		18	17	2	3	–	–	–	–	–	–	–	–	–	**Spiro[3H-indole-3,3′-[3H]pyrazol]-2(1H)-one, 2′,4′-dihydro-5′-(4-methoxyphenyl)-1-methyl-**	40526-77-0	**NI32327**
166	91	79	69	126	141	124	77	**307**	100	92	79	62	45	35	34	34		18	29	1	1	1	–	–	–	–	–	–	–	–	**Spiro[pyrrolo[1,2-b][1.4.2]oxathiazole-2,2′-tricyclo[3.3.1.1^{3,7}]decane], tetrahydro-5,5,7,7-tetramethyl-**	50455-49-7	**NI32328**
158	116	74	107	220	307	108	90	**307**	100	95	71	53	50	35	20	20		18	29	3	1	–	–	–	–	–	–	–	–	–	**L-Alanine, 3-(4-isopropoxyphenyl)-N-isopropyl-, isopropyl ester**	31552-13-3	**MS07134**
203	307	178	204	83	57	41	308	**307**	100	69	39	23	16	15	15	14		18	29	3	1	–	–	–	–	–	–	–	–	–	**Phenylpropanamide, 3,5-di-tert-butyl-β,4-dihydroxy-N-methyl-**		**IC03829**
70	83	97	126	112	140	154	182	**307**	100	68	47	40	39	33	30	28	**4.00**	18	33	1	3	–	–	–	–	–	–	–	–	–	**1,5-Diazacycloheptadecane-1-propanenitrile, 17-oxo-**	67171-89-5	**NI32329**
70	83	267	97	198	224	87	110	**307**	100	80	42	32	27	26	25	24	**22.00**	18	33	1	3	–	–	–	–	–	–	–	–	–	**Propanenitrile, 3-[[3-(2-oxoazacyclotridec-1-yl)propyl]amino]-**	67171-82-8	**NI32330**
230	259	231	307	248	260	290	291	**307**	100	58	19	16	15	13	9	7		19	17	1	1	1	–	–	–	–	–	–	–	–	**10-Propyl-6,7-benzophenothiazine-5-oxide**		**NI32331**
307	292	306	278	291	262	290	264	**307**	100	82	9	9	8	6	5	3		19	17	3	1	–	–	–	–	–	–	–	–	–	**Belemine**		**NI32332**
307	292	278	264	250	239	236	77	**307**	100	30	29	27	16	16	15	13		19	17	3	1	–	–	–	–	–	–	–	–	–	**Furo[2,3-c]acridin-6(2H)-one, 1,11-dihydro-5-hydroxy-11-methyl-2-(1-methylethenyl)-, (–)-**	17948-33-3	**NI32333**
307	292	77	278	264	239	236	51	**307**	100	38	24	22	22	17	15	15		19	17	3	1	–	–	–	–	–	–	–	–	–	**Furo[2,3-c]acridin-6(2H)-one, 1,11-dihydro-5-hydroxy-11-methyl-2-(1-methylethenyl)-, (–)-**	17948-33-3	**NI32334**
292	307	77	290	264	51	266	146	**307**	100	64	15	9	9	9	8	8		19	17	3	1	–	–	–	–	–	–	–	–	–	**Furo[3,2-b]acridin-5(2H)-one, 3,10-dihydro-4-hydroxy-10-methyl-2-(1-methylethenyl)-**		**NI32335**
201	307	173	202	130	308	186	82	**307**	100	69	57	33	33	32	28	27		19	17	3	1	–	–	–	–	–	–	–	–	–	**8-Methoxy-5-methyl-3-phenyl-1,4-dihydro-2H-pyrano[4,3-b]indol-1-one**		**DD01133**
77	202	104	105	185	51	82	184	**307**	100	72	55	53	36	33	29	22	**4.00**	19	17	3	1	–	–	–	–	–	–	–	–	–	**3-Phenyl-4-benzoyl-7-methyl-2,6-dioxa-3-azabicyclo[3.3.0]-7-octene**		**NI32336**
106	44	201	77	78	202	51	79	**307**	100	49	44	38	31	29	25	22	**6.70**	19	18	1	1	–	–	–	–	–	–	1	–	–	**Phosphinamide, diphenyl-N-benzyl-**		**MS07135**
169	168	75	167	147	193	170	84	**307**	100	51	42	34	27	15	15	14	**3.00**	19	21	1	1	–	–	–	–	–	1	–	–	–	**N,N-Diphenyl-2-(trimethylsilyl)-2,3-butadienamide**		**NI32337**
250	104	77	165	182	251	57	166	**307**	100	52	34	31	26	23	22	17	**13.00**	19	21	1	3	–	–	–	–	–	–	–	–	–	**4H-Imidazol-4-one, 2-amino-1-tert-butyl-1,5-dihydro-5,5-diphenyl-**	56634-13-0	**NI32338**
250	222	104	175	251	77	41	165	**307**	100	55	53	50	34	34	32	31	**12.99**	19	21	1	3	–	–	–	–	–	–	–	–	–	**4H-Imidazol-4-one, 2-amino-1-tert-butyl-1,5-dihydro-5,5-diphenyl-**	56634-13-0	**NI32339**
251	182	104	165	208	77	250	57	**307**	100	89	41	30	28	25	18	18	**18.00**	19	21	1	3	–	–	–	–	–	–	–	–	–	**4H-Imidazol-4-one, 2-amino-3-tert-butyl-3,5-dihydro-5,5-diphenyl-**	56954-65-5	**NI32340**
250	182	104	251	165	208	77	307	**307**	100	89	41	37	30	28	25	18		19	21	1	3	–	–	–	–	–	–	–	–	–	**4H-Imidazol-4-one, 2-amino-3-tert-butyl-3,5-dihydro-5,5-diphenyl-**	56954-65-5	**NI32341**
276	44	307	56	43	277	58	57	**307**	100	74	46	26	26	24	17	17		19	33	2	1	–	–	–	–	–	–	–	–	–	**Neosamandiol**		**LI06759**
44	278	307	43	72	41	56	55	**307**	100	24	13	12	8	8	7	7		19	33	2	1	–	–	–	–	–	–	–	–	–	**5,9,13-Pentadecatrien-2-one, 6,10,14-trimethyl-, (E,E)-**		**LI06760**
307	292	264	245	180	234	222	190	**307**	100	90	60	45	45	35	35	35		20	21	2	1	–	–	–	–	–	–	–	–	–	**Carbazole, 2-(3,3-dimethylallyl)-N-formyl-1-methoxy-3-methyl-**		**NI32342**
133	175	134	307	91	174	131	117	**307**	100	74	55	32	31	24	18	15		20	21	2	1	–	–	–	–	–	–	–	–	–	**1H-Dicyclopent[e,g]isoindole-1,3(2H)-dione, 3a,3b,4,5,6,7,8,9,9a,9b-decahydro-2-phenyl-, (3aα,3bα,9aα,9bα)-**	51624-48-7	**NI32343**
106	276	91	103	121	143	277	128	**307**	100	72	69	30	22	17	15	13	**0.00**	20	21	2	1	–	–	–	–	–	–	–	–	–	**(1R,2R)-N-(1-Phenyl-2-hydroxyethyl)-2-phenyl-5-hexynamide**		**DD01134**
106	276	67	128	103	77	91	104	**307**	100	23	23	19	16	16	15	14	**0.00**	20	21	2	1	–	–	–	–	–	–	–	–	–	**(1R,3R)-N-(1-Phenyl-2-hydroxyethyl)-3-phenyl-5-hexynamide**		**DD01135**
133	175	134	40	307	91	174	131	**307**	100	73	55	50	35	32	25	18		20	21	2	1	–	–	–	–	–	–	–	–	–	**N-Phenylmaleimide-1,1′-bicyclopentenyl endo adduct**		**LI06761**
207	71	236	73	234	219	205	72	**307**	100	96	88	73	70	48	39	26	**7.07**	20	25	–	3	–	–	–	–	–	–	–	–	–	**1H-Carbazole-4-carbonitrile, 2-[2-(dimethylamino)ethyl]-3-ethyl-2,3,4,9-tetrahydro-1-methylene-**	55320-33-7	**NI32344**
97	124	167	140	308				**307**	100	45	41	13	5				**0.00**	21	25	1	1	–	–	–	–	–	–	–	–	–	**8-Azabicyclo[3.2.1]octane, 3-(diphenylmethoxy)-8-methyl-, endo-**	86-13-5	**NI32345**
140	83	82	124	97	96	167	125	**307**	100	88	59	42	25	25	23	21	**6.41**	21	25	1	1	–	–	–	–	–	–	–	–	–	**8-Azabicyclo[3.2.1]octane, 3-(diphenylmethoxy)-8-methyl-, endo-**	86-13-5	**NI32346**
146	250	77	57	106	147	70	105	**307**	100	99	50	37	34	33	25	22	**8.50**	21	25	1	1	–	–	–	–	–	–	–	–	–	**Azetidine, 3-benzoyl-1-tert-butyl-2-phenyl-, cis-**	22592-78-5	**NI32347**
146	250	77	131	57	106	147	70	**307**	100	99	50	42	37	34	33	25	**8.50**	21	25	1	1	–	–	–	–	–	–	–	–	–	**Azetidine, 3-benzoyl-1-tert-butyl-2-phenyl-, cis-**	22592-78-5	**LI06762**
307	306	308	230	102	202	77	304	**307**	100	33	25	15	10	7	7	4		23	17	–	1	–	–	–	–	–	–	–	–	–	**Pyridine, 2,4,6-triphenyl-**	580-35-8	**NI32348**
59	124	154	94	89	188	248	278	**308**	100	82	69	62	10	8	5	4	**1.00**	–	–	4	4	–	–	2	–	–	–	–	–	2	**Cobalt, bis-, (di-μ-chloro)tetranitroso-**		**MS07136**
41	73	64	32	78	44	46	113	**308**	100	66	25	23	18	17	16	13	**0.00**	2	3	–	2	2	6	–	–	–	–	–	–	1	**5-Methyl-1,3,2,4-dithiadiazolium hexafluoroarsenate(v)**		**NI32349**
229	308	69	239	210	289	189		**308**	100	98	33	21	12	8	7			3	–	–	–	–	6	–	2	–	–	–	–	–	**Propane, 2,2-dibromo-1,1,1,3,3,3-hexafluoro-**	38568-21-7	**NI32350**
59	123	43	121	41	151	31	149	**308**	100	40	40	39	21	13	13	12	**0.00**	4	7	1	–	–	–	–	3	–	–	–	–	–	**2-Propanol, 1,1,1-tribromo-2-methyl-**	76-08-4	**NI32351**
275	277	273	240	310	238	312	85	**308**	100	64	62	55	43	42	34	33	**22.22**	5	–	–	–	–	2	6	–	–	–	–	–	–	**1,4-Pentadiene, 1,1,2,4,5,5-hexachloro-3,3-difluoro-**		**DO01091**
58	132	59	213	249	164	117	214	**308**	100	81	44	17	16	10	6	2		5	10	–	2	2	4	2	–	–	–	–	–	–	**Bis(chlorodifluoromethylsulphenyl)amine trimethylamine adduct**		**NI32352**
71	43	41	265	55	308	237	202	**308**	100	94	61	18	18	13	12	9		5	11	–	–	–	–	1	–	–	–	–	–	1	**Pentylmercuric chloride**		**LI06763**
39	41	67	54	97	56	95	153	**308**	100	87	55	29	18	15	10	9	**1.92**	6	5	3	–	–	–	–	–	1	–	–	–	1	**Iron, tricarbonyl-π-allyliodo-**		**MS07138**

MASS TO CHARGE RATIOS								M.W.	INTENSITIES								Parent	C	H	O	N	S	F	Cl	Br	I	Si	P	B	X	COMPOUND NAME	CAS Reg No	No
56	183	97	96	224	84	95	252	308	100	83	45	43	29	27	24	23	14.00	6	5	3	–	–	–	–	–	1	–	–	–	1	Iron, tricarbonyl-π-allyliodo-		MS07137
161	162	197	85	97	199	83	84	308	100	40	32	20	17	14	14	12	0.40	7	8	3	–	1	–	3	–	–	–	1	–	–	Phosphinic acid, 2-thienyl(2,2,2-trichloro-1-hydroxyethyl)-, methyl ester	1725-17-3	NI32353
228	181	211	212	227	308	213	199	308	100	89	73	66	60	55	47	46		8	6	–	–	–	5	–	–	–	–	1	–	1	Phosphine selenide, dimethyl(pentafluorophenyl)-	60575-50-0	NI32354
71	73	113	85	127	43	61	29	308	100	87	52	52	41	40	39	36	0.00	8	11	6	–	–	–	3	–	–	–	–	–	–	α-D-Glucofuranose, 1,2-O-(2,2,2-trichloroethylidene)-, (R)-	15879-93-3	NI32355
140	158	93	75	39	308	65	252	308	100	84	37	37	31	30	24	23		10	5	2	–	–	5	–	–	–	–	–	–	1	Iron, dicarbonyl-π-cyclopentadienyl(pentafluoropropenyl)-	12211-99-3	LI06764
140	158	93	75	39	308	65	56	308	100	83	37	37	31	30	24	23		10	5	2	–	–	5	–	–	–	–	–	–	1	Iron, dicarbonyl-π-cyclopentadienyl(pentafluoropropenyl)-	12211-99-3	NI32356
189	124	308	154	118	59	143	98	308	100	23	22	5	4	4	3	3		10	10	2	2	–	–	–	–	–	–	–	–	2	Cobalt, bis-, bis(cyclopentadienylnitrosyl)-		LI06765
206	194	278	250	128	66	161	222	308	100	41	38	37	29	23	22	20	0.00	10	10	3	–	1	–	–	–	–	–	–	–	1	π-Cyclopentadienyl-methylthiomethyl-tricarbonyl molybdenum		MS07139
174	56	176	230	308	310	232	145	308	100	88	56	53	50	33	32	24		10	10	3	2	1	–	2	–	–	–	–	–	–	Imidazolidinone, 1-(3′,4′-dichlorophenyl)-3-(methylsulphonyl)-		NI32357
127	95	41	55	67	59	43	73	308	100	49	32	30	27	24	20	18	0.00	10	13	3	–	–	3	2	–	–	–	–	–	–	Methyl 3-(2,2-dichloro-3,3,3-trifluoro-1-hydroxypropyl)-2,2-dimethylcyclopropanecarboxylate		DD01136
29	97	125	27	141	293	109	65	308	100	65	49	44	42	41	37	31	20.42	11	17	4	–	2	–	–	–	–	–	1	–	–	Phosphorothioic acid, O,O-diethyl O-[4-(methylsulphinyl)phenyl] ester	115-90-2	NI32358
293	97	125	141	308	109	153	65	308	100	66	54	51	37	36	25	23		11	17	4	–	2	–	–	–	–	–	1	–	–	Phosphorothioic acid, O,O-diethyl O-[4-(methylsulphinyl)phenyl] ester	115-90-2	NI32359
73	18	308	28	117	45	75	235	308	100	46	40	36	29	26	20	17		11	20	–	–	4	–	–	–	–	1	–	–	–	2,4,6,8-Tetrathiatricyclo[3.3.1.1^{3,7}]decane,9,10-dimethyl-1-(trimethylsilyl)-	57304-97-9	NI32360
73	147	191	45	75	221	192	148	308	100	60	59	16	12	10	10	10	0.00	11	28	4	–	–	–	–	–	–	3	–	–	–	Acetic acid, bis[(trimethylsilyl)oxyl]-, trimethylsilyl ester	55517-38-9	NI32361
154	98	106	96	138	69	308	198	308	100	52	48	48	42	30	26	18		12	8	4	2	2	–	–	–	–	–	–	–	–	Disulphide, bis(2-nitrophenyl)	1155-00-6	IC03830
96	154	98	69	70	106	138	78	308	100	97	94	67	62	55	48	38	19.50	12	8	4	2	2	–	–	–	–	–	–	–	–	Disulphide, bis(2-nitrophenyl)	1155-00-6	NI32362
126	156	279	309	140	127	280	157	308	100	72	69	45	45	36	30	24	8.91	12	8	4	2	2	–	–	–	–	–	–	–	–	Disulphide, bis(3-nitrophenyl)	537-91-7	NI32363
308	108	69	309	109	155	310	82	308	100	41	22	18	14	13	12	9		12	8	4	2	2	–	–	–	–	–	–	–	–	Disulphide, bis(3-nitrophenyl)	537-91-7	NI32364
308	108	69	82	154	96	63	171	308	100	90	48	29	20	18	16	15		12	8	4	2	2	–	–	–	–	–	–	–	–	Disulphide, bis(3-nitrophenyl)	537-91-7	NI32365
308	108	276	154	262	186	244	122	308	100	45	9	6	3	2	1	1		12	8	4	2	2	–	–	–	–	–	–	–	–	Disulphide, bis(4-nitrophenyl)		LI06766
308	109	309	138	310	82	140	124	308	100	29	16	12	11	8	6	6		12	8	4	2	2	–	–	–	–	–	–	–	–	Disulphide, bis(4-nitrophenyl)		IC03831
77	93	65	51	28	227	30	141	308	100	52	27	24	16	15	15	11	1.53	12	8	6	2	1	–	–	–	–	–	–	–	–	2,4-Dinitrodiphenylsulphone	896-80-0	NI32366
77	308	203	76	51	50	105	231	308	100	93	66	66	40	40	32	30		12	9	–	2	–	–	–	–	1	–	–	–	–	Azobenzene, 4-iodo-		MS07140
291	261	307	308	216	262	76	123	308	100	56	54	47	44	37	32	22		12	9	6	2	–	–	–	–	–	–	1	–	–	Phosphinic acid, di-3-nitrophenyl-		MS07141
135	43	136	164	174	60	108	178	308	100	81	80	69	68	34	31	30	4.00	12	16	4	6	–	–	–	–	–	–	–	–	–	Adenosine, 3′-acetamido-3′-deoxy-		LI06767
69	83	55	239	57	125	59	67	308	100	74	32	25	20	17	16	12	0.00	12	18	2	–	–	6	–	–	–	–	–	–	–	1,3-Dioxolane, 4-ethyl-5-pentyl-2,2-bis(trifluoromethyl)-, trans-	38363-94-9	MS07142
55	43	27	70	306	236	208	196	308	100	98	76	70	45	35	31	29		12	18	3	–	–	–	–	–	–	–	–	–	1	Molybdenum, tris[π-(3-buten-2-one)]		MS07143
131	235	234	236	250	178	308	205	308	100	27	22	20	15	14	11	10		12	20	2	–	–	–	–	–	–	2	–	–	1	Iron, dicarbonyl(η^5-2,4-cyclopentadien-1-yl)(pentamethyldisilanyl)-	53433-61-7	LI06768
131	235	234	236	250	178	179	191	308	100	27	22	20	15	14	12	11	11.00	12	20	2	–	–	–	–	–	–	2	–	–	1	Iron, dicarbonyl(η^5-2,4-cyclopentadien-1-yl)(pentamethyldisilanyl)-	53433-61-7	MS07144
73	147	205	117	103	45	74	75	308	100	35	22	21	21	13	9	8		12	32	3	–	–	–	–	–	–	3	–	–	–	Glycerol, tris[(trimethylsilyl)oxy]-	6787-10-6	NI32367
73	205	147	103	117	293	218	45	308	100	56	56	27	25	18	13	10		12	32	3	–	–	–	–	–	–	3	–	–	–	Glycerol, tris[(trimethylsilyl)oxy]-	6787-10-6	NI32368
73	147	205	103	117	45	133	59	308	100	50	38	30	29	21	13	12	0.00	12	32	3	–	–	–	–	–	–	3	–	–	–	Glycerol, tris[(trimethylsilyl)oxy]-	6787-10-6	NI32369
99	293	292	44	98	154	45	28	308	100	40	37	27	24	22	18	15	6.00	12	36	–	6	–	–	–	–	–	–	–	4	–	Tetraborane, hexakis(dimethylamino)-		LI06769
177	157	280	252	158	232	93	138	308	100	76	54	47	41	40	40	25	20.02	13	7	2	–	–	3	–	–	–	–	–	–	1	Iron, dicarbonyl(η^5-2,4-cyclopentadien-1-yl)(2,3,5-trifluorophenyl)-	33136-08-2	NI32370
157	177	158	232	280	252	93	138	308	100	99	88	69	63	63	57	44	44.04	13	7	2	–	–	3	–	–	–	–	–	–	1	Iron, dicarbonyl(η^5-2,4-cyclopentadien-1-yl)(2,3,6-trifluorophenyl)-	32759-06-1	NI32371
177	158	280	252	157	232	93	308	308	100	51	47	42	40	31	30	24		13	7	2	–	–	3	–	–	–	–	–	–	1	Iron, dicarbonyl-π-cyclopentadienyl(2,4,5-trifluorophenyl)-	12212-23-6	NI32372
186	132	177	157	158	280	252	232	308	100	72	68	60	54	48	45	42	30.76	13	7	2	–	–	3	–	–	–	–	–	–	1	Iron, dicarbonyl-π-cyclopentadienyl(2,4,5-trifluorophenyl)-	12212-23-6	MS07145
28	224	55	138	67	27	82	41	308	100	49	43	41	29	24	20	19	0.18	13	16	5	–	–	–	–	–	–	–	–	–	1	Iron, tricarbonyl[(1,2-η)-2,3-dimethyl-1,3-butadiene][(2,3-η)-methyl 2-propenoate]-	62265-42-3	NI32373
57	28	209	224	41	97	112	56	308	100	63	24	23	21	15	13	13	0.80	13	20	3	2	–	–	–	–	–	–	–	–	1	Iron, tricarbonyl[N,N′-1,2-ethanediylidenebis[2-methyl-2-propanamine]-N,N′]-	65036-35-3	NI32374
151	109	193	222	211	166	124	57	308	100	95	90	80	61	59	54	23	0.00	13	25	6	–	–	–	–	–	–	–	1	–	–	Dimethyl [3-(tert-butylacetoxy)-3-methyl-2-oxobutyl]phosphonate		NI32375
88	75	101	73	45	89	117	59	308	100	72	38	30	21	20	17	10		13	28	6	–	–	–	–	–	–	1	–	–	–	α-D-Glucopyranoside, methyl 2,3,4-tri-O-methyl-6-O-(trimethylsilyl)-	52430-37-2	NI32376
146	159	73	75	45	71	89	88	308	100	95	73	66	61	47	39	32		13	28	6	–	–	–	–	–	–	1	–	–	–	α-D-Glucopyranoside, methyl 3,4,6-tri-O-methyl-2-O-(trimethylsilyl)-	52430-36-1	NI32377
88	133	101	45	73	71	89	75	308	100	68	36	24	23	21	18	15		13	28	6	–	–	–	–	–	–	1	–	–	–	β-D-Glucopyranoside, methyl 2,3,6-tri-O-methyl-4-O-(trimethylsilyl)-	55028-67-6	NI32378
83	69	55	97	73	75	103	57	308	100	60	56	41	41	29	23	23	0.10	13	29	1	–	–	–	–	1	–	1	–	–	–	Silane, [(10-bromodecyl)oxy]trimethyl-	26306-02-5	NI32379
77	125	51	141	78	126	167	97	308	100	56	44	32	6	5	5	3	2.26	14	12	4	–	2	–	–	–	–	–	–	–	–	Benzene, 1,1′-[1,2-ethenediylbis(sulphonyl)]bis-, (Z)-	963-15-5	NI32380
28	307	227	151	228	77	105	231	308	100	80	80	75	61	57	45	29	18.15	14	12	6	–	1	–	–	–	–	–	–	–	–	Benzenesulphonic acid, 5-benzoyl-4-hydroxy-2-methoxy-	4065-45-6	NI32381
246	44	43	245	153	203	290	217	308	100	97	91	56	52	31	30	29	5.10	14	12	8	–	–	–	–	–	–	–	–	–	–	Fulvic acid	479-66-3	NI32382
308	209	248	210	178	191	249	272	308	100	81	76	33	26	16	15	11		14	12	8	–	–	–	–	–	–	–	–	–	–	2β,3β,4β,7-Tetrahydroxy-8,9-methylendioxy-1,2,3,4-tetrahydrodibenzo[b,d]pyran-6-one		NI32383
217	91	308	90	89	65	181	63	308	100	96	55	30	20	16	13	10		14	13	–	–	–	–	–	–	1	–	–	–	–	Benzene, 1-iodo-4-benzyl-	14310-25-9	NI32384
235	154	156	127	218	155	308	237	308	100	91	65	53	35	32	27	26		14	16	4	2	1	–	–	–	–	–	–	–	–	Ethyl α-sulphamoyl-1-naphthaleneacetate		DD01137
182	43	140	139	206	224	164	28	308	100	74	71	40	35	26	25	18	12.00	14	16	6	2	–	–	–	–	–	–	–	–	–	2,5-Diaminohydroquinone tetraacetate		NI32385
29	27	189	143	76	89	207	162	308	100	40	35	28	28	20	14	14	9.50	14	16	6	2	–	–	–	–	–	–	–	–	–	Diethyl 2-(4-nitroanilino)fumarate		MS07146

MASS TO CHARGE RATIOS	M.W.	INTENSITIES	Parent	C	H	O	N	S	F	Cl	Br	I	Si	P	B	X	COMPOUND NAME	CAS Reg No	No
277 44 308 233 132 117 156 90	**308**	100 86 69 48 34 34 23 23		14	16	6	2	–	–	–	–	–	–	–	–	–	**2-Oxazolidinone, 3,3'-(1,4-phenylene)bis[4-(hydroxymethyl)-**	69974-34-1	**NI32386**
79 43 108 107 77 51 91 28	**308**	100 74 69 54 53 32 30 29	**0.00**	14	20	4	4	–	–	–	–	–	–	–	–	–	**L-Arginine, N[2]-[(phenylmethoxy)carbonyl]-**	1234-35-1	**NI32387**
154 155 97 156 69 68 42 98	**308**	100 67 57 56 54 48 34 32	**10.47**	14	20	4	4	–	–	–	–	–	–	–	–	–	**Cyclobuta[1,2-d:4,3-d']dipyrimidine-2,4,5,7(3H,6H)-tetrone, hexahydro-1,3,4a,4b,6,8-hexamethyl-, (4aα,4bα,8aα,8bα)-**	7025-74-3	**NI32388**
154 155 97 69 68 156 42 98	**308**	100 84 70 68 62 43 41 38	**14.00**	14	20	4	4	–	–	–	–	–	–	–	–	–	**Cyclobuta[1,2-d:4,3-d']dipyrimidine-2,4,5,7(3H,6H)-tetrone, hexahydro-1,3,4a,4b,6,8-hexamethyl-, (4aα,4bα,8aα,8bα)-**	7025-74-3	**LI06770**
154 97 69 155 68 42 98 96	**308**	100 76 76 73 62 43 38 35	**1.40**	14	20	4	4	–	–	–	–	–	–	–	–	–	**Cyclobuta[1,2-d:4,3-d']dipyrimidine-2,4,5,7(3H,6H)-tetrone, hexahydro-1,3,4a,4b,6,8-hexamethyl-, (4aα,4bβ,8aβ,8bα)-**	19670-00-9	**LI06771**
154 97 69 155 68 42 98 96	**308**	100 76 76 72 64 44 39 37	**1.36**	14	20	4	4	–	–	–	–	–	–	–	–	–	**Cyclobuta[1,2-d:4,3-d']dipyrimidine-2,4,5,7(3H,6H)-tetrone, hexahydro-1,3,4a,4b,6,8-hexamethyl-, (4aα,4bβ,8aβ,8bα)-**	19670-00-9	**NI32389**
69 41 55 217 44 57 43 308	**308**	100 89 77 61 61 59 56 35		14	20	4	4	–	–	–	–	–	–	–	–	–	**2-Ethylhexanal (2,4-dinitrophenyl)hydrazone**		**LI06772**
111 41 43 69 55 42 178 29	**308**	100 86 84 71 58 40 27 20	**15.74**	14	20	4	4	–	–	–	–	–	–	–	–	–	**2-Octanone (2,4-dinitrophenyl)hydrazone**		**IC03832**
57 41 55 111 29 69 43 56	**308**	100 97 96 74 68 65 61 36	**20.00**	14	20	4	4	–	–	–	–	–	–	–	–	–	**3-Octanone (2,4-dinitrophenyl)hydrazone**		**IC03833**
89 195 83 55 54 96 73 59	**308**	100 66 53 50 50 45 42 42	**0.00**	14	28	7	–	–	–	–	–	–	–	–	–	–	**3-Bis(hydroxymethyl)-1,5,8,11,14-pentaoxacyclohexadecane**		**MS07147**
169 308 94 166 140 139 141 104	**308**	100 18 10 7 5 5 4 3		15	11	2	2	–	3	–	–	–	–	–	–	–	**N,N-Diphenyl-N'-(trifluoroacetyl)urea**		**DD01138**
178 262 220 263 151 78 308 307	**308**	100 99 99 60 54 53 50 37		15	12	2	6	–	–	–	–	–	–	–	–	–	**1,3,5-Triazine-2,4-diamine, 6-(2'-nitro[1,1'-biphenyl]yl)-**	29366-82-3	**NI32390**
308 307 261 178 85 265 260 177	**308**	100 72 63 37 37 26 26 26		15	12	2	6	–	–	–	–	–	–	–	–	–	**1,3,5-Triazine-2,4-diamine, 6-(3'-nitro[1,1'-biphenyl]yl)-**	55570-90-6	**NI32391**
308 307 85 177 265 151 224 194	**308**	100 98 55 43 35 32 27 27		15	12	2	6	–	–	–	–	–	–	–	–	–	**1,3,5-Triazine-2,4-diamine, 6-(4'-nitro[1,1'-biphenyl]yl)-**	55638-40-9	**NI32392**
273 140 125 275 106 77 274 89	**308**	100 63 56 37 29 25 20 15	**0.00**	15	14	1	2	–	–	2	–	–	–	–	–	–	**N,N'-Bis(2-chlorobenzyl)urea**	104569-03-1	**NI32393**
140 125 308 142 183 310 127 77	**308**	100 69 37 37 36 23 19 17		15	14	1	2	–	–	2	–	–	–	–	–	–	**N,N'-Bis(4-chlorobenzyl)urea**	92550-15-7	**NI32394**
141 143 106 140 308 77 28 142	**308**	100 33 26 17 16 15 14 13		15	14	1	2	–	–	2	–	–	–	–	–	–	**N-(2-Methyl-3-chlorophenyl)-N'-(2-chloro-3-methylphenyl)urea**		**MS07148**
141 143 106 140 306 142 77 310	**308**	100 33 22 16 14 13 11 9	**0.00**	15	14	1	2	–	–	2	–	–	–	–	–	–	**N,N'-Bis(2-methyl-3-chlorophenyl)urea**		**MS07149**
141 106 143 140 167 132 142 77	**308**	100 34 32 31 28 18 17 17	**15.90**	15	14	1	2	–	–	2	–	–	–	–	–	–	**N,N'-Bis(2-methyl-5-chlorophenyl)urea**		**MS07150**
141 273 106 143 140 77 104 275	**308**	100 49 49 33 22 21 18 16	**3.00**	15	14	1	2	–	–	2	–	–	–	–	–	–	**N,N'-Bis(2-methyl-6-chlorophenyl)urea**		**MS07151**
308 307 108 133 65 134 309 92	**308**	100 44 35 30 28 26 21 19		15	16	–	8	–	–	–	–	–	–	–	–	–	**1,3,5-Triazine-2,4,6-triamine, N,N'-bis(3-aminophenyl)-**	33933-64-1	**NI32395**
211 179 28 194 97 103 161 123	**308**	100 34 32 31 31 27 25 17	**2.55**	15	16	7	–	–	–	–	–	–	–	–	–	–	**Allamandin**	51820-82-7	**NI32396**
165 133 43 91 308 151 264 121	**308**	100 49 36 27 20 20 17 16		15	16	7	–	–	–	–	–	–	–	–	–	–	**1,5-Methanobenzo[1,2-c:3,4-c']difuran-4-carboxylic acid, decahydro-1,5-dimethyl-3,6,8-trioxo-, methyl ester**	32251-45-9	**MS07152**
308 279 235 261 233 249 205 247	**308**	100 69 55 48 38 27 25 24		15	16	7	–	–	–	–	–	–	–	–	–	–	**6-(Methoxycarbonylmethyl)-7-hydroxy-2-methyl-5,8-dimethoxychromone**		**NI32397**
215 250 94 45 77 59 308 249	**308**	100 35 26 26 20 20 16 12		15	17	5	–	–	–	–	–	–	–	1	–	–	**Phosphoric acid, 2-methoxyethyl diphenyl ester**	65444-10-2	**NI32398**
43 55 308 31 266 29 123 73	**308**	100 62 40 38 34 31 29 28		15	20	5	2	–	–	–	–	–	–	–	–	–	**N-Acetyl-N-(2-methoxycarbonylethyl)-3-acetamido-4-methoxyaniline**		**IC03834**
293 101 43 294 59 55 72 58	**308**	100 68 52 17 13 10 6 5	**0.70**	15	20	5	2	–	–	–	–	–	–	–	–	–	**3-C-Cyano-3-C-cyanomethyl-3-desoxy-1,2,5,6-di-O-isopropylidene-α-D-glucofuranose**		**NI32399**
293 101 43 235 55 294 175 59	**308**	100 36 35 30 20 16 12 7		15	20	5	2	–	–	–	–	–	–	–	–	–	**3-C-Cyano-3-C-cyanomethyl-1,2,5,6-di-O-isopropylidene-α-D-galactofuranose**		**NI32400**
43 93 193 119 92 220 235 174	**308**	100 76 73 61 53 28 27 27	**0.00**	15	20	5	2	–	–	–	–	–	–	–	–	–	**Diethyl 2-(acetylamino)-2-[(3-pyridyl)methyl]malonate**		**DD01139**
84 176 132 116 115 150 130 177	**308**	100 94 71 49 43 40 37 29	**4.00**	15	20	5	2	–	–	–	–	–	–	–	–	–	**Diethyl (1-hydroxy-2-indanyl)-bicarbamate**		**LI06773**
308 170 169 171 234 172 309 233	**308**	100 96 49 30 30 23 19 16		15	21	3	2	–	–	–	–	–	–	1	–	–	**3-(4-N,N-Dimethylaminophenyl)propenenitrile, 2-(diethoxyphosphinyl)-**	18896-69-0	**NI32401**
147 73 293 45 75 148 149 77	**308**	100 58 22 19 17 16 8 6	**3.20**	15	24	3	–	–	–	–	–	–	2	–	–	–	**Benzenepropanoic acid, α-oxo-β,β-bis(trimethylsilyl)-**	69782-88-3	**MS07153**
147 73 293 148 45 149 294 75	**308**	100 46 33 17 14 8 7 6	**3.70**	15	24	3	–	–	–	–	–	–	2	–	–	–	**Benzenepropanoic acid, α-oxo-β,β-bis(trimethylsilyl)-**	69782-88-3	**NI32402**
147 73 293 45 148 149 294 75	**308**	100 54 24 17 17 9 6 6	**2.87**	15	24	3	–	–	–	–	–	–	2	–	–	–	**Benzenepropanoic acid, α-oxo-β,β-bis(trimethylsilyl)-**	69782-88-3	**NI32403**
73 308 293 75 219 203 309 249	**308**	100 54 46 35 28 19 18 15		15	24	3	–	–	–	–	–	–	2	–	–	–	**Cinnamic acid, m-(trimethylsiloxy)-, trimethylsilyl ester**	32342-01-1	**NI32405**
73 293 308 219 203 249 75 309	**308**	100 96 93 46 44 44 31 26		15	24	3	–	–	–	–	–	–	2	–	–	–	**Cinnamic acid, m-(trimethylsiloxy)-, trimethylsilyl ester**	32342-01-1	**NI32404**
73 308 293 219 75 309 249 45	**308**	100 77 57 33 32 22 18 18		15	24	3	–	–	–	–	–	–	2	–	–	–	**Cinnamic acid, m-(trimethylsiloxy)-, trimethylsilyl ester**	32342-01-1	**MS07154**
73 147 293 45 308 161 148 75	**308**	100 75 22 22 18 13 13 11		15	24	3	–	–	–	–	–	–	2	–	–	–	**Cinnamic acid, o-(trimethylsiloxy)-, trimethylsilyl ester**	32426-62-3	**NI32406**
73 147 308 293 45 161 75 74	**308**	100 70 38 28 19 18 18 15		15	24	3	–	–	–	–	–	–	2	–	–	–	**Cinnamic acid, o-(trimethylsiloxy)-, trimethylsilyl ester**	32426-62-3	**MS07155**
73 147 308 293 75 45 161 74	**308**	100 67 32 22 15 15 14 12		15	24	3	–	–	–	–	–	–	2	–	–	–	**Cinnamic acid, o-(trimethylsiloxy)-, trimethylsilyl ester**	32426-62-3	**NI32407**
73 308 293 219 75 309 249 45	**308**	100 53 50 48 38 18 15 15		15	24	3	–	–	–	–	–	–	2	–	–	–	**Cinnamic acid, p-(trimethylsiloxy)-, trimethylsilyl ester**	10517-30-3	**NI32408**
293 308 219 249 309 294 73 179	**308**	100 98 83 42 27 22 21 20		15	24	3	–	–	–	–	–	–	2	–	–	–	**Cinnamic acid, p-(trimethylsiloxy)-, trimethylsilyl ester**	10517-30-3	**NI32409**
73 219 293 308 249 75 45 294	**308**	100 56 47 39 24 24 22 13		15	24	3	–	–	–	–	–	–	2	–	–	–	**Cinnamic acid, p-(trimethylsiloxy)-, trimethylsilyl ester**	10517-30-3	**NI32410**
136 293 73 137 108 79 75 173	**308**	100 30 12 11 11 11 10 9	**9.34**	15	24	3	2	–	–	–	–	–	1	–	–	–	**2,4,6(1H,3H,5H)-Pyrimidinetrione, 5-(1-cyclohexen-1-yl)-1,5-dimethyl-3-(trimethylsilyl)-**	55649-44-0	**NI32411**
223 56 224 121 167 252 168 186	**308**	100 98 87 72 65 59 57 47	**0.14**	16	12	3	–	–	–	–	–	–	–	–	–	1	**Iron, dicarbonyl-π-cyclopentadienyl(3-phenylacryloyl)-**		**MS07156**
308 250 249 263 276 307 275 78	**308**	100 52 39 37 30 29 28 25		16	12	3	4	–	–	–	–	–	–	–	–	–	**Methyl N-(3,5-dicyano-6-methoxy-4-phenylpyrimidinyl-2)carbamate**	82072-09-1	**NI32412**
91 132 308 77 309 64 92 76	**308**	100 69 58 18 11 10 8 8		16	12	3	4	–	–	–	–	–	–	–	–	–	**1-Phenyl-3-methyl-4-(p-nitrophenylimino)-2-pyrazolin-5-one**		**MS07157**
238 308 310 224 132 195 99 239	**308**	100 97 72 42 41 36 31 18		16	14	2	–	–	–	2	–	–	–	–	–	–	**Ethylene, 1,1-dichloro-2,2-bis(p-methoxyphenyl)-**	2132-70-9	**MS07158**

MASS TO CHARGE RATIOS								M.W.	INTENSITIES								Parent	C	H	O	N	S	F	Cl	Br	I	Si	P	B	X	COMPOUND NAME	CAS Reg No	No
166	44	77	79	167	45	135	41	**308**	100	83	60	53	49	48	46	40	**1.00**	16	20	4	–	1	–	–	–	–	–	–	–	–	Cyclohexenepropanoic acid, α-(phenylsulphonyl)-, methyl ester	74366-98-6	NI32413
164	43	206	308	266	180	249	224	**308**	100	20	19	7	5	4	3	3		16	20	6	–	–	–	–	–	–	–	–	–	–	2,5-Diacetoxy-3,4,6-trimethylbenzyl acetate		LI06774
163	149	128	164	181	77	74	96	**308**	100	22	17	16	13	11	7	7	**0.00**	16	20	6	–	–	–	–	–	–	–	–	–	–	Dimethyl 2-ethyl-3-carboxypropylphthalate		NI32414
147	133	119	249	148	120	146	164	**308**	100	41	30	23	20	20	16	15	**0.00**	16	20	6	–	–	–	–	–	–	–	–	–	–	Dimethyl (4R,5R)-2-(2,4,6-trimethylphenyl)-1,3-dioxolane-4,5-dicarboxylate		DD01140
99	43	159	105	91	248	77	308	**308**	100	97	51	43	35	29	23	19		16	20	6	–	–	–	–	–	–	–	–	–	–	Methyl 4,6-O-benzylidene-2-deoxy-α-D-arabino-hexopyranoside, 3-acetate		NI32415
43	91	92	113	65	59	129	142	**308**	100	77	21	15	15	11	7	4	**0.00**	16	20	6	–	–	–	–	–	–	–	–	–	–	[3aR-(3aα,5β,6α,6aα)]-Tetrahydro-2,2-dimethyl-5-acetoxy-6-(benzyloxy)furo[2,3-d]-1,3-dioxole	120520-94-7	DD01141
70	43	71	41	154	209	69	210	**308**	100	88	63	30	29	28	20	12	**2.00**	16	24	4	2	–	–	–	–	–	–	–	–	–	(3R,8aS)-2-(3-Methyl-2-oxobutyryl)-3-[(1R)-1-methylpropyl]-1,2,3,4,6,7,8,8a-octahydropyrrolo[1,2-a]pyrazine-1,4-dione		MS07159
237	181	293	236	68	41	125	238	**308**	100	47	40	39	33	30	29	28	**11.43**	16	38	–	4	–	–	–	–	–	–	–	2	–	β,β′-(tert-Butylamino)-N,N′-tert-butylborazine		IC03835
100	308	310	200	201	87	199	280	**308**	100	99	98	86	58	28	18	18		17	9	1	–	–	–	–	1	–	–	–	–	–	7H-Benz[de]anthracen-7-one, 3-bromo-	81-96-9	NI32416
308	279	273	204	77	102	307	281	**308**	100	86	68	58	46	38	38	32		17	13	–	4	–	–	1	–	–	–	–	–	–	Xanax (alprazolam)	28981-97-7	NI32417
308	106	83	78	202	309	79	51	**308**	100	40	37	35	27	24	23	16		17	16	2	4	–	–	–	–	–	–	–	–	–	Nicotinamide, N-antipyrinyl-	2139-47-1	NI32418
308	123	28	77	200	309	92	122	**308**	100	95	82	80	77	61	60	51		17	16	2	4	–	–	–	–	–	–	–	–	–	1-Phenyl-3-methyl-4-(2-methoxyphenylazo)-5-pyrazolone	15095-99-5	NI32419
76	308	201	309	28	123	107	91	**308**	100	99	69	65	36	35	23	21		17	16	2	4	–	–	–	–	–	–	–	–	–	1-Phenyl-3-methyl-4-(3-methoxyphenylazo)-5-pyrazolone	15095-98-4	NI32420
122	308	309	123	77	91	60	119	**308**	100	98	71	70	70	22	22	21		17	16	2	4	–	–	–	–	–	–	–	–	–	1-Phenyl-3-methyl-4-(4-methoxyphenylazo)-5-pyrazolone	15095-97-3	NI32421
265	77	188	104	220	51	266	91	**308**	100	51	32	27	24	17	16	15	**0.00**	17	20	–	6	–	–	–	–	–	–	–	–	–	4-(Dimethylamino)-1-methyl-3-(phenylamino)-5-(phenylimino)-Δ^2-1,2,4-triazoline	118631-36-0	DD01142
94	136	93	91	121	107	95	79	**308**	100	50	32	24	21	19	16	14	**2.00**	17	24	3	–	1	–	–	–	–	–	–	–	–	p-Tolyl 2-(methylcyclohexen-4-yl)propane sulphonate		MS07160
43	41	53	109	209	99	91	79	**308**	100	97	88	84	82	73	64	59	**2.48**	17	24	5	–	–	–	–	–	–	–	–	–	–	Benz[e]azulen-3(3aH)-one, 4,6a,7,8,9,10,10a,10b-octahydro-3a,8,10a-trihydroxy-5-(hydroxymethyl)-2,10-dimethyl-, [3aR-(3aα,6aα,8β,10β,10aβ,10bβ)]-	77573-30-9	NI32422
69	43	57	41	197	85	248	55	**308**	100	86	81	70	57	42	36	29	**4.00**	17	24	5	–	–	–	–	–	–	–	–	–	–	2-Cyclopenten-1-one, 4-acetyl-3,4-dihydroxy-5-(3-methyl-2-butenyl)-2-(3-methyl-1-oxobutyl)-	57195-46-7	NI32423
180	193	308	290	133	175	105	217	**308**	100	75	72	58	42	40	40	39		17	24	5	–	–	–	–	–	–	–	–	–	–	3β,20-Epoxy-3α,14α-dihydroxy-9β-podocarpan-19,6β-olide		NI32424
43	109	41	140	139	123	55	107	**308**	100	44	31	28	28	20	20	19	**0.80**	17	24	5	–	–	–	–	–	–	–	–	–	–	7-Hydroxyisotrichodermin		NI32425
43	123	109	41	122	108	107	95	**308**	100	51	37	33	28	16	16	16	**0.50**	17	24	5	–	–	–	–	–	–	–	–	–	–	8-Hydroxyisotrichodermin		NI32426
43	41	55	29	107	91	27	230	**308**	100	30	18	15	14	13	12	11	**0.00**	17	24	5	–	–	–	–	–	–	–	–	–	–	Naphtho[2,3-b]furan-2(3H)-one, 4-(acetyloxy)decahydro-8-hydroxy-3,8a-dimethyl-5-methylene-, [3R-(3α,3aα,4α,4aα,8β,8aβ,9aβ)]-	16822-17-6	NI32427
308	309	154	245	221	279	246	202	**308**	100	21	14	13	10	8	8	8		18	12	1	–	2	–	–	–	–	–	–	–	–	19-Oxa-20,21-dithiatetracyclo[14.2.1.1^{4,7}.1^{10,13}]heneicos-2,4,6,8,10,12,14,16,18-nonaene		LI06775
308	279	128	282	309	168	267	234	**308**	100	34	33	24	20	19	18	15		18	12	1	–	2	–	–	–	–	–	–	–	–	2-(5-Oxobicyclo[4.4.1]undeca-3,6,8,10-tetraen-2-ylidene)-1,3-benzodithiole		DD01143
308	307	291	115	171	165	144	309	**308**	100	78	57	52	20	19	19	18		18	12	5	–	–	–	–	–	–	–	–	–	–	trans-6-Carboxy-2-(4-hydroxystyryl)chromone		MS07161
280	308	307	309	281	282	91	310	**308**	100	97	81	46	41	37	37	34		18	13	1	2	–	–	1	–	–	–	–	–	–	Pinazepam		MS07162
293	250	308	251	294	28	222	139	**308**	100	58	51	45	21	19	14	12		18	16	3	2	–	–	–	–	–	–	–	–	–	6-Amino-2,2-dimethyl-1,2-dihydro-4H-anthra[1,2-d][1,3]oxazine-7,12-dione		NI32428
308	160	77	145	105	117	89	161	**308**	100	55	35	13	12	11	11	8		18	16	3	2	–	–	–	–	–	–	–	–	–	4-(4-Methoxybenzylidene)-2-methyl-1-phenyl-3,5-dioxopyrazolidine	106584-30-9	NI32429
223	222	57	224	290	225	84	308	**308**	100	78	64	37	36	26	17	15		18	16	3	2	–	–	–	–	–	–	–	–	–	3-(Methylamino)-N-(9,10-dioxaantracen-1-yl)propanamide		MS07163
138	217	248	146	276	144	153	205	**308**	100	88	49	46	43	38	33	28	**8.00**	18	28	4	–	–	–	–	–	–	–	–	–	–	Dimethyl 3-isopropyl-4-(3-methyl-1-butenyl)-1-cyclohexene-1,4-dicarboxylate		MS07164
91	72	187	101	185	129	83	139	**308**	100	92	46	42	33	24	22	20	**0.00**	18	28	4	–	–	–	–	–	–	–	–	–	–	Methyl (±)-2,4,6-trideoxy-2,4,6-trimethyl-7-O-(phenylmethyl)-D-glycero-α-D-ido-heptopyranoside		NI32430
91	72	101	69	73	57	92	107	**308**	100	97	52	47	42	39	35	33	**0.00**	18	28	4	–	–	–	–	–	–	–	–	–	–	Methyl (±)-2,4,6-trideoxy-2,4,6-trimethyl-7-O-(phenylmethyl)-D-glycero-α-D-talo-heptopyranoside		NI32431
165	125	55	95	43	138	81	41	**308**	100	78	77	76	73	53	52	50	**0.00**	18	28	4	–	–	–	–	–	–	–	–	–	–	Ostopanic acid		DD01144
43	248	55	139	124	41	81	230	**308**	100	80	53	47	41	40	37	34	**32.73**	18	28	4	–	–	–	–	–	–	–	–	–	–	1(2H)-Phenanthrenone, 7-(acetyloxy)dodecahydro-8a-hydroxy-2,4b-dimethyl-, [2S-(2α,4aα,4bβ,7β,8aα,10aβ)]-	41853-21-8	NI32432
43	163	149	41	91	135	191	55	**308**	100	37	34	32	29	28	27	25	**4.90**	18	28	4	–	–	–	–	–	–	–	–	–	–	Plucheinol acetonide		NI32433
129	128	87	73	130	59	131	75	**308**	100	26	26	26	11	6	4	4	**0.46**	18	37	1	–	–	–	–	–	–	1	–	1	–	Borinic acid, (2-cyclohexylidene-1,1-diethylpropyl)ethyl-, trimethylsilyl ester	74810-48-3	NI32434
197	308	111	139	115	141	310	75	**308**	100	56	36	35	28	24	22	22		19	13	2	–	–	–	1	–	–	–	–	–	–	4-Chloro-4′-(4-hydroxyphenyl)-benzophenone		IC03836
125	218	77	165	242	97	51	290	**308**	100	78	73	58	47	38	26	19	**14.66**	19	16	2	–	1	–	–	–	–	–	–	–	–	2-Benzyldiphenylsulphone		NI32435
265	308	121	266	187	103	145	251	**308**	100	29	28	26	26	12	11	10		19	16	4	–	–	–	–	–	–	–	–	–	–	2H-1-Benzopyran-2-one, 4-hydroxy-3-(3-oxo-1-phenylbutyl)-	81-81-2	LI06776
265	43	121	18	187	308	266	145	**308**	100	46	38	36	27	22	16	13		19	16	4	–	–	–	–	–	–	–	–	–	–	2H-1-Benzopyran-2-one, 4-hydroxy-3-(3-oxo-1-phenylbutyl)-	81-81-2	NI32436

MASS TO CHARGE RATIOS								M.W.	INTENSITIES								Parent	C	H	O	N	S	F	Cl	Br	I	Si	P	B	X	COMPOUND NAME	CAS Reg No	No
265	43	213	121	308	36	77	290	**308**	100	46	46	37	29	25	22	20		19	16	4	–	–	–	–	–	–	–	–	–	–	**2H-1-Benzopyran-2-one, 4-hydroxy-3-(3-oxo-1-phenylbutyl)-**	81-81-2	**NI32437**
308	293	280	265	309	199	238	209	**308**	100	34	34	30	26	21	17	16		19	16	4	–	–	–	–	–	–	–	–	–	–	**3,6-Dihydroxy-7,8-dimethyl-2-(2-propenyl)phenanthrene-1,4-dione**		**NI32438**
308	121	307	278	165	221	293	178	**308**	100	15	13	13	8	7	6	5		19	16	4	–	–	–	–	–	–	–	–	–	–	**3-Formyl-2-(4′-hydroxyphenyl)-7-methoxy-5-(E)-propenylbenzofuran**		**MS07165**
161	69	147	203	77	91	189	65	**308**	100	99	40	24	24	22	20	16	**3.30**	19	16	4	–	–	–	–	–	–	–	–	–	–	**1,3,4,6-Hexanetetrone, 1-(4-methylphenyl)-6-phenyl-**	58330-10-2	**NI32439**
189	249	59	190	308	218	178	163	**308**	100	30	26	22	19	11	10	9		19	16	4	–	–	–	–	–	–	–	–	–	–	**Spiro[cyclopropane-1,9′-[9H]fluorene]-2,3-dicarboxylic acid, dimethyl ester, cis-**	34296-55-4	**NI32440**
217	308	139	218	77	91	51	63	**308**	100	36	34	14	14	11	11	10		19	17	–	–	1	–	–	–	–	–	1	–	–	**Phosphine sulphide, diphenylbenzyl-**	15367-75-6	**NI32441**
202	308	91	201	203	155	77	167	**308**	100	37	29	15	13	11	11	10		19	17	2	–	–	–	–	–	–	–	1	–	–	**Benzyl diphenylphosphinate**		**MS07166**
308	215	307	309	165	197	166	290	**308**	100	44	40	20	17	14	13	12		19	17	2	–	–	–	–	–	–	–	1	–	–	**Phenyl phenyl-O-tolylphosphinate**		**MS07167**
306	122	307	43	198	41	154	96	**308**	100	25	24	21	20	20	16	15	**3.00**	19	20	2	2	–	–	–	–	–	–	–	–	–	**Aspidofractinine-3,8-dione, (2α,5α)-**	55724-70-4	**NI32442**
105	77	57	106	103	118	89	84	**308**	100	44	16	14	11	10	7	7	**5.00**	19	20	2	2	–	–	–	–	–	–	–	–	–	**1,2-Diazetidin-3-one, 1-benzoyl-2-tert-butyl-4-phenyl-**	39640-84-1	**NI32443**
238	308	239	237	55	307	209	210	**308**	100	91	34	33	33	29	28	25		19	20	2	2	–	–	–	–	–	–	–	–	–	**4,13-Dioxo-1,2,3,4,6,7,13,14,15,16-decahydropyrido[2,1-a]azepino[1,3-l,m] β-carboline**		**MS07168**
308	170	172	43	168	143	116	82	**308**	100	89	51	24	19	11	11	10		19	20	2	2	–	–	–	–	–	–	–	–	–	**12cα-Ethyl-1,2,3,3aα,4,6,7,12,12bα,12c-decahydrocyclopenta[1,2]indolizino[8,7-b]indol-3,4-dione**		**NI32444**
293	308	279	278	57	55	69	71	**308**	100	76	76	76	59	44	43	42		19	20	2	2	–	–	–	–	–	–	–	–	–	**(Z,Z)-4-Ethyl-3-methyl-5-[5-(furyl-2-methylene)-3,4-dimethyl-5H-pyrrolyl 2-methylene]-3-pyrrolin-2-one**		**MS07169**
180	308	279	208	77	251	165	104	**308**	100	80	78	61	50	27	26	21		19	20	2	2	–	–	–	–	–	–	–	–	–	**2,4-Imidazolidinedione, 1,3-diethyl-5,5-diphenyl-**	54964-77-1	**NI32445**
279	208	77	308	280	104	149	180	**308**	100	62	46	41	29	21	20	19		19	20	2	2	–	–	–	–	–	–	–	–	–	**4H-Imidazol-4-one, 2-ethoxy-3-ethyl-3,5-dihydro-5,5-diphenyl-**	54964-78-2	**NI32446**
123	157	131	185	158	186	156	132	**308**	100	76	38	35	19	13	9	4	**1.00**	19	20	2	2	–	–	–	–	–	–	–	–	–	**4-(4-Methoxyanilino)-2-(4-methylanilino)-cyclopent-2-enone**		**MS07170**
105	103	77	188	308	262	275	178	**308**	100	94	77	60	31	22	21	15		19	20	2	2	–	–	–	–	–	–	–	–	–	**(E)-1-Nitro-1,2-diphenyl-2-(1-piperidino)ethylene**	73013-90-8	**DD01145**
77	183	184	105	308	55	51	41	**308**	100	89	29	24	22	16	14	13		19	20	2	2	–	–	–	–	–	–	–	–	–	**3,5-Pyrazolidinedione, 4-butyl-1,2-diphenyl-**	50-33-9	**NI32447**
77	183	308	184	105	55	51	41	**308**	100	65	27	19	15	15	15	12		19	20	2	2	–	–	–	–	–	–	–	–	–	**3,5-Pyrazolidinedione, 4-butyl-1,2-diphenyl-**	50-33-9	**MS07171**
183	77	184	105	308	91	93	78	**308**	100	99	41	39	31	30	25	25		19	20	2	2	–	–	–	–	–	–	–	–	–	**3,5-Pyrazolidinedione, 4-butyl-1,2-diphenyl-**	50-33-9	**NI32448**
279	308	180	155	307	280	154	181	**308**	100	79	69	64	38	26	24	22		19	20	2	2	–	–	–	–	–	–	–	–	–	**Talbotine, 16-de(methoxycarbonyl)-17-deoxy-17-oxo-**	30809-24-6	**NI32449**
279	308	180	155	307	170	209	235	**308**	100	66	58	55	32	15	12	10		19	20	2	2	–	–	–	–	–	–	–	–	–	**Talbotine, 16-de(methoxycarbonyl)-17-deoxy-17-oxo-**	30809-24-6	**LI06777**
290	55	178	308	93	81	41	95	**308**	100	48	46	43	41	35	34	33		19	32	3	–	–	–	–	–	–	–	–	–	–	**Androstane-3,7,17-triol, (3α,5α,7α,17β)-**	19316-58-6	**NI32450**
290	55	41	81	95	43	178	93	**308**	100	76	56	53	49	49	47	46	**45.00**	19	32	3	–	–	–	–	–	–	–	–	–	–	**Androstane-3,7,17-triol, (3α,5α,7β,17β)-**	41843-62-3	**NI32451**
272	55	41	81	95	93	43	67	**308**	100	54	41	38	37	36	33	32	**9.00**	19	32	3	–	–	–	–	–	–	–	–	–	–	**Androstane-3,7,17-triol, (3α,5β,7α,17β)-**	55448-82-3	**NI32452**
272	290	55	43	41	95	81	93	**308**	100	92	76	66	58	52	51	49	**18.00**	19	32	3	–	–	–	–	–	–	–	–	–	–	**Androstane-3,7,17-triol, (3β,5β,7α,17β)-**	55448-83-4	**NI32453**
290	272	55	41	124	81	107	93	**308**	100	61	52	51	50	48	47	43	**35.00**	19	32	3	–	–	–	–	–	–	–	–	–	–	**Androstane-3,11,17-triol, (3α,5α,11α,17β)-**		**LI06778**
290	43	275	246	55	272	41	107	**308**	100	34	31	31	30	29	27	25	**16.00**	19	32	3	–	–	–	–	–	–	–	–	–	–	**Androstane-3,11,17-triol, (3α,5α,11β,17β)-**		**LI06779**
124	55	290	41	272	93	81	95	**308**	100	36	34	34	31	30	30	27	**2.00**	19	32	3	–	–	–	–	–	–	–	–	–	–	**Androstane-3,11,17-triol, (3α,5β,11α,17β)-**		**LI06780**
290	272	55	41	81	93	213	95	**308**	100	59	44	43	36	35	32	31	**14.00**	19	32	3	–	–	–	–	–	–	–	–	–	–	**Androstane-3,11,17-triol, (3α,5β,11β,17β)-**		**LI06781**
290	272	55	41	43	107	95	81	**308**	100	71	51	51	43	42	42	42	**32.00**	19	32	3	–	–	–	–	–	–	–	–	–	–	**Androstane-3,11,17-triol, (3β,5α,11α,17β)-**	32212-65-0	**LI06782**
290	55	41	272	81	246	107	93	**308**	100	47	47	36	36	32	32	31	**14.00**	19	32	3	–	–	–	–	–	–	–	–	–	–	**Androstane-3,11,17-triol, (3β,5α,11β,17β)-**	32212-65-0	**LI06783**
290	55	41	272	81	246	107	275	**308**	100	47	47	36	36	33	33	32	**13.70**	19	32	3	–	–	–	–	–	–	–	–	–	–	**Androstane-3,11,17-triol, (3β,5α,11β,17β)-**	32212-65-0	**NI32454**
124	164	290	55	41	108	107	93	**308**	100	39	36	36	36	31	30	30	**7.00**	19	32	3	–	–	–	–	–	–	–	–	–	–	**Androstane-3,11,17-triol, (3β,5β,11α,17β)-**		**LI06784**
290	41	55	272	81	107	93	95	**308**	100	48	46	41	39	38	37	34	**16.00**	19	32	3	–	–	–	–	–	–	–	–	–	–	**Androstane-3,11,17-triol, (3β,5β,11β,17β)-**		**LI06785**
290	275	272	55	41	81	43	93	**308**	100	90	66	57	51	48	45	34		19	32	3	–	–	–	–	–	–	–	–	–	–	**Androstane-3,12,17-triol, (3α,5β,12α,17β)-**	6656-28-6	**NI32455**
290	81	272	55	43	95	93	41	**308**	100	46	45	43	32	30	30	30	**3.33**	19	32	3	–	–	–	–	–	–	–	–	–	–	**Androstane-3,12,17-triol, (3α,5β,12β,17β)-**	53604-43-6	**NI32456**
275	290	55	81	41	43	107	57	**308**	100	91	36	32	29	28	27	27	**0.00**	19	32	3	–	–	–	–	–	–	–	–	–	–	**Androstane-3,12,17-triol, (3β,5α,12α,17β)-**	53604-41-4	**NI32457**
290	55	41	81	43	107	249	93	**308**	100	33	31	30	28	24	21	21	**3.03**	19	32	3	–	–	–	–	–	–	–	–	–	–	**Androstane-3,12,17-triol, (3β,5α,12β,17β)-**	53608-26-7	**NI32458**
179	194	55	137	95	69	81	41	**308**	100	67	17	16	15	15	14	12	**9.00**	19	32	3	–	–	–	–	–	–	–	–	–	–	**17-Nor-8-oxo-13β-labdan-15-oic acid**		**MS07172**
179	194	55	137	95	81	69	109	**308**	100	67	17	16	16	15	15	12	**9.00**	19	32	3	–	–	–	–	–	–	–	–	–	–	**17-Nor-8-oxo-13β-labdan-15-oic-acid**		**LI06786**
67	308	81	55	41	95	43	83	**308**	100	70	66	61	59	44	43	42		19	32	3	–	–	–	–	–	–	–	–	–	–	**8-Nonenoic acid, 9-(1,3-nonadienyloxy)-, methyl ester**	39692-46-1	**MS07173**
290	107	308	231	275	41	55	93	**308**	100	86	79	76	66	62	61	59		19	32	3	–	–	–	–	–	–	–	–	–	–	**19-Norkaurane-2,15,18-triol, (2β,4α,15β)-**	43059-71-8	**NI32459**
266	308	210	197	223	209	238	165	**308**	100	25	23	23	22	18	10	10		20	20	3	–	–	–	–	–	–	–	–	–	–	**Estra-1,3,5,7,9-pentaen-17-one, 3-(acetyloxy)-**	6030-91-7	**NI32460**
115	231	91	84	50	83	86	92	**308**	100	87	69	69	54	49	46	39	**4.00**	20	20	3	–	–	–	–	–	–	–	–	–	–	**Methyl 1,3-diphenyl-6-hydroxy-1-cyclohexene-5-carboxylate**		**MS07174**
222	207	235	206	308	223	69	204	**308**	100	43	38	23	22	22	19	18		20	24	1	2	–	–	–	–	–	–	–	–	–	**Acetamide, N-[2-(3-ethyl-1-methyl-9H-carbazol-2-yl)ethyl]-N-methyl-**	55320-31-5	**NI32461**
308	122	121	293	154	107	279		**308**	100	84	60	24	17	15	10			20	24	1	2	–	–	–	–	–	–	–	–	–	**Anhydrovobasindiol**		**LI06787**
188	308	280	107	41	43	78	56	**308**	100	55	29	25	23	20	19	16		20	24	1	2	–	–	–	–	–	–	–	–	–	**Aspidofractinine, 6,7-didehydro-17-methoxy-**	54965-83-2	**NI32462**
134	308	170	202	135	41	156	212	**308**	100	97	93	59	56	42	38	37		20	24	1	2	–	–	–	–	–	–	–	–	–	**2,20-Cycloaspidospermidine-3-methanol, 6,7-didehydro-, (2α,3α,5α,12β,19α,20R)-**	56270-84-9	**NI32463**
170	308	122	290	168	135	134	202	**308**	100	96	89	55	55	54	45	40		20	24	1	2	–	–	–	–	–	–	–	–	–	**2,20-Cycloaspidospermidine-3-methanol, 6,7-didehydro-, (3β,5α,12β,19α,20R)-**	52347-30-5	**NI32464**

MASS TO CHARGE RATIOS								M.W.	INTENSITIES								Parent	C	H	O	N	S	F	Cl	Br	I	Si	P	B	X	COMPOUND NAME	CAS Reg No	No
123	308	69	149	98	185	82	275	**308**	100	89	47	40	38	34	30	26		20	24	1	2	–	–	–	–	–	–	–	–	–	1H-Cyclobuta[jk]phenanthren-4-ol, 2,3,3a,4,4a,5,9b,9c-octahydro-4-(1H-imidazol-2-yl)-3a,9b-dimethyl-	69634-33-9	NI32465
308	183	144	182	157	131	277	291	**308**	100	36	22	18	16	10	7	4		20	24	1	2	–	–	–	–	–	–	–	–	–	12-Demethoxy-seredamine		LI06788
136	308	123	130	144	137	143	122	**308**	100	22	19	12	11	11	9	6		20	24	1	2	–	–	–	–	–	–	–	–	–	N-Methyl-17-methoxyaspidospermatidine		NI32466
308	68	293	265	96	166	131	197	**308**	100	87	79	75	68	57	51	43		20	24	1	2	–	–	–	–	–	–	–	–	–	Phenanthrene, 1-acetyl-1H-imidazol-2-yl-1,2,3,4,4a,9,10,10a-octahydro-1,4a-dimethyl-, [1R-(1α,4aα,10aα)]-	69634-31-7	NI32467
131	308	143	95	119	117	166	293	**308**	100	54	43	30	28	28	26	24		20	24	1	2	–	–	–	–	–	–	–	–	–	Phenanthrene, 1-acetyl-1H-imidazol-4-yl-1,2,3,4,4a,9,10,10a-octahydro-1,4a-dimethyl-, [1R-(1α,4aα,10aα)]-	69634-32-8	NI32468
237	166	238	165	167	151	239	181	**308**	100	21	18	15	10	9	6	3	**3.10**	20	36	–	–	1	–	–	–	–	–	–	–	–	Thiophene, 2,5-bis(1,1,3,3-tetramethylbutyl)-	54964-79-3	WI01399
55	41	43	69	98	196	83	42	**308**	100	93	57	45	36	32	30	26	**8.00**	20	36	2	–	–	–	–	–	–	–	–	–	–	1,3-Cyclododecanedione, 2-octyl-	29550-15-0	NI32469
81	41	95	109	55	93	43	107	**308**	100	87	85	77	73	72	72	66	**0.00**	20	36	2	–	–	–	–	–	–	–	–	–	–	Labd-14-ene-8,13-diol, (13R)-	515-03-7	NI32470
81	41	69	55	109	93	28	43	**308**	100	98	88	80	76	73	67	62	**0.00**	20	36	2	–	–	–	–	–	–	–	–	–	–	Labd-14-ene-8,13-diol, (13R)-	515-03-7	NI32471
67	81	55	95	41	68	82	54	**308**	100	83	67	59	52	48	46	46	**10.90**	20	36	2	–	–	–	–	–	–	–	–	–	–	Linoleic acid ethyl ester		NI32472
249	307	95	109	59	123	137	97	**308**	100	74	70	66	66	65	63	57	**24.80**	20	36	2	–	–	–	–	–	–	–	–	–	–	Linoleyl acetate		NI32473
81	43	95	41	55	67	29	69	**308**	100	76	69	67	64	47	42	38	**13.00**	20	36	2	–	–	–	–	–	–	–	–	–	–	Methyl 2-octylcyclopropene-1-octanoate	3220-60-8	NI32474
245	54	94	137	263	92	275	246	**308**	100	57	53	47	24	23	19	19	**0.00**	20	36	2	–	–	–	–	–	–	–	–	–	–	1H-Naphtho[2,1-b]pyran-3-ethanol, dodecahydro-3,4a,7,7,10a-pentamethyl-	56247-61-1	NI32475
71	276	68	82	96	277	110	124	**308**	100	8	7	5	4	2	2	1	**0.00**	20	36	2	–	–	–	–	–	–	–	–	–	–	9,12,15-Octadecatrienal, dimethyl acetal		LI06789
108	133	94	147	119	134	131	136	**308**	100	48	45	39	37	36	36	35	**0.00**	20	36	2	–	–	–	–	–	–	–	–	–	–	9,12,15-Octadecatrienal, dimethyl acetal	26574-38-9	NI32476
69	221	109	95	81	123	93	121	**308**	100	64	54	34	33	32	27	22	**0.10**	20	36	2	–	–	–	–	–	–	–	–	–	–	Peucelinendiol	72776-48-8	MS07175
85	55	56	41	84	101	67	57	**308**	100	41	27	26	21	18	18	18	**0.34**	20	36	2	–	–	–	–	–	–	–	–	–	–	2H-Pyran, tetrahydro-2-(12-pentadecynyloxy)-	56666-38-7	NI32477
84	97	139	98	43	41	111	55	**308**	100	74	38	38	27	22	21	21	**15.01**	20	40	–	2	–	–	–	–	–	–	–	–	–	1H-Imidazole, 2-heptadecyl-4,5-dihydro-	105-28-2	NI32478
84	97	139	98	43	41	111	307	**308**	100	74	38	38	27	22	21	20	**15.00**	20	40	–	2	–	–	–	–	–	–	–	–	–	1H-Imidazole, 2-heptadecyl-4,5-dihydro-	105-28-2	LI06790
131	91	130	129	118	117	105	115	**308**	100	34	32	22	21	20	20	19	**18.71**	21	24	2	–	–	–	–	–	–	–	–	–	–	Hydrocinnamic acid, 2-[(1,2,3,4-tetrahydro-2-naphthyl)methyl]-, methyl ester	23804-22-0	NI32479
43	175	308	171	159	173	174	135	**308**	100	42	30	21	18	17	16	16		21	24	2	–	–	–	–	–	–	–	–	–	–	Pregna-1,4,7,16-tetraene-3,20-dione	56145-33-6	NI32480
69	41	43	97	71	112	57	55	**308**	100	72	55	49	47	41	33	33	**3.82**	21	40	1	–	–	–	–	–	–	–	–	–	–	1-Cyclopentyl-1-hexadecanone		AI02026
308	252	309	250	280	251	253	307	**308**	100	27	23	22	17	9	6	5		22	12	2	–	–	–	–	–	–	–	–	–	–	Benzo[a]naphthacene-8,13-dione	65492-95-7	NI32481
280	308	281	250	126	125	252	113	**308**	100	31	27	24	22	19	14	11		22	12	2	–	–	–	–	–	–	–	–	–	–	Dibenzanthracene-5,6-quinone		NI32482
308	252	309	280	126	250	125	251	**308**	100	33	25	24	21	20	16	8		22	12	2	–	–	–	–	–	–	–	–	–	–	6,13-Pentacenedione	3029-32-1	NI32483
280	281	308	307	282	309	279	278	**308**	100	87	75	41	19	17	10	7		22	16	–	2	–	–	–	–	–	–	–	–	–	1-Naphthaldehyde azine	2144-00-5	LI06791
127	154	280	281	308	181	153	128	**308**	100	85	77	73	67	48	36	35		22	16	–	2	–	–	–	–	–	–	–	–	–	1-Naphthaldehyde azine	2144-00-5	NI32484
308	280	309	307	279	281	310	278	**308**	100	56	28	19	16	13	1	1		22	16	–	2	–	–	–	–	–	–	–	–	–	1-Naphthaldehyde azine	2144-00-5	LI06792
127	155	153	156	281	308	154	142	**308**	100	75	65	64	53	43	38	38		22	16	–	2	–	–	–	–	–	–	–	–	–	2-Naphthaldehyde azine	2144-02-7	LI06793
127	155	156	153	281	308	154	142	**308**	100	75	65	65	53	42	40	40		22	16	–	2	–	–	–	–	–	–	–	–	–	2-Naphthaldehyde azine	2144-02-7	NI32485
128	308	307	129	230	205	204	180	**308**	100	16	14	14	8	7	7	7		22	16	–	2	–	–	–	–	–	–	–	–	–	Pyridine, 3,3′-(1a,2,7,7a-tetrahydro-1,2,7-metheno-1H-cyclopropa[b]naphthalene-1,8-diyl)bis-	50559-66-5	NI32486
175	105	176	145	160	308	203	133	**308**	100	29	17	15	8	7	6	6		22	28	1	–	–	–	–	–	–	–	–	–	–	2-Butanone, 1-(4-tert-butyl-2,6-xylyl)-3-phenyl-	29549-18-6	LI06794
175	105	176	145	160	28	308	133	**308**	100	29	17	15	8	8	7	6		22	28	1	–	–	–	–	–	–	–	–	–	–	2-Butanone, 1-(4-tert-butyl-2,6-xylyl)-3-phenyl-	29549-18-6	NI32487
42	44	69	41	91	31	71	43	**308**	100	78	57	49	45	36	26	24	**2.00**	22	28	1	–	–	–	–	–	–	–	–	–	–	3,3-Dimethyl-6-benzyl-7-phenyl-hept-5-en-1-ol		LI06795
43	57	55	41	97	83	69	29	**308**	100	77	77	76	55	48	46	43	**8.71**	22	44	–	–	–	–	–	–	–	–	–	–	–	1-Docosene	1599-67-3	NI32488
308	202	203	279	307	278	276	138	**308**	100	32	22	21	17	16	16	16		23	16	1	–	–	–	–	–	–	–	–	–	–	9-(4-Formylstyryl)anthracene		LI06796
308	167	215	309	165	204	191	115	**308**	100	40	31	27	27	26	23	23		24	20	–	–	–	–	–	–	–	–	–	–	–	2-Phenyl-3-diphenylmethylenecyclopentene		LI06797
291	309	62	63	79	233	61	52	**309**	100	71	43	30	17	13	13	13		7	4	5	1	–	–	–	–	1	–	–	–	–	Salicylic acid, 5-iodo-3-nitro-	22621-43-8	NI32489
291	309	62	63	233	189	79	61	**309**	100	72	44	32	27	18	17	13		7	4	5	1	–	–	–	–	1	–	–	–	–	Salicylic acid, 5-iodo-3-nitro-	22621-43-8	MS07176
293	57	43	55	69	41	71	83	**309**	100	94	87	75	62	58	53	47	**1.00**	7	4	5	1	–	–	–	–	1	–	–	–	–	Salicylic acid, 5-iodo-3-nitro-	22621-43-8	LI06798
55	177	119	225	93	309	80	133	**309**	100	61	38	33	27	22	17	16		8	4	5	3	1	–	–	–	–	–	–	–	1	Tricarbonyl(η-azidosulphonylcyclopentadienyl)manganese		NI32490
18	162	164	45	63	17	27	98	**309**	100	99	67	59	35	33	28	23	**0.00**	8	8	5	–	1	–	2	–	–	–	–	–	1	Ethanol, 2-(2,4-dichlorophenoxy)-, hydrogen sulphate, sodium salt	136-78-7	NI32491
310	311	278	75	312	308	262	105	**309**	100	10	9	8	5	4	3	3	**0.80**	9	12	3	1	1	5	–	–	–	–	–	–	–	Methionine methyl ester, N-pentafluoropropionate		NI32492
43	170	85	230	127	45	79	41	**309**	100	50	46	25	22	22	21	21	**0.50**	9	15	7	3	1	–	–	–	–	–	–	–	–	Arabinopyranoside, methyl 4-azido-4-deoxy-, 3-acetate 2-methanesulphonate, β-L-	24905-23-5	NI32493
41	182	81	138	39	309	96	102	**309**	100	81	30	22	14	13	8	6		10	16	2	1	–	–	–	–	1	–	–	–	–	Allyl N-trans-2-iodocyclohexylcarbamate		LI06799
202	15	18	201	44	109	42	93	**309**	100	92	88	87	87	84	36	26	**5.50**	10	16	6	1	1	–	–	–	–	–	1	–	–	Phosphoric acid, [4-[(dimethylamino)sulphonyl]phenyl] dimethyl ester	960-25-8	NI32494
206	73	294	207	208	295	74	190	**309**	100	72	37	24	14	12	11	8	**0.91**	10	31	2	1	–	–	–	–	–	4	–	–	–	4,6-Dioxa-5-aza-2,3,7,8-tetrasilanonane, 2,2,3,3,7,7,8,8-octamethyl-	74793-29-6	NI32495

MASS TO CHARGE RATIOS								M.W.	INTENSITIES								Parent	C	H	O	N	S	F	Cl	Br	I	Si	P	B	X	COMPOUND NAME	CAS Reg No	No
135	136	43	164	174	60	178	108	**309**	100	81	81	69	68	34	32	32	**1.00**	11	15	4	7	–	–	–	–	–	–	–	–	–	3-Acetamido-3′-deoxyadenosine		MS07177
57	60	73	55	61	56	59	85	**309**	100	39	31	28	24	20	19	13	**0.00**	11	19	7	1	1	–	–	–	–	–	–	–	–	Desulphoprogoitrin		MS07178
56	73	128	75	45	130	74	247	**309**	100	89	71	35	25	18	17	13	**0.00**	11	27	3	1	1	–	–	–	–	2	–	–	–	Methionine sulphoxide, bis(trimethylsilyl)-		MS07179
84	85	42	73	83	41	75	45	**309**	100	7	6	5	3	3	2	2	**1.09**	11	28	3	1	–	–	–	–	–	2	1	–	–	Phosphonic acid, 1-[(isopropylideneamino)ethyl]-, bis(trimethylsilyl) ester	55108-64-0	NI32496
84	73	294	75	147	70	237	225	**309**	100	24	15	11	10	10	9	9	**6.00**	11	28	3	1	–	–	–	–	–	2	1	–	–	Phosphonic acid, 2-[(isopropylideneamino)ethyl]-, bis(trimethylsilyl) ester	55108-71-9	NI32497
255	253	57	257	41	311	92	309	**309**	100	80	56	49	39	18	15	14		12	11	–	1	–	–	4	–	–	–	–	–	–	2-tert-Butyl-4,5,6,7-tetrachloroisoindole		NI32498
124	43	69	41	126	184	68	55	**309**	100	94	51	36	32	28	26	20	**6.00**	12	15	5	5	–	–	–	–	–	–	–	–	–	5′-Azido-5′-deoxy-3′-O-acetylthymidine		MS07180
294	295	309	73	180	296	75	45	**309**	100	25	21	19	11	9	9	6		12	23	1	5	–	–	–	–	–	2	–	–	–	7H-Purin-2-amine, 7-methyl-N-(trimethylsilyl)-6-[(trimethylsilyl)oxy]-	32958-86-4	NI32499
252	254	217	216	84	45	56	309	**309**	100	84	48	46	41	40	39	30		13	9	–	3	1	–	2	–	–	–	–	–	–	Thiocyanic acid, 4-amino-2-chloro-5-[(4-chlorophenyl)amino]phenyl ester	23648-89-7	MS07181
311	58	309	274	276	139	111	197	**309**	100	90	84	81	77	65	52	48		13	9	1	1	–	–	1	1	–	–	–	–	–	5-Bromo-2′-chlorobenzophenone		MS07182
309	280	263	200	201	199	159	160	**309**	100	58	33	24	16	14	12	10		13	11	6	1	1	–	–	–	–	–	–	–	–	4-Methoxy-7-methyl-5-oxo-5H-furo[3,2-g][1]benzopyran-9-sulphonamide		MS07183
43	104	105	91	45	69	77	51	**309**	100	75	41	25	16	14	12	12	**0.00**	13	12	2	1	–	5	–	–	–	–	–	–	–	N-Acetyl-2-phenylethylamine, pentafluoropropionyl-		MS07184
136	309	163	43	44	91	235	137	**309**	100	84	81	78	71	60	51	18		13	15	6	3	–	–	–	–	–	–	–	–	–	2-(4-Nitrophenyl)hydrazonopropandioic acid, diethyl ester		NI32500
91	104	92	309	105	176	146	218	**309**	100	25	25	13	11	9	9	8		14	16	1	1	–	5	–	–	–	–	–	–	–	5-Phenylpentylamine, pentafluoropropionyl-		MS07185
73	147	192	45	74	75	149	148	**309**	100	35	14	13	9	8	6	6	**1.18**	14	23	3	1	–	–	–	–	–	2	–	–	–	Oxanilic acid, bis(trimethylsilyl)-		NI32501
173	175	145	309	147	311	109	177	**309**	100	61	21	19	13	12	12	10		15	13	2	1	–	–	2	–	–	–	–	–	–	3-Acetyl-1-(2,6-dichlorobenzoyl)-2,4-dimethylpyrrole	95883-30-0	NI32502
242	244	243	214	311	309	179	178	**309**	100	38	27	26	23	23	18	15		15	13	2	1	–	–	2	–	–	–	–	–	–	Benzoic acid, 2-[(2,6-dichloro-3-methylphenyl)amino]-, methyl ester	3254-79-3	NI32503
262	264	235	125	115	165	116	237	**309**	100	78	77	76	70	58	58	56	**3.10**	15	13	2	1	–	–	2	–	–	–	–	–	–	Propane, 1,1-bis(4-chlorophenyl)-2-nitro-	117-27-1	NI32504
91	236	155	65	139	80	43	41	**309**	100	87	68	20	12	11	10	10	**0.00**	15	19	4	1	1	–	–	–	–	–	–	–	–	Ethyl (E)-3-ethylidene-1-tosylazetidine-2-carboxylate		DD01146
91	236	155	65	80	256	53	39	**309**	100	80	80	17	10	8	7	7	**0.00**	15	19	4	1	1	–	–	–	–	–	–	–	–	Ethyl (Z)-3-ethylidene-1-tosylazetidine-2-carboxylate		DD01147
91	236	155	154	81	80	65	54	**309**	100	69	68	23	23	21	18	16	**0.00**	15	19	4	1	1	–	–	–	–	–	–	–	–	Ethyl (2SR,3RS)-1-tosyl-3-vinylazetidine-2-carboxylate		DD01148
43	153	73	165	30	207	208	28	**309**	100	40	28	26	26	23	22	20	**1.70**	15	19	6	1	–	–	–	–	–	–	–	–	–	Acetamide, N-[2-(acetyloxy)-2-[4-(acetyloxy)-3-methoxyphenyl]ethyl]-	54965-13-8	NI32505
87	44	181	43	86	139	309	45	**309**	100	83	53	48	30	25	24	23		15	19	6	1	–	–	–	–	–	–	–	–	–	Acetamide, N-[2-[3,4-bis(acetyloxy)phenyl]-2-hydroxyethyl]-N-methyl-	55712-68-0	NI32506
266	202	267	217	186	203	91	48	**309**	100	40	25	13	12	10	10	8	**1.00**	15	23	2	3	1	–	–	–	–	–	–	–	–	Benzenesulphonyl azide, 2,4,6-triisopropyl-	36982-84-0	NI32507
43	280	41	27	238	29	18	91	**309**	100	86	84	69	53	35	34	20	**5.90**	15	23	4	3	–	–	–	–	–	–	–	–	–	Aniline, 4-isopropyl-2,6-dinitro-N,N-dipropyl-	33820-53-0	NI32508
218	73	192	75	147	77	219	100	**309**	100	82	76	28	21	20	19	15	**0.00**	15	27	2	1	–	–	–	–	–	2	–	–	–	Phenylalanine, bis(trimethylsilyl)-	7364-51-4	MS07186
73	218	192	45	91	75	147	100	**309**	100	30	29	22	15	14	10	10	**0.00**	15	27	2	1	–	–	–	–	–	2	–	–	–	Phenylalanine, bis(trimethylsilyl)-	7364-51-4	NI32509
73	218	192	45	91	147	100	75	**309**	100	45	37	22	18	12	12	10	**0.00**	15	27	2	1	–	–	–	–	–	2	–	–	–	Phenylalanine, bis(trinethylsilyl)-	7364-51-4	MS07187
134	309	77	194	135	193	91	136	**309**	100	49	14	13	11	7	7	5		16	11	–	3	2	–	–	–	–	–	–	–	–	5-Phenylthiazolo[2,3-b]-1,3,4-thiadiazol-4-ium-2-phenylaminide		MS07188
309	217	216	263	262	30	140	189	**309**	100	45	43	41	37	35	33	28		16	11	4	3	–	–	–	–	–	–	–	–	–	Cyclopropanecarbonitrile, 1,2-bis(4-nitrophenyl)-	28752-28-5	NI32510
309	77	175	176	240	134	101	241	**309**	100	37	22	17	16	16	16	14		16	11	4	3	–	–	–	–	–	–	–	–	–	4-(4-Nitrobenzylidene)-1-phenyl-3,5-dioxopyrazolidine	4599-10-4	NI32511
309	132	133	90	18	279	104	89	**309**	100	88	39	32	32	26	23	22		16	11	4	3	–	–	–	–	–	–	–	–	–	2-Nitrodiftalone		MS07189
309	132	90	133	89	310	104	118	**309**	100	70	31	30	26	20	20	17		16	11	4	3	–	–	–	–	–	–	–	–	–	3-Nitrodiftalone		MS07190
309	90	119	192	310	102	92	221	**309**	100	30	25	22	21	18	14	12		16	15	2	5	–	–	–	–	–	–	–	–	–	2-Isopropyl-1,2,3-benzotriazinium 4-(2-nitroanilide)		LI06800
309	192	221	310	102	90	193	119	**309**	100	27	21	20	18	18	11	10		16	15	2	5	–	–	–	–	–	–	–	–	–	2-Propyl-1,2,3-benzotriazinium 4-(2-nitroanilide)		LI06801
120	105	274	228	104	77	155	121	**309**	100	63	34	32	17	16	15	14	**9.00**	16	20	3	1	–	–	1	–	–	–	–	–	–	Ethyl 3-chloro-1-[(1-phenylethyl)carbamoyl]cyclobutanecarboxylate	85318-08-7	NI32512
209	168	125	77	43	41	105	40	**309**	100	83	74	38	33	21	20	13	**0.00**	16	23	3	1	1	–	–	–	–	–	–	–	–	N,N-Diisopropyl-3-(phenylsulphonyl)-2-methylenepropanamide	110426-80-7	DD01149
151	250	83	85	43	164	47	162	**309**	100	75	62	43	24	22	15	14	**1.15**	16	23	5	1	–	–	–	–	–	–	–	–	–	N-Acetyl-3-(3-hydroxymethyl-4-methoxyphenyl)-L-alanine isopropyl ester	16825-26-6	NI32513
240	43	55	41	69	57	127	182	**309**	100	83	68	68	50	50	48	32	**2.40**	16	30	1	1	–	3	–	–	–	–	–	–	–	1-Aminotetradecane, trifluoroacetyl-		MS07191
130	73	58	131	45	59	132	74	**309**	100	51	16	13	7	5	4	4	**0.00**	16	31	1	1	–	–	–	–	–	2	–	–	–	Ephedrine, bis(trimethylsilyl)-		MS07192
130	73	131	147	132	45	74	59	**309**	100	38	13	4	4	4	3	3	**0.02**	16	31	1	1	–	–	–	–	–	2	–	–	–	Pseudoephedrine, bis(trimethylsilyl)-		MS07193
116	132	119	291	115	91	83	133	**309**	100	90	21	18	17	15	14	12	**2.00**	17	15	3	3	–	–	–	–	–	–	–	–	–	1-(1-Hydroxy-2-indanyl)-4-phenyl-1,2,4-triazolidine-3,5-dione		LI06802
237	252	309	169	223	171	253	170	**309**	100	78	33	28	25	25	19	17		17	19	1	5	–	–	–	–	–	–	–	–	–	1β-Azidomethyl-1α-ethyl-3-oxo-2,3,5,6,11,11bβ-hexahydro-1H-indolizino[8,7-b]indole		NI32514
58	59	84	43	42	41	309	44	**309**	100	4	2	2	2	2	1	1		17	27	4	1	–	–	–	–	–	–	–	–	–	1,4-Methano-2H-cyclopent[d]oxepin-2,5(4H)-dione, 6-[(dimethylamino)methyl]hexahydro-8a-hydroxy-5a-methyl-9-isopropyl-, [1R-(1α,4α,5aα,6β,8aα,9S*)]-	32562-94-0	NI32515
70	41	208	250	80	94	309		**309**	100	89	82	78	69	60	7			17	31	2	3	–	–	–	–	–	–	–	–	–	Palustrine		LI06803
105	237	265	160	106	77	116	115	**309**	100	24	19	13	13	13	11	10	**0.00**	18	15	4	1	–	–	–	–	–	–	–	–	–	Benzeneacetic acid, α-cyano-α-benzoyloxy-, ethyl ester		LI06804
263	308	105	44	77	235	186	130	**309**	100	98	93	93	87	79	79	73	**29.00**	18	15	4	1	–	–	–	–	–	–	–	–	–	5-Hydroxy-3-benzoyl-2-ethoxycarbonylindole		LI06805
217	308	139	309	183	63	152	185	**309**	100	86	46	20	16	10	5	4		18	16	–	1	1	–	–	–	–	–	1	–	–	N-Phenyl-P,P-diphenylphosphinothioic amide		MS07194
309	308	103	307	310	206	102	306	**309**	100	74	50	25	19	17	16	12		18	18	–	3	–	–	–	–	–	–	–	3	–	Borazine, 1,3,5-triphenyl-	976-29-4	NI32516
91	148	65	106	77	267	79	104	**309**	100	78	30	23	22	15	12	10	**9.60**	18	19	2	3	–	–	–	–	–	–	–	–	–	Acetamide, N-acetyl-N-[2-methyl-4-[(2-methylphenyl)azo]phenyl]-	83-63-6	NI32517

MASS TO CHARGE RATIOS								M.W.	INTENSITIES								Parent	C	H	O	N	S	F	Cl	Br	I	Si	P	B	X	COMPOUND NAME	CAS Reg No	No
77	264	195	44	51	105	91	64	309	100	63	36	28	27	26	14	13	0.00	18	19	2	3	–	–	–	–	–	–	–	–	–	3,5-Pyrazolidinedione, 4-[(dimethylamino)methyl]-1,2-diphenyl-	6136-32-9	NI32518
293	124	192	233	169	309	265	223	309	100	98	54	44	32	30	25	21		18	19	2	3	–	–	–	–	–	–	–	–	–	1-Pyrrolidino-3-oxo-11b-hydroxy-5,6,11,11b-tetrahydro-3H-indolo[3,2-g]indolizine		LI06806
309	254	294	279	224	241	266	140	309	100	55	37	26	14	13	10	6		19	19	3	1	–	–	–	–	–	–	–	–	–	1,4-Dihydro-7-methoxy-3-methyl-8-(3-methylbut-2-enyl)carbazole-1,4-dione		MS07195
168	167	165	169	166	152	265	212	309	100	99	84	82	79	77	76	66	49.00	19	19	3	1	–	–	–	–	–	–	–	–	–	Diphenylacetylproline		NI32519
120	105	129	224	143	131	121	235	309	100	80	22	19	13	13	13	12	0.00	19	19	3	1	–	–	–	–	–	–	–	–	–	Ethaneperoxoic acid, 1-cyano-1,4-diphenylbutyl ester	58422-76-7	NI32520
309	294	238	295	308	237	280	281	309	100	94	29	20	19	18	15	14		19	19	3	1	–	–	–	–	–	–	–	–	–	3-Ethoxy-1-(2,4-diethoxyphenyl)-isoquinoline		LI06807
309	241	225	254	242	310	226	213	309	100	99	45	44	25	23	17	14		19	19	3	1	–	–	–	–	–	–	–	–	–	6-Hydroxy-3,3,12-trimethyl-1,2-dihydropyrano[2,3-c]acridone		MS07196
218	176	77	105	43	91	219	174	309	100	57	38	32	31	28	14	14	0.00	19	19	3	1	–	–	–	–	–	–	–	–	–	Methyl (4S,5S)-4-benzyl-4,5-dihydro-5-methyl-2-phenyloxazole-4-carboxylate		DD01150
309	237	236	280	310	156	264	101	309	100	36	29	21	21	20	18	17		19	19	3	1	–	–	–	–	–	–	–	–	–	2-Propenoic acid, 3-[4-[[(4-methoxyphenyl)methylene]amino]phenyl]-, ethyl ester	6421-30-3	NI32521
252	98	253	126	43	280	58	56	309	100	43	22	22	20	18	18	18	6.00	19	35	2	1	–	–	–	–	–	–	–	–	–	Bicyclocassine	28111-16-2	LI06808
86	58	99	71	100	87	57	55	309	100	17	15	10	6	6	6	6	0.00	19	35	2	1	–	–	–	–	–	–	–	–	–	[1,1′-Bicyclohexyl]-1-carboxylic acid, 2-(diethylamino)ethyl ester	77-19-0	NI32522
266	308	230	231	268	189	267	310	309	100	57	46	39	34	29	27	22	16.00	20	20	–	1	–	–	1	–	–	–	–	–	–	3-Butyl-6-chloro-2-methyl-4-phenyl-quinoline		LI06809
266	230	308	268	231	267	189	251	309	100	43	39	35	35	24	22	16	14.00	20	20	–	1	–	–	1	–	–	–	–	–	–	3-Isobutyl-6-chloro-2-methyl-4-phenyl-quinoline		LI06810
309	254	294	41	224	278	127	211	309	100	45	22	20	10	9	9	8		20	23	2	1	–	–	–	–	–	–	–	–	–	1,7-Dimethoxy-3-methyl-8-(3-methylbut-2-enyl)-9H-carbazole		MS07197
215	229	230	207	248	178	216	202	309	100	49	44	43	28	26	23	23	0.00	20	23	2	1	–	–	–	–	–	–	–	–	–	10-Hydroxyamitriptyline N-oxide		MS07198
105	77	132	42	51	159	131	190	309	100	73	47	29	24	23	21	20	1.80	20	23	2	1	–	–	–	–	–	–	–	–	–	β-(4-Phenyl-4-hydroxypiperidino)propiophenone		MS07199
87	100	72	309	45	55	41	43	309	100	68	38	29	25	24	21	16		20	39	1	1	–	–	–	–	–	–	–	–	–	9-Octadecenamide, N,N-dimethyl-	3906-30-7	NI32523
309	310	206	156	231	308	282	311	309	100	24	6	5	4	3	3	3		21	15	–	3	–	–	–	–	–	–	–	–	–	2,6-Diphenyl-1,7-dihydrodipyrrolo[2,3-b:3′,2′-e]pyridine		NI32524
103	309	76	104	310	77	51	50	309	100	37	14	13	8	7	3	3		21	15	–	3	–	–	–	–	–	–	–	–	–	1,3,5-Triazine, 2,4,6-triphenyl-	493-77-6	NI32525
72	91	115	77	78	105	70	71	309	100	23	16	15	11	9	9	8	0.00	21	27	1	1	–	–	–	–	–	–	–	–	–	3-Heptanone, 6-(dimethylamino)-4,4-diphenyl-	76-99-3	NI32526
72	91	73	294	223	57	56	42	309	100	5	5	4	4	4	4	4	0.50	21	27	1	1	–	–	–	–	–	–	–	–	–	3-Heptanone, 6-(dimethylamino)-4,4-diphenyl-	76-99-3	NI32527
72	73	29	57	42	56	91	44	309	100	5	5	3	3	3	3	3	0.00	21	27	1	1	–	–	–	–	–	–	–	–	–	3-Heptanone, 6-(dimethylamino)-4,4-diphenyl-	76-99-3	NI32528
84	43	85	41	55	57	98	42	309	100	12	9	9	6	5	4	3	1.14	21	43	–	1	–	–	–	–	–	–	–	–	–	2-Hexadecyl-5-methylpyrrolidine		NI32529
308	309	218	217	310	204	203	191	309	100	76	36	19	18	17	17	16		23	19	–	1	–	–	–	–	–	–	–	–	–	2-Benzyl-2,3-dihydro-1H-dibenz[e,g]isoindole	110028-16-5	DD01151
309	91	218	310	217	219	65	189	309	100	81	77	26	25	14	13	10		23	19	–	1	–	–	–	–	–	–	–	–	–	1-Benzyl-2,4-diphenylpyrrole		MS07200
67	151	113	48	86	35	132	51	310	100	58	50	43	41	27	25	16	0.00	–	–	1	–	1	8	1	–	–	–	–	–	1	Chlorodifluoro-oxysulphonium hexafluoroarsenic		MS07201
69	127	183	163	310	114	128	113	310	100	66	54	48	47	33	30	23		4	2	–	–	–	7	–	–	1	–	–	–	–	1,1,1,2,4,4,4-Heptafluoro-2-iodobutane		LI06811
69	310	51	127	95	145	113	31	310	100	79	55	41	32	21	19	16		4	2	–	–	–	7	–	–	1	–	–	–	–	erythro-1,1,1,2,4,4,4-Heptafluoro-3-iodobutane		LI06812
310	69	127	95	145	51	128	113	310	100	79	66	34	20	16	15	13		4	2	–	–	–	7	–	–	1	–	–	–	–	threo-1,1,1,2,4,4,4-Heptafluoro-3-iodobutane		LI06813
69	145	127	310	159	114	183	51	310	100	82	70	59	59	58	55	53		4	2	–	–	–	7	–	–	1	–	–	–	–	1,1,1,3-Tetrafluoro-2-trifluoromethyl-3-iodopropane		LI06814
53	52	150	54	152	151	115	63	310	100	62	32	32	23	25	24	22	0.00	4	6	–	–	–	–	2	2	–	1	–	–	–	1,1-Dichloro-3,4-dibromo-1-silacyclopentane	75292-54-5	NI32530
183	55	155	127	27	39	29	141	310	100	50	18	16	11	7	7	6	0.60	4	8	–	–	–	–	–	–	2	–	–	–	–	Butane, 1,4-diiodo-	628-21-7	NI32531
183	55	128	127	310	254			310	100	94	35	19	11	5				4	8	–	–	–	–	–	–	2	–	–	–	–	Butane, 1,4-diiodo-	628-21-7	LI06815
265	263	267	264	45	59	262	202	310	100	70	56	55	47	46	43	26	1.06	4	9	1	–	–	–	1	–	–	–	–	–	1	Mercury, chloro(2-ethoxyethyl)-	124-01-6	NI32532
19	163	119	113	144	100	16	69	310	100	72	25	14	9	8	5	4	0.00	6	–	3	–	–	10	–	–	–	–	–	–	–	Perfluoropropionic anhydride (negative ion spectrum)		LI06816
119	69	147	44	28	31	100	97	310	100	63	51	21	21	11	10	10	0.00	6	–	3	–	–	10	–	–	–	–	–	–	–	Perfluoropropionic anhydride (positive ion spectrum)		LI06817
249	284	286	251	247	282	288	253	310	100	98	79	64	63	50	34	21	1.50	7	–	1	–	–	–	6	–	–	–	–	–	–	Bis(trichlorovinyl)cyclopropenone		LI06818
253	223	208	295	221	222	207	251	310	100	98	90	69	50	47	47	46	0.20	7	18	–	–	–	–	–	–	–	–	–	–	1	Plumbane, butyltrimethyl-	54964-75-9	NI32533
252	223	208	250	221	251	222	206	310	100	86	68	48	45	43	39	31	0.09	7	18	–	–	–	–	–	–	–	–	–	–	1	Plumbane, tert-butyltrimethyl-		AI02027
281	208	223	237	207	279	280	209	310	100	89	77	69	50	46	43	43	0.98	7	18	–	–	–	–	–	–	–	–	–	–	1	Plumbane, triethylmethyl-		AI02028
253	223	208	251	221	252	222	207	310	100	91	79	48	47	42	41	38	1.94	7	18	–	–	–	–	–	–	–	–	–	–	1	Plumbane, trimethylsec-butyl-		AI02029
277	279	275	242	240	170	281	244	310	100	60	57	32	22	17	12	11	0.00	8	4	–	–	–	–	6	–	–	–	–	–	–	Benzene, 1,3-bis(trichloromethyl)-	881-99-2	NI32534
277	279	275	242	240	170	85	244	310	100	71	64	62	51	31	30	28	1.19	8	4	–	–	–	–	6	–	–	–	–	–	–	Benzene, 1,4-bis(trichloromethyl)-	68-36-0	NI32535
277	279	275	242	240	244	170	281	310	100	67	65	61	48	28	27	22	0.00	8	4	–	–	–	–	6	–	–	–	–	–	–	Benzene, 1,4-bis(trichloromethyl)-	68-36-0	NI32536
277	275	279	242	240	170	172	244	310	100	70	62	51	41	39	25	25	11.75	8	4	–	–	–	–	6	–	–	–	–	–	–	Benzene, 1,2,4,5-tetrachloro-3,6-bis(chloromethyl)-	1079-17-0	NI32537
277	279	275	281	170	49	242	312	310	100	64	62	18	18	15	14	13	5.47	8	4	–	–	–	–	6	–	–	–	–	–	–	Benzene, 2,4,5,6-tetrachlro-1,3-bis(chloromethyl)-	1133-57-9	NI32538
140	125	310	77	51	97	231	94	310	100	92	59	53	46	40	17	11		8	8	1	–	1	–	–	2	–	–	–	–	–	Methylphenylsulphoxonium dibromomethylide		MS07202
29	232	69	55	41	209	57	43	310	100	99	99	99	90	80	60	40	0.00	8	10	1	2	–	6	1	–	–	–	–	1	–	2-Butyl-2-chloro-4,6-bis(trifluoromethyl)-1-oxa-3-azonia-5-aza-2-borata-3,5-cyclohexadiene		NI32539
45	97	99	169	61	28	62	27	310	100	93	61	21	19	18	16	16	0.00	8	10	4	–	–	–	4	–	–	–	–	–	–	Ethyl 1,2-bis(2,2-dichloropropionate)		DO01092

MASS TO CHARGE RATIOS								M.W.	INTENSITIES								Parent	C	H	O	N	S	F	Cl	Br	I	Si	P	B	X	COMPOUND NAME	CAS Reg No	No
281	282	283	253	225	40	197	126	**310**	100	34	28	23	15	14	14	13	**0.69**	8	22	5	–	–	–	–	–	–	4	–	–	–	1,3,5,7-Tetraethylbicyclo[3.3.1]tetrasiloxane	73420-21-0	NI32540
30	120	43	100	147	41	119	270	**310**	100	95	94	94	83	82	75	69	**28.30**	9	9	2	2	–	7	–	–	–	–	–	–	–	Propanamide, N-[3-(difluoromethyl)-2-oxo-3-piperidinyl]-2,2,3,3,3-pentafluoro-		NI32541
233	231	152	312	314	310	201	199	**310**	100	92	88	73	58	42	38	38		9	12	–	–	1	–	–	2	–	–	–	–	–	4-Bromo-2-bromomethyl-6-thiatricyclo[3.2.1.1^{3,8}]nonane		LI06819
59	250	117	18	217	185	131	45	**310**	100	61	46	35	30	26	25	21	**0.00**	9	15	3	–	4	–	–	–	–	–	–	1	–	2,4,6,8-Tetrathiatricyclo[3.3.1.1^{3,7}]decan-1-ol, 3,5,7-trimethyl-, monoester with boric acid	57274-36-9	NI32542
241	310	75	199	231	85	101	69	**310**	100	71	30	29	23	23	21	20		10	6	1	–	1	3	–	1	–	–	–	–	–	3-Buten-2-one, 4-(4-bromophenyl)-1,1,1-trifluoro-4-mercapto-	53833-48-0	NI32543
127	281	157	311	249	141	155	279	**310**	100	99	55	52	38	30	25	19	**1.48**	10	6	4	4	2	–	–	–	–	–	–	–	–	Bis(5-nitro-2-pyridyl) disulphide	2127-10-8	NI32544
310	123	82	170	311	156	187	64	**310**	100	24	17	16	14	13	11	11		10	6	4	4	2	–	–	–	–	–	–	–	–	Bis(5-nitro-2-pyridyl) disulphide	2127-10-8	NI32545
125	109	310	93	79	136	63	47	**310**	100	70	42	41	37	35	34	32		10	15	5	–	2	–	–	–	–	–	1	–	–	Phosphorothioic acid, O,O-dimethyl O-[3-methyl-4-(methylsulphonyl)phenyl] ester	3761-42-0	NI32546
207	73	295	208	209	296	191	297	**310**	100	97	83	46	28	24	24	15	**0.00**	10	30	3	–	–	–	–	–	–	4	–	–	–	Methyltris(trimethylsiloxy)silane	17928-28-8	NI32547
73	207	295	74	191	45	75	59	**310**	100	95	35	9	7	6	5	5	**0.00**	10	30	3	–	–	–	–	–	–	4	–	–	–	Tetrasiloxane, decamethyl-	141-62-8	MS07203
73	207	208	295	74	45	209	59	**310**	100	55	11	10	9	8	6	5	**0.00**	10	30	3	–	–	–	–	–	–	4	–	–	–	Tetrasiloxane, decamethyl-	141-62-8	NI32548
207	73	295	208	209	296	191	45	**310**	100	61	25	21	12	8	8	5	**0.00**	10	30	3	–	–	–	–	–	–	4	–	–	–	Tetrasiloxane, decamethyl-	141-62-8	NI32549
310	159	247	231	251	223	266	275	**310**	100	88	80	76	60	60	12	10		11	9	4	–	–	–	3	–	–	–	–	–	–	3,5-Methanobenzofuran-7a(2H)-carboxylic acid, 6,7,8-trichloro-3,3a,4,5-tetrahydro-2-oxo-, methyl ester	75420-32-5	NI32550
226	168	194	80	110	254	111	138	**310**	100	52	47	29	28	26	24	23	**0.28**	11	10	7	–	–	–	–	–	–	–	–	–	1	Iron, tricarbonyl(2,4-hexadienedioic acid)-, dimethyl ester	12153-21-8	NI32551
56	168	87	84	52	254	226	112	**310**	100	61	61	61	47	36	33	28	**3.30**	11	10	7	–	–	–	–	–	–	–	–	–	1	con-Trimethylenebis(methoxycarbonyl)iron tricarbonyl		NI32552
168	226	254	110	108	87	52	84	**310**	100	97	67	23	17	17	15	12	**0.00**	11	10	7	–	–	–	–	–	–	–	–	–	1	dis-Trimethylenebis(methoxycarbonyl)iron tricarbonyl		NI32553
57	127	81	109	141	209	255	254	**310**	100	86	76	30	14	13	5	2	**0.00**	11	19	2	–	–	–	–	–	1	–	–	–	–	tert-Butyl 3-(iodomethyl)cyclopentanecarboxylate		DD01152
41	251	253	57	249	195	197	193	**310**	100	63	57	57	52	36	35	33	**7.00**	11	24	–	–	–	–	–	–	–	–	–	–	2	1,2-Selenagermolane, 2,2-dibutyl-	18438-01-2	NI32554
290	310	291	224	193	271	311	222	**310**	100	74	59	21	20	16	15	15		12	5	1	–	–	6	–	–	–	–	1	–	–	Phosphine oxide, fluoro(pentafluorophenyl)phenyl-	53381-03-6	NI32555
152	76	231	233	312	151	75	310	**310**	100	60	53	51	32	26	22	20		12	8	–	–	–	–	–	2	–	–	–	–	–	1,1′-Biphenyl, 2,2′-dibromo-	13029-09-9	NI32556
152	231	233	76	151	312	75	150	**310**	100	42	41	27	26	25	20	16	**12.13**	12	8	–	–	–	–	–	2	–	–	–	–	–	1,1′-Biphenyl, 2,2′-dibromo-	13029-09-9	NI32557
152	312	76	310	314	151	75	63	**310**	100	98	73	43	42	23	20	17		12	8	–	–	–	–	–	2	–	–	–	–	–	1,1′-Biphenyl, 4,4′-dibromo-	92-86-4	NI32558
152	312	310	314	76	151	75	150	**310**	100	71	36	35	28	27	19	17		12	8	–	–	–	–	–	2	–	–	–	–	–	Biphenyl, 2,5-dibromo-	57422-77-2	NI32559
152	312	310	314	151	76	75	150	**310**	100	61	31	30	27	24	21	18		12	8	–	–	–	–	–	2	–	–	–	–	–	Biphenyl, 2,6-dibromo-	59080-32-9	NI32560
202	310	277	139	204	183	98	63	**310**	100	87	57	50	38	33	32	32		12	8	3	–	–	–	1	–	–	–	–	–	1	Phenoxarsine, 2-chloro-10-hydroxy-, 10-oxide	27796-65-2	NI32561
57	41	40	43	42	56	59	58	**310**	100	70	27	23	17	13	12	9	**4.00**	12	13	3	–	–	–	3	–	–	–	–	–	–	Acetic acid, (2,4,5-trichlorophenoxy)-, butyl ester	93-79-8	NI32562
57	41	29	42	43	196	209	198	**310**	100	41	17	12	11	10	9	9	**4.80**	12	13	3	–	–	–	3	–	–	–	–	–	–	Acetic acid, (2,4,5-trichlorophenoxy)-, 2-methylpropyl ester	4938-72-1	NI32563
295	296	297	310	249	238	223	277	**310**	100	25	18	13	11	8	8	7		12	14	2	4	2	–	–	–	–	–	–	–	–	2,2′-Dimethoxy-4′,6-bis(methylthio)-4,5′-bipyrimidine		NI32564
41	55	229	67	275	77	63	75	**310**	100	91	88	76	61	56	50	46	**25.00**	12	14	6	4	–	–	–	–	–	–	–	–	–	Adipaldehydic acid, (2,4-dinitrophenyl)hydrazone		IC03837
29	41	28	43	31	310	219	77	**310**	100	66	60	55	40	36	33	24		12	14	6	4	–	–	–	–	–	–	–	–	–	Ethyl acetoacetate 2,4-dinitrophenylhydrazone	4093-60-1	NI32565
69	41	55	97	45	115	73	39	**310**	100	79	55	46	36	30	27	26	**11.97**	12	22	4	–	–	–	–	–	–	–	–	–	1	Hexanoic acid, 6,6′-selenodi-	22676-16-0	NI32566
169	165	167	168	163	139	166	135	**310**	100	85	81	35	33	31	28	26	**0.00**	12	27	–	–	–	1	–	–	–	–	–	–	1	Tri-butyltin fluoride		MS07204
57	253	41	177	251	175	213	29	**310**	100	38	33	31	28	24	23	20	**9.50**	12	27	–	–	–	1	–	–	–	–	–	–	1	Triisobutyltin fluoride		MS07205
143	165	41	109	199	255	56	43	**310**	100	48	48	47	39	38	34	32	**0.00**	12	27	4	–	–	–	–	–	–	–	–	–	1	Arsenic acid, tributyl ester	15063-74-8	NI32568
143	165	41	109	255	56	43	55	**310**	100	47	47	45	37	34	32	31	**0.00**	12	27	4	–	–	–	–	–	–	–	–	–	1	Arsenic acid, tributyl ester	15063-74-8	NI32567
312	310	202	216	231	67	201	63	**310**	100	98	62	36	36	29	28	23		13	11	4	–	–	–	–	1	–	–	–	–	–	2-Bromo-5,8-dimethoxy-3-methyl-1,4-naphthalenedione		NI32569
137	177	310	97	179	113	139	267	**310**	100	87	60	49	47	42	41	35		13	14	7	2	–	–	–	–	–	–	–	–	–	6H-Furo[2′,3′:4,5]oxazolo[3,2-a]pyrimidin-6-one, 3-(acetyloxy)-2-[(acetyloxy)methyl]-2,3,3a,9a-tetrahydro-, [2R-(2α,3β,3aβ,9aβ)]-	28309-53-7	NI32570
114	113	99	297	295	59	39	115	**310**	100	67	59	43	42	34	32	32	**12.39**	13	15	–	2	1	–	–	1	–	–	–	–	–	4,4,6-Trimethyl-2-[4-bromophenylamino(imino)]-4H-1,3-thiazine	85298-51-7	NI32571
169	171	312	310	268	266	172	170	**310**	100	99	30	30	10	10	8	8		13	15	2	2	–	–	–	1	–	–	–	–	–	Δ^2-1,3,4-Oxadiazolin-5-one, 4-(4-bromophenyl)-2-pentyl-	28740-56-9	NI32572
310	210	208	182	211	168	206	157	**310**	100	48	45	34	28	22	18	15		13	19	2	–	–	–	–	–	–	–	–	–	1	Rhodium, [(1,2,5,6-η)-1,5-cyclooctadiene](2,4-pentanedionato-O,O′)-	12245-39-5	NI32573
310	157	211	210	103	208	182	168	**310**	100	79	77	45	39	38	32	29		13	19	2	–	–	–	–	–	–	–	–	–	1	Rhodium, (η^4-1,2-diethenylcyclobutane)(2,4-pentanedionato-O,O′)-, stereoisomer	74811-15-7	NI32574
73	117	31	180	75	295	131	130	**310**	100	58	22	17	17	13	10	9	**6.01**	13	22	3	4	–	–	–	–	–	1	–	–	–	7-[(2′-O-Trimethylsilyl)propyl]-1,3-dimethylxa nthine		NI32575
117	180	295	310	266	131	130	252	**310**	100	60	47	31	24	23	22	19		13	22	3	4	–	–	–	–	–	1	–	–	–	7-[(2′-O-Trimethylsilyl)propyl]-1,3-dimethylxanthine		NI32576
43	41	55	180	61	182	109	95	**310**	100	31	29	24	22	21	17	17	**0.00**	13	24	2	–	–	1	–	1	–	–	–	–	–	10-Bromo-11-fluoroundecan-1-yl acetate		DD01153
43	95	59	41	109	81	151	69	**310**	100	33	29	28	25	22	17	17	**0.00**	13	24	2	–	–	1	–	1	–	–	–	–	–	11-Bromo-10-fluoroundecan-1-yl acetate		DD01154
57	43	71	85	41	55	28	29	**310**	100	75	66	43	35	29	18	15	**0.34**	13	27	–	–	–	–	–	–	1	–	–	–	–	Tridecane, 1-iodo-	35599-77-0	NI32577
101	69	99	100	125	113	68	102	**310**	100	59	40	10	9	6	5	4	**0.00**	13	27	6	–	–	–	–	–	–	–	1	–	–	Dibutyl (3-methoxycarbonyl)propyl phosphate	89197-74-0	NI32578
310	312	184	275	256	284	282	254	**310**	100	97	53	49	48	46	46	44		14	5	2	–	–	–	3	–	–	–	–	–	–	1,4,5-Trichloroanthraquinone		IC03838
153	141	155	63	113	127	310	125	**310**	100	84	46	38	33	29	21	18		14	9	2	2	–	2	1	–	–	–	–	–	–	Benzamide, N-[[(4-chlorophenyl)amino]carbonyl]-2,6-difluoro-	35367-38-5	NI32579
153	141	155	63	113	125	57	44	**310**	100	76	34	33	29	23	23	23	**4.00**	14	9	2	2	–	2	1	–	–	–	–	–	–	Benzamide, N-[[(4-chlorophenyl)amino]carbonyl]-2,6-difluoro-	35367-38-5	NI32580

MASS TO CHARGE RATIOS	M.W.	INTENSITIES	Parent	C	H	O	N	S	F	Cl	Br	I	Si	P	B	X	COMPOUND NAME	CAS Reg No	No
250 308 115 116 96 280 125 224	310	100 19 19 18 17 16 15 12		14	12	2	–	–	–	–	–	–	–	–	–	1	Molybdenum, dicarbonyl-π-allyl-π-indenyl-		MS07206
252 91 308 198 96 224 174 161	310	100 17 16 14 14 12 8 8		14	12	2	–	–	–	–	–	–	–	–	–	1	Molybdenum, dicarbonyl-π-cyclopentadienyl-benzyl-		MS07207
252 91 78 96 39 65 174 66	310	100 28 27 22 20 15 14 14	10.00	14	12	2	–	–	–	–	–	–	–	–	–	1	Molybdenum, dicarbonyl-π-cyclopentadienyl-π-cycloheptatrienyl-		MS07208
91 105 104 92 77 65 45 51	310	100 20 20 8 7 7 4 3	0.00	14	14	4	–	2	–	–	–	–	–	–	–	–	Benzene, [[[benzylsulphonyl]methyl]sulphonyl]-	58751-71-6	NI32581
279 162 280 161 74 75 76 29	310	100 31 30 30 23 22 20 19	17.42	14	14	8	–	–	–	–	–	–	–	–	–	–	1,2,4,5-Benzenetetracarboxylic acid, tetramethyl ester	635-10-9	NI32582
192 62 91 27 30 77 193 43	310	100 95 91 85 64 62 59 59	0.00	14	14	8	–	–	–	–	–	–	–	–	–	–	Tetramethyl benzenetetracarboxylate		DO01093
231 79 91 93 77 65 147 232	310	100 28 27 24 24 24 19 17	7.00	14	15	1	–	1	–	–	1	–	–	–	–	–	3-[1-Bromo-2-(phenylthio)cycloprop-1-yl]cyclopentan-1-one		MS07209
91 139 121 173 171 174 172 110	310	100 91 90 82 80 77 77 77	5.00	14	15	1	–	1	–	–	1	–	–	–	–	–	1-[1-Bromo-2-(phenylthio)cycloprop-1-yl]cyclopent-2-en-1-ol(diastereoisomer 1)		MS07210
91 121 139 83 172 77 174 93	310	100 91 86 83 70 70 68 68	2.00	14	15	1	–	1	–	–	1	–	–	–	–	–	1-[1-Bromo-2-(phenylthio)cycloprop-1-yl]cyclopent-2-en-1-ol(diastereoisomer 2)		MS07211
277 246 310 245 169 278 63 140	310	100 48 33 20 18 18 18 15		14	15	2	–	2	–	–	–	–	–	1	–	–	Bis(4-methoxyphenyl)dithiophosphinic acid	68344-17-2	NI32583
205 241 310 243 106 167 141 159	310	100 91 88 56 55 48 31 20		14	15	4	2	–	–	1	–	–	–	–	–	–	Ethyl chloro-[[2-[(4-hydroxy-2-butynyl)oxy]phenyl]hydrazono]acetate		MS07212
155 200 310 63 77 91 156 109	310	100 52 21 13 12 11 8 8		14	16	–	–	2	–	–	–	–	–	2	–	–	1,2-Dimethyl-1,2-diphenyldiphosphane disulphide		MS07213
91 43 84 83 56 42 41 155	310	100 86 83 71 69 60 60 46	0.00	14	18	4	2	1	–	–	–	–	–	–	–	–	1-[N-Methyl-N-(4-toluenesulphonyl)aminocarbonyl]-4,4-dimethylazetidinone		NI32584
130 43 150 60 30 88 86 125	310	100 62 60 55 55 43 42 28	3.00	14	18	6	2	–	–	–	–	–	–	–	–	–	Acetamide, N-[3-hydroxy-3-methyl-4-[(4-nitrobenzoyl)oxy]butyl]-, (R)-	64018-50-4	NI32585
103 150 219 265 309	310	100 80 26 16 2	0.00	14	18	6	2	–	–	–	–	–	–	–	–	–	4-(4-Nitrobenzamido)butyraldehyde diethyl acetal		LI06820
57 177 137 85 69 310 113 97	310	100 99 99 71 71 69 58 58		14	18	6	2	–	–	–	–	–	–	–	–	–	Propanoic acid, 2,2-dimethyl-, 2,3,3a,9a-tetrahydro-2-(hydroxymethyl)-6-oxo-6H-furo[2',3':4,5]oxazolo[3,2-a]pyrimidin-3-yl ester, [2R-(2α,3β,3aβ,9aβ)]-	38642-35-2	NI32586
57 69 199 97 85 55 81 113	310	100 99 76 66 65 56 55 49	22.00	14	18	6	2	–	–	–	–	–	–	–	–	–	Propanoic acid, 2,2-dimethyl-, (2,3,3a,9a-tetrahydro-3-hydroxy-6-oxo-6H furo[2',3':4,5]oxazolo[3,2-a]pyrimidin-2-yl)methyl ester, [2R-(2α,3β,3aβ,9aβ)]-	40732-65-8	NI32587
56 69 84 41 55 42 39 312	310	100 67 51 43 28 19 17 17	16.99	14	19	1	2	–	–	–	1	–	–	–	–	–	3,4,4,6-Tetramethyl-2-(4-bromophenylimino)-tetrahydro-1,3-oxazine	53004-30-1	NI32588
84 185 187 41 212 214 56 186	310	100 58 51 48 42 39 39 34	15.99	14	19	1	2	–	–	–	1	–	–	–	–	–	4,4,6-Trimethyl-2-(N-methyl-4-bromophenylamino)-5,6-dihydro-4H-1,3-oxazine	53004-37-8	NI32589
147 73 295 148 45 149 221 76	310	100 66 22 17 16 9 7 7	3.00	14	22	4	–	–	–	–	–	–	2	–	–	–	1,2-Benzenedicarboxylic acid, bis(trimethylsilyl) ester	2078-22-0	NI32590
147 73 295 148 149 45 75 221	310	100 87 71 70 34 30 26 25	6.40	14	22	4	–	–	–	–	–	–	2	–	–	–	1,2-Benzenedicarboxylic acid, bis(trimethylsilyl) ester	2078-22-0	MS07214
147 73 45 295 148 149 141 75	310	100 63 17 16 16 9 7 7	1.99	14	22	4	–	–	–	–	–	–	2	–	–	–	1,2-Benzenedicarboxylic acid, bis(trimethylsilyl) ester	2078-22-0	NI32591
295 103 221 73 296 104 251 76	310	100 31 31 27 25 16 16 14	5.96	14	22	4	–	–	–	–	–	–	2	–	–	–	1,4-Benzenedicarboxylic acid, bis(trimethylsilyl) ester	4147-84-6	NI32592
82 81 110 30 43 238 194 28	310	100 77 66 54 40 26 24 24	18.45	14	22	4	4	–	–	–	–	–	–	–	–	–	N^2-(N-Acetylglycyl)histidine butyl ester		AI02030
221 237 85 177 83 235 251 236	310	100 95 71 70 70 60 49 49	3.00	14	30	–	–	–	–	–	–	–	4	–	–	–	Silane, 3,4-hexadien-1-yne-1,3,5,6-tetrayltetrakis[dimethyl-	61255-23-0	NI32593
310 213 186 311 170 198 309 184	310	100 70 46 34 26 24 20 18		15	13	2	2	–	3	–	–	–	–	–	–	–	Melatonin, trifluoroacetyl-		MS07215
236 278 45 221 162 279 246 263	310	100 96 89 62 51 50 36 34	19.00	15	18	7	–	–	–	–	–	–	–	–	–	–	Dimethyl 4,5-dihydro-4,5-epoxy-3-methoxymethyl-4-(2-oxopropyl)phthalate		NI32594
278 236 177 205 247 58 279 206	310	100 37 31 24 22 22 17 15	4.00	15	18	7	–	–	–	–	–	–	–	–	–	–	Dimethyl 6-hydroxy-4-methoxymethyl-3-(2-oxopropyl)phthalate		NI32595
278 207 310 251 69 219 59 77	310	100 92 60 36 30 29 29 26		15	18	7	–	–	–	–	–	–	–	–	–	–	Dimethyl 6-methoxy-8-methyl-8,9-dihydro-6H-pyrano[4,3-b]oxepine-4,5-dicarboxylate		NI32596
193 224 177 69 219 59 178 208	310	100 61 44 35 30 26 22 19	2.00	15	18	7	–	–	–	–	–	–	–	–	–	–	Dimethyl 11-methoxy-9-methyl-6,10-dioxapentacyclo[$5.4.0.0^{1,4}.0^{2,7}.0^{3,5}$]undeca-2,3-dicarboxylate		NI32597
193 69 124 178 219 249 279 59	310	100 96 71 69 68 60 58 56	2.00	15	18	7	–	–	–	–	–	–	–	–	–	–	Dimethyl 7-methoxy-9-methyl-8,11-dioxatricyclo[$6.2.1.0^{1,6}$]undeca-2,5-diene-2,3-dicarboxylate		NI32598
267 29 265 225 41 310 184 57	310	100 56 48 46 40 31 24 23		15	22	–	2	–	–	–	–	–	–	–	–	1	Selenocyanic acid, 4-(dibutylamino)phenyl ester	22037-12-3	NI32599
96 91 43 155 65 56 121 70	310	100 48 45 24 21 17 11 10	0.00	15	22	3	2	1	–	–	–	–	–	–	–	–	cis-1-Acetamido-2-(p-toluenesulphonamido)cyclohexane		MS07216
96 155 91 43 251 60 225 56	310	100 64 53 24 14 14 12 10	10.00	15	22	3	2	1	–	–	–	–	–	–	–	–	cis-1-Acetamido-3-(p-toluenesulphonamido)cyclohexane		MS07217
96 91 155 43 68 58 80 56	310	100 95 64 62 53 53 49 42	3.00	15	22	3	2	1	–	–	–	–	–	–	–	–	cis-1-Acetamido-4-(p-toluenesulphonamido)cyclohexane		MS07218
96 155 91 112 43 70 56 65	310	100 41 23 21 21 10 9 8	0.50	15	22	3	2	1	–	–	–	–	–	–	–	–	trans-1-Acetamido-2-(p-toluenesulphonamido)cyclohexane		MS07219
96 91 43 155 56 70 65 60	310	100 27 22 15 10 6 6 4	2.00	15	22	3	2	1	–	–	–	–	–	–	–	–	trans-1-Acetamido-3-(p-toluenesulphonamido)cyclohexane		MS07220
96 91 155 43 68 65 56 58	310	100 95 92 83 49 42 39 33	0.30	15	22	3	2	1	–	–	–	–	–	–	–	–	trans-1-Acetamido-4-(p-toluenesulphonamido)cyclohexane		MS07221
310 280 295 311 281 73 79 75	310	100 89 22 16 12 9 5 5		15	22	5	–	–	–	–	–	–	1	–	–	–	Cinnamic acid, 3,5-dimethoxy-4-(trimethylsiloxy)-, methyl ester	27798-72-7	NI32600
280 310 73 281 311 295 75 279	310	100 85 46 23 21 16 15 14		15	22	5	–	–	–	–	–	–	1	–	–	–	Cinnamic acid, 3,5-dimethoxy-4-(trimethylsiloxy)-, methyl ester	27798-72-7	NI32601
161 178 105 78 106 132 79 52	310	100 68 51 47 31 26 20 18	15.00	15	22	5	2	–	–	–	–	–	–	–	–	–	3-(2-Methoxy-5-acetamido-anilino)propionic acid, 2-methoxyethyl ester		IC03839
147 73 192 295 253 310 177 179	310	100 75 67 64 40 38 36 19		15	26	3	–	–	–	–	–	–	2	–	–	–	Benzenepropanoic acid, 2-[(trimethylsilyl)oxy]-, trimethylsilyl ester		MS07222
205 192 73 75 310 193 177 45	310	100 81 72 60 46 36 31 19		15	26	3	–	–	–	–	–	–	2	–	–	–	Benzenepropanoic acid, 3-[(trimethylsilyl)oxy]-, trimethylsilyl ester	55887-87-1	NI32602
73 205 75 192 310 177 206 179	310	100 86 86 76 32 28 14 10		15	26	3	–	–	–	–	–	–	2	–	–	–	Benzenepropanoic acid, 3-[(trimethylsilyl)oxy]-, trimethylsilyl ester	55887-87-1	NI32603

MASS TO CHARGE RATIOS								M.W.	INTENSITIES								Parent	C	H	O	N	S	F	Cl	Br	I	Si	P	B	X	COMPOUND NAME	CAS Reg No	No
179	192	310	193	180	93	73	177	**310**	100	99	30	19	16	14	13	10		15	26	3	–	–	–	–	–	–	2	–	–	–	Benzenepropanoic acid, 4-[(trimethylsilyl)oxy]-, trimethylsilyl ester	27750-62-5	NI32604
179	192	73	75	310	177	180	77	**310**	100	64	63	26	21	15	14	13		15	26	3	–	–	–	–	–	–	2	–	–	–	Benzenepropanoic acid, 4-[(trimethylsilyl)oxy]-, trimethylsilyl ester	27750-62-5	MS07223
179	73	192	75	310	180	177	45	**310**	100	67	64	22	21	17	15	14		15	26	3	–	–	–	–	–	–	2	–	–	–	Benzenepropanoic acid, 4-[(trimethylsilyl)oxy]-, trimethylsilyl ester	27750-62-5	NI32605
73	193	147	45	75	194	220	74	**310**	100	57	43	15	11	10	9	9	**0.00**	15	26	3	–	–	–	–	–	–	2	–	–	–	Benzenepropanoic acid, α-[(trimethylsilyl)oxy]-, trimethylsilyl ester	27750-45-4	NI32606
193	73	147	220	194	219	295	148	**310**	100	79	76	43	41	24	22	18	**0.00**	15	26	3	–	–	–	–	–	–	2	–	–	–	Benzenepropanoic acid, α-[(trimethylsilyl)]oxy-, trimethylsilyl ester		MS07224
73	148	147	75	89	120	45	77	**310**	100	96	50	50	48	40	35	34	**27.14**	15	26	3	–	–	–	–	–	–	2	–	–	–	4-Methoxyphenylketene bis(trimethylsilyl) acetal		MS07225
193	73	194	147	267	45	75	74	**310**	100	68	17	16	9	8	6	5	**0.00**	15	26	3	–	–	–	–	–	–	2	–	–	–	4-Methylmandelic acid, bis(trimethylsilyl)-		MS07226
193	73	194	147	75	267	45	74	**310**	100	68	22	20	11	8	8	6	**0.00**	15	26	3	–	–	–	–	–	–	2	–	–	–	α-Methylmandelic acid, bis(trimethylsilyl)-		MS07227
73	193	147	45	220	194	74	91	**310**	100	53	40	16	11	10	9	8	**0.00**	15	26	3	–	–	–	–	–	–	2	–	–	–	2-Phenyllactic acid, bis(trimethylsilyl) ester		NI32607
73	147	103	104	118	280	75	148	**310**	100	43	43	31	28	22	14	9	**1.64**	15	26	3	–	–	–	–	–	–	2	–	–	–	2-Phenyl-2-trimethylsiloxyacrylic acid, trimethylsilyl ester		NI32608
73	179	147	192	75	180	45	148	**310**	100	93	62	37	19	15	14	10	**3.44**	15	26	3	–	–	–	–	–	–	2	–	–	–	2-Phenyl-2-trimethylsiloxyacrylic acid, trimethylsilyl ester		NI32609
281	295	73	209	282	296	283	75	**310**	100	68	49	39	25	18	11	11	**5.94**	15	26	3	–	–	–	–	–	–	2	–	–	–	1-Propanone, 1-[2,4-bis[(trimethylsilyl)oxy]phenyl]-	55724-93-1	NI32610
115	143	30	189	63	94	144	116	**310**	100	89	40	30	20	16	14	14	**11.51**	16	10	5	2	–	–	–	–	–	–	–	–	–	Naphthalene, 1-(2,4-dinitrophenoxy)-	3761-15-7	NI32611
143	115	189	310	94	144	127	63	**310**	100	86	25	15	14	10	9	8		16	10	5	2	–	–	–	–	–	–	–	–	–	Naphthalene, 1-(2,4-dinitrophenoxy)-	3761-15-7	NI32612
79	106	107	78	80	204	105	144	**310**	100	99	69	55	50	48	44	41	**4.00**	16	14	1	4	1	–	–	–	–	–	–	–	–	5-Formyl-4-(1-pyridyl)thiazole-2-oxide α,α-methylphenylhydrazone		NI32613
121	56	104	91	105	78	103	77	**310**	100	82	74	55	50	45	42	36	**0.00**	16	14	3	–	–	–	–	–	–	–	–	–	1	Iron, cyclopentadienyl-(2-phenylpropanoyl)dicarbonyl		NI32614
251	139	253	111	141	252	255	75	**310**	100	80	64	33	26	15	12	12	**1.70**	16	16	2	–	–	–	2	–	–	–	–	–	–	Benzenemethanol, 4-chloro-α-(4-chlorophenyl)-α-(ethoxymethyl)-		MS07228
227	228	310	312	212	153	152	169	**310**	100	16	6	4	3	3	3	2		16	16	2	–	–	–	2	–	–	–	–	–	–	2,2-Dichloro-1,1-bis(4-methoxyphenyl)ethane	7388-31-0	MS07229
227	228	310	153	152	212	169	141	**310**	100	24	8	7	7	6	6	6		16	16	2	–	–	–	2	–	–	–	–	–	–	2,2-Dichloro-1,1-bis(4-methoxyphenyl)ethane	7388-31-0	NI32615
45	72	73	149	104	43	76	59	**310**	100	85	83	46	25	25	19	15	**0.00**	16	22	6	–	–	–	–	–	–	–	–	–	–	1,2-Benzenedicarboxylic acid, bis(2-ethoxyethyl) ester	605-54-9	NI32616
163	59	310	191	177	237	135	149	**310**	100	70	66	59	44	22	20	10		16	22	6	–	–	–	–	–	–	–	–	–	–	Diethyl α-(2,5-dimethoxyphenyl)-α-methylmalonate	83026-47-5	DD01155
163	59	310	191	43	149	237	177	**310**	100	50	41	40	30	26	20	20		16	22	6	–	–	–	–	–	–	–	–	–	–	Diethyl α-(3,4-dimethoxyphenyl)-α-methylmalonate		DD01156
151	310	75	73	295	164	152	177	**310**	100	84	68	51	45	44	32	25		16	26	4	–	–	–	–	–	–	1	–	–	–	3,4-Dimethoxyphenylpentanoic acid, trimethylsilyl-		NI32617
310	247	311	309	242	265	293	77	**310**	100	24	20	10	9	8	7	7		17	14	2	2	1	–	–	–	–	–	–	–	–	2,3,7,12-Tetrahydro-6-methylamino-7,12-dioxo-1H-anthra[2,1-b]thiazine		IC03840
43	268	191	105	267	77	162	103	**310**	100	43	40	39	36	28	26	15	**1.00**	17	14	4	2	–	–	–	–	–	–	–	–	–	3-(4-Nitrophenyl)-4-acetyl-5-phenyl-2-isoxazoline		NI32618
91	267	43	119	268	193	239	120	**310**	100	72	44	35	16	14	13	12	**4.00**	17	14	4	2	–	–	–	–	–	–	–	–	–	3-(4-Nitrophenyl)-4-phenyl-5-acetyl-2-isoxazoline		NI32619
119	42	253	161	310	252	92	146	**310**	100	72	67	60	55	55	43	41		17	14	4	2	–	–	–	–	–	–	–	–	–	Spiro[3H-1,4-benzodiazepine-3,2′-oxirane]-2,5(1H,4H)-dione, 3′-(3-hydroxyphenyl)-4-methyl-, cis-(-)-	20007-85-6	LI06821
132	119	42	253	161	310	252	92	**310**	100	86	60	56	51	46	46	36		17	14	4	2	–	–	–	–	–	–	–	–	–	Spiro[3H-1,4-benzodiazepine-3,2′-oxirane]-2,5(1H,4H)-dione, 3′-(3-hydroxyphenyl)-4-methyl-, cis-(-)-	20007-85-6	NI32620
254	255	208	180	310	196	152	311	**310**	100	16	16	16	12	5	5	2		17	14	4	2	–	–	–	–	–	–	–	–	–	1,3a,4,8b-Tetrahydro-8b-hydroxy-3a-(4-nitrophenyl)cyclopent[b]indol-3(2H)-one	90732-24-4	NI32621
148	267	133	310	91	77	29	119	**310**	100	72	67	46	41	32	31	30		17	18	–	4	1	–	–	–	–	–	–	–	–	N,N-Diethyl-4-(benzthiazol-2-ylazo)aniline		IC03841
310	311	239	295	267	308	224	175	**310**	100	20	10	8	7	6	5	4		17	18	2	–	–	–	–	–	–	–	–	–	1	1,4′-Diacetyl[3]ferrocenophane	41558-93-4	NI32622
310	224	311	267	308	225	162	239	**310**	100	24	22	8	7	4	4	3		17	18	2	–	–	–	–	–	–	–	–	–	1	2,2′-Diacetyl[3]ferrocenophane	41558-91-2	NI32623
310	311	267	224	308	162	239	56	**310**	100	21	13	13	6	4	3	3		17	18	2	–	–	–	–	–	–	–	–	–	1	2,3′-Diacetyl[3]ferrocenophane	38548-79-7	NI32625
310	311	267	224	308	268	239	162	**310**	100	23	14	14	6	3	3	3		17	18	2	–	–	–	–	–	–	–	–	–	1	2,3′-Diacetyl[3]ferrocenophane	38548-79-7	NI32624
310	311	267	239	308	224	312	295	**310**	100	21	9	9	6	5	3	3		17	18	2	–	–	–	–	–	–	–	–	–	1	3,3′-Diacetyl[3]ferrocenophane	41558-92-3	NI32626
268	310	58	115	211	70	282	253	**310**	100	43	28	13	9	8	7	4		17	18	2	4	–	–	–	–	–	–	–	–	–	Morphinan-3-ol, 8-azido-6,7-didehydro-4,5-epoxy-17-methyl-, (5α,8β)-	55781-29-8	NI32627
219	310	32	44	220	161	203	41	**310**	100	28	24	19	18	15	14	13		17	26	3	–	1	–	–	–	–	–	–	–	–	2,6-Di-tert-butyl-4-(2-thiobutyl-4-carboxyl)phenol		IC03842
43	149	277	235	217	55	83	41	**310**	100	38	34	20	12	12	11	11	**0.00**	17	26	5	–	–	–	–	–	–	–	–	–	–	3α-Acetylplucheinol		NI32628
43	213	41	190	195	203	218	55	**310**	100	88	41	38	34	30	28	28	**1.00**	17	26	5	–	–	–	–	–	–	–	–	–	–	Azuleno[5,6-c]furan-3(1H)-one, 8-ethoxy-4,4a,5,6,7,7a,8,9-octahydro-1,4-dihydroxy-6,6,8-trimethyl-	71305-64-1	NI32629
141	83	43	85	69	41	197	58	**310**	100	99	99	80	78	76	65	65	**25.00**	17	26	5	–	–	–	–	–	–	–	–	–	–	Portentol	21795-25-5	LI06822
141	224	123	267	211	283	240	125	**310**	100	85	59	41	36	24	24	22	**0.00**	17	26	5	–	–	–	–	–	–	–	–	–	–	3,6,10-Trimethyl-5-acetyloxybicyclo[7.3.0]-12-oxa-4,11-dodecadione	69782-92-9	NI32630
310	102	129	103	76	156	155	130	**310**	100	37	23	20	18	17	14	13		18	10	–	6	–	–	–	–	–	–	–	–	–	Quinoxalino[1″,2″:1′,2′]imidazo[4′,5′:4,5]imidazo[1,2-a]quinoxaline		MS07230
309	310	293	173	311	165	146	145	**310**	100	97	62	22	20	19	19	15		18	11	4	–	–	1	–	–	–	–	–	–	–	trans-6-Carboxy-2-(4-fluorostyryl)chromone		MS07231
310	295	311	234	121	202	312	296	**310**	100	52	22	22	22	16	15	11		18	14	1	–	2	–	–	–	–	–	–	–	–	3H-1,2-Dithiole, 5-phenyl-(1-phenyl-2-propanonyl)-	38489-98-4	NI32631
268	269	239	139	310	238	211	210	**310**	100	63	53	48	47	47	47	42		18	14	5	–	–	–	–	–	–	–	–	–	–	1-Acetoxy-3-methoxy-6-methylanthraquinone		NI32632
295	296	310	147.5	294	311	294	311	**310**	100	20	15	8	4	3	4	3		18	14	5	–	–	–	–	–	–	–	–	–	–	6-Deoxyisojacareubin	26486-92-0	LI06823
295	32	310	296	18	147.5	311	294	**310**	100	31	20	20	17	11	4	4		18	14	5	–	–	–	–	–	–	–	–	–	–	6-Deoxyisojacareubin	26486-92-0	LI06824
105	77	189	106	282	188	68	122	**310**	100	61	29	20	18	16	11	3	**1.00**	18	14	5	–	–	–	–	–	–	–	–	–	–	cis-2,5-Dibenzoxy-2,5-dihydrofuran		LI06825
310	202	267	265	311	281	237	268	**310**	100	99	86	42	19	16	16	14		18	14	5	–	–	–	–	–	–	–	–	–	–	8,10-Dimethoxy-7-oxo-6,12-dioxa-5,6,7,7a,11a,12-hexahydrobenz[a]anthracene		LI06826
295	296	310	147.5	43	311	294	297	**310**	100	28	23	19	7	5	5	4		18	14	5	–	–	–	–	–	–	–	–	–	–	Iso-osajaxanthone	26486-88-4	LI06827
295	18	28	296	310	147.5	57	311	**310**	100	26	24	20	19	11	6	4		18	14	5	–	–	–	–	–	–	–	–	–	–	Osajaxanthone	1043-08-9	LI06828

MASS TO CHARGE RATIOS								M.W.	INTENSITIES								Parent	C	H	O	N	S	F	Cl	Br	I	Si	P	B	X	COMPOUND NAME	CAS Reg No	No
94	44	65	39	66	40	55	38	**310**	100	38	25	23	21	12	8	8	**1.24**	18	15	3	–	–	–	–	–	–	–	1	–	–	Triphenyl phosphite	101-02-0	IC03843
94	66	39	77	217	65	51	40	**310**	100	55	51	50	49	43	23	22	**11.30**	18	15	3	–	–	–	–	–	–	–	1	–	–	Triphenyl phosphite	101-02-0	IC03844
217	77	310	51	153	218	152	65	**310**	100	80	27	23	22	13	12	11		18	15	3	–	–	–	–	–	–	–	1	–	–	Triphenyl phosphite	101-02-0	NI32633
295	296	265	297	238	277	165	249	**310**	100	23	7	6	4	4	4	3	**0.20**	18	18	3	–	–	–	–	–	–	1	–	–	–	1-Hydroxy-2-methylanthraquinone, trimethylsilyl-		NI32634
295	296	265	297	165	277	140	294	**310**	100	25	7	6	3	3	3	2	**0.30**	18	18	3	–	–	–	–	–	–	1	–	–	–	1-Hydroxy-3-methylanthraquinone, trimethylsilyl-	91701-13-2	NI32635
295	310	296	267	165	297	152	265	**310**	100	61	23	11	7	6	5	4		18	18	3	–	–	–	–	–	–	1	–	–	–	2-Hydroxy-3-methylanthraquinone, trimethylsilyl-	91701-15-4	NI32636
295	310	296	267	297	165	268	237	**310**	100	67	22	19	6	4	4	3		18	18	3	–	–	–	–	–	–	1	–	–	–	2-Hydroxy-4-methylanthraquinone, trimethylsilyl-	91701-14-3	NI32637
281	279	310	151	161	174	76	280	**310**	100	91	77	36	27	24	20	17		18	18	3	2	–	–	–	–	–	–	–	–	–	3-(2,6-Dimethoxyphenyl)-2-ethyl-4(3H)-quinazolinone		NI32638
310	224	203	233	118	148	311	253	**310**	100	95	46	40	36	29	21	18		18	18	3	2	–	–	–	–	–	–	–	–	–	N,N-Dimethyl-5-phenyl-5-(3-methoxyphenyl)hydantoin		NI32639
233	310	148	224	311	118	234	194	**310**	100	75	30	27	15	15	14	11		18	18	3	2	–	–	–	–	–	–	–	–	–	N,N-Dimethyl-5-phenyl-5-(4-methoxyphenyl)hydantoin		NI32640
77	105	176	84	310	119	266	207	**310**	100	90	88	70	62	20	10	10		18	18	3	2	–	–	–	–	–	–	–	–	–	5-Ethoxy-3-methyl-1,5-diphenyl-hydantoin	93879-71-1	DD01157
81	123	41	43	67	55	69	187	**310**	100	96	54	37	28	26	22	18	**2.14**	18	30	–	–	–	–	–	–	–	–	–	–	1	Zinc, bis[2,2-dimethyl-3-(2-methyl-2-propenyl)cyclopropyl]-	74842-24-3	NI32641
81	123	41	43	55	67	187	95	**310**	100	94	45	38	27	24	18	15	**3.30**	18	30	–	–	–	–	–	–	–	–	–	–	1	Zinc, bis[2,2-dimethyl-3-(2-methyl-2-propenyl)cyclopropyl]-, [cis(cis)]-	74810-57-4	NI32642
129	55	147	83	67	41	82	54	**310**	100	49	48	28	24	22	17	15	**0.30**	18	30	4	–	–	–	–	–	–	–	–	–	–	Dicyclohexyl adipate	849-99-0	IC03845
129	111	147	55	83	101	128	67	**310**	100	25	24	20	17	12	10	10	**0.00**	18	30	4	–	–	–	–	–	–	–	–	–	–	Dicyclohexyl adipate	849-99-0	NI32643
129	41	147	83	111	67	28	101	**310**	100	31	29	28	21	20	15	12	**0.12**	18	30	4	–	–	–	–	–	–	–	–	–	–	Dicyclohexyl adipate	849-99-0	IC03846
43	96	213	152	212	109	98	41	**310**	100	50	43	36	35	35	32	32	**0.50**	18	30	4	–	–	–	–	–	–	–	–	–	–	Dihydrodecarboxyportentol acetate		LI06829
139	112	310	197	154	209	179	182	**310**	100	92	20	17	15	12	11	6		18	30	4	–	–	–	–	–	–	–	–	–	–	9,12-Dioxo-octadec-trans-10-enoic acid		LI06830
169	99	129	125	41	168	141	143	**310**	100	71	68	67	57	44	40	39	**0.00**	18	30	4	–	–	–	–	–	–	–	–	–	–	4,4-Ethylenedioxy-2-hexadecen-15-olide		DD01158
85	67	101	84	55	135	93	79	**310**	100	17	16	15	15	14	14	12	**0.40**	18	30	4	–	–	–	–	–	–	–	–	–	–	Methyl 12-(tetrahydro-2-pyranyloxy)-4-dodecynoate		NI32644
310	171	170	81	96	118	295	215	**310**	100	83	60	50	47	34	23	17		19	18	4	–	–	–	–	–	–	–	–	–	–	1,3,5-Cyclohexanetrione, 6-(2H-1-benzopyran-2-ylidene)-2,2,4,4-tetramethyl-	65653-67-0	NI32645
310	295	267	255	239	213	283	282	**310**	100	73	27	20	11	11	10	8		19	18	4	–	–	–	–	–	–	–	–	–	–	Hortiolone	62395-58-8	NI32646
310	295	165	151	115	77	251	178	**310**	100	12	9	9	9	7	3	3		19	18	4	–	–	–	–	–	–	–	–	–	–	2-(2′-Hydroxy-4′-methoxyphenyl)-7-methoxy-5-(E)-propenylbenzofuran		MS07232
310	121	293	294	280	165	309	261	**310**	100	20	16	8	5	5	3	3		19	18	4	–	–	–	–	–	–	–	–	–	–	3-Hydroxymethyl-2-(4′-hydroxyphenyl)-7-methoxy-5-(E)-propenylbenzofuran		MS07233
191	184	237	178	38	310	165	238	**310**	100	88	84	72	54	47	40	25		19	18	4	–	–	–	–	–	–	–	–	–	–	9,9-Bis(methoxycarbonylmethyl)fluorene		IC03847
249	277	292	178	310	165	278	250	**310**	100	95	63	36	33	29	22	22		19	18	4	–	–	–	–	–	–	–	–	–	–	Phenanthro[1,2-b]furan-10,11-dione, 6,7,8,9-tetrahydro-1-(hydroxymethyl)-6,6-dimethyl-	76843-23-7	NI32647
121	174	77	160	105	175	148	263	**310**	100	99	94	88	73	69	33	27	**0.00**	19	18	4	–	–	–	–	–	–	–	–	–	–	1-Propanone, 3-methoxy-1-(4-methoxy-5-benzofuranyl)-3-phenyl-	64280-21-3	NI32648
279	261	233	280	165	178	251	278	**310**	100	90	66	52	50	46	43	42	**25.00**	19	18	4	–	–	–	–	–	–	–	–	–	–	Tanshinone II-b	17397-93-2	NI32649
310	249	277	177	292	165	295	267	**310**	100	99	73	46	39	28	26	25		19	18	4	–	–	–	–	–	–	–	–	–	–	Tanshinone II-b	17397-93-2	NI32650
310	295	267	251	241	225	277	253	**310**	100	33	13	10	10	10	9	9		19	18	4	–	–	–	–	–	–	–	–	–	–	Tetramethyldipyranocoumarin		NI32651
58	41	42	198	44	180	112	43	**310**	100	78	67	39	38	21	20	18	**8.01**	19	22	–	2	1	–	–	–	–	–	–	–	–	10H-Phenothiazine, 10-[(1-methyl-3-piperidinyl)methyl]-	60-89-9	NI32652
310	111	112	58	199	212	96	41	**310**	100	92	85	80	66	65	61	56		19	22	–	2	1	–	–	–	–	–	–	–	–	10H-Phenothiazine, 10-[(1-methyl-3-piperidinyl)methyl]-	60-89-9	NI32653
295	310	253	43	296	118	211	134	**310**	100	90	87	62	53	34	26	23		19	22	2	2	–	–	–	–	–	–	–	–	–	2,2-Bis(4-acetamidophenyl)propane		IC03848
310	185	71	309	279	69	311	184	**310**	100	53	53	33	33	26	25	25		19	22	2	2	–	–	–	–	–	–	–	–	–	Ajmalan-10,17-diol, 19,20-didehydro-, (17R,19E)-	70509-79-4	NI32654
105	77	188	51	134	135	106	176	**310**	100	47	15	11	10	9	9	8	**5.00**	19	22	2	2	–	–	–	–	–	–	–	–	–	N,N′-Dibenzoyl-1,5-diaminopentane		LI06831
281	180	310	309	208	155	154	169	**310**	100	69	44	31	19	11	11	9		19	22	2	2	–	–	–	–	–	–	–	–	–	19,20-Dihydrodeformyl-talbotine acid lactone		LI06832
310	143	129	154	157	130	128	111	**310**	100	67	56	44	40	33	24	23		19	22	2	2	–	–	–	–	–	–	–	–	–	3,10-Dioxoquebrachamine		LI06833
180	310	42	130	43	144	77	166	**310**	100	40	30	29	25	17	11	4		19	22	2	2	–	–	–	–	–	–	–	–	–	1-(7-Hydroxy-1,2,3,4,6,7-hexahydro-isoquinolin-6-one-3-yl)-2-(3-indolyl)-ethane		LI06834
84	310	43	42	118	253	167	185	**310**	100	63	47	38	22	18	18	17		19	22	2	2	–	–	–	–	–	–	–	–	–	N-Isopropyl-4-(N-phenylacetamido)-acetanilide		IC03849
98	82	96	97	310	110	143	171	**310**	100	91	62	48	40	23	17	16		19	22	2	2	–	–	–	–	–	–	–	–	–	Spiro[8-methyl-8-azabicyclo[3.2.1]octane-3,3′-(5-methyl-1,4-dihydro-1-oxo 2H-pyrano[4,3-b]indole)]		DD01159
310	309	225	186	311	308	253	198	**310**	100	29	23	22	21	21	19	18		19	23	3	–	–	–	–	–	–	–	–	1	–	2-Hydroxy-estrone methylboronate		NI32655
73	222	223	45	74	59	295	195	**310**	100	71	16	15	8	8	5	5	**4.26**	19	26	–	–	–	–	–	–	–	2	–	–	–	9H-Fluoren-9-ylidene, bis(trimethylsilyl)-	7351-45-3	NI32656
73	41	43	69	55	111	81	109	**310**	100	78	74	62	55	50	38	30	**0.00**	19	34	3	–	–	–	–	–	–	–	–	–	–	2,4-Dodecadienoic acid, 11-methoxy-3,7,11-trimethyl-, isopropyl ester, (E,E)-	40596-69-8	NI32657
73	110	43	69	81	109	41	95	**310**	100	33	27	26	25	22	19	18	**0.00**	19	34	3	–	–	–	–	–	–	–	–	–	–	2,4-Dodecadienoic acid, 11-methoxy-3,7,11-trimethyl-, isopropyl ester, (E,E)-	40596-69-8	NI32658
99	139	127	167	254	207	114	222	**310**	100	21	16	5	3	3	3	2	**0.00**	19	34	3	–	–	–	–	–	–	–	–	–	–	Methyl 9,10-methylene-12-oxo-heptadecanoate		LI06835
99	165	207	153	222	254	239	197	**310**	100	18	16	16	11	9	9	4	**0.00**	19	34	3	–	–	–	–	–	–	–	–	–	–	Methyl 10,11-methylene-12-oxo-heptadecanoate		LI06836
97	125	185	181	168	153	225	253	**310**	100	97	68	58	42	37	18	5	**0.00**	19	34	3	–	–	–	–	–	–	–	–	–	–	Methyl 11,12-methylene-9-oxo-heptadecanoate		LI06837
144	55	98	310	41	71	199	67	**310**	100	99	93	89	81	69	68	60		19	34	3	–	–	–	–	–	–	–	–	–	–	Oxacyclotetradec-10-en-2-one, 13-methoxy-14-pentyl-, [13S-(10Z,13R*,14R*)]-	75299-45-5	NI32659

MASS TO CHARGE RATIOS								M.W.	INTENSITIES								Parent	C	H	O	N	S	F	Cl	Br	I	Si	P	B	X	COMPOUND NAME	CAS Reg No	No
310	17	311	155	77	73	63	64	**310**	100	28	22	17	17	17	16	15		20	14	–	4	–	–	–	–	–	–	–	–	–	**1,4-Bis-(2-benzimidazolyl)benzene**		**IC03850**
310	311	155	269	202	309	44	114	**310**	100	22	13	9	7	6	6	5		20	14	–	4	–	–	–	–	–	–	–	–	–	**3,9-Dimethyldipyrido[3,2-a:3',2'-h]phenazine**		**NI32660**
310	309	155	311	65	44	77	308	**310**	100	72	22	21	7	7	6	5		20	14	–	4	–	–	–	–	–	–	–	–	–	**11H-Isoindolo[2,1-a]benzimidazole, 11-(2-aminophenylimino)-**		**IC03851**
267	268	282	310	185	283	141	165	**310**	100	69	61	25	25	13	10	9		20	22	3	–	–	–	–	–	–	–	–	–	–	**Aegyptinone A**	121704-42-5	**DD01160**
73	310	208	219	209	295	203	167	**310**	100	75	72	59	45	28	20	20		20	26	1	–	–	–	–	–	–	1	–	–	–	**trans-1,4,5,6,7,7a-Hexahydro-2-phenyl-7a-[2-(trimethylsilyl)ethenyl]-2H-inden-1-one**		**DD01161**
310	73	219	295	75	311	211	233	**310**	100	59	41	31	30	27	25	20		20	26	1	–	–	–	–	–	–	1	–	–	–	**(E,E)-1-(2-Styryl-1-cyclohexenyl)-3-(trimethylsilyl)-2-propen-1-one**		**DD01162**
43	310	166	144	44	157	41	311	**310**	100	96	87	73	73	59	42	29		20	26	1	2	–	–	–	–	–	–	–	–	–	**Ajmalan-17-ol, (2α,17R)-**	56142-96-2	**NI32661**
43	310	60	45	182	144	42	41	**310**	100	99	99	99	68	68	67	64		20	26	1	2	–	–	–	–	–	–	–	–	–	**Ajmalan-17-ol, (17R)-**	16641-67-1	**NI32662**
310	144	182	157	183	198	131	166	**310**	100	40	39	25	22	14	14	12		20	26	1	2	–	–	–	–	–	–	–	–	–	**Ajmalan-17-ol, (17R)-**	16641-67-1	**LI06838**
149	167	57	43	41	279	104	76	**310**	100	45	40	36	34	33	28	24	**0.00**	20	26	1	2	–	–	–	–	–	–	–	–	–	**Aspidofractinine-3-methanol, (2α,3β,5α)-**	2656-44-2	**NI32663**
109	282	310	124	188	108	41	187	**310**	100	59	39	39	29	19	19	14		20	26	1	2	–	–	–	–	–	–	–	–	–	**Aspidofractinine, 17-methoxy-**	2122-34-1	**NI32664**
109	282	310	124	188	283	41	108	**310**	100	65	40	38	31	22	21	15		20	26	1	2	–	–	–	–	–	–	–	–	–	**Aspidofractinine, 17-methoxy-**	2122-34-1	**NI32665**
310	240	281	124	311	238	125	241	**310**	100	79	46	35	20	17	17	15		20	26	1	2	–	–	–	–	–	–	–	–	–	**Aspidospermidine, 1,2-didehydro-17-methoxy-**	2912-26-7	**NI32666**
122	152	153	144	167	154	135	310	**310**	100	52	48	35	33	30	30	26		20	26	1	2	–	–	–	–	–	–	–	–	–	**Dihydro-anhydrovobasindiol**		**LI06839**
310	309	281	42	29	279	41	31	**310**	100	77	63	54	48	36	30	28		20	26	1	2	–	–	–	–	–	–	–	–	–	**Eburnamenine-14-methanol, 14,15-dihydro-, (3α,4β,16α)-**		**MS07234**
310	281	279	240	208				**310**	100	62	38	28	28					20	26	1	2	–	–	–	–	–	–	–	–	–	**Eburnamenine-14-methanol, 14,15-dihydro-, (3α,4β,16α)-**		**LI06840**
310	240	309	279	281	208	311	42	**310**	100	69	49	40	38	30	22	20		20	26	1	2	–	–	–	–	–	–	–	–	–	**Eburnamenine-14-methanol, 14,15-dihydro-, (3α,14α,16α)-**	23173-24-2	**MS07235**
310	309	281	42	29	279	41	31	**310**	100	77	63	59	49	36	31	30		20	26	1	2	–	–	–	–	–	–	–	–	–	**Eburnamenine-14-methanol, 14,15-dihydro-, (3α,14β,16α)-**	23173-27-5	**NI32667**
282	254	134	130	310	163	190	281	**310**	100	95	71	64	62	58	47	32		20	26	1	2	–	–	–	–	–	–	–	–	–	**13a-Ethyl-2,3,5,6,6a,11,12,13a,13b-decahydro-1H-cyclopenta[d,e]indolo[3,2-c]quinolizine-11-carbaldehyde**		**MS07236**
311	312	310	309	313	281	63	61	**310**	100	24	21	12	3	2	1	1		20	26	1	2	–	–	–	–	–	–	–	–	–	**Ibogamine, 12-methoxy-**	83-74-9	**NI32668**
136	135	310	149	122	225	155	41	**310**	100	73	53	47	38	29	18	18		20	26	1	2	–	–	–	–	–	–	–	–	–	**Ibogamine, 12-methoxy-**	83-74-9	**NI32669**
122	310	279	280	180	248	311	249	**310**	100	99	51	33	32	29	25	22		20	26	1	2	–	–	–	–	–	–	–	–	–	**N-Methyl-4,5-dihydro-cyano-pleiocarpaminol**		**LI06841**
55	74	69	278	41	83	43	97	**310**	100	99	99	96	93	89	88	69	**16.04**	20	38	2	–	–	–	–	–	–	–	–	–	–	**Cyclopropanedecanoic acid, 2-hexyl-, methyl ester**	5965-63-9	**NI32670**
55	69	74	41	83	97	43	87	**310**	100	85	68	68	66	59	58	49	**1.90**	20	38	2	–	–	–	–	–	–	–	–	–	–	**Cyclopropaneoctanoic acid, 2-octyl-, methyl ester**	10152-62-2	**NI32672**
55	41	69	43	74	278	83	97	**310**	100	89	66	62	61	58	48	41	**8.99**	20	38	2	–	–	–	–	–	–	–	–	–	–	**Cyclopropaneoctanoic acid, 2-octyl-, methyl ester**	10152-62-2	**NI32671**
55	41	69	43	278	74	83	97	**310**	100	89	66	62	58	58	48	41	**9.44**	20	38	2	–	–	–	–	–	–	–	–	–	–	**Cyclopropaneoctanoic acid, 2-octyl-, methyl ester, cis-**	3971-54-8	**NI32673**
55	41	69	43	279	74	83	97	**310**	100	74	61	55	46	45	44	38	**8.44**	20	38	2	–	–	–	–	–	–	–	–	–	–	**Cyclopropaneoctanoic acid, 2-octyl-, methyl ester, trans-**	5135-07-9	**NI32674**
41	55	43	69	278	83	97	96	**310**	100	93	82	58	50	49	44	44	**16.04**	20	38	2	–	–	–	–	–	–	–	–	–	–	**Cyclopropanepentanoic acid, 2-undecyl-, methyl ester**	10406-55-0	**NI32675**
41	43	55	29	74	57	69	27	**310**	100	88	60	50	35	28	25	23	**1.04**	20	38	2	–	–	–	–	–	–	–	–	–	–	**Cyclopropanepentanoic acid, 2-undecyl-, methyl ester, trans-**	42199-20-2	**NI32676**
57	43	71	239	267	55	85	41	**310**	100	74	72	70	50	49	41	35	**0.00**	20	38	2	–	–	–	–	–	–	–	–	–	–	**1,1,2,2-Ethanetetracarboxylic acid, tetraethyl ester**		**DO01094**
67	81	55	82	95	96	41	69	**310**	100	79	69	62	60	49	49	43	**2.14**	20	38	2	–	–	–	–	–	–	–	–	–	–	**Ethanol, 2-(9,12-octadecadienyloxy)-, (Z,Z)-**	17367-08-7	**NI32677**
311	279	97	277	111	143	171	157	**310**	100	82	61	51	47	40	39	38	**0.00**	20	38	2	–	–	–	–	–	–	–	–	–	–	**Methyl cis-9,10-methyleneoctadecanoate**		**NI32678**
109	189	69	153	95	136	123	55	**310**	100	95	89	82	77	59	53	53	**0.00**	20	38	2	–	–	–	–	–	–	–	–	–	–	**1-Naphthalenepropanol, α-ethyldecahydro-4-hydroxy-α,2,5,5,8a-pentamethyl-, [1S-[1α(R*),2α,4α,4aβ,8aα]]-**	57397-30-5	**NI32679**
55	41	69	43	74	83	97	87	**310**	100	76	65	59	58	51	41	39	**5.64**	20	38	2	–	–	–	–	–	–	–	–	–	–	**10-Nonadecenoic acid, methyl ester**	56599-83-8	**NI32680**
311	279	277	97	111	309	143	125	**310**	100	58	39	39	31	24	21	21	**0.00**	20	38	2	–	–	–	–	–	–	–	–	–	–	**10-Nonadecenoic acid, methyl ester**	56599-83-8	**NI32681**
311	312	279	313	309	277	213	171	**310**	100	23	4	3	2	2	1	1	**0.00**	20	38	2	–	–	–	–	–	–	–	–	–	–	**10-Nonadecenoic acid, methyl ester**	56599-83-8	**NI32682**
71	41	67	79	55	81	31	29	**310**	100	74	45	34	30	27	26	26	**0.00**	20	38	2	–	–	–	–	–	–	–	–	–	–	**9,12-Octadecadienal, dimethyl acetal**	1599-51-5	**NI32683**
71	68	278	82	96	110	279	124	**310**	100	10	8	8	6	4	2	2	**0.00**	20	38	2	–	–	–	–	–	–	–	–	–	–	**9,12-Octadecadienal, dimethyl acetal**	1599-51-5	**LI06842**
55	69	41	83	88	97	101	43	**310**	100	74	69	65	63	53	51	51	**13.91**	20	38	2	–	–	–	–	–	–	–	–	–	–	**9-Octadecenoic acid (Z)-, ethyl ester**	111-62-6	**WI01400**
43	55	69	264	97	88	83	41	**310**	100	68	67	59	59	53	52	50	**22.70**	20	38	2	–	–	–	–	–	–	–	–	–	–	**9-Octadecenoic acid (Z)-, ethyl ester**	111-62-6	**NI32684**
55	69	88	41	83	43	97	84	**310**	100	71	64	63	59	49	48	46	**4.50**	20	38	2	–	–	–	–	–	–	–	–	–	–	**9-Octadecenoic acid (Z)-, ethyl ester**	111-62-6	**NI32685**
97	311	111	60	125	89	251	83	**310**	100	96	96	91	75	70	67	65	**0.00**	20	38	2	–	–	–	–	–	–	–	–	–	–	**Oleyl acetate**		**NI32686**
267	83	173	81	95	55	69	116	**310**	100	91	90	59	58	48	40	37	**23.78**	20	39	–	–	–	–	–	–	–	–	1	–	–	**Phosphine, bis(5-methyl-2-isopropylcyclohexyl)-, [1α(1'R*,2'S*,5'R*,2β,5α)]-**	62238-19-1	**NI32687**
105	179	77	76	131	89	106	75	**310**	100	60	52	17	12	9	8	8	**7.01**	21	14	1	2	–	–	–	–	–	–	–	–	–	**Benzo[f]quinoline-2-carbonitrile, 1-benzoyl-1,2-dihydro-**	54964-81-7	**NI32688**
253	252	281	254	282	310	205	126	**310**	100	80	55	55	53	44	43	35		21	14	1	2	–	–	–	–	–	–	–	–	–	**2H-Inden-2-one, 3-diazo-1,3-dihydro-1,1-diphenyl-**	54964-80-6	**NI32689**
295	41	238	43	296	223	39	251	**310**	100	37	29	26	23	8	8	7	**6.02**	21	26	2	–	–	–	–	–	–	–	–	–	–	**6H-Dibenzo[b,d]pyran-1-ol, 6,6,9-trimethyl-3-pentyl-**	521-35-7	**NI32691**
295	296	238	310	239	223	297	224	**310**	100	23	13	10	6	6	4	3		21	26	2	–	–	–	–	–	–	–	–	–	–	**6H-Dibenzo[b,d]pyran-1-ol, 6,6,9-trimethyl-3-pentyl-**	521-35-7	**NI32690**
227	310	174	173	242	147	228	41	**310**	100	90	33	24	23	23	18	16		21	26	2	–	–	–	–	–	–	–	–	–	–	**19-Norpregna-1,3,5(10)-trien-20-yn-17-ol, 3-methoxy-, (17α)-**	72-33-3	**NI32694**
227	310	284	174	242	228	199	160	**310**	100	61	40	27	20	20	18	16		21	26	2	–	–	–	–	–	–	–	–	–	–	**19-Norpregna-1,3,5(10)-trien-20-yn-17-ol, 3-methoxy-, (17α)-**	72-33-3	**NI32693**
227	310	174	284	147	160	173	199	**310**	100	40	39	36	31	30	29	27		21	26	2	–	–	–	–	–	–	–	–	–	–	**19-Norpregna-1,3,5(10)-trien-20-yn-17-ol, 3-methoxy-, (17α)-**	72-33-3	**NI32692**
105	239	77	240	106	57	51	41	**310**	100	49	19	8	6	4	3	3	**1.14**	21	26	2	–	–	–	–	–	–	–	–	–	–	**4-Octylphenyl benzoate**		**IC03852**
295	296	310	223	311	297	237	224	**310**	100	40	26	14	11	9	6	6		21	26	2	–	–	–	–	–	–	–	–	–	–	**1-Pentyl-3-hydroxy-6,6,9-trimethyldibenzo[b,d]pyran**		**LI06843**

MASS TO CHARGE RATIOS								M.W.	INTENSITIES								Parent	C	H	O	N	S	F	Cl	Br	I	Si	P	B	X	COMPOUND NAME	CAS Reg No	No
43	311	310	159	171				310	100	30	28	26	23					21	26	2	–	–	–	–	–	–	–	–	–	–	Pregna-1,4,6-triene-3,20-dione	4192-93-2	NI32695
124	310	125	282	311	152	158	155	310	100	28	11	8	7	7	5	5		21	30	–	2	–	–	–	–	–	–	–	–	–	Aspidospermidine, 1-ethyl-	55162-76-0	NI32696
173	57	310	42	41	213	58	44	310	100	90	72	59	52	51	49	49		21	30	–	2	–	–	–	–	–	–	–	–	–	2,20-Cyclo-8,9-secoaspidospermidine, 3,9-dimethyl-, (2α,3β,5α,12β,19α,20R)-	56053-38-4	NI32697
169	43	51	71	58	41	56	59	310	100	84	60	54	50	40	37	27	12.00	21	42	1	–	–	–	–	–	–	–	–	–	–	11-Heneicosanone		LI06845
43	57	41	169	71	58	85	59	310	100	60	58	56	56	56	25	19	2.00	21	42	1	–	–	–	–	–	–	–	–	–	–	11-Heneicosanone		LI06844
43	57	41	71	58	169	55	29	310	100	63	58	57	57	48	42	42	2.88	21	42	1	–	–	–	–	–	–	–	–	–	–	11-Heneicosanone		AI02031
310	308	311	309	154	155	280	154.5	310	100	35	25	25	16	12	11	11		22	14	–	–	1	–	–	–	–	–	–	–	–	5-Phenylbenzo[b]naphtho[1,2-d]thiophene		LI06846
310	309	311	155	103	104	78	77	310	100	83	23	22	20	14	13	13		22	18	–	2	–	–	–	–	–	–	–	–	–	2,3-Diphenyl-6,7-dimethylquinoxaline	13362-56-6	NI32698
233	310	77	51	155	311	234	52	310	100	95	72	43	30	26	20	15		22	18	–	2	–	–	–	–	–	–	–	–	–	Pyridine, 1,4-dihydro-1-phenyl-4-(1-phenyl-4(1H)-pyridinylidene)-	27802-00-2	NI32699
310	207	91	121	206	282	120	165	310	100	53	24	23	22	20	20	12		22	18	–	2	–	–	–	–	–	–	–	–	–	1-(4-Tolyl)-4-phenyl-5H-2,3-benzodiazepine		NI32700
295	310	57	296	217	105	41	311	310	100	46	34	26	20	19	13	12		22	30	1	–	–	–	–	–	–	–	–	–	–	1-Phenyl-1-(4-hydroxy-2,6-di-tert-butylphenyl)-ethane		IC03853
57	43	71	85	41	55	99	56	310	100	79	74	57	43	32	19	18	1.90	22	46	–	–	–	–	–	–	–	–	–	–	–	Docosane	629-97-0	NI32702
43	57	71	41	85	55	56	69	310	100	91	54	48	39	29	15	13	1.47	22	46	–	–	–	–	–	–	–	–	–	–	–	Docosane	629-97-0	AI02032
310	85	71	57	311	43	127	113	310	100	23	23	23	22	10	8	8		22	46	–	–	–	–	–	–	–	–	–	–	–	Docosane	629-97-0	NI32701
71	43	57	70	85	55	41	267	310	100	95	86	55	43	29	28	22	0.43	22	46	–	–	–	–	–	–	–	–	–	–	–	Heneicosane, 4-methyl-		LI06847
43	57	85	71	84	41	55	56	310	100	87	79	49	45	27	24	19	0.32	22	46	–	–	–	–	–	–	–	–	–	–	–	Heneicosane, 5-methyl-		LI06848
57	71	85	84	43	41	55	69	310	100	74	69	60	58	39	36	28	0.40	22	46	–	–	–	–	–	–	–	–	–	–	–	Octadecane, 2,6,10,14-tetramethyl-	54964-82-8	LI06849
58	71	86	113	99	155	85	57	310	100	99	76	32	32	28	23	22	9.52	22	46	–	–	–	–	–	–	–	–	–	–	–	Octadecane, 2,6,10,14-tetramethyl-	54964-82-8	NI32703
295	18	296	147.5	310	155	132.5	17	310	100	35	24	23	18	9	7	6		23	18	1	–	–	–	–	–	–	–	–	–	–	13,13-Dimethyl-13H-dibenzo[b,i]xanthene		IC03854
310	233	309	205	203	204	202	191	310	100	56	45	32	18	16	15	12		23	18	1	–	–	–	–	–	–	–	–	–	–	4H-Pyran, 2,4,6-triphenyl-	801-06-9	NI32704
310	203	311	115	178	202	179	91	310	100	29	26	25	20	17	15	14		24	22	–	–	–	–	–	–	–	–	–	–	–	Benzene, 1,4-bis[2-(2-methylphenyl)vinyl]-	13280-61-0	NI32705
141	310	115	142	168	167	154	153	310	100	70	21	12	10	9	8	7		24	22	–	–	–	–	–	–	–	–	–	–	–	Naphthalene, 1,1′-(1,4-butanediyl)bis-	29571-17-3	NI32706
59	220	247	311	145	160	128	156	311	100	38	30	28	23	21	18	16		9	13	1	1	5	–	–	–	–	–	–	–	–	2,4,6,8,9-Pentathiatricyclo[3.3.1.1^{3,7}]decane-1-carboxamide, 3,5,7-trimethyl-	57289-33-5	NI32707
120	55	108	227	119	93	42	311	311	100	98	74	65	30	30	14	13		10	10	5	1	1	–	–	–	–	–	–	–	1	Tricarbonyl(η-N,N-dimethylsulphamoylcyclopentadienyl)manganese		NI32708
211	226	44	195	73	227	69	135	311	100	75	50	33	29	23	23	22	8.50	10	26	4	1	–	–	–	–	–	2	1	–	–	Phosphonic acid, [1-(acetylamino)ethyl]-, bis(trimethylsilyl) ester	55108-82-2	NI32709
211	226	73	227	225	296	135	133	311	100	60	43	33	24	23	23	20	18.51	10	26	4	1	–	–	–	–	–	2	1	–	–	Phosphonic acid, [2-(acetylamino)ethyl]-, bis(trimethylsilyl) ester	55108-88-8	NI32710
129	128	115	127	156	130	63	116	311	100	49	45	23	19	15	11	10	0.00	11	10	–	3	–	–	–	–	1	–	–	–	–	exo-5-Azido-anti-7-iodobenzonorbornene		LI06850
227	57	226	43	41	55	142	69	311	100	68	61	52	52	38	34	32	0.00	11	16	1	1	–	7	–	–	–	–	–	–	–	Heptylamine, heptafluorobutanoyl-		MS07237
59	60	84	56	55	57	85	58	311	100	38	25	19	15	13	11	9	0.00	11	21	7	1	1	–	–	–	–	–	–	–	–	Desulphoglucoconringiin		MS07238
86	44	211	73	75	43	135	45	311	100	65	28	26	12	12	11	9	1.60	11	30	3	1	–	–	–	–	–	2	1	–	–	Phosphonic acid, (1-aminopentyl)-, bis(trimethylsilyl) ester	53044-32-9	NI32711
86	211	73	75	135	195	147	133	311	100	28	26	12	11	9	9	8	1.60	11	30	3	1	–	–	–	–	–	2	1	–	–	Phosphonic acid, (1-aminopentyl)-, bis(trimethylsilyl) ester	53044-32-9	NI32712
86	211	73	75	43	135	195	45	311	100	28	26	12	12	11	9	9	1.60	11	30	3	1	–	–	–	–	–	2	1	–	–	Phosphonic acid, (1-aminopentyl)-, bis(trimethylsilyl) ester	53044-32-9	NI32713
170	172	77	91	313	311	51	141	311	100	98	70	47	46	46	29	20		12	10	2	1	1	–	–	1	–	–	–	–	–	Benzenesulphonamide, N-(4-bromophenyl)-	16468-97-6	NI32714
43	87	59	44	313	311	183	185	311	100	31	29	29	26	26	26	24		12	10	2	1	1	–	–	1	–	–	–	–	–	Thiazole, 2-(4-bromophenyl)-4-(hydroxymethyl)-5-acetyl-	67387-05-7	NI32715
135	77	107	136	92	64	78	63	311	100	36	26	21	18	9	5	5	0.00	12	10	3	1	–	5	–	–	–	–	–	–	–	2-(4-Methoxyphenyl)-2-ketoethylamine PFP		MS07239
183	185	111	155	157	128	311	313	311	100	99	20	16	15	12	11	10		12	10	4	1	–	–	–	1	–	–	–	–	–	α-(2-Bromobenzamido)-β-(hydroxymethyl)butenolide		LI06851
183	185	111	128	155	157	311	313	311	100	99	42	32	27	26	20	19		12	10	4	1	–	–	–	1	–	–	–	–	–	α-(3-Bromobenzamido)-β-(hydroxymethyl)butenolide		LI06852
183	185	111	155	157	311	313	128	311	100	99	26	20	19	14	13	13		12	10	4	1	–	–	–	1	–	–	–	–	–	α-(4-Bromobenzamido)-β-(hydroxymethyl)butenolide		LI06853
311	313	113	174	237	222	100	235	311	100	99	48	27	23	23	23	21		12	10	4	1	–	–	–	1	–	–	–	–	–	Naphthalene, 1-bromo-2,6-dimethoxy-5-nitro-	25314-99-2	LI06854
313	311	113	174	237	222	100	63	311	100	99	46	27	23	23	23	21		12	10	4	1	–	–	–	1	–	–	–	–	–	Naphthalene, 1-bromo-2,6-dimethoxy-5-nitro-	25314-99-2	NI32716
173	202	172	58	103	104	42	59	311	100	99	74	74	47	33	33	30	0.00	12	17	–	1	–	3	–	1	–	–	–	–	–	Benzeneethylaminium, N,N,N-trimethyl-4-(trifluoromethyl)-, bromide	26536-36-7	NI32717
136	251	250	135	208	190	148	104	311	100	55	40	30	13	12	10	10	2.00	12	17	3	5	1	–	–	–	–	–	–	–	–	D-Xylitol, 1-C-(6-amino-9H-purin-9-yl)-2,5-anhydro-1-S-ethyl-1-thio-	40031-68-3	NI32718
179	135	164	44	150	43	108	57	311	100	65	58	55	50	50	33	30	3.00	12	17	5	5	–	–	–	–	–	–	–	–	–	Guanosine, N,N-dimethyl-	2140-67-2	NI32719
28	44	179	27	135	150	43	42	311	100	76	64	50	47	32	27	26	0.00	12	17	5	5	–	–	–	–	–	–	–	–	–	Guanosine, N,N-dimethyl-	2140-67-2	NI32721
179	43	44	135	150	164	180	193	311	100	94	86	69	53	51	14	3	0.27	12	17	5	5	–	–	–	–	–	–	–	–	–	Guanosine, N,N-dimethyl-	2140-67-2	NI32720
60	218	43	158	102	135	45	219	311	100	77	75	47	47	42	27	25	0.00	12	25	4	1	2	–	–	–	–	–	–	–	–	L-Idose, 3-acetamido-3,6-dideoxy-, diethyl mercaptal	3509-33-9	NI32722
77	140	51	141	233	94	92	50	311	100	49	42	38	24	14	14	11	0.00	13	13	4	1	2	–	–	–	–	–	–	–	–	Benzenemethanesulphonamide, N-(phenylsulphonyl)-	54833-64-6	NI32723
134	91	121	119	190	135	65	93	311	100	90	87	59	43	25	12	11	0.00	13	14	2	1	–	5	–	–	–	–	–	–	–	2-(2-Methoxyphenyl)-N-methylethylamine, pentafluoropropionyl-		MS07240
134	190	121	135	119	140	91	42	311	100	40	16	13	12	11	9	7	6.40	13	14	2	1	–	5	–	–	–	–	–	–	–	2-(3-Methoxyphenyl)-N-methylethylamine, pentafluoropropionyl-		MS07241
121	134	135	119	122	190	91	77	311	100	85	13	9	8	7	6	6	2.40	13	14	2	1	–	5	–	–	–	–	–	–	–	2-(4-Methoxyphenyl)-N-methylethylamine, pentafluoropropionyl-		MS07242
121	134	135	122	119	44	190	91	311	100	81	13	9	8	7	6	6	2.10	13	14	2	1	–	5	–	–	–	–	–	–	–	2-(4-Methoxyphenyl)-N-methylethylamine, pentafluoropropionyl-		NI32724
121	91	311	148	119	135	122	105	311	100	93	66	44	27	23	20	16		13	14	2	1	–	5	–	–	–	–	–	–	–	3-(2-Methoxyphenyl)propylamine, pentafluoropropionyl-		MS07243

MASS TO CHARGE RATIOS								M.W.	INTENSITIES								Parent	C	H	O	N	S	F	Cl	Br	I	Si	P	B	X	COMPOUND NAME	CAS Reg No	No
122	135	311	121	91	105	148	77	**311**	100	54	38	24	20	12	11	10		13	14	2	1	–	5	–	–	–	–	–	–	–	3-(3-Methoxyphenyl)propylamine, pentafluoropropionyl-		MS07244
121	311	148	122	77	147	91	78	**311**	100	23	10	8	6	5	5	4		13	14	2	1	–	5	–	–	–	–	–	–	–	3-(4-Methoxyphenyl)propylamine, pentafluoropropionyl-		MS07245
130	144	143	311	131	115	117	77	**311**	100	43	22	10	10	10	5	5		14	17	2	1	–	–	–	–	–	–	–	–	1	Butyric acid, 4-[(2-indol-3-ylethyl)selenyl]-	1148-13-6	NI32725
311	83	280	85	77	236	141	170	**311**	100	83	61	53	38	35	32	30		14	17	5	1	1	–	–	–	–	–	–	–	–	2-Oxa-6-azatricyclo[3.3.1.1^{3,7}]decane-4,8-diol, 6-(phenylsulphonyl)-, (1α,3β,4β,5α,7β,8α)-	50267-34-0	NI32726
280	220	281	232	236	174	190	132	**311**	100	29	17	17	12	10	9	9	**5.20**	14	21	5	3	–	–	–	–	–	–	–	–	–	1-Butanol, 2-[[4-tert-butyl-2,6-dinitrophenyl]amino]-	55702-39-1	NI32727
90	280	220	225	155	212	281	65	**311**	100	95	30	25	24	21	18	12	**5.20**	14	21	5	3	–	–	–	–	–	–	–	–	–	2-Butanol, 2-[[4-tert-butyl-2,6-dinitrophenyl]amino]-	55702-40-4	NI32728
43	69	140	85	41	182	183	57	**311**	100	61	40	32	27	26	23	14	**0.00**	14	24	3	1	–	3	–	–	–	–	–	–	–	L-Leucine, N-(trifluoroacetyl)-, 1-methylpentyl ester	57983-55-8	NI32729
41	55	97	43	57	69	56	126	**311**	100	99	82	59	45	29	18	16	**0.00**	14	24	3	1	–	3	–	–	–	–	–	–	–	Octanoic acid, 2-[(trifluoroacetyl)amino]-, butyl ester	56051-55-9	NI32730
296	192	73	297	45	298	193	164	**311**	100	62	60	27	19	11	10	7	**6.40**	14	25	3	1	–	–	–	–	–	2	–	–	–	Benzoic acid, 2-[(trimethylsilyl)amino]-3-[(trimethylsilyl)oxy]-, methyl ester	57397-09-8	NI32731
73	222	296	75	45	310	147	311	**311**	100	40	30	28	26	22	18	16		14	25	3	1	–	–	–	–	–	2	–	–	–	Furo[3,4-c]pyridine, 1,3-dihydro-6-methyl-1,7-bis(trimethylsiloxy)-	14857-19-3	NI32732
296	73	192	297	45	193	298	176	**311**	100	76	72	27	24	13	10	9	**0.00**	14	25	3	1	–	–	–	–	–	2	–	–	–	3-Hydroxy-anthranilic acid, bis(trimethylsilyl)-, methyl ester		NI32733
73	222	296	75	45	310	147	74	**311**	100	42	32	24	24	19	17	12	**10.00**	14	25	3	1	–	–	–	–	–	2	–	–	–	Pyridoxal, bis(trimethylsilyl)-		LI06855
75	101	157	155	77	76	283	63	**311**	100	72	71	71	71	64	53	51	**4.21**	15	10	–	3	–	–	–	1	–	–	–	–	–	Quinoline, 2-[(4-bromophenyl)azo]-	25117-51-5	NI32734
93	157	155	94	95	185	183	261	**311**	100	99	99	99	68	55	55	40	**5.00**	15	10	–	3	–	–	–	1	–	–	–	–	–	Quinoline, 7-[(4-bromophenyl)azo]-	25117-52-6	NI32735
82	200	56	152	282	72	112	254	**311**	100	58	50	46	42	37	33	27	**0.00**	15	25	2	3	1	–	–	–	–	–	–	–	–	8-(Diethylamino)-1-(ethylthio)-2,6,7-trimethyl-2,6-diazabicyclo[2.2.2]oct-7 ene-3,5-dione		DD01163
102	73	103	45	59	210	104	74	**311**	100	46	10	7	6	4	4	4	**0.00**	15	29	2	1	–	–	–	–	–	2	–	–	–	Benzeneethylamine, 2-methoxy-N-(trimethylsilyl)-4[(trimethylsilyl)oxy]-	55887-47-3	NI32736
221	73	147	206	148	222	178	75	**311**	100	82	43	41	24	21	18	10	**0.36**	15	29	2	1	–	–	–	–	–	2	–	–	–	2,6-Dimethyl-3,4-bis(trimethylsiloxymethyl)pyridine		NI32737
102	73	147	75	103	210	207	45	**311**	100	70	34	34	15	11	11	10	**1.20**	15	29	2	1	–	–	–	–	–	2	–	–	–	3-Methoxytyramine, bis(trimethylsilyl)-		MS07246
101	296	312	224	297	223	240		**311**	100	70	36	22	18	16	15		**2.00**	15	29	2	1	–	–	–	–	–	2	–	–	–	3-Methoxytyramine, bis(trimethylsilyl)-		NI32738
102	73	103	45	59	210	104	74	**311**	100	46	10	7	6	4	4	4	**0.00**	15	29	2	1	–	–	–	–	–	2	–	–	–	3-Methoxytyramine, bis(trimethylsilyl)-		NI32739
311	230	276	229	310	313	102	264	**311**	100	83	74	68	40	36	35	34		16	10	2	3	–	–	1	–	–	–	–	–	–	2-Chloro-3-[(E)-2-(4-nitrophenyl)ethenyl]quinoxaline		MS07247
135	77	311	118	43	44	91	136	**311**	100	31	28	25	17	17	17	11		16	13	–	3	2	–	–	–	–	–	–	–	–	6-Anilino-2-(3-tolyl)-1,3,5-thiadiazine-4-thione	94014-52-5	NI32740
188	162	160	146	238	77	132	117	**311**	100	62	48	34	31	30	26	24	**0.00**	16	22	3	1	–	–	1	–	–	–	–	–	–	Glycine, N-(chloroacetyl)-N-(2,6-diethylphenyl)-, ethyl ester	38727-55-8	NI32741
97	69	218	311	152	190	247	232	**311**	100	40	29	12	9	8	7	7		16	25	3	1	1	–	–	–	–	–	–	–	–	N-[(E)-2-Methyl-2-pentenoyl]bornane-10,2-sultam		DD01164
268	266	238	222	311	194	240	43	**311**	100	89	88	81	57	37	30	27		16	25	5	1	–	–	–	–	–	–	–	–	–	1-Ethyl-3-acetyl-3,5-bis(ethoxycarbonyl)-6-methyl-1,2,3,4-tetrahydropyridine		IC03855
57	41	176	160	40	146	188	77	**311**	100	48	42	39	28	17	16	16	**1.00**	17	26	2	1	–	–	1	–	–	–	–	–	–	Acetamide, N-(butoxymethyl)-2-chloro-N-(2,6-diethylphenyl)-	23184-66-9	NI32742
57	176	160	188	41	146	162	77	**311**	100	68	59	29	27	18	16	15	**1.85**	17	26	2	1	–	–	1	–	–	–	–	–	–	Acetamide, N-(butoxymethyl)-2-chloro-N-(2,6-diethylphenyl)-	23184-66-9	NI32743
160	81	115	234	128	240	79	116	**311**	100	53	41	39	30	25	19	14	**0.00**	18	17	2	1	1	–	–	–	–	–	–	–	–	Cyclohexanespiro-2′-[4-(2-naphthylthio)-1,3-oxazol-5(2H)-one]		DD01165
151	131	311	104	255	103	123	163	**311**	100	81	47	40	37	17	11	10		18	17	2	1	1	–	–	–	–	–	–	–	–	2,3-Dihydro-2-phenyl-5-propionyl-1,5-benzothiazepin-4(5H)-one		MS07248
311	312	313	278	282	155	154	266	**311**	100	60	14	13	10	9	9	6		18	17	4	1	–	–	–	–	–	–	–	–	–	Actinodaphnine		LI06856
252	222	279	208	311	236			**311**	100	92	54	36	21	8				18	17	4	1	–	–	–	–	–	–	–	–	–	Methyl 2-methoxycarbonyl-5-methyl-benzo[e]indolizine-1-acetate		LI06857
278	279	190	307	292	264	249	188	**311**	100	23	22	20	11	7	7	7	**0.00**	18	17	4	1	–	–	–	–	–	–	–	–	–	Spiro[1,3-dioxolo[4,5-g]isoquinoline-5(6H),2′-[2H]indene]-1′,3′-diol, 1′,3′,7,8-tetrahydro-	56196-57-7	NI32744
91	135	93	119	79	134	41	178	**311**	100	63	60	45	44	43	34	29	**2.40**	18	21	2	3	–	–	–	–	–	–	–	–	–	1-(6,6-Dimethylbicyclo[3.1.1]hept-2-en-2-ylmethyl)-4-phenyl-1,2,4-triazolidine-3,5-dione	74402-29-2	NI32745
41	93	69	91	79	119	67	107	**311**	100	76	67	66	60	57	44	43	**41.10**	18	21	2	3	–	–	–	–	–	–	–	–	–	10,10-Dimethyl-N-phenyl-3,4-diazatricyclo[5.2.1.0^{1,5}]decane-3,4-dicarboximide		NI32746
73	168	154	57	153	149	311		**311**	100	44	41	33	32	25	2			18	21	2	3	–	–	–	–	–	–	–	–	–	2-Phenyl-3-isobutenyl-5,6-dimethyl-3,6-dihydro-1,2-oxazine		LI06858
267	70	238	198	87	84	101	210	**311**	100	75	70	70	70	70	65	50	**20.00**	18	37	1	3	–	–	–	–	–	–	–	–	–	Azacyclotridecan-2-one, 1-[3-[(3-aminopropyl)amino]propyl]-	64414-60-4	NI32747
70	281	268	112	84	98	87	139	**311**	100	99	80	80	55	50	45	38	**24.00**	18	37	1	3	–	–	–	–	–	–	–	–	–	1,5-Diazacycloheptadecan-6-one, 5-(3-aminopropyl)-	64414-57-9	NI32748
226	311	165	211	193	296	269	268	**311**	100	51	28	20	16	10	10	8		19	21	3	1	–	–	–	–	–	–	–	–	–	Apogalantamine monomethyl ether, acetate		LI06859
91	268	92	43	65	269	178	136	**311**	100	49	16	15	14	11	8	8	**5.20**	19	21	3	1	–	–	–	–	–	–	–	–	–	3-Benzyloxy-4-propionylacetanilide		IC03856
105	108	91	77	107	79	204	51	**311**	100	65	54	52	43	43	28	14	**0.00**	19	21	3	1	–	–	–	–	–	–	–	–	–	Benzyl (5-phenyl-5-oxopentyl)carbamate		DD01166
105	118	161	133	267	283	106	238	**311**	100	96	80	50	14	10	7	2	**0.00**	19	21	3	1	–	–	–	–	–	–	–	–	–	5-Ethoxy-2,3-dimethyl-2,5-diphenyl-4-oxazolidinone	116927-95-8	DD01167
105	94	206	162	190	311	282	267	**311**	100	23	19	18	8	6	3	3		19	21	3	1	–	–	–	–	–	–	–	–	–	2-Methylformyl-N-benzoylnuphamine		LI06860
311	241	188	312	41	310	81	294	**311**	100	22	22	21	18	14	13	11		19	21	3	1	–	–	–	–	–	–	–	–	–	Morphinan-3,6-diol, 7,8-didehydro-4,5-epoxy-17-allyl-, (5α,6α)-	62-67-9	NI32749
311	296	42	44	255	312	310	253	**311**	100	51	42	32	30	22	20	15		19	21	3	1	–	–	–	–	–	–	–	–	–	Morphinan, 6,7,8,14-tetradehydro-4,5-epoxy-3,6-dimethoxy-17-methyl-, (5α)-	115-37-7	NI32752
42	311	296	152	58	165	139	153	**311**	100	72	50	37	32	30	28	25		19	21	3	1	–	–	–	–	–	–	–	–	–	Morphinan, 6,7,8,14-tetradehydro-4,5-epoxy-3,6-dimethoxy-17-methyl-, (5α)-	115-37-7	NI32750
311	255	44	42	28	296	310	253	**311**	100	63	56	55	43	26	24	21		19	21	3	1	–	–	–	–	–	–	–	–	–	Morphinan, 6,7,8,14-tetradehydro-4,5-epoxy-3,6-dimethoxy-17-methyl-, (5α)-	115-37-7	NI32751

MASS TO CHARGE RATIOS								M.W.	INTENSITIES								Parent	C	H	O	N	S	F	Cl	Br	I	Si	P	B	X	COMPOUND NAME	CAS Reg No	No
311	310	282	268	312	42	225	283	**311**	100	41	35	31	21	12	9	7		19	21	3	1	–	–	–	–	–	–	–	–	–	**Pronuciferine**	2128-60-1	**MS07249**
311	310	282	268	312	28	42	225	**311**	100	40	34	30	20	15	12	8		19	21	3	1	–	–	–	–	–	–	–	–	–	**Pronuciferine**	2128-60-1	**LI06861**
311	310	282	268	312	28	42	225	**311**	100	41	35	33	20	15	13	8		19	21	3	1	–	–	–	–	–	–	–	–	–	**Pronuciferine**	2128-60-1	**NI32753**
105	77	180	51	153	50	179	106	**311**	100	55	27	21	15	11	8	8	**5.00**	20	13	1	3	–	–	–	–	–	–	–	–	–	**1,7-Phenanthroline-8-carbonitrile, 7-benzoyl-7,8-dihydro-**	29924-57-0	**NI32754**
205	240	125	91	242	186	129	84	**311**	100	54	28	26	25	20	17	17	**8.91**	20	22	–	1	–	–	1	–	–	–	–	–	–	**Pyrrolidine, 1-[4-(4-chlorophenyl)-3-phenyl-2-butenyl]-**	91-82-7	**NI32755**
310	311	282	296	280	283	268	188	**311**	100	99	92	78	23	27	27	25		20	25	2	1	–	–	–	–	–	–	–	–	–	**6-Azaestra-1,3,5(10),8(14)-tetraen-17-one, 6-ethyl-3-methoxy-**		**LI06862**
43	114	28	241	156	96	84	139	**311**	100	42	34	17	17	17	16	15	**0.00**	20	41	1	1	–	–	–	–	–	–	–	–	–	**Hexadecanamide, N-isobutyl-**	54794-72-8	**NI32756**
87	43	42	57	41	55	100	69	**311**	100	25	23	22	22	17	13	8	**1.33**	20	41	1	1	–	–	–	–	–	–	–	–	–	**2-Nonadecanone, O-methyloxime**	36379-39-2	**NI32757**
87	43	57	42	41	100	55	29	**311**	100	25	23	23	20	17	17	8	**1.40**	20	41	1	1	–	–	–	–	–	–	–	–	–	**2-Nonadecanone, O-methyloxime**	36379-39-2	**LI06863**
311	73	30	72	43	86	60	114	**311**	100	97	87	55	50	37	33	32		20	41	1	1	–	–	–	–	–	–	–	–	–	**Octadecyl phenyl ether**		**IC03857**
310	311	312	155	156	309	213	180	**311**	100	94	33	24	22	18	14	14		21	13	–	1	1	–	–	–	–	–	–	–	–	**2-(9-Phenanthryl)benzothiazole**		**NI32758**
311	310	312	309	155	313	156		**311**	100	89	25	21	8	7	1			21	13	–	1	1	–	–	–	–	–	–	–	–	**5-Phenylbenzo[4,5]thieno[2,3-c]isoquinoline**		**NI32759**
43	311	283	312	254	141.5	127	284	**311**	100	53	18	12	11	10	10	4		21	13	2	1	–	–	–	–	–	–	–	–	–	**1-Benzopyrano[6,7-b]indole, 3-phenyl-2-oxo-**		**IC03858**
283	282	206	178	179	207	165	280	**311**	100	91	74	34	24	15	13	12	**0.00**	21	17	–	3	–	–	–	–	–	–	–	–	–	**Benzene, 1,1′,1″-(1-azido-1-propene-1,2,3-triyl)tris-**	55823-11-5	**NI32760**
296	297	77	134	51	294	164	192	**311**	100	24	10	9	6	5	4	4	**0.70**	21	17	–	3	–	–	–	–	–	–	–	–	–	**2,6-Diphenyl-4,4-dimethyl-1,4-dihydropyridine-3,5-dicarbonitrile**	15513-21-0	**NI32761**
192	91	311	193	89	165	312	65	**311**	100	89	32	15	15	14	10	9		21	17	–	3	–	–	–	–	–	–	–	–	–	**1H-1,2,3-Triazole, 4,5-diphenyl-1-benzyl-**	33471-66-8	**NI32762**
192	91	311	89	193	165	312	67	**311**	100	84	37	16	15	14	10	9		21	17	–	3	–	–	–	–	–	–	–	–	–	**1H-1,2,3-Triazole, 4,5-diphenyl-1-benzyl-**	33471-66-8	**LI06864**
98	28	218	99	44	55	83	41	**311**	100	18	13	11	9	8	7	6	**1.03**	21	29	1	1	–	–	–	–	–	–	–	–	–	**1-Piperidinepropanol, α-5-norbornen-2-yl-α-phenyl-**	514-65-8	**NI32763**
226	142	44	128	57	43	41	58	**311**	100	11	9	7	6	6	5	4	**2.00**	21	45	–	1	–	–	–	–	–	–	–	–	–	**1-Heptylamine, N,N-diheptyl-**	2411-36-1	**LI06865**
226	227	142	44	30	128	57	43	**311**	100	17	11	9	8	7	6	6	**2.00**	21	45	–	1	–	–	–	–	–	–	–	–	–	**1-Heptylamine, N,N-diheptyl-**	2411-36-1	**NI32764**
178	72	44	114	59	74	86	152	**311**	100	69	67	35	21	17	12	11	**6.00**	22	17	1	1	–	–	–	–	–	–	–	–	–	**N-(4-Methoxyphenyl)-2,3-diphenylcyclopropenone imine**		**MS07250**
310	311	265	280	279	42	282	178	**311**	100	82	43	40	35	24	17	16		23	21	–	1	–	–	–	–	–	–	–	–	–	**Spiro[5H-dibenzo[a,d]cycloheptene-5,1′-[1H]isoindole], 2′,3′,10,11-tetrahydro-2′-methyl-**	35926-76-2	**NI32765**
127	47	89	103	20	150	149	70	**312**	100	90	64	39	33	33	32	31	**0.00**	1	2	1	2	2	10	–	–	–	–	–	–	–	**N,N′-Bis(pentafluorosulphanyl)urea**		**MS07251**
101	103	147	119	117	235	105	233	**312**	100	69	31	31	31	17	14	13	**0.00**	3	–	–	–	–	3	4	1	–	–	–	–	–	**1-Bromo-2,3,3-trifluoro-2,4,4,4-tetrachlorobutane**		**LI06866**
314	316	235	312	318	233	154	156	**312**	100	91	62	47	31	31	31	24		6	3	–	–	–	–	–	3	–	–	–	–	–	**Benzene, 1,2,4-tribromo-**	615-54-3	**NI32766**
314	316	74	235	75	312	73	233	**312**	100	98	85	53	52	33	28	26		6	3	–	–	–	–	–	3	–	–	–	–	–	**Benzene, 1,2,4-tribromo-**	615-54-3	**NI32767**
74	75	79	81	314	316	73	235	**312**	100	68	46	45	38	37	32	29	**13.10**	6	3	–	–	–	–	–	3	–	–	–	–	–	**Benzene, 1,3,5-tribromo-**	626-39-1	**NI32768**
44	58	28	18	29	86	114	70	**312**	100	89	66	63	49	31	26	18	**0.00**	6	16	14	–	–	–	–	–	–	–	–	–	–	**Cyclohexanehexone octahydrate**		**LI06867**
312	103	256	268	131	187	200	228	**312**	100	77	72	65	44	31	19	10		7	4	4	–	–	3	–	–	–	–	–	–	1	**Rhodium, dicarbonyl(1,1,1-trifluoro-2,4-pentanedionato-O,O′)-**	18517-13-0	**NI32769**
172	59	118	143	144	284	87	228	**312**	100	88	86	71	68	65	51	33	**29.00**	8	2	6	–	–	–	–	–	–	–	–	–	2	**Cobalt, hexacarbonyl(acetylene)di-**	12553-66-1	**NI32770**
172	200	59	118	143	144	284	87	**312**	100	89	88	86	71	68	65	51	**29.00**	8	2	6	–	–	–	–	–	–	–	–	–	2	**Cobalt, hexacarbonyl(acetylene)di-**	12553-66-1	**LI06868**
229	135	297	199	150	312	121	120	**312**	100	68	56	52	41	28	21	21		8	17	–	–	–	–	–	1	–	–	–	–	1	**Dimethylcyclohexyltin bromide**		**LI06869**
282	225	72	73	297	147	283	226	**312**	100	87	62	51	46	37	28	22	**0.00**	8	21	3	2	1	–	–	–	–	2	1	–	–	**Phosphonic acid, [(carbonothioylhydrazino)methyl]-, bis(trimethylsilyl) ester**	56051-78-6	**NI32771**
73	232	158	143	131	233	157	45	**312**	100	94	44	38	32	30	29	28	**7.71**	8	24	–	–	–	–	–	–	–	4	–	–	1	**2,3,4,5-Tetrakis(dimethyl)selenacyclopentasilane**	83295-91-4	**NI32772**
283	253	284	255	225	195	223	285	**312**	100	35	22	19	15	14	14	13	**0.59**	8	24	5	–	–	–	–	–	–	4	–	–	–	**1,3,5,7-Tetraethyl-1-oxycyclotetrasiloxane**	75221-54-4	**NI32773**
256	284	312	131	132	187	103	200	**312**	100	72	55	29	23	19	15	12		9	12	2	–	–	3	–	–	–	–	–	–	1	**Rhodium, bis(η^2-ethene)(1,1,1-trifluoro-2,4-pentanedionato-O,O′)-**	69372-77-6	**NI32774**
59	131	44	268	204	145	99	127	**312**	100	39	37	31	23	21	18	16	**0.46**	9	12	2	–	5	–	–	–	–	–	–	–	–	**2,4,6,8,9-Pentathiatricyclo[3.3.1.1^{3,7}]decane-1-carboxylic acid, 3,5,7-trimethyl-**	57274-34-7	**NI32775**
59	161	58	189	45	312	117	248	**312**	100	13	11	10	8	7	7	6		9	12	2	–	5	–	–	–	–	–	–	–	–	**2,4,6,8,9-Pentathiatricyclo[3.3.1.1^{3,7}]decane-1-carboxylic acid, 3,5,7-trimethyl-**	57274-34-7	**NI32776**
283	31	253	225	284	223	195	255	**312**	100	45	37	36	35	28	25	24	**0.39**	9	24	6	–	–	–	–	–	–	3	–	–	–	**1,3,5-Triethyl-1,3,5-trimethoxycyclotrisiloxane**		**NI32777**
202	139	63	277	204	113	76	110	**312**	100	93	47	37	33	22	22	20	**12.01**	12	7	1	–	–	–	2	–	–	–	–	–	1	**Phenoxarsine, 2,10-dichloro-**	27796-60-7	**NI32778**
253	312	90	59	43	254	313	225	**312**	100	71	20	19	16	14	10	10		12	9	4	–	–	5	–	–	–	–	–	–	–	**4-Hydroxyphenylacetic acid methyl ester PFP**		**NI32779**
312	314	258	256	75	286	284	214	**312**	100	99	24	24	19	16	16	16		12	9	5	–	–	–	–	1	–	–	–	–	–	**Ethyl 8-bromo-7-hydroxy-4-oxochromene-2-carboxylate**	30113-86-1	**NI32780**
297	299	219	314	312	69	234	67	**312**	100	99	53	49	49	27	14	11		12	9	5	–	–	–	–	1	–	–	–	–	–	**2-Methyl-5,7-dihydroxy-6-acetyl-8-bromochromone**		**LI06870**
153	76	152	235	233	154	63	77	**312**	100	88	83	77	76	33	24	13	**0.99**	12	10	–	–	–	–	–	2	–	–	–	–	–	**Naphthalene, 1,8-bis(bromomethyl)-**	2025-95-8	**NI32781**
312	279	251	314	313	284	69	256	**312**	100	40	27	20	18	13	13	12		12	12	–	2	4	–	–	–	–	–	–	–	–	**2,2-Diethyldithio-dithiazolo[4′,5′-1,2][4″,5″-5,4]benzene**		**IC03859**
91	246	65	156	92	139	278	172	**312**	100	70	43	30	27	21	20	15	**1.36**	12	12	4	2	2	–	–	–	–	–	–	–	–	**Benzenesulphonic acid, 2-(phenylsulphonyl)hydrazide**	6272-36-2	**NI32782**
93	92	65	66	39	128	219	127	**312**	100	84	31	30	23	16	15	11	**0.00**	12	13	–	2	–	–	–	–	1	–	–	–	–	**4-(4-Methylpyridinium-1-ylmethyl)pyridine iodide**		**LI06871**
254	196	284	226	168	255	283	252	**312**	100	81	47	47	45	35	32	32	**28.00**	12	21	1	–	–	–	–	–	–	1	–	–	1	**Triethylsilyl(hydrido)(carbonyl)cyclopentadienylrhodium(III)**		**NI32783**
135	75	43	29	87	47	28	27	**312**	100	30	28	27	25	24	19	19	**4.00**	12	24	5	–	2	–	–	–	–	–	–	–	–	**Galactose, 5,6-O-ethylidene-, diethyl mercaptal**	56781-36-3	**NI32784**
135	43	47	29	87	75	28	62	**312**	100	58	57	57	49	47	45	37	**7.01**	12	24	5	–	2	–	–	–	–	–	–	–	–	**D-Mannose, 5,6-O-ethylidene-, diethyl mercaptal**	72150-28-8	**NI32785**

MASS TO CHARGE RATIOS								M.W.	INTENSITIES								Parent	C	H	O	N	S	F	Cl	Br	I	Si	P	B	X	COMPOUND NAME	CAS Reg No	No
45	27	29	43	47	62	28	75	**312**	100	79	64	63	59	48	41	39	**0.00**	12	24	5	–	2	–	–	–	–	–	–	–	–	Mannose, 4,6-O-ethylidene-, diethyl mercaptal	56701-10-1	NI32786
75	85	145	101	127	163	133	103	**312**	100	4	3	3	2	1	1	1	**0.00**	12	24	9	–	–	–	–	–	–	–	–	–	–	(1RS)-1,3,3-Trimethoxypropyl β-D-glucopyranoside		MS07252
115	43	116	230	58	224	130	89	**312**	100	66	40	18	16	10	10	9	**1.10**	13	10	3	–	–	–	–	–	–	–	–	–	1	Molybdenum, π-indenylmethyltricarbonyl-		MS07253
232	231	233	82	80	121	217	60	**312**	100	90	24	21	21	18	16	16	**0.00**	13	13	–	–	2	–	–	1	–	–	–	–	–	1,2-Benzodithiol-1-ium, 4,5,6,7-tetrahydro-3-phenyl-, bromide	55836-85-6	NI32787
184	128	43	127	141	77	185	169	**312**	100	47	35	25	20	18	16	12	**0.00**	13	13	1	–	–	–	–	–	1	–	–	–	–	Pyrylium, 2,4-dimethyl-6-phenyl-, iodide	32339-28-9	NI32788
184	128	141	169	43	127	185	115	**312**	100	44	32	26	26	20	17	17	**0.00**	13	13	1	–	–	–	–	–	1	–	–	–	–	Pyrylium, 2,6-dimethyl-4-phenyl-, iodide		LI06872
184	128	141	169	43	127	185	115	**312**	100	44	30	25	25	20	18	18	**0.00**	13	13	1	–	–	–	–	–	1	–	–	–	–	Pyrylium, 2,6-dimethyl-4-phenyl-, iodide	32339-29-0	NI32789
43	141	183	87	42	41	140	28	**312**	100	49	28	24	23	17	10	9	**0.20**	13	16	5	2	1	–	–	–	–	–	–	–	–	L-Cysteine, N-acetyl-S-[5-(acetylamino)-2-hydroxyphenyl]-	52372-86-8	NI32790
116	100	41	60	61	42	171	314	**312**	100	98	83	56	51	49	42	40	**39.59**	13	17	–	2	1	–	–	1	–	–	–	–	–	4,4,6-Trimethyl-2-[4-bromophenylamino(imino)]-5,6-dihydro-4H-1,3-thiazine	53004-49-2	NI32791
165	269	227	297	135	105	77	150	**312**	100	95	68	58	41	33	33	18	**1.00**	13	20	1	–	–	–	–	–	–	–	–	–	1	Tin, (3-benzoylpropyl)trimethyl-		LI06873
81	73	96	312	75	129	145	297	**312**	100	77	32	26	21	18	11	9		13	20	5	2	–	–	–	–	–	1	–	–	–	3,6-Epoxy-2H,8H-pyrimido[6,1-b][1,3]oxazocine-8,10(9H)-dione, 3,4,5,6-tetrahydro-11-methyl-4-[(trimethylsilyl)oxy]-, [3R-(3α,4β,6α)]-	41547-76-6	NI32792
58	225	312	180	253	100	243	59	**312**	100	63	31	14	10	7	7	6		13	23	3	2	–	3	–	–	–	–	–	–	–	Propyl N-(trifluoroacetoxy)-N(ϵ),N(ϵ)-dimethyllysine	91795-79-8	NI32793
254	196	284	226	168	252	282	312	**312**	100	60	46	44	41	38	13	4		13	25	–	–	–	–	–	–	–	1	–	–	1	Triethylsilyl(hydrido)(ethene)cyclopentadienylrhodium(III)		NI32794
139	52	75	156	111	140	53	77	**312**	100	80	70	65	65	63	51	40	**0.00**	14	10	4	–	–	–	2	–	–	–	–	–	–	4-Chlorobenzaldehyde diperoxide		LI06874
134	200	312	77	193	113	97	92	**312**	100	55	50	45	38	29	28	27		14	11	1	2	1	3	–	–	–	–	–	–	–	cis-1-(Phenylcarbamoyl)-3-(3-thienyl)-2-(trifluoromethyl)aziridine		DD01168
112	139	296	111	80	79	59	141	**312**	100	36	34	32	32	29	20	18	**0.00**	14	16	4	–	2	–	–	–	–	–	–	–	–	4H-Thiopyran-4-one, 2,6-dimethyl-, 1-oxide, dimer	55649-34-8	NI32795
80	112	140	79	123	141	100	111	**312**	100	98	51	51	48	39	37	33	**3.00**	14	16	4	–	2	–	–	–	–	–	–	–	–	4H-Thiopyran-4-one, 2,6-dimethyl-, 1-oxide, dimer	55649-34-8	LI06875
183	43	225	226	167	312	211	267	**312**	100	62	31	17	16	10	9	7		14	16	8	–	–	–	–	–	–	–	–	–	–	Acetoxy-(4-acetoxy-2,6-dimethoxy)-phenylacetic acid		MS07254
139	228	123	121	43	210	193	178	**312**	100	56	53	42	42	38	33	24	**0.37**	14	16	8	–	–	–	–	–	–	–	–	–	–	Methyl ($\pm$)-2-hydroxy-3-(3,3-diacetoxy-4-oxo-cyclohexa-1,5-dienyl)-propionate		NI32796
185	91	129	115	143	128	77	39	**312**	100	83	48	31	28	22	20	20	**2.20**	14	17	–	–	–	–	–	–	1	–	–	–	–	Bicyclo[2.2.2]octane, 1-iodo-4-phenyl-	55044-15-0	NI32797
110	123	95	67	111	79	105	77	**312**	100	23	19	11	9	9	8	8	**2.00**	14	17	1	–	1	–	–	1	–	–	–	–	–	1-[1-Bromo-2-(phenylthio)cycloprop-1-yl]cyclopentan-1-ol		MS07255
43	185	183	41	99	71	39	55	**312**	100	67	64	29	24	20	15	14	**0.00**	14	17	3	–	–	–	–	1	–	–	–	–	–	4-Bromophenacyl hexanoate		NI32798
119	175	174	118	78	77	65	91	**312**	100	85	80	75	72	40	36	30	**5.00**	14	21	4	2	–	–	–	–	–	–	1	–	–	Phosphonic acid, [3-(aminocarbonyl)-2-(4-methylphenyl)-2-aziridinyl]-, diethyl ester	33895-62-4	NI32799
297	295	32	293	57	241	296	105	**312**	100	81	49	44	43	39	36	36	**1.00**	14	24	–	–	–	–	–	–	–	–	–	–	1	Tin, trimethyl(2,2-dimethylpropyl)-		MS07256
193	297	73	298	45	194	137	59	**312**	100	96	84	25	24	16	14	11	**5.00**	14	24	4	–	–	–	–	–	–	2	–	–	–	Benzoic acid, 2,3-bis(trimethylsiloxy)-, methyl ester	27798-54-5	NI32801
297	193	298	299	73	312	194	79	**312**	100	40	25	10	9	7	6	5		14	24	4	–	–	–	–	–	–	2	–	–	–	Benzoic acid, 2,3-bis(trimethylsiloxy)-, methyl ester	27798-54-5	NI32800
297	298	299	312	281	313	267	73	**312**	100	24	9	6	4	2	2	2		14	24	4	–	–	–	–	–	–	2	–	–	–	Benzoic acid, 2,4-bis(trimethylsiloxy)-, methyl ester	27798-55-6	NI32802
297	298	73	299	45	59	281	267	**312**	100	25	20	11	6	6	6	5	**3.81**	14	24	4	–	–	–	–	–	–	2	–	–	–	Benzoic acid, 2,4-bis(trimethylsiloxy)-, methyl ester	27798-55-6	NI32803
297	298	312	299	267	79	281	73	**312**	100	21	13	10	5	4	3	3		14	24	4	–	–	–	–	–	–	2	–	–	–	Benzoic acid, 2,5-bis(trimethylsiloxy)-, methyl ester	27798-56-7	NI32804
297	298	299	281	312	73	265	77	**312**	100	25	11	5	4	4	3	3		14	24	4	–	–	–	–	–	–	2	–	–	–	Benzoic acid, 2,6-bis(trimethylsiloxy)-, methyl ester	27798-57-8	NI32805
193	73	312	194	45	313	281	195	**312**	100	46	33	17	10	9	5	5		14	24	4	–	–	–	–	–	–	2	–	–	–	Benzoic acid, 3,4-bis(trimethylsiloxy)-, methyl ester	27798-58-9	NI32806
193	312	313	194	73	314	195	281	**312**	100	62	16	16	15	6	5	4		14	24	4	–	–	–	–	–	–	2	–	–	–	Benzoic acid, 3,4-bis(trimethylsiloxy)-, methyl ester	27798-58-9	NI32807
312	297	313	265	314	298	299	281	**312**	100	32	25	13	10	7	4	4		14	24	4	–	–	–	–	–	–	2	–	–	–	Benzoic acid, 3,5-bis(trimethylsiloxy)-, methyl ester	27798-59-0	NI32808
73	268	297	283	312	252	223	45	**312**	100	48	42	35	32	29	29	28		14	24	4	–	–	–	–	–	–	2	–	–	–	Benzoic acid, 3-methoxy-4-[(trimethylsilyl)oxy]-, trimethylsilyl ester	2078-15-1	NI32809
73	297	312	267	223	253	282	126	**312**	100	82	55	54	49	37	33	28		14	24	4	–	–	–	–	–	–	2	–	–	–	Benzoic acid, 3-methoxy-4-[(trimethylsilyl)oxy]-, trimethylsilyl ester	2078-15-1	NI32810
297	73	312	267	223	253	282	298	**312**	100	76	67	64	53	41	35	25		14	24	4	–	–	–	–	–	–	2	–	–	–	Benzoic acid, 3-methoxy-4-[(trimethylsilyl)oxy]-, trimethylsilyl ester	2078-15-1	MS07257
297	73	298	40	299	45	180	44	**312**	100	42	26	17	9	5	4	4	**3.60**	14	24	4	–	–	–	–	–	–	2	–	–	–	Benzoic acid, 5-methoxy-2-[(trimethylsilyl)oxy]-, trimethylsilyl ester	55517-47-0	NI32811
312	226	70	112	153	69	213	167	**312**	100	84	77	69	51	49	48	46		14	24	4	4	–	–	–	–	–	–	–	–	–	(2S-cis)-N,N′-[(3,6-Dioxo-2,5-piperazinediyl)di-3,1-propanediyl]bisacetamide	21028-09-1	NI32812
70	112	312	166	95	252	69	152	**312**	100	87	70	70	62	46	40	34		14	24	4	4	–	–	–	–	–	–	–	–	–	N[5]-Acetyl-L-ornithine anhydride		LI06876
264	218	277	77	115	312	130	129	**312**	100	95	70	58	52	45	32	28		15	12	4	4	–	–	–	–	–	–	–	–	–	Cinnamaldehyde, (2,4-dinitrophenyl)hydrazone	1237-69-0	NI32813
264	218	277	115	312	77	130	219	**312**	100	94	70	52	45	38	32	24		15	12	4	4	–	–	–	–	–	–	–	–	–	Cinnamaldehyde, (2,4-dinitrophenyl)hydrazone	1237-69-0	LI06877
110	130	103	76	132	182	108	131	**312**	100	84	56	49	40	33	33	31	**24.00**	15	16	2	6	–	–	–	–	–	–	–	–	–	Ethyl 5,5a,6,11,11a,12-hexahydroquinoxalino[2,3-g]pteridine-4-carboxylate		LI06878
99	140	59	74	86	141	100	312	**312**	100	47	20	15	12	6	5	4		15	20	5	–	1	–	–	–	–	–	–	–	–	1,4-Dioxaspiro[4.5]decan-8-ol, p-toluenesulphonate	23511-05-9	NI32814
205	177	44	42	159	91	79	77	**312**	100	97	78	78	68	54	51	49	**1.79**	15	20	7	–	–	–	–	–	–	–	–	–	–	Trichothec-9-en-8-one, 12,13-epoxy-3,4,7,15-tetrahydroxy-, (3α,4β,7α)-	23282-20-4	NI32815
177	205	159	187	77	91	161	79	**312**	100	89	57	47	45	41	37	36	**0.00**	15	20	7	–	–	–	–	–	–	–	–	–	–	Trichothec-9-en-8-one, 12,13-epoxy-3,4,7,15-tetrahydroxy-, (3α,4β,7α)-	23282-20-4	NI32816
183	102	131	239	211	103	267	29	**312**	100	60	59	50	48	47	40	40	**17.58**	15	21	5	–	–	–	–	–	–	–	1	–	–	3-Phenylpropenoic acid, 2-(diethoxyphosphinyl)-, ethyl ester	13507-49-8	NI32817
209	73	312	179	210	103	297	45	**312**	100	56	36	25	17	13	10	9		15	28	3	–	–	–	–	–	–	2	–	–	–	2-(3-Hydroxy-4-methoxyphenyl)ethanol bis(trimethylsilyl) ether		MS07258
209	73	312	210	179	76	77	103	**312**	100	83	19	19	17	16	15	13		15	28	3	–	–	–	–	–	–	2	–	–	–	2-(4-Hydroxy-3-methoxyphenyl)ethanol bis(trimethylsilyl) ether	56728-06-4	MS07259
209	73	312	210	179	297	45	75	**312**	100	54	26	18	12	10	8	7		15	28	3	–	–	–	–	–	–	2	–	–	–	2-(4-Hydroxy-3-methoxyphenyl)ethanol bis(trimethylsilyl) ether	56728-06-4	MS07260
297	283	298	73	100	75	55	41	**312**	100	35	23	23	19	18	16	15	**3.00**	15	28	3	2	–	–	–	–	–	1	–	–	–	N-Butyl-N′-trimethylsilylbarbital	51209-90-6	LI06879

MASS TO CHARGE RATIOS								M.W.	INTENSITIES								Parent	C	H	O	N	S	F	Cl	Br	I	Si	P	B	X	COMPOUND NAME	CAS Reg No	No
297	283	298	73	100	75	55	83	**312**	100	36	23	23	19	18	16	15	**2.51**	15	28	3	2	–	–	–	–	–	1	–	–	–	N-Butyl-N′-trimethylsilylbarbital	51209-90-6	NI32818
241	297	257	73	70	100	75	55	**312**	100	61	60	47	36	33	32	29	**8.75**	15	28	3	2	–	–	–	–	–	1	–	–	–	N-sec-Butyl-N′-trimethylsilylbarbital	51209-94-0	NI32819
297	241	257	283	73	100	55	298	**312**	100	49	43	34	34	24	22	21	**2.00**	15	28	3	2	–	–	–	–	–	1	–	–	–	N-Isobutyl-N′-trimethylsilylbarbital	51209-92-8	LI06880
167	297	166	168	241	69	257	68	**312**	100	99	72	61	49	47	43	41	**1.67**	15	28	3	2	–	–	–	–	–	1	–	–	–	N-Isobutyl-N′-trimethylsilylbarbital	51209-92-8	NI32820
150	233	176	312	314	178	75	151	**312**	100	87	63	60	56	43	42	35		16	9	2	–	–	–	–	1	–	–	–	–	–	2-Bromo-8H,9H-cycloocta[def]biphenylene-1,4-dione	119392-60-8	DD01169
232	116	90	89	115	231	312	233	**312**	100	80	35	32	31	19	16	16		16	12	–	2	–	–	–	–	–	–	–	–	1	3,3′-Selenodiindole	14293-12-0	LI06881
232	117	90	89	116	231	233	312	**312**	100	80	37	33	31	20	17	15		16	12	–	2	–	–	–	–	–	–	–	–	1	3,3′-Selenodiindole	14293-12-0	NI32821
312	314	201	77	313	91	67	127	**312**	100	99	99	99	88	67	65	60		16	13	1	4	–	–	1	–	–	–	–	–	–	1-Phenyl-3-methyl-4-(2-chlorophenylazo)-5-pyrazolone	15096-00-1	NI32822
77	312	314	201	91	127	99	313	**312**	100	98	75	73	52	41	40	35		16	13	1	4	–	–	1	–	–	–	–	–	–	1-Phenyl-3-methyl-4-(4-chlorophenylazo)-5-pyrazolone	6407-74-5	NI32823
77	118	268	92	104	312	136	195	**312**	100	29	24	13	12	10	8	4		16	16	1	4	1	–	–	–	–	–	–	–	–	2,5-Diphenylimino-3-hydroxyethyl-1,3,4-thiadiazolidine		LI06882
56	121	312	81	65	247	95	253	**312**	100	45	37	12	9	9	6	6		16	16	3	–	–	–	–	–	–	–	–	–	1	Cyclopentadienyl(4,5,6,7-tetrahydro-7-oxo-5-methoxycarbonylindenyl)iron		NI32824
221	91	145	77	312	175	146	117	**312**	100	31	26	7	5	5	5	5		16	16	3	4	–	–	–	–	–	–	–	–	–	7-[N′-(N′-Methyl-1′-phenyl-2′-propylamino)]-4-nitrobenzo-2-oxa-1,3-diazole		LI06883
312	311	142	313	157	170	158	102	**312**	100	31	26	19	12	11	11	10		16	16	3	4	–	–	–	–	–	–	–	–	–	1H-Pyrrolo[1,2-a]benzimidazolium, 2,3-dihydro-4-(1,2,3,4-tetrahydro-6-hydroxy-1,3-dimethyl-2,4-dioxo-5-pyrimidinyl)-	27356-02-1	NI32825
122	107	121	91	156	158			**312**	100	69	45	19	14	5			**0.00**	16	18	2	–	–	–	2	–	–	–	–	–	–	1,4-Etheno-2,8-dichloro-2,4,6,8-tetramethyl-octahydronaphthal-5-en-3,7-dione		LI06884
136	121	80	108	137	71	268	312	**312**	100	38	31	22	14	13	13	13		16	24	6	–	–	–	–	–	–	–	–	–	–	2,3-Benzo-1,4,7,10,13,16-hexaoxacyclooctadeca-2-ene		MS07261
45	177	117	145	189	176	71	236	**312**	100	78	78	39	35	28	28	27	**2.00**	16	24	6	–	–	–	–	–	–	–	–	–	–	Dimethyl 3-(methoxymethyl)-4-(3-methoxy-1-propenyl)-1-cyclohexene-1,4 dicarboxylate		MS07262
111	187	101	45	71	75	89	127	**312**	100	93	84	65	48	41	26	24	**0.00**	16	24	6	–	–	–	–	–	–	–	–	–	–	α-D-Glucopyranoside, phenyl 2,3,4,6-tetra-O-methyl-	3149-61-9	NI32826
107	135	79	82	121	199	139	59	**312**	100	96	85	80	60	50	47	40	**2.00**	16	24	6	–	–	–	–	–	–	–	–	–	–	3,7-Octadiene-1,1,8-tricarboxylic acid, 3,7-dimethyl-, trimethyl ester, (E,E)-	57683-64-4	NI32827
143	164	73	149	93	166	131	91	**312**	100	74	62	30	27	26	23	17	**1.60**	16	25	2	–	–	–	1	–	–	1	–	–	–	(2RS,3RS,4RS,6SR)-4-Chloro-2,4-dimethyl-6-phenyl-3-(trimethylsiloxy)tetrahydropyran		MS07263
72	123	151	139	211	141	242	183	**312**	100	24	8	8	7	7	6	5	**3.00**	16	28	4	2	–	–	–	–	–	–	–	–	–	N,N′,N′,O-Tetramethyl-2-(1′-methylbutyl)-2-allylalonurate		LI06885
312	311	279	313	121	178	235	203	**312**	100	56	32	28	25	21	19	17		17	12	–	–	3	–	–	–	–	–	–	–	–	[1,2]Dithiolo[1,5-b][1,2]dithiole-7-SIV, 2,3-diphenyl-	38443-38-8	NI32828
312	311	121	313	279	156	314	235	**312**	100	77	56	32	21	16	15	13		17	12	–	–	3	–	–	–	–	–	–	–	–	[1,2]Dithiolo[1,5-b][1,2]dithiole-7-SIV, 2,4-diphenyl-	14905-03-4	NI32829
121	312	43	311	58	235	313	248	**312**	100	67	62	34	28	24	20	17		17	12	–	–	3	–	–	–	–	–	–	–	–	[1,2]Dithiolo[1,5-b][1,2]dithiole-7-SIV, 2,5-diphenyl-	1033-90-5	NI32830
121	312	313	311	235	248	314	236	**312**	100	74	40	35	22	16	11	10		17	12	–	–	3	–	–	–	–	–	–	–	–	[1,2]Dithiolo[1,5-b][1,2]dithiole-7-SIV, 2,5-diphenyl-	1033-90-5	NI32831
312	29	227	284	228	256	39	283	**312**	100	41	32	31	29	28	27	25		17	12	6	–	–	–	–	–	–	–	–	–	–	Cyclopenta[c]furo[3′,2′:4,5]furo[2,3-h][1]benzopyran-1,11-dione, 2,3,6a,9a tetrahydro-4-methoxy-, (6aR-cis)-	1162-65-8	NI32832
312	313	284	283	228	226	241	213	**312**	100	16	12	12	8	8	6	6		17	12	6	–	–	–	–	–	–	–	–	–	–	Cyclopenta[c]furo[3′,2′:4,5]furo[2,3-h][1]benzopyran-1,11-dione, 2,3,6a,9a tetrahydro-4-methoxy-, (6aR-cis)-	1162-65-8	NI32833
77	267	312	57	56	314	269	232	**312**	100	78	75	45	42	40	39	38		17	13	2	2	–	–	1	–	–	–	–	–	–	1H-Pyrazole-4-acetic acid, 3-(4-chlorophenyl)-1-phenyl-	53808-88-1	NI32834
267	312	77	269	51	314	268	78	**312**	100	68	51	36	28	22	21	18		17	13	2	2	–	–	1	–	–	–	–	–	–	1H-Pyrazole-4-acetic acid, 5-(4-chlorophenyl)-1-phenyl-	32702-08-2	NI32835
224	312	165	239	225	121	313	166	**312**	100	52	38	28	18	13	10	9		17	16	–	2	2	–	–	–	–	–	–	–	–	4H-Imidazole, 2,5-bis(methylthio)-4,4-diphenyl-	2032-17-9	NI32836
193	61	312	165	194	118	207	210	**312**	100	94	68	42	19	19	14	13		17	16	–	2	2	–	–	–	–	–	–	–	–	2H-Imidazole-2-thione, 1,5-dihydro-1-methyl-4-(methylthio)-5,5-diphenyl-	16116-39-5	NI32837
207	312	99	102	165	265	313	208	**312**	100	91	55	33	26	25	20	19		17	16	–	2	2	–	–	–	–	–	–	–	–	4H-Imidazole-4-thione, 1,5-dihydro-1-methyl-2-(methylthio)-5,5-diphenyl-	22544-79-2	NI32838
224	312	165	239	225	121	313	166	**312**	100	56	31	19	17	17	13	8		17	16	–	2	2	–	–	–	–	–	–	–	–	4H-Imidazole-4-thione, 3,5-dihydro-3-methyl-2-(methylthio)-5,5-diphenyl-	22544-78-1	NI32839
312	210	165	313	118	239	211	314	**312**	100	70	36	20	15	12	11	10		17	16	–	2	2	–	–	–	–	–	–	–	–	2,4-Imidazolidinedithione, 1,3-dimethyl-5,5-diphenyl-	16116-40-8	NI32840
162	284	120	172	163	283	238	294	**312**	100	28	16	14	12	10	6	5	**3.00**	17	16	4	2	–	–	–	–	–	–	–	–	–	2aR*,3R*,8b S*-2a,8b-Dihydroxy-3-(4-nitrophenyl)-1,2,3,4-tetrahydrocyclobut[c]quinoline	90821-37-7	NI32841
298	220	268	300	297	299	240	270	**312**	100	86	51	34	31	28	27	25	**0.00**	17	17	–	4	–	–	1	–	–	–	–	–	–	7-Chloro-2-dimethylamino-3-methyl-5-phenyl-3H-1,3,4-benzotriazepine		MS07264
282	312	234	284	314	240	313	205	**312**	100	96	47	40	32	27	25	24		17	17	–	4	–	–	1	–	–	–	–	–	–	7-Chloro-2-ethylamino-3-methyl-5-phenyl-3H-1,3,4-benzotriazepine		MS07265
222	223	207	91	60	208	221	313	**312**	100	26	22	22	20	13	12	11	**7.20**	17	20	–	4	1	–	–	–	–	–	–	–	–	5-Methyl-2-N,N-dimethylaminobenzophenone thiosemicarbazone		MS07266
91	106	150	87	65	162			**312**	100	56	39	17	17	11			**0.00**	17	20	2	4	–	–	–	–	–	–	–	–	–	Methylene bis(3-benzylurea)		LI06886
123	312	59	58	70	115	184	171	**312**	100	67	22	15	10	7	6	6		17	20	2	4	–	–	–	–	–	–	–	–	–	Morphinan-3-ol, 6-azido-4,5-epoxy-17-methyl-, (5α,6β)-	22952-87-0	NI32842
36	185	198	210	211	186	32	199	**312**	100	90	88	53	27	27	23	19	**15.40**	17	20	2	4	–	–	–	–	–	–	–	–	–	3-[2-(2-Oxo)morpholinoethylamino]-4-methyl-6-phenylpyridazine	82246-39-7	NI32843
57	43	71	41	55	200	125	77	**312**	100	60	58	35	18	16	10	7	**5.90**	17	25	3	–	–	–	1	–	–	–	–	–	–	Acetic acid, (4-chloro-2-methylphenoxy)-, isooctyl ester	26544-20-7	NI32844
57	43	71	41	29	27	55	200	**312**	100	76	71	69	63	41	40	28	**9.90**	17	25	3	–	–	–	1	–	–	–	–	–	–	Acetic acid, (4-chloro-2-methylphenoxy)-, isooctyl ester	26544-20-7	NI32845
171	96	124	81	79	139	101	111	**312**	100	55	33	20	19	9	9	8	**0.00**	17	28	5	–	–	–	–	–	–	–	–	–	–	2,2-Dimethoxynorbornan-2-yl 2′-methoxy-2′-hydroxynorbornan-2-yl ether		NI32846
312	235	188	236	108	157	77	313	**312**	100	99	56	44	37	24	22	16		18	11	1	–	–	2	–	–	–	–	1	–	–	2,8-Difluoro-10-phenylphenoxaphosphine		LI06887
312	129	102	90	103	284	311	156	**312**	100	99	99	74	70	63	49	44		18	12	–	6	–	–	–	–	–	–	–	–	–	N-(2-Quinoxalinyl)imidazo[1,2-a]quinoxalin-2-amine		MS07267

MASS TO CHARGE RATIOS	M.W.	INTENSITIES	Parent	C	H	O	N	S	F	Cl	Br	I	Si	P	B	X	COMPOUND NAME	CAS Reg No	No
312 311 313 310 164 208 207 104	**312**	100 67 19 17 10 7 5 5		18	15	3	–	–	–	–	–	–	–	–	3	–	Boroxin, triphenyl-	3262-89-3	NI32847
75 44 84 45 47 43 73 41	**312**	100 66 53 21 19 17 15 15	**0.00**	18	15	3	–	–	–	–	–	–	–	–	3	–	Boroxin, triphenyl-	3262-89-3	NI32848
312 311 164 104 313 310 163 156	**312**	100 69 23 20 17 17 12 12		18	15	3	–	–	–	–	–	–	–	–	3	–	Boroxin, triphenyl-	3262-89-3	NI32849
128 312 91 65 57 77 41 129	**312**	100 75 73 72 65 60 57 56		18	16	3	–	1	–	–	–	–	–	–	–	–	4-Methyl-3-[(4-methylphenyl)sulphonyl]-2-phenylfuran	118356-46-0	DD01170
312 293 294 283 269 281	**312**	100 35 33 25 13 12		18	16	5	–	–	–	–	–	–	–	–	–	–	5,7-Dihydroxy-3-methoxy-6,8-dimethylflavone		NI32850
277 232 234 278 261 260 259 248	**312**	100 50 36 34 30 21 21 20	**0.00**	18	16	5	–	–	–	–	–	–	–	–	–	–	8,9-Epoxy-1α,2β,15ξ-trihydroxy-11-methyl-8,9-secogona-3,5,7,9,11,13-hexaen-17-one		NI32851
312 251 57 42 297 254 283 279	**312**	100 50 46 41 36 34 32 31		18	16	5	–	–	–	–	–	–	–	–	–	–	10-Hydroxy-2,5-dimethoxy-3,7-dimethyl-1,4-anthracenedione	84542-54-1	NI32852
312 57 42 55 69 40 71 266	**312**	100 93 87 70 63 62 56 44		18	16	5	–	–	–	–	–	–	–	–	–	–	10-Hydroxy-5,7-dimethoxy-2,3-dimethyl-1,4-anthracenedione	84542-59-6	NI32853
312 297 43 67 313 269 231 88	**312**	100 42 34 28 20 16 16 10		18	16	5	–	–	–	–	–	–	–	–	–	–	1H-Naphtho[2,1-b]pyran-1-one, 3-acetyl-7,8-dimethoxy-2-methyl-	55044-17-2	NI32854
297 312 187 43 67 215 251 104	**312**	100 55 37 35 33 27 25 17		18	16	5	–	–	–	–	–	–	–	–	–	–	4H-Naphtho[1,2-b]pyran-4-one, 3-acetyl-5,6-dimethoxy-2-methyl-	55044-16-1	NI32855
121 312 192 177 134 92 63 313	**312**	100 64 25 19 17 16 16 15		18	16	5	–	–	–	–	–	–	–	–	–	–	2′,3′,4′-Trimethoxyflavone		NI32856
151 312 65 91 63 79 76 51	**312**	100 31 20 18 17 13 13 13		18	16	5	–	–	–	–	–	–	–	–	–	–	2′,3′,7-Trimethoxyflavone	80710-39-0	NI32857
311 312 77 293 105 313 89 63	**312**	100 84 41 34 27 23 13 13		18	16	5	–	–	–	–	–	–	–	–	–	–	3,5,7-Trimethoxyflavone	26964-29-4	MS07268
313 314 77 293 105 281 142 69	**312**	100 60 50 45 38 21 20 20	**5.00**	18	16	5	–	–	–	–	–	–	–	–	–	–	3,5,7-Trimethoxyflavone	26964-29-4	NI32858
312 162 313 269 151 147 297 91	**312**	100 23 20 20 15 12 11 10		18	16	5	–	–	–	–	–	–	–	–	–	–	3′,4′,7-Trimethoxyflavone	22395-24-0	NI32859
44 57 56 87 312 162 115 297	**312**	100 27 11 9 2 2 2 1		18	16	5	–	–	–	–	–	–	–	–	–	–	3′,4′,7-Trimethoxyflavone	22395-24-0	LI06888
312 311 266 283 132 281 282 117	**312**	100 57 29 28 27 17 13 8		18	16	5	–	–	–	–	–	–	–	–	–	–	4′,5,7-Trimethoxyflavone	5631-70-9	MS07269
312 311 266 132 283 281 313 142	**312**	100 62 40 35 32 21 20 18		18	16	5	–	–	–	–	–	–	–	–	–	–	4′,5,7-Trimethoxyflavone	5631-70-9	NI32860
312 297 119 269 226 162 147 127	**312**	100 18 17 14 10 9 8 8		18	16	5	–	–	–	–	–	–	–	–	–	–	3′,4,7-Trimethoxyisoflavone		NI32861
312 105 269 284 313 139 152 163	**312**	100 56 47 22 21 12 11 10		18	16	5	–	–	–	–	–	–	–	–	–	–	4,5,7-Trimethoxy-3-phenylcoumarin	4222-03-1	NI32862
312 311 132 266 283 313 281 137	**312**	100 46 38 30 22 21 20 16		18	16	5	–	–	–	–	–	–	–	–	–	–	4′,5,7,-Trimeththoxyisoflavone		NI32863
127 206 312 311 313 186 178 207	**312**	100 60 46 31 29 29 26 25		18	17	1	2	–	–	1	–	–	–	–	–	–	2-(4-Chloroanilino)-4-(N-methylanilino)-cyclopent-2-enone		MS07270
241 243 242 269 270 227 312 271	**312**	100 62 55 50 35 22 21 19		18	17	1	2	–	–	1	–	–	–	–	–	–	7-Chloro-2,3-dihydro-3-isopropyl-5-phenyl-1H-1,4-benzodiazepin-2-one	51990-97-7	NI32864
284 311 242 256 269 312 283 257	**312**	100 89 85 79 76 75 67 63		18	17	1	2	–	–	1	–	–	–	–	–	–	6-(4-Chlorophenyl)-1,2,3,4-tetrahydro-1,8-dimethyl-1,5-benzodiazocin-2-one	63594-49-0	NI32865
148 179 178 312 149 43 28 77	**312**	100 27 15 12 11 10 10 7		18	20	3	2	–	–	–	–	–	–	–	–	–	Benzoic acid, 4-(dimethylamino)-, anhydride	7474-31-9	NI32866
225 226 98 55 312 43 224 167	**312**	100 16 8 7 6 5 4 4		18	20	3	2	–	–	–	–	–	–	–	–	–	Methyl 3-(3-oxo-1,2,3,5,6,11b-hexahydro-11h-pyrrolo[2,1-a]-β-carbolin-11b-yl) propionate		MS07271
312 297 73 313 298 314 195 299	**312**	100 51 35 30 15 11 6 6		18	24	1	–	–	–	–	–	–	2	–	–	–	1,3-Bis(trimethylsilyl)naphtho[1,2-c]furan		NI32867
241 243 135 57 73 60 242 55	**312**	100 37 35 20 17 17 14 12	**3.20**	18	29	2	–	–	–	1	–	–	–	–	–	–	Benzene, 1-[2-(2-chloroethoxy)ethoxy]-4-(1,1,3,3-tetramethylbutyl)-	65925-28-2	NI32868
55 41 98 67 248 280 249 281	**312**	100 72 65 61 25 21 20 19	**8.77**	18	32	4	–	–	–	–	–	–	–	–	–	–	Dimethyl hexadec-8-en-dioate		IC03860
114 184 113 169 141 171 242 127	**312**	100 74 74 50 49 44 43 38	**12.00**	18	32	4	–	–	–	–	–	–	–	–	–	–	9,12-Dioxostearic acid		LI06889
148 230 83 149 312 55 231 143	**312**	100 51 37 35 12 11 9 5		18	33	–	–	1	–	–	–	–	–	1	–	–	Phosphine sulphide, tricyclohexyl-	42201-98-9	NI32869
199 140 100 129 114 226 185 158	**312**	100 60 43 38 20 7 6 5	**5.00**	18	36	2	2	–	–	–	–	–	–	–	–	–	Heptanamide, N-[3-(acetylamino)propyl]-N-hexyl-	67138-91-4	NI32870
43 140 269 170 56 128 114 199	**312**	100 55 38 25 20 17 15 12	**5.00**	18	36	2	2	–	–	–	–	–	–	–	–	–	Heptanamide, N-[3-(acetylhexylamino)propyl]-	67139-03-1	NI32871
173 73 131 174 75 147 217 204	**312**	100 70 22 15 14 12 9 9	**0.00**	18	36	2	2	–	–	–	–	–	–	–	–	–	Tetradecanamide, N-(2-methylpropyl)-N-nitroso-	74420-89-6	NI32872
155 310 43 142 55 41 57 154	**312**	100 51 45 42 34 31 27 24	**1.50**	18	37	3	–	–	–	–	–	–	–	–	1	–	α-Hydroxymyristic acid butyl boronate		MS07272
152 153 43 55 57 69 217 41	**312**	100 70 38 36 34 32 30 28	**12.00**	19	17	2	–	–	–	1	–	–	–	–	–	–	cis-2-Acetoxy-3-chloro-endo-5,6-(1,8-naphthylene)norbornane		NI32873
57 43 55 69 41 71 152 312	**312**	100 95 90 70 63 59 32 23		19	17	2	–	–	–	1	–	–	–	–	–	–	cis-7-Acetoxy-2-chloro-endo-5,6-(1,8-naphthylene)norbornane		NI32874
312 152 153 217 314 313 216 165	**312**	100 95 40 39 33 18 18 18		19	17	2	–	–	–	1	–	–	–	–	–	–	cis-7-Acetoxy-2-chloro-exo-5,6-(1,8-naphthylene)norbornane		NI32875
198 165 199 312 121 166 57 73	**312**	100 64 40 37 36 24 20 18		19	20	2	–	1	–	–	–	–	–	–	–	–	1,3-Oxathiolan-5-one, 2,2-diethyl-4,4-diphenyl-	31061-72-0	NI32876
149 91 206 132 150 123 104 41	**312**	100 68 28 15 12 12 11 11	**6.91**	19	20	4	–	–	–	–	–	–	–	–	–	–	Butyl benzyl phthalate		IC03862
149 91 206 150 104 132 105 41	**312**	100 50 18 12 12 11 11 11	**3.00**	19	20	4	–	–	–	–	–	–	–	–	–	–	Butyl benzyl phthalate		IC03863
149 91 41 56 206 104 76 92	**312**	100 92 35 32 27 26 20 17	**0.01**	19	20	4	–	–	–	–	–	–	–	–	–	–	Butyl benzyl phthalate		IC03861
229 312 230 269 297 231 191 244	**312**	100 50 17 14 10 10 6 5		19	20	4	–	–	–	–	–	–	–	–	–	–	(±)-Deoxybruceol		LI06890
284 76 152 115 165 312 153 155	**312**	100 79 74 73 69 61 61 54		19	20	4	–	–	–	–	–	–	–	–	–	–	6,7-Dihydro-1,2,3-trimethoxybenzo[a]heptalen-10(5H)-one		MS07273
105 68 77 190 67 51 123 106	**312**	100 70 37 14 10 10 9 9	**0.50**	19	20	4	–	–	–	–	–	–	–	–	–	–	1,5-Pentanediol dibenzoate		NI32877
281 312 165 152 282 127 115 153	**312**	100 82 27 24 23 23 22 21		19	20	4	–	–	–	–	–	–	–	–	–	–	6,7,7b,10a-Tetrahydro-1,2,3-trimethoxybenzo[a]cyclopenta[3,4]cyclohepten-8(5H)-one		MS07274
241 297 295 312 77 257 242 127	**312**	100 68 41 31 19 18 18 15		19	20	4	–	–	–	–	–	–	–	–	–	–	Trachyphyllin		MS07275
100 101 197 72 312 84 212 179	**312**	100 7 4 4 3 3 2 2		19	24	–	2	1	–	–	–	–	–	–	–	–	10H-Phenothiazine-10-ethanamine, N,N-diethyl-α-methyl-	522-00-9	NI32878
100 44 101 72 56 42 41 70	**312**	100 46 39 24 9 8 8 7	**4.00**	19	24	–	2	1	–	–	–	–	–	–	–	–	10H-Phenothiazine-10-ethanamine, N,N-diethyl-α-methyl-	522-00-9	NI32879
100 101 198 72 56 98 70 42	**312**	100 7 4 4 4 3 3 3	**1.33**	19	24	–	2	1	–	–	–	–	–	–	–	–	10H-Phenothiazine-10-ethanamine, N,N-diethyl-α-methyl-	522-00-9	NI32880
112 312 311 140 313 43 294 130	**312**	100 85 46 33 20 18 16 13		19	24	2	2	–	–	–	–	–	–	–	–	–	Aspidodispermine, deoxy-	21446-30-0	NI32881
144 312 182 130 143 281 183 267	**312**	100 99 99 80 65 63 25 22		19	24	2	2	–	–	–	–	–	–	–	–	–	Curan-17,18-diol, 19,20-didehydro-, (19E)-	900-98-1	NI32882

MASS TO CHARGE RATIOS	M.W.	INTENSITIES	Parent	C	H	O	N	S	F	Cl	Br	I	Si	P	B	X	COMPOUND NAME	CAS Reg No	No
57 55 69 70 182 144 295 56	312	100 89 75 72 70 53 50 49	35.92	19	24	2	2	–	–	–	–	–	–	–	–	–	Curan-19,20-diol, 16,17-didehydro-, (19S)-	56192-68-8	NI32883
312 170 267 197 225 156 184 127	312	100 61 52 41 36 35 33 27		19	24	2	2	–	–	–	–	–	–	–	–	–	trans-5-Ethyl-6,6-(ethylenedioxy)-1,2,5,6,7,7a-hexahydro-4H-pyrido[1',2':1,2]pyrazino[4,3-a]indole		DD01171
49 84 51 69 55 43 41 312	312	100 83 35 29 25 25 25 9		19	24	2	2	–	–	–	–	–	–	–	–	–	trans-3-Ethyl-2,2-(ethylenedioxy)-1,2,3,4,6,7,12,12b-octahydroindolo[2,3-a]quinolizine		DD01172
312 156 172 130 146 159 140 186	312	100 59 55 36 34 33 33 29		19	24	2	2	–	–	–	–	–	–	–	–	–	2H-3,13-Methanooxireno[9,10]azacycloundecino[5,4-b]indol-5a(1aH)-ol, 13-ethyl-4,5,11,12,13,13a-hexahydro-, [1aR-(1aR*,5aR*,13S*,13aS*)]-	18269-16-4	NI32884
58 44 84 66 124 86 42 73	312	100 63 59 57 53 37 37 34	11.00	19	24	2	2	–	–	–	–	–	–	–	–	–	Methyl (Z)-5-ethylidene-2-indolyl-1-methylpiperidine-4-acetate		MS07276
44 55 155 41 43 69 81 57	312	100 98 91 86 80 56 50 49	1.00	19	36	3	–	–	–	–	–	–	–	–	–	–	2-Furanoctanoic acid, 5-hexyltetrahydro-, methyl ester	34724-76-0	NI32885
55 155 95 81 41 185 67 69	312	100 99 80 80 80 66 65 62	1.98	19	36	3	–	–	–	–	–	–	–	–	–	–	2-Furanpentanoic acid, tetrahydro-5-nonyl-, methyl ester	54725-58-5	NI32886
185 158 55 153 43 197 41 95	312	100 88 80 79 58 57 53 51	6.00	19	36	3	–	–	–	–	–	–	–	–	–	–	2-Furanpentanoic acid, tetrahydro-5-nonyl-, methyl ester	54725-58-5	NI32887
116 129 97 101 312 239 256 220	312	100 98 71 60 55 46 35 23		19	36	3	–	–	–	–	–	–	–	–	–	–	Heptadecanoic acid, 3-oxo-16-methyl-, methyl ester	87538-95-2	NI32888
55 81 67 116 220 94 95 82	312	100 94 88 85 81 81 79 75	29.17	19	36	3	–	–	–	–	–	–	–	–	–	–	Methyl 3-hydroxy-9-octadecanoate		NI32889
55 41 74 166 43 69 84 29	312	100 50 45 41 40 36 29 26	0.10	19	36	3	–	–	–	–	–	–	–	–	–	–	Methyl 12-hydroxy-9-octadecenoate		IC03864
74 69 97 84 67 124 96 81	312	100 99 98 97 97 96 95 95	0.00	19	36	3	–	–	–	–	–	–	–	–	–	–	Methyl 12-hydroxy-9-octadecenoate		LI06891
253 57 71 43 55 85 254 41	312	100 41 30 29 25 23 20 19	3.30	19	36	3	–	–	–	–	–	–	–	–	–	–	Octadecanoic acid, 2-oxo-, methyl ester	2380-18-9	MS07277
57 43 71 55 41 85 95 69	312	100 87 61 49 45 32 20 20	0.00	19	36	3	–	–	–	–	–	–	–	–	–	–	Octadecanoic acid, 2-oxo-, methyl ester	2380-18-9	NI32890
116 43 117 57 129 41 55 29	312	100 34 32 25 23 22 21 12	4.40	19	36	3	–	–	–	–	–	–	–	–	–	–	Octadecanoic acid, 3-oxo-, methyl ester	14531-34-1	NI32891
116 43 117 57 129 41 55 74	312	100 34 32 25 23 23 22 11	4.30	19	36	3	–	–	–	–	–	–	–	–	–	–	Octadecanoic acid, 3-oxo-, methyl ester	14531-34-1	WI01401
116 43 117 57 129 41 55 29	312	100 45 43 33 30 30 28 17	5.96	19	36	3	–	–	–	–	–	–	–	–	–	–	Octadecanoic acid, 3-oxo-, methyl ester	14531-34-1	NI32892
130 98 115 225 281 111 57 43	312	100 38 21 15 12 11 10 9	3.60	19	36	3	–	–	–	–	–	–	–	–	–	–	Octadecanoic acid, 4-oxo-, methyl ester		MS07278
158 126 55 43 111 57 197 41	312	100 67 47 41 37 30 27 27	4.00	19	36	3	–	–	–	–	–	–	–	–	–	–	Octadecanoic acid, 6-oxo-, methyl ester		MS07279
43 55 41 69 172 140 115 71	312	100 82 72 62 51 43 42 36	4.40	19	36	3	–	–	–	–	–	–	–	–	–	–	Octadecanoic acid, 7-oxo-, methyl ester		MS07280
186 184 129 171 55 169 43 154	312	100 93 92 82 70 62 61 56	9.00	19	36	3	–	–	–	–	–	–	–	–	–	–	Octadecanoic acid, 8-oxo-, methyl ester		MS07281
43 55 143 71 200 57 111 125	312	100 97 52 50 47 45 42 38	8.33	19	36	3	–	–	–	–	–	–	–	–	–	–	Octadecanoic acid, 9-oxo-, methyl ester	1842-70-2	NI32893
143 170 200 111 110 155 125 185	312	100 80 68 67 56 54 50 48	3.12	19	36	3	–	–	–	–	–	–	–	–	–	–	Octadecanoic acid, 9-oxo-, methyl ester	1842-70-2	NI32894
43 55 41 170 143 71 200 57	312	100 98 97 54 52 52 47 47	8.70	19	36	3	–	–	–	–	–	–	–	–	–	–	Octadecanoic acid, 9-oxo-, methyl ester	1842-70-2	MS07282
55 43 57 71 41 156 214 69	312	100 82 76 74 65 49 42 41	3.10	19	36	3	–	–	–	–	–	–	–	–	–	–	Octadecanoic acid, 10-oxo-, methyl ester	870-10-0	NI32895
43 41 113 69 58 128 85 98	312	100 54 33 31 29 28 28 27	1.40	19	36	3	–	–	–	–	–	–	–	–	–	–	Octadecanoic acid, 12-oxo-, methyl ester	2380-27-0	NI32896
43 55 71 41 99 114 59 58	312	100 62 48 48 36 26 24 24	4.10	19	36	3	–	–	–	–	–	–	–	–	–	–	Octadecanoic acid, 13-oxo-, methyl ester	2380-28-1	MS07283
99 114 98 256 167 126 82 199	312	100 75 74 42 42 42 38 37	5.86	19	36	3	–	–	–	–	–	–	–	–	–	–	Octadecanoic acid, 13-oxo-, methyl ester	2380-28-1	NI32897
43 71 41 55 269 69 58 74	312	100 68 58 50 34 25 23 20	11.00	19	36	3	–	–	–	–	–	–	–	–	–	–	Octadecanoic acid, 15-oxo-, methyl ester		MS07284
283 57 241 72 55 209 281 41	312	100 93 83 80 64 58 56 53	25.00	19	36	3	–	–	–	–	–	–	–	–	–	–	Octadecanoic acid, 16-oxo-, methyl ester		MS07285
43 223 58 55 255 222 41 73	312	100 82 73 58 53 53 52 51	8.60	19	36	3	–	–	–	–	–	–	–	–	–	–	Octadecanoic acid, 17-oxo-, methyl ester		MS07286
55 74 155 41 43 69 57 87	312	100 96 91 79 71 69 55 46	0.74	19	36	3	–	–	–	–	–	–	–	–	–	–	Oxiraneoctanoic acid, 3-octyl-, methyl ester, cis-	2566-91-8	NI32898
55 74 155 69 41 43 57 87	312	100 93 81 74 72 70 58 50	0.34	19	36	3	–	–	–	–	–	–	–	–	–	–	Oxiraneoctanoic acid, 3-octyl-, methyl ester, cis-	2566-91-8	NI32899
55 43 41 67 81 57 69 74	312	100 89 81 69 57 55 53 52	1.00	19	36	3	–	–	–	–	–	–	–	–	–	–	Oxiraneoctanoic acid, 3-octyl-, methyl ester, cis-	2566-91-8	NI32900
67 55 81 41 43 95 82 69	312	100 82 77 71 62 57 52 49	0.34	19	36	3	–	–	–	–	–	–	–	–	–	–	Oxiraneoctanoic acid, 3-octyl-, methyl ester, trans-	6084-76-0	NI32903
55 74 155 69 41 137 43 57	312	100 91 77 72 71 63 61 53	0.44	19	36	3	–	–	–	–	–	–	–	–	–	–	Oxiraneoctanoic acid, 3-octyl-, methyl ester, trans-	6084-76-0	NI32901
55 67 41 74 43 81 69 155	312	100 84 84 69 67 63 62 52	0.44	19	36	3	–	–	–	–	–	–	–	–	–	–	Oxiraneoctanoic acid, 3-octyl-, methyl ester, trans-	6084-76-0	NI32902
43 55 57 41 115 158 69 172	312	100 99 89 84 79 78 69 66	3.94	19	36	3	–	–	–	–	–	–	–	–	–	–	Oxiranepentanoic acid, 3-undecyl-, methyl ester, cis-	1041-25-4	NI32904
28 55 43 41 113 69 57 29	312	100 99 93 88 78 69 65 51	3.14	19	36	3	–	–	–	–	–	–	–	–	–	–	Oxiranepentanoic acid, 3-undecyl-, methyl ester, cis-	1041-25-4	NI32905
55 80 43 41 67 79 87 57	312	100 99 98 93 85 70 68 68	1.34	19	36	3	–	–	–	–	–	–	–	–	–	–	Oxiranepentanoic acid, 3-undecyl-, methyl ester, trans-	6175-11-7	NI32906
55 43 41 28 113 69 57 29	312	100 98 98 92 83 69 67 67	2.64	19	36	3	–	–	–	–	–	–	–	–	–	–	Oxiranepentanoic acid, 3-undecyl-, methyl ester, trans-	6175-11-7	NI32907
43 55 41 57 67 69 81 74	312	100 85 76 72 57 48 47 46	1.24	19	36	3	–	–	–	–	–	–	–	–	–	–	Oxiraneundecanoic acid, 3-pentyl-, methyl ester, cis-	38520-30-8	NI32908
55 41 43 74 69 57 29 87	312	100 81 71 55 52 48 45 39	0.54	19	36	3	–	–	–	–	–	–	–	–	–	–	Oxiraneundecanoic acid, 3-pentyl-, methyl ester, cis-	38520-30-8	NI32909
55 41 28 43 74 69 57 87	312	100 76 74 69 61 58 52 46	1.24	19	36	3	–	–	–	–	–	–	–	–	–	–	Oxiraneundecanoic acid, 3-pentyl-, methyl ester, trans-	38520-31-9	NI32910
67 55 81 41 95 82 43 69	312	100 78 74 70 58 52 51 46	0.24	19	36	3	–	–	–	–	–	–	–	–	–	–	Oxiraneundecanoic acid, 3-pentyl-, methyl ester, trans-	38520-31-9	NI32911
271 272 75 97 83 111 89 69	312	100 24 22 12 11 8 7 7	0.00	19	40	1	–	–	–	–	–	–	1	–	–	–	Allyldimethylsilyl tetradecyl ether		NI32912
312 313 156 44 255 311 128 114	312	100 23 14 11 10 9 7 5		20	12	2	2	–	–	–	–	–	–	–	–	–	Quinacridone		IC03865
312 313 311 255 156 55 44 28	312	100 23 10 9 9 9 8 5		20	12	2	2	–	–	–	–	–	–	–	–	–	Quinacridone		MS07287
77 312 92 194 180 51 65 311	312	100 93 55 51 40 33 26 23		20	16	–	4	–	–	–	–	–	–	–	–	–	4,5-Dihydro-1,4-diphenyl-3,5-phenylimino-1,2,4-triazole		IC03866
186 294 312 266 187 171 123 114	312	100 47 38 23 22 20 11 11		20	24	3	–	–	–	–	–	–	–	–	–	–	9,11-Dehydro-14α-hydroxy-D-homoestrone methyl ether		LI06892
312 186 241 294 187 159 171 313	312	100 76 55 32 27 25 20 18		20	24	3	–	–	–	–	–	–	–	–	–	–	9,11-Dehydro-14β-hydroxy-8α-D-homoisoestrone methyl ether		LI06893
270 185 271 146 172 312 213 186	312	100 21 20 12 11 7 6 6		20	24	3	–	–	–	–	–	–	–	–	–	–	Estra-1,3,5(10)-trien-17-one, 3-(acetyloxy)-	901-93-9	NI32913

MASS TO CHARGE RATIOS								M.W.	INTENSITIES								Parent	C	H	O	N	S	F	Cl	Br	I	Si	P	B	X	COMPOUND NAME	CAS Reg No	No
270	43	146	271	185	172	41	213	**312**	100	37	21	20	20	15	15	11	**10.80**	20	24	3	–	–	–	–	–	–	–	–	–	–	Estra-1,3,5(10)-trien-17-one, 3-(acetyloxy)-	901-93-9	NI32914
186	294	312	266	187	171	128	114	**312**	100	46	38	23	22	21	11	11		20	24	3	–	–	–	–	–	–	–	–	–	–	D-Homoestra-1,3,5(10),9(11)-tetraen-17a-one, 14-hydroxy-3-methoxy-	3940-07-6	NI32915
312	186	241	294	187	159	313	171	**312**	100	76	55	31	27	25	22	19		20	24	3	–	–	–	–	–	–	–	–	–	–	D-Homoestra-1,3,5(10),9(11)-tetraen-17a-one, 14-hydroxy-3-methoxy-, (8α,14β)-	17908-44-0	NI32916
213	312	214	137	311	105	57	41	**312**	100	58	51	39	26	26	21	18		20	24	3	–	–	–	–	–	–	–	–	–	–	2-Hydroxy-4-heptoxy-benzophenone		IC03867
107	91	105	280	133	36	147	175	**312**	100	44	42	25	24	15	12	4	**0.00**	20	24	3	–	–	–	–	–	–	–	–	–	–	5-Methoxy-7-(4-hydroxyphenyl)-1-phenyl-3-heptanone		DD01173
187	312	188	313	161	160	159	146	**312**	100	81	80	38	32	24	23	18		20	24	3	–	–	–	–	–	–	–	–	–	–	3-Methoxy-8,14-seco-14-oxo-estrone		LI06894
279	312	297	43	239	241	227	237	**312**	100	96	76	69	42	41	39	38		20	24	3	–	–	–	–	–	–	–	–	–	–	1,2,4,5,6,6a,7,11b-Octahydro-6-hydroxy-9-methoxy-6,11b-dimethyl-3H-benz[de]anthracen-3-one		NI32917
148	133	312	105	120	91	77	79	**312**	100	44	41	34	22	19	15	13		20	24	3	–	–	–	–	–	–	–	–	–	–	Rubifolide		MS07288
202	201	28	77	203	47	256	312	**312**	100	86	31	26	25	20	12	9		20	25	1	–	–	–	–	–	–	–	1	–	–	(E)-(1-Isobutylbut-2-enyl)diphenylphosphine oxide		MS07289
171	312	158	229	187	172	149	144	**312**	100	48	44	38	36	25	23	16		20	28	1	2	–	–	–	–	–	–	–	–	–	N-Acetyl-2-cyclohexylaminomethyl-3-propylindole		LI06895
124	312	125	284	160	152	311	41	**312**	100	25	10	9	6	6	5	5		20	28	1	2	–	–	–	–	–	–	–	–	–	Aspidospermidine, 17-methoxy-	2447-50-9	NI32918
212	312	213	297	283	227	158	184	**312**	100	67	50	40	38	31	24	19		20	28	1	2	–	–	–	–	–	–	–	–	–	4,6-Dimethyl-3,3-pentamethylene-3H-indole-2-carboxylic acid tert-butylamide		LI06896
212	312	213	297	283	78	227	158	**312**	100	67	50	40	38	38	31	24		20	28	1	2	–	–	–	–	–	–	–	–	–	4,6-Dimethyl-3,3-pentamethylene-3H-indole-2-carboxylic acid tert-butylamide		MS07290
311	312	170	169	197	185	267	171	**312**	100	82	57	45	44	27	21	20		20	28	1	2	–	–	–	–	–	–	–	–	–	1-Ethyl-1-(3-hydroxypropyl)-1,2,3,4,6,7,12,12b-octahydropyrido[2,1-a]-β-carboline		MS07291
138	137	136	174	139	132	122	131	**312**	100	21	15	10	10	6	6	5	**4.58**	20	28	1	2	–	–	–	–	–	–	–	–	–	Indole-2-ethanol, β-(1-ethyl-3-ethylidene-4-piperidyl)-3-methyl-	1850-31-3	NI32919
312	110	125	124	187	138	96	126	**312**	100	76	75	71	51	44	36	29		20	28	1	2	–	–	–	–	–	–	–	–	–	2H-3,7-Methanoazacycloundecino[5,4-b]indole, 7-ethyl-1,4,5,6,7,8,9,10-octahydro-11-methoxy-, (-)-	14358-58-8	NI32920
43	173	140	57	55	41	69	56	**312**	100	99	70	68	62	62	40	40	**5.93**	20	40	2	–	–	–	–	–	–	–	–	–	–	Decanoic acid, decyl ester	1654-86-0	NI32921
101	55	43	102	57	41	116	72	**312**	100	33	12	6	6	6	4	4	**0.80**	20	40	2	–	–	–	–	–	–	–	–	–	–	1,3-Dioxane, 4-methyl-2-pentadecyl-	54950-57-1	NI32922
101	55	43	71	41	102	73	57	**312**	100	34	16	14	9	6	6	6	**0.44**	20	40	2	–	–	–	–	–	–	–	–	–	–	1,3-Dioxepane, 2-pentadecyl-	41563-29-5	NI32923
101	73	55	43	102	41	57	29	**312**	100	16	14	12	6	6	5	4	**0.44**	20	40	2	–	–	–	–	–	–	–	–	–	–	1,3-Dioxolane, 4,5-dimethyl-2-pentadecyl-	56599-61-2	NI32924
101	55	43	102	41	57	56	29	**312**	100	34	10	6	6	4	3	3	**0.20**	20	40	2	–	–	–	–	–	–	–	–	–	–	1,3-Dioxolane, 4-ethyl-2-pentadecyl-	56599-60-1	NI32925
43	73	57	60	55	41	71	69	**312**	100	77	75	68	62	60	38	38	**31.14**	20	40	2	–	–	–	–	–	–	–	–	–	–	Eicosanoic acid	506-30-9	NI32926
82	96	55	83	81	69	95	67	**312**	100	79	75	58	58	58	57	57	**4.14**	20	40	2	–	–	–	–	–	–	–	–	–	–	Ethanol, 2-(9-octadecenyloxy)-, (E)-	17367-07-6	NI32927
55	82	69	96	83	67	81	41	**312**	100	97	77	71	67	66	63	62	**1.44**	20	40	2	–	–	–	–	–	–	–	–	–	–	Ethanol, 2-(9-octadecenyloxy)-, (Z)-	5353-25-3	NI32928
43	74	41	55	87	57	29	69	**312**	100	92	89	86	62	53	52	45	**44.00**	20	40	2	–	–	–	–	–	–	–	–	–	–	Heptadecanoic acid, 15-ethyl-, methyl ester		MS07292
88	101	312	57	157	55	43	89	**312**	100	68	63	28	27	25	22	19		20	40	2	–	–	–	–	–	–	–	–	–	–	Heptadecanoic acid, 15-methyl-, ethyl ester	57274-46-1	NI32929
73	257	69	239	129	71	116	83	**312**	100	88	64	56	56	54	52	42	**26.52**	20	40	2	–	–	–	–	–	–	–	–	–	–	Hexadecanoic acid, butyl ester	111-06-8	WI01402
56	57	41	43	55	29	73	257	**312**	100	82	66	56	38	37	33	29	**10.68**	20	40	2	–	–	–	–	–	–	–	–	–	–	Hexadecanoic acid, butyl ester	111-06-8	IC03868
116	43	41	55	29	101	57	129	**312**	100	52	42	37	34	33	26	21	**7.26**	20	40	2	–	–	–	–	–	–	–	–	–	–	Hexadecanoic acid, 2-propyl-, methyl ester		NI32930
115	116	88	225	255	143	117	111	**312**	100	49	30	19	14	6	6	6	**1.00**	20	40	2	–	–	–	–	–	–	–	–	–	–	Methyl 2,3,4-trimethylhexadecanoate		MS07293
74	87	312	75	143	43	55	57	**312**	100	67	65	36	33	30	21	19		20	40	2	–	–	–	–	–	–	–	–	–	–	Nonadecanoic acid, methyl ester	1731-94-8	NI32931
74	87	43	312	75	55	57	69	**312**	100	74	28	23	23	22	19	15		20	40	2	–	–	–	–	–	–	–	–	–	–	Nonadecanoic acid, methyl ester	1731-94-8	WI01403
74	43	55	41	75	57	69	143	**312**	100	58	31	29	22	20	17	11	**1.00**	20	40	2	–	–	–	–	–	–	–	–	–	–	Nonadecanoic acid, methyl ester	1731-94-8	NI32932
312	88	101	157	269	313	267	89	**312**	100	84	64	32	27	26	24	17		20	40	2	–	–	–	–	–	–	–	–	–	–	Octadecanoic acid, ethyl ester	111-61-5	WI01404
88	101	43	57	55	41	73	71	**312**	100	60	55	47	45	43	26	25	**9.51**	20	40	2	–	–	–	–	–	–	–	–	–	–	Octadecanoic acid, ethyl ester	111-61-5	NI32933
88	101	43	55	41	57	89	29	**312**	100	61	37	25	24	23	19	18	**13.60**	20	40	2	–	–	–	–	–	–	–	–	–	–	Octadecanoic acid, ethyl ester	111-61-5	NI32934
88	43	28	41	101	55	57	29	**312**	100	38	30	29	24	23	18	17	**12.04**	20	40	2	–	–	–	–	–	–	–	–	–	–	Octadecanoic acid, 2-methyl-, methyl ester	2490-22-4	NI32937
88	101	43	55	57	41	69	56	**312**	100	39	28	23	22	16	12	7	**2.00**	20	40	2	–	–	–	–	–	–	–	–	–	–	Octadecanoic acid, 2-methyl-, methyl ester	2490-22-4	NI32936
88	101	312	43	55	57	41	69	**312**	100	42	22	20	16	15	15	12		20	40	2	–	–	–	–	–	–	–	–	–	–	Octadecanoic acid, 2-methyl-, methyl ester	2490-22-4	NI32935
88	101	43	312	55	57	41	69	**312**	100	45	28	25	24	23	22	16		20	40	2	–	–	–	–	–	–	–	–	–	–	Octadecanoic acid, 2-methyl-, methyl ester, (–)-	57346-70-0	NI32938
87	74	57	255	43	241	55	71	**312**	100	65	47	45	38	37	31	30	**28.00**	20	40	2	–	–	–	–	–	–	–	–	–	–	Octadecanoic acid, 4-methyl-, methyl ester	2490-15-5	NI32939
236	74	87	55	75	115	83	43	**312**	100	93	55	26	25	24	24	23	**6.04**	20	40	2	–	–	–	–	–	–	–	–	–	–	Octadecanoic acid, 6-methyl-, methyl ester	2490-21-3	NI32940
236	74	87	115	75	83	237	111	**312**	100	98	56	25	24	22	19	19	**4.30**	20	40	2	–	–	–	–	–	–	–	–	–	–	Octadecanoic acid, 6-methyl-, methyl ester	2490-21-3	MS07294
74	87	157	312	129	125	97	75	**312**	100	89	68	62	34	34	29	28		20	40	2	–	–	–	–	–	–	–	–	–	–	Octadecanoic acid, 7-methyl-, methyl ester		MS07295
74	87	157	312	129	125	97	75	**312**	100	89	68	62	34	34	29	28		20	40	2	–	–	–	–	–	–	–	–	–	–	Octadecanoic acid, 7-methyl-, methyl ester, (±)-	57346-72-2	NI32941
143	87	74	43	55	57	41	312	**312**	100	69	61	51	48	43	34	27		20	40	2	–	–	–	–	–	–	–	–	–	–	Octadecanoic acid, 8-methyl-, methyl ester		MS07296
74	87	312	57	69	213	55	143	**312**	100	61	47	33	29	25	25	24		20	40	2	–	–	–	–	–	–	–	–	–	–	Octadecanoic acid, 9-methyl-, methyl ester		MS07297
74	312	87	43	55	57	143	41	**312**	100	94	79	62	58	50	48	41		20	40	2	–	–	–	–	–	–	–	–	–	–	Octadecanoic acid, 10-methyl-, methyl ester	2490-19-9	MS07298
74	312	143	199	269	213	172	157	**312**	100	59	46	20	17	12	10	10		20	40	2	–	–	–	–	–	–	–	–	–	–	Octadecanoic acid, 10-methyl-, methyl ester	2490-19-9	LI06897
74	87	143	312	75	57	71	85	**312**	100	65	39	36	32	21	16	15		20	40	2	–	–	–	–	–	–	–	–	–	–	Octadecanoic acid, 10-methyl-, methyl ester	2490-19-9	NI32942

MASS TO CHARGE RATIOS								M.W.	INTENSITIES								Parent	C	H	O	N	S	F	Cl	Br	I	Si	P	B	X	COMPOUND NAME	CAS Reg No	No
74	312	87	43	55	57	143	41	**312**	100	94	79	62	58	50	48	41		20	40	2	–	–	–	–	–	–	–	–	–	–	**Octadecanoic acid, 10-methyl-, methyl ester, (R)-**	55044-28-5	**NI32943**
312	74	87	43	143	55	269	313	**312**	100	79	64	44	37	28	25	24		20	40	2	–	–	–	–	–	–	–	–	–	–	**Octadecanoic acid, 14-methyl-, methyl ester**		**MS07299**
312	74	87	57	41	55	43	75	**312**	100	97	57	50	40	39	39	32		20	40	2	–	–	–	–	–	–	–	–	–	–	**Octadecanoic acid, 16-methyl-, methyl ester**	2490-16-6	**NI32944**
312	74	87	57	41	43	143	29	**312**	100	97	57	50	42	40	24	22		20	40	2	–	–	–	–	–	–	–	–	–	–	**Octadecanoic acid, 16-methyl-, methyl ester**	2490-16-6	**LI06898**
74	87	43	55	41	57	75	69	**312**	100	66	55	35	33	30	22	21	**18.00**	20	40	2	–	–	–	–	–	–	–	–	–	–	**Octadecanoic acid, 17-methyl-, methyl ester**		**MS07300**
60	89	253	311	313	251	97	111	**312**	100	61	41	25	23	23	21	15	**5.90**	20	40	2	–	–	–	–	–	–	–	–	–	–	**Octadecyl acetate**	822-23-1	**NI32945**
43	55	57	83	69	61	97	41	**312**	100	63	59	55	53	48	41	39	**0.94**	20	40	2	–	–	–	–	–	–	–	–	–	–	**Octadecyl acetate**	822-23-1	**IC03869**
43	57	83	55	69	97	61	70	**312**	100	81	75	65	63	60	55	41	**0.00**	20	40	2	–	–	–	–	–	–	–	–	–	–	**Octadecyl acetate**	822-23-1	**NI32946**
71	41	55	43	67	82	29	81	**312**	100	56	31	26	24	22	20	19	**0.00**	20	40	2	–	–	–	–	–	–	–	–	–	–	**Olealdehyde, dimethyl acetal**	15677-71-1	**NI32947**
71	82	96	68	110	138	124	152	**312**	100	28	21	18	8	5	5	4	**0.00**	20	40	2	–	–	–	–	–	–	–	–	–	–	**Olealdehyde, dimethyl acetal**	15677-71-1	**LI06899**
31	71	32	29	41	43	82	55	**312**	100	85	73	52	43	27	24	23	**0.00**	20	40	2	–	–	–	–	–	–	–	–	–	–	**Olealdehyde, dimethyl acetal**	15677-71-1	**NI32948**
88	101	157	97	57	222	129	71	**312**	100	45	17	9	9	6	6	6	**4.39**	20	40	2	–	–	–	–	–	–	–	–	–	–	**Pentadecanoic acid, 2,6,10,14-tetramethyl-, methyl ester**	1001-80-5	**NI32949**
88	312	101	55	57	41	69	43	**312**	100	30	28	26	24	23	17	14		20	40	2	–	–	–	–	–	–	–	–	–	–	**Pentadecanoic acid, 2,6,10,14-tetramethyl-, methyl ester**	1001-80-5	**LI06900**
43	41	88	55	57	29	69	27	**312**	100	75	69	53	39	39	24	19	**2.14**	20	40	2	–	–	–	–	–	–	–	–	–	–	**Pentadecanoic acid, 4,6,10,14-tetramethyl-, methyl ester**	56554-31-5	**NI32950**
174	269	243	312	201	118	43	69	**312**	100	65	64	63	48	44	43	41		20	41	–	–	–	–	–	–	–	–	1	–	–	**Phosphine, bis(2,2-dimethylpropyl)[2(or 5)-methyl-5(or 2)-isopropylcyclohexyl]-**	74753-31-4	**NI32951**
235	77	312	311	236	92	104	51	**312**	100	97	80	70	64	41	30	28		21	16	1	2	–	–	–	–	–	–	–	–	–	**3-Benzoyl-7-methyl-2-phenylpyrazolo[1,5-a]pyridine**	17408-40-1	**NI32952**
105	312	77	313	311	51			**312**	100	90	40	21	15	13				21	16	1	2	–	–	–	–	–	–	–	–	–	**2-(2-Hydroxyphenyl)-4,5-diphenyl-imidazole**		**IC03870**
312	220	180	295	235	209			**312**	100	56	42	30	16	8				21	16	1	2	–	–	–	–	–	–	–	–	–	**3-[(α-Phenylamino)benzylidene]indolin-2-one**		**LI06901**
169	312	144	89	115	116	143	168	**312**	100	29	25	23	13	8	7	5		21	16	1	2	–	–	–	–	–	–	–	–	–	**4′-Phenyl-indole-2-carboxyanilide**		**LI06902**
312	104	180	165	269	77	283	166	**312**	100	91	89	84	82	59	57	55		21	16	1	2	–	–	–	–	–	–	–	–	–	**2,5,5-Triphenyl-2-imidazolin-4-one**	37068-60-3	**NI32953**
312	180	165	268	193				**312**	100	27	11	5	2					21	16	1	2	–	–	–	–	–	–	–	–	–	**1,4,5-Triphenyl-imidazol-2-one**		**LI06903**
119	41	55	125	109	189	91	124	**312**	100	61	50	39	38	37	26	25	**1.00**	21	28	2	–	–	–	–	–	–	–	–	–	–	**2,5-Cyclohexadiene-1,4-dione, 2-[(1,4,4a,5,6,7,8,8a-octahydro-2,5,5,8a-tetramethyl-1-naphthalenyl)methyl]-, [1R-(1α,4aβ,8aα)]-**	39707-59-0	**NI32954**
41	55	69	123	189	137	95	81	**312**	100	80	79	75	67	64	63	63	**19.00**	21	28	2	–	–	–	–	–	–	–	–	–	–	**2,5-Cyclohexadiene-1,4-dione, 2-[(decahydro-5,5,8a-trimethyl-2-methylene-1-naphthalenyl)methyl]-, [1S-(1α,4aβ,8aα)]-**	39707-56-7	**NI32955**
279	277	294	292	280	105	91	79	**312**	100	46	45	44	22	21	21	15	**0.00**	21	28	2	–	–	–	–	–	–	–	–	–	–	**(+)-12β-Hydroxy-5α-pregna-2,7,16-trien-20-one**		**DD01174**
269	256	241	312	270	257	43	41	**312**	100	87	42	28	23	17	15	12		21	28	2	–	–	–	–	–	–	–	–	–	–	**s-Indacene-1,7-dione, 2,3,5,6-tetrahydro-3,3,5,5-tetramethyl-8-(3-methylbutyl)-**	40650-59-7	**NI32956**
227	312	294	313	213	279	225	171	**312**	100	83	30	25	25	17	16	14		21	28	2	–	–	–	–	–	–	–	–	–	–	**DL-3α-Methoxy-17aα-methyl-D-homoestra-1,3,5(10),8-tetraen-17a-ol**		**LI06904**
279	294	280	312	295	293	277	171	**312**	100	86	41	28	17	10	10	9		21	28	2	–	–	–	–	–	–	–	–	–	–	**DL-3α-Methoxy-17aβ-methyl-D-homoestra-1,3,5(10),8-tetraen-17a-ol**		**LI06905**
141	237	197	167	169	153	128	195	**312**	100	72	67	64	63	57	57	56	**53.00**	21	28	2	–	–	–	–	–	–	–	–	–	–	**Methyl 6,8,11,13-abietatetraen-18-oate**		**MS07301**
141	158	43	140	57	71	128	85	**312**	100	29	23	10	10	8	8	6	**4.50**	21	28	2	–	–	–	–	–	–	–	–	–	–	**1-Naphthalenemethyl decanoate**	71844-11-6	**NI32957**
227	174	312	242	173	147	240	228	**312**	100	51	44	38	31	29	19	19		21	28	2	–	–	–	–	–	–	–	–	–	–	**19-Norpregna-1,3,5(10),20-tetraen-17-ol, 3-methoxy-, (17α)-**	6885-48-9	**NI32958**
162	312	163	297	313	119	156	91	**312**	100	20	20	10	5	5	3	3		21	28	2	–	–	–	–	–	–	–	–	–	–	**2,2′,3,3′,5,5′,6,6′-Octamethyl-4,4′-dihydroxy-diphenylmethane**		**MS07302**
237	197	41	43	195	167	155	141	**312**	100	96	90	82	76	54	52	47	**18.00**	21	28	2	–	–	–	–	–	–	–	–	–	–	**1-Phenanthrenecarboxylic acid, 1,2,3,4,4a,10a-hexahydro-1,4a-dimethyl-7-isopropyl-, methyl ester, [1R-(1α,4aβ,10aα)]-**	18492-76-7	**NI32959**
43	312	297	227	91	55	105	41	**312**	100	81	57	49	24	18	16	16		21	28	2	–	–	–	–	–	–	–	–	–	–	**Pregna-4,7-diene-3,20-dione**	17398-60-6	**NI32960**
43	312	269	297	124	189	270	135	**312**	100	91	70	37	32	27	25	25		21	28	2	–	–	–	–	–	–	–	–	–	–	**Pregna-4,16-diene-3,20-dione**	1096-38-4	**NI32961**
312	43	136	160	159	297	269	174	**312**	100	72	43	40	32	28	28	28		21	28	2	–	–	–	–	–	–	–	–	–	–	**Pregna-4,16-diene-3,20-dione**	1096-38-4	**LI06906**
269	312	297	270	189	43	124	227	**312**	100	80	48	37	37	37	35	28		21	28	2	–	–	–	–	–	–	–	–	–	–	**Pregna-4,16-diene-3,20-dione**	1096-38-4	**LI06907**
124	189	312	43	147	135	269	227	**312**	100	88	74	64	48	28	24	23		21	28	2	–	–	–	–	–	–	–	–	–	–	**Pregna-4,16-diene-3,20-dione, (9β,10α)-**		**LI06908**
110	162	147	160	81	105	91	79	**312**	100	91	79	70	70	65	65	64	**0.00**	21	28	2	–	–	–	–	–	–	–	–	–	–	**Pregn-4-en-20-yn-3-one, 17-hydroxy-, (17α)-**	434-03-7	**NI32962**
43	312	57	55	41	149	71	85	**312**	100	76	40	35	30	29	26	25		21	28	2	–	–	–	–	–	–	–	–	–	–	**Progester-6-enone**		**MS07303**
178	134	312	210	313	179	311	235	**312**	100	34	30	10	7	7	5	4		22	16	–	–	1	–	–	–	–	–	–	–	–	**1,2-Diphenyl-2a,7b-dihydro-cyclobuta[b][1]benzothiophene**		**LI06909**
210	312	178	280	279	134	313	278	**312**	100	33	10	7	7	7	6	6		22	16	–	–	1	–	–	–	–	–	–	–	–	**2,2a-Diphenyl-2a,7b-dihydro-cyclobuta[b][1]benzothiophene**		**LI06910**
141	155	127	312	115	140	139	128	**312**	100	70	20	19	13	9	7	7		22	16	2	–	–	–	–	–	–	–	–	–	–	**1-Naphthalenemethyl naphthalene-2-carboxylate**	86328-63-4	**NI32963**
105	191	189	312	77	190	165	179	**312**	100	33	19	16	16	12	6	4		22	16	2	–	–	–	–	–	–	–	–	–	–	**9-Phenanthrenemethyl benzoate**	38418-18-7	**NI32964**
105	77	312	51	207	191	284	106	**312**	100	83	24	22	17	14	13	10		22	16	2	–	–	–	–	–	–	–	–	–	–	**1,3-Propanedione, 1,3-diphenyl-2-benzylidene-**	5398-64-1	**NI32965**
105	77	312	51	207	191	284	178	**312**	100	83	24	22	17	14	13	9		22	16	2	–	–	–	–	–	–	–	–	–	–	**1,3-Propanedione, 1,3-diphenyl-2-benzylidene-**	5398-64-1	**LI06911**
312	267	268	265	294	165	252	189	**312**	100	69	38	34	28	24	14	14		22	16	2	–	–	–	–	–	–	–	–	–	–	**2,4,4-Triphenyl-2,3-butadienoic acid**		**MS07304**
312	313	193	195	221	119	117	105	**312**	100	27	22	18	14	13	13	12		24	24	–	–	–	–	–	–	–	–	–	–	–	**Metacyclophane**	13612-55-0	**NI32966**
134	69	70	46	131	129	231	229	**313**	100	32	20	19	7	7	2	2		3	–	–	1	1	8	–	1	–	–	–	–	–	**N-(1,1,2,3,3,3-Hexafluoro-2-bromo-propyl)-thiodifluorimide**		**LI06912**
166	78	69	143	28	177			**313**	100	19	19	10	5	4			**0.00**	4	4	–	1	–	6	3	–	–	1	–	–	–	**2-[N,N-Bis(trifluoromethyl)amino]ethyltrichlorosilane**		28871-58-1	**LI06913**

MASS TO CHARGE RATIOS								M.W.	INTENSITIES								Parent	C	H	O	N	S	F	Cl	Br	I	Si	P	B	X	COMPOUND NAME	CAS Reg No	No
166	78	69	149	147	145	143	28	313	100	19	19	10	10	10	10	5	0.00	4	4	–	1	–	6	3	–	–	1	–	–	–	2-[N,N-Bis(trifluoromethyl)amino]ethyltrichlorosilane	28871-58-1	NI32967
43	273	245	30	315	243	271	142	313	100	99	99	94	89	86	84	78	70.72	8	3	2	3	–	–	4	–	–	–	–	–	–	1H-Benzotriazole, 1-acetoxy-4,5,6,7-tetrachloro-	18355-09-4	NI32968
43	226	227	41	169	69	58	42	313	100	67	57	39	23	21	12	9	0.00	9	10	3	1	–	7	–	–	–	–	–	–	–	N-(Heptafluorobutyryl)glycine propyl ester		MS07305
73	236	282	211	241	72	266	225	313	100	73	72	71	66	65	63	55	8.01	9	24	5	1	–	–	–	–	–	2	1	–	–	Phosphoric acid, 2-(methoxyimino)ethyl bis(trimethylsilyl) ester	55044-12-7	NI32969
110	58	278	63	43	64	42	41	313	100	65	40	40	33	29	27	27	0.42	10	14	2	1	1	–	2	–	–	–	1	–	–	Phosphoramidothioic acid, isopropyl-, O-(2,4-dichlorophenyl) O-methyl ester	299-85-4	NI32970
110	279	281	152	58	223	136	225	313	100	87	49	42	35	25	19	16	1.06	10	14	2	1	1	–	2	–	–	–	1	–	–	Phosphoramidothioic acid, isopropyl-, O-(2,4-dichlorophenyl) O-methyl ester	299-85-4	NI32971
110	278	280	58	152	223	136	63	313	100	84	43	36	35	25	18	15	0.00	10	14	2	1	1	–	2	–	–	–	1	–	–	Phosphoramidothioic acid, isopropyl-, O-(2,4-dichlorophenyl) O-methyl ester	299-85-4	NI32972
250	222	55	268	85	313			313	100	83	73	53	42	7				11	11	8	3	–	–	–	–	–	–	–	–	–	2,4-Dinitrophenyl-glutamic acid		LI06914
226	93	228	190	73	95	241	43	313	100	76	36	33	30	27	18	16	0.00	11	20	–	5	–	–	1	–	–	2	–	–	–	9H-Purin-2-amine, 6-chloro-N,9-bis(trimethylsilyl)-		MS07306
313	151	166	298	162	136	314	121	313	100	39	36	21	14	14	13	12		12	12	3	1	–	5	–	–	–	–	–	–	–	2,4-Dimethoxybenzylamine PFP		MS07307
313	166	151	139	314	263	124	77	313	100	80	22	16	13	12	9	8		12	12	3	1	–	5	–	–	–	–	–	–	–	3,5-Dimethoxybenzylamine PFP		MS07308
254	266	313	256	75	169			313	100	38	33	33	32	28				12	12	5	3	–	–	1	–	–	–	–	–	–	2-(5-Amino-4-chloro-3-oxo-2,3-dihydro-2-pyridazino)-cis-cis-muconic acid, dimethyl ester		NI32973
313	93	43	65	157	40	107	249	313	100	76	67	63	61	47	43	36		12	15	3	3	2	–	–	–	–	–	–	–	–	2-(4-Hydroxybenzenesulphonamido)-5-isobutyl-1,3,4-thiadiazine		LI06915
161	77	97	162	91	65	172	51	313	100	87	75	66	48	38	35	35	3.60	12	16	3	3	1	–	–	–	–	–	1	–	–	Phosphorothioic acid, O,O-diethyl O-(1-phenyl-1H-1,2,4-triazol-3-yl) ester	24017-47-8	NI32974
43	153	57	152	41	61	44	84	313	100	35	32	21	18	13	11	8	0.00	12	18	5	1	–	3	–	–	–	–	–	–	–	L-Threonine, N-(trifluoroacetyl)-, sec-butyl ester, acetate	57983-71-8	NI32975
207	208	192	164	178	248	313	278	313	100	50	50	45	33	31	28	28		12	20	1	7	–	–	1	–	–	–	–	–	–	N-(2-Chloro-1-ethoxyethyl)-N-cyano-N',N',N'',N''-tetramethyl-1,3,5-triazine-2,4,6-triamine	87166-16-3	NI32976
114	43	72	170	186	115	86	143	313	100	48	18	18	16	12	9	9	0.01	12	28	–	1	–	–	–	–	1	–	–	–	–	Tetra-propylammonium iodide		NI32977
125	171	281	313	209	208	59	97	313	100	95	28	24	15	14	12	11		13	15	6	1	1	–	–	–	–	–	–	–	–	2-Thiabicyclo[3.2.0]hepta-3,6-diene-4,6,7-tricarboxylic acid, 3-amino-, 4-ethyl 6,7-dimethyl ester	66599-28-8	NI32978
129	128	127	102	51	130	45	76	313	100	43	25	20	14	11	9	8	0.00	13	16	–	1	–	–	–	–	1	–	–	–	–	Quinoline butiodide		IC03871
129	57	41	102	128	183	51	39	313	100	73	44	29	18	16	16	15	0.00	13	16	–	1	–	–	–	–	1	–	–	–	–	Quinoline butiodide		IC03872
313	148	176	188	175	284	160	204	313	100	52	49	38	22	11	9	5		13	16	6	1	–	–	–	–	–	–	1	–	–	2,3,4,5-Tetrahydro-7,8-(methylenedioxy)-N-(dimethylphosphoryl)-3-benzazepin-1-one		NI32979
73	196	95	147	75	45	197	39	313	100	61	49	22	20	20	12	11	2.47	13	23	4	1	–	–	–	–	–	2	–	–	–	Glycine, N-(2-furanylcarbonyl)-N-(trimethylsilyl)-, trimethylsilyl ester	55556-83-7	NI32980
95	226	169	75	197	143	168	196	313	100	95	50	37	35	35	23	20	0.11	13	23	4	1	–	–	–	–	–	2	–	–	–	Glycine, N-(2-furanylcarbonyl)-N-(trimethylsilyl)-, trimethylsilyl ester	55556-83-7	NI32981
134	146	90	75	313	149	164	179	313	100	39	39	19	16	11	11	10		14	7	6	3	–	–	–	–	–	–	–	–	–	2-(2,4-Dinitrophenyl)-3,1-benzoxazin-4(4H)-one	68490-45-9	NI32982
146	313	90	75	269	149	223	74	313	100	57	38	10	7	1	1	1		14	7	6	3	–	–	–	–	–	–	–	–	–	2-(2,4-Dinitrophenyl)-3,1-benzoxazin-4(4H)-one	68490-45-9	NI32983
105	74	313	77	223	177	75	90	313	100	92	83	45	34	24	24	18		14	7	6	3	–	–	–	–	–	–	–	–	–	4,6-Dinitro-2-phenylisatogen		LI06916
313	298	314	252	267	75	51	44	313	100	26	17	14	9	7	5	5		14	11	4	5	–	–	–	–	–	–	–	–	–	2,3-Dimethoxy-8-nitro-4b,5,9b,10-tetraazaindeno[2,1-a]indene		IC03873
313	163	64	296	52	92	51	314	313	100	34	32	28	28	24	24	20		14	11	4	5	–	–	–	–	–	–	–	–	–	5-(2-Nitrophenylazo)-1-ethyl-3-cyano-6-hydroxypyrid-2-one		IC03874
181	166	135	313	72	151	107	182	313	100	29	28	22	19	17	17	16		14	19	7	1	–	–	–	–	–	–	–	–	–	17-Nitro-2,3-benzo-1,4,7,10,13-pentaoxacyclopentadeca-2-ene		MS07309
43	228	57	99	56	128	168	256	313	100	45	38	32	29	27	26	25	0.70	14	23	5	3	–	–	–	–	–	–	–	–	–	1-(4'-Acetoxybutyl)-3-methyl-1,3,4-triacetylguanidine		NI32984
85	198	91	155	65	41	58	43	313	100	70	70	56	16	15	12	12	0.00	15	23	4	1	1	–	–	–	–	–	–	–	–	(±)-N-[1-Methyl-2-[(tetrahydro-2H-pyran-2-yl)oxy]ethyl]-p-toluenesulphonamide		DD01175
298	101	43	72	240	98	59	299	313	100	63	44	30	18	18	18	17	0.00	15	23	6	1	–	–	–	–	–	–	–	–	–	3-C-Cyanomethyl-1,2,5,6-di-O-isopropylidene-3-O-methyl-α-D-glucofuranose		NI32985
222	223	91	28	181	203	180	65	313	100	12	11	8	6	5	5	4	0.30	16	12	–	1	–	5	–	–	–	–	–	–	–	1-Phenyl-N-(pentafluorobenzylidene)-propyl-2-amine		MS07310
312	285	286	313	266	238	283	284	313	100	100	67	63	48	43	40	33		16	12	3	3	–	1	–	–	–	–	–	–	–	2H-1,4-Benzodiazepin-2-one, 5-(2-fluorophenyl)-1,3-dihydro-1-methyl-7-nitro-	1622-62-4	NI32986
105	93	77	119	51	91	194	64	313	100	57	50	35	18	16	12	11	1.80	16	15	2	3	1	–	–	–	–	–	–	–	–	Carbamimidothioic acid, N-benzoyl-N'-[(phenylamino)carbonyl]-, methyl ester	39536-11-3	NI32987
135	163	120	103	254	285	92	194	313	100	37	25	25	8	7	7	1	0.60	16	15	4	3	–	–	–	–	–	–	–	–	–	Benzoic acid, 4,4'-(1-triazene-1,3-diyl)bis-, dimethyl ester	55842-28-9	NI32988
239	313	154	227	105	77	80	240	313	100	75	24	21	21	20	18	15		16	15	4	3	–	–	–	–	–	–	–	–	–	2-Phenyl-4-(ethoxycarbonyl)-6-methyl-3H-imidazo[1,5-b]pyridazine-5,7-dione	72481-76-6	NI32989
225	271	105	77	154	313	127	155	313	100	70	65	40	38	32	30	28		16	15	4	3	–	–	–	–	–	–	–	–	–	2-Phenyl-4-(isopropyloxycarbonyl)-3H-imidazo[1,5-b]pyridazine-5,7-(6H)-dione	76089-67-3	NI32990
225	313	271	105	227	77	154	43	313	100	54	44	33	32	31	24	24		16	15	4	3	–	–	–	–	–	–	–	–	–	2-Phenyl-4-(propoxycarbonyl)-3H-imidazo[1,5-b]pyridazine-5,7-(6H)-dione	76089-66-2	NI32991
220	234	235	41	55	207	221	43	313	100	69	29	22	21	20	17	17	5.00	16	27	3	1	1	–	–	–	–	–	–	–	–	1-(1-Hexahydroazepino)-5,5,7-trimethyl-8-oxa-3-thiatricyclo[5.2.0.0^{7,9}]nonane-3,3-dioxide	104196-21-6	NI32992

MASS TO CHARGE RATIOS								M.W.	INTENSITIES								Parent	C	H	O	N	S	F	Cl	Br	I	Si	P	B	X	COMPOUND NAME	CAS Reg No	No
138	93	59	139	137	94	43	80	**313**	100	53	43	37	18	18	15	10	**8.01**	16	27	5	1	–	–	–	–	–	–	–	–	–	Heliotrine	303-33-3	NI32993
138	93	59	139	137	80	156	313	**313**	100	54	44	37	18	12	10	8		16	27	5	1	–	–	–	–	–	–	–	–	–	Heliotrine	303-33-3	LI06917
158	128	129	313					**313**	100	99	56	26						17	15	3	1	1	–	–	–	–	–	–	–	–	3H-1-Benzazepin-3-one, 1,2-dihydro-1-(4-tolylsulphonyl)-	19673-37-1	NI32994
251	313	265	283	207	224	238		**313**	100	60	54	50	25	22	20			17	15	5	1	–	–	–	–	–	–	–	–	–	3-Methoxycarbonyl-4-carboxy-2′-methylcarbamoyl-biphenyl		LI06918
298	270	313	299	271	284	314	159	**313**	100	80	78	24	17	16	15	14		17	15	5	1	–	–	–	–	–	–	–	–	–	10-Methyl-1,2-dimethoxy-3,4-methylenedioxyacridone		MS07311
313	270	252	298	280	285	268	314	**313**	100	99	70	52	46	32	32	28		17	15	5	1	–	–	–	–	–	–	–	–	–	10-Methyl-1,4-dimethoxy-2,3-methylenedioxyacridone		MS07312
313	270	285	267	239	268	240	224	**313**	100	74	68	57	56	38	27	27		17	15	5	1	–	–	–	–	–	–	–	–	–	Tecleanthine		MS07313
298	300	230	242	105	313			**313**	100	37	25	21	21	16				17	16	1	3	–	–	1	–	–	–	–	–	–	5-Chloro-2-(2-isopropylidenehydrazomethyleneamino)benzophenone		LI06919
245	56	257	193	247	246	244	228	**313**	100	60	40	35	32	27	22	20	**8.00**	17	16	1	3	–	–	1	–	–	–	–	–	–	2-Chloro-11-(1-piperazinyl)dibenz[b,f]-1,4-oxazepine		MS07314
196	268	140	197	29	141	269	168	**313**	100	42	20	12	10	8	7	6	**2.20**	17	19	3	3	–	–	–	–	–	–	–	–	–	N-(2-Carboxyethyl)-N-(2-methoxyethyl)-3-methyl-4-(2,2-dicyanovinyl)-aniline		IC03875
82	110	313	295	125	83	204	186	**313**	100	40	30	30	30	30	28	27		17	19	3	3	–	–	–	–	–	–	–	–	–	Isolongistrobine		LI06920
254	59	114	130	73	174	100	86	**313**	100	74	64	46	35	33	26	24	**0.00**	17	35	2	1	–	–	–	–	–	1	–	–	–	(E)-1-[[[Bis(2-ethoxyethyl)amino]methyl]dimethylsilyl]-1,5-hexadiene		DD01176
254	59	130	67	174	86	100	208	**313**	100	73	53	45	32	28	24	11	**0.00**	17	35	2	1	–	–	–	–	–	1	–	–	–	(Z)-1-[[[Bis(2-ethoxyethyl)amino]methyl]dimethylsilyl]-1,5-hexadiene		DD01177
254	59	130	174	114	232	73	100	**313**	100	60	40	38	34	31	29	28	**0.00**	17	35	2	1	–	–	–	–	–	1	–	–	–	3-[[[Bis(2-ethoxyethyl)amino]methyl]dimethylsilyl]-1,5-hexadiene		DD01178
146	268	59	69	114	100	314	130	**313**	100	86	77	70	57	23	9	7	**0.00**	17	35	2	1	–	–	–	–	–	1	–	–	–	(E)-1-[[[Bis(2-methoxyethyl)amino]methyl]dimethylsilyl]-6-methyl-1,5-heptadiene		DD01179
146	204	268	59	314	282	160	100	**313**	100	88	71	60	44	30	21	19	**0.00**	17	35	2	1	–	–	–	–	–	1	–	–	–	3-[[[Bis(2-methoxyethyl)amino]methyl]dimethylsilyl]-6-methyl-1,5-heptadiene		DD01180
86	44	42	84	85	57	228	171	**313**	100	56	21	20	19	16	15	13	**0.30**	18	19	2	1	1	–	–	–	–	–	–	–	–	4-(4-Methylaminobutoxy)thioxanthen-9-one		MS07315
313	298	296	42	314	132	270	312	**313**	100	43	34	28	15	15	13	11		18	19	4	1	–	–	–	–	–	–	–	–	–	Claviculine	87035-67-4	NI32995
313	298	175	110	314	150	188	43	**313**	100	26	25	22	17	16	14	14		18	19	4	1	–	–	–	–	–	–	–	–	–	3,11-Didehydroxy-3,11-epoxycephalotaxine		NI32996
255	227	226	199	198	270	313		**313**	100	80	60	55	35	32	20			18	19	4	1	–	–	–	–	–	–	–	–	–	Erythratinone		LI06921
282	313	264	280	298	283	295	262	**313**	100	98	85	80	70	70	57	50		18	19	4	1	–	–	–	–	–	–	–	–	–	Erythrinan-11-ol, 1,2,6,7-tetradehydro-3-methoxy-15,16-[methylenebis(oxy)]-, (3β,11α)-	29306-29-4	MS07316
298	313	299	65	280	149	150	314	**313**	100	50	20	15	13	13	12	10		18	19	4	1	–	–	–	–	–	–	–	–	–	2-Hydroxy-4-diethylamino-2′-carboxybenzophenone		IC03876
313	241	225	242	314	43			**313**	100	92	84	76	42	30				18	19	4	1	–	–	–	–	–	–	–	–	–	7-Isoquinolinol, 2-acetyl-1,2,3,4-tetrahydro-4-(4-hydroxyphenyl)-6-methoxy-, (S)-	56771-98-3	NI32997
313	285	115	230	284	70	228	128	**313**	100	98	70	49	48	47	44	43		18	19	4	1	–	–	–	–	–	–	–	–	–	6-Oxo-3-methoxy-N-methyl-4,5,7,8-diepoxymorphine		MS07317
298	313	255	242	299	253	270	300	**313**	100	37	34	28	26	26	24	20		18	19	4	1	–	–	–	–	–	–	–	–	–	10-Methyl-2,3-dimethoxy-1-ethoxyacridone		MS07318
313	229	214	188	314	230	256	257	**313**	100	65	30	28	23	20	12	9		18	19	4	1	–	–	–	–	–	–	–	–	–	Morphinan-6-one, 7,8-didehydro-4,5-epoxy-14-hydroxy-3-methoxy-17-methyl-, (5α)-	508-54-3	NI32998
150	296	253	191	295	194	106	166	**313**	100	54	36	34	29	29	26	12	**0.00**	18	19	4	1	–	–	–	–	–	–	–	–	–	(±)-Oliverolin hydrate		DD01181
236	78	91	181	213	77	183	167	**313**	100	41	23	22	18	18	12	11	**0.80**	18	23	2	1	–	–	–	–	–	1	–	–	–	2,2-Diphenyl-6-ethyl-1,3-dioxa-6-aza-2-silacyclooctane		MS07319
313	242	223	71	44	210	119	238	**313**	100	83	50	50	46	43	40	36		18	23	2	3	–	–	–	–	–	–	–	–	–	Benzoic acid, 2-[(2-aminophenyl)[3-(methylamino)propyl]amino]-, methyl ester	13450-71-0	NI32999
188	67	121	68	161	97	173	209	**313**	100	98	80	74	33	16	11	6	**0.00**	18	23	2	3	–	–	–	–	–	–	–	–	–	N-Cyclohexyl-N′-(2,2-dimethyl-3-butynoyl)-N′-(3-pyridyl)urea		DD01182
57	86	212	41	113	240	154	130	**313**	100	92	79	65	56	52	50	49	**0.00**	18	35	3	1	–	–	–	–	–	–	–	–	–	tert-Butyl (7-oxotridecyl)carbamate	116437-40-2	DD01183
102	144	270	87	198	145	113	137	**313**	100	72	51	37	22	22	22	21	**7.00**	18	35	3	1	–	–	–	–	–	–	–	–	–	Methyl 3-acetylamino-12-methyltetradecanoate		NI33000
91	120	109	222	296	123	151	122	**313**	100	90	77	69	42	39	35	31	**13.00**	19	20	2	1	–	1	–	–	–	–	–	–	–	1-Benzyl-6-(4-fluorophenyl)-3a-hydroxy-3-methyl-4-piperidone	103022-00-0	NI33001
312	313	244	285	242	270	229	228	**313**	100	70	66	47	32	23	20	16		19	23	3	1	–	–	–	–	–	–	–	–	–	Bulbocodine	33792-78-8	NI33002
313	270	229	314	137	149	271	269	**313**	100	91	31	30	30	21	20	19		19	23	3	1	–	–	–	–	–	–	–	–	–	4-Diisopropylamino-5,6-dihydro-2H-pyrano[3,2-d]-1-benzoxepin-2-one	86896-96-0	NI33003
312	313	203	282	190	191	311	257	**313**	100	62	56	54	35	33	23	21		19	23	3	1	–	–	–	–	–	–	–	–	–	(±)-2,3-Dimethoxy-5,6,8,8a,9,10,13,13a-octahydro-11H-dibenzo[a,g]-quinolizin-11-one		MS07320
206	207	190	192	191	163	148	107	**313**	100	27	26	23	11	10	10	10	**0.10**	19	23	3	1	–	–	–	–	–	–	–	–	–	7,8-Dimethoxy-2-methyl-1-(4′-hydroxybenzyl)-1,2,3,4-tetrahydroisoquinoline		NI33004
239	270	313	135	165	208	151	224	**313**	100	47	25	17	10	8	8	7		19	23	3	1	–	–	–	–	–	–	–	–	–	(+)-6,7-Dimethoxy-2-methyl-4(R)-(4-methoxyphenyl)-1,2,3,4-tetrahydroisoquinoline	23367-60-4	LI06922
239	270	313	135	165	208	151	195	**313**	100	47	25	17	10	8	8	7		19	23	3	1	–	–	–	–	–	–	–	–	–	(+)-6,7-Dimethoxy-2-methyl-4(R)-(4-methoxyphenyl)-1,2,3,4-tetrahydroisoquinoline	23367-60-4	MS07321
282	298	313	280	234	242	255		**313**	100	44	40	17	14	11	7			19	23	3	1	–	–	–	–	–	–	–	–	–	Erysotrine	27740-43-8	LI06923
192	177	43	193	41	69	57	72	**313**	100	27	15	13	13	11	11	10	**0.10**	19	23	3	1	–	–	–	–	–	–	–	–	–	7-Hydroxy-8-methoxy-2-methyl-1-(4′-methoxybenzyl)-1,2,3,4-tetrahydroisoquinoline		NI33005
313	282	186	297	259	242	214	226	**313**	100	43	40	39	32	22	19	18		19	23	3	1	–	–	–	–	–	–	–	–	–	1-[6-Methoxy-7-(3-methylbut-2-enyl)indol-2-yl]-3-methylbutane-1,4-dione		MS07322
313	282	138	314	229	42	298	178	**313**	100	25	22	19	16	14	11	11		19	23	3	1	–	–	–	–	–	–	–	–	–	Morphinan, 7,8-didehydro-4,5-epoxy-3,6-dimethoxy-17-methyl-, (5α,6α)-	2859-16-7	LI06924
313	229	138	162	42	176	188	299	**313**	100	90	81	72	69	65	62	61		19	23	3	1	–	–	–	–	–	–	–	–	–	Morphinan, 7,8-didehydro-4,5-epoxy-3,6-dimethoxy-17-methyl-, (5α,6α)-	2859-16-7	NI33006

MASS TO CHARGE RATIOS								M.W.	INTENSITIES								Parent	C	H	O	N	S	F	Cl	Br	I	Si	P	B	X	COMPOUND NAME	CAS Reg No	No
313	162	314	284	243	124	214	42	**313**	100	24	22	20	16	14	11	11		19	23	3	1	–	–	–	–	–	–	–	–	–	**Morphinan-6-ol, 7,8-didehydro-4,5-epoxy-3-ethoxy-17-methyl-, (5α,6α)-**	76-58-4	**NI33007**
313	42	59	162	44	314	284	124	**313**	100	38	29	28	23	19	18	18		19	23	3	1	–	–	–	–	–	–	–	–	–	**Morphinan-6-ol, 7,8-didehydro-4,5-epoxy-3-ethoxy-17-methyl-, (5α,6α)-**	76-58-4	**NI33008**
313	298	42	314	299	270	59	44	**313**	100	85	32	24	24	22	21	18		19	23	3	1	–	–	–	–	–	–	–	–	–	**Morphinan-7-one, 5,6-didehydro-3,6-dimethoxy-17-methyl-**	55101-91-2	**NI33009**
313	298	299	270	59	42	314	192	**313**	100	70	27	24	24	24	22	20		19	23	3	1	–	–	–	–	–	–	–	–	–	**Morphinan-6-one, 7,8-didehydro-3,7-dimethoxy-17-methyl-, L-(+)-**	14910-53-3	**NI33010**
311	296	214	313	312	242	254	213	**313**	100	66	36	32	30	30	18	17		19	23	3	1	–	–	–	–	–	–	–	–	–	**Thebaine**	115-37-7	**LI06925**
144	198	112	56	43	41	199	55	**313**	100	90	27	25	21	15	14	14	**0.20**	19	39	2	1	–	–	–	–	–	–	–	–	–	**Octadecanoic acid, 6-amino-, methyl ester**	30616-17-2	**NI33011**
228	114	229	196	56	55	43	115	**313**	100	78	16	15	8	7	7	6	**1.20**	19	39	2	1	–	–	–	–	–	–	–	–	–	**Octadecanoic acid, 12-amino-, methyl ester**	30616-18-3	**NI33012**
77	180	51	181	313	78	314	312	**313**	100	99	90	43	41	11	9	4		20	15	1	3	–	–	–	–	–	–	–	–	–	**Anhydro-3-hydroxy-1,4,5-triphenyl-1,2,4-triazolium hydroxide**	17370-11-5	**NI33013**
161	162	313	285	160	284	256	312	**313**	100	21	13	11	9	8	8	7		20	27	2	1	–	–	–	–	–	–	–	–	–	**18-Ethyl-3-methoxy-8-azoestra-1,3,5(10)-trien-17-one**		**IC03877**
282	160	283	313	173	314	147	171	**313**	100	52	36	35	31	23	23	16		20	27	2	1	–	–	–	–	–	–	–	–	–	**1-Methyl-estrone-methyloxime**		**NI33014**
228	172	158	313	242	186	43	312	**313**	100	47	38	33	29	29	28	21		20	31	–	3	–	–	–	–	–	–	–	–	–	**4-(N,N-Di-hexylamino)quinazoline**		**MS07323**
313	314	312	236	315	237	157	316	**313**	100	21	18	18	5	3	3	1		21	15	–	1	1	–	–	–	–	–	–	–	–	**Benzo[b]thiophen-2-amine, 3-phenyl-N-benzylidene-**	51324-20-0	**NI33015**
313	236	312	314	165	237	235	207	**313**	100	60	30	24	12	11	9	9		21	15	–	1	1	–	–	–	–	–	–	–	–	**Benzo[b]thiophen-2-amine, 3-phenyl-N-benzylidene-**	51324-20-0	**NI33016**
91	313	106	312	65	314	296	77	**313**	100	91	91	27	27	23	18	18		21	15	2	1	–	–	–	–	–	–	–	–	–	**1-Benzylaminoanthraquinone**		**IC03878**
44	115	28	144	114	65	63	18	**313**	100	83	75	63	45	35	35	35	**9.00**	21	15	2	1	–	–	–	–	–	–	–	–	–	**2-Hydroxy-N-(2-hydroxy-1-naphthylimine)-1-naphthalene**		**LI06926**
313	298	312	314	299	156.5	269	240	**313**	100	75	37	22	16	16	8	8		21	15	2	1	–	–	–	–	–	–	–	–	–	**8-Methyl-14-oxo-8H,13H,14H-naphtho[1′,2′:5,6]pyrano[2,3-b]quinoline**		**MS07324**
254	282	313	253	255	252	283	251	**313**	100	94	47	37	30	28	26	17		21	15	2	1	–	–	–	–	–	–	–	–	–	**Methyl 2-(6-phenanthridyl)benzoate**		**NI33017**
106	191	189	313	192	78	179	190	**313**	100	92	46	44	25	24	22	19		21	15	2	1	–	–	–	–	–	–	–	–	–	**9-Phenanthrenemethyl nicotinate**	92174-45-3	**NI33018**
166	165	269	313	297				**313**	100	90	63	2	2					21	15	2	1	–	–	–	–	–	–	–	–	–	**4,5,5-Triphenyl-Δ^3-oxazol-2-one**		**LI06927**
282	313	281	266	158	145	124	283	**313**	100	48	37	30	25	25	25	24		21	31	1	1	–	–	–	–	–	–	–	–	–	**syn-Retinal methoxime**	54595-39-0	**NI33019**
105	222	91	313	77	167	90	106	**313**	100	43	42	36	35	21	9	8		22	19	1	1	–	–	–	–	–	–	–	–	–	**Aziridine, 3-phenyl-1-benzyl-2-phenyl ketone-, cis-**	6372-57-2	**NI33020**
105	222	77	313	91	167	106	223	**313**	100	47	33	29	28	21	9	8		22	19	1	1	–	–	–	–	–	–	–	–	–	**Aziridine, 3-phenyl-1-benzyl-2-phenyl ketone-, trans-**	6476-12-6	**NI33021**
208	313	296	105	77	160	167	165	**313**	100	98	90	90	66	51	47	33		22	19	1	1	–	–	–	–	–	–	–	–	–	**3-(1,1′-Biphenyl)-4-yl-1-methyl-2-aziridinyl-phenyl ketone-, trans-**	32044-32-9	**NI33022**
104	105	77	18	78	51	313	103	**313**	100	80	75	28	24	15	15	14		22	19	1	1	–	–	–	–	–	–	–	–	–	**2,4,6-Triphenyl-5,6-dihydro-4H-1,3-oxazine**	91875-67-1	**NI33023**
180	104	77	105	179	78	178	18	**313**	100	64	54	44	34	23	22	22	**2.70**	22	19	1	1	–	–	–	–	–	–	–	–	–	**2,5,6-Triphenyl-5,6-dihydro-4H-1,3-oxazine**	111664-27-8	**NI33024**
180	105	77	165	313	178	18	181	**313**	100	46	45	42	31	22	21	20		22	19	1	1	–	–	–	–	–	–	–	–	–	**2,6,6-Triphenyl-5,6-dihydro-4H-1,3-oxazine**		**NI33025**
313	312	283	311	310	298	284	257	**313**	100	80	74	67	44	22	21	20		23	23	–	1	–	–	–	–	–	–	–	–	–	**8,10,12-Triethylbenzo[a]acridine**		**MS07325**
44	145	58	127	100	95	72	172	**314**	100	93	87	80	40	33	30	29	**0.00**	4	3	2	2	1	9	–	–	–	–	–	–	–	**(3-Carbamoyl-2,2,3,3-tetrafluoropropionamido)sulphur pentafluoride**	91940-17-9	**NI33026**
64	314	100	82	131	69	63	145	**314**	100	88	45	43	41	41	40	25		5	–	–	–	2	10	–	–	–	–	–	–	–	**Perfluoro-1,2-dithiepane**		**MS07326**
77	51	50	74	314	112	78	75	**314**	100	73	42	11	9	7	7	7		6	5	–	–	–	–	1	–	–	–	–	–	1	**Mercury, chlorophenyl-**	100-56-1	**NI33029**
77	51	28	50	314	312	112	74	**314**	100	54	32	31	18	13	11	10		6	5	–	–	–	–	1	–	–	–	–	–	1	**Mercury, chlorophenyl-**	100-56-1	**NI33028**
77	314	312	313	311	316	310	279	**314**	100	56	40	29	26	23	14	10		6	5	–	–	–	–	1	–	–	–	–	–	1	**Mercury, chlorophenyl-**	100-56-1	**NI33027**
42	28	200	259	218	220	82	58	**314**	100	43	42	35	34	33	25	23	**3.00**	6	8	3	2	–	–	–	2	–	–	–	–	–	**Hydrouracil, 5,5-dibromo-6-hydroxy-1,3-dimethyl-**	27994-76-9	**NI33030**
42	28	200	259	220	218	82	202	**314**	100	43	42	35	34	34	24	23	**3.20**	6	8	3	2	–	–	–	2	–	–	–	–	–	**Hydrouracil, 5,5-dibromo-6-hydroxy-1,3-dimethyl-**	27994-76-9	**MS07327**
99	27	28	26	39	81	54	41	**314**	100	43	37	23	19	13	12	12	**0.00**	7	8	4	–	–	–	–	2	–	–	–	–	–	**Dibromopropyl maleate**		**IC03879**
214	212	89	133	85	215	213	107	**314**	100	99	39	23	19	10	10	8	**0.00**	9	8	–	–	1	4	–	1	–	–	–	1	–	**3-Bromo-1-methylbenzo[b]thiophenium tetrafluoroborate**		**LI06928**
29	169	241	242	69	28	45	214	**314**	100	61	51	41	39	24	23	21	**0.00**	9	9	4	–	–	7	–	–	–	–	–	–	–	**Ethyl lactate heptafluorobutyrate**		**NI33031**
139	299	134	136	297	169	167	295	**314**	100	96	69	64	59	47	37	34	**0.00**	9	10	–	–	–	4	–	–	–	–	–	–	1	**Stannane, trimethyl(2,3,5,6-tetrafluorophenyl)-**	23653-74-9	**NI33032**
157	108	158	143	159	125	93	75	**314**	100	53	50	38	33	30	23	23	**0.00**	9	12	2	–	3	–	1	–	–	–	1	–	–	**Phosphorodithioic acid, S-[[(4-chlorophenyl)thio]methyl] O,O-dimethyl ester**	953-17-3	**NI33033**
45	157	125	93	63	47	314	79	**314**	100	95	68	62	31	31	22	21		9	12	2	–	3	–	1	–	–	–	1	–	–	**Phosphorodithioic acid, S-[[(4-chlorophenyl)thio]methyl] O,O-dimethyl ester**	953-17-3	**NI33035**
157	125	45	93	159	314	63	47	**314**	100	58	53	49	39	37	26	26		9	12	2	–	3	–	1	–	–	–	1	–	–	**Phosphorodithioic acid, S-[[(4-chlorophenyl)thio]methyl] O,O-dimethyl ester**	953-17-3	**NI33034**
109	125	15	45	155	157	79	75	**314**	100	92	69	68	65	30	29	27	**0.00**	9	12	4	–	2	–	1	–	–	–	1	–	–	**Phosphorothioic acid, S-[[(4-chlorophenyl)sulphinyl]methyl] O,O-dimethyl ester**	62059-36-3	**NI33036**
284	282	314	286	258	312	256	235	**314**	100	46	42	38	34	25	20	11		9	13	–	–	–	–	–	–	–	–	–	–	1	**η^5-Cyclopentadienylbis(η^2-ethene)iridium(I)**	88411-52-3	**NI33037**
73	75	69	187	299	97	185	41	**314**	100	78	70	54	26	25	21	21	**3.10**	9	19	2	–	–	–	–	–	1	1	–	–	–	**Hexanoic acid, 6-iodo-, trimethylsilyl ester**	26305-95-3	**NI33038**
42	29	172	39	41	72	40	53	**314**	100	69	57	57	56	43	39	33	**0.00**	9	22	–	4	–	–	2	–	–	–	–	–	1	**Nickel, [1-chloro-N^3-[3-[(1-chloropropyl)amino]propyl]-1,1,3-propanetriamine]-**	56793-00-1	**NI33039**
299	314	300	73	301	315	75	316	**314**	100	28	25	15	12	7	4	3		9	27	4	–	–	–	–	–	–	3	1	–	–	**Silanol, trimethyl-, phosphate (3:1)**	10497-05-9	**NI33040**
73	45	299	133	75	74	59	147	**314**	100	92	61	43	34	27	24	23	**12.00**	9	27	4	–	–	–	–	–	–	3	1	–	–	**Silanol, trimethyl-, phosphate (3:1)**	10497-05-9	**MS07328**
299	73	300	314	301	45	133	207	**314**	100	45	25	22	14	11	7	5		9	27	4	–	–	–	–	–	–	3	1	–	–	**Silanol, trimethyl-, phosphate (3:1)**	10497-05-9	**NI33041**

MASS TO CHARGE RATIOS								M.W.	INTENSITIES								Parent	C	H	O	N	S	F	Cl	Br	I	Si	P	B	X	COMPOUND NAME	CAS Reg No	No
145	245	314	295	214	69	176	146	**314**	100	47	30	12	12	11	9	9		10	4	–	–	–	10	–	–	–	–	–	–	–	**2-Monofluoromethylene(E)-3-trifluoromethyl-1,4,7,7,8,8-hexafluorobicyclo[2.2.2]oct-5-ene**		**NI33042**
235	237	156	316	155	314	318	50	**314**	100	99	61	53	32	27	26	20		10	8	–	2	–	–	–	2	–	–	–	–	–	**Quinoxaline, 2,3-bis(bromomethyl)-**	3138-86-1	**NI33043**
235	237	156	316	155	318	314	50	**314**	100	98	62	55	31	26	26	20		10	8	–	2	–	–	–	2	–	–	–	–	–	**Quinoxaline, 2,3-bis(bromomethyl)-**	3138-86-1	**LI06929**
237	235	156	76	316	80	82	155	**314**	100	96	56	44	32	30	29	27	**19.00**	10	8	–	2	–	–	–	2	–	–	–	–	–	**Quinoxaline, 2,3-bis(bromomethyl)-**	3138-86-1	**NI33044**
29	27	18	97	28	17	63	15	**314**	100	58	47	25	14	14	13	13	**0.00**	10	13	3	–	1	–	2	–	–	–	1	–	–	**Phosphorothioic acid, O-(2,4-dichlorophenyl) O,O-diethyl ester**	97-17-6	**NI33045**
279	97	223	251	88	162	281	28	**314**	100	78	74	50	50	46	37	35	**4.77**	10	13	3	–	1	–	2	–	–	–	1	–	–	**Phosphorothioic acid, O-(2,4-dichlorophenyl) O,O-diethyl ester**	97-17-6	**NI33046**
279	223	281	162	109	29	164	17	**314**	100	50	42	42	29	23	22	21	**9.60**	10	13	3	–	1	–	2	–	–	–	1	–	–	**Phosphorothioic acid, O-(2,4-dichlorophenyl) O,O-diethyl ester**	97-17-6	**NI33047**
127	29	99	55	110	27	173	125	**314**	100	34	26	25	24	22	21	20	**9.00**	10	19	7	–	1	–	–	–	–	–	1	–	–	**Butanedioic acid, [(dimethoxyphosphinothioyl)oxy]-, diethyl ester**	52234-58-9	**NI33048**
127	29	99	55	110	269	27	173	**314**	100	34	26	25	24	22	22	21	**8.40**	10	19	7	–	1	–	–	–	–	–	1	–	–	**Butanedioic acid, [(dimethoxyphosphinyl)thio]-, diethyl ester**	1634-78-2	**NI33049**
127	99	55	109	125	79	110	142	**314**	100	43	29	28	22	21	16	14	**1.00**	10	19	7	–	1	–	–	–	–	–	1	–	–	**Butanedioic acid, [(dimethoxyphosphinyl)thio]-, diethyl ester**	1634-78-2	**NI33050**
127	99	109	142	125	110	79	55	**314**	100	61	40	34	31	29	27	27	**0.00**	10	19	7	–	1	–	–	–	–	–	1	–	–	**Butanedioic acid, [(dimethoxyphosphinyl)thio]-, diethyl ester**	1634-78-2	**NI33051**
69	41	133	314	71	68	55	56	**314**	100	65	58	28	24	17	16	12		10	20	5	–	1	–	–	–	–	–	2	–	–	**2,2′-Bis-1,3,2-dioxaphosphorinane, 5,5,5′,5′-tetramethyl-, 2-oxide 2′-sulphide**	16368-07-3	**NI33052**
69	229	247	68	299	67	314	110	**314**	100	37	30	27	13	10	8	7		10	20	7	–	–	–	–	–	–	–	2	–	–	**Oxybis(5,5-dimethyl-1,3,2-dioxaphosphorinane 2-oxide)**	4090-52-2	**NI33053**
151	191	179	148	227	229	221	149	**314**	100	57	57	48	43	41	41	41	**12.00**	11	15	2	4	–	–	–	1	–	–	–	–	–	**6-Acetyl-4-bromomethyl-4,6-dihydro-4-ethoxy-3-methyl-pyrazolo[3,2-c]-as-triazine**		**LI06930**
151	191	179	148	227	229	221	149	**314**	100	57	57	48	43	41	41	41	**12.01**	11	15	2	4	–	–	–	1	–	–	–	–	–	**Pyrazolo[5,1-c]-as-triazine, 6-acetyl-4-(bromomethyl)-4-ethoxy-4,6-dihydro-3-methyl-**	6578-50-3	**NI33054**
111	69	191	189	110	109	79	71	**314**	100	71	40	37	33	31	31	26	**0.00**	11	19	4	–	2	–	1	–	–	–	–	–	–	**S-(10-Chloro-2,9-dioxodecyl)methanesulphonothioate**	76078-76-7	**NI33055**
57	143	215	200	109	81	64	185	**314**	100	77	62	61	42	42	36	35	**0.00**	11	19	4	–	2	–	1	–	–	–	–	–	–	**1,2-Dithiol-1-ium, 3,5-bis(tert-butyl)-, perchlorate**	35610-83-4	**NI33056**
77	157	155	314	51	153	312	310	**314**	100	97	52	50	50	47	45	28		12	10	–	–	–	–	–	–	–	–	–	–	2	**Diselenide, diphenyl-**	1666-13-3	**LI06931**
77	157	155	314	51	154	312	310	**314**	100	98	52	51	50	46	45	28		12	10	–	–	–	–	–	–	–	–	–	–	2	**Diselenide, diphenyl-**	1666-13-3	**NI33057**
314	299	172	157	271	144	114	101	**314**	100	40	25	19	18	13	13	13		12	11	2	–	–	–	–	–	1	–	–	–	–	**Naphthalene, 1-iodo-2,6-dimethoxy-**	25315-11-1	**NI33058**
156	115	129	155	141	77	130	103	**314**	100	83	65	60	41	34	34	23	**0.00**	12	12	–	–	–	–	–	2	–	–	–	–	–	**4,4-Dibromo-1-methyl-1-phenyl-spiro[2.2]pentane**		**MS07329**
156	155	115	141	129	130	235	91	**314**	100	82	69	57	48	46	38	18	**0.00**	12	12	–	–	–	–	–	2	–	–	–	–	–	**4,4-Dibromo-1-(4-tolyl)-spiro[2.2]pentane**		**MS07330**
70	83	71	58	98	99	57	56	**314**	100	44	13	10	8	5	5	5	**2.60**	12	15	2	4	1	–	1	–	–	–	–	–	–	**7-Chloro-3-(4-methyl-1-piperazinyl)-4H-1,2,4-benzothiadiazine 1,1-dioxide**		**MS07331**
299	297	295	73	298	207	296	269	**314**	100	72	41	39	35	35	27	18	**1.20**	12	22	–	–	–	–	–	–	–	1	–	–	1	**Tin, 4-(trimethylsilyl)phenyl-trimethyl-**		**MS07332**
57	41	169	55	40	113	47	89	**314**	100	78	32	32	27	24	20	19	**1.70**	12	27	1	–	3	–	–	–	–	–	1	–	–	**Phosphorotrithioic acid, S,S,S-tributyl ester**	78-48-8	**NI33059**
57	41	169	202	170	56	113	55	**314**	100	75	62	36	36	32	31	30	**8.00**	12	27	1	–	3	–	–	–	–	–	1	–	–	**Phosphorotrithioic acid, S,S,S-tributyl ester**	78-48-8	**NI33060**
314	222	284	315	238	221	223	210	**314**	100	33	20	17	13	13	12	11		13	6	4	4	1	–	–	–	–	–	–	–	–	**Benzimidazo[2,1-b]benzothiazole, 7,9-dinitro-**		**IC03880**
230	138	55	112	231	258	314	232	**314**	100	95	32	26	19	12	9	8		13	7	4	–	1	–	–	–	–	–	–	–	1	**(2-Thenoyl)cymantrene**	65468-92-0	**NI33061**
105	77	51	50	29	314	316	106	**314**	100	81	52	27	26	25	22	21		13	9	1	2	–	–	3	–	–	–	–	–	–	**Benzoic acid, 2-(2,4,6-trichlorophenyl)hydrazide**	33422-33-2	**NI33062**
314	316	280	236	202	315	283	157	**314**	100	38	21	19	19	17	14	13		13	9	2	–	1	1	1	–	–	–	1	–	–	**10H-Phenothiaphosphine, 7-chloro-2-fluoro-10-methoxy-, 10-oxide**	55125-05-8	**NI33063**
59	235	191	51	89	77	91	117	**314**	100	99	88	33	30	30	20	19	**13.54**	13	11	7	–	–	–	1	–	–	–	–	–	–	**3,4-Bis(methoxycarbonyloxy)cinnamoyl chloride**	16180-95-3	**NI33064**
285	73	299	269	286	255	314	100	**314**	100	50	32	23	22	22	16	14		13	26	3	2	–	–	–	–	–	2	–	–	–	**Pyrimidine, 5-(ethoxymethyl)-2,4-bis[(trimethylsilyl)oxy]-**	56145-29-0	**NI33065**
186	77	39	65	121	154	51	66	**314**	100	43	36	32	30	26	26	25	**0.13**	14	10	3	–	1	–	–	–	–	–	–	–	1	**π-Cyclopentadienyl-benzoylthio-dicarbonyl iron**		**MS07333**
299	242	73	314	215	300	45	75	**314**	100	67	42	30	24	23	21	14		14	14	1	4	1	–	–	–	–	1	–	–	–	**1H-Benzimidazole, 5-isocyanato-2-(4-thiazolyl)-1-(trimethylsilyl)-**	55090-56-7	**NI33066**
79	78	80	77	39	51	52	50	**314**	100	71	56	52	24	21	20	15	**0.76**	14	16	2	–	–	–	–	–	–	–	–	–	1	**Molybdenum, dicarbonylbis(π-1,3-cyclohexadienyl)-**		**MS07334**
134	208	207	314	135	121	238	316	**314**	100	91	78	76	20	16	15	14		14	18	–	–	4	–	–	–	–	–	–	–	–	**1,2-Bis-(1,3-dithian-2-yl)-benzene**		**MS07335**
314	240	175	166	105	74	121	316	**314**	100	60	50	45	36	31	30	22		14	18	–	–	4	–	–	–	–	–	–	–	–	**1,4-Bis-(1,3-dithian-2-yl)-benzene**		**LI06932**
43	28	194	112	39	41	122	15	**314**	100	41	12	7	6	6	5	5	**0.00**	14	18	8	–	–	–	–	–	–	–	–	–	–	**1,4-Diacetoxybicyclo[2.2.2]octane-2,3-dicarboxylic acid**		**NI33067**
181	158	116	84	200	127	241	87	**314**	100	99	99	99	77	71	35	34	**1.00**	14	18	8	–	–	–	–	–	–	–	–	–	–	**meso-4,5-Diacetoxy-octa-2,6-dien-1,8-dioic acid, dimethyl ester**		**LI06933**
127	105	104	40	193	67	166	43	**314**	100	80	37	34	31	31	23	23	**0.10**	14	19	6	–	–	–	–	–	–	–	1	–	–	**2-Butenoic acid, 3-[(dimethoxyphosphinyl)oxy]-, 1-phenylethyl ester, (E)-**	7700-17-6	**NI33068**
127	105	193	104	121	179	43	166	**314**	100	96	34	31	30	26	26	24	**0.00**	14	19	6	–	–	–	–	–	–	–	1	–	–	**2-Butenoic acid, 3-[(dimethoxyphosphinyl)oxy]-, 1-phenylethyl ester, (Z)-**	29900-31-0	**NI33069**
57	314	72	54	106	82	240	149	**314**	100	75	49	33	30	30	23	23		14	22	6	2	–	–	–	–	–	–	–	–	–	**1-(tert-Butylamino)-3-(4,6-dihydroxy-3-methyl-2-nitrophenoxy)-2-propanol**		**MS07336**
197	73	75	152	151	103	198	72	**314**	100	84	49	21	20	16	12	12	**10.00**	14	26	4	2	–	–	–	–	–	1	–	–	–	**Fumarylacetone diethoxime TMS**		**NI33070**
192	108	65	314	80	63	121	77	**314**	100	87	87	43	43	43	41	39		15	10	2	2	2	–	–	–	–	–	–	–	–	**(3-Thio-2-benzo[b]thienylidene)-4-nitroaniline**	93575-32-7	**NI33071**
286	287	288	314	315	316	285	313	**314**	100	95	93	92	90	86	70	67		15	11	1	2	–	–	–	1	–	–	–	–	–	**7-Bromo-2,3-dihydro-5-phenyl-1H-1,4-benzodiazepin-2-one**	2894-61-3	**NI33072**
144	89	316	314	143	116	171	173	**314**	100	44	24	24	18	14	13	12		15	11	1	2	–	–	–	1	–	–	–	–	–	**4′-Bromo-indole-2-carboxanilide**		**LI06934**
91	77	159	132	158	104	93	141	**314**	100	53	29	28	23	10	10	9	**2.00**	15	14	2	4	1	–	–	–	–	–	–	–	–	**2-Benzyl-4-benzenesulphonylimino-1,2,4-triazolium-ylide**		**LI06935**
28	91	267	77	36	51	235	184	**314**	100	92	77	52	36	35	35	35	**6.99**	15	14	2	4	1	–	–	–	–	–	–	–	–	**N-[1-Benzyl-4-(1,2,4-triazolio)]benzolsulphonamidate**	32539-60-9	**NI33073**
91	314	117	132	43	65	135	281	**314**	100	95	82	74	58	52	42	34		15	14	4	4	–	–	–	–	–	–	–	–	–	**Acetophenone, 4′-methyl-, (2,4-dinitrophenyl)hydrazone**	1237-49-6	**NI33074**
314	315	135	281	78	279	136	43	**314**	100	20	18	10	4	3	3	2		15	14	4	4	–	–	–	–	–	–	–	–	–	**Acetophenone, 4′-methyl-, (2,4-dinitrophenyl)hydrazone**	1237-49-6	**NI33075**

MASS TO CHARGE RATIOS	M.W.	INTENSITIES	Parent	C	H	O	N	S	F	Cl	Br	I	Si	P	B	X	COMPOUND NAME	CAS Reg No	No
91 314 117 43 65 135 281 132	**314**	100 95 83 58 52 43 37 35		15	14	4	4	–	–	–	–	–	–	–	–	–	Acetophenone, 4'-methyl-, (2,4-dinitrophenyl)hydrazone	1237-49-6	LI06936
208 77 191 105 207 131 106 79	**314**	100 86 71 71 68 66 59 39	**0.00**	15	14	4	4	–	–	–	–	–	–	–	–	–	7-[N'-(2'-Amino-1'-hydroxy-1'-phenylpropanyl)]-4-nitrobenzo-2-oxa-1,3-diazole		LI06937
250 251 191 235 218 223 267 207	**314**	100 62 56 47 42 32 31 27	**1.00**	15	22	7	–	–	–	–	–	–	–	–	–	–	Dimethyl (1α,2α,5α)-3,7,7-trimethoxybicyclo[3.3.0]oct-3-en e-2,4-dicarboxylate		MS07337
135 91 103 145 85 163 127 77	**314**	100 56 38 28 25 21 16 10	**0.00**	15	22	7	–	–	–	–	–	–	–	–	–	–	(1RS)-1-Methoxy-2-phenylethyl β-D-glucopyranoside		MS07338
134 135 121 91 105 136 60 122	**314**	100 92 40 11 10 9 7 6	**1.00**	15	22	7	–	–	–	–	–	–	–	–	–	–	2-(4-Methoxyphenyl)ethyl β-D-glucopyranoside		NI33076
285 283 103 149 101 281 41 105	**314**	100 76 75 71 60 56 47 42	**1.00**	15	28	2	–	–	–	–	–	–	–	–	–	1	Germacyclododecane-6,7-dione, 1,1-diethyl-	22778-65-0	NI33077
123 95 314 75 149 124 69 50	**314**	100 80 37 16 8 8 6 4		16	8	3	2	–	2	–	–	–	–	–	–	–	Furazan, 3,4-bis(4-fluorobenzyl)-	55232-38-7	NI33078
235 314 260 236 313 259 287 118	**314**	100 34 20 16 15 13 12 12		16	10	–	8	–	–	–	–	–	–	–	–	–	5,5';4',4'';5'',5'''-Quaterpyrimidine		NI33079
149 225 65 121 122 166 314 76	**314**	100 36 29 21 14 10 9 9		16	10	7	–	–	–	–	–	–	–	–	–	–	2,4,5'-Benzophenonetricarboxylic acid	75144-29-5	NI33080
314 57 268 53 286 93 63 71	**314**	100 56 48 43 34 34 34 32		16	10	7	–	–	–	–	–	–	–	–	–	–	Majoronal	34425-62-2	LI06938
197 117 195 116 193 90 199 194	**314**	100 60 50 48 19 19 18 17	**17.00**	16	14	–	2	–	–	–	–	–	–	–	–	1	3,5-Bis(3-tolyl)-1,2,4-selenadiazole		MS07339
197 117 195 116 90 193 314 199	**314**	100 62 50 46 20 19 18 18		16	14	–	2	–	–	–	–	–	–	–	–	1	3,5-Bis(4-tolyl)-1,2,4-selenadiazole		MS07340
91 43 77 130 92 65 102 51	**314**	100 46 21 19 18 16 14 11	**0.00**	16	14	5	2	–	–	–	–	–	–	–	–	–	N-(α-Acetoxy-4-nitrobenzylidene)-O-benzylhydroxylamine		LI06939
91 43 92 150 255 272 314	**314**	100 12 10 5 3 2 1		16	14	5	2	–	–	–	–	–	–	–	–	–	N-(α-Acetoxy-4-nitrobenzylidene)-O-benzylhydroxylamine		LI06940
314 283 270 77 269 271 285 316	**314**	100 98 79 55 50 38 35 34		16	15	1	4	–	–	1	–	–	–	–	–	–	7-Chloro-2-(2'-hydroxyethyl)-5-phenyl-3H-1,3,4-benzotriazepine		MS07341
131 133 144 147 200 185 167 157	**314**	100 22 19 16 13 10 10 10	**5.00**	16	17	3	–	–	3	–	–	–	–	–	–	–	Acetic acid, trifluoro-, 5,6,7,8,9,10-hexahydro-7,7-dimethyl-10-oxo-5-benzocyclooctenyl ester	64129-26-6	NI33081
314 121 185 129 213 296 186 56	**314**	100 38 37 31 29 24 23 20		16	18	3	–	–	–	–	–	–	–	–	–	1	6-Ferrocenyl-6-oxohexanoic acid		NI33082
91 129 228 172 131 115 128 56	**314**	100 94 85 70 70 68 55 53	**27.66**	16	18	3	–	–	–	–	–	–	–	–	–	1	Iron, tricarbonyl[(1,2,3,4-η)-4a,4b,6,7,8,9,9a,9b-octahydro-5H-benzo[3,4]cyclobuta[1,2]cycloheptene]-	56784-34-0	MS07342
43 156 130 157 269 138 103 97	**314**	100 64 45 43 34 28 27 27	**0.00**	16	18	3	4	–	–	–	–	–	–	–	–	–	6-Ethoxycarbonyl-3-imino-5-isopropyl-1-methyl-4,7,7-tricyano-2-oxabicyclo[2.2.1]heptane		MS07343
43 130 156 131 146 97 55 103	**314**	100 23 19 17 16 14 14 12	**0.00**	16	18	3	4	–	–	–	–	–	–	–	–	–	6-Ethoxycarbonyl-3-imino-1-methyl-5-propyl-4,7,7-tricyano-2-oxabicyclo[2.2.1]heptane		MS07344
135 124 56 314 203 176 95 150	**314**	100 30 29 21 10 9 9 8		16	18	3	4	–	–	–	–	–	–	–	–	–	1-(3,4-Methylenedioxybenzyl)-4-(2-pyrimidinyl)-4-hydroxy-piperazine		NI33083
93 119 77 164 195 177 314 120	**314**	100 65 35 13 10 10 3 3		16	18	3	4	–	–	–	–	–	–	–	–	–	N,N'-Bis(phenylcarbamoyl) hydrazinoethanol		LI06941
135 120 108 57 122 121 55 106	**314**	100 15 13 11 10 9 9 8	**7.44**	16	18	3	4	–	–	–	–	–	–	–	–	–	Pyrimidine, 2-[4-(1,3-benzodioxol-5-ylmethyl)-1-piperazinyl]-, N-oxide	50602-52-3	NI33084
43 121 139 109 152 134 82 41	**314**	100 20 15 14 12 12 12 11	**0.00**	16	26	6	–	–	–	–	–	–	–	–	–	–	1,2-Cyclohexanedimethanol, 3-(acetyloxy)-1,2-dimethyl-, diacetate	55905-48-1	NI33085
229 257 69 55 41 299 97 243	**314**	100 36 34 20 19 17 12 11	**0.00**	16	26	6	–	–	–	–	–	–	–	–	–	–	Dimethyl (2R,3R,6S,9R)-6-isopropyl-9-methyl-1,4-dioxaspiro[4.5]decane-2,3-dicarboxylate		DD01184
45 77 120 31 94 43 65 39	**314**	100 21 15 15 13 12 10 10	**0.00**	16	26	6	–	–	–	–	–	–	–	–	–	–	Pentaethylene glycol monophenyl ether		IC03881
72 114 115 41 43 29 27 55	**314**	100 67 33 29 28 22 16 15	**2.38**	16	30	4	2	–	–	–	–	–	–	–	–	–	L-Valine, N-(N-acetyl-L-valyl)-, butyl ester	69833-62-1	NI33086
157 73 158 75 67 159 156 299	**314**	100 77 14 14 6 3 3 1	**1.00**	16	34	2	–	–	–	–	–	–	2	–	–	–	1,1'-Bicyclopentyl-1,1'-diol bis(trimethylsilyl) ether		MS07345
156 73 142 155 135 75 129 67	**314**	100 50 14 12 12 12 9 9	**1.00**	16	34	2	–	–	–	–	–	–	2	–	–	–	1,1'-Bicyclopentyl-2,2'-diol bis(trimethylsilyl) ether		MS07346
285 286 255 258 257 313 142 128	**314**	100 36 33 28 27 13 7 7	**0.00**	16	40	–	2	–	–	–	–	–	–	–	–	2	Di-aluminium, tetraethyl-bis[μ-(2-methyl-2-propanaminato)]-	55902-80-2	NI33087
278 215 314 316 280 279 315 216	**314**	100 71 58 40 36 27 23 22		17	12	–	–	–	–	2	–	–	1	–	–	–	7,7-Dichloro-7-sila-7,12-dihydropleiadene		NI33088
150 115 314 316 315 152 136 317	**314**	100 97 49 38 38 35 35 22		17	12	–	2	–	–	2	–	–	–	–	–	–	3,6-Bis(4-chlorophenyl)-4-methylpyridazine	87521-59-3	NI33089
254 212 199 314 241 272 271 230	**314**	100 99 40 28 2 1 1 1		17	14	4	–	1	–	–	–	–	–	–	–	–	5-(3,4-Diacetoxy-butyn-1-yl)-2-(penta-1,3-diynyl)-thiophene		LI06942
285 314 257 286 270 258 315 243	**314**	100 50 50 27 18 13 12 11		17	14	6	–	–	–	–	–	–	–	–	–	–	9,10-Anthracenedione, 1,3,8-trihydroxy-6-(1-hydroxypropyl)-	15979-74-5	NI33090
314 299 135 271 313 285 243 286	**314**	100 11 11 10 6 6 4 3		17	14	6	–	–	–	–	–	–	–	–	–	–	4H-1-Benzopyran-4-one, 3,5-dihydroxy-7-methoxy-2-(4-methoxyphenyl)-		MS07347
314 313 271 135 285 296 295 283	**314**	100 69 47 18 16 14 10 9		17	14	6	–	–	–	–	–	–	–	–	–	–	4H-1-Benzopyran-4-one, 5,7-dihydroxy-3-methoxy-2-(4-methoxyphenyl)-		MS07348
314 299 296 271 133 300 167 139	**314**	100 56 32 25 13 9 6 5		17	14	6	–	–	–	–	–	–	–	–	–	–	4H-1-Benzopyran-4-one, 5,7-dihydroxy-6-methoxy-2-(4-methoxyphenyl)-	520-12-7	NI33091
312 57 43 55 41 73 69 71	**314**	100 61 53 44 38 37 34 28	**3.52**	17	14	6	–	–	–	–	–	–	–	–	–	–	4H-1-Benzopyran-4-one, 2-(3,4-dimethoxyphenyl)-5,7-dihydroxy-	4712-12-3	NI33092
314 299 153 271 313 285 296 165	**314**	100 90 45 40 24 23 8 3		17	14	6	–	–	–	–	–	–	–	–	–	–	4H-1-Benzopyran-4-one, 2-(3,4-dimethoxyphenyl)-5,7-dihydroxy-	4712-12-3	NI33093
314 167 286 143 271 148 133 95	**314**	100 14 12 11 10 6 6 6		17	14	6	–	–	–	–	–	–	–	–	–	–	4H-1-Benzopyran-4-one, 5-hydroxy-2-(4-hydroxy-3-methoxyphenyl)-7-methoxy-	25739-41-7	NI33094
314 313 315 271 285 296 295 121	**314**	100 82 35 31 14 13 13 11		17	14	6	–	–	–	–	–	–	–	–	–	–	4H-1-Benzopyran-4-one, 5-hydroxy-2-(4-hydroxyphenyl)-3,7-dimethoxy-	3301-49-3	NI33095
299 314 271 153 285 69 95 81	**314**	100 99 55 45 43 37 33 32		17	14	6	–	–	–	–	–	–	–	–	–	–	4H-1-Benzopyran-4-one, 5-hydroxy-2-(4-hydroxyphenyl)-6,7-dimethoxy-	6601-62-3	NI33096
314 299 271 153 285 69 81 268	**314**	100 99 55 45 43 37 32 30		17	14	6	–	–	–	–	–	–	–	–	–	–	4H-1-Benzopyran-4-one, 5-hydroxy-2-(4-hydroxyphenyl)-6,7-dimethoxy-	6601-62-3	MS07349
314 315 285 271 167 284 143 243	**314**	100 18 15 11 10 8 8 7		17	14	6	–	–	–	–	–	–	–	–	–	–	4H-1-Benzopyran-4-one, 5-hydroxy-7-methoxy-2-(3-hydroxy-4-methoxyphenyl)-		MS07350
181 314 315 182 148 134 312 166	**314**	100 99 20 20 13 12 11 9		17	14	6	–	–	–	–	–	–	–	–	–	–	Coumarin, 4-hydroxy-3-(4-hydroxyphenyl)-5,7-dimethoxy-	4222-01-9	NI33097
314 271 315 285 286 272 269 215	**314**	100 25 17 11 4 4 3 3		17	14	6	–	–	–	–	–	–	–	–	–	–	Cyclopenta[c]furo[3',2':4,5]furo[2,3-h][1]benzopyran-1,11-dione, 2,3,6a,8,9,9a-hexahydro-4-methoxy-, (6aR-cis)-	7220-81-7	NI33098

MASS TO CHARGE RATIOS								M.W.	INTENSITIES								Parent	C	H	O	N	S	F	Cl	Br	I	Si	P	B	X	COMPOUND NAME	CAS Reg No	No
314	271	315	285	272	44	286	43	314	100	38	21	17	8	7	5	5		17	14	6	–	–	–	–	–	–	–	–	–	–	Cyclopenta[c]furo[3′,2′:4,5]furo[2,3-h][1]benzopyran-1,11-dione, 2,3,6a,8,9,9a-hexahydro-4-methoxy-, (6aR-cis)-	7220-81-7	MS07351
314	271	315	285	115	272	243	77	314	100	42	23	15	9	6	4	4		17	14	6	–	–	–	–	–	–	–	–	–	–	Cyclopenta[c]furo[3′,2′:4,5]furo[2,3-h][1]benzopyran-1,11-dione, 2,3,6a,8,9,9a-hexahydro-4-methoxy-, (6aR-cis)-	7220-81-7	NI33099
119	135	128	126	112	107	157	142	314	100	82	79	69	67	65	63	61	41.77	17	14	6	–	–	–	–	–	–	–	–	–	–	Cyclopenta[c]furo[3′,2′:4,5]furo[2,3-h][1]benzopyran-11(1H)-one, 2,3,6a,9a-tetrahydro-1-hydroxy-4-methoxy-, [1R-(1α,6aα,9aα)]-	29611-03-8	NI33101
314	257	313	297	115	296	77	69	314	100	48	45	45	44	43	41	40		17	14	6	–	–	–	–	–	–	–	–	–	–	Cyclopenta[c]furo[3′,2′:4,5]furo[2,3-h][1]benzopyran-11(1H)-one, 2,3,6a,9a-tetrahydro-1-hydroxy-4-methoxy-, [1R-(1α,6aα,9aα)]-	29611-03-8	NI33100
100	296	113	143	139	141	140	152	314	100	91	87	84	84	82	81	80	62.30	17	14	6	–	–	–	–	–	–	–	–	–	–	Cyclopenta[c]furo[3′,2′:4,5]furo[2,3-h][1]benzopyran-11(1H)-one, 2,3,6a,9a-tetrahydro-1-hydroxy-4-methoxy-, [1R-(1α,6aα,9aα)]-	29611-03-8	NI33102
314	291	271	115	200	297	109	296	314	100	91	20	19	18	16	16	15		17	14	6	–	–	–	–	–	–	–	–	–	–	1,7-Dihydroxy-6,8-dimethoxy-3-methylanthraquinone		LI06943
314	283	315	284	251	141	75	210	314	100	85	19	15	15	11	10	8		17	14	6	–	–	–	–	–	–	–	–	–	–	Dimethyl 3-methoxydibenzofuran-1,2-dicarboxylate		LI06944
314	299	162	313	149	271	177	164	314	100	28	23	10	8	6	6	6		17	14	6	–	–	–	–	–	–	–	–	–	–	6H-[1,3]Dioxolo[5,6]benzofuro[3,2-c][1]benzopyran-2-ol, 6a,12a-dihydro-3-methoxy-, (6aR-cis)-	30461-92-8	NI33103
314	167	133	299	183	131	132	182	314	100	99	48	36	32	28	26	25		17	14	6	–	–	–	–	–	–	–	–	–	–	Punctatin	27181-63-1	LI06945
255	314	40	256	51	77	283	316	314	100	92	76	51	50	49	48	30		17	15	2	2	–	–	1	–	–	–	–	–	–	3-Methylclobazam		NI33104
159	160	314	144	91	143	103	117	314	100	13	4	4	4	3	3	2		17	18	2	2	1	–	–	–	–	–	–	–	–	1,2-Dimethyl-3-(4-methylphenylsulphonamido)-indole		IC03883
159	160	314	144	158	143	117	91	314	100	13	7	5	4	4	4	4		17	18	2	2	1	–	–	–	–	–	–	–	–	1,2-Dimethyl-3-(4-methylphenylsulphonamido)-indole		IC03882
141	115	142	127	240	43	168	143	314	100	95	45	23	16	11	10	9	6.00	17	18	4	2	–	–	–	–	–	–	–	–	–	2-(2-Napthalenyl)hydrazonopropanedioic acid, diethyl ester		NI33105
252	43	192	297	238	254	237	42	314	100	72	71	63	54	42	30	23	3.70	17	18	4	2	–	–	–	–	–	–	–	–	–	4-(2-Nitrophenyl)-2,6-dimethyl-3,5-diacetyl-1,4-dihydropyridine	34551-63-8	NI33106
271	270	272	213	186	214	185	171	314	100	34	25	22	16	14	9	9	8.67	17	22	2	4	–	–	–	–	–	–	–	–	–	Benzo[g]pteridine-2,4(1H,3H)-dione, 5,10-dihydro-3,7,8,10-tetramethyl-5-isopropyl-	50387-39-8	NI33107
240	314	255	199	228	41	241	43	314	100	54	30	22	15	14	13	13		17	22	2	4	–	–	–	–	–	–	–	–	–	1H-Benzo[g]pyrrolo[2,1-e]pteridine-4,6(5H,7H)-dione, 2,3,7a,8-tetrahydro-5,8,10,11-tetramethyl-	55661-25-1	NI33108
100	113	82	80	201	94	202	79	314	100	56	55	36	35	29	21	19	2.00	17	22	2	4	–	–	–	–	–	–	–	–	–	3-(2-Morpholinoethylamino)-4-methyl-6-(2-hydroxyphenyl)pyridazine		NI33109
80	81	82	79	35	100	38	30	314	100	69	54	40	29	27	20	17	0.00	17	22	2	4	–	–	–	–	–	–	–	–	–	3-(2-Morpholinoethylamino)-4-methyl-6-(4-hydroxyphenyl)pyridazine		NI33110
41	67	55	54	81	95	43	29	314	100	98	98	93	92	58	54	53	0.14	17	31	–	–	–	–	–	1	–	–	–	–	–	8-Heptadecyne, 1-bromo-	56599-94-1	NI33111
91	158	77	65	129	43	128	51	314	100	47	46	44	40	37	33	30	0.00	18	18	3	–	1	–	–	–	–	–	–	–	–	4-Methyl-3-[(4-methylphenyl)sulphonyl]-2-phenyl-2,5-dihydrofuran	118356-41-5	DD01185
314	148	299	315	313	178	161	211	314	100	27	24	16	13	10	8	6		18	18	5	–	–	–	–	–	–	–	–	–	–	6H-Benzofuro[3,2-c][1]benzopyran, 6a,11a-dihydro-3,8,9-trimethoxy-	20390-12-9	NI33112
134	121	180	207	69	181	314	137	314	100	84	66	42	41	37	35	30		18	18	5	–	–	–	–	–	–	–	–	–	–	Chalcone, 2′-hydroxy-4,4′,6-trimethoxy-	109469-66-1	NI33113
212	240	43	213	314	272	169	115	314	100	50	34	28	23	23	22	13		18	18	5	–	–	–	–	–	–	–	–	–	–	1,3-Cyclohexadiene-1-carboxylic acid, 6-[2-(acetyloxy)phenyl]-4,6-dimethyl-5-oxo-, methyl ester	40801-44-3	NI33114
299	314	196	255	267	181	195	209	314	100	96	80	72	58	56	42	40		18	18	5	–	–	–	–	–	–	–	–	–	–	1,2-Dibenzofurandicarboxylic acid, 1,9b-dihydro-4,9b-dimethyl-, dimethyl ester	56247-80-4	NI33115
267	196	299	255	223	181	209	314	314	100	50	46	44	37	33	29	23		18	18	5	–	–	–	–	–	–	–	–	–	–	1,2-Dibenzofurandicarboxylic acid, 3,9b-dihydro-4,9b-dimethyl-, dimethyl ester	23911-60-6	NI33116
105	149	77	150	106	70	51	44	314	100	92	31	8	7	7	7	2	0.06	18	18	5	–	–	–	–	–	–	–	–	–	–	Diethylene glycol dibenzoate		IC03884
314	133	190	257	226	191	151	77	314	100	40	35	34	30	25	25	24		18	18	5	–	–	–	–	–	–	–	–	–	–	α-(2-Methoxyphenoxy)coniferaldehyde methyl ether		LI06946
314	269	283	151	285	313	239	163	314	100	79	56	50	43	39	36	36		18	18	5	–	–	–	–	–	–	–	–	–	–	6-Methyl-1,3-benzodioxol-5-yl 3-ethoxy-4-methoxyphenyl ketone	52806-41-4	NI33117
314	269	313	283	151	299	315	270	314	100	92	73	39	37	26	22	20		18	18	5	–	–	–	–	–	–	–	–	–	–	6-Methyl-1,3-benzodioxol-5-yl 4-ethoxy-3-methoxyphenyl ketone	52828-42-9	NI33118
314	195	210	167	237	313	103	211	314	100	33	27	19	13	11	10	7		18	18	5	–	–	–	–	–	–	–	–	–	–	2-Propen-1-one, 1-(2-hydroxy-3,4,6-trimethoxyphenyl)-3-phenyl-	6971-20-6	NI33119
121	134	181	55	69	314	95	207	314	100	74	72	59	53	50	36	32		18	18	5	–	–	–	–	–	–	–	–	–	–	4′,5,7-Trimethoxyflavanone	38302-15-7	NI33120
314	210	195	167	181	153	152	104	314	100	99	52	36	13	11	11	11		18	18	5	–	–	–	–	–	–	–	–	–	–	5,7,8-Trimethoxyflavone		NI33121
117	314	316	105	91	106	131	118	314	100	66	64	58	57	45	41	40		18	19	–	–	–	–	–	1	–	–	–	–	–	Tricyclo[10.2.2.2^{5,8}]octadeca-5,7,12,14,15,17-hexaene, 6-bromo-	24777-39-7	NI33122
58	86	314	229	185	228	42	85	314	100	38	31	16	11	10	9	7		18	22	1	2	1	–	–	–	–	–	–	–	–	10H-Phenothiazine-10-propanamine, 2-methoxy-N,N-dimethyl-	61-01-8	NI33123
58	86	229	228	185	42	268	85	314	100	22	13	10	9	9	8	7	5.78	18	22	1	2	1	–	–	–	–	–	–	–	–	10H-Phenothiazine-10-propanamine, 2-methoxy-N,N-dimethyl-	61-01-8	NI33125
58	86	314	185	42	228	229	44	314	100	96	60	43	40	36	35	30		18	22	1	2	1	–	–	–	–	–	–	–	–	10H-Phenothiazine-10-propanamine, 2-methoxy-N,N-dimethyl-	61-01-8	NI33124
73	75	314	77	74	104	45	270	314	100	16	9	9	9	8	8	7		18	22	3	–	–	–	–	–	–	1	–	–	–	Benzeneacetic acid, α-methyl-3-phenoxy-, trimethylsilyl ester, (±)-	74810-90-5	NI33126
167	282	154	221	223	314	184	284	314	100	50	43	39	29	28	28	27		18	22	3	2	–	–	–	–	–	–	–	–	–	Methyl (8α,10α)-10-methoxy-6-methylergoline-8-carboxylate		MS07352
167	55	314	154	184	57	284	69	314	100	47	43	37	36	35	33	32		18	22	3	2	–	–	–	–	–	–	–	–	–	Methyl (8α,10β)-10-methoxy-6-methylergoline-8-carboxylate		MS07353
167	314	154	284	184	283	168	282	314	100	41	35	34	30	29	29	27		18	22	3	2	–	–	–	–	–	–	–	–	–	Methyl (8β,10α)-10-methoxy-6-methylergoline-8-carboxylate		MS07354
167	154	184	314	168	284	299	283	314	100	55	48	45	41	39	35	33		18	22	3	2	–	–	–	–	–	–	–	–	–	Methyl (8β,10β)-10-methoxy-6-methylergoline-8-carboxylate		MS07355
69	125	41	83	55	57	43	70	314	100	83	39	38	29	24	19	13	2.20	18	34	–	–	–	–	–	–	–	–	–	–	1	Zinc, bis[2-tert-butyl-3,3-dimethylcyclopropyl]-, [1α(1R*,2R*,2β)]-	74793-36-5	NI33127
241	185	41	56	55	57	242	98	314	100	71	37	35	32	31	23	21	0.88	18	34	4	–	–	–	–	–	–	–	–	–	–	Decanedioic acid, dibutyl ester	109-43-3	NI33128
241	185	56	41	57	55	29	98	314	100	68	47	40	38	38	30	29		18	34	4	–	–	–	–	–	–	–	–	–	–	Decanedioic acid, dibutyl ester	109-43-3	DO01095
57	41	70	55	43	112	71	29	314	100	83	69	61	56	50	50	44	0.33	18	34	4	–	–	–	–	–	–	–	–	–	–	Decanedioic acid, dibutyl ester	109-43-3	IC03885

MASS TO CHARGE RATIOS								M.W.	INTENSITIES								Parent	C	H	O	N	S	F	Cl	Br	I	Si	P	B	X	COMPOUND NAME	CAS Reg No	No
129	85	43	84	213	147	111	69	**314**	100	29	23	17	14	11	10	10	**0.00**	18	34	4	–	–	–	–	–	–	–	–	–	–	Bis(1,3-dimethylbutyl) hexanedioate		DO01096
88	55	41	98	43	69	112	87	**314**	100	91	70	59	56	55	49	42	**0.14**	18	34	4	–	–	–	–	–	–	–	–	–	–	Dimethyl 2-methyl-1,15-pentadecandioate		IC03886
43	300	101	57	41	55	29	116	**314**	100	86	79	51	45	37	34	30	**0.00**	18	34	4	–	–	–	–	–	–	–	–	–	–	Dodecanoic acid, (2,2-dimethyl-1,3-dioxolan-4-yl)methyl ester	40630-75-9	NI33129
55	98	74	41	43	69	84	87	**314**	100	95	73	73	68	56	44	41	**0.73**	18	34	4	–	–	–	–	–	–	–	–	–	–	Hexadecanedioic acid, dimethyl ester	19102-90-0	IC03887
98	74	84	112	241	283	209	69	**314**	100	44	42	35	30	26	23	21	**8.00**	18	34	4	–	–	–	–	–	–	–	–	–	–	Hexadecanedioic acid, dimethyl ester	19102-90-0	NI33130
129	43	85	84	69	213	111	55	**314**	100	31	29	19	15	13	13	12	**0.00**	18	34	4	–	–	–	–	–	–	–	–	–	–	Hexanedioic acid, dihexyl ester	110-33-8	NI33131
98	55	84	41	43	69	112	60	**314**	100	65	58	51	44	36	34	32	**0.34**	18	34	4	–	–	–	–	–	–	–	–	–	–	Octadecanedioic acid	871-70-5	NI33132
73	147	293	308	254	161	148	190	**314**	100	91	33	25	21	20	20	11	**0.00**	18	38	2	–	–	–	–	–	–	1	–	–	–	Pentadecanoic acid, trimethylsilyl ester	74367-22-9	NI33133
157	115	59	73	87	229	187	271	**314**	100	75	69	49	36	24	24	23	**7.50**	18	42	–	–	–	–	–	–	–	2	–	–	–	Hexaisopropyldisilicon		LI06947
231	43	314	215	316	42	41	232	**314**	100	47	45	44	32	27	27	25		19	16	–	–	–	–	2	–	–	–	–	–	–	Naphthalene, 2-(2,2-dichloroethyl)-1-methyl-4-phenyl-	22242-71-3	NI33134
121	109	94	108	107	314	133	95	**314**	100	84	84	60	52	51	45	40		19	22	4	–	–	–	–	–	–	–	–	–	–	3H,7H-Benzo[1,2-c:3,4-c']dipyran, 9-(3-furanyl)decahydro-4-hydroxy-4a,10a-dimethyl-1,4-vinyl-, [1R(1α,4β,4aα,6aα,9β,10aβ,10bα)]-		NI33135
94	41	95	39	43	55	314	28	**314**	100	86	58	39	36	34	30	30		19	22	4	–	–	–	–	–	–	–	–	–	–	3H,7H-Benzo[1,2-c:3,4-c']dipyran, 9-(3-furanyl)decahydro-4-hydroxy-4a,10a-dimethyl-1,4-vinyl-, [1R(1α,4β,4aα,6aβ,9β,10aβ,10bα)]-		NI33136
299	314	267	300	269	253	249	142	**314**	100	28	23	20	13	9	8	8		19	22	4	–	–	–	–	–	–	–	–	–	–	2H,8H-Benzo[1,2-b:5,4-b']dipyran, 5-methoxy-2,2,8,8-tetramethyl-10-acetyl-	3818-99-3	NI33137
299	43	41	301	314	300	267	269	**314**	100	38	27	23	22	19	16	14		19	22	4	–	–	–	–	–	–	–	–	–	–	2H,8H-Benzo[1,2-b:5,4-b']dipyran, 5-methoxy-2,2,8,8-tetramethyl-10-acetyl-		MS07356
314	269	145	137	299	115	91	131	**314**	100	12	12	12	10	8	7	6		19	22	4	–	–	–	–	–	–	–	–	–	–	(2R,3R)-2,3-Dihydro-2-(4'-hydroxy-3'-methoxyphenyl)-5-(3''-hydroxypropyl)-3-methylbenzofuran		MS07357
192	162	147	145	164	314	177	163	**314**	100	79	65	41	37	36	30	20		19	22	4	–	–	–	–	–	–	–	–	–	–	2,3-trans-3,4-cis-4,5-cis-5-(4''-Hydroxyphenyl)-2-(4'-hydroxy-3'-methoxyphenyl)-3,4-dimethyltetrahydrofuran		MS07358
162	147	192	314	145	163	177	161	**314**	100	82	49	33	19	18	13	11		19	22	4	–	–	–	–	–	–	–	–	–	–	2,3-trans-3,4-cis-4,5-trans-5-(4''-Hydroxyphenyl)-2-(4'-hydroxy-3-methoxyphenyl)-3,4-dimethyltetrahydrofuran		MS07359
164	77	165	314	149	91	103	162	**314**	100	12	11	10	10	10	9	8		19	22	4	–	–	–	–	–	–	–	–	–	–	erythro-1-(4'-Hydroxyphenyl)-2-[2''-methoxy-4''-(E)-propenylphenoxy]propan-1-ol		MS07360
91	65	92	163	77	39	314	223	**314**	100	16	10	8	8	6	5	3		19	22	4	–	–	–	–	–	–	–	–	–	–	Methyl 4-benzyloxy-2-methoxy-3,5,6-trimethylbenzoate		MS07361
58	85	269	268	270	271	314	242	**314**	100	81	72	48	28	24	19	14		19	23	–	2	–	–	1	–	–	–	–	–	–	5H-Dibenz[b,f]azepine-5-propanamine, 3-chloro-10,11-dihydro-N,N-dimethyl-	303-49-1	NI33138
58	85	269	268	270	271	44	42	**314**	100	45	37	24	15	14	13	12	**10.00**	19	23	–	2	–	–	1	–	–	–	–	–	–	5H-Dibenz[b,f]azepine-5-propanamine, 3-chloro-10,11-dihydro-N,N-dimethyl-	303-49-1	MS07362
91	73	118	135	117	75	134	208	**314**	100	92	43	34	29	29	25	24	**0.00**	19	26	2	–	–	–	–	–	–	1	–	–	–	(E)-1-(6-Benzyloxy-1-cyclohexenyl)-3-(trimethylsilyl)-2-propen-1-one		DD01186
57	55	41	43	44	285	69	56	**314**	100	60	60	57	50	47	43	40	**0.00**	19	38	3	–	–	–	–	–	–	–	–	–	–	Hexadecanoic acid, 15,15-dimethyl-16-hydroxy-, methyl ester		IC03888
103	71	74	97	223	296	265	242	**314**	100	54	46	26	16	14	11	8	**3.00**	19	38	3	–	–	–	–	–	–	–	–	–	–	Methyl 3-hydroxy-16-methylheptadecanoate	87538-93-0	NI33139
43	55	41	69	155	74	57	87	**314**	100	99	88	76	59	59	58	52	**1.08**	19	38	3	–	–	–	–	–	–	–	–	–	–	Nonanoic acid, 9-(nonyloxy)-, methyl ester	39692-47-2	MS07363
43	314	55	41	57	297	97	87	**314**	100	39	17	16	15	13	11	11		19	38	3	–	–	–	–	–	–	–	–	–	–	Octadecanoic acid, 2-hydroxy-, methyl ester	2420-35-1	NI33140
255	314	43	57	18	83	55	69	**314**	100	91	90	75	75	73	73	70		19	38	3	–	–	–	–	–	–	–	–	–	–	Octadecanoic acid, 2-hydroxy-, methyl ester	2420-35-1	WI01405
43	41	55	57	69	83	97	90	**314**	100	75	72	59	54	43	39	30	**10.16**	19	38	3	–	–	–	–	–	–	–	–	–	–	Octadecanoic acid, 2-hydroxy-, methyl ester	2420-35-1	NI33141
255	314	43	57	83	55	69	90	**314**	100	92	91	76	73	73	71	65		19	38	3	–	–	–	–	–	–	–	–	–	–	Octadecanoic acid, 2-hydroxy-, methyl ester, (±)-	40617-55-8	NI33142
43	103	41	57	55	82	29	83	**314**	100	68	65	63	59	43	39	36	**0.00**	19	38	3	–	–	–	–	–	–	–	–	–	–	Octadecanoic acid, 3-hydroxy-, methyl ester	2420-36-2	MS07364
103	43	71	41	74	55	57	61	**314**	100	67	28	28	26	25	21	14	**0.00**	19	38	3	–	–	–	–	–	–	–	–	–	–	Octadecanoic acid, 3-hydroxy-, methyl ester	2420-36-2	NI33144
103	41	74	55	57	71	58	43	**314**	100	45	38	38	37	36	34	29	**0.10**	19	38	3	–	–	–	–	–	–	–	–	–	–	Octadecanoic acid, 3-hydroxy-, methyl ester	2420-36-2	NI33143
85	55	43	83	57	69	97	41	**314**	100	38	34	31	29	28	27	27	**0.00**	19	38	3	–	–	–	–	–	–	–	–	–	–	Octadecanoic acid, 4-hydroxy-, methyl ester		MS07365
85	31	55	84	29	41	32	43	**314**	100	85	82	81	67	62	55	52	**0.00**	19	38	3	–	–	–	–	–	–	–	–	–	–	Octadecanoic acid, 6-hydroxy-, methyl ester		MS07366
127	87	55	130	43	41	99	98	**314**	100	85	72	60	60	56	54	52	**0.00**	19	38	3	–	–	–	–	–	–	–	–	–	–	Octadecanoic acid, 7-hydroxy-, methyl ester		MS07367
141	87	144	173	55	43	101	41	**314**	100	56	53	43	38	38	33	33	**0.00**	19	38	3	–	–	–	–	–	–	–	–	–	–	Octadecanoic acid, 8-hydroxy-, methyl ester		MS07368
155	87	158	187	264	55	43	74	**314**	100	51	48	40	39	39	36	35	**1.00**	19	38	3	–	–	–	–	–	–	–	–	–	–	Octadecanoic acid, 9-hydroxy-, methyl ester	2447-53-2	NI33145
155	87	158	74	55	187	41	83	**314**	100	63	52	45	44	34	27	23	**0.00**	19	38	3	–	–	–	–	–	–	–	–	–	–	Octadecanoic acid, 9-hydroxy-, methyl ester	2447-53-2	NI33146
201	55	41	43	87	74	57	69	**314**	100	91	64	55	54	46	46	41	**0.00**	19	38	3	–	–	–	–	–	–	–	–	–	–	Octadecanoic acid, 10-hydroxy-, methyl ester, (±)-	55044-18-3	NI33147
201	55	41	43	87	74	57	69	**314**	100	92	65	55	54	47	47	41	**0.20**	19	38	3	–	–	–	–	–	–	–	–	–	–	Octadecanoic acid, 10-hydroxy-, methyl, ester		MS07369
297	298	295	229	265	313	200	199	**314**	100	20	14	10	7	6	4	4	**1.00**	19	38	3	–	–	–	–	–	–	–	–	–	–	Octadecanoic acid, 12-hydroxy-, methyl ester	141-23-1	NI33148
55	197	87	43	297	69	41	74	**314**	100	86	74	53	50	38	38	36	**0.00**	19	38	3	–	–	–	–	–	–	–	–	–	–	Octadecanoic acid, 12-hydroxy-, methyl ester	141-23-1	MS07370
55	87	74	83	211	41	88	214	**314**	100	95	93	66	64	50	28	26	**0.00**	19	38	3	–	–	–	–	–	–	–	–	–	–	Octadecanoic acid, 13-hydroxy-, methyl ester	2540-76-3	NI33149
211	214	143	243	175	193	171	215	**314**	100	45	39	30	30	28	28	16	**0.00**	19	38	3	–	–	–	–	–	–	–	–	–	–	Octadecanoic acid, 13-hydroxy-, methyl ester	2540-76-3	LI06948
97	83	57	81	111	96	85	74	**314**	100	86	75	71	69	66	64	64	**0.00**	19	38	3	–	–	–	–	–	–	–	–	–	–	Octadecanoic acid, 17-hydroxy-, methyl ester		LI06949
270	87	74	55	45	264	69	83	**314**	100	83	81	52	46	44	44	36	**0.30**	19	38	3	–	–	–	–	–	–	–	–	–	–	Octadecanoic acid, 17-hydroxy-, methyl ester		MS07371
74	98	75	83	84	87	97	82	**314**	100	86	52	45	39	36	34	34	**0.00**	19	38	3	–	–	–	–	–	–	–	–	–	–	Octadecanoic acid, 18-hydroxy-, methyl ester		LI06950
75	299	73	43	103	41	83	57	**314**	100	88	52	40	32	29	27	27	**0.00**	19	42	1	–	–	–	–	–	–	1	–	–	–	Silane, (hexadecyloxy)trimethyl-	6221-90-5	NI33150
299	75	73	57	300	69	55	43	**314**	100	77	29	29	26	26	26	22	**0.00**	19	42	1	–	–	–	–	–	–	1	–	–	–	Silane, (hexadecyloxy)trimethyl-	6221-90-5	MS07372
299	75	73	300	43	103	41	55	**314**	100	52	27	26	21	18	15	14	**0.42**	19	42	1	–	–	–	–	–	–	1	–	–	–	Silane, (hexadecyloxy)trimethyl-	6221-90-5	NI33151

MASS TO CHARGE RATIOS								M.W.	INTENSITIES								Parent	C	H	O	N	S	F	Cl	Br	I	Si	P	B	X	COMPOUND NAME	CAS Reg No	No
76	104	297	314	202	101	50	200	314	100	88	62	60	40	26	20	15		20	10	4	–	–	–	–	–	–	–	–	–	–	2,2′-Binaphthalene-1,1′,4,4′-tetrone	3408-13-7	NI33152
314	297	76	202	104	315	298	77	314	100	96	48	30	28	22	20	20		20	10	4	–	–	–	–	–	–	–	–	–	–	2,2′-Binaphthalene-1,1′,4,4′-tetrone	3408-13-7	NI33153
314	202	258	230	286				314	100	92	68	43	32					20	10	4	–	–	–	–	–	–	–	–	–	–	1,2-Phthaloyl-2a,3,8,8a-tetrahydro-3,8-diketonaphtho[b]cyclobutadiene		LI06951
313	314	143	172	115	315	140	127	314	100	70	50	38	20	18	12	11		20	14	–	2	1	–	–	–	–	–	–	–	–	1,1′-Azothionaphthalene		LI06952
314	151	113	161	178	285	133	259	314	100	55	35	30	29	23	20	15		20	14	2	2	–	–	–	–	–	–	–	–	–	5,6-Diphenyl-3-methylpyrazolo[1,5-a]pyridine-4,7-dione		NI33154
314	77	104	179	223	76	269	91	314	100	90	66	63	53	52	48	45		20	14	2	2	–	–	–	–	–	–	–	–	–	1,4-Phthalazinedione, 2,3-dihydro-2,3-diphenyl-	63546-88-3	NI33155
314	76	104	168	130	167	169	77	314	100	40	35	22	9	8	7	5		20	14	2	2	–	–	–	–	–	–	–	–	–	2-(4-Phthalimidobenzyl)pyridine		MS07373
208	65	107	209	92	165	207	39	314	100	43	29	24	24	14	13	13	7.00	20	18	–	4	–	–	–	–	–	–	–	–	–	Aniline, 2-[2-[2-[(2-aminophenyl)azo]phenyl]vinyl]-	69395-28-4	NI33156
314	171	299	285	312	297	315	169	314	100	61	41	22	19	19	13	12		20	18	–	4	–	–	–	–	–	–	–	–	–	1,2-Bis-(2-methylquinoxal-3-yl)-ethane		IC03889
227	229	228	296	215	240	281	243	314	100	74	25	18	17	15	9	9	7.00	20	26	3	–	–	–	–	–	–	–	–	–	–	Abieta-5,7,9(11),13-tetraen-12-one, 2α,11-dihydroxy-		NI33157
229	227	228	314	296	240	253	264	314	100	87	71	47	27	22	20	16		20	26	3	–	–	–	–	–	–	–	–	–	–	Abieta-5,7,9(11),13-tetraen-12-one, 11,19-dihydroxy-		NI33158
121	43	160	108	314	83	41	55	314	100	87	71	68	47	35	35	29		20	26	3	–	–	–	–	–	–	–	–	–	–	Bicyclo[2.2.1]heptan-2-ol, 3-[(4-methoxyphenyl)methylene]-1,7,7-trimethyl-, acetate, exo-	56816-64-9	MS07374
83	43	55	214	58	31	123	199	314	100	67	49	43	26	18	15	13	1.22	20	26	3	–	–	–	–	–	–	–	–	–	–	2-Butenoic acid, 2-methyl-, 4,4a,5,6,7,9-hexahydro-3,4a,5-trimethylnaphtho[2,3-b]furan-4-yl ester, [4S-[4α(Z),4aα,5α]]-	40072-63-7	NI33159
43	159	314	145	173	161	146	174	314	100	90	64	55	38	32	28	23		20	26	3	–	–	–	–	–	–	–	–	–	–	14-O-Deacetylverecynarmin A		DD01187
314	299	300	149	315	43	229	175	314	100	55	23	23	22	14	11	10		20	26	3	–	–	–	–	–	–	–	–	–	–	1,2-Epoxyhinokione		NI33160
314	315	97	229	187	283	190	61	314	100	23	16	14	11	10	10	9		20	26	3	–	–	–	–	–	–	–	–	–	–	Estra-1,3,5(10)-trien-17-one, 3,4-dimethoxy-	26624-38-4	NI33161
296	186	314	160	187	161	174	281	314	100	89	66	65	57	55	52	40		20	26	3	–	–	–	–	–	–	–	–	–	–	D-Homoestra-1,3,5(10)-trien-17a-one, 14-hydroxy-3-methoxy-	10003-04-0	NI33162
296	281	314	297	282	162	225	160	314	100	85	44	29	19	19	18	16		20	26	3	–	–	–	–	–	–	–	–	–	–	D-Homoestra-1,3,5(10)-trien-17a-one, 14-hydroxy-3-methoxy-, (8α,9β,14β)-	15908-92-6	NI33163
242	241	314	160	243	296	161	173	314	100	78	46	26	24	19	15	14		20	26	3	–	–	–	–	–	–	–	–	–	–	D-Homoestra-1,3,5(10)-trien-17a-one, 14-hydroxy-3-methoxy-, (8α,14β)-	10003-07-3	NI33164
161	314	160	159	174	147	173	172	314	100	92	63	39	33	31	27	26		20	26	3	–	–	–	–	–	–	–	–	–	–	D-Homoestra-1,3,5(10)-trien-17a-one, 14-hydroxy-3-methoxy-, (9β)-	10003-05-1	NI33165
296	186	314	187	161	174	160	127	314	100	89	66	57	55	53	43	40		20	26	3	–	–	–	–	–	–	–	–	–	–	D-Homoestra-1,3,5(10)-trien-17a-one, 14-hydroxy-3-methoxy-, (14α)-		LI06953
99	100	314	195	228	214	213	210	314	100	38	6	6	5	5	5	4		20	26	3	–	–	–	–	–	–	–	–	–	–	D-Homoestra-5(10),6,8-triene, 17,17-ethylenedioxy-3-hydroxy-, (3β)-		MS07375
121	43	160	108	314	83	41	55	314	100	87	71	68	47	35	35	29		20	26	3	–	–	–	–	–	–	–	–	–	–	Isobomeol, 3-(4-methoxybenzylidene)-, acetate		LI06954
137	314	109	147	315	139	119	107	314	100	64	45	14	12	12	10	10		20	26	3	–	–	–	–	–	–	–	–	–	–	Kaur-16-en-18-oic acid, 6-hydroxy-7-oxo-, γ-lactone, (4α,6α)-	19794-40-2	NI33166
165	314	166	257	91	77	105	137	314	100	82	22	13	13	8	8	7		20	26	3	–	–	–	–	–	–	–	–	–	–	4-Methoxyandrost-4-ene-3,17-dione		NI33167
314	203	243	270	43	202	244	216	314	100	96	94	85	83	80	71	68		20	26	3	–	–	–	–	–	–	–	–	–	–	9,10-Phenanthrenedione, 1,2,3,4,4a,10a-hexahydro-6-hydroxy-7-isopropyl 1,1,4a-trimethyl-, (4aS-cis)-	57377-89-6	NI33168
272	91	41	79	314	124	55	77	314	100	86	81	70	55	51	50	49		20	26	3	–	–	–	–	–	–	–	–	–	–	Pregn-4-ene-3,20-dione, 18-hydroxy-, γ-lactone		NI33169
176	69	160	81	164	95	190	314	314	100	59	57	57	53	48	41	34		20	26	3	–	–	–	–	–	–	–	–	–	–	Pseudojolkinolide A	97906-88-2	NI33170
89	83	71	69	97	82	67	68	314	100	99	99	98	92	57	57	55	0.20	20	42	2	–	–	–	–	–	–	–	–	–	–	1,4-Eicosanediol	16274-31-0	NI33171
57	43	71	55	41	85	63	45	314	100	80	64	53	47	43	43	39	0.14	20	42	2	–	–	–	–	–	–	–	–	–	–	Ethanol, 2-(octadecyloxy)-	2136-72-3	NI33172
71	75	82	41	43	96	68	29	314	100	46	33	31	25	17	17	16	0.14	20	42	2	–	–	–	–	–	–	–	–	–	–	Octadecane, 1,1-dimethoxy-	14620-55-4	NI33173
71	82	68	96	110	75	292	250	314	100	25	14	10	2	2	1	1	0.00	20	42	2	–	–	–	–	–	–	–	–	–	–	Octadecane, 1,1-dimethoxy-	14620-55-4	LI06955
71	41	43	82	29	31	55	68	314	100	50	35	25	20	19	18	14	0.00	20	42	2	–	–	–	–	–	–	–	–	–	–	Octadecane, 1,1-dimethoxy-	14620-55-4	NI33174
191	41	123	109	95	55	69	67	314	100	76	74	72	57	55	43	41	17.00	21	30	2	–	–	–	–	–	–	–	–	–	–	1,4-Benzenediol, 2-[(1,4,4a,5,6,7,8,8a-octahydro-2,5,5,8a-tetramethyl-1-naphthalenyl)methyl]-, [1R-(1α,4aβ,8aα)]-	39707-55-6	NI33175
191	123	314	95	109	69	161	121	314	100	90	52	49	43	32	29	29		21	30	2	–	–	–	–	–	–	–	–	–	–	1,4-Benzenediol, 2-[(decahydro-5,5,8a-trimethyl-2-methylene-1-naphthalenyl)methyl]-, [1R-(1α,4aβ,8aα)]-	39707-54-5	NI33176
231	314	299	232	233	271	258	193	314	100	22	17	16	10	6	5	5		21	30	2	–	–	–	–	–	–	–	–	–	–	Cannabichromene	20675-51-8	LI06956
231	232	174	314	299	233	187		314	100	21	10	8	5	3	3			21	30	2	–	–	–	–	–	–	–	–	–	–	Cannabicyclol	21366-63-2	LI06957
231	232	314	233	271	299	258	174	314	100	18	17	10	7	5	5	5		21	30	2	–	–	–	–	–	–	–	–	–	–	Δ^{8}-3,4-cis-Isotetrahydrocannabinol		LI06958
231	314	272	232	258	193	299	174	314	100	35	28	26	18	14	13	13		21	30	2	–	–	–	–	–	–	–	–	–	–	$\Delta^{4(8)}$-Isotetrahydrocannabinol		LI06959
42	231	314	43	299	91	67	271	314	100	90	80	80	75	55	55	51		21	30	2	–	–	–	–	–	–	–	–	–	–	Δ^{1}-Tetrahydrocannabinol	1972-08-3	NI33177
315	313	299	135	193				314	100	55	15	7	5				0.00	21	30	2	–	–	–	–	–	–	–	–	–	–	Δ^{1}-Tetrahydrocannabinol	1972-08-3	NI33178
41	43	299	314	231	55	69	67	314	100	99	51	50	48	35	32	32		21	30	2	–	–	–	–	–	–	–	–	–	–	Δ^{1}-Tetrahydrocannabinol	1972-08-3	NI33179
231	314	271	258	246	193	232	259	314	100	64	36	34	19	18	17	16		21	30	2	–	–	–	–	–	–	–	–	–	–	Δ^{8}-Tetrahydrocannabinol	5957-75-5	NI33181
231	314	271	258	43	193	41	246	314	100	80	44	43	42	31	31	23		21	30	2	–	–	–	–	–	–	–	–	–	–	Δ^{8}-Tetrahydrocannabinol	5957-75-5	NI33180
314	231	258	271	315	43	299	193	314	100	61	31	30	25	18	17	16		21	30	2	–	–	–	–	–	–	–	–	–	–	Δ^{8}-Tetrahydrocannabinol	5957-75-5	MS07376
315	193	313	135	231				314	100	50	44	35	17				0.00	21	30	2	–	–	–	–	–	–	–	–	–	–	$\Delta^{1(2)}$-trans-Cannabidiol	13956-29-1	NI33182
231	246	314	232	121	192	43	174	314	100	47	33	15	15	13	10	9		21	30	2	–	–	–	–	–	–	–	–	–	–	$\Delta^{1(2)}$-trans-Cannabidiol	13956-29-1	NI33183
231	41	67	43	232	68	91	55	314	100	29	19	19	18	16	13	13	2.95	21	30	2	–	–	–	–	–	–	–	–	–	–	$\Delta^{1(2)}$-trans-Cannabidiol	13956-29-1	NI33184
314	231	271	258	315	246	193	232	314	100	83	46	40	24	23	23	16		21	30	2	–	–	–	–	–	–	–	–	–	–	Dibenzo[b,d]pyran-3-ol, 6a,7,10,10a-tetrahydro-6,6,9-trimethyl-1-pentyl-		LI06961
231	314	175	246	271	258	232	201	314	100	73	26	25	23	22	18	18		21	30	2	–	–	–	–	–	–	–	–	–	–	Dibenzo[b,d]pyran-3-ol, 6a,7,10,10a-tetrahydro-6,6,9-trimethyl-1-pentyl-		LI06960

MASS TO CHARGE RATIOS	M.W.	INTENSITIES	Parent	C	H	O	N	S	F	Cl	Br	I	Si	P	B	X	COMPOUND NAME	CAS Reg No	No
314 43 110 148 231 91 79 55	**314**	100 95 51 49 36 35 28 27		21	30	2	–	–	–	–	–	–	–	–	–	–	**18,19-Dinorpregn-4-ene-3,20-dione, 13-ethyl-**	13934-52-6	**NI33185**
314 43 110 148 231 91 79 55	**314**	100 97 51 49 36 36 28 28		21	30	2	–	–	–	–	–	–	–	–	–	–	**18,19-Dinorpregn-4-ene-3,20-dione, 13-ethyl-**	13934-52-6	**NI33186**
247 229 246 288 244 231 314 215	**314**	100 91 84 56 45 42 40 39		21	30	2	–	–	–	–	–	–	–	–	–	–	**18,19-Dinorpregn-4-en-20-yn-3-one, 12-ethyl-17-hydroxy-, (5β,17α)-**		**NI33187**
314 200 91 254 106 260	**314**	100 59 33 29 28 8		21	30	2	–	–	–	–	–	–	–	–	–	–	**B-Homoestra-2,5(10)-dien-17-ol, (17β)-, acetate**		**LI06962**
231 232 314 174 299 271 187	**314**	100 20 10 8 7 3 3		21	30	2	–	–	–	–	–	–	–	–	–	–	**Isocannabichromene**	31508-73-3	**LI06963**
231 232 174 314 299 233 187	**314**	100 22 11 8 5 3 3		21	30	2	–	–	–	–	–	–	–	–	–	–	**Isocannabicyclol**	31508-74-4	**LI06964**
314 296 227 173 147 174 315 171	**314**	100 40 39 38 32 29 24 21		21	30	2	–	–	–	–	–	–	–	–	–	–	**19-Norpregna-1,3,5(10)-trien-17-ol, 3-methoxy-**	17550-02-6	**NI33188**
43 272 124 314 91 41 229 55	**314**	100 49 39 36 33 33 26 24		21	30	2	–	–	–	–	–	–	–	–	–	–	**19-Norpregn-4-ene-3,20-dione, 1-methyl-, (1α)-**	4050-25-3	**NI33190**
43 273 124 315 41 91 229 55	**314**	100 99 81 75 69 68 54 50	**0.00**	21	30	2	–	–	–	–	–	–	–	–	–	–	**19-Norpregn-4-ene-3,20-dione, 1-methyl-, (1α)-**	4050-25-3	**NI33189**
43 124 314 272 91 79 55 41	**314**	100 89 45 32 28 27 27 26		21	30	2	–	–	–	–	–	–	–	–	–	–	**19-Norpregn-4-ene-3,20-dione, 1-methyl-, (1β)-**	4183-69-1	**NI33191**
43 124 314 273 91 244 79 55	**314**	100 99 52 37 31 30 30 30		21	30	2	–	–	–	–	–	–	–	–	–	–	**19-Norpregn-4-ene-3,20-dione, 1-methyl-, (1β)-**	4183-69-1	**NI33192**
314 189 271 299 315 215 229 161	**314**	100 65 48 45 23 23 14 12		21	30	2	–	–	–	–	–	–	–	–	–	–	**4a(2H)-Phenanthrenecarboxaldehyde, 1,3,4,9,10,10a-hexahydro-7-isopropyl-6-methoxy-1,1-dimethyl-, (4aR-trans)-**	57397-33-8	**NI33193**
314 254 121 315 134 255 91 132	**314**	100 53 37 36 33 32 25 24		21	30	2	–	–	–	–	–	–	–	–	–	–	**Podocarpa-7,13,15-trien-15-oic acid, 13-isopropyl-, methyl ester**	54850-32-7	**NI33194**
239 240 299 314 43 197 141 129	**314**	100 22 19 14 10 8 8 8		21	30	2	–	–	–	–	–	–	–	–	–	–	**Podocarpa-8,11,13-trien-15-oic acid, 13-isopropyl-, methyl ester**	1235-74-1	**NI33195**
239 299 302 240 300 197 303 141	**314**	100 89 43 19 17 13 9 8	**0.00**	21	30	2	–	–	–	–	–	–	–	–	–	–	**Podocarpa-8,11,13-trien-15-oic acid, 13-isopropyl-, methyl ester**	1235-74-1	**NI33197**
239 240 141 173 197 129 155 128	**314**	100 25 24 21 19 19 18 17	**16.00**	21	30	2	–	–	–	–	–	–	–	–	–	–	**Podocarpa-8,11,13-trien-15-oic acid, 13-isopropyl-, methyl ester**	1235-74-1	**NI33196**
239 299 314 240 300 197 141 129	**314**	100 88 42 18 16 12 8 7		21	30	2	–	–	–	–	–	–	–	–	–	–	**Podocarpa-8,11,13-trien-16-oic acid, 13-isopropyl-, methyl ester**	24035-60-7	**LI06965**
239 299 314 240 300 197 315 141	**314**	100 90 43 18 17 13 9 8		21	30	2	–	–	–	–	–	–	–	–	–	–	**Podocarpa-8,11,13-trien-16-oic acid, 13-isopropyl-, methyl ester**	24035-60-7	**NI33198**
314 299 257 315 213 300 215 173	**314**	100 72 33 19 18 16 13 11		21	30	2	–	–	–	–	–	–	–	–	–	–	**Podocarpa-8,11,13-trien-3-one, 13-isopropyl-12-methoxy-**	18325-89-8	**NI33199**
314 299 257 315 213 300 215 41	**314**	100 72 33 19 18 16 14 10		21	30	2	–	–	–	–	–	–	–	–	–	–	**Podocarpa-8,11,13-trien-3-one, 13-isopropyl-12-methoxy-**	18325-89-8	**NI33200**
299 314 257 300 315 215 189 213	**314**	100 54 25 20 11 11 8 7		21	30	2	–	–	–	–	–	–	–	–	–	–	**Podocarpa-8,11,13-trien-3-one, 14-isopropyl-13-methoxy-**	18326-16-4	**LI06966**
299 314 257 300 315 215 189 173	**314**	100 54 25 19 11 11 8 7		21	30	2	–	–	–	–	–	–	–	–	–	–	**Podocarpa-8,11,13-trien-3-one, 14-isopropyl-13-methoxy-**	18326-16-4	**NI33201**
299 314 257 300 315 215 189 201	**314**	100 53 26 20 12 9 8 7		21	30	2	–	–	–	–	–	–	–	–	–	–	**Podocarpa-8,11,13-trien-3-one, 14-isopropyl-13-methoxy-**	18326-16-4	**NI33202**
296 281 267 297 314 282 270 268	**314**	100 39 28 24 23 9 8 8		21	30	2	–	–	–	–	–	–	–	–	–	–	**Pregna-4,6-dien-3-one, 20-hydroxy-**		**LI06967**
296 135 45 281 161 136 55 267	**314**	100 44 40 39 36 34 28 27	**22.00**	21	30	2	–	–	–	–	–	–	–	–	–	–	**Pregna-4,6-dien-3-one, 20-hydroxy-, (20α)-**		**MS07377**
43 314 91 145 105 203 299 107	**314**	100 41 23 20 20 19 18 18		21	30	2	–	–	–	–	–	–	–	–	–	–	**Pregna-5,16-dien-20-one, 3-hydroxy-, (3β)-**	1162-53-4	**NI33203**
296 91 105 77 281 145 121 79	**314**	100 65 47 39 38 36 33 33	**17.70**	21	30	2	–	–	–	–	–	–	–	–	–	–	**Pregna-5,16-dien-20-one, 3-hydroxy-, (3β)-**	1162-53-4	**MS07378**
314 79 124 91 93 95 105 67	**314**	100 63 60 54 52 50 49 48		21	30	2	–	–	–	–	–	–	–	–	–	–	**Pregn-5(6)-en-3,20-dione**		**LI06968**
163 137 41 136 55 43 80 81	**314**	100 82 76 61 58 46 44 42	**23.67**	21	30	2	–	–	–	–	–	–	–	–	–	–	**Pregn-4-ene-3,6-dione**	3510-11-0	**NI33204**
124 43 314 272 229 79 91 55	**314**	100 97 70 49 37 27 25 25		21	30	2	–	–	–	–	–	–	–	–	–	–	**Pregn-4-ene-3,20-dione**	57-83-0	**NI33206**
43 124 79 55 91 41 67 135	**314**	100 55 26 26 25 23 20 19	**4.02**	21	30	2	–	–	–	–	–	–	–	–	–	–	**Pregn-4-ene-3,20-dione**	57-83-0	**NI33205**
43 124 314 28 79 272 91 229	**314**	100 82 36 33 30 29 29 28		21	30	2	–	–	–	–	–	–	–	–	–	–	**Pregn-4-ene-3,20-dione**	57-83-0	**NI33207**
205 204 203 206 207 202 201 200	**314**	100 80 80 77 66 56 33 17	**0.00**	21	30	2	–	–	–	–	–	–	–	–	–	–	**Pregn-4-ene-3,20-dione, (8α)-**	5750-05-0	**NI33208**
196 195 181 180 179 197 194 198	**314**	100 90 85 75 50 41 34 32	**0.00**	21	30	2	–	–	–	–	–	–	–	–	–	–	**Pregn-4-ene-3,20-dione, (8α)-**	5750-05-0	**NI33210**
176 175 159 160 161 158 174 157	**314**	100 92 85 78 77 61 38 24	**0.00**	21	30	2	–	–	–	–	–	–	–	–	–	–	**Pregn-4-ene-3,20-dione, (8α)-**	5750-05-0	**NI33209**
124 43 314 91 79 55 107 93	**314**	100 82 44 24 24 22 18 18		21	30	2	–	–	–	–	–	–	–	–	–	–	**Pregn-4-ene-3,20-dione, (8α,10α)-**	3795-19-5	**NI33211**
124 43 314 91 79 55 147 93	**314**	100 82 43 25 24 23 20 19		21	30	2	–	–	–	–	–	–	–	–	–	–	**Pregn-4-ene-3,20-dione, (8α,10α)-**	3795-19-5	**NI33212**
124 314 91 147 191 133 105 93	**314**	100 27 20 19 17 16 13 13		21	30	2	–	–	–	–	–	–	–	–	–	–	**Pregn-4-ene-3,20-dione, (9β,10α)-**	2755-10-4	**NI33214**
124 43 147 91 314 133 79 55	**314**	100 68 20 17 16 16 16 15		21	30	2	–	–	–	–	–	–	–	–	–	–	**Pregn-4-ene-3,20-dione, (9β,10α)-**	2755-10-4	**NI33213**
124 43 314 91 229 55 107 79	**314**	100 63 28 17 16 16 14 14		21	30	2	–	–	–	–	–	–	–	–	–	–	**Pregn-4-ene-3,20-dione, (10α)-**	3562-13-8	**NI33215**
244 314 71 90 54 79 124 95	**314**	100 43 36 32 31 28 23 20		21	30	2	–	–	–	–	–	–	–	–	–	–	**Pregn-4-ene-3,20-dione, (17α)**		**LI06969**
43 314 105 229 55 91 41 107	**314**	100 39 33 29 28 25 22 18		21	30	2	–	–	–	–	–	–	–	–	–	–	**Pregn-11-ene-3,20-dione, (5β)-**	1096-39-5	**NI33216**
229 314 211 244 105 85 281 159	**314**	100 60 60 58 54 47 46 44		21	30	2	–	–	–	–	–	–	–	–	–	–	**Pregn-11-ene-3,20-dione, (5β)-**	1096-39-5	**NI33217**
314 159 41 105 119 91 55 69	**314**	100 93 90 81 74 72 64 59		21	30	2	–	–	–	–	–	–	–	–	–	–	**Retinoic acid, methyl ester**	339-16-2	**NI33219**
314 105 91 255 119 69 41 177	**314**	100 38 36 33 33 33 33 30		21	30	2	–	–	–	–	–	–	–	–	–	–	**Retinoic acid, methyl ester**	339-16-2	**NI33218**
314 299 231 41 271 43 315 243	**314**	100 69 53 43 37 32 25 19		21	30	2	–	–	–	–	–	–	–	–	–	–	**Δ^9-Tetrahydrocannabinol**	1972-08-3	**NI33220**
314 299 231 271 43 41 315 243	**314**	100 81 55 41 36 35 24 23		21	30	2	–	–	–	–	–	–	–	–	–	–	**Tetrahydrocannabinol**	33086-25-8	**NI33221**
314 299 231 271 243 258 43 41	**314**	100 95 80 50 32 30 28 28		21	30	2	–	–	–	–	–	–	–	–	–	–	**Tetrahydrocannabinol**	33086-25-8	**NI33222**
314 299 268 315 283 267 113 300	**314**	100 45 24 22 22 13 13 10		22	18	2	–	–	–	–	–	–	–	–	–	–	**Anthracene, 1,4-dimethoxy-9-phenyl-**		**MS07379**
254 253 252 126 113 125 250 255	**314**	100 97 60 54 27 26 21 20	**1.00**	22	18	2	–	–	–	–	–	–	–	–	–	–	**9H-Fluorene-9-methanol, α-phenyl-, acetate**	63839-89-4	**NI33223**
314 178 285 118 315 206 313 165	**314**	100 37 29 27 25 23 20 17		22	18	2	–	–	–	–	–	–	–	–	–	–	**2-Propenal, 2-(4-methoxyphenyl)-3,3-diphenyl-**		**IC03890**
314 285 315 313 108 286 165 178	**314**	100 51 20 17 12 11 10 8		22	18	2	–	–	–	–	–	–	–	–	–	–	**Styrene, β-formyl-β-(4-methoxyphenyl)-α-phenyl-**		**MS07380**
296 239 241 253 314 252 240 270	**314**	100 81 70 59 55 53 49 38		22	18	2	–	–	–	–	–	–	–	–	–	–	**cis-11,12,13,14-Tetrahydrobenzo[g]chrysene-11,12-diol**		**DD01188**
271 107 43 91 55 41 109 95	**314**	100 35 32 23 22 22 21 21	**7.07**	22	34	1	–	–	–	–	–	–	–	–	–	–	**Pregn-15-en-20-one, 17-methyl-, (5α,17ξ)-**	54411-97-1	**NI33224**
43 69 41 81 55 93 105 91	**314**	100 90 90 80 80 75 70 70	**55.00**	22	34	1	–	–	–	–	–	–	–	–	–	–	**Vitamin A, 4-ethoxyanhydro-**		**MS07381**

MASS TO CHARGE RATIOS	M.W.	INTENSITIES	Parent	C	H	O	N	S	F	Cl	Br	I	Si	P	B	X	COMPOUND NAME	CAS Reg No	No
314 167 203 91 243 258 115 286	**314**	100 41 38 36 30 23 22 18		23	22	1	–	–	–	–	–	–	–	–	–	–	**(±)-(1α,1aα,5aα,6aα)-1a,4,5,5a,6,6a-Hexahydro-5a-methyl-1,6a-diphenylcycloprop[a]inden-3(1H)-one**		**DD01189**
314 286 128 315 143 171 256 78	**314**	100 42 30 28 28 24 22 20		23	22	1	–	–	–	–	–	–	–	–	–	–	**1,4-Pentadiyn-3-one, 1,5-bis(2,4,6-trimethylphenyl)-**	51118-05-9	**MS07382**
314 165 243 91 210 257 195 286	**314**	100 27 23 22 20 19 19 17		23	22	1	–	–	–	–	–	–	–	–	–	–	**(±)-4,4a,5,6-Tetrahydro-4a-methyl-6,6-diphenyl-2(3H)-naphthalenone**		**DD01190**
243 165 244 241 166 314 242 239	**314**	100 30 26 6 6 5 5 4		24	26	–	–	–	–	–	–	–	–	–	–	–	**Butane, 3,3-dimethyl-1,1,1-triphenyl-**	24523-61-3	**NI33225**
243 165 244 41 166 57 29 39	**314**	100 37 28 15 9 9 9 7	**0.90**	24	26	–	–	–	–	–	–	–	–	–	–	–	**Butane, 3,3-dimethyl-1,1,1-triphenyl-**	24523-61-3	**WI01406**
314 271 145 215 165 229 129 115	**314**	100 40 22 18 17 16 15 15		24	26	–	–	–	–	–	–	–	–	–	–	–	**3,7-Decadien-5-yne, 2,9-dimethyl-4,7-diphenyl-**		**NI33226**
314 299 114 315 300 202 243 142	**314**	100 58 33 27 15 14 11 10		24	26	–	–	–	–	–	–	–	–	–	–	–	**Pyrene, 1,6-di-tert-butyl-**	55044-29-6	**WI01407**
314 299 114 315 300 202 243 142	**314**	100 58 33 27 15 14 11 10		24	26	–	–	–	–	–	–	–	–	–	–	–	**Pyrene, 3,8-di-tert-butyl-**		**AM00141**
127 146 100 170 126 124 128 150	**315**	100 36 31 21 17 12 11 10	**0.00**	4	2	3	1	1	9	–	–	–	–	–	–	–	**(3-Carboxy-2,2,3,3-tetrafluoropropionamido)sulphur pentafluoride**		**NI33227**
315 317 142 144 30 179 177 107	**315**	100 98 98 85 85 84 84 50		6	–	6	3	–	–	3	–	–	–	–	–	–	**Benzene, 1,3,5-trichloro-2,4,6-trinitro-**	2631-68-7	**LI06970**
142 144 30 315 177 317 179 46	**315**	100 99 91 71 70 69 62 56		6	–	6	3	–	–	3	–	–	–	–	–	–	**Benzene, 1,3,5-trichloro-2,4,6-trinitro-**	2631-68-7	**NI33228**
163 165 280 97 224 174 152 182	**315**	100 64 54 51 33 28 24 21	**5.00**	9	12	3	1	1	–	2	–	–	–	1	–	–	**Phosphorothioic acid, O,O-diethyl-, O-(3,6-dichloro-2-pyridyl) ester**		**LI06971**
114 142 315 113 88 102 63 101	**315**	100 74 73 62 23 21 16 14		10	6	3	1	–	–	–	–	1	–	–	–	–	**2-Iodo-4-nitro-1-naphthol**	53913-61-4	**NI33229**
316 317 155 154 153 95 82 318	**315**	100 11 3 2 2 2 2 1	**0.00**	10	10	3	3	–	5	–	–	–	–	–	–	–	**Histidine methyl ester, pentafluoropropionate**		**NI33230**
73 90 91 79 118 117 80 134	**315**	100 86 51 51 49 41 36 28	**4.90**	11	13	4	3	2	–	–	–	–	–	–	–	–	**3-((Methylthio)sulphonyl)propyl 4-azidobenzoate**	76091-05-9	**NI33231**
228 58 71 245 257 30 78 56	**315**	100 95 73 67 65 38 21 19	**15.00**	12	18	2	3	–	–	–	1	–	–	–	–	–	**Benzamide, 4-amino-5-bromo-2-methoxy-N-[(ethylamino)ethyl]-**		**MS07383**
315 149 222 316 137 193 99 269	**315**	100 20 18 16 15 13 11 10		13	5	7	3	–	–	–	–	–	–	–	–	–	**9H-Fluoren-9-one, 2,4,7-trinitro-**	129-79-3	**NI33232**
43 145 103 115 15 44 28 42	**315**	100 7 7 6 4 3 3 1	**0.00**	13	17	8	1	–	–	–	–	–	–	–	–	–	**D-Arabinononitrile, 2,3,4,5-tetraacetate**	34360-54-8	**MS07384**
43 115 103 145 15 157 98 44	**315**	100 8 5 4 4 3 3 2	**0.00**	13	17	8	1	–	–	–	–	–	–	–	–	–	**D-Lyxononitrile, 2,3,4,5-tetraacetate**	34360-56-0	**MS07385**
212 166 256 224 315 284	**315**	100 88 70 67 25 22		13	17	8	1	–	–	–	–	–	–	–	–	–	**1-Methyl-2,3-dihydro-pyrrole-2,3,4,5-tetracarboxylate tetramethyl ester**		**LI06972**
43 28 145 115 103 15 44 42	**315**	100 7 5 5 5 4 3 2	**0.00**	13	17	8	1	–	–	–	–	–	–	–	–	–	**D-Ribononitrile, 2,3,4,5-tetraacetate**	25546-50-3	**MS07386**
43 28 115 103 145 15 44 42	**315**	100 8 6 6 5 4 3 2	**0.00**	13	17	8	1	–	–	–	–	–	–	–	–	–	**D-Xylononitrile, 2,3,4,5-tetraacetate**	13501-95-6	**MS07387**
84 56 80 55 107 133 191 146	**315**	100 42 25 17 14 3 2 2	**0.00**	13	18	3	1	–	–	–	1	–	–	–	–	–	**threo-DL-2-(4-Hydroxyphenyl)-2-(2′-piperidyl)acetic acid hydrobromide**		**NI33233**
317 315 236 288 78 286 79 77	**315**	100 96 92 69 57 57 56 56		14	10	1	3	–	–	–	1	–	–	–	–	–	**2H-1,4-Benzodiazepin-2-one, 7-bromo-1,3-dihydro-5-(2-pyridinyl)-**	1812-30-2	**NI33234**
315 272 243 244 246 317 43 217	**315**	100 55 55 44 40 33 33 24		14	10	2	5	–	–	1	–	–	–	–	–	–	**7-(4-Chlorophenyl)-1,2,3,4-tetrahydro-1-methyl-8H-imidazo[2,1-f]purine-2,4-dione**		**NI33235**
91 60 59 239 93 240 92 77	**315**	100 43 6 6 6 5 4 4	**3.20**	14	13	2	5	1	–	–	–	–	–	–	–	–	**5-Nitro-2-aminobenzophenone thiosemicarbazone**		**MS07388**
120 42 77 28 79 106 30 105	**315**	100 86 45 38 32 31 31 24	**16.38**	14	13	4	5	–	–	–	–	–	–	–	–	–	**Aniline, 4-[(2,4-dinitrophenyl)azo]-N,N-dimethyl-**	35473-63-3	**NI33237**
120 42 315 148 77 105 79 28	**315**	100 48 31 27 22 21 18 13		14	13	4	5	–	–	–	–	–	–	–	–	–	**Aniline, 4-[(2,4-dinitrophenyl)azo]-N,N-dimethyl-**	35473-63-3	**NI33236**
135 242 199 270 228 70 116 104	**315**	100 80 65 43 25 25 22 22	**0.80**	14	21	5	1	1	–	–	–	–	–	–	–	–	**Benzenesulphonamide, 4-(methoxycarbonyl)-N-(2-hydroxypropyl)-N-propyl-**		**LI06974**
135 242 199 270 228 70 116 104	**315**	100 80 65 43 25 25 20 20	**0.00**	14	21	5	1	1	–	–	–	–	–	–	–	–	**Benzenesulphonamide, 4-(methoxycarbonyl)-N-(2-hydroxypropyl)-N-propyl-**		**LI06973**
135 270 199 72 70 42 116 74	**315**	100 92 61 38 38 38 21 21	**0.00**	14	21	5	1	1	–	–	–	–	–	–	–	–	**Benzenesulphonamide, 4-(methoxycarbonyl)-N-(3-hydroxypropyl)-N-propyl-**		**LI06976**
136 270 199 72 70 43 42	**315**	100 89 59 36 36 36 36	**0.00**	14	21	5	1	1	–	–	–	–	–	–	–	–	**Benzenesulphonamide, 4-(methoxycarbonyl)-N-(3-hydroxypropyl)-N-propyl-**		**LI06975**
104 195 91 167 105 65 105 65	**315**	100 76 27 14 9 9 9 9	**7.70**	15	10	1	1	–	5	–	–	–	–	–	–	–	**Phenyl ethylamine, N-(pentafluorobenzoyl)-**		**MS07390**
104 195 91 43 167 196 105 65	**315**	100 82 24 12 11 6 6 6	**2.40**	15	10	1	1	–	5	–	–	–	–	–	–	–	**Phenyl ethylamine, N-(pentafluorobenzoyl)-**		**MS07389**
315 317 76 152 208 90 104 151	**315**	100 84 57 49 41 33 21 19		15	10	2	1	–	–	–	1	–	–	–	–	–	**9,10-Anthracenedione, 1-amino-4-bromo-2-methyl-**	81-50-5	**NI33238**
287 315 314 288 286 316 289 317	**315**	100 92 63 45 44 38 28 24		15	10	3	3	–	–	1	–	–	–	–	–	–	**1H-1,4-Benzodiazepin-2-one, 5-(4-chlorophenyl)-2,3-dihydro-7-nitro-**	4928-04-5	**NI33239**
280 314 286 315 288 240 234 75	**315**	100 86 72 72 56 53 53 48		15	10	3	3	–	–	1	–	–	–	–	–	–	**2H-1,4-Benzodiazepin-2-one, 5-(2-chlorophenyl)-1,3-dihydro-7-nitro-**	1622-61-3	**NI33240**
125 154 127 156 126 89 36 90	**315**	100 54 35 19 10 10 10 5	**0.00**	15	16	–	1	–	–	3	–	–	–	–	–	–	**Phenethylamine, 4-chloro-N-(4-chlorobenzyl)-, hydrochloride**	13541-46-3	**NI33241**
86 201 110 185 213 315 243	**315**	100 64 40 36 16 8 8		15	29	4	3	–	–	–	–	–	–	–	–	–	**L-Alanine, N-(N-L-valyl-L-isoleucyl)-, methyl ester**	37580-54-4	**NI33242**
315 297 272 244 269 300 268 257	**315**	100 46 26 18 12 9 5 5		16	13	6	1	–	–	–	–	–	–	–	–	–	**8-Amino-1,5,7-trihydroxy-6-methoxy-3-methyl-anthraquinone**		**LI06977**
224 181 225 91 182 41 65 28	**315**	100 77 16 9 6 4 3 2	**0.13**	16	14	–	1	–	5	–	–	–	–	–	–	–	**Methamphetamine, N-pentafluorophenyl-**		**MS07391**
314 315 316 317 43 239 237 77	**315**	100 96 53 36 28 27 25 22		16	14	–	3	1	–	1	–	–	–	–	–	–	**7-Chloro-3-methyl-2-methylthio-5-phenyl-3H-1,3,4-benzothiazepine**		**MS07392**
88 237 315 77 89 118 241 90	**315**	100 13 11 9 6 5 5 5		16	14	–	3	1	–	1	–	–	–	–	–	–	**7-Chloro-1-methyl-2-methylthio-5-phenyl-3H-1,3,4-benzotriazepine**		**MS07393**
125 107 167 104 149 103 124 148	**315**	100 61 35 35 30 27 18 16	**7.00**	17	17	3	1	1	–	–	–	–	–	–	–	–	**3-[2-(Acetylamino)phenylthio]-3-phenylpropionic acid**		**MS07394**
160 315 145 286 155	**315**	100 4 4 1 1		17	17	3	1	1	–	–	–	–	–	–	–	–	**1H-3-Benzazepin-1-one, 2,3,4,5-tetrahydro-3-[(4-methylphenyl)sulphonyl]**	15218-07-2	**NI33243**
132 315 160 251 155 316 250 236	**315**	100 63 48 16 16 12 7 3		17	17	3	1	1	–	–	–	–	–	–	–	–	**5H-1-Benzazepin-5-one, 1,2,3,4-tetrahydro-1-[(4-methylphenyl)sulphonyl]**	24310-36-9	**NI33244**
132 315 160 251 155 250 236 222	**315**	100 63 48 16 16 7 3 2		17	17	3	1	1	–	–	–	–	–	–	–	–	**5H-1-Benzazepin-5-one, 1,2,3,4-tetrahydro-1-[(4-methylphenyl)sulphonyl]**	24310-36-9	**LI06978**
132 91 105 65 41 133 155 77	**315**	100 58 30 23 19 13 13 11	**0.00**	17	17	3	1	1	–	–	–	–	–	–	–	–	**2-Benzyl-1-(4-methylbenzenesulfonyl)azetidin-3-one**	82380-64-1	**NI33245**

MASS TO CHARGE RATIOS	M.W.	INTENSITIES	Parent	C	H	O	N	S	F	Cl	Br	I	Si	P	B	X	COMPOUND NAME	CAS Reg No	No
300 257 315 272 158 301 242 282	**315**	100 70 59 32 30 25 25 16		17	17	5	1	–	–	–	–	–	–	–	–	–	**Acridone, 10-methyl-1,2,3-trimethoxy-**		**MS07395**
300 315 284 158 257 242 301 316	**315**	100 94 93 51 48 36 31 24		17	17	5	1	–	–	–	–	–	–	–	–	–	**Acridone, 10-methyl-1,3,4-trimethoxy-**		**MS07396**
300 315 158 242 301 228 316 130	**315**	100 43 38 24 20 17 12 10		17	17	5	1	–	–	–	–	–	–	–	–	–	**Acridone, 10-methyl-2,3,4-trimethoxy-**		**MS07397**
125 42 96 124 126 162 41 44	**315**	100 23 19 9 8 6 5 4	**0.08**	17	17	5	1	–	–	–	–	–	–	–	–	–	**Lycorenan-7-one, 5-hydroxy-1-methyl-9,10-[methylenebis(oxy)]-, (5α)-**	477-17-8	**NI33246**
212 273 257 244 315 299 227 300	**315**	100 65 46 42 39 33 26 16		17	17	5	1	–	–	–	–	–	–	–	–	–	**5H-Pyrano[3,2-c]quinolin-5-one, 6-[(acetyloxy)methyl]-2,6-dihydro-7-hydroxy-2,2-dimethyl-**	65560-24-9	**NI33247**
315 44 73 42 43 28 104 30	**315**	100 78 58 49 44 42 24 24		18	13	1	5	–	–	–	–	–	–	–	–	–	**6,7-Diphenylpterin**		**MS07398**
315 317 316 314 318 319 111 102	**315**	100 65 32 21 16 13 12 10		18	15	–	1	–	–	2	–	–	–	–	–	–	**2,4-Dimethyl-1,3-bis(4-chlorophenyl)pyrrole**		**NI33248**
58 28 59 221 30 42	**315**	100 22 5 4 4 3	**0.00**	18	18	–	1	1	–	1	–	–	–	–	–	–	**1-Propylamine, 3-(2-chloro-9H-thioxanthen-9-ylidene)-N,N-dimethyl-, (Z)**	113-59-7	**NI33249**
316 257 314 259	**315**	100 20 10 7	**0.00**	18	18	–	1	1	–	1	–	–	–	–	–	–	**1-Propylamine, 3-(2-chloro-9H-thioxanthen-9-ylidene)-N,N-dimethyl-, (Z)**	113-59-7	**NI33251**
58 221 59 42 222 234 43 30	**315**	100 7 7 4 3 2 2 2	**0.44**	18	18	–	1	1	–	1	–	–	–	–	–	–	**1-Propylamine, 3-(2-chloro-9H-thioxanthen-9-ylidene)-N,N-dimethyl-, (Z)**	113-59-7	**NI33250**
315 258 94 259 135 79 93 272	**315**	100 39 35 29 14 14 12 9		18	21	4	1	–	–	–	–	–	–	–	–	–	**Benzoic acid, 3-(1-adamantyl)-5-nitro-, methyl ester**		**NI33252**
315 284 298 300 166 137 28 150	**315**	100 67 58 54 36 26 25 23		18	21	4	1	–	–	–	–	–	–	–	–	–	**Cephalotaxine**	24316-19-6	**MS07399**
284 298 166 315 300 137 214 150	**315**	100 90 83 70 70 55 44 43		18	21	4	1	–	–	–	–	–	–	–	–	–	**Cephalotaxine**	24316-19-6	**NI33253**
241 257 240 228 256 315 242 284	**315**	100 86 52 47 29 24 18 14		18	21	4	1	–	–	–	–	–	–	–	–	–	**Erythratine**	5550-20-9	**NI33254**
269 297 268 270 298 112 315 254	**315**	100 73 25 22 20 16 15 9		18	21	4	1	–	–	–	–	–	–	–	–	–	**Hasubanan-6,9-diol, 7,8-didehydro-4,5-epoxy-3-methoxy-17-methyl-, (5α,6α,9α,13β,14β)-**	20377-55-3	**NI33255**
269 297 268 270 298 112 315 254	**315**	100 67 23 21 20 15 14 10		18	21	4	1	–	–	–	–	–	–	–	–	–	**Hasubanan-6,9-diol, 7,8-didehydro-4,5-epoxy-3-methoxy-17-methyl-, (5α,6α,9α,13β,14β)-**	20377-55-3	**LI06979**
202 43 315 203 245 108 41 93	**315**	100 81 56 38 26 26 24 23		18	21	4	1	–	–	–	–	–	–	–	–	–	**1H-Isoindole-1,3(2H)-dione, 4-(acetyloxy)hexahydro-3a,7a-dimethyl-2-phenyl-, [3aS-(3aα,4β,7aα)]-**	54345-98-1	**NI33256**
315 112 44 43 42 107 58 316	**315**	100 75 73 66 32 24 22 18		18	21	4	1	–	–	–	–	–	–	–	–	–	**Morphinan-6,10-diol, 7,8-didehydro-4,5-epoxy-3-methoxy-17-methyl-, (5α,6α)-**		**LI06980**
315 229 316 230 188 175 214 189	**315**	100 33 21 21 20 17 13 10		18	21	4	1	–	–	–	–	–	–	–	–	–	**Morphinan-6,14-diol, 7,8-didehydro-4,5-epoxy-3-methoxy-17-methyl-, (5α,6α)-**	4829-46-3	**NI33257**
300 286 57 115 243 301 44 73	**315**	100 95 82 72 71 70 68 61	**50.00**	18	21	4	1	–	–	–	–	–	–	–	–	–	**Morphinan-7-one, 5,6-didehydro-4-hydroxy-3,6-dimethoxy-**		**LI06981**
300 286 57 243 301 44 74 132	**315**	100 95 82 72 71 63 62 60	**51.00**	18	21	4	1	–	–	–	–	–	–	–	–	–	**Morphinan-7-one, 5,6-didehydro-4-hydroxy-3,6-dimethoxy-**		**MS07400**
316 298	**315**	100 6	**0.00**	18	21	4	1	–	–	–	–	–	–	–	–	–	**Morphinan-6-one, 4,5-epoxy-14-hydroxy-3-methoxy-17-methyl-, (5α)-**	76-42-6	**NI33258**
315 257 284 229 316 226 199 223	**315**	100 76 38 32 20 16 14 13		18	21	4	1	–	–	–	–	–	–	–	–	–	**Pancracine, O,O-dimethyl-**	606-51-9	**NI33259**
315 257 284 229 316 226 199 258	**315**	100 75 39 33 20 16 14 13		18	21	4	1	–	–	–	–	–	–	–	–	–	**Pancracine, O,O-dimethyl-**	606-51-9	**LI06982**
224 91 315 208 162 149 178 97	**315**	100 88 63 26 22 21 16 16		18	21	4	1	–	–	–	–	–	–	–	–	–	**1H-Pyrrole-2,5-dicarboxylic acid, 3-ethyl-4-methyl-, 5-ethyl 2-(phenylmethyl) ester**	35889-90-8	**NI33260**
315 314 250 249 313 248 316 286	**315**	100 73 31 29 26 24 15 8		18	24	–	3	–	–	–	–	–	–	–	3	–	**Borazine, 2,4,6-tri-2,4-cyclopentadien-1-yl-1,3,5-trimethyl-**	33911-15-8	**NI33261**
297 139 91 92 266 209 298 267	**315**	100 80 70 57 51 34 21 19	**17.36**	19	25	3	1	–	–	–	–	–	–	–	–	–	**Estra-1,3,5(10)-trien-16-one, 3,17-dihydroxy-, O-methyloxime, (17α)-**	69833-59-6	**NI33262**
213 214 315 243 172 283 133 159	**315**	100 93 47 37 36 35 26 24		19	25	3	1	–	–	–	–	–	–	–	–	–	**Estra-1,3,5(10)-trien-16-one, 3,17-dihydroxy-, O-methyloxime, (17β)-**	69834-03-3	**NI33263**
91 92 297 157 139 96 266 65	**315**	100 77 18 13 11 11 10 10	**1.86**	19	25	3	1	–	–	–	–	–	–	–	–	–	**Estra-1,3,5(10)-trien-17-one, 3,7-dihydroxy-, O-methyloxime, (7α)-**	74299-31-3	**NI33264**
91 92 284 209 285 146 110 315	**315**	100 81 48 12 11 10 8 7		19	25	3	1	–	–	–	–	–	–	–	–	–	**Estra-1,3,5(10)-trien-17-one, 3,11-dihydroxy-, O-methyloxime, (11β)-**	69833-88-1	**NI33265**
91 92 297 266 160 298 282 65	**315**	100 77 42 28 17 14 13 11	**0.00**	19	25	3	1	–	–	–	–	–	–	–	–	–	**Estra-1,3,5(10)-trien-17-one, 3,14-dihydroxy-, O-methyloxime**	74299-29-9	**NI33266**
284 315 91 92 159 266 285 209	**315**	100 96 54 42 33 29 22 20		19	25	3	1	–	–	–	–	–	–	–	–	–	**Estra-1,3,5(10)-trien-17-one, 3,15-dihydroxy-, O-methyloxime, (15α)-**	69833-44-9	**NI33267**
91 92 266 157 159 146 133 158	**315**	100 75 37 23 22 22 19 17	**10.90**	19	25	3	1	–	–	–	–	–	–	–	–	–	**Estra-1,3,5(10)-trien-17-one, 3,15-dihydroxy-, O-methyloxime, (15β)-**	74312-89-3	**NI33268**
283 126 160 172 256 240 284 91	**315**	100 67 46 42 39 37 27 26	**5.67**	19	25	3	1	–	–	–	–	–	–	–	–	–	**Estra-1,3,5(10)-trien-17-one, 3,16-dihydroxy-, O-methyloxime, (16α)-**	69833-49-4	**NI33269**
105 77 51 315 106 42 41 210	**315**	100 38 10 8 8 8 8 6		19	25	3	1	–	–	–	–	–	–	–	–	–	**Ethyl 3-(1′-benzoyl-3′-allylpiperid-4-yl)-propionate**		**NI33270**
315 282 297 257 186 242 226 266	**315**	100 92 89 71 33 26 20 13		19	25	3	1	–	–	–	–	–	–	–	–	–	**4-Hydroxy-1-[6-methoxy-7-(3-methylbut-2-enyl)indol-2-yl]-3-methylbutan 1-one**		**MS07401**
106 181 139 107 147 83 315 120	**315**	100 34 32 26 25 24 17 16		19	25	3	1	–	–	–	–	–	–	–	–	–	**2-Methyl-3-methoxy-5-[4-(4-aminophenyl)-1,3-dimethyl-butyl]-pent-4,5-dien-5-olide**		**LI06983**
315 314 172 297 298 286 115 170	**315**	100 90 86 48 42 26 25 24		20	13	3	1	–	–	–	–	–	–	–	–	–	**Quinolin-2-ol, 3-(2-hydroxy-1-naphthoyl)-**		**MS07402**
315 245 91 316 105 217 54 117	**315**	100 35 28 24 20 18 17 16		20	29	2	1	–	–	–	–	–	–	–	–	–	**1H-Dicyclooct[e,g]isoindole-1,3(2H)-dione, 3a,3b,4,5,6,7,8,9,10,11,12,13,14,15,15a,15b-hexadecahydro-, (3aα,3bα,15aα,15bα)-**	51624-47-6	**NI33271**
315 244 91 316 105 40 217 118	**315**	100 37 30 25 20 20 18 17		20	29	2	1	–	–	–	–	–	–	–	–	–	**Maleimide-1,1′-bicyclooctenyl endo adduct**		**LI06984**
211 108 212 210 213 69 315 77	**315**	100 21 21 15 7 7 6 5		21	17	–	1	1	–	–	–	–	–	–	–	–	**2,4-Diphenyl-2,3-dihydro-1,5-benzothiazepine**	40358-31-4	**NI33272**
211 105 77 210 106 51 212 91	**315**	100 90 64 35 32 26 18 16	**1.00**	21	17	2	1	–	–	–	–	–	–	–	–	–	**Benzamide, N-benzoyl-N-phenyl-**		**MS07403**
254 256 255 315 282 253 180 241	**315**	100 42 36 25 25 25 19 18		21	17	2	1	–	–	–	–	–	–	–	–	–	**Carbazole, 9-methyl-1-(2-methoxycarbonylphenyl)-**		**LI06985**
130 141 315 115 77 40 129 103	**315**	100 44 12 9 8 6 5 5		21	17	2	1	–	–	–	–	–	–	–	–	–	**1-Naphthalenemethyl indole-2-acetate**	86328-62-3	**NI33273**
315 300 224 91 316 301 182 208	**315**	100 58 36 26 24 14 13 11		21	21	–	3	–	–	–	–	–	–	–	–	–	**4-Amino-3-(N,N-dimethylamino)-9-benzylcarbazole**	94127-28-3	**NI33274**

MASS TO CHARGE RATIOS								M.W.	INTENSITIES								Parent	C	H	O	N	S	F	Cl	Br	I	Si	P	B	X	COMPOUND NAME	CAS Reg No	No
232	30	233	315	130	84	117	314	315	100	80	63	61	55	47	37	27		21	21	–	3	–	–	–	–	–	–	–	–	–	Indoline, 2-(3-indolyl)-1-[2-(5-methyl-1-pyrrolinyl)]-		MS07404
43	130	45	144	69	99	315	117	315	100	80	73	70	58	56	50	38		21	21	–	3	–	–	–	–	–	–	–	–	–	1H-Pyrrolo[2,1-b][1,3]benzodiazepine, 2,3,5,6-tetrahydro-5-(3-indolyl)-3-methyl-		MS07405
104	105	77	106	210	78	315	65	315	100	85	33	11	9	6	5	5		21	21	–	3	–	–	–	–	–	–	–	–	–	s-Triazine, hexahydro-N,N′,N″-triphenyl-		MS07406
131	161	315	103	146	188	132	202	315	100	29	26	24	16	15	14	10		21	33	1	1	–	–	–	–	–	–	–	–	–	Phenylpropenamide, N-dodecyl-		TR00301
243	72	244	315	165	241	228	91	315	100	32	20	13	8	5	4	4		22	21	1	1	–	–	–	–	–	–	–	–	–	Acetamide, N,N-dimethyl-(4-phenyl)diphenyl-		NI33275
243	165	244	72	315	228	239	241	315	100	28	19	19	7	3	2	2		22	21	1	1	–	–	–	–	–	–	–	–	–	Acetamide, N,N-dimethyl-2,2,2-triphenyl-	66415-28-9	NI33276
109	234	235	110	206				315	100	68	10	9	2				0.00	22	21	1	1	–	–	–	–	–	–	–	–	–	3-Cyclohexen-1-amine, 6-(2-furanyl)-2,5-diphenyl-, (1α,2α,5α,6β)-	42715-18-4	NI33277
109	234	110	235	206				315	100	16	7	3	2				0.00	22	21	1	1	–	–	–	–	–	–	–	–	–	3-Cyclohexen-1-amine, 6-(2-furanyl)-2,5-diphenyl-, (1α,2β,5β,6β)-	42715-17-3	NI33278
208	77	209	315	214	179	104	199	315	100	29	25	21	16	15	13	12		22	21	1	1	–	–	–	–	–	–	–	–	–	Isquinoline, 1,2,3,4-tetrahydro-1-(4-methoxyphenyl)-2-phenyl-		MS07407
300	71	315	301	72	56	57	44	315	100	50	16	10	5	4	2	2		22	37	–	1	–	–	–	–	–	–	–	–	–	Conanine, (5α)-	468-33-7	NI33279
101	120	60	79	180	161	71	169	316	100	55	36	32	30	27	25	20	0.00	–	–	–	–	–	12	–	–	–	–	–	8	–	Octaborane(12), dodecafluoro-	55836-56-1	NI33280
69	71	109	39	159	89	65	50	316	100	73	28	28	25	19	12	12	0.00	3	4	–	–	–	8	–	–	–	–	4	–	–	1,1,2,2-Tetrakis(difluorophosphino)propane		DD01191
76	46	57	55	56	60	47	97	316	100	89	14	13	9	7	5	4	0.00	5	8	12	4	–	–	–	–	–	–	–	–	–	Pentaerythritol, tetranitrate	78-11-5	NI33281
76	46	57	55	56	60	47	43	316	100	88	13	13	9	7	4	4	0.00	5	8	12	4	–	–	–	–	–	–	–	–	–	Pentaerythritol, tetranitrate	78-11-5	MS07408
69	100	131	183	119	181	150	31	316	100	80	46	28	24	21	16	14	0.00	6	–	1	–	–	12	–	–	–	–	–	–	–	Hexanal, perfluoro-		IC03891
176	28	204	131	161	316	52	177	316	100	45	34	18	17	16	16	14		8	9	8	–	–	–	–	–	–	–	1	–	1	Chromium, pentacarbonyl(trimethyl phosphite-P)-	18461-34-2	NI33282
269	249	221	144	296	109	81	45	316	100	64	62	53	24	13	10	10	2.00	8	14	5	–	–	5	–	–	–	–	1	–	–	[2,2-Difluoro-2-methoxy-1-(trifluoromethyl)ethyl]diethylphosphate		MS07409
57	42	316	56	40	259	230	202	316	100	27	12	12	5	2	1	1		8	18	–	–	–	–	–	–	–	–	–	–	1	Mercury, dibutyl-		LI06986
243	241	245	108	274	215	213	272	316	100	79	46	23	20	20	15	15	0.00	9	4	4	–	–	–	4	–	–	–	–	–	–	1,4-Benzenedicarboxylic acid, 2,3,5,6-tetrachloro-, monomethyl ester	887-54-7	NI33283
287	285	289	142	107	36	144	59	316	100	73	53	42	34	30	27	26	16.40	9	4	4	–	–	–	4	–	–	–	–	–	–	1,4-Benzenedicarboxylic acid, 2,3,5,6-tetrachloro-, monomethyl ester	887-54-7	NI33284
77	180	230	154	210	135	39	161	316	100	16	10	6	5	5	5	4	4.37	9	5	3	–	–	3	–	–	–	–	–	–	1	π-Cyclopentadienyl-trifluoromethyl-tricarbonyl molybdenum	12152-60-2	MS07410
77	69	182	232	180	179	230	176	316	100	25	14	12	11	10	9	9	3.99	9	5	3	–	–	3	–	–	–	–	–	–	1	π-Cyclopentadienyl-trifluoromethyl-tricarbonyl molybdenum	12152-60-2	NI33285
148	316	149	53	52	317			316	100	21	9	9	7	2				9	9	1	4	–	–	–	–	1	–	–	–	–	Pyrazolo[5,1-c][1,2,4]triazine, 4,6-dihydro-6-(iodoacetyl)-3-methyl-4-methylene-	6763-71-9	NI33286
91	65	92	172	174	173	63	39	316	100	11	10	4	3	3	3	2	0.00	9	10	4	–	2	–	2	–	–	–	–	–	–	Benzene, [[[dichloro(methylsulphonyl)methyl]sulphonyl]methyl]-	56620-04-3	NI33287
281	283	125	109	218	93	266	79	316	100	38	35	25	17	15	14	14	6.47	9	11	4	–	1	–	2	–	–	–	1	–	–	Phosphorothioic acid, O-(4,5-dichloro-2-methoxyphenyl) O,O-dimethyl ester	55975-98-9	NI33288
318	102	50	101	209	75	211	51	316	100	66	64	62	61	60	58	53	52.42	10	6	2	–	–	–	–	2	–	–	–	–	–	1,5-Naphthalenediol, 2,6-dibromo-	84-59-3	NI33289
133	316	45	134	39	135	75	69	316	100	16	15	7	6	5	4	3		11	6	–	–	2	6	–	–	–	–	–	–	–	Thiophene, 2,2′-(1,1,2,2,3,3-hexafluoro-1,3-propanediyl)bis-	21250-11-3	NI33290
235	285	297	59	15	257	205	31	316	100	28	26	20	20	18	15	12	7.30	11	6	4	–	–	6	–	–	–	–	–	–	–	Dimethyl-1,2,3,4,7,7-hexafluorobicyclo[2.2.1]hepta-2,5-diene-5,6-dicarboxylate		MS07411
141	91	101	232	121	55	65	160	316	100	50	35	24	21	18	17	16	9.99	11	8	4	–	–	3	–	–	–	–	–	–	1	[1-Hydroxy-1-(trifluoromethyl)ethyl]cymantrene	33270-42-7	NI33291
68	55	79	69	67	54	82	41	316	100	94	79	69	65	63	61	61	0.00	11	24	6	–	2	–	–	–	–	–	–	–	–	1,9-Nonanediol, dimethanesulphonate	4248-77-5	NI33292
165	179	152	125	187	123	109	137	316	100	60	36	34	28	27	27	26	2.00	11	26	6	–	–	–	–	–	–	–	2	–	–	Tetraethyl propane-1,3-diphosphonate		MS07412
316	318	115	128	282	193	195	95	316	100	65	47	42	39	36	32	19		12	7	2	–	1	–	2	–	–	–	1	–	–	10H-Phenothiaphosphine, 2,7-dichloro-10-hydroxy-, 10-oxide	55125-04-7	NI33293
91	77	167	105	103	189	153	115	316	100	94	44	42	42	40	36	33	0.00	12	16	1	–	–	–	4	–	–	–	–	–	–	(1RS,4SR,5SR,6SR,7SR,8SR)-5,6,7,8-Tetrakis(chloromethyl)bicyclo[2.2.2]octan-2-one		DD01192
153	77	91	167	202	105	131	189	316	100	97	94	72	62	55	53	49	0.00	12	16	1	–	–	–	4	–	–	–	–	–	–	(1S,4R,5R,6R,7S,8S)-5,6,7,8-Tetrakis(chloromethyl)bicyclo[2.2.2]octan-2-one		DD01193
43	237	167	41	39	124	168	55	316	100	99	99	66	60	52	41	35	0.00	12	17	3	2	–	–	–	1	–	–	–	–	–	2,4,6(1H,3H,5H)-Pyrimidinetrione, 5-(2-bromoallyl)-5-(1-methylbutyl)-	1216-40-6	NI33294
169	73	74	75	57	100	71	69	316	100	68	42	25	19	12	11	10	1.90	12	20	6	2	–	–	–	–	–	1	–	–	–	Uridine, 5-(trimethylsilyl)-	51432-36-1	MS07413
131	101	99	71	146	72	129	142	316	100	62	56	54	50	49	46	43	12.67	12	20	6	4	–	–	–	–	–	–	–	–	–	1,10-Dioxa-4,7,13,16-tetraazacyclooctadecane-3,6,14,17-tetrone		NI33295
81	67	41	54	82	39	30	53	316	100	91	69	60	40	37	30	28	0.00	12	20	6	4	–	–	–	–	–	–	–	–	–	Bis-(2-nitro-nitrosocyclohexane)		IC03892
316	63	43	135	137	69	134	195	316	100	68	30	25	23	22	21	20		13	8	4	4	1	–	–	–	–	–	–	–	–	2-Thiophenecarboxaldehyde, 5-ethynyl-, (2,4-dinitrophenyl)hydrazone	56588-21-7	NI33296
316	121	317	224	120	318	109	63	316	100	24	18	16	10	8	7	7		13	8	4	4	1	–	–	–	–	–	–	–	–	2-Thiophenecarboxaldehyde, α-ethynyl-, (2,4-dinitrophenyl)hydrazone	56588-20-6	NI33297
139	111	85	197	75	141	230	228	316	100	79	67	56	37	36	31	26	2.00	13	13	3	–	2	–	1	–	–	–	–	–	–	Propanoic acid, 3-[[3-(4-chlorophenyl)-3-hydroxy-1-thioxo-2-propenyl]thio]-, methyl ester	72347-57-0	NI33298
238	222	252	253	206	205	119	316	316	100	84	40	36	36	36	28	24		13	16	5	–	2	–	–	–	–	–	–	–	–	2-(2′,3′-Dimethoxy-4′,5′-methylenedioxyphenyl)-1-methylsulphinyl-1-methylthioethylene	75629-02-6	NI33299
43	144	102	154	112	28	111	153	316	100	42	41	24	14	14	13	9	0.00	13	21	8	–	–	–	–	–	–	–	–	1	–	α-D-Glucopyranoside, methyl, cyclic 4,6-(ethylboronate) 2,3-diacetate	61553-57-9	NI33300
43	111	110	171	154	196	103	183	316	100	53	13	10	10	8	8	7	0.00	13	21	8	–	–	–	–	–	–	–	–	1	–	DL-Xylitol, cyclic 2,3-(ethylboronate) 1,4,5-triacetate	74793-01-4	NI33301
273	208	108	96	166	150	44	42	316	100	73	52	45	44	38	37	34	28.00	13	24	3	4	1	–	–	–	–	–	–	–	–	5-Butyl-2-ethylamino-6-methylpyrimidin-4-yl dimethylsulphamate (bupirimate)	41483-43-6	NI33302

MASS TO CHARGE RATIOS								M.W.	INTENSITIES								Parent	C	H	O	N	S	F	Cl	Br	I	Si	P	B	X	COMPOUND NAME	CAS Reg No	No
301	91	316	225	73	147	285	135	**316**	100	89	86	54	34	19	17	16		13	25	3	–	–	–	–	–	–	2	1	–	–	Phosphonic acid, benzyl-, bis(trimethylsilyl) ester	18406-56-9	NI33303
301	91	316	225	73	302	317	147	**316**	100	89	86	54	34	28	22	19		13	25	3	–	–	–	–	–	–	2	1	–	–	Phosphonic acid, benzyl-, bis(trimethylsilyl) ester	18406-56-9	NI33304
301	73	302	316	200	186	172	147	**316**	100	63	29	10	10	10	10	10		13	32	1	2	–	–	–	–	–	3	–	–	–	Butanoic acid, 2,4-diamino-, γ-lactam, tris(trimethylsilyl)-		LI06987
301	73	200	186	172	147	316	45	**316**	100	62	12	12	12	11	10	8		13	32	1	2	–	–	–	–	–	3	–	–	–	Butanoic acid, 2,4-diamino-, γ-lactam, tris(trimethylsilyl)-		MS07414
246	248	176	318	316	75	105	320	**316**	100	64	38	37	28	21	19	18		14	8	–	–	–	–	4	–	–	–	–	–	–	Benzene, 1-chloro-2-[2,2-dichloro-1-(4-chlorophenyl)vinyl]-	3424-82-6	NI33306
246	248	318	75	105	316	176	247	**316**	100	72	37	30	29	28	28	22		14	8	–	–	–	–	4	–	–	–	–	–	–	Benzene, 1-chloro-2-[2,2-dichloro-1-(4-chlorophenyl)vinyl]-	3424-82-6	NI33307
246	248	318	176	316	247	105	320	**316**	100	68	34	27	23	20	19	18		14	8	–	–	–	–	4	–	–	–	–	–	–	Benzene, 1-chloro-2-[2,2-dichloro-1-(4-chlorophenyl)vinyl]-	3424-82-6	NI33305
246	248	318	316	320	176	105	247	**316**	100	67	66	56	34	32	29	20		14	8	–	–	–	–	4	–	–	–	–	–	–	Benzene, 1,1′-(dichloroethenylidene)bis[4-chloro-	72-55-9	AI02033
246	318	316	248	320	176	247	319	**316**	100	95	69	64	48	34	16	14		14	8	–	–	–	–	4	–	–	–	–	–	–	Benzene, 1,1′-(dichloroethenylidene)bis[4-chloro-	72-55-9	NI33308
55	134	63	42	90	64	119	311	**316**	100	80	70	65	55	47	44	42	**0.00**	14	12	5	4	–	–	–	–	–	–	–	–	–	Acetophenone, 2′-hydroxy-, (2,4-dinitrophenyl)hydrazone	17744-50-2	NI33309
136	316	180	89	90	164	107	78	**316**	100	62	41	31	28	25	25	24		14	12	5	4	–	–	–	–	–	–	–	–	–	Azobenzene, 4-hydroxy-2,2′-dimethyl-5,5′-dinitro-		IC03893
316	119	91	137	77	51	107	63	**316**	100	86	44	43	38	23	22	19		14	12	5	4	–	–	–	–	–	–	–	–	–	Benzaldehyde, 2-methoxy-, (2,4-dinitrophenyl)hydrazone	1163-71-9	NI33310
316	119	137	91	77	107	51	92	**316**	100	85	44	44	39	24	24	18		14	12	5	4	–	–	–	–	–	–	–	–	–	Benzaldehyde, 2-methoxy-, (2,4-dinitrophenyl)hydrazone	1163-71-9	LI06988
316	135	136	137	77	92	152	317	**316**	100	61	38	37	30	23	19	18		14	12	5	4	–	–	–	–	–	–	–	–	–	Benzaldehyde, 4-methoxy-, (2,4-dinitrophenyl)hydrazone	1773-49-5	LI06989
316	135	136	137	77	92	317	152	**316**	100	60	35	34	30	20	18	18		14	12	5	4	–	–	–	–	–	–	–	–	–	Benzaldehyde, 4-methoxy-, (2,4-dinitrophenyl)hydrazone	1773-49-5	NI33311
121	316	149	77	90	91	103	89	**316**	100	41	35	32	17	16	14	10		14	12	5	4	–	–	–	–	–	–	–	–	–	Benzeneethanol, 4-[(2,4-dinitrophenyl)azo]-		NI33312
236	235	206	77	80	218	82	237	**316**	100	91	19	18	18	17	17	17	**0.00**	14	13	–	4	–	–	–	1	–	–	–	–	–	2-Imino-5-phenyl-1,2-dihydro-3H-1,3,4-benzotriazepine hydrobromide		MS07415
168	91	66	316	194	287	103	259	**316**	100	86	78	62	44	42	40	39		14	13	2	–	–	–	–	–	–	–	–	–	1	Rhodium, (2-formylnorbornadiene)(5-formylcyclopentadienyl)		NI33313
188	232	55	316	186	144	80	189	**316**	100	99	95	46	23	23	23	22		14	13	5	–	–	–	–	–	–	–	–	–	1	Manganese, tricarbonyl cyclopentadienyl (1-ethoxycarbonylisopropenyl)-		LI06990
261	263	243	45	55	245	73	316	**316**	100	62	50	48	32	27	24	15		14	14	4	–	–	–	2	–	–	–	–	–	–	Acetic acid, [2,3-dichloro-4-(2-methylene-1-oxobutyl)phenoxy]-, methyl ester	6463-21-4	NI33314
316	169	168	59	167	318	317	195	**316**	100	50	29	28	21	20	18	17		14	20	–	–	4	–	–	–	–	–	–	–	–	4H-1,3-Dithiin, 2,2,6-trimethyl-4-(2,2,6-trimethyl-4H-1,3-dithiin-4-ylidene)-	57274-44-9	NI33315
175	57	259	85	41	32	316	177	**316**	100	70	66	48	30	29	28	13		14	20	2	–	3	–	–	–	–	–	–	–	–	2-Butanone, 1,1′-(1,2,4-trithiolane-3,5-diylidene)bis[3,3-dimethyl-	19018-15-6	LI06991
175	57	259	85	316	41	29	176	**316**	100	70	66	48	29	29	29	12		14	20	2	–	3	–	–	–	–	–	–	–	–	2-Butanone, 1,1′-(1,2,4-trithiolane-3,5-diylidene)bis[3,3-dimethyl-	19018-15-6	NI33316
44	87	86	43	102	114	60	29	**316**	100	90	86	27	18	14	6	6	**0.41**	14	24	6	2	–	–	–	–	–	–	–	–	–	Serine, O-acetyl-N-(N-acetylalanyl)-, butyl ester		AI02034
259	257	203	41	201	29	145	255	**316**	100	75	75	60	56	54	50	45	**0.91**	14	28	–	–	–	–	–	–	–	–	–	–	1	Stannane, tributylethynyl-	994-89-8	NI33317
259	203	257	145	201	29	143	255	**316**	100	88	76	74	67	57	52	44	**0.82**	14	28	–	–	–	–	–	–	–	–	–	–	1	Stannane, tributylethynyl-	994-89-8	NI33318
73	171	75	183	301	172	43	45	**316**	100	94	39	37	22	14	13	12	**10.43**	14	28	4	–	–	–	–	–	–	2	–	–	–	4,6-Diketooctanoic acid, bis(trimethylsilyl)-		NI33319
73	109	75	103	153	181	55	58	**316**	100	38	37	34	31	31	25	17	**11.00**	14	28	4	2	–	–	–	–	–	1	–	–	–	Succinylacetone diethoxime monoTMS		NI33320
57	127	58	86	114	56	159	115	**316**	100	79	78	66	59	54	51	44	**0.66**	14	28	4	4	–	–	–	–	–	–	–	–	–	4,6,12,14-Tetramethyl-1,9-dioxa-4,6,12,14-tetraazahexadecane-5,13-dione	89913-94-0	NI33321
73	130	75	44	55	199	43	131	**316**	100	95	74	35	34	29	29	25	**0.00**	14	29	5	–	–	–	–	–	–	1	–	1	–	α-L-Galactopyranoside, methyl 6-deoxy-2-O-(trimethylsilyl)-, cyclic butylboronate	56196-84-0	NI33322
316	277	216	201	214	259	221	236	**316**	100	99	91	84	83	37	36	30		14	36	–	4	–	–	–	–	–	2	–	–	–	Cyclodisilazane-1,3-diamine, 2-butyl-2-isopropyl-N,N,N′,N′-4,4-hexamethyl-	66517-40-6	NI33323
135	181	91	119	136	65	44	79	**316**	100	22	20	16	9	9	6	5	**5.20**	15	9	2	–	–	5	–	–	–	–	–	–	–	Benzoic acid, 2-methyl-, pentafluorobenzyl ester		MS07416
119	181	91	316	65	271	44	272	**316**	100	65	54	28	25	22	21	15		15	9	2	–	–	5	–	–	–	–	–	–	–	Benzoic acid, 3-methyl-, pentafluorobenzyl ester		MS07417
119	181	91	316	65	271	44	120	**316**	100	44	34	32	14	12	11	10		15	9	2	–	–	5	–	–	–	–	–	–	–	Benzoic acid, 4-methyl-, pentafluorobenzyl ester		MS07418
91	181	316	92	65	161	89	63	**316**	100	36	14	12	11	3	3	3		15	9	2	–	–	5	–	–	–	–	–	–	–	Phenylacetic acid, pentafluorobenzyl ester		MS07419
105	237	209	104	77	76	213	211	**316**	100	63	50	23	22	19	5	5	**0.50**	15	9	3	–	–	–	–	1	–	–	–	–	–	3-Benzoyl-3-bromophthalide		LI06992
64	77	252	105	147	48	175	146	**316**	100	87	75	75	63	47	43	40	**0.00**	15	12	4	2	1	–	–	–	–	–	–	–	–	1H-2,1,3-Benzothiadiazin-4(3H)-one, 3-(2-oxo-2-phenylethyl)-, 2,2-dioxide	40467-23-0	NI33324
316	318	147	317	161	131	133	132	**316**	100	97	53	43	43	32	31	31		15	13	1	2	–	–	–	1	–	–	–	–	–	Benzimidazole, 2-(4-bromo-α-bromobenzyl)-1-methyl-		MS07420
56	121	199	316	28	78	272	79	**316**	100	93	65	64	59	29	28	26		15	16	4	–	–	–	–	–	–	–	–	–	1	2-(Ferrocenylmethyl)succinic acid		NI33325
44	255	316	272	256	273	258	43	**316**	100	60	52	47	35	30	30	30		15	16	4	4	–	–	–	–	–	–	–	–	–	Azirino[2′,3′:3,4]pyrrolo[1,2-a]indole-4,7-dione, 6-amino-8-[[(aminocarbonyl)oxy]methyl]-1,1a,2,8b-tetrahydro-1,5-dimethyl-	32633-56-0	NI33326
86	57	28	30	156	44	29	41	**316**	100	99	54	52	49	37	37	30	**10.10**	15	22	1	2	–	–	2	–	–	–	–	–	–	Urea, N-(3,4-dichlorophenyl)-N′,N′-dibutyl-		MS07421
316	298	272	240	283	280	317	285	**316**	100	54	50	45	21	21	19	15		16	12	5	–	1	–	–	–	–	–	–	–	–	Anthraquinone, 1,4-dihydroxy-2-[(2-hydroxyethyl)thio]-		IC03894
316	298	242	273	115	245	299	270	**316**	100	89	42	38	22	18	17	17		16	12	7	–	–	–	–	–	–	–	–	–	–	Anthraquinone, 1,2,6,7-tetrahydroxy-8-methoxy-3-methyl-		LI06993
316	270	245	269	298	273	268	287	**316**	100	38	25	19	17	10	7	4		16	12	7	–	–	–	–	–	–	–	–	–	–	Anthraquinone, 1,5,7,8-tetrahydroxy-6-methoxy-3-methyl-		LI06994
316	315	287	301	245	153	151	123	**316**	100	12	10	6	6	5	5	4		16	12	7	–	–	–	–	–	–	–	–	–	–	Flavone, 3,4′,5,7-tetrahydroxy-3′-methoxy-	480-19-3	MS07422
316	317	315	287	153	108	151	269	**316**	100	17	12	11	8	8	7	6		16	12	7	–	–	–	–	–	–	–	–	–	–	Flavone, 3,4′,5,7-tetrahydroxy-3′-methoxy-	480-19-3	NI33327
273	316	121	298	317	92	105	147	**316**	100	64	48	21	19	17	13	11		16	12	7	–	–	–	–	–	–	–	–	–	–	Flavone, 3,4′,5,7-tetrahydroxy-6-methoxy-	32520-55-1	MS07423
316	69	301	273	298	317	139	135	**316**	100	70	69	43	22	18	18	18		16	12	7	–	–	–	–	–	–	–	–	–	–	Flavone, 3′,4′,5,7-tetrahydroxy-6-methoxy-	520-11-6	MS07424
316	301	273	298	139	135	302	167	**316**	100	50	28	25	13	10	8	6		16	12	7	–	–	–	–	–	–	–	–	–	–	Flavone, 3′,4′,5,7-tetrahydroxy-6-methoxy-	520-11-6	NI33328
316	273	301	298	139	167	315	287	**316**	100	81	76	72	23	17	14	13		16	12	7	–	–	–	–	–	–	–	–	–	–	Flavone, 3′,4′,5,7-tetrahydroxy-6-methoxy-	520-11-6	NI33329

MASS TO CHARGE RATIOS								M.W.	INTENSITIES								Parent	C	H	O	N	S	F	Cl	Br	I	Si	P	B	X	COMPOUND NAME	CAS Reg No	No
316	298	315	301	183	155	121	119	316	100	27	26	18	10	8	6	5		16	12	7	–	–	–	–	–	–	–	–	–	–	Flavone, 5,6,8,4′-tetrahydroxy-7-methoxy-		NI33330
240	301	316	115	91	121	302	241	316	100	99	26	23	23	22	16	16		16	16	5	2	–	–	–	–	–	–	–	–	–	Benzene, 1-(2-tert-butylphenoxy)-2,4-dinitro-	32101-58-9	NI33331
58	70	193	42	161	192	316	318	316	100	73	71	47	30	29	27	27		16	17	–	2	–	–	–	1	–	–	–	–	–	N,N-Dimethyl-3-(3-pyridyl)-3-(4-bromophenyl)prop-2-enamine		NI33332
155	201	123	95	81	129	113	43	316	100	75	45	34	28	18	14	13	0.60	16	28	6	–	–	–	–	–	–	–	–	–	–	Myrothecic acid, hexahydro-, dimethyl ester	29953-53-5	NI33333
155	201	123	95	81	129	113	55	316	100	75	46	34	28	18	14	13	0.00	16	28	6	–	–	–	–	–	–	–	–	–	–	Myrothecic acid, hexahydro-, dimethyl ester	29953-53-5	LI06995
100	56	58	114	70	98	72	84	316	100	91	84	78	75	66	66	59	14.70	16	32	4	2	–	–	–	–	–	–	–	–	–	4,7,14,20-Tetraoxa-1,10-diazabicyclo[8.7.5]docosane		MS07425
100	72	114	58	88	70	84	98	316	100	92	83	83	75	71	67	63	45.80	16	32	4	2	–	–	–	–	–	–	–	–	–	5,8,15,20-Tetraoxa-1,2-diazabicyclo[10.5.5]docosane		MS07426
172	316	89	104	117	288	287	77	316	100	99	95	94	86	65	61	51		17	12	1	6	–	–	–	–	–	–	–	–	–	Bis(5-phenyl-1,2,3-triazol-4-yl) ketone		NI33334
91	316	225	284	65	92	285		316	100	18	16	11	9	8	6			17	16	4	–	1	–	–	–	–	–	–	–	–	Terephthalic acid, 2-(benzylthio)-, dimethyl ester		LI06996
107	316	210	195	167	183	182	301	316	100	73	51	42	14	10	7	4		17	16	6	–	–	–	–	–	–	–	–	–	–	4-Chromanone, 3,9-dihydro-5,7-dihydroxy-3-(4-hydroxybenzylidene)-6-methoxy-, (E)-		LI06997
257	316	213	198	258	256	212	211	316	100	70	70	21	17	17	16	15		17	16	6	–	–	–	–	–	–	–	–	–	–	1,2-Dibenzofurandicarboxylic acid, 1,9b-dihydro-7-methoxy-4,9b-dimethyl-	23911-62-8	NI33335
253	284	116	75	254	285	89	102	316	100	80	24	23	17	14	11	10	4.00	17	16	6	–	–	–	–	–	–	–	–	–	–	1,2-Dibenzofurandicarboxylic acid, 2,3-dihydro-3-methoxy-, dimethyl ester		LI06998
164	316	149	121	178	165	134	91	316	100	33	24	24	13	12	8	8		17	16	6	–	–	–	–	–	–	–	–	–	–	Isoflavanone, 5,7-dihydroxy-2′,4′-dimethoxy-		LI06999
180	122	181	107	137	136	152	287	316	100	99	93	28	25	23	20	17	5.00	17	16	6	–	–	–	–	–	–	–	–	–	–	(2S,3S)-3,7,4′-Trihydroxy-5-methoxy-6-methylflavanone	78417-24-0	NI33336
316	315	287	270	317	271	285	299	316	100	41	31	20	18	15	13	11		17	16	6	–	–	–	–	–	–	–	–	–	–	Xanthone, 1,3,5,6-tetramethoxy-		MS07427
316	301	259	317	216	273	226	217	316	100	74	25	21	18	17	17	16		17	20	2	2	1	–	–	–	–	–	–	–	–	4,4′-Diacetyl-3,5,3′,5′-tetramethyl-2,2′-dipyrrylthione		LI07000
192	193	134	124	106	78	138	285	316	100	79	62	53	43	41	38	15	3.00	17	20	4	2	–	–	–	–	–	–	–	–	–	6-O-Nicotinoyl-1,2,4a,7a-tetrahydrocantleyine		NI33337
154	155	227	152	229	228	151	316	316	100	88	76	76	71	59	41	35		17	21	1	–	–	–	–	–	–	–	–	–	1	(2-Hydroxypentyl)diphenylarsane		DD01194
259	154	155	145	229	228	169	260	316	100	99	90	69	63	55	36	33	0.00	17	21	1	–	–	–	–	–	–	–	–	–	1	Pentyldiphenylarsane oxide		DD01195
67	81	82	96	95	41	55	79	316	100	86	73	60	39	34	31	27	0.00	17	33	–	–	–	–	–	1	–	–	–	–	–	5-Heptadecene, 1-bromo-	56600-21-6	NI33338
316	126	219	221	317	125	201	75	316	100	62	44	22	17	13	10	7		18	12	–	–	–	3	–	–	–	–	1	–	–	Phosphine, tris(4-fluorophenyl)-	18437-78-0	NI33339
116	89	115	90	158	141	114	148	316	100	74	44	37	35	35	35	28	0.00	18	12	–	4	1	–	–	–	–	–	–	–	–	Thieno[2,3-b:4,5-b′]diquinoxaline, dimethyl-	38638-13-0	NI33340
316	196	301	39	120	65	94	317	316	100	89	71	56	45	37	26	22		18	12	2	4	–	–	–	–	–	–	–	–	–	1-Amino-2,6,6-tricyano-3,5-di-(2-furyl)-5-methylcyclohexa-1,3-diene	94644-68-5	NI33341
295	316	315	296	165	317	179	301	316	100	98	51	22	22	20	18	15		18	20	5	–	–	–	–	–	–	–	–	–	–	Benzophenone, 3,4,4′,5′-tetramethoxy-2′-methyl-	52806-37-8	NI33342
298	284	137	167	283	165	148	180	316	100	56	54	52	49	49	49	41	34.00	18	20	5	–	–	–	–	–	–	–	–	–	–	2H-1-Benzopyran, 3,4-dihydro-2-(4,6-dimethoxy-2-hydroxyphenyl)-7-methoxy-		LI07001
180	165	181	151	137	135	316	119	316	100	19	13	13	8	6	5	3		18	20	5	–	–	–	–	–	–	–	–	–	–	2H-1-Benzopyran-3,4-diol, 2-(3,4-dimethoxyphenyl)-3,4-dihydro-6-methyl, (2α,3α,4α)-	55125-21-8	NI33343
134	298	135	297	316	182	181	164	316	100	21	15	11	8	8	7	6		18	20	5	–	–	–	–	–	–	–	–	–	–	2H-1-Benzopyran-4-ol, 3,4-dihydro-5,7-dimethoxy-2-(4-methoxyphenyl)-	10493-01-3	NI33344
134	69	298	297	239	191	55	77	316	100	69	64	61	61	55	45	40	1.00	18	20	5	–	–	–	–	–	–	–	–	–	–	2H-1-Benzopyran-4-ol, 3,4-dihydro-5,7-dimethoxy-2-(4-methoxyphenyl)-	10493-01-3	NI33345
167	121	109	150	137	316	168	91	316	100	46	34	29	21	19	17	17		18	20	5	–	–	–	–	–	–	–	–	–	–	2H-1-Benzopyran-3-ol, 3,4-dihydro-5,7-dimethoxy-2-(4-methoxyphenyl)-, (2R-cis)-	3143-21-3	NI33346
166	107	77	105	43	79	150	51	316	100	90	80	65	45	42	34	20	0.00	18	20	5	–	–	–	–	–	–	–	–	–	–	Methyl 2-phenyl-2,3-dihydroxy-3-(phenoxymethyl)butanoate		DD01196
105	272	189	118	77	162	235	273	316	100	94	92	80	80	78	54	15	0.00	18	24	3	2	–	–	–	–	–	–	–	–	–	3-Cyclohexyl-5-ethoxy-1-methyl-5-phenyl-2,4-imidazolidinedione		DD01197
316	243	317	230	213	121	120	228	316	100	71	20	17	17	14	13	12		18	24	3	2	–	–	–	–	–	–	–	–	–	(3,5-Dimethyl-4-(2-carboxyethyl)-2-pyrrolyl)-(3-ethyl-4-methyl-5-oxo-2-pyrrolylidene)-methane		LI07002
316	301	84	257	86	241	227	287	316	100	80	70	65	45	25	25	22		18	24	3	2	–	–	–	–	–	–	–	–	–	(3-Ethyl-4-methyl-5-methoxy-2-pyrrolyl)-(3,5-dimethyl-4-(2-carboxyethyl)-2-pyrrolylidene)-methane		LI07003
316	150	301	255	123	193	149	108	316	100	95	75	45	45	35	35	35		18	24	3	2	–	–	–	–	–	–	–	–	–	3,5,3′,5′-Tetramethyl-4′-ethyl-4-(ethoxycarbonyl)-2,2′-bis-pyrrolyl-ketone		LI07004
299	43	120	57	71	55	61	95	316	100	68	60	60	40	40	39	36	0.00	18	36	2	–	1	–	–	–	–	–	–	–	–	1-(Methylsulphinyl)-2-heptadecanone	51840-11-0	DD01198
316	116	260	259	89	172	288	287	316	100	68	65	43	41	33	31	24		19	12	3	2	–	–	–	–	–	–	–	–	–	2-Indolinone, 3,3′-(oxopropanediylidene)di-	21905-77-1	NI33347
143	77	117	51	144	50	316	116	316	100	51	40	20	19	16	15	14		19	16	1	4	–	–	–	–	–	–	–	–	–	3-(1-Phenyl-3-methylpyrazol-5-yl)-2-methylquinazolin-4(3H)-one		MS07428
133	134	135	183	181	259	301	91	316	100	90	85	76	70	70	68	68	21.95	19	24	2	–	1	–	–	–	–	–	–	–	–	2,4,6-Trimethyl-2′-tert-butyldiphenylsulphone	40316-48-1	NI33348
237	316	298	195	238	194	119	301	316	100	92	91	69	65	62	60	59		19	24	2	–	1	–	–	–	–	–	–	–	–	2,4,6-Trimethyl-3′-tert-butyldiphenylsulphone	40316-47-0	NI33349
233	43	41	69	28	234	18	316	316	100	50	38	27	18	15	12	9		19	24	4	–	–	–	–	–	–	–	–	–	–	Benzopyran, 6-acetyl-5-hydroxy-8-methoxy-2-(4′-methylpent-3′-enyl)-2-methyl-		MS07429
280	269	316	174	298	91	159	171	316	100	74	49	45	44	36	35	34		19	24	4	–	–	–	–	–	–	–	–	–	–	Estra-1,3,5(10)-trien-17-one, 3,6,7-trihydroxy-1-methyl-, (6α,7α)-	25328-48-7	NI33350
272	229	227	203	91	159	204	55	316	100	98	62	61	61	50	49	45	6.00	19	24	4	–	–	–	–	–	–	–	–	–	–	Gibbane-1,10-dicarboxylic acid, 4a-hydroxy-1-methyl-8-methylene-, 1,4a-lactone, (1α,4aα,4bβ,10β)-	427-77-0	NI33351
272	259	81	80	316	189	97	107	316	100	62	59	55	35	32	22	20		19	24	4	–	–	–	–	–	–	–	–	–	–	(1R,2R,4S,8S,9S,11S,12S,15R,16R)-15-Hydroxy-2,16-dimethylpentacyclo[9.6.0.0^{2,9}.0^{4,8}.0^{12,16}]heptadeca-3,6,17-trione		DD01199
272	81	80	259	189	316	107	298	316	100	98	69	61	46	44	33	22		19	24	4	–	–	–	–	–	–	–	–	–	–	(1R,2R,4S,8S,9S,11S,12S,15S,16R)-15-Hydroxy-2,16-dimethylpentacyclo[9.6.0.0^{2,9}.0^{4,8}.0^{12,16}]heptadeca-3,6,17-trione		DD01200

MASS TO CHARGE RATIOS	M.W.	INTENSITIES	Parent	C	H	O	N	S	F	Cl	Br	I	Si	P	B	X	COMPOUND NAME	CAS Reg No	No
58 85 235 234 36 195 193 130	**316**	100 63 45 42 40 19 15 15	**0.00**	19	25	–	2	–	–	1	–	–	–	–	–	–	5H-Dibenz[b,f]azepine-5-propylamine, 10,11-dihydro-N,N-dimethyl-, monohydrochloride	113-52-0	NI33352
107 99 55 105 120 113 79 41	**316**	100 75 61 58 32 31 28 26	**1.00**	19	26	3	1	–	–	–	–	–	–	–	–	–	(1R,6R)-5,5-(Ethylenedioxy)-N-((S)-1-phenylethyl)bicyclo[4.2.0]octane-1-carboxamide		MS07430
107 99 55 105 120 113 41 79	**316**	100 65 54 52 31 31 31 23	**0.00**	19	26	3	1	–	–	–	–	–	–	–	–	–	(1S,6S)-5,5-(Ethylenedioxy)-N-((S)-1-phenylethyl)bicyclo[4.2.0]octane-1-carboxamide		MS07431
138 135 316 77 56 233 181 51	**316**	100 53 40 34 34 32 20 15		19	28	–	2	1	–	–	–	–	–	–	–	–	Thiourea, N,N-dicyclohexyl-N′-phenyl-	15093-49-9	NI33353
138 181 135 316 233 139 56 41	**316**	100 84 79 78 63 55 55 55		19	28	–	2	1	–	–	–	–	–	–	–	–	Thiourea, N,N-dicyclohexyl-N′-phenyl-	15093-49-9	NI33354
138 135 77 56 181 55 51 139	**316**	100 52 34 34 19 15 15 10	**1.00**	19	28	–	2	1	–	–	–	–	–	–	–	–	Thiourea, N,N-dicyclohexyl-N′-phenyl-	15093-49-9	LI07005
91 73 75 118 107 79 92 135	**316**	100 93 41 27 23 18 17 14	**0.00**	19	28	2	–	–	–	–	–	–	1	–	–	–	(E)-1-[6-(Benzyloxy)-1-cyclohexenyl]-3-(trimethylsilyl)-2-propen-1-ol		DD01201
273 100 188 41 160 44 29 57	**316**	100 72 57 43 41 34 30 21	**1.89**	19	28	2	2	–	–	–	–	–	–	–	–	–	Phthalimide, N-[3-(N,N-diisobutylamino)propyl]-		IC03895
193 316 136 123 194 147 108 120	**316**	100 48 45 22 20 18 18 17		19	28	2	2	–	–	–	–	–	–	–	–	–	1H-Pyrrole-2-carboxylic acid, 4-ethyl-5-[(4-ethyl-3,5-dimethyl-1H-pyrrol 2-yl)methyl]-3-methyl-, ethyl ester	15769-97-8	NI33355
193 316 136 194 123 147 120 108	**316**	100 49 45 21 21 18 18 18		19	28	2	2	–	–	–	–	–	–	–	–	–	4,3′,5′-Trimethyl-3,4′-diethyl-5-(ethoxycarbonyl)-2,2′-dipyrrolyl-methane		LI07006
316 271 317 202 76 31 289 272	**316**	100 50 26 17 16 16 14 14		20	12	4	–	–	–	–	–	–	–	–	–	–	Dibenzo[b,h]biphenylene-5,12-dione, 5a,11b-dihydro-6,11-dihydroxy-		LI07007
316 271 317 202 76 31 287 272	**316**	100 50 26 17 16 16 14 14		20	12	4	–	–	–	–	–	–	–	–	–	–	Dibenzo[b,h]biphenylene-6,11-dione, 5b,11a-dihydro-5,12-dihydroxy-		MS07432
316 76 105 317 271 272 231 158	**316**	100 30 28 20 16 15 15 14		20	12	4	–	–	–	–	–	–	–	–	–	–	Dibenzo[b,h]biphenylene-5,6,11,12-tetrone, 5a,5b,11a,11b-tetrahydro-, (5aα,5bα,11aα,11bα)-	14734-20-4	NI33356
316 76 104 159 102 50 130 75	**316**	100 86 80 67 56 44 40 24		20	12	4	–	–	–	–	–	–	–	–	–	–	Dibenzo[b,h]biphenylene-5,6,11,12-tetrone, 5a,5b,11a,11b-tetrahydro-, (5aα,5bβ,11aβ,11bα)-	14734-19-1	NI33357
316 76 104 158 102 50 130 317	**316**	100 85 80 66 55 44 40 24		20	12	4	–	–	–	–	–	–	–	–	–	–	Dibenzo[b,h]biphenylene-5,6,11,12-tetrone, 5a,5b,11a,11b-tetrahydro-, anti-		MS07433
316 76 104 159 102 50 130 202	**316**	100 86 82 66 56 44 40 24		20	12	4	–	–	–	–	–	–	–	–	–	–	Dibenzo[b,h]biphenylene-5,6,11,12-tetrone, 5a,5b,11a,11b-tetrahydro-, anti-		LI07008
316 317 76 104 272 271 231 158	**316**	100 39 30 29 15 15 14 13		20	12	4	–	–	–	–	–	–	–	–	–	–	Dibenzo[b,h]biphenylene-5,6,11,12-tetrone, 5a,5b,11a,11b-tetrahydro-, syn-		MS07434
316 77 104 271 231 159 101 50	**316**	100 30 29 15 15 14 11 11		20	12	4	–	–	–	–	–	–	–	–	–	–	Dibenzo[b,h]biphenylene-5,6,11,12-tetrone, 5a,5b,11a,11b-tetrahydro-, syn-		LI07009
316 318 317 281 315 282 280 158	**316**	100 42 29 23 20 18 15 15		20	13	–	2	–	–	1	–	–	–	–	–	–	13H-Benz[6,7]indolo[3,2-c]quinoline, 2-chloro-6-methyl-	4240-59-9	LI07010
316 318 317 280 315 281 158 279	**316**	100 43 30 23 20 18 17 16		20	13	–	2	–	–	1	–	–	–	–	–	–	13H-Benz[6,7]indolo[3,2-c]quinoline, 2-chloro-6-methyl-	4240-59-9	NI33358
161 127 126 316 163 318 252 317	**316**	100 98 94 68 33 24 24 16		20	13	–	2	–	–	1	–	–	–	–	–	–	Naphthalene, 8-chloro-2-(2-naphthylazo)-		IC03896
91 104 316 239 194 207 180 105	**316**	100 77 66 27 26 20 9 8		20	16	–	2	1	–	–	–	–	–	–	–	–	1,3,4-Thiadiazole, 2,3-dihydro-2,3,5-triphenyl-	13187-62-7	NI33359
91 104 316 239 194 207 317 103	**316**	100 77 66 28 27 21 19 11		20	16	–	2	1	–	–	–	–	–	–	–	–	1,3,4-Thiadiazole, 2,3-dihydro-2,3,5-triphenyl-	13187-62-7	NI33360
129 130 158 159 128 157 316	**316**	100 90 69 63 63 55 2		20	16	2	2	–	–	–	–	–	–	–	–	–	1,2-Ethanediol, 1,2-bis(2-quinolyl)-		LI07011
183 316 196 184 129 224 77 195	**316**	100 94 76 38 38 31 27 25		20	20	–	4	–	–	–	–	–	–	–	–	–	4H-Dipyridazino[1,6-a:4′,3′-c]quinoline, 1,4a,4b,5,6,13b-hexahydro-1-phenyl-, [4aα,4bα,13bα-(±)]-	34945-82-9	NI33361
183 316 196 184 129 77 195 224	**316**	100 32 32 22 17 16 13 11		20	20	–	4	–	–	–	–	–	–	–	–	–	4H-Dipyridazino[1,6-a:4′,3′-c]quinoline, 1,4a,4b,5,6,13b-hexahydro-1-phenyl-, [4aα,4bβ,13bα-(±)]-	34945-83-0	NI33362
273 316 255 256 288	**316**	100 71 63 57 5		20	28	3	–	–	–	–	–	–	–	–	–	–	Androstan-3,17,19-trione, 19a-methyl-, (5α)-		LI07012
255 316 237 197 212 147 158 213	**316**	100 56 53 45 43 31 30 29		20	28	3	–	–	–	–	–	–	–	–	–	–	Androst-11-en-17-one, 3-formoxy-, (3α,5α)-		NI33363
255 207 105 91 316 237 119 197	**316**	100 91 85 69 56 53 50 45		20	28	3	–	–	–	–	–	–	–	–	–	–	Androst-11-en-17-one, 3-formoxy-, (3α,5α)-		MS07435
252 144 255 105 91 145 237 119	**316**	100 47 40 39 34 30 26 25	**0.00**	20	28	3	–	–	–	–	–	–	–	–	–	–	Androst-11-en-17-one, 3-formoxy-, (3α,5β)-		MS07436
252 144 255 105 91 237 270 197	**316**	100 47 41 39 35 27 24 23	**0.70**	20	28	3	–	–	–	–	–	–	–	–	–	–	Androst-11-en-17-one, 3-formoxy-, (3α,5β)-		NI33364
41 316 91 107 93 113 109 95	**316**	100 85 60 38 38 34 34 30		20	28	3	–	–	–	–	–	–	–	–	–	–	Antiquorin		MS07437
219 203 316 243 185 159 157 175	**316**	100 46 18 14 13 12 11 10		20	28	3	–	–	–	–	–	–	–	–	–	–	Bacchotricuneatin D aldehyde alcohol		MS07438
316 191 41 55 247 317 83 81	**316**	100 52 52 47 35 25 19 17		20	28	3	–	–	–	–	–	–	–	–	–	–	Benzoic acid, 3,5-dicyclohexyl-4-hydroxy-, methyl ester		MS07439
121 93 81 95 215 91 217 105	**316**	100 88 86 81 78 78 70 69	**6.00**	20	28	3	–	–	–	–	–	–	–	–	–	–	Cleomeolide ketone		MS07440
123 150 121 93 81 41 55 108	**316**	100 32 32 22 21 21 20 18	**4.00**	20	28	3	–	–	–	–	–	–	–	–	–	–	Cyclopropanecarboxylic acid, 2,2-dimethyl-3-(2-methyl-1-propenyl)-, 3-(2 butenyl)-2-methyl-4-oxo-2-cyclopenten-1-yl ester		LI07013
123 150 121 93 81 41 43 85	**316**	100 41 23 16 16 15 14 13	**7.90**	20	28	3	–	–	–	–	–	–	–	–	–	–	Cyclopropanecarboxylic acid, 2,2-dimethyl-3-(2-methyl-1-propenyl)-, 3-(2 butenyl)-2-methyl-4-oxo-2-cyclopenten-1-yl ester, [1R-[1α[S*(Z)],3β]]-	25402-06-6	MS07441
123 81 43 41 93 121 29 79	**316**	100 41 39 35 32 24 24 23	**0.20**	20	28	3	–	–	–	–	–	–	–	–	–	–	Cyclopropanecarboxylic acid, 2,2-dimethyl-3-(2-methyl-1-propenyl)-, 3-(2 butenyl)-2-methyl-4-oxo-2-cyclopenten-1-yl ester, [1R-[1α[S*(Z)],3β]]-	25402-06-6	NI33365

MASS TO CHARGE RATIOS								M.W.	INTENSITIES								Parent	C	H	O	N	S	F	Cl	Br	I	Si	P	B	X	COMPOUND NAME	CAS Reg No	No
123	150	121	93	41	81	43	56	**316**	100	32	32	25	23	22	20	19	**1.48**	20	28	3	–	–	–	–	–	–	–	–	–	–	Cyclopropanecarboxylic acid, 2,2-dimethyl-3-(2-methyl-1-propenyl)-, 3-(2 butenyl)-2-methyl-4-oxo-2-cyclopenten-1-yl ester, [1R-[1α[S*(Z)],3β]]-	25402-06-6	NI33366
81	316	121	189	82	53	95	79	**316**	100	91	71	51	47	33	32	30		20	28	3	–	–	–	–	–	–	–	–	–	–	Damellic acid		NI33367
286	215	316	287	271	204	243	202	**316**	100	32	27	21	14	12	8	8		20	28	3	–	–	–	–	–	–	–	–	–	–	(-)-20-Deoxocarnosol		NI33368
87	99	316	86	91	254	145	79	**316**	100	70	46	39	26	25	15	15		20	28	3	–	–	–	–	–	–	–	–	–	–	Estra-5(10)-en-3,17-dione, cyclic 3-(1,2-ethanediyl acetal)	6193-99-3	MS07442
100	91	86	79	105	77	316	101	**316**	100	42	37	37	25	23	20	12		20	28	3	–	–	–	–	–	–	–	–	–	–	Estra-5-en-3,17-dione, cyclic 3-(1,2-ethanediyl acetal)	6193-98-2	MS07443
316	108	256	43	149	95	107	135	**316**	100	88	83	74	64	45	28	25		20	28	3	–	–	–	–	–	–	–	–	–	–	Estra-1-en-3-one, 17-acetoxy-, (5α)-		MS07444
316	43	256	108	274	149	107	95	**316**	100	78	72	53	40	39	28	24		20	28	3	–	–	–	–	–	–	–	–	–	–	Estra-1-en-3-one, 17-acetoxy-, (5β)-		MS07445
316	273	43	108	255	215	95	107	**316**	100	87	52	38	32	28	28	27		20	28	3	–	–	–	–	–	–	–	–	–	–	Estra-1-en-3-one, 17-acetyl-17-hydroxy-, (5α)-		MS07446
316	255	273	43	108	107	215	41	**316**	100	61	60	49	27	26	25	21		20	28	3	–	–	–	–	–	–	–	–	–	–	Estra-1-en-3-one, 17-acetyl-17-hydroxy-, (5β)-		MS07447
316	215	71	230	317	257	284	239	**316**	100	78	58	27	23	18	16	15		20	28	3	–	–	–	–	–	–	–	–	–	–	Estra-1,3,5(10)-trien-3-ol, 16,17-dimethoxy-, (16α,17β)-		NI33369
316	244	41	317	43	55	79	242	**316**	100	29	23	21	21	20	19	18		20	28	3	–	–	–	–	–	–	–	–	–	–	D-Homoandrosta-4,17-dien-3-one, 17,17a-dihydroxy-	30298-69-2	NI33370
109	316	317	110	298	259	108	137	**316**	100	83	13	13	8	8	8	7		20	28	3	–	–	–	–	–	–	–	–	–	–	Kaur-16-en-18-oic acid, 6,7-dihydroxy-, γ-lactone, (4α,6α,7α)-	15959-13-4	NI33371
298	137	109	299	270	145	163	107	**316**	100	41	35	20	14	11	10	10	**1.50**	20	28	3	–	–	–	–	–	–	–	–	–	–	Kaur-16-en-18-oic acid, 6,7-dihydroxy-, γ-lactone, (4α,6α,7β)-	5691-63-4	NI33372
81	121	82	316	53	41	95	91	**316**	100	57	50	44	40	35	31	30		20	28	3	–	–	–	–	–	–	–	–	–	–	Labda-8(20),13(16),14-trien-18-oic acid, 15,16-epoxy-	10267-14-8	NI33374
81	121	82	53	41	316	95	91	**316**	100	59	50	40	33	30	28	28		20	28	3	–	–	–	–	–	–	–	–	–	–	Labda-8(20),13(16),14-trien-18-oic acid, 15,16-epoxy-	10267-14-8	NI33373
81	316	121	189	82	53	95	79	**316**	100	90	71	51	47	37	32	30		20	28	3	–	–	–	–	–	–	–	–	–	–	Labda-8(20),13(16),14-trien-18-oic acid, 15,16-epoxy-, (5β,9βH,10α)-	1235-77-4	NI33375
82	81	41	111	95	192	43	109	**316**	100	62	41	32	31	26	26	25	**10.19**	20	28	3	–	–	–	–	–	–	–	–	–	–	1(4H)-Naphthalenone, 4-[2-(3-furanyl)ethyl]-4a,5,6,7,8,8a-hexahydro-4-hydroxy-3,4a,8,8-tetramethyl-	40883-07-6	NI33376
316	108	317	301	109	255	133	131	**316**	100	62	23	19	14	10	7	7		20	28	3	–	–	–	–	–	–	–	–	–	–	Phenanthro[3,2-b]furan-4-carboxylic acid, 1,2,3,4,4a,5,6,6a,7,11,11a,11b-dodecahydro-4,7,11b-trimethyl-, [4R-(4α,4aα,6aβ,7α,11aα,11bβ)]-	19941-61-8	NI33377
109	316	110	317	133	55	79	28	**316**	100	71	20	16	13	13	12	12		20	28	3	–	–	–	–	–	–	–	–	–	–	Phenanthro[3,2-b]furan-4-carboxylic acid, 1,2,3,4,4a,5,6,6a,7,11,11a,11b-dodecahydro-4,7,11b-trimethyl-, [4S-(4α,4aβ,6aα,7β,11aβ,11bα)]-	19941-59-4	NI33378
316	108	317	109	301	255	79	81	**316**	100	85	24	18	16	11	10	9		20	28	3	–	–	–	–	–	–	–	–	–	–	Phenanthro[3,2-b]furan-4-carboxylic acid, 1,2,3,4,4a,5,6,6a,7,11,11a,11b-dodecahydro-4,7,11b-trimethyl-, [4S-(4α,4aβ,6aα,7β,11aβ,11bα)]-	19941-59-4	NI33379
43	41	91	316	55	69	105	77	**316**	100	68	52	42	42	39	30	29		20	28	3	–	–	–	–	–	–	–	–	–	–	Retinoic acid, 5,6-epoxy-5,6-dihydro-	13100-69-1	NI33380
45	150	43	92	96	316	165	57	**316**	100	99	80	57	56	54	52	52		20	28	3	–	–	–	–	–	–	–	–	–	–	Retinoic acid, 5,8-epoxy-5,8-dihydro-	3012-76-8	NI33381
316	211	178	317	165	167	210	284	**316**	100	39	28	26	26	23	20	16		21	16	1	–	1	–	–	–	–	–	–	–	–	2H-1,3-Oxathiole, 2,4,5-triphenyl-		LI07014
181	152	77	182	153	105	44	316	**316**	100	18	18	17	16	15	13	9		21	16	3	–	–	–	–	–	–	–	–	–	–	Acetophenone, 2-hydroxy-4′-phenyl-, benozate	4347-61-9	NI33382
272	166	165	105	273	271	77	168	**316**	100	60	54	48	22	22	20	8	**0.00**	21	16	3	–	–	–	–	–	–	–	–	–	–	1,3-Dioxolan-4-one, 2,4,5-triphenyl-		LI07015
316	105	239	178	77	317	211	39	**316**	100	82	60	33	33	23	21	16		21	16	3	–	–	–	–	–	–	–	–	–	–	2,3-Diphenyl-5-methoxybenzo-1,4-dioxin	91201-56-8	NI33383
181	180	316	315	298	182	297	43	**316**	100	76	72	55	39	38	24	20		21	20	1	2	–	–	–	–	–	–	–	–	–	Acetamide, N-[2-[[2-(phenylamino)phenyl]methyl]phenyl]-	52812-75-6	NI33384
43	316	124	71	135	55	123	41	**316**	100	80	48	46	45	45	44	43		21	32	2	–	–	–	–	–	–	–	–	–	–	Androst-4-en-3-one, 17-hydroxy-7,17-dimethyl-, (7β,17β)-	17021-26-0	NI33385
124	316	79	91	55	95	93	67	**316**	100	53	40	38	35	34	33	32		21	32	2	–	–	–	–	–	–	–	–	–	–	Androsten-3-one, 17-(1-hydroxyethyl)-		LI07016
193	231	247	123	194	136	233	232	**316**	100	64	25	24	20	20	19	12	**11.60**	21	32	2	–	–	–	–	–	–	–	–	–	–	Cannabigerol	2808-33-5	NI33386
146	133	101	187	134	91	92	117	**316**	100	46	46	39	28	25	24	19	**15.00**	21	32	2	–	–	–	–	–	–	–	–	–	–	Cyclohexanecarboxylic acid, 1,3-dimethyl-2-[2′-(4-isopropylphenyl)ethyl]-, (1β,2α,3α)-, methyl ester		LI07017
146	133	101	187	134	91	92	123	**316**	100	46	46	39	28	25	24	19	**15.00**	21	32	2	–	–	–	–	–	–	–	–	–	–	Cyclohexanecarboxylic acid, 1,3-dimethyl-2-[2′-(4-isopropylphenyl)ethyl]-, (1β,2α,3α)-, methyl ester		MS07448
146	133	91	55	109	117	131	123	**316**	100	46	30	27	24	22	20	20	**7.50**	21	32	2	–	–	–	–	–	–	–	–	–	–	Cyclohexanecarboxylic acid, 1,3-dimethyl-2-[2′-(4-isopropylphenyl)ethyl]-, methyl ester		LI07018
146	101	133	134	147	187	116	92	**316**	100	74	63	62	52	48	42	41	**18.00**	21	32	2	–	–	–	–	–	–	–	–	–	–	Cyclohexanecarboxylic acid, 1,3-dimethyl-2-[2′-(4-isopropylphenyl)ethyl]-, (1R,2S,3S)-, methyl ester		MS07449
146	123	133	151	81	117	95	91	**316**	100	92	89	83	73	69	64	48	**5.00**	21	32	2	–	–	–	–	–	–	–	–	–	–	Cyclohexanecarboxylic acid, 2-(3-isopropylphenylethyl)-1,3-dimethyl-, methyl ester, [1R-(1α,2α,3α)]-	35482-10-1	NI33387
316	121	135	134	105	133	106	119	**316**	100	69	64	61	61	58	52	50		21	32	2	–	–	–	–	–	–	–	–	–	–	14,17-Cyclolabda-8(17),12-dien-18-oic acid, methyl ester, (20S)-		MS07450
273	193	260	316	231	233	274	136	**316**	100	84	79	56	39	32	28	27		21	32	2	–	–	–	–	–	–	–	–	–	–	6H-Dibenzo[b,d]pyran-1-ol, 6a,7,8,9,10,10a-hexahydro-6,6,9-trimethyl-3-pentyl-	6692-85-9	NI33388
110	316	189	95	317	207	274	298	**316**	100	82	22	20	19	19	14	10		21	32	2	–	–	–	–	–	–	–	–	–	–	19-Nor-D-homoandrost-4-en-3-one, 17a-hydroxy-16,16-dimethyl-, (17β)-		LI07019
110	316	189	298	187	207	109	256	**316**	100	72	56	53	32	23	21	20		21	32	2	–	–	–	–	–	–	–	–	–	–	19-Nor-D-homoandrost-4-en-3-one, 17-hydroxy-16,16-dimethyl-, (17α)-		LI07020
41	55	91	79	81	257	67	93	**316**	100	64	63	61	53	51	45	42	**37.71**	21	32	2	–	–	–	–	–	–	–	–	–	–	Kaur-16-en-18-oic acid, methyl ester, (4β)-	5524-25-4	NI33389
94	39	91	105	55	79	93	81	**316**	100	70	55	45	45	43	42	40	**18.00**	21	32	2	–	–	–	–	–	–	–	–	–	–	Kaur-16-en-18-oic acid, methyl ester, (4β)-	5524-25-4	MS07451
41	55	91	93	79	77	67	53	**316**	100	55	33	28	27	25	24	24	**0.00**	21	32	2	–	–	–	–	–	–	–	–	–	–	17-Nor-kaur-15-en-18-oic acid, 13-methyl-, (8β,13β)-, methyl ester		LI07021
81	93	79	121	91	119	105	175	**316**	100	89	80	79	65	63	60	57	**50.00**	21	32	2	–	–	–	–	–	–	–	–	–	–	Labda-8(17),12E,14-trien-19-oic acid, methyl ester		MS07452
121	316	135	134	119	133	161	317	**316**	100	87	87	32	31	28	26	25		21	32	2	–	–	–	–	–	–	–	–	–	–	Labda-8(20),12,14-trien-19-oic acid, methyl ester	1235-39-8	NI33390

MASS TO CHARGE RATIOS	M.W.	INTENSITIES	Parent	C	H	O	N	S	F	Cl	Br	I	Si	P	B	X	COMPOUND NAME	CAS Reg No	No
316 121 175 257 134 301 119 241	**316**	100 84 56 47 43 42 42 36		21	32	2	–	–	–	–	–	–	–	–	–	–	**Labda-8(20),12,14-trien-19-oic acid, methyl ester, isomer A**		**LI07022**
121 316 119 257 301 123 241 149	**316**	100 51 42 38 35 35 26 24		21	32	2	–	–	–	–	–	–	–	–	–	–	**Labda-8(20),12,14-trien-19-oic acid, methyl ester, isomer B**		**LI07023**
121 135 316 134 119 133 161 148	**316**	100 87 86 35 32 28 26 26		21	32	2	–	–	–	–	–	–	–	–	–	–	**Labda-8(20),12,14-trien-19-oic acid, methyl ester, isomer C**		**LI07024**
316 121 175 257 301 134 119 317	**316**	100 82 57 47 44 44 44 43		21	32	2	–	–	–	–	–	–	–	–	–	–	**Labda-8(20),12,14-trien-19-oic acid, methyl ester, [1S-(1α,4aα,5αE,8aβ)]**	15798-13-7	**NI33391**
121 175 316 119 81 105 93 107	**316**	100 88 76 66 50 49 49 45		21	32	2	–	–	–	–	–	–	–	–	–	–	**Labda-8(20),12,14-trien-19-oic acid, methyl ester, [1S-(1α,4aα,5αE,8aβ)]**	15798-13-7	**NI33392**
41 79 81 55 93 91 43 45	**316**	100 70 66 56 52 52 51 47	**5.00**	21	32	2	–	–	–	–	–	–	–	–	–	–	**Labda-8(20),12,14-trien-19-oic acid, methyl ester, [1S-(1α,4aα,5αZ,8aβ)]**	10178-35-5	**NI33393**
93 301 242 55 114 107 79 175	**316**	100 70 70 68 60 60 58 53	**35.00**	21	32	2	–	–	–	–	–	–	–	–	–	–	**Methyl abeoanticopalate**		**MS07453**
43 41 240 91 129 55 115 141	**316**	100 81 70 56 51 47 46 42	**3.11**	21	32	2	–	–	–	–	–	–	–	–	–	–	**Methyl abietate isomer**		**NI33394**
79 91 67 93 108 105 119 80	**316**	100 73 60 58 46 43 43 38	**1.20**	21	32	2	–	–	–	–	–	–	–	–	–	–	**Methyl eicosa-5,8,11,14,17-pentaenoate**		**NI33395**
241 105 121 107 106 257 91 316	**316**	100 96 85 83 80 79 77 75		21	32	2	–	–	–	–	–	–	–	–	–	–	**Methyl 7,15-isopimaradien-18-oate**		**MS07454**
241 257 316 301 107 91 105 121	**316**	100 76 74 74 62 60 58 55		21	32	2	–	–	–	–	–	–	–	–	–	–	**Methyl 8,15-isopimaradien-18-oate**		**MS07455**
146 151 133 123 95 55 187 117	**316**	100 92 88 65 65 44 39 35	**12.70**	21	32	2	–	–	–	–	–	–	–	–	–	–	**Methyl 9-(m-isopropylphenyl)-2,6-dimethyl-cis-6-nonenoate**		**LI07025**
93 107 55 114 67 79 135 147	**316**	100 93 82 68 67 65 55 38	**10.00**	21	32	2	–	–	–	–	–	–	–	–	–	–	**Methyl 3S,18-cyclolabda-8(17),E-13-dien-15-oate**		**MS07456**
121 316 119 257 123 301 135 133	**316**	100 51 43 38 35 31 28 28		21	32	2	–	–	–	–	–	–	–	–	–	–	**1-Naphthalenecarboxylic acid, decahydro-1,4a-dimethyl-6-methylene-5-(3 methylene-4-pentenyl)-, methyl ester, [1S-(1α,4aα,5α,8aβ)]-**	28720-17-4	**NI33396**
146 133 151 123 95 81 187 55	**316**	100 71 67 49 46 36 32 30	**6.50**	21	32	2	–	–	–	–	–	–	–	–	–	–	**6-Nonenoic acid, 9-(3-isopropylphenyl)-2,6-dimethyl-, methyl ester, (E)-**		**LI07026**
121 180 93 91 79 181 105 81	**316**	100 15 14 14 14 12 12 12	**6.62**	21	32	2	–	–	–	–	–	–	–	–	–	–	**1-Phenanthrenecarboxylic acid, 7-vinyl-1,2,3,4,4a,4b,5,6,7,9,10,10a-dodecahydro-1,4a,7-trimethyl-, methyl ester**	56051-68-4	**NI33397**
121 91 105 93 119 120 81 79	**316**	100 19 16 16 13 12 11 11	**5.00**	21	32	2	–	–	–	–	–	–	–	–	–	–	**1-Phenanthrenecarboxylic acid, 7-vinyl-1,2,3,4,4a,4b,5,6,7,9,10,10a-dodecahydro-1,4a,7-trimethyl-, methyl ester, [1R-(1α,4aβ,4bα,7α,10aα)]-**	1686-54-0	**NI33398**
316 283 301 317 241 284 213 215	**316**	100 46 31 25 13 10 9 8		21	32	2	–	–	–	–	–	–	–	–	–	–	**2,6-Phenanthrenediol, 1,2,3,4,4a,9,10,10a-octahydro-1,1,4a-trimethyl-7-isopropyl-, monomethyl ether, [2S-(2α,4aα,10aβ)]-**	72361-31-0	**NI33399**
316 283 301 317 241 284 215 213	**316**	100 35 24 22 11 8 8 8		21	32	2	–	–	–	–	–	–	–	–	–	–	**2,6-Phenanthrenediol, 1,2,3,4,4a,9,10,10a-octahydro-1,1,4a-trimethyl-7-isopropyl-, monomethyl ether, [2S-(2α,4aα,10aβ)]-**	72361-31-0	**NI33400**
316 301 317 189 283 241 163 161	**316**	100 28 24 17 12 11 10 9		21	32	2	–	–	–	–	–	–	–	–	–	–	**4a(2H)-Phenanthrenemethanol, 1,3,4,9,10,10a-hexahydro-7-isopropyl-6-methoxy-1,1-dimethyl-, (4aR-trans)-**	57397-36-1	**NI33401**
316 241 201 131 256 257 129 105	**316**	100 73 58 53 51 50 49 49		21	32	2	–	–	–	–	–	–	–	–	–	–	**Podocarpa-7,9(11)-dien-15-oic acid, 13-isopropyl-, methyl ester**		**MS07457**
316 121 91 256 105 136 93 79	**316**	100 69 69 67 67 60 42 42		21	32	2	–	–	–	–	–	–	–	–	–	–	**Podocarpa-7,13-dien-15-oic acid, 13-isopropyl-, methyl ester**	127-25-3	**NI33402**
256 121 241 105 316 91 213 131	**316**	100 70 69 61 57 51 46 44		21	32	2	–	–	–	–	–	–	–	–	–	–	**Podocarpa-7,13-dien-15-oic acid, 13-isopropyl-, methyl ester**	127-25-3	**NI33403**
121 105 91 131 136 185 256 93	**316**	100 83 73 62 59 51 49 46	**29.00**	21	32	2	–	–	–	–	–	–	–	–	–	–	**Podocarpa-7,13-dien-15-oic acid, 13-isopropyl-, methyl ester**	127-25-3	**NI33404**
146 101 92 91 121 134 133 105	**316**	100 75 75 73 70 54 53 49	**48.00**	21	32	2	–	–	–	–	–	–	–	–	–	–	**Podocarpa-8(14),12-dien-15-oic acid, 13-isopropyl-, methyl ester**	3513-69-7	**MS07458**
121 91 146 92 133 101 105 134	**316**	100 99 87 61 50 46 44 32	**20.00**	21	32	2	–	–	–	–	–	–	–	–	–	–	**Podocarpa-8(14),12-dien-15-oic acid, 13-isopropyl-, methyl ester**	3513-69-7	**NI33405**
135 316 148 121 91 136 134 93	**316**	100 63 59 56 56 46 42 40		21	32	2	–	–	–	–	–	–	–	–	–	–	**Podocarpa-8(14),13(15)-dien-15-oic acid, 13-isopropyl-, methyl ester**		**MS07459**
91 92 105 146 101 134 109 131	**316**	100 66 61 60 50 49 49 46	**34.00**	21	32	2	–	–	–	–	–	–	–	–	–	–	**Podocarpa-8,12-dien-15-oic acid, 13-isopropyl-, methyl ester**		**MS07460**
135 316 148 121 91 134 136 105	**316**	100 64 32 31 26 23 21 19		21	32	2	–	–	–	–	–	–	–	–	–	–	**Podocarpa-8,13(15)-dien-15-oic acid, 13-isopropyl-, methyl ester**		**MS07461**
43 91 41 105 55 107 149 106	**316**	100 90 87 77 64 62 56 53	**14.00**	21	32	2	–	–	–	–	–	–	–	–	–	–	**Podocarpa-8,13-dien-15-oic acid, 13-isopropyl-, methyl ester**		**NI33406**
91 113 121 105 151 316 95 131	**316**	100 99 82 81 71 63 62 57		21	32	2	–	–	–	–	–	–	–	–	–	–	**Podocarpa-8,13-dien-15-oic acid, 13-isopropyl-, methyl ester**		**MS07462**
241 316 301 257 119 149 105 91	**316**	100 70 68 50 49 45 41 41		21	32	2	–	–	–	–	–	–	–	–	–	–	**Podocarpa-8,15-dien-15-oic acid, 13-isopropyl-, methyl ester**		**MS07463**
135 316 148 121 91 136 134 93	**316**	100 63 59 56 56 46 42 40		21	32	2	–	–	–	–	–	–	–	–	–	–	**Podocarpa-8(14)-en-15-oic acid, 13-isopropylidene-, methyl ester**	3310-97-2	**NI33407**
135 148 316 136 134 119 91 93	**316**	100 21 14 14 14 9 8 7		21	32	2	–	–	–	–	–	–	–	–	–	–	**Podocarpa-8(14)-en-15-oic acid, 13-isopropylidene-, methyl ester**	3310-97-2	**NI33408**
241 91 119 105 121 79 187 107	**316**	100 98 90 90 68 67 56 55	**21.00**	21	32	2	–	–	–	–	–	–	–	–	–	–	**Podocarpa-7-en-15-oic acid, 13-methyl-13-vinyl-, methyl ester**	1686-62-0	**NI33409**
121 180 181 122 316 105 91 123	**316**	100 58 51 49 47 42 42 39		21	32	2	–	–	–	–	–	–	–	–	–	–	**Podocarpa-8(14)-en-15-oic acid, 13-methyl-13-vinyl-, methyl ester**	3730-56-1	**NI33410**
121 316 257 180 181 301 241 148	**316**	100 49 37 34 24 23 22 16		21	32	2	–	–	–	–	–	–	–	–	–	–	**Podocarpa-8(14)-en-15-oic acid, 13-methyl-13-vinyl-, methyl ester**	3730-56-1	**NI33411**
121 91 180 105 93 79 120 119	**316**	100 20 14 14 13 13 12 12	**0.00**	21	32	2	–	–	–	–	–	–	–	–	–	–	**Podocarpa-8(14)-en-15-oic acid, 13-methyl-13-vinyl-, methyl ester**	3730-56-1	**NI33412**
316 283 301 317 241 285 215 213	**316**	100 35 22 21 12 9 8 8		21	32	2	–	–	–	–	–	–	–	–	–	–	**Podocarpa-8,11,13-trien-3-ol, 13-isopropyl-12-methoxy-, (3β)-**	18326-13-1	**NI33413**
121 180 181 122 316 105 91 123	**316**	100 58 51 49 47 42 42 39		21	32	2	–	–	–	–	–	–	–	–	–	–	**Podocarp-8(14),15-dien-18-oic acid, 13β-ethyl-13-methyl-, methyl ester**		**MS07464**
121 181 316 122 105 93 91 180	**316**	100 45 40 39 38 38 38 36		21	32	2	–	–	–	–	–	–	–	–	–	–	**Podocarp-8(14)-15-oic acid, 13-isopropyl-, methyl ester**	1686-54-0	**NI33414**
121 301 316 257 91 181 105 93	**316**	100 26 25 22 22 19 17 16		21	32	2	–	–	–	–	–	–	–	–	–	–	**Podocarp-8(14)-15-oic acid, 13-isopropyl-, methyl ester**	1686-54-0	**NI33415**
316 247 55 81 41 260 109 301	**316**	100 87 42 37 37 29 26 25		21	32	2	–	–	–	–	–	–	–	–	–	–	**Pregnane-3,12-dione, (5β)-**	26991-52-6	**NI33416**
316 247 55 41 81 260 301 109	**316**	100 87 42 38 37 30 26 24		21	32	2	–	–	–	–	–	–	–	–	–	–	**Pregnane-3,12-dione, (5β)-**		**LI07027**
43 316 55 246 298 84 81 41	**316**	100 45 44 41 40 34 32 29		21	32	2	–	–	–	–	–	–	–	–	–	–	**Pregnane-3,20-dione**	7350-00-7	**NI33417**
43 316 84 55 81 231 298 95	**316**	100 60 53 42 34 28 27 27		21	32	2	–	–	–	–	–	–	–	–	–	–	**Pregnane-3,20-dione, (5α)-**	566-65-4	**LI07028**
43 316 84 55 81 41 67 95	**316**	100 59 52 40 34 29 28 27		21	32	2	–	–	–	–	–	–	–	–	–	–	**Pregnane-3,20-dione, (5α)-**	566-65-4	**NI33419**
43 84 316 55 231 81 41 67	**316**	100 61 55 44 38 38 38 33		21	32	2	–	–	–	–	–	–	–	–	–	–	**Pregnane-3,20-dione, (5α)-**	566-65-4	**NI33418**
316 246 298 213 231 272 228 255	**316**	100 66 60 23 19 17 16 13		21	32	2	–	–	–	–	–	–	–	–	–	–	**Pregnane-3,20-dione, (5β)-**	128-23-4	**MS07466**

MASS TO CHARGE RATIOS								M.W.	INTENSITIES								Parent	C	H	O	N	S	F	Cl	Br	I	Si	P	B	X	COMPOUND NAME	CAS Reg No	No
43	41	246	55	84	81	95	67	316	100	42	33	33	30	30	26	26	21.02	21	32	2	–	–	–	–	–	–	–	–	–	–	Pregnane-3,20-dione, (5β)-	128-23-4	MS07465
43	316	246	298	55	84	81	41	316	100	82	55	54	40	38	32	32		21	32	2	–	–	–	–	–	–	–	–	–	–	Pregnane-3,20-dione, (5β)-	128-23-4	NI33420
43	81	55	316	67	91	95	79	316	100	48	47	30	30	26	23	23		21	32	2	–	–	–	–	–	–	–	–	–	–	Pregnane-12,20-dione, (5α)-	6022-48-6	NI33421
81	55	67	316	109	79	301	95	316	100	63	54	44	42	40	37	35		21	32	2	–	–	–	–	–	–	–	–	–	–	Pregnane-12,20-dione, (5α)-	6022-48-6	NI33422
316	67	79	81	91	95	93	109	316	100	62	55	51	45	37	37	35		21	32	2	–	–	–	–	–	–	–	–	–	–	Pregnan-18-oic acid, 20-hydroxy-, γ-lactone, (5α)-	56143-34-1	NI33423
316	317	245	124	57	43	229	41	316	100	23	19	19	18	16	15	14		21	32	2	–	–	–	–	–	–	–	–	–	–	Pregn-4-en-3-one, 17-hydroxy-, (17α)-	1235-97-8	NI33424
124	316	43	91	44	45	230	110	316	100	72	53	42	37	34	31	30		21	32	2	–	–	–	–	–	–	–	–	–	–	Pregn-4-en-3-one, 20-hydroxy-, (20α)-		MS07467
124	45	91	79	149	55	316	41	316	100	41	34	33	32	31	30	30		21	32	2	–	–	–	–	–	–	–	–	–	–	Pregn-4-en-3-one, 20-hydroxy-, (20β)-	145-15-3	NI33425
124	45	316	79	91	41	55	149	316	100	43	41	35	34	31	29	28		21	32	2	–	–	–	–	–	–	–	–	–	–	Pregn-4-en-3-one, 20-hydroxy-, (20β)-	145-15-3	NI33426
124	316	175	149	274	91	79	93	316	100	60	24	23	21	21	21	20		21	32	2	–	–	–	–	–	–	–	–	–	–	Pregn-4-en-3-one, 20-hydroxy-, (20β)-	145-15-3	LI07029
43	316	298	107	85	105	91	145	316	100	40	30	30	27	26	24	23		21	32	2	–	–	–	–	–	–	–	–	–	–	Pregn-5-en-20-one, 3-hydroxy-, (3β)-	145-13-1	NI33427
316	91	55	105	298	107	283	67	316	100	85	74	71	70	68	55	55		21	32	2	–	–	–	–	–	–	–	–	–	–	Pregn-5-en-20-one, 3-hydroxy-, (3β)-	145-13-1	WI01408
43	316	298	231	283	213	255		316	100	40	26	24	23	14	9			21	32	2	–	–	–	–	–	–	–	–	–	–	Pregn-5-en-20-one, 3-hydroxy-, (3β)-	145-13-1	MS07468
43	316	213	105	91	85	71	55	316	100	96	43	31	31	30	30	28		21	32	2	–	–	–	–	–	–	–	–	–	–	Pregn-5-en-20-one, 3-hydroxy-, (3β,17α)-	566-63-2	NI33428
57	97	43	71	69	95	85	41	316	100	97	94	91	84	79	74	73	0.00	21	32	2	–	–	–	–	–	–	–	–	–	–	Pregn-7-en-20-one, 12-hydroxy-, (5α,12β)-		DD01202
121	272	301	257	81	273	160	231	316	100	77	53	50	24	18	18	17	14.00	21	32	2	–	–	–	–	–	–	–	–	–	–	Pregn-9(11)-en-12-one, 20-hydroxy-, (5α,20β)-	7704-94-1	NI33429
316	255	28	317	43	273	44	91	316	100	28	28	23	21	20	20	16		21	32	2	–	–	–	–	–	–	–	–	–	–	Pregn-14-en-20-one, 3-hydroxy-, (3β,5α)-		LI07030
316	317	255	43	273	28	91	105	316	100	24	24	20	19	17	14	12		21	32	2	–	–	–	–	–	–	–	–	–	–	Pregn-14-en-20-one, 3-hydroxy-, (3β,5α,17α)-		LI07031
283	298	91	93	255	79	77	105	316	100	96	91	75	71	68	64	56	48.00	21	32	2	–	–	–	–	–	–	–	–	–	–	Pregn-16-en-20-one, 3-hydroxy-, (3β,5β)-	566-59-6	MS07469
43	120	41	105	107	119	91	55	316	100	45	45	40	37	35	33	32	19.00	21	32	2	–	–	–	–	–	–	–	–	–	–	Retinol, acetate, (19β)-		NI33430
316	58	60	43	45	301	317	59	316	100	90	73	60	35	33	25	24		21	36	–	2	–	–	–	–	–	–	–	–	–	Androstan-17-one, (5α)-, N,N-dimethylhydrazone		LI07032
44	56	273	257	70	55	43	41	316	100	40	25	7	5	5	5	5	1.00	21	36	–	2	–	–	–	–	–	–	–	–	–	Pregn-5-ene-3,20-diamine, (3β,20S)-	3614-57-1	NI33431
206	207	129	91	28	191	165	59	316	100	75	52	31	23	22	18	15	0.00	22	20	–	–	1	–	–	–	–	–	–	–	–	Benzene, 1,1′-[2-methyl-2-(phenylthio)cyclopropylidene]bis-	56728-02-0	NI33432
105	77	211	316	106	212	103	51	316	100	26	21	14	8	5	5	4		22	20	2	–	–	–	–	–	–	–	–	–	–	Benzoic acid, 4-methyl-2-(1-phenylethyl)phenyl ester		MS07470
172	144	115	145	173	127	185	116	316	100	66	63	40	24	24	22	21	4.00	22	20	2	–	–	–	–	–	–	–	–	–	–	1,8:2,7-Dimethanodibenzo[a,e]cyclobuta[c]cycloocten-13-one, 1-ethoxy-1,2,2a,7,8,12b-hexahydro-	55649-66-6	NI33433
45	316	271	243	165	210	167	284	316	100	88	76	46	36	34	26	17		22	20	2	–	–	–	–	–	–	–	–	–	–	(E)-2-methoxymethoxy-4′-phenylstilbene		MS07471
45	316	271	243	165	210	167	284	316	100	45	38	26	23	18	14	9		22	20	2	–	–	–	–	–	–	–	–	–	–	(Z)-2-methoxymethoxy-4′-phenylstilbene		MS07472
121	105	316	211	122	77	317	91	316	100	30	18	13	10	9	5	4		22	20	2	–	–	–	–	–	–	–	–	–	–	1-Propanone, 1,2-diphenyl-3-(4-methoxyphenyl)-		MS07473
91	105	77	225	92	65	51	106	316	100	31	13	9	8	6	3	2	0.00	22	20	2	–	–	–	–	–	–	–	–	–	–	1-Propanone, 1-phenyl-3-[2-benzyloxyphenyl]-	56052-51-8	NI33434
316	317	165	301	158	195	55	101	316	100	24	14	13	9	8	7	6		22	20	2	–	–	–	–	–	–	–	–	–	–	Styrene, α,β-bis(4-methoxyphenyl)-		MS07474
158	117	91	56	129	225	316	144	316	100	75	48	48	32	30	18	9		22	24	–	2	–	–	–	–	–	–	–	–	–	3,8-Dimethyl-1,10-diphenyl-4,7-diazadeca-1,3,7,9-tetraene		NI33435
316	217	301	43	81	95	109	67	316	100	92	44	38	32	29	26	25		22	36	1	–	–	–	–	–	–	–	–	–	–	D-Homopregnan-20-one, (5α)-	32318-95-9	NI33436
317	299	315	259	257	109	318	135	316	100	62	48	40	33	25	22	19	18.00	22	36	1	–	–	–	–	–	–	–	–	–	–	Pregnane, 20,21-epoxy-20-methyl-		NI33437
98	217	55	95	81	109	67	149	316	100	66	65	64	64	63	60	55	25.25	22	36	1	–	–	–	–	–	–	–	–	–	–	Pregnane, 20,21-epoxy-20-methyl-, (5α)-	54411-89-1	NI33438
218	316	139	245	95	203	109	41	316	100	65	64	55	37	36	36	27		22	36	1	–	–	–	–	–	–	–	–	–	–	Pregnan-15-one, 20-methyl-, (5α)-	14012-16-9	NI33439
139	55	41	81	67	109	79	93	316	100	13	13	12	12	11	11	10	5.05	22	36	1	–	–	–	–	–	–	–	–	–	–	Pregnan-15-one, 20-methyl-, (5α,14β)-	14111-73-0	NI33440
217	216	109	121	107	301	106	135	316	100	96	57	38	37	32	31	30	24.25	22	36	1	–	–	–	–	–	–	–	–	–	–	Pregnan-16-one, 20-methyl-, (5α)-	56051-43-5	NI33441
316	317	301	255	298	273	302	229	316	100	27	24	21	15	11	7	6		22	36	1	–	–	–	–	–	–	–	–	–	–	Pregn-7-en-3-ol, 20-methyl-, (3β)-	57983-95-6	NI33442
316	41	81	82	107	67	55	93	316	100	46	45	44	43	42	42	40		22	36	1	–	–	–	–	–	–	–	–	–	–	Pregn-20-en-3-ol, 20-methyl-, (3β,5α)-	54411-87-9	NI33443
243	105	183	244	165	77	239	316	316	100	90	75	43	43	40	34	25		23	24	1	–	–	–	–	–	–	–	–	–	–	Butyl trityl ether	6226-44-4	LI07033
243	105	183	244	165	77	239	316	316	100	90	77	44	44	41	34	26		23	24	1	–	–	–	–	–	–	–	–	–	–	Butyl trityl ether	6226-44-4	NI33444
180	316	193	91	184	165	115	77	316	100	81	56	53	47	36	33	23		23	24	1	–	–	–	–	–	–	–	–	–	–	4,4a,5,6,7,8-Hexahydro-4a-methyl-6,6-diphenyl-2(3H)-naphthalenone		DD01203
243	165	244	91	166	228	245	241	316	100	29	18	5	4	3	2	2	0.31	23	24	1	–	–	–	–	–	–	–	–	–	–	1-Pentanol, 5,5,5-triphenyl-	2294-95-3	NI33445
301	316	105	91	223	77	18	115	316	100	82	71	66	44	42	37	36		23	24	1	–	–	–	–	–	–	–	–	–	–	Phenol, 2-methyl-4,6-bis(1-phenylethyl)-		IC03897
197	18	105	316	91	193	44	103	316	100	63	32	31	29	24	21	20		23	24	1	–	–	–	–	–	–	–	–	–	–	Phenol, 2-methyl-4,6-bis(1-phenylethyl)-		IC03898
105	316	106	43	41	317	57	55	316	100	26	26	22	17	7	7	7		23	40	–	–	–	–	–	–	–	–	–	–	–	Benzene, hexadecylmethyl-	72101-31-6	WI01409
105	106	91	118	131	43	104	41	316	100	15	11	10	7	7	6	5	4.00	23	40	–	–	–	–	–	–	–	–	–	–	–	Benzene, (1-methylhexadecyl)-	55125-25-2	WI01410
105	106	91	104	316	79	77	69	316	100	17	13	6	5	4	4	2		23	40	–	–	–	–	–	–	–	–	–	–	–	Benzene, (1-methylhexadecyl)-	55125-25-2	WI01411
259	57	316	41	203	44	29	147	316	100	52	28	28	23	19	17	13		23	40	–	–	–	–	–	–	–	–	–	–	–	Benzene, 1,3,5-tris(2,2-dimethylpropyl)-2,4-dimethyl-	40572-13-2	MS07475
259	316	301	245	229	91	57	215	316	100	41	25	18	16	16	14	10		24	28	–	–	–	–	–	–	–	–	–	–	–	Benzofulvene, 3,8-di-tert-butyl-8-phenyl-		NI33446
92	91	93	79	66	117	131	77	316	100	91	76	76	67	61	60	55	19.05	24	28	–	–	–	–	–	–	–	–	–	–	–	1,4:5,12:6,11:7,10-Tetramethanodibenzo[b,h]biphenylene, 1,4,4a,5,5a,5b,6,6a,7,10,10a,11,11a,11b,12,12a-hexadecahydro-	17829-32-2	NI33447
316	273	231	317	243	143	274	259	316	100	78	29	26	24	21	19	19		24	28	–	–	–	–	–	–	–	–	–	–	–	trans-1-(3S,2,2-Trimethyl-1-indanylidene)-3S,2,2-trimethylindan	96144-93-3	NI33448

MASS TO CHARGE RATIOS								M.W.	INTENSITIES								Parent	C	H	O	N	S	F	Cl	Br	I	Si	P	B	X	COMPOUND NAME	CAS Reg No	No
83	64	36	85	136	56	43	48	**317**	100	85	73	70	63	55	55	50	**0.00**	8	6	4	1	1	–	3	–	–	–	–	–	–	**Benzenemethanesulphonic acid, 4-nitro-, trichloromethyl ester**	15894-01-6	**NI33449**
305	150	307	316	318	317	152	115	**317**	100	38	32	30	17	17	12	12		9	6	1	1	–	–	1	–	1	–	–	1	–	**(5-Chloro-7-iodo-8-quinolinato-O,N)borane**	83210-21-3	**NI33450**
132	77	160	104	105	76	97	129	**317**	100	91	45	33	32	32	28	25	**0.00**	10	12	3	3	2	–	–	–	–	–	1	–	–	**Phosphorodithioic acid, O,O-dimethyl S-[(4-oxo-1,2,3-benzotriazin-3(4H) yl)methyl] ester**	86-50-0	**NI33451**
157	302	159	145	185	158	58	147	**317**	100	21	8	7	5	4	4	3	**0.00**	10	12	3	3	2	–	–	–	–	–	1	–	–	**Phosphorodithioic acid, O,O-dimethyl S-[(4-oxo-1,2,3-benzotriazin-3(4H) yl)methyl] ester**	86-50-0	**NI33453**
160	132	77	93	125	104	76	147	**317**	100	84	70	44	39	37	30	28	**0.00**	10	12	3	3	2	–	–	–	–	–	1	–	–	**Phosphorodithioic acid, O,O-dimethyl S-[(4-oxo-1,2,3-benzotriazin-3(4H) yl)methyl] ester**	86-50-0	**NI33452**
243	73	91	61	317	227	302	75	**317**	100	62	54	22	21	21	15	15		10	18	3	1	1	3	–	–	–	1	–	–	–	**L-Methionine, N-(trifluoroacetyl)-, trimethylsilyl ester**	52558-92-6	**NI33454**
160	161	77	93	76	104	47	63	**317**	100	22	22	19	17	13	10	9	**3.60**	11	12	4	1	2	–	–	–	–	–	1	–	–	**Phosphorodithioic acid, S-[(1,3-dihydro-1,3-dioxo-2H-isoindol-2-yl)methyl] O,O-dimethyl ester**	732-11-6	**NI33455**
157	185	177	159	225	227	145	81	**317**	100	15	13	13	4	3	3	3	**0.40**	11	12	4	1	2	–	–	–	–	–	1	–	–	**Phosphorodithioic acid, S-[(1,3-dihydro-1,3-dioxo-2H-isoindol-2-yl)methyl] O,O-dimethyl ester**	732-11-6	**NI33456**
160	61	76	77	133	104	15	161	**317**	100	18	17	16	15	15	14	13	**0.00**	11	12	4	1	2	–	–	–	–	–	1	–	–	**Phosphorodithioic acid, S-[(1,3-dihydro-1,3-dioxo-2H-isoindol-2-yl)methyl] O,O-dimethyl ester**	732-11-6	**NI33457**
317	319	236	50	76	173	287	238	**317**	100	93	91	75	54	53	51	50		12	6	3	1	–	–	3	–	–	–	–	–	–	**Benzene, 1,3,5-trichloro-2-(4-nitrophenoxy)-**	1836-77-7	**NI33458**
50	30	63	76	64	74	75	38	**317**	100	97	56	53	42	41	39	29	**18.21**	12	6	3	1	–	–	3	–	–	–	–	–	–	**Benzene, 1,3,5-trichloro-2-(4-nitrophenoxy)-**	1836-77-7	**NI33459**
148	120	317	105	77	69	91	79	**317**	100	77	45	41	38	26	26	17		12	10	1	1	–	7	–	–	–	–	–	–	–	**Butanamide, 2,2,3,3,4,4,4-heptafluoro-N-(2,4-dimethylphenyl)-**	90791-28-9	**NI33460**
148	317	91	77	120	105	79	104	**317**	100	29	20	20	17	14	14	13		12	10	1	1	–	7	–	–	–	–	–	–	–	**Butanamide, 2,2,3,3,4,4,4-heptafluoro-N-(2,6-dimethylphenyl)-**	101948-89-4	**NI33461**
104	91	105	169	65	69	226	92	**317**	100	47	19	8	7	6	5	5	**0.60**	12	10	1	1	–	7	–	–	–	–	–	–	–	**Butanamide, 2,2,3,3,4,4,4-heptafluoro-N-(2-phenylethyl)-**	29723-29-3	**NI33462**
104	91	105	69	65	28	169	77	**317**	100	52	21	12	11	11	7	5	**1.10**	12	10	1	1	–	7	–	–	–	–	–	–	–	**Butanamide, heptafluoro-N-(2-phenylethyl)-**		**MS07476**
104	91	105	69	65	28	92	77	**317**	100	52	20	11	10	10	5	5	**1.10**	12	10	1	1	–	7	–	–	–	–	–	–	–	**Butanamide, heptafluoro-N-(2-phenylethyl)-**		**LI07034**
50	30	63	64	75	76	236	317	**317**	100	79	70	58	49	48	40	36		13	7	3	1	–	3	1	–	–	–	–	–	–	**Benzene, 2-chloro-1-(4-nitrophenoxy)-4-(trifluoromethyl)-**	42874-01-1	**NI33463**
162	281	164	70	219	175	283	43	**317**	100	89	74	74	59	43	40	39	**8.00**	13	13	4	1	–	–	2	–	–	–	–	–	–	**L-Proline, 1-[(2,4-dichlorophenoxy)acetyl]-**	50648-98-1	**NI33464**
73	84	43	98	56	99	133	75	**317**	100	47	43	42	37	33	25	22	**0.00**	13	23	6	1	–	–	–	–	–	1	–	–	–	**Acetamide, N-[1-methoxy-2-oxo-7-[(trimethylsilyl)oxy]-3,9-dioxabicyclo[3.3.1]non-6-yl]-**	55622-39-4	**NI33465**
288	317	290	282	319	274	255	302	**317**	100	60	33	26	24	22	22	18		13	24	–	5	1	–	1	–	–	–	–	–	–	**6-(2-Chloroethylthio)-N,N,N',N'-tetramethyl-1,3,5-triazine-2,4-diamine**		**NI33466**
29	152	28	27	42	18	109	138	**317**	100	49	47	40	32	25	22	21	**5.20**	13	24	4	3	–	–	–	–	–	–	1	–	–	**Phosphoric acid, 2-(diethylamino)-6-methyl-4-pyrimidinyl diethyl ester**	36378-61-7	**NI33467**
198	188	129	67	317	93	77	119	**317**	100	69	65	64	56	47	41	25		14	14	2	1	1	3	–	–	–	–	–	–	–	**2-Oxo-N-phenyl-3-[(trifluoromethyl)thio]cyclohexanecarboxamide**		**DD01204**
43	162	91	131	69	15	65	103	**317**	100	60	44	23	13	13	10	9	**0.10**	14	14	4	1	–	3	–	–	–	–	–	–	–	**L-Alanine, N-acetyl-3-phenyl-N-(trifluoroacetyl)-, methyl ester**	33858-09-2	**NI33468**
83	84	43	249	41	60	139	134	**317**	100	85	83	82	52	45	41	32	**0.00**	14	19	2	7	–	–	–	–	–	–	–	–	–	**1H-Pyrrole-2-carboxamide, 4-[[(5-amino-3,4-dihydro-2H-pyrrol-2-yl)carbonyl]amino]-N-(3-amino-3-imino-1-propenyl)-1-methyl-**	37913-78-3	**NI33469**
198	257	81	139	127	199	165	166	**317**	100	33	33	17	15	14	14	10	**0.00**	14	23	5	1	1	–	–	–	–	–	–	–	–	**N-Acetyl-S-(trans-2-acetoxycyclohexyl)-L-cysteine methyl ester**		**NI33470**
145	73	75	130	147	69	41	55	**317**	100	80	68	49	44	39	21	20	**2.88**	14	31	3	1	–	–	–	–	–	2	–	–	–	**7-Ketooctanoic acid oxime, bis(trimethylsilyl)-**		**NI33471**
77	51	159	236	317	89	78	240	**317**	100	45	43	40	29	18	18	17		15	11	–	1	1	–	–	–	–	–	–	–	1	**(2-Selenyl-3-benzo[b]thienylidene)aniline**	92886-21-0	**NI33472**
318	319	244	216	320	262	204		**317**	100	16	3	3	1	1	1		**0.00**	15	18	3	1	–	3	–	–	–	–	–	–	–	**L-Phenylalanine, N-(trifluoroacetyl)-, butyl ester**	52574-47-7	**NI33474**
91	29	148	41	27	69	103	57	**317**	100	55	40	35	19	17	16	14	**0.00**	15	18	3	1	–	3	–	–	–	–	–	–	–	**L-Phenylalanine, N-(trifluoroacetyl)-, butyl ester**	52574-47-7	**NI33473**
91	176	131	65	103	77	92	148	**317**	100	42	21	14	13	12	11	10	**0.00**	15	18	3	1	–	3	–	–	–	–	–	–	–	**L-Phenylalanine, N-(trifluoroacetyl)-, sec-butyl ester**	58072-44-9	**NI33475**
87	43	202	101	230	42	116	203	**317**	100	62	46	12	11	7	6	5	**0.10**	15	27	6	1	–	–	–	–	–	–	–	–	–	**Tris(2-acetoxypropyl)-amine**		**IC03899**
162	77	65	91	63	92	317	163	**317**	100	21	20	17	16	13	13	13		16	15	4	1	1	–	–	–	–	–	–	–	–	**2,3-Dihydro-5-methoxy-1-(4-methylphenylsulphonyl)-3-oxoindole**		**NI33476**
115	168	253	141	140	139	197	41	**317**	100	99	95	94	82	77	76	76	**55.67**	16	15	4	1	1	–	–	–	–	–	–	–	–	**2-Pyrrolidinone, 1-(5,6-dihydronaphth[2,1-c][1,2]oxathiin-2-yl)-, S,S-dioxide**	40535-16-8	**MS07477**
59	51	162	77	106	91	163	65	**317**	100	68	64	21	14	12	9	9	**1.00**	16	19	2	3	1	–	–	–	–	–	–	–	–	**N(2),N(2)-Dimethyl-N(1)-tosylformohydrazide phenylimide**		**MS07478**
162	121	77	42	106	163	91	92	**317**	100	77	46	26	23	13	13	11	**7.00**	16	19	2	3	1	–	–	–	–	–	–	–	–	**N(2)-Methyl-N(2)-phenyl-N(1)-tosylformohydrazide methylimide**		**MS07479**
200	201	244	202	317	185	199	186	**317**	100	35	21	18	17	9	8	8		17	19	5	1	–	–	–	–	–	–	–	–	–	**β-O-Acetyldubinidine**		**LI07035**
268	285	227	209	284	224	256	226	**317**	100	84	37	23	12	12	11	7	**1.00**	17	19	5	1	–	–	–	–	–	–	–	–	–	**Crinan-6,11-diol, 1,2-didehydro-3-methoxy-, (3α,5α,6β,11R,13β,19α)-**	545-66-4	**NI33477**
268	284	227	269	43	209	285	44	**317**	100	54	17	13	12	10	9	7	**1.00**	17	19	5	1	–	–	–	–	–	–	–	–	–	**Crinan-6,11-diol, 1,2-didehydro-3-methoxy-, (3α,5α,6β,11R,13β,19α)-**	545-66-4	**NI33478**
227	226	225	258	257	284	228	199	**317**	100	82	60	40	30	27	25	24	**20.00**	17	19	5	1	–	–	–	–	–	–	–	–	–	**Crinan-6,11-diol, 1,2-didehydro-3-methoxy-, (3β,5α,6β,11R,13β,19α)-**	466-73-9	**NI33479**
227	317	284	43	268	225	41	45	**317**	100	81	52	47	38	36	36	33		17	19	5	1	–	–	–	–	–	–	–	–	–	**Crinan-6,11-diol, 1,2-didehydro-3-methoxy-, (3β,5α,6β,11R,13β,19α)-**	466-73-9	**NI33480**
227	226	225	258	268	257	284	199	**317**	100	83	58	38	34	28	26	26	**21.00**	17	19	5	1	–	–	–	–	–	–	–	–	–	**Crinan-6,11-diol, 1,2-didehydro-3-methoxy-, (3β,5α,6β,11R,13β,19α)-**	466-73-9	**LI07036**
215	59	216	317	200	43	172	44	**317**	100	36	32	24	18	17	8	8		17	19	5	1	–	–	–	–	–	–	–	–	–	**7-(2,3-Dihydroxy-3-methylbutoxy)-4-methoxyfuro[2,3-b]quinoline**		**NI33481**
83	96	82	317	97	84	57	42	**317**	100	52	25	16	7	7	6	5		17	19	5	1	–	–	–	–	–	–	–	–	–	**Lycorenan-7-one, 4,12-dihydro-5-hydroxy-1-methyl-9,10-[methylenebis(oxy)]-, (5α,12β)-**	19504-94-0	**NI33482**
83	96	42	82	41	43	57	55	**317**	100	59	36	30	22	21	16	16	**9.00**	17	19	5	1	–	–	–	–	–	–	–	–	–	**Lycorenan-7-one, 4,12-dihydro-5-hydroxy-1-methyl-9,10-[methylenebis(oxy)]-, (5α,12β,13β)-**		**LI07037**
268	299	212	214	242	250	317	224	**317**	100	54	50	48	40	30	24	24		17	19	5	1	–	–	–	–	–	–	–	–	–	**2-Methoxy-9,10-dioxolopluviine**		**LI07038**

MASS TO CHARGE RATIOS								M.W.	INTENSITIES								Parent	C	H	O	N	S	F	Cl	Br	I	Si	P	B	X	COMPOUND NAME	CAS Reg No	No
44	180	91	41	43	165	51	45	**317**	100	89	89	75	71	54	46	46	**15.71**	17	20	3	1	–	–	–	–	–	–	1	–	–	1,3,2-Dioxaphosphorinan-2-amine, N,N-dimethyl-5,5-diphenyl-, 2-oxide	31951-88-9	MS07480
73	317	45	318	74	319	75	43	**317**	100	50	22	14	8	5	5	4		17	27	1	1	–	–	–	–	–	2	–	–	–	1H-Indole, 1-(trimethylsilyl)-3-[2-[(trimethylsilyl)oxy]-1-propenyl]-	74367-61-6	NI33483
73	202	75	317	45	318	171	203	**317**	100	58	22	19	17	11	9	8		17	27	1	1	–	–	–	–	–	2	–	–	–	1H-Indole, 1-(trimethylsilyl)-3-[2-[(trimethylsilyl)oxy]-2-propenyl]-	74367-62-7	NI33484
73	318	231	304	319	288	232	289	**317**	100	93	70	26	25	17	14	11	**0.00**	17	27	1	1	–	–	–	–	–	2	–	–	–	1H-Indole, 1-(trimethylsilyl)-3-[2-[(trimethylsilyl)oxy]-2-propenyl]-	74367-62-7	NI33485
240	317	104	213	77	316	315	89	**317**	100	71	50	45	44	32	32	28		18	15	1	5	–	–	–	–	–	–	–	–	–	6,7-Diphenyl-5,6-dihydropterin		MS07481
315	43	104	195	314	77	64	103	**317**	100	95	75	50	41	36	36	32	**0.00**	18	15	1	5	–	–	–	–	–	–	–	–	–	6,7-Diphenyl-7,8-dihydropterin		MS07482
282	317	283	152	319	153	58	165	**317**	100	48	26	19	18	17	17	16		18	20	2	1	–	–	1	–	–	–	–	–	–	Hasubanan, 6-chloro-7,8-didehydro-4,5-epoxy-3-methoxy-17-methyl-, (5α,6β,13β,14β)-	2670-51-1	NI33486
282	317	284	42	319	152	58	281	**317**	100	48	25	19	18	16	16	14		18	20	2	1	–	–	1	–	–	–	–	–	–	Hasubanan, 6-chloro-7,8-didehydro-4,5-epoxy-3-methoxy-17-methyl-, (5α,6β,13β,14β)-	2670-51-1	NI33487
282	317	284	42	319	152	58	252	**317**	100	47	23	19	16	15	14	12		18	20	2	1	–	–	1	–	–	–	–	–	–	Hasubanan, 6-chloro-7,8-didehydro-4,5-epoxy-3-methoxy-17-methyl-, (5α,6β,13β,14β)-	2670-51-1	LI07039
229	286	230	285	202	228			**317**	100	93	72	63	59	55			**0.00**	18	23	4	1	–	–	–	–	–	–	–	–	–	Anhydroperforine		LI07040
316	243	286	318	300	242	317	229	**317**	100	81	72	66	54	54	45	28		18	23	4	1	–	–	–	–	–	–	–	–	–	Galanthan-1-ol, 3,12-didehydro-2,9,10-trimethoxy-, (1α,2β)-	517-78-2	NI33488
317	300	299	242	243	232	228	316	**317**	100	82	55	43	42	39	36	35		18	23	4	1	–	–	–	–	–	–	–	–	–	Hasubanan-6,9-diol, 4,5-epoxy-3-methoxy-17-methyl-, (5α,6α,9α,13β,14β)-	57237-96-4	NI33489
260	261	317	230	318	274	262	258	**317**	100	87	67	23	12	12	12	10		18	23	4	1	–	–	–	–	–	–	–	–	–	6α-Hydroxy-15,16-dimethoxy-cis-erythrinan-8-one		LI07041
260	261	317	258	230	318	299	274	**317**	100	79	70	18	15	13	13	13		18	23	4	1	–	–	–	–	–	–	–	–	–	6β-Hydroxy-15,16-dimethoxy-trans-erythrinan-8-one		LI07042
192	174	56	41	55	70	69	43	**317**	100	55	38	35	32	30	29	28	**0.00**	18	23	4	1	–	–	–	–	–	–	–	–	–	5-Isoindolinecarboxylic acid, 1,3-dioxo-, nonyl ester	33975-34-7	NI33490
192	174	55	41	54	70	69	43	**317**	100	55	39	36	33	31	30	30	**3.00**	18	23	4	1	–	–	–	–	–	–	–	–	–	5-Isoindolinecarboxylic acid, 1,3-dioxo-, nonyl ester	33975-34-7	NI33491
243	242	317	268	244	316	283	266	**317**	100	80	36	28	28	26	20	20		18	23	4	1	–	–	–	–	–	–	–	–	–	2-Methoxy-pluviine		LI07043
207	300	317	206	260	301	261	249	**317**	100	95	89	79	48	42	38	38		18	23	4	1	–	–	–	–	–	–	–	–	–	1,2-(1′-Spirocyclopentyl-3′-oxotrimethylene)-1-hydroxy-6′7-dimethoxy-1,2,3,4-tetrahydroisoquinoline		LI07044
317	201	174	202	203	175	259	176	**317**	100	43	27	24	20	20	16	14		19	11	4	1	–	–	–	–	–	–	–	–	–	[1,3]Benzodioxolo[5,6-c]-1,3-dioxolo[4,5-i]phenanthridine	522-30-5	NI33492
143	276	117	89	317	76	105	77	**317**	100	60	38	32	29	23	22	20		19	15	2	3	–	–	–	–	–	–	–	–	–	2-Methyl-3-(3-methyl-4-phenyl-5-isoxazolyl)-4(3H)-quinazolinone	86134-25-0	NI33493
219	317	276	90	220	91	76	318	**317**	100	65	33	27	20	20	17	15		19	15	2	3	–	–	–	–	–	–	–	–	–	2-(4-Tolyl)-3-(3-methyl-5-isoxazolyl)-4(3H)-quinazolinone	90059-43-1	NI33494
88	72	57	245	41	189	157	317	**317**	100	61	39	12	9	8	6	5		19	27	1	1	1	–	–	–	–	–	–	–	–	O-[4-tert-Butyl-2,6-bis(2-propenyloxy)phenyl] N,N-dimethylthiocarbamate		DD01205
72	57	245	189	41	157	317	115	**317**	100	49	10	9	9	6	4	4		19	27	1	1	1	–	–	–	–	–	–	–	–	S-[4-tert-Butyl-2,6-bis(2-propenyloxy)phenyl] N,N-dimethylthiocarbamate		DD01206
106	59	94	79	43	152	137	105	**317**	100	96	86	75	51	47	46	41	**0.00**	19	27	3	1	–	–	–	–	–	–	–	–	–	N-[(R)-1-Phenylethyl]-O-[(3R,4R)-8-hydroxy-1-p-menthen-3-yl]urethane		DD01207
195	105	77	196	167	51	106	317	**317**	100	71	40	18	9	9	7	6		20	15	3	1	–	–	–	–	–	–	–	–	–	N-Benzoyldiphenylamine-2-carboxylic acid		NI33495
105	77	195	117	119	121	106	51	**317**	100	63	56	42	40	13	13	11	**11.00**	20	15	3	1	–	–	–	–	–	–	–	–	–	N-Benzoyldiphenylamine-4-carboxylic acid		NI33496
222	91	95	194	223	193	103	77	**317**	100	83	57	42	17	14	11	11	**3.00**	20	15	3	1	–	–	–	–	–	–	–	–	–	3,4-Diphenyl-5α-furoyl-2-isoxazoline		NI33497
95	105	214	76	106	96	103	51	**317**	100	58	22	16	7	7	5	5	**1.00**	20	15	3	1	–	–	–	–	–	–	–	–	–	3-Phenyl-4α-furoyl-5-phenyl-2-isoxazoline		NI33498
317	125	137	153	151	286	318	138	**317**	100	94	75	69	37	33	22	21		20	31	2	1	–	–	–	–	–	–	–	–	–	Androst-4-en-3-one, 17-hydroxy-, O-methyloxime, (17β)-	3091-89-2	NI33499
84	32	219	43	317	98	234	151	**317**	100	82	52	47	42	40	38	38		20	35	–	3	–	–	–	–	–	–	–	–	–	Ormosanine, (18α)-	33792-80-2	MS07483
317	165	166	152	104	213	150	211	**317**	100	34	33	27	18	14	13	11		21	17	–	–	–	–	–	–	–	–	–	–	1	Titanium, (η^8-cyclooctatetraene)(η^5-fluorenyl)-	55672-85-0	NI33500
212	136	284	317	109	193	115	213	**317**	100	90	44	44	35	35	31	29		21	19	–	1	1	–	–	–	–	–	–	–	–	2,4-Diphenyl-2,3,4,5-tetrahydro-1,5-benzothiazepine	78031-25-1	NI33501
135	183	118	105	119	77	136	317	**317**	100	28	22	22	21	17	10	7		21	19	2	1	–	–	–	–	–	–	–	–	–	2-(2-Hydroxy-2,2-diphenyl-ethyl)-benzamide		MS07484
135	183	118	119	317	91	90	77	**317**	100	28	22	21	7	6	4	4		21	19	2	1	–	–	–	–	–	–	–	–	–	2-(2-Hydroxy-2,2-diphenyl-ethyl)-benzamide		LI07045
60	43	41	108	95	81	67	55	**317**	100	19	11	9	9	9	9	9	**3.70**	21	35	1	1	–	–	–	–	–	–	–	–	–	Acetamide, N-[(3α,5α)-androstan-3-yl]-	54411-43-7	NI33502
60	95	81	258	108	67	93	55	**317**	100	95	86	84	81	75	67	67	**29.31**	21	35	1	1	–	–	–	–	–	–	–	–	–	Acetamide, N-[(3α,5α)-androstan-3-yl]-	54411-43-7	NI33503
60	95	107	81	108	67	55	93	**317**	100	95	85	85	73	72	67	64	**36.20**	21	35	1	1	–	–	–	–	–	–	–	–	–	Acetamide, N-[(3β,5α)-androstan-3-yl]-	4642-61-9	NI33504
60	317	108	95	243	258	81	67	**317**	100	25	18	18	17	15	15	13		21	35	1	1	–	–	–	–	–	–	–	–	–	Acetamide, N-[(3β,5α)-androstan-3-yl]-	4642-61-9	NI33505
60	317	258	243	318	108	259	204	**317**	100	53	34	15	13	8	7	6		21	35	1	1	–	–	–	–	–	–	–	–	–	Acetamide, N-[(3β,5α)-androstan-3-yl]-	4642-61-9	NI33506
99	140	95	81	58	55	41	301	**317**	100	38	25	20	20	20	20	19	**15.00**	21	35	1	1	–	–	–	–	–	–	–	–	–	Androstan-3-one, 4,4-dimethyl-, oxime, (5α)-	56052-62-1	NI33507
41	112	55	317	81	67	246	79	**317**	100	99	97	70	65	59	57	55		21	35	1	1	–	–	–	–	–	–	–	–	–	5α-Pregnan-3-one, oxime	6845-59-6	LI07046
41	112	55	28	317	81	67	246	**317**	100	99	96	78	70	65	59	48		21	35	1	1	–	–	–	–	–	–	–	–	–	5α-Pregnan-3-one, oxime	6845-59-6	NI33509
41	112	55	317	81	67	246	69	**317**	100	99	96	69	65	59	47	45		21	35	1	1	–	–	–	–	–	–	–	–	–	5α-Pregnan-3-one, oxime	6845-59-6	NI33508
44	45	43	66	55	78	69	41	**317**	100	7	7	6	6	5	5	5	**0.00**	21	35	1	1	–	–	–	–	–	–	–	–	–	Pregn-5-en-3-ol, 20-amino-, (3β,20S)-	5035-10-9	NI33510
105	86	67	32	127	89	134	48	**318**	100	68	46	27	23	20	12	12	**0.00**	–	–	1	–	1	10	–	–	–	–	–	–	1	Pentafluoroseleniumoxypentafluorosulphur		LI07047
239	189	318	158	220	249	69		**318**	100	10	8	7	5	3	1			2	–	–	–	–	3	–	3	–	–	–	–	–	Ethane, 1,1,1-tribromo-2,2,2-trifluoro-	354-48-3	NI33511
69	249	70	234	46	51	146	50	**318**	100	78	69	32	29	18	13	6	**0.00**	3	–	–	2	2	10	–	–	–	–	–	–	–	N,N′-Hexafluoroisopropylidene-bis-(sulphurdifluoridimide)		LI07048

MASS TO CHARGE RATIOS								M.W.	INTENSITIES								Parent	C	H	O	N	S	F	Cl	Br	I	Si	P	B	X	COMPOUND NAME	CAS Reg No	No
85	167	169	151	87	101	153	163	**318**	100	46	42	35	33	28	23	21	**0.00**	4	–	–	–	–	5	5	–	–	–	–	–	–	1,1,2,3,4-Pentachloro-1,2,3,4,4-pentafluorobutane		DO01097
85	101	151	171	103	169	163	87	**318**	100	62	48	36	35	33	30	29	**0.00**	4	–	–	–	–	5	5	–	–	–	–	–	–	1,1,2,3,4-Pentachloro-1,2,3,4,4-pentafluorobutane		IC03900
93	178	109	143	31	147	180	29	**318**	100	27	27	20	15	11	9	7	**0.00**	4	–	–	–	–	5	5	–	–	–	–	–	–	Pentafluoropentachlorobutane		IC03901
241	81	80	243	239	37	320	44	**318**	100	95	90	57	55	53	47	45	**17.02**	4	1	–	–	1	–	–	3	–	–	–	–	–	Thiophene, 2,3,5-tribromo-	3141-24-0	NI33512
105	318	103	89	109	249	124	69	**318**	100	43	40	20	19	14	11	9		4	6	–	–	–	6	–	–	–	–	–	–	2	1,1-Bis(trifluoromethyl)-2,2-dimethyl-diarsine		LI07049
318	261	289	260	290	131	130	29	**318**	100	63	61	59	35	17	15	15		4	10	–	–	–	–	–	–	–	–	–	–	2	Tetraethyl-ditelluride		LI07050
283	219	75	285	36	221	120	99	**318**	100	69	54	35	34	25	25	19	**5.00**	7	4	6	–	2	–	2	–	–	–	–	–	–	3,5-Bis(chlorosulphonyl)-benzoic acid		IC03902
57	56	82	69	114	71	318	93	**318**	100	94	74	71	61	57	54	47		8	–	–	4	3	–	2	–	–	–	–	–	–	Bis(3-chloro-4-cyanoisothiazol-5-yl)sulphide		LI07051
66	253	208	273	223	65	238	132	**318**	100	99	99	96	37	11	8	5	**4.00**	8	14	–	–	–	–	–	–	–	–	–	–	1	Cyclopentadienyldimethyllead		MS07485
28	80	51	26	52	50	27	151	**318**	100	18	12	9	7	6	2	1	**0.14**	9	4	5	2	–	–	–	–	–	–	–	–	1	Molybdenum, pentacarbonyl(N1-pyridazine)-	67049-36-9	NI33513
57	69	74	43	93	45	219	119	**318**	100	60	26	25	23	18	10	8	**0.00**	9	7	2	–	–	9	–	–	–	–	–	–	–	1,2-Epoxy-3-(1-methylenenonafluoropentyloxy)propane		DD01208
119	117	115	121	118	123	43	200	**318**	100	76	56	40	29	24	21	20	**6.30**	9	9	–	–	1	5	–	–	–	–	–	–	1	Germane, trimethyl[(pentafluorophenyl)thio]-	22529-99-3	NI33514
91	155	139	69	318	65	92	64	**318**	100	94	51	29	28	20	8	8		9	9	5	–	2	3	–	–	–	–	–	–	–	Methanesulphonic acid, trifluoro-, [(4-methylphenyl)sulphonyl]methyl ester	37891-93-3	NI33515
265	303	280	119	43	262	231	123	**318**	100	94	87	26	25	24	23	22	**0.00**	9	10	1	–	–	3	–	–	1	–	–	–	–	1,1,1-Trifluoro-3,3-dimethyl-3H-1,2-benziodoxole		MS07486
167	100	226	154	152	211	123	140	**318**	100	70	42	24	24	19	18	17	**0.00**	11	14	7	2	1	–	–	–	–	–	–	–	–	Methyl 7-methoxycarbonylamino-3-methyl-3-cephem-4-carboxylate 1,1-dioxide	97609-87-5	NI33516
285	200	86	318	232	180	275	261	**318**	100	63	53	40	35	27	24	22		11	15	5	2	1	–	–	–	–	–	1	–	–	Phosphonothioic acid, morpholino-, O-methyl O-(4-nitrophenyl) ester	13351-40-1	NI33517
145	147	109	74	173	75	175	111	**318**	100	64	45	29	28	27	18	15	**7.54**	12	6	–	2	–	–	4	–	–	–	–	–	–	Diazene, bis(3,4-dichlorophenyl)-	14047-09-7	NI33518
54	67	80	39	41	79	66	77	**318**	100	85	44	42	21	20	20	19	**0.50**	12	12	4	–	–	–	–	–	–	–	–	–	1	Molybdenum, tetracarbonyl-π-1,5-cyclooctadienyl-		MS07487
91	153	202	168	105	204	139	251	**318**	100	57	49	46	40	35	31	13	**0.00**	12	18	1	–	–	–	4	–	–	–	–	–	–	(1RS,2SR,4SR,5SR,6SR,7RS,8RS)-5,6,7,8-Tetrakis(chloromethyl)bicyclo[2.2.2]octan-2-ol		DD01209
142	155	126	113	102	184	259	212	**318**	100	22	20	20	20	18	16	13	**6.00**	12	18	8	2	–	–	–	–	–	–	–	–	–	2,5-Dimethyl-3a,4a,6a-tris(methoxycarbonyl)-perhydroisoxazol[4,5-d]isoxazole		LI07052
225	153	59	193	167	257	197	226	**318**	100	50	34	32	23	15	15	11	**0.00**	12	18	8	2	–	–	–	–	–	–	–	–	–	2,3,5,6-Tetrakis(methoxycarbonyl)piperidine		NI33519
73	147	157	129	75	45	303	74	**318**	100	35	31	26	22	15	13	9	**0.00**	12	22	6	–	–	–	–	–	–	2	–	–	–	Isocitric lactone, bis(trimethylsilyl)-,	65143-62-6	NI33520
143	73	110	75	147	200	44	185	**318**	100	93	78	63	49	44	31	30	**0.00**	12	27	5	–	–	–	–	–	–	2	–	1	–	β-D-Xylofuranose, 1,2-bis-O-(trimethylsilyl)-, cyclic methylboronate	56196-82-8	NI33521
73	100	45	75	116	74	43	89	**318**	100	35	23	21	19	14	13	12	**0.00**	12	30	2	2	–	–	–	–	–	3	–	–	–	3-Isoxazolidinone, 4-[bis(trimethylsilyl)amino]-2-(trimethylsilyl)-	55517-42-5	NI33522
283	285	319	320	317	318	321	142	**318**	100	97	89	77	68	60	51	43		13	6	1	–	–	–	4	–	–	–	–	–	–	Tetrachloroxanthene		NI33523
318	139	137	63	39	135	65	79	**318**	100	45	36	33	30	29	25	23		13	10	6	4	–	–	–	–	–	–	–	–	–	Benzaldehyde, 2,4-dihydroxy-, (2,4-dinitrophenyl)hydrazone	1237-66-7	NI33524
216	234	55	144	172	318	174	72	**318**	100	81	55	35	30	20	20	20		13	11	6	–	–	–	–	–	–	–	–	–	1	4-Carboxybutanoyl-cyclopentadienylmanganese tricarbonyl		LI07053
51	84	56	93	67	77	111	160	**318**	100	79	63	59	55	40	29	24	**0.00**	13	16	1	–	–	–	–	–	–	–	–	–	1	8-(Phenyltelluro)-2-oxabicyclo[3.3.0]octane		DD01210
143	43	156	103	97	128	114	86	**318**	100	99	96	75	72	67	63	62	**0.00**	13	18	9	–	–	–	–	–	–	–	–	–	–	Acetyl 2,3,5-tri-O-acetyl-D-xylofuranoside	61248-15-5	NI33525
128	43	170	115	157	103	86	69	**318**	100	99	58	54	38	35	29	27	**0.00**	13	18	9	–	–	–	–	–	–	–	–	–	–	Acetyl 2,3,4-tri-O-acetyl-β-D-xylopyranoside	4049-33-6	NI33526
45	85	44	68	115	69	73	86	**318**	100	95	90	57	43	41	36	35	**0.00**	13	18	9	–	–	–	–	–	–	–	–	–	–	D-Arabinose, 2,3,4,5-tetraacetate	3891-58-5	NI33527
128	115	170	69	86	103	157	68	**318**	100	65	59	46	41	38	34	28	**0.00**	13	18	9	–	–	–	–	–	–	–	–	–	–	β-D-Ribopyranose, tetraacetate	4049-34-7	NI33528
43	156	103	143	97	69	145	84	**318**	100	96	85	77	69	59	53	53	**0.00**	13	18	9	–	–	–	–	–	–	–	–	–	–	D-Xylofuranose, tetraacetate	42927-46-8	NI33529
43	128	170	115	157	69	97	103	**318**	100	96	72	71	52	52	48	41	**0.00**	13	18	9	–	–	–	–	–	–	–	–	–	–	α-D-Xylopyranose, tetraacetate	4257-98-1	NI33530
128	43	170	115	157	97	103	86	**318**	100	99	75	61	48	41	34	34	**0.00**	13	18	9	–	–	–	–	–	–	–	–	–	–	α-D-Xylopyranose, tetraacetate	4257-98-1	NI33531
73	173	75	111	185	303	55	218	**318**	100	77	65	52	45	24	20	14	**0.00**	13	26	5	–	–	–	–	–	–	2	–	–	–	Heptanedioic acid, 4-oxo-, bis(trimethylsilyl) ester	55517-44-7	NI33532
235	237	165	236	199	239	238	88	**318**	100	64	38	15	12	11	9	9	**0.00**	14	10	–	–	–	–	4	–	–	–	–	–	–	Benzene, 1-chloro-2-[2,2-dichloro-1-(4-chlorophenyl)ethyl]-	53-19-0	AI02035
235	237	165	236	200	199	88	176	**318**	100	69	41	17	15	15	14	12	**3.06**	14	10	–	–	–	–	4	–	–	–	–	–	–	Benzene, 1-chloro-2-[2,2-dichloro-1-(4-chlorophenyl)ethyl]-	53-19-0	NI33534
235	237	165	199	75	236	176	200	**318**	100	67	58	23	21	16	16	15	**4.90**	14	10	–	–	–	–	4	–	–	–	–	–	–	Benzene, 1-chloro-2-[2,2-dichloro-1-(4-chlorophenyl)ethyl]-	53-19-0	NI33533
235	237	165	236	239	199	75	82	**318**	100	67	40	16	12	12	10	9	**3.07**	14	10	–	–	–	–	4	–	–	–	–	–	–	Benzene, 1-chloro-3-[2,2-dichloro-1-(4-chlorophenyl)ethyl]-	4329-12-8	NI33535
235	237	165	236	239	238	199	178	**318**	100	64	40	16	11	10	10	10	**0.00**	14	10	–	–	–	–	4	–	–	–	–	–	–	Benzene, 1,1′-(2,2-dichloroethylidene)bis[4-chloro-	72-54-8	AI02036
237	165	178	75	199	236	212	176	**318**	100	86	30	27	24	23	19	18	**1.79**	14	10	–	–	–	–	4	–	–	–	–	–	–	Benzene, 1,1′-(2,2-dichloroethylidene)bis[4-chloro-	72-54-8	NI33536
235	237	165	75	199	236	51	50	**318**	100	65	59	22	19	15	12	12	**3.70**	14	10	–	–	–	–	4	–	–	–	–	–	–	Benzene, 1,1′-(2,2-dichloroethylidene)bis[4-chloro-	72-54-8	NI33537
78	103	89	90	76	63	50	51	**318**	100	86	67	55	48	45	36	33	**1.00**	14	10	5	2	1	–	–	–	–	–	–	–	–	1,2-Benzisothiazole, 3-[(4-nitrophenyl)methoxy]-, 1,1-dioxide	55124-69-1	LI07054
78	103	89	90	76	63	50	64	**318**	100	85	67	53	48	45	35	25	**0.80**	14	10	5	2	1	–	–	–	–	–	–	–	–	1,2-Benzisothiazole, 3-[(4-nitrophenyl)methoxy]-, 1,1-dioxide	55124-69-1	NI33538
119	93	183	320	318	185	83	155	**318**	100	52	40	39	38	38	34	24		14	11	2	2	–	–	–	1	–	–	–	–	–	Benzamide, 3-bromo-N-[(phenylamino)carbonyl]-	56437-97-9	NI33539
303	301	299	302	300	307	305	304	**318**	100	82	54	46	46	27	27	27	**0.00**	14	14	1	–	–	–	–	–	–	–	–	–	1	10,10-Dimethylphenoxastannin		LI07055
134	162	56	190	28	290	262	91	**318**	100	58	50	16	15	11	11	11	**0.50**	14	14	5	–	–	–	–	–	–	–	–	–	1	Iron, tricarbonyl[(4,5,6,7-η)-3a,3b,7a,7b-tetrahydro-2,2-dimethylbenzo[3,4]cyclobuta[1,2-d]-1,3-dioxole]-	63444-27-9	NI33540
168	318	66	194	103	289	142	287	**318**	100	71	51	43	42	30	29	27		14	15	2	–	–	–	–	–	–	–	–	–	1	[eta[4]-2-(Hydroxymethyl)norbornadiene](η^5-formylcyclopentadienyl)rhodium	77906-36-6	NI33541
168	318	259	317	180	182	303	142	**318**	100	85	50	25	21	20	17	17		14	15	2	–	–	–	–	–	–	–	–	–	1	(eta[4]-2-Methoxycarbonylnorbornadiene)(η^5-cyclopentadienyl)rhodium	63533-26-6	NI33542

MASS TO CHARGE RATIOS	M.W.	INTENSITIES	Parent	C	H	O	N	S	F	Cl	Br	I	Si	P	B	X	COMPOUND NAME	CAS Reg No	No
127 99 29 199 173 142 227 171	318	100 77 69 65 53 51 44 43	0.00	14	22	8	–	–	–	–	–	–	–	–	–	–	Tetraethyl 1,1,2,2-ethanetetracarboxylate		DO01098
157 43 203 115 213 139 273 158	318	100 62 50 19 17 13 9 8	0.00	14	22	8	–	–	–	–	–	–	–	–	–	–	Triethyl acetylcitrate		IC03903
44 86 43 114 61 104 56 87	318	100 34 27 25 25 22 20 19	2.50	14	26	4	2	1	–	–	–	–	–	–	–	–	N-(N-Acetylalanyl)methionine butyl ester		AI02037
179 149 151 177 289 233 147 121	318	100 94 93 82 73 68 61 60	0.00	14	30	–	–	–	–	–	–	–	–	–	–	1	Butyldiethylcyclohexyltin		MS07488
135 41 137 55 133 179 219 275	318	100 84 82 73 68 52 51 48	0.00	14	30	–	–	–	–	–	–	–	–	–	–	1	Methylisopropylbutylcyclohexyltin		MS07489
147 175 149 145 231 173 205 229	318	100 83 70 70 68 59 57 48	0.00	14	30	–	–	–	–	–	–	–	–	–	–	1	Tributylvinyltin		MS07490
73 147 75 204 303 83 55 45	318	100 48 42 26 25 22 22 16	0.00	14	30	4	–	–	–	–	–	–	2	–	–	–	Glutaric acid, 2-propyl-, bis(trimethylsilyl) ester		MS07491
77 73 75 47 49 55 303 93	318	100 72 53 41 30 29 17 17	0.00	14	30	4	–	–	–	–	–	–	2	–	–	–	Octanedioic acid, bis(trimethylsilyl) ester	43199-48-0	MS07492
73 75 303 55 187 169 147 129	318	100 81 40 37 32 22 20 20	0.00	14	30	4	–	–	–	–	–	–	2	–	–	–	Octanedioic acid, bis(trimethylsilyl) ester	43199-48-0	NI33544
73 75 55 303 187 147 169 117	318	100 78 38 20 20 20 17 17	0.00	14	30	4	–	–	–	–	–	–	2	–	–	–	Octanedioic acid, bis(trimethylsilyl) ester	43199-48-0	NI33543
73 75 147 204 83 69 45 303	318	100 58 41 32 21 21 18 17	0.00	14	30	4	–	–	–	–	–	–	2	–	–	–	Pentanedioic acid, 2,2,3-trimethyl-, bis(trimethylsilyl) ester	55044-26-3	NI33545
318 69 317 319 137 150 136 77	318	100 24 23 18 12 11 11 8		15	10	8	–	–	–	–	–	–	–	–	–	–	4H-1-Benzopyran-4-one, 2-(3,4-dihydroxyphenyl)-3,5,6,7-tetrahydroxy-	90-18-6	NI33546
318 69 317 319 137 150 136 289	318	100 24 23 18 12 11 11 8		15	10	8	–	–	–	–	–	–	–	–	–	–	4H-1-Benzopyran-4-one, 2-(3,4-dihydroxyphenyl)-3,5,6,7-tetrahydroxy-	90-18-6	MS07493
202 174 230 318 59 134 79 77	318	100 50 24 22 20 17 12 11		15	10	8	–	–	–	–	–	–	–	–	–	–	5-Hydroxy-9-methoxy-2-methyl-4H,6H-pyrano[3,4-g]-1-benzopyran-4,6-dione-8-carboxylic acid	93752-78-4	NI33547
183 185 105 77 51 155 157 76	318	100 99 88 45 18 14 13 12	5.00	15	11	3	–	–	–	–	1	–	–	–	–	–	Acetophenone, 4′-bromo-2-hydroxy-, benzoate	7506-12-9	NI33548
105 77 118 185 183 106 51 76	318	100 20 13 12 12 10 8 5	3.00	15	11	3	–	–	–	–	1	–	–	–	–	–	Benzoic acid, 2-bromo-, 2-oxo-2-phenylethyl ester	55153-28-1	NI33549
105 77 51 118 76 106 75 50	318	100 24 10 9 9 8 7 7	3.00	15	11	3	–	–	–	–	1	–	–	–	–	–	Benzoic acid, 3-bromo-, 2-oxo-2-phenylethyl ester	55153-27-0	NI33550
105 77 106 185 183 76 51 75	318	100 17 8 7 7 6 6 5	1.00	15	11	3	–	–	–	–	1	–	–	–	–	–	Benzoic acid, 4-bromo-, 2-oxo-2-phenylethyl ester	55153-26-9	NI33551
86 318 42 36 58 30 29 320	318	100 88 75 52 42 41 35 30		15	11	4	2	–	–	1	–	–	–	–	–	–	2,3-Dimethoxy-6-nitro-9-chloroacridine	6628-92-8	NI33552
105 276 106 51 43 277 246 78	318	100 31 9 7 7 5 5 4	0.00	15	14	4	2	1	–	–	–	–	–	–	–	–	Benzenesulphonic acid, (α-hydroxybenzylidene)hydrazide, acetate	1666-15-5	NI33553
239 318 165 179 209 240 319 222	318	100 90 34 24 21 17 16 16		15	14	4	2	1	–	–	–	–	–	–	–	–	N-(9,10-Dihydro-3-nitro-2-phenanthryl)methanesulphonamide		MS07494
239 318 165 193 240 166 319 209	318	100 52 21 19 17 10 9 6		15	14	4	2	1	–	–	–	–	–	–	–	–	N-(9,10-Dihydro-7-nitro-2-phenanthryl)methanesulphonamide		MS07495
318 283 91 255 300 108 209	318	100 95 71 52 47 27 25		15	14	6	2	–	–	–	–	–	–	–	–	–	Bis(2-hydroxy-3-nitro-5-methyl-phenyl)-methane		LI07056
153 151 318 165 167 163 161 273	318	100 69 46 23 20 20 16 12		15	18	–	–	–	–	–	–	–	–	–	–	1	1-(Trimethylstannyl)-1,5-dihydro-s-indacene		MS07496
41 301 318 67 79 135 55 283	318	100 81 69 69 59 58 53 51		15	18	4	4	–	–	–	–	–	–	–	–	–	1-Formylcyclooct-1-ene, 2,4-dinitrophenylhydrazone		IC03904
318 303 43 40 41 122 94 55	318	100 97 34 33 28 23 20 20		15	18	4	4	–	–	–	–	–	–	–	–	–	2,5-Heptadien-4-one, 2,6-dimethyl-, 2,4-dinitrophenylhydrazone		IC03905
166 318 165 44 153 71 42 150	318	100 68 63 36 26 23 20 17		15	22	2	6	–	–	–	–	–	–	–	–	–	Bis[2-methyl-4-hydroxy-6-(N,N-dimethylamino)pyrimidinyl]methane		IC03906
45 187 99 69 87 72 70 42	318	100 19 17 14 13 12 12 12	0.04	15	26	7	–	–	–	–	–	–	–	–	–	–	1,4,7,10,13-Pentaoxacyclocosane-14,20-dione		MS07497
131 73 75 132 69 95 74 133	318	100 63 42 15 8 7 6 6	0.12	15	34	3	–	–	–	–	–	–	2	–	–	–	7-Trimethylsiloxy-7-methyloctanoic acid, trimethylsilyl ester		NI33554
318 320 111 151 115 125 43 89	318	100 63 48 36 31 30 25 22		16	12	1	2	–	–	2	–	–	–	–	–	–	2,4-Bis(4-chlorophenyl)-5-methyl-1,2-dihydro-3H-pyrazol-3-one	80510-63-0	NI33555
290 318 317 289 239 91 291 319	318	100 88 47 29 20 17 15 10		16	12	1	2	1	2	–	–	–	–	–	–	–	7-Difluoromethylthio-2,3-dihydro-5-phenyl-1H-1,4-benzodiazepin-2-one	55384-54-8	NI33556
209 128 129 318 77 102 109 103	318	100 93 40 36 34 29 21 15		16	14	–	–	1	–	–	–	–	–	–	–	1	(E)-β-Styryl-(Z)-2-(phenylthio)-1-ethenyl selenide		DD01211
128 208 77 109 318 209 115 319	318	100 80 55 32 23 15 15 14		16	14	–	–	1	–	–	–	–	–	–	–	1	(Z)-β-Styryl-(Z)-2-(phenylthio)-1-ethenyl selenide		DD01212
124 123 150 122 107 109 94 125	318	100 98 86 81 72 68 54 53	0.00	16	14	7	–	–	–	–	–	–	–	–	–	–	Benzoic acid, 2,4-dihydroxy-6-methyl-, 4-carboxy-3-hydroxy-5-methylphenyl ester	480-56-8	NI33557
303 274 318 59 289 235 53 93	318	100 79 74 41 40 39 36 32		16	14	7	–	–	–	–	–	–	–	–	–	–	2-(Methoxycarbonyl)-4,9-dimethoxy-7-methyl-5H-furo[3,2-g][1]benzopyran-5-one		NI33558
91 73 65 131 107 105 51 106	318	100 30 20 19 19 19 19 17	0.40	16	15	2	–	–	–	–	1	–	–	–	–	–	2-Butanone, 3-bromo-4-hydroxy-1,4-diphenyl-	34737-58-1	NI33559
45 165 320 318 275 273 166 194	318	100 33 24 24 24 24 17 12		16	15	2	–	–	–	–	1	–	–	–	–	–	(E)-2-Methoxymethoxy-4′-bromostilbene		MS07498
45 165 320 318 275 273 166 194	318	100 43 30 30 29 29 21 15		16	15	2	–	–	–	–	1	–	–	–	–	–	(Z)-2-Methoxymethoxy-4′-bromostilbene		MS07499
120 121 77 200 198 42 105 104	318	100 9 8 7 7 5 4 4	4.00	16	19	–	2	–	–	–	1	–	–	–	–	–	Ethylenediamine, N-(3-bromophenyl)-N,N′-dimethyl-N′-phenyl-	32857-46-8	NI33560
120 121 77 200 198 42 320 318	318	100 9 8 7 7 5 4 4		16	19	–	2	–	–	–	1	–	–	–	–	–	Ethylenediamine, N-(3-bromophenyl)-N,N′-dimethyl-N′-phenyl-	32857-46-8	LI07057
120 121 200 198 77 320 318 42	318	100 10 9 9 9 7 7 5		16	19	–	2	–	–	–	1	–	–	–	–	–	Ethylenediamine, N-(4-bromophenyl)-N,N′-dimethyl-N′-phenyl-	32857-44-6	NI33561
120 121 200 198 77 320 318 104	318	100 10 9 9 9 7 7 5		16	19	–	2	–	–	–	1	–	–	–	–	–	Ethylenediamine, N-(4-bromophenyl)-N,N′-dimethyl-N′-phenyl-	32857-44-6	LI07058
249 247 58 72 167 248 250 168	318	100 96 72 31 24 21 16 16	0.00	16	19	–	2	–	–	–	1	–	–	–	–	–	2-Pyridinepropanamine, γ-(4-bromophenyl)-N,N-dimethyl-	86-22-6	NI33562
58 249 247 72 167 248 168 42	318	100 79 74 39 34 22 22 21	0.30	16	19	–	2	–	–	–	1	–	–	–	–	–	2-Pyridinepropanamine, γ-(4-bromophenyl)-N,N-dimethyl-, (S)-	132-21-8	NI33563
146 303 100 117 118 75 73 304	318	100 69 31 26 19 16 16 15	11.76	16	22	3	2	–	–	–	–	–	1	–	–	–	2,4,6(1H,3H,5H)-Pyrimidinetrione, 5-ethyl-1-methyl-5-phenyl-3-(trimethylsilyl)-	52937-72-1	NI33564
221 145 83 318 117 81 303 146	318	100 14 14 8 4 4 3 3		16	22	3	4	–	–	–	–	–	–	–	–	–	7-(N′-(1′-Cyclohexyl-2′-methylamino-propyl))-4-nitrobenzo-2-oxa-1,3-diazole		LI07059
73 202 203 45 75 74 43 130	318	100 30 27 25 15 10 10 9	4.30	16	26	1	2	–	–	–	–	–	2	–	–	–	1H-Indole-3-acetamide, N,1-bis(trimethylsilyl)-	72101-39-4	NI33565
129 28 69 87 145 130 98 43	318	100 39 26 18 14 12 12 11	0.00	16	30	6	–	–	–	–	–	–	–	–	–	–	Octanoic acid, 6-(4-carboxy-3-methylbutoxy)-7-hydroxy-, dimethyl ester, (–)-	30244-30-5	NI33566
83 147 69 73 55 75 97 103	318	100 76 64 60 58 54 42 32	0.00	16	38	2	–	–	–	–	–	–	2	–	–	–	1,10-Decanediol bis(trimethylsilyl) ether	18546-99-1	NI33567
83 147 69 55 97 149 73 148	318	100 67 48 46 34 21 19 15	0.00	16	38	2	–	–	–	–	–	–	2	–	–	–	1,10-Decanediol bis(trimethylsilyl) ether	18546-99-1	NI33568

MASS TO CHARGE RATIOS								M.W.	INTENSITIES								Parent	C	H	O	N	S	F	Cl	Br	I	Si	P	B	X	COMPOUND NAME	CAS Reg No	No
91	115	77	65	131	128	105	89	**318**	100	73	65	54	43	32	30	30	**0.00**	17	18	4	–	1	–	–	–	–	–	–	–	–	**(E)-2-[(4-Methylphenyl)sulphonyl]-1-phenyl-2-butene-1,4-diol**	118356-24-4	**DD01213**
243	205	204	318	189	175	161	227	**318**	100	38	31	25	17	12	8	4		17	18	6	–	–	–	–	–	–	–	–	–	–	**(+)-trans-Acetyl-khellactone methyl ether**		**LI07060**
204	205	189	318	175	161	227	243	**318**	100	75	46	40	32	27	24	15		17	18	6	–	–	–	–	–	–	–	–	–	–	**(–)-cis-Acetyl-khellactone methyl ether**		**LI07061**
227	195	183	59	168	254	286	199	**318**	100	64	34	31	30	28	24	18	**8.00**	17	18	6	–	–	–	–	–	–	–	–	–	–	**Dimethyl 3-methoxy-1,2,3,11-tetrahydrobenzofuran-1,2-dicarboxylate**		**LI07062**
230	229	318	215	231	246	217	216	**318**	100	96	70	40	33	23	23	13		17	18	6	–	–	–	–	–	–	–	–	–	–	**7H-Furo[3,2-g][1]benzopyran-7-one, 4-(2,3-dihydroxy-3-methylbutyl)-9-methoxy-**	10523-56-5	**NI33569**
233	234	318	43	85	215	219	235	**318**	100	94	44	35	30	20	15	12		17	18	6	–	–	–	–	–	–	–	–	–	–	**2,4-Pentanedione, 1-(7-acetyl-4,6-dihydroxy-3,5-dimethyl-2-benzofuranyl)-**	21402-79-9	**NI33570**
287	318	138	276	181	250	106	69	**318**	100	99	87	75	73	72	65	64		17	18	6	–	–	–	–	–	–	–	–	–	–	**Spiro[benzofuran-2(3H),1′-[2]cyclohexene]-3,4′-dione, 2′,4,6-trimethoxy-6′-methyl-**	3680-32-8	**NI33572**
318	138	287	276	181	250	69	319	**318**	100	89	55	44	44	41	26	20		17	18	6	–	–	–	–	–	–	–	–	–	–	**Spiro[benzofuran-2(3H),1′-[2]cyclohexene]-3,4′-dione, 2′,4,6-trimethoxy-6′-methyl-**	3680-32-8	**NI33571**
58	86	85	42	43	44	57	59	**318**	100	71	33	27	20	17	16	15	**3.10**	17	19	–	2	1	–	1	–	–	–	–	–	–	**10H-Phenothiazine-10-propanamine, 2-chloro-N,N-dimethyl-**	50-53-3	**NI33574**
58	86	318	272	85	320	232	42	**318**	100	26	23	12	11	10	9	6		17	19	–	2	1	–	1	–	–	–	–	–	–	**10H-Phenothiazine-10-propanamine, 2-chloro-N,N-dimethyl-**	50-53-3	**NI33573**
58	318	86	85	320	272	42	319	**318**	100	27	26	11	10	6	6	5		17	19	–	2	1	–	1	–	–	–	–	–	–	**10H-Phenothiazine-10-propanamine, 2-chloro-N,N-dimethyl-**	50-53-3	**NI33575**
43	136	125	318	58	127	57	137	**318**	100	93	73	38	35	28	24	22		17	19	2	2	–	–	1	–	–	–	–	–	–	**1-(4-Chlorobenzyl)-1,3-dimethyl-3-(4-methoxyphenyl)urea**		**NI33576**
171	318	170	172	168	42	154	127	**318**	100	24	18	14	9	7	6	6		17	22	2	2	1	–	–	–	–	–	–	–	–	**1-Dimethylaminonaphthalene-5-sulphonylpiperidine**		**LI07063**
109	108	239	318	271	122	199	143	**318**	100	97	87	64	64	51	49	39		17	22	4	2	–	–	–	–	–	–	–	–	–	**4,3′,4′,5′-Tetramethyl-3,5-bis(methoxycarbonyl)-2,2′-dipyrrolylmethane**		**LI07064**
318	120	317	319	158	134	285	240	**318**	100	28	27	20	20	19	17	12		17	23	2	2	–	–	–	–	–	–	1	–	–	**Methyl bis(4-dimethylaminophenyl)phosphinate**		**LI07065**
57	43	70	55	69	71	84	41	**318**	100	64	56	45	44	41	39	37	**0.00**	17	35	–	–	–	–	–	1	–	–	–	–	–	**1-(3-Ethyl)amyl-4-ethyloctyl bromide**		**LI07066**
135	137	57	43	71	55	85	41	**318**	100	96	69	52	49	45	38	37	**2.00**	17	35	–	–	–	–	–	1	–	–	–	–	–	**Heptadecane, 1-bromo-**	3508-00-7	**NI33577**
57	43	71	135	137	55	70	85	**318**	100	76	58	52	51	37	36	35	**0.00**	17	35	–	–	–	–	–	1	–	–	–	–	–	**Heptadecane, 1-bromo-**	3508-00-7	**IC03907**
318	319	320	159	286	287	160	321	**318**	100	23	11	6	5	3	3	2		18	10	–	2	2	–	–	–	–	–	–	–	–	**Triphenodithiazine**		**IC03908**
318	319	159	63	320	289	226	159.5	**318**	100	22	10	8	7	3	3	3		18	10	2	2	1	–	–	–	–	–	–	–	–	**2,5-Bis(benzoxazol-2-yl)thiophene**		**IC03909**
93	121	65	318	77	94	198	39	**318**	100	14	12	8	8	7	6	6		18	14	2	4	–	–	–	–	–	–	–	–	–	**1,4-Bis(4-hydroxyphenylazo)benzene**		**IC03910**
318	89	117	174	62	103	77	38	**318**	100	77	61	42	38	23	14	14		18	14	2	4	–	–	–	–	–	–	–	–	–	**N-Methyl-4-(4-methyl-5-phenyl-2H-1,2,3-triazol-2-yl)phthalimide**	91289-97-3	**NI33578**
43	41	55	56	193	69	75	103	**318**	100	95	90	85	65	60	56	47	**2.00**	18	22	5	–	–	–	–	–	–	–	–	–	–	**1,2,4-Benzenetricarboxylic acid, cyclic 1,2-anhydride, nonyl ester**	33975-30-3	**NI33579**
193	43	175	194	44	56	41	126	**318**	100	82	63	55	55	52	50	48	**10.01**	18	22	5	–	–	–	–	–	–	–	–	–	–	**1,2,4-Benzenetricarboxylic acid, cyclic 1,2-anhydride, nonyl ester**	33975-30-3	**NI33580**
188	189	318	112	161	125	55	151	**318**	100	58	52	35	28	26	23	22		18	22	5	–	–	–	–	–	–	–	–	–	–	**1H-2-Benzoxacyclotetradecin-1,7(8H)-dione, 3,4,5,6,9,10-hexahydro-14,16-dihydroxy-3-methyl-, (E)-**	7645-23-0	**NI33581**
188	54	318	56	112	189	69	125	**318**	100	90	76	67	65	52	50	42		18	22	5	–	–	–	–	–	–	–	–	–	–	**1H-2-Benzoxacyclotetradecin-1,7(8H)-dione, 3,4,5,6,9,10-hexahydro-14,16-dihydroxy-3-methyl-, [S-(E)]-**	17924-92-4	**NI33582**
152	151	105	121	77	137	318	166	**318**	100	74	44	42	40	32	19	16		18	22	5	–	–	–	–	–	–	–	–	–	–	**Bis(3,4-dimethoxybenzyl) ether**		**LI07067**
228	185	41	55	83	229	91	231	**318**	100	90	75	71	70	57	57	51	**8.00**	18	22	5	–	–	–	–	–	–	–	–	–	–	**GA_9-17-norketone**		**NI33583**
300	150	163	137	301	177	124	148	**318**	100	98	50	41	20	19	16	15	**14.00**	18	22	5	–	–	–	–	–	–	–	–	–	–	**5-(2′-Hydroxy-4′-methoxyphenyl)-7aβ-methyl-1-oxo-2,3,3aα,4,5,6, 7,7a-octahydro-1H-indene-4ξ-carboxylic acid**		**MS07500**
167	149	165	151	318	121	108	109	**318**	100	85	40	39	38	25	18	15		18	26	3	2	–	–	–	–	–	–	–	–	–	**Cypholophine**		**LI07068**
218	318	275	147	259	217	120	108	**318**	100	87	64	60	53	50	35	33		18	26	3	2	–	–	–	–	–	–	–	–	–	**Lycocernuine, 14,15-didehydro-, acetate**	36283-10-0	**NI33584**
91	227	123	79	121	77	43	103	**318**	100	50	45	42	40	36	31	25	**0.00**	19	23	2	–	–	–	1	–	–	–	–	–	–	**Butane, 1-(benzyloxy)-2-[(benzyloxy)methyl]-4-chloro-**	33498-89-4	**NI33585**
91	121	227	92	209	211	106	103	**318**	100	6	5	4	2	1	1	1	**0.10**	19	23	2	–	–	–	1	–	–	–	–	–	–	**2-(2′-Chloroethyl)-1,3-dibenzyloxy-propane**		**LI07069**
69	318	249	250	123	221	71	207	**318**	100	65	61	38	21	20	20	16		19	26	4	–	–	–	–	–	–	–	–	–	–	**1,2,4-Cyclopentanetrione, 3,3-bis(3-methyl-2-butenyl)-5-(2-methyl-1-oxopropyl)-**	1891-34-5	**NI33586**
318	319	189	176	137	231	202	175	**318**	100	26	11	10	10	8	6	6		19	26	4	–	–	–	–	–	–	–	–	–	–	**Estra-1,3,5(10)-triene-3,16,17-triol, 2-methoxy-, (16α,17β)-**	1236-72-2	**NI33587**
105	303	271	300	318				**318**	100	10	10	3	1					19	26	4	–	–	–	–	–	–	–	–	–	–	**2-(5-Methoxycarbonylpentyl)-4-methyl-4-phenyl-4-pentanalide**		**LI07070**
43	28	55	163	41	318	91	59	**318**	100	37	22	19	16	12	12	11		19	26	4	–	–	–	–	–	–	–	–	–	–	**Methyl, 1,3-dimethyl-2-(3-methyl-7-oxo-1,3-octadienyl)-4-oxo-2-cyclohexene-1-carboxylate**		**MS07501**
260	41	55	79	93	91	245	81	**318**	100	57	56	46	41	40	35	35	**0.00**	19	26	4	–	–	–	–	–	–	–	–	–	–	**19-Norkauran-18-oic acid, 2,15-dioxo-, (4α)-**	51154-88-2	**NI33588**
263	318	247	207	303	43	264	191	**318**	100	72	56	56	35	23	19	18		19	26	4	–	–	–	–	–	–	–	–	–	–	**Tetrahydrofranklinone**		**MS07502**
318	253	175	127	254	115	126	319	**318**	100	84	76	70	40	30	24	23		20	14	2	–	1	–	–	–	–	–	–	–	–	**Binaphthyl sulphone**	32390-26-4	**IC03911**
127	175	115	126	318	253	147	107	**318**	100	85	63	46	33	24	23	21		20	14	2	–	1	–	–	–	–	–	–	–	–	**Binaphthyl sulphone**	32390-26-4	**NI33589**
105	77	51	106	52	318	78	27	**318**	100	40	9	8	3	3	3	2		20	14	4	–	–	–	–	–	–	–	–	–	–	**1,2-Bis(benzoyloxy)benzene**	643-94-7	**NI33590**
94	225	86	65	44	66	39	105	**318**	100	32	21	19	19	18	15	14	**0.00**	20	14	4	–	–	–	–	–	–	–	–	–	–	**Diphenyl phthalate**	84-62-8	**DO01099**
225	77	226	76	104	51	39	50	**318**	100	33	19	14	10	6	6	5	**0.00**	20	14	4	–	–	–	–	–	–	–	–	–	–	**Diphenyl phthalate**	84-62-8	**IC03912**
65	44	66	105	91	45	29	120	**318**	100	97	94	71	71	66	66	65	**0.00**	20	14	4	–	–	–	–	–	–	–	–	–	–	**Diphenyl phthalate**	84-62-8	**WI01412**
274	225	318	273	181	121	197	257	**318**	100	92	90	70	33	33	31	28		20	14	4	–	–	–	–	–	–	–	–	–	–	**1(3H)-Isobenzofuranone, 3,3-bis(4-hydroxyphenyl)-**	77-09-8	**LI07071**
318	274	225	273	121	257	319	181	**318**	100	95	90	65	31	29	25	21		20	14	4	–	–	–	–	–	–	–	–	–	–	**1(3H)-Isobenzofuranone, 3,3-bis(4-hydroxyphenyl)-**	77-09-8	**NI33591**
317	318	77	163	105	241	319	239	**318**	100	90	63	42	42	32	21	20		20	14	4	–	–	–	–	–	–	–	–	–	–	**Resorcinol dibenzoate**		**IC03913**

MASS TO CHARGE RATIOS								M.W.	INTENSITIES								Parent	C	H	O	N	S	F	Cl	Br	I	Si	P	B	X	COMPOUND NAME	CAS Reg No	No
318	140	317	139	290	168	151	319	318	100	37	32	30	28	28	25	22		20	14	4	–	–	–	–	–	–	–	–	–	–	Spiro[benzofuran-2(3H),2′-oxiran]-3-one, 6-methoxy-3′-(2-naphthalenyl)-	35405-28-8	LI07072
318	140	317	139	168	290	151	127	318	100	41	36	36	33	31	28	24		20	14	4	–	–	–	–	–	–	–	–	–	–	Spiro[benzofuran-2(3H),2′-oxiran]-3-one, 6-methoxy-3′-(2-naphthalenyl)-	35405-28-8	NI33592
317	289	318	241	183	152	159	199	318	100	73	64	21	19	17	15	14		20	15	2	–	–	–	–	–	–	–	1	–	–	Spiro[5H-dibenzophosphole-5,8′-[2H,3H][1,2]oxaphospholo[4,3,2-hi][2,1]benzoxaphosphole]	65930-87-2	NI33593
318	209	182	183	167	91	319	77	318	100	26	26	23	23	23	22	16		20	18	–	2	1	–	–	–	–	–	–	–	–	Carbanilide, N-benzylthio-	3053-39-2	NI33594
93	117	55	90	145	173	89	91	318	100	81	67	65	62	48	45	38	0.00	20	18	2	2	–	–	–	–	–	–	–	–	–	endo-9-Ethyl-1,2,3,4-tetrahydro-N-phenyl-1,4-iminonaphthalene-2,3-dicarboximide	123542-16-5	DD01214
213	318	303	301	105	289	290	197	318	100	40	33	28	27	25	23	17		20	18	2	2	–	–	–	–	–	–	–	–	–	1,2,3,10-Tetrahydro-4-benzoyl-5-methyl-pyrrolo[2,1-b]quinazoline-10-one		LI07073
45	199	41	241	43	117	119	77	318	100	43	27	25	23	19	18	15	0.20	20	22	–	2	–	–	–	–	–	1	–	–	–	Silacycloheptane-4-carbonitrile, 5-imino-7-methyl-1,1-diphenyl-	56437-94-6	NI33595
287	191	217	205	318	215	231	218	318	100	95	76	72	31	26	25	25		20	30	3	–	–	–	–	–	–	–	–	–	–	Abieta-8,11,13-trien-11,12,20-triol		MS07503
93	81	288	161	177	159	191	145	318	100	66	31	31	29	27	22	20	0.80	20	30	3	–	–	–	–	–	–	–	–	–	–	Bacchotricuneatin D	66563-31-3	MS07504
123	121	81	95	93	107	91	109	318	100	67	46	40	36	34	34	31	3.00	20	30	3	–	–	–	–	–	–	–	–	–	–	Cleomeolide	72188-81-9	MS07505
318	215	259	275	258	300	216	203	318	100	85	80	70	70	65	65	60		20	30	3	–	–	–	–	–	–	–	–	–	–	4,5-Deoxyneodolabellin		NI33596
275	257	300	43	249	185	318		318	100	73	58	57	50	39	2			20	30	3	–	–	–	–	–	–	–	–	–	–	3β,17β-Dihydroxy-19a-methyl-androst-5-en-19-one		LI07074
99	87	318	86	256	79	319	105	318	100	99	89	33	30	28	27	27		20	30	3	–	–	–	–	–	–	–	–	–	–	Estr-5(10)-en-3-one, 17-hydroxy-, cyclic 1,2-ethanediyl acetal, (17β)-	15342-09-3	MS07506
290	99	318	91	79	126	93	81	318	100	87	49	42	36	26	26	23		20	30	3	–	–	–	–	–	–	–	–	–	–	Estr-4-en-3-one, 17β-hydroxy-, cyclic ethylene acetal	13886-12-9	MS07507
100	318	91	86	79	105	93	77	318	100	52	38	34	28	23	17	17		20	30	3	–	–	–	–	–	–	–	–	–	–	Estr-5-en-3-one, 17β-hydroxy-, cyclic ethylene acetal	13963-24-1	MS07508
272	318	108	93	107	67	257	79	318	100	82	75	58	56	56	53	51		20	30	3	–	–	–	–	–	–	–	–	–	–	5α,3α-Formoxyandrostan-17-one		MS07509
272	318	108	93	107	257	147	201	318	100	82	75	58	56	53	47	45		20	30	3	–	–	–	–	–	–	–	–	–	–	5α-3α-Formoxyandrostan-17-one		NI33597
272	41	93	107	67	91	108	55	318	100	98	94	90	86	83	80	80	79.20	20	30	3	–	–	–	–	–	–	–	–	–	–	5β,3α-Formoxyandrostan-17-one		MS07510
272	93	107	91	108	318	257	149	318	100	94	83	83	80	79	67	60		20	30	3	–	–	–	–	–	–	–	–	–	–	5β-3α-Formoxyandrostan-17-one		NI33598
257	300	239	244	275				318	100	44	31	22	16				0.00	20	30	3	–	–	–	–	–	–	–	–	–	–	3β-Hydroxy-19a-methyl-5α-androstane-17,19-dione		LI07075
257	239	275	318	300				318	100	33	24	1	1					20	30	3	–	–	–	–	–	–	–	–	–	–	5α-Hydroxy-19a-methyl-5α-androstane-17,19-dione		LI07076
216	257	256	318	274	300			318	100	77	74	56	41	4				20	30	3	–	–	–	–	–	–	–	–	–	–	19R-Hydroxy-19a-methyl-5α-androstane-3,17-dione		LI07077
287	133	121	107	241	167	139	273	318	100	99	99	82	33	29	28	21	1.50	20	30	3	–	–	–	–	–	–	–	–	–	–	17-Hydroxyisopimara-8(14),15-dien-19-oic acid		MS07511
216	256	318	274	55	43	303	300	318	100	60	23	23	14	11	1	1		20	30	3	–	–	–	–	–	–	–	–	–	–	19R-Hydroxy-19-methyl-5α-androstane-3,17-dione		LI07078
300	318	134	121	133	285	59	105	318	100	99	97	97	86	77	67	64		20	30	3	–	–	–	–	–	–	–	–	–	–	Hydroxysandaracopimaric acid	1235-75-2	NI33599
258	318	304	259	121	159	243	125	318	100	83	68	54	46	44	36	34		20	30	3	–	–	–	–	–	–	–	–	–	–	2-Phenanthrenecarboxylic acid, 1,2,3,4,4a,4b,5,6,7,8,8a,9-dodecahydro-2,4b,8,8-tetramethyl-7-oxo-, methyl ester	57397-05-4	NI33600
159	160	158	220	105	219	243	91	318	100	68	44	39	33	32	31	27	25.00	20	30	3	–	–	–	–	–	–	–	–	–	–	2-Phenanthrenecarboxylic acid, 1,2,3,4,4a,4b,5,6,7,8,8a,9-dodecahydro-2,4b,8,8-tetramethyl-7-oxo-, methyl ester	57397-05-4	NI33601
287	302	227	159	288	242	303	243	318	100	70	60	51	23	22	20	17	1.70	20	30	3	–	–	–	–	–	–	–	–	–	–	1-Phenanthrenecarboxylic acid, 7-vinyl-1,2,3,4,4a,4b,5,6,7,8,10,10a-dodecahydro-3-hydroxy-4a,7-dimethyl-, methyl ester, [1R-(1α,3α,4aβ,4bα,7α,10aα)]-	57397-37-2	NI33602
305	121	287	320	261	159	105	145	318	100	62	50	40	40	40	40	33	0.00	20	30	3	–	–	–	–	–	–	–	–	–	–	1-Phenanthrenecarboxylic acid, 7-vinyl-1,2,3,4,4a,4b,5,6,7,8,10,10a-dodecahydro-3-hydroxy-4a,7-dimethyl-, methyl ester, [1R-(1α,3β,4aβ,4bα,7α,10aα)]-	57397-38-3	NI33603
121	41	43	55	134	81	105	99	318	100	95	91	83	75	64	61	58	8.47	20	30	3	–	–	–	–	–	–	–	–	–	–	1-Phenanthrenecarboxylic acid, 7-vinyl-1,2,3,4,4a,4b,5,6,7,9,10,10a-dodecahydro-6-hydroxy-1,4a,7-trimethyl-	56051-66-2	NI33604
301	192	70	162	111	215	285	271	318	100	69	62	60	45	35	30	29	0.00	20	30	3	–	–	–	–	–	–	–	–	–	–	Pisiferdiol	95263-32-4	DD01215
303	300	137	41	123	69	95	176	318	100	85	82	80	72	72	70	66	6.70	20	30	3	–	–	–	–	–	–	–	–	–	–	Spiro[furan-2(5H),2′(1′H)-naphtho[2,1-b]furan]-5-one, 3′a,4′,5′,5′a,6′,7′,8′,9′,9′a,9′b-decahydro-3,3′a,6′,6′,9′a-pentamethyl-, [2′R-(2′α,3′aβ,5′aα,9′aβ,9′bα)]-	1235-78-5	NI33605
303	43	41	304	137	123	69	95	318	100	22	21	20	18	18	18	17	6.00	20	30	3	–	–	–	–	–	–	–	–	–	–	Spiro[furan-2(5H),2′(1′H)-naphtho[2,1-b]furan]-5-one, 3′a,4′,5′,5′a,6′,7′,8′,9′,9′a,9′b-decahydro-3,3′a,6′,6′,9′a-pentamethyl-, [2′S-(2′α,3′aα,5′aβ,9′aα,9′bβ)]-	30987-48-5	NI33606
247	261	246	55	41	248	262	107	318	100	47	23	22	22	16	12	12	11.42	20	35	2	–	–	–	–	–	–	–	–	1	–	Spiro[4H-1,3,2-benzodioxaborin-4,1′-cyclohexane], 2-ethyl-5,6,7,8-tetrahydro-3′,3′,5,5′,7,7-hexamethyl-	74421-12-8	NI33607
160	159	144	115	158	116	161	149	318	100	73	29	22	20	13	12	12	9.00	21	18	3	–	–	–	–	–	–	–	–	–	–	1,8-(Epoxymethano)-2,7-methanodibenzo[a,e]cyclobuta[c]cycloocten-13-one, 1,2,2a,7,8,12b-hexahydro-2-methoxy-	52023-05-9	NI33608
303	183	318	317	165	152	277	201	318	100	24	15	14	13	7	6	5		21	19	1	–	–	–	–	–	–	–	1	–	–	Acetone, 1-(triphenylphosphoranylidene)-	1439-36-7	LI07079
303	318	317	304	183	277	165	152	318	100	39	36	26	23	15	11	9		21	19	1	–	–	–	–	–	–	–	1	–	–	Acetone, 1-(triphenylphosphoranylidene)-	1439-36-7	NI33609
222	318	220	193	317	194	104	178	318	100	39	15	15	13	12	12	9		21	22	1	2	–	–	–	–	–	–	–	–	–	1-(β-Amino-α-phenylcinnamoyl)-2,3-tetramethylene-aziridine		LI07080
91	318	132	106	93	319	77	227	318	100	99	88	49	37	25	25	23		21	22	1	2	–	–	–	–	–	–	–	–	–	cis-1-Benzyl-2-phenylimino-1H,4H,4′H,5H,8H,8′H-3,1-benzoxazine		MS07512
318	132	91	106	317	319	77	93	318	100	89	71	40	32	24	21	17		21	22	1	2	–	–	–	–	–	–	–	–	–	trans-1-Benzyl-2-phenylimino-1H,4H,4′H,5H,8H,8′H-3,1-benzoxazine		MS07513

MASS TO CHARGE RATIOS								M.W.	INTENSITIES								Parent	C	H	O	N	S	F	Cl	Br	I	Si	P	B	X	COMPOUND NAME	CAS Reg No	No
57	55	71	69	303	97	318	83	**318**	100	65	61	57	56	47	46	45		21	22	1	2	–	–	–	–	–	–	–	–	–	**(Z,Z)-4-Ethyl-3-methyl-5-[5-(phenyl-2-methylene)-3,4-dimethyl-5H-pyrrolyl-2-methylene]-3-pyrrolin-2-one**		**MS07514**
69	136	165	105	77	68	166	192	**318**	100	46	21	21	17	16	13	8	**1.20**	21	22	1	2	–	–	–	–	–	–	–	–	–	**Propanenitrile, 3-[(2,2-dimethyl-4,4-diphenyl-3-oxetanylidene)amino]-2-methyl-**	55044-22-9	**MS07515**
69	136	105	77	68	137	70	192	**318**	100	49	20	18	16	9	9	8	**4.69**	21	22	1	2	–	–	–	–	–	–	–	–	–	**Propanenitrile, 3-[(2,2-dimethyl-4,4-diphenyl-3-oxetanylidene)amino]-2-methyl-**	55044-22-9	**NI33610**
69	208	95	105	136	77	193	91	**318**	100	84	76	56	43	34	32	32	**4.35**	21	22	1	2	–	–	–	–	–	–	–	–	–	**Propanenitrile, 3-[(3,3-dimethyl-4,4-diphenyl-2-oxetanylidene)amino]-2-methyl-**	55044-24-1	**NI33611**
69	208	95	105	77	91	193	182	**318**	100	83	77	53	38	33	30	20	**0.60**	21	22	1	2	–	–	–	–	–	–	–	–	–	**Propanenitrile, 3-[(3,3-dimethyl-4,4-diphenyl-2-oxetanylidene)amino]-2-methyl-**	55044-24-1	**MS07516**
198	318	199	275	156	226	184	171	**318**	100	47	42	19	18	16	14	13		21	22	1	2	–	–	–	–	–	–	–	–	–	**3-Spiro-cycloheptane-3H-indole-2-carboxanilide**		**MS07517**
148	94	149	43	258	81	95	67	**318**	100	95	91	90	83	81	79	70	**1.19**	21	34	2	–	–	–	–	–	–	–	–	–	–	**Androstan-17-ol, acetate, (5α,17α)-**	22265-08-3	**NI33612**
148	149	243	95	94	43	258	109	**318**	100	91	68	65	64	64	63	60	**12.40**	21	34	2	–	–	–	–	–	–	–	–	–	–	**Androstan-17-ol, acetate, (5α,17β)-**	1236-49-3	**NI33613**
43	149	258	148	243	94	109	55	**318**	100	88	85	80	63	60	59	55	**20.02**	21	34	2	–	–	–	–	–	–	–	–	–	–	**Androstan-17-ol, acetate, (17β)-**	27736-62-5	**NI33614**
99	125	100	112	318	81	95	79	**318**	100	46	13	12	10	9	8	6		21	34	2	–	–	–	–	–	–	–	–	–	–	**Androstan-3-one, cyclic 1,2-ethanediyl acetal, (5α)-**	1434-14-6	**NI33616**
99	125	112	100	318	80	66	94	**318**	100	47	13	13	10	9	9	8		21	34	2	–	–	–	–	–	–	–	–	–	–	**Androstan-3-one, cyclic 1,2-ethanediyl acetal, (5α)-**	1434-14-6	**NI33615**
99	125	112	100	318	81	67	95	**318**	100	47	13	13	10	9	9	8		21	34	2	–	–	–	–	–	–	–	–	–	–	**Androstan-3-one, cyclic 1,2-ethanediyl acetal, (5α)-**	1434-14-6	**NI33617**
125	99	153	318	55	142	95	93	**318**	100	65	64	52	41	28	28	21		21	34	2	–	–	–	–	–	–	–	–	–	–	**Androstan-7-one, cyclic 1,2-ethanediyl acetal**	55252-96-5	**NI33618**
99	139	114	41	55	67	318	81	**318**	100	79	78	39	38	34	26	26		21	34	2	–	–	–	–	–	–	–	–	–	–	**Androstan-16-one, cyclic 1,2-ethanediyl acetal, (5α)-**	1778-87-6	**NI33619**
99	139	114	67	318	81	86	79	**318**	100	81	79	34	26	26	22	20		21	34	2	–	–	–	–	–	–	–	–	–	–	**Androstan-16-one, cyclic 1,2-ethanediyl acetal, (5α)-**	1778-87-6	**NI33620**
99	318	100	55	86	67	41	81	**318**	100	15	12	7	5	5	5	4		21	34	2	–	–	–	–	–	–	–	–	–	–	**Androstan-17-one, cyclic 1,2-ethanediyl acetal**	27736-67-0	**NI33621**
125	99	153	318	43	55	41	67	**318**	100	73	64	53	49	40	40	34		21	34	2	–	–	–	–	–	–	–	–	–	–	**Androstan-7-one, cyclic ethylene acetal, (5α)-**	1434-36-2	**NI33622**
91	93	79	81	105	55	67	107	**318**	100	90	87	81	75	74	70	63	**0.00**	21	34	2	–	–	–	–	–	–	–	–	–	–	**Androstan-17-one, 3-ethyl-3-hydroxy-, (5α)-**	57344-99-7	**NI33623**
43	41	81	55	67	69	79	93	**318**	100	66	62	61	59	54	51	50	**4.06**	21	34	2	–	–	–	–	–	–	–	–	–	–	**Androstan-3-one, 17-hydroxy-1,17-dimethyl-, (1α,5α,17β)-**	2881-21-2	**NI33624**
41	55	43	67	81	69	95	79	**318**	100	70	58	43	40	34	30	27	**17.00**	21	34	2	–	–	–	–	–	–	–	–	–	–	**Androstan-3-one, 17-hydroxy-2,4-dimethyl-, (2α,4α,5α,17β)-**	54550-08-2	**NI33625**
318	41	173	43	82	191	67	69	**318**	100	96	67	60	57	56	55	45		21	34	2	–	–	–	–	–	–	–	–	–	–	**Androstan-3-one, 17-hydroxy-4,4-dimethyl-, (5α,17β)-**	54550-06-0	**NI33626**
193	318	233	231	194	262			**318**	100	23	20	20	19	9				21	34	2	–	–	–	–	–	–	–	–	–	–	**1,3-Benzenediol, 2-[5-methyl-2-isopropylcyclohexyl]-5-pentyl-**	4460-20-2	**LI07081**
233	318	234	193	262	319	194	263	**318**	100	47	25	25	17	15	6	5		21	34	2	–	–	–	–	–	–	–	–	–	–	**1,3-Benzenediol, 2-[5-methyl-2-isopropylcyclohexyl]-5-pentyl-**	4460-20-2	**NI33627**
91	108	43	92	57	41	55	71	**318**	100	74	36	28	28	25	18	15	**0.00**	21	34	2	–	–	–	–	–	–	–	–	–	–	**Benzyl myristate**		**NI33628**
91	183	107	127	95	227	159	165	**318**	100	51	42	33	26	25	22	14	**0.00**	21	34	2	–	–	–	–	–	–	–	–	–	–	**(2R,7R)-2-[2-(Benzyloxy)ethyl]-7-hexyl-1-oxacycloheptane**		**DD01216**
79	91	106	150	93	80	119	105	**318**	100	87	74	57	55	52	51	47	**0.00**	21	34	2	–	–	–	–	–	–	–	–	–	–	**5,8,11,14-Eicosatetraenoic acid, methyl ester, (all-Z)-**	2566-89-4	**NI33629**
79	67	41	80	91	93	55	81	**318**	100	85	82	78	71	63	50	45	**14.04**	21	34	2	–	–	–	–	–	–	–	–	–	–	**5,8,11,14-Eicosatetraenoic acid, methyl ester, (all-Z)-**	2566-89-4	**NI33630**
79	67	41	93	91	55	81	106	**318**	100	91	87	66	64	61	54	39	**5.84**	21	34	2	–	–	–	–	–	–	–	–	–	–	**5,8,11,14-Eicosatetraenoic acid, methyl ester, (all-Z)-**	2566-89-4	**NI33631**
123	39	81	259	109	93	79	55	**318**	100	95	87	86	72	70	68	68	**26.00**	21	34	2	–	–	–	–	–	–	–	–	–	–	**Kauran-19-oic acid, methyl ester**	5282-19-9	**MS07518**
107	258	318	150	121	243	303	91	**318**	100	58	57	56	50	47	46	46		21	34	2	–	–	–	–	–	–	–	–	–	–	**Methyl 7-abieten-18-oate**		**MS07519**
275	215	318	121	181	81	107	91	**318**	100	85	69	66	52	47	45	38		21	34	2	–	–	–	–	–	–	–	–	–	–	**Methyl 8(14)-abieten-18-oate**		**MS07520**
243	100	303	91	79	318	107	105	**318**	100	55	50	49	48	47	44	44		21	34	2	–	–	–	–	–	–	–	–	–	–	**Methyl 8-abieten-18-oate**		**MS07521**
318	121	81	107	93	79	91	95	**318**	100	80	59	52	51	44	41	37		21	34	2	–	–	–	–	–	–	–	–	–	–	**Methyl 13(15)-abieten-18-oate**		**MS07522**
121	107	81	136	275	93	215	79	**318**	100	68	54	51	45	44	38	37	**19.00**	21	34	2	–	–	–	–	–	–	–	–	–	–	**Methyl 13-abieten-18-oate**		**MS07523**
107	91	79	121	258	93	81	150	**318**	100	58	58	56	55	53	51	48	**33.00**	21	34	2	–	–	–	–	–	–	–	–	–	–	**Methyl 13β-abiet-7-en-18-oate**		**MS07524**
275	215	121	318	181	136	107	81	**318**	100	97	92	73	58	58	58	45		21	34	2	–	–	–	–	–	–	–	–	–	–	**Methyl 13β-abiet-8(14)-en-18-oate**		**MS07525**
243	91	303	105	107	318	93	121	**318**	100	65	59	57	55	52	52	50		21	34	2	–	–	–	–	–	–	–	–	–	–	**Methyl 13β-abiet-8-en-18-oate**		**MS07526**
79	67	41	55	95	81	80	93	**318**	100	83	78	54	52	49	48	47	**2.00**	21	34	2	–	–	–	–	–	–	–	–	–	–	**Methyl (Z)-5,11,14,17-eicosatetraenoate**		**NI33632**
303	95	81	318	82	114	137	83	**318**	100	99	99	98	89	80	78	78		21	34	2	–	–	–	–	–	–	–	–	–	–	**Methyl 8E(17),13-labdadien-15-oate**		**MS07527**
114	137	81	69	95	55	109	123	**318**	100	84	81	73	66	59	53	46	**11.60**	21	34	2	–	–	–	–	–	–	–	–	–	–	**Methyl enantio-8(17),13-trans-labdadien-15-oate**		**MS07528**
41	69	28	55	81	95	303	135	**318**	100	87	81	78	75	65	61	51	**22.00**	21	34	2	–	–	–	–	–	–	–	–	–	–	**Methyl ent-D-norbeyerane-endo-15-carboxylate**		**MS07529**
41	81	69	137	55	123	95	318	**318**	100	94	90	84	83	75	75	38		21	34	2	–	–	–	–	–	–	–	–	–	–	**Methyl ent-D-norbeyerane-exo-15-carboxylate**		**MS07530**
121	243	258	318	259	91	107	109	**318**	100	83	79	78	77	71	67	65		21	34	2	–	–	–	–	–	–	–	–	–	–	**Methyl 7-isopimaren-18-oate**		**MS07531**
243	303	318	91	105	121	107	79	**318**	100	93	85	73	71	70	67	63		21	34	2	–	–	–	–	–	–	–	–	–	–	**Methyl 8(14)-isopimaren-18-oate**		**MS07533**
121	289	107	95	93	81	105	229	**318**	100	88	48	42	42	42	40	39	**23.00**	21	34	2	–	–	–	–	–	–	–	–	–	–	**Methyl 8(14)-isopimaren-18-oate**		**MS07532**
303	95	81	318	114	137	205	204	**318**	100	99	99	98	80	78	71	60		21	34	2	–	–	–	–	–	–	–	–	–	–	**Methyl labda-8(20),13-dien-15-oate**		**MS07534**
289	121	181	95	229	122	290	107	**318**	100	97	56	47	46	44	41	41	**30.00**	21	34	2	–	–	–	–	–	–	–	–	–	–	**Methyl 8(14)-pimaren-18-oate**		**MS07535**
243	303	318	105	259	91	121	107	**318**	100	75	65	56	53	52	50	45		21	34	2	–	–	–	–	–	–	–	–	–	–	**Methyl 8(14)-pimaren-18-oate**		**MS07536**
121	81	95	79	107	93	136	122	**318**	100	34	27	27	25	25	23	21	**17.00**	21	34	2	–	–	–	–	–	–	–	–	–	–	**Methyl 14S,17-cyclo-13ξ-labd-8(17)-en-18-oate**		**MS07537**

MASS TO CHARGE RATIOS								M.W.	INTENSITIES								Parent	C	H	O	N	S	F	Cl	Br	I	Si	P	B	X	COMPOUND NAME	CAS Reg No	No
289	121	229	290	181	95	107	81	**318**	100	81	32	23	23	18	16	13	**5.80**	21	34	2	–	–	–	–	–	–	–	–	–	–	1-Phenanthrenecarboxylic acid, 7-ethyl-1,2,3,4,4a,4b,5,6,7,9,10,10a-dodecahydro-1,4a,7-trimethyl-, methyl ester, [1R-(1α,4aβ,4bα,7α,10aα)]-	3867-54-7	NI33633
243	302	55	41	121	91	105	107	**318**	100	44	37	35	33	32	30	24	**3.80**	21	34	2	–	–	–	–	–	–	–	–	–	–	1-Phenanthrenecarboxylic acid, 7-ethyl-1,2,3,4,4a,4b,5,6,7,9,10,10a-dodecahydro-1,4a,7-trimethyl-, methyl ester, [1R-(1α,4aβ,4bα,7β,7aα)]-	3582-27-2	NI33635
121	95	107	91	289	93	105	81	**318**	100	18	15	15	14	14	13	13	**0.00**	21	34	2	–	–	–	–	–	–	–	–	–	–	1-Phenanthrenecarboxylic acid, 7-ethyl-1,2,3,4,4a,4b,5,6,7,9,10,10a-dodecahydro-1,4a,7-trimethyl-, methyl ester, [1R-(1α,4aβ,4bα,7β,7aα)]-	3582-27-2	NI33634
243	121	55	41	163	91	105	107	**318**	100	95	88	65	53	41	38	36	**14.60**	21	34	2	–	–	–	–	–	–	–	–	–	–	1-Phenanthrenecarboxylic acid, 7-ethyl-1,2,3,4,4a,4b,5,6,7,9,10,10a-dodecahydro-1,4a,7-trimethyl-, methyl ester, [1R-(1α,4aβ,4bα,7β,7aα)]-	3582-27-2	NI33636
177	121	123	81	69	192	95	107	**318**	100	68	65	65	63	62	62	60	**0.00**	21	34	2	–	–	–	–	–	–	–	–	–	–	8β-Podocarp-12-ene-14-carboxylic acid, 8,13-dimethyl-, methyl ester	15372-63-1	NI33637
289	247	57	318	55	85	81	43	**318**	100	60	47	45	44	42	36	34		21	34	2	–	–	–	–	–	–	–	–	–	–	Pregnan-3-one, 17-hydroxy-, (5α,17α)-	20112-30-5	NI33639
289	318	247	300	290	246	231	271	**318**	100	48	43	19	19	18	15	12		21	34	2	–	–	–	–	–	–	–	–	–	–	Pregnan-3-one, 17-hydroxy-, (5α,17α)-	20112-30-5	NI33638
233	110	274	81	149	109	67	55	**318**	100	32	31	30	23	23	22	21	**7.07**	21	34	2	–	–	–	–	–	–	–	–	–	–	Pregnan-12-one, 20-hydroxy-, (5α,20β)-	7704-91-8	NI33640
43	84	55	81	57	41	71	95	**318**	100	59	53	37	37	37	34	31	**18.80**	21	34	2	–	–	–	–	–	–	–	–	–	–	Pregnan-20-one, 3-hydroxy-		MS07538
43	300	215	318	230	257	242	246	**318**	100	84	43	40	16	15	14	11		21	34	2	–	–	–	–	–	–	–	–	–	–	Pregnan-20-one, 3-hydroxy-, (3α,5α)-		MS07539
84	215	300	318	107	285	233	135	**318**	100	82	74	51	38	33	31	31		21	34	2	–	–	–	–	–	–	–	–	–	–	Pregnan-20-one, 3-hydroxy-, (3α,5α)-		NI33641
300	246	215	108	285	301	84	257	**318**	100	44	39	35	32	29	29	22	**5.34**	21	34	2	–	–	–	–	–	–	–	–	–	–	Pregnan-20-one, 3-hydroxy-, (3α,5α)-		NI33642
84	81	79	67	93	107	91	95	**318**	100	96	94	85	84	78	78	77	**34.60**	21	34	2	–	–	–	–	–	–	–	–	–	–	Pregnan-20-one, 3-hydroxy-, (3α,5α)-		AM00142
43	300	84	318	81	107	93	95	**318**	100	52	41	32	32	31	30	28		21	34	2	–	–	–	–	–	–	–	–	–	–	Pregnan-20-one, 3-hydroxy-, (3β)-	4406-37-5	NI33643
81	79	67	300	93	95	215	107	**318**	100	95	95	89	89	87	81	79	**12.91**	21	34	2	–	–	–	–	–	–	–	–	–	–	Pregnan-20-one, 3-hydroxy-, (3β)-	4406-37-5	WI01413
318	84	215	107	300	233	260	303	**318**	100	79	53	51	44	33	25	18		21	34	2	–	–	–	–	–	–	–	–	–	–	Pregnan-20-one, 3-hydroxy-, (3β,5α)-	516-55-2	NI33644
43	55	41	81	67	242	109	95	**318**	100	43	41	34	32	27	24	23	**1.00**	21	34	2	–	–	–	–	–	–	–	–	–	–	Pregnan-20-one, 12-hydroxy-, (5α,12β)-	5618-22-4	NI33645
43	233	41	257	55	81	67	109	**318**	100	72	54	48	46	38	33	28	**5.00**	21	34	2	–	–	–	–	–	–	–	–	–	–	Pregnan-20-one, 12-hydroxy-, (12β)-	5618-23-5	NI33646
318	45	189	105	91	107	79	95	**318**	100	94	79	78	78	73	73	72		21	34	2	–	–	–	–	–	–	–	–	–	–	Pregn-5-ene-3,20-diol, (3β,20S)-	901-56-4	NI33647
57	91	41	318	138	180	124	55	**318**	100	83	54	47	45	40	39	39		21	35	–	–	–	–	–	–	–	–	1	–	–	Phosphine, tert-butyl 5-methyl-2-isopropylcyclohexyl(phenylmethyl)-	61141-92-2	NI33648
44	275	56	276	259	93	70	43	**318**	100	50	15	10	10	10	5	5	**1.00**	21	38	–	2	–	–	–	–	–	–	–	–	–	Pregnane-3,20-diamine, (3β,5α,20S)-	7050-25-1	NI33649
243	165	244	241	166	242	239	228	**318**	100	30	29	8	6	5	5	4	**0.00**	22	22	–	–	1	–	–	–	–	–	–	–	–	Propyl trityl sulphide	24523-63-5	NI33650
183	105	77	184	136	91	135	43	**318**	100	46	16	14	13	13	9	7	**6.00**	22	22	2	–	–	–	–	–	–	–	–	–	–	Benzenepropanol, 2-methoxy-α,α-diphenyl-	56052-54-1	NI33651
222	318	177	205	55	176	41	178	**318**	100	52	38	33	33	25	24	22		22	22	2	–	–	–	–	–	–	–	–	–	–	Cyclohexanemethyl anthracene-9-carboxylate	86338-72-9	NI33652
122	121	43	41	318	135	91	29	**318**	100	42	20	16	12	12	11	10		22	38	1	–	–	–	–	–	–	–	–	–	–	Benzene, 1-methoxy-3-pentadecyl-	15071-57-5	NI33653
247	57	248	41	318	175	191	161	**318**	100	19	18	6	5	5	3	3		22	38	1	–	–	–	–	–	–	–	–	–	–	Phenol, 2,4-dioctyl-	29988-16-7	NI33654
247	57	248	41	318	29	43	94	**318**	100	26	21	10	7	6	4	3		22	38	1	–	–	–	–	–	–	–	–	–	–	Phenol, 4-octyl-2,6-di-tert-butyl-		IC03914
318	303	275	132	144	138	288	124	**318**	100	84	8	8	7	7	5	5		23	14	–	2	–	–	–	–	–	–	–	–	–	1,8-Dicyano-13-methyltriptycene		NI33655
303	318	138	151	130	275	144	124	**318**	100	72	10	9	9	8	8	7		23	14	–	2	–	–	–	–	–	–	–	–	–	1,8-Dicyano-16-methyltriptycene		NI33656
206	167	318	91	249	290	41	55	**318**	100	85	75	59	52	48	41	29		23	26	1	–	–	–	–	–	–	–	–	–	–	Bicyclo[2.2.1]heptan-2-ol, 3-(diphenylmethylene)-1,7,7-dimethyl-, exo-	38765-75-2	MS07540
69	81	43	67	41	109	95	135	**318**	100	48	38	35	35	34	34	30	**8.01**	23	42	–	–	–	–	–	–	–	–	–	–	–	2,6,10-Nonadecatriene, 2,6,10,14-tetramethyl-, (E,E)-	31147-35-0	NI33657
191	69	55	95	81	123	41	137	**318**	100	62	46	46	42	35	32	31	**17.98**	23	42	–	–	–	–	–	–	–	–	–	–	–	15-Propyl-(13αH)-isocopalane		NI33658
318	241	319	239	317	242	240	237	**318**	100	42	29	29	20	9	9	6		25	18	–	–	–	–	–	–	–	–	–	–	–	9H-Fluorene, 9,9-diphenyl-	20302-14-1	NI33659
279	277	278	276	280	281	242	275	**319**	100	98	89	70	46	46	26	26	**9.63**	3	8	–	3	–	–	4	–	–	–	3	–	–	1-Isopropyl-1-hydridotetrachlorocyclotriphosphazene	71982-86-0	NI33660
278	276	280	242	279	277	240	244	**319**	100	82	54	17	11	11	11	10	**7.21**	3	8	–	3	–	–	4	–	–	–	3	–	–	1-Propyl-1-hydridotetrachlorocyclotriphosphazene	71982-85-9	NI33661
134	69	46	70	282	85	100	31	**319**	100	37	18	17	15	15	7	6	**0.00**	4	–	–	1	1	10	1	–	–	–	–	–	–	N-(1,1,2,2,3,3,4,4-Octafluoro-4-chloro-butyl)-thio-difluorimide		LI07082
43	41	197	74	75	277	62	120	**319**	100	70	17	15	12	9	7	5	**0.00**	9	9	8	3	1	–	–	–	–	–	–	–	–	1-Isopropylsulphonyl-2,4,6-trinitrobenzene		NI33662
321	319	323	199	197	242	87	240	**319**	100	62	48	46	46	37	36	27		10	4	2	1	–	–	2	1	–	–	–	–	–	N-(Bromophenyl)-2,3-dichloromaleicisoimide		MS07541
105	77	106	51	28	319	78	76	**319**	100	74	20	19	13	10	9	7		11	5	1	1	–	8	–	–	–	–	–	–	–	1-Benzoyl-2,2-difluoro-3,3-bis(trifluoromethyl)aziridine		LI07083
238	240	196	198	319	321	317	157	**319**	100	97	84	82	69	54	34	30		11	14	–	1	–	–	–	1	–	–	–	–	1	N-(4-Bromophenyl)selenopentamide		MS07542
69	43	190	232	233	41	70	203	**319**	100	71	52	46	22	21	10	8	**0.00**	12	18	3	1	–	5	–	–	–	–	–	–	–	N(O,S)-Pentafluoropropionyl-D-leucine isopropyl ester		MS07543
69	43	190	232	233	41	70	203	**319**	100	70	57	52	25	20	10	9	**0.00**	12	18	3	1	–	5	–	–	–	–	–	–	–	N(O,S)-Pentafluoropropionyl-L-leucine isopropyl ester		MS07544
73	75	147	89	55	198	157	45	**319**	100	52	27	26	19	16	15	13	**4.25**	12	25	5	1	–	–	–	–	–	2	–	–	–	Pentanedioic acid, 2-(methoxyimino)-, bis(trimethylsilyl) ester	60022-87-9	LI07084
73	75	147	198	55	304	156	74	**319**	100	30	17	13	12	10	10	10	**1.60**	12	25	5	1	–	–	–	–	–	2	–	–	–	Pentanedioic acid, 2-(methoxyimino)-, bis(trimethylsilyl) ester	60022-87-9	NI33663
73	75	158	45	147	304	56	102	**319**	100	36	31	20	14	13	11	10	**0.00**	12	25	5	1	–	–	–	–	–	2	–	–	–	Propanoic acid, 2-methyl-3-oxo-3-[[2-oxo-2-[(trimethylsilyl)oxy]ethyl](trimethylsilyl)amino]-	55557-20-5	NI33664

MASS TO CHARGE RATIOS								M.W.	INTENSITIES								Parent	C	H	O	N	S	F	Cl	Br	I	Si	P	B	X	COMPOUND NAME	CAS Reg No	No
73	75	158	45	304	147	102	56	**319**	100	35	35	24	15	14	12	10	**0.00**	12	25	5	1	–	–	–	–	–	2	–	–	–	Propanoic acid, 2-methyl-3-oxo-3-[[2-oxo-2-[(trimethylsilyl)oxy]ethyl](trimethylsilyl)amino]-	55557-20-5	NI33665
73	75	158	45	147	304	56	102	**319**	100	36	31	20	14	13	11	10	**0.00**	12	25	5	1	–	–	–	–	–	2	–	–	–	Propanoic acid, 3-[methyl[2-oxo-2-[(trimethylsilyl)oxy]ethyl]amino]-3-oxo-, trimethylsilyl ester	55887-45-1	NI33666
73	75	147	214	45	304	74	43	**319**	100	51	32	25	24	11	9	8	**2.53**	12	29	3	1	–	–	–	–	–	3	–	–	–	Monoamidomalonic acid, tris(trimethylsilyl)-		NI33667
122	80	43	29	196	44	150	304	**319**	100	51	36	28	22	22	21	20	**5.50**	13	22	4	1	1	–	–	–	–	–	1	–	–	Phosphoramidic acid, isopropyl-, ethyl 3-methyl-4-(methylsulphinyl)phenyl ester	31972-43-7	NI33668
174	73	304	147	75	59	45	86	**319**	100	51	28	24	12	9	9	8	**2.00**	13	33	2	1	–	–	–	–	–	3	–	–	–	Butanoic acid, 4-[bis(trimethylsilyl)amino]-, trimethylsilyl ester	39508-23-1	MS07545
73	174	59	75	147	86	74	100	**319**	100	34	24	22	18	13	10	9	**0.00**	13	33	2	1	–	–	–	–	–	3	–	–	–	Butanoic acid, 4-[bis(trimethylsilyl)amino]-, trimethylsilyl ester	39508-23-1	NI33669
73	174	75	147	45	304	175	59	**319**	100	96	30	28	21	18	17	16	**0.00**	13	33	2	1	–	–	–	–	–	3	–	–	–	Butanoic acid, 4-[bis(trimethylsilyl)amino]-, trimethylsilyl ester	39508-23-1	MS07546
73	174	59	133	45	248	86	175	**319**	100	74	29	26	26	21	20	19	**0.00**	13	33	2	1	–	–	–	–	–	3	–	–	–	Propanoic acid, 3-[bis(trimethylsilyl)amino]-2-methyl-, trimethylsilyl ester	55125-15-0	NI33670
83	84	302	43	249	41	149	42	**319**	100	80	43	37	26	24	21	21	**0.00**	14	21	2	7	–	–	–	–	–	–	–	–	–	1H-Pyrrole-2-carboxamide, 5-[[(5-amino-3,4-dihydro-2H-pyrrol-2-yl)carbonyl]amino]-N-(3-amino-3-iminopropyl)-1-methyl-	61233-55-4	NI33671
72	101	43	314	28	88	59	106	**319**	100	92	75	43	31	28	24	18	**0.00**	14	25	5	1	1	–	–	–	–	–	–	–	–	α-DL-Xylofuranose, 5-O-isopropyl-1,2-O-isopropylidene-, dimethylcarbamothioate	56678-63-8	NI33672
87	43	129	142	72	98	42	45	**319**	100	60	52	45	26	21	19	17	**0.14**	14	25	7	1	–	–	–	–	–	–	–	–	–	α-D-Galactopyranoside, methyl 2-(acetylmethylamino)-2-deoxy-3,4-di-O-methyl-, 6-acetate	36663-29-3	NI33673
115	43	142	98	73	101	71	45	**319**	100	85	70	50	47	43	35	35	**0.59**	14	25	7	1	–	–	–	–	–	–	–	–	–	α-D-Glucopuranoside, methyl 3-O-acetyl-2-deoxy-4,6-di-O-methyl-2-(N-methylacetamido)-	56389-83-4	NI33674
87	142	43	129	88	72	42	45	**319**	100	47	47	46	29	29	25	23	**0.17**	14	25	7	1	–	–	–	–	–	–	–	–	–	α-D-Glucopyranoside, methyl 2-(acetylmethylamino)-2-deoxy-3,4-di-O-methyl-, 6-acetate	36663-25-9	NI33675
87	142	129	43	98	88	72	100	**319**	100	75	60	55	25	24	21	21	**0.19**	14	25	7	1	–	–	–	–	–	–	–	–	–	α-D-Glucopyranoside methyl 6-O-acetyl-2-deoxy-3,4-di-O-methyl-2-(N-methylacetamido)-	56389-84-5	NI33676
87	75	98	45	129	72	141	128	**319**	100	70	60	60	50	25	20	20	**0.00**	14	25	7	1	–	–	–	–	–	–	–	–	–	α-D-Glucopyranoside, methyl 4-O-acetyl-2-deoxy-3,6-di-O-methyl-2-(N-methylacetamido)-	56341-49-2	NI33677
140	290	319	292	291	321	293	87	**319**	100	96	67	67	53	43	34	33		15	11	1	3	–	–	2	–	–	–	–	–	–	1-Ethoxy-4-(dichloro-s-triazinyl)naphthalene	21614-17-5	NI33678
319	290	321	292	140	291	293	29	**319**	100	75	67	54	51	48	31	20		15	11	1	3	–	–	2	–	–	–	–	–	–	1-Ethoxy-4-(4,6-dichlorotriaz-2-yl)naphthalene		IC03915
232	319	246	234	321	248	214	233	**319**	100	96	66	46	38	29	25	23		16	14	2	1	1	–	1	–	–	–	–	–	–	Methyl (2′-chlorophenothiazin-9-yl)propionate		LI07085
204	235	44	234	205	77	220	165	**319**	100	60	54	49	49	27	24	22	**0.00**	16	14	4	1	–	–	1	–	–	–	–	–	–	Isoquinolinium, 2-methyl-1-phenyl-, perchloroate	52806-09-4	NI33679
205	204	206	44	235	102	50	207	**319**	100	78	22	22	16	16	13	12	**0.00**	16	14	4	1	–	–	1	–	–	–	–	–	–	Quinolinium, 1-methyl-2-phenyl-, perchlorate	16192-31-7	NI33680
167	165	194	169	253	70	152	87	**319**	100	38	36	26	20	14	12	9	**7.00**	16	17	4	1	1	–	–	–	–	–	–	–	–	N-(Biphenyl-4-ylacetyl)taurine		MS07547
107	77	91	171	122	43	78	155	**319**	100	72	71	49	49	45	44	41	**0.00**	16	17	4	1	1	–	–	–	–	–	–	–	–	Carbamic acid, (p-tolylsulphonyl)-, α-methylbenzyl ester	32363-30-7	NI33681
107	91	171	155	197	108	260		**319**	100	71	49	41	31	15	2		**0.00**	16	17	4	1	1	–	–	–	–	–	–	–	–	Carbamic acid, (p-tolylsulphonyl)-, α-methylbenzyl ester	32363-30-7	LI07086
91	155	171	197	92	65	122	104	**319**	100	78	55	54	37	31	21	12	**0.00**	16	17	4	1	1	–	–	–	–	–	–	–	–	Carbamic acid, (p-tolylsulphonyl)-, phenethyl ester	18303-11-2	NI33682
91	155	171	197	107	108			**319**	100	78	55	54	11	8			**0.00**	16	17	4	1	1	–	–	–	–	–	–	–	–	Carbamic acid, (p-tolylsulphonyl)-, phenethyl ester	18303-11-2	LI07087
266	219	234	233	251	203	192	319	**319**	100	85	52	25	22	18	14	2		16	18	–	1	2	–	–	–	–	–	1	–	–	4-tert-Butyl-endo-5-cyano-2-phenyl-7-thia-1-phosphabicyclo[2.2.1]hept-2-ene 1-sulphide	93032-37-2	NI33683
106	319	107	149	162	148	108	77	**319**	100	76	56	39	37	22	20	17		16	21	2	3	1	–	–	–	–	–	–	–	–	1,4-Diamino-2-(N-phenyl-N-butylsulphonamido)-benzene		IC03916
202	73	319	203	320	304	75	45	**319**	100	57	29	19	8	8	7	7		16	25	2	1	–	–	–	–	–	2	–	–	–	1H-Indole-2-acetic acid, 1-(trimethylsilyl)-, trimethylsilyl ester	55125-13-8	NI33684
202	73	319	203	320	77	304	204	**319**	100	39	23	19	7	7	6	5		16	25	2	1	–	–	–	–	–	2	–	–	–	1H-Indole-3-acetic acid, 1-(trimethylsilyl)-, trimethylsilyl ester	56114-66-0	NI33685
202	73	203	319	45	75	74	204	**319**	100	77	19	18	15	8	7	5		16	25	2	1	–	–	–	–	–	2	–	–	–	1H-Indole-3-acetic acid, 1-(trimethylsilyl)-, trimethylsilyl ester	56114-66-0	NI33686
202	73	319	203	45	75	74	320	**319**	100	78	22	21	17	9	7	6		16	25	2	1	–	–	–	–	–	2	–	–	–	1H-Indole-3-acetic acid, 1-(trimethylsilyl)-, trimethylsilyl ester	56114-66-0	NI33687
289	202	272	201	288	224	217	173	**319**	100	20	17	10	9	9	9	7	**0.00**	17	21	5	1	–	–	–	–	–	–	–	–	–	Crinan-1,3-diol, 7-methoxy-, (1α,3α)-	4673-18-1	NI33688
319	44	43	57	41	55	232	72	**319**	100	40	31	23	23	21	20	19		17	21	5	1	–	–	–	–	–	–	–	–	–	1H,6H-5,11b-Ethano[1,3]dioxolo[4,5-j]phenanthridine-1,2-diol, 2,3,4,4a-tetrahydro-7-methoxy-	3660-65-9	NI33689
152	109	150	41	104	107	55	76	**319**	100	75	65	65	52	50	46	45	**0.00**	17	21	5	1	–	–	–	–	–	–	–	–	–	4-Isobenzofuranol, octahydro-3a,7a-dimethyl-, 4-nitrobenzoate, (3aα,4β,7aα)-(±)-	54346-04-2	NI33690
70	151	84	234	276	277	126	112	**319**	100	86	66	52	40	37	28	27	**18.00**	17	21	5	1	–	–	–	–	–	–	–	–	–	Pyrrolidine, 1-acetyl-2-[2-[4-(acetyloxy)-3-methoxyphenyl]-2-oxoethyl]-, (±)-	67257-67-4	NI33691
182	77	183	104	180	319	181	51	**319**	100	48	41	24	19	12	12	10		17	22	3	1	–	–	–	–	–	–	1	–	–	Phenylanilino(diethoxyphosphinyl)methane		IC03917
319	91	98	228	83	81	315	79	**319**	100	93	77	66	47	47	42	35		18	17	1	5	–	–	–	–	–	–	–	–	–	6,7-Diphenyl-5,6,7,8-tetrahydropterin		LI07088
319	91	98	228	95	109	83	81	**319**	100	93	77	66	60	49	47	47		18	17	1	5	–	–	–	–	–	–	–	–	–	6,7-Diphenyl-5,6,7,8-tetrahydropterin		MS07548
93	136	83	119	120	80	220	219	**319**	100	80	80	42	40	20	17	17	**10.00**	18	25	4	1	–	–	–	–	–	–	–	–	–	Retronecine, 7,9-bis-O-(2-methylbut-2-enoyl)-, (E,E)-		NI33692
232	150	165	97	123	291	248	233	**319**	100	50	39	30	29	28	18	15	**8.00**	18	25	4	1	–	–	–	–	–	–	–	–	–	Serratinine, 8-acetyl-13-dehydro-		LI07089
232	150	165	123	97	291	248	233	**319**	100	51	39	30	30	28	19	15	**8.36**	18	25	4	1	–	–	–	–	–	–	–	–	–	Serratinine, 14-deoxy-14-oxo-, acetate (ester)	7679-02-9	NI33693

MASS TO CHARGE RATIOS								M.W.	INTENSITIES								Parent	C	H	O	N	S	F	Cl	Br	I	Si	P	B	X	COMPOUND NAME	CAS Reg No	No
320	322	321	323	319	318			**319**	100	34	21	7	5	1				18	26	–	3	–	–	1	–	–	–	–	–	–	1,4-Pentanediamine, N[4]-(7-chloro-4-quinolinyl)-N[1],N[1]-diethyl-	54-05-7	NI33695
86	58	30	179	41	42	56	205	**319**	100	42	30	20	20	18	16	14	**4.80**	18	26	–	3	–	–	1	–	–	–	–	–	–	1,4-Pentanediamine, N[4]-(7-chloro-4-quinolinyl)-N[1],N[1]-diethyl-	54-05-7	IC03918
86	99	319	87	58	114	73	112	**319**	100	29	6	6	5	4	4	3		18	26	–	3	–	–	1	–	–	–	–	–	–	1,4-Pentanediamine, N[4]-(7-chloro-4-quinolinyl)-N[1],N[1]-diethyl-	54-05-7	NI33694
177	127	319	77	115	101	121	286	**319**	100	51	39	36	32	26	24	23		19	13	–	1	2	–	–	–	–	–	–	–	–	(2-Thio-3-benzo[b]thienylidene)-β-naphthylamine	18510-05-9	NI33696
319	127	286	128	192	143	159	77	**319**	100	64	49	38	37	32	28	28		19	13	–	1	2	–	–	–	–	–	–	–	–	(3-Thio-2-benzo[b]thienylidene)-β-naphthylamine	71643-97-5	NI33697
77	319	178	141	254	255	177	151	**319**	100	85	32	30	28	26	26	25		19	13	2	1	1	–	–	–	–	–	–	–	–	N-Fluorenylidene-S-phenylsulphonamide		LI07090
319	213	320	186	121	228	91	212	**319**	100	71	46	46	45	39	24	17		19	21	–	1	–	–	–	–	–	–	–	–	1	Ferrocenamine, N-benzyl-N-ethyl-	12148-88-8	MS07549
71	43	165	216	41	149	221	248	**319**	100	56	35	27	23	17	16	12	**6.00**	19	29	3	1	–	–	–	–	–	–	–	–	–	Benzoic acid 3-[(2,2-dimethyl-1-oxobutyl)amino]-5-(3-methylbutyl)-, methyl ester		MS07550
91	108	43	113	184	205	69	57	**319**	100	65	35	31	20	19	19	19	**0.00**	19	29	3	1	–	–	–	–	–	–	–	–	–	Benzyl 5-oxoundecylcarbamate		DD01217
276	41	55	67	303	81	288	95	**319**	100	50	47	46	44	41	37	36	**0.00**	19	29	3	1	–	–	–	–	–	–	–	–	–	17-Hydroxy-17-aza-D-homo-5α-androstane-16,17a-dione	123239-89-4	DD01218
83	141	55	271	104	105	79	41	**319**	100	80	24	22	21	21	19	18	**6.78**	19	29	3	1	–	–	–	–	–	–	–	–	–	N-Isobutyl-10-(1-isobutenylcarbonyloxy)(2E,6Z,8E)-decatrienamide		NI33698
105	77	135	41	134	55	246	162	**319**	100	31	19	13	12	12	10	9	**8.00**	19	29	3	1	–	–	–	–	–	–	–	–	–	Methyl N-benzoyl-11-aminoundecanoate		LI07091
55	81	67	109	95	41	121	163	**319**	100	90	90	83	82	75	38	30	**0.00**	19	29	3	1	–	–	–	–	–	–	–	–	–	15α-Nitro-5α-androstan-16-one	123239-83-8	DD01219
284	285	254	209	165	152	141	208	**319**	100	43	13	13	8	7	7	6	**4.00**	20	14	1	1	–	–	1	–	–	–	–	–	–	3-Chloro-2,3-diphenyl-2,3-dihydro-1-oxoisoindole		LI07092
106	120	304	171	121	115	57	77	**319**	100	79	75	60	54	53	46	44	**15.00**	20	17	3	1	–	–	–	–	–	–	–	–	–	N-Ethyl-N-phenyl-2-methyl-1,4-naphthoquinone-3-carboxamide		MS07551
120	121	304	91	115	171	45	65	**319**	100	83	45	35	32	26	21	19	**8.00**	20	17	3	1	–	–	–	–	–	–	–	–	–	N-Methyl-N-(p-tolyl)-2-methyl-1,4-naphthoquinone-3-carboxamide		MS07552
213	41	43	144	69	30	51	117	**319**	100	99	80	70	70	62	50	37	**3.00**	20	21	1	3	–	–	–	–	–	–	–	–	–	3-[2-(2-Aminophenyl)-1-(2-oxo-1-pyrrolidinyl)ethyl]indole		MS07553
126	319	70	113	61	260	288	87	**319**	100	30	26	24	24	21	20	20		20	33	2	1	–	–	–	–	–	–	–	–	–	Androstan-3-one, 17-hydroxy-, O-methyloxime, (5β,17β)-	69854-82-6	NI33699
288	289	96	87	270	213	286	107	**319**	100	27	24	22	12	10	9	9	**1.78**	20	33	2	1	–	–	–	–	–	–	–	–	–	Androstan-17-one, 3-hydroxy-, O-methyloxime, (3β,5α)-	5615-21-4	NI33700
164	319	146	107	58	148	302	320	**319**	100	85	65	50	40	32	20	19		20	33	2	1	–	–	–	–	–	–	–	–	–	1-(2-Hydroxy-5-dodecylphenyl)-1-ethanone (E)-oxime		DD01220
91	240	80	92	170	79	228	28	**319**	100	27	18	10	9	9	7	7	**1.18**	21	21	2	1	–	–	–	–	–	–	–	–	–	2-Azabicyclo[2.2.2]oct-5-ene-2-carboxylic acid, 3-phenyl-, benzyl ester, (1α,3β,4α)-	43044-10-6	MS07554
180	179	70	41	249	181	178	42	**319**	100	80	26	24	20	14	12	12	**0.00**	21	21	2	1	–	–	–	–	–	–	–	–	–	4,4a-Dihydro-1-isopropylidene-4,4-dimethyl-1,3-oxazino[3,4-f]phenanthridin-3-one		LI07093
262	105	120	214	319	109	183	91	**319**	100	97	96	82	72	52	43	30		21	22	–	1	–	–	–	–	–	–	1	–	–	Phosphinous amide, N-methyl-P,P-diphenyl-N-(1-phenylethyl)-, (S)-	52364-43-9	NI33701
123	43	319	108	59	41	42	136	**319**	100	32	29	28	20	16	14	13		21	37	1	1	–	–	–	–	–	–	–	–	–	2-(1-Oxohexadecyl)-5-methylpyrrole		NI33702
72	317	95	45	93	284	91	56	**319**	100	44	35	31	30	25	25	25	**3.16**	21	37	1	1	–	–	–	–	–	–	–	–	–	Pregnan-1-ol, 3-amino-, (1α,3α,5α)-	23931-02-4	NI33703
72	319	44	284	55	87	276	58	**319**	100	27	19	16	15	14	13	13		21	37	1	1	–	–	–	–	–	–	–	–	–	Pregnan-1-ol, 3-amino-, (1α,3α,5α)-	23931-02-4	LI07094
72	302	55	41	44	81	69	43	**319**	100	72	40	40	38	31	25	25	**2.00**	21	37	1	1	–	–	–	–	–	–	–	–	–	Pregnan-1-ol, 3-amino-, (1β,3α,5α)-	23931-05-7	LI07095
72	302	55	42	45	81	69	44	**319**	100	70	39	39	38	31	24	24	**0.00**	21	37	1	1	–	–	–	–	–	–	–	–	–	Pregnan-1-ol, 3-amino-, (1β,3α,5α)-	23931-05-7	NI33704
72	319	82	276	302				**319**	100	10	6	5	4					21	37	1	1	–	–	–	–	–	–	–	–	–	Pregnan-1-ol, 3-amino-, (1β,3β,5α)-		LI07096
181	301	222	219	138	152	56	223	**319**	100	93	93	64	62	54	50	41	**35.38**	22	25	1	1	–	–	–	–	–	–	–	–	–	cis-4-Biphenylyl 1-cyclohexyl-3-methyl-2-aziridinyl ketone	6372-59-4	NI33705
180	222	301	219	138	223	152	124	**319**	100	77	71	66	66	63	53	47	**20.39**	22	25	1	1	–	–	–	–	–	–	–	–	–	trans-4-Biphenylyl 1-cyclohexyl-3-methyl-2-aziridinyl ketone	32044-50-1	NI33706
119	319	91	118	181	200	236	320	**319**	100	72	60	34	30	29	28	16		22	25	1	1	–	–	–	–	–	–	–	–	–	cis-(1-Cyclohexyl-3-phenyl-2-aziridinyl) p-tolyl ketone	6476-39-7	NI33707
119	91	319	200	118	181	236	320	**319**	100	75	72	41	34	29	27	16		22	25	1	1	–	–	–	–	–	–	–	–	–	trans-(1-Cyclohexyl-3-phenyl-2-aziridinyl) p-tolyl ketone	6372-29-8	NI33708
228	91	319	158	249	229	104	65	**319**	100	87	40	34	22	18	12	10		23	29	–	1	–	–	–	–	–	–	–	–	–	Bicyclo[2.2.2]octan-1-amine, 4-methyl-N,N-dibenzyl-	55044-19-4	NI33709
319	320	318	317	241	315	316	321	**319**	100	27	17	15	11	8	6	5		24	17	–	1	–	–	–	–	–	–	–	–	–	1,4-Diphenylcarbazole		LI07097
127	47	66	193	320	301	174		**320**	100	71	44	31	30	1	1			–	–	–	–	–	2	–	–	2	1	–	–	–	Diiododifluorosilane		LI07098
59	243	293	291	245	241	322	324	**320**	100	25	15	15	15	15	9	8	**4.00**	4	3	2	–	–	–	–	3	–	–	–	–	–	Methyl tribromoacrylate		NI33710
277	275	273	247	43	135	276	245	**320**	100	75	45	35	31	30	28	26	**16.00**	5	13	–	–	–	–	–	–	1	–	–	–	1	Dimethylisopropyltin iodide		MS07555
69	119	51	169	113	201	131	31	**320**	100	49	32	28	23	20	6	5	**0.00**	6	1	–	–	–	13	–	–	–	–	–	–	–	Hexane, 1,1,1,2,2,3,4,4,5,5,6,6,6-tridecafluoro-		MS07556
69	151	51	119	113	169	131	93	**320**	100	35	24	13	11	10	5	4	**0.00**	6	1	–	–	–	13	–	–	–	–	–	–	–	Hexane, 1,1,1,2,3,3,4,4,5,5,6,6,6-tridecafluoro-		MS07557
51	69	101	131	119	169	151	100	**320**	100	58	20	15	13	8	7	5	**0.00**	6	1	–	–	–	13	–	–	–	–	–	–	–	Hexane, 1,1,2,2,3,3,4,4,5,5,6,6,6-tridecafluoro-		MS07558
85	91	75	282	263	135	213	160	**320**	100	56	37	33	33	33	26	20	**0.00**	6	2	–	–	–	10	–	–	–	2	–	–	–	4,5-Bis(trifluoromethyl)-1,1,2,2-tetrafluoro-1,2-dihydro-1,2-disilabenzene		LI07099
125	285	287	109	47	79	93	63	**320**	100	97	63	43	39	35	25	17	**0.00**	8	8	3	–	1	–	3	–	–	–	1	–	–	Phosphorothioic acid, O,O-dimethyl O-(2,4,5-trichlorophenyl) ester	299-84-3	AI02038
287	125	285	79	109	93	289	47	**320**	100	99	93	47	43	34	24	24	**1.33**	8	8	3	–	1	–	3	–	–	–	1	–	–	Phosphorothioic acid, O,O-dimethyl O-(2,4,5-trichlorophenyl) ester	299-84-3	NI33711
285	287	125	109	93	47	79	289	**320**	100	81	77	38	33	32	31	25	**7.89**	8	8	3	–	1	–	3	–	–	–	1	–	–	Phosphorothioic acid, O,O-dimethyl O-(2,4,5-trichlorophenyl) ester	299-84-3	NI33712
79	85	57	45	43	29	97	61	**320**	100	99	75	26	20	20	17	15	**0.11**	8	16	9	–	2	–	–	–	–	–	–	–	–	α-D-Xylopyranoside, methyl, 2,4-dimethanesulphonate	29709-44-2	NI33713
167	195	251	141	223	114	186	102	**320**	100	42	39	29	18	18	13	9	**0.00**	9	5	3	–	–	3	–	–	–	–	–	–	1	Ruthenium, dicarbonyl(η^5-2,4-cyclopentadien-1-yl)(trifluoroacetyl)-	55836-26-5	NI33714
307	309	305	113	115	93	311	63	**320**	100	67	62	57	40	24	23	19	**9.20**	9	9	–	–	–	–	5	–	–	1	–	–	–	(Pentachlorophenyl)trimethylsilane		MS07559
305	303	307	304	301	306	275	309	**320**	100	75	74	38	28	24	24	19	**4.50**	9	13	–	–	–	–	–	1	–	–	–	–	1	(4-Bromophenyl)trimethyltin		MS07560
43	279	277	281	287	285	167	166	**320**	100	75	68	43	34	30	27	24	**1.00**	9	21	–	–	–	–	2	–	–	–	–	–	1	Triisopropylantimony dichloride	58037-60-8	NI33715
278	246	63	43	218	247	279	91	**320**	100	93	29	24	22	19	10	10	**7.80**	10	9	4	–	–	–	–	–	1	–	–	–	–	Salicylic acid, 5-iodo-, methyl ester, acetate	22621-40-5	MS07561

MASS TO CHARGE RATIOS								M.W.	INTENSITIES								Parent	C	H	O	N	S	F	Cl	Br	I	Si	P	B	X	COMPOUND NAME	CAS Reg No	No
278	246	63	43	247	218	91	53	**320**	100	92	30	23	20	20	11	10	**8.01**	10	9	4	–	–	–	–	–	1	–	–	–	–	Salicylic acid, 5-iodo-, methyl ester, acetate	22621-40-5	NI33716
278	246	63	43	247	218	279	320	**320**	100	92	30	24	20	20	11	8		10	9	4	–	–	–	–	–	1	–	–	–	–	Salicylic acid, 5-iodo-, methyl ester, acetate	22621-40-5	LI07100
67	41	51	134	162	91	65	79	**320**	100	51	50	44	43	38	36	27	**5.12**	10	10	2	–	–	–	–	2	–	–	–	–	–	2,5-Cyclohexadiene-1,4-dione, 2,5-dibromo-3-methyl-6-(1-methylethyl)-	29096-93-3	NI33717
320	166	76	193	75	203	114	127	**320**	100	91	49	44	17	12	10	9		11	5	–	4	–	–	–	–	1	–	–	–	–	(3-Iodoanilino)ethenetricarbonitrile		DD01221
320	166	76	193	203	293	127	75	**320**	100	30	13	11	7	6	5	4		11	5	–	4	–	–	–	–	1	–	–	–	–	(4-Iodoanilino)ethenetricarbonitrile		DD01222
55	170	105	263	169	320	140	291	**320**	100	99	96	95	88	80	62	58		11	5	2	–	–	5	–	–	–	–	–	–	1	Iron, dicarbonyl(η^5-2,4-cyclopentadien-1-yl)(1,2,3,4,4-pentafluoro-2-cyclobuten-1-yl)-	56943-69-2	NI33718
320	180	206	207	136	321	138	67	**320**	100	33	23	22	15	13	12	12		11	11	4	4	–	3	–	–	–	–	–	–	–	7-[2-(Trifluoroacetyl)ethyl]-1,3-dimethylxanthine		NI33719
85	43	41	44	183	55	67	86	**320**	100	48	24	12	9	9	7	6	**0.00**	11	13	3	–	–	3	2	–	–	–	–	–	–	1-(1,1-Dichloro-2,2,2-trifluoroethyl)-3-methyl-2-butenyl 3-oxobutanoate	107686-52-2	DD01223
323	59	61	321	325	179	289	287	**320**	100	85	80	72	42	32	30	30	**0.00**	12	4	2	–	–	–	4	–	–	–	–	–	–	Dibenzo[b,e][1,4]dioxin, 1,2,3,4-tetrachloro-	30746-58-8	NI33720
322	320	324	257	50	259	194	161	**320**	100	73	45	21	20	19	15	15		12	4	2	–	–	–	4	–	–	–	–	–	–	Dibenzo[b,e][1,4]dioxin, 1,2,3,4-tetrachloro-	30746-58-8	AI02039
322	320	324	257	259	194	50	28	**320**	100	81	51	23	22	18	16	15		12	4	2	–	–	–	4	–	–	–	–	–	–	Dibenzo[b,e][1,4]dioxin, 1,2,3,4-tetrachloro-	30746-58-8	NI33721
322	320	324	257	259	194	161	321	**320**	100	82	62	30	29	28	22	20		12	4	2	–	–	–	4	–	–	–	–	–	–	Dibenzo[b,e][1,4]dioxin, 1,3,6,8-tetrachloro-	33423-92-6	NI33722
322	320	257	324	259	194	196	97	**320**	100	76	49	48	47	33	21	16		12	4	2	–	–	–	4	–	–	–	–	–	–	Dibenzo[b,e][1,4]dioxin, 1,3,6,8-tetrachloro-	33423-92-6	NI33723
322	320	324	257	259	194	323	196	**320**	100	78	50	31	30	21	14	14		12	4	2	–	–	–	4	–	–	–	–	–	–	Dibenzo[b,e][1,4]dioxin, 1,3,7,8-tetrachloro-	50585-46-1	NI33724
322	320	324	257	259	194	196	161	**320**	100	78	50	47	45	36	23	17		12	4	2	–	–	–	4	–	–	–	–	–	–	Dibenzo[b,e][1,4]dioxin, 1,3,7,9-tetrachloro-	62470-53-5	NI33725
322	320	324	257	259	194	323	161	**320**	100	79	48	30	29	20	13	13		12	4	2	–	–	–	4	–	–	–	–	–	–	Dibenzo[b,e][1,4]dioxin, 2,3,7,8-tetrachloro-	1746-01-6	NI33726
322	320	324	257	259	161	194	74	**320**	100	79	48	24	20	18	12	11		12	4	2	–	–	–	4	–	–	–	–	–	–	Dibenzo[b,e][1,4]dioxin, 2,3,7,8-tetrachloro-	1746-01-6	AI02040
320	264	206	292	236	234	208	103	**320**	100	78	70	53	32	32	32	32		12	9	4	–	–	–	–	–	–	–	–	–	1	Rhodium, dicarbonyl(1-phenyl-1,3-butanedionato)-	24151-55-1	NI33727
274	93	125	121	91	107	135	246	**320**	100	94	89	69	45	41	36	35	**12.21**	12	17	4	–	2	–	–	–	–	–	1	–	–	Benzeneacetic acid, α-[(dimethoxyphosphinothioyl)thio]-, ethyl ester	2597-03-7	NI33728
128	69	41	84	160	72	55	129	**320**	100	43	40	36	34	21	19	15	**7.81**	12	20	–	2	4	–	–	–	–	–	–	–	–	Piperidine, 1,1′-(dithiodicarbonothioyl)bis-	94-37-1	NI33729
142	114	117	118	85	86	56	72	**320**	100	75	70	67	52	50	47	37	**6.49**	12	20	6	2	1	–	–	–	–	–	–	–	–	1,4,10-Trioxa-13-thia-7,16-diazacyclooctadecane-8,12,15-trione	74229-41-7	NI33730
153	151	178	320	97	169	125	150	**320**	100	74	72	62	62	53	44	43		12	21	4	2	1	–	–	–	–	–	1	–	–	Phosphorothioic acid, O,O-diethyl O-[2-(1-hydroxyisopropyl)-6-methyl-4 pyrimidinyl] ester	29820-16-4	NI33731
153	151	178	320	97	169	125	305	**320**	100	74	73	63	63	53	48	40		12	21	4	2	1	–	–	–	–	–	1	–	–	Phosphorothioic acid, O,O-diethyl O-[2-(1-hydroxysiopropyl)-6-methyl-4 pyrimidinyl] ester	29820-16-4	LI07101
73	147	305	45	75	69	148	221	**320**	100	98	30	28	24	18	16	15	**0.00**	12	28	4	–	–	–	–	–	–	3	–	–	–	Propanedioic acid, (trimethylsilyl)-, bis(trimethylsilyl) ester	56051-48-0	NI33732
102	291	73	219	204	103	147	101	**320**	100	52	45	15	12	10	9	9	**2.00**	12	32	2	2	–	–	–	–	–	3	–	–	–	Propionic acid, 2,3-bis[(trimethylsilyl)amino]-, trimethylsilyl ester		MS07562
125	127	89	63	99	90	126	62	**320**	100	39	32	14	12	10	8	7	**1.00**	13	8	1	–	–	–	4	–	–	–	–	–	–	Benzene, trichloro[(chlorophenyl)methoxy]-	67332-31-4	NI33733
322	305	320	137	279	307	309	207	**320**	100	99	94	69	68	56	41	41		13	8	1	–	–	–	4	–	–	–	–	–	–	1,1′-Biphenyl, 2,3′,4,5′-tetrachloro-4′-methoxy-	74298-93-4	NI33734
91	55	93	28	29	27	121	41	**320**	100	57	33	22	11	8	7	7	**0.00**	13	12	4	4	1	–	–	–	–	–	–	–	–	2H-1,2,3-Benzothiadiazine, 2-(2,4-dinitrophenyl)-5,6,7,8-tetrahydro-	42141-19-5	NI33735
85	192	224	193	160	289	320		**320**	100	94	90	31	25	12	1			13	20	7	–	1	–	–	–	–	–	–	–	–	6-O-Methylsulphonyl-1-O-methyldihydroverbenalol		LI07102
85	192	224	160	193	320	289		**320**	100	54	30	29	22	20	10			13	20	7	–	1	–	–	–	–	–	–	–	–	5-O-Methylsulphonyl-1-O-methyl-5-epiloganin aglucone		LI07103
85	160	192	193	224	289	320		**320**	100	26	15	11	9	3	1			13	20	7	–	1	–	–	–	–	–	–	–	–	5-O-Methylsulphonyl-1-O-methylloganin aglucone		LI07104
130	43	129	75	87	74	158	101	**320**	100	99	69	62	51	36	26	23	**0.00**	13	20	9	–	–	–	–	–	–	–	–	–	–	Methyl(methyl 2,4-di-O-acetyl-3-O-methyl-α-D-galactopyranoside)uronate	87910-31-4	NI33736
129	43	87	117	74	85	127	99	**320**	100	99	78	69	53	42	26	26	**0.39**	13	20	9	–	–	–	–	–	–	–	–	–	–	Methyl(methyl 2,3-di-O-acetyl-4-O-methyl-α-D-mannopyranoside)uronate	87910-32-5	NI33737
74	43	129	127	87	116	158	75	**320**	100	99	59	45	45	36	32	22	**0.00**	13	20	9	–	–	–	–	–	–	–	–	–	–	Methyl(methyl 3,4-di-O-acetyl-2-O-methyl-α-D-mannopyranoside)uronate	87910-30-3	NI33738
278	320	108	43	92	44	28	140	**320**	100	71	60	45	38	35	29	29		14	12	5	2	1	–	–	–	–	–	–	–	–	4-Nitro-4′-acetamidodiphenylsulphone		NI33739
149	218	164	236	79	93	121	91	**320**	100	78	74	59	59	50	48	45	**0.00**	14	16	5	–	–	–	–	–	–	–	–	–	1	Tricarbonyl[(1,3,4,5-η)-5-(1-hydroxyisopropyl)-1-methyl-2-oxo-4-cycloheptene-1,3-diyl]iron		DD01224
57	29	41	56	45	85	27	42	**320**	100	82	73	56	39	31	28	25	**13.01**	14	18	4	–	–	–	2	–	–	–	–	–	–	Acetic acid, (2,4-dichlorophenoxy)-, 2-butoxyethyl ester	1929-73-3	NI33740
59	320	155	219	127	145	161	41	**320**	100	61	61	41	39	36	24	19		14	24	–	–	4	–	–	–	–	–	–	–	–	2,4,6,8-Tetrathiaadamantane, 1,3,5,7,9,9,10,10-octamethyl-	17749-62-1	NI33741
61	273	215	305	242	43	127	100	**320**	100	99	86	63	63	40	37	37	**9.00**	14	24	6	–	1	–	–	–	–	–	–	–	–	1,2:3,4-Di-O-isopropylidene-6-O-[(methylthio)methyl]-α-D-galactopyranose		LI07105
173	215	203	186	169	170	302	229	**320**	100	80	78	38	27	25	24	23	**5.13**	14	24	8	–	–	–	–	–	–	–	–	–	–	D-Allitol, 6-deoxy-3-C-methyl-2-O-methyl-, triacetate	56687-69-5	NI33742
117	43	101	89	233	127	173	99	**320**	100	99	41	32	11	11	11	11	**0.00**	14	24	8	–	–	–	–	–	–	–	–	–	–	D-Galactitol, 1,3,5-tri-O-acetyl-2,4-di-O-methyl-6-deoxy-	81703-39-1	NI33743
43	117	131	233	101	89	28	99	**320**	100	74	71	48	48	40	26	25	**5.00**	14	24	8	–	–	–	–	–	–	–	–	–	–	D-Mannitol, 1-deoxy-3,5-di-O-methyl-, triacetate	75363-16-5	NI33744
29	207	149	121	151	205	177	119	**320**	100	84	74	74	73	61	61	61	**0.00**	14	32	–	–	–	–	–	–	–	–	–	–	1	Stannane, tributylethyl-	19411-60-0	MS07563
207	121	151	179	263	291	235	177	**320**	100	81	68	57	52	33	26	25	**0.00**	14	32	–	–	–	–	–	–	–	–	–	–	1	Stannane, tributylethyl-	19411-60-0	NI33745
239	274	302	276	304	241	77	275	**320**	100	91	68	58	45	34	34	32	**0.00**	15	10	2	2	–	–	2	–	–	–	–	–	–	2H-1,4-Benzodiazepin-2-one, 7-chloro-5-(2-chlorophenyl)-1,3-dihydro-3-hydroxy-	846-49-1	NI33746
232	320	83	97	215	173	259	149	**320**	100	46	45	40	31	31	28	22		15	12	8	–	–	–	–	–	–	–	–	–	–	Biphenyl-2′-carboxylic acid, 3,4-dihydroxy-4′,5′-methylenedioxy-2,5-dioxy-2,3,4,5-tetrahydro-, methyl ester, (3R,4-cis)-		NI33747

MASS TO CHARGE RATIOS	M.W.	INTENSITIES	Parent	C	H	O	N	S	F	Cl	Br	I	Si	P	B	X	COMPOUND NAME	CAS Reg No	No
123 246 91 139 124 158 125 155	320	100 95 64 62 47 43 37 32	0.00	15	13	2	2	1	–	1	–	–	–	–	–	–	(1-Chloro-2-phenylethenyl)(4-methylphenylsulphonyl)diazene		DD01225
230 320 43 304 303 247 232 229	320	100 85 44 37 37 37 35 34		15	13	2	2	1	–	1	–	–	–	–	–	–	10-Methyl-3-chloro-7-acetylaminophenothiazine-5-oxide		NI33748
193 236 194 227 320	320	100 99 88 62 36		15	16	6	2	–	–	–	–	–	–	–	–	–	Acetamide, N-[3,4-bis(acetyloxy)-5-(2-furanyl)-1-methyl-1H-pyrrol-2-yl]-	50618-96-7	NI33749
43 159 172 158 171 160 105 173	320	100 66 31 15 8 6 6 4	4.01	15	17	7	–	–	–	–	–	–	–	–	1	–	α-D-Ribopyranose, cyclic 2,4-(phenylboronate) 1,3-diacetate	74825-23-3	NI33750
197 195 193 279 161 196 159 120	320	100 76 45 39 39 31 29 28	0.50	15	20	–	–	–	–	–	–	–	–	–	–	1	Triallylphenyltin		MS07564
89 88 43 86 129 85 130 59	320	100 83 83 55 51 48 45 42	0.00	15	28	7	–	–	–	–	–	–	–	–	–	–	Methyl 2-O-acetyl-3,4,6-tri-O-ethyl-α-D-galactopyranoside		NI33751
116 129 43 89 88 101 60 57	320	100 76 44 39 29 28 25 20	0.00	15	28	7	–	–	–	–	–	–	–	–	–	–	Methyl 6-O-acetyl-2,3,4-tri-O-ethyl-α-D-galactopyranoside		NI33752
88 129 43 130 85 101 102 103	320	100 79 75 70 53 28 26 25	0.00	15	28	7	–	–	–	–	–	–	–	–	–	–	Methyl 3-O-acetyl-2,4,6-tri-O-ethyl-α-D-mannopyranoside		NI33753
89 116 88 101 59 41 117 60	320	100 90 42 25 23 23 22 22	0.00	15	28	7	–	–	–	–	–	–	–	–	–	–	Methyl 4-O-acetyl-2,3,6-tri-O-ethyl-α-D-mannopyranoside		NI33754
55 73 69 60 43 57 83 41	320	100 79 78 77 61 60 57 49	9.39	15	29	2	–	–	–	–	1	–	–	–	–	–	Pentadecanoic acid, 14-bromo-	74685-41-9	NI33755
73 60 55 69 43 57 129 83	320	100 91 68 55 43 42 38 37	13.14	15	29	2	–	–	–	–	1	–	–	–	–	–	Pentadecanoic acid, 15-bromo-	56523-59-2	NI33756
258 61 100 131 83 132 81 111	320	100 63 29 27 25 23 22 15	0.33	15	32	1	2	2	–	–	–	–	–	–	–	–	Urea, N,N'-bis[6-(methylthio)hexyl]-		NI33757
320 206 90 149 176 144 162 102	320	100 38 30 28 17 16 15 12		16	8	2	4	1	–	–	–	–	–	–	–	–	12H,15H-[1,3,4]Thiadiazolo[2,3-b:5,4-b']diquinazoline-12,15-dione	81559-83-3	NI33758
319 320 292 291 293 91 321 235	320	100 97 94 70 17 14 8 7		16	11	2	2	–	3	–	–	–	–	–	–	–	2,3-Dihydro-5-phenyl-7-trifluoromethoxy-1H-1,4-benzodiazepin-2-one	36107-44-5	NI33759
302 320 29 304 303 322 291 290	320	100 65 40 37 28 23 20 19		16	13	5	–	–	–	1	–	–	–	–	–	–	7-Chloro-6-hydroxy-1,3-dimethoxy-8-methyl-9H-xanthen-9-one		MS07565
289 291 142 179 169 147 320 171	320	100 22 16 13 13 13 9 3		16	13	5	–	–	–	1	–	–	–	–	–	–	Methyl 3-(5-chloro-2-methoxybenzoyl)-4-hydroxybenzoate		NI33760
322 131 190 175 161 58 177 176	320	100 31 12 12 9 9 8 6		16	14	–	4	–	–	–	–	–	–	–	–	1	Nickel, N,N'-bis(2-aminobenzylidene)diaminato-		LI07106
222 91 152 221 133 36 223 65	320	100 54 33 26 18 18 16 16	3.90	16	14	1	2	–	–	2	–	–	–	–	–	–	N-(Chlorocarbonyl)-N,N'-di(p-tolyl)-α-chloroformamidine		IC03919
139 182 153 320 154 69 55 166	320	100 93 45 36 25 18 18 17		16	16	7	–	–	–	–	–	–	–	–	–	–	2H-1-Benzopyran-3,5,7-triol, 2-(3,4-dihydroxy-5-methoxyphenyl)-3,4-dihydro-, cis-(±)-	39689-32-2	MS07566
171 320 170 168 172 154 127 292	320	100 41 30 17 14 9 8 6		16	20	3	2	1	–	–	–	–	–	–	–	–	Dansyl-4-aminobutanal		NI33761
219 261 59 116 43 45 187 159	320	100 61 60 49 44 43 31 26	0.00	16	20	5	2	–	–	–	–	–	–	–	–	–	Dimethyl 1,2-dihydro-1-isopropoxy-4a,8a-methanophthalazine-1,4-dicarboxylate		DD01226
251 253 172 210 208 322 320 129	320	100 99 90 66 66 50 50 40		16	21	–	2	–	–	–	1	–	–	–	–	–	Dihydroflustramine C		NI33762
226 322 320 212 307 305 291 211	320	100 60 60 60 40 40 22 21		16	21	–	2	–	–	–	1	–	–	–	–	–	(3-Ethyl-4-methyl-5-bromo-2-pyrrolyl)(3,5-dimethyl-4-ethyl-2-pyrrolylidene)methane		LI07107
210 208 320 182 67 79 54 206	320	100 81 50 42 39 33 31 27		16	25	–	–	–	–	–	–	–	–	–	–	1	Rhodium, [(1,2,5,6-η)-1,5-cyclooctadiene][(1,2,3-η)-2-cycloocten-1-yl]-	31798-33-1	NI33763
60 73 144 45 143 255 320 176	320	100 53 29 26 24 14 8 4		16	32	–	–	3	–	–	–	–	–	–	–	–	3,5-Diheptyl-1,2,4-trithiolane		NI33764
179 291 79 235 135 193 151 292	320	100 60 37 31 30 19 14 13	0.00	16	36	4	–	–	–	–	–	–	1	–	–	–	Silicic acid (H4SiO4), tetrabutyl ester	4766-57-8	NI33765
247 277 57 79 55 41 43 203	320	100 53 33 31 29 28 25 22	0.89	16	36	4	–	–	–	–	–	–	1	–	–	–	Silicic acid (H4SiO4), tetrabutyl ester	4766-57-8	NI33766
303 105 305 320 77 287 304 224	320	100 88 76 57 46 33 29 24		17	14	2	–	–	–	2	–	–	–	–	–	–	5,6-exo-Dichloro-1,4-diphenyl-2,3-dioxabicyclo[2.2.1]hept-5-ene		NI33767
201 199 77 132 120 91 90 119	320	100 96 25 24 19 14 12 11	0.00	17	15	1	2	–	3	–	–	–	–	–	–	–	1-(Phenylcarbamoyl)-2-methyl-3-phenyl-2-(trifluoromethyl)aziridine		DD01227
320 121 186 56 212 318 227 161	320	100 45 37 31 24 18 18 11		17	16	1	2	–	–	–	–	–	–	–	–	1	1-[3-(2-Furyl)pyrazolin-5-yl]ferrocene		LI07108
115 43 131 116 187 160 188 103	320	100 78 67 52 42 39 33 27	0.00	17	20	6	–	–	–	–	–	–	–	–	–	–	Diethyl 3-acetoxyindan-1,1-dicarboxylate	118629-54-2	DD01228
167 154 65 93 137 39 122 148	320	100 88 49 48 37 26 25 22	2.00	17	20	6	–	–	–	–	–	–	–	–	–	–	3,5-Dimethoxy-4-hydroxybenzyl 2-methoxy-4-(hydroxymethyl)phenyl ether		NI33768
320 321 289 167 319 322 290 245	320	100 16 13 8 7 3 3 3		17	20	6	–	–	–	–	–	–	–	–	–	–	Bis(3,5-dimethoxy-4-hydroxyphenyl)methane		MS07567
167 154 57 43 71 41 55 28	320	100 80 68 61 38 38 35 30	2.00	17	20	6	–	–	–	–	–	–	–	–	–	–	2,2-(5-Methoxy-2-methyl-4-oxopyran-6-yl)propane		MS07568
44 75 247 207 169 245 320 302	320	100 75 46 46 29 21 20 14		17	20	6	–	–	–	–	–	–	–	–	–	–	Mycophenolic acid	24280-93-1	IC03920
277 180 151 91 161 165 189 76	320	100 53 35 33 28 26 23 23	3.03	17	20	6	–	–	–	–	–	–	–	–	–	–	Spiro[naphth[1,2-b]oxirene-5(1aH),2'-oxirane]-3'-carboxaldehyde, 2-(acetyloxy)-2,3,3a,4,6,7b-hexahydro-3,3',3a-trimethyl-6-oxo-, [1aR-[1aα,2β,3β,3aβ,5β(S*),7bα]]-		NI33769
226 194 227 198 199 284 193 228	320	100 24 14 8 8 7 5 4	0.00	17	21	–	2	1	–	1	–	–	–	–	–	–	N,N,β-Trimethyl-10H-phenothiazine-10-ethanamine hydrochloride		NI33770
222 178 122 136 320 192 166 208	320	100 82 71 53 45 43 38 34		17	28	2	4	–	–	–	–	–	–	–	–	–	1,5,7-Tributyloxipurinol		NI33771
255 320 128 89 101 192 161 160	320	100 76 57 30 27 25 23 14		18	12	–	2	2	–	–	–	–	–	–	–	–	Bis(isoquinolin-1-yl) disulphide		NI33772
90 320 145 118 89 132 161 160	320	100 92 90 74 74 40 14 9		18	12	4	2	–	–	–	–	–	–	–	–	–	2,2'-Biisoquinolinyl, 1,1',3,3'-tetraoxo-1,1',2,2',3,3',4,4'-octahydro-		LI07109
156 128 155 226 141 127 113 320	320	100 90 71 57 42 33 28 22		18	12	4	2	–	–	–	–	–	–	–	–	–	Naphthalene, 1-(2,4-dinitrostyryl)-		NI33773
320 322 165 243 89 321 115 77	320	100 36 32 26 26 20 19 17		18	13	–	4	–	–	1	–	–	–	–	–	–	2-Methyl-3-chloro-6,7-diphenylimidazo[1,2-b]-1,2,4-triazine		NI33774
91 320 229 77 128 273 117 102	320	100 30 15 15 11 10 10 10		18	16	–	4	1	–	–	–	–	–	–	–	–	1-Benzyl-2-(methylthio)-6-phenyl-1H-imidazo[1,2-b][1,2,4]triazole		MS07569
117 247 320 116 273 91 103 207	320	100 78 75 58 27 26 19 18		18	16	–	4	1	–	–	–	–	–	–	–	–	2-(Methylthio)-6-phenyl-1-(p-tolyl)-1H-imidazo[1,2-b][1,2,4]triazole		MS07570
320 255 225 321 197 141 139 256	320	100 66 53 26 16 16 13 11		18	16	2	–	–	–	–	–	–	–	–	–	1	Methyl 2-ferrocenylbenzoate		MS07571
320 321 255 225 197 141 139 54	320	100 66 59 47 14 14 13 12		18	16	2	–	–	–	–	–	–	–	–	–	1	Methyl 2-ferrocenylbenzoate		LI07110
320 255 121 56 129 128 185 184	320	100 31 31 25 20 15 8 7		18	16	2	–	–	–	–	–	–	–	–	–	1	1-[β-(5-Methylfur-2-yl)acryloyl]ferrocene		LI07111
105 77 292 68 51 320 293 172	320	100 29 8 6 6 1 1 1		18	16	2	4	–	–	–	–	–	–	–	–	–	4,5-Dimethyl-1-(α-benzoyloxybenzylideneimino)-1,2,3-triazole	19226-31-4	MS07572
105 77 51 292 68 293 145 104	320	100 47 12 11 5 2 1 1	0.15	18	16	2	4	–	–	–	–	–	–	–	–	–	4,5-Dimethyl-1α-dibenzoylamino-1,2,3-triazole	19226-35-8	MS07573

MASS TO CHARGE RATIOS	M.W.	INTENSITIES	Parent	C	H	O	N	S	F	Cl	Br	I	Si	P	B	X	COMPOUND NAME	CAS Reg No	No
91 132 320 77 321 64 131 92	**320**	100 72 63 28 14 11 10 9		18	16	2	4	–	–	–	–	–	–	–	–	–	**Formanilide, N-methyl-4′-[(3-methyl-5-oxo-1-phenyl-2-pyrazolin-4-ylidene)amino]-**	19362-42-6	**MS07574**
91 132 320 77 321 64 131 51	**320**	100 72 62 28 13 11 10 8		18	16	2	4	–	–	–	–	–	–	–	–	–	**Formanilide, N-methyl-4′-[(3-methyl-5-oxo-1-phenyl-2-pyrazolin-4-ylidene)amino]-**	19362-42-6	**NI33775**
260 77 28 289 44 259 43 31	**320**	100 91 75 74 66 48 41 41	**35.30**	18	16	2	4	–	–	–	–	–	–	–	–	–	**1,2-Propanediol, 1-(1-phenyl-1H-pyrazolo[3,4-b]quinoxalin-3-yl)-, (2S)-**	31504-92-4	**NI33776**
276 275 220 77 245 51 45 248	**320**	100 87 52 51 40 34 26 22	**0.70**	18	16	2	4	–	–	–	–	–	–	–	–	–	**1,2-Propanediol, 1-(1-phenyl-1H-pyrazolo[3,4-b]quinoxalin-3-yl)-, [S-(R*,S*)]-**	56804-89-8	**NI33777**
248 103 320 104 77 247 76 249	**320**	100 81 65 59 28 26 26 20		18	16	2	4	–	–	–	–	–	–	–	–	–	**s-Triazine-2-carbamic acid, 4,6-diphenyl-, ethyl ester**	29366-68-5	**NI33778**
91 265 174 104 320 65 92 119	**320**	100 96 93 88 76 60 45 42		18	20	–	6	–	–	–	–	–	–	–	–	–	**N-Benzyl-N-ethyl-4-(4-methyl-4H-1,2,4-triazol-3-yl)aniline**		**IC03921**
146 145 147 320 132 131 130 128	**320**	100 87 40 35 21 21 20 15		18	20	–	6	–	–	–	–	–	–	–	–	–	**3,6-Bis(4-dimethylaminophenyl)-1,2,4,5-tetrazine**	81258-54-0	**NI33779**
73 91 85 133 70 43 101 41	**320**	100 85 60 56 55 47 31 29	**1.50**	18	24	5	–	–	–	–	–	–	–	–	–	–	**4-Benzyl-2,2-dimethyl-3-(2-hydroxy-3-methylbutyryloxy)-4-butanolide**		**LI07112**
167 320 57 210 83 69 55 85	**320**	100 99 99 89 69 60 49 44		18	24	5	–	–	–	–	–	–	–	–	–	–	**7H-6,9a-Methanocycloocta[b]furan-7,10-dione, 2,3,3a,4,5,6-hexahydro-9-hydroxy-2,2-dimethyl-8-(3-methyl-1-oxobutyl)-**	55712-72-6	**NI33780**
73 211 195 251 121 109 320 285	**320**	100 74 51 50 49 29 17 11		18	28	1	–	1	–	–	–	–	1	–	–	–	**Silane, [[2,5-dimethyl-1-methylene-2-(phenylthio)-4-hexenyl]oxy]trimethyl-**	66032-87-9	**NI33781**
73 135 210 110 123 147 251 291	**320**	100 36 30 29 27 20 9 4	**2.00**	18	28	1	–	1	–	–	–	–	1	–	–	–	**Silane, [[2,5-dimethyl-1-[(phenylthio)methyl]-1,4-hexadienyl]oxy]trimethyl-**	75332-29-5	**NI33782**
209 320 179 223 210 305 277 290	**320**	100 42 33 30 25 13 10 5		18	28	3	–	–	–	–	–	–	1	–	–	–	**4-Octen-3-one, 1-[3-methoxy-4-[(trimethylsilyl)oxy]phenyl]-**	56700-90-4	**MS07575**
219 261 320 249 220 262 277 321	**320**	100 85 48 34 32 20 15 10		18	28	3	2	–	–	–	–	–	–	–	–	–	**Lycocernuine, acetate (ester)**	36101-26-5	**NI33783**
320 89 132 77 321 118 178 160	**320**	100 67 46 22 20 20 18 16		19	12	5	–	–	–	–	–	–	–	–	–	–	**2(3H)-Benzofuranone, 3-(3-methoxy-5-oxo-4-phenyl-2(5H)-furylidene)-**	23670-24-8	**NI33784**
320 89 132 76 321 117 176 160	**320**	100 68 45 23 21 21 19 16		19	12	5	–	–	–	–	–	–	–	–	–	–	**2(3H)-Benzofuranone, 3-(3-methoxy-5-oxo-4-phenyl-2(5H)-furylidene)-**	23670-24-8	**LI07113**
320 175 91 89 145 321 119 117	**320**	100 68 58 55 25 20 20 17		19	12	5	–	–	–	–	–	–	–	–	–	–	**Furo[3,2-b]furan-2,5-dione, 3-(2-methoxyphenyl)-6-phenyl-**	5099-87-6	**NI33785**
320 208 89 147 119 264 145 117	**320**	100 67 52 51 50 45 28 20		19	12	5	–	–	–	–	–	–	–	–	–	–	**Furo[3,2-b]furan-2,5-dione, 3-(4-methoxyphenyl)-6-phenyl-**	22628-17-7	**LI07114**
320 208 89 147 119 260 145 175	**320**	100 65 50 49 48 42 27 18		19	12	5	–	–	–	–	–	–	–	–	–	–	**Furo[3,2-b]furan-2,5-dione, 3-(4-methoxyphenyl)-6-phenyl-**	22628-17-7	**NI33786**
199 105 157 115 156 77 43	**320**	100 61 52 52 38 33 25	**0.00**	19	16	3	2	–	–	–	–	–	–	–	–	–	**N-Benzoyloxy-2-cyano-4,5-dimethyl-2-phenyl-2,3-dihydrooxazole**		**LI07115**
77 180 248 51 105 320 319 291	**320**	100 82 47 24 20 17 13 2		19	16	3	2	–	–	–	–	–	–	–	–	–	**Ethyl 3,4-dihydro-4-oxo-2,3-diphenylpyrimidine-6-carboxylate**		**NI33787**
320 91 251 159 321 160 77 187	**320**	100 57 35 29 22 19 18 17		19	20	1	4	–	–	–	–	–	–	–	–	–	**2-Pyrazolin-5-one, 4-[[4-(dimethylamino)-o-tolyl]imino]-3-methyl-1-phenyl-**	4719-47-5	**MS07576**
320 91 251 159 321 160 77 187	**320**	100 57 32 28 23 18 18 17		19	20	1	4	–	–	–	–	–	–	–	–	–	**2-Pyrazolin-5-one, 4-[[4-(dimethylamino)-o-tolyl]imino]-3-methyl-1-phenyl-**	4719-47-5	**NI33788**
320 160 321 186 173 159 115 171	**320**	100 60 21 21 21 20 18 14		19	22	2	–	–	2	–	–	–	–	–	–	–	**Estra-1,3,5(10)-trien-17-one, 16,16-difluoro-3-methoxy-**	2991-05-1	**NI33789**
320 160 115 173 159 186 128 129	**320**	100 91 37 35 32 30 27 24		19	22	2	–	–	2	–	–	–	–	–	–	–	**Estra-1,3,5(10)-trien-17-one, 16,16-difluoro-3-methoxy-**	2991-05-1	**NI33790**
320 160 173 159 186 115 91 171	**320**	100 45 18 16 15 12 12 11		19	22	2	–	–	2	–	–	–	–	–	–	–	**Estra-1,3,5(10)-trien-17-one, 16,16-difluoro-3-methoxy-**	2991-05-1	**NI33791**
276 300 277 91 79 137 97 258	**320**	100 30 18 14 13 13 12 11	**11.10**	19	25	3	–	–	1	–	–	–	–	–	–	–	**6-β-Fluoro-2-α-hydroxyandrost-4-ene-3,17-dione**	89561-81-9	**NI33792**
272 41 91 271 199 43 55 145	**320**	100 67 66 60 60 48 47 44	**0.00**	19	28	4	–	–	–	–	–	–	–	–	–	–	**Androst-5-en-17-one, 3,16,19-trihydro-, (3β,16α)-**		**LI07116**
185 260 200 143 145 131 187 157	**320**	100 69 58 50 48 44 37 34	**0.00**	19	28	4	–	–	–	–	–	–	–	–	–	–	**Coralloidin D**		**DD01229**
181 43 41 55 56 163 70 69	**320**	100 52 51 48 39 38 31 28	**3.00**	19	28	4	–	–	–	–	–	–	–	–	–	–	**Isophthalic acid, decyl methyl ester**	33975-27-8	**NI33793**
43 28 55 85 41 45 91 163	**320**	100 57 41 39 36 31 27 26	**19.00**	19	28	4	–	–	–	–	–	–	–	–	–	–	**Methyl trisporate C**	26055-31-2	**MS07577**
261 320 41 93 55 107 79 92	**320**	100 80 57 53 50 48 45 44		19	28	4	–	–	–	–	–	–	–	–	–	–	**19-Norkauran-18-oic acid, 2-hydroxy-15-oxo-, (2β,4α)-**	51107-83-6	**NI33794**
83 291 320 43 41 55 181 193	**320**	100 79 54 46 44 42 32 31		19	28	4	–	–	–	–	–	–	–	–	–	–	**Propionaldehyde dimedone**		**LI07117**
85 131 84 218 174 145 176 190	**320**	100 46 38 30 30 26 21 20	**0.60**	19	28	4	–	–	–	–	–	–	–	–	–	–	**Spiro[cyclobutane-1,2′(1′aH)-indeno[3a,4-b]oxiren]-2-one, hexahydro-4,4′a-dimethyl-5′-[(tetrahydro-2H-pyran-2-yl)oxy]-**	67884-37-1	**NI33795**
320 278 292 203 249 261 305 321	**320**	100 71 43 43 41 28 24 23		20	16	4	–	–	–	–	–	–	–	–	–	–	**Benz[a]anthracene-1,7,12(2H)-trione, 3,4-dihydro-8-methoxy-3-methyl-**	56630-78-5	**NI33796**
132 320 289 89 117 321 277 322	**320**	100 96 41 37 27 22 17 16		20	16	4	–	–	–	–	–	–	–	–	–	–	**2,5-Cyclohexadiene-1,4-dione, 2,5-bis(4-methoxyphenyl)-**	5333-03-9	**NI33797**
211 320 243 212 321 197 203 244	**320**	100 71 59 26 18 14 12 11		20	17	–	–	1	–	–	–	–	–	1	–	–	**10H-Phenothiaphosphine, 2,8-dimethyl-10-phenyl-**	23861-47-4	**LI07118**
211 320 243 212 321 197 203 165	**320**	100 71 59 26 18 14 12 11		20	17	–	–	1	–	–	–	–	–	1	–	–	**10H-Phenothiaphosphine, 2,8-dimethyl-10-phenyl-**	23861-47-4	**NI33798**
320 243 319 321 227 195 165 213	**320**	100 86 38 25 20 19 19 17		20	17	2	–	–	–	–	–	–	–	1	–	–	**2,8-Dimethyl-10-phenylphenoxaphosphine 10-oxide**		**LI07119**
319 320 291 165 289 183 201 152	**320**	100 87 36 18 13 7 3 3		20	17	2	–	–	–	–	–	–	–	1	–	–	**2H,6H-[1,2]Oxaphospholo[4,3,2-hi][2,1]benzoxaphosphole, 8,8-dihydro-8,8-diphenyl-**	51922-82-8	**NI33799**
320 183 107 305 319 77 152 51	**320**	100 98 39 35 28 26 22 22		20	18	–	–	–	–	–	–	–	–	2	–	–	**1,4-Benzodiphosphorin, 1,2,3,4-tetrahydro-1,4-diphenyl-**	56700-81-3	**MS07578**
58 57 42 59 44 43 246 228	**320**	100 5 4 4 3 2 1 1	**0.30**	20	20	–	2	1	–	–	–	–	–	–	–	–	**11-(3-Dimethylaminopropylidene)-6,11-dihydrodibenzo[b,e]thiepin-2-carbonitrile**		**MS07579**
320 228 77 292 291 188 51 321	**320**	100 93 83 44 39 35 29 24		20	20	–	2	1	–	–	–	–	–	–	–	–	**2,3,5,6,7,8-Hexahydro-1-phenyl-5-(phenylimino)-1H-cyclopenta[e][1,4]thiazepine**		**NI33800**
320 160 321 132 292 291 263 235	**320**	100 30 25 10 5 5 5 5		20	20	2	2	–	–	–	–	–	–	–	–	–	**Bis-6,6′-(8-formyl-1,2,3,4-tetrahydroquinolyl)**		**LI07120**

MASS TO CHARGE RATIOS	M.W.	INTENSITIES	Parent	C	H	O	N	S	F	Cl	Br	I	Si	P	B	X	COMPOUND NAME	CAS Reg No	No
229 230 143 320 228	320	100 15 12 6 6		20	20	2	2	–	–	–	–	–	–	–	–	–	1H-Indeno[1,2-d]pyrimidine-2,4(3H,4aH)-dione, 4aβ-benzyl-1,3-dimethyl 5,9bβ-dihydro-		LI07121
77 183 93 252 184 69 105 320	320	100 57 31 26 26 20 19 14		20	20	2	2	–	–	–	–	–	–	–	–	–	3,5-Pyrazolidinedione, 4-(3-methyl-2-butenyl)-1,2-diphenyl-	30748-29-9	NI33801
292 264 261 234 277 233 293 189	320	100 42 34 26 23 23 22 16	13.00	20	20	2	2	–	–	–	–	–	–	–	–	–	1,2,3,3a-Tetrahydro-6-methoxy-10-(3-methoxyphenyl)benzo[c]cyclopenta[f]-1,2-diazepine	56005-87-9	NI33802
78 262 183 108 51 263 77 52	320	100 82 56 28 23 22 21 18	17.00	20	21	–	2	–	–	–	–	–	–	1	–	–	N,N-Dimethyl-N′-(triphenylphosphoryanylidene)hydrazine		MS07580
320 260 146 147 131 219 163 165	320	100 99 47 36 36 30 15 12		20	29	2	–	–	1	–	–	–	–	–	–	–	Estr-5(10)-en-17-ol, 3-fluoro-, acetate, (3α,17β)-	22034-58-8	LI07122
43 320 260 245 91 146 147 131	320	100 99 99 68 63 45 36 36		20	29	2	–	–	1	–	–	–	–	–	–	–	Estr-5(10)-en-17-ol, 3-fluoro-, acetate, (3α,17β)-	22034-58-8	NI33803
260 320 146 147 131 219 321 261	320	100 99 47 36 36 30 22 20		20	29	2	–	–	1	–	–	–	–	–	–	–	Estr-5(10)-en-17-ol, 3-fluoro-, acetate, (3α,17β)-	22034-58-8	NI33804
320 245 219 147 146 131 165 163	320	100 50 24 20 20 20 18 10		20	29	2	–	–	1	–	–	–	–	–	–	–	Estr-5(10)-en-17-ol, 3-fluoro-, acetate, (3β,17β)-	22034-57-7	LI07123
320 245 219 321 147 146 131 165	320	100 50 24 22 20 20 20 18		20	29	2	–	–	1	–	–	–	–	–	–	–	Estr-5(10)-en-17-ol, 3-fluoro-, acetate, (3β,17β)-	22034-57-7	NI33805
320 43 260 245 91 131 321 147	320	100 82 64 50 36 23 22 22		20	29	2	–	–	1	–	–	–	–	–	–	–	Estr-5(10)-en-17-ol, 3-fluoro-, acetate, (3β,17β)-	22034-57-7	NI33806
41 55 115 93 91 81 79 107	320	100 81 74 74 67 67 65 64	0.00	20	32	1	–	1	–	–	–	–	–	–	–	–	5α-Androstan-17β-ol, 2β,3β-epithio-2-methyl-	5783-86-8	NI33807
152 121 320 153 71 57 55 43	320	100 53 45 45 26 26 22 19		20	32	3	–	–	–	–	–	–	–	–	–	–	Benzoic acid, 4-(dodecyloxy)-, methyl ester	40654-49-7	NI33808
152 135 153 320 154 136 107 77	320	100 30 29 6 3 3 3 3		20	32	3	–	–	–	–	–	–	–	–	–	–	Benzoic acid, 4-methoxy-, dodecyl ester	56954-80-4	NI33809
181 320 168 221 305 195 263 57	320	100 93 72 57 47 28 23 23		20	32	3	–	–	–	–	–	–	–	–	–	–	cis-4-tert-Butyl-1-methyl-1-(2,4,6-trimethoxyphenyl)cyclohexane		LI07124
168 320 181 221 109 195 57 305	320	100 33 30 16 14 11 10 4		20	32	3	–	–	–	–	–	–	–	–	–	–	trans-4-tert-Butyl-1-methyl-1-(2,4,6-trimethoxyphenyl)cyclohexane		LI07125
95 109 121 123 107 135 124 122	320	100 46 41 31 29 24 21 18	0.00	20	32	3	–	–	–	–	–	–	–	–	–	–	rel-(5S,8R,9S,10R)-2-Oxo-ent-3-cleroden-15-oic acid		MS07581
252 41 69 55 320 43 81 95	320	100 61 49 43 31 31 30 29		20	32	3	–	–	–	–	–	–	–	–	–	–	8β,12α-Dihydroxysandaracopimar-15-ene-11-one		NI33810
69 305 191 95 81 121 177 137	320	100 80 60 47 40 30 25 23	4.00	20	32	3	–	–	–	–	–	–	–	–	–	–	12(R)-8,12-Epoxylabdan-15(12)-olide		MS07582
69 81 95 109 305 191 121 125	320	100 98 85 59 57 53 46 40	8.00	20	32	3	–	–	–	–	–	–	–	–	–	–	12(S)-8,12-Epoxylabdan-15(12)-olide		MS07583
94 109 81 121 95 93 67 105	320	100 78 72 70 60 60 55 53	7.01	20	32	3	–	–	–	–	–	–	–	–	–	–	Kauran-18-oic acid, 16-hydroxy-, (4α)-	22376-06-3	NI33811
94 43 121 109 262 302 93 105	320	100 81 78 70 63 61 61 59	8.00	20	32	3	–	–	–	–	–	–	–	–	–	–	Kauran-19-oic acid, 16-hydroxy-, (16α)-(–)-		LI07126
249 259 205 260 201 55 203 41	320	100 60 59 51 33 29 28 26	5.71	20	32	3	–	–	–	–	–	–	–	–	–	–	5,9-Methanobenzocycloocten-4(1H)-one, 2,3,5,6,7,8,9,10-octahydro-5-(2-hydroxyethoxy)-2,2,7,7,9-pentamethyl-	15919-97-8	NI33812
305 43 286 55 81 67 41 306	320	100 94 68 63 46 41 39 36	0.00	20	32	3	–	–	–	–	–	–	–	–	–	–	9H-Naphtho[2,1-b]pyran-9-one, 3-vinyldodecahydro-7-(hydroxymethyl)-3,4a,7,10a-tetramethyl-, [3R-(3α,4aβ,6aα,7α,10aβ,10bα)]-	57495-57-5	NI33813
227 287 121 305 105 260 320 159	320	100 81 67 49 32 31 30 27		20	32	3	–	–	–	–	–	–	–	–	–	–	2-Phenanthrenecarboxylic acid, 1,2,3,4,4a,4b,5,6,7,8,8a,9-dodecahydro-7-hydroxy-2,4b,8,8-tetramethyl-, methyl ester	57397-06-5	NI33814
196 43 119 197 109 41 55 133	320	100 16 14 13 13 13 12 10	2.04	20	32	3	–	–	–	–	–	–	–	–	–	–	Spiro[furan-2(3H),1′(4′H)-naphthalene]-5-acetic acid, 4,4′a,5,5′,6′,7′,8′,8′a octahydro-2′,5,5′,5′,8′a-pentamethyl-, [1′R-[1′α(S*),4′aα,8′aβ]]-	1438-57-9	NI33815
320 321 305 277 287 211 43 303	320	100 25 17 10 8 6 6 5		21	20	3	–	–	–	–	–	–	–	–	–	–	6-(2′-Hydroxy-5′-methylphenyl)-4a,9b-dihydro-8,9b-dimethyl-3(4H)-dibenzofuranone		MS07584
320 321 305 277 287 211 302 212	320	100 26 17 11 9 7 6 6		21	20	3	–	–	–	–	–	–	–	–	–	–	6-(2′-Hydroxy-5′-methylphenyl)-4a,9b-dihydro-8,9b-dimethyl-3(4H)-dibenzofuranone		LI07127
320 305 91 321 159 306 185 277	320	100 92 27 26 23 22 19 17		21	20	3	–	–	–	–	–	–	–	–	–	–	2-(4-Methylphenoxy)-4a,9b-dihydro-8,9b-dimethyl-3(4H)-dibenzofuranone	53042-30-1	NI33816
53 55 320 70 52 51 291 319	320	100 75 42 40 28 26 22 16		21	24	1	2	–	–	–	–	–	–	–	–	–	Aspidofractinine-1-carboxaldehyde, 3-methylene-, (2α,5α)-	55724-71-5	NI33817
43 320 305 148 277 130 144 77	320	100 77 42 35 31 30 29 27		21	24	1	2	–	–	–	–	–	–	–	–	–	Curan, 1-acetyl-16,17,19,20-tetradehydro-	56053-16-8	NI33818
200 130 201 320	320	100 25 14 7		21	24	1	2	–	–	–	–	–	–	–	–	–	Spiro[cycloheptane-3-indoline]-2-carboxanilide		MS07585
28 67 79 41 80 93 81 55	320	100 62 56 54 53 45 39 38	8.34	21	36	2	–	–	–	–	–	–	–	–	–	–	5,8,11-Eicosatrienoic acid, methyl ester	2463-03-8	NI33819
67 80 79 41 55 81 93 94	320	100 92 91 70 67 64 61 44	18.74	21	36	2	–	–	–	–	–	–	–	–	–	–	7,10,13-Eicosatrienoic acid, methyl ester	30223-51-9	NI33820
67 80 79 81 93 95 55 94	320	100 94 91 70 66 53 53 51	22.04	21	36	2	–	–	–	–	–	–	–	–	–	–	7,10,13-Eicosatrienoic acid, methyl ester	30223-51-9	NI33821
74 55 41 67 81 43 79 69	320	100 91 90 81 76 71 59 57	16.04	21	36	2	–	–	–	–	–	–	–	–	–	–	8,11,14-Eicosatrienoic acid, methyl ester	17364-32-8	NI33822
79 41 67 55 95 108 80 81	320	100 74 63 56 55 42 40 37	0.00	21	36	2	–	–	–	–	–	–	–	–	–	–	11,14,17-Eicosatrienoic acid, methyl ester	55682-88-7	NI33823
43 41 42 67 55 79 30 57	320	100 82 62 48 47 41 39 36	1.14	21	36	2	–	–	–	–	–	–	–	–	–	–	8,11,14-Eicosatrienoic acid, methyl ester, (Z,Z,Z)-	21061-10-9	NI33824
69 41 81 109 136 95 67 68	320	100 45 40 31 21 19 18 17	5.79	21	36	2	–	–	–	–	–	–	–	–	–	–	6,10,14-Hexadecatrienoic acid, 3,7,11,15-tetramethyl-, methyl ester, [R-(E,E)]-	36237-72-6	NI33825
163 123 81 95 109 162 261 93	320	100 62 52 48 45 43 35 34	30.00	21	36	2	–	–	–	–	–	–	–	–	–	–	Methyl 8α,13β-abietan-18-oate		MS07586
163 164 261 123 320 162 81 109	320	100 32 30 29 28 25 21		21	36	2	–	–	–	–	–	–	–	–	–	–	Methyl 9β,13β-abietan-18-oate		MS07587
163 81 95 320 162 164 261 109	320	100 41 37 36 35 34 31 29		21	36	2	–	–	–	–	–	–	–	–	–	–	Methyl 13β-abietan-18-oate		MS07588
163 320 164 162 261 95 123 82	320	100 43 42 39 34 30 27 27		21	36	2	–	–	–	–	–	–	–	–	–	–	Methyl 18-abietanoate		MS07589
163 123 261 109 95 81 320 162	320	100 68 51 44 41 41 40 31		21	36	2	–	–	–	–	–	–	–	–	–	–	Methyl 8α-isopimaran-18-oate		MS07590
163 320 162 164 81 261 109 123	320	100 45 45 41 41 40 39 37		21	36	2	–	–	–	–	–	–	–	–	–	–	Methyl 18-isopimaranoate		MS07591
163 123 261 164 320 95 81 109	320	100 73 70 58 57 57 54 52		21	36	2	–	–	–	–	–	–	–	–	–	–	Methyl 8α-pimaran-18-oate		MS07592
163 320 261 81 95 162 164 123	320	100 46 46 39 36 35 33 30		21	36	2	–	–	–	–	–	–	–	–	–	–	Methyl 18-pimaranoate		MS07593
163 95 123 109 133 81 101 319	320	100 99 96 96 95 93 86 80	18.00	21	36	2	–	–	–	–	–	–	–	–	–	–	Methyl 14S,17-cyclo-13ξ-labdan-18-oate		MS07594

MASS TO CHARGE RATIOS								M.W.	INTENSITIES								Parent	C	H	O	N	S	F	Cl	Br	I	Si	P	B	X	COMPOUND NAME	CAS Reg No	No
191	122	57	109	55	43	41	69	**320**	100	96	78	76	69	68	65	62	**39.60**	21	36	2	–	–	–	–	–	–	–	–	–	–	1-Naphthalenepentanoic acid, 1,4,4a,5,6,7,8,8a-octahydro-β,2,5,5,8a-pentamethyl-, methyl ester, [1S-[1α(R*),4aβ,8aα]]-	1438-55-7	NI33826
137	81	177	69	41	95	55	305	**320**	100	62	60	57	57	56	50	49	**35.80**	21	36	2	–	–	–	–	–	–	–	–	–	–	1-Naphthalenepentanoic acid, decahydro-β,5,5,8a-tetramethyl-2-methylene-, methyl ester, [1R-[1α(S*),4aβ,8aα]]-	13008-80-5	MS07595
137	305	177	81	95	69	41	320	**320**	100	93	62	60	59	54	50	48		21	36	2	–	–	–	–	–	–	–	–	–	–	1-Naphthalenepentanoic acid, decahydro-β,5,5,8a-tetramethyl-2-methylene-, methyl ester, [1R-[1α(S*),4aβ,8aα]]-	13008-80-5	NI33827
163	55	81	261	123	162	41	95	**320**	100	23	22	21	20	19	19	18	**10.70**	21	36	2	–	–	–	–	–	–	–	–	–	–	1-Phenanthrenecarboxylic acid, 7-ethyltetradecahydro-1,4a,7-trimethyl-, methyl ester, [1R-(1α,4aβ,4bα,7α,8aβ,10aα)]-	33892-02-3	NI33828
163	55	41	81	95	123	109	67	**320**	100	32	28	27	23	21	19	18	**5.10**	21	36	2	–	–	–	–	–	–	–	–	–	–	1-Phenanthrenecarboxylic acid, 7-ethyltetradecahydro-1,4a,7-trimethyl-, methyl ester, [1R-(1α,4aβ,4bα,7β,8aβ,10aα)]-	4614-69-1	NI33829
216	234	81	93	45	95	67	55	**320**	100	92	59	56	54	52	51	51	**1.16**	21	36	2	–	–	–	–	–	–	–	–	–	–	Pregnane-3,20-diol, (3α,5α,20S)-	566-58-5	NI33830
216	234	81	95	45	93	41	107	**320**	100	73	62	58	58	54	54	50	**0.00**	21	36	2	–	–	–	–	–	–	–	–	–	–	Pregnane-3,20-diol, (3α,5β)-	55569-11-4	IC03922
234	216	302	233	215	217	162	235	**320**	100	81	62	40	30	18	18	16	**8.42**	21	36	2	–	–	–	–	–	–	–	–	–	–	Pregnane-3,20-diol, (3α,5β)-	55569-11-4	NI33831
285	286	301	303	283	302	287	165	**320**	100	23	22	17	14	7	6	5	**0.80**	21	36	2	–	–	–	–	–	–	–	–	–	–	Pregnane-3,20-diol, (3α,5β,20S)-	80-92-2	NI33832
234	233	302	216	235	215	164	287	**320**	100	37	34	22	17	14	10	8	**6.02**	21	36	2	–	–	–	–	–	–	–	–	–	–	Pregnane-3,20-diol, (3β,5α)-	38270-91-6	NI33833
285	303	301	286	319	165	302	259	**320**	100	56	49	37	21	18	14	13	**4.30**	21	36	2	–	–	–	–	–	–	–	–	–	–	Pregnane-3,20-diol, (3β,5α,20S)-	566-56-3	NI33834
67	81	95	79	69	109	149	91	**320**	100	95	67	65	64	62	48	47	**0.00**	21	36	2	–	–	–	–	–	–	–	–	–	–	Pregnane-18,20-diol, (5α)-	56053-10-2	NI33835
83	69	41	95	81	109	67	55	**320**	100	86	56	43	35	30	24	24	**7.53**	21	36	2	–	–	–	–	–	–	–	–	–	–	8,12-Tetradecadienoic acid, 5-vinyl-3,5,9,13-tetramethyl-, methyl ester	36237-73-7	NI33836
319	57	41	56	43	99	321	141	**320**	100	72	61	58	55	50	36	36	**25.00**	22	21	–	–	–	–	1	–	–	–	–	–	–	10-Phenyl-2-(3-chlorophenyl)bicyclo[4.4.0]dec-1(6),2-diene		LI07128
320	160	162	277	161	321	91	104	**320**	100	94	46	33	33	26	26	23		22	24	2	–	–	–	–	–	–	–	–	–	–	Benzo[b]triphenylene-9,14-dione, 1,2,3,4,5,6,7,8,8a,8b,14a,14b-dodecahydro-	31991-64-7	NI33837
178	223	177	55	320	41	118	107	**320**	100	86	81	67	28	26	23	23		22	24	2	–	–	–	–	–	–	–	–	–	–	Cyclohexanemethyl 2-phenylcinnamate		NI33838
206	193	180	167	165	115	179	178	**320**	100	81	65	60	43	35	33	32	**0.00**	22	24	2	–	–	–	–	–	–	–	–	–	–	Cyclohexanepropanal, 1-methyl-2-oxo-6,6-diphenyl-	50592-56-8	NI33839
160	159	161	145	129	115	144	117	**320**	100	18	14	8	7	7	6	4	**0.00**	22	24	2	–	–	–	–	–	–	–	–	–	–	Dibenzo[a,i]biphenylene, 2,11-dimethoxy-5,6,6a,6b,7,8,12b,12c-octahydro-		AI02041
79	80	214	93	167	91	77	179	**320**	100	98	89	72	52	41	41	39	**0.00**	22	24	2	–	–	–	–	–	–	–	–	–	–	4′,5′-Diphenylspiro[bicyclo[5.1.0]octane-2,2′-[1,3]dioxolane]		DD01230
145	320	305	277	185	146	147	107	**320**	100	33	23	23	15	13	12	10		22	24	2	–	–	–	–	–	–	–	–	–	–	3,9,9-Trimethyl-4,10-dioxa-5.6,11.12-dibenztricyclo[6.4.0.1[3,7]]tridecane		MS07596
175	117	145	320	176	115	57	105	**320**	100	48	27	21	17	15	11	10		23	28	1	–	–	–	–	–	–	–	–	–	–	3-Buten-2-one, 1-(4-tert-butyl-2,6-dimethylphenyl)-3-(2-methylphenyl)-		LI07129
175	117	145	320	176	115	28	57	**320**	100	48	27	21	17	15	14	11		23	28	1	–	–	–	–	–	–	–	–	–	–	3-Buten-2-one, 1-(4-tert-butyl-2,6-dimethylphenyl)-3-(3-methylphenyl)-	55823-02-4	NI33840
43	41	67	55	57	95	81	69	**320**	100	99	79	79	57	33	33	26	**1.20**	23	44	–	–	–	–	–	–	–	–	–	–	–	9-Tricosyne		NI33841
66	91	67	129	95	225	79	117	**320**	100	70	56	54	51	42	41	31	**28.28**	24	32	–	–	–	–	–	–	–	–	–	–	–	1,4:5,12:6,11:7,10-Tetramethanodibenzo[b,h]biphenylene, eicosahydro-	55101-93-4	NI33842
243	165	320	244	166	321	241	239	**320**	100	48	30	20	9	8	7	7		25	20	–	–	–	–	–	–	–	–	–	–	–	Tetraphenylmethane	630-76-2	NI33844
243	165	77	320	166	244	51	154	**320**	100	70	29	27	24	21	17	13		25	20	–	–	–	–	–	–	–	–	–	–	–	Tetraphenylmethane	630-76-2	NI33843
59	46	127	70	44	43	102	75	**321**	100	86	71	67	52	40	36	35	**0.30**	4	8	4	3	2	5	–	–	–	–	–	–	–	S,S-Bis(methoxyformamido)-N-pentafluorosulphanylsulphilimine		NI33845
30	87	323	28	168	166	263	61	**321**	100	27	14	14	11	11	9	8	**8.00**	6	1	3	3	–	–	–	2	–	–	–	–	–	Benzofurazan, 5,7-dibromo-4-nitro-	52120-98-6	NI33846
323	39	45	295	325	321	69	71	**321**	100	91	85	54	53	53	39	31		8	5	1	1	1	–	–	2	–	–	–	–	–	Thiazolo[3,2-a]pyridinium, 5,7-dibromo-8-hydroxy-3-methyl-, hydroxide, inner salt	30277-20-4	NI33847
306	117	115	116	304	323	308	51	**321**	100	96	90	55	52	50	50	40	**26.00**	9	9	2	1	–	–	–	2	–	–	–	–	–	Benzene, 1,3-dibromo-2,4,6-trimethyl-5-nitro-	40572-27-8	MS07597
272	63	274	273	77	65	210	91	**321**	100	36	32	12	11	10	7	6	**1.76**	11	13	4	3	–	–	2	–	–	–	–	–	–	Aniline, N,N-bis(2-chloroethyl)-4-methyl-2,6-dinitro-	26389-78-6	NI33848
59	191	131	159	99	321	98	133	**321**	100	53	45	31	24	23	20	18		11	15	–	1	5	–	–	–	–	–	–	–	–	2,4,6,8,9-Pentathiatricyclo[3.3.1.1[3,7]]decane-1-propanenitrile, 3,5,7-trimethyl-	57274-40-5	NI33849
43	58	44	42	162	164	98	63	**321**	100	71	57	44	42	29	26	26	**1.00**	12	13	5	1	–	–	2	–	–	–	–	–	–	L-Threonine, N-[(2,4-dichlorophenoxy)acetyl]-	50648-99-2	NI33850
43	88	29	42	85	27	60	45	**321**	100	21	18	16	16	11	10	8	**0.00**	12	18	6	–	–	–	–	–	–	–	–	–	1	Copper, bis(ethyl 3-oxobutanoato-O1′,O3)-	14284-06-1	NI33851
73	204	75	218	45	147	116	44	**321**	100	75	65	46	28	20	16	16	**0.00**	12	31	3	1	–	–	–	–	–	3	–	–	–	L-Serine, N,O-bis(trimethylsilyl)-, trimethylsilyl ester	7364-48-9	NI33852
204	73	218	205	147	116	132	100	**321**	100	62	61	21	17	14	13	11	**2.00**	12	31	3	1	–	–	–	–	–	3	–	–	–	L-Serine, N,O-bis(trimethylsilyl)-, trimethylsilyl ester	7364-48-9	MS07598
95	40	29	41	96	42	69	44	**321**	100	19	12	10	9	9	7	7	**0.60**	13	18	3	3	–	3	–	–	–	–	–	–	–	L-Histidine, 1-methyl-N-(trifluoroacetyl)-, butyl ester	55145-65-8	NI33853
179	178	142	180	127	152	76	151	**321**	100	20	16	15	9	7	7	6	**0.00**	14	12	–	1	–	–	–	–	1	–	–	–	–	Phenanthridinium, 5-methyl-, iodide	5412-06-6	NI33854
84	250	139	44	114	41	87	56	**321**	100	36	34	32	30	23	17	17	**6.00**	14	19	4	5	–	–	–	–	–	–	–	–	–	1H-Pyrrole-2-carboxamide, N-(3-amino-3-oxapropyl)-1-methyl-5-[[(5-oxo-2-pyrrolidinyl)carbonyl]amino]-	61233-57-6	NI33855
131	43	89	130	190	101	175	234	**321**	100	90	50	44	22	8	5	1	**0.00**	14	25	8	–	–	–	–	–	–	–	–	–	–	3,4-Di-O-methyl-1,2,5-tri-O-acetyl-fucitol		NI33856
91	68	122	223	155	166	65	159	**321**	100	82	52	48	39	27	27	22	**6.40**	15	15	5	1	1	–	–	–	–	–	–	–	–	Isobenzofuran-4,7-imine-1,3-dione, hexahydro-8-[(4-methylphenyl)sulphonyl]-, (3aα,4α,7α,7aα)-	17037-46-6	NI33857
252	321	223	77	76	222	69	50	**321**	100	36	17	7	7	5	5	5		16	10	1	1	1	3	–	–	–	–	–	–	–	2-Phenyl-3-[(trifluoromethyl)thio]-4(1H)-quinolinone		DD01231
321	252	234	69	190	165	223	222	**321**	100	97	79	36	26	24	23	15		16	10	1	1	1	3	–	–	–	–	–	–	–	3-[(Trifluoromethyl)thio]-4-phenyl-2(1H)-quinolinone		DD01232
17	224	270	27	30	181	44	26	**321**	100	33	18	18	13	11	8	8	**0.00**	16	13	2	1	–	–	2	–	–	–	–	–	–	2-Azetidinone, 3,4-dichloro-1-(4-methoxy[1,1′-biphenyl]-2-yl)-	56772-32-8	NI33858

MASS TO CHARGE RATIOS								M.W.	INTENSITIES								Parent	C	H	O	N	S	F	Cl	Br	I	Si	P	B	X	COMPOUND NAME	CAS Reg No	No
91	186	170	65	230	276	107	321	**321**	100	82	49	26	12	11	10	8		16	19	4	1	1	–	–	–	–	–	–	–	–	Ethyl 2-(2-benzyloxyethoxymethyl)thiazole-4-carboxylate		MS07599
189	248	228	217	272	16	230	190	**321**	100	59	30	25	21	20	15	15	**1.50**	16	19	6	1	–	–	–	–	–	–	–	–	–	3-Methyl-3-phenyl-1-(β-D-ribofuranosyl)-2,5-pyrrolidinedione		NI33859
142	57	100	143	185	86	84	136	**321**	100	28	22	17	11	6	4	3	**0.00**	16	36	–	1	–	–	–	1	–	–	–	–	–	Tetra-N-butylammonium bromide	1643-19-2	NI33860
153	209	191	29	57	265	124	125	**321**	100	44	38	38	32	20	16	10	**6.90**	17	23	5	1	–	–	–	–	–	–	–	–	–	Propanamide, N-[4-ethoxy-2-(1-oxopropoxy)phenyl]-N-(1-oxopropyl)-		NI33861
57	176	236	221	321	246	164	146	**321**	100	39	34	27	17	16	12	6		17	23	5	1	–	–	–	–	–	–	–	–	–	(3S)-1,2,3,4-Tetrahydro-6,7-dimethoxy-2-pivaloylisoquinoline-3-carboxylic acid		DD01233
130	77	107	131	192	103	51	78	**321**	100	33	30	29	21	16	10	8	**7.00**	18	15	1	3	1	–	–	–	–	–	–	–	–	4-Imidazolidinone, 5-(3-indolylmethyl)-3-phenyl-2-thioxo-		LI07130
321	322	219	276	206	234	220	205	**321**	100	19	18	15	14	11	11	11		18	15	3	3	–	–	–	–	–	–	–	–	–	1-(2-Oxocyclohexylidene)-2-nitro-1,5-dihydrophenazine		MS07600
93	43	77	61	45	51	119	211	**321**	100	42	37	27	23	21	17	14	**5.63**	18	15	3	3	–	–	–	–	–	–	–	–	–	4-Pyrimidinecarboxylic acid, 1,2,3,6-tetrahydro-6-oxo-3-phenyl-2-(phenylimino)-, methyl ester	16135-25-4	NI33862
321	106	322	252	146	134	41	147	**321**	100	72	22	20	18	16	16	15		18	19	1	5	–	–	–	–	–	–	–	–	–	2-Pyrazolin-5-one, 1-(4-aminophenyl)-4-[[4-(dimethylamino)phenyl]imino]3-methyl-	27808-00-0	NI33863
234	174	250	190	146	235	262	162	**321**	100	81	54	30	25	20	19	19	**11.90**	18	27	4	1	–	–	–	–	–	–	–	–	–	Lycofawcine, dehydro-	3327-34-2	NI33864
152	234	150	153	293	125	233	219	**321**	100	16	13	12	7	7	6	6	**2.39**	18	27	4	1	–	–	–	–	–	–	–	–	–	Serratinine, 8-acetate	5532-09-2	NI33865
152	234	150	153	233	193	124	218	**321**	100	16	13	12	6	6	6	5	**3.00**	18	27	4	1	–	–	–	–	–	–	–	–	–	Serratinine, 8-acetate	5532-09-2	LI07131
234	149	235	152	293	194	150	148	**321**	100	28	25	25	24	20	20	12	**3.00**	18	27	4	1	–	–	–	–	–	–	–	–	–	Serratinine, 13-acetate		LI07132
234	149	235	152	293	194	150	124	**321**	100	28	25	25	24	20	20	12	**2.42**	18	27	4	1	–	–	–	–	–	–	–	–	–	Serratinine, 14-acetate	5532-10-5	NI33866
321	216	244	322	230	91	176	306	**321**	100	53	28	27	19	17	13	12		19	15	2	1	1	–	–	–	–	–	–	–	–	1H-Benzo[d]pyrido[2,1-b]thiazol-1-one, 2-benzyl-3-methoxy-		LI07133
321	322	244	236	323	216	278	160.5	**321**	100	23	21	18	13	8	7	7		19	15	2	1	1	–	–	–	–	–	–	–	–	1H-Benzo[d]pyrido[2,1-b]thiazol-1-one, 4-benzyl-3-methoxy-		LI07134
321	306	322	220	263	278	248	291	**321**	100	39	23	20	19	17	13	11		19	15	4	1	–	–	–	–	–	–	–	–	–	7H-Dibenzo[de,g]quinolin-7-one, 1,2,3-trimethoxy-	5140-38-5	NI33867
218	139	185	140	183	219	217	124	**321**	100	23	20	20	13	12	12	12	**6.70**	19	16	–	1	1	–	–	–	–	–	1	–	–	Phosphinothioic amide, P,P-diphenyl-N-benzylidene-	58156-56-2	NI33868
49	43	69	291	84	57	41	321	**321**	100	94	83	80	63	61	61	34		19	19	2	3	–	–	–	–	–	–	–	–	–	1,2-Pentamethylene-3-(nitrophenylamino)-indolizine		MS07601
279	321	292	280	291	277	322	42	**321**	100	59	44	20	17	14	13	12		19	19	2	3	–	–	–	–	–	–	–	–	–	3-Pyridinecarboxylic acid, 5-(1,2,3,4,5,6-hexahydroazepino[4,5-b]indol-5-yl)-, methyl ester, (-)-	38940-72-6	NI33869
321	322	320	306	303	304	190	144	**321**	100	33	22	16	16	12	12	11		19	19	2	3	–	–	–	–	–	–	–	–	–	Pyrrolo[2,1-b]quinazolin-9(1H)-one, 2,3-dihydro-5-methoxy-3-(methylphenylamino)-	25662-85-5	NI33870
141	57	85	49	84	41	86	79	**321**	100	43	37	36	28	25	24	21	**4.07**	19	31	3	1	–	–	–	–	–	–	–	–	–	N-Isobutyl-10-(isobutylcarbonyloxy)(2E,6Z,8E)-decatrienamide		NI33871
185	286	189	285	287	234	277	319	**321**	100	72	64	45	42	34	33	31	**29.00**	20	19	1	1	1	–	–	–	–	–	–	–	–	8-Methyl-6,7,8,9,10,15-hexahydrobenzo[d][1]benzothieno[2,3-g]azecin-15-one		NI33872
105	77	51	41	55	106	216	43	**321**	100	61	16	12	8	8	8	8	**3.90**	20	19	3	1	–	–	–	–	–	–	–	–	–	2,N-Dibenzoyl-6-hexanelactam	102222-10-6	NI33873
105	77	188	106	51	54	321	95	**321**	100	38	19	8	7	6	5	4		20	19	3	1	–	–	–	–	–	–	–	–	–	O,N-Dibenzoyl-6-hexenelactam	102222-04-8	NI33874
201	321	173	77	130	202	105	322	**321**	100	67	58	41	40	37	31	29		20	19	3	1	–	–	–	–	–	–	–	–	–	3,5-Dimethyl-8-methoxy-3-phenyl-1,4-dihydro-2H-pyrano[4,3-b]indol-1-one		DD01234
306	321	292	307	263	293	322	308	**321**	100	61	43	26	23	18	15	10		20	19	3	1	–	–	–	–	–	–	–	–	–	7H-Pyrano[2,3-c]acridin-7-one, 6-methoxy-3,3,12-trimethyl-3,12-dihydro-	7008-42-6	LI07135
306	321	292	264	307	322	276	265	**321**	100	68	30	30	26	21	12	12		20	19	3	1	–	–	–	–	–	–	–	–	–	7H-Pyrano[2,3-c]acridin-7-one, 6-methoxy-3,3,12-trimethyl-3,12-dihydro-	7008-42-6	MS07602
120	201	306	77	202	51	121	47	**321**	100	58	34	25	12	10	9	9	**1.80**	20	20	1	1	–	–	–	–	–	–	1	–	–	N-(1-Methylbenzyl)diphenylphosphinamide		MS07603
235	72	236	250	86	234	249	146	**321**	100	57	20	17	13	6	5	5	**0.67**	20	23	1	3	–	–	–	–	–	–	–	–	–	2-(N,N-Diethylaminomethyl)-3-(o-tolyl)-4(3H)-quinazolinone		NI33875
265	250	196	118	165	77	266	104	**321**	100	61	51	36	26	25	19	15	**12.00**	20	23	1	3	–	–	–	–	–	–	–	–	–	4-Imidazolidinone, 3-tert-butyl-2-imino-1-methyl-5,5-diphenyl-	54508-07-5	NI33876
265	250	196	118	165	77	266	182	**321**	100	61	51	36	26	25	19	15	**11.99**	20	23	1	3	–	–	–	–	–	–	–	–	–	4-Imidazolidinone, 3-tert-butyl-2-imino-1-methyl-5,5-diphenyl-	54508-07-5	NI33877
58	292	44	321	110	57	43	70	**321**	100	46	34	26	17	14	10	8		20	35	2	1	–	–	–	–	–	–	–	–	–	N-Methylsamandiol		LI07136
203	91	43	136	41	204	92	29	**321**	100	42	26	21	19	18	17	16	**0.00**	21	23	2	1	–	–	–	–	–	–	–	–	–	Benzoic acid, 4-heptyl-, 4-cyanophenyl ester	38690-76-5	NI33878
219	232	191	203	217	204	91	220	**321**	100	72	40	36	30	30	28	24	**10.00**	21	23	2	1	–	–	–	–	–	–	–	–	–	Carbamic acid, [3-(10,11-dihydro-5H-dibenzo[a,d]cyclohepten-5-ylidene)propyl]-, ethyl ester	74810-80-3	NI33879
99	41	112	55	56	43	57	58	**321**	100	75	66	64	52	52	45	35	**19.04**	21	39	1	1	–	–	–	–	–	–	–	–	–	Azetidine, 1-(1-oxo-9-octadecenyl)-	56667-16-4	NI33880
57	99	41	112	58	55	80	43	**321**	100	46	39	36	36	34	21	18	**14.24**	21	39	1	1	–	–	–	–	–	–	–	–	–	9-Octadecenamide, N-cyclopropyl-	56630-45-6	NI33881
113	126	55	40	70	43	41	98	**321**	100	58	41	39	28	25	25	16	**15.04**	21	39	1	1	–	–	–	–	–	–	–	–	–	Pyrrolidine, 1-[8-(2-hexylcyclopropyl)-1-oxooctyl]-	56600-05-6	NI33882
137	131	136	274	291	275	321	263	**321**	100	81	52	38	32	29	27	22		22	11	2	1	–	–	–	–	–	–	–	–	–	1-Nitrobenzo(ghi)perylene	91259-17-5	NI33883
230	58	231	44	173	105	42	159	**321**	100	24	18	12	11	11	9	7	**1.00**	22	27	1	1	–	–	–	–	–	–	–	–	–	2,6-Methano-3-benzazocin-8-ol, 1,2,3,4,5,6-hexahydro-6,11-dimethyl-3-(2-phenylethyl)-	127-35-5	NI33884
230	231	58	174	105	44	173	42	**321**	100	17	12	8	7	7	6	6	**1.00**	22	27	1	1	–	–	–	–	–	–	–	–	–	2,6-Methano-3-benzazocin-8-ol, 1,2,3,4,5,6-hexahydro-6,11-dimethyl-3-(2-phenylethyl)-	127-35-5	NI33885
321	229	319	322	320	230			**321**	100	52	33	25	9	9				22	27	1	1	–	–	–	–	–	–	–	–	–	2,6-Methano-3-benzazocin-8-ol, 1,2,3,4,5,6-hexahydro-6,11-dimethyl-3-(2-phenylethyl)-	127-35-5	NI33886
321	159	173	160	227	214	115	147	**321**	100	44	41	41	40	37	27	24		22	27	1	1	–	–	–	–	–	–	–	–	–	19-Norpregna-1,3,5(10),17(20)-tetraene-20-carbonitrile, 3-methoxy-	60727-74-4	NI33887

MASS TO CHARGE RATIOS								M.W.	INTENSITIES								Parent	C	H	O	N	S	F	Cl	Br	I	Si	P	B	X	COMPOUND NAME	CAS Reg No	No
166	164	47	94	129	131	168	59	**322**	100	78	73	68	58	57	49	45	**0.00**	2	–	–	–	–	–	4	2	–	–	–	–	–	**Ethane, 1,2-dibromo-1,1,2,2-tetrachloro-**	630-25-1	**NI33888**
69	119	127	246	254	177	50	100	**322**	100	67	46	43	31	31	31	28	**0.00**	2	–	–	–	–	9	–	–	1	–	–	–	–	**Tetrafluoro(pentafluoroethyl)iodine**		**MS07604**
182	322	238	83	266	210	168.5	182.5	**322**	100	69	57	52	44	26	26	19		5	–	5	–	–	–	–	–	1	–	–	–	1	**Iodo-pentacarbonyl-manganese**		**MS07605**
131	133	95	135	157	97	117	119	**322**	100	99	99	91	85	74	59	56	**0.13**	5	4	3	–	–	–	6	–	–	–	–	–	–	**Bis(trichloroethyl) carbonate**		**MS07606**
214	212	216	53	133	135	31	27	**322**	100	52	50	43	34	30	20	17	**0.00**	5	9	1	–	–	–	–	3	–	–	–	–	–	**1-Propanol, 3-bromo-2,2-bis(bromomethyl)-**	1522-92-5	**NI33889**
253	251	255	145	143	217	83	85	**322**	100	79	75	51	50	47	46	42	**0.00**	6	5	–	–	–	–	7	–	–	–	–	–	–	**Cyclohexane, 1,1,2,3,4,5,6-heptachloro-**	707-55-1	**NI33890**
287	289	285	83	145	251	143	85	**322**	100	86	55	50	47	43	43	41	**1.46**	6	5	–	–	–	–	7	–	–	–	–	–	–	**Cyclohexane, 1,1,2,3,4,5,6-heptachloro-**	707-55-1	**NI33891**
277	178	101	175	129	79	51	76	**322**	100	73	9	6	6	6	5	4	**0.00**	6	6	13	–	–	–	–	–	–	–	–	–	4	**Beryllium, hexakis[μ-(formato-O:O′)]-μ⁴-oxotetra-**	42556-30-9	**NI33892**
44	179	19	16	30	27	14	46	**322**	100	62	57	53	33	28	25	18	**0.00**	6	6	13	–	–	–	–	–	–	–	–	–	4	**Beryllium, hexakis[μ-(formato-O:O′)]-μ⁴-oxotetra-**	42556-30-9	**MS07607**
322	202	97	266	65	121	238	174	**322**	100	53	48	39	37	36	34	32		8	20	5	–	2	–	–	–	–	–	2	–	–	**Thiodiphosphoric acid, tetraethyl ester**	368-02-5	**NI33894**
29	322	97	65	202	27	93	266	**322**	100	92	76	62	49	40	39	38		8	20	5	–	2	–	–	–	–	–	2	–	–	**Thiodiphosphoric acid, tetraethyl ester**	368-02-5	**LI07137**
29	322	97	65	202	93	27	121	**322**	100	93	78	63	50	40	40	36		8	20	5	–	2	–	–	–	–	–	2	–	–	**Thiodiphosphoric acid, tetraethyl ester**	368-02-5	**NI33893**
88	322	144	179	303	99	192	159	**322**	100	86	63	49	25	23	17	15		9	8	2	2	1	6	–	–	–	–	–	–	–	**2-Imidazolidinethione, 1-acetyl-3-[3,3,3-trifluoro-1-oxo-2-(trifluoromethyl)propyl]-**	61709-51-1	**NI33895**
88	179	144	322	91	99	159	192	**322**	100	53	50	46	33	30	22	21		9	8	4	2	–	6	–	–	–	–	–	–	–	**1-Imidazolidinecarboxylic acid, 2-oxo-3-[3,3,3-trifluoro-1-oxo-2-(trifluoromethyl)propyl]-, methyl ester**	61687-02-3	**NI33896**
139	137	309	311	109	307	148	107	**322**	100	98	59	35	31	29	27	21	**8.04**	9	12	1	–	–	–	–	2	–	1	–	–	–	**Silane, (2,4-dibromophenoxy)trimethyl-**	56002-67-6	**NI33897**
324	63	322	326	196	126	161	198	**322**	100	89	83	47	44	44	28	27		10	6	–	–	–	–	4	–	–	–	–	–	1	**Ferrocene, 1,1′,2,2′-tetrachloro-**	33306-54-6	**LI07138**
324	63	322	126	326	196	161	91	**322**	100	90	82	45	44	43	38	28		10	6	–	–	–	–	4	–	–	–	–	–	1	**Ferrocene, 1,1′,2,2′-tetrachloro-**	33306-54-6	**NI33898**
126	161	196	96	324	91	39	56	**322**	100	88	83	68	67	65	65	63	**55.05**	10	6	–	–	–	–	4	–	–	–	–	–	1	**Ferrocene, 1,2,3,4-tetrachloro-**	33306-63-7	**NI33899**
105	77	90	89	87	115	109	169	**322**	100	73	49	45	36	33	32	30	**0.00**	10	19	–	–	–	4	–	1	–	1	–	–	–	**Silane, (4-bromo-3,3,4,4-tetrafluorobutyl)triethyl-**	62185-59-5	**NI33900**
152	170	151	172	153	322	171	303	**322**	100	30	26	19	9	6	6	5		11	12	5	2	–	–	2	–	–	–	–	–	–	**Chloramphenicol**	56-75-7	**NI33901**
153	60	70	155	170	106	77	172	**322**	100	99	85	78	70	46	40	39	**0.00**	11	12	5	2	–	–	2	–	–	–	–	–	–	**Chloramphenicol**	56-75-7	**MS07608**
29	305	27	43	28	261	30	307	**322**	100	51	48	33	30	26	21	19	**4.00**	11	13	4	4	–	3	–	–	–	–	–	–	–	**1,3-Benzenediamine, N³,N³-diethyl-2,4-dinitro-6-(trifluoromethyl)-**	29091-05-2	**NI33902**
141	113	55	95	43	59	67	98	**322**	100	59	41	30	30	29	22	20	**0.00**	11	15	3	–	–	3	2	–	–	–	–	–	–	**Ethyl (1R*,3S*)-3-(2,2-dichloro-3,3,3-trifluoro-1-hydroxypropyl)-2,2-dimethylcyclopropanecarboxylate**		**DD01235**
17	252	324	254	322	326	108	28	**322**	100	88	78	64	55	36	32	32		12	6	–	–	1	–	4	–	–	–	–	–	–	**Benzene, 1,2,4-trichloro-5-[(4-chlorophenyl)thio]-**	2227-13-6	**NI33903**
324	322	223	326	225	53	73	325	**322**	100	71	55	31	27	8	4	3		12	6	2	–	–	–	4	–	–	–	–	–	–	**4,4′-Biphenyldiol, 3,3′,5,5′-tetrachloro-**	13049-13-3	**NI33904**
138	92	306	65	108	80	139	91	**322**	100	80	76	67	66	60	55	42	**2.48**	12	10	5	4	1	–	–	–	–	–	–	–	–	**3,3′-Dinitro-4,4′-diaminodiphenylsulphoxide**		**NI33905**
171	43	129	85	144	102	84	59	**322**	100	99	62	56	49	46	35	33	**0.00**	12	18	8	–	1	–	–	–	–	–	–	–	–	**Methyl 2,3-di-O-acetyl-4-S-acetyl-4-thio-α-D-xylopyranoside**		**LI07139**
86	87	118	119	90	74	103	147	**322**	100	74	60	27	25	17	17	15	**1.44**	12	18	8	–	1	–	–	–	–	–	–	–	–	**1,4,7,10,13-Pentaoxa-16-thiacyclooctadecane-3,14,17-trione**	79687-38-0	**NI33906**
86	87	118	119	90	74	103	147	**322**	100	74	60	27	25	17	17	15	**1.50**	12	18	8	–	1	–	–	–	–	–	–	–	–	**1,4,7,10,13-Pentaoxa-16-thiacyclooctadecane-3,14,17-trione**		**MS07609**
81	67	95	55	69	163	324	326	**322**	100	90	84	70	36	29	0	0	**0.22**	12	20	–	–	–	–	–	2	–	–	–	–	–	**(E,E)-1,12-Dibromo-2,10-dodecadiene**		**MS07610**
81	67	95	55	69	53	54	109	**322**	100	91	80	70	41	41	35	32	**0.21**	12	20	–	–	–	–	–	2	–	–	–	–	–	**(Z,Z)-1,12-Dibromo-2,10-dodecadiene**		**MS07611**
73	147	89	234	161	59	45	103	**322**	100	45	35	34	25	20	19	18	**0.00**	12	26	6	–	–	–	–	–	–	2	–	–	–	**Butanedioic acid, 2,3-bis[(trimethylsilyl)oxy]-, dimethyl ester, (R*,S*)-**	53319-85-0	**NI33907**
73	147	89	234	161	59	45	103	**322**	100	99	77	75	56	45	41	39	**0.50**	12	26	6	–	–	–	–	–	–	2	–	–	–	**Butanedioic acid, 2,3-bis[(trimethylsilyl)oxy]-, dimethyl ester, (R*,S*)-**	53319-85-0	**NI33908**
73	89	147	59	45	75	103	161	**322**	100	99	70	68	58	34	28	27	**0.09**	12	26	6	–	–	–	–	–	–	2	–	–	–	**Butanedioic acid, 2,3-bis[(trimethylsilyl)oxy]-, dimethyl ester, [S-(R*,R*)]-**	57456-93-6	**NI33909**
73	89	147	59	45	75	103	161	**322**	100	63	44	43	36	21	18	17	**0.00**	12	26	6	–	–	–	–	–	–	2	–	–	–	**3,6-Dioxa-2,7-disilanonane-4,5-dicarboxylic acid, 2,2,7,7-tetramethyl-, dimethyl ester, [R-(R*,R*)]-**	55659-11-5	**NI33910**
237	265	293	209	133	105	322	103	**322**	100	84	81	48	40	36	23	16		12	30	–	–	–	–	–	–	–	–	–	–	2	**Digermane, hexaethyl-**	993-62-4	**NI33911**
73	147	292	189	103	205	133	45	**322**	100	55	27	25	20	15	15	10	**1.20**	12	30	4	–	–	–	–	–	–	3	–	–	–	**Glyceric acid, tri-TMS**	38191-87-6	**MS07612**
73	147	189	292	103	205	102	45	**322**	100	60	37	31	19	14	12	9	**0.00**	12	30	4	–	–	–	–	–	–	3	–	–	–	**Glyceric acid, tri-TMS**	38191-87-6	**MS07613**
73	307	147	292	189	308	103	133	**322**	100	70	50	24	24	21	21	13	**8.63**	12	30	4	–	–	–	–	–	–	3	–	–	–	**Glyceric acid, tri-TMS**	38191-87-6	**NI33912**
243	245	324	163	164	326	81.5	82	**322**	100	98	77	73	51	47	47	39		13	8	–	–	–	–	–	2	–	–	–	–	–	**2,7-Dibromofluorene**		**AI02042**
155	123	44	43	45	125	77	69	**322**	100	45	42	32	32	31	30	29	**9.25**	13	10	6	2	1	–	–	–	–	–	–	–	–	**2,4-Dinitro-4′-methoxydiphenylsulphoxide**	1039-70-9	**NI33913**
107	91	241	165	139	79	108	155	**322**	100	72	42	40	37	36	35	35	**30.58**	13	10	6	2	1	–	–	–	–	–	–	–	–	**2,4-Dinitro-4′-methyldiphenylsulphone**	899-02-5	**NI33914**
76	200	154	305	50	275	292	184	**322**	100	68	59	43	43	32	30	30	**20.00**	13	11	6	2	–	–	–	–	–	–	1	–	–	**Methyl bis(4-nitrophenyl)phosphinate**		**MS07614**
43	203	103	102	28	100	145	101	**322**	100	99	63	62	34	33	31	30	**0.00**	13	19	8	–	–	1	–	–	–	–	–	–	–	**α-D-Galactopyranoside, methyl 4-deoxy-4-fluoro-, triacetate**	32934-08-0	**NI33915**
83	69	55	97	253	87	57	139	**322**	100	85	49	48	33	30	29	26	**0.00**	13	20	2	–	–	6	–	–	–	–	–	–	–	**1,3-Dioxolane, 4,5-dibutyl-2,2-bis(trifluoromethyl)-, cis-**	38274-70-3	**MS07615**
69	83	55	87	97	253	57	139	**322**	100	99	66	50	48	39	34	31	**0.00**	13	20	2	–	–	6	–	–	–	–	–	–	–	**1,3-Dioxolane, 4,5-dibutyl-2,2-bis(trifluoromethyl)-, trans-**	38274-71-4	**MS07616**
83	97	69	55	57	253	71	70	**322**	100	63	61	43	26	25	18	15	**0.00**	13	20	2	–	–	6	–	–	–	–	–	–	–	**1,3-Dioxolane, 4-ethyl-5-hexyl-2,2-bis(trifluoromethyl)-, cis-**	38363-97-2	**MS07617**
83	97	69	55	253	57	59	70	**322**	100	72	68	57	36	33	23	20	**0.00**	13	20	2	–	–	6	–	–	–	–	–	–	–	**1,3-Dioxolane, 4-ethyl-5-hexyl-2,2-bis(trifluoromethyl)-, trans-**	38274-67-8	**NI33916**
83	69	97	55	57	253	96	43	**322**	100	79	47	38	36	28	15	14	**0.00**	13	20	2	–	–	6	–	–	–	–	–	–	–	**1,3-Dioxolane, 4-heptyl-5-methyl-2,2-bis(trifluoromethyl)-, cis-**	38363-95-0	**MS07618**
83	69	55	97	253	57	56	96	**322**	100	99	53	48	47	37	20	19	**0.00**	13	20	2	–	–	6	–	–	–	–	–	–	–	**1,3-Dioxolane, 4-heptyl-5-methyl-2,2-bis(trifluoromethyl)-, trans-**	38363-96-1	**MS07619**
83	69	55	97	57	253	139	56	**322**	100	52	48	37	24	22	16	16	**0.00**	13	20	2	–	–	6	–	–	–	–	–	–	–	**1,3-Dioxolane, 4-pentyl-5-propyl-2,2-bis(trifluoromethyl)-, cis-**	38274-68-9	**MS07620**

MASS TO CHARGE RATIOS								M.W.	INTENSITIES								Parent	C	H	O	N	S	F	Cl	Br	I	Si	P	B	X	COMPOUND NAME	CAS Reg No	No
83	55	69	97	253	57	139	56	322	100	58	54	40	35	24	23	23	0.00	13	20	2	–	–	6	–	–	–	–	–	–	–	1,3-Dioxolane, 4-pentyl-5-propyl-2,2-bis(trifluoromethyl)-, trans-	38274-69-0	MS07621
41	56	44	57	86	55	69	124	322	100	42	36	24	17	15	12	1	1.00	13	23	4	–	–	–	–	1	–	–	–	–	–	Propanedioic acid, (bromomethyl)methyl-, di-tert-butyl ester	67587-04-6	NI33917
73	117	75	132	69	55	41	145	322	100	73	63	62	47	33	21	19	2.40	13	27	2	–	–	–	–	1	–	1	–	–	–	Decanoic acid, 10-bromo-, trimethylsilyl ester	34176-84-6	NI33918
103	73	219	147	129	104	74	45	322	100	89	28	18	13	10	8	8	0.00	13	34	3	–	–	–	–	–	–	3	–	–	–	Butane, 1,2,4-tris(trimethylsiloxy)-	33581-75-8	NI33919
103	73	219	147	129	104	45	75	322	100	69	24	24	11	10	8	7	0.00	13	34	3	–	–	–	–	–	–	3	–	–	–	Butane, 1,2,4-tris(trimethylsiloxy)-	33581-75-8	NI33920
322	165	167	152	166	195	194	177	322	100	41	19	16	15	14	13	11		14	11	1	–	–	–	–	–	1	–	–	–	–	(E)-2-Hydroxy-4′-iodostilbene		MS07622
102	103	91	77	119	324	322	178	322	100	56	55	51	46	31	31	28		14	11	2	–	1	–	–	1	–	–	–	–	–	Benzene, 1-bromo-4-(styrylsulphonyl)-	3406-64-2	LI07140
102	103	91	77	119	322	178	76	322	100	56	55	51	46	31	28	20		14	11	2	–	1	–	–	1	–	–	–	–	–	Benzene, 1-bromo-4-(styrylsulphonyl)-	3406-64-2	NI33921
169	153	96	154	166	112	97	69	322	100	46	42	22	19	12	10	10	0.00	14	18	5	4	–	–	–	–	–	–	–	–	–	Bis[1,3-dimethylthymin-5-yl] ether		NI33922
153	152	154	127	70	84	82	126	322	100	74	43	37	35	32	27	23	5.00	14	18	5	4	–	–	–	–	–	–	–	–	–	β-(1-Methyluracil-3-yl)ethyl ether		NI33923
43	101	45	117	145	129	161	71	322	100	36	33	30	22	21	15	9	0.00	14	26	8	–	–	–	–	–	–	–	–	–	–	D-Galactitol, 1,3,4,5-tetra-O-methyl-, diacetate	25521-30-6	NI33924
43	101	117	45	129	145	161	87	322	100	87	74	59	53	49	37	28	0.00	14	26	8	–	–	–	–	–	–	–	–	–	–	Galactitol, 1,3,4,5-tetra-O-methyl-, diacetate	19318-49-1	NI33925
43	45	101	117	145	129	161	87	322	100	38	35	29	26	25	16	13	0.00	14	26	8	–	–	–	–	–	–	–	–	–	–	Galactitol, 1,3,4,5-tetra-O-methyl-, diacetate	19318-49-1	NI33926
43	143	89	87	277	185	129	59	322	100	70	56	50	43	43	37	35	0.00	14	26	8	–	–	–	–	–	–	–	–	–	–	D-Glucitol, 1,2,5,6-tetra-O-methyl-, diacetate		MS07623
101	43	117	129	161	145	45	87	322	100	99	64	55	52	49	46	27	0.00	14	26	8	–	–	–	–	–	–	–	–	–	–	D-Glucitol, 2,3,4,6-tetra-O-methyl-, 1,5-diacetate	20316-14-7	NI33927
43	101	117	129	45	161	145	87	322	100	62	47	44	36	32	28	23	0.00	14	26	8	–	–	–	–	–	–	–	–	–	–	D-Mannitol, 1,3,4,5-tetra-O-methyl-, diacetate	19285-93-9	NI33928
101	161	117	129	43	145	87	71	322	100	80	72	61	53	48	23	16	0.00	14	26	8	–	–	–	–	–	–	–	–	–	–	Mannitol, 2,3,4,6-tetra-O-methyl-, 1,5-diacetate		NI33929
173	74	128	29	85	87	45	47	322	100	79	60	60	59	54	54	53	0.00	14	26	8	–	–	–	–	–	–	–	–	–	–	Marsectobiose	25153-15-5	LI07141
101	75	88	43	45	74	89	87	322	100	82	46	39	28	26	19	12	0.00	14	26	8	–	–	–	–	–	–	–	–	–	–	Methyl 2-O-acetyl-3,4,6,7-tetra-O-methyl-β-glycero-D-glucoheptopyranoside	84564-18-1	NI33930
101	75	74	43	45	89	88	115	322	100	46	32	30	26	21	18	15	0.00	14	26	8	–	–	–	–	–	–	–	–	–	–	Methyl 3-O-acetyl-2,4,6,7-tetra-O-methyl-α-glycero-D-glucoheptopyranoside		NI33931
88	75	45	43	87	101	89	73	322	100	92	29	18	15	13	13	11	0.00	14	26	8	–	–	–	–	–	–	–	–	–	–	Methyl 4-O-acetyl-2,3,6,7-tetra-O-methyl-β-glycero-D-glucoheptopyranoside	84564-16-9	NI33932
88	75	101	45	87	129	43	89	322	100	69	36	17	13	10	9	8	0.00	14	26	8	–	–	–	–	–	–	–	–	–	–	Methyl 6-O-acetyl-2,3,4,7-tetra-O-methyl-β-glycero-D-glucoheptopyranoside	84564-15-8	NI33933
88	75	101	43	117	73	45	221	322	100	73	67	23	12	12	12	9	0.00	14	26	8	–	–	–	–	–	–	–	–	–	–	Methyl 7-O-acetyl-2,3,4,6-tetra-O-methyl-β-glycero-D-glucoheptopyranoside		NI33934
88	101	75	45	89	43	59	263	322	100	71	54	41	24	23	19	9	0.00	14	26	8	–	–	–	–	–	–	–	–	–	–	Methyl 3-O-acetyl-2,4,6,7-tetra-O-methyl-α-L-glycero-D-mannoheptopyranoside		NI33935
88	101	75	43	89	45	117	221	322	100	95	91	29	22	18	16	13	0.00	14	26	8	–	–	–	–	–	–	–	–	–	–	Methyl 7-O-acetyl-2,3,4,6-tetra-O-methyl-α-L-glycero-D-mannoheptopyranoside	84564-14-7	NI33936
97	83	69	55	73	75	103	111	322	100	90	72	65	44	41	32	27	0.10	14	31	1	–	–	–	–	1	–	1	–	–	–	11-Bromoundecyl trimethylsilyl ether	26305-83-9	NI33937
152	151	154	209	322	51	282	77	322	100	96	79	67	65	47	44	40		15	10	–	–	–	3	–	–	–	–	–	–	1	Arsine, diphenyl(3,3,3-trifluoro-1-propynyl)-	33730-50-6	NI33938
322	307	43	277	323	289	308	279	322	100	79	25	19	18	15	13	10		15	14	8	–	–	–	–	–	–	–	–	–	–	7-Acetyl-2,3,6-trimethoxynaphthazarin		LI07142
300	322	307	43	277	323	289	308	322	100	99	80	24	18	17	14	12		15	14	8	–	–	–	–	–	–	–	–	–	–	1,4-Naphthoquinone, 2-acetyl-5,8-dihydroxy-3,6,7-trimethoxy-	14090-56-3	NI33939
322	307	43	323	289	308	279	261	322	100	79	26	19	15	13	10	8		15	14	8	–	–	–	–	–	–	–	–	–	–	1,4-Naphthoquinone, 2-acetyl-5,8-dihydroxy-3,6,7-trimethoxy-	14090-56-3	NI33940
181	133	105	322	119	77	92	134	322	100	54	54	50	35	35	22	21		15	16	2	2	–	–	1	–	–	–	1	–	–	2-(4-Chloro-3-methylphenoxy)-2,3-dihydro-1,3-dimethyl-1H-1,3,2-benzodiazaphosphole 2-oxide		NI33941
83	55	39	84	82	67	43	41	322	100	50	22	16	14	10	10	10	0.00	15	18	6	2	–	–	–	–	–	–	–	–	–	2-sec-Butyl-4,6-dinitrophenyl 3-methylcrotonate (binapacryl)	485-31-4	NI33942
139	217	63	250	218	183	107	77	322	100	99	74	59	59	59	59	59	37.00	15	19	–	–	2	–	–	–	–	1	1	–	–	Trimethylsilyl diphenyldithiophosphinate		NI33943
245	322	41	54	56	165	69	83	322	100	76	73	68	52	38	30	28		15	22	4	4	–	–	–	–	–	–	–	–	–	5-Nonanone, (2,4-dinitrophenyl)hydrazone	3657-08-7	NI33944
245	322	41	55	57	165	69	83	322	100	77	74	70	55	40	32	30		15	22	4	4	–	–	–	–	–	–	–	–	–	5-Nonanone, (2,4-dinitrophenyl)hydrazone	3657-08-7	NI33945
90	145	119	322	263	294	235	264	322	100	52	40	28	18	17	16	13		16	10	2	4	1	–	–	–	–	–	–	–	–	Bis(3,4-dihydro-4-oxo-2-quinazolinyl) sulphide		NI33946
322	307	323	238	237	236	308	153	322	100	20	19	16	16	9	4	4		16	10	4	4	–	–	–	–	–	–	–	–	–	5,12-Diamino-1,4,6,11-tetraoxo-1,2,3,4,6,11-hexahydronaphtho[2,3-g]phthalazine		IC03923
149	322	148	150	323	147	89	91	322	100	95	94	70	64	58	40	34		16	10	4	4	–	–	–	–	–	–	–	–	–	3,6-Bis(3′,4′-methylenedioxyphenyl)-1,2,4,5-tetrazine		NI33947
322	220	161	118	204	129	180	104	322	100	43	22	15	13	13	12	10		16	14	–	6	1	–	–	–	–	–	–	–	–	6-Phenylamino-7-phenyl-3-methylthio[1,2,4]triazolo[4,3-b][1,2,4]triazole		MS07624
238	239	322	240	241	223	307	225	322	100	92	60	44	37	35	29	24		16	15	5	–	–	–	1	–	–	–	–	–	–	Spiro[benzofuran-2(3H),1′-cyclohexane]-2′,3,4′-trione, 7-chloro-6-methoxy-4,6′-dimethyl-	26942-70-1	MS07625
214	322	216	215	324	78	280	140	322	100	65	37	29	24	19	16	16		16	15	5	–	–	–	1	–	–	–	–	–	–	Spiro[benzofuran-2(3H),1′-[2]cyclohexene]-3,4′-dione, 7-chloro-4,6-dimethoxy-6′-methyl-	17793-62-3	NI33948
254	322	294	225	135	279	256	239	322	100	75	48	43	43	37	34	32		16	15	5	–	–	–	1	–	–	–	–	–	–	Spiro[benzofuran-2(3H),1′-[3]cyclohexene]-2′,3-dione, 7-chloro-4,6-dimethoxy-6′-methyl-	26881-69-6	NI33949
98	225	68	69	185	172	198	322	322	100	42	40	39	28	28	22	21		16	15	5	–	–	–	1	–	–	–	–	–	–	Spiro[benzofuran-2(3H),1′-[3]cyclohexene]-2′,3-dione, 7-chloro-6-hydroxy 4′-methoxy-4,6′-dimethyl-	26881-78-7	NI33950

MASS TO CHARGE RATIOS								M.W.	INTENSITIES								Parent	C	H	O	N	S	F	Cl	Br	I	Si	P	B	X	COMPOUND NAME	CAS Reg No	No
259	322	260	261	262	231	247	290	**322**	100	26	21	19	18	16	13	12		16	18	7	–	–	–	–	–	–	–	–	–	–	trans-2,3,4,9-Tetramethoxy-2,3-dihydro-7-methyl-5H-furo[3,2-g]benzopyran-5-one		NI33951
173	104	158	72	175	71	103	55	**322**	100	66	55	54	36	35	28	22	**0.00**	16	22	1	2	2	–	–	–	–	–	–	–	–	6-exo-Amino-2-exo-isopropyl-2-endo-methyl-3-endo-(methylthio)-5-phenyl 4-thia-1-azabicyclo[3.2.0]heptan-7-one		NI33952
163	89	73	237	164	104	117	77	**322**	100	50	36	23	22	16	14	14	**0.00**	16	22	5	–	–	–	–	–	–	1	–	–	–	Phthalic acid, 4-[(trimethylsilyloxy)butyl]-		MS07626
73	307	322	308	323	45	74	309	**322**	100	80	54	20	15	13	8	8		16	26	3	–	–	–	–	–	–	2	–	–	–	6,7-Dihydroxy-1-oxotetrahydronaphthalene di-TMS		NI33953
143	68	115	89	55	107	125	69	**322**	100	63	60	55	41	40	38	38	**0.00**	16	34	6	–	–	–	–	–	–	–	–	–	–	1,3,5,7,11,13-Hexahydroxy-2,6,10-trimethyltridecane	85888-02-4	NI33954
278	103	75	175	74	322	150	206	**322**	100	87	86	37	37	22	21	20		17	6	7	–	–	–	–	–	–	–	–	–	–	Benzophenone-3,4,3′,4′-tetracarboxylic acid dianhydride	2421-28-5	NI33955
278	103	75	175	279	322	206	74	**322**	100	42	42	19	17	16	14	13		17	6	7	–	–	–	–	–	–	–	–	–	–	Benzophenone-3,5,3′,5′-tetracarboxylic acid dianhydride		IC03924
148	174	147	146	44	149	43	57	**322**	100	39	21	11	10	9	9	8	**7.30**	17	14	3	4	–	–	–	–	–	–	–	–	–	N,N′-Bis(3-isocyanato-4-methylphenyl)urea		IC03925
148	174	147	322	145	149	173	77	**322**	100	34	30	24	11	10	9	8		17	14	3	4	–	–	–	–	–	–	–	–	–	N,N′-Bis(4-isocyanato-2-methylphenyl)urea		IC03926
148	174	147	146	118	91	149	39	**322**	100	56	25	15	12	12	10	10	**4.80**	17	14	3	4	–	–	–	–	–	–	–	–	–	1-(2-Methyl-5-isocyanato-phenylcarbamoyl)-4-methyl-benzimidazolin-2-one		IC03927
77	322	91	67	275	303	65	51	**322**	100	50	20	18	17	13	13	12		17	14	3	4	–	–	–	–	–	–	–	–	–	1-Phenyl-3-methyl-5-hydroxy-4-(2-carboxyphenylazo)pyrazole		IC03928
280	93	77	281	65	92	91	105	**322**	100	88	78	36	24	22	21	16	**14.68**	17	14	3	4	–	–	–	–	–	–	–	–	–	1H-Pyrazole-4,5-dione, 3-(acetyloxy)-1-phenyl-, 4-(phenylhydrazone)	51471-67-1	NI33956
146	145	159	172	105	147	104	322	**322**	100	31	28	20	20	15	12	11		17	16	5	–	–	–	–	–	–	–	–	2	–	β-L-Arabinopyranose 1,2:3,4-bis(benzeneboronate)	52681-70-6	NI33957
146	159	172	105	145	322	104	158	**322**	100	65	51	32	24	17	17	16		17	16	5	–	–	–	–	–	–	–	–	2	–	α-D-Ribopyranose, cyclic 1,2:3,4-bis(phenylboronate)	74810-33-6	NI33958
159	146	322	105	147	91	321	104	**322**	100	83	70	56	40	38	36	34		17	16	5	–	–	–	–	–	–	–	–	2	–	α-D-Xylofuranose 1,2:3,5-bis(benzeneboronate)	52681-71-7	NI33959
43	189	161	216	93	83	55	246	**322**	100	26	24	22	20	20	19	18	**0.00**	17	22	6	–	–	–	–	–	–	–	–	–	–	Azuleno[4,5-b]furan-2,9-dione, 6-[(acetyloxy)methyl]decahydro-6a-hydroxy-9a-methyl-3-methylene-, [3aS-(3aα,6β,6aα,9aβ,9bα)]-	22621-72-3	NI33960
193	233	263	261	231	322	262	219	**322**	100	95	79	58	46	28	28	26		17	22	6	–	–	–	–	–	–	–	–	–	–	2,3-Dihydro-2-hydroxy-5-(methoxycarbonyl)-7-[1-(methoxycarbonyl)-1-isopropyl]-3,3-dimethylbenzofuran		MS07627
77	50	105	76	104	51	162	74	**322**	100	98	97	91	69	38	34	30	**0.00**	18	10	6	–	–	–	–	–	–	–	–	–	–	[2,2′-Bi-1H-indene]-1,1′,3,3′(2H,2′H)-tetrone, 2,2′-dihydroxy-	5103-42-4	NI33961
292	322	293	323	263	294	277	249	**322**	100	49	32	11	8	6	6	6		18	18	2	4	–	–	–	–	–	–	–	–	–	6-[(2-Aminoethyl)amino]-7,12-dioxo-1,2,3,4,7,12-hexahydronaphtho[2,3-f]quinoxaline		IC03929
312	322	262	263	157	77	277	173	**322**	100	99	90	66	23	23	20	16		18	18	2	4	–	–	–	–	–	–	–	–	–	Dianhydrohexulose phenylosazone		NI33962
290	292	288	200	93	262	201	171	**322**	100	85	48	14	13	12	12	10	**2.52**	18	18	2	4	–	–	–	–	–	–	–	–	–	3H-Pyrazolo[4,3-c]pyridazin-3-one, octahydro-7-hydroxy-6-methylene-2,5 diphenyl-	35426-80-3	NI33963
172	173	322	77	103	91	323	150	**322**	100	37	33	18	12	7	7	4		18	18	2	4	–	–	–	–	–	–	–	–	–	1,2,3-Triazole, 2-phenyl-4-[methyl(4-carboxyphenyl)aminomethyl]-5-methyl-		NI33964
173	165	174	175	99	81	322	247	**322**	100	23	11	10	10	10	7	7		18	26	1	–	2	–	–	–	–	–	–	–	–	2-(2′-Methylpropen-1′-yl)-1,3-dithiane		NI33965
149	263	57	104	77	41	133	76	**322**	100	29	22	15	15	14	11	9	**0.00**	18	26	5	–	–	–	–	–	–	–	–	–	–	1,2-Benzenedicarboxylic acid, 2-butoxyethyl butyl ester	33374-28-6	NI33966
161	290	291	162	121	148	41	91	**322**	100	34	30	26	19	17	17	16	**4.44**	18	26	5	–	–	–	–	–	–	–	–	–	–	Benzeneoctanoic acid, 3-methoxy-2-(methoxycarbonyl)-, methyl ester	56554-81-5	NI33967
168	150	163	81	235	322	99	55	**322**	100	61	57	46	44	43	43	34		18	26	5	–	–	–	–	–	–	–	–	–	–	1H-2-Benzoxacyclotetradecin-1-one, 3,4,5,6,7,8,9,10,11,12-decahydro-7,14,16-trihydroxy-3-methyl-, [3S-(3R*,7S*)]-	26538-44-3	NI33968
141	126	113	55	125	142	41	43	**322**	100	45	24	13	10	9	9	6	**2.10**	18	26	5	–	–	–	–	–	–	–	–	–	–	8,9-Diethoxy-4-hydroxytetracyclo[6.4.2.0^{1,9}.0^{4,14}]tetradeca-3,13-dione		NI33969
121	322	307	75	73	122	323	308	**322**	100	20	14	14	11	10	5	3		18	30	3	–	–	–	–	–	–	1	–	–	–	Octanoic acid, 4-(methoxyphenyl)-, trimethylsilyl ester		NI33970
320	171	322	100	258	159	115	102	**322**	100	68	50	35	18	18	18	18		19	11	3	–	–	–	1	–	–	–	–	–	–	2H-Naphth[1,8-bc]oxepin-2-one, 3-(4-chlorophenyl)-7-hydroxy-	27667-38-5	LI07143
320	171	322	100	255	161	120	102	**322**	100	68	51	35	18	18	18	18		19	11	3	–	–	–	1	–	–	–	–	–	–	2H-Naphth[1,8-bc]oxepin-2-one, 3-(4-chlorophenyl)-7-hydroxy-	27667-38-5	NI33971
290	145	322	89	291	117	323	234	**322**	100	85	51	45	20	18	11	11		19	14	5	–	–	–	–	–	–	–	–	–	–	Benzeneacetic acid, α-(3-hydroxy-5-oxo-4-phenyl-2(5H)-furanylidene)-, methyl ester, (E)-	521-52-8	NI33972
145	89	290	117	322	90	291	115	**322**	100	93	69	31	30	21	14	14		19	14	5	–	–	–	–	–	–	–	–	–	–	Benzeneacetic acid, α-(3-hydroxy-5-oxo-4-phenyl-2(5H)-furanylidene)-, methyl ester, (E)-	521-52-8	NI33973
145	89	290	322	117	77	63	90	**322**	100	90	64	29	26	22	22	20		19	14	5	–	–	–	–	–	–	–	–	–	–	Benzeneacetic acid, α-(3-hydroxy-5-oxo-4-phenyl-2(5H)-furanylidene)-, methyl ester, (E)-	521-52-8	NI33974
322	321	305	115	323	306	307	291	**322**	100	61	42	28	19	15	13	13		19	14	5	–	–	–	–	–	–	–	–	–	–	trans-6-Carboxy-2-(4-methoxystyryl)chromone		MS07628
322	121	279	71	83	97	85	81	**322**	100	54	25	23	20	16	16	15		19	14	5	–	–	–	–	–	–	–	–	–	–	2-Propenoic acid, 3-[2,3-dihydro-3-[(4-methoxyphenyl)methylene]-2-oxo-5 benzofuranyl]-	69721-63-7	NI33975
243	165	242	244	241	239	166	228	**322**	100	78	40	37	37	22	14	13	**0.00**	19	15	–	–	–	–	–	1	–	–	–	–	–	Trityl bromide	596-43-0	NI33976
322	211	323	229	77	213	324	289	**322**	100	33	26	10	10	9	6	5		19	18	3	–	–	–	–	–	–	1	–	–	–	Methyltriphenoxysilane	3439-97-2	NI33977
322	77	160	307	294	293	105	308	**322**	100	59	51	48	19	18	18	15		19	18	3	2	–	–	–	–	–	–	–	–	–	2-Ethyl-4-(4-methoxybenzylidene)-1-phenyl-3,5-dioxopyrazolidine	106584-31-0	NI33978
159	322	158	307	147	292	275	160	**322**	100	93	43	34	33	17	14	13		19	18	3	2	–	–	–	–	–	–	–	–	–	Spiro[2H-1-benzopyran-2,2′-[2H]indole], 1′,3′-dihydro-1′,3′,3′-trimethyl-6 nitro-	1498-88-0	NI33979
261	217	59	111	41	262	232	175	**322**	100	30	28	27	26	25	25	25	**2.00**	19	30	–	–	2	–	–	–	–	–	–	–	–	4,10,10-Trimethyl-7-(1′-methyl-2′,5′-dithiacyclopentyl)-tricyclo[4.4.0.2^{1,4}]dec-6-ene		MS07629
304	43	41	55	286	113	322	289	**322**	100	87	75	68	62	61	52	41		19	30	4	–	–	–	–	–	–	–	–	–	–	Androstan-17-one, 3,5,14-trihydroxy-, (3β,5α,14β)-	41853-34-3	NI33980

MASS TO CHARGE RATIOS	M.W.	INTENSITIES	Parent	C	H	O	N	S	F	Cl	Br	I	Si	P	B	X	COMPOUND NAME	CAS Reg No	No
187 202 262 159 133 119 107 108	322	100 82 70 63 42 41 38 28	0.00	19	30	4	–	–	–	–	–	–	–	–	–	–	3β,11-Diacetoxydrimene		MS07630
217 55 41 67 81 79 93 69	322	100 30 27 22 21 18 17 17	15.00	19	30	4	–	–	–	–	–	–	–	–	–	–	6β-Hydroxy-17-oxo-4,5-secoandrostan-4-oic acid		MS07631
71 43 234 279 322	322	100 99 29 3 2		19	30	4	–	–	–	–	–	–	–	–	–	–	4-Isobutyroyloxy-2-oxo-3,6,10-trimethyl-2,3,3a,4,5,8,9,10,11,11a-decahydrocyclodeca[b]furan		LI07144
155 139 81 293 156 137 41 89	322	100 25 20 13 12 11 11 9	0.00	19	31	1	–	–	1	–	–	–	1	–	–	–	Silane, ethyl(4-fluorophenyl)methyl[[5-methyl-2-isopropylcyclohexyl]oxy], (1α,2β,5α)-	74807-00-4	NI33981
322 323 307 324 161 321 146 145	322	100 24 20 11 9 8 6 5		20	18	–	–	2	–	–	–	–	–	–	–	–	1,2,7,8-Tetramethyl-4-phenylbenzo[1,2-b:4,3-b']dithiophene		LI07145
322 323 307 324 321 146 305 291	322	100 26 19 13 12 7 6 5		20	18	–	–	2	–	–	–	–	–	–	–	–	2,3,6,7-Tetramethyl-4-phenylbenzo[2,1-b:3,4-b']dithiophene		LI07146
322 294 76 266 238 104 210 105	322	100 58 55 52 40 37 28 21		20	18	4	–	–	–	–	–	–	–	–	–	–	2-(2',5'-Diethoxyphenyl)-1,4-naphthaquinone		MS07632
215 228 257 202 239 286 304 322	322	100 87 78 73 39 31 8 2		20	18	4	–	–	–	–	–	–	–	–	–	–	3,4-Dihydro-7,12-bis-dihydroxymethylbenz[a]anthracene-trans-3,4-diol		MS07633
202 215 226 257 244 239 288 286	322	100 95 68 62 52 46 43 16	16.00	20	18	4	–	–	–	–	–	–	–	–	–	–	8,9-Dihydro-7,12-bis-dihydroxymethylbenz[a]anthracene-trans-8,9-diol		MS07634
229 226 215 257 241 202 239 288	322	100 84 81 75 71 65 59 50	10.00	20	18	4	–	–	–	–	–	–	–	–	–	–	10,11-Dihydro-7,12-bis-dihydroxymethylbenz[a]anthracene-trans-10,11-diol		MS07635
43 55 41 57 69 203 81 307	322	100 89 85 79 73 63 53 52	51.00	20	18	4	–	–	–	–	–	–	–	–	–	–	Flemichapparin-A	32507-61-2	MS07636
119 65 91 219 53 322 39 202	322	100 35 34 30 20 18 11 9		20	18	4	–	–	–	–	–	–	–	–	–	–	3-Furoic acid, 2,5-dihydro-5-oxo-2,2-di-4-tolyl-, methyl ester	33545-33-4	MS07637
88 43 58 233 322 234 235 89	322	100 35 22 11 9 7 5 5		20	18	4	–	–	–	–	–	–	–	–	–	–	Indeno[1,2-b]pyran-5(2H)-one, 3,4-dihydro-2,2-dimethoxy-4-phenyl-	74685-19-1	NI33982
177 323 178 307 293 265 147	322	100 16 11 8 6 5 3	2.00	20	18	4	–	–	–	–	–	–	–	–	–	–	Warfarin methyl ether		NI33983
322 92 107 263 321 65 106 77	322	100 95 85 74 63 58 54 54		20	22	2	2	–	–	–	–	–	–	–	–	–	Akuammilan-17-oic acid, methyl ester	6475-05-4	NI33984
41 42 300 96 44 43 115 55	322	100 51 47 46 41 41 40 30	17.57	20	22	2	2	–	–	–	–	–	–	–	–	–	Aspidofractinine-1-carboxaldehyde, 3-oxo-, (2α,5α)-	55724-69-1	NI33985
322 121 252 235 42 56 41 263	322	100 84 79 72 71 61 54 52		20	22	2	2	–	–	–	–	–	–	–	–	–	Condyfolan-16-carboxylic acid, 2,14,16,19-tetradehydro-, methyl ester, (14E)-	4939-81-5	NI33986
84 86 60 83 61 85 88 71	322	100 59 24 20 19 14 11 11	4.40	20	22	2	2	–	–	–	–	–	–	–	–	–	1,16-Cyclocorynan-17-oic acid, 19,20-didehydro-, methyl ester, (16S,19E)	6393-66-4	NI33987
322 263 180 234 323 264 232 307	322	100 88 51 23 22 19 16 8		20	22	2	2	–	–	–	–	–	–	–	–	–	1,16-Cyclocorynan-17-oic acid, 19,20-didehydro-, methyl ester, (16S,19E)	6393-66-4	LI07147
136 145 322 147 161 146 119 77	322	100 68 42 42 28 28 23 23		20	22	2	2	–	–	–	–	–	–	–	–	–	1,4,2,5-Dioxadiazine, 3,6-bis(2,4,6-trimethylphenyl)-	20434-87-1	NI33988
130 145 147 161 146 119 91 105	322	100 68 42 28 27 23 23 22	1.40	20	22	2	2	–	–	–	–	–	–	–	–	–	1,4,2,5-Dioxadiazine, 3,6-bis(2,4,6-trimethylphenyl)-	20434-87-1	LI07148
108 322 279 251 134 120 323 77	322	100 40 39 14 14 14 10 10		20	22	2	2	–	–	–	–	–	–	–	–	–	Gelsemine	509-15-9	NI33989
322 321 184 169 41 156 170 55	322	100 76 68 56 44 38 28 27		20	22	2	2	–	–	–	–	–	–	–	–	–	Oxayohimban-19-one, 16-ethylidene-, (16Z)-	5545-94-8	NI33990
197 77 322 105 266 118 41 83	322	100 88 55 34 25 22 20 16		20	22	2	2	–	–	–	–	–	–	–	–	–	3,5-Pyrazolidinedione, 4-butyl-4-methyl-1,2-diphenyl-	26598-82-3	NI33991
323 120 122 117 121 129 324 104	322	100 94 47 32 30 29 28 25	14.00	20	22	2	2	–	–	–	–	–	–	–	–	–	3,5-Pyrazolidinedione, 4-butyl-4-methyl-1,2-diphenyl-	26598-82-3	NI33992
145 144 146 143 130 322 304 177	322	100 89 74 67 62 9 9 7		20	22	2	2	–	–	–	–	–	–	–	–	–	Spiro[2H-furo[3,2-b]indole-2,2'-[2H]indol]-3'-ol, 1',3,3',3a,4,7b-hexahydro 3a,3',7b-trimethyl-		LI07149
145 144 146 143 130 160 135 161	322	100 91 76 71 60 31 31 27	9.83	20	22	2	2	–	–	–	–	–	–	–	–	–	Spiro[2H-furo[3,2-b]indole-2,2'-[2H]indol]-3'-ol, 1',3,3',3a,4,8b-hexahydro 3',3a,8b-trimethyl-	24628-59-9	NI33993
280 322 321 307 281 265 291 323	322	100 49 28 22 22 18 14 11		20	26	–	4	–	–	–	–	–	–	–	–	–	5-tert-Butyl-N-(6-tert-butyl-2-pyridinyl)imidazo[1,2-a]pyridin-2-amine		MS07638
69 119 95 107 291 145 81 93	322	100 75 69 69 67 65 62 53	29.27	20	28	1	–	–	2	–	–	–	–	–	–	–	4,4-Difluororetinol (all-trans)	90660-21-2	NI33994
101 115 109 91 291 121 191 95	322	100 77 53 36 25 23 18 9	0.00	20	34	3	–	–	–	–	–	–	–	–	–	–	Ent-cleroda-4(18),11E-dien-13ξ,5,16-triol		NI33995
43 304 72 135 69 209 179 150	322	100 89 79 67 60 59 58 56	0.00	20	34	3	–	–	–	–	–	–	–	–	–	–	8α,13-Dihydroxy-14-labden-2-one		DD01236
71 81 41 55 69 109 95 68	322	100 94 67 63 62 61 60 56	0.00	20	34	3	–	–	–	–	–	–	–	–	–	–	8α,13-Dihydroxy-14-labden-3-one		DD01237
191 95 109 235 307 192 322	322	100 80 65 25 20 20 3		20	34	3	–	–	–	–	–	–	–	–	–	–	12(R)-8,12-epoxylabdan-15-oic acid		MS07639
191 95 81 109 307 137 121 123	322	100 60 55 50 40 40 40 30	4.00	20	34	3	–	–	–	–	–	–	–	–	–	–	12(S)-8,12-epoxylabdan-15-oic acid		MS07640
105 120 121 123 135 107 119 134	322	100 87 86 72 61 58 50 47	6.10	20	34	3	–	–	–	–	–	–	–	–	–	–	1,2-Ethanediol, 1-(2,3,4,4a,4b,5,6,7,8,8a,9,10-dodecahydro-7-hydroxy-2,4b,8,8-tetramethyl-2-phenanthrenyl)-, [2S-[2α(S*),4aβ,4bα,7α,8aβ]]-	5940-00-1	NI33996
167 109 95 123 121 177 251 192	322	100 85 53 51 42 36 32 31	4.00	20	34	3	–	–	–	–	–	–	–	–	–	–	12(R)-8-Hydroxylabdan-15(12)-olide		MS07641
125 124 109 123 177 121 223 235	322	100 82 61 60 34 31 30 25	10.00	20	34	3	–	–	–	–	–	–	–	–	–	–	12(S)-8-Hydroxylabdan-15(12)-olide		MS07642
95 175 121 136 191 161 206 270	322	100 33 29 23 22 10 4 2	0.00	20	34	3	–	–	–	–	–	–	–	–	–	–	Isolinaritriol	96888-13-0	NI33997
69 41 179 55 109 43 95 81	322	100 74 62 62 57 57 52 51	2.27	20	34	3	–	–	–	–	–	–	–	–	–	–	Sandaracopimar-15-ene-6β,8β,11α-triol	41756-31-4	NI33998
253 289 69 41 55 81 95 109	322	100 97 65 65 63 54 43 36	1.09	20	34	3	–	–	–	–	–	–	–	–	–	–	Sandaracopimar-15-ene-8β,11α-12β-triol	41756-16-5	NI33999
105 197 91 77 231 178 304 125	322	100 83 56 56 35 21 19 17	0.00	21	19	1	–	–	–	1	–	–	–	–	–	–	Benzeneethanol, α-benzyl-4-chloro-α-phenyl-	33966-10-8	NI34000
181 180 165 97 166 41 182 179	322	100 74 67 61 56 36 18 16	0.00	21	22	3	–	–	–	–	–	–	–	–	–	–	2-Methyl-3-phenylbenzyl 3-formyl-2,2-dimethylcyclopropanecarboxylate		DD01238
231 146 42 188 43 44 105 189	322	100 79 50 42 35 28 27 22	0.00	21	26	1	2	–	–	–	–	–	–	–	–	–	Acetanilide, N-(1-phenethyl-4-piperidyl)-	3258-84-2	NI34001
121 122 144 130 135 322 294 293	322	100 36 33 24 14 7 4 4		21	26	1	2	–	–	–	–	–	–	–	–	–	N-(α)-Acetyl-6,7-dehydro-aspidospermidine		LI07150
322 292 321 294 323 293 264 122	322	100 51 49 31 25 22 21 21		21	26	1	2	–	–	–	–	–	–	–	–	–	Aspidofractinine, 17-methoxy-3-methylene-, (2α,5α)-	55103-46-3	NI34002
184 322 41 202 42 170 182 122	322	100 55 52 42 42 39 36 35		21	26	1	2	–	–	–	–	–	–	–	–	–	2,20-Cycloaspidospermidine-3-methanol, 6,7-didehydro-1-methyl-, (3β,5α,12β,19α,20R)-	55856-76-3	NI34003
91 133 147 99 146 84 58 92	322	100 87 60 55 46 45 30 22	2.00	21	26	1	2	–	–	–	–	–	–	–	–	–	Diaziridinone, bis(1,1-dimethyl-2-phenylethyl)-	19694-14-5	NI34004

MASS TO CHARGE RATIOS								M.W.	INTENSITIES								Parent	C	H	O	N	S	F	Cl	Br	I	Si	P	B	X	COMPOUND NAME	CAS Reg No	No
281	134	253	130	322	294	252	282	**322**	100	87	81	50	41	36	27	24		21	26	1	2	–	–	–	–	–	–	–	–	–	13a-(1-Propenyl)-2,3,5,6,6a,11,12,13a,13b-decahydro-1H-cyclopenta[d,e]indolo[3,2-c]quinolizine-11-carbaldehyde		MS07643
73	45	43	41	55	59	67	81	**322**	100	96	95	66	60	52	50	47	**4.44**	21	38	2	–	–	–	–	–	–	–	–	–	–	[1,1′-Bicyclopropyl]-2-octanoic acid, 2′-hexyl-, methyl ester	56687-68-4	NI34005
149	94	81	67	55	41	43	54	**322**	100	99	97	94	94	93	82	78	**9.44**	21	38	2	–	–	–	–	–	–	–	–	–	–	Cyclopropanebutyric acid, 2-[(2-nonylcyclopropyl)methyl]-, methyl ester	10152-68-8	NI34006
55	149	41	67	81	69	54	82	**322**	100	99	80	74	69	60	60	57	**2.24**	21	38	2	–	–	–	–	–	–	–	–	–	–	Cyclopropanenonanoic acid, 2-[(2-butylcyclopropyl)methyl]-, methyl ester	10152-69-9	NI34007
55	81	67	41	59	95	31	69	**322**	100	93	89	86	81	75	71	65	**5.84**	21	38	2	–	–	–	–	–	–	–	–	–	–	Cyclopropaneoctanoic acid, 2-[(2-pentylcyclopropyl)methyl]-, methyl ester	56687-67-3	NI34009
41	55	81	67	95	69	82	54	**322**	100	99	78	78	54	54	52	51	**4.04**	21	38	2	–	–	–	–	–	–	–	–	–	–	Cyclopropaneoctanoic acid, 2-[(2-pentylcyclopropyl)methyl]-, methyl ester	56687-67-3	NI34008
149	55	81	67	95	41	69	54	**322**	100	96	95	93	92	92	91	87	**4.14**	21	38	2	–	–	–	–	–	–	–	–	–	–	Cyclopropaneoctanoic acid, 2-[(2-pentylcyclopropyl)methyl]-, methyl ester, trans,trans-	10152-66-6	NI34010
149	41	80	55	81	67	43	94	**322**	100	82	78	75	73	71	65	64	**4.04**	21	38	2	–	–	–	–	–	–	–	–	–	–	Cyclopropanepropionic acid, 2-[(2-decylcyclopropyl)methyl]-, methyl ester	10152-67-7	NI34011
67	81	55	41	95	82	68	96	**322**	100	85	71	68	63	58	48	47	**21.14**	21	38	2	–	–	–	–	–	–	–	–	–	–	8,11-Eicosadienoic acid, methyl ester	56599-56-5	NI34012
67	81	95	82	55	96	68	69	**322**	100	98	82	75	75	71	53	52	**35.64**	21	38	2	–	–	–	–	–	–	–	–	–	–	10,13-Eicosadienoic acid, methyl ester	30223-50-8	NI34013
67	81	55	41	95	82	68	96	**322**	100	81	71	64	60	56	47	45	**20.14**	21	38	2	–	–	–	–	–	–	–	–	–	–	11,13-Eicosadienoic acid, methyl ester	56599-57-6	NI34014
67	41	55	81	95	82	54	43	**322**	100	82	79	67	50	49	42	41	**1.00**	21	38	2	–	–	–	–	–	–	–	–	–	–	11,14-Eicosadienoic acid, methyl ester	2463-02-7	NI34015
126	55	127	41	124	44	43	322	**322**	100	10	8	5	3	3	3	2		21	42	–	2	–	–	–	–	–	–	–	–	–	1,5-Bis(dimethylpiperidyl)-2,2-dimethylpentane		IC03930
28	322	250	44	125	124	323	251	**322**	100	75	67	60	52	29	20	14		22	10	3	–	–	–	–	–	–	–	–	–	–	6H,8H-Benzo[10,11]chryseno[1,12-cd]pyran-6,8-dione	61142-11-8	NI34016
28	322	250	44	125	323	124	251	**322**	100	75	67	60	44	20	20	14		22	10	3	–	–	–	–	–	–	–	–	–	–	6H,8H-Benzo[10,11]chryseno[1,12-cd]pyran-6,8-dione	61142-11-8	NI34017
218	322	175	203	190	323	176	91	**322**	100	92	62	51	33	25	23	22		22	26	2	–	–	–	–	–	–	–	–	–	–	8-Oxatricyclo[5.4.1.0^{2,5}]dodeca-6,11-dien-12-one, 2,5,6,9,9-pentamethyl-4-phenyl-	55823-07-9	LI07151
218	322	175	203	307	190	323	176	**322**	100	92	63	51	49	33	26	23		22	26	2	–	–	–	–	–	–	–	–	–	–	8-Oxatricyclo[5.4.1.0^{2,5}]dodeca-6,11-dien-12-one, 2,5,6,9,9-pentamethyl-4-phenyl-	55823-07-9	NI34018
322	321	323	161	160.5	319	159.5	324	**322**	100	47	27	16	15	14	11	1		23	14	–	–	1	–	–	–	–	–	–	–	–	14H-Benzo[b]benzo[3,4]fluoreno[2,1-d]thiophene		LI07152
219	322	115	77	103	220	218	323	**322**	100	77	32	27	26	25	18	17		23	18	–	2	–	–	–	–	–	–	–	–	–	4H-1,2-Diazepine, 3,5,7-triphenyl-	25649-70-1	NI34019
175	119	176	145	91	322	160	28	**322**	100	45	19	15	10	9	8	7		23	30	1	–	–	–	–	–	–	–	–	–	–	2-Butanone, 1-(4-tert-butyl-2,6-xylyl)-3-(2-tolyl)-	29549-26-6	NI34020
175	119	176	145	91	322	160	120	**322**	100	45	19	15	10	9	8	6		23	30	1	–	–	–	–	–	–	–	–	–	–	2-Butanone, 1-(4-tert-butyl-2,6-xylyl)-3-(2-tolyl)-	29549-26-6	LI07153
55	43	41	57	83	82	71	69	**322**	100	95	88	85	78	67	51	39	**1.72**	23	46	–	–	–	–	–	–	–	–	–	–	–	9-Cyclohexylheptadecane		AI02043
55	57	83	82	71	69	97	85	**322**	100	90	81	71	52	40	38	35	**0.41**	23	46	–	–	–	–	–	–	–	–	–	–	–	9-Cyclohexylheptadecane		AI02044
55	43	57	83	69	97	41	70	**322**	100	94	89	87	86	81	79	45	**11.90**	23	46	–	–	–	–	–	–	–	–	–	–	–	9-Tricosene, (Z)-	27519-02-4	NI34021
41	43	55	57	69	71	83	111	**322**	100	98	81	78	32	31	24	15	**1.90**	23	46	–	–	–	–	–	–	–	–	–	–	–	9-Tricosene, (Z)-	27519-02-4	NI34022
43	41	55	57	83	69	97	56	**322**	100	77	74	68	51	48	38	35	**2.80**	23	46	–	–	–	–	–	–	–	–	–	–	–	11-Tricosene	52078-56-5	WI01414
57	91	105	41	161	160	117	104	**322**	100	46	33	14	13	12	12	12	**0.09**	24	34	–	–	–	–	–	–	–	–	–	–	–	Benzene, 1,1′-[1,2-bis(2,2-dimethylpropyl)-1,2-ethanediyl]bis-	29492-96-4	NI34023
95	161	67	226	91	79	66	228	**322**	100	86	82	63	52	50	41	40	**35.21**	24	34	–	–	–	–	–	–	–	–	–	–	–	2,2′-Bi-1,4:5,8-dimethanonaphthalene, eicosahydro-	15914-97-3	NI34024
167	181	322	168	195	182	165	323	**322**	100	47	31	16	13	13	10	8		24	34	–	–	–	–	–	–	–	–	–	–	–	Dodecylbiphenyl		MS07644
69	224	155	124	93	74	31	143	**323**	100	62	57	23	17	12	12	11	**0.50**	7	–	1	1	–	11	–	–	–	–	–	–	–	Perfluoro-4-methyl-3-oxa-4-azabicyclo[4.2.0]oct-1(6)-ene		MS07645
323	291	62	233	63	91	189	61	**323**	100	83	62	34	29	28	22	21		8	6	5	1	–	–	–	–	1	–	–	–	–	Salicylic acid, 5-iodo-3-nitro-, methyl ester	22621-44-9	MS07646
323	291	62	233	63	90	189	61	**323**	100	83	58	33	29	26	20	20		8	6	5	1	–	–	–	–	1	–	–	–	–	Salicylic acid, 5-iodo-3-nitro-, methyl ester	22621-44-9	LI07154
323	291	62	233	63	89	189	61	**323**	100	80	58	32	29	26	20	20		8	6	5	1	–	–	–	–	1	–	–	–	–	Salicylic acid, 5-iodo-3-nitro-, methyl ester	22621-44-9	NI34025
162	175	164	59	63	184	270	177	**323**	100	72	71	64	51	50	49	47	**18.00**	11	11	4	1	1	–	2	–	–	–	–	–	–	L-Cysteine, N-[(2,4-dichlorophenoxy)acetyl]-	50649-00-8	NI34027
162	175	164	59	184	270	220	177	**323**	100	72	71	64	50	49	47	47	**18.00**	11	11	4	1	1	–	2	–	–	–	–	–	–	L-Cysteine, N-[(2,4-dichlorophenoxy)acetyl]-	50649-00-8	NI34026
125	291	43	46	39	172	63	42	**323**	100	63	36	33	33	30	21	21	**0.00**	11	13	3	7	1	–	–	–	–	–	–	–	–	1H-Pyrazole-1-carboximidamide, 4-[[4-(aminosulphonyl)phenyl]hydrazono]-4,5-dihydro-3-methyl-5-oxo-	47220-10-0	LI07155
125	291	43	46	172	97	63	42	**323**	100	63	36	33	30	21	21	21	**0.00**	11	13	3	7	1	–	–	–	–	–	–	–	–	1H-Pyrazole-1-carboximidamide, 4-[[4-(aminosulphonyl)phenyl]hydrazono]-4,5-dihydro-3-methyl-5-oxo-	47220-10-0	NI34028
101	129	209	87	102	99	244	131	**323**	100	65	55	30	25	25	18	16	**0.00**	11	17	6	1	2	–	–	–	–	–	–	–	–	Succinimido 6-((methylsulphonyl)thio)hexanoate	76078-81-4	NI34029
93	323	285	248	193	44	295	117	**323**	100	82	38	27	26	20	16	16		12	–	2	1	–	7	–	–	–	–	–	–	–	Heptafluorophenoxazone-3		NI34030
322	324	320	165	242	164	244	138	**323**	100	52	52	43	41	40	39	19	**0.00**	12	7	–	1	–	–	–	2	–	–	–	–	–	9H-Carbazole, 3,6-dibromo-	6825-20-3	NI34031
325	164	165	327	323	244	246	138	**323**	100	67	63	53	48	37	35	30		12	7	–	1	–	–	–	2	–	–	–	–	–	9H-Carbazole, 3,6-dibromo-	6825-20-3	NI34032
323	57	71	170	73	75	153	28	**323**	100	48	46	43	41	41	37	35		12	9	6	3	1	–	–	–	–	–	–	–	–	3,3′-Dinitro-4′-aminodiphenylsulphone	40178-99-2	NI34033
323	153	65	44	91	139	138	30	**323**	100	52	47	46	46	44	43	42		12	9	6	3	1	–	–	–	–	–	–	–	–	3,4′-Dinitro-4-aminodiphenylsulphone		NI34034

MASS TO CHARGE RATIOS								M.W.	INTENSITIES								Parent	C	H	O	N	S	F	Cl	Br	I	Si	P	B	X	COMPOUND NAME	CAS Reg No	No
98	99	73	41	56	308	97	42	**323**	100	12	5	5	4	3	3	3	**1.00**	12	30	3	1	–	–	–	–	–	2	1	–	–	**Phosphonic acid, [1-(isopropylideneamino)propyl]-, bis(trimethylsilyl) ester**	55108-65-1	**NI34035**
84	73	225	308	75	147	98	41	**323**	100	36	32	25	21	20	16	16	**6.50**	12	30	3	1	–	–	–	–	–	2	1	–	–	**Phosphonic acid, [3-(isopropylideneamino)propyl]-, bis(trimethylsilyl) ester**	55108-72-0	**NI34036**
323	74	308	251	294	236	222	75	**323**	100	73	58	50	32	28	27	27		13	25	1	5	–	–	–	–	–	2	–	–	–	**6H-Purin-6-one, 2-(dimethylamino)-1,7-dihydro-, bis(trimethylsilyl)-**	72088-18-7	**NI34037**
323	185	157	77	169	141	63	110	**323**	100	98	76	67	65	56	56	42		14	14	4	1	1	–	–	–	–	–	1	–	–	**Phosphonothioic acid, phenyl-, O-ethyl O-(4-nitrophenyl) ester**	2104-64-5	**NI34038**
157	169	63	29	141	185	77	47	**323**	100	70	50	43	37	33	33	26	**9.50**	14	14	4	1	1	–	–	–	–	–	1	–	–	**Phosphonothioic acid, phenyl-, O-ethyl O-(4-nitrophenyl) ester**	2104-64-5	**NI34039**
63	157	77	47	169	51	140	185	**323**	100	97	73	57	56	48	43	31	**6.66**	14	14	4	1	1	–	–	–	–	–	1	–	–	**Phosphonothioic acid, phenyl-, O-ethyl O-(4-nitrophenyl) ester**	2104-64-5	**NI34040**
173	175	323	145	325	177	147	109	**323**	100	65	19	16	13	11	10	8		15	11	3	1	–	–	2	–	–	–	–	–	–	**Methyl 3-[1-(2,6-dichlorobenzoyl)pyrrol-2-yl]prop-2-enoate**		**NI34041**
195	101	129	167	123	277	179	249	**323**	100	94	71	60	56	54	28	20	**10.00**	15	14	5	1	–	–	1	–	–	–	–	–	–	**6-Chloro-3-(3-carbethoxypropionylamino)isocoumarin**		**LI07156**
91	92	43	41	57	71	105	323	**323**	100	42	39	22	20	18	12	12		15	18	1	1	–	5	–	–	–	–	–	–	–	**Phenylhexylamine, pentafluoropropionyl-**		**NI34042**
73	91	75	74	45	147	189	59	**323**	100	11	10	9	9	8	6	5	**1.69**	15	25	3	1	–	–	–	–	–	2	–	–	–	**Benzenepropanoic acid, α-[[(trimethylsilyl)oxy]imino]-, trimethylsilyl ester**	55530-64-8	**NI34043**
105	206	73	77	308	323	75	147	**323**	100	65	61	25	24	18	18	17		15	25	3	1	–	–	–	–	–	2	–	–	–	**Glycine, N-benzoyl-N-(trimethylsilyl)-, trimethylsilyl ester**	55133-85-2	**NI34045**
105	73	206	77	45	91	75	147	**323**	100	69	50	23	18	15	14	12	**8.13**	15	25	3	1	–	–	–	–	–	2	–	–	–	**Glycine, N-benzoyl-N-(trimethylsilyl)-, trimethylsilyl ester**	55133-85-2	**NI34044**
105	73	206	75	147	77	91	207	**323**	100	65	60	60	41	29	17	15	**7.20**	15	25	3	1	–	–	–	–	–	2	–	–	–	**Glycine, N-benzoyl-N-(trimethylsilyl)-, trimethylsilyl ester**	55133-85-2	**MS07647**
77	323	201	324	80	105	67	173	**323**	100	98	98	80	80	61	54	50		16	13	3	5	–	–	–	–	–	–	–	–	–	**1-Phenyl-3-methyl-4-(4-nitrophenylazo)-5-pyrazolone**		**NI34046**
125	235	241	276	165	237	127	278	**323**	100	80	59	56	55	54	50	41	**0.00**	16	15	2	1	–	–	2	–	–	–	–	–	–	**Benzene, 1,1′-(2-nitrobutylidene)bis[4-chloro-**	117-26-0	**NI34047**
43	15	44	42	179	56	45	28	**323**	100	36	30	21	16	16	9	9	**0.00**	16	21	6	1	–	–	–	–	–	–	–	–	–	**Acetamide, N-[2-(acetyloxy)-2-[4-(acetyloxy)-3-methoxyphenyl]ethyl]-N-methyl-**	55145-64-7	**NI34048**
119	117	118	234	116	120	143	126	**323**	100	99	82	53	40	37	29	28	**20.79**	16	21	6	1	–	–	–	–	–	–	–	–	–	**20-Norcrotalanan-11,15-dione, 3,8-didehydro-14,19-dihydro-12,13-dihydroxy-, (13α,14α)-**	23291-96-5	**NI34049**
59	168	98	91	169	65	55	110	**323**	100	73	39	30	20	16	16	14	**5.00**	16	25	2	3	1	–	–	–	–	–	–	–	–	**N[2],N[2]-Dimethyl-N[1]-tosylformohydrazide cyclohexylimide**		**MS07648**
222	91	134	221	233	131	193	104	**323**	100	71	50	25	16	15	10	10	**5.00**	17	13	–	3	2	–	–	–	–	–	–	–	–	**5-Phenylthiazolo[2,3-b]-1,3,4-thiadiazol-4-ium-2-(4-methylphenyl)aminide**		**MS07649**
323	77	175	105	51	254	101	201	**323**	100	79	24	20	15	13	13	7		17	13	4	3	–	–	–	–	–	–	–	–	–	**2-Methyl-4-(4-nitrobenzylidene)-1-phenyl-3,5-dioxopyrazolidine**	106584-29-6	**NI34050**
143	323	115	128	324	171	144	101	**323**	100	83	59	18	17	17	13	9		17	13	4	3	–	–	–	–	–	–	–	–	–	**Naphthalene, 2-hydroxy-1-(2-methoxy-4-nitrophenylazo)-**		**IC03931**
143	323	115	128	171	324	144	51	**323**	100	62	57	17	16	13	11	11		17	13	4	3	–	–	–	–	–	–	–	–	–	**Naphthalene, 2-hydroxy-1-(2-methoxy-5-nitrophenylazo)-**		**IC03932**
283	133	323	173	284	104	77	105	**323**	100	41	37	20	18	12	9	8		17	17	2	5	–	–	–	–	–	–	–	–	–	**N-Ethyl-N-(2-cyanoethyl)-4-(nitrophenylazo)-aniline**		**IC03933**
58	73	45	165	57	42	46	44	**323**	100	17	11	8	7	5	4	3	**0.01**	17	19	1	1	–	–	2	–	–	–	–	–	–	**N,N-Dimethyl-2-[α-(2-chlorophenyl)-3-chlorobenzyloxy]ethylamine**		**MS07650**
136	119	138	95	206	323	278	250	**323**	100	95	93	87	80	63	55	48		17	25	5	1	–	–	–	–	–	–	–	–	–	**Crotananine**		**MS07651**
323	91	92	324	65	290	239	325	**323**	100	95	38	37	28	13	13	11		18	13	3	1	1	–	–	–	–	–	–	–	–	**2-Benzylthio-6-methylpyrano[2,3-e]benzoxazol-8(H)-one**		**NI34051**
323	188	105	201	77	200	130	175	**323**	100	88	74	70	50	30	30	24		18	17	3	3	–	–	–	–	–	–	–	–	–	**4-(2-Benzoyloxyethylamino)-2-methyl-1-oxo-1,2-dihydrophthalazine**	73632-87-8	**NI34052**
65	157	138	186	131	185	91	92	**323**	100	98	70	50	48	46	46	35	**2.00**	18	17	3	3	–	–	–	–	–	–	–	–	–	**2-(4-methylanilino)-4-(4-nitroanilino)-cyclopent-2-enone**		**MS07652**
323	77	104	93	264	118	51	324	**323**	100	83	74	34	33	26	23	22		18	17	3	3	–	–	–	–	–	–	–	–	–	**4-Pyrimidinecarboxylic acid, 3,4,5,6-tetrahydro-6-oxo-3-phenyl-2-(phenylamino)-, methyl ester**	55124-78-2	**NI34053**
93	323	130	92	65	213	324		**323**	100	68	37	24	21	18	15			18	18	1	3	–	–	–	–	–	–	1	–	–	**N,N′,N″-Triphenylphosphoramide**		**IC03934**
262	41	263	43	55	42	244	174	**323**	100	30	18	17	15	13	11	9	**4.35**	18	29	2	1	1	–	–	–	–	–	–	–	–	**Annotine, 15-deoxo-15-(ethylthio)-2,3-dihydro-**	5096-61-7	**NI34054**
174	234	146	175	176	137	235	172	**323**	100	50	31	20	16	14	11	11	**3.15**	18	29	4	1	–	–	–	–	–	–	–	–	–	**Lycopodane-5,8,12-triol, 15-methyl-, 5-acetate, (5β,8R,15S)-**	3175-90-4	**NI34055**
266	323	248	305	210	267	324	29	**323**	100	60	36	17	15	14	13	12		18	29	4	1	–	–	–	–	–	–	–	–	–	**4-Morpholino-2,5-dibutoxyphenol**		**IC03935**
168	323	167	324	258	155			**323**	100	81	56	24	2	1				19	17	2	1	1	–	–	–	–	–	–	–	–	**Benzenesulphonamide, N-[1,1′-biphenyl]-2-yl-4-methyl-**	24310-30-3	**NI34056**
148	323	322	149	324	174	150	147	**323**	100	28	18	16	13	12	7	7		19	17	4	1	–	–	–	–	–	–	–	–	–	**4H-Bis[1,3]benzodioxolo[5,6-a:4′,5′-g]quinolizine, 6,7,12b,13-tetrahydro-, (±)-**	4312-32-7	**NI34057**
308	293	323	309	294	324	322	154	**323**	100	60	43	26	12	11	10	10		19	17	4	1	–	–	–	–	–	–	–	–	–	**7H-Pyrano[2,3-c]acridin-7-one, 3,12-dihydro-6,11-dihydroxy-3,3,12-trimethyl-**	27067-70-5	**MS07653**
148	323	322	174	91	163	77	308	**323**	100	70	40	29	21	20	11	4		19	17	4	1	–	–	–	–	–	–	–	–	–	**L-Stylopine**	84-39-9	**NI34058**
292	323	233	264	223				**323**	100	20	20	10	4					19	21	2	3	–	–	–	–	–	–	–	–	–	**1-Pyrrolidino-2-methoxy-3-oxo-5,6,11,11b-tetrahydro-3H-indolo[3,2-g]indolizine**		**LI07157**
194	209	306	68	222	181	54	81	**323**	100	92	55	44	38	36	35	32	**22.00**	19	33	3	1	–	–	–	–	–	–	–	–	–	**17-Norlabdan-15-oic acid, 8-oxo-, oxime**		**LI07158**
194	209	306	69	41	222	55	81	**323**	100	92	54	42	40	37	34	32	**22.10**	19	33	3	1	–	–	–	–	–	–	–	–	–	**17-Norlabdan-15-oic acid, 8-oxo-, oxime, (13β)-**		**MS07654**
91	290	286	92	291	287	146	99	**323**	100	42	37	29	27	17	12	11	**2.00**	20	21	1	1	1	–	–	–	–	–	–	–	–	**8-Methyl-6,7,8,9,10,15-hexahydrobenzo[d][1]benzothieno[2,3-g]azecin-15-ol**		**NI34059**
91	121	122	158	202	92	65	78	**323**	100	19	17	13	8	8	8	6	**0.00**	20	21	3	1	–	–	–	–	–	–	–	–	–	**1-Benzyloxycarbonyl-2-(4-methoxybenzyl)-2,5-dihydropyrrole**		**LI07159**
105	120	131	238	121	133	248	249	**323**	100	86	12	10	8	6	5	3	**0.00**	20	21	3	1	–	–	–	–	–	–	–	–	–	**Ethaneperoxoic acid, 1-cyano-1,4-diphenylpentyl ester**	58422-77-8	**NI34060**
323	83	85	91	324	293	87	217	**323**	100	97	84	57	47	45	27	18		20	21	3	1	–	–	–	–	–	–	–	–	–	**Ethyl 1-[(benzyloxy)methyl]-2-methylindole-3-carboxylate**		**DD01239**
221	323	222	223	207	196	180	41	**323**	100	96	55	42	41	29	27	27		20	25	1	3	–	–	–	–	–	–	–	–	–	**Ergoline-8-carboxamide, 9,10-didehydro-N,N-diethyl-6-methyl-, (8β)-**	50-37-3	**NI34062**
323	221	181	222	223	207	324	72	**323**	100	65	41	36	33	29	22	21		20	25	1	3	–	–	–	–	–	–	–	–	–	**Ergoline-8-carboxamide, 9,10-didehydro-N,N-diethyl-6-methyl-, (8β)-**	50-37-3	**NI34061**

MASS TO CHARGE RATIOS								M.W.	INTENSITIES								Parent	C	H	O	N	S	F	Cl	Br	I	Si	P	B	X	COMPOUND NAME	CAS Reg No	No
60	42	45	62	44	43	61	58	**323**	100	98	97	95	75	71	65	60	**4.12**	20	25	1	3	–	–	–	–	–	–	–	–	–	Ergoline-8-carboxamide, 9,10-didehydro-N,N-diethyl-6-methyl-, (8β)-	50-37-3	NI34063
323	266	324	269	308	322			**323**	100	60	36	17	11	10				20	25	1	3	–	–	–	–	–	–	–	–	–	Prodigiosin	82-89-3	LI07160
86	129	128	279	123	121	109	191	**323**	100	52	50	40	40	35	26	20	**5.00**	20	37	2	1	–	–	–	–	–	–	–	–	–	8-Hydroxylabdan-15-amide		MS07655
287	96	286	215	288	28	229	36	**323**	100	58	44	43	22	21	20	20	**0.00**	21	22	–	1	–	–	1	–	–	–	–	–	–	Cyproheptadine hydrochloride		MS07656
216	57	132	217	220	85	130	221	**323**	100	78	76	25	15	15	14	13	**0.00**	21	25	2	1	–	–	–	–	–	–	–	–	–	u-2-tert-Butoxy-1,2,3,4-tetrahydro-α-phenyl-1-isoquinolinemethanol		DD01240
180	250	43	181	179	251	71	152	**323**	100	40	23	21	9	7	6	2	**1.00**	21	25	2	1	–	–	–	–	–	–	–	–	–	5,6-Dihydro-6-(2-hydroxymethylprop-2-yl)-5-isobutyrylphenanthridine		LI07161
135	107	77	212	111	96	204	323	**323**	100	61	48	33	24	22	21	6		21	25	2	1	–	–	–	–	–	–	–	–	–	2-[2-Hydroxy-2-(4-methoxyphenyl)-2-phenylethyl]-4,4-dimethyl-1-pyrazoline		NI34064
169	95	81	143	137	141	140	67	**323**	100	80	65	58	43	34	28	22	**7.30**	21	25	2	1	–	–	–	–	–	–	–	–	–	Isoborneol naphthyl uretane		MS07657
101	43	114	44	41	55	49	57	**323**	100	99	89	89	81	69	59	33	**30.04**	21	41	1	1	–	–	–	–	–	–	–	–	–	9-Octadecenamide, N-isopropyl-	56630-50-3	NI34065
43	114	101	41	55	69	57	60	**323**	100	82	78	70	58	24	24	23	**21.04**	21	41	1	1	–	–	–	–	–	–	–	–	–	9-Octadecenamide, N-propyl-	56630-49-0	NI34066
113	43	55	41	140	98	70	71	**323**	100	31	26	21	19	14	14	11	**1.14**	21	41	1	1	–	–	–	–	–	–	–	–	–	Pyrrolidine, 1-(3,6-dimethyl-1-oxopentadecyl)-	56630-62-7	NI34067
113	43	55	41	140	70	98	71	**323**	100	33	24	19	16	14	12	12	**1.24**	21	41	1	1	–	–	–	–	–	–	–	–	–	Pyrrolidine, 1-(3-methyl-1-oxohexadecyl)-	56630-57-0	NI34068
113	43	55	41	70	126	71	57	**323**	100	42	28	23	19	18	14	12	**2.34**	21	41	1	1	–	–	–	–	–	–	–	–	–	Pyrrolidine, 1-(7-methyl-1-oxohexadecyl)-	56630-56-9	NI34069
113	43	49	55	41	70	126	57	**323**	100	30	29	25	24	20	19	19	**2.34**	21	41	1	1	–	–	–	–	–	–	–	–	–	Pyrrolidine, 1-(15-methyl-1-oxohexadecyl)-	56630-53-6	NI34070
113	43	55	41	126	70	71	57	**323**	100	38	24	22	19	17	13	9	**2.34**	21	41	1	1	–	–	–	–	–	–	–	–	–	Pyrrolidine, 1-(1-oxoheptadecyl)-	56630-54-7	NI34072
113	43	55	41	70	126	71	57	**323**	100	50	26	23	19	18	13	9	**2.04**	21	41	1	1	–	–	–	–	–	–	–	–	–	Pyrrolidine, 1-(1-oxoheptadecyl)-	56630-54-7	NI34071
113	43	55	140	41	98	70	71	**323**	100	42	24	19	19	13	13	12	**1.54**	21	41	1	1	–	–	–	–	–	–	–	–	–	Pyrrolidine, 1-(3,6,13-trimethyl-1-oxotetradecyl)-	56630-60-5	NI34073
179	115	180	323	116	178	126	151	**323**	100	34	28	17	17	15	14	12		23	17	1	1	–	–	–	–	–	–	–	–	–	Cyclobuta[a]naphthalen-4(3H)-one, 2a,8b-dihydro-2-phenyl-1-(3-pyridinyl)-	50559-65-4	NI34074
323	220	77	191	115	51	105	178	**323**	100	99	80	75	60	60	40	30		23	17	1	1	–	–	–	–	–	–	–	–	–	4,6-Diphenyl-2-(phenylimino)pyran		NI34075
307	306	308	78	230	202	77	41	**323**	100	30	24	17	13	8	8	7	**0.13**	23	17	1	1	–	–	–	–	–	–	–	–	–	Pyridine, 2,4,6-triphenyl-, 1-oxide	23022-74-4	NI34076
323	322	246	324	77	217	191	189	**323**	100	47	30	27	15	11	10	6		23	17	1	1	–	–	–	–	–	–	–	–	–	1H-Pyrrole, 2-benzoyl-3,5-diphenyl-	56588-16-0	NI34077
323	295	191	77	51	115	246	218	**323**	100	90	80	80	60	50	30	30		23	17	1	1	–	–	–	–	–	–	–	–	–	1,4,6-Triphenyl-2-pyridone	62257-63-0	NI34078
232	323	105	104	77	91	219	51	**323**	100	71	70	51	51	41	35	24		24	21	–	1	–	–	–	–	–	–	–	–	–	1H-Pyrrole, 2,4-diphenyl-1-phenylethyl-		MS07658
247	245	232	249	230	234	217	151	**324**	100	65	50	45	28	22	20	20	**0.00**	3	9	–	–	–	–	–	2	–	–	–	–	1	Trimethylantimony dibromide	24606-08-4	NI34079
309	307	75	294	279	93	292	277	**324**	100	61	33	24	22	16	14	13	**0.31**	3	9	4	–	–	–	–	–	–	1	–	–	1	Perrhenic acid (HReO4), trimethylsilyl ester	16687-12-0	MS07659
41	197	69	155	39	127	27	128	**324**	100	99	84	68	47	46	21	17	**7.10**	5	10	–	–	–	–	–	–	2	–	–	–	–	Pentane, 1,5-diiodo-	628-77-3	NI34080
41	69	197	128	324	127			**324**	100	98	88	40	36	21				5	10	–	–	–	–	–	–	2	–	–	–	–	Pentane, 1,5-diiodo-		LI07162
281	69	245	288	55	70	202	237	**324**	100	31	29	27	17	9	6	4	**0.60**	5	11	1	–	–	–	1	–	–	–	–	–	1	(2-Hydroxypentyl)mercuric chloride		LI07163
289	291	43	229	227	225	295	293	**324**	100	99	75	53	53	53	16	16	**11.00**	7	13	4	–	–	–	2	–	–	–	–	–	1	Niobium, dichlorodimethoxy(2,4-pentanedionato-O,O′)-	17099-32-0	NI34082
43	289	225	291	227	175	243	293	**324**	100	99	40	33	27	15	13	12	**7.50**	7	13	4	–	–	–	2	–	–	–	–	–	1	Niobium, dichlorodimethoxy(2,4-pentanedionato-O,O′)-	17099-32-0	NI34081
237	295	208	235	236	209	207	293	**324**	100	72	61	50	41	40	38	35	**0.27**	8	20	–	–	–	–	–	–	–	–	–	–	1	Tetraethyllead	78-00-2	AI02045
237	295	208	235	236	209	207	294	**324**	100	73	61	46	44	39	37	33	**0.26**	8	20	–	–	–	–	–	–	–	–	–	–	1	Tetraethyllead	78-00-2	NI34083
295	237	208	209	266	28	27	29	**324**	100	82	19	18	16	9	7	6	**1.51**	8	20	–	–	–	–	–	–	–	–	–	–	1	Tetraethyllead	78-00-2	MS07660
121	205	56	149	140	177	39	95	**324**	100	50	50	33	33	20	20	16	**0.50**	10	5	3	–	–	5	–	–	–	–	–	–	1	Iron, dicarbonyl(η^5-2,4-cyclopentadien-1-yl)(2,2,3,3,3-pentafluoro-1-oxopropyl)-	55836-25-4	NI34084
126	161	196	96	324	91	56	39	**324**	100	88	83	69	68	64	63	63		10	8	–	–	–	–	4	–	–	–	–	–	1	Ferrocene, 1,2,3,4-tetrachloro-		LI07164
130	86	162	324	56	76	131	60	**324**	100	52	39	15	15	14	13	12		10	16	2	2	4	–	–	–	–	–	–	–	–	Morpholine, 4,4′-(dithiodicarbonothioyl)bis-	729-46-4	NI34085
324	124	148	31	325	93	98	162	**324**	100	30	22	17	14	13	12	10		12	–	–	2	–	8	–	–	–	–	–	–	–	Phenazine, octafluoro-	18232-24-1	NI34086
254	256	326	328	324	291	127	128	**324**	100	97	86	58	57	57	56	54		12	5	–	–	–	–	5	–	–	–	–	–	–	1,1′-Biphenyl, 2,2′,3,3′,4-pentachloro-	52663-62-4	NI34087
326	254	256	127	291	128	328	109	**324**	100	93	92	70	68	66	61	60	**57.40**	12	5	–	–	–	–	5	–	–	–	–	–	–	1,1′-Biphenyl, 2,2′,3,3′,5-pentachloro-	60145-20-2	NI34088
254	256	326	291	328	324	127	128	**324**	100	98	93	64	59	58	58	52		12	5	–	–	–	–	5	–	–	–	–	–	–	1,1′-Biphenyl, 2,2′,3,3′,6-pentachloro-	52663-60-2	NI34089
326	127	254	256	128	328	324	184	**324**	100	88	83	81	79	63	63	52		12	5	–	–	–	–	5	–	–	–	–	–	–	1,1′-Biphenyl, 2,2′,3,4,5-pentachloro-	55312-69-1	NI34090
326	254	127	256	328	324	128	184	**324**	100	79	75	74	65	64	64	44		12	5	–	–	–	–	5	–	–	–	–	–	–	1,1′-Biphenyl, 2,2′,3,4,5′-pentachloro-	38380-02-8	NI34091
254	256	326	324	328	127	128	184	**324**	100	97	92	59	57	56	53	46		12	5	–	–	–	–	5	–	–	–	–	–	–	1,1′-Biphenyl, 2,2′,3,4,5′-pentachloro-	38380-02-8	NI34093
184	254	256	326	186	149	328	147	**324**	100	76	73	54	47	45	41	32	**30.76**	12	5	–	–	–	–	5	–	–	–	–	–	–	1,1′-Biphenyl, 2,2′,3,4,5′-pentachloro-	38380-02-8	NI34092
326	254	328	324	256	127	128	184	**324**	100	67	64	64	63	50	46	35		12	5	–	–	–	–	5	–	–	–	–	–	–	1,1′-Biphenyl, 2,2′,3,4,6-pentachloro-	55215-17-3	NI34094
184	256	186	254	149	145	147	146	**324**	100	77	70	67	67	49	45	41	**16.23**	12	5	–	–	–	–	5	–	–	–	–	–	–	1,1′-Biphenyl, 2,2′,3,4′,5′-pentachloro-	41464-51-1	NI34095
326	256	254	324	328	127	128	291	**324**	100	98	96	63	62	50	48	44		12	5	–	–	–	–	5	–	–	–	–	–	–	1,1′-Biphenyl, 2,2′,3,4′,5′-pentachloro-	41464-51-1	NI34096
326	254	328	324	256	127	128	109	**324**	100	68	64	63	60	58	51	43		12	5	–	–	–	–	5	–	–	–	–	–	–	1,1′-Biphenyl, 2,2′,3,4′,5′-pentachloro-	41464-51-1	NI34097
326	256	254	328	324	291	127	128	**324**	100	98	98	62	59	55	52	51		12	5	–	–	–	–	5	–	–	–	–	–	–	1,1′-Biphenyl, 2,2′,3,5,5′-pentachloro-	52663-61-3	NI34098
326	254	256	127	328	128	324	184	**324**	100	80	72	71	66	65	65	51		12	5	–	–	–	–	5	–	–	–	–	–	–	1,1′-Biphenyl, 2,2′,3,5,6-pentachloro-	73575-56-1	NI34099
184	254	256	326	328	149	145	147	**324**	100	96	89	70	49	41	40	39	**27.71**	12	5	–	–	–	–	5	–	–	–	–	–	–	1,1′-Biphenyl, 2,2′,3,5′,6-pentachloro-	38379-99-6	NI34100

MASS TO CHARGE RATIOS								M.W.	INTENSITIES								Parent	C	H	O	N	S	F	Cl	Br	I	Si	P	B	X	COMPOUND NAME	CAS Reg No	No
326	256	254	324	328	127	291	128	**324**	100	95	92	60	57	56	52	50		12	5	–	–	–	–	5	–	–	–	–	–	–	**1,1′-Biphenyl, 2,2′,3,5′,6-pentachloro-**	38379-99-6	**NI34101**
127	326	254	128	256	109	324	328	**324**	100	93	86	85	80	72	62	62		12	5	–	–	–	–	5	–	–	–	–	–	–	**1,1′-Biphenyl, 2,2′,3′,4,6-pentachloro-**	60233-25-2	**NI34102**
326	254	256	328	324	184	128	127	**324**	100	79	73	64	62	35	30	30		12	5	–	–	–	–	5	–	–	–	–	–	–	**1,1′-Biphenyl, 2,2′,4,4′,5-pentachloro-**	38380-01-7	**NI34103**
326	328	324	254	256	127	128	184	**324**	100	62	62	60	58	37	33	30		12	5	–	–	–	–	5	–	–	–	–	–	–	**1,1′-Biphenyl, 2,2′,4,4′,6-pentachloro-**	39485-83-1	**NI34104**
326	254	328	324	127	256	128	109	**324**	100	65	65	63	62	62	54	43		12	5	–	–	–	–	5	–	–	–	–	–	–	**1,1′-Biphenyl, 2,2′,4,5,5′-pentachloro-**	37680-73-2	**NI34105**
254	256	326	328	324	184	258	330	**324**	100	95	82	57	44	41	31	21		12	5	–	–	–	–	5	–	–	–	–	–	–	**1,1′-Biphenyl, 2,2′,4,5,5′-pentachloro-**	37680-73-2	**NI34107**
326	254	256	328	324	127	128	184	**324**	100	98	88	65	61	49	45	44		12	5	–	–	–	–	5	–	–	–	–	–	–	**1,1′-Biphenyl, 2,2′,4,5,5′-pentachloro-**	37680-73-2	**NI34106**
326	254	256	324	127	328	128	109	**324**	100	79	77	61	61	60	53	47		12	5	–	–	–	–	5	–	–	–	–	–	–	**1,1′-Biphenyl, 2,2′,4,5,6′-pentachloro-**	68194-06-9	**NI34108**
326	254	324	328	256	127	128	109	**324**	100	66	64	62	62	60	51	46		12	5	–	–	–	–	5	–	–	–	–	–	–	**1,1′-Biphenyl, 2,2′,4,5′,6-pentachloro-**	60145-21-3	**NI34109**
326	324	328	254	256	127	128	109	**324**	100	64	63	48	46	38	32	30		12	5	–	–	–	–	5	–	–	–	–	–	–	**1,1′-Biphenyl, 2,2′,4,6,6′-pentachloro-**	56558-16-8	**NI34110**
326	324	256	254	328	128	127	184	**324**	100	65	65	64	62	34	34	33		12	5	–	–	–	–	5	–	–	–	–	–	–	**1,1′-Biphenyl, 2,3,3′,4,4′-pentachloro-**	32598-14-4	**NI34111**
326	324	328	256	254	127	128	109	**324**	100	62	60	56	56	39	36	36		12	5	–	–	–	–	5	–	–	–	–	–	–	**1,1′-Biphenyl, 2,3,3′,4,4′-pentachloro-**	32598-14-4	**NI34112**
326	324	328	254	256	184	127	128	**324**	100	64	62	50	47	31	27	22		12	5	–	–	–	–	5	–	–	–	–	–	–	**1,1′-Biphenyl, 2,3,3′,4,6-pentachloro-**	74472-35-8	**NI34113**
326	254	256	328	324	184	127	109	**324**	100	75	71	66	63	33	30	28		12	5	–	–	–	–	5	–	–	–	–	–	–	**1,1′-Biphenyl, 2,3,3′,4′,6-pentachloro-**	38380-03-9	**NI34114**
326	328	324	254	127	256	128	109	**324**	100	65	63	42	41	40	34	26		12	5	–	–	–	–	5	–	–	–	–	–	–	**1,1′-Biphenyl, 2,3,4,4′,5-pentachloro-**	74472-37-0	**NI34115**
326	328	324	254	256	127	128	109	**324**	100	66	64	42	41	38	31	25		12	5	–	–	–	–	5	–	–	–	–	–	–	**1,1′-Biphenyl, 2,3,4,4′,6-pentachloro-**	74472-38-1	**NI34116**
326	328	324	254	256	147	135	221	**324**	100	66	63	62	60	59	46	43		12	5	–	–	–	–	5	–	–	–	–	–	–	**1,1′-Biphenyl, 2,3,4,5,6-pentachloro-**	18259-05-7	**NI34117**
326	324	328	254	256	127	128	184	**324**	100	64	64	52	49	41	34	29		12	5	–	–	–	–	5	–	–	–	–	–	–	**1,1′-Biphenyl, 2,3,4,5,6-pentachloro-**	18259-05-7	**NI34118**
326	328	324	254	256	127	184	128	**324**	100	66	64	43	43	28	25	25		12	5	–	–	–	–	5	–	–	–	–	–	–	**1,1′-Biphenyl, 2,3,4′,5,6-pentachloro-**	68194-11-6	**NI34119**
326	328	324	330	256	254	323	322	**324**	100	66	64	11	5	5	4	3		12	5	–	–	–	–	5	–	–	–	–	–	–	**1,1′-Biphenyl, 2,3′,4,4′,5-pentachloro-**	31508-00-6	**NI34120**
326	328	324	254	256	128	127	184	**324**	100	62	59	57	53	26	24	22		12	5	–	–	–	–	5	–	–	–	–	–	–	**1,1′-Biphenyl, 2,3′,4,4′,5-pentachloro-**	31508-00-6	**NI34121**
326	328	324	254	256	127	128	109	**324**	100	65	62	39	37	29	26	26		12	5	–	–	–	–	5	–	–	–	–	–	–	**1,1′-Biphenyl, 2,3′,4,4′,6-pentachloro-**	56558-17-9	**NI34122**
326	328	324	256	254	127	128	109	**324**	100	67	64	54	51	45	44	30		12	5	–	–	–	–	5	–	–	–	–	–	–	**1,1′-Biphenyl, 2,3′,4,5,5′-pentachloro-**	68194-12-7	**NI34123**
326	324	328	254	256	127	128	109	**324**	100	64	64	51	47	45	38	37		12	5	–	–	–	–	5	–	–	–	–	–	–	**1,1′-Biphenyl, 2,3′,4,5′,6-pentachloro-**	56558-18-0	**NI34124**
326	324	328	254	256	127	128	109	**324**	100	64	62	46	44	39	32	30		12	5	–	–	–	–	5	–	–	–	–	–	–	**1,1′-Biphenyl, 2′,3,4,5,5′-pentachloro-**	70424-70-3	**NI34125**
326	254	256	328	324	291	289	109	**324**	100	90	87	66	66	57	42	31		12	5	–	–	–	–	5	–	–	–	–	–	–	**1,1′-Biphenyl, pentachloro-**	25429-29-2	**NI34127**
127	109	128	184	74	256	110	73	**324**	100	93	89	76	74	73	67	62	**16.52**	12	5	–	–	–	–	5	–	–	–	–	–	–	**1,1′-Biphenyl, pentachloro-**	25429-29-2	**NI34126**
326	328	324	256	254	291	289	127	**324**	100	64	64	61	59	43	29	29		12	5	–	–	–	–	5	–	–	–	–	–	–	**1,1′-Biphenyl, pentachloro-**	25429-29-2	**NI34128**
324	296	279	77	266	219	278	75	**324**	100	99	32	28	22	20	16	15		12	8	9	2	–	–	–	–	–	–	–	–	–	**4H-1-Benzopyran-2-carboxylic acid, 7-hydroxy-6,8-dinitro-4-oxo-, ethyl ester**	30113-68-9	**LI07165**
296	324	279	77	266	219	238	278	**324**	100	99	32	28	22	20	17	16		12	8	9	2	–	–	–	–	–	–	–	–	–	**4H-1-Benzopyran-2-carboxylic acid, 7-hydroxy-6,8-dinitro-4-oxo-, ethyl ester**	30113-68-9	**NI34129**
252	36	254	126	91	127	77	154	**324**	100	75	62	50	47	43	42	34	**0.00**	12	12	–	2	–	–	4	–	–	–	–	–	–	**1,1′-Biphenyl-4,4′-diamine, 3,3′-dichloro-, dihydrochloride**	612-83-9	**NI34130**
176	268	52	161	144	131	91	28	**324**	100	42	27	22	15	15	15	15	**14.58**	12	17	5	–	–	–	–	–	–	–	1	–	1	**Chromium, tricarbonyl[(1,2,3,4,5,6-η)-1,3,5-cycloheptatriene](trimethyl phosphite-P)-**	61766-45-8	**NI34131**
207	43	193	309	209	208	225	191	**324**	100	31	28	27	25	22	20	15	**0.00**	12	32	4	–	–	–	–	–	–	3	–	–	–	**3,3-Diisopropoxy-1,1,1,5,5,5-hexamethyltrisiloxane**		**MS07661**
324	296	227	246	277	265	258	196	**324**	100	44	18	13	10	10	8	7		13	–	1	–	–	8	–	–	–	–	–	–	–	**9H-Fluorenone, octafluoro-**		**LI07166**
238	161	78	322	135	51	294	77	**324**	100	47	35	32	24	24	22	18		13	10	2	2	–	–	–	–	–	–	–	–	1	**π-Cyclopentadienyl-π-phenylazo-dicarbonyl molybdenum**		**LI07167**
238	161	78	322	135	51	294	212	**324**	100	46	35	32	24	23	21	18		13	10	2	2	–	–	–	–	–	–	–	–	1	**π-Cyclopentadienyl-π-phenylazo-dicarbonyl molybdenum**		**MS07662**
76	200	154	305	50	275	292	184	**324**	100	68	59	43	43	32	30	30	**0.00**	13	13	6	2	–	–	–	–	–	–	1	–	–	**Phosphinic acid, bis-(3-nitrophenyl)-, methyl ester**		**LI07168**
168	194	324	196	103	224	169	91	**324**	100	65	47	44	41	40	39	20		13	17	3	–	–	–	–	–	–	–	–	–	1	**Rhodium, (2-hydroxymethylnorbornadiene)acetylacetanato-**		**NI34132**
166	167	41	29	69	27	139	39	**324**	100	34	31	23	17	14	11	11	**6.50**	13	19	4	2	–	3	–	–	–	–	–	–	–	**L-Prolylglycine, N-trifluoroacetyl-, butyl ester**		**MS07663**
173	115	295	87	151	59	238	75	**324**	100	82	53	52	30	26	19	18	**0.50**	13	29	5	–	–	–	–	–	–	1	1	–	–	**Dimethyl [3-(triethylsiloxy)-3-methyl-2-oxobutyl]phosphonate**		**NI34133**
107	91	63	75	90	77	64	52	**324**	100	85	71	60	53	53	47	47	**5.00**	14	8	4	6	–	–	–	–	–	–	–	–	–	**Bis(2,4-dinitrosobenzal)-hydrazone**		**LI07169**
173	175	289	145	147	109	177	291	**324**	100	66	28	19	12	11	11	10	**8.16**	14	10	3	2	–	–	2	–	–	–	–	–	–	**Methyl 3-[1-(2,6-dichlorobenzoyl)pyrazol-4-yl]prop-2-enoate**	95883-05-9	**NI34134**
206	165	161	149	134	238	164	133	**324**	100	95	71	56	49	47	44	37	**21.12**	14	16	5	2	1	–	–	–	–	–	–	–	–	**1,3-Oxathiolane-4-carboxylic acid, 2-(carbamoylimino)-5-(4-methoxyphenyl)-, ethyl ester**	33242-92-1	**NI34135**
98	324	97	126	151	81	80	60	**324**	100	78	69	32	31	31	25	25		14	16	7	2	–	–	–	–	–	–	–	–	–	**3′,5′-Diacetyl-2,2′-anhydro-1-(ab-D-arabinofuranosyl)-thymine**		**NI34136**
98	324	97	126	113	151	81	60	**324**	100	77	69	32	32	31	31	26		14	16	7	2	–	–	–	–	–	–	–	–	–	**6H-Furo[2′,3′:4,5]oxazolo[3,2-c]pyrimidin-6-one, 3-(acetyloxy)-2-[(acetyloxy)methyl]-2,3,3a,9a-tetrahydro-7-methyl-, [2R-(2α,3β,3aβ,9aβ)]-**	55334-27-5	**NI34137**
111	99	225	125	110	57	43	141	**324**	100	41	39	33	26	25	22	19	**0.00**	14	27	6	–	–	–	–	–	–	–	–	3	–	**β-D-Fructopyranose, cyclic 2,3:4,5-bis(ethylboronate) 1-(diethylborinate)**	74779-79-6	**NI34138**
111	99	98	110	57	141	43	129	**324**	100	85	30	26	17	15	14	12	**0.00**	14	27	6	–	–	–	–	–	–	–	–	3	–	**β-L-Gulofuranose, cyclic 2,3:5,6-bis(ethylboronate) 1-(diethylboronate)**	74841-63-7	**NI34139**
111	99	110	125	181	112	57	113	**324**	100	29	24	23	10	9	9	8	**0.00**	14	27	6	–	–	–	–	–	–	–	–	3	–	**α-L-Sorbofuranose, cyclic 2,3:4,6-bis(ethylboronate) 1-(diethylborinate)**	74779-78-5	**NI34140**
87	89	59	58	73	133	175	57	**324**	100	84	84	60	48	40	36	34	**0.27**	14	28	8	–	–	–	–	–	–	–	–	–	–	**(18S,19S)-18,19-Dihydroxy-1,4,7,10,13,16-hexaoxocycloencosane**	77887-94-6	**NI34141**
57	71	43	85	41	55	69	29	**324**	100	67	63	44	28	25	12	12	**0.10**	14	29	–	–	–	–	–	–	1	–	–	–	–	**Tetradecyliodide**		**LI07170**

MASS TO CHARGE RATIOS								M.W.	INTENSITIES								Parent	C	H	O	N	S	F	Cl	Br	I	Si	P	B	X	COMPOUND NAME	CAS Reg No	No
115	99	153	155	125	116	127	114	**324**	100	81	19	17	10	8	6	6	**0.00**	14	29	6	–	–	–	–	–	–	–	1	–	–	Dibutyl 3-acetoxybutyl phosphate		NI34142
240	55	132	152	241	91	88	153	**324**	100	68	22	21	13	12	12	10	**0.49**	15	9	3	–	1	–	–	–	–	–	–	–	1	(Thiobenzoyl)cymantrene	70616-39-6	NI34143
324	145	30	146	144	119	92	90	**324**	100	74	73	63	58	42	25	23		15	12	3	6	–	–	–	–	–	–	–	–	–	2-(α-Carbamoyl)-[(2-nitrophenyl)hydrazonomethyl]-benzimidazole		IC03936
240	324	241	242	326	309	225	243	**324**	100	78	73	44	31	26	26	25		15	13	6	–	–	–	1	–	–	–	–	–	–	Spiro[benzofuran-2(3H),1′-cyclohexane]-2′,3,4′-trione, 7-chloro-4-hydroxy-6-methoxy-6′-methyl-	26891-77-0	NI34144
241	324	240	326	243	309	239	242	**324**	100	91	44	33	33	31	26	25		15	13	6	–	–	–	1	–	–	–	–	–	–	Spiro[benzofuran-2(3H),1′-cyclohexane]-2′,3,4′-trione, 7-chloro-6-hydroxy-4-methoxy-6′-methyl-	26891-76-9	NI34145
167	193	222	179	324	282	236	265	**324**	100	87	57	43	28	11	10	9		15	16	8	–	–	–	–	–	–	–	–	–	–	2,4-Diacetoxy-3a-hydroxy-6-methoxy-2,3,3a,8a-tetrahydrofuro[2,3-b]benzofuran		LI07171
120	147	324	121	220	43	119	90	**324**	100	15	13	9	8	5	4	4		15	16	8	–	–	–	–	–	–	–	–	–	–	1,7-Dioxaspiro[4.4]nonane-2,6-dione, 8-(1,2-dihydroxyethyl)-9-hydroxy-4-(4-hydroxyphenyl)-, [5S-[5α(R*),8α(R*),9β]]-	30358-74-8	NI34146
120	324	174	121	119	90	220	43	**324**	100	13	10	8	4	4	3	3		15	16	8	–	–	–	–	–	–	–	–	–	–	1,7-Dioxaspiro[4.4]nonane-2,6-dione, 8-(1,2-dihydroxyethyl)-9-hydroxy-4-(4-hydroxyphenyl)-, [5S-[5α(S*),8α(R*),9β]]-	14225-07-1	NI34147
138	43	180	139	240	282	69	181	**324**	100	98	93	79	74	45	43	34	**7.00**	15	16	8	–	–	–	–	–	–	–	–	–	–	Tetra-O-acetyl-2,3,5-trihydroxy-benzyl alcohol		MS07664
206	242	324	107	244	168	142	191	**324**	100	74	63	47	39	31	23	11		15	17	4	2	–	–	1	–	–	–	–	–	–	Ethyl chloro-[[2-[(4-methoxy-2-butynyl)oxy]phenyl]hydrazono]acetate		MS07665
232	103	175	257	169	275	77	194	**324**	100	80	48	42	38	36	36	33	**8.00**	15	17	4	2	–	–	1	–	–	–	–	–	–	2,4,6(1H,3H,5H)-Pyrimidinetrione, 5-(3-chloro-2-hydroxypropyl)-1,3-dimethyl-5-phenyl-	57396-64-2	NI34148
184	93	44	51	75	65	185	76	**324**	100	33	22	14	12	12	11	11	**0.00**	15	20	4	2	1	–	–	–	–	–	–	–	–	Benzenesulphonamide, 4-acetyl-N-[(cyclohexylamino)carbonyl]-	968-81-0	NI34149
183	110	324	77	225	57	84	82	**324**	100	66	50	30	22	21	18	17		15	20	4	2	1	–	–	–	–	–	–	–	–	2,6-Diazatricyclo[3.3.1.1^{3,7}]decane-4,8-diol, 2-methyl-6-(phenylsulphonyl)-, (1α,3β,4β,5α,7β,8α)-	50267-41-9	NI34150
265	43	109	193	108	191	119	44	**324**	100	98	67	37	35	34	32	25	**0.00**	15	20	6	2	–	–	–	–	–	–	–	–	–	Diethyl 2-(acetylamino)-2-[(N-oxido-4-pyridyl)methyl]malonate		DD01241
98	244	146	80	82	97	68	96	**324**	100	31	21	15	14	13	12	11	**0.00**	15	21	1	2	–	–	–	1	–	–	–	–	–	Anagyrine, monohydrobromide	33478-03-4	NI34151
225	226	324	141	167	43	143	142	**324**	100	60	40	40	10	10	6	6		15	21	6	–	–	–	–	–	–	–	–	–	1	Aluminium, tri(acetylacetonato)-	13963-57-0	NI34152
225	324	226	141	43	147	143	325	**324**	100	11	11	5	4	3	3	2		15	21	6	–	–	–	–	–	–	–	–	–	1	Aluminium, tri(acetylacetonato)-	13963-57-0	IC03937
225	226	141	43	28	324	143	167	**324**	100	10	8	7	6	5	4	2		15	21	6	–	–	–	–	–	–	–	–	–	1	Aluminium, tri(acetylacetonato)-	13963-57-0	NI34153
184	185	324	43	41	277	186	142	**324**	100	52	33	26	26	23	19	14		15	24	–	4	2	–	–	–	–	–	–	–	–	Thiazolo[5,4-d]pyrimidine, 5-amino-2-(decylthio)-	19844-44-1	NI34154
44	82	110	81	238	43	86	28	**324**	100	82	67	48	38	38	29	25	**24.28**	15	24	4	4	–	–	–	–	–	–	–	–	–	N[2]-(N-Acetylalanyl)histidine butyl ester		AI02046
324	191	164	58	147	132	177	77	**324**	100	24	23	17	15	15	8	6		16	14	2	2	–	–	–	–	–	–	–	–	1	Nickel, [[2,2′-[1,2-ethanediylbis(nitrilomethylidyne)]bis[phenolato]](2-)-N,N′,O,O′]-	14167-20-5	LI07172
324	326	191	164	325	58	147	132	**324**	100	43	25	25	22	18	15	15		16	14	2	2	–	–	–	–	–	–	–	–	1	Nickel, [[2,2′-[1,2-ethanediylbis(nitrilomethylidyne)]bis[phenolato]](2-)-N,N′,O,O′]-	14167-20-5	NI34155
139	251	253	111	141	75	113	252	**324**	100	79	54	49	35	20	17	13	**0.80**	16	14	3	–	–	–	2	–	–	–	–	–	–	Benzeneacetic acid, 4-chloro-α-(4-chlorophenyl)-α-hydroxy-, ethyl ester	510-15-6	NI34156
139	251	253	111	141	113	252	75	**324**	100	78	54	48	35	16	13	10	**0.00**	16	14	3	–	–	–	2	–	–	–	–	–	–	Benzeneacetic acid, 4-chloro-α-(4-chlorophenyl)-α-hydroxy-, ethyl ester	510-15-6	NI34157
139	251	111	253	141	75	113	29	**324**	100	50	49	34	32	19	16	14	**0.00**	16	14	3	–	–	–	2	–	–	–	–	–	–	Benzeneacetic acid, 4-chloro-α-(4-chlorophenyl)-α-hydroxy-, ethyl ester	510-15-6	NI34158
94	324	120	67	239	78	264	110	**324**	100	62	27	24	23	21	20	20		16	16	2	6	–	–	–	–	–	–	–	–	–	Dianhydrohexulose 2-pyridylosazone		NI34159
268	254	324	281	270	255	296	295	**324**	100	69	49	47	35	35	28	27		16	17	5	–	–	–	1	–	–	–	–	–	–	Spiro[benzofuran-2(3H),1′-cyclohexane]-2′,3-dione, 7-chloro-4,6-dimethoxy-6′-methyl-	26891-86-1	NI34160
324	241	326	255	215	243	309	254	**324**	100	66	36	36	24	22	19	19		16	17	5	–	–	–	1	–	–	–	–	–	–	Spiro[benzofuran-2(3H),1′-cyclohexane]-3,4′-dione, 7-chloro-4,6-dimethoxy-6′-methyl-	17793-65-6	NI34161
43	217	235	208	249	193	250	264	**324**	100	90	83	81	79	60	53	52	**0.00**	16	20	7	–	–	–	–	–	–	–	–	–	–	Bicyclo[2.2.2]oct-2-ene-2,3-dicarboxylic acid, 1-hydroxy-8,8-dimethyl-5-oxo-, dimethyl ester, acetate	25864-64-6	MS07666
43	217	235	208	249	193	250	254	**324**	100	90	83	81	79	60	53	52	**0.00**	16	20	7	–	–	–	–	–	–	–	–	–	–	Dimethyl 1-acetoxy-8,8-dimethyl-3-oxobicyclo[2.2.2]oct-5-ene-5,6-dicarboxylate		LI07173
292	261	245	323	277	45	293	217	**324**	100	86	77	51	44	38	32	28	**8.00**	16	20	7	–	–	–	–	–	–	–	–	–	–	Dimethyl 6-hydroxy-4-methoxymethyl-3-(2-methoxy-1-propenyl)-phthalate		NI34162
324	278	251	268	250	239	235	207	**324**	100	94	94	75	75	70	69	68		16	20	7	–	–	–	–	–	–	–	–	–	–	Ethyl 2-(2′,3′-dimethoxy-4′,5′-methylenedioxy-6′-propionylphenyl)acetate		NI34163
140	169	91	152	126	114	112	124	**324**	100	84	38	35	13	13	12	9	**0.00**	16	24	3	2	1	–	–	–	–	–	–	–	–	1,5-Diazacycloundecan-6-one, 1-[(4-methylphenyl)sulphonyl]-	67370-70-1	NI34164
122	98	121	83	147	245	324	281	**324**	100	99	85	80	42	38	25	15		16	24	3	2	1	–	–	–	–	–	–	–	–	trans-2-Isopropyl-3-methyl-N-[N-(methylsulphonyl)(2,6-dimethylphenoxy)carbimidoyl]aziridine		DD01242
163	117	73	164	129	103	89	144	**324**	100	45	44	40	37	37	37	32	**0.00**	16	24	5	–	–	–	–	–	–	1	–	–	–	Methyl 3-hydroxybutyl phthalate trimethylsilyl ether		MS07667
192	309	73	324	147	177	193	41	**324**	100	42	40	36	36	28	23	21		16	28	3	–	–	–	–	–	–	2	–	–	–	(2-Hydroxyphenyl)butyric acid di-TMS		NI34165
192	73	75	324	309	179	193	177	**324**	100	38	23	21	20	19	19	11		16	28	3	–	–	–	–	–	–	2	–	–	–	(3-Hydroxyphenyl)butyric acid di-TMS		NI34166
192	73	75	193	309	179	324	177	**324**	100	45	32	30	26	26	24	23		16	28	3	–	–	–	–	–	–	2	–	–	–	(4-Hydroxyphenyl)butyric acid di-TMS		MS07668
147	133	119	146	105	131	115	126	**324**	100	83	48	37	34	29	25	23	**14.00**	17	18	2	–	–	–	2	–	–	–	–	–	–	5,7a-Dichloro-1a,3,3a,7a,7b,8,9,9a-octahydro-1a,7b-dimethyl-1H-cyclopropa[a]phenanthrene-4,7-dione		MS07669
147	159	146	91	105	324	104	160	**324**	100	47	41	31	29	24	23	20		17	18	5	–	–	–	–	–	–	–	–	2	–	DL-Xylitol, cyclic 1,3:2,4-bis(phenylboronate)	74867-29-1	NI34167

MASS TO CHARGE RATIOS								M.W.	INTENSITIES								Parent	C	H	O	N	S	F	Cl	Br	I	Si	P	B	X	COMPOUND NAME	CAS Reg No	No
109	78	167	57	95	55	158	41	324	100	78	71	70	68	67	63	61	3.00	17	24	1	–	–	–	–	–	–	–	–	–	1	Cyclopropanol, 1-[2,6-dimethyl-1-(phenylseleno)cyclohexyl]-	74281-18-8	NI34168
43	91	147	135	134	177	115	105	324	100	93	56	50	43	30	25	25	3.00	17	24	4	–	1	–	–	–	–	–	–	–	–	Benzenepropanoic acid, α-(acetyloxy)-β-butyl-α-(methylthio)-, methyl ester	63608-59-3	NI34169
123	85	43	264	161	55	95	246	324	100	78	71	58	58	56	48	45	0.05	17	24	6	–	–	–	–	–	–	–	–	–	–	Azuleno[4,5-b]furan-2(3H)-one, 9a-[(acetyloxy)methyl]decahydro-6a,9-dihydroxy-6-methyl-3-methylene-, [3aS-(3aα,6β,6aα,9β,9aβ,9bα)]-	28587-47-5	NI34170
45	105	89	107	175	77	91	83	324	100	99	95	93	75	42	36	36	7.00	17	24	6	–	–	–	–	–	–	–	–	–	–	2-(2-Hydroxyethoxy)ethyl 4,6-O-benzylidene-2,3-dideoxy-β-D-erythro-hexopyranoside		NI34171
43	124	55	91	105	205	53	106	324	100	77	59	58	48	47	39	38	0.00	17	24	6	–	–	–	–	–	–	–	–	–	–	Monoacetoxyscirpenol		NI34172
151	324	75	152	309	73	325	294	324	100	39	16	15	13	12	10	6		17	28	4	–	–	–	–	–	–	1	–	–	–	(3,4-Dimethoxyphenyl)hexanoic acid TMS		NI34173
73	129	75	45	130	74	147	59	324	100	95	20	17	12	8	7	5	0.10	17	32	2	–	–	–	–	–	–	2	–	–	–	1,8-Undecadien-5-yne 3,7-bis-trimethylsilyl ether, (Z)-		MS07670
239	295	183	309	253	73	57	167	324	100	86	79	20	20	20	17	12	4.90	17	42	–	2	–	–	–	–	–	1	–	2	–	[tert-Butyl(diethylboryl)amino][tert-butyl(trimethylsilyl)amino]ethylborane		MS07671
324	325	326	258	162	290	245	213	324	100	21	16	12	9	8	7	4		18	12	–	–	3	–	–	–	–	–	–	–	–	19,20,21-Trithiatetracyclo[14.12.1.1^{4,7}.1^{10,13}]heneicosa-2,4,6,8,10,12,14,16,18-nonaene		LI07174
324	105	77	163	325	44	247	32	324	100	60	32	25	23	15	14	14		18	12	2	–	2	–	–	–	–	–	–	–	–	2,4-Bis(benzoylmethylidene)-1,3-dithietane		LI07175
324	134	105	51	145	77	107	44	324	100	47	42	37	31	30	27	25		18	12	6	–	–	–	–	–	–	–	–	–	–	2,5-Cyclohexadiene-1,4-dione, 2,5-dihydroxy-3,6-bis(4-hydroxyphenyl)-	519-67-5	NI34174
293	324	265	294	249	75	208	131	324	100	36	25	18	16	16	15	15		18	12	6	–	–	–	–	–	–	–	–	–	–	Dimethyl anthraquinone-2,6-dicarboxylate		IC03938
324	294	265	325	75	209	279	206	324	100	36	23	19	18	13	11	11		18	12	6	–	–	–	–	–	–	–	–	–	–	Dimethyl anthraquinone-2,6-dicarboxylate		NI34175
134	324	223	106	105	107	77	78	324	100	91	42	40	33	31	28	25		18	12	6	–	–	–	–	–	–	–	–	–	–	3-Hydroxy-6-(4-hydroxybenzylidene)-4-(4-hydroxyphenyl)-2H-pyran-2,5(6H)-dione		NI34176
324	306	295	265	325	278	307	63	324	100	49	33	24	19	17	12	10		18	12	6	–	–	–	–	–	–	–	–	–	–	Isosterigmatocystin	55334-14-0	MS07672
324	296	307	325	297	265	308	278	324	100	83	64	35	31	31	25	23		18	12	6	–	–	–	–	–	–	–	–	–	–	Sterigmatocystin	10048-13-2	NI34177
324	295	306	325	265	296	307	63	324	100	87	28	19	19	14	10	10		18	12	6	–	–	–	–	–	–	–	–	–	–	Sterigmatocystin	10048-13-2	IC03939
324	295	306	325	265	296	307	278	324	100	76	28	20	15	14	10	8		18	12	6	–	–	–	–	–	–	–	–	–	–	Sterigmatocystin	10048-13-2	MS07673
295	257	297	77	324	296	239	104	324	100	42	35	33	29	21	16	16		18	13	2	2	–	–	1	–	–	–	–	–	–	3-Hydroxypinazepam		MS07674
245	244	246	243	230	242	82	80	324	100	33	19	11	11	8	8	8	0.00	18	15	–	1	–	–	–	1	–	–	–	–	–	6,7-Dihydro-13-methyldibenzo[a,f]quinolizinium bromide		NI34178
291	292	324	322	290	275	321	276	324	100	63	55	55	52	50	28	28		18	16	–	2	2	–	–	–	–	–	–	–	–	N,N′-Dimethyl-3,3′-bis(2-indolinethione)		LI07176
130	129	102	131	103	77	160	161	324	100	27	17	12	10	10	9	7	0.31	18	16	–	2	2	–	–	–	–	–	–	–	–	1H-Indole, 3,3′-[dithiobis(methylene)]bis-	17004-43-2	NI34179
204	281	296	324	307	176	232	323	324	100	84	73	65	60	38	34	22		18	16	4	2	–	–	–	–	–	–	–	–	–	Cinchoninanilide, 1,2-dihydro-6,7-dimethoxy-2-oxo-	22765-49-7	NI34180
293	324	133	132	294	265	148	130	324	100	30	22	20	19	19	17	17		18	16	4	2	–	–	–	–	–	–	–	–	–	7,14-Dimethoxydiftalone	56632-15-6	MS07675
233	324	91	161	234	188	325	160	324	100	54	30	20	15	13	11	10		18	16	4	2	–	–	–	–	–	–	–	–	–	1H-Indole-3-acetamide, 5-methoxy-α-oxo-6-benzyloxy-	4790-22-1	NI34181
324	109	255	325	173	146	95	150	324	100	75	52	23	21	17	13	12		18	17	1	4	–	1	–	–	–	–	–	–	–	2-Pyrazolin-5-one, 4-[[4-(dimethylamino)phenyl]imino]-1-(4-fluorophenyl) 3-methyl-	27808-01-1	NI34182
324	109	255	145	325	173	146	95	324	100	75	51	30	22	21	17	13		18	17	1	4	–	1	–	–	–	–	–	–	–	2-Pyrazolin-5-one, 4-[[4-(dimethylamino)phenyl]imino]-1-(4-fluorophenyl) 3-methyl-	27808-01-1	MS07676
265	30	324	139	257	168	140	167	324	100	99	77	77	55	48	41	35		18	20	2	4	–	–	–	–	–	–	–	–	–	1,4-Bis(2-aminoethylamino)-anthraquinone		IC03940
282	58	324	115	70	296	225	224	324	100	49	42	16	9	7	7	6		18	20	2	4	–	–	–	–	–	–	–	–	–	Morphinan, 8-azido-6,7-didehydro-4,5-epoxy-3-methoxy-17-methyl-, (5α,8β)-	25650-95-7	NI34183
176	149	162	163	324	295	150	109	324	100	80	72	36	32	20	18	15		18	24	–	6	–	–	–	–	–	–	–	–	–	1,4-Bis(4′-amino-5′,6′-trimethylene-2′-pyrimidinyl)butane		NI34184
162	41	29	43	27	135	163	105	324	100	85	82	68	59	41	34	31	0.00	18	28	3	–	1	–	–	–	–	–	–	–	–	1,3-Benzodioxole, 5-[2-(octylsulphinyl)propyl]-	120-62-7	NI34185
123	173	108	151	109	145	107	95	324	100	62	50	37	29	25	20	20	0.00	18	28	5	–	–	–	–	–	–	–	–	–	–	Diethyl (2-(3,3-dimethyl-2-norbornyl)-2-oxoethyl)propanedioate	79861-76-0	NI34186
213	324	139	326	214	111	325	157	324	100	55	20	16	13	13	10	10		19	13	3	–	–	–	1	–	–	–	–	–	–	4-Chloro-4′-(4-hydroxyphenoxy)-benzophenone		IC03941
240	282	239	146	121	145	222	223	324	100	61	29	21	21	17	15	13	10.00	19	16	5	–	–	–	–	–	–	–	–	–	–	4,4′-Diacetoxychalcone		MS07677
177	69	77	147	189	135	219	178	324	100	60	30	25	24	23	21	17	9.90	19	16	5	–	–	–	–	–	–	–	–	–	–	1,3,4,6-Hexanetetrone, 1-(4-methoxyphenyl)-6-phenyl-	58330-11-3	NI34187
281	178	103	136	131	324	137	265	324	100	85	38	36	31	29	29	26		19	16	5	–	–	–	–	–	–	–	–	–	–	3-(α-(2-Hydroxypropyl)benzyl)-4,5-dihydroxycoumarin		LI07177
281	187	121	43	324	282	147	120	324	100	63	39	36	32	22	19	15		19	16	5	–	–	–	–	–	–	–	–	–	–	4′-Hydroxywarfarin		LI07178
281	324	282	137	203	136	103	77	324	100	21	20	20	15	15	12	9		19	16	5	–	–	–	–	–	–	–	–	–	–	6-Hydroxywarfarin		LI07179
281	103	131	145	137	43	77	69	324	100	46	42	40	40	30	27	22	0.00	19	16	5	–	–	–	–	–	–	–	–	–	–	7-Hydroxywarfarin		LI07180
281	136	137	324	203	43	282	145	324	100	31	30	25	24	24	22	17		19	16	5	–	–	–	–	–	–	–	–	–	–	8-Hydroxywarfarin		LI07181
269	254	324	241	283	323	295	296	324	100	81	76	75	67	63	56	54		19	17	1	2	–	–	1	–	–	–	–	–	–	1-Allyl-8-chloro-1,2,3,4-tetrahydro-5-phenyl-1,5-benzodiazocin-2-one	63563-57-5	NI34188
185	115	218	217	144	77	169	289	324	100	79	72	67	59	58	40	39	10.00	19	17	1	2	–	–	1	–	–	–	–	–	–	2-Chloroacetyl-1-phenyl-1,2,3,4-tetrahydro-β-carboline		MS07678
309	324	294	295	57	83	69	55	324	100	82	46	40	39	36	30	30		19	20	1	2	1	–	–	–	–	–	–	–	–	(Z,Z)-4-Ethyl-3-methyl-5-[5-(thienyl-2-methylene)-3,4-dimethyl-5H-pyrrolyl-2-methylene]-3-pyrrolin-2-one		MS07679
93	109	199	77	119	162	55	324	324	100	79	34	31	28	25	24	21		19	20	3	2	–	–	–	–	–	–	–	–	–	3,5-Pyrazolidinedione, 4-butyl-1-(4-hydroxyphenyl)-2-phenyl-	129-20-4	NI34189
199	324	93	77	65	55	121	135	324	100	78	78	74	42	35	26	24		19	20	3	2	–	–	–	–	–	–	–	–	–	3,5-Pyrazolidinedione, 4-butyl-2-(4-hydroxyphenyl)-1-phenyl-		MS07680
216	201	77	202	217	199	51	124	324	100	88	62	54	49	45	35	26	4.70	19	21	1	2	–	–	–	–	–	–	1	–	–	N-(1-Cyanocyclohexyl)diphenylphosphinamide		MS07681

MASS TO CHARGE RATIOS								M.W.	INTENSITIES								Parent	C	H	O	N	S	F	Cl	Br	I	Si	P	B	X	COMPOUND NAME	CAS Reg No	No
181	55	111	324	41	43	97	67	**324**	100	67	64	57	57	53	38	38		19	32	2	–	1	–	–	–	–	–	–	–	–	Methyl 9,12-epithio-9,11-octadecanoate		MS07682
103	55	41	31	91	67	121	324	**324**	100	80	63	58	40	35	34	18		19	32	4	–	–	–	–	–	–	–	–	–	–	(+)-(S)-3-[((1Z,5Z)-16-Hydroxyhexadeca-1,5-dien-3-ynyl)oxy]-1,2-propanediol		MS07683
109	137	124	190	121	119	249	175	**324**	100	66	55	53	48	45	44	42	**0.00**	19	32	4	–	–	–	–	–	–	–	–	–	–	1-Naphthaleneacetic acid, 1α,2,3,4,4a,5,6,7,8,8a-decahydro-2α-hydroxy-2,5,5,8aβ-tetramethyl-, methyl ester, acetate, (-)-	19954-81-5	NI34190
43	55	71	141	113	97	193	123	**324**	100	87	76	40	40	38	32	24	**0.00**	19	32	4	–	–	–	–	–	–	–	–	–	–	4,8,12,16-Nonadecanetetrone	55110-15-1	NI34191
83	141	127	81	43	55	41	69	**324**	100	89	86	68	65	55	55	45	**0.50**	19	32	4	–	–	–	–	–	–	–	–	–	–	2H-Pyran-6-carboxylic acid, 2-propoxy-5,6-dihydro-, 1-menthyl ester		MS07684
324	269	147	199	239	255	307	279	**324**	100	99	50	35	20	10	5	5		20	20	4	–	–	–	–	–	–	–	–	–	–	[3,6′-Bi-2H-1-benzopyran]-7,7′-diol, 3′,4′-dihydro-2′,2′-dimethyl-	41347-48-2	MS07685
105	80	77	57	56	44	41	51	**324**	100	87	35	24	15	15	14	12	**0.00**	20	20	4	–	–	–	–	–	–	–	–	–	–	1,4-Cyclohexanediol, dibenzoate	19150-32-4	NI34192
240	282	241	324	225	283	199	198	**324**	100	44	18	15	13	9	8	8		20	20	4	–	–	–	–	–	–	–	–	–	–	2,3-Diacetoxy-1,4-diallylnaphthalene		NI34193
240	199	197	181	198	282	241	211	**324**	100	67	56	51	49	45	32	18	**15.70**	20	20	4	–	–	–	–	–	–	–	–	–	–	2,7-Diacetoxy-1,8-diallylnaphthalene		NI34194
116	115	43	134	105	77	175	117	**324**	100	56	51	34	29	25	21	11	**1.84**	20	20	4	–	–	–	–	–	–	–	–	–	–	2,3-Dihydro-2-acetoxy-2,5-dimethyl-3,6-diphenyl-1,4-dioxin		NI34195
167	306	165	288	152	91	324	233	**324**	100	24	11	9	7	7	5	5		20	20	4	–	–	–	–	–	–	–	–	–	–	cis,cis-2-(Diphenylmethyl)-1,3-cyclopentanedicarboxylic acid		DD01243
143	115	288	306	246	160	324		**324**	100	29	16	5	5	4	1			20	20	4	–	–	–	–	–	–	–	–	–	–	4-(2-Formylphenyl)-1-(2-carboxyethylphenyl)butan-2-one		LI07182
324	238	77	267	202	152	209	135	**324**	100	55	33	29	26	25	24	24		20	20	4	–	–	–	–	–	–	–	–	–	–	Naphtho[1,2-d]-1,3-dioxole, 9-(1,3-benzodioxol-5-yl)-6,7,8,9-tetrahydro-7,8-dimethyl-, [7S-(7α,8β,9α)]-	3738-01-0	NI34196
324	309	162						**324**	100	18	12							20	20	4	–	–	–	–	–	–	–	–	–	–	2,3,8,9-Tetramethoxy-5,6-didehydro-11,12-dihydrodibenzo[a,e]cyclooctyne		LI07183
307	293	177	325	308	321	294		**324**	100	91	49	35	22	20	19		**4.50**	20	20	4	–	–	–	–	–	–	–	–	–	–	Warfarin alcohol, O-methyl-		NI34197
324	199	198	173	184	293	160	200	**324**	100	32	20	18	16	15	14	13		20	24	2	2	–	–	–	–	–	–	–	–	–	Ajmalan-17-ol, 19,20-didehydro-1-demethyl-12-methoxy-, (17R,19E)-	30171-06-3	NI34198
144	198	324	145	183	182	150	157	**324**	100	88	84	36	30	25	24	19		20	24	2	2	–	–	–	–	–	–	–	–	–	Ajmalan-17-one, 21-hydroxy-, (21α)-	639-30-5	NI34199
194	324	144	139	251	143	130	79	**324**	100	55	47	38	28	27	26	24		20	24	2	2	–	–	–	–	–	–	–	–	–	Akuammicine, 2β,16β-dihydro-	34441-83-3	NI34200
43	41	55	57	42	69	44	56	**324**	100	88	61	49	37	28	21	20	**2.00**	20	24	2	2	–	–	–	–	–	–	–	–	–	Akuammilan-17-ol, 10-methoxy-	56259-10-0	NI34201
324	323	325	267	268	239	96	201	**324**	100	27	23	22	20	20	20	17		20	24	2	2	–	–	–	–	–	–	–	–	–	Aspidofractinin-3-one, 17-methoxy-, (2α,5α)-	55724-66-8	NI34202
123	174	324	138	42	186	188	41	**324**	100	52	30	26	25	24	22	22		20	24	2	2	–	–	–	–	–	–	–	–	–	Aspidofractinin-6-one, 17-methoxy-	54965-82-1	NI34203
162	28	324	41	134	39	163	55	**324**	100	92	79	73	51	46	43	42		20	24	2	2	–	–	–	–	–	–	–	–	–	Bicyclo[4.1.0]hept-3-ene-2,5-dione, 3,7,7-trimethyl-, 5-[(4,7,7-trimethyl-5-oxobicyclo[4.1.0]hept-3-en-2-ylidene)hydrazone]	55334-15-1	MS07686
71	95	229	84	180	167	324	181	**324**	100	64	38	29	28	25	23	16		20	24	2	2	–	–	–	–	–	–	–	–	–	Condyfolan-16-carboxylic acid, 2,16-didehydro-, methyl ester	6711-69-9	NI34204
105	77	190	51	106	202	148	134	**324**	100	50	14	12	9	7	7	7	**3.00**	20	24	2	2	–	–	–	–	–	–	–	–	–	N,N′-Dibenzoyl-1,6-diaminohexane		LI07184
162	91	119	250	324	179	58	163	**324**	100	48	43	26	25	18	16	13		20	24	2	2	–	–	–	–	–	–	–	–	–	2,2′-Bis[(dimethylamino)methyl]benzil		LI07185
145	162	147	144	130	288	306	149	**324**	100	99	99	99	53	39	23	23	**1.50**	20	24	2	2	–	–	–	–	–	–	–	–	–	2,3-Dimethyl-3-hydroxy-2-[(3-hydroxy-3-methylindolin-2-yl)methyl]indoline		LI07186
93	232	94	324	55	41	190	120	**324**	100	23	23	13	12	12	11	11		20	24	2	2	–	–	–	–	–	–	–	–	–	N,N′-Diphenyl-2-methylpimelamide		IC03942
324	323	295	254	325	267	322	162	**324**	100	72	23	14	11	11	10	10		20	24	2	2	–	–	–	–	–	–	–	–	–	Eburnamonine, 11-methoxy-	4800-93-5	NI34205
324	323	295	254	267	325	162	296	**324**	100	70	22	15	11	10	9	7		20	24	2	2	–	–	–	–	–	–	–	–	–	Eburnamonine, 11-methoxy-	4800-93-5	LI07187
198	199	306	323	324	293	212		**324**	100	84	59	57	56	36	16			20	24	2	2	–	–	–	–	–	–	–	–	–	Gardnerine	23172-92-1	LI07188
110	166	324	295	296	82	122	55	**324**	100	69	57	39	33	33	28	27		20	24	2	2	–	–	–	–	–	–	–	–	–	Gelsemine, 18,19-dihydro-	29576-86-1	NI34206
144	162	147	145	146	143	163	130	**324**	100	99	99	99	93	85	77	52	**1.71**	20	24	2	2	–	–	–	–	–	–	–	–	–	3-Indolinol, 2,3,3′-trimethyl-2,2′-methylenedi-	24644-80-2	NI34207
281	180	324	323	181	222	197	182	**324**	100	54	31	22	22	16	15	14		20	24	2	2	–	–	–	–	–	–	–	–	–	4-N-Methyl-19,20-dihydrodeformyltalbotine acid lactone		LI07189
324	323	168	250	248	155	143	115	**324**	100	86	75	74	46	42	32	31		20	24	2	2	–	–	–	–	–	–	–	–	–	Methyl [[2R-(2α,3Z,12bβ)]-3-ethylidene-1,2,3,4,6,7,12,12b-octahydroindolo[2,3-a]quinolizin-2-yl]acetate		NI34208
83	214	242	82	103	241	131	84	**324**	100	97	91	55	48	19	18	18	**13.00**	20	24	2	2	–	–	–	–	–	–	–	–	–	Pyridine, 1,1′-[1,2-phenylenebis(1-oxo-2,1-ethanediyl)]bis[1,2,3,4-tetrahydro-	52881-80-8	NI34209
136	81	324	189	55	137	42	41	**324**	100	23	20	18	18	15	13	13		20	24	2	2	–	–	–	–	–	–	–	–	–	Quinidine	56-54-2	NI34210
136	158	159	186	81	77	173	82	**324**	100	65	61	55	48	46	42	40	**23.20**	20	24	2	2	–	–	–	–	–	–	–	–	–	Quinidine	56-54-2	AM00143
136	189	81	324	137	173	82	79	**324**	100	15	15	14	13	11	7	6		20	24	2	2	–	–	–	–	–	–	–	–	–	Quinidine	56-54-2	NI34211
325	136	326	327	307	137	308	63	**324**	100	26	24	9	9	4	2	2	**2.36**	20	24	2	2	–	–	–	–	–	–	–	–	–	Quinine	130-95-0	NI34212
136	158	159	186	81	77	173	82	**324**	100	65	61	55	48	46	42	40	**23.22**	20	24	2	2	–	–	–	–	–	–	–	–	–	Quinine	130-95-0	WI01415
136	28	41	137	42	81	55	39	**324**	100	15	13	12	12	10	9	5	**1.00**	20	24	2	2	–	–	–	–	–	–	–	–	–	Quinine	130-95-0	NI34213
183	324	184	182	196	170	181	168	**324**	100	75	40	33	30	20	16	15		20	24	2	2	–	–	–	–	–	–	–	–	–	Sarpagan-21-ol, 17,19-epoxy-19,20-dihydro-1-methyl-, (21α)-	38990-06-6	NI34214
149	216	164	121	120	28	188	175	**324**	100	64	56	39	36	36	33	33	**3.83**	20	28	–	4	–	–	–	–	–	–	–	–	–	4,4′-Bis(diethylamino)azobenzene		MS07687
57	162	91	71	118	175	117	163	**324**	100	40	39	25	18	17	12	11	**3.00**	20	28	–	4	–	–	–	–	–	–	–	–	–	1,2,4,5-Tetrazine, hexahydro-1,4-bis(3-phenylpropyl)-	35035-71-3	NI34215
57	162	91	71	175	118	163	117	**324**	100	42	42	27	19	19	10	10	**3.00**	20	28	–	4	–	–	–	–	–	–	–	–	–	1,2,4,5-Tetrazine, hexahydro-1,4-bis(3-phenylpropyl)-	35035-71-3	LI07190
251	103	111	183	132	75	95	83	**324**	100	71	42	40	33	30	29	25	**1.39**	20	36	3	–	–	–	–	–	–	–	–	–	–	4-Hexadecynoic acid, 2-ethoxy-, ethyl ester	40924-18-3	NI34216
113	153	141	254	181	207	128	239	**324**	100	24	20	6	5	4	3	2	**0.00**	20	36	3	–	–	–	–	–	–	–	–	–	–	Octadecanoic acid, 9,10-methylene-12-oxo-, methyl ester		LI07191
127	57	150	41	43	55	81	109	**324**	100	72	42	42	40	37	25	21	**10.00**	20	36	3	–	–	–	–	–	–	–	–	–	–	Octadecanoic acid, 9,10-methylene-8-oxo-, methyl ester, cis-		LI07192

MASS TO CHARGE RATIOS								M.W.	INTENSITIES								Parent	C	H	O	N	S	F	Cl	Br	I	Si	P	B	X	COMPOUND NAME	CAS Reg No	No
55	41	83	69	97	43	171	81	324	100	60	56	55	42	41	37	37	19.00	20	36	3	–	–	–	–	–	–	–	–	–	–	Octadecanoic acid, 9,10-methylene-8-oxo-, methyl ester, cis-		LI07193
109	124	324	296	202	41	154	323	324	100	92	60	46	37	24	22	21		21	28	1	2	–	–	–	–	–	–	–	–	–	Aspidofractinin-17-ol, 1-ethyl-	55724-54-4	NI34217
124	43	41	324	125	296	152	42	324	100	10	8	7	7	5	5	4		21	28	1	2	–	–	–	–	–	–	–	–	–	Aspidospermidine, 1-acetyl-	2111-81-1	NI34218
124	68	79	67	69	91	94	77	324	100	50	23	22	20	19	16	15	6.00	21	28	1	2	–	–	–	–	–	–	–	–	–	Aspidospermidine, 1-acetyl-	2111-81-1	NI34220
124	324	296	125	323	325	43	152	324	100	31	20	13	8	7	7	5		21	28	1	2	–	–	–	–	–	–	–	–	–	Aspidospermidine, 1-acetyl-	2111-81-1	NI34219
309	324	132	310	265	147	237	29	324	100	33	25	23	19	19	11	10		21	28	1	2	–	–	–	–	–	–	–	–	–	Benzophenone, 4,4′-bis(diethylamino)-	90-93-7	NI34221
309	324	310	294	279	295	323	281	324	100	57	50	15	15	14	10	10		21	28	1	2	–	–	–	–	–	–	–	–	–	2-(5,7-Di-tert-butyl-3,3-dimethyl-2-indolinylidene)-2-formylacetonitrile		DD01244
184	215	41	43	136	42	324	123	324	100	80	60	54	45	45	35	30		21	28	1	2	–	–	–	–	–	–	–	–	–	2,20-Cycloaspidospermidine-3-methanol, 1-methyl-, (2α,3α,5α,12β,19α,20R)-	56053-42-0	NI34222
184	124	215	136	324	123	122	168	324	100	84	58	58	57	31	26	24		21	28	1	2	–	–	–	–	–	–	–	–	–	N-Methyldihydrovindolininol	55856-76-3	NI34223
55	69	41	43	74	83	292	97	324	100	73	72	63	57	56	52	46	9.00	21	40	2	–	–	–	–	–	–	–	–	–	–	11-Eicosenoic acid, methyl ester	3946-08-5	NI34224
43	41	55	74	67	57	81	84	324	100	79	76	72	45	44	41	40	0.00	21	40	2	–	–	–	–	–	–	–	–	–	–	Eicosenoic acid, methyl ester	27070-40-2	NI34225
85	82	43	55	41	67	28	96	324	100	58	55	53	50	46	41	38	0.30	21	40	2	–	–	–	–	–	–	–	–	–	–	Furan, 2-(5-heptadecenyloxy)tetrahydro-		NI34226
45	88	87	103	58	71	59	15	324	100	28	27	21	17	14	14	14	0.00	21	40	2	–	–	–	–	–	–	–	–	–	–	Methyl 3,3-dioctylmethacrylate		DO01100
43	41	57	100	55	113	71	69	324	100	99	90	85	70	59	51	43	5.60	21	40	2	–	–	–	–	–	–	–	–	–	–	Octadecanoic acid, 2-propenyl ester	6289-31-2	NI34227
55	281	41	186	81	43	69	95	324	100	87	77	62	52	50	48	44	22.66	21	41	–	–	–	–	–	–	–	–	1	–	–	Phosphine, methylbis(methylisopropylcyclohexyl)-	74764-60-6	NI34228
324	221	220	144	193	165	325	222	324	100	85	50	30	28	26	17	12		22	16	1	2	–	–	–	–	–	–	–	–	–	3-Benzylidene-1-(benzylideneamino)-2-indolinone		AI02047
324	208	295	247	78	325	323	77	324	100	36	31	21	20	18	16	13		22	16	1	2	–	–	–	–	–	–	–	–	–	3-Benzylidene-2,3-dihydro-5-phenyl-1H-1,4-benzodiazepin-2-one	62642-81-3	NI34229
324	51	220	77	246	191	178	203	324	100	80	60	60	50	50	50	30		22	16	1	2	–	–	–	–	–	–	–	–	–	4,6-Diphenyl-2-(2-pyridylimino)pyran	62219-24-3	NI34230
324	77	296	51	191	115	218	203	324	100	80	70	60	40	40	35	30		22	16	1	2	–	–	–	–	–	–	–	–	–	4,6-Diphenyl-1-(2-pyridyl)-2-pyridone		NI34231
324	293	307	77	204	325	294	295	324	100	40	32	31	28	27	22	20		22	16	1	2	–	–	–	–	–	–	–	–	–	2,4,5-Triphenyl-3-nitrosopyrrole		NI34232
195	324	180	165	326	288	273	231	324	100	13	8	8	5	2	1	1		22	25	–	–	–	–	1	–	–	–	–	–	–	exo-2-Chloro-syn-7-[bis(4-methylphenyl)methyl]norbornane		DD01245
309	187	310	324	147	172	134	203	324	100	31	23	22	11	10	8	7		22	28	2	–	–	–	–	–	–	–	–	–	–	5-Methoxy-1-(3-methoxy-4-methylphenyl)-1,3,3,6-tetramethylindan		NI34233
163	121	164	55	91	135	77	109	324	100	53	28	21	10	8	7	6	3.90	22	28	2	–	–	–	–	–	–	–	–	–	–	3,5-Bis(4-methoxyphenyl)-5-ethylhept-2-ene		LI07194
162	91	163	28	29	79	118	104	324	100	14	13	12	7	6	5	5	0.00	22	32	–	2	–	–	–	–	–	–	–	–	–	meso-N,N,N′,N′-Tetraethyl-1,2-diphenylethylenediamine		DD01246
162	163	91	28	79	29	134	133	324	100	40	27	22	16	15	13	13	0.00	22	32	–	2	–	–	–	–	–	–	–	–	–	(±)-N,N,N′,N′-Tetraethyl-1,2-diphenylethylenediamine		DD01247
324	325	246	191	233	215	323	247	324	100	25	15	15	11	11	9	9		23	17	–	–	–	–	–	–	–	–	1	–	–	Phosphorin, 2,4,6-triphenyl-	13497-36-4	NI34234
71	43	57	70	85	55	41	281	324	100	99	91	53	43	31	31	23	0.41	23	48	–	–	–	–	–	–	–	–	–	–	–	Docosane, 4-methyl-		LI07195
43	57	85	71	84	41	55	56	324	100	84	73	46	44	27	25	18	0.35	23	48	–	–	–	–	–	–	–	–	–	–	–	Docosane, 5-methyl-		LI07196
57	43	71	98	41	55	85	56	324	100	59	32	31	30	25	22	17	0.70	23	48	–	–	–	–	–	–	–	–	–	–	–	Docosane, 6-methyl-		AI02048
71	85	210	211	99	83	239	113	324	100	79	29	26	21	20	14	14	0.05	23	48	–	–	–	–	–	–	–	–	–	–	–	Heptadecane, 9-hexyl-		AI02049
57	71	43	85	55	69	113	41	324	100	69	60	44	29	21	20	20	1.60	23	48	–	–	–	–	–	–	–	–	–	–	–	Nonadecane, 2,6,10,14-tetramethyl-	55124-80-6	LI07197
57	71	85	113	99	56	55	169	324	100	99	58	33	32	25	23	22	7.30	23	48	–	–	–	–	–	–	–	–	–	–	–	Nonadecane, 2,6,10,14-tetramethyl-	55124-80-6	NI34235
57	43	71	85	41	55	29	56	324	100	91	61	40	36	25	24	14	1.26	23	48	–	–	–	–	–	–	–	–	–	–	–	Tricosane	638-67-5	AI02051
70	57	71	43	85	55	41	99	324	100	58	55	53	46	22	21	15	4.70	23	48	–	–	–	–	–	–	–	–	–	–	–	Tricosane	638-67-5	WI01416
43	57	71	41	85	55	56	69	324	100	93	55	46	39	29	15	14	2.17	23	48	–	–	–	–	–	–	–	–	–	–	–	Tricosane	638-67-5	AI02050
309	77	105	310	247	219	204	202	324	100	33	32	25	18	13	11	10	8.16	24	20	1	–	–	–	–	–	–	–	–	–	–	2,4-Hexadien-1-one, 1,3,5-triphenyl-	52588-88-2	NI34236
292	290	294	327	329	325	312	296	325	100	81	50	34	22	22	11	11		1	3	–	3	–	–	5	–	–	–	3	–	–	1-Methylpentachlorocyclotriphosphazene	71332-21-3	NI34237
69	327	325	308	306	239	237	218	325	100	60	60	20	20	20	20	17		5	1	–	1	–	9	–	1	–	–	–	–	–	Vinylamine, 2-bromo-1,N,N-tris(trifluoromethyl)-, cis-		LI07198
69	327	325	239	237	220	218	306	325	100	48	48	33	33	16	16	11		5	1	–	1	–	9	–	1	–	–	–	–	–	Vinylamine, 2-bromo-1,N,N-tris(trifluoromethyl)-, trans-		LI07199
245	247	82	80	110	51	81	79	325	100	99	75	75	63	54	50	50	0.00	8	9	1	1	1	–	–	2	–	–	–	–	–	Thiazolo[3,2-a]pyridinium, 7-bromo-2,3-dihydro-8-hydroxy-5-methyl-, bromide	22989-63-5	NI34238
218	18	93	125	44	15	109	79	325	100	74	68	61	45	37	29	29	1.00	10	16	5	1	2	–	–	–	–	–	1	–	–	Phosphorothioic acid, O-[4-[(dimethylamino)sulphonyl]phenyl] O,O-dimethyl ester	52-85-7	NI34239
218	93	125	28	217	44	29	109	325	100	53	51	44	34	33	24	23	2.08	10	16	5	1	2	–	–	–	–	–	1	–	–	Phosphorothioic acid, O-[4-[(dimethylamino)sulphonyl]phenyl] O,O-dimethyl ester	52-85-7	NI34240
81	53	82	51	52	50	327	110	325	100	14	5	2	2	1	1	1	0.58	11	7	2	1	–	–	4	–	–	–	–	–	–	Pyridine, 2-chloro-6-(2-furanylmethoxy)-4-(trichloromethyl)-	70166-48-2	NI34241
290	292	130	88	102	56	198	196	325	100	65	48	43	37	21	17	17	8.00	11	10	4	1	–	–	3	–	–	–	–	–	–	Glycine, N-[(2,4,5-trichlorophenoxy)acetyl]-, methyl ester	66789-85-3	NI34242
211	226	58	73	195	100	43	227	325	100	84	81	39	30	27	27	25	9.50	11	28	4	1	–	–	–	–	–	2	1	–	–	Phosphonic acid, [1-(acetylamino)propyl]-, bis(trimethylsilyl) ester	55108-81-1	NI34243
240	269	73	225	211	226	253	135	325	100	99	81	74	62	60	59	38	6.00	11	28	4	1	–	–	–	–	–	2	1	–	–	Phosphonic acid, [3-(acetylamino)propyl]-, bis(trimethylsilyl) ester	55108-89-9	NI34244
269	103	297	115	325	242	134	193	325	100	99	19	12	10	7	6	5		12	4	2	3	–	–	–	–	–	–	–	–	1	eta^5-(Tricyanovinylcyclopentadienyl)bis(carbonyl)rhodium(I)		NI34245
227	41	43	226	156	57	69	55	325	100	58	52	50	45	45	44	40	0.00	12	18	1	1	–	7	–	–	–	–	–	–	–	1-Aminooctane, heptafluorobutanoyl-		MS07688
101	118	67	252	88	246	102	159	325	100	96	70	46	43	28	17	16	1.90	12	23	5	1	2	–	–	–	–	–	–	–	–	1-(6-((Methylsulphonyl)thio)hexanamido)ethyl propanoate	76078-74-5	NI34246
135	77	136	107	92	119	64	42	325	100	10	8	8	6	4	2	2	0.00	13	12	3	1	–	5	–	–	–	–	–	–	–	Benzenepropylamine, α-keto-4-methoxy-N-(pentafluoropropionyl)-		MS07689

MASS TO CHARGE RATIOS								M.W.	INTENSITIES								Parent	C	H	O	N	S	F	Cl	Br	I	Si	P	B	X	COMPOUND NAME	CAS Reg No	No
135	77	136	92	119	42	190	51	**325**	100	16	9	7	5	4	3	2	**0.00**	13	12	3	1	–	5	–	–	–	–	–	–	–	2′-Methoxy-N-methyl-2-oxo-2-phenylethylamine PFP		NI34247
135	107	77	136	92	119	190	42	**325**	100	21	12	8	8	7	7	5	**2.60**	13	12	3	1	–	5	–	–	–	–	–	–	–	3′-Methoxy-N-methyl-2-oxo-2-phenylethylamine PFP		NI34248
135	77	136	92	107	119	190	64	**325**	100	14	12	10	9	6	5	4	**0.30**	13	12	3	1	–	5	–	–	–	–	–	–	–	4′-Methoxy-N-methyl-2-oxo-2-phenylethylamine PFP		NI34249
158	293	295	265	267	246	294	266	**325**	100	79	76	35	34	28	14	13	**7.00**	13	12	4	1	–	–	–	1	–	–	–	–	–	(±)-Methyl 2-bromo-2-[2-(2,5-dioxo-1-pyrrolidinyl)phenyl]acetate		NI34250
106	78	307	202	51				**325**	100	58	41	33	22				**0.00**	14	10	3	3	–	3	–	–	–	–	–	–	–	Isonicotinaldehyde, (1,1,1-trifluoro-3-furoylisopropylidene)hydrazide		LI07200
199	198	158	79	114	200	100	31	**325**	100	31	30	28	26	23	15	14	**0.00**	14	15	4	1	2	–	–	–	–	–	–	–	–	Naphtho[1,2-d]thiazolium, 1,2-dimethyl-, methyl sulphate	64415-17-4	NI34251
294	295	209	151	266	59	237	234	**325**	100	98	73	52	35	28	25	23	**2.91**	14	15	8	1	–	–	–	–	–	–	–	–	–	2,3,4,5-Tetramethoxycarbonyl-6-methylpyridine		NI34252
121	325	122	119	91	78	77	41	**325**	100	19	9	4	4	4	4	3		14	16	2	1	–	5	–	–	–	–	–	–	–	Benzenebutylamine, 4-methoxy-N-(pentafluoropropionyl)-4-(4-methoxyphenyl)-		MS07690
174	135	325	282	130	107			**325**	100	52	21	14	4	2				14	19	6	1	–	–	–	–	–	1	–	–	–	1-(4-Methoxybenzoyl)-2,8,9-trioxa-5-aza-1-silatricyclo[3.3.3.0^{1,5}]undecane		MS07691
79	91	138	105	155	107	93	139	**325**	100	57	37	21	17	16	16	13	**0.00**	15	16	3	1	1	–	1	–	–	–	–	–	–	(1RS,2RS,4SR,5RS,6RS)-6-exo-Chloro-5-endo-(2-nitrophenylthio)spiro[bicyclo[2.2.2]octane-2,2′-oxirane]		DD01248
79	138	91	171	93	105	139	107	**325**	100	49	44	32	25	21	19	14	**0.00**	15	16	3	1	1	–	1	–	–	–	–	–	–	(1RS,2RS,4SR,5SR,6SR)-5-exo-Chloro-6-endo-(2-nitrophenylthio)spiro[bicyclo[2.2.2]octane-2,2′-oxirane]		DD01249
79	91	105	93	107	138	155	106	**325**	100	56	20	20	18	16	12	12	**0.00**	15	16	3	1	1	–	1	–	–	–	–	–	–	(1RS,2SR,4SR,5RS,6RS)-6-exo-Chloro-5-endo-(2-nitrophenylthio)spiro[bicyclo[2.2.2]octane-2,2′-oxirane]		DD01250
79	91	93	105	138	107	171	106	**325**	100	59	24	23	22	19	14	13	**0.00**	15	16	3	1	1	–	1	–	–	–	–	–	–	(1RS,2SR,4SR,5SR,6SR)-5-exo-Chloro-6-endo-(2-nitrophenylthio)spiro[bicyclo[2.2.2]octane-2,2′-oxirane]		DD01251
79	91	93	84	105	80	107	138	**325**	100	59	32	21	18	18	15	14	**0.00**	15	16	3	1	1	–	1	–	–	–	–	–	–	(1RS,2SR,4SR,5RS,6RS)-5-endo-Chloro-6-exo-(2-nitrophenylthio)spiro[bicyclo[2.2.2]octane-2,2′-oxirane]		DD01252
130	144	143	115	325	55	117	77	**325**	100	47	23	10	8	6	5	5		15	19	2	1	–	–	–	–	–	–	–	–	1	Valeric acid, 5-[(2-indol-3-ylethyl)selenyl]-	1919-98-8	LI07201
130	144	143	115	325	55	145	117	**325**	100	46	22	10	8	6	5	5		15	19	2	1	–	–	–	–	–	–	–	–	1	Valeric acid, 5-[(2-indol-3-ylethyl)selenyl]-	1919-98-8	NI34253
73	179	75	253	210	223	209	235	**325**	100	67	63	58	44	38	32	24	**6.79**	15	27	3	1	–	–	–	–	–	2	–	–	–	Benzeneacetamide, 3-methoxy-N-(trimethylsilyl)-4-[(trimethylsilyl)oxy]-	72361-06-9	NI34254
119	120	236	136	93	237	137	94	**325**	100	95	72	48	41	23	20	20	**0.00**	16	23	6	1	–	–	–	–	–	–	–	–	–	20-Norcrotalanan-11,15-dione, 14,19-dihydro-12,13-dihydroxy-, (13α,14α)-	315-22-0	NI34255
96	83	56	193	325	112	268	296	**325**	100	48	38	28	24	13	12	8		16	27	2	3	1	–	–	–	–	–	–	–	–	8-(Diethylamino)-1-(ethylthio)-2,4,6,7-tetramethyl-2,6-diazabicyclo[2.2.2]oct-7-ene-3,5-dione		DD01253
116	73	44	210	117	180	310	209	**325**	100	76	44	22	20	15	11	9	**0.00**	16	31	2	1	–	–	–	–	–	2	–	–	–	Tyramine, 3-methoxy-N-methyl-N,O-bis(trimethylsilyl)-		MS07693
116	73	147	75	44	77	207	85	**325**	100	95	83	61	52	46	26	23	**0.00**	16	31	2	1	–	–	–	–	–	2	–	–	–	Tyramine, 3-methoxy-N-methyl-N,O-bis(trimethylsilyl)-		MS07692
290	292	264	236	308	326	55	291	**325**	100	33	32	29	26	22	20	19	**3.45**	17	24	3	1	–	–	1	–	–	–	–	–	–	Cyclohexanol, 1,1′-(4-chloro-2,6-pyridinediyl)bis-, N-oxide	34277-38-8	MS07694
57	114	268	325	88	59	58	70	**325**	100	95	67	53	52	49	49	42		17	27	5	1	–	–	–	–	–	–	–	–	–	17-(1-Hydroxyethyl)-2,3-benzo-10-methyl-1,4,7,13-tetraoxa-10-azacyclopentadeca-2-ene		MS07695
93	73	167	252	43	111	86	310	**325**	100	52	26	25	18	16	15	14	**0.00**	17	31	3	1	–	–	–	–	–	1	–	–	–	N,N-Diisopropyl-4-(methoxycarbonyl)-4-(trimethylsilyl)cyclopent-1-enecarboxamide	110426-84-1	DD01254
252	325	253	326	254	139	79	196	**325**	100	35	17	8	8	8	7	6		18	15	5	1	–	–	–	–	–	–	–	–	–	Anthraquinone, 1-hydroxy-4-(methoxycarbonylethylamino)-		IC03943
325	94	78	65	92	326	168	231	**325**	100	35	34	23	20	19	18	15		18	16	3	1	–	–	–	–	–	–	1	–	–	Phosphoramidic acid, phenyl-, diphenyl ester	3848-51-9	NI34256
86	173	153	174	325	255			**325**	100	94	39	36	25	16				18	31	4	1	–	–	–	–	–	–	–	–	–	N-Dihydrojasmonylisoleucine		LI07202
310	325	324	152	282	280	311	165	**325**	100	79	48	30	28	22	20	20		19	19	4	1	–	–	–	–	–	–	–	–	–	5H-Benzo[g]-1,3-benzodioxolo[6,5,4-de]quinolin-12-ol, 6,7,7a,8-tetrahydro-11-methoxy-7-methyl-, (S)-	298-45-3	NI34257
165	234	91	77	28	51	79	65	**325**	100	97	63	46	33	31	31	29	**4.69**	19	19	4	1	–	–	–	–	–	–	–	–	–	4-Benzyl-2-(3,4-dimethoxyphenyl)-5,6-dihydro-4H-1,3-oxazin-5-one		NI34258
148	325	149	324	57	91	176	43	**325**	100	34	18	16	13	13	12	12		19	19	4	1	–	–	–	–	–	–	–	–	–	(±)-Cheilanthifoline	483-44-3	NI34259
269	270	325	226	254	211	326	210	**325**	100	19	16	16	6	5	3	3		19	19	4	1	–	–	–	–	–	–	–	–	–	8-Methoxy-2-(4-methoxyphenyl)-1,2,4,5-tetrahydro-1-benzazocine-3,6-dione	90732-26-6	NI34260
190	325	324	282	191	239	206	192	**325**	100	49	46	40	13	12	12	10		19	19	4	1	–	–	–	–	–	–	–	–	–	Reframoline	21238-92-6	LI07203
148	325	149	324	176	91	89	326	**325**	100	33	18	16	15	14	9	7		19	19	4	1	–	–	–	–	–	–	–	–	–	(±)-Tetrahydrogroenlandicine		NI34261
255	131	254	125	257	256	57	58	**325**	100	71	54	39	35	32	16	14	**1.40**	19	20	–	3	–	–	1	–	–	–	–	–	–	1H-Benzimidazole, 1-[(4-chlorophenyl)methyl]-2-(1-pyrrolidinylmethyl)-	442-52-4	NI34262
326	256	324	255					**325**	100	80	28	25					**0.00**	19	20	–	3	–	–	1	–	–	–	–	–	–	1H-Benzimidazole, 1-[(4-chlorophenyl)methyl]-2-(1-pyrrolidinylmethyl)-	442-52-4	NI34263
234	73	235	325	91	45	74	206	**325**	100	80	18	17	17	10	6	5		19	23	2	1	–	–	–	–	–	1	–	–	–	1H-Indole, 5-methoxy-6-benzyloxy-1-(trimethylsilyl)-	74367-56-9	NI34264
73	105	262	176	206	159	77	75	**325**	100	76	47	46	43	34	34	29	**0.00**	19	23	2	1	–	–	–	–	–	1	–	–	–	1H-Indole, 5-methoxy-6-benzyloxy-1-(trimethylsilyl)-	74367-56-9	NI34265
73	217	147	205	218	103	231	204	**325**	100	78	29	21	19	18	11	10	**0.00**	19	23	2	1	–	–	–	–	–	1	–	–	–	1H-Indole, 6-methoxy-5-benzyloxy-1-(trimethylsilyl)-	74367-55-8	NI34266
234	73	325	235	91	45	206	74	**325**	100	77	22	19	17	10	9	6		19	23	2	1	–	–	–	–	–	1	–	–	–	1H-Indole, 6-methoxy-5-benzyloxy-1-(trimethylsilyl)-	74367-55-8	NI34267
58	268	70	281	84	255	99	238	**325**	100	90	70	50	31	30	28	24	**4.00**	19	39	1	3	–	–	–	–	–	–	–	–	–	1,5-Diazacycloheptadecan-6-one, 5-[3-(methylamino)propyl]-	67370-90-5	NI34268
58	70	84	98	127	253	267	196	**325**	100	65	45	40	30	21	20	20	**20.00**	19	39	1	3	–	–	–	–	–	–	–	–	–	1,5,9-Triazacycloheneicosan-10-one, 9-methyl-	67370-91-6	NI34269
325	279	267	201	326	139	175	280	**325**	100	65	23	21	20	15	13	12		20	11	2	3	–	–	–	–	–	–	–	–	–	11-Nitrodibenzo(a,c)phenazine	4661-57-8	NI34270
167	168	169	326	166	165	153	327	**325**	100	99	90	81	80	78	73	64	**0.00**	20	23	3	1	–	–	–	–	–	–	–	–	–	Leucine, diphenylacetyl-	65707-79-1	NI34271
165	137	177	109	131	103	77	94	**325**	100	96	80	65	64	60	57	48	**17.00**	20	23	3	1	–	–	–	–	–	–	–	–	–	2-Propenamide, N-[2-ethoxy-2-(4-methoxyphenyl)ethyl]-3-phenyl-	75363-20-1	NI34272

MASS TO CHARGE RATIOS								M.W.	INTENSITIES								Parent	C	H	O	N	S	F	Cl	Br	I	Si	P	B	X	COMPOUND NAME	CAS Reg No	No
325	297	253	36	281	269	296	282	**325**	100	98	45	41	36	34	25	24		20	27	1	3	–	–	–	–	–	–	–	–	–	3H-Phenoxazine, 3,7-bis(diethylamino)-		IC03945
30	325	44	297	281	253	326	269	**325**	100	73	69	45	22	21	16	16		20	27	1	3	–	–	–	–	–	–	–	–	–	3H-Phenoxazine, 3,7-bis(diethylamino)-		IC03944
325	326	240	120.5	269	241	134.5	106.5	**325**	100	23	10	8	7	7	4	4		21	11	3	1	–	–	–	–	–	–	–	–	–	5,8,13,14-Tetrahydro-5,8,14-trioxonaphtha[2,3-c]acridine		IC03946
105	192	77	89	297	193	51	165	**325**	100	65	53	22	19	11	9	8	**7.00**	21	15	1	3	–	–	–	–	–	–	–	–	–	1H-1,2,3-Triazole, 1-benzoyl-4,5-diphenyl-	33471-67-9	NI34273
105	192	77	89	297	193	165	106	**325**	100	65	53	23	21	12	8	8	**8.00**	21	15	1	3	–	–	–	–	–	–	–	–	–	1H-1,2,3-Triazole, 1-benzoyl-4,5-diphenyl-	33471-67-9	LI07204
151	107	149	175	325	299	105	157	**325**	100	53	52	50	48	39	33	32		21	27	2	1	–	–	–	–	–	–	–	–	–	Estra-4,6-dien-3-one, 17α-ethylnyl-17β-hydroxy-, methyloxime		NI34274
282	325	91	268	324	248	281	326	**325**	100	94	50	44	28	28	25	21		21	32	1	1	–	–	–	–	–	–	–	1	–	2H-1,3,2-Benzoxazaborine, 3-cyclohexyl-2-ethyloctahydro-8a-phenyl-	74810-36-9	NI34275
204	73	205	147	206	103	217	74	**325**	100	46	21	9	8	6	5	4	**0.00**	21	43	1	1	–	–	–	–	–	–	–	–	–	Hexadecanamide, N-pentyl-	73392-18-4	NI34276
89	325	192	77	63	105	51	193	**325**	100	98	77	43	25	19	15	14		22	15	2	1	–	–	–	–	–	–	–	–	–	5-Benzoyl-2,4-diphenyloxazole		LI07205
310	311	77	294	51	141	140	154	**325**	100	25	10	6	5	5	5	4	**2.00**	22	19	–	3	–	–	–	–	–	–	–	–	–	1,4,4-Trimethyl-2,6-diphenyl-1,4-dihydropyridine-3,5-dicarbonitrile	67439-19-4	NI34277
135	93	91	39	136	79	105	53	**325**	100	25	24	24	23	23	20	20	**0.80**	22	31	1	1	–	–	–	–	–	–	–	–	–	Aziridinone, 1,3-bis(tricyclo[3.3.1.1^{3,7}]dec-1-yl)-	17385-51-2	NI34278
91	217	218	198	204	233	216	203	**325**	100	41	32	31	27	22	22	16	**0.00**	23	19	1	1	–	–	–	–	–	–	–	–	–	2-Benzyl-2,3-dihydro-1H-dibenz[e,g]isoindole N-oxide	110028-18-7	DD01255
169	168	141	129	157	128	127	167	**325**	100	99	48	31	20	19	17	14	**0.40**	23	19	1	1	–	–	–	–	–	–	–	–	–	2-(N-β-Hydroxy-β-naphth-2-ylethyliminomethyl)naphthalene		IC03947
310	325	205	324	191	203	202	165	**325**	100	97	90	48	44	38	34	30		23	19	1	1	–	–	–	–	–	–	–	–	–	2-Methyl-4,4-diphenyl-2,3-butadienanilide		MS07696
206	255	191	325	254	256	207	192	**325**	100	91	70	69	28	27	25	17		23	19	1	1	–	–	–	–	–	–	–	–	–	Spiro[fluorene-9,2′-oxetane]-4′-imine, 3′,3′-dimethyl-N-phenyl-	15183-48-9	NI34279
206	255	191	325	119	165	254	91	**325**	100	94	71	68	68	43	31	28		23	19	1	1	–	–	–	–	–	–	–	–	–	Spiro[fluorene-9,2′-oxetane]-4′-imine, 3′,3′-dimethyl-N-phenyl-	15183-48-9	MS07697
119	91	206	234	118	120	115	128	**325**	100	22	21	14	10	9	8	7	**0.00**	24	23	–	1	–	–	–	–	–	–	–	–	–	3-Cyclohexen-1-amine, 2,5,6-triphenyl-, (1α,2α,5α,6β)-	33441-42-8	NI34280
119	206	91	234	120	118	128	115	**325**	100	25	22	15	9	9	7	7	**0.00**	24	23	–	1	–	–	–	–	–	–	–	–	–	3-Cyclohexen-1-amine, 2,5,6-triphenyl-, (1α,2α,5α,6β)-	33441-42-8	NI34281
119	206	234	120	207	235			**325**	100	27	14	9	5	3			**0.00**	24	23	–	1	–	–	–	–	–	–	–	–	–	3-Cyclohexen-1-amine, 2,5,6-triphenyl-, (1α,2α,5α,6β)-	33441-42-8	NI34282
119	206	91	120	118	205	128	115	**325**	100	25	17	9	8	5	5	5	**0.00**	24	23	–	1	–	–	–	–	–	–	–	–	–	3-Cyclohexen-1-amine, 2,5,6-triphenyl-, (1α,2β,5β,6β)-	33512-00-4	NI34283
119	206	120	207	234				**325**	100	31	9	6	3				**0.00**	24	23	–	1	–	–	–	–	–	–	–	–	–	3-Cyclohexen-1-amine, 2,5,6-triphenyl-, (1α,2β,5β,6β)-	33512-00-4	NI34285
119	206	91	120	118	207	205	128	**325**	100	25	17	9	8	5	5	5	**0.00**	24	23	–	1	–	–	–	–	–	–	–	–	–	3-Cyclohexen-1-amine, 2,5,6-triphenyl-, (1α,2β,5β,6β)-	33512-00-4	NI34284
296	236	326	191	266	221	206	144	**326**	100	78	71	71	57	36	36	28		2	6	4	4	2	–	–	–	–	–	–	–	2	Diiron, di-μ-methanethiolatotetranitroso-		MS07698
117	82	165	83	145	130	111	95	**326**	100	54	52	26	21	17	17	7	**0.00**	4	1	2	–	–	–	7	–	–	–	–	–	–	1,2,2,2-Tetrachloroethyl trichloroacetate		LI07206
142	152	127	60	59	45	75	64	**326**	100	75	67	58	58	58	50	30	**0.00**	5	11	–	–	4	–	–	–	1	–	–	–	–	1,2,5-Trithiepane-5-methylsulphonium iodide		LI07207
57	73	43	55	60	69	71	83	**326**	100	60	60	55	53	50	38	36	**5.00**	6	–	–	–	–	–	5	1	–	–	–	–	–	Pentachlorobromobenzene		IC03948
81	47	96	69	257	169	119	116	**326**	100	92	59	27	15	9	9	7	**2.00**	8	2	–	–	–	12	–	–	–	–	–	–	–	Cyclobutene, 1,2,3,4-tetrakis(trifluoromethyl)-	74367-21-8	NI34286
242	163	216	137	191	270	298	98	**326**	100	87	77	43	41	35	34	31	**29.20**	8	5	3	–	–	–	–	1	–	–	–	–	1	Molybdenum, bromotricarbonyl-π-cyclopentadienyl-	12079-79-7	MS07699
242	244	241	240	238	135	134	163	**326**	100	83	67	65	58	53	50	48	**8.59**	8	5	3	–	–	–	–	1	–	–	–	–	1	Molybdenum, bromotricarbonyl-π-cyclopentadienyl-	12079-79-7	NI34287
57	41	242	29	86	243	39	27	**326**	100	40	22	20	19	18	15	13	**0.00**	9	12	3	–	–	–	–	2	–	–	–	–	–	(2R)-5-Bromo-6-(bromomethyl)-2-tert-butyl-2H,4H-1,3-dioxin-4-one		DD01256
41	39	76	38	99	36	28	27	**326**	100	64	33	18	17	15	15	14	**0.00**	9	18	4	–	–	–	3	–	–	–	1	–	–	Tris(chloropropyl) phosphate		IC03949
225	73	311	296	237	268	226	86	**326**	100	41	29	29	27	26	19	18	**0.00**	9	23	3	2	1	–	–	–	–	2	1	–	–	Phosphonic acid, [1-(carbonothioylhydrazino)ethyl]-, bis(trimethylsilyl) ester	56051-82-2	NI34288
268	73	211	226	296	311	195	190	**326**	100	74	68	57	42	41	35	34	**0.00**	9	23	3	2	1	–	–	–	–	2	1	–	–	Phosphonic acid, [2-(carbonothioylhydrazino)ethyl]-, bis(trimethylsilyl) ester	56051-81-1	NI34289
297	298	299	211	267	183	44	269	**326**	100	28	18	6	5	5	5	5	**0.69**	9	26	5	–	–	–	–	–	–	4	–	–	–	1,3,5,7-Tetraethyl-1-methoxycyclotetrasiloxane	75221-55-5	NI34290
116	240	210	238	268	237	234	208	**326**	100	54	43	38	36	36	36	30	**22.00**	10	16	4	2	–	–	–	–	–	–	–	–	1	Molybdenum, tetracarbonyl(N,N,N′,N′-tetramethylethylenediamino)-		LI07208
220	218	140	221	217	222	216	219	**326**	100	54	49	24	22	20	20	16	**3.00**	11	6	–	2	–	–	–	–	–	–	–	–	2	Selenocyanic acid, (3-cyanobenzo[b]selenophene-2-yl)methyl ester	39828-50-7	NI34291
45	109	29	81	75	111	157	27	**326**	100	82	72	49	40	38	33	33	**4.90**	11	16	3	–	2	–	1	–	–	–	1	–	–	Phosphorothioic acid, S-[[(4-chlorophenyl)thio]methyl] O,O-diethyl ester	7173-84-4	NI34292
59	28	143	43	117	175	235	58	**326**	100	38	23	18	15	11	10	8	**5.93**	11	18	1	–	5	–	–	–	–	–	–	–	–	2,4,6,8,9-Pentathiaadamantane-3-methanol, α,α,1,5,7-pentamethyl-	21404-63-7	NI34293
99	157	211	141	169	57	71	183	**326**	100	60	39	16	15	15	9	5	**0.00**	11	19	3	–	–	–	–	–	1	–	–	–	–	3-(tert-Butylacetoxy)-1-iodo-3-methylbutan-2-one		NI34294
127	326	77	51	109	69	217	128	**326**	100	41	23	18	14	13	6	6		12	5	–	–	1	6	–	–	–	–	1	–	–	Phosphinothioic fluoride, (pentafluorophenyl)phenyl-	53327-27-8	NI34295
227	159	181	228	216	199	127	161	**326**	100	54	46	43	42	41	40	40	**12.44**	12	7	5	2	1	–	1	–	–	–	–	–	–	2,4-Dinitro-4′-chlorodiphenylsulphoxyde	966-72-3	NI34296
170	188	76	50	94	75	326	30	**326**	100	95	67	58	51	51	39	38		12	7	6	2	1	1	–	–	–	–	–	–	–	Benzene, 1-fluoro-2-nitro-4-[(3-nitrophenyl)sulphonyl]-	383-21-1	NI34297
297	327	267	124	140	281	142	94	**326**	100	97	30	29	26	20	19	17	**0.00**	12	7	6	2	1	1	–	–	–	–	–	–	–	Benzene, 1-fluoro-2-nitro-4-[(3-nitrophenyl)sulphonyl]-	383-21-1	NI34298
139	70	168	328	84	326	330	140	**326**	100	66	64	60	37	33	27	25		12	8	1	–	–	–	–	2	–	–	–	–	–	[1,1′-Biphenyl]-2-ol, 3,5-dibromo-	55815-20-8	NI34299
83	55	41	81	244	242	84	67	**326**	100	51	22	13	12	10	8	8	**6.00**	12	22	–	–	–	–	–	–	–	–	–	–	2	Dicyclohexyl diselenide		MS07700
106	326	43	73	133	107	134	61	**326**	100	82	59	53	48	36	31	30		12	22	2	–	4	–	–	–	–	–	–	–	–	6-Acetoxy-1,4,8,11-tetrathiacyclotetradecane		NI34300
41	43	57	55	69	71	83	39	**326**	100	95	84	84	57	44	36	33	**0.00**	12	24	–	–	–	–	–	2	–	–	–	–	–	Dodecane, 1,2-dibromo-		MS07701
55	69	41	57	43	135	137	71	**326**	100	72	71	61	44	42	41	34	**0.00**	12	24	–	–	–	–	–	2	–	–	–	–	–	Dodecane, 1,12-dibromo-	3344-70-5	NI34301
41	43	69	57	55	56	80	82	**326**	100	80	78	71	69	52	43	42	**0.00**	12	24	–	–	–	–	–	2	–	–	–	–	–	Undecane, 1,2-dibromo-2-methyl-		MS07702
57	85	86	28	149	326	206	29	**326**	100	66	46	30	29	17	17	16		12	26	–	2	4	–	–	–	–	–	–	–	–	Piperazine, 1,4-bis(tert-butyl)dithio-	37004-94-7	MS07703
269	267	29	41	57	271	265	268	**326**	100	71	62	56	54	36	36	33	**0.81**	12	27	–	–	–	–	1	–	–	–	–	–	1	Tributyltin chloride	1461-22-9	NI34302
270	57	29	268	41	155	266	272	**326**	100	77	77	73	69	35	33	31	**0.00**	12	27	–	–	–	–	1	–	–	–	–	–	1	Tributyltin chloride	1461-22-9	IC03950

MASS TO CHARGE RATIOS								M.W.	INTENSITIES								Parent	C	H	O	N	S	F	Cl	Br	I	Si	P	B	X	COMPOUND NAME	CAS Reg No	No
28	41	177	179	57	56	269	29	**326**	100	97	88	77	73	70	67	63	**0.00**	12	27	–	–	–	–	1	–	–	–	–	–	1	Tributyltin chloride	1461-22-9	MS07704
269	267	213	41	211	268	265	271	**326**	100	71	49	45	39	37	37	36	**0.00**	12	27	–	–	–	–	1	–	–	–	–	–	1	Triisobutyltin chloride		MS07705
250	175	176	298	140	270	195	326	**326**	100	88	85	63	60	58	58	52		13	6	2	–	–	4	–	–	–	–	–	–	1	Iron, dicarbonyl-π-cyclopentadienyl(2,3,5,6-tetrafluorophenyl)-	12110-35-9	NI34303
186	150	250	175	176	80	298	270	**326**	100	82	51	46	44	37	35	34	**24.13**	13	6	2	–	–	4	–	–	–	–	–	–	1	Iron, dicarbonyl-π-cyclopentadienyl(2,3,5,6-tetrafluorophenyl)-	12110-35-9	MS07706
155	154	157	156	99	63	127	326	**326**	100	56	31	29	23	21	16	13		13	8	4	2	–	–	2	–	–	–	–	–	–	Benzamide, 5-chloro-N-(2-chloro-4-nitrophenyl)-2-hydroxy-	50-65-7	NI34304
261	291	297	327	299	329	165	263	**326**	100	89	70	68	51	42	42	38	**4.74**	13	8	4	2	–	–	2	–	–	–	–	–	–	Benzamide, 5-chloro-N-(2-chloro-4-nitrophenyl)-2-hydroxy-	50-65-7	NI34305
166	165	326	325	249	247	245	169	**326**	100	98	90	38	35	33	28	25		13	10	–	–	–	–	–	–	–	–	–	–	2	2-Phenyl-1,3-benzodiselenole		DD01257
253	225	326	119	90	78	327	44	**326**	100	20	18	12	11	8	6	4		13	11	4	–	–	5	–	–	–	–	–	–	–	4-Hydroxyphenylacetic acid ethyl ester PFP		MS07707
148	178	135	149	175	162	108	176	**326**	100	96	58	52	48	36	36	25	**24.00**	13	14	1	10	–	–	–	–	–	–	–	–	–	9-[3-(Aden-9-yl)propyl]guanine		LI07209
42	262	55	73	233	56	43	57	**326**	100	99	59	38	30	29	29	28	**13.30**	13	14	4	2	2	–	–	–	–	–	–	–	–	10H-3,10a-Epidithiopyrazino[1,2-a]indole-1,4-dione, 2,3,5a,6-tetrahydro-6 hydroxy-3-(hydroxymethyl)-2-methyl-, [3R-(3α,5aβ,6β,10aα)]-	67-99-2	NI34306
128	81	112	97	129	326	153	98	**326**	100	97	86	83	82	72	71	69		13	14	6	2	1	–	–	–	–	–	–	–	–	6H-Furo[2',3':4,5]oxazolo[3,2-a]pyrimidine-6-thione, 3-(acetyloxy)-2-[(acetyloxy)methyl]-2,3,3a,9a-tetrahydro-, [2R-(2α,3β,3aβ,9aβ)]-		NI34307
149	177	211	231	298	213	326	151	**326**	100	95	60	54	45	38	36	36		13	20	1	–	2	–	2	–	–	–	–	–	–	Bis(2-chlorocyclohexyl) S,S-dithiocarbonate	105214-29-7	NI34308
132	104	105	160	77	131	78	133	**326**	100	54	51	35	34	21	18	16	**0.00**	13	22	4	6	–	–	–	–	–	–	–	–	–	Ethyl 5,6,7,8-tetrahydro-6,7-bis-(2-hydroxyethylamino)pteridine-4-carboxylate		LI07210
99	101	69	125	113	100	102	153	**326**	100	84	63	14	14	13	10	2	**0.00**	13	27	7	–	–	–	–	–	–	–	1	–	–	Butyl 3-hydroxybutyl (3-methoxycarbonyl)propyl phosphate		NI34309
326	328	327	329	330	69	291	325	**326**	100	68	38	23	15	15	13	8		14	8	5	–	–	–	2	–	–	–	–	–	–	5,7-Dichloro-1,3,6-trihydroxy-8-methyl-9H-xanthen-9-one		MS07708
326	234	90	327	250	296	280	233	**326**	100	33	21	18	16	15	11	11		14	10	4	6	–	–	–	–	–	–	–	–	–	2-[2-(4-Nitroanilino)]-4-amino-6-nitroquinazoline		IC03951
141	155	91	51	65	139	296	39	**326**	100	99	76	40	33	30	27	18	**0.29**	14	14	5	–	2	–	–	–	–	–	–	–	–	Methanol, [(4-methylphenyl)sulphonyl]-, benzenesulphonate	31081-09-1	NI34310
155	91	77	65	296	51	156	92	**326**	100	97	26	23	15	15	9	9	**0.10**	14	14	5	–	2	–	–	–	–	–	–	–	–	Methanol, (phenylsulphonyl)-, 4-methylbenzenesulphonate	31081-06-8	NI34311
232	43	41	326	328	56	57	42	**326**	100	98	83	79	79	74	68	56		14	15	4	–	–	–	–	1	–	–	–	–	–	2-Bromo-3-methyl-4,5,8-trimethoxy-1-naphthol	101459-31-8	NI34312
43	211	41	163	27	240	205	147	**326**	100	45	15	9	9	6	6	6	**0.00**	14	18	7	2	–	–	–	–	–	–	–	–	–	Carbonic acid, isopropyl 2-sec-butyl-4,6-dinitrophenyl ester	973-21-7	NI34313
81	43	201	126	127	326	150	110	**326**	100	55	6	4	2	1	1	1		14	18	7	2	–	–	–	–	–	–	–	–	–	Deoxy-2'-α-D-ribofuranosyl-1-thyminediacetyl		LI07211
81	43	201	126	326	127	193	110	**326**	100	48	9	3	2	2	1	1		14	18	7	2	–	–	–	–	–	–	–	–	–	Deoxy-2'-β-D-ribofuranosyl-1-thyminediacetyl		LI07212
81	43	141	201	126	127	326	110	**326**	100	79	15	9	5	3	1	1		14	18	7	2	–	–	–	–	–	–	–	–	–	Deoxy-2'-α-D-ribopyranosyl-1-thyminediacetyl		LI07213
97	100	41	327	325	326	328	313	**326**	100	56	51	26	26	26	24	23		14	19	–	2	1	–	–	1	–	–	–	–	–	4,4,6-Trimethyl-2-(N-methyl-4-bromophenylamino)-5,6-dihydro-4H-1,3-thiazine	53004-55-0	NI34314
209	152	210	77	153	151	105	76	**326**	100	24	18	11	9	8	8	7	**0.00**	15	9	2	–	–	–	3	–	–	–	–	–	–	10-Hydroxy-10-(trichloromethyl)anthrone		IC03952
86	326	240	105	213	77	56	172	**326**	100	90	40	35	21	21	17	15		15	13	3	2	–	3	–	–	–	–	–	–	–	2-Phenyl-4-(trifluoroacetyl)-5-(4-morpholino)oxazole	51770-10-6	NI34315
86	326	240	105	213	77	172	257	**326**	100	91	40	36	21	21	16	10		15	13	3	2	–	3	–	–	–	–	–	–	–	2-Phenyl-4-(trifluoroacetyl)-5-(4-morpholino)oxazole	51770-10-6	NI34316
98	81	326	253	254	174	307	219	**326**	100	80	52	41	36	25	23	19		15	13	4	–	–	2	–	–	–	–	1	–	–	Diphenyl 2,2-difluoro-1-methylethenyl phosphate		DD01258
176	326	311	296	327	42	325	43	**326**	100	99	57	25	20	20	17	17		15	14	3	6	–	–	–	–	–	–	–	–	–	Terpyrimidine		NI34317
57	58	129	41	196	43	270	178	**326**	100	43	35	30	20	19	17	14	**2.01**	15	18	8	–	–	–	–	–	–	–	–	–	–	4H,5aH,9H-Furo[2,3-b]furo[3',2':2,3]cyclopenta[1,2-c]furan-2,4,7(3H,8H) trione, 9-tert-butyl-10,10a-dihydro-8,9-dihydroxy-, [5aS-(3aR*,5aα,8β,8aR*,9α,10aα)]-	33570-04-6	NI34318
197	195	283	241	281	199	239	193	**326**	100	71	69	58	57	46	45	35	**0.80**	15	26	–	–	–	–	–	–	–	–	–	–	1	Phenyltriisopropyltin		MS07709
78	163	43	251	207	161	205	249	**326**	100	69	67	61	53	48	43	37	**0.38**	15	26	–	–	–	–	–	–	–	–	–	–	1	Phenyltripropyltin		MS07710
326	327	253	194	311	267	328	73	**326**	100	25	16	16	15	10	9	9		15	26	4	–	–	–	–	–	–	2	–	–	–	Acetic acid, [2,5-bis(trimethylsiloxy)phenyl]-, methyl ester	27798-63-6	NI34319
179	73	326	45	267	180	75	74	**326**	100	83	29	26	19	14	7	7		15	26	4	–	–	–	–	–	–	2	–	–	–	Benzeneacetic acid, 3,4-bis[(trimethylsilyl)oxy]-, methyl ester	27798-64-7	MS07711
326	179	267	327	73	180	328	268	**326**	100	85	28	26	17	12	10	6		15	26	4	–	–	–	–	–	–	2	–	–	–	Benzeneacetic acid, 3,4-bis[(trimethylsilyl)oxy]-, methyl ester	27798-64-7	NI34320
179	73	326	267	180	327	45	268	**326**	100	66	53	32	16	14	13	7		15	26	4	–	–	–	–	–	–	2	–	–	–	Benzeneacetic acid, 3,4-bis[(trimethylsilyl)oxy]-, methyl ester	27798-64-7	NI34321
73	267	268	89	45	179	59	269	**326**	100	67	24	22	19	13	12	11	**0.00**	15	26	4	–	–	–	–	–	–	2	–	–	–	Benzeneacetic acid, α,4-bis[(trimethylsilyl)oxy]-, methyl ester	55334-40-2	NI34322
267	268	269	73	311	283	75	89	**326**	100	23	10	8	5	3	3	2	**0.50**	15	26	4	–	–	–	–	–	–	2	–	–	–	Benzeneacetic acid, α,4-bis[(trimethylsilyl)oxy]-, methyl ester	55334-40-2	NI34323
209	73	210	147	45	75	283	74	**326**	100	83	18	18	15	9	8	7	**0.00**	15	26	4	–	–	–	–	–	–	2	–	–	–	Benzeneacetic acid, 2-methoxy-α-[(trimethylsilyl)oxy]-, trimethylsilyl ester	55530-66-0	NI34324
209	73	210	147	283	45	211	135	**326**	100	70	28	21	16	9	8	8	**0.00**	15	26	4	–	–	–	–	–	–	2	–	–	–	Benzeneacetic acid, 2-methoxy-α-[(trimethylsilyl)oxy]-, trimethylsilyl ester	55530-66-0	MS07712
73	326	209	75	267	179	311	45	**326**	100	34	32	27	20	20	17	16		15	26	4	–	–	–	–	–	–	2	–	–	–	Benzeneacetic acid, 3-methoxy-4-[(trimethylsilyl)oxy]-, trimethylsilyl ester	37148-61-1	MS07713
73	209	326	75	179	45	311	267	**326**	100	27	22	16	15	14	13	11		15	26	4	–	–	–	–	–	–	2	–	–	–	Benzeneacetic acid, 3-methoxy-4-[(trimethylsilyl)oxy]-, trimethylsilyl ester	37148-61-1	NI34325
326	209	267	311	179	73	327	296	**326**	100	74	47	45	40	34	25	12		15	26	4	–	–	–	–	–	–	2	–	–	–	Benzeneacetic acid, 3-methoxy-4-[(trimethylsilyl)oxy]-, trimethylsilyl ester	37148-61-1	NI34326
209	73	147	210	45	75	74	283	**326**	100	99	21	18	14	9	9	7	**1.22**	15	26	4	–	–	–	–	–	–	2	–	–	–	Benzeneacetic acid, 3-methoxy-α-[(trimethylsilyl)oxy]-, trimethylsilyl ester	55530-67-1	NI34327

MASS TO CHARGE RATIOS								M.W.	INTENSITIES								Parent	C	H	O	N	S	F	Cl	Br	I	Si	P	B	X	COMPOUND NAME	CAS Reg No	No
209	73	147	283	267	178	179	163	**326**	100	70	68	24	11	10	8	7	**0.80**	15	26	4	–	–	–	–	–	–	2	–	–	–	**Benzeneacetic acid, 3-methoxy-α-[(trimethylsilyl)oxy]-, trimethylsilyl ester**	55530-67-1	**NI34328**
73	326	179	267	209	75	311	45	**326**	100	32	29	25	21	15	13	12		15	26	4	–	–	–	–	–	–	2	–	–	–	**Benzeneacetic acid, 4-methoxy-3-[(trimethylsilyl)oxy]-, trimethylsilyl ester**		**MS07714**
209	73	283	147	135	173	153	169	**326**	100	40	11	10	5	2	2	1	**0.02**	15	26	4	–	–	–	–	–	–	2	–	–	–	**Benzeneacetic acid, 4-methoxy-α-[(trimethylsilyl)oxy]-, trimethylsilyl ester**	55556-95-1	**NI34329**
209	73	210	147	75	283	77	45	**326**	100	50	17	14	12	8	8	6	**0.00**	15	26	4	–	–	–	–	–	–	2	–	–	–	**Benzeneacetic acid, 4-methoxy-α-[(trimethylsilyl)oxy]-, trimethylsilyl ester**	55556-95-1	**MS07715**
75	73	44	117	97	45	43	103	**326**	100	43	42	35	28	23	18	17	**0.00**	15	28	6	–	–	–	–	–	–	–	–	2	–	**α-D-Mannopyranoside, methyl, cyclic 2,3:4,6-bis(butylboronate)**	54400-84-9	**NI34330**
326	327	293	280	247	328	325	279	**326**	100	21	15	15	12	7	6	5		16	14	2	4	1	–	–	–	–	–	–	–	–	**N,N-Dimethyl-4-(6-nitro-2-benzothien-1-ylazo)aniline**		**IC03953**
104	326	150	177	108	122	74	134	**326**	100	74	74	54	30	25	24	21		16	14	2	4	1	–	–	–	–	–	–	–	–	**Urea, N-(4-methoxyphenyl)-N'-(5-phenyl-1,2,4-thiadiazol-3-yl)-**	42053-83-8	**NI34331**
278	232	326	291	309	279	233	231	**326**	100	59	47	43	32	28	14	13		16	14	4	4	–	–	–	–	–	–	–	–	–	**3-Buten-2-one, 4-phenyl-, (2,4-dinitrophenyl)hydrazone**	976-25-0	**NI34332**
118	102	130	76	51	129	90	103	**326**	100	46	44	43	40	38	35	30	**13.00**	16	14	4	4	–	–	–	–	–	–	–	–	–	**2,2'-Diformyloxanilide dioxime**		**LI07214**
147	148	91	164	163	326			**326**	100	42	18	13	11	4				16	22	7	–	–	–	–	–	–	–	–	–	–	**β-D-Glucopyranoside, 3-(4-methoxyphenyl)-2-propenyl-**		**MS07716**
115	197	145	253	143	225	281	116	**326**	100	85	55	49	46	46	42	39	**32.76**	16	23	5	–	–	–	–	–	–	–	1	–	–	**Propenoic acid, 3-(3-methylphenyl)-2-diethoxyphosphinyl-, ethyl ester**	107846-67-3	**NI34333**
73	253	151	75	136	237	326	311	**326**	100	61	53	36	34	28	18	16		16	26	5	–	–	–	–	–	–	1	–	–	–	**2,3-Dimethoxyphenyllactic acid ethyl ester TMS**		**MS07717**
151	73	75	236	152	326	253	45	**326**	100	35	14	8	8	7	6	6		16	26	5	–	–	–	–	–	–	1	–	–	–	**3-(3,4-Dimethoxyphenyl)lactic acid ethyl ester TMS**		**MS07718**
253	73	225	254	166	197	75	223	**326**	100	22	20	19	12	9	9	8	**4.40**	16	26	5	–	–	–	–	–	–	1	–	–	–	**1-(4-Hydroxy-3-methoxyphenyl)-1-ethoxyacetic acid ethyl ester TMS**		**MS07719**
206	73	326	205	236	179	207	209	**326**	100	78	47	46	35	29	24	23		16	30	3	–	–	–	–	–	–	2	–	–	–	**3-Vanillylpropanol, bis(trimethylsilyl)-**		**NI34334**
206	73	326	205	236	179	209	311	**326**	100	80	70	49	39	30	26	25		16	30	3	–	–	–	–	–	–	2	–	–	–	**3-Vanillylpropanol, bis(trimethylsilyl)-**		**NI34335**
326	328	189	197	269	175	267	297	**326**	100	95	65	32	30	28	25	24		17	11	2	–	–	–	–	1	–	–	–	–	–	**Spiro[4H-cyclopenta[def]phenanthrene-4,2'-[1,3]dioxolane], 8-bromo-**	72361-07-0	**NI34336**
235	91	159	131	117	103	77	326	**326**	100	32	15	8	6	5	4	3		17	18	3	4	–	–	–	–	–	–	–	–	–	**7-[N'-(N'-Ethyl-1'-phenyl-2'-propylamino)]-4-nitrobenzo-2,1,3-oxadiazole**		**LI07215**
326	171	325	155	327	172	156	102	**326**	100	82	49	20	19	16	16	16		17	18	3	4	–	–	–	–	–	–	–	–	–	**Pyrido[1,2-a]benzimidazolium, 1,2,3,4-tetrahydro-5-(1,2,3,4-tetrahydro-6-hydroxy-1,3-dimethyl-2,4-dioxo-5-pyrimidinyl)-**	27356-04-3	**NI34337**
70	112	113	140	41	43	29	28	**326**	100	71	27	12	10	9	8	5	**2.19**	17	30	4	2	–	–	–	–	–	–	–	–	–	**N-(N-Acetylprolyl)isoleucine butyl ester**		**AI02052**
70	112	113	140	43	41	71	29	**326**	100	80	34	15	9	9	5	5	**3.04**	17	30	4	2	–	–	–	–	–	–	–	–	–	**N-(N-Acetylprolyl)leucine butyl ester**		**AI02053**
73	129	75	131	45	130	74	157	**326**	100	67	21	18	13	8	8	7	**0.00**	17	34	2	–	–	–	–	–	–	2	–	–	–	**1,5-cis,8-cis-Undecatriene-3,7-diol bis(trimethylsilyl)ether**		**MS07720**
326	325	309	127	189	165	162	291	**326**	100	93	67	52	41	41	41	40		18	11	4	–	–	–	1	–	–	–	–	–	–	**trans-6-Carboxy-2-(4-chlorostyryl)chromone**		**MS07721**
326	135	121	325	327	262	295	328	**326**	100	65	61	46	29	22	20	19		18	14	–	–	3	–	–	–	–	–	–	–	–	**[1,2]Dithiolo[1,5-b][1,2]dithiole-7-S(IV), 2-(4-methylphenyl)-5-phenyl-**	38443-42-4	**NI34338**
326	170	171	141	169	142	149	77	**326**	100	94	54	53	51	51	49	45		18	14	4	–	1	–	–	–	–	–	–	–	–	**2-Hydroxy-4'-phenoxydiphenylsulphone**	58194-85-7	**NI34339**
326	186	217	185	141	77	129	218	**326**	100	78	77	73	65	52	51	49		18	14	4	–	1	–	–	–	–	–	–	–	–	**4-Hydroxy-4'-phenoxydiphenylsulphone**		**NI34340**
242	44	28	243	42	41	284	326	**326**	100	98	28	19	19	11	10	2		18	14	6	–	–	–	–	–	–	–	–	–	–	**6H-Dibenzo[b,d]pyran-6-one, 3,7-diacetoxy-4-methoxy-1-methyl-**		**LI07216**
143	144	135	103	76	295	104	75	**326**	100	11	8	7	5	4	4	3	**1.09**	18	14	6	–	–	–	–	–	–	–	–	–	–	**4,4'-Bis(methoxycarbonyl)benzil**		**IC03955**
163	135	103	164	295	76	77	75	**326**	100	15	11	8	3	3	2	2	**0.00**	18	14	6	–	–	–	–	–	–	–	–	–	–	**4,4'-Bis(methoxycarbonyl)benzil**		**IC03954**
163	135	105	164	103	76	104	295	**326**	100	14	11	9	9	6	5	4	**0.20**	18	14	6	–	–	–	–	–	–	–	–	–	–	**4,4'-Bis(methoxycarbonyl)benzil**		**IC03956**
163	267	295	326	103	164	75	132	**326**	100	68	30	27	27	21	18	16		18	14	6	–	–	–	–	–	–	–	–	–	–	**Methyl 1,3-dihydro-3-[4-(methoxycarbonyl)phenyl]-1-oxoisobenzofuran-6 carboxylate**		**NI34341**
311	326	295	312	310	158.5	94	56	**326**	100	20	20	19	11	9	6	4		18	14	6	–	–	–	–	–	–	–	–	–	–	**2H,6H-Pyrano[3,2-b]xanthen-6-one, 5,9,10-trihydroxy-2,2-dimethyl-**		**MS07722**
145	132	146	104	193	144	117	195	**326**	100	99	35	25	20	20	18	7	**1.00**	18	15	2	2	–	–	1	–	–	–	–	–	–	**3-Chloro-N-[(1,2,3,4-tetrahydroisoquinolin-2-yl)methyl]phthalimide**	96919-06-1	**NI34342**
326	77	325	65	94	51	170	39	**326**	100	80	75	50	33	31	30	28		18	15	4	–	–	–	–	–	–	–	1	–	–	**Triphenyl phosphate**	115-86-6	**IC03957**
326	325	77	65	170	94	233	327	**326**	100	71	57	29	25	22	20	19		18	15	4	–	–	–	–	–	–	–	1	–	–	**Triphenyl phosphate**	115-86-6	**IC03958**
326	325	77	65	170	94	51	215	**326**	100	84	61	35	22	20	19	19		18	15	4	–	–	–	–	–	–	–	1	–	–	**Triphenyl phosphate**	115-86-6	**NI34343**
40	326	325	327	257	191	311	190	**326**	100	67	27	15	15	14	11	11		18	18	2	2	1	–	–	–	–	–	–	–	–	**(E)-9,10-Dimethoxy-4,5,6,7-tetrahydrothieno[3,2-e][3]benzazecine-5-carbonitrile**		**MS07723**
326	105	281	180	132	282	221	225	**326**	100	99	44	44	40	24	10	8		18	18	2	2	1	–	–	–	–	–	–	–	–	**5-Ethoxy-3-methyl-1,5-diphenyl-2-thioxo-4-imidazolidinone**	116927-98-1	**DD01259**
282	105	180	326	121	148	162	249	**326**	100	60	20	16	12	10	8	6		18	18	2	2	1	–	–	–	–	–	–	–	–	**5-Ethoxy-3-methyl-1,5-diphenyl-4-thioxo-2-imidazolidinone**	116928-03-1	**DD01260**
326	170	168	327	154	171	217	127	**326**	100	89	19	17	14	13	12	12		18	18	2	2	1	–	–	–	–	–	–	–	–	**1-Naphthalenesulphonamide, 5-(dimethylamino)-N-phenyl-**	34532-47-3	**NI34344**
326	170	327	168	171	169	328	174	**326**	100	87	20	20	15	9	8	3		18	18	2	2	1	–	–	–	–	–	–	–	–	**1-Naphthalenesulphonamide, 5-(dimethylamino)-N-phenyl-**	34532-47-3	**NI34345**
297	326	298	241	279	148.5	270	29	**326**	100	57	20	18	17	17	15	14		18	18	4	2	–	–	–	–	–	–	–	–	–	**5-Amino-1-sec-butylamino-4,8-dihydroxyanthraquinone**		**IC03959**
326	311	325	294	268	293	312	163	**326**	100	52	20	16	13	11	10	10		18	18	4	2	–	–	–	–	–	–	–	–	–	**1,4-Dihydroxy-5,8-bis(ethylamino)anthraquinone**		**IC03960**
167	168	152	165	169	166	194	212	**326**	100	94	72	71	63	63	55	53	**0.00**	18	18	4	2	–	–	–	–	–	–	–	–	–	**Diphenylacetylglycylglycine**	65707-76-8	**NI34346**
326	228	269	283	310	270	77	325	**326**	100	48	47	46	40	38	37	36		18	19	–	4	–	–	1	–	–	–	–	–	–	**7-Chloro-2-butylamino-5-phenyl-3H-1,3,4-benzotriazepine**		**MS07724**
326	283	121	327	56	185	213	255	**326**	100	60	23	20	18	12	11	10		18	22	2	–	–	–	–	–	–	–	–	–	1	**1,1'-Dibutrylferrocene**		**IC03961**
123	326	59	298	198	70	185	115	**326**	100	58	33	16	12	12	10	10		18	22	2	4	–	–	–	–	–	–	–	–	–	**Morphinan, 6-azido-4,5-epoxy-3-methoxy-17-methyl-, (5α,6β)-**	22958-08-3	**NI34347**
169	142	57	43	41	29	326	171	**326**	100	85	61	54	54	48	44	40		18	27	3	–	–	–	1	–	–	–	–	–	–	**Propanoic acid, 2-(4-chloro-2-methylphenoxy)-, isooctyl ester**	28473-03-2	**NI34348**

MASS TO CHARGE RATIOS								M.W.	INTENSITIES								Parent	C	H	O	N	S	F	Cl	Br	I	Si	P	B	X	COMPOUND NAME	CAS Reg No	No
121	326	193	161	133	134	311	297	**326**	100	45	40	40	35	25	18	18		19	18	5	–	–	–	–	–	–	–	–	–	–	**Benzaldehyde, 2,4-dimethoxy-5-[3-(4-methoxyphenyl)-1-oxo-2-propenyl]-, (E)-**	65621-11-6	**NI34349**
282	325	283	326	281	141	140	267	**326**	100	72	21	17	17	17	10	8		19	18	5	–	–	–	–	–	–	–	–	–	–	**5-Benzofuranpropanol, 2-(1,3-benzodioxol-5-yl)-7-methoxy-**		**MS07725**
242	284	326	181	84	58	55	69	**326**	100	36	11	8	7	7	7	5		19	18	5	–	–	–	–	–	–	–	–	–	–	**4′,5′-Diacetoxy-3′-methoxystilbene**		**NI34350**
121	263	65	292	189	265	131	77	**326**	100	34	30	27	24	23	23	20	**0.00**	19	18	5	–	–	–	–	–	–	–	–	–	–	**3-[α-(2-Hydroxypropyl)benzyl]-4-hydroxycoumarin**		**LI07217**
95	176	189	256	326	282	91	77	**326**	100	57	52	30	23	20	20	18		19	18	5	–	–	–	–	–	–	–	–	–	–	**Spiro[furan-3(2H),6′-[6H]naphtho[1,8-bc]furan]-2,2′(4′H)-dione, 5-(3-furanyl)-3′,5′,5′a,7′,8′,8′a-hexahydro-7′-methyl-, [5′aS-(5′aα,6′β,7′α,8′aα)]-**	66322-19-8	**NI34351**
269	270	91	271	272	241	57	242	**326**	100	47	44	39	15	15	15	12	**3.00**	19	19	1	2	–	–	1	–	–	–	–	–	–	**1-tert-Butyl-7-chloro-1,3-dihydro-5-phenyl-2H-1,4-benzodiazepin-2-one**		**MS07726**
298	256	241	271	269	243	255	283	**326**	100	99	61	57	56	50	49	44	**40.49**	19	19	1	2	–	–	1	–	–	–	–	–	–	**8-Chloro-1,2,3,4-tetrahydro-6-phenyl-1-propyl-1,5-benzodiazocin-2-one**	63563-56-4	**NI34352**
103	131	104	136	109	91	326	178	**326**	100	97	92	83	77	75	24	20		19	22	1	2	1	–	–	–	–	–	–	–	–	**1,5-Benzothiazepin-4(5H)-one, 5-[2-(dimethylamino)ethyl]-2,3-dihydro-2-phenyl-**	5845-26-1	**NI34353**
71	72	58	59	42	131	103	57	**326**	100	50	50	21	16	11	11	11	**1.27**	19	22	1	2	1	–	–	–	–	–	–	–	–	**1,5-Benzothiazepin-4(5H)-one, 5-[2-(dimethylamino)ethyl]-2,3-dihydro-2-phenyl-**	5845-26-1	**NI34354**
58	197	86	42	43	44	196	85	**326**	100	17	16	13	11	8	7	7	**5.00**	19	22	1	2	1	–	–	–	–	–	–	–	–	**10H-Phenothiazine, 2-acetyl-10-[3-(dimethylamino)propyl]-**	61-00-7	**NI34356**
327	89	61	326	86	328	84	85	**326**	100	76	71	39	37	27	11	10		19	22	1	2	1	–	–	–	–	–	–	–	–	**10H-Phenothiazine, 2-acetyl-10-[3-(dimethylamino)propyl]-**	61-00-7	**NI34355**
58	86	42	43	85	326	44	197	**326**	100	14	13	10	8	7	5	4		19	22	1	2	1	–	–	–	–	–	–	–	–	**10H-Phenothiazine, 2-acetyl-10-[3-(dimethylamino)propyl]-**	61-00-7	**NI34357**
283	60	326	208	284	196	154	169	**326**	100	33	25	23	22	18	18	14		19	22	3	2	–	–	–	–	–	–	–	–	–	**Azecino[4,5,6-cd]indole-11-carboxylic acid, 6-acetyl-2,6,7,10,11,12-hexahydro-8-methyl-, methyl ester**	55702-36-8	**NI34358**
326	327	222	249	251	264	168	167	**326**	100	21	17	15	12	10	9	9		19	22	3	2	–	–	–	–	–	–	–	–	–	**1H-12a,5-(Epoxymethano)pyrido[3′,4′:5,6]cyclohept[1,2-b]indole-14-one, 2,3,4,4a,5,6,11,12-octahydro-5-(hydroxymethyl)-12-methyl-**	3329-95-1	**NI34359**
127	55	284	142	115	170	183	99	**326**	100	98	77	72	64	63	60	55	**0.00**	19	22	3	2	–	–	–	–	–	–	–	–	–	**trans-5-Ethyl-6,6-(ethylenedioxy)-2-oxo-1,2,5,6,7,7a-hexahydro-4H-pyrido[1′,2′:1,2]pyrazino[4,3-a]indole**		**DD01261**
178	326	241	121	122	270	324	226	**326**	100	76	57	53	44	27	19	19		19	22	3	2	–	–	–	–	–	–	–	–	–	**2,4-Bis(p-methoxyphenyl)-5,5-dimethylpyrazolidin-3-one**	79481-64-4	**NI34360**
239	240	55	155	154	144	105	41	**326**	100	20	16	4	4	4	4	4	**3.00**	19	22	3	2	–	–	–	–	–	–	–	–	–	**4-(4-Oxo-1,2,3,4,6,7,12,12b-octahydropyrido[2,1-a]-β-carbolin-12b-yl)butanoic acid**		**MS07727**
155	187	166	194	156	211	295	199	**326**	100	35	32	20	18	10	7	7	**0.00**	19	34	4	–	–	–	–	–	–	–	–	–	–	**9,10-12,13-Diepoxy-octadecanoic acid methyl ester**		**LI07218**
171	181	141	203	139	111	228	113	**326**	100	84	58	49	41	35	32	27	**22.00**	19	34	4	–	–	–	–	–	–	–	–	–	–	**8,10-Diketomalvalic acid methyl ester**		**LI07219**
55	41	67	81	294	262	295	276	**326**	100	67	59	55	26	26	19	13	**7.77**	19	34	4	–	–	–	–	–	–	–	–	–	–	**Dimethyl heptadec-8-enedioate**		**IC03962**
73	75	117	311	55	129	41	145	**326**	100	94	64	53	51	50	40	29	**7.55**	19	38	2	–	–	–	–	–	–	1	–	–	–	**Palmitelaidic acid, trimethylsilyl ester**		**NI34361**
73	75	55	117	41	129	43	311	**326**	100	93	56	54	47	38	30	28	**3.90**	19	38	2	–	–	–	–	–	–	1	–	–	–	**Palmitoleic acid, trimethylsilyl ester**		**NI34362**
282	310	326	311	283	281	241	240	**326**	100	96	47	29	22	16	16	16		20	14	1	4	–	–	–	–	–	–	–	–	–	**Pyridazino[4,3-f:5,6-f′]biquinaldine-3-oxide**		**NI34363**
266	251	282	326	235				**326**	100	18	12	9	5					20	22	4	–	–	–	–	–	–	–	–	–	–	**1-Acetoxy-4-(4-methoxyphenyl)-6-methoxy-1,2,3,4-tetrahydronaphthalene**		**LI07220**
311	326	312	327	325	313	253	283	**326**	100	34	22	8	4	4	4	3		20	22	4	–	–	–	–	–	–	–	–	–	–	**2H,8H-Benzo[1,2-b:5,4-b′]dipyran-2-one, 5-methoxy-8,8-dimethyl-10-(3-methyl-2-butenyl)-**		**MS07728**
191	149	204	326	271	123	260	135	**326**	100	99	90	73	27	27	14	9		20	22	4	–	–	–	–	–	–	–	–	–	–	**[3,6′-Bi-2H-1-benzopyran]-7,7′-diol, 3,3′,4,4′-tetrahydro-2′,2′-dimethyl-**	41347-51-7	**MS07729**
297	121	131	298	122	237	282	95	**326**	100	79	68	55	48	42	36	36	**28.78**	20	22	4	–	–	–	–	–	–	–	–	–	–	**1-Dehydroaldosterone-γ-lactone**		**NI34364**
326	164	327	149	77	103	137	91	**326**	100	61	23	21	19	17	15	14		20	22	4	–	–	–	–	–	–	–	–	–	–	**Dehydrodieugenol**	4433-08-3	**MS07730**
326	164	327	165	299	311	283	297	**326**	100	62	21	11	9	8	7	6		20	22	4	–	–	–	–	–	–	–	–	–	–	**Dehydrodieugenol**	4433-08-3	**NI34365**
326	327	311	203	137	149	151	164	**326**	100	23	9	8	8	7	6	5		20	22	4	–	–	–	–	–	–	–	–	–	–	**Dehydrodiisoeugenol**	2680-81-1	**MS07731**
326	309	327	151	311	137	202	149	**326**	100	18	17	11	10	10	8	7		20	22	4	–	–	–	–	–	–	–	–	–	–	**Dehydrodiisoeugenol**	2680-81-1	**NI34366**
326	327	151	137	311	202	149		**326**	100	17	11	10	9	8	6			20	22	4	–	–	–	–	–	–	–	–	–	–	**Dehydrodiisoeugenol**	2680-81-1	**LI07221**
326	311	137	202	163	151	189	283	**326**	100	6	6	5	4	4	2	1		20	22	4	–	–	–	–	–	–	–	–	–	–	**(2R,3R)-2,3-Dihydro-2-(4′-hydroxy-3′-methoxyphenyl)-7-methoxy-3-methyl-5-(E)-propenylbenzofuran**		**MS07732**
193	294	326	207	178	208	295	165	**326**	100	65	61	45	44	41	35	28		20	22	4	–	–	–	–	–	–	–	–	–	–	**(±)-3,3-Dimethoxy-9,11,13,15-tetramethyl-4-oxatricyclo[8.5.0.0^{2,6}]pentadeca-2(6),7,9,11,13,15-hexaen-5-one**		**DD01262**
326	235	236	267	207	165	251	208	**326**	100	79	58	46	46	46	45	41		20	22	4	–	–	–	–	–	–	–	–	–	–	**(±)-5,5-Dimethoxy-9,11,13,15-tetramethyl-4-oxatricyclo[8.5.0.0^{2,6}]pentadeca-2(6),7,9,11,13,15-hexaen-3-one**		**DD01263**
284	212	173	213	172	285	266	211	**326**	100	50	41	31	29	25	25	20	**14.28**	20	22	4	–	–	–	–	–	–	–	–	–	–	**14α,15α-Epoxyestrone-3-acetate**		**NI34367**
105	79	107	163	164	325	220	106	**326**	100	77	66	50	48	30	28	26	**19.00**	20	22	4	–	–	–	–	–	–	–	–	–	–	**Galactitol, 2,3:4,5-di-O-benzylidene-1,6-dideoxy-**	19342-56-4	**LI07222**
163	91	105	107	79	164	325	220	**326**	100	55	45	30	18	11	7	6	**4.30**	20	22	4	–	–	–	–	–	–	–	–	–	–	**Galactitol, 2,3:4,5-di-O-benzylidene-1,6-dideoxy-**	19342-56-4	**NI34368**
105	82	77	54	67	123	51	106	**326**	100	43	35	21	16	9	9	9	**0.80**	20	22	4	–	–	–	–	–	–	–	–	–	–	**1,6-Hexanediol dibenzoate**	22915-73-7	**NI34369**
311	326	293	173	280	265	281	121	**326**	100	85	72	52	49	49	31	28		20	22	4	–	–	–	–	–	–	–	–	–	–	**3,3-Bis(3-methoxyphenyl)-4,4-dimethylcyclopropanecarboxylic acid**		**NI34370**
135	326	191	163	164	162	77	107	**326**	100	54	39	29	26	16	14	10		20	22	4	–	–	–	–	–	–	–	–	–	–	**1-(4′-Methoxyphenyl)-2-[2″-methoxy-4″-(E)-propenylphenoxy]propan-1-one**		**MS07733**
284	285	240	227	228	270	173	186	**326**	100	20	11	11	9	8	7	6	**6.03**	20	22	4	–	–	–	–	–	–	–	–	–	–	**6-Oxoestrone-3-acetate**		**NI34371**
175	132	123	326	165	95	188	138	**326**	100	54	40	33	20	18	8	6		20	23	1	2	–	1	–	–	–	–	–	–	–	**1-Butanone, 1-(4-fluorophenyl)-4-(4-phenyl-1-piperazinyl)-**	2354-61-2	**NI34373**

MASS TO CHARGE RATIOS	M.W.	INTENSITIES	Parent	C	H	O	N	S	F	Cl	Br	I	Si	P	B	X	COMPOUND NAME	CAS Reg No	No
175 188 132 123 70 326 189 56	326	100 76 53 40 34 33 22 21		20	23	1	2	–	1	–	–	–	–	–	–	–	1-Butanone, 1-(4-fluorophenyl)-4-(4-phenyl-1-piperazinyl)-	2354-61-2	NI34372
175 188 132 123 326 70 105 104	326	100 75 52 40 34 34 32 23		20	23	1	2	–	1	–	–	–	–	–	–	–	1-Butanone, 1-(4-fluorophenyl)-4-(4-phenyl-1-piperazinyl)-	2354-61-2	LI07223
144 182 183 145 326 131 157 158	326	100 64 54 53 47 41 38 36		20	26	2	2	–	–	–	–	–	–	–	–	–	Ajmalan-17,21-diol, (17R,21α)-	4360-12-7	NI34374
110 42 41 83 47 43 85 55	326	100 91 82 81 53 43 38 35	13.00	20	26	2	2	–	–	–	–	–	–	–	–	–	Aspidofractinin-3-ol, 17-methoxy-, (2α,5α)-	55724-65-7	NI34375
144 143 69 130 326 168 44 41	326	100 91 91 83 75 74 72 72		20	26	2	2	–	–	–	–	–	–	–	–	–	Burnamicine	2134-96-5	NI34376
326 325 199 281 186 200 69 71	326	100 98 98 81 75 74 55 39		20	26	2	2	–	–	–	–	–	–	–	–	–	Corynan-17-ol, 18,19-didehydro-10-methoxy-	56053-12-4	NI34377
196 144 130 197 326 143 199 133	326	100 60 48 39 38 32 28 10		20	26	2	2	–	–	–	–	–	–	–	–	–	Curan-17-oic acid, methyl ester, (16α)-	7267-97-2	NI34378
166 326 160 174 173 309 295 281	326	100 84 35 29 27 9 8 4		20	26	2	2	–	–	–	–	–	–	–	–	–	Curan-17-ol, 19,20-didehydro-10-methoxy-, (19E)-	59630-35-2	NI34379
69 73 83 81 71 60 124 97	326	100 52 44 43 41 35 34 34	4.00	20	26	2	2	–	–	–	–	–	–	–	–	–	Dasycarpidan-1-methanol, acetate	55724-48-6	NI34380
279 238 308 31 280 307 29 42	326	100 64 55 24 23 22 22 21	18.20	20	26	2	2	–	–	–	–	–	–	–	–	–	Epivincaminol		MS07734
279 238 308 326 267 295	326	100 64 58 18 14 8		20	26	2	2	–	–	–	–	–	–	–	–	–	Epivincaminol		LI07224
326 309 138 202 122 311 327 189	326	100 48 40 32 30 25 23 21		20	26	2	2	–	–	–	–	–	–	–	–	–	9H-Ibogaine, 9-hydroxy-	20454-39-1	NI34381
326 309 138 202 122 311 327 203	326	100 47 40 32 30 25 22 21		20	26	2	2	–	–	–	–	–	–	–	–	–	9H-Ibogaine, 9-hydroxy-	20454-39-1	LI07225
326 150 122 138 327 151 108 109	326	100 46 42 35 24 23 22 17		20	26	2	2	–	–	–	–	–	–	–	–	–	Iboluteine	468-11-1	NI34382
326 150 122 138 327 151 136 108	326	100 45 41 34 23 22 22 21		20	26	2	2	–	–	–	–	–	–	–	–	–	Iboluteine	468-11-1	LI07226
225 226 43 46 224 112 105 223	326	100 18 12 7 6 6 6 5	1.00	20	26	2	2	–	–	–	–	–	–	–	–	–	Methyl 4-(1,2,3,4,6,7,12,12b-octahydropyrido[2,1-a]-β-carbolin-12b-yl)butanoate		MS07735
184 326 325 156 170 169 185 171	326	100 58 47 28 23 20 17 17		20	26	2	2	–	–	–	–	–	–	–	–	–	18,19-Secoyohimban-16-methanol, 16,17-didehydro-19-hydroxy-, (15β,16Z,20ξ)-	5523-47-7	NI34383
279 238 308 326 267 42 325 208	326	100 66 55 54 51 31 29 25		20	26	2	2	–	–	–	–	–	–	–	–	–	Vincaminol	3382-95-4	MS07736
279 238 326 308 267 295	326	100 68 56 55 54 18		20	26	2	2	–	–	–	–	–	–	–	–	–	Vincaminol	3382-95-4	LI07227
163 164 108 70 119 92 135 80	326	100 12 5 5 4 4 2 2	0.00	20	30	–	4	–	–	–	–	–	–	–	–	–	meso-N,N,N',N'-Tetraethyl-1,2-bis(2-pyridyl)ethylenediamine		DD01264
163 164 92 70 119 108 80 42	326	100 14 7 7 6 6 4 4	0.00	20	30	–	4	–	–	–	–	–	–	–	–	–	(±)-N,N,N',N'-Tetraethyl-1,2-bis(2-pyridyl)ethylenediamine		DD01265
267 95 69 81 55 83 90 59	326	100 97 90 84 79 78 66 66	50.80	20	38	3	–	–	–	–	–	–	–	–	–	–	Cyclopropanedecanoic acid, 2-hexyl-α-hydroxy-, methyl ester		MS07737
155 43 55 57 98 29 71 60	326	100 78 68 62 58 58 44 44	0.00	20	38	3	–	–	–	–	–	–	–	–	–	–	Dodecanoic anhydride		LI07228
191 95 109 177 123 235 121 137	326	100 90 70 60 55 50 50 40	0.00	20	38	3	–	–	–	–	–	–	–	–	–	–	12(R)-8,12,15-labdanetriol		MS07738
191 95 177 109 123 235 121 137	326	100 85 66 61 51 46 44 39	0.00	20	38	3	–	–	–	–	–	–	–	–	–	–	12(S)-8,12,15-labdanetriol		MS07739
116 117 43 129 57 55 41 69	326	100 34 31 23 23 21 19 12	4.80	20	38	3	–	–	–	–	–	–	–	–	–	–	Nonadecanoic acid, 3-oxo-, methyl ester	55334-37-7	WI01417
43 55 41 74 58 69 98 57	326	100 73 53 52 52 48 45 40	12.00	20	38	3	–	–	–	–	–	–	–	–	–	–	Nonadecanoic acid, 18-oxo-, methyl ester	54725-53-0	NI34384
55 41 43 69 264 67 83 57	326	100 88 76 57 47 42 40 38	5.04	20	38	3	–	–	–	–	–	–	–	–	–	–	9-Octadecenoic acid, 2-hydroxyethyl ester, (Z)-	4500-01-0	NI34385
183 326 185 108 325 152 262 165	326	100 47 37 15 13 12 3 3		21	15	–	2	–	–	–	–	–	–	1	–	–	(Dicyanomethylene)triphenylphosphorane		LI07229
189 190 105 91 117 131 115 77	326	100 13 9 9 3 2 2 2	0.00	21	26	1	–	1	–	–	–	–	–	–	–	–	2,4,6-Triethylbenzothioic acid, S-[2-(1-phenylpropyl)] ester		NI34386
284 269 326 285 173 174 160 187	326	100 36 34 25 18 15 14 11		21	26	3	–	–	–	–	–	–	–	–	–	–	Estra-1,3,5(10),15-tetraen-17β-ol, 3-methoxy-, acetate	19590-59-1	NI34387
213 326 214 137 325 105 43 41	326	100 71 60 37 28 28 25 21		21	26	3	–	–	–	–	–	–	–	–	–	–	2-Hydroxy-4-octoxybenzophenone	1843-05-6	IC03964
213 214 137 43 326 41 105 77	326	100 47 36 29 28 26 25 20		21	26	3	–	–	–	–	–	–	–	–	–	–	2-Hydroxy-4-octoxybenzophenone	1843-05-6	IC03963
213 214 137 105 326 325 43 41	326	100 39 38 24 19 14 11 10		21	26	3	–	–	–	–	–	–	–	–	–	–	2-Hydroxy-4-octoxybenzophenone	1843-05-6	NI34388
105 110 77 43 41 106 57 55	326	100 15 14 9 9 7 6 6	4.39	21	26	3	–	–	–	–	–	–	–	–	–	–	3-Octoxyphenyl benzoate		IC03965
243 135 84 69 242 268 41 43	326	100 96 53 45 37 27 26 22	3.00	21	26	3	–	–	–	–	–	–	–	–	–	–	Oxetane, 2,2-bis(4-methoxyphenyl)-3,3,4,4-tetramethyl-	56440-24-5	NI34389
41 91 77 53 69 79 43 39	326	100 95 60 57 51 45 40 33	30.03	21	26	3	–	–	–	–	–	–	–	–	–	–	Podocarpa-1,12-diene-δ^{14},α-acetic acid, 7-hydroxy-8,13-dimethyl-3-oxo-, δ-lactone	5195-83-5	NI34390
43 326 121 239 160 159 147 41	326	100 45 22 18 17 17 14 14		21	26	3	–	–	–	–	–	–	–	–	–	–	Pregna-1,4-diene-3,11,20-trione	4368-11-0	NI34391
135 121 255 136 57 256 326 41	326	100 95 58 13 13 12 11 10		21	26	3	–	–	–	–	–	–	–	–	–	–	4-(1,1,3,3-Tetramethylbutyl)octyl salicylate		IC03966
326 327 218 231 311 244 205 93	326	100 29 28 26 19 16 12 12		21	30	1	–	–	–	–	–	–	1	–	–	–	Estra-1,3,5(10),16-tetraene, 3-[(trimethylsilyl)oxy]-	69688-27-3	NI34392
124 326 125 174 152 327 163 159	326	100 24 9 6 6 5 4 3		21	30	1	2	–	–	–	–	–	–	–	–	–	Aspidospermidine, 17-methoxy-1-methyl-	55400-14-1	NI34393
121 168 308 172 158 227 122 41	326	100 57 33 28 19 15 14 14	9.00	21	30	1	2	–	–	–	–	–	–	–	–	–	Curan-17-ol, 1-ethyl-, (16α)-	56053-23-7	NI34394
182 98 42 183 144 56 41 213	326	100 37 21 15 12 12 12 10	2.82	21	30	1	2	–	–	–	–	–	–	–	–	–	4,21-Secoajmalan-17-ol, 4-methyl-, (17R)-	56009-13-3	NI34395
103 57 43 85 55 83 69 41	326	100 75 55 49 39 36 35 33	2.20	21	42	2	–	–	–	–	–	–	–	–	–	–	Butanoic acid, 3-methyl-, hexadecyl ester		MS07740
115 28 69 43 41 57 55 45	326	100 52 44 29 25 21 19 18	0.94	21	42	2	–	–	–	–	–	–	–	–	–	–	1,3-Dioxane, 4,5-dimethyl-2-pentadecyl-	56599-32-7	NI34396
115 69 43 116 57 55 45 41	326	100 29 9 7 6 6 6 6	0.74	21	42	2	–	–	–	–	–	–	–	–	–	–	1,3-Dioxane, 4,6-dimethyl-2-pentadecyl-	56599-77-0	NI34397
115 69 43 57 45 116 55 41	326	100 29 13 8 8 7 7 7	0.84	21	42	2	–	–	–	–	–	–	–	–	–	–	1,3-Dioxane, 4,6-dimethyl-2-pentadecyl-	56599-77-0	NI34398
115 69 85 116 56 87 41 57	326	100 31 14 8 5 4 4 3	0.24	21	42	2	–	–	–	–	–	–	–	–	–	–	1,3-Dioxepane, 5-methyl-2-pentadecyl-	56599-34-9	NI34399
115 69 85 41 43 57 55 28	326	100 85 42 36 33 22 22 20	0.24	21	42	2	–	–	–	–	–	–	–	–	–	–	1,3-Dioxocane, 2-pentadecyl-	41583-11-3	NI34400
115 69 57 43 41 55 71 45	326	100 56 18 17 16 13 8 7	0.34	21	42	2	–	–	–	–	–	–	–	–	–	–	1,3-Dioxolane, 4-ethyl-4-methyl-2-pentadecyl-	56599-78-1	NI34401
115 69 116 43 57 55 41 70	326	100 30 8 4 3 3 3 2	0.54	21	42	2	–	–	–	–	–	–	–	–	–	–	1,3-Dioxolane, 4-isopropyl-2-pentadecyl-	56599-35-0	NI34402
115 87 69 43 116 41 57 55	326	100 25 10 8 7 6 5 5	0.24	21	42	2	–	–	–	–	–	–	–	–	–	–	1,3-Dioxolane, 4,4,5-trimethyl-2-pentadecyl-	56599-79-2	NI34403
74 87 43 75 326 55 57 41	326	100 71 29 28 23 22 19 16		21	42	2	–	–	–	–	–	–	–	–	–	–	Eicosanoic acid, methyl ester	1120-28-1	WI01418

MASS TO CHARGE RATIOS								M.W.	INTENSITIES								Parent	C	H	O	N	S	F	Cl	Br	I	Si	P	B	X	COMPOUND NAME	CAS Reg No	No
74	43	87	41	55	57	75	69	**326**	100	64	56	37	33	30	27	19	**1.00**	21	42	2	–	–	–	–	–	–	–	–	–	–	Eicosanoic acid, methyl ester	1120-28-1	NI34404
74	87	43	75	55	326	41	57	**326**	100	67	37	27	24	23	21	21		21	42	2	–	–	–	–	–	–	–	–	–	–	Eicosanoic acid, methyl ester	1120-28-1	NI34405
88	101	326	43	157	57	55	41	**326**	100	79	55	48	33	33	31	26		21	42	2	–	–	–	–	–	–	–	–	–	–	Ethyl 9-methyloctadecanoate		MS07741
326	88	101	157	283	327	213	43	**326**	100	78	74	54	28	26	20	18		21	42	2	–	–	–	–	–	–	–	–	–	–	Ethyl 10-methyloctadecanoate		MS07742
43	57	73	60	55	41	71	69	**326**	100	78	74	64	64	61	38	37	**35.14**	21	42	2	–	–	–	–	–	–	–	–	–	–	Heneicosanoic acid	2363-71-5	NI34406
101	74	43	57	75	55	69	41	**326**	100	58	32	29	22	22	19	16	**5.51**	21	42	2	–	–	–	–	–	–	–	–	–	–	Hexadecanoic acid, 3,7,11,15-tetramethyl-, methyl ester	1118-77-0	NI34407
101	74	43	57	55	41	75	69	**326**	100	80	67	48	38	34	29	28	**0.79**	21	42	2	–	–	–	–	–	–	–	–	–	–	Hexadecanoic acid, 3,7,11,15-tetramethyl-, methyl ester	1118-77-0	NI34408
217	73	202	14	218	45			**326**	100	70	35	20	15	10			**0.00**	21	42	2	–	–	–	–	–	–	–	–	–	–	Hexadecanoic acid, 3,7,11,15-tetramethyl-, methyl ester	1118-77-0	NI34409
70	43	71	55	41	57	239	257	**326**	100	55	35	23	21	13	11	10	**8.00**	21	42	2	–	–	–	–	–	–	–	–	–	–	3-Methylbutyl hexadecanoate	81974-61-0	NI34410
74	87	43	57	326	55	143	41	**326**	100	81	70	54	53	52	49	43		21	42	2	–	–	–	–	–	–	–	–	–	–	Methyl 11-DL-methylnonadecanoate		LI07230
74	87	326	143	57	43	75	55	**326**	100	71	59	40	38	33	30	26		21	42	2	–	–	–	–	–	–	–	–	–	–	Methyl 10-methylnonadecanoate		MS07743
74	87	43	57	326	55	143	41	**326**	100	83	71	55	53	53	49	43		21	42	2	–	–	–	–	–	–	–	–	–	–	Methyl 11-methylnonadecanoate		MS07744
88	43	41	101	55	57	69	70	**326**	100	82	60	52	44	40	24	21	**0.00**	21	42	2	–	–	–	–	–	–	–	–	–	–	Nonadecanoic acid, ethyl ester	18281-04-4	NI34411
43	28	61	41	57	55	102	326	**326**	100	78	60	56	48	44	42	36		21	42	2	–	–	–	–	–	–	–	–	–	–	Octadecanoic acid, propyl ester	3634-92-2	MS07745
61	43	102	57	73	115	60	55	**326**	100	75	58	42	41	40	39	36	**35.34**	21	42	2	–	–	–	–	–	–	–	–	–	–	Octadecanoic acid, propyl ester	3634-92-2	NI34412
73	87	43	74	55	41	142	57	**326**	100	42	28	22	18	17	16	14	**0.22**	21	42	2	–	–	–	–	–	–	–	–	–	–	1-Propanol, 3-(1-octadecenyloxy)-	72347-46-7	NI34413
326	297	77	105	249	298	325	327	**326**	100	89	81	79	51	28	26	24		22	14	3	–	–	–	–	–	–	–	–	–	–	3-Benzoyl-4-phenylcoumarin		MS07746
326	297	77	105	249	298	325	221	**326**	100	88	82	80	52	28	27	24		22	14	3	–	–	–	–	–	–	–	–	–	–	3-Benzoyl-4-phenylcoumarin		LI07231
326	134	192	191	135	327	91	193	**326**	100	86	69	38	35	25	22	18		22	18	1	2	–	–	–	–	–	–	–	–	–	2H-1,5-Benzodiazepin-2-one, 1,3-dihydro-3-phenyl-4-benzyl-	56771-67-6	MS07747
118	326	77	165	269	166	327	119	**326**	100	49	31	29	24	18	15	11		22	18	1	2	–	–	–	–	–	–	–	–	–	1-Methyl-2,4,4-triphenyl-2-imidazolin-5-one	37927-96-1	NI34414
311	77	326	208	165	118	51	312	**326**	100	90	88	71	64	55	27	26		22	18	1	2	–	–	–	–	–	–	–	–	–	1-Methyl-2,5,5-triphenyl-2-imidazolin-4-one		NI34415
326	180	165	77	327	269	166	311	**326**	100	56	52	31	29	25	21	14		22	18	1	2	–	–	–	–	–	–	–	–	–	2,5,5-Triphenyl-4-methoxyimidazole	24133-93-5	NI34416
249	77	206	104	326	91			**326**	100	99	82	38	24	18				22	18	1	2	–	–	–	–	–	–	–	–	–	3,4,6-Triphenyl-2-oxo-1,2,3,4-tetrahydropyrimidine		LI07232
127	326	144	327	325	115	128	242	**326**	100	79	25	16	16	16	14	10		22	19	2	–	–	–	–	–	–	–	–	1	–	Boronic acid, ethyl-, di-2-naphthalenyl ester	61233-76-9	NI34417
326	57	311	120	327	41	148	312	**326**	100	48	36	29	25	19	13	10		22	30	2	–	–	–	–	–	–	–	–	–	–	3,3′-Di-tert-butyl-4,4′-dihydroxy-5,5′-dimethylbiphenyl		IC03967
326	311	57	327	312	134	283	148	**326**	100	74	42	23	17	13	12	10		22	30	2	–	–	–	–	–	–	–	–	–	–	5,5′-Di-tert-butyl-2,2′-dimethyldi-p-phenylol		IC03968
43	91	326	283	311	79	124	105	**326**	100	33	30	25	23	23	22	22		22	30	2	–	–	–	–	–	–	–	–	–	–	Cyclopropa[16,17]pregn-4-ene-3,20-dione, 3′,16-dihydro-, (16β)-	2205-92-7	NI34418
326	311	269	57	312	255	100	176	**326**	100	99	25	25	24	23	15	12		22	30	2	–	–	–	–	–	–	–	–	–	–	Di-O-isopropylidene-L-sorbose		LI07233
279	308	293	280	326	227	309	226	**326**	100	57	39	32	24	20	14	11		22	30	2	–	–	–	–	–	–	–	–	–	–	cis-DL-3-Methoxy-17a-ethyl-D-homooestra-1,3,5(10),8-tetraen-17a-ol		LI07234
227	326	213	327	226	228	225	308	**326**	100	80	44	23	21	17	17	14		22	30	2	–	–	–	–	–	–	–	–	–	–	trans-DL-3-Methoxy-17a-ethyl-D-homooestra-1,3,5(10),8-tetraen-17a-ol		LI07235
43	326	91	147	136	283	79	41	**326**	100	31	31	25	22	21	21	21		22	30	2	–	–	–	–	–	–	–	–	–	–	19-Norpregn-4-ene-3,20-dione, 10-vinyl-	13258-85-0	NI34419
202	284	213	199	43	200	326	185	**326**	100	88	61	54	48	44	35	35		22	30	2	–	–	–	–	–	–	–	–	–	–	Podocarpa-6,8,11,13-tetraen-12-ol, 13-isopropyl-, acetate	22160-86-7	NI34420
43	311	283	91	326	41	79	77	**326**	100	34	23	23	17	17	15	15		22	30	2	–	–	–	–	–	–	–	–	–	–	Pregna-4,16-diene-3,20-dione, 16-methyl-	13485-43-3	NI34421
70	257	258	326	71	57	42	41	**326**	100	65	14	5	4	4	3	3		22	34	–	2	–	–	–	–	–	–	–	–	–	23-Norcona-5,18(22)-dienin-3-amine, N-methyl-, (3β)-	6877-20-9	NI34422
83	82	69	97	96	71	68	95	**326**	100	95	90	75	66	61	52	46	**0.00**	22	46	1	–	–	–	–	–	–	–	–	–	–	1-Docosanol	661-19-8	WI01419
43	57	55	83	69	41	97	82	**326**	100	79	79	64	60	59	47	37	**0.00**	22	46	1	–	–	–	–	–	–	–	–	–	–	1-Docosanol	661-19-8	NI34423
43	55	57	41	97	83	69	71	**326**	100	89	84	77	76	71	61	43	**0.00**	22	46	1	–	–	–	–	–	–	–	–	–	–	1-Docosanol	661-19-8	WI01420
97	57	129	71	43	85	83	113	**326**	100	66	64	44	43	26	18	10	**0.00**	22	46	1	–	–	–	–	–	–	–	–	–	–	1-Methoxy-5,5,8,11,15-pentamethyl-hexadecane		LI07236
326	249	325	215	247				**326**	100	79	53	13	12					23	18	–	–	1	–	–	–	–	–	–	–	–	4H-Thiopyran, 2,4,6-triphenyl-	7584-36-3	NI34424
221	105	77	115	51	222	78	91	**326**	100	90	76	62	49	32	18	17	**4.00**	23	18	2	–	–	–	–	–	–	–	–	–	–	1,2-Dibenzoyl-3-phenylcyclopropane	29128-64-1	NI34425
141	115	326	139	40	127	128	155	**326**	100	7	5	3	2	1	1	1		23	18	2	–	–	–	–	–	–	–	–	–	–	1-Naphthalenemethyl naphthalene-2-acetate	86328-64-5	NI34426
191	91	192	189	326	165	208	190	**326**	100	47	45	39	36	26	23	22		23	18	2	–	–	–	–	–	–	–	–	–	–	9-Phenanthrenemethyl phenylacetate	92174-43-1	NI34427
205	167	326	165	77	65	51	327	**326**	100	50	40	20	15	12	6	5		23	18	2	–	–	–	–	–	–	–	–	–	–	Phenyl 2-methyl-4,4-diphenyl-2,3-butadienoate		MS07748
105	77	221	203	51	106	115	204	**326**	100	44	13	12	11	8	6	5	**3.70**	23	18	2	–	–	–	–	–	–	–	–	–	–	1,3,5-Triphenylpent-2-ene-1,5-dione	23800-57-9	NI34428
133	194	105	77	193	113	91	55	**326**	100	76	41	35	28	25	19	15	**2.00**	24	22	1	–	–	–	–	–	–	–	–	–	–	2-Benzoyl-1,1-diphenylcyclopentane		LI07237
171	157	326	172	167	143	129	128	**326**	100	44	35	25	14	14	14	14		24	38	–	–	–	–	–	–	–	–	–	–	–	Acenaphthylene, 6-dodecyl-1,2,2a,3,4,5-hexahydro-	55334-56-0	WI01421
135	136	190	134	79	163	93	107	**326**	100	13	12	12	11	10	10	6	**1.00**	24	38	–	–	–	–	–	–	–	–	–	–	–	DL-2,3-Di-1-adamantylbutane		NI34429
135	204	205	79	136	93	67	41	**326**	100	41	16	14	12	12	6	6	**0.00**	24	38	–	–	–	–	–	–	–	–	–	–	–	meso-2,3-Di-1-adamantylbutane		NI34430
91	105	117	92	65	77	129	128	**326**	100	35	15	15	15	14	13	13	**3.20**	25	26	–	–	–	–	–	–	–	–	–	–	–	Benzene, 1,1′-[3-(2-phenylethylidene)-1,5-pentanediyl]bis-	55334-57-1	WI01422
243	165	244	166	241	228	326	239	**326**	100	28	22	5	4	4	3	3		25	26	–	–	–	–	–	–	–	–	–	–	–	Cyclohexyltriphenylmethane	13619-64-2	NI34431
82	63	31	44	50	76	85	117	**327**	100	60	31	17	14	3	2	1	**0.00**	3	–	–	1	3	4	3	–	–	–	–	–	–	N-(Chlorodifluoromethylthio)(chlorodifluoromethyldithio)chloromethimine		NI34432
60	165	163	130	128	312	133	98	**327**	100	99	99	70	70	42	42	27	**0.00**	4	12	–	1	1	–	2	–	–	–	1	–	1	Phosphoramidothioic dichloride, methyl(trimethylstannyl)-	41006-32-0	NI34433

MASS TO CHARGE RATIOS								M.W.	INTENSITIES								Parent	C	H	O	N	S	F	Cl	Br	I	Si	P	B	X	COMPOUND NAME	CAS Reg No	No
166	69	78	177	175	96	95	113	327	100	64	30	26	26	10	10	9	0.00	5	3	–	1	–	9	–	1	–	–	–	–	–	3-[Bis(trifluoromethyl)]amino-2-bromo-1,1,1-trifluoropropane		LI07238
294	292	221	187	223	127	99	327	327	100	80	52	52	49	26	20	6		6	3	–	3	–	–	6	–	–	–	–	–	–	3,5,6,7,7,8-Hexachloro-5,6,7,8-tetrahydro-s-triazolo[4,3-a]pyridine		LI07239
294	108	296	292	110	298	119	117	327	100	71	64	62	46	21	17	17	4.00	6	3	–	3	–	–	6	–	–	–	–	–	–	1,3,5-Triazine, 2-methyl-4,6-bis(trichloromethyl)-	949-42-8	NI34434
331	329	90	327	170	168	333	250	327	100	99	44	34	33	33	30	29		6	4	–	1	–	–	–	3	–	–	–	–	–	Aniline, 2,4,6-tribromo-	147-82-0	NI34435
329	331	327	332	250	170	168	90	327	100	98	36	32	24	21	19	13		6	4	–	1	–	–	–	3	–	–	–	–	–	Aniline, 2,4,6-tribromo-	147-82-0	NI34436
94	96	31	36	29	30			327	100	98	97	46	43	40			2.62	7	4	2	1	–	–	1	2	–	–	–	–	–	Benzene, 2-chloro-1-(dibromomethyl)-4-nitro-	55956-19-9	NI34437
39	332	330	328	316	312	300	297	327	100	3	3	3	3	3	3	3	2.58	8	13	2	2	–	–	–	2	–	–	–	–	–	1H-Imidazol-1-yloxy, 4-(dibromomethyl)-2,5-dihydro-2,2,5,5-tetramethyl-, 3-oxide	51973-28-5	NI34438
80	82	67	39	79	81	123	219	327	100	98	57	45	43	40	31	28	5.10	8	13	2	2	–	–	–	2	–	–	–	–	–	1H-Imidazol-1-yloxy, 4-(dibromomethyl)-2,5-dihydro-2,2,5,5-tetramethyl-, 3-oxide	51973-28-5	NI34439
211	73	85	133	243	45	86	75	327	100	92	70	50	39	38	36	21	1.00	9	22	4	1	1	–	–	–	–	2	1	–	–	Phosphoric acid, 2-isothiocyanatoethyl bis(trimethylsilyl) ester	56051-85-5	NI34440
43	55	268	239	98	69	226	42	327	100	84	69	45	18	17	16	16	1.99	10	12	3	1	–	7	–	–	–	–	–	–	–	N-(Heptafluorobutyryl)-β-alanine propyl ester		MS07749
240	241	45	69	220	70	47	88	327	100	27	13	6	4	4	4	3	0.00	10	12	3	1	–	7	–	–	–	–	–	–	–	N-(Heptafluorobutyryl)-alanine propyl ester		MS07750
211	282	73	236	266	225	227	133	327	100	81	76	68	62	51	50	50	6.51	10	26	5	1	–	–	–	–	–	2	1	–	–	Phosphoric acid, 2-(ethoxyimino)ethyl bis(trimethylsilyl) ester	55334-45-7	NI34441
102	73	298	147	312	211	207	75	327	100	64	40	12	10	10	9	7	1.50	10	30	3	1	–	–	–	–	–	3	1	–	–	Phosphonic acid, [[(trimethylsilyl)amino]methyl]-, bis(trimethylsilyl) ester	53044-36-3	NI34442
102	73	298	147	312	211	103	45	327	100	64	40	12	10	10	10	9	1.50	10	30	3	1	–	–	–	–	–	3	1	–	–	Phosphonic acid, [[(trimethylsilyl)amino]methyl]-, bis(trimethylsilyl) ester	53044-36-3	NI34443
135	60	177	43	117	192	102	104	327	100	74	60	57	25	22	21	19	9.44	12	25	5	1	2	–	–	–	–	–	–	–	–	D-Galactose, 2-(acetylamino)-2-deoxy-, 1-(diethyl mercaptal)	42890-21-1	NI34444
106	65	77	39	327				327	100	37	35	25	5					13	14	2	1	1	–	–	1	–	–	–	–	–	2-Methyl-N-(4-bromobenzenesulphonyl)azepine		LI07240
164	136	91	327	149	151	165	121	327	100	71	40	38	32	22	16	14		13	14	3	1	–	5	–	–	–	–	–	–	–	2-(2,3-Dimethoxyphenyl)-N-(pentafluoropropionyl)ethylamine		MS07751
151	121	164	327	152	91	149	77	327	100	41	22	12	10	9	7	6		13	14	3	1	–	5	–	–	–	–	–	–	–	2-(2,4-Dimethoxyphenyl)-N-(pentafluoropropionyl)ethylamine		MS07752
164	327	121	151	149	165	91	77	327	100	69	61	48	26	18	18	11		13	14	3	1	–	5	–	–	–	–	–	–	–	2-(2,5-Dimethoxyphenyl)-N-(pentafluoropropionyl)ethylamine		MS07753
164	327	151	165	135	91	77	121	327	100	34	28	17	12	8	8	7		13	14	3	1	–	5	–	–	–	–	–	–	–	2-(3,5-Dimethoxyphenyl)-N-(pentafluoropropionyl)ethylamine		MS07754
84	98	42	43	327	97	55	56	327	100	44	32	25	18	17	17	16		13	17	1	3	3	–	–	–	–	–	–	–	–	5-Acetyl-4-amino-3-(2-N-pyrrolidinylethylthio)thieno[2,3-c]isothiazole	97090-70-5	NI34445
84	97	42	43	55	70	98	85	327	100	55	26	21	12	11	10	10	4.00	13	17	1	3	3	–	–	–	–	–	–	–	–	5-Acetyl-4-amino-3-(2-N-pyrrolidinylethylthio)thieno[3,2-d]isothiazole	92090-77-2	NI34446
327	107	77	43	171	121	92	263	327	100	79	73	69	67	58	58	55		13	17	3	3	2	–	–	–	–	–	–	–	–	Benzenesulphonamide, 4-methoxy-N-[5-isobutyl-1,3,4-thiadiazol-2-yl]-	3567-08-6	NI34447
327	107	77	43	171	121	92	263	327	100	77	72	68	67	59	57	53		13	17	3	3	2	–	–	–	–	–	–	–	–	Benzenesulphonamide, 4-methoxy-N-[5-isobutyl-1,3,4-thiadiazol-2-yl]-	3567-08-6	LI07241
162	161	56	163	120	247	77	91	327	100	81	52	29	20	15	15	11	0.00	13	18	2	3	–	–	–	1	–	–	–	–	–	Acetamide, N-[2-[2-(phenylimino)-3-oxazolidinyl]ethyl]-, monohydrobromide	23008-60-8	NI34448
327	297	89	209	328	56	83	117	327	100	21	19	18	17	13	12	11		14	9	5	5	–	–	–	–	–	–	–	–	–	1,2,3,4-Tetrahydro-1-methyl-7-(4-nitrophenyl)oxazolo[2,3-f]purine-2,4-dione		NI34449
156	94	182	77	172	174	173	93	327	100	89	56	32	28	21	21	21	0.00	14	11	2	1	1	–	2	–	–	–	–	–	–	Sulphoximine, N-(2,6-dichlorobenzoyl)-S-methyl-S-phenyl-	54090-92-5	NI34450
180	181	267	87	208	170	327	268	327	100	26	15	13	7	6	2	2		14	26	3	2	–	3	–	–	–	–	–	–	–	Propyl N-trifluoroacetoxy-N(ε),N(ε),N(ε)-trimethylammoniolysine	91795-80-1	NI34451
128	100	70	129	127	227	327	199	327	100	68	16	9	8	7	6	6		15	25	5	3	–	–	–	–	–	–	–	–	–	Acmeala-pro-megly-ome		NI34452
128	100	70	225	116	129	114	469	327	100	33	16	14	13	11	11	9	0.00	15	25	5	3	–	–	–	–	–	–	–	–	–	Acmegly-pro-meala-ome		NI34453
207	116	195	236	168	105	208	117	327	100	74	47	37	29	21	20	18	3.20	16	10	1	1	–	5	–	–	–	–	–	–	–	N-(Pentafluorobenzoyl)tranylcypromine		MS07755
143	115	327	171	128	63	329	144	327	100	79	36	20	14	13	12	12		16	10	3	3	–	–	1	–	–	–	–	–	–	2-Hydroxy-1-(2-chloro-4-nitrophenylazo)naphthalene		IC03969
291	216	217	272	214	327	290	189	327	100	60	33	30	13	12	12	11		16	11	–	1	–	–	1	–	–	–	–	–	1	Benzo[a]phenarsazine, 12-chloro-7,12-dihydro-	13493-35-1	NI34454
291	292	146	216	214	289	215	290	327	100	62	32	26	25	22	22	15	0.00	16	11	–	1	–	–	1	–	–	–	–	–	1	Benzo[c]phenarsazine, 7-chloro-7,12-dihydro-	10352-43-9	NI34455
327	93	79	67	135	91	247	105	327	100	38	31	22	21	18	16	14		17	17	4	3	–	–	–	–	–	–	–	–	–	1-(1-Adamantyl)-2-cyano-3,5-dinitrobenzene		NI34456
94	327	93	79	135	107	105	67	327	100	90	51	32	25	17	17	17		17	17	4	3	–	–	–	–	–	–	–	–	–	1-(1-Adamantyl)-4-cyano-3,5-dinitrobenzene		NI34457
239	77	154	327	128	51	43	105	327	100	41	38	25	24	24	24	23		17	17	4	3	–	–	–	–	–	–	–	–	–	2-Phenyl-4-(isopropyloxycarbonyl)-6-methyl-3H-imidazo[1,5-b]pyridazine-5,7-dione		NI34458
85	113	135	271	55	69	152	207	327	100	31	14	12	12	10	5	3	0.00	17	29	3	1	1	–	–	–	–	–	–	–	–	N-[(2R)-2,4-Dimethylpentanoyl]bornane-10,2-sultam		DD01266
85	135	55	113	271	93	152	272	327	100	51	50	47	39	24	15	6	0.00	17	29	3	1	1	–	–	–	–	–	–	–	–	N-[(2R,3R)-2,3-Dimethylpentanoyl]bornane-10,2-sultam		DD01267
85	113	91	55	271	135	152	284	327	100	29	20	20	12	12	7	3	0.00	17	29	3	1	1	–	–	–	–	–	–	–	–	N-[(2R)-2-Methylhexanoyl]bornane-10,2-sultam		DD01268
124	83	135	268	183	327	270	313	327	100	49	40	30	13	12	11	9		17	29	5	1	–	–	–	–	–	–	–	–	–	Cornucervine		LI07242
327	77	28	132	210	250	46	188	327	100	78	42	41	27	18	18	14		18	13	–	7	–	–	–	–	–	–	–	–	–	3-Methyl-8-(nitrophenyl)-1-phenyl-1H-pyrazolo[4,3-e][1,2,4]triazolo[4,3-a][1,2,4]triazine		NI34459
170	327	171	328	155	154	127	186	327	100	58	14	12	9	9	7	6		18	17	3	1	1	–	–	–	–	–	–	–	–	O-[1-(Dimethylamino)naphthalene-5-sulphonyl]phenol		NI34460
170	327	171	328	329	169	168	167	327	100	58	15	12	5	5	2	2		18	17	3	1	1	–	–	–	–	–	–	–	–	1-Naphthalenesulphonic acid, 5-(dimethylamino)-, phenyl ester	55837-12-2	NI34461
167	212	168	165	152	166	213	195	327	100	91	88	86	77	76	57	54	0.00	18	17	5	1	–	–	–	–	–	–	–	–	–	Diphenylacetylaspartic acid	78490-35-4	NI34462
121	65	39	93	190	77	63	51	327	100	50	30	19	14	10	10	9	8.00	18	17	5	1	–	–	–	–	–	–	–	–	–	Fontaphillin	22667-62-5	NI34463
312	284	270	327	254	170	158	169	327	100	68	65	62	32	24	24	23		18	17	5	1	–	–	–	–	–	–	–	–	–	10-Methyl-1-ethoxy-2-methoxy-3,4-methylenedioxyacridone		MS07756
294	312	327	284	270	268	254	158	327	100	88	83	78	57	53	53	40		18	17	5	1	–	–	–	–	–	–	–	–	–	10-Methyl-1-ethoxy-4-methoxy-2,3-methylenedioxyacridone		MS07757

MASS TO CHARGE RATIOS								M.W.	INTENSITIES								Parent	C	H	O	N	S	F	Cl	Br	I	Si	P	B	X	COMPOUND NAME	CAS Reg No	No
58	86	59	71	327	84	57	44	327	100	7	4	3	2	2	2	2		19	21	2	1	1	–	–	–	–	–	–	–	–	4-(3-Dimethylaminopropoxy)-10,11-dihydrodibenzo[b,f]thiepin-11-one		MS07758
43	181	85	44	29	100	42	166	327	100	87	76	75	65	59	55	47	0.00	19	21	2	1	1	–	–	–	–	–	–	–	–	Spiro[9H-fluorene-9,2′(3′H)-thiophen]-4-amine, 4′,5′-dihydro-N,N,3′-trimethyl-, 1′,1′-dioxide	56666-86-5	NI34464
135	136	134	206	161	234	137	163	327	100	59	29	25	22	16	14	13	0.39	19	21	4	1	–	–	–	–	–	–	–	–	–	2-(1-Adamantyl)-2-phenoxyoxazolidine-4,5-dione		LI07243
326	327	284	312	328	253	310	283	327	100	97	30	27	18	13	11	9		19	21	4	1	–	–	–	–	–	–	–	–	–	4H-Dibenzo[de,g]quinoline-1,9-diol, 5,6,6a,7-tetrahydro-2,10-dimethoxy-6 methyl-, (S)-	3019-51-0	NI34465
326	327	284	312	310	253	252	282	327	100	81	28	27	16	14	11	10		19	21	4	1	–	–	–	–	–	–	–	–	–	4H-Dibenzo[de,g]quinoline-1,9-diol, 5,6,6a,7-tetrahydro-2,10-dimethoxy-6 methyl-, (S)-	3019-51-0	NI34466
326	327	312	296	328	284	313	269	327	100	82	34	25	20	19	11	10		19	21	4	1	–	–	–	–	–	–	–	–	–	4H-Dibenzo[de,g]quinoline-2,9-diol, 5,6,6a,7-tetrahydro-1,10-dimethoxy-6 methyl-, (S)-	476-70-0	NI34467
178	150	327	151	177	107	135	176	327	100	78	36	23	22	21	15	14		19	21	4	1	–	–	–	–	–	–	–	–	–	6H-Dibenzo[a,g]quinazoline-1,10-diol, 5,8,13,13a-tetrahydro-2,11-dimethoxy-, (S)-	27498-04-0	LI07244
178	327	150	326	176	135	106	328	327	100	71	57	43	29	27	19	16		19	21	4	1	–	–	–	–	–	–	–	–	–	6H-Dibenzo[a,g]quinolizine-2,9-diol, 5,8,13,13a-tetrahydro-3,10-dimethoxy-, (±)-	6451-72-5	NI34468
178	150	327	326	135	312	310	107	327	100	80	56	53	32	17	16	15		19	21	4	1	–	–	–	–	–	–	–	–	–	6H-Dibenzo[a,g]quinolizine-2,9-diol, 5,8,13,13a-tetrahydro-3,10-dimethoxy-, (S)-	6451-73-6	NI34469
192	210	267	239	299	298	327	266	327	100	74	65	58	53	42	40	26		19	21	4	1	–	–	–	–	–	–	–	–	–	2aR*,3R*,8bS*-2a,8b-Dihydroxy-7-methoxy-3-(4-methoxyphenyl)-1,2,3,4-tetrahydrocyclobut[c]quinoline	90732-21-1	NI34470
178	327	150	326	176	135	106	328	327	100	71	58	44	30	27	21	18		19	21	4	1	–	–	–	–	–	–	–	–	–	Discoulerine		LI07245
295	296	294	280	327	297	312		327	100	90	55	32	24	22	16			19	21	4	1	–	–	–	–	–	–	–	–	–	11-Methoxyerythraline		LI07246
206	327	326	190					327	100	22	15	6						19	21	4	1	–	–	–	–	–	–	–	–	–	1-(3,4-Methylenedioxyphenyl)-2-methyl-6,7-dimethoxy-1,2,3,4-tetrahydroisoquinoline		LI07247
271	254	327	242	226	208	152	180	327	100	63	31	23	23	20	20	18		19	21	4	1	–	–	–	–	–	–	–	–	–	2-Propenoic acid, 2-cyano-3-[4-[3-(1-methylpropoxy)-3-oxo-1-propenyl]phenyl]-, ethyl ester, [S-(E,E)]-	64666-23-5	NI34471
327	284	31	28	312	299	45	328	327	100	60	53	39	33	31	24	23		19	21	4	1	–	–	–	–	–	–	–	–	–	Salutaridine	1936-18-1	LI07248
327	312	328	174	281	326	310	284	327	100	51	44	40	36	29	29	24		19	21	4	1	–	–	–	–	–	–	–	–	–	Sarcocapnidine	87069-33-8	NI34472
280	250	281	266	192	251	327	292	327	100	78	31	27	22	16	15	11		19	21	4	1	–	–	–	–	–	–	–	–	–	Spiro[indan-2,1′(2′H)-isoquinoline]-1,3-diol, 3′,4′-dihydro-6′,7′-dimethoxy-	33863-49-9	NI34473
107	165	123	95	83	79	121	78	327	100	20	20	20	15	15	13	12	5.96	19	22	1	3	–	1	–	–	–	–	–	–	–	1-Butanone, 1-(4-fluorophenyl)-4-[4-(2-pyridinyl)-1-piperazinyl]-	1649-18-9	NI34474
198	58	115	70	199	42	71	72	327	100	55	33	31	19	18	15	13	0.44	19	25	–	3	1	–	–	–	–	–	–	–	–	Phenothiazine, 10-[2,3-bis(dimethylamino)propyl]-	58-37-7	NI34476
198	70	58	115	71	269	72	56	327	100	62	59	34	21	12	12	12	2.90	19	25	–	3	1	–	–	–	–	–	–	–	–	Phenothiazine, 10-[2,3-bis(dimethylamino)propyl]-	58-37-7	NI34475
250	91	41	137	298	55	54	43	327	100	63	44	40	38	29	26	26	1.00	19	25	2	1	–	–	–	–	–	1	–	–	–	2,2-Diphenyl-6-propyl-1,3-dioxa-6-aza-2-silacyclooctane		MS07759
327	58	36	85	242	119	43	223	327	100	78	72	71	50	40	33	27		19	25	2	3	–	–	–	–	–	–	–	–	–	Benzoic acid, 2-[(2-aminophenyl)[3-(dimethylamino)propyl]amino]-, methyl ester	13450-59-4	NI34477
327	58	85	242	43	223	151	131	327	100	79	71	51	33	27	23	22		19	25	2	3	–	–	–	–	–	–	–	–	–	Benzoic acid, 2-[(2-aminophenyl)[3-(dimethylamino)propyl]amino]-, methyl ester	13450-59-4	NI34478
222	44	77	327	73	30			327	100	84	54	49	48	30				20	13	2	3	–	–	–	–	–	–	–	–	–	4-Phthalimidoazobenzene		IC03970
105	102	97	43	145	60	58	182	327	100	99	70	50	43	30	30	23	7.00	20	25	1	1	1	–	–	–	–	–	–	–	–	2-Hydroxy-N,N-diisopropyl-2,2-diphenylthioacetamide		MS07760
191	163	121	161	204	77	190	132	327	100	79	75	65	54	54	53	52	0.00	20	25	3	1	–	–	–	–	–	–	–	–	–	Isoquinoline, 1,2,3,4-tetrahydro-6,7-dimethoxy-1-[(4-methoxyphenyl)methyl]-2-methyl-, (S)-	3423-02-7	NI34479
206	207	191	121	78	326	205	176	327	100	15	11	8	6	5	5	5	0.76	20	25	3	1	–	–	–	–	–	–	–	–	–	7-Isoquinolinol, 1,2,3,4-tetrahydro-6-methoxy-1-[(4-methoxyphenyl)methyl]-2,8-dimethyl-	16767-25-2	NI34480
226	327	42	240	326	282	87	241	327	100	84	47	36	30	28	28	23		20	25	3	1	–	–	–	–	–	–	–	–	–	Morphinan-6-one, 8,14-didehydro-3-methoxy-17-methyl-, cyclic ethylene acetal	23185-95-7	NI34481
58	327	59	121	328	177	206	165	327	100	16	9	5	4	4	2	2		20	25	3	1	–	–	–	–	–	–	–	–	–	Phenol, 3-[2-(dimethylamino)ethyl]-6-methoxy-2-[2-(4-methoxyphenyl)ethenyl]-, (E)-	2609-29-2	MS07761
161	131	103	91	77	327	154	228	327	100	72	40	15	13	10	8	7		20	25	3	1	–	–	–	–	–	–	–	–	–	Retrofractamide-A	94079-67-1	NI34482
327	298	328	299	270	163.5	326	101	327	100	64	24	13	7	6	5	4		21	13	3	1	–	–	–	–	–	–	–	–	–	12,13-Dihydro-12-methyl-13,14-dioxo-14H-naphtho[1′,2′:5,6]pyrano[3,2-c]quinoline		MS07762
327	312	328	299	326	134.5	310	313	327	100	37	27	20	15	14	13	12		21	13	3	1	–	–	–	–	–	–	–	–	–	8-Methyl-13,14-dioxo-8H,13H,14H-naphtho[1′,2′:5,6]pyrano[2,3-b]quinoline		MS07763
250	105	327	77	222	76	251	104	327	100	63	59	57	35	30	20	15		21	13	3	1	–	–	–	–	–	–	–	–	–	2-Phthalimidobenzophenone		IC03971
178	327	179	176	328	177	152	151	327	100	19	16	8	5	4	4	4		21	14	–	3	–	1	–	–	–	–	–	–	–	as-Triazine, 3-(4-fluorophenyl)-5,6-diphenyl-	22158-35-6	MS07764
298	77	119	299	327	182	165	104	327	100	89	78	60	59	52	37	37		21	17	1	3	–	–	–	–	–	–	–	–	–	2-Imino-3,5,5-triphenyl-4-imidazolidinone		NI34483
327	180	77	258	257	165	328	51	327	100	60	55	53	53	47	25	15		21	17	1	3	–	–	–	–	–	–	–	–	–	4-Imino-1,5,5-triphenyl-2-imidazolidinone		NI34484
326	327	119	180	104	77	165	118	327	100	56	50	37	28	24	16	12		21	17	1	3	–	–	–	–	–	–	–	–	–	5-Imino-1,4,4-triphenyl-2-imidazolidinone	52460-96-5	NI34485
192	91	193	89	105	165	178	77	327	100	62	32	20	19	16	15	12	0.00	21	17	1	3	–	–	–	–	–	–	–	–	–	1H-1,2,3-Triazole, 4,5-diphenyl-1-(phenylmethoxy)-	56701-36-1	NI34486

MASS TO CHARGE RATIOS								M.W.	INTENSITIES								Parent	C	H	O	N	S	F	Cl	Br	I	Si	P	B	X	COMPOUND NAME	CAS Reg No	No
219	220	327	203	312	57	41	328	**327**	100	18	13	12	11	10	7	4		21	29	2	1	–	–	–	–	–	–	–	–	–	2-Cyano-2-(3,5-di-tert-butyl-4-hydroxybenzyl)cyclopentanone		IC03972
105	103	77	69	67	81	82	95	**327**	100	58	54	48	40	35	27	22	**0.19**	21	29	2	1	–	–	–	–	–	–	–	–	–	4,5-Dihydro-4,4-undecamethylene-2-phenyl-1,3-oxazin-6-one		NI34487
301	91	92	137	138	270	57	302	**327**	100	42	37	36	30	29	26	21	**19.98**	21	29	2	1	–	–	–	–	–	–	–	–	–	19-Norpregn-4-en-20-yn-3-one, 17-hydroxy-, O-methyloxime, (17α)-	58001-83-5	NI34488
222	77	105	327	89	51	223	78	**327**	100	64	50	21	21	21	13	7		22	17	2	1	–	–	–	–	–	–	–	–	–	1-Anilino-1,2-dibenzoylethylene		MS07765
105	77	327	106	206	51			**327**	100	45	9	9	8	8				22	17	2	1	–	–	–	–	–	–	–	–	–	cis-α-Benzamidobenzalacetophenone		LI07249
105	77	327	206	106	78	222	40	**327**	100	44	19	10	10	9	8	8		22	17	2	1	–	–	–	–	–	–	–	–	–	trans-α-Benzamido-benzalacetophenone		LI07250
105	77	180	51	103	42	106	178	**327**	100	69	25	20	12	10	10	8	**3.19**	22	17	2	1	–	–	–	–	–	–	–	–	–	4,5-Dihydro-2,4,4-triphenyl-1,3-oxazin-6-one	77074-12-5	NI34489
327	76	310	328	104	255	226	152	**327**	100	28	27	23	18	11	11	11		22	17	2	1	–	–	–	–	–	–	–	–	–	2-[4-(Dimethylamino)-1-naphthyl]naphthoquinone		MS07766
105	222	77	91	194	103	223	106	**327**	100	62	41	37	22	14	10	8	**2.00**	22	17	2	1	–	–	–	–	–	–	–	–	–	3,4-Diphenyl-5-benzoyl-2-isoxazoline		NI34490
105	77	222	327	106	167	90	89	**327**	100	72	55	35	19	18	14	13		22	17	2	1	–	–	–	–	–	–	–	–	–	1,4-Diphenyl-4-benzoyloxy-2-azabutadiene		LI07251
135	327	210	328	77	78	138	211	**327**	100	95	53	24	15	13	10	8		22	17	2	1	–	–	–	–	–	–	–	–	–	3,4-Diphenyl-5-(4-methoxyphenyl)isoxazole		LI07252
105	77	224	106	51	223	103	78	**327**	100	24	4	3	3	2	2	2	**2.00**	22	17	2	1	–	–	–	–	–	–	–	–	–	3-Phenyl-4-benzoyl-5-phenyl-2-isoxazoline		NI34491
228	200	327	284	214	172	118	270	**327**	100	49	35	17	16	14	12	9		22	33	1	1	–	–	–	–	–	–	–	–	–	1′,4′-Dihydro-2′-butyl-4-tert-butylspiro[cyclohexane-1,3′(2H)-isoquinolin]-1′-one		DD01269
208	119	257	91	77	165	193	129	**327**	100	73	49	38	36	31	26	19	**12.36**	23	21	1	1	–	–	–	–	–	–	–	–	–	Aniline, N-(3,3-dimethyl-4,4-diphenyl-2-oxetanylidene)-	14251-66-2	NI34492
91	327	236	220	65	92	121	296	**327**	100	63	32	17	14	12	11	5		23	21	1	1	–	–	–	–	–	–	–	–	–	1-Benzyl-3-p-methoxybenzylindole		LI07253
327	236	213	228	220	91	206	219	**327**	100	75	37	30	26	26	21	20		23	21	1	1	–	–	–	–	–	–	–	–	–	2-Benzyl-3-p-methoxybenzylindole		LI07254
327	91	206	121	77	218	143	236	**327**	100	72	70	61	61	51	51	50		23	21	1	1	–	–	–	–	–	–	–	–	–	3-Benzyl-2-p-methoxybenzylindole		LI07255
105	327	222	310	167	77	165	174	**327**	100	93	82	69	68	49	42	38		23	21	1	1	–	–	–	–	–	–	–	–	–	trans-(3-[1,1′-Biphenyl]-4-yl-1-ethyl-2-aziridinyl)phenylmethanone	32044-34-1	NI34493
131	327	103	91	236	77	328	132	**327**	100	41	27	25	21	13	11	11		23	21	1	1	–	–	–	–	–	–	–	–	–	N,N-Dibenzyl-3-phenylpropenamide		TR00302
208	257	91	77	165	193	180	64	**327**	100	49	39	36	30	26	26	17	**11.00**	23	21	1	1	–	–	–	–	–	–	–	–	–	2-Phenylimino-3,3-dimethyl-4,4-diphenyloxetane		MS07767
145	165	77	144	146	130	327	166	**327**	100	16	16	15	11	9	8	8		23	21	1	1	–	–	–	–	–	–	–	–	–	3-Phenylimino-2,2-dimethyl-4,4-diphenyloxetane	55470-95-6	NI34494
145	165	144	77	146	130	166	327	**327**	100	20	16	16	11	10	9	8		23	21	1	1	–	–	–	–	–	–	–	–	–	3-Phenylimino-2,2-dimethyl-4,4-diphenyloxetane	55470-95-6	MS07768
270	250	271	165	180	327	251		**327**	100	78	46	36	32	27	21			24	25	–	1	–	–	–	–	–	–	–	–	–	2H-2-Benzazepine, 3,4,5,5a-tetrahydro-5,5-dimethyl-1,2-diphenyl-	64308-83-4	NI34495
146	147	312	327	131	165	182	313	**327**	100	56	39	29	20	12	12	10		24	25	–	1	–	–	–	–	–	–	–	–	–	N-Benzhydrylidene-1-(2,4,6-trimethylphenyl)ethylamine	75326-11-3	NI34496
270	271	180	165	250	327	195		**327**	100	44	24	20	10	8	7			24	25	–	1	–	–	–	–	–	–	–	–	–	Pyrrolidine, 3,3-dimethyl-1,2,2-triphenyl-	64278-16-6	NI34497
251	253	80	82	79	81	249	173	**328**	100	97	54	52	52	50	36	36	**0.00**	1	–	–	–	–	–	–	4	–	–	–	–	–	Carbon tetrabromide	558-13-4	NI34498
69	328	309	226	76	245	264	152	**328**	100	25	8	6	6	5	4	4		6	–	–	2	–	12	–	–	–	–	–	–	–	1,2-Bis[bis(trifluoromethyl)amino]acetylene		LI07256
330	332	328	334	62	143	141	61	**328**	100	96	32	31	26	17	17	15		6	3	1	–	–	–	–	3	–	–	–	–	–	Phenol, 2,4,6-tribromo-	118-79-6	IC03973
62	61	330	332	81	79	141	63	**328**	100	76	65	64	51	51	47	45	**22.60**	6	3	1	–	–	–	–	3	–	–	–	–	–	Phenol, 2,4,6-tribromo-	118-79-6	NI34499
202	200	328	199	326	201	327	325	**328**	100	77	71	56	51	43	38	32		7	7	–	–	–	–	1	–	–	–	–	–	1	Benzyl(chloro)mercury	2117-39-7	NI34500
91	328	65	39	89	313	51	50	**328**	100	71	30	19	13	11	10	9		7	7	–	–	–	–	1	–	–	–	–	–	1	o-Tolylmercuric chloride		LI07257
91	328	63	89	51	50	293	62	**328**	100	81	13	12	7	7	5	5		7	7	–	–	–	–	1	–	–	–	–	–	1	p-Tolylmercuric chloride		LI07258
110	191	189	43	109	166	154	233	**328**	100	99	96	94	62	52	52	29	**0.00**	8	10	4	–	–	–	–	2	–	–	–	–	–	(1R,4S,5R,6S)-4-Acetoxy-2,6-dibromo-1,5-dihydroxycyclohexene		DD01270
74	67	82	80	56	39	81	79	**328**	100	57	51	40	24	16	15	15	**0.20**	8	14	2	2	–	–	–	2	–	–	–	–	–	1-Hydroxy-4-dibromomethyl-2,2,5,5-tetramethyl-3-imidazoline-3-oxide	51973-30-9	NI34501
285	283	241	287	163	199	43	161	**328**	100	76	72	70	56	54	54	41	**3.31**	9	21	–	–	–	–	–	1	–	–	–	–	1	Tri-propyltin bromide		MS07769
214	212	89	133	47	215	213	107	**328**	100	99	25	15	15	13	12	10	**0.00**	10	10	–	–	1	4	–	1	–	–	–	1	–	3-Bromo-1-ethylbenzo[b]thiophenium tetrafluoroborate		LI07259
87	45	164	59	328	263	71	145	**328**	100	34	32	13	7	7	7	6		10	16	–	–	6	–	–	–	–	–	–	–	–	2,4,6,8,9,10-Hexathiatricyclo[3.3.1.1^{3,7}]decane, 1,3-dipropyl-	57274-50-7	NI34502
209	140	210	114	119	219	169	69	**328**	100	40	9	9	8	5	5	5	**2.00**	11	6	–	–	–	10	–	–	–	–	–	–	–	1,3,5-Cycloheptatriene, 7,7-bis(pentafluoroethyl)-	17838-67-4	LI07260
209	140	210	114	119	169	159	69	**328**	100	41	10	10	9	6	6	6	**2.00**	11	6	–	–	–	10	–	–	–	–	–	–	–	1,3,5-Cycloheptatriene, 7,7-bis(pentafluoroethyl)-	17838-67-4	NI34503
69	57	41	152	140	198	56	168	**328**	100	76	64	52	41	35	25	22	**0.00**	11	19	4	4	–	3	–	–	–	–	–	–	–	L-Homoserine, O-(aminoiminomethyl)amino-N-(trifluoroacetyl)-, butyl ester	55521-06-7	NI34504
95	55	151	81	41	109	123	83	**328**	100	97	96	87	85	70	63	56	**0.00**	11	22	1	–	–	–	–	2	–	–	–	–	–	10,11-Dibromoundecan-1-ol		DD01271
328	329	164	297	330	259	234	296	**328**	100	15	14	8	6	4	4	3		12	–	–	–	1	8	–	–	–	–	–	–	–	Dibenzothiophene, octafluoro-	7136-57-4	NI34505
328	164	297	330	265	234	259	296	**328**	100	13	6	5	5	4	3	2		12	–	–	–	1	8	–	–	–	–	–	–	–	Dibenzothiophene, octafluoro-	7136-57-4	NI34506
328	164	329	327	262	300	284	261	**328**	100	24	18	6	5	4	4	4		12	4	–	2	–	8	–	–	–	–	–	–	–	[1,1′-Biphenyl]-4,4′-diamine, 2,2′,3,3′,5,5′,6,6′-octafluoro-	1038-66-0	NI34507
328	164	329	44	327	289	262	261	**328**	100	24	14	8	6	5	4	4		12	4	–	2	–	8	–	–	–	–	–	–	–	[1,1′-Biphenyl]-4,4′-diamine, 2,2′,3,3′,5,5′,6,6′-octafluoro-	1038-66-0	NI34508
64	48	250	44	170	141	77	169	**328**	100	33	26	23	16	13	13	10	**2.60**	12	12	5	2	2	–	–	–	–	–	–	–	–	4,4′-Di(sulphonamido)diphenyl oxide		IC03974
106	84	112	94	217	237	293	265	**328**	100	56	52	14	12	6	5	3	**0.00**	12	16	4	–	–	2	1	–	–	–	1	–	–	Diethyl 2-chloro-2,2-difluoro-1-hydroxy-1-phenylethanephosphonate		DD01272
194	18	43	28	310	107	176	148	**328**	100	95	82	67	65	55	50	50	**0.00**	12	16	7	4	–	–	–	–	–	–	–	–	–	6-Methyl-8-(2,3,4,5-tetrahydroxypentyl)pterid-2,4,7-trione		LI07261
121	253	45	91	29	209	77	135	**328**	100	84	72	28	22	20	19	18	**0.00**	12	28	8	–	–	–	–	–	–	1	–	–	–	Tetra(methoxyethoxy)silane	2157-45-1	NI34509
147	73	45	273	148	75	149	198	**328**	100	92	19	16	15	15	9	9	**0.00**	12	48	5	–	–	–	–	–	–	2	–	–	–	Dimethylmaleic acid, bis(trimethylsilyl) ester		NI34510
190	64	63	30	126	75	50	162	**328**	100	30	30	30	29	27	25	24	**16.60**	13	7	5	2	–	3	–	–	–	–	–	–	–	Benzene, 2-nitro-1-(4-nitrophenoxy)-4-(trifluoromethyl)-	15457-05-3	NI34511

MASS TO CHARGE RATIOS								M.W.	INTENSITIES								Parent	C	H	O	N	S	F	Cl	Br	I	Si	P	B	X	COMPOUND NAME	CAS Reg No	No
328	201	230	184	160	330	200	299	**328**	100	89	57	36	34	33	30	28		13	9	6	–	1	–	1	–	–	–	–	–	–	**4-Methoxy-7-methyl-5-oxo-5H-furo[3,2-g][1]benzopyran-9-sulphonyl chloride**		**MS07770**
70	83	56	58	71	258	98	57	**328**	100	45	13	10	9	6	6	4	**0.80**	13	17	2	4	1	–	1	–	–	–	–	–	–	**7-Chloro-2-methyl-3-(4-methyl-1-piperazinyl)-1,2,4-benzothiadiazine 1,1-dioxide**		**MS07771**
70	83	71	58	98	258	56	140	**328**	100	46	19	11	7	6	6	3	**0.70**	13	17	2	4	1	–	1	–	–	–	–	–	–	**7-Chloro-4-methyl-3-(4-methyl-1-piperazinyl)-1,2,4-benzothiadiazine 1,1-dioxide**		**MS07772**
132	104	105	160	77	131	133	194	**328**	100	43	42	33	19	18	17	15	**3.00**	13	20	6	4	–	–	–	–	–	–	–	–	–	**Ethyl 5,6,7,8-tetrahydro-6,7-bis(2-hydroxyethoxy)pteridine-4-carboxylate**		**LI07262**
163	164	75	43	121	45	103	87	**328**	100	35	35	28	23	18	15	15	**9.86**	13	28	5	–	2	–	–	–	–	–	–	–	–	**D-Glucose, 3-O-methyl-, dipropyl mercaptal**	3806-83-5	**NI34512**
105	77	75	328	270	106	63	103	**328**	100	46	22	20	20	16	10	4		14	8	6	4	–	–	–	–	–	–	–	–	–	**Sydnone, 3-(2,4-dinitrophenyl)-4-phenyl-**	69978-05-8	**NI34513**
61	328	56	102	79	211	93	121	**328**	100	36	22	14	14	9	9	9		14	20	3	2	2	–	–	–	–	–	–	–	–	**5,5-Diallyl-1,3-bis(methylthiomethyl)-2,4,6-pyrimidinetrione**		**MS07773**
158	112	214	199	328	167	227	114	**328**	100	33	25	24	17	13	9	9		14	20	5	2	1	–	–	–	–	–	–	–	–	**t-Butyl 3-methyl-7-carbomethoxyamino-2-cephem-4-carboxylate**		**NI34514**
58	91	116	155	42	197	84	255	**328**	100	78	42	35	17	15	14	13	**0.17**	14	20	5	2	1	–	–	–	–	–	–	–	–	**Methyl 3-methyl-N-(4-toluenesulphonylaminocarbonyl)-3-aminobutanoate**		**NI34515**
88	55	168	141	328	131	101	84	**328**	100	95	69	68	62	58	55	51		14	24	5	4	–	–	–	–	–	–	–	–	–	**1-Oxa-4,7,14,17-tetraazacyclononadecane-5,8,13,16-tetrone**		**NI34516**
198	196	157	117	77	195	183	118	**328**	100	49	33	31	30	25	25	23	**0.00**	15	20	3	–	–	–	–	–	–	–	–	–	1	**(2R,6R)-2-tert-Butyl-6-methyl-5-(phenylselenyl)-1,3-dioxan-4-one**		**DD01273**
43	106	78	124	107	85	127	77	**328**	100	44	31	27	26	20	19	16	**0.10**	15	20	8	–	–	–	–	–	–	–	–	–	–	**Fragilin**		**MS07774**
186	100	125	55	113	69	118	141	**328**	100	71	56	55	50	40	35	35	**6.44**	15	24	6	2	–	–	–	–	–	–	–	–	–	**3,3-Dimethyl-1,5-dioxa-10,13-diazacycloheptadecane-6,9,14,17-tetrone**	79688-10-1	**NI34517**
70	55	69	56	113	100	84	101	**328**	100	80	75	47	42	34	33	30	**3.67**	15	24	6	2	–	–	–	–	–	–	–	–	–	**1,5-Dioxa-10,15-diazacyclononadecane-6,9,16,19-tetrone**	79688-11-2	**NI34518**
328	45	283	69	329	330	93	251	**328**	100	78	37	29	24	22	20	18		16	8	–	–	4	–	–	–	–	–	–	–	–	**Cycloocta[1,2-c:3,4-c′:5,6-c″:7,8-c‴]tetrathiophene**		**NI34519**
77	135	93	195	106	118	177	178	**328**	100	62	27	21	14	6	5	4	**0.00**	16	16	–	4	2	–	–	–	–	–	–	–	–	**3-(N-Phenylthiocarbamoylamino)-2-(phenylimino)thiazolidine**		**LI07263**
328	130	158	129	173	102	272	329	**328**	100	54	46	41	30	24	22	20		16	16	4	4	–	–	–	–	–	–	–	–	–	**5-(Dihydro-2H-1,4-oxazino[4,3-b]benzimidazolio-1)-1,3-dimethyl-4-barbiturate**		**LI07264**
328	130	158	129	173	102	272	145	**328**	100	54	46	41	30	24	22	20		16	16	4	4	–	–	–	–	–	–	–	–	–	**5-(Dihydro-2H-1,4-oxazino[4,3-b]benzimidazolio-1)-1,3-dimethyl-4-barbiturate**		**NI34520**
221	222	145	205	77	105	106	146	**328**	100	69	41	39	36	26	19	14	**1.00**	16	16	4	4	–	–	–	–	–	–	–	–	–	**7-[N′-(2′-Methylamino-1′-hydroxy-1′-phenylpropanyl)]-4-nitrobenzo-2-oxa 1,3-diazole**		**LI07265**
166	237	194	118	91	69	139	96	**328**	100	62	40	27	27	16	11	9	**0.00**	16	19	2	2	–	3	–	–	–	–	–	–	–	**N-(trifluoroacetyl)prolylamphetamine**		**NI34521**
130	120	163	328	77	31	164	132	**328**	100	83	50	42	38	35	25	24		16	20	2	6	–	–	–	–	–	–	–	–	–	**1,2-Di(2′-carboxyphenylamino)ethane dihydrazide**		**LI07266**
128	69	115	44	41	169	43	87	**328**	100	89	82	79	75	69	52	45	**14.32**	16	21	5	–	–	–	1	–	–	–	–	–	–	**Ethyl 2,4-dimethyl-2-(4-chloro-2-formylphenoxy)-4-hydroxypentanoate**		**IC03975**
281	205	265	264	237	297	149	296	**328**	100	89	80	80	77	49	46	43	**15.00**	16	24	7	–	–	–	–	–	–	–	–	–	–	**Dimethyl (1α,2β,5α)-3,7,7-trimethoxy-2-methylbicyclo[3.3.0]oct-3-ene-2,4-dicarboxylate**		**MS07775**
118	91	90	79	107	164	199	244	**328**	100	34	34	28	25	22	12	7	**0.00**	16	25	5	–	–	–	–	–	–	–	1	–	–	**Ethyl α-(diisopropoxyphosphinyl)phenylacetate**		**DD01274**
227	56	41	43	199	42	55	57	**328**	100	72	58	36	34	30	17	16	**11.20**	16	28	5	2	–	–	–	–	–	–	–	–	–	**1,4-Piperazinediacetic acid, 2-oxo-, dibutyl ester**	56701-25-8	**MS07776**
328	329	227	44	256	241	213	63	**328**	100	21	15	12	11	10	10	10		17	12	7	–	–	–	–	–	–	–	–	–	–	**Aflatoxin G1**	1165-39-5	**NI34522**
328	329	227	300	299	255	241	228	**328**	100	19	8	6	5	5	5	5		17	12	7	–	–	–	–	–	–	–	–	–	–	**Aflatoxin G1**	1165-39-5	**NI34523**
328	299	271	28	243	329	273	300	**328**	100	67	53	29	26	19	13	12		17	12	7	–	–	–	–	–	–	–	–	–	–	**Aflatoxin M1**	6795-23-9	**NI34524**
328	205	329	282	254	43	243	221	**328**	100	26	20	17	16	16	13	13		17	12	7	–	–	–	–	–	–	–	–	–	–	**Aflatoxin Q1**	52819-96-2	**NI34525**
284	282	310	328	238	266	255	254	**328**	100	68	51	32	24	22	20	16		17	12	7	–	–	–	–	–	–	–	–	–	–	**1,6-Dihydroxy-γ-methoxy-3-methylanthraquinone-2-carboxylic acid**		**LI07267**
191	189	179	207	192	208	190	178	**328**	100	37	25	24	24	21	17	17	**16.00**	17	13	2	–	–	–	–	1	–	–	–	–	–	**9-Phenanthrenemethyl monobromoacetate**	92174-38-4	**NI34526**
91	244	186	272	129	242	143	79	**328**	100	88	88	85	76	65	51	51	**48.82**	17	20	3	–	–	–	–	–	–	–	–	–	1	**Iron, tricarbonyl[(4a,4b,10a,10b-η)-1,2,3,4,5,6,7,8,9,10-decahydrobenzo[3,4]cyclobuta[1,2]cyclooctene]-**	38959-33-0	**MS07777**
244	242	272	116	56	240	135	134	**328**	100	85	63	63	62	54	41	37	**32.35**	17	20	3	–	–	–	–	–	–	–	–	–	1	**Iron, tricarbonyl[(4a,4b,10a,10b-η)-1,2,3,4,5,6,7,8,9,10-decahydrobenzo[3,4]cyclobuta[1,2]cyclooctene]-**	38959-33-0	**MS07778**
190	105	133	242	112	119	53	188	**328**	100	98	82	63	52	48	48	47	**15.09**	17	20	3	–	–	–	–	–	–	–	–	–	1	**Iron, tricarbonyl[(6a,6b,12a,12b-η)-1,2,3,4,5,6,7,8,9,10,11,12-dodecahydrocyclobuta[1,2:3,4]dicyclooctene]-**	69815-46-9	**MS07779**
328	70	216	58	187	115	286	271	**328**	100	42	30	15	11	6	5	3		17	20	3	4	–	–	–	–	–	–	–	–	–	**Morphinan-3,14-diol, 6-azido-4,5-epoxy-17-methyl-, (5α,6β)-**	54301-19-8	**NI34527**
119	292	174	328	262	267	188	158	**328**	100	76	66	43	42	37	33	33		17	20	3	4	–	–	–	–	–	–	–	–	–	**D-erythro-Pentose phenylosazone**		**LI07268**
55	269	241	115	41	209	181	43	**328**	100	82	67	41	23	22	21	17	**0.00**	17	28	6	–	–	–	–	–	–	–	–	–	–	**[S-(R*,R*)]-Methyl α-octyltetrahydro-2-methoxycarbonyl-5-oxo-2-furanacetate**		**NI34528**
83	138	81	95	57	103	69	55	**328**	100	66	61	60	33	27	27	27	**0.10**	17	28	6	–	–	–	–	–	–	–	–	–	–	**Propanoic acid, 2,3-bis(acetyloxy)-, 5-methyl-2-isopropylcyclohexyl ester**	69502-99-4	**NI34529**
73	83	95	138	55	81	75	41	**328**	100	73	62	56	51	48	44	35	**0.00**	17	32	4	–	–	–	–	–	–	1	–	–	–	**Menthyl trimethylsilylmethylmalonate**		**NI34530**
114	72	86	115	43	41	100	55	**328**	100	79	78	62	10	9	8	8	**1.40**	17	32	4	2	–	–	–	–	–	–	–	–	–	**L-Leucine, N-(N-acetyl-L-valyl)-, butyl ester**	62167-68-4	**NI34531**
128	86	72	129	28	100	43	41	**328**	100	65	64	39	18	13	13	11	**1.75**	17	32	4	2	–	–	–	–	–	–	–	–	–	**L-Valine, N-(N-acetyl-L-isoleucyl)-, butyl ester**	69833-63-2	**NI34532**
86	128	72	129	44	43	41	30	**328**	100	81	77	40	32	15	12	11	**0.82**	17	32	4	2	–	–	–	–	–	–	–	–	–	**L-Valine, N-(N-acetyl-L-leucyl)-, butyl ester**	62167-67-3	**NI34533**
147	69	181	77	189	51	223	105	**328**	100	93	31	26	23	16	15	15	**1.90**	18	13	4	–	–	–	1	–	–	–	–	–	–	**1,3,4,6-Hexanetetrone, 1-(4-chlorophenyl)-6-phenyl-**	58330-12-4	**NI34534**
328	329	296	313	151	89	132	237	**328**	100	22	22	20	18	15	14	13		18	16	2	–	2	–	–	–	–	–	–	–	–	**1,4-Dithiin, 2,5-bis(4-methoxyphenyl)-**	37989-51-8	**NI34535**
173	145	105	128	185	117	149	91	**328**	100	53	45	38	36	35	29	28	**0.00**	18	16	4	–	1	–	–	–	–	–	–	–	–	**2-Methyl-3-[(4-methylphenyl)sulphonyl]-4-phenyl-2(5H)-furanone**	118356-52-8	**DD01275**

MASS TO CHARGE RATIOS								M.W.	INTENSITIES								Parent	C	H	O	N	S	F	Cl	Br	I	Si	P	B	X	COMPOUND NAME	CAS Reg No	No
43	123	286	51	94	77	164	106	328	100	84	70	60	50	39	31	27	3.20	18	16	6	–	–	–	–	–	–	–	–	–	–	Benzoylgentisyl alcohol diacetate		MS07780
297	328	253	133	298	329	59	151	328	100	88	20	19	18	17	15	12		18	16	6	–	–	–	–	–	–	–	–	–	–	Biphenyl-2,4′,5-tricarboxylic acid, trimethyl ester		NI34536
328	313	329	149	162	281	282	175	328	100	20	16	16	15	12	6	6		18	16	6	–	–	–	–	–	–	–	–	–	–	6H-[1,3]Dioxolo[5,6]benzofuro[3,2-c][1]benzopyran, 6a,12a-dihydro-3,4-dimethoxy-, (6aR-cis)-	2694-70-4	NI34537
328	162	78	83	313	311	85	298	328	100	70	47	42	40	30	28	24		18	16	6	–	–	–	–	–	–	–	–	–	–	6H-[1,3]Dioxolo[5,6]benzofuro[3,2-c][1]benzopyran, 2,3-dimethoxy-	55334-50-4	NI34538
157	158	129	201	202	298	115	229	328	100	42	36	33	29	28	24	21	11.00	18	16	6	–	–	–	–	–	–	–	–	–	–	3b,9a[3′,4′]-Furano-as-indaceno[4,5-c]furan-1,3,11,13-tetrone, 3a,4,5,6,7,8,9,9b,10,14-decahydro-	32304-26-0	MS07781
43	127	202	203	174	89	85	269	328	100	75	60	18	18	15	15	1	0.30	18	16	6	–	–	–	–	–	–	–	–	–	–	7H-Furo[3,2-g][1]benzopyran-7-one, 9-[[4-(acetyloxy)-3-methyl-2-butenyl]oxy]-, (E)-	65853-07-8	NI34539
282	328	283	255	239	310	285	254	328	100	68	35	21	21	18	17	17		18	16	6	–	–	–	–	–	–	–	–	–	–	3-Hydroxy-4′,5,7-trimethoxyflavone	1098-92-6	NI34540
328	167	297	164	161	162	310	313	328	100	40	37	29	23	20	9	8		18	16	6	–	–	–	–	–	–	–	–	–	–	5-Hydroxy-7,2′,4′-trimethoxyisoflavone		NI34541
163	164	135	136	165	77	104	103	328	100	34	20	15	12	10	9	9	0.00	18	16	6	–	–	–	–	–	–	–	–	–	–	4,4′-Bis(methoxycarbonyl)benzoin		IC03976
44	43	28	105	77	93	41	39	328	100	46	39	33	26	23	21	20	0.32	18	16	6	–	–	–	–	–	–	–	–	–	–	Methyl 4-methoxycarbonylbenzyl terephthalate		MS07782
163	328	297	149	104	90	76	121	328	100	42	21	20	16	15	15	14		18	16	6	–	–	–	–	–	–	–	–	–	–	Methyl 4-methoxycarbonylbenzyl terephthalate		IC03977
163	149	328	133	297	164	89	135	328	100	17	16	16	14	12	7	6		18	16	6	–	–	–	–	–	–	–	–	–	–	Methyl 4-methoxycarbonylbenzyl terephthalate		IC03978
148	328	153	176	137	268	123	296	328	100	96	79	77	56	54	52	32		18	16	6	–	–	–	–	–	–	–	–	–	–	4-O-Methyl-4′,5′-O,O-methylenemopanol		MS07783
328	167	146	313	183	182	168		328	100	66	60	38	22	21	10			18	16	6	–	–	–	–	–	–	–	–	–	–	4′-O-Methylpunctatin		LI07269
313	328	149	298	296	281	314	269	328	100	85	43	37	25	23	20	18		18	16	6	–	–	–	–	–	–	–	–	–	–	1,4,5,8-Tetramethoxyanthraquinone		IC03979
330	328	105	331	329	106	226	225	328	100	99	27	20	20	14	11	11		18	17	1	–	–	–	–	1	–	–	–	–	–	5-Acetyl-11-bromotricyclo[8.2.2.$2^{4,7}$]hexadeca-4,6,10,12,13,15-hexaene	18931-44-7	NI34542
147	146	330	328	148	195	185	197	328	100	60	20	20	13	9	8	6		18	17	1	–	–	–	–	1	–	–	–	–	–	5-Acetyl-11-bromotricyclo[8.2.2.$2^{4,7}$]hexadeca-4,6,10,12,13,15-hexaene, stereoisomer	24417-97-8	NI34543
330	328	104	331	329	105	226	224	328	100	99	25	20	20	16	11	11		18	17	1	–	–	–	–	1	–	–	–	–	–	p-Bromoacetyl[2,2]paracyclophane		LI07270
146	145	330	328	147	245	182		328	100	60	20	20	12	10	8			18	17	1	–	–	–	–	1	–	–	–	–	–	pseudo-p-Bromoacetyl[2,2]paracyclophane		LI07271
123	177	151	205	179	178	206	207	328	100	37	22	17	12	10	8	7	1.00	18	17	2	2	–	–	1	–	–	–	–	–	–	2-(4-Chloroanilino)-4-(4-methoxyanilino)-cyclopent-2-enone		MS07784
198	292	296	250	164	330	236	328	328	100	33	30	11	11	10	10	1		18	19	1	3	–	–	1	–	–	–	–	–	–	2-Chloro-1-pyrrolidino-3-oxo-5,6,11,11b-tetrahydro-3H-indolo[3,2-g]indolizine		LI07272
198	292	296	250	164	330	236	327	328	100	33	30	11	11	10	10	1		18	19	1	3	–	–	1	–	–	–	–	–	–	2-Chloro-1-pyrrolidino-3-oxo-5,6,11,11b-tetrahydro-3H-indolo[3,2-g]indolizine		LI07273
70	97	125	164	124	123	135	328	328	100	99	62	45	40	31	22	18		18	20	–	2	2	–	–	–	–	–	–	–	–	3,8-Dimethyl-1,10-bis(2-thienyl)-4,7-diazadeca-1,3,7,9-tetraene		NI34544
328	191	163	82	165	76	164	84	328	100	60	56	44	37	36	31	29		18	20	4	2	–	–	–	–	–	–	–	–	–	Benzaldehyde, 3,4-dimethoxy-, [(3,4-dimethoxyphenyl)methylene]hydrazone	17745-86-7	NI34545
107	87	77	30	108	195	78	134	328	100	94	72	52	42	39	35	32	14.00	18	20	4	2	–	–	–	–	–	–	–	–	–	1,2-Bis[(4-hydroxyphenyl)acetamido]ethane		IC03980
132	165	164	77	328	133	105		328	100	50	50	20	12	12	10			18	20	4	2	–	–	–	–	–	–	–	–	–	1,2-Bis[(2-methoxycarbonylphenyl)amino]ethane		LI07274
328	313	149	77	329	300	165	98	328	100	90	82	81	74	71	68	64		18	24	2	4	–	–	–	–	–	–	–	–	–	5H,10H-Diimidazo[1,5-a:1′,5′-d]pyrazine-5,10-dione, 3,8-diethyl-1,6-diisopropyl-	56701-37-2	NI34546
42	328	41	58	43	121	164	68	328	100	55	49	36	32	25	19	18		18	28	–	6	–	–	–	–	–	–	–	–	–	1,2-Bis(5-methyl-1,3-diazaadamant-6-ylidene)hydrazine		MS07785
57	257	272	41	259	274	313	258	328	100	96	89	51	34	31	23	20	16.50	18	30	1	–	–	–	1	–	–	–	1	–	–	Phosphinic chloride, [(3,5-di-tert-butylphenyl)phenyl]tert-butyl-	29634-14-8	NI34547
79	80	67	81	41	93	91	28	328	100	82	80	75	67	55	40	38	0.94	18	32	3	–	1	–	–	–	–	–	–	–	–	Sulphuric acid, 5,8,11-heptadecatrienyl methyl ester	56554-67-7	NI34548
55	41	84	43	102	81	95	57	328	100	97	75	50	40	37	35	31	0.20	18	32	5	–	–	–	–	–	–	–	–	–	–	Oxacyclooctadec-3-en-2-one, 5,6,7-trihydroxy-18-methyl-	52461-05-9	NI34549
154	119	155	140	164	209	127	153	328	100	55	35	35	30	20	20	14	14.00	19	12	2	4	–	–	–	–	–	–	–	–	–	Pyrrolo[1,2-a]-1,3,5-triazine-8-carbonitrile, 1,2,3,4-tetrahydro-2,4-dioxo-3,7-diphenyl-	54450-44-1	NI34550
150	328	180	145	181	191	179	221	328	100	50	50	45	44	43	43	32		19	20	5	–	–	–	–	–	–	–	–	–	–	1,3:2,4-Di-O-benzylideneribitol	4148-60-1	LI07275
179	149	150	180	328	191	181	145	328	100	98	23	12	11	10	10	10		19	20	5	–	–	–	–	–	–	–	–	–	–	1,3:2,4-Di-O-benzylideneribitol	4148-60-1	NI34551
179	149	328	327	205	180	162	161	328	100	97	39	30	16	15	15	14		19	20	5	–	–	–	–	–	–	–	–	–	–	1,3:2,4-Di-O-benzylidenexylitol	55448-52-7	NI34552
328	327	205	180	162	161	297	222	328	100	79	41	40	40	36	32	31		19	20	5	–	–	–	–	–	–	–	–	–	–	1,3:2,4-Di-O-benzylidenexylitol	55448-52-7	LI07276
83	213	55	228	214	84	77	91	328	100	33	26	12	6	5	3	2	2.31	19	20	5	–	–	–	–	–	–	–	–	–	–	2-Butenoic acid, 3-methyl-, 4a,5,8,9-tetrahydro-9,9-dimethyl-5-oxobenzo[1,2-b:4,3-b′]dipyran-8-yl ester	55334-28-6	NI34553
83	213	55	228	214	328	84	229	328	100	83	70	29	12	7	7	6		19	20	5	–	–	–	–	–	–	–	–	–	–	2-Butenoic acid, 2-methyl-, 9,10-dihydro-8,8-dimethyl-2-oxo-2H,8H-benzo[1,2-b:3,4-b′]dipyran-9-yl ester	1165-60-2	NI34554
213	328	83	55	228	329	214	229	328	100	87	79	55	35	20	13	5		19	20	5	–	–	–	–	–	–	–	–	–	–	2-Butenoic acid, 2-methyl-, 9,10-dihydro-8,8-dimethyl-2-oxo-2H,8H-benzo[1,2-b:3,4-b′]dipyran-9-yl ester, [S-(Z)]-	19427-82-8	NI34555
78	83	55	77	228	57	56	51	328	100	79	53	26	22	18	16	15	5.00	19	20	5	–	–	–	–	–	–	–	–	–	–	2-Butenoic acid, 2-methyl-, 9,10-dihydro-8,8-dimethyl-2-oxo-2H,8H-benzo[1,2-b:3,4-b′]dipyran-9-yl ester, [S-(Z)]-	19427-82-8	LI07277
213	83	55	228	214	229	328	187	328	100	61	46	30	14	8	7	6		19	20	5	–	–	–	–	–	–	–	–	–	–	Columbianadin	5058-13-9	LI07278
244	43	60	45	286	44	245	243	328	100	58	37	34	28	19	15	13	8.00	19	20	5	–	–	–	–	–	–	–	–	–	–	4,5-Diacetoxy-2,3-dihydro-2,3,3-trimethylnaphtho[1,2-b]furan		LI07279
328	299	151	243	215	271	329	217	328	100	52	37	29	28	27	22	17		19	20	5	–	–	–	–	–	–	–	–	–	–	3,8-Dihydroxy-1,9-dimethyl-6-sec-butyldibenzo[b,e][1,4]-dioxepin-11-one		NI34556

MASS TO CHARGE RATIOS								M.W.	INTENSITIES								Parent	C	H	O	N	S	F	Cl	Br	I	Si	P	B	X	COMPOUND NAME	CAS Reg No	No
115	171	128	152	328	127	153	165	**328**	100	85	80	67	61	59	49	47		19	20	5	–	–	–	–	–	–	–	–	–	–	9-Hydroxy-6,7-dihydro-1,2,3-trimethoxybenzo[a]heptalen-10(5H)-one		MS07786
115	128	152	165	127	153	171	141	**328**	100	74	65	53	50	48	43	43	**27.00**	19	20	5	–	–	–	–	–	–	–	–	–	–	11-Hydroxy-6,7-dihydro-1,2,3-trimethoxybenzo[a]heptalen-10(5H)-one		MS07787
328	103	300	313	131	225	167	195	**328**	100	44	31	22	22	19	19	12		19	20	5	–	–	–	–	–	–	–	–	–	–	2-Propen-1-one, 3-phenyl-1-(2,3,4,6-tetramethoxyphenyl)-	63878-53-5	NI34557
83	213	228	187	328				**328**	100	85	45	24	15					19	20	5	–	–	–	–	–	–	–	–	–	–	O-Senecionyl dihydrooroselol		LI07280
300	328	195	150	121	161	180	285	**328**	100	54	45	31	30	18	17	15		19	20	5	–	–	–	–	–	–	–	–	–	–	2′,4,4′,6′-Tetramethoxychalcone		MS07788
58	328	100	42	59	228	185	43	**328**	100	14	10	7	6	5	5	5		19	24	1	2	1	–	–	–	–	–	–	–	–	10H-Phenothiazine-10-propanamine, 2-methoxy-N,N,β-trimethyl-, (–)-	60-99-1	NI34560
58	328	100	59	135	42	229	228	**328**	100	14	8	7	5	5	4	4		19	24	1	2	1	–	–	–	–	–	–	–	–	10H-Phenothiazine-10-propanamine, 2-methoxy-N,N,β-trimethyl-, (–)-	60-99-1	NI34558
58	328	100	229	228	185	329	242	**328**	100	36	15	9	9	9	8	8		19	24	1	2	1	–	–	–	–	–	–	–	–	10H-Phenothiazine-10-propanamine, 2-methoxy-N,N,β-trimethyl-, (–)-	60-99-1	NI34559
328	112	310	327	140	43	286	329	**328**	100	91	50	48	27	23	21	19		19	24	3	2	–	–	–	–	–	–	–	–	–	20,21-Dinoraspidospermidine-5,17-diol, 1-acetyl-	21451-67-2	NI34561
152	84	328	297	122	56	299	67	**328**	100	93	38	23	19	18	15	15		19	24	3	2	–	–	–	–	–	–	–	–	–	Gelsedine	7096-96-0	NI34562
171	87	55	211	41	69	97	67	**328**	100	71	70	64	46	44	43	43	**32.09**	19	36	2	–	1	–	–	–	–	–	–	–	–	Methyl 9,12-epithiostearate		MS07789
57	70	43	112	55	83	71	41	**328**	100	76	71	37	35	33	32	31	**0.00**	19	36	4	–	–	–	–	–	–	–	–	–	–	Bis(2-ethylhexyl) malonate		DO01101
43	41	45	55	57	87	97	69	**328**	100	40	28	25	18	13	12	12	**0.00**	19	36	4	–	–	–	–	–	–	–	–	–	–	Methyl 2-acetoxyhexadecanoate	21987-12-2	NI34563
167	157	225	257	199	214	207		**328**	100	40	20	11	11	9	6		**0.00**	19	36	4	–	–	–	–	–	–	–	–	–	–	Methyl 13-hydroxy-10-oxooctadecanoate		LI07281
98	74	112	101	297	69	55	83	**328**	100	94	87	59	58	56	55	45	**0.90**	19	36	4	–	–	–	–	–	–	–	–	–	–	Methyl 3-methylhexadecane-1,16-dioate		MS07790
73	117	313	75	132	43	129	145	**328**	100	74	72	71	40	32	31	29	**6.40**	19	40	2	–	–	–	–	–	–	1	–	–	–	Hexadecanoic acid, trimethylsilyl ester	55520-89-3	MS07791
73	117	75	313	132	129	43	145	**328**	100	77	73	64	49	33	33	32	**7.97**	19	40	2	–	–	–	–	–	–	1	–	–	–	Hexadecanoic acid, trimethylsilyl ester	55520-89-3	NI34565
73	75	117	43	132	41	129	55	**328**	100	87	70	52	46	37	35	34	**4.78**	19	40	2	–	–	–	–	–	–	1	–	–	–	Hexadecanoic acid, trimethylsilyl ester	55520-89-3	NI34564
105	77	173	183	251	145	51	78	**328**	100	58	34	31	25	25	23	14	**11.61**	20	15	1	–	–	3	–	–	–	–	–	–	–	[(4-Trifluoromethyl)phenyl]diphenylmethanol	21856-96-2	NI34566
328	327	329	164	77	169	299	168	**328**	100	28	24	18	12	10	6	6		20	16	1	4	–	–	–	–	–	–	–	–	–	2,5-Bis(1-methylbenzimidazol-2-yl)furan		IC03981
298	328	297	31	41	267	39	299	**328**	100	72	58	50	38	37	29	28		20	24	4	–	–	–	–	–	–	–	–	–	–	Aegyptinone B	121704-43-6	DD01276
284	91	79	40	42	105	77	54	**328**	100	91	79	67	61	50	50	50	**35.00**	20	24	4	–	–	–	–	–	–	–	–	–	–	Aldosterone γ-lactone	2507-89-3	LI07282
284	286	328	266	269	131	300	242	**328**	100	62	44	40	35	35	34	31		20	24	4	–	–	–	–	–	–	–	–	–	–	Aldosterone γ-lactone	2507-89-3	NI34567
216	255	171	328	199	272	217	29	**328**	100	43	34	30	28	26	17	15		20	24	4	–	–	–	–	–	–	–	–	–	–	2,6-Bis(butoxycarbonyl)naphthalene		IC03982
326	164	165	299	311	283			**328**	100	62	11	9	7	7			**0.00**	20	24	4	–	–	–	–	–	–	–	–	–	–	Dehydrodieugenol		LI07283
328	329	299	149	151	137	91	115	**328**	100	22	15	12	11	9	9	8		20	24	4	–	–	–	–	–	–	–	–	–	–	Dehydrodihydrodiisoeugenol		MS07792
151	82	135	178	67	79	91	53	**328**	100	95	55	50	48	45	40	39	**9.00**	20	24	4	–	–	–	–	–	–	–	–	–	–	Epilophodione		MS07793
91	82	151	105	68	178	53	135	**328**	100	86	85	79	64	62	62	52	**31.00**	20	24	4	–	–	–	–	–	–	–	–	–	–	Gersemolide		MS07794
242	328	229	243	115	227	128	211	**328**	100	67	57	35	30	27	24	21		20	24	4	–	–	–	–	–	–	–	–	–	–	(4bS,7S,8aR)-4b,5,6,7,8,8a-Hexahydro-3,7-dihydroxy-2-(2-propenyl)-4b,8,8-trimethylphenanthrene-1,4-dione		NI34568
122	328	123	300	329	121	285	135	**328**	100	38	21	15	9	9	8	8		20	24	4	–	–	–	–	–	–	–	–	–	–	18-Hydroxy-11-dehydrocorticosterone γ-lactone	4731-87-7	NI34569
137	148	91	105	328	180	77	310	**328**	100	61	50	47	33	33	15	10		20	24	4	–	–	–	–	–	–	–	–	–	–	5-Hydroxy-7-(4-hydroxy-3-methoxyphenyl)-1-phenyl-3-heptanone		DD01277
160	57	161	244	85	41	131	328	**328**	100	14	13	10	6	6	4	3		20	24	4	–	–	–	–	–	–	–	–	–	–	Pentanoic acid, 2,7-naphthalenediyl ester	55724-89-5	NI34570
244	43	286	175	136	229	55	328	**328**	100	90	22	15	15	14	10	7		20	24	4	–	–	–	–	–	–	–	–	–	–	5,6,7,8-Tetrahydro-7-isopropenyl-6-methyl-6-vinylnaphthalene-1,4-diol diacetate	77880-94-5	NI34571
91	73	75	237	41	148	65	45	**328**	100	90	51	23	13	11	11	11	**0.00**	20	28	2	–	–	–	–	–	–	1	–	–	–	(E)-1-[2-(Benzyloxymethyl)-1-cyclohexenyl]-3-(trimethylsilyl)-2-propen-1-one		DD01278
140	43	83	58	198	144	270	41	**328**	100	96	30	27	24	23	20	18	**6.00**	20	28	2	2	–	–	–	–	–	–	–	–	–	Aspidospermidine-3-methanol, 20-hydroxy-	56192-91-7	NI34572
140	328	141	300	269	168	329	160	**328**	100	20	11	8	7	6	5	4		20	28	2	2	–	–	–	–	–	–	–	–	–	Aspidospermidin-21-ol, 17-methoxy-	28189-98-2	NI34573
41	43	328	141	300	110	42	160	**328**	100	97	83	80	63	60	60	53		20	28	2	2	–	–	–	–	–	–	–	–	–	Aspidospermidin-21-ol, 17-methoxy-	28189-98-2	NI34574
78	328	143	58	77	51	52	185	**328**	100	99	91	80	49	47	45	42		20	28	2	2	–	–	–	–	–	–	–	–	–	8H-Azecino[5,4-b]indol-8-one, 5-ethyl-1,2,3,4,5,6,7,9-octahydro-6-(2-hydroxyethyl)-3-methyl-, [5R-(5R*,6S*)]-	2246-31-3	NI34575
328	327	43	200	41	199	42	255	**328**	100	93	74	65	65	60	57	51		20	28	2	2	–	–	–	–	–	–	–	–	–	Corynan-17-ol, 10-methoxy-	15266-55-4	NI34576
327	328	200	199	329	255	186	41	**328**	100	85	25	22	19	18	14	12		20	28	2	2	–	–	–	–	–	–	–	–	–	Corynan-17-ol, 10-methoxy-	15266-55-4	NI34577
328	327	200	199	60	326	329	168	**328**	100	90	87	46	34	27	22	22		20	28	2	2	–	–	–	–	–	–	–	–	–	Corynan-17-ol, 10-methoxy-	15266-55-4	NI34578
122	110	130	70	136	124	123	108	**328**	100	18	17	17	15	14	13	13	**1.10**	20	28	2	2	–	–	–	–	–	–	–	–	–	Dasycarpidan-8(16H)-ethanol, 3,18-didehydro-1-(hydroxymethyl)-, (2ξ,4ξ)-	55724-45-3	NI34579
84	112	215	69	328	216	41	243	**328**	100	90	66	41	39	38	33	32		20	28	2	2	–	–	–	–	–	–	–	–	–	Piperidine, 1,1′-[1,4-phenylenebis(1-oxo-2,1-ethanediyl)]bis-	55723-85-8	NI34580
124	123	122	158	125	144	204	130	**328**	100	27	16	14	14	13	12	12	**8.71**	20	28	2	2	–	–	–	–	–	–	–	–	–	1,3-Propanediol, 2-(3-ethylidene-1-methyl-4-piperidinyl)-2-(3-methylindol-2-yl)-	1850-32-4	NI34581
43	57	313	41	55	71	29	85	**328**	100	78	50	48	44	39	24	23	**0.00**	20	40	3	–	–	–	–	–	–	–	–	–	–	1,3-Dioxolane, 2,2-dimethyl-4-[(tetradecyloxy)methyl]-	13879-87-3	NI34582
43	104	55	98	57	41	117	69	**328**	100	95	64	60	59	56	44	35	**10.04**	20	40	3	–	–	–	–	–	–	–	–	–	–	2-Hydroxyethyl octadecanoate	111-60-4	NI34584
43	99	55	41	57	32	29	98	**328**	100	82	77	67	63	47	46	44	**4.04**	20	40	3	–	–	–	–	–	–	–	–	–	–	2-Hydroxyethyl octadecanoate	111-60-4	NI34583
71	43	41	55	102	57	69	83	**328**	100	97	45	42	37	36	33	21	**0.10**	20	40	3	–	–	–	–	–	–	–	–	–	–	2-Hydroxyhexadecyl butanoate		MS07795
115	74	69	55	73	45	83	57	**328**	100	83	71	66	62	62	52	40	**0.00**	20	40	3	–	–	–	–	–	–	–	–	–	–	Methyl 16-hydroxy-3,3-dimethylheptadecanoate		IC03983
229	143	69	55	41	43	71	45	**328**	100	72	54	37	31	24	21	21	**0.32**	20	40	3	–	–	–	–	–	–	–	–	–	–	Methyl (±)-11-methoxyoctadecanoate	55334-49-1	NI34585

MASS TO CHARGE RATIOS	M.W.	INTENSITIES	Parent	C	H	O	N	S	F	Cl	Br	I	Si	P	B	X	COMPOUND NAME	CAS Reg No	No
75 117 43 41 55 29 313 57	328	100 61 47 34 24 22 21 20	0.50	20	40	3	–	–	–	–	–	–	–	–	–	–	Methyl 3-methoxyoctadecanoate		MS07796
131 71 241 43 41 55 83 69	328	100 37 22 21 19 14 12 12	0.00	20	40	3	–	–	–	–	–	–	–	–	–	–	Methyl 4-methoxyoctadecanoate		MS07797
229 143 69 55 41 43 71 45	328	100 70 54 37 31 24 21 21	0.40	20	40	3	–	–	–	–	–	–	–	–	–	–	Methyl 11-methoxyoctadecanoate		MS07798
59 313 298 281 264 297 74 55	328	100 31 16 14 12 8 7 7	1.10	20	40	3	–	–	–	–	–	–	–	–	–	–	Methyl 17-methoxyoctadecanoate		MS07799
45 55 74 296 87 98 83 69	328	100 77 75 53 52 46 45 45	8.60	20	40	3	–	–	–	–	–	–	–	–	–	–	Methyl 18-methoxyoctadecanoate		MS07800
271 75 272 83 97 69 57 68	328	100 41 25 23 19 16 15 13	0.00	20	44	1	–	–	–	–	–	–	1	–	–	–	1-(tert-Butyldimethylsilyloxy)tetradecane		NI34586
328 311 76 104 215 329 271 312	328	100 81 52 31 25 23 21 19		21	12	4	–	–	–	–	–	–	–	–	–	–	3-Methyl-2,2′-binaphthalene-1,1′,4,4′-tetrone	72872-49-2	NI34587
165 269 166 296 328 270 295 164	328	100 90 58 31 28 28 16 11		21	16	–	2	1	–	–	–	–	–	–	–	–	2,5,5-Triphenyl-2-imidazolin-4-thione	24134-04-1	NI34588
105 77 103 328 89 165 104 78	328	100 65 9 8 5 4 3 3		21	16	2	2	–	–	–	–	–	–	–	–	–	Benzaldehyde dibenzoylhydrazone		LI07284
328 257 180 165 329 77 256 258	328	100 54 50 26 24 20 11 10		21	16	2	2	–	–	–	–	–	–	–	–	–	1,5,5-Triphenyl-2,4-imidazolidinedione	52460-88-5	NI34589
180 328 181 104 329 209 165 77	328	100 71 54 36 16 16 16 16		21	16	2	2	–	–	–	–	–	–	–	–	–	3,5,5-Triphenyl-2,4-imidazolidinedione	52461-02-6	NI34590
256 133 257 255 298 268 147 272	328	100 99 77 39 22 20 18 12	4.20	21	28	3	–	–	–	–	–	–	–	–	–	–	17-Acetoxy-19-norisopimara-4(18),8(14),15-trien-7-one		MS07801
328 284 256 257 313 271 269 295	328	100 39 26 25 14 12 12 7		21	28	3	–	–	–	–	–	–	–	–	–	–	B-Bishomo-5α,7α-cycloandrostan-3,7a,17-trione		LI07285
285 328 258 257 214 43 286 270	328	100 98 56 47 42 32 27 25		21	28	3	–	–	–	–	–	–	–	–	–	–	Cannabicoumaronone	70474-97-4	MS07802
328 297 44 32 269 329 83 36	328	100 53 49 39 26 25 25 25		21	28	3	–	–	–	–	–	–	–	–	–	–	Cortex-6-enone		MS07803
328 43 121 147 329 159 173 91	328	100 62 31 28 23 23 19 19		21	28	3	–	–	–	–	–	–	–	–	–	–	1,5-Cycloandrost-3-en-2-one, 17-(acetyloxy)-, (1α,5β,10α,17β)-	2363-60-2	NI34591
296 328 221 43 236 195 297 253	328	100 62 29 29 26 26 23 21		21	28	3	–	–	–	–	–	–	–	–	–	–	1H-Cyclopenta[a]phenanthren-17-ol, 2,3,4,11,12,13,14,15,16,17-decahydro 3-methoxy-13-methyl-, acetate	56771-80-3	MS07804
123 161 133 41 162 81 91 43	328	100 100 95 42 31 29 26 25	11.00	21	28	3	–	–	–	–	–	–	–	–	–	–	Cyclopropanecarboxylic acid, 2,2-dimethyl-3-(2-methyl-1-propenyl)-, 2-methyl-4-oxo-3-(1,3-pentadienyl)-2-cyclopenten-1-yl ester	5768-81-0	NI34592
123 45 162 80 133 91 56 47	328	100 26 20 20 19 16 15 14	5.00	21	28	3	–	–	–	–	–	–	–	–	–	–	3-[2,2-Dimethyl-3-(2-methylprop-1-enyl]cyclopropylcarbonyloxy)-4-methyl-5-(penta-2,4-dienyl)cyclopent-4-enone		LI07286
43 216 295 310 267 277 201 311	328	100 90 71 64 36 24 18 16	0.00	21	28	3	–	–	–	–	–	–	–	–	–	–	(+)-7α,8α-Epoxy-12β-hydroxy-5α-pregna-2,16-dien-20-one		DD01279
328 43 295 310 329 84 91 313	328	100 84 77 65 23 22 21 10		21	28	3	–	–	–	–	–	–	–	–	–	–	3α,4α-Epoxy-12β-hydroxy-5α-pregna-7,16-dien-20-one		DD01280
43 216 310 108 277 91 295 267	328	100 62 50 34 31 31 30 28	0.00	21	28	3	–	–	–	–	–	–	–	–	–	–	7α,8α-Epoxy-12β-hydroxy-5α-pregna-3,16-dien-20-one		DD01281
328 147 160 173 329 186 159 174	328	100 41 40 30 23 21 20 19		21	28	3	–	–	–	–	–	–	–	–	–	–	Estra-1,3,5(10)-triene-3,17-diol, 1-methyl-, 17-acetate, (17β)-	2456-11-3	NI34593
229 230 302 310 281 265 110 242	328	100 89 59 58 56 53 42 39	16.31	21	28	3	–	–	–	–	–	–	–	–	–	–	DL-16β-Hydroxynorgestrel	40915-03-5	NI34594
217 310 141 161 295 253 228 328	328	100 52 26 24 20 20 16 12		21	28	3	–	–	–	–	–	–	–	–	–	–	15β-Hydroxypregna-1,4-diene-3,20-dione	70019-87-3	NI34595
285 267 268 286 143 310 225 328	328	100 54 27 20 16 7 7 1		21	28	3	–	–	–	–	–	–	–	–	–	–	17-Methyl-18-norpregna-4,13-dien-16-ol-3,20-dione		LI07287
171 227 328 228 172 141 185 184	328	100 88 17 16 15 7 6 6		21	28	3	–	–	–	–	–	–	–	–	–	–	2-Naphthalenepropanoic acid, β-butyl-6-methoxy-α,α-dimethyl-, methyl ester	57274-57-4	NI34596
328 285 243 313 257 201 173 329	328	100 93 65 50 46 25 24 23		21	28	3	–	–	–	–	–	–	–	–	–	–	2,4(1H,3H)-Phenanthrenedione, 4a,9,10,10a-tetrahydro-7-methoxy-1,1,4a-trimethyl-8-isopropyl-	55334-48-0	NI34597
328 243 285 257 173 313 201 216	328	100 70 64 48 39 38 35 30		21	28	3	–	–	–	–	–	–	–	–	–	–	Podocarpa-8,11,13-triene-1,3-dione, 13-isopropyl-12-methoxy-	18326-18-6	NI34598
243 284 257 173 215 328 298 244	328	100 69 46 35 31 26 20 20		21	28	3	–	–	–	–	–	–	–	–	–	–	Podocarpa-8,11,13-triene-1,3-dione, 14-isopropyl-13-methoxy-	15372-53-9	NI34599
243 285 257 173 215 328 300 286	328	100 69 46 34 31 26 23 22		21	28	3	–	–	–	–	–	–	–	–	–	–	Podocarpa-8,11,13-triene-1,3-dione, 14-isopropyl-13-methoxy-	15372-53-9	NI34600
328 243 43 91 41 105 93 55	328	100 75 62 48 31 24 22 21		21	28	3	–	–	–	–	–	–	–	–	–	–	Pregn-9(11)-ene-3,6,20-trione, (5α)-	37717-04-7	NI34601
243 43 328 105 85 91 242 41	328	100 47 45 36 27 25 22 20		21	28	3	–	–	–	–	–	–	–	–	–	–	Pregn-9(11)-ene-3,6,20-trione, (5α,17α)-	56362-36-8	NI34602
123 108 93 79 41 136 81 39	328	100 30 26 26 23 20 19 19	2.46	21	28	3	–	–	–	–	–	–	–	–	–	–	Pyrethrin I	121-21-1	NI34603
123 41 81 43 91 55 105 67	328	100 50 38 32 31 28 26 20	0.20	21	28	3	–	–	–	–	–	–	–	–	–	–	Pyrethrin I	121-21-1	NI34604
123 41 43 162 81 133 91 55	328	100 25 21 20 20 19 16 14	3.20	21	28	3	–	–	–	–	–	–	–	–	–	–	Pyrethrin I	121-21-1	MS07805
328 329 231 257 232 330 229 218	328	100 28 20 13 8 7 7 7		21	32	1	–	–	–	–	–	–	1	–	–	–	Estra-1,3,5(10)-triene, 3-[(trimethylsilyl)oxy]-	29755-24-6	NI34605
77 89 75 57 71 43 55 41	328	100 81 77 73 67 66 48 40	3.64	21	44	2	–	–	–	–	–	–	–	–	–	–	1-Propanol, 3-(octadecyloxy)-	17367-36-1	NI34606
223 105 77 183 121 224 328 93	328	100 67 43 31 22 16 11 10		22	16	3	–	–	–	–	–	–	–	–	–	–	2-(Diphenylhydroxymethyl)chromone	20924-64-5	NI34607
223 105 77 183 121 224 328 251	328	100 66 43 30 22 16 11 10		22	16	3	–	–	–	–	–	–	–	–	–	–	2-(Diphenylhydroxymethyl)chromone	20924-64-5	LI07288
328 43 28 18 313 41 253 255	328	100 99 94 90 76 68 66 40		22	32	2	–	–	–	–	–	–	–	–	–	–	8-Acetoxymethyl-4b,8-dimethyl-12-methylidene-2,3,4,4a,4b,5,6,7,8,10-decahydro-1H-2,10a-ethanophenanthrene		LI07289
41 328 68 91 67 137 55 213	328	100 84 71 64 63 60 59 55		22	32	2	–	–	–	–	–	–	–	–	–	–	Androst-5-en-17-one, 3-hydroxy-16-isopropylidene-, (3β)-	3509-94-2	NI34608
328 137 294 213 310 217 313 243	328	100 30 24 22 21 17 16 9		22	32	2	–	–	–	–	–	–	–	–	–	–	Androst-5-en-17-one, 3-hydroxy-16-isopropylidene-, (3β)-	3509-94-2	LI07290
137 69 41 175 191 121 81 177	328	100 23 20 18 14 12 12 11	11.00	22	32	2	–	–	–	–	–	–	–	–	–	–	1,3-Benzenediol, 5-methyl-2-(3,7,11-trimethyl-2,6,10-dodecatrienyl)-, (E,E)-	6903-07-7	NI34609
43 41 57 149 55 71 70 39	328	100 70 50 36 31 22 18 17	0.00	22	32	2	–	–	–	–	–	–	–	–	–	–	6H-Dibenzo[b,d]pyran-1-ol, 3-hexyl-6a,7,8,10a-tetrahydro-6,6,9-trimethyl , (6aR-trans)-	36482-24-3	NI34610
329 328 271 81 330 327 311 285	328	100 34 31 30 29 22 18 16		22	32	2	–	–	–	–	–	–	–	–	–	–	3-Hexyl-Δ^9-tetrahydrocannibinol		NI34611
43 328 124 84 91 79 55 41	328	100 56 40 34 25 25 24 21		22	32	2	–	–	–	–	–	–	–	–	–	–	18-Norpregn-4-ene-3,20-dione, 13-ethyl-	5895-84-1	NI34612
43 328 124 84 91 79 55 41	328	100 99 72 61 45 45 44 37		22	32	2	–	–	–	–	–	–	–	–	–	–	18-Norpregn-4-ene-3,20-dione, 13-ethyl-	5895-84-1	NI34613

MASS TO CHARGE RATIOS	M.W.	INTENSITIES	Parent	C	H	O	N	S	F	Cl	Br	I	Si	P	B	X	COMPOUND NAME	CAS Reg No	No
43 328 138 286 55 41 79 91	**328**	100 40 30 24 23 23 22 21		22	32	2	–	–	–	–	–	–	–	–	–	–	19-Norpregn-4-ene-3,20-dione, 10-ethyl-	4781-53-7	NI34614
43 328 138 286 41 55 79 91	**328**	100 99 75 59 58 57 56 52		22	32	2	–	–	–	–	–	–	–	–	–	–	19-Norpregn-4-ene-3,20-dione, 10-ethyl-	4781-53-7	NI34615
271 286 175 272 201 43 328 287	**328**	100 66 35 21 21 21 16 14		22	32	2	–	–	–	–	–	–	–	–	–	–	2-Phenanthrenol, 4b,5,6,7,8,8a,9,10-octahydro-4b,8,8-trimethyl-1-isopropyl-, acetate, (4bS-trans)-	15340-82-6	NI34616
286 271 287 328 189 272 175 43	**328**	100 49 21 17 11 10 10 10		22	32	2	–	–	–	–	–	–	–	–	–	–	3-Phenanthrenol, 4b,5,6,7,8,8a,9,10-octahydro-4b,8,8-trimethyl-2-isopropyl-, acetate, (4bS-trans)-	15340-79-1	NI34617
310 295 91 105 77 159 107 145	**328**	100 87 84 60 47 44 42 41	**25.72**	22	32	2	–	–	–	–	–	–	–	–	–	–	Pregna-5,16-dien-20-one, 3-hydroxy-16-methyl-, (3β)-	1808-63-5	WI01423
138 43 328 137 243 286 55 41	**328**	100 83 56 29 27 26 23 22		22	32	2	–	–	–	–	–	–	–	–	–	–	Pregn-4-ene-3,20-dione, 6-methyl-, (6α)-	903-71-9	NI34618
138 43 137 109 147 93 191 41	**328**	100 76 36 19 18 18 17 16	**11.71**	22	32	2	–	–	–	–	–	–	–	–	–	–	Pregn-4-ene-3,20-dione, 6-methyl-, (6α,9β,10α)-	4255-80-5	NI34619
138 43 328 137 91 55 93 67	**328**	100 58 30 23 15 15 12 12		22	32	2	–	–	–	–	–	–	–	–	–	–	Pregn-4-ene-3,20-dione, 6-methyl-, (6α,10α)-	55220-88-7	NI34620
138 43 137 91 55 93 109 67	**328**	100 58 24 16 15 13 12 12	**0.00**	22	32	2	–	–	–	–	–	–	–	–	–	–	Pregn-4-ene-3,20-dione, 6-methyl-, (6α,10α)-	55220-88-7	NI34621
328 43 138 286 137 244 329 91	**328**	100 54 41 39 34 28 25 20		22	32	2	–	–	–	–	–	–	–	–	–	–	Pregn-4-ene-3,20-dione, 6-methyl-, (6β)-	2300-06-3	NI34622
43 328 124 286 55 41 79 95	**328**	100 99 99 59 44 38 35 34		22	32	2	–	–	–	–	–	–	–	–	–	–	Pregn-4-ene-3,20-dione, 7-methyl-, (7α)-	2640-71-3	NI34623
43 328 124 286 55 41 95 79	**328**	100 60 60 36 27 23 21 21		22	32	2	–	–	–	–	–	–	–	–	–	–	Pregn-4-ene-3,20-dione, 7-methyl-, (7α)-	2640-71-3	NI34624
43 124 328 149 41 95 55 79	**328**	100 78 45 33 31 28 28 26		22	32	2	–	–	–	–	–	–	–	–	–	–	Pregn-4-ene-3,20-dione, 14-methyl-	55162-96-4	NI34625
43 124 328 149 41 95 55 79	**328**	100 99 58 41 40 36 36 33		22	32	2	–	–	–	–	–	–	–	–	–	–	Pregn-4-ene-3,20-dione, 14-methyl-	55162-96-4	NI34626
327 43 124 286 243 244 107 55	**328**	100 76 69 41 25 24 20 20	**0.00**	22	32	2	–	–	–	–	–	–	–	–	–	–	Pregn-4-ene-3,20-dione, 16-methyl-, (16α)-	1239-79-8	NI34627
122 124 286 54 243 85 244 107	**328**	100 84 25 25 24 24 23 21	**20.00**	22	32	2	–	–	–	–	–	–	–	–	–	–	Pregn-4-ene-3,20-dione, 16-methyl-, (16β)-	1424-09-5	NI34628
43 124 328 286 91 55 79 41	**328**	100 71 67 34 26 26 25 25		22	32	2	–	–	–	–	–	–	–	–	–	–	Pregn-4-ene-3,20-dione, 16-methyl-, (16β)-	1424-09-5	NI34629
43 124 328 286 91 55 41 79	**328**	100 99 93 47 36 36 35 34		22	32	2	–	–	–	–	–	–	–	–	–	–	Pregn-4-ene-3,20-dione, 16-methyl-, (16β)-	1424-09-5	NI34630
224 85 245 91 121 79 67 104	**328**	100 33 18 12 11 11 10 9	**8.00**	22	32	2	–	–	–	–	–	–	–	–	–	–	Pregn-4-ene-3,20-dione, 16-methyl-, (16β,17α)-		LI07291
244 43 85 41 124 121 55 79	**328**	100 49 45 15 14 14 13 12	**2.87**	22	32	2	–	–	–	–	–	–	–	–	–	–	Pregn-4-ene-3,20-dione, 16-methyl-, (16β,17α)-	1922-34-5	NI34631
244 128 245 91 122 79 124 67	**328**	100 21 18 13 11 11 10 10	**8.16**	22	32	2	–	–	–	–	–	–	–	–	–	–	Pregn-4-ene-3,20-dione, 16-methyl-, (16β,17α)-	1922-34-5	NI34632
244 43 328 41 55 91 229 85	**328**	100 97 44 41 35 33 28 28		22	32	2	–	–	–	–	–	–	–	–	–	–	Pregn-4-ene-3,20-dione, 17-methyl-	1046-28-2	NI34633
43 244 328 41 55 91 85 79	**328**	100 99 44 41 36 34 29 28		22	32	2	–	–	–	–	–	–	–	–	–	–	Pregn-4-ene-3,20-dione, 17-methyl-	1046-28-2	NI34634
43 112 119 105 41 145 91 55	**328**	100 37 34 29 29 28 27 24	**14.00**	22	32	2	–	–	–	–	–	–	–	–	–	–	Retinol, acetate	127-47-9	LI07292
43 41 105 91 55 119 69 145	**328**	100 58 42 41 38 36 34 33	**5.04**	22	32	2	–	–	–	–	–	–	–	–	–	–	Retinol, acetate	127-47-9	NI34636
43 112 119 105 145 41 91 55	**328**	100 39 34 31 29 29 28 26	**14.00**	22	32	2	–	–	–	–	–	–	–	–	–	–	Retinol, acetate	127-47-9	NI34635
41 43 55 69 57 45 91 79	**328**	100 99 76 61 52 45 38 36	**4.00**	22	32	2	–	–	–	–	–	–	–	–	–	–	Spiro[androst-5-ene-17,1′-cyclobutan]-2′-one, 3-hydroxy-, (3β,17β)-	60534-16-9	NI34637
70 259 57 260 56 71 58 55	**328**	100 60 40 13 10 8 8 8	**5.00**	22	36	–	2	–	–	–	–	–	–	–	–	–	23-Norcon-5-enin-3-amine, N-methyl-, (3β)-	468-41-7	NI34638
70 328 259 96 244 329 91 67	**328**	100 51 35 34 20 14 11 10		22	36	–	2	–	–	–	–	–	–	–	–	–	23-Norcon-18(22)-enin-3-amine, N-methyl-, (3β)-	56282-40-7	NI34639
70 328 259 96 55 57 244 329	**328**	100 51 35 34 32 24 20 16		22	36	–	2	–	–	–	–	–	–	–	–	–	23-Norcon-20-enin-3-amine, N-methyl-, (3β,5α)-		LI07293
259 260 181 105 183 272 182 261	**328**	100 24 18 14 10 8 7 6	**4.43**	23	24	–	–	–	–	–	–	–	1	–	–	–	Silane, (2-methyl-2-butenyl)triphenyl-	74630-82-3	NI34640
104 91 105 117 77 131 78 208	**328**	100 31 23 21 15 9 9 8	**1.30**	24	24	1	–	–	–	–	–	–	–	–	–	–	cis-4-Benzyl-2,6-diphenyltetrahydropyran		MS07806
243 165 244 183 105 77 41 166	**328**	100 25 24 16 16 7 5 4	**1.90**	24	24	1	–	–	–	–	–	–	–	–	–	–	Cyclopentyl trityl ether	1241-40-3	NI34642
243 165 244 105 183 228 215 77	**328**	100 25 24 16 14 10 10 6	**3.00**	24	24	1	–	–	–	–	–	–	–	–	–	–	Cyclopentyl trityl ether	1241-40-3	NI34641
78 105 77 149 298 55 57 328	**328**	100 66 65 45 37 37 35 31		24	24	1	–	–	–	–	–	–	–	–	–	–	Isobenzofuran, 4,7-dihydro-4,5,6,7-tetramethyl-1,3-diphenyl-	51566-92-8	NI34643
328 161 55 232 95 105 81 215	**328**	100 87 81 76 68 65 64 62		24	40	–	–	–	–	–	–	–	–	–	–	–	Chol-7-ene, (5β)-	54411-75-5	NI34644
55 41 81 95 109 67 287 217	**328**	100 94 81 80 75 72 71 64	**14.00**	24	40	–	–	–	–	–	–	–	–	–	–	–	Chol-23-ene, (5β)-	54411-82-4	NI34645
328 271 91 313 179 105 193 257	**328**	100 77 70 68 26 22 20 19		25	28	–	–	–	–	–	–	–	–	–	–	–	3-[(E)-2,2-Dimethylpropylidene]-2-[(Z)-(4-methylphenyl)methylene]-1-[(E) phenylmethylene]cyclopentane		MS07807
328 327 162 163 326 329 164 150	**328**	100 55 40 33 30 26 17 17		26	16	–	–	–	–	–	–	–	–	–	–	–	9H-Fluorene, 9-(9H-fluoren-9-ylidene)-	746-47-4	LI07294
328 300 329 301 162 327 326 324	**328**	100 42 27 20 20 18 18 18		26	16	–	–	–	–	–	–	–	–	–	–	–	Hexahelicene	187-83-7	NI34646
328 300 329 301 324 162 327 326	**328**	100 46 29 21 20 20 18 18		26	16	–	–	–	–	–	–	–	–	–	–	–	Hexahelicene	187-83-7	LI07295
328 278 329 279 164 139 326 276	**328**	100 83 31 20 19 16 13 12		26	16	–	–	–	–	–	–	–	–	–	–	–	Naphtho[1,2-b]triphenylene		NI34647
59 329 238 150 206 174 118 164	**329**	100 19 19 13 11 8 7 6		8	11	1	1	6	–	–	–	–	–	–	–	–	2,4,6,8,9,10-Hexathiatricyclo[3.3.1.1^{3,7}]decane-1-carboxamide, 3,5,7-trimethyl-	57289-12-0	NI34648
329 288 197 77 152 314 114 149	**329**	100 87 69 57 24 21 20 18		9	10	–	3	–	4	–	–	–	–	3	–	–	2-(1-Propenyl)-2-phenyltetrafluorocyclotriphosphazene		MS07808
185 173 159 187 172 157 186 175	**329**	100 34 13 9 7 7 6 6	**0.00**	10	20	5	1	2	–	–	–	–	–	1	–	–	7-Oxa-5-thia-2-aza-6-phosphanonanoic acid, 6-ethoxy-2-methyl-3-oxo-, ethyl ester, 6-sulphide	2595-54-2	NI34649
43 44 70 105 331 45 69 316	**329**	100 26 22 14 13 11 10 9	**9.00**	12	9	1	1	1	–	1	1	–	–	–	–	–	Thiazole, 5-acetyl-2-(4-bromophenyl)-4-(chloromethyl)-	41981-07-1	NI34650
216 126 203 109 69 147 217 78	**329**	100 50 27 22 22 19 14 14	**1.20**	12	9	3	1	–	6	–	–	–	–	–	–	–	2-(2-Hydroxyphenyl)ethylamine DITFA		MS07809
203 216 69 105 175 109 77 126	**329**	100 68 53 44 37 25 20 18	**1.20**	12	9	3	1	–	6	–	–	–	–	–	–	–	2-Hydroxy-2-phenylethylamine DITFA		MS07810
266 264 265 267 111 234 154 155	**329**	100 98 66 21 11 7 6 4	**0.60**	12	12	4	3	1	–	1	–	–	–	–	–	–	2-(4-Chlorobenzenesulphonamido)-4,6-dimethoxypyrimidine		IC03984

MASS TO CHARGE RATIOS								M.W.	INTENSITIES								Parent	C	H	O	N	S	F	Cl	Br	I	Si	P	B	X	COMPOUND NAME	CAS Reg No	No
116	115	117	69	91	238	105	132	**329**	100	42	36	12	12	11	11	10	**5.10**	13	10	1	1	–	7	–	–	–	–	–	–	–	trans-2-Phenylcyclopropylamine, heptafluorobutanoyl-		NI34651
56	43	58	314	41	187	44	70	**329**	100	93	61	41	38	26	25	24	**1.79**	13	13	3	3	–	–	2	–	–	–	–	–	–	1-Imidazolidinecarboxamide, 3-(3,5-dichlorophenyl)-N-(1-methylethyl)-2,4-dioxo-	36734-19-7	NI34652
43	56	40	58	44	70	41	42	**329**	100	94	75	60	42	40	40	35	**0.00**	13	13	3	3	–	–	2	–	–	–	–	–	–	1-Imidazolidinecarboxamide, 3-(3,5-dichlorophenyl)-N-(1-methylethyl)-2,4-dioxo-	36734-19-7	MS07811
127	43	40	142	56	99	42	44	**329**	100	98	85	75	72	62	52	50	**7.19**	13	13	3	3	–	–	2	–	–	–	–	–	–	1-Imidazolidnecarboxamide, 3-(1-methylethyl)-N-(3,5-dichlorophenyl)-2,4-dioxo-		MS07812
73	115	75	41	143	329	113	40	**329**	100	82	44	44	41	29	29	25		13	31	1	3	–	–	–	–	–	3	–	–	–	Creatinine TRITMS		MS07813
42	91	59	58	92	65	55	178	**329**	100	41	30	30	29	24	21	19	**12.12**	14	19	4	1	2	–	–	–	–	–	–	–	–	Propionic acid, 3-[[2-(carboxyamino)ethyl]dithio]-, N-benzyl methyl ester	26599-16-6	NI34653
117	91	90	116	89	51	63	39	**329**	100	52	45	38	27	19	17	15	**0.00**	14	19	6	1	1	–	–	–	–	–	–	–	–	Desulphoglucotropaeolin		MS07814
84	83	150	85	149	81	86	79	**329**	100	67	35	35	15	15	13	13	**7.00**	14	19	8	1	–	–	–	–	–	–	–	–	–	5H-Pyrano[3,2-d]oxazole-6,7-diol, 5-[(acetyloxy)methyl]-3a,6,7,7a-tetrahydro-2-methyl-, diacetate, [3aR-(3aα,5α,6β,7α,7aα)]-	35954-65-5	NI34654
84	196	85	83	96	150	97	149	**329**	100	62	55	50	33	25	25	23	**7.50**	14	19	8	1	–	–	–	–	–	–	–	–	–	5H-Pyrano[3,2-d]oxazole-6,7-diol, 5-[(acetyloxy)methyl]-3a,6,7,7a-tetrahydro-2-methyl-, diacetate, [3aS-(3aα,5β,6α,7β,7aα)]-	20757-53-3	NI34655
145	103	129	141	117	159	87	99	**329**	100	56	42	33	20	18	17	12	**0.00**	14	19	8	1	–	–	–	–	–	–	–	–	–	2,3,4,5-Tetra-O-acetyl-6-deoxy-D-galactonitrile	6298-84-6	NI34656
301	303	329	331	139	166	222	167	**329**	100	99	98	97	48	29	28	27		15	8	3	1	–	–	–	1	–	–	–	–	–	1-Amino-4-bromoanthraquinone-2-carboxaldehyde		NI34657
329	96	164	183	69	165	77	242	**329**	100	84	59	57	53	43	19	13		15	10	–	1	–	6	–	–	–	–	–	1	–	(Hexafluoroisopropylidenimino)diphenylborane		LI07296
169	171	331	329	332	330	212	210	**329**	100	99	84	84	13	13	10	10		15	12	1	3	–	–	–	1	–	–	–	–	–	Δ^2-1,2,4-Triazolin-5-one, 1-(4-bromophenyl)-3-methyl-4-phenyl-	32589-66-5	NI34658
329	331	302	330	250	300	18	303	**329**	100	97	60	51	47	45	37	36		15	12	1	3	–	–	–	1	–	–	–	–	–	N-Methyl bromoazepam		NI34659
330	314	240						**329**	100	80	47						**0.00**	15	31	3	1	–	–	–	–	–	2	–	–	–	8-Azabicyclo[3.2.1]octane-2-carboxylic acid, 8-methyl-3-[(trimethylsilyl)oxy]-, trimethylsilyl ester, [1R-(exo,exo)]-	74367-12-7	NI34660
195	43	238	118	44	91	167	45	**329**	100	57	46	30	28	16	12	8	**0.00**	16	12	1	1	–	5	–	–	–	–	–	–	–	1-Methyl-2-phenylethylamine pentafluorobenzoyl ester		MS07815
195	238	118	91	167	196	239	117	**329**	100	45	34	12	11	7	5	5	**0.46**	16	12	1	1	–	5	–	–	–	–	–	–	–	1-Methyl-2-phenylethylamine pentafluorobenzoyl ester		MS07816
118	195	105	207	117	43	119	77	**329**	100	42	32	18	14	14	8	7	**1.20**	16	12	1	1	–	5	–	–	–	–	–	–	–	2-(4-Methylphenyl)ethylamine pentafluorobenzoyl ester		MS07817
104	30	181	105	91	329			**329**	100	42	35	23	20	14				16	12	1	1	–	5	–	–	–	–	–	–	–	N-Pentafluorophenylacetyl-2-phenylethylamine		MS07818
237	218	222	176	238	132	117	91	**329**	100	43	39	17	13	13	12	12	**0.00**	16	12	1	1	–	5	–	–	–	–	–	–	–	2-Pentafluorophenyl-3-methyl-5-phenyl-1,3-oxazolidine		MS07819
112	41	44	55	42	56	28	18	**329**	100	38	33	32	28	22	17	15	**0.30**	16	21	2	1	–	–	2	–	–	–	–	–	–	Benzoic acid, 3,4-dichloro-, 3-(2-methyl-1-piperidinyl)propyl ester	3478-94-2	NI34661
245	230	217	44	36	216	244	231	**329**	100	80	70	67	67	47	33	24	**0.00**	17	12	4	1	–	–	1	–	–	–	–	–	–	6,7-Dihydro[a,f]quinolizium perchlorate	71721-61-4	NI34662
285	254	284	239	311	240	286	329	**329**	100	52	49	26	23	21	19	14		17	15	6	1	–	–	–	–	–	–	–	–	–	1-Carboxymethyl-2-methoxycarbonyl-5,7-dimethylcyclo[3.2.2]azine-4-carboxylic acid		LI07297
270	297	329	271	211	254	153	152	**329**	100	42	34	17	11	10	10	7		17	15	6	1	–	–	–	–	–	–	–	–	–	Diethyl (1-methoxycarbonyl)methylcyclo[3.2.2]azine-2,4-dicarboxylate		LI07298
157	59	43	69	71	46	55	41	**329**	100	56	39	35	32	32	27	27	**12.00**	17	31	5	1	–	–	–	–	–	–	–	–	–	Isoxazolidine, 5-hexyl-2-[2,3-O-isopropylidene-β-D-ribofuranosyl]-	69575-62-8	NI34663
79	139	106	78	111	51	75	141	**329**	100	72	70	54	38	33	25	21	**9.60**	18	13	1	1	–	–	2	–	–	–	–	–	–	3-Pyridinemethanol, α,α-bis(4-chlorophenyl)-	17781-31-6	NI34664
130	329	157	91	156	131	65	77	**329**	100	40	32	26	19	16	12	11		18	19	3	1	1	–	–	–	–	–	–	–	–	3-(3-Indolyl)propyl 4-toluenesulphonate		LI07299
214	297	59	72	298	254	255	168	**329**	100	69	45	30	29	23	21	19	**12.00**	18	19	5	1	–	–	–	–	–	–	–	–	–	1-Acetyl-2,5-dimethyl-3,4-bis(methoxycarbonyl)-6,7-benzo-1H-azepine		NI34665
269	181	240	211	224	225	251	300	**329**	100	73	37	26	24	22	9	5	**1.00**	18	19	5	1	–	–	–	–	–	–	–	–	–	3-O-Acetylhamayne		NI34666
314	329	256	315	270	286	271	228	**329**	100	50	33	25	23	19	17	17		18	19	5	1	–	–	–	–	–	–	–	–	–	9(10H)-Acridinone, 1,2,3,4-tetramethoxy-10-methyl-	517-73-7	NI34667
314	329	273	274	245	315	330	287	**329**	100	85	70	23	22	20	17	15		18	19	5	1	–	–	–	–	–	–	–	–	–	Ethyl 8-oxo-3-tert-butylpyrano[2,3-f]-1,4-benzoxazine-6-carboxylate		LI07300
329	139	150	70	286	296	330	268	**329**	100	77	45	26	25	22	20	19		18	19	5	1	–	–	–	–	–	–	–	–	–	11-Hydroxycephalotaxinone hemiketal	49686-61-5	NI34668
214	145	329	144	255	298	254	256	**329**	100	54	36	30	17	15	14	10		18	19	5	1	–	–	–	–	–	–	–	–	–	6,7-Bis(methoxycarbonyl)-1,5-dimethyl-2-acetyl-3,4-benzo-2-azabicyclo[3.2.0]hepta-3,6-diene		NI34669
270	242	226	314	329	227			**329**	100	55	40	36	30	18				18	19	5	1	–	–	–	–	–	–	–	–	–	5H-Pyrano[3,2-c]quinolin-5-one, 6-[(acetyloxy)methyl]-2,6-dihydro-7-methoxy-2,2-dimethyl-	65560-23-8	NI34670
298	91	118	206	56	74	42	45	**329**	100	93	40	21	21	17	13	10	**0.80**	18	23	3	1	–	–	–	–	–	1	–	–	–	2,2-Diphenyl-6-(2′-hydroxyethyl)-1,3-dioxa-6-aza-2-silacyclooctane		MS07820
329	73	314	330	149.5	157	142.5	141.5	**329**	100	44	38	32	19	3	2	2		18	27	1	1	–	–	–	–	–	2	–	–	–	Silylamine, 1,1,1-trimethyl-N-[4′-(trimethylsiloxy)-4-biphenylyl]-	31396-30-2	LI07301
329	73	314	330	331	315	150	316	**329**	100	44	36	32	12	11	5	4		18	27	1	1	–	–	–	–	–	2	–	–	–	Silylamine, 1,1,1-trimethyl-N-[4′-(trimethylsiloxy)-4-biphenylyl]-	31396-30-2	NI34671
270	114	59	130	69	174	100	160	**329**	100	54	53	35	33	29	16	15	**0.00**	18	39	2	1	–	–	–	–	–	1	–	–	–	(E)-1-[[[Bis(2-ethoxyethyl)amino]methyl]dimethylsilyl]-1-heptene		DD01282
270	69	59	130	232	114	174	100	**329**	100	62	62	37	26	26	25	18	**0.00**	18	39	2	1	–	–	–	–	–	1	–	–	–	3-[[[Bis(2-ethoxyethyl)amino]methyl]dimethylsilyl]-1-heptene		DD01283
284	146	59	114	89	204	69	100	**329**	100	89	80	26	26	25	25	14	**0.00**	18	39	2	1	–	–	–	–	–	1	–	–	–	3-[[[Bis(2-methoxyethyl)amino]methyl]dimethylsilyl]-1-nonene		DD01284
103	329	178	111	77	76	119	286	**329**	100	78	78	65	60	56	42	31		19	15	1	5	–	–	–	–	–	–	–	–	–	2-Acetylamino-6,7-diphenylpyrazolo[5,1-c][1,2,4]triazine		NI34672
256	91	137	31	71	210	192	238	**329**	100	97	62	20	9	8	8	5	**3.00**	19	23	2	1	1	–	–	–	–	–	–	–	–	Ethyl N-(3-methylthiobenzyl)benzylaminoethanoate		MS07821
314	146	329	192	82	178	59	244	**329**	100	40	36	35	34	28	28	21		19	23	4	1	–	–	–	–	–	–	–	–	–	8,14-Dihydrosalutaridine	15444-27-6	MS07822
329	314	298	42	330	112	44	300	**329**	100	53	31	27	22	22	22	16		19	23	4	1	–	–	–	–	–	–	–	–	–	7a,9c-(Iminoethano)phenanthro[4,5-bcd]furan-5-ol, 4aα,5α,8,9-tetrahydro-3,8β-dimethoxy-12-methyl-	19742-01-9	NI34673
314	329	44	243	330	315	192	42	**329**	100	99	39	20	18	18	13	11		19	23	4	1	–	–	–	–	–	–	–	–	–	Isosinomenine	510-42-9	LI07302

MASS TO CHARGE RATIOS								M.W.	INTENSITIES								Parent	C	H	O	N	S	F	Cl	Br	I	Si	P	B	X	COMPOUND NAME	CAS Reg No	No
252	253	196	256	224	329	284	228	**329**	100	18	12	9	9	8	7	4		19	23	4	1	–	–	–	–	–	–	–	–	–	**3,5-Pyridinedicarboxylic acid, 1,4-dihydro-2,6-dimethyl-4-phenyl-, diethyl ester**	1165-06-6	**LI07303**
252	253	196	329	256	224	284	300	**329**	100	17	12	8	8	8	6	4		19	23	4	1	–	–	–	–	–	–	–	–	–	**3,5-Pyridinedicarboxylic acid, 1,4-dihydro-2,6-dimethyl-4-phenyl-, diethyl ester**	1165-06-6	**NI34674**
252	196	224	253	256	29	284	42	**329**	100	24	23	15	9	8	6	6	**5.00**	19	23	4	1	–	–	–	–	–	–	–	–	–	**3,5-Pyridinedicarboxylic acid, 1,4-dihydro-2,6-dimethyl-4-phenyl-, diethyl ester**	1165-06-6	**NI34675**
314	329	192	78	330	178	44	315	**329**	100	99	36	23	20	20	20	19		19	23	4	1	–	–	–	–	–	–	–	–	–	**Sinomenine**		**LI07304**
83	59	85	329	84	185	115	211	**329**	100	96	66	64	40	18	15	14		19	23	4	1	–	–	–	–	–	–	–	–	–	**Tetrahydroamurine**		**LI07305**
301	91	240	135	192	256	117	119	**329**	100	58	46	37	36	34	33	32	**11.00**	19	27	2	3	–	–	–	–	–	–	–	–	–	**4H-Imidazole-2-acetic acid, 5-(dimethylamino)-α-ethyl-4,4-dimethyl-α-phenyl-, ethyl ester**	69315-96-4	**NI34676**
314	329	268	57	212	315	330	28	**329**	100	69	53	40	28	21	17	15		20	27	3	1	–	–	–	–	–	–	–	–	–	**Cinnamic acid, 3,5-di-tert-butyl-α-cyano-4-hydroxy-, ethyl ester**	3293-92-3	**NI34677**
257	297	186	268	239	298	258	266	**329**	100	94	33	24	21	20	19	17	**0.00**	20	27	3	1	–	–	–	–	–	–	–	–	–	**Estra-1,3,5(10)-trien-16-one, 17-hydroxy-3-methoxy-, O-methyloxime, (17β)-**	69833-87-0	**NI34678**
329	298	91	92	330	299	189	187	**329**	100	53	52	42	23	20	19	19		20	27	3	1	–	–	–	–	–	–	–	–	–	**Estra-1,3,5(10)-trien-17-one, 2-hydroxy-3-methoxy-, O-methyloxime**	74299-30-2	**NI34679**
329	96	299	298	300	330	189	187	**329**	100	57	44	42	27	23	14	10		20	27	3	1	–	–	–	–	–	–	–	–	–	**Estra-1,3,5(10)-trien-17-one, 3-hydroxy-2-methoxy-, O-methyloxime**	69833-94-9	**NI34680**
157	297	266	209	91	92	210	144	**329**	100	89	85	70	69	55	53	38	**24.82**	20	27	3	1	–	–	–	–	–	–	–	–	–	**Estra-1,3,5(10)-trien-17-one, 3-hydroxy-6-methoxy-, O-methyloxime, (6β)-**	74299-26-6	**NI34681**
297	157	266	91	92	298	126	96	**329**	100	46	34	30	24	23	23	22	**1.20**	20	27	3	1	–	–	–	–	–	–	–	–	–	**Estra-1,3,5(10)-trien-17-one, 3-hydroxy-6-methoxy-, O-methyloxime, (6β)-**	74299-26-6	**NI34682**
91	92	266	298	329	158	133	65	**329**	100	80	30	15	12	10	10	9		20	27	3	1	–	–	–	–	–	–	–	–	–	**Estra-1,3,5(10)-trien-17-one, 3-hydroxy-15-methoxy-, O-methyloxime, (15β)-**	74299-13-1	**NI34683**
329	298	91	92	299	330	189	241	**329**	100	85	42	33	26	23	20	16		20	27	3	1	–	–	–	–	–	–	–	–	–	**Estra-1,3,5(10)-trien-17-one, 4-hydroxy-3-methoxy-, O-methyloxime**	74299-28-8	**NI34684**
135	214	329	77	152	57	55	141	**329**	100	31	29	29	28	27	26	23		20	27	3	1	–	–	–	–	–	–	–	–	–	**Pipercallosine [(2E,4E)-N-isobutyl-9-(3,4-methylenedioxyphenyl)nona-2,4 dienamide]**	83029-39-4	**NI34685**
43	175	93	193	133	135	119	109	**329**	100	95	85	78	76	74	73	58	**0.00**	20	27	3	1	–	–	–	–	–	–	–	–	–	**(1RS,2RS,3RS,7RS)-1,2,6,6-Tetramethyl-10-oxatricyclo[5.2.1.0^{2,7}]dec-3-yl N-phenylcarbamate**		**DD01285**
79	59	73	52	329	75	150	51	**329**	100	100	60	52	52	51	43	28		20	31	1	1	–	–	–	–	–	1	–	–	–	**Levorphanol, TMS**		**NI34686**
105	329	77	330	223	106	331	224	**329**	100	77	22	17	8	7	5	4		21	15	1	1	1	–	–	–	–	–	–	–	–	**Benzamide, N-(3-phenylbenzo[b]thien-2-yl)-**	51324-21-1	**NI34688**
105	329	77	330	223	106	331	224	**329**	100	81	23	18	9	7	6	4		21	15	1	1	1	–	–	–	–	–	–	–	–	**Benzamide, N-(3-phenylbenzo[b]thien-2-yl)-**	51324-21-1	**NI34687**
180	77	105	329	83	181	85	106	**329**	100	50	25	20	19	18	18	7		21	15	1	1	1	–	–	–	–	–	–	–	–	**Thiazolium, 4-hydroxy-2,3,5-triphenyl-**	18100-80-6	**NI34689**
186	329	328	159	115	89	300	130	**329**	100	82	50	30	30	30	22	20		21	15	3	1	–	–	–	–	–	–	–	–	–	**1-Oxo-2-formyl-3-(methylphenylamino)-9-oxy-1H-phenalene**		**MS07823**
193	105	90	77	43	104	89	57	**329**	100	78	73	67	64	62	58	57	**0.00**	21	19	1	3	–	–	–	–	–	–	–	–	–	**Benzil hydrazone O-benzyloxime**	56667-02-8	**NI34690**
252	178	224	179	329	79	300	273	**329**	100	51	48	43	27	22	20	18		21	19	1	3	–	–	–	–	–	–	–	–	–	**3-(4-Nitrophenylamino)-1,2-diphenylindolizine**		**MS07824**
303	329	298	230	246	272	261	248	**329**	100	85	63	47	42	35	31	29		21	31	2	1	–	–	–	–	–	–	–	–	–	**17α-Ethynyl-17β-hydroxy-5α-estran-3-one-methyloxime**		**NI34691**
262	303	329	298	272	281	230	255	**329**	100	76	62	50	39	36	29	27		21	31	2	1	–	–	–	–	–	–	–	–	–	**17α-Ethynyl-17β-hydroxy-5β-estran-3-one-methyloxime**		**NI34692**
224	146	105	221	329	160	249	172	**329**	100	91	87	78	74	49	39	35		22	19	2	1	–	–	–	–	–	–	–	–	–	**trans-2-Benzoylbicyclo[2.2.2.]oct-5-ene-3-spiro-3′-(1′-acetyl-2′-indolinone)**		**LI07306**
284	77	329	152	105	224	165	153	**329**	100	45	42	18	15	11	11	9		22	19	2	1	–	–	–	–	–	–	–	–	–	**3-Ethoxy-2,3-diphenyl-2,3-dihydro-1-oxoisoindole**		**LI07307**
41	42	130	69	43	113	58	217	**329**	100	36	26	23	21	20	20	13	**11.00**	22	23	–	3	–	–	–	–	–	–	–	–	–	**2,3,5,6-Tetrahydro-5-(3-indolyl)-3,3-dimethyl-1H-pyrrolinyl[2,1-b]-1,3-benzodiazepine**		**MS07825**
113	41	55	43	70	126	72	98	**329**	100	70	61	58	52	51	33	32	**8.60**	22	35	1	1	–	–	–	–	–	–	–	–	–	**N-Octadeca-6,9,12,15-tetraenoylpyrrolidide**		**NI34693**
243	165	244	166	228	239	105	91	**329**	100	62	34	11	8	6	6	6	**1.36**	23	23	1	1	–	–	–	–	–	–	–	–	–	**Morpholine, 4-(triphenylmethyl)-**	1420-06-0	**NI34694**
117	79	64	47	76	114	44	295	**330**	100	36	30	14	13	10	7	6	**1.70**	2	–	–	–	3	–	6	–	–	–	–	–	–	**Bis(trichloromethyl) trisulphide**		**LI07308**
79	117	47	82	48	44	64	114	**330**	100	61	56	55	30	28	20	9	**0.10**	2	–	2	–	2	–	6	–	–	–	–	–	–	**Trichloromethyl trichloromethanethiosulphonate**		**LI07309**
69	169	119	117	48	64	147	67	**330**	100	26	22	16	14	13	7	7	**0.00**	5	–	3	–	1	10	–	–	–	–	–	–	–	**Trifluoromethylsulphinyl heptafluorobutyrate**		**LI07310**
69	330	78	96	222	310	260	128	**330**	100	30	20	16	13	11	6	6		6	2	–	2	–	12	–	–	–	–	–	–	–	**Bis[bis(trifluoromethyl)amino]ethylene**		**LI07311**
178	69	330	90	109	166	159	64	**330**	100	49	41	36	27	12	9	8		6	2	–	2	–	12	–	–	–	–	–	–	–	**Bis[bis(trifluoromethyl)amino]ethylene**		**LI07312**
69	330	78	311	223	166	96	173	**330**	100	32	19	16	14	13	13	10		6	2	–	2	–	12	–	–	–	–	–	–	–	**cis-1,2-Bis[bis(trifluoromethyl)amino]ethylene**		**LI07313**
330	203	76	50	331	127	75	74	**330**	100	30	27	13	7	5	5	5		6	4	–	–	–	–	–	–	2	–	–	–	–	**1,2-Diiodobenzene**		**MS07826**
330	203	76	50	74	331	165	75	**330**	100	33	33	15	9	7	6	6		6	4	–	–	–	–	–	–	2	–	–	–	–	**1,3-Diiodobenzene**		**MS07827**
330	203	76	50	74	75	331	165	**330**	100	32	32	19	10	8	7	5		6	4	–	–	–	–	–	–	2	–	–	–	–	**1,4-Diiodobenzene**		**MS07828**
330	92	63	65	295	64	93	294	**330**	100	60	46	43	42	40	24	21		6	5	1	–	–	–	1	–	–	–	–	–	1	**(2-Hydroxyphenyl)mercuric chloride**		**LI07314**
121	330	241	153	209	150	119	60	**330**	100	99	99	99	98	97	93	93		6	18	–	6	1	–	–	–	–	–	4	–	–	**2,4,6,8,9,10-Hexaaza-1,3,5,7-tetraphosphatricyclo[3.3.1.1^{3,7}]decane, 2,4,6,8,9,10-hexamethyl-, 1-sulphide**	38448-57-6	**NI34695**

MASS TO CHARGE RATIOS								M.W.	INTENSITIES								Parent	C	H	O	N	S	F	Cl	Br	I	Si	P	B	X	COMPOUND NAME	CAS Reg No	No
88	210	73	42	315	212	107	44	330	100	56	9	6	5	5	5	5	0.00	7	15	–	2	4	–	–	–	–	–	–	–	1	Carbamodithioic acid, dimethyl-, bis(anhydrosulphide) with methylarsonodithious acid	2445-07-0	NI34696
167	165	86	121	42	123	43	87	330	100	96	55	30	29	28	15	10	0.00	8	12	4	–	–	–	–	2	–	–	–	–	–	Bis(2-bromomethyl-1,3-dioxolan-2-yl)		LI07315
167	165	121	123	43	42	87	45	330	100	97	61	56	39	24	22	21	2.00	8	12	4	–	–	–	–	2	–	–	–	–	–	4a,8a-Bis(bromomethyl)-1,4,5,8-tetraoxadecalin		LI07316
43	190	192	99	70	250	252	69	330	100	14	13	11	10	9	8	6	0.00	8	12	4	–	–	–	–	2	–	–	–	–	–	1,4-Diacetoxy-2,3-dibromo-butane		IC03985
29	99	127	179	151	181	153	55	330	100	42	35	19	15	15	15	11	0.00	8	12	4	–	–	–	–	2	–	–	–	–	–	DL-Diethyl 2,3-dibromosuccinate	608-82-2	NI34697
99	127	26	179	55	151	153	181	330	100	88	39	35	31	31	29	29	0.00	8	12	4	–	–	–	–	2	–	–	–	–	–	meso-Diethyl 2,3-dibromosuccinate	1114-30-3	NI34698
301	303	302	273	217	245	304	35	330	100	51	28	24	16	15	12	11	0.99	8	23	4	–	–	–	1	–	–	4	–	–	–	1,3,5,7-Tetraethyl-1-chlorocyclotetrasilane	73420-23-2	NI34699
315	317	316	150	319	318	130	151	330	100	99	58	47	44	44	33	32	2.20	8	24	–	2	–	–	2	–	–	4	–	–	–	Cyclodisilazane, 1,3-bis(chlorodimethylsilyl)-2,2,4,4-tetramethyl-	2329-10-4	NI34700
315	317	316	150	319	318	151	130	330	100	80	29	23	22	21	16	16	1.00	8	24	–	2	–	–	2	–	–	4	–	–	–	Cyclodisilazane, 1,3-bis(chlorodimethylsilyl)-2,2,4,4-tetramethyl-	2329-10-4	LI07317
59	45	31	315	259	47	43	129	330	100	65	36	27	19	16	16	15	2.40	9	15	5	–	–	–	–	–	1	–	–	–	–	α-D-Glucofuranose, 5-deoxy-5-iodo-1,2-O-isopropylidene-	35810-92-5	MS07829
206	255	283	227	181	45	81	109	330	100	81	78	56	38	38	30	18	0.00	9	16	5	–	–	5	–	–	–	–	1	–	–	[2,2-Difluoro-2-ethoxy-1-(trifluoromethyl)ethyl]diethylphosphate		MS07830
301	299	303	150	332	330	142	221	330	100	78	49	38	32	25	18	16		10	6	4	–	–	–	4	–	–	–	–	–	–	1,4-Benzenedicarboxylic acid, 2,3,5,6-tetrachloro-, dimethyl ester	1861-32-1	NI34701
301	299	303	332	221	330	223	142	330	100	75	49	26	23	20	19	19		10	6	4	–	–	–	4	–	–	–	–	–	–	1,4-Benzenedicarboxylic acid, 2,3,5,6-tetrachloro-, dimethyl ester	1861-32-1	NI34702
301	299	303	332	45	330	221	334	330	100	80	48	32	30	24	19	17		10	6	4	–	–	–	4	–	–	–	–	–	–	1,4-Benzenedicarboxylic acid, 2,3,5,6-tetrachloro-, dimethyl ester	1861-32-1	NI34703
43	42	332	330	157	155	76	75	330	100	45	30	30	20	20	18	18		10	11	2	4	1	–	–	1	–	–	–	–	–	5-(4-Bromophenylsulphonylimino)-1,2-dimethyl-Δ^3-1,2,3-triazoline		MS07831
42	43	82	111	187	185	76	75	330	100	60	46	26	22	22	18	18	6.00	10	11	2	4	1	–	–	1	–	–	–	–	–	5-[(4-Bromophenylsulphonyl)methylamino]-1-methyl-1,2,3-triazole		MS07832
43	111	332	330	56	76	75	155	330	100	61	15	15	13	11	11	10		10	11	2	4	1	–	–	1	–	–	–	–	–	5-[(4-Bromophenylsulphonyl)methylamino]-2-methyl-1,2,3-triazole		MS07833
125	173	93	127	29	158	99	55	330	100	98	95	83	49	48	28	19	1.60	10	19	6	–	2	–	–	–	–	–	1	–	–	Malathion	121-75-5	AI02054
173	127	125	93	158	99	29	143	330	100	99	99	96	54	44	32	24	1.40	10	19	6	–	2	–	–	–	–	–	1	–	–	Malathion	121-75-5	NI34704
248	130	166	165	295	150	273	110	330	100	60	40	40	25	22	9	6	1.00	10	20	–	2	–	–	4	–	–	–	–	2	–	tert-Butylaldiminodichloroborane dimer		LI07318
69	70	133	41	134	330	55	43	330	100	76	66	48	43	13	9	8		10	20	4	–	2	–	–	–	–	–	2	–	–	2,2′-Bi-1,3,2-dioxaphosphorinane, 5,5,5′,5′-tetramethyl-, 2,2′-disulphide	16368-08-4	NI34705
288	290	268	332	289	287	118	117	330	100	40	40	30	15	10	10	10	10.00	11	12	–	2	–	–	–	2	–	–	–	–	–	5,7-Dibromo-3-[(dimethylamino)methyl]indole		NI34706
193	99	127	155	148	137	120	109	330	100	98	84	78	49	39	28	26	0.00	11	24	7	–	–	–	–	–	–	–	2	–	–	[2-(Diethoxyphosphinyl)-1-methyleneethyl] diethyl phosphate		DD01286
232	330	168	230	328	139	215	332	330	100	91	61	52	47	46	43	41		12	8	2	–	–	–	1	–	–	–	1	–	1	O,O′-(2,2′-Biphenylylene)selenophosphoric acid chloride	77671-18-2	NI34707
81	53	330	96	82	332	64	44	330	100	11	9	7	6	4	3	3		12	11	5	2	1	–	1	–	–	–	–	–	–	Benzoic acid, 5-(aminosulphonyl)-4-chloro-2-[(2-furanylmethyl)amino]-	54-31-9	NI34708
81	44	53	330	96	82	332	331	330	100	18	13	13	7	6	5	2		12	11	5	2	1	–	1	–	–	–	–	–	–	Benzoic acid, 5-(aminosulphonyl)-4-chloro-2-[(2-furanylmethyl)amino]-	54-31-9	NI34709
251	172	146	171	103	134	157	119	330	100	90	51	44	24	23	22	12	0.00	12	12	1	–	–	–	–	2	–	–	–	–	–	4,4-Dibromo-1-(4-methoxyphenyl)-spiro[2.2]pentane		MS07834
70	57	71	83	58	60	84	56	330	100	61	36	27	26	24	21	16	13.90	12	15	3	4	1	–	1	–	–	–	–	–	–	7-Chloro-3-(4-methyl-1-piperazinyl)-4H-1,2,4-benzothiadiazine 1,1,4′-trioxide		MS07835
43	85	41	55	69	57	29	83	330	100	89	33	27	25	20	12	11	5.00	12	26	–	–	–	–	–	–	–	–	–	–	2	Bis(2-ethylbutyl) diselenide		MS07836
173	247	165	109	201	166	65	219	330	100	96	52	52	44	30	23	22	0.50	12	28	6	–	–	–	–	–	–	–	2	–	–	2,2-Diethyl-1,1-diisopropyl-ethane-1,2-diphosphonate		MS07837
165	179	193	137	166	201	55	138	330	100	84	83	53	41	29	28	20	0.50	12	28	6	–	–	–	–	–	–	–	2	–	–	Tetraethyl butane-1,4-diphosphonate		MS07838
143	115	102	128	60	116	57	58	330	100	24	19	17	16	15	13	12	2.92	13	12	2	–	1	–	–	–	–	–	–	–	1	(π-Indenyl)(π-methylthiomethyl)dicarbonyl molybdenum		MS07839
332	330	317	315	334	319	333	281	330	100	75	74	56	25	19	14	13		13	12	3	–	–	–	1	1	–	–	–	–	–	2-Bromo-4-chloro-5,8-dimethoxy-3-methyl-1-naphthol	104506-11-8	NI34710
111	198	139	170	169	77	197	138	330	100	91	79	58	50	40	34	23	0.00	13	15	4	2	1	–	1	–	–	–	–	–	–	2H-1,2,3-Benzothiadiazine, 5,6,7,8-tetrahydro-2-phenyl-, monoperchlorate	42141-12-8	NI34711
297	200	110	298	232	330	55	98	330	100	23	18	17	12	10	10	9		13	19	4	2	1	–	–	–	–	–	1	–	–	Methyl (4-nitrophenyl)-N-cyclohexylphosphoramidothioate		LI07319
43	41	57	55	71	56	69	124	330	100	57	53	27	21	16	15	11	0.00	13	22	–	–	–	2	1	1	–	–	–	–	–	1-Bromo-2-chloro-1,1-difluoro-2-tridecene		DD01287
330	150	77	151	149	44	121	51	330	100	47	47	43	38	36	34	24		14	10	4	4	1	–	–	–	–	–	–	–	–	2-Thiophenecarboxaldehyde, 5-(1-propynyl)-, (2,4-dinitrophenyl)hydrazone	56588-22-8	NI34712
149	150	330	63	121	65	76	91	330	100	98	82	64	37	35	33	30		14	10	6	4	–	–	–	–	–	–	–	–	–	1,3-Benzodioxole-5-carboxaldehyde, (2,4-dinitrophenyl)hydrazone	14400-81-8	NI34713
330	58	164	136	194	158	150	65	330	100	34	30	30	23	14	14	10		14	12	4	2	–	–	–	–	–	–	–	–	1	Nickel, bis(4-methyl-o-benzoquinone 2-oximato)-	30993-30-7	NI34714
45	97	126	108	98	169	185	127	330	100	61	43	34	28	19	14	12	0.00	14	18	9	–	–	–	–	–	–	–	–	–	–	(1RS,3SR,4RS,7RS)-3-exo,6,6,7-syn-Tetraacetyl-2-oxabicyclo[2.2.1]heptane		DD01288
126	97	69	144	113	185	139	60	330	100	93	37	33	33	21	21	21	2.66	14	18	9	–	–	–	–	–	–	–	–	–	–	Tetra-O-acetyl-1-deoxy-D-arabino-hex-1-enopyranose		MS07840
98	97	109	228	60	69	169	144	330	100	80	45	35	35	32	30	25	0.00	14	18	9	–	–	–	–	–	–	–	–	–	–	Tetra-O-acetyl-3-deoxy-α-D-erythro-hex-2-enopyranose		MS07841
98	97	109	228	60	69	169	126	330	100	80	45	35	35	32	30	25	0.00	14	18	9	–	–	–	–	–	–	–	–	–	–	Tetra-O-acetyl-3-deoxy-α-D-erythro-hex-2-enopyranose		LI07320
61	330	267	315	95	56	123	109	330	100	38	20	16	15	13	11	10		14	22	3	2	2	–	–	–	–	–	–	–	–	5-Allyl-1,3-bis(methylthiomethyl)-5-isopropyl-2,4,6-pyrimidinetrione		MS07842
138	180	152	154	139	183	199	198	330	100	33	24	23	23	13	11	11	4.00	14	22	7	2	–	–	–	–	–	–	–	–	–	Uridine, 5-(4-hydroxypentyl)-	68972-49-6	NI34715
43	125	124	113	183	28	155	57	330	100	12	6	5	4	4	3	3	0.00	14	23	8	–	–	–	–	–	–	–	–	1	–	L-Mannitol, 1-deoxy-, cyclic 3,4-(ethylboronate) 2,5,6-triacetate	74779-70-7	NI34716
332	330	331	223	329	77	333	91	330	100	98	67	46	36	31	27	23		15	11	–	2	1	–	–	1	–	–	–	–	–	7-Bromo-2,3-dihydro-5-phenyl-1H-1,4-benzodiazepin-2-thione	34099-70-2	NI34717
330	183	154	115	331	156	128	155	330	100	53	45	21	17	16	15	14		15	11	1	2	–	5	–	–	–	–	–	–	–	N-Acetyltryptamine, pentafluoropropionyl-		MS07843
104	177	329	127	74	135	331	77	330	100	71	60	46	46	23	20	19	16.00	15	11	1	4	1	–	1	–	–	–	–	–	–	Urea, N-(4-chlorophenyl)-N′-(5-phenyl-1,2,4-thiadiazol-3-yl)-	42053-82-7	NI34718
91	156	155	159	92	175	108	65	330	100	90	83	67	67	58	57	28	10.00	15	13	3	–	1	3	–	–	–	–	–	–	–	2,2,2-Trifluoro-1-phenylethyl 4-toluenesulphonate		MS07844
330	133	77	148	151	43	135	134	330	100	97	46	44	30	30	26	25		15	14	5	4	–	–	–	–	–	–	–	–	–	4-Methoxyacetophenone, (2,4-dinitrophenyl)hydrazone	854-04-6	LI07321

MASS TO CHARGE RATIOS								M.W.	INTENSITIES								Parent	C	H	O	N	S	F	Cl	Br	I	Si	P	B	X	COMPOUND NAME	CAS Reg No	No
330	133	77	148	151	43	135	92	330	100	98	45	43	30	30	27	25		15	14	5	4	–	–	–	–	–	–	–	–	–	4-Methoxyacetophenone, (2,4-dinitrophenyl)hydrazone	854-04-6	NI34719
250	249	206	220	207	222	80	251	330	100	74	31	27	24	19	18	18	0.00	15	15	–	4	–	–	–	1	–	–	–	–	–	2-Imino-3-methyl-5-phenyl-1,2-dihydro-3H-1,3,4-benzotriazepine hydrobromide		MS07845
106	172	27	147	41	31	28	39	330	100	82	75	70	48	48	48	47	3.00	15	15	6	2	–	–	–	–	–	–	–	1	–	Uridine, cyclic 2′,3′-(phenylboronate)	19455-04-0	NI34720
201	229	149	120	81	148	147	257	330	100	63	62	54	41	41	41	40	13.95	15	20	5	–	–	1	–	–	–	–	1	–	–	3-(3-Fluorophenyl)propenoic acid, 2-(diethoxyphosphinyl)-, ethyl ester		NI34721
201	120	149	229	81	29	148	257	330	100	69	59	54	43	42	41	40	9.90	15	20	5	–	–	1	–	–	–	–	1	–	–	2-Propenoic acid, 2-(diethoxyphosphinyl)-3-(4-fluorophenyl)-, ethyl ester	76386-69-1	NI34722
135	107	106	134	73	330	74	196	330	100	36	16	12	7	6	5	3		15	22	–	–	4	–	–	–	–	–	–	–	–	13,14-Benzo-1,4,8,11-tetrathiacyclopentadecane	25676-64-6	NI34723
330	135	107	270	196	302	106	134	330	100	87	58	50	40	32	27	21		15	22	–	–	4	–	–	–	–	–	–	–	–	1,4,8,11-Tetrathiametacyclophane		NI34724
91	155	43	99	69	65	45	113	330	100	46	36	33	32	31	29	28	0.00	15	22	6	–	1	–	–	–	–	–	–	–	–	(+)-(4S,5S)-2,2-Dimethyl-4-hydroxymethyl-5-methoxymethyl-1,3-dioxolane-4-toluenesulphonate	83481-29-2	NI34725
43	210	85	126	100	129	228	197	330	100	84	67	58	47	46	39	29	0.00	15	22	8	–	–	–	–	–	–	–	–	–	–	2-Furanhexanoic acid, tetrahydro-β,δ-dihydroxy-5-oxo-, methyl ester, diacetate	18142-16-0	NI34726
43	97	69	139	72	137	85	70	330	100	91	86	81	59	45	38	36	0.00	15	22	8	–	–	–	–	–	–	–	–	–	–	D-Glycero-L-gulo-octitol, 2,6:5,7-dianhydro-8-deoxy-7-C-methyl-, triacetate	32970-01-7	NI34727
45	167	253	255	137	91	29	121	330	100	94	76	64	29	28	24	18	0.00	15	26	6	–	–	–	–	–	–	1	–	–	–	Silane, tris(2-methoxyethoxy)phenyl-	17903-05-8	NI34728
139	175	141	176	111	177	75	50	330	100	71	65	46	35	30	30	21	15.00	16	8	2	2	–	–	2	–	–	–	–	–	–	α-(4-Chloro-benzoyloxy)-α-(4-chlorophenyl)malonitrile		NI34729
134	166	121	69	45	165	89	63	330	100	74	24	19	15	13	13	12	0.00	16	10	–	–	4	–	–	–	–	–	–	–	–	1,2-Bis(5-benzo[b]thienyl)-1,2-dithiaethane		AI02055
302	125	69	286	149	123	256	273	330	100	79	67	65	62	61	56	53	0.00	16	10	8	–	–	–	–	–	–	–	–	–	–	Phenanthrenecarboxylic acid, 1,4,9,10-tetrahydro-5,8,10-trihydroxy-10-methyl-1,4,9-trioxo-		LI07322
107	149	181	91	105	79	77	104	330	100	44	24	23	17	14	11	10	2.40	16	11	2	–	–	5	–	–	–	–	–	–	–	Phenylpropionic acid, pentafluorobenzyl ester		MS07846
91	105	77	124	51	59	147	65	330	100	62	52	27	25	24	23	13	9.20	16	14	2	2	2	–	–	–	–	–	–	–	–	Carbamodithioic acid, [(benzoylamino)carbonyl]-, benzyl ester	74793-42-3	NI34730
41	30	39	43	79	91	51	29	330	100	97	95	88	68	66	60	60	5.00	16	14	6	2	–	–	–	–	–	–	–	–	–	Benzofuran, 7-(2,4-dinitrophenoxy)-2,3-dihydro-2,2-dimethyl-	62059-44-3	NI34731
330	135	118	195	91	90	149	107	330	100	94	70	66	57	51	39	37		16	14	6	2	–	–	–	–	–	–	–	–	–	anti-3-Nortricyclanol, 6-methyl-5-methylidene-, 3,5-dinitrobenzoate		NI34732
91	43	70	71	42	41	92	139	330	100	53	35	30	10	8	7	4	1.00	16	14	6	2	–	–	–	–	–	–	–	–	–	Styrene, 4-(benzyloxy)-5-methoxy-β,2-dinitro-	2426-89-3	NI34733
233	330	332	314	147	300	234	302	330	100	75	69	39	36	28	28	26		16	15	1	2	–	–	–	1	–	–	–	–	–	2-(4-Bromo-α-hydroxybenzyl)-5,6-dimethylbenzimidazole		LI07323
43	41	56	57	121	60	73	286	330	100	81	71	45	36	22	21	20	0.00	16	18	4	–	–	–	–	–	–	–	–	–	1	2-(Ferrocenylmethyl)glutaric acid		NI34734
289	330	91	107	41	53	79	105	330	100	30	25	20	20	19	18	16		16	18	4	4	–	–	–	–	–	–	–	–	–	Carvone (2,4-dinitrophenyl)hydrazone		NI34735
165	152	42	166	164	150	330	69	330	100	94	46	38	38	36	32	28		16	22	2	6	–	–	–	–	–	–	–	–	–	1,4-Bis(3,5-dimethyl-4-oxo-3,4-dihydropyrimidyl-2)piperazine		NI34736
43	101	315	59	113	139	85	72	330	100	29	28	23	15	13	13	12	0.00	16	26	7	–	–	–	–	–	–	–	–	–	–	1,2:3,4:6,7-Tri-O-isopropylidene-D-manno-heptulofuranose (major isomer)		NI34737
101	315	126	59	72	141	113	98	330	100	47	36	33	32	29	22	21	0.00	16	26	7	–	–	–	–	–	–	–	–	–	–	1,2:3,4:6,7-Tri-O-isopropylidene-D-manno-heptulofuranose (minor isomer)		NI34738
113	69	114	112	86	87	70	41	330	100	32	8	8	8	4	3	3	0.00	16	26	7	–	–	–	–	–	–	–	–	–	–	Tetraethylene glycol dimethacrylate	109-17-1	NI34739
107	253	173	251	330	185	175	332	330	100	98	79	78	62	59	51	42		17	12	1	2	–	–	2	–	–	–	–	–	–	5-Pyrimidinemethanol, α-(2,4-dichlorophenyl)-α-phenyl	26766-27-8	IC03986
107	173	77	175	79	253	251	105	330	100	58	44	38	33	28	24	23	17.45	17	12	1	2	–	–	2	–	–	–	–	–	–	5-Pyrimidinemethanol, α-(2,4-dichlorophenyl)-α-phenyl-	26766-27-8	NI34740
43	41	55	39	57	69	44	45	330	100	54	42	32	29	24	22	19	6.14	17	14	7	–	–	–	–	–	–	–	–	–	–	Aflatoxicol H1	55446-27-0	NI34741
330	28	44	43	57	51	331	71	330	100	84	55	27	24	22	20	20		17	14	7	–	–	–	–	–	–	–	–	–	–	Aflatoxin G2	7241-98-7	MS07847
330	331	287	328	329	301	286	243	330	100	20	8	7	5	5	5	4		17	14	7	–	–	–	–	–	–	–	–	–	–	Aflatoxin G2	7241-98-7	NI34742
330	287	331	243	301	286	269	69	330	100	27	20	14	12	12	11	11		17	14	7	–	–	–	–	–	–	–	–	–	–	Aflatoxin G2	7241-98-7	NI34743
330	273	246	274	301	331	284	271	330	100	64	35	27	22	17	17	16		17	14	7	–	–	–	–	–	–	–	–	–	–	Aflatoxin M2	6885-57-0	NI34744
330	315	153	181	287	301	284	331	330	100	89	71	42	26	24	21	19		17	14	7	–	–	–	–	–	–	–	–	–	–	Flavone, 2′,5,5′-trihydroxy-6,7-dimethoxy-		DD01289
315	330	181	153	316	158	331	125	330	100	56	35	27	18	14	9	8		17	14	7	–	–	–	–	–	–	–	–	–	–	Flavone, 2′,5,5′-trihydroxy-7,8-dimethoxy-		DD01290
330	331	151	329	287	149	69	301	330	100	22	13	10	10	8	8	6		17	14	7	–	–	–	–	–	–	–	–	–	–	Flavone, 3,4′,5-trihydroxy-3′,7-dimethoxy-	552-54-5	NI34745
330	329	287	331	137	167	312	151	330	100	78	41	29	26	16	15	13		17	14	7	–	–	–	–	–	–	–	–	–	–	Flavone, 3′,4′,5-trihydroxy-3,7-dimethoxy-	2068-02-2	NI34746
330	315	287	312	181	153	299	137	330	100	72	24	12	10	10	7	4		17	14	7	–	–	–	–	–	–	–	–	–	–	Flavone, 3′,4′,7-trihydroxy-5,6-dimethoxy-	88153-47-3	NI34747
330	287	329	315	331	151	153	69	330	100	53	51	47	26	17	14	14		17	14	7	–	–	–	–	–	–	–	–	–	–	Flavone, 4′,5,7-trihydroxy-3,3′-dimethoxy-	4382-17-6	NI34748
315	330	297	197	329	169	287	119	330	100	69	10	10	5	5	2	2		17	14	7	–	–	–	–	–	–	–	–	–	–	Flavone, 5,6,4′-trihydroxy-7,8-dimethoxy-		NI34749
330	315	312	287	139	167	329	301	330	100	63	53	47	24	19	8	8		17	14	7	–	–	–	–	–	–	–	–	–	–	Flavone, 5,7,4′-trihydroxy-5,6-dimethoxy-	18085-97-7	NI34750
330	315	297	197	169	329	312	118	330	100	97	23	21	10	9	7	6		17	14	7	–	–	–	–	–	–	–	–	–	–	Flavone, 5,8,4′-trihydroxy-6,7-dimethoxy-	98755-25-0	NI34751
247	203	289	204	330	118	246	248	330	100	72	53	51	31	24	19	17		17	18	5	2	–	–	–	–	–	–	–	–	–	Acetamide, N-[3,4-bis(acetyloxy)-1-methyl-5-phenyl-1H-pyrrol-2-yl]-	50618-98-9	NI34752
207	192	124	106	78	148	178	298	330	100	66	58	53	43	28	14	8	2.00	17	18	5	2	–	–	–	–	–	–	–	–	–	6-(O-Nicotinoyl)strychnovoline		NI34753
43	73	45	155	29	41	27	55	330	100	68	61	33	29	20	18	16	0.00	17	30	6	–	–	–	–	–	–	–	–	–	–	Decanoic acid, 2-(acetyloxy)-1-[(acetyloxy)methyl]ethyl ester	55124-86-2	WI01424
73	155	158	159	257	130	115	106	330	100	53	26	25	22	11	8	8	0.00	17	30	6	–	–	–	–	–	–	–	–	–	–	Decanoic acid, 2-(acetyloxy)-1-[(acetyloxy)methyl]ethyl ester	55124-86-2	WI01425
330	40	189	29	240	332	300	205	330	100	93	73	59	57	39	38	37		18	15	2	–	1	–	1	–	–	–	–	–	–	[4-(4-Chlorophenyl)-1-hydroxy-2-naphthalenyl]dimethylsulphoxonium hydroxide inner salt	92608-99-6	NI34754
234	208	43	164	270	219	203	191	330	100	97	97	89	88	88	88	76	0.10	18	18	4	–	1	–	–	–	–	–	–	–	–	anti-11-(Methylsulpho)-2-methoxy-9,10-dihydro-9,10-ethanoanthracene		NI34755

MASS TO CHARGE RATIOS								M.W.	INTENSITIES								Parent	C	H	O	N	S	F	Cl	Br	I	Si	P	B	X	COMPOUND NAME	CAS Reg No	No
208	164	209	235	330	278	236	331	**330**	100	69	68	39	31	26	20	8		18	18	4	–	1	–	–	–	–	–	–	–	–	anti-12-(Methylsulpho)-2-methoxy-9,10-dihydro-9,10-ethanoanthracene		NI34756
208	164	209	330	235	234	236	331	**330**	100	70	69	35	29	13	9	8		18	18	4	–	1	–	–	–	–	–	–	–	–	syn-12-(Methylsulpho)-2-methoxy-9,10-dihydro-9,10-ethanoanthracene		NI34757
246	137	288	110	136	43	109	81	**330**	100	59	45	29	28	19	17	12	**3.50**	18	18	6	–	–	–	–	–	–	–	–	–	–	1,2-Bis(4-acetoxyphenoxy)-ethane		IC03987
270	43	255	69	219	41	220	45	**330**	100	76	60	42	39	27	23	21	**2.00**	18	18	6	–	–	–	–	–	–	–	–	–	–	Acetyl shikonin		LI07324
41	273	75	233	274	115	103	188	**330**	100	98	25	24	15	14	10	9	**6.23**	18	18	6	–	–	–	–	–	–	–	–	–	–	1,3,5-Benzenetricarboxylic acid, tri-2-propenyl ester	17832-16-5	NI34758
105	77	165	106	123	51	269	177	**330**	100	20	8	7	5	5	3	3	**0.00**	18	18	6	–	–	–	–	–	–	–	–	–	–	1,2,3,4-Butanetetrol, 1,4-dibenzoate, (R*,R*)-	74793-37-6	NI34759
105	165	77	123	106	195	166	86	**330**	100	20	16	10	7	3	2	2	**0.00**	18	18	6	–	–	–	–	–	–	–	–	–	–	1,2,3,4-Butanetetrol, 1,4-dibenzoate, (R*,S*)-	74793-33-2	NI34760
140	112	270	190	266	243	211	330	**330**	100	58	26	26	19	19	18	17		18	18	6	–	–	–	–	–	–	–	–	–	–	1,3-Cyclohexadiene-1,2-dicarboxylic acid, 6-(2-hydroxyphenyl)-4,6-dimethyl-5-oxo-, dimethyl ester	40801-36-3	NI34761
270	184	330	271	243	238	255	183	**330**	100	47	42	24	15	15	13	12		18	18	6	–	–	–	–	–	–	–	–	–	–	1,2-Dibenzofurandicarboxylic acid, 1,9b-dihydro-1-hydroxy-4,9b-dimethyl-, dimethyl ester	40801-37-4	NI34762
238	239	298	139	285	269	330	198	**330**	100	94	71	68	62	59	45	38		18	18	6	–	–	–	–	–	–	–	–	–	–	1,2-Dicarbomethoxy-3-carboethoxy-4-methylnaphthalene		MS07848
329	330	181	207	150	137	313	180	**330**	100	96	96	73	45	44	22	19		18	18	6	–	–	–	–	–	–	–	–	–	–	2′,3-Dihydroxy-4,4′,6′-trimethoxychalcone	38186-71-9	NI34763
312	181	311	313	326	151	150	314	**330**	100	75	43	30	16	16	16	13	**6.00**	18	18	6	–	–	–	–	–	–	–	–	–	–	Flavanone, 2′-hydroxy-4′,5,7-trimethoxy-		LI07325
180	135	165	151	314	147	107	181	**330**	100	93	44	40	33	15	15	13	**0.00**	18	18	6	–	–	–	–	–	–	–	–	–	–	Flavanone, 3-hydroxy-3′,4′,6-trimethoxy-	55125-20-7	NI34764
181	121	150	301	193	77	182	122	**330**	100	54	34	23	16	15	12	10	**3.00**	18	18	6	–	–	–	–	–	–	–	–	–	–	Flavanone, 3-hydroxy-4′,5,7-trimethoxy-	47335-95-5	NI34765
312	181	330	152	135	150	149	124	**330**	100	63	42	27	23	22	12	11		18	18	6	–	–	–	–	–	–	–	–	–	–	Flavanone, 2′,5,7-trihydroxy-4′-methoxy-6,8-dimethoxy-, (2R)-		NI34766
150	204	330	160	145	118	302	228	**330**	100	69	43	31	20	20	15	15		18	18	6	–	–	–	–	–	–	–	–	–	–	3H-4,9a[3′,4′]-Furanonaphtho[1,2-c]furan-1,3,11,13(4H)-tetrone, 3a,6,7,8,9,9b,10,14-octahydro-5,8-dimethyl-	32251-38-0	MS07849
262	190	41	233	69	44	330	191	**330**	100	62	53	33	32	29	23	18		18	18	6	–	–	–	–	–	–	–	–	–	–	7H-Furo[3,2-g][1]benzopyran-7-one, 5-[(1,1-dimethyl-2-propenyl)oxy]-4,6-dimethoxy-	28988-27-4	NI34767
262	190	69	41	233	44	330	191	**330**	100	63	58	54	34	30	24	18		18	18	6	–	–	–	–	–	–	–	–	–	–	7H-Furo[3,2-g][1]benzopyran-7-one, 5-[(1,1-dimethyl-2-propenyl)oxy]-4,6-dimethoxy-		LI07326
330	179	134	178	121	135	299	147	**330**	100	92	87	61	44	34	32	31		18	18	6	–	–	–	–	–	–	–	–	–	–	1,2-Bis(4-methoxycarbonylphenoxy)ethane		IC03989
330	179	134	178	135	147	299	119	**330**	100	94	67	60	35	33	30	26		18	18	6	–	–	–	–	–	–	–	–	–	–	1,2-Bis(4-methoxycarbonylphenoxy)ethane		IC03988
69	41	176	245	330	148	146	177	**330**	100	66	17	10	6	4	4	2		18	18	6	–	–	–	–	–	–	–	–	–	–	3-Methoxy-4-methacrylyloxycinnamic methacrylic anhydride		NI34768
330	315	329	147	208	221			**330**	100	31	10	10	3	2				18	18	6	–	–	–	–	–	–	–	–	–	–	9-O-Methylphilenopteran		LI07327
210	330	195	167	211	181	153	152	**330**	100	85	66	43	21	20	19	14		18	18	6	–	–	–	–	–	–	–	–	–	–	2-Propen-1-one, 3-(4-hydroxyphenyl)-1-(2-hydroxy-3,4,6-trimethoxyphenyl)-	69616-74-6	NI34769
58	59	42	43	86	272	180	152	**330**	100	11	7	4	3	2	2	2	**2.00**	18	22	2	2	1	–	–	–	–	–	–	–	–	10H-Phenothiazine-10-propanamine, N,N,β-trimethyl-, 5,5-dioxide	3689-50-7	LI07328
58	59	42	43	330	86	57	51	**330**	100	11	7	3	2	2	2	2		18	22	2	2	1	–	–	–	–	–	–	–	–	10H-Phenothiazine-10-propanamine, N,N,β-trimethyl-, 5,5-dioxide	3689-50-7	NI34770
165	44	107	55	71	79	121	77	**330**	100	76	70	52	45	30	28	28	**0.00**	18	22	4	2	–	–	–	–	–	–	–	–	–	1H-Azepine-1-carboxylic acid, 3-methyl-, methyl ester, dimer	56909-29-6	NI34771
287	105	241	330	169	213	288	242	**330**	100	51	43	33	30	29	27	26		18	22	4	2	–	–	–	–	–	–	–	–	–	3,9-Diaza-3-methyl-1,1-bis(ethoxycarbonyl)-1,2,3,4-tetrahydrofluorene		MS07850
330	73	331	315	150	332	316	74	**330**	100	61	30	22	18	12	7	5		18	26	2	–	–	–	–	–	–	2	–	–	–	4,4′-Bis(trimethylsilyloxy)-biphenyl	1094-86-6	NI34772
330	73	331	315	150	332	150.5	135	**330**	100	61	31	22	17	12	5	3		18	26	2	–	–	–	–	–	–	2	–	–	–	4,4′-Bis(trimethylsilyloxy)-biphenyl	1094-86-6	LI07329
82	67	81	41	55	83	69	54	**330**	100	33	30	12	11	10	8	7	**5.00**	18	33	3	–	–	–	–	–	–	–	–	3	–	Boroxin, tricyclohexyl-	28289-86-3	NI34773
116	98	41	95	43	117	115	99	**330**	100	75	34	22	22	21	21	19	**1.69**	18	34	5	–	–	–	–	–	–	–	–	–	–	Oxacyclooctadecan-2-one, 5,6,7-trihydroxy-18-methyl-	52461-08-2	NI34774
73	271	89	315	103	83	69	43	**330**	100	83	42	29	24	23	22	21	**0.00**	18	38	3	–	–	–	–	–	–	1	–	–	–	Methylmyristate, 2-trimethylsiloxy-		NI34775
43	56	55	57	41	29	42	118	**330**	100	76	65	51	43	39	37	34	**0.01**	18	39	–	–	–	–	–	–	–	–	–	–	1	Arsine, trihexyl-	5852-60-8	NI34776
288	330	331	43	56	55	41	57	**330**	100	80	20	6	5	4	3	2		18	39	–	–	–	–	–	–	–	–	–	–	1	Arsine, trihexyl-	5852-60-8	NI34777
330	91	119	331	92	120			**330**	100	51	26	19	4	3				19	14	2	4	–	–	–	–	–	–	–	–	–	Δ^2-1,2,4-Triazoline-3-carboxamide, N-1-naphthyl-5-oxo-1-phenyl-	32589-63-2	NI34778
185	77	105	183	165	262	243	242	**330**	100	78	76	70	70	60	21	20	**0.00**	19	15	–	–	–	4	–	–	–	–	–	1	–	Tritylium tetrafluoroborate	341-02-6	NI34779
194	330	179	149	182	181	151	121	**330**	100	52	52	41	27	21	21	19		19	22	5	–	–	–	–	–	–	–	–	–	–	2H-1-Benzopyran, 3,4-dihydro-7-methoxy-3-(2,3,4-trimethoxyphenyl)-	60478-76-4	NI34780
194	179	300	182	181	151	167		**330**	100	58	55	40	26	19	10		**0.00**	19	22	5	–	–	–	–	–	–	–	–	–	–	2H-1-Benzopyran, 3,4-dihydro-7-methoxy-3-(2,3,4-trimethoxyphenyl)-	60478-76-4	LI07330
83	330	161	85	271	43	41	47	**330**	100	99	70	60	50	40	34	27		19	22	5	–	–	–	–	–	–	–	–	–	–	1H-Cyclopenta[7,8]naphtho[2,3-b]furan-7-carboxylic acid, 2,3,3a,4,5,7,8,10b-octahydro-10-hydroxy-3a,6-dimethyl-3-oxo-, methyl ester	40625-49-8	NI34781
245	260	43	205	71	330	246	217	**330**	100	47	42	40	27	23	16	16		19	22	5	–	–	–	–	–	–	–	–	–	–	7-Demethylpinnarin butyrate		LI07331
253	43	301	330	257	197	285	255	**330**	100	41	32	24	19	18	15	7		19	22	5	–	–	–	–	–	–	–	–	–	–	Diethyl 2,6-dimethyl-4-phenyl-4H-pyran-3,5-dicarboxylate		LI07332
183	330	257	43	292	331	211	212	**330**	100	85	71	30	22	21	15	11		19	22	5	–	–	–	–	–	–	–	–	–	–	Diethyl α-(2-methoxy-8-naphthyl)-α-methylmalonate	118647-59-9	DD01291
183	257	330	43	115	141	229	331	**330**	100	90	39	32	30	28	12	8		19	22	5	–	–	–	–	–	–	–	–	–	–	Diethyl α-(4-methoxy-1-naphthyl)-α-methylmalonate	118647-72-6	DD01292
213	215	228	176	175	230	187	83	**330**	100	42	24	24	18	9	8	7	**4.50**	19	22	5	–	–	–	–	–	–	–	–	–	–	Dihydroselenidin	23662-03-5	LI07333
43	91	41	297	56	65	254	98	**330**	100	90	90	34	29	25	20	13	**10.00**	19	26	1	2	1	–	–	–	–	–	–	–	–	1-Benzyl-3-isopropyl-2-(2-oxo-4-thioxopentan-3-ylidene)hexahydropyrimidine		MS07851
156	115	55	69	143	299	83	99	**330**	100	88	36	34	27	19	19	16	**0.00**	19	26	3	2	–	–	–	–	–	–	–	–	–	trans-5-Ethyl-1-(2-hydroxyethyl)-2-(2-indolyl)-4-piperidone ethylene acetal		DD01293

MASS TO CHARGE RATIOS								M.W.	INTENSITIES								Parent	C	H	O	N	S	F	Cl	Br	I	Si	P	B	X	COMPOUND NAME	CAS Reg No	No
125	55	96	41	124	43	97	81	330	100	79	34	34	29	25	19	19	0.00	19	35	2	–	–	–	1	–	–	–	–	–	–	2-Octene, 1-[chloro(2-octenyloxy)propoxy]-	56600-22-7	NI34782
256	134	98	239	299	257	270	238	330	100	94	75	68	65	41	34	32	9.01	19	38	4	–	–	–	–	–	–	–	–	–	–	Hexadecanoic acid, 2,3-dihydroxypropyl ester	542-44-9	WI01427
43	57	41	55	98	29	74	73	330	100	69	67	62	44	44	35	32	0.00	19	38	4	–	–	–	–	–	–	–	–	–	–	Hexadecanoic acid, 2,3-dihydroxypropyl ester	542-44-9	WI01426
331	332	313	333	314	239	329	159	330	100	22	8	3	2	2	1	1	0.00	19	38	4	–	–	–	–	–	–	–	–	–	–	Hexadecanoic acid, 2,3-dihydroxypropyl ester	542-44-9	NI34783
201	129	97	137	169	70	213	128	330	100	45	35	28	25	19	17	17	0.00	19	38	4	–	–	–	–	–	–	–	–	–	–	Hexadecanoic acid, 9,10-dimethoxy-, methyl ester	55124-85-1	NI34784
134	239	98	74	299	256	84	257	330	100	92	74	50	42	39	33	29	1.00	19	38	4	–	–	–	–	–	–	–	–	–	–	Hexadecanoic acid, 2-hydroxy-1-(hydroxymethyl)ethyl ester	23470-00-0	WI01428
43	57	41	98	55	74	84	29	330	100	85	74	71	69	63	48	48	1.00	19	38	4	–	–	–	–	–	–	–	–	–	–	Hexadecanoic acid, 2-hydroxy-1-(hydroxymethyl)ethyl ester	23470-00-0	WI01429
313	331	314	332	280				330	100	36	15	3	3				0.00	19	38	4	–	–	–	–	–	–	–	–	–	–	Hexadecanoic acid, 2-hydroxy-1-(hydroxymethyl)ethyl ester	23470-00-0	NI34785
185	101	315	131	152	73	189	153	330	100	56	16	14	8	7	1	1	0.00	19	38	4	–	–	–	–	–	–	–	–	–	–	1-O-(2-Methoxydodecyl)-2,3-O-isopropylideneglycerol		NI34786
55	43	41	155	74	57	30		330	100	93	91	74	72	72	64		0.00	19	38	4	–	–	–	–	–	–	–	–	–	–	Octadecanoic acid, 9,10-dihydroxy-, methyl ester		MS07852
155	187	55	74	69	87	138	57	330	100	99	89	87	76	75	72	67	0.00	19	38	4	–	–	–	–	–	–	–	–	–	–	Octadecanoic acid, 9,10-dihydroxy-, methyl ester, (R*,R*)-	3639-31-4	NI34787
77	330	120	105	51	64	92	63	330	100	82	81	36	18	16	14	13		20	14	3	2	–	–	–	–	–	–	–	–	–	3,5-Pyrazolidinedione, 4-(2-furanylmethylene)-1,2-diphenyl-	2652-80-4	NI34788
77	157	330	130	76	103	51	131	330	100	88	82	67	62	33	31	25		20	18	1	4	–	–	–	–	–	–	–	–	–	3-(1-Phenyl-3-methylpyrazol-5-yl)-2-ethylquinazolin-4(3H)-one		MS07853
91	41	330	198	197	57	44	38	330	100	57	21	20	20	15	15	15		20	26	–	–	2	–	–	–	–	–	–	–	–	1-[(2-Butylphenyl)thio]-2-(butylthio)benzene		DD01294
83	230	174	201	215	301	53	107	330	100	95	43	38	30	28	23	13	3.00	20	26	4	–	–	–	–	–	–	–	–	–	–	3β-Angeloyloxy-9,10-didehydro-8-epieremophilenolide		NI34789
315	330	316	285	331	127	43	41	330	100	22	22	6	5	5	3	3		20	26	4	–	–	–	–	–	–	–	–	–	–	2H,8H-Benzo[1,2-b:5,4-b′]dipyran-10-propanol, 5-methoxy-2,2,8,8-tetramethyl-	26535-37-5	NI34790
133	158	330	83	230	55	43	105	330	100	99	93	93	81	79	68	62		20	26	4	–	–	–	–	–	–	–	–	–	–	2-Butenoic acid, 2-methyl-, 2,3,3a,4,5,7,9a,9b-octahydro-3,6,9-trimethyl-2 oxoazuleno[4,5-b]furan-3-yl ester, [3R-[3α(Z),3aβ,9aβ,9bα]]-	33439-65-5	NI34791
83	55	43	41	230	39	201	53	330	100	69	32	32	17	17	12	11	0.87	20	26	4	–	–	–	–	–	–	–	–	–	–	2-Butenoic acid, 2-methyl-, 1a,2,4,4a,5,9-hexahydro-4,4a,6-trimethyl-3H-oxireno[8,8a]naphtho[2,3-b]furan-5-yl ester	56246-42-5	NI34792
55	29	83	41	91	43	53	159	330	100	45	38	31	26	26	25	24	10.00	20	26	4	–	–	–	–	–	–	–	–	–	–	2-Butenoic acid, 2-methyl-, 4,5,6,6a,7,10b-hexahydro-3,10-dimethyl-2H,3H-oxeto[2′,3′:4,4a]naphtho[2,3-b]furan-4-yl ester	56298-87-4	NI34793
149	41	167	150	83	55	249	67	330	100	12	10	9	9	9	5	5	0.00	20	26	4	–	–	–	–	–	–	–	–	–	–	Dicyclohexyl phthalate	84-61-7	MS07854
149	167	55	150	41	83	249	67	330	100	24	14	10	10	7	6	4	0.04	20	26	4	–	–	–	–	–	–	–	–	–	–	Dicyclohexyl phthalate	84-61-7	IC03990
149	57	31	45	41	167	43	55	330	100	27	26	24	20	17	16	15	0.00	20	26	4	–	–	–	–	–	–	–	–	–	–	Dicyclohexyl phthalate	84-61-7	WI01430
286	106	257	258	330	163	231	256	330	100	49	32	31	29	29	28	27		20	26	4	–	–	–	–	–	–	–	–	–	–	5α-Dihydroaldosterone γ-lactone		NI34794
302	286	257	256	231	233	232	122	330	100	38	30	30	29	26	25	24	11.86	20	26	4	–	–	–	–	–	–	–	–	–	–	5β-Dihydroaldosterone γ-lactone		NI34795
298	270	41	91	227	226	243	55	330	100	79	54	51	49	48	43	39	7.00	20	26	4	–	–	–	–	–	–	–	–	–	–	Gibberellin A9 methyl ester	2112-08-5	MS07855
286	215	271	284	44	204	41	287	330	100	74	53	33	29	23	23	21	13.81	20	26	4	–	–	–	–	–	–	–	–	–	–	Isocarnosol	92519-82-9	NI34796
124	41	43	121	317	109	110	29	330	100	51	47	44	39	31	30	24	0.00	20	26	4	–	–	–	–	–	–	–	–	–	–	1-Phenanthrenecarboxylic acid, 7-vinyl-1,2,4a,4b,5,6,7,8,10,10a-decahydro-3-hydroxy-4a,7-dimethyl-2-oxo-, methyl ester, [1S-(1α,4aβ,4bα,7α,10aα)]-	57397-04-3	NI34797
151	137	81	95	123	109	330	161	330	100	77	70	49	44	33	21	21		20	26	4	–	–	–	–	–	–	–	–	–	–	Pseudojolkinolide B		NI34798
91	73	31	104	32	75	149	43	330	100	97	56	41	39	29	28	22	0.00	20	30	2	–	–	–	–	–	–	1	–	–	–	(E)-1-[2-(Benzyloxymethyl)-1-cyclohexenyl]-3-(trimethylsilyl)-2-propen-1-ol		DD01295
201	202	69	186	214	200	158	79	330	100	15	10	5	4	4	4	4	4.00	20	30	2	2	–	–	–	–	–	–	–	–	–	1H-Pyrido[3,4-b]indole-1-butanol, γ-sec-butyl-2,3,4,9-tetrahydro-6-methoxy-	14426-66-5	NI34799
172	130	173	330	87	158	243	116	330	100	83	65	51	22	21	20	18		20	30	2	2	–	–	–	–	–	–	–	–	–	L-Tryptophan, N,1-diisopropyl-, isopropyl ester	56805-35-7	MS07856
330	240	43	312	45	331	313	60	330	100	47	30	28	28	25	21	21		21	14	4	–	–	–	–	–	–	–	–	–	–	2-Benzylquinizarin		MS07857
330	91	331	252	74	115	329	253	330	100	80	30	29	25	14	13	13		21	14	4	–	–	–	–	–	–	–	–	–	–	2-Benzylxanthopurpurin		LI07334
105	77	210	209	106	91	51	122	330	100	30	10	10	9	8	7	3	0.50	21	18	2	2	–	–	–	–	–	–	–	–	–	Benzoic acid, 2-benzoyl-1-benzylhydrazide	24664-22-0	MS07858
315	55	76	104	56	330	57	69	330	100	87	66	57	48	45	41	39		21	18	2	2	–	–	–	–	–	–	–	–	–	1,4-Naphthoquinone, 2-cyano-3-[4(dimethylamino)phenyl]-		MS07859
43	124	147	228	55	91	41	79	330	100	82	70	30	28	27	25	24	16.01	21	30	3	–	–	–	–	–	–	–	–	–	–	Androst-4-en-3-one, 17-(acetyloxy)-, (17β)-	1045-69-8	NI34800
124	147	228	288	330	146	185	148	330	100	60	39	32	26	22	17	16		21	30	3	–	–	–	–	–	–	–	–	–	–	Androst-4-en-3-one, 17-(acetyloxy)-, (17β)-	1045-69-8	NI34801
43	270	41	91	55	159	79	67	330	100	41	33	30	23	22	20	19	0.00	21	30	3	–	–	–	–	–	–	–	–	–	–	Androst-5-en-7-one, 3-(acetyloxy)-, (3β)-	25845-92-5	NI34802
302	43	330	41	134	55	69	81	330	100	75	67	65	63	56	46	41		21	30	3	–	–	–	–	–	–	–	–	–	–	Androst-14-en-17-one, 3-hydroxy-, acetate, (3β,5α)-	1239-32-3	NI34803
205	330	247	204	148	147	43	108	330	100	62	52	39	30	29	26	22		21	30	3	–	–	–	–	–	–	–	–	–	–	Cannabielsoin I	54002-78-7	NI34804
299	271	253	147	55	41	91	79	330	100	58	40	40	39	26	24	24	3.28	21	30	3	–	–	–	–	–	–	–	–	–	–	Corticosterone, 11-deoxy-	64-85-7	NI34805
299	271	300	32	253	147	330	93	330	100	30	19	16	15	13	10	8		21	30	3	–	–	–	–	–	–	–	–	–	–	Corticosterone, 11-deoxy-	64-85-7	MS07860
299	18	330	300	59	28	43	331	330	100	49	26	21	20	20	19	13		21	30	3	–	–	–	–	–	–	–	–	–	–	Δ^9-Tetrahydrocannabinol, 11-hydroxy-		MS07861
299	300	330	43	231	217	193	41	330	100	22	16	10	9	9	9	9		21	30	3	–	–	–	–	–	–	–	–	–	–	Δ^9-Tetrahydrocannabinol, 11-hydroxy-	36557-05-8	NI34806
297	330	149	259	312	193	231	269	330	100	69	69	55	30	29	28	23		21	30	3	–	–	–	–	–	–	–	–	–	–	6H-Dibenzo[b,d]pyran-1,8-diol, 6a,7,8,9,10,10a-hexahydro-6,6-dimethyl-9 methylene-3-pentyl-, (6aα,8β,10aβ)-(-)-	27279-12-5	NI34807
330	149	231	259	297	271	193	41	330	100	69	63	45	43	43	39	36		21	30	3	–	–	–	–	–	–	–	–	–	–	6H-Dibenzo[b,d]pyran-1,8-diol, 6a,7,8,9,10,10a-hexahydro-6,6-dimethyl-9 methylene-3-pentyl-, [6aR-(6aα,8α,10aβ)]-	52522-55-1	NI34808

MASS TO CHARGE RATIOS	M.W.	INTENSITIES	Parent	C	H	O	N	S	F	Cl	Br	I	Si	P	B	X	COMPOUND NAME	CAS Reg No	No
297 214 312 43 330 256 193 69	**330**	100 45 29 29 28 26 24 18		21	30	3	–	–	–	–	–	–	–	–	–	–	**6H-Dibenzo[b,d]pyran-1,9-diol, 6a,9,10,10a-tetrahydro-6,6,9-trimethyl-3-pentyl-, [6aR-(6aα,9β,10aβ)]-**	52493-13-7	**NI34809**
330 231 331 193 43 274 41 269	**330**	100 36 27 21 21 19 19 15		21	30	3	–	–	–	–	–	–	–	–	–	–	**6H-Dibenzo[b,d]pyran-9-methanol, 6a,7,10,10a-tetrahydro-1-hydroxy-6,6-dimethyl-3-pentyl-, (6aR-trans)-**	28646-40-4	**NI34810**
123 164 41 56 81 93 43 135	**330**	100 27 24 23 20 19 19 17	**1.97**	21	30	3	–	–	–	–	–	–	–	–	–	–	**Jasmolin I**	4466-14-2	**NI34811**
123 43 55 41 164 69 81 57	**330**	100 33 31 31 27 25 24 21	**4.80**	21	30	3	–	–	–	–	–	–	–	–	–	–	**Jasmolin I**	4466-14-2	**MS07862**
123 43 41 81 55 93 79 29	**330**	100 42 41 31 31 24 19 19	**0.08**	21	30	3	–	–	–	–	–	–	–	–	–	–	**Jasmolin I**	4466-14-2	**NI34812**
237 162 330 255 238 195 163 253	**330**	100 55 26 22 20 20 14 12		21	30	3	–	–	–	–	–	–	–	–	–	–	**Methyl 7β-hydroxyabieta-8,11,13-trien-18-oate**	1802-09-1	**NI34813**
81 139 41 82 235 53 203 55	**330**	100 74 74 54 44 44 42 41	**10.47**	21	30	3	–	–	–	–	–	–	–	–	–	–	**1-Naphthalenecarboxylic acid, 5-[2-(3-furanyl)ethyl]-3,4,4a,5,6,7,8,8a-octahydro-5,6,8a-trimethyl-, methyl ester, [4aS-(4aα,5α,6β,8aβ)]-**	56630-98-9	**NI34814**
121 330 81 189 107 82 95 331	**330**	100 81 58 44 24 20 19 18		21	30	3	–	–	–	–	–	–	–	–	–	–	**1-Naphthalenecarboxylic acid, 5-[2-(3-furanyl)ethyl]decahydro-1,4a-dimethyl-6-methylene-, methyl ester, [1R-(1α,4aα,5α,8aβ)]-**	1238-98-8	**NI34815**
121 81 330 189 82 149 107 95	**330**	100 52 50 28 24 18 18 18		21	30	3	–	–	–	–	–	–	–	–	–	–	**1-Naphthalenecarboxylic acid, 5-[2-(3-furanyl)ethyl]decahydro-1,4a-dimethyl-6-methylene-, methyl ester, [1R-(1α,4aβ,5β,8aα)]-**	10267-15-9	**NI34816**
81 121 41 53 55 82 330 27	**330**	100 86 52 43 35 33 31 26		21	30	3	–	–	–	–	–	–	–	–	–	–	**1-Naphthalenecarboxylic acid, 5-[2-(3-furanyl)ethyl]decahydro-1,4a-dimethyl-6-methylene-, methyl ester, [1S-(1α,4aα,5α,8aβ)]-**	4966-14-7	**NI34817**
330 160 173 45 161 159 199 81	**330**	100 66 29 19 17 17 16 16		21	30	3	–	–	–	–	–	–	–	–	–	–	**19-Norpregna-1,3,5(10)-triene-11,20-diol, 3-methoxy-, (11α,20R)-**	56145-32-5	**NI34818**
287 43 269 229 330 91 41 55	**330**	100 96 65 64 50 48 42 30		21	30	3	–	–	–	–	–	–	–	–	–	–	**19-Norpregn-4-ene-3,20-dione, 17-hydroxy-1-methyl-**	56246-44-7	**NI34819**
108 330 109 331 161 131 133 55	**330**	100 80 20 19 16 14 12 12		21	30	3	–	–	–	–	–	–	–	–	–	–	**Phenanthro[3,2-b]furan-4-carboxylic acid, 1,2,3,4,4a,5,6,6a,7,11,11a,11b-dodecahydro-4,7,11b-trimethyl-, methyl ester, (4S,4aR,6aS,7R,11aS,11bR)-**	4614-50-0	**NI34820**
43 84 55 330 41 245 79 71	**330**	100 42 37 34 30 26 25 25		21	30	3	–	–	–	–	–	–	–	–	–	–	**Pregnane-3,6,20-trione**	30802-24-5	**NI34821**
43 330 84 55 312 81 331 245	**330**	100 88 35 34 30 23 20 20		21	30	3	–	–	–	–	–	–	–	–	–	–	**Pregnane-3,6,20-trione, (5α)-**	1923-27-9	**NI34822**
84 43 330 93 85 81 91 79	**330**	100 98 88 46 39 39 37 37		21	30	3	–	–	–	–	–	–	–	–	–	–	**Pregnane-3,6,20-trione, (5α)-**	1923-27-9	**NI34823**
330 43 55 84 312 81 234 79	**330**	100 98 38 35 31 27 26 26		21	30	3	–	–	–	–	–	–	–	–	–	–	**Pregnane-3,6,20-trione, (5β)-**	1239-92-5	**NI34824**
330 43 55 84 312 81 234 67	**330**	100 98 38 35 32 27 26 25		21	30	3	–	–	–	–	–	–	–	–	–	–	**Pregnane-3,6,20-trione, (5β)-**	1239-92-5	**LI07335**
43 330 192 312 55 245 41 260	**330**	100 81 73 64 52 51 49 47		21	30	3	–	–	–	–	–	–	–	–	–	–	**Pregnane-3,7,20-trione, (5α)-**	26991-60-6	**NI34825**
330 43 55 81 331 95 41 259	**330**	100 69 26 23 22 22 22 18		21	30	3	–	–	–	–	–	–	–	–	–	–	**Pregnane-3,11,20-trione, (5α)-**	2089-06-7	**NI34826**
43 330 109 124 122 85 81 55	**330**	100 59 45 43 43 41 37 35		21	30	3	–	–	–	–	–	–	–	–	–	–	**Pregnane-3,11,20-trione, (5β)-**	1474-68-6	**NI34828**
330 43 55 41 315 81 122 124	**330**	100 94 35 34 28 27 26 25		21	30	3	–	–	–	–	–	–	–	–	–	–	**Pregnane-3,11,20-trione, (5β)-**	1474-68-6	**NI34827**
43 330 260 95 55 109 85 81	**330**	100 77 49 47 37 34 33 31		21	30	3	–	–	–	–	–	–	–	–	–	–	**Pregnane-3,11,20-trione, (5β)-**	1474-68-6	**NI34829**
330 43 287 55 217 81 232 41	**330**	100 97 62 59 55 51 46 42		21	30	3	–	–	–	–	–	–	–	–	–	–	**Pregnane-3,15,20-trione, (5α)-**	7755-32-0	**NI34830**
330 43 55 41 81 79 259 162	**330**	100 99 56 44 43 33 31 30		21	30	3	–	–	–	–	–	–	–	–	–	–	**Pregnane-3,15,20-trione, (5β)-**	19953-81-2	**NI34831**
330 43 55 81 41 259 287 162	**330**	100 98 55 43 43 32 30 30		21	30	3	–	–	–	–	–	–	–	–	–	–	**Pregnane-3,15,20-trione, (5β)-**	19953-81-2	**LI07336**
330 245 84 312 71 260 137 85	**330**	100 75 75 59 42 40 38 34		21	30	3	–	–	–	–	–	–	–	–	–	–	**Pregn-4-ene-3,20-dione, 6-hydroxy-, (6α)-**		**NI34832**
330 84 312 245 137 85 159 71	**330**	100 88 79 72 41 39 37 36		21	30	3	–	–	–	–	–	–	–	–	–	–	**Pregn-4-ene-3,20-dione, 6-hydroxy-, (6β)-**	604-19-3	**NI34833**
43 330 152 55 41 95 79 67	**330**	100 52 29 26 20 19 19 19		21	30	3	–	–	–	–	–	–	–	–	–	–	**Pregn-4-ene-3,20-dione, 6-hydroxy-, (6β)-**	604-19-3	**NI34834**
330 124 43 109 331 55 194 93	**330**	100 85 43 22 19 19 18 15		21	30	3	–	–	–	–	–	–	–	–	–	–	**Pregn-4-ene-3,20-dione, 9-hydroxy-**	15981-54-1	**NI34835**
330 124 163 43 123 312 41 55	**330**	100 77 60 50 47 36 31 28		21	30	3	–	–	–	–	–	–	–	–	–	–	**Pregn-4-ene-3,20-dione, 11-hydroxy-, (11α)-**	80-75-1	**NI34836**
124 163 123 330 91 312 55 79	**330**	100 81 60 59 44 40 37 35		21	30	3	–	–	–	–	–	–	–	–	–	–	**Pregn-4-ene-3,20-dione, 11-hydroxy-, (11α)-**	80-75-1	**NI34837**
43 124 163 123 55 137 91 122	**330**	100 61 38 33 24 23 22 21	**14.74**	21	30	3	–	–	–	–	–	–	–	–	–	–	**Pregn-4-ene-3,20-dione, 11-hydroxy-, (11α)-**	80-75-1	**NI34838**
312 163 124 43 123 91 79 313	**330**	100 47 44 41 32 26 25 22	**13.92**	21	30	3	–	–	–	–	–	–	–	–	–	–	**Pregn-4-ene-3,20-dione, 11-hydroxy-, (11β)-**	600-57-7	**NI34841**
43 124 163 312 91 55 41 123	**330**	100 38 33 32 26 22 22 20	**4.73**	21	30	3	–	–	–	–	–	–	–	–	–	–	**Pregn-4-ene-3,20-dione, 11-hydroxy-, (11β)-**	600-57-7	**NI34839**
312 124 163 55 123 91 81 79	**330**	100 67 55 49 43 40 35 33	**16.82**	21	30	3	–	–	–	–	–	–	–	–	–	–	**Pregn-4-ene-3,20-dione, 11-hydroxy-, (11β)-**	600-57-7	**NI34840**
330 245 91 79 121 105 124 81	**330**	100 79 59 48 44 41 37 36		21	30	3	–	–	–	–	–	–	–	–	–	–	**Pregn-4-ene-3,20-dione, 12-hydroxy-**		**LI07337**
330 312 124 43 123 313 269 105	**330**	100 68 63 63 30 26 26 26		21	30	3	–	–	–	–	–	–	–	–	–	–	**Pregn-4-ene-3,20-dione, 14-hydroxy-**	16031-66-6	**NI34843**
43 312 91 55 145 124 93 41	**330**	100 64 28 24 19 19 19 19	**11.94**	21	30	3	–	–	–	–	–	–	–	–	–	–	**Pregn-4-ene-3,20-dione, 14-hydroxy-**	16031-66-6	**NI34842**
330 124 331 288 55 43 107 91	**330**	100 23 21 20 13 13 10 10		21	30	3	–	–	–	–	–	–	–	–	–	–	**Pregn-4-ene-3,20-dione, 15-hydroxy-, (15α)-**	600-73-7	**NI34845**
330 124 91 123 288 79 71 55	**330**	100 76 43 40 37 37 36 36		21	30	3	–	–	–	–	–	–	–	–	–	–	**Pregn-4-ene-3,20-dione, 15-hydroxy-, (15α)-**	600-73-7	**NI34844**
312 231 91 55 79 124 107 105	**330**	100 60 58 57 52 50 42 39	**19.00**	21	30	3	–	–	–	–	–	–	–	–	–	–	**Pregn-4-ene-3,20-dione, 15-hydroxy-, (15β)-**	600-72-6	**LI07338**
312 91 55 231 79 124 107 71	**330**	100 62 62 61 54 49 45 42	**20.11**	21	30	3	–	–	–	–	–	–	–	–	–	–	**Pregn-4-ene-3,20-dione, 15-hydroxy-, (15β)-**	600-72-6	**NI34846**
43 231 100 124 91 79 41 269	**330**	100 60 47 25 25 24 23 22	**0.57**	21	30	3	–	–	–	–	–	–	–	–	–	–	**Pregn-4-ene-3,20-dione, 16-hydroxy-, (16α)-**	438-07-3	**NI34847**
231 100 55 312 91 269 124 79	**330**	100 70 58 56 55 50 42 41	**5.00**	21	30	3	–	–	–	–	–	–	–	–	–	–	**Pregn-4-ene-3,20-dione, 16-hydroxy-, (16α)-**	438-07-3	**LI07339**
269 312 231 297 91 124 56 79	**330**	100 88 61 54 52 46 46 41	**1.76**	21	30	3	–	–	–	–	–	–	–	–	–	–	**Pregn-4-ene-3,20-dione, 16-hydroxy-, (16α)-**	438-07-3	**NI34848**
287 229 330 90 78 56 302 124	**330**	100 78 70 59 57 57 55 53		21	30	3	–	–	–	–	–	–	–	–	–	–	**Pregn-4-ene-3,20-dione, 16β-hydroxy-**	17779-71-4	**NI34849**
287 43 330 229 269 145 55 123	**330**	100 54 48 46 20 19 19 18		21	30	3	–	–	–	–	–	–	–	–	–	–	**Pregn-4-ene-3,20-dione, 17-hydroxy-**	68-96-2	**NI34851**

MASS TO CHARGE RATIOS								M.W.	INTENSITIES								Parent	C	H	O	N	S	F	Cl	Br	I	Si	P	B	X	COMPOUND NAME	CAS Reg No	No
287	43	330	229	269	145	244	55	330	100	53	47	45	21	20	19	19		21	30	3	–	–	–	–	–	–	–	–	–	–	Pregn-4-ene-3,20-dione, 17-hydroxy-	68-96-2	NI34850
287	43	330	91	79	41	229	55	330	100	99	77	73	71	64	63	62		21	30	3	–	–	–	–	–	–	–	–	–	–	Pregn-4-ene-3,20-dione, 17-hydroxy-	68-96-2	NI34852
43	300	91	330	110	41	79	55	330	100	71	19	18	18	17	15	15		21	30	3	–	–	–	–	–	–	–	–	–	–	Pregn-4-ene-3,20-dione, 19-hydroxy-	596-63-4	NI34853
91	105	269	77	79	55	159	107	330	100	74	71	63	60	59	56	53	27.80	21	30	3	–	–	–	–	–	–	–	–	–	–	Pregn-5-en-20-one, 16,17-epoxy-3-hydroxy-, (3β,16α)-	974-23-2	MS07863
159	330	255	55	91	108	106	79	330	100	58	56	55	52	50	48	44		21	30	3	–	–	–	–	–	–	–	–	–	–	Pregn-5-en-20-one, 16,17-epoxy-3-hydroxy-, (3β,16α)-	974-23-2	LI07340
330	149	271	43	315	91	69	107	330	100	58	52	46	45	36	34	33		21	30	3	–	–	–	–	–	–	–	–	–	–	Retinoic acid, 5,6-epoxy-5,6-dihydro-, methyl ester	7432-30-6	NI34854
330	149	271	43	315	107	164	95	330	100	56	51	44	43	36	32	31		21	30	3	–	–	–	–	–	–	–	–	–	–	Retinoic acid, 5,8-epoxy-5,8-dihydro-, methyl ester	50876-25-0	NI34855
97	174	234	159	133	131	147	145	330	100	94	47	42	36	33	30	27	0.00	21	30	3	–	–	–	–	–	–	–	–	–	–	Spiro[bicyclo[3.1.0]hexane-6,3′(2′H)-as-indacen]-2-one, decahydro-6′-hydroxy-4,5′a-dimethyl-, acetate	2383-65-5	NI34856
97	174	234	133	159	132	131	119	330	100	68	42	42	41	32	31	28	0.00	21	30	3	–	–	–	–	–	–	–	–	–	–	Spiro[bicyclo[3.1.0]hexane-6,3′(2′H)-as-indacen]-2-one, decahydro-6′-hydroxy-4,5′a-dimethyl-, acetate (stereoisomer)	2383-76-8	NI34857
97	174	133	159	147	148	145	146	330	100	22	20	17	14	13	13	11	0.62	21	30	3	–	–	–	–	–	–	–	–	–	–	Spiro[bicyclo[3.1.0]hexane-6,3′(2′H)-as-indacen]-2-one, decahydro-6′-hydroxy-5,5′a-dimethyl-, acetate (stereoisomer)	2587-18-0	NI34858
273	57	274	41	29	246	28	18	330	100	57	29	22	20	14	14	9	0.70	21	30	3	–	–	–	–	–	–	–	–	–	–	1,3,5-Tripivaloylbenzene		LI07341
275	184	197	330	329	156	223	169	330	100	98	96	91	42	31	29	29		22	22	1	2	–	–	–	–	–	–	–	–	–	2-Benzoyloctahydroindolo[2,3-a]quinolizine		LI07342
225	330	184	197	329	210	156	169	330	100	83	66	59	47	32	27	24		22	22	1	2	–	–	–	–	–	–	–	–	–	2-Benzoyloctahydroindolo[2,3-a]quinolizine		LI07343
105	144	77	183	145	225	130	330	330	100	52	43	37	31	30	26	19		22	22	1	2	–	–	–	–	–	–	–	–	–	4-Benzoyl-1,2,3,4-tetrahydro-1-[2-(indol-3-yl)ethyl]pyridine		LI07344
200	330	225	201	105	144	77	143	330	100	20	18	16	16	13	13	8		22	22	1	2	–	–	–	–	–	–	–	–	–	4-Benzoyl-1,2,3,6-tetrahydro-1-[2-(indol-3-yl)ethyl]pyridine		LI07345
271	286	330	297	244	328	272	270	330	100	89	68	63	41	40	38	34		22	22	1	2	–	–	–	–	–	–	–	–	–	1H-Indeno[4,5-g]quinazolin-3-ol, 2,3β,3a,4,5,11bβ-hexahydro-3aα-methyl 8-phenyl-	17305-66-7	NI34859
118	91	106	107	223	77	65	120	330	100	81	54	48	33	25	20	14	0.00	22	22	1	2	–	–	–	–	–	–	–	–	–	2,2-Bis(4-methylanilino)acetophenone	79866-34-5	NI34860
55	41	81	91	105	43	67	79	330	100	97	90	81	77	77	71	68	35.08	22	34	2	–	–	–	–	–	–	–	–	–	–	Androst-5-en-17-one, 3-methoxy-16,16-dimethyl-, (3α)-	55837-02-0	NI34861
55	41	298	81	330	121	91	69	330	100	81	80	77	75	72	69	69		22	34	2	–	–	–	–	–	–	–	–	–	–	Androst-5-en-17-one, 3-methoxy-16,16-dimethyl-, (3β)-	55837-03-1	NI34862
313	331	314	255	329	295	151	273	330	100	35	23	15	14	14	9	8	8.00	22	34	2	–	–	–	–	–	–	–	–	–	–	3β-Hydroxy-20-methyl-20,21-epoxy-Δ^5-pregnane		NI34863
43	69	41	81	93	55	91	67	330	100	63	60	48	43	43	35	33	15.89	22	34	2	–	–	–	–	–	–	–	–	–	–	Kauren-18-ol, acetate, (4β)-	72150-74-4	NI34864
310	295	91	105	77	159	107	145	330	100	87	84	60	47	44	42	41	0.00	22	34	2	–	–	–	–	–	–	–	–	–	–	16-Methyl-16-dehydropregnenolone		MS07864
330	91	55	107	105	85	312	79	330	100	89	84	81	75	71	70	70		22	34	2	–	–	–	–	–	–	–	–	–	–	16α-Methylpregnenolone		MS07865
43	41	55	91	93	105	81	135	330	100	64	38	33	28	26	24	21	5.00	22	34	2	–	–	–	–	–	–	–	–	–	–	17-Norkaur-15-ene-18-ol, 13-methyl-, acetate, (4α,8β,13β)-	16394-71-1	LI07346
43	44	239	123	29	69	97	41	330	100	51	47	35	34	27	23	22	2.20	22	34	2	–	–	–	–	–	–	–	–	–	–	1-Phenanthrenecarboxylic acid, 1,2,3,4,4a,4b,5,9,10,10a-decahydro-1,4a-dimethyl-7-isopropyl-, ethyl ester, [1R-(1α,4aβ,4bα,10aα)]-	57274-59-6	NI34865
43	29	44	121	123	41	42	69	330	100	53	45	44	28	26	20	19	8.50	22	34	2	–	–	–	–	–	–	–	–	–	–	1-Phenanthrenecarboxylic acid, 7-vinyl-1,2,3,4,4a,4b,5,6,7,9,10,10a-dodecahydro-1,4a,7-trimethyl-, ethyl ester, [1R-(1α,4aβ,4bα,7α,10aα)]-	57274-61-0	NI34866
121	55	91	81	41	120	93	257	330	100	20	19	17	17	16	16	15	8.00	22	34	2	–	–	–	–	–	–	–	–	–	–	Pimaric acid, ethyl ester	57274-58-5	NI34867
99	86	301	245	330	100	55	41	330	100	41	27	18	15	13	11	11		22	34	2	–	–	–	–	–	–	–	–	–	–	Pregn-4-en-3-one, 17-methoxy-, (17α)-	20112-33-8	NI34868
57	315	70	274	316	96	330	58	330	100	55	43	35	15	15	13	13		22	38	–	2	–	–	–	–	–	–	–	–	–	Norconanine, 3β-(methylamino)-	6869-86-9	NI34869
70	44	261	174	71	330	262	57	330	100	27	20	7	6	5	5	3		22	38	–	2	–	–	–	–	–	–	–	–	–	Pregn-5-ene-3,20-diamine, N^3-methyl-, (3β,20S)-	3734-04-1	NI34870
91	115	27	117	41	212	128	330	330	100	45	38	33	31	26	23	21		23	22	2	–	–	–	–	–	–	–	–	–	–	anti-8-Hydroxy-syn-8-isopropenyl-2,4-diphenylbicyclo[3.2.1]oct-6-en-3-one		DD01296
273	274	196	271	165	330	290	270	330	100	69	60	57	57	54	48	45		23	23	–	–	–	–	–	–	–	–	1	–	–	10-tert-Butyl-9,10-dihydro-9-phenyl-9-phosphaanthracene		MS07866
130	159	172	330	131	300			330	100	35	25	14	12	4				23	26	–	2	–	–	–	–	–	–	–	–	–	1,1′-Methylenebis(3-propylindole)		LI07347
217	259	301	109	216	121	135	107	330	100	76	52	37	30	27	24	22	6.41	23	38	1	–	–	–	–	–	–	–	–	–	–	24-Norcholan-16-one, (5α,20ξ)-	69831-76-1	NI34871
95	121	330	148	93	81	258	79	330	100	76	57	40	39	39	38	34		23	38	1	–	–	–	–	–	–	–	–	–	–	Pregn-2-en-17-ol, 4,4-dimethyl-, (5α,17ξ)-	53286-40-1	NI34872
330	331	207	165	73	329	281	126	330	100	30	11	7	2	1	0	0		24	14	–	2	–	–	–	–	–	–	–	–	–	Tribenzo[a,c,h]phenazine	215-29-2	NI34873
330	287	105	315	119	145	278	302	330	100	37	34	33	32	30	25	24		24	26	1	–	–	–	–	–	–	–	–	–	–	1,1-Dimethyl-2-(4-methylphenyl)-5-[(E)-(4-methylphenyl)methylene]spiro[2.4]heptan-4-one		MS07867
243	244	330	253	331	245	254	105	330	100	93	70	42	23	17	10	5		24	26	1	–	–	–	–	–	–	–	–	–	–	Pentyl trityl ether	20705-39-9	NI34874
243	105	244	183	165	330	253	77	330	100	52	44	41	38	24	24	16		24	26	1	–	–	–	–	–	–	–	–	–	–	Pentyl trityl ether	20705-39-9	NI34876
243	105	244	183	165	253	330	77	330	100	53	44	40	37	24	22	16		24	26	1	–	–	–	–	–	–	–	–	–	–	Pentyl trityl ether	20705-39-9	NI34875
119	330	43	120	41	91	105	57	330	100	31	27	21	21	15	13	9		24	42	–	–	–	–	–	–	–	–	–	–	–	Benzene, ethylhexadecyl-	72101-27-0	WI01431
301	330	302	43	331	175	217	287	330	100	55	24	20	14	12	11	8		24	42	–	–	–	–	–	–	–	–	–	–	–	Benzene, hexapropyl-	2456-68-0	NI34877
119	105	91	175	133	71	120	118	330	100	44	32	15	14	13	11	9	0.70	24	42	–	–	–	–	–	–	–	–	–	–	–	Benzene, octadecyl-	4445-07-2	WI01432
105	106	91	43	330	41	104	29	330	100	13	10	7	6	5	4	3		24	42	–	–	–	–	–	–	–	–	–	–	–	2-Phenyloctadecane		AI02056
91	203	105	217	104	43	92	41	330	100	18	17	16	10	10	9	8	7.49	24	42	–	–	–	–	–	–	–	–	–	–	–	9-Phenyloctadecane		AI02057
330	331	144	251	329	326	327	252	330	100	24	17	14	10	8	7	7		26	18	–	–	–	–	–	–	–	–	–	–	–	Anthracene, 9,10-diphenyl-	1499-10-1	NI34878
165	166	163	330	139	115	331	162	330	100	10	6	6	2	1	1	1		26	18	–	–	–	–	–	–	–	–	–	–	–	9,9′-Bi-9H-fluorene	1530-12-7	NI34879

MASS TO CHARGE RATIOS								M.W.	INTENSITIES								Parent	C	H	O	N	S	F	Cl	Br	I	Si	P	B	X	COMPOUND NAME	CAS Reg No	No
330	253	252	331	329	165	327	250	330	100	36	35	28	27	11	10	10		26	18	–	–	–	–	–	–	–	–	–	–	–	9H-Fluorene, 9-(diphenylmethylene)-	4709-68-6	NI34880
330	253	252	331	329	327	165	251	330	100	37	35	28	26	10	10	9		26	18	–	–	–	–	–	–	–	–	–	–	–	9H-Fluorene, 9-(diphenylmethylene)-	4709-68-6	LI07348
330	252	253	157	329	163	331	150	330	100	78	76	37	32	25	23	21		26	18	–	–	–	–	–	–	–	–	–	–	–	9H-Fluorene, 9-(diphenylmethylene)-	4709-68-6	NI34881
330	331	252	253	329	326	150	39	330	100	28	20	17	15	9	9	8		26	18	–	–	–	–	–	–	–	–	–	–	–	Phenanthrene, 9,10-diphenyl-	602-15-3	WI01433
330	329	252	253					330	100	33	33	28						26	18	–	–	–	–	–	–	–	–	–	–	–	9-Phenylanthracene benzyne centre ring adduct		LI07349
330	252	253	329					330	100	82	51	40						26	18	–	–	–	–	–	–	–	–	–	–	–	9-Phenylanthracene benzyne end ring adduct		LI07350
219	142	176	119	76	275	247	100	331	100	89	84	57	54	41	39	28	18.00	10	18	4	3	–	–	–	–	–	–	1	–	1	Tris(dimethylamino)phosphine-tetracarbonyl iron		MS07868
104	41	103	69	252	254	77	39	331	100	41	33	33	25	24	20	17	0.00	12	15	–	1	–	–	–	2	–	–	–	–	–	Benzenecarboximidoyl bromide, N-(5-bromopentyl)-	41182-89-2	NI34882
69	103	41	104	76	39	152	77	331	100	96	69	43	34	18	16	16	0.00	12	15	–	1	–	–	–	2	–	–	–	–	–	Piperidinium, 1-(bromophenylmethylene)-, bromide	41182-92-7	NI34883
296	298	300	178	149	333	91	180	331	100	95	28	22	22	19	18	16	0.00	12	17	1	1	–	–	4	–	–	–	–	–	–	(1RS,5SR,6SR,7RS,8RS,9SR)-6,7,8,9-Tetrakis(chloromethyl)-2-azabicyclo[3.2.2]nonan-3-one		DD01297
296	298	107	282	300	260	158	192	331	100	91	77	28	27	15	15	8	0.00	12	17	1	1	–	–	4	–	–	–	–	–	–	(1RS,4SR,5SR,6SR,7SR,8SR)-5,6,7,8-Tetrakis(chloromethyl)bicyclo[2.2.2]octan-2-one oxime		DD01298
331	51	63	76	77	79	151	75	331	100	52	50	48	47	44	40	40		13	9	6	5	–	–	–	–	–	–	–	–	–	Benzaldehyde, 3-nitro-, (2,4-dinitrophenyl)hydrazone	2571-09-7	LI07351
331	51	63	76	77	79	151	75	331	100	52	50	46	45	42	40	40		13	9	6	5	–	–	–	–	–	–	–	–	–	Benzaldehyde, 3-nitro-, (2,4-dinitrophenyl)hydrazone	2571-09-7	NI34884
240	118	91	119	169	241	69	41	331	100	66	25	12	11	8	8	8	0.15	13	12	1	1	–	7	–	–	–	–	–	–	–	Heptafluorobutyrylamphetamine	1545-26-2	MS07869
282	240	313	114	331				331	100	60	42	14	1					13	17	9	1	–	–	–	–	–	–	–	–	–	Dimethyl 2-(methylamino)-5-oxo-3,4-bis(methoxycarbonyl)-2-hexenedicarboxylate		LI07352
272	258	114	240	331				331	100	14	14	5	1					13	17	9	1	–	–	–	–	–	–	–	–	–	2-Methyl-3,4,5-tris(methoxycarbonyl)-2,3-dihydroisoxazole-3-acetic acid		LI07353
75	73	44	45	117	43	77	47	331	100	89	43	33	28	25	21	21	2.30	13	26	6	1	–	–	–	–	–	1	–	1	–	α-D-Glucopyranoside, methyl 2-(acetylamino)-2-deoxy-3-O-(trimethylsilyl)-, cyclic methylboronate	54477-01-9	NI34885
120	150	28	331	42	211	134	119	331	100	75	61	51	37	24	19	19		14	13	5	5	–	–	–	–	–	–	–	–	–	Aniline, 4-[(2,4-dinitrophenyl)-ONN-azoxy]-N,N-dimethyl-	61142-09-4	NI34886
43	101	42	54	45	71	129	41	331	100	55	36	27	27	25	21	20	0.00	14	21	8	1	–	–	–	–	–	–	–	–	–	2,3,5-Tri-O-acetyl-4,6-di-O-methyl-D-mannonitrile	58720-12-0	MS07870
101	129	112	45	161	214	154	87	331	100	43	36	34	24	22	20	13	0.00	14	21	8	1	–	–	–	–	–	–	–	–	–	2,3,5-Tri-O-acetyl-4,6-di-O-methyl-D-mannonitrile	58720-12-0	NI34887
87	45	129	113	131	99	142	83	331	100	83	79	46	36	31	30	21	0.00	14	21	8	1	–	–	–	–	–	–	–	–	–	2,4,5-Tri-O-acetyl-3,6-di-O-methyl-D-mannonitrile		MS07871
129	43	87	189	126	99	130	142	331	100	95	49	34	19	15	8	7	0.00	14	21	8	1	–	–	–	–	–	–	–	–	–	2,5,6-Tri-O-acetyl-3,4-di-O-methyl-D-mannonitrile	59061-09-5	MS07872
129	189	87	126	99	88	45	71	331	100	34	26	10	8	6	6	5	0.00	14	21	8	1	–	–	–	–	–	–	–	–	–	2,5,6-Tri-O-acetyl-3,4-di-O-methyl-D-mannonitrile	59061-09-5	NI34888
43	87	45	159	169	115	129	117	331	100	20	19	14	13	12	11	9	0.00	14	21	8	1	–	–	–	–	–	–	–	–	–	3,4,5-Tri-O-acetyl-2,6-di-O-methyl-D-mannonitrile		MS07873
43	129	87	186	112	189	154	159	331	100	52	28	23	21	20	18	15	0.00	14	21	8	1	–	–	–	–	–	–	–	–	–	3,5,6-Tri-O-acetyl-2,4-di-O-methyl-D-mannonitrile		MS07874
43	127	85	99	133	115	88	87	331	100	12	12	11	10	6	4	4	0.00	14	21	8	1	–	–	–	–	–	–	–	–	–	4,5,6-Tri-O-acetyl-2,3-di-O-methyl-D-mannonitrile	59061-07-3	MS07875
85	127	99	115	87	100	45	88	331	100	96	96	68	54	45	44	41	0.00	14	21	8	1	–	–	–	–	–	–	–	–	–	4,5,6-Tri-O-acetyl-2,3-di-O-methyl-D-mannonitrile	59061-07-3	NI34889
58	204	104	117	115	205	105	91	331	100	51	14	13	12	11	10	10	0.40	14	22	–	1	–	–	–	–	1	–	–	–	–	2,2-Dimethyl-2,3,4,5,6,7-hexahydro-1H-2-benzazoninium iodide		NI34890
73	182	214	75	45	89	300	55	331	100	69	46	44	29	20	19	18	12.40	14	29	4	1	–	–	–	–	–	2	–	–	–	4,6-Dioxoheptanoic acid methoxime, bis(trimethylsilyl)-		NI34891
331	333	28	234	315	313	151	252	331	100	98	57	43	42	41	26	22		15	10	3	1	–	–	–	1	–	–	–	–	–	1-Amino-4-bromo-2-(hydroxymethyl)anthraquinone		NI34892
91	240	194	178	166	331			331	100	24	13	13	13	10				15	13	6	3	–	–	–	–	–	–	–	–	–	2,4-(Dinitrophenyl)phenylalanine		LI07354
212	81	93	143	331	77	202	55	331	100	69	63	50	48	45	39	39		15	16	2	1	1	3	–	–	–	–	–	–	–	2-Oxo-N-phenyl-3-[(trifluoromethyl)thio]cycloheptanecarboxamide		DD01299
129	43	187	45	69	101	59	55	331	100	92	62	60	57	38	35	35	18.00	15	25	7	1	–	–	–	–	–	–	–	–	–	5-Isoxazolidinecarboxylic acid, 5-methyl-2-[5-O-methyl-2,3-O-isopropylidene-β-D-ribofuranosyl]-, methyl ester	64018-34-4	NI34893
212	43	55	177	41	218	69	57	331	100	97	82	80	79	60	60	56	5.20	15	26	1	1	–	5	–	–	–	–	–	–	–	1-Aminododecane, pentafluoropropionyl-		MS07876
77	159	51	236	316	314	78	238	331	100	67	63	51	43	22	21	20	5.00	16	13	–	1	1	–	–	–	–	–	–	–	1	(3-Methylseleno-2-benzo[b]thienylidene)aniline	71740-03-9	NI34894
91	159	65	250	331	89	240	145	331	100	84	84	65	61	40	32	29		16	13	–	1	1	–	–	–	–	–	–	–	1	(2-Selenyl-3-benzo[b]thienylidene)-4-methylaniline	92886-20-9	NI34895
273	316	331	44	287	42	288	271	331	100	50	42	25	20	19	14	14		16	17	5	3	–	–	–	–	–	–	–	–	–	Azirino[2′,3′:3,4]pyrrolo[1,2-a]indole-4,7-dione, 8-[[(aminocarbonyl)oxy]methyl]-1,1a,2,8b-tetrahydro-6-methoxy-1,5-dimethyl-	15973-07-6	NI34896
91	148	43	216	149	103	218	71	331	100	98	88	27	27	25	17	16	0.00	16	20	3	1	–	3	–	–	–	–	–	–	–	L-Phenylalanine, N-(trifluoroacetyl)-, 1-methylbutyl ester	58072-48-3	NI34897
84	241	156	82	92.5	187	331	170	331	100	39	34	30	18	16	14	10		16	33	4	3	–	–	–	–	–	–	–	–	–	Lysinonorleucine diethyl ester		LI07355
73	187	144	59	158	74	88	143	331	100	59	36	33	26	17	17	14	0.00	16	41	–	3	–	–	–	–	–	2	–	–	–	N(1),N(3)-Bis(dimethyl-tert-butylsilyl)diethylenetriamine		NI34898
260	259	288	287	261	289	262	290	331	100	91	82	70	59	38	36	29	0.00	17	15	1	3	–	1	1	–	–	–	–	–	–	2H-1,4-Benzodiazepin-2-one, 1-(2-aminoethyl)-7-chloro-5-(2-fluorophenyl)-1,3-dihydro-		NI34899
284	299	286	254	158	301	258	243	331	100	84	61	43	34	30	28	25	11.00	17	17	6	1	–	–	–	–	–	–	–	–	–	10-Methyl-2,4-dimethoxy-3,4-dihydro-1-hydroxy-3,4-methylenedioxy-acridone		MS07877
43	331	121	286	244	287	56	316	331	100	85	78	60	48	35	35	25		17	25	–	1	1	–	–	–	–	–	–	–	1	(R,S)-1-(1-(Dimethylamino)ethyl)-2-(isopropylthio)ferrocene		MS07878
331	286	41	287	43	316	121	288	331	100	71	62	48	38	36	17	13		17	25	–	1	1	–	–	–	–	–	–	–	1	(R,S)-1-(1-(Dimethylamino)ethyl)-2-(propylthio)ferrocene		MS07879

MASS TO CHARGE RATIOS	M.W.	INTENSITIES	Parent	C	H	O	N	S	F	Cl	Br	I	Si	P	B	X	COMPOUND NAME	CAS Reg No	No
331 242 73 332 316 215 243 75	**331**	100 51 44 28 23 18 14 12		17	25	2	1	–	–	–	–	–	2	–	–	–	2-Propenoic acid, 3-[1-(trimethylsilyl)-1H-indol-3-yl]-, trimethylsilyl ester	55124-88-4	NI34900
73 331 242 332 45 316 75 215	**331**	100 60 37 17 15 13 10 8		17	25	2	1	–	–	–	–	–	2	–	–	–	2-Propenoic acid, 3-[1-(trimethylsilyl)-1H-indol-3-yl]-, trimethylsilyl ester	55124-88-4	NI34901
84 141 98 107 331 85 41 190	**331**	100 67 30 15 14 10 10 8		17	25	2	5	–	–	–	–	–	–	–	–	–	5-Amino-2,4-dihydro-4-[3-[3-(piperidinomethyl)phenoxy]propyl]-3H-1,2,4-triazol-3-one		MS07880
84 98 57 41 59 42 107 141	**331**	100 37 26 24 22 21 20 15	**3.00**	17	25	2	5	–	–	–	–	–	–	–	–	–	2-Amino-5-[3-[3-(piperidinomethyl)phenoxy]propylamino]-1,3,4-oxadiazole		MS07881
110 150 221 178 179 107 135 109	**331**	100 68 63 39 31 13 12 11	**4.00**	18	21	3	1	1	–	–	–	–	–	–	–	–	9,10-Dimethoxy-5-methyl-4,5,6,7,12,13-hexahydrothieno[3,2-e][3]benzazecin-12-one		MS07882
142 331 190 161 141 138 83 110	**331**	100 88 84 73 48 38 36 31		18	21	5	1	–	–	–	–	–	–	–	–	–	Cephalotaxine, 2,11-epoxy-1,2-dihydro-, (2α,11α)-	49686-57-9	NI34902
142 331 190 161 141 154 138 83	**331**	100 86 84 73 46 41 36 34		18	21	5	1	–	–	–	–	–	–	–	–	–	Cephalotaxine, 2,11-epoxy-1,2-dihydro-, (2α,11α)-	49686-57-9	NI34903
142 161 190 138 331 110 141 131	**331**	100 92 85 74 50 49 46 42		18	21	5	1	–	–	–	–	–	–	–	–	–	Cephalotaxine, 2,11-epoxy-1,2-dihydro-, (2α,11α)-	49686-57-9	NI34904
331 44 314 270 313 298 295 28	**331**	100 92 61 59 43 39 36 29		18	21	5	1	–	–	–	–	–	–	–	–	–	Cephalotaxine, 11-hydroxy-	24268-95-9	NI34905
331 314 298 313 255 270 300 110	**331**	100 99 89 84 70 69 57 55		18	21	5	1	–	–	–	–	–	–	–	–	–	Cephalotaxine, 11-hydroxy-	24268-95-9	NI34906
331 287 260 257 255 299 241 211	**331**	100 52 47 35 30 29 27 24		18	21	5	1	–	–	–	–	–	–	–	–	–	Crinan-11-ol, 1,2-didehydro-3,7-dimethoxy-, (3α,11S)-	3660-62-6	NI34907
331 287 260 257 299 241 211 302	**331**	100 68 58 44 34 30 27 24		18	21	5	1	–	–	–	–	–	–	–	–	–	Crinan-11-ol, 1,2-didehydro-3,7-dimethoxy-, (3α,11S)-	3660-62-6	LI07356
201 43 115 331 172 173 143 135	**331**	100 60 54 46 30 26 17 15		18	21	5	1	–	–	–	–	–	–	–	–	–	2,4-Pentadienamide, N-[2-(Acetyloxy)-2-methylpropyl]-5-(1,3-benzodioxol 5-yl)-, (E,E)-	79097-20-4	NI34908
201 115 43 173 172 331 135 271	**331**	100 43 43 38 27 20 16 15		18	21	5	1	–	–	–	–	–	–	–	–	–	2,4-Pentadienamine, N-[2-(acetyloxy)-2-methylpropyl]-5-(1,3-benzodioxol 5-yl)-, (Z,Z)-	79097-21-5	NI34909
231 229 230 331 228 215 214 216	**331**	100 28 24 14 9 6 5 4		18	21	5	1	–	–	–	–	–	–	–	–	–	Stephabyssine	36871-84-8	NI34910
247 331 71 70 46 45 298 316	**331**	100 65 36 30 22 22 20 19		18	21	5	1	–	–	–	–	–	–	–	–	–	Tazettine	507-79-9	NI34911
247 314 331 330 332 230 300 181	**331**	100 80 52 46 41 39 31 28		18	21	5	1	–	–	–	–	–	–	–	–	–	Tazettine	507-79-9	NI34912
71 331 301 70 45 44 260 199	**331**	100 59 56 52 48 36 32 29		18	21	5	1	–	–	–	–	–	–	–	–	–	Tazettine, (3α)-	545-16-4	NI34913
295 280 264 235 249	**331**	100 29 25 12 5	**0.00**	18	22	1	3	–	–	1	–	–	–	–	–	–	2,4-Dimethyl-3-ethyl-6-methoxyprodigiosene hydrochloride		LI07357
331 105 77 288 332 203 204 51	**331**	100 65 55 34 24 19 18 17		19	13	3	3	–	–	–	–	–	–	–	–	–	2-Phenyl-4-benzoyl-3H-imidazo[1,5-b]pyridazine-5,7-(6H)-dione	7248-17-1	NI34914
105 331 316 43 77 254 57 41	**331**	100 70 52 48 44 40 29 29		19	17	1	5	–	–	–	–	–	–	–	–	–	N(5)-Methyl-6,7-diphenyl-5,6-dihydropterin		MS07883
254 73 43 44 331 57 41 28	**331**	100 89 67 50 48 48 48 46		19	17	1	5	–	–	–	–	–	–	–	–	–	N(8)-Methyl-6,7-diphenyl-7,8-dihydropterin		MS07884
151 152 180	**331**	100 19 9	**0.00**	19	25	4	1	–	–	–	–	–	–	–	–	–	N-Benzyl-β-(4-chlorophenyl)ethylamine		LI07358
164 41 123 79 81 39 77 43	**331**	100 52 47 45 32 26 25 24	**0.10**	19	25	4	1	–	–	–	–	–	–	–	–	–	Cyclopropanecarboxylic acid, 2,2-dimethyl-3-(2-methyl-1-propenyl)-, (1,3,4,5,6,7-hexahydro-1,3-dioxo-2H-isoindol-2-yl)methyl ester	7696-12-0	NI34916
164 123 165 81 41 107 43 79	**331**	100 45 16 13 10 9 9 8	**0.00**	19	25	4	1	–	–	–	–	–	–	–	–	–	Cyclopropanecarboxylic acid, 2,2-dimethyl-3-(2-methyl-1-propenyl)-, (1,3,4,5,6,7-hexahydro-1,3-dioxo-2H-isoindol-2-yl)methyl ester	7696-12-0	NI34915
286 314 332 123 164 151 135 333	**331**	100 50 40 21 18 11 10 9	**0.00**	19	25	4	1	–	–	–	–	–	–	–	–	–	Cyclopropanecarboxylic acid, 2,2-dimethyl-3-(2-methyl-1-propenyl)-, (1,3,4,5,6,7-hexahydro-1,3-dioxo-2H-isoindol-2-yl)methyl ester	7696-12-0	NI34917
192 169 55 56 70 43 103 69	**331**	100 58 55 46 43 43 40 38	**0.00**	19	25	4	1	–	–	–	–	–	–	–	–	–	5-Isoindolinecarboxylic acid, 1,3-dioxo-, decyl ester	33975-35-8	NI34918
331 77 332 51 226 115 105 127	**331**	100 24 23 17 13 12 12 11		20	13	4	1	–	–	–	–	–	–	–	–	–	1-Amino-2-phenoxy-4-hydroxyanthraquinone		IC03991
259 331 302 286 181	**331**	100 99 21 15 13		20	17	2	1	–	–	–	–	–	1	–	–	–	1-Oxa-3-aza-2-silacyclopentan-5-one, 2,2,3-triphenyl-	53268-81-8	NI34919
105 77 196 331 107 65 51 332	**331**	100 87 60 44 29 13 11 10		20	17	2	3	–	–	–	–	–	–	–	–	–	1-(Benzoylamino)-4-(4-hydroxytolylazo)benzene		IC03992
93 119 91 64 66 65 212 39	**331**	100 81 30 21 14 14 12 12	**0.80**	20	17	2	3	–	–	–	–	–	–	–	–	–	1,3,5-Triphenylbiuret	2645-39-8	MS07885
170 331 316 330 332 77 43 105	**331**	100 84 57 43 20 20 18 15		20	18	4	–	–	–	–	–	–	–	–	–	1	Beryllium, bis(1-phenyl-1,3-butanedionato)-	14128-75-7	NI34920
170 331 316 330 77 105 92 128	**331**	100 84 56 42 19 14 11 10		20	18	4	–	–	–	–	–	–	–	–	–	1	Beryllium, bis(1-phenyl-1,3-butanedionato)-	14128-75-7	LI07359
332 69 263 264 233 333 123 207	**331**	100 89 83 55 26 24 23 20	**0.00**	20	27	4	–	–	–	–	–	–	–	–	–	–	Cyclopenta[b]furylium, 3a,4-dihydro-3a-(3-methyl-2-butenyl)-6-(2-methyl-1-oxobutyl)-4-oxo-5-propoxy-	57174-10-4	NI34921
331 95 81 41 127 137 67 55	**331**	100 76 72 67 65 63 60 60		20	33	1	3	–	–	–	–	–	–	–	–	–	Androstan-3-one, (aminocarbonyl)hydrazone, (5α)-	54725-03-0	NI34922
139 331 214 333 141 111 332 165	**331**	100 97 54 35 32 30 25 19		21	14	1	1	–	–	1	–	–	–	–	–	–	3,4-Diphenyl-5-(4-chlorophenyl)-isoxazole		LI07360
213 180 179 136 165 214 181 331	**331**	100 90 70 59 41 33 27 16		21	17	1	1	1	–	–	–	–	–	–	–	–	trans-2,3-Dihydro-2,3-diphenyl-1,5-benzothiazepin-4(5H)-one		NI34923
105 77 51 331 106 78 50 332	**331**	100 63 18 12 9 5 5 3		21	17	3	1	–	–	–	–	–	–	–	–	–	2-Benzoxy-5-benzamidotoluene		IC03993
209 105 77 210 180 331 51 106	**331**	100 52 30 17 12 7 6 5		21	17	3	1	–	–	–	–	–	–	–	–	–	N-Benzoyl-N-phenyl-N[6]-methylphenylamine-2-carboxylic acid		NI34924
180 77 181 105 51 331 78 92	**331**	100 35 17 8 8 6 6 4		21	17	3	1	–	–	–	–	–	–	–	–	–	2-Methoxycarbonylphenyl-N-phenylbenzimidate		NI34925
180 77 181 51 105 78 76 50	**331**	100 37 16 11 5 4 3 3	**3.00**	21	17	3	1	–	–	–	–	–	–	–	–	–	4-Methoxycarbonylphenyl-N-phenylbenzimidate		NI34926
247 58 84 248 260 85 231 273	**331**	100 87 22 19 16 12 11 10	**5.00**	21	21	1	3	–	–	–	–	–	–	–	–	–	5,6-Dihydro-5-oxo-6-(3-dimethylaminopropyl)dibenzo[b,f]-1,8-naphthyridine		IC03994
313 282 284 298 121 314 331 137	**331**	100 43 38 28 27 24 21 19		21	33	2	1	–	–	–	–	–	–	–	–	–	19-Norpregn-4-en-3-one, 17-hydroxy-, O-methyloxime, (17α)-	69833-65-4	NI34927
149 119 148 118 77 150 105 91	**331**	100 37 13 13 12 11 11 9	**1.20**	22	21	2	1	–	–	–	–	–	–	–	–	–	N-Methyl-2-(β,β-diphenyl-β-hydroxyethyl)benzamide		MS07886

MASS TO CHARGE RATIOS								M.W.	INTENSITIES								Parent	C	H	O	N	S	F	Cl	Br	I	Si	P	B	X	COMPOUND NAME	CAS Reg No	No
100	300	87	70	55	67	81	41	331	100	62	50	24	21	20	18	18	5.00	22	37	1	1	–	–	–	–	–	–	–	–	–	5α-Pregnan-20-one, methyloxime		LI07361
113	126	70	55	72	98	41	331	331	100	41	35	33	29	25	24	23		22	37	1	1	–	–	–	–	–	–	–	–	–	Pyrrolidine, 1-(1-oxo-6,9,12-octadecatrienyl)-	56600-02-3	NI34928
113	55	41	70	43	126	67	72	331	100	54	49	46	43	42	34	34	18.40	22	37	1	1	–	–	–	–	–	–	–	–	–	Pyrrolidine, 1-(1-oxo-6,9,12-octadecatrienyl)-	56600-02-3	NI34929
113	126	70	55	331	72	41	98	331	100	62	51	44	40	37	31	25		22	37	1	1	–	–	–	–	–	–	–	–	–	Pyrrolidine, 1-(1-oxo-9,12,15-octadecatrienyl)-	56600-08-9	NI34930
302	220	301	303	55	125	136	83	331	100	40	24	22	22	16	13	11	7.26	22	42	–	1	–	–	–	–	–	–	–	1	–	Boranamine, N-1-cyclododecen-1-yl-N-cyclohexyl-1,1-diethyl-	74793-30-9	NI34931
69	206	191	165	151	207	189	180	331	100	81	55	49	25	20	20	17	1.60	23	25	1	1	–	–	–	–	–	–	–	–	–	Spiro[fluorene-9,2′-oxetan]-4′-imine, N-cyclohexyl-3′,3′-dimethyl-	15183-47-8	MS07887
69	206	191	165	207	189	83	180	331	100	48	48	43	17	16	15	14	2.60	23	25	1	1	–	–	–	–	–	–	–	–	–	Spiro[fluorene-9,2′-oxetan]-4′-imine, N-cyclohexyl-3′,3′-dimethyl-	15183-47-8	NI34932
84	110	97	71	55	96	57	54	331	100	48	13	13	7	6	6	6	5.00	23	41	–	1	–	–	–	–	–	–	–	–	–	Pregnan-3-amine, N,N-dimethyl-, (3β,5α)-	20840-32-8	NI34933
72	73	331	56	54				331	100	3	1	1	1					23	41	–	1	–	–	–	–	–	–	–	–	–	Pregnan-20-amine, N,N-dimethyl-, (5α,20R)-	55156-14-4	NI34934
253	223	79	208	251	252	221	222	332	100	70	60	48	45	43	38	36	0.00	4	12	2	–	1	–	–	–	–	–	–	–	1	Plumbane, trimethyl[(methylsulphinyl)oxy]-	44657-41-2	LI07362
253	223	79	208	251	252	221	222	332	100	72	62	49	44	43	36	35	0.00	4	12	2	–	1	–	–	–	–	–	–	–	1	Plumbane, trimethyl[(methylsulphinyl)oxy]-	44657-41-2	MS07888
131	63	100	69	332	82	113	163	332	100	41	30	27	26	20	17	9		6	–	–	–	1	12	–	–	–	–	–	–	–	Perfluorothiepane		MS07889
69	97	263	85	235	313	219	131	332	100	85	74	19	17	13	13	13	0.00	6	–	2	–	–	12	–	–	–	–	–	–	–	Acetic acid, trifluoro-, 2,2,2-trifluoro-1,1-bis(trifluoromethyl)ethyl ester	24165-10-4	NI34935
272	274	237	270	239	235	276	299	332	100	80	58	47	38	36	35	29	3.74	7	3	–	–	–	–	7	–	–	–	–	–	–	Bicyclo[2.2.1]hept-2-ene, 1,2,3,4,5,7,7-heptachloro-	5202-36-8	NI34936
122	177	175	174	172	162	248	192	332	100	8	6	5	5	5	4	4	3.35	8	10	4	–	2	–	–	–	–	–	–	–	1	Molybdenum, (2,5-dithiahexane)tetracarbonyl-		LI07363
59	227	197	165	211	121	135	317	332	100	92	33	33	30	26	19	16	3.51	8	24	8	–	–	–	–	–	–	3	–	–	–	Trisilane, octamethoxy-	49582-74-3	NI34937
227	59	197	165	211	121	135	317	332	100	99	33	33	30	26	19	16	4.00	8	24	8	–	–	–	–	–	–	3	–	–	–	Trisilane, octamethoxy-	49582-74-3	NI34938
43	91	92	58	39	65	27	332	332	100	95	79	22	21	15	15	12		9	9	–	–	–	5	–	–	–	–	–	–	1	Stannane, trimethyl(pentafluorophenyl)-	1015-53-8	NI34939
123	77	167	92	44	224	91	42	332	100	44	34	33	29	28	28	23	6.33	9	11	2	2	2	1	2	–	–	–	–	–	–	Methanesulphenamide, 1,1-dichloro-N-[(dimethylamino)sulphonyl]-1-fluoro-N-phenyl-	1085-98-9	NI34940
56	112	84	28	193	165	221	137	332	100	51	37	27	14	13	7	7	4.00	10	4	6	–	–	–	–	–	–	–	–	–	2	Diiron, (π-butatrienyl)hexacarbonyl-		MS07890
192	112	164	220	304	56	52	332	332	100	75	50	48	43	33	28	25		10	4	6	–	–	–	–	–	–	–	–	–	2	Diiron, (π-butatrienyl)hexacarbonyl-		MS07891
334	332	336	216	214	256	255	254	332	100	53	50	16	16	12	12	12		10	6	3	–	–	–	–	2	–	–	–	–	–	Chromone, 3,5-dibromo-6-hydroxy-2-methyl-	30095-77-3	NI34941
195	75	120	121	149	118	119	74	332	100	80	78	49	42	42	30	26	7.00	10	8	6	2	–	–	–	–	–	–	–	–	1	3-(3,5-Dinitrobenzoyloxy)-selenetane		MS07892
109	297	269	299	28	271	137	93	332	100	54	39	38	37	27	23	23	0.00	10	12	2	–	1	–	3	–	–	–	1	–	–	Phosphonothioic acid, ethyl-, O-ethyl O-(2,4,5-trichlorophenyl) ester	327-98-0	NI34942
194	192	278	179	190	276	193	177	332	100	81	56	56	52	46	46	42	27.86	11	21	–	–	–	–	–	–	–	–	–	–	2	Cobalt, bis(η^4-1,3-butadiene)(trimethylstannyl)-	74811-06-6	NI34943
179	166	98	69	82	42	110	167	332	100	99	99	99	98	91	79	69	4.49	12	14	2	2	–	6	–	–	–	–	–	–	–	N,N′-Ethylenebis(5,5,5-trifluoro-4-oxopentan-2-imine)	1870-12-8	NI34944
166	98	179	42	82	97	69	68	332	100	55	50	35	25	20	20	20	7.00	12	14	2	2	–	6	–	–	–	–	–	–	–	2-Pentanone, 4,4′-(1,2-ethanediyldinitrilo)bis[1,1,1-trifluoro-	433-30-7	NI34945
98	166	110	82	179	96	67	68	332	100	60	34	33	26	22	20	18	5.00	12	14	2	2	–	6	–	–	–	–	–	–	–	2-Pentanone, 4,4′-(1,2-ethanediyldinitrilo)bis[1,1,1-trifluoro-	433-30-7	NI34946
192	205	178	235	79	41	69	42	332	100	64	59	39	37	36	34	24	22.00	12	14	2	2	–	6	–	–	–	–	–	–	–	cis-2,9-Bis(trifluoroacetyl)-2,9-diazabicyclo[4.4.0]decane		NI34947
178	192	235	41	79	69	332	205	332	100	69	44	38	31	29	23	23		12	14	2	2	–	6	–	–	–	–	–	–	–	trans-2,9-Bis(trifluoroacetyl)-2,9-diazabicyclo[4.4.0]decane		NI34948
102	92	239	241	126	203	205	153	332	100	73	67	65	52	50	32	27	0.00	12	16	2	–	–	–	4	–	–	–	–	–	–	(1RS,5SR,6SR,7SR,8SR,9SR)-6,7,8,9-Tetrakis(chloromethyl)-3-oxabicyclo[3.2.2]nonan-2-one		DD01300
275	263	218	41	44	43	55	42	332	100	75	65	50	40	35	20	20	0.00	12	19	1	2	–	6	–	–	–	–	–	1	–	2,2,2-Trifluro-N-(2,2,2-trifluoro-1-iminoethyl)acetic acid imide, dibutylboryl ester		NI34949
193	195	158	194	334	197	196	160	332	100	98	50	43	37	34	33	33	30.91	13	8	–	2	–	–	4	–	–	–	–	–	–	Benzenecarbohydrazonoyl chloride, N-(2,4,6-trichlorophenyl)-	25939-05-3	NI34950
171	63	77	51	109	297	47	15	332	100	91	81	74	51	49	38	33	0.00	13	11	2	–	1	–	2	–	–	–	1	–	–	Phosphonothioic acid, phenyl-, O-(2,5-dichlorophenyl) O-methyl ester	53490-78-1	NI34951
136	218	135	43	204	118	85	59	332	100	97	82	33	32	27	26	24	0.40	13	16	3	8	–	–	–	–	–	–	–	–	–	Adenosine, 5′-azido-5′-deoxy-2′,3′-O-isopropylidene-	34245-48-2	NI34952
73	75	130	147	157	44	115	45	332	100	97	96	40	39	38	31	29	0.00	13	29	5	–	–	–	–	–	–	2	–	1	–	L-Galactopyranose, 6-deoxy-1,2-bis-O-(trimethylsilyl)-, cyclic methylboronate	56196-83-9	NI34953
167	69	109	108	123	63	91	64	332	100	22	18	16	15	14	13	13	0.00	14	8	–	2	4	–	–	–	–	–	–	–	–	Benzothiazole, 2,2′-dithiobis-	120-78-5	NI34954
151	96	69	64	50	152	63	76	332	100	61	39	35	30	25	25	21	0.20	14	8	2	2	3	–	–	–	–	–	–	–	–	Bis(2,3-dihydro-3-oxo-1,2-benzisothiazol-2-yl)sulphide		IC03995
332	152	151	63	270	52	153	51	332	100	74	63	30	25	24	23	22		14	12	6	4	–	–	–	–	–	–	–	–	–	Benzaldehyde, 4-hydroxy-3-methoxy-, (2,4-dinitrophenyl)hydrazone	1166-13-8	NI34955
332	135	150	333	107	42	136	43	332	100	60	25	17	13	13	11	11		14	12	6	4	–	–	–	–	–	–	–	–	–	2,4-Dihydroxyacetophenone (2,4-dinitrophenyl)hydrazone		IC03996
274	272	77	180	275	273	209	260	332	100	92	60	50	48	44	36	35	29.90	14	13	1	4	–	–	–	1	–	–	–	–	–	5-Bromo-2-aminobenzophenone semicarbazone		MS07893
226	332	317	151	152	240	302	165	332	100	51	49	28	25	17	16	10		14	14	–	–	–	–	–	–	–	–	–	–	2	Arsanthrene, 5,10-dihydro-5,10-dimethyl-	20193-91-3	MS07894
115	157	184	142	103	111	73	83	332	100	72	25	23	22	21	20	19	0.00	14	20	9	–	–	–	–	–	–	–	–	–	–	Acetyl 2,3,4-tri-O-acetyl-6-deoxy-α-D-glucopyranoside	7404-35-5	NI34956
115	157	142	184	103	72	100	83	332	100	73	27	24	24	19	17	17	0.00	14	20	9	–	–	–	–	–	–	–	–	–	–	Acetyl 2,3,4-tri-O-acetyl-6-deoxy-β-D-glucopyranoside	17081-04-8	NI34957
115	157	83	73	111	103	142	100	332	100	58	40	28	22	22	20	18	0.00	14	20	9	–	–	–	–	–	–	–	–	–	–	α-D-arabino-Hexopyranose, 2-deoxy-, tetraacetate	16750-06-4	NI34958
68	97	103	110	145	111	69	170	332	100	99	58	43	40	38	36	33	0.00	14	20	9	–	–	–	–	–	–	–	–	–	–	β-D-arabino-Hexopyranose, 2-deoxy-, tetraacetate	16750-07-5	NI34959
86	45	44	128	110	60	55	103	332	100	77	32	29	23	19	18	17	0.00	14	20	9	–	–	–	–	–	–	–	–	–	–	D-arabino-Hexose, 2-deoxy-, 3,4,5,6-tetraacetate	3891-61-0	NI34960
115	157	99	44	45	73	71	100	332	100	73	65	54	49	36	32	31	0.00	14	20	9	–	–	–	–	–	–	–	–	–	–	Fucose, 2,3,4,5-tetraacetate, L-	30571-58-5	NI34961
126	71	169	211	115	84	114	85	332	100	73	52	28	25	24	23	21	0.00	14	20	9	–	–	–	–	–	–	–	–	–	–	α-D-Glucofuranose, 1,2-O-1,2-ethanediyl-, triacetate	28069-78-5	NI34962
97	110	139	182	98	69	183	111	332	100	93	70	45	18	17	13	12	0.00	14	20	9	–	–	–	–	–	–	–	–	–	–	Tetraacetyl 2,5-anhydro-D-mannitol	65729-88-6	NI34963

MASS TO CHARGE RATIOS	M.W.	INTENSITIES	Parent	C	H	O	N	S	F	Cl	Br	I	Si	P	B	X	COMPOUND NAME	CAS Reg No	No
202 167 204 241 141 142 168 203	332	100 78 40 30 24 20 19 14	0.00	14	21	5	2	–	–	1	–	–	–	–	–	–	Methyl 5-(2-acetoxy-3-chloropropyl)-5-(1-methylbutyl)barbiturate		LI07364
57 115 116 99 58 85 114 86	332	100 79 74 69 69 67 57 48	1.51	14	24	7	2	–	–	–	–	–	–	–	–	–	7,16-Dimethyl-1,4,10,13-tetraoxa-7,16-diazacyclooctadecane-2,6,17-trione	87989-13-7	NI34964
77 141 332 105 104 78 51 133	332	100 88 51 32 28 20 18 17		15	12	5	2	1	–	–	–	–	–	–	–	–	2,4(1H,3H)-Quinazolinedione, 1-methyl-3-[(phenylsulphonyl)oxy]-	41120-15-4	NI34965
332 334 333 331 231 229 77 28	332	100 94 79 63 56 52 51 28		15	13	–	2	1	–	–	1	–	–	–	–	–	2-(4-Bromophenylimino)-3-phenylthiazolidine		NI34966
332 94 39 333 106 53 66 52	332	100 30 25 20 16 16 15 15		15	16	5	4	–	–	–	–	–	–	–	–	–	1-Ethyl-4-methyl-5-(4-methoxy-2-nitrophenylazo)-6-hydroxypyrid-2-one		IC03997
126 43 186 194 84 83 85 127	332	100 92 61 40 40 38 33 31	1.01	15	24	8	–	–	–	–	–	–	–	–	–	–	2-Furanhexanoic acid, tetrahydro-β,δ,5-trihydroxy-, methyl ester, β,δ-diacetate	18142-15-9	NI34967
44 32 45 31 115 173 127 133	332	100 87 66 65 40 35 29 25	0.00	15	24	8	–	–	–	–	–	–	–	–	–	–	Propane-1,1,3,3-tetracarboxylic acid, tetraethyl ester		IC03998
193 191 135 137 28 194 41 249	332	100 78 74 69 69 65 50 49	2.00	15	32	–	–	–	–	–	–	–	–	–	–	1	Dibutylmethylcyclohexyltin		MS07895
151 149 121 233 119 147 193 207	332	100 93 80 78 65 61 60 58	0.00	15	32	–	–	–	–	–	–	–	–	–	–	1	Ethylisopropylbutylcyclohexyltin		MS07896
73 75 55 317 117 201 147 129	332	100 76 40 24 20 18 18 18	0.00	15	32	4	–	–	–	–	–	–	2	–	–	–	Azelaic acid, bis(trimethylsilyl) ester	17906-08-0	NI34968
73 75 55 317 117 147 201 129	332	100 76 40 24 20 18 18 18	0.00	15	32	4	–	–	–	–	–	–	2	–	–	–	Azelaic acid, bis(trimethylsilyl) ester	17906-08-0	NI34969
73 147 245 83 55 75 45 317	332	100 52 33 31 25 15 12 11	0.00	15	32	4	–	–	–	–	–	–	2	–	–	–	Butylethylmalonic acid, bis(trimethylsilyl) ester		NI34970
259 73 97 317 261 260 69 288	332	100 97 64 41 33 29 14 14	0.00	15	32	4	–	–	–	–	–	–	2	–	–	–	Dipropylmalonic acid, bis(trimethylsilyl) ester		NI34971
73 147 55 217 75 248 148 45	332	100 96 38 33 29 17 16 14	0.00	15	32	4	–	–	–	–	–	–	2	–	–	–	Hexylmalonic acid, bis(trimethylsilyl) ester		NI34972
130 132 104 36 262 131 76 263	332	100 83 50 48 44 40 39 35	2.00	16	10	2	2	–	–	2	–	–	–	–	–	–	7,14-Dichlorodiftalone		MS07897
123 95 314 140 124 75 69 50	332	100 50 12 10 8 8 5 5	0.00	16	10	4	2	–	2	–	–	–	–	–	–	–	Butanetetrone, bis(4-fluorophenyl)-, 2,3-dioxime	55232-37-6	NI34973
332 314 331 317 197 183 134 155	332	100 36 35 14 12 5 5 4		16	12	8	–	–	–	–	–	–	–	–	–	–	5,6,8,3′,4′-Pentahydroxy-7-methoxyflavone	103633-25-6	NI34974
105 77 222 194 165 89 253 51	332	100 37 25 25 19 12 9 7	0.50	16	13	3	–	–	–	–	1	–	–	–	–	–	Phenyl 2-(α-bromo-2-methoxycarbonyl)benzyl ketone		LI07365
178 206 332 179 238 120 78 333	332	100 16 15 11 9 8 8 3		16	16	6	2	–	–	–	–	–	–	–	–	–	Dimethyl α,β-bis(2-pyridonyl)succinate		LI07366
105 69 68 77 106 113 112 57	332	100 90 47 31 29 28 28 28	3.00	16	16	6	2	–	–	–	–	–	–	–	–	–	Uridine, 2′,3′-O-benzylidene-	3257-71-4	NI34975
241 211 332 301 257 242 331 227	332	100 80 72 30 28 26 24 24		16	20	4	–	–	–	–	–	–	2	–	–	–	9,9,10,10-Tetramethoxy-9,10-disila-9,10-dihydroanthracene		NI34976
332 41 55 79 69 67 93 107	332	100 90 49 46 43 41 40 36		16	20	4	4	–	–	–	–	–	–	–	–	–	Camphor dinitrophenylhydrazone		NI34977
113 175 177 111 221 163 223 165	332	100 78 52 52 49 34 31 22	3.20	16	22	3	–	–	–	2	–	–	–	–	–	–	Acetic acid, (2,4-dichlorophenoxy)-, 2-ethylhexyl ester	1928-43-4	NI34979
57 43 71 41 70 55 29 220	332	100 73 51 32 29 22 22 21	6.09	16	22	3	–	–	–	2	–	–	–	–	–	–	Acetic acid, (2,4-dichlorophenoxy)-, 2-ethylhexyl ester	1928-43-4	AI02058
220 222 175 162 145 111 147 71	332	100 62 35 32 30 28 27 26	6.00	16	22	3	–	–	–	2	–	–	–	–	–	–	Acetic acid, (2,4-dichlorophenoxy)-, 2-ethylhexyl ester	1928-43-4	NI34978
57 43 41 29 71 55 70 27	332	100 86 67 52 47 33 27 24	3.50	16	22	3	–	–	–	2	–	–	–	–	–	–	Acetic acid, (2,4-dichlorophenoxy)-, isooctyl ester	25168-26-7	NI34980
57 43 55 71 41 69 56 70	332	100 72 31 30 30 20 20 17	6.47	16	22	3	–	–	–	2	–	–	–	–	–	–	Acetic acid, (2,4-dichlorophenoxy)-, isooctyl ester	25168-26-7	AI02059
57 43 69 71 41 55 84 29	332	100 99 54 50 48 46 31 23	9.79	16	22	3	–	–	–	2	–	–	–	–	–	–	Acetic acid, (2,4-dichlorophenoxy)-, isooctyl ester	25168-26-7	AI02060
73 147 233 201 75 69 317 74	332	100 50 31 24 20 13 11 9	0.00	16	36	3	–	–	–	–	–	–	2	–	–	–	3-Trimethylsiloxycapric acid, trimethylsilyl ester		NI34981
73 147 233 201 317 75 275 189	332	100 67 56 41 27 20 17 13	0.00	16	36	3	–	–	–	–	–	–	2	–	–	–	3-Trimethylsiloxycapric acid, trimethylsilyl ester		NI34982
233 201 147 73 317 275 234 189	332	100 67 66 61 48 25 21 19	0.00	16	36	3	–	–	–	–	–	–	2	–	–	–	3-Trimethylsiloxycapric acid, trimethylsilyl ester		NI34983
73 75 317 147 69 55 149 227	332	100 89 76 40 40 36 22 22	0.00	16	36	3	–	–	–	–	–	–	2	–	–	–	Trimethylsiloxycapric acid, trimethylsilyl ester		NI34984
117 73 75 317 118 69 55 147	332	100 70 40 12 11 11 10 10	0.00	16	36	3	–	–	–	–	–	–	2	–	–	–	9-Trimethylsiloxydecanoic acid, trimethylsilyl ester		NI34985
141 332 142 91 333 140 101	332	100 11 9 4 2 2 2		17	14	–	–	–	6	–	–	–	–	–	–	–	1,3-Bis(3-methylphenyl)hexafluoropropane		LI07367
151 150 269 300 209 272 268 122	332	100 63 52 43 34 31 29 29	14.01	17	16	7	–	–	–	–	–	–	–	–	–	–	Benzoic acid, 2-(2,6-dihydroxy-4-methylbenzoyl)-5-hydroxy-3-methoxy-, methyl ester	519-57-3	NI34986
332 153 133 165 180 39 69 51	332	100 61 57 43 40 32 26 26		17	16	7	–	–	–	–	–	–	–	–	–	–	4H-1-Benzopyran-4-one, 2,3-dihydro-5,7-dihydroxy-3-(4-hydroxy-2-methoxyphenyl)-3-methoxy-	56909-13-8	MS07898
332 258 44 314 259 243 300 257	332	100 70 68 55 50 49 49 47		17	16	7	–	–	–	–	–	–	–	–	–	–	Methyl 6,9-dihydroxy-1-methyl-5,10-dioxonaphtho[2,3-c]pyran-3-acetate	107298-06-6	NI34987
55 119 332 213 137 214 77 105	332	100 60 32 26 25 22 16 13		17	17	5	–	–	–	–	–	–	–	1	–	–	2-(3-Hydroxy-4-benzoylphenoxy)-methyl-1,3,2-dioxophosphinane		IC03999
289 260 154 155 227 290 246 170	332	100 51 51 39 36 33 31 31	0.00	17	21	2	–	–	–	–	–	–	–	–	–	1	(2-Hydroxypentyl)diphenylarsane oxide		DD01301
98 317 213 83 332 100 73 55	332	100 69 66 42 28 26 19 18		17	24	3	2	–	–	–	–	–	1	–	–	–	2,4,6(1H,3H,5H)-Pyrimidinetrione, 5,5-diethyl-1-phenyl-3-(trimethylsilyl)-	51209-95-1	NI34988
98 317 213 83 100 332 73 55	332	100 68 66 42 26 18 18 18		17	24	3	2	–	–	–	–	–	1	–	–	–	2,4,6(1H,3H,5H)-Pyrimidinetrione, 5,5-diethyl-1-phenyl-3-(trimethylsilyl)-	51209-95-1	LI07368
55 97 133 163 69 177 175 149	332	100 67 26 15 14 12 10 8	0.00	17	40	2	–	–	–	–	–	–	2	–	–	–	Heptyl 1,7-bis(propyldimethylsilyl) ether		NI34989
69 147 189 149 148 73 41 275	332	100 51 18 9 8 6 6 5	0.00	17	40	2	–	–	–	–	–	–	2	–	–	–	Pentyl 1,5-bis(tert-butyldimethylsilyl) ether		NI34990
75 73 147 83 97 69 55 103	332	100 89 85 83 77 74 68 47	0.42	17	40	2	–	–	–	–	–	–	2	–	–	–	1,11-Bis(trimethylsiloxy)undecane	60671-18-3	NI34991
332 334 336 333 226 335 262 224	332	100 98 32 22 21 20 20 8		18	11	–	–	–	–	3	–	–	–	–	–	–	1,1′:4′,1″-Terphenyl, 2,4,6-trichloro-	57346-61-9	NI34992
262 226 264 113 224 263 332 334	332	100 59 30 19 18 18 14 13		18	11	–	–	–	–	3	–	–	–	–	–	–	1,1′:4′,1″-Terphenyl, 2,4,6-trichloro-	57346-61-9	NI34993
332 334 226 262 336 113 333 335	332	100 96 50 33 27 15 14 13		18	11	–	–	–	–	3	–	–	–	–	–	–	Terphenyl, 2,5,4″-trichloro-	61576-93-0	NI34994
144 91 77 67 105 65 39 51	332	100 77 76 69 47 43 42 33	0.00	18	20	4	–	1	–	–	–	–	–	–	–	–	(E)-3-Methyl-2-[(4-methylphenyl)sulphonyl]-1-phenyl-2-butene-1,4-diol	118356-32-4	DD01302
195 332 167 135 196 77 224 303	332	100 64 57 35 16 16 11 9		18	20	6	–	–	–	–	–	–	–	–	–	–	Benzophenone, 2-hydroxy-4,4′,6-trimethoxy-3-ethoxy-		MS07899
332 165 333 195 317 301	332	100 37 27 17 12 8		18	20	6	–	–	–	–	–	–	–	–	–	–	Benzophenone, 3,3′,4,4′,5-pentamethoxy-	22699-97-4	NI34995
43 125 140 182 290 248 69 110	332	100 40 35 19 18 16 15 12	0.15	18	20	6	–	–	–	–	–	–	–	–	–	–	1,3-Diacetoxy-10,10-dimethyl-1,4,4a,8a-tetrahydro-1,4-ethanonaphthalene-5,8-dione		LI07369
180 151 165 153 286 332 137 314	332	100 26 16 12 11 8 6 5		18	20	6	–	–	–	–	–	–	–	–	–	–	3′,4′,7-Trimethoxy-2,3-cis-flavan-3,4-cis-diol		LI07370

MASS TO CHARGE RATIOS	M.W.	INTENSITIES	Parent	C	H	O	N	S	F	Cl	Br	I	Si	P	B	X	COMPOUND NAME	CAS Reg No	No
150 121 182 183 332 135 286 167	332	100 22 21 19 12 7 6 6		18	20	6	–	–	–	–	–	–	–	–	–	–	4′,7,8-Trimethoxy-2,3-cis-flavan-3,4-cis-diol		LI07371
180 286 151 165 314 332 277 137	332	100 26 18 15 11 8 8 6		18	20	6	–	–	–	–	–	–	–	–	–	–	3′,4′,7-Trimethoxy-2,3-trans-flavan-3,4-cis-diol		LI07372
180 151 165 286 137 153 152 314	332	100 30 26 21 14 13 11 10	8.00	18	20	6	–	–	–	–	–	–	–	–	–	–	3′,4′,7-Trimethoxy-2,3-trans-flavan-3,4-trans-diol		LI07373
77 105 81 51 91 165 109 152	332	100 85 38 37 36 30 23 18	0.00	18	21	4	–	–	–	–	–	–	–	1	–	–	Diethyl (2-oxo-1,2-diphenylethyl)phosphonate		DD01303
73 332 317 289 333 277 318 45	332	100 80 77 57 20 20 19 19		18	24	4	–	–	–	–	–	–	1	–	–	–	2H-1-Benzopyran-2-one, 3-(1′,1′-dimethylallyl)-7-[(trimethylsilyl)oxy]-6-methoxy-		MS07900
83 98 146 90 69 126 163 118	332	100 99 73 35 35 26 23 20	4.00	18	24	4	2	–	–	–	–	–	–	–	–	–	Benzoic acid, 4-(tetrahydro-5,5,7,7-tetramethyl-2-oxopyrrolo[1,2-b][1,2,4]oxadiazol-1(2H)-yl)-, ethyl ester	39931-32-3	NI34996
271 317 332 225 272 179 135 318	332	100 70 60 50 17 15 14 13		18	24	4	2	–	–	–	–	–	–	–	–	–	1-(4-Methyl-5-ethoxycarbonyl-2-pyrrolyl)-1-(4-methyl-5-ethoxycarbonyl-3 pyrrolyl)ethane		LI07374
332 303 275 304 190 276 333 134	332	100 82 65 50 22 21 18 18		18	24	4	2	–	–	–	–	–	–	–	–	–	2,4,6(1H,3H,5H)-Pyrimidinetrione, 5-(4-ethoxyphenyl)-1,3,5-triethyl-	55133-83-0	NI34997
332 285 271 253 272 239 300 123	332	100 90 64 63 55 48 43 34		18	24	4	2	–	–	–	–	–	–	–	–	–	4,3′,5′-Trimethyl-4′-ethyl-3,5-bis(ethoxycarbonyl)-2,2′-dipyrrolylmethane		LI07375
146 317 261 161 156 57 73 106	332	100 46 41 33 33 21 15 13	8.80	18	33	1	2	–	–	–	–	–	1	–	1	–	3-tert-Butyl-2-[tert-butyl(trimethylsilyl)amino]-4-phenyl-1,3,2-oxazaboretidine		MS07901
57 136 137 252 97 83 95 81	332	100 85 84 65 59 57 44 40	1.20	18	36	3	–	1	–	–	–	–	–	–	–	–	1-(Methylsulphonyl)-2-heptadecanone	76078-78-9	NI34998
57 137 135 43 55 71 41 85	332	100 98 97 79 74 73 63 59	0.30	18	37	–	–	–	–	–	1	–	–	–	–	–	Octadecane, 1-bromo-	112-89-0	NI34999
57 43 71 135 55 85 69 41	332	100 64 60 54 46 35 31 30	0.00	18	37	–	–	–	–	–	1	–	–	–	–	–	Octadecane, 1-bromo-	112-89-0	LI07376
57 56 41 71 55 43 69 83	332	100 17 9 8 7 7 6 5	0.00	18	37	–	–	–	–	–	1	–	–	–	–	–	2-(2,2,4-Trimethyl)butyl-5,7,7-trimethyloctyl bromide		LI07377
91 78 28 245 304 51 65 272	332	100 57 29 18 16 14 13 12	1.42	19	16	2	4	–	–	–	–	–	–	–	–	–	Methyl 2-azido-3-(1′-benzylindol-3-yl)prop-2-enoate	102244-90-6	NI35000
43 55 193 41 57 56 69 44	332	100 72 71 66 61 59 55 52	0.00	19	24	5	–	–	–	–	–	–	–	–	–	–	1,2,4-Benzenetricarboxylic acid, cyclic 1,2-anhydride, decyl ester	33975-31-4	NI35001
162 43 153 71 81 59 134 171	332	100 95 91 66 47 46 26 21	5.00	19	24	5	–	–	–	–	–	–	–	–	–	–	6,7-Dihydroxy-3,7-dimethyl-2-octenyloxycoumarin		MS07902
69 31 82 93 196 168 136 67	332	100 52 37 28 17 14 14 14	1.00	19	24	5	–	–	–	–	–	–	–	–	–	–	4-Geranyloxy-3-hydroxy-5-methoxyphthalaldehyde		NI35002
69 332 41 109 91 43 135 79	332	100 23 23 14 12 12 11 10		19	24	5	–	–	–	–	–	–	–	–	–	–	Trichothec-9-en-4-ol, 7,8:12,13-diepoxy-, 2-butenoate, [4β(Z),7β,8β]-	21284-11-7	NI35003
69 41 43 218 79 39 175 109	332	100 51 17 14 14 14 12 11	2.97	19	24	5	–	–	–	–	–	–	–	–	–	–	Trichothec-9-en-8-one, 12,13-epoxy-4-[(1-oxo-2-butenyl)oxy]-, (4β)-	6379-69-7	NI35004
105 218 106 120 58 77 41 104	332	100 85 59 58 55 34 34 30	0.00	19	28	3	2	–	–	–	–	–	–	–	–	–	tert-Butylammonium 3-methyl-1-[(1-phenylethyl)carbamoyl]cyclobutanecarboxylate		NI35005
287 160 117 145 131 288 188 220	332	100 43 32 27 25 19 17 11	0.00	19	28	3	2	–	–	–	–	–	–	–	–	–	(R,S)-1-[[3-Phenyl-4-(carboxymethyl)-1-methylbutylidene]amino]-2-(methoxymethyl)pyrrolidine		DD01304
108 43 57 55 87 41 110 121	332	100 84 53 52 51 49 36 28	2.80	19	37	2	–	–	–	1	–	–	–	–	–	–	Octadecanoic acid, 2-chloro-, methyl ester	41753-99-5	WI01434
160 118 332 161 234 115 147 131	332	100 19 18 14 10 10 7 7		20	16	1	2	1	–	–	–	–	–	–	–	–	15-Oxo-6,7,8,9,10,15-hexahydrobenzo[d][1]benzothieno[2,3-g]azecine-8-carbonitrile		NI35006
216 199 77 28 124 217 51 140	332	100 66 58 48 47 46 46 36	0.43	20	17	1	2	–	–	–	–	–	–	1	–	–	N-(1-Cyanobenzyl)diphenylphosphinamide		MS07903
332 333 314 121 330 313 228 166	332	100 24 16 10 8 7 7 6		20	20	1	–	–	–	–	–	–	–	–	–	1	Iron, (η^5-2,4-cyclopentadien-1-yl)[(1,2,3,3a,7a-η)-4,5,6,7-tetrahydro-4-hydroxy-5-phenyl-1H-inden-1-yl]-, stereoisomer	12149-22-3	NI35007
332 194 247 314 330 249 191 178	332	100 20 12 10 8 8 8 8		20	20	1	–	–	–	–	–	–	–	–	–	1	Iron, (η^5-2,4-cyclopentadien-1-yl)[(1,2,3,3a,7a-η)-4,5,6,7-tetrahydro-4-hydroxy-5-phenyl-1H-inden-1-yl]-, stereoisomer	12149-23-4	NI35008
317 255 318 256 195 319 257 197	332	100 92 30 21 20 10 9 7	6.00	20	20	1	–	–	–	–	–	–	2	–	–	–	1,3-Diphenyl-1,3-dimethyl-1,3-disila-2-oxaindan		NI35009
122 312 160 268 108 284 242 134	332	100 72 32 18 15 13 12 12	6.10	20	25	3	–	–	1	–	–	–	–	–	–	–	9α-Fluoro-11β-hydroxy-16α-methyl-1,4-androstadiene-3,17-dione		MS07904
122 312 160 121 313 268 123 108	332	100 72 32 32 24 18 18 15	6.10	20	25	3	–	–	1	–	–	–	–	–	–	–	9α-Fluoro-11β-hydroxy-16α-methyl-1,4-androstadiene-3,17-dione		MS07905
83 55 41 69 157 57 139 95	332	100 46 34 33 31 26 25 21	19.87	20	26	–	–	–	–	1	–	–	–	1	–	–	Phosphinous chloride, (methylisopropyl)-1-naphthalenyl-	74645-89-9	NI35010
83 55 109 31 41 108 43 44	332	100 79 76 74 68 63 61 34	5.79	20	28	4	–	–	–	–	–	–	–	–	–	–	2-Butenoic acid, 2-methyl-, 4,4a,5,6,7,8,8a,9-octahydro-8a-hydroxy-3,4a,5 trimethylnaphtho[2,3-b]furan-6-yl ester	56298-88-5	NI35011
332 263 69 264 57 289 235 333	332	100 82 72 55 31 23 22 21		20	28	4	–	–	–	–	–	–	–	–	–	–	1,3-Cyclopentanedione, 4-hydroxy-2,2-bis(3-methyl-2-butenyl)-5-(3-methyl-1-oxobutyl)-	38574-31-1	NI35012
69 41 57 43 235 263 249 304	332	100 98 38 24 22 20 18 17	13.67	20	28	4	–	–	–	–	–	–	–	–	–	–	1,2,4-Cyclopentanetrione, 3,3-bis(3-methyl-2-butenyl)-5-(3-methyl-1-oxobutyl)-	468-62-2	NI35013
332 333 167 283 153 330 187 205	332	100 69 26 22 21 19 18 17		20	28	4	–	–	–	–	–	–	–	–	–	–	Estra-1,3,5(10)-trien-2,17-diol, 3,4-dimethoxy-, (17β)-	65969-01-9	NI35014
332 333 167 330 301 283 232 206	332	100 23 10 6 5 5 5 5		20	28	4	–	–	–	–	–	–	–	–	–	–	Estra-1,3,5(10)-trien-3,17-diol, 2,4-dimethoxy-, (17β)-	65968-99-2	NI35015
332 333 153 167 330 205 317 203	332	100 38 23 17 12 11 10 9		20	28	4	–	–	–	–	–	–	–	–	–	–	Estra-1,3,5(10)-trien-4,17-diol, 2,3-dimethoxy-, (17β)-	65968-98-1	NI35016
202 187 121 188 287 332	332	100 72 66 50 19 10		20	28	4	–	–	–	–	–	–	–	–	–	–	Ethyl 2-acetyl-8-(3-methoxyphenyl)-5-methyl-oct-5-enoate		LI07378
81 284 31 41 43 203 95 55	332	100 99 88 59 51 44 41 41	7.35	20	28	4	–	–	–	–	–	–	–	–	–	–	1α-(Hydroxymethyl)-7α,8α-dimethyl-7-[2-(3-furyl)ethyl]bicyclo[4.4.0]dec 2-ene 2-carboxylic acid		NI35017
332 93 40 163 55 317 82 79	332	100 91 86 67 54 51 50 41		20	28	4	–	–	–	–	–	–	–	–	–	–	6β-Hydroxyrosenonolactone		LI07379
109 314 296 268 299 286 253 137	332	100 70 34 33 27 25 25 25	10.00	20	28	4	–	–	–	–	–	–	–	–	–	–	Kaur-16-en-18-oic acid, 6,7,13-trihydroxy-, γ-lactone, (4α,6α,7β)-	19018-46-3	NI35018
314 268 94 95 315 284 93 107	332	100 37 30 24 23 21 21 19	2.00	20	28	4	–	–	–	–	–	–	–	–	–	–	Kaur-16-en-18-oic acid, 6,7,19-trihydroxy-, γ-lactone, (4α,6α,7β)-	7758-47-6	NI35019
314 268 94 315 93 91 79 41	332	100 14 9 8 6 6 6 5	0.00	20	28	4	–	–	–	–	–	–	–	–	–	–	Kaur-16-en-18-oic acid, 6,17,18-trihydroxy-, γ-lactone, (4α,6α)-		MS07906

MASS TO CHARGE RATIOS	M.W.	INTENSITIES	Parent	C	H	O	N	S	F	Cl	Br	I	Si	P	B	X	COMPOUND NAME	CAS Reg No	No
129 91 157 43 115 171 141 117	332	100 71 65 65 53 45 37 35	0.00	20	28	4	–	–	–	–	–	–	–	–	–	–	9,17-Octadecadiene-12,14-diyne-1,11,16-triol, 1-acetate	52061-42-4	NI35020
195 332 314 123 299 261 317	332	100 99 70 60 43 28 26		20	28	4	–	–	–	–	–	–	–	–	–	–	1,4-Phenanthrenedione, 4b,5,6,7,8,8a,9,10-octahydro-3,7-dihydroxy-4b,8,8 trimethyl-2-isopropyl-	69494-10-6	NI35021
284 147 226 332 166 197 242 314	332	100 68 66 64 60 56 52 46		20	28	4	–	–	–	–	–	–	–	–	–	–	Sordaricin	51493-69-7	LI07380
165 198 121 288 287 166 332 289	332	100 90 84 82 38 30 26 16		21	16	2	–	1	–	–	–	–	–	–	–	–	5-Oxo-2,4,4-triphenyl-1,3-oxathiolane		LI07381
202 230 332 229 201 246 203 43	332	100 73 51 43 29 21 18 18		21	16	4	–	–	–	–	–	–	–	–	–	–	1,3-Dioxane-4,6-dione, 5-(9-anthracenylmethylene)-2,2-dimethyl-	62337-84-2	NI35022
332 106 241 135 91 77 197 136	332	100 78 55 50 50 40 38 35		21	20	–	2	1	–	–	–	–	–	–	–	–	Thiourea, N'-phenyl-N,N-dibenzyl-	15093-53-5	NI35023
331 332 169 333 317 144 162 163	332	100 99 44 24 23 21 18 12		21	20	2	2	–	–	–	–	–	–	–	–	–	Benz[g]indolo[2,3-a]quinolizine-1-carboxylic acid, 5,7,8,13,13b,14-hexahydro-, methyl ester, (S)-	25425-11-0	NI35024
117 187 90 130 89 63 116 77	332	100 87 48 41 40 25 24 23	0.00	21	20	2	2	–	–	–	–	–	–	–	–	–	endo-9-Ethyl-1,2,3,4-tetrahydro-N-(4-methylphenyl)-1,4-iminonaphthalene 2,3-dicarboximide	123542-17-6	DD01305
159 117 160 128 317 129 173 332	332	100 37 15 7 4 4 3 1		21	20	2	2	–	–	–	–	–	–	–	–	–	exo-1,2,3,4-Tetrahydro-9-isopropyl-N-phenyl-1,4-iminonaphthalene-2,3-dicarboximide	123620-17-7	DD01306
104 221 187 105 77 222 220 130	332	100 70 54 30 24 22 21 21	0.00	21	20	2	2	–	–	–	–	–	–	–	–	–	endo-1,2,3,4-Tetrahydro-N-methyl-9-(β-phenylethyl)-1,4-iminonaphthalene-2,3-dicarboximide	123542-32-5	DD01307
124 111 41 55 137 93 79 332	332	100 43 25 23 18 16 14 13		21	32	3	–	–	–	–	–	–	–	–	–	–	5β-Androstane-17β-carboxylic acid, 1-oxo-, methyl ester	15173-62-3	NI35026
124 111 56 137 94 79 332 81	332	100 44 24 18 17 15 13 13		21	32	3	–	–	–	–	–	–	–	–	–	–	5β-Androstane-17β-carboxylic acid, 1-oxo-, methyl ester	15173-62-3	NI35025
231 332 246 260 317 300 261 230	332	100 64 51 22 16 16 14 6		21	32	3	–	–	–	–	–	–	–	–	–	–	Androstane-17-carboxylic acid, 3-oxo-, methyl ester, (5α,17β)-	4139-88-2	NI35027
246 230 261 229 332 317 300 299	332	100 38 36 33 30 18 15 15		21	32	3	–	–	–	–	–	–	–	–	–	–	Androstane-17-carboxylic acid, 3-oxo-, methyl ester, (5β,17β)-	50305-74-3	LI07382
246 261 230 229 332 317 299 300	332	100 36 36 32 29 22 15 14		21	32	3	–	–	–	–	–	–	–	–	–	–	Androstane-17-carboxylic acid, 3-oxo-, methyl ester, (5β,17β)-	50305-74-3	NI35028
125 318 99 113 126 112 86 153	332	100 63 60 40 36 35 35 27	0.00	21	32	3	–	–	–	–	–	–	–	–	–	–	Androstane-3,16-dione, cyclic 16-(1,2-ethanediyl acetal)	47328-49-4	NI35029
149 257 231 161 94 148 332 273	332	100 85 44 38 38 35 32 32		21	32	3	–	–	–	–	–	–	–	–	–	–	Androstan-3-one, 17-(acetyloxy)-, (5α,17β)-	1164-91-6	IC04000
79 91 94 93 41 55 95 107	332	100 90 79 73 72 66 65 59	0.00	21	32	3	–	–	–	–	–	–	–	–	–	–	Androstan-3-one, 17-(acetyloxy)-, (5α,17β)-	1164-91-6	NI35030
43 272 257 95 55 93 67 79	332	100 67 60 53 46 45 38 33	1.33	21	32	3	–	–	–	–	–	–	–	–	–	–	Androstan-6-one, 3-(acetyloxy)-, (3β,5α)-	20835-29-4	NI35031
43 41 55 178 95 93 81 79	332	100 86 69 46 25 24 22 21	4.16	21	32	3	–	–	–	–	–	–	–	–	–	–	Androstan-7-one, 3-(acetyloxy)-, (3β,5α)-	54594-47-7	NI35032
43 41 241 95 55 67 91 135	332	100 42 34 34 31 30 29 23	0.00	21	32	3	–	–	–	–	–	–	–	–	–	–	Androst-5-ene-3,19-diol, 3-acetate, (3β)-	55320-48-4	NI35033
193 231 332 43 271 276 233 234	332	100 97 81 81 53 51 51 32		21	32	3	–	–	–	–	–	–	–	–	–	–	6H-Dibenzo[b,d]pyran-1,8-diol, 6a,7,8,9,10,10a-hexahydro-6,6,9-trimethyl 3-pentyl-, [6aR-(6aα,8α,9α,10aβ)]-	36403-92-6	NI35034
299 314 271 231 43 332 258 41	332	100 92 63 56 53 51 37 35		21	32	3	–	–	–	–	–	–	–	–	–	–	6H-Dibenzo[b,d]pyran-1,9-diol, 6a,7,8,9,10,10a-hexahydro-6,6,9-trimethyl 3-pentyl-, [6aR-(6aα,9β,10aβ)]-	52522-56-2	NI35035
43 41 55 94 81 67 79 93	332	100 28 19 13 13 13 12 10	2.00	21	32	3	–	–	–	–	–	–	–	–	–	–	Estran-3-one, 17-(acetyloxy)-2-methyl-, (2α,5α,17β)-	54550-10-6	NI35036
91 69 143 41 115 97 81 55	332	100 71 66 44 42 42 36 33	0.00	21	32	3	–	–	–	–	–	–	–	–	–	–	Ethyl 7-oxa-7-(2-benzyl-2-methylcyclopentyl)heptanoate		NI35037
70 41 55 81 43 95 69 83	332	100 99 72 45 37 35 35 25	2.00	21	32	3	–	–	–	–	–	–	–	–	–	–	(3R,2E)-2-(Hexadec-15-ynylidene)-3-hydroxy-4-methylenebutanolide	71339-49-6	NI35038
70 41 55 43 81 69 95 83	332	100 99 69 50 40 32 30 28	2.00	21	32	3	–	–	–	–	–	–	–	–	–	–	(3R,2Z)-2-(Hexadec-15-ynylidene)-3-hydroxy-4-methylenebutanolide	71339-48-5	NI35039
43 91 105 121 131 255 274 314	332	100 83 67 61 56 33 28 22	0.00	21	32	3	–	–	–	–	–	–	–	–	–	–	Kaur-11-enoic acid, 16α-hydroxy-, methyl ester	70901-54-1	NI35040
121 332 107 109 81 55 91 95	332	100 43 33 32 31 31 29 27		21	32	3	–	–	–	–	–	–	–	–	–	–	Kaur-16-en-18-oic acid, 13-hydroxy-, methyl ester, (4α)-(±)-	29444-14-2	NI35041
314 206 176 191 110 70 299 332	332	100 63 35 33 26 26 21 10		21	32	3	–	–	–	–	–	–	–	–	–	–	7-Methoxypisiferdiol		DD01308
272 41 123 55 81 79 332 93	332	100 55 54 53 52 50 43 41		21	32	3	–	–	–	–	–	–	–	–	–	–	Methyl 7-keto-16β-(–)-kauran-18-oate		MS07907
55 258 114 95 121 107 133 109	332	100 92 74 59 52 50 46 43	32.00	21	32	3	–	–	–	–	–	–	–	–	–	–	Methyl 3-oxo-8(17),E-13-labdadien-15-oate		MS07908
43 314 81 55 41 95 271 79	332	100 69 37 35 31 28 25 25	13.68	21	32	3	–	–	–	–	–	–	–	–	–	–	Pregnane-3,20-dione, 11-hydroxy-, (5α,11α)-	565-96-8	NI35042
43 314 271 55 81 41 71 315	332	100 99 46 31 26 25 23 22	1.57	21	32	3	–	–	–	–	–	–	–	–	–	–	Pregnane-3,20-dione, 11-hydroxy-, (5α,11β)-	565-94-6	NI35043
43 122 314 81 55 41 332 95	332	100 74 70 34 31 27 25 22		21	32	3	–	–	–	–	–	–	–	–	–	–	Pregnane-3,20-dione, 11-hydroxy-, (5β,11α)-	565-93-5	NI35044
43 314 55 41 244 95 271 71	332	100 99 36 30 28 24 23 23	1.56	21	32	3	–	–	–	–	–	–	–	–	–	–	Pregnane-3,20-dione, 11-hydroxy-, (11β)-	565-95-7	NI35045
314 271 244 247 315 248 272 229	332	100 75 48 43 23 21 16 13	8.30	21	32	3	–	–	–	–	–	–	–	–	–	–	Pregnane-3,20-dione, 14-hydroxy-, (5α,14β,17α)-		LI07383
43 81 55 41 289 79 147 95	332	100 41 34 34 33 32 31 31	0.00	21	32	3	–	–	–	–	–	–	–	–	–	–	Pregnane-11,20-dione, 3-hydroxy-, (3α,5α)-	23930-19-0	NI35046
43 260 219 85 332 81 93 314	332	100 50 42 41 40 34 31 29		21	32	3	–	–	–	–	–	–	–	–	–	–	Pregnane-11,20-dione, 3-hydroxy-, (3α,5β)-	565-99-1	NI35047
43 332 219 260 314 85 81 93	332	100 68 63 59 41 38 33 31		21	32	3	–	–	–	–	–	–	–	–	–	–	Pregnane-11,20-dione, 3-hydroxy-, (3α,5β)-	565-99-1	NI35048
332 43 147 81 333 55 289 95	332	100 80 34 27 23 23 22 22		21	32	3	–	–	–	–	–	–	–	–	–	–	Pregnane-11,20-dione, 3-hydroxy-, (3β,5α)-	600-59-9	NI35049
43 219 332 108 81 93 107 85	332	100 78 43 41 36 33 32 31		21	32	3	–	–	–	–	–	–	–	–	–	–	Pregnane-11,20-dione, 3-hydroxy-, (3β,5β)-	566-00-7	NI35050
55 91 79 93 95 67 314 81	332	100 91 89 82 76 70 68 67	39.74	21	32	3	–	–	–	–	–	–	–	–	–	–	Pregnan-20-one, 5,6-epoxy-3-hydroxy-, (3β,5α,6α)-	2193-00-2	WI01435
55 79 67 91 81 314 261 53	332	100 72 69 68 63 55 55 55	49.64	21	32	3	–	–	–	–	–	–	–	–	–	–	Pregnan-20-one, 5,6-epoxy-3-hydroxy-, (3β,5β,6β)-	6585-70-2	WI01436
43 271 253 332 91 213 105 81	332	100 63 55 45 43 42 40 38		21	32	3	–	–	–	–	–	–	–	–	–	–	Pregn-5-en-20-one, 3,17-dihydroxy-, (3β)-	387-79-1	NI35051
91 105 55 79 77 145 314 107	332	100 79 66 63 58 57 54 49	31.00	21	32	3	–	–	–	–	–	–	–	–	–	–	Pregn-5-en-20-one, 3,17-dihydroxy-, (3β)-	387-79-1	MS07909
43 95 230 211 105 85 91 55	332	100 97 59 43 40 37 36 31	2.80	21	32	3	–	–	–	–	–	–	–	–	–	–	Pregn-9(11)-en-20-one, 3,6-dihydroxy-, (3β,5α,6α)-	37717-02-5	MS07910
95 230 43 85 44 81 105 91	332	100 63 58 48 45 41 36 33	16.16	21	32	3	–	–	–	–	–	–	–	–	–	–	Pregn-9(11)-en-20-one, 3,6-dihydroxy-, (3β,5α,6α)-	37717-02-5	NI35052
43 95 230 81 85 105 211 42	332	100 77 45 27 26 23 22 22	10.00	21	32	3	–	–	–	–	–	–	–	–	–	–	Pregn-9(11)-en-20-one, 3,6-dihydroxy-, (3β,5α,6α,17α)-	56362-34-6	NI35053

MASS TO CHARGE RATIOS								M.W.	INTENSITIES								Parent	C	H	O	N	S	F	Cl	Br	I	Si	P	B	X	COMPOUND NAME	CAS Reg No	No
332	43	271	95	333	93	55	41	332	100	64	49	29	24	23	23	22		21	32	3	–	–	–	–	–	–	–	–	–	–	Pregn-14-en-20-one, 3,6-dihydroxy-, (3β,5α,6α)-	40148-15-0	NI35054
242	129	146	332	243	201	147	185	332	100	54	31	21	19	17	15	14		21	36	1	–	–	–	–	–	–	1	–	–	–	Androst-4-ene-17-ol, [17-(trimethylsilyl)oxy]-, (17β)-	56793-02-3	MS07911
44	56	289	92	258	79	57	41	332	100	75	22	10	8	7	7	6	2.00	21	36	1	2	–	–	–	–	–	–	–	–	–	Pregn-5-en-18-ol, 3,20-diamino-, (3β,20S)-	468-31-5	NI35055
332	105	333	289	317	227	77	334	332	100	49	46	46	39	34	30	17		22	20	1	–	1	–	–	–	–	–	–	–	–	2-Benzoyl-5-isopropyl-7-methyl-8H-azuleno[1,8-bc]thiophene		LI07384
91	65	92	181	43	39	89	63	332	100	18	8	4	4	4	3	3	2.00	22	20	3	–	–	–	–	–	–	–	–	–	–	Acetophenone, 3,5-dibenzyloxy-	28924-21-2	NI35056
105	77	227	314	332	106	228	51	332	100	25	23	12	10	8	4	3		22	20	3	–	–	–	–	–	–	–	–	–	–	2-(2-Hydroxy-5-methylbenzyl)-4-methylphenyl benzoate		NI35057
183	105	77	43	165	272	332	256	332	100	52	24	14	5	2	1	1		22	20	3	–	–	–	–	–	–	–	–	–	–	(R)-1,1,2-Triphenyl-1,2-ethanediol-2-acetate	95061-47-5	DD01309
135	317	95	255	137	136	105	43	332	100	34	29	25	17	14	13	12	0.00	22	24	1	–	–	–	–	–	–	1	–	–	–	(Z)-1-(1-Cyclopentenyl)-3-(dimethylphenylsilyl)-2-phenyl-2-propen-1-one		DD01310
317	83	332	69	91	149	119	302	332	100	99	95	92	70	66	59	54		22	24	1	2	–	–	–	–	–	–	–	–	–	(Z,Z)-4-Ethyl-3-methyl-5-[5-(4-methylphenyl-2-methylene)-3,4-dimethyl-5H-pyrrolyl-2-methylene]-3-pyrrolin-one		MS07912
225	332	331	226	184	223	169	156	332	100	78	64	19	19	14	14	13		22	24	1	2	–	–	–	–	–	–	–	–	–	2-(α-Hydroxybenzyl)octahydroindolo[2,3-a]quinolizine		LI07385
225	332	331	226	184	223	169	156	332	100	84	66	20	18	14	14	12		22	24	1	2	–	–	–	–	–	–	–	–	–	2-(α-Hydroxybenzyl)octahydroindolo[2,3-a]quinolizine		LI07386
198	332	17	144	184	157	330	289	332	100	81	62	55	40	27	26	18		22	24	1	2	–	–	–	–	–	–	–	–	–	7-Methyl-3-spirocyclohexane-3H-indole-2-2′-methylcarboxanilide		MS07913
202	173	105	42	203	77	130	144	332	100	23	19	16	15	12	10	9	2.00	22	24	1	2	–	–	–	–	–	–	–	–	–	1,2,3,6-Tetrahydro-4-(α-hydroxybenzyl)-1-[(2-indol-3-yl)ethyl]pyridine		LI07387
108	43	95	93	81	55	67	107	332	100	84	56	37	35	32	29	28	0.00	22	36	2	–	–	–	–	–	–	–	–	–	–	Androstan-3-ol, 9-methyl-, acetate, (3β,5α)-	54550-12-8	NI35058
108	41	248	107	216	332	55	81	332	100	95	78	78	77	74	69	67		22	36	2	–	–	–	–	–	–	–	–	–	–	Androstan-17-one, 3-methoxy-16,16-dimethyl-, (3α,5α)-	55836-75-4	NI35059
41	55	108	107	81	43	67	216	332	100	97	90	80	74	70	69	62	57.25	22	36	2	–	–	–	–	–	–	–	–	–	–	Androstan-17-one, 3-methoxy-16,16-dimethyl-, (3β,5α)-	55837-01-9	NI35060
79	67	80	91	41	93	55	81	332	100	79	73	72	64	63	48	45	1.35	22	36	2	–	–	–	–	–	–	–	–	–	–	Ethyl arachidonate	1808-26-0	MS07914
79	91	67	80	93	81	41	55	332	100	74	74	69	64	48	48	43	10.00	22	36	2	–	–	–	–	–	–	–	–	–	–	Ethyl arachidonate	1808-26-0	NI35061
205	276	57	41	43	191	206	55	332	100	27	26	17	17	16	14	7	0.30	22	36	2	–	–	–	–	–	–	–	–	–	–	2,5-Di-tert-octyl-4-benzoquinone	7511-47-9	NI35062
41	43	55	69	81	67	79	91	332	100	99	80	60	58	50	40	34	2.00	22	36	2	–	–	–	–	–	–	–	–	–	–	8a(2H)-Phenanthrenol, 7-vinyldodecahydro-1,1,4a,7-tetramethyl-, acetate, [4aS-(4aα,4bβ,7β,8aα,10aβ)]-	41756-14-3	NI35063
99	86	303	100	55	304	87	41	332	100	55	44	13	9	8	8	8	8.11	22	36	2	–	–	–	–	–	–	–	–	–	–	Pregnan-3-one, 17-methoxy-, (5α,17α)-	20112-32-7	NI35064
330	91	55	107	105	85	312	79	332	100	89	84	81	75	71	70	70	2.80	22	36	2	–	–	–	–	–	–	–	–	–	–	Pregn-5-en-3β,20-diol, 16α-methyl-	31320-91-9	WI01437
41	216	55	67	91	81	93	79	332	100	85	68	65	60	60	59	58	30.00	22	36	2	–	–	–	–	–	–	–	–	–	–	16,16′-O-Trimethylethiocholanolone		NI35065
247	332	81	43	29	41	27	333	332	100	86	63	41	40	37	26	21		22	40	–	2	–	–	–	–	–	–	–	–	–	1,4-Benzenediamine, N,N′-dioctyl-	1241-28-7	WI01438
44	70	96	289	332	273	56	263	332	100	70	43	35	22	22	15	12		22	40	–	2	–	–	–	–	–	–	–	–	–	5α-Pregnane-3β,20α-diamine, N[3]-methyl-	27741-50-0	NI35066
191	208	55	192	189	97	332	190	332	100	45	42	26	21	17	15	13		23	24	2	–	–	–	–	–	–	–	–	–	–	9-Phenanthrenemethyl cyclohexaneacetate	92174-44-2	NI35067
247	261	57	275	248	303	233	317	332	100	50	32	24	22	19	18	14	13.64	23	40	1	–	–	–	–	–	–	–	–	–	–	2,6-Di-tert-butyl-4-nonylphenol		IC04001
43	69	41	81	31	136	122	56	332	100	95	90	70	50	49	49	49	20.02	23	40	1	–	–	–	–	–	–	–	–	–	–	9,13,17-Nonadecatrien-2-one, 6,10,14,18-tetramethyl-	55124-84-0	NI35068
332	333	166	165	330	334	331	164	332	100	24	22	12	11	7	5	3		24	12	–	–	1	–	–	–	–	–	–	–	–	Diacenaphtho[1,2-b:1′,2′-d]thiophene	203-42-9	NI35069
332	333	166	330	28	165	334	166.5	332	100	27	22	12	11	10	7	6		24	12	–	–	1	–	–	–	–	–	–	–	–	Diacenaphtho[1,2-b:1′,2′-d]thiophene	203-42-9	MS07915
332	166	333	165	330	334	331	144	332	100	25	24	14	11	7	5	4		24	12	–	–	1	–	–	–	–	–	–	–	–	Diacenaphtho[1,2-b:1′,2′-d]thiophene	203-42-9	NI35070
166	332	140	139	333	167	331	138	332	100	41	19	15	11	9	7	5		24	16	–	2	–	–	–	–	–	–	–	–	–	9,9′-Biscarbazole	1914-12-1	NI35071
332	331	166	333	165	329	330	167	332	100	75	32	30	26	20	12	11		24	16	–	2	–	–	–	–	–	–	–	–	–	7H-Dibenzo[d,f]-4,5-benzisoquino[2,1-a]-1,3-diazepine		LI07388
332	333	165	166	331	330	329	164	332	100	26	19	18	16	7	7	6		24	16	–	2	–	–	–	–	–	–	–	–	–	16H-Dibenzo[c,e]perimidino[1,2-a]azepine		LI07389
332	331	333	329	77	255	330	51	332	100	52	26	18	15	13	8	6		24	16	–	2	–	–	–	–	–	–	–	–	–	1,10-Phenanthroline, 4,7-diphenyl-	1662-01-7	NI35072
191	69	95	81	55	123	109	137	332	100	47	40	34	31	30	25	23	17.56	24	44	–	–	–	–	–	–	–	–	–	–	–	15-Isobutyl-(13αH)-isocopalane		NI35073
332	289	144.5	166	143.5	151	150	149	332	100	44	35	20	17	9	8	7		25	16	1	–	–	–	–	–	–	–	–	–	–	9-Methoxynaphtho[2,3-a]pyrene		LI07390
332	333	255	254	334	78	329	253	332	100	50	20	19	15	15	14	14		26	20	–	–	–	–	–	–	–	–	–	–	–	Anthracene, 9,10-dihydro-9,10-diphenyl-	803-58-7	NI35074
204	203	332	165	202	152	141	205	332	100	65	38	26	24	20	19	18		26	20	–	–	–	–	–	–	–	–	–	–	–	1-(1-Azulenyl)-6-(6-azulenyl)hexa-1,3,5-triene	94154-67-3	NI35075
332	303	302	333	304	50	151	151.5	332	100	54	30	28	17	16	13	9		26	20	–	–	–	–	–	–	–	–	–	–	–	Spiro[7H-benzo[c]fluorene-7,1′(2H)-naphthalene], 3′,4′-dihydro-	40548-42-3	AI02061
105	183	184	182	312	106	165	77	332	100	46	12	8	7	7	4	4	0.46	26	20	–	–	–	–	–	–	–	–	–	–	–	Tetraphenylethylene	632-51-9	NI35076
332	74	28	30	56	333	253	252	332	100	57	55	44	35	27	19	17		26	20	–	–	–	–	–	–	–	–	–	–	–	Tetraphenylethylene	632-51-9	IC04002
332	253	252	254	255	241	165	239	332	100	26	20	19	17	9	9	8		26	20	–	–	–	–	–	–	–	–	–	–	–	Tetraphenylethylene	632-51-9	NI35077
278	276	280	216	293	242	41	291	333	100	81	49	44	18	18	15	15	9.64	4	10	–	3	–	–	4	–	–	–	3	–	–	1-Butyl-1-hydridotetrachlorocyclotriphosphazene	71982-87-1	NI35078
278	276	280	293	242	41	291	240	333	100	81	49	18	18	15	15	12	9.64	4	10	–	3	–	–	4	–	–	–	3	–	–	1-Butyl-1-hydridotetrachlorocyclotriphosphazene	71982-87-1	NI35079
57	279	278	277	276	41	280	281	333	100	32	30	27	23	18	17	16	3.83	4	10	–	3	–	–	4	–	–	–	3	–	–	1-tert-Butyl-1-hydridotetrachlorocyclotriphosphazene	71982-89-3	NI35080
69	114	164	314	119	226	169	145	333	100	58	17	15	10	9	8	6	0.00	6	–	–	1	–	13	–	–	–	–	–	–	–	Perfluoro(2-azahept-2-ene)		MS07916
29	18	27	81	109	28	15	197	333	100	54	51	38	37	27	19	17	0.60	9	11	4	1	–	–	3	–	–	–	1	–	–	Phosphoric acid, diethyl 3,5,6-trichloro-2-pyridinyl ester	5598-15-2	NI35081
235	293	265	321	245	103	180	118	333	100	63	49	43	36	32	10	8	0.00	13	12	3	1	–	–	–	–	–	–	–	–	1	Rhodium, dicarbonyl[4-(phenylimino)-2-pentanonato-N,O]-	15228-18-9	NI35082
43	58	42	162	44	164	220	41	333	100	76	12	9	7	6	4	4	1.00	13	13	5	1	–	–	2	–	–	–	–	–	–	L-Proline, 1-[(2,4-dichlorophenoxy)acetyl]-4-hydroxy-, trans-	50649-02-0	NI35083
55	43	276	56	41	316	292	69	333	100	33	24	24	20	19	11	11	6.88	13	14	4	3	–	3	–	–	–	–	–	–	–	Benzenamine, N-ethyl-N-(2-methyl-2-propenyl)-2,6-dinitro-4-(trifluoromethyl)-	55283-68-6	NI35084

MASS TO CHARGE RATIOS								M.W.	INTENSITIES								Parent	C	H	O	N	S	F	Cl	Br	I	Si	P	B	X	COMPOUND NAME	CAS Reg No	No
56	217	123	83	215	64	42	119	333	100	78	72	63	53	52	52	44	0.00	13	16	4	3	1	–	–	–	–	–	–	–	1	Sulpyrine	68-89-3	LI07391
40	168	72	109	44	67	57	42	333	100	25	24	23	23	22	22	22	13.21	13	24	3	3	1	–	–	–	–	–	1	–	–	Phosphorothioic acid, O-[2-(diethylamino)-6-methyl-4-pyrimidinyl] O,O-diethyl ester	23505-41-1	NI35085
333	318	57	304	168	180	71	166	333	100	82	60	57	50	47	40	34		13	24	3	3	1	–	–	–	–	–	1	–	–	Phosphorothioic acid, O-[2-(diethylamino)-6-methyl-4-pyrimidinyl] O,O-diethyl ester	23505-41-1	NI35086
75	73	84	158	216	156	43	45	333	100	92	67	62	49	47	42	26	0.00	13	27	5	1	–	–	–	–	–	2	–	–	–	L-Glutamic acid, N-acetyl-, bis(trimethylsilyl)-	56272-70-9	NI35087
73	75	55	147	302	258	45	89	333	100	77	28	25	20	13	13	12	0.00	13	27	5	1	–	–	–	–	–	2	–	–	–	Hexanedioic acid, 2-(methoxyimino)-, bis(trimethylsilyl) ester	55494-14-9	NI35088
145	173	305	160	131	115	104	216	333	100	48	16	4	4	4	4	1	1.00	14	9	–	3	–	6	–	–	–	–	–	–	–	1-Triazene, 1,3-bis[2-(trifluoromethyl)phenyl]-	7639-93-2	NI35089
168	333	103	169	142	259	233	285	333	100	41	38	35	27	26	26	25		14	16	2	1	–	–	–	–	–	–	–	–	1	[η^4-2-(2-Nitroethyl)norbornadiene](η^5-cyclopentadienyl)rhodium	90752-89-9	NI35090
298	133	162	135	184	41	333	69	333	100	94	88	65	57	57	53	49		14	17	4	1	–	–	2	–	–	–	–	–	–	L-Isoleucine, N-[(2,4-dichlorophenoxy)acetyl]-	7407-67-2	NI35091
162	298	175	164	145	177	128	147	333	100	99	67	67	64	61	61	49	5.00	14	17	4	1	–	–	2	–	–	–	–	–	–	L-Leucine, N-[(2,4-dichlorophenoxy)acetyl]-	2752-54-7	NI35092
43	58	42	162	44	164	220	41	333	100	76	12	9	7	6	4	4	1.00	14	17	4	1	–	–	2	–	–	–	–	–	–	L-Proline, hydroxy-, N-[(2,4-dichlorophenoxy)acetyl]-		NI35093
220	205	104	91	318	221	206	135	333	100	53	41	37	23	18	9	5	2.30	14	18	3	1	–	3	–	–	–	1	–	–	–	L-Phenylalanine, N-(trifluoroacetyl)-, trimethylsilyl ester	52558-84-6	NI35094
43	128	115	87	73	116	86	72	333	100	33	28	16	14	10	10	10	0.00	14	23	8	1	–	–	–	–	–	–	–	–	–	Galactopyranose, 2-acetamido-2-deoxy-3,4-di-O-methyl-, 1,6-diacetate, α D-	17296-17-2	MS07917
43	45	114	73	115	74	87	84	333	100	33	29	29	26	15	14	13	0.33	14	23	8	1	–	–	–	–	–	–	–	–	–	Galactopyranose, 2-acetamido-2-deoxy-3,6-di-O-methyl-, 1,4-diacetate, α D-	17296-16-1	MS07918
43	101	59	45	143	112	74	214	333	100	83	40	35	23	19	17	14	0.00	14	23	8	1	–	–	–	–	–	–	–	–	–	Galactopyranoside, methyl 2-acetamido-2-deoxy-3,4-diacetyl-6-O-methyl-, α-D-	17429-93-5	NI35095
43	101	59	45	143	112	74	140	333	100	83	40	34	22	19	15	11	0.00	14	23	8	1	–	–	–	–	–	–	–	–	–	Galactopyranoside, methyl 2-acetamido-2-deoxy-3,4-diacetyl-6-O-methyl-, α-D-	17429-93-5	MS07919
101	59	45	58	143	112	74	114	333	100	50	38	28	26	22	18	14	0.00	14	23	8	1	–	–	–	–	–	–	–	–	–	Galactopyranoside, methyl 2-acetamido-2-deoxy-3,4-diacetyl-6-O-methyl-, α-D-	17429-93-5	NI35096
115	73	43	75	74	84	58	45	333	100	94	91	88	15	10	10	9	0.00	14	23	8	1	–	–	–	–	–	–	–	–	–	Galactopyranoside, methyl 2-acetamido-2-deoxy-3-O-methyl-4,6-di-O-acetyl-, α-D-	17296-12-7	MS07920
115	73	75	74	116	84	58	98	333	100	93	82	15	14	10	10	9	0.00	14	23	8	1	–	–	–	–	–	–	–	–	–	Galactopyranoside, methyl 2-acetamido-2-deoxy-3-O-methyl-4,6-di-O-acetyl-, α-D-	17296-12-7	NI35097
101	71	143	128	59	98	87	73	333	100	43	40	36	21	18	18	11	0.00	14	23	8	1	–	–	–	–	–	–	–	–	–	Glucopyranoside, methyl 2-acetamido-2-deoxy-4-O-methyl-3,6-diacetyl-, α-D-		NI35098
174	318	73	147	75	200	244	86	333	100	21	21	10	7	5	4	4	2.00	14	35	2	1	–	–	–	–	–	3	–	–	–	Pentanoic acid, 5-[bis(trimethylsilyl)amino]-, trimethylsilyl ester	55191-54-3	MS07921
174	73	175	318	40	147	82	176	333	100	39	22	17	16	15	12	10	1.40	14	35	2	1	–	–	–	–	–	3	–	–	–	Pentanoic acid, 5-[bis(trimethylsilyl)amino]-, trimethylsilyl ester	55191-54-3	NI35099
102	101	144	318	72	202	174	156	333	100	56	25	16	11	5	4	3	0.01	15	27	7	1	–	–	–	–	–	–	–	–	–	1-Acetamido-2,3-5,6-di-O-isopropylidene-1-O-methyl-D-glucitol	83937-96-6	NI35100
100	88	72	59	185	161	86	217	333	100	75	42	35	34	30	28	27	0.00	15	27	7	1	–	–	–	–	–	–	–	–	–	1-O-[(2-(Acetylamino)propyl)]-2-O-acetyl-3,4-di-O-methyl-L-rhamanoside		LI07392
43	114	100	202	156	216	45	142	333	100	53	51	43	38	31	30	27	0.00	15	27	7	1	–	–	–	–	–	–	–	–	–	1,5-Di-O-acetyl-3,6-dideoxy-2,4-di-O-methyl-3-(N-methylacetamido)-L-glucitol	82497-76-5	NI35101
216	73	77	217	44	75	318	144	333	100	32	28	19	15	12	9	9	5.60	16	23	3	1	–	–	–	–	–	2	–	–	–	1H-Indole-3-acetic acid, α-oxo-1-(trimethylsilyl)-, trimethylsilyl ester	55191-46-3	NI35102
318	231	73	304	319	288	289	232	333	100	68	42	32	26	21	15	14	12.00	16	23	3	1	–	–	–	–	–	2	–	–	–	2-Quinolinecarboxylic acid, 4-[(trimethylsilyl)oxy]-, trimethylsilyl ester	36972-84-6	NI35103
193	165	234	235	194	179	204	192	333	100	84	68	50	47	35	30	30	0.00	17	16	4	1	–	–	1	–	–	–	–	–	–	5H-Dibenz[c,e]azepine, 6-allyl-, perchlorate		MS07922
178	91	155	150	210	134	65	55	333	100	70	23	21	19	19	17	16	15.70	17	19	4	1	1	–	–	–	–	–	–	–	–	Benzenesulphonamide, N-(4,8-dioxotricyclo[3.3.1.1^{3,7}]dec-2-yl)-4-methyl-	56781-97-6	MS07923
58	73	45	57	44	43	167	165	333	100	25	9	8	7	7	6	5	0.00	17	20	1	1	–	–	–	1	–	–	–	–	–	Ethanamine, 2-[(4-bromophenyl)phenylmethoxy]-N,N-dimethyl-	118-23-0	NI35104
245	247	334	165	332				333	100	98	9	9	5				0.00	17	20	1	1	–	–	–	1	–	–	–	–	–	Ethanamine, 2-[(4-bromophenyl)phenylmethoxy]-N,N-dimethyl-	118-23-0	NI35105
109	168	167	81	77	51	169	196	333	100	69	65	62	60	46	45	41	32.00	17	20	4	1	–	–	–	–	–	–	1	–	–	N-(Diethylphosphonatocarbonyl)diphenylamine		IC04003
161	219	235	75	176	57	250	118	333	100	84	34	21	20	20	16	16	0.00	17	27	2	1	–	–	–	–	–	2	–	–	–	1H-Indole-3-acetic acid, 5-methyl-1-(trimethylsilyl)-, trimethylsilyl ester	74367-50-3	NI35106
216	73	333	217	75	45	334	318	333	100	51	24	18	8	8	7	6		17	27	2	1	–	–	–	–	–	2	–	–	–	1H-Indole-3-acetic acid, 5-methyl-1-(trimethylsilyl)-, trimethylsilyl ester	74367-50-3	NI35107
318	73	319	333	320	147	45	244	333	100	35	25	14	9	7	7	6		17	27	2	1	–	–	–	–	–	2	–	–	–	1H-Indole-2-carboxylic acid, 5-ethyl-1-(trimethylsilyl)-, trimethylsilyl ester	74367-45-6	NI35108
235	161	204	191	250	73	75	133	333	100	88	77	71	45	44	42	19	0.00	17	27	2	1	–	–	–	–	–	2	–	–	–	1H-Indole-2-carboxylic acid, 5-ethyl-1-(trimethylsilyl)-, trimethylsilyl ester	74367-45-6	NI35109
202	333	73	203	334	77	204	200	333	100	32	28	20	9	8	6	6		17	27	2	1	–	–	–	–	–	2	–	–	–	1H-Indole-3-propanoic acid, 1-(trimethylsilyl)-, trimethylsilyl ester	55191-57-6	NI35110
202	73	333	203	45	334	75	200	333	100	68	27	17	11	8	7	6		17	27	2	1	–	–	–	–	–	2	–	–	–	1H-Indole-3-propanoic acid, 1-(trimethylsilyl)-, trimethylsilyl ester	55191-57-6	NI35111
171	333	318	332	43	78	334	129	333	100	69	58	25	20	19	15	13		18	16	4	2	–	–	–	–	–	–	–	–	1	Beryllium, bis[1-(3-pyridinyl)-1,3-butanedionato-O,O′]-	30983-53-0	NI35112
261	263	135	139	262	72	111	141	333	100	34	31	28	16	10	9	8	0.00	18	20	3	1	–	–	1	–	–	–	–	–	–	Benzeneacetamide, 4-chloro-α-methoxy-α-(4-methoxyphenyl)-N,N-dimethyl-	63953-35-5	NI35113
112	78	41	67	55	51	164	136	333	100	55	38	26	22	22	21	20	6.00	18	23	3	1	1	–	–	–	–	–	–	–	–	Cyclopentaneacetic acid, 3-oxo-2-(2-pentenyl)-4-(2-pyridinylthio)-, methyl ester	67012-87-7	NI35114

MASS TO CHARGE RATIOS								M.W.	INTENSITIES								Parent	C	H	O	N	S	F	Cl	Br	I	Si	P	B	X	COMPOUND NAME	CAS Reg No	No
110	204	191	190	257	301	315	300	333	100	64	49	36	35	33	31	26	3.00	18	23	3	1	1	–	–	–	–	–	–	–	–	9,10-Dimethoxy-5-methyl-4,5,6,7,12,13-hexahydrothieno[3,2-e][3]benzazecin-12-ol		MS07924
164	165	315	149	150	314	301	272	333	100	20	17	16	14	13	11	11	0.00	18	23	3	1	1	–	–	–	–	–	–	–	–	4-(2-Hydroxymethyl-4,5-dimethoxybenzyl)-5-methyl-4,5,6,7-tetrahydrothieno[3,2-c]pyridine		MS07925
57	233	232	277	174	162	333	260	333	100	55	25	19	13	12	6	5		18	23	5	1	–	–	–	–	–	–	–	–	–	tert-Butyl 2-[2-(2,5-dioxo-1-pyrrolidinyl)-5-methoxy-4-methylphenyl]acetate		NI35115
194	318	195	152	333	290	167	262	333	100	90	90	83	73	69	45	43		18	23	5	1	–	–	–	–	–	–	–	–	–	(+)-Limalongine		DD01311
120	119	43	93	94	136	95	138	333	100	80	76	74	69	59	52	43	2.60	18	23	5	1	–	–	–	–	–	–	–	–	–	Seneciphylline	480-81-9	NI35116
231	230	333	229	232	215	228	258	333	100	36	23	17	16	9	9	8		18	23	5	1	–	–	–	–	–	–	–	–	–	Stephaboline	36871-86-0	NI35117
192	41	179	333	55	69	95	79	333	100	17	10	7	7	5	4	4		18	30	1	1	–	3	–	–	–	–	–	–	–	4,5-Dihydro-2-methyl-1-trifluoroacetyl-3-undecylpyrrole		NI35118
333	331	335	334	330	329	253	252	333	100	52	23	23	23	20	20	20		19	11	–	1	–	–	–	–	–	–	–	–	1	Benzo[h]selenopheno[3,2-a]acridine		MS07926
333	331	335	334	330	329	332	166.5	333	100	52	24	24	22	22	16	14		19	11	–	1	–	–	–	–	–	–	–	–	1	Benzo[j]selenopheno[3,2-a]acridine		MS07927
333	227	121	91	56	202	268	334	333	100	53	48	48	34	33	31	30		19	19	1	1	–	–	–	–	–	–	–	–	1	Acetamide, N-benzyl-N-ferrocenyl-	34738-73-3	MS07928
333	91	242	120	180	77	256	318	333	100	77	63	51	40	37	16	6		19	19	1	5	–	–	–	–	–	–	–	–	–	N[5]-Methyl-6,7-diphenyl-5,6,7,8-tetrahydropterin		MS07929
333	242	334	91	254	43	40	256	333	100	26	23	21	16	14	11	10		19	19	1	5	–	–	–	–	–	–	–	–	–	N[8]-Methyl-6,7-diphenyl-5,6,7,8-tetrahydropterin		MS07930
196	333	44	302	42	274	334	260	333	100	100	32	30	29	28	21	21		19	27	4	1	–	–	–	–	–	–	–	–	–	6-Epioreobeiline	97400-78-7	NI35119
333	196	302	274	334	260	244	44	333	100	87	33	29	21	20	18	14		19	27	4	1	–	–	–	–	–	–	–	–	–	Oreobeiline	97400-76-5	NI35120
196	333	332	59	334	197	318	44	333	100	76	26	24	22	15	11	10		19	27	4	1	–	–	–	–	–	–	–	–	–	Tetrahydrosalutaridinol		MS07931
196	333	332	334	59	197	318	274	333	100	97	26	22	22	14	12	10		19	27	4	1	–	–	–	–	–	–	–	–	–	Tetrahydrosalutaridinol		LI07393
91	333	155	269	177	178	151	268	333	100	60	41	37	16	14	14	13		20	15	2	1	1	–	–	–	–	–	–	–	–	N-Fluorenylidene-S-p-tolylsulphonamide		LI07394
333	290	318	275	317	291	232	164	333	100	49	28	19	13	10	8	7		20	15	4	1	–	–	–	–	–	–	–	–	–	[1,3]Benzodioxolo[5,6-c]phenanthridine, 1,2-dimethoxy-	6900-99-8	NI35121
274	333	275	302					333	100	68	24	15						20	15	4	1	–	–	–	–	–	–	–	–	–	Methyl trans-3-(1-benzoyl-4-oxo-4H-quinolizin-3-yl) acrylate		LI07395
134	333	200	185	199	184	173	155	333	100	27	27	21	12	12	11	11		20	19	2	3	–	–	–	–	–	–	–	–	–	9-Methyl-1-(2-methylaminobenzoyl)-4-oxo-2,3,4,9-tetrahydro-1H-pyrrolo[2,3-b]quinoline		LI07396
284	300	261	333	315	302	220	109	333	100	63	60	48	48	43	37	31		20	31	3	1	–	–	–	–	–	–	–	–	–	Androstane-11,17-dione, 3-hydroxy-, 17-O-methyloxime, $(3\alpha,5\beta)$-	3091-91-6	NI35122
125	91	333	137	92	153	151	138	333	100	93	76	76	70	59	36	23		20	31	3	1	–	–	–	–	–	–	–	–	–	Androst-4-en-3-one, 16,17-dihydroxy-, O-methyloxime, $(16\alpha,17\beta)$-	69688-33-1	NI35123
91	92	283	274	224	268	239	256	333	100	82	79	63	63	52	52	38	4.39	20	31	3	1	–	–	–	–	–	–	–	–	–	Androst-5-en-17-one, 3,16-dihydroxy-, O-methyloxime, $(3\beta,16\alpha)$-	69688-30-8	NI35124
113	334	129	144	100				333	100	30	27	18	15				0.00	20	31	3	1	–	–	–	–	–	–	–	–	–	Cyclopentanecarboxylic acid, 1-phenyl-, 2-[2-(diethylamino)ethoxy]ethyl ester	77-23-6	NI35125
333	115	128	135	60	198	74	261	333	100	80	47	29	20	19	17	11		20	31	3	1	–	–	–	–	–	–	–	–	–	N-Isobutyl-9-(3,4-methylenedioxyphenyl)nonanamide	83029-42-9	NI35126
230	193	89	232	195	90	201	77	333	100	45	36	34	31	30	27	17	2.00	21	16	1	1	–	–	1	–	–	–	–	–	–	Oxazole, 5-(4-chlorophenyl)-2,5-dihydro-2,4-diphenyl-	55955-59-4	NI35127
107	106	57	71	77	55	69	79	333	100	99	99	94	90	84	82	78	31.00	21	23	1	3	–	–	–	–	–	–	–	–	–	(Z,Z)-4-Ethyl-3-methyl-5-[5-(4-aminophenyl-2-methylene)-3,4-dimethyl-5H-pyrrolyl-2-methylene]-3-pyrrolin-2-one		MS07932
55	69	81	67	95	70	332	57	333	100	95	65	61	60	55	48	46	11.00	21	35	–	1	1	–	–	–	–	–	–	–	–	Spiro[androstane-3,2'-thiazolidine], (5α)-	4642-50-6	NI35128
57	260	204	41	29	318	277	261	333	100	80	42	29	24	21	21	16	5.50	21	35	2	1	–	–	–	–	–	–	–	–	–	Benzene, 1,3,5-tris(2,2-dimethylpropyl)-2-nitro-	40572-19-8	MS07933
318	164	70	333	319	71	107	81	333	100	44	33	27	27	20	16	16		21	35	2	1	–	–	–	–	–	–	–	–	–	D-Bishomo-17b-aza-androstane-17a-one, $(3\beta,5\alpha)$-		LI07397
86	118	117	216	44	190	42	248	333	100	31	26	18	14	13	13	12	4.00	21	35	2	1	–	–	–	–	–	–	–	–	–	D-Norandrostane, 16-acetamido-3-methoxy-, $(3\beta,16\alpha)$-		LI07398
318	164	333	70	319	71	79	72	333	100	44	35	34	26	20	16	16		21	35	2	1	–	–	–	–	–	–	–	–	–	2H-Phenanthro[2,1-b]azepin-2-one, 1,3,4,5,5aα,5bβ,6,7,7aα,8,9,10,11,11a,11bα,12,13,13a-octadecahydro-9β-methoxy-11aβ,13aβ-dimethyl-	21003-00-9	NI35129
82	55	41	83	68	56	42	54	333	100	89	55	49	26	24	23	20	0.00	21	39	–	3	–	–	–	–	–	–	–	–	–	N,N,N-Tricyclohexylhexahydrotriazine		MS07934
318	69	333	179	180	264	70	248	333	100	96	29	21	13	8	8	7		22	23	2	1	–	–	–	–	–	–	–	–	–	9,10-Dihydro-10-(1-methacryloyloxy-2-methylpropenyl)-9-methylacridine		LI07399
193	234	70	263	194	69	318	180	333	100	57	39	27	23	19	11	10	6.00	22	23	2	1	–	–	–	–	–	–	–	–	–	2-Isopropylidene-5,5,6-trimethyldibenzo-3-oxa-1-azabicyclo[4.2.2]deca-7,9-dien-4-one		LI07400
105	134	108	183	185	109	214	262	333	100	46	33	28	27	25	14	13	3.00	22	24	–	1	–	–	–	–	–	–	1	–	–	Phosphinous amide, N-ethyl-P,P-diphenyl-N-(1-phenylethyl)-, (S)-	61937-85-7	NI35130
235	234	333	227	193	276	334	318	333	100	99	80	60	60	30	20	5		22	27	–	3	–	–	–	–	–	–	–	–	–	2-(tert-Butylamino)-4,5-diphenyl-1-isopropylimidazole	118515-07-4	DD01312
58	286	55	80	69	66	90	78	333	100	15	12	10	10	10	8	8	2.00	22	39	1	1	–	–	–	–	–	–	–	–	–	Pregnan-18-ol, 20-amino-20-methyl-	55659-10-4	NI35131
86	59	333	58	290	96	56	82	333	100	16	14	12	11	11	8	7		22	39	1	1	–	–	–	–	–	–	–	–	–	Pregnan-1-ol, 3-(methylamino)-, $(1\alpha,3\alpha,5\alpha)$-	23925-82-8	LI07401
86	58	333	290	96	57	79	55	333	100	16	14	12	12	12	8	8		22	39	1	1	–	–	–	–	–	–	–	–	–	Pregnan-1-ol, 3-(methylamino)-, $(1\alpha,3\alpha,5\alpha)$-	23925-82-8	NI35132
86	96	290	333					333	100	30	13	10						22	39	1	1	–	–	–	–	–	–	–	–	–	Pregnan-1-ol, 3-(methylamino)-, $(1\alpha,3\beta,5\alpha)$-		LI07402
86	57	302	55	96	81	58	70	333	100	29	20	15	14	12	12	11	7.72	22	39	1	1	–	–	–	–	–	–	–	–	–	Pregnan-1-ol, 3-(methylamino)-, $(1\beta,3\alpha,5\alpha)$-	23931-06-8	NI35133
86	96	302	333	290				333	100	9	7	5	3					22	39	1	1	–	–	–	–	–	–	–	–	–	Pregnan-1-ol, 3-(methylamino)-, $(1\beta,3\beta,5\alpha)$-	23925-82-8	LI07403
43	41	70	55	113	71	42	39	333	100	47	27	27	26	23	21	18	1.94	22	39	1	1	–	–	–	–	–	–	–	–	–	Pyrrolidine, 1-(1-oxo-2,5-octadecadienyl)-	56666-48-9	NI35134
113	98	55	333	70	43	85	41	333	100	19	18	16	16	14	11	11		22	39	1	1	–	–	–	–	–	–	–	–	–	Pyrrolidine, 1-(1-oxo-4,7-octadecadienyl)-	56600-14-7	NI35135
113	70	55	43	126	85	333	114	333	100	11	11	11	10	10	9	7		22	39	1	1	–	–	–	–	–	–	–	–	–	Pyrrolidine, 1-(1-oxo-5,8-octadecadienyl)-	56600-16-9	NI35136
113	126	70	55	333	72	43	41	333	100	50	32	31	27	27	21	20		22	39	1	1	–	–	–	–	–	–	–	–	–	Pyrrolidine, 1-(1-oxo-6,9-octadecadienyl)-	56600-18-1	NI35137

MASS TO CHARGE RATIOS	M.W.	INTENSITIES	Parent	C	H	O	N	S	F	Cl	Br	I	Si	P	B	X	COMPOUND NAME	CAS Reg No	No
43 113 70 55 41 126 72 98	**333**	100 92 67 65 59 40 30 28	**12.04**	22	39	1	1	–	–	–	–	–	–	–	–	–	**Pyrrolidine, 1-(1-oxo-7,10-octadecadienyl)-**	56599-76-9	**NI35138**
113 126 70 55 333 72 43 98	**333**	100 51 39 33 28 27 27 22		22	39	1	1	–	–	–	–	–	–	–	–	–	**Pyrrolidine, 1-(1-oxo-7,10-octadecadienyl)-**	56599-76-9	**NI35139**
83 113 85 43 70 55 41 47	**333**	100 70 68 68 59 39 36 35	**23.04**	22	39	1	1	–	–	–	–	–	–	–	–	–	**Pyrrolidine, 1-(1-oxo-8,10-octadecadienyl)-, (E,E)-**	56666-47-8	**NI35140**
113 126 70 55 333 72 41 43	**333**	100 72 47 46 43 30 28 25		22	39	1	1	–	–	–	–	–	–	–	–	–	**Pyrrolidine, 1-(1-oxo-8,11-octadecadienyl)-**	56630-35-4	**NI35141**
83 85 47 113 35 48 126 41	**333**	100 69 43 30 29 22 20 19	**10.04**	22	39	1	1	–	–	–	–	–	–	–	–	–	**Pyrrolidine, 1-(1-oxo-9,11-octadecadienyl)-, (E,E)-**	56666-45-6	**NI35142**
113 126 55 70 72 333 43 41	**333**	100 64 50 47 45 40 31 31		22	39	1	1	–	–	–	–	–	–	–	–	–	**Pyrrolidine, 1-(1-oxo-9,11-octadecadienyl)-, (Z,Z)-**	56051-64-0	**NI35143**
113 126 333 55 70 72 41 67	**333**	100 68 47 44 40 28 26 25		22	39	1	1	–	–	–	–	–	–	–	–	–	**Pyrrolidine, 1-(1-oxo-9,12-octadecadienyl)-**	56630-36-5	**NI35144**
113 41 43 55 126 70 42 39	**333**	100 99 86 74 61 61 34 31	**19.04**	22	39	1	1	–	–	–	–	–	–	–	–	–	**Pyrrolidine, 1-(1-oxo-9,15-octadecadienyl)-**	56630-65-0	**NI35145**
43 70 113 83 41 55 85 67	**333**	100 80 76 60 59 57 41 37	**23.04**	22	39	1	1	–	–	–	–	–	–	–	–	–	**Pyrrolidine, 1-(1-oxo-10,12-octadecadienyl)-, (E,E)-**	56666-46-7	**NI35146**
113 126 55 70 333 67 41 72	**333**	100 58 45 41 40 27 26 25		22	39	1	1	–	–	–	–	–	–	–	–	–	**Pyrrolidine, 1-(1-oxo-10,13-octadecadienyl)-**	56630-38-7	**NI35147**
113 126 333 70 55 67 72 41	**333**	100 56 41 41 39 28 23 23		22	39	1	1	–	–	–	–	–	–	–	–	–	**Pyrrolidine, 1-(1-oxo-11,14-octadecadienyl)-**	56630-46-7	**NI35148**
113 126 333 70 55 67 41 72	**333**	100 56 45 41 40 27 22 21		22	39	1	1	–	–	–	–	–	–	–	–	–	**Pyrrolidine, 1-(1-oxo-12,15-octadecadienyl)-**	56630-44-5	**NI35149**
113 126 333 55 70 41 72 43	**333**	100 51 43 43 42 21 19 19		22	39	1	1	–	–	–	–	–	–	–	–	–	**Pyrrolidine, 1-(1-oxo-13,16-octadecadienyl)-**	56600-15-8	**NI35150**
113 126 70 333 55 41 43 72	**333**	100 46 37 35 33 20 17 16		22	39	1	1	–	–	–	–	–	–	–	–	–	**Pyrrolidine, 1-(1-oxo-14,17-octadecadienyl)-**	56600-13-6	**NI35151**
113 98 70 55 43 41 126 71	**333**	100 38 31 31 22 17 16 16	**8.44**	22	39	1	1	–	–	–	–	–	–	–	–	–	**Pyrrolidine, 1-(1-oxo-6-octadecynyl)-**	56600-12-5	**NI35152**
113 55 41 43 70 67 126 81	**333**	100 47 35 31 28 28 27 27	**2.94**	22	39	1	1	–	–	–	–	–	–	–	–	–	**Pyrrolidine, 1-(1-oxo-9-octadecynyl)-**	56630-92-3	**NI35153**
113 55 43 41 70 126 67 81	**333**	100 43 32 31 30 26 22 19	**3.74**	22	39	1	1	–	–	–	–	–	–	–	–	–	**Pyrrolidine, 1-(1-oxo-10-octadecynyl)-**	56630-91-2	**NI35154**
113 55 43 70 41 126 67 72	**333**	100 40 33 32 28 25 17 15	**7.74**	22	39	1	1	–	–	–	–	–	–	–	–	–	**Pyrrolidine, 1-(1-oxo-11-octadecynyl)-**	56666-39-8	**NI35155**
113 55 81 41 67 95 43 70	**333**	100 66 54 52 51 38 37 33	**5.74**	22	39	1	1	–	–	–	–	–	–	–	–	–	**Pyrrolidine, 1-(1-oxo-12-octadecynyl)-**	56630-90-1	**NI35156**
208 69 151 105 209 193 55 115	**333**	100 84 41 38 20 20 20 17	**2.00**	23	27	1	1	–	–	–	–	–	–	–	–	–	**2-Cyclohexylimino-3,3-dimethyl-4,4-diphenyl-oxetane**		**MS07935**
333 59 150 332 265 276 42 77	**333**	100 84 65 47 36 32 32 27		23	27	1	1	–	–	–	–	–	–	–	–	–	**3-Phenoxy-N-methylmorphinan**	67562-76-9	**NI35157**
256 152 333 165 178 166 257 77	**333**	100 55 44 31 30 25 22 21		25	19	–	1	–	–	–	–	–	–	–	–	–	**[1,1′-Biphenyl]-2-amine, N-(diphenylmethylene)-**	51677-35-1	**NI35158**
101 103 151 169 85 153 167 117	**334**	100 63 30 21 21 19 15 15	**0.00**	4	–	–	–	–	4	6	–	–	–	–	–	–	**1,1,2,3,4,4-Hexachloro-1,2,3 4-tetrafluorobutane**		**DO01102**
101 103 163 109 194 159 44 165	**334**	100 61 47 39 33 33 32 30	**0.00**	4	–	–	–	–	4	6	–	–	–	–	–	–	**Tetrafluorohexachlorobutane**		**IC04004**
257 305 255 259 303 149 199 147	**334**	100 86 68 64 49 48 43 36	**0.00**	6	15	–	–	–	–	–	–	1	–	–	–	1	**Triethyltin iodide**		**MS07936**
250 306 334 125 249 222 278 221	**334**	100 46 43 30 28 28 26 23		8	6	3	–	–	–	–	–	–	–	–	–	1	**Tungsten, (π-cyclopentadienyl)hydridotricarbonyl-**		**MS07937**
250 334 306 278 249 305 277 333	**334**	100 58 50 42 33 23 18 8		8	6	3	–	–	–	–	–	–	–	–	–	1	**Tungsten, (π-cyclopentadienyl)hydridotricarbonyl-**		**LI07404**
299 301 263 208 36 265 210 133	**334**	100 98 65 62 58 42 41 36	**0.06**	8	10	2	4	–	–	4	–	–	–	–	–	–	**1,3,4,6-Tetrakis(chloromethyl)-2,5-dioxoperhydroimidazo[4,5-d]imidazole**		**IC04005**
87 43 45 41 39 27 69 42	**334**	100 64 52 40 16 16 8 7	**2.62**	8	14	4	–	–	–	–	–	–	–	–	–	2	**Diselenodibutyric acid**	14362-48-2	**NI35159**
217 219 172 174 46 161 189 163	**334**	100 71 69 48 42 29 19 19	**4.58**	9	7	1	2	–	–	5	–	–	–	–	–	–	**Formamide, N-[2,2,2-trichloro-1-[(3,4-dichlorophenyl)amino]ethyl]-**	20856-57-9	**NI35160**
319 75 320 273 288 272 289 321	**334**	100 30 24 23 20 20 19 13	**0.00**	9	27	4	–	–	–	–	–	–	3	–	–	1	**Vanadic acid (H3VO4), tris(trimethylsilyl) ester**	5590-56-7	**NI35161**
192 257 271 336 178 206 113 166	**334**	100 96 92 36 24 20 13 12	**0.00**	10	8	–	–	–	–	–	2	–	–	–	–	1	**Titanium, bis(π-cyclopentadienyl)dibromide**		**LI07405**
69 41 45 55 31 43 29 70	**334**	100 38 34 27 17 16 16 11	**6.82**	10	22	2	–	–	–	–	–	–	–	–	–	2	**3,3′-Diselenobis(2,2-dimethyl-1-propanol)**	23243-51-8	**NI35162**
175 55 64 41 87 69 176 45	**334**	100 98 95 95 76 71 59 59	**0.00**	10	22	4	–	4	–	–	–	–	–	–	–	–	**Methanesulphonothioic acid 1,8-octanediyl ester**		**MS07938**
334 265 335 266 315 296 234 167	**334**	100 32 13 13 10 10 7 6		12	–	–	–	–	10	–	–	–	–	–	–	–	**1,1′-Biphenyl, 2,2′,3,3′,4,4′,5,5′,6,6′-decafluoro-**	434-90-2	**IC04006**
334 265 335 315 296 234 167 266	**334**	100 33 18 15 12 10 9 5		12	–	–	–	–	10	–	–	–	–	–	–	–	**1,1′-Biphenyl, 2,2′,3,3′,4,4′,5,5′,6,6′-decafluoro-**	434-90-2	**IC04007**
334 265 93 117 167 141 234 165	**334**	100 96 80 63 45 40 37 35		12	–	–	–	–	10	–	–	–	–	–	–	–	**1,1′-Biphenyl, 2,2′,3,3′,4,4′,5,5′,6,6′-decafluoro-**	434-90-2	**NI35163**
145 147 124 109 159 173 161 175	**334**	100 77 51 34 29 26 24 20	**15.00**	12	6	1	2	–	–	4	–	–	–	–	–	–	**Diazene, bis(3,4-dichlorophenyl)-, 1-oxide**	21232-47-3	**NI35164**
42 204 66 161 65 39 248 304	**334**	100 80 30 25 17 16 15 14		12	12	5	–	–	–	–	–	–	–	–	–	1	**Molybdenum, (π-cyclopentadienyl)(ethoxycarbonylmethyl)dicarbonyl-**		**MS07939**
334 180 193 221 237 220 335 136	**334**	100 81 32 20 19 19 18 16		12	13	4	4	–	3	–	–	–	–	–	–	–	**7-[(2′-O-Trifluoroacetyl)propyl]-1,3-dimethylxanthine**		**NI35165**
207 132 54 69 209 36 257 255	**334**	100 90 73 68 65 50 46 46	**0.00**	12	22	–	2	–	–	4	–	–	–	–	–	–	**Hexanimidamide, 2-chloro-N-(1,2-dichloro-1-hexenyl)-, monohydrochloride**	41027-16-1	**NI35166**
83 111 155 127 99 129 298 205	**334**	100 63 21 15 15 5 1 1	**0.00**	12	22	4	–	–	2	1	–	–	–	1	–	–	**Diethyl 2-chloro-1-cyclohexyl-2,2-difluoro-1-hydroxyethanephosphonate**		**DD01313**
102 73 147 217 319 146 100 75	**334**	100 86 24 23 16 13 12 12	**1.49**	12	30	3	2	–	–	–	–	–	3	–	–	–	**Hydantoic acid, tris(trimethylsilyl)-**		**NI35167**
128 158 140 254 222 220 255 334	**334**	100 67 67 11 11 10 9 5		13	10	2	–	–	3	–	1	–	–	–	–	–	**4-Bromo-4-trifluoroacetoxytetracyclo[4.3.2.0^{2,4}.0^{8,9}]undeca-7,10-diene**		**NI35168**
173 147 219 191 129 218 148 174	**334**	100 95 24 24 22 18 16 13	**0.00**	13	26	6	–	–	–	–	–	–	2	–	–	–	**Succinyl lactate, bis(trimethylsilyl)-**		**NI35169**
147 72 82 45 319 55 148 113	**334**	100 95 50 26 25 19 16 15	**1.75**	13	30	4	–	–	–	–	–	–	3	–	–	–	**Methylmalonic acid, tris(trimethylsilyl)-**		**NI35170**
147 40 73 75 83 45 319 148	**334**	100 80 71 26 25 18 18 16	**2.01**	13	30	4	–	–	–	–	–	–	3	–	–	–	**Methylmalonic acid, tris(trimethylsilyl)-**		**NI35171**
147 73 83 45 319 55 148 113	**334**	100 95 50 26 25 19 16 15	**1.75**	13	30	4	–	–	–	–	–	–	3	–	–	–	**Trimethylsilyl 2-methyl-3,3-bis(trimethylsiloxy)acrylate**		**NI35172**
334 151 76 179 75 150 74 90	**334**	100 49 27 24 23 21 9 8		14	7	2	–	–	–	–	–	1	–	–	–	–	**1-Iodoanthraquinone**		**IC04008**
176 88 336 87 334 174 338 175	**334**	100 64 40 28 22 19 19 18		14	8	–	–	–	–	–	2	–	–	–	–	–	**Anthracene, 9,10-dibromo-**	523-27-3	**NI35173**
181 290 112 55 110 82 335 43	**334**	100 86 82 80 71 64 60 46	**0.00**	14	9	3	1	–	5	–	–	–	–	–	–	–	**5-Methyl-3-isoxazolepropanoic acid, pentafluoropropionyl-**		**NI35174**
56 75 91 147 104 162 111 188	**334**	100 53 28 16 14 13 13 12	**0.00**	14	14	6	–	–	–	–	–	–	–	–	–	1	**(1RS,4RS,5SR,6RS)-C,5,6,C-η-[1-(Dimethoxymethyl)-5,6-dimethylidene-7-oxabicyclo[2.2.1]hept-2-ene]iron tricarbonyl**		**DD01314**

MASS TO CHARGE RATIOS								M.W.	INTENSITIES								Parent	C	H	O	N	S	F	Cl	Br	I	Si	P	B	X	COMPOUND NAME	CAS Reg No	No
232	188	250	55	160	70	72	233	334	100	61	51	49	26	26	24	18	4.00	14	15	6	–	–	–	–	–	–	–	–	–	1	Manganese, (1-ethoxycarbonyl-2-cyclopentadienyl)tricarbonyl-		LI07406
55	70	162	73	160	120	93	232	334	100	69	65	43	30	27	16	12	0.79	14	19	4	–	–	–	–	–	–	1	–	–	1	(1-Hydroxy-1-methyl-2-trimethylsilylethyl)cymantrene	12147-45-4	NI35175
43	118	127	85	93	59	261	201	334	100	91	12	12	10	8	5	5	0.00	14	22	9	–	–	–	–	–	–	–	–	–	–	2-O-Methyl-1,3,4,5-tetra-O-acetyl-arabinitol	78037-60-2	NI35176
117	43	127	99	85	159	261	58	334	100	30	14	12	10	7	6	5	0.00	14	22	9	–	–	–	–	–	–	–	–	–	–	2-O-Methyl-1,3,4,5-tetra-O-acetyl-arabinitol	78037-60-2	NI35177
43	129	130	87	189	88	190	71	334	100	86	72	42	32	30	26	8	0.00	14	22	9	–	–	–	–	–	–	–	–	–	–	3-O-Methyl-1,2,4,5-tetra-O-acetyl-arabinitol		NI35178
87	43	112	129	113	44	101	115	334	100	99	96	92	85	52	35	31	0.00	14	22	9	–	–	–	–	–	–	–	–	–	–	Methyl 3,5,6-tri-O-acetyl-2-O-methyl-α-D-galactofuranoside		NI35179
113	43	87	144	141	112	157	98	334	100	99	89	54	54	54	51	51	0.00	14	22	9	–	–	–	–	–	–	–	–	–	–	Methyl 2,3,4-tri-O-acetyl-6-O-methyl-α-D-galactopyranoside	72945-53-0	NI35180
87	43	129	74	71	130	112	116	334	100	99	79	65	56	36	24	22	0.00	14	22	9	–	–	–	–	–	–	–	–	–	–	Methyl 2,3,6-tri-O-acetyl-4-O-methyl-α-D-mannopyranoside	53098-96-7	NI35181
75	43	74	116	117	101	69	112	334	100	99	96	53	18	15	14	12	0.00	14	22	9	–	–	–	–	–	–	–	–	–	–	Methyl 2,4,6-tri-O-acetyl-3-O-methyl-α-D-mannopyranoside	53098-97-8	NI35182
74	43	112	113	116	129	87	84	334	100	99	68	48	43	35	34	33	0.00	14	22	9	–	–	–	–	–	–	–	–	–	–	Methyl 3,4,6-tri-O-acetyl-2-O-methyl-α-D-mannopyranoside	20672-76-8	NI35183
254	183	197	59	169	80	144	155	334	100	99	91	85	78	73	71	69	0.00	14	23	–	–	2	–	–	1	–	–	–	–	–	Cyclododeca-1,2-dithiol-1-ium, 4,5,6,7,8,9,10,11,12,13-decahydro-3-methyl-, bromide	55836-90-3	NI35184
73	143	117	69	75	43	159	161	334	100	77	58	58	46	33	22	16	0.00	14	30	5	–	–	–	–	–	–	2	–	–	–	Dimeric 3-hydroxybutyrate, bis(trimethylsilyl)-		NI35185
255	256	121	191	257	223	77	179	334	100	27	24	13	12	11	11	9	0.00	15	11	–	–	2	–	–	1	–	–	–	–	–	1,2-Dithiol-1-ium, 3,5-diphenyl-, bromide	55836-88-9	NI35186
334	333	131	76	103	77	207	231	334	100	86	42	42	41	36	34	29		15	11	1	–	–	–	–	–	1	–	–	–	–	2-Propen-1-one, 1-(4-iodophenyl)-3-phenyl-	7466-60-6	NI35187
334	207	77	130	105	102	333	51	334	100	69	69	57	52	48	35	24		15	11	1	–	–	–	–	–	1	–	–	–	–	2-Propen-1-one, 3-(4-iodophenyl)-1-phenyl-	14548-42-6	NI35188
142	334	143	335	65	92	125	114	334	100	46	8	7	7	4	3	3		15	12	–	2	–	6	–	–	–	–	–	–	–	1,3-Bis(3-aminophenyl)hexafluoropropane		LI07407
44	302	242	273	243	258	241	303	334	100	99	99	75	35	22	22	19	12.17	15	18	5	4	–	–	–	–	–	–	–	–	–	Mitomycin C	50-07-7	NI35189
242	273	241	258	230	228	302	54	334	100	65	22	19	12	11	10	10	1.00	15	18	5	4	–	–	–	–	–	–	–	–	–	Mitomycin C	50-07-7	LI07408
57	58	41	59	42	175	43	114	334	100	58	34	17	14	13	13	10	1.00	15	20	4	–	–	–	2	–	–	–	–	–	–	Acetic acid, (2,4-dichlorophenoxy)-, 2-butoxypropyl ester	10463-14-6	NI35190
58	41	29	43	31	27	45	114	334	100	64	42	28	20	20	19	18	0.50	15	20	4	–	–	–	2	–	–	–	–	–	–	Acetic acid, (2,4-dichlorophenoxy)-, 3-butoxypropyl ester	1928-45-6	NI35191
191	57	185	63	41	27	135	59	334	100	89	88	84	63	61	54	49	27.32	15	23	4	–	1	–	1	–	–	–	–	–	–	Sulphurous acid, 2-chloroethyl 2-(4-tert-butylphenoxy)-1-isopropyl ester	140-57-8	NI35192
81	121	93	135	119	95	80	109	334	100	78	47	21	15	14	14	13	0.00	15	24	1	–	–	–	1	1	–	–	–	–	–	Oxirane, 2-bromo-3-chloro-2-(1,5,9-trimethyl-4,8-decadienyl)-	56772-06-6	NI35193
101	143	85	86	129	88	87	130	334	100	82	72	69	68	64	45	44	0.00	15	26	8	–	–	–	–	–	–	–	–	–	–	Methyl 2,3-di-O-acetyl-4,6-di-O-ethyl-α-D-mannopyranoside		NI35194
88	43	89	117	127	130	101	59	334	100	99	87	59	47	38	30	29	0.00	15	26	8	–	–	–	–	–	–	–	–	–	–	Methyl 2,4-di-O-acetyl-3,6-di-O-ethyl-α-D-mannopyranoside		NI35195
89	43	88	101	116	85	115	130	334	100	99	85	62	49	36	27	26	0.00	15	26	8	–	–	–	–	–	–	–	–	–	–	Methyl 2,6-di-O-acetyl-3,4-di-O-ethyl-α-D-mannopyranoside		NI35196
88	43	101	143	130	59	112	127	334	100	99	57	46	41	39	33	29	0.00	15	26	8	–	–	–	–	–	–	–	–	–	–	Methyl 3,4-di-O-acetyl-2,6-di-O-ethyl-α-D-mannopyranoside		NI35197
88	43	130	129	85	101	102	103	334	100	83	69	60	38	32	29	23	0.00	15	26	8	–	–	–	–	–	–	–	–	–	–	Methyl 3,6-di-O-acetyl-2,4-di-O-ethyl-α-D-mannopyranoside		NI35198
116	43	89	88	60	102	74	61	334	100	99	74	55	30	19	17	17	0.00	15	26	8	–	–	–	–	–	–	–	–	–	–	Methyl 4,6-di-O-acetyl-2,3-di-O-ethyl-α-D-mannopyranoside		NI35199
137	98	136	97	234	193	150	84	334	100	84	37	36	27	24	24	18	0.00	15	27	4	2	–	–	1	–	–	–	–	–	–	Sparteine monoperchlorate	14427-91-9	NI35200
121	177	179	29	119	221	41	175	334	100	98	89	85	78	73	66	65	0.00	15	34	–	–	–	–	–	–	–	–	–	–	1	Dibutylsec-butylisopropyltin		MS07940
57	41	43	121	29	55	71	85	334	100	89	85	73	57	56	44	41	0.00	15	34	–	–	–	–	–	–	–	–	–	–	1	Isopropyltributyltin		MS07941
177	121	29	179	119	221	175	277	334	100	99	97	95	79	72	68	67	0.00	15	34	–	–	–	–	–	–	–	–	–	–	1	Propyltributyltin		MS07942
183	303	76	334	104	50	152	136	334	100	34	24	18	16	16	14	14		16	14	6	–	1	–	–	–	–	–	–	–	–	4,4′-Sulphonylbisbenzoic acid, dimethyl ester	3965-53-5	NI35201
165	334	163	164	254	166	127	55	334	100	74	45	40	25	25	25	24		16	15	–	–	–	–	–	–	1	–	–	–	–	Naphthalene, 1-(1-hexynyl)-8-iodo-	22360-84-5	NI35202
152	318	317	166	319	320	334	89	334	100	87	71	63	54	51	49	35		16	16	–	4	–	–	2	–	–	–	–	–	–	3-Chloro-4-methylbenzamide azine		MS07943
171	334	170	168	172	154	169	155	334	100	31	30	21	14	12	8	8		16	18	4	2	1	–	–	–	–	–	–	–	–	Homoserine lactone 1-(dimethylamino)naphthalene-5-sulphonamide		MS07944
217	139	63	58	334	183	107	185	334	100	68	27	26	22	20	18	13		16	19	–	2	2	–	–	–	–	–	1	–	–	1-Diphenylphosphinothioyl-3-propylthiourea		MS07945
66	39	65	132	96	40	186	130	334	100	21	19	8	8	8	7	7	0.41	16	20	1	–	–	–	–	–	–	–	–	–	1	Palladium, (π-cyclopentadienyl)(9-methoxytricyclo[5.2.1.0]dec-4-ene)-		MS07946
81	41	291	108	134	69	55	91	334	100	65	62	50	46	45	45	44	28.00	16	22	4	4	–	–	–	–	–	–	–	–	–	Menthone (2,4-dinitrophenyl)hydrazone	27853-78-7	NI35203
81	291	334	41	55	111	108	69	334	100	86	57	56	45	44	40	35		16	22	4	4	–	–	–	–	–	–	–	–	–	Menthone (2,4-dinitrophenyl)hydrazone	27853-78-7	NI35204
159	41	134	187	107	135	232	69	334	100	73	67	50	42	36	35	35	3.16	16	22	4	4	–	–	–	–	–	–	–	–	–	7H-Pyrrolo[2,3-d]pyrimidin-4-amine, N-(3-methyl-2-butenyl)-7-β-D-ribofuranosyl-	23589-13-1	NI35205
55	109	81	197	91	179	165	123	334	100	80	78	38	35	32	28	25	0.00	16	31	5	–	–	–	–	–	–	–	1	–	–	3-(Diethoxyphosphinyl)-1-oxacyclotridecan-2-one		DD01315
264	223	97	141	55	140	70	41	334	100	42	40	36	32	22	20	19	1.00	16	32	3	–	–	–	–	–	–	–	2	–	–	1,1′-Bis(2,2,3,4,4-pentamethylphosphetan 1-oxide) ether		NI35206
233	58	114	100	234	71	72	56	334	100	85	54	50	37	30	27	27	2.40	16	34	5	2	–	–	–	–	–	–	–	–	–	13-(5-Dimethylamino-3-oxapentyl)-1,4,7,10-tetraoxa-13-azacyclopentadecane		MS07947
250	55	158	56	132	103	93	251	334	100	99	66	24	23	17	13	12	0.99	17	11	4	–	–	–	–	–	–	–	–	–	1	Cinnamoilcymantrene	12184-59-7	NI35207
55	250	77	132	80	251	165	144	334	100	67	17	16	7	6	6	5	0.99	17	11	4	–	–	–	–	–	–	–	–	–	1	(3-Oxo-3-phenyl-1-propenyl)cymantrene		NI35208
142	77	97	95	115	124	215	201	334	100	34	29	28	22	14	10	10	5.00	17	15	–	4	–	–	–	–	–	–	–	–	1	Cobalt, π-cyclopentadienyl(1,4-diphenyltetraazenediyl)-		LI07409
250	220	110	235	292	252	222	334	334	100	62	53	45	35	35	22	21		17	15	5	–	–	–	1	–	–	–	–	–	–	[1,1′-Biphenyl]-3,5-diol, 4′-chloro-4-methoxy-, diacetate	57346-60-8	NI35209
152	228	151	78	334	107	106	153	334	100	17	16	15	13	9	9	8		17	15	5	–	–	–	1	–	–	–	–	–	–	2-[3-(Chloroacetoxy)-4-methoxyphenyl]-1,3-benzodioxane		MS07948
316	334	318	317	336	335	273	319	334	100	64	37	28	22	14	14	13		17	15	5	–	–	–	1	–	–	–	–	–	–	2-Chloro-1,3,6-trimethoxy-8-methyl-9H-xanthen-9-one		MS07949
183	334	185	336	184	147	126	125	334	100	38	35	24	13	13	13	13		17	16	1	2	–	–	2	–	–	–	–	–	–	2,4-Bis(4-chlorophenyl)-5,5-dimethylpyrazolidin-3-one	104908-83-0	NI35210
196	167	334	139	168	197	55	69	334	100	42	35	28	20	14	13	10		17	18	7	–	–	–	–	–	–	–	–	–	–	2H-1-Benzopyran-3,5,7-triol, 3,4-dihydro-2-(4-hydroxy-3,5-dimethoxyphenyl)-, cis-(±)-	39689-30-0	MS07950

MASS TO CHARGE RATIOS								M.W.	INTENSITIES								Parent	C	H	O	N	S	F	Cl	Br	I	Si	P	B	X	COMPOUND NAME	CAS Reg No	No
232	217	202	59	189	174	334	43	**334**	100	73	68	26	17	17	14	14		17	18	7	–	–	–	–	–	–	–	–	–	–	**7H-Furo[3,2-g][1]benzopyran-7-one, 9-(2,3-dihydroxy-3-methylbutoxy)-4-methoxy-**	88548-17-8	**NI35211**
232	59	217	43	31	233	42	189	**334**	100	74	71	40	29	21	15	14	**8.00**	17	18	7	–	–	–	–	–	–	–	–	–	–	**7H-Furo[3,2-g][1]benzopyran-7-one, 9-(2,3-dihydroxy-3-methylbutoxy)-4-methoxy-, (R)-**	482-25-7	**LI07410**
43	135	163	162	149	139	334	177	**334**	100	62	25	23	21	17	15	13		17	18	7	–	–	–	–	–	–	–	–	–	–	**Zeylanidine**	13761-06-3	**MS07951**
58	86	42	334	220	87	185	59	**334**	100	74	11	6	6	6	5	5		17	19	1	2	1	–	1	–	–	–	–	–	–	**3-Hydroxychlorpromazine**	3930-47-0	**NI35212**
58	86	42	334	220	87	59	44	**334**	100	78	12	6	5	5	5	4		17	19	1	2	1	–	1	–	–	–	–	–	–	**3-Hydroxychlorpromazine**	3930-47-0	**NI35213**
58	86	334	42	185	248	220	41	**334**	100	78	47	27	18	17	17	17		17	19	1	2	1	–	1	–	–	–	–	–	–	**3-Hydroxychlorpromazine**	3930-47-0	**NI35214**
58	86	334	85	44	336	248	42	**334**	100	17	11	8	6	5	4	4		17	19	1	2	1	–	1	–	–	–	–	–	–	**6-Hydroxychlorpromazine**	3926-65-6	**NI35215**
58	86	334	42	248	85	44	59	**334**	100	45	25	19	14	11	11	10		17	19	1	2	1	–	1	–	–	–	–	–	–	**7-Hydroxychlorpromazine**	2095-62-7	**NI35216**
58	86	334	78	85	42	59	44	**334**	100	28	12	9	8	6	4	4		17	19	1	2	1	–	1	–	–	–	–	–	–	**8-Hydroxychlorpromazine**	3926-67-8	**NI35219**
58	86	334	42	85	59	248	44	**334**	100	42	26	18	11	10	9	9		17	19	1	2	1	–	1	–	–	–	–	–	–	**8-Hydroxychlorpromazine**	3926-67-8	**NI35217**
58	86	334	42	336	85	44	243	**334**	100	45	28	18	11	11	10	9		17	19	1	2	1	–	1	–	–	–	–	–	–	**8-Hydroxychlorpromazine**	3926-67-8	**NI35218**
141	158	140	115	129	69	334	336	**334**	100	64	26	12	11	11	10	10		17	19	2	–	–	–	–	1	–	–	–	–	–	**1-Naphthalenemethyl 6-bromohexanoate**	86328-58-7	**NI35220**
41	43	56	57	42	127	199	45	**334**	100	98	93	79	73	67	62	59	**6.00**	17	19	3	2	–	–	1	–	–	–	–	–	–	**4′-Chloro-3-[2-(diethylamino)-4-oxo-1-cyclobutenyl]-3-oxopropananilide**		**NI35221**
171	334	170	168	172	154	306	316	**334**	100	45	33	15	14	8	5	3		17	22	3	2	1	–	–	–	–	–	–	–	–	**Dansyl-5-aminopentanal**		**NI35222**
73	102	233	232	45	234	39	334	**334**	100	76	63	62	14	13	7	6		17	30	1	2	–	–	–	–	–	2	–	–	–	**1-(Trimethylsilyl)-3-[(trimethylsilyl)aminoethyl]-5-methoxyindole**		**NI35223**
41	299	288	253	289	301	334	255	**334**	100	15	12	10	8	7	5	4		18	16	2	–	–	–	2	–	–	–	–	–	–	**2,2-Bis(4-chlorophenyl)-3,3-dimethylcyclopropanecarboxylic acid**		**NI35224**
99	56	165	42	44	58	43	262	**334**	100	82	51	43	40	34	34	31	**14.00**	18	20	–	2	–	–	2	–	–	–	–	–	–	**1-[α-(2-Chlorophenyl)-3-chlorobenzyl]-4-methylpiperazine**		**MS07952**
188	161	55	189	204	162	110	176	**334**	100	56	42	30	24	21	19	14	**0.50**	18	22	6	–	–	–	–	–	–	–	–	–	–	**1H-2-Benzoxacyclotetradecin-1,7(8H)-dione, 3,4,5,6,9,10-hexahydro-5,14,16-trihydroxy-3-methyl-, [3S-(3R*,5R*,11E)]-**	40785-65-7	**NI35225**
188	161	189	204	176	162	110	71	**334**	100	51	37	30	23	23	20	16	**6.90**	18	22	6	–	–	–	–	–	–	–	–	–	–	**1H-2-Benzoxacyclotetradecin-1,7(8H)-dione, 3,4,5,6,9,10-hexahydro-5,14,16-trihydroxy-3-methyl-, [3S-(3R*,5R*,11E)]-**	40785-65-7	**NI35226**
331	316	28	332	167	317	152	137	**334**	100	46	23	20	20	9	6	5	**0.00**	18	22	6	–	–	–	–	–	–	–	–	–	–	**1,1′-Biphenyl, 3,3′,4,4′,5,5′-hexamethoxy-**	56772-00-0	**NI35227**
137	109	136	334	110	65	81	121	**334**	100	64	56	48	29	25	24	18		18	22	6	–	–	–	–	–	–	–	–	–	–	**1,8-Bis(2-hydroxyphenoxy)-3,6-dioxaoctane**		**MS07953**
83	192	177	119	105	203	161	149	**334**	100	21	18	9	9	8	8	8	**4.50**	18	22	6	–	–	–	–	–	–	–	–	–	–	**6-Methoxy-10-acetoxy-8,9-epoxythymol 3-angelate**		**NI35228**
204	334	177	99	70	183	115	42	**334**	100	53	43	31	29	26	24	23		18	23	2	2	–	–	1	–	–	–	–	–	–	**trans-2-(3-Chloro-2-indolyl)-5-ethyl-1-methyl-4-piperidone ethylene acetal**		**DD01316**
334	225	335	269	139	223	197	141	**334**	100	39	35	24	13	10	9	8		19	18	2	–	–	–	–	–	–	–	–	–	1	**Ethyl 2-ferrocenylbenzoate**		**MS07954**
43	41	55	91	316	119	288	105	**334**	100	80	78	56	50	42	41	39	**1.00**	19	26	5	–	–	–	–	–	–	–	–	–	–	**2β,3β,14α-Trihydroxy-5α-androst-7-ene-6,17-dione**		**MS07955**
316	132	117	288	160	91	41	55	**334**	100	64	56	47	45	40	39	34	**16.40**	19	26	5	–	–	–	–	–	–	–	–	–	–	**2β,3β,14α-Trihydroxy-5β-androst-7-ene-6,17-dione**		**MS07956**
209	334	179	277	210	319	223	304	**334**	100	37	24	16	13	12	9	7		19	30	3	–	–	–	–	–	–	1	–	–	–	**4-Nonen-3-one, 1-[3-methoxy-4-[(trimethylsilyl)oxy]phenyl]-**	56700-91-5	**MS07957**
248	235	264	189	276	191	292	176	**334**	100	67	44	40	37	35	32	28	**26.00**	20	14	5	–	–	–	–	–	–	–	–	–	–	**O-Acetylprzewaquinone B**		**NI35229**
255	253	254	336	334	252	114	126	**334**	100	63	60	51	51	50	47	45		20	15	–	–	–	–	–	1	–	–	–	–	–	**2-(2-Bromophenyl)-1,1-diphenylethylene**		**MS07958**
255	254	253	334	336	252	126	239	**334**	100	89	89	61	60	60	49	29		20	15	–	–	–	–	–	1	–	–	–	–	–	**2-(2-Bromophenyl)-1,1-diphenylethylene**		**LI07411**
334	336	255	253	254	252	178	126	**334**	100	99	69	60	59	57	46	36		20	15	–	–	–	–	–	1	–	–	–	–	–	**2-(3-Bromophenyl)-1,1-diphenylethylene**		**LI07412**
334	336	255	253	254	252	178	126	**334**	100	99	68	59	58	49	43	36		20	15	–	–	–	–	–	1	–	–	–	–	–	**2-(3-Bromophenyl)-1,1-diphenylethylene**		**MS07959**
334	336	255	253	254	252	178	126	**334**	100	98	65	59	56	48	43	36		20	15	–	–	–	–	–	1	–	–	–	–	–	**2-(4-Bromophenyl)-1,1-diphenylethylene**		**MS07960**
255	126	252	334	253	239	120	336	**334**	100	42	34	33	33	33	33	30		20	15	–	–	–	–	–	1	–	–	–	–	–	**Bromotriphenylethylene**	1607-57-4	**NI35230**
290	291	160	334	147	115	292	234	**334**	100	80	61	48	32	25	22	21		20	18	1	2	1	–	–	–	–	–	–	–	–	**15-Hydroxy-6,7,8,9,10,15-hexahydro[d][1]benzothieno[2,3-g]azecine-8-carbonitrile**		**NI35231**
317	301	334	315	318	237	316	101	**334**	100	53	40	33	26	25	22	22		20	18	1	2	1	–	–	–	–	–	–	–	–	**2-(Methylthio)-4-(4-methylphenyl)-6-phenylpyridine-3-carboxamide**	90732-67-5	**NI35232**
334	319	91	77	335	131	320	159	**334**	100	91	69	32	22	21	20	18		20	22	1	4	–	–	–	–	–	–	–	–	–	**3H-Pyrazol-3-one, 4-[[4-(diethylamino)phenyl]imino]-2,4-dihydro-5-methyl-2-phenyl-**	4595-01-1	**MS07961**
334	319	91	77	335	131	320	250	**334**	100	90	69	32	23	21	20	16		20	22	1	4	–	–	–	–	–	–	–	–	–	**3H-Pyrazol-3-one, 4-[[4-(diethylamino)phenyl]imino]-2,4-dihydro-5-methyl-2-phenyl-**	4595-01-1	**NI35233**
334	292	127	274	179	178	132	117	**334**	100	79	56	46	32	31	28	27		20	27	3	–	–	1	–	–	–	–	–	–	–	**Estra-5(10)-en-6-one, 3β-fluoro-17β-hydroxy-, acetate**	33547-52-3	**LI07413**
334	43	292	127	274	179	132	233	**334**	100	82	79	56	46	32	28	27		20	27	3	–	–	1	–	–	–	–	–	–	–	**Estra-5(10)-en-6-one, 3β-fluoro-17β-hydroxy-, acetate**	33547-52-3	**NI35234**
334	274	292	128	147	272	215	133	**334**	100	99	87	75	35	27	25	21		20	27	3	–	–	1	–	–	–	–	–	–	–	**Estr-4-en-3-one, 6α-fluoro-17β-hydroxy-, acetate**	3801-28-3	**LI07414**
274	334	128	292	214	147	213	133	**334**	100	55	53	36	32	23	13	12		20	27	3	–	–	1	–	–	–	–	–	–	–	**Estr-4-en-3-one, 6α-fluoro-17β-hydroxy-, acetate**	3801-28-3	**NI35235**
43	334	274	128	292	83	147	41	**334**	100	99	99	75	68	38	35	28		20	27	3	–	–	1	–	–	–	–	–	–	–	**Estr-4-en-3-one, 6α-fluoro-17β-hydroxy-, acetate**	3801-28-3	**NI35236**
274	334	292	128	147	272	215	133	**334**	100	99	87	75	35	27	25	21		20	27	3	–	–	1	–	–	–	–	–	–	–	**Estr-4-en-3-one, 6β-fluoro-17β-hydroxy-, acetate**	18119-95-4	**NI35238**
274	43	334	128	292	314	147	91	**334**	100	71	56	53	36	32	23	19		20	27	3	–	–	1	–	–	–	–	–	–	–	**Estr-4-en-3-one, 6β-fluoro-17β-hydroxy-, acetate**	18119-95-4	**NI35237**
274	334	128	292	314	147	213	249	**334**	100	55	53	36	32	23	13	10		20	27	3	–	–	1	–	–	–	–	–	–	–	**Estr-4-en-3-one, 6β-fluoro-17β-hydroxy-, acetate**	18119-95-4	**LI07415**
316	43	41	334	55	317	79	69	**334**	100	41	38	33	27	22	16	16		20	30	4	–	–	–	–	–	–	–	–	–	–	**Acetic acid, (dodecahydro-7-hydroxy-1,4b,8,8-tetramethyl-10-oxo-2(1H)-phenanthrenylidene)-, [1R-(1α,2E,4aα,4bβ,7β,8aα,10aβ)]-**	468-74-6	**NI35239**
107	255	270	301	316	213	334	199	**334**	100	93	80	65	64	57	56	38		20	30	4	–	–	–	–	–	–	–	–	–	–	**Androstan-11-ol-17-one, 3-formoxy-, (3α,5α,11β)-**		**NI35240**

MASS TO CHARGE RATIOS								M.W.	INTENSITIES								Parent	C	H	O	N	S	F	Cl	Br	I	Si	P	B	X	COMPOUND NAME	CAS Reg No	No
41	334	270	255	301	213	226	316	**334**	100	74	65	53	52	51	42	38		20	30	4	–	–	–	–	–	–	–	–	–	–	Androstan-11-ol-17-one, 3-formoxy-, (3α,5β,11β)-		NI35241
334	279	289	233	335	280	41	177	**334**	100	89	40	29	24	17	14	13		20	30	4	–	–	–	–	–	–	–	–	–	–	2H,6H-Benzo[1,2-b:5,4-b′]dipyran-10-propanol, 3,4,7,8-tetrahydro-5-methoxy-2,2,8,8-tetramethyl-	26537-39-3	NI35242
334	279	289	233	335	280	41	177	**334**	100	90	40	29	23	15	14	12		20	30	4	–	–	–	–	–	–	–	–	–	–	2H,6H-Benzo[1,2-b:5,4-b′]dipyran-10-propanol, 3,4,7,8-tetrahydro-5-methoxy-2,2,8,8-tetramethyl-	26537-39-3	NI35243
334	279	289	233	335	205	41	177	**334**	100	72	65	33	25	25	22	15		20	30	4	–	–	–	–	–	–	–	–	–	–	2H,8H-Benzo[1,2-b:3,4-b′]dipyran-6-propanol, 3,4,9,10-tetrahydro-5-methoxy-2,2,8,8-tetramethyl-	26537-40-6	NI35245
334	279	289	233	335	205	41	177	**334**	100	72	65	34	24	22	22	15		20	30	4	–	–	–	–	–	–	–	–	–	–	2H,8H-Benzo[1,2-b:3,4-b′]dipyran-6-propanol, 3,4,9,10-tetrahydro-5-methoxy-2,2,8,8-tetramethyl-	26537-40-6	NI35244
41	69	57	43	247	264	29	55	**334**	100	91	41	34	29	23	23	22	**13.47**	20	30	4	–	–	–	–	–	–	–	–	–	–	1,2,4-Cyclopentanetrione, 3-isopentyl-5-isovaleryl-3-(3-methyl-2-butenyl)-	1891-39-0	NI35246
105	155	43	57	333	41	71	100	**334**	100	86	56	45	44	41	39	37	**27.64**	20	30	4	–	–	–	–	–	–	–	–	–	–	Decanoic acid, 2-phenyl-1,3-dioxan-5-yl ester	56630-72-9	NI35247
124	316	137	301	92	203	79	81	**334**	100	50	35	29	29	27	23	22	**0.00**	20	30	4	–	–	–	–	–	–	–	–	–	–	1-Naphthalenecarboxylic acid, 5-(4-carboxy-3-methyl-3-butenyl)-3,4,4a,5,6,7,8,8a-octahydro-5,6,8a-trimethyl-	55252-88-5	NI35248
44	121	81	275	41	107	93	91	**334**	100	93	83	64	53	49	48	47	**0.41**	20	30	4	–	–	–	–	–	–	–	–	–	–	1-Naphthalenecarboxylic acid, 5-(4-carboxy-3-methyl-3-butenyl)decahydro-1,4a-dimethyl-6-methylene-, [1S-[1α,4aα,5α(E),8aβ]]-	640-28-8	NI35249
149	43	85	84	150	57	55	44	**334**	100	28	22	13	11	9	9	9	**0.00**	20	30	4	–	–	–	–	–	–	–	–	–	–	Phthalic acid, bis(2-ethylbutyl) ester	7299-89-0	LI07416
149	43	85	167	84	150	41	44	**334**	100	28	22	15	13	11	10	9	**0.00**	20	30	4	–	–	–	–	–	–	–	–	–	–	Phthalic acid, bis(2-ethylbutyl) ester	7299-89-0	WI01439
149	43	28	85	167	84	150	41	**334**	100	28	28	22	15	13	11	10	**0.00**	20	30	4	–	–	–	–	–	–	–	–	–	–	Phthalic acid, bis(2-ethylbutyl) ester	7299-89-0	DO01103
149	41	43	29	150	223	57	55	**334**	100	17	10	10	9	7	6	5	**0.20**	20	30	4	–	–	–	–	–	–	–	–	–	–	Phthalic acid, butyl 2-ethylhexyl ester	85-69-8	NI35250
149	41	150	76	104	56	65	43	**334**	100	18	10	10	8	5	4	4	**0.00**	20	30	4	–	–	–	–	–	–	–	–	–	–	Phthalic acid, butyl 2-ethylhexyl ester	85-69-8	NI35251
149	41	150	43	223	57	29	55	**334**	100	16	9	9	8	7	7	6	**0.40**	20	30	4	–	–	–	–	–	–	–	–	–	–	Phthalic acid, butyl octyl ester	84-78-6	NI35252
149	41	43	150	57	55	56	76	**334**	100	22	15	10	8	7	6	5	**0.00**	20	30	4	–	–	–	–	–	–	–	–	–	–	Phthalic acid, butyl octyl ester	84-78-6	NI35253
104	149	76	41	43	39	50	27	**334**	100	96	91	79	68	55	53	45	**1.00**	20	30	4	–	–	–	–	–	–	–	–	–	–	Phthalic acid, dihexyl ester	84-75-3	NI35254
149	43	85	150	41	251	56	84	**334**	100	21	19	9	8	5	5	5	**0.00**	20	30	4	–	–	–	–	–	–	–	–	–	–	Phthalic acid, diisohexyl ester	146-50-9	NI35255
190	185	245	316	191	217	246	183	**334**	100	29	23	21	20	17	16	13	**1.87**	20	30	4	–	–	–	–	–	–	–	–	–	–	Prosta-5,10,13-trien-1-oic acid, 15-hydroxy-9-oxo-, (5Z,13E,15S)-	13345-50-1	NI35256
195	177	71	150	85	57	43	55	**334**	100	16	16	15	15	15	14	12	**2.00**	20	30	4	–	–	–	–	–	–	–	–	–	–	Terephthalic acid, decyl ethyl ester	54699-33-1	NI35257
334	335	84	336	85	29	249	292	**334**	100	42	38	8	6	6	3	2		20	30	4	–	–	–	–	–	–	–	–	–	–	Terephthalic acid, dihexyl ester		LI07418
43	55	41	56	91	42	29	28	**334**	100	79	70	69	65	47	37	35	**0.00**	20	30	4	–	–	–	–	–	–	–	–	–	–	Terephthalic acid, dihexyl ester		DO01104
43	55	41	56	91	42	29	85	**334**	100	79	70	69	65	47	37	26	**0.10**	20	30	4	–	–	–	–	–	–	–	–	–	–	Terephthalic acid, dihexyl ester		LI07417
91	59	69	107	179	92	98	83	**334**	100	31	19	17	12	10	8	7	**0.00**	20	30	4	–	–	–	–	–	–	–	–	–	–	(±)-2,4,6-Trideoxy-2,4,6-trimethyl-3,5-O-(1-methylethylidene)-7-O-(phenylmethyl)-D-glycero-D-talo-heptose		NI35258
137	305	167	91	306	121	151	179	**334**	100	99	44	27	25	23	21	18	**1.13**	20	34	2	–	–	–	–	–	–	1	–	–	–	Silane, ethyl(2-methoxyphenyl)methyl[[5-methyl-2-isopropylcyclohexyl]oxy]-, (1α,2β,5α)-	74842-22-1	NI35259
149	105	77	150	51	334	106	115	**334**	100	42	23	10	4	3	3	2		21	18	4	–	–	–	–	–	–	–	–	–	–	2-(4-Phenoxyphenoxy)ethyl benzoate		DO01105
320	243	211	321	212	244	160	197	**334**	100	83	52	25	17	16	11	7		21	19	–	–	1	–	–	–	–	–	1	–	–	10H-Phenothiaphosphine, 2,8-dimethyl-10-(4-methylphenyl)-	23861-40-7	LI07419
334	211	243	335	290	212	244	227	**334**	100	58	57	25	16	16	12	9		21	19	–	–	1	–	–	–	–	–	1	–	–	10H-Phenothiaphosphine, 2,8-dimethyl-10-(4-methylphenyl)-	23861-40-7	NI35260
277	278	199	201	77	51	183	152	**334**	100	35	18	12	11	8	7	7	**0.00**	21	19	2	–	–	–	–	–	–	–	1	–	–	Acetic acid, (triphenylphosphoranylidene)-, methyl ester	2605-67-6	NI35261
301	183	185	165	277	77	51	302	**334**	100	72	54	54	47	42	42	38	**10.00**	21	19	2	–	–	–	–	–	–	–	1	–	–	Acetic acid, (triphenylphosphoranylidene)-, methyl ester	2605-67-6	LI07420
306	334	77	242	305	247	273	335	**334**	100	97	77	38	38	28	26	26		21	22	–	2	1	–	–	–	–	–	–	–	–	2,3,5,6,7,8,9-Heptahydro-1-phenyl-5-(phenylimino)-1H-benzo[e][1,4]thiazepine		NI35262
28	54	43	57	84	41	55	71	**334**	100	80	44	37	25	24	22	17	**3.00**	21	22	2	2	–	–	–	–	–	–	–	–	–	2,2-Bis[4-(2-cyano)phenyl]propane		IC04009
57	55	319	69	71	67	83	121	**334**	100	89	67	61	56	46	42	39	**31.00**	21	22	2	2	–	–	–	–	–	–	–	–	–	(Z,Z)-4-Ethyl-3-methyl-5-[5-(4-hydroxyphenyl-2-methylene)-3,4-dimethyl-5H-pyrrolyl-2-methylene]-3-pyrrolin-2-one		MS07962
334	44	120	77	41	107	55	144	**334**	100	44	37	35	34	33	33	32		21	22	2	2	–	–	–	–	–	–	–	–	–	Strychnine	57-24-9	NI35263
334	120	130	107	162	335	143	161	**334**	100	32	30	27	26	25	23	21		21	22	2	2	–	–	–	–	–	–	–	–	–	Strychnine	57-24-9	MS07963
335	334	336	71	69	61	87	220	**334**	100	21	19	11	10	9	8	5		21	22	2	2	–	–	–	–	–	–	–	–	–	Strychnine	57-24-9	NI35264
44	81	91	227	67	274	55	60	**334**	100	68	64	48	45	41	40	37	**1.00**	21	34	3	–	–	–	–	–	–	–	–	–	–	13α-Acetoxy-3β-methoxy-C-homo-D-dinorandrostane		LI07421
274	43	148	107	134	79	91	242	**334**	100	40	30	30	25	14	11	10	**0.00**	21	34	3	–	–	–	–	–	–	–	–	–	–	8-(2-Acetoxypropyl)-2-methoxy-4a-methyl-1,2,3,4,4a,4b,5,6,8a,9,10,10a-dodecahydrophenanthrene		LI07422
316	334	215	317	301	319	233	303	**334**	100	38	26	23	20	14	14	10		21	34	3	–	–	–	–	–	–	–	–	–	–	Androstane-17-carboxylic acid, 3-hydroxy-, methyl ester, (3β,5α,17β)-	10002-85-4	NI35265
334	233	215	335	319	248	208	316	**334**	100	41	35	22	22	14	14	10		21	34	3	–	–	–	–	–	–	–	–	–	–	Androstane-17-carboxylic acid, 3-hydroxy-, methyl ester, (3β,5β,17β)-	10002-84-3	NI35266
101	85	84	127	301	303	41	79	**334**	100	36	32	26	24	14	11	9	**7.00**	21	34	3	–	–	–	–	–	–	–	–	–	–	5α-Androstane-3,17-dione, 3-(dimethylacetal)	3591-19-3	NI35267
141	99	334	230	156	86	113	67	**334**	100	67	40	40	29	29	24	22		21	34	3	–	–	–	–	–	–	–	–	–	–	Androstan-7-one, 3-hydroxy-, cyclic 1,2-ethanediyl acetal, (3β,5α)-	54498-55-4	NI35268
141	99	41	334	230	55	156	86	**334**	100	67	40	39	39	32	29	29		21	34	3	–	–	–	–	–	–	–	–	–	–	Androstan-7-one, 3-hydroxy-, cyclic 1,2-ethanediyl acetal, (3β,5α)-	54498-55-4	NI35269
272	45	108	334	273	107	93	81	**334**	100	93	34	28	28	28	26	26		21	34	3	–	–	–	–	–	–	–	–	–	–	Androstan-17-one, 3-(methoxymethoxy)-, (3α)-	57156-78-2	NI35270

MASS TO CHARGE RATIOS								M.W.	INTENSITIES								Parent	C	H	O	N	S	F	Cl	Br	I	Si	P	B	X	COMPOUND NAME	CAS Reg No	No
45	122	272	108	107	41	67	55	**334**	100	61	47	22	22	22	21	21	**18.00**	21	34	3	–	–	–	–	–	–	–	–	–	–	Androstan-17-one, 3-(methoxymethoxy)-, (3β)-	57156-81-7	NI35271
152	153	135	334	107	154	136	109	**334**	100	30	23	17	6	4	4	4		21	34	3	–	–	–	–	–	–	–	–	–	–	Benzoic acid, 3-methoxy-, tridecyl ester	69833-37-0	NI35272
168	336	139	167	210	152	122	151	**334**	100	78	63	54	20	14	13	12	**0.00**	21	34	3	–	–	–	–	–	–	–	–	–	–	α-Dodecyl-3,5-dimethoxybenzyl alcohol		LI07423
70	41	55	43	81	95	57	69	**334**	100	83	72	61	55	43	38	33	**3.00**	21	34	3	–	–	–	–	–	–	–	–	–	–	2(3H)-Furanone, 3-(15-hexadecynylidene)dihydro-4-hydroxy-5-methyl-, [4R-(3E,4α,5β)]-	71339-54-3	NI35273
70	41	43	81	69	55	95	57	**334**	100	70	60	40	30	30	25	22	**3.00**	21	34	3	–	–	–	–	–	–	–	–	–	–	2(3H)-Furanone, 3-(15-hexadecynylidene)dihydro-4-hydroxy-5-methyl-, [4R-(3Z,4α,5β)]-	71339-53-2	NI35274
70	41	55	43	81	57	95	69	**334**	100	99	76	74	68	65	51	45	**6.00**	21	34	3	–	–	–	–	–	–	–	–	–	–	2(3H)-Furanone, 3-(15-hexadecynylidene)dihydro-4-hydroxy-5-methyl-, [4S-(3E,4α,5α)]-	71325-96-7	NI35275
70	43	41	57	55	81	69	95	**334**	100	75	70	60	52	50	28	26	**4.00**	21	34	3	–	–	–	–	–	–	–	–	–	–	2(3H)-Furanone, 3-(15-hexadecynylidene)dihydro-4-hydroxy-5-methyl-, [4S-(3Z,4α,5α)]-	71325-95-6	NI35276
70	57	43	41	81	83	69	55	**334**	100	91	81	77	64	61	58	58	**3.00**	21	34	3	–	–	–	–	–	–	–	–	–	–	(3R,2E)-2-(Hexadec-15-enylidene)-3-hydroxy-4-methylenebutanolide		NI35277
70	41	55	43	69	57	71	83	**334**	100	99	83	53	52	45	33	23	**2.00**	21	34	3	–	–	–	–	–	–	–	–	–	–	(3R,2Z)-2-(Hexadec-15-enylidene)-3-hydroxy-4-methylenebutanolide		NI35278
121	316	109	81	123	107	92	93	**334**	100	71	60	60	55	50	47	43	**18.02**	21	34	3	–	–	–	–	–	–	–	–	–	–	Kauran-18-oic acid, 16-hydroxy-, methyl ester, (4α)-	22376-08-5	NI35279
43	94	121	257	316	241	187	273	**334**	100	98	81	61	60	56	54	40	**4.00**	21	34	3	–	–	–	–	–	–	–	–	–	–	Kauranoic acid, 16α-hydroxy-, methyl ester	94992-34-4	NI35280
257	275	196	107	183	121	95	109	**334**	100	76	76	64	63	61	59	54	**27.32**	21	34	3	–	–	–	–	–	–	–	–	–	–	Methyl 1,11-epoxy-abietanoate		NI35281
43	121	58	44	41	55	289	39	**334**	100	33	18	17	15	11	10	9	**0.00**	21	34	3	–	–	–	–	–	–	–	–	–	–	Methyl 15-hydroxypimerate		NI35282
121	99	161	109	81	177	107	95	**334**	100	96	74	61	61	57	57	57	**4.34**	21	34	3	–	–	–	–	–	–	–	–	–	–	Methyl 14-oxo-13,14-dihydrophotolevopimerate	69745-80-8	NI35283
82	205	161	163	189	206	119	95	**334**	100	62	37	36	25	23	21	21	**4.00**	21	34	3	–	–	–	–	–	–	–	–	–	–	Methyl rel-(5S,8R,9S,10R)-2-oxo-ent-cleroda-3,13-dien-15-oate		MS07964
334	109	219	335	135	95	178	129	**334**	100	36	30	25	22	20	19	17		21	34	3	–	–	–	–	–	–	–	–	–	–	1-Naphthalenepentanoic acid, 1,4,4a,5,6,7,8,8a-octahydro-β,2,5,5,8a-pentamethyl-4-oxo-, methyl ester, [1S-[1α(R*),4aβ,8aα]]-	1438-56-8	NI35284
55	81	319	28	121	67	29	95	**334**	100	86	75	71	50	48	44	38	**1.20**	21	34	3	–	–	–	–	–	–	–	–	–	–	1H-Naphtho[2,1-b]pyran-7-carboxylic acid, 3-vinyldodecahydro-3,4a,7,10a-tetramethyl-, methyl ester, [3R-(3α,4aβ,6aα,7β,10aβ,10bα)]-	56687-75-3	NI35285
43	319	55	81	41	67	121	236	**334**	100	64	64	62	58	36	30	28	**0.70**	21	34	3	–	–	–	–	–	–	–	–	–	–	1H-Naphtho[2,1-b]pyran-7-carboxylic acid, 3-vinyldodecahydro-3,4a,7,10a-tetramethyl-, methyl ester, [3S-(3α,4aα,6aβ,7α,10aα,10bβ)]-	56687-76-4	NI35286
114	303	95	81	107	121	109	123	**334**	100	81	67	67	65	59	58	57	**17.55**	21	34	3	–	–	–	–	–	–	–	–	–	–	2-Pentenoic acid, 5-[decahydro-5-(hydroxymethyl)-5,8a-dimethyl-2-methylene-1-naphthalenyl]-3-methyl-, methyl ester, [1S-[1α(E),4aβ,5α,8aα]]-	1757-87-5	NI35287
121	239	289	91	109	123	79	77	**334**	100	58	56	43	41	31	31	31	**0.00**	21	34	3	–	–	–	–	–	–	–	–	–	–	1-Phenanthrenecarboxylic acid, 1,2,3,4,4a,4b,5,6,8a,9,10,10a-dodecahydro 8a-hydroxy-1,4a-dimethyl-7-isopropyl-, methyl ester, [1R-(1α,4aβ,4bα,8aα,10aα)]-	69765-29-3	NI35288
247	56	43	109	95	81	124	68	**334**	100	87	82	73	68	66	53	52	**8.00**	21	34	3	–	–	–	–	–	–	–	–	–	–	1-Phenanthrenecarboxylic acid, 7-ethyl-1,4a,7-trimethyl-9-oxoperhydro-, methyl ester		LI07424
146	223	187	121	134	123	163	133	**334**	100	54	47	44	32	30	29	28	**3.20**	21	34	3	–	–	–	–	–	–	–	–	–	–	8α-Podocarp-12-en-15-oic acid, 8-hydroxy-13-isopropyl-, methyl ester	25859-61-4	NI35289
69	83	316	85	71	81	60	73	**334**	100	78	77	56	56	55	49	45	**0.00**	21	34	3	–	–	–	–	–	–	–	–	–	–	Pregnan-18-oic acid, 20-hydroxy-, (5α)-	56143-33-0	NI35290
221	316	262	301	208	334	175	107	**334**	100	77	73	69	50	48	46	46		21	34	3	–	–	–	–	–	–	–	–	–	–	Pregnan-11-one, 3,20-dihydroxy-, (3α,5β,20β)-		NI35291
298	213	316	301	95	283	159	255	**334**	100	62	57	51	50	48	44	39	**0.00**	21	34	3	–	–	–	–	–	–	–	–	–	–	Pregnan-20-one, 3,6-dihydroxy-, (3α,5β,6α)-		NI35292
43	95	41	55	81	67	79	71	**334**	100	46	41	39	32	29	23	23	**0.50**	21	34	3	–	–	–	–	–	–	–	–	–	–	Pregnan-20-one, 3,6-dihydroxy-, (3α,5β,6α)-		IC04010
43	316	81	55	95	41	107	67	**334**	100	38	32	30	26	24	21	20	**0.00**	21	34	3	–	–	–	–	–	–	–	–	–	–	Pregnan-20-one, 3,11-dihydroxy-, (3α,5α,11α)-	38398-44-6	NI35293
316	43	81	55	107	301	41	95	**334**	100	84	26	25	23	22	22	20	**0.60**	21	34	3	–	–	–	–	–	–	–	–	–	–	Pregnan-20-one, 3,11-dihydroxy-, (3α,5α,11β)-	23930-29-2	NI35294
43	124	316	81	106	298	107	95	**334**	100	95	60	35	32	31	29	27	**4.53**	21	34	3	–	–	–	–	–	–	–	–	–	–	Pregnan-20-one, 3,11-dihydroxy-, (3α,5β,11α)-	600-52-2	NI35295
298	213	283	255	85	147	145	107	**334**	100	62	51	50	37	33	32	32	**0.00**	21	34	3	–	–	–	–	–	–	–	–	–	–	Pregnan-20-one, 3,11-dihydroxy-, (3α,5β,11β)-	565-92-4	NI35297
43	298	81	316	55	41	95	93	**334**	100	71	37	33	30	26	24	22	**0.75**	21	34	3	–	–	–	–	–	–	–	–	–	–	Pregnan-20-one, 3,11-dihydroxy-, (3α,5β,11β)-	565-92-4	NI35296
43	316	298	81	107	55	41	95	**334**	100	64	42	34	28	28	28	26	**0.00**	21	34	3	–	–	–	–	–	–	–	–	–	–	Pregnan-20-one, 3,11-dihydroxy-, (3β,5α,11α)-	565-91-3	NI35298
316	43	81	55	107	41	93	95	**334**	100	77	31	29	27	26	25	24	**0.46**	21	34	3	–	–	–	–	–	–	–	–	–	–	Pregnan-20-one, 3,11-dihydroxy-, (3β,5α,11β)-	565-89-9	NI35299
43	124	31	298	81	41	107	55	**334**	100	65	41	29	29	25	24	24	**1.76**	21	34	3	–	–	–	–	–	–	–	–	–	–	Pregnan-20-one, 3,11-dihydroxy-, (3β,5β,11α)-	52340-98-4	NI35300
43	316	81	41	107	93	95	301	**334**	100	86	34	29	27	26	24	22	**0.35**	21	34	3	–	–	–	–	–	–	–	–	–	–	Pregnan-20-one, 3,11-dihydroxy-, (3β,5β,11β)-	48200-75-5	NI35301
316	273	249	246	317	250	155	274	**334**	100	68	47	35	23	23	15	14	**13.40**	21	34	3	–	–	–	–	–	–	–	–	–	–	Pregnan-20-one, 3,14-dihydroxy-, (3β,5α,14β)-		LI07425
306	230	248	43	316	249	98	56	**334**	100	83	79	76	64	52	33	33	**30.00**	21	34	3	–	–	–	–	–	–	–	–	–	–	Pregnan-20-one, 3,14-dihydroxy-, (3β,14β)-		LI07426
43	255	55	229	215	107	81	93	**334**	100	59	47	45	44	41	40	37	**21.68**	21	34	3	–	–	–	–	–	–	–	–	–	–	Pregnan-20-one, 3,17-dihydroxy-, (3α)-	6609-97-8	NI35302
230	215	255	147	121	135	107	273	**334**	100	89	71	71	64	63	62	61	**28.41**	21	34	3	–	–	–	–	–	–	–	–	–	–	Pregnan-20-one, 3,17-dihydroxy-, (3α,5α,17α)-		NI35303
230	215	231	135	255	121	217	107	**334**	100	59	45	42	37	36	32	30	**10.76**	21	34	3	–	–	–	–	–	–	–	–	–	–	Pregnan-20-one, 3,17-dihydroxy-, (3α,5β,17α)-		NI35304
91	79	93	55	67	107	81	230	**334**	100	99	98	94	83	80	76	71	**12.70**	21	34	3	–	–	–	–	–	–	–	–	–	–	Pregnan-20-one, 3,17-dihydroxy-, (3β,5β)-	570-53-6	MS07965
255	298	213	147	135	215	107	145	**334**	100	92	84	82	78	77	72	67	**25.50**	21	34	3	–	–	–	–	–	–	–	–	–	–	Pregnan-20-one, 3,21-dihydroxy-, (3α,5α)-		NI35305

MASS TO CHARGE RATIOS	M.W.	INTENSITIES	Parent	C	H	O	N	S	F	Cl	Br	I	Si	P	B	X	COMPOUND NAME	CAS Reg No	No
257 95 303 81 147 149 161 107	**334**	100 40 37 37 35 32 30 30	**0.00**	21	34	3	–	–	–	–	–	–	–	–	–	–	Pregnan-20-one, 3,21-dihydroxy-, (3α,5β)-		NI35306
257 303 147 81 161 135 258 107	**334**	100 49 31 30 29 27 26 26	**0.00**	21	34	3	–	–	–	–	–	–	–	–	–	–	Pregnan-20-one, 3,21-dihydroxy-, (3β,5β)-		NI35307
210 211 109 136 82 69 55 43	**334**	100 13 10 8 4 4 4 4	**1.00**	21	34	3	–	–	–	–	–	–	–	–	–	–	Spiro[furan-2(5H),1'(4'H)-naphthalene]-5-acetic acid, 3,4,4'a,5',6',7',8',8'a octahydro-2',5,5',5',8'a-pentamethyl-, methyl ester	1438-58-0	NI35308
43 298 213 255 81 55 93 283	**334**	100 91 48 45 45 44 41 37	**1.00**	21	34	3	–	–	–	–	–	–	–	–	–	–	Tetrahydro-21-deoxycorticosterone		MS07966
110 334 82 69 95 55 137 123	**334**	100 32 29 17 9 9 8 8		21	38	1	2	–	–	–	–	–	–	–	–	–	Cyclobutanol, 1-(1H-imidazol-2-yl)-2-tetradecyl-	69393-17-5	NI35309
110 133 147 316 95 123 161 55	**334**	100 69 35 24 23 16 15 12	**4.00**	21	38	1	2	–	–	–	–	–	–	–	–	–	Cyclobutanol, 1-(1H-imidazol-4-yl)-2-tetradecyl-	69393-21-1	NI35310
44 291 260 56 287 81 261 57	**334**	100 35 30 25 15 12 10 10	**3.00**	21	38	1	2	–	–	–	–	–	–	–	–	–	5α-Pregnan-18-ol, 3β,20α-diamino-	6869-28-9	NI35311
185 216 150 91 141 153 115 186	**334**	100 80 69 47 41 27 27 26	**0.00**	22	22	3	–	–	–	–	–	–	–	–	–	–	8,8-Dimethoxy-2,4-diphenylbicyclo[3.2.1]oct-6-en-3-one		DD01317
43 186 334 170 162 134 185 135	**334**	100 51 45 40 40 39 34 25		22	26	1	2	–	–	–	–	–	–	–	–	–	2,20-Cycloaspidospermidine, 1-acetyl-6,7-didehydro-3-methyl-, (2α,3β,5α,12β,19α,20R)-	56053-36-2	NI35312
41 55 67 81 95 54 79 69	**334**	100 93 89 76 55 48 43 43	**0.94**	22	38	2	–	–	–	–	–	–	–	–	–	–	Cyclopropaneoctanoic acid, 2-[[2-[(2-ethylcyclopropyl)methyl]cyclopropyl]methyl]-, methyl ester	10152-71-3	NI35313
263 264 334 57 191 151 41 265	**334**	100 19 7 6 3 3 3 2		22	38	2	–	–	–	–	–	–	–	–	–	–	4,6-Di-tert-octylresorcinol		IC04011
206 180 167 165 193 179 115 91	**334**	100 68 61 30 23 23 22 20	**0.00**	23	26	2	–	–	–	–	–	–	–	–	–	–	Cyclohexanone, 2-methyl-2-(3-oxobutyl)-6,6-diphenyl-	50592-55-7	NI35314
319 320 334 262 247 335 263 131	**334**	100 24 22 16 8 5 4 4		23	26	2	–	–	–	–	–	–	–	–	–	–	4-Pentyl-7,7,10-trimethylfuro[1,2-a]dibenzopyran		NI35315
191 208 189 334 190 179 57 165	**334**	100 42 22 20 16 12 11 9		23	26	2	–	–	–	–	–	–	–	–	–	–	9-Phenanthrenemethyl octanoate	92174-42-0	NI35316
334 333 276 138 305 167 278 306	**334**	100 49 16 13 7 7 3 2		24	14	2	–	–	–	–	–	–	–	–	–	–	1H-Indene-1,3(2H)-dione, 2-(2,3-diphenyl-2-cyclopropen-1-ylidene)-	19164-54-6	NI35317
257 334 154 258 180 335 152 255	**334**	100 58 32 23 20 19 18 16		24	18	–	–	–	–	–	–	–	1	–	–	–	9H-9-Silafluorene, 9,9-diphenyl-	5550-08-3	NI35318
153 152 334 333 181 154 151	**334**	100 76 74 28 20 18 18		24	18	–	2	–	–	–	–	–	–	–	–	–	2-(2-Biphenylazo)biphenyl		LI07427
318 130 217 76 334 304 51 103	**334**	100 95 52 49 36 30 28 26		24	18	–	2	–	–	–	–	–	–	–	–	–	Tribenzo[c,g,k]-1,2-diazacyclotetradecane		NI35319
166 57 97 167 111 110 83 55	**334**	100 91 69 37 35 32 24 23	**3.00**	24	46	–	–	–	–	–	–	–	–	–	–	–	meso-4,5-Dicyclohexyl-2,2,7,7-tetramethyloctane		NI35320
91 92 41 105 55 104 334 69	**334**	100 99 17 15 13 11 10 10		25	34	–	–	–	–	–	–	–	–	–	–	–	1,5-Diphenyl-3-(3'-cyclopentylpropyl)-pentane		AI02062
92 91 41 105 55 104 93 69	**334**	100 81 19 15 12 10 8 8	**8.20**	25	34	–	–	–	–	–	–	–	–	–	–	–	1,5-Diphenyl-3-(3'-cyclopentylpropyl)-pentane		AI02063
257 334 243 165 335 179 258 241	**334**	100 95 44 38 27 24 22 15		26	22	–	–	–	–	–	–	–	–	–	–	–	Benzene, 1-methyl-2-(triphenylmethyl)-	24523-62-4	NI35321
243 165 244 241 239 228 166 91	**334**	100 40 22 5 5 5 5 5	**0.30**	26	22	–	–	–	–	–	–	–	–	–	–	–	Ethane, 1,1,1,2-tetraphenyl-	2294-94-2	NI35322
167 165 168 152 166 183 105 43	**334**	100 17 13 10 6 3 1 1	**0.00**	26	22	–	–	–	–	–	–	–	–	–	–	–	Ethane, 1,1,2,2-tetraphenyl-	632-50-8	NI35323
243 165 244 241 239 228 166 91	**334**	100 40 22 5 5 5 5 5	**0.29**	26	22	–	–	–	–	–	–	–	–	–	–	–	Ethane, 1,1,2,2-tetraphenyl-	632-50-8	LI07428
91 217 334 243 165 227 215 105	**334**	100 79 75 65 25 24 23 23		26	22	–	–	–	–	–	–	–	–	–	–	–	1,2,3-Tris[(E)-phenylmethylene]cyclopentane		MS07967
291 289 337 293 335 339 319 317	**335**	100 68 25 25 19 7 5 4		10	7	3	1	1	–	1	1	–	–	–	–	–	(6-Bromo-7-chloro-3,4-dihydro-3-oxo-2H-1,4-benzothiazin-2-yl)acetic acid		LI07429
162 164 43 300 63 98 220 175	**335**	100 95 49 37 29 27 25 25	**1.00**	12	11	6	1	–	–	2	–	–	–	–	–	–	L-Aspartic acid, N-[(2,4-dichlorophenoxy)acetyl]-	35144-55-9	NI35324
73 147 218 320 174 219 45 86	**335**	100 73 72 38 16 15 14 13	**0.00**	12	29	4	1	–	–	–	–	–	3	–	–	–	Aminomalonic acid, tris(trimethylsilyl)-	27762-05-6	NI35325
306 264 43 41 307 290 335 248	**335**	100 72 47 15 12 12 10 10		13	16	4	3	–	3	–	–	–	–	–	–	–	Aniline, 2,6-dinitro-N,N-dipropyl-4-(trifluoromethyl)-	1582-09-8	NI35327
43 306 264 17 41 290 248 307	**335**	100 99 97 67 35 17 17 15	**13.50**	13	16	4	3	–	3	–	–	–	–	–	–	–	Aniline, 2,6-dinitro-N,N-dipropyl-4-(trifluoromethyl)-	1582-09-8	NI35326
43 41 306 264 42 57 39 40	**335**	100 82 32 31 20 13 9 8	**2.58**	13	16	4	3	–	3	–	–	–	–	–	–	–	Aniline, 2,6-dinitro-N,N-dipropyl-4-(trifluoromethyl)-	1582-09-8	NI35328
294 32 266 42 43 278 295 58	**335**	100 30 27 25 16 15 12 12	**0.00**	13	16	4	3	–	3	–	–	–	–	–	–	–	Aniline, N-butyl-N-ethyl-2,6-dinitro-4-(trifluoromethyl)-	1861-40-1	NI35329
292 264 276 293 160 57 318 145	**335**	100 24 14 13 7 7 6 6	**6.20**	13	16	4	3	–	3	–	–	–	–	–	–	–	Aniline, N-butyl-N-ethyl-2,6-dinitro-4-(trifluoromethyl)-	1861-40-1	NI35330
292 41 43 264 57 105 293 42	**335**	100 39 38 29 20 14 11 11	**6.40**	13	16	4	3	–	3	–	–	–	–	–	–	–	Aniline, N-butyl-N-ethyl-2,6-dinitro-4-(trifluoromethyl)-	1861-40-1	NI35331
44 80 58 320 43 212 122 29	**335**	100 89 86 78 77 68 54 50	**5.80**	13	22	5	1	1	–	–	–	–	–	1	–	–	Phosphoramidic acid, isopropyl-, ethyl 3-methyl-4-(methylsulphonyl)phenyl ester	31972-44-8	NI35332
73 61 75 45 43 116 147 261	**335**	100 44 27 26 23 18 14 13	**0.61**	13	29	3	1	1	–	–	–	–	2	–	–	–	N-Acetylmethionine, bis(trimethylsilyl)-		NI35333
218 73 103 219 128 147 220 75	**335**	100 57 26 20 20 12 8 8	**0.80**	13	33	3	1	–	–	–	–	–	3	–	–	–	Homoserine, tris(trimethylsilyl)-		MS07968
73 117 218 45 219 101 75 57	**335**	100 25 21 19 18 9 9 9	**0.00**	13	33	3	1	–	–	–	–	–	3	–	–	–	L-Threonine, N,O-bis(trimethylsilyl)-, trimethylsilyl ester	7537-02-2	MS07969
73 75 77 147 218 219 117 44	**335**	100 86 48 35 31 26 26 23	**0.00**	13	33	3	1	–	–	–	–	–	3	–	–	–	L-Threonine, N,O-bis(trimethylsilyl)-, trimethylsilyl ester	7537-02-2	MS07970
73 117 218 219 45 147 75 101	**335**	100 25 23 21 18 15 15 14	**0.00**	13	33	3	1	–	–	–	–	–	3	–	–	–	L-Threonine, N,O-bis(trimethylsilyl)-, trimethylsilyl ester	7537-02-2	NI35334
166 122 167 337 335 46	**335**	100 10 8 1 1 1		14	10	4	1	–	–	–	1	–	–	–	–	–	Benzenemethanol, 3-bromo-, 4-nitrobenzoate	53218-04-5	NI35335
166 167 337 335 122 46	**335**	100 8 1 1 1 1		14	10	4	1	–	–	–	1	–	–	–	–	–	Benzenemethanol, 4-bromo-, 4-nitrobenzoate	53218-06-7	NI35336
77 137 44 39 93 78 51 66	**335**	100 65 62 56 50 50 50 46	**8.00**	14	12	4	2	–	–	–	–	–	–	–	–	1	Bis(4-methyl-1-quinone-2-oximato)copper(II)		MS07971
57 41 281 279 283 217 337 219	**335**	100 92 86 66 43 28 27 26	**20.00**	14	13	–	1	–	–	4	–	–	–	–	–	–	1-tert-Butyl-4,5,6,7-tetrachloro-3a,7b-dihydrobenzo[3,4]cyclobuta[1,2-b]pyrrole		NI35337
41 322 57 320 324 255 99 253	**335**	100 90 90 70 45 43 43 38	**14.00**	14	13	–	1	–	–	4	–	–	–	–	–	–	5,6,7,8-Tetrachloro-1,4-dihydro-1,4-tert-butyliminonaphthalene		NI35338
73 209 179 192 335 222 45 210	**335**	100 88 34 26 24 17 16 15		14	20	3	1	–	3	–	–	–	1	–	–	–	3-Methoxy-N-(trifluoroacetyl)-O-(trimethylsilyl)tyramine		MS07972
320 335 73 321 336 75 322 147	**335**	100 49 45 26 14 12 11 10		14	25	1	5	–	–	–	–	–	2	–	–	–	2-Pteridinamine, 6,7-dimethyl-N-(trimethylsilyl)-4-[(trimethylsilyl)oxy]-	36972-94-8	NI35339

MASS TO CHARGE RATIOS								M.W.	INTENSITIES								Parent	C	H	O	N	S	F	Cl	Br	I	Si	P	B	X	COMPOUND NAME	CAS Reg No	No
320	335	86	147	73	321	336	93	**335**	100	68	44	35	30	28	18	13		14	25	1	5	–	–	–	–	–	2	–	–	–	4(1H)-Pteridinone, 2-amino-6,7-dimethyl-, bis(trimethylsilyl)-	56272-94-7	NI35340
115	73	75	128	89	88	116	43	**335**	100	94	46	23	20	16	15	15	**0.00**	14	29	6	1	–	–	–	–	–	1	–	–	–	α-D-Galactopyranoside, methyl 2-(acetylamino)-2-deoxy-3,4-di-O-methyl-6-O-(trimethylsilyl)-	56211-02-0	NI35341
115	73	186	75	89	146	45	116	**335**	100	74	44	21	18	14	14	12	**0.00**	14	29	6	1	–	–	–	–	–	1	–	–	–	α-D-Galactopyranoside, methyl 2-(acetylamino)-2-deoxy-3,6-di-O-methyl-4-O-(trimethylsilyl)-	56211-05-3	NI35342
115	73	75	69	128	89	116	88	**335**	100	75	47	22	20	15	13	13	**0.00**	14	29	6	1	–	–	–	–	–	1	–	–	–	α-D-Galactopyranoside, methyl 2-(acetylamino)-2-deoxy-3,6-di-O-methyl-4-O-(trimethylsilyl)-	56211-05-3	NI35343
173	73	131	75	233	45	174	43	**335**	100	44	41	33	27	24	17	15	**0.00**	14	29	6	1	–	–	–	–	–	1	–	–	–	α-D-Galactopyranoside, methyl 2-(acetylamino)-2-deoxy-4,6-di-O-methyl-3-O-(trimethylsilyl)-	56211-03-1	NI35344
115	73	186	75	89	69	45	146	**335**	100	81	45	30	19	17	16	15	**0.00**	14	29	6	1	–	–	–	–	–	1	–	–	–	α-D-Glucopyranoside, methyl 2-(acetylamino)-2-deoxy-3,6-di-O-methyl-4 O-(trimethylsilyl)-	56227-35-1	NI35345
173	131	73	175	69	45	174	146	**335**	100	54	39	21	20	18	17	14	**0.00**	14	29	6	1	–	–	–	–	–	1	–	–	–	α-D-Glucopyranoside, methyl 2-(acetylamino)-2-deoxy-4,6-di-O-methyl-3 O-(trimethylsilyl)-	56211-04-2	NI35346
173	131	73	175	174	146	69	45	**335**	100	43	30	19	18	15	15	14	**0.00**	14	29	6	1	–	–	–	–	–	1	–	–	–	α-D-Glucopyranoside, methyl 2-(acetylamino)-2-deoxy-4,6-di-O-methyl-3 O-(trimethylsilyl)-	56211-04-2	NI35347
128	77	101	204	205	75	51	333	**335**	100	56	46	41	31	16	16	16	**15.40**	15	11	–	1	–	–	–	–	–	–	–	–	1	2-Quinolyl phenyl telluride	87803-46-1	NI35348
207	179	165	128	206	208	178	193	**335**	100	27	27	23	19	18	17	15	**0.00**	15	14	–	1	–	–	–	–	1	–	–	–	–	Phenanthridinium, 5,6-dimethyl-, iodide	16511-48-1	NI35349
160	136	203	335	188	41	137	202	**335**	100	83	76	66	57	46	36	32		15	21	4	5	–	–	–	–	–	–	–	–	–	Adenosine, N-(3-methyl-2-butenyl)-	7724-76-7	NI35350
148	30	28	335	135	36	41	29	**335**	100	33	15	14	12	12	10	10		15	21	4	5	–	–	–	–	–	–	–	–	–	Adenosine, N-(3-methyl-3-butenyl)-	16465-37-5	NI35351
164	41	232	69	43	45	218	55	**335**	100	74	67	53	35	26	22	21	**7.09**	15	21	4	5	–	–	–	–	–	–	–	–	–	1H-Pyrazolo[4,3-d]pyrimidin-7-amine, N-(3-methyl-2-butenyl)-3β-D-ribofuranosyl-	33987-26-7	NI35352
106	212	78	51	317	164	143	288	**335**	100	96	64	32	24	19	19	16	**0.00**	16	12	2	3	–	3	–	–	–	–	–	–	–	1,1,1-Trifluoro-3-benzoylisopropylideneisonicotinic acid hydrazide		LI07430
300	284	302	286	301	285	299	283	**335**	100	72	35	24	22	16	9	7	**0.00**	16	15	1	3	–	–	2	–	–	–	–	–	–	3H-1,4-Benzodiazepin-2-amine, 7-chloro-N-methyl-5-phenyl-, 4-oxide, monohydrochloride	438-41-5	NI35353
105	77	149	123	51	186	170	291	**335**	100	80	79	28	28	15	14	3	**0.00**	16	17	5	1	1	–	–	–	–	–	–	–	–	Ethyl 2-(2-benzoyloxyethoxymethyl)thiazole-4-carboxylate		MS07973
166	180	179	91	150	167	142	120	**335**	100	35	33	23	10	9	9	9	**0.00**	16	17	5	1	1	–	–	–	–	–	–	–	–	1-[[2-Methoxy-2-(4-nitrophenyl)ethyl]sulphonyl]-4-methylbenzene		DD01318
86	217	335	302	226	94	77	56	**335**	100	97	82	49	48	47	40	21		16	18	3	1	1	–	–	–	–	–	1	–	–	Phosphonothioic acid, morpholino-, O,O-diphenyl ester	22077-43-6	LI07431
86	217	335	302	226	94	77	292	**335**	100	97	82	50	48	47	40	27		16	18	3	1	1	–	–	–	–	–	1	–	–	Phosphonothioic acid, morpholino-, O,O-diphenyl ester	22077-43-6	NI35354
218	73	335	219	75	336	320	45	**335**	100	50	38	20	13	11	11	10		16	25	3	1	–	–	–	–	–	2	–	–	–	5-Trimethylsiloxyindoleacetic acid, trimethylsilyl ester	69937-41-3	NI35355
151	152	164	155	335	184	336		**335**	100	45	37	36	24	12	5			17	21	4	1	1	–	–	–	–	–	–	–	–	Benzenesulphonamide, N-[2-(3,4-dimethoxyphenyl)ethyl]-4-methyl-	14165-67-4	NI35356
207	164	181	110	192	191	169	208	**335**	100	76	64	63	61	56	53	50	**4.00**	17	21	4	1	1	–	–	–	–	–	–	–	–	N-2-(2-Hydroxymethyl-4,5-dimethoxyphenyl)-[2-(2-thienyl)ethyl]ethanamide		MS07974
68	91	180	223	124	222	155	65	**335**	100	44	42	23	23	20	13	13	**2.30**	17	21	4	1	1	–	–	–	–	–	–	–	–	Isobenzofuran-4,7-imin-1(3H)-one, hexahydro-3,3-dimethyl-8-(4-tolylsulphonyl)-	17037-50-2	NI35357
59	94	84	241	79	114	135	100	**335**	100	13	12	12	7	5	4	4	**0.00**	17	38	–	1	–	–	–	1	–	–	–	–	–	N-Tetradecyltrimethylammonium bromide	1119-97-7	NI35358
141	126	83	82	142	131	41	68	**335**	100	21	14	13	10	10	9	9	**0.10**	18	22	3	1	–	–	1	–	–	–	–	–	–	4-(4-Chlorophenyl)-5,5-dimethyl-6-(N-morpholinyl)-(4H,5H,6H)-oxin-3-carboxaldehyde		NI35359
136	120	93	119	121	138	95	94	**335**	100	90	89	88	82	71	68	64	**6.40**	18	25	5	1	–	–	–	–	–	–	–	–	–	Doronenine	74217-57-5	NI35360
119	120	93	95	136	121	43	94	**335**	100	83	83	65	64	62	55	51	**2.10**	18	25	5	1	–	–	–	–	–	–	–	–	–	Integerrimine	480-79-5	NI35361
136	93	83	55	120	94	220	237	**335**	100	86	59	49	46	41	36	33	**6.55**	18	25	5	1	–	–	–	–	–	–	–	–	–	Neotriangularine	87392-67-4	NI35362
120	119	136	93	94	43	95	121	**335**	100	86	86	81	64	60	56	48	**4.20**	18	25	5	1	–	–	–	–	–	–	–	–	–	Senecionine	130-01-8	NI35363
57	203	206	160	190	234	250	335	**335**	100	30	20	13	11	7	5	4		18	25	5	1	–	–	–	–	–	–	–	–	–	(3S)-1,2,3,4-Tetrahydro-6,7-dimethoxy-3-methyl-2-pivaloylisoquinoline-3-carboxylic acid		DD01319
136	93	28	83	120	220	55	94	**335**	100	84	54	53	46	42	41	40	**5.53**	18	25	5	1	–	–	–	–	–	–	–	–	–	Triangularine	87340-27-0	NI35364
91	73	92	77	75	265	41	45	**335**	100	17	7	7	7	5	5	4	**0.00**	18	29	3	1	–	–	–	–	–	1	–	–	–	Octanoic acid, 2-[(phenylmethoxy)imino]-, trimethylsilyl ester	55520-99-5	NI35365
334	335	292	106	293	276	120	187	**335**	100	64	29	27	10	9	9	7		19	17	3	3	–	–	–	–	–	–	–	–	–	4-[(4-Acetylaminophenyl)methyl]-2-methyl-6,7-methylenedioxyquinazoline		NI35366
300	318	274	336	317	335	246	301	**335**	100	72	37	32	28	24	23	21		19	29	4	1	–	–	–	–	–	–	–	–	–	1,1′-Bis(4-ethoxy-2,6-pyridinediyl)cyclohexanol, N-oxide	17117-07-6	MS07975
335	334	336	176	333	304	189	177	**335**	100	80	21	15	10	10	10	8		20	17	4	1	–	–	–	–	–	–	–	–	–	2,3:8,9-Bis(ethylendioxy)-5,5a,6,7-tetrahydro-6-methylbenz[e]indeno[2,1-b]azepine		NI35367
242	335	77	199	51	83	111	227	**335**	100	73	56	38	24	22	16	14		20	17	4	1	–	–	–	–	–	–	–	–	–	Methyl 1,2-dihydro-3-methyl-2-oxo-6-phenoxy-1-phenyl-5-pyridinecarboxylate		DD01320
136	122	129	57	137	41	43	55	**335**	100	87	85	71	67	50	47	44	**15.00**	20	17	4	1	–	–	–	–	–	–	–	–	–	N-Methyl-N-(4-anisyl)-2-methyl-1,4-naphthoquinone-3-carboxamide		MS07976
335	334	171	150.5	277	276	136.5	246	**335**	100	78	28	21	14	14	14	12		20	21	2	3	–	–	–	–	–	–	–	–	–	Benzoic acid, 2-(methylamino)-5-(1,2,3,9-tetrahydropyrrolo[2,1-b]quinazolin-3-yl)-, methyl ester		LI07432
115	77	130	91	104	63	89	65	**335**	100	90	69	64	54	44	39	36	**0.36**	20	21	2	3	–	–	–	–	–	–	–	–	–	2,4-Dimethyl-1-(2-oximino-1-phenylpropyl)-5-phenylimidazole 3-oxide		LI07433

MASS TO CHARGE RATIOS	M.W.	INTENSITIES	Parent	C	H	O	N	S	F	Cl	Br	I	Si	P	B	X	COMPOUND NAME	CAS Reg No	No
335 320 144 304 303 166 176 157	335	100 61 42 24 21 13 12 12		20	21	2	3	–	–	–	–	–	–	–	–	–	Pyrrolo[2,1-b]quinazolin-9(1H)-one, 3-[2-(dimethylamino)phenyl]-2,3-dihydro-5-methoxy-	16688-22-5	NI35368
125 167 208 83 81 58 55 43	335	100 60 45 45 37 30 29 28	0.00	20	33	3	1	–	–	–	–	–	–	–	–	–	N,N-Diisopropyl-4-[(cyclohexyloxy)carbonyl]-4-methylcyclopent-1-enecarboxamide	110426-81-8	DD01321
92 120 189 104 134 77 162 147	335	100 40 37 36 26 25 24 22	0.00	20	34	1	1	–	–	–	–	–	–	1	–	–	3-Phenyl-5,5,5-tributyl-1,2,5-oxazaphosphol-2-ine		LI07434
91 171 335 336 170 214 128 247	335	100 86 85 59 58 49 48 46		21	21	3	1	–	–	–	–	–	–	–	–	–	4-[(Benzyloxy)methyl]-3,3-dimethyl-1,4-dihydro-2H-pyrano[4,3-b]indol-1-one		DD01322
180 179 266 221 220 335 207 191	335	100 66 48 48 33 24 24 9		21	21	3	1	–	–	–	–	–	–	–	–	–	2,3,4,5-Tetrahydro-3,3,5,5-tetramethyl-5αH-phenanthridino[5,6-b]-1,2-oxazepine-2,4-dione		LI07435
189 307 147 118 146 335 199 201	335	100 35 27 26 23 18 13 10		21	25	1	3	–	–	–	–	–	–	–	–	–	8-Benzyl-3-methyl-1-phenyl-1,3,8-triazaspiro[4.5]decan-4-one		MS07977
44 183 320 335 184 67 43 198	335	100 98 89 86 64 64 60 50		21	25	1	3	–	–	–	–	–	–	–	–	–	(Z,Z)-4-Ethyl-3-methyl-5-[5-(3,4-dimethyl-1H-pyrrolyl-2-methylene)-3,4-dimethyl-5H-pyrrolyl-2-methylene]-3-pyrrolin-2-one		MS07978
208 320 335 77 209 165 57 321	335	100 40 31 27 21 13 12 10		21	25	1	3	–	–	–	–	–	–	–	–	–	4-Imidazolidinone, 1-(1,1-dimethylethyl)-3-ethyl-2-imino-5,5-diphenyl-	56954-66-6	NI35369
208 320 71 335 77 209 165 57	335	100 40 33 31 27 16 13 12		21	25	1	3	–	–	–	–	–	–	–	–	–	4-Imidazolidinone, 1-(1,1-dimethylethyl)-3-ethyl-2-imino-5,5-diphenyl-	56954-66-6	NI35370
250 279 104 280 251 335 208 182	335	100 70 34 25 21 17 16 16		21	25	1	3	–	–	–	–	–	–	–	–	–	4-Imidazolidinone, 3-(1,1-dimethylethyl)-1-ethyl-2-imino-5,5-diphenyl-	54508-08-6	NI35371
250 279 104 251 335 182 165 77	335	100 70 34 22 17 16 16 16		21	25	1	3	–	–	–	–	–	–	–	–	–	4-Imidazolidinone, 3-(1,1-dimethylethyl)-1-ethyl-2-imino-5,5-diphenyl-	54508-08-6	NI35372
214 91 82 83 42 77 215 105	335	100 33 33 32 22 21 18 18	17.78	22	25	2	1	–	–	–	–	–	–	–	–	–	8-Azabicyclo[3.2.1]octan-3-ol, 8-methyl-2-benzyl-, benzoate	55925-25-2	NI35373
161 335 175 162 145 91 174 131	335	100 63 63 49 35 35 31 25		22	25	2	1	–	–	–	–	–	–	–	–	–	1H-Dibenz[e,g]isoindole-1,3(2H)-dione, 3a,3b,4,5,6,7,8,9,10,11,11a,11b-dodecahydro-2-phenyl-, (3aα,3bα,11aα,11bα)-	51624-49-8	NI35374
161 335 175 162 145 91 174 131	335	100 63 63 50 35 35 33 25		22	25	2	1	–	–	–	–	–	–	–	–	–	N-Phenylmaleimide-1,1'-bicyclohexenyl endo adduct		LI07436
113 166 98 55 43 70 41 335	335	100 65 53 51 47 36 31 27		22	41	1	1	–	–	–	–	–	–	–	–	–	Pyrrolidine, 1-(1-oxo-4-octadecenyl)-	56599-68-9	NI35375
113 166 98 55 43 70 41 335	335	100 56 40 40 28 23 23 17		22	41	1	1	–	–	–	–	–	–	–	–	–	Pyrrolidine, 1-(1-oxo-4-octadecenyl)-	56599-68-9	NI35377
113 43 166 55 41 98 70 57	335	100 88 68 65 63 56 36 26	15.04	22	41	1	1	–	–	–	–	–	–	–	–	–	Pyrrolidine, 1-(1-oxo-4-octadecenyl)-	56599-68-9	NI35376
113 43 55 85 70 114 41 98	335	100 15 13 10 10 9 9 7	5.24	22	41	1	1	–	–	–	–	–	–	–	–	–	Pyrrolidine, 1-(1-oxo-5-octadecenyl)-	56599-67-8	NI35378
113 126 55 43 70 127 98 41	335	100 63 33 28 27 25 21 21	16.14	22	41	1	1	–	–	–	–	–	–	–	–	–	Pyrrolidine, 1-(1-oxo-6-octadecenyl)-	52380-34-4	NI35379
113 126 51 54 43 70 98 41	335	100 58 53 34 29 27 20 19	17.04	22	41	1	1	–	–	–	–	–	–	–	–	–	Pyrrolidine, 1-(1-oxo-7-octadecenyl)-	56600-01-2	NI35380
113 43 70 126 55 41 335 85	335	100 67 60 52 33 26 18 18		22	41	1	1	–	–	–	–	–	–	–	–	–	Pyrrolidine, 1-(1-oxo-8-octadecenyl)-	56599-75-8	NI35381
113 126 55 70 43 335 41 98	335	100 61 41 29 28 25 25 18		22	41	1	1	–	–	–	–	–	–	–	–	–	Pyrrolidine, 1-(1-oxo-9-octadecenyl)-	52380-36-6	NI35382
113 126 55 307 70 41 43 98	335	100 64 37 28 26 23 20 16	0.00	22	41	1	1	–	–	–	–	–	–	–	–	–	Pyrrolidine, 1-(1-oxo-9-octadecenyl)-	52380-36-6	NI35383
113 126 55 70 335 43 41 98	335	100 64 41 28 25 25 23 18		22	41	1	1	–	–	–	–	–	–	–	–	–	Pyrrolidine, 1-(1-oxo-9-octadecenyl)-, (E)-	52380-42-4	NI35384
113 126 55 43 70 335 57 41	335	100 60 44 33 29 27 27 27		22	41	1	1	–	–	–	–	–	–	–	–	–	Pyrrolidine, 1-(1-oxo-9-octadecenyl)-, (Z)-	4637-54-1	NI35385
43 41 55 98 113 78 57 70	335	100 96 93 71 46 45 43 34	8.04	22	41	1	1	–	–	–	–	–	–	–	–	–	Pyrrolidine, 1-(1-oxo-10-octadecenyl)-	56600-00-1	NI35386
113 126 55 335 70 43 41 98	335	100 46 36 30 28 24 19 15		22	41	1	1	–	–	–	–	–	–	–	–	–	Pyrrolidine, 1-(1-oxo-10-octadecenyl)-	56600-00-1	NI35387
113 126 55 335 70 43 41 69	335	100 41 35 27 22 20 18 11		22	41	1	1	–	–	–	–	–	–	–	–	–	Pyrrolidine, 1-(1-oxo-11-octadecenyl)-	52380-35-5	NI35388
113 126 55 335 70 41 43 69	335	100 41 39 33 27 22 17 14		22	41	1	1	–	–	–	–	–	–	–	–	–	Pyrrolidine, 1-(1-oxo-12-octadecenyl)-	56599-74-7	NI35389
113 126 55 335 70 43 41 69	335	100 47 41 35 29 29 27 16		22	41	1	1	–	–	–	–	–	–	–	–	–	Pyrrolidine, 1-(1-oxo-12-octadecenyl)-	56599-74-7	NI35390
113 126 55 335 70 41 43 71	335	100 45 42 33 24 19 16 13		22	41	1	1	–	–	–	–	–	–	–	–	–	Pyrrolidine, 1-(1-oxo-13-octadecenyl)-	56599-73-6	NI35391
113 55 126 335 70 43 41 69	335	100 38 36 30 24 16 16 11		22	41	1	1	–	–	–	–	–	–	–	–	–	Pyrrolidine, 1-(1-oxo-13-octadecenyl)-	56599-73-6	NI35392
113 126 335 55 70 41 43 40	335	100 41 40 33 22 17 14 12		22	41	1	1	–	–	–	–	–	–	–	–	–	Pyrrolidine, 1-(1-oxo-14-octadecenyl)-	56599-72-5	NI35393
113 126 55 335 70 41 43 69	335	100 40 33 31 27 22 18 16		22	41	1	1	–	–	–	–	–	–	–	–	–	Pyrrolidine, 1-(1-oxo-15-octadecenyl)-	56599-71-4	NI35394
113 126 335 55 70 41 43 69	335	100 39 34 28 22 18 15 13		22	41	1	1	–	–	–	–	–	–	–	–	–	Pyrrolidine, 1-(1-oxo-15-octadecenyl)-	56599-71-4	NI35395
113 126 55 335 70 43 41 71	335	100 34 33 32 21 15 15 12		22	41	1	1	–	–	–	–	–	–	–	–	–	Pyrrolidine, 1-(1-oxo-16-octadecenyl)-	56599-70-3	NI35397
113 126 55 335 70 41 43 71	335	100 33 33 26 19 15 14 11		22	41	1	1	–	–	–	–	–	–	–	–	–	Pyrrolidine, 1-(1-oxo-16-octadecenyl)-	56599-70-3	NI35396
113 55 126 70 43 41 71 335	335	100 22 19 14 12 11 10 8		22	41	1	1	–	–	–	–	–	–	–	–	–	Pyrrolidine, 1-(1-oxo-17-octadecenyl)-	52380-50-4	NI35399
113 55 126 70 43 41 71 335	335	100 23 22 19 15 13 11 10		22	41	1	1	–	–	–	–	–	–	–	–	–	Pyrrolidine, 1-(1-oxo-17-octadecenyl)-	52380-50-4	NI35398
98 99 86 112 41 70 55 43	335	100 58 28 20 20 18 17 13	0.10	22	41	1	1	–	–	–	–	–	–	–	–	–	2-Pyrrolidinone, 1-(9-octadecenyl)-	69494-15-1	NI35400
320 335 321 291 290 146 295 105	335	100 36 27 18 18 18 15 13		24	17	1	1	–	–	–	–	–	–	–	–	–	Benzo[f]naphtho[2',1':3,4]cyclobut[1,2-h]isoquinoline, 8c,14b-dihydro-14b methoxy-	50559-62-1	NI35401
320 335 321 291 290 304 292 146	335	100 33 29 17 16 13 13 10		24	17	1	1	–	–	–	–	–	–	–	–	–	Benzo[h]naphtho[1',2':3,4]cyclobut[1,2-f]isoquinoline, 8c,14b-dihydro-14b methoxy-	50559-50-7	NI35402
243 165 244 166 77 335 228 245	335	100 31 27 6 6 5 5 4		25	21	–	1	–	–	–	–	–	–	–	–	–	Aniline, N-trityl-	4471-22-1	NI35403
336 82 127 81 209 45 38 37	336	100 58 35 17 16 15 14 12		4	2	–	–	1	–	–	–	2	–	–	–	–	Thiophene, 2,3-diiodo-	19259-05-3	NI35404
336 82 127 38 81 209 57 37	336	100 56 24 16 14 11 10 9		4	2	–	–	1	–	–	–	2	–	–	–	–	Thiophene, 2,5-diiodo-	625-88-7	NI35405
81 82 67 41 55 99 54 39	336	100 46 35 32 24 20 19 19	0.64	6	11	1	–	–	–	1	–	–	–	–	–	1	(2-Hydroxycyclohexyl)mercuric chloride		LI07437

MASS TO CHARGE RATIOS								M.W.	INTENSITIES								Parent	C	H	O	N	S	F	Cl	Br	I	Si	P	B	X	COMPOUND NAME	CAS Reg No	No
280	336	252	308	224	250	226	225	336	100	55	50	20	20	18	14	12		8	5	3	–	–	–	–	–	–	–	–	–	1	Rhenium, π-cyclopentadienyl-		MS07979
230	232	303	169	171	338	305	301	336	100	91	85	80	74	61	61	57	35.41	10	6	–	–	–	–	6	–	–	–	–	–	–	γ-Chlordene	56641-38-4	NI35406
303	230	232	305	301	36	267	196	336	100	96	93	64	64	57	43	32	14.24	10	6	–	–	–	–	6	–	–	–	–	–	–	1,4-Ethenopentalene, 1,2,3,5,7,8-hexachloro-1,3a,4,5,6,6a-hexahydro-	69743-74-4	NI35407
232	230	303	27	305	301	267	39	336	100	90	67	48	43	40	40	39	6.10	10	6	–	–	–	–	6	–	–	–	–	–	–	1,4-Ethenopentalene, 1,2,3,5,7,8-hexachloro-1,3a,4,5,6,6a-hexahydro-, (1α,3aα,4β,5α,6aα)-	56534-02-2	NI35408
27	51	39	49	85	50	230	75	336	100	86	85	78	64	62	59	57	17.21	10	6	–	–	–	–	6	–	–	–	–	–	–	1,6-Methano-1H-indene, 2,3,3a,4,5,7-hexachloro-3a,6,7,7a-tetrahydro-, (1α,3aβ,6α,7α,7aβ)-	56534-03-3	NI35409
230	232	36	169	171	303	267	98	336	100	97	88	66	61	56	47	43	18.63	10	6	–	–	–	–	6	–	–	–	–	–	–	1,6-Methano-1H-indene, 2,3,3a,4,5,8-hexachloro-3a,6,7,7a-tetrahydro-	69743-73-3	NI35410
66	101	80	67	65	239	235	103	336	100	8	8	8	6	3	3	3	1.00	10	6	–	–	–	–	6	–	–	–	–	–	–	4,7-Methano-1H-indene, 4,5,6,7,8,8-hexachloro-3a,4,7,7a-tetrahydro-	3734-48-3	NI35411
66	101	80	67	39	65	237	103	336	100	8	7	7	7	5	4	3	1.00	10	6	–	–	–	–	6	–	–	–	–	–	–	4,7-Methano-1H-indene, 4,5,6,7,8,8-hexachloro-3a,4,7,7a-tetrahydro-	3734-48-3	NI35412
66	65	101	67	39	103	100	40	336	100	7	5	5	4	2	2	2	0.00	10	6	–	–	–	–	6	–	–	–	–	–	–	4,7-Methano-1H-indene, 4,5,6,7,8,8-hexachloro-3a,4,7,7a-tetrahydro-	3734-48-3	NI35413
230	232	36	39	169	171	51	75	336	100	98	86	62	56	51	46	44	2.81	10	6	–	–	–	–	6	–	–	–	–	–	–	1,3,5-Metheno-1H-cyclopropa[a]pentalene, 1,1a,4,5,5a,6-hexachlorooctahydro-	69743-71-1	NI35414
232	230	196	36	267	39	231	198	336	100	99	77	76	56	51	50	48	0.41	10	6	–	–	–	–	6	–	–	–	–	–	–	1,4,5-Metheno-1H-cyclopropa[a]pentalene, 1,1a,3,5,5a,6-hexachlorooctahydro-	69743-77-7	NI35415
338	336	340	202	238	339	203	240	336	100	73	52	35	27	24	22	19		11	8	–	–	–	–	4	–	–	2	–	–	–	1,1,3,3-Tetrachloro-1,3-disilaphenalane	32538-49-1	NI35416
336	338	337	339	209	250	223	252	336	100	98	17	15	12	11	10	9		12	9	3	4	–	–	–	1	–	–	–	–	–	5-(3-Methyl-6-bromo-1-benzimidazolio)-4-barbiturate		LI07438
112	113	143	85	122	114	157	144	336	100	34	24	24	22	8	6	3	0.00	12	18	1	4	1	–	2	–	–	–	–	–	–	Thiamine dihydrochloride		LI07439
86	103	128	336	102	60	69	57	336	100	81	70	67	61	50	50	39		12	24	3	4	2	–	–	–	–	–	–	–	–	1,4,7-Trioxa-10,12,15,17-tetraazacyclononadecane-11,16-dithione		NI35417
86	147	60	103	69	129	130	128	336	100	50	50	42	42	41	39	37	36.81	12	24	3	4	2	–	–	–	–	–	–	–	–	1,4,12-Trioxa-7,9,15,17-tetraazacyclononadecane-8,16-dithione		NI35418
113	111	85	114	99	155	173	244	336	100	59	39	18	12	11	3	1	0.00	12	24	4	–	–	2	1	–	–	–	1	–	–	Diethyl 1-(chlorodifluoromethyl)-1-hydroxyheptanephosphonate		DD01323
73	147	292	102	293	133	45	148	336	100	58	20	16	13	11	11	9	0.00	12	28	5	–	–	–	–	–	–	3	–	–	–	Propanedioic acid, [(trimethylsilyl)oxy]-, bis(trimethylsilyl) ester	38165-93-4	NI35420
73	147	102	45	292	293	133	74	336	100	52	22	16	14	11	11	10	0.00	12	28	5	–	–	–	–	–	–	3	–	–	–	Propanedioic acid, [(trimethylsilyl)oxy]-, bis(trimethylsilyl) ester	38165-93-4	NI35419
73	147	45	102	133	74	292	75	336	100	29	20	16	11	9	8	7	0.00	12	28	5	–	–	–	–	–	–	3	–	–	–	Propanedioic acid, [(trimethylsilyl)oxy]-, bis(trimethylsilyl) ester	38165-93-4	NI35421
43	55	108	85	99	81	63	77	336	100	55	32	31	24	24	24	20	0.00	13	18	2	–	–	–	–	–	–	–	–	–	1	2-[[(4-Methoxyphenyl)telluro]methyl]tetrahydropyran		DD01324
73	147	217	205	103	117	204	189	336	100	63	52	38	38	26	16	16	0.00	13	32	4	–	–	–	–	–	–	3	–	–	–	Butanal, 2,3,4-tris[(trimethylsilyl)oxy]-, (R*,R*)-	56297-94-0	NI35422
73	117	147	292	102	130	75	220	336	100	58	35	24	20	14	13	12	0.00	13	32	4	–	–	–	–	–	–	3	–	–	–	Butanoic acid, 2,3-bis[(trimethylsilyl)oxy]-, trimethylsilyl ester	71294-30-9	NI35423
73	117	292	147	220	75	102	293	336	100	73	38	36	14	14	14	12	0.00	13	32	4	–	–	–	–	–	–	3	–	–	–	Butanoic acid, 2,3-bis[(trimethylsilyl)oxy]-, trimethylsilyl ester, (R*,R*)-		NI35424
73	117	147	292	75	220	102	293	336	100	70	36	36	14	14	13	12	0.00	13	32	4	–	–	–	–	–	–	3	–	–	–	Butanoic acid, 2,3-bis[(trimethylsilyl)oxy]-, trimethylsilyl ester, (R*,S*)-		NI35425
103	73	219	147	220	104	75	45	336	100	76	34	30	11	9	9	9	0.00	13	32	4	–	–	–	–	–	–	3	–	–	–	Butanoic acid, 2,4-bis[(trimethylsilyl)oxy]-, trimethylsilyl ester	55191-52-1	NI35426
73	233	147	189	231	75	74	321	336	100	26	25	18	9	9	9	8	0.00	13	32	4	–	–	–	–	–	–	3	–	–	–	Butanoic acid, 3,4-bis[(trimethylsilyl)oxy]-, trimethylsilyl ester		NI35427
73	147	233	189	231	75	117	321	336	100	46	42	34	17	14	13	12	0.00	13	32	4	–	–	–	–	–	–	3	–	–	–	Butanoic acid, 3,4-bis[(trimethylsilyl)oxy]-, trimethylsilyl ester, (R*,S*)-		NI35428
73	147	233	189	231	117	246	321	336	100	47	40	31	29	20	15	10	0.00	13	32	4	–	–	–	–	–	–	3	–	–	–	2-Deoxytetronic acid, tris(trimethylsilyl)-		NI35429
73	147	218	191	217	45	75	148	336	100	65	36	22	21	11	11	11	0.00	13	32	4	–	–	–	–	–	–	3	–	–	–	D-Erythrose, tris(trimethylsilyl)-	73745-84-3	NI35430
73	219	147	129	233	75	43	131	336	100	52	49	22	18	18	18	15	0.14	13	32	4	–	–	–	–	–	–	3	–	–	–	Propanoic acid, 2-methyl-2,3-bis[(trimethylsilyl)oxy]-, trimethylsilyl ester	38166-00-6	NI35431
73	219	147	129	233	75	44	31	336	100	51	47	22	17	17	16	14	0.20	13	32	4	–	–	–	–	–	–	3	–	–	–	Propanoic acid, 2-methyl-2,3-bis[(trimethylsilyl)oxy]-, trimethylsilyl ester	38166-00-6	LI07440
336	334	256	176	166	254	167	174	336	100	91	33	27	25	18	16	10		14	8	–	–	–	–	–	–	–	–	–	–	2	Benzoselenopheno[2,3-b]benzoselenophene		LI07441
178	338	89	179	176	76	336	340	336	100	18	14	13	11	10	9	8		14	10	–	–	–	–	–	2	–	–	–	–	–	Benzene, 1,1′-(1,2-dibromo-1,2-ethenediyl)bis-	32047-17-9	NI35432
89	76	209	75	211	50	63	103	336	100	96	84	79	76	63	51	45	4.00	14	10	–	–	–	–	–	2	–	–	–	–	–	Benzene, 1,1′-(1,2-ethenediyl)bis[2-bromo-, (E)-	56667-12-0	NI35433
338	178	336	340	339	176			336	100	80	53	45	15	15				14	10	–	–	–	–	–	2	–	–	–	–	–	Benzene, 1,1′-(1,2-ethenediyl)bis[2-bromo-, (Z)-	56667-11-9	NI35434
176	336	334	338	177	174	175	150	336	100	80	40	38	23	21	19	15		14	10	–	–	–	–	–	2	–	–	–	–	–	9,10-Dibromo-phenanthrene		LI07442
95	336	306	141	238	251	307	225	336	100	70	65	61	45	40	27	25		14	16	6	4	–	–	–	–	–	–	–	–	–	N-(Cyclooctylidine)-2,4,6-trinitroaniline		IC04012
150	122	29	75	76	67	92	18	336	100	99	89	82	78	45	39	31	12.70	14	16	6	4	–	–	–	–	–	–	–	–	–	Diethyl 2-(4-nitrophenyltriazeno)fumarate	55721-93-2	MS07980
81	99	96	127	139	126	55	41	336	100	90	42	20	19	18	18	18	0.50	14	16	6	4	–	–	–	–	–	–	–	–	–	Uridine, 2′,5′-dideoxy-5′-(3,4-dihydro-5-methyl-2,4-dioxo-1(2H)-pyrimidinyl)-	35959-40-1	NI35435
166	128	70	142	96	170	85	129	336	100	95	70	45	25	15	15	14	4.00	14	19	4	2	–	3	–	–	–	–	–	–	–	Methyl N-trifluoroacetyl-(S)-(-)-prolyl-2-pyrrolidineacetate		NI35436
84	190	56	107	82	41	180	81	336	100	32	32	15	13	13	10	10	0.20	14	20	4	6	–	–	–	–	–	–	–	–	–	L-Histidinamide, 5-oxo-L-prolyl-L-alanyl-	37666-99-2	NI35437
73	117	75	132	55	69	83	41	336	100	74	70	60	43	34	33	31	1.00	14	29	2	–	–	–	–	1	–	1	–	–	–	Undecanoic acid, 11-bromo-, trimethylsilyl ester	34176-85-7	NI35438
103	73	117	219	147	75	45	104	336	100	90	55	33	23	11	10	8	0.00	14	36	3	–	–	–	–	–	–	3	–	–	–	3,8-Dioxa-2,9-disiladecane, 2,2,4,9,9-pentamethyl-5-[(trimethylsilyl)oxy]-	74685-21-5	NI35439
73	143	71	147	144	75	85	74	336	100	99	25	23	13	12	8	8	0.00	14	36	3	–	–	–	–	–	–	3	–	–	–	3,9-Dioxa-2,10-disilaundecane, 2,2,10,10-tetramethyl-5-[(trimethylsilyl)oxy]-	74685-22-6	NI35440
103	73	147	156	75	219	104	74	336	100	90	31	22	21	16	8	8	0.00	14	36	3	–	–	–	–	–	–	3	–	–	–	3,9-Dioxa-2,10-disilaundecane, 2,2,10,10-tetramethyl-6-[(trimethylsilyl)oxy]-	74685-20-4	NI35441
146	336	144	95	65	39	337	114	336	100	19	7	5	5	4	3	3		15	10	2	–	–	6	–	–	–	–	–	–	–	1,3-Bis(4-hydroxyphenyl)hexafluoropropane		LI07443

MASS TO CHARGE RATIOS								M.W.	INTENSITIES								Parent	C	H	O	N	S	F	Cl	Br	I	Si	P	B	X	COMPOUND NAME	CAS Reg No	No
108	123	255	122	239	254	256	336	336	100	84	69	66	48	44	29	27		15	16	2	2	–	–	–	–	–	–	–	–	1	N,N'-Bis(4-methoxyphenyl)selenourea		MS07981
182	254	28	208	308	110	56	29	336	100	74	68	53	49	48	48	38	5.76	15	20	5	–	–	–	–	–	–	–	–	–	1	Iron, (η^4-1,3-butadiene)carbonyl[(2,3,4,5-η)-diethyl 2,4-hexadienedioate]-	62278-90-4	NI35442
73	99	41	43	75	68	87	59	336	100	45	40	38	33	30	29	24	0.41	15	24	1	6	1	–	–	–	–	–	–	–	–	7,9-Diethyl-2,4-bis(dimethylamino)-10-imino-8-thio-1,7,9-triazaspiro[4.5]-1,3-decadiene-6,8-dione		NI35443
74	173	101	59	85	45	127	87	336	100	70	38	38	37	25	23	23	0.30	15	28	8	–	–	–	–	–	–	–	–	–	–	D-Arabino-hexopyranoside, methyl 2,6-dideoxy-4-O-(6-deoxy-3-O-methyl β-D-allopyranosyl)-3-O-methyl-	55902-83-5	NI35444
152	99	125	179	157	250	194	165	336	100	37	26	24	16	14	13	10	0.00	15	29	6	–	–	–	–	–	–	–	1	–	–	Diethyl [3-(tert-butylacetoxy)-3-methyl-2-oxobutyl]phosphonate		NI35445
175	83	100	274	164	210	147	98	336	100	95	75	63	56	46	37	33	2.50	15	32	2	2	2	–	–	–	–	–	–	–	–	Urea, N-[6-(methylsulphinyl)hexyl]-N'-[6-(methylthio)hexyl]-, (±)-		NI35446
153	43	57	151	321	41	135	165	336	100	51	49	42	40	37	35	31	0.00	15	36	4	–	–	–	–	–	–	2	–	–	–	1,1,1-Tributoxy-3,3,3-trimethyldisiloxane		MS07982
336	338	155	266	225	268	136	337	336	100	75	54	41	33	27	24	23		16	10	–	–	2	–	2	–	–	–	–	–	–	1,4-Dithiin, 2,5-bis(4-chlorophenyl)-	2244-77-1	NI35447
104	77	147	161	175	43	293	51	336	100	61	31	29	26	25	22	18	6.40	16	14	2	2	–	–	2	–	–	–	–	–	–	2-Anilino-2',5'-dichloro-3-hydroxycrotonanilide		MS07983
318	301	239	112	122	199	336		336	100	95	74	65	53	45	10			16	17	3	–	–	–	–	1	–	–	–	–	–	Bis-(2-hydroxy-3-bromo-5-methyl-phenyl)-methane		LI07444
336	93	94	135	71	167	79	338	336	100	97	80	62	60	56	48	36		16	17	4	2	–	–	1	–	–	–	–	–	–	1-(1-Adamantyl)-2-chloro-3,5-dinitrobenzene		NI35448
336	94	93	79	338	280	279	67	336	100	95	47	39	37	24	24	20		16	17	4	2	–	–	1	–	–	–	–	–	–	1-(1-Adamantyl)-4-chloro-3,5-dinitrobenzene		NI35449
336	206	223	208	222	179	224	207	336	100	93	76	70	46	42	35	35		16	20	6	2	–	–	–	–	–	–	–	–	–	3-Methyl-4-(N-acetylalanyl)amino-5-hydroxy-8-methoxy-3,4-dihydroisocoumarin		LI07445
251	253	211	209	279	277	322	320	336	100	99	76	76	44	44	34	34	17.00	16	21	1	2	–	–	–	1	–	–	–	–	–	Dihydroflustramine C N-oxide		NI35450
336	127	301	337	153	243	242	154	336	100	33	27	22	17	16	16	16		17	12	4	4	–	–	–	–	–	–	–	–	–	2-Naphthaldehyde, (2,4-dinitrophenyl)hydrazone	1773-52-0	NI35452
336	127	301	288	156	337	139	242	336	100	42	30	27	26	21	20	18		17	12	4	4	–	–	–	–	–	–	–	–	–	2-Naphthaldehyde, (2,4-dinitrophenyl)hydrazone	1773-52-0	NI35451
336	301	337	242	243	255	289	216	336	100	26	22	18	17	16	14	11		17	12	4	4	–	–	–	–	–	–	–	–	–	1-Naphthalenecarboxaldehyde, (2,4-dinitrophenyl)hydrazone	1773-51-9	NI35454
336	127	301	337	128	255	153	154	336	100	53	30	20	20	19	19	18		17	12	4	4	–	–	–	–	–	–	–	–	–	1-Naphthalenecarboxaldehyde, (2,4-dinitrophenyl)hydrazone	1773-51-9	NI35453
58	71	70	57	75	42	111	59	336	100	21	10	10	7	6	5	3	0.00	17	18	1	2	–	–	2	–	–	–	–	–	–	O-(2-Di(methylamino)ethyl)-2-chlorophenyl-3-chlorophenylketoxime		MS07984
336	335	321	337	133	292	93	92	336	100	25	23	22	19	18	15	15		17	20	–	8	–	–	–	–	–	–	–	–	–	1,3,5-Triazine-2,4,6-triamine, N',N''-bis(3-aminophenyl)-N,N-dimethyl-	33933-65-2	NI35455
173	200	215	159	185	336	338	257	336	100	27	26	15	11	9	8	7		17	21	2	–	–	–	–	1	–	–	–	–	–	Cupalaurenol acetate		NI35456
226	228	174	257	215	268	270	173	336	100	98	85	58	57	40	38	37	3.50	17	21	2	–	–	–	–	1	–	–	–	–	–	Cyclolaurenol acetate		NI35457
278	293	200	43	199	237	115	58	336	100	75	45	40	35	33	30	28	0.00	17	21	2	–	–	–	–	1	–	–	–	–	–	Isolaurinterol acetate		LI07446
204	275	243	336	176	205	189	276	336	100	64	41	33	15	15	13	12		17	24	1	2	2	–	–	–	–	–	–	–	–	5,14-Diazagona-8-ene-2,11-dione-2-ethylenethioketal		NI35458
91	162	206	92	336	207	163	291	336	100	92	71	10	7	6	6	4		17	24	5	2	–	–	–	–	–	–	–	–	–	Benzyloxycarbonylvalylglycine methyl ester		LI07447
91	176	220	92	177	221	107	88	336	100	55	35	13	9	8	8	8	4.00	17	24	5	2	–	–	–	–	–	–	–	–	–	Glycine, N-[N-[(phenylmethoxy)carbonyl]-L-leucyl]-, methyl ester	5084-98-0	NI35459
116	73	221	117	44	220	45	74	336	100	41	10	10	7	4	4	3	0.00	17	29	–	2	–	1	–	–	–	2	–	–	–	1H-Indole-3-ethylamine, 5-fluoro-α-methyl-N,1-bis(trimethylsilyl)-	77590-54-6	NI35460
116	73	117	45	221	100	74	59	336	100	48	11	6	4	4	4	4	0.10	17	29	–	2	–	1	–	–	–	2	–	–	–	1H-Indole-3-ethylamine, 6-fluoro-α-methyl-N,1-bis(trimethylsilyl)-	77590-55-7	NI35461
179	73	308	147	180	281	309	45	336	100	64	28	23	20	8	7	7	0.02	17	29	–	2	–	1	–	–	–	2	–	–	–	1H-Indole-3-ethylamine, 5-fluoro-α-methyl-N,N(or N,1)-bis(trimethylsilyl)-	74367-10-5	NI35462
289	215	213	290	243	189	214	202	336	100	64	48	18	18	16	16	12	0.00	18	12	5	2	–	–	–	–	–	–	–	–	–	2-(2,4-Dinitrostyryl)-1-naphthol		NI35463
133	103	336	263	248	116	77	289	336	100	56	54	43	19	15	15	13		18	16	1	4	1	–	–	–	–	–	–	–	–	1-(4-Methoxyphenyl)-2-(methylthio)-6-phenyl-1H-imidazo[1,2-b][1,2,4]triazole		MS07985
336	77	201	337	151	96	67	173	336	100	43	34	23	14	13	10	9		18	16	3	4	–	–	–	–	–	–	–	–	–	1-Phenyl-3-methyl-4,5-dihydro-5-oxo-4-(4-carbomethoxyphenylazo)pyrazole		IC04013
275	276	247	220	77	245	51	277	336	100	55	32	28	22	21	10	8	3.00	18	16	3	4	–	–	–	–	–	–	–	–	–	1,2,3-Propanetriol, 1-(1-phenyl-1H-pyrazolo[3,4-b]quinoxalin-3-yl)-, [S-(R*,S*)]-	17460-16-1	NI35464
275	276	247	220	77	245	51	277	336	100	56	32	28	24	21	11	9	2.10	18	16	3	4	–	–	–	–	–	–	–	–	–	1,2,3-Propanetriol, 1-(1-phenyl-1H-pyrazolo[3,4-b]quinoxalin-3-yl)-, [S-(R*,S*)]-	17460-16-1	NI35465
171	335	93	276	61	105	43	172	336	100	98	65	55	50	39	32	28	20.00	18	16	3	4	–	–	–	–	–	–	–	–	–	1H-Pyrazole-4,5-dione, 3-[(acetyloxy)methyl]-1-phenyl-, 4-(phenylhydrazone)	35426-93-8	NI35466
152	181	210	336	338	280	282	308	336	100	88	79	47	29	23	15	9		18	18	2	–	–	–	2	–	–	–	–	–	–	2,2-Dichloro-1,1-bis(4-ethoxyphenyl)ethylene	2132-71-0	NI35467
146	145	55	57	43	41	69	105	336	100	27	27	22	21	18	18	15	5.14	18	18	5	–	–	–	–	–	–	–	–	2	–	α-L-Fucopyranose 1,2:3,4-bis(benzeneboronate)	102281-26-5	NI35468
149	150	41	57	56	205	104	76	336	100	13	10	7	6	5	5	5	0.00	18	24	6	–	–	–	–	–	–	–	–	–	–	Butyl carbobutoxymethylphthalate	85-70-1	NI35469
149	263	150	104	57	133	77	41	336	100	24	10	10	10	9	8	7	1.50	18	24	6	–	–	–	–	–	–	–	–	–	–	Butyl carbobutoxymethylphthalate	85-70-1	NI35470
149	150	41	57	56	223	205	104	336	100	13	10	7	6	5	5	5	0.00	18	24	6	–	–	–	–	–	–	–	–	–	–	Butyl carbobutoxymethylphthalate	85-70-1	MS07986
163	149	164	181	77	55	82	156	336	100	26	20	19	16	14	11	8	0.00	18	24	6	–	–	–	–	–	–	–	–	–	–	Dimethyl 2-carboxymethylhexylphthalate		NI35471
163	149	181	55	156	164	124	82	336	100	44	31	25	22	21	21	20	0.00	18	24	6	–	–	–	–	–	–	–	–	–	–	Dimethyl 2-ethyl-5-carboxypentylphthalate		NI35472
43	304	272	234	187	245	261	305	336	100	27	21	20	20	19	13	5	2.00	18	24	6	–	–	–	–	–	–	–	–	–	–	Dimethyl 4-(4-oxo-1-pentenyl)-3-(2-oxopropyl)-cyclohex-1-ene-1,4-dicarboxylate		MS07987
192	179	164	252	163	149	193	336	336	100	58	28	19	17	17	16	14		18	24	6	–	–	–	–	–	–	–	–	–	–	6-Methoxy-10-acetoxy-8,9-epoxythymol 3-isovalerate		NI35473
265	43	318	247	266	69	67	57	336	100	31	25	16	15	8	7	6	0.00	18	24	6	–	–	–	–	–	–	–	–	–	–	Sessiliflorene		DD01325
336	321	59	207	277	279	303	278	336	100	62	39	36	34	32	30	24		18	24	6	–	–	–	–	–	–	–	–	–	–	Sessiliflorol A		DD01326

MASS TO CHARGE RATIOS								M.W.	INTENSITIES								Parent	C	H	O	N	S	F	Cl	Br	I	Si	P	B	X	COMPOUND NAME	CAS Reg No	No
336	279	321	207	59	303	57	319	336	100	80	37	34	31	22	20	16		18	24	6	–	–	–	–	–	–	–	–	–	–	Sessiliflorol B		DD01327
91	300	209	165	123	181	110	301	336	100	91	57	53	27	26	23	22	0.00	18	25	2	2	–	–	1	–	–	–	–	–	–	1(2H)-Pyridinecarboxylic acid, 3,4-dihydro-5-(2-piperidinyl)-, benzyl ester, monohydrochloride	54966-07-3	NI35474
168	169	28	170	97	85	56	29	336	100	74	37	33	32	25	25	23	0.00	18	28	–	2	2	–	–	–	–	–	–	–	–	meso-N,N,N',N'-Tetraethyl-1,2-bis(2-thienyl)ethylenediamine		DD01328
168	169	28	192	170	97	56	29	336	100	25	13	11	11	10	9	8	0.00	18	28	–	2	2	–	–	–	–	–	–	–	–	(±)-N,N,N',N'-Tetraethyl-1,2-bis(2-thienyl)ethylenediamine		DD01329
112	42	250	56	41	238	43	100	336	100	79	66	66	40	39	28	25	1.20	18	28	4	2	–	–	–	–	–	–	–	–	–	4-(2-Morpholinylmethyl)-2-(2-ethoxyphenoxymethyl)morpholine		IC04014
59	87	336	234	206	73	337	221	336	100	49	45	42	27	24	16	16		18	36	–	–	–	–	–	–	–	3	–	–	–	1-Methyl-1,2-pentadien-4-yne-1,3,5-tris(ethyldimethylsilane)	61227-83-6	NI35475
336	89	308	265	237	280	191	249	336	100	40	32	20	18	15	12	10		19	12	6	–	–	–	–	–	–	–	–	–	–	2H-1-Benzopyran-2-one, 6-methoxy-7-[(2-oxo-2H-1-benzopyran-7-yl)oxy]	33621-49-7	NI35476
113	91	105	120	79	77	336	55	336	100	74	72	61	55	55	54	48		19	28	5	–	–	–	–	–	–	–	–	–	–	Androst-5-en-17-one, 3,8,12,14-tetrahydroxy-, (3β,12β,14β)-	6785-38-2	NI35477
113	91	105	120	336	79	77	55	336	100	74	72	61	53	53	53	48		19	28	5	–	–	–	–	–	–	–	–	–	–	Androst-5-en-17-one, 3,8,12,14-tetrahydroxy-, (3β,12β,14β)-	6785-38-2	LI07448
43	249	41	318	91	55	231	42	336	100	40	27	23	19	19	18	18	2.50	19	28	5	–	–	–	–	–	–	–	–	–	–	Androst-7-en-6-one, 2,3,14,17-tetrahydroxy-, (2β,3β,14α,17β)-		MS07988
71	43	336	293	248				336	100	75	11	9	4					19	28	5	–	–	–	–	–	–	–	–	–	–	3,10-Dimethyl-6-formyl-4-isobutyroyloxy-2-oxo-2,3,3a,4,5,8,9,10,11,11a-decahydro-cyclodeca[b]furan		LI07449
121	336	321	207	75	73	122	337	336	100	31	18	15	13	12	9	8		19	32	3	–	–	–	–	–	–	1	–	–	–	(4-Methoxyphenyl)nonanoic acid TMS		NI35478
121	336	75	321	147	122	73	77	336	100	26	22	16	10	10	10	9		19	32	3	–	–	–	–	–	–	1	–	–	–	(4-Methoxyphenyl)nonanoic acid TMS		NI35479
241	123	95	336	75	242	199	65	336	100	60	52	50	18	15	15	15		20	13	4	–	–	1	–	–	–	–	–	–	–	4-Fluoro-4'-(4-carboxy-phenoxy)-benzophenone		IC04015
321	336	229	337	322	105	77	305	336	100	95	28	22	22	22	20	9		20	16	5	–	–	–	–	–	–	–	–	–	–	2-Benzoyl-3-methyl-4,8-dimethoxybenzo[1,2-b:5,4-b]difuran		NI35480
89	221	336	63	77	104	117	249	336	100	43	28	23	18	17	16	15		20	16	5	–	–	–	–	–	–	–	–	–	–	$\Delta^2(^5H)$,α-Furanacetic acid, 3-methoxy-5-oxo-α,4-diphenyl-, methyl ester, (E)-	22628-22-4	NI35482
336	337	221	249	219	89	305	145	336	100	21	10	6	6	6	5	5		20	16	5	–	–	–	–	–	–	–	–	–	–	$\Delta^2(^5H)$,α-Furanacetic acid, 3-methoxy-5-oxo-α,4-diphenyl-, methyl ester, (E)-	22628-22-4	NI35481
280	336	281	337	319	282	139	91	336	100	89	73	25	15	14	14	13		20	16	5	–	–	–	–	–	–	–	–	–	–	1H,7H-Furo[3,2-c:5,4-f']bis[1]benzopyran-7-one, 2,3-dihydro-10-hydroxy-3,3-dimethyl-	18979-00-5	NI35483
336	281	280	282	337	252	253	197	336	100	99	90	66	60	40	33	16		20	16	5	–	–	–	–	–	–	–	–	–	–	Psoralidin	18642-23-4	LI07450
227	336	217	105	139	125	77	183	336	100	94	73	67	55	42	42	33		20	17	1	–	1	–	–	–	–	–	1	–	–	Acetophenone, 2-(diphenylphosphinothioyl)-	58156-55-1	NI35484
91	166	143	182	115	293	144	65	336	100	70	69	53	41	26	25	22	6.75	20	20	1	2	1	–	–	–	–	–	–	–	–	Propanamide, 2-amino-N-2-(naphthalenyl)-3-(benzylthio)-, (R)-	7436-63-7	NI35485
86	70	57	87	80	331	82	223	336	100	30	28	12	12	11	11	10	0.00	20	20	3	2	–	–	–	–	–	–	–	–	–	2-(tert-Butylamino)-N-(9,10-dioxaanthracen-2-yl)-ethanamide		MS07989
86	58	87	57	71	149	72	56	336	100	37	27	17	12	12	10	10	0.00	20	20	3	2	–	–	–	–	–	–	–	–	–	2-(Dimethylamino)-N-(9,10-dioxaanthracen-2-yl)-ethanamide		MS07990
43	119	174	77	336	202	51	293	336	100	99	98	97	63	58	54	47		20	20	3	2	–	–	–	–	–	–	–	–	–	4-Methyl-4-(3-oxobutyl)-1,2-diphenyl-3,5-pyrazolidinedione		MS07991
33	182	154	70	196	181	336	155	336	100	56	33	31	29	29	26	21		20	20	3	2	–	–	–	–	–	–	–	–	–	9H-Pyrrolo[1',2':2,3]isoindolo[4,5,6-cd]indol-9-one, 10-acetyl-2,6,6a,7,11a,11b-hexahydro-11-hydroxy-7,7-dimethyl-, (6aα,11aβ,11bα)-	18172-33-3	NI35486
182	336	181	196	70	155	154	337	336	100	75	56	49	46	43	41	20		20	20	3	2	–	–	–	–	–	–	–	–	–	9H-Pyrrolo[1',2':2,3]isoindolo[4,5,6-cd]indol-9-one, 10-acetyl-2,6,6a,7,11a,11b-hexahydro-11-hydroxy-7,7-dimethyl-, (6aα,11aβ,11bα)-	18172-33-3	NI35487
336	337	307	335	279	308	291	293	336	100	22	22	18	11	9	9	8		20	20	3	2	–	–	–	–	–	–	–	–	–	Schizozygine	2047-63-4	LI07451
73	163	111	139	97	131	91	84	336	100	99	88	85	73	66	61	60	10.00	20	26	2	–	–	2	–	–	–	–	–	–	–	4,4-Difluororetinoic acid (all-trans)	90660-20-1	NI35488
124	44	123	45	55	41	43	137	336	100	80	31	28	22	21	19	14	10.00	20	32	4	–	–	–	–	–	–	–	–	–	–	2H-1-Benzoxacyclohexadecin-16(18aH)-one, 3,4,5,6,7,8,9,10,11,12,13,14-dodecahydro-18,18a-dihydroxy-2-methyl-	65580-01-0	NI35489
124	123	45	125	137	41	55	43	336	100	25	22	17	15	15	14	8	6.00	20	32	4	–	–	–	–	–	–	–	–	–	–	2,5-Cyclohexadiene-1,4-dione, 2-hydroxy-6-(13-hydroxytetradecyl)-	65579-91-1	NI35490
220	57	189	248	153	168	245	196	336	100	78	77	72	65	45	43	42	39.00	20	32	4	–	–	–	–	–	–	–	–	–	–	Dimethyl 3-tert-butyl-4-(3,3-dimethyl-1-butenyl)-1-cyclohexene-1,4-dicarboxylate		MS07992
125	165	138	171	95	113	55	223	336	100	78	55	43	41	39	33	18	0.00	20	32	4	–	–	–	–	–	–	–	–	–	–	Ethyl ostopanate		DD01330
235	55	217	41	81	79	93	67	336	100	30	28	25	23	23	22	22	2.00	20	32	4	–	–	–	–	–	–	–	–	–	–	17β-Hydroxy-6-oxo-4,5-secoandrostan-4-oic acid, methyl ester		MS07993
277	121	278	259	41	93	43	55	336	100	22	21	21	21	19	17	16	7.90	20	32	4	–	–	–	–	–	–	–	–	–	–	2-Phenanthreneacetic acid, tetradecahydro-7-hydroxy-1,4b,8,8-tetramethyl-10-oxo-	57289-50-6	NI35491
219	336	237	235	220	247	289	191	336	100	39	29	22	22	20	10	10		20	32	4	–	–	–	–	–	–	–	–	–	–	Prostaglandin B1	13345-51-2	LI07452
155	199	198	95	55	41	77	139	336	100	79	44	19	18	18	15	14	1.86	20	33	2	–	–	–	–	–	–	–	1	–	–	Phosphinous acid, (4-methoxyphenyl)isopropyl-, 5-methyl-2-isopropylcyclohexyl ester	74685-37-3	NI35492
154	153	152	155	151	115	270	44	336	100	70	69	54	51	41	38	37	4.58	21	20	2	–	1	–	–	–	–	–	–	–	–	2,4,6-Trimethyl-4'-phenyldiphenylsulphone	3406-48-2	NI35493
162	147	161	146	336	148	175	159	336	100	99	84	80	76	44	36	36		21	20	4	–	–	–	–	–	–	–	–	–	–	Benzo[1,2-b:3,4-b']bisbenzofuran-7(6H)-one, 5a,7a,12a,12b-tetrahydro-12a-hydroxy-2,9,12b-trimethyl-	21272-75-3	NI35494
162	135	336	161	146	190	40	148	336	100	64	62	52	51	44	44	36		21	20	4	–	–	–	–	–	–	–	–	–	–	Benzo[1,2-b:5,4-b']bisbenzofuran-12(6H)-one, 5a,6a,11b,12a-tetrahydro-5a-hydroxy-2,10,11b-trimethyl-	21272-74-2	NI35495
162	147	161	146	336	148	175	159	336	100	98	84	80	75	44	36	36		21	20	4	–	–	–	–	–	–	–	–	–	–	Benzo[1,2-b:3,4-b']bisbenzofuran-12(11H)-one, 5a,5b,10a,12a-tetrahydro-5a-hydroxy-2,5b,7-trimethyl-		MS07994

MASS TO CHARGE RATIOS								M.W.	INTENSITIES								Parent	C	H	O	N	S	F	Cl	Br	I	Si	P	B	X	COMPOUND NAME	CAS Reg No	No
162	135	336	161	146	190	40	148	336	100	62	60	52	50	44	44	36		21	20	4	–	–	–	–	–	–	–	–	–	–	Benzo[1,2-b:5,4-b']bisbenzofuran-12(6H)-one, 5a,5b,11b,12a-tetrahydro-5a-hydroxy-2,10,11b-trimethyl-		LI07453
43	294	336	293	178	235	251	189	336	100	62	27	10	5	2	1	1		21	20	4	–	–	–	–	–	–	–	–	–	–	Benzofuran, 2-(4'-acetylphenyl)-7-methoxy-3-methyl-5-(E)-propenyl-		MS07995
321	336	149	322	160	167	153	185	336	100	78	76	38	30	27	26	23		21	20	4	–	–	–	–	–	–	–	–	–	–	2H,6H-Benzofuro[3,2-c]pyrano[2,3-h][1]benzopyran, 6a,11a-dihydro-9-methoxy-2,2'-dimethyl-, (6aS-cis)-	66446-92-2	NI35496
336	308	249	337	78	233	198	189	336	100	30	28	24	13	10	9	9		21	20	4	–	–	–	–	–	–	–	–	–	–	11H-Cyclopenta[a]phenanthrene-15-carboxylic acid, 12,13,16,17-tetrahydro-3-methoxy-13-methyl-17-oxo-, methyl ester, (S)-	56588-56-8	NI35497
217	321	336	218					336	100	38	36	12						21	20	4	–	–	–	–	–	–	–	–	–	–	Obovatin methyl ether	69640-78-4	NI35498
189	29	263	27	218	190	165	336	336	100	86	47	31	24	23	23	19		21	20	4	–	–	–	–	–	–	–	–	–	–	Spiro[cyclopropane-1,9'-[9H]fluorene]-2,3-dicarboxylic acid, diethyl ester, cis-	41411-72-7	NI35499
301	91	121	119	105	302	223	131	336	100	92	79	72	64	57	55	39	0.00	21	21	–	–	–	–	1	–	–	1	–	–	–	Silane, chlorotris(phenylmethyl)-	18740-59-5	NI35500
170	197	181	336	98	149	337	212	336	100	73	48	33	19	18	17	11		21	24	2	2	–	–	–	–	–	–	–	–	–	Alstophyllan-19-one	25921-22-6	NI35501
170	70	197	181	62	91	182	171	336	100	80	74	49	31	26	24	18	12.40	21	24	2	2	–	–	–	–	–	–	–	–	–	Alstophyllan-19-one	25921-22-6	MS07996
307	266	336	42	308	29	267	264	336	100	94	32	26	22	20	18	16		21	24	2	2	–	–	–	–	–	–	–	–	–	Apovincamine	4880-92-6	MS07997
307	266	336	308	267	306	264	251	336	100	90	31	21	18	14	14	11		21	24	2	2	–	–	–	–	–	–	–	–	–	Apovincamine	4880-92-6	NI35502
92	107	135	106	44	65	168	197	336	100	99	49	48	42	40	36	35	5.00	21	24	2	2	–	–	–	–	–	–	–	–	–	Aspidospermidine-3-carboxylic acid, 2,3,6,7-tetradehydro-, methyl ester, (5α,12β,19α)-	4429-63-4	NI35503
208	321	77	209	57	322	165	180	336	100	45	19	16	11	10	9	7	4.99	21	24	2	2	–	–	–	–	–	–	–	–	–	1-(tert-Butyl)-3-ethyl-5,5-diphenyl-2,4-imidazolidinedione	52531-80-3	NI35504
134	170	84	336	135	86	156	230	336	100	77	65	63	48	43	38	35		21	24	2	2	–	–	–	–	–	–	–	–	–	2,20-Cycloaspidospermidine-3-carboxylic acid, 6,7-didehydro-, methyl ester, (2α,3α,5α,12β,19α,20R)-	6822-38-4	NI35505
336	134	170	135	156	230	337	120	336	100	80	62	39	30	28	24	23		21	24	2	2	–	–	–	–	–	–	–	–	–	2,20-Cycloaspidospermidine-3-carboxylic acid, 6,7-didehydro-, methyl ester, (2α,3α,5α,12β,19α,20R)-	6822-38-4	NI35506
170	336	135	134	215	169	168	122	336	100	98	87	87	39	36	32	31		21	24	2	2	–	–	–	–	–	–	–	–	–	2,20-Cycloaspidospermidine-3-carboxylic acid, 6,7-didehydro-, methyl ester, (2α,3β,5α,12β,19α,20R)-	5980-02-9	NI35507
336	182	183	335	337	168	277	321	336	100	79	71	63	41	34	32	23		21	24	2	2	–	–	–	–	–	–	–	–	–	Deshydroxymethylvoachalotine	2912-15-4	MS07998
208	321	209	77	57	165	322	104	336	100	45	20	19	11	10	8	7	4.50	21	24	2	2	–	–	–	–	–	–	–	–	–	2,4-Imidazolidinedione, 1-tert-butyl-3-ethyl-5,5-diphenyl-	56954-70-2	NI35508
336	174	307	213	212	200	187	293	336	100	40	24	12	12	11	10	9		21	24	2	2	–	–	–	–	–	–	–	–	–	Purpeline	2246-33-5	LI07454
169	336	335	277	43	168	263	41	336	100	92	80	79	75	60	20	19		21	24	2	2	–	–	–	–	–	–	–	–	–	Sarpagan-17-ol, acetate	3986-01-4	NI35509
43	95	192	109	235	191	217	137	336	100	82	55	53	47	44	37	25	2.00	21	36	3	–	–	–	–	–	–	–	–	–	–	12(R),13(S)-14-Acetoxy-8,12-epoxy-15-norlabdane		MS07999
43	95	191	144	218	276	123	293	336	100	36	33	30	18	15	15	13	2.00	21	36	3	–	–	–	–	–	–	–	–	–	–	12(R,S)-13-Acetoxy-8,12-epoxy-15-norlabdane		MS08000
43	191	109	95	81	192	137	276	336	100	50	30	30	30	25	20	18	3.00	21	36	3	–	–	–	–	–	–	–	–	–	–	12(S),13(S)-14-Acetoxy-8,12-epoxy-15-norlabdane		MS08001
279	280	57	156	41	87	69	43	336	100	20	14	9	8	7	6	5	0.20	21	36	3	–	–	–	–	–	–	–	–	–	–	1,3,5-Tris(2,2-dimethyl-1-hydroxy-propyl)-benzene		LI07455
70	41	55	43	57	69	140	95	336	100	68	60	54	40	32	30	30	12.00	21	36	3	–	–	–	–	–	–	–	–	–	–	(3R,4S,2E)-2-(Hexadec-15-enylidene)-3-hydroxy-4-methylbutanolide		NI35510
70	43	41	57	55	69	81	95	336	100	66	60	50	45	36	35	32	5.00	21	36	3	–	–	–	–	–	–	–	–	–	–	(3R,4S,2Z)-2-(Hexadec-15-enylidene)-3-hydroxy-4-methylbutanolide		NI35511
70	41	55	43	57	81	95	69	336	100	87	78	68	50	44	40	40	7.00	21	36	3	–	–	–	–	–	–	–	–	–	–	(3S,4S,2E)-2-(Hexadec-15-enylidene)-3-hydroxy-4-methylbutanolide		NI35512
70	57	43	41	95	83	55	81	336	100	80	62	55	50	47	40	30	5.00	21	36	3	–	–	–	–	–	–	–	–	–	–	(3S,4S,2Z)-2-(Hexadec-15-enylidene)-3-hydroxy-4-methylbutanolide		NI35513
135	109	43	95	123	117	69	226	336	100	90	87	82	80	75	73	70	1.00	21	36	3	–	–	–	–	–	–	–	–	–	–	Methyl 3,4,4a,5,6,7,8,8a-octahydro-β-hydroxy-β,2,5,5,8a-pentamethyl-1-naphthalenepentanoate		NI35514
43	81	95	137	122	123	136	121	336	100	55	45	40	37	35	30	27	2.00	21	36	3	–	–	–	–	–	–	–	–	–	–	Methyl 3(S),6(R,S)-3,6-dimethyl-6-[3,3-dimethyl-2-(3-oxobutyl)-1-cyclohexenyl]-hexanoate		MS08002
191	144	81	95	109	137	123	121	336	100	90	77	76	70	46	45	40	5.00	21	36	3	–	–	–	–	–	–	–	–	–	–	Methyl 12(R)-8,12-epoxylabdan-15-oate		MS08003
191	321	95	81	109	101	144	137	336	100	76	65	60	47	45	40	36	2.00	21	36	3	–	–	–	–	–	–	–	–	–	–	Methyl 12(S)-8,12-epoxylabdan-15-oate		MS08004
321	54	245	137	177	135	322	303	336	100	79	68	65	25	25	24	24	0.00	21	36	3	–	–	–	–	–	–	–	–	–	–	1H-Naphtho[2,1-b]pyran-3-acetic acid, dodecahydro-3,4a,7,7,10a-pentamethyl-, methyl ester	56247-60-0	NI35515
43	55	41	307	81	121	29	67	336	100	48	48	43	33	31	31	23	0.53	21	36	3	–	–	–	–	–	–	–	–	–	–	1H-Naphtho[2,1-b]pyran-7-carboxylic acid, 3-ethyldodecahydro-3,4a,7,10a-tetramethyl-, methyl ester, [3S-(3α,4aβ,6aα,7β,10aβ,10bα)]-	56630-93-4	NI35516
98	121	82	153	95	111	107	318	336	100	85	43	40	40	38	31	29	25.00	21	40	1	2	–	–	–	–	–	–	–	–	–	1H-Imidazole-2-methanol, α-heptadecyl-	69393-16-4	NI35517
97	121	107	318	82	149	55	69	336	100	37	37	28	27	14	13	12	4.00	21	40	1	2	–	–	–	–	–	–	–	–	–	1H-Imidazole-4-methanol, α-heptadecyl-	69393-20-0	NI35518
321	336	77	322	104	103	337	206	336	100	58	33	21	21	15	15	14		22	16	–	4	–	–	–	–	–	–	–	–	–	1-Amino-2,6,6-tricyano-3,5-diphenyl-5-methylcyclohexa-1,3-diene	71908-64-0	NI35519
201	202	336	173	245	121	158	129	336	100	40	37	17	12	12	11	11		22	24	3	–	–	–	–	–	–	–	–	–	–	(S)-4-(4-Methoxyphenyl)-4-[2-(benzyloxy)ethyl]cyclohex-2-en-1-one		DD01331
146	245	189	42	105	57	44	246	336	100	97	66	37	27	22	19	18	0.00	22	28	1	2	–	–	–	–	–	–	–	–	–	Propanamide, N-phenyl-N-[1-(2-phenylethyl)-4-piperidinyl]-	437-38-7	NI35520
57	41	29	61	188	30	132	43	336	100	55	38	36	30	26	25	23	3.50	22	37	–	–	–	–	1	–	–	–	–	–	–	Benzene, 2-(chloromethyl)-1,3,5-tris(2,2-dimethylpropyl)-	40572-18-7	MS08005
43	55	41	57	182	71	98	69	336	100	87	67	57	53	52	47	38	12.00	22	40	2	–	–	–	–	–	–	–	–	–	–	Cyclododecanone, 2-decanoyl-	29653-08-5	NI35521
55	41	224	28	43	98	83	69	336	100	78	70	67	57	41	32	32	6.00	22	40	2	–	–	–	–	–	–	–	–	–	–	1,3-Cyclotetradecanedione, 2-octyl-	29653-06-3	NI35522
55	85	79	41	84	97	67	83	336	100	82	67	62	61	54	48	45	0.00	22	40	2	–	–	–	–	–	–	–	–	–	–	2H-Pyran, 2-(2-heptadecynyloxy)tetrahydro-	69502-96-1	NI35523

MASS TO CHARGE RATIOS								M.W.	INTENSITIES								Parent	C	H	O	N	S	F	Cl	Br	I	Si	P	B	X	COMPOUND NAME	CAS Reg No	No
55	85	41	43	81	84	97	95	**336**	100	78	75	58	53	49	47	47	**6.24**	22	40	2	–	–	–	–	–	–	–	–	–	–	2H-Pyran, 2-(3-heptadecynyloxy)tetrahydro-	56666-94-5	NI35524
91	77	329	79	155	46	78	62	**336**	100	41	35	22	21	19	18	18	**0.00**	22	40	2	–	–	–	–	–	–	–	–	–	–	2H-Pyran, 2-(5-heptadecynyloxy)tetrahydro-	56666-93-4	NI35525
55	41	43	67	81	91	79	95	**336**	100	83	76	75	74	70	54	52	**0.00**	22	40	2	–	–	–	–	–	–	–	–	–	–	2H-Pyran, 2-(7-heptadecynyloxy)tetrahydro-	56599-50-9	NI35526
55	41	83	198	81	117	254	115	**336**	100	67	50	49	49	44	42	32	**18.58**	22	41	–	–	–	–	–	–	–	–	1	–	–	Phosphine, dicyclohexyl[2(or 5)-methyl-5(or 2)-isopropylcyclohexyl]-	74764-13-9	NI35527
336	337	308	306	168	154	153	338	**336**	100	26	16	13	12	12	10	8		23	12	1	–	1	–	–	–	–	–	–	–	–	14H-Benzo[b]benzo[3,4]fluoreno[2,1-d]thiophen-14-one		LI07456
91	83	171	117	129	115	55	65	**336**	100	22	14	12	10	10	10	7	**5.00**	23	28	2	–	–	–	–	–	–	–	–	–	–	1,1-Dimethyl-4,4-dibenzyl-cyclohexan-3a-methoxy-5-one		LI07457
97	83	69	95	336	98	81	96	**336**	100	99	87	84	81	77	68	65		23	44	1	–	–	–	–	–	–	–	–	–	–	Cyclopropaneundecanal, 2-nonyl-	56196-17-9	NI35528
336	259	231	105	337	77	202	307	**336**	100	88	54	41	25	25	19	18		24	16	2	–	–	–	–	–	–	–	–	–	–	2,3-Diphenylnaphtho[2,3-b][1,4]dioxin	91201-57-9	NI35529
202	231	289	291	336				**336**	100	62	15	11	2					24	16	2	–	–	–	–	–	–	–	–	–	–	7,12-Epoxy-7,12-dihydro-12-phenylpleiaden-7-ol		LI07458
259	182	336	181	105	260	183	258	**336**	100	90	39	35	27	22	18	17		24	20	–	–	–	–	–	–	–	1	–	–	–	Tetraphenylsilane	1048-08-4	NI35530
259	182	181	336	260	105	180	258	**336**	100	67	42	39	25	20	16	15		24	20	–	–	–	–	–	–	–	1	–	–	–	Tetraphenylsilane	1048-08-4	NI35531
78	182	259	184	130	75	105	147	**336**	100	96	90	70	65	64	53	52	**30.00**	24	20	–	–	–	–	–	–	–	1	–	–	–	Tetraphenylsilane	1048-08-4	LI07459
336	168	167	337	44	77	43	166	**336**	100	49	27	26	26	23	20	12		24	20	–	2	–	–	–	–	–	–	–	–	–	[1,1′-Biphenyl]-4,4′-diamine, N,N′-diphenyl-	531-91-9	NI35532
233	336	234	335	337	77	232	115	**336**	100	64	21	20	18	17	16	16		24	20	–	2	–	–	–	–	–	–	–	–	–	4H-1,2-Diazepine, 5-(4-methylphenyl)-3,7-diphenyl-	25649-71-2	NI35533
169	168	167	28	336	78	77	51	**336**	100	77	39	24	17	15	15	15		24	20	–	2	–	–	–	–	–	–	–	–	–	Tetraphenylhydrazine	632-52-0	MS08006
169	168	167	336	78	77	51	170	**336**	100	77	39	17	15	15	15	13		24	20	–	2	–	–	–	–	–	–	–	–	–	Tetraphenylhydrazine	632-52-0	LI07460
168	169	167	236	77	237	51	170	**336**	100	46	37	25	9	8	8	7		24	20	–	2	–	–	–	–	–	–	–	–	–	Tetraphenylhydrazine	632-52-0	IC04016
83	97	55	69	57	43	41	71	**336**	100	97	90	84	76	70	56	53	**43.20**	24	48	–	–	–	–	–	–	–	–	–	–	–	Cyclotetracosane	297-03-0	NI35534
259	165	336	181	260	152	337	166	**336**	100	41	33	24	20	11	9	8		25	20	1	–	–	–	–	–	–	–	–	–	–	Phenol, 4-(triphenylmethyl)-	978-86-9	NI35535
91	92	105	41	336	43	104	29	**336**	100	93	15	11	10	10	8	8		25	36	–	–	–	–	–	–	–	–	–	–	–	1-Phenyl-3-(2-phenylethyl)undecane		AI02065
92	91	105	43	41	336	104	93	**336**	100	80	14	10	10	8	8	8		25	36	–	–	–	–	–	–	–	–	–	–	–	1-Phenyl-3-(2-phenylethyl)undecane		AI02064
321	336	243	91	291	229	103	215	**336**	100	90	71	31	28	24	24	21		26	24	–	–	–	–	–	–	–	–	–	–	–	8-tert-Butyl-3,8-diphenylbenzofulvene		NI35536
265	336	266	337	263	334	43	41	**336**	100	71	25	19	16	10	9	8		26	24	–	–	–	–	–	–	–	–	–	–	–	3-Hexylperylene		AI02066
113	151	36	132	46	32	94	38	**337**	100	90	62	56	54	24	22	22	**0.00**	–	–	–	1	2	6	2	–	–	–	–	–	1	Dichlorodithionitronium hexafluoroarsenate(V)	60563-08-8	NI35537
69	260	258	179	64	96	27	91	**337**	100	38	38	33	25	23	23	9	**0.00**	4	3	–	1	–	6	–	2	–	–	–	–	–	1-Bis(trifluoromethyl)amino-1,2-dibromoethane		LI07461
210	121	128	80	337	138	94	41	**337**	100	34	22	21	17	11	11	10		11	16	3	1	–	–	–	–	1	–	–	–	–	2,5-Methanofuro[3,2-b]pyridine-4(2H)-carboxylic acid, hexahydro-8-iodo-, ethyl ester, (2α,3aβ,5α,7aβ,8S*)-	49656-59-9	NI35538
210	121	80	128	43	56	41	55	**337**	100	49	31	30	25	19	17	16	**16.00**	11	16	3	1	–	–	–	–	1	–	–	–	–	2,5-Methanofuro[3,2-b]pyridine-4(2H)-carboxylic acid, hexahydro-8-iodo-, ethyl ester, (2α,3aβ,5α,7aβ,8R*)-	49656-58-8	NI35539
61	75	221	43	203	263	337	62	**337**	100	72	34	32	30	20	18	15		11	16	3	1	1	5	–	–	–	–	–	–	–	N(O,S)-Pentafluoropropionyl-D-methionine isopropyl ester		MS08007
61	75	221	43	203	263	62	337	**337**	100	72	34	33	30	19	16	15		11	16	3	1	1	5	–	–	–	–	–	–	–	N(O,S)-Pentafluoropropionyl-L-methionine isopropyl ester		MS08008
336	338	337	339	209	250	223	210	**337**	100	98	17	15	12	11	10	9		12	10	3	4	–	–	–	1	–	–	–	–	–	1H-Benzimidazolium, 5-bromo-1-(hexahydro-2,4,6-trioxo-5-pyrimidinyl)-3-methyl-	56247-90-6	NI35540
112	113	143	85	125	144	114	124	**337**	100	50	39	32	22	16	15	10	**0.00**	12	17	2	3	1	–	2	–	–	–	–	–	–	Oxythiamine dihydrochloride		LI07462
73	220	218	100	45	221	219	147	**337**	100	61	55	18	16	12	11	11	**0.00**	12	31	2	1	1	–	–	–	–	3	–	–	–	L-Cysteine, N,S-bis(trimethylsilyl)-, trimethylsilyl ester	7364-50-3	NI35541
73	220	218	75	221	219	147	100	**337**	100	89	79	22	18	16	15	15	**0.00**	12	31	2	1	1	–	–	–	–	3	–	–	–	L-Cysteine, tris(trimethylsilyl)-	56272-69-6	NI35542
73	220	218	147	75	45	100	59	**337**	100	42	39	23	23	17	14	10	**2.00**	12	31	2	1	1	–	–	–	–	3	–	–	–	L-Cysteine, tris(trimethylsilyl)-	56272-69-6	MS08009
220	218	73	45	100	147	40	75	**337**	100	99	99	86	73	42	41	38	**2.80**	12	31	2	1	1	–	–	–	–	3	–	–	–	L-Cysteine, tris(trimethylsilyl)-	56272-69-6	MS08010
112	226	254	44	73	41	211	69	**337**	100	49	42	26	25	20	16	15	**0.00**	13	32	3	1	–	–	–	–	–	2	1	–	–	Phosphonic acid, [1-[(2,2-dimethylpropylidene)amino]ethyl]-, bis(trimethylsilyl) ester	55108-73-1	NI35543
226	73	225	254	41	322	98	240	**337**	100	98	77	65	57	46	39	38	**2.00**	13	32	3	1	–	–	–	–	–	2	1	–	–	Phosphonic acid, [2-[(2,2-dimethylpropylidene)amino]ethyl]-, bis(trimethylsilyl) ester	55108-78-6	NI35544
112	70	113	73	58	55	42	41	**337**	100	9	8	7	4	4	4	4	**0.00**	13	32	3	1	–	–	–	–	–	2	1	–	–	Phosphonic acid, [1-[isopropylideneamino]butyl]-, bis(trimethylsilyl) ester	55108-66-2	NI35545
112	73	55	113	98	42	41	63	**337**	100	10	9	8	8	8	6	5	**0.00**	13	32	3	1	–	–	–	–	–	2	1	–	–	Phosphonic acid, [2-methyl-1-[isopropylideneamino]propyl]-, bis(trimethylsilyl) ester	55108-68-4	NI35546
147	73	161	102	45	89	101	75	**337**	100	88	49	40	40	34	22	20	**0.00**	14	31	6	1	–	–	–	–	–	1	–	–	–	D-Galactose, 2,3,4,6-tetra-O-methyl-5-O-(trimethylsilyl)-, O-methyloxime	56196-38-4	NI35547
171	107	77	75	123	166	111	92	**337**	100	55	49	42	27	25	24	24	**13.69**	15	12	4	1	1	–	1	–	–	–	–	–	–	5-Chloro-2,3-dihydro-1-(4-methoxyphenylsulphonyl)-3-oxoindole	83372-83-2	NI35548
162	43	77	41	57	55	63	69	**337**	100	52	36	34	29	24	23	19	**3.35**	15	12	4	1	1	–	1	–	–	–	–	–	–	1-(4-Chlorophenylsulphonyl)-2,3-dihydro-5-methoxy-3-oxoindole	83372-82-1	NI35549
148	149	234	135	162	206	248	136	**337**	100	82	79	79	66	60	31	25	**16.00**	15	23	4	5	–	–	–	–	–	–	–	–	–	6-(3-Methylbutylamino)-9β-D-ribofuranosylpurine		LI07463
146	43	44	262	277	55	57	41	**337**	100	93	87	85	83	79	78	75	**51.00**	16	7	6	3	–	–	–	–	–	–	–	–	–	Trinitropyrene		IC04017
173	175	145	147	337	109	177	174	**337**	100	65	18	12	11	11	11	8		16	13	3	1	–	–	2	–	–	–	–	–	–	Ethyl 3-[1-(2,6-dichlorobenzoyl)pyrrol-2-yl]prop-2-enoate	95883-17-3	NI35550
152	109	91	65	182	110	151	111	**337**	100	88	81	64	34	34	32	22	**0.00**	16	19	3	1	2	–	–	–	–	–	–	–	–	(±)-N-(2-Hydroxyisopropyl)-N-(phenylsulphenyl)-p-toluenesulphonamide		DD01332

MASS TO CHARGE RATIOS								M.W.	INTENSITIES								Parent	C	H	O	N	S	F	Cl	Br	I	Si	P	B	X	COMPOUND NAME	CAS Reg No	No
77	220	94	337	141	79	154	81	337	100	90	82	73	66	62	56	49		16	19	5	1	1	–	–	–	–	–	–	–	–	2-Oxa-6-azatricyclo[3.3.1.1^{3,7}]decan-4-ol, 6-(phenylsulphonyl)-, acetate, (1α,3β,4β,5α,7β)-	50267-33-9	NI35551
43	151	193	28	235	15	150	149	337	100	72	48	32	23	16	15	10	0.00	16	19	7	1	–	–	–	–	–	–	–	–	–	Acetamide, N-[2-(acetyloxy)-2-[3,4-bis(acetyloxy)phenyl]ethyl]-	55191-56-5	MS08011
151	193	235	43	277	49	236	194	337	100	97	79	76	25	22	19	13	0.00	16	19	7	1	–	–	–	–	–	–	–	–	–	Acetamide, N-[2-(acetyloxy)-2-[3,4-bis(acetyloxy)phenyl]ethyl]-	55191-56-5	MS08012
208	236	156	128	127	264	29	81	337	100	84	72	69	55	54	52	46	26.49	16	20	5	1	–	–	–	–	–	–	1	–	–	Propenoic acid, 3-(3-cyanophenyl)-2-(diethoxyphosphinyl)-, ethyl ester	107846-71-9	NI35552
208	236	128	264	156	127	29	81	337	100	69	59	56	55	48	48	46	14.87	16	20	5	1	–	–	–	–	–	–	1	–	–	Propenoic acid, 3-(4-cyanophenyl)-2-(diethoxyphosphinyl)-, ethyl ester	107846-65-1	NI35553
87	41	43	57	55	29	56	28	337	100	95	82	79	73	71	68	56	0.00	16	35	4	1	1	–	–	–	–	–	–	–	–	1-Dodecanesulphonamide, N,N-bis(2-hydroxyethyl)-	27854-50-8	WI01440
293	321	337	322	294	151			337	100	55	40	37	20	16				17	11	5	3	–	–	–	–	–	–	–	–	–	2,4-Dimethoxy-6-(1,3-dioxo-1H,3H-naphtho[1,8-cd]pyran-6-yl)-s-triazine		IC04018
72	125	265	267	127	337	165	152	337	100	60	45	30	18	8	8	7		17	17	2	1	–	–	2	–	–	–	–	–	–	Acetamide, N,N-dimethyl-2-(4-chlorophenyl)-2-(2-methoxy-4-chlorophenyl)-	63953-34-4	NI35554
265	267	139	111	72	266	141	269	337	100	65	42	19	16	15	15	11	0.00	17	17	2	1	–	–	2	–	–	–	–	–	–	Benzeneacetamide, 4-chloro-α-(4-chlorophenyl)-α-methoxy-N,N-dimethyl-	63953-29-7	NI35555
168	140	126	337	84	43	238	69	337	100	84	63	58	54	51	31	29		17	23	6	1	–	–	–	–	–	–	–	–	–	Methyl 2-hydroxy-4-oxo-4-(4-hexylamino-6-methyl-2-pyrone-3-yl)-2-butenoate	86488-08-6	NI35556
112	43	30	196	154	252	214	55	337	100	62	42	40	37	33	33	32	21.10	17	27	4	3	–	–	–	–	–	–	–	–	–	Acetamide, N-[4-(1-acetyl-2,5-dioxo-4-imidazolidinyl)butyl]-N-cyclohexyl-, (±)-	77828-56-9	NI35557
266	250	264	322	267	338	294	251	337	100	48	26	23	23	13	13	13	0.00	17	31	2	1	–	–	–	–	–	2	–	–	–	Salsoline, di-TMS		NI35558
261	76	239	50	90	262	54	337	337	100	42	40	29	28	22	22	21		18	12	2	3	–	–	1	–	–	–	–	–	–	2-(2-Chlorophenyl)-3-(3-methyl-5-isoxazolyl)-4(3H)-quinazolinone	90059-39-5	NI35559
239	337	296	241	76	339	90	240	337	100	59	44	38	27	22	20	19		18	12	2	3	–	–	1	–	–	–	–	–	–	2-(4-Chlorophenyl)-3-(3-methyl-5-isoxazolyl)-4(3H)-quinazolinone	90059-40-8	NI35560
145	144	132	146	103	320	104	75	337	100	90	88	85	70	60	45	45	10.00	18	15	4	3	–	–	–	–	–	–	–	–	–	3-Nitro-N-[(1,2,3,4-tetrahydroisoquinolin-2-yl)methyl]phthalimide	96919-08-3	NI35561
145	132	146	144	105	104	75	117	337	100	99	55	50	43	40	20	15	5.00	18	15	4	3	–	–	–	–	–	–	–	–	–	4-Nitro-N-[(1,2,3,4-tetrahydroisoquinolin-2-yl)methyl]phthalimide	96919-07-2	NI35562
140	82	122	123	211	138	106	337	337	100	90	59	53	49	47	39	33		18	27	5	1	–	–	–	–	–	–	–	–	–	Hastacine		LI07464
82	140	122	123	211	138	337	96	337	100	96	65	55	53	52	27	26		18	27	5	1	–	–	–	–	–	–	–	–	–	Neoplatyphylline		LI07465
140	82	138	122	211	123	96	337	337	100	98	64	63	60	55	29	27		18	27	5	1	–	–	–	–	–	–	–	–	–	Platyphylline	480-78-4	LI07466
140	82	122	138	123	211	96	43	337	100	67	65	61	48	40	27	25	10.00	18	27	5	1	–	–	–	–	–	–	–	–	–	Platyphylline	480-78-4	NI35563
278	41	55	70	80	199	94	337	337	100	99	70	63	44	33	30	1		18	31	3	3	–	–	–	–	–	–	–	–	–	Palustridine		LI07467
218	91	337	92	69	338	65	89	337	100	99	98	97	72	33	31	28		19	15	3	1	1	–	–	–	–	–	–	–	–	2-Benzylthio-7,8-dimethylpyrano[2,3-e]benzoxazol-6(H)-one	88973-19-7	NI35564
337	294	251	322	338	236	279	208	337	100	40	34	30	22	21	20	18		19	15	5	1	–	–	–	–	–	–	–	–	–	9H-Azuleno[1,2,3-ij]isoquinolin-9-one, 10-hydroxy-4,5,6-trimethoxy-	74631-22-4	NI35565
337	322	338	294	323	262	307	290	337	100	24	23	17	11	11	10	9		19	15	5	1	–	–	–	–	–	–	–	–	–	7H-Dibenzo[de,g]quinolin-7-one, 9-hydroxy-1,2,10-trimethoxy-	5140-35-2	NI35566
337	294	255	45	338	41	295	240	337	100	69	46	42	23	17	14	14		20	19	4	1	–	–	–	–	–	–	–	–	–	1,4-Dihydroxy-2-cyclohexylaminoanthraquinone		IC04019
322	337	323	278	308	338	307	277	337	100	51	21	21	16	13	10	7		20	19	4	1	–	–	–	–	–	–	–	–	–	11-Hydroxyacronine		LI07468
306	307	337	262	308	338	292	261	337	100	23	21	18	10	7	7	7		20	19	4	1	–	–	–	–	–	–	–	–	–	7H-Pyrano[2,3-c]acridine, 3,12-dimethyl-6-methoxy-7-oxo-3-hydroxymethyl-		LI07469
337	207	281	208	266	248	322	235	337	100	90	70	35	30	30	25	25		20	23	–	3	1	–	–	–	–	–	–	–	–	2-(tert-Butylamino)-5,5-diphenyl-4-(methylthio)-5-isoimidazole	118515-00-7	DD01333
180	43	250	181	179	71	41	152	337	100	42	18	15	9	9	7	6	0.10	21	23	3	1	–	–	–	–	–	–	–	–	–	6-(1-Carboxyisopropyl)-5,6-dihydro-5-isobutyrylphenanthridine		LI07470
161	71	43	77	337	105	231	146	337	100	59	54	34	33	21	19	16		21	23	3	1	–	–	–	–	–	–	–	–	–	α-(3,3-Dimethyl-2-oxoindol-1-yl)benzyl isobutyrate		LI07471
91	337	231	158	338	307	262	292	337	100	66	53	49	22	20	15	14		21	23	3	1	–	–	–	–	–	–	–	–	–	Ethyl 1-[(benzyloxy)methyl]-2-ethylindole-3-carboxylate		DD01334
337	235	195	57	237	236	221	338	337	100	52	48	39	38	32	28	22		21	27	1	3	–	–	–	–	–	–	–	–	–	1-Methyllysergic acid diethylamide		MS08013
105	268	107	232	337	94			337	100	52	23	3	2	1				22	27	2	1	–	–	–	–	–	–	–	–	–	N-Benzoylnuphenine		LI07472
132	216	57	217	43	85	41	77	337	100	98	97	81	60	34	28	23	0.00	22	27	2	1	–	–	–	–	–	–	–	–	–	u-2-tert-Butoxy-1,2,3,4-tetrahydro-α-methyl-α-phenyl-1-isoquinolinemethanol		DD01335
59	72	55	41	43	69	57	83	337	100	61	54	44	41	34	22	16	8.00	22	43	1	1	–	–	–	–	–	–	–	–	–	Erucamide	112-84-5	IC04020
128	41	115	55	57	43	73	69	337	100	89	78	78	77	59	29	29	27.34	22	43	1	1	–	–	–	–	–	–	–	–	–	9-Octadecenamide, N-butyl-	56630-51-4	NI35567
105	77	122	51	173	106	231	50	337	100	97	58	50	31	29	26	17	3.93	22	43	1	1	–	–	–	–	–	–	–	–	–	9-Octadecenamide, N-isobutyl-, (Z)-	74420-96-5	NI35568
115	128	58	337	100	72	55	41	337	100	68	48	40	38	31	31	23		22	43	1	1	–	–	–	–	–	–	–	–	–	9-Octadecenamide, N,N-diethyl-	56630-39-8	NI35569
113	43	55	126	41	70	71	57	337	100	32	21	18	17	14	12	10	2.44	22	43	1	1	–	–	–	–	–	–	–	–	–	Pyrrolidine, 1-(1-oxooctadecyl)-	33707-76-5	NI35570
113	126	43	71	70	114	41	72	337	100	17	16	10	10	9	8	6	5.59	22	43	1	1	–	–	–	–	–	–	–	–	–	Pyrrolidine, 1-(1-oxooctadecyl)-	33707-76-5	NI35571
98	99	43	70	41	112	126	55	337	100	73	21	18	18	17	15	15	4.00	22	43	1	1	–	–	–	–	–	–	–	–	–	2-Pyrrolidinone, 1-octadecyl-	7425-87-8	NI35572
337	336	338	160	167	161	322	335	337	100	25	24	10	9	7	6	6		23	19	–	3	–	–	–	–	–	–	–	–	–	2,3-Dimethyl-5,6-diphenyl-1,7-dihydrodipyrrolo[2,3-b:3′,2′-e]pyridine		NI35573
308	336	337	309	77	270	294	306	337	100	66	31	25	15	12	10	8		23	19	–	3	–	–	–	–	–	–	–	–	–	2,6-Diphenyl-4,4-tetramethylene-1,4-dihydropyridine-3,5-dicarbonitrile	52831-28-4	NI35574
337	220	119	51	178	115	77	105	337	100	99	70	60	50	50	40	30		24	19	1	1	–	–	–	–	–	–	–	–	–	4,6-Diphenyl-2-(4-tolylimino)pyran	62219-22-1	NI35575
337	77	309	191	51	115	178	103	337	100	80	70	70	60	50	30	30		24	19	1	1	–	–	–	–	–	–	–	–	–	4,6-Diphenyl-1-(4-tolyl)-2-pyridone	86605-68-7	NI35576
337	191	338	77	309	206	103	308	337	100	32	27	24	23	16	16	15		24	19	1	1	–	–	–	–	–	–	–	–	–	N-(4-Methylbenzylidene)-3,5-diphenyl-2-furanamine		NI35577
158	115	159	143	116	179	144	128	337	100	23	13	10	4	3	2	2	1.00	24	19	1	1	–	–	–	–	–	–	–	–	–	Pyridine, 3-(1a,2,7,7a-tetrahydro-2-methoxy-1-phenyl-1,2,7-metheno-1H-cyclopropa[b]naphthalen-8-yl)-	50559-57-4	NI35578
158	115	159	143	116	179	126	322	337	100	50	27	23	10	6	6	4	1.00	24	19	1	1	–	–	–	–	–	–	–	–	–	Pyridine, 3-(1a,2,7,7a-tetrahydro-2-methoxy-8-phenyl-1,2,7-metheno-1H-cyclopropa[b]naphthalen-1-yl)-	50559-53-0	NI35579

MASS TO CHARGE RATIOS	M.W.	INTENSITIES	Parent	C	H	O	N	S	F	Cl	Br	I	Si	P	B	X	COMPOUND NAME	CAS Reg No	No
158 115 143 116 322 128 126 127	337	100 78 46 14 9 8 8 7	4.00	24	19	1	1	–	–	–	–	–	–	–	–	–	Pyridine, 4-(1a,2,7,7a-tetrahydro-2-methoxy-8-phenyl-1,2,7-metheno-1H-cyclopropa[b]naphthalen-1-yl)-	50559-47-2	NI35580
158 322 115 159 337 143 323 338	337	100 53 44 29 28 23 14 8		24	19	1	1	–	–	–	–	–	–	–	–	–	Pyridine, 4-(2a,8b-dihydro-8b-methoxy-2-phenylcyclobuta[a]naphthalen-1 yl)-	50559-49-4	NI35581
69 119 169 131 100 31 219 231	338	100 35 23 13 8 6 4 3	0.04	6	–	–	–	–	14	–	–	–	–	–	–	–	Hexane, perfluoro-		LI07473
69 119 169 131 100 31 219 93	338	100 28 19 14 13 11 4 4	0.05	6	–	–	–	–	14	–	–	–	–	–	–	–	Hexane, perfluoro-		AI02067
69 119 169 131 100 219 31 319	338	100 27 22 13 7 5 5 3	0.00	6	–	–	–	–	14	–	–	–	–	–	–	–	Hexane, perfluoro-		IC04021
19 169 219 319 338 269 119 69	338	100 76 34 24 15 10 4 2		6	–	–	–	–	14	–	–	–	–	–	–	–	Hexane, perfluoro- (negative ion spectrum)		LI07474
69 119 169 131 219 231 100 181	338	100 13 10 10 5 4 4 3	0.00	6	–	–	–	–	14	–	–	–	–	–	–	–	Pentane, 2-methylperfluoro-		LI07475
69 131 119 169 31 100 93 181	338	100 13 11 9 8 7 5 4	0.00	6	–	–	–	–	14	–	–	–	–	–	–	–	Pentane, 2-methylperfluoro-		AI02068
169 338 319 19 219 269 262 119	338	100 25 23 14 13 5 2 1		6	–	–	–	–	14	–	–	–	–	–	–	–	Pentane, 2-methylperfluoro- (negative ion spectrum)		LI07476
55 83 211 338 128 127	338	100 83 73 54 40 21		6	12	–	–	–	–	–	–	2	–	–	–	–	1,6-Diiodohexane		LI07477
214 56 282 183 127 87 338 97	338	100 86 80 80 80 80 50 40		7	7	4	–	–	–	–	–	1	–	–	–	1	Iron, 2-methoxy-π-allyl-, tricarbonyl iodide		MS08014
77 51 50 279 78 277 28 276	338	100 62 28 15 15 12 12 8	0.20	8	8	2	–	–	–	–	–	–	–	–	–	1	Mercury, (acetato-O)phenyl-	62-38-4	NI35582
77 279 277 78 276 278 202 200	338	100 17 13 11 10 9 8 7	2.63	8	8	2	–	–	–	–	–	–	–	–	–	1	Mercury, (acetato-O)phenyl-	62-38-4	NI35583
121 119 79 103 105 95 93 101	338	100 96 90 77 67 52 50 45	18.43	8	8	2	–	–	–	–	–	–	–	–	–	1	Mercury, (acetato-O)phenyl-	62-38-4	NI35584
73 338 198 105 55 254 226 86	338	100 99 99 99 90 67 47 47		9	–	5	–	–	5	–	–	–	–	–	–	1	Manganese, pentacarbonyl(2,3,3,4,4-pentafluoro-1-cyclobuten-1-yl)-	15043-65-9	NI35585
338 198 105 74 28 55 254 86	338	100 99 99 99 99 89 65 48		9	–	5	–	–	5	–	–	–	–	–	–	1	Manganese, pentacarbonyl(2,3,3,4,4-pentafluoro-1-cyclobuten-1-yl)-	15043-65-9	LI07478
69 72 88 338 159 249 179 102	338	100 83 64 39 29 21 20 18		9	8	3	2	1	6	–	–	–	–	–	–	–	1-Imidazolidinecarboxylic acid, 2-thioxo-3-[3,3,3-trifluoro-1-oxo-2-(trifluoromethyl)propyl]-, methyl ester	61687-03-4	NI35586
282 280 308 284 306 310 201 203	338	100 66 59 59 36 32 32 27	9.00	9	8	4	–	–	–	–	2	–	–	–	–	–	Benzoic acid, 3,5-dibromo-2,4-dihydroxy-6-methyl-, methyl ester	715-33-3	NI35587
139 181 69 153 111 70 84 196	338	100 40 36 33 33 30 27 26	0.00	9	8	5	2	–	6	–	–	–	–	–	–	–	N,N-Bis(trifluoroacetyl)glutamine		NI35588
252 209 253 295 338 281 239 210	338	100 19 7 6 4 1 1 1		9	21	–	–	–	–	–	–	–	–	–	–	1	Bismuthine, tripropyl-	3692-82-8	NI35589
340 338 305 303 342 341 307 339	338	100 73 54 54 52 21 18 16		10	6	1	–	–	–	4	–	–	2	–	–	–	1,1,3,3-Tetrachloro-1,3-disila-2-oxaphenalane	54321-29-8	NI35590
340 338 170 205 342 207 139 277	338	100 57 56 55 53 46 40 36		12	3	1	–	–	–	5	–	–	–	–	–	–	Dibenzofuran, 1,2,3,7,8-pentachloro-	57117-41-6	NI35591
321 319 323 331 327	338	100 80 45 38 34	0.00	12	3	1	–	–	–	5	–	–	–	–	–	–	Dibenzofuran, 1,2,4,7,8-pentachloro-	58802-15-6	NI35592
75 127 50 143 74 108 99 111	338	100 98 98 85 78 65 59 58	8.20	12	6	1	–	1	–	4	–	–	–	–	–	–	Benzene, 1,2,4-trichloro-5-[(4-chlorophenyl)sulphinyl]-	35850-29-4	NI35593
155 157 185 183 76 75 340 50	338	100 94 61 61 47 46 36 30	18.23	12	8	–	2	–	–	–	2	–	–	–	–	–	Diazene, bis(2-bromophenyl)-	15426-16-1	NI35594
155 157 76 185 183 75 340 28	338	100 97 62 39 39 37 31 28	14.95	12	8	–	2	–	–	–	2	–	–	–	–	–	Diazene, bis(3-bromophenyl)-	15426-17-2	NI35595
155 157 183 76 185 75 340 338	338	100 95 52 52 49 40 37 19		12	8	–	2	–	–	–	2	–	–	–	–	–	Diazene, bis(4-bromophenyl)-	1601-98-5	NI35596
282 284 281 283 338 39 336 27	338	100 96 72 70 67 67 54 44		12	14	–	–	–	–	–	–	–	–	–	–	1	Hafnium, (η^4-1,3-butadiene)(η^8-1,3,5,7-cyclooctatetraene)-	74779-90-1	NI35597
172 44 28 140 112 45 68 42	338	100 43 32 31 30 22 17 16	2.40	12	14	4	6	1	–	–	–	–	–	–	–	–	Methyl 7-(1H-tetrazol-1-ylacetamido)deacetoxycephalosporanate		MS08015
43 141 296 113 281 155 41 157	338	100 86 61 54 50 42 41 41	17.37	12	19	3	–	3	–	–	–	–	–	1	–	–	Bolstar sulphoxide		NI35598
123 338 137 124 120 155 107 139	338	100 57 43 42 42 40 40 40		13	10	7	2	1	–	–	–	–	–	–	–	–	2,4-Dinitro-4′-methoxydiphenylsulphone	972-48-5	NI35599
169 168 167 338 170 215 318 210	338	100 54 26 20 11 11 10 6		13	11	1	2	1	5	–	–	–	–	–	–	–	(3,3-Diphenylureido)sulphur pentafluoride	90598-10-0	NI35600
135 60 143 126 144 59 101 185	338	100 71 65 62 55 50 48 47	3.60	13	26	4	2	2	–	–	–	–	–	–	–	–	L-Xylose, 4,5-diacetamido-4,5-dideoxy-, diethyl mercaptal	3509-40-8	NI35601
338 340 339 342 341 169 171 343	338	100 66 17 12 11 8 5 2		14	8	4	2	–	–	2	–	–	–	–	–	–	1,4-Diamino-2,3-dichloro-5,8-dihydroxyanthraquinone		IC04022
44 57 69 338 219 215 279 187	338	100 25 12 7 6 6 5 5		14	10	4	–	3	–	–	–	–	–	–	–	–	1-Hydroxy-9-[(methoxycarbonyl)dithio]phenoxathiin	120577-46-0	DD01336
180 179 261 259 178 165 89 181	338	100 51 50 49 38 25 23 15	0.74	14	12	–	–	–	–	–	2	–	–	–	–	–	Benzene, 1,1′-(1,2-dibromo-1,2-ethanediyl)bis-	5789-30-0	NI35602
238 161 322 78 65 39 294 66	338	100 15 10 8 8 8 7 7	0.00	14	12	2	2	–	–	–	–	–	–	–	–	1	Molybdenum, 4-tolylazo-π-cyclopentadienyldicarbonyl-		LI07479
252 161 336 92 65 39 308 66	338	100 15 10 8 8 8 7 7		14	12	2	2	–	–	–	–	–	–	–	–	1	Molybdenum, 4-tolylazo-π-cyclopentadienyldicarbonyl-		MS08016
41 55 81 95 123 79 69 206	338	100 97 75 70 62 52 51 47	21.30	14	18	6	4	–	–	–	–	–	–	–	–	–	Suberic hemialdehyde-2,4-dinitrophenylhydrazone		IC04023
238 168 65 338 236 220 210 91	338	100 63 28 26 25 22 22 22		14	19	3	–	–	–	–	–	–	–	–	–	1	[η^4-2-(1-Hydroxyethyl)norbornadiene]acetylacetonatorhodium	78392-06-0	NI35603
338 208 168 238 103 182 207 209	338	100 58 51 50 39 38 32 30		14	19	3	–	–	–	–	–	–	–	–	–	1	[η^4-2-(Methoxymethyl)norbornadiene]acetylacetonatorhodium	114366-68-6	NI35604
166 167 41 28 139 29 55 69	338	100 90 30 30 21 21 15 14	4.70	14	21	4	2	–	3	–	–	–	–	–	–	–	β-Alanine, N-[(trifluoroacetyl)-L-prolyl]- butyl ester		MS08017
166 167 41 237 29 194 139 28	338	100 61 21 18 18 17 16 15	4.20	14	21	4	2	–	3	–	–	–	–	–	–	–	Alanine, N-[(trifluoroacetyl)-L-prolyl]- butyl ester		MS08018
31 166 167 69 72 139 44 70	338	100 80 55 39 35 30 30 27	0.00	14	21	4	2	–	3	–	–	–	–	–	–	–	L-Isoleucine, N-[1-(trifluoroacetyl)-L-prolyl]-, methyl ester	52183-94-5	NI35605
55 41 83 69 111 45 81 43	338	100 88 80 66 59 34 33 33	20.00	14	26	4	–	–	–	–	–	–	–	–	–	1	Heptanoic acid, 7,7′-selenodi-	22676-17-1	LI07480
55 41 83 69 111 45 81 43	338	100 80 79 65 60 33 32 31	20.79	14	26	4	–	–	–	–	–	–	–	–	–	1	Heptanoic acid, 7,7′-selenodi-	22676-17-1	NI35606
173 175 303 145 147 109 305 177	338	100 63 32 22 14 13 11 10	9.19	15	12	3	2	–	–	2	–	–	–	–	–	–	Ethyl 3-[1-(2,6-dichlorobenzoyl)pyrazol-4-yl]prop-2-enoate	95883-06-0	NI35607
173 175 303 145 165 147 338 109	338	100 64 25 24 19 16 14 14		15	12	3	2	–	–	2	–	–	–	–	–	–	Methyl 3-[1-(2,6-dichlorobenzoyl)-4-methylpyrazol-5-yl]prop-2-enoate	95883-09-3	NI35608
338 208 234 236 232 206 238 103	338	100 71 57 52 48 26 24 22		15	23	2	–	–	–	–	–	–	–	–	–	1	Rhodium, (η^2-1,2-divinylcyclohexane)(2,4-pentanedionato-O,O′)-	74842-26-5	NI35609
338 234 232 236 208 103 339 230	338	100 67 67 44 26 21 17 17		15	23	2	–	–	–	–	–	–	–	–	–	1	Rhodium, (η^4-1,2-divinylcyclohexane)(2,4-pentanedionato-O,O′)-	74810-98-3	NI35610
43 93 65 92 199 77 338 39	338	100 70 43 41 38 32 31 17		16	13	3	2	–	3	–	–	–	–	–	–	–	N-Phenyl-N′-(4-acetoxyphenyl)-N′-trifluoroacetyl-hydrazine		LI07481

MASS TO CHARGE RATIOS								M.W.	INTENSITIES								Parent	C	H	O	N	S	F	Cl	Br	I	Si	P	B	X	COMPOUND NAME	CAS Reg No	No
338	255	254	340	323	257	253	256	338	100	99	40	37	33	33	27	26		16	15	6	–	–	–	1	–	–	–	–	–	–	Spiro[benzofuran-2(3H),1′-cyclohexane]-2′,3,4′-trione, 7-chloro-4,6-dimethoxy-6′-methyl-	26891-73-6	NI35611
98	68	338	240	69	201	165	40	338	100	46	35	27	26	20	20	17		16	15	6	–	–	–	1	–	–	–	–	–	–	Spiro[benzofuran-2(3H),1′-[3]cyclohexene]-2′,3-dione, 7-chloro-6-hydroxy 4,4′-dimethoxy-6′-methyl-	26881-74-3	NI35612
138	338	296	69	201	340	307	298	338	100	64	44	36	28	24	23	17		16	15	6	–	–	–	1	–	–	–	–	–	–	Spiro[benzofuran-2(3H),1′-[2]cyclohexene]-3,4′-dione, 7-chloro-6-hydroxy 2′,4-dimethoxy-6′-methyl-, (1′S-trans)-	20168-88-1	NI35613
98	68	336	238	199	69	223	40	338	100	30	26	22	15	15	12	11	10.01	16	15	6	–	–	–	1	–	–	–	–	–	–	Spiro[benzofuran-2(3H),1′-[3]cyclohexene]-2′,3-dione, 7-chloro-4,4′,6-trimethoxy-	2855-89-2	NI35614
155	338	111	156	339	183	340	109	338	100	20	13	10	3	3	2	2		16	18	4	–	2	–	–	–	–	–	–	–	–	1,2-Bis(2-hydroxy-5-methoxyphenylthio)ethane	93885-80-4	NI35615
91	247	338	169	65	137	92	248	338	100	94	62	62	38	30	28	24		16	20	–	–	2	–	–	–	–	–	2	–	–	1,2-Dimethyl-1,2-dibenzyldiphosphane disulphide		MS08019
91	44	79	108	100	56	101	77	338	100	33	32	30	27	27	26	25	0.10	16	22	6	2	–	–	–	–	–	–	–	–	–	L-Alanine, N-[N-(benzyloxycarbonyl)-L-threonyl]-, methyl ester	2483-53-6	NI35616
73	219	338	339	45	75	220	191	338	100	80	53	20	20	18	16	11		16	26	4	–	–	–	–	–	–	2	–	–	–	Cinnamic acid, 3,4-bis(trimethylsiloxy)-, methyl ester	22020-29-7	NI35617
338	219	339	73	220	77	340	75	338	100	99	30	27	18	17	12	6		16	26	4	–	–	–	–	–	–	2	–	–	–	Cinnamic acid, 3,4-bis(trimethylsiloxy)-, methyl ester	22020-29-7	NI35618
73	219	338	45	339	75	220	191	338	100	80	64	21	20	18	16	15		16	26	4	–	–	–	–	–	–	2	–	–	–	Cinnamic acid, 3,4-bis(trimethylsiloxy)-, methyl ester	22020-29-7	LI07482
323	338	206	324	339	308	325	89	338	100	49	41	26	25	16	11	10		16	26	4	–	–	–	–	–	–	2	–	–	–	Cinnamic acid, 4,α-bis(trimethylsiloxy)-, methyl ester	29372-28-9	NI35620
323	73	206	338	89	308	324	45	338	100	83	46	42	34	30	28	20		16	26	4	–	–	–	–	–	–	2	–	–	–	Cinnamic acid, 4,α-bis(trimethylsiloxy)-, methyl ester	29372-28-9	NI35619
73	311	77	338	75	308	323	312	338	100	90	82	59	57	37	32	28		16	26	4	–	–	–	–	–	–	2	–	–	–	Cinnamic acid, 4-methoxy-3-(trimethylsiloxy)-, trimethylsilyl ester	32342-04-4	MS08020
338	73	323	249	339	75	308	45	338	100	86	43	43	28	27	25	18		16	26	4	–	–	–	–	–	–	2	–	–	–	Cinnamic acid, 4-methoxy-3-(trimethylsiloxy)-, trimethylsilyl ester	32342-04-4	NI35621
338	73	249	323	339	308	75	45	338	100	87	47	43	29	27	27	18		16	26	4	–	–	–	–	–	–	2	–	–	–	Cinnamic acid, 4-methoxy-3-(trimethylsiloxy)-, trimethylsilyl ester	32342-04-4	LI07483
338	323	308	73	249	293	339	324	338	100	73	67	52	48	40	28	22		16	26	4	–	–	–	–	–	–	2	–	–	–	Cinnamic acid, 3-methoxy-4-(trimethylsilyloxy)-, trimethylsilyl ester	10517-09-6	MS08021
338	73	323	308	249	339	75	219	338	100	82	41	40	40	29	28	22		16	26	4	–	–	–	–	–	–	2	–	–	–	Cinnamic acid, 3-methoxy-4-(trimethylsilyloxy)-, trimethylsilyl ester	10517-09-6	NI35623
338	308	323	249	73	293	339	324	338	100	70	69	46	44	39	33	21		16	26	4	–	–	–	–	–	–	2	–	–	–	Cinnamic acid, 3-methoxy-4-(trimethylsilyloxy)-, trimethylsilyl ester	10517-09-6	NI35622
73	338	308	323	249	75	45	293	338	100	67	41	38	36	25	25	21		16	26	4	–	–	–	–	–	–	2	–	–	–	Ferulic acid, trimethylsiloxy, trimethylsilyl ester	67460-51-9	NI35624
73	338	308	249	323	293	219	279	338	100	52	46	38	28	22	14	8		16	26	4	–	–	–	–	–	–	2	–	–	–	Ferulic acid, trimethylsiloxy, trimethylsilyl ester, trans-		NI35625
147	73	323	75	148	267	45	149	338	100	48	36	20	16	14	14	10	8.40	16	26	4	–	–	–	–	–	–	2	–	–	–	Pyruvic acid, (2-methoxyphenyl)-trimethylsilyl-, trmethylsilyl ester		MS08023
147	73	148	267	75	149	77	44	338	100	50	16	12	9	8	7	6	0.00	16	26	4	–	–	–	–	–	–	2	–	–	–	Pyruvic acid, (2-methoxyphenyl)-trimethylsilyl-, trmethylsilyl ester		MS08022
147	73	323	148	75	149	45	324	338	100	40	32	20	11	10	9	8	4.40	16	26	4	–	–	–	–	–	–	2	–	–	–	Pyruvic acid, (3-methoxyphenyl)-trimethylsilyl-, trmethylsilyl ester		MS08024
147	73	323	75	267	148	45	44	338	100	68	38	32	25	20	16	12	9.20	16	26	4	–	–	–	–	–	–	2	–	–	–	Pyruvic acid, (4-methoxyphenyl)-trimethylsilyl-, trmethylsilyl ester		MS08025
169	73	170	75	249	233	171	79	338	100	47	20	19	13	11	8	8	0.50	16	26	4	2	–	–	–	–	–	1	–	–	–	2,4,6(1H,3H,5H)-Pyrimidinetrione, 1,3,5-trimethyl-5-[3-[(trimethylsilyl)oxy]-1-cyclohexen-1-yl]-	55299-22-4	NI35626
337	77	338	144	103	298	135	202	338	100	95	55	51	29	20	20	19		17	14	–	4	2	–	–	–	–	–	–	–	–	2-Phenylimino-3-phenyl-4-thioxothiazolo[3,2-a]tetrahydro-s-triazine		LI07484
306	241	56	129	121	211	213	95	338	100	92	24	22	21	15	14	12	0.00	17	14	4	–	–	–	–	–	–	–	–	–	1	Ferrocene, 1-(β-fur-2-oylvinyl)-		LI07485
105	77	51	106	122	216	69	295	338	100	100	52	49	4	3	3	2	1.06	17	14	4	4	–	–	–	–	–	–	–	–	–	5-[1(R,S),2-Dibenzoyloxyethyl]tetrazole		NI35627
43	338	202	196	65	221	264	193	338	100	23	22	18	18	16	14	14		17	14	4	4	–	–	–	–	–	–	–	–	–	6-Ethoxycarbonyl-5-(2-furyl)-3-imino-1-methyl-4,7,7-tricyano-2-oxabicyclo[2.2.1]heptane		MS08026
236	235	234	86	237	84	49	221	338	100	67	45	22	21	20	20	15	0.00	17	15	–	–	1	4	–	–	–	–	–	1	–	1-Methyl-trans-2-styrylbenzo[b]thiophenium tetrafluoroborate		LI07486
139	251	111	43	253	141	75	41	338	100	53	50	42	35	32	23	18	0.36	17	16	3	–	–	–	2	–	–	–	–	–	–	Benzeneacetic acid, 4-chloro-α-(4-chlorophenyl)-α-hydroxy-, isopropyl ester	5836-10-2	NI35628
321	139	323	251	185	169	253	301	338	100	76	69	67	54	44	41	29	0.00	17	16	3	–	–	–	2	–	–	–	–	–	–	Benzeneacetic acid, 4-chloro-α-(4-chlorophenyl)-α-hydroxy-, isopropyl ester	5836-10-2	NI35630
139	251	111	253	43	141	75	41	338	100	51	48	35	35	33	18	15	0.00	17	16	3	–	–	–	2	–	–	–	–	–	–	Benzeneacetic acid, 4-chloro-α-(4-chlorophenyl)-α-hydroxy-, isopropyl ester	5836-10-2	NI35629
98	79	55	77	69	53	107	91	338	100	89	87	65	63	58	57	56	2.80	17	22	7	–	–	–	–	–	–	–	–	–	–	7-Acetyldeoxynivalenol		NI35631
43	41	98	79	249	137	107	55	338	100	34	28	20	20	19	18	16	0.80	17	22	7	–	–	–	–	–	–	–	–	–	–	15-Acetyldeoxynivalenol		NI35632
44	173	83	85	43	41	47	39	338	100	95	75	68	64	64	60	59	12.16	17	22	7	–	–	–	–	–	–	–	–	–	–	Deoxynivalenyl acetate		NI35633
105	143	69	151	98	91	123	59	338	100	52	46	35	28	26	25	25	24.00	17	22	7	–	–	–	–	–	–	–	–	–	–	6′-Methoxy-exo-spiro[cyclopropan-1,8′-tricyclo[3.2.1.0$^{2'',4''}$]octane]tricarbonic acid (3′-endo-6′-endo-7′-exo) trimethyl ester		LI07487
105	91	143	59	187	215	115	77	338	100	54	50	33	24	22	22	22	7.00	17	22	7	–	–	–	–	–	–	–	–	–	–	6′-Methoxy-exo-spiro[cyclopropan-1,8′-tricyclo[3.2.1.0$^{2'',4''}$]octane]tricarbonic acid (3′-endo-6′-exo-7′-endo) trimethyl ester		LI07488
43	67	109	81	41	79	53	40	338	100	83	54	39	34	33	25	25	2.00	17	23	2	–	–	–	–	1	–	–	–	–	–	Laurencin		LI07489
73	75	147	179	323	338	248	41	338	100	96	60	55	37	36	31	31		17	30	3	–	–	–	–	–	–	2	–	–	–	Pentanoic acid, (2-hydroxyphenyl)-, bis(trimethylsilyl)-		NI35634
73	179	75	180	192	45	338	323	338	100	92	50	16	15	13	11	9		17	30	3	–	–	–	–	–	–	2	–	–	–	Pentanoic acid, (3-hydroxyphenyl)-, bis(trimethylsilyl)-		NI35635
179	73	75	93	77	338	323	192	338	100	60	36	32	26	23	19	18		17	30	3	–	–	–	–	–	–	2	–	–	–	Pentanoic acid, (4-hydroxyphenyl)-, bis(trimethylsilyl)-		MS08027

MASS TO CHARGE RATIOS								M.W.	INTENSITIES								Parent	C	H	O	N	S	F	Cl	Br	I	Si	P	B	X	COMPOUND NAME	CAS Reg No	No
123	152	75	338	141	59	294	146	**338**	100	52	48	46	36	31	28	19		17	34	1	4	–	–	–	–	–	1	–	–	–	3-Dimethylhydrazono-1-[3-(dimethylhydrazono)-1-propenyl]-2-ethyl-5-(trimethylsilyloxy)cyclopentane		NI35636
152	151	136	338	134	321	149	237	**338**	100	62	41	34	28	25	25	20		18	10	7	–	–	–	–	–	–	–	–	–	–	Anhydrogrevillin D		NI35637
310	338	309	339	311	279	280	340	**338**	100	94	93	19	19	19	7	6		18	10	7	–	–	–	–	–	–	–	–	–	–	Versicolorin A		NI35638
338	194	77	116	91	85	104	86	**338**	100	67	37	15	14	11	10	6		18	14	3	2	1	–	–	–	–	–	–	–	–	2-Phenylimino-3-phenyl-6-methoxycarbonyl-2,3-dihydro-1,3-thiazin-4-one		MS08028
338	279	307	339	138	323	337	294	**338**	100	68	64	36	36	32	19	14		18	14	5	2	–	–	–	–	–	–	–	–	–	Methyl 1,2-dihydro-4-[4-(methoxycarbonyl)phenyl]-1-oxophthalazine-6-carboxylate		NI35639
245	80	244	91	109	141	137	154	**338**	100	63	55	44	41	38	38	35	**1.70**	18	15	3	2	–	–	–	–	–	–	1	–	–	2-Phenoxy-2,2′-(3H,3′H)-spiro[1.3.2-benzoxazaphosphole]	52961-94-1	NI35640
69	41	253	70	68	67	87	81	**338**	100	19	5	5	5	3	2	2	**0.00**	18	26	6	–	–	–	–	–	–	–	–	–	–	2-Propenoic acid, 2-methyl-, 2-ethyl-2-[[(2-methyl-1-oxo-2-propenyl)oxy]methyl]-1,3-propanediyl ester	3290-92-4	NI35641
151	338	152	75	323	73	339	308	**338**	100	50	17	15	14	13	12	8		18	30	4	–	–	–	–	–	–	1	–	–	–	Heptanoic acid, (3,4-dimethoxyphenyl)-, trimethylsilyl ester		NI35642
83	114	56	170	55	123	154	70	**338**	100	43	28	28	21	14	13	10	**0.00**	18	30	4	2	–	–	–	–	–	–	–	–	–	N,N′-Bis(2,6-dimethyl-6-nitrosohept-1-en-4-one)	94514-28-0	NI35643
83	55	29	39	84	27	41	56	**338**	100	39	10	8	7	5	5	5	**0.00**	18	30	4	2	–	–	–	–	–	–	–	–	–	N,N′-Bis(2,6-dimethyl-6-nitrosohept-2-en-4-one)	66737-12-0	NI35644
58	98	42	183	338	99	339	212	**338**	100	52	44	40	36	36	26	23		18	34	2	4	–	–	–	–	–	–	–	–	–	1,2-Bis[1′-(2′-oxo-3′,3′,5′,5′-tetramethylpiperazine)]ethane	71029-16-8	NI35645
112	223	168	140	27	84	195	110	**338**	100	78	43	42	36	28	15	15	**0.00**	18	40	–	2	–	–	–	–	–	–	–	–	2	Aluminium, tetraethyldi-μ-1-piperidinyldi-	42777-01-5	MS08029
254	253	296	152	255	236	237	225	**338**	100	51	41	22	18	18	14	14	**1.30**	19	14	6	–	–	–	–	–	–	–	–	–	–	1-Acetoxy-2-acetoxymethylanthraquinone		NI35646
307	175	149	160	320	148	122	132	**338**	100	99	99	52	34	25	23	16	**0.00**	19	14	6	–	–	–	–	–	–	–	–	–	–	1,3-Propanedione, 1-(1,3-benzodioxol-5-yl)-3-(4-methoxy-5-benzofuranyl)-	64280-22-4	NI35647
77	103	233	338	105	70	133	95	**338**	100	55	51	47	45	36	33	22		19	18	4	2	–	–	–	–	–	–	–	–	–	Benzoic acid, 4-(phenylazo)-, 2-(tetrahydro-2-oxo-3-furanyl)ethyl ester	36763-88-9	MS08030
84	322	338	324	340	323	294	77	**338**	100	72	56	25	19	17	17	16		19	19	–	4	–	–	1	–	–	–	–	–	–	7-Chloro-2-(1′-piperidinyl)-5-phenyl-3H-1,3,4-benzotriazeyine		MS08031
165	73	166	176	135	77	281	253	**338**	100	91	76	67	65	62	58	34	**10.01**	19	22	2	2	–	–	–	–	–	1	–	–	–	2,4-Imidazolidinedione, 3-methyl-5,5-diphenyl-1-(trimethylsilyl)-	57326-20-2	NI35648
323	338	73	91	321	65	324	45	**338**	100	64	55	54	38	30	28	26		19	22	2	2	–	–	–	–	–	1	–	–	–	4(3H)-Quinazolinone, 2-methyl-3-(2-methylphenyl)-6-[(trimethylsilyl)oxy]	55683-26-6	NI35649
323	73	338	324	321	45	143	77	**338**	100	67	29	25	22	17	16	16		19	22	2	2	–	–	–	–	–	1	–	–	–	4(3H)-Quinazolinone, 2-methyl-3-[2-methyl-3-[(trimethylsilyl)oxy]phenyl]	55683-28-8	NI35650
73	323	338	321	143	45	77	154	**338**	100	75	36	30	27	24	20	19		19	22	2	2	–	–	–	–	–	1	–	–	–	4(3H)-Quinazolinone, 2-methyl-3-[2-methyl-4-[(trimethylsilyl)oxy]phenyl]	55683-27-7	NI35651
73	338	247	323	179	77	75	235	**338**	100	70	66	55	49	40	33	29		19	22	2	2	–	–	–	–	–	1	–	–	–	4(3H)-Quinazolinone, 2-methyl-3-[2-[[(trimethylsilyl)oxy]methyl]phenyl]-	55683-29-9	NI35652
230	84	79	91	229	203	338	231	**338**	100	48	38	27	25	22	20	19		19	22	2	4	–	–	–	–	–	–	–	–	–	N-(exo-Bicyclo[4.1.0]hept-3-en-7-yl)-N′-(1-phenyl-2,3-dimethylpyrazol-5-on-4-yl)urea	90013-17-5	NI35653
56	230	84	79	91	57	229	203	**338**	100	89	43	34	24	24	22	20	**18.00**	19	22	2	4	–	–	–	–	–	–	–	–	–	N-(exo-Bicyclo[4.1.0]hept-3-en-7-yl)-N′-(1-phenyl-2,3-dimethylpyrazol-5-on-4-yl)urea	85156-52-1	NI35654
45	57	135	176	41	79	149	77	**338**	100	91	88	80	78	57	44	44	**3.92**	19	30	5	–	–	–	–	–	–	–	–	–	–	1,3-Benzodioxole, 5-[1-[2-(2-butoxyethoxy)ethoxy]butyl]-	23473-61-2	NI35655
176	177	194	57	165	107	45	149	**338**	100	41	26	19	17	15	15	14	**7.70**	19	30	5	–	–	–	–	–	–	–	–	–	–	1,3-Benzodioxole, 5-[[2-(2-butoxyethoxy)ethoxy]methyl]-6-propyl-	51-03-6	NI35656
176	177	57	45	41	149	29	119	**338**	100	36	22	18	15	13	12	9	**6.00**	19	30	5	–	–	–	–	–	–	–	–	–	–	1,3-Benzodioxole, 5-[[2-(2-butoxyethoxy)ethoxy]methyl]-6-propyl-	51-03-6	NI35657
109	69	173	95	123	55	155	197	**338**	100	54	51	39	35	26	23	21	**0.00**	19	30	5	–	–	–	–	–	–	–	–	–	–	Methyl grindelistrictate		NI35658
43	113	55	97	71	85	183	165	**338**	100	80	72	50	48	44	40	40	**0.00**	19	30	5	–	–	–	–	–	–	–	–	–	–	2,6,10,14,18-Nonadecanepentone	55110-17-3	NI35659
337	338	323	51	79	207	246	259	**338**	100	83	63	39	29	24	23	21		20	14	–	6	–	–	–	–	–	–	–	–	–	1-Amino-2,6,6-tricyano-3,5-di-(3-pyridyl)-5-methylcyclohexa-1,3-diene	94638-77-4	NI35660
338	263	149	322	321	291			**338**	100	20	20	13	13	13				20	18	5	–	–	–	–	–	–	–	–	–	–	5-Allyl-3-hydroxymethyl-7-methoxy-2-(1,3-benzodioxol-5-yl)-benzofuran		LI07490
338	265	237	121	278	194	165	138	**338**	100	36	21	8	6	6	6	4		20	18	5	–	–	–	–	–	–	–	–	–	–	5-Benzofuranpropanoic acid, 2,3-dihydro-3-[(4-methoxyphenyl)methylene]-2-oxo-, methyl ester	75311-78-3	NI35661
283	295	338	296	321	284	323	339	**338**	100	67	55	18	16	16	15	10		20	18	5	–	–	–	–	–	–	–	–	–	–	4H-1-Benzopyran-4-one, 5,7-dihydroxy-3-(4-hydroxyphenyl)-6-(3-methylbutenyl)-	75024-15-6	NI35662
279	338	278	261	251	165	178	339	**338**	100	93	90	79	52	29	24	19		20	18	5	–	–	–	–	–	–	–	–	–	–	Phenanthro[1,2-b]furan-6-carboxylic acid, 6,7,8,9,10,11-hexahydro-1,6-dimethyl-10,11-dioxo-, methyl ester	18887-19-9	NI35664
279	338	278	261	251	165	178	179	**338**	100	92	90	79	52	28	23	20		20	18	5	–	–	–	–	–	–	–	–	–	–	Phenanthro[1,2-b]furan-6-carboxylic acid, 6,7,8,9,10,11-hexahydro-1,6-dimethyl-10,11-dioxo-, methyl ester	18887-19-9	NI35663
323	338	270	283	324	339	141	165	**338**	100	80	48	37	23	19	18	11		20	18	5	–	–	–	–	–	–	–	–	–	–	5,7,4′-Trihydroxy-8-(3,3-dimethylallyl)isoflavone (lupiwighteone)		NI35665
247	188	91	338	176	160	146	266	**338**	100	27	26	22	20	12	8	5		20	22	3	2	–	–	–	–	–	–	–	–	–	3-(N-Acetylaminoethyl)-6-benzyloxy-5-methoxyindole		LI07491
124	214	338	154	167	180	281	125	**338**	100	25	22	18	15	13	11	7		20	22	3	2	–	–	–	–	–	–	–	–	–	19-Demethyl-20-oxoiboxyphylline		DD01337
247	91	159	92	169				**338**	100	46	19	11	10				**0.50**	20	22	3	2	–	–	–	–	–	–	–	–	–	5,5-Dibenzyl-6-hydroxy-1,3-dimethylhydro-uracil		LI07492
338	295	120	296	319	43	174	280	**338**	100	87	64	47	31	29	26	24		20	22	3	2	–	–	–	–	–	–	–	–	–	20,21-Dinoraspidospermidin-10-one, 1-acetyl-5,19-didehydro-17-methoxy-	36458-94-3	NI35666
185	171	186	156	183	338	198	144	**338**	100	18	15	13	10	9	8	8		20	22	3	2	–	–	–	–	–	–	–	–	–	2(5H)-Furanone, 3-(1,2,3,3a,4,5,6,7-octahydro-7-hydroxy-3-methyl-3,7a-diazacyclohepta[jk]fluoren-5-yl)-, [3aS-(3aα,5β,7α)]-	75667-85-5	NI35667
189	122	174	123	108	207	149	95	**338**	100	89	70	65	45	33	21	20	**11.00**	20	22	3	2	–	–	–	–	–	–	–	–	–	5-Methoxy-3-methyl-3-ethyl-3H-indole-2-carboxylic acid-(4′-methoxy-anilide)		MS08032
338	188	177	216	323	148	134	77	**338**	100	96	76	75	57	51	42	40		20	22	3	2	–	–	–	–	–	–	–	–	–	2-(4-Methoxy-phenylimino)-3-methyl-pent-3-enoic-(4-methoxy-anilide)		LI07493
265	338	279	158	134	130	307	266	**338**	100	77	46	22	22	17	16	15		20	22	3	2	–	–	–	–	–	–	–	–	–	Polyneuridine oxindole, de(hydroxymethyl)-	32303-70-1	NI35668
198	185	156	154	338	155			**338**	100	50	25	20	10	8				20	22	3	2	–	–	–	–	–	–	–	–	–	Bis(secodehydrocyclopiazonic acid)		LI07494
320	239	261	168	182	338	279		**338**	100	64	43	37	27	12	10			20	22	3	2	–	–	–	–	–	–	–	–	–	Vincaridin	27501-23-1	LI07495

MASS TO CHARGE RATIOS								M.W.	INTENSITIES								Parent	C	H	O	N	S	F	Cl	Br	I	Si	P	B	X	COMPOUND NAME	CAS Reg No	No
95	137	41	69	57	81	43	67	338	100	72	57	45	36	31	31	28	3.41	20	34	–	–	–	–	–	–	–	–	–	–	1	Zinc, bis[2-(1,1-dimethyl-2-propenyl)-3,3-dimethylcyclopropyl]-, [1α(1R*,2R*,2α)]-	74793-76-3	NI35669
93	320	107	338	178	160	264	55	338	100	60	52	50	45	42	40	40		20	34	4	–	–	–	–	–	–	–	–	–	–	Diethyl 2-[(E)-2-isopropenyldec-2-enyl]propanedioate		MS08033
267	89	45	99	133	268	112	57	338	100	30	27	26	22	18	18	14	2.40	20	34	4	–	–	–	–	–	–	–	–	–	–	Ethanol, 2-[2-[2-[4-(1,1,3,3-tetramethylbutyl)phenoxy]ethoxy]ethoxy]-	2315-62-0	NI35670
67	55	41	81	79	80	93	69	338	100	89	88	79	72	54	45	33	2.00	20	34	4	–	–	–	–	–	–	–	–	–	–	Methyl octadeca-9,12-diene-1,18-dioate		LI07496
95	221	68	113	67	55	209	81	338	100	93	83	80	80	79	79	74	0.00	20	34	4	–	–	–	–	–	–	–	–	–	–	21-Normontanol		NI35671
123	124	55	96	95	43	41	289	338	100	65	64	57	55	55	52	50	2.00	20	34	4	–	–	–	–	–	–	–	–	–	–	Portulol	34756-98-4	MS08034
83	55	69	139	57	41	170	30	338	100	58	43	36	35	29	22	20	1.00	20	38	2	2	–	–	–	–	–	–	–	–	–	Diazene, bis[5-methyl-2-isopropylcyclohexyl]-, 1,2-dioxide	56052-59-6	NI35672
338	339	207	310	77	179	169	104	338	100	38	27	25	12	11	11	10		21	14	1	4	–	–	–	–	–	–	–	–	–	Pyrimido[5,4-a]phenazin-5-ol, 1-methyl-3-phenyl-	61416-96-4	NI35673
91	65	92	181	338	51	89	63	338	100	12	7	5	3	2	1	1		21	19	2	–	–	–	1	–	–	–	–	–	–	Benzene, 4-(chloromethyl)-1,2-bis(phenylmethoxy)-	1699-59-8	NI35674
245	152	168	184	338	261			338	100	96	73	63	16	10				21	22	2	–	1	–	–	–	–	–	–	–	–	4-Phenoxy-2-phenoxy-6-thiatricyclo[3.2.1.1^{3,8}]nonane		LI07497
338	339	263	307	340	235	171	165	338	100	24	8	6	5	4	4	4		21	22	4	–	–	–	–	–	–	–	–	–	–	11H-Cyclopenta[a]phenanthrene-15-carboxylic acid, 12,13,16,17-tetrahydro-17-hydroxy-3-methoxy-13-methyl-, methyl ester, (13S-cis)-	56588-55-7	NI35675
296	338	297	137	281	123	115	172	338	100	41	21	17	15	15	10	9		21	22	4	–	–	–	–	–	–	–	–	–	–	(2R,3R)-2,3-Dihydro-2-(4′-acetyl-3′-methoxyphenyl)-3-methyl-5-(E)-propenylbenzofuran		MS08035
296	338	202	107	121	281	119	115	338	100	71	13	12	8	7	7	7		21	22	4	–	–	–	–	–	–	–	–	–	–	(2R,3R)-2,3-Dihydro-2-(4′-acetylphenyl)-7-methoxy-3-methyl-5-(E)-propenylbenzofuran		MS08036
74	55	45	100	73	28	27	101	338	100	96	84	75	71	53	47	27	0.58	21	22	4	–	–	–	–	–	–	–	–	–	–	Methyl 3-carboxymethyl-4,5-diphenyl-hex-3-enoate		NI35676
201	309	338	77	310	269	256	215	338	100	57	42	38	28	19	11	9		21	23	2	–	–	–	–	–	–	–	1	–	–	(1RS,4RS,5Rs)-2-(Diphenylphosphinoyl)-4-methylbicyclo[3.2.1]oct-2-en-1 ol		MS08037
201	202	309	310	269	338	287	295	338	100	53	48	43	43	35	28	15		21	23	2	–	–	–	–	–	–	–	1	–	–	(1RS,4SR,5RS,S(S)R(S))-2-(Diphenylphosphinoyl)-4-methylbicyclo[3.2.1]oct-2-en-1-ol		MS08038
202	201	215	310	203	216	125	338	338	100	60	42	40	20	18	18	15		21	23	2	–	–	–	–	–	–	–	1	–	–	exo-5-Methyl-endo-6-[(diphenylphosphinoyl)methyl]bicyclo[2.2.1]heptan-2 one		MS08039
202	215	201	257	216	338	310	125	338	100	92	88	40	40	32	28	28		21	23	2	–	–	–	–	–	–	–	1	–	–	endo-5-Methyl-exo-6-[(diphenylphosphinoyl)methyl]bicyclo[2.2.1]heptan-2 one		MS08040
338	339	295	282	269	310	281	323	338	100	30	25	19	16	15	13	11		21	26	2	–	–	–	–	–	–	1	–	–	–	Estra-1,3,5,7,9-pentaen-17-one, 3-[(trimethylsilyl)oxy]-	69833-75-6	NI35677
338	339	282	323	340	283	281	309	338	100	28	20	10	8	7	7	6		21	26	2	–	–	–	–	–	–	1	–	–	–	Estra-1,3,5,7,9-pentaen-17-one, 3-[(trimethylsilyl)oxy]-, (14β)-	75330-99-3	NI35678
109	43	338	124	310	295	337	339	338	100	79	70	30	28	24	22	16		21	26	2	2	–	–	–	–	–	–	–	–	–	Aspidofiline	6871-24-5	NI35679
109	43	338	124	310	295	337	339	338	100	78	69	29	26	23	22	15		21	26	2	2	–	–	–	–	–	–	–	–	–	Aspidofiline	6871-24-5	NI35680
109	338	310	124	139	110	41	136	338	100	35	29	23	10	10	10	9		21	26	2	2	–	–	–	–	–	–	–	–	–	Aspidofractinine-3-carboxylic acid, methyl ester, (2α,3α,5α)-	28161-78-6	NI35681
124	109	125	338	110	310	136	130	338	100	78	11	10	10	5	5	4		21	26	2	2	–	–	–	–	–	–	–	–	–	Aspidofractinine-3-carboxylic acid, methyl ester, (2α,3β,5α)-	559-51-3	NI35682
136	338	123	174	137	160	339	122	338	100	33	15	12	11	10	8	7		21	26	2	2	–	–	–	–	–	–	–	–	–	Aspidospermatine	5794-14-9	NI35683
310	338	138	43	294	160	311	55	338	100	45	38	34	28	23	20	19		21	26	2	2	–	–	–	–	–	–	–	–	–	Aspidospermidine, 1-acetyl-19,21-epoxy-	2671-45-6	NI35684
124	338	125	41	339	254	44	42	338	100	11	10	5	3	2	2	2		21	26	2	2	–	–	–	–	–	–	–	–	–	Aspidospermidine-3-carboxylic acid, 2,3-didehydro-, methyl ester, (5α,12β,19α)-	3247-10-7	NI35685
124	55	46	41	57	69	81	95	338	100	63	63	63	50	44	30	28	6.00	21	26	2	2	–	–	–	–	–	–	–	–	–	Aspidospermidine-3-carboxylic acid, 2,3-didehydro-, methyl ester, (5α,12β,19α)-	3247-10-7	NI35687
124	125	41	338	57	56	42	55	338	100	11	10	7	5	5	5	4		21	26	2	2	–	–	–	–	–	–	–	–	–	Aspidospermidine-3-carboxylic acid, 2,3-didehydro-, methyl ester, (5α,12β,19α)-	3247-10-7	NI35686
135	121	122	144	107	252	93	41	338	100	77	47	42	29	23	20	18	18.00	21	26	2	2	–	–	–	–	–	–	–	–	–	Aspidospermidine-3-carboxylic acid, 6,7-didehydro-, methyl ester, (2β,3β,5α,12β,19α)-	50868-35-4	NI35688
135	121	122	144	107	252	93	41	338	100	78	49	43	29	24	21	19	19.02	21	26	2	2	–	–	–	–	–	–	–	–	–	Aspidospermidine-3-carboxylic acid, 6,7-didehydro-, methyl ester, (2β,3β,5α,12β,19α)-	50868-35-4	NI35689
229	170	123	110	122	43	339	137	338	100	96	94	91	65	51	47	46	0.00	21	26	2	2	–	–	–	–	–	–	–	–	–	2,20-Cycloaspidospermidine-3-carboxylic acid, methyl ester, (2α,3α,5α,12β,19α,20R)-	25343-00-4	NI35690
124	338	136	229	170	123	84	86	338	100	97	78	52	51	43	38	25		21	26	2	2	–	–	–	–	–	–	–	–	–	2,20-Cycloaspidospermidine-3-carboxylic acid, methyl ester, (2α,3β,5α,12β,19α,20R)-	17172-16-6	NI35691
338	337	208	309	268	42	279	339	338	100	49	49	44	38	26	25	23		21	26	2	2	–	–	–	–	–	–	–	–	–	Deoxyvincamine	21019-23-8	MS08041
338	208	309	268	279	252	249		338	100	48	42	38	24	20	18			21	26	2	2	–	–	–	–	–	–	–	–	–	Deoxyvincamine	21019-23-8	LI07498
105	77	204	51	106	148	134	135	338	100	46	11	11	8	7	7	6	3.00	21	26	2	2	–	–	–	–	–	–	–	–	–	N,N′-Dibenzoyl-1,7-diaminoheptane		LI07499
338	309	208	279	252	249	268		338	100	48	34	28	10	10	8			21	26	2	2	–	–	–	–	–	–	–	–	–	Epideoxyvincamine		LI07500
338	337	309	208	279	268	339	249	338	100	62	47	34	25	25	23	23		21	26	2	2	–	–	–	–	–	–	–	–	–	Epideoxyvincamine		MS08042
338	337	182	169	309	183	170	339	338	100	85	48	45	39	28	25	24		21	26	2	2	–	–	–	–	–	–	–	–	–	(E,3R,12bS)-3-Ethyl-1,2,3,4,6,7,12,12b-octahydroindolo[2,3-a]quinolizine-2-ylidene acetic acid ethyl ester		MS08043

MASS TO CHARGE RATIOS								M.W.	INTENSITIES								Parent	C	H	O	N	S	F	Cl	Br	I	Si	P	B	X	COMPOUND NAME	CAS Reg No	No
197	191	338	320	170	251	182	144	**338**	100	95	84	52	51	33	32	10		21	26	2	2	–	–	–	–	–	–	–	–	–	Macroline	2269-93-4	LI07501
138	340	210	170	139	124	341	137	**338**	100	48	48	16	15	15	14	13	**2.00**	21	26	2	2	–	–	–	–	–	–	–	–	–	2H-3,7-Methanoazacycloundecino[5,4-b]indole-9-carboxylic acid, 5-ethyl-1,4,7,8,9,10-hexahydro-, methyl ester, [7R-(7R*,9S*)]-	35478-77-4	NI35692
138	141	340	210	215	170	82	341	**338**	100	86	63	63	31	22	18	16	**3.00**	21	26	2	2	–	–	–	–	–	–	–	–	–	2H-3,7-Methanoazacycloundecino[5,4-b]indole-9-carboxylic acid, 5-ethyl-1,4,7,8,9,10-hexahydro-, methyl ester, [7R-(7R*,9R*)]-	26251-91-2	NI35693
223	209	224	210	338	207	279	208	**338**	100	22	20	16	12	5	4	4		21	26	2	2	–	–	–	–	–	–	–	–	–	Methyl 1,2,3,4,4a,5,6,8,9,9a,14,14a-dodecahydroindolo[7a,1-a]-β-carbolin-1-yl acetate		MS08044
124	338	125	339	57	169	168	167	**338**	100	11	6	3	3	2	2	2		21	26	2	2	–	–	–	–	–	–	–	–	–	Methyl 3a-ethyl-2,3,3a,4,6,11,12,13a-octahydro-1H-indolizino[8,1-cd]carbazole-5-carboxylate		MS08045
338	213	174	108	197	187	323	307	**338**	100	20	15	11	8	8	6	6		21	26	2	2	–	–	–	–	–	–	–	–	–	Seredamine	3911-20-4	LI07502
70	57	61	281	55	69	60	73	**338**	100	88	81	64	58	45	43	40	**16.88**	21	26	2	2	–	–	–	–	–	–	–	–	–	Strychane, 1-acetyl-20α-hydroxy-16-methylene-	2111-98-0	NI35694
197	338	181	225	182	170	196	183	**338**	100	64	28	27	27	24	20	20		21	26	2	2	–	–	–	–	–	–	–	–	–	Tarcarpin		NI35695
70	41	43	55	57	69	142	83	**338**	100	84	72	40	30	18	15	15	**1.00**	21	38	3	–	–	–	–	–	–	–	–	–	–	(3R,4S,2E)-2-Hexadecanylidene-3-hydroxy-4-methylbutanolide		NI35696
70	41	43	55	57	140	69	142	**338**	100	90	82	52	39	31	24	19	**3.00**	21	38	3	–	–	–	–	–	–	–	–	–	–	(3S,4S,2E)-2-Hexadecanylidene-3-hydroxy-4-methylbutanolide	71358-20-8	NI35697
85	81	41	55	67	97	109	123	**338**	100	97	97	65	49	43	35	34	**0.00**	21	38	3	–	–	–	–	–	–	–	–	–	–	1-Methoxy-5-[(3,4,5,6-tetrahydro-2H-pyranyl)oxy]-1,3-pentadecadiene		DD01338
144	125	124	101	109	123	177	137	**338**	100	93	93	66	64	49	45	43	**17.00**	21	38	3	–	–	–	–	–	–	–	–	–	–	1-Naphthalenepentanoic acid, decahydro-2-hydroxy-β,2,5,5,8a-pentamethyl-, methyl ester, [1R-[1α(S*),2β,4aβ,8aα]]-	10267-25-1	NI35698
305	320	191	69	95	81	109	137	**338**	100	71	54	54	52	50	48	45	**0.21**	21	38	3	–	–	–	–	–	–	–	–	–	–	1-Naphthalenepentanoic acid, decahydro-2-hydroxy-β,2,5,5,8a-pentamethyl-, methyl ester, [1R-[1α(S*,2β,4aβ,8aα)]]-	10267-25-1	NI35699
125	144	124	109	101	123	185	177	**338**	100	90	89	78	72	56	32	30	**8.00**	21	38	3	–	–	–	–	–	–	–	–	–	–	1-Naphthalenepentanoic acid, decahydro-2-hydroxy-β,2,5,5,8a-pentamethyl-, methyl ester, [1S-[1α(S*,2β,4aβ,8aα)]]-	3950-05-8	NI35700
338	207	339	219	220	238	44	91	**338**	100	44	27	22	20	19	16	15		22	18	–	4	–	–	–	–	–	–	–	–	–	Benzenamine, 4,4′-(6-phenyl-2,4-pyrimidinediyl)bis-	85489-55-0	NI35701
338	117	339	118	234	89	90	337	**338**	100	47	26	12	9	8	7	7		22	18	–	4	–	–	–	–	–	–	–	–	–	2-Phenyl-4-(3-aminophenyl)-6-(4-aminophenyl)pyrimidine		NI35702
123	171	143	81	128	91	172	41	**338**	100	49	35	31	29	16	15	14	**4.99**	22	26	3	–	–	–	–	–	–	–	–	–	–	Cyclopropanecarboxylic acid, 2,2-dimethyl-3-(2-methyl-1-propenyl)-, [5-benzyl-3-furanyl]methyl ester	10453-86-8	NI35703
123	171	143	41	81	43	128	91	**338**	100	50	38	37	34	34	30	24	**4.40**	22	26	3	–	–	–	–	–	–	–	–	–	–	Cyclopropanecarboxylic acid, 2,2-dimethyl-3-(2-methyl-1-propenyl)-, [5-benzyl-3-furanyl]methyl ester	10453-86-8	NI35704
124	125	57	338	41	310	152	55	**338**	100	31	31	22	20	18	16	16		22	30	1	2	–	–	–	–	–	–	–	–	–	Aspidospermidine, 1-(1-oxopropyl)-	17472-57-0	NI35705
55	43	41	69	83	74	97	57	**338**	100	81	68	64	51	47	42	41	**5.74**	22	42	2	–	–	–	–	–	–	–	–	–	–	Cyclopropanedecanoic acid, 2-octyl-, methyl ester	10152-64-4	NI35706
43	100	41	113	56	85	170	155	**338**	100	83	69	60	46	33	31	31	**3.64**	22	42	2	–	–	–	–	–	–	–	–	–	–	7,9-Docosanedione	52262-76-7	NI35707
55	69	41	83	43	97	57	320	**338**	100	71	67	61	59	48	42	34	**5.00**	22	42	2	–	–	–	–	–	–	–	–	–	–	Docos-13-enoic acid		IC04024
101	88	43	306	338	41	55	307	**338**	100	81	76	69	65	58	42	39		22	42	2	–	–	–	–	–	–	–	–	–	–	2-Eicosenoic acid, 2-methyl-, methyl ester	57346-69-7	NI35708
82	55	43	57	83	257	67	41	**338**	100	71	60	56	55	54	44	44	**0.00**	22	42	2	–	–	–	–	–	–	–	–	–	–	Hexadecanoic acid, cyclohexyl ester	1673-08-1	NI35709
84	41	55	57	43	97	83	29	**338**	100	81	60	51	51	45	42	34	**0.00**	22	42	2	–	–	–	–	–	–	–	–	–	–	Palmitaldehyde, diallyl acetal	18302-67-5	NI35710
337	338	77	104	339	281	206	189	**338**	100	65	18	17	15	8	8	5		23	18	1	2	–	–	–	–	–	–	–	–	–	5-Benzyl-4,6-diphenyl-2-pyrimidole		LI07503
338	222	309	339	261	337	91	310	**338**	100	35	29	22	21	10	8	7		23	18	1	2	–	–	–	–	–	–	–	–	–	3-Benzylidene-2,3-dihydro-7-methyl-5-phenyl-1H-1,4-benzodiazepin-2-one	55056-31-0	NI35711
338	220	51	191	178	77	115	105	**338**	100	99	60	50	40	40	30	30		23	18	1	2	–	–	–	–	–	–	–	–	–	4,6-Diphenyl-2-(5-methyl-2-pyridylimino)pyran		NI35712
338	77	51	310	115	191	203	178	**338**	100	80	60	50	50	40	30	30		23	18	1	2	–	–	–	–	–	–	–	–	–	4,6-Diphenyl-1-(5-methyl-2-pyridyl)-2-pyridone		NI35713
338	231	232	339	58	84	107	105	**338**	100	73	66	26	13	12	10	10		23	18	1	2	–	–	–	–	–	–	–	–	–	1H-Phenanthro[9,10-d]imidazole-6-ethanol, α-phenyl-	69707-19-3	NI35714
158	115	159	143	144	116	149	127	**338**	100	24	13	11	5	3	2	2	**1.00**	23	18	1	2	–	–	–	–	–	–	–	–	–	Pyridine, 3,3′-(1a,2,7,7a-tetrahydro-2-methoxy-1,2,7-metheno-1H-cyclopropa[b]naphthalene-1,8-diyl)bis-	50559-68-7	NI35715
338	307	321	339	204	308	189	309	**338**	100	31	29	28	26	21	16	13		23	18	1	2	–	–	–	–	–	–	–	–	–	2-(4-Tolyl)-4,5-diphenyl-3-nitrosopyrrole		NI35716
204	338	307	203	339	177	176	178	**338**	100	72	35	33	20	20	14	10		23	18	1	2	–	–	–	–	–	–	–	–	–	2,4,5-Triphenyl-3(H)-pyrrol-3-one, O-methyl oxime		NI35717
338	187	171	173	185	107	91	55	**338**	100	91	67	62	56	52	51	50		23	30	2	–	–	–	–	–	–	–	–	–	–	Pregn-1,4,6-triene-3,20-dione, 6,16-dimethyl-, (16α)-	55191-51-0	WI01441
338	187	171	173	185	107	91	85	**338**	100	91	67	62	56	52	51	49		23	30	2	–	–	–	–	–	–	–	–	–	–	Pregn-1,4,6-triene-3,20-dione, 6,16-dimethyl-, (16α)-	55191-51-0	AM00144
71	57	58	43	183	55	85	59	**338**	100	99	77	77	71	43	40	33	**0.10**	23	46	1	–	–	–	–	–	–	–	–	–	–	12-Tricosanone	540-09-0	NI35718
320	291	303	219	191	178	202	165	**338**	100	66	61	56	36	28	27	23	**8.00**	24	18	2	–	–	–	–	–	–	–	–	–	–	Benzaldehyde, 2,2′-(1,2-phenylenedi-2,1-ethenediyl)bis-	69395-30-8	NI35719
338	295	252	126	339	169	296	147.5	**338**	100	39	34	32	26	11	10	10		24	18	2	–	–	–	–	–	–	–	–	–	–	5,12-Dimethoxydibenz[a,h]anthracene		AI02069
338	252	126	295	339	169	323	147.5	**338**	100	49	45	43	27	23	19	14		24	18	2	–	–	–	–	–	–	–	–	–	–	6,13-Dimethoxydibenz[a,h]anthracene		AI02070
246	45	29	338	46	27	43	44	**338**	100	90	86	76	38	29	27	22		24	18	2	–	–	–	–	–	–	–	–	–	–	2,2′-Diphenoxydiphenyl		MS08046
131	191	189	103	338	165	178	179	**338**	100	76	28	22	17	14	10	8		24	18	2	–	–	–	–	–	–	–	–	–	–	9-Phenanthrenemethyl cinnamate	92174-48-6	NI35720
211	28	43	98	281	126	197	112	**338**	100	91	64	61	35	34	33	33	**0.00**	24	34	1	–	–	–	–	–	–	–	–	–	–	Bis[ar-(1,1-dimethylbutyl)phenyl]ether		DO01106
323	338	324	95	57	339	67	202	**338**	100	70	26	26	22	18	14	13		24	34	1	–	–	–	–	–	–	–	–	–	–	2,6-Dinorborn-2-yl-4-tert-butylhydroxybenzene		MS08047
211	197	183	213	198	225	184	77	**338**	100	75	36	18	14	13	6	5	**0.00**	24	34	1	–	–	–	–	–	–	–	–	–	–	4-Dodecylphenyl phenyl ether		DO01107
309	269	268	281	280	338	310	307	**338**	100	76	69	48	46	32	29	24		24	50	–	–	–	–	–	–	–	–	–	–	–	Docosane, 3,5-dimethyl-	13897-12-6	NI35721
57	71	43	85	55	113	56	69	**338**	100	75	60	46	30	24	23	21	**0.80**	24	50	–	–	–	–	–	–	–	–	–	–	–	Nonadecane, 2,6,10,14,18-pentamethyl-	55191-61-2	LI07504

MASS TO CHARGE RATIOS								M.W.	INTENSITIES								Parent	C	H	O	N	S	F	Cl	Br	I	Si	P	B	X	COMPOUND NAME	CAS Reg No	No
58	71	86	113	183	99	112	128	338	100	99	57	41	34	29	28	24	10.87	24	50	–	–	–	–	–	–	–	–	–	–	–	Nonadecane, 2,6,10,14,18-pentamethyl-	55191-61-2	NI35722
57	43	71	85	41	55	29	56	338	100	86	58	40	33	27	22	16	7.36	24	50	–	–	–	–	–	–	–	–	–	–	–	Tetracosane		AI02071
57	43	71	85	41	55	29	69	338	100	84	71	59	33	28	20	16	4.44	24	50	–	–	–	–	–	–	–	–	–	–	–	Tetracosane		AI02072
85	43	57	71	41	55	29	56	338	100	45	38	20	16	11	9	7	0.33	24	50	–	–	–	–	–	–	–	–	–	–	–	Tricosane, 2-methyl-		AI02073
43	57	71	55	41	85	29	56	338	100	94	49	34	34	33	22	21	0.66	24	50	–	–	–	–	–	–	–	–	–	–	–	Tricosane, 2-methyl-		AI02074
43	57	71	41	85	55	56	69	338	100	79	43	36	30	25	17	14	0.71	24	50	–	–	–	–	–	–	–	–	–	–	–	Tricosane, 2-methyl-		AI02075
43	57	85	84	71	55	41	56	338	100	90	77	47	47	28	28	21	0.29	24	50	–	–	–	–	–	–	–	–	–	–	–	Tricosane, 5-methyl-		LI07505
141	142	338	339	43	41	166	154	338	100	12	9	2	2	2	1	1		25	38	–	–	–	–	–	–	–	–	–	–	–	1-Pentadecylnaphthalene		AI02077
141	338	142	43	41	29	339	27	338	100	53	30	24	19	18	15	8		25	38	–	–	–	–	–	–	–	–	–	–	–	1-Pentadecylnaphthalene		AI02078
141	338	142	43	41	339	29	154	338	100	68	29	26	20	19	18	8		25	38	–	–	–	–	–	–	–	–	–	–	–	1-Pentadecylnaphthalene		AI02076
104	118	117	91	338	234	207	105	338	100	77	73	52	50	29	29	20		26	26	–	–	–	–	–	–	–	–	–	–	–	Bicyclo[8.2.2]tetradeca-5,10,12,13-tetraene, 4,7-diphenyl-, (4R*,5E,7R*)-	50703-40-7	NI35723
278	313	276	315	311	280	306	304	339	100	88	79	51	50	43	33	26	8.90	2	5	–	3	–	–	5	–	–	–	3	–	–	1-Ethylpentachlorocyclotriphosphazene	71332-23-5	NI35724
341	339	343	337	260	181	262	258	339	100	88	51	38	37	31	28	18		7	3	–	1	–	–	–	2	–	–	–	–	1	Selenolo[2,3-b]pyridine, 2,3-dibromo-	55108-60-6	NI35725
183	128	211	185	127	213	184	148	339	100	67	60	38	31	23	16	16	0.00	10	11	–	1	1	–	1	–	1	–	–	–	–	Benzothiazolium, 5-chloro-3-ethyl-2-methyl-, iodide	33599-35-8	LI07506
183	128	211	185	127	213	45	184	339	100	65	59	38	31	22	20	17	0.00	10	11	–	1	1	–	1	–	1	–	–	–	–	Benzothiazolium, 5-chloro-3-ethyl-2-methyl-, iodide	33599-35-8	NI35726
294	274	82	180	295	275	277	167	339	100	38	33	25	22	19	18	17	1.30	10	11	5	1	–	6	–	–	–	–	–	–	–	Leucine, η-hydroxy-O,N-bis(trifluoroacetyl)-		LI07507
59	150	339	118	209	58	117	98	339	100	22	17	16	11	8	7	7		10	13	–	1	6	–	–	–	–	–	–	–	–	2,4,6,8,9,10-Hexathiatricyclo[3.3.1.1^{3,7}]decane-1-propanenitrile, 3,5,7-trimethyl-	57274-37-0	NI35727
83	55	41	43	297	295	84	299	339	100	52	24	20	15	11	11	9	0.50	10	20	–	–	–	–	–	1	–	–	–	–	1	Methylisopropylcyclohexyltin bromide		MS08048
303	282	280	305	56	102	209	144	339	100	75	75	68	47	32	27	27	10.00	12	12	4	1	–	–	3	–	–	–	–	–	–	L-Alanine, N-[(2,4,5-trichlorophenoxy)acetyl]-, methyl ester	66789-86-4	NI35728
43	31	210	59	42	87	41	60	339	100	77	30	30	30	27	23	20	5.40	12	17	3	7	1	–	–	–	–	–	–	–	–	N-Acetyl-S-[4-amino-6-(1-methyl-1-cyanoethylamino)-s-triazinyl-2]-L-cysteine		LI07508
226	72	211	73	114	227	195	324	339	100	93	88	39	35	33	31	30	12.01	12	30	4	1	–	–	–	–	–	2	1	–	–	Phosphonic acid, [1-(acetylamino)butyl]-, bis(trimethylsilyl) ester	55108-83-3	NI35729
211	226	72	73	114	227	195	43	339	100	89	74	53	42	36	35	24	3.50	12	30	4	1	–	–	–	–	–	2	1	–	–	Phosphonic acid, [1-(acetylamino)-2-methylpropyl]-, bis(trimethylsilyl) ester	55108-85-5	NI35730
77	119	246	51	65	141	39	150	339	100	48	31	18	18	14	12	9	1.60	13	10	2	1	1	5	–	–	–	–	–	–	–	(Diphenoxymethylenimino)sulphur pentafluoride	81439-22-7	NI35731
105	77	106	50	339	175	76	49	339	100	26	8	8	3	2	2	2		13	10	4	1	–	5	–	–	–	–	–	–	–	Glycine, N-benzoyl-N-(2,2,3,3,3-pentafluoro-1-oxopropyl)-, methyl ester	72347-42-3	NI35732
105	77	106	339	175				339	100	26	8	3	2					13	10	4	1	–	5	–	–	–	–	–	–	–	Glycine, N-benzoyl-N-(2,2,3,3,3-pentafluoro-1-oxopropyl)-, methyl ester	72347-42-3	NI35733
105	77	106	50	339	78	76	49	339	100	26	8	8	3	2	2	2		13	10	4	1	–	5	–	–	–	–	–	–	–	Glycine, N-benzoyl-N-(2,2,3,3,3-pentafluoro-1-oxopropyl)-, methyl ester	72347-42-3	MS08049
149	150	108	104	107	105	133	122	339	100	68	38	36	35	24	22	17	0.00	13	17	6	5	–	–	–	–	–	–	–	–	–	Methyl 2-amino-3,4-dihydro-4-oxo-7-(2-amino-2-deoxy-β-arabinofuranosyl)-7H-pyrrolo[2,3-d]pyrimidine-5-carboxylate		NI35734
227	43	170	41	55	226	69	56	339	100	65	55	55	51	49	45	33	0.00	13	20	1	1	–	7	–	–	–	–	–	–	–	Nonane, 1-amino-, heptaflurobutionyl-		MS08050
253	161	93	226	200	309	92	175	339	100	37	33	29	24	22	21	20	1.60	14	11	3	1	–	–	–	–	–	–	–	–	1	π-Cyclopentadienyl-π-2-methylpyridyl-tricarbonyl molybdenum		MS08051
339	177	164	308	338	289	340	324	339	100	40	36	22	20	18	16	16		14	14	3	1	–	5	–	–	–	–	–	–	–	Isoquinoline, 1,2,3,4-tetrahydro-6,7-dimethoxy-, pentafluoropropionyl-		MS08052
136	135	251	250	203	190	180	204	339	100	98	90	83	42	38	36	35	2.00	14	21	3	5	1	–	–	–	–	–	–	–	–	D-Xylitol, 1-C-(6-amino-9H-purin-9-yl)-2,5-anhydro-1-S-(2-methylpropyl)1-thio-	40031-69-4	NI35735
246	60	43	102	163	172	158	247	339	100	85	78	52	41	33	27	25	0.00	14	29	4	1	2	–	–	–	–	–	–	–	–	L-Idose, 3-acetamido-3,6-dideoxy-, dipropyl mercaptal	3509-34-0	NI35736
121	43	339	71	57	41	122	56	339	100	20	17	11	10	10	10	7		15	18	2	1	–	5	–	–	–	–	–	–	–	Pentane, 1-amino-(3-methoxyphenyl)-, pentafluoropropionyl-		NI35737
267	73	339	268	179	193	45	269	339	100	67	50	45	41	37	21	15		15	25	2	1	1	–	–	–	–	2	–	–	–	4-(2-Isothiocyanatoethyl)benzene, 1,2-bis[(trimethylsilyl)oxy]-	35424-97-6	MS08053
267	339	193	268	340	341	269	73	339	100	77	27	24	23	13	9	6		15	25	2	1	1	–	–	–	–	2	–	–	–	4-(2-Isothiocyanotoethyl)benzene, 1,2-bis[(trimethylsilyl)oxy]-	35424-97-6	NI35738
73	324	206	193	75	45	325	194	339	100	90	83	82	34	26	24	15	4.91	15	25	4	1	–	–	–	–	–	2	–	–	–	Glycine, N-[2-[(trimethylsilyl)oxy]benzoyl]-, trimethylsilyl ester	55887-56-4	NI35739
324	73	206	193	75	325	45	207	339	100	95	88	84	30	27	25	16	5.53	15	25	4	1	–	–	–	–	–	2	–	–	–	Glycine, N-[2-[(trimethylsilyl)oxy]benzoyl]-, trimethylsilyl ester	55887-56-4	NI35740
193	294	75	73	295	194	339	324	339	100	80	57	56	45	44	38	30		15	25	4	1	–	–	–	–	–	2	–	–	–	Glycine, N-[4-[(trimethylsilyl)oxy]benzoyl]-, trimethylsilyl ester	55622-53-2	NI35741
324	206	193	325	294	249	73	326	339	100	57	33	27	18	15	13	11	3.83	15	25	4	1	–	–	–	–	–	2	–	–	–	Glycine, N-[4-[(trimethylsilyl)oxy]benzoyl]-, trimethylsilyl ester	55622-53-2	NI35742
73	294	193	295	75	339	324	194	339	100	81	69	24	21	18	16	12		15	25	4	1	–	–	–	–	–	2	–	–	–	Hippuric acid, 3-(trimethylsiloxy)-, trimethylsilyl ester		NI35743
73	324	206	193	75	45	325	91	339	100	92	68	57	43	29	23	15	6.08	15	25	4	1	–	–	–	–	–	2	–	–	–	Salicyluric acid, bis(trimethylsilyl)-	71428-95-0	NI35744
93	105	77	339	51	65	45	151	339	100	57	48	17	17	11	7	6		16	12	2	1	1	3	–	–	–	–	–	–	–	α-[(Trifluoromethyl)thio]benzoylacetanilide		MS08054
77	339	340	119	91	187	159	153	339	100	73	15	13	13	10	9	8		16	13	4	5	–	–	–	–	–	–	–	–	–	1-Phenyl-3-methyl-5-hydroxy-4-(2-hydroxy-4-nitrophenylazo)pyrazole		IC04025
218	73	75	222	150	147	219	100	339	100	99	59	58	43	23	22	22	0.00	16	29	3	1	–	–	–	–	–	2	–	–	–	2-Methoxyphenylalanine, bis(trimethylsilyl)-		MS08055
73	218	222	75	150	147	219	45	339	100	97	67	49	24	24	20	15	0.00	16	29	3	1	–	–	–	–	–	2	–	–	–	3-Methoxyphenylalanine, bis(trimethylsilyl)-		MS08056
71	72	70	192	39	59	57	193	339	100	85	48	35	22	18	18	17	10.99	16	40	–	4	–	–	–	–	–	–	–	–	1	Tetra(bisethylamino)vanadium		NI35745
342	340	341	339	343	135	145	130	339	100	98	43	24	23	19	18	14		18	16	–	2	–	–	–	1	–	–	–	–	–	Cyclopropane, 1-(4-bromophenyl)-1-cyano-2-[4-(dimethylamino)phenyl]-		LI07509
97	125	83	124	55	67	135	107	339	100	99	98	97	97	92	90	65	0.00	18	29	3	1	1	–	–	–	–	–	–	–	–	(E)-N-[(3R)-3-Ethyl-4-hexenoyl]bornane-10,2-sultam		DD01339
125	55	126	69	82	135	95	218	339	100	53	22	22	13	11	11	10	0.00	18	29	3	1	1	–	–	–	–	–	–	–	–	N-[(E)-2-Methyl-2-heptenoyl]bornane-10,2-sultam		DD01340

MASS TO CHARGE RATIOS								M.W.	INTENSITIES								Parent	C	H	O	N	S	F	Cl	Br	I	Si	P	B	X	COMPOUND NAME	CAS Reg No	No
77	104	81	51	105	53	39	78	**339**	100	81	63	56	45	44	27	25	**0.00**	19	17	3	1	1	–	–	–	–	–	–	–	–	3-Phenyl-4-benzoyl-7-mercaptomethyl-2,6-dioxa-3-azabicyclo[3.3.0]-7-octene		NI35746
179	194	339	159	154	144	308		**339**	100	54	42	26	22	10	8			19	21	3	3	–	–	–	–	–	–	–	–	–	7-Pyrrolidino-7a-methoxy-5,2′-dioxo-2,3,5,7a,2′,3′-hexahydro-1H-pyrrolizin-1-spiro-3′-indole		LI07510
294	146	59	93	188	41	100	75	**339**	100	97	97	56	36	35	32	32	**0.00**	19	37	2	1	–	–	–	–	–	1	–	–	–	1-[[[Bis(2-methoxyethyl)amino]methyl]dimethylsilyl]-2-(4-methyl-3-cyclohexenyl)-2-propene		DD01341
324	339	326	69	325	176	341	296	**339**	100	43	37	25	21	20	18	15		20	18	2	1	–	–	1	–	–	–	–	–	–	2-Chloro-3-[4-(diethylaminophenyl)]naphthoquinone		MS08057
339	297	340	296	43	267	268	28	**339**	100	48	22	22	19	18	11	10		20	21	4	1	–	–	–	–	–	–	–	–	–	N-Acetyl-O-methyl-crotonosine		MS08058
339	297	340	296	267	43	268	338	**339**	100	37	23	23	18	18	12	10		20	21	4	1	–	–	–	–	–	–	–	–	–	N-Acetyl-O-methyl-crotonosine		LI07511
338	339	308	324	125	337	152	265	**339**	100	66	47	41	34	33	33	29		20	21	4	1	–	–	–	–	–	–	–	–	–	4H-Benzo[de][1,3]benzodioxolo[5,6-g]quinoline, 5,6,6a,7-tetrahydro-1,2-dimethoxy-6-methyl-, (S)-	2565-01-7	NI35747
338	339	296	340	265	307	324	322	**339**	100	71	28	25	21	15	10	10		20	21	4	1	–	–	–	–	–	–	–	–	–	5H-Benzo[g][1,3]benzodioxolo[6,5,4-de]quinoline, 6,7,7a,8-tetrahydro-10,11-dimethoxy-7-methyl-, (S)-	517-66-8	NI35748
338	324	339	308	328	293	154	340	**339**	100	99	63	24	22	16	16	14		20	21	4	1	–	–	–	–	–	–	–	–	–	5H-Benzo[g][1,3]benzodioxolo[6,5,4-de]quinoline, 6,7,7a,8-tetrahydro-10,11-dimethoxy-7-methyl-, (S)-	517-66-8	LI07512
164	149	339	174	338	165	121	308	**339**	100	72	68	26	24	24	23	21		20	21	4	1	–	–	–	–	–	–	–	–	–	6H-Benzo[g][1,3]benzodioxolo[5,6-a]quinolizine, 5,8,13,13a-tetrahydro-9,10,dimethoxy-, (±)-	29074-38-2	NI35749
91	57	43	55	121	41	69	77	**339**	100	37	30	28	25	25	18	16	**2.00**	20	21	4	1	–	–	–	–	–	–	–	–	–	1-Benzyloxycarbonyl-3,4-epoxy-2-(4-methoxybenzyl)-pyrrolidine		LI07514
91	121	92	47	44	77	69	65	**339**	100	11	11	11	10	9	9	9	**0.00**	20	21	4	1	–	–	–	–	–	–	–	–	–	1-Benzyloxycarbonyl-3,4-epoxy-2-(4-methoxybenzyl)-pyrrolidine		LI07513
339	324	169.5	281	296	266	309	294	**339**	100	65	33	12	7	5	4	3		20	21	4	1	–	–	–	–	–	–	–	–	–	Cryptaustoline	26754-21-2	LI07515
324	212	149	213	339	83	167	227	**339**	100	94	78	75	72	66	64	60		20	21	4	1	–	–	–	–	–	–	–	–	–	Haplophylline	93710-20-4	NI35750
324	339	338	325	340	308	293	280	**339**	100	96	90	24	21	21	12	8		20	21	4	1	–	–	–	–	–	–	–	–	–	Isoquinoline, 1-[(3,4-dimethoxyphenyl)methyl]-6,7-dimethoxy-	58-74-2	LI07516
324	338	339	308	325	293	220	340	**339**	100	94	75	28	22	22	18	16		20	21	4	1	–	–	–	–	–	–	–	–	–	Isoquinoline, 1-[(3,4-dimethoxyphenyl)methyl]-6,7-dimethoxy-	58-74-2	NI35752
31	45	324	339	46	338	27	28	**339**	100	61	30	28	27	26	15	14		20	21	4	1	–	–	–	–	–	–	–	–	–	Isoquinoline, 1-[(3,4-dimethoxyphenyl)methyl]-6,7-dimethoxy-	58-74-2	NI35751
190	339	150	338	324	188	135	202	**339**	100	72	40	35	22	18	18	17		20	21	4	1	–	–	–	–	–	–	–	–	–	7,8-Isoquinolinediol, 3-(6-vinyl-4-methyl-1,3-benzodioxol-5-yl)-1,2,3,4-tetrahydro-2-methyl-	55299-21-3	NI35753
204	339	338	296	42	205	188	153	**339**	100	28	27	27	20	13	13	12		20	21	4	1	–	–	–	–	–	–	–	–	–	Reframine	21305-35-1	LI07517
148	339	190	338	91	324	177	119	**339**	100	90	81	76	42	25	22	11		20	21	4	1	–	–	–	–	–	–	–	–	–	L-Sinactine	22554-51-4	NI35754
339	297	340	296	43	267	268	338	**339**	100	38	23	23	20	18	11	10		20	21	4	1	–	–	–	–	–	–	–	–	–	Spiro[2,5-cyclohexadiene-1,7′(1′H)-cyclopent[ij]isoquinolin]-4-one, 1′-acetyl-2′,3′,8′,8′a-tetrahydro-5′,6′-dimethoxy-, (R)-	4880-87-9	NI35755
339	206	204	340	324	294	220	308	**339**	100	58	54	26	22	22	22	12		20	21	4	1	–	–	–	–	–	–	–	–	–	Spiro[2H-indene-2,1′(2′H)-isoquinolin]-1(3H)-one, 3′,4′-dihydro-3-hydroxy-7′,8′-dimethoxy-	55700-35-1	NI35756
103	339	75	47	223	222	207	219	**339**	100	70	70	68	60	55	28	21		20	25	2	3	–	–	–	–	–	–	–	–	–	9-(2,2-Diethoxyethyl)-2,8,9-triaza-3,4,6,7-dibenzbicyclo[3.3.1]nonane		LI07518
339	221	196	207	181	223	222	340	**339**	100	85	49	46	44	42	40	22		20	25	2	3	–	–	–	–	–	–	–	–	–	Ergoline-8-carboxamide, 9,10-didehydro-N-[1-(hydroxymethyl)propyl]-6-methyl-, [8β(S)]-	113-42-8	NI35757
43	85	41	99	55	57	60	72	**339**	100	62	45	40	40	39	27	24	**22.62**	20	37	3	1	–	–	–	–	–	–	–	–	–	2-Tridecyl-5-(acetylamino)tetrahydro-γ-pyrone		MS08059
339	340	117	221	220	118	338	238	**339**	100	28	16	16	15	15	10	7		21	17	–	5	–	–	–	–	–	–	–	–	–	2,4-Bis(4-aminophenyl)-6-(pyridyl)pyrimidine		NI35758
339	118	340	220	221	119	91	338	**339**	100	30	30	13	11	11	10	9		21	17	–	5	–	–	–	–	–	–	–	–	–	2,4-Bis(4-aminophenyl)-6-(pyridyl)pyrimidine		NI35759
339	340	221	220	238	78	143	338	**339**	100	28	21	15	9	8	8	8		21	17	–	5	–	–	–	–	–	–	–	–	–	2,4-Bis(4-aminophenyl)-6-(pyridyl)pyrimidine		NI35760
195	212	194	193	167	239	196	213	**339**	100	78	59	56	43	26	24	22	**1.14**	21	29	1	3	–	–	–	–	–	–	–	–	–	2-Pyridineacetamide, α-[2-(diisopropylamino)ethyl]-α-phenyl-	3737-09-5	NI35761
58	57	91	59	115	105	77	42	**339**	100	5	4	4	3	3	3	3	**0.00**	22	29	2	1	–	–	–	–	–	–	–	–	–	Benzeneethanol, α-[2-(dimethylamino)-1-isopropyl]-α-phenyl-, propanoate (ester), [S-(R*,S*)]-	469-62-5	NI35762
58	83	91	117	115	208	85	57	**339**	100	8	7	6	6	5	5	5	**0.00**	22	29	2	1	–	–	–	–	–	–	–	–	–	Benzeneethanol, α-[2-(dimethylamino)-1-isopropyl]-α-phenyl-, propanoate (ester), [S-(R*,S*)]-	469-62-5	NI35763
142	184	268	43	44	41	170	114	**339**	100	39	38	37	22	14	13	13	**2.00**	22	45	1	1	–	–	–	–	–	–	–	–	–	Hexanamide, N,N-dioctyl-		MS08060
115	128	57	73	43	41	55	30	**339**	100	23	23	20	18	14	13	13	**2.00**	22	45	1	1	–	–	–	–	–	–	–	–	–	Octadecanamide, N-butyl-		MS08061
217	73	132	204	218	103	147	116	**339**	100	83	32	26	23	20	17	15	**0.00**	22	45	1	1	–	–	–	–	–	–	–	–	–	Octadecanamide, N-isobutyl-	54794-73-9	NI35764
75	56	55	74	83	61	47	163	**339**	100	75	50	45	39	39	37	29	**0.00**	22	45	1	1	–	–	–	–	–	–	–	–	–	Octadecanamide, N-isobutyl-	54794-73-9	NI35765
173	186	187	144	310	174	339	208	**339**	100	92	25	17	16	15	12	11		23	33	1	1	–	–	–	–	–	–	–	–	–	4(1H)-Quinolone, 1-methyl-2-(8-tridecenyl)-	15266-38-3	NI35766
262	132	58	91	261	263	184	131	**339**	100	58	46	34	24	19	17	15	**9.54**	24	26	–	1	–	–	–	–	–	–	–	1	–	2-Propen-1-amine, 3-(diphenylboryl)-N,N,2-trimethyl-3-phenyl-, (E)-	74685-77-1	NI35767
133	134	234	206					**339**	100	12	6	3					**0.00**	25	25	–	1	–	–	–	–	–	–	–	–	–	3-Cyclohexen-1-amine, 6-(4-methylphenyl)-2,5-diphenyl-, (1α,2α,5α,6β)-	42715-14-0	NI35768
133	134	206						**339**	100	12	3						**0.00**	25	25	–	1	–	–	–	–	–	–	–	–	–	3-Cyclohexen-1-amine, 6-(4-methylphenyl)-2,5-diphenyl-, (1α,2β,5β,6β)-	42715-13-9	NI35769
69	202	271	200	269	199	31	201	**340**	100	24	20	19	16	14	14	11	**0.75**	2	–	–	–	–	6	–	–	–	–	–	–	1	Mercury, bis(trifluoromethyl)-	371-76-6	AI02079
69	202	271	200	269	199	31	268	**340**	100	24	20	19	16	14	14	11	**0.80**	2	–	–	–	–	6	–	–	–	–	–	–	1	Mercury, bis(trifluoromethyl)-	371-76-6	NI35770

MASS TO CHARGE RATIOS	M.W.	INTENSITIES	Parent	C	H	O	N	S	F	Cl	Br	I	Si	P	B	X	COMPOUND NAME	CAS Reg No	No
127 170 124 89 104 129 70 150	**340**	100 36 15 10 10 5 3 2	**0.10**	2	2	2	2	2	10	–	–	–	–	–	–	–	**Dioxoethylenebis(iminosulphur pentafluoride)**	80409-45-6	**NI35771**
95 237 307 309 272 119 60 36	**340**	100 99 93 91 64 58 58 58	**0.00**	5	–	–	–	–	–	8	–	–	–	–	–	–	**Cyclopentene, octachloro-**	706-78-5	**NI35772**
69 28 44 41 241 79 307 239	**340**	100 24 12 12 11 11 9 9	**1.79**	9	6	1	–	–	–	6	–	–	–	–	–	–	**4,7-Methanoisobenzofuran, 4,5,6,7,8,8-hexachloro-1,3,3a,4,7,7a-hexahydro-**	3369-52-6	**NI35773**
121 186 140 312 69 56 340 167	**340**	100 86 52 51 45 42 29 29		10	6	1	2	–	6	–	–	–	–	–	–	1	**π-Cyclopentadienyl-trifluoroacetimino-trifluoroacetylnitrilo-monocarbonyl-iron**		**MS08062**
54 200 118 59 53 172 228 312	**340**	100 90 75 68 50 44 40 33	**11.00**	10	6	6	–	–	–	–	–	–	–	–	–	2	**Cobalt, hexacarbonyl(2-butyne)di-**	37726-81-1	**NI35774**
73 147 75 55 217 45 245 129	**340**	100 54 28 26 17 12 12 10	**0.00**	10	21	4	–	–	–	–	1	–	2	–	–	–	**Bis(trimethylsilyl)bromosuccinate**		**NI35775**
225 73 310 251 325 226 241 147	**340**	100 36 27 25 21 20 15 15	**0.00**	10	25	3	2	1	–	–	–	–	2	1	–	–	**Phosphonic acid, [1-(carbonothioylhydrazino)propyl]-, bis(trimethylsilyl) ester**	56051-87-7	**NI35776**
225 310 240 73 251 325 147 268	**340**	100 63 49 48 39 34 31 27	**0.00**	10	25	3	2	1	–	–	–	–	2	1	–	–	**Phosphonic acid, [3-(carbonothioylhydrazino)propyl]-, bis(trimethylsilyl) ester**	56051-86-6	**NI35777**
59 73 217 189 276 249 117 45	**340**	100 99 47 29 27 22 18 15	**6.47**	11	20	–	–	5	–	–	–	–	1	–	–	–	**2,4,6,8,9-Pentathiatricyclo[3.3.1.1^{3,7}]decane, 1,5,7-trimethyl-3-[(trimethylsilyl)oxy]-**	57379-39-2	**NI35778**
281 73 325 282 207 193 45 283	**340**	100 87 32 27 25 19 19 16	**0.00**	11	32	4	–	–	–	–	–	–	4	–	–	–	**3-Ethoxy-1,1,1,5,5,5-hexamethyl-3-(trimethylsiloxy)trisiloxane**		**MS08063**
272 270 342 344 340 207 306 274	**340**	100 98 43 19 18 16 14 13		12	5	1	–	–	–	5	–	–	–	–	–	–	**Phenol, 2-phenyl-2′,3′,4′,5,5′-pentachloro-**	67651-36-9	**NI35779**
270 272 342 306 120 344 340 171	**340**	100 97 76 52 51 47 46 46		12	5	1	–	–	–	5	–	–	–	–	–	–	**Phenol, 4-phenyl-2′,3,3′,4′,5′-pentachloro-**	67651-35-8	**NI35780**
340 186 36 64 44 57 55 341	**340**	100 42 21 20 19 19 19 18		12	8	8	2	1	–	–	–	–	–	–	–	–	**3,3′-Dinitro-4,4′-dihydroxydiphenylsulphone**	7149-20-4	**NI35781**
264 227 262 229 266 249 231 78	**340**	100 81 79 79 48 42 36 32	**0.00**	13	12	2	–	–	–	4	–	–	–	–	–	–	**1,8,9,10-Tetrachloro-11,11-dimethoxy-endo-tricyclo[6.2.1.0^{2,7}]undeca-3,5,9 triene**		**MS08064**
297 174 340 150 130 166	**340**	100 86 69 10 5 4		13	16	7	2	–	–	–	–	–	1	–	–	–	**1-(4-Nitrobenzoyl)-2,8,9-trioxa-5-aza-1-silatricyclo[3.3.3.0^{1,5}]undecane**		**MS08065**
266 253 269 267 119 340 103 77	**340**	100 57 53 25 22 20 18 16		14	13	4	–	–	5	–	–	–	–	–	–	–	**4-Hydroxyphenylpropionic acid ethyl ester PFP**		**MS08066**
92 65 156 93 139 172 107 63	**340**	100 43 30 26 21 15 15 13	**1.00**	14	16	4	2	2	–	–	–	–	–	–	–	–	**Hydrazine, N,N′-bis(4-toluenesulphonyl)-**		**LI07519**
73 235 137 171 325 75 187 103	**340**	100 72 63 48 45 43 41 41	**12.00**	14	20	6	2	–	–	–	–	–	1	–	–	–	**6H-Furo[2′,3′:4,5]oxazolo[3,2-a]pyrimidin-6-one, 3-[(acetyloxy)methyl]-2,3,3a,9a-tetrahydro-2-[[(trimethylsilyl)oxy]methyl]-, [2R-(2α,3β,3aβ,9aβ)]-**	40732-58-9	**NI35782**
73 169 75 137 209 117 283 57	**340**	100 78 49 48 45 38 32 32	**16.00**	14	20	6	2	–	–	–	–	–	1	–	–	–	**6H-Furo[2′,3′:4,5]oxazolo[3,2-a]pyrimidin-6-one, 2-[(acetyloxy)methyl]-2,3,3a,9a-tetrahydro-3-[(trimethylsilyl)oxy]-, [2R-(2α,3β,3aβ,9aβ)]-**	40732-60-3	**NI35783**
155 91 310 65 139 156 92 39	**340**	100 88 25 19 15 12 8 7	**0.06**	15	16	5	–	2	–	–	–	–	–	–	–	–	**Methanol, [(4-methylphenyl)sulphonyl]-, 4-methylbenzenesulphonate**	14894-58-7	**NI35784**
178 150 60 69 73 179 51 76	**340**	100 92 19 17 13 13 12 10	**0.00**	15	16	9	–	–	–	–	–	–	–	–	–	–	**2H-1-Benzopyran-2-one, 6-(β-D-glucopyranosyloxy)-7-hydroxy-**	531-75-9	**NI35785**
223 73 224 147 45 297 77 75	**340**	100 59 16 13 8 6 6 6	**0.00**	15	24	5	–	–	–	–	–	–	2	–	–	–	**3,4-Methylenedioxymandelic acid di-TMS**		**MS08067**
86 340 119 254 227 91 186 77	**340**	100 71 36 28 20 11 8 6		16	15	3	2	–	3	–	–	–	–	–	–	–	**Oxazole, 4-(α,α,α-trifluoroacetyl)-2-(4-tolyl)-5-(4-morpholinyl)-**	51770-11-7	**NI35786**
86 340 119 254 227 341 91 56	**340**	100 71 36 27 20 13 11 11		16	15	3	2	–	3	–	–	–	–	–	–	–	**Oxazole, 4-(α,α,α-trifluoroacetyl)-2-(4-tolyl)-5-(4-morpholinyl)-**	51770-11-7	**NI35787**
341 87 86 114 150 131 228 340	**340**	100 91 91 87 82 74 72 66		16	24	6	2	–	–	–	–	–	–	–	–	–	**N-Acetyl-alanylactinobicyclone-O-acetate**		**LI07520**
211 209 207 297 163 213 295 120	**340**	100 71 61 60 49 47 45 39	**2.60**	16	28	–	–	–	–	–	–	–	–	–	–	1	**Benzyltriisopropyltin**		**NI35788**
340 73 269 194 75 341 297 267	**340**	100 67 40 32 30 28 25 24		16	28	4	–	–	–	–	–	–	2	–	–	–	**Benzeneacetic acid, 2,5-bis[(trimethylsilyl)oxy]-, ethyl ester**		**MS08068**
179 340 73 267 341 180 207 268	**340**	100 63 58 54 15 15 14 13		16	28	4	–	–	–	–	–	–	2	–	–	–	**Benzeneacetic acid, 3,4-bis[(trimethylsilyl)oxy]-, ethyl ester**		**MS08069**
267 73 268 179 75 269 147 45	**340**	100 60 25 14 14 11 9 9	**0.00**	16	28	4	–	–	–	–	–	–	2	–	–	–	**Benzeneacetic acid, α,2-bis[(trimethylsilyl)oxy]-, ethyl ester**		**MS08070**
267 73 268 75 77 269 179 45	**340**	100 53 25 18 13 12 6 6	**1.20**	16	28	4	–	–	–	–	–	–	2	–	–	–	**Benzeneacetic acid, α,3-bis[(trimethylsilyl)oxy]-, ethyl ester**		**MS08071**
267 73 268 75 147 269 45 74	**340**	100 85 23 23 12 11 9 8	**0.00**	16	28	4	–	–	–	–	–	–	2	–	–	–	**Benzeneacetic acid, α,3-bis[(trimethylsilyl)oxy]-, ethyl ester**		**MS08072**
267 73 268 75 77 269 193 44	**340**	100 47 26 20 15 12 7 6	**0.00**	16	28	4	–	–	–	–	–	–	2	–	–	–	**Benzeneacetic acid, α,4-bis[(trimethylsilyl)oxy]-, ethyl ester**		**MS08073**
179 73 180 45 89 75 59 181	**340**	100 49 17 8 6 6 6 5	**3.50**	16	28	4	–	–	–	–	–	–	2	–	–	–	**Benzeneacetic acid, α-methyl-α,4-bis[(trimethylsilyl)oxy]-, methyl ester**	57397-10-1	**NI35789**
73 209 340 192 179 75 310 325	**340**	100 98 67 49 33 30 27 25		16	28	4	–	–	–	–	–	–	2	–	–	–	**Benzenepropanoic acid, 3-methoxy-4-[(trimethylsilyl)oxy]-, trimethylsilyl ester**	56051-49-1	**NI35790**
73 75 209 192 179 340 310 222	**340**	100 36 30 30 26 19 9 8		16	28	4	–	–	–	–	–	–	2	–	–	–	**Benzenepropanoic acid, 3-methoxy-4-[(trimethylsilyl)oxy]-, trimethylsilyl ester**	56051-49-1	**NI35791**
73 223 147 121 219 250 224 207	**340**	100 97 42 32 28 20 19 18	**0.00**	16	28	4	–	–	–	–	–	–	2	–	–	–	**Benzeneprpanoic acid, 2-methoxy-, β-[(trimethylsilyl)oxy]- trimethylsilyl ester**		**MS08074**
73 223 147 121 219 250 224 207	**340**	100 97 42 32 28 20 19 18	**0.00**	16	28	4	–	–	–	–	–	–	2	–	–	–	**Benzeneprpanoic acid, 3-methoxy-, β-[(trimethylsilyl)oxy]- trimethylsilyl ester**		**MS08075**
121 73 147 250 75 223 235 45	**340**	100 79 30 29 24 16 11 11	**0.00**	16	28	4	–	–	–	–	–	–	2	–	–	–	**Benzeneprpanoic acid, 4-methoxy-, β-[(trimethylsilyl)oxy]- trimethylsilyl ester**		**MS08076**
179 340 267 73 341 180 342 268	**340**	100 85 32 25 23 15 8 7		16	28	4	–	–	–	–	–	–	2	–	–	–	**Hydrocinnamic acid, 3,4-bis(trimethylsiloxy)-, methyl ester**	27798-75-0	**NI35792**
137 102 232 103 111 129 77 340	**340**	100 41 35 27 18 14 14 12		17	13	–	4	1	–	1	–	–	–	–	–	–	**1-(4-Chlorophenyl)-2-(methylthio)-6-phenyl-1H-imidazo[1,2-b][1,2,4]triazole**		**MS08077**

MASS TO CHARGE RATIOS								M.W.	INTENSITIES								Parent	C	H	O	N	S	F	Cl	Br	I	Si	P	B	X	COMPOUND NAME	CAS Reg No	No
145	291	77	290	47	51	48	292	**340**	100	67	52	19	18	16	14	12	**0.00**	17	16	2	4	1	–	–	–	–	–	–	–	–	1H-1,3,5-Triazepine-6,7-dione, 2,3-dihydro-2-(methylthio)-3-phenyl-2-(phenylamino)-	74793-07-0	NI35793
93	43	340	77	94	65	341	51	**340**	100	28	20	14	9	7	5	5		17	16	4	4	–	–	–	–	–	–	–	–	–	2-(4-Methyl-2-nitrophenylazo)-acetoacetanilide		IC04026
206	105	77	323	91	340	281	117	**340**	100	69	66	64	51	49	38	34		17	16	4	4	–	–	–	–	–	–	–	–	–	7-[N′-(Tetrahydro-3-methyl-2-phenyl-1,4-oxazinyl)]-4-nitrobenzo-2-oxa-1,3-diazole		LI07521
147	238	148	91	43	77	149	58	**340**	100	25	11	9	9	7	6	6	**0.00**	17	17	–	–	1	4	–	–	–	–	–	1	–	1-Methyl-2-phenethylbenzo[b]thiophenium tetrafluoroborate		LI07522
207	133	208	189	92	164	340	322	**340**	100	57	38	16	13	12	3	2		17	24	3	–	2	–	–	–	–	–	–	–	–	1,2,3,4-Tetrahydro-2-hydroxy-5,8-dimethoxy-2-(1-methyl-1,3-dithian-1-yl)naphthalene		NI35794
340	134	121	150	239	123	105	122	**340**	100	50	48	47	38	35	33	31		18	12	7	–	–	–	–	–	–	–	–	–	–	6-(3,4-Dihydroxybenzylidene)-3-hydroxy-4-(4-hydroxyphenyl)-2H-pyran-2,5(6H)-dione		NI35795
297	340	325	311	69	298	341	312	**340**	100	93	54	36	21	20	19	19		18	12	7	–	–	–	–	–	–	–	–	–	–	Versicolorin B	4331-22-0	NI35796
297	340	325	28	29	311	43	69	**340**	100	73	41	39	33	30	28	24		18	12	7	–	–	–	–	–	–	–	–	–	–	Versicolorin C	16049-49-3	NI35797
342	340	263	265	341	105	343	43	**340**	100	99	24	23	22	20	19	19		18	13	2	–	–	–	–	1	–	–	–	–	–	6-Benzoyl-1-bromo-2-methoxy-naphthalene		MS08078
260	261	217	262	246	263	94	218	**340**	100	45	45	41	24	16	13	12	**0.00**	18	15	1	1	–	–	–	1	–	–	–	–	–	6,7-Dihydro-2-methoxydibenzo[a,f]quinolizinium bromide		NI35798
260	130	261	245	340	338	122	129	**340**	100	34	32	27	24	12	11	10		18	16	–	2	–	–	–	–	–	–	–	–	1	Indole, 3,3′-selenobis[1-methyl-	21001-19-4	NI35799
129	130	102	76	51	128	103	131	**340**	100	91	50	20	19	18	16	14	**0.31**	18	16	–	2	–	–	–	–	–	–	–	–	1	Indole, 3,3′-(selenodimethylene)di-	21903-68-4	NI35800
340	77	194	104	118	59	91	119	**340**	100	67	42	37	20	20	18	16		18	16	3	2	1	–	–	–	–	–	–	–	–	2-Phenylimino-3-phenyl-6-methoxycarbonyl-perhydro-1,3-thiazin-4-one		MS08079
100	131	121	158	116	189	89	134	**340**	100	95	67	59	46	37	22	20	**6.01**	18	16	5	2	–	–	–	–	–	–	–	–	–	2-Propenoic acid, 2-[methyl(2-nitrobenzoyl)amino]-3-phenyl-, methyl ester	55299-23-5	NI35801
100	131	121	158	116	189	89	150	**340**	100	98	68	61	47	37	23	19	**6.00**	18	16	5	2	–	–	–	–	–	–	–	–	–	2-Propenoic acid, 2-[methyl(2-nitrobenzoyl)amino]-3-phenyl-, methyl ester	55299-23-5	LI07523
340	179	195	181	178	150	177	323	**340**	100	80	55	40	33	22	20	18		18	16	5	2	–	–	–	–	–	–	–	–	–	2,4,6-Trimethoxy-2′-nitro-4′-cyano-stilbene		LI07524
342	340	341	339	134	343	146	130	**340**	100	99	41	23	23	22	17	13		18	17	–	2	–	–	–	1	–	–	–	–	–	Cyclopropanecarbonitrile, 1-(4-bromophenyl)-2-[4-(dimethylamino)phenyl]-	32589-49-4	NI35802
340	342	341	339	343	146	130	107	**340**	100	99	39	22	20	17	14	13		18	17	–	2	–	–	–	1	–	–	–	–	–	Cyclopropanecarbonitrile, 1-(4-bromophenyl)-2-[4-(dimethylamino)phenyl]-	32589-49-4	NI35803
91	340	132	342	77	271	179	341	**340**	100	86	40	30	30	26	21	20		18	17	1	4	–	–	1	–	–	–	–	–	–	1-Phenyl-3-methyl-4-(2-chloro-4-(dimethylamino)phenylimino)-2-pyrazolin-5-one		MS08080
123	218	105	77	140	124	54	56	**340**	100	89	78	60	50	22	20	18	**0.00**	18	20	1	4	1	–	–	–	–	–	–	–	–	2-[α-(4-Amino-2-methyl-5-pyrimidinylmethyl)-α-hydroxybenzyl]-4,5-dimethyl-thiazole		LI07525
254	340	255	339	341	311	269	312	**340**	100	24	18	7	6	5	2	1		18	20	3	4	–	–	–	–	–	–	–	–	–	N,N-Ethylpropyl-4′-nitroazobenzene-carbamide		MS08081
93	221	28	87	73	119	77	91	**340**	100	99	96	65	59	53	49	39	**5.16**	18	20	3	4	–	–	–	–	–	–	–	–	–	4-Hydroxy-5,5-dimethyl-3-phenyl-1-(3-phenylureido)imidazolidin-2-one		NI35804
298	340	70	115	58	312			**340**	100	16	4	3	3	2				18	20	3	4	–	–	–	–	–	–	–	–	–	Morphinan-14-ol, 6-azido-7,8-didehydro-4,5-epoxy-3-methoxy-17-methyl-, (5α,6β)-	38211-22-2	NI35805
298	340	115	58	70	312	283		**340**	100	18	12	10	7	5	3			18	20	3	4	–	–	–	–	–	–	–	–	–	Morphinan-14-ol, 8-azido-6,7-didehydro-4,5-epoxy-3-methoxy-17-methyl-, (5α,8β)-	38211-23-3	NI35806
340	322	188	92	93	77	65	174	**340**	100	41	19	17	16	15	15	13		18	20	3	4	–	–	–	–	–	–	–	–	–	D-Ribo-hexos-2-ulose, 3,6-anhydro-, bis(phenylhydrazone)	35522-78-2	NI35807
226	180	57	211	210	197	182	228	**340**	100	58	58	56	51	34	31	28	**0.00**	18	28	6	–	–	–	–	–	–	–	–	–	–	Di-tert-butyl 3-[(hydroxycarbonyl)methyl]-1-cyclohexene-1,5-dicarboxylate		MS08082
256	241	298	340	257	255	227	242	**340**	100	53	27	15	14	10	10	8		19	16	6	–	–	–	–	–	–	–	–	–	–	1,3-Benzenediol, 4-(6-methoxy-2-benzofuranyl)-, diacetate	67699-39-2	NI35808
256	298	340	257	255	213	241	212	**340**	100	31	19	18	16	10	7	7		19	16	6	–	–	–	–	–	–	–	–	–	–	6-Benzofuranol, 2-[4-(acetyloxy)-2-methoxyphenyl]-, acetate	67685-31-8	NI35809
340	325	297	283	312				**340**	100	77	45	25	23					19	16	6	–	–	–	–	–	–	–	–	–	–	1,3-Dihydroxy-2,9-dimethyl-4-acetyl-10,10a-dihydro-10-oxobenzo[b]furano[d-1,2]benzo[3,4-2′,3′]furan		LI07526
44	50	60	57	55	71	69	70	**340**	100	20	16	16	14	7	7	6	**0.50**	19	16	6	–	–	–	–	–	–	–	–	–	–	4′,7-Dimethoxy-3′-acetoxy-flavone		LI07527
171	340	170	172	168	91	341	105	**340**	100	22	18	12	9	7	5	3		19	20	2	2	1	–	–	–	–	–	–	–	–	1-Naphthalenesulphonamide, 5-(dimethylamino)-N-benzyl-	34532-49-5	NI35811
171	340	170	172	168	154	91	127	**340**	100	21	18	13	9	6	6	5		19	20	2	2	1	–	–	–	–	–	–	–	–	1-Naphthalenesulphonamide, 5-(dimethylamino)-N-benzyl-	34532-49-5	NI35810
325	326	282	327	155	140	163	283	**340**	100	22	10	8	5	4	4	3	**0.30**	19	20	4	–	–	–	–	–	–	1	–	–	–	Anthraquinone, 1-[(trimethylsilyl)oxy]-3-methoxy-6-methyl-		NI35812
238	340	59	239	183	77	237	341	**340**	100	59	41	24	23	23	20	12		19	20	4	2	–	–	–	–	–	–	–	–	–	2,3-Butanediol, 3-methyl-1-[4-[2-(3-pyridinyl)-4-oxazolyl]phenoxy]-	21059-65-4	NI35813
84	195	82	166	152	102	115	85	**340**	100	30	30	29	28	26	25	23	**8.00**	19	20	4	2	–	–	–	–	–	–	–	–	–	N[2]-(Biphenyl-4-ylacetyl)glutamine		MS08083
44	167	43	340	266	143	168	194	**340**	100	65	50	28	20	20	19	7		19	20	4	2	–	–	–	–	–	–	–	–	–	2-(4-Phenylphenyl)hydrazonopropandioic acid, diethyl ester		NI35814
91	150	175	92	174	83	65	132	**340**	100	39	34	24	22	16	15	11	**1.68**	19	24	2	4	–	–	–	–	–	–	–	–	–	5-Amino-2-methoxybenzoic acid (4-benzyl-1-piperazinyl)amide		MS08084
123	340	59	283	212	312	115	70	**340**	100	62	30	17	13	10	10	9		19	24	2	4	–	–	–	–	–	–	–	–	–	Morphinan, 6-azido-4,5-epoxy-3-ethoxy-17-methyl-, (5α,6β)-	54301-21-2	NI35815
173	91	55	41	43	69	83	70	**340**	100	78	68	60	58	55	45	42	**0.00**	19	32	3	–	1	–	–	–	–	–	–	–	–	Dodecyl 4-methylbenzenesulphonate	10157-76-3	NI35816
185	213	199	214	241	227	255	311	**340**	100	81	41	39	19	19	18	11	**8.00**	19	32	3	–	1	–	–	–	–	–	–	–	–	Methyl 4-(1-ethyldecyl)benzenesulphonate		MS08085
199	200	185	340	213	227	255	241	**340**	100	40	8	5	3	2	1	1		19	32	3	–	1	–	–	–	–	–	–	–	–	Methyl 4-(1-methylundecyl)benzenesulphonate		MS08086

MASS TO CHARGE RATIOS								M.W.	INTENSITIES								Parent	C	H	O	N	S	F	Cl	Br	I	Si	P	B	X	COMPOUND NAME	CAS Reg No	No
190	145	162	175	117	149	135	340	**340**	100	55	35	33	21	20	20	15		20	20	5	–	–	–	–	–	–	–	–	–	–	**1,3-Benzodioxole, 5,5′-(tetrahydro-3,4-dimethyl-2,5-furandiyl)bis-, [2S-(2α,3β,4α,5β)]-**	528-64-3	**LI07528**
190	145	142	162	175	117	149	135	**340**	100	55	45	35	33	21	20	20	**15.01**	20	20	5	–	–	–	–	–	–	–	–	–	–	**1,3-Benzodioxole, 5,5′-(tetrahydro-3,4-dimethyl-2,5-furandiyl)bis-, [2S-(2α,3β,4α,5β)]-**	528-64-3	**NI35817**
340	299	135	341	149	178	239	271	**340**	100	71	38	25	19	18	17	15		20	20	5	–	–	–	–	–	–	–	–	–	–	**Benzofuran, 5-allyl-5-methoxy-3-methyl-2-(3′,4′-methylenedioxyphenyl)-2,3,5,6-tetrahydro-6-oxo-, (2S,3S,5S)-**		**NI35818**
340	311	281	323					**340**	100	16	13	7						20	20	5	–	–	–	–	–	–	–	–	–	–	**Benzofuran, 5-propyl-3-hydroxymethyl-7-methoxy-2-(1,3-benzodioxol-5-yl)-**		**LI07529**
310	44	91	77	229	39	95	115	**340**	100	82	80	58	54	54	48	47	**33.57**	20	20	5	–	–	–	–	–	–	–	–	–	–	**5(S),9(S),10(S)-15,16-Epoxycleroda-3,8,13(16),14-tetraene-19,18:20,12(S)-diolactone (swassin)**	69749-03-7	**NI35819**
163	135	340	162	77	78	147	76	**340**	100	78	69	60	40	36	35	32		20	20	5	–	–	–	–	–	–	–	–	–	–	**Futoenone**	19913-01-0	**LI07530**
167	168	165	91	342	340	152	157	**340**	100	14	11	10	9	9	6	5		20	21	–	–	–	–	–	1	–	–	–	–	–	**exo-2-Bromo-syn-7-(diphenylmethyl)norbornane**		**DD01342**
72	44	73	43	57	41	55	42	**340**	100	22	12	9	8	8	7	6	**2.00**	20	24	1	2	1	–	–	–	–	–	–	–	–	**10-(2-Dimethylaminopropyl)-2-propionyl-phenothiazine**	362-29-8	**NI35820**
72	149	269	255	197	73	340	254	**340**	100	17	5	5	4	4	3	3		20	24	1	2	1	–	–	–	–	–	–	–	–	**10-(2-Dimethylaminopropyl)-2-propionyl-phenothiazine**	362-29-8	**NI35821**
72	70	58	71	73	56	42	57	**340**	100	62	59	21	12	12	11	10	**2.00**	20	24	1	2	1	–	–	–	–	–	–	–	–	**10-(2-Dimethylaminopropyl)-2-propionyl-phenothiazine**	362-29-8	**LI07531**
108	340	109	106	174	180	144	130	**340**	100	13	12	12	9	8	7	7		20	24	3	2	–	–	–	–	–	–	–	–	–	**Aspidodasycarpine, de(hydroxymethyl)-**	3620-79-9	**NI35822**
297	222	340	298	210	237	168	181	**340**	100	32	31	22	21	19	19	15		20	24	3	2	–	–	–	–	–	–	–	–	–	**Azecino[4,5,6-cd]indole-11-carboxylic acid, 6-acetyl-2,6,7,10,11,12-hexahydro-1,8-dimethyl-, methyl ester**	55702-35-7	**NI35823**
136	323	340	158	186	159	173	324	**340**	100	66	38	34	29	27	23	18		20	24	3	2	–	–	–	–	–	–	–	–	–	**Cinchonan, 6′-methoxy-, 1,1′-dioxide, (8α,9R)-**	56588-18-2	**NI35824**
71	84	95	229	180	167	87	182	**340**	100	68	62	40	33	32	28	25	**6.00**	20	24	3	2	–	–	–	–	–	–	–	–	–	**Condyfolan-16-carboxylic acid, 2,16-didehydro-, methyl ester, 4-oxide**	40169-69-5	**NI35825**
44	92	107	94	43	42	180	157	**340**	100	18	17	14	14	12	11	11	**1.50**	20	24	3	2	–	–	–	–	–	–	–	–	–	**Curan-17-oic acid, 2,16-didehydro-19-hydroxy-, methyl ester, (20ξ)-**	38681-90-2	**NI35826**
44	264	123	107	92	41	156	168	**340**	100	43	39	26	23	23	21	19	**6.00**	20	24	3	2	–	–	–	–	–	–	–	–	–	**Curan-17-oic acid, 2,16-didehydro-20-hydroxy-, methyl ester, (20α)-**	5980-01-8	**NI35827**
210	300	144	340	139	130	79	143	**340**	100	99	69	57	57	36	30	27		20	24	3	2	–	–	–	–	–	–	–	–	–	**Curan-17-oic acid, 19,20-didehydro-14-hydroxy-, methyl ester, (16α,19E)**	56114-73-9	**NI35828**
340	244	160	108	149	184	341	245	**340**	100	97	52	48	45	26	24	24		20	24	3	2	–	–	–	–	–	–	–	–	–	**Desethylroacangine**		**LI07532**
91	231	133	236	42	172	115	69	**340**	100	60	50	40	38	26	23	23	**5.00**	20	24	3	2	–	–	–	–	–	–	–	–	–	**5,5-Dibenzyl-tetrahydro-cis-4,6-dihydroxy-1,3-dimethyl-2(1H)-pyrimidinone**		**LI07533**
91	231	236	133	172	42	263	115	**340**	100	64	54	54	36	35	29	25	**4.00**	20	24	3	2	–	–	–	–	–	–	–	–	–	**5,5-Dibenzyl-tetrahydro-trans-4,6-dihydroxy-1,3-dimethyl-2(1H)-pyrimidinone**		**LI07534**
144	57	200	340	283				**340**	100	85	70	18	4					20	24	3	2	–	–	–	–	–	–	–	–	–	**10,11-Dihydrobissecodehydrocyclopiazonic acid**		**LI07535**
136	340	81	137	95	205			**340**	100	28	27	25	13	8				20	24	3	2	–	–	–	–	–	–	–	–	–	**1,2-Dihydroquinidine-2-one**		**LI07536**
58	340	310	309	57	43	41	44	**340**	100	99	71	47	40	36	35	32		20	24	3	2	–	–	–	–	–	–	–	–	–	**2,5-Ethano-2H-azocino[4,3-b]indole-6-carboxylic acid, 4-ethylidene-1,3,4,5,6,7-hexahydro-6-(hydroxymethyl)-, methyl ester**	56324-22-2	**NI35829**
201	200	322	202	253	340	172	187	**340**	100	27	17	16	14	11	11	10		20	24	3	2	–	–	–	–	–	–	–	–	–	**10-Hydroxyakagerine**	74765-89-2	**NI35830**
184	340	170	252	183	311	156	180	**340**	100	69	67	58	58	43	42	35		20	24	3	2	–	–	–	–	–	–	–	–	–	**1H-Indolo[3,2,1-de][1,5]naphthyridine-6-carboxylic acid, 2,3,3a,4,5,6-hexahydro-5-[1-(hydroxymethyl)-1-propenyl]-, methyl ester, [3aS-[3aα,5β(E),6α]]-**	30809-33-7	**NI35831**
340	184	170	252	183	180	308	242	**340**	100	95	83	64	62	53	40	39		20	24	3	2	–	–	–	–	–	–	–	–	–	**1H-Indolo[3,2,1-de][1,5]naphthyridine-6-carboxylic acid, 2,3,3a,4,5,6-hexahydro-5-[1-(hydroxymethyl)-1-propenyl]-, methyl ester, [3aS-[3aα,5β(E),6α]]-**	30809-33-7	**NI35832**
98	82	97	96	340	110	259	201	**340**	100	74	35	34	12	12	10	8		20	24	3	2	–	–	–	–	–	–	–	–	–	**Spiro[8-methyl-8-azabicyclo[3.2.1]octane-3,3′-(8-methoxy-5-methyl-1,4-dihydro-1-oxo-2H-pyrano[4,3-b]indole)]**		**DD01343**
57	70	41	112	71	55	43	117	**340**	100	90	71	62	62	62	60	58	**0.33**	20	36	4	–	–	–	–	–	–	–	–	–	–	**Dioctyl maleate**		**IC04027**
112	70	57	55	100	71	43	83	**340**	100	96	80	63	61	56	47	46	**0.20**	20	36	4	–	–	–	–	–	–	–	–	–	–	**Bis(2-ethylhexyl) fumarate**		**IC04028**
57	70	43	85	41	55	83	56	**340**	100	39	36	32	31	26	21	21	**0.00**	20	36	4	–	–	–	–	–	–	–	–	–	–	**Bis(2-ethylhexyl) fumarate**		**DO01108**
183	141	185	195	113	255	211	125	**340**	100	86	76	39	36	25	24	23	**17.00**	20	36	4	–	–	–	–	–	–	–	–	–	–	**Methyl 9,11-diketosterculate**		**LI07537**
67	81	69	74	95	98	59	43	**340**	100	99	80	64	60	59	58	57	**8.00**	20	36	4	–	–	–	–	–	–	–	–	–	–	**Methyl octadec-1-ene-1,18-dioate**		**LI07538**
55	41	81	67	276	308	309	290	**340**	100	74	61	61	32	28	21	18	**11.22**	20	36	4	–	–	–	–	–	–	–	–	–	–	**Methyl octadec-9-ene-1,18-dioate**		**IC04029**
71	97	227	41	55	67	43	45	**340**	100	86	62	57	40	30	30	27	**0.00**	20	36	4	–	–	–	–	–	–	–	–	–	–	**Methyl (11R,12R,13S)-(Z)-12,13-epoxy-11-methoxy-9-octadecenoate**		**NI35833**
73	186	217	147	132	103	319	205	**340**	100	29	25	24	23	19	18	18	**0.00**	20	40	2	2	–	–	–	–	–	–	–	–	–	**Hexadecanamide, N-isobutyl-N-nitroso-**	74420-92-1	**NI35834**
198	57	30	115	240	166	128	74	**340**	100	61	59	54	40	40	40	39	**2.96**	20	40	2	2	–	–	–	–	–	–	–	–	–	**Sebacamide, N,N-dibutyl-**		**IC04030**
118	104	340	77	341	220	119	91	**340**	100	92	65	31	16	16	13	13		21	16	1	4	–	–	–	–	–	–	–	–	–	**4H-Imidazol-4-one, 1,5-dihydro-1-phenyl-2-(phenylamino)-5-(phenylimino)-**	68822-96-8	**NI35835**
198	165	199	121	340	166	43	200	**340**	100	41	36	20	18	17	11	9		21	24	2	–	1	–	–	–	–	–	–	–	–	**1,3-Oxathiolan-5-one, 4,4-diphenyl-2,2-dipropyl-**	31061-73-1	**NI35836**
340	164	177	176	341	165	299	298	**340**	100	72	35	24	22	10	8	8		21	24	4	–	–	–	–	–	–	–	–	–	–	**2,2′-Dihydroxy-3,3′-dimethoxy-6,6′-diallyl-diphenylmethane**		**MS08087**
253	340	254	341	283	187	112	202	**340**	100	55	21	15	15	13	12	8		21	24	4	–	–	–	–	–	–	–	–	–	–	**Galcatin**	9002-62-4	**LI07539**
325	340	326	31	29	57	269	27	**340**	100	23	21	18	18	17	10	6		21	24	4	–	–	–	–	–	–	–	–	–	–	**Bis(4-glycidyloxyphenyl)dimethylmethane**	1675-54-3	**NI35837**

MASS TO CHARGE RATIOS								M.W.	INTENSITIES								Parent	C	H	O	N	S	F	Cl	Br	I	Si	P	B	X	COMPOUND NAME	CAS Reg No	No
149	191	150	137	340	135	167	216	**340**	100	72	62	41	39	37	31	28		21	24	4	–	–	–	–	–	–	–	–	–	–	2′-Hydroxy-4′-methoxy-3″,4″-dihydro-2″,2″-dimethylpyrano[6″,5″:7,8]isoflavan		NI35838
91	340	307	279	267	193	322	145	**340**	100	42	40	33	29	29	22	22		21	24	4	–	–	–	–	–	–	–	–	–	–	2-Hydroxy-4-methoxy-3-(3-methylbut-2-enyl)-6-(2-phenylethyl)benzoic acid		NI35839
340	253	254	283	341	284	135	149	**340**	100	44	33	25	23	21	10	7		21	24	4	–	–	–	–	–	–	–	–	–	–	Isogalcatin	24150-38-7	LI07540
308	340	309	341	293	272	310	325	**340**	100	93	75	21	18	17	16	13		21	24	4	–	–	–	–	–	–	–	–	–	–	(±)-12-Isopropyl-3,3-dimethoxy-9,15-dimethyl-4-oxatricyclo[8.5.0.0^{2,6}]pentadeca-2(6),7,9,11,13,15-hexaen-5-one		DD01344
340	255	281	221	325	179	309	207	**340**	100	89	58	50	48	40	27	27		21	24	4	–	–	–	–	–	–	–	–	–	–	(±)-12-Isopropyl-5,5-dimethoxy-9,15-dimethyl-4-oxatricyclo[8.5.0.0^{2,6}]pentadeca-2(6),7,9,11,13,15-hexaen-3-one		DD01345
340	341	242	255	216	73	229	57	**340**	100	67	67	60	30	27	23	23		21	28	2	–	–	–	–	–	–	1	–	–	–	Estra-1,3,5(10),6-tetraen-17-one, 3-[(trimethylsilyl)oxy]-	69688-26-2	NI35840
340	341	342	338	242	194	283	216	**340**	100	30	14	12	12	12	11	10		21	28	2	–	–	–	–	–	–	1	–	–	–	Estra-1,3,5(10),7-tetraen-17-one, 3-[(trimethylsilyl)oxy]-	69688-22-8	NI35841
340	325	57	341	282	73	283	323	**340**	100	54	54	29	29	24	22	17		21	28	2	–	–	–	–	–	–	1	–	–	–	Estra-1,3,5(10),9(11)-tetraen-17-one, 3-[(trimethylsilyl)oxy]-	69833-95-0	NI35842
340	341	218	73	148	312	244	283	**340**	100	34	21	20	19	18	18	17		21	28	2	–	–	–	–	–	–	1	–	–	–	Estra-1,3,5(10),15-tetraen-17-one-, 3-[(trimethylsilyl)oxy]-, 14α-		NI35843
244	340	245	341	217	73	232	243	**340**	100	78	63	23	21	21	20	19		21	28	2	–	–	–	–	–	–	1	–	–	–	Estra-1,3,5(10),15-tetraen-17-one-, 3-[(trimethylsilyl)oxy]-, 14β-		NI35844
222	235	207	223	308	71	69	206	**340**	100	38	19	17	16	12	12	11	**0.00**	21	28	2	2	–	–	–	–	–	–	–	–	–	Acetamide, N-[2-(3-ethyl-2,3,4,9-tetrahydro-4-methoxy-1-methylene-1H-carbazol-2-yl)ethyl]-N-methyl-	55320-30-4	NI35845
158	144	157	196	166	143	145	115	**340**	100	90	59	45	36	28	22	22	**21.02**	21	28	2	2	–	–	–	–	–	–	–	–	–	Akuammilan-16-methanol, 1,2-dihydro-17-hydroxy-1-methyl-	7385-60-6	NI35846
154	109	41	42	340	139	188	312	**340**	100	90	72	69	68	56	47	44		21	28	2	2	–	–	–	–	–	–	–	–	–	Aspidofractinine, 6,17-dimethoxy-, (6β)-	54965-81-0	NI35847
109	340	312	124	110	86	282	41	**340**	100	50	37	34	33	25	24	21		21	28	2	2	–	–	–	–	–	–	–	–	–	Aspidofractinine-3-methanol, 17-methoxy-, (2α,3α,5α)-	55103-44-1	NI35848
124	109	41	340	39	312	196	172	**340**	100	40	33	25	24	19	19	17		21	28	2	2	–	–	–	–	–	–	–	–	–	Aspidofractinine-3-methanol, 17-methoxy-, (2α,3β,5α)-	55103-42-9	NI35849
124	254	125	41	340	130	144	42	**340**	100	10	10	9	5	5	4	4		21	28	2	2	–	–	–	–	–	–	–	–	–	Aspidospermidine-3-carboxylic acid, methyl ester, (2β,3α,5α,12β,19α)-	55700-36-2	NI35850
124	254	125	340	143	55	41	144	**340**	100	20	18	12	9	8	8	7		21	28	2	2	–	–	–	–	–	–	–	–	–	Aspidospermidine-3-carboxylic acid, methyl ester, (2β,3β,5α,12β,19α)-	55781-30-1	NI35851
124	254	125	41	340	55	130	42	**340**	100	50	50	45	26	26	24	22		21	28	2	2	–	–	–	–	–	–	–	–	–	Aspidospermidine-3-carboxylic acid, methyl ester, (2β,5α,12β,19α)-	55514-94-8	NI35852
124	340	339	41	59	57	45	43	**340**	100	31	18	11	10	10	10	10		21	28	2	2	–	–	–	–	–	–	–	–	–	Aspidospermidin-17-ol, 1-acetyl-	2122-21-6	NI35853
138	340	210	170	139	124	341	137	**340**	100	47	47	17	17	16	14	13		21	28	2	2	–	–	–	–	–	–	–	–	–	18α-Carbomethoxy-4α-dihydrocleavamine		LI07541
138	57	210	149	69	340	71	83	**340**	100	76	65	57	57	52	45	43		21	28	2	2	–	–	–	–	–	–	–	–	–	18α-Carbomethoxy-4β-dihydrocleavamine		LI07542
138	340	210	215	124	170	137	82	**340**	100	63	63	32	32	22	22	18		21	28	2	2	–	–	–	–	–	–	–	–	–	18β-Carbomethoxy-4β-dihydrocleavamine		LI07543
340	215	138	284	341	299	216	186	**340**	100	58	42	39	28	28	28	28		21	28	2	2	–	–	–	–	–	–	–	–	–	Condyfolan, 1-acetyl-11-methoxy-	56143-38-5	NI35854
117	119	121	82	340	47	36	297	**340**	100	85	38	30	17	14	13	12		21	28	2	2	–	–	–	–	–	–	–	–	–	Cyclohexane, 1-(4-methylamino-3-methoxyphenyl)-1-(4-amino-3-methoxyphenyl)-		IC04031
202	41	43	340	138	42	203	171	**340**	100	90	70	61	60	50	42	37		21	28	2	2	–	–	–	–	–	–	–	–	–	2,20-Cyclo-8,9-secoaspidospermidine-3-carboxylic acid, methyl ester, (2α,3β,5α,12β,19α,20R)-	56053-31-7	NI35855
339	225	340	197	223	42	55	43	**340**	100	95	86	52	45	44	39	39		21	28	2	2	–	–	–	–	–	–	–	–	–	10-Deoxy-18,19-dihydro-15-epi-hunterburnine acetate		MS08088
339	340	170	251	225	169	341	311	**340**	100	97	39	31	28	28	25	22		21	28	2	2	–	–	–	–	–	–	–	–	–	3-Ethyl-1,2,3,4,6,7,12,12b-octahydroindolo[2,3-a]quinolizine-2-yl acetic acid ethyl ester		MS08089
138	57	210	149	69	340	81	71	**340**	100	76	65	56	56	52	51	43		21	28	2	2	–	–	–	–	–	–	–	–	–	2H-3,7-Methanoazacycloundecino[5,4-b]indole-9-carboxylic acid, 5-ethyl-1,4,5,6,7,8,9,10-octahydro-, methyl ester, [5S-(5R*,7R*,9S*)]-	26251-90-1	NI35857
138	340	215	210	124	126	341	170	**340**	100	85	54	48	35	30	25	25		21	28	2	2	–	–	–	–	–	–	–	–	–	2H-3,7-Methanoazacycloundecino[5,4-b]indole-9-carboxylic acid, 5-ethyl-1,4,5,6,7,8,9,10-octahydro-, methyl ester, [5S-(5R*,7R*,9S*)]-	26251-90-1	NI35856
138	340	215	210	124	126	170	82	**340**	100	83	54	47	34	30	28	20		21	28	2	2	–	–	–	–	–	–	–	–	–	2H-3,7-Methanoazacycloundecino[5,4-b]indole-9-carboxylic acid, 5-ethyl-1,4,5,6,7,8,9,10-octahydro-, methyl ester, [5S-(5R*,7R*,9S*)]-	26251-90-1	LI07544
197	70	338	310	225	182	170	196	**340**	100	96	59	53	27	27	23	20	**0.00**	21	28	2	2	–	–	–	–	–	–	–	–	–	18-Noralstophyllan-19-ol, 20,21-dihydro-21-methyl-	16101-12-5	NI35858
197	338	310	225	182	170	196	181	**340**	100	65	34	27	24	22	21	20	**1.53**	21	28	2	2	–	–	–	–	–	–	–	–	–	18-Noralstophyllan-19-ol, 20,21-dihydro-21-methyl-	16101-12-5	NI35859
340	339	44	283	43	253	69	41	**340**	100	77	32	31	31	25	22	22		21	28	2	2	–	–	–	–	–	–	–	–	–	18,19-Secoyohimban-19-oic acid, 16-methyl-, methyl ester, (15α,20ξ)-	56259-09-7	NI35860
170	43	228	55	98	69	71	57	**340**	100	89	76	66	62	62	57	51	**8.00**	21	40	3	–	–	–	–	–	–	–	–	–	–	Eicosanoic acid, 11-oxo-, methyl ester	2388-76-3	NI35861
170	139	228	171	155	138	111	110	**340**	100	62	55	53	50	39	31	31	**0.00**	21	40	3	–	–	–	–	–	–	–	–	–	–	Eicosanoic acid, 11-oxo-, methyl ester	2388-76-3	NI35862
170	139	228	171	155	138	213	135	**340**	100	63	56	54	51	39	30	30	**0.00**	21	40	3	–	–	–	–	–	–	–	–	–	–	Eicosanoic acid, 11-oxo-, methyl ester	2388-76-3	LI07545
43	99	71	98	114	149	195	59	**340**	100	81	77	54	51	35	29	29	**0.00**	21	40	3	–	–	–	–	–	–	–	–	–	–	Eicosanoic acid, 15-oxo-, methyl ester	19271-79-5	LI07546
43	99	71	98	114	97	149	115	**340**	100	82	76	55	50	36	35	32	**0.26**	21	40	3	–	–	–	–	–	–	–	–	–	–	Eicosanoic acid, 15-oxo-, methyl ester	19271-79-5	NI35863
43	41	57	116	55	129	322	69	**340**	100	90	63	60	56	40	23	23	**11.00**	21	40	3	–	–	–	–	–	–	–	–	–	–	(2R,3R,4S)-2-Hexadecanyl-3-hydroxy-4-methylbutanolide		NI35864
55	41	43	264	69	98	67	31	**340**	100	95	64	60	57	53	47	43	**3.04**	21	40	3	–	–	–	–	–	–	–	–	–	–	Oleic acid, 3-hydroxypropyl ester	821-17-0	NI35865
299	75	97	300	83	89	57	69	**340**	100	52	27	26	26	17	16	14	**1.10**	21	44	1	–	–	–	–	–	–	1	–	–	–	Allyldimethylsilyl hexadecyl ether		NI35866
180	77	340	181	105	51	341	43	**340**	100	47	32	15	11	11	8	6		22	16	2	2	–	–	–	–	–	–	–	–	–	1,5-Diphenyl-3-benzoyl-4-hydroxypyrazole	95884-33-6	NI35867
340	341	140	309	280	281	279	170	**340**	100	20	11	10	10	8	6	5		22	16	2	2	–	–	–	–	–	–	–	–	–	Methyl 4-(indol-3-yl)carbazole-1-carboxylate		NI35868
339	340	178	77	89	51	165	176	**340**	100	69	15	15	8	7	7	5		22	16	2	2	–	–	–	–	–	–	–	–	–	4-Phenoxy-5,6-diphenyl-2H-pyridazin-3-one	77891-97-5	NI35869
340	208	263	194	165	132	341	209	**340**	100	82	70	42	36	30	26	20		22	16	2	2	–	–	–	–	–	–	–	–	–	7-Phenyldiftalone		MS08090

MASS TO CHARGE RATIOS	M.W.	INTENSITIES	Parent	C	H	O	N	S	F	Cl	Br	I	Si	P	B	X	COMPOUND NAME	CAS Reg No	No
340 77 130 105 102 271 51 91	340	100 83 30 27 23 20 14 6		22	16	2	2	–	–	–	–	–	–	–	–	–	3,5-Pyrazolidinedione, 1,2-diphenyl-4-benzylidene-	2652-78-0	NI35870
104 340 77 180 339 181 105 341	340	100 86 73 65 35 24 22 19		22	16	2	2	–	–	–	–	–	–	–	–	–	2(1H)-Pyrimidinone, 5-phenoxy-4,6-diphenyl-	28567-85-3	NI35871
286 325 287 271 263 222 326 262	340	100 51 28 19 17 17 16 16	0.80	22	20	–	2	–	–	–	–	–	1	–	–	–	10-Methyl-10-phenyl-9-(β-cyanoethyl)-9,10-dihydro-10-sila-2-azaanthracene	92973-64-3	NI35872
267 55 107 91 136 79 67 77	340	100 98 89 86 65 58 57 52	50.80	22	28	3	–	–	–	–	–	–	–	–	–	–	Canrenone		MS08091
340 159 199 133 157 160 212 213	340	100 82 81 73 66 62 53 48		22	28	3	–	–	–	–	–	–	–	–	–	–	19-Norbisnorchola-1,3,5(10),17(20)-tetraen-22-carboxylic acid, 3-hydroxy , methyl ester		LI07547
340 199 157 133 155 158 212 213	340	100 84 82 73 65 62 53 48		22	28	3	–	–	–	–	–	–	–	–	–	–	19-Norpregna-1,3,5(10),17(20)-tetraene-20-carboxylic acid, 3-hydroxy-, methyl ester	55399-33-2	NI35873
124 340 341 125 170 325 188 152	340	100 42 11 10 7 6 6 6		22	32	1	2	–	–	–	–	–	–	–	–	–	Aspidospermidine, 1-ethyl-17-methoxy-	5523-58-0	NI35874
43 41 98 55 57 73 60 69	340	100 71 69 68 61 47 47 39	6.61	22	44	2	–	–	–	–	–	–	–	–	–	–	Docosanoic acid	112-85-6	WI01442
43 57 73 60 55 41 340 69	340	100 70 63 55 54 54 44 33		22	44	2	–	–	–	–	–	–	–	–	–	–	Docosanoic acid	112-85-6	NI35875
43 57 73 60 55 71 41 69	340	100 84 79 68 57 43 43 40	11.00	22	44	2	–	–	–	–	–	–	–	–	–	–	Docosanoic acid	112-85-6	NI35876
74 101 43 41 55 75 57 69	340	100 74 52 39 37 36 29 23	9.50	22	44	2	–	–	–	–	–	–	–	–	–	–	Eicosanoic acid, 3-methyl-, methyl ester		MS08092
74 340 199 143 157 297 227 195	340	100 40 38 30 17 10 10 5		22	44	2	–	–	–	–	–	–	–	–	–	–	Eicosanoic acid, 12-methyl-, methyl ester		LI07548
43 57 83 55 44 69 97 41	340	100 62 58 53 47 46 42 41	1.76	22	44	2	–	–	–	–	–	–	–	–	–	–	Eicosyl acetate		IC04032
74 87 340 43 75 57 55 18	340	100 63 40 35 28 26 26 23		22	44	2	–	–	–	–	–	–	–	–	–	–	Heneicosanoic acid, methyl ester	6064-90-0	WI01443
74 87 43 57 55 41 69 83	340	100 77 77 67 57 51 42 33	19.20	22	44	2	–	–	–	–	–	–	–	–	–	–	Heneicosanoic acid, methyl ester	6064-90-0	MS08093
74 87 43 55 41 57 75 69	340	100 64 55 36 32 30 20 17	1.00	22	44	2	–	–	–	–	–	–	–	–	–	–	Heneicosanoic acid, methyl ester	6064-90-0	NI35877
57 43 83 69 55 97 71 56	340	100 97 85 82 76 71 53 49	0.10	22	44	2	–	–	–	–	–	–	–	–	–	–	1-Heneicosyl formate		MS08094
87 44 340 57 43 55 75 41	340	100 70 52 48 48 47 46 39		22	44	2	–	–	–	–	–	–	–	–	–	–	Heptanoic acid, 4,8,12,16-tetramethyl-, methyl ester		LI07549
73 257 69 129 116 71 239 285	340	100 85 70 63 58 57 54 51	24.02	22	44	2	–	–	–	–	–	–	–	–	–	–	Octadecanoic acid, butyl ester	123-95-5	WI01444
285 340 56 267 284 129 57 341	340	100 86 53 45 39 32 30 27		22	44	2	–	–	–	–	–	–	–	–	–	–	Octadecanoic acid, butyl ester	123-95-5	WI01445
56 57 285 43 73 41 340 55	340	100 54 44 40 28 28 26 26		22	44	2	–	–	–	–	–	–	–	–	–	–	Octadecanoic acid, butyl ester	123-95-5	NI35878
57 56 41 43 55 29 285 284	340	100 81 66 64 37 29 26 23	5.60	22	44	2	–	–	–	–	–	–	–	–	–	–	Octadecanoic acid, isobutyl ester	646-13-9	NI35879
87 43 55 57 74 69 41 71	340	100 38 36 35 33 28 23 21	15.00	22	44	2	–	–	–	–	–	–	–	–	–	–	Octadecanoic acid, 4,8,12-trimethyl-, methyl ester		MS08095
127 57 43 41 55 69 56 83	340	100 65 57 44 41 21 18 17	0.00	22	44	2	–	–	–	–	–	–	–	–	–	–	Octanoic acid, 1-methyltridecyl ester	55193-79-8	NI35880
173 167 340 165 168 28 89 152	340	100 89 57 40 28 27 26 15		23	16	3	–	–	–	–	–	–	–	–	–	–	1H-Indene-1,3(2H)-dione, 2-(diphenylacetyl)-	82-66-6	NI35881
339 340 222 220 104 193 194 178	340	100 46 45 20 17 16 14 14		23	20	1	2	–	–	–	–	–	–	–	–	–	β-Amino-α-phenylcinnamic acid (2-phenylaziridide)		LI07550
233 340 341 77 107 103 79 178	340	100 88 23 18 15 15 15 8		23	20	1	2	–	–	–	–	–	–	–	–	–	1H-Imidazole-1-ethanol, α,4,5-triphenyl-	58275-51-7	NI35882
249 263 91 104 77 250 115 340	340	100 66 52 40 31 18 18 17		23	20	1	2	–	–	–	–	–	–	–	–	–	Pyrimidine, 5-benzyl-2-oxo-4,6-diphenyl-1,2,3,4-tetrahydro-		LI07551
206 263 91 77 340 104	340	100 48 32 31 24 24		23	20	1	2	–	–	–	–	–	–	–	–	–	Pyrimidine, 3-tolyl-4,6-diphenyl-2-oxo-1,2,3,4-tetrahydro-		LI07552
177 161 164 340 149 57 121 127	340	100 74 58 47 42 29 27 18		23	32	2	–	–	–	–	–	–	–	–	–	–	Bis(3-tert-butyl-2-hydroxy-5-methylphenyl)methane	119-47-1	IC04033
177 161 164 340 149 178 155 121	340	100 71 48 36 28 14 14 14		23	32	2	–	–	–	–	–	–	–	–	–	–	Bis(3-tert-butyl-2-hydroxy-5-methylphenyl)methane	119-47-1	IC04034
177 340 164 149 57 121 127 341	340	100 79 61 39 23 21 20 19		23	32	2	–	–	–	–	–	–	–	–	–	–	Bis(3-tert-butyl-2-hydroxy-5-methylphenyl)methane	119-47-1	NI35883
325 340 283 341 326 177 161 127	340	100 95 33 24 24 24 14 11		23	32	2	–	–	–	–	–	–	–	–	–	–	Bis(3-tert-butyl-4-hydroxy-5-methylphenyl)methane		IC04035
340 325 283 341 326 177 127 155	340	100 77 28 26 20 20 14 10		23	32	2	–	–	–	–	–	–	–	–	–	–	Bis(3-tert-butyl-4-hydroxy-5-methylphenyl)methane		MS08096
325 171 161 340 326 176 283 127	340	100 40 34 30 24 24 21 16		23	32	2	–	–	–	–	–	–	–	–	–	–	Bis(3-tert-butyl-4-hydroxy-6-methylphenyl)methane		IC04036
340 325 135 341 326 283 57 148.5	340	100 98 34 28 27 16 7 5		23	32	2	–	–	–	–	–	–	–	–	–	–	3,5-Dimethyl-3′,5′-di-tert-butyl-4,4′-dihydroxy-diphenylmethane		MS08097
285 173 340 147 171 267 322 227	340	100 85 81 66 64 61 51 47		23	32	2	–	–	–	–	–	–	–	–	–	–	Estra-1,3,5(10)-trien-17α-ol, 3-methoxy-17-isobutenyl-	17550-04-8	NI35884
285 173 340 147 171 267 322 227	340	100 85 81 68 63 61 52 48		23	32	2	–	–	–	–	–	–	–	–	–	–	Estra-1,3,5(10)-trien-17α-ol, 3-methoxy-17β-isobutenyl-		LI07553
340 43 297 109 189 175 187 41	340	100 94 59 29 27 26 24 20		23	32	2	–	–	–	–	–	–	–	–	–	–	Pregna-4,6-diene-3,20-dione, 6,17-dimethyl-	977-79-7	NI35885
340 91 55 325 85 256 77 67	340	100 51 50 42 42 33 33 30		23	32	2	–	–	–	–	–	–	–	–	–	–	Pregn-4-ene-3,20-dione, 16-methyl-6-methylene-, (16α)-	55334-58-2	WI01446
340 91 55 325 85 256 105 79	340	100 51 50 42 42 33 30 30		23	32	2	–	–	–	–	–	–	–	–	–	–	Pregn-4-ene-3,20-dione, 16-methyl-6-methylene-, (16α)-	55334-58-2	AM00145
43 55 57 41 97 69 83 56	340	100 91 85 78 68 61 59 43	0.00	23	48	1	–	–	–	–	–	–	–	–	–	–	1-Tricosanol		MS08098
105 267 311 340 165 265 312 294	340	100 97 82 67 40 36 28 26		24	20	2	–	–	–	–	–	–	–	–	–	–	Benzeneacetic acid, α-(diphenylvinylidene)-, ethyl ester	36858-84-1	LI07554
105 267 311 340 165 265 312 294	340	100 97 84 67 40 36 28 26		24	20	2	–	–	–	–	–	–	–	–	–	–	Benzeneacetic acid, α-(diphenylvinylidene)-, ethyl ester	36858-84-1	NI35886
340 312 311 267 165 105 341 265	340	100 70 64 55 50 38 30 30		24	20	2	–	–	–	–	–	–	–	–	–	–	3-Ethoxy-2,4,4-triphenyl-2-cyclobuten-1-one		DD01346
267 105 311 340 265 165 312 294	340	100 85 80 60 40 40 25 25		24	20	2	–	–	–	–	–	–	–	–	–	–	Ethyl 2,4,4-triphenyl-2,3-butadienoate		MS08099
191 133 105 189 340 165 190 103	340	100 48 13 12 10 8 7 3		24	20	2	–	–	–	–	–	–	–	–	–	–	9-Phenanthrenemethyl 2,6-dimethylbenzoate	92174-46-4	NI35887
340 341 55 83 271 133 41 189	340	100 27 14 11 8 8 6 5		24	36	1	–	–	–	–	–	–	–	–	–	–	2,4,6-Tricyclohexylphenol		MS08100
92 91 41 55 69 83 97 93	340	100 37 24 20 17 15 10 10	1.36	25	40	–	–	–	–	–	–	–	–	–	–	–	1,7-Dicyclopentyl-4-(2′-phenylethyl)-heptane		AI02080
69 41 94 81 107 55 109 121	340	100 61 60 36 34 30 27 26	4.00	25	40	–	–	–	–	–	–	–	–	–	–	–	2,6,12,17-Pentamethyl-9-methylenenonadeca-2,6,13,17-tetraene		LI07555
92 91 55 41 83 69 97 67	340	100 48 38 25 23 18 12 10	4.36	25	40	–	–	–	–	–	–	–	–	–	–	–	1-Phenyl-3-(2′-cyclohexylethyl)-6-cyclopentylhexane		AI02081
92 91 55 41 83 69 97 67	340	100 38 35 31 23 16 15 9	1.07	25	40	–	–	–	–	–	–	–	–	–	–	–	1-Phenyl-3-(2′-cyclohexylethyl)-6-cyclopentylhexane		AI02082
340 241 239 341 242 240 57 43	340	100 94 33 28 26 18 9 9		26	28	–	–	–	–	–	–	–	–	–	–	–	6-Octylchrysene		AI02083

MASS TO CHARGE RATIOS								M.W.	INTENSITIES								Parent	C	H	O	N	S	F	Cl	Br	I	Si	P	B	X	COMPOUND NAME	CAS Reg No	No
241	340	239	242	341	240	41	43	**340**	100	76	30	28	22	15	15	13		26	28	–	–	–	–	–	–	–	–	–	–	–	6-Octylchrysene		AI02084
241	340	239	242	341	240	41	43	**340**	100	77	32	30	22	17	13	11		26	28	–	–	–	–	–	–	–	–	–	–	–	6-Octylchrysene		AI02085
43	71	226	69	169	272	119	298	**341**	100	49	32	14	10	5	3	1	**0.10**	11	14	3	1	–	7	–	–	–	–	–	–	–	3-Methylbutyl N-heptafluorobutyrylglycinate		NI35888
282	43	41	61	254	102	226	60	**341**	100	74	63	57	46	40	36	35	**6.66**	11	14	3	1	–	7	–	–	–	–	–	–	–	Propyl N-(heptafluorobutyryl)-γ-aminobutyrate		MS08101
224	73	128	222	147	220	226	218	**341**	100	93	69	49	30	22	21	21	**6.80**	11	27	2	1	–	–	–	–	–	2	–	–	1	N,O-Bis(trimethylsilyl)selenomethionine		NI35889
116	73	298	147	117	299	59	45	**341**	100	50	29	10	10	8	7	7	**1.30**	11	32	3	1	–	–	–	–	–	3	1	–	–	Phosphonic acid, [1-[(trimethylsilyl)amino]ethyl]-, bis(trimethylsilyl) ester	53044-37-4	NI35890
116	73	298	147	207	211	100	75	**341**	100	50	29	10	7	5	5	5	**1.30**	11	32	3	1	–	–	–	–	–	3	1	–	–	Phosphonic acid, [1-[(trimethylsilyl)amino]ethyl]-, bis(trimethylsilyl) ester	53044-37-4	NI35891
73	211	226	225	44	45	75	254	**341**	100	70	67	59	58	48	45	39	**0.00**	11	32	3	1	–	–	–	–	–	3	1	–	–	Phosphonic acid, [2-[(trimethylsilyl)amino]ethyl]-, bis(trimethylsilyl) ester	56051-90-2	NI35892
70	234	176	193	280	222			**341**	100	44	36	26	19	15			**0.00**	12	15	7	5	–	–	–	–	–	–	–	–	–	2,4-Dinitrophenyl-citrulline		LI07556
61	55	129	82	60	54	87	73	**341**	100	64	58	55	40	27	22	20	**0.00**	12	23	6	1	2	–	–	–	–	–	–	–	–	Desulphoglucoerucin		MS08102
183	185	241	243	41	157	155	54	**341**	100	99	65	63	56	43	42	42	**8.00**	13	12	1	1	2	–	–	1	–	–	–	–	–	Benzenepropanedithioic acid, 4-bromo-β-hydroxy-, 3-cyanopropyl ester	72347-61-6	NI35893
15	30	75	341	343	63	173	74	**341**	100	93	86	66	53	53	39	39		14	9	5	1	–	–	2	–	–	–	–	–	–	Benzoic acid, 5-(2,4-dichlorophenoxy)-2-nitro-, methyl ester	42576-02-3	NI35894
110	106	78	69	218	51	153	323	**341**	100	99	94	64	57	57	51	45	**0.00**	14	10	2	3	1	3	–	–	–	–	–	–	–	1,1,1-Trifluoro-3-thienoylisopropylidene isonicotinohydrazide		LI07557
151	341	152	342	165	119	107	91	**341**	100	68	18	11	10	7	6	6		14	16	3	1	–	5	–	–	–	–	–	–	–	Propylamine, 3-(3,4-dimethoxyphenyl)-, pentafluoropropionyl-		MS08103
343	341	224	222	28	29	31	27	**341**	100	99	99	99	88	68	48	35		14	16	4	1	–	–	–	1	–	–	–	–	–	Diethyl 2-(4-bromoanilino)fumarate		MS08104
98	112	43	55	41	42	99	56	**341**	100	42	33	27	27	25	16	14	**13.00**	14	19	1	3	3	–	–	–	–	–	–	–	–	5-Acetyl-4-amino-3-(2-N-piperidinylethylthio)thieno[2,3-c]isothiazole	97090-69-2	NI35895
98	111	43	42	41	55	56	99	**341**	100	24	24	22	20	15	9	8	**3.00**	14	19	1	3	3	–	–	–	–	–	–	–	–	5-Acetyl-4-amino-3-(2-N-piperidinylethylthio)thieno[3,2-d]isothiazole	97090-76-1	NI35896
342	343	214	212	344	268	240	184	**341**	100	16	7	5	2	2	1	1	**0.00**	14	22	5	1	–	3	–	–	–	–	–	–	–	L-Aspartic acid, N-(trifluoroacetyl)-, dibutyl ester	2926-77-4	NI35897
57	41	240	139	184	140	212	167	**341**	100	74	45	43	38	28	24	19	**0.00**	14	22	5	1	–	3	–	–	–	–	–	–	–	L-Aspartic acid, N-(trifluoroacetyl)-, dibutyl ester	2926-77-4	NI35898
57	212	184	185	41	139	43	45	**341**	100	59	55	48	43	34	15	13	**0.00**	14	22	5	1	–	3	–	–	–	–	–	–	–	L-Aspartic acid, N-(trifluoroacetyl)-, di-sec-butyl ester	57983-73-0	NI35899
341	166	148	188	176	256	138	313	**341**	100	73	52	39	38	29	26	20		15	20	6	1	–	–	–	–	–	–	1	–	–	2,3,4,5-Tetrahydro-7,8-(methylenedioxy)-N-(diethylphosphoryl)-3-benzazepin-1-one		NI35900
211	163	147	117	77	53	89	63	**341**	100	51	34	32	25	25	24	22	**0.00**	16	27	5	3	–	–	–	–	–	–	–	–	–	Phenol, 2-sec-butyl-4,6-dinitro-, compd. with triethylamine (1:1)	31931-88-1	NI35901
118	18	30	74	56	29	27	45	**341**	100	93	86	84	83	65	58	56	**0.00**	16	27	5	3	–	–	–	–	–	–	–	–	–	Phenol, 2-sec-butyl-4,6-dinitro-, compd. with triethylamine (1:1)	31931-88-1	NI35902
75	73	297	44	45	77	68	40	**341**	100	71	58	31	20	19	18	16	**0.00**	16	31	3	1	–	–	–	–	–	2	–	–	–	Phenethylamine, 3-methoxy-N-methyl-β,4-bis(trimethylsiloxy)-	10538-87-1	MS08105
297	73	298	299	75	267	326	223	**341**	100	43	28	10	8	5	4	4	**0.00**	16	31	3	1	–	–	–	–	–	2	–	–	–	Phenethylamine, 3-methoxy-N-methyl-β,4-bis(trimethylsiloxy)-	10538-87-1	NI35903
295	75	296	74	53	341	63	69	**341**	100	52	36	36	36	35	35	31		17	11	7	1	–	–	–	–	–	–	–	–	–	1-Carboxy-3,4-methylenedioxy-8-methoxy-10-nitro-phenanthrene		MS08106
305	230	306	231	229	228	202	307	**341**	100	46	34	11	11	9	6	5	**2.50**	17	13	–	1	–	–	1	–	–	–	–	–	1	Benzo[c]phenarsazine, 7-chloro-7,12-dihydro-9-methyl-	13493-36-2	NI35904
154	156	42	115	63	56	144	158	**341**	100	61	18	17	16	14	13	10	**2.30**	17	21	2	1	–	–	2	–	–	–	–	–	–	1-(1-Naphthyloxy)-3-[bis(2-chloroethyl)amino]-2-propanol		NI35905
44	149	135	190	296	162	176	122	**341**	100	34	30	29	28	26	16	14	**0.00**	17	23	1	7	–	–	–	–	–	–	–	–	–	1-[6-(Aden-9-yl)hexyl]-3-carbamoyl-1,4-dihydropyridine		LI07558
138	93	43	94	80	139	136	137	**341**	100	74	48	35	17	15	11	10	**2.59**	17	27	6	1	–	–	–	–	–	–	–	–	–	3′-Acetyllycopsamine	73544-48-6	NI35906
341	310	311	238	252	193	165	164	**341**	100	90	19	15	14	14	8	7		18	15	6	1	–	–	–	–	–	–	–	–	–	Trimethyl benzo[g]indolizine-1,2,3-tricarboxylate		LI07559
119	91	280	65	120	29	43	62	**341**	100	37	24	13	9	8	7	6	**0.69**	18	19	2	3	1	–	–	–	–	–	–	–	–	Carbamothioic acid, [[(3-methylbenzoyl)amino](phenylamino)methylene]-, S-ethyl ester	74793-49-0	NI35907
134	299	284	341	65	138	135	30	**341**	100	67	58	51	32	27	26	25		18	23	2	5	–	–	–	–	–	–	–	–	–	N-Ethyl-N-(3-aminopropyl)-3-methyl-4-(4-nitrophenylazo)-aniline		IC04037
169	70	126	112	298	100	157	98	**341**	100	94	91	89	73	56	54	54	**8.50**	18	35	3	3	–	–	–	–	–	–	–	–	–	1,6,10-Triacetyl-1,6,10-triazapentadecane	55521-02-3	NI35908
169	70	126	112	298	157	100	143	**341**	100	94	90	88	73	54	54	53	**8.50**	18	35	3	3	–	–	–	–	–	–	–	–	–	1,6,10-Triacetyl-1,6,10-triazapentadecane	55521-02-3	LI07560
214	216	111	75	215	77	113	51	**341**	100	34	19	19	17	14	7	7	**1.00**	19	13	1	1	–	–	2	–	–	–	–	–	–	4-Chloro-2-methoxycarbonylphenyl-N-(2′-chlorophenyl)benzimidate		NI35909
341	340	311	312	153	326	266	277	**341**	100	90	89	76	54	46	34	32		19	19	5	1	–	–	–	–	–	–	–	–	–	Cassifiline		LI07561
299	256	228	341	257	270	282	242	**341**	100	84	61	60	35	32	30	29		19	19	5	1	–	–	–	–	–	–	–	–	–	4H-Cyclopenta[a][1,3]dioxolo[4,5-H]pyrrolo[2,1-b][3]benazepin-2(3H)-one, 1-(acetyloxy)-5,6,7,8-9-tetrahydro-, (S)-	50908-90-2	NI35910
167	166	165	212	194	168	152	153	**341**	100	97	97	96	96	96	96	81	**0.00**	19	19	5	1	–	–	–	–	–	–	–	–	–	Diphenylacetylglutamic acid	78490-36-5	NI35911
204	323	70	205	91	324	202	135	**341**	100	43	32	13	12	11	9	9	**2.60**	19	23	3	3	–	–	–	–	–	–	–	–	–	Azacyclol L-proline-L-phenylalanine-L-proline		NI35912
97	124	326	309	341	311			**341**	100	55	15	4	3	3				19	23	3	3	–	–	–	–	–	–	–	–	–	7-Pyrrolidino-7a-methoxy-5,2-dioxo-2,3,5,6,7a,2,3-octahydro-1H-pyrrolizin-1-spiro-3-indole		LI07562
282	69	59	130	114	174	100	160	**341**	100	82	60	51	47	39	30	21	**0.00**	19	39	2	1	–	–	–	–	–	1	–	–	–	(E)-1-[[[Bis(2-ethoxyethyl)amino]methyl]dimethylsilyl]-6-methyl-1,5-heptadiene		DD01347
282	69	114	59	130	174	100	160	**341**	100	92	76	68	57	50	29	24	**0.00**	19	39	2	1	–	–	–	–	–	1	–	–	–	(Z)-1-[[[Bis(2-ethoxyethyl)amino]methyl]dimethylsilyl]-6-methyl-1,5-heptadiene		DD01348
282	59	130	174	232	81	100	342	**341**	100	51	49	43	40	34	33	17	**0.00**	19	39	2	1	–	–	–	–	–	1	–	–	–	3-[[[Bis(2-ethoxyethyl)amino]methyl]dimethylsilyl]-6-methyl-1,5-heptadiene		DD01349

MASS TO CHARGE RATIOS								M.W.	INTENSITIES								Parent	C	H	O	N	S	F	Cl	Br	I	Si	P	B	X	COMPOUND NAME	CAS Reg No	No
105	77	106	178	78	76	297	281	**341**	100	27	7	3	3	3	2	2	**1.20**	20	11	3	3	–	–	–	–	–	–	–	–	–	3-Benzoyl-1,3-dihydro-1-ketoisoxazolo[4,3-a]phenazine		LI07563
326	341	327	342	324	174	145	115	**341**	100	26	21	7	7	7	7	5		20	23	4	1	–	–	–	–	–	–	–	–	–	1H-[1]Benzoxepino[2,3,4-ij]isoquinoline, 2,3,12,12a-tetrahydro-6,9,10-trimethoxy-1-methyl-, (S)-	479-39-0	NI35913
204	341	190	340	170.5				**341**	100	33	30	22	3					20	23	4	1	–	–	–	–	–	–	–	–	–	Dibenzo[a,e]cycloocten-5,11-imin-4-ol, 5,6,11,12-tetrahydro-3,8,9-trimethoxy-13-methyl-, (5S)-	18826-68-1	LI07564
341	340	267	298	326	310	236	283	**341**	100	92	57	50	36	28	21	11		20	23	4	1	–	–	–	–	–	–	–	–	–	4H-Dibenzo[de,g]quinolin-1-ol, 5,6,6a,7-tetrahydro-2,9,10-trimethoxy-6-methyl-, (S)-	5083-88-5	LI07565
341	310	326	324	340	342	311	298	**341**	100	97	73	45	42	23	20	15		20	23	4	1	–	–	–	–	–	–	–	–	–	4H-Dibenzo[de,g]quinolin-1-ol, 5,6,6a,7-tetrahydro-2,10,11-trimethoxy-6-methyl-, (S)-	476-69-7	NI35914
339	340	341	324	326	325	310	266	**341**	100	99	83	71	40	28	24	19		20	23	4	1	–	–	–	–	–	–	–	–	–	4H-Dibenzo[de,g]quinolin-2-ol, 5,6,6a,7-tetrahydro-1,9,10-trimethoxy-6-methyl-		LI07566
164	165	341	121	342	176	166	149	**341**	100	40	29	12	10	10	10	8		20	23	4	1	–	–	–	–	–	–	–	–	–	6H-Dibenzo[a,g]quinolizin-1-ol, 5,8,13,13a-tetrahydro-2,10,11-trimethoxy, (S)-	34413-12-2	LI07567
164	165	341	121	176	166	342	149	**341**	100	41	29	10	9	9	8	7		20	23	4	1	–	–	–	–	–	–	–	–	–	6H-Dibenzo[a,g]quinolizin-1-ol, 5,8,13,13a-tetrahydro-2,10,11-trimethoxy, (S)-	34413-12-2	NI35915
192	341	190	150	135	149	310	326	**341**	100	58	27	23	23	19	12	8		20	23	4	1	–	–	–	–	–	–	–	–	–	6H-Dibenzo[a,g]quinolizin-10-ol, 5,8,13,13a-tetrahydro-2,3,9-trimethoxy-, (S)-	30413-84-4	LI07568
341	226	286	186	210	214	156	242	**341**	100	51	46	35	19	17	15	13		20	23	4	1	–	–	–	–	–	–	–	–	–	4,5-Dihydro-3-[6-methoxy-7-(3-methylbut-2-enyl)indol-2-ylcarbonyl]-4-methylfuran-2-one		MS08107
190	191	188	151	91	175	132	77	**341**	100	64	12	10	10	8	8	7	**1.40**	20	23	4	1	–	–	–	–	–	–	–	–	–	1,3-Dioxolo[4,5-g]isoquinoline, 5-[(3,4-dimethoxyphenyl)methyl]-5,6,7,8-tetrahydro-6-methyl-, (R)-	5544-49-0	NI35916
312	190	341	340	313	269	281	156	**341**	100	72	44	36	25	18	11	11		20	23	4	1	–	–	–	–	–	–	–	–	–	Isopavine	25727-46-2	NI35917
341	282	342	229	43	59	204	42	**341**	100	38	21	17	15	13	12	11		20	23	4	1	–	–	–	–	–	–	–	–	–	Morphinan-6-ol, 7,8-didehydro-4,5-epoxy-3-methoxy-17-methyl-, acetate, (5α,6α)-	6703-27-1	LI07569
282	43	42	229	59	204	81	124	**341**	100	68	59	52	40	29	26	24	**0.00**	20	23	4	1	–	–	–	–	–	–	–	–	–	Morphinan-6-ol, 7,8-didehydro-4,5-epoxy-3-methoxy-17-methyl-, acetate, (5α,6α)-	6703-27-1	NI35918
341	282	229	204	59	124	214	81	**341**	100	71	61	56	32	27	25	25		20	23	4	1	–	–	–	–	–	–	–	–	–	Morphinan-6-ol, 7,8-didehydro-4,5-epoxy-3-methoxy-17-methyl-, acetate, (5α,6α)-	6703-27-1	NI35919
341	340	342	31	44	227	268	30	**341**	100	45	22	20	19	17	13	11		21	15	2	3	–	–	–	–	–	–	–	–	–	1-(4,6-Dimethoxy-s-triazin-2-yl)pyrene		IC04038
121	323	311	326	341	221	57	324	**341**	100	95	82	69	50	39	37	26		21	27	3	1	–	–	–	–	–	–	–	–	–	3,5-Di-tert-butyl-4,2′-dihydroxydiphenylformaldehyde, oxime		IC04039
341	146	152	190	150	192	296	312	**341**	100	93	85	55	55	48	44	35		21	27	3	1	–	–	–	–	–	–	–	–	–	Diethylmorphine		LI07570
268	269	211	159	212	84	299	133	**341**	100	24	24	24	16	16	15	15	**0.00**	21	27	3	1	–	–	–	–	–	–	–	–	–	Estrone-3-acetate-17-methyloxime		NI35920
190	177	192	162	191	121	91	161	**341**	100	56	55	55	52	46	44	42	**0.80**	21	27	3	1	–	–	–	–	–	–	–	–	–	Isoquinoline, 7-ethoxy-1,2,3,4-tetrahydro-6-methoxy-1-(4-methoxybenzyl)2-methyl-	3423-06-1	NI35921
69	43	41	57	55	81	73	60	**341**	100	68	66	60	53	41	41	38	**0.80**	21	29	2	2	–	–	–	–	–	–	–	–	–	10-Methoxy-nb-α-methylcorynantheol	55322-92-4	NI35922
172	226	227	173	85	310	72	44	**341**	100	90	15	11	8	6	3	3	**2.00**	21	43	2	1	–	–	–	–	–	–	–	–	–	Octadecanoic acid, 6-(dimethylamino)-, methyl ester	56817-90-4	NI35923
230	341							**341**	100	15								22	15	3	1	–	–	–	–	–	–	–	–	–	1,3-Dioxolo[4,5-c]quinolin-4(5H)-one, 2,2-diphenyl-	38274-11-2	NI35924
238	104	76	119	221	105	91	222	**341**	100	91	68	55	36	25	22	21	**2.40**	22	15	3	1	–	–	–	–	–	–	–	–	–	3,3a-Diphenylindano[2,1-d]oxazolidine-2,4-dione	98294-00-9	NI35925
221	205	249	92	222	76	89	155	**341**	100	28	28	23	20	11	9	8	**2.30**	22	15	3	1	–	–	–	–	–	–	–	–	–	2-Phenyl-2-(phenylcarbamoyl)indan-1,3-dione	88702-68-5	NI35926
281	280	43	282	221	297	77	165	**341**	100	81	71	57	47	25	22	19	**4.00**	22	19	1	3	–	–	–	–	–	–	–	–	–	Acrylophenone, 3,3-diphenyl-, semicarbazone	16983-75-8	NI35927
222	105	77	223	120	51	106	78	**341**	100	57	35	17	7	6	4	4	**0.03**	23	19	2	1	–	–	–	–	–	–	–	–	–	Acetophenone, 2-(3,5-diphenyl-2-isoxazolin-5-yl)-	19363-81-6	NI35928
341	310	342	135	233	206	77	340	**341**	100	44	28	27	24	17	17	16		23	19	2	1	–	–	–	–	–	–	–	–	–	Indole, 3-benzyl-2-(4-methoxybenzoyl)-		LI07571
128	156	157	185	129	127	158	102	**341**	100	75	62	50	34	33	18	18	**2.02**	23	19	2	1	–	–	–	–	–	–	–	–	–	2-Naphthalenol, 1-(1H-naphth[1,2-e][1,3]oxazin-2(3H)-ylmethyl)-	50617-01-1	NI35929
341	236	192	210	191	342	312	264	**341**	100	49	32	30	25	24	22	17		23	19	2	1	–	–	–	–	–	–	–	–	–	2H-Pyrrol-2-one, 1,5-dihydro-1-(4-methoxyphenyl)-5,5-diphenyl-	53774-23-5	NI35930
105	77	236	165	298	342	194	42	**341**	100	42	21	21	18	12	11	11	**4.79**	24	23	1	1	–	–	–	–	–	–	–	–	–	Aziridine, 2-benzoyl-2-isopropyl-3-[1,1′-biphenyl]-4-yl-, cis-	32044-35-2	NI35931
105	341	77	165	298	236	342	167	**341**	100	48	33	26	19	19	12	11		24	23	1	1	–	–	–	–	–	–	–	–	–	Aziridine, 2-benzoyl-2-isopropyl-3-[1,1′-biphenyl]-4-yl-, trans-	32044-36-3	NI35932
195	146	147	131	178	165	196	179	**341**	100	52	37	33	30	19	17	16	**5.91**	24	23	1	1	–	–	–	–	–	–	–	–	–	2-[1-(2,4,6-Trimethylphenyl)ethyl]spiro[9H-fluorene-9,3′-oxaziridine]	89839-56-5	NI35933
243	341	98	165	244	55	342	56	**341**	100	37	33	18	17	10	4	3		24	23	1	1	–	–	–	–	–	–	–	–	–	Triphenylacetylpyrrolidine	70008-36-5	NI35934
243	244	165	264	258	167	242	166	**341**	100	45	44	29	17	13	12	12	**2.50**	25	27	–	1	–	–	–	–	–	–	–	–	–	Cyclohexylamine, N-trityl-	20360-17-2	MS08108
265	267	186	26	263	269	105	107	**342**	100	97	38	35	34	32	29	27	**1.60**	2	2	–	–	–	–	–	4	–	–	–	–	–	1,1,2,2-Tetrabromoethane		IC04040
265	267	186	263	269	105	107	184	**342**	100	93	40	38	35	34	31	21	**2.25**	2	2	–	–	–	–	–	4	–	–	–	–	–	1,1,2,2-Tetrabromoethane		IC04041
265	267	263	269	186	346	184	188	**342**	100	98	35	33	24	14	13	11	**3.38**	2	2	–	–	–	–	–	4	–	–	–	–	–	1,1,2,2-Tetrabromoethane		IC04042
131	133	61	31	28	96	64	48	**342**	100	95	75	66	57	46	43	35	**0.02**	4	4	3	–	1	–	6	–	–	–	–	–	–	Di-2,2,2-trichloroethyl sulphite		IC04043
346	344	329	331	141	303	301	143	**342**	100	99	98	95	52	50	49	45	**36.34**	7	5	1	–	–	–	–	3	–	–	–	–	–	Benzene, 1,3,5-tribromo-2-methoxy-	607-99-8	NI35935

MASS TO CHARGE RATIOS								M.W.	INTENSITIES								Parent	C	H	O	N	S	F	Cl	Br	I	Si	P	B	X	COMPOUND NAME	CAS Reg No	No
202	314	174	286	118	87	115	59	**342**	100	42	37	28	27	26	23	22	**12.50**	8	–	8	–	–	–	–	–	–	–	–	–	2	Di-μ-Carbonyl-hexacarbonyl-dicobalt		MS08109
207	208	209	73	133	105	75	191	**342**	100	20	10	10	8	8	7	5	**0.00**	9	27	3	–	–	–	–	–	–	3	–	–	1	Arsenous acid, tris(trimethylsilyl) ester	55429-29-3	NI35936
281	73	327	207	193	282	283	267	**342**	100	69	36	34	30	29	20	18	**0.00**	9	30	4	–	–	–	–	–	–	5	–	–	–	1,1,1,3,5,7,9,9,9-Nonamethylpentasiloxane	84409-41-6	NI35937
157	45	97	121	159	125	153	342	**342**	100	57	55	48	42	41	40	30		11	16	2	–	3	–	1	–	–	–	1	–	–	Phosphorodithioic acid, S-[[(4-chlorophenyl)thio]methyl] O,O-diethyl ester	786-19-6	NI35938
157	45	121	65	153	97	342	159	**342**	100	64	62	52	44	40	39	39		11	16	2	–	3	–	1	–	–	–	1	–	–	Phosphorodithioic acid, S-[[(4-chlorophenyl)thio]methyl] O,O-diethyl ester	786-19-6	NI35939
157	159	145	155	127	158	147	173	**342**	100	70	55	38	23	19	19	18	**0.00**	11	16	2	–	3	–	1	–	–	–	1	–	–	Phosphorodithioic acid, S-[[(4-chlorophenyl)thio]methyl] O,O-diethyl ester	786-19-6	NI35940
127	342	111	99	243	175	75	215	**342**	100	76	74	58	51	51	50	49		12	7	6	2	1	–	1	–	–	–	–	–	–	2,4-Dinitro-4′-chlorodiphenylsulphone	899-03-6	NI35941
342	170	204	75	76	344	122	110	**342**	100	91	84	37	34	33	33	32		12	7	6	2	1	–	1	–	–	–	–	–	–	3,3′-Dinitro-4′-chlorodiphenylsulphone	38040-20-9	NI35942
99	155	57	41	211	29	56	55	**342**	100	25	18	17	11	11	6	6	**0.00**	12	12	–	2	–	–	–	2	–	–	–	–	–	9,10-Dihydro-8a,10a-diazoniaphenanthrenedibromide		IC04044
15	191	59	73	159	192	29	86	**342**	100	43	34	31	27	25	24	22	**4.50**	12	14	4	4	2	–	–	–	–	–	–	–	–	Carbamic acid, [1,2-phenylenebis(iminocarbonothioyl)]bis-, dimethyl ester	23564-05-8	NI35943
192	160	342	59	150	177	105	209	**342**	100	83	39	38	35	28	24	23		12	14	4	4	2	–	–	–	–	–	–	–	–	Carbamic acid, [1,2-phenylenebis(iminocarbonothioyl)]bis-, dimethyl ester	23564-05-8	NI35944
325	343	163						**342**	100	82	45						**0.00**	12	22	11	–	–	–	–	–	–	–	–	–	–	D-Fructose, 4-O-β-D-galactopyranosyl-	4618-18-2	NI35945
343	163	325	181					**342**	100	41	20	12					**0.00**	12	22	11	–	–	–	–	–	–	–	–	–	–	D-Fructose, 6-O-α-D-glucopyranosyl-	13718-94-0	NI35946
73	29	43	60	44	31	61	57	**342**	100	76	73	68	64	62	47	47	**0.00**	12	22	11	–	–	–	–	–	–	–	–	–	–	β-D-Glucopyranose, 4-O-β-D-galactopyranosyl-	5965-66-2	NI35947
343	163	325						**342**	100	57	14						**0.00**	12	22	11	–	–	–	–	–	–	–	–	–	–	α-D-Glucopyranoside, β-D-fructofuranosyl-	57-50-1	NI35949
73	57	31	43	60	61	44	71	**342**	100	95	75	57	53	37	28	23	**0.00**	12	22	11	–	–	–	–	–	–	–	–	–	–	α-D-Glucopyranoside, β-D-fructofuranosyl-	57-50-1	NI35948
73	60	44	57	43	61	71	85	**342**	100	64	35	34	31	24	20	17	**0.00**	12	22	11	–	–	–	–	–	–	–	–	–	–	α-D-Glucopyranoside, α-D-glucopyranosyl	99-20-7	NI35950
343	325	163						**342**	100	26	12						**0.00**	12	22	11	–	–	–	–	–	–	–	–	–	–	α-D-Glucopyranoside, α-D-glucopyranosyl-	99-20-7	NI35951
73	60	57	43	44	61	71	97	**342**	100	75	40	35	32	32	28	21	**0.00**	12	22	11	–	–	–	–	–	–	–	–	–	–	D-Glucose, 4-O-β-D-galactopyranosyl-	63-42-3	NI35952
73	60	43	44	29	31	61	57	**342**	100	73	67	54	47	46	37	32	**0.00**	12	22	11	–	–	–	–	–	–	–	–	–	–	D-Glucose, 6-O-α-D-galactopyranosyl-	585-99-9	NI35953
73	60	57	43	97	44	61	69	**342**	100	74	46	43	36	35	32	31	**0.00**	12	22	11	–	–	–	–	–	–	–	–	–	–	D-Glucose, 4-O-α-D-glucopyranosyl-	69-79-4	NI35954
343	163	307	325	145	310	127	73	**342**	100	67	33	17	13	12	9	8	**0.00**	12	22	11	–	–	–	–	–	–	–	–	–	–	D-Glucose, 4-O-β-D-glucopyranosyl-		LI07572
325	343	163						**342**	100	30	19						**0.00**	12	22	11	–	–	–	–	–	–	–	–	–	–	D-Glucose, 6-O-β-D-glucopyranosyl-	554-91-6	NI35955
342	283	59	107	195	119	65	69	**342**	100	99	53	52	44	35	27	26		13	11	5	–	–	5	–	–	–	–	–	–	–	Benzeneacetic acid, 3-methoxy-4-(2,2,3,3,3-pentafluoro-1-oxopropoxy)-, methyl ester	55683-23-3	MS08110
342	283	195	107	343	151	59	284	**342**	100	73	27	17	15	14	12	10		13	11	5	–	–	5	–	–	–	–	–	–	–	Benzeneacetic acid, 3-methoxy-4-(2,2,3,3,3-pentafluoro-1-oxopropoxy)-, methyl ester	55683-23-3	NI35956
283	342	284	107	65	119	59	69	**342**	100	27	14	11	11	10	10	9		13	11	5	–	–	5	–	–	–	–	–	–	–	Benzeneacetic acid, 4-methoxy-3-(2,2,3,3,3-pentafluoro-1-oxopropoxy)-, methyl ester		MS08111
133	89	195	169	137	156	99	117	**342**	100	55	49	47	47	40	34	31	**0.29**	13	22	5	2	–	–	–	–	–	2	–	–	–	6H-Furo[2′,3′:4,5]oxazolo[3,2-a]pyrimidin-6-one, 3-[(dimethylsilyl)oxy]-2-[[(dimethylsilyl)oxy]methyl]-2,3,3a,9a-tetrahydro-, [2R-(2α,3β,3aβ,9aβ)]-	55429-30-6	NI35957
73	327	237	45	328	75	252	74	**342**	100	58	31	21	17	16	16	11	**8.84**	13	30	1	4	–	–	–	–	–	3	–	–	–	4-Amino-5-imidazole carboxamide, tris(trimethylsilyl)- deriv.		NI35958
91	65	92	39	181	63	342	89	**342**	100	30	16	11	9	7	6	6		14	14	–	–	–	–	–	–	–	–	–	–	2	Dibenzyl diselenide	1482-82-2	NI35959
91	181	342	340	338	182	92	65	**342**	100	45	14	13	8	8	8	6		14	14	–	–	–	–	–	–	–	–	–	–	2	Dibenzyl diselenide	1482-82-2	MS08112
91	171	169	342	340	65	39	89	**342**	100	66	41	37	34	27	23	21		14	14	–	–	–	–	–	–	–	–	–	–	2	Di-p-tolyl diselenide		LI07573
117	183	79	119	91	77	51	105	**342**	100	73	53	51	45	45	45	38	**0.60**	14	16	–	–	–	–	–	2	–	–	–	–	–	4,7-Methano-2,3,8-methenocyclopent[a]indene, 5,10-dibromododecahydro	55780-43-3	NI35960
43	240	155	156	180	198	103	181	**342**	100	95	80	56	51	45	33	28	**0.00**	15	18	9	–	–	–	–	–	–	–	–	–	–	Furan-2-carboxylic acid, 5-(triacetyloxypropyl)-, D-erythro-		LI07574
114	131	212	57	171	194	170	165	**342**	100	94	93	93	82	81	73	67	**11.00**	15	22	7	2	–	–	–	–	–	–	–	–	–	N-Acetylactinobolin		LI07575
79	73	93	75	77	327	52	342	**342**	100	87	73	72	68	56	51	44		15	26	5	–	–	–	–	–	–	2	–	–	–	Benzoic acid, 3,5-dimethoxy-4-[(trimethylsilyl)oxy]-, trimethylsilyl ester	10517-29-0	NI35961
239	73	342	240	343	45	241	74	**342**	100	95	58	21	20	12	10	9		15	30	3	–	–	–	–	–	–	3	–	–	–	1,2,3-Tris(trimethylsiloxy)benzene	17864-23-2	MS08113
73	239	342	240	343	45	133	74	**342**	100	99	54	22	19	19	13	12		15	30	3	–	–	–	–	–	–	3	–	–	–	1,2,3-Tris(trimethylsiloxy)benzene	17864-23-2	NI35962
73	239	342	45	343	240	344	74	**342**	100	62	42	16	14	14	11	9		15	30	3	–	–	–	–	–	–	3	–	–	–	1,2,3-Tris(trimethylsiloxy)benzene	17864-23-2	NI35963
342	73	327	343	328	147	344	45	**342**	100	80	71	34	25	20	15	13		15	30	3	–	–	–	–	–	–	3	–	–	–	1,3,5-Tris(trimethylsiloxy)benzene	10586-12-6	NI35964
342	327	73	343	328	344	147	329	**342**	100	75	66	33	24	15	15	11		15	30	3	–	–	–	–	–	–	3	–	–	–	1,3,5-Tris(trimethylsiloxy)benzene	10586-12-6	NI35965
327	285	73	75	300	100	328	97	**342**	100	70	68	36	33	29	27	21	**4.73**	15	30	3	2	–	–	–	–	–	2	–	–	–	2,4,6(1H,3H,5H)-Pyrimidinetrione, 5-ethyl-5-isopropyl-1,3-bis(trimethylsilyl)-	52937-65-2	NI35966
213	159	161	29	269	241	102	160	**342**	100	58	50	47	42	42	41	35	**23.92**	16	23	6	–	–	–	–	–	–	–	1	–	–	Propenoic acid, 3-(3-methoxyphenyl)-2-(diethoxyphosphinyl)-, ethyl ester	107846-66-2	NI35967
213	160	250	132	297	269	161	159	**342**	100	74	69	55	52	49	49	41	**35.99**	16	23	6	–	–	–	–	–	–	–	1	–	–	Propenoic acid, 3-(4-Methoxyphenyl)-2-(diethoxyphosphinyl)-, ethyl ester	14656-25-8	NI35968

MASS TO CHARGE RATIOS								M.W.	INTENSITIES								Parent	C	H	O	N	S	F	Cl	Br	I	Si	P	B	X	COMPOUND NAME	CAS Reg No	No
117	73	327	75	143	256	118	69	342	100	50	40	26	18	16	13	12	0.10	16	30	4	2	–	–	–	–	–	1	–	–	–	Barbituric acid, 5-ethyl-1,3-dimethyl-5-[1-methyl-3-(trimethylsiloxy)butyl]-	29836-74-6	NI35969
131	327	73	143	75	132	328	169	342	100	32	29	19	17	8	5	4	0.10	16	30	4	2	–	–	–	–	–	1	–	–	–	Barbituric acid, 5-ethyl-1,3-dimethyl-5-[1-methyl-3-(trimethylsiloxy)butyl]-	29836-74-6	NI35970
313	342	315	238	344	203	75	137	342	100	79	65	64	55	29	28	27		17	12	–	4	–	–	2	–	–	–	–	–	–	Halcion (triazolam)	28911-01-5	NI35971
254	171	157	41	185	39	255	90	342	100	39	37	18	17	15	13	13	0.00	17	14	4	2	1	–	–	–	–	–	–	–	–	1-(4-Sulphophenyl)-3-methyl-4-benzylidenepyrazol-5-one		IC04045
343	110	55	341	315	77	299	314	342	100	98	94	89	80	74	73	72	61.59	17	15	1	2	–	–	–	1	–	–	–	–	–	8-Bromo-1,2,3,4-tetrahydro-1-methyl-6-phenyl-1,5-benzodiazocin-2-one	63563-58-6	NI35972
327	325	323	297	326	295	324	293	342	100	75	43	43	37	33	30	21	15.80	17	18	–	–	–	–	–	–	–	–	–	–	1	9-Phenanthryltrimethyltin		MS08114
58	166	251	91	69	119	96	194	342	100	86	65	34	14	11	9	8	0.00	17	21	2	2	–	3	–	–	–	–	–	–	–	N-(Trifluoroacetyl)prolylmethamphetamine		NI35973
342	151	121	341	343	278	344	310	342	100	67	55	41	26	18	17	17		18	14	1	–	3	–	–	–	–	–	–	–	–	[1,2]Dithiolo[1,5-b][1,2]dithiole-7-SIV, 2-(4-methoxyphenyl)-5-phenyl-	2204-32-2	NI35974
311	342	283	312	194	343	166	155	342	100	49	17	16	10	9	8	5		18	14	7	–	–	–	–	–	–	–	–	–	–	Trimethyl dibenzofuran-1,2,3-tricarboxylate		LI07576
311	342	312	343	193	252	140	224	342	100	75	18	14	9	5	5	3		18	14	7	–	–	–	–	–	–	–	–	–	–	Trimethyl dibenzofuran-1,2,4-tricarboxylate		LI07577
307	322	91	327	308	147	117	116	342	100	63	33	32	32	30	30	27	23.00	18	15	1	–	–	5	–	–	–	–	–	–	–	Octafluoroanthraquinone		LI07578
105	173	201	133	306	146	237	289	342	100	64	58	50	39	33	24	17	6.00	18	15	3	2	–	–	1	–	–	–	–	–	–	N-[2-[(4-Hydroxy-2-butynyl)oxy]phenyl]-α-oxobenzeneethanehydrazonyl chloride		MS08115
43	77	15	51	65	92	39	105	342	100	77	29	27	20	17	17	13	4.41	18	18	5	2	–	–	–	–	–	–	–	–	–	Methyl (±)-2-acetoxy-3-[4-(phenylazo)-5-hydroxyphenyl]propionate		NI35975
73	341	342	343	45	344	327	91	342	100	90	64	44	25	20	17	16		18	19	1	2	–	–	1	–	–	1	–	–	–	2H-1,4-Benzodiazepin-2-one, 7-chloro-1,3-dihydro-5-phenyl-1-(trimethylsilyl)-	55299-24-6	NI35976
157	342	111	55	129	128	121	185	342	100	93	48	47	32	32	30	26		18	22	3	–	–	–	–	–	–	–	–	–	1	Ethyl 6-ferrocenyl-6-oxohexanoate		NI35977
91	200	131	129	105	117	256	204	342	100	46	44	43	40	36	34	33	10.00	18	22	3	–	–	–	–	–	–	–	–	–	1	Iron, tricarbonyl[(5a,5b,11a,11b-η)-2,3,4,5,6,7,8,9,10,11-decahydro-1H-cyclohepta[3,4]cyclobuta[1,2]cyclooctene]-	69815-45-8	MS08116
262	306	342	119	174	267	188	249	342	100	99	59	55	44	39	32	25		18	22	3	4	–	–	–	–	–	–	–	–	–	6-Deoxy-L-arabino-hexose phenylosazone		LI07579
342	70	230	201	314	115	58	300	342	100	28	20	11	8	5	5	3		18	22	3	4	–	–	–	–	–	–	–	–	–	Morphinan-14-ol, 6-azido-4,5-epoxy-3-methoxy-17-methyl-, (5α,6β)-	38211-26-6	NI35978
114	228	81	115	342	196	71	79	342	100	41	32	23	17	11	10	7		18	30	–	–	3	–	–	–	–	–	–	–	–	7,14,21-Trithiatrispiro[5.1.5.1.5.1]heneicosane	177-58-2	NI35979
114	228	81	115	342	196	71	79	342	100	41	36	23	17	11	10	7		18	30	–	–	3	–	–	–	–	–	–	–	–	7,14,21-Trithiatrispiro[5.1.5.1.5.1]heneicosane		AI02086
55	283	255	115	43	41	223	195	342	100	71	58	42	28	26	24	20	0.00	18	30	6	–	–	–	–	–	–	–	–	–	–	[S-(R*,R*)]-Methyl α-nonyltetrahydro-2-methoxycarbonyl-5-oxo-2-furanacetate		NI35980
57	112	41	157	29	156	56	43	342	100	84	73	68	60	52	43	43	5.00	18	30	6	–	–	–	–	–	–	–	–	–	–	Tributyl aconitate		IC04046
112	157	57	156	139	41	29	213	342	100	94	76	64	44	41	41	29	0.83	18	30	6	–	–	–	–	–	–	–	–	–	–	Tributyl aconitate		DO01109
57	28	112	29	157	41	156	139	342	100	96	84	71	68	67	42	30	0.91	18	30	6	–	–	–	–	–	–	–	–	–	–	Tributyl aconitate		IC04047
86	128	129	44	43	41	57	30	342	100	46	28	14	9	8	7	6	0.45	18	34	4	2	–	–	–	–	–	–	–	–	–	L-Leucine, N-(N-acetyl-L-leucyl)-, butyl ester	55712-43-1	NI35981
41	256	120	89	242	28	76	55	342	100	79	71	61	60	60	56	55	24.79	18	37	–	–	–	–	–	–	–	–	1	–	1	Nickel, [dimethyl(5-methyl-2-isopropylcyclohexyl)phosphine]methyl[(1,2,3-η)-2-pentenyl]-	74421-54-8	NI35982
170	73	111	171	81	75	342	74	342	100	34	20	16	9	9	5	4		18	38	2	–	–	–	–	–	–	2	–	–	–	1,1′-Bicyclohexyl-2,2′-diol bis(trimethylsilyl) ether		MS08117
42	113	43	44	72	58	70	41	342	100	71	70	56	51	44	35	26	0.00	18	42	–	6	–	–	–	–	–	–	–	–	–	1,3,5-Tris-(3-dimethylaminopropyl)-hexahydro-s-triazine		IC04048
343	163	345	86	191	115	344	181	342	100	52	34	32	25	22	21	17	0.00	19	15	4	–	–	–	1	–	–	–	–	–	–	2H-1-Benzopyran-2-one, 3-[1-(4-chlorophenyl)-3-oxobutyl]-4-hydroxy-	81-82-3	NI35984
299	43	121	301	342	187	300	120	342	100	46	41	34	23	23	20	15		19	15	4	–	–	–	1	–	–	–	–	–	–	2H-1-Benzopyran-2-one, 3-[1-(4-chlorophenyl)-3-oxobutyl]-4-hydroxy-	81-82-3	NI35983
161	69	91	203	119	181	162	65	342	100	99	21	20	17	15	14	13	2.10	19	15	4	–	–	–	1	–	–	–	–	–	–	1,3,4,6-Hexanetetrone, 1-(4-chlorophenyl)-6-(4-methylphenyl)-	58330-14-6	NI35985
300	230	202	272	187	244	257	229	342	100	80	39	18	17	15	10	5	3.00	19	18	6	–	–	–	–	–	–	–	–	–	–	Anhydromethyldihydrousnic acid monoacetate		LI07580
107	134	240	282	108	300	258	176	342	100	60	54	48	40	33	29	24	5.00	19	18	6	–	–	–	–	–	–	–	–	–	–	Benzoic acid, 2,3-bis(acetyloxy)-6-(2-phenylethyl)-	75299-43-3	NI35986
192	191	342	177	193	343	149	121	342	100	36	34	26	16	9	7	7		19	18	6	–	–	–	–	–	–	–	–	–	–	[1]Benzopyrano[3,4-b][1]benzopyran-12(6H)-one, 6a,12a-dihydro-2,3,9-trimethoxy-, (6aS-cis)-	3564-85-0	NI35987
342	299	171	157	256	271	149.5	163.5	342	100	11	8	6	5	3	3	2		19	18	6	–	–	–	–	–	–	–	–	–	–	Coumarin, 5,6,7,4′-tetramethoxy-4-phenyl-		MS08118
342	327	299	271	157	256	171	149.5	342	100	48	46	15	15	9	8	3		19	18	6	–	–	–	–	–	–	–	–	–	–	Coumarin, 5,7,8,4′-tetramethoxy-4-phenyl-		MS08119
342	299	343	181	135	300	171	157	342	100	30	23	10	10	7	7	7		19	18	6	–	–	–	–	–	–	–	–	–	–	Coumarin, 4,5,7-trimethoxy-3-(4-methoxyphenyl)-	4222-04-2	NI35988
298	297	296	134	120	269	253	77	342	100	96	87	85	85	82	82	82	70.99	19	18	6	–	–	–	–	–	–	–	–	–	–	Cyclopenta[c]furo[3′,2′:4,5]furo[2,3-h][1]benzopyran-11(1H)-one, 1-ethoxy-2,3,6a,9a-tetrahydro-4-methoxy-	69831-77-2	NI35989
342	327	171	341	328	299			342	100	43	15	13	10	6				19	18	6	–	–	–	–	–	–	–	–	–	–	3,4,8,9-Tetramethoxy-6a,11a-dehydropterocarpan		LI07581
327	342	328	137	165	343	313	70	342	100	57	20	20	15	14	9	9		19	18	6	–	–	–	–	–	–	–	–	–	–	2′,5,5,6′-Tetramethoxyflavone	14813-19-5	NI35990
327	342	328	137	166	44	43	343	342	100	57	21	20	16	14	14	13		19	18	6	–	–	–	–	–	–	–	–	–	–	2′,5,6,6′-Tetramethoxyflavone		MS08120
342	341	327	299	311	343	151	77	342	100	85	70	45	28	20	19	19		19	18	6	–	–	–	–	–	–	–	–	–	–	3,3′,4′,7-Tetramethoxyflavone		MS08121
341	342	323	135	157	327	313	311	342	100	91	39	28	27	16	14	14		19	18	6	–	–	–	–	–	–	–	–	–	–	3,4′,5,7-Tetramethoxyflavone		MS08122
341	342	323	135	157	327	343	311	342	100	78	25	16	13	12	11	9		19	18	6	–	–	–	–	–	–	–	–	–	–	3,4′,5,7-Tetramethoxyflavone		MS08123
341	342	323	135	327	343	311	313	342	100	78	25	16	12	11	9	7		19	18	6	–	–	–	–	–	–	–	–	–	–	3,4′,5,7-Tetramethoxyflavone		MS08124
311	342	341	181	293	77	312	91	342	100	89	59	37	24	23	21	20		19	18	6	–	–	–	–	–	–	–	–	–	–	3,5,7,2′-Tetramethoxyflavone		MS08125
342	341	323	135	77	343	157	119	342	100	99	38	33	24	20	20	20		19	18	6	–	–	–	–	–	–	–	–	–	–	3,5,7,4′-Tetramethoxyflavone		MS08126

MASS TO CHARGE RATIOS								M.W.	INTENSITIES								Parent	C	H	O	N	S	F	Cl	Br	I	Si	P	B	X	COMPOUND NAME	CAS Reg No	No
342	341	296	313	83	85	312	311	**342**	100	55	29	27	24	15	13	13		19	18	6	–	–	–	–	–	–	–	–	–	–	**3′,4′,5,7-Tetramethoxyflavone**	855-97-0	**NI35991**
342	341	296	313	325	297	157	311	**342**	100	67	30	28	18	17	14	13		19	18	6	–	–	–	–	–	–	–	–	–	–	**3′,4′,5,7-Tetramethoxyflavone**	855-97-0	**NI35992**
342	327	151	157	142	69	149	127	**342**	100	44	20	18	18	14	11	10		19	18	6	–	–	–	–	–	–	–	–	–	–	**3′,4′,5′,7-Tetramethoxyflavone**	3044-57-3	**NI35993**
327	342	328	167	299	343	311	296	**342**	100	26	22	12	8	6	6	6		19	18	6	–	–	–	–	–	–	–	–	–	–	**4′,5,6,7-Tetramethoxyflavone**	1168-42-9	**MS08127**
327	342	328	167	299	95	81	69	**342**	100	26	22	12	8	6	6	6		19	18	6	–	–	–	–	–	–	–	–	–	–	**4′,5,6,7-Tetramethoxyflavone**	1168-42-9	**NI35994**
327	342	298	167	297	139	328	157	**342**	100	84	29	24	19	19	18	18		19	18	6	–	–	–	–	–	–	–	–	–	–	**4′,5,7,8-Tetramethoxyflavone**	6601-66-7	**NI35995**
342	343	327	171	341	142	299	256	**342**	100	22	12	10	7	5	4	4		19	18	6	–	–	–	–	–	–	–	–	–	–	**3′,4′,6,7-Tetramethoxyisoflavone**	24126-93-0	**NI35996**
342	311	161	341	43	181	343	162	**342**	100	68	27	26	25	24	21	20		19	18	6	–	–	–	–	–	–	–	–	–	–	**Trimethylbarpisoflavone A**	101691-27-4	**NI35997**
342	58	86	60	85	343	214	59	**342**	100	71	66	30	21	20	7	7		19	22	2	2	1	–	–	–	–	–	–	–	–	**10H-Phenothiazin-3-ol, 10-[3-(dimethylamino)propyl]-, acetate**	56438-23-4	**NI35998**
324	107	135	122	121	325	77	161	**342**	100	47	45	28	23	20	20	17	**1.25**	19	22	4	2	–	–	–	–	–	–	–	–	–	**2-(4-Methoxyphenyl)-3-methyl-3-(4-methoxyphenylazo)-butanoic acid**		**NI35999**
150	192	193	137	134	178	77	311	**342**	100	89	80	80	52	49	19	17	**6.00**	19	22	4	2	–	–	–	–	–	–	–	–	–	**6-O-(5-Vinylnicotinoyl)-1,2,4a,7a-tetrahydrocantleyine**		**NI36000**
55	200	83	125	97	185	143	41	**342**	100	50	50	47	40	38	37	36	**0.70**	19	34	5	–	–	–	–	–	–	–	–	–	–	**Heptadecanedioic acid, 9-oxo-, dimethyl ester**	29263-73-8	**NI36001**
73	75	117	129	309	132	93	145	**342**	100	89	79	45	42	32	28	25	**1.00**	19	38	3	–	–	–	–	–	–	1	–	–	–	**Trimethylsilyl 16-oxohexadecanoate**		**MS08128**
104	77	76	51	50	91	28	195	**342**	100	83	53	28	23	15	14	12	**2.00**	20	14	2	4	–	–	–	–	–	–	–	–	–	**cis-2,3-Diphenyl-1-phthalimido-azimine**		**NI36002**
104	77	76	51	50	105	91	28	**342**	100	83	51	28	23	15	15	14	**2.00**	20	14	2	4	–	–	–	–	–	–	–	–	–	**trans-2,3-Diphenyl-1-phthalimido-azimine**		**NI36003**
165	314	166	192	315	268	190	164	**342**	100	93	28	26	22	15	13	12	**5.00**	20	14	2	4	–	–	–	–	–	–	–	–	–	**1H-1,2,3-Triazole, 1-(3-nitrophenyl)-4,5-diphenyl-**	33471-64-6	**NI36005**
165	314	166	315	76	268	163	190	**342**	100	94	24	22	20	16	16	13	**4.00**	20	14	2	4	–	–	–	–	–	–	–	–	–	**1H-1,2,3-Triazole, 1-(3-nitrophenyl)-4,5-diphenyl-**	33471-64-6	**NI36004**
165	314	166	192	315	268	190	164	**342**	100	94	30	28	23	17	14	12	**5.00**	20	14	2	4	–	–	–	–	–	–	–	–	–	**1H-1,2,3-Triazole, 1-(3-nitrophenyl)-4,5-diphenyl-**	33471-64-6	**LI07582**
342	190	43	267	243	299	241	93	**342**	100	76	56	39	28	25	25	22		20	22	5	–	–	–	–	–	–	–	–	–	–	**(±)-17β-Acetoxy-3-methoxy-6-oxaestra-1,3,5(10),8(9)-tetraen-7-one**		**MS08129**
310	296	179	237	252	193	165	59	**342**	100	95	75	59	51	43	39	33	**0.00**	20	22	5	–	–	–	–	–	–	–	–	–	–	**3-Acetyl-1,2,3,4,5,8-hexahydro-7-(methoxycarbonyl)anthracene-9-acetic acid**		**DD01350**
342	164	144	152	271	165	343	178	**342**	100	56	46	41	40	40	34	32		20	22	5	–	–	–	–	–	–	–	–	–	–	**(1R,5R,6R,7R)-1-Allyl-6-(4′-hydroxy-3′-methoxyphenyl)-3-methoxy-7-methyl-4,8-dioxobicyclo[3.2.1]oct-2-ene**		**NI36006**
312	95	94	145	131	189	81	342	**342**	100	98	90	41	27	21	20	19		20	22	5	–	–	–	–	–	–	–	–	–	–	**Bacchotricuneatin A**	65596-25-0	**MS08130**
312	95	94	190	145	172	105	81	**342**	100	40	32	18	17	15	14	14	**0.40**	20	22	5	–	–	–	–	–	–	–	–	–	–	**Bacchotricuneatin B**	65596-26-1	**MS08131**
71	83	69	112	81	95	104	97	**342**	100	97	97	84	82	78	67	63	**2.00**	20	22	5	–	–	–	–	–	–	–	–	–	–	**Bacchotricuneatin C**	66563-30-2	**MS08132**
175	181	162	257	311	284	342	214	**342**	100	85	64	58	48	48	45	25		20	22	5	–	–	–	–	–	–	–	–	–	–	**6(2H)-Benzofuranone, 2-(1,3-benzodioxol-5-yl)-3,3a,4,5-tetrahydro-5-methoxy-3-methyl-3a-allyl-, [2S-(2α,3β,3aα,5β)]-**	70522-39-3	**NI36007**
324	269	220	165	205	147	342	123	**342**	100	99	75	75	65	33	25	23		20	22	5	–	–	–	–	–	–	–	–	–	–	**[3,6′-Bi-2H-1-Benzopyran]-3,7,7′(4H)-triol, 3′,4′-dihydro-2′,2′-dimethyl-**	56771-79-0	**MS08133**
342	298	343	149	299	212	211	210	**342**	100	75	22	18	16	14	14	14		20	22	5	–	–	–	–	–	–	–	–	–	–	**Coumarone, 7-methoxy-5-(γ-hydroxypropyl)-2-(3′,4′-dimethoxyphenyl)-**		**MS08134**
342	258	164	285	151	343	314	257	**342**	100	71	28	26	24	23	22	20		20	22	5	–	–	–	–	–	–	–	–	–	–	**Dibenzo[b,e][1,4]dioxepin-11-one, 3-hydroxy-8-methoxy-1,9-dimethyl-6-sec-butyl-**		**NI36008**
99	100	241	40	242	256	243	342	**342**	100	65	27	27	25	18	17	6		20	22	5	–	–	–	–	–	–	–	–	–	–	**(±)-17,17-Ethylenedioxy-3-methoxy-6-oxaestra-1,3,5(10),8(9)-tetraen-7-one**		**MS08135**
125	127	106	77	107	134	41	342	**342**	100	33	13	13	12	11	10	8		20	23	1	2	–	–	1	–	–	–	–	–	–	**1-(4-Chlorobenzyl)-1-cyclopentyl-3-methyl-3-phenylurea**		**NI36009**
43	41	29	84	77	55	57	86	**342**	100	87	71	68	61	56	42	39	**5.00**	20	26	3	2	–	–	–	–	–	–	–	–	–	**2-Acetyl-3-phenyl-4-ethoxycarbonyl-2,3-diaza[3.3.3]propellane**		**NI36010**
342	112	300	341	140	343	178	301	**342**	100	74	45	28	24	23	11	10		20	26	3	2	–	–	–	–	–	–	–	–	–	**Aspidodispermine, O-methyl-**	21451-69-4	**NI36011**
148	269	270	44	43	56	55	41	**342**	100	95	19	19	12	11	11	11	**10.00**	20	26	3	2	–	–	–	–	–	–	–	–	–	**Carbethoxy-hydroxy-bis(4-dimethylaminophenyl)-methane**		**IC04049**
311	91	342	92	96	312	202	279	**342**	100	96	90	70	56	42	23	22		20	26	3	2	–	–	–	–	–	–	–	–	–	**Estra-1,3,5(10)-triene-6,17-dione, 3-hydroxy-, bis(O-methyloxime)**	69833-90-5	**NI36012**
83	139	55	69	95	138	81	57	**342**	100	55	48	42	40	35	34	32	**2.98**	20	38	–	–	–	–	–	–	–	–	–	–	1	**Zinc, bis(5-methyl-2-isopropylcyclohexyl)-, [1α(1R*,2R*,5R*),2β,5α]-**	74792-89-5	**NI36013**
102	55	69	103	41	44	43	283	**342**	100	25	23	20	20	19	19	16	**0.78**	20	38	4	–	–	–	–	–	–	–	–	–	–	**Dimethyl 15,15-dimethyl hexadecanoate**		**IC04050**
88	74	98	112	43	214	255	126	**342**	100	95	76	65	48	27	24	21	**6.00**	20	38	4	–	–	–	–	–	–	–	–	–	–	**Dimethyl 2-methylheptadecane-1,17-dioate**		**LI07583**
102	55	43	69	41	57	74	83	**342**	100	43	40	33	33	25	20	18	**0.00**	20	38	4	–	–	–	–	–	–	–	–	–	–	**Dimethyl 2-methylpentadecane 2,5-dicarboxylate**		**IC04051**
57	119	112	101	70	71	43	113	**342**	100	78	76	71	68	66	54	52	**0.30**	20	38	4	–	–	–	–	–	–	–	–	–	–	**Bis(2-ethyl hexyl) succinate**		**IC04052**
43	55	41	29	215	57	69	272	**342**	100	88	81	64	53	49	42	31	**0.00**	20	38	4	–	–	–	–	–	–	–	–	–	–	**Ethyl pentyl 2,2-dipentylmalonate**		**IC04053**
282	264	222	151	125	196	239	180	**342**	100	87	87	65	52	45	28	26	**9.80**	20	38	4	–	–	–	–	–	–	–	–	–	–	**Heptadecanoic acid, 3-acetoxy-16-methyl-**		**NI36014**
129	227	57	55	87	149	43	41	**342**	100	73	66	30	29	28	28	23	**1.90**	20	38	4	–	–	–	–	–	–	–	–	–	–	**Di-heptyl adipate**		**IC04054**
222	43	57	82	96	224	83	55	**342**	100	99	71	55	50	47	45	45	**0.00**	20	38	4	–	–	–	–	–	–	–	–	–	–	**1,1-Hexadecanediol, diacetate**	56438-11-0	**NI36015**
55	126	74	269	69	41	43	59	**342**	100	90	72	70	66	63	53	51	**4.60**	20	38	4	–	–	–	–	–	–	–	–	–	–	**Methyl 3,6,9,12-tetramethyltetradecane-1,14-dioate**		**MS08136**
74	98	55	43	41	69	84	87	**342**	100	97	81	55	55	48	42	41	**2.90**	20	38	4	–	–	–	–	–	–	–	–	–	–	**Octadecanedioic acid, dimethyl ester**	1472-93-1	**MS08137**
98	74	55	43	41	84	69	87	**342**	100	85	77	54	50	47	45	40	**0.94**	20	38	4	–	–	–	–	–	–	–	–	–	–	**Octadecanedioic acid, dimethyl ester**	1472-93-1	**NI36016**
115	195	83	227	145	166	95	74	**342**	100	80	68	66	52	43	20	15	**0.00**	20	38	4	–	–	–	–	–	–	–	–	–	–	**9-Octadecenoic acid, 12-hydroxy-13-methoxy-, methyl ester**	75299-48-8	**NI36017**
145	95	71	241	210	69	81	83	**342**	100	54	40	29	26	25	22	16	**1.00**	20	38	4	–	–	–	–	–	–	–	–	–	–	**9-Octadecenoic acid, 13-hydroxy-12-methoxy-, methyl ester**	75299-47-7	**NI36018**
57	70	101	71	41	111	118	55	**342**	100	90	76	76	67	64	62	62	**0.24**	20	38	4	–	–	–	–	–	–	–	–	–	–	**Di-octyl succinate**		**IC04055**
85	55	41	56	29	84	83	69	**342**	100	82	52	35	33	31	31	30	**0.00**	20	38	4	–	–	–	–	–	–	–	–	–	–	**2H-Pyran, 2,2′-[1,10-decanediylbis(oxy)]bis[tetrahydro-**	56599-49-6	**NI36019**
327	43	101	41	57	55	116	328	**342**	100	54	53	28	27	24	23	22	**0.00**	20	38	4	–	–	–	–	–	–	–	–	–	–	**Tetradecanoic acid, (2,2-dimethyl-1,3-dioxolan-4-yl)methyl ester**	56630-71-8	**NI36020**

MASS TO CHARGE RATIOS	M.W.	INTENSITIES	Parent	C	H	O	N	S	F	Cl	Br	I	Si	P	B	X	COMPOUND NAME	CAS Reg No	No
73 75 117 43 132 55 327 41	342	100 88 61 53 37 36 35 35	4.90	20	42	2	–	–	–	–	–	–	1	–	–	–	Heptadecanoic acid, trimethylsilyl ester	55517-58-3	NI36022
121 134 75 73 252 91 77 135	342	100 81 24 21 15 14 13 10	0.00	20	42	2	–	–	–	–	–	–	1	–	–	–	Heptadecanoic acid, trimethylsilyl ester	55517-58-3	NI36021
73 117 75 132 327 43 145 129	342	100 71 70 48 41 38 30 29	6.65	20	42	2	–	–	–	–	–	–	1	–	–	–	Heptadecanoic acid, trimethylsilyl ester	55517-58-3	NI36023
105 77 342 106 343 344 341 327	342	100 28 17 5 4 1 0 0		21	14	3	2	–	–	–	–	–	–	–	–	–	Benzamide, N-(5-amino-9,10-dihydro-9,10-dioxo-1-anthracenyl)-	117-06-6	NI36024
77 105 342 106 51 162 163 343	342	100 97 90 31 23 19 17 9		21	14	3	2	–	–	–	–	–	–	–	–	–	Benzamide, N-[(2-nitrofluoren-9-ylidene)methyl]-	27398-55-6	NI36025
300 152 151 149 105 91 107 79	342	100 40 39 16 10 9 8 5	0.00	21	26	4	–	–	–	–	–	–	–	–	–	–	4-Acetoxyandrosta-4,6-diene-3,17-dione	89593-30-6	NI36026
311 326 312 327 241 309 215 97	342	100 64 26 19 9 8 7 7	0.00	21	26	4	–	–	–	–	–	–	–	–	–	–	2-Acetoxy-3-methoxy-estrone		NI36027
121 110 137 213 222 325 342 55	342	100 45 16 15 15 13 11 9		21	26	4	–	–	–	–	–	–	–	–	–	–	Benzophenone, 2,2'-dihydroxy-4-(octyloxy)-	85-24-5	NI36028
342 327 285 283 232 219 328 57	342	100 94 36 35 34 26 21 19		21	26	4	–	–	–	–	–	–	–	–	–	–	(±)-2-tert-Butyl-3,17β-dihydroxy-6-oxaestra-1,3,5(10),8(9)-tetraen-7-one		MS08138
296 268 297 264 342 311 282 298	342	100 99 60 40 20 14 13 9		21	26	4	–	–	–	–	–	–	–	–	–	–	Methyl 3-carbethoxy-3-(1',2',3',7',8',9',10',10'a-octahydrocyclohepta[de]naphthalen-3'-ylidene)-propionate		LI07584
342 218 324 232 244 343 230 231	342	100 46 43 34 30 28 19 15		21	30	2	–	–	–	–	–	–	1	–	–	–	Estra-1,3,5(10)-triene, 16,17-epoxy-3-[(trimethylsilyl)oxy]-, (16α,17α)-	74312-41-7	NI36029
342 244 343 218 232 299 257 245	342	100 35 28 23 21 13 12 10		21	30	2	–	–	–	–	–	–	1	–	–	–	Estra-1,3,5(10)-triene, 16,17-epoxy-3-[(trimethylsilyl)oxy]-, (16β,17β)-	74299-44-8	NI36030
342 244 343 245 231 327 344 341	342	100 36 32 17 13 10 9 9		21	30	2	–	–	–	–	–	–	1	–	–	–	Estra-1,3,5(10)-trien-16-one, 3-[(trimethylsilyl)oxy]-		NI36031
342 73 218 257 343 244 219 258	342	100 93 51 48 31 22 17 16		21	30	2	–	–	–	–	–	–	1	–	–	–	Estra-1,3,5(10)-trien-17-one, 3-[(trimethylsilyl)oxy]-	1839-54-9	NI36032
342 257 343 218 244 258 57 327	342	100 40 28 26 15 12 10 8		21	30	2	–	–	–	–	–	–	1	–	–	–	Estra-1,3,5(10)-trien-17-one, 3-[(trimethylsilyl)oxy]-	1839-54-9	NI36033
140 57 342 56 41 141 42 43	342	100 36 28 15 13 11 9 8		21	30	2	2	–	–	–	–	–	–	–	–	–	Aspidospermidin-21-ol, 17-methoxy-1-methyl-	54658-04-7	NI36034
126 127 342 156 124 170 157 143	342	100 12 8 7 5 4 4 2		21	30	2	2	–	–	–	–	–	–	–	–	–	16,17,15,20-Tetrahydro-secondine		LI07585
90 57 46 71 55 69 44 85	342	100 87 69 60 45 44 38 35	22.00	21	42	3	–	–	–	–	–	–	–	–	–	–	Hexadecanoic acid, 2-hydroxy-3,7,11,15-tetramethyl-, methyl ester	24257-07-6	LI07587
90 57 71 46 69 55 84 44	342	100 92 65 65 44 44 38 33	17.00	21	42	3	–	–	–	–	–	–	–	–	–	–	Hexadecanoic acid, 2-hydroxy-3,7,11,15-tetramethyl-, methyl ester	24257-07-6	LI07586
90 57 55 84 117 82 97 283	342	100 91 44 38 29 29 24 19	0.00	21	42	3	–	–	–	–	–	–	–	–	–	–	Hexadecanoic acid, 2-hydroxy-3,7,11,15-tetramethyl-, methyl ester	24257-07-6	NI36035
69 55 101 74 57 41 83 97	342	100 99 81 79 62 45 42 34	0.77	21	42	3	–	–	–	–	–	–	–	–	–	–	Hexadecanoic acid, 16-hydroxy-3,7,11,15-tetramethyl-, methyl ester		IC04056
43 98 58 57 55 41 118 84	342	100 96 85 67 67 61 58 58	1.04	21	42	3	–	–	–	–	–	–	–	–	–	–	Octadecanoic acid, 3-hydroxypropyl ester	10108-23-3	NI36036
342 252 253 250 127 126 171 343	342	100 99 90 44 38 33 26 25		22	14	4	–	–	–	–	–	–	–	–	–	–	8,8'-Dicarboxy-1,1'-binaphthyl		IC04057
342 325 90 89 343 118 170 326	342	100 65 32 25 24 18 16 16		22	14	4	–	–	–	–	–	–	–	–	–	–	5,5'-Dimethyl-2,2'-binaphthalene-1,1',4,4'-tetrone	97308-78-6	NI36037
342 325 90 89 343 118 326 170	342	100 74 45 35 24 24 19 16		22	14	4	–	–	–	–	–	–	–	–	–	–	8,8'-Dimethyl-2,2'-binaphthalene-1,1',4,4'-tetrone	97308-81-1	NI36038
313 314 285 267 226 18 239 228	342	100 25 21 18 15 14 11 10	0.00	22	14	4	–	–	–	–	–	–	–	–	–	–	1,1':4',1''-Terphenyl-2,2',2'',5'-tetracarboxaldehyde		AI02087
269 165 166 342 270 343 121 118	342	100 84 45 44 25 14 12 12		22	18	–	2	1	–	–	–	–	–	–	–	–	1-Methyl-2,4,4-triphenyl-2-imidazolin-5-thione	24134-07-4	NI36039
284 165 342 285 224 166 327 310	342	100 61 46 26 24 23 22 16		22	18	–	2	1	–	–	–	–	–	–	–	–	1-Methyl-2,5,5-triphenyl-2-imidazolin-4-thione	37927-98-3	NI36040
269 165 166 342 270 309 343 164	342	100 94 64 40 25 16 12 11		22	18	–	2	1	–	–	–	–	–	–	–	–	2,5,5-Triphenyl-4-(methylthio)imidazole	24133-97-9	NI36041
342 105 104 237 91 76 193 297	342	100 93 55 40 38 35 27 25		22	18	2	2	–	–	–	–	–	–	–	–	–	1,4-Phthalazinedione, 2,3-dihydro-2,3-bis(4-methylphenyl)-	63546-89-4	NI36042
104 265 249 77 342 105 78 343	342	100 93 84 84 69 28 22 19		22	18	2	2	–	–	–	–	–	–	–	–	–	Pyrimidine, 4,6-diphenyl-5-phenoxy-2-oxo-1,2,3,4-tetrahydro-		LI07588
327 183 342 145 328 168 102 169	342	100 90 46 38 26 22 22 20		22	22	–	4	–	–	–	–	–	–	–	–	–	Bis(1',3'-dimethyl-1',2'-dihydroquinoxaline)-1,2-ethane diylidene		LI07589
43 342 314 91 135 105 93 55	342	100 77 75 33 27 26 24 24		22	30	3	–	–	–	–	–	–	–	–	–	–	1,5-Cycloandrost-3-ene-2-one, 17-(acetyloxy)-1-methyl-, (1α,1β,5β,10α,17β)-		NI36043
108 177 91 104 80 94 66 107	342	100 80 70 67 60 50 44 40	17.00	22	30	3	–	–	–	–	–	–	–	–	–	–	Dihydronardosinondiol benzaldehyde acetal		LI07590
342 213 160 159 146 133 145 132	342	100 54 36 31 21 21 17 16		22	30	3	–	–	–	–	–	–	–	–	–	–	19-Norpregna-1,3,5(10)-triene-20α-carboxylic acid, 3-hydroxy-, methyl ester	20248-21-9	NI36044
267 342 268 343 211 41 161 213	342	100 81 22 21 18 16 13 12		22	30	3	–	–	–	–	–	–	–	–	–	–	4a(2H)-Phenanthrenecarboxylic acid, 1,3,4,9,10,10a-hexahydro-6-methoxy-1,1-dimethyl-7-isopropenyl-, methyl ester, (4aR-trans)-	57397-32-7	NI36045
300 43 285 243 301 147 41 125	342	100 63 35 25 23 16 14 13	13.00	22	30	3	–	–	–	–	–	–	–	–	–	–	2(1H)-Phenanthrenone, 6-(acetyloxy)-3,4,4a,9,10,10a-hexahydro-1,1,4a-trimethyl-7-isopropyl-, (4aS-trans)-	18385-56-3	LI07591
300 43 244 285 243 301 147 41	342	100 62 37 36 25 23 16 15	13.21	22	30	3	–	–	–	–	–	–	–	–	–	–	2(1H)-Phenanthrenone, 6-(acetyloxy)-3,4,4a,9,10,10a-hexahydro-1,1,4a-trimethyl-7-isopropyl-, (4aS-trans)-	18385-56-3	NI36046
300 43 267 342 285 197 215 203	342	100 85 54 53 32 26 25 21		22	30	3	–	–	–	–	–	–	–	–	–	–	4(1H)-Phenanthrenone, 7-(acetyloxy)-2,3,4a,9,10,10a-hexahydro-1,1,4a-trimethyl-8-isopropyl-, (4aS-trans)-	57397-35-0	NI36047
260 327 342 243 285 43 41 257	342	100 83 69 61 42 38 37 32		22	30	3	–	–	–	–	–	–	–	–	–	–	Podocarpa-1,8,11,13-tetraen-3-one, 14-isopropyl-1,13-dimethoxy-	18326-20-0	NI36048
216 173 217 201 342 126 299 159	342	100 14 12 9 7 7 5 4		22	30	3	–	–	–	–	–	–	–	–	–	–	Podocarpa-2,8,11,13-tetraen-1-one, 14-isopropyl-3,13-dimethoxy-	18326-21-1	NI36049
285 300 243 286 342 301 43 201	342	100 56 30 18 13 10 10 9		22	30	3	–	–	–	–	–	–	–	–	–	–	Podocarpa-8,11,13-trien-3-one, 13-hydroxy-14-isopropyl-, acetate	18468-20-7	NI36050
285 300 243 286 342 301 43 201	342	100 55 20 18 14 10 10 9		22	30	3	–	–	–	–	–	–	–	–	–	–	Podocarpa-8,11,13-trien-3-one, 13-hydroxy-14-isopropyl-, acetate	18468-20-7	NI36051
57 43 71 63 55 85 83 41	342	100 80 69 57 51 49 43 41	0.14	22	46	2	–	–	–	–	–	–	–	–	–	–	Ethanol, 2-(eicosyloxy)-	2136-73-4	NI36052
57 43 45 82 41 55 96 68	342	100 76 59 53 44 32 30 28	0.00	22	46	2	–	–	–	–	–	–	–	–	–	–	Palmitaldehyde, diisopropyl acetal	18302-66-4	NI36053
57 43 99 82 31 41 55 29	342	100 64 54 40 38 36 22 19	0.00	22	46	2	–	–	–	–	–	–	–	–	–	–	Palmitaldehyde, dipropyl acetal	22140-81-4	NI36054
270 240 269 298 342 241 271 299	342	100 74 73 51 34 30 26 18		23	18	3	–	–	–	–	–	–	–	–	–	–	Acenaphthene-2-one-1-spiro-(2'-ethoxycarbonyl-3'-phenyl)-cyclopropane		LI07592
167 165 342 152 175 91 147 115	342	100 12 7 7 6 3 2 2		23	18	3	–	–	–	–	–	–	–	–	–	–	2H-1-Benzopyran-2-one, 8-(2,2-diphenylethyl)-7-hydroxy-	31231-48-8	MS08139
105 342 237 77 183 121 343 160	342	100 70 67 61 42 25 18 18		23	18	3	–	–	–	–	–	–	–	–	–	–	2H-1-Benzopyran-2-one, 2-(hydroxydiphenylmethyl)-3-methyl-	20924-67-8	LI07593

MASS TO CHARGE RATIOS								M.W.	INTENSITIES								Parent	C	H	O	N	S	F	Cl	Br	I	Si	P	B	X	COMPOUND NAME	CAS Reg No	No
191	192	107	105	77	189	79	165	**342**	100	54	47	22	22	20	18	14	**11.00**	23	18	3	–	–	–	–	–	–	–	–	–	–	9-Phenanthrenemethyl mandelate	92174-50-0	NI36055
135	191	189	192	92	165	342	190	**342**	100	86	30	21	19	16	15	15		23	18	3	–	–	–	–	–	–	–	–	–	–	9-Phenanthrenemethyl 3-methoxybenzoate		NI36056
77	194	105	165	166	180	195	43	**342**	100	89	47	42	36	17	16	12	**3.10**	23	22	1	2	–	–	–	–	–	–	–	–	–	2-Isopropyl-1,4,4-triphenyl-1,2-diazetidin-3-one		MS08140
342	151	231	213	309	327	324	229	**342**	100	36	35	30	22	21	18	14		23	34	2	–	–	–	–	–	–	–	–	–	–	5-Androsten-17-one, 3β-hydroxy-16-isobutylidene		LI07594
274	221	43	273	243	342	275	173	**342**	100	46	24	22	21	19	19	15		23	34	2	–	–	–	–	–	–	–	–	–	–	Benzene, 1,3-dimethoxy-2-[3-methyl-6-isopropenyl-2-cyclohexen-1-yl]-5-pentyl-, (1R-trans)-	1242-67-7	NI36058
208	43	221	274	152	55	97	109	**342**	100	99	94	63	62	58	57	54	**0.00**	23	34	2	–	–	–	–	–	–	–	–	–	–	Benzene, 1,3-dimethoxy-2-[3-methyl-6-isopropenyl-2-cyclohexen-1-yl]-5-pentyl-, (1R-trans)-	1242-67-7	NI36057
221	274	287	342	43	243	234	275	**342**	100	93	63	29	25	24	24	19		23	34	2	–	–	–	–	–	–	–	–	–	–	Benzene, 1,3-dimethoxy-2-[3-methyl-6-isopropenyl-3-cyclohexen-1-yl]-5-pentyl-, (1R-trans)-	55332-72-4	NI36059
245	208	221	152	43	246	344	41	**342**	100	48	34	34	30	19	17	13	**0.00**	23	34	2	–	–	–	–	–	–	–	–	–	–	Benzene, 1,3-dimethoxy-2-[3-methyl-6-isopropenyl-3-cyclohexen-1-yl]-5-pentyl-, (1R-trans)-	55332-72-4	NI36060
151	69	41	205	121	191	189	190	**342**	100	28	26	21	15	12	11	7	**7.00**	23	34	2	–	–	–	–	–	–	–	–	–	–	Phenol, 3-methoxy-5-methyl-2-(3,7,11-trimethyl-2,6,10-dodecatrienyl)-, (E,E)-	64432-04-8	NI36061
43	286	342	272	91	229	107	55	**342**	100	99	63	36	22	21	21	21		23	34	2	–	–	–	–	–	–	–	–	–	–	Pregn-4-ene-3,20-dione, 2,2-dimethyl-	56196-80-6	NI36062
138	342	55	137	91	258	93	67	**342**	100	85	61	50	46	42	39	39		23	34	2	–	–	–	–	–	–	–	–	–	–	Pregn-4-ene-3,20-dione, 6,16-dimethyl-, (6α,16α)-	1816-78-0	WI01447
138	43	342	137	300	55	109	258	**342**	100	99	56	28	23	23	21	19		23	34	2	–	–	–	–	–	–	–	–	–	–	Pregn-4-ene-3,20-dione, 6,16-dimethyl-, (6α,16α)-	1816-78-0	NI36063
138	258	342	55	137	91	85	81	**342**	100	87	64	59	51	47	46	37		23	34	2	–	–	–	–	–	–	–	–	–	–	Pregn-4-ene-3,20-dione, 6,16-dimethyl-, (6α,16β)-	55400-13-0	WI01448
43	342	138	137	85	55	41	300	**342**	100	57	40	30	24	23	22	21		23	34	2	–	–	–	–	–	–	–	–	–	–	Pregn-4-ene-3,20-dione, 6,16-dimethyl-, (6β,16α)-	1816-79-1	NI36064
43	342	138	137	85	55	41	300	**342**	100	99	70	53	42	40	38	37		23	34	2	–	–	–	–	–	–	–	–	–	–	Pregn-4-ene-3,20-dione, 6,16-dimethyl-, (6β,16α)-	1816-79-1	NI36065
258	43	137	342	138	41	121	55	**342**	100	72	45	42	31	26	25	24		23	34	2	–	–	–	–	–	–	–	–	–	–	Pregn-4-ene-3,20-dione, 6,17-dimethyl-, (6α)-	5087-55-8	NI36066
43	244	55	41	91	342	299	79	**342**	100	96	36	31	30	29	29	26		23	34	2	–	–	–	–	–	–	–	–	–	–	Pregn-4-ene-3,20-dione, 17-ethyl-	1048-01-7	NI36067
43	244	55	91	41	342	299	229	**342**	100	97	36	31	31	30	30	27		23	34	2	–	–	–	–	–	–	–	–	–	–	Pregn-4-ene-3,20-dione, 17-ethyl-	1048-01-7	NI36068
71	327	70	342	273	328	341	56	**342**	100	85	85	28	27	22	13	12		23	38	–	2	–	–	–	–	–	–	–	–	–	Con-5-enin-3-amine, N-methyl-, (3β)-	468-36-0	NI36069
129	172	128	171	43	127	155	342	**342**	100	58	52	42	42	22	20	18		24	22	2	–	–	–	–	–	–	–	–	–	–	5-Acetyl-1,2,4,5-tetrahydro-2-methyl-2-(1-naphthyl)-1,4-methano-3-benzoxepin		MS08141
259	260	181	183	105	182	342	272	**342**	100	25	20	15	15	8	7	6		24	26	–	–	–	–	–	–	–	1	–	–	–	Silane, (2,3-dimethyl-2-butenyl)triphenyl-	74630-81-2	NI36070
57	77	147	131	109	69	53	43	**342**	100	18	17	16	15	14	14	14	**0.00**	24	38	1	–	–	–	–	–	–	–	–	–	–	Androst-5-en-3-ol, 17-(2-pentenyl)-, [3β,17(Z)]-	57983-06-9	NI36071
41	255	55	95	91	107	67	105	**342**	100	92	64	53	51	50	49	48	**44.06**	24	38	1	–	–	–	–	–	–	–	–	–	–	Chol-5,22-dien-3-ol, (3β,22E)-	57597-08-7	NI36072
69	43	105	107	95	67	55	91	**342**	100	39	33	32	32	32	32	29	**1.08**	24	38	1	–	–	–	–	–	–	–	–	–	–	Chol-5,22-dien-3-ol, (3β,22Z)-	57597-14-5	NI36073
231	342	55	81	41	135	43	232	**342**	100	60	35	28	28	24	20	19		24	38	1	–	–	–	–	–	–	–	–	–	–	Chol-7-en-12-one, (5β)-	54411-74-4	NI36074
243	183	105	244	165	77	215	154	**342**	100	45	40	29	29	19	9	9	**2.00**	25	26	1	–	–	–	–	–	–	–	–	–	–	Cyclohexyl triphenylmethyl ether	20705-40-2	LI07595
243	183	105	244	165	77	154	155	**342**	100	45	40	29	29	19	9	7	**2.00**	25	26	1	–	–	–	–	–	–	–	–	–	–	Cyclohexyl triphenylmethyl ether	20705-40-2	NI36075
117	43	118	41	130	116	55	91	**342**	100	13	11	9	6	5	4	3	**2.02**	25	42	–	–	–	–	–	–	–	–	–	–	–	1-Hexadecylindan		AI02090
117	118	43	41	116	29	342	115	**342**	100	10	9	6	5	5	4	4		25	42	–	–	–	–	–	–	–	–	–	–	–	1-Hexadecylindan		AI02089
117	118	43	41	116	29	57	55	**342**	100	10	8	6	5	5	4	4	**3.49**	25	42	–	–	–	–	–	–	–	–	–	–	–	1-Hexadecylindan		AI02088
117	342	43	104	131	41	118	116	**342**	100	90	68	57	42	40	35	27		25	42	–	–	–	–	–	–	–	–	–	–	–	2-Hexadecylindan		AI02092
342	117	43	104	131	41	343	116	**342**	100	73	47	40	30	29	27	25		25	42	–	–	–	–	–	–	–	–	–	–	–	2-Hexadecylindan		AI02091
145	146	43	131	342	41	57	29	**342**	100	55	53	43	34	33	30	27		25	42	–	–	–	–	–	–	–	–	–	–	–	1,2,3,4-Tetrahydro-5-pentadecyl-naphthalene		AI02093
145	43	146	342	131	41	57	29	**342**	100	55	51	46	40	34	30	27		25	42	–	–	–	–	–	–	–	–	–	–	–	1,2,3,4-Tetrahydro-5-pentadecyl-naphthalene		AI02094
215	342	343	216	213	228	227	202	**342**	100	71	20	20	6	5	4	4		26	30	–	–	–	–	–	–	–	–	–	–	–	3-Decylpyrene		AI02095
91	342	223	131	104	117	105	92	**342**	100	40	40	26	20	17	14	13		26	30	–	–	–	–	–	–	–	–	–	–	–	1,4-Bis(4'-phenylbutyl)-benzene		AI02096
91	342	223	131	104	117	105	92	**342**	100	37	34	25	20	17	14	13		26	30	–	–	–	–	–	–	–	–	–	–	–	1,4-Bis(4'-phenylbutyl)-benzene		AI02097
310	312	308	345	347	314	275	249	**343**	100	80	51	35	34	34	30	23	**15.04**	7	–	–	1	–	–	7	–	–	–	–	–	–	Aniline, 2,3,4,5,6-pentachloro-N-(dichloromethylene)-	2666-66-2	NI36076
64	41	61	68	132	63	60	45	**343**	100	80	62	52	49	32	27	27	**0.00**	11	21	7	1	2	–	–	–	–	–	–	–	–	Desulphoglucoiberin		MS08142
44	45	28	298	172	42	184	71	**343**	100	57	38	37	35	26	23	21	**0.00**	12	26	4	7	–	–	–	–	–	–	–	1	–	Dimethylammonium bis(N,N',N''-trimethylbiuret-1,5-diyl)borate(1–)		NI36077
65	229	117	116	91	68	114	92	**343**	100	97	97	97	96	95	78	74	**0.00**	13	11	3	1	–	6	–	–	–	–	–	–	–	Acetic acid, trifluoro-, 3-phenyl-2-[(trifluoroacetyl)amino]propyl ester, (S)-	59080-42-1	NI36078
343	312	196	181	328	166	344	192	**343**	100	82	52	29	26	24	21	20		13	14	4	1	–	5	–	–	–	–	–	–	–	2,3,4-Trimethoxybenzylamine, pentafluoropropionyl-		MS08143
343	328	181	207	44	344	69	300	**343**	100	45	24	20	16	11	10	9		13	14	4	1	–	5	–	–	–	–	–	–	–	2,4,5-Trimethoxybenzylamine, pentafluoropropionyl-		MS08144
343	328	181	207	344	138	44	153	**343**	100	36	18	14	13	8	8	7		13	14	4	1	–	5	–	–	–	–	–	–	–	3,4,5-Trimethoxybenzylamine, pentafluoropropionyl-		MS08145
100	114	56	43	42	70	55	343	**343**	100	48	25	25	20	16	15	10		13	17	2	3	3	–	–	–	–	–	–	–	–	5-Acetyl-4-amino-3-(2-N-morpholinylethylthio)thieno[2,3-c]isothiazole	97090-71-6	NI36079
100	113	43	42	56	70	55	84	**343**	100	30	30	28	25	16	15	13	**2.00**	13	17	2	3	3	–	–	–	–	–	–	–	–	5-Acetyl-4-amino-3-(2-N-morpholinylethylthio)thieno[3,2-d]isothiazole	97090-78-3	NI36080

MASS TO CHARGE RATIOS								M.W.	INTENSITIES								Parent	C	H	O	N	S	F	Cl	Br	I	Si	P	B	X	COMPOUND NAME	CAS Reg No	No
43	41	171	57	54	56	107	58	343	100	53	40	31	30	29	28	21	11.00	13	17	4	3	2	–	–	–	–	–	–	–	–	2-(4-Methoxybenzenesulphonamido)-5-(1′-hydroxy-2′-methylpropyl)-1,3,4-thiadiazine		LI07596
114	59	43	285	107	171	44	77	343	100	97	86	55	41	39	33	30	9.50	13	17	4	3	2	–	–	–	–	–	–	–	–	2-(4-Methoxybenzenesulphonamido)-5-(2′-hydroxy-2′-methylpropyl)-1,3,4-thiadiazine		LI07597
171	77	107	40	92	54	63	122	343	100	71	62	47	45	33	31	26	21.00	13	17	4	3	2	–	–	–	–	–	–	–	–	2-(4-Methoxybenzenesulphonamido)-5-(3′-hydroxy-2′-methylpropyl)-1,3,4-thiadiazine		LI07598
146	127	196	96	343	89	197	216	343	100	54	20	19	13	13	12	11		14	10	–	–	–	6	–	–	–	–	–	–	1	Vanadium, bis(benzotrifluoride)-	53966-11-3	NI36081
152	181	165	73	60	43	164	57	343	100	72	23	22	20	20	19	17	14.50	14	17	9	1	–	–	–	–	–	–	–	–	–	2-(2,7-Dihydroxy-1,4-2H-benzoxazin-3(4H)-one)-β-D-gluco-pyranoside		MS08146
187	84	70	161	162	160	145	55	343	100	37	35	32	29	26	26	23	0.00	14	21	5	3	1	–	–	–	–	–	–	–	–	8-[(4-Amino-1-methylbutyl)amino]-6-hydroxyquinoline sulphate		NI36082
86	99	58	30	228	42	245	149	343	100	91	47	47	42	36	14	12	1.30	14	22	2	3	–	–	–	1	–	–	–	–	–	N-(Diethylaminoethyl)-2-methoxy-4-amino-5-bromobenzamide		MS08147
118	147	148	161	149	77	343	101	343	100	96	62	45	45	45	40	37		15	13	5	5	–	–	–	–	–	–	–	–	–	Formamide, N-[2-[[(2,4-dinitrophenyl)hydrazono]methyl]phenyl]-N-methyl-	52381-17-6	NI36083
91	131	60	73	65	61	92	43	343	100	26	9	8	8	8	7	5	0.00	15	21	6	1	1	–	–	–	–	–	–	–	–	Desulphogluconasturtiin		MS08148
81	91	100	72	155	271	171	343	343	100	28	26	26	15	6	5	2		15	21	6	1	1	–	–	–	–	–	–	–	–	1-(N,N-Dimethoxycarboxamido)-2-deoxy-5-O-tosyl-α-L-erythro-pentofuranose		NI36084
81	91	72	155	100	271	171	343	343	100	38	30	22	9	7	6	3		15	21	6	1	1	–	–	–	–	–	–	–	–	1-(N,N-Dimethoxycarboxamido)-2-deoxy-5-O-tosyl-β-L-erythro-pentofuranose		NI36085
81	72	91	100	155	171	271	343	343	100	24	23	22	13	8	7	2		15	21	6	1	1	–	–	–	–	–	–	–	–	1-(N,N-Dimethylcarboxamido)-2-deoxy-5-O-tosyl-α-D-erythro-pentofuranose		NI36086
81	91	72	172	155	271	100	343	343	100	40	21	18	14	6	6	5		15	21	6	1	1	–	–	–	–	–	–	–	–	1-(N,N-Dimethylcarboxamido)-2-deoxy-5-O-tosyl-β-D-erythro-pentofuranose		NI36087
134	311	125	111	152	310	102	127	343	100	39	36	31	26	25	25	20	6.00	16	10	–	3	2	–	1	–	–	–	–	–	–	5-Phenylthiazolo[2,3-b]-1,3,4-thiadiazol-4-ium-2-(4-chlorophenyl)aminide		MS08149
343	345	316	317	314	344	315	78	343	100	97	74	49	49	48	47	45		16	14	1	3	–	–	–	1	–	–	–	–	–	Amido-N-ethylbromoazepam		NI36088
343	328	176	260	43	134	344	329	343	100	88	82	50	47	42	38	23		16	14	8	–	–	–	–	–	–	–	–	–	1	Bis(dehydroacetato)beryllium		NI36089
120	105	262	308	28	159	104	77	343	100	69	50	34	31	18	18	18	2.50	16	19	3	1	–	–	2	–	–	–	–	–	–	Ethyl 3,3-dichloro-1-[(1-phenylethyl)carbamoyl]cyclobutanecarboxylate	85291-82-3	NI36090
72	86	44	43	128	227	255	269	343	100	75	51	50	49	43	26	18	4.00	16	29	5	3	–	–	–	–	–	–	–	–	–	N-Acetyl-leucyl-valyl-alanine		LI07599
126	84	30	43	72	73	127	56	343	100	46	42	24	13	12	11	8	6.51	16	29	5	3	–	–	–	–	–	–	–	–	–	N^2-N-Acetylglycyl-N^6-acetyllysine, butyl ester		AI02098
236	222	221	204	203	111	104	264	343	100	64	61	52	44	40	32	31	27.00	17	14	–	1	1	–	–	1	–	–	–	–	–	11-(2-Bromoethyl)-11-cyano-6,11-dihydrodibenzo[b,e]thiepine		MS08150
195	252	44	167	91	118	42	41	343	100	75	21	9	7	6	4	4	0.00	17	14	1	1	–	5	–	–	–	–	–	–	–	1,N-Dimethyl-2-phenylethylamine pentafluorobenzoyl ester		MS08151
43	105	77	162	224	226	302	161	343	100	56	44	42	38	34	29	27	4.00	17	14	2	1	–	–	–	1	–	–	–	–	–	3-(4-Bromophenyl)-4-acetyl-5-phenyl-2-isoxazoline		NI36091
91	43	193	119	302	300	90	89	343	100	23	19	16	14	14	13	13	3.00	17	14	2	1	–	–	–	1	–	–	–	–	–	3-(4-Bromophenyl)-4-phenyl-5-acetyl-2-isoxazoline		NI36092
138	177	151	205	179	153	207	178	343	100	90	51	38	29	17	13	12	0.00	17	14	3	3	–	–	1	–	–	–	–	–	–	2-(4-Chloroanilino)-4-(4-nitroanilino)-cyclopent-2-enone		MS08152
284	286	299	252	158	258	104	300	343	100	86	84	42	40	30	28	25	2.00	18	17	6	1	–	–	–	–	–	–	–	–	–	Acridone, 10-methyl-4-ethoxy-2-methoxy-3,4-methylenedioxy-1-oxo-		MS08153
260	217	343	232	83	261	85	84	343	100	70	62	60	54	47	47	45		18	17	6	1	–	–	–	–	–	–	–	–	–	1,3(2H,9bH)-Dibenzofurandione, 6-acetyl-2-(1-aminoethylidene)-7,9-dihydroxy-8,9b-dimethyl-	55721-24-9	NI36093
264	179	178	86	345	343	77	260	343	100	79	64	64	44	44	32	1		18	18	1	1	–	–	–	1	–	–	–	–	–	1-Bromo-2-(4-morpholino)-1,2-diphenyl-ethane		LI07600
43	114	156	328	58	84	96	72	343	100	70	65	44	43	38	33	33	5.00	18	33	5	1	–	–	–	–	–	–	–	–	–	Acetamide, N,N-bis[2-(2,2,4-trimethyl-1,3-dioxolan-4-yl)ethyl]-, [S-(R*,R*)]-	64061-18-3	NI36094
135	136	206	161	137	134	235	219	343	100	70	6	6	6	5	3	2	0.19	19	21	3	1	1	–	–	–	–	–	–	–	–	2-(1′-Adamantyl)-2-phenylthio-oxazolidine-4,5-dione		LI07601
167	118	168	269	165	166	194	152	343	100	95	94	81	66	61	59	52	41.00	19	21	3	1	1	–	–	–	–	–	–	–	–	Diphenylacetylmethionine	65707-81-5	NI36095
135	125	167	105	131	177	343	124	343	100	49	43	43	38	24	18	18		19	21	3	1	1	–	–	–	–	–	–	–	–	Ethyl 3-[2-(acetylamino)phenylthio]-3-phenylpropionate		MS08154
328	314	343	256	270	315	284	329	343	100	80	63	47	45	31	30	26		19	21	5	1	–	–	–	–	–	–	–	–	–	9-Acridanone, 4-ethoxy-1,2,3-trimethoxy-10-methyl-	17014-68-5	NI36096
328	300	343	286	242	158	329	284	343	100	55	51	46	27	26	22	22		19	21	5	1	–	–	–	–	–	–	–	–	–	Acridone, 1-ethoxy-2,3,4-trimethoxy-10-methyl-		MS08155
292	307	306	293	248	276	249	290	343	100	38	36	19	13	10	10	9	0.00	20	22	2	1	–	–	1	–	–	–	–	–	–	Isoquinoline, 3-ethyl-6,7-dimethoxy-1-benzyl-, hydrochloride	1163-37-7	NI36097
206	343	342	269					343	100	36	26	20						20	25	4	1	–	–	–	–	–	–	–	–	–	6,7-Dimethoxy-2-methyl-1-(3,4-dimethoxyphenyl)-1,2,3,4-tetrahydro-isoquinoline		LI07602
311	312	310	343	313	296	328		343	100	96	78	31	31	31	20			20	25	4	1	–	–	–	–	–	–	–	–	–	Erythristemine	28619-41-2	LI07603
328	343	312	283	300	256	344	329	343	100	96	46	30	25	24	23	21		20	25	4	1	–	–	–	–	–	–	–	–	–	C-Homoerythrinan, 1,2-didehydro-6,7-epoxy-3,15,16-trimethoxy-, (3α,6ξ)	39024-12-9	NI36098
328	283	343	312	256	240	300	268	343	100	46	42	35	34	30	28	26		20	25	4	1	–	–	–	–	–	–	–	–	–	C-Homoerythrinan, 1,2-didehydro-6,7-epoxy-3,15,16-trimethoxy-, (3β,6ξ)	39024-15-2	NI36099
328	343	312	256	283	300	284	244	343	100	56	42	37	36	26	25	18		20	25	4	1	–	–	–	–	–	–	–	–	–	Homoerythrinane alkaloid 7		LI07604
270	266	343	226	298	238	210	242	343	100	43	29	23	15	14	14	10		20	25	4	1	–	–	–	–	–	–	–	–	–	3,5-Pyridinedicarboxylic acid, 1,4-dihydro-1,2,6-trimethyl-4-phenyl-, diethyl ester	3274-34-8	NI36100
344	100	342	271					343	100	14	12	4					0.00	20	29	2	3	–	–	–	–	–	–	–	–	–	4-Quinolinecarboxamide, 2-butoxy-N-[2-(diethylamino)ethyl]-	85-79-0	NI36101
86	87	58	116	99	57	28	56	343	100	73	47	34	32	24	21	20	6.64	20	29	2	3	–	–	–	–	–	–	–	–	–	4-Quinolinecarboxamide, 2-butoxy-N-[2-(diethylamino)ethyl]-	85-79-0	NI36102
86	87	99	58	116	30	343	29	343	100	95	48	48	38	33	29	26		20	29	2	3	–	–	–	–	–	–	–	–	–	4-Quinolinecarboxamide, 2-butoxy-N-[2-(diethylamino)ethyl]-	85-79-0	NI36103
343	151	165	77	344	342	193	180	343	100	79	41	25	24	20	18	13		21	17	–	3	1	–	–	–	–	–	–	–	–	4-Imino-1,5,5-triphenyl-2-imidazolinethione		NI36104

MASS TO CHARGE RATIOS								M.W.	INTENSITIES								Parent	C	H	O	N	S	F	Cl	Br	I	Si	P	B	X	COMPOUND NAME	CAS Reg No	No
343	119	165	342	77	151	344	104	**343**	100	61	48	42	36	35	26	26		21	17	–	3	1	–	–	–	–	–	–	–	–	5-Imino-1,4,4-triphenyl-2-imidazolinethione	52460-97-6	NI36105
328	130	266	102	329	105	76	149	**343**	100	52	39	32	29	14	13	11	**4.79**	21	17	2	1	–	–	–	–	–	1	–	–	–	Phthalimide, N-(methyldiphenylsilyl)-	31684-03-4	NI36106
91	92	343	312	190	65	344	313	**343**	100	79	32	22	10	10	7	7		21	29	3	1	–	–	–	–	–	–	–	–	–	Estra-1,3,5(10)-trien-17-one, 3,4-dimethoxy-, O-methyloxime	74299-27-7	NI36107
343	284	328	240	268	200	226	57	**343**	100	70	57	50	45	45	43	43		21	29	3	1	–	–	–	–	–	–	–	–	–	Glyoxylic acid, (5,7-di-tert-butyl-3,3-dimethyl-3H-indol-2-yl)-, methyl ester		DD01351
127	135	84	140	131	103	343	112	**343**	100	53	52	50	45	36	33	33		21	29	3	1	–	–	–	–	–	–	–	–	–	Piperidine, 1-[9-(1,3-benzodioxol-5-yl)-1-oxo-8-nonenyl]-, (E)-	30505-89-6	LI07605
109	196	43	242	343	97	96	209	**343**	100	88	83	80	71	71	66	59		21	29	3	1	–	–	–	–	–	–	–	–	–	1H-Pyrrolo[2,1,5-de]quinazoline, 3-acetyldecahydro-4,6-dihydroxy-4,7-dimethyl-6-phenyl-, [2aS-(2aα,3α,4α,5aα,6α,7α,8aβ)]-	34443-74-8	NI36108
343	171	314	213	185	184			**343**	100	17	5	5	5	5				22	17	1	1	1	–	–	–	–	–	–	–	–	4,6-Bisbenzylidene-4,5-dihydro-2-methylbenzothiazol-7(6H)-one		LI07606
105	221	106	51	50	222	165	52	**343**	100	16	14	9	4	3	2	2	**1.00**	22	17	3	1	–	–	–	–	–	–	–	–	–	N-1-Phenyl-1-phenylglyoxylmethylbenzamide		LI07607
328	57	55	84	343	69	71	86	**343**	100	83	58	50	49	48	46	36		22	21	1	3	–	–	–	–	–	–	–	–	–	(Z,Z)-4-Ethyl-3-methyl-5-[5-(4-cyanophenyl-2-methylene)-3,4-dimethyl-5H-pyrrolyl-2-methylene]-3-pyrrolin-2-one		MS08156
276	312	343	295	317	277	244	286	**343**	100	35	33	27	23	23	20	15		22	33	2	1	–	–	–	–	–	–	–	–	–	18,19-Dinorpregn-4-en-20-yn-3-one, 13-ethyl-dihydro-17-hydroxy, O-methyloxime, (5β)-		NI36109
125	343	137	153	151	312	138	93	**343**	100	99	77	63	41	30	21	21		22	33	2	1	–	–	–	–	–	–	–	–	–	Pregn-4-en-3,20-dione, 3-O-methyloxime		LI07608
100	87	312	257	70	244	93	201	**343**	100	71	59	37	36	22	17	14	**12.00**	22	33	2	1	–	–	–	–	–	–	–	–	–	Pregn-4-en-3,20-dione, 20-O-methyloxime		LI07609
170	242	199	271	272	128	227	184	**343**	100	60	50	45	42	41	38	36	**7.00**	22	37	–	3	–	–	–	–	–	–	–	–	–	Tricyclo[5.2.1.0^{4,10}]deca-2,5,8-triene, 1,4,7-tris(diethylamino)-		NI36110
343	266	344	342	345	341	268	217	**343**	100	79	63	47	36	36	28	22		23	18	–	1	–	–	1	–	–	–	–	–	–	Quinoline, 6-chloro-2-(2-phenylethyl)-4-phenyl-		LI07610
43	70	82	343	54	41	314	139	**343**	100	60	51	46	43	43	38	34		23	37	1	1	–	–	–	–	–	–	–	–	–	Acetamide, N-5α-androst-2-en-3-yl-N-ethyl-	4642-51-7	NI36111
147	146	148	165	182	145	343	180	**343**	100	25	14	4	4	3	2	1		24	25	1	1	–	–	–	–	–	–	–	–	–	N-Benzhydrylidene-1-(2,4,6-trimethylphenyl)ethylamine N-oxide	89839-57-6	NI36112
147	105	77	148	131	91	165	119	**343**	100	14	13	13	8	6	6	5	**0.90**	24	25	1	1	–	–	–	–	–	–	–	–	–	3,3-Diphenyl-2-[1-(2,4,6-trimethylphenyl)ethyl]oxaziridine	75326-12-4	NI36113
142	344	329	342	327	127	217	341	**344**	100	88	80	70	67	59	57	51		1	3	–	–	–	–	–	–	1	–	–	–	1	Mercury, iodomethyl-	143-36-2	NI36114
344	329	342	341	327	343	326	328	**344**	100	82	80	63	63	53	48	42		1	3	–	–	–	–	–	–	1	–	–	–	1	Mercury, iodomethyl-	143-36-2	NI36115
270	255	225	232	165	150	217	240	**344**	100	82	36	26	24	19	17	4	**0.00**	1	3	–	–	–	–	–	–	2	–	–	–	1	Arsine, diiodomethyl-	58299-04-0	NI36116
231	69	114	248	259	344	212	50	**344**	100	17	15	14	9	8	8	5		1	3	1	4	–	7	–	–	–	–	4	–	–	1,3,5,2,4,6-Triazatriphosphorine, 2-[(difluorophosphinyl)methylamino]-2,2,4,4,6,6-pentafluoro-2,2,4,4,6,6-hexahydro-	41006-36-4	NI36117
69	164	92	202	114	180	145	76	**344**	100	48	11	7	7	6	2	2	**0.00**	6	–	1	2	–	12	–	–	–	–	–	–	–	Propanamide, 3,3,3-trifluoro-N,N-bis(trifluoromethyl)-2-(trifluoromethylimino)-		LI07611
346	348	77	50	53	49	344	157	**344**	100	95	82	44	39	36	35	34		6	3	2	–	–	–	–	3	–	–	–	–	–	1,3-Benzenediol, 2,4,6-tribromo-	2437-49-2	NI36118
33	244	163	64	83	69	319	320	**344**	100	87	60	59	54	46	43	40	**0.40**	6	6	4	–	–	9	–	–	–	–	1	–	–	Ethanol, 2,2,2-trifluoro-, phosphate (3:1)	358-63-4	WI01449
329	165	135	120	269	121	299	255	**344**	100	54	34	9	7	6	5	5	**0.00**	7	20	–	–	–	–	–	–	–	–	–	–	2	Stannane, methylenebis[trimethyl-	16812-43-4	NI36119
213	73	45	75	43	211	74	59	**344**	100	99	88	70	63	62	56	56	**3.70**	8	24	–	–	1	–	–	–	–	4	–	–	1	2,2,3,3,5,5,6,6-Octamethyl-1-selena-4-thiacyclohexasilane	85263-62-3	NI36120
77	180	230	314	161	154	135	258	**344**	100	20	10	7	7	7	7	5	**0.03**	10	5	4	–	–	3	–	–	–	–	–	–	1	π-Cyclopentadienyl-trifluoroacetyl-tricarbonyl molybdenum		MS08157
346	344	348	265	267	102	51	237	**344**	100	51	49	43	43	34	25	23		11	6	3	–	–	–	–	2	–	–	–	–	–	2,6-Dibromo-5-hydroxy-7-methylnaphthalene-1,4-dione	104506-28-7	NI36121
81	208	253	210	109	45	344	346	**344**	100	94	91	64	60	54	52	35		11	15	4	–	1	–	2	–	–	–	1	–	–	Chlorothiophos o-analog 2,5-isomer		NI36122
68	73	140	147	41	98	125	241	**344**	100	97	88	82	57	54	45	44	**0.00**	11	29	6	–	–	–	–	–	–	2	1	–	–	Phosphoric acid, 2,3-bis(trimethylsiloxy)propyl dimethyl ester	18151-73-0	NI36123
188	94	140	142	93	344	82	50	**344**	100	46	17	17	11	11	10	10		12	6	6	2	1	2	–	–	–	–	–	–	–	Benzene, 1,1′-sulphonylbis[4-fluoro-3-nitro-	312-30-1	NI36124
229	227	81	311	82	231	309	79	**344**	100	78	75	63	63	51	49	43	**6.20**	12	9	1	–	–	–	5	–	–	–	–	–	–	2,7:3,6-Dimethanonaphth[2,3-b]oxirene, 3,4,6,9,9-pentachloro-1a,2,2a,3,6,6a,7,7a-octahydro-, (1aα,2β,2aβ,3β,6β,6aβ,7β,7aα)-	33487-96-6	MS08158
311	309	79	313	283	67	175	281	**344**	100	75	73	51	45	41	38	36	**13.50**	12	9	1	–	–	–	5	–	–	–	–	–	–	2,5,7-Metheno-3H-cyclopenta[a]pentalen-3-one, 4,5,6,6,6a-pentachlorodecahydro-, (2α,3aβ,3bβ,4β,5β,6aβ,7α,7aβ,8R*)-	24557-95-7	MS08159
346	348	344	331	303	333	329	301	**344**	100	50	50	45	43	20	20	20		12	10	2	–	–	–	–	2	–	–	–	–	–	Naphthalene, 1,5-dibromo-2,6-dimethoxy-	25315-06-4	LI07612
346	344	348	331	303	329	301	113	**344**	100	51	50	44	44	22	22	22		12	10	2	–	–	–	–	2	–	–	–	–	–	Naphthalene, 1,5-dibromo-2,6-dimethoxy-	25315-06-4	NI36125
80	156	155	81	78	128	51	129	**344**	100	97	37	37	25	19	18	16	**0.00**	12	14	–	2	–	–	–	2	–	–	–	–	–	1,1′-Dimethyl-2,2′-dipyridylium dibromide		IC04058
224	103	97	102	182	85	225	170	**344**	100	96	71	53	44	30	27	27	**0.00**	12	15	8	–	–	3	–	–	–	–	–	–	–	Methyl 2,3-di-O-acetyl-4-O-trifluoroacetyl-β-D-xylopyranoside		MS08160
97	224	95	103	171	86	114	182	**344**	100	77	76	61	59	43	42	40	**0.00**	12	15	8	–	–	3	–	–	–	–	–	–	–	Methyl 2,4-di-O-acetyl-3-O-trifluoroacetyl-β-D-xylopyranoside		MS08161
224	182	97	225	95	86	211	87	**344**	100	80	75	50	47	45	43	41	**0.00**	12	15	8	–	–	3	–	–	–	–	–	–	–	Methyl 3,4-di-O-acetyl-2-O-trifluoroacetyl-β-D-xylopyranoside		MS08162
186	168	268	121	193	194	316	140	**344**	100	67	65	47	44	43	40	40	**25.23**	13	5	2	–	–	5	–	–	–	–	–	–	1	Iron, dicarbonyl(η^5-2,4-cyclopentadien-1-yl)(pentafluorophenyl)-	12176-60-2	MS08163
175	268	140	195	194	156	316	75	**344**	100	83	58	52	48	46	44	43	**41.14**	13	5	2	–	–	5	–	–	–	–	–	–	1	Iron, dicarbonyl(η^5-2,4-cyclopentadien-1-yl)(pentafluorophenyl)-	12176-60-2	NI36126
127	75	217	111	137	263	129	126	**344**	100	44	38	36	35	33	32	23	**18.00**	13	10	–	2	–	–	2	–	–	–	–	–	1	N,N′-Bis(4-chlorophenyl)selenourea		MS08164
309	311	308	310	313	241	99	237	**344**	100	99	40	39	33	29	23	20	**5.00**	13	16	2	–	–	–	4	–	–	–	–	–	–	Bicyclo[2.2.1]hept-2-ene, 1,4,5,6-tetrachloro-7,7-dimethoxy-2-(1-methyl-1-propenyl)-	55493-64-6	NI36127
70	83	258	56	71	98	97	58	**344**	100	55	13	12	9	8	8	7	**1.40**	13	17	3	4	1	–	1	–	–	–	–	–	–	1,2,4-Benzothiadiazine, 7-chloro-2-methyl-3-(4-methyl-1-piperazinyl)-, 1,1,4′-trioxide		MS08165

MASS TO CHARGE RATIOS								M.W.	INTENSITIES								Parent	C	H	O	N	S	F	Cl	Br	I	Si	P	B	X	COMPOUND NAME	CAS Reg No	No
73	329	45	344	330	147	343	241	344	100	43	23	15	13	11	9	9		13	28	3	2	–	–	–	–	–	3	–	–	–	Pyrimidine, 2,4,5-tris[(trimethylsilyl)oxy]-	32865-92-2	NI36128
73	147	329	344	343	45	330	345	344	100	53	49	31	30	23	15	12		13	28	3	2	–	–	–	–	–	3	–	–	–	Pyrimidine, 2,4,6-tris[(trimethylsilyl)oxy]-	31111-39-4	NI36129
193	207	165	179	152	137	151	215	344	100	89	49	40	25	24	22	19	0.50	13	30	6	–	–	–	–	–	–	–	2	–	–	Tetraethyl pentane-1,5-diphosphonate		MS08166
346	344	318	348	218	316	311	309	344	100	76	52	45	43	41	35	35		14	4	2	–	–	–	4	–	–	–	–	–	–	1,4,5,8-Tetrachloroanthraquinone		IC04059
246	248	346	63	344	348	65	247	344	100	68	58	54	44	30	19	16		14	8	–	–	–	–	4	–	–	1	–	–	–	1,1,4,5-Tetrachloro-1-sila-2,3:6,7-dibenzocycloheptatriene		NI36130
127	198	52	344	146	108	77	89	344	100	44	33	22	18	18	18	16		14	10	–	–	–	6	–	–	–	–	–	–	1	Chromium, bis[(1,2,3,4,5,6-η)-(trifluoromethyl)benzene]-	53966-07-7	NI36131
182	344	143	162	139	202	204	187	344	100	89	79	44	38	35	22	22		14	16	–	2	–	4	–	–	–	2	–	–	–	Hydrazine, 1,1-bis(difluorophenylsilyl)-2,2-dimethyl-	66436-29-1	NI36132
70	86	68	60	85	69	112	226	344	100	87	85	81	80	79	54	38	3.00	14	24	6	4	–	–	–	–	–	–	–	–	–	Acetamide, N,N'-[(3,6-dioxo-2,5-piperazinediyl)di-3,1-propanediyl]bis[N-hydroxy-, (2S-cis)-	18928-00-2	NI36133
60	69	67	70	91	97	95	71	344	100	91	70	64	60	58	56	56	4.00	14	24	6	4	–	–	–	–	–	–	–	–	–	2,5-Piperazinediacetamide, N,N'-dihydroxy-3,6-dioxo-N,N'-dipropyl-	69833-34-7	NI36134
183	185	119	77	155	346	344	157	344	100	97	63	40	39	39	39	38		15	9	3	2	–	–	–	1	–	–	–	–	–	5-(4'-Bromophenyl)-3-phenyl-4-nitroisoxazole	96751-50-7	NI36135
344	72	58	345	130	217	230	69	344	100	46	36	24	15	13	12	9		15	16	4	6	–	–	–	–	–	–	–	–	–	1-(4',6'-Dimethoxy-1',3',5'-triazin-2'-yl)-5-(4',6'-dimethoxy-1',2'-dihydro-1',3',5'-triazin-2'-ylidene)-cyclopenta-1,3-diene		MS08167
139	141	113	111	199	73	85	75	344	100	37	37	37	34	23	20	19	5.00	15	17	3	–	2	–	1	–	–	–	–	–	–	Propanoic acid, 3-[[3-(4-chlorophenyl)-3-hydroxy-1-thioxo-2-propenyl]thio]-, isopropyl ester	72347-56-9	NI36136
41	43	175	57	27	177	29	258	344	100	71	60	59	48	42	38	25	7.20	15	18	3	2	–	–	2	–	–	–	–	–	–	1,3,4-Oxadiazol-2(3H)-one, 3-[2,4-dichloro-5-isopropoxyphenyl]-5-tert-butyl-	19666-30-9	NI36137
205	203	149	147	124	247	245	206	344	100	99	14	14	10	9	9	8	7.20	15	21	4	–	–	–	–	1	–	–	–	–	–	3-Bromo-4-methoxy-6-nonanoyl-2H-pyran-2-one		MS08168
126	55	113	125	127	158	98	56	344	100	45	41	36	32	26	23	22	0.77	15	24	7	2	–	–	–	–	–	–	–	–	–	1,5,13-Trioxa-10,16-diazacycloeicosane-6,9,17,20-tetrone	79688-15-6	NI36138
126	55	125	127	158	98	100	56	344	100	45	36	32	26	23	21	20	0.00	15	24	7	2	–	–	–	–	–	–	–	–	–	1,5,13-Trioxa-10,16-diazacycloeicosane-6,9,17,20-tetrone		MS08169
41	149	39	109	30	177	43	79	344	100	95	51	44	43	40	40	33	7.30	16	12	7	2	–	–	–	–	–	–	–	–	–	3(2H)-Benzofuranone, 7-(2,4-dinitrophenoxy)-2,2-dimethyl-	62059-45-4	NI36139
121	227	122	114	228	152	77	63	344	100	48	10	9	7	7	6	6	0.60	16	15	2	–	–	–	3	–	–	–	–	–	–	Benzene, 1-methoxy-2-[2,2,2-trichloro-1-(4-methoxyphenyl)ethyl]-	30667-99-3	NI36140
238	223	227	152	239	119	308	310	344	100	64	61	44	40	39	28	26		16	15	2	–	–	–	3	–	–	–	–	–	–	Benzene, 1,1'-(2,2,2-trichloroethylidene)bis[4-methoxy-	72-43-5	NI36141
109	239	121	241	237	133	135	275	344	100	91	61	57	53	52	50	31	0.00	16	15	2	–	–	–	3	–	–	–	–	–	–	Benzene, 1,1'-(2,2,2-trichloroethylidene)bis[4-methoxy-	72-43-5	NI36142
227	228	212	169	152	346	344	141	344	100	15	8	7	7	6	6	5		16	15	2	–	–	–	3	–	–	–	–	–	–	Benzene, 1,1'-(2,2,2-trichloroethylidene)bis[4-methoxy-	72-43-5	NI36143
166	238	194	237	69	77	98	96	344	100	77	24	21	20	14	13	11	0.00	16	19	3	2	–	3	–	–	–	–	–	–	–	N-(Trifluoroacetyl)prolylnorephedrine		NI36144
43	139	181	109	235	121	95	65	344	100	47	42	37	36	23	22	20	0.00	16	21	4	–	1	–	1	–	–	–	–	–	–	(±)-1-Methyl-5-exo-methoxy-5-endo-(phenylthio)-6-exo-chloro-7-oxabicyclo[2.2.1]heptane-2-exo,3-exo-dimethanol	117626-65-0	DD01352
107	108	172	93	91	79	109	97	344	100	68	60	50	42	30	17	15	0.80	16	24	–	–	4	–	–	–	–	–	–	–	–	1,2,3,4,5,8,9,10,11,12,13,16-Dodecahydrodibenzo[d,j][1,2,7,8]tetrathiacyclododecin		MS08170
144	43	186	41	85	187	57	70	344	100	72	59	41	36	35	34	26	0.37	16	28	6	2	–	–	–	–	–	–	–	–	–	N-(N-Acetyl-O-butyl-α-aspartyl)glycine, butyl ester		AI02099
197	158	329	141	215	271	176	143	344	100	65	58	58	50	47	39	26	12.00	16	36	4	–	–	–	–	–	–	–	–	–	1	Chromium, tetra-tert-butoxy-		LI07613
197	158	329	141	215	271	176	140	344	100	65	58	58	50	47	39	26	0.00	16	36	4	–	–	–	–	–	–	–	–	–	1	2-Propanol, 2-methyl-, chromium(4+) salt	10585-25-8	NI36145
117	163	91	181	104	145	65	121	344	100	79	68	49	30	17	15	11	2.20	17	13	2	–	–	5	–	–	–	–	–	–	–	Phenylbutyric acid, pentafluorobenzyl ester		NI36146
220	164	266	272	274	344	181	346	344	100	52	16	13	12	7	7	6		17	13	3	–	–	–	–	1	–	–	–	–	–	9-Phenanthrenecarboxylic acid, 8-bromo-9,10-dihydro-10-oxo-, ethyl ester	56666-61-6	MS08171
202	203	155	127	77	186	128	156	344	100	34	23	12	8	7	6	5	4.30	17	14	2	2	–	–	1	–	–	–	1	–	–	1H-Naphtho[1,8-de]-1,3,2-diazaphosphorine, 2-(4-chloro-3-methylphenoxy)-2,3-dihydro-, 2-oxide		NI36147
148	149	91	162	133	105	79	41	344	100	90	50	44	41	35	35	31	28.00	17	20	4	4	–	–	–	–	–	–	–	–	–	2,3,3a,5,6,7-Hexahydro-3a-methylazulen-4(1H)-one dinitrophenylhydrazone		NI36148
344	185	346	157	345	139	115	217	344	100	65	38	27	23	23	21	18		18	13	3	–	1	–	1	–	–	–	–	–	–	4-Chlorophenyl 4-(4-methoxyphenyl)phenyl sulphone		IC04060
195	111	43	83	77	255	344	284	344	100	51	10	4	4	3	2	2		18	16	3	–	2	–	–	–	–	–	–	–	–	(R)-2-Phenyl-1,1-di-2-thienyl-1,2-ethanediol-2-acetate	110773-03-0	DD01353
344	345	329	57	153	55	43	41	344	100	21	20	20	17	15	15	15		18	16	7	–	–	–	–	–	–	–	–	–	–	4H-1-Benzopyran-4-one, 5,7-dihydroxy-2-(3,4,5-trimethoxyphenyl)-	18103-42-9	NI36149
329	344	330	105	301	343	311	197	344	100	73	19	12	9	8	8	6		18	16	7	–	–	–	–	–	–	–	–	–	–	4H-1-Benzopyran-4-one, 5,7-dihydroxy-3,6,8-trimethoxy-2-phenyl-	71479-92-0	NI36150
57	73	71	55	43	69	85	83	344	100	98	75	68	68	61	45	42	12.64	18	16	7	–	–	–	–	–	–	–	–	–	–	4H-1-Benzopyran-4-one, 2-(3,4-dimethoxyphenyl)-3,5-dihydroxy-7-methoxy-	6068-80-0	NI36151
344	329	326	301	139	167	315	343	344	100	70	59	52	22	14	10	8		18	16	7	–	–	–	–	–	–	–	–	–	–	4H-1-Benzopyran-4-one, 2-(3,4-dimethyoxyphenyl)-5,7-dihydroxy-6-methoxy-	22368-21-4	NI36152
344	152	329	129	148	183	234	300	344	100	95	90	90	65	60	50	47		18	16	7	–	–	–	–	–	–	–	–	–	–	4H-1-Benzopyran-4-one, 5-hydroxy-2-(4-hydroxy-3-methoxyphenyl)-6,7-dimethoxy-	41365-32-6	NI36153
299	44	244	216	243	229	344	91	344	100	72	39	34	28	28	26	22		18	16	7	–	–	–	–	–	–	–	–	–	–	11H-Dibenzo[b,e][1,4]dioxepin-7-carboxylic acid, 3,8-dihydroxy-1,4,6,9-tetramethyl-11-oxo-	4665-02-5	NI36154
300	244	216	243	229	344	91	136	344	100	43	37	30	30	27	24	22		18	16	7	–	–	–	–	–	–	–	–	–	–	11H-Dibenzo[b,e][1,4]dioxepin-7-carboxylic acid, 3,8-dihydroxy-1,4,6,9-tetramethyl-11-oxo-	4665-02-5	LI07614
44	344	300	233	179	258	260	243	344	100	99	99	66	62	49	42	38		18	16	7	–	–	–	–	–	–	–	–	–	–	11H-Dibenzo[b,e][1,4]dioxepin-7-carboxylic acid, 8-hydroxy-3-methoxy-1,4,6-trimethyl-11-oxo-	38636-86-1	NI36155

MASS TO CHARGE RATIOS								M.W.	INTENSITIES								Parent	C	H	O	N	S	F	Cl	Br	I	Si	P	B	X	COMPOUND NAME	CAS Reg No	No
233	260	43	344	217	55	234	69	344	100	61	57	50	20	18	16	13		18	16	7	–	–	–	–	–	–	–	–	–	–	1,3(2H,9bH)-Dibenzofurandione, 2,6-diacetyl-7,9-dihydroxy-8,9b-dimethyl-	125-46-2	NI36158
233	260	344	234	217	261	232	345	344	100	97	89	74	73	67	64	63		18	16	7	–	–	–	–	–	–	–	–	–	–	1,3(2H,9bH)-Dibenzofurandione, 2,6-diacetyl-7,9-dihydroxy-8,9b-dimethyl-	125-46-2	NI36157
233	260	344	44	261	232	234	217	344	100	50	44	35	27	26	24	23		18	16	7	–	–	–	–	–	–	–	–	–	–	1,3(2H,9bH)-Dibenzofurandione, 2,6-diacetyl-7,9-dihydroxy-8,9b-dimethyl-	125-46-2	NI36156
260	233	344	217	261	345	234	55	344	100	90	73	45	19	15	15	15		18	16	7	–	–	–	–	–	–	–	–	–	–	1(9bH)-Dibenzofuranone, 2,8-diacetyl-3,7,9-trihydroxy-6,9b-dimethyl-, (+)-	18058-86-1	NI36159
242	241	203	227	302	225	243	199	344	100	50	50	47	37	29	28	28	0.00	18	16	7	–	–	–	–	–	–	–	–	–	–	2H-Naphtho[2,3-b]pyran-5,10-dione, 3,4-bis(acetyloxy)-3,4-dihydro-2-methyl-, [2R-(2α,3β,4β)]-	75420-29-0	NI36160
242	227	203	241	302	173	105	199	344	100	46	43	37	35	27	27	20	0.00	18	16	7	–	–	–	–	–	–	–	–	–	–	2H-Naphtho[2,3-b]pyran-5,10-dione, 3,4-bis(acetyloxy)-3,4-dihydro-2-methyl-, [2S-(2α,3α,4α)]-	75420-30-3	NI36161
260	216	302	202	261	105	146	174	344	100	84	64	56	46	36	32	25	20.00	18	16	7	–	–	–	–	–	–	–	–	–	–	4H-Naphtho[2,3-b]pyran-4-one, 5,10-bis(acetyloxy)-2,3-dihydro-3-hydroxy-2-methyl-, trans-	63755-82-8	NI36162
110	217	191	192	176	190	326	301	344	100	46	31	21	18	17	14	13	10.00	18	20	3	2	1	–	–	–	–	–	–	–	–	12-Hydroxy-9,10-dimethoxy-4,5,6,7,12,13-hexahydrothieno[3,2-e][3]benzazecine-5-carbonitrile		MS08172
163	181	164	138	153	165	122	344	344	100	99	35	19	16	14	14	11		18	20	3	2	1	–	–	–	–	–	–	–	–	4-(2-Hydroxymethyl-4,5-dimethoxybenzyl)-4,5,6,7-tetrahydrothieno[3,2-c]pyridine-5-carbonitrile		MS08173
329	344	239	224	346	331	209	207	344	100	96	93	81	37	36	30	30		18	21	1	2	–	–	1	–	–	1	–	–	–	9-(4-Chlorobutanamido)-10,10-dimethyl-9,10-dihydro-10-sila-2-azaanthracene	84841-79-2	NI36163
206	261	344	70	55	189	205	175	344	100	19	11	10	9	7	6	6		18	25	3	–	–	–	–	–	–	–	–	–	1	Pentaethylcymantrene	36180-27-5	NI36164
344	329	345	178	330	297	163	115	344	100	30	19	16	6	6	5	4		19	20	6	–	–	–	–	–	–	–	–	–	–	6H-Benzofuro[3,2-c][1]benzopyran, 6a,11a-dihydro-3,4,8,9-tetramethoxy-	20390-16-3	NI36165
285	326	286	254	270	284	253	239	344	100	36	19	18	14	10	10	10	4.00	19	20	6	–	–	–	–	–	–	–	–	–	–	Benzoic acid, 2,4-dimethoxy-6-[(4-methoxyphenyl)methyl]-, anhydride with acetic acid	55429-38-4	NI36166
137	180	208	344	193	179	285	163	344	100	97	96	53	42	40	32	28		19	20	6	–	–	–	–	–	–	–	–	–	–	[2]Benzopyrano[4,3-b][1]benzopyran-6a(7H)-ol, 5,12a-dihydro-2,3,10-trimethoxy-	14991-62-9	NI36167
192	193	177	164	153	191	121	77	344	100	14	14	14	11	9	9	9	7.00	19	20	6	–	–	–	–	–	–	–	–	–	–	[2]Benzopyrano[4,3-b][1]benzopyran-7-ol, 5,6a,7,12a-tetrahydro-2,3,10-trimethoxy-, [6aS-(6aα,7α,12aβ)]-	16877-49-9	NI36168
151	164	344	343	207	181	137	327	344	100	58	50	42	29	21	9	8		19	20	6	–	–	–	–	–	–	–	–	–	–	Chalcone, 2′-hydroxy-3,4,4′,6′-tetramethoxy-	10496-67-0	NI36169
181	121	237	344	329	313	135	314	344	100	84	81	80	79	79	74	67		19	20	6	–	–	–	–	–	–	–	–	–	–	Chalcone, 2′-hydroxy-α,4,4′,6′-tetramethoxy-, cis-	57601-13-5	NI36170
297	329	344	226	270	285	59	147	344	100	93	74	64	50	49	35	32		19	20	6	–	–	–	–	–	–	–	–	–	–	1,2-Dibenzofurandicarboxylic acid, 1,9b-dihydro-7-methoxy-4,9b-dimethyl-, dimethyl ester	56247-79-1	NI36171
287	329	270	285	253	226	344	298	344	100	60	29	24	24	23	21	20		19	20	6	–	–	–	–	–	–	–	–	–	–	1,2-Dibenzofurandicarboxylic acid, 3,9b-dihydro-7-methoxy-4,9b-dimethyl-, dimethyl ester	23911-61-7	NI36172
344	329	161	343	208	221			344	100	33	14	10	4	2				19	20	6	–	–	–	–	–	–	–	–	–	–	Philenopteran dimethyl ether		LI07615
195	148	180	149	120	166	152	139	344	100	99	40	39	38	33	27	24	13.60	19	20	6	–	–	–	–	–	–	–	–	–	–	1,2-Propandione, 3-(4-methoxyphenyl)-1-(2,4,6-trimethoxyphenyl)-	54764-73-7	NI36173
317	193	151	302	319	195	158	304	344	100	80	52	48	38	31	17	16	4.00	19	21	2	2	–	–	1	–	–	–	–	–	–	1-Cyano-N,N-dimethyl-1-(2-chlorophenyl)-2-(3,4-dimethoxyphenyl)ethylamine		LI07616
108	122	167	109	181	123	120	121	344	100	60	29	17	14	14	12	11	7.00	19	24	4	2	–	–	–	–	–	–	–	–	–	[3,5-Dimethyl-4-(2-carboxyethyl)-2-pyrrolyl][3,5-dimethyl-4-(2-carboxyethyl)-2-pyrrolidene]methane		LI07617
273	344	256	329	300	255	299	345	344	100	85	31	30	30	30	29	20		19	24	4	2	–	–	–	–	–	–	–	–	–	[3,5-Dimethyl-4-ethoxycarbonyl-2-pyrrolyl][3,5-dimethyl-4-ethoxycarbonyl-2-pyrrolylidene]methane		LI07618
344	73	179	345	157	343	346	329	344	100	85	42	34	24	22	12	11		19	28	2	–	–	–	–	–	–	2	–	–	–	Silane, [methylenebis(p-phenyleneoxy)]bis[trimethyl-	18641-44-6	NI36174
346	73	179	347	157	329	180	74	344	100	85	42	28	24	11	7	7	0.00	19	28	2	–	–	–	–	–	–	2	–	–	–	Silane, [methylenebis(p-phenyleneoxy)]bis[trimethyl-	18641-44-6	NI36175
56	69	74	44	87	83	70	98	344	100	84	78	73	70	70	69	68	0.00	19	36	5	–	–	–	–	–	–	–	–	–	–	1,2,4-Trioxolane-2-octanoic acid, 5-octyl-, methyl ester	55398-23-7	NI36176
100	142	112	216	128	287	86	97	344	100	56	28	27	25	22	22	17	2.00	19	41	1	2	–	–	–	–	–	–	1	–	–	1-[1-Amino-3-(dibutylamino)propane]-2,2,3,4,4-pentamethylphosphetane 1 oxide		NI36177
195	221	205	105	77	207	196	151	344	100	83	58	55	28	24	23	23	1.00	20	24	5	–	–	–	–	–	–	–	–	–	–	Benzene, 5-benzoyloxy-1,3-dimethoxy-4-methyl-6-(1-methyl-3-hydroxypropyl)-		LI07619
329	330	344	311	345	157	77	43	344	100	22	19	5	4	4	3	3		20	24	5	–	–	–	–	–	–	–	–	–	–	2H,8H-Benzo[1,2-b:3,4-b′]dipyran-6-propanoic acid, 5-methoxy-2,2,8,8-tetramethyl-	26535-36-4	NI36178
329	330	344	311	345	157	127	43	344	100	22	20	5	4	4	3	3		20	24	5	–	–	–	–	–	–	–	–	–	–	2H,8H-Benzo[1,2-b:3,4-b′]dipyran-6-propanoic acid, 5-methoxy-2,2,8,8-tetramethyl-	26535-36-4	NI36179
329	330	344	311	345	157	43	269	344	100	22	19	6	5	5	4	3		20	24	5	–	–	–	–	–	–	–	–	–	–	2H,8H-Benzo[1,2-b:3,4-b′]dipyran-6-propanoic acid, 5-methoxy-2,2,8,8-tetramethyl-	26535-36-4	NI36180

MASS TO CHARGE RATIOS								M.W.	INTENSITIES								Parent	C	H	O	N	S	F	Cl	Br	I	Si	P	B	X	COMPOUND NAME	CAS Reg No	No
329	344	299	142	157	127	120	69	**344**	100	22	3	3	2	2	2	2		20	24	5	–	–	–	–	–	–	–	–	–	–	2H,8H-Benzo[1,2-b:5,4-b′]dipyran-10-propanoic acid, 5-methoxy-2,2,8,8-tetramethyl-	26535-35-3	MS08174
329	344	330	345	126	142	78	43	**344**	100	30	20	7	5	4	4	4		20	24	5	–	–	–	–	–	–	–	–	–	–	2H,8H-Benzo[1,2-b:5,4-b′]dipyran-10-propanoic acid, 5-methoxy-2,2,8,8-tetramethyl-	26535-35-3	NI36181
329	344	330	345	299	269	142	127	**344**	100	30	22	8	4	4	4	4		20	24	5	–	–	–	–	–	–	–	–	–	–	2H,8H-Benzo[1,2-b:5,4-b′]dipyran-10-propanoic acid, 5-methoxy-2,2,8,8-tetramethyl-	26535-35-3	NI36182
197	344	225	182	57	183	141	211	**344**	100	54	30	18	18	16	12	8		20	24	5	–	–	–	–	–	–	–	–	–	–	Diethyl α-(4-methoxy-1-naphthyl)-α-ethylmalonate	118657-04-8	DD01354
284	43	302	41	55	186	161	147	**344**	100	95	81	34	33	32	28	27	**22.78**	20	24	5	–	–	–	–	–	–	–	–	–	–	Equilin acetate glycol		NI36183
344	177	43	137	256	202	150	241	**344**	100	74	62	32	24	18	17	16		20	24	5	–	–	–	–	–	–	–	–	–	–	Estra-1,3,5(10)-trien-7-one, 17β-acetoxy-3-methoxy-6-oxa-8α-, (±)-		MS08175
99	100	344	300	69	55	86	57	**344**	100	34	14	14	11	11	10	9		20	24	5	–	–	–	–	–	–	–	–	–	–	Estra-1,3,5(10)-trien-7-one, 17,17-ethylenedioxy-3-methoxy-6-oxa-, (±)-		MS08176
105	121	156	129	128	136	115	77	**344**	100	79	61	61	61	57	57	54	**3.20**	20	24	5	–	–	–	–	–	–	–	–	–	–	Gibberellin A5 methyl ester	15355-45-0	IC04061
240	43	105	28	41	156	242	300	**344**	100	86	73	70	59	56	55	50	**15.00**	20	24	5	–	–	–	–	–	–	–	–	–	–	Gibberellin A5 methyl ester	15355-45-0	MS08177
281	222	221	223	91	282	41	155	**344**	100	62	51	43	43	36	35	34	**3.00**	20	24	5	–	–	–	–	–	–	–	–	–	–	Gibberellin A7 methyl ester	5508-47-4	MS08178
344	312	41	91	187	79	55	284	**344**	100	97	95	82	55	47	47	42		20	24	5	–	–	–	–	–	–	–	–	–	–	Gibberellin A11 methyl ester	55093-68-0	MS08179
213	43	228	176	71	214	99	229	**344**	100	34	29	20	14	13	13	12	**10.00**	20	24	5	–	–	–	–	–	–	–	–	–	–	(+)-Hexanoyllomatin		LI07620
249	95	81	41	344	43	55	91	**344**	100	94	88	72	65	65	59	55		20	24	5	–	–	–	–	–	–	–	–	–	–	4β-Hydroxyisobacchasmacranone		DD01355
344	41	298	297	283	43	39	53	**344**	100	88	67	62	56	48	45	44		20	24	5	–	–	–	–	–	–	–	–	–	–	Phenanthrene-1,4-dione, 4b,5,6,7,8,8a,9,10-octahydro-3,9,10-trihydroxy-4b,7-dimethyl-8-methylidene-2-(2-propenyl)-, (4bS,7R,8aS,9S,10S)-		DD01356
344	301	346	267	310	158	331	282	**344**	100	60	53	36	30	20	17	12		21	13	1	2	–	–	1	–	–	–	–	–	–	Naphtho[1,2-b]phenazin-5(8H)-one, 6-chloro-8-methyl-	52736-89-7	NI36184
344	194	165	166	77	345	180	343	**344**	100	74	73	49	25	24	21	20		21	16	1	2	1	–	–	–	–	–	–	–	–	1,5,5-Triphenyl-2-thioxo-4-imidazolidinone	52460-92-1	NI36185
344	135	180	225	165	103	77	345	**344**	100	69	57	33	33	33	33	24		21	16	1	2	1	–	–	–	–	–	–	–	–	3,5,5-Triphenyl-2-thioxo-4-imidazolidinone	52460-98-7	NI36186
344	345	220	343	327	329	152	316	**344**	100	24	12	9	9	5	5	4		21	16	3	2	–	–	–	–	–	–	–	–	–	1-Methylamino-4-(4-hydroxyanilino)anthraquinone		IC04062
343	344	105	77	297	193	165	51	**344**	100	73	49	48	18	13	12	11		21	16	3	2	–	–	–	–	–	–	–	–	–	3-(4-Nitrophenylamino)-1,2-diphenylpropen-1-one		MS08180
77	299	344	300	345	170	41	168	**344**	100	80	70	29	24	18	12	9		21	16	3	2	–	–	–	–	–	–	–	–	–	1H-Pyrazole-4-acetic acid, 5-(2-furanyl)-1,3-diphenyl-	36664-52-5	NI36187
344	209	194	77	208	224	93	251	**344**	100	96	80	64	60	48	44	36		21	20	1	4	–	–	–	–	–	–	–	–	–	5-Methyl-2-aminobenzophenone 4-phenyl-semicarbazone		MS08181
344	43	83	314	271	55	326	105	**344**	100	41	34	25	15	13	12	11		21	28	4	–	–	–	–	–	–	–	–	–	–	Anhydrohirundigenin	20853-99-0	LI07621
69	41	81	136	193	121	344	94	**344**	100	43	29	22	14	14	9	7		21	28	4	–	–	–	–	–	–	–	–	–	–	1,4-Benzoquinone, 2,4-dihydroxy-3-farnesyl-	34198-83-9	LI07622
344	345	202	176	137	255	189	346	**344**	100	25	9	7	7	5	5	4		21	28	4	–	–	–	–	–	–	–	–	–	–	Estradiol, 2-methoxy-, 17-acetate, (17β)-		NI36188
239	284	41	91	79	240	344	195	**344**	100	50	43	39	27	24	23	23		21	28	4	–	–	–	–	–	–	–	–	–	–	Gibberellin A15 methyl ester	13744-19-9	MS08182
260	259	302	344					**344**	100	96	10	7						21	28	4	–	–	–	–	–	–	–	–	–	–	Heptadeca-1,9-diene-4,6-diyne, 3,8-diacetoxy-	20853-99-0	LI07623
123	161	133	41	81	162	43	91	**344**	100	78	67	32	28	24	22	20	**0.90**	21	28	4	–	–	–	–	–	–	–	–	–	–	Monooxopyrethrin I		MS08183
344	329	301	313	262	247	261	316	**344**	100	55	38	36	36	34	33	26		21	28	4	–	–	–	–	–	–	–	–	–	–	3,9-Phenanthrenedione, 4b,5,6,7,8,8a-hexahydro-4-hydroxy-1-methoxy-4b,8,8-trimethyl-2-isopropyl-, (4bS-trans)-	60371-71-3	NI36189
137	108	43	150	41	149	135	109	**344**	100	51	40	38	38	37	32	30	**17.02**	21	28	4	–	–	–	–	–	–	–	–	–	–	Picrasa-1,13-diene-3,16-dione, 15-hydroxy-4-methyl-	6072-93-1	NI36190
216	43	326	310	201	283	308	295	**344**	100	62	39	32	23	22	18	10	**0.00**	21	28	4	–	–	–	–	–	–	–	–	–	–	Pregn-16-en-20-one, (+)-2α,3α:7α,8α-diepoxy-12β-hydroxy-, (5α)-		DD01357
257	344	43	259	301	201	231	345	**344**	100	95	67	63	47	34	32	26		21	28	4	–	–	–	–	–	–	–	–	–	–	Pregn-4-en-3,15,20-trione, 17-hydroxy-, (17α)-		MS08184
313	344	77	91	314	315	282	186	**344**	100	70	32	31	25	24	22	22		21	32	2	2	–	–	–	–	–	–	–	–	–	Gon-4-ene, 13β-ethyl-3,17-dimethoximino-		NI36191
215	186	69	174	185	108	81	67	**344**	100	7	7	6	5	4	4	4	**1.67**	21	32	2	2	–	–	–	–	–	–	–	–	–	1H-Pyrido[3,4-b]indole-1-butanol, γ-sec-butyl-2,3,4,9-tetrahydro-6-methoxy-2-methyl-	14358-60-2	NI36192
58	43	89	57	45	71	72	41	**344**	100	65	60	59	58	56	46	40	**8.58**	21	44	3	–	–	–	–	–	–	–	–	–	–	Chimilether		NI36193
344	329	314	330	345	76	202	133	**344**	100	95	27	24	23	17	14	12		22	16	4	–	–	–	–	–	–	–	–	–	–	2,2′-Binaphthalene, 1′4′-dimethoxy-1,4-dioxo-	54808-07-0	NI36194
313	344	329	314	345	202	172	165	**344**	100	87	53	32	20	17	12	12		22	16	4	–	–	–	–	–	–	–	–	–	–	2,2′-Binaphthylidene, 4,4′-dimethoxy-1,1′-dioxo-1,1′,2,2′-tetrahydro-	1244-92-4	NI36195
173	344	105	77	174	144	115	106	**344**	100	31	28	27	22	21	15	13		22	16	4	–	–	–	–	–	–	–	–	–	–	Dibenzo[b,h]biphenylene-5,6,11,12-tetrone, 5a,5b,11a,11b-tetrahydro-5a,11a-dimethyl-	55721-23-8	NI36196
172	40	344	104	76	173	144	115	**344**	100	60	33	28	27	23	20	15		22	16	4	–	–	–	–	–	–	–	–	–	–	Dibenzo[b,h]biphenylene-5,6,11,12-tetrone, 5a,5b,11a,11b-tetrahydro-5a,11a-dimethyl-	55721-23-8	MS08185
43	260	202	203	200	201	231	302	**344**	100	89	70	24	23	19	18	8	**6.00**	22	16	4	–	–	–	–	–	–	–	–	–	–	syn-Dibenzofulvalene, 3,3′-diacetoxy-		LI07624
251	183	77	37	252	102	155	65	**344**	100	77	40	20	19	18	17	13	**0.50**	22	16	4	–	–	–	–	–	–	–	–	–	–	Propanedioic acid, benzylidene-, diphenyl ester	25601-04-1	NI36197
251	183	77	252	155	102	36		**344**	100	77	39	20	17	17	17		**0.50**	22	16	4	–	–	–	–	–	–	–	–	–	–	Propanedioic acid, benzylidene-, diphenyl ester	25601-04-1	LI07625
108	119	91	344	225	226	345	107	**344**	100	50	44	43	31	17	13	13		22	20	2	2	–	–	–	–	–	–	–	–	–	N,N′-Bis[(2-methoxyphenyl)methylene]-1,4-benzenediamine	24588-85-0	NI36198
168	285	344	167	286	283	169	132	**344**	100	50	30	10	7	7	7	5		22	20	2	2	–	–	–	–	–	–	–	–	–	Methyl 1,2,3,4-tetrahydro-4-indol-3-ylcarbazole-1-carboxylate		NI36199
344	238	251	224	345	327	77	343	**344**	100	53	35	31	27	14	14	13		22	20	2	2	–	–	–	–	–	–	–	–	–	N,N′-Bis(salicylidene)-4,5-dimethyl-O-phenylenediamine	91585-59-0	NI36200
147	209	344	223	198	329	183	157	**344**	100	63	51	29	22	19	16	15		22	24	–	4	–	–	–	–	–	–	–	–	–	6,12-Dihydro-6a,7,13a,14-tetramethylquinoxalino[2,3-b]phenazine		LI07626
172	344	301	173	345	302	130	303	**344**	100	93	63	45	44	44	43	40		22	24	–	4	–	–	–	–	–	–	–	–	–	Hodgkinsine	18210-71-4	MS08186
271	270	189	175	69	201	344	272	**344**	100	27	23	23	22	18	16	14		22	32	3	–	–	–	–	–	–	–	–	–	–	Ferniginol, 15-ethoxycarbonyl-	34323-09-6	LI07627
344	147	43	123	104	91	284	93	**344**	100	98	86	78	65	62	61	60		22	32	3	–	–	–	–	–	–	–	–	–	–	A-Norandrostan-3-one, 17β-acetoxy-5α-vinyl-		MS08187

MASS TO CHARGE RATIOS	M.W.	INTENSITIES	Parent	C	H	O	N	S	F	Cl	Br	I	Si	P	B	X	COMPOUND NAME	CAS Reg No	No
43 344 55 243 79 229 91 123	**344**	100 63 55 50 43 38 37 33		22	32	3	–	–	–	–	–	–	–	–	–	–	18-Norpregn-4-ene-3,20-dione, 13-ethyl-17-hydroxy-	6030-25-7	NI36201
43 344 55 301 243 79 229 91	**344**	100 63 56 51 50 44 38 38		22	32	3	–	–	–	–	–	–	–	–	–	–	18-Norpregn-4-ene-3,20-dione, 13-ethyl-17-hydroxy-	6030-25-7	NI36202
344 269 329 227 345 270 161 330	**344**	100 69 39 22 17 11 8 7		22	32	3	–	–	–	–	–	–	–	–	–	–	4a(2H)-Phenanthrenecarboxylic acid, 1,3,4,9,10,10a-hexahydro-6-methoxy-1,1-dimethyl-7-isopropyl-, methyl ester, (4aR-trans)-	57397-34-9	NI36203
302 269 303 43 344 147 199 41	**344**	100 44 22 19 18 15 13 13		22	32	3	–	–	–	–	–	–	–	–	–	–	2,6-Phenanthrenediol, 1,2,3,4,4a,9,10,10a-octahydro-1,1,4a-trimethyl-7-isopropyl-, 6-acetate, [2S-(2α,4aα,10aβ)]-	18326-14-2	NI36204
201 284 43 41 268 55 188 69	**344**	100 94 84 58 50 46 37 30	**0.80**	22	32	3	–	–	–	–	–	–	–	–	–	–	4(1H)-Phenanthrenone, 9-(acetyloxy)-2-vinyl-2,3,4b,5,6,7,8,8a,9,10-decahydro-2,4b,8,8-tetramethyl-, [2R-(2α,4bβ,8aα,9β)]-	41756-33-6	NI36205
302 269 303 43 344 147 199 41	**344**	100 45 23 20 19 15 14 14		22	32	3	–	–	–	–	–	–	–	–	–	–	Podocarpa-8,11,13-triene-3β,12-diol, 13-isopropyl-, monoacetate	27216-72-4	NI36206
271 270 189 175 69 201 344 272	**344**	100 27 24 23 23 19 16 14		22	32	3	–	–	–	–	–	–	–	–	–	–	Podocarpa-8,11,13-trien-17-oic acid, 12-hydroxy-13-isopropyl-, ethyl ester	24067-43-4	NI36207
43 344 55 109 85 233 329 207	**344**	100 78 48 40 38 22 19 13		22	32	3	–	–	–	–	–	–	–	–	–	–	Pregnane-3,11,20-trione, 16-methyl-, (5α,16α)-		LI07628
43 55 85 344 67 260 329 246	**344**	100 37 33 28 28 15 12 11		22	32	3	–	–	–	–	–	–	–	–	–	–	Pregnane-3,11,20-trione, 16-methyl-, (5β,16α)-		LI07629
326 311 55 316 114 91 231 123	**344**	100 70 70 60 57 57 53 50	**23.00**	22	32	3	–	–	–	–	–	–	–	–	–	–	Pregn-4-ene-3,20-dione, 16-hydroxy-17-methyl-, (16α,17α)-		LI07630
43 301 243 41 344 91 55 283	**344**	100 96 42 34 31 31 28 27		22	32	3	–	–	–	–	–	–	–	–	–	–	Pregn-4-ene-3,20-dione, 17-hydroxy-2-methyl-, (2α)-	56145-30-3	NI36208
43 301 243 344 137 283 55 91	**344**	100 81 61 49 43 36 36 31		22	32	3	–	–	–	–	–	–	–	–	–	–	Pregn-4-ene-3,20-dione, 17-hydroxy-6-methyl-, (6α)-	520-85-4	NI36209
138 137 316 243 344 136 109 71	**344**	100 67 66 56 45 32 31 31		22	32	3	–	–	–	–	–	–	–	–	–	–	Pregn-4-ene-3,20-dione, 17-hydroxy-6-methyl-, (6α)-	520-85-4	NI36210
43 301 243 344 137 283 55 91	**344**	100 82 62 50 44 38 36 31		22	32	3	–	–	–	–	–	–	–	–	–	–	Pregn-4-ene-3,20-dione, 17-hydroxy-6-methyl-, (6α)-	520-85-4	NI36211
55 91 79 344 77 67 53 284	**344**	100 94 86 76 72 63 63 61		22	32	3	–	–	–	–	–	–	–	–	–	–	Pregn-4-ene-3,20-dione, 17-hydroxy-16-methyl-, (16α)-	2868-02-2	WI01450
55 91 79 344 77 67 53 284	**344**	100 94 86 76 72 63 63 61		22	32	3	–	–	–	–	–	–	–	–	–	–	Pregn-4-ene-3,20-dione, 17-hydroxy-16-methyl-, (16α,17α)-		AM00146
269 301 145 43 91 41 55 302	**344**	100 63 50 36 20 20 17 15	**1.26**	22	32	3	–	–	–	–	–	–	–	–	–	–	Pregn-4-ene-3,20-dione, 17-methoxy-	13254-82-5	NI36212
269 301 145 43 91 41 55 79	**344**	100 64 50 36 21 20 18 16	**1.04**	22	32	3	–	–	–	–	–	–	–	–	–	–	Pregn-4-ene-3,20-dione, 17-methoxy-	13254-82-5	NI36213
43 284 129 149 165 95 41 55	**344**	100 76 67 56 55 44 44 43	**26.00**	22	32	3	–	–	–	–	–	–	–	–	–	–	Retinol, 5,6-epoxy-5,6-dihydro-, acetate	801-72-9	NI36214
43 284 165 149 269 95 121 344	**344**	100 99 80 57 53 51 48 47		22	32	3	–	–	–	–	–	–	–	–	–	–	Retinol, 5,8-epoxy-5,8-dihydro-, acetate	744-98-9	NI36215
111 174 234 159 133 147 131 145	**344**	100 66 52 34 28 19 19 17	**1.87**	22	32	3	–	–	–	–	–	–	–	–	–	–	Spiro[bicyclo[3.1.0]hexane-6,3′(2′H)-as-indacen]-2-one, decahydro-6′-hydroxy-1,4,5′a-trimethyl-, acetate	6390-15-4	NI36216
147 344 145 149 108 160 284 133	**344**	100 82 60 50 32 31 28 13		22	32	3	–	–	–	–	–	–	–	–	–	–	Vouacapane, 7β-acetoxy-	18210-71-4	NI36217
172 91 344 173 170 82 130 65	**344**	100 80 17 11 9 8 5 5		23	24	1	2	–	–	–	–	–	–	–	–	–	3-Benzyl-2,3,4,5,6,7,8,9-octahydro-2,6-methano-1H-azecino[5,4-b]indol-8-one		LI07631
344 215 329 233 313 326 301 287	**344**	100 35 34 16 12 11 10 9		23	36	2	–	–	–	–	–	–	–	–	–	–	Androstan-17-one, 3-hydroxy-16-isobutylidene-, (3α,5α)-		LI07632
43 41 95 269 55 135 67 81	**344**	100 30 26 25 25 20 19 17	**3.00**	23	36	2	–	–	–	–	–	–	–	–	–	–	Androst-5-en-3β-ol, 4,4-dimethyl-, acetate	7673-18-9	NI36218
41 57 55 91 311 105 43 79	**344**	100 69 63 42 39 36 36 35	**31.25**	23	36	2	–	–	–	–	–	–	–	–	–	–	Androst-5-en-17-one, 16-tert-butyl-3-hydroxy-, (3β,16α)-	55162-58-8	NI36219
41 289 329 312 290 344 286 330	**344**	100 95 54 44 20 18 15 13		23	36	2	–	–	–	–	–	–	–	–	–	–	Cyclopregnane-20-carbaldehyde, 6-methoxy-, 20S-(3α,5α,6β)-		MS08188
91 261 81 95 79 55 93 43	**344**	100 77 68 66 63 61 60 60	**19.00**	23	36	2	–	–	–	–	–	–	–	–	–	–	3,5-Cyclopregnane-20-carboxaldehyde, 6-methoxy-, (3β,5α,6β)-	56298-02-3	NI36220
57 316 301 317 28 260 41 29	**344**	100 95 23 21 21 12 11 11	**1.00**	23	36	2	–	–	–	–	–	–	–	–	–	–	1,4-Epoxynaphthalene-1(2H)-methanol, 4,5,7-tri-tert-butyl-3,4-dihydro-	56771-86-9	MS08189
91 119 92 131 105 327 117 175	**344**	100 22 9 8 6 4 4 3	**0.30**	23	36	2	–	–	–	–	–	–	–	–	–	–	Palmitoleic acid, benzyl ester	77509-01-4	NI36221
91 55 69 41 43 83 97 67	**344**	100 44 36 36 27 24 17 15	**0.00**	23	36	2	–	–	–	–	–	–	–	–	–	–	Palmitoleic acid, benzyl ester	77509-01-4	NI36222
58 70 287 44 272 288 83 59	**344**	100 65 27 15 10 7 7 5	**1.00**	23	40	–	2	–	–	–	–	–	–	–	–	–	Pregn-5-ene-3,20-diamine, N,N′-dimethyl-, (3β,20S)-	21632-03-1	NI36223
344 172 138 315 290 288 289 166	**344**	100 12 8 7 4 4 2 2		24	12	1	2	–	–	–	–	–	–	–	–	–	7H-Acenaphth[1′,2′:4,5]imidazo[1,2-b]benz[d,e]isoquinolinone		LI07633
243 259 165 244 260 167 166 241	**344**	100 97 62 57 43 20 18 10	**5.00**	24	24	2	–	–	–	–	–	–	–	–	–	–	Pivalic acid, trityl ester	24523-65-7	NI36224
255 256 326 55 132 91 43 41	**344**	100 21 15 9 5 5 5 5	**1.01**	24	40	1	–	–	–	–	–	–	–	–	–	–	Chol-7-en-12-ol, (5β,12α)-	54411-73-3	NI36225
344 55 41 257 329 215 109 95	**344**	100 45 42 41 38 36 35 34		24	40	1	–	–	–	–	–	–	–	–	–	–	Chol-8(14)-en-24-ol, (5β)-	54411-88-0	NI36226
259 260 217 109 121 135 111 107	**344**	100 21 19 16 9 8 7 7	**0.50**	24	40	1	–	–	–	–	–	–	–	–	–	–	24-Norcholan-16-one, 22-methyl-, (5α)-	54498-41-8	NI36227
344 259 345 55 217 57 43 41	**344**	100 28 26 26 24 23 23 21		24	40	1	–	–	–	–	–	–	–	–	–	–	1-Pentanone, 1-[(5α)-androstan-17-yl]-	54482-52-9	NI36228
344 259 57 287 326 217 149 41	**344**	100 50 33 31 28 25 23 23		24	40	1	–	–	–	–	–	–	–	–	–	–	1-Pentanone, 1-[(5α,17β)-androstan-17-yl]-	54411-84-6	NI36229
91 329 105 344 129 143 157 185	**344**	100 96 92 84 57 50 38 28		25	28	1	–	–	–	–	–	–	–	–	–	–	syn-2,2,7,7-Tetramethyl-1,6-diphenyldispiro[2.1.2.2]nonan-4-one		MS08190
329 344 133 157 330 158 345 211	**344**	100 61 56 44 27 22 18 17		25	28	1	–	–	–	–	–	–	–	–	–	–	(E)-2,5-Bis(2,4,6-trimethylphenylmethylene)cyclopentanone		MS08191
133 329 157 344 158 142 330 211	**344**	100 87 62 51 39 27 24 20		25	28	1	–	–	–	–	–	–	–	–	–	–	(E,Z)-2,5-Bis(2,4,6-trimethylphenylmethylene)cyclopentanone		MS08192
329 133 344 157 330 158 211 159	**344**	100 39 37 34 27 17 10 8		25	28	1	–	–	–	–	–	–	–	–	–	–	(Z,Z)-2,5-Bis(2,4,6-trimethylphenylmethylene)cyclopentanone		MS08193
135 344 231 95 67 81 43 149	**344**	100 63 61 55 48 42 40 31		25	44	–	–	–	–	–	–	–	–	–	–	–	Benz[de]anthracene, 6-octylhexadecahydro-		AI02100
135 231 95 344 81 121 79 189	**344**	100 62 55 47 46 31 30 29		25	44	–	–	–	–	–	–	–	–	–	–	–	Benz[de]anthracene, 6-octylhexadecahydro-		AI02101
133 105 43 344 41 91 134 104	**344**	100 50 47 46 30 29 19 16		25	44	–	–	–	–	–	–	–	–	–	–	–	Benzene, hexadecylpropyl-	72101-17-8	WI01451
92 91 43 57 41 55 71 93	**344**	100 48 30 19 19 13 10 8	**2.08**	25	44	–	–	–	–	–	–	–	–	–	–	–	Heptadecane, 9-(2-phenylethyl)-		AI02102
92 91 43 41 57 55 71 44	**344**	100 47 32 22 21 15 10 9	**1.94**	25	44	–	–	–	–	–	–	–	–	–	–	–	Heptadecane, 9-(2-phenylethyl)-		AI02103
91 90 57 43 71 92 85 55	**344**	100 33 18 15 11 7 7 7	**1.00**	25	44	–	–	–	–	–	–	–	–	–	–	–	Heptadecane, 9-(2-phenylethyl)-		AI02104
91 92 344 41 57 29 55 105	**344**	100 99 30 27 22 18 16 13		25	44	–	–	–	–	–	–	–	–	–	–	–	Phenylnonadecane		MS08194

MASS TO CHARGE RATIOS								M.W.	INTENSITIES								Parent	C	H	O	N	S	F	Cl	Br	I	Si	P	B	X	COMPOUND NAME	CAS Reg No	No
245	344	241	41	229	228	243	27	**344**	100	69	34	29	27	27	25	24		26	32	–	–	–	–	–	–	–	–	–	–	–	9-Octyl-1,2,3,4-tetrahydronaphthacene		AI02105
344	267	345	265	165	343	266	252	**344**	100	45	30	27	15	12	10	10		27	20	–	–	–	–	–	–	–	–	–	–	–	Tetraphenylpropyne	20143-13-9	NI36230
344	267	345	265	165	343	268	266	**344**	100	45	29	28	15	11	10	10		27	20	–	–	–	–	–	–	–	–	–	–	–	Tetraphenylpropyne	20143-13-9	LI07634
344	342	265	267	163	78			**344**	100	57	43	39	9	8				27	20	–	–	–	–	–	–	–	–	–	–	–	1,1,3-Triphenylindene		LI07635
344	342	267	265	163	78			**344**	100	55	42	40	10	8				27	20	–	–	–	–	–	–	–	–	–	–	–	1,2,3-Triphenylindene		LI07636
310	345	146	73	240	111	216	76	**345**	100	26	12	11	8	4	3	3		–	–	–	3	–	–	6	–	–	–	3	–	–	1,3,5,2,4,6-Triazatriphosphorine, 2,2,4,4,6,6-hexachloro-2,2,4,4,6,6-hexahydro-	940-71-6	NI36231
79	206	70	41	163	74	73	71	**345**	100	49	45	45	31	30	23	23	**0.28**	8	15	8	3	2	–	–	–	–	–	–	–	–	α-D-Xylopyranoside, methyl-4-azido-4-deoxy-, 2,3-dimethanesulphonate	6160-92-5	NI36232
103	73	212	315	147	211	330	240	**345**	100	73	44	36	33	28	27	27	**0.00**	11	22	4	1	–	3	–	–	–	2	–	–	–	L-Serine, N-(trifluoroacetyl)-O-(trimethylsilyl)-, trimethylsilyl ester	52558-12-0	NI36233
132	77	160	105	29	76	65	104	**345**	100	88	79	31	29	27	26	25	**0.00**	12	16	3	3	2	–	–	–	–	–	1	–	–	Phosphorodithioic acid, O,O-diethyl S-[(4-oxo-1,2,3-benzotriazin-3(4H)-yl)methyl] ester	2642-71-9	NI36234
134	285	289	317	274	136	164	78	**345**	100	26	23	22	21	18	14	13	**5.00**	13	18	2	1	–	2	–	–	–	–	1	–	1	Iron, acetylcarbonyl(η^5-2,4-cyclopentadien-1-yl)(1-piperidinylphosphonous difluoride-P)-	37725-98-7	MS08195
43	41	274	316	63	57	42	39	**345**	100	64	45	39	17	13	13	13	**2.25**	13	19	6	3	1	–	–	–	–	–	–	–	–	Aniline, 4-(methylsulphonyl)-2,6-dinitro-N,N-dipropyl-	4726-14-1	NI36235
316	39	274	38	27	317	75	61	**345**	100	81	74	51	33	14	14	14	**7.05**	13	19	6	3	1	–	–	–	–	–	–	–	–	Aniline, 4-(methylsulphonyl)-2,6-dinitro-N,N-dipropyl-	4726-14-1	NI36236
176	147	345	91	132	119	148	69	**345**	100	45	29	23	21	21	17	15		14	14	1	1	–	7	–	–	–	–	–	–	–	N-Heptafluorobutyryl-2,6-diethylaniline		NI36237
107	133	69	132	78	73	77	60	**345**	100	93	56	47	42	38	34	32	**0.00**	14	19	7	1	1	–	–	–	–	–	–	–	–	Desulphoglucosinalbin		MS08196
120	254	73	211	75	147	91	135	**345**	100	65	61	52	38	24	22	15	**1.30**	14	28	3	1	–	–	–	–	–	2	1	–	–	Phosphonic acid, (1-amino-2-phenylethyl)-, bis(trimethylsilyl) ester	53044-34-1	NI36238
120	254	73	211	75	147	119	91	**345**	100	65	61	52	38	24	24	22	**1.30**	14	28	3	1	–	–	–	–	–	2	1	–	–	Phosphonic acid, (1-amino-2-phenylethyl)-, bis(trimethylsilyl) ester	53044-34-1	NI36239
345	245	163	246	263	164	145	127	**345**	100	80	64	55	37	36	18	17		15	21	6	–	–	–	–	–	–	–	–	–	1	Tris(acetylacetonato)titanium		LI07637
84	158	58	142	203	71	143	115	**345**	100	74	47	42	30	28	26	26	**0.00**	15	24	–	1	–	–	–	–	1	–	–	–	–	Cyclohexanaminium, N,N,N-trimethyl-2-phenyl-, iodide, cis-	52182-68-0	NI36240
84	142	203	58	85	42	158	56	**345**	100	39	23	15	8	7	5	5	**0.00**	15	24	–	1	–	–	–	–	1	–	–	–	–	Cyclohexanaminium, N,N,N-trimethyl-2-phenyl-, iodide, trans-	52182-67-9	NI36241
58	218	219	59	119	117	91	115	**345**	100	40	9	7	6	5	4	4	**0.00**	15	24	–	1	–	–	–	–	1	–	–	–	–	2,2,10-Trimethyl-2,3,4,5,6,7-hexahydro-1H-2-benzazoninium iodide		NI36242
128	40	142	56	199	42	41	44	**345**	100	55	50	46	35	34	34	32	**8.50**	15	27	6	3	–	–	–	–	–	–	–	–	–	Pentanedioic acid, 2-amino-2-[[3-(2-methoxycarbonylazetidin-1-yl)propyl]amino]-, dimethyl ester		LI07638
84	56	41	100	57	94	122	55	**345**	100	78	77	59	43	42	40	40	**0.00**	15	27	6	3	–	–	–	–	–	–	–	–	–	Trimethyl hexahydro-1,3,5-triazine-2,4,6-tripropanoate	77526-16-0	NI36243
345	302	29	41	57	242	196	256	**345**	100	71	53	36	32	28	28	25		16	15	4	3	1	–	–	–	–	–	–	–	–	10H-Phenothiazine, 10-butyl-3,7-dinitro-	35076-83-6	LI07639
345	302	29	41	57	346	196	242	**345**	100	67	50	33	31	27	27	25		16	15	4	3	1	–	–	–	–	–	–	–	–	10H-Phenothiazine, 10-butyl-3,7-dinitro-	35076-83-6	NI36244
148	91	43	149	103	216	147	65	**345**	100	83	54	22	22	18	14	14	**0.00**	17	22	3	1	–	3	–	–	–	–	–	–	–	L-Phenylalanine, N-(trifluoroacetyl)-, 1-methylpentyl ester	58072-49-4	NI36245
330	231	73	175	98	57	114	345	**345**	100	25	18	14	9	8	7	2		17	32	–	5	–	–	–	–	–	1	–	1	–	1-tert-Butyl-5-[tert-butyl(trimethylsilyl)amino]-4-phenyl-2,3-dehydrotetrazaboroline		MS08197
253	190	345	252	346	254	176	224	**345**	100	85	83	57	18	17	11	9		18	19	4	1	1	–	–	–	–	–	–	–	–	5H-2-Benzazepin-5-one, 1,2,3,4-tetrahydro-8-methoxy-2-[(4-methylphenyl)sulphonyl]-	24310-35-8	NI36246
190	345	147	346	176	175	281	280	**345**	100	25	11	5	5	4	1	1		18	19	4	1	1	–	–	–	–	–	–	–	–	4(1H)-Quinolone, 2,3-dihydro-6-methoxy-2-methyl-1-(p-tolylsulphonyl)-	24310-39-2	NI36247
316	317	231	301	259	345			**345**	100	78	55	16	9	2				18	19	6	1	–	–	–	–	–	–	–	–	–	Crinan-11-one, 1,2-epoxy-3,7-dimethoxy-, (1β,2β,3α)-	58189-39-2	NI36248
144	115	89	231	116	345	87	200	**345**	100	85	43	28	20	15	15	13		18	19	6	1	–	–	–	–	–	–	–	–	–	Indole, 1,3-bis(2-methoxycarbonylpropionyl)-		LI07640
125	42	96	41	126	124	44	53	**345**	100	31	24	9	8	8	8	6	**0.04**	18	19	6	1	–	–	–	–	–	–	–	–	–	Lycorenan-7-one, 5-hydroxy-11-methoxy-1-methyl-9,10-[methylenebis(oxy)]-, (5α)-	1167-58-4	NI36249
72	345	56	300	301	121	44	330	**345**	100	80	48	42	38	36	34	28		18	27	–	1	1	–	–	–	–	–	–	–	1	(R,S)-1-((1-Dimethylamino)ethyl)-2-(isobutylthio)ferrocene		MS08198
84	98	107	108	56	70	155	42	**345**	100	76	61	58	57	56	38	33	**11.00**	18	27	2	5	–	–	–	–	–	–	–	–	–	(±)-2-Amino-5-[3-[3-(piperidinomethyl)phenoxy]-2-methylpropylamino]-3,4-oxadiazole		MS08199
28	77	38	151	39	51	91	95	**345**	100	93	87	86	82	82	79	76	**0.69**	19	18	3	–	–	–	–	–	–	–	–	–	1	Methylvanadium triphenoxide		NI36250
245	213	345	230	243	244	242	196	**345**	100	60	50	34	32	31	18	12		19	23	5	1	–	–	–	–	–	–	–	–	–	Hasubanan-7-one, 8,10-epoxy-8-hydroxy-3,4-dimethoxy-17-methyl-, (8β,10β)-	1805-86-3	NI36251
109	108	110	107	94	82	81	42	**345**	100	10	7	1	1	1	1	1	**0.04**	19	23	5	1	–	–	–	–	–	–	–	–	–	Lycorenan-7-one, 8,9,10-trimethoxy-1-methyl-	668-63-3	NI36252
135	226	198	345	346	276	300	299	**345**	100	76	27	18	7	6	3	3		19	23	5	1	–	–	–	–	–	–	–	–	–	Pyrrole-2-carboxylic acid, 3-(4-methoxyphenyl)-4-(methoxycarbonylethyl)-5-methyl-, ethyl ester		LI07641
345	105	77	346	288	51	231	239	**345**	100	65	60	25	17	15	14	12		20	15	3	3	–	–	–	–	–	–	–	–	–	2-Phenyl-4-benzoyl-6-methyl-3H-imidazo[1,5-b]pyridazine-5,7-dione		NI36253
286	287	270	194	272	256	132	314	**345**	100	94	73	56	55	30	27	20	**16.00**	20	27	4	1	–	–	–	–	–	–	–	–	–	Dihydroepiwilsonine		NI36254
194	165	151	195	166	43			**345**	100	91	16	9	9	9			**0.00**	20	27	4	1	–	–	–	–	–	–	–	–	–	2,2'-Bis(3,4-dimethoxyphenoxy)diethylamine		LI07642
121	242	345	312	243	77	241	346	**345**	100	66	39	20	10	10	9	8		21	15	–	1	2	–	–	–	–	–	–	–	–	Benzenecarbothioamide, N-(3-phenylbenzo[b]thien-2-yl)-	40532-76-1	NI36255
345	330	118	181	183	259	182	275	**345**	100	95	65	57	43	33	27	19		21	19	2	1	–	–	–	–	–	1	–	–	–	1-Oxa-3-aza-2-silacyclopentan-5-one, 4-methyl-2,2,3-triphenyl-	21654-64-8	NI36256
345	193	91	328	208	346	254	299	**345**	100	70	62	40	34	24	20	18		21	19	2	3	–	–	–	–	–	–	–	–	–	3-(N,N-Dimethylamino)-4-nitro-9-benzylcarbazole	94127-26-1	NI36257
219	220	203	345	191	57	128	41	**345**	100	17	12	7	5	5	4	3		21	31	3	1	–	–	–	–	–	–	–	–	–	Cyclopentanone, 2-carbamyl-2-(3,5-di-tert-butyl-4-hydroxybenzyl)-		IC04063

MASS TO CHARGE RATIOS								M.W.	INTENSITIES								Parent	C	H	O	N	S	F	Cl	Br	I	Si	P	B	X	COMPOUND NAME	CAS Reg No	No
85	56	57	345	87	55	317	46	**345**	100	43	33	27	19	15	14	14		21	31	3	1	–	–	–	–	–	–	–	–	–	3-Aza-A-homopregnan-16-ol, 1,4-epoxy-20-methyl-, acetate (ester), (1α,4α,5β,16β)-	24206-15-3	LI07643
165	257	166	285	77	180	256	258	**345**	100	99	38	30	30	25	23	20	**2.50**	22	19	3	1	–	–	–	–	–	–	–	–	–	4-Oxazolidinone, 2-methoxy-3,5,5-triphenyl-	56440-41-6	NI36258
239	91	240	273	72	167	165	77	**345**	100	30	19	12	11	10	10	9	**2.60**	22	23	1	3	–	–	–	–	–	–	–	–	–	N,N-Dimethyl-2,2-diphenyl-2-phenylhydrazinoacetamide	72572-06-6	NI36259
100	296	312	327	91	66	105	87	**345**	100	75	67	61	60	54	52	44	**2.00**	22	35	2	1	–	–	–	–	–	–	–	–	–	Pregn-5-en-20-one, 3-hydroxy-, O-methyloxime, (3β)-		LI07644
117	91	208	345					**345**	100	47	30	10						23	23	2	1	–	–	–	–	–	–	–	–	–	Carbamic acid, N-phenyl, 2,2-dibenzylethyl ester		LI07645
89	43	41	55	95	82	67	85	**345**	100	33	19	15	14	14	14	13	**9.33**	23	39	1	1	–	–	–	–	–	–	–	–	–	Acetamide, N-[(3β,5α)-androstan-3-yl]-N-ethyl-	55320-49-5	NI36260
114	101	300	70	55	67	81	41	**345**	100	64	56	18	18	17	16	16	**7.00**	23	39	1	1	–	–	–	–	–	–	–	–	–	5α-Pregnan-2-one, ethyloxime		LI07646
299	345	298	287	149	148	300	296	**345**	100	84	66	66	59	35	25	22		24	11	2	1	–	–	–	–	–	–	–	–	–	1-Nitrocoronene	81316-84-9	NI36261
345	346	315	344	343	316	347		**345**	100	28	19	18	12	12	9			25	15	1	1	–	–	–	–	–	–	–	–	–	7-Phenylindolo[3,2,1-de]phenanthridin-13-one		LI07647
268	267	269	266	265	239	345	77	**345**	100	26	21	14	5	5	4	4		26	19	–	1	–	–	–	–	–	–	–	–	–	10,14b-Dihydro-14b-phenylisoindolo[2,1-f]phenanthridine		NI36262
230	311	313	276	350	348	346		**346**	100	47	17	2	1	1	1			–	–	–	4	1	5	2	–	–	–	3	–	–	N-(2,4,4,6,6-Pentafluoro-1,3,5,2,4,6-triazaphosphorin-2-yl) sulphur dichloride imide		LI07648
46	30	29	28	44	45	43	31	**346**	100	78	8	8	6	3	3	3	**0.00**	6	10	13	4	–	–	–	–	–	–	–	–	–	Diglycerol tetranitrate		MS08200
107	81	331	77	91	69	73	43	**346**	100	84	81	61	49	33	23	21	**0.00**	8	13	4	–	–	6	–	–	–	1	1	–	–	2-Oxo-2-[2,2,2-trifluoro-1-(trifluoromethyl)-1-(trimethylsiloxy)ethyl]-1,3,2$\lambda^5\sigma^4$-dioxaphospholane		MS08201
311	313	229	227	225	267	265	346	**346**	100	99	76	76	76	31	31	28		9	11	4	–	–	–	2	–	–	–	–	–	1	Niobium, dichloro(2-hydroxybenzaldehydato-O,O')dimethoxy-	52690-26-3	NI36263
311	225	227	313	265	77	346	51	**346**	100	57	38	33	31	30	21	20		9	11	4	–	–	–	2	–	–	–	–	–	1	Niobium, dichloro(2-hydroxybenzaldehydato-O,O')dimethoxy-	52690-26-3	NI36264
301	299	303	221	28	223	142	317	**346**	100	95	63	23	21	20	20	19	**5.73**	10	6	3	–	1	–	4	–	–	–	–	–	–	Benzoic acid, 2,3,5,6-tetrachloro-4-[(methylthio)carbonyl]-, methyl ester	3765-57-9	NI36265
288	318	346	290	260	249	223	262	**346**	100	25	20	20	20	10	6	4		10	10	2	–	–	–	–	–	–	–	–	–	1	(π-Allyl)dicarbonylcyclopentadienyltungsten		NI36266
29	131	226	100	27	245	178	45	**346**	100	58	57	30	27	22	13	12	**0.00**	10	10	4	–	–	8	–	–	–	–	–	–	–	Hexanedioic acid, octafluoro-, diethyl ester	376-50-1	NI36267
137	40	238	181	37	240	44	138	**346**	100	36	32	31	21	20	19	15	**4.73**	10	13	2	2	2	1	2	–	–	–	–	–	–	Sulphamide, N-[(dichlorofluoromethyl)thio]-N',N'-dimethyl-N-o-tolyl-	1087-94-1	NI36268
61	73	133	106	119	134	303	107	**346**	100	82	73	65	43	38	32	32	**6.00**	10	19	–	–	4	–	–	1	–	–	–	–	–	6-Bromo-1,4,8,11-tetrathiacyclotetradecane	91375-66-5	NI36269
69	346	133	68	165	67	261	313	**346**	100	60	29	16	13	7	2	1		10	20	5	–	2	–	–	–	–	–	2	–	–	Oxybis(5,5-dimethyl-1,3,2-dioxaphosphorinane 2-sulphide)	4090-51-1	NI36270
142	141	115	205	69	138	203	89	**346**	100	95	62	23	18	15	10	8	**5.00**	11	9	–	–	–	–	–	–	–	–	–	–	1	eta[5]-Phenylcyclopentadienylthallium(I)	90510-42-2	NI36271
82	305	304	306	349	303	307	346	**346**	100	87	69	58	53	50	47	40		11	12	1	2	–	–	–	2	–	–	–	–	–	3,5-Dibromcytisine		LI07649
43	41	27	317	275	29	28	39	**346**	100	66	55	38	27	23	13	12	**3.30**	12	18	6	4	1	–	–	–	–	–	–	–	–	Benzenesulphonamide, 4-(dipropylamino)-3,5-dinitro-	19044-88-3	NI36272
180	98	30	42	41	167	104	70	**346**	100	86	74	47	45	44	39	36	**11.50**	13	16	2	2	–	6	–	–	–	–	–	–	–	N,N'-Propylenebis(5,5,5-trifluoro-4-oxopentan-2-imine)	2730-31-6	NI36273
234	70	236	75	262	101	290	264	**346**	100	91	88	36	33	24	16	11	**0.00**	14	8	5	–	–	–	1	–	–	–	–	–	1	Tetracarbonyl[1-(4-chlorophenyl)-3-oxo-1-butenyl]manganese (OC-6-23)	122967-40-2	DD01358
346	86	29	274	68	347	102	316	**346**	100	48	28	24	24	17	14	12		14	10	5	4	1	–	–	–	–	–	–	–	–	Ethyl 2-(4-nitrophenyl)-7-oxo-7H-1,3,4-thiadiazolo[3,2-a]pyrimidine-5-carboxylate	82077-88-1	NI36274
110	121	111	93	149	77	147	91	**346**	100	15	9	9	8	8	7	7	**2.00**	14	16	1	–	1	–	1	1	–	–	–	–	–	1-[1-Bromo-2-(phenylthio)cycloprop-1-yl]-2-chlorocyclopentan-1-ol		MS08202
29	81	27	109	28	89	210	182	**346**	100	50	40	32	30	26	23	19	**15.71**	14	16	6	–	–	–	1	–	–	–	1	–	–	Phosphoric acid, 3-chloro-4-methyl-2-oxo-2H-1-benzopyran-7-yl diethyl ester	321-54-0	NI36275
81	89	109	346	91	210	63	182	**346**	100	57	55	50	45	44	35	34		14	16	6	–	–	–	1	–	–	–	1	–	–	Phosphoric acid, 3-chloro-4-methyl-2-oxo-2H-1-benzopyran-7-yl diethyl ester	321-54-0	NI36276
114	156	96	139	84	241	72	126	**346**	100	34	27	23	22	21	21	13	**3.00**	14	20	9	1	–	–	–	–	–	–	–	–	–	2-Acetamido-1,3,4,6-tetra-O-acetyl-2-deoxy-α-D-glucopyranose		LI07650
73	147	45	303	67	346	331	229	**346**	100	62	21	19	16	15	15	14		14	30	4	–	–	–	–	–	–	3	–	–	–	Acetopyruvic acid, tris(trimethylsilyl)-		NI36277
73	147	156	95	346	45	215	75	**346**	100	60	23	23	19	17	11	10		14	30	4	–	–	–	–	–	–	3	–	–	–	Glutaconic acid, tris(trimethylsilyl)- deriv.		NI36278
91	131	146	30	233	103	232	216	**346**	100	82	72	67	43	42	37	28	**20.30**	15	17	4	2	–	3	–	–	–	–	–	–	–	Glycine, N-[3-phenyl-N-(trifluoroacetyl)-L-alanyl]-, ethyl ester	57274-60-9	NI36279
162	88	120	91	140	131	163	161	**346**	100	42	29	28	21	19	14	11	**2.00**	15	17	4	2	–	3	–	–	–	–	–	–	–	DL-Phenylalanine, N-[N-(trifluoroacetyl)-L-alanyl]-, methyl ester		LI07651
162	140	120	91	131	88	171	163	**346**	100	34	26	26	19	19	15	12	**0.75**	15	17	4	2	–	3	–	–	–	–	–	–	–	L-Phenylalanine, N-[N-(trifluoroacetyl)-L-alanyl]-, methyl ester	329-36-2	NI36280
162	88	120	91	140	131	163	161	**346**	100	43	30	28	22	20	15	12	**2.00**	15	17	4	2	–	3	–	–	–	–	–	–	–	L-Phenylalanine, N-[N-(trifluoroacetyl)-L-alanyl]-, methyl ester	329-36-2	NI36281
88	91	162	120	140	69	55	70	**346**	100	82	65	50	30	30	26	25	**0.00**	15	17	4	2	–	3	–	–	–	–	–	–	–	L-Phenylalanine, N-[N-(trifluoroacetyl)-L-alanyl]-, methyl ester	329-36-2	NI36282
95	121	56	81	67	139	109	207	**346**	100	91	72	61	56	22	18	11	**0.00**	15	20	1	–	–	–	–	–	–	–	–	–	1	3-(Phenyltelluro)-1-oxaspiro[4.5]decane		DD01359
217	165	29	136	245	81	101	164	**346**	100	74	65	64	62	55	49	43	**17.02**	15	20	5	–	–	–	1	–	–	–	1	–	–	2-Propenoic acid, 3-(3-chlorophenyl)-2-(diethoxyphosphinyl)-, ethyl ester		NI36283
217	165	136	29	245	81	273	164	**346**	100	68	63	54	53	48	46	42	**14.46**	15	20	5	–	–	–	1	–	–	–	1	–	–	2-Propenoic acid, 3-(4-chlorophenyl)-2-(diethoxyphosphinyl)-, ethyl ester	13507-50-1	NI36284
43	331	28	143	168	115	128	109	**346**	100	99	45	41	38	35	28	28	**0.00**	15	22	9	–	–	–	–	–	–	–	–	–	–	α-D-Glucofuranose, 1,2-O-isopropylidene-, triacetate	29364-56-5	NI36285
184	111	100	124	93	69	83	153	**346**	100	88	77	66	57	55	53	33	**0.00**	15	22	9	–	–	–	–	–	–	–	–	–	–	D-Gluco-heptitol, 1,4,5,7-tetra-O-acetyl-2,6-anhydro-3-deoxy-		MS08203
111	69	83	124	100	71	153	81	**346**	100	99	60	50	40	40	35	30	**0.00**	15	22	9	–	–	–	–	–	–	–	–	–	–	D-Manno-heptitol, 1,4,5,7-tetra-O-acetyl-2,6-anhydro-3-deoxy-		MS08204
121	91	346	181	65	122	347	78	**346**	100	85	44	33	11	8	7	7		16	11	3	–	–	5	–	–	–	–	–	–	–	2-Methoxyphenylacetic acid, pentafluorobenzyl ester		MS08205
121	346	181	91	122	78	347	77	**346**	100	55	43	16	11	11	8	8		16	11	3	–	–	5	–	–	–	–	–	–	–	3-Methoxyphenylacetic acid, pentafluorobenzyl ester		MS08206
121	346	181	122	78	77	347	51	**346**	100	23	15	9	6	5	4	3		16	11	3	–	–	5	–	–	–	–	–	–	–	4-Methoxyphenylacetic acid, pentafluorobenzyl ester		MS08207
121	91	346	181	65	347	122	78	**346**	100	80	56	32	10	9	9	7		16	11	3	–	–	5	–	–	–	–	–	–	–	Phenylacetic acid, 2-methoxy-, pentafluorobenzyl ester		NI36286

MASS TO CHARGE RATIOS								M.W.	INTENSITIES								Parent	C	H	O	N	S	F	Cl	Br	I	Si	P	B	X	COMPOUND NAME	CAS Reg No	No
121	346	181	91	122	78	347	77	**346**	100	45	34	18	13	9	8	6		16	11	3	–	–	5	–	–	–	–	–	–	–	Phenylacetic acid, 3-methoxy-, pentafluorobenzyl ester		NI36287
121	346	181	122	347	78	77	135	**346**	100	33	16	9	6	5	5	4		16	11	3	–	–	5	–	–	–	–	–	–	–	Phenylacetic acid, 4-methoxy-, pentafluorobenzyl ester		NI36288
133	213	211	346	103	90	209	134	**346**	100	57	28	17	13	12	11	10		16	14	2	2	–	–	–	–	–	–	–	–	1	3,5-Bis(4-methoxyphenyl)-1,2,4-selenadiazole		MS08208
317	315	313	316	314	170	168	321	**346**	100	76	45	38	33	31	25	19	**0.00**	16	18	1	–	–	–	–	–	–	–	–	–	1	10,10-Diethylphenoxastannin		LI07652
155	127	174	301	99	229	227	199	**346**	100	85	50	26	18	16	15	14	**0.00**	16	26	8	–	–	–	–	–	–	–	–	–	–	Diethyl 2,3-dimethyl-2,3-bis(ethoxycarbonyl)succinate	49846-73-3	DD01360
174	127	173	160	301	255	55	99	**346**	100	90	81	44	32	30	30	28	**0.00**	16	26	8	–	–	–	–	–	–	–	–	–	–	Diethyl 2-methyl-2,4-bis(ethoxycarbonyl)pentanedioate	17696-77-4	DD01361
43	101	69	72	41	315	59	45	**346**	100	78	29	22	20	18	18	14	**0.00**	16	26	8	–	–	–	–	–	–	–	–	–	–	2-Furanhexanoic acid, tetrahydro-β,δ-dihydroxy-5-methoxy-, methyl ester, diacetate	18142-14-8	NI36289
101	153	111	273	58	85	83	115	**346**	100	68	52	23	21	17	17	14	**0.00**	16	26	8	–	–	–	–	–	–	–	–	–	–	1,2:3,4-Di-O-isopropylidene-5,6-di-O-acetyl-D-glucitol		NI36290
101	153	85	111	187	58	57	127	**346**	100	71	64	53	43	42	42	35	**0.00**	16	26	8	–	–	–	–	–	–	–	–	–	–	1,2:5,6-Di-O-isopropylidene-3,4-di-O-acetyl-D-glucitol	71696-36-1	NI36291
101	153	111	72	273	83	85	331	**346**	100	58	50	22	20	20	18	15	**0.00**	16	26	8	–	–	–	–	–	–	–	–	–	–	1,3:5,6-Di-O-isopropylidene-2,4-di-O-acetyl-D-glucitol		NI36292
153	101	115	111	58	85	331	273	**346**	100	97	92	82	40	36	31	27	**0.00**	16	26	8	–	–	–	–	–	–	–	–	–	–	2,3:5,6-Di-O-isopropylidene-1,4-di-O-acetyl-D-glucitol	73745-57-0	NI36293
101	85	153	58	111	127	143	331	**346**	100	83	59	47	46	44	40	29	**0.00**	16	26	8	–	–	–	–	–	–	–	–	–	–	3,4:5,6-Di-O-isopropylidene-1,2-di-O-acetyl-D-glucitol	109680-93-5	NI36294
101	85	111	153	59	143	187	127	**346**	100	86	53	50	50	47	46	40	**0.00**	16	26	8	–	–	–	–	–	–	–	–	–	–	1,2:3,4-Di-O-isopropylidene-5,6-di-O-acetyl-D-mannitol	62819-11-8	NI36295
85	115	101	59	111	153	127	187	**346**	100	91	79	79	71	56	47	32	**0.00**	16	26	8	–	–	–	–	–	–	–	–	–	–	1,2:3,6-Di-O-isopropylidene-4,5-di-O-acetyl-D-mannitol	92175-34-3	NI36296
101	115	153	111	59	85	157	273	**346**	100	85	71	66	33	36	20	19	**0.00**	16	26	8	–	–	–	–	–	–	–	–	–	–	1,2:4,5-Di-O-isopropylidene-3,6-di-O-acetyl-D-mannitol	91739-76-3	NI36297
101	111	115	153	59	85	72	273	**346**	100	59	55	52	42	32	26	12	**0.00**	16	26	8	–	–	–	–	–	–	–	–	–	–	1,2:4,6-Di-O-isopropylidene-3,5-di-O-acetyl-D-mannitol	93375-26-9	NI36298
101	153	111	85	59	273	185	171	**346**	100	58	55	22	22	13	13	13	**0.00**	16	26	8	–	–	–	–	–	–	–	–	–	–	1,2:5,6-Di-O-isopropylidene-3,4-di-O-acetyl-D-mannitol	76880-55-2	NI36299
114	135	134	93	121	95	107	109	**346**	100	81	77	45	38	35	34	29	**0.00**	16	27	3	–	–	–	–	1	–	–	–	–	–	2,6-Dodecadienoic acid, 10-bromo-11-hydroxy-3,7,11-trimethyl-, methyl ester	70038-03-8	NI36300
207	205	151	149	203	121	261	119	**346**	100	88	72	64	57	56	48	47	**1.00**	16	34	–	–	–	–	–	–	–	–	–	–	1	Dibutylethylcyclohexyltin		MS08209
331	73	215	117	116	217	75	204	**346**	100	87	70	39	33	32	32	28	**1.67**	16	34	4	–	–	–	–	–	–	2	–	–	–	Sebacic acid, bis(trimethylsilyl) ester	18408-42-9	NI36301
73	75	55	331	117	129	215	147	**346**	100	76	34	25	21	19	16	16	**0.00**	16	34	4	–	–	–	–	–	–	2	–	–	–	Sebacic acid, bis(trimethylsilyl) ester	18408-42-9	NI36302
73	75	55	129	331	117	69	215	**346**	100	91	31	21	19	13	13	12	**0.10**	16	34	4	–	–	–	–	–	–	2	–	–	–	Sebacic acid, bis(trimethylsilyl) ester	18408-42-9	NI36303
150	122	197	272	151	69	53	39	**346**	100	30	23	15	12	12	11	9	**4.00**	17	14	8	–	–	–	–	–	–	–	–	–	–	Spiro[1,3-benzodioxan-2,1'-[2,5]cyclohexadiene]-2'-carboxylic acid, 5-hydroxy-6'-methoxy-7-methyl-4,4'-dioxo-, methyl ester	3405-53-6	NI36304
315	346	329	303	179	153	194	167	**346**	100	88	45	29	23	22	15	12		17	14	8	–	–	–	–	–	–	–	–	–	–	2',5,5',7-Tetrahydroxy-3,4'-dimethoxyflavone		MS08210
346	153	192	315	164	165	300	303	**346**	100	98	70	39	39	30	26	22		17	14	8	–	–	–	–	–	–	–	–	–	–	3,3',5,7-Tetrahydroxy-2',4'-dimethoxyflavone		DD01362
346	331	328	345	183	155	197	148	**346**	100	30	30	29	9	9	8	7		17	14	8	–	–	–	–	–	–	–	–	–	–	5,6,8,4'-Tetrahydroxy-7,3'-dimethoxyflavone	103633-26-7	NI36305
346	331	328	303	345	167	139	111	**346**	100	72	62	44	19	16	15	14		17	14	8	–	–	–	–	–	–	–	–	–	–	5,7,3',4'-Tetrahydroxy-6,5'-dimethoxyflavone	79154-47-5	NI36306
346	196	135	319	283	332	123	304	**346**	100	97	90	28	25	24	24	22		17	15	2	–	–	5	–	–	–	–	–	–	–	3-Pentafluorobenzoylcamphor		NI36307
234	77	233	235	51	103	117	118	**346**	100	29	22	17	13	12	11	10	**0.00**	17	18	4	2	1	–	–	–	–	–	–	–	–	Desmethyl avenge		NI36308
329	284	224	268	330	270	42	225	**346**	100	83	68	39	20	18	16	13	**7.20**	17	18	6	2	–	–	–	–	–	–	–	–	–	Dimethyl 1,4-dihydro-2,6-dimethyl-4-(2-nitrophenyl)pyridine-3,5-dicarboxylate	21829-25-4	NI36309
224	192	287	225	315	346	164	149	**346**	100	7	6	4	4	4	4	4		17	18	6	2	–	–	–	–	–	–	–	–	–	Dimethyl 1,4-dihydro-2,6-dimethyl-4-(3-nitrophenyl)pyridine-3,5-dicarboxylate		NI36310
224	225	192	287	164	331	149	42	**346**	100	14	8	6	5	4	4	4	**3.80**	17	18	6	2	–	–	–	–	–	–	–	–	–	Dimethyl 1,4-dihydro-2,6-dimethyl-4-(4-nitrophenyl)pyridine-3,5-dicarboxylate		NI36311
186	145	69	41	55	143	97	112	**346**	100	92	80	74	58	57	56	53	**35.00**	17	30	3	–	2	–	–	–	–	–	–	–	–	Ethyl trans-7-oxa-7-(2-methyl-6,10-dithiaspiro[4.5]decan-1-yl)heptanoate	92640-74-9	NI36312
45	99	89	87	42	69	215	72	**346**	100	21	17	9	6	4	3	2	**0.04**	17	30	7	–	–	–	–	–	–	–	–	–	–	1,4,7,10,13-Pentaoxacyclodocosane-14,22-dione		MS08211
233	215	147	73	331	289	234	189	**346**	100	58	53	49	48	21	21	17	**0.00**	17	38	3	–	–	–	–	–	–	2	–	–	–	3-Trimethylsiloxy(trimethylsilyl)undecanoate		NI36313
346	311	151	348	317	132	302	272	**346**	100	50	46	32	25	17	14	7		18	15	5	–	–	–	1	–	–	–	–	–	–	6-Chloro-5,7,4'-trimethoxyflavone		NI36314
346	345	317	132	348	151	302	184	**346**	100	72	64	60	54	54	27	25		18	15	5	–	–	–	1	–	–	–	–	–	–	8-Chloro-5,7,4'-trimethoxyflavone	83516-32-9	NI36315
151	164	136	196	79	77	53	107	**346**	100	98	93	74	47	45	45	37	**16.00**	18	18	7	–	–	–	–	–	–	–	–	–	–	Benzoic acid, 2,4-dihydroxy-3,6-dimethyl-, 3-hydroxy-4-(methoxycarbonyl)-5-methylphenyl ester	75422-05-8	NI36316
138	165	44	39	109	107	108	77	**346**	100	78	73	52	50	48	30	26	**0.00**	18	18	7	–	–	–	–	–	–	–	–	–	–	Benzoic acid, 2,4-dimethoxy-6-methyl-, 4-carboxy-3-hydroxy-5-methylphenyl ester	75422-04-7	NI36317
165	61	43	40	346	57	69	41	**346**	100	83	48	46	21	21	18	17		18	18	7	–	–	–	–	–	–	–	–	–	–	Benzoic acid, 2-hydroxy-4-[(2-hydroxy-4-methoxy-6-methylbenzoyl)oxy]-6-methyl-, methyl ester	3542-22-1	NI36318
165	150	122	182	151	69	66	65	**346**	100	42	27	15	15	12	11	11	**0.00**	18	18	7	–	–	–	–	–	–	–	–	–	–	Benzoic acid, 2-hydroxy-4-[(2-hydroxy-4-methoxy-6-methylbenzoyl)oxy]-6-methyl-, methyl ester	3542-22-1	NI36319
165	346	166	150	109	136	123	122	**346**	100	70	40	26	16	12	12	12		18	18	7	–	–	–	–	–	–	–	–	–	–	Benzoic acid, 2-hydroxy-4-[(2-hydroxy-4-methoxy-6-methylbenzoyl)oxy]-6-methyl-, methyl ester	3542-22-1	LI07653
165	164	346	136	166	182	43	149	**346**	100	69	63	54	46	37	23	17		18	18	7	–	–	–	–	–	–	–	–	–	–	Benzoic acid, 2-hydroxy-4-methoxy-3,6-dimethyl-, 4-carboxy-3-hydroxy-5-methylphenyl ester	500-37-8	NI36320

MASS TO CHARGE RATIOS								M.W.	INTENSITIES								Parent	C	H	O	N	S	F	Cl	Br	I	Si	P	B	X	COMPOUND NAME	CAS Reg No	No
330	164	151	133	165	150	97	88	**346**	100	99	86	43	3	2	2	2	**0.00**	18	18	7	–	–	–	–	–	–	–	–	–	–	4H-1-Benzopyran-4-one, 2-(3,4-dimethoxyphenyl)-2,3-dihydro-5,7-dihydroxy-6-methoxy-	56847-13-3	MS08212
105	43	141	77	140	163	99	122	**346**	100	84	40	40	35	32	30	28	**2.00**	18	18	7	–	–	–	–	–	–	–	–	–	–	Cyclohex-4-ene, 1-benzoyloxymethyl-2,3-diacetoxy-1,6-epoxy-, (1S,2S,3R,6S)-		LI07654
346	287	243	234	44	219	331	257	**346**	100	63	55	52	31	30	26	25		18	18	7	–	–	–	–	–	–	–	–	–	–	1,2-Dibenzofurandicarboxylic acid, 1,9b-dihydro-7,9-dimethoxy-4,9b-dimethyl-	55937-84-3	NI36321
346	220	233	205	221	347	57	55	**346**	100	95	68	50	40	26	26	26		18	18	7	–	–	–	–	–	–	–	–	–	–	1(4H)-Dibenzofuranone, 2,6-diacetyl-4a,9b-dihydro-3,7,9-trihydroxy-8,9b-dimethyl-	18058-88-3	NI36322
43	85	202	145	103	59	328	174	**346**	100	69	62	31	19	16	11	10	**7.00**	18	18	7	–	–	–	–	–	–	–	–	–	–	7H-Furo[3,2-g][1]benzopyran-7-one, R-(+)-4-(2,3-dihydroxy-3-methylbutoxy)-, monoacetate	40693-09-2	NI36323
194	179	177	195	153	151	328	165	**346**	100	61	22	21	18	14	13	8	**2.70**	18	18	7	–	–	–	–	–	–	–	–	–	–	3,4-Isoflavandiol, 2′,7-dimethoxy-4′,5′-(methylenedioxy)-	3207-46-3	NI36324
43	45	229	244	187	227	346	201	**346**	100	46	34	16	15	14	8	3		18	18	7	–	–	–	–	–	–	–	–	–	–	Oroselol, 9-acetoxy-O-acetyldihydro-		LI07655
71	216	43	303	159	57	131	115	**346**	100	64	55	45	43	37	22	20	**0.00**	18	34	6	–	–	–	–	–	–	–	–	–	–	Ethyl 2-hydroxy-2-[2,2-diisopropyl-5-(2-methylpropoxy)-1,3-dioxolan-4-yl]propanoate		DD01363
43	57	41	55	29	83	28	69	**346**	100	92	82	72	42	41	40	39	**1.50**	18	35	1	–	–	–	–	1	–	–	–	–	–	Octadecanal, 2-bromo-	56599-95-2	NI36325
75	73	147	69	83	97	55	149	**346**	100	90	71	70	69	57	56	50	**0.49**	18	42	2	–	–	–	–	–	–	2	–	–	–	1,12-Bis(trimethylsiloxy)dodecane		NI36326
346	345	331	316	179	347	315	297	**346**	100	27	21	20	20	19	14	13		19	22	6	–	–	–	–	–	–	–	–	–	–	Benzophenone, 3,3′,4,4′,5-pentamethoxy-2-methyl-	56890-08-5	NI36327
164	328	165	191	327	329	346	149	**346**	100	48	17	15	13	9	8	8		19	22	6	–	–	–	–	–	–	–	–	–	–	2H-1-Benzopyran-4-ol, 2-(3,4-dimethoxyphenyl)-3,4-dihydro-5,7-dimethoxy-	10496-69-2	NI36328
167	180	346	151	166	91	107	95	**346**	100	30	23	20	16	16	14	14		19	22	6	–	–	–	–	–	–	–	–	–	–	2H-1-Benzopyran-3-ol, 2-(3,4-dimethoxyphenyl)-3,4-dihydro-5,7-dimethoxy-, (2R-cis)-	51196-02-2	NI36329
167	180	151	346	137	168	109	152	**346**	100	32	24	22	14	12	11	9		19	22	6	–	–	–	–	–	–	–	–	–	–	2H-1-Benzopyran-3-ol, 2-(3,4-dimethoxyphenyl)-3,4-dihydro-5,7-dimethoxy-, (2R-trans)-	51079-25-5	NI36330
210	197	167	195	346	163	137	198	**346**	100	82	64	57	55	38	28	26		19	22	6	–	–	–	–	–	–	–	–	–	–	2H-Benzopyran, 3-(3,4,6-trimethoxy-2-hydroxy-phenyl)-3,4-dihydro-7-methoxy-		LI07656
167	180	346	168	151	165	137	181	**346**	100	52	27	25	22	13	12	10		19	22	6	–	–	–	–	–	–	–	–	–	–	3-Flavanol, 3′,4,4′,7-tetramethoxy-, trans-2,3,cis-2,4-, (+)-	30655-11-9	MS08213
44	136	284	121	91	300	155	195	**346**	100	69	47	47	35	33	32	30	**13.95**	19	22	6	–	–	–	–	–	–	–	–	–	–	Gibb-3-ene-1,10-carboxylic acid, 2,4a,7-trihydroxy-1-methyl-8-methylene-1,4-lactone		NI36331
284	136	155	121	195	223	209	238	**346**	100	99	70	66	54	53	45	44	**6.00**	19	22	6	–	–	–	–	–	–	–	–	–	–	Gibb-3-ene-1,10-dicarboxylic acid, 2,4a,7-trihydroxy-1-methyl-8-methylene-, 1,4a-lactone, (1α,2β,4aα,4bβ,10β)-	77-06-5	NI36332
121	136	346	109	110	137	149	80	**346**	100	91	87	87	78	74	70	70		19	22	6	–	–	–	–	–	–	–	–	–	–	6-Hydroxy-2,3,9,10-dibenzo-1,4,8,11,14-pentaoxacyclohexadeca-2,9-diene		MS08214
55	69	43	301	97	204	149	127	**346**	100	96	84	40	37	33	29	26	**0.00**	19	23	2	2	–	–	1	–	–	–	–	–	–	trans-8-Chloro-5-ethyl-6,6-(ethylenedioxy)-1,2,5,6,7,7a-hexahydro-4H-pyrido[1′,2′:1,2]pyrazino[4,3-a]indole		DD01364
41	91	43	59	65	346	313	98	**346**	100	90	70	40	27	20	17	15		19	26	–	2	2	–	–	–	–	–	–	–	–	1-Benzyl-3-isopropyl-2-(2,4-dithioxopentan-3-ylidene)hexahydropyrimidine		MS08215
57	103	41	159	58	55	131	47	**346**	100	17	15	14	3	2	2	1	**0.00**	19	38	5	–	–	–	–	–	–	–	–	–	–	2-Propanone, 1,1,3,3-tetrabutoxy-		NI36333
346	289	262	236	347	264	291	290	**346**	100	73	62	49	36	29	23	23		20	10	6	–	–	–	–	–	–	–	–	–	–	[1,1′-Binaphthalene]-5,5′,8,8′-tetrone, 4,4′-dihydroxy-	101459-03-4	NI36334
346	92	329	347	91	262	289	290	**346**	100	36	31	22	21	21	21	16		20	10	6	–	–	–	–	–	–	–	–	–	–	[2,2′-Binaphthalene]-1,1′,4,4′-tetrone, 5,5′-dihydroxy-	61836-42-8	NI36335
346	329	92	347	289	262	290	91	**346**	100	31	26	21	18	15	15	13		20	10	6	–	–	–	–	–	–	–	–	–	–	[2,2′-Binaphthalene]-1,1′,4,4′-tetrone, 5,8′-dihydroxy-	50838-56-7	NI36336
91	106	346	44	78	105	77	51	**346**	100	59	47	37	35	26	19	17		20	10	6	–	–	–	–	–	–	–	–	–	–	[2,2′-Binaphthalene]-5,5′,8,8′-tetrone, 1,1′-dihydroxy-	89475-02-5	NI36337
346	77	136	347	105	108	51	348	**346**	100	66	42	32	26	14	10	9		20	14	2	2	1	–	–	–	–	–	–	–	–	3,5-Pyrazolidinedione, 1,2-diphenyl-4-(2-thienylmethylene)-	2652-81-5	NI36338
346	299	300	301	167	105			**346**	100	55	40	32	17	13				20	14	4	2	–	–	–	–	–	–	–	–	–	4-Carboxy-2,3,4,5-tetrahydrospiro[2,2′-β-carbolineindan]-1′,3′-dione		LI07657
346	347	77	93	69	327	329	71	**346**	100	24	12	10	10	9	7	7		20	14	4	2	–	–	–	–	–	–	–	–	–	1,8-Dihydroxy-4-anilino-5-aminoanthraquinone		IC04064
77	346	173	347	213	241	185	254	**346**	100	77	22	20	18	9	7	5		20	18	2	4	–	–	–	–	–	–	–	–	–	3,3′-Dimethyl-5,5′-dioxo-1,1′-diphenylbipyrazol-4-yl		IC04065
346	209	106	73	107	75	208	330	**346**	100	71	63	63	53	47	41	36		20	18	2	4	–	–	–	–	–	–	–	–	–	5-Methyl-2-aminobenzophenone 4-nitrophenylhydrazone		MS08216
77	93	121	346	65	122	105	51	**346**	100	96	58	52	50	40	40	24		20	18	2	4	–	–	–	–	–	–	–	–	–	2-Phenylazo-4-methoxy-5-(4-hydroxyphenylazo)toluene		IC04066
290	291	261	289	332	331	262	92	**346**	100	83	73	72	71	71	31	31	**0.00**	20	18	2	4	–	–	–	–	–	–	–	–	–	3H-Pyrazolo[4,3-c]pyridazin-3-one, 4-acetyl-1,2,3a,4,5,6-hexahydro-6-methylene-2,5-diphenyl-	35426-82-5	NI36339
55	83	107	246	119	213	140	228	**346**	100	58	50	31	15	14	12	11	**0.00**	20	26	5	–	–	–	–	–	–	–	–	–	–	1β-Angeloyloxy-9α-hydroxy-α-cyclocostunolide		MS08217
262	244	304	136	229	175	161	286	**346**	100	38	25	23	20	16	13	6	**4.10**	20	26	5	–	–	–	–	–	–	–	–	–	–	1,4,8a(6H)Anthracenetriol, 5,7,8,9,10,10a-hexahydro-5,5-dimethyl-, 1,4-diacetate	77880-98-9	NI36340
217	189	55	151	41	207	115	204	**346**	100	95	58	45	44	36	32	29	**12.00**	20	26	5	–	–	–	–	–	–	–	–	–	–	1H-2-Benzoxacyclotetradecin-1,7(8H)-dione, 3,4,5,6,9,10-hexahydro-14,16-dimethoxy-3-methyl-, [S-(E)]-	10497-40-2	NI36341
162	230	215	346	161	175	99	41	**346**	100	92	58	47	47	36	32	25		20	26	5	–	–	–	–	–	–	–	–	–	–	2-Butenoic acid, 2-(hydroxymethyl)-, 4,5,6,7,8,9-hexahydro-3,4a,5-trimethyl-9-oxonaphtho[2,3-b]furan-8a(4aH)-yl ester, [4aR-(4aα,5α,8aα)]-	56196-60-2	NI36342

MASS TO CHARGE RATIOS	M.W.	INTENSITIES	Parent	C	H	O	N	S	F	Cl	Br	I	Si	P	B	X	COMPOUND NAME	CAS Reg No	No
53 83 55 263 82 84 39 41	**346**	100 96 85 77 72 64 64 60	**0.00**	20	26	5	–	–	–	–	–	–	–	–	–	–	**2-Butenoic acid, 2-methyl-, dodecahydro-3a,8c-dimethyl-6-methylene-7-oxooxireno[7,8]naphtho[1,2-b]furan-3-yl ester, [1aR-[1aα,3β(Z),3aα,5aβ,8aα,8bβ,8cα]]-**	56196-59-9	**NI36343**
55 83 41 39 27 53 228 43	**346**	100 88 32 21 20 17 16 15	**0.00**	20	26	5	–	–	–	–	–	–	–	–	–	–	**2-Butenoic acid, 2-methyl-, dodecahydro-8-hydroxy-8a-methyl-3,5-bis(methylene)-2-oxonaphtho[2,3-b]furan-4-yl ester, [3aR-[3aα,4α(Z),4aα,8β,8aβ,9aβ]]-**	34226-89-6	**NI36344**
219 346 304 261 289	**346**	100 28 17 15 10		20	26	5	–	–	–	–	–	–	–	–	–	–	**Dihydroisodesbenzylidenerubranine acetate**		**LI07658**
314 224 28 284 91 41 225 43	**346**	100 93 85 59 56 54 50 39	**7.00**	20	26	5	–	–	–	–	–	–	–	–	–	–	**Gibbane-1,10-dicarboxylic acid, 2,4a-dihydroxy-1-methyl-8-methylene-, 1,4a-lactone,10-methyl ester, (1α,2β,4aα,4bβ,10β)-**	19124-90-4	**MS08218**
314 55 346 303 286 41 43 315	**346**	100 52 50 49 46 45 42 32		20	26	5	–	–	–	–	–	–	–	–	–	–	**Gibbane-1,10-dicarboxylic acid, 4a,7-dihydroxy-1-methyl-8-methylene-, 1,4a-lactone,10-methyl ester, (1α,4aα,4bβ,10β)-**	15355-41-6	**MS08219**
346 302 233 41 231 205 232 43	**346**	100 82 81 68 62 60 56 51		20	26	5	–	–	–	–	–	–	–	–	–	–	**6α-Hydroxycarnosol**		**NI36345**
149 137 123 175 299 177 109 163	**346**	100 89 56 48 40 40 35 26	**11.00**	20	26	5	–	–	–	–	–	–	–	–	–	–	**16-Hydroxypseudojolkinolide B**		**NI36346**
346 285 41 328 231 43 300 258	**346**	100 96 94 85 64 63 62 60		20	26	5	–	–	–	–	–	–	–	–	–	–	**6aH-Isobenzofuro[1,7a-b]benzofuran-11-carboxaldehyde, 1,2,3,4,4a,5-hexahydro-5,8-dihydroxy-4,4-dimethyl-9-(1-methylethyl)-, [4aS-(4aα,5β,6aβ,11bS*)]-**	98941-40-3	**NI36347**
69 41 121 328 43 107	**346**	100 77 72 69 64 55	**0.01**	20	26	5	–	–	–	–	–	–	–	–	–	–	**Phorbobutanone**	18326-01-7	**LI07659**
74 43 57 56 55 41 69 83	**346**	100 27 20 20 18 15 14 12	**0.00**	20	42	4	–	–	–	–	–	–	–	–	–	–	**1,2(S),3(R),4(R)-Icosanetetrol**		**MS08220**
93 95 331 121 57 71 43 333	**346**	100 35 23 19 11 10 10 9	**0.40**	20	43	–	–	–	–	1	–	–	1	–	–	–	**Silane, chlorodimethyloctadecyl-**	18643-08-8	**NI36348**
267 268 265 189 132 263 165 126	**346**	100 24 13 6 6 5 5 4	**0.00**	21	15	–	–	–	–	–	1	–	–	–	–	–	**1,2,3-Triphenylcyclopropenyl bromide**		**LI07660**
346 105 226 28 77 121 56 65	**346**	100 44 40 24 20 12 8 3		21	22	1	–	–	–	–	–	–	–	–	–	1	**1-Benzoyl-2-tert-butylferrocene**		**NI36349**
346 91 347 277 77 213 185 129	**346**	100 48 25 25 24 16 15 10		21	22	1	4	–	–	–	–	–	–	–	–	–	**2-Pyrazolin-5-one, 3-methyl-1-phenyl-4-[(4-piperidinophenyl)imino]-**	4719-53-3	**NI36350**
346 91 347 277 77 213 185 132	**346**	100 48 25 25 24 17 15 10		21	22	1	4	–	–	–	–	–	–	–	–	–	**2-Pyrazolin-5-one, 3-methyl-1-phenyl-4-[(4-piperidinophenyl)imino]-**	4719-53-3	**MS08221**
329 174 202 105 104 58 346 31	**346**	100 68 64 60 60 56 46 46		21	22	1	4	–	–	–	–	–	–	–	–	–	**1,2,3-Triazole, 4-[2-phenyl-2-(diethylamino)propenoyl]-5-phenyl-**		**NI36351**
133 286 304 346 154 271 151 300	**346**	100 56 46 41 39 22 22 14		21	27	3	–	–	1	–	–	–	–	–	–	–	**Androsta-4,6-dien-3-one, 2β-fluoro-17β-hydroxy-, acetate**	32307-43-0	**LI07661**
133 43 286 304 346 154 271 151	**346**	100 65 57 46 41 39 22 22		21	27	3	–	–	1	–	–	–	–	–	–	–	**Androsta-4,6-dien-3-one, 2β-fluoro-17β-hydroxy-, acetate**	32307-43-0	**NI36352**
346 229 271 208 347 257 303 43	**346**	100 32 26 26 24 13 12 12		21	30	4	–	–	–	–	–	–	–	–	–	–	**Abieta-8,11,13-triene, 12-methoxy-3-oxo-7,11-dihydroxy-, (17β)-**		**NI36353**
43 346 55 317 79 41 148 93	**346**	100 29 28 26 24 24 22 22		21	30	4	–	–	–	–	–	–	–	–	–	–	**Androstane-3,6-dione, 17-acetoxy-, (5α,17β)-**		**MS08222**
99 112 154 55 41 86 81 79	**346**	100 94 89 84 84 79 58 58	**9.00**	21	30	4	–	–	–	–	–	–	–	–	–	–	**Androstane-3,7,17-trione, cyclic 7-(1,2-ethanediyl acetal), (5α)-**	54498-68-9	**NI36354**
112 154 99 54 346 167 139 91	**346**	100 99 84 40 19 18 18 18		21	30	4	–	–	–	–	–	–	–	–	–	–	**Androstane-3,7,17-trione, cyclic 7-(1,2-ethanediyl acetal), (5α)-**	54498-68-9	**NI36355**
287 288 165 314 346 154 91 105	**346**	100 20 19 14 13 8 7 6		21	30	4	–	–	–	–	–	–	–	–	–	–	**Androst-6-en-3,17-dione, 4,4-dimethoxy-**		**NI36356**
268 91 105 79 55 143 145 67	**346**	100 85 79 58 58 46 44 42	**0.00**	21	30	4	–	–	–	–	–	–	–	–	–	–	**Androst-5-en-17-one, 3-(acetyloxy)-14-hydroxy-, (3β)-**	1443-89-6	**NI36357**
43 255 41 55 91 67 79 81	**346**	100 36 36 30 26 21 18 14	**0.00**	21	30	4	–	–	–	–	–	–	–	–	–	–	**Androst-5-en-17-one, 3-(acetyloxy)-19-hydroxy-, (3β)-**	2857-42-3	**NI36358**
81 43 41 57 31 45 284 105	**346**	100 97 94 88 85 78 74 70	**1.85**	21	30	4	–	–	–	–	–	–	–	–	–	–	**Bicyclo[4.4.0]dec-2-ene-2-carboxylic acid, 1α-(hydroxymethyl)-7α,8α-dimethyl-7-[(2-fur-3-yl)ethyl]-, methyl ester**		**NI36359**
191 221 346 43 83 328 91 316	**346**	100 75 59 22 16 8 8 7		21	30	4	–	–	–	–	–	–	–	–	–	–	**Dihydroanydrohinundigenin**		**LI07662**
346 347 167 344 315 297 181 201	**346**	100 24 16 6 6 6 6 5		21	30	4	–	–	–	–	–	–	–	–	–	–	**Estra-1,3,5(10)-trien-17-ol, 2,3,4-trimethoxy-, (17β)-**	4314-54-9	**NI36360**
43 138 123 151 55 79 91 81	**346**	100 45 36 34 28 20 18 18	**4.05**	21	30	4	–	–	–	–	–	–	–	–	–	–	**Gonane-12,17-dione, 3-(acetyloxy)-5,14-dimethyl-, (3β,5β,8α,9β,10α,13ξ)**	55515-20-3	**NI36361**
44 138 151 123 55 67 148 139	**346**	100 99 78 78 33 27 21 21	**8.97**	21	30	4	–	–	–	–	–	–	–	–	–	–	**Gonane-12,17-dione, 3-(acetyloxy)-5,14-dimethyl-, (3β,5β,8α,9β,10α,13ξ,14β)-**	55429-37-3	**NI36362**
97 143 69 91 41 55 115 159	**346**	100 72 63 60 46 45 31 9	**0.00**	21	30	4	–	–	–	–	–	–	–	–	–	–	**Heptanoic acid, 7-oxa-7-(2-benzyl-2-methyl-5-oxocyclopentyl)-, ethyl ester**		**NI36363**
108 122 28 149 95 316 79 43	**346**	100 95 82 73 69 65 59 51	**27.00**	21	30	4	–	–	–	–	–	–	–	–	–	–	**Isodihydroanhydrohirundigenin**		**LI07663**
43 149 41 150 55 121 302 137	**346**	100 90 87 63 63 59 50 48	**36.03**	21	30	4	–	–	–	–	–	–	–	–	–	–	**Podocarp-13-ene-14-glycolic acid, 7-hydroxy-8,13-dimethyl-3-oxo-, δ-lactone**	31858-08-9	**NI36364**
43 109 41 79 91 110 105 81	**346**	100 99 93 83 80 78 71 71	**0.00**	21	30	4	–	–	–	–	–	–	–	–	–	–	**Pregn-5-ene-3,12-diol, 14,20:18,20-diepoxy-, (3β,12β,14β,20S)-**	41410-50-8	**NI36365**
315 287 269 316 81 79 121 67	**346**	100 53 26 25 20 18 17 16	**2.18**	21	30	4	–	–	–	–	–	–	–	–	–	–	**Pregn-4-ene-3,20-dione, 2,21-dihydroxy-, (2α)-**	56196-78-2	**NI36366**
43 285 55 41 227 269 91 79	**346**	100 62 38 32 27 26 26 25	**15.00**	21	30	4	–	–	–	–	–	–	–	–	–	–	**Pregn-4-ene-3,20-dione, 6,17-dihydroxy-, (6β)-**	604-03-5	**NI36367**
43 229 267 124 310 247 100 328	**346**	100 57 45 32 21 21 21 13	**2.60**	21	30	4	–	–	–	–	–	–	–	–	–	–	**Pregn-4-ene-3,20-dione, 7,15-dihydroxy-, (7β,15β)-**	601-90-1	**NI36368**
43 124 247 163 346 310 229 328	**346**	100 95 69 57 38 31 19 10		21	30	4	–	–	–	–	–	–	–	–	–	–	**Pregn-4-ene-3,20-dione, 11,15-dihydroxy-, (11α,15β)-**	82538-36-1	**NI36369**
247 124 163 56 100 123 91 122	**346**	100 42 40 33 31 27 26 23	**14.55**	21	30	4	–	–	–	–	–	–	–	–	–	–	**Pregn-4-ene-3,20-dione, 11,16-dihydroxy-, (11α,16α)-**	55622-61-2	**NI36370**
315 269 316 145 227 55 81 79	**346**	100 61 23 21 18 18 16 15	**0.00**	21	30	4	–	–	–	–	–	–	–	–	–	–	**Pregn-4-ene-3,20-dione, 11,21-dihydroxy-, (11α)-**	600-67-9	**NI36371**
315 269 55 41 91 316 145 227	**346**	100 66 30 25 23 21 21 20	**1.64**	21	30	4	–	–	–	–	–	–	–	–	–	–	**Pregn-4-ene-3,20-dione, 11,21-dihydroxy-, (11β)-**	50-22-6	**NI36372**
315 269 145 91 43 41 55 81	**346**	100 71 45 45 44 43 41 37	**2.05**	21	30	4	–	–	–	–	–	–	–	–	–	–	**Pregn-4-ene-3,20-dione, 11,21-dihydroxy-, (11β)-**	50-22-6	**NI36373**
285 43 55 97 91 121 105 79	**346**	100 31 12 8 8 7 7 7	**0.00**	21	30	4	–	–	–	–	–	–	–	–	–	–	**Pregn-4-ene-3,20-dione, 14,17-dihydroxy-**	14226-13-2	**NI36374**
285 43 55 41 91 79 105 97	**346**	100 25 9 9 7 7 6 6	**0.00**	21	30	4	–	–	–	–	–	–	–	–	–	–	**Pregn-4-ene-3,20-dione, 14,17-dihydroxy-**	14226-13-2	**NI36375**

MASS TO CHARGE RATIOS	M.W.	INTENSITIES	Parent	C	H	O	N	S	F	Cl	Br	I	Si	P	B	X	COMPOUND NAME	CAS Reg No	No
297 315 55 43 91 227 79 81	**346**	100 37 23 13 12 10 10 8	**0.00**	21	30	4	–	–	–	–	–	–	–	–	–	–	Pregn-4-ene-3,20-dione, 14,21-dihydroxy-	595-71-1	NI36376
303 43 285 229 55 267 91 346	**346**	100 56 36 28 22 20 19 18		21	30	4	–	–	–	–	–	–	–	–	–	–	Pregn-4-ene-3,20-dione, 16,17-dihydroxy-	56193-65-8	NI36377
97 229 123 145 121 105 287 43	**346**	100 88 77 73 66 61 60 60	**8.97**	21	30	4	–	–	–	–	–	–	–	–	–	–	Pregn-4-ene-3,20-dione, 17,21-dihydroxy-	152-58-9	NI36378
287 229 346 43 55 91 41 79	**346**	100 47 37 36 34 33 33 32		21	30	4	–	–	–	–	–	–	–	–	–	–	Pregn-4-ene-3,20-dione, 17,21-dihydroxy-	152-58-9	NI36379
299 91 79 41 300 67 95 93	**346**	100 30 30 30 24 22 20 20	**0.00**	21	30	4	–	–	–	–	–	–	–	–	–	–	Pregn-4-ene-3,20-dione, 18,21-dihydroxy-	379-68-0	NI36380
299 91 79 41 300 67 315 105	**346**	100 30 30 30 24 20 19 19	**9.00**	21	30	4	–	–	–	–	–	–	–	–	–	–	Pregn-4-ene-3,20-dione, 18,21-dihydroxy-	379-68-0	LI07664
315 257 316 55 91 41 269 79	**346**	100 94 54 52 44 43 30 30	**12.10**	21	30	4	–	–	–	–	–	–	–	–	–	–	Pregn-4-ene-3,20-dione, 19,21-dihydroxy-	2394-23-2	NI36381
284 346 316 180 328 147 256 296	**346**	100 60 50 36 30 30 20 14		21	30	4	–	–	–	–	–	–	–	–	–	–	Sordaricin methyl ester		LI07665
129 256 346 331 160 215 130 57	**346**	100 69 50 23 22 19 18 15		21	34	2	–	–	–	–	–	–	1	–	–	–	Estr-4-en-3-one, 17-[(trimethylsilyl)oxy]-, (17β)-	6717-72-2	NI36382
256 129 346 215 199 213 257 147	**346**	100 94 39 33 24 23 20 19		21	34	2	–	–	–	–	–	–	1	–	–	–	Estr-5(10)-en-3-one, 17-[(trimethylsilyl)oxy]-, (17β)-	59384-55-3	NI36383
315 316 260 96 87 126 83 285	**346**	100 36 21 15 9 8 6 5	**2.48**	21	34	2	2	–	–	–	–	–	–	–	–	–	Androstane-3,17-dione, bis(O-methyloxime), (5α)-	3091-35-8	NI36384
315 316 283 260 284 285 214 216	**346**	100 27 17 17 7 6 5 4	**3.18**	21	34	2	2	–	–	–	–	–	–	–	–	–	Androstane-3,17-dione, bis(O-methyloxime), (5β)-	56210-80-1	NI36385
346 218 317 131 318 214 77 265	**346**	100 45 27 27 20 12 12 10		22	10	1	4	–	–	–	–	–	–	–	–	–	9H-Indeno[1,2-b]pyrazine-2,3-dicarbonitrile, 9-(2,3-dihydro-2-oxo-1H-inden-1-ylidene)-	52197-19-0	NI36386
91 107 149 65 92 77 105 79	**346**	100 84 65 22 20 11 7 7	**0.20**	22	18	4	–	–	–	–	–	–	–	–	–	–	1,2-Benzenedicarboxylic acid, dibenzyl ester	523-31-9	NI36387
149 107 91 92 65 150 167 108	**346**	100 99 99 26 17 11 9 9	**0.80**	22	18	4	–	–	–	–	–	–	–	–	–	–	1,2-Benzenedicarboxylic acid, dibenzyl ester	523-31-9	IC04067
346 271 302 239 301 347 272 303	**346**	100 91 85 41 40 27 22 20		22	18	4	–	–	–	–	–	–	–	–	–	–	1(3H)-Isobenzofuranone, 3,3-bis(4-methoxyphenyl)-	6315-80-6	NI36388
346 271 302 211 195 135 239 301	**346**	100 88 85 65 63 58 43 42		22	18	4	–	–	–	–	–	–	–	–	–	–	1(3H)-Isobenzofuranone, 3,3-bis(4-methoxyphenyl)-	6315-80-6	NI36389
346 271 302 301 239 347 272 303	**346**	100 88 84 43 42 32 22 20		22	18	4	–	–	–	–	–	–	–	–	–	–	1(3H)-Isobenzofuranone, 3,3-bis(4-methoxyphenyl)-	6315-80-6	LI07666
86 58 346 214 71 85 87 84	**346**	100 63 7 6 6 5 4 3		22	22	2	2	–	–	–	–	–	–	–	–	–	5-[3-(N,N-Dimethylamino)propoxy]-7H-benzo[c]fluoren-7-one oxime		MS08223
172 173 130 174 171 131 117 103	**346**	100 71 48 28 22 14 9 7	**3.00**	22	26	–	4	–	–	–	–	–	–	–	–	–	Calycanthidine, 1-demethyl-	5545-89-1	NI36390
172 173 130 346 157 144 347	**346**	100 33 26 12 2 2 1		22	26	–	4	–	–	–	–	–	–	–	–	–	Calycanthidine, 1-demethyl-, (±)-	4147-36-8	MS08224
172 173 130 346 157 144	**346**	100 25 10 7 2 2		22	26	–	4	–	–	–	–	–	–	–	–	–	Calycanthidine, 1-demethyl-, meso-	4147-37-9	MS08225
346 231 347 232 288 302 245 172	**346**	100 34 28 20 18 15 15 12		22	26	–	4	–	–	–	–	–	–	–	–	–	Calycanthine	595-05-1	MS08226
346 231 288 232 302 245 172 83	**346**	100 45 19 19 16 16 12 12		22	26	–	4	–	–	–	–	–	–	–	–	–	Calycanthine	595-05-1	NI36391
347 63 348 346 61 91 316 345	**346**	100 27 25 16 14 8 7 6		22	26	–	4	–	–	–	–	–	–	–	–	–	Calycanthine	595-05-1	NI36392
149 57 55 101 167 71 79 69	**346**	100 36 31 26 18 14 12 11	**0.00**	22	34	3	–	–	–	–	–	–	–	–	–	–	3-Oxo-18-nor-ent-ros-4-ene 15β,16-acetonide	119769-57-2	DD01365
43 331 93 55 41 82 107 95	**346**	100 38 35 35 35 33 29 27	**7.00**	22	34	3	–	–	–	–	–	–	–	–	–	–	Androstan-17-ol, 2,3-epoxy-1-methyl-, acetate, (1α,2α,3α,5α,17β)-	21623-55-2	NI36393
43 55 41 93 81 107 95 79	**346**	100 39 38 37 34 32 31 28	**15.00**	22	34	3	–	–	–	–	–	–	–	–	–	–	Androstan-17-ol, 2,3-epoxy-1-methyl-, acetate, (1α,2β,3β,5α,17β)-	21623-48-3	NI36394
43 346 41 93 55 105 81 95	**346**	100 46 33 29 29 27 27 26		22	34	3	–	–	–	–	–	–	–	–	–	–	Androstan-17-ol, 2,3-epoxy-3-methyl-, acetate, (2α,3α,5α,17β)-	16321-36-1	NI36395
43 93 41 81 55 346 95 79	**346**	100 30 30 28 28 27 25 22		22	34	3	–	–	–	–	–	–	–	–	–	–	Androstan-17-ol, 2,3-epoxy-3-methyl-, acetate, (2β,3β,5α,17β)-	16321-28-1	NI36396
43 81 95 41 201 93 55 67	**346**	100 20 19 19 18 18 18 17	**13.00**	22	34	3	–	–	–	–	–	–	–	–	–	–	Androstan-3-one, 17-(acetyloxy)-2-methyl-, (2β,5β,17β)-	54550-09-3	NI36397
157 286 268 186 328 197 225 201	**346**	100 52 40 38 21 21 12 12	**5.00**	22	34	3	–	–	–	–	–	–	–	–	–	–	4,7-Azulenediol, 5-(1,5-dimethyl-4-hexenyl)-1,3a,4,5,6,7,8,8a-octahydro-3-methyl-8-methylene-, 7-acetate, [3aS-[3aα,4α,5α(S*),7α,8aβ]]-	63908-64-5	NI36398
43 41 81 55 67 123 79 91	**346**	100 53 50 50 41 39 36 33	**1.11**	22	34	3	–	–	–	–	–	–	–	–	–	–	Kauran-18-al, 17-(acetyloxy)-, (4β)-	55902-84-6	NI36399
331 73 41 43 91 55 332 314	**346**	100 72 46 35 34 30 25 25	**24.72**	22	34	3	–	–	–	–	–	–	–	–	–	–	1-Phenanthrenecarboxylic acid, 1,2,3,4,4a,4b,5,6,10,10-decahydro-7-(1-methoxy-1-methylethyl)-1,4a-dimethyl-, methyl ester, [1R-(1α,4aβ,4bα,10aα)]-	26548-91-4	MS08227
143 142 346 127 189 317 202 75	**346**	100 71 50 38 22 19 16 16		22	38	1	–	–	–	–	–	–	1	–	–	–	Androst-3-ene, 3-[(trimethylsilyl)oxy]-, (5α)-	69833-86-9	NI36400
217 129 256 135 241 121 95 107	**346**	100 86 45 37 34 27 23 18	**17.00**	22	38	1	–	–	–	–	–	–	1	–	–	–	Androst-5-ene, 3-[(trimethylsilyl)oxy]-, (3β)-	49774-79-0	NI36401
149 135 69 331 136 71 150 137	**346**	100 94 70 61 58 55 44 43	**7.30**	22	38	1	2	–	–	–	–	–	–	–	–	–	Nonadec-1-ene, 1,1-dicyano-2-methoxy-		LI07667
70 44 277 303 272 57 42 41	**346**	100 50 20 10 10 8 8 7	**3.00**	22	38	1	2	–	–	–	–	–	–	–	–	–	Pregn-5-en-18-ol, 20-amino-3-(methylamino)-, (3β,20S)-	27741-48-6	NI36402
289 288 183 290 211 241 56 228	**346**	100 33 24 23 19 18 15 11	**5.00**	23	23	1	–	–	–	–	–	–	–	1	–	–	10-tert-Butyl-9,10-dihydro-10-hydroxy-9-phenyl-9-phosphaanthracene		MS08228
172 91 185 186 346 82 173 170	**346**	100 99 71 62 40 40 29 20		23	26	1	2	–	–	–	–	–	–	–	–	–	1H-Azecino[5,4-b]indol-8-ol, 3-benzyl-2,3,4,5,6,7,8,9-octahydro-2,6-methano-		LI07668
329 148 253 328 269 225 346 330	**346**	100 83 76 64 51 48 44 24		23	26	1	2	–	–	–	–	–	–	–	–	–	Benzenemethanol, 4-(dimethylamino)-α-[4-(dimethylamino)phenyl]-α-phenyl-	510-13-4	IC04068
269 329 346 330 148 253 105 77	**346**	100 72 68 62 62 47 35 23		23	26	1	2	–	–	–	–	–	–	–	–	–	Benzenemethanol, 4-(dimethylamino)-α-[4-(dimethylamino)phenyl]-α-phenyl-	510-13-4	NI36403
317 274 77 198 91 104 346 255	**346**	100 36 31 20 19 18 17 7		23	26	1	2	–	–	–	–	–	–	–	–	–	2(1H)-Pyridinone, 3-benzyl-4-(diethylamino)-5-methyl-1-phenyl-		DD01366
41 43 95 55 81 69 67 231	**346**	100 99 68 56 48 42 42 32	**12.40**	23	38	2	–	–	–	–	–	–	–	–	–	–	5α-Androstan-3β-ol, 4,4-dimethyl-, acetate		NI36404
41 43 67 79 55 93 81 80	**346**	100 93 91 90 87 81 73 62	**6.04**	23	38	2	–	–	–	–	–	–	–	–	–	–	6,9,12,15-Docosatetraenoic acid, methyl ester	17364-34-0	NI36405
255 119 327 256 131 108 239 328	**346**	100 56 45 17 16 16 13 9	**4.00**	23	38	2	–	–	–	–	–	–	–	–	–	–	Palmitic acid, benzyl ester		NI36406
91 108 43 57 41 55 92 71	**346**	100 72 56 50 29 27 26 24	**0.00**	23	38	2	–	–	–	–	–	–	–	–	–	–	Palmitic acid, benzyl ester		NI36407
124 346 347 137 136 125 123 41	**346**	100 87 76 72 65 54 44 30		23	38	2	–	–	–	–	–	–	–	–	–	–	Phenol, 5-[2(Z)-heptadecenyl]-3-hydroxy-		NI36408
243 165 244 42 241 242 166 41	**346**	100 37 29 12 11 9 8 7	**0.00**	24	26	–	–	1	–	–	–	–	–	–	–	–	Pentyl trityl sulphide	20705-42-4	NI36409

MASS TO CHARGE RATIOS								M.W.	INTENSITIES								Parent	C	H	O	N	S	F	Cl	Br	I	Si	P	B	X	COMPOUND NAME	CAS Reg No	No
243	165	244	228	215	42	226	202	**346**	100	38	29	20	15	12	10	10	**0.00**	24	26	–	–	1	–	–	–	–	–	–	–	–	**Pentyl trityl sulphide**	20705-42-4	LI07669
243	165	244	42	241	242	166	55	**346**	100	37	28	11	10	9	8	8	**0.00**	24	26	–	–	1	–	–	–	–	–	–	–	–	**Pentyl trityl sulphide**	20705-42-4	NI36410
149	91	240	167	150	148	107	105	**346**	100	33	29	21	18	17	14	14	**0.00**	24	26	2	–	–	–	–	–	–	–	–	–	–	**Spiro[azulene-1(2H),2′-[1,3]dioxolane], 3,4,5,6,7,8-hexahydro-4′,5′-diphenyl-**	117583-55-8	DD01367
122	121	43	41	135	346	55	29	**346**	100	47	26	17	12	11	11	10		24	42	1	–	–	–	–	–	–	–	–	–	–	**Anisole, 3-heptadecyl-**	22165-13-5	NI36411
346	215	328	55	216	81	95	43	**346**	100	97	76	74	59	55	53	52		24	42	1	–	–	–	–	–	–	–	–	–	–	**Cholan-3-ol, (3α,5β)-**	5352-77-2	NI36412
346	215	328	216	233	217	165	347	**346**	100	96	75	59	38	35	34	28		24	42	1	–	–	–	–	–	–	–	–	–	–	**Cholan-3-ol, (3α,5β)-**	5352-77-2	NI36413
346	233	215	149	234	331	217	347	**346**	100	52	48	42	37	33	29	28		24	42	1	–	–	–	–	–	–	–	–	–	–	**Cholan-3-ol, (3β,5β)-**	42921-44-8	NI36414
346	55	233	215	95	81	43	107	**346**	100	65	51	47	47	47	47	42		24	42	1	–	–	–	–	–	–	–	–	–	–	**Cholan-3-ol, (3β,5β)-**	42921-44-8	NI36415
261	43	275	247	41	57	71	55	**346**	100	49	47	44	36	34	22	21	**2.60**	24	42	1	–	–	–	–	–	–	–	–	–	–	**Phenol, 2,4-dinonyl-**	137-99-5	IC04069
261	275	262	43	289	247	57	317	**346**	100	44	19	18	17	15	11	10	**5.00**	24	42	1	–	–	–	–	–	–	–	–	–	–	**Phenol, 2,4-dinonyl-**	137-99-5	NI36416
261	275	83	57	43	71	141	55	**346**	100	71	51	43	43	40	37	34	**11.00**	24	42	1	–	–	–	–	–	–	–	–	–	–	**Phenol, 2,6-dinonyl-**	137-99-5	IC04070
180	181	346	167	152	140	347		**346**	100	22	16	14	14	10	4			25	18	–	2	–	–	–	–	–	–	–	–	–	**Bis(9-carbazyl)methane**	6510-63-0	LI07670
180	181	152	346	167	166	151	347	**346**	100	15	14	10	8	4	4	3		25	18	–	2	–	–	–	–	–	–	–	–	–	**Bis(9-carbazyl)methane**	6510-63-0	NI36417
55	83	41	97	235	69	234	67	**346**	100	82	55	44	36	27	26	21	**0.45**	25	46	–	–	–	–	–	–	–	–	–	–	–	**1,5-Dicyclohexyl-3-(2-cyclohexylethyl)pentane**		AI02106
55	83	97	235	69	234	67	96	**346**	100	87	47	32	30	22	21	20	**0.15**	25	46	–	–	–	–	–	–	–	–	–	–	–	**1,5-Dicyclohexyl-3-(2-cyclohexylethyl)pentane**		AI02107
55	83	41	97	235	69	234	67	**346**	100	74	56	48	31	29	24	23	**0.58**	25	46	–	–	–	–	–	–	–	–	–	–	–	**1,5-Dicyclohexyl-3-(2-cyclohexylethyl)pentane**		AI02108
55	83	41	97	69	67	111	96	**346**	100	95	69	62	48	27	22	20	**0.23**	25	46	–	–	–	–	–	–	–	–	–	–	–	**1,5-Dicyclohexyl-3-(3-cyclopentylpropyl)pentane**		AI02109
55	83	97	41	69	67	111	96	**346**	100	93	67	64	48	29	25	22	**0.59**	25	46	–	–	–	–	–	–	–	–	–	–	–	**1,5-Dicyclohexyl-3-(3-cyclopentylpropyl)pentane**		AI02110
55	83	41	97	69	67	111	82	**346**	100	94	78	69	65	34	25	23	**0.09**	25	46	–	–	–	–	–	–	–	–	–	–	–	**1,7-Dicyclopentyl-4-(2-cyclohexylethyl)heptane**		AI02111
83	55	41	97	69	67	234	111	**346**	100	99	78	73	67	36	32	29	**0.40**	25	46	–	–	–	–	–	–	–	–	–	–	–	**1,7-Dicyclopentyl-4-(2-cyclohexylethyl)heptane**		AI02112
69	83	55	41	97	67	235	111	**346**	100	89	85	80	61	39	31	24	**0.00**	25	46	–	–	–	–	–	–	–	–	–	–	–	**1,7-Dicyclopentyl-4-(3-cyclopentylpropyl)heptane**		AI02113
83	41	55	69	79	67	82	111	**346**	100	90	89	85	70	38	28	26	**0.12**	25	46	–	–	–	–	–	–	–	–	–	–	–	**1,7-Dicyclopentyl-4-(3-cyclopentylpropyl)heptane**		AI02114
191	69	95	81	55	123	109	41	**346**	100	50	44	38	37	31	28	28	**16.01**	25	46	–	–	–	–	–	–	–	–	–	–	–	**15-(2-Methylbutyl)-(13αH)-isocopalane**		NI36418
105	77	346	239	241	269	268	51	**346**	100	49	35	24	20	15	15	13		26	18	1	–	–	–	–	–	–	–	–	–	–	**Anthracene, 9,10-epoxy-9,10-diphenyl-9,10-dihydro-**		LI07671
346	269	239	347	270	268	240	252	**346**	100	98	45	33	23	17	14	10		26	18	1	–	–	–	–	–	–	–	–	–	–	**9(10H)-Anthracenone, 10,10-diphenyl-**	3216-03-3	NI36419
219	346	43	41	220	347	205	91	**346**	100	96	48	35	34	27	26	22		26	34	–	–	–	–	–	–	–	–	–	–	–	**3-Decyl-4b,5,9b,10-tetrahydroindeno[2,1-a]indene**		AI02115
191	346	347	192	189	215	342	178	**346**	100	82	23	18	10	9	7	6		26	34	–	–	–	–	–	–	–	–	–	–	–	**9-Dodecylanthracene**		AI02116
191	346	192	29	347	43	41	189	**346**	100	46	18	14	13	13	12	7		26	34	–	–	–	–	–	–	–	–	–	–	–	**9-Dodecylanthracene**		AI02117
191	346	192	347	209	43	41	189	**346**	100	39	17	11	11	9	8	6		26	34	–	–	–	–	–	–	–	–	–	–	–	**9-Dodecylanthracene**		AI02118
346	191	192	347	189	267	204	190	**346**	100	96	38	28	14	10	9	9		26	34	–	–	–	–	–	–	–	–	–	–	–	**2-Dodecylphenanthrene**		AI02119
191	346	192	43	347	41	189	204	**346**	100	81	54	27	23	23	14	10		26	34	–	–	–	–	–	–	–	–	–	–	–	**2-Dodecylphenanthrene**		AI02120
191	346	192	254	44	330	347	204	**346**	100	74	52	43	27	25	22	15		26	34	–	–	–	–	–	–	–	–	–	–	–	**2-Dodecylphenanthrene**		AI02121
191	346	192	347	254	44	330	228	**346**	100	70	60	31	21	18	16	13		26	34	–	–	–	–	–	–	–	–	–	–	–	**9-Dodecylphenanthrene**		AI02122
191	346	192	43	41	347	217	344	**346**	100	64	49	24	21	19	14	11		26	34	–	–	–	–	–	–	–	–	–	–	–	**9-Dodecylphenanthrene**		AI02123
191	346	192	347	189	344	217	165	**346**	100	80	43	22	11	10	10	9		26	34	–	–	–	–	–	–	–	–	–	–	–	**9-Dodecylphenanthrene**		AI02124
167	268	165	346	255	191	179	269	**346**	100	98	77	75	70	48	39	34		27	22	–	–	–	–	–	–	–	–	–	–	–	**Cyclopropane, 1,1,2,2-tetraphenyl-**	1053-23-2	NI36420
347	127	66	220	44	93	193	53	**347**	100	16	10	7	6	5	3	2		5	3	1	1	–	–	–	–	2	–	–	–	–	**4(1H)-Pyridinone, 3,5-diiodo-**	5579-93-1	NI36421
108	109	82	80	110	83	81	111	**347**	100	52	51	41	33	28	22	12	**0.00**	7	12	–	1	–	–	1	–	–	–	–	–	1	**Mercury, chloro(hexahydro-1H-pyrrolizin-1-yl)-, trans-**	75751-69-8	NI36422
79	80	77	78	39	107	70	51	**347**	100	27	25	21	11	10	7	7	**1.17**	10	9	2	1	1	–	4	–	–	–	–	–	–	**1H-Isoindole-1,3(2H)-dione, 3a,4,7,7a-tetrahydro-2-[(1,1,2,2-tetrachloroethyl)thio]-**	2425-06-1	NI36423
180	314	312	295	152	293	261	263	**347**	100	88	75	46	45	37	32	31	**0.00**	10	9	2	1	1	–	4	–	–	–	–	–	–	**1H-Isoindole-1,3(2H)-dione, 3a,4,7,7a-tetrahydro-2-[(1,1,2,2-tetrachloroethyl)thio]-**	2425-06-1	NI36424
79	80	78	77	144	107	150	149	**347**	100	32	22	21	20	19	12	11	**2.05**	10	9	2	1	1	–	4	–	–	–	–	–	–	**1H-Isoindole-1,3(2H)-dione, 3a,4,7,7a-tetrahydro-2-[(1,1,2,2-tetrachloroethyl)thio]-**	2425-06-1	NI36425
79	77	80	78	39	27	107	51	**347**	100	46	42	30	27	25	15	14	**0.40**	10	9	2	1	1	–	4	–	–	–	–	–	–	**1H-Isoindole-1,3(2H)-dione, 3a,4,7,7a-tetrahydro-2-[(1,1,2,2-tetrachloroethyl)thio]-, cis-**	2939-80-2	NI36426
135	108	28	81	44	54	43	53	**347**	100	35	20	15	15	15	14	11	**0.00**	10	14	7	5	–	–	–	–	–	–	1	–	–	**Adenosine 5′-monophosphate**	61-19-8	NI36427
172	347	174	149	349	102	185	119	**347**	100	90	65	59	56	48	48	44		12	8	1	1	–	5	2	–	–	–	–	–	–	**Isoquinoline, 7,8-dichloro-tetrahydro-, pentafluoropropionyl-**		NI36428
145	220	227	104	255	229	87	89	**347**	100	86	68	52	51	45	44	40	**10.00**	14	15	1	1	2	–	2	–	–	–	–	–	–	**Bicyclo[3.2.0]heptan-7-one, 6,6-dichloro-2,2-dimethyl-3-endo-(methylthio) 5-phenyl-4-thia-1-aza-**		NI36429
318	330	69	264	347	319	248	145	**347**	100	54	28	21	18	15	9	9		14	16	4	3	–	3	–	–	–	–	–	–	–	**Aniline, N-(cyclopropylmethyl)-2,6-dinitro-N-propyl-4-(trifluoromethyl)-**	26399-36-0	NI36430
157	111	144	112	44	43	143	42	**347**	100	98	71	70	69	59	53	51	**4.30**	14	25	5	3	1	–	–	–	–	–	–	–	–	**S-[L-2-(Acetylamino)-2-carboxyethyl]-N-(dimethylaminomethylene)-L-homocysteine methyl ester**		NI36431

MASS TO CHARGE RATIOS								M.W.	INTENSITIES								Parent	C	H	O	N	S	F	Cl	Br	I	Si	P	B	X	COMPOUND NAME	CAS Reg No	No
143	217	44	43	42	56	99	144	347	100	51	50	35	22	22	20	19	3.90	14	25	5	3	1	–	–	–	–	–	–	–	–	S-[L-3-(Acetylamino)-3-carboxypropyl]-N-(dimethylaminomethylene)-L-cysteine methyl ester		NI36432
73	258	116	158	214	103	117	97	347	100	78	70	49	47	46	40	39	0.00	14	29	5	1	–	–	–	–	–	2	–	–	–	α-D-Glucopyranose, 2-(acetylamino)-3,6-anhydro-2-deoxy-1,4-bis-O-(trimethylsilyl)-	56196-86-2	NI36433
73	258	160	159	101	117	97	116	347	100	68	44	44	35	32	30	27	0.00	14	29	5	1	–	–	–	–	–	2	–	–	–	β-D-Glucopyranose, 2-(acetylamino)-3,6-anhydro-2-deoxy-1,4-bis-O-(trimethylsilyl)-	56196-85-1	NI36434
73	75	140	55	45	147	89	108	347	100	64	57	29	19	18	17	15	5.80	14	29	5	1	–	–	–	–	–	2	–	–	–	Heptanedioic acid, 4-(methoxyimino)-, bis(trimethylsilyl) ester	55520-92-8	NI36435
73	75	147	40	45	59	171	103	347	100	29	26	24	13	11	11	10	0.00	14	33	3	1	–	–	–	–	–	3	–	–	–	Monoamidoethylmalonic acid, tris(trimethylsilyl)- deriv.		NI36436
73	230	140	45	75	232	74	147	347	100	50	21	19	13	10	7	6	0.00	14	33	3	1	–	–	–	–	–	3	–	–	–	L-Proline, 1-(trimethylsilyl)-3-[(trimethylsilyl)oxy]-, trimethylsilyl ester		MS08229
230	73	140	231	75	147	45	69	347	100	70	40	23	21	18	13	11	0.00	14	33	3	1	–	–	–	–	–	3	–	–	–	L-Proline, 1-(trimethylsilyl)-4-[(trimethylsilyl)oxy]-, trimethylsilyl ester, trans-	55429-66-8	NI36437
230	73	140	231	45	147	232	75	347	100	75	34	25	11	11	10	10	0.00	14	33	3	1	–	–	–	–	–	3	–	–	–	L-Proline, 1-(trimethylsilyl)-4-[(trimethylsilyl)oxy]-, trimethylsilyl ester, trans-	55429-66-8	NI36438
230	73	140	231	45	147	75	232	347	100	78	27	22	11	9	9	9	0.00	14	33	3	1	–	–	–	–	–	3	–	–	–	L-Proline, 1-(trimethylsilyl)-4-[(trimethylsilyl)oxy]-, trimethylsilyl ester, trans-	55429-66-8	NI36439
105	347	77	51	89	348	220	106	347	100	48	42	14	10	9	9	9		15	10	1	1	–	–	–	–	1	–	–	–	–	Isoxazole, 3,5-diphenyl-4-iodo-		LI07672
69	41	82	70	231	261	53	81	347	100	98	84	80	62	58	53	48	35.00	15	13	7	3	–	–	–	–	–	–	–	–	–	1H-Isoindole-1,3(2H)-dione, 4,7-epoxy-2-(2,4-dinitrophenyl)hexahydro-3a methyl-, [3aR-(3aα,4β,7β,7aα)]-	54346-09-7	NI36440
167	273	107	79	91	183	240		347	100	56	50	36	23	17	6		0.00	15	13	7	3	–	–	–	–	–	–	–	–	–	Tyrosine, N-(2,4-dinitrophenyl)-		LI07673
246	248	290	139	111	247	44	292	347	100	33	24	22	20	17	9	9	0.00	15	15	5	1	–	–	1	–	–	–	–	–	1	Sodium 5-(4-chlorobenzoyl)-1,4-dimethyl-1H-pyrrole-2-acetatedihydrate		NI36441
215	81	186	347	44	214	216	53	347	100	47	41	34	20	19	17	13		15	17	5	5	–	–	–	–	–	–	–	–	–	Adenosine, N-(2-furanylmethyl)-	4338-47-0	NI36442
168	347	233	259	103	287	169	246	347	100	92	48	43	36	32	28	26		15	18	2	1	–	–	–	–	–	–	–	–	1	[η^4-2-(1-Nitro-2-propyl)norbornadiene](η^5-cyclopentadienyl)rhodium	90752-91-3	NI36443
250	150	177	176	220	246	251	136	347	100	68	61	51	23	17	16	14	6.99	15	20	3	3	–	3	–	–	–	–	–	–	–	Butyl [(4,6-dimethylpyrimidin-2-yl)(trifluoroacetyl)amino]propionate		MS08230
43	169	72	28	184	242	97	59	347	100	75	40	38	34	26	21	19	17.02	15	25	6	1	1	–	–	–	–	–	–	–	–	α-D-Galactopyranose, 1,2:3,4-bis-O-isopropylidene-, dimethylcarbamothioate	16795-64-5	NI36444
115	98	73	45	112	72	154	87	347	100	75	70	40	28	26	25	20	0.00	15	25	8	1	–	–	–	–	–	–	–	–	–	Methyl 3,4-di-O-acetyl-2-deoxy-6-O-methyl-2-(N-methylacetamido)-α-D-glucopyranoside	56341-54-9	NI36445
115	43	142	98	73	87	274	242	347	100	70	61	48	42	27	21	21	0.00	15	25	8	1	–	–	–	–	–	–	–	–	–	Methyl 3,6-di-O-acetyl-2-deoxy-4-O-methyl-2-(N-methylacetamido)-α-D-glucopyranoside	56341-53-8	NI36446
87	129	98	43	115	75	142	128	347	100	76	69	60	53	40	34	30	0.79	15	25	8	1	–	–	–	–	–	–	–	–	–	Methyl 4,6-di-O-acetyl-2-deoxy-3-O-methyl-2-(N-methylacetamido)-α-D-glucopyranoside	56341-51-6	NI36447
87	43	143	84	128	86	126	100	347	100	66	42	32	24	21	17	16	3.49	15	25	8	1	–	–	–	–	–	–	–	–	–	Methyl 3,4,6-tri-O-acetyl-2-deoxy-2-dimethylamino-α-D-glucopyranoside	81093-28-9	NI36448
161	71	129	101	246	75	42	186	347	100	45	36	26	25	23	22	21	0.30	15	25	8	1	–	–	–	–	–	–	–	–	–	L-Proline, 4-methoxy-1-methyl-5-oxo-4-(tetrahydro-2,3,4-trimethoxy-2-furanyl)-, methyl ester	56817-95-9	MS08231
174	73	332	147	96	86	59	45	347	100	42	16	13	10	10	9	5	2.00	15	37	2	1	–	–	–	–	–	3	–	–	–	Hexanoic acid, 6-[bis(trimethylsilyl)amino]-, trimethylsilyl ester	25688-76-0	MS08232
174	73	332	147	96	75	86	188	347	100	43	17	14	12	12	10	2	2.00	15	37	2	1	–	–	–	–	–	3	–	–	–	Hexanoic acid, 6-[bis(trimethylsilyl)amino]-, trimethylsilyl ester	25688-76-0	LI07674
174	73	332	147	96	75	86	59	347	100	42	15	12	10	10	9	8	1.80	15	37	2	1	–	–	–	–	–	3	–	–	–	Hexanoic acid, 6-[bis(trimethylsilyl)amino]-, trimethylsilyl ester	25688-76-0	NI36449
188	347	220	54	26	349	75	160	347	100	75	71	46	38	29	29	23		16	10	4	1	1	–	1	–	–	–	–	–	–	4-Chlorophenyl 4-(N-maleimido)phenyl sulphone		IC04071
204	205	203	102	206	176	88	51	347	100	46	12	11	7	6	5	3	0.00	16	14	–	1	–	–	–	–	1	–	–	–	–	Isoquinolinium, 2-methyl-1-phenyl-, iodide	52806-07-2	NI36450
205	204	206	102	203	76	75	51	347	100	75	16	12	11	6	5	5	0.00	16	14	–	1	–	–	–	–	1	–	–	–	–	Quinolinium, 1-methyl-2-phenyl-, iodide	14886-84-1	NI36451
43	230	170	188	60	128	56	41	347	100	70	55	45	40	35	35	30	0.00	16	29	5	1	1	–	–	–	–	–	–	–	–	3-O-Acetyl-6-acetylamino-4,6-dideoxy-1-hexylthio-α-DL-lyxo-hexopyranoside		NI36452
268	253	269	42	132	165	152	58	347	100	43	20	18	15	13	13	12	5.92	17	18	2	1	–	–	–	1	–	–	–	–	–	Hasubanan-3-ol, 6-bromo-7,8-didehydro-4,5-epoxy-17-methyl-, (5α,6β,13β,14β)-	2670-52-2	NI36453
131	89	201	347	218	213	174	315	347	100	28	20	3	3	3	3	2		17	21	5	3	–	–	–	–	–	–	–	–	–	1,3-Butanediol, 2-methoxy-1-(2-phenyl-2H-1,2,3-triazol-4-yl)-, diacetate (ester), [1S-(1R*,2R*,3R*)]-	35405-78-8	NI36454
130	258	131	347	129	170	259	231	347	100	96	76	43	39	36	34	31		17	21	5	3	–	–	–	–	–	–	–	–	–	Glycine, N-(N-carboxy-L-tryptophyl)-, N-ethyl methyl ester	21026-89-1	NI36455
91	107	92	65	108	79	39	77	347	100	16	8	8	5	4	4	3	0.01	18	18	4	1	–	–	1	–	–	–	–	–	–	L-Alanine, 3-chloro-N-(benzyloxycarbonyl)-, benzyl ester	55822-82-7	MS08233
192	91	347	193	42	164	163	155	347	100	22	13	13	11	10	9	8		18	21	4	1	1	–	–	–	–	–	–	–	–	Benzenesulphonamide, N-(4,8-dioxotricyclo[3.3.1.1^{3,7}]dec-2-yl)-N,4-dimethyl-	56781-98-7	MS08234
91	159	146	158	92	65	347	68	347	100	18	12	10	10	8	6	5		18	21	6	1	–	–	–	–	–	–	–	–	–	7-Azabicyclo[2.2.1]heptane-2,3,7-tricarboxylic acid, 7-benzyl dimethyl ester, trans-	17037-56-8	NI36456
245	347	246	230	244	288	43	59	347	100	32	25	23	5	3	3	2		18	21	6	1	–	–	–	–	–	–	–	–	–	2,3-Butanediol, 1-[(4,7-dimethoxyfuro[2,3-b]quinolin-6-yl)oxy]-3-methyl,-(–)-	77145-74-5	NI36457
227	244	59	347	245	228	216	230	347	100	54	44	40	36	26	26	25		18	21	6	1	–	–	–	–	–	–	–	–	–	2,3-Butanediol, 1-[(4,8-dimethoxyfuro[2,3-b]quinolin-7-yl)oxy]-3-methyl-	522-11-2	NI36458
318	205	347	274	273				347	100	40	33	32	15					18	21	6	1	–	–	–	–	–	–	–	–	–	Crinan-11-ol, 1,2-epoxy-3,7-dimethoxy-, (1β,2β,3α,11R)-	58189-38-1	NI36459

MASS TO CHARGE RATIOS								M.W.	INTENSITIES								Parent	C	H	O	N	S	F	Cl	Br	I	Si	P	B	X	COMPOUND NAME	CAS Reg No	No
125	329	96	311	313	330	282	94	**347**	100	96	29	24	22	20	17	16	**1.19**	18	21	6	1	–	–	–	–	–	–	–	–	–	**Lycorenan-5,7-diol, 11-methoxy-1-methyl-9,10-[methylenebis(oxy)]-, (5α,7α)-**	905-37-3	**NI36460**
83	96	57	97	82	69	43	71	**347**	100	69	24	22	22	19	19	16	**13.89**	18	21	6	1	–	–	–	–	–	–	–	–	–	**Neronine, 4β,5-dihydro-**	19483-30-8	**NI36461**
101	105	332	146	77	290	145	104	**347**	100	97	33	27	21	16	12	11	**5.56**	18	21	6	1	–	–	–	–	–	–	–	–	–	**2-Oxazoline, 4,5-(3-O-acetyl-2-deoxy-5,6-O-isopropylidene-α-D-glucofurano)-2-phenyl-**		**NI36462**
202	73	347	215	203	45	75	348	**347**	100	88	36	30	21	13	12	11		18	29	2	1	–	–	–	–	–	2	–	–	–	**1H-Indole-3-butanoic acid, 1-(trimethylsilyl)-, trimethylsilyl ester**	55429-34-0	**NI36463**
202	73	347	215	203	348	332	75	**347**	100	42	38	34	23	13	9	8		18	29	2	1	–	–	–	–	–	2	–	–	–	**1H-Indole-3-butanoic acid, 1-(trimethylsilyl)-, trimethylsilyl ester**	55429-34-0	**NI36464**
149	105	150	205	185	151	161	159	**347**	100	28	10	2	2	2	1	1	**0.00**	19	25	5	1	–	–	–	–	–	–	–	–	–	**Ethaneperoxoic acid, 1-cyano-1-[2-(2-phenyl-1,3-dioxolan-2-yl)ethyl]pentyl ester**	58422-92-7	**NI36465**
221	220	41	222	43	69	55	125	**347**	100	28	21	17	13	11	11	10	**3.00**	19	30	4	1	–	–	–	–	–	–	–	1	–	**2,4-Dimethyl-6-[2-(piperidine-2,6-dion-4-yl)acetyl]cyclohex-1(or 6)-enol butyl boronate**		**MS08235**
347	289	260	330	318				**347**	100	19	18	10	10					20	13	5	1	–	–	–	–	–	–	–	–	–	**12-Methoxycarbonyl-5-methyl-5-azachrysine-1,4,6-trione**		**LI07675**
347	346	44	318	289	69	159	203	**347**	100	62	34	25	14	12	11	9		20	13	5	1	–	–	–	–	–	–	–	–	–	**Oxysanguinarine**	548-30-1	**NI36466**
347	80	349	348	57	55	105	77	**347**	100	37	29	21	15	12	11	10		20	14	1	3	–	–	1	–	–	–	–	–	–	**7-Chloro-2,3-dihydro-5-phenyl-3-[(pyrrolyl-2)methylene]-1H-1,4-benzodiazepin-2-one**	55056-40-1	**NI36467**
86	99	167	87	165	58	36	30	**347**	100	9	8	7	5	5	5	5	**0.00**	20	26	2	1	–	–	1	–	–	–	–	–	–	**Benzeneacetic acid, α-phenyl-, 2-(diethylamino)ethyl ester, hydrochloride**	50-42-0	**NI36468**
105	139	178	77	103	286	123	152	**347**	100	88	58	42	30	23	18	15	**0.00**	21	17	2	1	1	–	–	–	–	–	–	–	–	**(E)-1-[(4-Methylphenyl)thio]-2-nitro-1,2-diphenylethylene**	73013-89-5	**DD01368**
288	228	315	347	230	229	272	244	**347**	100	34	31	26	20	12	8	8		21	17	4	1	–	–	–	–	–	–	–	–	–	**Methyl 2-(methoxycarbonyl)pyrrolo[1,2-f]phenanthridine-1-acetate**		**LI07676**
98	138	112	136	194	180	152	212	**347**	100	49	20	11	10	8	8	6	**0.00**	21	33	3	1	–	–	–	–	–	–	–	–	–	**rel-[2S,6S,2(3R)]-2-[3-[[(Benzyloxy)carbonyl]oxy]heptyl]-6-methylpiperidine**		**DD01369**
59	297	50	43	71	84	29	18	**347**	100	11	8	4	3	2	2	2	**0.00**	21	46	–	1	–	–	1	–	–	–	–	–	–	**Stearyltrimethylammonium chloride**	112-03-8	**NI36469**
217	174	105	218	143	175	121	77	**347**	100	78	43	17	11	10	6	6	**0.00**	22	25	1	3	–	–	–	–	–	–	–	–	–	**Benzamide, N-[1-[2-(1H-indol-3-yl)ethyl]-4-piperidinyl]-**	26844-12-2	**NI36470**
274	57	218	291	41	275	29	332	**347**	100	85	58	30	29	21	19	18	**8.10**	22	37	2	1	–	–	–	–	–	–	–	–	–	**Benzene, 1,3,5-tris(2,2-dimethylpropyl)-2-methyl-4-nitro-**	40572-20-1	**MS08236**
100	87	316	70	317	113	88	74	**347**	100	55	51	26	16	9	9	7	**5.02**	22	37	2	1	–	–	–	–	–	–	–	–	–	**Pregnan-20-one, 3-hydroxy-, O-methyloxime, (3α,5α)-**		**NI36471**
100	87	316	70	88	317	298	347	**347**	100	57	30	28	14	12	11	10		22	37	2	1	–	–	–	–	–	–	–	–	–	**Pregnan-20-one, 3-hydroxy-, O-methyloxime, (3α,5β)-**		**NI36472**
100	87	316	70	91	92	317	113	**347**	100	68	45	23	13	12	11	7	**4.72**	22	37	2	1	–	–	–	–	–	–	–	–	–	**Pregnan-20-one, 3-hydroxy-, O-methyloxime, (3β,5α)-**	3091-40-5	**NI36473**
100	87	316	70	92	91	317	298	**347**	100	62	45	26	17	16	12	8	**4.65**	22	37	2	1	–	–	–	–	–	–	–	–	–	**Pregnan-20-one, 3-hydroxy-, O-methyloxime, (3β,5β)-**	75112-90-2	**NI36474**
70	192	277	208	207	193	278	262	**347**	100	75	73	48	38	18	16	14	**2.00**	23	25	2	1	–	–	–	–	–	–	–	–	–	**6-Ethyl-2-isopropylidene-5,5-dimethyldibenzo-3-oxa-1-azabicyclo[4.2.2]deca-7,9-dien-4-one**		**LI07677**
100	304	110	347	71	84	72	59	**347**	100	37	35	26	17	12	10	10		23	41	1	1	–	–	–	–	–	–	–	–	–	**Pregnan-1-ol, 3-(dimethylamino)-, (1α,3α,5α)-**	17808-81-0	**LI07678**
100	304	110	347	71	83	72	56	**347**	100	36	34	27	17	11	10	9		23	41	1	1	–	–	–	–	–	–	–	–	–	**Pregnan-1-ol, 3-(dimethylamino)-, (1α,3α,5α)-**	17808-81-0	**NI36475**
100	110	71	347	304				**347**	100	47	29	20	5					23	41	1	1	–	–	–	–	–	–	–	–	–	**Pregnan-1-ol, 3-(dimethylamino)-, (1α,3β,5α)-**		**LI07679**
100	71	110	347	59	56	72	101	**347**	100	27	15	11	10	9	8	7		23	41	1	1	–	–	–	–	–	–	–	–	–	**Pregnan-1-ol, 3-(dimethylamino)-, (1β,3α,5α)-**	23931-07-9	**LI07680**
100	71	110	347	56	53	101	72	**347**	100	25	15	11	10	9	8	8		23	41	1	1	–	–	–	–	–	–	–	–	–	**Pregnan-1-ol, 3-(dimethylamino)-, (1β,3α,5α)-**	23931-07-9	**NI36476**
100	71	110	347	302				**347**	100	29	18	7	4					23	41	1	1	–	–	–	–	–	–	–	–	–	**Pregnan-1-ol, 3-(dimethylamino)-, (1β,3β,5α)-**		**LI07681**
347	98	260	58	115	71	348	84	**347**	100	75	31	31	29	25	24	16		23	41	1	1	–	–	–	–	–	–	–	–	–	**Pregnan-3-ol, 2-(dimethylamino)-, (2β,3α)-**	69830-83-7	**NI36477**
72	73	55	41	66	58	56	43	**347**	100	13	5	5	3	3	3	3	**0.10**	23	41	1	1	–	–	–	–	–	–	–	–	–	**Pregnan-18-ol, 20-methyl-20-(methylamino)-, (5α)-**	32253-20-6	**NI36478**
347	77	167	166	180	270	89	103	**347**	100	59	53	43	17	10	7	6		26	21	–	1	–	–	–	–	–	–	–	–	–	**Benzenemethanamine, N-(diphenylmethylene)-α-phenyl-**	5350-59-4	**LI07682**
167	347	165	168	166	152	346	348	**347**	100	27	27	13	11	11	9	7		26	21	–	1	–	–	–	–	–	–	–	–	–	**Benzenemethanamine, N-(diphenylmethylene)-α-phenyl-**	5350-59-4	**NI36479**
69	281	48	181	243	67	231	262	**348**	100	48	48	36	28	24	16	14	**0.00**	6	–	1	–	1	12	–	–	–	–	–	–	–	**2,2,3,3-Tetrakis(trifluoromethyl)thiirane-1-oxide**		**DD01370**
348	111	75	50	313	78	113	74	**348**	100	44	34	23	15	15	14	14		6	4	–	–	–	–	2	–	–	–	–	–	1	**(4-Chlorophenyl)mercuric chloride**		**LI07683**
350	352	348	179	252	250	354	215	**348**	100	72	60	31	27	27	24	21		8	5	1	–	–	–	5	–	–	2	–	–	–	**1-Oxa-2-silanaphthalene, 2,2-dichloro-3-trichlorosilyl-**	73063-01-1	**NI36480**
147	333	31	89	73	334	256	182	**348**	100	94	30	27	20	12	10	4	**0.01**	8	15	4	–	–	6	–	–	–	1	1	–	–	**Dimethyl [2,2,2-trifluoro-1-(trifluoromethyl)-1-(trimethylsiloxy)ethyl]phosphonate**		**MS08237**
262	264	260	249	247	292	290	266	**348**	100	89	67	60	60	49	47	47	**25.32**	9	8	3	–	–	–	–	–	–	–	–	–	1	**Tungsten, methyltricarbonyl-π-cyclopentadienyl-**	12082-27-8	**NI36481**
174	279	348	124	69	229	155	93	**348**	100	41	31	28	23	20	17	14		10	–	–	–	–	12	–	–	–	–	–	–	–	**4,7-Methano-1H-indene, 1,1,2,3,3a,4,5,6,7,7a,8,8-dodecafluoro-3a,4,7,7a-tetrahydro-**	1482-08-2	**NI36482**
112	56	208	180	152	292	236	320	**348**	100	98	96	96	95	94	94	89	**0.00**	10	4	7	–	–	–	–	–	–	–	–	–	2	**Iron, heptacarbonyl-μ-[(1-3-η³:2-η)-2-propene-1,2-diyl]di-**	12300-81-1	**DD01371**
236	208	180	56	154	348	128	112	**348**	100	76	74	73	72	47	39	36		10	4	7	–	–	–	–	–	–	–	–	–	2	**Iron, hexacarbonyl[μ-[(1,2-η:3,4-η)-1-hydroxy-1,3-butadiene-1,4-diyl]]di-, (Fe-Fe)**	74381-48-9	**NI36483**
250	350	252	348	215	352	315	313	**348**	100	83	65	58	48	46	33	33		12	8	–	–	–	–	4	–	–	2	–	–	–	**9,10-Disilaanthracene, 9,9,10,10-tetrachloro-9,10-dihydro-**	32962-41-7	**NI36484**
348	333	350	305	206	113	148	99	**348**	100	43	32	25	21	18	17	16		12	10	2	–	–	–	1	–	1	–	–	–	–	**Naphthalene, 1-chloro-5-iodo-2,6-dimethoxy-**	25315-07-5	**NI36485**
146	248	348	102	88	70	76	29	**348**	100	81	46	24	22	19	18	18		12	12	4	–	4	–	–	–	–	–	–	–	–	**Cyclohexanecarboxylic acid, 1-hydroxy-**		**MS08238**

MASS TO CHARGE RATIOS								M.W.	INTENSITIES								Parent	C	H	O	N	S	F	Cl	Br	I	Si	P	B	X	COMPOUND NAME	CAS Reg No	No
197	224	256	107	165	211	225	192	**348**	100	33	27	27	25	17	16	16	**0.30**	12	16	8	2	1	–	–	–	–	–	–	–	–	Methyl 7-(methoxycarbonylamino)-3-(methoxymethyl)-3-cephem-4-carboxylate 1,1-dioxide	97609-95-5	NI36486
301	303	302	304	300	299	305	261	**348**	100	96	80	77	77	71	46	44	**15.53**	12	20	–	–	–	–	–	–	–	–	–	–	1	Tungsten, tetrakis(η^3-2-propenyl)-	11077-54-6	NI36487
174	348	73	159	217	246	225	116	**348**	100	90	90	36	25	22	22	22		12	36	–	4	–	–	–	–	–	4	–	–	–	Dodecamethyl-1,2,5,6-tetrasila-3,4,7,8-tetrazacyclooctane		MS08239
181	348	329	328	180	279	298	278	**348**	100	95	48	48	21	19	15	14		13	2	–	–	–	10	–	–	–	–	–	–	–	Benzene, 1,1′-methylenebis[2,3,4,5,6-pentafluoro-	5736-46-9	NI36488
161	163	187	189	165	124	162	99	**348**	100	73	32	19	12	12	11	9	**7.56**	13	8	1	2	–	–	4	–	–	–	–	–	–	N,N′-Bis(2,3-dichlorophenyl)urea		IC04072
161	163	187	189	28	124	90	126	**348**	100	63	56	37	29	28	15	14	**3.70**	13	8	1	2	–	–	4	–	–	–	–	–	–	N,N′-Bis(2,3-dichlorophenyl)urea		MS08240
161	163	187	189	124	126	28	63	**348**	100	62	49	32	27	13	13	12	**3.30**	13	8	1	2	–	–	4	–	–	–	–	–	–	N,N′-Bis(2,4-dichlorophenyl)urea		MS08241
161	163	187	162	189	124	28	63	**348**	100	62	43	31	27	21	14	12	**4.60**	13	8	1	2	–	–	4	–	–	–	–	–	–	N,N′-Bis(2,5-dichlorophenyl)urea		MS08242
166	187	168	189	44	124	43	170	**348**	100	48	34	30	21	17	17	9	**5.00**	13	8	1	2	–	–	4	–	–	–	–	–	–	N,N′-Bis(3,4-dichlorophenyl)urea		IC04073
161	163	187	124	63	189	62	126	**348**	100	59	53	38	35	35	24	21	**0.61**	13	8	1	2	–	–	4	–	–	–	–	–	–	N,N′-Bis(3,5-dichlorophenyl)urea	73439-19-7	NI36489
28	52	156	104	236	208	78	103	**348**	100	17	10	7	6	6	6	4	**0.80**	13	8	5	–	–	–	–	–	–	–	–	–	2	Chromium, pentacarbonyl(1,3,5,7-cyclooctatetraene)di-	61178-32-3	NI36490
73	147	45	333	75	171	69	148	**348**	100	77	25	20	16	13	13	12	**0.00**	13	28	5	–	–	–	–	–	–	3	–	–	–	Butanedioic acid, oxo(trimethylsilyl)-, bis(trimethylsilyl) ester	55517-46-9	NI36491
147	73	333	75	200	171	148	221	**348**	100	71	26	26	20	15	15	14	**0.00**	13	28	5	–	–	–	–	–	–	3	–	–	–	Butanedioic acid, oxo(trimethylsilyl)-, bis(trimethylsilyl) ester	55517-46-9	MS08243
73	147	75	333	45	171	44	148	**348**	100	81	46	39	26	14	14	12	**0.30**	13	28	5	–	–	–	–	–	–	3	–	–	–	Butanedioic acid, oxo(trimethylsilyl)-, bis(trimethylsilyl) ester	55517-46-9	NI36492
73	44	116	231	159	75	132	147	**348**	100	69	58	45	38	35	25	24	**4.90**	13	32	3	2	–	–	–	–	–	3	–	–	–	L-Asparagine, N,N^2-bis(trimethylsilyl)-, trimethylsilyl ester	55649-62-2	NI36493
73	116	231	132	75	188	147	45	**348**	100	56	29	25	20	14	14	14	**1.80**	13	32	3	2	–	–	–	–	–	3	–	–	–	L-Asparagine, N,N^2-bis(trimethylsilyl)-, trimethylsilyl ester	55649-62-2	NI36494
174	73	175	333	176	102	86	147	**348**	100	36	19	18	13	11	9	8	**0.72**	13	32	3	2	–	–	–	–	–	3	–	–	–	Glycylglycine, tris(trimethylsilyl)-		NI36495
274	273	275	272	194	77	276	91	**348**	100	95	92	85	54	33	20	19	**13.10**	14	13	–	4	1	–	–	1	–	–	–	–	–	5-Bromo-2-aminobenzophenone thiosemicarbazone		MS08244
120	42	105	148	77	350	348	79	**348**	100	42	22	21	20	15	15	15		14	13	2	4	–	–	–	1	–	–	–	–	–	Aniline, 4-[(2-bromo-4-nitrophenyl)azo]-N,N-dimethyl-	74421-32-2	NI36496
55	333	83	69	348	67	95	305	**348**	100	98	82	59	42	42	41	40		14	15	2	–	–	7	–	–	–	–	–	–	–	3-Pinanone, 4-(heptafluorobutyryl)-		NI36497
83	333	149	69	67	291	95	305	**348**	100	83	75	70	39	38	37	36	**33.00**	14	15	2	–	–	7	–	–	–	–	–	–	–	4-Pinanone, 3-(heptafluorobutyryl)-		NI36498
348	122	123	291	277	305	333	320	**348**	100	97	82	63	58	57	55	55		14	15	2	–	–	7	–	–	–	–	–	–	–	3-Thujanone, 2-(heptafluorobutyryl)-		NI36499
55	67	93	108	77	63	137	348	**348**	100	43	31	30	29	21	6	5		14	18	2	–	–	–	–	–	–	–	–	–	1	2-Oxabicyclo[3.3.0]octane, 8-[(4-methoxyphenyl)telluro]-		DD01372
43	45	97	84	83	77	55	41	**348**	100	77	67	63	55	55	47	47	**3.00**	14	18	4	2	–	–	2	–	–	–	–	–	–	L-Lysine, N^2-[(2,4-dichlorophenoxy)acetyl]-	50649-08-6	NI36500
43	127	115	157	169	116	103	144	**348**	100	99	67	51	38	33	33	30	**0.00**	14	20	10	–	–	–	–	–	–	–	–	–	–	Methyl (methyl 2,3,4-tri-O-acetyl-α-D-mannopyranoside)uronate	87936-90-1	NI36501
43	97	115	45	98	60	157	103	**348**	100	99	85	72	66	57	42	40	**0.00**	14	20	10	–	–	–	–	–	–	–	–	–	–	2,3,4,6-Tetra-O-acetyl-D-glucopyranose	10343-06-3	NI36502
73	216	117	147	205	89	217	59	**348**	100	80	42	40	27	15	12	12	**1.00**	14	28	6	–	–	–	–	–	–	2	–	–	–	L-Ascorbic acid, 2,3-di-O-methyl-5,6-bis-O-(trimethylsilyl)-	57397-26-9	NI36503
73	216	117	147	205	89	45	59	**348**	100	99	88	73	42	32	32	31	**0.79**	14	28	6	–	–	–	–	–	–	2	–	–	–	D-erythro-Hex-2-enonic acid, 2,3-di-O-methyl-5,6-bis-O-(trimethylsilyl)-, γ-lactone	57397-28-1	NI36504
73	216	117	147	205	89	59	45	**348**	100	42	37	30	18	13	13	13	**0.00**	14	28	6	–	–	–	–	–	–	2	–	–	–	D-erythro-Hex-2-enonic acid, di-O-methylbis-O-(trimethylsilyl)-, γ-lactone	56211-35-9	NI36505
147	73	333	45	171	148	97	127	**348**	100	87	24	21	18	18	16	15	**2.09**	14	32	4	–	–	–	–	–	–	3	–	–	–	Propanedioic acid, ethyl-, tris(trimethylsilyl)-		NI36506
147	73	142	70	43	128	75	115	**348**	100	99	79	68	49	45	44	34	**26.00**	14	36	2	2	–	–	–	–	–	3	–	–	–	N,N′-Tris(trimethylsilyl)-2,5-diaminovalerolactam		MS08245
350	321	348	349	319	322	315	313	**348**	100	79	77	71	52	48	47	47		15	10	1	2	–	–	1	1	–	–	–	–	–	1H-1,4-Benzodiazepin-2-one, 7-bromo-5-(2-chlorophenyl)-2,3-dihydro-	51753-57-2	NI36507
269	304	275	277	306	350	305	303	**348**	100	72	59	43	42	38	36	35	**31.00**	15	10	1	2	–	–	1	1	–	–	–	–	–	1H-1,4-Benzodiazepin-2-one, 5-(2-bromophenyl)-7-chloro-2,3-dihydro-	63574-83-4	NI36508
125	77	51	127	89	193	141	50	**348**	100	77	31	27	26	22	16	15	**0.49**	15	13	2	4	1	–	1	–	–	–	–	–	–	Benzenesulphonamidate, N-[1-(4-chlorobenzyl)-4-(1,2,4-triazolio)]-	32539-61-0	NI36509
125	167	153	95	111	138	180	68	**348**	100	84	65	40	37	35	30	29	**19.00**	15	17	1	6	1	1	–	–	–	–	–	–	–	1,3,5-Triazine, 2-acetamido-4-[3-(4-fluorophenyl)thioureido]-6-propyl-		NI36510
135	348	198	197	73	271	275	256	**348**	100	20	20	19	12	3	2	1		15	19	–	–	–	–	–	–	–	1	–	–	1	Diphenyl(trimethylsilyl)stibine		MS08246
249	67	348	166	167	183	150	265	**348**	100	34	32	26	10	7	5	4		15	21	6	–	–	–	–	–	–	–	–	–	1	Tris(acetylacetonato)vanadium		LI07684
43	101	88	143	85	127	144	87	**348**	100	99	63	58	45	40	39	34	**0.00**	15	24	9	–	–	–	–	–	–	–	–	–	–	α-D-Galactopyranoside, methyl 2,3,6-tri-O-acetyl-4-O-ethyl-		NI36511
127	101	99	85	59	115	55	289	**348**	100	42	40	33	31	30	28	27	**0.00**	15	24	9	–	–	–	–	–	–	–	–	–	–	α-D-Mannopyranoside, methyl 2,3,4-tri-O-acetyl-6-O-ethyl-		NI36512
43	88	89	130	115	131	74	99	**348**	100	99	81	32	17	16	16	12	**0.00**	15	24	9	–	–	–	–	–	–	–	–	–	–	α-D-Mannopyranoside, methyl 2,4,6-tri-O-acetyl-3-O-ethyl-		NI36513
43	88	112	127	101	143	130	70	**348**	100	99	40	32	30	28	25	14	**0.00**	15	24	9	–	–	–	–	–	–	–	–	–	–	α-D-Mannopyranoside, methyl 3,4,6-tri-O-acetyl-2-O-ethyl-		NI36514
350	348	246	244	218	216	104	103	**348**	100	99	99	99	55	55	40	40		16	13	4	–	–	–	–	1	–	–	–	–	–	4H-1-Benzopyran-4-one, 6-bromo-5,7-dihydroxy-8-methyl-2-phenyl-2,3-dihydro-, (S)-(–)-	84314-27-2	NI36515
246	244	350	348	103	273	271	104	**348**	100	99	75	75	40	30	30	30		16	13	4	–	–	–	–	1	–	–	–	–	–	4H-1-Benzopyran-4-one, 8-bromo-5,7-dihydroxy-6-methyl-2-phenyl-2,3-dihydro-, (±)-	84314-28-3	NI36516
317	172	319	90	133	174	318	348	**348**	100	54	34	34	20	19	18	17		16	17	3	4	–	–	1	–	–	–	–	–	–	N-Ethyl-N-(2-hydroxyethyl)-4-(2-chloro-4-nitrophenylazo)aniline		IC04074
173	158	104	55	131	174	72	56	**348**	100	57	51	23	17	14	13	13	**0.00**	16	20	1	4	2	–	–	–	–	–	–	–	–	4-Thia-1-azabicyclo[3.2.0]heptan-7-one, 6-endo-azido-2-exo-isopropyl-2-endo-methyl-3-exo-(methylthio)-5-phenyl-		NI36517
173	158	104	131	119	174	55	72	**348**	100	51	44	17	16	15	15	12	**0.00**	16	20	1	4	2	–	–	–	–	–	–	–	–	4-Thia-1-azabicyclo[3.2.0]heptan-7-one, 6-exo-azido-2-exo-isopropyl-2-endo-methyl-3-endo-(methylthio)-5-phenyl-		NI36518
42	256	272	287	68	70	255	316	**348**	100	60	59	51	41	25	20	14	**1.00**	16	20	5	4	–	–	–	–	–	–	–	–	–	Azirino[2′,3′:3,4]pyrrolo[1,2-a]indole-4,7-dione, 6-amino-8-[[(aminocarbonyl)oxy]methyl]-1,1a,2,8,8a,8b-hexahydro-8a-methoxy 1,5-dimethyl-, [1aR-(1aα,8β,8aα,8bα)]-	801-52-5	LI07685

MASS TO CHARGE RATIOS								M.W.	INTENSITIES								Parent	C	H	O	N	S	F	Cl	Br	I	Si	P	B	X	COMPOUND NAME	CAS Reg No	No
42	272	44	256	287	68	70	43	**348**	100	56	56	55	48	41	24	24	**8.01**	16	20	5	4	–	–	–	–	–	–	–	–	–	**Azirino[2′,3′:3,4]pyrrolo[1,2-a]indole-4,7-dione, 6-amino-8-[[(aminocarbonyl)oxy]methyl]-1,1a,2,8,8a,8b-hexahydro-8a-methoxy 1,5-dimethyl-, [1aR-(1aα,8β,8aα,8bα)]-**	801-52-5	**NI36519**
259	168	348	89	61	169	317	319	**348**	100	70	55	55	36	32	22	21		16	21	2	–	–	–	–	–	–	–	–	–	1	**Rhodium, (η^4-2-(methoxyethoxymethyl)norbornadiene)(η^5-cyclopentadienyl)-**	97373-72-3	**NI36520**
41	57	29	45	87	101	69	27	**348**	100	91	74	71	56	49	47	43	**0.00**	16	22	4	–	–	–	2	–	–	–	–	–	–	**Butanoic acid, 4-(2,4-dichlorophenoxy)-, 2-(2-methylpropoxy)ethyl ester**	62059-41-0	**NI36521**
100	118	56	88	58	114	248	102	**348**	100	73	73	68	68	64	59	59	**25.00**	16	32	6	2	–	–	–	–	–	–	–	–	–	**3,10-Dihydroxy-5,8,15,20-tetraoxa-1,12-diazabicyclo[10,5,5]docosane**		**MS08247**
100	56	114	58	88	70	144	72	**348**	100	99	94	83	78	72	61	61	**11.10**	16	32	6	2	–	–	–	–	–	–	–	–	–	**12,16-Dihydroxy-4,7,14,20-tetraoxa-1,10-diazabicyclo[8.7.5]docosane**		**MS08248**
56	58	100	70	57	72	86	144	**348**	100	86	85	55	55	53	43	42	**0.00**	16	32	6	2	–	–	–	–	–	–	–	–	–	**20,21-Dihydroxy-4,7,13,16-tetraoxa-1,10-diazabicyclo[8.8.4]docosane**		**MS08249**
179	121	235	291	177	120	123	135	**348**	100	76	71	68	24	23	20	6	**0.00**	16	36	–	–	–	–	–	–	–	–	–	–	1	**Stannane, tetrabutyl-**	1461-25-2	**NI36522**
189	187	235	291	233	185	289	121	**348**	100	93	88	69	65	59	52	43	**0.00**	16	36	–	–	–	–	–	–	–	–	–	–	1	**Stannane, tetrabutyl-**	1461-25-2	**MS08250**
179	235	291	121	177	123	120	135	**348**	100	87	66	33	22	14	9	3	**0.10**	16	36	–	–	–	–	–	–	–	–	–	–	1	**Stannane, tetrabutyl-**	1461-25-2	**LI07686**
235	179	291	121	135	177	120	123	**348**	100	77	40	31	15	12	11	6	**0.00**	16	36	–	–	–	–	–	–	–	–	–	–	1	**Stannane, tetraisobutyl-**		**LI07687**
235	179	233	231	43	175	41	234	**348**	100	81	76	45	43	42	38	35	**0.00**	16	36	–	–	–	–	–	–	–	–	–	–	1	**Stannane, tetraisobutyl-**		**MS08251**
348	307	121	316	317	284	245	288	**348**	100	80	59	55	46	35	32	30		17	16	4	–	2	–	–	–	–	–	–	–	–	**2-Allylthio-5-phenyl-3,4-thiophenedicarboxylic acid, dimethyl ester**		**NI36523**
333	348	305	318	290	289			**348**	100	90	6	3	3	2				17	16	8	–	–	–	–	–	–	–	–	–	–	**Xanthone, 1,8-dihydroxy-3,5,6,7-tetramethoxy-**		**NI36524**
156	91	118	197	72	348	234	274	**348**	100	80	40	20	18	14	14	12		17	20	4	2	1	–	–	–	–	–	–	–	–	**Methyl benzylpenillinate**		**LI07688**
95	136	121	93	195	110	149	75	**348**	100	79	56	54	43	32	19	16	**0.00**	17	20	6	2	–	–	–	–	–	–	–	–	–	**Isoborneol 3,5-dinitrobenzoate**		**MS08252**
348	217	72	139	183	63	185	107	**348**	100	82	82	68	36	29	22	18		17	21	–	2	2	–	–	–	–	–	1	–	–	**1-Butyl-3-(diphenylphosphinothioyl)thiourea**		**LI07689**
43	83	163	125	124	97	164	151	**348**	100	96	40	38	37	35	23	20	**0.10**	17	24	–	4	2	–	–	–	–	–	–	–	–	**Imidazo[1,2-e]1,3,5-thiadiazine-4-thione, 2-(cyclohexylimino)-3-cyclohexyl-**		**NI36525**
232	333	73	117	233	75	175	276	**348**	100	47	32	23	19	19	17	15	**5.90**	17	24	4	2	–	–	–	–	–	1	–	–	–	**2,4,6(1H,3H,5H)-Pyrimidinetrione, 1,3-dimethyl-5-phenyl-5-[1-[(trimethylsilyl)oxy]ethyl]-**	57396-65-3	**NI36526**
73	319	291	348	206	320	45	333	**348**	100	98	81	74	62	31	29	25		17	24	4	2	–	–	–	–	–	1	–	–	–	**2,4,6(1H,3H,5H)-Pyrimidinetrione, 5-ethyl-1,3-dimethyl-5-[4-[(trimethylsilyl)oxy]phenyl]-**	55268-55-8	**NI36527**
245	73	348	259	147	333	104	230	**348**	100	56	25	25	20	14	9	4		17	28	2	2	–	–	–	–	–	2	–	–	–	**Quinoxaline, 2-methyl-3-[1(S),2-bis(trimethylsilyloxy)ethyl]-**		**NI36528**
245	259	73	169	144	147	333	205	**348**	100	98	48	34	33	30	25	22	**0.00**	17	28	2	2	–	–	–	–	–	2	–	–	–	**Quinoxaline, 2-[2(S),3-bis(trimethylsilyloxy)propyl]-**		**NI36529**
202	73	203	75	45	204	74	130	**348**	100	51	19	8	8	6	6	5	**1.42**	17	28	2	2	–	–	–	–	–	2	–	–	–	**Tryptophan, bis(trimethylsilyl)-**	61445-27-0	**NI36530**
348	212	333	136	109	45	39	349	**348**	100	97	63	60	45	34	29	25		18	12	–	4	2	–	–	–	–	–	–	–	–	**1,3-Cyclohexadiene-2,6,6-tricarbonitrile, 1-amino-3,5-di-(2-thienyl)-5-methyl-**	94638-76-3	**NI36531**
116	89	148	174	348	173	206	115	**348**	100	56	53	44	43	29	27	25		18	12	–	4	2	–	–	–	–	–	–	–	–	**[1,4]Dithiino[2,3-b:5,6-b′]diquinoxaline, 1,9(or 1,10)-dimethyl-**	38638-14-1	**NI36532**
207	348	116	148	349	115	206	174	**348**	100	32	20	10	8	8	7	7		18	12	–	4	2	–	–	–	–	–	–	–	–	**[1,4]Dithiino[2,3-b:5,6-b′]diquinoxaline, 2,9(or 2,10)-dimethyl-**	38638-14-1	**NI36533**
129	103	187	77	161	51	75	93	**348**	100	87	51	44	43	43	28	17	**0.00**	18	12	4	4	–	–	–	–	–	–	–	–	–	**5(4H)-Isoxazolone, 4,4′-azobis[3-phenyl-**	56666-60-5	**MS08253**
261	307	348	104	76	90	82	102	**348**	100	64	58	45	45	42	34	29		18	12	4	4	–	–	–	–	–	–	–	–	–	**4(3H)-Quinazolinone, 2-(2-nitrophenyl)-3-(3-methyl-5-isoxazolyl)-**	90059-42-0	**NI36534**
250	76	307	348	54	82	90	50	**348**	100	60	55	44	37	28	27	27		18	12	4	4	–	–	–	–	–	–	–	–	–	**4(3H)-Quinazolinone, 2-(4-nitrophenyl)-3-(3-methyl-5-isoxazolyl)-**	90059-41-9	**NI36535**
308	348	309	133	104	349	105	77	**348**	100	33	19	10	10	7	7	7		18	16	–	6	1	–	–	–	–	–	–	–	–	**N-Ethyl-N-(2-cyanoethyl)-4-(3,5-dicyano-4-methylthien-2-ylazo)aniline**		**IC04075**
348	216	289	103	77	349	119	232	**348**	100	46	36	34	32	26	19	16		18	16	–	6	1	–	–	–	–	–	–	–	–	**[1,2,4]Triazolo[1,5-d][1,2,4]triazinylium-2-phenylaminide, 6,8-dimethyl-1-phenyl-5-thioxo**		**MS08254**
348	305	275	320	121	350	135	134	**348**	100	61	55	47	44	33	22	19		18	17	5	–	–	–	1	–	–	–	–	–	–	**4H-1-Benzopyran-4-one, 8-chloro-2,3-dihydro-5,7-dimethoxy-2-(4-methoxyphenyl)-**	98752-29-5	**NI36536**
214	121	347	134	161	348	171	133	**348**	100	68	60	60	25	22	18	15		18	17	5	–	–	–	1	–	–	–	–	–	–	**Chalcone, 2′-hydroxy-3′-chloro-4′,6,4-trimethoxy-**	83516-31-8	**NI36537**
43	111	75	208	115	50	44	41	**348**	100	4	4	2	2	2	2	2	**0.30**	18	18	1	2	–	–	2	–	–	–	–	–	–	**2-Pentanone, 3-(4-chlorophenyl)-4-methyl-4-(4-chlorophenylazo)-**	75478-75-0	**NI36538**
233	69	119	348	131	234	215	51	**348**	100	82	21	20	18	14	14	14		18	20	7	–	–	–	–	–	–	–	–	–	–	**2-Butenoic acid, 4-(7-acetyl-4,6-dihydroxy-3,5-dimethyl-2-benzofuranyl)-3-hydroxy-, ethyl ester**	55649-37-1	**NI36539**
44	58	43	42	286	110	348	45	**348**	100	90	44	41	29	25	21	20		18	24	5	2	–	–	–	–	–	–	–	–	–	**1,2-Benzenediol, 4-[1-(3,4-dihydroxyphenyl)-2-(methylamino)ethyl]-5-[1-hydroxy-2-(methylamino)ethyl]-**	25349-47-7	**NI36540**
58	290	348	291	349	59	292	73	**348**	100	30	12	9	4	4	3	2		18	32	1	2	–	–	–	–	–	2	–	–	–	**1H-Indole-3-ethanamine, N,N-dimethyl-1-(trimethylsilyl)-4-[(trimethylsilyl)oxy]-**	55760-24-2	**NI36541**
179	227	209	116	313	331	131	198	**348**	100	41	37	31	27	24	24	20	**0.00**	18	36	6	–	–	–	–	–	–	–	–	–	–	**Dihydroaspicilolic acid**		**NI36542**
304	348	44	104	76	330	146	149	**348**	100	96	43	35	33	29	27	26		19	12	5	2	–	–	–	–	–	–	–	–	–	**Dioxindole-3-(α-phthalimido)propionic acid lactone**		**NI36543**
275	348	289	276	77	245	220	349	**348**	100	46	22	22	21	16	12	10		19	16	3	4	–	–	–	–	–	–	–	–	–	**3-β-D-erythro-Furanosyl-1-phenylpyrazolo[3,4-b]quinoxaline**		**MS08255**
43	348	131	274	303	275	212	206	**348**	100	36	36	28	26	26	26	26		19	16	3	4	–	–	–	–	–	–	–	–	–	**2-Oxabicyclo[2.2.1]heptane-6-carboxylic acid, 3-imino-1-methyl-5-phenyl 4,7,7-tricyano-, ethyl ester**		**MS08256**
211	131	257	185	43	177	31	29	**348**	100	48	36	31	24	21	18	17	**0.85**	19	24	6	–	–	–	–	–	–	–	–	–	–	**Diethyl α,α′-diacetyl-β-phenylglutarate**		**LI07690**

MASS TO CHARGE RATIOS								M.W.	INTENSITIES								Parent	C	H	O	N	S	F	Cl	Br	I	Si	P	B	X	COMPOUND NAME	CAS Reg No	No
119	228	118	43	213	120	229	199	**348**	100	61	17	14	12	11	10	10	**0.10**	19	24	6	–	–	–	–	–	–	–	–	–	–	Naphtho[2,3-b]furan-2(3H)-one, 4,8-bis(acetyloxy)-3a,4,4a,7,8,8a,9,9a-octahydro-5,8a-dimethyl-3-methylene-, [3aR-(3aα,4α,4aα,8β,8aβ,9aβ)]-	13962-21-5	MS08257
228	229	183	43	200	184	213	185	**348**	100	28	27	24	19	18	16	13	**0.10**	19	24	6	–	–	–	–	–	–	–	–	–	–	Naphtho[2,3-b]furan-2(3H)-one, 4,8-bis(acetyloxy)decahydro-8a-methyl-3,5-bis(methylene)-, [3aS-(3aα,4α,4aα,8β,8aβ,9aβ)]-	16822-14-3	MS08258
181	137	167	151	348	316	165	256	**348**	100	41	29	22	21	19	18	7		19	24	6	–	–	–	–	–	–	–	–	–	–	2-Propanol, 1-(3,4-dimethoxyphenyl)-3-(2-hydroxy-4-methoxyphenyl)-1-methoxy-		NI36544
40	348	44	181	151	258	167	136	**348**	100	50	27	24	18	10	6	6		19	24	6	–	–	–	–	–	–	–	–	–	–	Bis(3,4,5-trimethoxyphenyl)methane		NI36545
191	84	157	57	137	192	163	55	**348**	100	48	46	26	24	22	20	15	**14.00**	19	28	4	2	–	–	–	–	–	–	–	–	–	2-Propenamide, N-[4-(acetylmethylamino)butyl]-3-(3,4-dimethoxyphenyl)-N-methyl-	55760-31-1	NI36546
348	319	88	174	291	176	233	261	**348**	100	19	10	9	7	7	6	5		20	12	6	–	–	–	–	–	–	–	–	–	–	Furo[3′,4′:6,7]naphtho[1,2-d]-1,3-dioxol-7(9H)-one, 10-(1,3-benzodioxol-5-yl)-	18920-47-3	NI36547
348	248	247	304	289	219	218	276	**348**	100	72	66	63	61	59	48	45		20	16	4	2	–	–	–	–	–	–	–	–	–	1H-Pyrano[3′,4′:6,7]indolizino[1,2-b]quinoline-3,14(4H,12H)-dione, 4-ethyl-4-hydroxy-, (S)-	7689-03-4	NI36548
225	348	241	223	139	226	283	56	**348**	100	80	58	58	38	36	32	26		20	20	2	–	–	–	–	–	–	–	–	–	1	Propyl 2-ferrocenylbenzoate		MS08259
348	91	132	77	349	279	215	130	**348**	100	79	32	32	23	21	17	17		20	20	2	4	–	–	–	–	–	–	–	–	–	2-Pyrazolin-5-one, 3-methyl-4-[(4-morpholinophenyl)imino]-1-phenyl-	4722-21-8	MS08260
348	91	77	349	279	215	130	129	**348**	100	79	32	24	22	18	18	16		20	20	2	4	–	–	–	–	–	–	–	–	–	2-Pyrazolin-5-one, 3-methyl-4-[(4-morpholinophenyl)imino]-1-phenyl-	4722-21-8	NI36549
43	305	239	152	98	55	169	288	**348**	100	63	63	63	18	18	16	9	**9.00**	20	28	3	–	1	–	–	–	–	–	–	–	–	Cyclododecanone, 2-(acetyloxy)-2-(phenylthio)-	63608-51-5	NI36550
245	136	137	43	41	165	164	123	**348**	100	92	85	77	77	74	74	71	**2.53**	20	28	5	–	–	–	–	–	–	–	–	–	–	11,12-Anhydroingol		NI36551
293	348	289	233	275	349	294	177	**348**	100	99	48	47	27	25	21	16		20	28	5	–	–	–	–	–	–	–	–	–	–	2H,6H-Benzo[1,2-b:5,4-b′]dipyran-10-propionic acid, 3,4,7,8-tetrahydro-5-methoxy-2,2,8,8-tetramethyl-	26535-38-6	NI36552
293	348	233	289	275	349	294	177	**348**	100	98	45	30	25	24	20	17		20	28	5	–	–	–	–	–	–	–	–	–	–	2H,6H-Benzo[1,2-b:5,4-b′]dipyran-10-propionic acid, 3,4,7,8-tetrahydro-5-methoxy-2,2,8,8-tetramethyl-	26535-38-6	NI36553
199	217	200	83	216	229	201	41	**348**	100	94	79	76	59	47	44	41	**0.94**	20	28	5	–	–	–	–	–	–	–	–	–	–	4-Desoxy-4α-phorbol		NI36554
43	259	41	91	55	243	245	105	**348**	100	40	34	31	30	26	24	21	**1.00**	20	28	5	–	–	–	–	–	–	–	–	–	–	Gibberellin A_{10} methyl ester		MS08261
57	85	145	246	233	147	348	263	**348**	100	57	23	7	7	3	1	1		20	28	5	–	–	–	–	–	–	–	–	–	–	Thymol, 8,9-epoxy-10-(3-methylbutoxy)-, 3-isovalerate		NI36555
209	348	277	223	333	210	179	318	**348**	100	51	28	25	20	10	10	7		20	32	3	–	–	–	–	–	–	1	–	–	–	4-Decen-3-one, 1-[3-methoxy-4-[(trimethylsilyl)oxy]phenyl]-	56700-92-6	MS08262
348	333	350	335	213	121	349	334	**348**	100	99	36	34	31	28	26	25		21	13	3	–	–	–	1	–	–	–	–	–	–	Dinaphtho[1,2-b:2′,1′-d]furan, 6-chloro-8-hydroxy-5-methoxy-		NI36556
348	333	350	349	213	335	106	174	**348**	100	56	35	25	25	19	17	16		21	13	3	–	–	–	1	–	–	–	–	–	–	Dinaphtho[1,2-b:1′,2′-d]furan-5-ol, 6-chloro-8-methoxy-	91913-11-0	NI36557
117	145	90	89	203	116	64	160	**348**	100	81	76	54	52	41	41	28	**0.00**	21	20	3	2	–	–	–	–	–	–	–	–	–	Naphthalene-1,4-dicarboximide, 1,4-imino-endo-9-ethyl-1,2,3,4-tetrahydro-N-(4-methoxyphenyl)-	123542-18-7	DD01373
93	56	121	66	290	65	348	77	**348**	100	71	60	54	41	38	31	30		21	24	1	–	–	–	–	–	–	–	–	–	1	Ferrocene, (1-phenoxypentyl)-		NI36558
43	312	313	91	158	93	41	55	**348**	100	61	26	24	23	18	18	17	**11.87**	21	29	2	–	–	–	1	–	–	–	–	–	–	Pregn-4-ene-3,20-dione, 4-chloro-	1164-81-4	NI36559
43	312	313	91	158	93	41	79	**348**	100	62	26	24	23	18	18	17	**12.00**	21	29	2	–	–	–	1	–	–	–	–	–	–	Pregn-4-ene-3,20-dione, 4-chloro-	1164-81-4	NI36560
259	319	163	260	320	328	299	239	**348**	100	58	55	18	12	11	8	8	**2.00**	21	29	3	–	–	1	–	–	–	–	–	–	–	Androst-5-en-19-al, 17-(acetyloxy)-3-fluoro-, (3β,17β)-	28344-68-5	NI36561
259	319	163	328	299	239	348		**348**	100	58	55	11	8	8	2			21	29	3	–	–	1	–	–	–	–	–	–	–	Androst-5-en-19-al, 17-(acetyloxy)-3-fluoro-, (3β,17β)-	28344-68-5	LI07691
259	319	163	43	277	260	90	320	**348**	100	60	52	47	31	27	23	21	**3.10**	21	29	3	–	–	1	–	–	–	–	–	–	–	Androst-5-en-19-al, 17-(acetyloxy)-3-fluoro-, (3β,17β)-	28344-68-5	NI36562
239	259	163	43	319	240	260	91	**348**	100	30	29	19	18	11	8	7	**1.50**	21	29	3	–	–	1	–	–	–	–	–	–	–	Androst-5-en-19-al, 3α-fluoro-17β-hydroxy-, acetate	28344-49-2	NI36563
319	259	163	320	239	260	164	240	**348**	100	31	29	20	10	5	3	2	**1.00**	21	29	3	–	–	1	–	–	–	–	–	–	–	Androst-5-en-19-al, 3α-fluoro-17β-hydroxy-, acetate	28344-49-2	NI36564
147	348	306	288	142	246	133	141	**348**	100	91	79	68	63	43	32	25		21	29	3	–	–	1	–	–	–	–	–	–	–	Androst-4-en-3-one, 17-(acetyloxy)-6-fluoro-, (6α,17β)-	855-55-0	NI36565
43	147	246	348	306	288	142	91	**348**	100	99	93	91	80	68	64	36		21	29	3	–	–	1	–	–	–	–	–	–	–	Androst-4-en-3-one, 17-(acetyloxy)-6-fluoro-, (6β,17β)-	2627-94-3	NI36566
147	328	348	330	288	124	302	133	**348**	100	66	53	47	46	36	31	28		21	29	3	–	–	1	–	–	–	–	–	–	–	Androst-4-en-3-one, 17-(acetyloxy)-6-fluoro-, (6β,17β)-	2627-94-3	NI36567
348	43	140	97	91	330	94	93	**348**	100	89	34	33	32	28	27	27		21	29	3	–	–	1	–	–	–	–	–	–	–	Androst-4-en-6-one, 3β-fluoro-17β-hydroxy-, acetate	31756-75-9	NI36568
348	140	330	141	133	121	328	313	**348**	100	34	28	26	26	19	14	13		21	29	3	–	–	1	–	–	–	–	–	–	–	Androst-4-en-6-one, 3β-fluoro-17β-hydroxy-, acetate	31756-75-9	LI07692
348	140	330	141	133	349	121	328	**348**	100	34	28	26	26	24	19	14		21	29	3	–	–	1	–	–	–	–	–	–	–	Androst-4-en-6-one, 3β-fluoro-17β-hydroxy-, acetate	31756-75-9	NI36569
43	348	147	91	330	133	260	273	**348**	100	99	76	63	57	49	47	44		21	29	3	–	–	1	–	–	–	–	–	–	–	Androst-5-en-4-one, 3β-fluoro-17β-hydroxy-, acetate	31756-72-6	NI36570
348	147	330	133	260	273	119	288	**348**	100	76	57	49	47	44	39	37		21	29	3	–	–	1	–	–	–	–	–	–	–	Androst-5-en-4-one, 3β-fluoro-17β-hydroxy-, acetate	31756-72-6	LI07693
288	348	156	273	194	141	181	155	**348**	100	90	88	53	40	34	25	23		21	29	3	–	–	1	–	–	–	–	–	–	–	Androst-5-en-7-one, 3β-fluoro-17β-hydroxy-, acetate	31756-73-7	LI07694
288	348	156	273	289	194	141	181	**348**	100	91	88	53	50	40	34	25		21	29	3	–	–	1	–	–	–	–	–	–	–	Androst-5-en-7-one, 3β-fluoro-17β-hydroxy-, acetate	31756-73-7	NI36571
348	195	190	105	121	91	135	134	**348**	100	98	93	92	80	72	66	57		21	32	4	–	–	–	–	–	–	–	–	–	–	allo-Cassaic acid, methyl ester		NI36572
302	215	287	241	108	93	256	147	**348**	100	54	53	53	48	48	47	45	**0.40**	21	32	4	–	–	–	–	–	–	–	–	–	–	Androstane, 3,17-diformyloxy-, (3α,5α,17β)-		NI36573
256	302	215	93	241	107	81	147	**348**	100	91	83	63	52	51	50	41	**0.10**	21	32	4	–	–	–	–	–	–	–	–	–	–	Androstane, 3,17-diformyloxy-, (3α,5β,17β)-		NI36574
278	43	279	348	55	41	105	95	**348**	100	54	18	15	14	14	13	13		21	32	4	–	–	–	–	–	–	–	–	–	–	Androstan-4-one, 17-acetoxy-5-hydroxy-, (5α,17β)-		MS08263
43	95	288	55	81	67	135	93	**348**	100	56	46	44	40	35	34	33	**0.00**	21	32	4	–	–	–	–	–	–	–	–	–	–	Androstan-6-one, 3-(acetyloxy)-5-hydroxy-, (3β,5α)-	4725-53-5	NI36575
234	43	270	348	55	81	41	67	**348**	100	63	61	50	29	27	24	19		21	32	4	–	–	–	–	–	–	–	–	–	–	Androstan-17-one, 3-(acetyloxy)-5-hydroxy-, (3β,5α)-	41853-36-5	NI36576
47	55	83	43	85	41	67	81	**348**	100	82	81	80	73	73	55	54	**23.23**	21	32	4	–	–	–	–	–	–	–	–	–	–	Androstan-17-one, 3-(acetyloxy)-14-hydroxy-, (3β,5α)-	1241-30-1	NI36577

MASS TO CHARGE RATIOS								M.W.	INTENSITIES								Parent	C	H	O	N	S	F	Cl	Br	I	Si	P	B	X	COMPOUND NAME	CAS Reg No	No
43	348	273	288	215	107	55	113	348	100	71	64	58	57	45	42	40		21	32	4	–	–	–	–	–	–	–	–	–	–	Androstan-17-one, 3-(acetyloxy)-14-hydroxy-, (3β,5α,14β)-	38632-10-9	NI36578
88	89	58	61	118	70	79	81	348	100	96	13	11	10	9	6	5	0.10	21	32	4	–	–	–	–	–	–	–	–	–	–	Androst-5-en-17-one, 3-hydroxy-16,16-dimethoxy-, (3β)-	63608-60-6	NI36579
107	330	159	201	123	248	230	219	348	100	84	74	53	50	47	26	13	2.00	21	32	4	–	–	–	–	–	–	–	–	–	–	5-Cyclohexen-1-one, 2,5-dihydroxy-6-[[1-(1,5-dimethylhex-4-enyl)-4-hydroxy-4-methylcyclopent-1-en-5-yl]methyl]-		NI36580
263	43	320	97	81	55	105	57	348	100	86	39	34	34	33	25	23	11.00	21	32	4	–	–	–	–	–	–	–	–	–	–	Digipurpurogenine I		LI07695
81	287	269	43	330	105	251	145	348	100	62	56	56	38	33	30	28	9.00	21	32	4	–	–	–	–	–	–	–	–	–	–	Digipurpurogenine II		LI07696
181	163	43	69	182	55	70	41	348	100	23	13	11	10	10	9	9	3.00	21	32	4	–	–	–	–	–	–	–	–	–	–	Isophthalic acid, dodecyl methyl ester	33975-26-7	NI36581
125	106	348	165	159	183	191	205	348	100	80	62	43	43	38	25	18		21	32	4	–	–	–	–	–	–	–	–	–	–	Methyl all-trans-5,6-erythro-5,6-dihydroxy-5,6-dihydroretinoate		NI36582
109	127	161	348	221	106	189	178	348	100	70	50	40	34	34	23	13		21	32	4	–	–	–	–	–	–	–	–	–	–	Methyl all-trans-5,6-threo-5,6-dihydroxy-5,6-dihydroretinoate		NI36583
105	43	157	97	55	41	69	139	348	100	81	50	46	42	42	37	30	0.00	21	32	4	–	–	–	–	–	–	–	–	–	–	Octanoic acid, 4-[(R)-2-phenylpropionyloxy]-3-methyl-, isopropyl ester, (3R,4R)- or (3S,4S)-		NI36584
43	41	55	105	42	157	199	44	348	100	34	24	21	17	14	9	8	0.00	21	32	4	–	–	–	–	–	–	–	–	–	–	Octanoic acid, 4-[(R)-2-phenylpropionyloxy]-3-methyl-, isopropyl ester, (3S,4R)- or (3R,4S)-		NI36585
320	223	163	121	305	138	123	321	348	100	72	55	45	28	26	26	24	3.58	21	32	4	–	–	–	–	–	–	–	–	–	–	1-Phenanthrenecarboxylic acid, 1,2,3,4,4a,4b,5,8,8a,9,10,10a-dodecahydro 8a-hydroxy-1,4a-dimethyl-7-isopropyl-8-oxo-, methyl ester, [1-(1α,4aβ,4bα,8aα,10aα)]-	32111-53-8	NI36586
149	330	270	137	150	271	255	135	348	100	49	38	37	36	35	35	30	4.78	21	32	4	–	–	–	–	–	–	–	–	–	–	1-Phenanthrenecarboxylic acid, 1,2,3,4,4a,4b,5,6,8a,9,10,10a-dodecahydro 8a-hydroxy-1,4a-dimethyl-7-isopropyl-6-oxo-, methyl ester, [1R-(1α,4aβ,4bα,8aα,10aα)]-	3787-85-7	NI36587
243	288	348	330	43	229	147	107	348	100	54	41	39	32	24	23	23		21	32	4	–	–	–	–	–	–	–	–	–	–	Pregnane-11,20-dione, 3,17-dihydroxy-	56193-64-7	MS08264
43	348	121	261	55	107	41	93	348	100	46	45	44	43	41	38	36		21	32	4	–	–	–	–	–	–	–	–	–	–	Pregnane-11,20-dione, 3,17-dihydroxy-	56193-64-7	NI36588
217	247	55	57	315	71	249	215	348	100	80	62	58	51	44	43	41	9.00	21	32	4	–	–	–	–	–	–	–	–	–	–	Prosta-5,8(12),13-trien-1-oic acid, 15-hydroxy-9-oxo-, methyl ester, (5Z,13E,15S)-	55760-05-9	NI36589
348	81	151	109	95	69	67	123	348	100	43	42	41	39	35	33	31		21	32	4	–	–	–	–	–	–	–	–	–	–	Strobanoic acid, 6-oxo-8,13-epoxy-, methyl ester, (8α,13α,14α)-		NI36590
43	348	55	109	151	123	81	69	348	100	84	40	36	32	32	32	28		21	32	4	–	–	–	–	–	–	–	–	–	–	Strobanoic acid, 6-oxo-8,13-epoxy-, methyl ester, (8α,13α,14β)-		NI36591
266	83	305	300	288	330	95	107	348	100	99	79	69	65	64	58	57	50.00	21	32	4	–	–	–	–	–	–	–	–	–	–	Tetrahydroanhydrohirundigenin		LI07697
266	330	83	248	318	43	236	230	348	100	85	81	50	33	30	28	23	5.00	21	32	4	–	–	–	–	–	–	–	–	–	–	Tetrahydroanhydrohirundigenin		LI07698
272	257	108	215	348	216	273	107	348	100	37	33	31	30	21	20	20		21	36	2	–	–	–	–	–	–	1	–	–	–	Androstan-17-one, 5-[(dimethylsilyl)oxy]-, (5α)-		NI36592
95	348	310	295	319	230	309	125	348	100	69	27	27	16	16	15	15		22	20	4	–	–	–	–	–	–	–	–	–	–	Androtephrostachin, trans-		NI36593
275	276	183	303	165	29	185	347	348	100	66	60	54	42	26	26	24	19.30	22	21	2	–	–	–	–	–	–	–	1	–	–	Acetic acid, (triphenylphosphoranylidene)-, ethyl ester	1099-45-2	NI36594
320	348	15	77	31	319	91	349	348	100	94	79	61	31	30	26	25		22	24	–	2	1	–	–	–	–	–	–	–	–	1H-Benzo[e][1,4]thiazepine, 2,3,5,6,7,8,9-heptahydro-1-phenyl-5-(4-tolylimino)-		NI36595
261	77	348	320	256	130	262	287	348	100	74	40	36	32	24	23	22		22	24	–	2	1	–	–	–	–	–	–	–	–	1H-Cyclohepta[e][1,4]thiazepine, 2,3,5,6,7,8,9,10-octahydro-1-phenyl-5-(phenylimino)-		NI36596
348	262	144	212	230	120	132	160	348	100	69	68	48	45	40	38	27		22	24	2	2	–	–	–	–	–	–	–	–	–	1H-1-Benzazepine-3,4-dimethanol, 2,3-dihydro-1-methyl-2-(1-methyl-1H-indol-3-yl)-	34221-67-5	NI36597
174	187	91	105	77	82	28	158	348	100	63	58	46	40	32	25	24	10.44	22	24	2	2	–	–	–	–	–	–	–	–	–	1-Butanone, 3,3′-(1,2-ethanediyldinitrilo)bis[1-phenyl-	16087-30-2	NI36598
105	77	243	106	213	122	51	244	348	100	41	35	9	8	7	7	6	3.00	22	24	2	2	–	–	–	–	–	–	–	–	–	2,9-Diazabicyclo[4.4.0]decane, 2,9-dibenzoyl-, cis-		NI36599
105	77	243	122	106	51	244	213	348	100	45	39	10	8	8	7	5	3.00	22	24	2	2	–	–	–	–	–	–	–	–	–	2,9-Diazabicyclo[4.4.0]decane, 2,9-dibenzoyl-, trans-		NI36600
333	348	57	71	97	69	318	319	348	100	88	58	45	42	39	38	37		22	24	2	2	–	–	–	–	–	–	–	–	–	3-Pyrrolin-2-one, 4-ethyl-3-methyl-5-[[5-(4-methoxybenzylidene)-3,4-dimethyl-5H-pyrrol-2-yl]methylene]-, (Z,Z)-		MS08265
125	348	113	112	99	349	126	73	348	100	64	63	20	17	16	12	10		22	36	3	–	–	–	–	–	–	–	–	–	–	Androstan-3-one, 17-hydroxy-2-methyl-, cyclic 1,2-ethanediyl acetal, (2α,5α,17α)-	54498-59-8	NI36601
125	348	113	112	99	349	126	81	348	100	59	59	22	19	12	10	9		22	36	3	–	–	–	–	–	–	–	–	–	–	Androstan-3-one, 17-hydroxy-2-methyl-, cyclic 1,2-ethanediyl acetal, (2β,5β,17β)-	54498-60-1	NI36602
125	348	113	112	43	41	99	55	348	100	59	58	26	25	24	18	17		22	36	3	–	–	–	–	–	–	–	–	–	–	Androstan-3-one, 17-hydroxy-2-methyl-, cyclic 1,2-ethanediyl acetal, (2β,5β,17β)-	54498-60-1	NI36603
99	100	55	41	43	139	81	67	348	100	16	10	8	6	5	5	4	3.90	22	36	3	–	–	–	–	–	–	–	–	–	–	Androstan-3-one, 17-hydroxy-4-methyl-, cyclic 1,2-ethanediyl acetal, (4α,5α,17β)-	3347-76-0	NI36604
115	57	91	79	123	135	290	149	348	100	89	17	15	6	4	3	3	0.00	22	36	3	–	–	–	–	–	–	–	–	–	–	11,15-Hexadecadien-13-yne, 16-[(2,3-isopropylidenedioxy)propoxy]-		MS08266
330	123	81	94	109	69	257	135	348	100	68	63	61	60	60	58	58	7.01	22	36	3	–	–	–	–	–	–	–	–	–	–	Kaurane-16,18-diol, 18-acetate	30320-70-8	NI36605
121	109	81	288	122	107	93	289	348	100	51	47	45	45	38	34	32	19.00	22	36	3	–	–	–	–	–	–	–	–	–	–	8(17)-Labden-19-oic acid, 15-methyl-15-oxo-, methyl ester		MS08267
43	270	189	41	69	55	81	288	348	100	55	52	49	48	47	37	33	1.03	22	36	3	–	–	–	–	–	–	–	–	–	–	8a,10(2H)-Phenanthrenediol, 7-ethenyldodecahydro-1,1,4a,7-tetramethyl-, 10-acetate, [4aS-(4aα,4bβ,7β,8aα,10α,10aβ)]-	41756-26-7	NI36606
43	270	41	69	55	81	137	109	348	100	99	82	81	72	56	49	48	3.33	22	36	3	–	–	–	–	–	–	–	–	–	–	4,10a(1H)-Phenanthrenediol, 2-vinyldodecahydro-2,4b,8,8-tetramethyl-, 4 acetate, [2S-(2α,4α,4aα,4bβ,8aα,10aβ)]-	41756-38-1	NI36607

MASS TO CHARGE RATIOS								M.W.	INTENSITIES								Parent	C	H	O	N	S	F	Cl	Br	I	Si	P	B	X	COMPOUND NAME	CAS Reg No	No
312	105	297	91	119	159	55	79	**348**	100	64	61	56	43	42	39	36	**0.00**	22	36	3	–	–	–	–	–	–	–	–	–	–	Pregnan-20-one, 3,5-dihydroxy-6-methyl-, (3β,5α,6β)-	56630-86-5	MS08268
55	79	93	91	107	81	67	269	**348**	100	84	82	81	75	75	75	55	**22.80**	22	36	3	–	–	–	–	–	–	–	–	–	–	Pregnan-20-one, 3,17-dihydroxy-16-methyl-, (3β,5β,16α)-	56630-85-4	MS08269
74	258	95	243	81	162	148	73	**348**	100	99	90	83	68	65	51	51	**2.50**	22	40	1	–	–	–	–	–	–	1	–	–	–	Androstane, 3-(trimethylsiloxy)-, (3α,5β)-	57397-25-8	NI36608
75	333	258	95	81	334	73	142	**348**	100	91	38	35	28	26	24	21	**14.14**	22	40	1	–	–	–	–	–	–	1	–	–	–	Androstane, 3-(trimethylsiloxy)-, (3β,5α)-	18899-44-0	NI36609
258	75	95	108	81	73	67	55	**348**	100	90	58	57	48	45	34	32	**0.50**	22	40	1	–	–	–	–	–	–	1	–	–	–	Androstane, 3-(trimethylsiloxy)-, (5β,5β)-	18899-46-2	NI36610
144	75	258	73	243	257	217	333	**348**	100	80	75	61	58	56	56	41	**32.50**	22	40	1	–	–	–	–	–	–	1	–	–	–	Androstane, 16-(trimethylsiloxy)-, (5α,16β)-	37977-23-4	NI36611
348	243	258	73	333	75	95	67	**348**	100	69	64	55	39	37	35	33		22	40	1	–	–	–	–	–	–	1	–	–	–	Androstane, 17-(trimethylsiloxy)-, (5α,17β)-,	7604-82-2	NI36612
348	243	258	73	41	55	333	75	**348**	100	67	62	54	43	42	40	37		22	40	1	–	–	–	–	–	–	1	–	–	–	Androstane, 17-(trimethylsiloxy)-, (5α,17β)-	7604-82-2	NI36613
129	73	243	258	149	109	75	67	**348**	100	73	72	51	48	43	43	39	**34.00**	22	40	1	–	–	–	–	–	–	1	–	–	–	Androstane, 17-(trimethylsiloxy)-, (5α,17β)-	7604-82-2	NI36614
348	347	349	255	241	270	78	158	**348**	100	30	25	24	19	17	17	12		23	16	–	4	–	–	–	–	–	–	–	–	–	Spiro[9H-fluorene-9,3′(2′H)-[1,2,4]triazolo[4,3-a]pyridine], 2′-(2-pyridinyl)-	16672-26-7	NI36615
348	39	51	77	165	244	63	50	**348**	100	70	64	47	32	30	24	22		23	16	–	4	–	–	–	–	–	–	–	–	–	s-Triazolo[4,3-a]pyrazine, 3,5,6-triphenyl-	33590-26-0	LI07699
348	349	39	51	77	165	244	63	**348**	100	75	70	64	47	32	30	24		23	16	–	4	–	–	–	–	–	–	–	–	–	s-Triazolo[4,3-a]pyrazine, 3,5,6-triphenyl-	33590-26-0	NI36616
91	123	43	155	171	115	226	118	**348**	100	50	50	44	42	31	26	25	**0.00**	23	24	3	–	–	–	–	–	–	–	–	–	–	Bicyclo[3.2.1]oct-6-en-3-one, anti-8-hydroxy-syn-8-(1-hydroxyisopropyl)-2,4-diphenyl-		DD01374
348	240	228	225	223	121	108	90	**348**	100	95	87	83	73	60	55	42		23	24	3	–	–	–	–	–	–	–	–	–	–	Phenol, 2,6-bis(2-hydroxy-5-methylbenzyl)-4-methyl-		LI07700
348	240	228	225	223	121	108	91	**348**	100	95	87	81	73	59	54	41		23	24	3	–	–	–	–	–	–	–	–	–	–	Phenol, 2,6-bis(2-hydroxy-5-methylbenzyl)-4-methyl-		LI07701
200	105	77	348	333				**348**	100	21	20	16	11					23	28	1	2	–	–	–	–	–	–	–	–	–	2-(2,6-Dimethylphenylimino)-3,4-dimethyl-3-pentenoic acid 2,6-dimethylanilide		LI07702
200	105	77	201	41	348	79	333	**348**	100	21	19	17	17	16	14	11		23	28	1	2	–	–	–	–	–	–	–	–	–	2-(2,6-Dimethylphenylimino)-3,4-dimethyl-3-pentenoic acid 2,6-dimethylanilide		MS08270
219	220	292	236	190	348	191	165	**348**	100	58	57	41	31	30	21	16		23	28	1	2	–	–	–	–	–	–	–	–	–	Indole-2-carboxamide, N,1-di-tert-butyl-3-phenyl-	17537-55-2	NI36617
219	292	220	236	190	348	191	165	**348**	100	58	51	40	31	30	21	16		23	28	1	2	–	–	–	–	–	–	–	–	–	Indole-2-carboxamide, N,1-di-tert-butyl-3-phenyl-	17537-55-2	LI07703
333	348	161	159	291	305	43	133	**348**	100	90	23	19	17	11	6	5		23	28	1	2	–	–	–	–	–	–	–	–	–	Phenanthrene, 4b,5,6,7,9,10-hexahydro-4b,8-dimethyl-2-isopropyl-9-(1H-imidazolyl-4-ylcarbonyl)-, (4bS)-	75601-29-5	NI36618
58	247	305	131	91	42	104	43	**348**	100	36	26	25	21	19	18	18	**10.00**	23	28	1	2	–	–	–	–	–	–	–	–	–	2-Propen-1-one, 1-[hexahydro-1,3-dimethyl-5-(2,6-xylyl)-5-pyrimidinyl]-2-phenyl-	29549-15-3	LI07704
55	67	41	81	79	43	80	95	**348**	100	65	65	56	46	46	42	41	**13.04**	23	40	2	–	–	–	–	–	–	–	–	–	–	8,11,14-Docosatrienoic acid, methyl ester	56847-02-0	NI36619
333	348	334	276	261	349	277	138	**348**	100	28	25	14	10	8	4	4		24	28	2	–	–	–	–	–	–	–	–	–	–	Furo[1,2-a]dibenzopyran, 2,7,7,10-tetramethyl-4-pentyl-	87734-66-5	NI36620
105	124	76	109	106	41	51	174	**348**	100	35	27	6	6	6	5	4	**1.05**	24	28	2	–	–	–	–	–	–	–	–	–	–	1,5-Pentanedione, 3-isopropyl-2-(2-methyl-1-propenyl)-1,5-diphenyl-, (R*,R*)-	36597-10-1	MS08271
105	124	76	106	41	51	175	69	**348**	100	41	36	11	11	7	5	5	**1.58**	24	28	2	–	–	–	–	–	–	–	–	–	–	1,5-Pentanedione, 3-isopropyl-2-(2-methyl-1-propenyl)-1,5-diphenyl-, (R*,S*)-	36597-11-2	MS08272
174	91	175	41	180	179	28	178	**348**	100	18	14	13	4	4	4	3	**0.00**	24	32	–	2	–	–	–	–	–	–	–	–	–	1,2-Diphenyl-1,2-bis(1-piperidyl)ethane, (±)-		DD01375
174	91	41	175	69	42	28	104	**348**	100	21	15	14	4	4	4	3	**0.00**	24	32	–	2	–	–	–	–	–	–	–	–	–	1,2-Diphenyl-1,2-bis(1-piperidyl)ethane, meso-		DD01376
55	177	95	138	192	191	93	137	**348**	100	93	77	64	61	46	31	30	**0.00**	24	44	1	–	–	–	–	–	–	–	–	–	–	1H-Naphtho[2,1-b]pyran, 3-hexyldodecahydro-3,4a,7,7,10a-pentamethyl-	56247-59-7	NI36621
348	138	276	305	150	174	133		**348**	100	39	29	10	6	3	1			25	16	2	–	–	–	–	–	–	–	–	–	–	Naphtho[2,1,8-qra]naphthacen-12-one, 7,12-dihydro-9-methoxy-		LI07705
83	55	97	41	69	43	111	96	**348**	100	81	55	51	37	29	22	22	**0.14**	25	48	–	–	–	–	–	–	–	–	–	–	–	1,5-Dicyclohexyl-3-octylpentane		AI02125
55	83	97	41	69	43	57	96	**348**	100	82	61	61	39	35	31	25	**0.06**	25	48	–	–	–	–	–	–	–	–	–	–	–	1,5-Dicyclohexyl-3-octylpentane		AI02126
55	83	41	97	69	43	57	96	**348**	100	81	66	55	36	36	27	24	**0.50**	25	48	–	–	–	–	–	–	–	–	–	–	–	1,5-Dicyclohexyl-3-octylpentane		AI02127
83	41	55	69	97	43	57	67	**348**	100	95	94	86	78	57	51	36	**0.09**	25	48	–	–	–	–	–	–	–	–	–	–	–	1,7-Dicyclopentyl-4-octylheptane		AI02128
96	81	67	43	41	55	97	123	**348**	100	57	47	47	40	39	37	31	**20.00**	25	48	–	–	–	–	–	–	–	–	–	–	–	7-Hexadecylbicyclo[4.3.0]nonane		AI02129
96	81	67	55	43	41	97	123	**348**	100	66	57	49	46	41	38	33	**19.84**	25	48	–	–	–	–	–	–	–	–	–	–	–	7-Hexadecylbicyclo[4.3.0]nonane		AI02130
96	81	97	123	82	95	83	43	**348**	100	66	37	31	30	23	18	18	**18.16**	25	48	–	–	–	–	–	–	–	–	–	–	–	7-Hexadecylbicyclo[4.3.0]nonane		AI02131
123	96	81	43	41	67	55	97	**348**	100	99	96	81	70	65	64	48	**16.98**	25	48	–	–	–	–	–	–	–	–	–	–	–	8-Hexadecylbicyclo[4.3.0]nonane		AI02132
137	81	95	55	43	41	67	69	**348**	100	47	42	35	31	30	27	22	**10.50**	25	48	–	–	–	–	–	–	–	–	–	–	–	2-Pentadecylbicyclo[4.4.0]decane		AI02133
137	81	95	43	41	55	67	136	**348**	100	52	41	37	33	30	26	23	**8.82**	25	48	–	–	–	–	–	–	–	–	–	–	–	2-Pentadecylbicyclo[4.4.0]decane		AI02134
137	81	95	43	41	55	67	69	**348**	100	49	42	41	39	34	27	22	**4.71**	25	48	–	–	–	–	–	–	–	–	–	–	–	2-Pentadecylbicyclo[4.4.0]decane		AI02135
271	193	165	345	270	269	194	57	**348**	100	27	20	17	17	16	15	14	**0.00**	26	20	1	–	–	–	–	–	–	–	–	–	–	1,3,3-Triphenylphthalane		MS08273
348	349	221	207	205	222	178	193	**348**	100	28	27	9	8	7	6	5		26	36	–	–	–	–	–	–	–	–	–	–	–	4-Decyl-1,2,3,6,7,8-hexahydropyrene		AI02136
348	221	43	207	41	29	222	205	**348**	100	45	18	17	14	14	12	11		26	36	–	–	–	–	–	–	–	–	–	–	–	4-Decyl-1,2,3,6,7,8-hexahydropyrene		AI02137
348	193	349	194	178	191	192	179	**348**	100	93	29	17	11	9	6	6		26	36	–	–	–	–	–	–	–	–	–	–	–	2-Dodecyl-9,10-dihydrophenanthrene		AI02138
193	348	43	41	194	178	191	179	**348**	100	71	24	19	18	12	9	7		26	36	–	–	–	–	–	–	–	–	–	–	–	2-Dodecyl-9,10-dihydrophenanthrene		AI02139
193	348	191	192	194	346	349	178	**348**	100	42	36	20	19	13	12	11		26	36	–	–	–	–	–	–	–	–	–	–	–	2-Dodecyl-9,10-dihydrophenanthrene		AI02140
243	165	244	41	55	166	91	228	**348**	100	31	23	14	12	5	5	4	**1.00**	27	24	–	–	–	–	–	–	–	–	–	–	–	Benzene, 1,1′,1″,1‴-(1-propanyl-3-ylidyne)tetrakis-	54966-03-9	NI36622

MASS TO CHARGE RATIOS								M.W.	INTENSITIES								Parent	C	H	O	N	S	F	Cl	Br	I	Si	P	B	X	COMPOUND NAME	CAS Reg No	No
141	142	128	115	129	155	184	143	**348**	100	44	44	29	22	20	15	11	**0.00**	27	24	–	–	–	–	–	–	–	–	–	–	–	Benzo[3,4]cyclobuta[1,2-a]cyclobuta[b]cycloheptene, 2a,3,4,5,6,6a-hexahydro-1,2-diphenyl-	56978-65-5	NI36623
259	260	91	77	105	115	348	231	**348**	100	79	76	72	65	42	28	27		27	24	–	–	–	–	–	–	–	–	–	–	–	Cyclopentane, 2-(4-methylbenzylidene)-1,3-bis[(E)-benzylidene]-		MS08274
348	349	174	347	350	151	91	39	**348**	100	30	12	5	4	4	4	4		27	24	–	–	–	–	–	–	–	–	–	–	–	1,1′:3′,1″-Terphenyl, 4,4″-dimethyl-5′-(4-methylphenyl)-	50446-43-0	WI01452
69	92	47	114	180	100	131	119	**349**	100	27	19	12	10	6	5	4	**0.00**	6	–	1	1	–	13	–	–	–	–	–	–	–	N-(Nonafluorobutyl)-N-(trifluoromethyl)carbamoyl fluoride		DD01377
69	119	92	169	47	114	100	164	**349**	100	51	37	35	34	13	10	9	**0.00**	6	–	1	1	–	13	–	–	–	–	–	–	–	N-(Pentafluoroethyl)-N-(heptafluoropropyl)carbamoyl fluoride		DD01378
69	99	64	186	44	76	319	122	**349**	100	65	60	34	33	24	22	22	**5.90**	8	6	7	1	2	3	–	–	–	–	–	–	–	Methanesulphonic acid, trifluoro-, [(4-nitrophenyl)sulphonyl]methyl ester	56177-36-7	NI36624
199	197	97	314	258	316	201	260	**349**	100	97	89	48	38	33	28	26	**0.00**	9	11	3	1	1	–	3	–	–	–	1	–	–	Phosphorothioic acid, O,O-diethyl O-(3,5,6-trichloro-2-pyridinyl) ester	2921-88-2	NI36625
97	32	197	199	314	316	125	208	**349**	100	99	97	94	64	45	39	31	**7.80**	9	11	3	1	1	–	3	–	–	–	1	–	–	Phosphorothioic acid, O,O-diethyl O-(3,5,6-trichloro-2-pyridinyl) ester	2921-88-2	NI36626
97	197	199	65	314	125	258	47	**349**	100	54	48	32	26	24	20	19	**2.25**	9	11	3	1	1	–	3	–	–	–	1	–	–	Phosphorothioic acid, O,O-diethyl O-(3,5,6-trichloro-2-pyridinyl) ester	2921-88-2	NI36627
349	290	248	261	175	307			**349**	100	8	8	4	3	1				10	18	–	6	2	–	–	–	–	–	–	–	1	Copper, [1,1′-(1-methyl-2-pentylethylene)bis[3-thiosemicarbazidato]^2-]-	24457-74-7	LI07706
349	160	351	186	59	162	105	18	**349**	100	60	58	34	34	30	25	25		10	18	–	6	2	–	–	–	–	–	–	–	1	Copper, [1,1′-(1-methyl-2-pentylethylene)bis[3-thiosemicarbazidato]^2-]-	24457-74-7	NI36628
349	310	330	155	280	117	350	279	**349**	100	40	31	31	16	16	14	14		12	1	–	1	–	10	–	–	–	–	–	–	–	Aniline, 2,3,4,5,6-pentafluoro-N-(pentafluorophenyl)-	1535-92-8	NI36629
349	30	165	164	76	64	63	28	**349**	100	28	24	24	19	18	17	16		12	7	8	5	–	–	–	–	–	–	–	–	–	Aniline, 2,3,6-trinitro-N-(4-nitrophenyl)-	56830-87-6	MS08275
349	165	164	30	350	76	63	28	**349**	100	25	22	20	16	14	12	12		12	7	8	5	–	–	–	–	–	–	–	–	–	Aniline, 2,4,6-trinitro-N-(4-nitrophenyl)-	38417-97-9	MS08276
73	232	75	45	100	147	74	43	**349**	100	33	23	20	19	14	11	9	**0.10**	13	31	4	1	–	–	–	–	–	3	–	–	–	L-Aspartic acid, N-(trimethylsilyl)-, bis(trimethylsilyl) ester	55268-53-6	NI36630
73	232	100	45	75	147	74	43	**349**	100	42	18	18	15	11	11	9	**0.00**	13	31	4	1	–	–	–	–	–	3	–	–	–	L-Aspartic acid, N-(trimethylsilyl)-, bis(trimethylsilyl) ester	55268-53-6	MS08277
232	73	218	160	147	93	100	75	**349**	100	42	17	13	13	13	12	10	**2.00**	13	31	4	1	–	–	–	–	–	3	–	–	–	L-Aspartic acid, N-(trimethylsilyl)-, bis(trimethylsilyl) ester	55268-53-6	MS08278
247	192	177	235	205	234	149	206	**349**	100	57	30	28	28	26	24	19	**1.00**	14	15	6	5	–	–	–	–	–	–	–	–	–	2-Acetylamino-7-(1,2-diacetoxyethyl)-3,4-dihydropteridin-4-one		LI07707
73	232	142	75	147	233	70	89	**349**	100	99	60	25	24	22	22	20	**5.50**	14	35	3	1	–	–	–	–	–	3	–	–	–	L-Norvaline, N-(trimethylsilyl)-5-[(trimethylsilyl)oxy]-, trimethylsilyl ester	53261-79-3	MS08279
131	349	103	219	77	132	350	51	**349**	100	41	26	19	16	10	7	5		15	12	1	1	–	–	–	–	1	–	–	–	–	4′-Iodo-3-phenylpropenanilide		TR00303
276	129	349	226	277	102	299	119	**349**	100	35	33	19	13	11	7	7		15	12	3	1	–	5	–	–	–	–	–	–	–	Indole-3-acetic acid, 1-pentafluoropropionyl-, ethyl ester		MS08280
181	110	82	122	168	97	94	85	**349**	100	64	39	33	28	22	16	16	**11.00**	15	12	3	1	–	5	–	–	–	–	–	–	–	3-Isoxazolebutanoic acid, 5-methyl- pentafluorobenzyl ester		NI36631
250	251	349	151	247	252	235	43	**349**	100	19	17	12	6	4	4	4		15	21	6	–	–	–	–	–	–	–	–	–	1	Chromium, tris(2,4-pentanedionato)-	21679-31-2	NI36632
250	151	349	168	235	136	133	52	**349**	100	45	23	12	5	4	3	2		15	21	6	–	–	–	–	–	–	–	–	–	1	Chromium, tris(2,4-pentanedionato)-	21679-31-2	LI07708
251	28	350	151	31	252	160	236	**349**	100	99	53	50	47	42	17	8	**0.00**	15	21	6	–	–	–	–	–	–	–	–	–	1	Chromium, tris(2,4-pentanedionato)-	21679-31-2	NI36633
195	275	349	273	347	276	274	272	**349**	100	48	41	25	20	16	15	13		16	15	1	1	1	–	–	–	–	–	–	–	1	Benzamide, 2-[(allylthio)seleno]-N-phenyl-	118398-38-2	DD01379
197	183	140	198	126	153	154	81	**349**	100	33	32	29	27	25	23	17	**0.08**	16	16	6	1	–	–	–	–	–	–	1	–	–	1H-Benz[de]isoquinoline-1,3(2H)-dione, 2-[(diethoxyphosphinyl)oxy]-	1491-41-4	NI36634
44	288	257	273	243	302	317	54	**349**	100	50	23	10	10	8	7	7	**1.00**	16	19	6	3	–	–	–	–	–	–	–	–	–	Azirino[2′,3′:3,4]pyrrolo[1,2-a]indole-4,7-dione, 8-[[(aminocarbonyl)oxy]methyl]-1,1a,2,8,8a,8b-hexahydro-6,8a-dimethoxy-5-methyl-, [1aR-(1aα,8β,8aα,8bα)]-	4055-39-4	LI07709
44	288	258	260	349	243	273	228	**349**	100	45	24	17	11	11	10	10		16	19	6	3	–	–	–	–	–	–	–	–	–	Azirino[2′,3′:3,4]pyrrolo[1,2-a]indole-4,7-dione, 8-[[(aminocarbonyl)oxy]methyl]-1,1a,2,8,8a,8b-hexahydro-6,8a-dimethoxy-5-methyl-, [1aR-(1aα,8β,8aα,8bα)]-	4055-39-4	NI36635
70	44	288	316	331	42	349	273	**349**	100	53	52	41	34	23	19	17		16	19	6	3	–	–	–	–	–	–	–	–	–	Azirino[2′,3′:3,4]pyrrolo[1,2-a]indole-4,7-dione, 8-[[(aminocarbonyl)oxy]methyl]-1,1a,2,8,8a,8b-hexahydro-8a-hydroxy 6-methoxy-1,5-dimethyl-, [1aR-(1aα,8α,8aα,8bα)]-	4055-40-7	NI36636
70	288	42	273	68	245	243	271	**349**	100	51	23	16	12	8	8	7	**2.00**	16	19	6	3	–	–	–	–	–	–	–	–	–	Azirino[2′,3′:3,4]pyrrolo[1,2-a]indole-4,7-dione, 8-[[(aminocarbonyl)oxy]methyl]-1,1a,2,8,8a,8b-hexahydro-8a-hydroxy 6-methoxy-1,5-dimethyl-, [1aR-(1aα,8α,8aα,8bα)]-	4055-40-7	LI07710
73	232	45	233	349	75	74	28	**349**	100	77	21	16	15	10	9	5		17	27	3	1	–	–	–	–	–	2	–	–	–	1H-Indole-3-acetic acid, 5-methoxy-1-(trimethylsilyl)-, trimethylsilyl ester	55887-55-3	NI36637
84	43	126	121	44	228	186	168	**349**	100	70	43	29	13	10	8	8	**2.00**	18	23	6	1	–	–	–	–	–	–	–	–	–	1-Acetyl-3,4-diacetoxy-2-(4-methoxybenzyl)pyrrolidine		LI07711
138	84	58	55	268	98	349	152	**349**	100	88	75	69	67	67	65	65		18	23	6	1	–	–	–	–	–	–	–	–	–	2-Butenoic acid, 2-hydroxy-4-oxo-4-(4-cyclohexylamino-6-methyl-2-oxo-2H-pyran-3-yl)-, ethyl ester	86488-06-4	NI36638
120	136	93	119	94	349	80	138	**349**	100	90	74	63	60	44	40	36		18	23	6	1	–	–	–	–	–	–	–	–	–	Norsenecionan-11,16-dione, 13,19-epoxy-12-hydroxy-14-methyl-, (14β)-		NI36639
120	43	93	94	119	80	41	95	**349**	100	85	69	63	56	42	41	34	**1.50**	18	23	6	1	–	–	–	–	–	–	–	–	–	Senecionan-11,16-dione, 13,19-didehydro-15,20-epoxy-15,20-dihydro-12-hydroxy-, (15α,20S)-	5532-23-0	LI07712
119	120	93	94	95	43	136	138	**349**	100	98	85	61	56	47	28	26	**1.60**	18	23	6	1	–	–	–	–	–	–	–	–	–	Senecionan-11,16-dione, 13,19-didehydro-15,20-epoxy-15,20-dihydro-12-hydroxy-, (15α,20S)-	5532-23-0	NI36640
270	214	351	349	306	228	308	200	**349**	100	33	18	18	18	16	10	10		18	24	1	1	–	–	–	1	–	–	–	–	–	Spiro[cyclohexane-1,3′(2′H)-isoquinolin]-1′-one, 4′-bromo-1′,4′-dihydro-2′butyl-	120637-32-3	DD01380

MASS TO CHARGE RATIOS								M.W.	INTENSITIES								Parent	C	H	O	N	S	F	Cl	Br	I	Si	P	B	X	COMPOUND NAME	CAS Reg No	No
73	129	128	114	334	349	115	130	**349**	100	88	23	22	21	16	13	12		18	31	2	1	–	–	–	–	–	2	–	–	–	**4-Piperidinecarboxylic acid, 4-phenyl-1-(trimethylsilyl)-, trimethylsilyl ester**	55268-57-0	**NI36641**
349	334	174	146	321	350	348	320	**349**	100	37	33	33	32	23	22	22		19	15	2	3	1	–	–	–	–	–	–	–	–	**5H-1,3,4-Thiadiazolo[3,2-a]pyrimidin-5-one, 2-(2-ethoxyphenyl)-7-phenyl-**	82077-85-8	**NI36642**
105	77	43	202	174	106	307	200	**349**	100	45	12	8	8	8	7	7	**4.00**	19	15	4	3	–	–	–	–	–	–	–	–	–	**1,4-Naphthoquinone, 2-(acetylamino)-3-(N'-benzoylhydrazido)-**		**NI36643**
120	85	42	77	347	131	118	43	**349**	100	95	67	39	36	36	32	32	**0.00**	19	19	2	5	–	–	–	–	–	–	–	–	–	**6,7-Diphenyl-10-hydroxy-2-imino-5-methyl-4-oxo-2,3,4,5,6,7,8,10-octahydropteridine**		**MS08281**
107	208	86	108	58	334	306	230	**349**	100	29	11	9	9	8	6	5	**0.00**	19	27	3	1	1	–	–	–	–	–	–	–	–	**1-Cyclopentene-1-carboxamide, N,N-diisopropyl-4-(phenylsulphonyl)-4-methyl-**	119039-06-4	**DD01381**
110	28	239	66	109	238	77	178	**349**	100	37	34	34	28	23	23	21	**0.00**	20	15	1	1	2	–	–	–	–	–	–	–	–	**3H-Indol-3-one, 1,2-dihydro-2,2-bis(phenylthio)-**	38168-16-0	**MS08282**
307	292	349	308	293	221	220	264	**349**	100	51	29	18	11	11	10	9		20	15	5	1	–	–	–	–	–	–	–	–	–	**Noraporphin-7-one, 4,5,6,6a-tetradehydro-2-hydroxy-1,3-dimethoxy-, acetate (ester)**	5140-37-4	**NI36644**
349	78	316	185	77	72	350	83	**349**	100	24	14	11	9	9	6	6		20	19	3	3	–	–	–	–	–	–	–	–	–	**Benzoic acid, 2-(methylamino)-5-(1,2,3,9-tetrahydro-9-oxopyrrolo[2,1-b]quinazolin-3-yl)-, methyl ester**	16688-19-0	**NI36645**
349	275	350	183	274	302	200	276	**349**	100	42	24	18	15	12	12	11		20	19	3	3	–	–	–	–	–	–	–	–	–	**Benzoic acid, 2-[(1,2,3,9-tetrahydro-9-oxopyrrolo[2,1-b]quinazolin-3-yl)amino]-, ethyl ester**	16688-20-3	**NI36646**
182	224	210	153	83	321	198	196	**349**	100	85	56	47	37	35	35	35	**23.00**	20	35	2	3	–	–	–	–	–	–	–	–	–	**Acetamide, N-(2-cyanoethyl)-N-[3-(2-oxoazacyclotridec-1-yl)propyl]-**	67171-83-9	**NI36647**
227	105	77	228	198	78	51	106	**349**	100	78	56	18	16	12	10	8	**4.00**	21	16	3	1	–	1	–	–	–	–	–	–	–	**Benzoic acid, 2-[[N-(2-fluorophenyl)-N-benzoyl]amino]-3-methyl-**		**NI36648**
348	349	333	350	57	334	55	95	**349**	100	83	42	20	19	17	16	13		21	19	4	1	–	–	–	–	–	–	–	–	–	**[1,3]Benzodioxolo[5,6-c]phenanthridine, 12,13-dihydro-2,3-dimethoxy-12-methyl-**	13063-06-4	**MS08283**
117	39	118	349	76	206	43	317	**349**	100	61	35	18	16	14	7	6		21	19	4	1	–	–	–	–	–	–	–	–	–	**6,7-Benzo-1H-azepine, 1-methyl-2-phenyl-3,4-bis(methoxycarbonyl)-**		**NI36649**
77	43	331	51	78	349	332	65	**349**	100	72	36	20	16	11	10	10		21	19	4	1	–	–	–	–	–	–	–	–	–	**1,3-Butanedione, 1-(5-benzoyl-3-oxo-1-phenyl-2-pyrrolidinyl)-**	35307-17-6	**NI36650**
349	258	231	289	290	230	350	259	**349**	100	68	45	35	30	26	25	18		21	19	4	1	–	–	–	–	–	–	–	–	–	**2-Butenedioic acid, 2-(1-methyl-2-phenylindol-3-yl)-, dimethyl ester, (E)-**		**NI36651**
349	258	232	289	290	231	350	216	**349**	100	81	55	47	33	30	23	18		21	19	4	1	–	–	–	–	–	–	–	–	–	**2-Butenedioic acid, 2-(1-methyl-2-phenylindol-3-yl)-, dimethyl ester, (Z)-**		**NI36652**
136	150	151	108	41	57	115	43	**349**	100	90	45	39	39	37	34	31	**11.00**	21	19	4	1	–	–	–	–	–	–	–	–	–	**N-Ethyl-N-(4-anisyl)-2-methyl-1,4-naphthoquinone-3-carboxamide**		**MS08284**
217	188	146	133	145	161	189	231	**349**	100	96	93	81	73	58	49	27	**24.00**	21	23	–	3	1	–	–	–	–	–	–	–	–	**1-[(2-Benzyl-4-thiazolyl)methyl]-4-phenylpiperazine**		**MS08285**
198	43	83	55	41	183	199	182	**349**	100	35	29	29	21	19	17	17	**2.26**	21	23	2	3	–	–	–	–	–	–	–	–	–	**Dehydrodeoxybrevianamide E**		**NI36653**
201	72	230	187	103	349	91	175	**349**	100	36	23	20	15	9	9	8		21	23	2	3	–	–	–	–	–	–	–	–	–	**1-Pyrimidinium, 2-(diethylamino)-3,6-dihydro-5-methyl-6-oxo-1,3-diphenyl-4-olate**		**DD01382**
334	335	197	135	336	28	318	256	**349**	100	28	8	8	6	6	4	4	**1.59**	21	24	–	1	–	–	–	–	–	1	1	–	–	**Silanamine, 1,1,1-trimethyl-N-(triphenylphosphoranylidene)-**	13892-06-3	**NI36654**
334	335	135	197	336	28	160	318	**349**	100	28	9	8	6	6	5	4	**1.59**	21	24	–	1	–	–	–	–	–	1	1	–	–	**Silanamine, 1,1,1-trimethyl-N-(triphenylphosphoranylidene)-**	13892-06-3	**NI36655**
349	208	209	152	166	196	306	180	**349**	100	90	45	35	30	20	15	15		21	35	3	1	–	–	–	–	–	–	–	–	–	**5-(Butylamino)-2-hydroxy-3-undecyl-1,4-benzoquinone**		**NI36656**
108	248	107	55	101	216	190	40	**349**	100	70	65	64	45	44	35	32	**2.00**	21	35	3	1	–	–	–	–	–	–	–	–	–	**Methyl 3-β-methoxy-D-norandrostane-16α-carbamate**		**LI07713**
178	105	45	77	44	43	122	51	**349**	100	64	40	37	36	31	19	19	**9.00**	22	14	–	1	–	3	–	–	–	–	–	–	–	**Cyclopropenone imine, N-[4-(trifluoromethyl)phenyl]-2,3-diphenyl-**		**MS08286**
139	258	141	260	111	259	43	140	**349**	100	48	33	16	15	9	9	8	**7.00**	22	20	1	1	–	–	1	–	–	–	–	–	–	**Benzamide, 4-chloro-N-(1,2-diphenylethyl)-N-methyl-**		**LI07714**
105	77	51	244	106	99	122	41	**349**	100	56	11	10	10	7	6	5	**0.80**	22	23	3	1	–	–	–	–	–	–	–	–	–	**2,N-Dibenzoyl-8-octanelactam**	102222-11-7	**NI36657**
91	349	241	184	304	168	350	228	**349**	100	73	66	27	26	24	21	20		22	23	3	1	–	–	–	–	–	–	–	–	–	**1H-Indole-3-carboxylic acid, 1-[(benzyloxy)methyl]-2-(1,2-dimethyl-1-propenyl)-**		**DD01383**
91	349	171	247	128	115	350	64	**349**	100	52	46	44	37	33	32	32		22	23	3	1	–	–	–	–	–	–	–	–	–	**2H-Pyrano[4,3-b]indol-1-one, 4-[(benzyloxy)methyl]-3-propyl-1,4-dihydro**		**DD01384**
125	43	41	29	55	80	27	94	**349**	100	76	73	55	39	27	25	23	**7.44**	22	39	–	1	1	–	–	–	–	–	–	–	–	**Aniline, 3-(hexadecylthio)-**	57946-69-7	**NI36658**
83	349	82	125	55	141	350	126	**349**	100	79	38	37	25	22	20	15		22	39	2	1	–	–	–	–	–	–	–	–	–	**1H-Pyrrole-2,5-dione, 3-methyl-4-(5,9,13-trimethyltetradecyl)-**	55268-59-2	**NI36659**
217	216	57	132	85	133	131	130	**349**	100	93	91	87	58	45	43	41	**0.00**	23	27	2	1	–	–	–	–	–	–	–	–	–	**1-Isoquinolinemethanol, u-2-tert-butoxy-1,2,3,4-tetrahydro-α-styryl-**		**DD01385**
59	349	150	348	42	292	350	281	**349**	100	95	80	47	31	27	25	22		23	27	2	1	–	–	–	–	–	–	–	–	–	**Morphinan, 3-(4-hydroxyphenoxy)-N-methyl-**	67562-61-2	**NI36660**
306	167	55	252	41	349	152	115	**349**	100	60	43	31	27	22	13	11		23	27	2	1	–	–	–	–	–	–	–	–	–	**Oxetane, 2,2-diphenyl-3-(N-morpholino)-3,4-tetramethylene-**		**LI07715**
127	140	84	349	55	69	41	43	**349**	100	58	34	32	27	25	25	18		23	43	1	1	–	–	–	–	–	–	–	–	–	**Piperidine, 1-(1-oxo-9-octadecenyl)-**	56630-42-3	**NI36661**
113	55	43	49	126	41	70	84	**349**	100	58	52	49	48	40	34	25	**10.04**	23	43	1	1	–	–	–	–	–	–	–	–	–	**Pyrrolidine, 1-[8-(2-octylcyclopropyl)-1-oxooctyl]-**	56630-59-2	**NI36662**
113	126	55	70	43	208	98	41	**349**	100	56	27	22	21	17	17	16	**7.34**	23	43	1	1	–	–	–	–	–	–	–	–	–	**Pyrrolidine, 1-[1-oxo-5-(2-undecylcyclopropyl)pentyl]-**	56600-03-4	**NI36663**
272	243	91	258	244	165	273	106	**349**	100	60	50	40	39	37	25	14	**2.00**	26	23	–	1	–	–	–	–	–	–	–	–	–	**Benzylamine, N-trityl-**	3378-73-2	**NI36664**
179	350	127	254	44	28	128	278	**350**	100	63	39	28	26	15	14	6		4	–	3	–	–	–	–	–	2	–	–	–	–	**Diiodomaleic anhydride**		**MS08287**
131	69	181	93	31	119	100	162	**350**	100	58	14	13	13	10	6	5	**1.20**	7	–	–	–	–	14	–	–	–	–	–	–	–	**1-Heptene, 1,1,2,3,3,4,4,5,5,6,6,7,7,7-tetradecafluoro-**	355-63-5	**WI01453**
69	131	100	181	31	150	93	119	**350**	100	16	15	10	10	9	7	4	**0.00**	7	–	–	–	–	14	–	–	–	–	–	–	–	**Perfluoro-1,1-dimethylcyclopentane**		**LI07716**
69	131	181	100	93	231	162	119	**350**	100	50	27	13	8	6	6	6	**0.00**	7	–	–	–	–	14	–	–	–	–	–	–	–	**Tetradecafluoromethylcyclohexane**		**IC04076**
69	131	181	100	93	31	119	162	**350**	100	44	20	15	11	10	5	4	**0.00**	7	–	–	–	–	14	–	–	–	–	–	–	–	**Tetradecafluoromethylcyclohexane**		**AI02141**
81	71	113	82	67	55	318	54	**350**	100	34	18	14	10	7	6	6	**4.03**	7	13	1	–	–	–	1	–	–	–	–	–	1	**2-Methoxycyclohexylmercuric chloride**		**LI07717**

MASS TO CHARGE RATIOS								M.W.	INTENSITIES								Parent	C	H	O	N	S	F	Cl	Br	I	Si	P	B	X	COMPOUND NAME	CAS Reg No	No
59	221	81	147	85	271	93	109	**350**	100	21	10	9	9	8	8	5	**0.00**	8	6	4	–	–	6	2	–	–	–	–	–	–	Butanedioic acid, 2,3-difluoro-2,3-bis(chlorodifluoromethyl)-, dimethyl ester		NI36665
175	99	129	56	83	72	84	55	**350**	100	87	60	50	43	18	17	17	**0.00**	8	14	8	8	–	–	–	–	–	–	–	–	–	2,2′-Bis(1,3-dinitrohexahydropyrimidine)		DD01386
294	350	264	292	266	348	262	28	**350**	100	76	68	63	56	45	34	27		9	7	3	–	–	–	–	–	–	–	–	–	1	Tricarbonylmethylcyclopentadienylrhenium		IC04077
150	108	250	181	69	350	113	206	**350**	100	20	14	13	12	10	7	6		10	2	–	–	–	12	–	–	–	–	–	–	–	2H,5H-Dodecafluorotricyclo[4.2.2.0^{2,5}]dec-7-ene		NI36666
250	231	181	100	331	200	251	212	**350**	100	37	37	22	19	18	11	4	**0.00**	10	2	–	–	–	12	–	–	–	–	–	–	–	2,3-Bis(trifluoromethyl)-1,4,7,7,8,8-hexafluorobicyclo[2.2.2]octa-2,5-diene		NI36667
250	231	100	181	200	69	331	251	**350**	100	46	46	33	25	25	17	13	**0.00**	10	2	–	–	–	12	–	–	–	–	–	–	–	2,3-Bis(trifluoromethyl)-1,6,7,7,8,8-hexafluorobicyclo[2.2.2]octa-2,5-diene		NI36668
28	93	266	208	180	165	55	135	**350**	100	87	84	70	52	45	45	33	**2.62**	10	15	8	–	–	–	–	–	–	–	1	–	1	Iron, tricarbonyl[(2,3-η)-methyl 2-propenoate](trimethyl phosphite-P)-	62199-95-5	NI36669
350	183	155	117	272	331	167	303	**350**	100	58	57	20	17	10	10	9		12	–	1	–	–	10	–	–	–	–	–	–	–	Benzene, 1,1′-oxybis[2,3,4,5,6-pentafluoro-	1800-30-2	LI07718
350	183	155	117	272	351	331	167	**350**	100	58	57	20	17	13	10	10		12	–	1	–	–	10	–	–	–	–	–	–	–	Benzene, 1,1′-oxybis[2,3,4,5,6-pentafluoro-	1800-30-2	NI36670
217	315	317	201	202	351	219	218	**350**	100	63	28	27	21	19	16	16	**1.61**	12	8	4	–	2	–	2	–	–	–	–	–	–	[1,1′-Biphenyl]-4,4′-disulphonyl dichloride	3406-84-6	NI36671
152	76	251	315	151	63	153	64	**350**	100	57	55	52	39	27	23	23	**22.45**	12	8	4	–	2	–	2	–	–	–	–	–	–	[1,1′-Biphenyl]-4,4′-disulphonyl dichloride	3406-84-6	NI36672
250	249	251	231	350	223	200	229	**350**	100	28	15	12	8	7	7	5		13	4	–	–	–	10	–	–	–	–	–	–	–	4H-Decafluoro-5-methyltricyclo[6.2.2.0^{2,7}]dodeca-2,4,6,9-triene		NI36673
43	41	27	29	28	18	42	39	**350**	100	78	73	31	21	13	12	12	**0.40**	13	17	4	4	–	3	–	–	–	–	–	–	–	1,3-Benzenediamine, 2,4-dinitro-N^3,N^3-dipropyl-6-(trifluoromethyl)-	29091-21-2	NI36674
73	147	233	75	74	245	190	148	**350**	100	40	20	11	9	8	7	7	**0.00**	13	30	5	–	–	–	–	–	–	3	–	–	–	Butanedioic acid, [(trimethylsilyl)oxy]-, bis(trimethylsilyl) ester	38166-11-9	NI36675
73	147	233	75	45	74	55	133	**350**	100	37	13	12	11	9	9	7	**0.00**	13	30	5	–	–	–	–	–	–	3	–	–	–	Butanedioic acid, [(trimethylsilyl)oxy]-, bis(trimethylsilyl) ester	38166-11-9	NI36676
73	147	233	75	245	74	45	56	**350**	100	45	16	12	9	9	9	8	**0.20**	13	30	5	–	–	–	–	–	–	3	–	–	–	Butanedioic acid, [(trimethylsilyl)oxy]-, bis(trimethylsilyl) ester	38166-11-9	NI36677
350	334	284	145	319	349	235	63	**350**	100	66	63	47	30	29	23	23		14	8	2	2	–	6	–	–	–	–	–	–	–	4-Nitroso-2,3′-bis(trifluoromethyl)-N-hydroxydiphenylamine		IC04078
270	269	271	272	206	80	82	234	**350**	100	86	44	33	27	26	25	23	**0.00**	14	12	–	4	–	–	1	1	–	–	–	–	–	3H-1,3,4-Benzotriazepine, 7-chloro-2-imino-5-phenyl-1,2-dihydro-, hydrobromide		MS08288
168	103	43	115	145	116	126	117	**350**	100	99	99	88	83	75	67	46	**0.00**	14	19	9	–	–	1	–	–	–	–	–	–	–	β-D-Glucopyranose, 1-deoxy-1-fluoro-, tetraacetate	2823-46-3	NI36678
43	145	103	188	160	130	117	115	**350**	100	12	11	6	4	4	4	4	**0.00**	14	19	9	–	–	1	–	–	–	–	–	–	–	β-D-Glucopyranose, 2-deoxy-2-fluoro-, tetraacetate	31077-89-1	NI36679
43	115	98	157	140	200	103	73	**350**	100	13	10	7	4	3	3	3	**0.00**	14	19	9	–	–	1	–	–	–	–	–	–	–	β-D-Glucopyranose, 3-deoxy-3-fluoro-, tetraacetate	7226-43-9	NI36680
43	103	100	160	145	202	99	44	**350**	100	13	10	9	7	5	2	2	**0.00**	14	19	9	–	–	1	–	–	–	–	–	–	–	β-D-Glucopyranose, 4-deoxy-4-fluoro-, tetraacetate	27108-07-2	NI36681
43	160	115	202	157	118	103	73	**350**	100	12	12	11	11	7	6	3	**0.00**	14	19	9	–	–	1	–	–	–	–	–	–	–	β-D-Glucopyranose, 6-deoxy-6-fluoro-, tetraacetate	33557-29-8	NI36682
95	55	109	67	83	81	69	97	**350**	100	92	75	73	69	63	47	41	**0.00**	14	24	–	–	–	–	–	2	–	–	–	–	–	1,14-Dibromo-2,12-tetradecadiene, (E,E)-		MS08289
95	67	81	109	83	69	97	273	**350**	100	85	71	65	58	44	32	1	**0.00**	14	24	–	–	–	–	–	2	–	–	–	–	–	1,14-dibromo-2,12-tetradecadiene, (Z,Z)-		MS08290
146	28	73	217	114	303	204	75	**350**	100	75	33	25	25	22	21	21	**15.00**	14	30	6	–	–	–	–	–	–	2	–	–	–	2,6-Dioxabicyclo[3.3.0]octane, 4,8-bis(trimethylsilyloxy)-3,5-dimethoxy-		LI07719
73	146	75	131	89	45	59	147	**350**	100	65	38	24	23	21	15	13	**0.00**	14	30	6	–	–	–	–	–	–	2	–	–	–	D-Gluco-Hexodialdodifuranoside, dimethyl 2,5-bis-O-(trimethylsilyl)-	74685-72-6	NI36683
73	27	75	248	147	133	103	175	**350**	100	64	44	20	17	15	15	14	**0.00**	14	30	6	–	–	–	–	–	–	2	–	–	–	Tartaric acid, bis(trimethylsilyl)-, diethyl ester	73639-61-9	NI36684
73	75	147	129	218	191	103	81	**350**	100	72	49	45	35	35	33	30	**0.00**	14	34	4	–	–	–	–	–	–	3	–	–	–	D-erythro-Pentofuranose, 2-deoxy-1,3,5-tris-O-(trimethylsilyl)-	56227-33-9	NI36685
73	147	116	75	101	217	103	191	**350**	100	59	59	58	46	37	28	27	**0.00**	14	34	4	–	–	–	–	–	–	3	–	–	–	D-erythro-Pentopyranose, 2-deoxy-1,3,4-tris-O-(trimethylsilyl)-	56227-34-0	NI36686
73	116	101	147	103	75	191	217	**350**	100	78	39	35	24	20	19	17	**0.00**	14	34	4	–	–	–	–	–	–	3	–	–	–	D-erythro-Pentopyranose, 2-deoxy-1,3,4-tris-O-(trimethylsilyl)-	56227-34-0	NI36687
73	103	147	170	143	69	75	260	**350**	100	36	32	13	13	13	12	8	**0.00**	14	34	4	–	–	–	–	–	–	3	–	–	–	Propanoic acid, 2,2-bis(trimethylsiloxymethyl)-, trimethylsilyl ester		NI36688
77	105	78	51	301	350	181	237	**350**	100	47	45	41	36	27	26	25		15	11	4	2	1	–	1	–	–	–	–	–	–	Imidazo[1,2-a]pyridinium, 3-(chloroacetyl)-2,3-dihydro-2-oxo-1-(phenylsulphonyl)-, hydroxide, inner salt	11066-38-9	NI36689
78	79	135	217	189	133	108	77	**350**	100	32	20	15	15	15	15	15	**1.00**	15	18	6	4	–	–	–	–	–	–	–	–	–	Anhydro-3-(1,2-dicarbethoxyhydrazino)-2-hydroxy-1-methyl-4-oxopyrido[1,2-a]pyrimidinium hydroxide		LI07720
59	58	261	116	130	115	89	129	**350**	100	98	69	67	30	26	26	19	**0.00**	15	18	6	4	–	–	–	–	–	–	–	–	–	4a,8a-Methanophthalazine-1,4-dicarboxylic acid, 1,2-dihydro-1-[N′-(methoxycarbonyl)hydrazino]-, dimethyl ester		DD01387
81	126	55	53	28	27	206	54	**350**	100	27	25	17	15	14	12	11	**0.50**	15	18	6	4	–	–	–	–	–	–	–	–	–	Thymidine, 5′-deoxy-5′-(3,4-dihydro-5-methyl-2,4-dioxo-1(2H)-pyrimidinyl)-	35959-35-4	NI36690
105	77	175	140	218	253	163	121	**350**	100	66	51	45	43	38	38	36	**28.00**	15	22	4	6	–	–	–	–	–	–	–	–	–	2-Amino-6-(3-methyl-2-butenylamino)-9-β-ribofuranosylpurine		LI07721
69	97	83	111	57	55	71	281	**350**	100	77	67	56	55	49	31	27	**0.00**	15	24	2	–	–	6	–	–	–	–	–	–	–	1,3-Dioxolane, 4-ethyl-5-octyl-2,2-bis(trifluoromethyl)-, cis-	38274-72-5	MS08291
69	97	83	57	111	55	281	59	**350**	100	68	65	48	47	44	30	25	**0.00**	15	24	2	–	–	6	–	–	–	–	–	–	–	1,3-Dioxolane, 4-ethyl-5-octyl-2,2-bis(trifluoromethyl)-, trans-	38274-73-6	NI36691
43	117	101	99	129	87	45	71	**350**	100	54	52	42	35	32	21	13	**0.00**	15	26	9	–	–	–	–	–	–	–	–	–	–	D-Galactitol, 2,3,4-tri-O-methyl-, triacetate	25521-27-1	NI36692
43	117	129	45	161	101	87	160	**350**	100	55	32	26	17	17	12	10	**0.00**	15	26	9	–	–	–	–	–	–	–	–	–	–	Galactitol, 1,3,5-tri-O-methyl-, triacetate	19318-48-0	NI36693
43	117	129	45	101	161	87	71	**350**	100	85	69	42	38	30	24	17	**0.00**	15	26	9	–	–	–	–	–	–	–	–	–	–	Galactitol, 1,3,5-tri-O-methyl-, triacetate	19318-48-0	NI36694
43	101	117	129	99	87	70	161	**350**	100	99	80	60	55	42	33	32	**0.00**	15	26	9	–	–	–	–	–	–	–	–	–	–	D-Glucitol, 2,3,4-tri-O-methyl-, triacetate	24905-18-8	NI36695
43	45	117	42	87	101	99	233	**350**	100	31	26	20	13	11	10	9	**0.00**	15	26	9	–	–	–	–	–	–	–	–	–	–	Glucitol, 2,3,6-tri-O-methyl-, triacetate		NI36696
117	113	99	43	233	173	101	129	**350**	100	53	45	39	33	32	31	23	**0.00**	15	26	9	–	–	–	–	–	–	–	–	–	–	Glucitol, 2,3,6-tri-O-methyl-, triacetate	20250-41-3	NI36697
101	43	74	117	75	116	88	69	**350**	100	50	46	42	37	33	21	10	**0.00**	15	26	9	–	–	–	–	–	–	–	–	–	–	α-glycero-D-Glucoheptopyranoside, methyl 3,7-di-O-acetyl-2,4,6-tri-O-methyl-	84564-10-3	NI36698
101	43	129	45	87	75	89	74	**350**	100	71	36	28	27	27	25	22	**0.00**	15	26	9	–	–	–	–	–	–	–	–	–	–	β-glycero-D-Glucoheptopyranoside, methyl 2,3-di-O-acetyl-4,6,7-tri-O-methyl-	84564-07-8	NI36699

MASS TO CHARGE RATIOS	M.W.	INTENSITIES	Parent	C	H	O	N	S	F	Cl	Br	I	Si	P	B	X	COMPOUND NAME	CAS Reg No	No
43 75 74 87 45 116 129 115	**350**	100 99 77 54 54 35 28 21	**0.00**	15	26	9	–	–	–	–	–	–	–	–	–	–	β-glycero-D-Glucoheptopyranoside, methyl 2,4-di-O-acetyl-3,6,7-tri-O-methyl-	84564-06-7	NI36700
75 101 43 74 117 88 116 45	**350**	100 99 73 47 37 33 18 17	**0.00**	15	26	9	–	–	–	–	–	–	–	–	–	–	β-glycero-D-Glucoheptopyranoside, methyl 2,7-di-O-acetyl-3,4,6-tri-O-methyl-	84564-12-5	NI36701
43 74 116 87 185 89 75 129	**350**	100 99 67 57 48 48 48 40	**0.00**	15	26	9	–	–	–	–	–	–	–	–	–	–	β-glycero-D-Glucoheptopyranoside, methyl 3,4-di-O-acetyl-2,6,7-tri-O-methyl-	84564-05-6	NI36702
43 75 74 115 101 87 116 45	**350**	100 99 92 72 70 60 56 46	**0.00**	15	26	9	–	–	–	–	–	–	–	–	–	–	β-glycero-D-Glucoheptopyranoside, methyl 3,6-di-O-acetyl-2,4,7-tri-O-methyl-	84564-11-4	NI36703
88 75 43 45 73 101 74 115	**350**	100 94 38 22 11 9 9 8	**0.00**	15	26	9	–	–	–	–	–	–	–	–	–	–	β-glycero-D-Glucoheptopyranoside, methyl 4,6-di-O-acetyl-2,3,7-tri-O-methyl-	84564-09-0	NI36704
75 88 43 87 117 101 45 85	**350**	100 91 37 13 11 11 10 8	**0.00**	15	26	9	–	–	–	–	–	–	–	–	–	–	β-glycero-D-Glucoheptopyranoside, methyl 4,7-di-O-acetyl-2,3,6-tri-O-methyl-	84564-08-9	NI36705
88 75 101 43 45 73 87 89	**350**	100 51 46 44 13 12 11 8	**0.00**	15	26	9	–	–	–	–	–	–	–	–	–	–	β-glycero-D-Glucoheptopyranoside, methyl 6,7-di-O-acetyl-2,3,4-tri-O-methyl-	84564-13-6	NI36706
43 74 87 88 75 45 89 185	**350**	100 99 92 87 82 82 74 66	**0.00**	15	26	9	–	–	–	–	–	–	–	–	–	–	α-L-glycero-D-Mannoheptopyranoside, methyl 3,4-di-O-acetyl-2,6,7-tri-O-methyl-		NI36707
100 43 75 117 74 88 45 116	**350**	100 62 56 44 44 31 24 21	**0.00**	15	26	9	–	–	–	–	–	–	–	–	–	–	α-L-glycero-D-Mannoheptopyranoside, methyl 3,7-di-O-acetyl-2,4,6-tri-O-methyl-		NI36708
43 129 45 87 161 101 71 99	**350**	100 98 54 47 26 18 16 12	**0.00**	15	26	9	–	–	–	–	–	–	–	–	–	–	D-Mannitol, 1,3,4-tri-o-methyl-, triacetate	19285-92-8	NI36709
43 117 42 45 87 101 113 233	**350**	100 26 25 23 16 15 13 11	**0.00**	15	26	9	–	–	–	–	–	–	–	–	–	–	Mannitol, 2,3,6-tri-O-methyl-, triacetate		NI36710
129 161 189 87 43 99 101 145	**350**	100 45 28 20 12 11 10 6	**0.00**	15	26	9	–	–	–	–	–	–	–	–	–	–	Mannitol, 3,4,6-tri-O-methyl-, triacetate		NI36711
101 117 129 99 161 189 43 87	**350**	100 81 72 51 47 35 30 29	**0.00**	15	26	9	–	–	–	–	–	–	–	–	–	–	Mannitol, 2,3,4-tri-O-methyl-1,5,6-tri-O-acetyl-	72002-73-4	NI36712
177 293 179 235 137 291 251 181	**350**	100 99 70 60 35 20 20 10	**0.00**	15	34	1	–	–	–	–	–	–	–	–	–	1	1-Propanol, 3-(tributylstannyl)-	29346-30-3	NI36713
322 349 350 234 206 207 323 205	**350**	100 87 85 72 68 50 40 40		16	12	3	2	1	2	–	–	–	–	–	–	–	1H-1,4-Benzodiazepin-2-one, 7-(difluoromethylsulphonyl)-2,3-dihydro-5-phenyl-	55384-55-9	NI36714
333 29 259 335 41 334 350 213	**350**	100 67 45 34 22 20 19 18		16	15	3	2	1	–	1	–	–	–	–	–	–	10H-Phenothiazine, 10-butyl-3-chloro-7-nitro-, 5-oxide	35076-82-5	NI36715
333 29 259 335 41 350 213 231	**350**	100 69 49 38 24 23 20 19		16	15	3	2	1	–	1	–	–	–	–	–	–	10H-Phenothiazine, 10-butyl-7-chloro-3-nitro-, 5-oxide		LI07722
77 107 94 79 100 163 108 209	**350**	100 67 47 40 33 27 27 23	**0.00**	16	18	5	2	1	–	–	–	–	–	–	–	–	DL-Valine, 3-mercapto-N-[[5-oxo-2-(phenoxymethyl)-4(5H)-oxazolylidene]methyl]-	51703-49-2	NI36716
28 182 254 97 67 208 68 53	**350**	100 68 61 42 38 35 33 27	**2.50**	16	22	5	–	–	–	–	–	–	–	–	–	1	Iron, carbonyl[(2,3,4,5-η)-diethyl 2,4-hexadienedioate][(1,2,3,4-η)-2-methyl-1,3-butadiene]-	62265-43-4	NI36717
73 87 89 86 59 71 57 99	**350**	100 74 58 42 42 39 32 31	**3.50**	16	30	8	–	–	–	–	–	–	–	–	–	–	(1S,14S)-Bicyclo[12.10.0]-3,6,9,12,15,18,21,24-octaoxatetracosane		MS08292
87 73 89 57 59 71 96 58	**350**	100 97 58 54 51 46 38 36	**12.90**	16	30	8	–	–	–	–	–	–	–	–	–	–	(2S,2′S)-2,2′-Bis[1,4,7,10-tetraoxacyclododecane]		MS08293
87 73 89 57 59 71 187 86	**350**	100 97 58 54 51 46 42 38	**12.89**	16	30	8	–	–	–	–	–	–	–	–	–	–	2,2′-Bis[1,4,7,10-tetraoxacyclododecane], (2S,2′S)-	95637-56-2	NI36718
73 163 91 245 178 55 74 350	**350**	100 33 18 15 10 10 9 8		16	38	–	–	2	–	–	–	–	2	–	–	–	3,14-Dithia-2,15-disilahexadecane, 2,2,15,15-tetramethyl-	22396-25-4	NI36719
350 102 322 351 116 77 144 349	**350**	100 66 34 23 20 16 15 13		17	10	3	4	1	–	–	–	–	–	–	–	–	7H-1,3,4-Thiadiazolo[3,2-a]pyrimidin-7-one, 2-(4-nitrophenyl)-5-phenyl-	82077-83-6	NI36720
350 197 352 168 140 199 198 351	**350**	100 47 38 26 26 22 21 20		17	15	6	–	–	–	1	–	–	–	–	–	–	Spiro[benzofuran-2(3H),1′-[2,5]cyclohexadiene]-3,4′-dione, 7-chloro-2′,4,6 trimethoxy-6′-methyl-, (S)-	3573-90-8	NI36721
104 43 161 77 162 350 134 352	**350**	100 60 39 39 26 23 15 14		17	16	2	2	–	–	2	–	–	–	–	–	–	2-Anilino-2′,5′-dichloro-3-methoxycrotonanilide		MS08294
59 139 111 125 43 141 127 75	**350**	100 72 70 65 39 38 27 27	**3.90**	17	16	2	2	–	–	2	–	–	–	–	–	–	Butanmoic acid, 2-(4-chlorophenyl)-3-methyl-3-(4-chlorophenylazo)-		NI36722
118 161 91 189 163 307 120 65	**350**	100 73 65 38 32 30 25 25	**11.80**	17	16	2	2	–	–	2	–	–	–	–	–	–	2′,5′-Dichloro-3-hydroxy-2-(4-methylanilino)crotonanilide		MS08295
57 72 124 91 350 73 44 42	**350**	100 92 42 42 36 23 23 20		17	22	2	2	2	–	–	–	–	–	–	–	–	2-(Tosylamido)benzenesulphenyldiethylamide		LI07723
131 160 178 174 91 173 72 87	**350**	100 48 35 30 23 22 19 18	**1.00**	17	22	6	2	–	–	–	–	–	–	–	–	–	15-Benzyl-1,4,10-trioxa-7,13-diazacyclopentadecane-2,6,14-trione		MS08296
219 222 193 207 178 221 208 237	**350**	100 97 89 81 77 76 76 64	**24.00**	17	22	6	2	–	–	–	–	–	–	–	–	–	3-Methyl-4-(N-acetylalanyl)amino-5,8-dimethoxy-3,4-dihydroisocoumarin		LI07724
131 160 178 174 91 173 72 135	**350**	100 48 35 30 23 22 19 18	**1.02**	17	22	6	2	–	–	–	–	–	–	–	–	–	1,4,10-Trioxa-7,13-diazacyclopentadecane-2,6,14-trione, 15-benzyl-	74229-37-1	NI36723
350 352 321 305 256 226 319 322	**350**	100 99 74 72 64 59 56 54		17	23	1	2	–	–	–	1	–	–	–	–	–	(3-Ethyl-4-methyl-5-bromo-2-pyrrolyl)(3-methyl-4-ethyl-5-methoxymethyl 2-pyrrolylidene)methane		LI07725
55 43 245 111 69 57 97 83	**350**	100 87 64 60 56 54 52 51	**45.00**	17	26	4	4	–	–	–	–	–	–	–	–	–	Undecan-5-one, 2,4-dinitrophenylhydrazone		LI07726
235 220 73 204 75 145 236 130	**350**	100 60 50 38 38 22 18 14	**0.00**	17	30	2	2	–	–	–	–	–	2	–	–	–	Propanediamide, 2-ethyl-2-phenyl-N,N′-bis(trimethylsilyl)-	55268-56-9	NI36724
132 91 118 218 131 143 175 350	**350**	100 82 72 49 41 38 37 36		18	18	–	6	1	–	–	–	–	–	–	–	–	6-(4-Methylphenyl)amino-7-(4-methylphenyl)-3-methylthio[1,2,4]triazolo[4,3-b][1,2,4]triazole		MS08297
217 234 145 161 307 291 308 189	**350**	100 98 41 40 27 23 20 20	**0.00**	18	22	7	–	–	–	–	–	–	–	–	–	–	Indan-1,1-dicarboxylic acid, 3-acetoxy-6-methoxy-, diethyl ester	118598-65-5	DD01388
187 146 205 350 132 305	**350**	100 98 88 73 73 69		19	14	5	2	–	–	–	–	–	–	–	–	–	α-Phthalimidooxindole-3-propionic acid		LI07727
146 187 205 350 132 305 104 130	**350**	100 99 89 74 73 70 57 33		19	14	5	2	–	–	–	–	–	–	–	–	–	N-Phthaloyl-2-oxotryptophan		NI36725
135 81 173 39 18 41 57 28	**350**	100 63 36 31 30 29 27 26	**13.43**	19	26	4	–	1	–	–	–	–	–	–	–	–	Sulphurous acid, 2-[4-tert-butylphenoxy]cyclohexyl 2-propynyl ester	2312-35-8	NI36726
43 109 111 138 248 137 119 186	**350**	100 62 56 33 26 20 20 17	**0.60**	19	26	6	–	–	–	–	–	–	–	–	–	–	Bakkenolide B diacetate		LI07728

MASS TO CHARGE RATIOS	M.W.	INTENSITIES	Parent	C	H	O	N	S	F	Cl	Br	I	Si	P	B	X	COMPOUND NAME	CAS Reg No	No
262 290 96 291 159 125 119 67	350	100 97 84 82 81 81 80 80	3.03	19	26	6	–	–	–	–	–	–	–	–	–	–	Calonectrin		NI36727
43 91 290 262 105 123 122 41	350	100 27 25 23 22 21 20 18	0.00	19	26	6	–	–	–	–	–	–	–	–	–	–	Calonectrin		NI36728
71 201 172 91 173 105 262 307	350	100 41 31 26 23 21 16 11	0.00	19	26	6	–	–	–	–	–	–	–	–	–	–	Marasmene, 3α,15-diacetoxy-		MS08298
43 55 93 41 232 91 105 131	350	100 84 74 70 68 66 65 64	0.00	19	26	6	–	–	–	–	–	–	–	–	–	–	Propanoic acid, 2-methyl-, (decahydro-6a-hydroxy-9a-methyl-3-methylene-2,9-dioxoazuleno[4,5-b]furan-6-yl)methyl ester, [3aS-(3aα,6β,6aα,9aβ,9bα)]-	33649-17-1	NI36729
232 43 91 105 79 55 93 41	350	100 62 51 43 41 41 39 39	0.00	19	26	6	–	–	–	–	–	–	–	–	–	–	Propanoic acid, 2-methyl-, (decahydro-6a-hydroxy-9a-methyl-3-methylene-2,9-dioxoazuleno[4,5-b]furan-6-yl)methyl ester, [3aS-(3aα,6β,6aα,9aβ,9bα)]-	33649-17-1	NI36730
43 82 124 125 167 166 350 83	350	100 83 36 19 18 18 16 13		19	26	6	–	–	–	–	–	–	–	–	–	–	Sambucinol, diacetyl-		NI36731
43 55 109 111 138 41 44 110	350	100 82 63 56 34 34 29 28	0.73	19	26	6	–	–	–	–	–	–	–	–	–	–	Spiro[furan-3(2H),2′-[2H]inden]-2-one, 1′,7′-bis(acetyloxy)decahydro-3′a,4′-dimethyl-4-methylene-, [1′R-(1′α,2′α,3′aα,4′α,7′β,7′aα)]-	19903-94-7	NI36732
99 70 56 58 101 278 279 217	350	100 76 55 50 47 15 12 10	5.00	19	27	–	2	1	–	1	–	–	–	–	–	–	cis-8-Chloro-1,2,3,4,4a,10,11,11a-octahydro-10-(4-methylpiperazin-1-yl)dibenzo[b,f]thiepin		LI07729
99 70 58 100 56 217 278 294	350	100 85 76 65 55 30 18 17	8.00	19	27	–	2	1	–	1	–	–	–	–	–	–	trans-8-Chloro-1,2,3,4,4a,10,11,11a-octahydro-10-(4-methylpiperazin-1-yl)dibenzo[b,f]thiepin		LI07730
195 167 91 166 196 168 350 65	350	100 44 28 20 18 17 16 14		20	18	2	2	1	–	–	–	–	–	–	–	–	Carbazole, 3-(N-methyl-N-tosylamino)-		NI36733
350 307 249 335 292 264 221 351	350	100 48 38 36 36 27 26 23		20	18	4	2	–	–	–	–	–	–	–	–	–	Isoquinoline, 1-(2-cyano-4-methoxyphenyl)-5,6,7-trimethoxy-		DD01389
105 229 145 186 187 77 43 350	350	100 61 46 23 20 19 13 10		20	18	4	2	–	–	–	–	–	–	–	–	–	Oxazole, 2-anisyl-N-benzoyloxy-2-cyano-4,5-dimethyl-2,3-dihydro-		LI07731
52 145 229 173 146 92 53 67	350	100 55 17 11 11 11 11 6	0.00	20	19	–	4	–	–	1	–	–	–	–	–	–	Phenazinium, 3,7-diamino-2,8-dimethyl-5-phenyl-, chloride	477-73-6	NI36734
105 184 350 77 107 227 152 154	350	100 42 35 33 27 24 24 22		20	19	1	–	–	–	–	–	–	–	–	–	1	Arsane, (2-hydroxy-2-phenylethyl)diphenyl-		DD01390
152 246 228 229 227 153 181 154	350	100 97 96 95 95 92 87 87	0.00	20	19	1	–	–	–	–	–	–	–	–	–	1	Arsane oxide, phenethyldiphenyl-		DD01391
161 318 319 162 175 55 43 41	350	100 39 32 28 17 15 15 15	5.74	20	30	5	–	–	–	–	–	–	–	–	–	–	Benzenedecanoic acid, 3-methoxy-2-(methoxycarbonyl)-, methyl ester	56554-82-6	NI36735
293 307 43 41 350 235 203 249	350	100 79 71 57 53 40 25 23		20	30	5	–	–	–	–	–	–	–	–	–	–	Benzo[b]furan-4,6-diol, 2-(1-hydroxyisopropyl)-7-isobutyryl-5-isopentyl-2,3-dihydro-		LI07732
41 249 57 278 282 44 165 179	350	100 84 79 78 69 68 61 58	0.00	20	30	5	–	–	–	–	–	–	–	–	–	–	Tricyclo[3.3.1.1^{3,7}]decane-1,3-dicarboxylic acid, 4-oxo-, dibutyl ester	55724-15-7	NI36736
350 241 257 117 166 94 75 73	350	100 69 54 44 37 31 30 27		20	34	3	–	–	–	–	–	–	1	–	–	–	Undecanoic acid, 11-phenoxy-, trimethylsilyl ester	21273-14-3	NI36737
308 350 43 309 307 121 165 293	350	100 46 35 18 11 7 6 5		21	18	5	–	–	–	–	–	–	–	–	–	–	Benzofuran-3-carboxaldehyde, 2-(4-acetylphenyl)-7-methoxy-5(E)-propenyl-		MS08299
350 295 351 294 296 335 91 251	350	100 73 48 29 22 7 7 6		21	18	5	–	–	–	–	–	–	–	–	–	–	1H,7H-Benzofuro[3,2-c]pyrano[3,2-g][1]benzopyran-7-one, 2,3-dihydro-10-methoxy-3,3-dimethyl-	72060-16-3	NI36738
350 295 351 294 296 335 91 175	350	100 74 48 31 20 7 7 6		21	18	5	–	–	–	–	–	–	–	–	–	–	Isopsoralidine monomethyl ether		LI07733
59 60 45 44 41 107 74 43	350	100 94 93 60 60 58 54 52	0.00	21	19	3	–	–	–	–	–	–	–	1	–	–	1,3,2-Dioxaphosphorinane, 2,5,5-triphenyl-, 2-oxide	34284-46-3	MS08300
350 168 169 182 349 351 319 291	350	100 92 54 36 26 24 24 21		21	22	3	2	–	–	–	–	–	–	–	–	–	Sarpagan-16-carboxylic acid, 17-oxo-, methyl ester, (16R)-	2520-44-7	NI36739
108 107 350 121 109 120 201 200	350	100 46 19 19 11 9 6 6		21	22	3	2	–	–	–	–	–	–	–	–	–	Schizozyginemethine		LI07734
78 77 51 52 50 79 76 74	350	100 15 15 14 11 6 4 4	3.37	21	22	3	2	–	–	–	–	–	–	–	–	–	18,19-Secoyohimban-19-oic acid, 16,17,20,21-tetradehydro-16-formyl-, methyl ester, (15β,16E)-	5523-37-5	NI36740
350 180 305 307 155 279 183 154	350	100 52 43 31 22 19 19 17		21	22	3	2	–	–	–	–	–	–	–	–	–	Talbotine, 4-acetyl-16-de(methoxycarbonyl)-17-deoxy-17-oxo-	30809-26-8	NI36741
350 291 198 184 351 277 169 199	350	100 60 52 27 25 25 19 15		21	22	3	2	–	–	–	–	–	–	–	–	–	Voachalotine, 16-de(hydroxymethyl)-6-oxo-	32303-68-7	NI36742
56 57 55 63 78 83 69 61	350	100 62 42 41 29 24 24 23	13.33	21	22	3	2	–	–	–	–	–	–	–	–	–	Yohimban-16-carboxylic acid, 19,20-didehydro-17-oxo-, methyl ester, (16α)-	54725-63-2	NI36743
125 177 149 69 137 153 57 159	350	100 91 89 58 50 49 48 44	10.53	21	28	2	–	–	2	–	–	–	–	–	–	–	Retinoic acid, 4,4-difluoro-, methyl ester (all-trans)-	90660-19-8	NI36744
231 137 175 215 271 259 124 119	350	100 71 64 49 48 46 46 33	19.00	21	31	2	–	–	–	1	–	–	–	–	–	–	1-Amyl-3-hydroxy-6,6,9-trimethyl-9-chloro-6a,10a-trans-hexahydrodibenzo[b,d]pyran		LI07735
314 231 271 258 315 299 193 259	350	100 93 47 42 26 26 26 25	5.00	21	31	2	–	–	–	1	–	–	–	–	–	–	3-Amyl-1-hydroxy-6,6,9-trimethyl-9-chloro-6a,10a,-trans-hexahydrodibenzo[b,d]pyran		LI07736
43 148 93 350 133 91 79 105	350	100 84 84 78 78 68 60 59		21	31	3	–	–	1	–	–	–	–	–	–	–	Androstan-17-ol, 5,6-epoxy-3-fluoro-, acetate, (3α,5α,6α,17β)-	28344-40-3	NI36745
43 350 93 41 133 91 67 55	350	100 99 45 39 36 32 30 29		21	31	3	–	–	1	–	–	–	–	–	–	–	Androstan-17-ol, 5,6-epoxy-3-fluoro-, acetate, (3α,5β,6β,17β)-	28344-39-0	MS08301
43 148 350 133 93 142 55 81	350	100 99 89 74 73 67 61 59		21	31	3	–	–	1	–	–	–	–	–	–	–	Androstan-17-ol, 5,6-epoxy-3-fluoro-, acetate, (3β,5α,6α,17β)-	28344-60-7	MS08302
43 350 93 133 134 91 41 79	350	100 58 49 39 36 36 34 31		21	31	3	–	–	1	–	–	–	–	–	–	–	Androstan-17-ol, 5,6-epoxy-3-fluoro-, acetate, (3β,5β,6β,17β)-	28526-86-5	MS08303
259 319 330 163 299 239 350 290	350	100 43 40 40 20 6 5 4		21	31	3	–	–	1	–	–	–	–	–	–	–	Androst-5-ene-17,19-diol, 3-fluoro-, 17-acetate, (3β,17β)-	28344-65-2	LI07737
259 319 330 163 299 260 331 320	350	100 43 40 40 20 18 8 8	5.00	21	31	3	–	–	1	–	–	–	–	–	–	–	Androst-5-ene-17,19-diol, 3-fluoro-, 17-acetate, (3β,17β)-	28344-65-2	NI36746
163 259 319 350 330 345 239 290	350	100 80 60 32 20 12 8 5		21	31	3	–	–	1	–	–	–	–	–	–	–	Androst-5-ene-17β,19-diol, 3α-fluoro-, 17-acetate	28344-47-0	LI07738
163 259 319 350 330 260 335 320	350	100 80 60 32 20 15 12 11		21	31	3	–	–	1	–	–	–	–	–	–	–	Androst-5-ene-17β,19-diol, 3α-fluoro-, 17-acetate	28344-47-0	NI36747
163 259 43 319 277 350 91 93	350	100 83 83 58 37 31 29 23		21	31	3	–	–	1	–	–	–	–	–	–	–	Androst-5-ene-17β,19-diol, 3α-fluoro-, 17-acetate	28344-47-0	NI36748
259 163 43 257 319 277 350 290	350	100 81 78 75 47 41 38 34		21	31	3	–	–	1	–	–	–	–	–	–	–	Estr-5(10)-ene-6-methanol, 17-(acetyloxy)-3-fluoro-, (3α,6β,16β)-	28344-54-9	NI36749

MASS TO CHARGE RATIOS								M.W.	INTENSITIES								Parent	C	H	O	N	S	F	Cl	Br	I	Si	P	B	X	COMPOUND NAME	CAS Reg No	No
259	163	350	277	290	131	145	133	**350**	100	85	56	40	35	20	15	12		21	31	3	–	–	1	–	–	–	–	–	–	–	**Estr-5(10)-ene-6-methanol, 17-(acetyloxy)-3-fluoro-, (3α,6β,17β)-**	28344-54-9	**LI07739**
259	163	350	277	290	131	260	145	**350**	100	85	56	40	35	20	18	15		21	31	3	–	–	1	–	–	–	–	–	–	–	**Estr-5(10)-ene-6-methanol, 17-(acetyloxy)-3-fluoro-, (3α,6β,17β)-**	28344-54-9	**NI36750**
259	43	163	330	277	91	260	319	**350**	100	85	78	63	37	30	25	21	**9.90**	21	31	3	–	–	1	–	–	–	–	–	–	–	**Estr-5(10)-ene-6-methanol, 17-(acetyloxy)-3-fluoro-, (3β,6β,17β)-**	28344-74-3	**NI36751**
259	163	330	277	319	145	133	131	**350**	100	90	82	46	27	25	23	23	**14.01**	21	31	3	–	–	1	–	–	–	–	–	–	–	**Estr-5(10)-ene-6-methanol, 17-(acetyloxy)-3-fluoro-, (3β,6β,17β)-**	28344-74-3	**NI36752**
81	41	350	55	95	44	40	67	**350**	100	75	71	64	57	52	46	43		21	34	–	–	2	–	–	–	–	–	–	–	–	**Androstan-3-one, cyclic 1,2-ethanediyl mercaptole, (5α)-**	2791-42-6	**NI36753**
44	41	55	67	43	81	45	131	**350**	100	71	63	54	49	42	42	38	**20.62**	21	34	–	–	2	–	–	–	–	–	–	–	–	**Androstan-12-one, cyclic ethylene mercaptole, (5α)-**	25597-02-8	**NI36754**
55	43	67	45	53	61	81	42	**350**	100	70	63	61	39	38	28	26	**0.57**	21	34	–	–	2	–	–	–	–	–	–	–	–	**Androstan-16-one, cyclic ethylene mercaptole, (5α)-**	2759-86-6	**NI36755**
278	332	55	28	41	93	81	43	**350**	100	99	80	64	63	61	55	53	**2.06**	21	34	4	–	–	–	–	–	–	–	–	–	–	**Androstane-17-carboxylic acid, 3,5-dihydroxy-, methyl ester, (3β,5α,17β)-**	55955-57-2	**NI36756**
225	43	227	116	134	232	133	205	**350**	100	58	41	32	29	24	19	18	**7.00**	21	34	4	–	–	–	–	–	–	–	–	–	–	**1-Naphthalenecarboxylic acid, 3-oxo-3,4,4a,5,6,7,8,8a-octahydro-β-hydroxy-β,2,5,5,8a-pentamethyl-, methyl ester**		**NI36757**
179	177	180	151	133	332	207	178	**350**	100	40	15	9	9	8	7	7	**4.00**	21	34	4	–	–	–	–	–	–	–	–	–	–	**2-Pentyl-3-(3-hydroxy-9-methoxycarbonyl-1-nonenyl)cyclopent-2-enone**		**LI07740**
277	210	121	350	93	43	95	55	**350**	100	60	58	42	42	37	36	36		21	34	4	–	–	–	–	–	–	–	–	–	–	**2-Phenanthreneacetic acid, tetradecahydro-7-hydroxy-1,4b,8,8-tetramethyl-10-oxo-, methyl ester**	57289-54-0	**NI36758**
112	350	251	239	223	121	163	97	**350**	100	67	47	41	39	39	38	29		21	34	4	–	–	–	–	–	–	–	–	–	–	**1-Phenanthrenecarboxylic acid, 1,2,3,4,4a,4b,5,8,8a,9,10,10a-dodecahydro 8,8a-dihydroxy-1,4a-dimethyl-7-isopropyl-, methyl ester, [1R-(1α,4aβ,4bα,8α,8aα,10aα)]-**	32111-52-7	**NI36759**
112	121	239	223	251	163	91	107	**350**	100	93	81	72	63	56	56	49	**11.01**	21	34	4	–	–	–	–	–	–	–	–	–	–	**1-Phenanthrenecarboxylic acid, 1,2,3,4,4a,4b,5,8,8a,9,10,10a-dodecahydro 8,8a-dihydroxy-1,4a-dimethyl-7-isopropyl-, methyl ester, [1R-(1α,4aβ,4bα,8β,8aα,10aα)]-**	34217-14-6	**NI36760**
307	121	239	107	229	251	112	109	**350**	100	74	59	53	41	37	35	32	**5.60**	21	34	4	–	–	–	–	–	–	–	–	–	–	**1-Phenanthrenecarboxylic acid, 1,2,3,4,4a,4b,5,6,7,9,10,10a-dodecahydro-6,7-dihydroxy-1,4a-dimethyl-7-isopropyl-, methyl ester, [1R-(1α,4aβ,4bα,6α,7α,10aα)]-**	34217-11-3	**NI36761**
220	121	161	307	123	221	175	119	**350**	100	88	81	46	36	29	25	24	**3.09**	21	34	4	–	–	–	–	–	–	–	–	–	–	**8α-Podocarpan-15-oic acid, 12α,13α-epoxy-8-hydroxy-13-isopropyl-, methyl ester**	25859-62-5	**NI36762**
239	121	314	289	146	299	181	332	**350**	100	63	49	48	32	23	23	21	**0.00**	21	34	4	–	–	–	–	–	–	–	–	–	–	**8α-Podocarp-13-en-15-oic acid, 8,12α-dihydroxy-13-isopropyl-, methyl ester**	14029-35-7	**NI36763**
43	332	57	55	69	41	71	81	**350**	100	99	93	93	84	82	76	73	**1.80**	21	34	4	–	–	–	–	–	–	–	–	–	–	**Pregnan-20-one, 3,11,12-trihydroxy-, (3β,5α,11α,12β)-**	13384-56-0	**NI36764**
75	73	79	95	67	41	93	55	**350**	100	95	77	53	51	46	39	39	**7.71**	21	38	2	–	–	–	–	–	–	1	–	–	–	**α-Linolenic acid, trimethylsilyl ester**	97844-13-8	**NI36765**
350	107	106	352	315	77	165	239	**350**	100	41	39	35	31	11	6	5		22	19	2	–	–	–	1	–	–	–	–	–	–	**cis-1,2-Dianisyl-2-phenylvinyl chloride**		**LI07741**
350	107	106	352	315	77	165	239	**350**	100	41	39	35	30	14	9	5		22	19	2	–	–	–	1	–	–	–	–	–	–	**trans-1,2-Dianisyl-2-phenylvinyl chloride**		**LI07742**
350	197	351	218	184	198	185	165	**350**	100	38	23	23	15	13	8	7		22	22	–	–	2	–	–	–	–	–	–	–	–	**1-[(2-Butylphenyl)thio]-2-(phenylthio)benzene**		**DD01392**
350	349	351	184	169	170	156	307	**350**	100	97	22	21	21	18	15	9		22	26	2	2	–	–	–	–	–	–	–	–	–	**3β,20α,19β-Acetonylyohimb-17-one**		**LI07743**
43	321	78	350	322	41	42	51	**350**	100	72	72	70	52	46	33	31		22	26	2	2	–	–	–	–	–	–	–	–	–	**Aspidofractinine-1-carboxaldehyde, 17-methoxy-3-methylene-, (2α,5α)-**	55724-68-0	**NI36766**
170	291	350	184	43	135	28	168	**350**	100	76	62	56	54	48	44	33		22	26	2	2	–	–	–	–	–	–	–	–	–	**2,20-Cycloaspidospermidine-3-methanol, 6,7-didehydro-, acetate (ester), (2α,3β,5α,12β,19α,20R)-**	56053-25-9	**NI36767**
136	78	350	137	77	58	55	52	**350**	100	19	12	9	5	5	4	4		22	26	2	2	–	–	–	–	–	–	–	–	–	**1H-Indolizino[8,1-cd]carbazole, 3a-(2-propenyl)-2,3,3a,4,6,11,12,13a-octahydro-**		**MS08304**
350	351	349	144	291	183	182	157	**350**	100	22	20	11	9	9	6	6		22	26	2	2	–	–	–	–	–	–	–	–	–	**Tetraphyllicine, acetate (ester)**	25926-60-7	**NI36768**
247	45	57	89	206	350	191	41	**350**	100	97	96	34	32	31	20	20		22	38	3	–	–	–	–	–	–	–	–	–	–	**2-[2-(2,4,6-Tri-tert-butylphenoxy)ethoxy]ethanol**		**IC04079**
273	350	274	78	170	165	351	257	**350**	100	36	20	19	12	12	9	8		23	18	–	4	–	–	–	–	–	–	–	–	–	**1,2,4-Triazolo[4,3-a]pyridine, 2,3-dihydro-3,3-diphenyl-2-(2-pyridinyl)-**	16672-25-6	**NI36769**
193	115	350	91	178	293	165	167	**350**	100	34	19	19	12	11	9	6		23	26	1	–	1	–	–	–	–	–	–	–	–	**Cyclohexanone, 6-[(butylthio)methylene]-2,2-diphenyl-**	50592-51-3	**NI36770**
119	78	43	59	239	253	350	149	**350**	100	74	61	36	34	26	21	18		23	30	1	2	–	–	–	–	–	–	–	–	–	**Phenanthrene, 1,2,3,4,4a,9,10,10a-octahydro-1-(1H-imidazol-2-ylcarbonyl)-1,4a-dimethyl-7-isopropyl-, [1R-(1α,4aβ,10aα)]-**	69634-26-0	**NI36771**
350	335	95	197	55	43	69	173	**350**	100	75	67	25	19	19	18	16		23	30	1	2	–	–	–	–	–	–	–	–	–	**Phenanthrene, 4b,5,6,7,8,8a,9,10-octahydro-9-(1H-imidazol-2-ylcarbonyl)-4b,8-dimethyl-2-isopropyl-, [4bS-(4bα,8α,8β)]-**	75601-48-8	**NI36772**
68	96	322	78	43	350	50	109	**350**	100	72	61	59	24	23	22	20		23	30	1	2	–	–	–	–	–	–	–	–	–	**Phenanthrene, 4b,5,6,7,8,8a,9,10-octahydro-9-(1H-imidazol-2-ylcarbonyl)-4b,8-dimethyl-2-isopropyl-, [4bS-(4bα,8α,8β)]-**	75601-27-3	**NI36773**
273	350	135	54	274	197	351	152	**350**	100	46	25	21	17	17	16	15		24	18	1	–	–	–	–	–	–	1	–	–	–	**10H-Phenoxasilin, 10,10-diphenyl-**	18733-65-8	**NI36774**
273	350	135	55	274	197	351	152	**350**	100	91	49	41	34	34	32	31		24	18	1	–	–	–	–	–	–	1	–	–	–	**10H-Phenoxasilin, 10,10-diphenyl-**	18733-65-8	**LI07744**
105	284	273	178	77	66	179	350	**350**	100	51	46	46	41	34	24	22		24	18	1	2	–	–	–	–	–	–	–	–	–	**4H-Pyran-3-carbonitrile, 2-amino-4,5,6-triphenyl-**	86606-31-7	**NI36775**
335	132	43	133	91	336	350	176	**350**	100	41	38	30	30	26	25	22		24	34	–	2	–	–	–	–	–	–	–	–	–	**Diazene, bis[2,6-diisopropylphenyl]-**	74685-83-9	**NI36776**
350	273	77	105	152	351	51	141	**350**	100	79	75	45	36	29	25	15		25	18	2	–	–	–	–	–	–	–	–	–	–	**4-(4-Phenylphenoxy)benzophenone**		**IC04080**
182	77	183	104	169	51	168	167	**350**	100	41	15	15	12	9	5	4	**3.42**	25	22	–	2	–	–	–	–	–	–	–	–	–	**Methanediamine, N,N,N′,N′-tetraphenyl-**	21905-92-0	**NI36777**
256	255	165	257	241	154	179	178	**350**	100	35	25	23	21	20	16	16	**0.30**	25	22	–	2	–	–	–	–	–	–	–	–	–	**3,8-Azo-4,7-methanocyclobuta[b]naphthalene, 2a,3,3a,4,7,7a,8,8a-octahydro-3,8-diphenyl-, (2aα,3β,3aα,4β,7β,7aα,8β,8aα)-**	40229-71-8	**NI36778**

MASS TO CHARGE RATIOS								M.W.	INTENSITIES								Parent	C	H	O	N	S	F	Cl	Br	I	Si	P	B	X	COMPOUND NAME	CAS Reg No	No
43	57	41	55	83	69	97	71	**350**	100	84	84	72	64	63	50	48	**0.42**	25	50	–	–	–	–	–	–	–	–	–	–	–	**Dodecane, 1-cyclopentyl-4-octyl-**		**AI02142**
43	57	41	55	83	69	71	97	**350**	100	99	81	74	66	61	57	54	**0.05**	25	50	–	–	–	–	–	–	–	–	–	–	–	**Dodecane, 1-cyclopentyl-4-octyl-**		**AI02143**
55	43	57	41	83	97	71	69	**350**	100	96	82	79	74	49	45	41	**0.07**	25	50	–	–	–	–	–	–	–	–	–	–	–	**Heptadecane, 9-(2-cyclohexylethyl)-**		**AI02144**
43	55	57	69	41	83	56	97	**350**	100	93	84	79	73	70	58	51	**9.62**	25	50	–	–	–	–	–	–	–	–	–	–	–	**Heptadecane, 9-octyl-**		**AI02145**
83	97	71	69	85	111	237	96	**350**	100	72	64	53	44	37	35	28	**0.10**	25	50	–	–	–	–	–	–	–	–	–	–	–	**Undecane, 1-cyclohexyl-3-octyl-**		**AI02146**
91	200	244	243	105	165	77	245	**350**	100	99	82	76	49	38	19	16	**0.00**	26	22	1	–	–	–	–	–	–	–	–	–	–	**Benzene, 1,1′,1″-[benzyloxymethylidyne]tris-**	5333-62-0	**NI36779**
91	244	243	105	165	77	245	167	**350**	100	82	76	48	38	20	16	12	**0.00**	26	22	1	–	–	–	–	–	–	–	–	–	–	**Benzene, 1,1′,1″-[benzyloxymethylidyne]tris-**	5333-62-0	**NI36780**
91	244	243	105	165	228	77	245	**350**	100	82	75	48	39	20	20	16	**0.00**	26	22	1	–	–	–	–	–	–	–	–	–	–	**Benzene, 1,1′,1″-[benzyloxymethylidyne]tris-**	5333-62-0	**LI07745**
273	350	274	165	351	195	228	229	**350**	100	44	21	19	11	8	7	6		26	22	1	–	–	–	–	–	–	–	–	–	–	**Benzene, 1-methoxy-4-(triphenylmethyl)-**	7402-89-3	**NI36781**
167	350	351	168	152	165			**350**	100	48	15	12	9	8				26	38	–	–	–	–	–	–	–	–	–	–	–	**Acenaphthene, 5-(4-pentylnonyl)-**		**LI07746**
167	350	351	168	152	165			**350**	100	49	15	12	9	6				26	38	–	–	–	–	–	–	–	–	–	–	–	**Acenaphthene, 5-tetradecyl-**		**LI07747**
350	225	351	279	155	226	141	167	**350**	100	42	24	20	18	16	15	11		26	38	–	–	–	–	–	–	–	–	–	–	–	**2-(4-Cyclohexylbutyl)-6-hexylnaphthalene**		**LI07748**
195	196	165	105	43	180	41	29	**350**	100	17	6	6	6	5	5	5	**3.61**	26	38	–	–	–	–	–	–	–	–	–	–	–	**Dodecane, 1,1-bis(4-methylphenyl)-**		**AI02147**
195	196	43	105	41	180	165	179	**350**	100	17	9	8	8	7	7	3	**1.28**	26	38	–	–	–	–	–	–	–	–	–	–	–	**Dodecane, 1,1-bis(4-methylphenyl)-**		**AI02148**
181	350	351	182	165	153	179		**350**	100	62	20	20	13	5	3			26	38	–	–	–	–	–	–	–	–	–	–	–	**Phenalene, 6-tridecyl-2,3-dihydro-**		**LI07749**
167	168	43	40	91	165	57	55	**350**	100	15	13	10	8	6	4	4	**1.66**	26	38	–	–	–	–	–	–	–	–	–	–	–	**Tetradecane, 1,1-diphenyl-**		**AI02149**
167	168	165	43	91	41	29	152	**350**	100	15	8	8	7	7	7	4	**2.99**	26	38	–	–	–	–	–	–	–	–	–	–	–	**Tetradecane, 1,1-diphenyl-**		**AI02150**
43	309	291	62	351	292	205	63	**351**	100	98	54	14	13	10	10	8		9	6	6	1	–	–	–	–	1	–	–	–	–	**Salicylic acid, 5-iodo-3-nitro-, acetate (ester)**	22884-73-7	**NI36782**
309	43	291	62	351	205	292	310	**351**	100	92	56	17	16	12	10	9		9	6	6	1	–	–	–	–	1	–	–	–	–	**Salicylic acid, 5-iodo-3-nitro-, acetate (ester)**	22884-73-7	**MS08305**
309	291	43	351	62	205	292	310	**351**	100	55	50	15	14	11	10	9		9	6	6	1	–	–	–	–	1	–	–	–	–	**Salicylic acid, 5-iodo-3-nitro-, acetate (ester)**	22884-73-7	**LI07750**
135	103	200	177	29	137	46	136	**351**	100	70	67	65	64	58	55	46	**0.00**	11	20	4	1	2	3	–	–	–	–	–	–	–	**2-Trifluoromethylacetamido-1,1-bis(ethylthio)fructose**		**NI36783**
121	137	77	104	105	103	51	76	**351**	100	22	22	20	17	15	12	11	**10.00**	13	10	2	3	1	–	–	1	–	–	–	–	–	**2-(α-Mercaptobenzylidenehydrazino)-5-bromo-1-nitrobenzene**		**IC04081**
61	162	277	164	279	116	75	56	**351**	100	99	68	66	46	46	46	37	**5.00**	13	15	4	1	1	–	2	–	–	–	–	–	–	**L-Methionine, N-[(2,4-dichlorophenoxy)acetyl]-**	50834-39-4	**NI36784**
179	105	180	195	177	178	127	196	**351**	100	79	45	44	44	36	36	18	**0.00**	14	10	2	1	–	–	–	–	1	–	–	–	–	**Ethylene, 1-iodo-2-nitro-1,2-diphenyl-, (E)-**	55902-54-0	**DD01393**
105	177	153	152	179	195	180	178	**351**	100	99	98	98	97	96	96	84	**0.00**	14	10	2	1	–	–	–	–	1	–	–	–	–	**Ethylene, 1-iodo-2-nitro-1,2-diphenyl-, (Z)-**	73013-87-3	**DD01394**
351	178	352	176	177	305	47	165	**351**	100	37	16	15	10	9	9	7		14	10	2	1	–	–	–	–	1	–	–	–	–	**Phenanthrene, 2-iodo-3-nitro-9,10-dihydrophenanthrene**		**MS08306**
198	67	129	127	153	351	222	55	**351**	100	70	65	54	47	46	46	37		14	13	2	1	1	3	1	–	–	–	–	–	–	**Cyclohexanecarboxamide, N-(4-chlorophenyl)-2-oxo-3-[(trifluoromethyl)thio]-**		**DD01395**
292	308	136	148	266	149	178	218	**351**	100	97	83	75	72	59	48	45	**28.00**	14	17	6	5	–	–	–	–	–	–	–	–	–	**Lentysine diacetate methyl ester**		**MS08307**
73	103	88	147	102	89	249	131	**351**	100	49	32	30	21	21	19	19	**0.00**	14	33	5	1	–	–	–	–	–	2	–	–	–	**D-Ribose, 2,3-di-O-methyl-4,5-bis-O-(trimethylsilyl)-, O-methyloxime**	56196-37-3	**NI36785**
126	226	268	58	73	41	127	69	**351**	100	24	23	17	15	14	9	9	**0.00**	14	34	3	1	–	–	–	–	–	2	1	–	–	**Phosphonic acid, [1-[(2,2-dimethylpropylidene)amino]-isopropyl]-, bis(trimethylsilyl) ester**	56196-70-4	**NI36786**
126	226	268	253	73	41	58	69	**351**	100	49	25	23	22	21	19	15	**0.00**	14	34	3	1	–	–	–	–	–	2	1	–	–	**Phosphonic acid, [1-[(2,2-dimethylpropylidene)amino]propyl]-, bis(trimethylsilyl) ester**	55108-74-2	**NI36787**
73	336	240	226	225	41	268	267	**351**	100	87	83	80	72	52	40	35	**6.52**	14	34	3	1	–	–	–	–	–	2	1	–	–	**Phosphonic acid, [3-[(2,2-dimethylpropylidene)amino]propyl]-, bis(trimethylsilyl) ester**	55108-79-7	**NI36788**
126	127	73	41	70	69	58	42	**351**	100	10	7	7	6	5	5	4	**0.00**	14	34	3	1	–	–	–	–	–	2	1	–	–	**Phosphonic acid, [1-[isopropylideneamino]pentyl]-, bis(trimethylsilyl) ester**	55108-67-3	**NI36789**
202	188	57	136	28	43	135	160	**351**	100	93	87	76	70	64	60	55	**5.93**	15	21	5	5	–	–	–	–	–	–	–	–	–	**Adenosine, N-(4-hydroxy-3-methyl-2-butenyl)-, (E)-**	6025-53-2	**NI36790**
136	202	201	188	43	135	29	28	**351**	100	88	75	74	66	55	53	51	**16.00**	15	21	5	5	–	–	–	–	–	–	–	–	–	**Purine, 6-(cis-4-hydroxy-3-methylbut-2-enylamino)-9-β-D-ribofuranosyl-**		**LI07751**
173	175	145	147	109	177	351	353	**351**	100	62	20	13	12	10	6	4		16	11	4	1	–	–	2	–	–	–	–	–	–	**2-Propenoic acid, 3-[1-(2,6-dichlorobenzoyl)-5-formylpyrrol-2-yl]-, methyl ester**	95883-21-9	**NI36791**
139	92	91	65	40	39	156	196	**351**	100	83	46	24	14	14	11	9	**1.00**	16	17	4	1	2	–	–	–	–	–	–	–	–	**Aziridine, 2,3-bis(4-toluenesulphonyl)-**		**LI07752**
43	111	213	153	81	44	109	39	**351**	100	90	53	47	29	20	17	16	**0.00**	16	17	8	1	–	–	–	–	–	–	–	–	–	**2H-Pyran-2-methanol, 3-(acetyloxy)-3,6-dihydro-6-(4-nitrophenoxy)-, acetate (ester)**	56196-13-5	**NI36792**
351	308	287	234	233	352	218	206	**351**	100	29	23	23	22	21	18	18		17	12	4	1	–	3	–	–	–	–	–	–	–	**Dimethyl 4,5,6-trifluoro-1-methyl-1H-benzo[de]quinoline-2,3-dicarboxylate**		**NI36793**
351	121	56	211	239	210	319	133	**351**	100	58	37	26	23	18	14	14		17	13	4	1	–	–	–	–	–	–	–	–	1	**1-[β-(5-Nitrofur-2-oyl)vinyl]ferrocene**		**LI07753**
173	175	351	145	353	177	174	147	**351**	100	65	17	12	11	11	8	8		17	15	3	1	–	–	2	–	–	–	–	–	–	**2-Propenoic acid, 3-[1-(2,6-dichlorobenzoyl)-2-methylpyrrol-3-yl]-, methyl ester**	95883-23-1	**NI36794**
91	68	65	155	130	196	222	67	**351**	100	63	33	30	26	23	19	18	**2.90**	17	21	5	1	1	–	–	–	–	–	–	–	–	**7-Azabicyclo[2.2.1]heptane-2-carboxylic acid, 3-acetyl-7-(4-tolylsulphonyl)-, methyl ester, 2-endo,3-endo-**	17037-47-7	**NI36795**
91	68	155	130	222	223	196	308	**351**	100	59	34	28	26	24	23	22	**10.80**	17	21	5	1	1	–	–	–	–	–	–	–	–	**7-Azabicyclo[2.2.1]heptane-2-carboxylic acid, 3-acetyl-7-(4-tolylsulphonyl)-, methyl ester, 2-endo,3-exo-**	17037-48-8	**NI36796**

MASS TO CHARGE RATIOS	M.W.	INTENSITIES	Parent	C	H	O	N	S	F	Cl	Br	I	Si	P	B	X	COMPOUND NAME	CAS Reg No	No
43 107 165 15 55 207 110 77	**351**	100 44 31 30 28 23 22 21	**0.00**	17	21	7	1	–	–	–	–	–	–	–	–	–	**Acetamide, N-[2-(acetyloxy)-2-[3,4-bis(acetyloxy)phenyl]ethyl]-N-methyl-**	55268-54-7	**NI36797**
165 249 207 291 56 250 164 292	**351**	100 99 88 45 37 21 20 19	**0.00**	17	21	7	1	–	–	–	–	–	–	–	–	–	**Acetamide, N-[2-(acetyloxy)-2-[3,4-bis(acetyloxy)phenyl]ethyl]-N-methyl-**	55268-54-7	**MS08308**
43 44 86 207 165 45 87 181	**351**	100 77 70 62 57 54 52 47	**31.59**	17	21	7	1	–	–	–	–	–	–	–	–	–	**Acetamide, N-[2-(acetyloxy)-2-[3,4-bis(acetyloxy)phenyl]ethyl]-N-methyl-, (R)-**	28371-31-5	**NI36798**
91 92 43 351 41 55 105 97	**351**	100 66 26 19 19 17 14 12		17	22	1	1	–	5	–	–	–	–	–	–	–	**Benzeneoctanamine, N-(pentafluoropropionyl)-**		**NI36799**
73 260 234 218 91 75 147 43	**351**	100 63 59 51 51 43 36 26	**4.40**	17	29	3	1	–	–	–	–	–	2	–	–	–	**Phenylalanine, N-acetyl-N-(trimethylsilyl)-, trimethylsilyl ester**		**MS08309**
145 207 117 351 63 352 209 89	**351**	100 71 61 42 39 37 32 30		18	12	2	2	–	–	–	–	–	–	–	–	1	**Copper, bis(8-quinolinolato-N[1],O[8])-**	10380-28-6	**NI36800**
145 146 173 185 144 170 230 78	**351**	100 10 4 3 3 2 1 1	**0.00**	18	12	2	2	–	–	–	–	–	–	–	–	1	**Copper, bis(8-quinolinolato-N[1],O[8])-**	10380-28-6	**NI36801**
105 77 351 45 106 46 57 43	**351**	100 27 21 15 8 7 6 6		18	13	5	3	–	–	–	–	–	–	–	–	–	**6-(2,6-Dihydroxy-3-pyridylazo)resorcinol 3-benzoate**		**MS08310**
216 77 105 119 100 144 121 51	**351**	100 33 32 29 29 26 21 21	**0.00**	18	16	1	1	1	3	–	–	–	–	–	–	–	**1,3-Oxazole, 2,5-dihydro-2,2-dimethyl-5-phenyl-4-(phenylthio)-5-(trifluoromethyl)-**		**DD01396**
351 282 145 146 352 173 90 136	**351**	100 32 28 25 23 18 12 7		18	17	3	5	–	–	–	–	–	–	–	–	–	**3H-Pyrazol-3-one, 4-[[4-(dimethylamino)phenyl]imino]-2,4-dihydro-5-methyl-2-(4-nitrophenyl)-**	27808-02-2	**NI36802**
43 105 101 31 59 60 107 45	**351**	100 79 56 46 45 33 32 31	**0.00**	18	25	6	1	–	–	–	–	–	–	–	–	–	**α-D-Altroyranoside, methyl 3-acetamido-3-deoxy-4,6-(phenylisopropylidene)-**		**NI36803**
351 140 168 126 84 43 252 194	**351**	100 99 91 76 65 57 43 30		18	25	6	1	–	–	–	–	–	–	–	–	–	**2-Butenoic acid, 2-hydroxy-4-oxo-4-(4-hexylamino-6-methyl-2-oxo-2H-pyran-3-yl)-, ethyl ester**		**NI36804**
93 136 94 120 253 137 138 80	**351**	100 77 49 37 36 33 27 25	**8.70**	18	25	6	1	–	–	–	–	–	–	–	–	–	**Doriasenine**		**NI36805**
120 95 93 119 94 121 43 122	**351**	100 97 74 73 54 47 33 26	**1.60**	18	25	6	1	–	–	–	–	–	–	–	–	–	**Senecionan-11,16-dione, 15,20-15,20-epoxy-15,20-dihydro-12-hydroxy-, (15α,20S)-**	6870-67-3	**NI36806**
105 351 305 69 77 181 169 174	**351**	100 60 40 30 25 20 18 16		19	13	4	1	1	–	–	–	–	–	–	–	–	**Pyrano[2,3-e]benzoxazol-8(H)-one, 2-[(benzoylmethyl)thio]-6-methyl-**		**NI36807**
305 41 29 40 84 351 68 30	**351**	100 94 75 60 58 51 31 30		19	13	6	1	–	–	–	–	–	–	–	–	–	**Anthra[1,2-b]furan-6,11-dione, 2-amino-3-ethoxycarbonyl-5-hydroxy-**		**IC04082**
308 311 351 269 54 309 255 158	**351**	100 88 72 32 28 24 24 23		19	21	2	5	–	–	–	–	–	–	–	–	–	**N-Butyl-N-(cyanoethyl)-4-(4-nitrophenylazo)aniline**		**IC04083**
120 180 118 131 90 347 77 91	**351**	100 76 41 22 21 18 18 14	**0.00**	19	21	2	5	–	–	–	–	–	–	–	–	–	**N[5]-Methyl-6,7-diphenyl-10-hydroxy-5,6,7,8,9,10-hexahydropterin**		**MS08311**
105 77 51 122 50 39 78 38	**351**	100 88 63 60 42 16 14 13	**0.74**	19	26	3	1	–	–	1	–	–	–	–	–	–	**Ethyl 3-[1-benzoyl-3-(2-chloroethyl)piperid-4-yl]propionate**		**NI36808**
292 43 213 70 80 50 351	**351**	100 45 30 26 21 18 6		19	33	3	3	–	–	–	–	–	–	–	–	–	**N[6]-Acetylpalustrine**		**LI07754**
105 77 229 231 194 51 166 106	**351**	100 58 52 18 12 12 10 10	**3.00**	20	14	3	1	–	–	1	–	–	–	–	–	–	**Benzoic acid, 2-(benzoylphenylamino)-**		**NI36809**
351 287 204 288 352 232 320 304	**351**	100 96 35 28 25 25 22 22		20	17	5	1	–	–	–	–	–	–	–	–	–	**3-Benzyl-4-oxo-4H-quinolizine-1,2-dicarboxylic acid, dimethyl ester**		**NI36810**
322 351 323 135 336 264 175 176	**351**	100 38 27 18 14 14 13 11		20	17	5	1	–	–	–	–	–	–	–	–	–	**Spiro[1,3-dioxolo[4,5-g]isoquinoline-5(6H),7′-[7H]indeno[4,5-d][1,3]dioxol]-8′(6′H)-one, 7,8-dihydro-6-methyl-, (S)-**	20411-03-4	**NI36811**
322 323 351 336 264 135 175 190	**351**	100 78 71 61 45 37 31 12		20	17	5	1	–	–	–	–	–	–	–	–	–	**Spiro[1,3-dioxolo[4,5-g]isoquinoline-5(6H),7′-[7H]indeno[4,5-d][1,3]dioxol]-8′(6′H)-one, 7,8-dihydro-6-methyl-, (S)-**	20411-03-4	**NI36812**
351 352 336 308 350 320 337 306	**351**	100 25 21 17 14 14 11 8		20	17	5	1	–	–	–	–	–	–	–	–	–	**1,2,9,10-Tetramethoxy-7-oxodibenzo[de,g]quinoline**		**MS08312**
161 191 318 351 303 132 77 160	**351**	100 61 39 37 37 30 24 21		20	21	3	3	–	–	–	–	–	–	–	–	–	**Pyrrolo[2,1-b]quinazolin-9(1H)-one, 3-[2-(dimethylamino)phenyl]-2,3-dihydro-3-hydroxy-5-methoxy-**	29048-81-5	**NI36813**
336 351 337 322 352 334 306 292	**351**	100 65 25 21 16 12 12 12		21	21	4	1	–	–	–	–	–	–	–	–	–	**9-Methoxyacronine**		**LI07755**
351 350 349 148 320 352 336 322	**351**	100 50 48 34 26 24 24 24		21	21	4	1	–	–	–	–	–	–	–	–	–	**Ochotensine**		**LI07756**
198 51 50 52 183 182 199 70	**351**	100 38 36 31 17 17 16 15	**8.93**	21	25	2	3	–	–	–	–	–	–	–	–	–	**Deoxybrevianamide E**		**NI36814**
103 351 233 116 175 118 77 91	**351**	100 87 82 15 12 12 10 5		22	17	–	5	–	–	–	–	–	–	–	–	–	**3H-Imidazo[1,2-b][1,2,4]triazole, 3,5-diphenyl-2-(phenylamino)-**		**MS08313**
112 94 81 105 42 91 77 82	**351**	100 75 43 40 36 35 34 27	**12.77**	22	25	3	1	–	–	–	–	–	–	–	–	–	**8-Azabicyclo[3.2.1]octane-3,6-diol, 8-methyl-2-benzyl-, 6-benzoate**	55925-26-3	**NI36815**
105 351 184 204 144 120 129 104	**351**	100 47 41 36 35 35 26 25		22	25	3	1	–	–	–	–	–	–	–	–	–	**Cyclobutanecarboxylic acid, 3-phenyl-1-[(1-phenylethyl)carbamoyl]-, ethyl ester**	85318-03-2	**NI36816**
240 282 41 252 69 351 228 283	**351**	100 64 32 28 22 20 20 16		22	25	3	1	–	–	–	–	–	–	–	–	–	**Furo[2,3-b]quinoline, 7-(3-methyl-2-butenoxy)-8-(3-methyl-2-butenyl)-4-methoxy-**	85802-30-8	**NI36817**
180 250 43 181 179 71 152	**351**	100 22 19 14 8 7 5	**0.15**	22	25	3	1	–	–	–	–	–	–	–	–	–	**6-(1-Methoxycarbonylisopropyl)-5,6-dihydro-5-isobutylphenanthridine**		**LI07757**
351 336 59 323 55 81 162 124	**351**	100 98 82 62 53 44 42 38		22	25	3	1	–	–	–	–	–	–	–	–	–	**6′H-Morphin-7-eno[3,2-b]pyran-6-ol, 4,5-epoxy-6′,6′,17-trimethyl-, (5α,6β)-**		**NI36818**
131 163 136 69 162 160 103 161	**351**	100 74 53 52 51 51 50 49	**3.00**	22	25	3	1	–	–	–	–	–	–	–	–	–	**2-Propenamide, 3-phenyl-N-[2-hydroxy-2-[4-(3,3-dimethylallyloxy)phenyl]ethyl]-**		**NI36819**
175 93 41 57 71 91 55 69	**351**	100 46 21 20 18 17 17 16	**2.00**	22	25	3	1	–	–	–	–	–	–	–	–	–	**α-(3,3,7-Trimethyl-2-oxoindol-1-yl)benzyl isobutyrate**		**LI07758**
200 57 186 242 309 281 336 257	**351**	100 65 62 59 55 41 34 31	**0.00**	22	29	1	3	–	–	–	–	–	–	–	–	–	**3a,2-(Epoxyethano)-2H-indole-9,10-dicarbonitrile, 5,7-di-tert-butyl-3,3a-dihydro-3,3-dimethyl-, rel-(2R,3aS,9R,10S)-**		**DD01397**
351 200 57 281 258 218 336 84	**351**	100 87 87 78 78 78 70 70		22	29	1	3	–	–	–	–	–	–	–	–	–	**3a,2-(Epoxyethano)-2H-indole-9,10-dicarbonitrile, 5,7-di-tert-butyl-3,3a-dihydro-3,3-dimethyl-, rel-(2R,3aS,9S,10S)-**		**DD01398**
98 286 99 70 315 287 300 72	**351**	100 42 10 10 9 9 3 3	**0.00**	22	38	–	1	–	–	1	–	–	–	–	–	–	**Androst-2-en-17-amine, N-propyl-, hydrochloride, (5α)-**	53286-41-2	**NI36820**
43 57 41 55 83 71 113 84	**351**	100 74 69 49 41 40 33 33	**0.84**	22	41	2	1	–	–	–	–	–	–	–	–	–	**Pyrrolidine, 1-(1,6-dioxooctadecyl)-**	56630-89-8	**NI36821**

MASS TO CHARGE RATIOS								M.W.	INTENSITIES								Parent	C	H	O	N	S	F	Cl	Br	I	Si	P	B	X	COMPOUND NAME	CAS Reg No	No
83	85	47	113	48	43	55	49	**351**	100	63	41	29	22	16	15	14	**1.04**	22	41	2	1	–	–	–	–	–	–	–	–	–	Pyrrolidine, 1-[8-(3-octyloxiranyl)-1-oxooctyl]-	56630-37-6	NI36822
113	70	126	43	55	49	84	72	**351**	100	30	29	27	26	23	21	16	**3.84**	22	41	2	1	–	–	–	–	–	–	–	–	–	Pyrrolidine, 1-[8-(3-octyloxiranyl)-1-oxooctyl]-	56630-37-6	NI36823
83	85	47	113	43	70	35	55	**351**	100	67	46	43	41	34	31	27	**1.44**	22	41	2	1	–	–	–	–	–	–	–	–	–	Pyrrolidine, 1-[8-(3-octyloxiranyl)-1-oxooctyl]-	56630-37-6	NI36824
113	55	43	70	41	126	72	98	**351**	100	58	55	53	49	36	22	19	**1.84**	22	41	2	1	–	–	–	–	–	–	–	–	–	Pyrrolidine, 1-[1-oxo-5-(3-undecyloxiranyl)pentyl]-	56666-43-4	NI36825
196	168	167	197	155	351	129	51	**351**	100	41	22	16	15	14	11	11		23	17	1	3	–	–	–	–	–	–	–	–	–	2(1H)-Isoquinolinecarboxamide, 1-cyano-N,N-diphenyl-	4053-46-7	NI36826
255	351	256	295	131	309	308	159	**351**	100	88	32	17	14	12	12	12		23	29	2	1	–	–	–	–	–	–	–	–	–	3,6-Dispirocyclohexyl-1,2,3,4,5,6,7,8-octahydro-1,8-acridinedione		NI36827
113	43	55	41	140	70	98	71	**351**	100	36	23	21	15	14	13	13	**0.00**	23	45	1	1	–	–	–	–	–	–	–	–	–	Pyrrolidine, 1-(2-methyl-1-oxooctadecyl)-	56630-58-1	NI36828
113	43	70	126	55	71	114	41	**351**	100	21	18	15	14	9	8	8	**2.84**	23	45	1	1	–	–	–	–	–	–	–	–	–	Pyrrolidine, 1-(6-methyl-1-oxooctadecyl)-	56600-07-8	NI36829
351	229	303	128	304	228	302	352	**351**	100	81	43	43	38	35	32	26		24	17	2	1	–	–	–	–	–	–	–	–	–	1,4-Methanonaphthalene, 1,4-dihydro-9-[α-(4-nitrophenyl)benzylidene]-	116972-02-2	DD01399
191	90	189	165	351	64	178	116	**351**	100	32	14	13	10	10	8	7		24	17	2	1	–	–	–	–	–	–	–	–	–	9-Phenanthrenemethyl indole-2-carboxylate	92174-52-2	NI36830
308	309	350	351	294	295	77	51	**351**	100	26	20	19	16	12	12	6		24	21	–	3	–	–	–	–	–	–	–	–	–	3,5-Pyridinedicarbonitrile, 4,4-pentamethylene-1,4-dihydro-2,6-diphenyl-	52831-29-5	NI36831
322	323	351	350	284	77	91	186	**351**	100	27	23	23	23	17	16	12		24	21	–	3	–	–	–	–	–	–	–	–	–	3,5-Pyridinedicarbonitrile, 4,4-tetramethylene-1,4-dihydro-1-methyl-2,6-diphenyl-	83078-30-2	NI36832
119	351	266	91	280	118	120	107	**351**	100	56	52	46	29	29	28	24		24	33	1	1	–	–	–	–	–	–	–	–	–	2H-1,3-Benzoxazine, 3,4-dihydro-6-nonyl-3-(4-tolyl)-		IC04084
333	276	295	314	257	238	157	138	**352**	100	31	27	26	16	8	3	3	**0.38**	–	–	–	–	–	6	–	–	–	–	–	–	1	Uranium hexafluoride	7783-81-5	NI36833
333	276	295	314	257	238	128.5	138	**352**	100	31	27	26	16	8	4	3	**0.38**	–	–	–	–	–	6	–	–	–	–	–	–	1	Uranium hexafluoride	7783-81-5	AM00147
127	89	333	70	352	103	46	51	**352**	100	28	9	9	6	6	6	2		–	–	–	2	3	12	–	–	–	–	–	–	–	Bis(pentafluorosulphanylimino)sulphur difluoride		LI07759
198	232	168	292	124	138	322	154	**352**	100	72	67	57	57	46	43	43	**14.00**	–	–	4	4	–	–	1	1	–	–	–	–	2	ξ-Bromo-ξ-chlorotetranitrosodicobalt		MS08314
59	147	87	69	175	28	88	176	**352**	100	63	48	38	34	25	18	17	**0.00**	1	1	1	–	–	9	–	–	–	–	3	–	1	Tris(trifluorophosphine)carbonylcobalt hydride		MS08315
268	212	184	240	296	352	196	224	**352**	100	65	64	57	52	47	14	12		6	–	6	–	–	–	–	–	–	–	–	–	1	Tungsten hexacarbonyl	14040-11-0	MS08316
28	268	270	266	212	214	210	184	**352**	100	19	17	16	10	9	9	9	**7.46**	6	–	6	–	–	–	–	–	–	–	–	–	1	Tungsten hexacarbonyl	14040-11-0	NI36834
107	109	73	231	27	261	233	229	**352**	100	98	65	39	36	28	20	20	**0.50**	6	11	2	–	–	–	–	3	–	–	–	–	–	1,1-Bis(2-bromoethoxy)-2-bromoethane		IC04085
59	83	69	72	55	309	73	67	**352**	100	73	64	35	34	8	7	7	**0.71**	7	15	1	–	–	–	1	–	–	–	–	–	1	(2-Methoxy-4-methylpentane)-3-mercuric chloride		LI07760
77	279	277	276	278	78	275	74	**352**	100	26	21	15	13	11	9	5	**0.00**	9	10	2	–	–	–	–	–	–	–	–	–	1	Mercury, phenyl(propanoato-O)-	103-27-5	NI36835
317	43	243	319	253	225	245	197	**352**	100	97	80	33	33	33	27	26	**9.00**	9	17	4	–	–	–	2	–	–	–	–	–	1	Niobium, dichlorodiethoxy(2,4-pentanedionato-O,O')-	17035-43-7	NI36836
317	43	243	319	253	225	245	255	**352**	100	97	80	33	33	33	27	21	**9.00**	9	17	4	–	–	–	2	–	–	–	–	–	1	Niobium, dichlorodiethoxy(2,4-pentanedionato-O,O')-	17035-43-7	LI07761
317	319	245	243	43	257	255	253	**352**	100	99	80	80	73	44	44	44	**11.90**	9	17	4	–	–	–	2	–	–	–	–	–	1	Niobium, dichlorodiethoxy(2,4-pentanedionato-O,O')-	17035-43-7	NI36837
183	181	73	75	109	36	39	185	**352**	100	79	59	57	45	45	42	41	**1.69**	10	6	1	–	–	–	6	–	–	–	–	–	–	1,4-Ethanopentalen-8-one, 1,2,3,5,7,7-hexachloro-1,3a,4,5,6,6a-hexahydro-	69653-75-4	NI36838
36	217	183	38	63	39	219	73	**352**	100	37	37	37	34	34	33	32	**1.62**	10	6	1	–	–	–	6	–	–	–	–	–	–	1,2,3,5,7,9-Hexachloro-1,3a,4,5,6,6a-hexahydro-1,4-endo-oxiranopentalene		NI36839
337	339	335	283	341	281	285	249	**352**	100	83	53	43	35	33	24	18	**0.00**	10	6	1	–	–	–	6	–	–	–	–	–	–	4,7-Methano-1H-inden-1-ol, 4,5,6,7,8,8-hexachloro-3a,4,7,7a-tetrahydro-	2597-11-7	NI36840
81	80	217	219	319	37	27	47	**352**	100	92	27	25	19	17	16	15	**0.00**	10	6	1	–	–	–	6	–	–	–	–	–	–	4,7-Methano-1H-inden-1-ol, 4,5,6,7,8,8-hexachloro-3a,4,7,7a-tetrahydro-	2597-11-7	NI36841
181	172	215	193	281	253	352	289	**352**	100	42	33	16	6	4	3	2		10	6	1	–	–	–	6	–	–	–	–	–	–	4,6-Methano-4H-indeno[1,2-b]oxirene, 1a,1b,2,3,6a,7-hexachloro-1a,1b,5,5a,6,6a-hexahydro-	57160-08-4	NI36842
308	310	306	354	309	352	307	356	**352**	100	50	50	20	18	10	10	9		10	10	4	–	–	–	–	2	–	–	–	–	–	Benzoic acid, 2,4-dihydroxy-3,5-dibromo-6-methyl-, ethyl ester	21855-46-9	NI36843
43	164	162	41	39	163	123	161	**352**	100	17	9	9	8	7	7	6	**0.00**	12	16	2	–	–	–	–	–	–	–	–	–	2	2,4-Diselenatricyclo[3.3.1.1^{3,7}]decane-6,8-dione, 1,3,5,7-tetramethyl-	57289-42-6	NI36844
41	67	55	82	81	268	352	109	**352**	100	94	94	90	89	80	78	72		12	16	2	–	–	–	–	–	–	–	–	–	2	3(2H)-Selenophenone, 2-(dihydro-4,4-dimethyl-3-oxoselenophene-2(3H)-ylidene)dihydro-4,4-dimethyl-	56009-21-3	NI36845
245	243	36	247	244	246	28	38	**352**	100	79	75	49	37	32	30	22	**0.00**	12	24	–	4	–	–	2	–	–	–	–	–	1	Nickel, [6,14-dichloro-1,5,9,13-tetraazacyclohexadecanato(2-)-N1,N5,N9,N15]-	69815-43-6	NI36846
57	41	293	179	177	291	175	295	**352**	100	98	95	85	73	72	49	44	**0.00**	12	24	4	–	–	–	–	–	–	–	–	–	1	Dibutyl tin diacetate		IC04086
56	83	42	217	28	123	64	57	**352**	100	26	22	19	17	15	12	11	**0.00**	13	19	5	3	1	–	–	–	–	–	–	–	1	Methanesulphonic acid, [(2,3-dihydro-1,5-dimethyl-3-oxo-2-phenyl-1H-pyrazol-4-yl)methylamino]-, sodium salt, monohydrate	5907-38-0	NI36847
235	237	165	236	212	239	199	238	**352**	100	68	38	16	13	12	11	10	**1.20**	14	9	–	–	–	–	5	–	–	–	–	–	–	Benzene, 1,1'-(2,2,2-trichloroethylidene)bis[4-chloro-	50-29-3	NI36848
235	237	36	165	236	38	239	82	**352**	100	66	51	40	18	17	13	12	**5.56**	14	9	–	–	–	–	5	–	–	–	–	–	–	Benzene, 1,1'-(2,2,2-trichloroethylidene)bis[4-chloro-	50-29-3	NI36849
212	235	237	282	284	165	176	88	**352**	100	86	59	47	45	40	38	35	**0.00**	14	9	–	–	–	–	5	–	–	–	–	–	–	Benzene, 1,1'-(2,2,2-trichloroethylidene)bis[4-chloro-	50-29-3	AI02151
235	237	165	75	199	176	236	50	**352**	100	66	59	23	22	18	15	12	**2.60**	14	9	–	–	–	–	5	–	–	–	–	–	–	Ethane, 1,1,1-trichloro-2-(2-chlorophenyl)-2-(4-chlorophenyl)-	789-02-6	MS08317
235	237	165	236	176	212	199	238	**352**	100	62	37	16	16	15	14	11	**1.83**	14	9	–	–	–	–	5	–	–	–	–	–	–	Ethane, 1,1,1-trichloro-2-(2-chlorophenyl)-2-(4-chlorophenyl)-	789-02-6	NI36850
235	237	219	165	199	236	176	239	**352**	100	72	61	46	18	16	16	12	**3.23**	14	9	–	–	–	–	5	–	–	–	–	–	–	Ethane, 1,1,1-trichloro-2-(2-chlorophenyl)-2-(4-chlorophenyl)-	789-02-6	NI36851
294	104	77	105	78	69	283	51	**352**	100	65	65	60	60	55	50	45	**3.00**	14	15	1	2	–	6	–	–	–	–	–	1	–	2,2,2-Trifluoro-N-(2,2,2-trifluoro-1-iminoethyl)acetic acid imide, butylphenylboryl ester		NI36852
178	120	122	119	105	92	352	176	**352**	100	76	46	43	25	21	10	7		14	16	5	4	1	–	–	–	–	–	–	–	–	Bicarbamic acid, [(3,4-dihydro-4-oxo-2-quinazolinyl)thio]-, diethyl ester	81559-85-5	NI36853
252	352	194	169	223	168	195	91	**352**	100	78	61	41	35	35	33	26		14	17	4	–	–	–	–	–	–	–	–	–	1	Rhodium, [η^4-2-(methoxycarbonyl)norbornadiene]acetylacetonato-	85150-73-8	NI36854
43	41	102	176	58	144	60	92	**352**	100	59	50	49	34	32	30	29	**2.20**	14	28	–	2	4	–	–	–	–	–	–	–	–	Thioperoxydicarbonic diamide, tetraisopropyl-	4136-91-8	NI36855

MASS TO CHARGE RATIOS								M.W.	INTENSITIES								Parent	C	H	O	N	S	F	Cl	Br	I	Si	P	B	X	COMPOUND NAME	CAS Reg No	No
43	144	176	41	102	58	60	177	**352**	100	86	69	59	54	34	30	27	**18.92**	14	28	–	2	4	–	–	–	–	–	–	–	–	Thioperoxydicarbonic diamide, tetrapropyl-	2556-42-5	NI36856
207	57	41	209	337	193	225	208	**352**	100	61	44	42	31	31	27	24	**0.00**	14	36	4	–	–	–	–	–	–	3	–	–	–	Trisiloxane, 3,3-dibutoxy-1,1,1,5,5,5-hexamethyl-		MS08318
57	211	27	43	41	55	29	56	**352**	100	96	73	56	47	45	42	35	**0.00**	14	36	4	–	–	–	–	–	–	3	–	–	–	Trisiloxane, 1,3,5-triethyl-1,5-dibutoxy-		NI36857
43	309	155	51	77	307	41	153	**352**	100	77	73	73	69	64	59	55	**1.00**	15	17	–	–	–	–	1	–	–	–	–	–	1	Tin, chlorodiphenyl(1-methylethyl)-	70335-31-8	NI36858
149	84	95	86	82	107	108	121	**352**	100	74	44	42	24	22	21	19	**0.00**	15	19	3	–	–	–	3	–	–	–	–	–	–	Acetic acid, trichloro-, 3-(cis-3a,4,5,6,7,7a-hexahydro-5-methyl-1-oxo-1H inden-5-yl)propyl ester		DD01400
253	154	238	352	211	139	136		**352**	100	70	32	16	6	3	3			15	21	6	–	–	–	–	–	–	–	–	–	1	Manganese, tris(acetylacetonato)-		LI07762
166	167	58	41	251	29	139	194	**352**	100	67	31	25	24	21	16	14	**3.50**	15	23	4	2	–	3	–	–	–	–	–	–	–	Butanoic acid, 2-amino-N-(trifluoroacetyl)-L-prolyl-, butyl ester		MS08319
58	167	166	41	29	42	251	27	**352**	100	89	73	60	43	40	33	26	**0.00**	15	23	4	2	–	3	–	–	–	–	–	–	–	Butanoic acid, 2-amino-N-(trifluoroacetyl)-L-prolyl-, butyl ester		MS08320
167	166	41	139	69	29	42	70	**352**	100	56	26	20	18	15	12	11	**1.80**	15	23	4	2	–	3	–	–	–	–	–	–	–	Butanoic acid, 3-amino-N-(trifluoroacetyl)-L-prolyl-, butyl ester		MS08321
167	166	41	69	139	29	28	42	**352**	100	84	34	25	21	19	18	12	**4.30**	15	23	4	2	–	3	–	–	–	–	–	–	–	Butanoic acid, 3-amino-N-(trifluoroacetyl)-L-prolyl-, butyl ester		MS08322
167	166	41	139	86	29	28	69	**352**	100	92	37	23	17	17	17	16	**3.30**	15	23	4	2	–	3	–	–	–	–	–	–	–	Butanoic acid, 4-amino-N-(trifluoroacetyl)-L-prolyl-, butyl ester		MS08323
88	177	101	75	45	117	71	89	**352**	100	79	49	37	25	19	16	13	**0.00**	15	28	9	–	–	–	–	–	–	–	–	–	–	Methyl 3-O-methyl-2-O-(2′,3′,4′,6′-tetra-O-methyl-α-D-mannopyranosyl)-D-glycerate		NI36859
190	100	117	164	175	126	162	61	**352**	100	69	57	45	31	29	25	24	**0.33**	15	32	3	2	2	–	–	–	–	–	–	–	–	Urea, N,N′-bis[6-(methylsulphinyl)hexyl]-, (±)-		NI36860
42	55	41	143	43	70	69	44	**352**	100	95	95	56	52	47	24	23	**0.00**	15	33	4	–	–	–	–	–	–	–	–	–	1	Arsenic acid, tripentyl ester	15063-75-9	NI36861
352	121	193	353	56	39	122	173	**352**	100	43	27	18	16	8	4	3		16	9	–	–	–	5	–	–	–	–	–	–	1	Ferrocene, (pentafluorophenyl)-	55282-20-7	NI36862
173	175	317	145	319	147	177	109	**352**	100	65	49	24	17	14	11	11	**10.00**	16	14	3	2	–	–	2	–	–	–	–	–	–	2-Propenoic acid, 3-[1-(2,6-dichlorobenzoyl)-3-methylpyrazol-4-yl]-, methyl ester	95883-07-1	NI36863
166	268	226	292	250	352	208	43	**352**	100	71	55	21	15	11	11	11		16	16	9	–	–	–	–	–	–	–	–	–	–	Benzaldehyde, 2,3,4-triacetoxy-6-(acetoxymethyl)-	84018-93-9	NI36864
135	204	149	148	136	352	108	190	**352**	100	83	77	61	47	46	41	40		16	20	–	10	–	–	–	–	–	–	–	–	–	Adenine, 9,9′-dimethylenebis-		LI07763
61	304	352	147	117	91	305	174	**352**	100	93	63	53	41	28	26	22		16	20	3	2	2	–	–	–	–	–	–	–	–	5-Ethyl-1,3-bis(methylthiomethyl)-5-phenyl-2,4,6-pyrimidinetrione		MS08324
57	193	137	267	179	293	337	69	**352**	100	58	50	48	44	42	33	33	**32.00**	16	20	7	2	–	–	–	–	–	–	–	–	–	Propanoic acid, 2,2-dimethyl-, [3-(acetyloxy)-2,3,3a,9a-tetrahydro-6-oxo-6H-furo[2′,3′:4,5]oxazolo[3,2-a]pyrimidin-2-yl]methyl ester, [2R-(2α,3β,3aβ,9aβ)]-	40732-64-7	NI36865
57	177	352	85	137	113	96	97	**352**	100	67	56	40	39	25	25	21		16	20	7	2	–	–	–	–	–	–	–	–	–	Propanoic acid, 2,2-dimethyl-, 2-[(acetyloxy)methyl]-2,3,3a,9a-tetrahydro 6-oxo-6H-furo[2′,3′:4,5]oxazolo[3,2-a]pyrimidin-3-yl ester, [2R-(2α,3β,3aβ,9aβ)]-	40773-09-9	NI36866
147	73	75	337	281	45	148	44	**352**	100	78	44	34	21	19	18	18	**10.40**	16	24	5	–	–	–	–	–	–	2	–	–	–	2-Propenoic acid, 3-(1,3-benzodioxol-5-yl)-2-[(trimethylsilyl)oxy]-, trimethylsilyl ester		MS08325
353	281	209	337					**352**	100	90	85	25					**0.00**	16	28	3	2	–	–	–	–	–	2	–	–	–	2,4,6(1H,3H,5H)-Pyrimidinetrione, 5,5-di-2-propenyl-1,3-bis(trimethylsilyl)-	74367-13-8	NI36867
69	177	55	111	41	43	57	113	**352**	100	70	62	53	51	40	30	24	**0.39**	16	32	4	–	2	–	–	–	–	–	–	–	–	1,4-Dithiane, 2,5-dihexyl-, 1,1,4,4-tetraoxide		IC04087
89	133	83	73	59	58	57	55	**352**	100	42	39	38	37	37	37	37	**0.00**	16	32	8	–	–	–	–	–	–	–	–	–	–	3-Bis(hydroxymethyl)-1,5,8,11,14,17-hexaoxacyclononadecane		MS08326
57	71	43	85	28	55	41	99	**352**	100	76	61	55	42	23	23	15	**0.00**	16	33	–	–	–	–	–	–	1	–	–	–	–	Hexadecane, 1-iodo-		DO01110
57	71	43	85	55	41	69	29	**352**	100	68	66	45	31	30	21	12	**0.00**	16	33	–	–	–	–	–	–	1	–	–	–	–	Hexadecane, 1-iodo-		LI07764
93	118	240	260	77	352	131	130	**352**	100	91	82	74	62	54	51	44		17	15	1	2	1	3	–	–	–	–	–	–	–	2-Butenamide, N-phenyl-2-[(trifluoromethyl)thio]-3-(phenylamino)-, (E)-		DD01401
254	98	352	68	256	215	165	225	**352**	100	68	48	43	34	28	28	26		17	17	6	–	–	–	1	–	–	–	–	–	–	Spiro[benzofuran-2(3H),1′-[3]cyclohexene]-2′,3-dione, 7-chloro-4,4′,6-trimethoxy-6′-methyl-	17818-12-1	NI36868
255	97	352	68	257	214	165	225	**352**	100	68	49	44	36	28	28	26		17	17	6	–	–	–	1	–	–	–	–	–	–	Spiro[benzofuran-2(3H),1′-[3]cyclohexene]-2′,3-dione, 7-chloro-4,4′,6-trimethoxy-6′-methyl-	17818-12-1	NI36869
138	352	310	215	214	321	354	284	**352**	100	93	59	53	50	47	32	23		17	17	6	–	–	–	1	–	–	–	–	–	–	Spiro[benzofuran-2(3H),1′-[2]cyclohexene]-3,4′-dione, 7-chloro-2′,4,6-trimethoxy-6′-methyl-, (1′S-trans)-	126-07-8	NI36870
138	352	310	214	354	215	69	284	**352**	100	93	57	39	35	34	30	25		17	17	6	–	–	–	1	–	–	–	–	–	–	Spiro[benzofuran-2(3H),1′-[2]cyclohexene]-3,4′-dione, 7-chloro-2′,4,6-trimethoxy-6′-methyl-, (1′S-trans)-	126-07-8	NI36871
138	352	310	214	215	69	321	354	**352**	100	66	50	40	37	33	25	23		17	17	6	–	–	–	1	–	–	–	–	–	–	Spiro[benzofuran-2(3H),1′-[2]cyclohexene]-3,4′-dione, 7-chloro-2′,4,6-trimethoxy-6′-methyl-, (1′S-trans)-	126-07-8	NI36872
182	334	184	336	125	126	167	166	**352**	100	44	32	30	20	16	12	12	**0.10**	17	18	2	2	–	–	2	–	–	–	–	–	–	Butanoic acid, 2-(4-chlorophenyl)-3-methyl-3-(4-chlorophenylhydrazo)-		NI36873
131	161	91	148	101	175	160	147	**352**	100	96	63	54	48	39	38	37	**5.75**	17	20	8	–	–	–	–	–	–	–	–	–	–	1,4,7,10,13-Pentaoxacyclopentadecane-2,5,9-trione, 3-benzyl-	79687-34-6	NI36874
91	138	244	96	137	101	179	60	**352**	100	42	29	23	20	20	17	16	**0.00**	17	24	6	2	–	–	–	–	–	–	–	–	–	α-D-Glucopyranoside, benzyl 2,3-diacetamido-2,3-dideoxy-	27539-64-6	NI36875
73	147	103	45	308	293	117	146	**352**	100	34	33	16	15	15	15	14	**0.00**	17	28	4	–	–	–	–	–	–	2	–	–	–	Propanedioic acid, (2-phenylethyl)-, bis(trimethylsilyl) ester	55887-54-2	NI36876
73	264	352	45	265	155	97	74	**352**	100	84	32	29	23	15	15	15		17	36	–	–	–	–	–	–	–	4	–	–	–	Silane, 1,2-pentadien-4-yne-1,1,3,5-tetrayltetrakis[trimethyl-	30314-58-0	NI36877
352	32	154	183	353	108	51	127	**352**	100	69	36	29	21	20	18	16		18	10	–	–	–	5	–	–	–	–	1	–	–	Phosphine, diphenylpentafluoro-		IC04088
302	154	183	353	108	301	127	255	**352**	100	39	32	21	21	16	16	14	**0.00**	18	10	–	–	–	5	–	–	–	–	1	–	–	Phosphine, diphenylpentafluoro-		MS08327
198	200	154	77	153	352	275	152	**352**	100	80	68	33	15	14	13	13		18	15	–	–	–	–	–	–	–	–	–	–	1	Stibine, triphenyl-	603-36-1	MS08328
198	200	154	77	153	352	275	152	**352**	100	73	67	26	18	13	13	12		18	15	–	–	–	–	–	–	–	–	–	–	1	Stibine, triphenyl-	603-36-1	NI36878

MASS TO CHARGE RATIOS								M.W.	INTENSITIES								Parent	C	H	O	N	S	F	Cl	Br	I	Si	P	B	X	COMPOUND NAME	CAS Reg No	No
198	200	153	154	152	199	275	352	**352**	100	75	21	17	16	13	12	11		18	15	–	–	–	–	–	–	–	–	–	–	1	Stibine, triphenyl-	603-36-1	MS08329
236	235	234	237	202	221	117	118	**352**	100	62	30	20	13	12	10	9	**0.00**	18	17	–	–	1	4	–	–	–	–	–	1	–	1-Ethyl-trans-2-styrylbenzo[b]thiophenium tetrafluoroborate		LI07765
352	354	179	219	337	353	181	58	**352**	100	99	38	30	27	22	18	12		18	18	2	2	–	–	–	–	–	–	–	–	1	Nickel, [[2,2'-[(1,1-dimethyl-1,2-ethanediyl)bis(nitrilomethylidyne)]bis[phenolato]](2-)-N,N',O,O']-	32593-86-5	NI36879
352	179	219	337	58	322	177	176	**352**	100	38	29	28	12	9	9	9		18	18	2	2	–	–	–	–	–	–	–	–	1	Nickel, isobutylenediamino-N,N'-bis(salicylidene)-		LI07766
105	321	159	173	320	146	104	39	**352**	100	91	61	52	45	43	43	41	**0.00**	18	18	6	–	–	–	–	–	–	–	–	2	–	β-D-Fructopyranose 2,3:4,5-bis(benzeneboronate)	53829-58-6	NI36880
159	147	146	175	352	158	104	91	**352**	100	85	64	30	27	26	26	22		18	18	6	–	–	–	–	–	–	–	–	2	–	α-D-Glucofuranose 1,2:3,5-bis(benzeneboronate)	20229-53-2	NI36881
147	105	91	104	159	146	160	77	**352**	100	79	76	69	62	48	41	29	**2.20**	18	18	6	–	–	–	–	–	–	–	–	2	–	DL-Glyceraldehyde benzeneboronate		NI36882
352	353	280	282					**352**	100	72	5	4						18	19	–	2	1	3	–	–	–	–	–	–	–	10H-Phenothiazine-10-propylamine, N,N-dimethyl-2-(trifluoromethyl)-	146-54-3	NI36883
58	44	57	41	43	55	72	85	**352**	100	32	29	24	23	20	16	15	**8.01**	18	19	–	2	1	3	–	–	–	–	–	–	–	10H-Phenothiazine-10-propylamine, N,N-dimethyl-2-(trifluoromethyl)-	146-54-3	NI36884
58	86	352	85	306	266	59	248	**352**	100	32	25	19	10	6	6	5		18	19	–	2	1	3	–	–	–	–	–	–	–	10H-Phenothiazine-10-propylamine, N,N-dimethyl-2-(trifluoromethyl)-	146-54-3	NI36885
52	202	352	135	130	280	208	176	**352**	100	73	46	19	2	1	1	1		18	28	–	–	–	–	–	–	–	2	–	–	1	Chromium, bis(trimethylsilylphenyl)-		LI07767
239	295	240	352	182	41	181	57	**352**	100	49	36	35	32	32	29	22		18	28	5	2	–	–	–	–	–	–	–	–	–	Morpholine, 4-(2,5-dibutoxy-4-nitrophenyl)-	86-15-7	NI36886
179	73	352	337	75	147	247	207	**352**	100	84	58	46	33	22	19	18		18	32	3	–	–	–	–	–	–	2	–	–	–	Hexanoic acid, 6-[2-(trimethylsiloxy)phenyl]-, trimethylsilyl ester		NI36887
179	352	73	337	180	353	75	338	**352**	100	40	23	20	15	11	9	6		18	32	3	–	–	–	–	–	–	2	–	–	–	Hexanoic acid, 6-[3-(trimethylsiloxy)phenyl]-, trimethylsilyl ester		NI36888
352	179	308	267	159	151	324	337	**352**	100	52	30	30	30	22	12	8		19	12	7	–	–	–	–	–	–	–	–	–	–	2H-1-Benzopyran-2-one, 7-hydroxy-6-methoxy-3-[(2-oxo-2H-1-benzopyran-7-yl)oxy]-	2034-69-7	NI36889
160	148	188	205	91	161	147	77	**352**	100	50	23	19	15	12	8	7	**2.30**	19	16	5	2	–	–	–	–	–	–	–	–	–	L-Alanine, 3-phenyl-N-(phthalimidoacetyl)-	1170-07-6	NI36890
352	324	353	321	264	292	163	146	**352**	100	40	21	18	16	11	11	9		19	16	5	2	–	–	–	–	–	–	–	–	–	7-Phthalazinecarboxylic acid, 3,4-dihydro-1-[4-(methoxycarbonyl)phenyl] 3-methyl-4-oxo-, methyl ester		NI36891
94	352	353	120	67	239	78	264	**352**	100	88	29	27	24	22	21	20		19	20	3	4	–	–	–	–	–	–	–	–	–	Dianhydroheptulose, phenylosazone		NI36892
43	83	141	152	239	179	352	181	**352**	100	41	22	20	14	12	10	9		19	28	6	–	–	–	–	–	–	–	–	–	–	Acetylportentol		LI07768
273	274	163	55	255	177	109	41	**352**	100	24	20	17	15	15	14	12	**4.00**	19	29	1	–	–	–	–	1	–	–	–	–	–	Androstan-15-one, 14-bromo-, (5α,14β)-	4004-59-5	NI36893
151	352	152	75	73	337	353	322	**352**	100	45	17	16	14	13	11	7		19	32	4	–	–	–	–	–	–	1	–	–	–	Octanoic acid, 8-(3,4-dimethoxyphenyl)-, trimethylsilyl ester		NI36894
320	352	208	89	147	264	145	119	**352**	100	84	57	57	52	42	41	39		20	16	6	–	–	–	–	–	–	–	–	–	–	Benzeneacetic acid, α-[3-hydroxy-4-(4-methoxyphenyl)-5-oxo-2(5H)-furanylidene]-, methyl ester, (E)-	481-64-1	NI36895
145	320	89	352	290	208	264	147	**352**	100	88	73	67	67	59	38	38		20	16	6	–	–	–	–	–	–	–	–	–	–	Benzeneacetic acid, α-[3-hydroxy-4-(4-methoxyphenyl)-5-oxo-2(5H)-furanylidene]-, methyl ester, (E)-	481-64-1	NI36896
320	175	352	145	91	89	119	321	**352**	100	78	60	54	51	42	22	21		20	16	6	–	–	–	–	–	–	–	–	–	–	Benzeneacetic acid, α-(3-hydroxy-5-oxo-4-phenyl-2(5H)-furanylidene)-2-methoxy-, methyl ester, (E)-	481-59-4	NI36897
145	320	89	290	352	208	264	147	**352**	100	90	72	70	67	58	38	38		20	16	6	–	–	–	–	–	–	–	–	–	–	Benzeneacetic acid, α-(3-hydroxy-5-oxo-4-phenyl-2(5H)-furanylidene)-2-methoxy-, methyl ester, (E)-	481-59-4	LI07769
320	175	352	145	91	89	119	117	**352**	100	79	62	57	54	46	25	20		20	16	6	–	–	–	–	–	–	–	–	–	–	Benzeneacetic acid, α-(3-hydroxy-5-oxo-4-phenyl-2(5H)-furanylidene)-2-methoxy-, methyl ester, (E)-	481-59-4	LI07770
105	352	44	79	51	253	46	354	**352**	100	89	84	63	60	57	57	56		20	16	6	–	–	–	–	–	–	–	–	–	–	2,5-Cyclohexadiene-1,4-dione, 2,5-bis(4-hydroxyphenyl)-3,6-dimethoxy-	51860-93-6	NI36898
352	148	121	324	119	353	159	133	**352**	100	33	31	30	30	24	22	22		20	16	6	–	–	–	–	–	–	–	–	–	–	2,5-Cyclohexadiene-1,4-dione, 2,5-dihydroxy-3,6-bis(4-methoxyphenyl)-	38558-72-4	NI36899
60	206	205	177	174	250	196	207	**352**	100	19	11	9	9	8	8	7	**0.00**	20	16	6	–	–	–	–	–	–	–	–	–	–	Diacetyldiphenylmethylenemalonate		LI07771
352	337	105	77	351	177	321	353	**352**	100	94	53	41	32	30	25	24		20	16	6	–	–	–	–	–	–	–	–	–	–	4H-Furo[2,3-h]-1-benzopyran-4-one, 3,5,6-trimethoxy-2-phenyl-	77970-07-1	NI36900
278	250	279	352	222	165	251	193	**352**	100	52	26	19	17	15	12	7		20	16	6	–	–	–	–	–	–	–	–	–	–	Viridin		MS08330
265	323	352	307	324	266	247	206	**352**	100	70	24	24	21	19	19	19		20	20	4	2	–	–	–	–	–	–	–	–	–	6,21-Cyclo-4,5-secoakuammilan-17-oic acid, 4,5-epoxy-5-hydroxy-, methyl ester, (5S,6α)-	63950-46-9	NI36901
133	307	352	279	308	148	105	134	**352**	100	80	39	29	19	16	13	11		20	20	4	2	–	–	–	–	–	–	–	–	–	7,14-Diethoxydiftalone		MS08331
105	77	294	205	206	189	178	190	**352**	100	85	36	16	13	12	10	8	**2.00**	20	20	4	2	–	–	–	–	–	–	–	–	–	8,9-Dimethoxy-11-benzoyl-1,2,5,6-tetrahydro-1,5-imino-3-benzazocin-4(3H)-one		LI07772
105	77	294	106	189	146	78	76	**352**	100	48	18	7	5	5	4	4	**1.00**	20	20	4	2	–	–	–	–	–	–	–	–	–	8,9-Dimethoxy-11-benzoyl-1,2,5,6-tetrahydro-2,6-imino-3-benzazocin-4(3H)-one		LI07773
132	144	207	146	145	220	102	117	**352**	100	53	44	29	22	16	12	11	**0.00**	20	20	4	2	–	–	–	–	–	–	–	–	–	Phthalimide, 4,5-dimethoxy-N-[(1,2,3,4-tetrahydroisoquinolin-2-yl)methyl]-		NI36902
158	159	352	144	174	337	353	143	**352**	100	93	63	54	49	27	22	22		20	20	4	2	–	–	–	–	–	–	–	–	–	Spiro[2H-1-benzopyran-2,2'-[2H]indole], 1',3'-dihydro-8-methoxy-1',3',3'-trimethyl-6-nitro-	1498-89-1	NI36903
265	352	266	252	309	353	277	251	**352**	100	98	49	39	37	26	23	23		20	24	2	4	–	–	–	–	–	–	–	–	–	Anthraquinone, 1,4-bis[2-(methylamino)ethylamino]-		IC04089
161	77	160	118	203	145	104	91	**352**	100	78	33	18	16	16	7	6	**0.00**	20	24	2	4	–	–	–	–	–	–	–	–	–	Pyrazolidine, 1-acetyl-2-phenyl-3-methyl-5-(1-phenyl-2-acetylhydrazino)-		NI36904
316	298	334	105	148	135	107	109	**352**	100	88	57	53	48	48	45	42	**15.00**	20	32	5	–	–	–	–	–	–	–	–	–	–	$\Delta^{10}(^{18})$-Andromedenol		LI07774
85	210	84	109	211	131	174	86	**352**	100	47	10	9	8	8	7	7	**0.00**	20	32	5	–	–	–	–	–	–	–	–	–	–	1H-Indene-5-propanoic acid, 2,3,3a,6,7,7a-hexahydro-3a-hydroxy-β,7a-dimethyl-1-[(tetrahydro-2H-pyran-2-yl)oxy]-, methyl ester	67884-42-8	NI36905
224	206	135	173	107	119	187	217	**352**	100	80	80	69	53	38	31	25	**2.00**	20	32	5	–	–	–	–	–	–	–	–	–	–	5,10-Seco-ent-kaur-1(10)-en-5-one, 3,6,14,16-tetrahydroxy-, (3S,6R,14R)-		NI36906

MASS TO CHARGE RATIOS								M.W.	INTENSITIES								Parent	C	H	O	N	S	F	Cl	Br	I	Si	P	B	X	COMPOUND NAME	CAS Reg No	No
190	217	316	245	191	199	207	229	**352**	100	42	41	31	22	18	17	15	**0.00**	20	32	5	–	–	–	–	–	–	–	–	–	–	Prosta-5,13-dien-1-oic acid, 11,15-dihydroxy-9-oxo-, (5Z,11α,13E,15S)-	363-24-6	NI36907
73	89	299	42	103	82	75	56	**352**	100	37	35	31	21	20	19	19	**0.00**	20	42	–	3	–	–	–	–	–	1	–	–	–	Hexadecanoic acid, 2-[(trimethylsilyl)oxy]-, methyl ester		LI07775
309	161	115	297	310	352	267	337	**352**	100	37	33	30	22	13	11	9		21	20	5	–	–	–	–	–	–	–	–	–	–	9,10-Anthracenedione, 1,8-dihydroxy-3-methoxy-6-methyl-2-(3-methyl-1-butenyl)-, (E)-	73340-48-4	NI36908
309	352	297	337	294	267	323	295	**352**	100	33	28	13	7	7	6	6		21	20	5	–	–	–	–	–	–	–	–	–	–	9,10-Anthracenedione, 1,8-dihydroxy-3-methoxy-6-methyl-2-(3-methyl-1-butenyl)-, (E)-	73340-48-4	NI36909
141	195	352	115	139	140	40	53	**352**	100	65	18	18	9	8	7	6		21	20	5	–	–	–	–	–	–	–	–	–	–	Benzoic acid, 3,4,5-trimethoxy-, 1-naphthalenemethyl ester	86338-69-4	NI36910
352	267	351	279	217	295	353	268	**352**	100	42	38	29	29	27	25	22		21	20	5	–	–	–	–	–	–	–	–	–	–	4H-1-Benzopyran-4-one, 5-(acetyloxy)-7-hydroxy-3-(4-hydroxyphenyl)-8-(3-methyl-2-butenyl)-		NI36911
337	173	352	187	164	163	151	188	**352**	100	61	60	43	43	38	37	32		21	20	5	–	–	–	–	–	–	–	–	–	–	Leiocin		NI36912
292	277	249	293	352	278	178	275	**352**	100	89	58	25	22	22	13	7		21	20	5	–	–	–	–	–	–	–	–	–	–	Phenanthro[1,2-b]furan-10,11-dione, 1-(acetyloxymethyl)-6,7,8,9-tetrahydro-6,6-dimethyl-		NI36913
279	292	261	352	251	178	165	152	**352**	100	90	42	32	32	30	28	18		21	20	5	–	–	–	–	–	–	–	–	–	–	Phenanthro[1,2-b]furan-10,11-dione, 6-[(acetyloxy)methyl]-6,7,8,9-tetrahydro-1,6-dimethyl-	30860-36-7	LI07776
279	292	276	60	261	352	251	178	**352**	100	90	50	45	42	32	32	30		21	20	5	–	–	–	–	–	–	–	–	–	–	Phenanthro[1,2-b]furan-10,11-dione, 6-[(acetyloxy)methyl]-6,7,8,9-tetrahydro-1,6-dimethyl-	30860-36-7	NI36914
279	292	261	352	251	178	165	280	**352**	100	90	42	32	32	30	28	20		21	20	5	–	–	–	–	–	–	–	–	–	–	Phenanthro[1,2-b]furan-10,11-dione, 6-[(acetyloxy)methyl]-6,7,8,9-tetrahydro-1,6-dimethyl-	30860-36-7	NI36915
337	139	57	352	140	322	323	55	**352**	100	99	85	84	70	66	64	64		21	21	1	2	–	–	1	–	–	–	–	–	–	3-Pyrrolin-2-one, 4-ethyl-3-methyl-5-[[5-(4-chlorobenzylidene)-3,4-dimethyl-5H-pyrrol-2-yl]-2-methylene]-, (Z,Z)-		MS08332
138	352	353	245	139	199	214	137	**352**	100	99	23	19	12	12	12	11		21	21	3	–	–	–	–	–	–	–	1	–	–	Phosphine, tris(4-methoxyphenyl)-	855-38-9	NI36916
190	352	222	143	130	131	162	78	**352**	100	96	80	70	63	36	33	31		21	24	3	2	–	–	–	–	–	–	–	–	–	Ajmalan-16-carboxylic acid, 19,20-didehydro-1-demethyl-17-hydroxy-, methyl ester, (2α,17S,19E)-	4835-69-2	MS08333
352	130	44	190	143	222	131	41	**352**	100	83	80	73	71	56	35	32		21	24	3	2	–	–	–	–	–	–	–	–	–	Ajmalan-16-carboxylic acid, 19,20-didehydro-1-demethyl-17-hydroxy-, methyl ester, (2α,17S,19E)-	4835-69-2	NI36917
78	45	44	77	92	107	52	51	**352**	100	26	26	25	17	16	15	15	**2.33**	21	24	3	2	–	–	–	–	–	–	–	–	–	Akuammilan-17-oic acid, 10-methoxy-, methyl ester	38734-64-4	NI36918
170	181	352	197	267	224	239	144	**352**	100	92	88	81	20	20	18	17		21	24	3	2	–	–	–	–	–	–	–	–	–	Alstophyllan-19-one, 4-oxide	67497-75-0	NI36919
324	352	78	323	325	267	268	207	**352**	100	33	32	31	22	21	20	18		21	24	3	2	–	–	–	–	–	–	–	–	–	Aspidofractinine-1-carboxaldehyde, 17-methoxy-3-oxo-, (2α,5α)-	55724-67-9	NI36920
322	252	235	263	323	307	279	249	**352**	100	36	24	23	22	21	15	14	**1.00**	21	24	3	2	–	–	–	–	–	–	–	–	–	Condyfolan-16-carboxylic acid, 1,2,14,19-tetradehydro-16-(hydroxymethyl)-, methyl ester	2779-17-1	NI36921
169	156	251	352	119	129	323	237	**352**	100	67	59	45	45	35	29	24		21	24	3	2	–	–	–	–	–	–	–	–	–	Corynan-16-carboxylic acid, 16,17,19,20-tetradehydro-17-hydroxy-, methyl ester, (16E,19Z)-		NI36922
352	83	85	172	92	91	44	118	**352**	100	75	48	43	38	34	32	31		21	24	3	2	–	–	–	–	–	–	–	–	–	1,16-Cyclocorynan-17-oic acid, 19,20-didehydro-10-methoxy-, methyl ester, (16ξ,19E)-	55724-50-0	NI36923
324	108	352	199	198	283	160	227	**352**	100	57	52	37	34	27	24	22		21	24	3	2	–	–	–	–	–	–	–	–	–	Demethylseredamine, Na-formyl-		LI07777
352	263	235	138	167	214	295	168	**352**	100	68	47	43	30	29	28	26		21	24	3	2	–	–	–	–	–	–	–	–	–	3,7-Epiibophyllidine, deethyl-15-(1-oxopropyl)-, (15β)-		DD01402
138	352	214	180	167	295	194	154	**352**	100	50	20	20	18	14	13	13		21	24	3	2	–	–	–	–	–	–	–	–	–	3,7-Epiiboxyphylline, 3,7-epi-20-oxo-, 19-ξ-		DD01403
138	352	214	180	139	167	154	295	**352**	100	42	22	16	15	13	11	6		21	24	3	2	–	–	–	–	–	–	–	–	–	19-Epiiboxyphylline, 19-ξ-3,7-epi-20-oxo-		DD01404
138	352	214	182	295	139	180	167	**352**	100	48	19	18	13	12	10	10		21	24	3	2	–	–	–	–	–	–	–	–	–	19-Epiiboxyphylline, 20-oxo-		DD01405
108	321	352	309	291	322	120	70	**352**	100	60	39	25	15	14	14	13		21	24	3	2	–	–	–	–	–	–	–	–	–	Gelsemine, 1-methoxy-	38990-03-3	NI36924
263	295	352	138	235	168	167	353	**352**	100	87	60	55	45	33	23	22		21	24	3	2	–	–	–	–	–	–	–	–	–	Ibophyllidine, deethyl-15-(1-oxopropyl)-, 15-α-		DD01406
138	352	214	180	139	182	167	154	**352**	100	31	19	14	12	11	11	11		21	24	3	2	–	–	–	–	–	–	–	–	–	Iboxyphylline, 20-oxo-		DD01407
203	188	189	186	352	97	247		**352**	100	78	44	44	25	19	12			21	24	3	2	–	–	–	–	–	–	–	–	–	3H-Indole-2-carboxanilide, 3,3-diethyl-4′,5-dimethoxy-		LI07778
203	188	189	186	352	204	97	247	**352**	100	78	44	44	25	21	19	12		21	24	3	2	–	–	–	–	–	–	–	–	–	3H-Indole-2-carboxanilide, 3,3-diethyl-4′,5-dimethoxy-		MS08334
188	203	352	187	204	123	148	189	**352**	100	60	18	12	10	10	9	8		21	24	3	2	–	–	–	–	–	–	–	–	–	3H-Indole-2-carboxanilide, 3-isopropyl-4′,5-dimethoxy-3-methyl-		MS08335
188	203	352	337					**352**	100	60	18	4						21	24	3	2	–	–	–	–	–	–	–	–	–	3H-Indole-2-carboxanilide, 3-isopropyl-4′,5-dimethoxy-3-methyl-		MS08336
321	334	139	130	122	352	335	144	**352**	100	73	62	42	41	40	39	37		21	24	3	2	–	–	–	–	–	–	–	–	–	Isostrychnine, 12,13-dihydro-12-hydroxy-, (12α,12β,13α)-		NI36925
352	156	351	184	169	353	40	170	**352**	100	71	69	51	25	23	23	13		21	24	3	2	–	–	–	–	–	–	–	–	–	Oxayohimban-16-carboxylic acid, 16,17-didehydro-19-methyl-, methyl ester, (19α)-	483-04-5	NI36926
352	156	351	184	169	353	170	157	**352**	100	74	71	52	28	24	14	14		21	24	3	2	–	–	–	–	–	–	–	–	–	Oxayohimban-16-carboxylic acid, 16,17-didehydro-19-methyl-, methyl ester, (19α)-	483-04-5	NI36927
352	351	156	337	169	223	353	170	**352**	100	86	71	36	32	26	23	21		21	24	3	2	–	–	–	–	–	–	–	–	–	Oxayohimban-16-carboxylic acid, 16,17-didehydro-19-methyl-, methyl ester, (19α,20α)-	6474-90-4	LI07779
352	351	156	337	169	353	223	170	**352**	100	81	69	29	29	23	20	19		21	24	3	2	–	–	–	–	–	–	–	–	–	Oxayohimban-16-carboxylic acid, 16,17-didehydro-19-methyl-, methyl ester, (19α,20α)-	6474-90-4	NI36928
120	165	150	122	353	100	124	101	**352**	100	89	79	45	38	38	34	33	**4.00**	21	24	3	2	–	–	–	–	–	–	–	–	–	Oxyphenylbutazone, dimethyl ester		NI36929

MASS TO CHARGE RATIOS	M.W.	INTENSITIES	Parent	C	H	O	N	S	F	Cl	Br	I	Si	P	B	X	COMPOUND NAME	CAS Reg No	No
323 352 202 230 191 176 148 337	352	100 66 36 29 18 17 17 11		21	24	3	2	–	–	–	–	–	–	–	–	–	3-Pentenoxanilide, 3-ethyl-2-(4-methoxyphenylimino)-		LI07780
352 351 337 156 223 169 321 170	352	100 65 20 13 8 6 4 4		21	24	3	2	–	–	–	–	–	–	–	–	–	Rauniticine		LI07781
352 351 249 169 321 168 353 69	352	100 73 64 58 45 42 39 19		21	24	3	2	–	–	–	–	–	–	–	–	–	Sarpagan-16-carboxylic acid, 17-hydroxy-, methyl ester	639-36-1	MS08337
352 169 249 168 351 321 353 293	352	100 71 64 64 60 48 24 22		21	24	3	2	–	–	–	–	–	–	–	–	–	Sarpagan-16-carboxylic acid, 17-hydroxy-, methyl ester	639-36-1	NI36930
352 169 168 249 351 321 353 293	352	100 74 64 62 57 46 29 24		21	24	3	2	–	–	–	–	–	–	–	–	–	Sarpagan-16-carboxylic acid, 17-hydroxy-, methyl ester	639-36-1	NI36931
168 169 352 249 334 321 351 41	352	100 80 72 59 48 41 40 31		21	24	3	2	–	–	–	–	–	–	–	–	–	Sarpagan-16-carboxylic acid, 17-hydroxy-, methyl ester, (16R)-	6872-44-2	NI36932
57 43 55 59 69 71 45 95	352	100 95 91 78 77 65 64 52	7.67	21	24	3	2	–	–	–	–	–	–	–	–	–	18,19-Secoyohimban-19-oic acid, 16,17,20,21-tetradehydro-16-(hydroxymethyl)-, methyl ester, (15β,16E)-	5523-49-9	NI36933
265 352 159 171 185 41 324 183	352	100 87 58 50 49 46 42 42		21	24	3	2	–	–	–	–	–	–	–	–	–	Vincoridine	21290-53-9	NI36934
265 352 171 185 324 183 144 158	352	100 87 51 48 43 42 32 24		21	24	3	2	–	–	–	–	–	–	–	–	–	Vincoridine	21290-53-9	LI07782
279 352 172 130 132 134 293 280	352	100 51 41 40 38 32 26 21		21	24	3	2	–	–	–	–	–	–	–	–	–	Voachalotine oxindole, 16-de(hydroxymethyl)-	26126-87-4	NI36935
180 179 352 181 122 120 178 121	352	100 20 14 14 10 8 6 6		21	24	3	2	–	–	–	–	–	–	–	–	–	Vobasan-17-oic acid, 3-oxo-, methyl ester	2134-83-0	NI36936
352 351 156 169 170 144 353 143	352	100 93 70 56 39 35 30 30		21	24	3	2	–	–	–	–	–	–	–	–	–	Yohimban-16-carboxylic acid, 19,20-didehydro-17-hydroxy-, methyl ester, (16α,17α)-	54725-26-7	NI36937
352 351 353 169 170 184 156 44	352	100 73 19 18 14 13 12 10		21	24	3	2	–	–	–	–	–	–	–	–	–	Yohimban-16-carboxylic acid, 17-oxo-, methyl ester, (16α)-	2671-57-0	NI36938
179 177 180 151 133 332 207 178	352	100 40 15 9 9 8 7 7	0.39	21	36	4	–	–	–	–	–	–	–	–	–	–	1-Cyclopentene-1-decanoic acid, η-hydroxy-3-oxo-2-pentyl-, methyl ester	55521-10-3	NI36939
109 43 219 350 352 210 300 315	352	100 60 50 32 30 28 21 20		21	36	4	–	–	–	–	–	–	–	–	–	–	1-Naphthalenepentanoic acid, 3,4,4a,5,6,7,8,8a-octahydro-β,3-dihydroxy-β,2,5,5,8a-pentamethyl-, methyl ester, [3S-(1R*,3α,4aβ,8aα)]-		NI36940
98 43 139 83 41 55 69 71	352	100 60 55 50 40 37 35 34	0.30	21	36	4	–	–	–	–	–	–	–	–	–	–	1-Naphthalenepentanoic acid, decahydro-2-hydroxy-β,2,5,5,8a-pentamethyl-6-oxo-, methyl ester, [1S-[1α(S*),2β,4aβ,8aα]]-	52567-63-2	NI36941
79 95 67 55 41 81 93 108	352	100 90 86 82 80 70 60 55	15.01	21	36	4	–	–	–	–	–	–	–	–	–	–	Nonanoic acid, 9-(3-hexenylidenecyclopropylidene)-, 2-hydroxy-1-(hydroxymethyl)ethyl ester, (Z,Z,Z)-	55268-58-1	WI01454
352 353 109 296 278 261 323 123	352	100 25 24 12 11 8 6 5		21	36	4	–	–	–	–	–	–	–	–	–	–	Nonanoic acid, 9-(3-hexenylidenecyclopropylidene)-, 2-hydroxy-1-(hydroxymethyl)ethyl ester, (Z,Z,Z)-	55268-58-1	WI01455
79 41 67 43 95 81 29 57	352	100 89 79 67 58 58 58 56	2.00	21	36	4	–	–	–	–	–	–	–	–	–	–	9,12,15-Octadecatrienoic acid, 2,3-dihydroxypropyl ester, (Z,Z,Z)-	18465-99-1	WI01456
103 91 45 77 57 352 121 31	352	100 80 60 30 25 23 20 20		21	36	4	–	–	–	–	–	–	–	–	–	–	1,2-Propanediol, [(17-hydroxy-1,5-octadecadien-3-ynyl)oxy]-, (Z,Z)-(+)-		DD01408
75 73 67 55 81 41 79 54	352	100 90 60 50 44 40 23 23	2.50	21	40	2	–	–	–	–	–	–	1	–	–	–	9,12-Octadecadienoic acid (Z,Z)-, trimethylsilyl ester	56259-07-5	NI36942
73 75 67 55 81 41 95 337	352	100 99 65 51 50 47 32 29	3.96	21	40	2	–	–	–	–	–	–	1	–	–	–	9,12-Octadecadienoic acid (Z,Z)-, trimethylsilyl ester	56259-07-5	NI36943
352 77 93 126 245 121 247 353	352	100 55 36 26 25 24 20 18		22	16	1	4	–	–	–	–	–	–	–	–	–	Naphthalene, 1-(4-hydroxyphenylazo)-2-phenylazo-		IC04090
143 247 115 352 171 76 105 248	352	100 98 68 58 29 18 17 17		22	16	1	4	–	–	–	–	–	–	–	–	–	2-Naphthalenol, 1-[[4-(phenylazo)phenyl]azo]-	85-86-9	NI36944
196 352 77 167 233 168 169 197	352	100 29 28 26 19 18 17 16		22	16	1	4	–	–	–	–	–	–	–	–	–	2(1H)-Phthalazinecarboxamide, 1-cyano-N,N-diphenyl-	21415-94-1	NI36945
148 105 77 176 161 352 133 120	352	100 69 69 44 26 13 12 6		22	24	4	–	–	–	–	–	–	–	–	–	–	Di-o-thymotide		MS08338
352 351 293 144 182 193 160 157	352	100 32 32 23 22 14 12 12		22	28	2	2	–	–	–	–	–	–	–	–	–	Ajmaline, O-acetyl-21-desoxy-		LI07783
109 43 352 324 124 308 353 41	352	100 58 55 43 36 24 13 13		22	28	2	2	–	–	–	–	–	–	–	–	–	Aspidofractinine, 1-acetyl-17-methoxy-	55724-53-3	NI36946
265 109 43 352 170 264 324 291	352	100 64 61 53 36 35 30 24		22	28	2	2	–	–	–	–	–	–	–	–	–	Aspidofractinine-3-methanol, acetate, (2α,3α,5α)-	55103-47-4	NI36947
124 125 352 69 57 55 168 56	352	100 10 5 4 4 4 3 3		22	28	2	2	–	–	–	–	–	–	–	–	–	Aspidospermidine-3-carboxylic acid, 2,3-didehydro-1-methyl-, methyl ester, (5α,12β,19α)-	19074-77-2	NI36948
105 77 218 247 135 106 134 51	352	100 44 12 10 10 9 8 8	5.00	22	28	2	2	–	–	–	–	–	–	–	–	–	1,8-Octanediamine, N,N'-dibenzoyl-		LI07784
122 135 195 352 292 148 209 146	352	100 91 46 38 37 23 20 16		22	28	2	2	–	–	–	–	–	–	–	–	–	Picralinol, triacetate		LI07785
246 247 42 120 218 172 106 91	352	100 18 12 9 7 7 7 6	0.50	22	28	2	2	–	–	–	–	–	–	–	–	–	4-Piperidinecarboxylic acid, 1-[2-(4-aminophenyl)ethyl]-4-phenyl-, ethyl ester	144-14-9	NI36949
91 309 120 160 43 92 106 70	352	100 47 20 18 18 12 11 10	3.00	22	28	2	2	–	–	–	–	–	–	–	–	–	Putrescine, N,N'-diacetyl-N,N'-dibenzoyl-		LI07786
43 55 97 83 57 41 69 29	352	100 71 56 55 53 52 49 26	0.00	22	40	3	–	–	–	–	–	–	–	–	–	–	2,5-Furandione, dihydro-3-octadecyl-	47458-32-2	NI36950
43 223 117 265 211 83 95 251	352	100 81 59 45 34 26 25 24	1.74	22	40	3	–	–	–	–	–	–	–	–	–	–	4-Hexadecynoic acid, 2-propoxy-, propyl ester	40924-19-4	NI36951
85 95 177 109 191 192 278 337	352	100 35 25 23 16 5 4 3	2.00	22	40	3	–	–	–	–	–	–	–	–	–	–	1-Naphthalenepentanoic acid, decahydro-2-hydroxy-8-methoxy-β,2,5,5,8a pentamethyl-, methyl ester, [1R-[1α(S*),2β,4aβ,8aα]]-		MS08339
352 77 275 105 323 104 351 353	352	100 77 64 63 55 52 40 27		23	16	2	2	–	–	–	–	–	–	–	–	–	2(1H)-Pyrimidinone, 5-benzoyl-4,6-diphenyl-	42919-51-7	NI36952
352 77 275 105 323 104 351 353	352	100 78 66 63 57 53 40 29		23	16	2	2	–	–	–	–	–	–	–	–	–	2(1H)-Pyrimidinone, 5-benzoyl-4,6-diphenyl-	42919-51-7	LI07787
257 180 258 78 206 117 168 130	352	100 58 55 35 30 26 22 21	0.30	23	20	–	4	–	–	–	–	–	–	–	–	–	3,8-Azo-4,7-methanocyclobuta[b]naphthalene, 2a,3,3a,4,7,7a,8,8a-octahydro-3,8-di-2-pyridinyl-, (2aα,3β,3aα,4β,7β,7aα,8β,8aα)-	40229-68-3	NI36953
180 206 91 167 165 233 115 179	352	100 99 55 50 50 46 44 35	34.10	23	25	1	–	–	–	1	–	–	–	–	–	–	Cyclohexanone, 2-(3-chloro-2-butenyl)-2-methyl-6,6-diphenyl-	50592-54-6	NI36954
43 352 227 197 185 184 170 41	352	100 42 37 33 33 29 21 18		23	32	1	2	–	–	–	–	–	–	–	–	–	2,20-Cyclo-8,9-secoaspidospermidine, 9-acetyl-1,3-dimethyl-, (2α,3β,5α,12β,19α,20R)-	56053-39-5	NI36955
43 41 55 29 69 74 27 57	352	100 72 62 34 31 30 27 24	1.34	23	44	2	–	–	–	–	–	–	–	–	–	–	13-Docosenoic acid, methyl ester	56630-69-4	NI36956
55 69 41 43 74 83 57 97	352	100 70 66 56 55 49 37 33	3.86	23	44	2	–	–	–	–	–	–	–	–	–	–	13-Docosenoic acid, methyl ester, (Z)-	1120-34-9	NI36957
55 43 69 41 83 74 57 67	352	100 73 66 61 46 41 32 27	0.00	23	44	2	–	–	–	–	–	–	–	–	–	–	13-Docosenoic acid, methyl ester, (Z)-	1120-34-9	NI36958

MASS TO CHARGE RATIOS								M.W.	INTENSITIES								Parent	C	H	O	N	S	F	Cl	Br	I	Si	P	B	X	COMPOUND NAME	CAS Reg No	No
100	43	85	41	55	57	101	113	352	100	54	48	21	19	18	17	15	1.33	23	44	2	–	–	–	–	–	–	–	–	–	–	Tricosane-2,4-dione	65351-36-2	NI36959
309	55	214	41	283	81	43	83	352	100	71	60	53	47	44	43	42	30.24	23	45	–	–	–	–	–	–	–	–	1	–	–	Phosphine, bis(isopropylmethylcyclohexyl)propyl-	74645-95-7	NI36960
309	214	283	83	172	55	69	215	352	100	85	51	51	48	46	45	44	43.89	23	45	–	–	–	–	–	–	–	–	1	–	–	Phosphine, isopropylbis(isopropylmethylcyclohexyl)-	74779-92-3	NI36961
202	352	351	203	335	201	189	200	352	100	81	42	18	16	9	9	7		24	16	3	–	–	–	–	–	–	–	–	–	–	3(2H)-Benzofuranone, 6-methoxy-2-(9-anthrylmethylene)-	77764-99-9	NI36962
249	352	234	250	77	353	351	103	352	100	77	40	21	21	20	18	12		24	20	1	2	–	–	–	–	–	–	–	–	–	4H-1,2-Diazepine, 5-(4-methoxyphenyl)-3,7-diphenyl-	25649-72-3	NI36963
352	337	309	353	338	324	354	43	352	100	54	32	30	16	16	14	14		24	32	2	–	–	–	–	–	–	–	–	–	–	4,4′-Dioxo-3,3′,5,5′-tetraisopropyl-bicyclohexylidene		IC04091
243	165	244	241	166	242	239	167	352	100	43	38	13	12	11	9	8	0.10	25	20	–	–	1	–	–	–	–	–	–	–	–	Phenyl trityl sulphide	16928-73-7	NI36964
57	71	85	113	127	99	56	55	352	100	81	51	26	25	23	18	17	11.59	25	52	–	–	–	–	–	–	–	–	–	–	–	Eicosane, 2,6,10,14,18-pentamethyl-	51794-16-2	NI36965
57	71	85	43	69	113	127	55	352	100	76	50	50	29	27	25	25	0.60	25	52	–	–	–	–	–	–	–	–	–	–	–	Eicosane, 2,6,10,14,18-pentamethyl-	51794-16-2	LI07788
57	71	86	113	98	127	56	55	352	100	99	63	33	32	28	23	23	6.59	25	52	–	–	–	–	–	–	–	–	–	–	–	Eicosane, 2,6,10,14,19-pentamethyl-	55268-60-5	NI36966
57	43	71	85	41	55	69	99	352	100	92	63	43	42	31	17	14	0.02	25	52	–	–	–	–	–	–	–	–	–	–	–	Heptadecane, 9-octyl-		AI02152
43	57	71	41	85	55	69	56	352	100	93	55	43	39	29	15	14	1.67	25	52	–	–	–	–	–	–	–	–	–	–	–	Pentacosane		AI02153
352	43	253	239	41	353	309	55	352	100	46	42	39	38	31	17	13		26	40	–	–	–	–	–	–	–	–	–	–	–	Chrysene, 1,2,3,4,4a,7,8,9,10,11,12,12a-dodecahydro-6-octyl-	55281-94-2	AI02154
352	43	41	253	353	239	309	55	352	100	46	38	37	32	32	15	13		26	40	–	–	–	–	–	–	–	–	–	–	–	Chrysene, 1,2,3,4,4a,7,8,9,10,11,12,12a-dodecahydro-6-octyl-	55281-94-2	NI36967
352	253	43	239	41	353	309	55	352	100	46	46	45	42	28	20	15		26	40	–	–	–	–	–	–	–	–	–	–	–	Chrysene, 1,2,3,4,5,6,6a,7,8,9,10a-dodecahydro-12-octyl-		AI02155
352	176	353	175	350	162	78	324	352	100	40	30	27	18	10	7	2		28	16	–	–	–	–	–	–	–	–	–	–	–	Anthra[1,2-e]acephenanthrylene	28512-55-2	LI07789
352	176	353	175	350	173	177	162	352	100	40	30	27	17	14	10	10		28	16	–	–	–	–	–	–	–	–	–	–	–	Anthra[1,2-e]acephenanthrylene	28512-55-2	NI36968
352	350	353	351	348	175	174	149	352	100	40	31	22	18	13	12	10		28	16	–	–	–	–	–	–	–	–	–	–	–	Anthracene, 9,9′,10,10′-tetrahydrobis-		MS08340
352	176	353	174	350	172	177	170	352	100	40	34	30	20	20	13	10		28	16	–	–	–	–	–	–	–	–	–	–	–	Benz[a]indeno[1,2,3-de]naphthacene	27983-38-6	NI36969
352	176	353	175	350	162	78	324	352	100	39	35	31	20	10	5	2		28	16	–	–	–	–	–	–	–	–	–	–	–	Benz[a]indeno[1,2,3-de]naphthacene	27983-38-6	LI07790
352	174	175	174.5	176	351	349	175.5	352	100	99	51	48	36	35	27	27		28	16	–	–	–	–	–	–	–	–	–	–	–	1,16:4,5:8,9:12,13-Tetramethanotetraphenylene		LI07791
313	278	276	315	311	280	43	317	353	100	95	75	65	64	47	26	22	8.29	3	7	–	3	–	–	5	–	–	–	3	–	–	1-Isopropylpentachlorocyclotriphosphazene	75155-05-4	NI36970
327	329	325	278	276	313	280	331	353	100	64	63	62	48	32	31	21	11.10	3	7	–	3	–	–	5	–	–	–	3	–	–	1-Propylpentachlorotriphosphazene	75132-80-8	NI36971
270	271	269	311	312	313	39	61	353	100	86	85	80	78	61	59	44	15.19	9	19	–	–	–	–	–	–	–	–	1	–	1	Platinum, bis(η^3-2-propenyl)(trimethylphosphine)-	74811-16-8	NI36972
184	240	298	114	115	354	139	166	353	100	86	83	74	50	43	33	30	0.00	11	13	5	1	–	6	–	–	–	–	–	–	–	L-Serine, N-(trifluoroacetyl)-, butyl ester, trifluoroacetate (ester)	5282-96-2	NI36973
354	240	257	355	185	101	241	298	353	100	58	19	15	10	10	8	3	0.00	11	13	5	1	–	6	–	–	–	–	–	–	–	L-Serine, N-(trifluoroacetyl)-, butyl ester, trifluoroacetate (ester)	5282-96-2	NI36974
57	41	138	139	69	49	84	86	353	100	49	36	35	28	17	11	9	0.00	11	13	5	1	–	6	–	–	–	–	–	–	–	L-Serine, N-(trifluoroacetyl)-, sec-butyl ester, trifluoroacetate (ester)	57983-19-4	NI36975
59	197	202	230	198	117	353	175	353	100	41	28	24	20	16	13	11		11	15	–	1	6	–	–	–	–	–	–	–	–	Propanenitrile, 3-[(1,5,7-trimethyl-2,4,6,8,9-pentathiatricyclo[3.3.1.1^{3,7}]dec-3-yl)thio]-	57274-41-6	NI36976
353	310	138	355	202	312	203	139	353	100	86	50	40	40	38	30	30		11	16	4	3	2	–	1	–	–	–	–	–	–	6-Dimethylsulphamoyl-4,4-dioxo-1,3-dimethyl-7-chloro-1,3-diaza-4-thia-1,2,3,4-tetrahydronaphthalene		MS08341
215	272	122	77	216	30	186	217	353	100	47	47	44	40	40	37	33	2.28	12	7	8	3	1	–	–	–	–	–	–	–	–	2,4,4′-Trinitrodiphenyl sulphone	1098-16-4	NI36977
266	69	41	267	169	71	43	68	353	100	31	21	20	16	15	15	8	6.92	12	14	3	1	–	7	–	–	–	–	–	–	–	Proline, N-(heptafluorobutanoyl)-, propyl ester		MS08342
43	235	42	69	101	210	68	143	353	100	99	97	95	59	56	54	51	29.00	13	19	3	7	1	–	–	–	–	–	–	–	–	2-Amino-4-[2-acetamido-2-(methoxycarbonyl)ethylthio]-6-(1-cyano-1-methylethylamino)-s-triazine		LI07792
86	226	211	73	128	227	195	43	353	100	82	78	40	35	32	27	21	9.00	13	32	4	1	–	–	–	–	–	2	1	–	–	Phosphonic acid, [1-(acetylamino)pentyl]-, bis(trimethylsilyl) ester	55108-84-4	NI36978
105	77	106	51	45	69	161	100	353	100	68	24	19	12	11	9	9	1.20	14	12	4	1	–	5	–	–	–	–	–	–	–	Hippuric acid, N-(pentafluoropropanoyl)-, ethyl ester		MS08343
227	43	184	55	41	226	57	69	353	100	66	62	54	54	44	42	41	2.40	14	22	1	1	–	7	–	–	–	–	–	–	–	1-Decylamine, N-(heptafluorobutanoyl)-		MS08344
288	286	290	287	285	353	351	355	353	100	86	80	66	58	26	22	21		15	15	–	–	–	–	–	–	–	–	–	–	1	Gadolinium, tris(η^5-2,4-cyclopentadien-1-yl)-	1272-21-5	NI36979
91	190	148	103	266	206	43	147	353	100	80	68	30	28	28	25	20	0.00	15	16	3	1	–	5	–	–	–	–	–	–	–	D-Phenylalanine, N-pentafluoropropionyl-, isopropyl ester		MS08345
254	353	239	154	255	252	71	43	353	100	19	19	19	13	8	8	6		15	21	6	–	–	–	–	–	–	–	–	–	1	Iron, tris(2,4-pentanedionato-O,O′)-	14024-18-1	NI36980
43	85	155	100	254	239	16	27	353	100	90	87	82	70	54	28	23	0.00	15	21	6	–	–	–	–	–	–	–	–	–	1	Iron, tris(2,4-pentanedionato-O,O′)-	14024-18-1	NI36981
254	155	239	353	172	56			353	100	74	47	16	8	5				15	21	6	–	–	–	–	–	–	–	–	–	1	Iron, tris(2,4-pentanedionato-O,O′)-	14024-18-1	LI07793
148	162	136	250	44	43	163	36	353	100	98	58	47	38	33	30	30	6.67	15	23	5	5	–	–	–	–	–	–	–	–	–	Adenosine, N-(3-hydroxy-3-methylbutyl)-	23477-26-1	NI36982
162	148	250	135	43	37	57	30	353	100	79	63	54	42	33	32	32	12.00	15	23	5	5	–	–	–	–	–	–	–	–	–	Adenosine, N-(3-hydroxy-3-methylbutyl)-	23477-26-1	LI07794
28	148	135	162	149	250	136	33	353	100	98	51	50	41	35	27	24	5.40	15	23	5	5	–	–	–	–	–	–	–	–	–	Adenosine, N-(4-hydroxy-3-methylbutyl)-	22663-55-4	NI36983
295	353	321	309	59	165	190	31	353	100	61	58	40	36	34	29	29		16	11	5	5	–	–	–	–	–	–	–	–	–	Carbamic acid, N-[3,5-dicyano-6-methoxy-4-(3-nitrophenyl)pyridin-2-yl]-, methyl ester	82072-10-4	NI36984
209	73	210	41	179	353	192	45	353	100	44	16	16	14	7	7	7		16	23	4	1	1	–	–	–	–	1	–	–	–	Propanoic acid, 3-[4-(trimethylsilyl)oxy-3-methoxyphenyl]- 2-isothiocyanato-, ethyl ester		MS08346
73	190	277	75	44	74	45	338	353	100	42	16	15	14	11	10	9	5.10	16	27	4	1	–	–	–	–	–	2	–	–	–	Benzenepropanoic acid, α-(methoxyimino)-4-[(trimethylsilyl)oxy]-, trimethylsilyl ester	55520-97-3	NI36985
73	190	277	278	75	338	179	40	353	100	50	21	14	13	11	11	9	6.00	16	27	4	1	–	–	–	–	–	2	–	–	–	Benzenepropanoic acid, α-(methoxyimino)-4-[(trimethylsilyl)oxy]-, trimethylsilyl ester		MS08347

MASS TO CHARGE RATIOS								M.W.	INTENSITIES								Parent	C	H	O	N	S	F	Cl	Br	I	Si	P	B	X	COMPOUND NAME	CAS Reg No	No
75	73	135	77	93	147	45	263	353	100	66	61	40	30	18	15	13	0.00	16	27	4	1	–	–	–	–	–	2	–	–	–	Glycine, N-(4-methoxybenzoyl)-, bis(trimethylsilyl) ester		MS08348
218	73	75	135	147	236	219	45	353	100	83	38	36	22	19	19	16	0.00	16	27	4	1	–	–	–	–	–	2	–	–	–	3,4-Methylenedioxyphenylalanine, N-(trimethylsilyl)oxy-, trimethylsilyl ester		MS08349
153	99	98	73	85	353	75	83	353	100	64	58	44	34	30	30	28		16	31	2	7	–	–	–	–	–	–	–	–	–	Arginine, N-tris(dimethylaminomethyl)-, methyl ester		NI36986
88	116	165	235	237	56	201	199	353	100	77	62	43	29	23	21	20	1.20	17	17	3	1	–	–	2	–	–	–	–	–	–	Glycine, N,N-bis[(4-chlorophenyl)acetyl]-, methyl ester		NI36987
174	73	175	86	75	176	147	45	353	100	82	30	22	20	18	13	12	0.00	17	35	1	1	–	–	–	–	–	3	–	–	–	Phenylethanolamine, tris(trimethylsilyl)-		MS08350
174	73	175	86	59	45	176	338	353	100	62	22	16	14	12	10	8	0.00	17	35	1	1	–	–	–	–	–	3	–	–	–	Tyramine, tris(trimethylsilyl)-		MS08351
353	180	179	59	177	178	176.5	338	353	100	15	13	10	9	8	8	6		18	18	2	2	–	–	–	–	–	–	–	–	1	Cobalt, [[2,2′-[(1,1-dimethyl-1,2-ethanediyl)bis(nitrilomethylidyne)]bis[phenolato]](2-)-N,N′,O,O′]-	32593-87-6	LI07795
353	354	180	179	297	59	194	177	353	100	22	16	12	10	10	9	9		18	18	2	2	–	–	–	–	–	–	–	–	1	Cobalt, [[2,2′-[(1,1-dimethyl-1,2-ethanediyl)bis(nitrilomethylidyne)]bis[phenolato]](2-)-N,N′,O,O′]-	32593-87-6	NI36988
267	86	338	268	263	339	269	264	353	100	66	28	26	24	11	9	9	4.00	18	35	2	1	–	–	–	–	–	2	–	–	–	Benzeneethanamine, N-butyl-β,4-bis[(trimethylsilyl)oxy]-	40629-66-1	NI36989
353	111	178	177	151	175	355	289	353	100	84	74	59	48	47	40	37		19	12	2	1	1	–	1	–	–	–	–	–	–	N-Fluorenylidene-S-(4-chlorophenyl)sulphonamide		LI07796
353	322	324	271	336	266	354	352	353	100	53	30	30	28	28	23	23		20	19	5	1	–	–	–	–	–	–	–	–	–	Acronine, 11-hydroxy-3-(hydroxymethyl)-		LI07797
332	333	304	335	176	303	334	162	353	100	85	67	48	40	38	34	31	9.70	20	19	5	1	–	–	–	–	–	–	–	–	–	Chelidonine	476-32-4	NI36990
148	178	335	176	149	334	69	304	353	100	32	18	18	18	14	14	11	3.00	20	19	5	1	–	–	–	–	–	–	–	–	–	Dihydroribasine	88151-42-2	NI36991
190	338	353	105	122	191	339	322	353	100	40	30	25	19	13	10	10		20	19	5	1	–	–	–	–	–	–	–	–	–	Fumariline, dihydro-	24181-80-4	NI36992
145	105	91	146	77	149	104	78	353	100	81	66	51	42	38	27	19	10.00	20	19	5	1	–	–	–	–	–	–	–	–	–	Furo[3,2-d]oxazol-6-ol, 3a,5,6,6a-tetrahydro-2-phenyl-5-(2-phenyl-1,3-dioxolan-4-yl)-	23626-69-9	NI36993
135	57	83	71	43	353	77	338	353	100	98	71	62	60	47	47	45		20	19	5	1	–	–	–	–	–	–	–	–	–	Isoquinoline, 1-(4-methoxybenzoyl)-5,6,7-trimethoxy-	84716-73-4	NI36994
145	105	91	146	77	149	175	352	353	100	78	64	50	42	38	12	9	8.00	20	19	5	1	–	–	–	–	–	–	–	–	–	2-Oxazoline, 2-phenyl-4,5-(3,4,6-tri-O-acetyl-2-deoxy-α-D-glucopyrano)-		LI07798
324	353	338	177	192	162	77	42	353	100	43	18	14	8	7	4	3		20	19	5	1	–	–	–	–	–	–	–	–	–	D-Parfumine	28230-70-8	NI36995
148	163	190	91	134	353	77	338	353	100	19	11	8	7	6	5	2		20	19	5	1	–	–	–	–	–	–	–	–	–	Protopine	130-86-9	NI36996
190	176	353	325	149	177	162	148	353	100	56	50	44	30	27	23	23		20	19	5	1	–	–	–	–	–	–	–	–	–	Rheadan, 16-methyl-2,3:10,11-bis[methylenebis(oxy)]-	30833-13-7	NI36997
148	163	190	267	281	252	209	177	353	100	21	9	6	4	4	3	3	3.00	20	19	5	1	–	–	–	–	–	–	–	–	–	7,13a-Secoberbin-13a-ene, 7-methyl-2,3:9,10-bis(methylenedioxy)-	130-86-9	LI07799
148	354	55	355	149	352	353	53	353	100	82	20	18	15	14	11	11		20	19	5	1	–	–	–	–	–	–	–	–	–	7,13a-Secoberbin-13a-ene, 7-methyl-2,3:9,10-bis(methylenedioxy)-	130-86-9	NI36998
146	59	308	114	67	93	309	107	353	100	75	58	33	25	16	9	9	0.00	20	39	2	1	–	–	–	–	–	1	–	–	–	1-Butene, 1-[[[bis(2-methoxyethyl)amino]methyl]dimethylsilyl]-2-(4-methyl-3-cyclohexenyl)-, (E)-		DD01409
98	99	55	41	44	42	70	112	353	100	8	6	5	4	4	3	2	1.00	21	23	2	1	1	–	–	–	–	–	–	–	–	Dibenzo[b,f]thiepin-11-one, 4-(1-ethyl-3-piperidyloxy)-10,11-dihydro-		MS08352
72	217	353	215	281	338	202	218	353	100	82	55	43	40	33	31	29		21	23	2	1	1	–	–	–	–	–	–	–	–	Naphthalene, 4-benzyl-1-(N,N-diethylsulphonamido)-		IC04092
91	181	77	86	71	180	96	41	353	100	90	74	63	53	51	29	26	2.00	21	23	2	1	1	–	–	–	–	–	–	–	–	5-Oxa-8-thia-6-azaspiro[3.4]octan-2-one, 1,1,3,3-tetramethyl-6,7-diphenyl-	50455-60-2	NI36999
325	283	353	282	43	311	326	253	353	100	63	54	43	43	29	22	21		21	23	4	1	–	–	–	–	–	–	–	–	–	Crotonosine, N-acetyl-O-ethyl-		MS08353
325	283	353	282	43	311	324	326	353	100	63	53	43	42	29	23	22		21	23	4	1	–	–	–	–	–	–	–	–	–	Crotonosine, N-acetyl-O-ethyl-		LI07800
247	218	175	201	248	79	232	353	353	100	65	57	51	42	36	24	16		21	23	4	1	–	–	–	–	–	–	–	–	–	1H-Indole-3-carboxylic acid, 2-(2-hydroxy-2-phenylethyl)-5-methoxy-, ethyl ester		DD01410
165	353	150	352	338	204	187	354	353	100	97	75	55	30	30	27	23		21	23	4	1	–	–	–	–	–	–	–	–	–	Isoquinoline, 3-(6-vinyl-1,3-benzodioxol-5-yl)-1,2,3,4-tetrahydro-7,8-dimethoxy-2-methyl-	29079-32-1	NI37000
325	283	353	282	43	311	324	326	353	100	63	54	43	43	30	23	21		21	23	4	1	–	–	–	–	–	–	–	–	–	Spiro[2,5-cyclohexadiene-1,7′(1′H)-cyclopent[ij]isoquinolin]-4-one, 1′-acetyl-2′,3′,8′,8′a-tetrahydro-5′-ethoxy-6′-methoxy-, (R)-	10214-74-1	NI37001
235	210	335	236	353	126	195	221	353	100	88	82	61	57	37	31	30		21	27	2	3	–	–	–	–	–	–	–	–	–	Ergoline-8-carboxamide, 9,10-didehydro-N-[1-(hydroxymethyl)propyl]-1,6 dimethyl-, (8β)-	361-37-5	MS08354
353	235	210	195	221	237	236	354	353	100	50	47	39	36	34	26	24		21	27	2	3	–	–	–	–	–	–	–	–	–	Ergoline-8-carboxamide, 9,10-didehydro-N-[1-(hydroxymethyl)propyl]-1,6 dimethyl-, (8β)-	361-37-5	NI37002
43	336	105	41	214	162	77	230	353	100	84	83	76	61	56	52	51	32.80	22	27	3	1	–	–	–	–	–	–	–	–	–	Annotinol, phenyl-	5096-62-8	NI37003
353	322	28	338	161	133	354	323	353	100	79	63	44	33	31	26	26		22	31	1	3	–	–	–	–	–	–	–	–	–	4,4-Bis(diethylamino)benzophenone oxime, O-methyl ether		MS08355
113	43	55	41	70	126	184	57	353	100	35	30	24	23	19	15	14	0.94	22	43	2	1	–	–	–	–	–	–	–	–	–	Pyrrolidine, 1-(6-hydroxy-1-oxooctadecyl)-	56666-42-3	NI37004
49	43	113	84	70	55	86	41	353	100	88	87	59	54	38	37	36	1.14	22	43	2	1	–	–	–	–	–	–	–	–	–	Pyrrolidine, 1-(12-hydroxy-1-oxooctadecyl)-	56666-49-0	NI37005
227	353	212	226	354	210	228	198	353	100	88	73	34	23	20	18	15		23	31	2	1	–	–	–	–	–	–	–	–	–	Monascaminone methyl ether		LI07801
76	150	50	104	92	75	57	41	353	100	99	98	91	65	52	51	49	0.00	23	47	1	1	–	–	–	–	–	–	–	–	–	Octadecanamide, N-pentyl-	74420-94-3	NI37006
353	352	183	122	262	107	185	108	353	100	77	59	24	15	15	14	13		24	20	–	1	–	–	–	–	–	–	1	–	–	Phosphine imide, tetraphenyl-	2325-27-1	LI07802
353	352	183	354	122	262	107	185	353	100	77	59	25	24	15	15	14		24	20	–	1	–	–	–	–	–	–	1	–	–	Phosphine imide, tetraphenyl-	2325-27-1	NI37007
198	44	199	43	30	57	55	41	353	100	31	16	15	10	8	7	7	1.80	24	51	–	1	–	–	–	–	–	–	–	–	–	Didodecylamine		MS08356
254	255	112	156	98	154	143	84	353	100	20	6	4	4	3	3	3	1.70	24	51	–	1	–	–	–	–	–	–	–	–	–	Trioctylamine	1116-76-3	NI37008
233	338	353	218	103	129	119	105	353	100	65	62	60	20	15	14	13		25	28	–	1	–	–	–	–	–	–	–	1	–	Borane, (benzylideneamino)dimesityl-	29098-27-9	NI37009
147	148							353	100	13							0.00	26	27	–	1	–	–	–	–	–	–	–	–	–	3-Cyclohexen-1-amine, N,N-dimethyl-2,5,6-triphenyl-, (1α,2α,5α,6β)-	42715-10-6	NI37010
147	148	206						353	100	13	2						0.00	26	27	–	1	–	–	–	–	–	–	–	–	–	3-Cyclohexen-1-amine, N,N-dimethyl-2,5,6-triphenyl-, (1α,2β,5β,6β)-	42715-09-3	NI37011

MASS TO CHARGE RATIOS								M.W.	INTENSITIES								Parent	C	H	O	N	S	F	Cl	Br	I	Si	P	B	X	COMPOUND NAME	CAS Reg No	No
77	353	91	276	104	180			**353**	100	60	58	52	44	37				26	27	–	1	–	–	–	–	–	–	–	–	–	**1H-Isoindole, octahydro-1,2,3-triphenyl-**	55268-61-6	**NI37012**
222	354	100	254	127	31	177	208	**354**	100	92	87	85	82	78	44	32		2	–	–	–	–	4	–	–	2	–	–	–	–	**Ethane, 1,1,2,2-tetrafluoro-1,2-diiodo-**		**IC04093**
127	185	89	212	69	128	170	103	**354**	100	38	37	30	21	15	14	13	**2.00**	3	4	2	2	2	10	–	–	–	–	–	–	–	**1,3-Dioxo-1,3-propylenebis(imino)bis(sulphur pentafluoride)**	80409-46-7	**NI37013**
324	236	177	354	266	208	178	176	**354**	100	77	77	71	71	71	71	63		4	10	4	4	2	–	–	–	–	–	–	–	2	**Iron, di-μ-ethanethiolato(tetranitroso)bis-**		**MS08357**
69	247	159	85	61	107	81	50	**354**	100	32	9	9	8	2	2	2	**0.00**	5	–	1	2	–	9	–	1	–	–	–	–	–	**2,5-Diaza-2-hexen-4-one, 3-bromo-5-(trifluoromethyl)hexafluoro-**		**LI07803**
69	174	180	92	123	54	202	275	**354**	100	44	31	14	8	5	4	3	**0.00**	5	–	1	2	–	9	–	1	–	–	–	–	–	**2,5-Diaza-2-hexen-4-one, 3-bromo-5-(trifluoromethyl)hexafluoro-**		**LI07804**
354	339	292	248	277	324	262	230	**354**	100	80	26	22	17	6	6	5		5	15	–	–	–	–	–	–	–	–	9	–	–	**Nonaphosphane(5), pentamethyl-**	79180-88-4	**NI37014**
277	321	239	237	279	275	241	272	**354**	100	77	76	74	65	64	61	55	**9.62**	9	4	2	–	–	–	6	–	–	–	–	–	–	**4,7-Methanoisobenzofuran-1(3H)-one, 4,5,6,7,8,8-hexachloro-3a,4,7,7a-tetrahydro-**	3868-61-9	**NI37015**
36	302	44	38	28	89	133	260	**354**	100	90	42	36	32	29	16	15	**0.00**	9	8	1	2	1	–	1	–	1	–	–	–	–	**2(3H)-Thiazolimine, 3-hydroxy-4-(4-iodophenyl)-, monohydrochloride**	58275-68-6	**NI37016**
117	238	73	195	43	99	45	168	**354**	100	35	25	23	22	20	16	16	**2.40**	9	11	5	2	–	–	–	–	1	–	–	–	–	**Uridine, 2′-deoxy-5-iodo-**	54-42-2	**NI37017**
186	167	354	298	326	166	158	232	**354**	100	40	30	19	17	17	16	15		10	5	2	–	–	5	–	–	–	–	–	–	1	**Ruthenium, dicarbonyl(η^5-2,4-cyclopentadien-1-yl)(1,2,3,3,3-pentafluoro-1-propenyl)-**	33195-20-9	**NI37018**
64	113	77	100	177	95	69	51	**354**	100	25	14	10	5	4	4	4	**0.01**	10	6	–	–	–	12	–	–	–	–	–	–	–	**Cyclobutane, 1,1′-(1,1,2,2-tetrafluoro-1,2-ethanediyl)bis[2,2,3,3-tetrafluoro-**	35207-94-4	**MS08358**
39	65	66	117	52	63	185	264	**354**	100	93	76	66	60	30	26	19	**7.33**	10	10	4	4	–	–	–	–	–	–	–	–	2	**Chromium, bis(η^5-2,4-cyclopentadien-1-yl)bis-μ-(nitrosyl-N:N-)dinitrosyldi-, (Cr-Cr)**	50277-98-0	**MS08359**
39	65	66	117	52	128	304	244	**354**	100	93	76	67	61	38	33	32	**7.33**	10	10	4	4	–	–	–	–	–	–	–	–	2	**Chromium, bis(η^5-2,4-cyclopentadien-1-yl)bis[μ-(nitrosyl-N:N)]dinitrosyldi-, (Cr-Cr)**	50277-98-0	**NI37019**
39	65	66	117	52	304	244	63	**354**	100	93	76	67	61	33	32	31	**7.00**	10	10	4	4	–	–	–	–	–	–	–	–	2	**Chromium, bis(η^5-2,4-cyclopentadien-1-yl)bis[μ-(nitrosyl-N:N)]dinitrosyldi-, (Cr-Cr)**	50277-98-0	**LI07805**
354	356	355	358	357	360	359	148	**354**	100	59	16	15	11	3	3	1		10	20	–	2	4	–	–	–	–	–	–	–	1	**Nickel, bis(diethylcarbamodithioato-S,S′)-**	52610-81-8	**NI37020**
116	88	29	60	354	27	356	72	**354**	100	84	84	63	53	53	32	32		10	20	–	2	4	–	–	–	–	–	–	–	1	**Nickel, bis(diethylcarbamodithioato-S,S′)-**		**IC04094**
242	354	241	75	244	280	356	76	**354**	100	65	61	61	55	54	50	50		11	23	–	–	–	–	1	–	–	–	2	–	1	**Ruthenium, chloro(η^5-2,4-cyclopentadien-1-yl)bis(trimethylphosphine)-**	74558-74-0	**NI37021**
225	73	324	227	339	241	226	147	**354**	100	39	34	25	22	20	20	20	**0.00**	11	27	3	2	1	–	–	–	–	2	1	–	–	**Phosphonic acid, [1-(carbonothioylhydrazino)-2-methylpropyl]-, bis(trimethylsilyl) ester**	56051-89-9	**NI37022**
356	293	354	230	228	358	291	178	**354**	100	66	64	62	57	54	51	48		12	3	2	–	–	–	5	–	–	–	–	–	–	**Dibenzo[b,e][1,4]dioxin, 1,2,3,7,8-pentachloro-**	40321-76-4	**NI37024**
356	358	354	293	291	360	230	228	**354**	100	66	62	27	23	22	16	16		12	3	2	–	–	–	5	–	–	–	–	–	–	**Dibenzo[b,e][1,4]dioxin, 1,2,3,7,8-pentachloro-**	40321-76-4	**NI37023**
356	358	354	293	291	360	228	230	**354**	100	64	63	27	23	21	17	16		12	3	2	–	–	–	5	–	–	–	–	–	–	**Dibenzo[b,e][1,4]dioxin, 1,2,4,7,8-pentachloro-**	58802-08-7	**NI37025**
193	195	145	109	354	352	147	75	**354**	100	70	35	33	32	29	28	26		12	6	2	–	1	–	4	–	–	–	–	–	–	**Benzene, 1,2-dichloro-4-[(3,4-dichlorophenyl)sulphonyl]-**		**IC04095**
159	111	227	229	75	356	161	127	**354**	100	87	59	58	54	44	44	44	**32.27**	12	6	2	–	1	–	4	–	–	–	–	–	–	**Benzene, 1,2,3-trichloro-5-[(4-chlorophenyl)sulphonyl]-**	16485-36-2	**NI37026**
111	159	75	127	229	74	50	227	**354**	100	96	80	39	38	37	37	36	**15.31**	12	6	2	–	1	–	4	–	–	–	–	–	–	**Benzene, 1,2,4-trichloro-5-[(4-chlorophenyl)sulphonyl]-**	116-29-0	**NI37027**
111	75	159	50	74	127	229	227	**354**	100	91	88	42	39	38	37	37	**12.10**	12	6	2	–	1	–	4	–	–	–	–	–	–	**Benzene, 1,2,4-trichloro-5-[(4-chlorophenyl)sulphonyl]-**	116-29-0	**NI37028**
354	199	121	56	228	129	185	183	**354**	100	70	59	34	18	12	9	7		12	11	1	–	–	–	–	–	1	–	–	–	1	**Ferrocene, (iodoacetyl)-**	54804-01-2	**NI37029**
43	188	113	312	41	63	141	172	**354**	100	99	51	40	40	35	31	26	**6.56**	12	19	4	–	3	–	–	–	–	–	1	–	–	**Bolstar sulphone**		**NI37030**
73	281	43	207	339	282	45	41	**354**	100	78	54	43	30	21	18	13	**0.00**	12	34	4	–	–	–	–	–	–	4	–	–	–	**Trisiloxane, 3-isopropoxy-1,1,1,5,5,5-hexamethyl-3-(trimethylsilyl)oxy-**		**MS08360**
159	161	160	163	63	62	124	162	**354**	100	67	11	10	9	7	6	5	**1.20**	13	7	1	–	–	–	5	–	–	–	–	–	–	**Benzene, trichloro[(dichlorophenyl)methoxy]-**	74421-51-5	**NI37031**
217	137	215	219	102	139	213	214	**354**	100	69	48	44	26	23	17	14	**13.00**	14	8	–	2	–	–	2	–	–	–	–	–	1	**1,2,4-Selenadiazole, 3,5-bis(4-chlorophenyl)-**		**MS08361**
186	242	121	149	56	354	298	270	**354**	100	68	40	32	30	29	26	22		14	10	4	–	–	–	–	–	–	–	–	–	2	**Iron, di-μ-carbonyldicarbonylbis(η^5-2,4-cyclopentadien-1-yl)di-, (Fe-Fe)-,**		**MS08362**
121	28	186	56	149	177	95	242	**354**	100	78	59	59	23	22	15	11	**3.50**	14	10	4	–	–	–	–	–	–	–	–	–	2	**Iron, di-μ-carbonyldicarbonylbis(η^5-2,4-cyclopentadien-1-yl)di-, (Fe-Fe), cis-**	33221-55-5	**NI37032**
121	28	186	56	149	177	95	242	**354**	100	82	65	56	22	21	15	10	**3.10**	14	10	4	–	–	–	–	–	–	–	–	–	2	**Iron, di-μ-carbonyldicarbonylbis(η^5-2,4-cyclopentadien-1-yl)di-, (Fe-Fe), trans-**	32757-46-3	**NI37033**
101	197	199	283	281	209	163	165	**354**	100	40	39	34	34	18	18	15	**0.00**	14	17	4	–	–	–	3	–	–	–	–	–	–	**Acetic acid, (2,4,5-trichlorophenoxy)-, 2-butoxyethyl ester**	2545-59-7	**NI37034**
254	256	57	85	209	211	181	145	**354**	100	96	92	88	72	68	68	68	**8.00**	14	17	4	–	–	–	3	–	–	–	–	–	–	**Acetic acid, (2,4,5-trichlorophenoxy)-, 2-butoxyethyl ester**	2545-59-7	**NI37035**
57	29	41	56	18	45	27	42	**354**	100	72	50	44	33	25	20	19	**1.10**	14	17	4	–	–	–	3	–	–	–	–	–	–	**Acetic acid, (2,4,5-trichlorophenoxy)-, 2-butoxyethyl ester**	2545-59-7	**NI37036**
57	29	41	56	43	42	27	45	**354**	100	87	85	59	44	36	36	35	**2.90**	14	17	4	–	–	–	3	–	–	–	–	–	–	**Acetic acid, (2,4,5-trichlorophenoxy)-, 2-(isobutoxy)ethyl ester**	62059-40-9	**NI37037**
91	155	65	171	39	183			**354**	100	33	25	18	15	4			**0.00**	15	18	4	2	2	–	–	–	–	–	–	–	–	**Bis(4-tolylsulphonamido)methane**		**LI07806**
171	354	128	222	170	183			**354**	100	57	40	33	22	20				16	10	4	4	1	–	–	–	–	–	–	–	–	**3-Indoleacetonitrile, 2-(2,4-dinitrophenylthio)-**		**LI07807**
354	177	44	18	72	178	28	89	**354**	100	49	33	31	25	22	22	21		16	10	6	4	–	–	–	–	–	–	–	–	–	**3,10-Dinitrodiftalone**		**MS08363**
354	356	355	311	358	357	313	290	**354**	100	66	20	15	13	13	10	10		16	12	5	–	–	–	2	–	–	–	–	–	–	**9H-Xanthen-9-one, 2,7-dichloro-1-hydroxy-3,6-dimethoxy-8-methyl-**	22346-58-3	**NI37038**
147	77	119	91	131	154	115	207	**354**	100	47	31	31	29	27	21	10	**0.00**	16	16	1	–	–	–	–	–	–	–	–	–	1	**Benzofuran, 2,3-dihydro-7-methyl-2-[(phenyltelluro)methyl]-**		**DD01411**
163	180	162	134	43	44	89	29	**354**	100	82	64	26	24	19	18	17	**14.00**	16	18	9	–	–	–	–	–	–	–	–	–	–	**Cyclohexanecarboxylic acid, 3-[[3-(3,4-dihydroxyphenyl)-1-oxo-2-propenyl]oxy]-1,4,5-trihydroxy-, [1S-(1α,3β,4α,5α)]-**	327-97-9	**NI37039**

MASS TO CHARGE RATIOS								M.W.	INTENSITIES								Parent	C	H	O	N	S	F	Cl	Br	I	Si	P	B	X	COMPOUND NAME	CAS Reg No	No
43	109	312	126	81	281	73	69	**354**	100	90	60	53	40	30	30	30	**7.00**	16	18	9	–	–	–	–	–	–	–	–	–	–	β-D-Galactopyranose, 2-O-acetyl-1,6-anhydro-3,4-O-[5-(acetoxymethyl)-2-furfurylidene]-	95881-41-7	NI37040
73	135	264	147	237	75	249	77	**354**	100	94	52	46	24	21	16	14	**4.40**	16	26	5	–	–	–	–	–	–	2	–	–	–	Propanoic acid, 3-(1,3-benzodioxol-5-yl)-2-[(trimethylsilyl)oxy]-, trimethylsilyl ester		MS08364
43	139	168	297	210	181	83	167	**354**	100	29	29	21	15	12	10	10	**0.00**	16	28	7	–	–	–	–	–	–	–	–	2	–	α-D-Glucofuranose cyclic 1,2:3,5-bis(butylboronate)-6-acetate	61501-06-2	NI37041
43	139	168	83	127	181	297	210	**354**	100	88	75	39	35	27	20	20	**0.00**	16	28	7	–	–	–	–	–	–	–	–	2	–	α-D-Glucofuranose cyclic 1,2:3,5-bis(butylboronate)-6-acetate	61501-06-2	NI37042
177	297	87	55	354	298	241	57	**354**	100	66	66	45	36	28	25	25		16	36	–	–	2	–	–	–	–	–	2	–	–	Tetrabutyldiphosphane disulphide		MS08365
254	294	322	268	225	279	255	135	**354**	100	60	56	47	40	39	36	36	**23.02**	17	19	6	–	–	–	1	–	–	–	–	–	–	Spiro[benzofuran-2(3H),1'-cyclohexane]-2',3-dione, 7-chloro-4,4',6-trimethoxy-6'-methyl-	26891-85-0	NI37043
255	354	254	257	256	100	356	69	**354**	100	44	37	33	26	20	16	15		17	19	6	–	–	–	1	–	–	–	–	–	–	Spiro[benzofuran-2(3H),1'-cyclohexane]-3,4'-dione, 7-chloro-2',4,6-trimethoxy-6'-methyl-	17793-64-5	NI37044
144	112	58	90	145	164	322	132	**354**	100	77	50	33	30	14	13	13	**3.00**	17	22	8	–	–	–	–	–	–	–	–	–	–	3,6-Methanobenzofuran-4,5,8-tricarboxylic acid, octahydro-6,7a-dimethyl 2-oxo-, trimethyl ester	32251-36-8	MS08366
43	179	41	29	151	98	95	79	**354**	100	72	61	59	58	58	58	58	**19.13**	17	22	8	–	–	–	–	–	–	–	–	–	–	Trichothec-9-en-8-one, 4-(acetyloxy)-12,13-epoxy-3,7,15-trihydroxy-, (3α,4β,7β)-	23255-69-8	NI37045
43	67	109	81	41	79	53	40	**354**	100	83	52	40	34	32	25	25	**2.47**	17	23	3	–	–	–	–	1	–	–	–	–	–	2H-Oxocin-2-methanol, 7-bromo-8-ethyl-3,6,7,8-tetrahydro-α-2-penten-4-ynyl-, acetate, [2R-[2α[R*(E)],7β,8α]]-	3442-58-8	NI37046
43	67	109	81	41	79	40	53	**354**	100	80	52	39	33	26	25	13	**2.00**	17	23	3	–	–	–	–	1	–	–	–	–	–	2H-Oxocin-2-methanol, 7-bromo-8-ethyl-3,6,7,8-tetrahydro-α-2-penten-4-ynyl-, acetate, [2R-[2α[R*(E)],7β,8α]]-		LI07808
354	86	196	157	224	182	236	164	**354**	100	95	47	44	30	26	24	24		17	26	2	2	2	–	–	–	–	–	–	–	–	Spiro[3.3]heptane, 2,6-bis(morpholinothiocarbonyl)-		LI07809
179	354	73	267	355	180	268	45	**354**	100	86	85	56	26	18	14	14		17	30	4	–	–	–	–	–	–	2	–	–	–	Phenylpropanoic acid, 3,4-bis[(trimethylsilyl)oxy]-, ethyl ester		MS08367
179	207	73	77	180	75	281	264	**354**	100	44	36	16	15	15	12	11	**4.40**	17	30	4	–	–	–	–	–	–	2	–	–	–	Phenylpropanoic acid, α,4-bis[(trimethylsilyl)oxy]-, ethyl ester		MS08368
267	73	75	354	268	339	269	147	**354**	100	81	39	29	29	13	13	13		17	30	4	–	–	–	–	–	–	2	–	–	–	Phenylpropanoic acid, β,3-bis[(trimethylsilyl)oxy]-, ethyl ester		MS08369
243	168	354	319	139	356	244	186	**354**	100	53	47	26	18	17	14	13		18	12	1	–	–	–	1	–	–	–	–	–	1	10H-Phenoxarsine, 10-(4-chlorophenyl)-	27796-61-8	NI37047
177	149	233	60	131	234	219	162	**354**	100	55	54	53	51	40	39	34	**7.80**	18	14	2	2	2	–	–	–	–	–	–	–	–	2H-3,1-Benzothiazin-2-one, 1,4-dihydro-4-[(4-methyl-2-oxo-1,4-dihydro-2H-3,1-benzothiazin-4-yl)methylene]-		MS08370
167	261	354	65	77	140	94	51	**354**	100	95	58	35	31	25	24	17		18	15	3	–	–	–	–	–	–	–	–	–	1	Arsenous acid, triphenyl ester	1529-86-8	NI37048
93	292	77	92	353	65	291	91	**354**	100	81	47	41	40	21	20	16	**13.75**	18	18	4	4	–	–	–	–	–	–	–	–	–	1H-Pyrazole-4,5-dione, 1-phenyl-3-(1,2,3-trihydroxypropyl)-, 4-(phenylhydrazone), [S-(R*,R*)]-	55191-37-2	NI37049
163	148	92	120	164	149	102	354	**354**	100	52	39	36	32	22	19	15		18	18	4	4	–	–	–	–	–	–	–	–	–	1,2,4,5-Tetrazine, 3,6-bis(3,4-dimethoxyphenyl)-	77355-32-9	NI37050
147	69	57	55	238	83	71	43	**354**	100	32	29	28	26	23	23	23	**0.00**	18	19	–	–	1	4	–	–	–	–	–	1	–	1-Ethyl-2-phenethylbenzo[b]thiophenium tetrafluoroborate		LI07810
181	354	29	182	235	355	77	148	**354**	100	44	35	21	15	9	9	8		18	26	7	–	–	–	–	–	–	–	–	–	–	Benzene, 1-(2,2-dicarboethoxypropyl)-3,4,5-trimethoxy-		NI37051
283	285	204	205	275	187	233	121	**354**	100	99	92	62	40	29	21	19	**6.67**	18	27	2	–	–	–	–	1	–	–	–	–	–	5,9-Methanobenzocycloocten-4(1H)-one, 10-bromo-2,3,5,6,7,8,9,10-octahydro-5-hydroxy-2,2,7,7,9-pentamethyl-	6244-20-8	NI37052
191	163	206	89	67	65	63	148	**354**	100	90	85	59	59	55	54	36	**35.00**	19	14	7	–	–	–	–	–	–	–	–	–	–	4H-Furo[2,3-h]-1-benzopyran-4-one, 5-hydroxy-6-methoxy-2-(3,4-methylenedioxyphenyl)-		NI37053
354	355	339	325	265	353	279	336	**354**	100	20	13	10	8	7	7	6		19	14	7	–	–	–	–	–	–	–	–	–	–	7H-Furo[3',2':4,5]furo[2,3-c]xanthen-7-one, 3a,12c-dihydro-12c-hydroxy-6,8-dimethoxy-	20797-80-2	NI37054
354	44	339	338	310	309	355	325	**354**	100	52	38	38	38	36	33	25		19	14	7	–	–	–	–	–	–	–	–	–	–	7H-Furo[3',2':4,5]furo[2,3-c]xanthen-7-one, 3a,12c-dihydro-12c-hydroxy-6,8-dimethoxy-	20797-80-2	NI37055
339	354	325	295	355	340	321	155	**354**	100	89	30	29	21	21	18	18		19	14	7	–	–	–	–	–	–	–	–	–	–	7H-Furo[3',2':4,5]furo[2,3-c]xanthen-7-one, 3a,12c-dihydro-8-hydroxy-6,11-dimethoxy-, (3aR-cis)-	22897-08-1	NI37056
275	274	245	276	260	231	217	82	**354**	100	82	21	20	18	13	13	13	**0.00**	19	17	1	1	–	–	–	1	–	–	–	–	–	Dibenzo[a,f]quinolizinium bromide, 6,7-dihydro-2-methoxy-13-methyl-		NI37057
354	235	339	233	217	220	355	218	**354**	100	92	52	42	38	36	24	22		19	18	5	2	–	–	–	–	–	–	–	–	–	2-Quinolinecarboxamide, 4-hydroxy-5,6,8-trimethoxy-N-phenyl-	55191-29-2	NI37058
354	73	104	268	325	355	282	77	**354**	100	50	43	38	37	27	27	24		19	22	3	2	–	–	–	–	–	1	–	–	–	2,4-Imidazolidinedione, 3-methyl-5-phenyl-5-[3-[(trimethylsilyl)oxy]phenyl]-	55191-27-0	NI37059
354	325	277	268	104	73	269	355	**354**	100	85	85	65	65	60	30	26		19	22	3	2	–	–	–	–	–	1	–	–	–	2,4-Imidazolidinedione, 3-methyl-5-phenyl-5-[4-[(trimethylsilyl)oxy]phenyl]-	55191-28-1	NI37060
294	43	266	156	224	123	41	353	**354**	100	98	86	59	51	49	41	39	**0.00**	19	30	2	–	2	–	–	–	–	–	–	–	–	Spiro[2H-cyclopenta[a]pentalene-2,2'-[1,3]dithiolan]-7-ol, decahydro-3,4,4,6a-tetramethyl-, acetate	57984-02-8	NI37061
149	135	150	161	354	122	203	178	**354**	100	46	41	37	29	28	21	14		20	18	6	–	–	–	–	–	–	–	–	–	–	Asarinin, (+)-	133-03-9	MS08371
149	354	150	135	160	203	122	131	**354**	100	51	34	20	19	14	13	11		20	18	6	–	–	–	–	–	–	–	–	–	–	Asarinin, (+)-	133-03-9	NI37062
149	135	150	161	122	131	354	203	**354**	100	60	40	33	30	26	23	23		20	18	6	–	–	–	–	–	–	–	–	–	–	Asarinin, (+)-	133-03-9	LI07811
317	318	336	321	354	319	339	311	**354**	100	89	38	38	25	25	21	17		20	18	6	–	–	–	–	–	–	–	–	–	–	6H-Benzofuro[3,2-c]furo[3,2-g][1]benzopyran-6a,9(11aH)-diol, 2-(1-hydroxy-1-methylethyl)-, (6aS-cis)-	78873-52-6	NI37063

MASS TO CHARGE RATIOS								M.W.	INTENSITIES								Parent	C	H	O	N	S	F	Cl	Br	I	Si	P	B	X	COMPOUND NAME	CAS Reg No	No
137	354	298	281	299	189	138	311	**354**	100	51	34	11	10	9	9	8		20	18	6	–	–	–	–	–	–	–	–	–	–	4H-1-Benzopyran-4-one, 3-(3,4-dihydro-6,8-dihydroxy-2,2-dimethyl-2H-1-benzopyran-5-yl)-7-hydroxy-		MS08372
321	43	283	284	336	55	322	295	**354**	100	78	63	48	36	28	22	22	**4.48**	20	18	6	–	–	–	–	–	–	–	–	–	–	4H-1-Benzopyran-4-one, 5,7-dihydroxy-3-(4-hydroxyphenyl)-6-[3-(hydroxymethyl)-2-butenyl]-		NI37064
354	137	326	124	151	336	55	165	**354**	100	56	42	38	29	28	24	23		20	18	6	–	–	–	–	–	–	–	–	–	–	α,3′-Bicinnamaldehyde, 4,4′-dihydroxy-3,5′-dimethoxy-	32666-21-0	NI37065
137	83	354	202	85	124	230	265	**354**	100	44	41	27	27	23	20	19		20	18	6	–	–	–	–	–	–	–	–	–	–	Butanedial, bis[(3-hydroxy-4-methoxyphenyl)methylene]-	56247-62-2	NI37066
354	137	326	124	151	336	152	355	**354**	100	54	40	36	30	27	27	22		20	18	6	–	–	–	–	–	–	–	–	–	–	Cinnamaldehyde, α-[5-(2-formylvinyl)-2-hydroxy-3-methoxyphenyl]-4-hydroxy-3-methoxy-		LI07812
354	177	89	176	297	137	164	77	**354**	100	49	45	43	37	37	36	36		20	18	6	–	–	–	–	–	–	–	–	–	–	Cinnamaldehyde, α-[4-(2-formylvinyl)-2-methoxyphenoxy]-4-hydroxy-3-methoxy-	32022-24-5	NI37067
354	177	89	51	297	164	137	77	**354**	100	47	45	37	36	36	36	36		20	18	6	–	–	–	–	–	–	–	–	–	–	Cinnamaldehyde, α-[4-(2-formylvinyl)-2-methoxyphenoxy]-4-hydroxy-3-methoxy-	32022-24-5	LI07813
135	77	136	354	50	51	105	43	**354**	100	22	20	16	16	12	10	7		20	18	6	–	–	–	–	–	–	–	–	–	–	2(3H)-Furanone, 3,4-bis(1,3-benzodioxol-5-ylmethyl)dihydro-, (3R-trans)-	26543-89-5	NI37068
299	354	311	298	153	300	312	147	**354**	100	99	70	44	29	20	14	11		20	18	6	–	–	–	–	–	–	–	–	–	–	Isoflavone, 5,7,2′,4′-tetrahydroxy-3′-(3,3-dimethylallyl)-		NI37069
57	121	43	55	354	266	253	152	**354**	100	43	41	32	25	24	22	22		20	18	6	–	–	–	–	–	–	–	–	–	–	5,12-Naphthacenedione, 8-ethyl-7,8,9,10-tetrahydro-1,6,8,11-tetrahydroxy	4877-81-0	NI37070
279	254	307	336	354	318	280	282	**354**	100	57	50	46	45	32	30	14		20	18	6	–	–	–	–	–	–	–	–	–	–	5,12-Naphthacenedione, 8β-ethyl-1,6,7β,8α-tetrahydroxy-7,8,9,10-tetrahydro-	79036-55-8	NI37071
354	228	77	262	356	326	355	51	**354**	100	84	63	43	40	34	30	23		20	19	–	2	1	–	1	–	–	–	–	–	–	1H-Cyclopenta[e][1,4]thiazepine, 2,3,5,6,7,8-hexahydro-1-phenyl-5-(4-chlorophenylimino)-		NI37072
171	354	170	172	168	91	120	154	**354**	100	19	15	14	8	8	7	6		20	22	2	2	1	–	–	–	–	–	–	–	–	1-Naphthalenesulphonamide, 5-(dimethylamino)-N-methyl-N-benzyl-	55837-11-1	NI37073
171	354	170	172	168	91	355	169	**354**	100	21	15	14	9	9	6	4		20	22	2	2	1	–	–	–	–	–	–	–	–	1-Naphthalenesulphonamide, 5-(dimethylamino)-N-methyl-N-benzyl-	55837-11-1	NI37074
171	170	354	172	168	105	169	355	**354**	100	16	15	15	10	9	8	5		20	22	2	2	1	–	–	–	–	–	–	–	–	1-Naphthalenesulphonamide, 5-(dimethylamino)-N-(1-phenylethyl)-	55837-10-0	NI37075
171	170	354	172	168	105	169	154	**354**	100	15	14	13	11	8	7	6		20	22	2	2	1	–	–	–	–	–	–	–	–	1-Naphthalenesulphonamide, 5-(dimethylamino)-N-(1-phenylethyl)-	55837-10-0	NI37076
170	171	354	168	91	355	263	172	**354**	100	81	45	16	11	9	9	9		20	22	2	2	1	–	–	–	–	–	–	–	–	1-Naphthalenesulphonamide, 5-(dimethylamino)-N-(2-phenylethyl)-	5282-81-5	NI37077
170	171	354	168	91	154	234	127	**354**	100	71	43	15	12	11	10	9		20	22	2	2	1	–	–	–	–	–	–	–	–	1-Naphthalenesulphonamide, 5-(dimethylamino)-N-(2-phenylethyl)-	5282-81-5	NI37078
170	172	352	354	205	185	262	330	**354**	100	64	39	6	6	6	5	4		20	22	2	2	1	–	–	–	–	–	–	–	–	1-Naphthalenesulphonamide, 5-(dimethylamino)-N-(2-phenylethyl)-	5282-81-5	LI07814
254	339	213	151	58	184	281	146	**354**	100	95	66	33	29	25	23	15	**7.90**	20	22	2	2	1	–	–	–	–	–	–	–	–	1,1′-Spirobi[1,2-benzisothiazole]-3,3′(2H,2′H)-dione, 2,2′-diisopropyl-	63744-02-5	NI37079
69	167	283	226	129	209	71	194	**354**	100	72	70	66	64	55	52	47	**23.19**	20	22	4	2	–	–	–	–	–	–	–	–	–	Curan-17-oic acid, 2,16-didehydro-20-hydroxy-19-oxo-, methyl ester	56053-15-7	NI37080
306	246	295	354	323	307	107	206	**354**	100	66	65	45	42	40	36	27		20	22	4	2	–	–	–	–	–	–	–	–	–	6,21-Cyclo-4,5-secoakuammilan-17-oic acid, 4,5-dihydroxy-, methyl ester, (6α)-	63944-63-8	NI37081
205	297	354	298	206	57	41	121	**354**	100	77	54	20	18	18	18	16		20	26	2	–	–	–	–	–	–	–	–	–	1	Ferrocene, 1,1′-dipentanoyl-		IC04096
171	113	129	73	105	79	297	143	**354**	100	90	89	51	40	31	24	23	**0.00**	20	38	3	–	–	–	–	–	–	1	–	–	–	1,7-Dioxaspiro[5.5]undecane, 2-allyl-8,9-dimethyl-4-(dimethyl-tert-butylsiloxy)-, (2R,4S,6R,8R,9S)-		MS08373
91	43	185	152	157	203	354	294	**354**	100	96	46	45	33	24	23	5		21	22	3	–	1	–	–	–	–	–	–	–	–	Cyclohexanone, 2-(acetyloxy)-6-benzyl-2-(phenylthio)-	63608-53-7	NI37082
354	298	283	339	123	297	311	213	**354**	100	50	49	43	30	23	19	19		21	22	5	–	–	–	–	–	–	–	–	–	–	6H-Benzofuro[3,2-c][1]benzopyran-3,9-diol, 6a,11a-dihydro-7-(3,3-dimethylallyl)-8-methoxy-		MS08374
354	179	177	234	219	191	311	205	**354**	100	96	56	40	38	35	34	31		21	22	5	–	–	–	–	–	–	–	–	–	–	4H-1-Benzopyran-4-one, 2,3-dihydro-7-hydroxy-2-(4-hydroxyphenyl)-5-methoxy-8-(3-methyl-2-butenyl)-	521-48-2	NI37083
174	354	228	175	213	159	355	212	**354**	100	81	73	63	23	23	19	19		21	26	3	2	–	–	–	–	–	–	–	–	–	Ajmalan-17-one, 21-hydroxy-12-methoxy-, (21α)-	639-28-1	NI37084
138	326	310	160	139	354	309	110	**354**	100	28	28	17	11	10	9	8		21	26	3	2	–	–	–	–	–	–	–	–	–	Aspidoalbidine, 1-formyl-17-methoxy-	2671-44-5	NI37085
294	295	251	293	43	160	130	112	**354**	100	26	20	16	16	14	9	8	**0.00**	21	26	3	2	–	–	–	–	–	–	–	–	–	Aspidodispermine, 7-deoxy-, acetate (ester)	21451-70-7	NI37086
125	41	140	42	268	296	278	158	**354**	100	62	55	52	47	38	34	33	**0.00**	21	26	3	2	–	–	–	–	–	–	–	–	–	Aspidofractinine-3-carboxylic acid, 6-hydroxy-, methyl ester, (2α,3β,5α,6β)-	55724-58-8	NI37087
83	122	140	44	47	85	84	50	**354**	100	90	87	71	68	64	57	40	**7.21**	21	26	3	2	–	–	–	–	–	–	–	–	–	Aspidospermidine-3-carboxylic acid, 2,3-didehydro-20-hydroxy-, methyl ester, (5α,12β,19α,20R)-	6801-25-8	NI37088
138	82	55	77	83	57	94	81	**354**	100	56	39	31	28	27	25	22	**10.00**	21	26	3	2	–	–	–	–	–	–	–	–	–	Aspidospermidin-17-ol, 1-acetyl-19,21-epoxy-	2671-46-7	NI37089
251	169	354	353	252	171	249	170	**354**	100	62	43	36	35	34	33	33		21	26	3	2	–	–	–	–	–	–	–	–	–	Corynan-16-carboxylic acid, 19,20-didehydro-17-hydroxy-, methyl ester, (16R,19Z)-		NI37090
251	169	249	354	353	170	171	115	**354**	100	61	44	40	39	33	31	31		21	26	3	2	–	–	–	–	–	–	–	–	–	Corynan-16-carboxylic acid, 19,20-didehydro-17-hydroxy-, methyl ester, (16S,19Z)-		NI37091
144	336	323	182	130	143	291	293	**354**	100	90	62	48	43	33	29	28	**8.00**	21	26	3	2	–	–	–	–	–	–	–	–	–	Curan-17,18-diol, 1-acetyl-19,20-didehydro-, (19E)-	13013-60-0	NI37092
144	354	295	224	143	130	164	296	**354**	100	99	99	98	51	46	36	30		21	26	3	2	–	–	–	–	–	–	–	–	–	Curan-17,18-diol, 19,20-didehydro-, 17-acetate, (19E)-	59630-33-0	NI37093
300	306	71	262	276	200	77	107	**354**	100	99	68	58	54	51	51	50	**0.00**	21	26	3	2	–	–	–	–	–	–	–	–	–	1,16-Cyclocorynan-16-methanol, 19,20-didehydro-17-hydroxy-10-methoxy, (19E)-	55724-51-1	NI37094
354	171	309	355	144	172	168	204	**354**	100	84	27	26	15	14	9	8		21	26	3	2	–	–	–	–	–	–	–	–	–	Deformylcorynine		LI07815
252	354	353	307	224	253	295	167	**354**	100	89	61	32	30	27	26	22		21	26	3	2	–	–	–	–	–	–	–	–	–	14-Epivincamine	6835-99-0	NI37095

MASS TO CHARGE RATIOS								M.W.	INTENSITIES								Parent	C	H	O	N	S	F	Cl	Br	I	Si	P	B	X	COMPOUND NAME	CAS Reg No	No
252	354	353	307	237	224	253	295	**354**	100	89	60	31	30	30	27	25		21	26	3	2	–	–	–	–	–	–	–	–	–	14-Epivincamine	6835-99-0	MS08375
252	354	307	224	295	237	267	265	**354**	100	90	30	30	26	24	8	6		21	26	3	2	–	–	–	–	–	–	–	–	–	14-Epivincamine	6835-99-0	LI07816
71	259	297	354	84	227	224	210	**354**	100	99	40	35	25	21	19	18		21	26	3	2	–	–	–	–	–	–	–	–	–	3,5-Ethano-3H-pyrrolo[2,3-d]carbazole-6-carboxylic acid, 4-ethyl-1,2,3a,4,5,7-hexahydro-11-methoxy-, methyl ester, stereoisomer	2122-32-9	NI37096
323	110	293	354	166	324	82	70	**354**	100	50	33	25	25	23	17	14		21	26	3	2	–	–	–	–	–	–	–	–	–	Gelsemine, 18,19-dihydro-1-methoxy-	41478-35-7	NI37097
109	310	354	311	355	353	110	140	**354**	100	62	49	15	12	11	11	9		21	26	3	2	–	–	–	–	–	–	–	–	–	3-Isokopsinine, 14-hydroxy-, (14α)-		LI07817
354	156	169	43	83	57	108	130	**354**	100	88	86	76	70	64	61	60		21	26	3	2	–	–	–	–	–	–	–	–	–	4-Piperidineacetic acid, 5-ethylidene-2-[3-(2-hydroxyethyl)-1H-indol-2-yl] α-methylene-, methyl ester, [2S-(2α,4α,5E)]-	3668-17-5	NI37098
158	354	144	145	143	157	153	115	**354**	100	95	95	60	60	45	40	35		21	26	3	2	–	–	–	–	–	–	–	–	–	3,4-Secoajmalicine		LI07818
123	354	122	108	124	355	224	81	**354**	100	48	20	17	15	11	10	8		21	26	3	2	–	–	–	–	–	–	–	–	–	3,7-Secocuran-16-carboxylic acid, 2,7,19,20-tetradehydro-17-hydroxy-, methyl ester, (19E)-	10012-73-4	NI37099
247	248	232	235	249	137	116	336	**354**	100	64	59	54	46	39	38	37	**3.45**	21	26	3	2	–	–	–	–	–	–	–	–	–	17,18-Secoyohimban-16-carboxylic acid, 19,20-didehydro-17-hydroxy-, methyl ester, (16S,19E)-	6519-27-3	NI37100
41	43	184	57	322	321	42	56	**354**	100	77	67	61	57	47	45	32	**10.00**	21	26	3	2	–	–	–	–	–	–	–	–	–	18,19-Secoyohimban-19-oic acid, 16,17-didehydro-16-(hydroxymethyl)-, methyl ester, (15β,16Z,20ξ)-	5523-48-8	NI37101
354	252	267	224	295	237	307	284	**354**	100	70	50	48	40	28	16	12		21	26	3	2	–	–	–	–	–	–	–	–	–	Vincamine		LI07819
354	252	267	224	353	42	295	167	**354**	100	70	52	47	44	42	38	32		21	26	3	2	–	–	–	–	–	–	–	–	–	Vincamine		MS08376
182	42	322	183	354	164	122	152	**354**	100	19	15	15	14	13	13	12		21	26	3	2	–	–	–	–	–	–	–	–	–	Vobasan-17-oic acid, 19,20-dihydro-3-oxo-, methyl ester, (20α)-	2299-26-5	MS08377
124	296	138	122	125	98	94	123	**354**	100	32	29	15	12	10	10	9	**0.00**	21	26	3	2	–	–	–	–	–	–	–	–	–	Vobasan-17-oic acid, 19,20-dihydro-3-oxo-, methyl ester, (20β)-	2134-98-7	NI37102
182	122	180	354	181	130	179	174	**354**	100	74	59	45	21	17	16	16		21	26	3	2	–	–	–	–	–	–	–	–	–	Vobasan-17-oic acid, 3-hydroxy-, methyl ester, (3β)-	7168-77-6	NI37103
182	122	180	354	181	130	194	174	**354**	100	74	60	45	21	18	17	17		21	26	3	2	–	–	–	–	–	–	–	–	–	Vobasan-17-oic acid, 3-hydroxy-, methyl ester, (3β)-	7168-77-6	NI37104
281	322	156	282	354	221	144	321	**354**	100	25	22	21	18	18	15	11		21	26	3	2	–	–	–	–	–	–	–	–	–	18,19-Seco-15β-yohimban-19-oic acid, 20,21-didehydro-16β-(hydroxymethyl)-, methyl ester	5552-25-0	NI37105
355	61	63	356	337	353	354	257	**354**	100	76	46	24	24	22	18	15		21	26	3	2	–	–	–	–	–	–	–	–	–	Yohimbine	146-48-5	NI37106
354	353	43	355	47	83	169	57	**354**	100	80	24	23	22	21	19	19		21	26	3	2	–	–	–	–	–	–	–	–	–	Yohimbine	146-48-5	NI37107
353	354	169	170	41	156	355	184	**354**	100	91	38	28	24	23	20	20		21	26	3	2	–	–	–	–	–	–	–	–	–	Yohimbine	146-48-5	NI37108
354	353	169	170	156	355	184	41	**354**	100	99	50	32	32	25	25	24		21	26	3	2	–	–	–	–	–	–	–	–	–	Yohimbine, β-	549-84-8	NI37109
43	316	91	107	105	79	95	298	**354**	100	38	26	24	24	23	22	21	**0.00**	21	32	2	–	–	2	–	–	–	–	–	–	–	Pregnan-20-one, 5,6-difluoro-3-hydroxy-, (3β)-	56193-66-9	NI37110
115	213	187	309	200	354	85	69	**354**	100	49	21	15	15	13	13	11		21	38	4	–	–	–	–	–	–	–	–	–	–	2-Cyclopentadecen-1-one, 1,4-di-O-methyl-D-threitol ketal, (E)-		NI37111
115	213	187	200	85	69	354	81	**354**	100	38	20	14	14	12	11	10		21	38	4	–	–	–	–	–	–	–	–	–	–	2-Cyclopentadecen-1-one, 1,4-di-O-methyl-D-threitol ketal, (Z)-		NI37112
67	81	55	41	95	262	79	69	**354**	100	82	73	58	56	55	42	40	**3.00**	21	38	4	–	–	–	–	–	–	–	–	–	–	9,12-Octadecadienoic acid (Z,Z)-, 2,3-dihydroxypropyl ester	2277-28-3	WI01457
262	280	354	263	278	352	281	123	**354**	100	41	32	22	17	10	10	9		21	38	4	–	–	–	–	–	–	–	–	–	–	9,12-Octadecadienoic acid (Z,Z)-, 2,3-dihydroxypropyl ester	2277-28-3	WI01458
67	81	55	41	95	79	262	69	**354**	100	84	78	69	49	42	39	39	**1.00**	21	38	4	–	–	–	–	–	–	–	–	–	–	9,12-Octadecadienoic acid (Z,Z)-, 2-hydroxy-1-(hydroxymethyl)ethyl ester	3443-82-1	WI01459
262	354	280	263	355	336	264	281	**354**	100	34	25	21	9	9	6	5		21	38	4	–	–	–	–	–	–	–	–	–	–	9,12-Octadecadienoic acid (Z,Z)-, 2-hydroxy-1-(hydroxymethyl)ethyl ester	3443-82-1	WI01460
74	87	55	43	41	143	75	57	**354**	100	70	29	25	24	16	15	15	**0.00**	21	38	4	–	–	–	–	–	–	–	–	–	–	9-Octadecenoic acid, 12-(acetyloxy)-, methyl ester, [R-(Z)]-	140-03-4	NI37113
73	117	75	55	43	41	129	28	**354**	100	87	81	45	45	42	37	33	**6.00**	21	42	2	–	–	–	–	–	–	1	–	–	–	6-Octadecenoic acid, trimethylsilyl ester, (Z)-	96851-53-5	NI37114
73	75	117	55	41	129	43	28	**354**	100	87	71	52	44	42	40	34	**4.30**	21	42	2	–	–	–	–	–	–	1	–	–	–	9-Octadecenoic acid, trimethylsilyl ester, (E)-	96851-47-7	NI37115
73	297	267	312	223	217	253	282	**354**	100	53	43	40	33	30	24	23	**0.00**	21	42	2	–	–	–	–	–	–	1	–	–	–	9-Octadecenoic acid, trimethylsilyl ester, (Z)-	21556-26-3	NI37116
73	75	117	55	129	41	339	145	**354**	100	99	63	58	47	42	31	31	**6.50**	21	42	2	–	–	–	–	–	–	1	–	–	–	9-Octadecenoic acid, trimethylsilyl ester, (Z)-	21556-26-3	NI37117
73	75	117	55	129	339	41	43	**354**	100	88	63	55	45	44	40	31	**6.92**	21	42	2	–	–	–	–	–	–	1	–	–	–	9-Octadecenoic acid, trimethylsilyl ester, (Z)-	21556-26-3	NI37118
73	75	55	339	117	129	69	145	**354**	100	88	54	47	47	45	29	28	**8.75**	21	42	2	–	–	–	–	–	–	1	–	–	–	11-Octadecenoic acid, trimethylsilyl ester, (E)-		NI37119
73	75	55	28	117	41	129	339	**354**	100	95	61	52	50	47	45	40	**5.63**	21	42	2	–	–	–	–	–	–	1	–	–	–	11-Octadecenoic acid, trimethylsilyl ester, (Z)-		NI37120
354	91	194	221	355	77	89	223	**354**	100	45	40	30	27	13	12	10		22	18	1	4	–	–	–	–	–	–	–	–	–	2-Pyrazolin-5-one, 4-[[4-(methylamino)phenyl]imino]-1,3-diphenyl-	13617-71-5	NI37121
91	354	307	267	322	279			**354**	100	61	61	45	42	39				22	26	4	–	–	–	–	–	–	–	–	–	–	Benzoic acid, 2-hydroxy-4-methoxy-3-(3-methylbut-2-enyl)-6-(2-phenylethyl)-, methyl ester		NI37122
205	163	218	354	299	177	267	283	**354**	100	99	95	55	15	15	8	5		22	26	4	–	–	–	–	–	–	–	–	–	–	[3,6'-Bi-2H-1-Benzopyran]-7,7'-diol, 3,3',4,4'-tetrahydro-7,7'-dimethoxy-2',2'-dimethyl-	41347-52-8	MS08378
354	297	241	57	31	29	355	298	**354**	100	88	41	40	37	31	24	21		22	26	4	–	–	–	–	–	–	–	–	–	–	4,4'-Diglycidyl-3,3',5,5'-tetramethylbiphenyl		IC04097
312	158	354	313	252	170	157	237	**354**	100	28	27	23	21	15	15	14		22	26	4	–	–	–	–	–	–	–	–	–	–	Estra-1,3,5(10),6-tetraene-3,17-diol, diacetate, (17β)-	1971-65-9	NI37123
43	252	312	237	224	210	354	223	**354**	100	94	90	50	34	33	32	32		22	26	4	–	–	–	–	–	–	–	–	–	–	Estra-1,3,5(10),9(11)-tetraene-3,17-diol, diacetate, (17β)-	1169-54-6	NI37124
270	312	255	271	160	159	313	173	**354**	100	34	30	19	9	9	8	7	**6.59**	22	26	4	–	–	–	–	–	–	–	–	–	–	Estra-1,3,5(10),16-tetraene-3,17-diol, diacetate	20592-42-1	NI37125
124	354	125	326	353	152	160	122	**354**	100	14	10	7	6	6	4	3		22	30	2	2	–	–	–	–	–	–	–	–	–	Aspidospermidine, 1-acetyl-17-methoxy-	466-49-9	NI37126
124	354	326	353	152	355	311	160	**354**	100	88	42	34	24	22	22	16		22	30	2	2	–	–	–	–	–	–	–	–	–	Aspidospermidine, 1-acetyl-17-methoxy-	466-49-9	NI37127
354	178	122	83	176	177	69	85	**354**	100	94	77	66	59	53	53	44		22	30	2	2	–	–	–	–	–	–	–	–	–	2,3,11,12-Dibenzo-1,13-dioxa-4,10-diazacyclooctadeca-2,11,diene		MS08379

MASS TO CHARGE RATIOS								M.W.	INTENSITIES								Parent	C	H	O	N	S	F	Cl	Br	I	Si	P	B	X	COMPOUND NAME	CAS Reg No	No
58	354	296	126	309	266	227	96	**354**	100	28	27	19	14	14	9	9		22	30	2	2	–	–	–	–	–	–	–	–	–	Schizozygine, 14-deoxotetrahydro-N,N-dimethyl-	14937-07-6	NI37128
110	109	354	91	79	111	41	124	**354**	100	28	19	13	12	10	10	8		22	31	3	–	–	–	–	–	–	–	–	1	–	Pregn-4-ene-3,20-dione, 21-hydroxy-, methyl boronate		MS08380
110	109	354	111	124	105	355	353	**354**	100	28	19	10	8	8	5	5		22	31	3	–	–	–	–	–	–	–	–	1	–	Pregn-4-ene-3,20-dione, 21-hydroxy-, methyl boronate		LI07820
43	54	58	41	74	98	69	71	**354**	100	46	43	31	30	27	27	26	**11.07**	22	42	3	–	–	–	–	–	–	–	–	–	–	Heneicosanoic acid, 20-oxo-, methyl ester	56247-70-2	NI37129
354	339	355	340	356	226	76	163	**354**	100	56	27	15	13	13	9	8		23	14	4	–	–	–	–	–	–	–	–	–	–	Dinaphth[1,2-b:2',3'-f]oxepin-9,14-dione, 5-methoxy-	84101-93-9	NI37130
353	354	77	178	177	176	89	283	**354**	100	68	15	13	7	7	7	5		23	18	2	2	–	–	–	–	–	–	–	–	–	2H-Pyridazin-3-one, 2-methyl-4-phenoxy-5,6-diphenyl-	79369-61-2	NI37131
150	249	354	149	353	277	355		**354**	100	64	55	42	40	34	15			23	18	2	2	–	–	–	–	–	–	–	–	–	2(1H)-Pyrimidinone, 4-benzoyl-1,2,3,4-tetrahydro-3,5-diphenyl-		LI07821
354	77	118	104	353	180	355	194	**354**	100	44	42	40	39	34	25	24		23	18	2	2	–	–	–	–	–	–	–	–	–	2(1H)-Pyrimidinone, 4-(4-tolyl)-5-phenoxy-6-phenyl-	42919-55-1	NI37132
213	214	43	137	41	105	55	57	**354**	100	61	52	45	42	40	32	29	**2.90**	23	30	3	–	–	–	–	–	–	–	–	–	–	Benzophenone, 4-(decyloxy)-2-hydroxy-		IC04098
354	237	297	252	279	339	312	294	**354**	100	59	43	28	25	6	6	4		23	30	3	–	–	–	–	–	–	–	–	–	–	B-Bishomo-4,6-cyclo-5,7b-cycloandrosta-4,7-diene-3-one, (4β,5α,6α,7α)-		LI07822
339	340	354	309	268	325	294	284	**354**	100	23	13	6	6	5	3	3		23	30	3	–	–	–	–	–	–	–	–	–	–	Cannabinol, 7-methoxy-O-methyl-		LI07823
279	294	295	174	280	171	121	172	**354**	100	90	22	21	20	15	14	12	**0.14**	23	30	3	–	–	–	–	–	–	–	–	–	–	D-Homoestra-1,3,5(10),8-tetraene, 17a-acetoxy-3-methoxy-17a-methyl-, (17α,17β)-, (E,E)-		LI07824
294	279	354	172	295	173	171	186	**354**	100	58	32	32	28	28	26	20		23	30	3	–	–	–	–	–	–	–	–	–	–	D-Homoestra-1,3,5(10),8-tetraene, 17a-acetoxy-3-methoxy-17a-methyl-, (17α,17β)-, (E,E)-		LI07825
339	59	97	83	340	43	297	57	**354**	100	75	30	30	25	25	23	20	**0.00**	23	46	2	–	–	–	–	–	–	–	–	–	–	1,3-Dioxolane, 4-butyl-2,2-dimethyl-5-tetradecyl-, cis-	55622-60-1	NI37133
339	86	43	340	59	58	97	83	**354**	100	47	28	26	25	23	21	17	**0.00**	23	46	2	–	–	–	–	–	–	–	–	–	–	1,3-Dioxolane, 4-heptadecyl-2,2,5-trimethyl-, cis-	55622-59-8	NI37134
74	87	75	43	57	55	143	69	**354**	100	98	89	71	65	53	45	45	**4.84**	23	46	2	–	–	–	–	–	–	–	–	–	–	Docosanoic acid, methyl ester	929-77-1	NI37135
74	87	43	41	55	57	75	69	**354**	100	71	71	38	32	28	26	18	**4.00**	23	46	2	–	–	–	–	–	–	–	–	–	–	Docosanoic acid, methyl ester	929-77-1	NI37136
74	87	354	43	75	57	55	143	**354**	100	71	45	45	36	30	28	22		23	46	2	–	–	–	–	–	–	–	–	–	–	Docosanoic acid, methyl ester	929-77-1	MS08381
97	83	111	57	69	308	82	71	**354**	100	86	64	62	58	50	45	45	**2.20**	23	46	2	–	–	–	–	–	–	–	–	–	–	1-Docosanol, formate	15155-62-1	WI01461
43	57	55	41	201	69	71	83	**354**	100	78	72	72	61	56	33	32	**4.00**	23	46	2	–	–	–	–	–	–	–	–	–	–	Dodecanoic acid, undecyl ester		IC04099
105	77	249	106	233	250	78	115	**354**	100	35	27	8	7	5	3	2	**1.00**	24	18	3	–	–	–	–	–	–	–	–	–	–	Cyclopropane, 1,2,3-tribenzoyl-, trans-	2892-42-4	NI37137
105	77	249	106	51	233	78	115	**354**	100	39	26	12	12	8	8	6	**2.40**	24	18	3	–	–	–	–	–	–	–	–	–	–	Cyclopropane, 1,2,3-tribenzoyl-, trans-	2892-42-4	MS08382
354	77	168	141	115	233	51	64	**354**	100	11	10	8	7	5	5	3		24	18	3	–	–	–	–	–	–	–	–	–	–	Bis(4-phenoxyphenyl) ether		DO01111
57	71	85	43	83	44	97	69	**354**	100	80	56	55	44	42	41	41	**0.00**	24	50	1	–	–	–	–	–	–	–	–	–	–	Didodecyl ether	4542-57-8	IC04100
57	71	43	85	86	83	55	97	**354**	100	58	55	35	33	30	29	27	**0.00**	24	50	1	–	–	–	–	–	–	–	–	–	–	Didodecyl ether	4542-57-8	IC04101
57	169	43	71	55	85	168	69	**354**	100	88	88	55	45	42	38	38	**3.03**	24	50	1	–	–	–	–	–	–	–	–	–	–	Didodecyl ether	4542-57-8	NI37138
43	57	55	41	83	69	29	71	**354**	100	76	76	72	47	47	39	34	**0.00**	24	50	1	–	–	–	–	–	–	–	–	–	–	Tetracosanol		MS08383
157	158	354	130	129	115			**354**	100	15	9	9	9	5				26	42	–	–	–	–	–	–	–	–	–	–	–	Acenaphthylene, 5-tetradecyl-1,2,2a,3,4,5-hexahydro-		LI07826
199	265	297	200	117	354	129	69	**354**	100	25	18	15	11	10	8	7		26	42	–	–	–	–	–	–	–	–	–	–	–	Anthracene, 6-(1-butyloctyl)-1,2,3,4,4a,9,9a,10-octahydro-		LI07827
199	354	355	200	117	185	105	131	**354**	100	94	47	41	25	24	20	11		26	42	–	–	–	–	–	–	–	–	–	–	–	Anthracene, 6-dodecyl-1,2,3,4,4a,9,9a,10-octahydro-		LI07828
185	186	143	354	129				**354**	100	16	10	6	4					26	42	–	–	–	–	–	–	–	–	–	–	–	Anthracene, 1-dodecyl-1,2,3,4,5,6,7,8-octahydro-		LI07829
199	354	185	200	355	157	186	143	**354**	100	90	85	41	25	18	15	10		26	42	–	–	–	–	–	–	–	–	–	–	–	Anthracene, 9-dodecyl-1,2,3,4,5,6,7,8-octahydro-		LI07830
339	340	43	354	283	41	29	297	**354**	100	28	21	20	17	15	15	10		26	42	–	–	–	–	–	–	–	–	–	–	–	s-Hydrindacene, 1,1,5,5-tetramethyl-4,8-diisopentyl-		AI02156
57	41	129	29	241	297	185	43	**354**	100	54	51	50	45	39	30	23	**4.05**	26	42	–	–	–	–	–	–	–	–	–	–	–	Naphthalene, tetrabutyldihydro-	74646-41-6	NI37139
57	241	297	29	129	41	185	141	**354**	100	57	56	54	52	52	34	19	**0.00**	26	42	–	–	–	–	–	–	–	–	–	–	–	Naphthalene, tetrabutyldihydro-	74646-41-6	NI37140
199	354	185	200	355	157	186	143	**354**	100	82	57	36	22	14	10	10		26	42	–	–	–	–	–	–	–	–	–	–	–	Phenanthrene, 9-dodecyl-1,2,3,4,5,6,7,8-octahydro-		LI07831
354	39	178	355	351	350	175	168.5	**354**	100	65	40	32	27	26	23	19		28	18	–	–	–	–	–	–	–	–	–	–	–	9,9'-Bianthracene		AI02157
69	278	276	109	197	145	143	96	**355**	100	36	36	16	14	10	10	8	**0.00**	4	2	–	1	–	7	–	2	–	–	–	–	–	1-Fluoro-1,2-dibromo-2-(perfluorodimethylamino)ethane, erythro-		LI07832
69	278	276	109	197	145	143	96	**355**	100	36	36	14	10	10	10	8	**0.00**	4	2	–	1	–	7	–	2	–	–	–	–	–	1-Fluoro-1,2-dibromo-2-(perfluorodimethylamino)ethane, threo-		LI07833
32	73	75	136	69	43	44	31	**355**	100	99	75	67	63	44	41	39	**0.00**	7	10	1	1	1	9	–	–	–	1	–	–	–	1-Butanesulphinamide, 1,1,2,2,3,3,4,4,4-nonafluoro-N-(trimethylsilyl)-	51735-80-9	NI37141
215	76	119	172	52	243	355	327	**355**	100	51	43	39	20	16	15	15		11	18	5	3	–	–	–	–	–	–	1	–	1	Chromium, pentacarbonyltris(dimethylamino)phosphino-		LI07834
215	95	172	355	243	317	139	128	**355**	100	44	36	17	16	11	10	10		11	18	5	3	–	–	–	–	–	–	1	–	1	Chromium, pentacarbonyltris(dimethylamino)phosphino-		MS08384
76	119	215	172	95	163	60	243	**355**	100	94	55	32	22	21	20	13	**8.33**	11	18	5	3	–	–	–	–	–	–	1	–	1	Chromium, pentacarbonyltris(dimethylamino)phosphino-		MS08385
54	302	114	82	304	196	198	96	**355**	100	87	85	62	58	43	42	30	**0.00**	12	12	5	1	–	–	3	–	–	–	–	–	–	L-Serine, N-[(2,4,5-trichlorophenoxy)acetyl]-, methyl ester	66789-87-5	NI37142
63	43	27	306	41	326	65	264	**355**	100	92	81	52	46	42	33	32	**2.32**	12	13	4	3	–	3	1	–	–	–	–	–	–	Aniline, N-(2-chloroethyl)-2,6-dinitro-N-propyl-4-(trifluoromethyl)-	33245-39-5	NI37143
240	43	71	69	169	119	286		**355**	100	80	36	17	10	4	2		**0.00**	12	16	3	1	–	7	–	–	–	–	–	–	–	Alanine, N-(heptafluorobutanoyl)-, 3-methylbutyl ester		NI37144
268	43	226	294	69	55	269	169	**355**	100	98	30	28	27	27	25	24	**4.41**	12	16	3	1	–	7	–	–	–	–	–	–	–	Valine, N-(heptafluorobutanoyl)-, propyl ester		MS08386
268	55	43	253	41	269	214	69	**355**	100	75	44	25	19	13	9	8	**0.00**	12	16	3	1	–	7	–	–	–	–	–	–	–	Valine, N-(heptafluorobutanoyl)-, propyl ester		MS08387
130	73	298	131	299	147	59	45	**355**	100	44	24	13	6	6	6	6	**1.50**	12	34	3	1	–	–	–	–	–	3	1	–	–	Phosphonic acid, [1-[(trimethylsilyl)amino]propyl]-, bis(trimethylsilyl) ester	53044-38-5	NI37145
130	73	298	147	100	75	135	133	**355**	100	44	24	6	4	4	3	3	**1.30**	12	34	3	1	–	–	–	–	–	3	1	–	–	Phosphonic acid, [1-[(trimethylsilyl)amino]propyl]-, bis(trimethylsilyl) ester	53044-38-5	NI37146

MASS TO CHARGE RATIOS								M.W.	INTENSITIES								Parent	C	H	O	N	S	F	Cl	Br	I	Si	P	B	X	COMPOUND NAME	CAS Reg No	No
73	225	135	240	195	56	268	226	355	100	71	56	53	46	44	42	41	0.00	12	34	3	1	–	–	–	–	–	3	1	–	–	Phosphonic acid, [3-[(trimethylsilyl)amino]propyl]-, bis(trimethylsilyl) ester	56051-92-4	NI37147
44	296	30	187	226	168	55	113	355	100	59	48	34	29	22	21	19	3.00	13	20	5	3	–	3	–	–	–	–	–	–	–	L-Alanine, N-[N-[N-(trifluoroacetyl)-L-valyl]glycyl]-, methyl ester	55191-31-6	NI37148
44	296	187	30	127	226	168	55	355	100	64	39	38	36	34	26	24	2.80	13	20	5	3	–	3	–	–	–	–	–	–	–	L-Alanine, N-[N-[N-(trifluoroacetyl)-L-valyl]glycyl]-, methyl ester	55191-31-6	NI37149
31	29	45	30	42	18	15	86	355	100	95	79	57	42	42	39	38	0.00	14	20	3	1	–	–	3	–	–	–	–	–	–	Acetic acid, (2,4,5-trichlorophenoxy)-, compd. with N,N-diethylethanamine (1:1)	2008-46-0	NI37150
57	198	226	41	42	154	84	153	355	100	96	86	67	60	42	28	25	1.04	15	24	5	1	–	3	–	–	–	–	–	–	–	L-Aspartic acid, N-methyl-N-(trifluoroacetyl)-, diisopropyl ester	57983-69-4	NI37151
29	41	57	27	180	56	152	198	355	100	77	42	24	22	21	20	16	0.00	15	24	5	1	–	3	–	–	–	–	–	–	–	L-Glutamic acid, N-(trifluoroacetyl)-, dibutyl ester	816-59-1	NI37152
57	41	198	226	180	152	85	45	355	100	83	77	68	59	50	40	29	0.00	15	24	5	1	–	3	–	–	–	–	–	–	–	L-Glutamic acid, N-(trifluoroacetyl)-, diisopropyl ester	57983-23-0	NI37153
240	73	239	241	75	209	147	45	355	100	67	32	19	9	8	7	7	0.00	16	29	4	1	–	–	–	–	–	2	–	–	–	Benzeneacetamide, 3,5-dimethoxy-N-(trimethylsilyl)-α-[(trimethylsilyl)oxy]-		MS08388
355	145	173	144	172	310	29	174	355	100	98	65	59	42	33	30	24		17	26	5	1	–	–	–	–	–	–	1	–	–	2-Propenoic acid, 3-[4-(dimethylamino)phenyl]-2-(diethoxyphosphinyl)-, ethyl ester	66564-08-7	NI37154
77	65	51	39	94	93	50	76	355	100	39	31	25	16	12	12	11	1.00	18	14	5	1	–	–	–	–	–	–	1	–	–	Diphenyl 3-nitrophenylphosphonate		IC04102
161	44	43	162	133	55	77	41	355	100	92	59	56	48	40	36	34	2.50	18	17	5	3	–	–	–	–	–	–	–	–	–	Hydroxylamine, bis[3-(2-hydroxyphenyl)isoxazolin-5-yl]-		NI37155
119	93	91	39	85	64	77	41	355	100	95	91	83	79	63	62	62	2.35	18	21	3	5	–	–	–	–	–	–	–	–	–	2-Imidazolidinone, 4-(hydroxyamino)-5,5-dimethyl-3-phenyl-1-(3-phenylureido)-	85601-76-9	NI37156
148	320	319	91	174	355	77	305	355	100	74	42	20	12	10	10	9		19	14	4	1	–	–	1	–	–	–	–	–	–	Bis[1,3]benzodioxolo[5,6-a:4′,5′-g]quinolizinium, 6,7-dihydro-	3486-66-6	NI37157
170	355	171	168	356	154	155	127	355	100	64	16	16	14	12	11	10		19	17	4	1	1	–	–	–	–	–	–	–	–	1-Naphthalenamine, 5-[[(2-formylphenyl)sulphonyl]oxy]-		LI07835
170	355	356	171	168	154	155	127	355	100	66	15	15	15	12	10	9		19	17	4	1	1	–	–	–	–	–	–	–	–	1-Naphthalenamine, 5-[[(4-formylphenyl)sulphonyl]oxy]-		LI07836
57	71	135	257	107	93	151	258	355	100	61	60	25	11	11	7	3	0.00	19	33	3	1	1	–	–	–	–	–	–	–	–	Bornane-10,2-sultam, N-[(3R)-3-ethylheptanoyl]-		DD01412
77	93	51	149	355	57	55	105	355	100	52	37	35	33	31	28	26		20	13	2	5	–	–	–	–	–	–	–	–	–	Pyridazino[4,5:2′,3′]pyrrolo[4′,5′-d]pyridazine-1,6-dione, 2,7-diphenyl-	91757-06-1	NI37158
192	340	355	58	177	43	77	164	355	100	42	26	10	9	9	6	5		20	21	5	1	–	–	–	–	–	–	–	–	–	Fumaritine	24181-78-0	NI37159
192	340	355	193	110	177	324	190	355	100	34	20	15	12	10	8	6		20	21	5	1	–	–	–	–	–	–	–	–	–	Fumaritine	24181-78-0	NI37160
355	337	163	340	354	192	178	326	355	100	44	42	29	25	18	18	16		20	21	5	1	–	–	–	–	–	–	–	–	–	5H-Isoindolo[1,2-b][3]benzazepin-5-one,7,8,13,13a-tetrahydro-10-hydroxy 3,4,12-trimethoxy-		NI37161
176	149	57	337	322	119	307	278	355	100	66	16	6	3	3	2	2	0.00	20	21	5	1	–	–	–	–	–	–	–	–	–	(±)-Ophiocarpine		DD01413
84	223	104	337	111	86	77	41	355	100	60	56	49	42	38	38	37	19.00	20	22	1	3	–	–	1	–	–	–	–	–	–	1H-Imidazol-4-ol, 1-(4-chlorophenyl)-2-phenyl-5-piperidino-4,5-dihydro-	80365-97-5	NI37162
84	117	214	97	68	77	75	42	355	100	92	72	70	69	68	59	59	10.00	20	22	1	3	–	–	1	–	–	–	–	–	–	1H-Imidazol-5-ol, 1-(4-chlorophenyl)-2-phenyl-4-piperidino-4,5-dihydro-	80365-98-6	NI37163
84	228	171	172	128	42	69	200	355	100	73	50	29	28	28	26	24	0.20	21	25	2	1	1	–	–	–	–	–	–	–	–	Thioxanthen-9-one, 4-[4-(butylamino)butoxy]-		MS08389
354	355	340	324	281	356	312	341	355	100	97	45	26	21	20	16	12		21	25	4	1	–	–	–	–	–	–	–	–	–	Aporphine, 1,2,9,10-tetramethoxy-, (6aα)-	475-81-0	NI37164
356	357	355	354	63	61	147	106	355	100	24	19	13	11	10	5	4		21	25	4	1	–	–	–	–	–	–	–	–	–	Aporphine, 1,2,9,10-tetramethoxy-, (6aα)-	475-81-0	NI37165
58	355	356	204	59	151	152	42	355	100	42	10	9	8	4	3	3		21	25	4	1	–	–	–	–	–	–	–	–	–	1,3-Benzodioxole-5-ethanamine, 6-[2-(3,4-dimethoxyphenyl)vinyl]-N,N-dimethyl-, (E)-	50657-26-6	NI37166
58	356	204	59	151	42	161	152	355	100	21	18	17	7	6	5	4	0.00	21	25	4	1	–	–	–	–	–	–	–	–	–	1,3-Benzodioxole-5-ethanamine, 6-[2-(3,4-dimethoxyphenyl)vinyl]-N,N-dimethyl-, (E)-	50657-26-6	NI37167
58	355	356	204	59	151	152	161	355	100	40	12	10	10	5	4	3		21	25	4	1	–	–	–	–	–	–	–	–	–	1,3-Benzodioxole-5-ethylamine, 6-[2-(3,4-dimethoxyphenyl)vinyl]-N,N-dimethyl-, (E)-	50657-26-6	LI07837
164	355	165	354	149	121	356	190	355	100	30	19	13	9	9	7	7		21	25	4	1	–	–	–	–	–	–	–	–	–	Berbine, 1,2,10,11-tetramethoxy-, (13aα)-		LI07838
164	149	355	354	190	165	121	150	355	100	76	72	40	32	28	28	17		21	25	4	1	–	–	–	–	–	–	–	–	–	Berbine, 2,3,9,10-tetramethoxy-, (±)-	2934-97-6	LI07839
164	149	355	354	190	165	121	104	355	100	76	69	40	30	28	28	17		21	25	4	1	–	–	–	–	–	–	–	–	–	Berbine, 2,3,9,10-tetramethoxy-, (±)-	2934-97-6	NI37168
164	355	165	354	190	149	166	121	355	100	43	34	17	17	14	10	10		21	25	4	1	–	–	–	–	–	–	–	–	–	Berbine, 2,3,10,11-tetramethoxy-, (13aα)-	523-02-4	NI37169
164	165	355	121	190	166	354	149	355	100	30	22	15	13	11	10	9		21	25	4	1	–	–	–	–	–	–	–	–	–	Berbine, 2,3,10,11-tetramethoxy-, (13aα)-	523-02-4	NI37170
164	121	356	165	191	149	166	103	355	100	68	50	40	30	21	18	15	7.28	21	25	4	1	–	–	–	–	–	–	–	–	–	Berbine, 2,3,10,11-tetramethoxy-, (13aβ)-	6872-27-1	NI37171
110	313	282	212	254	173	96	108	355	100	94	94	63	41	38	26	25	0.00	21	25	4	1	–	–	–	–	–	–	–	–	–	Estra-17-one, 14,15-epoxy-3-acetoxy-, O-methyloxime, (14α,15α)-		NI37172
42	204	217	123	186	172	43	95	355	100	96	62	52	43	36	34	26	2.44	22	26	2	1	–	1	–	–	–	–	–	–	–	Butanophenone, 4′-fluoro-4-[4-hydroxy-4-(4-tolyl)piperidino]-	1050-79-9	NI37173
204	217	123	186	95	205	91	165	355	100	73	40	26	21	17	17	16	1.70	22	26	2	1	–	1	–	–	–	–	–	–	–	Butanophenone, 4′-fluoro-4-[4-hydroxy-4-(4-tolyl)piperidino]-	1050-79-9	NI37174
135	131	103	57	220	152	161	355	355	100	68	47	41	38	37	30	28		22	29	3	1	–	–	–	–	–	–	–	–	–	N-Isobutyl-11-(3,4-methylenedioxyphenyl)-2E,4E,10E-undecatrienoic amide		MS08390
105	77	91	197	106	132	198	104	355	100	95	44	40	22	14	13	10	2.00	23	17	3	1	–	–	–	–	–	–	–	–	–	5-Benzyl-6-oxo-2,3-diphenyl-6H-1,3-oxazin-3-ium-4-olate		NI37175
105	77	250	129	251	106	355	232	355	100	36	32	12	11	9	8	6		23	17	3	1	–	–	–	–	–	–	–	–	–	5H-Pyrrol-2-one, 4-benzoyl-5-hydroxy-3,5-diphenyl-		NI37176
355	356	251	250	105	357	100	91	355	100	35	19	14	9	5	4	3		23	17	3	1	–	–	–	–	–	–	–	–	–	2,4(1H,3H)-Quinolinedione, 3-benzoyl-3-benzyl-		NI37177
355	182	127	223	340	65	264	154	355	100	91	31	21	17	11	8	7		23	21	1	3	–	–	–	–	–	–	–	–	–	2-(Phenylmethylamino)-4-methyl-5-(1-oxo-1,2-dihydro-2-naphthylmethylene)imidazole		LI07840
105	77	286	70	355	42	51	41	355	100	93	57	33	26	24	13	12		24	21	2	1	–	–	–	–	–	–	–	–	–	2-Azetidinone, 1-benzoyl-3,3-dimethyl-4,4-diphenyl-	24549-63-1	NI37178
206	307	325	308	207	326	309	193	355	100	43	40	40	19	11	11	11	0.00	24	21	2	1	–	–	–	–	–	–	–	–	–	Benzene, 1,1′,1″-(3-nitro-5-cyclohexene-1,2,4-triyl)tris-, (1α,2α,3β,4α)-	32581-60-5	NI37179

MASS TO CHARGE RATIOS	M.W.	INTENSITIES	Parent	C	H	O	N	S	F	Cl	Br	I	Si	P	B	X	COMPOUND NAME	CAS Reg No	No
167 309 231 310 168 232 143 233	**355**	100 57 27 16 15 7 5 3	**0.00**	24	21	2	1	–	–	–	–	–	–	–	–	–	**Benzene, 1,1′,1″-(3-nitro-5-cyclohexene-1,2,4-triyl)tris-, (1α,2α,3β,4α)-**	32581-60-5	**NI37180**
206 193 207 194 307 167	**355**	100 54 17 9 2 2	**0.00**	24	21	2	1	–	–	–	–	–	–	–	–	–	**Benzene, 1,1′,1″-(3-nitro-5-cyclohexene-1,2,4-triyl)tris-, (1α,2β,3α,4α)-**	32581-61-6	**NI37181**
105 355 298 299 77 165 282 300	**355**	100 55 52 45 35 34 26 21		25	25	1	1	–	–	–	–	–	–	–	–	–	**Benzene, 4-acetyl-2-aziridinyl-1-tert-butyl-3-(1,1′-biphenyl)-**	38751-82-5	**NI37182**
203 57 220 202 45 55 84 127	**355**	100 82 74 71 40 35 30 29	**0.00**	25	25	1	1	–	–	–	–	–	–	–	–	–	**3-Cyclohexen-1-amine, 6-(4-methoxyphenyl)-2,5-diphenyl-, (1α,2β,5β,6β)**	42715-19-5	**NI37183**
174 175 340 355 159 341 165 356	**355**	100 27 21 17 16 6 6 5		26	29	–	1	–	–	–	–	–	–	–	–	–	**Ethylamine, N-(diphenylmethylene)-1-(pentamethylphenyl)-**	75326-13-5	**NI37184**
147 145 191 279 132 119 117 281	**356**	100 73 60 53 52 50 50 45	**0.00**	3	–	–	–	–	3	3	2	–	–	–	–	–	**Propane, 1,1-dibromo-3,3,3-trichloro-1,2,2-trifluoro-**		**LI07841**
69 119 31 180 111 130 80 50	**356**	100 72 23 20 19 11 6 6	**0.00**	4	–	–	–	–	12	–	–	–	–	–	–	1	**Bis(pentafluoroethyl)selenium difluoride**		**NI37185**
69 198 110 168 277 279 129 179	**356**	100 48 44 37 31 29 26 20	**7.50**	6	–	–	–	–	5	–	2	–	–	1	–	–	**Phosphine, pentafluorophenyldibromo-**		**MS08391**
140 99 39 41 142 101 358 58	**356**	100 99 65 41 39 39 25 25	**8.38**	6	10	–	–	–	–	–	2	–	–	–	–	2	**Nickel, di-μ-bromobis(η³-2-propenyl)di-**	12012-90-7	**NI37186**
140 99 39 41 142 101 358 360	**356**	100 90 52 40 36 35 30 28	**8.65**	6	10	–	–	–	–	–	2	–	–	–	–	2	**Nickel, di-μ-bromobis(η³-2-propenyl)di-**	12012-90-7	**NI37187**
358 360 341 343 356 74 362 339	**356**	100 98 67 64 34 32 31 22		7	3	2	–	–	–	–	3	–	–	–	–	–	**Benzoic acid, 2,4,6-tribromo-**	633-12-5	**NI37188**
156 56 158 121 91 139 300 109	**356**	100 44 32 31 24 20 18 18	**1.68**	7	5	2	–	–	–	3	–	–	–	–	–	2	**Iron, (trichlorogermyl)dicarbonyl-π-cyclopentadienyl-**	12107-06-1	**NI37189**
69 119 64 48 31 169 101 50	**356**	100 18 18 18 18 17 9 7	**0.00**	7	9	2	–	1	9	–	–	–	1	–	–	–	**1-Butanesulphinic acid, 1,1,2,2,3,3,4,4,4-nonafluoro-, trimethylsilyl ester**	51735-77-4	**NI37190**
341 73 249 229 233 342 147 77	**356**	100 75 43 33 21 20 12 11	**1.50**	9	18	3	–	–	5	–	–	–	2	1	–	–	**Bis(trimethylsilyl) 2,2-difluoro-1-(trifluoromethyl)ethenylphosphonate**		**MS08392**
179 75 45 150 81 100	**356**	100 82 68 58 23 18	**7.58**	10	2	4	–	–	6	–	–	–	–	–	–	1	**Iron, tricarbonylbis(trifluoromethyl)oxocyclopentadienyl-**		**LI07842**
358 230 117 119 232 322 160 201	**356**	100 84 80 78 76 75 75 70	**66.67**	10	5	–	–	–	–	5	–	–	–	–	–	1	**Ferrocene, 1,2,3,4,5-pentachloro-**	33306-53-5	**NI37191**
358 230 322 232 160 360 356 195	**356**	100 83 75 75 75 68 68 62		10	5	–	–	–	–	5	–	–	–	–	–	1	**Ferrocene, 1,2,3,4,5-pentachloro-**	33306-53-5	**NI37192**
358 230 232 322 160 195 130 95	**356**	100 84 76 75 75 61 40 34	**0.00**	10	5	–	–	–	–	5	–	–	–	–	–	1	**Ferrocene, 1,2,3,4,5-pentachloro-**	33306-53-5	**LI07843**
41 29 199 43 313 27 57 177	**356**	100 99 71 71 66 66 64 54	**0.00**	11	25	–	–	–	–	–	1	–	–	–	–	1	**Dibutylisopropyltin bromide**		**MS08393**
341 73 251 75 299 325 342 356	**356**	100 96 36 30 24 23 22 21		11	29	5	–	–	–	–	–	–	3	1	–	–	**Acetic acid, [bis[(trimethylsilyl)oxy]phosphinyl]-, trimethylsilyl ester**	53044-27-2	**NI37193**
341 73 297 251 75 299 325 342	**356**	100 96 40 36 30 24 23 22	**21.02**	11	29	5	–	–	–	–	–	–	3	1	–	–	**Acetic acid, [bis[(trimethylsilyl)oxy]phosphinyl]-, trimethylsilyl ester**	53044-27-2	**NI37194**
341 73 297 251 75 299 325 356	**356**	100 96 40 36 30 24 23 21		11	29	5	–	–	–	–	–	–	3	1	–	–	**Acetic acid, [bis[(trimethylsilyl)oxy]phosphinyl]-, trimethylsilyl ester**	53044-27-2	**NI37195**
77 51 356 354 353 78 355 279	**356**	100 21 20 14 11 10 9 9		12	10	–	–	–	–	–	–	–	–	–	–	1	**Mercury, diphenyl-**	587-85-9	**NI37196**
154 202 200 199 153 201 152 198	**356**	100 53 42 31 27 25 24 19	**0.00**	12	10	–	–	–	–	–	–	–	–	–	–	1	**Mercury, diphenyl-**	587-85-9	**NI37197**
77 51 50 356 78 354 279 76	**356**	100 40 13 8 8 6 6 6		12	10	–	–	–	–	–	–	–	–	–	–	1	**Mercury, diphenyl-**	587-85-9	**NI37198**
45 57 101 79 69 42 40 41	**356**	100 83 61 47 45 42 40 34	**0.45**	12	15	4	–	–	7	–	–	–	–	–	–	–	**Butanoic acid, heptafluoro-, 1-(butoxycarbonyl)propyl ester**	55649-48-4	**NI37199**
59 45 31 69 119 40 150 100	**356**	100 63 54 46 39 38 37 35	**0.47**	12	15	4	–	–	7	–	–	–	–	–	–	–	**Butanoic acid, heptafluoro-, 3-butoxy-1-methyl-3-oxopropyl ester**	55649-47-3	**NI37200**
45 69 31 119 150 57 59 40	**356**	100 91 64 54 51 50 49 49	**0.55**	12	15	4	–	–	7	–	–	–	–	–	–	–	**Butanoic acid, heptafluoro-, 4-butoxy-4-oxobutyl ester**	55649-46-2	**NI37201**
182 296 326 297 327 281 183 185	**356**	100 87 78 29 28 28 24 18	**2.00**	12	16	4	2	–	–	–	–	–	–	–	–	2	**Chromium, μ-methoxybis(nitrosylcyclopentadienyl)di-**		**LI07844**
339 341 356 204 89 340 309 158	**356**	100 36 27 24 19 18 18 18		13	9	6	2	1	–	1	–	–	–	–	–	–	**3,3′-Dinitro-4-methyl-4′-chlorodiphenyl sulphone**		**NI37202**
172 90 184 63 126 75 174 142	**356**	100 71 54 37 34 34 33 30	**0.00**	13	10	4	4	–	–	2	–	–	–	–	–	–	**Methane, bis(2-chloro-5-nitroanilino)-**		**IC04103**
83 69 75 55 73 97 103 41	**356**	100 63 57 57 49 43 28 27	**0.40**	13	29	1	–	–	–	–	–	1	1	–	–	–	**Silane, [(10-iododecyl)oxy]trimethyl-**	26305-80-6	**NI37203**
283 356 137 284 107 255 119 105	**356**	100 47 16 15 11 10 8 6		14	13	5	–	–	5	–	–	–	–	–	–	–	**Benzeneacetic acid, 3-methoxy-4-[(pentafluoropropionyl)oxy]-, ethyl ester**		**MS08394**
79 93 121 91 98 43 164 55	**356**	100 53 50 43 37 37 28 28	**0.00**	14	28	6	–	2	–	–	–	–	–	–	–	–	**2,2-Dipropylcyclohexane-trans-1,3-diol dimethosulphate**		**LI07845**
327 356 358 329 27 41 314 357	**356**	100 94 69 68 33 29 27 26		15	14	2	2	1	–	2	–	–	–	–	–	–	**Phenothiazine, 3-amino-7,9-dichloro-10-propyl-, S,S-dioxide**		**LI07846**
125 356 77 51 97 44 156 141	**356**	100 37 35 11 6 6 2 2		15	16	4	–	3	–	–	–	–	–	–	–	–	**Dimethylsulphonio bis(phenylsulphonyl)methanide**		**LI07847**
171 78 107 77 326 91 155 123	**356**	100 34 18 15 10 9 8 7	**1.90**	15	16	6	–	2	–	–	–	–	–	–	–	–	**Benzenesulphonic acid, 4-methoxy-, [(4-methylphenyl)sulphonyl]methyl ester**	31081-10-4	**NI37204**
155 91 171 65 77 107 92 262	**356**	100 76 53 21 16 15 14 11	**1.20**	15	16	6	–	2	–	–	–	–	–	–	–	–	**Methanol, [(4-methoxyphenyl)sulphonyl]-, 4-methylbenzenesulphonate**	31081-05-7	**NI37205**
295 178 185 191 183 179 213 177	**356**	100 66 54 40 40 40 36 36	**0.00**	15	16	10	–	–	–	–	–	–	–	–	–	–	**2H-1-Benzopyran-2-one, 7-(β-D-glucopyranosyloxy)-8-hydroxy-**	486-55-5	**NI37206**
257 158 242 59 356 74 143 84	**356**	100 92 66 20 17 17 15 13		15	21	6	–	–	–	–	–	–	–	–	–	1	**Cobalt, tris(2,4-pentanedionato-O,O′)-**	21679-46-9	**NI37207**
257 158 242 356 143 299 341 314	**356**	100 66 47 29 7 6 4 2		15	21	6	–	–	–	–	–	–	–	–	–	1	**Cobalt, tris(2,4-pentanedionato-O,O′)-**	21679-46-9	**LI07848**
356 257 158 241 43 357 59 74	**356**	100 56 51 38 22 17 11 10		15	21	6	–	–	–	–	–	–	–	–	–	1	**Cobalt, tris(2,4-pentanedionato-O,O′)-**	21679-46-9	**NI37208**
114 131 86 87 57 208 225 72	**356**	100 98 79 72 72 56 33 29	**0.00**	15	24	6	4	–	–	–	–	–	–	–	–	–	**Actinobolin-5-pyrazolone, N-acetyl-**		**LI07849**
214 356 216 358 321 69 215 140	**356**	100 39 35 27 23 21 20 16		16	14	5	–	–	–	2	–	–	–	–	–	–	**Spiro[benzofuran-2(3H),1′-[2]cyclohexene]-3,4′-dione, 2′,7-dichloro-4,6-dimethoxy-6′-methyl-**	26881-56-1	**NI37209**
254 356 358 225 256 328 293 188	**356**	100 62 41 41 37 32 32 31		16	14	5	–	–	–	2	–	–	–	–	–	–	**Spiro[benzofuran-2(3H),1′-[3]cyclohexene]-2′,3-dione, 4′,7-dichloro-4,6-dimethoxy-6′-methyl-**	1235-50-3	**NI37210**
86 356 135 270 243 357 242 119	**356**	100 69 53 23 17 14 9 9		16	15	4	2	–	3	–	–	–	–	–	–	–	**Oxazole, 2-(4-methoxyphenyl)-4-(trifluoroacetyl)-5-(4-morpholino)-**	51770-12-8	**NI37211**
52 210 272 129 114 211 240 184	**356**	100 69 55 26 25 17 16 14	**9.10**	16	16	6	–	–	–	–	–	–	–	–	–	1	**Chromium, tricarbonyl-π-(2-methoxy-3-methyl-2-phenyl-1-methoxycarbonylcyclopropyl)-**		**LI07850**
297 61 265 356 296 237 59 60	**356**	100 91 73 69 49 47 40 36		16	20	5	–	2	–	–	–	–	–	–	–	–	**Spiro[adamantane-2,2′-[1,3]dithiolane]-1,5-dicarboxylic acid, 6-oxo-**	19930-84-8	**NI37212**

MASS TO CHARGE RATIOS	M.W.	INTENSITIES	Parent	C	H	O	N	S	F	Cl	Br	I	Si	P	B	X	COMPOUND NAME	CAS Reg No	No
81 77 189 218 125 85 83 311	356	100 97 56 46 24 21 17 16	0.00	16	20	7	–	1	–	–	–	–	–	–	–	–	2H-Pyran-2-methanol, 3-(benzoyloxy)-6-ethoxy-3,6-dihydro-, methanesulphonate	56248-13-6	NI37213
179 169 57 114 86 226 297 139	356	100 98 51 46 38 37 35 26	4.00	16	24	7	2	–	–	–	–	–	–	–	–	–	Actinobolin, N-acetyl-, methyl enol ether		LI07851
128 145 57 100 101 357 338 356	356	100 98 72 67 10 7 6 2		16	24	7	2	–	–	–	–	–	–	–	–	–	Actinobolin, N-propanoyl-		LI07852
227 228 212 169 346 344 152 141	356	100 15 8 7 6 6 6 5	0.83	16	27	2	–	–	–	3	–	–	–	–	–	–	Ethane, 1,1,1-trichloro-2,2-bis(4-methoxyphenyl)-		NI37214
239 73 240 147 313 209 241 45	356	100 40 18 10 9 7 5 5	3.20	16	28	5	–	–	–	–	–	–	2	–	–	–	Benzeneacetic acid, 2,5-dimethoxy-α-[(trimethylsilyl)oxy]-, trimethylsilyl ester		MS08395
239 73 240 313 147 241 165 45	356	100 37 17 8 8 5 4 3	0.80	16	28	5	–	–	–	–	–	–	2	–	–	–	Benzeneacetic acid, 3,4-dimethoxy-α-[(trimethylsilyl)oxy]-, trimethylsilyl ester		MS08396
239 73 240 147 45 313 75 74	356	100 72 18 16 8 7 7 6	3.60	16	28	5	–	–	–	–	–	–	2	–	–	–	Benzeneacetic acid, 3,5-dimethoxy-α-[(trimethylsilyl)oxy]-, trimethylsilyl ester		MS08397
73 297 45 298 89 59 74 299	356	100 80 20 19 11 11 8 6	3.10	16	28	5	–	–	–	–	–	–	2	–	–	–	Benzeneacetic acid, 3-methoxy-α,4-bis[(trimethylsilyl)oxy]-, methyl ester	3223-45-8	MS08398
297 298 73 299 341 356 267 75	356	100 28 15 12 7 5 5 5		16	28	5	–	–	–	–	–	–	2	–	–	–	Benzeneacetic acid, 3-methoxy-α,4-bis[(trimethylsilyl)oxy]-, methyl ester	3223-45-8	NI37215
297 73 298 299 341 267 89 45	356	100 52 28 11 8 6 6 6	5.10	16	28	5	–	–	–	–	–	–	2	–	–	–	Benzeneacetic acid, 3-methoxy-α,4-bis[(trimethylsilyl)oxy]-, methyl ester, (±)-	57397-11-2	NI37216
73 297 298 45 89 59 299 194	356	100 78 20 19 11 11 9 9	3.20	16	28	5	–	–	–	–	–	–	2	–	–	–	Benzeneacetic acid, 4-methoxy-α,3-bis[(trimethylsilyl)oxy]-, methyl ester		MS08399
297 73 298 299 296 341 267 194	356	100 45 28 11 8 7 6 6	0.00	16	28	5	–	–	–	–	–	–	2	–	–	–	Benzeneacetic acid, 4-methoxy-α,3-bis[(trimethylsilyl]oxy)-, methyl ester		NI37217
73 149 181 147 75 166 123 77	356	100 59 31 31 25 24 20 18	18.00	16	28	5	–	–	–	–	–	–	2	–	–	–	Propanoic acid, 3-(2-methoxyphenoxy)-2-[(trimethylsilyl)oxy]-, trimethylsilyl ester		MS08400
73 356 194 147 45 357 267 74	356	100 37 28 15 15 13 13 9		16	32	3	–	–	–	–	–	–	3	–	–	–	Benzene, 2,5-bis(trimethylsiloxy)-1-[(trimethylsiloxy)methyl]-		NI37218
73 267 179 356 45 147 268 357	356	100 50 40 26 17 15 13 9		16	32	3	–	–	–	–	–	–	3	–	–	–	Benzene, 3,4-bis(trimethylsiloxy)-1-[(trimethylsiloxy)methyl]-	68595-79-9	NI37219
341 73 300 342 100 75 55 147	356	100 85 38 35 31 24 24 20	7.35	16	32	3	2	–	–	–	–	–	2	–	–	–	2,4,6(1H,3H,5H)-Pyrimidinetrione, 5-butyl-5-ethyl-1,3-bis(trimethylsilyl)	52988-92-8	NI37220
285 341 300 73 75 286 100 97	356	100 69 62 57 32 24 24 24	2.94	16	32	3	2	–	–	–	–	–	2	–	–	–	2,4,6(1H,3H,5H)-Pyrimidinetrione, 5-ethyl-5-sec-butyl-1,3-bis(trimethylsilyl)-	52937-66-3	NI37221
43 313 108 267 192 121 314 76	356	100 24 23 20 18 14 11 11	3.20	17	12	3	2	2	–	–	–	–	–	–	–	–	Aniline, (3-acetylthio-2-benzo[b]thienylidene)-4-nitro-		NI37222
256 356 258 257 358 255 215 229	356	100 46 37 20 17 16 15 12		17	21	6	–	–	–	1	–	–	–	–	–	–	Spiro[benzofuran-2(3H),1′-cyclohexan]-3-one, 7-chloro-4′-hydroxy-2′,4,6-trimethoxy-6′-methyl-	26891-87-2	NI37223
89 99 45 131 263 72 57 90	356	100 96 94 46 43 31 29 27	2.00	17	24	8	–	–	–	–	–	–	–	–	–	–	7-Oxabicyclo[2.2.0]hepta-2,5-diene-2,3-dicarboxylic acid, 1-(2,2-dimethoxypropyl)-6-(methoxymethyl)-, dimethyl ester	106710-25-2	NI37224
89 99 131 45 71 194 163 59	356	100 15 14 6 5 4 4 3	0.00	17	24	8	–	–	–	–	–	–	–	–	–	–	3-Oxatetracyclo[3.2.0.0^{2,7}.0^{4,6}]heptane-1,5-dicarboxylic acid, 2-(2,2-dimethoxypropyl)-7-(methoxymethyl)-, dimethyl ester	106710-30-9	NI37225
89 90 45 261 293 59 324 356	356	100 12 10 9 7 6 2 1		17	24	8	–	–	–	–	–	–	–	–	–	–	Oxepin-4,5-dicarboxylic acid, 2-(2,2-dimethoxypropyl)-3-(methoxymethyl)-, dimethyl ester	106710-38-7	NI37226
55 56 97 100 113 168 84 101	356	100 95 93 86 78 69 66 51	1.07	17	28	6	2	–	–	–	–	–	–	–	–	–	1,5-Dioxa-10,17-diazacycloheneicosane-6,9,18,21-tetrone	79688-13-4	NI37227
55 56 97 100 113 114 168 84	356	100 95 94 86 79 77 69 66	0.00	17	28	6	2	–	–	–	–	–	–	–	–	–	1,5-Dioxa-10,17-diazacycloheneicosane-6,9,18,21-tetrone	79688-13-4	MS08401
70 100 140 84 55 113 127 153	356	100 60 58 48 44 42 35 33	8.61	17	28	6	2	–	–	–	–	–	–	–	–	–	1,5-Dioxa-10,15-diazacyclononadecane-6,9,16,19-tetrone, 3,3-dimethyl-	79688-12-3	NI37228
103 75 73 47 266 209 237 179	356	100 90 79 50 45 29 28 28	0.00	17	32	4	–	–	–	–	–	–	2	–	–	–	Benzeneethanol, β-ethoxy-3′-methoxy-α,4′-bis[(trimethylsilyl)oxy]-		MS08402
105 77 38 110 85 51 356 106	356	100 40 36 13 12 12 10 8		18	12	2	–	3	–	–	–	–	–	–	–	–	1,2,4-Trithiolane, 3,5-(benzoylmethylidene)-		LI07853
43 272 28 314 69 45 216 189	356	100 87 50 42 20 20 18 18	1.00	18	12	8	–	–	–	–	–	–	–	–	–	–	4H-Naphtho[2,3-b]pyran-4,6,9-trione, 5,8-bis(acetyloxy)-2-methyl-	62681-87-2	NI37229
310 123 338 161 104 150 122 294	356	100 89 72 67 50 35 35 30	13.00	18	12	8	–	–	–	–	–	–	–	–	–	–	2H-Pyran-2,5(6H)-dione, 4-(2,5-dihydroxyphenyl)-6-(3,4-dihydroxybenzylidene)-3-hydroxy-		NI37230
123 150 122 121 312 135 149 134	356	100 66 34 27 25 21 18 18	12.00	18	12	8	–	–	–	–	–	–	–	–	–	–	2H-Pyran-2,5(6H)-dione, 4-(3,4-dihydroxyphenyl)-6-(3,4-dihydroxybenzylidene)-3-hydroxy-		NI37231
247 217 172 139 107 63 215 185	356	100 99 97 81 18 18 15 15	0.97	18	17	–	2	2	–	–	–	–	–	1	–	–	4,6-Dimethylpyrimidin-2-yl diphenylphosphinodithioate		MS08403
313 315 328 330 77 286 287 329	356	100 98 78 76 75 62 60 59	0.00	18	17	1	2	–	–	–	1	–	–	–	–	–	1,5-Benzodiazocin-2-one, 8-bromo-1-ethyl-1,2,3,4-tetrahydro-6-phenyl-	63563-59-7	NI37232
328 330 301 357 329 300 355 302	356	100 99 89 87 84 78 77 72	69.00	18	17	1	2	–	–	–	1	–	–	–	–	–	1,5-Benzodiazocin-2-one, 6-(3-bromophenyl)-1,2,3,4-tetrahydro-1,8-dimethyl-	76061-55-7	NI37233
197 195 193 196 120 194 279 118	356	100 73 43 31 30 27 23 22	1.20	18	20	–	–	–	–	–	–	–	–	–	–	1	Tin, diallyldiphenyl-		MS08404
356 313 254 298 283 240 269 296	356	100 73 56 32 28 25 19 13		18	28	–	4	–	–	–	–	–	2	–	–	–	Cyclodisilazane-1,3-diamine, N,N,N′,N′,2,4-hexamethyl-2,4-diphenyl-	66436-35-9	NI37234
356 341 236 285 313 298	356	100 33 13 10 9 8		19	16	7	–	–	–	–	–	–	–	–	–	–	6H-Benzofuran[3,2-c][1]benzopyran-6-one, 3,4,8,9-tetramethoxy-		LI07854
221 163 356 325 147 222 135 103	356	100 72 40 32 24 18 13 12		19	16	7	–	–	–	–	–	–	–	–	–	–	Benzophenone-2,4′,5-tricarboxylic acid, trimethyl ester		NI37235
314 162 356 299 315 83 164 229	356	100 61 44 27 17 17 10 9		19	16	7	–	–	–	–	–	–	–	–	–	–	6H-[1,3]Dioxolo[5,6]benzofuro[3,2-c][1]benzopyran-2-ol, 6a,12a-dihydro-3-methoxy-, acetate, (6aR-cis)-	55556-76-8	NI37236
83 311 151 356 205 177 206 178	356	100 56 56 50 33 33 14 14		19	16	7	–	–	–	–	–	–	–	–	–	–	Flavonol, 2′-formyl-4′,5′,7-trimethoxy-		LI07855
356 327 357 325 323 310 341 328	356	100 45 22 22 18 16 9 9		19	16	7	–	–	–	–	–	–	–	–	–	–	Isobenzofuro[5,6-b]benzofuran-8-carboxylic acid, 1,3-dihydro-7,10-dimethoxy-9-methyl-1-oxo-, methyl ester	54725-11-0	NI37237

MASS TO CHARGE RATIOS								M.W.	INTENSITIES								Parent	C	H	O	N	S	F	Cl	Br	I	Si	P	B	X	COMPOUND NAME	CAS Reg No	No
270	43	356	295	298	83	296	271	**356**	100	68	62	52	37	26	19	18		19	16	7	–	–	–	–	–	–	–	–	–	–	**5,12-Naphthacenedione, 7,8,9,10-tetrahydro-1,6,7,8,11-pentahydroxy-8-methyl-**	6824-61-9	**NI37238**
245	217	235	264	42	236	147	175	**356**	100	78	69	67	63	45	43	32	**0.00**	19	17	3	2	–	–	1	–	–	–	–	–	–	**5H-Pyrrolo[1,2-d][1,4]benzodiazepine-3,6-dione, 10-chloro-1,2,7,11b-tetrahydro-(2R)-hydroxy-trans-2-methyl-trans-11b-phenyl-**	89839-35-0	**NI37239**
91	165	356	92	120	108	55	357	**356**	100	48	28	11	10	7	7	6		19	20	5	2	–	–	–	–	–	–	–	–	–	**Benzoic acid, N-(benzyloxycarbonyl)glycyl-4-amino-, ethyl ester**		**DD01414**
297	356	298	357	147	118	91	55	**356**	100	39	18	9	9	7	6	6		19	24	3	4	–	–	–	–	–	–	–	–	–	**Aniline, N-ethyl-N-(hydroxybutyl)-3-methyl-4-(4-nitrophenylazo)-**		**IC04104**
356	192	357	73	341	163	355	358	**356**	100	75	33	32	30	18	14	12		19	28	1	2	–	–	–	–	–	2	–	–	–	**Benzophenone, 4,4′-bis[(trimethylsilyl)amino]-**	31396-45-9	**NI37240**
55	297	269	115	43	41	237	209	**356**	100	73	54	36	24	22	18	17	**0.00**	19	32	6	–	–	–	–	–	–	–	–	–	–	**2-Furanacetic acid, α-decyltetrahydro-2-(methoxycarbonyl)-5-oxo-, methyl ester, [S-(R*,R*)]-**		**NI37241**
129	75	101	173	281	145	95	255	**356**	100	68	50	43	41	39	39	37	**0.00**	19	36	4	–	–	–	–	–	–	1	–	–	–	**1,7-Dioxaspiro[5.5]undecane, 2-(formylmethyl)-4-(dimethyl-tert-butylsiloxy)-8,9-dimethyl-, (2S,4S,6R,8R,9S)-**		**MS08405**
44	199	72	43	87	198	55	100	**356**	100	50	35	28	23	22	18	17	**2.40**	19	36	4	2	–	–	–	–	–	–	–	–	–	**L-Valine, N-[N-(1-oxodecyl)-L-alanyl]-, methyl ester**	28415-50-1	**NI37242**
233	340	341	248	247	356	139	234	**356**	100	72	59	56	33	30	29	28		20	17	3	–	1	1	–	–	–	–	–	–	–	**1H-Indene-3-acetic acid, 5-fluoro-2-methyl-1-[[4-(methylsulphinyl)phenyl]methylene]-, (Z)-**	38194-50-2	**NI37243**
354	352	356	66	128	65	350		**356**	100	99	55	30	19	11	5			20	20	–	–	–	–	–	–	–	–	–	–	2	**Bis(titanocene)**		**LI07856**
150	151	314	273	356	271	315	279	**356**	100	11	10	9	7	3	2	2		20	20	6	–	–	–	–	–	–	–	–	–	–	**2H-1-Benzopyran-7-ol, 3-[2-(acetyloxy)-4-methoxyphenyl]-3,4-dihydro-, acetate**	38822-06-9	**NI37244**
283	323	44	356	338	270	284	165	**356**	100	42	26	24	24	20	19	19		20	20	6	–	–	–	–	–	–	–	–	–	–	**4H-1-Benzopyran-4-one, 5,7-dihydroxy-3-(4-hydroxyphenyl)-8-(3-hydroxy 3-methylbutyl)-**		**NI37245**
356	327	313	299	271	328	285	178	**356**	100	70	31	16	12	6	6	6		20	20	6	–	–	–	–	–	–	–	–	–	–	**2H-1-Benzopyran-2-one, 5,7-dimethoxy-8-ethoxy-4-(4-methoxyphenyl)-**		**MS08406**
205	204	59	121	173	145	146	147	**356**	100	62	49	47	44	39	28	21	**14.80**	20	20	6	–	–	–	–	–	–	–	–	–	–	**2-Butene, 1,4-bis[4-(methoxycarbonyl)phenoxy]-, cis-**		**IC04105**
205	204	121	173	145	256	147	146	**356**	100	94	94	43	38	27	27	27	**26.30**	20	20	6	–	–	–	–	–	–	–	–	–	–	**2-Butene, 1,4-bis[4-(methoxycarbonyl)phenoxy]-, trans-**		**IC04106**
115	116	59	91	175	143	117	131	**356**	100	37	29	25	18	16	15	13	**1.26**	20	20	6	–	–	–	–	–	–	–	–	–	–	**2,5-Cyclohexadien-1-one, 4-(methoxycarbonyl)-4-[2-[(2-methoxycarbonyl)vinyl]benzyl]-3-methoxy-**		**NI37246**
137	356	135	138	136	106	162	194	**356**	100	95	37	12	12	11	9	6		20	20	6	–	–	–	–	–	–	–	–	–	–	**2(3H)-Furanone, dihydro-4-piperonyl-3-vanillyl-**		**LI07857**
356	241	357	255	271	272	232	175	**356**	100	26	24	16	14	9	8	8		20	20	6	–	–	–	–	–	–	–	–	–	–	**Naphtho[2,3-c]furan-1(3H)-one, 3a,4,9,9a-tetrahydro-6-hydroxy-4-(4-hydroxy-3-methoxyphenyl)-7-methoxy-, [3aR-(3aα,4α,9aβ)]-**	518-55-8	**LI07858**
356	241	357	255	271	272	137	232	**356**	100	25	22	16	15	10	8	7		20	20	6	–	–	–	–	–	–	–	–	–	–	**Naphtho[2,3-c]furan-1(3H)-one, 3a,4,9,9a-tetrahydro-6-hydroxy-4-(4-hydroxy-3-methoxyphenyl)-7-methoxy-, [3aR-(3aα,4α,9aβ)]-**	518-55-8	**MS08407**
356	241	357	255	271	272	137	175	**356**	100	26	24	15	14	9	9	8		20	20	6	–	–	–	–	–	–	–	–	–	–	**Naphtho[2,3-c]furan-1(3H)-one, 3a,4,9,9a-tetrahydro-6-hydroxy-4-(4-hydroxy-3-methoxyphenyl)-7-methoxy-, [3aR-(3aα,4α,9aβ)]-**	518-55-8	**NI37247**
356	149	152	135	150	163	161	203	**356**	100	53	27	24	20	15	14	13		20	20	6	–	–	–	–	–	–	–	–	–	–	**Phenol, 2-methoxy-4-[4-[3,4-(methylenedioxy)phenyl]-1H,3H-furo[3,4-c]furan-1-yl]-**		**LI07859**
226	167	225	209	194	168	180	283	**356**	100	94	80	58	57	56	48	47	**38.38**	20	24	4	2	–	–	–	–	–	–	–	–	–	**Curan-17-oic acid, 2,16-didehydro-19,20-dihydroxy-, methyl ester, (19S)-**	2270-73-7	**NI37248**
339	106	130	108	180	143	120	188	**356**	100	36	29	28	19	18	17	15	**3.00**	20	24	4	2	–	–	–	–	–	–	–	–	–	**6,21-Cyclo-4,5-secoakuammilan-17-oic acid, 1,2-dihydro-4,5-dihydroxy-, methyl ester, (2ξ,6α)-**	63944-66-1	**NI37249**
43	57	73	356	166	42	165	147	**356**	100	27	16	6	6	6	4	4		20	28	2	–	–	–	–	–	–	2	–	–	–	**Diphenylketene bis(trimethylsilyl) acetal**		**MS08408**
285	287	135	286	63	107	57	151	**356**	100	38	37	17	14	12	12	9	**1.50**	20	33	3	–	–	–	1	–	–	–	–	–	–	**Benzene, 1-[2-[2-(2-chloroethoxy)ethoxy]ethoxy]-4-(1,1,3,3-tetramethylbutyl)-**	67025-22-3	**NI37250**
116	55	117	74	69	41	43	98	**356**	100	30	28	25	22	22	20	13	**0.10**	20	36	5	–	–	–	–	–	–	–	–	–	–	**Octadecanedioic acid, 3-oxo-, dimethyl ester**	55191-33-8	**WI01462**
55	200	214	125	97	83	41	43	**356**	100	67	62	56	48	44	44	42	**3.80**	20	36	5	–	–	–	–	–	–	–	–	–	–	**Octadecanedioic acid, 9-oxo-, dimethyl ester**	40393-43-9	**NI37251**
253	225	271	201	117	130	183	142	**356**	100	51	26	23	21	20	18	18	**4.40**	20	38	4	1	–	–	–	–	–	–	–	–	–	**3-Oxazolidine, 2-(3-carboxypropyl)-4-dimethyl-2-undecyl-**		**MS08409**
179	325	237	193	59	165	296	279	**356**	100	75	38	35	33	29	25	23	**0.00**	21	24	5	–	–	–	–	–	–	–	–	–	–	**9-Anthraceneacetic acid, 3-acetyl-1,2,3,4,5,8-hexahydro-7-(methoxycarbonyl)-, methyl ester**		**DD01415**
180	134	356	179	254	45	253	43	**356**	100	99	96	94	85	46	44	41		21	24	5	–	–	–	–	–	–	–	–	–	–	**4H-1-Benzopyran-4-one, 2,3-dihydro-7-hydroxy-2-(4-hydroxyphenyl)-5-methoxy-8-(3-methylbutyl)-**	72060-15-2	**NI37252**
295	41	310	43	296	91	185	241	**356**	100	42	34	25	21	21	14	13	**0.00**	21	24	5	–	–	–	–	–	–	–	–	–	–	**1,4-Phenanthrenedione, 7-formyloxy-4b,5,6,7,8,8a-hexahydro-3-hydroxy-4b,8,8-trimethyl-2-(2-propenyl)-, (4bS,7R,8aR)-**		**DD01416**
205	218	123	356	219	70	162	190	**356**	100	63	47	46	41	32	29	24		21	25	2	2	–	1	–	–	–	–	–	–	–	**1-Butanone, 1-(4-fluorophenyl)-4-[4-(2-methoxyphenyl)-1-piperazinyl]-**	1480-19-9	**LI07860**
205	218	123	356	219	70	162	56	**356**	100	63	48	47	42	31	30	30		21	25	2	2	–	1	–	–	–	–	–	–	–	**1-Butanone, 1-(4-fluorophenyl)-4-[4-(2-methoxyphenyl)-1-piperazinyl]-**	1480-19-9	**NI37253**
356	341	357	342	299	258	245	232	**356**	100	43	29	12	10	10	9	8		21	28	3	–	–	–	–	–	–	1	–	–	–	**Estra-1,3,5(10)-triene-6,17-dione, 3-(trimethylsilyloxy)-**	2786-99-4	**NI37254**
286	356	232	287	357	258	231	218	**356**	100	87	26	25	24	17	13	12		21	28	3	–	–	–	–	–	–	1	–	–	–	**Estra-1,3,5(10)-triene-6,17-dione, 3-(trimethylsilyloxy)-**	2786-99-4	**NI37255**
356	357	232	233	341	358	271	281	**356**	100	28	16	12	10	9	6	5		21	28	3	–	–	–	–	–	–	1	–	–	–	**Estra-1,3,5(10)-triene-7,17-dione, 3-(trimethylsilyloxy)-**		**NI37256**
166	196	56	356	174	43	164	326	**356**	100	92	82	74	71	61	60	50		21	28	3	2	–	–	–	–	–	–	–	–	–	**Akuammilan-16-methanol, 1,2-dihydro-17-hydroxy-10-methoxy-, (2α)-**	40169-73-1	**NI37257**
168	356	282	57	41	263	169	55	**356**	100	43	31	18	13	12	12	11		21	28	3	2	–	–	–	–	–	–	–	–	–	**Aspidospermidin-21-oic acid, 17-methoxy-, methyl ester**	22222-79-3	**NI37258**
168	356	282	169	357	283	196	325	**356**	100	31	23	11	9	8	8	6		21	28	3	2	–	–	–	–	–	–	–	–	–	**Aspidospermidin-21-oic acid, 17-methoxy-, methyl ester**	22222-79-3	**NI37259**

MASS TO CHARGE RATIOS								M.W.	INTENSITIES								Parent	C	H	O	N	S	F	Cl	Br	I	Si	P	B	X	COMPOUND NAME	CAS Reg No	No
356	328	325	355	126	127	357	43	**356**	100	33	32	30	28	23	21	20		21	28	3	2	–	–	–	–	–	–	–	–	–	**21-Noraspidospermidin-20-ol, 1-acetyl-17-methoxy-**	36458-98-7	**NI37260**
356	216	202	174	112	111	357	188	**356**	100	90	87	69	38	38	25	25		21	28	3	2	–	–	–	–	–	–	–	–	–	**Schizozygine, hexahydro-N-methyl-**	14509-83-2	**NI37261**
356	216	285	270	204	190	44	124	**356**	100	82	70	42	34	32	32	22		21	28	3	2	–	–	–	–	–	–	–	–	–	**3,4-Secocondyfolan-3-one, 14,19-didehydro-10,11-dimethoxy-4-methyl-, (2β,7β,14E,15α)-**	36954-74-2	**NI37262**
341	199	119	277	243	356	121	111	**356**	100	77	51	46	41	40	34	29		21	40	2	–	1	–	–	–	–	–	–	–	–	**9,11-Methanethiol adduct of methyl sterculeate**		**LI07861**
146	43	55	41	114	29	57	69	**356**	100	88	68	67	53	42	33	32	**5.20**	21	40	4	–	–	–	–	–	–	–	–	–	–	**Butanedioic acid, pentadecyl-, dimethyl ester**	41240-29-3	**MS08410**
145	43	117	28	57	146	55	41	**356**	100	47	22	14	13	7	6	6	**0.64**	21	40	4	–	–	–	–	–	–	–	–	–	–	**1,3-Dioxan-5-ol, 2-pentadecyl-, acetate, cis-**	30889-23-7	**NI37263**
145	43	28	117	57	41	55	29	**356**	100	82	27	25	20	14	11	9	**0.54**	21	40	4	–	–	–	–	–	–	–	–	–	–	**1,3-Dioxan-5-ol, 2-pentadecyl-, acetate, cis-**	30889-23-7	**NI37264**
145	43	117	57	41	55	146	29	**356**	100	63	25	14	9	8	7	6	**0.44**	21	40	4	–	–	–	–	–	–	–	–	–	–	**1,3-Dioxan-5-ol, 2-pentadecyl-, acetate, trans-**	30889-26-0	**NI37265**
145	43	117	28	57	146	41	55	**356**	100	45	21	13	12	7	7	6	**0.70**	21	40	4	–	–	–	–	–	–	–	–	–	–	**1,3-Dioxan-5-ol, 2-pentadecyl-, acetate, trans-**	30889-26-0	**NI37266**
145	43	117	28	57	41	55	29	**356**	100	97	30	25	22	16	12	10	**0.44**	21	40	4	–	–	–	–	–	–	–	–	–	–	**1,3-Dioxan-5-ol, 2-pentadecyl-, acetate, trans-**	30889-26-0	**NI37267**
327	28	57	43	55	101	71	328	**356**	100	63	48	33	30	29	27	24	**0.00**	21	40	4	–	–	–	–	–	–	–	–	–	–	**1,3-Dioxolane-4-methanol, 2-pentadecyl-, acetate**	56599-54-3	**NI37268**
145	43	28	57	41	146	55	29	**356**	100	44	18	13	10	7	7	6	**0.24**	21	40	4	–	–	–	–	–	–	–	–	–	–	**1,3-Dioxolane-4-methanol, 2-pentadecyl-, acetate, cis-**	30889-29-3	**NI37269**
145	43	57	41	28	146	55	29	**356**	100	43	12	9	8	7	7	6	**0.14**	21	40	4	–	–	–	–	–	–	–	–	–	–	**1,3-Dioxolane-4-methanol, 2-pentadecyl-, acetate, trans-**	30889-32-8	**NI37270**
145	28	43	31	57	69	55	41	**356**	100	87	36	24	18	15	14	14	**0.54**	21	40	4	–	–	–	–	–	–	–	–	–	–	**1,3-Dioxolane-4-methanol, 2-pentadecyl-, acetate, trans-**	30889-32-8	**NI37271**
102	55	87	98	41	126	69	43	**356**	100	46	33	32	31	30	27	26	**1.20**	21	40	4	–	–	–	–	–	–	–	–	–	–	**Heptadecanedioic acid, 2-ethyl-, dimethyl ester**	41240-35-1	**MS08411**
297	264	313	325	240	222	128	356	**356**	100	94	90	74	61	56	43	31		21	40	4	–	–	–	–	–	–	–	–	–	–	**Heptadecanoic acid, 3-acetoxy-16-methyl-, methyl ester**		**NI37272**
116	87	98	168	55	43	69	41	**356**	100	51	43	41	34	22	16	16	**1.40**	21	40	4	–	–	–	–	–	–	–	–	–	–	**Hexadecanedioic acid, 2-propyl-, dimethyl ester**	41240-34-0	**MS08412**
116	98	174	97	125	113	325	117	**356**	100	44	24	12	11	11	9	8	**1.29**	21	40	4	–	–	–	–	–	–	–	–	–	–	**Hexadecanedioic acid, 2-propyl-, dimethyl ester**	41240-34-0	**NI37273**
174	114	142	55	43	87	41	69	**356**	100	49	43	37	37	32	32	25	**3.60**	21	40	4	–	–	–	–	–	–	–	–	–	–	**Hexanedioic acid, 2-tridecyl-, dimethyl ester**	41240-31-7	**MS08413**
43	56	85	55	41	42	28	69	**356**	100	78	53	45	37	30	30	26	**0.00**	21	40	4	–	–	–	–	–	–	–	–	–	–	**Nonanedioic acid, dihexyl ester**		**DO01112**
88	112	98	55	69	41	74	43	**356**	100	51	49	41	31	30	29	27	**2.40**	21	40	4	–	–	–	–	–	–	–	–	–	–	**Octadecanedioic acid, 2-methyl-, dimethyl ester**	41240-36-2	**MS08414**
55	41	69	43	83	98	67	57	**356**	100	72	64	56	48	46	43	42	**2.00**	21	40	4	–	–	–	–	–	–	–	–	–	–	**9-Octadecenoic acid (Z)-, 2,3-dihydroxypropyl ester**	111-03-5	**WI01463**
264	265	356	282	338	357	97	280	**356**	100	20	18	15	8	5	5	4		21	40	4	–	–	–	–	–	–	–	–	–	–	**9-Octadecenoic acid (Z)-, 2,3-dihydroxypropyl ester**	111-03-5	**WI01464**
357	358	339	265	340	359	255	241	**356**	100	26	23	7	4	3	2	2	**0.00**	21	40	4	–	–	–	–	–	–	–	–	–	–	**9-Octadecenoic acid (Z)-, 2,3-dihydroxypropyl ester**	111-03-5	**NI37274**
55	41	69	43	67	81	264	83	**356**	100	90	69	66	65	57	51	51	**1.00**	21	40	4	–	–	–	–	–	–	–	–	–	–	**9-Octadecenoic acid (Z)-, 2-hydroxy-1-(hydroxymethyl)ethyl ester**	3443-84-3	**WI01465**
264	262	265	356	280	338	263	282	**356**	100	28	24	10	9	8	8	6		21	40	4	–	–	–	–	–	–	–	–	–	–	**9-Octadecenoic acid (Z)-, 2-hydroxy-1-(hydroxymethyl)ethyl ester**	3443-84-3	**WI01466**
55	87	43	41	202	228	138	69	**356**	100	93	92	83	72	64	62	57	**3.40**	21	40	4	–	–	–	–	–	–	–	–	–	–	**Octanedioic acid, 2-undecyl-, dimethyl ester**	41240-32-8	**MS08415**
160	43	55	41	128	74	57	69	**356**	100	65	59	53	48	33	32	31	**5.70**	21	40	4	–	–	–	–	–	–	–	–	–	–	**Pentanedioic acid, 2-tetradecyl-, dimethyl ester**	29238-07-1	**MS08416**
132	145	43	55	41	87	57	133	**356**	100	93	37	32	26	17	17	16	**4.30**	21	40	4	–	–	–	–	–	–	–	–	–	–	**Propanedioic acid, hexadecyl-, dimethyl ester**	23130-42-9	**MS08417**
87	158	55	98	41	43	69	74	**356**	100	84	81	71	63	60	50	38	**0.10**	21	40	4	–	–	–	–	–	–	–	–	–	–	**Tridecanedioic acid, 2-hexyl-, dimethyl ester**	25751-93-3	**MS08418**
73	117	75	132	43	145	129	55	**356**	100	79	75	65	43	33	31	31	**4.72**	21	44	2	–	–	–	–	–	–	1	–	–	–	**Octadecanoic acid, trimethylsilyl ester**	18748-91-9	**NI37275**
73	117	75	132	43	341	145	129	**356**	100	74	70	50	43	42	33	30	**7.24**	21	44	2	–	–	–	–	–	–	1	–	–	–	**Octadecanoic acid, trimethylsilyl ester**	18748-91-9	**NI37276**
117	132	341	145	73	313	342	356	**356**	100	95	89	62	62	51	24	20		21	44	2	–	–	–	–	–	–	1	–	–	–	**Octadecanoic acid, trimethylsilyl ester**	18748-91-9	**NI37277**
341	75	73	55	83	69	130	342	**356**	100	75	63	51	41	37	30	27	**8.00**	21	44	2	–	–	–	–	–	–	1	–	–	–	**12-Octadecanone, 1-[(trimethylsilyl)oxy]-**		**NI37278**
99	98	70	56	57	259	161	69	**356**	100	57	55	48	34	29	27	20	**9.00**	21	45	3	–	–	–	–	–	–	–	–	1	–	**Boric acid (H3BO3), triheptyl ester**	2938-83-2	**NI37279**
189	190	221	91	79	117	107	77	**356**	100	14	3	3	3	2	2	2	**0.00**	22	28	2	–	1	–	–	–	–	–	–	–	–	**Benzenecarbothioic acid, 2,4,6-triethyl-, S-(2-hydroxy-1-methyl-2-phenylethyl) ester**	67902-81-2	**NI37280**
314	315	172	254	225	146	160	226	**356**	100	23	14	11	10	8	7	6	**6.11**	22	28	4	–	–	–	–	–	–	–	–	–	–	**Estra-1,3,5(10)-triene-3,17-diol (17β)-, diacetate**	3434-88-6	**NI37281**
314	43	315	172	158	44	41	133	**356**	100	54	23	18	17	17	16	15	**5.00**	22	28	4	–	–	–	–	–	–	–	–	–	–	**Estra-1,3,5(10)-triene-3,17-diol (17β)-, diacetate**	3434-88-6	**NI37282**
99	100	252	43	294	55	253	41	**356**	100	25	14	10	7	7	5	4	**4.00**	22	28	4	–	–	–	–	–	–	–	–	–	–	**Estra-1,3,5(10)-trien-17-one, 3-(acetyloxy)-, cyclic 1,2-ethanediyl acetal**	27736-68-1	**NI37283**
100	91	120	85	43	341	283	178	**356**	100	89	85	37	29	26	10	9	**0.50**	22	28	4	–	–	–	–	–	–	–	–	–	–	**Hexacyclo[8.4.2.0^{2,9}.0^{3,5}.0^{4,8}.0^{11,14}]hexadec-15-ene, 6,7:12,13-bis(isopropylidenedioxy)-**	75765-44-5	**NI37284**
187	188	186	161	128	115	82	356	**356**	100	20	15	15	13	13	13	10		22	28	4	–	–	–	–	–	–	–	–	–	–	**8,14-seco-D-Homoestra-1,3,5(10),9(11)-tetraen-14-one, 17a-acetoxy-3-methoxy-, (17α)-**		**LI07862**
269	356	178	151	150	270	357	234	**356**	100	68	45	35	27	20	19	14		22	28	4	–	–	–	–	–	–	–	–	–	–	**Naphthalene, 1-(3,4-dimethoxyphenyl)-1,2,3,4-tetrahydro-6,7-dimethoxy-2,3-dimethyl-, [1S-(1α,2β,3α)]-**	521-54-0	**NI37285**
356	269	357	270	299	285	212	238	**356**	100	73	24	12	10	7	7	6		22	28	4	–	–	–	–	–	–	–	–	–	–	**Naphthalene, 1-(3,4-dimethoxyphenyl)-1,2,3,4-tetrahydro-6,7-dimethoxy-2,3-dimethyl-, [1S-(1α,2β,3α)]-**	521-54-0	**NI37286**
145	159	173	161	296	281	356	280	**356**	100	86	79	72	28	17	12	5		22	28	4	–	–	–	–	–	–	–	–	–	–	**Verecynarmin A**		**DD01417**
265	91	266	120	65	174	92	87	**356**	100	41	16	7	6	4	4	4	**0.06**	22	30	–	–	–	–	–	–	–	–	2	–	–	**1,6-Diphosphecane, 1,6-dibenzyl-**	61142-49-2	**NI37287**
186	169	183	73	266	199	214	75	**356**	100	50	32	27	24	24	20	20	**19.63**	22	32	2	–	–	–	–	–	–	1	–	–	–	**Estra-1,3,5(10),8-tetraene, 3-methoxy-17-(trimethylsilyloxy)-, (17α)-**	55191-32-7	**NI37288**
356	357	232	271	258	272	341	231	**356**	100	30	23	22	22	18	13	13		22	32	2	–	–	–	–	–	–	1	–	–	–	**Estra-1,3,5(10)-trien-17-one, 1-methyl-3-(trimethylsilyloxy)-**		**NI37289**
140	356	141	188	297	168	202	41	**356**	100	23	13	9	8	7	6	5		22	32	2	2	–	–	–	–	–	–	–	–	–	**Aspidospermidin-21-ol, 1-ethyl-17-methoxy-**	54751-76-7	**NI37290**
135	77	136	41	69	107	43	55	**356**	100	10	9	8	7	6	6	4	**0.30**	22	32	2	2	–	–	–	–	–	–	–	–	–	**1-Dodecanone, 2-(imidazol-1-yl)-1-(4-methoxyphenyl)-**		**MS08419**
110	43	109	91	356	105	79	41	**356**	100	70	33	32	25	25	25	23		22	33	3	–	–	–	–	–	–	–	–	1	–	**Pregna-5,17(20)-dien-3-ol, 20,21-[(methylborylene)bis(oxy)]-, (3β)-**	30881-79-9	**MS08420**

MASS TO CHARGE RATIOS	M.W.	INTENSITIES	Parent	C	H	O	N	S	F	Cl	Br	I	Si	P	B	X	COMPOUND NAME	CAS Reg No	No
110 109 356 105 229 213 107 145	**356**	100 33 25 25 19 19 18 17		22	33	3	–	–	–	–	–	–	–	–	1	–	Pregna-5,17(20)-dien-3-ol, 20,21-[(methylborylene)bis(oxy)]-, (3β)-	30881-79-9	NI37291
110 109 356 124 111 55 81 91	**356**	100 28 15 13 10 10 8 7		22	33	3	–	–	–	–	–	–	–	–	1	–	Pregn-17(20)-en-3-one, 20,21-[(methylborylene)bis(oxy)]-, (5α)-	31012-60-9	LI07863
110 109 356 124 111 55 81 67	**356**	100 27 14 11 10 10 8 7		22	33	3	–	–	–	–	–	–	–	–	1	–	Pregn-17(20)-en-3-one, 20,21-[(methylborylene)bis(oxy)]-, (5α)-	31012-60-9	NI37292
124 314 85 356 79 91 271 173	**356**	100 24 22 20 19 18 16 16		22	33	3	–	–	–	–	–	–	–	–	1	–	Pregn-4-en-3-one, 20,21-[(methyleneborylene)bis(oxy)]-, (20R)-	30882-65-6	LI07864
124 314 85 356 91 79 95 93	**356**	100 23 20 18 18 18 17 16		22	33	3	–	–	–	–	–	–	–	–	1	–	Pregn-4-en-3-one, 20,21-[(methyleneborylene)bis(oxy)]-, (20R)-	30882-65-6	NI37293
124 314 85 356 79 91 95 271	**356**	100 23 20 18 18 17 16 15		22	33	3	–	–	–	–	–	–	–	–	1	–	Pregn-4-en-3-one, 20,21-[(methyleneborylene)bis(oxy)]-, (20R)-	30882-65-6	MS08421
43 104 57 55 98 41 117 69	**356**	100 70 62 61 53 52 35 34	**5.04**	22	44	3	–	–	–	–	–	–	–	–	–	–	Eicosanoic acid, 2-hydroxyethyl ester	26158-80-5	NI37294
57 101 341 71 85 149 356 195	**356**	100 93 51 48 42 11 2 2		22	44	3	–	–	–	–	–	–	–	–	–	–	Hexadecane, 16-[(2,3-(isopropylidenedioxy)propyl)oxy]-		MS08422
73 43 55 41 57 71 69 29	**356**	100 61 59 35 32 26 18 18	**0.34**	22	44	3	–	–	–	–	–	–	–	–	–	–	Octadecanoic acid, 4-hydroxybutyl ester	15337-64-1	NI37295
73 43 55 57 72 41 45 71	**356**	100 88 66 62 58 49 36 31	**1.14**	22	44	3	–	–	–	–	–	–	–	–	–	–	Octadecanoic acid, 2-hydroxy-1-sec-butyl ester	14251-39-9	NI37296
299 300 75 97 83 111 89 301	**356**	100 28 23 14 14 8 8 7	**0.00**	22	48	1	–	–	–	–	–	–	1	–	–	–	Hexadecyl tert-butyldimethylsilyl ether		NI37297
291 292 289 356 215 290 144.5 145	**356**	100 67 52 43 32 29 24 21		23	16	2	–	1	–	–	–	–	–	–	–	–	2H-Naphtho[2,3-b]thiete, 3,8-diphenyl-, 1,1-dioxide		LI07865
195 180 196 152 83 165 85 76	**356**	100 91 83 78 78 70 70 70	**6.50**	23	16	4	–	–	–	–	–	–	–	–	–	–	1H-Indene-1,3(2H)-dione, 2-hydroxy-2-(9-methoxy-9H-fluoren-9-yl)-	50616-96-1	NI37298
253 356 255 115 77 254 217 358	**356**	100 51 35 28 27 26 20 19		23	17	–	2	–	–	1	–	–	–	–	–	–	4H-1,2-Diazepine, 5-(4-chlorophenyl)-3,7-diphenyl-	25649-73-4	NI37299
267 296 254 268 266 355 326 338	**356**	100 93 63 33 33 30 28 27	**6.00**	23	18	3	1	–	–	–	–	–	–	–	–	–	5-(1-Methoxycarbonyl-2-oxidovinyl)-6-phenylphenanthridinine		LI07866
249 279 356 104 77 91	**356**	100 96 88 56 30 20		23	20	2	2	–	–	–	–	–	–	–	–	–	2-Oxo-4,6-diphenyl-5-tolyloxy-1,2,3,4-tetrahydropyrimidine		LI07867
279 263 356 118 91 77 104	**356**	100 90 82 44 22 20 17		23	20	2	2	–	–	–	–	–	–	–	–	–	2-Oxo-4-phenyl-5-phenoxy-6-tolyl-1,2,3,4-tetrahydropyrimidine		LI07868
263 356 265 104 77 118 91	**356**	100 62 56 42 26 10 10		23	20	2	2	–	–	–	–	–	–	–	–	–	2-Oxo-4-tolyl-5-phenoxy-6-phenyl-1,2,3,4-tetrahydropyrimidine		LI07869
338 55 93 356 91 241 53 173	**356**	100 87 81 75 69 58 52 49		23	32	3	–	–	–	–	–	–	–	–	–	–	Pregna-4,6-diene-3,20-dione, 17-hydroxy-6,16-dimethyl-, (16α)-	5244-58-6	WI01467
296 91 105 281 297 145 77 121	**356**	100 40 29 26 24 23 22 20	**0.00**	23	32	3	–	–	–	–	–	–	–	–	–	–	Pregna-5,16-dien-20-one, 3-(acetyloxy)-, (3β)-	979-02-2	WI01468
43 296 297 145 121 91 281 107	**356**	100 90 20 19 18 18 17 16	**0.00**	23	32	3	–	–	–	–	–	–	–	–	–	–	Pregna-5,16-dien-20-one, 3-(acetyloxy)-, (3β)-	979-02-2	NI37300
91 296 79 105 67 93 81 77	**356**	100 99 83 67 64 63 63 61	**0.00**	23	32	3	–	–	–	–	–	–	–	–	–	–	Pregna-5,16-dien-20-one, 3-(acetyloxy)-, (3β)-	979-02-2	NI37301
341 209 342 73 155 153 140 167	**356**	100 45 30 26 23 22 22 21	**9.89**	23	36	1	–	–	–	–	–	–	1	–	–	–	19-Norpregn-4-en-20-yne, 17-[(trimethylsilyl)oxy]-, (17α)-	56771-62-1	NI37302
341 342 209 356 153 140 155 133	**356**	100 30 21 15 10 10 9 9		23	36	1	–	–	–	–	–	–	1	–	–	–	19-Norpregn-4-en-20-yne, 17-[(trimethylsilyl)oxy]-, (17α)-	56771-62-1	MS08423
341 342 356 340 339 311 357 178	**356**	100 26 20 7 7 7 6 5		24	20	3	–	–	–	–	–	–	–	–	–	–	7H-Dibenzo[c,h]xanthene-5,9-diol, 6,7,8-trimethyl-	99893-92-2	NI37303
258 43 78 77 69 71 259 51	**356**	100 65 48 32 32 27 21 18	**0.00**	24	24	1	2	–	–	–	–	–	–	–	–	–	Acetamide, α-anilino-α,α-diphenyl-N-pyrrolidino-		LI07870
236 237 91 165	**356**	100 19 14 7	**0.00**	24	24	1	2	–	–	–	–	–	–	–	–	–	Acetamide, α,α,N-triphenyl-α-(N-pyrrolidino)-		LI07871
91 84 86 106 196 107 77 251	**356**	100 96 65 38 28 28 27 17	**17.00**	24	24	1	2	–	–	–	–	–	–	–	–	–	Acetic acid, 2-methyl-2-phenyl-1-[2-phenyl-1-benzylvinyl]hydrazide	67134-53-6	NI37304
356 286 285 43 55 99 211 41	**356**	100 68 31 30 29 28 27 25		24	36	2	–	–	–	–	–	–	–	–	–	–	Androst-4-ene-3,17-dione, 6-pentyl-, (6α)-		MS08424
356 341 213 165 323 245 231 229	**356**	100 21 21 16 9 9 9 7		24	36	2	–	–	–	–	–	–	–	–	–	–	Androst-5-en-17-one, 3-hydroxy-16-(2,2-dimethylpropylidene)-, (3β)-		LI07872
258 43 137 55 41 356 313 91	**356**	100 86 36 34 27 23 23 23		24	36	2	–	–	–	–	–	–	–	–	–	–	Pregn-4-ene-3,20-dione, 17-ethyl-6-methyl-, (6α)-	978-80-3	NI37305
118 136 55 59 91 81 57 67	**356**	100 99 72 63 60 49 47 45	**13.33**	24	36	2	–	–	–	–	–	–	–	–	–	–	9,10-Secochola-5,7,10(19)-trien-24-al, 3-hydroxy-, (3β,5Z,7E)-	40013-88-5	NI37306
84 71 356 341 58 56 70 44	**356**	100 9 3 3 2 2 1 1		24	40	–	2	–	–	–	–	–	–	–	–	–	Con-5-enin-3-amine, N,N-dimethyl-, (3β)-	546-06-5	NI37307
55 83 81 255 67 107 356 79	**356**	100 89 48 40 36 34 33 32		25	40	1	–	–	–	–	–	–	–	–	–	–	26,27-Dinorcholesta-5,22-dien-3-ol, (3β,22E)-	57597-10-1	NI37308
243 244 183 165 105 55 77 41	**356**	100 26 23 22 20 10 5 4	**3.00**	26	28	1	–	–	–	–	–	–	–	–	–	–	4-Methylcyclohexyl trityl ether	20705-41-3	NI37309
243 244 183 165 105 55 134 77	**356**	100 26 23 21 19 10 5 5	**2.40**	26	28	1	–	–	–	–	–	–	–	–	–	–	4-Methylcyclohexyl trityl ether	20705-41-3	NI37310
243 244 183 165 105 228 55 215	**356**	100 26 23 21 20 10 10 5	**2.00**	26	28	1	–	–	–	–	–	–	–	–	–	–	4-Methylcyclohexyl trityl ether	20705-41-3	LI07873
257 81 55 41 271 95 258 69	**356**	100 33 30 30 25 25 24 24	**6.00**	26	44	–	–	–	–	–	–	–	–	–	–	–	21-Norcholest-24-ene, (5α)-	14949-19-0	NI37311
257 81 55 41 95 271 69 67	**356**	100 33 30 30 25 24 24 23	**5.00**	26	44	–	–	–	–	–	–	–	–	–	–	–	21-Norcholest-24-ene, (5α)-	14949-19-0	LI07874
133 341 221 356 327 207 147 183	**356**	100 94 44 39 26 26 22 20		27	32	–	–	–	–	–	–	–	–	–	–	–	Cyclopentane, 2-ethylidene-1,3-bis[(E)-2,4,6-trimethylbenzylidene]-		MS08425
133 223 356 208 209 157 224 193	**356**	100 68 52 18 17 17 16 14		27	32	–	–	–	–	–	–	–	–	–	–	–	Cyclopentene, 1-(2,4,6-trimethylbenzyl)-3-[(E)-2,4,6-trimethylbenzylidene]-2-vinyl-		MS08426
356 357 278 276 139 277 163 358	**356**	100 31 9 9 9 6 6 5		28	20	–	–	–	–	–	–	–	–	–	–	–	Azulene, 1,2,3-triphenyl-		AI02158
178 152 151 179 279 356 278 149	**356**	100 30 29 20 11 10 8 8		28	20	–	–	–	–	–	–	–	–	–	–	–	2:3,4:5-Dibenzobicyclo[4,2,0]octa-2,4,7-triene		LI07875
356 357 178 165 194 279 166 278	**356**	100 29 28 28 25 20 18 15		28	20	–	–	–	–	–	–	–	–	–	–	–	5H-Dibenzo[a,d]cycloheptene, 5-(diphenylmethylene)-	18916-68-2	NI37312
356 280 279 339 277 278 326	**356**	100 12 9 8 8 6 5		28	20	–	–	–	–	–	–	–	–	–	–	–	Naphthalene, 1,2,3-triphenyl-		DO01113
251 359 250 253 361 357 266 252	**357**	100 60 52 50 31 30 30 24		8	9	3	1	1	–	–	2	–	–	–	–	–	2,6-Dibromo-4-(hydroxyethylsulphonyl)aniline		IC04107
107 77 92 171 120 40 42 63	**357**	100 86 82 74 57 55 50 45	**15.00**	13	15	5	3	2	–	–	–	–	–	–	–	–	2-(4-Methoxybenzenesulphonamido)-5-(2-carboxypropyl)-1,3,4-thiadiazine		LI07876
43 58 162 164 220 184 59 45	**357**	100 40 36 25 17 17 17 14	**1.00**	14	13	4	3	–	–	2	–	–	–	–	–	–	L-Histidine, N-[(2,4-dichlorophenoxy)acetyl]-		NI37313
181 151 194 182 136 179 57 43	**357**	100 33 15 10 7 6 6 4	**0.00**	14	16	4	1	–	5	–	–	–	–	–	–	–	Benzeneethanamine, 2,4,5-trimethoxy-N-(pentafluoropropionyl)-		MS08427
181 121 182 136 194 179 151 123	**357**	100 55 39 30 18 15 12 11	**0.00**	14	16	4	1	–	5	–	–	–	–	–	–	–	Benzeneethanamine, 2,4,6-trimethoxy-N-(pentafluoropropionyl)-		MS08428
359 357 361 265 343 341 345 263	**357**	100 56 55 54 38 21 20 19		14	17	–	1	–	–	–	2	–	–	–	–	–	Carbazole, 6,8-dibromo-1,2,3,4,4a,9a-hexahydro-4a,9-dimethyl-, trans-		LI07877
339 225 247 105 357 293 309	**357**	100 42 30 30 14 14 13		15	11	6	5	–	–	–	–	–	–	–	–	–	1H-Indole, 1-methyl-2-(picrylamino)-		LI07878
166 195 357 196 28 167 138 110	**357**	100 82 15 15 13 12 12 8		15	19	9	1	–	–	–	–	–	–	–	–	–	2H-1,4-Benzoxazin-3(4H)-one, 2-(β-D-glucopyranosyloxy)-7-methoxy-	17622-26-3	NI37314

MASS TO CHARGE RATIOS								M.W.	INTENSITIES								Parent	C	H	O	N	S	F	Cl	Br	I	Si	P	B	X	COMPOUND NAME	CAS Reg No	No
153	315	195	255	213	170	212	169	**357**	100	47	33	28	16	16	13	13	**0.00**	15	19	9	1	–	–	–	–	–	–	–	–	–	2-Oxabicyclo[2.2.1]heptane, 6-endo-(N-acetylcarbamoyl)-3-exo,6-exo,7-syn-triacetyl-, (1RS,3SR,4RS,6RS,7RS)-		DD01418
228	256	284	29	81	312	101	238	**357**	100	89	67	57	48	47	45	44	**26.00**	15	20	7	1	–	–	–	–	–	–	1	–	–	2-Propenoic acid, 2-(diethoxyphosphinyl)-3-(3-nitrophenyl)-, ethyl ester		NI37315
228	256	284	312	202	249	229	283	**357**	100	85	67	41	29	28	25	24	**20.00**	15	20	7	1	–	–	–	–	–	–	1	–	–	2-Propenoic acid, 2-(diethoxyphosphinyl)-3-(4-nitrophenyl)-, ethyl ester	76386-70-4	NI37316
77	106	105	78	112	96	329	122	**357**	100	96	19	13	5	5	1	1	**0.10**	16	15	5	5	–	–	–	–	–	–	–	–	–	Uridine, 5′-acetyl-5′-deoxy-2′,3′-O-benzylidene-	35837-31-1	NI37317
242	210	357	57	238	41	178	301	**357**	100	82	40	25	18	18	15	14		16	23	8	1	–	–	–	–	–	–	–	–	–	1,3-Butadiene-1,2,3,4-tetracarboxylic acid, 1-(tert-butylamino)-, tetramethyl ester		NI37318
298	300	191	357	359	190	220	299	**357**	100	99	80	64	62	34	26	19		17	12	3	1	–	–	–	1	–	–	–	–	–	4H-Quinolizine-3-carboxylic acid, 1-bromo-4-oxo-2-phenyl-, methyl ester		LI07879
317	357	133	319	173	318	359	54	**357**	100	56	47	38	30	20	19	19		17	16	2	5	–	–	1	–	–	–	–	–	–	Aniline, N-(cyanoethyl)-N-ethyl-4-(2-chloro-4-nitrophenylazo)-		IC04108
317	104	133	173	357	318	105	77	**357**	100	79	66	45	40	18	15	14		17	16	2	5	–	–	1	–	–	–	–	–	–	Aniline, N-(cyanoethyl)-N-ethyl-4-(2-chloro-4-nitrophenylazo)-		IC04109
162	91	118	77	79	133	65	134	**357**	100	61	40	39	38	28	28	27	**2.00**	17	19	4	5	–	–	–	–	–	–	–	–	–	Aniline, N,N-diethyl-3-methyl-4-(2,4-dinitrophenylazo)-		IC04110
115	342	135	116	216	151	97	79	**357**	100	15	11	7	5	5	5	5	**0.00**	17	27	5	1	1	–	–	–	–	–	–	–	–	Bornane-10,2-sultam, N-[(4R,5S)-2,2,5-trimethyl-1,3-dioxolane-4-carbonyl]-		DD01419
124	268	226	43	108	57	41	106	**357**	100	45	37	32	17	11	10	9	**1.00**	17	27	5	1	1	–	–	–	–	–	–	–	–	Pyridine, 1-acetyl-2,6-diacetoxy-3,5-dimethyl-3-(tert-butylmercapto)-1,2,3,6-tetrahydro-		MS08429
231	230	242	105	273	315	357		**357**	100	69	50	50	42	12	5			18	16	3	3	–	–	1	–	–	–	–	–	–	2′-Benzoyl-4′-chloro-N-(2-acetylhydrazono)methylacetanilide		LI07880
357	145	224	120	63	132	184	183	**357**	100	78	64	24	20	19	11	11		18	18	2	2	–	–	–	–	–	–	–	–	1	Copper, isobutylenediamino-N,N′-bis(salicylidenyl)-	32593-88-7	LI07881
357	145	224	359	226	358	120	63	**357**	100	77	64	50	30	25	25	20		18	18	2	2	–	–	–	–	–	–	–	–	1	Copper, isobutylenediamino-N,N′-bis(salicylidenyl)-	32593-88-7	NI37319
43	133	41	55	57	134	69	60	**357**	100	89	89	70	53	31	27	24	**6.00**	18	35	–	3	2	–	–	–	–	–	–	–	–	1,3,4-Thiadiazole, 2-amino-5-(hexadecylthio)-	33313-11-0	NI37320
44	28	29	31	43	60	73	41	**357**	100	89	82	81	65	60	56	52	**8.50**	19	16	4	1	–	–	1	–	–	–	–	–	–	1H-Indole-3-acetic acid, 1-(4-chlorobenzoyl)-5-methoxy-2-methyl-	53-86-1	NI37321
139	141	357	111	140	359	113	75	**357**	100	37	26	23	10	9	8	8		19	16	4	1	–	–	1	–	–	–	–	–	–	Indomethacin	53-86-1	NI37322
91	146	155	357	202	157	174	130	**357**	100	57	54	35	34	34	31	29		19	19	4	1	1	–	–	–	–	–	–	–	–	1H-Indole-3-carboxylic acid, 1-(4-tolylsulphonyl)-2-methyl-, ethyl ester		DD01420
103	47	75	357	252	180			**357**	100	53	48	13	12	12				19	23	4	3	–	–	–	–	–	–	–	–	–	2-[1-(4,4-Diethoxy-2-azabut-1-enyl)]-2′-nitrodiphenylamine		LI07882
266	193	268	342	77	357	147	135	**357**	100	98	81	61	42	32	28	27		19	27	2	1	–	–	–	–	–	2	–	–	–	Benzamide, N-phenyl-N-(trimethylsilyl)-2-[(trimethylsilyl)oxy]-		LI07883
322	323	320	306	321	307	50	308	**357**	100	51	46	42	38	22	18	12	**0.00**	20	20	3	1	–	–	1	–	–	–	–	–	–	5,6-Dihydro-2,3,11-trimethoxydibenzo[a,g]quinolizinium chloride		MS08430
135	357	125	181	105	131	177	124	**357**	100	65	52	44	26	19	16	12		20	23	3	1	1	–	–	–	–	–	–	–	–	Benzenepropanoic acid, γ-[2-(propionylamino)phenylthio]-, ethyl ester		MS08431
130	185	92	156	143	357	131	117	**357**	100	30	29	25	24	21	21	21		20	23	3	1	1	–	–	–	–	–	–	–	–	5-(3-Indolyl)pentyl-1-toluene-4-sulphonate		LI07884
208	341	206	135	150	149	326		**357**	100	52	29	24	16	16	12		**0.00**	20	23	5	1	–	–	–	–	–	–	–	–	–	Capaurimine		LI07885
208	357	206	135	150	149	326	342	**357**	100	52	29	24	16	16	10	5		20	23	5	1	–	–	–	–	–	–	–	–	–	Capaurimine		LI07886
208	357	356	206	360	326	209	135	**357**	100	73	41	34	14	14	14	14		20	23	5	1	–	–	–	–	–	–	–	–	–	Capaurimine		LI07887
298	266	150	357	314	137	326	214	**357**	100	48	47	32	23	23	19	18		20	23	5	1	–	–	–	–	–	–	–	–	–	Cephalotaxine, acetate (ester)	24274-60-0	NI37323
298	357	28	266	43	314	150	299	**357**	100	58	38	31	29	27	22	21		20	23	5	1	–	–	–	–	–	–	–	–	–	Cephalotaxine, 6-acetyl-		MS08432
314	357	283	271	329	257	270	244	**357**	100	97	76	65	60	58	50	50		20	23	5	1	–	–	–	–	–	–	–	–	–	1,7-Cycloerythrinan-2,8-dione, 15,16-dimethoxy-, cyclic 2-(1,2-ethanediyl acetal)	75444-10-9	NI37324
339	324	340	296	341	337	281	310	**357**	100	80	43	25	20	20	20	14	**1.00**	20	23	5	1	–	–	–	–	–	–	–	–	–	4H-Dibenzo[de,g]quinolin-9-ol, 5,6,6a,7-tetrahydro-1,2,10-trimethoxy-N-methyl-, β-N-oxide, (+)-		NI37325
298	357	314	299	358	43	297	268	**357**	100	93	35	24	22	20	17	16		20	23	5	1	–	–	–	–	–	–	–	–	–	7a,9c-(Iminoethano)phenanthro[4,5-bcd]furan-5,8-diol, 4aα,5α,8α,9-tetrahydro-3-methoxy-12-methyl-, 8-acetate, (+)-	33064-79-8	NI37326
44	70	42	43	55	58	115	41	**357**	100	95	80	48	32	27	26	22	**14.01**	20	23	5	1	–	–	–	–	–	–	–	–	–	Morphinan-6-one, 7,8-didehydro-4,5α-epoxy-14-hydroxy-3-methoxy-17-methyl-, cyclic ethylene acetal	7415-43-2	NI37327
208	83	70	205	136	135	148	120	**357**	100	99	99	80	70	65	64	57	**36.00**	20	27	3	3	–	–	–	–	–	–	–	–	–	2,3,11,12-Dibenzo-1,7,13-thioxa-4,10,16-triazacyclooctadeca-2,11-diene		MS08433
298	114	55	130	174	299	258	100	**357**	100	83	73	34	32	18	18	18	**0.00**	20	43	2	1	–	–	–	–	–	1	–	–	–	1-Nonene, 1-[[[bis(2-ethoxyethyl)amino]methyl]dimethylsilyl]-, (E)-		DD01421
298	114	55	174	130	299	69	100	**357**	100	55	37	22	22	18	12	11	**0.00**	20	43	2	1	–	–	–	–	–	1	–	–	–	1-Nonene, 1-[[[bis(2-ethoxyethyl)amino]methyl]dimethylsilyl]-, (Z)-		DD01422
298	55	130	114	174	69	299	232	**357**	100	71	25	24	22	21	18	17	**0.00**	20	43	2	1	–	–	–	–	–	1	–	–	–	1-Nonene, 3-[[[bis(2-ethoxyethyl)amino]methyl]dimethylsilyl]-		DD01423
119	357	91	41	43	120	64	93	**357**	100	36	15	13	12	8	7	6		21	15	3	3	–	–	–	–	–	–	–	–	–	1,3,5-Triazine-2,4,6(1H,3H,5H)-trione, 1,3,5-triphenyl-	1785-02-0	NI37328
77	264	51	65	121	238	70	265	**357**	100	90	42	37	31	28	18	16	**1.80**	21	15	3	3	–	–	–	–	–	–	–	–	–	1,3,5-Triazine, 2,4,6-triphenoxy-	1919-48-8	NI37329
357	41	57	69	43	55	73	56	**357**	100	79	69	60	57	52	49	43		21	15	3	3	–	–	–	–	–	–	–	–	–	1,3,5-Triazine, 2,4,6-triphenoxy-	1919-48-8	IC04111
357	79	356	80	278	78	342	277	**357**	100	95	75	69	33	28	24	22		21	30	–	3	–	–	–	–	–	–	–	3	–	1,3,5,2,4,6-Triazatriborine, 1,3,5-trimethyl-2,4,6-tris(1-methyl-2,4-cyclopentadien-1-yl)-	72088-05-2	NI37330
55	57	41	43	69	71	91	85	**357**	100	92	90	88	53	52	49	29	**0.00**	21	40	1	1	–	–	1	–	–	–	–	–	–	Pyridinium, 1-hexadecyl-, chloride, monohydrate	6004-24-6	NI37331
356	357	77	269	258	165	310	180	**357**	100	94	81	56	53	51	44	43		22	19	–	3	1	–	–	–	–	–	–	–	–	2-Imidazoline, 4-imino-2-methylthio-1,5,5-triphenyl-		NI37332
224	357	165	77	239	356	358	225	**357**	100	61	32	26	19	17	16	16		22	19	–	3	1	–	–	–	–	–	–	–	–	2-Imidazoline, 5-imino-2-methylthio-1,4,4-triphenyl-		NI37333
91	77	196	209	118	266	161	106	**357**	100	83	79	73	50	48	36	10	**3.00**	22	19	2	3	–	–	–	–	–	–	–	–	–	2,3-Pyridinedicarboxamide, N-[(dibenzylamino)methyl]-	96919-16-3	NI37334
266	91	196	118	161	209	210	208	**357**	100	69	48	20	18	15	10	10	**1.00**	22	19	2	3	–	–	–	–	–	–	–	–	–	3,4-Pyridinedicarboxamide, N-[(dibenzylamino)methyl]-	96919-17-4	NI37335
357	258	331	339	137	259	308	272	**357**	100	87	79	77	61	60	54	53		22	31	3	1	–	–	–	–	–	–	–	–	–	DL-18,19-Dinorpregn-4-en-20-yn-3-one, 13-ethyl-16,17-hydroxy-, (16β,17α)-, O-methyl oxime		NI37336

MASS TO CHARGE RATIOS								M.W.	INTENSITIES								Parent	C	H	O	N	S	F	Cl	Br	I	Si	P	B	X	COMPOUND NAME	CAS Reg No	No
210	260	77	100	105	242	91	58	**357**	100	80	77	53	34	28	27	24	**7.00**	22	31	3	1	–	–	–	–	–	–	–	–	–	**3-Hexen-2-one, 6-(6-acetonyl-5-hydroxy-1,4-dimethyl-5-phenyl-2-piperidyl)-**	34443-89-5	**NI37337**
357	168	76	121	159	135	122	157	**357**	100	71	54	45	44	40	38	37		22	31	3	1	–	–	–	–	–	–	–	–	–	**N-Retinoylglycine (all-trans)**	71407-30-2	**NI37338**
357	328	298	246	314	299	270	180	**357**	100	16	16	13	10	7	7	6		22	31	3	1	–	–	–	–	–	–	–	–	–	**Sangorine**		**LI07888**
91	210	120	160	92	77	65	107	**357**	100	17	16	8	8	7	7	6	**3.00**	23	19	3	1	–	–	–	–	–	–	–	–	–	**Benzeneethanamine, 4-benzyloxy-N-phthaloyl-**		**NI37339**
252	105	357	77	253	358	92	106	**357**	100	52	34	30	18	8	5	4		23	19	3	1	–	–	–	–	–	–	–	–	–	**1-(4-Methoxyanilino)-1,2-dibenzoylethylene**		**MS08434**
106	276	91	103	121	143	277	128	**357**	100	72	69	30	22	17	15	13	**0.00**	24	23	2	1	–	–	–	–	–	–	–	–	–	**5-Hexynamide, N-(1-phenyl-2-hydroxyethyl)-2-(1-naphthyl)-, (1R,2R)-**		**DD01424**
106	276	67	128	103	77	91	104	**357**	100	23	23	19	16	16	15	14	**0.00**	24	23	2	1	–	–	–	–	–	–	–	–	–	**5-Hexynamide, N-(1-phenyl-2-hydroxyethyl)-3-(1-naphthyl)-, (1R,3R)-**		**DD01425**
113	55	70	85	72	126	41	98	**357**	100	19	16	14	14	13	13	12	**6.54**	24	39	1	1	–	–	–	–	–	–	–	–	–	**Pyrrolidine, 1-(1-oxo-5,8,11,14-eicosatetraenyl)-**	56599-69-0	**NI37340**
358	231	202	357	329	203	127	330	**358**	100	66	65	50	48	47	26	19		2	5	–	–	–	–	–	–	2	–	–	–	1	**Arsine, diiodoethyl-**	20738-28-7	**NI37341**
272	303	302	300	271	244	243	358	**358**	100	99	99	99	99	86	86	59		5	3	5	–	–	–	–	–	–	–	1	–	1	**Tungsten, pentacarbonylphosphinyl-**		**LI07889**
358	301	273	243	271	150	122	121	**358**	100	88	73	71	40	21	19	16		8	20	–	–	–	–	–	–	–	–	–	–	2	**Distibine, tetraethyl-**		**LI07890**
162	143	93	136	163	66	71	69	**358**	100	43	31	25	18	17	13	13	**2.00**	9	5	2	2	–	7	–	–	–	–	–	–	1	**Chromium, dinitrosocyclopentadienylheptafluorobut-1-enyl-**		**NI37342**
69	31	229	259	277	237	257	261	**358**	100	69	27	24	22	22	21	19	**1.57**	9	8	2	–	–	–	6	–	–	–	–	–	–	**Bicyclo[2.2.1]hept-5-ene-2,3-dimethanol, 1,4,5,6,7,7-hexachloro-**	2157-19-9	**NI37343**
343	193	358	344	207	345	75	269	**358**	100	36	33	23	15	13	13	12		9	27	4	–	–	–	–	–	–	3	–	–	1	**Silanol, trimethyl-, triester with arsenic acid (H3AsO4)**	17921-76-5	**NI37344**
343	193	358	75	73	344	207	299	**358**	100	36	35	33	27	23	21	17		9	27	4	–	–	–	–	–	–	3	–	–	1	**Silanol, trimethyl-, triester with arsenic acid (H3AsO4)**	17921-76-5	**NI37345**
59	117	118	150	176	61	55	208	**358**	100	24	16	12	7	7	7	6	**3.17**	10	14	2	–	6	–	–	–	–	–	–	–	–	**2,4,6,8,9,10-Hexathiatricyclo[3.3.1.1^{3,7}]decane-1-propanoic acid, 3,5,7-trimethyl-**	57274-38-1	**NI37346**
44	41	67	60	45	69	64	43	**358**	100	84	55	31	31	28	22	21	**0.00**	10	16	9	1	2	–	–	–	–	–	–	–	–	**β-D-Glucopyranose, 1-thio-, 1-(3-butenohydroximate)NO-(sulphonic acid)**		**MS08435**
281	279	251	253	74	362	97	75	**358**	100	99	72	70	69	56	51	51	**1.00**	11	4	4	–	–	–	–	2	–	–	–	–	–	**1,2-Naphthaquinone-7-carboxylic acid, 3,5-dibromo-**		**LI07891**
283	328	284	182	52	217	329	285	**358**	100	43	35	31	17	16	15	15	**13.00**	11	14	3	2	1	–	–	–	–	–	–	–	2	**Chromium, μ-(methylthio)-μ-hydroxy-(nitrosocyclopentadienyl)bis-**		**LI07892**
97	153	125	29	199	45	65	27	**358**	100	83	83	60	51	44	37	27	**0.00**	11	16	3	–	3	–	1	–	–	–	1	–	–	**Phosphorodithioic acid, S-[[(4-chlorophenyl)sulphinyl]methyl] O,O-diethyl ester**	17297-40-4	**NI37347**
109	29	75	81	183	139	111	27	**358**	100	79	74	61	59	41	38	33	**0.00**	11	16	5	–	2	–	1	–	–	–	1	–	–	**Phosphorothioic acid, S-[[(4-chlorophenyl)sulphonyl]methyl] O,O-diethyl ester**	16662-87-6	**NI37348**
299	73	343	315	211	300	103	227	**358**	100	94	60	45	30	25	20	18	**0.00**	11	31	5	–	–	–	–	–	–	3	1	–	–	**Phosphoric acid, bis(trimethylsilyl) 2-[(trimethylsilyl)oxy]ethyl ester**	55887-73-5	**LI07893**
299	73	343	315	211	300	133	344	**358**	100	93	65	42	29	25	18	17	**4.07**	11	31	5	–	–	–	–	–	–	3	1	–	–	**Phosphoric acid, bis(trimethylsilyl) 2-[(trimethylsilyl)oxy]ethyl ester**	55887-73-5	**NI37349**
360	362	290	358	288	329	364	292	**358**	100	82	67	53	51	47	37	29		12	4	–	–	–	–	6	–	–	–	–	–	–	**1,1′-Biphenyl, 2,2′,3,3′,4,4′-hexachloro-**	38380-07-3	**NI37350**
360	218	362	220	358	182	290	162	**358**	100	92	81	60	54	52	48	46		12	4	–	–	–	–	6	–	–	–	–	–	–	**1,1′-Biphenyl, 2,2′,3,3′,4,4′-hexachloro-**	38380-07-3	**NI37351**
360	362	145	290	144	288	358	109	**358**	100	81	80	77	68	61	54	53		12	4	–	–	–	–	6	–	–	–	–	–	–	**1,1′-Biphenyl, 2,2′,3,3′,4,4′-hexachloro-**	38380-07-3	**NI37352**
360	145	362	290	144	288	358	109	**358**	100	87	82	82	74	65	54	52		12	4	–	–	–	–	6	–	–	–	–	–	–	**1,1′-Biphenyl, 2,2′,3,3′,4,5-hexachloro-**	55215-18-4	**NI37353**
360	362	145	290	358	144	288	325	**358**	100	78	61	55	52	50	44	40		12	4	–	–	–	–	6	–	–	–	–	–	–	**1,1′-Biphenyl, 2,2′,3,3′,4,5-hexachloro-**	55215-18-4	**NI37354**
145	360	290	36	109	144	288	362	**358**	100	85	83	82	80	75	71	69	**34.20**	12	4	–	–	–	–	6	–	–	–	–	–	–	**1,1′-Biphenyl, 2,2′,3,3′,4,5′-hexachloro-**	52663-66-8	**NI37355**
145	360	144	290	362	288	109	146	**358**	100	94	85	83	72	66	63	58	**49.00**	12	4	–	–	–	–	6	–	–	–	–	–	–	**1,1′-Biphenyl, 2,2′,3,3′,4,6-hexachloro-**	61798-70-7	**NI37356**
360	290	362	288	145	144	358	218	**358**	100	94	83	77	70	58	50	48		12	4	–	–	–	–	6	–	–	–	–	–	–	**1,1′-Biphenyl, 2,2′,3,3′,4,6′-hexachloro-**	38380-05-1	**NI37357**
360	362	145	290	144	288	358	109	**358**	100	79	77	72	69	57	54	54		12	4	–	–	–	–	6	–	–	–	–	–	–	**1,1′-Biphenyl, 2,2′,3,3′,5,5′-hexachloro-**	35694-04-3	**NI37358**
290	360	362	288	325	218	358	292	**358**	100	99	84	78	60	56	54	45		12	4	–	–	–	–	6	–	–	–	–	–	–	**1,1′-Biphenyl, 2,2′,3,3′,5,6′-hexachloro-**	52744-13-5	**NI37359**
360	362	218	180	181	358	182	220	**358**	100	87	71	55	53	52	48	44		12	4	–	–	–	–	6	–	–	–	–	–	–	**1,1′-Biphenyl, 2,2′,3,3′,6,6′-hexachloro-**	38411-22-2	**NI37360**
360	290	362	288	145	358	218	144	**358**	100	80	78	62	57	52	48	46		12	4	–	–	–	–	6	–	–	–	–	–	–	**1,1′-Biphenyl, 2,2′,3,3′,6,6′-hexachloro-**	38411-22-2	**NI37361**
360	362	290	145	144	288	358	109	**358**	100	78	72	68	61	57	52	42		12	4	–	–	–	–	6	–	–	–	–	–	–	**1,1′-Biphenyl, 2,2′,3,4,4′,5-hexachloro-**	35694-06-5	**NI37362**
360	362	290	358	288	364	292	145	**358**	100	82	33	28	26	19	18	16		12	4	–	–	–	–	6	–	–	–	–	–	–	**1,1′-Biphenyl, 2,2′,3,4,4′,5′-hexachloro-**	35065-28-2	**NI37363**
360	362	290	145	144	288	358	109	**358**	100	79	69	66	56	55	54	43		12	4	–	–	–	–	6	–	–	–	–	–	–	**1,1′-Biphenyl, 2,2′,3,4,4′,6-hexachloro-**	56030-56-9	**NI37364**
360	362	290	145	358	288	144	218	**358**	100	79	60	53	53	46	40	36		12	4	–	–	–	–	6	–	–	–	–	–	–	**1,1′-Biphenyl, 2,2′,3,4,5,5′-hexachloro-**	52712-04-6	**NI37365**
360	362	218	290	358	288	364	220	**358**	100	83	54	52	51	43	38	36		12	4	–	–	–	–	6	–	–	–	–	–	–	**1,1′-Biphenyl, 2,2′,3,4,5,5′-hexachloro-**	52712-04-6	**NI37366**
360	362	290	358	288	292	364	218	**358**	100	77	72	54	54	19	18	17		12	4	–	–	–	–	6	–	–	–	–	–	–	**1,1′-Biphenyl, 2,2′,3,4,5,5′-hexachloro-**	52712-04-6	**NI37367**
360	362	290	145	144	288	358	109	**358**	100	78	75	73	61	60	53	47		12	4	–	–	–	–	6	–	–	–	–	–	–	**1,1′-Biphenyl, 2,2′,3,4,5,6′-hexachloro-**	68194-15-0	**NI37368**
360	362	290	145	288	144	358	109	**358**	100	79	74	72	56	55	48	43		12	4	–	–	–	–	6	–	–	–	–	–	–	**1,1′-Biphenyl, 2,2′,3,4′,5,5′-hexachloro-**	51908-16-8	**NI37369**
360	290	362	288	145	144	218	358	**358**	100	87	75	71	63	53	48	47		12	4	–	–	–	–	6	–	–	–	–	–	–	**1,1′-Biphenyl, 2,2′,3,4′,5′,6-hexachloro-**	38380-04-0	**NI37370**
145	360	290	144	109	36	288	362	**358**	100	92	86	81	74	70	67	64	**40.90**	12	4	–	–	–	–	6	–	–	–	–	–	–	**1,1′-Biphenyl, 2,2′,3,5,5′,6-hexachloro-**	52663-63-5	**NI37371**
360	362	218	358	288	180	220	181	**358**	100	80	66	52	45	44	42	40		12	4	–	–	–	–	6	–	–	–	–	–	–	**1,1′-Biphenyl, 2,2′,3,5,5′,6-hexachloro-**	52663-63-5	**NI37372**
360	362	290	288	358	145	218	144	**358**	100	83	76	61	52	51	39	39		12	4	–	–	–	–	6	–	–	–	–	–	–	**1,1′-Biphenyl, 2,2′,4,4′,5,5′-hexachloro-**	35065-27-1	**NI37373**
360	358	290	288	364	292	361	363	**358**	100	55	48	39	37	25	13	11		12	4	–	–	–	–	6	–	–	–	–	–	–	**1,1′-Biphenyl, 2,2′,4,4′,5,5′-hexachloro-**	35065-27-1	**NI37374**
360	362	145	144	290	358	288	109	**358**	100	92	86	77	73	71	60	57		12	4	–	–	–	–	6	–	–	–	–	–	–	**1,1′-Biphenyl, 2,2′,4,4′,5,5′-hexachloro-**	35065-27-1	**NI37375**

MASS TO CHARGE RATIOS								M.W.	INTENSITIES								Parent	C	H	O	N	S	F	Cl	Br	I	Si	P	B	X	COMPOUND NAME	CAS Reg No	No
360	362	145	290	144	358	288	109	**358**	100	77	74	68	64	54	54	54		12	4	–	–	–	–	6	–	–	–	–	–	–	1,1′-Biphenyl, 2,2′,4,4′,5′,6-hexachloro-	60145-22-4	NI37376
360	362	358	290	145	144	288	364	**358**	100	79	54	48	45	38	38	34		12	4	–	–	–	–	6	–	–	–	–	–	–	1,1′-Biphenyl, 2,2′,4,4′,6,6′-hexachloro-	33979-03-2	NI37377
360	362	290	358	288	364	292	218	**358**	100	78	60	56	46	17	16	16		12	4	–	–	–	–	6	–	–	–	–	–	–	1,1′-Biphenyl, 2,2′,4,4′,6,6′-hexachloro-	33979-03-2	NI37378
360	362	290	358	145	288	144	218	**358**	100	77	60	53	53	49	41	36		12	4	–	–	–	–	6	–	–	–	–	–	–	1,1′-Biphenyl, 2,3,3′,4,4′,5-hexachloro-	38380-08-4	NI37379
360	362	145	290	358	144	109	127	**358**	100	79	70	55	55	53	51	46		12	4	–	–	–	–	6	–	–	–	–	–	–	1,1′-Biphenyl, 2,3,3′,4,4′,5-hexachloro-	38380-08-4	NI37380
360	362	358	290	144	145	109	288	**358**	100	71	55	55	45	41	40	37		12	4	–	–	–	–	6	–	–	–	–	–	–	1,1′-Biphenyl, 2,3,3′,4,4′,5′-hexachloro-	69782-90-7	NI37381
360	362	145	290	144	358	288	109	**358**	100	81	68	63	62	55	53	50		12	4	–	–	–	–	6	–	–	–	–	–	–	1,1′-Biphenyl, 2,3,3′,4,5,6-hexachloro-	41411-62-5	NI37382
360	362	290	358	288	145	144	364	**358**	100	83	59	51	47	46	38	34		12	4	–	–	–	–	6	–	–	–	–	–	–	1,1′-Biphenyl, 2,3′,4,4′,5,5′-hexachloro-	52663-72-6	NI37383
360	362	358	290	145	288	364	144	**358**	100	79	51	45	37	34	32	31		12	4	–	–	–	–	6	–	–	–	–	–	–	1,1′-Biphenyl, 3,3′,4,4′,5,5′-hexachloro-	32774-16-6	NI37384
75	109	145	127	128	74	144	97	**358**	100	62	41	39	35	33	31	29	**6.40**	12	4	–	–	–	–	6	–	–	–	–	–	–	1,1′-Biphenyl, hexachloro-	26601-64-9	NI37385
360	290	362	288	358	145	292	144	**358**	100	85	80	66	52	44	41	34		12	4	–	–	–	–	6	–	–	–	–	–	–	1,1′-Biphenyl, hexachloro-	26601-64-9	NI37386
360	326	290	362	288	328	324	358	**358**	100	88	83	80	64	56	56	52		12	4	–	–	–	–	6	–	–	–	–	–	–	1,1′-Biphenyl, hexachloro-	26601-64-9	NI37387
358	342	326	343	341	359	299	31	**358**	100	84	38	18	18	17	17	17		12	6	–	4	–	8	–	–	–	–	–	–	–	Hydrazine, 1,1′-(2,2′,3,3′,5,5′,6,6′-octafluoro[1,1′-biphenyl]-4,4′-diyl)bis-	2200-68-2	NI37388
168	358	356	360	359	357	354	362	**358**	100	58	39	36	22	22	22	19		12	8	1	–	–	–	2	–	–	–	–	–	1	Phenoxastannin, 10,10-dichloro-		LI07894
267	323	269	325	81	109	295	170	**358**	100	76	63	50	49	35	30	21	**2.55**	12	14	4	–	–	–	3	–	–	–	1	–	–	Phosphoric acid, 2-chloro-1-(2,4-dichlorophenyl)vinyl diethyl ester	470-90-6	NI37389
81	109	267	269	91	99	123	323	**358**	100	57	34	24	24	18	17	16	**0.00**	12	14	4	–	–	–	3	–	–	–	1	–	–	Phosphoric acid, 2-chloro-1-(2,4-dichlorophenyl)vinyl diethyl ester	470-90-6	NI37390
267	323	269	81	325	109	295	170	**358**	100	66	66	54	50	36	29	22	**2.63**	12	14	4	–	–	–	3	–	–	–	1	–	–	Phosphoric acid, 2-chloro-1-(2,4-dichlorophenyl)vinyl diethyl ester	470-90-6	NI37391
29	81	27	109	28	267	91	269	**358**	100	62	45	33	18	17	15	12	**0.10**	12	14	4	–	–	–	3	–	–	–	1	–	–	Phosphoric acid, 2-chloro-1-(2,4-dichlorophenyl)vinyl diethyl ester, (E)-	18708-86-6	NI37392
29	81	109	27	267	28	269	91	**358**	100	68	36	36	28	21	17	16	**0.00**	12	14	4	–	–	–	3	–	–	–	1	–	–	Phosphoric acid, 2-chloro-1-(2,4-dichlorophenyl)vinyl diethyl ester, (Z)-	18708-87-7	NI37393
103	75	227	211	73	47	313	61	**358**	100	70	63	55	49	29	28	21	**0.00**	12	31	6	–	–	–	–	–	–	2	1	–	–	Phosphoric acid, 2,2-diethoxyethyl bis(trimethylsilyl) ester	55191-35-0	NI37394
246	274	302	215	358	176	98	203	**358**	100	83	74	70	55	52	36	25		13	12	4	2	–	–	–	–	–	–	–	–	1	Molybdenum, tetracarbonyl[N-(2-pyridinylmethylene)-2-propanamine-N,N′]-	36100-15-9	MS08436
58	288	72	290	286	128	127	360	**358**	100	70	50	35	35	13	9	8	**4.00**	13	16	–	2	–	–	–	2	–	–	–	–	–	Diethylamine, N-[(5,7-dibromoindol-3-yl)methyl]-		NI37395
56	161	77	132	105	104	358	51	**358**	100	81	60	56	55	50	43	37		14	13	1	2	–	7	–	–	–	–	–	–	–	Piperazine, 1-phenyl-4-heptafluorobutyryl-		NI37396
340	243	223	341	69	41	55	79	**358**	100	29	23	15	13	13	12	11	**9.00**	14	16	2	2	–	6	–	–	–	–	–	–	–	Pyridine, 1,2,3,4-tetrahydro-1-(trifluoroacetyl)-5-[1-(trifluoroacetyl)-2-piperidinyl]-	54966-08-4	NI37397
73	358	75	79	343	77	147	45	**358**	100	95	92	63	54	53	49	48		14	30	3	2	–	–	–	–	–	3	–	–	–	Pyrimidine, 2,4-bis[(trimethylsilyl)oxy]-5-[[(trimethylsilyl)oxy]methyl]-	31517-04-1	MS08437
73	315	100	259	244	186	147	116	**358**	100	56	19	16	11	7	6	6	**0.00**	15	30	4	2	–	–	–	–	–	2	–	–	–	2,4-Imidazolidinedione, 5-acetylbis(2,5-trimethylsilyloxy)-3-sec-butyl-		NI37398
358	41	64	359	266	44	120	221	**358**	100	64	51	20	18	18	15	14		16	10	4	2	2	–	–	–	–	–	–	–	–	1,4-Dithiin, 2,5-bis(4-nitrophenyl)-	38172-60-0	NI37399
44	73	30	42	43	58	41	40	**358**	100	54	41	25	13	11	11	10	**3.00**	16	14	6	4	–	–	–	–	–	–	–	–	–	1,2-Ethanediamine, N,N′-bis(2-formyl-4-nitrophenyl)-		LI07895
163	28	103	133	91	18	77	90	**358**	100	39	30	26	23	23	21	16	**12.85**	16	14	6	4	–	–	–	–	–	–	–	–	–	Propanoic acid, 3-[4-(2,4-dinitrophenylazo)phenyl]-, methyl ester		NI37400
163	28	133	103	77	91	266	326	**358**	100	42	26	26	24	18	14	13	**11.57**	16	14	6	4	–	–	–	–	–	–	–	–	–	Propanoic acid, 3-[4-(2,4-dinitrophenylazo)phenyl]-, methyl ester		NI37401
103	43	131	163	77	298	90	104	**358**	100	54	51	40	28	19	12	11	**0.34**	16	14	6	4	–	–	–	–	–	–	–	–	–	Propanoic acid, 3-[4-(2,4-dinitrophenylazo)phenyl]-, methyl ester		NI37402
127	44	85	73	70	179	197	128	**358**	100	48	41	35	33	32	23	23	**0.00**	16	22	9	–	–	–	–	–	–	–	–	–	–	1H,3H-Pyrano[3,4-c]pyran-1-one, 5-vinyl-6-(β-D-glucopyranosyloxy)-4,4a,5,6-tetrahydro-, [4aS-(4aα,5β,6α)]-	14215-86-2	MS08438
267	87	86	114	85	57	72	282	**358**	100	54	47	43	33	31	22	6	**0.00**	16	26	7	2	–	–	–	–	–	–	–	–	–	Dihydro-N-acetyllactinobolin methyl enol ether		LI07896
165	358	119	166	359	55	118	105	**358**	100	22	21	9	5	4	3	3		17	18	5	4	–	–	–	–	–	–	–	–	–	Propionanilide, 3-(N′-methyl-4-nitroanilino)-4′-nitro-N-methyl-		IC04112
58	166	251	252	69	77	96	194	**358**	100	64	19	17	16	14	10	7	**0.00**	17	21	3	2	–	3	–	–	–	–	–	–	–	Ephedrine, N-[(trifluoroacetyl)prolyl]-		NI37403
158	30	73	84	43	29	41	72	**358**	100	66	64	54	31	31	30	25	**1.18**	17	30	6	2	–	–	–	–	–	–	–	–	–	Dibutyl N-(N-acetylglycyl)glutamate		AI02159
77	51	144	89	178	358	39	323	**358**	100	47	46	32	19	17	17	14		18	12	2	2	–	–	2	–	–	–	–	–	–	Benzoquinone, 2,5-bis(anilino)-3,6-dichloro-		IC04113
203	91	160	155	130	77	105	83	**358**	100	78	53	34	17	12	11	10	**0.30**	18	18	4	2	1	–	–	–	–	–	–	–	–	Benzenesulphonamide, N-[3-(1,3-dihydro-1,3-dioxo-2H-isoindol-2-yl)propyl]-4-methyl-	38353-77-4	NI37404
343	345	73	323	111	152	252	154	**358**	100	33	10	8	8	5	4	4	**1.00**	18	19	2	2	–	–	1	–	–	1	–	–	–	4(3H)-Quinazolinone, 3-(2-chlorophenyl)-2-methyl-5-(trimethylsiloxy)-		NI37405
343	345	73	251	111	323	252	307	**358**	100	33	14	9	9	8	8	3	**0.00**	18	19	2	2	–	–	1	–	–	1	–	–	–	4(3H)-Quinazolinone, 3-(2-chlorophenyl)-2-methyl-6-(trimethylsiloxy)-		NI37406
323	358	73	343	154	360	111	152	**358**	100	50	41	24	22	17	15	13		18	19	2	2	–	–	1	–	–	1	–	–	–	4(3H)-Quinazolinone, 3-(2-chlorophenyl)-2-methyl-7-(trimethylsiloxy)-		NI37407
343	152	345	154	73	111	358	113	**358**	100	41	33	33	30	25	8	8		18	19	2	2	–	–	1	–	–	1	–	–	–	4(3H)-Quinazolinone, 3-(2-chlorophenyl)-2-methyl-8-(trimethylsiloxy)-		NI37408
323	73	343	251	143	345	117	116	**358**	100	25	10	4	4	3	3	3	**1.00**	18	19	2	2	–	–	1	–	–	1	–	–	–	4(3H)-Quinazolinone, 3-[2-chloro-3-(trimethylsiloxy)phenyl]-2-methyl-		NI37409
323	73	143	343	285	358	164	116	**358**	100	25	11	6	5	3	3	3		18	19	2	2	–	–	1	–	–	1	–	–	–	4(3H)-Quinazolinone, 3-[2-chloro-4-(trimethylsiloxy)phenyl]-2-methyl-		NI37410
323	73	143	343	116	165	164	117	**358**	100	30	6	4	3	2	2	2	**0.00**	18	19	2	2	–	–	1	–	–	1	–	–	–	4(3H)-Quinazolinone, 3-[2-chloro-5-(trimethylsiloxy)phenyl]-2-methyl-		NI37411
323	343	73	358	345	143	360	307	**358**	100	26	21	16	9	7	6	6		18	19	2	2	–	–	1	–	–	1	–	–	–	4(3H)-Quinazolinone, 3-[2-chloro-6-(trimethylsiloxy)phenyl]-2-methyl-		NI37412
340	188	174	358	322	119	158	249	**358**	100	56	55	38	34	28	27	23		18	22	4	4	–	–	–	–	–	–	–	–	–	D-Arabino-hexose, phenylosazone		LI07897
93	92	77	65	358	340	322	94	**358**	100	69	63	52	40	28	26	24		18	22	4	4	–	–	–	–	–	–	–	–	–	D-Arabino-hexose, phenylosazone		MS08439
358	81	359	105	119	160	288	162	**358**	100	33	29	23	12	10	9	9		18	22	4	4	–	–	–	–	–	–	–	–	–	Bicyclo[4.4.0]dec-1-en-3-one, 6,10-dimethyl-, 2,4-dinitrophenylhydrazone, (6S,10R)-(+)-		MS08440
274	275	55	43	70	358	244	273	**358**	100	18	16	16	9	8	5	3		18	23	4	–	–	–	–	–	–	–	–	–	1	Manganese, (1-acetyl-2,3,4,5-tetraethylcyclopentadienyl)tricarbonyl-	12267-19-5	NI37413

MASS TO CHARGE RATIOS								M.W.	INTENSITIES								Parent	C	H	O	N	S	F	Cl	Br	I	Si	P	B	X	COMPOUND NAME	CAS Reg No	No
227	209	264	327	309	296	322	308	**358**	100	25	17	5	5	5	4	4	**1.50**	18	30	7	–	–	–	–	–	–	–	–	–	–	Methyl 4-[5S-(2S,3S-epoxy-5S-hydroxy-4S-methylhexyl)-3R,4R-dihydroxytetrahydropyran-2S-yl]-3-methylbut-2(E)-enoate		NI37414
227	210	209	256	278	277	267	291	**358**	100	65	21	11	8	6	5	3	**0.50**	18	30	7	–	–	–	–	–	–	–	–	–	–	Methyl 4-[5S-(2S,3S-epoxy-5S-hydroxy-4S-methylhexyl)-3R,4R-dihydroxytetrahydropyran-2S-yl]-3-methylbut-2(Z)-enoate		NI37415
177	69	135	109	223	219	178	77	**358**	100	47	24	17	15	14	14	13	**4.00**	19	15	5	–	–	–	1	–	–	–	–	–	–	1,3,4,6-Hexanetetrone, 1-(4-chlorophenyl)-6-(4-methoxyphenyl)-	58368-97-1	NI37416
193	165	121	120	194	93	65	92	**358**	100	68	58	22	13	12	9	8	**0.90**	19	18	7	–	–	–	–	–	–	–	–	–	–	Benzoic acid, 2,2′-[carbonylbis(oxy)]bis-, diethyl ester	118-27-4	NI37417
358	315	357	343	359	165	327	316	**358**	100	62	50	41	25	24	23	13		19	18	7	–	–	–	–	–	–	–	–	–	–	4H-1-Benzopyran-4-one, 2-(3,4-dimethoxyphenyl)-5-hydroxy-3,7-dimethoxy-	1245-15-4	NI37418
358	343	357	153	312	315	329	344	**358**	100	82	40	40	26	25	24	21		19	18	7	–	–	–	–	–	–	–	–	–	–	4H-1-Benzopyran-4-one, 5-hydroxy-6,7-dimethoxy-2-(3,4-dimethoxyphenyl)-	21763-80-4	NI37419
44	314	258	91	65	243	43	257	**358**	100	99	41	34	27	25	22	21	**9.00**	19	18	7	–	–	–	–	–	–	–	–	–	–	11H-Dibenzo[b,e][1,4]dioxepin-7-carboxylic acid, 3,8-dihydroxy-1,4,6,9-tetramethyl-11-oxo-, methyl ester	4665-03-6	NI37420
358	284	326	298	285	149	180	359	**358**	100	72	66	66	38	27	25	22		19	18	7	–	–	–	–	–	–	–	–	–	–	11H-Dibenzo[b,e][1,4]dioxepin-7-carboxylic acid, 8-hydroxy-3-methoxy-1,4,6-trimethyl-11-oxo-, methyl ester	38629-30-0	NI37421
151	208	207	193	311	358	340		**358**	100	61	51	36	29	25	8			19	18	7	–	–	–	–	–	–	–	–	–	–	Isochromano[4′,3′:2,3]chroman-4-one, 3(3′)-hydroxy-7,6′,7′-trimethoxy-, (±)-		LI07898
120	358	121	92	164	65	239	300	**358**	100	9	8	8	5	4	3	2		19	22	5	2	–	–	–	–	–	–	–	–	–	Benzamide, 4-amino-N-[4-(2-(2-methoxyethoxy)ethoxy)-carbonylphenyl]-		IC04114
358	239	343	224	345	360	359	344	**358**	100	99	97	60	41	39	30	27		19	23	1	2	–	–	1	–	–	1	–	–	–	10-Sila-2-azaanthracene, 9-(5-chloropentanamido)-10,10-dimethyl-9,10-dihydro-	84841-80-5	NI37422
169	315	154	259	155	227	229	91	**358**	100	89	85	84	67	61	51	41	**0.00**	19	23	2	–	–	–	–	–	–	–	–	–	1	Arsane oxide, [(1-hydroxycyclohexyl)methyl]diphenyl-		DD01426
43	131	137	173	113	215	71	83	**358**	100	60	50	49	48	47	46	37	**0.50**	19	34	6	–	–	–	–	–	–	–	–	–	–	1,3,10-Decanetriol, 2,4,6-trimethyl-, triacetate		NI37423
73	94	220	180	174	74	108	99	**358**	100	29	6	6	6	6	5	5	**0.00**	19	34	6	–	–	–	–	–	–	–	–	–	–	Dodecanoic acid, 2-(acetyloxy)-1-[(acetyloxy)methyl]ethyl ester	55191-43-0	WI01469
73	45	43	41	57	55	29	149	**358**	100	70	34	17	15	15	14	13	**0.00**	19	34	6	–	–	–	–	–	–	–	–	–	–	Dodecanoic acid, 2-(acetyloxy)-1-[(acetyloxy)methyl]ethyl ester	55191-43-0	WI01470
73	218	125	74	182	176	162	108	**358**	100	5	5	5	4	4	4	4	**0.00**	19	34	6	–	–	–	–	–	–	–	–	–	–	Dodecanoic acid, 2,3-bis(acetyloxy)propyl ester	55191-44-1	WI01471
73	45	43	41	74	55	57	29	**358**	100	62	40	20	18	18	15	15	**0.00**	19	34	6	–	–	–	–	–	–	–	–	–	–	Dodecanoic acid, 2,3-bis(acetyloxy)propyl ester	55191-44-1	WI01472
280	281	297	265	311	326	343	282	**358**	100	43	26	17	12	8	6	6	**0.00**	19	34	6	–	–	–	–	–	–	–	–	–	–	Ethyl 7-(2-(ethoxycarbonyl)-3α,5β-dimethoxycyclopentyl)heptanoate		NI37424
70	358	254	150	74	102	178	256	**358**	100	67	18	15	15	11	5	4		20	6	–	8	–	–	–	–	–	–	–	–	–	6,6′-Biquinoxaline, 2,2′,3,3′-tetracyano-		MS08441
137	151	138	152	77	65	122	91	**358**	100	41	22	20	18	16	15	15	**11.80**	20	22	6	–	–	–	–	–	–	–	–	–	–	3-Benzofuranmethanol, 2,3-dihydro-2-(4-hydroxy-3-methoxyphenyl)-5-(3-hydroxy-1-propenyl)-7-methoxy-	4263-87-0	NI37425
137	151	152	138	77	65	122	91	**358**	100	41	27	26	18	17	15	15	**12.80**	20	22	6	–	–	–	–	–	–	–	–	–	–	3-Benzofuranmethanol, 2,3-dihydro-2-(4-hydroxy-3-methoxyphenyl)-5-(3-hydroxy-1-propenyl)-7-methoxy-	4263-87-0	MS08442
121	221	122	109	93	120	81	29	**358**	100	13	8	6	5	5	5	5	**1.80**	20	22	6	–	–	–	–	–	–	–	–	–	–	Benzoic acid, 4-[(butoxycarbonyl)oxy]-, 4-ethoxyphenyl ester	16494-24-9	NI37426
358	192	209	359	137	287	165	179	**358**	100	79	56	25	21	20	19	17		20	22	6	–	–	–	–	–	–	–	–	–	–	Bicyclo[3.2.1]oct-2-ene, 1-allyl-3,8-dihydroxy-6-(3-methoxy-4,5-methylenedioxyphenyl)-7-methyl-4-oxo-, (1S,5S,6R,7R,8S)-		NI37427
94	95	81	91	55	105	77	119	**358**	100	82	57	39	37	36	30	21	**4.00**	20	22	6	–	–	–	–	–	–	–	–	–	–	Chamaedroxide, 2-deoxy-		MS08443
94	109	81	44	121	108	79	95	**358**	100	77	69	67	65	59	52	45	**4.20**	20	22	6	–	–	–	–	–	–	–	–	–	–	1,4-Etheno-3H,7H-benzo[1,2-c:3,4-c′]dipyran-3,7-dione, 9-(3-furanyl)decahydro-4-hydroxy-4a,10a-dimethyl-, [1R-(1α,4β,4aα,6aα,9β,10aβ,10bα)]-	471-54-5	NI37428
44	91	79	77	94	204	28	95	**358**	100	23	22	18	17	16	16	15	**0.00**	20	22	6	–	–	–	–	–	–	–	–	–	–	1,4-Etheno-3H,7H-benzo[1,2-c:3,4-c′]dipyran-3,7-dione, 9-(3-furanyl)decahydro-4-hydroxy-4a,10a-dimethyl-, [1R-(1α,4β,4aα,6aβ,9β,10aβ,10bα)]-	546-97-4	NI37429
94	109	81	44	121	108	79	95	**358**	100	77	69	67	65	59	52	45	**4.20**	20	22	6	–	–	–	–	–	–	–	–	–	–	1,4-Etheno-3H,7H-benzo[1,2-c:3,4-c′]dipyran-3,7-dione, 9-(3-furanyl)decahydro-4-hydroxy-4a,10a-dimethyl-, [1R-(1α,4β,4aα,6aβ,9β,10aβ,10bα)]-	546-97-4	AM00148
94	95	44	81	93	121	107	91	**358**	100	80	63	51	44	41	41	38	**1.40**	20	22	6	–	–	–	–	–	–	–	–	–	–	1,4-Etheno-3H,7H-benzo[1,2-c:3,4-c′]dipyran-3,7-dione, 9-(3-furanyl)decahydro-4-hydroxy-4a,10a-dimethyl-, [1R-(1α,4β,4aα,6aβ,9β,10aβ,10bα)]-	546-97-4	NI37430
240	239	358	257	256	298			**358**	100	50	45	43	43	28				20	22	6	–	–	–	–	–	–	–	–	–	–	Ethylene glycol, 1,2-bis(4-methoxyphenyl)-		LI07899
137	358	138	122	94	41	43	55	**358**	100	45	20	18	14	13	12	10		20	22	6	–	–	–	–	–	–	–	–	–	–	2(3H)-Furanone, dihydro-3,4-bis[(4-hydroxy-3-methoxyphenyl)methyl]-, (3R-trans)-	580-72-3	NI37431
185	328	128	143	186	129	115	229	**358**	100	39	28	27	26	22	22	21	**0.00**	20	22	6	–	–	–	–	–	–	–	–	–	–	3b,11a[3′,4′]-Furanophenanthro[9,10-c]furan-1,3,13,15-tetrone, tetradecahydro-	56830-86-5	MS08444
179	107	106	108	105	358	357	180	**358**	100	82	76	73	64	41	39	35		20	22	6	–	–	–	–	–	–	–	–	–	–	Galactitol, 2,3:4,5-di-O-benzylidene-	19342-55-3	LI07900
179	107	105	106	108	358	357	180	**358**	100	83	63	37	36	21	20	15		20	22	6	–	–	–	–	–	–	–	–	–	–	Galactitol, 2,3:4,5-di-O-benzylidene-	19342-55-3	NI37432
179	192	149	191	358	210	123	357	**358**	100	77	76	68	55	47	46	40		20	22	6	–	–	–	–	–	–	–	–	–	–	D-Glucitol, 2,4:3,5-di-O-benzylidene-	13265-76-4	LI07901
105	107	179	192	149	191	358	123	**358**	100	85	30	23	23	20	16	14		20	22	6	–	–	–	–	–	–	–	–	–	–	D-Glucitol, 2,4:3,5-di-O-benzylidene-	13265-76-4	NI37433

MASS TO CHARGE RATIOS								M.W.	INTENSITIES								Parent	C	H	O	N	S	F	Cl	Br	I	Si	P	B	X	COMPOUND NAME	CAS Reg No	No
105	179	191	149	108	192	180	107	**358**	100	68	35	14	14	12	10	9	**5.43**	20	22	6	–	–	–	–	–	–	–	–	–	–	D-Mannitol, 2,4:3,5-di-O-benzylidene-	19342-60-0	NI37434
191	179	149	192	180	105	107	161	**358**	100	98	40	34	30	28	27	25	**15.00**	20	22	6	–	–	–	–	–	–	–	–	–	–	D-Mannitol, 2,4:3,5-di-O-benzylidene-	19342-60-0	LI07902
243	83	227	258	358	326			**358**	100	40	35	14	3	3				20	22	6	–	–	–	–	–	–	–	–	–	–	Oroselol, dihydro-9-methoxy-O-senecionyl-		LI07903
325	358	326	297	299	298	115	340	**358**	100	29	27	16	15	13	9	8		20	22	6	–	–	–	–	–	–	–	–	–	–	3,9-Phenanthrenedione, 4,4a-dihydro-5,6,8-trihydroxy-7-(2-hydroxypropyl)-1,2,4a-trimethyl-, (2'S,4aS)-		NI37435
199	84	200	143	56	144	81	58	**358**	100	18	10	7	6	5	5	5	**2.00**	20	26	4	2	–	–	–	–	–	–	–	–	–	β-Carboline-1-butanoic acid, 2-ethyl-1,2,3,4-tetrahydro-α-(methoxycarbonyl)-, methyl ester		NI37436
57	144	228	60	55	69	70	71	**358**	100	95	89	85	80	65	63	61	**7.89**	20	26	4	2	–	–	–	–	–	–	–	–	–	Curan-17-oic acid, 19,20-dihydroxy-, methyl ester, (19S)-	2111-90-2	NI37437
328	327	358	270	87	271	96	329	**358**	100	70	45	28	23	22	18	15		20	26	4	2	–	–	–	–	–	–	–	–	–	Estrone, 3-methoxy-4-nitro-, methyloxime		NI37438
152	84	358	122	153	56	55	80	**358**	100	23	18	15	14	12	9	8		20	26	4	2	–	–	–	–	–	–	–	–	–	Gelsedine, 11-methoxy-	6887-28-1	NI37439
301	358	189	258	43	57	55	69	**358**	100	88	79	69	39	31	28	27		20	26	4	2	–	–	–	–	–	–	–	–	–	3,4-Secocondylfolan-3-one, 16,19-epoxy-2-hydroxy-11-methoxy-4-methyl-, (14β,16β,19R)-	55283-41-5	NI37440
83	139	55	69	41	57	81	43	**358**	100	67	37	34	31	30	29	26	**0.03**	20	38	3	–	1	–	–	–	–	–	–	–	–	Cyclohexanol, 5-methyl-2-isopropyl-, sulphite, [1R-[1α(1R*,2S*,5R*),2β,5α]]-	26510-92-9	NI37441
75	311	55	73	343	69	43	103	**358**	100	87	80	75	58	48	44	43	**2.00**	20	42	3	–	–	–	–	–	–	1	–	–	–	Hexadecanoic acid, 16-(trimethylsiloxy)-, methyl ester	21987-14-4	NI37442
75	311	54	73	343	69	74	43	**358**	100	86	80	74	57	46	45	36	**2.00**	20	42	3	–	–	–	–	–	–	1	–	–	–	Hexadecanoic acid, 16-(trimethylsiloxy)-, methyl ester	21987-14-4	LI07904
73	45	299	89	44	84	103	75	**358**	100	50	47	39	28	23	21	21	**0.23**	20	42	3	–	–	–	–	–	–	1	–	–	–	Hexadecanoic acid, 2-[(trimethylsilyl)oxy]-, methyl ester	21987-11-1	LI07905
73	89	299	43	83	75	57	55	**358**	100	40	36	30	20	20	20	20	**0.00**	20	42	3	–	–	–	–	–	–	1	–	–	–	Hexadecanoic acid, 2-[(trimethylsilyl)oxy]-, methyl ester	21987-11-1	NI37443
73	299	89	43	129	103	83	55	**358**	100	46	41	25	23	23	23	22	**1.00**	20	42	3	–	–	–	–	–	–	1	–	–	–	Hexadecanoic acid, 2-[(trimethylsilyl)oxy]-, methyl ester	21987-11-1	MS08445
73	299	89	343	43	103	83	300	**358**	100	85	41	30	29	24	23	22	**0.43**	20	42	3	–	–	–	–	–	–	1	–	–	–	Hexadecanoic acid, 2-[(trimethylsilyl)oxy]-, methyl ester	21987-11-1	NI37444
73	201	259	44	75	69	55	155	**358**	100	69	60	51	47	35	35	22	**0.23**	20	42	3	–	–	–	–	–	–	1	–	–	–	Hexadecanoic acid, 9-[(trimethylsilyl)oxy]-, methyl ester		LI07906
117	73	75	314	118	74	311	55	**358**	100	29	14	11	10	8	7	7	**0.40**	20	42	3	–	–	–	–	–	–	1	–	–	–	Hexadecanoic acid, 15-[(trimethylsilyl)oxy]-, methyl ester	56196-04-4	NI37445
217	147	287	69	103	43	41	89	**358**	100	77	33	18	12	12	11	8	**0.00**	20	44	–	–	–	–	–	–	–	–	–	–	1	Tetrapentylgermane		MS08446
169	76	358	104	316	288	141	43	**358**	100	80	78	62	49	43	40	35		21	14	4	2	–	–	–	–	–	–	–	–	–	8-Acetyl-1-oxo-2-phthalimidocyclopenta[b]indole		LI07907
169	358	76	104	316	288	141	140	**358**	100	79	79	61	50	43	40	37		21	14	4	2	–	–	–	–	–	–	–	–	–	4-Acetyl-2-phthalimido-1,2-dihydro-3-oxocyclopent[b]indole		NI37446
77	358	91	51	65	359	39	118	**358**	100	49	30	24	17	12	12	11		21	18	2	4	–	–	–	–	–	–	–	–	–	Bis(3-methyl-5-oxo-1-phenyl-2-pyrazolinyl)methane	4174-09-8	IC04115
358	77	91	359	341	118	266	51	**358**	100	40	27	26	23	11	10	10		21	18	2	4	–	–	–	–	–	–	–	–	–	Bis(3-methyl-5-oxo-1-phenyl-2-pyrazolinyl)methane	4174-09-8	NI37447
343	358	344	327	313	311	43	41	**358**	100	27	23	7	7	6	6	6		21	26	5	–	–	–	–	–	–	–	–	–	–	2H,8H-Benzo[1,2-b:5,4-b']dipyran-10-propanoic acid, 5-methoxy-2,2,8,8-tetramethyl-, methyl ester	26574-30-1	NI37448
343	358	344	313	328	269	43	41	**358**	100	26	22	7	6	6	6	6		21	26	5	–	–	–	–	–	–	–	–	–	–	2H,8H-Benzo[1,2-b:5,4-b']dipyran-10-propanoic acid, 5-methoxy-2,2,8,8-tetramethyl-, methyl ester	26574-30-1	NI37449
229	312	230	231	358	269	297	191	**358**	100	30	15	13	10	7	6	6		21	26	5	–	–	–	–	–	–	–	–	–	–	12-Methyl-15-(1'-methyl-1'-ethoxyethyl)-5,11-dioxatetracyclo[8.6.0.1^{12,16}.0^{4,9}]heptadeca-1,3,9-trien-6-one		LI07908
162	358	163	134	359	226	147	287	**358**	100	20	13	11	5	3	3	2		21	26	5	–	–	–	–	–	–	–	–	–	–	Monascoflavin		LI07909
43	29	55	91	39	27	44	41	**358**	100	93	80	74	67	66	62	57	**0.00**	21	26	5	–	–	–	–	–	–	–	–	–	–	Pregna-1,4-diene-3,11,20-trione, 17,21-dihydroxy-	53-03-2	NI37450
202	200	199	326	201	232	101	204	**358**	100	78	53	44	39	28	26	21	**12.20**	21	26	5	–	–	–	–	–	–	–	–	–	–	Sonderianin	79405-82-6	NI37451
358	342	257	218	343	359	179	244	**358**	100	71	42	40	32	28	23	17		21	30	3	–	–	–	–	–	–	1	–	–	–	Estra-1,3,5(10)-trien-17-one, 4-hydroxy-3-[(trimethylsilyl)oxy]-	74299-18-6	NI37452
340	341	284	283	312	342	285	73	**358**	100	29	19	17	15	8	6	6	**5.63**	21	30	3	–	–	–	–	–	–	1	–	–	–	Estra-1,3,5(10)-trien-17-one, 14-hydroxy-3-[(trimethylsilyl)oxy]-	69597-46-2	NI37453
105	176	148	70	84	57	343	77	**358**	100	60	56	53	52	52	46	42	**0.00**	21	30	3	2	–	–	–	–	–	–	–	–	–	2,4-Imidazolidinedione, 1-tert-butyl-3-cyclohexyl-5-ethoxy-5-phenyl-		DD01427
72	310	71	82	358	69	68	67	**358**	100	83	65	63	59	53	50	50		21	30	3	2	–	–	–	–	–	–	–	–	–	Pyrido[3',4':5,6]cyclohept[1,2-b]indole-5,5(1H)-dimethanol, 2-ethyl-2,3,4,4a,6,11,12,12a-octahydro-12a-hydroxy-12-methyl-	3368-88-5	NI37454
71	213	101	343	187	157	143	87	**358**	100	26	5	3	1	1	1	1	**0.10**	21	42	4	–	–	–	–	–	–	–	–	–	–	1-O-(2-Isopentyloxydecyl)-2,3-O-isopropylideneglycerol		NI37455
359	341	360	342	262	260	358	357	**358**	100	80	23	21	6	6	4	4		21	42	4	–	–	–	–	–	–	–	–	–	–	Octadecanoic acid, 2,3-dihydroxypropyl ester	123-94-4	NI37456
284	267	285	134	327	266	98	298	**358**	100	28	27	21	19	15	14	11	**1.00**	21	42	4	–	–	–	–	–	–	–	–	–	–	Octadecanoic acid, 2,3-dihydroxypropyl ester	123-94-4	WI01473
43	73	57	55	60	41	101	98	**358**	100	87	83	77	71	70	57	53	**0.00**	21	42	4	–	–	–	–	–	–	–	–	–	–	Octadecanoic acid, 2,3-dihydroxypropyl ester	123-94-4	NI37457
201	169	127	241	95	137	173	209	**358**	100	72	72	44	30	16	10	6	**0.00**	21	42	4	–	–	–	–	–	–	–	–	–	–	Octadecanoic acid, 9,12-dimethoxy-, methyl ester		LI07910
43	57	98	55	41	74	73	69	**358**	100	85	71	71	62	45	45	43	**1.00**	21	42	4	–	–	–	–	–	–	–	–	–	–	Octadecanoic acid, 2-hydroxy-1-(hydroxymethyl)ethyl ester	621-61-4	WI01474
98	43	57	41	55	74	84	69	**358**	100	97	92	72	69	59	55	47	**0.00**	21	42	4	–	–	–	–	–	–	–	–	–	–	Octadecanoic acid, 2-hydroxy-1-(hydroxymethyl)ethyl ester	621-61-4	IC04116
284	134	267	327	98	285	266	298	**358**	100	56	46	44	34	32	30	22	**8.01**	21	42	4	–	–	–	–	–	–	–	–	–	–	Octadecanoic acid, 2-hydroxy-1-(hydroxymethyl)ethyl ester	621-61-4	WI01475
358	329	360	359	242	281	77	178	**358**	100	38	28	24	23	17	16	15		22	15	1	2	–	–	1	–	–	–	–	–	–	1H-1,4-Benzodiazepin-2-one, 3-benzylidene-7-chloro-2,3-dihydro-5-phenyl	55056-30-9	NI37458
358	301	267	324	360	331	134	164	**358**	100	79	54	37	29	21	14	12		22	15	1	2	–	–	1	–	–	–	–	–	–	Naphtho[1,2-b]phenazin-5(8H)-one, 6-chloro-8-ethyl-	52736-85-3	NI37459
357	358	359	360	281	190	189	225	**358**	100	65	46	24	13	11	10	9		22	15	1	2	–	–	1	–	–	–	–	–	–	Pyridazine, 3-chloro-4-phenoxy-5,6-diphenyl-	77892-03-6	NI37460
311	358	145	77	165	312	180	359	**358**	100	77	69	48	39	23	23	19		22	18	1	2	1	–	–	–	–	–	–	–	–	2-Imidazolin-4-one, 2-methylthio-1,5,5-triphenyl-		NI37461
283	77	358	330	284	165	180	224	**358**	100	36	32	31	25	19	14	12		22	18	1	2	1	–	–	–	–	–	–	–	–	2-Imidazolin-5-one, 2-methylthio-1,4,4-triphenyl-		NI37462
281	358	249	77	104	247	110	359	**358**	100	85	79	54	49	28	28	22		22	18	1	2	1	–	–	–	–	–	–	–	–	2(1H)-Pyrimidinone, 3,4-dihydro-4,6-diphenyl-5-phenylthio-		LI07911
105	103	77	76	133	131	121	119	**358**	100	57	53	22	20	20	12	9	**7.00**	22	18	3	2	–	–	–	–	–	–	–	–	–	Benzaldehyde benzoyl phenoxyacetyl hydrazone		LI07912

MASS TO CHARGE RATIOS								M.W.	INTENSITIES								Parent	C	H	O	N	S	F	Cl	Br	I	Si	P	B	X	COMPOUND NAME	CAS Reg No	No
358	343	359	179	344	122	163	77	**358**	100	32	25	10	8	7	5	4		22	18	3	2	–	–	–	–	–	–	–	–	–	1-Methylamino-4-(4-methoxyanilino)anthraquinone		IC04117
358	164	149	359	136	57	195	146	**358**	100	75	41	25	16	14	12	12		22	30	2	–	1	–	–	–	–	–	–	–	–	Phenol, 2,2′-thiobis[6-tert-butyl-4-methyl-	90-66-4	NI37463
358	164	149	136	57	195	121	91	**358**	100	76	40	14	14	12	8	8		22	30	2	–	1	–	–	–	–	–	–	–	–	Phenol, 2,2′-thiobis[6-tert-butyl-4-methyl-	90-66-4	LI07913
358	359	343	136	164	28	57		**358**	100	34	33	25	20	17	16			22	30	2	–	1	–	–	–	–	–	–	–	–	Phenol, 4,4′-thiobis[2-tert-butyl-5-methyl-	96-69-5	IC04118
358	343	136	359	179	139	128	57	**358**	100	55	36	26	23	13	13	13		22	30	2	–	1	–	–	–	–	–	–	–	–	Phenol, 4,4′-thiobis[2-tert-butyl-5-methyl-	96-69-5	NI37464
358	359	136	343	195	30	360	57	**358**	100	36	24	18	15	10	9	9		22	30	2	–	1	–	–	–	–	–	–	–	–	Phenol, 4,4′-thiobis[2-tert-butyl-6-methyl-	96-66-2	NI37465
358	343	136	195	359	41	179	57	**358**	100	31	30	26	25	20	17	15		22	30	2	–	1	–	–	–	–	–	–	–	–	Phenol, 4,4′-thiobis[2-tert-butyl-6-methyl-	96-66-2	NI37466
358	343	136	164	359	179	150	57	**358**	100	62	31	26	23	18	17	14		22	30	2	–	1	–	–	–	–	–	–	–	–	Phenol, 4,4′-thiobis[2-tert-butyl-6-methyl-	96-66-2	IC04119
203	81	95	119	107	93	201	145	**358**	100	57	41	30	23	23	21	12	**0.30**	22	30	4	–	–	–	–	–	–	–	–	–	–	Bacchotricuneatin D monoacetate ketone		MS08447
358	359	267	216	151	171	360	81	**358**	100	28	7	6	6	5	4	4		22	30	4	–	–	–	–	–	–	–	–	–	–	2,3-Dimethoxy-estradiol 17β-acetate		NI37467
298	137	109	270	299	147	316	105	**358**	100	79	59	33	26	21	19	16	**10.00**	22	30	4	–	–	–	–	–	–	–	–	–	–	Kaur-16-en-18-oic acid, 7-(acetyloxy)-6-hydroxy-, γ-lactone, (4α,6α,7β)-	52474-66-5	NI37468
137	298	109	270	316	145	107	105	**358**	100	86	80	45	24	22	20	19	**5.00**	22	30	4	–	–	–	–	–	–	–	–	–	–	Kaur-16-en-18-oic acid, 6α,7α-dihydroxy-, γ-lactone, acetate	31560-87-9	NI37469
327	328	128	358	231	295	329	267	**358**	100	60	46	40	33	30	26	23		22	30	4	–	–	–	–	–	–	–	–	–	–	Methyl pregn-4-en-3-one-17,20-epoxy-20α-carboxylate		LI07914
43	269	287	107	329	230	91	105	**358**	100	84	25	17	14	13	10	9	**0.00**	22	30	4	–	–	–	–	–	–	–	–	–	–	19-Norpregn-4-ene-3,20-dione, 17-(acetyloxy)-	31981-44-9	NI37470
227	358	129	268	174	359	228	226	**358**	100	88	45	37	27	26	19	19		22	34	2	–	–	–	–	–	–	1	–	–	–	Estra-1,3,5(10)-trien-17-ol, 3-methoxy-3-[(trimethylsilyl)oxy]-, (17β)-	18880-67-6	NI37471
270	271	43	55	69	241	83	57	**358**	100	23	17	16	14	12	12	12	**1.00**	22	34	2	2	–	–	–	–	–	–	–	–	–	2-(1,1-Dimethylpropyl)-6-isopentyl-3-(2-amino-2-methoxycarbonylethyl)indole		MS08448
110	109	358	124	81	79	67	93	**358**	100	30	19	19	16	15	15	14		22	35	3	–	–	–	–	–	–	–	–	1	–	3α,21-Dihydroxy-5α-pregnan-20-one methyl boronate		MS08449
110	109	358	124	107	215	111	105	**358**	100	30	19	19	13	11	11	10		22	35	3	–	–	–	–	–	–	–	–	1	–	3α,21-Dihydroxy-5α-pregnan-20-one methyl boronate		LI07915
358	343	359	344	327	328	311	312	**358**	100	58	26	15	15	12	8	6		23	18	4	–	–	–	–	–	–	–	–	–	–	2,2′-Binaphthalene-1,4-dione, 1-ethoxy-4′-methoxy-		NI37472
358	343	359	328	285	344	300	179	**358**	100	65	32	20	18	16	14	11		23	18	4	–	–	–	–	–	–	–	–	–	–	Dinaphtho[1,2-b:2′,1′-d]furan, 5,6,8-trimethoxy-	91913-09-6	NI37473
172	91	186	173	358	130	187	129	**358**	100	97	80	70	65	38	23	21		23	22	2	2	–	–	–	–	–	–	–	–	–	3-Benzyl-2,3,4,5,6,7,8,9-octahydro-2,6-methano-1H-azecino[5,4-b]indole-4,8-dione		LI07916
43	41	283	55	67	81	95	135	**358**	100	65	56	45	35	29	28	26	**15.00**	23	34	3	–	–	–	–	–	–	–	–	–	–	Androst-5-en-7-one, 3-(acetyloxy)-4,4-dimethyl-, (3β)-	54549-89-2	NI37474
299	217	94	157	204	300	82	131	**358**	100	36	36	28	27	24	20	18	**10.46**	23	34	3	–	–	–	–	–	–	–	–	–	–	4-Hexadecynoic acid, 2-phenoxy-, methyl ester	40924-22-9	NI37475
358	276	343	275	274	327			**358**	100	70	65	60	60	10				23	34	3	–	–	–	–	–	–	–	–	–	–	Phenanthrene, 1,2,3,4,4a,9-hexahydro-5,6,8-trimethoxy-1,1,4a-trimethyl-7 isopropyl-, (S)-	60371-74-6	NI37476
315	340	55	316	91	85	109	67	**358**	100	44	27	23	23	23	18	17	**6.60**	23	34	3	–	–	–	–	–	–	–	–	–	–	Pregn-4-ene-3,20-dione, 6-hydroxy-6,16-dimethyl-, (6β,16α)-	56630-84-3	MS08450
55	358	137	91	53	79	93	77	**358**	100	91	80	76	61	58	57	53		23	34	3	–	–	–	–	–	–	–	–	–	–	Pregn-4-ene-3,20-dione, 17-hydroxy-6,16-dimethyl-, (6α,16α)-	2738-39-8	WI01476
43	297	315	137	243	55	41	358	**358**	100	89	87	48	41	37	36	31		23	34	3	–	–	–	–	–	–	–	–	–	–	Pregn-4-ene-3,20-dione, 17-hydroxy-6,16-dimethyl-, (6α,16α)-	2738-39-8	NI37477
43	297	315	137	243	55	41	91	**358**	100	87	85	47	38	36	36	28	**28.28**	23	34	3	–	–	–	–	–	–	–	–	–	–	Pregn-4-ene-3,20-dione, 17-hydroxy-6,16-dimethyl-, (6α,16α)-	2738-39-8	NI37478
43	297	315	243	137	55	41	101	**358**	100	71	61	44	35	35	30	27	**27.00**	23	34	3	–	–	–	–	–	–	–	–	–	–	Pregn-4-ene-3,20-dione, 17-hydroxy-6,16-dimethyl-, (6β,16α)-	14486-40-9	NI37479
255	283	43	55	298	159	256	145	**358**	100	52	51	33	30	29	23	22	**0.00**	23	34	3	–	–	–	–	–	–	–	–	–	–	Pregn-7-en-20-one, 12-acetoxy-, (5α,7β)-		DD01428
298	299	107	91	105	121	283	147	**358**	100	23	23	22	20	19	17	17	**0.00**	23	34	3	–	–	–	–	–	–	–	–	–	–	Pregn-5-en-20-one, 3-(acetyloxy)-, (3β)-	1778-02-5	WI01477
121	315	231	43	18	298	81	190	**358**	100	52	45	39	36	29	25	20	**11.00**	23	34	3	–	–	–	–	–	–	–	–	–	–	Pregn-9(11)-en-12-one, 20-(acetyloxy)-, (5α,20β)-	7704-93-0	NI37480
43	255	283	358	93	315	107	81	**358**	100	55	35	34	22	16	15	15		23	34	3	–	–	–	–	–	–	–	–	–	–	Pregn-16-en-20-one, 3-(acetyloxy)-, (3α)-	56193-61-4	NI37481
298	283	255	91	93	79	77	107	**358**	100	86	64	60	56	49	45	39	**14.90**	23	34	3	–	–	–	–	–	–	–	–	–	–	Pregn-16-en-20-one, 3-(acetyloxy)-, (3β,5β)-		MS08451
298	55	67	149	81	255	283	109	**358**	100	75	54	51	49	44	42	37	**1.01**	23	34	3	–	–	–	–	–	–	–	–	–	–	Pregn-16-en-20-one, 12-(acetyloxy)-, (5α,12β)-	5618-21-3	NI37482
73	358	95	103	133	169	121	119	**358**	100	56	39	29	26	22	22	20		23	38	1	–	–	–	–	–	–	1	–	–	–	Silane, [[3,7-dimethyl-9-(2,6,6-trimethyl-2-cyclohexen-1-ylidene)-3,5,7-nonatrienyl]oxy]trimethyl-	16729-20-7	NI37483
73	130	168	69	41	55	75	119	**358**	100	74	38	31	26	25	24	23	**13.69**	23	38	1	–	–	–	–	–	–	1	–	–	–	Silane, [[3,7-dimethyl-9-(2,6,6-trimethyl-1-cyclohexen-1-yl)-2,4,6,8-nonatetraenyl]oxy]trimethyl-, all-(E)-	16729-17-2	NI37484
73	255	168	358	119	69	75	105	**358**	100	99	98	56	54	51	50	42		23	38	1	–	–	–	–	–	–	1	–	–	–	Silane, [[3,7-dimethyl-9-(2,6,6-trimethyl-1-cyclohexen-1-yl)-2,4,6,8-nonatetraenyl]oxy]trimethyl-, all-(E)-	16729-17-2	NI37485
57	343	302	344	58	358	60	56	**358**	100	95	55	23	13	11	10	10		23	38	1	2	–	–	–	–	–	–	–	–	–	Formamide, N-methyl-N-norconanin-3β-yl-	27769-09-1	NI37486
119	358	105	239	253				**358**	100	56	23	9	5					24	22	3	–	–	–	–	–	–	–	–	–	–	1-Benzoyl-1-(4-methoxyphenyl)-2-(4-methylbenzoyl)ethane		LI07917
135	358	119	239	223				**358**	100	41	24	12	12					24	22	3	–	–	–	–	–	–	–	–	–	–	1-Phenyl-1-(4-methoxybenzoyl)-2-(4-methylbenzoyl)ethane		LI07918
172	91	358	186	185	173	144	170	**358**	100	70	33	31	23	19	15	13		24	26	1	2	–	–	–	–	–	–	–	–	–	3-Benzyl-9-methyl-2,3,4,5,6,7,8,9-octahydro-2,6-methano-1H-azecino[5,4-b]indol-8-one		LI07919
77	255	360	115	256	128	361	51	**358**	100	88	62	20	19	19	18	14	**2.00**	24	26	1	2	–	–	–	–	–	–	–	–	–	3-Phenylazo-Δ-1,3,5(10)-estratrien-17-one		NI37487
121	229	358	81	55	147	41	43	**358**	100	76	42	28	26	24	23	22		24	38	2	–	–	–	–	–	–	–	–	–	–	Chol-9(11)-en-12-one, 3-hydroxy-, (3α,5β)-	42921-48-2	NI37488
84	71	343	110	358				**358**	100	64	55	40	31					24	42	–	2	–	–	–	–	–	–	–	–	–	Conanin-3-amine, N,N-dimethyl-, (3β,5α)-	14278-81-0	NI37489
84	58	85	358	301	71	44	42	**358**	100	30	6	4	2	2	2	2		24	42	–	2	–	–	–	–	–	–	–	–	–	Pregn-5-ene-3,20-diamine, N^3,N^3,N^{20}-trimethyl-, (3β,20S)-	22953-27-1	NI37490
72	70	73	287	71	57	56	55	**358**	100	15	8	2	1	1	1	1	**0.70**	24	42	–	2	–	–	–	–	–	–	–	–	–	Pregn-5-ene-3,20-diamine, N^3,N^{20},N^{20}-trimethyl-, (3β,20S)-	27769-13-7	NI37491
358	267	357	240	359	268	343	77	**358**	100	99	77	32	26	21	15	14		25	30	–	2	–	–	–	–	–	–	–	–	–	Aniline, N-(3,4,5,6-tetraethyl-1-phenyl-2(1H)-pyridinylidene)-	71704-81-9	NI37492
91	160	267	92	65	210	162	70	**358**	100	33	27	10	10	9	8	8	**1.00**	25	30	–	2	–	–	–	–	–	–	–	–	–	N,N,N′-Tribenzylputrescine		LI07920

MASS TO CHARGE RATIOS	M.W.	INTENSITIES	Parent	C	H	O	N	S	F	Cl	Br	I	Si	P	B	X	COMPOUND NAME	CAS Reg No	No
343 259 217 344 301 260 121 218	**358**	100 60 25 23 11 11 7 4	**4.38**	25	42	1	–	–	–	–	–	–	–	–	–	–	**Cholan-16-one, 23-methyl-, (5α,20ξ)-**	56143-17-0	**NI37493**
217 43 95 109 81 55 358 108	**358**	100 82 56 54 54 53 49 45		25	42	1	–	–	–	–	–	–	–	–	–	–	**26,27-Dinor-5β-cholestan-24-one**	14949-16-7	**NI37494**
358 359 179 191 192 178 180 176	**358**	100 37 26 18 16 16 10 10		26	18	–	2	–	–	–	–	–	–	–	–	–	**1,2-Dicarbazol-9-ylethene**		**LI07921**
358 181 357 359 178 165 191 190	**358**	100 33 30 28 16 16 12 10		26	18	–	2	–	–	–	–	–	–	–	–	–	**2,3:4,5:8,9:10,11-Tetrabenzo-1,6-diaza[12]annulene**		**LI07922**
358 130 91 129 359 117 158 131	**358**	100 71 39 36 29 28 22 22		26	30	1	–	–	–	–	–	–	–	–	–	–	**Cyclododecanone, 2,12-dibenzylidene-**	49709-11-7	**NI37495**
245 244 105 183 165 281 358 228	**358**	100 45 45 37 33 19 15 15		26	30	1	–	–	–	–	–	–	–	–	–	–	**Heptyl trityl ether**	16519-22-5	**LI07923**
243 244 105 183 165 281 358 41	**358**	100 46 45 37 33 19 15 12		26	30	1	–	–	–	–	–	–	–	–	–	–	**Heptyl trityl ether**	16519-22-5	**NI37496**
147 105 358 43 41 91 57 104	**358**	100 68 57 56 36 32 21 20		26	46	–	–	–	–	–	–	–	–	–	–	–	**Benzene, butylhexadecyl-**	72101-16-7	**WI01478**
105 43 358 41 106 91 57 232	**358**	100 63 39 39 27 26 22 18		26	46	–	–	–	–	–	–	–	–	–	–	–	**Benzene, 1,3-didecyl-**		**AI02160**
105 106 43 91 57 231 92 41	**358**	100 43 20 17 16 15 13 9	**8.50**	26	46	–	–	–	–	–	–	–	–	–	–	–	**Benzene, 1,3-didecyl-**		**AI02161**
105 43 358 41 29 106 91 57	**358**	100 60 47 35 31 26 26 24		26	46	–	–	–	–	–	–	–	–	–	–	–	**Benzene, 1,3-didecyl-**		**AI02162**
105 358 43 41 104 29 91 359	**358**	100 66 52 29 28 26 23 19		26	46	–	–	–	–	–	–	–	–	–	–	–	**Benzene, 1,4-didecyl-**		**AI02163**
105 358 43 104 91 41 29 57	**358**	100 45 33 26 22 19 19 17		26	46	–	–	–	–	–	–	–	–	–	–	–	**Benzene, 1,4-didecyl-**		**AI02164**
105 43 92 106 358 91 104 41	**358**	100 16 15 13 12 12 10 7		26	46	–	–	–	–	–	–	–	–	–	–	–	**Benzene, 1,4-didecyl-**		**AI02165**
119 358 120 43 41 359 57 55	**358**	100 47 44 34 23 14 12 11		26	46	–	–	–	–	–	–	–	–	–	–	–	**Benzene, 1,4-dimethyl-3-octadecyl-**		**AI02166**
119 120 358 43 41 359 105 57	**358**	100 54 42 29 20 12 10 10		26	46	–	–	–	–	–	–	–	–	–	–	–	**Benzene, 1,4-dimethyl-3-octadecyl-**		**AI02167**
121 122 43 41 67 81 261 135	**358**	100 59 54 54 49 41 35 35	**20.00**	26	46	–	–	–	–	–	–	–	–	–	–	–	**3-Decylperhydroindeno[2,1-a]indene**		**AI02168**
217 135 121 43 41 28 95 67	**358**	100 79 77 53 45 43 41 39	**37.00**	26	46	–	–	–	–	–	–	–	–	–	–	–	**3-Decylperhydropyrene**		**AI02169**
121 122 41 67 81 261 55 79	**358**	100 60 59 52 46 38 38 36	**33.70**	26	46	–	–	–	–	–	–	–	–	–	–	–	**Indeno[2,1-a]indene, 2-decylhexadecahydro-**	55191-42-9	**NI37497**
121 122 41 67 81 55 261 135	**358**	100 59 59 47 44 38 37 34	**28.10**	26	46	–	–	–	–	–	–	–	–	–	–	–	**Indeno[2,1-a]indene, 2-decylhexadecahydro-**	55191-42-9	**AI02170**
105 203 259 43 106 41 29 55	**358**	100 20 14 11 9 9 8 6	**3.53**	26	46	–	–	–	–	–	–	–	–	–	–	–	**8-(4-Methylphenyl)nonadecane**		**AI02171**
105 203 41 106 259 55 118 57	**358**	100 14 12 9 7 6 4 4	**2.29**	26	46	–	–	–	–	–	–	–	–	–	–	–	**8-(4-Methylphenyl)nonadecane**		**AI02172**
245 41 95 67 81 149 43 55	**358**	100 68 65 61 59 53 53 52	**15.50**	26	46	–	–	–	–	–	–	–	–	–	–	–	**9-Octyloctadecahydronaphthacene**		**AI02173**
245 41 149 67 95 43 81 55	**358**	100 58 52 51 46 46 43 42	**19.90**	26	46	–	–	–	–	–	–	–	–	–	–	–	**6-Octylperhydrochrysene**		**AI02174**
245 149 41 67 95 81 43 135	**358**	100 51 50 48 47 46 44 37	**22.89**	26	46	–	–	–	–	–	–	–	–	–	–	–	**6-Octylperhydrochrysene**		**AI02175**
92 91 358 43 57 41 105 55	**358**	100 90 39 33 18 18 13 12		26	46	–	–	–	–	–	–	–	–	–	–	–	**1-Phenyleicosane**		**AI02176**
92 91 43 57 358 133 105 93	**358**	100 47 15 13 9 7 7 7		26	46	–	–	–	–	–	–	–	–	–	–	–	**1-Phenyleicosane**		**AI02177**
92 91 43 41 57 55 105 358	**358**	100 78 50 30 25 19 13 12		26	46	–	–	–	–	–	–	–	–	–	–	–	**1-Phenyleicosane**		**AI02178**
105 106 91 358 43 104 41 57	**358**	100 13 9 7 7 4 4 3		26	46	–	–	–	–	–	–	–	–	–	–	–	**2-Phenyleicosane**		**AI02179**
105 106 91 117 43 104 57 41	**358**	100 13 6 5 5 4 3 3	**2.30**	26	46	–	–	–	–	–	–	–	–	–	–	–	**2-Phenyleicosane**		**AI02180**
105 106 43 91 358 41 104 57	**358**	100 14 10 9 8 6 4 4		26	46	–	–	–	–	–	–	–	–	–	–	–	**2-Phenyleicosane**		**AI02181**
91 119 43 41 330 57 120 55	**358**	100 85 26 20 17 14 12 11	**10.38**	26	46	–	–	–	–	–	–	–	–	–	–	–	**3-Phenyleicosane**		**AI02182**
91 119 43 41 57 55 120 105	**358**	100 86 40 24 20 15 13 11	**5.10**	26	46	–	–	–	–	–	–	–	–	–	–	–	**3-Phenyleicosane**		**AI02183**
119 91 57 105 43 120 41 71	**358**	100 89 17 16 16 13 11 10	**3.80**	26	46	–	–	–	–	–	–	–	–	–	–	–	**3-Phenyleicosane**		**AI02184**
91 133 43 41 315 92 57 55	**358**	100 47 18 11 10 9 9 8	**7.35**	26	46	–	–	–	–	–	–	–	–	–	–	–	**4-Phenyleicosane**		**AI02185**
91 133 57 43 141 315 105 71	**358**	100 54 15 15 11 10 9 9	**5.80**	26	46	–	–	–	–	–	–	–	–	–	–	–	**4-Phenyleicosane**		**AI02186**
91 133 43 41 57 92 55 105	**358**	100 43 23 16 13 10 10 9	**1.53**	26	46	–	–	–	–	–	–	–	–	–	–	–	**4-Phenyleicosane**		**AI02187**
91 147 105 43 301 92 57 41	**358**	100 44 14 11 9 9 9 8	**4.90**	26	46	–	–	–	–	–	–	–	–	–	–	–	**5-Phenyleicosane**		**AI02188**
91 147 43 41 105 57 29 92	**358**	100 32 23 16 12 12 11 9	**2.05**	26	46	–	–	–	–	–	–	–	–	–	–	–	**5-Phenyleicosane**		**AI02189**
91 147 43 301 105 41 92 104	**358**	100 43 15 11 11 11 9 8	**7.34**	26	46	–	–	–	–	–	–	–	–	–	–	–	**5-Phenyleicosane**		**AI02190**
91 175 43 105 273 41 104 29	**358**	100 30 19 15 12 12 10 9	**7.67**	26	46	–	–	–	–	–	–	–	–	–	–	–	**7-Phenyleicosane**		**AI02191**
91 175 105 43 119 273 92 57	**358**	100 30 22 13 11 10 9 9	**5.50**	26	46	–	–	–	–	–	–	–	–	–	–	–	**7-Phenyleicosane**		**AI02192**
91 175 105 43 273 104 92 41	**358**	100 28 14 14 11 10 9 9	**6.67**	26	46	–	–	–	–	–	–	–	–	–	–	–	**7-Phenyleicosane**		**AI02193**
91 43 105 41 203 57 92 55	**358**	100 25 16 16 12 12 9 9	**3.48**	26	46	–	–	–	–	–	–	–	–	–	–	–	**9-Phenyleicosane**		**AI02194**
91 105 203 245 119 43 133 57	**358**	100 22 21 13 13 12 10 10	**5.60**	26	46	–	–	–	–	–	–	–	–	–	–	–	**9-Phenyleicosane**		**AI02195**
91 203 245 105 43 104 41 92	**358**	100 23 16 16 15 11 10 9	**8.21**	26	46	–	–	–	–	–	–	–	–	–	–	–	**9-Phenyleicosane**		**AI02196**
119 120 358 359 43 105 41 106	**358**	100 67 58 17 14 10 9 6		26	46	–	–	–	–	–	–	–	–	–	–	–	**1-(2,5-Xylyl)octadecane**		**AI02197**
358 281 252 359 157 250 163 357	**358**	100 72 41 28 21 18 18 17		27	18	1	–	–	–	–	–	–	–	–	–	–	**9(10H)-Anthracenone, 10-benzhydrylidene-**	667-91-4	**NI37498**
358 141 280 179 165 359 202 152	**358**	100 92 51 43 32 30 29 27		28	22	–	–	–	–	–	–	–	–	–	–	–	**1,8-Bis(1-azulenyl)octa-1,3,5,7-tetraene**	94154-54-8	**NI37499**
267 358 167 281 268 280 265 279	**358**	100 72 50 30 23 22 20 18		28	22	–	–	–	–	–	–	–	–	–	–	–	**1,2,3,4-Tetraphenyl-1,3-butadiene**	806-71-3	**DO01114**
358 267 167 359 191 281 280 268	**358**	100 84 52 30 27 26 25 20		28	22	–	–	–	–	–	–	–	–	–	–	–	**1,2,3,4-Tetraphenyl-1,3-butadiene**	806-71-3	**NI37500**
358 267 167 359 281 280 265 178	**358**	100 88 52 29 27 26 21 21		28	22	–	–	–	–	–	–	–	–	–	–	–	**1,2,3,4-Tetraphenyl-1,3-butadiene**	806-71-3	**NI37501**
69 247 159 180 92	**359**	100 68 34 28 18	**0.00**	6	1	1	3	–	12	–	–	–	–	–	–	–	**3-(Trifluoromethylamino)perfluoro-5-methyl-2,5-diaza-hex-2-en-4-one**		**LI07924**
326 108 328 324 47 110 36 122	**359**	100 73 69 59 50 49 45 41	**6.78**	6	3	–	3	1	–	6	–	–	–	–	–	–	**2-Methylthio-4,6-bis(trichloromethyl)-s-triazine**		**MS08452**

MASS TO CHARGE RATIOS								M.W.	INTENSITIES								Parent	C	H	O	N	S	F	Cl	Br	I	Si	P	B	X	COMPOUND NAME	CAS Reg No	No
69	119	195	290	145	109	220	40	**359**	100	62	40	32	32	18	9	9	**0.15**	8	2	–	1	–	13	–	–	–	–	–	–	–	Butanenitrile, 3,3-bis(pentafluoroethyl)-4,4,4-trifluoro-		IC04120
275	277	273	274	220	249	222	331	**359**	100	89	83	70	39	37	36	35	**25.00**	9	5	3	1	–	–	–	–	–	–	–	–	1	Tungsten, cyanotricarbonyl-π-cyclopentadienyl-	33272-03-6	NI37502
310	64	36	28	312	42	43	62	**359**	100	58	54	44	43	37	30	26	**4.10**	9	11	4	3	2	–	2	–	–	–	–	–	–	2H-1,2,4-Benzothiadiazine-7-sulphonamide, 6-chloro-3-(chloromethyl)-3,4 dihydro-2-methyl-, 1,1-dioxide	135-07-9	NI37503
301	299	303	330	305	15	328	221	**359**	100	76	46	14	11	11	10	8	**2.49**	11	9	4	1	–	–	4	–	–	–	–	–	–	Benzoic acid, 2,3,5,6-tetrachloro-4-[(methoxymethylamino)carbonyl]-, methyl ester	14419-01-3	NI37504
68	41	148	81	67	80	79	44	**359**	100	70	27	27	23	22	20	17	**0.00**	11	21	8	1	2	–	–	–	–	–	–	–	–	Desulphoglucoheirolin		MS08453
266	65	93	77	139	85	94	238	**359**	100	72	56	52	41	33	30	20	**15.00**	12	10	2	1	2	5	–	–	–	–	–	–	–	N-(Pentafluorosulphanyl)-S,S-diphenoxysulphilimine		NI37505
359	168	218	77	125	110	123	109	**359**	100	90	84	82	78	75	71	71		12	10	3	3	3	1	–	–	–	–	–	–	–	Diphenyl sulphanuric fluoride		LI07925
359	113	283	100	126	268	360	115	**359**	100	30	19	18	14	13	12	11		12	10	4	1	–	–	–	–	1	–	–	–	–	Naphthalene, 1-iodo-2,6-dimethoxy-5-nitro-	25315-00-8	NI37506
359	113	283	100	126	268	360	298	**359**	100	30	20	18	14	13	12	10		12	10	4	1	–	–	–	–	1	–	–	–	–	Naphthalene, 1-iodo-2,6-dimethoxy-5-nitro-	25315-00-8	LI07926
117	73	300	147	225	118	315	226	**359**	100	35	13	13	11	11	9	9	**0.00**	12	24	4	1	–	3	–	–	–	2	–	–	–	L-Threonine, N-(trifluoroacetyl)-O-(trimethylsilyl)-, trimethylsilyl ester	52558-13-1	NI37507
246	149	43	233	77	69	41	57	**359**	100	42	34	28	25	25	24	19	**9.20**	13	11	4	1	–	6	–	–	–	–	–	–	–	Tyramine, 3-methoxybis(trifluoroacetyl)-		MS08454
107	133	79	119	78	56	359	69	**359**	100	57	57	40	33	32	26	25		13	12	1	3	–	7	–	–	–	–	–	–	–	Piperazine, 1-(2-pyridyl)-4-(heptafluorobutyryl)-		NI37508
361	359	289	287	318	316	43	290	**359**	100	98	41	35	32	32	26	25		14	10	2	5	–	–	–	1	–	–	–	–	–	8H-Imidazo[2,1-f]purine-2,4-dione, 7-(4-bromophenyl)-1,2,3,4-tetrahydro-1-methyl-		NI37509
360	362	331	333	332	361	330	359	**359**	100	99	90	82	69	68	54	53		15	10	3	3	–	–	–	1	–	–	–	–	–	1H-1,4-Benzodiazepin-2-one, 7-bromo-2,3-dihydro-5-(3-nitrophenyl)-	65247-15-6	NI37510
107	79	77	105	78	106	51	44	**359**	100	69	55	22	14	11	11	8	**0.00**	15	21	7	1	1	–	–	–	–	–	–	–	–	Desulphoglucobarbarin		MS08455
187	145	115	103	85	184	127	157	**359**	100	87	83	67	56	50	43	28	**0.00**	15	21	9	1	–	–	–	–	–	–	–	–	–	D-Galactonitrile, 3,4,5,6-tetra-O-acetyl-2-O-methyl-	84710-70-3	NI37511
45	87	197	159	129	115	117	212	**359**	100	62	46	38	38	34	32	28	**0.00**	15	21	9	1	–	–	–	–	–	–	–	–	–	D-Gluconitrile, 2,3,4,5-tetra-O-acetyl-6-O-methyl-	35439-76-0	NI37512
129	112	87	214	154	189	113	44	**359**	100	64	60	40	27	26	23	20	**0.00**	15	21	9	1	–	–	–	–	–	–	–	–	–	D-Gluconitrile, 2,3,5,6-tetra-O-acetyl-4-O-methyl-	63953-04-8	NI37513
142	115	83	85	127	99	84	217	**359**	100	70	69	66	60	54	45	36	**0.00**	15	21	9	1	–	–	–	–	–	–	–	–	–	D-Gluconitrile, 2,4,5,6-tetra-O-acetyl-3-O-methyl-	35439-74-8	NI37514
43	180	115	138	157	96	170	128	**359**	100	47	39	27	24	20	15	13	**0.07**	15	21	9	1	–	–	–	–	–	–	–	–	–	D-Xylopyranose, 5-acetamido-5-deoxy-, tetraacetate		MS08456
148	135	224	166	138	359	110	196	**359**	100	46	43	16	8	7	7	4		15	22	7	1	–	–	–	–	–	–	1	–	–	Glycine, N-[2-[(3,4-methylenedioxy)phenyl]ethyl]-N-(diethylphosphoryl)-	92241-52-6	NI37515
328	78	359	235	146	329	281	236	**359**	100	67	61	30	28	24	16	16		16	13	5	3	1	–	–	–	–	–	–	–	–	2-(2,4-Dinitrophenylthio)-3-indoleethanol		LI07927
151	359	152	208	107	181	106	78	**359**	100	17	11	9	7	6	5	4		17	14	2	1	–	5	–	–	–	–	–	–	–	2-(3,4-Dimethoxyphenyl)-N-(pentafluorophenylmethylene)ethylamine		MS08457
154	156	115	63	106	41	144	42	**359**	100	66	32	27	22	22	20	20	**4.40**	17	20	1	1	–	–	3	–	–	–	–	–	–	1-Propanamine, 2-chloro-N,N-bis(2-chloroethyl)-3-(1-naphthyloxy)-		NI37516
57	111	41	43	112	39	168	80	**359**	100	62	62	49	38	34	29	24	**3.00**	17	29	3	1	2	–	–	–	–	–	–	–	–	1-Acetyl-2-acetoxy-3,6-di-tert-butylmercapto-1,2,3,6-tetrahydropyridine		MS08458
240	43	177	55	41	69	57	218	**359**	100	71	60	53	47	44	44	38	**4.40**	17	30	1	1	–	5	–	–	–	–	–	–	–	1-Aminotetradecane, N-(pentafluoropropionyl)-		MS08459
200	226	201	359	244	202	129	185	**359**	100	39	25	17	16	15	11	10		19	21	6	1	–	–	–	–	–	–	–	–	–	Di-O-acetyldubinidine		LI07928
84	169	276	98	83	107	359	189	**359**	100	75	52	44	44	33	30	30		19	29	2	5	–	–	–	–	–	–	–	–	–	1,3,4-Oxadiazol-5-amine, 2-(dimethylamino)-N-[3-[3-(piperidinomethyl)phenoxy]propyl]-		MS08460
156	56	85	84	44	128	71	72	**359**	100	93	73	71	69	58	55	49	**2.00**	19	29	2	5	–	–	–	–	–	–	–	–	–	1,3,4-Oxadiazol-5-amine, 2-(ethylamino)-N-[3-[3-(piperidinomethyl)phenoxy]propyl]-		MS08461
359	17	360	76	312	325	104	165	**359**	100	33	23	19	15	14	13	6		20	13	4	3	–	–	–	–	–	–	–	–	–	4-Phthalimido-1-nitrodiphenylamine		IC04121
359	91	77	92	360	144	132	103	**359**	100	66	42	32	25	24	22	12		20	17	2	5	–	–	–	–	–	–	–	–	–	2-Pyrazolin-5-one, 4-(4-amino-5-oxo-1-phenyl-3-methyl-2-pyrazolinyl)-1-phenyl-3-methyl-	16394-90-4	NI37517
91	77	359	192	132	144	51	65	**359**	100	76	66	44	30	22	18	16		20	17	2	5	–	–	–	–	–	–	–	–	–	Rubazonic acid		LI07929
327	328	344	42	329	44	312	284	**359**	100	51	23	19	15	15	14	14	**8.64**	20	25	5	1	–	–	–	–	–	–	–	–	–	Hasubanan-9-ol, 7,8-didehydro-4,5-epoxy-3,6,6-trimethoxy-17-methyl-, (5α,9α,13β,14β)-	40135-50-0	NI37518
301	359	244	229	302	216	270	214	**359**	100	65	43	29	27	25	22	22		20	25	5	1	–	–	–	–	–	–	–	–	–	Hasubanan-6-one, 7,8-didehydro-3-hydroxy-4,7,8-trimethoxy-17-methyl-	2689-15-8	NI37519
44	42	99	70	55	100	86	115	**359**	100	92	83	71	55	35	33	29	**27.02**	20	25	5	1	–	–	–	–	–	–	–	–	–	Morphinan-6-one, 4,5-epoxy-14-hydroxy-3-methoxy-17-methyl-, cyclic 1,2-ethanediyl acetal, (5α)-	2859-02-1	NI37520
151	57	41	303	84	73	56	58	**359**	100	47	41	34	28	19	19	12	**0.23**	20	29	3	3	–	–	–	–	–	–	–	–	–	5-tert-Butyl-2-(tert-butylcyanomethylene)-5-cyano-6-methoxy-3-oxa-1-azabicyclo[4.2.0]octan-4-one		MS08462
192	41	218	55	179	69	95	79	**359**	100	29	17	16	12	12	10	9	**8.00**	20	32	1	1	–	3	–	–	–	–	–	–	–	1H-Pyrrole, 4,5-dihydro-2-methyl-1-trifluoroacetyl-3-[(5Z)-tridecenyl]-		NI37521
69	232	91	217	219	204	97	95	**359**	100	86	66	38	37	37	34	33	**3.10**	21	20	1	1	–	3	–	–	–	–	–	–	–	Acetamide, N-[3-(10,11-dihydro-5H-dibenzo[a,d]cyclohepten-5-ylidene)propyl]-2,2,2-trifluoro-N-methyl-	55334-12-8	NI37522
58	165	255	359	166	73	199	45	**359**	100	11	10	9	8	7	4	4		21	29	–	1	2	–	–	–	–	–	–	–	–	Ethanamine, 2-[[[4-(butylthio)phenyl]benzyl]thio]-N,N-dimethyl-	486-17-9	NI37523
328	301	359	178	300	286	165	146	**359**	100	96	60	47	41	30	13	9		21	29	4	1	–	–	–	–	–	–	–	–	–	Homo-erythrinane alkaloid 5		LI07930
192	55	174	43	69	56	83	41	**359**	100	60	56	55	51	41	38	38	**8.00**	21	29	4	1	–	–	–	–	–	–	–	–	–	1H-Isoindole-5-carboxylic acid, 2,3-dihydro-1,3-dioxo-, dodecyl ester	33975-36-9	NI37524
312	359	313	360	361	314	234	310	**359**	100	71	29	21	8	7	7	4		22	17	–	1	2	–	–	–	–	–	–	–	–	Benzenecarboximidothioic acid, N-(3-phenylbenzo[b]thien-2-yi)-, methyl ester	51324-18-6	NI37525
312	359	313	360	361	314	234	235	**359**	100	69	28	20	8	7	7	3		22	17	–	1	2	–	–	–	–	–	–	–	–	Benzenecarboximidothioic acid, N-(3-phenylbenzo[b]thien-2-yl)-, methyl ester	51324-18-6	NI37526
131	105	77	359	254	255	211	236	**359**	100	99	97	38	37	33	23	18		22	17	2	1	1	–	–	–	–	–	–	–	–	1,5-Benzothiazepin-4(5H)-one, 5-benzoyl-2,3-dihydro-2-phenyl-		MS08463

MASS TO CHARGE RATIOS								M.W.	INTENSITIES								Parent	C	H	O	N	S	F	Cl	Br	I	Si	P	B	X	COMPOUND NAME	CAS Reg No	No
359	185	183	358	262	201	108	165	**359**	100	53	47	43	9	9	8	7		22	18	2	1	–	–	–	–	–	–	1	–	–	Phosphorane, cyanocarbomethoxymethylene-, triphenyl-		LI07931
91	194	250	315	147	121	109	92	**359**	100	45	37	25	15	13	12	10	**7.39**	22	21	–	3	1	–	–	–	–	–	–	–	–	1,3,4-Thiadiazol-2-amine, 2,3-dihydro-N,N-dimethyl-2,3,5-triphenyl-	53067-49-5	NI37527
91	194	250	315	147	121	109	103	**359**	100	45	35	24	15	13	10	9	**7.00**	22	21	–	3	1	–	–	–	–	–	–	–	–	1,3,4-Thiadiazol-2-amine, 2,3-dihydro-N,N-dimethyl-2,3,5-triphenyl-	53067-49-5	NI37528
330	359	331	152	181	154	77	183	**359**	100	30	25	16	9	9	9	8		22	21	2	1	–	–	–	–	–	1	–	–	–	1-Oxa-3-aza-2-silacyclopentan-5-one, 4-ethyl-2,2,3-triphenyl-	53268-82-9	NI37529
159	359	185	158	36	160	157	144	**359**	100	18	18	17	17	11	10	10		22	21	2	3	–	–	–	–	–	–	–	–	–	4-Methyl-2-oxo-7-(1,3,3-trimethylindol-2-ylidenemethylazo)chroman		IC04122
181	182	165	166	128	119	127	91	**359**	100	18	18	14	10	9	4	4	**0.00**	22	21	2	3	–	–	–	–	–	–	–	–	–	1,2,4-Triazolidine-3,5-dione, 4-phenyl-1-(1,2,3,4-tetrahydro-9-(2-propenyl) 1,4-methanonaphthalen-9-yl)-	56009-18-8	NI37530
359	184	344	43	360	358	345	185	**359**	100	85	66	26	24	18	15	10		22	22	4	–	–	–	–	–	–	–	–	–	1	Beryllium, bis(3-phenyl-2,4-pentanedionato)-	21264-49-3	NI37531
359	184	344	43	358	142	317	180	**359**	100	85	66	25	18	8	5	3		22	22	4	–	–	–	–	–	–	–	–	–	1	Beryllium, bis(3-phenyl-2,4-pentanedionato)-	21264-49-3	LI07932
91	232	162	245	107	106	92	93	**359**	100	99	79	69	69	69	69	63	**32.69**	22	33	3	1	–	–	–	–	–	–	–	–	–	11-Azapregn-16-en-20-one, 11-acetyl-3-hydroxy-, (3β)-	56782-27-5	NI37532
105	77	106	121	32	327	206		**359**	100	43	33	25	19	16	12		**0.00**	23	21	3	1	–	–	–	–	–	–	–	–	–	2-Benzamido-1,3-diphenyl-3-methoxypropanone		LI07933
194	91	195	65	105	92	77	51	**359**	100	21	16	10	5	5	5	5	**5.00**	23	21	3	1	–	–	–	–	–	–	–	–	–	6-Methyl-2-(methoxycarbonyl)phenyl-N-(2-methylphenyl)benzimidate		NI37533
168	167	169	91	359	165	166	119	**359**	100	99	91	91	86	79	75	73		23	21	3	1	–	–	–	–	–	–	–	–	–	Phenylalanine, diphenylacetyl-	65707-84-8	NI37534
359	272	344	274	258	286	316	230	**359**	100	78	77	42	39	30	28	28		23	37	2	1	–	–	–	–	–	–	–	–	–	Daphnialcohol acetate		LI07934
114	87	328	43	344	91	41	105	**359**	100	49	48	28	27	27	27	24	**11.00**	23	37	2	1	–	–	–	–	–	–	–	–	–	3β-Hydroxy-16α-methyl-pregn-5-en-20-one, methyl oxime		LI07935
114	101	91	71	296	105	326	344	**359**	100	73	50	47	42	41	38	35	**2.00**	23	37	2	1	–	–	–	–	–	–	–	–	–	3β-Hydroxy-pregn-5-ene-20-one, ethyl oxime		LI07936
206	153	234	155	207	235	154	156	**359**	100	97	48	33	19	9	9	3	**0.00**	24	22	–	1	–	–	1	–	–	–	–	–	–	3-Cyclohexen-1-amine, 6-(4-chlorophenyl)-2,5-diphenyl-, (1α,2α,5α,6β)-	42715-16-2	NI37535
206	153	155	207	154	234	156		**359**	100	92	31	17	9	5	3		**0.00**	24	22	–	1	–	–	1	–	–	–	–	–	–	3-Cyclohexen-1-amine, 6-(4-chlorophenyl)-2,5-diphenyl-, (1α,2β,5β,6β)-	42715-15-1	NI37536
344	44	84	58	345	85	55	41	**359**	100	67	54	43	27	15	14	14	**4.00**	24	41	1	1	–	–	–	–	–	–	–	–	–	Pregnan-3-ol, 20-[isopropylideneamino]-, (3β)-	55448-51-6	LI07937
344	44	84	58	345	41	55	67	**359**	100	70	55	40	32	18	17	15	**5.00**	24	41	1	1	–	–	–	–	–	–	–	–	–	Pregnan-3-ol, 20-[isopropylideneamino]-, (3β)-	55448-51-6	NI37537
85	58	44	84	55	67	41	81	**359**	100	52	50	30	13	12	12	10	**8.00**	24	41	1	1	–	–	–	–	–	–	–	–	–	Pregnan-18-ol, 20-[isopropylideneamino]-	55400-12-9	NI37538
85	58	44	84	55	41			**359**	100	50	50	28	12	10			**0.00**	24	41	1	1	–	–	–	–	–	–	–	–	–	Pregnan-18-ol, 20-[isopropylideneamino]-	55400-12-9	LI07938
58	359	344	45	41	81	55	67	**359**	100	40	38	20	18	17	16	15		24	41	1	1	–	–	–	–	–	–	–	–	–	Pregnan-20-ol, 3-[isopropylideneamino]-, (3β)-	55400-11-8	NI37539
113	70	126	55	359	72	67	43	**359**	100	62	60	47	38	37	37	35		24	41	1	1	–	–	–	–	–	–	–	–	–	Pyrrolidine, 1-(1-oxo-8,11,14-eicosatrienyl)-	56600-10-3	NI37540
359	360	358	255	77	51	357	179.5	**359**	100	43	40	21	21	20	19	15		25	17	–	3	–	–	–	–	–	–	–	–	–	1,2-Diphenyl-2,6-dihydropyrazolo[3,4-c]carbazole		LI07939
181	359	182	180	77	344	275	91	**359**	100	24	18	18	17	10	10	10		26	33	–	1	–	–	–	–	–	–	–	–	–	3-Azabicyclo[4.3.0]non-1(6)-ene, 7,7,8,8,9,9-hexamethyl-3,4-diphenyl-	123187-18-8	DD01429
359	360	199	228	358	343	341	252	**359**	100	28	11	7	6	5	5	5		27	21	–	1	–	–	–	–	–	–	–	–	–	9,10-Diphenyl-1-(methylamino)anthracene		MS08464
118	184	300	242	150	330	360	183	**360**	100	75	70	70	65	60	50	50		4	10	4	4	2	–	–	–	–	–	–	–	2	Di-μ-ethanethiolatotetranitrosodicobalt		MS08465
129	66	128	229	63	227	131	130	**360**	100	54	48	43	43	41	39	30	**11.00**	5	6	–	–	–	–	6	–	–	3	–	–	–	2,4,8-Trisilabicyclo[3.2.1]-6-octene, 2,2,4,4,8,8-hexachloro-		NI37541
69	31	119	131	181	179	50	93	**360**	100	80	42	31	24	20	20	15	**0.00**	6	–	–	–	–	11	–	1	–	–	–	–	–	Perfluoro-trans-2-bromohex-2-ene		LI07940
69	75	163	44	28	241	119	341	**360**	100	42	29	28	27	21	18	15	**7.00**	8	1	1	–	–	13	–	–	–	–	–	–	–	Perfluoro(3-methylpent-3-yl) ketene		IC04123
91	92	276	274	65	78	39	360	**360**	100	45	42	25	21	16	16	15		10	8	3		–	–	–	–	–	–	–	–	1	π-Cycloheptatrienetricarbonyltungsten		MS08466
92	91	276	304	274	65	360	39	**360**	100	99	27	24	24	19	17	16		10	8	3	–	–	–	–	–	–	–	–	–	1	π-Toluenetricarbonyltungsten		MS08467
69	43	264	100	56	120	147	30	**360**	100	88	84	70	68	49	47	44	**3.90**	10	9	2	2	–	9	–	–	–	–	–	–	–	Butanamide, N-[3-(difluoromethyl)-2-oxo-3-piperidinyl]-2,2,3,3,4,4,4-heptafluoro-		NI37542
57	56	41	280	278	29	282	58	**360**	100	24	17	13	10	10	6	6	**0.00**	10	18	–	2	–	–	4	–	–	–	–	–	2	Aluminum, di-μ-chlorodichlorobis(2-isocyano-2-methylpropane)di-	62431-10-1	NI37543
260	241	210	141	360	291	117	93	**360**	100	67	27	12	11	10	8	7		11	–	–	–	–	12	–	–	–	–	–	–	–	Perfluorotricyclo[5.2.2.0^{2,6}]undeca-2,3,8-triene		NI37544
260	241	210	360	291	261	141	117	**360**	100	61	21	18	12	10	9	9		11	–	–	–	–	12	–	–	–	–	–	–	–	Perfluorotricyclo[5.2.2.0^{2,6}]undeca-2,5,8-triene		NI37545
362	53	83	360	364	281	283	67	**360**	100	79	54	52	51	48	45	40		11	6	4	–	–	–	–	2	–	–	–	–	–	1,4-Naphthalenedione, 2,6-dibromo-5,8-dihydroxy-3-methyl-	104506-33-4	NI37546
362	364	360	283	281	53	67	255	**360**	100	53	52	34	34	30	17	15		11	6	4	–	–	–	–	2	–	–	–	–	–	1,4-Naphthalenedione, 3,6-dibromo-5,8-dihydroxy-2-methyl-	104506-35-6	NI37547
72	40	81	35	83	28	42	82	**360**	100	93	63	52	41	32	31	28	**0.00**	11	12	3	2	–	–	4	–	–	–	–	–	–	N′-(4-Chlorophenyl)-N,N-dimethylurea trichloroacetic acid	140-41-0	NI37548
244	199	227	246	201	270	229	245	**360**	100	67	36	32	22	16	12	11	**0.00**	11	12	3	2	–	–	4	–	–	–	–	–	–	N′-(4-Chlorophenyl)-N,N-dimethylurea trichloroacetic acid	140-41-0	NI37549
72	44	82	45	47	42	84	63	**360**	100	77	61	59	50	42	40	31	**0.00**	11	12	3	2	–	–	4	–	–	–	–	–	–	N′-(4-Chlorophenyl)-N,N-dimethylurea trichloroacetic acid	140-41-0	NI37550
97	29	27	45	109	65	125	269	**360**	100	93	38	32	30	28	26	18	**3.50**	11	15	3	–	2	–	2	–	–	–	1	–	–	Phosphorothioic acid, O-[2,5-dichloro-4-(methylthio)phenyl] O,O-diethyl ester	21923-23-9	NI37551
90	30	43	45	174	184	247	149	**360**	100	84	72	54	45	41	39	37	**24.75**	11	15	6	2	1	3	–	–	–	–	–	–	–	Methyl-S-carboxymethyl(trifluoroacetyl)cysteinylglycine		LI07941
328	360	297	87	329	322	361	63	**360**	100	89	22	20	17	14	13	10		12	–	–	–	2	8	–	–	–	–	–	–	–	Thianthrene, octafluoro-	16012-83-2	NI37552
360	328	296	87	329	111	361	63	**360**	100	69	24	23	18	15	14	11		12	–	–	–	2	8	–	–	–	–	–	–	–	Thianthrene, octafluoro-	16012-83-2	NI37553
328	360	297	87	111	361	117	168	**360**	100	99	24	21	13	11	8	6		12	–	–	–	2	8	–	–	–	–	–	–	–	Thianthrene, octafluoro-	16012-83-2	LI07942
56	28	112	84	220	190	248	134	**360**	100	63	59	40	33	20	19	13	**7.00**	12	8	6	–	–	–	–	–	–	–	–	–	2	1,4-Dimethylbutatriene(hexacarbonyl)diiron		MS08468
220	112	192	248	332	56	152	134	**360**	100	72	54	39	35	31	22	22	**0.00**	12	8	6	–	–	–	–	–	–	–	–	–	2	Iron, hexacarbonyl-μ-[(1,2,2α-η^3:3,3α,4-η^3)-2,3-bis(methylene)-1,4-butanediyl]di-	18662-05-0	DD01430
248	112	220	192	56	360	131	134	**360**	100	91	89	74	65	48	47	40		12	8	6	–	–	–	–	–	–	–	–	–	2	Iron, hexacarbonyl[μ-[(1,2-η:3,4-η)-1-ethyl-1,3-butadiene-1,4-diyl]]di-	74381-49-0	NI37554
78	360	108	134	359	191	163	80	**360**	100	68	34	29	17	15	15	14		12	11	1	4	–	7	–	–	–	–	–	–	–	Piperazine, 1-(2-pyrimidyl)-4-(heptafluorobutyryl)-		NI37555

MASS TO CHARGE RATIOS								M.W.	INTENSITIES								Parent	C	H	O	N	S	F	Cl	Br	I	Si	P	B	X	COMPOUND NAME	CAS Reg No	No
156	80	82	128	155	81	78	79	**360**	100	87	85	56	52	52	51	49	**0.00**	12	14	1	2	–	–	–	2	–	–	–	–	–	Dipyrido[1,2-a:2',1'-c]pyrazinediium, 6,7-dihydro-, dibromide, monohydrate	6385-62-2	NI37556
41	42	39	27	43	273	272	270	**360**	100	81	64	42	41	29	29	27	**10.21**	12	20	–	–	–	–	–	–	–	–	–	–	2	Molybdenum, tetrakis(η^3-2-propenyl)di-	12318-77-3	NI37557
238	314	360	136	192	164	110	135	**360**	100	82	39	32	29	26	22	21		13	4	9	4	–	–	–	–	–	–	–	–	–	9H-Fluoren-9-one, 2,4,5,7-tetranitro-	746-53-2	NI37558
314	238	28	192	164	256	136	360	**360**	100	93	71	54	31	29	28	26		13	4	9	4	–	–	–	–	–	–	–	–	–	9H-Fluoren-9-one, 2,4,5,7-tetranitro-	746-53-2	NI37559
230	198	59	131	98	99	58	197	**360**	100	83	80	76	74	59	29	25	**22.83**	13	16	–	2	5	–	–	–	–	–	–	–	–	2,4,6,8,9-Pentathiatricyclo[3.3.1.1^{3,7}]decane-1,5-dipropanenitrile, 3,7-dimethyl-	57274-43-8	NI37560
360	362	364	361	363	77	75	87	**360**	100	99	32	21	17	12	10	9		14	7	5	–	–	–	3	–	–	–	–	–	–	9H-Xanthen-9-one, 2,4,5-trichloro-1,3,6-trihydroxy-8-methyl-	20716-96-5	NI37561
194	30	44	63	52	75	39	119	**360**	100	47	44	42	36	25	24	16	**11.00**	14	8	8	4	–	–	–	–	–	–	–	–	–	4,4'-Dicarboxy-2,2'-dinitroazobenzene		IC04124
155	91	175	111	139	65	177	75	**360**	100	69	67	57	38	26	24	21	**0.23**	14	13	5	–	2	–	1	–	–	–	–	–	–	Benzenesulphonic acid, 4-chloro-, [(4-methylphenyl)sulphonyl]methyl ester	31081-11-5	NI37562
155	91	65	330	156	111	92	75	**360**	100	60	12	9	9	5	5	5	**0.04**	14	13	5	–	2	–	1	–	–	–	–	–	–	Methanol, [(3-chlorophenyl)sulphonyl]-, 4-methylbenzenesulphonate	31081-07-9	NI37563
164	135	136	43	70	246	204	218	**360**	100	98	96	95	62	54	54	40	**0.50**	14	16	4	8	–	–	–	–	–	–	–	–	–	Adenosine, 5'-azido-5'-deoxy-N-formyl-2',3'-O-isopropylidene-	25030-10-8	NI37564
138	207	110	111	42	208	206	41	**360**	100	62	47	29	24	21	19	17	**4.45**	14	18	2	2	–	6	–	–	–	–	–	–	–	N,N'-Butylenebis(5,5,5-trifluoro-4-oxopentan-2-imine)	87961-37-3	NI37565
88	131	157	143	102	72	58	101	**360**	100	69	52	50	48	48	47	44	**36.04**	14	24	7	4	–	–	–	–	–	–	–	–	–	1,4,13-Trioxa-7,10,16,19-tetraazacycloheneicosane-6,9,17,20-tetrone		NI37566
107	132	133	28	77	176	44	78	**360**	100	59	33	30	21	20	19	17	**0.03**	15	12	7	4	–	–	–	–	–	–	–	–	–	4-Hydroxyphenylpyruvic acid 2,4-dinitrophenylhydrazone		NI37567
138	156	198	111	110	153	180	139	**360**	100	66	64	54	43	36	24	21	**0.10**	15	20	10	–	–	–	–	–	–	–	–	–	–	Cyclohexanecarboxylic acid, 1,3,4,5-tetrakis(acetyloxy)-, (1α,3α,4α,5β)-(–)-	32469-24-2	MS08469
100	141	139	241	73	101	58	181	**360**	100	86	55	43	34	25	24	23	**0.00**	15	24	8	2	–	–	–	–	–	–	–	–	–	α-D-Glucopyranoside, methyl 4,6-di-O-acetyl-2,3-diacetamido-2,3-dideoxy-	27851-40-7	NI37568
100	142	139	241	101	58	84	73	**360**	100	81	50	28	26	25	21	21	**0.00**	15	24	8	2	–	–	–	–	–	–	–	–	–	β-D-Glucopyranoside, methyl 4,6-di-O-acetyl-2,3-diacetamido-2,3-dideoxy-	7721-85-9	NI37569
73	345	149	221	41	311	101	75	**360**	100	46	41	24	24	23	21	21	**0.46**	15	25	4	2	–	–	1	–	–	1	–	–	–	2,4,6(1H,3H,5H)-Pyrimidinetrione, 5-[3-chloro-2-[(trimethylsilyl)oxy]propyl]-1,3-dimethyl-5-(2-propenyl)-	55836-74-3	NI37570
73	147	360	170	361	345	45	229	**360**	100	61	45	41	15	15	12	11		15	32	4	–	–	–	–	–	–	3	–	–	–	2-Pentenedioic acid, 2-methyl-, tris(trimethylsilyl)-		NI37571
73	147	170	229	45	360	75	109	**360**	100	82	28	26	24	21	17	16		15	32	4	–	–	–	–	–	–	3	–	–	–	2-Pentenedioic acid, 3-methyl-, tris(trimethylsilyl)-		NI37572
360	213	186	361	198	184	170	54	**360**	100	57	33	26	19	18	18	12		16	13	2	2	–	5	–	–	–	–	–	–	–	Melatonin, (pentafluoropropionyl)-		MS08470
241	360	242	207	120	361	184	169	**360**	100	38	15	14	11	9	6	6		16	13	2	2	–	5	–	–	–	–	–	–	–	6-Methoxyharmalan, (pentafluoropropionyl)-		MS08471
121	95	135	67	77	56	139	207	**360**	100	78	64	56	47	46	31	22	**0.00**	16	22	1	–	–	–	–	–	–	–	–	–	1	2-Oxabicyclo[4.4.0]decane, 3-[(phenyltelluro)methyl]-		DD01431
121	95	135	67	77	56	207	360	**360**	100	96	95	58	48	45	25	11		16	22	1	–	–	–	–	–	–	–	–	–	1	2-Oxabicyclo[5.4.0]undecane, 4-(phenyltelluro)-		DD01432
121	56	95	135	77	67	139	207	**360**	100	99	94	88	54	54	38	26	**0.00**	16	22	1	–	–	–	–	–	–	–	–	–	1	1-Oxaspiro[4.5]decane, 2-[(phenyltelluro)methyl]-		DD01433
28	276	82	177	55	95	83	274	**360**	100	70	51	48	45	39	37	33	**3.25**	17	24	3	2	–	–	–	–	–	–	–	–	1	Iron, tricarbonyl[N,N'-1,2-ethanediylidenebis[cyclohexanamine]-N,N']-	54446-63-8	NI37573
201	173	269	160	29	167	109	127	**360**	100	93	87	74	69	41	36	34	**0.90**	17	28	8	–	–	–	–	–	–	–	–	–	–	1,1,5,5-Pentanetetracarboxylic acid, tetraethyl ester	3779-30-4	NI37574
86	61	286	188	43	56	104	41	**360**	100	55	49	47	44	39	38	26	**1.69**	17	32	4	2	1	–	–	–	–	–	–	–	–	N-(N-Acetylmethionyl)leucine butyl ester		AI02198
41	55	121	177	119	179	29	317	**360**	100	85	77	66	62	58	53	50	**0.00**	17	36	–	–	–	–	–	–	–	–	–	–	1	Dibutylisopropylcyclohexyltin		MS08472
217	197	227	287	243	191	345	189	**360**	100	91	80	60	51	51	49	46	**11.40**	17	36	4	–	–	–	–	–	–	2	–	–	–	Cyclopentaneacetic acid, 2-propyl-3,5-bis[(trimethylsilyl)oxy]-, methyl ester, (1α,2α,3α,5β)-	61177-10-4	NI37575
191	217	197	227	243	218	345	196	**360**	100	52	34	33	21	12	11	10	**0.00**	17	36	4	–	–	–	–	–	–	2	–	–	–	Cyclopentaneacetic acid, 2-propyl-3,5-bis[(trimethylsilyl)oxy]-, methyl ester, (1α,2α,3β,5β)-	61177-09-1	NI37576
73	75	345	55	229	217	117	129	**360**	100	60	44	21	19	18	18	16	**0.00**	17	36	4	–	–	–	–	–	–	2	–	–	–	Undecanedioic acid, bis(trimethylsilyl) ester		MS08473
73	75	345	55	117	229	147	129	**360**	100	68	50	28	22	21	18	17	**0.00**	17	36	4	–	–	–	–	–	–	2	–	–	–	Undecanedioic acid, bis(trimethylsilyl) ester		MS08474
73	75	345	229	55	129	117	217	**360**	100	82	55	27	27	24	23	17	**2.29**	17	36	4	–	–	–	–	–	–	2	–	–	–	Undecanedioic acid, bis(trimethylsilyl) ester		NI37577
360	152	201	362	361	173	139	75	**360**	100	74	55	44	22	22	22	22		18	13	2	–	2	–	1	–	–	–	–	–	–	4-Chloro-4'-(4-mercaptophenyl)diphenyl sulphone		IC04125
145	132	146	104	117	228	230	232	**360**	100	70	45	35	25	11	7	1	**0.00**	18	14	2	2	–	–	2	–	–	–	–	–	–	Phthalimide, 4,5-dichloro-N-[(1,2,3,4-tetrahydroisoquinolin-2-yl)methyl]-	96919-04-9	NI37578
137	115	180	109	217	181	253	138	**360**	100	32	30	21	20	19	17	14	**4.15**	18	16	4	–	2	–	–	–	–	–	–	–	–	2-Methylthianaphthene-1,1-dioxide anti-head-to-head photodimer		MS08475
360	359	361	317	311	343	344	325	**360**	100	66	20	16	16	12	11	9		18	16	8	–	–	–	–	–	–	–	–	–	–	4H-1-Benzopyran-4-one, 5,6-dihydroxy-2-(3-hydroxy-4-methoxyphenyl)-3,7-dimethoxy-	548-74-3	NI37579
360	345	359	361	317	342	346	299	**360**	100	63	25	21	21	14	12	12		18	16	8	–	–	–	–	–	–	–	–	–	–	4H-1-Benzopyran-4-one, 5,7-dihydroxy-2-(4-hydroxy-3-methoxyphenyl)-3,6-dimethoxy-	10173-01-0	NI37580
29	234	31	32	45	77	95	249	**360**	100	99	98	74	54	45	44	41	**0.00**	18	20	4	2	1	–	–	–	–	–	–	–	–	1H-Pyrazolium, 1,2-dimethyl-3,5-diphenyl-, methyl sulphate	43222-48-6	NI37581
44	69	43	55	57	41	60	73	**360**	100	97	88	83	74	61	60	58	**0.49**	18	20	6	2	–	–	–	–	–	–	–	–	–	Benzaldehyde, 4-hydroxy-3,5-dimethoxy-, [(4-hydroxy-3,5-dimethoxyphenyl)methylene]hydrazone	14414-32-5	NI37582
105	77	122	201	166	67	51	134	**360**	100	78	53	52	39	24	19	18	**6.55**	18	20	6	2	–	–	–	–	–	–	–	–	–	2-Butenedioic acid, 2-[[1-methyl-3-oxo-3-(benzoylmethylamino)-1-propenyl]amino]-, dimethyl ester		NI37583
211	209	269	149	147	267	120	207	**360**	100	78	63	63	49	47	45	40	**0.60**	18	24	–	–	–	–	–	–	–	–	–	–	1	Dibenzyldiethyltin		NI37584
87	57	43	41	71	69	42	55	**360**	100	68	62	61	58	42	26	25	**0.10**	18	26	3	–	–	–	2	–	–	–	–	–	–	Butanoic acid, 4-(2,4-dichlorophenoxy)-, isooctyl ester	1320-15-6	NI37585

MASS TO CHARGE RATIOS								M.W.	INTENSITIES								Parent	C	H	O	N	S	F	Cl	Br	I	Si	P	B	X	COMPOUND NAME	CAS Reg No	No
145	200	97	69	41	143	106	55	**360**	100	50	34	34	32	20	17	17	**15.00**	18	32	3	–	2	–	–	–	–	–	–	–	–	Hexanoic acid, 6-[(2,2-dimethyl-6,10-dithiaspiro[4.5]decan-1-yl)oxy]-, ethyl ester	92640-75-0	NI37586
259	185	361	129	260	213	231	133	**360**	100	67	52	18	16	14	13	12	**0.20**	18	32	7	–	–	–	–	–	–	–	–	–	–	Citric acid, tri-butyl-		NI37587
185	129	41	29	259	57	56	27	**360**	100	74	48	44	41	41	16	10	**0.00**	18	32	7	–	–	–	–	–	–	–	–	–	–	Citric acid, tri-butyl-		IC04126
100	114	101	58	70	56	84	113	**360**	100	97	89	89	86	80	74	71	**16.60**	18	36	5	2	–	–	–	–	–	–	–	–	–	4,7,13,16,22-Pentaoxa-1,10-diazabicyclo[8.8.7]pentacosane		MS08476
100	72	54	79	88	128	114	84	**360**	100	90	88	77	71	66	64	64	**19.10**	18	36	5	2	–	–	–	–	–	–	–	–	–	5,8,11,18,23-Pentaoxa-1,15-diazabicyclo[13.5.5]pentacosane		MS08477
58	79	72	70	114	56	84	57	**360**	100	97	95	81	77	75	64	62	**10.10**	18	36	5	2	–	–	–	–	–	–	–	–	–	5,8,15,18,23-Pentaoxa-1,12-diazabicyclo[10.8.5]pentacosane		MS08478
233	229	345	147	73	234	303	189	**360**	100	50	49	45	37	21	19	15	**0.00**	18	40	3	–	–	–	–	–	–	2	–	–	–	Dodecanoic acid, 3-(trimethylsiloxy)-, trimethylsilyl ester		NI37588
73	233	147	229	345	75	303	234	**360**	100	67	65	38	33	18	16	14	**0.00**	18	40	3	–	–	–	–	–	–	2	–	–	–	Dodecanoic acid, 3-(trimethylsiloxy)-, trimethylsilyl ester		NI37589
181	180	151	165	360	182	152	195	**360**	100	35	16	15	13	13	13	11		19	20	7	–	–	–	–	–	–	–	–	–	–	4H-1-Benzopyran-4-one, 2-(3,4-dimethoxyphenyl)-2,3-dihydro-3-hydroxy-7,8-dimethoxy-	35683-21-7	NI37590
181	331	180	360	151	332	193	165	**360**	100	64	38	19	15	13	12	12		19	20	7	–	–	–	–	–	–	–	–	–	–	4H-1-Benzopyran-4-one, 2-(3,4-dimethoxyphenyl)-2,3-dihydro-3-hydroxy-5,7-dimethoxy-, (2R-trans)-	6563-36-6	NI37591
167	59	74	31	43	209	168	29	**360**	100	21	19	16	12	11	11	10	**1.70**	19	20	7	–	–	–	–	–	–	–	–	–	–	3,5-Dimethoxy-4-O-acetylbenzyl 2-methoxy-4-formylphenyl ether		NI37592
121	360	147	135	42	91	257	199	**360**	100	75	58	46	28	24	18	14		19	20	7	–	–	–	–	–	–	–	–	–	–	Icacenone	99624-07-4	NI37593
360	167	345	165	194	166	95	179	**360**	100	52	38	38	27	24	20	19		19	20	7	–	–	–	–	–	–	–	–	–	–	(3S)-6,7,2′,3′-Tetramethoxyisoflavanquinone		NI37594
193	299	148	360	194	167	165	345	**360**	100	70	70	65	40	40	40	35		19	24	5	2	–	–	–	–	–	–	–	–	–	Bis(3,5-dimethyl-4-ethoxycarbonyl-2-pyrrolyl) ketone		LI07943
243	73	360	244	345	361	228	242	**360**	100	75	43	22	18	14	12	11		19	28	3	–	–	–	–	–	–	2	–	–	–	Naphthaleneacetic acid, α-methyl-6-(trimethylsiloxy)-, trimethylsilyl ester		NI37595
101	83	129	157	88	201	139	81	**360**	100	33	19	16	16	13	13	12	**0.00**	19	36	6	–	–	–	–	–	–	–	–	–	–	L-Menthyl-2,3,4-tri-O-methyl-5-methoxy-D-xyloside		LI07944
55	97	147	189	69	149	191	163	**360**	100	88	53	29	16	15	14	9	**0.00**	19	44	2	–	–	–	–	–	–	2	–	–	–	Heptane, 1,7-bis[(tert-butyldimethylsilyl)oxy]-		NI37596
343	159	344	131	130	160	103	327	**360**	100	46	22	17	13	6	6	5	**0.00**	20	17	3	2	–	–	–	–	–	–	–	–	1	Bis(2-methyl-8-hydroxyquinolinolinato)aluminium hydroxide		LI07945
85	105	77	259	84	93	67	79	**360**	100	98	87	85	55	45	36	31	**2.00**	20	24	6	–	–	–	–	–	–	–	–	–	–	1H-Cyclopenta[c]furan-1-one, 4-[(benzoyloxy)methyl]hexahydro-5-[(tetrahydro-2H-pyran-2-yl)oxy]-, [3aR-(3aα,4α,5β,6aα)]-		NI37597
137	151	138	152	77	165	122	180	**360**	100	42	22	20	18	16	15	14	**12.00**	20	24	6	–	–	–	–	–	–	–	–	–	–	Dehydrodiconiferyl alcohol		LI07946
136	121	137	43	80	45	109	52	**360**	100	95	35	26	25	25	17	14	**2.42**	20	24	6	–	–	–	–	–	–	–	–	–	–	Dibenzo[b,k][1,4,7,10,13,16]hexaoxacyclooctadecin, 6,7,9,10,17,18,20,21-octahydro-	14187-32-7	NI37598
136	121	137	80	360	109	45	65	**360**	100	91	29	26	24	20	15	14		20	24	6	–	–	–	–	–	–	–	–	–	–	Dibenzo[b,k][1,4,7,10,13,16]hexaoxacyclooctadecin, 6,7,9,10,17,18,20,21-octahydro-	14187-32-7	NI37599
94	109	95	121	28	81	107	108	**360**	100	66	66	65	57	54	49	43	**28.63**	20	24	6	–	–	–	–	–	–	–	–	–	–	Dihydrocolumbin	10207-58-6	NI37600
360	94	41	28	133	55	43	44	**360**	100	78	78	76	73	60	55	54		20	24	6	–	–	–	–	–	–	–	–	–	–	Dihydroisocolumbin	10413-81-7	NI37601
211	210	181	360	286	121	167	183	**360**	100	82	61	22	19	16	15	13		20	24	6	–	–	–	–	–	–	–	–	–	–	3-Flavanol, 4-ethoxy-4′,7,8-trimethoxy-, trans-2,3,cis-3,4-	16821-15-1	NI37602
137	43	297	29	55	237	59	41	**360**	100	99	94	88	82	70	68	64	**10.00**	20	24	6	–	–	–	–	–	–	–	–	–	–	Gibberellin A_3 methyl ester		MS08479
328	28	43	41	55	163	91	301	**360**	100	55	42	42	41	40	32	31	**28.00**	20	24	6	–	–	–	–	–	–	–	–	–	–	Gibberellin A_6 methyl ester		MS08480
328	155	91	238	55	41	270	77	**360**	100	78	72	56	55	52	51	50	**3.00**	20	24	6	–	–	–	–	–	–	–	–	–	–	Gibberellin A_{22} methyl ester		MS08481
43	41	69	91	77	53	55	345	**360**	100	93	80	50	40	38	34	16	**14.00**	20	24	6	–	–	–	–	–	–	–	–	–	–	1,1,2,3,12,13-Hexadehydrochaparrolide		LI07947
137	360	151	153	194	138	180	150	**360**	100	39	37	25	19	19	12	12		20	24	6	–	–	–	–	–	–	–	–	–	–	(+)-Lariciresinol	27003-73-2	NI37603
224	360	211	209	212	180	197	181	**360**	100	84	76	46	22	10	9	6		20	24	6	–	–	–	–	–	–	–	–	–	–	Lonchocarpan dimethyl ether		LI07948
209	224	211	166	360	151	181	136	**360**	100	72	57	32	30	24	20	19		20	24	6	–	–	–	–	–	–	–	–	–	–	2′,3′,4′,6′,7-Pentamethoxyisoflavan		LI07949
187	147	345	203	55	83	360		**360**	100	58	21	19	14	12	2			20	24	6	–	–	–	–	–	–	–	–	–	–	Phenol, 2-senecionyl-4-(1-acetylsarracinoyloxyethyl)-		NI37604
41	83	69	301	247	91	245	77	**360**	100	62	50	49	41	32	31	31	**0.00**	20	24	6	–	–	–	–	–	–	–	–	–	–	Sanguinone A		DD01434
360	329	298	328	59	127	314	99	**360**	100	92	33	26	26	25	23	20		20	24	6	–	–	–	–	–	–	–	–	–	–	1,2-Bis(3,4,5-trimethoxyphenyl)ethylene	61240-22-0	NI37605
299	360	345	194	300	361	193	148	**360**	100	75	48	26	20	17	17	15		20	28	4	2	–	–	–	–	–	–	–	–	–	1,1′-Bis[3,4-dimethyl-5-(ethoxycarbonyl)-2-pyrrolyl]ethane		LI07950
114	247	97	70	99	175	218	248	**360**	100	76	44	34	23	19	18	13	**0.00**	20	28	4	2	–	–	–	–	–	–	–	–	–	1H-Indole-3-carboxylic acid, 2-[(1-hydroxy-4-piperidinyl)methyl]-5-methoxy-, ethyl ester		DD01435
285	299	286	253	314	238	267	121	**360**	100	77	55	55	45	45	43	43	**21.59**	20	28	4	2	–	–	–	–	–	–	–	–	–	Pyrrole-2,4-dicarboxylic acid, 5-[(4-ethyl-3,5-dimethylpyrrol-2-yl)methyl]-3-methyl-, diethyl ester	15845-55-3	NI37606
120	92	240	360	64	196	76	39	**360**	100	97	90	53	19	9	9	8		21	12	6	–	–	–	–	–	–	–	–	–	–	Trisalicylide		MS08482
360	316	288	361	75	103	317	289	**360**	100	48	48	23	15	14	10	10		21	16	4	2	–	–	–	–	–	–	–	–	–	5-Dimethylcarbamyl-1,3-dioxo-2,3-dihydro-2-(3-hydroxyquinol-2-yl)indene		IC04127
92	65	77	267	94	93	91	105	**360**	100	78	75	65	50	23	20	13	**11.00**	21	16	4	2	–	–	–	–	–	–	–	–	–	2-Phenylhydrazonopropanedioic acid, diphenyl ester		NI37607
91	131	42	283	77	105	161	79	**360**	100	70	69	60	60	54	44	44	**2.65**	21	28	5	–	–	–	–	–	–	–	–	–	–	Aldosterone	52-39-1	NI37608
91	79	131	42	283	77	105	255	**360**	100	71	70	70	59	59	54	46	**2.00**	21	28	5	–	–	–	–	–	–	–	–	–	–	Aldosterone	52-39-1	LI07951
122	121	179	207	208	310	123	328	**360**	100	92	60	52	50	49	44	29	**15.00**	21	28	5	–	–	–	–	–	–	–	–	–	–	Crotophorbolon-20-methyl ether		LI07952

MASS TO CHARGE RATIOS								M.W.	INTENSITIES								Parent	C	H	O	N	S	F	Cl	Br	I	Si	P	B	X	COMPOUND NAME	CAS Reg No	No
107	41	167	55	121	44	149	134	**360**	100	90	81	78	75	74	69	62	**22.90**	21	28	5	–	–	–	–	–	–	–	–	–	–	**Cyclopropanecarboxylic acid, 3-(3-methoxy-2-methyl-3-oxo-1-propenyl)-2,2-dimethyl-, 3-(2-butenyl)-2-methyl-4-oxo-2-cyclopenten-1-yl ester, [1R-[1α[S*(Z)],3β(E)]]-**	121-20-0	**MS08483**
107	121	93	91	79	43	55	29	**360**	100	62	61	40	33	33	32	32	**0.00**	21	28	5	–	–	–	–	–	–	–	–	–	–	**Cyclopropanecarboxylic acid, 3-(3-methoxy-2-methyl-3-oxo-1-propenyl)-2,2-dimethyl-, 3-(2-butenyl)-2-methyl-4-oxo-2-cyclopenten-1-yl ester, [1R-[1α[S*(Z)],3β(E)]]-**	121-20-0	**NI37609**
121	167	149	93	107	150	91	80	**360**	100	99	99	97	96	93	82	61	**10.68**	21	28	5	–	–	–	–	–	–	–	–	–	–	**Cyclopropanecarboxylic acid, 3-(3-methoxy-2-methyl-3-oxo-1-propenyl)-2,2-dimethyl-, 3-(2-butenyl)-2-methyl-4-oxo-2-cyclopenten-1-yl ester, [1R-[1α[S*(Z)],3β(E)]]-**	121-20-0	**NI37610**
91	147	175	79	107	148	190	176	**360**	100	93	82	64	54	37	35	32	**0.00**	21	28	5	–	–	–	–	–	–	–	–	–	–	**Dendryphiellin A**		**DD01436**
314	316	245	231	299	232	301	273	**360**	100	98	55	55	52	50	33	28	**0.00**	21	28	5	–	–	–	–	–	–	–	–	–	–	**7-O-Formylhorminone**		**DD01437**
193	194	175	43	176	97	55	111	**360**	100	66	18	18	17	15	14	13	**0.00**	21	28	5	–	–	–	–	–	–	–	–	–	–	**5-Isobenzofurancarboxylic acid, 1,3-dihydro-1,3-dioxo-, dodecyl ester**	22485-51-4	**NI37611**
332	43	263	137	55	262	41	125	**360**	100	95	89	81	61	57	48	47	**14.00**	21	28	5	–	–	–	–	–	–	–	–	–	–	**Pregnan-3,12,15,20-tetrone, 14-hydroxy-, (5ξ,14β)-**		**MS08484**
122	43	55	41	69	57	121	45	**360**	100	95	79	76	68	55	48	47	**32.43**	21	28	5	–	–	–	–	–	–	–	–	–	–	**Pregn-4-ene-3,11,20-trione, 17,21-dihydroxy-**	53-06-5	**NI37612**
360	301	41	91	122	43	258	121	**360**	100	90	70	69	63	63	60	58		21	28	5	–	–	–	–	–	–	–	–	–	–	**Pregn-4-ene-3,11,20-trione, 17,21-dihydroxy-**	53-06-5	**NI37613**
360	301	91	258	41	121	43	55	**360**	100	76	51	48	48	47	47	43		21	28	5	–	–	–	–	–	–	–	–	–	–	**Pregn-4-ene-3,11,20-trione, 17,21-dihydroxy-**	53-06-5	**NI37614**
360	125	137	126	361	329	345	138	**360**	100	66	37	32	24	23	13	13		21	32	3	2	–	–	–	–	–	–	–	–	–	**Androst-4-ene-3,17-dione, 12-hydroxy-, bis(O-methyloxime), (12β)-**	69688-31-9	**NI37615**
267	362	360	252	281	265			**360**	100	43	43	30	21	18				22	17	–	–	–	–	–	1	–	–	–	–	–	**2-[(10-Phenylanthracen-9-yl)]ethyl bromide**		**MS08485**
91	194	105	251	360	315	121	104	**360**	100	44	27	23	13	13	12	12		22	20	1	2	1	–	–	–	–	–	–	–	–	**1,3,4-Thiadiazole, 2-ethoxy-2,3-dihydro-2,3,5-triphenyl-**	42141-07-1	**NI37616**
106	94	360	66	65	39	133	239	**360**	100	53	18	10	10	10	7	6		22	20	3	2	–	–	–	–	–	–	–	–	–	**Phenyl 4-[(4-aminobenzyl)carbonylamino]phenylacetate**		**IC04128**
180	181	109	28	152	216	29	124	**360**	100	65	26	22	20	17	17	13	**0.00**	22	30	–	2	–	2	–	–	–	–	–	–	–	**Ethylenediamine, N,N,N′,N′-tetraethyl-1,2-bis(4-fluorophenyl)-, (±)-**		**DD01438**
180	181	28	152	109	29	97	122	**360**	100	20	8	6	6	5	4	3	**0.00**	22	30	–	2	–	2	–	–	–	–	–	–	–	**Ethylenediamine, N,N,N′,N′-tetraethyl-1,2-bis(4-fluorophenyl)-, meso-**		**DD01439**
257	43	273	300	360	342			**360**	100	87	79	61	41	15				22	32	4	–	–	–	–	–	–	–	–	–	–	**19R-Acetoxy-19a-methyl-5α-androstane-3,17-dione**		**LI07953**
300	273	256	216	360				**360**	100	64	63	38	19					22	32	4	–	–	–	–	–	–	–	–	–	–	**19S-Acetoxy-19a-methyl-5α-androstane-3,17-dione**		**LI07954**
300	273	256	360	43	216	87	317	**360**	100	98	96	89	44	35	19	9		22	32	4	–	–	–	–	–	–	–	–	–	–	**19R-Acetoxy-19-methyl-5α-androstane-3,17-dione**		**LI07955**
43	94	93	95	107	79	41	55	**360**	100	99	99	91	78	74	66	58	**0.16**	22	32	4	–	–	–	–	–	–	–	–	–	–	**5β,14β-Androstane-17β-carboxylic acid, 3β,14-dihydroxy-, γ-lactone, acetate**	10124-02-4	**NI37617**
136	121	123	81	93	91	107	105	**360**	100	91	78	70	67	66	63	59	**8.00**	22	32	4	–	–	–	–	–	–	–	–	–	–	**Cleomeolide acetate**		**MS08486**
32	60	84	96	191	136	271	257	**360**	100	76	75	65	19	15	2	2	**0.00**	22	32	4	–	–	–	–	–	–	–	–	–	–	**ent-Cleroda-4(18),12-dien-16-al, 11,15-epoxy-15-acetoxy-**		**NI37618**
99	87	86	360	91	298	93	79	**360**	100	80	52	45	20	18	13	13		22	32	4	–	–	–	–	–	–	–	–	–	–	**Estr-5(10)-en-3-one, 17-(acetyloxy)-, cyclic 1,2-ethanediyl acetal, (17β)-**	18367-54-9	**MS08487**
99	332	360	333	91	110	126	105	**360**	100	98	51	22	17	15	14	13		22	32	4	–	–	–	–	–	–	–	–	–	–	**Estr-4-en-3-one, 17β-hydroxy-, cyclic ethylene acetal, acetate**	24811-26-5	**MS08488**
100	360	91	79	105	86	93	81	**360**	100	32	32	24	23	23	17	17		22	32	4	–	–	–	–	–	–	–	–	–	–	**Estr-5-en-3-one, 17β-hydroxy-, cyclic ethylene acetal, acetate**	21528-89-2	**MS08489**
300	41	240	55	301	43	239	328	**360**	100	33	32	30	24	24	23	22	**2.00**	22	32	4	–	–	–	–	–	–	–	–	–	–	**Gibberellin A_{12} methyl ester**		**MS08490**
149	81	91	233	43	55	45	41	**360**	100	92	88	80	80	79	77	69	**16.92**	22	32	4	–	–	–	–	–	–	–	–	–	–	**Naphthalenecarboxylic acid, 1-[2-(3-furanyl)ethyl]-1,2,3,4,4a,7,8,8a-octahydro-4a-(methoxymethyl)-1,2-dimethyl-, methyl ester**	72361-24-1	**NI37619**
142	127	360	143	75	288	331	361	**360**	100	55	52	45	23	18	17	15		22	36	2	–	–	–	–	–	–	1	–	–	–	**Androst-3-en-17-one, 3-[(trimethylsilyl)oxy]-, (5α)-**	69833-85-8	**NI37620**
142	127	143	360	75	361	291	201	**360**	100	51	50	48	21	14	12	12		22	36	2	–	–	–	–	–	–	1	–	–	–	**Androst-3-en-17-one, 3-[(trimethylsilyl)oxy]-, (5β)-**	69833-84-7	**NI37621**
129	73	147	75	91	360	79	130	**360**	100	68	39	37	32	31	30	26		22	36	2	–	–	–	–	–	–	1	–	–	–	**Androst-4-en-3-one, 17-[(trimethylsilyl)oxy]-, (17α)-**		**NI37622**
360	270	129	226	361	345	147	304	**360**	100	78	65	35	33	31	26	24		22	36	2	–	–	–	–	–	–	1	–	–	–	**Androst-4-en-3-one, 17-[(trimethylsilyl)oxy]-, (17β)-**	5055-42-5	**NI37623**
129	73	147	75	152	91	130	105	**360**	100	73	36	35	34	31	29	29	**28.00**	22	36	2	–	–	–	–	–	–	1	–	–	–	**Androst-4-en-3-one, 17-[(trimethylsilyl)oxy]-, (17β)-**	5055-42-5	**NI37624**
142	360	143	137	75	361	73	281	**360**	100	55	40	23	22	14	14	12		22	36	2	–	–	–	–	–	–	1	–	–	–	**Androst-4-en-17-one, 3-[(trimethylsilyl)oxy]-, (3α)-**		**LI07956**
129	304	270	231	130	73	360	255	**360**	100	17	12	12	12	8	7	7		22	36	2	–	–	–	–	–	–	1	–	–	–	**Androst-5-en-16-one, 3-[(trimethylsilyl)oxy]-, (3β)-**	49774-91-6	**NI37625**
129	304	130	231	270	360	255	230	**360**	100	15	15	13	12	10	7	5		22	36	2	–	–	–	–	–	–	1	–	–	–	**Androst-5-en-17-one, 3-[(trimethylsilyl)oxy]-, (3α)-**		**LI07957**
129	304	360	231	145	117	270	132	**360**	100	16	14	14	13	13	10	10		22	36	2	–	–	–	–	–	–	1	–	–	–	**Androst-5-en-17-one, 3-[(trimethylsilyl)oxy]-, (3α)-**		**LI07958**
304	231	270	129	360	73	255	213	**360**	100	90	74	74	71	51	46	41		22	36	2	–	–	–	–	–	–	1	–	–	–	**Androst-5-en-17-one, 3-[(trimethylsilyl)oxy]-, (3β)-**	3747-91-9	**NI37626**
129	304	360	270	231	255	130	230	**360**	100	32	25	25	21	10	10	8		22	36	2	–	–	–	–	–	–	1	–	–	–	**Androst-5-en-17-one, 3-[(trimethylsilyl)oxy]-, (3β)-**	3747-91-9	**NI37627**
129	73	75	231	130	304	91	105	**360**	100	50	26	15	15	14	14	13	**12.50**	22	36	2	–	–	–	–	–	–	1	–	–	–	**Androst-5-en-17-one, 3-[(trimethylsilyl)oxy]-, (3β)-**	3747-91-9	**NI37628**
73	255	75	105	91	41	55	79	**360**	100	99	99	91	90	75	69	55	**30.96**	22	36	2	–	–	–	–	–	–	1	–	–	–	**Androst-9(11)-en-17-one, 3-[(trimethylsilyl)oxy]-, (3α,5α)-**	57305-04-1	**NI37629**
85	91	81	79	93	107	105	95	**360**	100	91	90	89	88	80	76	75	**2.00**	22	37	3	–	–	–	–	–	–	–	–	1	–	**Pregnane-3,11,20,21-tetrol, methyl boronate, (3α,5α,11β,20β)-**		**MS08491**
342	215	55	81	67	217	79	93	**360**	100	85	77	71	70	68	67	62	**13.01**	22	37	3	–	–	–	–	–	–	–	–	1	–	**Pregnane-3,17,20-triol, cyclic 17,20-(methylboronate), (3α,20R)-**	31012-66-5	**NI37630**
342	215	124	107	217	111	105	125	**360**	100	77	68	61	60	52	51	49	**10.01**	22	37	3	–	–	–	–	–	–	–	–	1	–	**Pregnane-3,17,20-triol, cyclic 17,20-(methylboronate), (3α,20R)-**	31012-66-5	**NI37631**
342	215	55	81	67	217	79	107	**360**	100	85	76	70	69	67	66	62	**13.00**	22	37	3	–	–	–	–	–	–	–	–	1	–	**Pregnane-3,17,20-triol, methyl boronate, (3α,5β,17α,20α)-**		**MS08492**
342	55	43	41	67	215	81	79	**360**	100	89	84	81	79	77	77	74	**10.00**	22	37	3	–	–	–	–	–	–	–	–	1	–	**Pregnane-3,17,20-triol, methyl boronate, (3α,5β,17α,20β)-**		**MS08493**
359	360	105	77	283	343			**360**	100	96	83	51	38	14				23	20	4	–	–	–	–	–	–	–	–	–	–	**2,6-Dibenzoyl-3,5-dimethoxytoluene**		**LI07959**
221	220	222	70	191	262	290	360	**360**	100	41	18	18	17	10	9	7		23	24	2	2	–	–	–	–	–	–	–	–	–	**4-Isopropylidene-1,1,6-trimethyl-7-phenyl-4H[1,3,4]oxadiazino[4,3-a]cinnolin-2(1H)-one**		**LI07960**

MASS TO CHARGE RATIOS								M.W.	INTENSITIES								Parent	C	H	O	N	S	F	Cl	Br	I	Si	P	B	X	COMPOUND NAME	CAS Reg No	No
172	186	187	173	144	360	130	145	**360**	100	74	32	32	26	22	21	16		23	28	–	4	–	–	–	–	–	–	–	–	–	Calycanthidine		MS08494
43	55	41	95	94	93	81	67	**360**	100	28	27	18	17	16	16	16	**14.00**	23	36	3	–	–	–	–	–	–	–	–	–	–	Androstan-3-one, 17-(acetyloxy)-2,4-dimethyl-, (2α,4α,5α,17β)-	54594-48-8	NI37632
273	55	81	274	93	91	67	41	**360**	100	34	30	28	27	27	26	26	**17.00**	23	36	3	–	–	–	–	–	–	–	–	–	–	5-(2-Butenyl)-17-oxo-4-nor-3,5-seco-5α-androstan-3-oic acid, methyl ester		MS08495
231	314	232	233	360	299	193	315	**360**	100	18	18	12	10	5	5	4		23	36	3	–	–	–	–	–	–	–	–	–	–	8-Ethoxy-iso-tetrahydrocannabinol		LI07961
55	360	287	41	67	81	43	93	**360**	100	86	84	84	69	64	60	50		23	36	3	–	–	–	–	–	–	–	–	–	–	6α-Pentyl-4-oxa-5β-androstane-3,17-dione		MS08496
360	99	43	255	81	41	55	361	**360**	100	71	56	33	33	32	30	28		23	36	3	–	–	–	–	–	–	–	–	–	–	Pregnane-12,20-dione, cyclic 12-(ethylene acetal), (5α)-	5618-27-9	NI37633
360	99	255	81	361	67	86	79	**360**	100	72	33	33	28	20	16	16		23	36	3	–	–	–	–	–	–	–	–	–	–	Pregnane-12,20-dione, cyclic 12-(ethylene acetal), (5α)-	5618-27-9	NI37634
99	43	360	40	41	81	55	255	**360**	100	92	54	48	47	46	41	31		23	36	3	–	–	–	–	–	–	–	–	–	–	Pregnane-12,20-dione, cyclic 12-(ethylene acetal), (5α)-	5618-27-9	NI37635
317	43	300	233	81	360	109	55	**360**	100	78	66	54	41	33	33	33		23	36	3	–	–	–	–	–	–	–	–	–	–	Pregnan-12-one, 20-(acetyloxy)-, (5α,20β)-	7704-92-9	NI37636
43	300	81	93	215	95	79	107	**360**	100	51	27	26	25	24	24	22	**0.63**	23	36	3	–	–	–	–	–	–	–	–	–	–	Pregnan-20-one, 3-(acetyloxy)-, (3α,5β)-		MS08497
300	43	215	230	242	257	246	282	**360**	100	57	28	22	13	11	8	7	**4.51**	23	36	3	–	–	–	–	–	–	–	–	–	–	Pregnan-20-one, 3-(acetyloxy)-, (3α,5β)-		MS08498
43	79	91	93	81	67	105	55	**360**	100	42	40	33	30	29	25	25	**0.00**	23	36	3	–	–	–	–	–	–	–	–	–	–	Pregnan-20-one, 3-(acetyloxy)-, (3β,5α)-	906-83-2	NI37637
300	257	55	81	67	242	109	95	**360**	100	93	88	71	69	54	50	45	**0.00**	23	36	3	–	–	–	–	–	–	–	–	–	–	Pregnan-20-one, 12-(acetyloxy)-, (5α,12β)-	3002-93-5	NI37638
143	270	130	144	345	131	228	145	**360**	100	23	21	18	10	9	5	5	**4.00**	23	40	1	–	–	–	–	–	–	1	–	–	–	Androst-3-en-17-ol, 17-methyl-17-[(trimethylsilyl)oxy]-, (5β,17β)-	56771-60-9	MS08499
157	241	270	144	145	158	331	242	**360**	100	55	50	36	25	20	18	11	**2.70**	23	40	1	–	–	–	–	–	–	1	–	–	–	19-Norpregn-14-en-17-ol, 17-[(trimethylsilyl)oxy]-, (17α)-	56771-61-0	MS08500
303	304	305	360	275	262			**360**	100	22	3	2	2	1				24	25	1	–	–	–	–	–	–	–	1	–	–	2-Butanone, 3,3-dimethyl-1-(triphenylphosphoranylidene)-	26487-93-4	NI37639
303	275	276	304	318	262	360	317	**360**	100	33	30	21	10	9	8	8		24	25	1	–	–	–	–	–	–	–	1	–	–	2-Pentanone, 4-methyl-1-(triphenylphosphoranylidene)-	27653-95-8	NI37640
240	360	212	241	361	105	158	132	**360**	100	47	23	19	13	13	10	10		24	28	1	2	–	–	–	–	–	–	–	–	–	2-Cyclohexenyl-2-(2,6-dimethylanil)glyoxylic acid 2,6-dimethylanilide		MS08501
212	213	360	158	198	317	184	361	**360**	100	95	80	62	37	35	35	30		24	28	1	2	–	–	–	–	–	–	–	–	–	4,6-Dimethyl-3,3-pentamethylene-3H-indole-2-carboxylic acid (3,5-dimethylanilide)		MS08502
212	213	360	158	198	317	184	306	**360**	100	95	80	62	37	35	35	20		24	28	1	2	–	–	–	–	–	–	–	–	–	4,6-Dimethyl-3,3-pentamethylene-3H-indole-2-carboxylic acid (3,5-dimethylanilide)		LI07962
185	105	170	77	360	184	43	44	**360**	100	67	42	37	34	32	31	30		24	28	1	2	–	–	–	–	–	–	–	–	–	2-Phenyl-4,4-dimethyl-8-ethyl-8-benzamido-3-azabicyclo[3.3.0]oct-2-ene		LI07963
57	244	188	41	187	300	43	243	**360**	100	50	45	33	22	21	21	19	**1.90**	24	40	2	–	–	–	–	–	–	–	–	–	–	Benzenemethanol, 2,4,6-tris(2,2-dimethylpropyl)-, acetate	41080-93-7	MS08503
360	361	231	81	55	121	249	41	**360**	100	27	27	24	22	19	17	14		24	40	2	–	–	–	–	–	–	–	–	–	–	Cholan-12-one, 3-hydroxy-, (3α,5β)-	54411-58-4	NI37641
255	43	41	342	256	228	55	57	**360**	100	28	26	24	21	19	17	13	**1.01**	24	40	2	–	–	–	–	–	–	–	–	–	–	Chol-7-ene-12,24-diol, (5β,12α)-	54411-70-0	NI37642
43	57	267	94	71	55	41	69	**360**	100	88	83	77	51	49	48	30	**5.50**	24	40	2	–	–	–	–	–	–	–	–	–	–	Phenyl octadecanoate		IC04129
85	259	95	67	41	149	55	81	**360**	100	43	24	20	20	19	17	16	**0.00**	24	40	2	–	–	–	–	–	–	–	–	–	–	2H-Pyran, 2-[[(3β,5α)-androstan-3-yl]oxy]tetrahydro-	55162-80-6	NI37643
85	84	41	67	55	81	57	43	**360**	100	26	15	13	13	8	7	6	**0.00**	24	40	2	–	–	–	–	–	–	–	–	–	–	2H-Pyran, 2-[[(5α,17β)-androstan-17-yl]oxy]tetrahydro-	55162-81-7	NI37644
58	84	110	303	288	360	70	289	**360**	100	80	25	23	15	13	12	7		24	44	–	2	–	–	–	–	–	–	–	–	–	Pregnane-3,20-diamine, N³,N³,N²⁰-trimethyl-, (3β,5α,20S)-	15112-47-7	NI37645
288	303	84	289	360	304	58	71	**360**	100	51	29	23	11	11	10	9		24	44	–	2	–	–	–	–	–	–	–	–	–	Pregnane-3,20-diamine, N³,N³,N²⁰-trimethyl-, (3β,5α,20S)-	15112-47-7	NI37646
243	165	244	241	43	56	242	41	**360**	100	29	24	8	8	7	6	5	**0.00**	25	28	–	–	1	–	–	–	–	–	–	–	–	Hexyl trityl sulphide	20705-43-5	NI37647
318	43	109	360	290	167	91	243	**360**	100	96	84	80	72	66	62	60		25	28	2	–	–	–	–	–	–	–	–	–	–	Bicyclo[2.2.1]heptan-2-ol, 3-(diphenylmethylene)-1,7,7-trimethyl-, acetate, exo-	38761-80-7	MS08504
318	43	109	360	290	167	243	206	**360**	100	96	84	80	72	66	60	50		25	28	2	–	–	–	–	–	–	–	–	–	–	Bicyclo[2.2.1]heptan-2-ol, 3-(diphenylmethylene)-1,7,7-trimethyl-, acetate, exo-	38761-80-7	LI07964
180	77	51	360	181	361	78	50	**360**	100	52	18	17	16	5	5	2		26	20	–	2	–	–	–	–	–	–	–	–	–	Aniline, N,N′-(1,2-diphenyl-1,2-ethanediylidene)bis-	7510-33-0	NI37648
283	77	180	165	360	359	284	51	**360**	100	88	52	50	47	28	23	23		26	20	–	2	–	–	–	–	–	–	–	–	–	Benzophenone azine	983-79-9	NI37649
283	360	180	165	359	284	256	257	**360**	100	58	50	40	28	25	17	16		26	20	–	2	–	–	–	–	–	–	–	–	–	Benzophenone azine	983-79-9	NI37650
360	283	165	359	180	361	77	284	**360**	100	99	47	43	41	28	25	23		26	20	–	2	–	–	–	–	–	–	–	–	–	Benzophenone azine	983-79-9	TR00304
360	361	345	267	282				**360**	100	38	35	20	19					26	20	–	2	–	–	–	–	–	–	–	–	–	9,10-Dianilinophenanthrene		LI07965
360	361	217	180	115	233	362	218	**360**	100	30	13	13	7	5	4	4		26	20	–	2	–	–	–	–	–	–	–	–	–	1,4-Bis(naphth-1-ylamino)benzene		IC04130
360	361	359	180	172	158	171	77	**360**	100	27	15	12	11	9	5	5		26	20	–	2	–	–	–	–	–	–	–	–	–	1,10-Phenanthroline, 2,9-dimethyl-4,7-diphenyl-	4733-39-5	NI37651
83	55	97	41	69	82	67	81	**360**	100	86	47	42	35	32	26	24	**1.53**	26	48	–	–	–	–	–	–	–	–	–	–	–	1,4-Bis(4-cyclohexylbutyl)cyclohexane		AI02199
191	95	109	43	41	81	67	55	**360**	100	71	45	41	37	33	33	31	**6.45**	26	48	–	–	–	–	–	–	–	–	–	–	–	9-Dodecyltetradecahydroanthracene		AI02200
191	95	43	109	41	55	81	67	**360**	100	88	55	53	51	42	41	41	**4.55**	26	48	–	–	–	–	–	–	–	–	–	–	–	9-Dodecyltetradecahydroanthracene		AI02201
95	43	41	81	67	191	55	109	**360**	100	79	72	68	67	61	57	51	**35.90**	26	48	–	–	–	–	–	–	–	–	–	–	–	2-Dodecyltetradecahydrophenanthrene		AI02202
191	95	109	43	41	67	81	55	**360**	100	79	53	49	48	40	38	37	**10.09**	26	48	–	–	–	–	–	–	–	–	–	–	–	9-Dodecyltetradecahydrophenanthrene		AI02203
191	95	109	55	69	41	81	43	**360**	100	72	47	41	39	39	37	35	**13.65**	26	48	–	–	–	–	–	–	–	–	–	–	–	9-Dodecyltetradecahydrophenanthrene		AI02204
191	95	109	41	43	67	81	55	**360**	100	78	55	44	41	40	39	39	**10.20**	26	48	–	–	–	–	–	–	–	–	–	–	–	9-Dodecyltetradecahydrophenanthrene		AI02205
89	154	69	70	169	277	192	100	**361**	100	44	42	15	10	9	9	9	**0.00**	3	–	–	1	2	13	–	–	–	–	–	–	–	cis-Heptafluoroisopropyltetrafluorosulphur-imino-difluorosulphur		MS08505
89	154	69	192	70	277	169	100	**361**	100	46	39	16	14	12	12	7	**0.00**	3	–	–	1	2	13	–	–	–	–	–	–	–	trans-Heptafluoroisopropyltetrafluorosulphur-imino-difluorosulphur		MS08506
328	330	326	108	293	332	110	295	**361**	100	80	52	48	47	34	32	30	**1.70**	6	2	–	3	–	–	7	–	–	–	–	–	–	s-Triazine, 2-(chloromethyl)-4,6-bis(trichloromethyl)-	30361-97-8	LI07966

MASS TO CHARGE RATIOS								M.W.	INTENSITIES								Parent	C	H	O	N	S	F	Cl	Br	I	Si	P	B	X	COMPOUND NAME	CAS Reg No	No
328	330	326	108	293	332	110	291	**361**	100	80	52	48	47	34	32	30	**1.70**	6	2	–	3	–	–	7	–	–	–	–	–	–	s-Triazine, 2-(chloromethyl)-4,6-bis(trichloromethyl)-	30361-97-8	NI37652
103	45	57	74	44	192	191	73	**361**	100	49	44	39	37	31	30	30	**0.00**	13	20	5	5	–	–	1	–	–	–	–	–	–	2-Amino-6,7-dimethyl-4-oxo-8-(1′-D-ribityl)-4,8-dihydropteridine hydrochloride		LI07967
346	361	347	100	147	341	348	151	**361**	100	33	28	20	19	18	15	11		13	28	2	3	–	1	–	–	–	3	–	–	–	2(1H)-Pyrimidinone, 4-[bis(trimethylsilyl)amino]-5-fluoro-6-[(trimethylsilyl)oxy]-	75365-79-6	NI37653
296	91	298	119	92	300	104	120	**361**	100	98	97	94	76	54	52	43	**14.70**	14	10	2	1	1	–	3	–	–	–	–	–	–	5,6-Benzo-2-imino-N-(2,4,6-trichlorophenyl)-1,3-dioxa-2-thiacycloheptane		MS08507
64	245	149	105	117	172	199	121	**361**	100	63	51	35	25	16	11	9	**4.00**	15	11	2	3	3	–	–	–	–	–	–	–	–	Thiazolo[3,2-a]thiazolo[5,4-d]pyrimidinium hydroxide, anhydro-1-hydroxy-2-phenyl-4-methyl-7-(methylthio)-5-oxo-	86999-01-1	NI37654
248	105	249	77	172	51	176	63	**361**	100	65	46	40	18	14	11	9	**0.00**	15	12	1	3	1	–	–	1	–	–	–	–	–	Pyrido[2,3-d]thiazolo[3,2-b]pyridazin-4-ium, 2,3-dihydro-3-hydroxy-3-phenyl-, bromide	18592-61-5	NI37655
101	131	102	115	258	241	139	84	**361**	100	62	53	44	41	41	37	37	**0.00**	15	23	9	1	–	–	–	–	–	–	–	–	–	α-D-Allopyranoside, methyl 3-acetamido-3-deoxy-, 2,4,6-triacetate	41624-03-7	NI37656
114	96	139	98	97	84	140	156	**361**	100	70	54	49	48	45	41	38	**0.00**	15	23	9	1	–	–	–	–	–	–	–	–	–	α-D-Altrofuranoside, methyl 2-acetamido-2-deoxy-, 3,5,6-triacetate		NI37657
128	86	114	127	156	60	85	144	**361**	100	62	51	38	35	28	27	24	**0.00**	15	23	9	1	–	–	–	–	–	–	–	–	–	D-Altrofuranoside, methyl 3-acetamido-3-deoxy-, 2,5,6-triacetate	100645-83-8	NI37658
101	138	102	139	60	59	258	115	**361**	100	55	44	39	35	33	31	31	**0.00**	15	23	9	1	–	–	–	–	–	–	–	–	–	α-D-Altropyranoside, methyl 3-acetamido-3-deoxy-, 2,4,6-triacetate	72523-30-9	NI37659
43	114	110	73	74	241	115	156	**361**	100	99	90	73	63	53	53	49	**0.00**	15	23	9	1	–	–	–	–	–	–	–	–	–	α-D-Galactopyranose, 2-acetamido-2-deoxy-3-O-methyl-, 1,4,6-triacetate	17429-94-6	NI37660
43	114	110	73	241	156	96	111	**361**	100	18	17	13	10	9	8	7	**0.30**	15	23	9	1	–	–	–	–	–	–	–	–	–	α-D-Galactopyranose, 2-acetamido-2-deoxy-3-O-methyl-, 1,4,6-triacetate	17429-94-6	MS08508
43	101	71	128	59	80	88	86	**361**	100	45	33	30	30	22	14	14	**0.00**	15	23	9	1	–	–	–	–	–	–	–	–	–	α-D-Galactopyranose, 2-acetamido-2-deoxy-4-O-methyl-, 1,3,6-triacetate	17296-15-0	MS08509
43	101	71	59	128	80	86	41	**361**	100	99	73	68	67	48	33	27	**0.00**	15	23	9	1	–	–	–	–	–	–	–	–	–	α-D-Galactopyranose, 2-acetamido-2-deoxy-4-O-methyl-, 1,3,6-triacetate	17296-15-0	NI37661
43	114	45	96	139	72	213	59	**361**	100	31	20	11	10	10	8	8	**0.00**	15	23	9	1	–	–	–	–	–	–	–	–	–	α-D-Galactopyranose, 2-acetamido-2-deoxy-6-O-methyl-, 1,3,4-triacetate	17429-95-7	MS08510
43	114	45	259	302	301	318	258	**361**	100	99	67	65	63	47	40	40	**0.00**	15	23	9	1	–	–	–	–	–	–	–	–	–	α-D-Galactopyranose, 2-acetamido-2-deoxy-6-O-methyl-, 1,3,4-triacetate	17429-95-7	NI37662
43	101	143	114	59	140	98	139	**361**	100	88	28	18	17	16	16	15	**0.00**	15	23	9	1	–	–	–	–	–	–	–	–	–	α-D-Galactopyranoside, methyl 2-acetamido-2-deoxy-, 3,4,6-triacetate	17296-10-5	MS08511
43	101	143	114	140	139	98	97	**361**	100	99	32	20	18	18	18	14	**0.00**	15	23	9	1	–	–	–	–	–	–	–	–	–	α-D-Galactopyranoside, methyl 2-acetamido-2-deoxy-, 3,4,6-triacetate	17296-10-5	NI37663
101	143	59	140	114	98	139	112	**361**	100	28	22	21	19	18	16	15	**0.00**	15	23	9	1	–	–	–	–	–	–	–	–	–	α-D-Galactopyranoside, methyl 2-acetamido-2-deoxy-, 3,4,6-triacetate	17296-10-5	NI37664
101	143	140	59	98	139	302	114	**361**	100	29	28	23	22	19	17	16	**0.00**	15	23	9	1	–	–	–	–	–	–	–	–	–	β-D-Galactopyranoside, methyl 3,4,6-tri-O-acetyl-2-acetamido-2-deoxy-, 3,4,6-triacetate	50605-02-2	NI37665
101	143	59	114	98	140	139	102	**361**	100	30	24	19	18	17	16	15	**0.00**	15	23	9	1	–	–	–	–	–	–	–	–	–	α-D-Glucopyranoside, methyl 2-acetamido-2-deoxy-, 3,4,6-triacetate	2595-39-3	NI37666
101	138	102	84	60	144	59	258	**361**	100	49	45	40	38	35	34	32	**0.00**	15	23	9	1	–	–	–	–	–	–	–	–	–	α-D-Glucopyranoside, methyl 3-acetamido-3-deoxy-, 2,4,6-triacetate	2595-38-2	NI37667
127	140	85	128	115	72	200	169	**361**	100	52	45	42	41	39	38	38	**0.00**	15	23	9	1	–	–	–	–	–	–	–	–	–	α-D-Glucopyranoside, methyl 6-acetamido-6-deoxy-, 2,3,4-triacetate		NI37668
101	143	59	140	139	102	98	114	**361**	100	27	24	23	22	19	19	18	**0.00**	15	23	9	1	–	–	–	–	–	–	–	–	–	β-D-Glucopyranoside, methyl 3,4,6-tri-O-acetyl-2-acetamido-2-deoxy-, 3,4,6-triacetate	2771-48-4	NI37669
244	73	154	245	97	75	147	246	**361**	100	64	35	23	12	11	10	9	**0.00**	15	35	3	1	–	–	–	–	–	3	–	–	–	2-Piperidinecarboxylic acid, 1-(trimethylsilyl)-5-[(trimethylsilyl)oxy]-, trimethylsilyl ester	55517-48-1	NI37670
344	29	270	345	41	302	288	224	**361**	100	48	37	28	27	18	15	15	**15.00**	16	15	5	3	1	–	–	–	–	–	–	–	–	10H-Phenothiazine, 10-butyl-3,7-dinitro-, 5-oxide	35076-81-4	LI07968
344	29	270	345	41	57	302	361	**361**	100	48	35	27	27	18	17	16		16	15	5	3	1	–	–	–	–	–	–	–	–	10H-Phenothiazine, 10-butyl-3,7-dinitro-, 5-oxide	35076-81-4	NI37671
264	177	176	150	162	246	265	190	**361**	100	61	51	38	23	19	17	15	**7.00**	16	22	3	3	–	3	–	–	–	–	–	–	–	Butyl [(4,6-dimethylpyrimidin-2-yl)(trifluoroacetyl)amino]butyrate		MS08512
142	100	244	43	117	184	98	202	**361**	100	52	43	35	24	12	10	9	**0.00**	16	27	8	1	–	–	–	–	–	–	–	–	–	L-Glucitol, 3,6-dideoxy-2-O-methyl-3-(N-methylacetamido)-, 1,4,5-triacetate	82486-49-5	NI37672
128	230	86	43	114	131	87	130	**361**	100	77	44	44	38	31	15	14	**0.00**	16	27	8	1	–	–	–	–	–	–	–	–	–	L-Glucitol, 3,6-dideoxy-4-O-methyl-3-(N-methylacetamido)-, 1,2,5-triacetate	82486-50-8	NI37673
174	259	199	216	301	103	217	145	**361**	100	76	50	35	22	21	19	14	**13.81**	17	19	6	3	–	–	–	–	–	–	–	–	–	1,2,3-Propanetriol, 1-(1-phenyl-1H-1,2,3-triazol-4-yl)-, triacetate, D-threo	34279-95-3	NI37674
315	361	269	273	227	245	200		**361**	100	61	48	45	29	23	10			18	19	7	1	–	–	–	–	–	–	–	–	–	Ethyl 2-[2-(ethoxycarbonyl)-5,6-(methylenedioxy)indol-3-yl]acetoacetate		LI07969
67	103	77	200	51	161	174	91	**361**	100	91	88	54	43	41	36	34	**0.00**	19	15	3	5	–	–	–	–	–	–	–	–	–	5(4H)-Isoxazolone, 4-[(4,5-dihydro-3-methyl-5-oxo-1-phenyl-1H-pyrazol-4-yl)azo]-3-phenyl-	56666-64-9	MS08513
101	83	129	157	88	201	139	81	**361**	100	33	19	16	16	13	13	12	**0.00**	19	37	6	–	–	–	–	–	–	–	–	–	–	1-p-Menthyl-2,3,4-tri-O-methyl-5-methoxyxylopyranose		NI37675
361	142	128	328	288	77	118	196	**361**	100	61	59	43	17	15	14	12		20	19	–	5	1	–	–	–	–	–	–	–	–	Imidazo[5,1-f][1,2,4]triazine, 4-(methylamino)-7-(methylthio)-5-phenyl-2-(4-tolyl)-		MS08514
285	287	256	286	210	272	270	238	**361**	100	94	46	38	31	12	12	9	**8.00**	20	27	5	1	–	–	–	–	–	–	–	–	–	C-Homoerythrinan-2,7-diol, 1,6-didehydro-3,15,16-trimethoxy-, (2β,3β)-	68156-55-8	NI37676
186	200	145	201	171	113	226	361	**361**	100	48	37	33	28	17	16	14		20	27	5	1	–	–	–	–	–	–	–	–	–	Isoxazolo[2,3-a]indole-2,3-dicarboxylic acid, 7-tert-butyl-2,3,3a,4-tetrahydro-4,4-dimethyl-, dimethyl ester, rel-(2R,3S,3aR)-		DD01440
186	200	113	145	201	171	361	286	**361**	100	72	52	38	37	28	24	20		20	27	5	1	–	–	–	–	–	–	–	–	–	Isoxazolo[2,3-a]indole-2,3-dicarboxylic acid, 7-tert-butyl-2,3,3a,4-tetrahydro-4,4-dimethyl-, dimethyl ester, rel-(2S,3S,3aR)-		DD01441
124	91	83	55	82	288	140	361	**361**	100	24	23	15	13	6	2	1		20	27	5	1	–	–	–	–	–	–	–	–	–	Phalaenopsine T	23412-97-7	LI07970
99	344	150	256	258	190	163	100	**361**	100	75	61	41	36	30	30	30	**23.00**	20	27	5	1	–	–	–	–	–	–	–	–	–	Spiro[1,3-dioxolane-2,1′-[1H]indene]-3a′-carboxamide, octahydro-5′-(2-hydroxy-4-methoxyphenyl)-7a′-methyl-, (±)-(3′α,5′β,7a′β)-		MS08515
124	83	82	73	94	125	96	42	**361**	100	19	17	15	12	10	10	9	**6.01**	20	31	3	1	–	–	–	–	–	1	–	–	–	Benzeneacetic acid, α-[[(trimethylsilyl)oxy]methyl]-, 8-methyl-8-azabicyclo[3.2.1]oct-3-yl ester, endo-(±)-	55334-03-7	NI37677

MASS TO CHARGE RATIOS								M.W.	INTENSITIES								Parent	C	H	O	N	S	F	Cl	Br	I	Si	P	B	X	COMPOUND NAME	CAS Reg No	No
192	41	179	55	361	69	95	79	**361**	100	14	12	8	7	6	5	5		20	34	1	1	–	3	–	–	–	–	–	–	–	Pyrrole, 4,5-dihydro-2-methyl-1-(trifluoroacetyl)-3-tridecyl-		NI37678
105	77	361	106	51	363	164	76	**361**	100	43	10	7	7	3	3	3		21	12	3	1	–	–	1	–	–	–	–	–	–	Benzamide, N-(4-chloro-9,10-dihydro-9,10-dioxo-1-anthracenyl)-	81-45-8	NI37679
251	236	36	293	180	42	234	278	**361**	100	94	66	57	54	53	44	41	**0.00**	21	30	1	2	–	–	1	–	–	–	–	–	–	η^2-Dihydromavacurin-methochloride		LI07971
74	288	260	108	43	245	91	57	**361**	100	92	67	27	26	18	18	18	**5.50**	21	31	4	1	–	–	–	–	–	–	–	–	–	Icasamine		NI37680
346	318	361	288	302	272	330		**361**	100	56	55	22	17	15	12			22	19	4	1	–	–	–	–	–	–	–	–	–	Dimethyl 9a-methyl-9aH-pyrido[1,2-f]phenanthridine-7,9-dicarboxylate		LI07972
284	105	77	91	361	104	41	55	**361**	100	48	38	32	29	27	23	20		22	23	2	1	–	–	–	–	–	1	–	–	–	2,2,6-Triphenyl-1,3-dioxa-6-aza-2-silacyclooctane		MS08516
312	91	41	343	105	79	116	42	**361**	100	95	95	77	75	70	68	68	**20.00**	22	35	3	1	–	–	–	–	–	–	–	–	–	$3\beta,17\alpha$-Dihydroxy-pregn-5-en-20-one, methyl oxime		LI07973
41	147	55	69	71	43	91	79	**361**	100	55	37	24	22	22	21	20	**2.92**	22	35	3	1	–	–	–	–	–	–	–	–	–	4'H,5'H-Mutilano[13,14-d]oxazol-11-one, 3-methoxy-		NI37681
346	361	69	57	55	70	82	96	**361**	100	78	78	78	78	61	46	37		23	27	1	3	–	–	–	–	–	–	–	–	–	3-Pyrrolin-2-one, 4-ethyl-3-methyl-5-[5-[4-(dimethylamino)phenyl-2-methylene]-3,4-dimethyl-5H-pyrrolyl-2-methylene]-, (Z,Z)-		MS08517
57	232	288	41	27	289	233	176	**361**	100	74	72	44	27	17	13	12	**9.80**	23	39	2	1	–	–	–	–	–	–	–	–	–	Benzene, 1,3,5-tris(2,2-dimethylpropyl)-2,4-dimethyl-6-nitro-	40572-22-3	MS08518
249	250	232	204	112	70	360	105	**361**	100	68	57	52	51	48	34	29	**15.60**	24	27	2	1	–	–	–	–	–	–	–	–	–	1,1-Diphenyl-2-cyano-2-(octyloxycarbonyl)acetylene		IC04131
55	113	41	70	67	43	126	361	**361**	100	99	79	67	60	57	52	35		24	43	1	1	–	–	–	–	–	–	–	–	–	Pyrrolidine, 1-(1-oxo-11,14-eicosadienyl)-	56666-44-5	NI37682
113	49	84	70	40	43	126	55	**361**	100	74	62	58	57	50	49	45	**6.94**	24	43	1	1	–	–	–	–	–	–	–	–	–	Pyrrolidine, 1-[1-oxo-8-[2-[(2-pentylcyclopropyl)methyl]cyclopropyl]octyl]-	56600-06-7	NI37683
241	318	319	242	239	317	165	77	**361**	100	57	37	37	27	26	23	19	**0.00**	25	19	–	3	–	–	–	–	–	–	–	–	–	1,1'-Biphenyl, 2-(azidodiphenylmethyl)-	32366-27-1	NI37684
117	82	48	47	76	44	81	64	**362**	100	39	26	25	24	23	21	21	**0.00**	2	–	2	–	3	–	6	–	–	–	–	–	–	1,3-Bis(trichloromethyl)trisulphane, 1,1-dioxide		LI07974
229	231	115	233	364	116	117	366	**362**	100	99	72	36	31	30	28	26	**15.00**	4	4	1	–	–	–	6	–	–	3	–	–	–	3-Oxa-2,4,8-trisilabicyclo[3.2.1]oct-6-ene, 2,2,4,4,8,8-hexachloro-		NI37685
278	362	334	250	222	187	306	139	**362**	100	78	78	58	57	54	49	27		5	–	5	–	–	–	1	–	–	–	–	–	1	Chloropentacarbonylrhenium		MS08519
364	366	349	347	81	80	368	161	**362**	100	98	63	63	53	53	43	43	**38.03**	5	1	2	–	1	–	–	3	–	–	–	–	–	3-Thiophenecarboxylic acid, 2,4,5-tribromo-	16694-22-7	NI37686
160	366	364	349	347	81	80	368	**362**	100	42	42	27	27	22	22	18	**16.30**	5	1	2	–	1	–	–	3	–	–	–	–	–	3-Thiophenecarboxylic acid, 2,4,5-tribromo-	16694-22-7	MS08520
60	241	153	362	121	31	150	89	**362**	100	99	73	60	43	38	29	25		6	18	–	6	2	–	–	–	–	–	4	–	–	2,4,6,8,9,10-Hexaaza-1,3,5,7-tetraphosphatricyclo[3.3.1.1^{3,7}]decane, 2,4,6,8,9,10-hexamethyl-, 1,3-disulphide	38448-56-5	NI37687
241	46	30	75	240	76	74	62	**362**	100	99	78	42	41	29	19	16	**0.00**	8	6	11	6	–	–	–	–	–	–	–	–	–	N-(2-Hydroxyethyl)-N-(2,4,6-trinitrophenyl)nitramine nitrate		MS08521
235	79	107	93	108	91	39	77	**362**	100	32	30	19	17	16	14	12	**10.80**	8	12	–	–	–	–	–	–	2	–	–	–	–	Bicyclo[2.2.2]octane, 1,4-diiodo-	10364-05-3	NI37688
91	315	118	362	103	300	197	150	**362**	100	83	62	52	47	39	39	34		10	18	–	–	7	–	–	–	–	–	–	–	–	Bis[1,2,2-tris(methylthio)vinyl] sulphide	82766-69-6	NI37689
101	43	41	29	39	27	45	54	**362**	100	68	48	44	32	20	14	12	**2.81**	10	18	4	–	–	–	–	–	–	–	–	–	2	Butyric acid, 3,3'-diselenobis[3-methyl-	1599-27-5	NI37690
101	43	41	29	39	27	45	54	**362**	100	69	48	44	33	20	15	12	**3.00**	10	18	4	–	–	–	–	–	–	–	–	–	2	Butyric acid, 3,3'-diselenobis[3-methyl-	1599-27-5	LI07975
101	55	83	59	43	29	41	27	**362**	100	89	47	29	26	21	15	15	**3.62**	10	18	4	–	–	–	–	–	–	–	–	–	2	Pentanoic acid, 5,5'-diselenobis-	18285-95-5	NI37691
43	85	223	127	79	97	45	69	**362**	100	99	65	57	57	31	28	26	**0.23**	10	18	10	–	2	–	–	–	–	–	–	–	–	α-D-Xylopyranoside, methyl, 3-acetate 2,4-dimethanesulphonate	29709-43-1	NI37692
250	194	222	334	362	306	168	278	**362**	100	91	88	50	48	36	36	30		11	6	7	–	–	–	–	–	–	–	–	–	2	Iron, hexacarbonyl[μ-[(1,2-η:3,4-η)-1-methoxy-1,3-butadiene-1,4-diyl]]di-	74381-47-8	NI37693
167	117	362	195	93	98	168	363	**362**	100	44	29	28	12	7	6	4		12	–	–	2	–	10	–	–	–	–	–	–	–	Diazene, bis(pentafluorophenyl)-	2285-06-5	NI37694
167	117	362	195	93	31	98	168	**362**	100	43	31	28	16	10	8	7		12	–	–	2	–	10	–	–	–	–	–	–	–	Diazene, bis(pentafluorophenyl)-	2285-06-5	NI37695
66	263	265	261	91	65	79	39	**362**	100	93	61	59	54	34	28	26	**1.20**	12	8	–	–	–	–	6	–	–	–	–	–	–	1,4:5,8-Dimethanonaphthalene, 1,2,3,4,10,10-hexachloro-1,4,4a,5,8,8a-hexahydro-, ($1\alpha,4\alpha,4a\beta,5\alpha,8\alpha,8a\beta$)-	309-00-2	NI37696
66	263	79	91	101	265	261	65	**362**	100	45	42	36	34	28	28	25	**3.22**	12	8	–	–	–	–	6	–	–	–	–	–	–	1,4:5,8-Dimethanonaphthalene, 1,2,3,4,10,10-hexachloro-1,4,4a,5,8,8a-hexahydro-, ($1\alpha,4\alpha,4a\beta,5\alpha,8\alpha,8a\beta$)-	309-00-2	NI37697
66	263	265	261	91	65	79	293	**362**	100	93	61	59	54	34	28	27	**0.87**	12	8	–	–	–	–	6	–	–	–	–	–	–	1,4:5,8-Dimethanonaphthalene, 1,2,3,4,10,10-hexachloro-1,4,4a,5,8,8a-hexahydro-, ($1\alpha,4\alpha,4a\beta,5\alpha,8\alpha,8a\beta$)-	309-00-2	NI37698
193	195	263	66	261	265	197	147	**362**	100	89	46	38	30	29	29	26	**0.00**	12	8	–	–	–	–	6	–	–	–	–	–	–	1,4:5,8-Dimethanonaphthalene, 1,2,3,4,10,10-hexachloro-1,4,4a,5,8,8a-hexahydro-, ($1\alpha,4\alpha,4a\beta,5\beta,8\beta,8a\beta$)-	465-73-6	NI37699
66	193	195	39	65	147	91	263	**362**	100	52	43	30	26	22	22	16	**0.00**	12	8	–	–	–	–	6	–	–	–	–	–	–	1,4:5,8-Dimethanonaphthalene, 1,2,3,4,10,10-hexachloro-1,4,4a,5,8,8a-hexahydro-, ($1\alpha,4\alpha,4a\beta,5\beta,8\beta,8a\beta$)-	465-73-6	NI37700
193	195	66	263	147	40	197	44	**362**	100	80	80	37	34	33	32	32	**12.31**	12	8	–	–	–	–	6	–	–	–	–	–	–	1,4:5,8-Dimethanonaphthalene, 1,2,3,4,10,10-hexachloro-1,4,4a,5,8,8a-hexahydro-, ($1\alpha,4\alpha,4a\beta,5\beta,8\beta,8a\beta$)-	465-73-6	NI37701
55	41	64	87	45	203	67	204	**362**	100	89	87	77	77	56	46	45	**0.00**	12	26	4	–	4	–	–	–	–	–	–	–	–	Methanesulphonothioic acid, 1,10-decanediyl ester		MS08522
195	167	117	362	196	92	98	78	**362**	100	32	25	20	9	7	6	4		13	–	1	–	–	10	–	–	–	–	–	–	–	Decafluorobenzophenone		LI07976
195	167	117	362	93	196	98	31	**362**	100	62	57	34	15	12	7	7		13	–	1	–	–	10	–	–	–	–	–	–	–	Decafluorobenzophenone		MS08523
317	92	121	69	120	29	95	169	**362**	100	77	70	67	41	31	25	24	**18.00**	13	9	4	–	–	7	–	–	–	–	–	–	–	Benzoic acid, 2-[(heptafluorobutyryl)oxy]-, ethyl ester		NI37702
262	160	362	29	116	84	264	75	**362**	100	86	70	23	21	19	17	17		13	14	4	–	4	–	–	–	–	–	–	–	–	2-Methyl-6,7-bis(ethoxycarbonyl)tetrathiafulvalene		MS08524
181	161	362	31	131	69	155	81	**362**	100	8	6	4	3	3	2	2		14	4	–	–	–	10	–	–	–	–	–	–	–	1,2-Bis(pentafluorophenyl)ethane		LI07977
109	97	65	226	125	362	210	89	**362**	100	83	41	36	33	28	28	22		14	16	5	–	1	–	1	–	–	–	1	–	–	Phosphorothioic acid, O-(3-chloro-4-methyl-2-oxo-2H-1-benzopyran-7-yl) O,O-diethyl ester	56-72-4	NI37703

MASS TO CHARGE RATIOS								M.W.	INTENSITIES								Parent	C	H	O	N	S	F	Cl	Br	I	Si	P	B	X	COMPOUND NAME	CAS Reg No	No
109	228	364	97	65	334	29	212	362	100	75	54	48	40	32	29	27	12.82	14	16	5	–	1	–	1	–	–	–	1	–	–	Phosphorothioic acid, O-(3-chloro-4-methyl-2-oxo-2H-1-benzopyran-7-yl) O,O-diethyl ester	56-72-4	NI37704
362	109	226	97	210	364	28	334	362	100	95	78	74	56	45	33	30		14	16	5	–	1	–	1	–	–	–	1	–	–	Phosphorothioic acid, O-(3-chloro-4-methyl-2-oxo-2H-1-benzopyran-7-yl) O,O-diethyl ester	56-72-4	NI37705
253	309	362	347	294				362	100	71	65	13	10					14	30	–	–	–	–	–	–	–	–	2	–	1	Bis(2-methylallyl)-1,2-bis(dimethylphosphino)ethaneruthenium		LI07978
73	147	347	75	319	55	45	148	362	100	80	34	31	17	17	17	14	2.60	14	30	5	–	–	–	–	–	–	3	–	–	–	2-Pentenedioic acid, 2-[(trimethylsilyl)oxy]-, bis(trimethylsilyl) ester	55590-97-1	NI37706
73	147	347	55	75	318	45	128	362	100	42	26	25	19	15	15	10	1.90	14	30	5	–	–	–	–	–	–	3	–	–	–	2-Pentenedioic acid, 2-[(trimethylsilyl)oxy]-, bis(trimethylsilyl) ester	55590-97-1	NI37707
73	147	77	75	347	318	348	319	362	100	25	21	16	3	2	1	1	0.00	14	30	5	–	–	–	–	–	–	3	–	–	–	2-Pentenedioic acid, 2-[(trimethylsilyl)oxy]-, bis(trimethylsilyl) ester	55590-97-1	NI37708
73	147	347	75	45	257	67	231	362	100	40	20	19	19	13	12	10	2.25	14	30	5	–	–	–	–	–	–	3	–	–	–	2-Pentenedioic acid, 3-[(trimethylsilyl)oxy]-, bis(trimethylsilyl) ester	72361-20-7	NI37709
73	147	347	75	257	67	244	231	362	100	35	23	17	15	11	10	9	3.66	14	30	5	–	–	–	–	–	–	3	–	–	–	3-Pentenedioic acid, 3-[(trimethylsilyl)oxy]-, bis(trimethylsilyl) ester	72361-20-7	NI37710
75	44	45	73	47	43	77	55	362	100	45	24	19	18	17	14	14	0.00	14	31	6	–	–	–	–	–	–	2	–	1	–	α-D-Galactopyranoside, methyl 2,3-bis-O-(trimethylsilyl)-, cyclic methylboronate	54400-88-3	NI37711
73	133	204	147	75	45	189	116	362	100	96	92	67	48	37	35	31	0.60	14	31	6	–	–	–	–	–	–	2	–	1	–	α-D-Galactopyranoside, methyl 2,3-bis-O-(trimethylsilyl)-, cyclic methylboronate	54400-88-3	NI37712
73	115	75	147	218	191	204	246	362	100	65	50	46	45	44	43	37	0.00	14	31	6	–	–	–	–	–	–	2	–	1	–	α-D-Galactopyranoside, methyl 2,6-bis-O-(trimethylsilyl)-, cyclic methylboronate	54400-89-4	NI37713
73	75	204	45	44	147	117	103	362	100	98	46	43	42	40	39	36	0.00	14	31	6	–	–	–	–	–	–	2	–	1	–	β-D-Galactopyranoside, methyl 2,3-bis-O-(trimethylsilyl)-, cyclic methylboronate	56211-08-6	NI37714
73	75	117	43	129	44	191	115	362	100	98	53	51	49	42	41	41	0.00	14	31	6	–	–	–	–	–	–	2	–	1	–	β-D-Galactopyranoside, methyl 2,6-bis-O-(trimethylsilyl)-, cyclic methylboronate	56211-06-4	NI37715
75	117	73	44	103	129	43	139	362	100	69	58	49	42	30	24	23	0.00	14	31	6	–	–	–	–	–	–	2	–	1	–	α-D-Glucopyranoside, methyl 2,3-bis-O-(trimethylsilyl)-, cyclic methylboronate	54400-90-7	NI37716
204	73	147	75	45	189	205	44	362	100	97	45	42	31	30	27	25	0.00	14	31	6	–	–	–	–	–	–	2	–	1	–	α-D-Glucopyranoside, methyl 2,3-bis-O-(trimethylsilyl)-, cyclic methylboronate	54400-90-7	NI37717
75	117	44	73	55	41	43	103	362	100	84	52	46	29	23	22	21	0.00	14	31	6	–	–	–	–	–	–	2	–	1	–	β-D-Glucopyranoside, methyl 2,3-bis-O-(trimethylsilyl)-, cyclic methylboronate	56211-07-5	NI37718
73	156	75	155	45	245	147	74	362	100	47	28	23	17	16	13	11	0.00	14	34	3	2	–	–	–	–	–	3	–	–	–	L-Glutamine, tris(trimethylsilyl)-	56145-13-2	MS08525
73	156	155	75	245	147	157	74	362	100	88	86	45	29	27	14	11	2.52	14	34	3	2	–	–	–	–	–	3	–	–	–	L-Glutamine, tris(trimethylsilyl)-	56145-13-2	NI37719
364	362	106	79	78	242	151	363	362	100	99	99	35	35	34	33	24		15	15	–	4	1	–	–	1	–	–	–	–	–	5,7-Dimethyl-3-(5-bromo-4,6-dimethylpyrid-2-ylimino)-3H-[1,2,4]thiadiazolo[4,3-a]pyridine		LI07979
79	75	207	73	44	52	204	51	362	100	64	53	42	40	37	34	34	0.00	15	22	10	–	–	–	–	–	–	–	–	–	–	Arabinitol, pentaacetate	26674-23-7	NI37720
115	217	85	98	157	145	69	112	362	100	33	30	24	21	18	18	17	0.00	15	22	10	–	–	–	–	–	–	–	–	–	–	D-Galactofuranoside, methyl, pentaacetate	39598-81-7	NI37721
43	28	115	99	98	81	87	74	362	100	15	8	6	6	6	5	5	0.00	15	22	10	–	–	–	–	–	–	–	–	–	–	α-D-Glucopyranoside, methyl, tetraacetate	604-70-6	NI37722
157	115	81	74	103	112	98	145	362	100	98	81	81	80	76	76	69	0.00	15	22	10	–	–	–	–	–	–	–	–	–	–	α-D-Glucopyranoside, methyl, tetraacetate	604-70-6	NI37723
115	81	157	103	98	102	145	112	362	100	95	88	82	80	76	69	66	0.00	15	22	10	–	–	–	–	–	–	–	–	–	–	β-D-Glucopyranoside, methyl, tetraacetate	4860-85-9	NI37724
43	115	157	98	103	81	145	44	362	100	12	8	7	6	6	5	4	0.00	15	22	10	–	–	–	–	–	–	–	–	–	–	Glucose methylglycoside, tetraacetate		MS08526
43	81	115	103	102	98	157	145	362	100	10	8	8	8	8	6	6	0.00	15	22	10	–	–	–	–	–	–	–	–	–	–	Glucose methylglycoside, tetraacetate		MS08527
43	115	157	81	103	98	56	74	362	100	10	9	9	8	8	8	7	0.00	15	22	10	–	–	–	–	–	–	–	–	–	–	α-D-Mannopyranoside, methyl, tetraacetate		MS08528
43	81	98	115	102	103	85	69	362	100	10	9	8	8	7	7	6	0.00	15	22	10	–	–	–	–	–	–	–	–	–	–	β-D-Mannopyranoside, methyl, tetraacetate		MS08529
149	79	207	75	52	73	177	217	362	100	67	49	43	40	35	34	29	0.00	15	22	10	–	–	–	–	–	–	–	–	–	–	Ribitol, pentaacetate	7208-42-6	NI37725
79	75	135	207	73	52	45	197	362	100	93	90	88	58	48	46	42	0.00	15	22	10	–	–	–	–	–	–	–	–	–	–	D-Xylitol, pentaacetate	13437-68-8	NI37726
115	43	145	103	217	127	85	187	362	100	99	64	57	42	35	32	31	0.00	15	22	10	–	–	–	–	–	–	–	–	–	–	D-Xylitol, pentaacetate	13437-68-8	NI37727
73	84	156	362	93	102	128	230	362	100	98	47	28	28	25	20	18		15	38	2	2	–	–	–	–	–	3	–	–	–	L-Lysine, N^2,N^6-bis(trimethylsilyl)-, trimethylsilyl ester	24595-69-5	MS08530
73	84	147	75	156	199	93	102	362	100	82	70	68	40	30	28	23	15.00	15	38	2	2	–	–	–	–	–	3	–	–	–	L-Lysine, N^2,N^6-bis(trimethylsilyl)-, trimethylsilyl ester	24595-69-5	NI37728
84	73	156	102	74	15	128	59	362	100	65	38	16	14	11	9	8	5.00	15	38	2	2	–	–	–	–	–	3	–	–	–	L-Lysine, N^2,N^6-bis(trimethylsilyl)-, trimethylsilyl ester	24595-69-5	NI37729
73	84	156	75	93	362	102	41	362	100	97	46	32	30	27	25	20		15	38	2	2	–	–	–	–	–	3	–	–	–	L-Lysine, tris(trimethylsilyl)-	25737-19-3	LI07980
73	84	156	75	93	362	102	128	362	100	97	47	32	28	27	24	18		15	38	2	2	–	–	–	–	–	3	–	–	–	L-Lysine, tris(trimethylsilyl)-	25737-19-3	LI07981
73	84	156	75	93	362	102	41	362	100	97	46	30	28	27	24	20		15	38	2	2	–	–	–	–	–	3	–	–	–	L-Lysine, tris(trimethylsilyl)-	25737-19-3	NI37730
174	73	75	142	175	200	128	86	362	100	44	30	25	19	13	13	10	1.00	15	38	2	2	–	–	–	–	–	3	–	–	–	Tris(trimethylsilyl)ornithine methyl ester		MS08531
104	76	50	74	148	75	77	105	362	100	95	43	22	18	12	9	8	0.00	16	14	6	2	1	–	–	–	–	–	–	–	–	Benzoic acid, 2-[[[4-[(acetylamino)sulphonyl]phenyl]amino]carbonyl]-	131-69-1	NI37731
91	125	193	155	171	77	140	166	362	100	76	71	66	64	56	50	40	1.05	16	15	2	4	1	–	1	–	–	–	–	–	–	N-[1-(4-Chlorobenzyl)-4-(1,2,4-triazolio)]-4-toluenesulphonamidate	32539-62-1	NI37732
84	82	70	154	81	234	153	122	362	100	88	71	43	27	23	20	14	3.00	16	22	4	6	–	–	–	–	–	–	–	–	–	L-Prolinamide, 5-oxo-L-prolyl-L-histidyl-	24305-27-9	MS08532
103	43	75	47	104	155	197	101	362	100	46	34	32	6	5	3	3	0.00	16	26	9	–	–	–	–	–	–	–	–	–	–	3,4,6-Tri-O-acetyl-2,5-anhydro-D-mannose diethyl acetal		MS08533
145	99	71	73	243	117	289	127	362	100	81	73	63	54	44	40	37	0.00	16	26	9	–	–	–	–	–	–	–	–	–	–	Butanedioic acid, 2,2′-oxybis-, tetraethyl ester, (R,S)-		NI37733
145	99	71	243	73	174	128	289	362	100	68	63	62	62	60	49	47	0.00	16	26	9	–	–	–	–	–	–	–	–	–	–	Butanedioic acid, 2,2′-oxybis-, tetraethyl ester, (R,R or S,S)-	98870-52-1	NI37734

MASS TO CHARGE RATIOS	M.W.	INTENSITIES	Parent	C	H	O	N	S	F	Cl	Br	I	Si	P	B	X	COMPOUND NAME	CAS Reg No	No
89 66 79 169 151 227 257 215	362	100 19 10 8 7 4 3 3	0.00	16	26	9	–	–	–	–	–	–	–	–	–	–	1H-Cyclopenta[c]furan-4-carboxylic acid, hexahydro-5-[(2-methoxyethoxy)methoxy]-1-oxo-, 2-(methoxyethoxy)methyl ester, [3aS-(3aα,4α,5β,6aα)]-		NI37735
268 15 269 94 96 134 77 121	362	100 28 13 11 10 9 9 8	0.00	17	19	–	2	1	–	–	1	–	–	–	–	–	Benzothiazolium, 2-[4-(dimethylamino)phenyl]-3,6-dimethyl-, bromide	47070-04-2	NI37736
269 270 271 297 255 283 267	362	100 16 5 2 2 1 1	0.00	17	19	–	2	1	–	–	1	–	–	–	–	–	Benzothiazolium, 2-[4-(dimethylamino)phenyl]-3,6-dimethyl-, bromide	47070-04-2	NI37737
73 127 57 259 147 43 41 75	362	100 99 95 80 44 31 26 24	0.00	17	38	4	–	–	–	–	–	–	2	–	–	–	Monocaprylin, bis(trimethylsilyl)-		NI37738
122 362 178 333 334 77 364 319	362	100 84 51 44 36 33 31 31		18	15	6	–	–	–	1	–	–	–	–	–	–	11H-Dibenzo[b,e][1,4]dioxepin-7-carboxaldehyde, 2-chloro-6-hydroxy-3-methoxy-1,4,8-trimethyl-11-oxo-	489-30-5	NI37739
362 298 331 363 270 299 315 165	362	100 88 27 22 20 19 16 15		18	18	4	–	2	–	–	–	–	–	–	–	–	4,5-Dicarbomethoxy-2,3,6,7-tetramethylbenzo[2,1-b:3,4-b′]dithiophene		MS08534
347 362 304 319 344 333 303 361	362	100 79 9 8 7 6 5 2		18	18	8	–	–	–	–	–	–	–	–	–	–	Xanthone, 1-hydroxy-3,5,6,7,8-pentamethoxy-		NI37740
91 156 87 216 288 229 329 183	362	100 25 23 21 15 15 12 12	8.00	18	22	4	2	1	–	–	–	–	–	–	–	–	Dimethyl 7-benzylperillate		LI07982
201 229 362 203 202 247 256 273	362	100 33 27 27 22 15 14 11		18	22	6	2	–	–	–	–	–	–	–	–	–	Diethyl 5-acetoxy-4,5-dihydro-4-methyl-1H-1,2-benzodiazepine-3,4-dicarboxylate		NI37741
209 142 362 154 141 345 222 223	362	100 92 83 57 54 16 14 9		18	26	4	4	–	–	–	–	–	–	–	–	–	3,3′-(1,6-Hexanediyl)bis[1,6-dimethyl-2,4(1H,3H)-pyrimidinedione]		MS08535
290 73 303 291 304 362 292 305	362	100 73 72 27 20 19 10 8		18	30	2	2	–	–	–	–	–	2	–	–	–	1H-Indol-5-ol, 1-acetyl-3-(2-aminoethyl)-, bis(trimethylsilyl)-	56145-10-9	NI37742
146 73 334 232 247 246 117 147	362	100 60 47 47 34 21 21 20	15.51	18	30	2	2	–	–	–	–	–	2	–	–	–	4,6(1H,5H)-Pyrimidinedione, 5-ethyldihydro-5-phenyl-1,3-bis(trimethylsilyl)-	55334-06-0	NI37743
259 273 73 144 147 347 184 129	362	100 38 34 23 16 9 9 9	0.00	18	30	2	2	–	–	–	–	–	2	–	–	–	Quinoxaline, 2-[3(R),4-bis[(trimethylsilyl)oxy]butyl]-,		NI37744
74 87 364 362 283 41 43 363	362	100 56 52 49 49 40 36 33		18	35	2	–	–	–	–	1	–	–	–	–	–	Heptadecanoic acid, 7-bromo-, methyl ester	57289-62-0	NI37745
77 362 180 165 51 105 363 166	362	100 56 35 30 30 27 18 15		19	14	4	4	–	–	–	–	–	–	–	–	–	Benzophenone 2,4-dinitrophenylhydrazone	1733-62-6	NI37746
77 362 180 165 51 105 327 166	362	100 57 36 31 30 27 24 15		19	14	4	4	–	–	–	–	–	–	–	–	–	Benzophenone 2,4-dinitrophenylhydrazone	1733-62-6	LI07983
362 121 43 320 56 115 141 255	362	100 54 44 40 30 14 12 10		19	18	2	2	–	–	–	–	–	–	–	–	1	1-(1-Acetyl-5-fur-2-yl-2-pyrazolin-3-yl)ferrocene		LI07984
91 95 81 242 149 121 260 214	362	100 91 87 59 54 53 51 45	4.00	19	22	7	–	–	–	–	–	–	–	–	–	–	Anhydroverlotorin, 8α,15-diacetoxy-		MS08536
195 194 168 362 196 167	362	100 63 13 10 10 9		19	22	7	–	–	–	–	–	–	–	–	–	–	Benzophenone, 2,2′-dihydroxy-4,4′,6,6′-tetramethoxy-3,3′-dimethyl-	78417-12-6	NI37747
180 181 165 151 183 182 149 316	362	100 18 18 14 11 11 11 9	4.00	19	22	7	–	–	–	–	–	–	–	–	–	–	2H-1-Benzopyran-3,4-diol, 2-(3,4-dimethoxyphenyl)-3,4-dihydro-7,8-dimethoxy-, [2R-(2α,3α,4α)]-	22425-59-8	NI37748
180 181 165 183 182 151 362 167	362	100 21 13 12 11 11 5 5		19	22	7	–	–	–	–	–	–	–	–	–	–	2H-1-Benzopyran-3,4-diol, 2-(3,4-dimethoxyphenyl)-3,4-dihydro-7,8-dimethoxy-, [2R-(2α,3α,4α)]-	22425-59-8	LI07985
210 195 362 181 151 153 301 137	362	100 42 17 15 11 10 8 8		19	22	7	–	–	–	–	–	–	–	–	–	–	3′,4′,5′,7-Tetramethoxy-2,3-cis-flavan-3,4-cis-diol		LI07986
210 195 301 316 181 344 151 153	362	100 50 25 22 22 19 15 14	13.00	19	22	7	–	–	–	–	–	–	–	–	–	–	3′,4′,5′,7-Tetramethoxy-2,3-trans-flavan-3,4-trans-diol		LI07987
364 362 160 173 365 283 228 159	362	100 98 33 21 19 19 17 15		19	23	2	–	–	–	–	1	–	–	–	–	–	Estra-1,3,5(10)-trien-17-one, 16-bromo-3-methoxy-	51262-23-8	NI37749
362 364 160 173 363 365 283 228	362	100 99 17 11 10 9 9 8		19	23	2	–	–	–	–	1	–	–	–	–	–	Estra-1,3,5(10)-trien-17-one, 16-bromo-3-methoxy-	51262-23-8	NI37750
127 55 115 313 43 171 77 89	362	100 69 53 44 43 40 40 32	0.00	19	23	3	2	–	–	1	–	–	–	–	–	–	4-Piperidone, 1-(chloroacetyl)-2-(2-indolyl)-, ethylene acetal, trans-		DD01442
84 83 131 57 70 91 304 215	362	100 69 46 44 31 21 19 18	6.40	19	26	5	2	–	–	–	–	–	–	–	–	–	1,4-Dioxa-7,12-diazacyclotetradecane-2,6,13-trione, 14-benzyl-7,12-dimethyl-	91234-76-3	NI37751
362 92 306 120 363 44 63 57	362	100 35 22 22 22 21 20 18		20	10	7	–	–	–	–	–	–	–	–	–	–	[2,2′-Binaphthalene]-1,1′,4,4′-tetrone, 3,8,8′-trihydroxy-	88381-93-5	NI37752
334 105 77 51 335 362 192 76	362	100 59 53 32 24 18 13 9		20	14	3	2	1	–	–	–	–	–	–	–	–	Benzamide, N-1,2-benzisothiazol-3-yl-N-phenyl-, S,S-dioxide	21387-30-4	NI37753
180 77 362 51 181 105 104 50	362	100 57 27 23 15 13 12 12		20	14	3	2	1	–	–	–	–	–	–	–	–	Benzophenone, O-1,2-benzisothiazol-3-yloxime S,S-dioxide	7668-32-8	NI37754
56 362 36 43 41 54 55 17	362	100 93 59 52 50 47 46 46		20	14	5	2	–	–	–	–	–	–	–	–	–	1,5-Diamino-4-hydroxy-7-(4-hydroxyphenyl)anthraquinone		IC04132
137 59 44 77 215 69 362 78	362	100 57 49 39 34 31 30 26		20	17	3	–	–	3	–	–	–	–	–	–	–	Estra-1,3,5,7,9-pentaen-17-one, 3-[(trifluoroacetyl)oxy]-	56588-12-6	NI37755
77 93 362 121 65 122 106 105	362	100 92 65 62 60 34 29 25		20	18	3	4	–	–	–	–	–	–	–	–	–	1-Phenylazo-4-(4-hydroxyphenylazo)-2,6-dimethoxybenzene		IC04133
300 257 239 285 203 268 269 267	362	100 60 60 50 45 40 35 35	5.00	20	26	6	–	–	–	–	–	–	–	–	–	–	17,19-Bis(abeo)-6β,7α,16ξ-trihydroxyroyleanone		NI37756
137 138 189 362 194 206 122 163	362	100 29 12 9 8 6 6 5		20	26	6	–	–	–	–	–	–	–	–	–	–	1,4-Butanediol, 2,3-bis[(4-hydroxy-3-methoxyphenyl)methyl]-, [R-(R*,R*)]-	29388-59-8	NI37757
260 219 217 190 202 218 189 261	362	100 88 67 40 32 25 23 20	6.00	20	26	6	–	–	–	–	–	–	–	–	–	–	Butanoic acid, 3-methyl-, 2-hydroxy-1-[(7-methoxy-2-oxo-2H-1-benzopyran-8-yl)methyl]-2-methylpropyl ester, (S)-	88640-83-9	NI37758
181 182 151 137 180 121 166 165	362	100 76 70 67 53 45 38 35	15.90	20	26	6	–	–	–	–	–	–	–	–	–	–	3-(2,4-Dimethoxyphenyl)-1-(3,4-dimethoxyphenyl)-1-methoxypropan-2-ol		NI37759
97 55 83 95 81 69 71 109	362	100 90 89 87 87 87 83 82	7.00	20	26	6	–	–	–	–	–	–	–	–	–	–	Germacra-1(10),11(13)-dien-6,12-olide, 4,5-epoxy-, 8-(4-hydroxysenecioate), (4α,5β,8α)-		NI37760
83 43 55 84 28 69 41 57	362	100 80 64 54 48 47 46 43	5.00	20	26	6	–	–	–	–	–	–	–	–	–	–	Germacra-1(10),11(13)-dien-6,12-olide, 4,5-epoxy-9-hydroxy-, 8-senecioate, (4α,5β,8α,9α)-		NI37761
330 55 43 44 41 77 303 137	362	100 76 70 68 58 45 44 44	42.00	20	26	6	–	–	–	–	–	–	–	–	–	–	Gibberellin A_1 methyl ester	4747-53-9	MS08537
330 303 284 362 300 312 344 256	362	100 29 19 15 15 11 9 9		20	26	6	–	–	–	–	–	–	–	–	–	–	Gibberellin A_1 methyl ester	4747-53-9	NI37762
330 303 284 322 275 239 362 312	362	100 37 32 27 25 22 21 21		20	26	6	–	–	–	–	–	–	–	–	–	–	Gibberellin A_1 methyl ester	4747-53-9	NI37763
302 330 137 91 77 59 41 228	362	100 95 95 92 81 79 67 66	31.00	20	26	6	–	–	–	–	–	–	–	–	–	–	Gibberellin A_{16} methyl ester		MS08538
316 362 344 257 302 288 284 298	362	100 62 58 42 40 30 28 26		20	26	6	–	–	–	–	–	–	–	–	–	–	Gibberellin C, methyl ester	15004-61-2	NI37764

MASS TO CHARGE RATIOS								M.W.	INTENSITIES								Parent	C	H	O	N	S	F	Cl	Br	I	Si	P	B	X	COMPOUND NAME	CAS Reg No	No
316	344	288	257	298	362	302	284	**362**	100	71	66	66	58	54	46	36		20	26	6	–	–	–	–	–	–	–	–	–	–	Gibberellin C, methyl ester	15004-61-2	NI37765
221	43	270	41	69	125	79	53	**362**	100	90	66	57	53	49	44	43	**0.60**	20	26	6	–	–	–	–	–	–	–	–	–	–	4-Methoxy-15,16,17-trinorcrotophorbolone-20-acetate		NI37766
55	83	262	263	163	217	245		**362**	100	83	9	8	8	5	4			20	26	6	–	–	–	–	–	–	–	–	–	–	Oxireno[7,8]naphtho[1,2-b]furan-7(2H)-one, decahydro-3-hydroxy-3a,8c-dimethyl-6-methylene, 9-(angeloyloxy)-, [1aR-(1aα,3β,3aα,5aβ,8aα,8bβ,8cα,9α)]-		MS08539
329	43	275	362	41	55	330	45	**362**	100	48	43	35	34	23	22	18		20	26	6	–	–	–	–	–	–	–	–	–	–	9,10-Phenanthrenedione, 1,2,3,4,4a,10a-hexahydro-5,6,8-trihydroxy-7-(2-hydroxypropyl)-1,1,4a-trimethyl-, (2′-ζ,4aS,10aS)-		DD01443
329	41	275	55	43	362	69	330	**362**	100	49	35	34	26	25	25	23		20	26	6	–	–	–	–	–	–	–	–	–	–	9(1H)-Phenanthrenone, 2,3,4,4a-tetrahydro-5,6,8,10-tetrahydroxy-7-(2-hydroxypropyl)-1,1,4a-trimethyl-, (2′ζ,4aR)-		DD01444
41	43	55	91	53	77	79	362	**362**	100	82	50	47	36	31	30	28		20	26	6	–	–	–	–	–	–	–	–	–	–	1,1,12,13-Tetradehydrochaparrolide		LI07988
251	94	55	250	41	70	28	43	**362**	100	60	58	56	51	44	39	37	**13.00**	20	27	4	–	–	–	–	–	–	–	1	–	–	Phosphoric acid, 2-ethylhexyl diphenyl ester	1241-94-7	IC04134
251	94	250	55	41	70	77	43	**362**	100	50	33	26	24	23	16	16	**4.00**	20	27	4	–	–	–	–	–	–	–	1	–	–	Phosphoric acid, 2-ethylhexyl diphenyl ester	1241-94-7	NI37767
94	55	70	251	41	38	27	29	**362**	100	98	90	73	70	61	44	36	**7.01**	20	27	4	–	–	–	–	–	–	–	1	–	–	Phosphoric acid, octyl diphenyl ester	115-88-8	WI01479
362	363	347	139	255	171	77	348	**362**	100	25	25	25	21	18	10	6		21	14	4	–	1	–	–	–	–	–	–	–	–	1-Hydroxy-4-(4-methoxyphenylthio)anthraquinone		IC04135
117	118	334	107	146	275	173	201	**362**	100	75	63	35	25	22	22	16	**11.00**	21	22	2	4	–	–	–	–	–	–	–	–	–	17β-Hydroxy-5-(1-phenyl-1H-tetrazol-5-yloxy)de-A-estra-5,7,9-triene		NI37768
331	43	303	332	91	79	55	95	**362**	100	48	28	22	11	10	9	8	**5.11**	21	27	4	–	–	1	–	–	–	–	–	–	–	Pregna-4,16-diene-3,20-dione, 9-fluoro-11β,21-dihydroxy-	802-71-1	NI37769
93	69	135	55	107	77	109	79	**362**	100	88	68	67	60	50	48	43	**0.00**	21	30	3	–	1	–	–	–	–	–	–	–	–	2,6,10-Dodecatrien-1-ol, 3,7,11-trimethyl-9-(phenylsulphonyl)-, (E,E)-	57683-67-7	NI37770
302	216	201	287	248	276	274	261	**362**	100	66	37	27	20	12	5	4	**0.00**	21	30	5	–	–	–	–	–	–	–	–	–	–	3α-Acetoxy-D-homo-17-oxo-5α-androstane-16,17a-dione		LI07989
362	307	275	233	289	363	308	177	**362**	100	82	49	45	38	25	18	18		21	30	5	–	–	–	–	–	–	–	–	–	–	2H,6H-Benzo[1,2-b:5,4-b′]dipyran-10-propionic acid, 3,4,7,8-tetrahydro-5 methoxy-2,2,8,8-tetramethyl-, methyl ester	26535-39-7	NI37771
362	307	275	233	289	363	177	41	**362**	100	80	50	45	38	25	18	18		21	30	5	–	–	–	–	–	–	–	–	–	–	2H,6H-Benzo[1,2-b:5,4-b′]dipyran-10-propionic acid, 3,4,7,8-tetrahydro-5 methoxy-2,2,8,8-tetramethyl-, methyl ester	26535-39-7	NI37772
69	41	182	43	181	238	294	223	**362**	100	99	71	49	47	43	41	41	**2.70**	21	30	5	–	–	–	–	–	–	–	–	–	–	2,4-Cyclohexadien-1-one, 3,5,6-trihydroxy-4,6-bis(3-methyl-2-butenyl)-2-(3-methyl-1-oxobutyl)-, (R)-	26472-41-3	NI37773
69	41	70	39	67	55	53	43	**362**	100	68	20	17	14	14	13	13	**0.20**	21	30	5	–	–	–	–	–	–	–	–	–	–	2,4-Cyclohexadien-1-one, 3,5,6-trihydroxy-4,6-bis(3-methyl-2-butenyl)-2-(3-methyl-1-oxobutyl)-, (R)-	26472-41-3	NI37774
41	69	97	57	197	39	85	55	**362**	100	69	67	28	25	19	15	14	**1.00**	21	30	5	–	–	–	–	–	–	–	–	–	–	2-Cyclopenten-1-one, 3,4-dihydroxy-5-(3-methyl-2-butenyl)-2-(3-methyl-1-oxobutyl)-4-(4-methyl-1-oxo-2-pentenyl)-	57195-47-8	NI37775
69	41	197	43	44	60	18	55	**362**	100	64	62	44	42	35	30	21	**3.00**	21	30	5	–	–	–	–	–	–	–	–	–	–	2-Cyclopenten-1-one, 3,4-dihydroxy-5-(3-methyl-2-butenyl)-2-(3-methyl-1-oxobutyl)-4-(4-methyl-1-oxo-3-pentenyl)-	25522-96-7	NI37776
69	197	41	57	266	293	97	43	**362**	100	82	77	25	23	22	18	18	**8.00**	21	30	5	–	–	–	–	–	–	–	–	–	–	2-Cyclopenten-1-one, 3,4-dihydroxy-5-(3-methyl-2-butenyl)-2-(3-methyl-1-oxobutyl)-4-(4-methyl-1-oxo-3-pentenyl)-	25522-96-7	NI37777
91	255	270	316	213	226	237	301	**362**	100	94	90	46	43	38	33	21	**17.00**	21	30	5	–	–	–	–	–	–	–	–	–	–	5α,3α,11β-Diformoxyandrostan-17-one		NI37778
106	270	226	255	213	252	237	301	**362**	100	93	77	53	51	40	36	29	**26.00**	21	30	5	–	–	–	–	–	–	–	–	–	–	5β,3α,11β-Diformoxyandrostan-17-one		NI37779
344	55	83	57	149	109	43	91	**362**	100	52	47	45	38	35	35	29	**5.00**	21	30	5	–	–	–	–	–	–	–	–	–	–	Hirundigenin		LI07990
69	267	131	285	41	137	326	55	**362**	100	98	85	68	65	53	50	49	**45.00**	21	30	5	–	–	–	–	–	–	–	–	–	–	Methyl 6α,7β-dihydroxyvouacapan-17β-oate		NI37780
320	109	260	55	107	41	95	67	**362**	100	73	53	46	44	40	39	39	**0.00**	21	30	5	–	–	–	–	–	–	–	–	–	–	Norketocassamic acid	2225-69-6	NI37781
28	41	57	69	97	197	43	44	**362**	100	99	76	72	47	40	36	33	**1.84**	21	30	5	–	–	–	–	–	–	–	–	–	–	1-Oxaspiro[4.4]non-8-ene-4,7-dione, 9-hydroxy-6-(3-methyl-2-butenyl)-2-isopropyl-8-(3-methyl-1-oxobutyl)-	31319-12-7	NI37782
43	59	41	44	57	85	91	105	**362**	100	47	42	27	24	23	22	21	**2.73**	21	30	5	–	–	–	–	–	–	–	–	–	–	1-Oxaspiro[4.4]non-8-ene-4,7-dione, 9-hydroxy-6-(3-methyl-2-butenyl)-2-isopropyl-8-(3-methyl-1-oxobutyl)-	31319-12-7	NI37783
57	56	41	43	86	235	222	362	**362**	100	38	25	21	15	12	12	10		21	30	5	–	–	–	–	–	–	–	–	–	–	1-Oxaspiro[4.4]non-8-ene-4,7-dione, 9-hydroxy-6-(3-methyl-2-butenyl)-2-isopropyl-8-(3-methyl-1-oxobutyl)-	31319-12-7	NI37784
258	318	44	286	107	91	109	93	**362**	100	99	66	58	52	51	49	46	**0.00**	21	30	5	–	–	–	–	–	–	–	–	–	–	1-Phenanthrenecarboxylic acid, 7-(carboxymethylene)tetradecahydro-1,4a,8-trimethyl-9-oxo-, 1-methyl ester, [1S-(1α,4aα,4bβ,8β,8aα,10aβ)]-	471-70-5	NI37785
288	107	147	260	165	230	332	106	**362**	100	89	79	43	43	40	39	37	**0.00**	21	30	5	–	–	–	–	–	–	–	–	–	–	Pregnan-18-oic acid, 3,11,21-trihydroxy-20-oxo-, γ-lactone, (3β,5α,11β)-	72101-51-0	NI37786
41	55	57	43	69	163	123	91	**362**	100	85	84	69	59	58	55	53	**2.11**	21	30	5	–	–	–	–	–	–	–	–	–	–	Pregn-4-ene-3,20-dione, 11,17,21-trihydroxy-, (11β)-	50-23-7	NI37787
285	43	362	41	91	227	55	79	**362**	100	97	88	88	78	77	74	66		21	30	5	–	–	–	–	–	–	–	–	–	–	Pregn-4-ene-3,20-dione, 11,17,21-trihydroxy-, (11β)-	50-23-7	NI37788
362	285	43	242	227	55	41	91	**362**	100	72	57	50	47	43	40	39		21	30	5	–	–	–	–	–	–	–	–	–	–	Pregn-4-ene-3,20-dione, 11,17,21-trihydroxy-, (11β)-	50-23-7	NI37789
43	109	318	44	41	145	91	105	**362**	100	99	79	70	54	53	52	47	**2.20**	21	30	5	–	–	–	–	–	–	–	–	–	–	Pregn-5-ene-12,20-dione, 3,14,15-trihydroxy-, (3β,14β,15α)-	10005-77-3	NI37790
362	316	331	363	138	317	137	332	**362**	100	76	43	28	22	21	16	13		21	34	3	2	–	–	–	–	–	–	–	–	–	17β-Hydroxy-5α-androstane-3,6-dione, bis(methyloxime)		NI37791
347	105	77	362	263	331	348	115	**362**	100	48	45	27	27	26	25	22		22	22	1	2	1	–	–	–	–	–	–	–	–	1H-[1]Benzothieno[3,2-d]azonine-3-carbonitrile, 2,3,4,5,6,7-hexahydro-7-methoxy-7-phenyl-	99659-38-8	NI37792
117	159	203	90	188	160	204	189	**362**	100	77	73	31	27	26	9	4	**0.00**	22	22	3	2	–	–	–	–	–	–	–	–	–	2,3-Naphthalenedicarboximide, 1,4-diimino-1,2,3,4-tetrahydro-9-isopropyl N-(4-methoxyphenyl)-, endo-	123542-21-2	DD01445

MASS TO CHARGE RATIOS								M.W.	INTENSITIES								Parent	C	H	O	N	S	F	Cl	Br	I	Si	P	B	X	COMPOUND NAME	CAS Reg No	No
57	302	55	362	53	69	71	289	**362**	100	96	92	81	53	40	32	31		22	22	3	2	–	–	–	–	–	–	–	–	–	3-Pyrrolin-2-one, 4-ethyl-3-methyl-5-[5-(2-carboxyphenyl-2-methylene)-3,4-dimethyl-5H-pyrrolyl-2-methylene]-, (Z,E)-		MS08540
302	257	275	239	284	185			**362**	100	96	88	66	52	38			**0.00**	22	34	4	–	–	–	–	–	–	–	–	–	–	19S-Acetoxy-3β,17β-dihydroxy-19a-methylandrost-5-ene		LI07991
149	41	43	29	223	150	57	55	**362**	100	19	13	12	9	9	6	6	**0.00**	22	34	4	–	–	–	–	–	–	–	–	–	–	1,2-Benzenedicarboxylic acid, butyl decyl ester	89-19-0	NI37793
149	57	150	41	265	43	55	29	**362**	100	10	9	7	5	5	4	3	**0.00**	22	34	4	–	–	–	–	–	–	–	–	–	–	1,2-Benzenedicarboxylic acid, diheptyl ester	3648-21-3	NI37794
149	57	265	43	99	41	56	55	**362**	100	59	32	27	24	21	20	16	**1.51**	22	34	4	–	–	–	–	–	–	–	–	–	–	1,2-Benzenedicarboxylic acid, diiso-heptyl ester		IC04136
195	43	177	165	149	196	55	41	**362**	100	15	14	12	11	10	9	9	**2.00**	22	34	4	–	–	–	–	–	–	–	–	–	–	1,4-Benzenedicarboxylic acid, dodecyl ethyl ester	54699-32-0	NI37795
168	169	362	167	336	153	139	364	**362**	100	65	52	18	14	14	14	12		22	34	4	–	–	–	–	–	–	–	–	–	–	1,4-Benzoquinone, 2-hydroxy-5-methoxy-3-(pentadecenyl)-	26915-11-7	NI37796
125	140	223	153	44	56	265	362	**362**	100	78	42	41	33	25	23	22		22	34	4	–	–	–	–	–	–	–	–	–	–	1,10-Decanedione, 1,10-bis(2-oxocyclohexyl)-		LI07992
125	140	223	153	47	55	265	362	**362**	100	78	42	41	33	26	24	22		22	34	4	–	–	–	–	–	–	–	–	–	–	1,10-Decanedione, 1,10-bis(2-oxocyclohexyl)-	17343-93-0	NI37797
43	139	123	181	167	287	259	217	**362**	100	93	11	10	6	5	5	4	**0.00**	22	34	4	–	–	–	–	–	–	–	–	–	–	4,8-Dodecadiyne-2,11-dione, 6,7-dihydroxy-3,6,7,10-tetramethyl-3,10-diisopropyl-	63922-56-5	NI37798
105	43	57	41	183	55	29	100	**362**	100	87	86	65	58	58	43	36	**25.04**	22	34	4	–	–	–	–	–	–	–	–	–	–	Dodecanoic acid, 2-phenyl-1,3-dioxan-5-yl ester	56630-70-7	NI37799
43	123	164	109	149	124	139	165	**362**	100	35	28	21	18	15	12	10	**0.00**	22	34	4	–	–	–	–	–	–	–	–	–	–	1,5-Hexadiyne-3,4-diol, 1,6-bis[3,3-dimethyl-2-isopropyloxiranyl]-3,4-dimethyl-	63922-54-3	NI37800
271	43	329	231	59	272	229	203	**362**	100	84	64	24	22	21	21	20	**0.00**	22	34	4	–	–	–	–	–	–	–	–	–	–	2,5-Hexanedione, 3,4-bis[3-hydroxy-3-methyl-2-isopropyl-1-butenylidene]	69651-45-2	NI37801
115	121	189	302	81	303	114	107	**362**	100	59	35	28	24	20	18	17	**7.41**	22	34	4	–	–	–	–	–	–	–	–	–	–	1-Naphthalenecarboxylic acid, decahydro-5-(5-methoxy-3-methyl-5-oxo-3 pentenyl)-1,4a-dimethyl-6-methylene-, methyl ester, [1S-[1α,4aα,5α(E),8aβ]]-	1757-85-3	NI37802
121	81	41	55	82	189	67	79	**362**	100	81	71	57	37	35	35	34	**2.36**	22	34	4	–	–	–	–	–	–	–	–	–	–	1-Naphthalenecarboxylic acid, decahydro-5-(5-methoxy-3-methyl-5-oxo-3 pentenyl)-1,4a-dimethyl-6-methylene-, methyl ester, [1S-[1α,4aα,5α(E),8aβ]]-	1757-85-3	NI37803
121	189	302	81	303	114	107	249	**362**	100	60	47	41	34	31	29	28	**12.60**	22	34	4	–	–	–	–	–	–	–	–	–	–	1-Naphthalenecarboxylic acid, decahydro-5-(5-methoxy-3-methyl-5-oxo-3 pentenyl)-1,4a-dimethyl-6-methylene-, methyl ester, [1S-[1α,4aα,5α(E),8aβ]]-	1757-85-3	NI37804
43	252	82	69	55	41	79	93	**362**	100	68	63	63	56	54	44	37	**11.58**	22	34	4	–	–	–	–	–	–	–	–	–	–	4(1H)-Phenanthrenone, 3-(acetyloxy)-2-ethenyldodecahydro-10a-hydroxy-2,4b,8,8-tetramethyl-, [2S-(2α,3β,4aα,4bβ,8aα,10aβ)]-	41756-17-6	NI37805
55	91	79	67	95	81	93	53	**362**	100	55	54	53	51	49	47	46	**33.00**	22	34	4	–	–	–	–	–	–	–	–	–	–	Pregnan-20-one, 5,6-epoxy-3,17-dihydroxy-16-methyl-, (3β,5α,6α,16α)-	56630-87-6	MS08541
129	73	130	257	116	75	272	149	**362**	100	37	28	25	23	22	21	21	**8.00**	22	38	2	–	–	–	–	–	–	1	–	–	–	Androstan-3-one, 17-[(trimethylsilyl)oxy]-, (5α,17β)-	18899-43-9	NI37806
129	272	347	257	149	116	130	273	**362**	100	63	38	37	24	24	23	13	**8.82**	22	38	2	–	–	–	–	–	–	1	–	–	–	Androstan-3-one, 17-[(trimethylsilyl)oxy]-, (5β,17β)-	18880-43-8	NI37807
129	73	75	130	116	79	101	81	**362**	100	55	33	29	26	25	22	22	**11.00**	22	38	2	–	–	–	–	–	–	1	–	–	–	Androstan-3-one, 17-[(trimethylsilyl)oxy]-, (5β,17β)-	18880-43-8	NI37808
75	272	73	271	347	81	55	93	**362**	100	99	80	75	56	50	42	41	**30.96**	22	38	2	–	–	–	–	–	–	1	–	–	–	Androstan-17-one, 3-[(trimethylsilyl)oxy]-, (3α,5α)-	7337-91-9	NI37809
272	271	347	75	155	215	129	108	**362**	100	71	56	30	26	25	24	23	**20.23**	22	38	2	–	–	–	–	–	–	1	–	–	–	Androstan-17-one, 3-[(trimethylsilyl)oxy]-, (3α,5α)-	7337-91-9	NI37810
272	271	347	215	257	362	155	75	**362**	100	64	57	25	23	22	21	20		22	38	2	–	–	–	–	–	–	1	–	–	–	Androstan-17-one, 3-[(trimethylsilyl)oxy]-, (3α,5α)-	7337-91-9	NI37811
75	272	73	244	108	257	93	55	**362**	100	99	93	56	47	43	42	41	**9.50**	22	38	2	–	–	–	–	–	–	1	–	–	–	Androstan-17-one, 3-[(trimethylsilyl)oxy]-, (3α,5β)-	4867-14-5	NI37812
272	244	257	108	75	176	147	215	**362**	100	64	48	41	29	28	23	22	**6.06**	22	38	2	–	–	–	–	–	–	1	–	–	–	Androstan-17-one, 3-[(trimethylsilyl)oxy]-, (3α,5β)-	4867-14-5	NI37813
272	244	257	108	176	273	215	75	**362**	100	66	48	36	25	23	22	22	**6.08**	22	38	2	–	–	–	–	–	–	1	–	–	–	Androstan-17-one, 3-[(trimethylsilyl)oxy]-, (3α,5β)-	4867-14-5	NI37814
347	75	348	272	71	362	155	271	**362**	100	43	25	23	19	18	18	17		22	38	2	–	–	–	–	–	–	1	–	–	–	Androstan-17-one, 3-[(trimethylsilyl)oxy]-, (3β,5α)-	10426-95-6	NI37815
347	75	272	348	271	215	155	130	**362**	100	37	32	29	21	15	15	15	**12.81**	22	38	2	–	–	–	–	–	–	1	–	–	–	Androstan-17-one, 3-[(trimethylsilyl)oxy]-, (3β,5α)-	10426-95-6	NI37816
347	75	73	348	155	129	107	81	**362**	100	80	40	26	25	23	23	22	**20.50**	22	38	2	–	–	–	–	–	–	1	–	–	–	Androstan-17-one, 3-[(trimethylsilyl)oxy]-, (3β,5α)-	10426-95-6	NI37817
272	257	108	273	244	215	271	75	**362**	100	36	34	23	20	19	18	15	**1.88**	22	38	2	–	–	–	–	–	–	1	–	–	–	Androstan-17-one, 3-[(trimethylsilyl)oxy]-, (3β,5β)-	10426-96-7	NI37818
272	75	73	108	257	81	273	271	**362**	100	54	34	30	26	24	22	22	**1.50**	22	38	2	–	–	–	–	–	–	1	–	–	–	Androstan-17-one, 3-[(trimethylsilyl)oxy]-, (3β,5β)-	10426-96-7	NI37819
272	75	257	108	73	79	93	155	**362**	100	56	42	37	30	28	26	25	**2.00**	22	38	2	–	–	–	–	–	–	1	–	–	–	Androstan-17-one, 3-[(trimethylsilyl)oxy]-, (3β,5β)-	10426-96-7	NI37820
143	73	270	75	144	130	91	115	**362**	100	50	21	21	15	7	7	6	**0.00**	22	38	2	–	–	–	–	–	–	1	–	–	–	19-Nortestosterone, 17-methyl(trimethylsilyl)-, (17α)-		NI37821
262	362	183	361	277	185	108	201	**362**	100	91	77	61	60	35	30	25		23	23	2	–	–	–	–	–	–	–	1	–	–	Carboethoxy methyl methylene triphenyl phosphorane		LI07993
275	77	362	334	91	276	301	270	**362**	100	42	38	34	25	23	22	22		23	26	–	2	1	–	–	–	–	–	–	–	–	1H-Cyclohepta[e][1,4]thiazepine, 2,3,5,6,7,8,9,10-octahydro-1-phenyl-5-(4 tolylimino)-		NI37822
362	291	363	238	41	305	55	43	**362**	100	29	28	27	15	12	11	8		23	26	2	2	–	–	–	–	–	–	–	–	–	1-Amino-4-(trimethylcyclohexylamino)anthraquinone		IC04137
54	135	111	83	82	79	93	67	**362**	100	81	80	56	51	35	34	28	**0.00**	23	26	2	2	–	–	–	–	–	–	–	–	–	2,3-Naphthalenedicarboximide, 1,4-imino-9-adamantyl-1,2,3,4-tetrahydro-N-methyl-, endo-	123542-33-6	DD01446
152	153	135	107	154	109	136	77	**362**	100	34	25	6	5	5	4	4	**0.00**	23	38	3	–	–	–	–	–	–	–	–	–	–	Benzoic acid, 3-methoxy-, pentadecyl ester	56954-75-7	NI37823
152	153	135	154	136	107	77	92	**362**	100	31	30	3	3	3	3	2	**0.00**	23	38	3	–	–	–	–	–	–	–	–	–	–	Benzoic acid, 4-methoxy-, pentadecyl ester	56954-74-6	NI37824
55	41	81	93	105	79	67	43	**362**	100	85	84	75	72	72	72	70	**32.00**	23	38	3	–	–	–	–	–	–	–	–	–	–	17β-Hydroxy-6α-pentyl-4-oxa-5β-androstan-3-one		MS08542
275	257	161	109	55	81	107	93	**362**	100	86	54	54	53	49	48	48	**16.00**	23	38	3	–	–	–	–	–	–	–	–	–	–	17β-Hydroxy-4-propyl-3,4-seco-5-androsten-3-oic acid, methyl ester		MS08543
289	271	362	290	55	233	81	41	**362**	100	48	26	25	25	23	22	18		23	38	3	–	–	–	–	–	–	–	–	–	–	17-Oxo-6α-pentyl-4-nor-3,5-secoandrostan-3-oic acid		MS08544
289	271	55	41	81	67	95	233	**362**	100	72	47	45	44	41	34	31	**22.00**	23	38	3	–	–	–	–	–	–	–	–	–	–	17-Oxo-6β-pentyl-4-nor-3,5-secoandrostan-3-oic acid		MS08545

MASS TO CHARGE RATIOS	M.W.	INTENSITIES	Parent	C	H	O	N	S	F	Cl	Br	I	Si	P	B	X	COMPOUND NAME	CAS Reg No	No
326 105 311 91 159 121 119 55	**362**	100 54 46 45 36 35 35 34	**0.00**	23	38	3	–	–	–	–	–	–	–	–	–	–	Pregnan-20-one, 3,5-dihydroxy-6,16-dimethyl-, (3β,5α,6β,16α)-	55334-07-1	WI01480
45 43 84 300 81 95 55 67	**362**	100 79 56 51 49 45 42 41	**26.00**	23	38	3	–	–	–	–	–	–	–	–	–	–	Pregnan-20-one, 3-(methoxymethoxy)-, (3β,5α)-	57156-80-6	NI37825
275 257 55 276 81 95 67 41	**362**	100 34 32 29 27 21 20 20	**13.00**	23	38	3	–	–	–	–	–	–	–	–	–	–	17-Oxo-4-propyl-3,4-seco-5α-androstan-3-oic acid, methyl ester		MS08546
98 278 41 118 95 55 106 105	**362**	100 77 73 51 50 50 45 44	**14.28**	23	38	3	–	–	–	–	–	–	–	–	–	–	16,16′-O-Trimethyl-11β-methoxyetiocholanolone		NI37826
302 362 124 330 138 151 347 275	**362**	100 83 35 30 29 21 20 16		24	14	2	2	–	–	–	–	–	–	–	–	–	Triptycene-13-carboxylic acid, 1,8-dicyano-, methyl ester		NI37827
77 152 257 362 51 180 105 153	**362**	100 77 58 40 24 20 20 12		24	18	–	4	–	–	–	–	–	–	–	–	–	Bis(4-azobenzene)		MS08547
180 182 181 183 153 154 51 50	**362**	100 84 67 12 6 5 4 4	**0.00**	24	18	–	4	–	–	–	–	–	–	–	–	–	Phenazine, 5,10-dihydro-,	4733-23-7	NI37828
362 319 161 334 347 173 265 253	**362**	100 26 23 19 16 11 7 5		24	26	3	–	–	–	–	–	–	–	–	–	–	Spiro[2.4]heptan-4-one, 2-(4-methoxyphenyl)-5-[(4-methoxyphenyl)methylene]-1,1-dimethyl-, (E)-		MS08548
255 55 81 326 273 43 95 41	**362**	100 64 57 56 51 43 40 38	**0.00**	24	42	2	–	–	–	–	–	–	–	–	–	–	Cholane-3,12-diol, (3α,5β,12α)-	32726-81-1	NI37829
289 55 41 81 67 290 93 43	**362**	100 45 29 22 22 21 21 21	**6.00**	24	42	2	–	–	–	–	–	–	–	–	–	–	6-Pentylidene-4,5-secoandrostane-4,17β-diol		MS08549
362 254 257 252 105 181 255 363	**362**	100 65 52 43 37 34 31 29		26	18	2	–	–	–	–	–	–	–	–	–	–	Benzo[b]benzo[3,4]cyclobuta[1,2-e][1,4]dioxin, 4b,10a-dihydro-4b,10a-diphenyl-	53486-89-8	NI37830
362 363 319 289 347 290 276 320	**362**	100 32 27 27 13 13 13 10		26	18	2	–	–	–	–	–	–	–	–	–	–	9-Methoxy-10-methyl-12-oxo-7,12-dihydronaphtho[2,1,8-qra]naphthacene		LI07994
141 40 178 73 362 115 176 177	**362**	100 17 14 14 13 11 9 8		26	18	2	–	–	–	–	–	–	–	–	–	–	1-Naphthalenemethyl anthracene-9-carboxylate	86328-66-7	NI37831
191 155 192 189 127 178 165 190	**362**	100 43 39 22 17 15 15 12	**11.00**	26	18	2	–	–	–	–	–	–	–	–	–	–	9-Phenanthrenemethyl naphthalene-2-carboxylate	92174-49-7	NI37832
184 179 167 362 185 178 183 168	**362**	100 34 23 20 20 18 16 14		26	22	–	2	–	–	–	–	–	–	–	–	–	9,10,19,20-Tetrahydrotetrabenzo[b,d,l,j][1,6]diazacyclododecine		LI07995
184 179 362 167 178 183 168 185	**362**	100 38 25 23 20 19 17 13		26	22	–	2	–	–	–	–	–	–	–	–	–	1,6,7,12-Tetrahydro-2,3:4,5:8,9:10,11-tetrabenzo-1,6-diaza[12]annulene		LI07996
67 55 82 69 83 57 81 137	**362**	100 99 90 86 81 70 63 61	**19.20**	26	50	–	–	–	–	–	–	–	–	–	–	–	1-Cyclopentyl-2-hexadecylcyclopentane		AI02206
55 83 41 97 43 96 57 69	**362**	100 83 73 60 54 36 35 34	**0.33**	26	50	–	–	–	–	–	–	–	–	–	–	–	1,1-Dicyclohexyltetradecane		AI02207
83 41 97 43 82 57 96 69	**362**	100 79 76 61 49 49 47 44	**0.03**	26	50	–	–	–	–	–	–	–	–	–	–	–	1,1-Dicyclohexyltetradecane		AI02208
41 83 43 69 55 57 97 68	**362**	100 95 87 85 80 60 59 53	**0.24**	26	50	–	–	–	–	–	–	–	–	–	–	–	1,1-Dicyclopentylhexadecane		AI02209
81 82 43 96 95 67 55 83	**362**	100 99 98 95 90 80 78 65	**22.02**	26	50	–	–	–	–	–	–	–	–	–	–	–	11-Hexacosyne	34291-69-5	NI37833
81 82 43 96 95 67 55 83	**362**	100 99 99 97 90 82 79 68	**24.00**	26	50	–	–	–	–	–	–	–	–	–	–	–	11-Hexacosyne	34291-69-5	LI07997
82 96 81 95 43 67 55 83	**362**	100 96 96 88 84 78 70 63	**23.00**	26	50	–	–	–	–	–	–	–	–	–	–	–	13-Hexacosyne	34291-68-4	LI07998
82 96 81 95 43 67 55 83	**362**	100 98 98 87 83 78 70 62	**20.02**	26	50	–	–	–	–	–	–	–	–	–	–	–	13-Hexacosyne	34291-68-4	NI37834
137 95 81 55 67 43 304 57	**362**	100 66 66 47 45 45 43 32	**9.21**	26	50	–	–	–	–	–	–	–	–	–	–	–	7-Hexadecylspiro[4.5]decane		AI02210
55 97 41 43 69 264 83 96	**362**	100 52 34 29 24 23 23 20	**0.08**	26	50	–	–	–	–	–	–	–	–	–	–	–	1,1-Bis(4-methylcyclohexyl)dodecane		AI02211
55 97 41 43 69 83 111 81	**362**	100 54 39 33 27 26 24 21	**0.04**	26	50	–	–	–	–	–	–	–	–	–	–	–	1,1-Bis(4-methylcyclohexyl)dodecane		AI02212
43 57 67 41 55 81 109 71	**362**	100 81 77 72 62 57 46 44	**5.23**	26	50	–	–	–	–	–	–	–	–	–	–	–	2-(2-Octyldecyl)-cis-bicyclo[3.3.0]octane		AI02213
43 57 67 41 55 81 109 71	**362**	100 85 74 67 63 57 49 45	**3.80**	26	50	–	–	–	–	–	–	–	–	–	–	–	2-(2-Octyldecyl)-cis-bicyclo[3.3.0]octane		AI02214
317 240 318 362 242 285 333	**362**	100 99 82 78 78 50 19		27	22	1	–	–	–	–	–	–	–	–	–	–	1-Ethoxy-2-(diphenylmethyl)acenaphthene		LI07999
167 362 168 165 152 166 363 180	**362**	100 18 14 12 8 7 6 5		28	26	–	–	–	–	–	–	–	–	–	–	–	Benzene, 1,1′,1″,1‴-(1,4-butanediylidene)tetrakis-	1483-64-3	NI37835
193 362 347 317 363 332 331 194	**362**	100 57 43 22 19 19 19 19		28	26	–	–	–	–	–	–	–	–	–	–	–	1,3-Butadiene, 1,4-bis(6,8-dimethyl-4-azulenyl)-	94154-58-2	NI37836
364 235 234 363 236 366 233 222	**363**	100 73 71 64 64 61 60 57		6	10	1	–	–	–	2	–	–	–	–	–	1	Platinum, dichloro[(1,1′,2,2′-η)-3,3′-oxybis[propene]]-	12012-94-1	NI37837
139 141 111 27 75 156 28 113	**363**	100 39 29 19 14 12 12 9	**0.00**	10	4	2	1	–	–	1	2	–	–	–	–	–	Benzoic acid, 4-chloro-, 2,2-dibromo-1-cyanovinyl ester	55976-00-6	NI37838
363 149 271 43 57 179 279 167	**363**	100 48 43 28 27 19 18 18		12	5	9	5	–	–	–	–	–	–	–	–	–	10H-Phenoxazine, 1,3,7,9-tetranitro-	24050-28-0	MS08550
43 234 262 235 189 190 41 216	**363**	100 58 41 33 30 21 20 18	**0.00**	13	18	5	1	–	5	–	–	–	–	–	–	–	D-Asparaginic acid, N-pentafluoropropionyl-, isopropyl ester		MS08551
43 234 262 235 189 190 41 216	**363**	100 59 41 34 30 20 20 19	**0.00**	13	18	5	1	–	5	–	–	–	–	–	–	–	L-Asparaginic acid, N-pentafluoropropionyl-, isopropyl ester		MS08552
328 330 304 329 175 306 177 160	**363**	100 59 33 29 27 21 16 14	**4.40**	14	15	6	1	–	–	2	–	–	–	–	–	–	2,4-Dichlorophenoxyacetylaspartate		MS08553
115 98 103 149 158 200 140 116	**363**	100 99 91 83 72 69 63 63	**0.00**	14	21	10	1	–	–	–	–	–	–	–	–	–	Methyl 2,4,6-tri-O-acetyl-3-O-carbamoyl-α-D-mannopyranoside		MS08554
119 76 163 60 181 116 66 279	**363**	100 99 33 25 13 12 4 3	**0.72**	14	23	3	3	–	–	–	–	–	–	1	–	1	Tris(dimethylamino)phoshine-π-cyclopentadienyl-tricarbonylvanadium		MS08555
73 156 45 147 75 157 230 74	**363**	100 77 26 25 24 11 10 10	**0.00**	14	33	4	1	–	–	–	–	–	3	–	–	–	Glutamic acid, N,O,O′-tris(trimethylsilyl)-	15985-07-6	MS08556
73 246 156 147 93 75 232 218	**363**	100 61 55 30 28 28 25 23	**3.00**	14	33	4	1	–	–	–	–	–	3	–	–	–	Glutamic acid, N,O,O′-tris(trimethylsilyl)-	15985-07-6	MS08557
73 246 75 45 128 147 84 74	**363**	100 53 24 20 18 15 11 11	**0.00**	14	33	4	1	–	–	–	–	–	3	–	–	–	Glutamic acid, N,O,O′-tris(trimethylsilyl)-	15985-07-6	MS08558
134 263 262 89 90 44 264 77	**363**	100 82 64 28 21 19 17 15	**0.00**	15	10	4	3	1	–	1	–	–	–	–	–	–	Pyrido[2,3-d]thiazolo[3,2-b]pyridazin-4-ium, 3-phenyl-, perchlorate	18592-55-7	NI37839
134 262 263 89 90 135 44 77	**363**	100 72 44 32 25 24 22 18	**0.00**	15	10	4	3	1	–	1	–	–	–	–	–	–	Pyrido[3,2-d]thiazolo[3,2-b]pyridazin-4-ium, 3-phenyl-, perchlorate	18592-54-6	NI37840
235 261 192 231 333 277 193 234	**363**	100 58 47 36 25 25 25 22	**3.00**	15	17	6	5	–	–	–	–	–	–	–	–	–	2-Acetylamino-7-(2,3-diacetoxypropyl)-3,4-dihydropteridin-4-one		LI08000
306 307 192 308 348 193 165 363	**363**	100 30 30 10 8 8 8 4		15	25	2	5	–	–	–	–	–	2	–	–	–	Adenine, N,N′-bis[(tert-butyldimethylsilyl)oxy]-		NI37841
363 364 242 198 30 201 317 257	**363**	100 20 10 7 7 5 4 3		16	5	6	5	–	–	–	–	–	–	–	–	–	Propanedinitrile, (2,4,7-trinitro-9H-fluoren-9-ylidene)-	1172-02-7	NI37842
58 72 77 105 42 56 91 363	**363**	100 70 45 22 20 12 11 10		16	17	1	3	3	–	–	–	–	–	–	–	–	Thieno[2,3-c]isothiazol-4-amine, 5-benzoyl-3-[2-(dimethylamino)ethylthio]	97090-75-0	NI37843
58 71 77 42 72 59 44 43	**363**	100 30 27 15 13 8 6 6	**1.00**	16	17	1	3	3	–	–	–	–	–	–	–	–	Thieno[3,2-d]isothiazol-4-amine, 5-benzoyl-3-[2-(dimethylamino)ethylthio]-	97090-79-4	NI37844

MASS TO CHARGE RATIOS								M.W.	INTENSITIES								Parent	C	H	O	N	S	F	Cl	Br	I	Si	P	B	X	COMPOUND NAME	CAS Reg No	No
91	218	189	290	78	363	119	65	363	100	33	27	19	17	15	15	15		16	21	5	5	–	–	–	–	–	–	–	–	–	Ethyl 7-methyl-5-oxo-6-(4-tolyl)-1,2,3,6,7-pentaazabicyclo[2.2.1]heptane-2,3-dicarboxylate		LI08001
167	166	168	83	139	140			363	100	15	12	10	9	7			0.00	16	29	8	1	–	–	–	–	–	–	–	–	–	D-Galactitol, 2-(acetylmethylamino)-2-deoxy-3,4,6-tri-O-methyl-, 1,5-diacetate	74410-42-7	NI37845
116	43	158	98	129	145	74	45	363	100	46	38	38	35	32	30	30	0.00	16	29	8	1	–	–	–	–	–	–	–	–	–	Glucitol, 2-(acetylmethylamino)-2-deoxy-3,4,6-tri-O-methyl-, 1,5-diacetate		NI37846
42	273	287	68	302	70	271	57	363	100	43	40	35	29	24	16	12	3.66	17	21	6	3	–	–	–	–	–	–	–	–	–	Azirino[2',3':3,4]pyrrolo[1,2-a]indole-4,7-dione, 8-[[(aminocarbonyl)oxy]methyl]-1,1a,2,8,8a,8b-hexahydro-6,8a-dimethoxy-1,5-dimethyl-, [1aR-(1aα,8β,8aα,8bα)]-	18209-14-8	NI37847
42	287	68	302	70	271	316	257	363	100	40	35	29	24	16	10	9	0.40	17	21	6	3	–	–	–	–	–	–	–	–	–	Azirino[2',3':3,4]pyrrolo[1,2-a]indole-4,7-dione, 8-[[(aminocarbonyl)oxy]methyl]-1,1a,2,8,8a,8b-hexahydro-6,8a-dimethoxy-1,5-dimethyl-, [1aR-(1aα,8β,8aα,8bα)]-	18209-14-8	LI08002
73	45	245	40	75	246	362	44	363	100	35	33	23	22	20	19	18	12.10	17	25	4	1	–	–	–	–	–	2	–	–	–	2-Quinolinecarboxylic acid, 8-methoxy-4-[(trimethylsilyl)oxy]-, trimethylsilyl ester	55517-50-5	NI37848
105	199	77	201	200	106	50	184	363	100	65	29	21	8	8	5	4	0.00	18	18	5	1	–	–	1	–	–	–	–	–	–	Benzoic acid, anhydride with 3-chloro-N-ethoxy-2,6-dimethoxybenzenecarboximidic acid	29104-30-1	NI37849
105	199	77	201	106	200	51	184	363	100	74	31	25	11	8	6	4	0.00	18	18	5	1	–	–	1	–	–	–	–	–	–	Benzoic acid, anhydride with 3-chloro-N-ethoxy-2,6-dimethoxybenzenecarboximidic acid	29104-30-1	NI37850
199	201	364	242	260	200	244	105	363	100	33	17	17	12	11	6	6	0.00	18	18	5	1	–	–	1	–	–	–	–	–	–	Benzoic acid, anhydride with 3-chloro-N-ethoxy-2,6-dimethoxybenzenecarboximidic acid	29104-30-1	NI37851
84	98	275	273	191	58	107	363	363	100	42	27	24	22	20	18	13		18	29	3	5	–	–	–	–	–	–	–	–	–	Methyl 2-[3-[3-(piperidinomethyl)phenoxy]propylaminocarbonyl]hydrazinecarboximidate		MS08559
180	77	51	181	363	63	104	78	363	100	88	31	15	10	10	7	6		19	13	5	3	–	–	–	–	–	–	–	–	–	Benzophenone, O-(2,4-dinitrophenyl)oxime	17188-67-9	NI37852
180	77	51	181	363	63	104	107	363	100	88	31	15	10	10	7	3		19	13	5	3	–	–	–	–	–	–	–	–	–	Benzophenone, O-(2,4-dinitrophenyl)oxime	17188-67-9	LI08003
321	200	241	94	201	257	79	93	363	100	43	41	38	26	25	25	23	8.00	19	25	4	1	1	–	–	–	–	–	–	–	–	1-(1-Adamantyl)-4-isopropylsulphonyl-3-nitrobenzene		NI37853
284	200	228	214	365	363	348	346	363	100	28	21	16	10	10	10	10		19	26	1	1	–	–	–	1	–	–	–	–	–	Spiro[cycloheptane-1,3'(2'H)-isoquinolin]-1'-one, 4'-bromo-2'-butyl-1',4'-dihydro-	120637-34-5	DD01447
292	73	334	348	293	335	248	363	363	100	99	89	59	29	28	22	10		19	33	2	1	–	–	–	–	–	2	–	–	–	Benzamide, N,N-diethyl-2-formyl-6-[bis(trimethylsilyl)methyl]-	85370-81-6	DD01448
58	59	57	55	45	84	56	269	363	100	8	7	6	5	3	2	2	0.00	19	42	–	1	–	–	–	1	–	–	–	–	–	1-Hexadecanaminium, N,N,N-trimethyl-, bromide	57-09-0	NI37854
319	262	234	263	206	363	304	290	363	100	78	70	68	55	50	50	45		20	17	4	3	–	–	–	–	–	–	–	–	–	12-Aminocamptothecin	58546-28-4	NI37855
86	105	77	87	99	183	30	28	363	100	12	10	8	7	6	5	5	0.00	20	26	3	1	–	–	1	–	–	–	–	–	–	Benzeneacetic acid, α-hydroxy-α-phenyl-, 2-(diethylamino)ethyl ester, hydrochloride	57-37-4	NI37856
86	105	77	99	87	183	58	44	363	100	72	60	56	44	39	30	30	0.00	20	26	3	1	–	–	1	–	–	–	–	–	–	Benzeneacetic acid, α-hydroxy-α-phenyl-, 2-(diethylamino)ethyl ester, hydrochloride	57-37-4	NI37857
180	208	178	179	152	335	363	155	363	100	72	67	51	34	20	16	1		21	17	3	1	1	–	–	–	–	–	–	–	–	7H-Dibenz[b,d]azepin-7-one, 5,6-dihydro-5-(4-tolylsulphonyl)-	24127-35-3	NI37858
208	363	180	299	298	284	270		363	100	50	24	7	5	4	1			21	17	3	1	1	–	–	–	–	–	–	–	–	11-Morphanthridinone, 5,6-dihydro-5-(4-tolylsulphonyl)-	23145-74-6	NI37859
332	57	317	71	149	97	83	111	363	100	55	39	35	34	33	27	23	10.00	21	17	5	1	–	–	–	–	–	–	–	–	–	Sanguinarine, dihydro-8-methoxy-, (–)-	72401-54-8	NI37860
119	91	77	244	93	125	246	127	363	100	30	27	14	10	7	5	2	0.00	21	18	1	3	–	–	1	–	–	–	–	–	–	N-(Anilinocarbonyl)-N'-phenyl-(4-chlorophenyl)acetamidine		LI08004
105	135	118	55	51	77	363	83	363	100	29	21	16	16	12	10	6		21	21	1	3	1	–	–	–	–	–	–	–	–	1,3,5-Oxadiazine-4-thione, 2-cyclohexylimino-3,6-diphenyl-	80772-90-3	NI37861
203	218	40	363	94	67	198	68	363	100	46	24	20	20	19	17	16		21	21	3	3	–	–	–	–	–	–	–	–	–	Austamide		NI37862
363	362	169	184	183	170	156	364	363	100	87	55	48	43	39	32	23		21	21	3	3	–	–	–	–	–	–	–	–	–	6H-Pyrido[4'',3'':4',5']azepino[1',2':1,2]pyrido[3,4-b]indole-1-carboxylic acid, 5,8,9,14,14a,15-hexahydro-5-hydroxy-, methyl ester	38940-73-7	MS08560
348	363	333	346	334	349	302	288	363	100	83	79	72	65	60	40	31		21	21	3	3	–	–	–	–	–	–	–	–	–	3-Pyrrolin-2-one, 4-ethyl-3-methyl-5-[5-(4-nitrophenyl-2-methylene)-3,4-dimethyl-5H-pyrrolyl-2-methylene]-, (Z,Z)-		MS08561
203	363	55	57	128	30	41	219	363	100	76	43	42	41	37	36	28		21	33	4	1	–	–	–	–	–	–	–	–	–	2,6-Di-tert-butyl-4-[5-(carboxypentanamido)methyl]phenol		IC04138
167	243	75	244	79	193	168	183	363	100	33	20	19	18	13	13	10	0.00	22	21	2	1	1	–	–	–	–	–	–	–	–	L-Cysteine, S-(triphenylmethyl)-	2799-07-7	NI37863
105	118	43	183	77	91	154	58	363	100	78	73	67	64	43	23	20	0.00	22	21	2	1	1	–	–	–	–	–	–	–	–	L-Cysteine, S-(triphenylmethyl)-	2799-07-7	NI37864
363	348	364	349	305	277	320	332	363	100	54	27	12	8	5	4	2		22	21	4	1	–	–	–	–	–	–	–	–	–	6,7-Dimethoxy-(5,6-dimethoxyinden-1-yl)isoquinoline		LI08005
363	265	348	335	240	266	320	292	363	100	88	86	62	61	57	45	44		22	25	2	3	–	–	–	–	–	–	–	–	–	2,4-Pyrrolidinedione, 3,3-dimethyl-5-[5-[4-(dimethylamino)phenyl-2-methylene]-3,4-dimethyl-5H-pyrrolyl-2-methylene]-, (Z,Z)-		MS08562
314	332	116	86	330	70	296	107	363	100	97	70	42	39	33	31	30	14.63	22	37	3	1	–	–	–	–	–	–	–	–	–	5α-Pregnane-3α,17α-diol-20-one, methyloxime		NI37865
100	87	70	332	314	330	88	239	363	100	55	25	22	19	10	10	8	2.31	22	37	3	1	–	–	–	–	–	–	–	–	–	5β-Pregnane-3α,6α-diol-20-one, methyloxime		NI37866
116	332	314	296	86	330	297	345	363	100	89	86	76	48	33	30	27	17.86	22	37	3	1	–	–	–	–	–	–	–	–	–	5β-Pregnane-3α,17α-diol-20-one, methyloxime		NI37867
363	364	180	77	362	129	272	231	363	100	82	74	41	39	38	34	24		23	17	–	5	–	–	–	–	–	–	–	–	–	1,5-Diphenyl-3-(5-phenyl-1,2,3-triazol-4-yl)pyrazole		NI37868

MASS TO CHARGE RATIOS								M.W.	INTENSITIES								Parent	C	H	O	N	S	F	Cl	Br	I	Si	P	B	X	COMPOUND NAME	CAS Reg No	No
294	363	165	208	279	251	189	220	**363**	100	90	25	20	18	10	9	8		23	25	3	1	–	–	–	–	–	–	–	–	–	Dibenzo[f,h]pyrrolo[1,2-b]isoquinoline, 9,11,12,13,13a,14-hexahydro-3,6,7 trimethoxy-, (S)-	54781-87-2	NI37869
251	207	104	363	307	77	250	208	**363**	100	71	63	52	51	41	37	36		23	29	1	3	–	–	–	–	–	–	–	–	–	4-Imidazolidinone, 1,3-di-tert-butyl-2-imino-5,5-diphenyl-	54508-12-2	NI37870
208	363	252	77	57	251	209	104	**363**	100	42	32	31	27	22	21	19		23	29	1	3	–	–	–	–	–	–	–	–	–	4-Imidazolidinone, 1,3-di-tert-butyl-2-imino-5,5-diphenyl-	54508-12-2	NI37871
250	104	251	182	208	57	264	165	**363**	100	31	22	17	16	16	15	15	**3.10**	23	29	1	3	–	–	–	–	–	–	–	–	–	4-Imidazolidinone, 1,3-di-tert-butyl-2-imino-5,5-diphenyl-	54508-12-2	NI37872
208	292	363	252	77	57	251	250	**363**	100	63	38	36	31	27	22	19		23	29	1	3	–	–	–	–	–	–	–	–	–	4-Imidazolidinone, 1-(tert-butyl)-2-(tert-butylimino)-5,5-diphenyl-		NI37873
251	207	104	182	363	307	208	77	**363**	100	71	63	57	52	51	43	41		23	29	1	3	–	–	–	–	–	–	–	–	–	4-Imidazolidinone, 3-(tert-butyl)-2-(tert-butylimino)-5,5-diphenyl-		NI37874
97	363	206	143	70	207	193	232	**363**	100	99	74	65	44	40	35	34		23	29	1	3	–	–	–	–	–	–	–	–	–	1-Piperazineethanol, 4-[3-(5H-dibenz[b,f]azepin-5-yl)propyl]-	315-72-0	NI37875
363	189	175	159	364	92	160	104	**363**	100	96	44	33	30	30	23	22		24	29	2	1	–	–	–	–	–	–	–	–	–	1H-Dicyclohept[e,g]isoindole-1,3(2H)-dione, 3a,3b,4,5,6,7,8,9,10,11,12,13,13a,13b-tetradecahydro-2-phenyl-, (3aα,3bα,13aα,13bα)-	51624-50-1	NI37876
150	363	362	59	364	266	148	42	**363**	100	86	37	35	24	18	17	14		24	29	2	1	–	–	–	–	–	–	–	–	–	Morphinan, 3-methoxy-17-methyl-4-phenoxy-	47523-05-7	NI37877
363	189	175	159	364	91	160	105	**363**	100	93	45	35	30	30	23	23		24	29	2	1	–	–	–	–	–	–	–	–	–	N-Phenylmaleimide-1,1′-bicycloheptenyl endo adduct		LI08006
55	136	67	149	122	83	79	77	**363**	100	98	81	77	62	60	60	60	**32.00**	24	33	–	3	–	–	–	–	–	–	–	–	–	Prodigiosene, 2,7,12-triethyl-1,3,8,11,13-pentamethyl-		MS08563
141	154	55	363	126	41	43	98	**363**	100	99	69	58	57	43	35	32		24	45	1	1	–	–	–	–	–	–	–	–	–	1H-Azepine, hexahydro-1-(1-oxo-9-octadecenyl)-	56630-43-4	NI37878
113	70	43	126	55	363	41	69	**363**	100	45	43	41	39	27	23	16		24	45	1	1	–	–	–	–	–	–	–	–	–	Pyrrolidine, 1-(1-oxo-11-eicosenyl)-	56650-77-2	NI37879
99	98	55	112	70	41	71	86	**363**	100	80	23	21	20	20	17	15	**4.60**	24	45	1	1	–	–	–	–	–	–	–	–	–	2-Pyrrolidinone, 1-[3,7-dimethyl-9-(2,2,6-trimethylcyclohexyl)nonyl]-	63913-39-3	NI37880
128	191	178	127	189	157	192	165	**363**	100	35	21	18	17	17	13	13	**3.80**	25	17	2	1	–	–	–	–	–	–	–	–	–	9-Phenanthrenemethyl quinaldate	92147-54-1	NI37881
147	364	288	247	106	329	323	253	**364**	100	94	65	30	14	13	12	12		6	10	–	–	–	–	2	–	–	–	–	–	2	Di-μ-chlorodiallyldipalladium		LI08007
41	36	76	67	54	49	147	53	**364**	100	99	96	26	14	12	4	3	**2.40**	6	10	–	–	–	–	2	–	–	–	–	–	2	Di-μ-chlorodiallyldipalladium		MS08564
147	149	146	366	290	368	145	76	**364**	100	88	78	56	53	52	48	47	**43.00**	6	10	–	–	–	–	2	–	–	–	–	–	2	Di-μ-chlorodiallyldipalladium		LI08008
331	125	329	47	79	109	93	333	**364**	100	79	75	42	38	36	30	28	**0.00**	8	8	3	–	1	–	2	1	–	–	1	–	–	Phosphorothioic acid, O-(4-bromo-2,5-dichlorophenyl) O,O-dimethyl ester	2104-96-3	NI37882
331	329	125	333	93	109	332	62	**364**	100	70	52	30	16	13	11	8	**2.02**	8	8	3	–	1	–	2	1	–	–	1	–	–	Phosphorothioic acid, O-(4-bromo-2,5-dichlorophenyl) O,O-dimethyl ester	2104-96-3	NI37883
81	71	41	82	113	67	45	54	**364**	100	35	22	18	17	17	16	10	**0.30**	8	15	1	–	–	–	1	–	–	–	–	–	1	(2-Hydroxycyclooctyl)mercuric chloride		LI08009
84	56	196	112	140				**364**	100	67	31	25	16				**0.00**	9	–	9	–	–	–	–	–	–	–	–	–	2	Tri-μ-carbonyl-hexacarbonyldiiron		MS08565
329	109	331	333	15	79	240	332	**364**	100	95	85	33	29	20	13	12	**0.66**	10	9	4	–	–	–	4	–	–	–	1	–	–	Phosphoric acid, 2-chloro-1-(2,4,5-trichlorophenyl)vinyl dimethyl ester	961-11-5	NI37884
15	109	79	329	331	93	47	31	**364**	100	80	40	26	23	17	14	14	**0.20**	10	9	4	–	–	–	4	–	–	–	1	–	–	Phosphoric acid, 2-chloro-1-(2,4,5-trichlorophenyl)vinyl dimethyl ester, (Z)-	22248-79-9	NI37885
109	329	331	79	333	93	74	47	**364**	100	35	32	20	10	9	5	4	**0.00**	10	9	4	–	–	–	4	–	–	–	1	–	–	Phosphoric acid, 2-chloro-1-(2,4,5-trichlorophenyl)vinyl dimethyl ester, (Z)-	22248-79-9	NI37886
139	162	137	121	135	138	177	164	**364**	100	71	71	56	55	50	49	46	**0.00**	10	12	2	–	–	–	2	–	–	–	–	–	2	Iron, (dichloropropylgermyl)dicarbonyl-π-cyclopentadienyl-	31811-22-0	NI37887
220	305	347	203	93	178	161	101	**364**	100	85	73	19	11	9	8	8	**0.00**	10	16	12	–	–	–	–	–	–	–	–	–	4	Beryllium, pentakis[μ-(acetato-O:O′)]-μ-hydroxy-$μ^4$-oxotetra-	38999-30-3	NI37888
169	97	171	45	196	198	99	27	**364**	100	74	68	64	57	55	43	33	**2.83**	11	9	3	–	–	–	5	–	–	–	–	–	–	Propanoic acid, 2,2-dichloro-, 2-(2,4,5-trichlorophenoxy)ethyl ester	136-25-4	NI37889
285	156	158	366	287	103	206	368	**364**	100	69	65	37	34	34	23	17	**17.00**	12	6	–	4	–	–	–	2	–	–	–	–	–	Pyrido[1″,2″:1′,2′]imidazo[4′,5′:4,5]imidazo[1,2-a]pyridine, 1,10-dibromo-		MS08566
69	55	41	140	142	56	43	83	**364**	100	98	75	49	37	31	27	25	**0.00**	12	20	2	2	–	–	4	–	–	–	–	–	–	Acetamide, N,N′-1,8-octanediylbis[2,2-dichloro-	1477-57-2	NI37890
197	195	364	28	196	117	99	155	**364**	100	60	48	48	30	22	22	21		13	2	1	–	–	10	–	–	–	–	–	–	–	Benzenemethanol, 2,3,4,5,6-pentafluoro-α-(pentafluorophenyl)-	1766-76-3	LI08010
197	195	364	28	196	117	99	167	**364**	100	61	48	48	31	22	22	20		13	2	1	–	–	10	–	–	–	–	–	–	–	Benzenemethanol, 2,3,4,5,6-pentafluoro-α-(pentafluorophenyl)-	1766-76-3	NI37891
181	155	31	69	161	93	117	182	**364**	100	17	17	15	12	11	10	9	**0.20**	13	2	1	–	–	10	–	–	–	–	–	–	–	Benzene, pentafluoro[(pentafluorophenoxy)methyl]-	38795-53-8	NI37892
364	366	185	187	89	79	62	102	**364**	100	99	86	64	57	55	48	43		13	9	4	4	–	–	–	1	–	–	–	–	–	Benzaldehyde, 2-bromo-, (2,4-dinitrophenyl)hydrazone	34158-85-5	LI08011
366	364	185	187	89	79	62	75	**364**	100	99	84	63	55	54	46	42		13	9	4	4	–	–	–	1	–	–	–	–	–	Benzaldehyde, 2-bromo-, (2,4-dinitrophenyl)hydrazone	34158-85-5	NI37893
364	366	185	187	89	63	164	180	**364**	100	99	76	64	31	30	26	25		13	9	4	4	–	–	–	1	–	–	–	–	–	Benzaldehyde, 4-bromo-, (2,4-dinitrophenyl)hydrazone	2087-20-9	NI37894
69	364	96	137	177	157	95	97	**364**	100	93	81	78	50	43	30	25		13	11	7	2	–	3	–	–	–	–	–	–	–	Acetic acid, trifluoro-, [3-(acetyloxy)-2,3,3a,9a-tetrahydro-6-oxo-6H-furo[2′,3′:4,5]oxazolo[3,2-a]pyrimidin-2-yl]methyl ester, [2R-(2α,3β,3aβ,9aβ)]-	40732-62-5	NI37895
193	364	321	69	179	113	97	57	**364**	100	66	51	51	44	44	43	42		13	11	7	2	–	3	–	–	–	–	–	–	–	Acetic acid, trifluoro-, 2-[(acetyloxy)methyl]-2,3,3a,9a-tetrahydro-6-oxo-6H-furo[2′,3′:4,5]oxazolo[3,2-a]pyrimidin-3-yl ester, [2R-(2α,3β,3aβ,9aβ)]-	40732-63-6	NI37896
69	364	137	177	262	193	191	113	**364**	100	93	78	50	18	12	12	12		13	11	7	2	–	3	–	–	–	–	–	–	–	3′-O-Acetyl-5′-O-trifluoroacetyl-2,2′-anhydrouridine		LI08012
69	193	364	321	179	113	137	177	**364**	100	47	31	24	20	20	18	17		13	11	7	2	–	3	–	–	–	–	–	–	–	3′-O-Trifluoroacetyl-5′-O-acetyl-2,2′-anhydrouridine		LI08013
364	211	210	208	182	157	168	207	**364**	100	52	45	33	27	21	18	17		13	16	2	–	–	3	–	–	–	–	–	–	1	Rhodium, [(1,2,5,6-η)-1,5-cyclooctadiene](1,1,1-trifluoro-2,4-pentanedionato-O,O′)-	32610-46-1	NI37897

MASS TO CHARGE RATIOS								M.W.	INTENSITIES								Parent	C	H	O	N	S	F	Cl	Br	I	Si	P	B	X	COMPOUND NAME	CAS Reg No	No
211	157	364	103	43	210	155	182	**364**	100	90	61	35	34	27	23	16		13	16	2	–	–	3	–	–	–	–	–	–	1	Rhodium, (η^4-1,2-divinylcyclobutane)(1,1,1-trifluoro-2,4-pentanedionato-O,O')-	74810-94-9	NI37898
264	249	364	265	263	345	242	223	**364**	100	24	14	14	13	4	3	3		14	6	–	–	–	10	–	–	–	–	–	–	–	Decafluoro-4,5-dimethyltricyclo[6.2.2.0^{2,7}]dodeca-2,4,6,9-tetraene		NI37899
61	41	76	59	284	286	56	39	**364**	100	84	66	63	54	47	36	36	**23.91**	14	32	–	–	–	–	–	–	–	–	2	–	1	Ruthenium, bis[(1,2,3-η)-2-methyl-2-propenyl]bis(trimethylphosphine)-	74792-85-1	NI37900
73	117	75	147	45	102	74	130	**364**	100	20	18	15	10	9	8	7	**0.87**	14	32	5	–	–	–	–	–	–	3	–	–	–	Arabinonic acid, 2,3,5-tris-O-(trimethylsilyl)-, γ-lactone	74742-31-7	NI37901
73	217	147	75	45	220	218	74	**364**	100	42	24	16	15	11	11	10	**2.20**	14	32	5	–	–	–	–	–	–	3	–	–	–	D-Arabinonic acid, 2,3,4-tris-O-(trimethylsilyl)-, lactone	32384-58-0	NI37902
73	117	147	231	217	189	75	45	**364**	100	24	16	12	12	12	10	9	**7.31**	14	32	5	–	–	–	–	–	–	3	–	–	–	D-Arabinonic acid, 2,3,5-tris-O-(trimethylsilyl)-, γ-lactone	32384-55-7	NI37903
73	129	147	247	75	45	203	85	**364**	100	69	40	35	27	17	13	12	**0.00**	14	32	5	–	–	–	–	–	–	3	–	–	–	Pentanedioic acid, 2-[(trimethylsilyl)oxy]-, bis(trimethylsilyl) ester	55530-62-6	NI37904
147	247	73	320	115	349	259	133	**364**	100	68	48	32	24	23	21	20	**0.00**	14	32	5	–	–	–	–	–	–	3	–	–	–	Propanedioic acid, ethyl[(trimethylsilyl)oxy]-, bis(trimethylsilyl) ester		MS08567
73	147	117	217	75	103	74	45	**364**	100	24	19	11	10	9	9	8	**5.20**	14	32	5	–	–	–	–	–	–	3	–	–	–	D-Ribonic acid, 2,3,5-tris-O-(trimethylsilyl)-, γ-lactone	10589-34-1	NI37905
73	117	147	45	75	74	189	103	**364**	100	16	11	10	8	8	6	6	**1.37**	14	32	5	–	–	–	–	–	–	3	–	–	–	Ribonic acid, 2,3,5-tris-O-(trimethylsilyl)-, γ-lactone	74742-38-4	NI37906
73	147	117	217	75	103	74	45	**364**	100	24	19	11	10	9	9	8	**5.20**	14	32	5	–	–	–	–	–	–	3	–	–	–	Ribonic acid, 2,3,5-tris-O-(trimethylsilyl)-, γ-lactone	74742-38-4	MS08568
73	217	147	75	220	218	45	74	**364**	100	43	25	14	12	11	11	9	**2.40**	14	32	5	–	–	–	–	–	–	3	–	–	–	D-Xylonic acid, 2,3,4-tris-O-(trimethylsilyl)-, δ-lactone	32384-59-1	NI37907
73	117	217	147	244	75	103	45	**364**	100	17	14	14	11	10	9	9	**2.70**	14	32	5	–	–	–	–	–	–	3	–	–	–	D-Xylonic acid, 2,3,5-tris-O-(trimethylsilyl)-, γ-lactone	10589-36-3	NI37908
73	147	117	217	75	45	74	102	**364**	100	23	18	16	11	10	9	7	**5.91**	14	32	5	–	–	–	–	–	–	3	–	–	–	D-Xylonic acid, 2,3,5-tris-O-(trimethylsilyl)-, γ-lactone	10589-36-3	NI37909
73	117	217	147	244	75	189	103	**364**	100	17	14	14	10	10	9	9	**2.70**	14	32	5	–	–	–	–	–	–	3	–	–	–	Xylonic acid, 2,3,5-tris-O-(trimethylsilyl)-, γ-lactone	10589-36-3	MS08569
51	77	364	237	197	88	76	50	**364**	100	67	60	55	49	44	44	44		15	9	3	–	–	–	–	–	1	–	–	–	–	2H-1-Benzopyran-2-one, 3-iodo-4-phenoxy-	68903-74-2	NI37910
351	349	187	152	353	189	75	63	**364**	100	81	55	47	46	35	35	32	**16.80**	15	12	2	–	–	–	4	–	–	–	–	–	–	Phenol, 4,4'-(1-methylethylidene)bis[2,6-dichloro-	79-95-8	NI37911
351	349	353	187	366	189	364	352	**364**	100	82	53	33	30	22	22	18		15	12	2	–	–	–	4	–	–	–	–	–	–	Phenol, 4,4'-(1-methylethylidene)bis[2,6-dichloro-	79-95-8	NI37912
220	228	284	283	80	82	241	285	**364**	100	85	73	47	35	34	30	28	**0.00**	15	14	–	4	–	–	1	1	–	–	–	–	–	1H-1,3,4-Benzotriazepine, 7-chloro-2,3-dihydro-2-imino-1-methyl-5-phenyl-, hydrobromide		MS08570
284	283	206	285	286	80	82	205	**364**	100	69	53	39	34	27	26	17	**0.00**	15	14	–	4	–	–	1	1	–	–	–	–	–	3H-1,3,4-Benzotriazepine, 7-chloro-1,2-dihydro-2-imino-3-methyl-5-phenyl-, hydrobromide		MS08571
69	84	96	281	239	83	97	364	**364**	100	90	42	40	15	15	13	12		15	19	2	–	–	7	–	–	–	–	–	–	–	(Heptafluorobutyryl)camphorylmethane		NI37913
73	147	103	247	75	233	143	45	**364**	100	26	26	24	21	17	13	10	**0.00**	15	36	4	–	–	–	–	–	–	3	–	–	–	Pentanoic acid, 3,5-bis(trimethylsiloxy)-3-methyl-, trimethylsilyl ester		NI37914
73	103	147	247	75	233	143	74	**364**	100	25	22	17	16	13	12	11	**0.00**	15	36	4	–	–	–	–	–	–	3	–	–	–	Pentanoic acid, 3,5-bis(trimethylsiloxy)-3-methyl-, trimethylsilyl ester		NI37915
364	362	202	334	332	269			**364**	100	94	15	10	9	5				16	12	–	–	–	–	–	–	–	–	–	–	2	3,8-Dimethylbenzoselenopheno[2,3-b]benzoselenophene		LI08014
191	304	263	158	192	147	121	115	**364**	100	58	58	24	22	22	18	18	**0.00**	16	12	4	–	3	–	–	–	–	–	–	–	–	1-Allylthio-3-oxo-4-phenyl-2,7-dithiabicyclo[2.2.1]heptan-5-endo,6-endo-dicarboxylic acid, anhydride		NI37916
150	364	182	109	106	181	77	151	**364**	100	96	56	46	45	44	42	40		16	16	–	2	4	–	–	–	–	–	–	–	–	Bis(methylphenylthiocarbamoyl) disulphide	10591-84-1	NI37917
121	170	69	164	308	235	234	122	**364**	100	43	40	38	36	36	31	28	**0.40**	16	20	6	–	–	–	–	–	–	–	–	–	1	Iron, [2,2-bis(ethoxycarbonyl)propyl]cyclopentadienyldicarbonyl-		NI37918
70	28	166	41	170	114	263	29	**364**	100	86	61	23	19	17	15	14	**0.00**	16	23	4	2	–	3	–	–	–	–	–	–	–	Proline, N-(trifluoroacetyl)-L-prolyl-, butyl ester		MS08572
84	72	82	81	44	41	43	42	**364**	100	65	53	53	35	33	28	23	**1.00**	16	24	4	6	–	–	–	–	–	–	–	–	–	L-Norvalinamide, 5-oxo-L-prolyl-L-histidyl-	37666-91-4	NI37919
84	82	81	120	121	110	56	41	**364**	100	56	42	36	29	20	18	18	**0.40**	16	24	4	6	–	–	–	–	–	–	–	–	–	L-Valinamide, 5-oxo-L-prolyl-L-histidyl-	27058-71-5	NI37920
56	100	114	146	57	88	70	58	**364**	100	90	70	57	57	53	47	43	**3.50**	16	32	5	2	1	–	–	–	–	–	–	–	–	12,16-Dihydroxy-4,7,20-trioxa-14-thia-1,10-diazabicyclo[8.7.5]docosane		MS08573
307	251	177	137	234	235	195	291	**364**	100	80	58	40	25	15	14	10	**0.00**	16	36	1	–	–	–	–	–	–	–	–	–	1	2-Butanol, 4-(tributylstannyl)-	30988-55-7	NI37921
307	177	137	251	291	235	195	193	**364**	100	24	19	17	5	5	5	5	**0.00**	16	36	1	–	–	–	–	–	–	–	–	–	1	1-Propanol, 2-methyl-3-(tributylstannyl)-	31123-79-2	NI37922
364	121	349	365	56	162	39	122	**364**	100	53	27	18	11	9	6	5		17	12	1	–	–	4	–	–	–	–	–	–	1	Ferrocene, (2,3,5,6-tetrafluoro-4-methoxyphenyl)-	55334-25-3	NI37923
157	364	158	127	114				**364**	100	28	9	7	5					17	14	2	–	–	6	–	–	–	–	–	–	–	1,3-Bis(4-methoxyphenyl)hexafluoropropane		LI08015
103	77	246	130	105	131	51	104	**364**	100	94	68	63	56	52	40	33	**0.00**	17	17	5	2	–	–	1	–	–	–	–	–	–	Isoquinolinium, 2-(2-hydroxyimino-2-phenylethyl)-3,4-dihydro-, perchlorate, (E)-		NI37924
145	103	246	264	130	247	131	117	**364**	100	54	50	46	46	42	38	38	**0.00**	17	17	5	2	–	–	1	–	–	–	–	–	–	Isoquinolinium, 2-(2-hydroxyimino-2-phenylethyl)-3,4-dihydro-, perchlorate, (Z)-		NI37925
111	110	95	83	206	192	122	299	**364**	100	98	66	66	39	31	30	26	**0.00**	17	18	3	4	–	2	–	–	–	–	–	–	–	D-erythro-Pentos-2-ulose, bis[(2-fluorophenyl)hydrazone]	51306-41-3	NI37926
108	137	109	119	283	253	282	284	**364**	100	70	58	51	45	42	35	26	**13.00**	17	20	2	2	–	–	–	–	–	–	–	–	1	N,N'-Bis(4-ethoxyphenyl)selenourea		MS08574
211	366	364	270	291	305	225	293	**364**	100	70	70	69	30	27	27	26		17	21	2	2	–	–	–	1	–	–	–	–	–	[4-Methyl-5-bromo-3-(2-carboxyethyl)-2-pyrrolyl](3,5-dimethyl-4-ethyl-2-pyrrolylidene)methane		LI08016
212	213	291	293	366	364	223	305	**364**	100	22	21	19	18	18	13	12		17	21	2	2	–	–	–	1	–	–	–	–	–	[4-Methyl-5-bromo-3-(ethoxycarbonyl)-2-pyrrolyl](3,5-dimethyl-4-ethyl-2-pyrrolylidene)methane		LI08017
43	159	244	231	158	202	201	160	**364**	100	40	20	9	9	8	6	6	**0.52**	17	21	8	–	–	–	–	–	–	–	–	1	–	DL-Xylitol, cyclic 2,3-(phenylboronate) 1,4,5-triacetate	74793-44-5	NI37927
254	182	208	28	336	110	209	138	**364**	100	88	46	32	30	29	24	24	**5.54**	17	24	5	–	–	–	–	–	–	–	–	–	1	Iron, carbonyl[(2,3,4,5-η)-diethyl 2,4-hexadienedioate][(1,2,3,4-η)-2,3-dimethyl-1,3-butadiene]-	62337-83-1	NI37928
364	262	264	260	265	263	103	261	**364**	100	68	60	38	31	26	25	24		17	25	2	–	–	–	–	–	–	–	–	–	1	Rhodium, bis[5,6-bis(η^2-vinyl)cyclooctene](2,4-pentanedionato-O,O')-	74810-99-4	NI37929
364	262	260	264	234	258	41	79	**364**	100	51	38	31	28	24	23	22		17	25	2	–	–	–	–	–	–	–	–	–	1	Rhodium, [1,2-bis(η^2-vinyl)-4-vinylcyclohexane](2,4-pentanedionato-O,O')	74825-25-5	NI37930
318	288	58	57	43	73	69	41	**364**	100	91	60	60	56	38	35	33	**0.00**	18	8	7	2	–	–	–	–	–	–	–	–	–	3,5-Dinitro-1-hydroxynaphthacenequinone		IC04139
364	366	142	107	368	253	183	144	**364**	100	97	81	34	33	28	27	27		18	12	–	–	–	–	3	–	–	–	1	–	–	Tri(4-chlorophenyl)phosphine		IC04140

MASS TO CHARGE RATIOS								M.W.	INTENSITIES								Parent	C	H	O	N	S	F	Cl	Br	I	Si	P	B	X	COMPOUND NAME	CAS Reg No	No
333	54	56	364	158	334	324	105	**364**	100	57	32	23	22	21	20	19		18	16	3	6	–	–	–	–	–	–	–	–	–	N-Cyanoethyl-N-hydroxyethyl-4-(2-cyano-4-nitrophenylazo)aniline		IC04141
184	65	39	121	69	41	94	55	**364**	100	81	68	44	39	38	35	32	**8.10**	18	17	6	–	–	–	1	–	–	–	–	–	–	6H-Oxireno[e][2]benzoxacyclotetradecin-6,12(7H)-dione, 8-chloro-1a,14,15,15a-tetrahydro-9,11-dihydroxy-14-methyl-	12772-57-5	NI37931
118	175	43	91	364	366	176	119	**364**	100	44	38	37	23	16	16	16		18	18	2	2	–	–	2	–	–	–	–	–	–	2′,5′-Dichloro-3-methoxy-2-(4-methylanilino)crotonanilide		MS08575
161	285	41	160	257	162	146	43	**364**	100	19	19	17	16	16	12	11	**0.00**	18	23	2	1	–	–	–	1	–	–	–	–	–	8-Azaestra-1,3,5(10)-trien-17-one, 3-methoxy-, hydrobromide	57983-81-0	NI37932
284	161	228	178	188	160	162	41	**364**	100	63	35	29	28	24	20	18	**0.00**	18	23	2	1	–	–	–	1	–	–	–	–	–	8-Azaestra-1,3,5(10)-trien-17-one, 3-methoxy-, hydrobromide, (9β,14β)-	58072-50-7	NI37933
170	181	172	314	358	360	316	183	**364**	100	98	97	76	40	16	8	7	**0.00**	18	24	4	2	1	–	–	–	–	–	–	–	–	Dansylleucine		LI08018
69	41	39	29	43	27	70	18	**364**	100	46	17	13	10	8	7	5	**0.00**	18	24	6	2	–	–	–	–	–	–	–	–	–	2-Butenoic acid, 2(or 4)-isooctyl-4,6(or 2,6)-dinitrophenyl ester	39300-45-3	NI37934
57	252	239	137	136	97	83	95	**364**	100	96	82	75	74	71	58	54	**0.50**	18	36	3	–	2	–	–	–	–	–	–	–	–	S-(2-Oxoheptadecyl) methanesulphonothioate	76078-83-6	NI37935
259	168	364	77	260	103	233	115	**364**	100	74	58	34	29	28	27	26		19	17	1	–	–	–	–	–	–	–	–	–	1	Rhodium, (2-benzoylnorbornadiene)(5-cyclopentadienyl)-		NI37936
42	56	165	58	235	88	129	74	**364**	100	92	91	86	59	46	42	38	**7.00**	19	22	1	2	–	–	2	–	–	–	–	–	–	1-Piperazineethanol, [4-[α-(2-chlorophenyl)-3-chlorobenzyl]-		MS08576
245	149	364	305	233	193	263	261	**364**	100	69	52	48	39	36	25	22		19	24	7	–	–	–	–	–	–	–	–	–	–	Benzofuran, 2-acetoxy-2,3-dihydro-5-(methoxycarbonyl)-7-[1-(methoxycarbonyl)-1-isopropyl]-3,3-dimethyl-		MS08577
91	95	151	198	226	244	322	304	**364**	100	51	45	29	27	22	13	9	**0.00**	19	24	7	–	–	–	–	–	–	–	–	–	–	Dentatin b, 15-acetoxy-		MS08578
57	75	91	55	69	79	77	105	**364**	100	75	57	57	53	37	36	34	**7.00**	19	24	7	–	–	–	–	–	–	–	–	–	–	Guaia-4(15),10(14),11(13)-trien-6,12-olide, 3β-hydroxy-, 8-(α,β-dihydroxybutyrate)		NI37937
366	364	240	307	305	238	146	251	**364**	100	95	44	32	32	25	24	22		19	25	2	–	–	–	–	1	–	–	–	–	–	Estra-1,3,5(10)-triene-3,17-diol, 2-bromo-1-methyl-, (17β)-	16373-36-7	NI37938
364	319	320	88	318	290	348	176	**364**	100	36	35	30	28	28	27	27		20	12	7	–	–	–	–	–	–	–	–	–	–	Furo[3′,4′:6,7]naphtho[1,2-d]-1,3-dioxol-7(9H)-one, 10-(7-hydroxy-1,3-benzodioxol-5-yl)-	75340-41-9	NI37939
364	321	349	365	293	348	152	43	**364**	100	38	30	25	23	15	10	10		20	12	7	–	–	–	–	–	–	–	–	–	–	5,12-Naphthacenedione, 8-acetyl-1,6,10,11-tetrahydroxy-	20982-42-7	MS08579
364	366	365	97	77	152	151	367	**364**	100	37	24	22	9	8	8	7		20	13	1	2	1	–	1	–	–	–	–	–	–	1H-1,4-Benzodiazepin-2-one, 7-chloro-2,3-dihydro-5-phenyl-3-[(thienyl-2)methylene]-	55056-39-8	NI37940
320	234	264	263	235	305	292	364	**364**	100	92	85	71	71	56	47	41		20	16	5	2	–	–	–	–	–	–	–	–	–	Camptothecin, 10-hydroxy-	67656-30-8	NI37941
320	364	264	263	235	305	335	234	**364**	100	84	69	57	50	47	25	24		20	16	5	2	–	–	–	–	–	–	–	–	–	Camptothecin, 10-hydroxy-	67656-30-8	NI37942
264	305	263	235	291	292	320	364	**364**	100	99	99	99	70	68	58	51		20	16	5	2	–	–	–	–	–	–	–	–	–	Camptothecin, 11-hydroxy-	68426-53-9	NI37943
235	305	264	263	291	292	320	364	**364**	100	88	81	81	66	64	55	48		20	16	5	2	–	–	–	–	–	–	–	–	–	Camptothecin, 11-hydroxy-	68426-53-9	NI37944
364	320	264	263	235	305	291	292	**364**	100	99	78	68	63	52	43	42		20	16	5	2	–	–	–	–	–	–	–	–	–	Camptothecin, 12-hydroxy-	58546-29-5	NI37945
220	364	180	258	257	221	291	152	**364**	100	39	31	26	26	26	25	20		20	16	5	2	–	–	–	–	–	–	–	–	–	4′,5′-Bis(ethoxycarbonyl)spiroacenaphthenone-2,3′(3′H)-pyrazole		LI08019
322	364	303	93	313	77	92	292	**364**	100	41	36	23	22	22	18	14		20	20	3	4	–	–	–	–	–	–	–	–	–	Acetic acid, [[5-(1,2-dihydroxyethyl)-1-phenyl-1H-pyrazol-3-yl]methylene]phenylhydrazide	55649-78-0	NI37946
44	292	364	293	365	290	305	263	**364**	100	61	27	13	6	4	3	3		20	20	3	4	–	–	–	–	–	–	–	–	–	6-(Acetylaminoethylamino)-1,2,3,4-tetrahydro[1,2]pyrazinoanthraquinone		IC04142
308	307	364	89	117	57	62	38	**364**	100	18	17	15	14	13	6	3		20	20	3	4	–	–	–	–	–	–	–	–	–	Phthalimide, 4-(5-butoxy-6-methyl-2H-benzotriazol-2-yl)-N-methyl-		NI37947
138	55	28	41	42	364	227	110	**364**	100	63	44	33	28	26	26	22		20	23	1	2	–	3	–	–	–	–	–	–	–	Cinchonidine, dihydro-7′-(trifluoromethyl)-	38199-24-5	NI37948
61	289	135	364	41	187	91	44	**364**	100	30	29	27	25	24	22	16		20	28	2	–	2	–	–	–	–	–	–	–	–	2,6-Dimethyl-6-thiomethoxymethylcyclohexa-2,4-dien-1-one dimer		MS08580
207	239	221	333	208	56	55	43	**364**	100	30	27	13	13	12	12	12	**2.00**	20	28	6	–	–	–	–	–	–	–	–	–	–	1,2,4-Benzenetricarboxylic acid, 1,2-dimethyl nonyl ester	33975-28-9	NI37949
207	239	221	333	208	56	55	41	**364**	100	29	26	13	12	12	12	11	**1.00**	20	28	6	–	–	–	–	–	–	–	–	–	–	1,2,4-Benzenetricarboxylic acid, 1,2-dimethyl nonyl ester	33975-28-9	NI37950
346	209	41	43	331	313	364	328	**364**	100	40	38	37	18	12	5	2		20	28	6	–	–	–	–	–	–	–	–	–	–	Castelanolide		LI08020
207	41	43	53	179	69	79	55	**364**	100	94	88	76	65	65	55	50	**0.00**	20	28	6	–	–	–	–	–	–	–	–	–	–	15,17-Dihydro-15-hydroxycrotophorbolone		NI37951
113	105	29	91	43	55	138	318	**364**	100	65	57	54	53	52	51	49	**20.00**	20	28	6	–	–	–	–	–	–	–	–	–	–	12-O-Formyl-3β,8β,12β,14β-tetrahydroxy-androst-5-en-20-one		LI08021
43	112	275	55	41	28	91	29	**364**	100	32	31	28	27	24	23	21	**1.00**	20	28	6	–	–	–	–	–	–	–	–	–	–	Gibberellin A_2 methyl ester		MS08581
43	41	77	91	121	93	53	55	**364**	100	82	72	71	54	54	48	44	**0.22**	20	28	6	–	–	–	–	–	–	–	–	–	–	16-Hydroxyingenol		NI37952
57	83	85	262	163	320			**364**	100	78	74	41	18	3			**0.00**	20	28	6	–	–	–	–	–	–	–	–	–	–	Ludovicin A, 9α-(isovaleryloxy)-		MS08582
43	41	45	57	91	105	107	79	**364**	100	81	53	34	30	21	18	18	**1.00**	20	28	6	–	–	–	–	–	–	–	–	–	–	Schkuhridin A		MS08583
85	135	237	153	136	147	187	164	**364**	100	55	45	31	28	26	21	16	**1.67**	20	28	6	–	–	–	–	–	–	–	–	–	–	Tetrahydrosubacaulin		NI37953
135	207	206	163	234	136	79	72	**364**	100	50	30	19	13	13	11	10	**3.00**	20	32	4	2	–	–	–	–	–	–	–	–	–	L-Valine, N-[N-(1-adamantylcarbonyl)-L-alanyl]-, methyl ester	28417-03-0	NI37954
209	210	364	181	180	167	91	152	**364**	100	17	15	15	11	10	7	6		21	20	2	2	1	–	–	–	–	–	–	–	–	3-Carbazolamine, N,9-dimethyl-N-tosyl-		NI37955
209	210	364	193	78	194	166	365	**364**	100	19	16	8	8	5	5	4		21	20	2	2	1	–	–	–	–	–	–	–	–	3-Carbazolamine, N,N-dimethyl-9-tosyl-	94127-17-0	NI37956
259	205	364	260	206	243	305	219	**364**	100	39	25	20	14	11	10	8		21	20	4	2	–	–	–	–	–	–	–	–	–	Talbotine, 3,4,5,6-tetradehydro-	30809-25-7	NI37957
131	364	304	91	105	95	81	79	**364**	100	46	45	31	30	29	28	28		21	32	1	–	2	–	–	–	–	–	–	–	–	Androstane-2,11-dione, cyclic 2-(1,2-ethanediyl mercaptole), (5α)-	54498-87-2	NI37958
97	346	364	263	156	348	41	55	**364**	100	99	85	82	52	50	45	43		21	32	5	–	–	–	–	–	–	–	–	–	–	Androstane-17-carboxylic acid, 3,4-dihydroxy-11-oxo-, methyl ester, (3β,4β,5α,17β)-	55282-39-8	NI37959
43	286	55	268	81	107	93	67	**364**	100	63	51	50	35	28	27	25	**0.00**	21	32	5	–	–	–	–	–	–	–	–	–	–	Androstan-17-one, 3-(acetyloxy)-5,14-dihydroxy-, (3β,5α,14β)-	41853-32-1	NI37960
55	300	213	285	254	239	259	228	**364**	100	38	22	15	13	11	6	4	**0.10**	21	32	5	–	–	–	–	–	–	–	–	–	–	5α,3α,17β-Diformoxyandrostan-11β-ol		NI37961
55	300	213	228	254	239	274	259	**364**	100	30	23	14	12	10	8	8	**0.10**	21	32	5	–	–	–	–	–	–	–	–	–	–	5β-3α,17β-Diformoxyandrostan-11β-ol		NI37962
303	109	43	91	83	145	55	346	**364**	100	42	37	32	31	23	21	18	**13.00**	21	32	5	–	–	–	–	–	–	–	–	–	–	Dihydro-14-seco-hirundigenin		LI08022

MASS TO CHARGE RATIOS								M.W.	INTENSITIES								Parent	C	H	O	N	S	F	Cl	Br	I	Si	P	B	X	COMPOUND NAME	CAS Reg No	No
121	223	113	126	123	237	163	127	**364**	100	95	83	80	69	65	61	47	**7.88**	21	32	5	–	–	–	–	–	–	–	–	–	–	1H-3,10a-Epoxyphenanthrene-8-carboxylic acid, dodecahydro-2-hydroxy-4a,8-dimethyl-2-isopropyl-1-oxo-, methyl ester, [2S-(2α,3α,4aα,4bβ,8α,8aα,10aα)]-	34226-18-1	NI37963
42	69	197	56	54	85	109	81	**364**	100	70	57	39	26	17	15	14	**2.06**	21	32	5	–	–	–	–	–	–	–	–	–	–	ρ-Isohumulone A_1		MS08584
197	69	41	198	267	210	85	56	**364**	100	31	30	15	11	11	11	11	**0.94**	21	32	5	–	–	–	–	–	–	–	–	–	–	ρ-Isohumulone A_2		MS08585
197	69	42	266	198	57	210	85	**364**	100	27	26	19	14	11	10	8	**0.00**	21	32	5	–	–	–	–	–	–	–	–	–	–	ρ-Isohumulone B_1		MS08586
197	69	41	198	266	346	211	57	**364**	100	31	28	15	13	10	10	10	**0.63**	21	32	5	–	–	–	–	–	–	–	–	–	–	ρ-Isohumulone B_2		MS08587
321	135	69	91	81	107	93	71	**364**	100	45	44	43	40	39	36	36	**24.72**	21	32	5	–	–	–	–	–	–	–	–	–	–	Methyl-12-hydroxy-8α,9α,13α,14α-diepoxyabietanoate		NI37964
304	322	191	305	273	203	221	189	**364**	100	30	28	24	24	19	18	15	**3.00**	21	32	5	–	–	–	–	–	–	–	–	–	–	Methyl prostaglandin A_1		LI08023
303	276	243	247	275	233	287	215	**364**	100	57	44	39	37	34	32	31	**2.49**	21	32	5	–	–	–	–	–	–	–	–	–	–	Phenanthro[1,2-b]oxirene-4-carboxylic acid, tetradecahydro-1b-hydroxy-4,7a-dimethyl-9a-isopropyl-9-oxo-, methyl ester, [1aR-(1aα,1bβ,3aβ,4β,7aα,7bβ,9aα)]-	34217-21-5	NI37965
226	120	138	147	109	107	227	105	**364**	100	73	30	29	28	28	25	24	**10.74**	21	32	5	–	–	–	–	–	–	–	–	–	–	Pregn-5-en-20-one, 3,8,12,14-tetrahydroxy-, (3β,12β,14β)-	7102-32-1	NI37966
43	226	120	28	105	97	79	41	**364**	100	96	95	52	48	48	48	48	**20.42**	21	32	5	–	–	–	–	–	–	–	–	–	–	Pregn-5-en-20-one, 3,8,12,14-tetrahydroxy-, (3β,12β,14β)-	7102-32-1	NI37967
346	180	147	123	120	313	105	328	**364**	100	52	49	46	45	42	42	40	**0.60**	21	32	5	–	–	–	–	–	–	–	–	–	–	Pregn-5-en-20-one, 3,8,12,14-tetrahydroxy-, (3β,12β,14β,17α)-	6869-50-7	NI37968
43	346	120	147	97	41	123	105	**364**	100	98	90	76	76	72	64	64	**1.67**	21	32	5	–	–	–	–	–	–	–	–	–	–	Pregn-5-en-20-one, 3,8,12,14-tetrahydroxy-, (3β,12β,14β,17α)-	6869-50-7	NI37969
73	59	89	104	117	131	143	169	**364**	100	99	57	37	8	7	6	4	**0.00**	21	36	3	–	–	–	–	–	–	1	–	–	–	4-Octene, 8-[(2-methoxyethoxy)methoxy]-1-phenyl-4-(trimethylsilyl)-, (Z)-	123186-20-9	DD01449
70	267	97	84	83	238	198	111	**364**	100	80	80	60	60	53	47	38	**15.00**	21	40	1	4	–	–	–	–	–	–	–	–	–	Propanenitrile, 3-[[3-[[3-(2-oxoazacyclotridec-1-yl)propyl]amino]propyl]amino]-	67171-91-9	NI37970
45	323	277	165	291	324	253	151	**364**	100	64	45	14	13	12	6	6	**0.00**	22	20	5	–	–	–	–	–	–	–	–	–	–	Anthraquinone, 1-hydroxy-2-(1,5-hexadien-3-yl)-3-(methoxymethoxy)-	120022-34-6	DD01450
45	284	293	139	79	254	323	167	**364**	100	50	21	16	16	15	8	6	**0.00**	22	20	5	–	–	–	–	–	–	–	–	–	–	Anthraquinone, 3-(methoxymethoxy)-1-(2,5-hexadien-1-yloxy)-	120022-33-5	DD01451
349	364	350	174	149	217	222	320	**364**	100	35	24	17	16	12	9	8		22	20	5	–	–	–	–	–	–	–	–	–	–	Derrone, methyl-5-O-methyl-		NI37971
364	170	183	181	336	197	225	226	**364**	100	98	62	58	53	48	36	31		22	24	3	2	–	–	–	–	–	–	–	–	–	Alstophyllan-19-one, 4-acetyl-4-demethyl-	67497-76-1	NI37972
215	364	214	123	186	200	160	335	**364**	100	85	51	37	31	21	19	14		22	24	3	2	–	–	–	–	–	–	–	–	–	5-Methoxy-3-spiro[cyclohexane-3H-indole-2-(4-methoxycarboxanilide)]		MS08588
215	364	214	200	149	186	160	335	**364**	100	50	46	36	25	18	16	14		22	24	3	2	–	–	–	–	–	–	–	–	–	6-Methoxy-3-spiro[cyclohexane-3H-indole-2-(3-methoxycarboxanilide)]		MS08589
305	364	130	290	57	126	168	143	**364**	100	84	30	26	25	22	21	21		22	24	3	2	–	–	–	–	–	–	–	–	–	16,19-Secostrychnidine-10,16-dione, 19-methyl-	5525-31-5	NI37973
94	145	364	132	304	289	144	131	**364**	100	63	48	46	32	22	17	15		22	30	2	–	–	2	–	–	–	–	–	–	–	Retinol, 4,4-difluoro-, acetate, (all-trans)-	90660-22-3	NI37974
121	41	81	95	122	109	93	79	**364**	100	23	19	18	17	16	16	15	**3.00**	22	36	4	–	–	–	–	–	–	–	–	–	–	5β,9βH,10α-Labd-8(20)-ene-15,19-dioic acid, dimethyl ester	13902-83-5	MS08590
304	121	305	364	161	122	221	181	**364**	100	96	63	28	23	22	21	21		22	36	4	–	–	–	–	–	–	–	–	–	–	5β,9βH,10α-Labd-8(20)-ene-15,19-dioic acid, dimethyl ester	13902-83-5	NI37975
43	91	115	117	105	131	161	145	**364**	100	83	73	44	43	24	20	15	**2.00**	22	36	4	–	–	–	–	–	–	–	–	–	–	11,15-Hexadecadien-13-yn-1-ol, 16-[2,3-(isopropylidenedioxy)propoxy]-		MS08591
273	55	41	43	67	81	69	291	**364**	100	93	57	42	40	31	31	26	**15.00**	22	36	4	–	–	–	–	–	–	–	–	–	–	6-Hydroxy-17-oxo-4-propyl-3,4-seco-5β-androstan-3-oic acid		MS08592
273	55	274	217	81	41	346	97	**364**	100	27	25	19	18	18	17	16	**1.00**	22	36	4	–	–	–	–	–	–	–	–	–	–	6α-Hydroxy-17-oxo-4-propyl-3,4-seco-5α-androstan-3-oic acid		MS08593
273	291	55	41	81	274	67	79	**364**	100	38	37	25	23	22	20	18	**10.00**	22	36	4	–	–	–	–	–	–	–	–	–	–	6β-Hydroxy-17-oxo-4-propyl-3,4-seco-5α-androstan-3-oic acid		MS08594
121	304	305	55	81	41	161	109	**364**	100	32	25	24	20	19	18	18	**7.01**	22	36	4	–	–	–	–	–	–	–	–	–	–	1-Naphthalenepentanoic acid, decahydro-5-(methoxycarbonyl)-β,5,8a-trimethyl-2-methylene-, methyl ester, [1S-[1α(S*),4aβ,5β,8aα]]-	13346-06-0	NI37976
121	41	55	304	81	59	67	109	**364**	100	44	38	31	29	27	21	20	**6.00**	22	36	4	–	–	–	–	–	–	–	–	–	–	1-Naphthalenepentanoic acid, decahydro-5-(methoxycarbonyl)-β,5,8a-trimethyl-2-methylene-, methyl ester, [1S-[1α(S*),4aβ,5β,8aα]]-	13346-06-0	NI37977
83	81	194	140	96	67	193	95	**364**	100	98	81	78	73	68	66	62	**10.00**	22	36	4	–	–	–	–	–	–	–	–	–	–	Pentadecanoic acid, 9,10-bis(3-oxo-1-propenyl)-, methyl ester	95452-49-6	NI37978
43	69	81	304	55	41	95	289	**364**	100	70	60	58	54	51	50	43	**7.92**	22	36	4	–	–	–	–	–	–	–	–	–	–	3,4,10a(1H)-Phenanthrenetriol, 2-vinyldodecahydro-2,4b,8,8-tetramethyl-, 3-acetate, [2S-(2α,3β,4α,4aα,4bβ,8aα,10aβ)]-	41756-15-4	NI37979
30	218	81	43	108	41	67	55	**364**	100	83	78	66	62	52	46	46	**7.69**	22	36	4	–	–	–	–	–	–	–	–	–	–	3,4,10a(1H)-Phenanthrenetriol, 2-vinyldodecahydro-2,4b,8,8-tetramethyl-, 4-acetate, [2S-(2α,3β,4α,4aα,4bβ,8aα,10aβ)]-	41756-22-3	NI37980
349	99	167	58	350	98	126	55	**364**	100	52	32	32	23	17	11	11	**0.00**	22	40	2	2	–	–	–	–	–	–	–	–	–	3,20-Diaza-7-oxadispiro[5.1.11.2]heneicosan-21-one, 2,2,4,4-tetramethyl-	64338-16-5	NI37981
69	41	349	364	350	70	39	40	**364**	100	34	19	5	5	5	4	2		23	24	4	–	–	–	–	–	–	–	–	–	–	Bisphenol A, dimethacrylate	3253-39-2	NI37982
280	281	364	322	279	193	192	265	**364**	100	21	10	10	10	5	5	4		23	24	4	–	–	–	–	–	–	–	–	–	–	Phenanthrene, 3,4-diacetyl-7,8-dimethyl-2-isopropyl-		MS08595
253	165	126	210	209	330	118	166	**364**	100	88	68	37	34	32	31	26	**0.00**	23	25	–	2	–	–	1	–	–	–	–	–	–	Methanaminium, N-[4-[[4-(dimethylamino)phenyl]phenylmethylene]-2,5-cyclohexadien-1-ylidene]-N-methyl-, chloride	569-64-2	NI37983
292	132	44	18	117	42	175	146	**364**	100	58	49	30	27	25	23	23	**15.00**	23	28	2	2	–	–	–	–	–	–	–	–	–	Acetaldehyde, [[[5-(2,6-dimethylphenyl)-1-methyl-4-oxo-3-phenyl-3-piperidinyl]methyl]amino]-	55823-01-3	NI37984
105	273	72	151	110	259	96	140	**364**	100	40	28	27	24	21	12	11	**3.00**	23	28	2	2	–	–	–	–	–	–	–	–	–	cis-1-Acetamido-2-(N-phenethylbenzamido)cyclohexane		MS08596
105	273	110	77	151	231	81	91	**364**	100	57	31	24	23	7	4	3	**1.00**	23	28	2	2	–	–	–	–	–	–	–	–	–	cis-1-Acetamido-3-(N-phenethylbenzamido)cyclohexane		MS08597
105	273	110	151	231	77	91	81	**364**	100	38	35	26	21	21	10	8	**1.00**	23	28	2	2	–	–	–	–	–	–	–	–	–	cis-1-Acetamido-4-(N-phenethylbenzamido)cyclohexane		MS08598
105	151	110	273	77	259	140	81	**364**	100	46	42	41	23	15	11	10	**2.00**	23	28	2	2	–	–	–	–	–	–	–	–	–	trans-1-Acetamido-2-(N-phenethylbenzamido)cyclohexane		MS08599
105	140	77	273	200	81	98	60	**364**	100	40	30	15	7	7	6	6	**1.00**	23	28	2	2	–	–	–	–	–	–	–	–	–	trans-1-Acetamido-3-(N-phenethylbenzamido)cyclohexane		MS08600
105	273	140	77	110	60	81	96	**364**	100	26	21	15	5	5	4	3	**2.00**	23	28	2	2	–	–	–	–	–	–	–	–	–	trans-1-Acetamido-4-(N-phenethylbenzamido)cyclohexane		MS08601

MASS TO CHARGE RATIOS								M.W.	INTENSITIES								Parent	C	H	O	N	S	F	Cl	Br	I	Si	P	B	X	COMPOUND NAME	CAS Reg No	No
184	198	43	42	305	41	364	183	**364**	100	42	35	35	34	30	29	23		23	28	2	2	–	–	–	–	–	–	–	–	–	2,20-Cycloaspidospermidine-3-methanol, 6,7-didehydro-1-methyl-, acetate, (2α,3β,5α,12β,19α,20R)-	56053-30-6	NI37985
120	105	217	106	121	336	202	79	**364**	100	84	16	15	14	12	11	11	**4.50**	23	28	2	2	–	–	–	–	–	–	–	–	–	Cyclobutane, 3-methyl-1,1-bis[(1-phenylethyl)carbamoyl]-	85291-65-2	NI37986
208	251	209	265	57	77	165	105	**364**	100	21	20	16	16	13	7	7	**2.10**	23	28	2	2	–	–	–	–	–	–	–	–	–	2,4-Imidazolidinedione, 1,3-di-tert-butyl-5,5-diphenyl-	53000-03-6	NI37987
208	251	265	209	57	77	183	165	**364**	100	21	16	16	16	13	8	7	**0.00**	23	28	2	2	–	–	–	–	–	–	–	–	–	2,4-Imidazolidinedione, 1,3-di-tert-butyl-5,5-diphenyl-	88700-41-8	NI37988
91	108	79	136	107	77	123	122	**364**	100	83	60	46	46	40	27	25	**18.00**	23	28	2	2	–	–	–	–	–	–	–	–	–	3,4,3′,5′-Tetramethyl-4′-ethyl-5-(benzyloxycarbonyl)-2,2′-dipyrrolylmethane		LI08024
70	41	43	55	292	69	57	83	**364**	100	60	57	39	35	26	26	22	**1.00**	23	40	3	–	–	–	–	–	–	–	–	–	–	2(5H)-Furanone, dihydro-4-hydroxy-5-methylene-3-octadecanylidene-, (3R,2E)-	78420-65-2	NI37989
196	364	152	365	168	197	153	151	**364**	100	96	41	29	25	19	16	12		24	16	2	–	–	–	–	–	–	1	–	–	–	3,8-Dioxasilaspiro[5.5]undecane, 1:2,4:5,6:7,9:10-tetrabenzo-		NI37990
335	141	233	102	205	77	130	364	**364**	100	80	50	49	35	30	11	1		24	16	2	2	–	–	–	–	–	–	–	–	–	3H-[1,3]Diazepino[2,1-a]isoindol-7-one, 3,11b-epoxy-7,11b-dihydro-2,5-diphenyl-		NI37991
364	365	77	309	105	89	182	88	**364**	100	29	25	13	13	13	12	11		24	16	2	2	–	–	–	–	–	–	–	–	–	1,4-Bis(5-phenyloxazol-2-yl)benzene		IC04143
181	182	364	183	128	271	154	273	**364**	100	74	10	10	9	8	5	4		24	20	–	4	–	–	–	–	–	–	–	–	–	Cyclobutane, 1,2-bis(3-pyridazinyl)-3,4-diphenyl-, (1α,2α,3β,4β)-		MS08602
181	182	149	97	105	183	364	271	**364**	100	33	18	14	12	6	3	2		24	20	–	4	–	–	–	–	–	–	–	–	–	Cyclobutane, 1,3-bis(3-pyridazinyl)-2,4-diphenyl-, (1α,2α,3β,4β)-		MS08603
364	91	137	273	163	161	150	255	**364**	100	58	57	51	50	30	23	12		24	28	3	–	–	–	–	–	–	–	–	–	–	3-Methoxy-7-phenyl-6-oxaestra-1,3,5(10)-trien-17β-ol		MS08604
82	336	110	364	321	141	96	349	**364**	100	64	28	18	18	12	10	9		24	32	1	2	–	–	–	–	–	–	–	–	–	1H-Imidazole, 1-methyl-2-acetyl-, [4b,5,6,7,8,8a,9,10-octahydro-4b,8-dimethyl-2-isopropyl-9-phenanthrenyl]-, [4bS-(4bα,8α,8aβ)]-	75601-28-4	NI37992
196	364	168	77	365	197	167	51	**364**	100	55	29	25	17	17	17	9		25	20	1	2	–	–	–	–	–	–	–	–	–	Urea, tetraphenyl-	632-89-3	NI37993
364	157	163	366	252	162	326	365	**364**	100	44	41	25	24	20	19	14		26	17	–	–	–	–	1	–	–	–	–	–	–	Anthracene, 1-chloro-9,10-diphenyl-	43217-27-2	NI37994
259	243	165	105	244	260	242	241	**364**	100	91	64	62	50	21	21	20	**6.61**	26	20	2	–	–	–	–	–	–	–	–	–	–	Benzenemethanol, α,α-diphenyl-, benzoate	17714-77-1	NI37995
364	305	304	365	306	202	302	226	**364**	100	96	48	30	23	22	21	21		26	20	2	–	–	–	–	–	–	–	–	–	–	Benzo[a]cyclopropa[cd]pentalene-2a(2bH)-carboxylic acid, 6b,6c-dihydro-1,6c-diphenyl-, methyl ester	40897-25-4	NI37996
364	305	304	365	119	306	303	302	**364**	100	85	40	35	35	23	14	14		26	20	2	–	–	–	–	–	–	–	–	–	–	Benzo[a]cyclopropa[cd]pentalene-1-carboxylic acid, 2a,2b,6b,6c-tetrahydro-2a,6c-diphenyl-, methyl ester	40897-24-3	NI37997
105	364	77	155	183	181	209	154	**364**	100	94	88	75	68	50	49	34		26	20	2	–	–	–	–	–	–	–	–	–	–	4-Benzoyl-4′-phenylbenzhydrol		NI37998
105	259	364	181	77	285	79	183	**364**	100	98	96	88	68	30	23	19		26	20	2	–	–	–	–	–	–	–	–	–	–	4-Benzoyl-4′-phenylbenzhydrol		NI37999
287	208	152	180	288	364	209	270	**364**	100	28	25	24	21	18	18	17		26	20	2	–	–	–	–	–	–	–	–	–	–	9,10-Diphenyl-9,10-dihydroxy-9,10-dihydroanthracene		MS08605
204	305	364	304	306	202	205	203	**364**	100	73	46	33	22	22	20	16		26	20	2	–	–	–	–	–	–	–	–	–	–	1,2,7-Metheno-1H-cyclopropa[b]naphthalene-1-carboxylic acid, 1a,2,7,7a-tetrahydro-1a,8-diphenyl-, methyl ester	40897-22-1	NI38000
141	179	178	115	318	364	229	207	**364**	100	50	31	16	7	6	6	5		26	20	2	–	–	–	–	–	–	–	–	–	–	1-Naphthalenemethyl 2-phenylcinnamate	86328-65-6	NI38001
269	347	239	270	348	268	240	252	**364**	100	96	78	39	37	26	26	21	**0.00**	26	20	2	–	–	–	–	–	–	–	–	–	–	1,3,3-Triphenyl-1-oxyphthalane		MS08606
83	82	55	43	41	57	69	97	**364**	100	92	41	38	31	25	15	11	**3.79**	26	52	–	–	–	–	–	–	–	–	–	–	–	1-Cyclohexyleicosane		AI02215
83	82	55	43	41	57	69	29	**364**	100	77	40	35	26	22	14	11	**2.11**	26	52	–	–	–	–	–	–	–	–	–	–	–	1-Cyclohexyleicosane		AI02216
83	82	43	55	41	57	69	56	**364**	100	81	42	41	31	24	14	9	**0.99**	26	52	–	–	–	–	–	–	–	–	–	–	–	1-Cyclohexyleicosane		AI02217
82	83	55	43	41	57	111	69	**364**	100	76	60	55	46	40	35	32	**0.29**	26	52	–	–	–	–	–	–	–	–	–	–	–	2-Cyclohexyleicosane		AI02218
83	55	43	82	41	57	69	71	**364**	100	94	91	90	77	71	54	36	**0.10**	26	52	–	–	–	–	–	–	–	–	–	–	–	3-Cyclohexyleicosane		AI02219
83	82	69	71	97	85	67	139	**364**	100	74	41	37	30	25	23	22	**0.07**	26	52	–	–	–	–	–	–	–	–	–	–	–	4-Cyclohexyleicosane		AI02220
83	82	71	97	69	85	67	111	**364**	100	80	56	49	47	35	27	19	**0.17**	26	52	–	–	–	–	–	–	–	–	–	–	–	5-Cyclohexyleicosane		AI02221
55	83	43	82	57	41	69	71	**364**	100	98	97	92	90	88	49	47	**1.53**	26	52	–	–	–	–	–	–	–	–	–	–	–	5-Cyclohexyleicosane		AI02222
43	57	55	83	82	41	71	97	**364**	100	97	96	90	90	62	59	50	**0.25**	26	52	–	–	–	–	–	–	–	–	–	–	–	5-Cyclohexyleicosane		AI02223
83	82	71	97	85	67	111	70	**364**	100	82	63	53	44	31	27	19	**0.11**	26	52	–	–	–	–	–	–	–	–	–	–	–	7-Cyclohexyleicosane		AI02224
83	82	71	97	69	85	67	111	**364**	100	82	69	54	51	46	31	27	**0.13**	26	52	–	–	–	–	–	–	–	–	–	–	–	9-Cyclohexyleicosane		AI02225
43	68	69	41	55	57	83	82	**364**	100	99	93	75	71	69	68	33	**4.81**	26	52	–	–	–	–	–	–	–	–	–	–	–	1-Cyclopentylheneicosane		AI02226
69	83	97	71	68	85	111	67	**364**	100	92	89	82	63	55	42	32	**0.18**	26	52	–	–	–	–	–	–	–	–	–	–	–	11-Cyclopentylheneicosane		AI02227
69	83	71	97	68	85	111	67	**364**	100	89	79	75	63	52	37	32	**0.30**	26	52	–	–	–	–	–	–	–	–	–	–	–	11-Cyclopentylheneicosane		AI02228
43	83	97	55	57	41	69	111	**364**	100	87	83	68	63	55	54	39	**4.48**	26	52	–	–	–	–	–	–	–	–	–	–	–	1,3-Didecylcyclohexane		AI02229
43	83	55	57	97	41	69	223	**364**	100	81	78	71	70	64	56	50	**7.87**	26	52	–	–	–	–	–	–	–	–	–	–	–	1,3-Didecylcyclohexane		AI02230
43	83	55	97	57	41	69	71	**364**	100	83	77	75	65	64	56	33	**2.23**	26	52	–	–	–	–	–	–	–	–	–	–	–	1,4-Didecylcyclohexane		AI02231
43	83	97	55	41	57	69	223	**364**	100	71	69	67	62	60	48	35	**8.93**	26	52	–	–	–	–	–	–	–	–	–	–	–	1,4-Didecylcyclohexane		AI02232
43	83	55	97	57	69	41	29	**364**	100	96	93	87	83	63	63	38	**5.03**	26	52	–	–	–	–	–	–	–	–	–	–	–	1,4-Didecylcyclohexane		AI02233
111	69	55	43	41	110	57	112	**364**	100	39	33	28	21	20	17	9	**0.54**	26	52	–	–	–	–	–	–	–	–	–	–	–	1,4-Dimethyl-3-octadecylcyclohexane		AI02234
43	57	55	41	97	69	83	71	**364**	100	82	79	65	58	48	47	38	**7.11**	26	52	–	–	–	–	–	–	–	–	–	–	–	1-Hexacosene	18835-33-1	NI38002
167	364	43	41	29	365	168	165	**364**	100	59	32	24	22	18	16	13		27	40	–	–	–	–	–	–	–	–	–	–	–	1-(5-Acenaphthenyl)pentadecane		AI02235
364	167	365	168	165	152	166	153	**364**	100	92	29	15	11	9	6	5		27	40	–	–	–	–	–	–	–	–	–	–	–	1-(5-Acenaphthenyl)pentadecane		AI02236
167	364	168	365	165	152	57	55	**364**	100	62	16	15	9	9	8	7		27	40	–	–	–	–	–	–	–	–	–	–	–	1-(5-Acenaphthenyl)pentadecane		AI02237

MASS TO CHARGE RATIOS								M.W.	INTENSITIES								Parent	C	H	O	N	S	F	Cl	Br	I	Si	P	B	X	COMPOUND NAME	CAS Reg No	No
364	334	365	319	335	179	165	349	**364**	100	74	29	26	25	23	18	15		28	28	–	–	–	–	–	–	–	–	–	–	–	1,2-Bis(4,6,8-trimethyl-1-azulenyl)ethylene	94154-50-4	NI38003
127	100	150	176	119	131	170	112	**365**	100	44	43	23	22	19	19	10	**0.00**	5	2	3	1	1	11	–	–	–	–	–	–	–	(2,2,3,3,4,4-Hexafluoro-4-carboxybutyramido)sulphur pentafluoride		NI38004
323	291	62	43	292	189	61	365	**365**	100	52	33	21	13	12	12	11		10	8	6	1	–	–	–	–	1	–	–	–	–	Salicylic acid, 5-iodo-3-nitro-, methyl ester, acetate	22621-46-1	MS08607
323	291	62	43	292	189	233	59	**365**	100	50	32	20	12	12	11	11	**10.01**	10	8	6	1	–	–	–	–	1	–	–	–	–	Salicylic acid, 5-iodo-3-nitro-, methyl ester, acetate	22621-46-1	NI38005
323	291	62	43	292	189	365	233	**365**	100	51	34	21	13	12	11	11		10	8	6	1	–	–	–	–	1	–	–	–	–	Salicylic acid, 5-iodo-3-nitro-, methyl ester, acetate	22621-46-1	LI08025
367	64	288	286	287	289	365	369	**365**	100	90	75	71	55	53	52	50		11	5	–	5	–	–	–	2	–	–	–	–	–	7-Bromo-2-bromomethyl-4-cyano-1,3,6-triazacycl[3.3.3]azine		NI38006
288	286	367	206	287	365	289	369	**365**	100	96	92	84	49	47	46	45		11	5	–	5	–	–	–	2	–	–	–	–	–	4-Cyano-2-dibromomethyl-1,3,6-triazacycl[3.3.3]azine		NI38007
367	369	365	18	245	247	368	142	**365**	100	52	52	25	16	15	14	11		11	5	–	5	–	–	–	2	–	–	–	–	–	1,3,6,9b-Tetraazaphenalene-4-carbonitrile, 7,9-dibromo-2-methyl-	37160-11-5	NI38008
367	286	288	164	365	369	179	178	**365**	100	83	82	58	54	41	41	37		14	9	1	1	–	–	–	2	–	–	–	–	–	10-Methyl-2,7-dibromoacridone		MS08608
69	70	330	83	332	306	308	128	**365**	100	35	31	24	21	13	12	11	**0.18**	14	14	4	1	–	–	3	–	–	–	–	–	–	L-Proline, 1-[(2,4,5-trichlorophenoxy)acetyl]-, methyl ester	66789-88-6	NI38009
67	68	53	83	142	169	82	55	**365**	100	79	76	73	53	39	31	25	**0.00**	14	24	2	1	–	–	–	–	1	–	–	–	–	Anodendrine, iodide, methyl ester	27510-47-0	LI08026
67	53	68	83	142	169	82	55	**365**	100	79	77	74	54	40	32	25	**0.00**	14	24	2	1	–	–	–	–	1	–	–	–	–	Anodendrine, iodide, methyl ester	27510-47-0	NI38010
73	147	45	218	75	74	291	59	**365**	100	20	17	14	9	8	6	6	**0.00**	14	35	2	1	1	–	–	–	–	3	–	–	–	Penicillamine, tris(trimethylsilyl)-		MS08609
73	147	103	205	117	75	74	45	**365**	100	25	17	16	15	9	9	9	**1.43**	14	35	4	1	–	–	–	–	–	3	–	–	–	Butanal, 2,3,4-tris[(trimethylsilyl)oxy]-, O-methyloxime, [R-(R*,R*)]-	56196-36-2	NI38011
91	75	65	76	212	210	50	51	**365**	100	46	44	40	30	30	26	22	**14.00**	15	12	3	1	1	–	–	1	–	–	–	–	–	3H-Indol-3-one, 5-bromo-1,2-dihydro-1-(4-tolylsulphonyl)-	83372-85-4	NI38012
127	212	81	55	365	143	153	236	**365**	100	96	92	70	52	52	42	38		15	15	2	1	1	3	1	–	–	–	–	–	–	Cycloheptanecarboxamide, N-(4-chlorophenyl)-2-oxo-3-[(trifluoromethyl)thio]-		DD01452
304	30	163	44	145	148	91	57	**365**	100	59	49	46	44	35	32	29	**18.00**	15	19	4	5	1	–	–	–	–	–	–	–	–	N-Ethyl-N-(2,4-dihydroxybutyl)-3-methyl-4-(4-nitrothiaz-2-ylazo)aniline		IC04144
128	66	59	127	55	95	123	81	**365**	100	79	74	50	42	24	21	21	**0.00**	15	28	1	1	–	–	–	–	1	–	–	–	–	[3-(3,3-Dimethyl-2-norbornyl)-3-oxopropyl]trimethylammonium iodide		NI38013
140	253	73	41	240	226	282	69	**365**	100	62	30	23	17	17	16	15	**0.00**	15	36	3	1	–	–	–	–	–	2	1	–	–	Phosphonic acid, [1-[(2,2-dimethylpropylidene)amino]butyl]-, bis(trimethylsilyl) ester	55108-75-3	NI38014
140	226	267	73	55	41	282	141	**365**	100	37	32	26	20	17	16	11	**0.00**	15	36	3	1	–	–	–	–	–	2	1	–	–	Phosphonic acid, [1-[(2,2-dimethylpropylidene)amino]-isobutyl]-, bis(trimethylsilyl) ester	55108-77-5	NI38015
262	73	75	263	130	59	45	350	**365**	100	43	18	16	7	7	6	5	**0.50**	15	39	3	1	–	–	–	–	–	3	–	–	–	Triethanolamine, tris(trimethylsilyl)-		IC04145
173	175	145	177	147	174	365	197	**365**	100	64	13	11	9	8	6	6		17	13	4	1	–	–	2	–	–	–	–	–	–	2-Propenoic acid, 3-[1-(2,6-dichlorobenzoyl)-5-formylpyrrol-2-yl]-, ethyl ester	95883-22-0	NI38016
365	121	186	56	212	115	141	147	**365**	100	43	38	29	24	16	9	8		17	15	3	3	–	–	–	–	–	–	–	–	1	1-[3-(5-Nitrofur-2-yl)-2-pyrazolin-5-yl]ferrocene		LI08027
73	217	205	333	292	147	319	75	**365**	100	61	47	42	40	39	26	17	**0.26**	17	31	2	1	–	–	–	–	–	3	–	–	–	1H-Indole, 1-(trimethylsilyl)-2,5-bis[(trimethylsilyl)oxy]-	74381-41-2	NI38017
365	73	366	367	45	350	74	277	**365**	100	81	30	14	11	10	6	4		17	31	2	1	–	–	–	–	–	3	–	–	–	1H-Indole, 1-(trimethylsilyl)-2,5-bis[(trimethylsilyl)oxy]-	74381-41-2	NI38018
105	77	106	51	29	292	294	145	**365**	100	39	11	6	6	5	4	3	**2.10**	18	17	3	1	–	–	2	–	–	–	–	–	–	DL-Alanine, N-benzoyl-N-(3,4-dichlorophenyl)-, ethyl ester	22212-55-1	NI38019
105	77	106	292	143	127	75	32	**365**	100	30	10	8	8	7	7	7	**2.54**	18	17	3	1	–	–	2	–	–	–	–	–	–	DL-Alanine, N-benzoyl-N-(3,4-dichlorophenyl)-, ethyl ester	22212-55-1	NI38020
102	130	165	70	235	44	199	59	**365**	100	38	35	20	13	12	11	11	**0.20**	18	17	3	1	–	–	2	–	–	–	–	–	–	DL-Alanine, N,N-bis(4-chlorophenyl)acetyl-, methyl ester		NI38021
173	175	365	145	367	177	147	174	**365**	100	65	17	14	12	11	9	8		18	17	3	1	–	–	2	–	–	–	–	–	–	2-Propenoic acid, 3-[1-(2,6-dichlorobenzoyl)-2,5-dimethylpyrrol-3-yl]-, ethyl ester		NI38022
105	365	77	89	366	106	247	219	**365**	100	31	20	14	8	8	5	5		20	15	4	1	1	–	–	–	–	–	–	–	–	Pyrano[3,2-f]benzoxazol-6(H)-one, 4,8-dimethyl-2-[(benzoylmethyl)thio]-		NI38023
365	188	364	189	178	320	216	187	**365**	100	86	42	38	36	15	13	13		20	19	4	3	–	–	–	–	–	–	–	–	–	Benzoic acid, 4-[N-[2-methyl-6,7-(methylenedioxy)quinazolin-4-ylmethyl]amino]-, ethyl ester	85591-11-3	NI38024
209	73	365	75	179	210	224	149	**365**	100	93	50	35	25	24	22	22		20	35	3	1	–	–	–	–	–	1	–	–	–	Nonanamide, N-[[3-methoxy-4-[(trimethylsilyl)oxy]phenyl]methyl]-	56784-18-0	MS08610
105	330	77	331	106	51	166	78	**365**	100	52	48	13	10	6	4	4	**2.00**	21	16	3	1	–	–	1	–	–	–	–	–	–	N-Benzoyl-N-(2-chlorophenyl)-2-(methoxycarbonyl)phenylamine		NI38025
214	216	105	111	215	77	75	76	**365**	100	40	24	20	18	18	13	8	**6.00**	21	16	3	1	–	–	1	–	–	–	–	–	–	N-Benzoyl-N-(4-chlorophenyl)-2-(methoxycarbonyl)phenylamine		NI38026
105	77	365	238	106	367	166	51	**365**	100	33	16	10	10	7	7	5		21	16	3	1	–	–	1	–	–	–	–	–	–	N-Benzoyl-N-(4-chlorophenyl)-2-(methoxycarbonyl)phenylamine		NI38027
105	77	243	208	245	330	51	106	**365**	100	59	56	30	18	16	10	9	**1.00**	21	16	3	1	–	–	1	–	–	–	–	–	–	N-Benzoyl-N-(2-chlorophenyl)-6-methylphenylamine-2-carboxylic acid		NI38028
210	182	365	180	195	179	178	350	**365**	100	40	31	28	21	4	4	1		21	19	3	1	1	–	–	–	–	–	–	–	–	5H-Dibenz[b,d]azepin-7-ol, 6,7-dihydro-5-(4-tolylsulphonyl)-	24310-34-7	NI38029
155	105	104	365	260	209	106		**365**	100	43	34	23	19	11	9			21	19	3	1	1	–	–	–	–	–	–	–	–	4-Toluenesulphonanilide, N-phenacyl-	24310-48-3	NI38030
365	366	350	322	337	336	294	262	**365**	100	23	11	11	6	6	5	3		21	19	5	1	–	–	–	–	–	–	–	–	–	Atheroline, O-ethyl-	11036-98-9	NI38031
365	336	308	337	366	322	306	277	**365**	100	70	36	25	24	11	10	10		21	19	5	1	–	–	–	–	–	–	–	–	–	7H-Dibenzo[de,g]quinolin-7-one, 1-ethoxy-2,9,10-trimethoxy-	15358-02-8	NI38032
365	366	322	350	336	294	320	364	**365**	100	24	24	20	11	10	8	7		21	19	5	1	–	–	–	–	–	–	–	–	–	7H-Dibenzo[de,g]quinolin-7-one, 10-ethoxy-1,2,9-trimethoxy-		MS08611
77	105	104	51	43	81	200	122	**365**	100	79	51	40	33	32	24	19	**1.00**	21	19	5	1	–	–	–	–	–	–	–	–	–	2,6-Dioxa-3-azabicyclo[3.3.0]-7-octene, 7-(acetoxymethyl)-4-benzoyl-3-phenyl-		NI38033
365	176	148	147	366	206	188	336	**365**	100	91	66	27	23	21	11	7		21	19	5	1	–	–	–	–	–	–	–	–	–	1(2H)-Isoquinolone, N-methyl-3-(5-vinyl-1,3-benzodioxol-6-yl)-4-methyl-7,8-(methylenedioxy)-3,4-dihydro-, (±)-trans-	118514-50-4	DD01453
135	198	187	136	104	77	92	183	**365**	100	18	6	6	5	5	4	3	**1.90**	21	19	5	1	–	–	–	–	–	–	–	–	–	2-Norbornen-7-ol, 7-(4-methoxyphenyl)-, 4-nitrobenzoate, anti-	19719-68-7	NI38034
135	198	150	136	77	76	104	92	**365**	100	24	14	9	8	8	7	6	**2.70**	21	19	5	1	–	–	–	–	–	–	–	–	–	2-Norbornen-7-ol, 7-(4-methoxyphenyl)-, 4-nitrobenzoate, anti-	19719-68-7	NI38035
135	198	150	365	77	136	92	215	**365**	100	30	18	9	9	8	8	6		21	19	5	1	–	–	–	–	–	–	–	–	–	2-Norbornen-7-ol, 7-(4-methoxyphenyl)-, 4-nitrobenzoate, syn-	27999-80-0	NI38036

MASS TO CHARGE RATIOS								M.W.	INTENSITIES								Parent	C	H	O	N	S	F	Cl	Br	I	Si	P	B	X	COMPOUND NAME	CAS Reg No	No
135	198	150	104	77	76	365	92	**365**	100	37	23	15	12	12	11	10		21	19	5	1	–	–	–	–	–	–	–	–	–	2-Norbornen-7-ol, 7-(4-methoxyphenyl)-, 4-nitrobenzoate, syn-	27999-80-0	NI38037
114	44	142	365	42	115	263	128	**365**	100	91	86	79	47	43	36	35		21	23	1	3	1	–	–	–	–	–	–	–	–	10H-Phenothiazine-2-carbonitrile, 10-[3-(4-hydroxy-1-piperidinyl)propyl]	2622-26-6	NI38038
296	365	297	133	366	165	268	146	**365**	100	99	50	37	36	35	34	26		21	23	3	3	–	–	–	–	–	–	–	–	–	Spiro[5H,6H-5a,9a-(iminomethano)-1H-cyclopent[f]indolizine-7(8H),2'-[2H]indole]-3',5,10(1'H)-trione, 2,3,8a,9-tetrahydro-8,8-dimethyl-		NI38039
83	85	47	296	365	48	42	87	**365**	100	64	26	14	13	13	12	11		21	23	3	3	–	–	–	–	–	–	–	–	–	Spiro[5H,6H-5a,9a-(iminomethano)-1H-cyclopent[f]indolizine-7(8H),2'-[2H]indole]-3',5,10(1'H)-trione, 2,3,8a,9-tetrahydro-8,8-dimethyl-	23402-09-7	NI38040
116	124	278	123	208	158	165	365	**365**	100	99	77	55	53	50	49	31		21	35	4	1	–	–	–	–	–	–	–	–	–	L-Tyrosine, 3-(1-methylethoxy)-N,O-bisisopropyl-, isopropyl ester	56771-68-7	MS08612
365	85	364	366	83	350	336	148	**365**	100	50	40	24	24	18	18	16		22	23	4	1	–	–	–	–	–	–	–	–	–	Ochotensimine		LI08028
91	323	200	92	65	365	291	260	**365**	100	23	20	17	17	9	9	6		22	23	4	1	–	–	–	–	–	–	–	–	–	3-Quinolinecarboxylic acid, 6-butyl-1,4-dihydro-4-oxo-7-(phenylmethoxy), methyl ester	13997-19-8	NI38041
296	70	365	152	267	165	181	225	**365**	100	55	22	17	16	16	15	13		22	23	4	1	–	–	–	–	–	–	–	–	–	Tylophorinidine	32523-69-6	NI38042
296	70	295	365	152	165	181	225	**365**	100	56	28	25	18	17	15	13		22	23	4	1	–	–	–	–	–	–	–	–	–	Tylophorinidine	32523-69-6	MS08613
207	208	104	365	264	267	165	366	**365**	100	65	58	36	34	28	19	13		22	27	–	3	1	–	–	–	–	–	–	–	–	2-Imidazoline-5-thione, 2-(tert-butylamino)-4,4-diphenyl-1-isopropyl-	118515-05-2	DD01454
365	263	221	264	223	29	366	72	**365**	100	41	33	28	24	24	23	22		22	27	2	3	–	–	–	–	–	–	–	–	–	1-Acetyllysergic acid diethylamide		MS08614
121	57	55	71	69	83	70	365	**365**	100	95	72	53	53	41	38	37		22	27	2	3	–	–	–	–	–	–	–	–	–	2-Pyrrolidinone, 3,3-dimethyl-5-[5-[4-(dimethylamino)phenyl-2-methylene]-3,4-dimethyl-5H-pyrrolyl-2-methylene]-4-hydroxy-, (Z,Z)-		MS08615
91	43	232	274	65	41	39	204	**365**	100	89	68	36	29	12	12	11	**3.00**	23	27	3	1	–	–	–	–	–	–	–	–	–	Carbazole, 9-acetyl-6-benzyloxy-7-methoxy-cis-4a-methyl-1,2,3,4,4a,9a-hexahydro-		MS08616
91	84	191	82	365	274	110	250	**365**	100	43	32	32	26	24	18	14		23	27	3	1	–	–	–	–	–	–	–	–	–	2-Keto-4-(3-benzyloxy-4-methoxyphenyl)(E)-trans-quinolizidine		LI08029
86	43	41	58	55	100	44	322	**365**	100	67	46	42	39	38	29	20	**16.04**	24	47	1	1	–	–	–	–	–	–	–	–	–	9-Octadecenamide, N,N-diisopropyl-	5831-78-7	NI38043
72	43	156	365	114	143	55	41	**365**	100	33	32	29	29	27	24	19		24	47	1	1	–	–	–	–	–	–	–	–	–	9-Octadecenamide, N,N-dipropyl-	56630-40-1	NI38044
113	43	55	41	140	57	71	70	**365**	100	36	26	20	16	16	15	13	**1.34**	24	47	1	1	–	–	–	–	–	–	–	–	–	Pyrrolidine, 1-(3,7,11,15-tetramethyl-1-oxohexadecyl)-	56630-63-8	NI38045
98	99	112	43	70	41	86	57	**365**	100	62	25	22	16	15	13	13	**2.70**	24	47	1	1	–	–	–	–	–	–	–	–	–	2-Pyrrolidinone, 1-(3,7,11,15-tetramethylhexadecyl)-	63913-38-2	NI38046
222	144	221	223	143	145	219	77	**365**	100	54	23	17	9	7	5	5	**3.00**	25	23	–	3	–	–	–	–	–	–	–	–	–	1H-Indole, 3,3'-[1-(4-pyridinyl)-1,2-ethanediyl]bis[2-methyl-	37707-65-6	NI38047
322	323	308	77	309	365	364	366	**365**	100	28	19	13	11	8	6	5		25	23	–	3	–	–	–	–	–	–	–	–	–	3,5-Pyridinedicarbonitrile, 4,4-pentamethylene-1,4-dihydro-1-methyl-2,6-diphenyl-		NI38048
44	30	322	56	365	55	43	364	**365**	100	71	49	49	48	45	45	37		25	51	–	1	–	–	–	–	–	–	–	–	–	Azacyclohexacosane		MS08617
115	134	137	175	153	118	96	99	**366**	100	99	92	50	28	26	19	12	**0.00**	–	–	1	–	–	10	–	–	–	–	–	–	2	Bis(pentafluoroselenium) oxide		LI08030
69	215	113	151	67	51	132	127	**366**	100	85	29	22	12	10	6	6	**0.00**	6	2	2	–	1	12	–	–	–	–	–	–	–	Bis(1,1,1,3,3,3-hexafluoroisopropyl) sulphone		DD01455
175	177	207	205	111	113	69	27	**366**	100	95	92	92	36	35	28	20	**0.00**	6	6	1	–	–	6	–	2	–	–	–	–	–	Bis(2-bromo-3,3,3-trifluoropropyl) ether		IC04146
131	366	103	310	338	241	213	282	**366**	100	94	88	75	63	59	26	9		7	1	4	–	–	6	–	–	–	–	–	–	1	Rhodium, dicarbonyl(1,1,1,5,5,5-hexafluoro-2,4-pentanedionato-O,O')-	18517-12-9	NI38049
109	93	173	107	44	339	91	189	**366**	100	52	50	39	31	30	28	23	**0.00**	7	10	6	–	–	6	–	–	–	–	2	–	–	1,3,2-Dioxaphospholane, 2-[1-(dimethoxyphosphinyl)-2,2,2-trifluoro-1-(trifluoromethyl)ethoxy]-	111557-97-2	DD01456
253	223	351	208	251	252	221	29	**366**	100	64	61	58	50	43	35	33	**6.00**	7	14	2	2	–	–	–	–	–	–	–	–	1	(Ethoxycarbonyl)(trimethyllead)diazomethane		LI08031
176	190	177	41	119	204	43	72	**366**	100	73	66	60	58	53	35	32	**5.20**	9	8	2	2	–	10	–	–	–	–	–	–	–	1,3-Diaminopropane, N,N'-bis(pentafluoropropionyl)-		MS08618
366	199	347	297	367	368	339	200	**366**	100	41	17	17	13	5	3	3		12	–	–	–	1	10	–	–	–	–	–	–	–	Benzene, 1,1'-thiobis[2,3,4,5,6-pentafluoro-	1043-50-1	NI38050
174	172	173	175	368	145	147	366	**366**	100	99	47	43	25	21	20	12		12	8	–	4	–	–	–	2	–	–	–	–	–	Imidazo[1,2-a]pyridin-2-amine, 5-bromo-N-(6-bromo-2-pyridinyl)-		MS08619
331	203	333	36	332	68	135	366	**366**	100	59	32	20	16	14	13	12		12	8	2	8	–	–	2	–	–	–	–	–	–	1,4-Bis(2-amino-4-chloro-s-triaz-6-yloxy)benzene		IC04147
366	365	363	57	55	97	69	337	**366**	100	41	41	33	28	22	22	21		13	17	–	–	–	–	–	–	–	–	–	–	1	π-Cyclopentadienyl-π-cyclo-octa-1,5-dieneiridium		MS08620
73	147	75	55	45	74	18	148	**366**	100	47	21	18	17	9	8	8	**1.49**	13	30	4	–	1	–	–	–	–	3	–	–	–	Mercaptosuccinic acid, tris(trimethylsilyl) ester		NI38051
147	73	75	44	56	45	43	88	**366**	100	96	83	70	67	41	30	27	**3.60**	13	30	4	2	1	–	–	–	–	2	–	–	–	4-Thia-2,7-diaminooctanedioate, bis(trimethylsilyl)-		MS08621
115	218	157	176	103	145	220	98	**366**	100	80	78	50	50	35	27	25	**0.00**	14	19	9	–	–	–	1	–	–	–	–	–	–	α-D-Glucopyranose, 6-chloro-6-deoxy-, tetraacetate	35816-31-0	NI38052
67	189	131	116	57	41	235	115	**366**	100	50	29	27	24	19	14	14	**5.50**	14	26	2	–	–	–	–	–	–	3	–	–	1	Iron, dicarbonyl(η^5-2,4-cyclopentadien-1-yl)(heptamethyltrisilanyl)-	56784-37-3	MS08622
189	131	116	57	41	265	235	71	**366**	100	57	54	48	37	27	27	25	**11.00**	14	26	2	–	–	–	–	–	–	3	–	–	1	Iron, dicarbonyl(η^5-2,4-cyclopentadien-1-yl)(heptamethyltrisilanyl)-	56784-37-3	LI08032
84	182	168	99	351	69	282	152	**366**	100	69	64	59	55	28	10	10	**0.50**	14	32	–	2	–	–	–	–	–	–	–	–	2	Gallium, tetramethyldi-μ-1-piperidinyldi-	36351-56-1	MS08623
105	366	77	231	76	106	51	50	**366**	100	21	20	17	12	11	8	7		15	11	3	–	–	–	–	–	1	–	–	–	–	Benzoic acid, 2-iodo-, 2-oxo-2-phenylethyl ester	55153-31-6	NI38053
105	77	76	231	106	203	51	50	**366**	100	15	10	7	7	5	5	5	**5.00**	15	11	3	–	–	–	–	–	1	–	–	–	–	Benzoic acid, 3-iodo-, 2-oxo-2-phenylethyl ester	55153-30-5	NI38054
105	77	76	231	106	50	203	51	**366**	100	16	11	7	7	6	5	5	**5.00**	15	11	3	–	–	–	–	–	1	–	–	–	–	Benzoic acid, 4-iodo-, 2-oxo-2-phenylethyl ester	55153-29-2	NI38055
127	129	128	126	63	78	99	77	**366**	100	31	11	11	10	9	8	8	**0.11**	15	12	3	4	–	–	2	–	–	–	–	–	–	N-(Chlorophenyl)-2-(4-chloro-2-nitrophenylazo)propionamide		IC04148
84	83	81	82	44	56	41	42	**366**	100	40	26	23	22	19	13	10	**0.00**	15	22	5	6	–	–	–	–	–	–	–	–	–	L-Threoninamide, 5-oxo-L-prolyl-L-histidyl-	33217-48-0	NI38056
88	133	73	101	89	117	75	45	**366**	100	95	45	33	26	16	16	11	**0.00**	15	34	6	–	–	–	–	–	–	2	–	–	–	D-Glucopyranose, 2,3,4-tri-O-methyl-1,6-bis-O-(trimethylsilyl)-	32388-38-8	NI38057
88	133	159	73	89	45	71	59	**366**	100	77	47	44	28	19	15	11	**0.00**	15	34	6	–	–	–	–	–	–	2	–	–	–	D-Glucopyranose, 2,3,6-tri-O-methyl-1,4-bis-O-(trimethylsilyl)-	32388-39-9	NI38058
73	204	133	45	71	146	89	75	**366**	100	87	43	33	24	21	21	21	**0.00**	15	34	6	–	–	–	–	–	–	2	–	–	–	α-D-Glucopyranoside, methyl 3,4-di-O-methyl-2,6-bis-O-(trimethylsilyl)-	52430-39-4	NI38059

MASS TO CHARGE RATIOS								M.W.	INTENSITIES								Parent	C	H	O	N	S	F	Cl	Br	I	Si	P	B	X	COMPOUND NAME	CAS Reg No	No
146	159	73	75	89	88	59	71	**366**	100	93	89	88	44	28	24	23	**0.00**	15	34	6	–	–	–	–	–	–	2	–	–	–	α-D-Glucopyranoside, methyl 3,4-di-O-methyl-2,6-bis-O-(trimethylsilyl)-	52430-39-4	NI38060
146	159	73	75	89	88	71	59	**366**	100	94	90	88	46	27	26	25	**0.00**	15	34	6	–	–	–	–	–	–	2	–	–	–	α-D-Glucopyranoside, methyl 3,4-di-O-methyl-2,6-bis-O-(trimethylsilyl)-	52430-39-4	NI38061
73	204	133	45	71	159	89	75	**366**	100	87	42	35	26	22	22	22	**0.00**	15	34	6	–	–	–	–	–	–	2	–	–	–	α-D-Glucopyranoside, methyl 4,6-di-O-methyl-2,3-bis-O-(trimethylsilyl)-		NI38062
128	77	102	129	103	366	209	367	**366**	100	55	46	35	24	16	14	10		16	14	–	–	–	–	–	–	–	–	–	–	2	β-Styryl 2-(phenylselenyl)-1-ethenyl selenide, (Z,Z)-		DD01457
366	45	321	165	260	166	194	293	**366**	100	74	52	29	13	13	12	11		16	15	2	–	–	–	–	–	1	–	–	–	–	Stilbene, 2-(methoxymethoxy)-4′-iodo-, (E)-		MS08624
45	366	321	165	166	194	260	293	**366**	100	96	63	56	22	18	17	16		16	15	2	–	–	–	–	–	1	–	–	–	–	Stilbene, 2-(methoxymethoxy)-4′-iodo-, (Z)-		MS08625
29	133	27	317	39	43	182	41	**366**	100	57	52	30	29	26	25	24	**8.50**	16	16	2	4	–	–	2	–	–	–	–	–	–	N-Ethyl-N-(2-chloroethyl)-4-(2-chloro-4-nitrophenylazo)aniline		IC04149
57	43	71	41	70	55	256	254	**366**	100	76	54	51	41	28	16	16	**4.00**	16	21	3	–	–	–	3	–	–	–	–	–	–	Acetic acid, (2,4,5-trichlorophenoxy)-, 2-ethylhexyl ester	1928-47-8	NI38063
57	43	71	70	41	55	29	254	**366**	100	68	53	37	32	21	21	13	**3.60**	16	21	3	–	–	–	3	–	–	–	–	–	–	Acetic acid, (2,4,5-trichlorophenoxy)-, 2-ethylhexyl ester	1928-47-8	AI02238
57	43	40	41	87	71	55	69	**366**	100	88	78	63	58	44	44	42	**2.90**	16	21	3	–	–	–	3	–	–	–	–	–	–	Isooctyl (2,4,5-trichlorophenoxy)acetate	25168-15-4	NI38064
57	43	69	41	55	71	84	29	**366**	100	96	63	53	52	50	31	25	**6.91**	16	21	3	–	–	–	3	–	–	–	–	–	–	Isooctyl (2,4,5-trichlorophenoxy)acetate	25168-15-4	AI02239
57	43	71	55	41	69	56	70	**366**	100	66	33	30	29	22	22	18	**5.17**	16	21	3	–	–	–	3	–	–	–	–	–	–	Isooctyl (2,4,5-trichlorophenoxy)acetate	25168-15-4	AI02240
167	166	72	41	29	265	139	42	**366**	100	92	84	35	35	34	23	20	**0.00**	16	25	4	2	–	3	–	–	–	–	–	–	–	Butanoic acid, N-(trifluoroacetyl)-L-prolyl-2-methyl-2-amino-, methyl ester		MS08626
166	167	72	28	265	29	41	139	**366**	100	68	41	39	29	26	25	17	**2.90**	16	25	4	2	–	3	–	–	–	–	–	–	–	Norvaline, N-(trifluoroacetyl)-L-prolyl-, butyl ester		MS08627
166	167	28	72	265	41	29	139	**366**	100	63	43	41	40	27	21	15	**1.70**	16	25	4	2	–	3	–	–	–	–	–	–	–	Valine, N-(trifluoroacetyl)-L-prolyl-, butyl ester		MS08628
55	41	125	83	97	43	81	29	**366**	100	60	44	29	27	26	23	20	**14.64**	16	30	4	–	–	–	–	–	–	–	–	–	1	Octanoic acid, 8,8′-selenodi-	22676-18-2	NI38065
55	41	125	83	97	43	81	69	**366**	100	59	44	30	27	27	23	19	**14.00**	16	30	4	–	–	–	–	–	–	–	–	–	1	Octanoic acid, 8,8′-selenodi-	22676-18-2	LI08033
101	43	89	117	205	75	59	45	**366**	100	89	66	63	43	29	25	24	**0.00**	16	30	9	–	–	–	–	–	–	–	–	–	–	L-Glycero-D-mannoheptitol, 1,5-di-O-acetyl-2,3,4,6,7-penta-O-methyl-	103489-00-5	NI38066
88	101	143	175	75	115	71	99	**366**	100	67	62	42	38	22	20	18	**0.00**	16	30	9	–	–	–	–	–	–	–	–	–	–	α-D-Xylopyranoside, methyl 2,3-di-O-methyl-4-O-(2,3,4-tri-O-methyl-α-D-xylopyranosyl)-		NI38067
235	101	115	88	75	175	143	71	**366**	100	90	79	75	65	39	32	13	**0.00**	16	30	9	–	–	–	–	–	–	–	–	–	–	α-D-Xylopyranoside, methyl 2,4-di-O-methyl-3-O-(2,3,4-tri-O-methyl-α-D-xylopyranosyl)-		NI38068
175	143	88	101	235	75	115	99	**366**	100	62	62	57	50	19	18	18	**0.00**	16	30	9	–	–	–	–	–	–	–	–	–	–	α-D-Xylopyranoside, methyl 3,4-di-O-methyl-2-O-(2,3,4-tri-O-methyl-α-D-xylopyranosyl)-	74405-62-2	NI38069
101	88	175	143	75	235	115	71	**366**	100	90	65	63	52	32	22	19	**0.00**	16	30	9	–	–	–	–	–	–	–	–	–	–	β-D-Xylopyranoside, methyl 2,3-di-O-methyl-4-O-(2,3,4-tri-O-methyl-β-D-xylopyranosyl)-	74405-72-4	NI38070
235	101	115	75	88	175	143	114	**366**	100	77	68	52	50	37	24	14	**0.00**	16	30	9	–	–	–	–	–	–	–	–	–	–	β-D-Xylopyranoside, methyl 2,4-di-O-methyl-3-O-(2,3,4-tri-O-methyl-β-D-xylopyranosyl)-	74405-69-9	NI38071
175	101	88	235	143	115	75	99	**366**	100	59	58	51	45	17	17	15	**0.00**	16	30	9	–	–	–	–	–	–	–	–	–	–	β-D-Xylopyranoside, methyl 3,4-di-O-methyl-2-O-(2,3,4-tri-O-methyl-β-D-xylopyranosyl)-	74405-63-3	NI38072
101	143	175	99	103	115	75	114	**366**	100	51	46	30	20	16	15	12	**0.00**	16	30	9	–	–	–	–	–	–	–	–	–	–	β-D-Xylopyranoside, 2,3,4-tri-O-methyl-(2,3,4-tri-O-methyl-α-D-xylopyranosyl)-		NI38073
366	206	367	121	347	56	225	75	**366**	100	30	18	17	13	13	5	4		17	11	–	–	–	5	–	–	–	–	–	–	1	Ferrocene, [(pentafluorophenyl)methyl]-	55334-24-2	NI38074
173	331	175	333	145	332	366	147	**366**	100	69	65	25	19	15	14	12		17	16	3	2	–	–	2	–	–	–	–	–	–	2-Propenoic acid, 3-[1-(2,6-dichlorobenzoyl)-3,5-dimethylpyrazol-4-yl]-, ethyl ester	95883-08-2	NI38075
110	152	194	82	43	81	195	44	**366**	100	93	88	59	59	30	28	28	**25.43**	17	26	5	4	–	–	–	–	–	–	–	–	–	N-(N^2,N-Diacetylhistidyl)alanine butyl ester		AI02241
368	366	226	370	296	224	260	298	**366**	100	72	41	40	27	26	24	14		18	10	–	–	–	–	4	–	–	–	–	–	–	1,1′:4′,1″-Terphenyl, 2,4,4″,6-tetrachloro-	61576-97-4	NI38076
108	186	55	240	282	107	78	366	**366**	100	60	40	34	33	30	23	10		18	16	3	–	–	–	–	–	–	–	1	–	1	Allyldiphenylphosphinemanganese tricarbonyl		IC04150
296	254	41	366	69	225	298	255	**366**	100	94	64	52	43	39	38	37		18	19	6	–	–	–	1	–	–	–	–	–	–	Spiro[benzofuran-2(3H),1′-cyclohexane]-2′,3,4′-trione, 7-chloro-4,6-dimethoxy-3′,3′,6′-trimethyl-	26891-78-1	NI38077
152	366	335	215	324	337	214	368	**366**	100	50	48	45	22	20	20	18		18	19	6	–	–	–	1	–	–	–	–	–	–	Spiro[benzofuran-2(3H),1′-[2]cyclohexene]-3,4′-dione, 7-chloro-2′,4,6-trimethoxy-3′,6′-dimethyl-	26881-59-4	NI38078
112	215	179	366	43	178	54	254	**366**	100	48	45	35	33	27	24	21		18	19	6	–	–	–	1	–	–	–	–	–	–	Spiro[benzofuran-2(3H),1′-[3]cyclohexene]-2′,3-dione, 7-chloro-4,4′,6-trimethoxy-3′,6′-dimethyl-	26881-71-0	NI38079
194	43	41	226	268	56	69	195	**366**	100	96	87	85	57	56	52	47	**0.00**	18	22	8	–	–	–	–	–	–	–	–	–	–	Bicyclo[2.2.2]octa-2,5-diene-2,3-dicarboxylic acid, 1,5-dihydroxy-8,8-dimethyl-, dimethyl ester, diacetate	25864-63-5	MS08629
222	162	264	43	324	366	282	307	**366**	100	80	30	30	17	14	8	5		18	22	8	–	–	–	–	–	–	–	–	–	–	2,3-Diacetoxy-6-acetoxymethyl-5,6-dimethylbenzyl acetate		LI08034
111	153	273	83	94	171	129	69	**366**	100	74	47	47	35	19	12	9	**0.00**	18	22	8	–	–	–	–	–	–	–	–	–	–	β-L-Rhamnopyranoside, phenyl 2,3,4-tri-O-acetyl-	4468-73-9	NI38080
103	75	273	47	321	217	293	94	**366**	100	45	40	31	15	7	6	6	**0.24**	18	23	6	–	–	–	–	–	–	–	1	–	–	Phosphoric acid, 2,2-diethoxyethyl diphenyl ester	74498-93-4	NI38081
73	351	335	366	263	352	353	336	**366**	100	90	47	44	44	36	19	19		18	38	–	–	–	–	–	–	–	4	–	–	–	1,2,4,5-Tetrakis(trimethylsilyl)benzene	17156-61-5	NI38082
275	276	245	274	56	247	77	220	**366**	100	60	29	24	22	21	19	14	**8.00**	19	18	4	4	–	–	–	–	–	–	–	–	–	1,2,3,4-Butanetetrol, 1-(1-phenyl-1H-pyrazolo[3,4-b]quinoxalin-3-yl)-, [1R-(1R*,2S*,3R*)]-	31504-90-2	MS08630
275	276	247	245	220	77	274	277	**366**	100	57	30	19	16	13	10	9	**4.10**	19	18	4	4	–	–	–	–	–	–	–	–	–	1,2,3,4-Butanetetrol, 1-(1-phenyl-1H-pyrazolo[3,4-b]quinoxalin-3-yl)-, [1R-(1R*,2S*,3R*)]-	31504-90-2	NI38083
275	276	247	245	77	277	220	222	**366**	100	79	35	18	18	16	14	8	**3.00**	19	18	4	4	–	–	–	–	–	–	–	–	–	1-Phenyl-3-(D-lyxo-tetritol-1-yl)pyrazolo[3,4-b]quinoxaline		MS08631

MASS TO CHARGE RATIOS	M.W.	INTENSITIES	Parent	C	H	O	N	S	F	Cl	Br	I	Si	P	B	X	COMPOUND NAME	CAS Reg No	No
326 366 133 173 328 368 104 367	**366**	100 92 81 80 37 33 25 24		19	19	–	6	–	–	1	–	–	–	–	–	–	**N-Ethyl-N-(2-cyanoethyl)-4-(1-methyl-chlorobenzimidaz-2-ylazo)aniline**		**IC04151**
259 168 366 233 103 77 260 261	**366**	100 32 28 16 15 14 13 11		19	19	1	–	–	–	–	–	–	–	–	–	1	**Rhodium, [η^4-2-(α-hydroxybenzyl)norbornadiene](η^5-cyclopentadienyl)-**	72950-49-3	**NI38084**
146 159 160 145 312 105 104 311	**366**	100 62 54 32 29 24 23 19	**0.00**	19	20	6	–	–	–	–	–	–	–	–	2	–	**α-D-Mannopyranoside, methyl, cyclic 2,3:4,6-bis(phenylboronate)**	41356-07-4	**NI38085**
146 160 159 145 104 105 147 103	**366**	100 40 29 24 18 16 13 9	**0.10**	19	20	6	–	–	–	–	–	–	–	–	2	–	**α-D-Mannopyranoside, methyl, cyclic 2,3:4,6-bis(phenylboronate)**	41356-07-4	**NI38086**
43 159 160 147 105 146 104 91	**366**	100 87 67 62 33 23 23 22	**13.28**	19	20	6	–	–	–	–	–	–	–	–	2	–	**DL-Xylitol, cyclic 1,3:2,4-bis(phenylboronate) 5-acetate**	74793-45-6	**NI38087**
83 85 84 47 45 43 87 86	**366**	100 69 18 14 14 13 12 9	**0.00**	19	26	7	–	–	–	–	–	–	–	–	–	–	**Azuleno[4,5-b]furan-2(3H)-one, 9-(acetyloxy)-9a-[(acetyloxy)methyl]decahydro-6a-hydroxy-6-methyl-3-methylene-, [3aS-(3aα,6β,6aα,9β,9aβ,9bα)]-**	28587-46-4	**NI38088**
83 85 43 47 84 45 86 49	**366**	100 72 24 19 17 16 12 10	**0.00**	19	26	7	–	–	–	–	–	–	–	–	–	–	**Azuleno[4,5-b]furan-2(3H)-one, 9-(acetyloxy)-6-[(acetyloxy)methyl]decahydro-6a-hydroxy-9a-methyl-3-methylene-, [3aS-(3aα,6β,6aα,9β,9aβ,9bα)]-**	25383-32-8	**NI38089**
43 161 246 228 162 202 85 189	**366**	100 52 51 39 38 37 31 29	**0.00**	19	26	7	–	–	–	–	–	–	–	–	–	–	**Azuleno[4,5-b]furan-2(3H)-one, 9-(acetyloxy)-6-[(acetyloxy)methyl]decahydro-6a-hydroxy-9a-methyl-3-methylene-, [3aS-(3aα,6β,6aα,9β,9aβ,9bα)]-**	25383-32-8	**NI38090**
306 235 95 107 108 71 137 264	**366**	100 98 97 95 85 80 76 75	**0.00**	19	26	7	–	–	–	–	–	–	–	–	–	–	**Azuleno[6,5-b]furan-2(3H)-one, 6,7-bis(acetyloxy)decahydro-5-hydroxy-4a,8-dimethyl-3-methylene-, [3aR-(3aα,4aβ,5β,6α,7α,7aα,8α,9aα)]-**	51292-55-8	**NI38091**
121 105 109 107 122 79 91 95	**366**	100 88 72 72 66 61 59 56	**0.30**	19	26	7	–	–	–	–	–	–	–	–	–	–	**Trichothec-9-ene-3,15-diol, 12,13-epoxy-7-hydroxy-, 3,15-diacetate, (3α)-**		**NI38092**
43 121 122 41 109 107 105 138	**366**	100 39 14 14 12 12 11 10	**0.00**	19	26	7	–	–	–	–	–	–	–	–	–	–	**Trichothec-9-ene-3,15-diol, 12,13-epoxy-8-hydroxy-, 3,15-diacetate, (3α)-**		**NI38093**
74 55 41 87 43 69 81 29	**366**	100 31 30 27 26 18 15 13	**0.00**	19	36	2	–	–	–	2	–	–	–	–	–	–	**Octadecanoic acid, 9,10-dichloro-, methyl ester**	33094-27-8	**NI38094**
74 55 41 87 43 69 75 81	**366**	100 31 30 27 25 18 17 15	**0.14**	19	36	2	–	–	–	2	–	–	–	–	–	–	**Octadecanoic acid, 9,10-dichloro-, methyl ester**	33094-27-8	**NI38095**
366 305 265 323 249 163 351 251	**366**	100 11 11 10 6 6 5 5		20	14	7	–	–	–	–	–	–	–	–	–	–	**Naphtho[2,3-c]furan-1(3H)-one, 9-(1,3-benzodioxol-5-yl)-7,8-dihydroxy-4-methoxy-**		**NI38096**
91 275 366 131 117 221 85 86	**366**	100 93 30 28 5 4 4 3		20	18	3	2	1	–	–	–	–	–	–	–	–	**2-Benzylimino-3-benzyl-6-methoxycarbonyl-2,3-dihydro-1,3-thiazin-4-one**		**MS08632**
259 227 154 229 321 152 169 155	**366**	100 48 45 42 25 25 20 20	**0.00**	20	19	2	–	–	–	–	–	–	–	–	–	1	**(2-Hydroxy-2-phenylethyl)diphenylarsane oxide**		**DD01458**
57 104 43 87 105 41 190 69	**366**	100 79 60 33 27 27 22 22	**0.00**	20	30	4	–	1	–	–	–	–	–	–	–	–	**1,3-Dioxolane-4-carbothioate, 2-tert-butyl-4-(α-hydroxy-α-methylbenzyl)-, S-tert-butyl ester, (2R,4R)-**		**DD01459**
58 136 149 73 81 121 59 57	**366**	100 20 9 8 8 6 6 5	**3.30**	20	30	6	–	–	–	–	–	–	–	–	–	–	**2,3-Benzo-11,12-cyclohexano-1,4,7,10,13,16-hexaoxacyclooctadecane**		**MS08633**
71 91 149 107 233 235 217 175	**366**	100 47 33 32 17 10 10 10	**0.00**	20	30	6	–	–	–	–	–	–	–	–	–	–	**Eudesman-8-one, 6,7-didehydro-3α-(2-methyl-2,3-epoxybutyryloxy)-4α,11 dihydroxy-**		**NI38097**
149 71 91 107 233 333 217 175	**366**	100 81 49 47 28 24 21 21	**0.00**	20	30	6	–	–	–	–	–	–	–	–	–	–	**Eudesman-8-one, 6,7-didehydro-3α-(2-methyl-2,3-epoxybutyryloxy)-4α,11 dihydroxy-**		**NI38098**
57 41 149 45 56 43 39 101	**366**	100 92 88 83 75 71 71 58	**1.20**	20	30	6	–	–	–	–	–	–	–	–	–	–	**Phthalic acid, di(2-butoxyethyl) ester**		**MS08634**
57 56 149 45 101 29 85 50	**366**	100 75 65 64 45 44 42 42	**0.63**	20	30	6	–	–	–	–	–	–	–	–	–	–	**Phthalic acid, di(2-butoxyethyl) ester**		**IC04152**
41 43 55 107 91 119 67 93	**366**	100 90 88 47 41 40 40 36	**12.00**	20	30	6	–	–	–	–	–	–	–	–	–	–	**Picrasane-11,16-dione, 1,2,12-trihydroxy-, (1β,2α,12β)-**		**LI08035**
73 223 366 276 224 221 277 75	**366**	100 56 43 40 30 24 17 17		20	38	2	–	–	–	–	–	–	2	–	–	–	**Tetracyclo[4.4.4.0.0^{2,7}]tetradecane, 2,7-bis[(trimethylsilyl)oxy]-**		**NI38099**
121 122 186 180 78 77 123 45	**366**	100 87 23 22 17 15 12 11	**0.30**	21	18	–	–	3	–	–	–	–	–	–	–	–	**1,3,5-Trithiane, 2,4,6-triphenyl-**		**AI02242**
366 219 89 251 191 145 91 367	**366**	100 59 53 32 32 27 26 22		21	18	6	–	–	–	–	–	–	–	–	–	–	**Benzeneacetic acid, 2-methoxy-α-(3-methoxy-5-oxo-4-phenyl-2(5H)-furanylidene)-, methyl ester, (E)-**	22628-25-7	**NI38100**
366 307 251 89 367 279 308 219	**366**	100 69 51 30 22 22 15 13		21	18	6	–	–	–	–	–	–	–	–	–	–	**Benzeneacetic acid, 4-methoxy-α-(3-methoxy-5-oxo-4-phenyl-2(5H)-furanylidene)-, methyl ester, (E)-**	22628-24-6	**NI38101**
366 251 119 367 279 147 84 89	**366**	100 41 31 22 21 20 16 14		21	18	6	–	–	–	–	–	–	–	–	–	–	**Benzeneacetic acid, α-[3-methoxy-4-(4-methoxyphenyl)-5-oxo-2(5H)-furanylidene]-, methyl ester, (E)-**	22736-30-7	**NI38102**
366 311 283 168 351 297 137 309	**366**	100 44 34 26 25 24 23 22		21	18	6	–	–	–	–	–	–	–	–	–	–	**11H-Benzofuro[2,3-b][1]benzopyran-11-one, 3,8-dihydroxy-10-(3,3-dimethylallyl)-9-methoxy-**		**MS08635**
202 97 174 201 41 89 145 69	**366**	100 17 10 8 7 6 4 4	**3.00**	21	18	6	–	–	–	–	–	–	–	–	–	–	**7H-Furo[3,2-g][1]benzopyran-7-one, 9-[[4-(2,5-dihydro-4-methyl-5-oxo-2-furanyl)-3-methyl-2-butenyl]oxy]-, (Z)-**	89824-26-0	**NI38103**
180 105 198 91 165 181 77 167	**366**	100 57 32 29 21 18 14 10	**0.00**	21	19	2	–	1	–	–	–	–	–	1	–	–	**1,3,2-Dioxaphosphorinane, 2,5,5-triphenyl-, 2-sulphide**	39879-97-5	**MS08636**
336 366 307 244 97 237 232 204	**366**	100 54 31 29 27 23 23 23		21	22	4	2	–	–	–	–	–	–	–	–	–	**6,21-Cyclo-4,5-secoakuammilan-17-oic acid, 4,5-epoxy-5-methoxy-, methyl ester, (5S,6α)-**	63964-36-3	**NI38104**
366 236 149 293 367 167 222 69	**366**	100 35 35 24 23 23 22 21		21	22	4	2	–	–	–	–	–	–	–	–	–	**1H-12a,5-(Epoxymethano)pyrido[3',4':5,6]cyclohept[1,2-b]indole-5(6H)-carboxaldehyde, 2-acetyl-2,3,4,4a,11,12-hexahydro-12-methyl-14-oxo-**	3329-93-9	**NI38105**
366 367 307 239 159 222 197 184	**366**	100 40 20 20 17 9 8 8		21	22	4	2	–	–	–	–	–	–	–	–	–	**Voachalotine pseudoindoxyl, 1-demethyl-17-deoxy-6,17-epoxy-, (6β)-**	32326-32-2	**NI38106**
43 257 41 163 91 79 55 167	**366**	100 41 30 22 22 20 19 18	**0.60**	21	31	4	–	–	1	–	–	–	–	–	–	–	**Androstane-17,19-diol, 5,6-epoxy-3-fluoro-, 17-acetate, (3β,5α,6α,17β)-**	40242-94-2	**MS08637**
43 336 142 133 93 91 134 81	**366**	100 79 72 36 36 36 32 32	**0.20**	21	31	4	–	–	1	–	–	–	–	–	–	–	**Androstane-17,19-diol, 5,6-epoxy-3-fluoro-, 17-acetate, (3β,5β,6β,17β)-**	40242-93-1	**MS08638**
264 366 130 133 141 249 119 149	**366**	100 89 47 45 22 21 21 15		21	31	4	–	–	1	–	–	–	–	–	–	–	**Androstan-6-one, 17-(acetyloxy)-3-fluoro-5-hydroxy-, (3β,5α,17β)-**		**LI08036**

MASS TO CHARGE RATIOS								M.W.	INTENSITIES								Parent	C	H	O	N	S	F	Cl	Br	I	Si	P	B	X	COMPOUND NAME	CAS Reg No	No
264	366	130	133	141	249	119	367	**366**	100	89	47	45	22	21	21	19		21	31	4	–	–	1	–	–	–	–	–	–	–	Androstan-6-one, 17-(acetyloxy)-3-fluoro-5-hydroxy-, (3β,5β,17β)-	55723-87-0	NI38107
348	43	57	41	69	366			**366**	100	61	54	37	36	34				21	34	5	–	–	–	–	–	–	–	–	–	–	Androstane-17-carboxylic acid, 3,11,12-dihydroxy-, methyl ester, (3β,5α,11β,12β,17β)-	10005-91-1	NI38108
348	366	43	57	55	38	71	42	**366**	100	87	82	59	49	36	27	26		21	34	5	–	–	–	–	–	–	–	–	–	–	Androstane-17-carboxylic acid, 3,11,12-trihydroxy-, methyl ester (3β,5α,11α,12β,17β)-	10005-94-4	NI38109
223	163	121	123	348	195	224	161	**366**	100	42	37	32	25	23	20	19	**0.80**	21	34	5	–	–	–	–	–	–	–	–	–	–	Methyl 8α,12α-epoxy-13α,14-dihydroxyabietanoate		NI38110
69	67	71	121	79	81	77	93	**366**	100	75	71	70	70	69	67	64	**0.00**	21	34	5	–	–	–	–	–	–	–	–	–	–	Phenanthro[1,2-b]oxirene-4-carboxylic acid, tetradecahydro-1b,9-dihydroxy-4,7a-dimethyl-9a-isopropyl-, methyl ester	34217-20-4	NI38111
223	121	123	95	115	163	113	109	**366**	100	81	65	55	51	48	47	42	**0.00**	21	34	5	–	–	–	–	–	–	–	–	–	–	8α-Podocarpan-15-oic acid, 8,12α-epoxy-13α,14α-dihydroxy-13-isopropyl , methyl ester	25594-18-7	NI38112
301	319	283	337	302	299	320	317	**366**	100	39	38	33	21	10	8	8	**0.50**	21	34	5	–	–	–	–	–	–	–	–	–	–	Pregnan-20-one, 3,11,17,21-tetrahydroxy-, (3α,5β,11β,17α)-		NI38113
31	59	43	29	45	74	321	81	**366**	100	93	73	73	72	57	49	43	**1.06**	21	34	5	–	–	–	–	–	–	–	–	–	–	Pregn-5-ene-3,12,14,17,20-pentol, (3β,12β,14β,17α,20S)-	28417-32-5	NI38114
91	321	105	286	145	285	113	304	**366**	100	99	99	93	93	85	76	68	**0.00**	21	34	5	–	–	–	–	–	–	–	–	–	–	Pregn-5-ene-3,12,14,17,20-pentol, (3β,12β,14β,17α,20S)-	28417-32-5	LI08037
67	319	43	109	305	95	81	122	**366**	100	98	98	95	89	88	88	82	**0.00**	21	34	5	–	–	–	–	–	–	–	–	–	–	Tetrahydro-14-seco-hirundigenin		LI08038
57	254	256	200	198	41	43	310	**366**	100	28	27	25	25	19	16	14	**11.60**	21	35	–	–	–	–	–	1	–	–	–	–	–	Benzene, 2-bromo-1,3,5-tris(2,2-dimethylpropyl)-	25347-03-9	MS08639
286	365	288	366	153	151	127	144	**366**	100	94	94	77	23	21	20	19		22	21	–	–	–	–	–	–	–	–	–	2	1	Cobalt, [(2,3,5,6-η)-1,4-dihydro-1-phenylborin][(1,2,3,4,5,6-η)-1-phenylboratabenzene]-	70181-49-6	NI38115
365	286	288	366	59	151	153	127	**366**	100	62	57	51	37	31	29	27		22	21	–	–	–	–	–	–	–	–	–	2	1	Cobalt, [(3,4,5,6-η)-1,2-dihydro-1-phenylborin][(1,2,3,4,5,6-η)-1-phenylboratabenzene]-	70181-50-9	NI38116
323	311	366	324	351	312	367	281	**366**	100	93	48	23	18	17	11	9		22	22	5	–	–	–	–	–	–	–	–	–	–	7,4′-Dimethoxy-5-hydroxy-6-isopentenylisoflavone		NI38117
311	366	323	312	367	281	132	335	**366**	100	76	32	23	19	15	15	10		22	22	5	–	–	–	–	–	–	–	–	–	–	5,6-Dimethylchroman-7,4′-dimethoxyisoflavone		NI38118
351	176	178	175	165	366	173	174	**366**	100	63	61	61	61	60	57	53		22	22	5	–	–	–	–	–	–	–	–	–	–	2′-O-Methylisoleiocin		NI38119
351	173	178	366	165	133	166	135	**366**	100	82	76	70	70	65	40	40		22	22	5	–	–	–	–	–	–	–	–	–	–	2′-O-Methylleiocin		NI38120
348	319	349	287	317	320	366	225	**366**	100	81	38	31	27	21	20	13		22	22	5	–	–	–	–	–	–	–	–	–	–	Tephrostachin, trans-	80377-43-1	NI38121
366	199	307	198	323	184	160	173	**366**	100	29	25	20	8	8	8	7		22	26	3	2	–	–	–	–	–	–	–	–	–	O-Acetyl-Na-demethylseredamine		LI08039
366	109	365	124	367	41	29	351	**366**	100	79	59	39	22	18	17	16		22	26	3	2	–	–	–	–	–	–	–	–	–	Aspidofractine	2348-67-6	NI38122
366	199	160	198	307	184	173	185	**366**	100	21	19	15	14	11	10	9		22	26	3	2	–	–	–	–	–	–	–	–	–	O′-Demethyl-O-acetylseredamine		LI08040
109	366	124	365	310	125	110	136	**366**	100	36	29	28	19	15	12	8		22	26	3	2	–	–	–	–	–	–	–	–	–	21-Isoaspidofractine	6870-43-5	NI38123
216	217	160	366	364				**366**	100	16	12	9	2					22	26	3	2	–	–	–	–	–	–	–	–	–	5-Methoxy-3-spiro[cyclohexane-indoline-2-(4-methoxycarboxanilide)]		MS08640
323	366	324	199	198	297	160	173	**366**	100	89	66	34	34	30	22	21		22	26	3	2	–	–	–	–	–	–	–	–	–	Na-Acetyldemethylseredamine		LI08041
135	366	259	107	121	122	43	367	**366**	100	85	40	35	32	28	25	19		22	26	3	2	–	–	–	–	–	–	–	–	–	Na,Nb-Diacetyl-8,9-dihydro-8,9-cyanoisotuboxenine		LI08042
143	144	223	130	236	150	176	55	**366**	100	30	22	16	11	9	8	8	**6.00**	22	26	3	2	–	–	–	–	–	–	–	–	–	Propanoic acid, 3-(6-oxo-1,2,3,3a,4,5,6,8,9,14-decahydro-2,3-cyclopentapyrido[2,1-a]-β-carbolin-1-yl)-, methyl ester		MS08641
366	181	183	170	168	225	196	322	**366**	100	62	60	25	23	20	15	10		22	26	3	2	–	–	–	–	–	–	–	–	–	Talpinine-Nb-acetamide		NI38124
157	144	190	366	222	145	158	162	**366**	100	72	57	55	38	23	22	17		22	26	3	2	–	–	–	–	–	–	–	–	–	Vincamajine	2506-26-5	NI38125
366	263	183	182	365	335	367	264	**366**	100	73	71	51	50	50	20	20		22	26	3	2	–	–	–	–	–	–	–	–	–	Voachalotine	664-25-5	NI38126
183	366	182	263	365	335	168	364	**366**	100	84	81	68	59	44	39	37		22	26	3	2	–	–	–	–	–	–	–	–	–	Voachalotine	664-25-5	MS08642
194	222	366	180	172	195	193	77	**366**	100	80	54	30	26	20	20	20		23	18	1	4	–	–	–	–	–	–	–	–	–	1,2,3-Triazole, 4-[γ-(phenylamino)cinnamoyl]-5-phenyl-		NI38127
366	159	43	147	183	160	181	197	**366**	100	42	31	16	14	12	10	8		23	30	2	2	–	–	–	–	–	–	–	–	–	Androsta-4,6-dieno[3,2-c]pyrazole, 17β-acetoxy-17α-methyl-		MS08643
105	77	261	232	135	106	51	134	**366**	100	40	11	10	10	9	9	8	**6.00**	23	30	2	2	–	–	–	–	–	–	–	–	–	N,N′-Dibenzoyl-1,9-diaminononane		LI08043
366	323	264	185	144	130	252	213	**366**	100	74	40	40	39	39	36	33		23	30	2	2	–	–	–	–	–	–	–	–	–	Na,Nb-Diacetyl-8,9-dihydro-8,9-chano-isotuboxenin		LI08044
246	366	106	247	133	260	234	367	**366**	100	53	24	19	18	17	16	15		23	30	2	2	–	–	–	–	–	–	–	–	–	4-Piperidinecarboxylic acid, 4-phenyl-1-[3-(phenylamino)propyl]-, ethyl ester	13495-09-5	NI38128
41	55	289	271	43	81	85	67	**366**	100	95	82	79	78	67	61	60	**18.00**	23	42	3	–	–	–	–	–	–	–	–	–	–	6-Pentyl-4-nor-3,5-secoandrostane-3,5β,17β-triol		MS08644
288	197	271	152	287	273	289	366	**366**	100	62	41	41	37	36	32	25		24	18	2	–	–	–	–	–	–	1	–	–	–	9-Oxa-10-silaanthracene, 9,10-dihydro-10-(2-phenoxyphenyl)-		NI38129
105	77	366	106	51	367	78	261	**366**	100	48	21	8	6	5	3	2		24	18	2	2	–	–	–	–	–	–	–	–	–	1,4-Dibenzamidonaphthalene		IC04153
105	77	366	106	367	78	51	261	**366**	100	35	26	8	7	3	3	2		24	18	2	2	–	–	–	–	–	–	–	–	–	1,4-Dibenzamidonaphthalene		IC04154
236	206	115	219	130	91	366	104	**366**	100	99	78	69	66	56	55	55		24	18	2	2	–	–	–	–	–	–	–	–	–	Pyridazino[1,2-b]phthalazine-6,11-dione, 1,4,6,11-tetrahydro-1,4-diphenyl		NI38130
91	209	182	92	152	121	118	366	**366**	100	52	51	49	42	40	28	23		24	22	–	4	–	–	–	–	–	–	–	–	–	1,8-Naphthalenediamine, N[1]-methyl-N[2]-[(1-methyl-2-perimidinyl)methyl]-	93767-82-9	NI38131
41	55	43	29	28	69	74	57	**366**	100	84	66	55	49	38	32	30	**0.84**	24	46	2	–	–	–	–	–	–	–	–	–	–	Cyclopropanedodecanoic acid, 2-octyl-, methyl ester	10152-65-5	NI38132
43	55	41	29	69	56	84	57	**366**	100	60	58	32	31	31	25	24	**0.50**	24	46	2	–	–	–	–	–	–	–	–	–	–	9-Octadecenoic acid (Z)-, hexyl ester	20290-84-0	WI01481
55	82	96	69	83	43	97	57	**366**	100	93	80	76	72	62	52	52	**0.10**	24	46	2	–	–	–	–	–	–	–	–	–	–	14-Tricosenyl formate, (Z)-		MS08645
366	289	77	105	367	152	184	260	**366**	100	55	51	43	29	25	24	16		25	18	1	–	1	–	–	–	–	–	–	–	–	4-Phenyl-4′-benzoyldiphenyl sulphide		IC04155
197	198	44	104	28	141	91	77	**366**	100	15	13	12	12	9	9	8	**0.00**	25	18	3	–	–	–	–	–	–	–	–	–	–	4,4′-Diphenoxybenzophenone		DO01115
366	337	367	338	309	365	368	351	**366**	100	50	38	13	13	12	5	5		25	38	–	2	–	–	–	–	–	–	–	–	–	Bis(4-amino-3-isobutyl-5-ethylphenyl)methane		IC04156
85	57	43	59	58	55	41	56	**366**	100	80	58	54	38	27	25	24	**4.15**	25	50	1	–	–	–	–	–	–	–	–	–	–	2-Docosanone, 4,21,21-trimethyl-, L-(–)-	18607-44-8	NI38133

MASS TO CHARGE RATIOS								M.W.	INTENSITIES								Parent	C	H	O	N	S	F	Cl	Br	I	Si	P	B	X	COMPOUND NAME	CAS Reg No	No
55	82	67	96	110	124	138	348	**366**	100	99	82	80	26	19	11	5	**0.00**	25	50	1	–	–	–	–	–	–	–	–	–	–	4-Pentacosen-1-ol		DD01460
55	82	69	96	110	124	138	348	**366**	100	64	63	47	18	11	7	4	**0.00**	25	50	1	–	–	–	–	–	–	–	–	–	–	18-Pentacosen-1-ol, (Z)-		DD01461
243	165	244	91	241	242	214	166	**366**	100	66	55	49	26	25	23	19	**0.00**	26	22	–	–	1	–	–	–	–	–	–	–	–	Benzyl trityl sulphide	6622-14-6	MS08646
243	165	244	124	109	242	241	166	**366**	100	61	32	29	27	20	18	13	**0.00**	26	22	2	–	–	–	–	–	–	–	–	–	–	Benzene, 1-methoxy-4-(trityloxy)-	20705-45-7	NI38134
105	183	77	184	51	182	106	78	**366**	100	93	65	17	13	8	8	8	**0.00**	26	22	2	–	–	–	–	–	–	–	–	–	–	1,2-Ethanediol, 1,1,2,2-tetraphenyl-	464-72-2	NI38135
183	105	77	184	182	165	106	181	**366**	100	75	23	19	8	7	6	4	**0.00**	26	22	2	–	–	–	–	–	–	–	–	–	–	1,2-Ethanediol, 1,1,2,2-tetraphenyl-	464-72-2	NI38136
188	178	159	366	189	44	77	191	**366**	100	80	66	35	16	16	15	14		26	22	2	–	–	–	–	–	–	–	–	–	–	2,2′-Bis[2-(4-methyl-3-furyl)vinyl]biphenyl		MS08647
188	178	159	366	189	191	179	145	**366**	100	80	66	35	16	14	13	11		26	22	2	–	–	–	–	–	–	–	–	–	–	cis,trans-2,2′-Bis[(4-methyl-3-furyl)vinyl]biphenyl		LI08045
188	178	159	366	191	189	179	160	**366**	100	85	84	38	24	19	13	13		26	22	2	–	–	–	–	–	–	–	–	–	–	trans,trans-2,2′-Bis[(4-methyl-3-furyl)]vinylbiphenyl		LI08046
170	366	196	197	152	153	169	141	**366**	100	79	75	70	51	37	35	34		26	22	2	–	–	–	–	–	–	–	–	–	–	1,2-Bis(4-phenylphenoxy)ethane		IC04157
211	366	212	157	158	226	184	225	**366**	100	82	38	31	26	10	7	6		26	38	1	–	–	–	–	–	–	–	–	–	–	19-Norcholesta-1,3,5(10)-trien-6-one	19454-79-6	NI38137
57	43	71	85	41	55	69	56	**366**	100	95	74	47	40	34	18	15	**0.03**	26	54	–	–	–	–	–	–	–	–	–	–	–	5-Butyldocosane		AI02243
43	57	71	41	85	55	126	69	**366**	100	90	64	47	42	38	22	20	**0.06**	26	54	–	–	–	–	–	–	–	–	–	–	–	5-Butyldocosane		AI02244
43	57	71	85	41	55	69	56	**366**	100	94	54	41	41	35	18	15	**0.02**	26	54	–	–	–	–	–	–	–	–	–	–	–	7-Butyldocosane		AI02245
43	57	71	41	85	55	69	56	**366**	100	96	58	47	40	37	20	17	**0.09**	26	54	–	–	–	–	–	–	–	–	–	–	–	9-Butyldocosane		AI02246
57	43	71	85	41	55	69	56	**366**	100	87	63	43	37	31	16	14	**0.03**	26	54	–	–	–	–	–	–	–	–	–	–	–	9-Butyldocosane		AI02247
43	57	71	41	85	55	69	56	**366**	100	95	57	47	40	37	20	17	**0.08**	26	54	–	–	–	–	–	–	–	–	–	–	–	11-Butyldocosane		AI02248
71	85	69	99	83	113	70	97	**366**	100	71	30	23	19	16	15	14	**0.03**	26	54	–	–	–	–	–	–	–	–	–	–	–	11-Butyldocosane		AI02249
71	85	69	126	83	70	127	99	**366**	100	65	27	23	18	15	14	14	**0.03**	26	54	–	–	–	–	–	–	–	–	–	–	–	5,14-Dibutyloctadecane		AI02250
57	56	41	43	55	29	71	27	**366**	100	55	36	33	21	16	15	15	**0.02**	26	54	–	–	–	–	–	–	–	–	–	–	–	11-(2,2-Dimethylpropyl)heneicosane		AI02251
57	43	71	41	85	29	295	27	**366**	100	85	57	55	41	31	30	29	**0.11**	26	54	–	–	–	–	–	–	–	–	–	–	–	6,11-Dipentylhexadecane		AI02252
71	85	69	99	97	83	84	70	**366**	100	83	38	28	28	28	27	25	**0.00**	26	54	–	–	–	–	–	–	–	–	–	–	–	3-Ethyl-5-(2-ethylbutyl)octadecane		AI02253
43	57	71	41	85	55	69	70	**366**	100	61	47	35	34	32	18	13	**0.00**	26	54	–	–	–	–	–	–	–	–	–	–	–	3-Ethyl-5-(2-ethylbutyl)octadecane		AI02254
43	57	71	41	70	55	29	85	**366**	100	74	63	35	33	33	22	21	**0.02**	26	54	–	–	–	–	–	–	–	–	–	–	–	3-Ethyltetracosane		AI02255
43	57	71	41	70	55	337	85	**366**	100	80	65	47	43	43	31	30	**0.12**	26	54	–	–	–	–	–	–	–	–	–	–	–	3-Ethyltetracosane		AI02256
43	57	71	41	70	55	29	85	**366**	100	69	61	39	34	34	30	26	**0.03**	26	54	–	–	–	–	–	–	–	–	–	–	–	3-Ethyltetracosane		AI02257
57	43	71	41	85	55	29	69	**366**	100	97	59	36	29	29	23	15	**1.87**	26	54	–	–	–	–	–	–	–	–	–	–	–	Hexacosane	630-01-3	AI02258
57	43	71	85	41	55	56	99	**366**	100	77	55	35	26	22	12	10	**1.00**	26	54	–	–	–	–	–	–	–	–	–	–	–	Hexacosane	630-01-3	NI38138
57	43	71	85	41	55	56	99	**366**	100	77	55	35	26	22	17	10	**1.80**	26	54	–	–	–	–	–	–	–	–	–	–	–	Hexacosane	630-01-3	AI02259
57	43	71	85	41	55	69	56	**366**	100	93	61	40	39	31	16	14	**0.02**	26	54	–	–	–	–	–	–	–	–	–	–	–	7-Hexyleicosane		AI02260
43	57	71	85	70	41	55	69	**366**	100	87	62	39	39	37	35	18	**0.00**	26	54	–	–	–	–	–	–	–	–	–	–	–	11-(3-Pentyl)heneicosane		AI02261
57	43	71	85	41	55	69	56	**366**	100	92	63	43	39	32	17	14	**0.02**	26	54	–	–	–	–	–	–	–	–	–	–	–	11-Pentylheneicosane		AI02262
43	57	71	85	55	69	224	56	**366**	100	97	56	40	40	20	19	17	**0.00**	26	54	–	–	–	–	–	–	–	–	–	–	–	11-Pentylheneicosane		AI02263
43	57	71	41	85	55	69	224	**366**	100	91	55	45	39	36	19	18	**0.12**	26	54	–	–	–	–	–	–	–	–	–	–	–	11-Pentylheneicosane		AI02264
327	329	325	278	276	331	313	280	**367**	100	64	62	32	26	22	16	16	**4.30**	4	9	–	3	–	–	5	–	–	–	3	–	–	1-Butylpentachlorocyclotriphosphazene	75132-82-0	NI38139
313	57	315	311	278	41	276	280	**367**	100	91	64	62	53	48	42	26	**13.89**	4	9	–	3	–	–	5	–	–	–	3	–	–	1-tert-Butylpentachlorocyclotriphosphazene	75155-06-5	NI38140
100	127	69	150	131	31	70	64	**367**	100	27	27	26	26	22	16	11	**0.00**	5	1	2	1	1	12	–	–	–	–	–	–	–	[2,2,3,3,4,4-Hexafluoro-4-(fluoroformyl)butyramido]sulphur pentafluoride	91940-19-1	NI38141
354	352	77	356	369	367	371	358	**367**	100	78	59	48	27	21	13	10		7	8	–	3	–	–	4	–	–	–	3	–	–	1-Methyl-1-phenyltetrachlorocyclotriphosphazene	84811-29-0	NI38142
280	252	43	41	69	28	57	281	**367**	100	99	69	42	40	40	33	29	**0.00**	12	12	4	1	–	7	–	–	–	–	–	–	–	Propyl N-(heptafluorobutyryl)pyroglutamate		MS08648
182	121	184	154	367	97	241	58	**367**	100	64	54	38	36	34	27	24		12	15	4	1	2	–	1	–	–	–	1	–	–	Phosphorodithioic acid, S-[(6-chloro-2-oxo-3(2H)-benzoxazolyl)methyl] O,O-diethyl ester	2310-17-0	NI38143
182	121	97	65	184	111	154	125	**367**	100	66	49	38	34	32	28	22	**17.57**	12	15	4	1	2	–	1	–	–	–	1	–	–	Phosphorodithioic acid, S-[(6-chloro-2-oxo-3(2H)-benzoxazolyl)methyl] O,O-diethyl ester	2310-17-0	NI38144
184	182	183	155	186	170	185	127	**367**	100	49	38	38	26	25	18	16	**0.00**	12	15	4	1	2	–	1	–	–	–	1	–	–	Phosphorodithioic acid, S-[(6-chloro-2-oxo-3(2H)-benzoxazolyl)methyl] O,O-diethyl ester	2310-17-0	NI38145
57	41	29	153	69	152	56	27	**367**	100	74	73	53	38	31	18	16	**0.00**	12	15	5	1	–	6	–	–	–	–	–	–	–	L-Threonine, N,N-bis(trifluoroacetyl)-, butyl ester	55282-41-2	NI38146
57	41	29	153	69	152	56	27	**367**	100	74	73	53	38	31	18	16	**0.00**	12	15	5	1	–	6	–	–	–	–	–	–	–	Threonine, N,O-bis(trifluoroacetyl)-, butyl ester		MS08649
57	153	41	152	69	84	49	56	**367**	100	50	45	37	24	16	10	9	**0.00**	12	15	5	1	–	6	–	–	–	–	–	–	–	L-Threonine, N,O-bis(trifluoroacetyl)-, sec-butyl ester	57983-18-3	NI38147
368	240	254	369	255	312	241	370	**367**	100	48	36	10	7	5	3	1	**0.00**	12	15	5	1	–	6	–	–	–	–	–	–	–	L-Threonine, N-(trifluoroacetyl)-, butyl ester, trifluoroacetate (ester)	5282-97-3	NI38148
44	28	136	153	52	39	31	29	**367**	100	34	31	27	25	25	25	25	**4.50**	14	12	6	2	–	–	–	–	–	–	–	–	1	Bis(3-methoxy-1-quinone-2-oximato)copper(II)		MS08650
196	198	83	97	55	332	308	72	**367**	100	96	84	47	46	43	43	39	**5.00**	14	16	4	1	–	–	3	–	–	–	–	–	–	L-Valine, N-[(2,4,5-trichlorophenoxy)acetyl]-, methyl ester	66789-89-7	NI38149
73	352	367	75	353	77	147	45	**367**	100	48	32	24	16	14	12	12		14	29	1	5	–	–	–	–	–	3	–	–	–	9H-Purin-6-amine, N,9-bis(trimethylsilyl)-8-[(trimethylsilyl)oxy]-	56145-36-9	NI38150
367	245	93	53	192	67	348	271	**367**	100	22	20	20	19	19	17	13		15	6	1	5	–	5	–	–	–	–	–	–	–	10(9H)-Pyrimido[1,2-a]purinone, 7-(pentafluorophenyl)-1-methyl-		NI38151
170	367	111	352	74	126	128	139	**367**	100	94	88	66	48	42	40	38		15	17	8	3	–	–	–	–	–	–	–	–	–	N^6-Acetyl-2′,3′-O-diacetyl-O^6,5′-cyclocytidine		NI38152

MASS TO CHARGE RATIOS	M.W.	INTENSITIES	Parent	C	H	O	N	S	F	Cl	Br	I	Si	P	B	X	COMPOUND NAME	CAS Reg No	No
213 129 121 185 56 367 211 138	**367**	100 93 89 70 62 54 43 41		16	13	4	3	–	–	–	–	–	–	–	–	1	1-(5-Nitrofurfurylidenehydrazinocarbonyl)ferrocene		LI08047
77 367 200 107 66 321 105 172	**367**	100 50 37 29 23 15 15 14		16	13	4	7	–	–	–	–	–	–	–	–	–	1-Phenyl-3-methyl-4-(2,4-dinitrophenylazo)-5-aminopyrazole		IC04158
91 184 219 175 148 147 65 186	**367**	100 59 31 31 28 28 28 25	2.00	17	15	4	1	–	–	2	–	–	–	–	–	–	L-Phenylalanine, N-[(2,4-dichlorophenoxy)acetyl]-	2752-55-8	NI38153
242 244 243 73 365 214 369 277	**367**	100 42 34 21 19 17 15 15	0.00	17	19	2	1	–	–	2	–	–	1	–	–	–	Benzoic acid, 2-[(2,6-dichloro-3-methylphenyl)amino]-, trimethylsilyl ester	74810-88-1	NI38154
91 68 65 223 212 155 92 152	**367**	100 51 33 32 22 19 17 14	6.10	17	21	6	1	1	–	–	–	–	–	–	–	–	7-Azabicyclo[2.2.1]heptane-2,3-dicarboxylic acid, 7-(4-tolylsulphonyl)-, dimethyl ester, trans-	17037-57-9	NI38155
352 320 111 260 336 284 367	**367**	100 81 17 15 14 10 1		17	21	8	1	–	–	–	–	–	–	–	–	–	5-tert-Butylimino-2-methoxy-2-(methoxycarbonylethynyl)-3,4-bis(methoxycarbonyl)-2,5-dihydrofuran		LI08048
121 367 122 207 368 147 119 91	**367**	100 14 9 4 3 2 2 2		17	22	2	1	–	5	–	–	–	–	–	–	–	Benzeneheptanamine, 4-methoxy-N-(pentafluoropropionyl)-		NI38156
223 179 73 224 310 41 192 75	**367**	100 44 41 18 14 12 10 10	9.20	17	25	4	1	1	–	–	–	–	1	–	–	–	Benzenepropanoic acid, 3-ethoxy-4-[(trimethylsilyl)oxy]-α-isothiocyanato, ethyl ester		MS08651
130 73 247 75 232 131 203 77	**367**	100 39 22 10 8 8 6 5	0.00	17	26	2	1	–	–	1	–	–	2	–	–	–	1H-Indole-3-acetic acid, 5-chloro-2-methyl-1-(trimethylsilyl)-, trimethylsilyl ester	74367-57-0	NI38157
73 250 252 367 251 45 75 74	**367**	100 68 26 19 15 14 10 9		17	26	2	1	–	–	1	–	–	2	–	–	–	1H-Indole-3-acetic acid, 5-chloro-2-methyl-1-(trimethylsilyl)-, trimethylsilyl ester	74367-57-0	NI38158
179 73 308 75 43 293 180 147	**367**	100 90 50 30 24 18 18 18	0.00	17	29	4	1	–	–	–	–	–	2	–	–	–	Tyrosine, N-acetylbis(trimethylsilyl)-		NI38159
369 367 323 325 204 77 102 88	**367**	100 99 85 80 47 47 38 38		18	10	3	1	–	–	–	1	–	–	–	–	–	2-(1,3-Dioxoindan-2-yl)3-hydroxybromoquinoline		IC04159
367 241 225 273 254 242 211 271	**367**	100 65 39 33 31 31 29 27		18	16	4	1	–	3	–	–	–	–	–	–	–	7-Isoquinolinol, 1,2,3,4-tetrahydro-4-(4-hydroxyphenyl)-6-methoxy-2-(trifluoroacetyl)-, (S)-	56771-97-2	NI38160
115 143 127 116 286 159 89 128	**367**	100 96 52 40 36 28 26 25	11.00	19	13	–	1	1	–	–	–	–	–	–	–	1	2-Naphthalenamine, (2-selenyl-3-benzo[b]thienylidene)-	71643-93-1	NI38161
59 309 291 281 44 139 367 350	**367**	100 85 54 35 22 21 16 12		19	13	7	1	–	–	–	–	–	–	–	–	–	Ekatetrone		MS08652
151 152 180 179 107 106 36 65	**367**	100 17 10 5 5 4 4 2	0.00	19	26	4	1	–	–	1	–	–	–	–	–	–	Benzeneethanamine, N-[(3,4-dimethoxyphenyl)methyl]-3,4-dimethoxy-, hydrochloride	55124-87-3	NI38162
190 131 177 77 91 117 160 149	**367**	100 77 67 62 57 45 37 36	7.00	20	17	6	1	–	–	–	–	–	–	–	–	–	D-Bicuculline	485-49-4	NI38163
367 352 183 324 368 184 353 279	**367**	100 49 26 24 20 16 12 9		20	17	6	1	–	–	–	–	–	–	–	–	–	5H-Isoindolo[1,2-b][3]benzazepin-5-one, 7,8-dihydro-3,12-dimethoxy-4-hydroxy-10,11-methylenedioxy-		MS08653
192 65 63 77 89 188 176 163	**367**	100 94 88 79 71 50 50 35	21.00	20	17	6	1	–	–	–	–	–	–	–	–	–	Ribasidine	88151-44-4	NI38164
367 322 190 204 352 338 368 336	**367**	100 59 50 33 30 25 23 20		20	17	6	1	–	–	–	–	–	–	–	–	–	Sibiricine	24181-66-6	NI38165
307 107 323 308 167 367 106 247	**367**	100 79 72 65 64 63 58 52		20	21	4	3	–	–	–	–	–	–	–	–	–	6,21-Cyclo-4,5-secoakuammilan-17-oic acid, 4-hydroxy-5-(hydroxyimino)-, methyl ester, (6α)-	63944-64-9	NI38166
207 179 167 43 119 267 367 41	**367**	100 69 47 40 33 28 25 23		20	33	5	1	–	–	–	–	–	–	–	–	–	1-Cycloheptene-1-carboxamide, N,N-diisopropyl-4,6-bis(methoxycarbonyl)-4,6-dimethyl-	119039-14-4	DD01462
367 43 283 325 282 368 28 254	**367**	100 79 67 39 30 24 23 20		21	21	5	1	–	–	–	–	–	–	–	–	–	Crotonosine, 6-acetyl-, acetate	10214-67-2	MS08654
367 43 283 325 282 368 254 253	**367**	100 90 68 40 30 24 21 18		21	21	5	1	–	–	–	–	–	–	–	–	–	Crotonosine, 6-acetyl-, acetate	10214-67-2	LI08049
43 222 165 30 27 216 115 41	**367**	100 9 7 7 7 6 6 6	0.00	21	21	5	1	–	–	–	–	–	–	–	–	–	Crotonosine, 6-acetyl-, acetate	10214-67-2	NI38167
367 338 162 336 337 366 352 368	**367**	100 76 54 50 32 28 28 25		22	25	4	1	–	–	–	–	–	–	–	–	–	Dihydroochotensimine		LI08050
57 132 216 217 85 130 65 117	**367**	100 66 61 56 15 13 13 12	0.00	22	25	4	1	–	–	–	–	–	–	–	–	–	1-Isoquinolinemethanol, u-2-tert-butoxy-1,2,3,4-tetrahydro-α-(1,3-benzodioxol-5-yl)-		DD01463
367 368 336 279 352 282 229 337	**367**	100 31 20 20 16 14 11 9		22	29	2	1	–	–	–	–	–	1	–	–	–	Estra-1,3,5,7,9-pentaen-17-one, 3-[(trimethylsilyl)oxy]-, O-methyloxime	69833-43-8	NI38168
264 367 217 77 218 368 265 115	**367**	100 87 40 32 28 23 22 17		23	17	2	3	–	–	–	–	–	–	–	–	–	4H-1,2-Diazepine, 5-(3-nitrophenyl)-3,7-diphenyl-	25649-76-7	NI38169
264 367 217 77 218 78 368 103	**367**	100 91 51 41 37 34 25 23		23	17	2	3	–	–	–	–	–	–	–	–	–	4H-1,2-Diazepine, 5-(4-nitrophenyl)-3,7-diphenyl-	25649-75-6	NI38170
367 130 158 91 248 352 368 103	**367**	100 67 39 34 32 30 28 13		23	21	–	5	–	–	–	–	–	–	–	–	–	2-Methyl-3-[4-(2-tolyazo)-3-methylphenylazo]indole		IC04160
105 245 262 77 219 246 367 178	**367**	100 59 55 50 38 30 26 26		23	29	3	1	–	–	–	–	–	–	–	–	–	4,10-Cyclolycopodane-2,5-diol, 15-methyl-, 2-benzoate, (2α,5β,15R)-		LI08051
91 154 276 367 84 110 82	**367**	100 38 37 36 27 22 20		23	29	3	1	–	–	–	–	–	–	–	–	–	2-Hydroxy-4-(3-benzyloxy-4-methoxyphenyl)-trans-quinolizidine		LI08052
105 245 77 122 51 246 83 50	**367**	100 60 55 43 17 13 10 10	5.00	24	17	3	1	–	–	–	–	–	–	–	–	–	2-(N-Benzoyl-1-naphthylamino)benzoic acid		NI38171
105 77 310 194 339 180 367 283	**367**	100 82 61 61 54 49 36 18		24	17	3	1	–	–	–	–	–	–	–	–	–	2,3-Furandione, 4-[α-(4-methylphenylimino)benzyl]-5-phenyl-	90140-28-6	NI38172
227 43 57 30 367 240 73 41	**367**	100 79 56 55 53 47 40 39		24	49	1	1	–	–	–	–	–	–	–	–	–	Dodecanamide, N-dodecyl-	33422-43-4	NI38173
263 367 131 264 158 132 133 339	**367**	100 31 26 21 16 7 6 4		25	21	2	1	–	–	–	–	–	–	–	–	–	1,8-Acridinedione, 3,6-diphenyl-1,2,3,4,5,6,7,8-octahydro-		NI38174
367 366 183 185 262 136 290 108	**367**	100 58 36 9 8 8 7 6		25	22	–	1	–	–	–	–	–	–	1	–	–	Aniline, 4-methyl-N-(triphenylphosphoranylidene)-	2327-67-5	LI08053
367 366 183 368 277 136 185 77	**367**	100 55 35 25 12 10 9 8		25	22	–	1	–	–	–	–	–	–	1	–	–	Aniline, 4-methyl-N-(triphenylphosphoranylidene)-	2327-67-5	NI38175
367 309 91 310 289 366 368 275	**367**	100 94 54 45 44 34 29 20		25	37	1	1	–	–	–	–	–	–	–	–	–	6-Azaandrostan-17β-ol, 6-benzyl-	16373-54-9	NI38176
93 109 79 369 47 353 97 337	**368**	100 53 17 16 13 10 10 9	0.00	7	12	6	–	–	6	–	–	–	–	2	–	–	[1-(Dimethoxyphosphino)-2,2,2-trifluoro-1-(trifluoromethyl)ethyl] dimethyl phosphate	111557-95-0	DD01464

MASS TO CHARGE RATIOS								M.W.	INTENSITIES								Parent	C	H	O	N	S	F	Cl	Br	I	Si	P	B	X	COMPOUND NAME	CAS Reg No	No
284	286	282	258	260	256	283	312	**368**	100	96	73	61	49	49	45	43	**9.51**	8	5	3	–	–	–	1	–	–	–	–	–	1	**Tungsten, chloro(η^5-2,4-cyclopentadien-1-yl)tricarbonyl-**	12128-24-4	**NI38177**
350	36	284	247	221	249	222	277	**368**	100	78	47	47	47	42	42	40	**8.09**	8	5	3	–	–	–	1	–	–	–	–	–	1	**Tungsten, chloro(π-cyclopentadienyl)tricarbonyl-**		**MS08655**
36	263	44	80	82	265	261	38	**368**	100	54	53	44	41	36	36	34	**5.00**	9	2	3	–	–	–	6	–	–	–	–	–	–	**4,7-Methanoisobenzofuran-1,3-dione, 4,5,6,7,8,8-hexachloro-3a,4,7,7a-tetrahydro-**	115-27-5	**NI38178**
153	89	370	368	372	152	63	56	**368**	100	50	43	24	22	22	20	17		12	10	–	–	–	–	–	2	–	–	–	–	1	**1,1-Dibromo-2-ferrocenylethylene**		**MS08656**
225	227	73	226	279	338	147	135	**368**	100	59	44	40	38	33	30	17	**0.00**	12	29	3	2	1	–	–	–	–	2	1	–	–	**Phosphonic acid, [1-(carbonothioylhydrazino)pentyl]-, bis(trimethylsilyl) ester**	56051-91-3	**NI38179**
281	283	282	253	309	43	284	267	**368**	100	43	27	19	14	14	13	10	**0.00**	12	32	5	–	–	–	–	–	–	4	–	–	–	**Cyclotetrasiloxane, 1,3,5,7-tetraethyl-1-butoxy-**	73420-27-6	**NI38180**
281	57	282	283	59	325	58	85	**368**	100	98	57	56	34	25	17	13	**0.00**	12	32	5	–	–	–	–	–	–	4	–	–	–	**Cyclotrisiloxane, 1,3,5-triethyl-1-(ethylbutoxysiloxy)-**	73420-25-4	**NI38181**
305	307	172	309	207	59	209	269	**368**	100	95	50	33	31	24	21	18	**4.00**	13	12	2	2	–	–	4	–	–	–	–	–	–	**Pentacyclo[5.4.0.0^{2,6}.0^{3,10}.0^{5,9}]undecane, 8,11-diazenediyl-4,4-dimethoxy-2,3,5,6-tetrachloro-**		**MS08657**
145	28	44	108	164	146	30	54	**368**	100	44	34	27	24	20	19	18	**0.00**	13	16	7	6	–	–	–	–	–	–	–	–	–	**N-[9-(β-D-ribofuranosyl)purin-6-ylcarbamoyl]glycine**		**LI08054**
57	73	41	281	207	45	282	353	**368**	100	69	58	37	22	12	11	9	**0.00**	13	36	4	–	–	–	–	–	–	4	–	–	–	**3-Butoxy-1,1,1,5,5,5-hexamethyl-3-(trimethylsiloxy)trisiloxane**		**MS08658**
208	180	75	288	304	224	139	150	**368**	100	80	73	66	65	56	38	32	**2.00**	14	8	8	–	2	–	–	–	–	–	–	–	–	**1,5-Disulphoanthraquinone**		**IC04161**
139	111	75	50	141	76	251	51	**368**	100	83	74	47	44	35	31	27	**0.00**	14	9	1	–	–	–	5	–	–	–	–	–	–	**Benzenemethanol, 2-chloro-α-(4-chlorophenyl)-α-(trichloromethyl)-**	10606-46-9	**NI38182**
139	251	111	253	141	75	113	50	**368**	100	61	46	40	34	30	17	13	**0.30**	14	9	1	–	–	–	5	–	–	–	–	–	–	**Benzenemethanol, 4-chloro-α-(4-chlorophenyl)-α-(trichloromethyl)-**	115-32-2	**NI38183**
139	251	253	111	141	75	252	113	**368**	100	74	48	35	30	14	13	11	**0.00**	14	9	1	–	–	–	5	–	–	–	–	–	–	**Benzenemethanol, 4-chloro-α-(4-chlorophenyl)-α-(trichloromethyl)-**	115-32-2	**NI38184**
139	251	253	111	141	246	75	252	**368**	100	98	64	33	30	18	16	15	**1.13**	14	9	1	–	–	–	5	–	–	–	–	–	–	**Benzenemethanol, 4-chloro-α-(4-chlorophenyl)-α-(trichloromethyl)-**	115-32-2	**NI38185**
368	270	280	324	282	326	372	369	**368**	100	90	40	36	36	34	24	18		14	10	4	–	–	–	–	–	–	–	–	–	2	**Bis(copper benzoate)**		**LI08055**
28	256	55	184	27	227	200	140	**368**	100	73	57	24	18	14	14	14	**0.00**	14	16	8	–	–	–	–	–	–	–	–	–	1	**Iron, [1,4-bis(ethoxycarbonyl)-1,4-butanediyl]tetracarbonyl-**	62265-45-6	**NI38186**
71	169	170	200	368	201	82	141	**368**	100	89	78	51	44	44	41	31		14	20	4	6	1	–	–	–	–	–	–	–	–	**Bis(1,3-dimethyl-2,4-dioxo-6-methylamino-1,2,3,4-tetrahydro-pyrimid-5-yl) sulphide**		**LI08056**
73	353	368	325	45	354	171	369	**368**	100	58	33	22	19	18	15	11		14	28	2	4	–	–	–	–	–	3	–	–	–	**8-Azaguanine, N^2,O^6,9-tris(trimethylsilyl)-**		**NI38187**
73	353	368	147	354	75	40	45	**368**	100	84	57	39	29	23	23	19		14	28	2	4	–	–	–	–	–	3	–	–	–	**9H-Purine, 9-(trimethylsilyl)-2,6-bis[(trimethylsilyl)oxy]-**	18551-03-6	**NI38188**
43	135	60	84	158	138	126	72	**368**	100	87	69	42	39	27	26	22	**0.42**	14	28	5	2	2	–	–	–	–	–	–	–	–	**D-Altrose, 3,6-bis(acetylamino)-3,6-dideoxy-, 1-(diethylmercaptal)**	3509-39-5	**NI38189**
135	43	60	132	174	72	177	102	**368**	100	67	46	36	25	24	23	23	**2.94**	14	28	5	2	2	–	–	–	–	–	–	–	–	**L-Idose, 2,6-diacetamido-2,6-dideoxy-, diethyl mercaptal**	3509-38-4	**NI38190**
77	149	65	117	51	151	118	91	**368**	100	91	63	52	43	42	39	27	**1.00**	15	12	–	8	2	–	–	–	–	–	–	–	–	**1H-Tetrazole, 5,5′-[methylenebis(thio)]bis[1-phenyl-**	25746-68-3	**NI38191**
57	41	58	29	43	42	59	114	**368**	100	63	55	46	31	26	23	14	**0.50**	15	19	4	–	–	–	3	–	–	–	–	–	–	**Acetic acid, (2,4,5-trichlorophenoxy)-, 2-butoxypropyl ester**	3084-62-6	**NI38192**
155	156	130	44	34	60	43	76	**368**	100	76	58	51	50	32	26	18	**0.00**	16	20	6	2	1	–	–	–	–	–	–	–	–	**Desulphoglucobrassicin**		**MS08659**
28	310	215	110	82	27	55	45	**368**	100	68	57	57	57	54	50	50	**5.00**	16	20	8	2	–	–	–	–	–	–	–	–	–	**Uridine, 2′,3′-O-isopropylidene-, 4,5′-diacetate**	29015-22-3	**NI38193**
368	154	326	321	370	214	69	215	**368**	100	81	54	48	41	35	34	31		17	17	5	–	1	–	1	–	–	–	–	–	–	**Spiro[benzofuran-2(3H),1′-[2]cyclohexene]-3,4′-dione, 7-chloro-4,6-dimethoxy-6′-methyl-2′-(methylthio)-**	26881-62-9	**NI38194**
254	114	67	340	321	368	215	256	**368**	100	56	52	49	43	42	40	35		17	17	5	–	1	–	1	–	–	–	–	–	–	**Spiro[benzofuran-2(3H),1′-[3]cyclohexene]-2′,3-dione, 7-chloro-4,6-dimethoxy-6′-methyl-4′-(methylthio)-**	26881-76-5	**NI38195**
215	368	310	180	214	370	312	217	**368**	100	79	62	42	38	37	31	30		17	17	7	–	–	–	1	–	–	–	–	–	–	**Spiro[benzofuran-2(3H),1′-[2]cyclohexene]-3,4′-dione, 7-chloro-5′-hydroxy-2′,4,6-trimethoxy-6′-methyl-, [1′S-(1′α,5′β,6′β)]-**	52745-96-7	**MS08660**
263	353	368	251	73	277	262	249	**368**	100	74	38	29	24	21	21	20		17	19	2	2	–	3	–	–	–	1	–	–	–	**Flanixin, trimethylsilyl ester**		**NI38196**
113	145	224	59	192	165	337	277	**368**	100	80	42	31	29	27	21	18	**0.30**	17	20	9	–	–	–	–	–	–	–	–	–	–	**Tricyclo[5.2.0.0^{2,5}]nonane-3,4,8,9-tetracarboxylic acid, 6-oxo-, tetramethyl ester**	74742-36-2	**NI38197**
216	368	97	125	77	105	115	215	**368**	100	90	89	76	36	32	25	22		17	21	5	–	1	–	–	–	–	–	1	–	–	**Phosphorothioic acid, O,O-diethyl O-(7,8,9,10-tetrahydro-6-oxo-6H-dibenzo[b,d]pyran-3-yl) ester**	572-48-5	**NI38198**
368	353	338	73	369	75	370	69	**368**	100	29	29	29	24	15	10	10		17	28	5	–	–	–	–	–	–	2	–	–	–	**Cinnamic acid, 3,5-dimethoxy-4-(trimethylsiloxy)-, trimethylsilyl ester**	27750-80-7	**LI08057**
368	353	338	73	369	75	370	249	**368**	100	30	30	28	26	15	13	12		17	28	5	–	–	–	–	–	–	2	–	–	–	**Cinnamic acid, 3,5-dimethoxy-4-(trimethylsiloxy)-, trimethylsilyl ester**	27750-80-7	**NI38199**
147	73	353	148	45	75	368	149	**368**	100	60	38	15	12	10	8	8		17	28	5	–	–	–	–	–	–	2	–	–	–	**2-Propenoic acid, 3-(2,3-dimethoxyphenyl)-2-[(trimethylsilyl)oxy]-, trimethylsilyl ester**		**MS08661**
147	73	77	148	75	149	44	310	**368**	100	56	36	16	11	10	8	7	**0.00**	17	28	5	–	–	–	–	–	–	2	–	–	–	**2-Propenoic acid, 3-(2,3-dimethoxyphenyl)-2-[(trimethylsilyl)oxy]-, trimethylsilyl ester**		**MS08662**
147	73	297	353	368	282	354	298	**368**	100	71	66	65	26	26	20	17		17	28	5	–	–	–	–	–	–	2	–	–	–	**2-Propenoic acid, 3-(2,4-dimethoxyphenyl)-2-[(trimethylsilyl)oxy]-, trimethylsilyl ester**		**MS08663**
147	73	353	368	148	77	354	75	**368**	100	58	36	18	16	13	10	8		17	28	5	–	–	–	–	–	–	2	–	–	–	**2-Propenoic acid, 3-(2,5-dimethoxyphenyl)-2-[(trimethylsilyl)oxy]-, trimethylsilyl ester**		**MS08664**
73	353	236	45	368	59	206	89	**368**	100	33	24	24	17	15	13	13		17	28	5	–	–	–	–	–	–	2	–	–	–	**2-Propenoic acid, 3-[3-methoxy-4-[(trimethylsilyl)oxy]phenyl]-2-[(trimethylsilyl)oxy]-, methyl ester**		**MS08665**
353	73	312	354	325	100	311	75	**368**	100	77	49	30	29	26	21	21	**3.16**	17	32	3	2	–	–	–	–	–	2	–	–	–	**2,4,6(1H,3H,5H)-Pyrimidinetrione, 5-isobutyl-5-(2-propenyl)-1,3-bis(trimethylsilyl)-**	52937-70-9	**NI38200**

MASS TO CHARGE RATIOS								M.W.	INTENSITIES								Parent	C	H	O	N	S	F	Cl	Br	I	Si	P	B	X	COMPOUND NAME	CAS Reg No	No
73	297	353	312	339	75	100	109	368	100	99	73	70	52	47	40	34	7.06	17	32	3	2	–	–	–	–	–	2	–	–	–	2,4,6(1H,3H,5H)-Pyrimidinetrione, 5-sec-butyl-5-(2-propenyl)-1,3-bis(trimethylsilyl)-	52937-69-6	NI38201
351	78	349	274	347	350	272	197	368	100	92	76	44	43	39	35	34	0.19	18	16	1	–	–	–	–	–	–	–	–	–	1	Stannane, hydroxytriphenyl-	76-87-9	NI38202
351	120	154	78	197	349	118	77	368	100	97	95	91	85	77	72	70	0.00	18	16	1	–	–	–	–	–	–	–	–	–	1	Stannane, hydroxytriphenyl-	76-87-9	NI38203
167	122	123	166	229	228	368	185	368	100	56	50	34	20	20	19	19		18	24	8	–	–	–	–	–	–	–	–	–	–	Malonic acid, isobutylidene-, bimol. cyclic ethylene ester	21620-32-6	NI38204
197	311	199	309	255	195	253	196	368	100	88	76	59	56	54	43	29	0.00	18	32	–	–	–	–	–	–	–	–	–	–	1	Phenyltributyltin		MS08666
309	281	368	291	139	322	252	253	368	100	48	25	24	9	5	5	4		19	12	8	–	–	–	–	–	–	–	–	–	–	Ekatetrone acid		MS08667
93	217	139	275	63	107	183	152	368	100	68	61	36	35	34	23	9	2.00	19	17	–	2	2	–	–	–	–	–	1	–	–	1-Diphenylphosphinothioyl-3-phenylthiourea		MS08668
176	161	91	175	189	177	124	280	368	100	80	58	46	30	20	17	16	0.00	19	21	–	–	1	4	–	–	–	–	–	1	–	1-Phenethyl-2,3,5-trimethylbenzo[b]thiophenium tetrafluoroborate		LI08058
325	368	353	339	369	310	326	354	368	100	96	85	24	23	21	20	17		20	16	7	–	–	–	–	–	–	–	–	–	–	Aversin		NI38205
284	326	43	368	283	44	238	266	368	100	66	53	46	39	31	24	23		20	16	7	–	–	–	–	–	–	–	–	–	–	4H-1-Benzopyran-4-one, 7-(acetyloxy)-5-methoxy-3-[4-(acetyloxy)phenyl]		NI38206
310	325	368	286	297	285	350	311	368	100	97	96	93	92	92	85	85		20	16	7	–	–	–	–	–	–	–	–	–	–	2,6-Epoxy-2H-anthra[2,3-b]oxocin-8,13-dione, 3,4,5,6-tetrahydro-7,9,11-trihydroxy-2-methyl-	14016-29-6	NI38207
368	367	275	369	276	256	227	257	368	100	84	66	24	13	12	11	10		20	17	3	–	1	–	–	–	–	–	1	–	–	Phenothiaphosphine, 2,8-dimethyl-10-phenyl-, 5,5,10-trioxide	23861-49-6	NI38208
91	85	149	132	114	59	87	368	368	100	31	15	15	12	12	9	6		20	20	3	2	1	–	–	–	–	–	–	–	–	2-Benzylimino-3-benzyl-6-(methoxycarbonyl)perhydro-1,3-thiazin-4-one		MS08669
263	120	222	323	204	250	368	178	368	100	64	45	29	28	23	22	19		20	20	5	2	–	–	–	–	–	–	–	–	–	Oxonareline	63999-23-5	NI38209
105	124	111	77	123	110	106	125	368	100	69	28	19	17	7	7	5	0.00	20	21	6	–	–	–	–	–	–	–	–	1	–	1,3,2-Dioxaborinane-4-methanol, 5-(benzoyloxy)-2-ethyl-, benzoate, trans	74793-13-8	NI38210
105	124	111	123	77	110	106	125	368	100	90	51	22	20	13	8	7	0.00	20	21	6	–	–	–	–	–	–	–	–	1	–	1,3,2-Dioxaborolane-4,5-dimethanol, 2-ethyl-, dibenzoate, (4S-trans)-	74793-25-2	NI38211
324	368	104	224	196	180	295	253	368	100	48	38	36	34	31	24	20		20	24	3	2	–	–	–	–	–	1	–	–	–	2,4-Imidazolidinedione, 1,3-dimethyl-5-phenyl-5-[3-[(trimethylsilyl)oxy]phenyl]-	57326-23-5	NI38212
324	368	104	224	196	180	295	253	368	100	48	38	36	34	31	24	20		20	24	3	2	–	–	–	–	–	1	–	–	–	2,4-Imidazolidinedione, 3-ethyl-5-phenyl-5-[3-[(trimethylsilyl)oxy]phenyl]	57326-24-6	NI38213
324	247	295	224	180	196	225	325	368	100	66	49	42	33	29	27	22	5.50	20	24	3	2	–	–	–	–	–	1	–	–	–	2,4-Imidazolidinedione, 3-ethyl-5-phenyl-5-[4-[(trimethylsilyl)oxy]phenyl]	57326-27-9	NI38214
73	311	368	176	165	206	268	296	368	100	99	94	84	77	62	54	51		20	24	3	2	–	–	–	–	–	1	–	–	–	2,4-Imidazolidinedione, 5-(3-methoxyphenyl)-3-methyl-5-phenyl-1-(trimethylsilyl)-	57326-22-4	NI38215
73	368	196	165	296	206	219	283	368	100	99	97	61	58	46	43	41		20	24	3	2	–	–	–	–	–	1	–	–	–	2,4-Imidazolidinedione, 5-(4-methoxyphenyl)-3-methyl-5-phenyl-1-(trimethylsilyl)-	57346-57-3	NI38216
267	296	219	210	134	77	180	297	368	100	92	90	56	48	39	34	27	17.02	20	24	3	2	–	–	–	–	–	1	–	–	–	2,4-Imidazolidinedione, 5-(4-methoxyphenyl)-3-methyl-5-phenyl-1-(trimethylsilyl)-	57346-57-3	NI38217
254	368	255	325	283	369	269	367	368	100	25	15	7	7	6	6	5		20	24	3	4	–	–	–	–	–	–	–	–	–	N-Propyl-N-butyl-4'-nitroazobenzenecarboxamide		MS08670
60	85	164	41	57	43	135	45	368	100	43	40	40	36	35	17	15	0.00	20	32	6	–	–	–	–	–	–	–	–	–	–	11-Deacyloxytetrahydrovaltrate		MS08671
98	85	36	84	185	99	42	41	368	100	15	13	12	7	7	5	5	0.10	20	36	4	2	–	–	–	–	–	–	–	–	–	2,3-Bis(ethoxycarbonyl)-1,4-bis(piperidino)butane		NI38218
336	121	368	293	295	305	205	135	368	100	16	10	10	8	7	7	5		21	20	6	–	–	–	–	–	–	–	–	–	–	Benzeneacetic acid, 2-hydroxy-5-(3-methoxy-3-oxo-1-propenyl)-α-[(4-methoxyphenyl)methylene]-, methyl ester	69682-23-1	NI38219
368	233	179	369	191	219	335	313	368	100	43	37	12	12	7	5	4		21	20	6	–	–	–	–	–	–	–	–	–	–	4H-1-Benzopyran-4-one, 3-(2,4-dihydroxyphenyl)-7-hydroxy-5-methoxy-8(3-methyl-2-butenyl)-	104703-88-0	NI38220
137	368	232	169	149	138	231	243	368	100	30	14	10	8	8	7	6		21	20	6	–	–	–	–	–	–	–	–	–	–	4H-1-Benzopyran-4-one, 7-hydroxy-3-[2,4-dihydroxy-5-methoxy-6-(3-methyl-2-butenyl)phenyl]-		MS08672
179	368	335	134	369	313	295	233	368	100	87	33	32	22	20	15	13		21	20	6	–	–	–	–	–	–	–	–	–	–	Cyclobarpisoflavone B		NI38221
253	43	105	267	309	295	77	337	368	100	75	48	40	32	26	26	18	2.00	21	20	6	–	–	–	–	–	–	–	–	–	–	2,4-Diacetoxy-5-methyl-6-methoxychalcone		MS08673
151	152	368	25	217	135			368	100	10	5	5	2	1				21	20	6	–	–	–	–	–	–	–	–	–	–	4-(3,4-Dimethoxybenzyl)-3-(3,4-methylenedioxybenzylidene)tetrahydrofuran-2-one		LI08059
105	77	51	178	263	190	69	50	368	100	32	8	3	2	2	2	2	0.00	21	20	6	–	–	–	–	–	–	–	–	–	–	Dimethyl (α,α'-dibenzoyl)glutarate		LI08060
177	191	150	145	350	137	272	232	368	100	22	20	16	11	11	10	7	5.00	21	20	6	–	–	–	–	–	–	–	–	–	–	1,6-Heptadiene-3,5-dione, 1,7-bis(4-hydroxy-3-methoxyphenyl)-	458-37-7	NI38222
297	368	312	325	269				368	100	64	54	35	17					21	20	6	–	–	–	–	–	–	–	–	–	–	Phenanthrenequinone, 1,4,8-trihydroxy-2-methyl-3-(4-methylvaleryl)-	24329-93-9	LI08061
297	368	312	325	298	269	369	270	368	100	62	51	35	26	17	15	14		21	20	6	–	–	–	–	–	–	–	–	–	–	Phenanthrenequinone, 1,4,8-trihydroxy-2-methyl-3-(4-methylvaleryl)-	24329-93-9	NI38223
105	77	44	122	106	204	78	41	368	100	55	19	9	6	5	5	4	0.00	21	20	6	–	–	–	–	–	–	–	–	–	–	2H-Pyran-2-methanol, 3-(benzoyloxy)-3,6-dihydro-6-methoxy-, benzoate	56248-18-1	NI38224
77	340	368	51	111	342	341	370	368	100	58	42	31	24	23	19	18		21	21	–	2	1	–	1	–	–	–	–	–	–	1H-Benzo[e][1,4]thiazepine, 2,3,5,6,7,8,9-heptahydro-1-phenyl-5-(4-chlorophenylimino)-		NI38225
368	91	108	107	77	65	179	367	368	100	90	70	70	70	70	50	40		21	21	4	–	–	–	–	–	–	–	1	–	–	Tricresyl phosphate		IC04162
368	367	369	91	165	108	243	65	368	100	41	24	24	20	16	14	14		21	21	4	–	–	–	–	–	–	–	1	–	–	Tricresyl phosphate		IC04163
107	108	122	77	79	44	121	39	368	100	73	24	23	19	18	11	10	0.00	21	21	4	–	–	–	–	–	–	–	1	–	–	Tricresyl phosphate		IC04164
170	368	171	277	234	168	369	91	368	100	35	35	22	20	11	9	7		21	24	2	2	1	–	–	–	–	–	–	–	–	1-Naphthalenesulphonamide, 5-(dimethylamino)-N-(1-methyl-2-phenylethyl)-, (±)-	51581-19-2	NI38226
170	368	171	277	234	168	369	154	368	100	35	34	22	18	10	8	7		21	24	2	2	1	–	–	–	–	–	–	–	–	1-Naphthalenesulphonamide, 5-(dimethylamino)-N-(1-methyl-2-phenylethyl)-, (±)-	51581-19-2	NI38227
171	368	170	168	203	169	172	369	368	100	32	25	20	13	13	12	9		21	24	2	2	1	–	–	–	–	–	–	–	–	1-Naphthalenesulphonamide, 5-(dimethylamino)-N-(3-phenylpropyl)-	55837-08-6	NI38228

MASS TO CHARGE RATIOS	M.W.	INTENSITIES	Parent	C	H	O	N	S	F	Cl	Br	I	Si	P	B	X	COMPOUND NAME	CAS Reg No	No
171 368 170 168 169 115 203 128	**368**	100 31 24 19 13 13 12 12		21	24	2	2	1	–	–	–	–	–	–	–	–	1-Naphthalenesulphonamide, 5-(dimethylamino)-N-(3-phenylpropyl)-	55837-08-6	NI38229
368 168 130 167 154 180 158 69	**368**	100 33 33 32 26 25 25 23		21	24	4	2	–	–	–	–	–	–	–	–	–	1H-12a,5-(Epoxymethano)pyrido[3',4':5,6]cyclohept[1,2-b]indol-14-one, 2-acetyl-2,3,4,4a,5,6,11,12-octahydro-5-(hydroxymethyl)-12-methyl-	3329-92-8	NI38230
223 368 69 130 55 42 41 117	**368**	100 55 41 36 31 28 26 21		21	24	4	2	–	–	–	–	–	–	–	–	–	Formosanan-16-carboxylic acid, 19-methyl-2-oxo-, methyl ester, (19α)-	509-80-8	NI38231
69 70 368 130 223 180 337 267	**368**	100 37 29 22 18 15 12 12		21	24	4	2	–	–	–	–	–	–	–	–	–	Isopteropodine	5171-37-9	NI38232
337 239 368 338 157 180 240 43	**368**	100 51 25 25 24 22 17 16		21	24	4	2	–	–	–	–	–	–	–	–	–	Picraline, deacetyl-	6808-68-0	NI38233
368 293 184 238 180 339 322 280	**368**	100 57 51 47 31 26 18 15		21	24	4	2	–	–	–	–	–	–	–	–	–	Talbotine	30809-15-5	LI08062
368 293 184 238 266 234 143 180	**368**	100 57 51 48 43 38 32 31		21	24	4	2	–	–	–	–	–	–	–	–	–	Talbotine	30809-15-5	NI38234
350 269 368 199 291 336 309 337	**368**	100 77 71 27 24 17 17 12		21	24	4	2	–	–	–	–	–	–	–	–	–	Vincaricine		LI08063
333 115 315 351 103 279 183 334	**368**	100 87 53 45 37 36 30 23	**0.50**	21	36	5	–	–	–	–	–	–	–	–	–	–	Prosta-5,13-dien-1-oic acid, 9,11,15-trihydroxy-, methyl ester, (5Z,9α,11α,13E,15S)-	33854-16-9	NI38235
247 332 300 190 301 219 279 192	**368**	100 92 83 78 73 70 63 59	**0.00**	21	36	5	–	–	–	–	–	–	–	–	–	–	Prost-13-en-1-oic acid, 11,15-dihydroxy-9-oxo-, methyl ester, (11α,13E,15S)-	3434-33-1	NI38236
368 73 98 202 75 187 155 129	**368**	100 78 76 54 43 27 27 27		21	40	3	–	–	–	–	–	–	1	–	–	–	Oxacyclotetradec-10-en-2-one, 14-pentyl-13-[(trimethylsilyl)oxy]-, [13S-(10Z,13R*,14R*)]-	75299-44-4	NI38237
84 238 70 267 98 198 112 210	**368**	100 90 90 65 52 45 36 35	**7.00**	21	44	1	4	–	–	–	–	–	–	–	–	–	Azacyclotridecan-2-one, 1-[3-[[3-[(3-aminopropyl)amino]propyl]amino]propyl]-	67473-75-0	NI38238
70 84 98 112 267 253 255 273	**368**	100 70 50 32 25 24 20 19	**6.00**	21	44	1	4	–	–	–	–	–	–	–	–	–	1,5,9,13-Tetraazacyclopentacosan-14-one	67171-93-1	NI38239
105 77 106 368 218 51 117 78	**368**	100 25 9 4 4 4 2 2		22	24	5	–	–	–	–	–	–	–	–	–	–	Ethyl 2-benzoyloxy-2-isobutylbenzoylacetate		MS08674
91 105 231 77 204 92 43 161	**368**	100 73 14 13 12 9 7 5	**0.00**	22	24	5	–	–	–	–	–	–	–	–	–	–	Furo[2,3-d]-1,3-dioxole, tetrahydro-2,2-dimethyl-5-(benzoylmethyl)-6-(benzyloxy)-, [3aR-(3aα,5α,6α,6aα)]-	120417-99-4	DD01465
91 105 203 248 77 92 231 120	**368**	100 90 50 31 31 18 17 15	**0.00**	22	24	5	–	–	–	–	–	–	–	–	–	–	Furo[2,3-d]-1,3-dioxole, tetrahydro-2,2-dimethyl-5-(benzoylmethyl)-6-(benzyloxy)-, [3aR-(3aα,5β,6α,6aα)]-	120417-98-3	DD01466
91 105 231 103 77 43 92 165	**368**	100 46 14 10 10 10 8 6	**0.00**	22	24	5	–	–	–	–	–	–	–	–	–	–	Furo[2,3-d]-1,3-dioxole, tetrahydro-2,2-dimethyl-5-(benzoylmethyl)-6-(benzyloxy)-, [3aR-(3aα,5β,6β,6aα)]-	120417-95-0	DD01467
125 140 310 172 282 158 41 42	**368**	100 85 72 39 36 32 23 22	**0.00**	22	28	3	2	–	–	–	–	–	–	–	–	–	Aspidofractinine-3-carboxylic acid, 6-hydroxy-1-methyl-, methyl ester, (2α,3β,5α,6β)-	55724-63-5	NI38240
282 278 154 310 41 139 42 109	**368**	100 95 85 83 81 75 72 64	**0.00**	22	28	3	2	–	–	–	–	–	–	–	–	–	Aspidofractinine-3-carboxylic acid, 6-methoxy-, methyl ester, (2α,3β,5α,6β)-	55724-57-7	NI38241
124 109 368 336 340 125 77 154	**368**	100 53 30 15 11 11 8 7		22	28	3	2	–	–	–	–	–	–	–	–	–	Aspidofractinine-3-carboxylic acid, 17-methoxy-, methyl ester, (2α,3β,5α)-	55103-43-0	NI38242
325 368 326 138 369 310 160 353	**368**	100 47 21 15 11 8 8 6		22	28	3	2	–	–	–	–	–	–	–	–	–	Aspidospermidin-20-one, 1-acetyl-17-methoxy-	36458-96-5	NI38243
367 353 184 368 169 170 156 225	**368**	100 99 97 83 58 56 56 36		22	28	3	2	–	–	–	–	–	–	–	–	–	Corynan-16-carboxylic acid, 16,17-didehydro-17-methoxy-, methyl ester, (3β,16E)-	7729-23-9	NI38244
184 368 353 367 369 239 169 156	**368**	100 55 55 45 27 27 23 23		22	28	3	2	–	–	–	–	–	–	–	–	–	Corynan-16-carboxylic acid, 16,17-didehydro-17-methoxy-, methyl ester, (3β,16E,20β)-	7729-22-8	NI38245
368 367 184 353 169 170 156 168	**368**	100 88 75 62 48 44 37 34		22	28	3	2	–	–	–	–	–	–	–	–	–	Corynan-16-carboxylic acid, 16,17-didehydro-17-methoxy-, methyl ester, (16E)-	50439-68-4	NI38246
184 368 367 170 169 353 156 168	**368**	100 88 81 66 66 58 52 37		22	28	3	2	–	–	–	–	–	–	–	–	–	Corynan-16-carboxylic acid, 16,17-didehydro-17-methoxy-, methyl ester, (16E)-	50439-68-4	NI38247
184 368 367 353 169 170 156 239	**368**	100 53 47 47 33 29 29 23		22	28	3	2	–	–	–	–	–	–	–	–	–	Corynan-16-carboxylic acid, 16,17-didehydro-17-methoxy-, methyl ester, (16E,20β)-	23407-35-4	NI38248
69 71 67 81 83 97 79 77	**368**	100 82 77 64 61 44 44 44	**32.00**	22	28	3	2	–	–	–	–	–	–	–	–	–	Corynan-17-ol, 18,19-didehydro-10-methoxy-, acetate	56053-13-5	NI38249
168 294 295 368 169 196 43 144	**368**	100 72 26 21 11 7 7 6		22	28	3	2	–	–	–	–	–	–	–	–	–	Cylindrocarine, 1-acetyl-17-demethoxy-	22226-35-3	NI38250
368 309 160 198 199 176 325 173	**368**	100 14 13 11 7 7 6 6		22	28	3	2	–	–	–	–	–	–	–	–	–	O'-Demethyl-O-acetyl-19,20-dihydroseredamine		LI08064
183 182 368 265 184 238 168 237	**368**	100 74 51 44 26 24 23 21		22	28	3	2	–	–	–	–	–	–	–	–	–	Dihydrovoachalotine		MS08675
138 300 340 324 368 341 325 160	**368**	100 99 49 46 25 17 16 16		22	28	3	2	–	–	–	–	–	–	–	–	–	13a,3a-(Epoxyethano)-1H-indolizino[8,1-cd]carbazole, 6-acetyl-2,3,4,5,5a,6,11,12-octahydro-7-methoxy-	2671-43-4	NI38251
138 340 324 160 368 323 325 139	**368**	100 38 23 16 13 12 10 10		22	28	3	2	–	–	–	–	–	–	–	–	–	13a,3a-(Epoxyethano)-1H-indolizino[8,1-cd]carbazole, 6-acetyl-2,3,4,5,5a,6,11,12-octahydro-7-methoxy-	2671-43-4	NI38252
58 125 368 154 96 200 84 55	**368**	100 62 19 17 15 14 12 12		22	28	3	2	–	–	–	–	–	–	–	–	–	Schizozygine, hexahydro-N,N-dimethyl-	2283-33-2	NI38253
325 368 143 234 210 262 265 184	**368**	100 96 68 34 34 32 29 23		22	28	3	2	–	–	–	–	–	–	–	–	–	Talbotine, 16-di(methoxycarbonyl)-16-(hydroxymethyl)-O,4-dimethyl-	30809-22-4	NI38254
368 136 369 184 124 244 135 122	**368**	100 41 25 20 15 14 14 13		22	28	3	2	–	–	–	–	–	–	–	–	–	Voacangine	510-22-5	NI38255
136 124 122 135 184 368 160 137	**368**	100 40 35 33 30 22 21 16		22	28	3	2	–	–	–	–	–	–	–	–	–	Voacangine	510-22-5	NI38256
368 136 184 124 122 244 160 353	**368**	100 99 41 40 38 34 25 23		22	28	3	2	–	–	–	–	–	–	–	–	–	Voacangine	510-22-5	LI08065

MASS TO CHARGE RATIOS								M.W.	INTENSITIES								Parent	C	H	O	N	S	F	Cl	Br	I	Si	P	B	X	COMPOUND NAME	CAS Reg No	No
115	187	84	174	85	325	67	69	**368**	100	74	20	19	16	15	14	11	**5.00**	22	40	4	–	–	–	–	–	–	–	–	–	–	**Bicyclo[13.1.0]hexadecan-2-one, 1,4-di-O-methyl-D-threitolketal, (1S,15S)-**		**NI38257**
101	175	117	88	353	59	205	223	**368**	100	44	42	34	13	12	2	1	**0.70**	22	40	4	–	–	–	–	–	–	–	–	–	–	**1-O-(2-Hydroxy-4-hexadecynyl)-2,3-O-isopropylideneglycerol**		**NI38258**
55	149	101	69	81	95	57	67	**368**	100	99	73	60	52	50	47	45	**5.08**	22	40	4	–	–	–	–	–	–	–	–	–	–	**Methyl cis-11,12-methylene-2-acetoxyoctadecanoate**		**MS08676**
84	43	97	41	30	29	44	55	**368**	100	51	44	26	20	19	18	14	**0.00**	22	44	2	2	–	–	–	–	–	–	–	–	–	**1H-Imidazole, 2-heptadecyl-4,5-dihydro-, monoacetate**	556-22-9	**NI38259**
203	73	204	147	205	217	216	75	**368**	100	55	21	12	10	7	6	6	**0.00**	22	44	2	2	–	–	–	–	–	–	–	–	–	**Octadecanamide, N-isobutyl-N-nitroso-**	74420-93-2	**NI38260**
368	235	91	369	194	237	145	89	**368**	100	45	40	29	27	20	12	12		23	20	1	4	–	–	–	–	–	–	–	–	–	**3H-Pyrazol-3-one, 4-[[4-(dimethylamino)phenyl]imino]-2,4-dihydro-2,5-diphenyl-**	13617-68-0	**MS08677**
368	235	91	369	194	237	145	77	**368**	100	44	40	28	26	20	12	11		23	20	1	4	–	–	–	–	–	–	–	–	–	**3H-Pyrazol-3-one, 4-[[4-(dimethylamino)phenyl]imino]-2,4-dihydro-2,5-diphenyl-**	13617-68-0	**NI38261**
227	83	228	368	56	55	41	43	**368**	100	94	72	60	57	56	46	32		23	28	4	–	–	–	–	–	–	–	–	–	–	**Benzaldehyde dimedone**		**LI08066**
228	227	368	229	140	102	241	369	**368**	100	92	67	33	33	33	30	17		23	28	4	–	–	–	–	–	–	–	–	–	–	**Benzaldehyde dimedone**		**IC04165**
227	107	284	55	105	114	228	242	**368**	100	60	38	25	23	21	18	17	**4.00**	23	28	4	–	–	–	–	–	–	–	–	–	–	**Pentanoic acid, 4-[[4-hydroxy-3-(1-oxopentyl)phenyl]methyl]phenyl ester**	55724-87-3	**NI38262**
200	107	57	284	201	85	41	183	**368**	100	36	17	15	15	8	8	6	**1.00**	23	28	4	–	–	–	–	–	–	–	–	–	–	**Pentanoic acid, methylenedi-4,1-phenylene ester**	55724-88-4	**NI38263**
311	368	339	127	191	81	312	44	**368**	100	67	39	28	25	22	21	18		23	28	4	–	–	–	–	–	–	–	–	–	–	**Bis(1-pentanone), 1,1′-[methylenebis(6-hydroxy-3,1-phenylene)]**	55282-09-2	**NI38264**
311	368	339	127	191	312	81	369	**368**	100	67	30	28	25	23	22	19		23	28	4	–	–	–	–	–	–	–	–	–	–	**Bis(1-pentanone), 1,1′-[methylenebis(6-hydroxy-3,1-phenylene)]**	55282-09-2	**NI38265**
73	269	75	352	353	57	297	159	**368**	100	53	49	45	43	20	19	19	**0.00**	23	32	2	–	–	–	–	–	–	1	–	–	–	**2,5-Cyclohexadiene-1,4-dione, 2,3-dibutyl-5-phenyl-6-(trimethylsilyl)-**	122949-73-9	**DD01468**
73	75	353	129	57	55	297	91	**368**	100	27	17	15	12	12	10	9	**0.00**	23	32	2	–	–	–	–	–	–	1	–	–	–	**4-Cyclopentene-1,3-dione, 4,5-dibutyl-2-[phenyl(trimethylsilyl)methylene]**	122967-67-3	**DD01469**
353	73	354	86	75	125	311	209	**368**	100	33	29	20	20	15	13	12	**9.94**	23	32	2	–	–	–	–	–	–	1	–	–	–	**Estra-4,6-dien-3-one, 17α-ethynyl-17β-[(trimethylsilyl)oxy]-**		**NI38266**
83	41	43	85	55	57	368	69	**368**	100	71	69	63	61	43	38	37		23	32	2	2	–	–	–	–	–	–	–	–	–	**4,25-Secoobscurinervan, 21-deoxy-16-methoxy-22-methyl-, (22α)-**	6872-63-5	**NI38267**
124	368	41	125	43	55	190	152	**368**	100	22	15	12	10	9	8	7		23	32	2	2	–	–	–	–	–	–	–	–	–	**4,25-Secoobscurinervan, 21-deoxy-16-methoxy-22-methyl-, (22β)-**	6855-59-0	**NI38268**
124	368	369	125	190	142	367	184	**368**	100	74	20	10	9	8	7	7		23	32	2	2	–	–	–	–	–	–	–	–	–	**4,25-Secoobscurinervan, 21-deoxy-16-methoxy-22-methyl-, (22β)-**	6855-59-0	**NI38269**
124	190	368	125	184	340	83	41	**368**	100	36	35	29	27	22	22	19		23	32	2	2	–	–	–	–	–	–	–	–	–	**4,25-Secoobscurinervan, 21-deoxy-16-methoxy-23-methyl-**	54725-08-5	**NI38270**
55	43	256	170	71	69	57	126	**368**	100	98	86	86	76	69	56	54	**10.00**	23	44	3	–	–	–	–	–	–	–	–	–	–	**Docosanoic acid, 13-oxo-, methyl ester**	2388-99-0	**NI38271**
114	115	116	312	255	222	223	111	**368**	100	59	56	44	42	41	40	34	**4.04**	23	44	3	–	–	–	–	–	–	–	–	–	–	**Docosanoic acid, 17-oxo-, methyl ester**	19271-80-8	**NI38272**
325	368	55	230	81	41	95	43	**368**	100	64	52	43	37	37	34	31		23	45	1	–	–	–	–	–	–	–	1	–	–	**Phosphine oxide, bis[methylisopropylcyclohexyl]propyl-**	74645-96-8	**NI38273**
143	73	75	144	123	81	43	57	**368**	100	23	21	14	12	7	7	6	**0.24**	23	48	1	–	–	–	–	–	–	1	–	–	–	**Silane, [(3,7,11,15-tetramethyl-2-hexadecenyl)oxy]trimethyl-**	57397-39-4	**NI38274**
143	73	75	144	123	43	57	145	**368**	100	29	23	13	12	8	6	4	**0.44**	23	48	1	–	–	–	–	–	–	1	–	–	–	**Silane, [(3,7,11,15-tetramethyl-2-hexadecenyl)oxy]trimethyl-**	57397-39-4	**NI38275**
150	263	368	367	149	277	369	264	**368**	100	81	67	47	36	34	22	14		24	20	2	2	–	–	–	–	–	–	–	–	–	**2(1H)-Pyrimidinone, 4-benzoyl-3,4-dihydro-3-phenyl-6-tolyl-**		**LI08067**
368	91	214	350	232	105	79	335	**368**	100	94	93	78	78	78	74	53		24	32	3	–	–	–	–	–	–	–	–	–	–	**Bufa-14,20,22-trienolide, 3-hydroxy-, (3β,5β)-**	7439-77-2	**LI08068**
368	91	214	350	232	105	79	93	**368**	100	93	92	80	78	77	72	56		24	32	3	–	–	–	–	–	–	–	–	–	–	**Bufa-14,20,22-trienolide, 3-hydroxy-, (3β,5β)-**	7439-77-2	**NI38276**
74	368	143	255	325	228	223		**368**	100	74	30	12	10	9	4			24	48	2	–	–	–	–	–	–	–	–	–	–	**Docosanoic acid, 14-methyl-, methyl ester**		**LI08069**
43	83	57	97	61	69	55	82	**368**	100	83	79	77	67	62	58	46	**0.70**	24	48	2	–	–	–	–	–	–	–	–	–	–	**1-Docosanol, acetate**	822-26-4	**WI01482**
116	43	57	41	55	71	69	29	**368**	100	64	57	46	45	28	24	21	**2.50**	24	48	2	–	–	–	–	–	–	–	–	–	–	**Eicosanoic acid, 2-methyl-2-ethyl-, methyl ester**		**MS08678**
88	115	368	311	369	69	83	57	**368**	100	49	38	18	11	10	9	8		24	48	2	–	–	–	–	–	–	–	–	–	–	**Heneicosanoic acid, 2,3-dimethyl-, methyl ester**		**MS08679**
88	101	69	55	43	57	41	745	**368**	100	68	39	38	36	33	29	21	**9.40**	24	48	2	–	–	–	–	–	–	–	–	–	–	**Heneicosanoic acid, 2,4-dimethyl-, methyl ester**		**MS08680**
88	101	368	55	43	311	57	83	**368**	100	32	27	22	22	20	20	15		24	48	2	–	–	–	–	–	–	–	–	–	–	**Heneicosanoic acid, 2,5-dimethyl-, methyl ester**	55282-05-8	**NI38277**
115	74	295	57	73	43	71	55	**368**	100	78	65	45	42	33	32	28	**15.00**	24	48	2	–	–	–	–	–	–	–	–	–	–	**Heneicosanoic acid, 3,3-dimethyl-, methyl ester**	55282-07-0	**MS08681**
115	74	295	57	73	43	71	69	**368**	100	78	65	45	42	33	32	28	**15.00**	24	48	2	–	–	–	–	–	–	–	–	–	–	**Heneicosanoic acid, 3,3-dimethyl-, methyl ester**	55282-07-0	**NI38278**
84	43	285	61	57	56	85	55	**368**	100	93	78	56	46	41	35	35	**27.04**	24	48	2	–	–	–	–	–	–	–	–	–	–	**Octadecanoic acid, hexyl ester**	3460-37-5	**NI38279**
112	57	70	71	43	55	41	111	**368**	100	90	67	64	62	39	36	31	**2.50**	24	48	2	–	–	–	–	–	–	–	–	–	–	**Palmitic acid, 2-ethyl-, hexyl ester**		**IC04166**
368	43	57	73	55	60	71	41	**368**	100	76	72	54	42	40	36	30		24	48	2	–	–	–	–	–	–	–	–	–	–	**Tetracosanoic acid**		**IC04167**
74	87	368	43	75	57	55	143	**368**	100	71	41	40	35	30	27	22		24	48	2	–	–	–	–	–	–	–	–	–	–	**Tricosanoic acid, methyl ester**	2433-97-8	**WI01483**
74	43	87	41	55	57	75	143	**368**	100	76	69	39	35	31	27	19	**4.00**	24	48	2	–	–	–	–	–	–	–	–	–	–	**Tricosanoic acid, methyl ester**	2433-97-8	**NI38280**
74	87	43	57	55	41	69	75	**368**	100	73	64	61	52	44	36	28	**24.40**	24	48	2	–	–	–	–	–	–	–	–	–	–	**Tricosanoic acid, methyl ester**	2433-97-8	**MS08682**
368	231	353	213	177	257	229	299	**368**	100	19	18	18	18	10	9	8		25	36	2	–	–	–	–	–	–	–	–	–	–	**Androst-5-en-17-one, 3-hydroxy-16-cyclopentylmethylene-, (3β)-**		**LI08070**
191	178	368	175	163	312	57	135	**368**	100	70	68	66	55	32	31	30		25	36	2	–	–	–	–	–	–	–	–	–	–	**Bis(3-tert-butyl-5-ethyl-2-hydroxyphenyl)methane**	88-24-4	**IC04168**
191	175	178	368	163	57	135	169	**368**	100	68	65	54	39	27	22	19		25	36	2	–	–	–	–	–	–	–	–	–	–	**Bis(3-tert-butyl-5-ethyl-2-hydroxyphenyl)methane**	88-24-4	**IC04169**
191	368	178	175	163	57	169	369	**368**	100	77	69	60	35	34	23	22		25	36	2	–	–	–	–	–	–	–	–	–	–	**Bis(3-tert-butyl-5-ethyl-2-hydroxyphenyl)methane**	88-24-4	**IC04170**
255	135	368	256	57	175	369	136	**368**	100	58	25	23	19	13	8	7		25	36	2	–	–	–	–	–	–	–	–	–	–	**1,1-Bis(2-hydroxy-3,5-dimethylphenyl)-6,6-dimethylheptane**		**IC04171**
91	67	55	41	43	79	69	57	**368**	100	31	31	31	27	21	15	15	**0.00**	25	36	2	–	–	–	–	–	–	–	–	–	–	**Linolenic acid, γ-, benzyl ester**		**NI38281**
91	79	41	67	55	93	95	39	**368**	100	21	21	18	17	12	11	11	**0.00**	25	36	2	–	–	–	–	–	–	–	–	–	–	**Linolenic acid, benzyl ester**		**NI38282**
91	119	117	109	145	131	95	92	**368**	100	17	14	11	9	8	8	8	**0.00**	25	36	2	–	–	–	–	–	–	–	–	–	–	**Linolenic acid, benzyl ester**		**NI38283**
83	97	55	43	69	111	125	71	**368**	100	99	97	88	82	73	58	54	**0.00**	25	52	1	–	–	–	–	–	–	–	–	–	–	**Pentacosanol**		**DD01470**
57	71	43	83	97	196	113	154	**368**	100	90	53	48	37	17	16	10	**0.00**	25	52	1	–	–	–	–	–	–	–	–	–	–	**Perhydroisomoenocinol**		**LI08071**
57	71	43	97	83	171	127	111	**368**	100	70	58	51	47	27	16	16	**0.00**	25	52	1	–	–	–	–	–	–	–	–	–	–	**Perhydromoenocinol**		**LI08072**

MASS TO CHARGE RATIOS	M.W.	INTENSITIES	Parent	C	H	O	N	S	F	Cl	Br	I	Si	P	B	X	COMPOUND NAME	CAS Reg No	No
192 176 91 191 115 193 368 117	**368**	100 59 36 32 32 30 28 27		26	24	2	–	–	–	–	–	–	–	–	–	–	**Cyclopent[a]indene-1-carboxylic acid, 1,2,3,3a,8,8a-hexahydro-3,8a-diphenyl-, methyl ester, (1α,3α,3aα,8aα)-**	40897-32-3	**NI38284**
57 43 41 105 55 211 81 91	**368**	100 84 70 63 57 56 54 47	**32.01**	27	44	–	–	–	–	–	–	–	–	–	–	–	**Cholesta-2,4-diene**	4117-50-4	**NI38285**
43 57 41 55 44 105 93 81	**368**	100 92 75 48 45 37 37 37	**20.17**	27	44	–	–	–	–	–	–	–	–	–	–	–	**Cholesta-3,5-diene**	747-90-0	**NI38286**
368 369 43 81 55 145 105 41	**368**	100 30 25 23 19 17 17 17		27	44	–	–	–	–	–	–	–	–	–	–	–	**Cholesta-3,5-diene**	747-90-0	**NI38287**
81 368 43 41 55 105 145 107	**368**	100 99 99 89 87 72 71 71		27	44	–	–	–	–	–	–	–	–	–	–	–	**Cholesta-3,5-diene**	747-90-0	**NI38288**
368 255 353 43 55 369 41 145	**368**	100 62 54 34 32 30 29 22		27	44	–	–	–	–	–	–	–	–	–	–	–	**Cholesta-7,14-diene, (5α)-**	54482-39-2	**NI38289**
368 255 369 43 256 241 353 41	**368**	100 69 31 27 20 20 19 17		27	44	–	–	–	–	–	–	–	–	–	–	–	**Cholesta-7,14-diene, (5α)-**	54482-39-2	**NI38290**
91 104 117 105 131 92 160 145	**368**	100 57 21 21 13 11 10 10	**0.36**	28	32	–	–	–	–	–	–	–	–	–	–	–	**1,7-Diphenyl-4-(3-phenylpropyl)-3-heptene**		**AI02265**
292 290 294 146 296 220 242 148	**369**	100 88 43 9 9 7 5 5	**0.50**	1	3	–	3	–	–	4	1	–	–	3	–	–	**1-Methyl-1-bromotetrachlorocyclotriphosphazene**	77589-25-4	**NI38291**
135 69 85 137 87 284 50 196	**369**	100 83 77 33 26 17 15 11	**0.00**	5	–	1	1	–	11	2	–	–	–	–	–	–	**Undecafluoro[O-2-chloroethyl-N-methyl-N-(2-chloroethyl)]hydroxylamine**		**LI08073**
73 240 139 77 75 354 121 211	**369**	100 52 32 30 19 15 12 11	**0.00**	10	13	5	1	–	6	–	–	–	1	–	–	–	**L-Serine, N-(trifluoroacetyl)-, trimethylsilyl ester, trifluoroacetate**	52558-89-1	**NI38292**
314 370 140 114 131 184 240 115	**369**	100 87 58 54 48 44 39 30	**0.00**	11	13	4	1	1	6	–	–	–	–	–	–	–	**L-Cysteine, N-(trifluoroacetyl)-, butyl ester, trifluoroacetate**	5282-99-5	**NI38293**
41 69 57 140 56 55 45 43	**369**	100 99 99 43 43 35 32 27	**0.00**	11	13	4	1	1	6	–	–	–	–	–	–	–	**L-Cysteine, N-(trifluoroacetyl)-, butyl ester, trifluoroacetate**	5282-99-5	**NI38294**
370 240 371 258 372 314 241 242	**369**	100 45 10 8 6 5 5 3	**0.00**	11	13	4	1	1	6	–	–	–	–	–	–	–	**L-Cysteine, N-(trifluoroacetyl)-, butyl ester, trifluoroacetate**	5282-99-5	**NI38295**
285 55 191 176 189 119 221 369	**369**	100 96 84 66 37 34 30 26		12	12	7	1	1	–	–	–	–	–	–	–	1	**Manganese, η-[1-(N,N-dimethylsulphamoyl)-2-carbomethoxycyclopentadienyl]tricarbonyl-**		**NI38296**
204 77 51 205 74 76 75 78	**369**	100 84 29 7 7 6 6 5	**0.20**	13	8	4	1	–	–	–	–	1	–	–	–	–	**5-Nitro-2-phenyliodonio benzoate**		**LI08074**
171 173 17 63 91 62 16 92	**369**	100 76 58 35 26 22 21 16	**0.10**	13	9	2	1	–	–	–	2	–	–	–	–	–	**Benzamide, 3,5-dibromo-2-hydroxy-N-phenyl-**	2577-72-2	**NI38297**
290 292 88 130 102 196 198 97	**369**	100 67 45 42 39 37 36 25	**0.00**	13	14	5	1	–	–	3	–	–	–	–	–	–	**L-Threonine, N-[(2,4,5-trichlorophenoxy)acetyl]-, methyl ester**	66789-90-0	**NI38298**
43 78 55 59 83 148 77 58	**369**	100 66 51 32 27 26 24 24	**6.80**	13	16	6	1	1	–	–	–	–	–	–	–	1	**Manganese, η-[1-(N,N-dimethylsulphamoyl)]-2-[(2-hydroxy-2-propyl)cyclopentadienyl]tricarbonyl-**		**NI38299**
69 282 41 43 253 57 29 313	**369**	100 56 37 35 29 14 13 10	**0.00**	13	18	3	1	–	7	–	–	–	–	–	–	–	**N-(Heptafluorobutyryl)isoleucine propyl ester**		**MS08683**
69 43 282 240 41 253 214 283	**369**	100 61 42 33 23 7 7 6	**0.00**	13	18	3	1	–	7	–	–	–	–	–	–	–	**N-(Heptafluorobutyryl)leucine propyl ester**		**MS08684**
69 282 43 41 226 214 283 70	**369**	100 82 41 37 23 13 11 11	**0.00**	13	18	3	1	–	7	–	–	–	–	–	–	–	**N-(Heptafluorobutyryl)norleucine propyl ester**		**MS08685**
144 73 298 145 147 72 45 41	**369**	100 58 23 15 10 8 8 8	**1.10**	13	36	3	1	–	–	–	–	–	3	1	–	–	**Phosphonic acid, [2-methyl-1-[(trimethylsilyl)amino]propyl]-, bis(trimethylsilyl) ester**	53044-41-0	**NI38300**
144 73 298 147 227 211 100 75	**369**	100 58 23 10 7 7 6 6	**1.10**	13	36	3	1	–	–	–	–	–	3	1	–	–	**Phosphonic acid, [2-methyl-1-[(trimethylsilyl)amino]propyl]-, bis(trimethylsilyl) ester**	53044-41-0	**NI38301**
144 73 298 147 207 211 75 100	**369**	100 53 26 10 6 5 5 4	**1.60**	13	36	3	1	–	–	–	–	–	3	1	–	–	**Phosphonic acid, [1-[(trimethylsilyl)amino]butyl]-, bis(trimethylsilyl) ester**	53044-39-6	**NI38302**
144 73 298 145 147 299 45 72	**369**	100 53 26 14 10 8 8 7	**1.60**	13	36	3	1	–	–	–	–	–	3	1	–	–	**Phosphonic acid, [1-[(trimethylsilyl)amino]butyl]-, bis(trimethylsilyl) ester**	53044-39-6	**NI38303**
155 157 185 183 76 29 75 68	**369**	100 95 62 62 29 26 22 20	**2.90**	14	16	4	3	–	–	–	1	–	–	–	–	–	**Diethyl 2-(4-bromophenyltriazeno)fumarate**		**MS08686**
290 202 230 203 337 176 339 231	**369**	100 97 58 49 36 36 32 28	**17.00**	15	16	5	1	–	–	–	1	–	–	–	–	–	**Benzeneacetic acid, α-bromo-2-(2,5-dioxo-1-pyrrolidinyl)-5-methoxy-4-methyl-, methyl ester, (±)-**		**NI38304**
148 178 149 310 135 43 59 136	**369**	100 98 69 49 44 34 29 28	**4.50**	15	23	6	5	–	–	–	–	–	–	–	–	–	**Adenosine, N-(2,3-dihydroxy-3-methylbutyl)-**	17179-59-8	**NI38305**
28 27 43 148 136 44 41 29	**369**	100 93 67 46 39 25 23 23	**0.00**	15	23	6	5	–	–	–	–	–	–	–	–	–	**6-(2,3-Dihydroxy-3-methyl-butylamino)-9β-D-ribofuranosylpurine**		**LI08075**
27 29 43 28 148 39 41 30	**369**	100 95 87 81 73 50 43 40	**0.00**	15	23	6	5	–	–	–	–	–	–	–	–	–	**6-(3,4-Dihydroxy-3-methyl-butylamino)-9β-D-ribofuranosylpurine**		**LI08076**
199 73 200 75 280 81 369 184	**369**	100 79 70 67 45 31 23 19		15	27	4	3	–	–	–	–	–	2	–	–	–	**3,6-Epoxy-2H,8H-pyrimido[6,1-b][1,3]oxazocin-8-one, 3,4,5,6,9,10-hexahydro-10-[(trimethylsilyl)imino]-4-[(trimethylsilyl)oxy]-, [3R-(3α,4α,6α)]-**	41569-25-9	**NI38306**
43 71 185 184 212 139 41 55	**369**	100 64 31 29 24 20 16 15	**0.00**	16	26	5	1	–	3	–	–	–	–	–	–	–	**L-Aspartic acid, N-(trifluoroacetyl)-, bis(1-methylbutyl) ester**	57983-44-5	**NI38307**
194 57 41 212 152 240 55 44	**369**	100 89 69 60 57 48 45 32	**0.00**	16	26	5	1	–	3	–	–	–	–	–	–	–	**Hexanedioic acid, 2-[(trifluoroacetyl)amino]-, bis-sec-butyl ester**	57983-10-5	**NI38308**
73 354 355 192 356 74 75 147	**369**	100 82 27 19 13 10 9 8	**2.30**	16	31	3	1	–	–	–	–	–	3	–	–	–	**Benzoic acid, 2-amino-3-hydroxy-, tris(trimethylsilyl)-**	56272-78-7	**NI38309**
354 355 73 356 280 281 207 222	**369**	100 36 31 16 14 10 9 8	**8.40**	16	31	3	1	–	–	–	–	–	3	–	–	–	**Benzoic acid, 4-amino-2-hydroxy-, tris(trimethylsilyl)-**	56272-79-8	**MS08687**
354 73 355 282 356 369 75 45	**369**	100 46 31 18 15 13 13 9		16	31	3	1	–	–	–	–	–	3	–	–	–	**Benzoic acid, 4-amino-2-hydroxy-, tris(trimethylsilyl)-**	56272-79-8	**NI38310**
142 100 57 242 143 128 185 184	**369**	100 26 19 13 12 7 6 4	**0.00**	16	36	–	1	–	–	–	–	1	–	–	–	–	**Tetrabutylammonium iodide**	311-28-4	**NI38311**
371 369 370 373 372 368 299 333	**369**	100 83 65 48 43 43 28 23		17	11	–	1	–	–	4	–	–	–	–	–	–	**1,2,3,4-Tetrachloro-6a,11a-dihydro-11-methylbenzo[a]carbazole**		**NI38312**
82 111 369 309 96 280 110 83	**369**	100 75 53 50 50 38 38 37		17	11	4	1	–	4	–	–	–	–	–	–	–	**Dimethyl 6,7,8,9-tetrafluoro-1-methylbenz[g]indole-2,3-dicarboxylate**		**NI38313**
224 225 192 42 338 165 15 369	**369**	100 13 10 5 4 4 4 3		17	17	4	1	–	–	2	–	–	–	–	–	–	**3,5-Pyridinedicarboxylic acid, 4-(2,6-dichlorophenyl)-1,4-dihydro-2,6-dimethyl-, dimethyl ester**		**NI38314**

MASS TO CHARGE RATIOS								M.W.	INTENSITIES								Parent	C	H	O	N	S	F	Cl	Br	I	Si	P	B	X	COMPOUND NAME	CAS Reg No	No
369	298	313	234	138	338	101	111	**369**	100	99	87	78	42	32	32	26		17	23	8	1	–	–	–	–	–	–	–	–	–	**5-tert-Butylamino-2-(1-methoxy-2-methoxycarbonylvinyl)-3,4-bis-methoxycarbonylfuran**		**LI08077**
176	188	285	148	327	110	369	204	**369**	100	61	60	50	30	27	21	15		17	24	6	1	–	–	–	–	–	–	1	–	–	**3-Benzazepin-1-one, 7,8-(methylenedioxy)-2,3,4,5-tetrahydro-N-(diisopropylphosphoryl)-**	87212-50-8	**NI38315**
218	283	145	125	81	67	40	117	**369**	100	87	87	87	87	87	87	86	**0.00**	17	31	4	1	–	–	–	–	–	2	–	–	–	**3,4-Dimethoxyphenylalanine, bis(trimethylsilyl)-**		**MS08688**
102	73	103	45	104	74	268	59	**369**	100	51	11	7	5	5	4	4	**0.65**	17	35	2	1	–	–	–	–	–	3	–	–	–	**Benzeneethanamine, 3,4-dihydroxy-, tris(trimethylsilyl)-**	68595-54-0	**NI38316**
267	102	73	268	269	75	103	147	**369**	100	65	45	22	10	10	9	8	**0.09**	17	35	2	1	–	–	–	–	–	3	–	–	–	**Benzeneethanamine, N-(trimethylsilyl)-β,4-bis[(trimethylsilyl)oxy]-**	68595-60-8	**NI38317**
267	102	73	268	269	147	103	74	**369**	100	69	64	24	9	8	7	6	**0.00**	17	35	2	1	–	–	–	–	–	3	–	–	–	**Octopamine, tris(trimethylsilyl)-**	56145-08-5	**NI38318**
73	102	267	45	268	103	74	55	**369**	100	93	41	15	9	9	8	6	**0.00**	17	35	2	1	–	–	–	–	–	3	–	–	–	**Octopamine, tris(trimethylsilyl)-**	56145-08-5	**MS08689**
75	77	73	267	47	102	45	147	**369**	100	64	41	34	22	21	17	13	**0.00**	17	35	2	1	–	–	–	–	–	3	–	–	–	**Octopamine, tris(trimethylsilyl)-**	56145-08-5	**MS08690**
120	119	93	43	80	95	94	121	**369**	100	82	56	55	41	37	35	34	**0.50**	18	27	7	1	–	–	–	–	–	–	–	–	–	**Senecionan-11,16-dione, 15,20-dihydro-12,15,20-trihydroxy-, (15α,20R)-**	480-76-2	**NI38319**
190	188	162	191	148	149	178	175	**369**	100	28	21	19	18	14	5	4	**0.00**	20	19	6	1	–	–	–	–	–	–	–	–	–	**Egenine**	6883-44-9	**NI38320**
369	354	192	284	179	370	162	326	**369**	100	56	34	27	27	25	20	19		20	19	6	1	–	–	–	–	–	–	–	–	–	**Fumarofine**	34114-84-6	**NI38321**
369	43	28	69	354	71	192	179	**369**	100	78	60	59	42	42	40	35		20	19	6	1	–	–	–	–	–	–	–	–	–	**Fumarofine**	34114-84-6	**NI38322**
176	57	84	149	43	71	86	135	**369**	100	56	39	32	32	31	25	11	**0.00**	20	19	6	1	–	–	–	–	–	–	–	–	–	**Norhydrastine, (±)-β-**		**DD01471**
322	351	190	323	336	293	292	352	**369**	100	43	27	24	20	16	15	11	**2.74**	20	19	6	1	–	–	–	–	–	–	–	–	–	**Ochrobirine**	24181-64-4	**NI38323**
177	354	369	178	176	91	338	149	**369**	100	97	80	22	20	20	19	18		20	19	6	1	–	–	–	–	–	–	–	–	–	**Papaverrubine E**	6807-95-0	**NI38324**
205	221	187	120	103	206	159	325	**369**	100	58	31	18	16	14	14	13	**6.30**	20	19	6	1	–	–	–	–	–	–	–	–	–	**L-Phenylalanine, N-[(3,4-dihydro-8-hydroxy-3-methyl-1-oxo-1H-2-benzopyran-7-yl)carbonyl]-, (R)-**	4825-86-9	**NI38325**
206	163	192	162	369	190	207	177	**369**	100	81	57	19	18	18	12	12		20	19	6	1	–	–	–	–	–	–	–	–	–	**Rheagenine**	5574-77-6	**NI38326**
58	311	313	147	312	127	118	142	**369**	100	24	9	6	5	5	5	4	**3.00**	20	24	–	5	–	–	1	–	–	–	–	–	–	**N-Ethyl-N-(2-dimethylaminoethyl)-3-methyl-4-[chloro-(cyanophenylazo)]aniline**		**IC04172**
71	57	135	85	271	127	93	155	**369**	100	81	51	49	34	30	15	14	**0.00**	20	35	3	1	1	–	–	–	–	–	–	–	–	**Bornane-10,2-sultam, N-[(2R,3R)-3-ethyl-2-methylheptanoyl]-**		**DD01472**
369	133	370	368	134	92	237	104	**369**	100	30	27	27	16	10	9	9		21	19	–	7	–	–	–	–	–	–	–	–	–	**1,3,5-Triazine-2,4-diamine, N,N'-bis(3-aminophenyl)-6-phenyl-**	33933-67-4	**NI38327**
164	206	369	163	338	354	205	43	**369**	100	8	5	5	2	2	2	1		21	23	5	1	–	–	–	–	–	–	–	–	–	**β-Allocryptopine**		**NI38328**
148	179	190	91	150	77	135	369	**369**	100	34	12	12	10	9	6	5		21	23	5	1	–	–	–	–	–	–	–	–	–	**Cryptopine**	482-74-6	**NI38329**
206	354	369	338	355	207	142	164	**369**	100	45	32	14	13	12	12	9		21	23	5	1	–	–	–	–	–	–	–	–	–	**Fumaricine**	24181-77-9	**NI38330**
206	354	369	148	338	190	322	163	**369**	100	52	35	19	18	18	12	1		21	23	5	1	–	–	–	–	–	–	–	–	–	**Fumaricine**	24181-77-9	**NI38331**
43	42	44	268	81	70	59	327	**369**	100	55	40	14	14	11	11	9	**2.35**	21	23	5	1	–	–	–	–	–	–	–	–	–	**Morphinan-3,6-diol, 7,8-didehydro-4,5-epoxy-17-methyl- (5α,6α)-, diacetate**	561-27-3	**NI38332**
215	204	81	124	70	146	162	59	**369**	100	62	56	45	45	41	36	34	**0.00**	21	23	5	1	–	–	–	–	–	–	–	–	–	**Morphinan-3,6-diol, 7,8-didehydro-4,5-epoxy-17-methyl- (5α,6α)-, diacetate**	561-27-3	**NI38333**
327	369	268	43	310	215	42	204	**369**	100	65	54	49	47	30	28	25		21	23	5	1	–	–	–	–	–	–	–	–	–	**Morphinan-3,6-diol, 7,8-didehydro-4,5-epoxy-17-methyl- (5α,6α)-, diacetate**	561-27-3	**NI38334**
368	369	326	338	354	183	339	152	**369**	100	84	68	60	35	35	28	28		21	23	5	1	–	–	–	–	–	–	–	–	–	**Ocoteine**		**LI08078**
369	220	178	192	221	191	251	354	**369**	100	99	99	59	44	27	19	18		21	23	5	1	–	–	–	–	–	–	–	–	–	**L-Palmatine, tetrahydro-8-oxo-**		**NI38335**
105	163	83	77	177	206	149	109	**369**	100	99	99	99	61	48	44	22	**1.30**	21	27	1	3	1	–	–	–	–	–	–	–	–	**2-(Benzoylimino)-3-cyclohexyl-4-(cyclohexylimino)-1,3-thiazetidine**		**MS08691**
105	83	77	163	125	149	177	206	**369**	100	99	99	78	73	38	29	26	**1.30**	21	27	1	3	1	–	–	–	–	–	–	–	–	**3-Cyclohexyl-2-(cyclohexylimino)-6-phenyl-1,3,5-oxadiazine-4-thione**		**MS08692**
178	163	369	192	368	135	354	338	**369**	100	89	80	79	74	70	64	57		22	27	4	1	–	–	–	–	–	–	–	–	–	**D-Corydaline**	518-69-4	**NI38336**
250	296	266	210	238	369	324	268	**369**	100	37	35	23	19	18	10	10		22	27	4	1	–	–	–	–	–	–	–	–	–	**3,5-Pyridinedicarboxylic acid, 1,4-dihydro-1,2,6-trimethyl-4-(2-phenylethenyl)-, diethyl ester**	42972-38-3	**NI38337**
251	185	58	156	153	169	248	354	**369**	100	72	59	53	52	50	40	34	**3.00**	22	29	3	2	–	–	–	–	–	–	–	–	–	**Corynanium, 19,20-didehydro-16-(hydroxymethyl)-17-methoxy-4-methyl-17-oxo-, (4α,19E)-**	69306-89-4	**NI38338**
369	370	338	229	216	57	354	126	**369**	100	31	26	24	21	19	15	12		22	31	2	1	–	–	–	–	–	1	–	–	–	**Estra-1,3,5(10),6-tetraen-17-one, 3-[(trimethylsilyl)oxy]-, O-methyloxime**	69833-99-4	**NI38339**
369	370	338	216	354	139	126	229	**369**	100	31	19	18	15	14	13	12		22	31	2	1	–	–	–	–	–	1	–	–	–	**Estra-1,3,5(10),7-tetraen-17-one, 3-[(trimethylsilyl)oxy]-, O-methyloxime**	69833-42-7	**NI38340**
244	338	245	339	94	369	57	73	**369**	100	44	29	26	23	14	14	12		22	31	2	1	–	–	–	–	–	1	–	–	–	**Estra-1,3,5(10),15-tetraen-17-one, 3-[(trimethylsilyl)oxy]-, O-methyloxime**		**NI38341**
369	328	370	103	260	129	165	368	**369**	100	59	32	32	19	15	14	13		23	23	–	5	–	–	–	–	–	–	–	–	–	**Imidazo[1,2-b]-1,2,4-triazine, 2-methyl-6,7-diphenyl-3-(1-piperidino)-**		**NI38342**
267	266	268	325	326	280	278	281	**369**	100	54	49	39	36	36	33	27	**3.00**	24	19	3	1	–	–	–	–	–	–	–	–	–	**5-(1-Methoxycarbonyl-2-oxidovinyl)-6-(2-tolyl)phenanthridinium**		**LI08079**
267	325	326	268	266	295	280	354	**369**	100	80	70	65	39	35	30	27	**11.00**	24	19	3	1	–	–	–	–	–	–	–	–	–	**5-(2-Methoxy-1-methoxycarbonylvinyl)-6-(2-tolyl)phenanthridinium**		**LI08080**
105	103	77	98	69	81	82	83	**369**	100	91	56	52	52	40	39	36	**0.29**	24	35	2	1	–	–	–	–	–	–	–	–	–	**1,3-Oxazin-6-one, 4,5-dihydro-4,4-tetradecamethylene-2-phenyl-**		**NI38343**
181	312	152	117	146	243	153	57	**369**	100	80	37	36	35	26	26	20	**14.70**	26	27	1	1	–	–	–	–	–	–	–	–	–	**4-Biphenylyl 1-tert-butyl-2-phenyl-3-azetidinyl-, cis-**	13871-55-1	**NI38344**
181	117	312	152	57	153	104	70	**369**	100	72	51	32	30	25	24	24	**10.23**	26	27	1	1	–	–	–	–	–	–	–	–	–	**4-Biphenylyl 1-tert-butyl-2-phenyl-3-azetidinyl-, trans-**	13871-53-9	**NI38345**
213	293	295	27	53	133	215	211	**370**	100	74	68	68	65	55	52	52	**0.20**	4	6	–	–	–	–	–	4	–	–	–	–	–	**1,2,3,4-Tetrabromobutane**		**IC04173**

MASS TO CHARGE RATIOS								M.W.	INTENSITIES								Parent	C	H	O	N	S	F	Cl	Br	I	Si	P	B	X	COMPOUND NAME	CAS Reg No	No
238	195	168	254	239	152	57	127	370	100	37	19	14	11	10	10	8	1.07	9	11	6	2	–	–	–	–	1	–	–	–	–	Uridine, 5-iodo-	1024-99-3	NI38346
100	65	102	103	38	99	63	272	370	100	70	35	33	21	17	16	15	0.25	10	5	–	–	–	–	7	–	–	–	–	–	–	4,7-Methano-1H-indene, 1,4,5,6,7,8,8-heptachloro-3a,4,7,7a-tetrahydro-	76-44-8	NI38347
100	272	274	65	270	102	36	238	370	100	60	46	42	35	34	29	28	4.70	10	5	–	–	–	–	7	–	–	–	–	–	–	4,7-Methano-1H-indene, 1,4,5,6,7,8,8-heptachloro-3a,4,7,7a-tetrahydro-	76-44-8	NI38348
73	355	267	356	357	268	269	74	370	100	60	52	22	14	13	9	8	0.00	10	30	5	–	–	–	–	–	–	5	–	–	–	Cyclopentasiloxane, decamethyl-	541-02-6	NI38349
73	355	267	45	59	341	193		370	100	50	37	5	4	3	3		0.01	10	30	5	–	–	–	–	–	–	5	–	–	–	Cyclopentasiloxane, decamethyl-	541-02-6	LI08081
341	87	59	342	281	253	343	44	370	100	55	43	32	30	27	19	14	0.49	10	30	5	–	–	–	–	–	–	5	–	–	–	Cyclopentasiloxane, 1,3,5,7,9-pentaethyl-	17995-44-7	NI38350
370	235	77	63	184	156	214	79	370	100	80	80	70	60	59	40	28		12	6	4	2	–	–	2	–	–	–	–	–	1	Nickel, bis(4-chloro-1,2-dioxo-3,5-cyclohexadienyl 2-oximato)-	30993-27-2	NI38351
227	203	205	181	199	228	171	173	370	100	87	84	81	70	66	66	64	11.13	12	7	5	2	1	–	–	1	–	–	–	–	–	2,4-Dinitro-4′-bromodiphenyl sulphoxide	966-70-1	NI38352
313	43	311	315	57	58	312	257	370	100	92	66	60	45	41	32	29	0.81	12	27	–	–	–	–	–	1	–	–	–	–	1	Tributyltin bromide		MS08693
57	41	313	180	43	29	105	269	370	100	60	43	39	38	34	33	32	0.60	12	27	–	–	–	–	–	1	–	–	–	–	1	Triisobutyltin bromide		MS08694
355	73	253	356	226	357	211	55	370	100	39	29	24	16	15	15	13	7.00	12	31	5	–	–	–	–	–	–	3	1	–	–	Propanoic acid, 3-[bis[(trimethylsilyl)oxy]phosphinyl]-, trimethylsilyl ester	53044-28-3	NI38353
355	73	253	226	211	195	147	75	370	100	39	29	16	15	12	10	9	7.00	12	31	5	–	–	–	–	–	–	3	1	–	–	Propanoic acid, 3-[bis[(trimethylsilyl)oxy]phosphinyl]-, trimethylsilyl ester	53044-28-3	NI38354
184	354	64	370	91	109	79	77	370	100	69	69	62	61	57	54	52		13	10	7	2	2	–	–	–	–	–	–	–	–	2,4-Dinitro-4′-(sulphonylmethyl)diphenyl sulphoxide	75317-04-3	NI38355
176	161	175	107	189	177	109	49	370	100	75	40	16	15	15	14	13	0.00	13	16	–	–	1	4	–	1	–	–	–	1	–	1-(2-Bromoethyl)-2,3,5-trimethylbenzo[b]thiophenium tetrafluoroborate		LI08082
87	206	45	174	219	59	370	164	370	100	24	20	13	10	10	9	9		13	22	–	–	6	–	–	–	–	–	–	–	–	2,4,6,8,9,10-Hexathiatricyclo[3.3.1.1^{3,7}]decane, 1,3,5-tripropyl-	57274-48-3	NI38356
73	75	69	355	117	243	55	41	370	100	97	45	38	38	36	35	30	4.90	13	27	2	–	–	–	–	–	1	1	–	–	–	Decanoic acid, 10-iodo-, trimethylsilyl ester	26305-96-4	NI38357
121	180	122	179	178	77	78	165	370	100	86	72	56	35	24	22	21	0.00	14	12	–	–	1	–	–	2	–	–	–	–	–	α,α′-Dibromodibenzyl sulphide	31247-20-8	NI38358
198	291	293	199	197	277	279	292	370	100	23	21	18	15	13	12	10	0.00	14	16	–	2	–	–	–	2	–	–	–	–	–	(3,4-Dimethyl-5-bromo-2-pyrrolyl)(3,4-dimethyl-5-bromomethyl-2-pyrrolylidene)methane		LI08083
29	133	59	205	159	86	27	34	370	100	83	74	51	48	34	34	33	0.00	14	18	4	4	2	–	–	–	–	–	–	–	–	Carbamic acid, [1,2-phenylenebis(iminocarbonothioyl)]bis-, diethyl ester	23564-06-9	NI38359
206	29	133	150	370	160	134	178	370	100	78	69	56	43	42	39	34		14	18	4	4	2	–	–	–	–	–	–	–	–	Carbamic acid, [1,2-phenylenebis(iminocarbonothioyl)]bis-, diethyl ester	23564-06-9	NI38360
166	167	41	69	57	29	139	28	370	100	28	25	17	15	13	12	10	1.70	14	21	4	2	1	3	–	–	–	–	–	–	–	Cysteine, N-(trifluoroacetyl)-L-prolyl-, butyl ester		MS08695
97	83	75	69	55	73	103	355	370	100	96	75	73	65	58	40	36	0.70	14	31	1	–	–	–	–	–	1	1	–	–	–	Silane, [(11-iodoundecyl)oxy]trimethyl-	26305-84-0	NI38361
103	259	73	209	217	137	189	75	370	100	99	99	98	97	52	50	34	14.37	15	26	5	2	–	–	–	–	–	2	–	–	–	6H-Furo[2′,3′:4,5]oxazolo[3,2-a]pyrimidin-6-one, 2,3,3a,9a-tetrahydro-3-[(trimethylsilyl)oxy]-2-[[(trimethylsilyl)oxy]methyl]-, [2R-(2α,3β,4aβ,9aβ)]-	32414-34-9	NI38362
221	105	91	77	253	177	255	151	370	100	89	84	68	63	24	5	5	4.00	16	9	–	–	2	–	3	–	–	–	–	–	–	Trichloromethyl 9-anthracenecarbodithioate	91631-91-3	NI38363
138	140	43	111	181	327	329	209	370	100	46	40	33	30	24	23	22	8.20	16	13	2	2	–	–	3	–	–	–	–	–	–	2′,5′-Dichloro-2-(4-chloroanilino)-3-hydroxycrotonanilide		MS08696
370	372	252	287	290	253	69	289	370	100	46	43	37	33	29	29	28		16	15	4	–	2	–	1	–	–	–	–	–	–	Spiro[benzofuran-2(3H),1′-cyclohexane]-2′,3,4′-trione, 7-chloro-6′-methyl-4,6-bis(methylthio)-	26891-74-7	NI38364
327	356	358	329	27	41	357	314	370	100	96	73	72	32	29	27	27	0.00	16	16	2	2	1	–	2	–	–	–	–	–	–	10H-Phenothiazin-3-amine, 10-butyl-7,9-dichloro-, 5,5-dioxide	55191-34-9	NI38365
105	209	106	77	79	370	368	103	370	100	11	11	9	7	5	5	5		16	18	–	–	–	–	–	–	–	–	–	–	2	4,4′-Bis(4-methylbenzyl) diselenide		MS08697
73	355	356	45	357	74	75	147	370	100	75	24	20	12	9	8	6	2.71	16	30	4	–	–	–	–	–	–	3	–	–	–	Benzoic acid, 2,5-bis(trimethylsiloxy)-, trimethylsilyl ester	3618-20-0	NI38366
355	73	356	357	267	147	74	75	370	100	50	31	15	4	4	4	3	3.45	16	30	4	–	–	–	–	–	–	3	–	–	–	Benzoic acid, 2,5-bis(trimethylsiloxy)-, trimethylsilyl ester	3618-20-0	NI38367
355	73	356	357	45	75	147	74	370	100	75	32	15	12	7	6	6	3.20	16	30	4	–	–	–	–	–	–	3	–	–	–	Benzoic acid, 2,5-bis(trimethylsiloxy)-, trimethylsilyl ester	3618-20-0	MS08698
355	73	356	45	357	281	74	147	370	100	89	34	18	17	9	8	7	1.16	16	30	4	–	–	–	–	–	–	3	–	–	–	Benzoic acid, 2,4-bis[(trimethylsilyl)oxy]-, trimethylsilyl ester	10586-16-0	NI38368
355	73	356	251	18	89	59	357	370	100	30	30	28	20	18	18	11	3.51	16	30	4	–	–	–	–	–	–	3	–	–	–	Benzoic acid, 2,4-bis[(trimethylsilyl)oxy]-, trimethylsilyl ester	10586-16-0	NI38369
73	355	45	356	74	357	75	133	370	100	43	25	16	9	8	6	5	0.69	16	30	4	–	–	–	–	–	–	3	–	–	–	Benzoic acid, 2,4-bis[(trimethylsilyl)oxy]-, trimethylsilyl ester	10586-16-0	NI38370
73	355	356	267	147	357	74	265	370	100	94	31	21	17	15	9	7	1.25	16	30	4	–	–	–	–	–	–	3	–	–	–	Benzoic acid, 2,6-bis[(trimethylsilyl)oxy]-, trimethylsilyl ester	3782-85-2	NI38371
73	355	45	356	267	147	357	74	370	100	70	27	22	17	14	11	9	0.00	16	30	4	–	–	–	–	–	–	3	–	–	–	Benzoic acid, 2,6-bis[(trimethylsilyl)oxy]-, trimethylsilyl ester	3782-85-2	NI38372
193	73	370	355	311	194	371	281	370	100	47	42	26	15	15	13	9		16	30	4	–	–	–	–	–	–	3	–	–	–	Benzoic acid, 3,4-bis[(trimethylsilyl)oxy]-, trimethylsilyl ester	2347-40-2	NI38373
73	193	370	355	45	371	194	311	370	100	85	38	22	22	13	13	11		16	30	4	–	–	–	–	–	–	3	–	–	–	Benzoic acid, 3,4-bis[(trimethylsilyl)oxy]-, trimethylsilyl ester	2347-40-2	NI38374
193	73	370	355	371	311	194	45	370	100	59	55	35	20	20	14	12		16	30	4	–	–	–	–	–	–	3	–	–	–	Benzoic acid, 3,4-[bis(trimethylsilyloxy)]-, trimethylsilyl ester	2347-40-2	MS08699
73	370	355	371	356	147	372	281	370	100	58	47	19	16	14	10	10		16	30	4	–	–	–	–	–	–	3	–	–	–	Benzoic acid, 3,5-bis[(trimethylsilyl)oxy]-, trimethylsilyl ester	79314-27-5	NI38375
73	355	370	158	156	68	212	75	370	100	84	77	76	63	61	60	54		16	30	4	2	–	–	–	–	–	2	–	–	–	5H,10H-Dipyrrolo[1,2-a:1′,2′-d]pyrazine-5,10-dione, octahydro-2,7-bis[(trimethylsilyl)oxy]-	56817-96-0	MS08700
104	76	132	187	248	77	105	225	370	100	74	70	54	27	27	26	24	0.80	17	11	3	2	–	–	–	1	–	–	–	–	–	2-Azetidinone, 4-(2-bromophenyl)-3-phthalimido-, (Z)-		NI38376
130	43	103	77	58	370			370	100	24	9	9	6	1				17	14	6	4	–	–	–	–	–	–	–	–	–	2,4-(Dinitrophenyl)tryptophan		LI08084
194	77	93	195	370	51	45	65	370	100	36	34	29	26	26	25	18		17	18	–	6	2	–	–	–	–	–	–	–	–	1H-1,2,4-Triazole, 4,5-dihydro-4-[(N,N′-diphenyl)guanidino]-3,5-bis(methylthio)-		MS08701
297	73	298	299	75	45	355	267	370	100	49	26	12	7	6	5	5	3.68	17	30	5	–	–	–	–	–	–	2	–	–	–	Benzeneacetic acid, 3-methoxy-α,4-bis[(trimethylsilyl)oxy]-, ethyl ester	55590-98-2	NI38377
297	73	298	299	75	267	355	45	370	100	96	94	39	20	18	17	16	10.40	17	30	5	–	–	–	–	–	–	2	–	–	–	Benzeneacetic acid, 3-methoxy-α,4-bis[(trimethylsilyl)oxy]-, ethyl ester	55590-98-2	MS08702
253	73	147	237	151	75	254	249	370	100	98	35	26	22	20	19	19	13.20	17	30	5	–	–	–	–	–	–	2	–	–	–	Benzenepropanoic acid, 2,3-dimethoxy-α-[(trimethylsilyl)oxy]-, trimethylsilyl ester		MS08703

MASS TO CHARGE RATIOS	M.W.	INTENSITIES	Parent	C	H	O	N	S	F	Cl	Br	I	Si	P	B	X	COMPOUND NAME	CAS Reg No	No
73 253 151 147 280 370 249 237	370	100 63 53 40 28 27 20 18		17	30	5	–	–	–	–	–	–	2	–	–	–	Benzenepropanoic acid, 2,5-dimethoxy-α-[(trimethylsilyl)oxy]-, trimethylsilyl ester		MS08704
151 73 280 147 253 370 75 152	370	100 79 47 38 28 17 16 15		17	30	5	–	–	–	–	–	–	2	–	–	–	Benzenepropanoic acid, 3,4-dimethoxy-α-[(trimethylsilyl)oxy]-, trimethylsilyl ester		MS08705
209 73 210 179 45 59 75 89	370	100 65 18 12 11 8 6 5	5.60	17	30	5	–	–	–	–	–	–	2	–	–	–	Benzenepropanoic acid, 3-methoxy-α,4-bis[(trimethylsilyl)oxy]-, methyl ester		MS08706
267 370 73 268 193 371 179 269	370	100 45 44 25 19 15 11 10		17	34	3	–	–	–	–	–	–	3	–	–	–	Benzene, 3,4-bis[(trimethylsilyl)oxy]-1-[2-[(trimethylsilyl)oxy]ethyl]-		MS08707
73 267 370 268 193 45 75 371	370	100 98 37 25 20 18 13 12		17	34	3	–	–	–	–	–	–	3	–	–	–	Benzene, 3,4-bis[(trimethylsilyl)oxy]-1-[2-[(trimethylsilyl)oxy]ethyl]-		MS08708
267 73 268 147 75 269 45 74	370	100 67 24 22 15 9 9 6	0.00	17	34	3	–	–	–	–	–	–	3	–	–	–	Benzene, 4-[(trimethylsilyl)oxy]-1-[1,2-bis[(trimethylsilyl)oxy]ethyl]-		MS08709
267 73 268 77 269 193 75 179	370	100 42 24 20 11 10 10 7	0.00	17	34	3	–	–	–	–	–	–	3	–	–	–	Benzene, 4-[(trimethylsilyl)oxy]-1-[1,2-bis[(trimethylsilyl)oxy]ethyl]-		MS08710
285 300 355 73 75 286 100 97	370	100 80 67 64 37 28 24 22	2.67	17	34	3	2	–	–	–	–	–	2	–	–	–	2,4,6(1H,3H,5H)-Pyrimidinetrione, 5-ethyl-5-(1-methylbutyl)-1,3-bis(trimethylsilyl)-	52937-68-5	NI38378
355 73 300 356 100 285 75 55	370	100 74 46 37 26 23 22 17	3.67	17	34	3	2	–	–	–	–	–	2	–	–	–	2,4,6(1H,3H,5H)-Pyrimidinetrione, 5-ethyl-5-(3-methylbutyl)-1,3-bis(trimethylsilyl)-	52937-67-4	NI38379
243 168 244 128 139 130 66 64	370	100 30 17 17 10 7 7 5	4.00	18	12	2	–	–	–	1	–	–	–	–	–	1	Phenoxarsine, 10-(4-chlorophenoxy)-	27796-63-0	NI38380
185 77 292 59 81 178 186 63	370	100 61 35 22 19 18 16 16	2.00	18	22	–	–	–	4	–	–	–	2	–	–	–	1,2-Bis[3-(dimethylsilyl)phenyl]tetrafluoroethane	37601-86-8	NI38381
147 295 264 221 370 190 263 119	370	100 88 49 40 36 35 30 27		18	26	–	–	4	–	–	–	–	–	–	–	–	1-(2-Butyl-1,3-dithian-2-yl)-2-(1,3-dithian-2-yl)benzene		MS08711
106 91 133 238 178 147 307 187	370	100 69 67 36 35 31 16 14	0.00	18	26	8	–	–	–	–	–	–	–	–	–	–	Propanedioic acid, [[7-[bis(methoxycarbonyl)methyl]-1-cyclohepten-1-yl]methyl]-, dimethyl ester		DD01473
100 84 56 70 129 98 72 58	370	100 62 51 46 41 31 31 30	2.30	18	34	4	4	–	–	–	–	–	–	–	–	–	N,N′,N″,N‴-Tetraacetylspermine	40563-82-4	NI38382
100 327 84 268 129 70 256 98	370	100 60 56 55 51 45 41 33	2.73	18	34	4	4	–	–	–	–	–	–	–	–	–	N,N′,N″,N‴-Tetraacetylspermine	40563-82-4	NI38383
93 92 370 57 77 69 65 55	370	100 61 58 57 56 39 39 34		19	22	4	4	–	–	–	–	–	–	–	–	–	3,6-Anhydro-D-galacto-heptulose phenylosazone		MS08712
370 371 188 92 174 352 77 93	370	100 23 21 20 17 16 14 13		19	22	4	4	–	–	–	–	–	–	–	–	–	3,6-Anhydro-D-manno-heptos-2-ulose, 3,6-anhydro-, bis(phenylhydrazone)	21238-55-1	NI38384
93 92 77 57 370 55 69 65	370	100 58 53 49 48 40 37 37		19	22	4	4	–	–	–	–	–	–	–	–	–	3,6-Anhydro-D-talo-2-heptulose phenylosazone		MS08713
234 252 206 235 202 203 224 124	370	100 70 59 55 48 40 31 31	0.00	19	30	7	–	–	–	–	–	–	–	–	–	–	Cyclopentaneheptanoic acid, 2-acetyl-3,5-bis(acetyloxy)-, methyl ester	55334-66-2	NI38385
342 78 74 370 102 194 343 51	370	100 84 68 46 34 32 25 23		20	10	4	4	–	–	–	–	–	–	–	–	–	N,N′-Bis(2-pyridyl)pyromellitimide	31663-81-7	NI38386
370 78 64 74 178 51 371 77	370	100 70 64 38 32 32 30 24		20	10	4	4	–	–	–	–	–	–	–	–	–	N,N′-Bis(3-pyridyl)pyromellitimide	31663-82-8	NI38387
370 78 74 178 51 171 102 222	370	100 70 38 32 32 30 24 16		20	10	4	4	–	–	–	–	–	–	–	–	–	N,N′-Bis(3-pyridyl)pyromellitimide	31663-82-8	LI08085
370 78 74 178 51 371 102 77	370	100 70 38 32 32 30 24 24		20	10	4	4	–	–	–	–	–	–	–	–	–	N,N′-Bis(3-pyridyl)pyromellitimide	31663-82-8	MS08714
77 51 90 194 368 341 41 39	370	100 34 31 28 26 14 13 12	8.00	20	14	2	6	–	–	–	–	–	–	–	–	–	Pyridazino[4,5-b]pyridazino[4,5-e]pyrazine, 1,6-dihydroxy-2,7-diphenyl-		NI38388
136 139 116 138 151 150 203 117	370	100 40 29 23 15 14 12 11	2.00	20	18	7	–	–	–	–	–	–	–	–	–	–	1,3-Benzodioxole, 5-[4-(1,3-benzodioxol-5-yloxy)tetrahydro-1H-furo[3,4-c]furan-1-yl]-		NI38389
370 299 165 134 371 298 300 237	370	100 58 35 31 16 12 4 3		20	18	7	–	–	–	–	–	–	–	–	–	–	2H,6H-Benzo[1,2-b:5,4-b′]dipyran-6-one, 3,4-dihydro-3,5-dihydroxy-7-(2,4-dihydroxyphenyl)-2,2-dimethyl-	91681-65-1	NI38390
297 352 337 298 370 284 69 353	370	100 40 30 19 16 14 11 10		20	18	7	–	–	–	–	–	–	–	–	–	–	11H-Benzofuro[2,3-b]benzopyran-11-one, 1,3,8-trihydroxy-4-(3-hydroxy-3-methylbutyl)-	104691-85-2	NI38391
370 327 44 29 356 341 41 43	370	100 65 50 48 41 34 32 31		20	18	7	–	–	–	–	–	–	–	–	–	–	[2]Benzopyrano[4,3-b][1]benzopyran-7(5H)-one, 2,3,8,10-tetramethoxy-		MS08715
29 356 357 43 41 44 39 69	370	100 74 68 68 61 51 50 45	13.83	20	18	7	–	–	–	–	–	–	–	–	–	–	[2]Benzopyrano[4,3-b][1]benzopyran-7(5H)-one, 3,4,8,10-tetramethoxy-		MS08716
299 370 298 300 153 43 371 311	370	100 86 53 49 28 23 13 13		20	18	7	–	–	–	–	–	–	–	–	–	–	4H-1-Benzopyran-4-one, 5,7-dihydroxy-3-(2,3-dihydro-2,2-dimethyl-3,5-dihydroxychroman-6-yl)-		NI38392
299 370 300 298 153 311 57 55	370	100 63 32 27 27 7 6 6		20	18	7	–	–	–	–	–	–	–	–	–	–	4H-1-Benzopyran-4-one, 5,7-dihydroxy-3-(2,3-dihydro-2,2-dimethyl-3,5-dihydroxychroman-8-yl)-		NI38393
311 312 370 59 153 44 337 160	370	100 52 51 24 14 14 9 5		20	18	7	–	–	–	–	–	–	–	–	–	–	4H-1-Benzopyran-4-one, 5,7-dihydroxy-3-[2,3-dihydro-4-hydroxy-2-(2-hydroxyisopropyl)benzofuran-5-yl]-		NI38394
311 312 44 313 153 284 59 283	370	100 62 5 4 4 3 3 2	0.67	20	18	7	–	–	–	–	–	–	–	–	–	–	4H-1-Benzopyran-4-one, 5,7-dihydroxy-3-[2,3-dihydro-4-hydroxy-2-(2-hydroxyisopropyl)benzofuran-7-yl]-		NI38395
370 311 59 312 337 371 179 177	370	100 76 51 42 21 12 11 11		20	18	7	–	–	–	–	–	–	–	–	–	–	5H-Furo[3,2-g][1]benzopyran-5-one, 2,3-dihydro-4-hydroxy-2-(1-hydroxy-1-methylethyl)-6-(2,4-dihydroxyphenyl)-	91681-64-0	NI38396
163 43 165 164 135 136 103 76	370	100 41 18 17 8 5 5 4	0.10	20	18	7	–	–	–	–	–	–	–	–	–	–	4,4′-Bis(methoxycarbonyl)benzoin acetate		IC04174
299 327 370 314 300 352	370	100 35 31 18 18 7		20	18	7	–	–	–	–	–	–	–	–	–	–	Norsolorinic acid		LI08086
270 295 298 370 352 296 334 323	370	100 51 37 31 31 25 15 15		20	18	7	–	–	–	–	–	–	–	–	–	–	β-Rhodomycinone, 10-deoxy-		NI38397
328 370 330 329 327 372 371 331	370	100 44 36 27 22 15 11 7		20	19	3	2	–	–	1	–	–	–	–	–	–	3H-2,3-Benzodiazepine, 3-acetyl-1-(3-chlorophenyl)-4,5-dihydro-7,8-dimethoxy-4-methylene-	90140-67-3	NI38398
328 370 330 327 329 287 372 371	370	100 71 35 34 31 31 27 17		20	19	3	2	–	–	1	–	–	–	–	–	–	3H-2,3-Benzodiazepine, 3-acetyl-1-(3-chlorophenyl)-7,8-dimethoxy-4-methyl-	90140-68-4	NI38399

MASS TO CHARGE RATIOS								M.W.	INTENSITIES								Parent	C	H	O	N	S	F	Cl	Br	I	Si	P	B	X	COMPOUND NAME	CAS Reg No	No
86	57	70	87	355	257	164	71	370	100	22	21	10	8	4	4	4	0.00	20	19	3	2	–	–	1	–	–	–	–	–	–	2-(tert-Butylamino)-N-(9,10-dioxaanthracen-2-yl)ethanamide		MS08717
336	259	249	278	250	251	231	194	370	100	99	99	93	90	80	80	50	0.00	20	19	3	2	–	–	1	–	–	–	–	–	–	5H-Pyrrolo[1,2-d][1,4]benzodiazepine-3,6-dione, 10-chloro-1,2,7,11b-tetrahydro-R-2-hydroxy-trans-2,7-dimethyl-trans-11b-phenyl-	84980-51-8	NI38400
170	171	235	370	168	263	172	154	370	100	79	49	38	14	13	9	9		20	22	3	2	1	–	–	–	–	–	–	–	–	N-(1-Dimethylaminonaphthalene-5-sulphonyl)-2-hydroxy-2-phenylethylamine		NI38401
170	171	235	370	263	168	371	172	370	100	80	48	35	14	14	10	10		20	22	3	2	1	–	–	–	–	–	–	–	–	1-Naphthalenesulphonic acid, 5-(dimethylamino)-, 2-amino-1-phenylethyl ester	55837-09-7	NI38402
91	146	79	217	77	108	107	192	370	100	47	41	36	29	28	22	21	0.00	20	22	5	2	–	–	–	–	–	–	–	–	–	Benzoic acid, N-[(benzyloxycarbonyl)-L-alanyl]-4-amino-, ethyl ester		DD01474
57	160	161	79	81	221	137	67	370	100	68	50	37	35	30	30	30	0.00	20	34	6	–	–	–	–	–	–	–	–	–	–	Diethyl rel-1-tert-butyl-cis-4-carboethoxy-cis-3-cyclohexylmalonate		NI38403
160	161	211	165	137	266	240	221	370	100	50	30	28	23	22	18	18	0.30	20	34	6	–	–	–	–	–	–	–	–	–	–	Diethyl rel-1-tert-butyl-trans-4-carboethoxy-cis-3-cyclohexylmalonate		NI38404
160	161	210	326	211	57	370	41	370	100	52	42	34	32	26	19	17		20	34	6	–	–	–	–	–	–	–	–	–	–	Diethyl rel-1-tert-butyl-trans-4-carboethoxy-trans-3-cyclohexylmalonate		NI38405
255	59	119	129	43	69	101	163	370	100	95	90	71	65	58	51	27	10.00	20	34	6	–	–	–	–	–	–	–	–	–	–	Octadecanedioic acid, 3,16-dioxo-, dimethyl ester	51414-51-8	NI38406
43	370	126	154	183	283	114	112	370	100	77	68	66	63	50	44	43		20	38	4	2	–	–	–	–	–	–	–	–	–	1,4-Dioxa-7,10-diazacyclododecane-6,11-dione, 7,10-dihexyl-	67456-23-9	NI38407
370	192	205	166	191	371	125	165	370	100	50	38	33	29	26	23	21		21	22	6	–	–	–	–	–	–	–	–	–	–	(1S,5S,6R,7R)-1-Allyl-3-methoxy-6-(3-methoxy-4,5-methylenedioxyphenyl)-7-methyl-4,8-dioxobicyclo[3.2.1]oct-2-ene		NI38408
338	265	121	237	370	165	278	194	370	100	34	26	23	11	7	6	6		21	22	6	–	–	–	–	–	–	–	–	–	–	Benzenepropanoic acid, 4-hydroxy-3-[1-(methoxycarbonyl)-2-(4-methoxyphenyl)vinyl]-, methyl ester, (E)-	69721-91-1	NI38409
149	57	43	41	180	179	370	71	370	100	48	48	35	26	24	21	20		21	22	6	–	–	–	–	–	–	–	–	–	–	1,3-Benzodioxole, 6-[2,3-dihydro-5-methoxy-3-methyl-6-(2-propenyloxy)-2 benzofuranyl]-4-methoxy-, (2R-cis)-	64272-51-1	NI38410
370	177	329	41	371	165	121	106	370	100	60	28	26	25	17	14	14		21	22	6	–	–	–	–	–	–	–	–	–	–	1,3-Benzodioxole, 6-[2,3-dihydro-5-methoxy-3-methyl-6-(2-propenyloxy)-2 benzofuranyl]-4-methoxy-, (2S-trans)-	64272-50-0	NI38411
43	59	370	165	149	41	329	57	370	100	47	39	30	27	26	22	20		21	22	6	–	–	–	–	–	–	–	–	–	–	6(2H)-Benzofuranone, 3,5-dihydro-5-methoxy-2-(7-methoxy-1,3-benzodioxol-5-yl)-3-methyl-5-(2-propenyl)-, [2R-(2α,3α,5α)]-	64272-49-7	NI38412
370	325	240	312	296	297	281	268	370	100	85	67	48	48	43	43	33		21	22	6	–	–	–	–	–	–	–	–	–	–	11-Desacetoxywortmannin		NI38413
355	370	356	369	326	371	298	255	370	100	57	23	18	18	17	17	17		21	22	6	–	–	–	–	–	–	–	–	–	–	3′,5-Diethoxy-4′,7-dimethoxyflavone		MS08718
296	297	281	307	370	325	268	267	370	100	31	18	12	11	10	10	10		21	22	6	–	–	–	–	–	–	–	–	–	–	8,9-Dihydro-9,11-dihydrowortmannin		NI38414
236	221	121	370	371	237	193	91	370	100	89	60	40	31	24	23	23		21	22	6	–	–	–	–	–	–	–	–	–	–	Furano[7,8:2″,3″]flavan, 3,4,5,6-tetramethoxy-	77970-11-7	NI38415
137	177	150	219	191	370	122	233	370	100	88	32	28	28	14	12	6		21	22	6	–	–	–	–	–	–	–	–	–	–	6-Heptene-3,5-dione, 1,7-bis(4-hydroxy-3-methoxyphenyl)-, (E)-	76474-56-1	NI38416
83	270	255	100	229	271	272	108	370	100	31	24	16	7	6	6	5	0.08	21	22	6	–	–	–	–	–	–	–	–	–	–	Shikonin β, β-dimethylacrylate	24502-79-2	NI38417
83	270	255	100	69	41	187	55	370	100	68	38	27	21	21	19	16	1.00	21	22	6	–	–	–	–	–	–	–	–	–	–	Shikonin, β,β-dimethylacrylate	24502-79-2	LI08087
108	107	336	121	350	308	120	109	370	100	45	20	18	14	10	10	10	0.00	21	23	3	2	–	1	–	–	–	–	–	–	–	Schizozygine, trifluoromethyl-		LI08088
83	45	85	121	47	98	43	44	370	100	79	75	64	40	38	37	25	10.59	21	26	–	2	2	–	–	–	–	–	–	–	–	10H-Phenothiazine, 10-[2-(1-methyl-2-piperidinyl)ethyl]-2-(methylthio)-	50-52-2	NI38418
98	70	370	126	99	185	244	125	370	100	14	9	8	8	5	4	3		21	26	–	2	2	–	–	–	–	–	–	–	–	10H-Phenothiazine, 10-[2-(1-methyl-2-piperidinyl)ethyl]-2-(methylthio)-	50-52-2	NI38419
98	370	70	126	42	99	371	185	370	100	35	19	11	11	9	8	6		21	26	–	2	2	–	–	–	–	–	–	–	–	10H-Phenothiazine, 10-[2-(1-methyl-2-piperidinyl)ethyl]-2-(methylthio)-	50-52-2	NI38420
108	370	263	109	83	69	267	81	370	100	22	12	12	12	12	9	8		21	26	4	2	–	–	–	–	–	–	–	–	–	Aspidodasycarpine	2744-47-0	NI38421
108	370	109	263	69	83	267	60	370	100	21	13	12	12	11	10	9		21	26	4	2	–	–	–	–	–	–	–	–	–	Aspidodasycarpine	2744-47-0	NI38422
112	370	328	310	327	43	140	286	370	100	87	67	38	36	35	27	23		21	26	4	2	–	–	–	–	–	–	–	–	–	Aspidodispermine, monoacetate	21451-68-3	NI38423
370	184	341	282	180	311	309	208	370	100	90	77	75	47	24	24	24		21	26	4	2	–	–	–	–	–	–	–	–	–	19,20-Dihydrotalbotine		LI08089
108	370	144	156	130	115	232	159	370	100	15	13	6	5	5	4	4		21	26	4	2	–	–	–	–	–	–	–	–	–	Lonicerine	26988-11-4	LI08090
108	370	109	130	122	107	106	156	370	100	19	14	8	8	8	8	7		21	26	4	2	–	–	–	–	–	–	–	–	–	Lonicerine	26988-11-4	NI38424
57	196	81	197	370	171	143	226	370	100	83	77	67	61	58	40	33		21	26	4	2	–	–	–	–	–	–	–	–	–	Spiro[1-(tert-butoxycarbonyl)piperidine-4,3′-(5-methyl-1,4-dihydro-1-oxo-2H-pyrano[4,3-b]indole)]		DD01475
201	138	144	143	182	370	130	202	370	100	42	41	39	28	22	22	20		21	26	4	2	–	–	–	–	–	–	–	–	–	Splendoline		LI08091
369	370	169	170	156	184	144	143	370	100	99	16	14	12	9	9	7		21	26	4	2	–	–	–	–	–	–	–	–	–	Yohimban-16-carboxylic acid, 17,18-dihydroxy-, methyl ester, (16α,17α,18α)-	69862-18-6	NI38425
224	225	148	370	130	69	77	94	370	100	43	23	17	16	16	15	14		21	26	4	2	–	–	–	–	–	–	–	–	–	Yohimbine pseudoindoxyl	6872-98-6	NI38426
256	135	95	107	241	148	257	91	370	100	67	51	35	34	29	28	28	0.99	21	29	2	–	–	3	–	–	–	–	–	–	–	Androst-5-en-3-ol, trifluoroacetate, (3β)-	56438-15-4	NI38427
126	95	280	109	308	263	151	182	370	100	73	36	35	30	26	20	18	0.00	21	38	5	–	–	–	–	–	–	–	–	–	–	Butanedioic acid, (15-hydroxyhexadecyl)methylene-	62722-96-7	NI38428
335	309	317	163	353	336	319	280	370	100	65	45	40	35	22	17	17	0.00	21	38	5	–	–	–	–	–	–	–	–	–	–	Prost-13-en-1-oic acid, 9,11,15-trihydroxy-, methyl ester, (9α,11α,13E,15S)-	13227-94-6	NI38429
370	73	371	44	342	43	105	159	370	100	30	28	25	24	24	20	16		22	10	6	–	–	–	–	–	–	–	–	–	–	5,8,13,14-Pentaphenetetrone, 6,7-dihydroxy-		NI38430
370	337	371	338	369	305	185	336	370	100	91	42	40	32	28	27	23		22	14	–	2	2	–	–	–	–	–	–	–	–	5,13-Dihydrophenothiazino[2,1-a]phenothiazine		MS08719
370	371	337	306	168	305	369	372	370	100	27	23	18	18	14	13	12		22	14	–	2	2	–	–	–	–	–	–	–	–	8,16-Dihydrophenothiazino[4,3-c]phenothiazine		MS08720
91	104	76	105	223	65	50	28	370	100	75	58	56	47	42	35	33	2.00	22	18	2	4	–	–	–	–	–	–	–	–	–	cis-2,3-Bis(4-tolyl)-1-phthalimidoazimine		NI38431
91	104	76	105	223	65	342	50	370	100	83	75	56	47	42	41	35	2.00	22	18	2	4	–	–	–	–	–	–	–	–	–	trans-2,3-Bis(4-tolyl)phthalimidoazimine		NI38432

MASS TO CHARGE RATIOS								M.W.	INTENSITIES								Parent	C	H	O	N	S	F	Cl	Br	I	Si	P	B	X	COMPOUND NAME	CAS Reg No	No
156	370	372	371	329	355	342	327	370	100	77	52	25	24	23	22	22		22	20	1	–	–	–	2	–	–	–	–	–	–	Spiro[2.4]heptan-4-one, 2-(4-chlorophenyl)-5-[(E)-4-chlorobenzylidene]-1,1-dimethyl-		MS08721
50	370	52	355	371	185	44	356	370	100	52	30	23	13	13	13	6		22	22	–	6	–	–	–	–	–	–	–	–	–	2-Amino-5,12-dimethyl-3-dimethylamino-5,12-dihydroquinoxalino[2,3-b]phenazine		IC04175
370	355	185	181	167	190	192	165	370	100	43	23	17	10	8	7	7		22	26	5	–	–	–	–	–	–	–	–	–	–	Benzofuran, 2,3-dihydro-7-methoxy-3-methyl-5-(1-propenyl)-2-(3,4,5-trimethoxyphenyl)-, [2R-[2α,3β,5(E)]]-	60297-83-8	NI38433
178	356	316	165	315	167	166	151	370	100	99	71	60	59	48	48	42	0.00	22	26	5	–	–	–	–	–	–	–	–	–	–	6(2H)-Benzofuranone, 2-[(3,4-dimethoxyphenyl)methyl]-3,3a-dihydro-5-methoxy-3-methyl-3a-(2-propenyl)-, [2S-(2α,3α,3aα)]-	75299-49-9	NI38434
178	356	342	182	166	165	167	149	370	100	99	62	46	42	42	38	38	0.00	22	26	5	–	–	–	–	–	–	–	–	–	–	6(2H)-Benzofuranone, 2-[(3,4-dimethoxyphenyl)methyl]-3,5-dihydro-5-methoxy-3-methyl-5-(2-propenyl)-, [2S-(2α,3α,5β)]-	75299-50-2	NI38435
370	201	227	202	187	371	186	129	370	100	25	22	17	16	12	11	11		22	26	5	–	–	–	–	–	–	–	–	–	–	Cyclopentaneacetic acid, 2-[(4,6-dimethoxy-1-naphthyl)methyl]-1-methyl-5-oxo-, methyl ester	15384-83-5	NI38436
370	214	201	371	215	202	190	188	370	100	70	41	25	11	6	4	3		22	26	5	–	–	–	–	–	–	–	–	–	–	8,14-seco-13ξ-Estra-1,3,5,7,9-pentaene-17-carboxylic acid, 3,6-dimethoxy-14-oxo-, methyl ester	23569-27-9	NI38437
328	214	213	329	172	61	268	146	370	100	38	26	23	23	14	13	13	6.18	22	26	5	–	–	–	–	–	–	–	–	–	–	Estra-1,3,5(10)-triene-3,17-diol, 16,17-epoxy-, diacetate, (16α,17β)-	69744-63-4	NI38438
328	329	174	227	173	268	286	160	370	100	23	16	15	13	12	9	5	2.32	22	26	5	–	–	–	–	–	–	–	–	–	–	Estra-1,3,5(10)-trien-6-one, 3,17-bis(acetyloxy)-, (17β)-	3434-45-5	NI38439
268	310	269	286	170	311	157	144	370	100	22	22	12	12	8	7	7	1.04	22	26	5	–	–	–	–	–	–	–	–	–	–	Estra-1,3,5(10)-trien-17-one, 3,6-bis(acetyloxy)-, (6β)-		NI38440
328	268	329	253	370	269	210	310	370	100	67	40	38	35	31	27	24		22	26	5	–	–	–	–	–	–	–	–	–	–	Estra-1,3,5(10)-trien-17-one, 3,11-bis(acetyloxy)-, (11β)-		NI38441
268	328	240	269	310	172	211	198	370	100	28	27	23	21	13	11	9	3.45	22	26	5	–	–	–	–	–	–	–	–	–	–	Estra-1,3,5(10)-trien-17-one, 3,15-bis(acetyloxy)-, (15α)-		NI38442
328	214	329	213	172	268	146	159	370	100	34	23	18	15	10	10	9	7.75	22	26	5	–	–	–	–	–	–	–	–	–	–	Estra-1,3,5(10)-trien-17-one, 3,16-bis(acetyloxy)-, (16β)-	1247-70-7	NI38443
328	214	329	213	172	268	146	159	370	100	38	25	19	16	10	10	9	8.60	22	26	5	–	–	–	–	–	–	–	–	–	–	Estra-1,3,5(10)-trien-17-one, 3,16-bis(acetyloxy)-, (16β)-	1247-70-7	NI38444
213	228	55	176	229	370	214	187	370	100	39	31	20	18	17	16	9		22	26	5	–	–	–	–	–	–	–	–	–	–	3'(R)-(+)-3'-cis-4-Octenoyloxy-3',4'-dihydroseselin		LI08092
328	172	96	310	268	173	215	329	370	100	89	83	43	42	40	33	30	16.56	22	26	5	–	–	–	–	–	–	–	–	–	–	15-Oxoestradiol-3,17β-diacetate		NI38445
328	329	370	213	172	286	285	173	370	100	24	10	10	7	6	4	4		22	26	5	–	–	–	–	–	–	–	–	–	–	16-Oxoestradiol-3,17β-diacetate		NI38446
370	270	165	152	115	127	128	141	370	100	94	88	70	62	61	55	48		22	26	5	–	–	–	–	–	–	–	–	–	–	Spiro[benzo[a]cyclopenta[3,4]cyclobuta[1,2-c]cycloheptene-8(5H),2'-[1,3]dioxane], 6,7,7b,10a-tetrahydro-1,2,3-trimethoxy-		MS08722
284	269	339	165	285	370	238	127	370	100	32	29	24	23	22	22	22		22	26	5	–	–	–	–	–	–	–	–	–	–	Spiro[3H-benzo[e]cyclobut[c]azulene-1(11aH),2'-[1,3]dioxane], 4,5-dihydro-7,8,9-trimethoxy-		MS08723
204	355	354	205	340	190	324		370	100	54	36	22	6	6	5		0.00	22	28	4	1	–	–	–	–	–	–	–	–	–	(–)-Argemonine methohydroxide		LI08093
41	91	337	43	42	70	65	98	370	100	95	60	40	40	35	24	18	8.00	22	30	1	2	1	–	–	–	–	–	–	–	–	Pyrimidine, 1-benzylhexahydro-3-isopropyl-2-(4,4-dimethyl-2-oxo-6-thioxocyclohexylidene)-		MS08724
124	370	168	154	326	125	327	174	370	100	52	32	26	22	22	20	18		22	30	3	2	–	–	–	–	–	–	–	–	–	Aspidospermidin-3-ol, 1-acetyl-17-methoxy-, (3α)-	54725-60-9	NI38447
55	124	69	71	67	99	57	83	370	100	58	54	47	47	43	40	37	6.00	22	30	3	2	–	–	–	–	–	–	–	–	–	Aspidospermidin-17-ol, 1-acetyl-16-methoxy-	466-45-5	NI38448
370	369	371	125	43	342	69	57	370	100	42	27	23	19	18	15	15		22	30	3	2	–	–	–	–	–	–	–	–	–	Aspidospermidin-17-ol, 1-acetyl-16-methoxy-	466-45-5	NI38449
370	325	342	140	141	168	371	326	370	100	71	57	44	30	27	25	25		22	30	3	2	–	–	–	–	–	–	–	–	–	Aspidospermidin-20-ol, 1-acetyl-17-methoxy-	36458-97-6	NI38450
168	370	371	169	296	196	174	339	370	100	55	14	12	10	9	8	6		22	30	3	2	–	–	–	–	–	–	–	–	–	Cylindrocarine, 1-methyl-	22226-36-4	NI38451
140	370	342	141	369	168	43	311	370	100	22	13	12	7	7	7	6		22	30	3	2	–	–	–	–	–	–	–	–	–	Cylindrocarpinol, N-acetyl-	22226-34-2	NI38452
58	126	154	370	174	96	112	83	370	100	94	28	24	23	15	14	10		22	30	3	2	–	–	–	–	–	–	–	–	–	Schizozygine, hexahydro-N,N-dimethyl-	2283-33-2	NI38453
370	124	182	297	211	215	173		370	100	53	41	21	20	18	13			22	30	3	2	–	–	–	–	–	–	–	–	–	Vincatine		LI08094
244	370	229	91	79	55	77	93	370	100	75	50	50	49	43	31	30		22	31	4	–	–	–	–	–	–	–	–	1	–	Pregn-4-ene-3,20-dione, 17,21-[(methylborylene)bis(oxy)]-	30888-52-9	NI38454
244	370	229	91	79	55	105	77	370	100	73	49	49	48	42	32	30		22	31	4	–	–	–	–	–	–	–	–	1	–	Pregn-4-ene-3,20-dione, 17α,21-[(methylborylene)bis(oxy)]-		MS08725
55	98	74	42	339	40	297	69	370	100	96	87	87	73	64	60	55	8.00	22	42	4	–	–	–	–	–	–	–	–	–	–	Eicosanedioic acid, dimethyl ester		LI08095
43	355	101	57	41	55	116	29	370	100	74	64	49	44	38	29	27	0.00	22	42	4	–	–	–	–	–	–	–	–	–	–	Hexadecanoic acid, (2,2-dimethyl-1,3-dioxolan-4-yl)methyl ester	18418-21-8	NI38455
88	43	57	59	55	41	371	58	370	100	82	52	48	33	32	28	23	0.43	22	42	4	–	–	–	–	–	–	–	–	–	–	2-Hexadecanone, 1-[(2,2-dimethyl-1,3-dioxolan-4-yl)methoxy]-	39033-39-1	NI38456
43	57	41	370	55	28	71	87	370	100	53	38	37	29	26	25	24		22	42	4	–	–	–	–	–	–	–	–	–	–	2-Hexadecanone, 1-[(2,2-dimethyl-1,3-dioxolan-4-yl)methoxy]-	39033-39-1	NI38457
129	57	70	55	71	112	43	41	370	100	53	45	40	36	33	30	27	0.48	22	42	4	–	–	–	–	–	–	–	–	–	–	Hexanedioic acid, bis(2-ethylhexyl) ester	103-23-1	IC04176
41	57	55	43	27	29	70	39	370	100	90	73	62	57	51	50	50	0.50	22	42	4	–	–	–	–	–	–	–	–	–	–	Hexanedioic acid, bis(2-ethylhexyl) ester	103-23-1	NI38458
57	129	71	43	55	41	56	69	370	100	74	54	48	44	38	31	25	0.84	22	42	4	–	–	–	–	–	–	–	–	–	–	Hexanedioic acid, bis(6-methylheptyl) ester		IC04177
129	57	70	112	71	55	43	41	370	100	53	39	34	32	30	26	22	0.39	22	42	4	–	–	–	–	–	–	–	–	–	–	Hexanedioic acid, dioctyl ester	123-79-5	IC04178
57	41	70	55	43	129	112	259	370	100	77	73	71	68	54	50	14	2.70	22	42	4	–	–	–	–	–	–	–	–	–	–	Hexanedioic acid, dioctyl ester	123-79-5	IC04179
129	57	55	70	43	41	71	112	370	100	75	56	47	44	41	38	33	0.10	22	42	4	–	–	–	–	–	–	–	–	–	–	Hexanedioic acid, dioctyl ester	123-79-5	WI01484
117	175	101	59	355	352	353	225	370	100	77	54	26	13	2	1	1	0.20	22	42	4	–	–	–	–	–	–	–	–	–	–	1-O-(2-Hydroxy-4-hexadecenyl)-2,3-O-isopropylideneglycerol		NI38459
325	283	98	88	55	112	69	326	370	100	84	58	33	32	23	23	22	14.90	22	42	4	–	–	–	–	–	–	–	–	–	–	Octadecanedioic acid, diethyl ester	1472-90-8	NI38460
73	217	147	103	205	189	219	74	370	100	44	23	15	10	10	8	8	0.00	22	46	2	–	–	–	–	–	–	1	–	–	–	Nonadecanoic acid, trimethylsilyl ester	74367-35-4	NI38461
73	75	117	132	43	355	145	55	370	100	66	66	48	46	32	31	30	6.92	22	46	2	–	–	–	–	–	–	1	–	–	–	Nonadecanoic acid, trimethylsilyl ester	74367-35-4	NI38462

MASS TO CHARGE RATIOS								M.W.	INTENSITIES								Parent	C	H	O	N	S	F	Cl	Br	I	Si	P	B	X	COMPOUND NAME	CAS Reg No	No
272	271	273	270	239	135	268	240	370	100	28	22	21	10	10	5	4	0.70	23	15	3	–	–	–	–	–	–	–	1	–	–	9,10-Dihydro-10-phenyl-9,10-ethano-9-phosphaanthracene-11,12-dicarboxylic anhydride		MS08726
83	265	194	342	118	370	91	119	370	100	40	37	35	30	23	11	10		23	18	3	2	–	–	–	–	–	–	–	–	–	1-Pyrimidinium-4-olate, 3,6-dihydro-5-methyl-6-oxo-2-phenoxy-1,3-diphenyl-		DD01476
132	190	189	206	188	164	135	160	370	100	99	99	85	49	38	34	25	12.00	23	30	2	–	1	–	–	–	–	–	–	–	–	Benzenecarbothioic acid, 2,4,6-triethyl-, S-(2-hydroxy-2-phenylbutyl) ester	67902-79-8	NI38463
178	192	179	191	91	177	77	79	370	100	42	35	18	16	13	12	11	9.00	23	30	4	–	–	–	–	–	–	–	–	–	–	2H-1-Benzopyran, 2-(2,5-dimethoxy-3-methylphenyl)-3,4-dihydro-6-methoxy-2,4,4,8-tetramethyl-	38102-52-2	MS08727
192	179	370	178	193	177	163	91	370	100	31	27	23	15	14	8	7		23	30	4	–	–	–	–	–	–	–	–	–	–	2H-1-Benzopyran, 2-(2,5-dimethoxy-4-methylphenyl)-3,4-dihydro-6-methoxy-2,4,4,7-tetramethyl-	38102-49-7	MS08728
192	178	179	370	191	177	193	163	370	100	79	73	49	24	24	15	15		23	30	4	–	–	–	–	–	–	–	–	–	–	2H-1-Benzopyran, 2-(2,5-dimethoxy-4-methylphenyl)-3,4-dihydro-6-methoxy-2,4,4,8-tetramethyl-	38102-51-1	MS08729
55	41	44	43	91	81	79	105	370	100	97	95	95	88	75	73	67	7.67	23	30	4	–	–	–	–	–	–	–	–	–	–	D(17a)-Homo-C,18-dinorcard-20(22)-enolide, 14-hydroxy-17a-methylene-3-oxo-, (5β)-	56993-82-9	NI38464
370	310	157	371	81	57	53	175	370	100	61	54	45	44	43	38	37		23	30	4	–	–	–	–	–	–	–	–	–	–	Pregna-4,6-diene-20α-carboxylic acid, 3,12-dioxo-, methyl ester	20314-98-1	NI38465
243	370	225	338	81	202	55	283	370	100	90	54	38	38	36	35	33		23	30	4	–	–	–	–	–	–	–	–	–	–	Pregna-4,9(11)-diene-3,12-dione, 20-(acetyloxy)-	55282-22-9	NI38466
143	159	144	280	265	211	130	73	370	100	23	17	16	11	7	7	7	2.00	23	34	2	–	–	–	–	–	–	1	–	–	–	1,4,6-Androstatrien-3-one, 17α-methyl-17β-[(trimethylsilyl)oxy]-		NI38467
355	356	140	125	153	209	303	73	370	100	29	26	26	24	19	15	15	7.60	23	34	2	–	–	–	–	–	–	1	–	–	–	19-Norpregn-4-en-20-yn-3-one, 17-[(trimethylsilyl)oxy]-, (17α)-	28426-43-9	NI38468
355	196	356	209	153	75	140	73	370	100	33	29	26	24	24	19	18	8.20	23	34	2	–	–	–	–	–	–	1	–	–	–	19-Norpregn-5(10)-en-20-yn-3-one, 17-[(trimethylsilyl)oxy]-, (17α)-	38631-87-7	NI38469
101	43	57	40	59	55	41	71	370	100	31	28	23	12	12	12	11	0.54	23	46	3	–	–	–	–	–	–	–	–	–	–	1,3-Dioxolane, 2,2-dimethyl-4-[[(1-methylhexadecyl)oxy]methyl]-	56599-62-3	NI38470
101	43	57	355	55	71	41	59	370	100	28	27	12	11	10	10	9	0.00	23	46	3	–	–	–	–	–	–	–	–	–	–	1,3-Dioxolane, 2,2-dimethyl-4-[[(1-methylhexadecyl)oxy]methyl]-	56599-62-3	NI38471
101	43	57	55	41	71	354	58	370	100	55	50	28	27	25	22	17	0.64	23	46	3	–	–	–	–	–	–	–	–	–	–	1,3-Dioxolane, 2,2-dimethyl-4-[[(2-methylhexadecyl)oxy]methyl]-	56600-20-5	NI38472
57	370	55	311	69	83	97	90	370	100	88	81	56	52	50	48	38		23	46	3	–	–	–	–	–	–	–	–	–	–	Methyl 2-hydroxydocosanoate		MS08730
97	57	43	369	83	111	311	69	370	100	86	78	74	74	64	60	56	23.00	23	46	3	–	–	–	–	–	–	–	–	–	–	Methyl 2-hydroxydocosanoate		MS08731
114	113	112	312	255	223	222	297	370	100	60	57	44	42	40	40	33	1.00	23	46	3	–	–	–	–	–	–	–	–	–	–	Methyl 17-oxodocosanoate		LI08096
89	55	59	45	121	83	41	58	370	100	74	68	65	62	58	55	50	31.16	23	46	3	–	–	–	–	–	–	–	–	–	–	9-Octadecene, 1-(2,3-dimethoxypropoxy)-, (Z)-(±)-	55282-66-1	NI38473
327	75	328	97	83	57	111	89	370	100	35	28	27	25	20	16	14	0.00	23	50	1	–	–	–	–	–	–	1	–	–	–	Octadecyl propyldimethylsilyl ether		NI38474
310	295	91	105	267	311	77	107	370	100	76	46	34	31	25	24	23	0.00	24	34	3	–	–	–	–	–	–	–	–	–	–	Pregna-5,16-dien-20-one, 3-(acetyloxy)-16-methyl-, (3β)-	982-06-9	WI01485
43	55	41	57	201	69	83	61	370	100	67	67	53	49	42	26	19	3.55	24	50	–	–	1	–	–	–	–	–	–	–	–	Didodecyl sulphide		MS08732
56	43	41	55	69	70	42	57	370	100	87	87	86	62	60	57	55	0.01	24	50	2	–	–	–	–	–	–	–	–	–	–	1,1-Dioctyloxyoctane		IC04180
103	85	68	84	71	83	97	70	370	100	56	44	26	25	22	21	18	0.00	24	50	2	–	–	–	–	–	–	–	–	–	–	2,4-Docosanediol, 3,5-dimethyl-	56324-81-3	NI38475
57	41	43	82	29	56	113	55	370	100	50	40	37	35	34	31	23	0.00	24	50	2	–	–	–	–	–	–	–	–	–	–	Palmitaldehyde, dibutyl acetal	18302-68-6	NI38476
289	55	43	41	91	79	93	105	370	100	98	96	95	89	84	79	78	11.88	25	38	2	–	–	–	–	–	–	–	–	–	–	3,5-Cyclochol-23-yn-22-ol, 6-methoxy-, (3β,5α)-	58072-57-4	NI38477
91	67	41	81	43	55	95	69	370	100	48	37	35	34	31	21	17	0.00	25	38	2	–	–	–	–	–	–	–	–	–	–	Linoleic acid, benzyl ester		NI38478
91	119	117	131	92	201	187	105	370	100	19	19	10	8	6	6	6	0.00	25	38	2	–	–	–	–	–	–	–	–	–	–	Linoleic acid, benzyl ester		NI38479
310	295	91	105	267	311	77	107	370	100	76	46	34	31	25	24	23	0.00	25	38	2	–	–	–	–	–	–	–	–	–	–	16-Methyl-16-dehydropregnenolone acetate		AM00149
43	327	138	370	137	248	259	55	370	100	99	90	55	48	42	41	40		25	38	2	–	–	–	–	–	–	–	–	–	–	Pregn-4-ene-3,20-dione, 6-methyl-16-isopropyl-, (6α,16α)-	1922-28-7	NI38480
43	327	138	370	137	258	55	91	370	100	56	51	32	27	24	23	20		25	38	2	–	–	–	–	–	–	–	–	–	–	Pregn-4-ene-3,20-dione, 6-methyl-16-isopropyl-, (6α,16α)-	1922-28-7	NI38481
370	369	176	266	371	265	165	264	370	100	61	51	36	27	23	21	19		26	20	1	–	–	–	–	–	–	–	–	2	–	Bis(9,10-dihydro-9-bora-9-anthranyl)oxide		LI08097
370	372	340	355	339	371	177	236	370	100	82	65	60	37	31	27	26		26	30	–	2	–	–	–	–	–	–	–	–	–	Aniline, N,N′-(2,5-dimethyl-2,5-cyclohexadiene-1,4-diylidene)bis[2,3,5-trimethyl-	54245-92-0	NI38482
69	111	55	95	41	370	109	81	370	100	71	61	60	60	58	57	56		26	42	1	–	–	–	–	–	–	–	–	–	–	21-Nor-5α-cholest-24-en-20-one	14949-18-9	NI38483
370	371	215	162	95	79	43	217	370	100	33	31	31	27	27	24	23		26	42	1	–	–	–	–	–	–	–	–	–	–	3,5-Cyclo-b-nor-5-cholestan-6-one, (3α,5α)-		NI38484
315	355	338	370	283	201	213	323	370	100	65	50	45	36	15	14	12		26	42	1	–	–	–	–	–	–	–	–	–	–	3,5-Cyclo-26,27-dinorcholest-23-ene, 6-methoxy-, (E)-(3α,6β)-		MS08733
315	338	283	355	370	217	215	201	370	100	68	44	42	40	31	17	17		26	42	1	–	–	–	–	–	–	–	–	–	–	3,5-Cyclo-26,27-dinorcholest-23-ene, 6-methoxy-, (Z)-(3α,6β)-		MS08734
105	300	273	79	161	259	121	213	370	100	95	73	73	71	64	57	54	2.32	26	42	1	–	–	–	–	–	–	–	–	–	–	26,27-Dinorergosta-5,22-dien-3-ol, (3β,22E)-	38788-81-7	NI38485
55	370	97	255	300	81	159	107	370	100	67	66	57	49	42	32	30		26	42	1	–	–	–	–	–	–	–	–	–	–	26,27-Dinorergosta-5,22-dien-3-ol, (3β,22Z)-	52745-87-6	NI38486
55	69	81	41	133	95	105	271	370	100	95	86	75	68	62	51	49	2.02	26	42	1	–	–	–	–	–	–	–	–	–	–	26,27-Dinorergosta-5,23-dien-3-ol, (3β)-	35882-88-3	NI38487
271	314	55	107	81	145	105	95	370	100	65	63	59	59	54	54	51	20.00	26	42	1	–	–	–	–	–	–	–	–	–	–	26,27-Dinorergosta-5,24-dien-3-ol, (3β)-	56588-49-9	NI38488
55	81	91	271	314	145	213	355	370	100	71	58	51	43	38	22	11	0.00	26	42	1	–	–	–	–	–	–	–	–	–	–	26,27-Dinorergosta-5,24-dien-3-ol, (3β)-	56588-49-9	DD01477
271	55	355	213	300	337	370	352	370	100	99	35	27	22	20	14	12		26	42	1	–	–	–	–	–	–	–	–	–	–	27-Norcholesta-5,24(Z)-dien-3β-ol		DD01478
69	370	111	259	149	109	355	95	370	100	98	89	84	70	68	61	61		26	42	1	–	–	–	–	–	–	–	–	–	–	21-Norcholest-24-en-20-one, (5α,17ξ)-	54482-51-8	NI38489
370	371	369	266	190	165	77	51	370	100	28	25	14	12	12	7	7		27	18	–	2	–	–	–	–	–	–	–	–	–	1H-Phenanthro[9,10-d]imidazole, 1,2-diphenyl-	16408-28-9	NI38490
370	369	266	190	165	267	77	51	370	100	25	14	12	12	7	7	7		27	18	–	2	–	–	–	–	–	–	–	–	–	1H-Phenanthro[9,10-d]imidazole, 1,2-diphenyl-	16408-28-9	LI08098
370	371	215	355	108	81	95	93	370	100	29	23	22	22	18	17	16		27	46	–	–	–	–	–	–	–	–	–	–	–	Cholest-3-ene, (5α)-	28338-69-4	NI38491
370	81	108	55	95	43	107	93	370	100	51	48	43	42	42	41	38		27	46	–	–	–	–	–	–	–	–	–	–	–	Cholest-3-ene, (5α)-	28338-69-4	NI38492

MASS TO CHARGE RATIOS								M.W.	INTENSITIES								Parent	C	H	O	N	S	F	Cl	Br	I	Si	P	B	X	COMPOUND NAME	CAS Reg No	No
370	108	43	109	55	355	81	95	370	100	92	39	37	37	36	36	33		27	46	–	–	–	–	–	–	–	–	–	–	–	Cholest-4-ene	16732-86-8	NI38493
370	57	108	55	41	81	43	95	370	100	62	60	51	46	42	42	41		27	46	–	–	–	–	–	–	–	–	–	–	–	Cholest-4-ene	16732-86-8	NI38494
370	355	55	215	43	109	95	81	370	100	53	44	42	42	39	35	35		27	46	–	–	–	–	–	–	–	–	–	–	–	Cholest-5-ene	570-74-1	NI38495
370	109	355	111	215	135	257	123	370	100	76	55	54	53	42	41	41		27	46	–	–	–	–	–	–	–	–	–	–	–	Cholest-5-ene	570-74-1	LI08099
370	257	215	81	95	67	109	355	370	100	83	69	55	53	49	48	46		27	46	–	–	–	–	–	–	–	–	–	–	–	Cholest-7-ene, (5α)-	40071-65-6	NI38496
370	257	55	215	355	43	41	95	370	100	56	45	38	36	36	35	33		27	46	–	–	–	–	–	–	–	–	–	–	–	Cholest-7-ene, (5α)-	40071-65-6	NI38497
370	355	120	55	95	121	57	43	370	100	89	67	59	58	56	55	54		27	46	–	–	–	–	–	–	–	–	–	–	–	Cholest-7-ene, (5α,14β)-	40071-68-9	NI38498
370	355	109	215	135	371	147	257	370	100	42	37	34	31	29	29	26		27	46	–	–	–	–	–	–	–	–	–	–	–	Cholest-8(14)-ene, (5α)-	54725-42-7	NI38499
81	286	355	95	257	107	105	109	370	100	99	72	60	57	46	45	43	**43.00**	27	46	–	–	–	–	–	–	–	–	–	–	–	Cholest-9(11)-ene, (5α,15β)-		MS08735
257	258	93	81	94	95	67	147	370	100	26	26	26	24	23	23	21	**4.00**	27	46	–	–	–	–	–	–	–	–	–	–	–	Cholest-14-ene, (5α)-	54725-04-1	NI38500
149	55	109	95	259	81	163	286	370	100	98	90	86	85	81	73	61	**22.13**	27	46	–	–	–	–	–	–	–	–	–	–	–	Cholest-22-ene, (5α)-	55282-65-0	NI38501
55	149	109	95	81	259	163	286	370	100	82	76	71	71	68	60	54	**22.00**	27	46	–	–	–	–	–	–	–	–	–	–	–	Cholest-22-ene, (5α,20ξ)-	54514-99-7	NI38502
286	149	109	95	56	259	257	81	370	100	93	85	82	82	78	77	74	**61.00**	27	46	–	–	–	–	–	–	–	–	–	–	–	Cholest-22-ene, (5α,22E)-	54514-98-6	NI38503
55	81	149	95	109	259	67	163	370	100	79	77	75	74	70	58	57	**22.00**	27	46	–	–	–	–	–	–	–	–	–	–	–	Cholest-22-ene, (5α,22Z)-	15076-93-4	NI38504
55	109	95	81	287	257	67	41	370	100	86	80	80	63	61	55	50	**4.68**	27	46	–	–	–	–	–	–	–	–	–	–	–	Cholest-23-ene, (5β)-	30658-62-9	NI38505
287	257	123	147	163	177	165	121	370	100	98	63	62	56	54	48	48	**6.67**	27	46	–	–	–	–	–	–	–	–	–	–	–	Cholest-23-ene, (5β)-	30658-62-9	NI38506
55	109	95	81	287	257	67	41	370	100	86	80	80	64	62	54	49	**4.04**	27	46	–	–	–	–	–	–	–	–	–	–	–	Cholest-23-ene, (5β,23Z)-	14949-12-3	NI38507
259	69	109	258	55	41	95	81	370	100	52	45	43	43	38	36	35	**1.01**	27	46	–	–	–	–	–	–	–	–	–	–	–	Cholest-24-ene, (5α,20ξ)-	54482-49-4	NI38508
257	55	95	81	69	109	41	67	370	100	62	61	56	55	52	50	45	**11.57**	27	46	–	–	–	–	–	–	–	–	–	–	–	Cholest-24-ene, (5β)-	14949-23-6	NI38509
257	217	258	121	355	149	123	135	370	100	36	30	27	25	25	23	21	**10.89**	27	46	–	–	–	–	–	–	–	–	–	–	–	Cholest-24-ene, (5β)-	14949-23-6	NI38510
257	355	258	69	95	217	81	55	370	100	34	31	27	24	22	20	20	**16.83**	27	46	–	–	–	–	–	–	–	–	–	–	–	Cholest-24-ene, (5β)-	14949-23-6	NI38511
91	28	43	41	105	120	55	29	370	100	98	89	85	84	64	56	55	**2.84**	27	46	–	–	–	–	–	–	–	–	–	–	–	11-Phenyl-10-heneicosene		AI02266
118	43	91	41	117	57	55	105	370	100	68	65	55	44	27	26	21	**0.00**	27	46	–	–	–	–	–	–	–	–	–	–	–	11-Phenyl-10-heneicosene		AI02267
355	249	370	356	133	237	371	234	370	100	57	47	30	30	17	14	13		28	34	–	–	–	–	–	–	–	–	–	–	–	Benzene, 1,4-bis[1-(dimethylphenyl)ethyl]dimethyl-	72121-37-0	NI38512
147	133	119	355	327	207	370	341	370	100	55	20	18	16	14	11	10		28	34	–	–	–	–	–	–	–	–	–	–	–	Cyclopentane, 2-propylidene-1,3-bis[(E)-(2,4,6-trimethylbenzylidene)]-		MS08736
370	133	237	157	238	207	355	195	370	100	84	35	19	18	18	13	12		28	34	–	–	–	–	–	–	–	–	–	–	–	Cyclopentene, 2-[(E)-propenyl]-1-(2,4,6-trimethylbenzyl)-3-[(E)-2,4,6-trimethylbenzylidene]-		MS08737
91	92	104	105	370	117	41	131	370	100	35	15	14	13	10	7	6		28	34	–	–	–	–	–	–	–	–	–	–	–	1,7-Diphenyl-4-(3-phenylpropyl)heptane		AI02268
91	92	104	370	105	117	65	77	370	100	29	13	12	11	9	6	5		28	34	–	–	–	–	–	–	–	–	–	–	–	1,7-Diphenyl-4-(3-phenylpropyl)heptane		AI02269
41	118	57	135	43	55	370	27	370	100	95	95	92	87	83	72	72		28	34	–	–	–	–	–	–	–	–	–	–	–	1,4-Bis(tolylethyl)durene		IC04181
192	370	178	176	189	190	371	165	370	100	45	27	23	20	15	13	11		29	22	–	–	–	–	–	–	–	–	–	–	–	Methyl 2,3,4,5-dibenzobicyclo[4.2.0]octa-2,4,7-triene		LI08100
370	293	78	291	215	289	265	77	370	100	27	18	17	15	10	9	8		29	22	–	–	–	–	–	–	–	–	–	–	–	9,10,11-Triphenylbicyclo[6,3,0]undecapenta-2,4,6,8,10-ene		LI08101
371	117	88	62	89	61	53	127	371	100	62	21	19	16	11	10	8		7	3	1	1	–	–	–	–	2	–	–	–	–	Benzonitrile, 4-hydroxy-3,5-diiodo-	1689-83-4	NI38513
371	117	28	87	59	127	88	58	371	100	67	36	25	22	17	16	13		7	3	1	1	–	–	–	–	2	–	–	–	–	Benzonitrile, 4-hydroxy-3,5-diiodo-	1689-83-4	NI38514
57	127	41	43	29	88	55	243	371	100	70	51	43	36	31	31	28	**7.19**	7	3	1	1	–	–	–	–	2	–	–	–	–	Benzonitrile, 4-hydroxy-3,5-diiodo-	1689-83-4	NI38515
299	73	282	300	301	211	236	75	371	100	99	60	35	27	24	22	21	**3.50**	11	30	5	1	–	–	–	–	–	3	1	–	–	Acetaldehyde, [(trimethylsilyl)oxy]-, O-[bis[(trimethylsilyl)oxy]phosphinyl]oxime	55282-61-6	NI38516
299	73	282	356	236	75	300	211	371	100	90	75	35	34	31	29	28	**3.50**	11	30	5	1	–	–	–	–	–	3	1	–	–	Acetaldehyde, [(trimethylsilyl)oxy]-, O-[bis[(trimethylsilyl)oxy]phosphinyl]oxime	55282-61-6	NI38517
254	327	356	73	164	116	328	211	371	100	92	73	54	48	25	24	20	**1.80**	13	24	4	1	–	3	–	–	–	2	–	–	–	L-Proline, 1-(trifluoroacetyl)-4-[(trimethylsilyl)oxy]-, trimethylsilyl ester, trans-	52669-44-0	NI38518
155	91	139	186	122	341	65	76	371	100	59	31	29	17	16	11	9	**0.26**	14	13	7	1	2	–	–	–	–	–	–	–	–	Benzenesulphonic acid, 4-nitro-, [(4-methylphenyl)sulphonyl]methyl ester	31081-12-6	NI38519
155	91	185	156	65	186	92	40	371	100	33	20	14	4	3	3	2	**0.04**	14	13	7	1	2	–	–	–	–	–	–	–	–	Methanol, [(4-nitrophenyl)sulphonyl]-, 4-methylbenzenesulphonate	31081-08-0	NI38520
371	86	285	217	302	150	70	341	371	100	64	45	22	20	19	17	14		15	12	5	3	–	3	–	–	–	–	–	–	–	Oxazole, 4-trifluoroacetyl-5-(4-morpholinyl)-2-(4-nitrophenyl)-	52044-90-3	NI38521
371	86	285	56	217	302	150	372	371	100	64	45	22	21	20	19	18		15	12	5	3	–	3	–	–	–	–	–	–	–	Oxazole, 4-trifluoroacetyl-5-(4-morpholinyl)-2-(4-nitrophenyl)-	52044-90-3	NI38522
73	154	45	254	155	74	75	43	371	100	74	21	17	14	9	7	7	**0.00**	15	33	2	3	–	–	–	–	–	3	–	–	–	Histamine, tris(trimethylsilyl)-		MS08738
73	154	147	82	182	58	110	75	371	100	72	62	51	49	35	26	25	**0.20**	15	33	2	3	–	–	–	–	–	3	–	–	–	L-Histidine, N,1-bis(trimethylsilyl)-, trimethylsilyl ester	17908-25-7	NI38523
154	73	254	155	182	45	74	218	371	100	68	21	18	14	11	8	7	**0.62**	15	33	2	3	–	–	–	–	–	3	–	–	–	L-Histidine, N,1-bis(trimethylsilyl)-, trimethylsilyl ester	17908-25-7	NI38524
337	312	355	340	307	323	293	245	371	100	54	40	34	31	23	22	22	**4.00**	16	13	6	5	–	–	–	–	–	–	–	–	–	1,3-Dimethyl-2-(picrylamino)indole		LI08102
371	311	312	212	280	200	253	340	371	100	52	48	41	39	39	35	26		17	13	4	1	–	4	–	–	–	–	–	–	–	Dimethyl 6,7,8,9-tetrafluoro-3a,9b-dihydro-1-methylbenz[g]indole-2,3-dicarboxylate		NI38525
312	253	56	100	229	271	371	238	371	100	53	42	40	27	25	23	18		17	13	4	1	–	4	–	–	–	–	–	–	–	Dimethyl (5,6,7,8-tetrafluoro-N-methyl-1-naphthylamino)maleate		NI38526
328	79	371	235	329	281	236		371	100	59	36	30	20	15	14			17	13	5	3	1	–	–	–	–	–	–	–	–	2-(2,4-Dinitrophenylthio)-3-indoleacetone		LI08103
43	83	109	174	67	55	125	85	371	100	71	68	53	39	39	35	27	**0.00**	17	25	8	1	–	–	–	–	–	–	–	–	–	4-Acetoxy-3-nitropent-2-yl camphanoate, (2R,3R,4S)-		DD01479

MASS TO CHARGE RATIOS								M.W.	INTENSITIES								Parent	C	H	O	N	S	F	Cl	Br	I	Si	P	B	X	COMPOUND NAME	CAS Reg No	No
217	139	371	183	63	107	152	215	**371**	100	36	12	10	7	5	4	2		18	14	2	1	2	–	–	–	–	–	1	–	–	**4-Nitrophenyl diphenylphosphinodithioate**		**MS08739**
358	314	316	151	357	373	371	359	**371**	100	32	29	23	21	20	20	20		18	14	3	1	–	–	–	1	–	–	–	–	–	**4H-Anthra[1,2-d][1,3]oxazine-7,12-dione, 6-bromo-1,2-dihydro-2,2-dimethyl-**	61487-32-9	**NI38527**
44	269	55	371	79	72	284	29	**371**	100	40	32	31	30	29	25	19		18	17	6	3	–	–	–	–	–	–	–	–	–	**9-Acridinamine, 2,3-dimethoxy-6-nitro-N-(2-carboxyethyl)-**	80648-48-2	**NI38528**
331	187	371	333	147	332	118	373	**371**	100	60	46	35	26	20	20	16		18	18	2	5	–	–	1	–	–	–	–	–	–	**N-Ethyl-N-(2-cyanoethyl)-3-methyl-4-(2-chloro-4-nitrophenylazo)aniline**		**IC04182**
198	199	226	155	259	200	227	173	**371**	100	99	30	26	14	14	8	8	**3.00**	18	33	5	3	–	–	–	–	–	–	–	–	–	**Glycine, N-[N-[N-(1-oxodecyl)-L-alanyl]glycyl]-, methyl ester**	35146-55-5	**NI38529**
139	141	129	111	142	158	140	87	**371**	100	32	26	23	12	11	10	9	**6.90**	20	18	4	1	–	–	1	–	–	–	–	–	–	**1H-Indole-3-acetic acid, 1-(4-chlorobenzoyl)-5-methoxy-2-methyl-, methyl ester**	1601-18-9	**NI38530**
139	141	111	371	140	113	75	158	**371**	100	32	24	16	8	8	8	7		20	18	4	1	–	–	1	–	–	–	–	–	–	**1H-Indole-3-acetic acid, 1-(4-chlorobenzoyl)-5-methoxy-2-methyl-, methyl ester**	1601-18-9	**NI38531**
252	250	311	253	371	226	251	227	**371**	100	78	26	26	25	24	19	15		20	21	6	1	–	–	–	–	–	–	–	–	–	**1,2-Diacetoxy-9,10-dioxolopluviine**		**LI08104**
197	169	168	198	167	170	371	196	**371**	100	44	31	18	9	6	5	4		20	21	6	1	–	–	–	–	–	–	–	–	–	**1,3-Dioxolane-2-carboxylic acid, 2-ethoxy-4-oxo-5-[(Z)-2,3,4,9-tetrahydro 1H-carbazol-1-ylidene]-, ethyl ester, (±)-**	86657-66-1	**NI38532**
92	77	120	189	134	147	162	161	**371**	100	73	35	30	22	20	19	17	**0.00**	20	35	1	1	–	–	1	–	–	–	1	–	–	**2-Oximinophenacyltributylphosphonium chloride**		**LI08105**
371	312	43	297	281	372	298	313	**371**	100	86	43	43	29	26	23	21		21	25	5	1	–	–	–	–	–	–	–	–	–	**N-Acetylcolchinol methyl ether**	65967-01-3	**NI38533**
204	355	354	205	190	369	371		**371**	100	45	18	10	8	3	1			21	25	5	1	–	–	–	–	–	–	–	–	–	**(–)-Argemonine N-oxide**		**LI08106**
207	156	208	371	312	127	152	126	**371**	100	17	16	14	14	14	12	12		21	25	5	1	–	–	–	–	–	–	–	–	–	**Benzo[a]heptalen-9(5H)-one, 6,7-dihydro-1,2,3,10-tetramethoxy-7-(methylamino)-, (S)-**	477-30-5	**NI38534**
164	149	371	206	340	356	208		**371**	100	96	83	46	25	12	12			21	25	5	1	–	–	–	–	–	–	–	–	–	**Capaurine**	478-14-8	**LI08107**
164	149	371	206	370	165	121	135	**371**	100	91	77	46	36	33	31	23		21	25	5	1	–	–	–	–	–	–	–	–	–	**Capaurine**	478-14-8	**NI38535**
282	372	313	356					**371**	100	64	33	15					**0.00**	21	29	3	1	–	–	–	–	–	1	–	–	–	**Morphinan-6-ol, 7,8-didehydro-4,5-epoxy-3-methoxy-17-methyl-6-(trimethylsilyl)-, (5α,6α)-**	74367-14-9	**NI38536**
73	371	178	196	42	372	70	59	**371**	100	76	41	39	29	23	21	21		21	29	3	1	–	–	–	–	–	1	–	–	–	**Morphinan-6-ol, 7,8-didehydro-4,5-epoxy-3-methoxy-17-methyl-6-(trimethylsilyl)-, (5α,6α)-**	74367-14-9	**NI38537**
254	75	43	56	255	73	326	41	**371**	100	29	27	25	21	20	15	11	**0.00**	21	45	2	1	–	–	–	–	–	1	–	–	–	**Octadecanoic acid, 2-amino-, trimethylsilyl-**	56272-83-4	**NI38538**
371	330	372	103	66	259	165	260	**371**	100	35	33	31	23	18	17	15		22	21	1	5	–	–	–	–	–	–	–	–	–	**Imidazo[1,2-b]-1,2,4-triazine, 2-methyl-3-(4-morpholino)-6,7-diphenyl-**		**NI38539**
371	244	372	341	369	340	231	245	**371**	100	56	39	33	30	25	25	21		22	33	2	1	–	–	–	–	–	1	–	–	–	**Estra-1,3,5(10)-trien-16-one, 3-[(trimethylsilyl)oxy]-, O-methyloxime**		**NI38540**
371	340	341	96	218	372	231	193	**371**	100	89	73	39	36	32	26	22		22	33	2	1	–	–	–	–	–	1	–	–	–	**Estra-1,3,5(10)-trien-17-one, 3-[(trimethylsilyl)oxy]-, O-methyloxime**	69688-21-7	**NI38541**
236	266	371						**371**	100	65	20							23	17	4	1	–	–	–	–	–	–	–	–	–	**1,3-Dioxolo[4,5-c]quinolin-4(5H)-one, 2-(4-methoxyphenyl)-2-phenyl-**	54616-40-9	**NI38542**
133	118	294	119	132	77	91	104	**371**	100	80	61	59	58	57	50	41	**40.29**	23	21	2	3	–	–	–	–	–	–	–	–	–	**1,3,5-Triazine-2,4(1H,3H)-dione, 1-ethyldihydro-3,5,6-triphenyl-**	13787-70-7	**NI38543**
249	221	220	105	144	291	116	146	**371**	100	74	54	48	39	22	22	17	**4.00**	24	21	3	1	–	–	–	–	–	–	–	–	–	**cis-2-Benzoylbicyclo[2.2.2]oct-5-ene-3-spiro-3′-(1′-acetyl)-2′-indolinone**		**LI08108**
249	105	221	266	220	250	224	146	**371**	100	90	63	53	53	37	32	32	**21.00**	24	21	3	1	–	–	–	–	–	–	–	–	–	**trans-2-Benzoylbicyclo[2.2.2]oct-5-ene-3-spiro-3′-(1′-acetyl)-2′-indolinone**		**LI08109**
266	254	296	370	294	265	267	255	**371**	100	86	70	67	44	38	33	30	**22.00**	24	21	3	1	–	–	–	–	–	–	–	–	–	**5-(2-Methoxy-1-methoxycarbonylvinyl)-6-phenylphenanthridinium**		**LI08110**
105	104	77	258	130	122	69	67	**371**	100	45	35	30	22	20	20	12	**3.00**	25	41	1	1	–	–	–	–	–	–	–	–	–	**2-Oxazoline, 4,5-dioctyl-2-phenyl-, cis-**	26015-58-7	**NI38544**
105	258	104	77	69	130	122	67	**371**	100	65	60	45	30	18	18	17	**2.00**	25	41	1	1	–	–	–	–	–	–	–	–	–	**2-Oxazoline, 4,5-dioctyl-2-phenyl-, trans-**	25943-12-8	**NI38545**
58	72	44	371	59	42	40	68	**371**	100	89	18	14	14	14	10	9		26	29	1	1	–	–	–	–	–	–	–	–	–	**1-(1,2-Diphenyl-2-ethylvinyl)-4-(2-dimethylaminoethoxy)benzene**		**IC04183**
175	174	182	188	356	189	183	340	**371**	100	80	20	4	3	3	3	2	**1.00**	26	29	1	1	–	–	–	–	–	–	–	–	–	**Oxaziridine, 2-[1-(pentamethylphenyl)ethyl]-3,3-diphenyl-**	75326-14-6	**NI38546**
69	296	100	127	254	150	31	131	**372**	100	41	40	26	24	22	17	14	**0.00**	3	–	–	–	–	11	–	–	1	–	–	–	–	**Tetrafluoro(heptafluoroisopropyl)iodine**		**MS08740**
172	174	64	79	52	26	78	76	**372**	100	73	56	41	39	38	36	36	**5.60**	4	–	–	4	4	–	4	–	–	–	–	–	–	**2,3,7,8-Tetrachloro-5,10,11,12-tetrathia-1,4,6,9-tetraazatricyclo[5.3.1.1^{2,6}]dodeca-3,8-diene**		**LI08111**
176	232	204	316	372	260	144	344	**372**	100	40	37	29	24	24	18	17		8	4	6	–	2	–	–	–	–	–	–	–	2	**Diiron, μ-(π-ethylenedithio)hexacarbonyl-**		**MS08741**
372	118	117	245	91	51	115	39	**372**	100	41	31	28	16	14	12	12		9	10	–	–	–	–	–	–	2	–	–	–	–	**Benzene, 2,4-diiodo-1,3,5-trimethyl-**	53779-84-3	**MS08742**
248	124	189	372	59	155	341	218	**372**	100	97	38	36	13	12	11	8		11	23	6	–	–	–	–	–	–	–	2	–	1	**Cobalt, π-cyclopentadienylbis(trimethylphosphite)**		**LI08112**
246	274	171	156	28	289	216	73	**372**	100	76	62	41	39	37	25	25	**0.00**	11	29	6	–	–	–	–	–	–	3	1	–	–	**Acetic acid, [[bis[(trimethylsilyl)oxy]phosphinyl]oxy]-, trimethylsilyl ester**	56009-10-0	**NI38547**
357	73	299	328	147	358	45	133	**372**	100	93	36	29	28	27	19	16	**2.00**	11	29	6	–	–	–	–	–	–	3	1	–	–	**Acetic acid, [[bis[(trimethylsilyl)oxy]phosphinyl]oxy]-, trimethylsilyl ester**	56009-10-0	**NI38548**
355	357	353	321	319	359	323	356	**372**	100	70	67	55	45	25	25	15	**1.35**	12	2	1	–	–	–	6	–	–	–	–	–	–	**Dibenzofuran, 1,2,4,6,7,9-hexachloro-**	75627-02-0	**NI38549**
374	376	239	155	241	372	187	120	**372**	100	87	61	61	58	52	50	50		12	2	1	–	–	–	6	–	–	–	–	–	–	**Dibenzofuran, 2,3,4,6,7,8-hexachloro-**	60851-34-5	**NI38550**
202	109	372	154	77	51	293	186	**372**	100	72	70	64	46	45	30	24		13	10	1	–	1	–	–	2	–	–	–	–	–	**Diphenylsulphoxonium dibromomethylide**		**MS08743**
182	155	153	151	157	246	154	190	**372**	100	73	49	26	25	20	19	18	**3.80**	14	14	–	–	–	–	2	–	–	–	–	–	1	**Dibenzyltin dichloride**		**NI38551**
254	357	73	255	358	147	359	256	**372**	100	79	59	28	26	16	13	11	**10.54**	14	28	4	2	–	–	–	–	–	3	–	–	–	**4-Pyrimidinecarboxylic acid, 2,6-bis[(trimethylsilyl)oxy]-, trimethylsilyl ester**	31111-36-1	**NI38552**

MASS TO CHARGE RATIOS								M.W.	INTENSITIES								Parent	C	H	O	N	S	F	Cl	Br	I	Si	P	B	X	COMPOUND NAME	CAS Reg No	No
73	75	147	45	77	357	255	79	**372**	100	92	74	60	52	51	47	46	**7.90**	14	28	4	2	–	–	–	–	–	3	–	–	–	5-Pyrimidinecarboxylic acid, 2,4-bis[(trimethylsilyl)oxy]-, trimethylsilyl ester	56804-96-7	MS08744
254	357	255	358	73	147	256	359	**372**	100	87	47	35	33	22	18	17	**14.36**	14	28	4	2	–	–	–	–	–	3	–	–	–	4-Pyrimidinecarboxylic acid, 1,2,3,6-tetrahydro-2,6-dioxo-1,3-bis(trimethylsilyl)-, trimethylsilyl ester	55530-63-7	NI38553
254	73	357	255	358	45	147	256	**372**	100	65	61	26	17	16	15	10	**1.91**	14	28	4	2	–	–	–	–	–	3	–	–	–	4-Pyrimidinecarboxylic acid, 1,2,3,6-tetrahydro-2,6-dioxo-1,3-bis(trimethylsilyl)-, trimethylsilyl ester	55530-63-7	NI38554
73	254	357	255	45	147	358	100	**372**	100	87	46	23	22	21	14	13	**4.80**	14	28	4	2	–	–	–	–	–	3	–	–	–	4-Pyrimidinecarboxylic acid, 1,2,3,6-tetrahydro-2,6-dioxo-1,3-bis(trimethylsilyl)-, trimethylsilyl ester	55530-63-7	NI38555
56	118	175	146	91	69	119	132	**372**	100	91	82	69	60	56	53	50	**0.00**	15	15	1	2	–	7	–	–	–	–	–	–	–	Piperazine, 1-(2-tolyl)-4-(heptafluorobutyryl)-		NI38556
92	98	142	70	84	74	80	44	**372**	100	75	66	60	53	53	43	38	**0.00**	15	17	6	–	–	–	–	1	–	–	–	–	–	α-D-Altropyranoside, methyl 3-bromo-3-deoxy-4,6-O-benzylidene-, acetate	56701-38-3	NI38557
255	372	214	338	215	188	69	374	**372**	100	61	57	54	53	53	45	41		16	14	6	–	–	–	2	–	–	–	–	–	–	Spiro[benzofuran-2(3H),1′-cyclohexane]-2′,3,4′-trione, 3′,7-dichloro-4,6-dimethoxy-6′-methyl-	26891-72-5	NI38558
73	251	340	339	372	75	193	147	**372**	100	57	21	21	20	20	16	12		16	28	6	–	–	–	–	–	–	2	–	–	–	4,5-Bis(trimethylsilyloxy)-1,4-cyclohexadiene-1,2-dicarboxylic acid, dimethyl ester		NI38559
255	73	256	45	147	257	239	75	**372**	100	90	50	28	24	21	17	17	**4.00**	16	32	4	–	–	–	–	–	–	3	–	–	–	2,4,6-Heptatrienoic acid, 4,6-bis[(trimethylsilyl)oxy]-, trimethylsilyl ester		NI38560
270	234	272	271	269	235	255	284	**372**	100	60	40	30	30	23	16	8	**0.00**	17	14	–	–	1	4	1	–	–	–	–	1	–	trans-5-Chloro-1-methyl-2-styrylbenzo[b]thiophenium tetrafluoroborate		LI08113
162	166	194	135	372	96	77	69	**372**	100	76	26	16	13	9	9	8		17	19	4	2	–	3	–	–	–	–	–	–	–	Amphetamine, 3,4-methylenedioxy-N-[(trifluoroacetyl)prolyl]-		NI38561
372	77	46	326	132	186	228	373	**372**	100	83	52	26	23	16	14	12		18	12	2	8	–	–	–	–	–	–	–	–	–	1H-Pyrazolo[4,3-e][1,2,4]triazolo[4,3-a][1,2,4]triazine, 3-methyl-1,8-diphenyl-		NI38562
299	354	325	343	297	326	300	298	**372**	100	68	61	33	32	24	24	23	**18.26**	18	12	9	–	–	–	–	–	–	–	–	–	–	Anthra[2,3-b]furo[3,2-d]furan-5,10-dione, 2,3,3a,12a-tetrahydro-2,3a,4,6,9 pentahydroxy-, [2R-(2α,3aβ,12aβ)]-	31456-72-1	NI38563
372	44	83	354	179	177	327	326	**372**	100	71	67	61	50	48	47	39		18	12	9	–	–	–	–	–	–	–	–	–	–	7H-Isobenzofuro[4,5-b][1,4]benzodioxepin-11-carboxaldehyde, 1,3-dihydro-1,4,10-trihydroxy-5,8-dimethyl-3,7-dioxo-	571-67-5	NI38564
147	69	227	225	189	105	185	183	**372**	100	80	40	40	35	31	26	26	**0.70**	18	13	4	–	–	–	–	1	–	–	–	–	–	1,3,4,6-Hexanetetrone, 1-(4-bromophenyl)-6-phenyl-	58330-13-5	NI38565
372	374	295	373	297	296	335	375	**372**	100	68	54	39	36	24	22	21		18	14	1	–	–	–	2	–	–	2	–	–	–	1,3-Disila-2-oxaindan, 1,3-dichloro-1,3-diphenyl-	58194-85-7	NI38566
306	193	147	180	107	105	108	121	**372**	100	80	80	78	60	60	50	41	**13.00**	18	16	7	2	–	–	–	–	–	–	–	–	–	Pyrrolo[1,2-a]quinoxaline-1,2,3-tricarboxylic acid, 4,5-dihydro-5-methyl-4-oxo-, trimethyl ester	37922-02-4	NI38567
187	56	101	55	287	87	188	119	**372**	100	83	79	43	28	28	17	17	**0.26**	18	28	8	–	–	–	–	–	–	–	–	–	–	1,5,10,14-Tetraoxacyclooctadecane-6,9,15,18-tetrone, 3,3,12,12-tetramethyl-	68671-56-7	NI38568
84	44	200	158	228	201	43	57	**372**	100	78	76	71	50	47	43	33	**2.37**	18	32	6	2	–	–	–	–	–	–	–	–	–	N-(N-Acetyl-O-butyl-α-glutamyl)alanine butyl ester		AI02270
87	44	158	86	84	43	41	114	**372**	100	92	55	54	20	17	17	14	**1.58**	18	32	6	2	–	–	–	–	–	–	–	–	–	Dibutyl N-(N-acetylalanyl)glutamate		AI02271
256	283	284	285	257	255	258	286	**372**	100	86	76	40	38	38	36	30	**0.00**	18	32	6	2	–	–	–	–	–	–	–	–	–	D-Glucitol, 1-[(10-amino-1,10-dioxodecyl)amino]-1,6-dideoxy-2,4:3,5-di-O methylene-	6613-85-0	NI38569
156	144	143	316	200	260	101	140	**372**	100	82	82	76	76	70	70	57	**13.63**	18	32	6	2	–	–	–	–	–	–	–	–	–	L-Serine, N-[N-(1-oxohexyl)glycyl]-, methyl ester, hexanoate	56272-46-9	NI38570
357	372	358	373	149	371	153	69	**372**	100	60	22	14	11	8	8	6		19	16	8	–	–	–	–	–	–	–	–	–	–	5-Hydroxy-3,7,8-trimethoxy-2-(3,4-methylenedioxyphenyl)-1-benzopyran-4-one		NI38571
372	272	312	258	340	285	284	256	**372**	100	81	77	68	61	37	27	24		19	16	8	–	–	–	–	–	–	–	–	–	–	Methyl 4-formyl-3,8-dihydroxy-1,6,9-trimethyl-11-oxo-11H-dibenzo[b,e][1,4]dioxepin-7-carboxylate		NI38572
73	343	257	272	256	345	283	45	**372**	100	77	41	40	32	31	29	27	**0.00**	19	21	2	2	–	–	1	–	–	1	–	–	–	2H-1,4-Benzodiazepin-2-one, 7-chloro-1,3-dihydro-1-methyl-5-phenyl-3-[(trimethylsilyl)oxy]-	35147-95-6	NI38573
343	372	357	345	257	256	344	73	**372**	100	59	39	39	35	26	25	22		19	21	2	2	–	–	1	–	–	1	–	–	–	2H-1,4-Benzodiazepin-2-one, 7-chloro-1,3-dihydro-1-methyl-5-phenyl-3-[(trimethylsilyl)oxy]-	35147-95-6	NI38574
344	73	372	371	346	345	373	374	**372**	100	78	48	39	33	31	26	20		19	21	2	2	–	–	1	–	–	1	–	–	–	2H-1,4-Benzodiazepin-2-one, 7-chloro-1,3-dihydro-1-methyl-5-[4-[(trimethylsilyl)oxy]phenyl]-	55282-59-2	NI38575
93	357	134	41	91	107	316	77	**372**	100	98	81	81	79	74	72	68	**56.14**	19	24	4	4	–	–	–	–	–	–	–	–	–	α-Ionone dinitrophenylhydrazone		NI38576
227	209	224	264	270	245	244	309	**372**	100	35	30	13	11	10	10	7	**2.00**	19	32	7	–	–	–	–	–	–	–	–	–	–	Ethyl 4-[5S-(2S,3S-epoxy-5S-hydroxy-4S-methylhexyl)-3R,4R-dihydroxytetrahydropyran-2S-yl]-3-methylbut-2(E)-enoate		NI38577
227	224	209	291	309	336	327	354	**372**	100	69	23	9	4	2	2	1	**0.50**	19	32	7	–	–	–	–	–	–	–	–	–	–	Ethyl 4-[5S-(2S,3S-epoxy-5S-hydroxy-4S-methylhexyl)-3R,4R-dihydroxytetrahydropyran-2S-yl]-3-methylbut-2(Z)-enoate		NI38578
245	262	372	227	247	261	163	264	**372**	100	65	55	36	35	26	26	21		20	17	5	–	–	–	1	–	–	–	–	–	–	2,6-Bis(2,4-dihydroxyphenylmethyl)-4-chlorophenol		IC04184
339	372	354	326	39	115	43	357	**372**	100	15	13	13	12	11	10	9		20	20	7	–	–	–	–	–	–	–	–	–	–	1H-Benzo[g]cyclopenta[de][1]benzopyran-4(2H),6(10bH)-dione, 7,9,10-trihydroxy-8-(2-hydroxyisopropyl)-3,10b-dimethyl-, (1′S,10bS)-		DD01480
166	151	167	178	165	372	77	39	**372**	100	20	12	7	7	6	4	4		20	20	7	–	–	–	–	–	–	–	–	–	–	Benzoic acid, 3,5-diformyl-2,4-dihydroxy-6-methyl-, 3-methoxy-2,5,6-trimethylphenyl ester	23004-60-6	MS08745
273	315	251	181	43	134	271	135	**372**	100	76	43	40	35	30	28	20	**5.00**	20	20	7	–	–	–	–	–	–	–	–	–	–	Benzophenone, 2,4′-dihydroxy-4,6-dimethoxy-2′-methyl-, diacetate	4650-76-4	NI38579

MASS TO CHARGE RATIOS								M.W.	INTENSITIES								Parent	C	H	O	N	S	F	Cl	Br	I	Si	P	B	X	COMPOUND NAME	CAS Reg No	No
299	165	354	298	311	134	69	55	**372**	100	36	19	17	16	8	8	8	**6.00**	20	20	7	–	–	–	–	–	–	–	–	–	–	4H-1-Benzopyran-4-one, 5,7-dihydroxy-6-(3-hydroxy-3-methylbutyl)-3-(2,4-dihydroxyphenyl)-	91681-62-8	NI38580
341	372	371	75	342	181	69	172	**372**	100	74	50	26	25	25	23	21		20	20	7	–	–	–	–	–	–	–	–	–	–	4H-1-Benzopyran-4-one, 2-(2,4-dimethoxyphenyl)-3,5,7-trimethoxy-	7555-80-8	NI38581
341	372	371	75	342	181	69	323	**372**	100	74	50	26	25	25	23	21		20	20	7	–	–	–	–	–	–	–	–	–	–	4H-1-Benzopyran-4-one, 2-(2,4-dimethoxyphenyl)-3,5,7-trimethoxy-	7555-80-8	MS08746
372	341	168	342	313	195	179	314	**372**	100	6	6	2	2	2	2	1		20	20	7	–	–	–	–	–	–	–	–	–	–	4H-1-Benzopyran-4-one, 2-(3,4-dimethoxyphenyl)-5,6,7-trimethoxy-	2306-27-6	NI38582
357	372	167	328	139	329	358	327	**372**	100	56	35	32	25	21	20	19		20	20	7	–	–	–	–	–	–	–	–	–	–	4H-1-Benzopyran-4-one, 2-(3,4-dimethoxyphenyl)-5,7,8-trimethoxy-	17290-70-9	NI38583
372	371	357	172	353	341	343	329	**372**	100	74	57	23	20	20	10	10		20	20	7	–	–	–	–	–	–	–	–	–	–	4H-1-Benzopyran-4-one, 2-(3,4-dimethoxyphenyl)-3,5,7-trimethoxyflavone		MS08747
357	358	372	182	314	197	83	296	**372**	100	22	17	17	16	14	14	9		20	20	7	–	–	–	–	–	–	–	–	–	–	4H-1-Benzopyran-4-one, 2-(4-methoxyphenyl)-5,6,7,8-tetramethoxy-	481-53-8	NI38584
372	341	311	357	329	151	243	299	**372**	100	90	48	39	27	26	20	19		20	20	7	–	–	–	–	–	–	–	–	–	–	4H-1-Benzopyran-4-one, 3-(2,3,4,6-tetramethoxyphenyl)-7-methoxy-		LI08114
372	371	326	151	373	343	167	195	**372**	100	66	45	36	32	32	20	19		20	20	7	–	–	–	–	–	–	–	–	–	–	4H-1-Benzopyran-4-one, 2-(2,3,4-trimethoxyphenyl)-5,7-dimethoxy-	89121-55-1	NI38585
372	371	326	343	341	327	342	355	**372**	100	63	33	31	21	19	18	12		20	20	7	–	–	–	–	–	–	–	–	–	–	4H-1-Benzopyran-4-one, 2-(3,4,5-trimethoxyphenyl)-5,7-dimethoxy-	53350-26-8	NI38586
372	341	181	191	273	149	357	157	**372**	100	56	34	24	23	20	18	14		20	20	7	–	–	–	–	–	–	–	–	–	–	4H-1-Benzopyran-4-one, 3-(2,4,5-trimethoxyphenyl)-6,7-dimethoxy-	24203-68-7	NI38587
372	312	258	340	272	285	156	83	**372**	100	91	43	39	39	37	29	23		20	20	7	–	–	–	–	–	–	–	–	–	–	11H-Dibenzo[b,e][1,4]dioxepin-7-carboxylic acid, 8-hydroxy-3-methoxy-1,4,6,9-tetramethyl-11-oxo-, methyl ester	6161-70-2	NI38588
330	372	315						**372**	100	86	57							20	20	7	–	–	–	–	–	–	–	–	–	–	9-O-Methylphilenopteran acetate		LI08115
194	49	213	151	84	168	152	210	**372**	100	99	88	85	77	68	60	53	**7.40**	20	24	3	2	1	–	–	–	–	–	–	–	–	Benzamide, 2,2′-sulphinylbis[N-isopropyl-	65838-71-3	NI38589
311	77	372	341	268	313	309	105	**372**	100	65	56	55	53	38	31	29		20	24	5	2	–	–	–	–	–	–	–	–	–	4,5-Pyridinedicarboxylic acid, 6-(diethylamino)-1,2-dihydro-3-methyl-2-oxo-1-phenyl-, dimethyl ester		DD01481
45	81	99	89	143	98	73	82	**372**	100	97	83	67	57	39	39	35	**1.65**	20	36	6	–	–	–	–	–	–	–	–	–	–	Dicyclohexyl-18-crown-6	16069-36-6	NI38590
81	45	99	89	73	143	43	41	**372**	100	96	77	55	49	40	37	36	**0.00**	20	36	6	–	–	–	–	–	–	–	–	–	–	Dicyclohexyl-18-crown-6	16069-36-6	NI38591
45	81	43	99	41	89	73	57	**372**	100	43	38	34	34	23	23	22	**0.00**	20	36	6	–	–	–	–	–	–	–	–	–	–	Dicyclohexyl-18-crown-6, cis-anti-cis-	15128-66-2	NI38592
45	81	89	99	73	41	143	43	**372**	100	86	56	54	42	41	36	34	**0.00**	20	36	6	–	–	–	–	–	–	–	–	–	–	Dicyclohexyl-18-crown-6, cis-syn-cis-	54383-26-5	NI38593
45	81	99	73	89	41	143	55	**372**	100	88	46	40	37	37	31	29	**0.00**	20	36	6	–	–	–	–	–	–	–	–	–	–	Dicyclohexyl-18-crown-6, trans-syn-trans-	15128-65-1	NI38594
121	372	340	77	159	146	107	91	**372**	100	2	2	2	1	1	1	1		21	24	6	–	–	–	–	–	–	–	–	–	–	Benzenepropanoic acid, 4-hydroxy-3-[2-methoxy-1-[(4-methoxyphenyl)methyl]-2-oxoethyl]-, methyl ester	69721-65-9	NI38595
303	135	331	77	162	104	115	91	**372**	100	96	87	87	86	82	62	60	**30.00**	21	24	6	–	–	–	–	–	–	–	–	–	–	6(2H)-Benzofuranone, 2-(1,3-benzodioxol-5-yl)-3,3a,4,5-tetrahydro-5,7-dimethoxy-3-methyl-3a-(2-propenyl)-, [2S-(2α,3β,3aα,5β)]-	70561-33-0	NI38596
164	167	206	165	168	163	372	149	**372**	100	97	19	14	11	10	8	5		21	24	6	–	–	–	–	–	–	–	–	–	–	3,5-Dimethoxy-4-O-acetylbenzyl 2-methoxy-4-(1-propenyl) ether		NI38597
151	137	165	177	152	166	138	372	**372**	100	90	75	60	60	55	45	25		21	24	6	–	–	–	–	–	–	–	–	–	–	Phenol, 4-[4-(3,4-dimethoxyphenyl)tetrahydro-1H,3H-furo[3,4-c]furan-1-yl]-2-methoxy-, [1S-(1α,3aα,4β,6aα)]-	487-39-8	NI38598
122	336	372	180	313	282	277	94	**372**	100	99	97	97	62	39	35	34		21	25	2	2	–	–	1	–	–	–	–	–	–	5-Acetoxy-9-(2-chloroethyl)-10-methyl-12-ethylidine-6,10-diazatetracyclo[4.3.0.$3^{2,4}$.$4^{7,8}$]-hexadeca-1(9),7,13,15-tetraene		LI08116
322	263	180	234	44	323	264	232	**372**	100	72	30	26	25	24	23	21	**4.00**	21	25	2	2	–	–	1	–	–	–	–	–	–	Pleiocarpamine methyl chloride		LI08117
122	336	372	180	313	282	277	374	**372**	100	99	98	98	62	39	35	33		21	25	2	2	–	–	1	–	–	–	–	–	–	Pleiocarpamine methyl chloride		LI08118
125	136	127	372	137	41	124	122	**372**	100	66	33	17	16	11	9	9		21	25	2	2	–	–	1	–	–	–	–	–	–	Urea, 1-(4-chlorobenzyl)-1-cyclopentyl-3-methyl-3-(4-methoxyphenyl)-		NI38599
176	372	87	168	122	180	149	57	**372**	100	63	61	50	44	33	33	29		21	28	2	2	1	–	–	–	–	–	–	–	–	5,6,14,15-Dibenzo-4,16-dioxa-1-thia-7,13-diazacyclooctadeca-5,14-diene	79688-17-8	MS08748
184	372	285	283	282	58	212	373	**372**	100	85	71	35	30	26	24	22		21	28	4	2	–	–	–	–	–	–	–	–	–	Aspidospermidin-21-oic acid, 20-hydroxy-17-methoxy-, methyl ester	36459-02-6	NI38600
184	372	285	283	282	212	373	284	**372**	100	68	67	35	28	22	17	17		21	28	4	2	–	–	–	–	–	–	–	–	–	Aspidospermidin-21-oic acid, 20-hydroxy-17-methoxy-, methyl ester	36459-02-6	NI38601
73	299	45	103	75	44	84	57	**372**	100	65	38	33	33	26	20	20	**0.15**	21	44	3	–	–	–	–	–	–	1	–	–	–	Hexadecanoic acid, 2-(trimethylsiloxy)ethyl ester	22613-62-3	LI08119
239	73	116	57	43	357	75	55	**372**	100	94	89	80	75	72	64	58	**1.00**	21	44	3	–	–	–	–	–	–	1	–	–	–	Hexadecanoic acid, 2-(trimethylsiloxy)ethyl ester	22613-62-3	NI38602
73	57	43	116	239	75	71	55	**372**	100	95	88	80	57	55	54	53	**0.79**	21	44	3	–	–	–	–	–	–	1	–	–	–	Hexadecanoic acid, 2-(trimethylsiloxy)ethyl ester	22613-62-3	NI38603
173	287	75	55	89	197	59	95	**372**	100	60	39	34	31	20	20	17	**0.00**	21	44	3	–	–	–	–	–	–	1	–	–	–	Octadecanoic acid, 12-[(dimethylsilyl)oxy]-, methyl ester	38429-53-7	NI38604
372	373	186	213	374	356	298	105	**372**	100	25	8	5	4	3	3	3		22	12	6	–	–	–	–	–	–	–	–	–	–	5,7,12,14-Tetrahydroxy-6,13-dioxopentacene		IC04185
64	306	308	305	48	307	153	309	**372**	100	99	92	88	70	39	29	26	**0.00**	22	16	2	2	1	–	–	–	–	–	–	–	–	1,3-Diphenyl-1,3-dihydrothieno[3,4-b]quinoxaline 2,2-dioxide		LI08120
372	329	330	373	331	139	43	374	**372**	100	34	28	25	18	7	6	4		22	16	4	2	–	–	–	–	–	–	–	–	–	1-Hydroxy-4-(4-acetamidoanilino)anthraquinone		IC04186
267	105	77	268	372	220	76	78	**372**	100	65	52	43	27	27	13	12		22	16	4	2	–	–	–	–	–	–	–	–	–	1-(4-Nitroanilino)-1,2-dibenzoylethylene		MS08749
186	371	372	224	187	314	345	225	**372**	100	70	26	23	21	14	13	10		22	20	2	4	–	–	–	–	–	–	–	–	–	Nipecotamide, 3,5-dicyano-N,1-dimethyl-6-oxo-2,4-diphenyl-		LI08121
372	178	295	149	283	371	179	167	**372**	100	66	62	48	38	36	28	27		22	20	2	4	–	–	–	–	–	–	–	–	–	1H-Purine-2,6-dione, 8-(1,2-diphenylvinyl)-3,7-dihydro-1,3,7-trimethyl-, (Z)-		NI38605
286	43	123	87	330	372	258	257	**372**	100	15	9	6	3	2	2	2		22	28	5	–	–	–	–	–	–	–	–	–	–	Androst-4-ene-3,6,17-trione, 19R-acetoxy-19-methyl-		LI08122
357	372	358	43	373	359	297	141	**372**	100	25	25	9	6	5	5	4		22	28	5	–	–	–	–	–	–	–	–	–	–	2H,8H-Benzo[1,2-b:3,4-b′]dipyran-6-propanol, 5-methoxy-2,2,8,8-tetramethyl-, acetate	56336-20-0	NI38606
133	160	107	161	167	91	105	28	**372**	100	85	82	80	71	69	50	50	**1.47**	22	28	5	–	–	–	–	–	–	–	–	–	–	Cyclopropanecarboxylic acid, 3-(3-methoxy-2-methyl-3-oxo-1-propenyl)-2,2-dimethyl-, 2-methyl-4-oxo-3-(2,4-pentadienyl)-2-cyclopenten-1-yl ester, [1R-[1α[S*(Z)],3β(E)]]-	121-29-9	NI38607

MASS TO CHARGE RATIOS								M.W.	INTENSITIES								Parent	C	H	O	N	S	F	Cl	Br	I	Si	P	B	X	COMPOUND NAME	CAS Reg No	No
107	41	91	105	133	55	161	160	**372**	100	94	93	80	61	49	38	36	**0.00**	22	28	5	–	–	–	–	–	–	–	–	–	–	Cyclopropanecarboxylic acid, 3-(3-methoxy-2-methyl-3-oxo-1-propenyl)-2,2-dimethyl-, 2-methyl-4-oxo-3-(2,4-pentadienyl)-2-cyclopenten-1-yl ester, [1R-[1α[S*(Z)],3β(E)]]-	121-29-9	NI38608
161	133	107	91	41	105	162	93	**372**	100	92	23	22	19	18	17	15	**7.50**	22	28	5	–	–	–	–	–	–	–	–	–	–	Cyclopropanecarboxylic acid, 3-(3-methoxy-2-methyl-3-oxo-1-propenyl)-2,2-dimethyl-, 2-methyl-4-oxo-3-(1,3-pentadienyl)-2-cyclopenten-1-yl ester, [1S-[1α[1R*,4(1E,3E)],2β(E)]]-	64655-46-5	NI38609
133	45	106	90	158	162	56	105	**372**	100	96	83	74	56	53	53	51	**5.00**	22	28	5	–	–	–	–	–	–	–	–	–	–	3-[2,2-Dimethyl-3-[2-(methoxycarbonyl)prop-1-enyl]cyclopropylcarbonyloxy]-4-methyl-5-(penta-2,4-dienyl)cyclopent-4-enone		LI08123
330	331	252	294	172	372	160	328	**372**	100	49	35	30	23	22	19	15		22	28	5	–	–	–	–	–	–	–	–	–	–	Estra-3,17-diol, 3,17-diacetoxy-, (3β,15β,17β)-		NI38610
206	191	175	207	178	151	138	165	**372**	100	48	40	16	14	13	12	11	**7.91**	22	28	5	–	–	–	–	–	–	–	–	–	–	Furan, 2,5-bis(3,4-dimethoxyphenyl)tetrahydro-3,4-dimethyl-, [2R-(2α,3β,4β,5α)]-	528-63-2	NI38611
206	191	175	372	207	165	160	178	**372**	100	57	50	34	15	15	14	13		22	28	5	–	–	–	–	–	–	–	–	–	–	Furan, 2,5-bis(3,4-dimethoxyphenyl)tetrahydro-3,4-dimethyl-, [2R-(2α,3β,4β,5α)]-	528-63-2	NI38612
206	191	175	178	207	194	165	372	**372**	100	58	48	36	16	15	14	11		22	28	5	–	–	–	–	–	–	–	–	–	–	Furan, 2,5-bis(3,4-dimethoxyphenyl)tetrahydro-3,4-dimethyl-, [2S-(2α,3β,4α,5β)]-	10569-12-7	NI38613
213	228	57	176	43	214	229	372	**372**	100	42	25	19	18	16	15	12		22	28	5	–	–	–	–	–	–	–	–	–	–	(+)-Octanoyllomatin		LI08124
372	129	227	143	254	187	200	373	**372**	100	96	56	44	40	40	34	30		22	32	3	–	–	–	–	–	–	1	–	–	–	Estra-1,3,5(10)-trien-16-one, 3-methoxy-17-[(trimethylsilyl)oxy]-, (17β)-	69688-17-1	NI38614
342	372	343	373	357	287	248	218	**372**	100	73	30	22	22	16	14	14		22	32	3	–	–	–	–	–	–	1	–	–	–	Estra-1,3,5(10)-trien-17-one, 2-methoxy-3-[(trimethylsilyl)oxy]-	29825-40-9	NI38615
372	342	373	343	374	370	57	357	**372**	100	31	30	10	8	8	8	7		22	32	3	–	–	–	–	–	–	1	–	–	–	Estra-1,3,5(10)-trien-17-one, 3-methoxy-2-[(trimethylsilyl)oxy]-	29825-41-0	NI38616
372	342	373	357	343	374	358	300	**372**	100	57	30	24	17	9	8	8		22	32	3	–	–	–	–	–	–	1	–	–	–	Estra-1,3,5(10)-trien-17-one, 3-methoxy-4-[(trimethylsilyl)oxy]-	51497-41-7	NI38617
340	341	242	342	255	243	283	229	**372**	100	29	12	9	8	7	6	6	**4.19**	22	32	3	–	–	–	–	–	–	1	–	–	–	Estra-1,3,5(10)-trien-17-one, 6-methoxy-3-[(trimethylsilyl)oxy]-, (6α)-	74299-20-0	NI38618
372	340	373	341	342	218	217	192	**372**	100	34	30	21	17	10	10	10		22	32	3	–	–	–	–	–	–	1	–	–	–	Estra-1,3,5(10)-trien-17-one, 6-methoxy-3-[(trimethylsilyl)oxy]-, (6β)-	74299-21-1	NI38619
372	218	373	286	374	219	340	258	**372**	100	32	30	20	8	8	7	7		22	32	3	–	–	–	–	–	–	1	–	–	–	Estra-1,3,5(10)-trien-17-one, 15-methoxy-3-[(trimethylsilyl)oxy]-, (15β)-	74299-22-2	NI38620
246	56	80	41	82	68	92	372	**372**	100	86	57	56	53	53	48	45		22	33	4	–	–	–	–	–	–	–	–	1	–	17α,21-Dihydroxy-5β-pregnane-3,20-dione methyl boronate		MS08750
287	269	91	79	124	55	105	145	**372**	100	99	50	40	35	35	34	31	**30.00**	22	33	4	–	–	–	–	–	–	–	–	1	–	17α,20α,21-Trihydroxypregn-4-en-3-one methyl boronate		MS08751
269	287	91	79	124	105	55	145	**372**	100	99	52	42	36	36	36	33	**30.00**	22	33	4	–	–	–	–	–	–	–	–	1	–	17α,20α,21-Trihydroxypregn-4-en-3-one methyl boronate		LI08125
287	269	91	79	43	55	124	105	**372**	100	47	45	43	43	36	34	32	**26.00**	22	33	4	–	–	–	–	–	–	–	–	1	–	17α,20β,21-Trihydroxypregn-4-en-3-one methyl boronate		MS08752
287	269	124	105	107	229	372	145	**372**	100	47	34	32	29	28	26	26		22	33	4	–	–	–	–	–	–	–	–	1	–	17α,20β,21-Trihydroxypregn-4-en-3-one methyl boronate		LI08126
101	88	43	59	57	55	41	357	**372**	100	55	45	30	30	18	17	15	**0.00**	22	44	4	–	–	–	–	–	–	–	–	–	–	1-Hexadecanol, 1-[(2,2-dimethyl-1,3-dioxolan-4-yl)methoxy]-	56599-81-6	NI38621
101	88	43	28	59	57	41	55	**372**	100	64	50	49	32	30	22	21	**0.00**	22	44	4	–	–	–	–	–	–	–	–	–	–	2-Hexadecanol, 1-[(2,2-dimethyl-1,3-dioxolan-4-yl)methoxy]-	13879-81-7	NI38622
105	372	77	373	295	121	355	367	**372**	100	70	45	20	19	18	17	10		23	16	1	–	2	–	–	–	–	–	–	–	–	Acetophenone, 2-(4,5-diphenyl-3H-1,2-dithiol-3-ylidene)-	56196-68-0	NI38623
105	372	77	373	374	295	355	267	**372**	100	93	61	27	14	13	11	11		23	16	1	–	2	–	–	–	–	–	–	–	–	Acetophenone, 2-phenyl-2-(4-phenyl-3H-1,2-dithiol-3-ylidene)-	5245-04-5	NI38624
372	105	77	373	121	374	295	234	**372**	100	43	43	27	17	13	12	8		23	16	1	–	2	–	–	–	–	–	–	–	–	Acetophenone, 2-phenyl-2-(5-phenyl-3H-1,2-dithiol-3-ylidene)-	38490-06-1	NI38625
121	355	344	77	343	310	178	51	**372**	100	58	38	36	32	27	16	15	**0.00**	23	16	1	–	2	–	–	–	–	–	–	–	–	Benzeneacetaldehyde, α-(4,5-diphenyl-3H-1,2-dithiol-3-ylidene)-	32783-38-3	NI38626
330	331	315	372	299	202	43	76	**372**	100	24	19	12	11	9	9	8		23	16	5	–	–	–	–	–	–	–	–	–	–	2,2′-Binaphthalene-1,4-dione, 1′-acetoxy-4′-methoxy-	72971-98-3	NI38627
372	194	104	212	122	373	91	371	**372**	100	36	34	30	28	27	23	21		23	17	2	2	–	1	–	–	–	–	–	–	–	2(1H)-Pyrimidinone, 4-(4-fluorophenyl)-5-(4-methylphenoxy)-6-phenyl-	42919-57-3	NI38628
264	77	91	104	119	117	180	105	**372**	100	46	35	29	20	16	15	13	**2.00**	23	20	3	2	–	–	–	–	–	–	–	–	–	1,8-Dioxa-2,6-diazaspiro[4.4]nonan-7-one, 2,3,6-triphenyl-	52512-28-4	NI38629
263	77	91	93	180	119	118	182	**372**	100	70	61	19	18	17	17	14	**7.00**	23	20	3	2	–	–	–	–	–	–	–	–	–	Oxazolo[5,4-d]isoxazol-5(2H)-one, tetrahydro-6a-methyl-2,3,6-triphenyl-	55780-94-4	NI38630
372	121	140	357	141	330	297	179	**372**	100	31	27	26	25	22	18	18		23	29	3	–	–	1	–	–	–	–	–	–	–	17α-Pregn-4-en-20-yn-6-one, 3β-fluoro-17-hydroxy-, acetate	33547-25-0	NI38631
243	313	271	372	201	228	256	312	**372**	100	94	75	60	50	36	35	24		23	32	4	–	–	–	–	–	–	–	–	–	–	Abieta-6,8,11,13-tetraen-20-oic acid, 11,12-dimethoxy-, methyl ester		MS08753
69	168	81	167	205	41	204	119	**372**	100	90	62	55	52	44	35	35	**4.00**	23	32	4	–	–	–	–	–	–	–	–	–	–	Benzaldehyde, 3-farnesyloxy-2-hydroxy-4-methoxy-		NI38632
313	357	372	314	207	358	231	245	**372**	100	56	32	22	18	14	13	12		23	32	4	–	–	–	–	–	–	–	–	–	–	11-Nor-Δ^9-tetrahydrocannabinol-9-carboxylic acid, 1-O-methyl-, methyl ester	52762-27-3	NI38633
312	149	372	297	231	269	256	69	**372**	100	86	82	79	54	52	35	33		23	32	4	–	–	–	–	–	–	–	–	–	–	6H-Dibenzo[b,d]pyran-9-methanol, 6a,7,10,10a-tetrahydro-1-hydroxy-6,6-dimethyl-3-pentyl-, 9-acetate, (6aR-trans)-	36871-18-8	NI38634
372	357	373	358	167	315	178.5	374	**372**	100	50	26	13	11	5	5	4		23	32	4	–	–	–	–	–	–	–	–	–	–	3,5-Dimethoxy-3′,5′-di-tert-butyl-4,4′-dihydroxydiphenylmethane		MS08754
43	269	287	107	329	230	91	312	**372**	100	83	26	17	14	13	11	9	**0.00**	23	32	4	–	–	–	–	–	–	–	–	–	–	19-Norpregn-4-ene-3,20-dione, 17-(acetyloxy)-16-methyl-, (16α)-	56196-79-3	NI38635
330	372	231	43	271	274	313	193	**372**	100	99	68	52	43	42	41	32		23	32	4	–	–	–	–	–	–	–	–	–	–	6H-Oxireno[4,5]benzo[1,2-c][1]benzopyran-1-ol, 6a,7,7a,8a,9,9a-hexahydro-6,6,8a-trimethyl-3-pentyl-, acetate, (6aα,7aβ,8aβ,9aβ)-	34984-80-0	NI38636
313	271	372	315	330	231	43	312	**372**	100	84	82	73	64	42	40	38		23	32	4	–	–	–	–	–	–	–	–	–	–	6H-Oxireno[4,5]benzo[1,2-c][1]benzopyran-1-ol, 6a,7,7a,8a,9,9a-hexahydro-6,6,8a-trimethyl-3-pentyl-, acetate, [6aR-(6aα,7aα,8aα,9aβ)]-	52613-01-1	NI38637
330	372	43	28	312	244	259	274	**372**	100	96	77	72	61	59	52	50		23	32	4	–	–	–	–	–	–	–	–	–	–	3H-Oxireno[3,4]benzo[1,2-c][1]benzopyran-9-ol, 1a,2,3a,4,9b,9c-hexahydro-1a,4,4-trimethyl-7-pentyl-, acetate, [1aS-(1aα,3aα,9bβ,9cα)]-	38215-39-3	NI38638

MASS TO CHARGE RATIOS								M.W.	INTENSITIES								Parent	C	H	O	N	S	F	Cl	Br	I	Si	P	B	X	COMPOUND NAME	CAS Reg No	No
285	312	107	372	124	245	284	81	**372**	100	53	53	48	44	43	36	36		23	32	4	–	–	–	–	–	–	–	–	–	–	**Pregn-4-ene-20-carboxylic acid, 3,12-dioxo-, methyl ester, (20S)-**	24341-26-2	**NI38639**
43	122	286	121	55	147	91	41	**372**	100	96	19	17	14	13	12	12	**0.00**	23	32	4	–	–	–	–	–	–	–	–	–	–	**Pregn-4-ene-3,20-dione, 6-(acetyloxy)-, (6β)-**	2076-31-5	**NI38640**
43	330	91	331	55	93	79	41	**372**	100	64	18	15	15	13	13	13	**9.59**	23	32	4	–	–	–	–	–	–	–	–	–	–	**Pregn-4-ene-3,20-dione, 6-(acetyloxy)-, (6β)-**	1675-95-2	**NI38641**
43	269	287	270	330	229	145	79	**372**	100	89	43	22	14	14	12	12	**2.54**	23	32	4	–	–	–	–	–	–	–	–	–	–	**Pregn-4-ene-3,20-dione, 17-(acetyloxy)-**	302-23-8	**NI38642**
43	269	287	330	329	312	229	91	**372**	100	99	43	17	15	12	12	11	**4.51**	23	32	4	–	–	–	–	–	–	–	–	–	–	**Pregn-4-ene-3,20-dione, 17-(acetyloxy)-**	302-23-8	**NI38643**
43	147	312	135	149	93	91	105	**372**	100	41	27	24	19	18	17	16	**13.76**	23	32	4	–	–	–	–	–	–	–	–	–	–	**Pregn-16-ene-11,20-dione, 3-(acetyloxy)-**	56193-63-6	**NI38644**
328	91	55	79	77	105	93	53	**372**	100	73	58	55	51	50	47	41	**0.00**	23	32	4	–	–	–	–	–	–	–	–	–	–	**Pregn-5-en-20-one, 3-(acetyloxy)-16,17-epoxy-, (3β,16α)-**	34209-81-9	**NI38645**
312	91	269	251	105	145	55	77	**372**	100	54	52	46	44	33	31	30	**0.00**	23	32	4	–	–	–	–	–	–	–	–	–	–	**Pregn-5-en-20-one, 3-(acetyloxy)-16,17-epoxy-, (3β,16α)-**	34209-81-9	**NI38646**
43	312	269	105	91	107	296	145	**372**	100	68	22	20	20	18	17	17	**0.00**	23	32	4	–	–	–	–	–	–	–	–	–	–	**Pregn-5-en-20-one, 3-(acetyloxy)-16,17-epoxy-, (16α)-**	56193-62-5	**NI38647**
312	91	77	79	297	93	105	294	**372**	100	64	51	50	47	43	38	33	**3.00**	23	32	4	–	–	–	–	–	–	–	–	–	–	**Pregn-16-en-20-one, 3-(acetyloxy)-5,6-epoxy-, (3β,5α,6α)-**	14279-42-6	**WI01486**
91	28	105	44	336	201	79	145	**372**	100	83	75	68	66	64	64	59	**4.00**	23	32	4	–	–	–	–	–	–	–	–	–	–	**Xysmalogenin**		**LI08127**
189	147	43	41	372	29	190	55	**372**	100	55	54	40	33	30	27	14		23	36	–	2	1	–	–	–	–	–	–	–	–	**2-Thiazolamine, 4-phenyl-5-tetradecyl-**	64415-14-1	**NI38648**
239	73	240	43	173	357	255	171	**372**	100	38	21	18	12	11	11	9	**7.00**	23	36	2	–	–	–	–	–	–	1	–	–	–	**Abietic acid, dehydro-, trimethylsilyl ester**		**NI38649**
239	73	240	43	171	357	173	75	**372**	100	54	21	16	14	12	12	12	**11.00**	23	36	2	–	–	–	–	–	–	1	–	–	–	**Abietic acid, dehydro-, trimethylsilyl ester**		**NI38650**
143	282	161	144	122	219	194	283	**372**	100	35	21	18	16	8	8	7	**0.51**	23	36	2	–	–	–	–	–	–	1	–	–	–	**Androsta-1,4-dien-3-one, 17β-methyl-17α-[(trimethylsilyl)oxy]-**		**NI38651**
143	282	161	122	283	144	219	194	**372**	100	73	29	23	16	16	13	10	**5.14**	23	36	2	–	–	–	–	–	–	1	–	–	–	**Androsta-1,4-dien-4-one, 17α-methyl-17β-[(trimethylsilyl)oxy]-**		**NI38652**
372	357	329	301	373	289	330	316	**372**	100	79	59	59	34	33	25	25		23	36	2	–	–	–	–	–	–	1	–	–	–	**Δ^1-Tetrahydrocannabinol dimethylsilyl ether**		**NI38653**
357	358	140	372	153	343	154	125	**372**	100	45	40	29	29	15	14	14		23	36	2	–	–	–	–	–	–	1	–	–	–	**17α-Ethynyl-17β-[(trimethylsilyl)oxy]-5α-estran-3-one**		**NI38654**
357	140	358	372	153	196	125	155	**372**	100	36	31	19	19	18	13	11		23	36	2	–	–	–	–	–	–	1	–	–	–	**17α-Ethynyl-17β-[(trimethylsilyl)oxy]-5β-estran-3-one**		**NI38655**
241	372	73	174	129	173	282	75	**372**	100	73	65	41	40	35	25	25		23	36	2	–	–	–	–	–	–	1	–	–	–	**3,5,10-Gonatriene, 13-ethyl-3-methoxy-17-[(trimethylsilyl)oxy]-, (13β,17β)-**		**NI38656**
73	372	255	183	75	69	159	105	**372**	100	27	25	20	14	14	13	13		23	36	2	–	–	–	–	–	–	1	–	–	–	**2,4,6,8-Nonatetraenoic acid, 3,7-dimethyl-9-(2,6,6-trimethyl-1-cyclohexen-1-yl)-, trimethylsilyl ester, (all-E)-**	56227-27-1	**NI38657**
100	372	153	341	91	273	93	79	**372**	100	80	68	60	51	35	32	32		23	36	2	2	–	–	–	–	–	–	–	–	–	**Pregn-4-ene-3,20-dione, bis(O-methyloxime)**	26432-00-8	**LI08128**
372	341	125	273	153	100	342	286	**372**	100	82	43	41	40	38	30	29		23	36	2	2	–	–	–	–	–	–	–	–	–	**Pregn-4-ene-3,20-dione, bis(O-methyloxime)**	26432-00-8	**NI38658**
372	373	44	314	43	42	58	45	**372**	100	37	24	22	21	20	17	16		23	40	–	4	–	–	–	–	–	–	–	–	–	**Androstane-3,17-dione, bis(dimethylhydrazone), (5α)-**	54966-64-2	**NI38659**
194	372	371	176	374	195	97	166	**372**	100	74	72	30	25	23	22	17		24	18	1	2	–	–	–	–	–	–	–	2	–	**Bis(10,9-borazaro-10-phenanthryl) ether**		**LI08129**
372	195	186	152	151	150	356	344	**372**	100	14	5	2	2	2	1	1		24	18	1	2	–	–	–	–	–	–	–	2	–	**Bis(10,9-borazaro-10-phenanthryl) ether**		**LI08130**
186	91	372	187	144	145	143	147	**372**	100	97	67	63	40	27	25	17		24	24	2	2	–	–	–	–	–	–	–	–	–	**3-Benzyl-9-methyl-2,3,4,5,6,7,8,9-octahydro-2,6-methanoazecino[5,4-b]indole-4,8-dione**		**LI08131**
91	217	231	105	266	265	126	140	**372**	100	71	47	30	16	15	15	14	**3.06**	24	36	3	–	–	–	–	–	–	–	–	–	–	**4-Hexadecynoic acid, 2-(phenylmethoxy)-, methyl ester**	40924-21-8	**NI38660**
312	107	91	105	121	55	85	313	**372**	100	28	27	25	24	20	19	18	**0.00**	24	36	3	–	–	–	–	–	–	–	–	–	–	**16α-Methylpregnenolone acetate**		**MS08755**
313	295	314	225	312	311	296	131	**372**	100	28	23	12	11	10	7	3	**0.00**	24	36	3	–	–	–	–	–	–	–	–	–	–	**Pregn-5-en-3-ol, 3-(acetyloxy)-20,21-epoxy-20-methyl-, (3β)-**		**NI38661**
312	91	269	251	105	145	55	77	**372**	100	54	52	46	44	33	31	30	**0.00**	24	36	3	–	–	–	–	–	–	–	–	–	–	**Pregn-5-en-20-one, 3-(acetyloxy)-16,17-epoxy-, (3β,16α,17α)-**		**AM00150**
312	91	77	79	297	93	105	294	**372**	100	64	51	50	47	43	38	33	**3.00**	24	36	3	–	–	–	–	–	–	–	–	–	–	**Pregn-16-en-20-one, 3-(acetyloxy)-5,6-epoxy-, (3β,5α,6α)-**		**AM00151**
312	107	91	105	121	55	85	313	**372**	100	28	27	25	24	20	19	18	**0.00**	24	36	3	–	–	–	–	–	–	–	–	–	–	**Pregn-5-en-20-one, 3-(acetyloxy)-16-methyl-, (3β,16α)-**	1863-41-8	**WI01487**
169	241	331	282	73	332	242	156	**372**	100	99	71	54	30	20	20	20	**15.00**	24	40	1	–	–	–	–	–	–	1	–	–	–	**Estr-4-en-17-one, 17-(2-propenyl)-17-[(trimethylsilyl)oxy]-, (17β)-**	56771-58-5	**MS08756**
117	73	75	118	119	372	145	143	**372**	100	40	12	11	6	3	3	3		24	40	1	–	–	–	–	–	–	1	–	–	–	**Pregna-3,5-diene, 20α-[(trimethylsilyl)oxy]-**		**NI38662**
117	73	118	75	119	372	174	161	**372**	100	32	11	10	6	4	3	3		24	40	1	–	–	–	–	–	–	1	–	–	–	**Pregna-3,5-diene, 20β-[(trimethylsilyl)oxy]-**		**NI38663**
372	329	286	301	357	312	373	300	**372**	100	85	67	50	33	33	32	32		24	40	1	2	–	–	–	–	–	–	–	–	–	**19-Nor-5β-con-9-enin-6α-ol, 3β-(dimethylamino)-5-methyl-**	5880-30-8	**NI38664**
372	287	357	310	301	373	300	329	**372**	100	70	59	48	41	30	23	20		24	40	1	2	–	–	–	–	–	–	–	–	–	**19-Nor-5β-con-9-enin-6β-ol, 3β-(dimethylamino)-5-methyl-**	5874-52-2	**NI38665**
372	118	84	249	281	91	263	353	**372**	100	72	57	56	47	45	38	31		25	24	3	–	–	–	–	–	–	–	–	–	–	**9-Anthracene-6-carboxylic acid, 8a-styryl-7,8,8a,9,10,10a-hexahydro-, methyl ester**		**MS08757**
372	317	340	81	357	95	107	93	**372**	100	95	86	56	54	53	50	48		25	40	2	–	–	–	–	–	–	–	–	–	–	**3,5-Cyclochol-22-en-24-ol, 6-methoxy-, (3β,5α,6β,22E)-**	56259-13-3	**NI38666**
283	55	317	81	121	95	105	57	**372**	100	45	38	35	32	30	29	29	**19.00**	25	40	2	–	–	–	–	–	–	–	–	–	–	**3,5-Cyclochol-23-en-22-ol, 6-methoxy-, (3β,5α)-**	57984-04-0	**NI38667**
91	117	118	55	69	131	215	41	**372**	100	83	44	43	32	30	28	27	**5.73**	25	40	2	–	–	–	–	–	–	–	–	–	–	**4-Hexyl-1-(7-methoxycarbonylheptyl)bicyclo[4.4.0]deca-2,5,7-triene**		**MS08758**
91	43	41	55	57	69	83	42	**372**	100	65	50	43	37	34	30	26	**0.00**	25	40	2	–	–	–	–	–	–	–	–	–	–	**Oleic acid, benzyl ester**	55130-16-0	**NI38668**
91	119	131	92	105	355	117	97	**372**	100	22	8	8	6	4	4	4	**0.30**	25	40	2	–	–	–	–	–	–	–	–	–	–	**Oleic acid, benzyl ester**	55130-16-0	**NI38669**
72	84	73	301	286	85	70	58	**372**	100	30	7	3	3	3	2	2	**1.20**	25	44	–	2	–	–	–	–	–	–	–	–	–	**Pregn-5-ene-3,20-diamine, N,N,N',N'-tetramethyl-, (3β,20S)-**	6869-45-0	**NI38670**
97	55	81	302	257	273	95	107	**372**	100	84	59	54	54	51	48	45	**26.12**	26	44	1	–	–	–	–	–	–	–	–	–	–	**24-Nor-cholest-22-ene**		**MS08759**
357	259	217	358	301	260	121	135	**372**	100	78	50	28	24	15	13	11	**7.63**	26	44	1	–	–	–	–	–	–	–	–	–	–	**26,27-Dinorergostan-16-one, (5α)-**	54548-11-7	**NI38671**
372	354	261	81	55	95	43	107	**372**	100	84	62	62	62	60	57	54		26	44	1	–	–	–	–	–	–	–	–	–	–	**26,27-Dinorergost-5-en-3-ol, (3β)-**	38819-44-2	**NI38672**
372	261	57	55	95	81	69	287	**372**	100	51	46	41	39	39	33	32		26	44	1	–	–	–	–	–	–	–	–	–	–	**26,27-Dinorergost-5-en-3-ol, (3β)-**	38819-44-2	**NI38673**
243	288	259	372	147	260	69	95	**372**	100	66	48	47	34	27	25	24		26	44	1	–	–	–	–	–	–	–	–	–	–	**19-Norcholest-22-en-3-ol, (3β,5α,22E)-**	56362-44-8	**NI38674**
372	261	354	287	339	213	357	255	**372**	100	95	77	59	54	52	50	46		26	44	1	–	–	–	–	–	–	–	–	–	–	**27-Norcholest-5-en-3-ol, (3β)-**	4420-91-1	**NI38675**
372	178	149	111	109	254	179	105	**372**	100	37	33	33	30	20	20	20		27	20	–	2	–	–	–	–	–	–	–	–	–	**Pyrrolo[2,3-c]carbazole, 1-(3-methylphenyl)-2-phenyl-**		**NI38676**

MASS TO CHARGE RATIOS								M.W.	INTENSITIES								Parent	C	H	O	N	S	F	Cl	Br	I	Si	P	B	X	COMPOUND NAME	CAS Reg No	No
372	91	254	356	371	186	253	149	**372**	100	99	17	13	12	12	9	9		27	20	–	2	–	–	–	–	–	–	–	–	–	**Pyrrolo[2,3-c]carbazole, 1-(4-methylphenyl)-2-phenyl-**		**NI38677**
161	105	43	372	41	91	104	57	**372**	100	88	68	65	41	36	27	26		27	48	–	–	–	–	–	–	–	–	–	–	–	**Benzene, hexadecylpentyl-**	72101-15-6	**WI01488**
217	43	55	41	218	149	109	81	**372**	100	93	71	62	60	57	54	52	**40.70**	27	48	–	–	–	–	–	–	–	–	–	–	–	**Cholestane**	14982-53-7	**AI02272**
217	218	149	357	372	109	219	43	**372**	100	48	34	31	29	19	17	16		27	48	–	–	–	–	–	–	–	–	–	–	–	**Cholestane**	14982-53-7	**NI38678**
217	218	149	109	372	357	121	108	**372**	100	60	54	48	42	35	21	20		27	48	–	–	–	–	–	–	–	–	–	–	–	**Cholestane**	14982-53-7	**NI38679**
217	372	218	55	357	43	149	81	**372**	100	62	61	57	47	45	44	42		27	48	–	–	–	–	–	–	–	–	–	–	–	**Cholestane, (5α)-**	481-21-0	**NI38680**
217	372	218	357	149	219	123	121	**372**	100	61	60	48	44	20	18	18		27	48	–	–	–	–	–	–	–	–	–	–	–	**Cholestane, (5α)-**	481-21-0	**NI38681**
217	43	55	218	41	149	109	95	**372**	100	93	71	62	62	55	52	52	**40.74**	27	48	–	–	–	–	–	–	–	–	–	–	–	**Cholestane, (5α)-**	481-21-0	**WI01489**
149	217	372	218	357	203	232	151	**372**	100	99	53	53	38	21	20	17		27	48	–	–	–	–	–	–	–	–	–	–	–	**Cholestane, (5α,8α,9β,14β)-**		**MS08760**
217	218	149	372	357	203	232	219	**372**	100	76	67	37	32	23	18	18		27	48	–	–	–	–	–	–	–	–	–	–	–	**Cholestane, (5α,8α,14β)-**		**MS08761**
218	217	219	372	149	357	203	163	**372**	100	58	42	36	36	20	17	16		27	48	–	–	–	–	–	–	–	–	–	–	–	**Cholestane, (5α,14β)-**	40071-70-3	**MS08762**
218	217	149	109	95	219	123	372	**372**	100	51	43	40	39	38	36	29		27	48	–	–	–	–	–	–	–	–	–	–	–	**Cholestane, (5α,14β)-**	40071-70-3	**NI38682**
217	372	218	55	95	81	43	109	**372**	100	78	57	51	48	47	43	39		27	48	–	–	–	–	–	–	–	–	–	–	–	**Cholestane, (5β)-**	481-20-9	**NI38683**
91	43	231	105	41	57	92	55	**372**	100	27	17	16	16	12	9	9	**3.66**	27	48	–	–	–	–	–	–	–	–	–	–	–	**11-Phenylheneicosane**		**AI02273**
91	231	105	43	119	133	57	92	**372**	100	31	21	14	13	10	10	9	**5.00**	27	48	–	–	–	–	–	–	–	–	–	–	–	**11-Phenylheneicosane**		**AI02274**
91	231	43	41	105	57	104	55	**372**	100	31	31	21	17	13	12	10	**5.25**	27	48	–	–	–	–	–	–	–	–	–	–	–	**11-Phenylheneicosane**		**AI02275**
133	134	44	372	43	120	55	41	**372**	100	9	5	3	3	2	2	2		27	48	–	–	–	–	–	–	–	–	–	–	–	**1,3,5-Trimethyl-2-octadecylbenzene**		**AI02276**
133	43	372	134	41	57	55	373	**372**	100	18	16	12	12	7	6	5		27	48	–	–	–	–	–	–	–	–	–	–	–	**1,3,5-Trimethyl-2-octadecylbenzene**		**AI02277**
133	372	43	134	41	373	57	55	**372**	100	17	15	13	10	5	5	5		27	48	–	–	–	–	–	–	–	–	–	–	–	**1,3,5-Trimethyl-2-octadecylbenzene**		**AI02278**
372	267	105	77	265	165	263	51	**372**	100	36	31	22	12	12	5	4		28	20	1	–	–	–	–	–	–	–	–	–	–	**Furan, tetraphenyl-**	1056-77-5	**MS08763**
372	281	373	91	203	295	115	296	**372**	100	65	32	28	15	14	12	11		29	24	–	–	–	–	–	–	–	–	–	–	–	**1,2,3,5-Tetraphenylcyclopentene**		**LI08132**
69	246	221	158	127	177	83	31	**373**	100	61	24	10	9	4	4	3	**0.00**	5	1	–	1	–	9	–	–	1	–	–	–	–	**3,3,3-Trifluoro-1-bis(trifluoromethyl)amino-1-iodo-1-propene**		**LI08133**
69	246	221	158	127	177	134	89	**373**	100	76	26	14	10	6	4	3	**0.00**	5	1	–	1	–	9	–	–	1	–	–	–	–	**cis-3,3,3-Trifluoro-1-bis(trifluoromethyl)amino-1-iodo-1-propene**		**LI08134**
340	342	338	36	29	108	344	27	**373**	100	69	59	44	37	27	23	22	**3.77**	7	5	–	3	1	–	6	–	–	–	–	–	–	**2-Ethylthio-4,6-bis(trichloromethyl)-s-triazine**		**MS08764**
55	82	123	81	79	54	94	57	**373**	100	78	56	51	46	45	34	24	**0.00**	12	23	8	1	2	–	–	–	–	–	–	–	–	**Desulphoglucoerysolin**		**MS08765**
174	185	183	375	373	332	330	157	**373**	100	40	40	32	32	32	32	20		13	16	5	1	–	–	–	1	–	1	–	–	–	**2,8,9-Trioxa-5-aza-1-silatricyclo[3.3.3.0[1,5]]undecane, 1-(4-bromobenzoyl)-**		**MS08766**
43	114	85	101	103	69	68	44	**373**	100	99	90	89	79	69	60	47	**0.00**	14	19	9	3	–	–	–	–	–	–	–	–	–	**α-D-Galactopyranose, 1,3,4,6-tetra-O-acetyl-2-azido-2-deoxy-**	67817-30-5	**NI38684**
221	97	232	373	91	29	237	65	**373**	100	33	31	22	21	19	16	16		14	20	5	3	1	–	–	–	–	–	1	–	–	**Pyrazolo[1,5-a]pyrimidine-6-carboxylic acid, 2-[(diethoxyphosphinothioyl)oxy]-5-methyl-, ethyl ester**	13457-18-6	**NI38685**
148	106	131	211	73	373	226	147	**373**	100	96	47	44	36	24	21	20		15	28	4	1	–	–	–	–	–	2	1	–	–	**Phosphonic acid, (acetylamino)benzyl-, bis(trimethylsilyl) ester**	56196-73-7	**NI38686**
127	139	44	75	55	41	126	43	**373**	100	81	74	57	57	57	53	52	**0.00**	16	32	6	1	–	–	–	–	–	1	–	1	–	**α-D-Glucopyranoside, methyl 2-(acetylamino)-2-deoxy-3-O-(trimethylsilyl)-, cyclic butylboronate**	56196-88-4	**NI38687**
242	182	95	173	131	155	181	123	**373**	100	50	40	30	28	27	23	16	**8.00**	17	27	8	1	–	–	–	–	–	–	–	–	–	**5-tert-Butylamino-2-methoxy-2-(2-methoxycarbonylethyl)-3,4-bis(methoxycarbonyl)-2,5-dihydrofuran**		**LI08135**
168	134	121	252	210	110	373	152	**373**	100	42	41	40	35	16	3	3		17	28	6	1	–	–	–	–	–	–	1	–	–	**Glycine, N-[2-[(4-methoxy)phenyl]ethyl]-N-(diisopropylphosphoryl)-**	87212-46-2	**NI38688**
172	173	243	144	270	271	229	174	**373**	100	58	24	19	17	15	15	15	**2.20**	17	31	6	3	–	–	–	–	–	–	–	–	–	**1-Isobutoxypermethylglycyl-α-aminobutyrylglycine**		**LI08136**
91	294	337	266	246	258	282	238	**373**	100	28	27	20	15	11	10	10	**8.00**	18	25	3	1	–	–	2	–	–	–	–	–	–	**1-Piperidinecarboxylic acid, 4-(3,3-dichloro-2-hydroxypropyl)-3-ethyl-, benzyl ester, [3α,4α(S*)]-(±)-**	66087-83-0	**NI38689**
57	41	125	43	93	39	110	126	**373**	100	58	56	48	35	25	22	18	**1.00**	18	31	3	1	2	–	–	–	–	–	–	–	–	**1-Acetyl-2-acetoxy-4-methyl-3,6-di-tert-butylmercapto-1,2,3,6-tetrahydropyridine**		**MS08767**
106	373	148	149	162	225	161	107	**373**	100	63	53	47	35	34	32	27		19	23	3	3	1	–	–	–	–	–	–	–	–	**Aniline, 4-(acrylylamino)-3-(N-butyl-N-phenylsulphonamido)-**		**IC04187**
94	165	293	80	151	236	132	278	**373**	100	50	47	46	42	34	13	8	**0.00**	19	24	–	3	–	–	–	1	–	–	–	–	–	**2-Methyl-3-pentylprodigiosene hydrobromide**		**LI08137**
225	148	70	120	182	134	56	82	**373**	100	38	32	28	27	21	19	16	**6.60**	19	27	3	5	–	–	–	–	–	–	–	–	–	**Uracil, 6-[3-[4-(2-hydroxyphenyl)piperazinyl]propylamino]-1,3-dimethyl-**	91453-03-1	**NI38690**
211	163	162	56	70	150	149	148	**373**	100	66	43	33	32	31	28	27	**25.00**	19	27	3	5	–	–	–	–	–	–	–	–	–	**Uracil, 6-[3-[4-(2-methoxyphenyl)piperazinyl]propylamino]-1-methyl-**	88719-11-3	**NI38691**
211	163	162	70	148	150	182	56	**373**	100	86	66	47	44	43	41	38	**13.20**	19	27	3	5	–	–	–	–	–	–	–	–	–	**Uracil, 6-[3-[4-(2-methoxyphenyl)piperazinyl]propylamino]-3-methyl-**		**NI38692**
314	91	373	330	121	315	282	65	**373**	100	74	48	26	25	21	16	15		20	23	4	1	1	–	–	–	–	–	–	–	–	**Dimethyl 5-(benzylthio)spiro[2H-pyrrole-2,1′-cyclohexane]-3,4-dicarboxylate**		**DD01482**
314	373	161	172	43	315	374	154	**373**	100	77	36	33	27	22	20	20		20	23	6	1	–	–	–	–	–	–	–	–	–	**3-O-Acetyldrupacine**		**NI38693**
113	70	373	141	43	42	127	71	**373**	100	78	47	44	36	30	21	21		20	24	–	3	1	–	1	–	–	–	–	–	–	**10H-Phenothiazine, 2-chloro-10-[3-(4-methyl-1-piperazinyl)propyl]-**	58-38-8	**NI38694**
113	70	373	141	43	42	375	44	**373**	100	90	59	49	44	44	21	21		20	24	–	3	1	–	1	–	–	–	–	–	–	**10H-Phenothiazine, 2-chloro-10-[3-(4-methyl-1-piperazinyl)propyl]-**	58-38-8	**NI38695**
70	113	43	42	232	141	214	71	**373**	100	99	97	92	80	76	65	55	**37.03**	20	24	–	3	1	–	1	–	–	–	–	–	–	**10H-Phenothiazine, 2-chloro-10-[3-(4-methyl-1-piperazinyl)propyl]-**	58-38-8	**NI38696**
130	314	170	44	43	131	159	41	**373**	100	19	17	13	12	11	9	5	**3.05**	20	27	4	3	–	–	–	–	–	–	–	–	–	**N-(N^2-Acetyltryptophyl)alanine butyl ester**		**AI02279**
130	44	43	243	131	170	159	86	**373**	100	18	14	11	11	4	4	4	**3.64**	20	27	4	3	–	–	–	–	–	–	–	–	–	**N^2-(N-Acetylalanyl)tryptophan butyl ester**		**AI02280**

MASS TO CHARGE RATIOS								M.W.	INTENSITIES								Parent	C	H	O	N	S	F	Cl	Br	I	Si	P	B	X	COMPOUND NAME	CAS Reg No	No
84	43	70	100	98	42	184	41	373	100	55	45	41	41	36	32	32	**5.00**	20	31	2	5	–	–	–	–	–	–	–	–	–	1,3,4-Oxadiazol-2-amine, N-isopropyl-5-[3-[3-(piperidinomethyl)phenoxy]propylamino]-		MS08768
56	91	77	84	42	40	55	373	373	100	30	29	23	22	18	13	12		21	19	2	5	–	–	–	–	–	–	–	–	–	Methylrubazonic acid		LI08138
338	204	353	151	339	352	354	165	373	100	73	70	53	34	24	18	17	**0.00**	21	24	4	1	–	1	–	–	–	–	–	–	–	N-Methylpapaverine fluoride		LI08139
206	373	372	190	299				373	100	53	22	16	15					21	27	5	1	–	–	–	–	–	–	–	–	–	6,7-Dimethoxy-2-methyl-1-(3,4,5-trimethoxyphenyl)-1,2,3,4-tetrahydro-isoquinoline		LI08140
315	314	373	258	284	316	298	245	373	100	64	56	27	24	22	16	15		21	27	5	1	–	–	–	–	–	–	–	–	–	Hasubanonine	1805-85-2	NI38697
300	266	256	210	164	373	328	238	373	100	25	20	17	11	10	8	8		21	27	5	1	–	–	–	–	–	–	–	–	–	3,5-Pyridinedicarboxylic acid, 1,4-dihydro-4-(4-methoxyphenyl)-1,2,6-trimethyl-, diethyl ester	42972-34-9	NI38698
358	373	75	73	359	374	91	79	373	100	99	67	47	32	31	27	24		22	35	2	1	–	–	–	–	–	1	–	–	–	Gon-4-en-17-one, 13β-ethyl-3-[(trimethylsilyl)oxyimino]-		NI38699
131	151	255	91	373	272	104	103	373	100	93	80	80	68	48	42	21		23	19	2	1	1	–	–	–	–	–	–	–	–	1,5-Benzothiazepine-4(5H)-one, 2,3-dihydro-2-phenyl-5-phenylacetyl-		MS08769
373	372	277	183	185	262	278	152	373	100	30	27	23	20	17	12	4		23	20	2	1	–	–	–	–	–	–	1	–	–	[(Ethoxycarbonyl)cyanomethylene]triphenylphosphorane		LI08141
282	45	140	283	105	59	373	87	373	100	21	20	18	17	14	13	9		23	23	2	3	–	–	–	–	–	–	–	–	–	N-(2-Phenylethyl)-N-(2-methoxycarbonylethyl)-3-methyl-4-(2,2-dicyanovinyl)aniline		IC04188
105	77	135	45	327	106			373	100	45	19	19	14	9			**0.00**	24	23	3	1	–	–	–	–	–	–	–	–	–	2-Benzamido-1,3-diphenyl-3-ethoxypropanone		LI08142
250	292	251	373	331	293	268		373	100	36	29	21	18	14	14			24	23	3	1	–	–	–	–	–	–	–	–	–	2-Benzobicyclo[2,2,2]octane-3-spiro-3′-(1′-acetyl-2′-indolinone)		LI08143
58	328	373	310	295	358			373	100	28	20	12	9	1				24	39	2	1	–	–	–	–	–	–	–	–	–	Pregn-5-en-20-one, 3-hydroxy-21-(dimethylamino)methyl-, (3β)-		LI08144
128	101	328	91	115	42	41	55	373	100	44	33	31	30	30	30	28	**12.00**	24	39	2	1	–	–	–	–	–	–	–	–	–	Pregn-5-en-20-one, 3-hydroxy-16-methyl-, O-ethyloxime, (3β,16α)-		LI08145
242	164	165	163	318	373	345		373	100	97	85	78	7	1	1			25	20	–	3	–	–	–	–	–	–	–	1	–	N-(Phenylazo)triphenylnitrilium borate		LI08146
253	252	237	373	126	165	372	208	373	100	72	59	57	30	25	23	22		25	31	–	3	–	–	–	–	–	–	–	–	–	Benzenamine, 4,4′,4″-methylidynetris[N,N-dimethyl-	603-48-5	NI38700
120	373	316	372	374	358	317	330	373	100	27	18	14	12	9	7	5		26	47	–	1	–	–	–	–	–	–	–	–	–	6-Azacholestane	16359-52-7	NI38701
130	132	79	83	85	134	109	81	374	100	99	97	68	43	36	32	32	**0.00**	6	6	1	–	–	–	8	–	–	–	–	–	–	Bis(2,3,3,3-tetrachloropropyl) ether	127-90-2	NI38702
47	206	62	191	48	176	234	318	374	100	71	62	45	45	42	27	25	**24.28**	8	6	6	–	2	–	–	–	–	–	–	–	2	Diiron, anti-[di-μ-(π-methanethiolato)]hexacarbonyl-		MS08770
206	135	234	176	318	290	374	346	374	100	51	43	41	31	30	29	26		8	6	6	–	2	–	–	–	–	–	–	–	2	Diiron, (di-μ-methanethiolato)hexacarbonyl-		MS08771
206	176	191	234	290	318	144	346	374	100	53	51	40	32	27	26	25	**22.00**	8	6	6	–	2	–	–	–	–	–	–	–	2	Diiron, syn-[di-μ-(π-methanethiolato)]hexacarbonyl-		MS08772
77	177	127	227	197	113	51	69	374	100	99	71	63	63	62	36	33	**16.50**	10	7	–	–	–	13	–	–	–	–	–	–	–	4-Decene, 1,1,1,2,2,5,6,6,9,9,10,10,10-tridecafluoro-	40723-77-1	NI38703
94	233	112	289	113	120	97	83	374	100	34	11	10	10	9	6	5	**3.98**	10	17	1	2	–	4	–	–	–	–	2	–	1	Manganese, carbonyl(η^5-2,4-cyclopentadien-1-yl)bis(dimethylphosphoramidous difluoride-P)-	37725-92-1	MS08773
121	196	127	56	149	205	140	177	374	100	74	67	53	41	32	30	25	**0.20**	11	5	3	–	–	7	–	–	–	–	–	–	1	Iron, (π-cyclopentadienyl)heptafluorobutyryldicarbonyl-		MS08774
288	318	346	260	290	328	249	262	374	100	26	22	19	16	14	9	3	**0.00**	11	10	3	–	–	–	–	–	–	–	–	–	1	Tungsten, (σ-allyl)(cyclopentadienyl)tricarbonyl-		NI38704
78	91	106	79	77	39	288	105	374	100	58	35	32	25	25	21	20	**9.66**	11	10	3	–	–	–	–	–	–	–	–	–	1	Tungsten, π-(1,3,5-cyclooctatrienyl)tricarbonyl-		MS08775
91	106	105	77	51	39	79	78	374	100	96	42	19	16	14	11	10	**3.61**	11	10	3	–	–	–	–	–	–	–	–	–	1	Tungsten, π-(1,4-dimethylphenyl)tricarbonyl-		MS08776
97	153	125	29	199	65	159	75	374	100	98	89	76	52	47	37	34	**0.40**	11	16	4	–	3	–	1	–	–	–	1	–	–	Phosphorodithioic acid, S-[[(4-chlorophenyl)sulphonyl]methyl] O,O-diethyl ester	16662-85-4	NI38705
376	374	378	295	297	53	67	77	374	100	52	49	42	40	38	35	26		12	8	4	–	–	–	–	2	–	–	–	–	–	1,4-Naphthalenedione, 2,6-dibromo-8-hydroxy-5-methoxy-3-methyl-		NI38706
376	378	374	347	53	267	67	63	374	100	54	50	26	24	22	19	17		12	8	4	–	–	–	–	2	–	–	–	–	–	1,4-Naphthalenedione, 3,6-dibromo-8-hydroxy-5-methoxy-2-methyl-	104506-36-7	NI38707
97	339	283	29	341	285	125	65	374	100	75	67	60	51	48	31	28	**5.17**	12	14	3	–	1	–	3	–	–	–	1	–	–	Phosphorothioic acid, O-[2-chloro-1-(2,5-dichlorophenyl)vinyl] O,O-diethyl ester	1757-18-2	NI38708
376	374	377	375	107	91	218	345	374	100	92	22	21	20	18	16	13		13	9	2	–	1	–	1	1	–	–	1	–	–	Phenothiaphosphine, 7-bromo-2-chloro-10-methoxy-, 10-oxide	30427-90-8	NI38709
376	361	374	363	378	359	377	362	374	100	96	55	52	50	49	15	12		13	12	3	–	–	–	–	2	–	–	–	–	–	1-Naphthalenol, 2,6-dibromo-5,8-dimethoxy-3-methyl-	104506-12-9	NI38710
67	166	109	79	41	112	80	168	374	100	99	78	62	44	39	38	37	**26.63**	13	14	2	–	–	6	–	–	–	–	–	–	1	Nickel, [(1,4,5-η)-4-cycloocten-1-yl](1,1,1,5,5,5-hexafluoro-2,4-pentanedionato-O,O′)-	71900-00-0	NI38711
253	255	59	86	57	257	207	209	374	100	98	67	60	55	30	26	23	**0.10**	13	14	4	–	–	–	4	–	–	–	–	–	–	Tricyclo[6.2.1.0^{2,7}]undeca-4,9-diene-3,6-diol, 1,8,9,10-tetrachloro-11,11-dimethoxy-, endo-		MS08777
178	241	63	376	276	374	176	278	374	100	50	42	27	27	21	20	18		14	10	–	–	–	–	4	–	–	2	–	–	–	1,2-Disilacyclo-3-butene, 1,1,2,2-tetrachloro-3,4-diphenyl-		NI38712
206	168	190	208	63	28	231	170	374	100	63	53	37	29	29	26	22	**4.72**	14	13	2	2	1	–	2	–	–	–	1	–	–	Phosphoramidothioic acid, (1-iminoethyl)-, O,O-bis(4-chlorophenyl) ester	4104-14-7	NI38713
125	63	69	108	374	109	82	45	374	100	80	42	32	29	23	22	22		14	14	4	–	4	–	–	–	–	–	–	–	–	Bis[(4-methylsulphonyl)phenyl] disulphide		MS08778
183	185	243	241	85	157	162	77	374	100	89	66	65	40	36	34	32	**3.00**	14	15	3	–	2	–	–	1	–	–	–	–	–	Propanoic acid, 3-[[3-(4-bromophenyl)-3-hydroxy-1-thioxo-2-propenyl]thio]-, ethyl ester	56248-41-0	NI38714
57	284	283	239	282	212	286	247	374	100	23	23	19	14	14	8	8	**3.00**	14	16	4	4	–	–	2	–	–	–	–	–	–	5-Chloro-6-methyl-(3′,5′)-5′-chloro-6′-methyl-5′,6′-dihydro-6′-hydroxy-6′,2 anhydro-3′-tert-butyluracilyl-uracil		NI38715
264	112	293	104	262	305	266	185	374	100	90	85	69	52	49	44	42	**6.50**	14	20	–	2	–	–	–	2	–	–	–	–	–	Benzenemethanamine, 2-amino-3,5-dibromo-N-cyclohexyl-N-methyl-	3572-43-8	NI38716
69	254	97	62	256	234	88		374	100	88	70	65	62	62	61		**0.00**	15	7	2	2	–	3	2	–	–	–	–	–	–	1H-Benzimidazole-1-carboxylic acid, 5,6-dichloro-2-(trifluoromethyl)-, phenyl ester	14255-88-0	NI38717

MASS TO CHARGE RATIOS								M.W.	INTENSITIES								Parent	C	H	O	N	S	F	Cl	Br	I	Si	P	B	X	COMPOUND NAME	CAS Reg No	No
43	41	175	27	177	18	112	39	**374**	100	95	69	61	42	40	31	28	**3.20**	15	16	5	2	–	–	2	–	–	–	–	–	–	1,3,4-Oxadiazole-2-acetic acid, 4-[2,4-dichloro-5-(1-methylethoxy)phenyl] 4,5-dihydro-α,α-dimethyl-5-oxo-	57198-84-2	NI38718
61	76	59	319	45	241	298	243	**374**	100	70	66	38	34	33	30	30	**0.00**	15	30	–	–	–	–	–	–	–	–	2	–	1	Ruthenium, (η^5-2,4-cyclopentadien-1-yl)(2-methyl-2-propenyl)bis(trimethylphosphine)-	74573-19-6	NI38719
193	253	91	194	315	285	221	175	**374**	100	22	15	13	11	11	8	8	**0.48**	16	14	7	4	–	–	–	–	–	–	–	–	–	Methyl (±)-2-hydroxy-3-[4-(2,4-dinitrophenylazo)phenyl]propionate		NI38720
187	104	189	78	161	83	103	77	**374**	100	45	39	38	30	30	29	20	**6.60**	16	16	–	–	–	–	2	–	–	–	–	–	2	Titanium, di-μ-chlorobis(η^8-1,3,5,7-cyclooctatetraene)di-	12566-82-4	NI38721
187	189	83	161	188	185	78	85	**374**	100	37	31	23	18	13	13	10	**0.00**	16	16	–	–	–	–	2	–	–	–	–	–	2	Titanium, di-μ-chlorobis(η^8-1,3,5,7-cyclooctatetraene)di-	12566-82-4	NI38722
330	331	108	94	65	93	357	39	**374**	100	69	32	20	19	18	16	16	**8.00**	16	18	5	6	–	–	–	–	–	–	–	–	–	1-Ethyl-4-methyl-5-carbamylaminomethyl-3-(2-nitrophenylazo)pyrid-2-one		IC04189
153	111	152	171	170	212	213	195	**374**	100	77	67	40	39	34	30	29	**0.00**	16	22	10	–	–	–	–	–	–	–	–	–	–	Cyclohexanecarboxylic acid, 1,3,4,5-tetrakis(acetyloxy)-, methyl ester, [1R-(1α,3α,4α,5β)]-	50605-64-6	NI38723
126	125	127	55	99	144	98	100	**374**	100	38	33	32	23	21	21	17	**0.47**	16	26	8	2	–	–	–	–	–	–	–	–	–	1,4,7,15-Tetraoxa-12,18-diazacyclodocosane-8,11,19,22-tetrone	79688-17-8	NI38724
126	125	127	55	99	143	98	100	**374**	100	38	33	32	23	21	21	17	**0.00**	16	26	8	2	–	–	–	–	–	–	–	–	–	1,4,7,15-Tetraoxa-12,18-diazacyclodocosane-8,11,19,22-tetrone		MS08779
73	257	258	147	45	75	259	74	**374**	100	90	21	15	14	13	9	9	**7.00**	16	34	4	–	–	–	–	–	–	3	–	–	–	3,5-Heptadienoic acid, 4,6-bis[(trimethylsilyl)oxy]-, trimethylsilyl ester		NI38725
189	190	131	219	202	77	217	175	**374**	100	65	53	41	22	20	17	15	**6.00**	17	17	6	–	–	3	–	–	–	–	–	–	–	2H-1-Benzopyran-2-one, 7-methoxy-8-(2-trifluoroacetoxy-3-hydroxy-3'-methylbutyl)-	93525-01-0	NI38726
129	128	69	130	70	236	127	235	**374**	100	96	60	58	47	38	38	35	**0.00**	18	14	9	–	–	–	–	–	–	–	–	–	–	2,3',4,5',6-Pentahydroxy-4'-(3,5-dihydroxyphenoxy)diphenyl ether		NI38727
246	128	247	77	127	105	245	115	**374**	100	43	22	22	20	20	14	10	**0.00**	18	15	1	–	–	–	–	–	1	–	–	–	–	Pyrylium, 2-methyl-4,6-diphenyl-, iodide	26105-53-3	NI38728
246	128	77	247	105	127	245	123	**374**	100	30	21	20	16	13	8	8	**0.00**	18	15	1	–	–	–	–	–	1	–	–	–	–	Pyrylium, 4-methyl-2,6-diphenyl-, iodide	26105-56-6	NI38729
29	43	41	27	133	79	39	30	**374**	100	69	47	45	35	31	31	25	**1.80**	18	18	7	2	–	–	–	–	–	–	–	–	–	Benzofuran, 7-(2,4-dinitrophenoxy)-3-ethoxy-2,3-dihydro-2-dimethyl-	62059-46-5	NI38730
147	306	195	89	375	222	97	55	**374**	100	89	75	60	28	28	18	15	**0.00**	18	36	2	2	–	–	–	–	–	–	2	–	–	N,N'-Bis[1-(2,2,3,4,4-pentamethylphosphetan 1-oxide)]diaminoethane		NI38731
41	55	179	177	121	29	119	175	**374**	100	91	71	70	61	60	52	45	**0.00**	18	38	–	–	–	–	–	–	–	–	–	–	1	Cyclohexyltributyltin		MS08780
359	73	75	243	217	360	117	55	**374**	100	88	59	38	32	28	26	24	**0.00**	18	38	4	–	–	–	–	–	–	2	–	–	–	1,12-Dodecanedioic acid, bis(trimethylsilyl) ester		MS08781
73	99	144	187	158	59	44	230	**374**	100	63	36	33	31	25	22	19	**0.00**	18	46	–	4	–	–	–	–	–	2	–	–	–	Triethylenetetramine, N[1],N[4]-bis(dimethyl-tert-butylsilyl)-		NI38732
164	196	136	179	165	53	150	44	**374**	100	60	60	53	24	16	14	13	**5.00**	19	18	8	–	–	–	–	–	–	–	–	–	–	Benzoic acid, 3-formyl-2,4-dihydroxy-6-methyl-, 3-hydroxy-4-(methoxycarbonyl)-2,5-dimethylphenyl ester	479-20-9	NI38733
150	178	69	374	327	163	223	342	**374**	100	72	63	56	41	41	31	23		19	18	8	–	–	–	–	–	–	–	–	–	–	Benzoic acid, 2-[[5-hydroxy-4-oxo-6-(1-propenyl)-4H-pyran-3-yl]carbonyl]-3,5-dimethoxy-, methyl ester		LI08147
374	359	331	373	356	165	345	167	**374**	100	69	50	31	27	11	3	3		19	18	8	–	–	–	–	–	–	–	–	–	–	4H-1-Benzopyran-4-one, 5,7-dihydroxy-3,6-dimethoxy-2-(3,4-dimethoxyphenyl)-	35688-42-7	NI38734
359	374	329	43	344	58	88	360	**374**	100	77	58	57	47	47	45	22		19	18	8	–	–	–	–	–	–	–	–	–	–	4H-1-Benzopyran-4-one, 5,6-dihydroxy-2-(3,4-dimethoxyphenyl)-7,8-dimethoxy-	13509-93-8	NI38735
374	359	331	135	187	345	273	161	**374**	100	90	26	9	7	5	3	3		19	18	8	–	–	–	–	–	–	–	–	–	–	4H-1-Benzopyran-4-one, 3,8-dihydroxy-5,6,7-trimethoxy-2-(4-methoxyphenyl)-	40522-42-7	MS08782
374	373	359	331	299	343	167	271	**374**	100	80	35	35	7	6	6	5		19	18	8	–	–	–	–	–	–	–	–	–	–	4H-1-Benzopyran-4-one, 5-hydroxy-2-(4-hydroxy-3,5-dimethoxyphenyl)-3,7-dimethoxy-		NI38736
216	184	217	218	185	56	171	152	**374**	100	46	24	11	11	11	7	5	**0.50**	19	22	2	2	2	–	–	–	–	–	–	–	–	1-Piperazinecarboxylic acid, 4-(4,9-dihydrobenzo[e]thieno[2,3-b]thiepin-4 yl)-, ethyl ester		MS08783
357	298	252	196	29	284	282	270	**374**	100	51	43	32	32	26	25	25	**7.50**	19	22	6	2	–	–	–	–	–	–	–	–	–	3,5-Pyridinedicarboxylic acid, 4-(2-nitrophenyl)-2,6-dimethyl-1,4-dihydro, diethyl ester	21829-26-5	NI38737
73	374	359	346	375	318	45	75	**374**	100	51	19	19	18	18	15	13		19	26	4	–	–	–	–	–	–	2	–	–	–	2H-Pyran-2-one, (E)-4-[(trimethylsilyl)oxy]-6-[2-[4-[(trimethylsilyl)oxy]phenyl]vinyl]-		MS08784
73	81	135	194	114	99	93	121	**374**	100	34	20	7	6	6	6	4	**0.30**	19	31	5	–	–	–	1	–	–	–	–	–	–	Methyl 10-(chloroacetoxy)-11-methoxy-3,7,11-trimethyl-2,6-dodecadienoate		MS08785
145	214	69	41	143	97	55	106	**374**	100	33	28	26	19	19	19	11	**8.00**	19	34	3	–	2	–	–	–	–	–	–	–	–	Hexanoic acid, 6-[(2-ethyl-2-methyl-6,10-dithiaspiro[4.5]decan-1-yl)oxy]-, ethyl ester	92640-70-5	NI38738
374	193	249	124	235	218	318	275	**374**	100	42	27	26	23	19	13	9		19	37	–	2	–	–	–	–	–	–	–	2	1	Cobalt, η^4-(1,3-di-tert-butyl-2,4-diisopropyl-1,3,2,4-diazadiboretidine)-eta^5 cyclopentadienyl-		NI38739
374	338	376	375	373	337	339	340	**374**	100	64	60	48	46	33	25	23		20	14	–	2	–	–	2	–	–	–	–	2	–	Naphtho[2,1-e]-1,3,2,4-diazadiborine, 2,4-dichloro-1,2,3,4-tetrahydro-3-(1-naphthalenyl)-	35704-34-8	NI38740
222	165	41	57	77	91	357	281	**374**	100	81	74	69	67	61	60	59	**5.00**	20	14	4	4	–	–	–	–	–	–	–	–	–	Benzene, 1-[(2-nitrophenyl)azo]-2-[2-(2-nitrophenyl)vinyl]-	69395-32-0	NI38741
67	174	77	91	200	105	51	78	**374**	100	87	86	67	56	39	27	16	**1.00**	20	18	2	6	–	–	–	–	–	–	–	–	–	2-Pyrazolin-5-one, 4,4'-azobis[3-methyl-1-phenyl-	13615-29-7	MS08786
180	222	297	151	314	374	181	165	**374**	100	39	26	20	17	11	11	9		20	22	7	–	–	–	–	–	–	–	–	–	–	(+)-2,3-trans-3,4-trans-3β-Acetoxy-3',4',7-trimethoxyflavan-4α-ol		MS08787
43	55	83	248	138	274	374	292	**374**	100	85	80	50	50	42	25	20		20	22	7	–	–	–	–	–	–	–	–	–	–	Budlein A	59481-48-0	NI38742
181	374	194	179	373	207	331	193	**374**	100	50	18	18	15	14	12	11		20	22	7	–	–	–	–	–	–	–	–	–	–	Chalcone, 2'-hydroxy-3,4,4',5,6'-pentamethoxy-	80496-71-5	NI38743

MASS TO CHARGE RATIOS								M.W.	INTENSITIES								Parent	C	H	O	N	S	F	Cl	Br	I	Si	P	B	X	COMPOUND NAME	CAS Reg No	No
374	300	315	256	359	327	283	375	**374**	100	81	71	68	61	59	25	23		20	22	7	–	–	–	–	–	–	–	–	–	–	**1,2-Dibenzofurandicarboxylic acid, 1,9b-dihydro-7,9-dimethoxy-4,9b-dimethyl-, dimethyl ester**	56247-77-9	**NI38744**
327	359	300	315	256	328	269	283	**374**	100	49	38	30	25	19	18	17	**10.00**	20	22	7	–	–	–	–	–	–	–	–	–	–	**1,2-Dibenzofurandicarboxylic acid, 3,9b-dihydro-7,9-dimethoxy-4,9b-dimethyl-, dimethyl ester**	40801-29-4	**NI38745**
181	223	43	182	151	91	152	77	**374**	100	14	13	12	10	7	6	5	**0.20**	20	22	7	–	–	–	–	–	–	–	–	–	–	**1-(3,5-Dimethoxy-4-acetoxyphenyl)ethyl 2-methoxy-4-formylphenyl ether**		**NI38746**
152	223	180	121	181	224	374	209	**374**	100	58	34	34	23	13	11	10		20	22	7	–	–	–	–	–	–	–	–	–	–	**(±)-2-(α,4-Dimethoxybenzyl)-2,4,6-trimethoxybenzo[b]furan-3(2H)-one**		**NI38747**
137	138	374	122	94	77	56	55	**374**	100	13	8	6	5	3	3	3		20	22	7	–	–	–	–	–	–	–	–	–	–	**2(3H)-Furanone, dihydro-3-hydroxy-3,4-bis(4-hydroxy-3-methoxybenzyl)-, (3S-cis)-**	34444-37-6	**NI38748**
224	209	374	166	151	122	181	150	**374**	100	91	29	26	23	20	16	15		20	22	7	–	–	–	–	–	–	–	–	–	–	**2′,3′,4′,6′,7-Pentamethoxyisoflavanone**		**LI08148**
185	259	374	43	285	157	127	314	**374**	100	66	28	28	8	7	4	3		20	22	7	–	–	–	–	–	–	–	–	–	–	**Propanedioic acid, α-(4-methoxy-1-naphthyl)-α-acetoxy-, diethyl ester**	118657-08-2	**DD01483**
168	84	374	56	58	342	80	70	**374**	100	38	30	25	24	21	21	20		20	26	5	2	–	–	–	–	–	–	–	–	–	**Gelsedine, 14-hydroxy-11-methoxy-**	41478-09-5	**NI38749**
43	374	255	300	166	148	299	167	**374**	100	82	53	51	46	36	35	33		20	26	5	2	–	–	–	–	–	–	–	–	–	**4,3′,5′-Trimethyl-5,4′-bis(ethoxycarbonyl)-3-acetyl-2,2′-dipyrrolylmethane**		**LI08149**
136	121	137	109	80	65	77	110	**374**	100	96	78	57	57	30	30	28	**24.80**	21	26	6	–	–	–	–	–	–	–	–	–	–	**6-Methyl-2,3,11,12-dibenzo-1,4,7,10,13,16-hexaoxacyclooctadeca-2,11-diene**		**MS08788**
201	31	203	45	197	166	165	132	**374**	100	43	31	28	27	26	24	18	**6.08**	21	27	2	2	–	–	1	–	–	–	–	–	–	**Ethanol, 2-[2-[4-[(4-chlorophenyl)phenylmethyl]-1-piperazinyl]ethoxy]-**	68-88-2	**NI38750**
201	203	45	166	165	299	56	132	**374**	100	35	24	22	21	20	18	16	**5.61**	21	27	2	2	–	–	1	–	–	–	–	–	–	**Ethanol, 2-[2-[4-[(4-chlorophenyl)phenylmethyl]-1-piperazinyl]ethoxy]-**	68-88-2	**NI38751**
255	301	256	328	180	240	374	241	**374**	100	34	31	27	20	19	18	17		21	30	4	2	–	–	–	–	–	–	–	–	–	**Bis(2-ethoxycarbonyl-4-methyl-5-ethyl-3-pyrrolyl)methane**		**LI08150**
193	374	194	180	134	147	181	120	**374**	100	88	32	24	18	17	15	13		21	30	4	2	–	–	–	–	–	–	–	–	–	**Bis(5-ethoxycarbonyl-4-methyl-3-ethyl-2-pyrrolyl)methane**		**LI08151**
374	49	303	84	86	44	278	153	**374**	100	60	49	47	31	27	26	25		22	14	6	–	–	–	–	–	–	–	–	–	–	**1,1′-Binaphthalene-5,5′,8,8′-tetrone, 4,4′-dihydroxy-6,6′-dimethyl-**		**NI38752**
374	359	357	375	360	92	358	373	**374**	100	40	28	24	10	10	7	6		22	14	6	–	–	–	–	–	–	–	–	–	–	**2,2′-Binaphthalene-1,1′,4,4′-tetrone, 5,5-dihydroxy-3,3′-dimethyl-**		**NI38753**
91	92	106	77	51	63	374	65	**374**	100	73	48	35	32	31	29	25		22	14	6	–	–	–	–	–	–	–	–	–	–	**2,2′-Binaphthalene-1,1′,4,4′-tetrone, 5,5′-dihydroxy-7,7′-dimethyl-**	24456-79-9	**NI38754**
374	359	92	63	120	331	303	317	**374**	100	66	30	21	18	13	8	5		22	14	6	–	–	–	–	–	–	–	–	–	–	**2,2′-Binaphthalene-1,1′,4,4′-tetrone, 8,8′-dihydroxy-3,3′-dimethyl-**		**LI08152**
374	106	357	77	290	375	318	91	**374**	100	49	38	28	25	24	24	24		22	14	6	–	–	–	–	–	–	–	–	–	–	**2,2′-Binaphthalene-1,1′,4,4′-tetrone, 8,8′-dihydroxy-6,6′-dimethyl-**	17734-93-9	**NI38755**
374	106	134	135	77	359	163	78	**374**	100	45	28	25	25	15	15	12		22	14	6	–	–	–	–	–	–	–	–	–	–	**2,2′-Binaphthalene-1,4,5′,8′-tetrone, 1′,5-dihydroxy-3′,6-dimethyl-**		**MS08789**
374	375	92	359	120	346	121	63	**374**	100	25	19	15	11	10	10	10		22	14	6	–	–	–	–	–	–	–	–	–	–	**2,2′-Binaphthalene-1,4,5′,8′-tetrone, 1′,8-dihydroxy-3,6′-dimethyl-**	58274-95-6	**NI38756**
374	375	331	278	41	346	153	357	**374**	100	30	14	14	13	12	12	11		22	14	6	–	–	–	–	–	–	–	–	–	–	**2,2′-Binaphthalene-5,5′,8,8′-tetrone, 1,1′-dihydroxy-6,6′-dimethyl-**	20175-85-3	**NI38757**
359	374	331	288	76	360	375	232	**374**	100	80	38	32	29	24	19	16		22	14	6	–	–	–	–	–	–	–	–	–	–	**2,2′-Binaphthalene-1,1′,4,4′-tetrone, 3,3′-dimethoxy-**		**NI38758**
374	76	44	357	375	327	104	101	**374**	100	89	50	36	26	26	26	24		22	14	6	–	–	–	–	–	–	–	–	–	–	**2,2′-Binaphthalene-1,1′,4,4′-tetrone, 5,5′-dimethoxy-**	61836-44-0	**NI38759**
374	357	76	375	108	358	78	104	**374**	100	60	38	26	23	15	12	11		22	14	6	–	–	–	–	–	–	–	–	–	–	**2,2′-Binaphthalene-1,1′,4,4′-tetrone, 5,8′-dimethoxy-**		**NI38760**
374	375	357	329	359	44	301	119	**374**	100	30	22	15	12	12	11	10		22	14	6	–	–	–	–	–	–	–	–	–	–	**Dinaphtho[1,2-b:2′,3′-d]furan-7,12-dione, 5-hydroxy-4,11-dimethoxy-**	75446-03-6	**NI38761**
319	374	320	375	132	251	69	55	**374**	100	62	23	18	17	14	14	12		22	18	4	2	–	–	–	–	–	–	–	–	–	**9,10-Anthracenedione, 2,6-bis(2-oxo-1-pyrrolidinyl)-**		**MS08790**
178	372	371	179	373	194	295	283	**374**	100	56	32	17	15	9	5	4	**2.00**	22	22	2	4	–	–	–	–	–	–	–	–	–	**1H-Purine-2,6-dione, 8-(1,2-diphenylethyl)-3,7-dihydro-1,3,7-trimethyl-**		**NI38762**
374	43	133	57	376	27	31	375	**374**	100	44	43	40	35	33	24	23		22	27	3	–	–	–	1	–	–	–	–	–	–	**3-Chloro-17β-acetoxy-2-formyl-2,4,6-androstatriene**		**MS08791**
374	328	215	55	69	300	231	115	**374**	100	87	68	51	40	29	24	24		22	30	5	–	–	–	–	–	–	–	–	–	–	**7-Ethoxyrosmanol**		**MS08792**
314	226	225	286	227	285	91	342	**374**	100	87	78	77	70	64	44	43	**4.00**	22	30	5	–	–	–	–	–	–	–	–	–	–	**Gibberellin A_{24} methyl ester**		**MS08793**
43	219	95	81	279	283	301	314	**374**	100	87	71	62	52	38	31	27	**0.00**	22	30	5	–	–	–	–	–	–	–	–	–	–	**Hautriwaic acid acetate**		**DD01484**
163	107	134	167	121	93	55	29	**374**	100	88	72	70	53	51	51	51	**3.98**	22	30	5	–	–	–	–	–	–	–	–	–	–	**Jasmolin II**	1172-63-0	**NI38763**
163	58	107	167	135	55	121	161	**374**	100	75	69	58	58	48	46	40	**7.80**	22	30	5	–	–	–	–	–	–	–	–	–	–	**Jasmolin II**	1172-63-0	**MS08794**
107	93	55	43	41	91	79	29	**374**	100	67	63	59	56	50	37	36	**0.00**	22	30	5	–	–	–	–	–	–	–	–	–	–	**Jasmolin II**	1172-63-0	**NI38764**
286	230	218	243	271	328	374	332	**374**	100	48	20	19	11	9	2	2		22	30	5	–	–	–	–	–	–	–	–	–	–	**Podocarpa-8,11,13-trien-17-oic acid, 11-acetoxy-12-hydroxy-13-isopropyl-**		**MS08795**
43	359	45	58	87	299	42	374	**374**	100	75	62	34	26	23	19	16		22	30	5	–	–	–	–	–	–	–	–	–	–	**2β,3β,14α-Trihydroxy-5β-androst-7-ene-6,17-dione-2,3-acetonide**		**MS08796**
359	360	258	169	243	361	181	216	**374**	100	29	15	11	10	8	7	6	**0.80**	22	34	3	–	–	–	–	–	–	1	–	–	–	**Androst-4-ene-3,17-dione, 2-[(trimethylsilyl)oxy]-, (2β)-**	69688-13-7	**NI38765**
359	318	374	360	319	73	317	375	**374**	100	68	41	28	18	14	13	12		22	34	3	–	–	–	–	–	–	1	–	–	–	**Androst-4-ene-3,17-dione, 6-[(trimethylsilyl)oxy]-, (6β)-**	69833-67-6	**NI38766**
359	374	360	284	224	375	75	251	**374**	100	50	29	23	19	13	13	12		22	34	3	–	–	–	–	–	–	1	–	–	–	**Androst-4-ene-3,17-dione, 7-[(trimethylsilyl)oxy]-, (7α)-**	61103-05-7	**NI38767**
359	360	267	143	361	145	225	268	**374**	100	32	31	13	9	9	8	6	**0.78**	22	34	3	–	–	–	–	–	–	1	–	–	–	**Androst-4-ene-3,17-dione, 12-[(trimethylsilyl)oxy]-, (12β)-**	69688-28-4	**NI38768**
129	318	130	73	131	303	245	75	**374**	100	15	13	8	4	3	3	2	**0.59**	22	34	3	–	–	–	–	–	–	1	–	–	–	**Androst-5-ene-11,17-dione, 3-[(trimethylsilyl)oxy]-, (3β)-**	49774-80-3	**NI38769**
301	302	43	359	176	187	41	28	**374**	100	23	10	6	5	4	4	3	**3.00**	22	34	3	2	–	–	–	–	–	–	–	–	–	**17β-Hydroxy-4-propyl-3,4-seco-5-androstene[4,5,6-cd]pyrazol-3-oic acid**		**MS08797**
229	230	215	107	105	147	356	121	**374**	100	45	35	35	33	30	25	24	**3.00**	22	35	4	–	–	–	–	–	–	–	–	1	–	**Pregnan-20-one, 3-hydroxy-17,21-[(methylborylene)bis(oxy)], (3α,5β)-**	30888-60-9	**NI38770**
241	45	71	83	97	69	57	43	**374**	100	90	87	81	76	61	50	46	**0.00**	22	46	4	–	–	–	–	–	–	–	–	–	–	**Propane, 1,2-dimethoxy-3-[(2-methoxyhexadecyl)oxy]-**	16725-44-3	**LI08153**
241	45	83	97	71	69	57	43	**374**	100	90	81	75	67	61	50	46	**0.00**	22	46	4	–	–	–	–	–	–	–	–	–	–	**Propane, 1,2-dimethoxy-3-[(2-methoxyhexadecyl)oxy]-**	16725-44-3	**NI38771**
173	374	165	174	201	166	89	167	**374**	100	19	19	18	11	11	11	9		23	15	3	–	–	–	1	–	–	–	–	–	–	**1H-Indene-1,3(2H)-dione, 2-[(4-chlorophenyl)phenylacetyl]-**	3691-35-8	**NI38772**
338	91	356	79	105	107	93	77	**374**	100	81	67	66	60	45	44	41	**1.93**	23	34	4	–	–	–	–	–	–	–	–	–	–	**Card-20(22)-enolide, 3,14-dihydroxy-, (3β,5α)-**	466-09-1	**NI38773**
203	111	147	107	93	91	79	85	**374**	100	36	30	30	29	28	28	27	**4.00**	23	34	4	–	–	–	–	–	–	–	–	–	–	**Card-20(22)-enolide, 3,14-dihydroxy-, (3β,5β)-**	143-62-4	**NI38774**
203	43	356	246	79	95	162	81	**374**	100	46	37	34	27	21	20	20	**0.00**	23	34	4	–	–	–	–	–	–	–	–	–	–	**Card-20(22)-enolide, 3,14-dihydroxy-, (3β,5β)-**	143-62-4	**MS08798**
183	356	55	111	79	92	67	90	**374**	100	81	62	60	60	58	57	56	**22.86**	23	34	4	–	–	–	–	–	–	–	–	–	–	**Card-20(22)-enolide, 3,14-dihydroxy-, (3β,5β)-**	143-62-4	**NI38775**
344	215	257	173	159	345	231	197	**374**	100	70	42	33	32	24	23	22	**0.00**	23	34	4	–	–	–	–	–	–	–	–	–	–	**Methyl 3-oxo-19-hydroxybisnorchol-4-en-22-oate**		**LI08154**

MASS TO CHARGE RATIOS								M.W.	INTENSITIES								Parent	C	H	O	N	S	F	Cl	Br	I	Si	P	B	X	COMPOUND NAME	CAS Reg No	No
374	331	359	249					**374**	100	19	13	6						23	34	4	–	–	–	–	–	–	–	–	–	–	**9(5H)-Phenanthrenone, 4b,6,7,8,8a,10-hexahydro-1,3,4-trimethoxy-4b,8,8-trimethyl-2-isopropyl-, (4bS-trans)-**	60371-72-4	**NI38776**
43	219	314	108	85	81	299	93	**374**	100	46	36	26	26	24	23	23	**15.53**	23	34	4	–	–	–	–	–	–	–	–	–	–	**Pregnane-11,20-dione, 3-(acetyloxy)-, (3α)-**	56193-60-3	**NI38777**
314	93	55	91	95	79	81	67	**374**	100	51	51	48	47	46	42	38	**4.10**	23	34	4	–	–	–	–	–	–	–	–	–	–	**Pregnan-20-one, 3-(acetyloxy)-5,6-epoxy-, (3β,5α,6α)-**	14148-09-5	**WI01490**
314	91	93	79	55	285	286	95	**374**	100	76	73	73	71	64	62	62	**3.40**	23	34	4	–	–	–	–	–	–	–	–	–	–	**Pregnan-20-one, 3-(acetyloxy)-5,6-epoxy-, (3β,5β,6β)-**	6661-94-5	**WI01491**
344	215	257	174	160	345	232	197	**374**	100	70	41	32	32	25	23	23	**0.00**	23	34	4	–	–	–	–	–	–	–	–	–	–	**Pregn-4-ene-20α-carboxylic acid, 19-hydroxy-3-oxo-, methyl ester**	20248-19-5	**NI38778**
296	91	105	79	314	107	77	145	**374**	100	58	48	34	33	32	31	30	**0.20**	23	34	4	–	–	–	–	–	–	–	–	–	–	**Pregn-5-en-20-one, 3-(acetyloxy)-17-hydroxy-, (3β)-**	1863-39-4	**WI01492**
43	301	255	81	55	145	159	107	**374**	100	86	86	39	35	33	32	32	**26.33**	23	34	4	–	–	–	–	–	–	–	–	–	–	**Pregn-5-en-20-one, 21-(acetyloxy)-3-hydroxy-, (3β)-**	566-78-9	**NI38779**
43	314	81	271	253	213	91	105	**374**	100	37	35	29	28	28	27	26	**3.88**	23	34	4	–	–	–	–	–	–	–	–	–	–	**Pregn-5-en-20-one, 3,17-dihydroxy-, 3-acetate**	10343-51-8	**NI38780**
296	91	105	79	314	107	77	145	**374**	100	58	48	34	33	32	31	30	**0.20**	23	34	4	–	–	–	–	–	–	–	–	–	–	**Pregn-5-en-20-one, 3,17-dihydroxy-, monoacetate, (3β)-**	56273-07-5	**NI38781**
256	73	241	257	374	32	213	185	**374**	100	42	35	25	19	18	16	16		23	38	2	–	–	–	–	–	–	1	–	–	–	**Abietic acid, trimethylsilyl ester**		**NI38782**
73	374	122	129	121	75	134	91	**374**	100	90	84	53	48	40	39	35		23	38	2	–	–	–	–	–	–	1	–	–	–	**Gona-3,5-diene, 13-ethyl-3-methoxy-17-[(trimethylsilyl)oxy]-, (13β,17β)-**		**NI38783**
73	256	241	257	75	55	41	81	**374**	100	73	61	30	27	27	27	25	**6.06**	23	38	2	–	–	–	–	–	–	1	–	–	–	**Isopimaric acid, trimethylsilyl ester**		**NI38784**
73	121	120	257	75	81	55	41	**374**	100	63	30	24	21	18	18	18	**10.78**	23	38	2	–	–	–	–	–	–	1	–	–	–	**Pimaric acid, trimethylsilyl ester**		**NI38785**
157	303	345	144	158	284	255	290	**374**	100	45	39	33	20	15	15	13	**5.62**	23	38	2	–	–	–	–	–	–	1	–	–	–	**19-Nor-17α-pregn-4-en-3-one, 17-(trimethylsiloxy)-**	6689-89-0	**NI38786**
100	343	288	87	42	79	70	41	**374**	100	99	64	58	57	48	47	47	**19.00**	23	38	2	2	–	–	–	–	–	–	–	–	–	**5β-Pregnane-3,20-dione, bis(O-methyloxime)**	3091-37-0	**NI38787**
304	303	339	374	302	128	151	228	**374**	100	73	71	46	45	39	37	34		24	16	–	–	–	–	2	–	–	–	–	–	–	**1,4-Methanonaphthalene, 9-[bis(4-chlorophenyl)methylene]-1,4-dihydro-**	116972-01-1	**DD01485**
195	151	374	194	150	196	373	140	**374**	100	32	25	25	23	17	15	15		24	16	3	–	–	–	–	–	–	–	–	2	–	**Bis(10,9-boroxaro-10-phenanthryl) ether**		**LI08155**
187	188	176	159	84	131	116	29	**374**	100	14	5	5	5	4	4	4	**0.00**	24	30	–	4	–	–	–	–	–	–	–	–	–	**Ethylenediamine, 1,2-bis(4-cyanophenyl)-N,N,N′,N′-tetraethyl-, (±)-**		**DD01486**
187	188	159	29	131	116	28	171	**374**	100	94	29	28	26	22	17	14	**0.00**	24	30	–	4	–	–	–	–	–	–	–	–	–	**Ethylenediamine, 1,2-bis(4-cyanophenyl)-N,N,N′,N′-tetraethyl-, meso-**		**DD01487**
186	187	144	374	145	188	185	375	**374**	100	52	43	32	19	14	14	12		24	30	–	4	–	–	–	–	–	–	–	–	–	**Folicanthine**		**MS08799**
85	272	273	57	67	41	86	43	**374**	100	84	19	9	7	6	5	5	**0.00**	24	38	3	–	–	–	–	–	–	–	–	–	–	**Androst-5-en-17-ol, 3-[(tetrahydro-2H-pyran-2-yl)oxy]-, (3β,17β)-**	5419-51-2	**MS08800**
288	85	275	289	374	373	58	273	**374**	100	82	43	38	18	11	9	8		24	38	3	–	–	–	–	–	–	–	–	–	–	**Androst-5-en-17-ol, 3-[(tetrahydro-2H-pyran-2-yl)oxy]-, (3β,17β)-**	5419-51-2	**MS08801**
85	288	289	272	86	374			**374**	100	12	4	4	3	2				24	38	3	–	–	–	–	–	–	–	–	–	–	**Androst-5-en-17-ol, 3-[(tetrahydro-2H-pyran-2-yl)oxy]-, (3β,17β)-**	5419-51-2	**NI38788**
161	180	55	41	374	175	121	162	**374**	100	61	45	44	40	31	31	29		24	38	3	–	–	–	–	–	–	–	–	–	–	**Benzoic acid, 2-methoxy-6-(8-pentadecenyl)-, methyl ester**	56554-66-6	**NI38789**
287	288	55	105	43	41	91	81	**374**	100	23	19	17	17	17	14	14	**4.00**	24	38	3	–	–	–	–	–	–	–	–	–	–	**17-Oxo-6-pentyl-4-nor-3,5-seco-5-androsten-3-oic acid methyl ester**		**MS08802**
245	129	73	95	246	75	331	135	**374**	100	36	29	21	19	18	15	14	**3.03**	24	42	1	–	–	–	–	–	–	1	–	–	–	**Androst-5-ene, 4,4-dimethyl-3-[(trimethylsilyl)oxy]-, (3β)-**	7604-81-1	**NI38790**
129	245	163	121	107	95	123	130	**374**	100	40	28	24	18	16	14	13	**3.70**	24	42	1	–	–	–	–	–	–	1	–	–	–	**Pregn-5-en-3β-ol trimethylsilyl ether**		**MS08803**
128	246	231	297	374	286	270	129	**374**	100	77	75	62	60	39	26	24		25	18	–	4	–	–	–	–	–	–	–	–	–	**[1,2,4]Triazolo[4,3-a]quinoline, 1,2-dihydro-1-phenyl-2-(2-quinolinyl)-**	51659-05-3	**NI38791**
374	217	55	81	67	359	95	79	**374**	100	67	62	59	52	44	41	35		25	42	2	–	–	–	–	–	–	–	–	–	–	**Cholan-24-oic acid, methyl ester, (5β)-**	2204-14-0	**NI38792**
74	55	41	67	81	87	43	69	**374**	100	97	92	81	74	70	70	60	**0.00**	25	42	2	–	–	–	–	–	–	–	–	–	–	**Cyclopropanebutanoic acid, 2-[[2-[[2-[(2-pentylcyclopropyl)methyl]cyclopropyl]methyl]cyclopropyl]methyl]-, methyl ester**	56051-53-7	**NI38793**
99	139	100	55	81	95	69	41	**374**	100	99	99	52	36	35	30	29	**1.79**	25	42	2	–	–	–	–	–	–	–	–	–	–	**D-Homo-5α-androstan-3-one, 5,8,17aβ-trimethyl-, cyclic ethylene acetal**	20078-53-9	**NI38794**
283	355	119	284	130	108	281	131	**374**	100	78	60	18	18	14	13	11	**7.00**	25	42	2	–	–	–	–	–	–	–	–	–	–	**Stearic acid, benzyl ester**		**NI38795**
91	108	57	43	92	41	55	71	**374**	100	56	47	45	27	25	23	22	**0.00**	25	42	2	–	–	–	–	–	–	–	–	–	–	**Stearic acid, benzyl ester**		**NI38796**
72	84	303	73	110	71	58	374	**374**	100	12	5	5	3	3	3	2		25	46	–	2	–	–	–	–	–	–	–	–	–	**5α-Pregnane-3β,20α-diamine, N,N,N′,N′-tetramethyl-**	1172-08-3	**NI38797**
181	374	152	153	179	41	375	29	**374**	100	99	68	52	48	36	32	32		26	18	1	2	–	–	–	–	–	–	–	–	–	**2,5-Bis(4-biphenylyl)-1,3,4-oxadiazole**		**IC04190**
259	115	357	355	260	245	137	303	**374**	100	75	50	28	26	23	23	20	**2.00**	26	46	1	–	–	–	–	–	–	–	–	–	–	**20-Hydroxy-26-norcholestane**		**NI38798**
105	374							**374**	100	32								27	18	2	–	–	–	–	–	–	–	–	–	–	**Phenanthro[9,10-d]-1,3-dioxole, 2,2-diphenyl-**	17529-29-2	**NI38799**
41	67	95	43	55	69	81	83	**374**	100	95	85	73	69	64	60	53	**4.14**	27	50	–	–	–	–	–	–	–	–	–	–	–	**1,3-Dicyclopentyl-2-dodecylcyclopentane**		**AI02281**
121	163	43	81	41	67	122	55	**374**	100	88	66	62	50	49	42	40	**19.10**	27	50	–	–	–	–	–	–	–	–	–	–	–	**5-Pentadecyltricyclo[6.3.1.0^{4,12}]dodecane**		**AI02282**
374	373	375	165	179	218	178	152	**374**	100	83	29	29	18	13	12	12		28	22	1	–	–	–	–	–	–	–	–	–	–	**2,6-Bis(α-naphthylidene)cyclohexanone**		**MS08804**
131	374	117	132	91	375	129	116	**374**	100	38	33	26	13	12	11	9		28	38	–	–	–	–	–	–	–	–	–	–	–	**1,10-Bis(5-indanyl)decane**		**AI02283**
91	194	180	207	167	374	270		**374**	100	33	26	17	15	7	7			29	26	–	–	–	–	–	–	–	–	–	–	–	**1,1-Diphenyl-2-(2,2-diphenylethyl)cyclopropane**		**LI08156**
69	239	96	218	126	83	277	158	**375**	100	96	47	27	15	13	11	8	**0.00**	5	3	–	1	–	9	–	–	1	–	–	–	–	**3-Bis(trifluoromethyl)amino-3-iodo-1,1,1-trifluoropropane**		**LI08157**
44	60	219	157	73	63	62	221	**375**	100	62	55	55	55	55	55	45	**17.24**	12	6	4	2	–	–	2	–	–	–	–	–	1	**Copper, bis(4-chloro-O-benzoquinone 2-oximato)-**	30993-28-3	**MS08805**
99	219	44	63	189	161	60	62	**375**	100	33	29	24	20	20	18	16	**14.67**	12	6	4	2	–	–	2	–	–	–	–	–	1	**Copper, bis(4-chloro-O-benzoquinone 2-oximato)-**	30993-28-3	**NI38800**
202	77	218	200	199	186	154	112	**375**	100	97	96	96	90	90	90	87	**0.00**	12	10	3	3	3	–	1	–	–	–	–	–	–	**1-Chloro-3,5-diphenyl-1,3,5-triaza-2,4,6-trithiacyclohexane-2,4,6-trioxide**		**LI08158**
216	296	217	201	375	202	189	294	**375**	100	81	38	38	25	25	25	16		15	21	6	1	2	–	–	–	–	–	–	–	–	**Pyrrolidine, 2-[2-[3-methoxy-4-[(methylsulphonyl)oxy]phenyl]ethenyl]-1-(methylsulphonyl)-, (E)-(±)-**	67257-64-1	**NI38801**
129	173	255	248	104	130	115	257	**375**	100	66	56	54	54	41	40	38	**13.00**	16	19	1	1	2	–	2	–	–	–	–	–	–	**4-Thia-1-azabicyclo[3.2.0]heptan-7-one, 6,6-dichloro-2-exo-isopropyl-2-endo-methyl-3-endo-(methylthio)-5-phenyl-**		**NI38802**

MASS TO CHARGE RATIOS								M.W.	INTENSITIES								Parent	C	H	O	N	S	F	Cl	Br	I	Si	P	B	X	COMPOUND NAME	CAS Reg No	No
129	255	248	257	283	104	55	173	**375**	100	94	81	61	57	55	49	48	**4.00**	16	19	1	1	2	–	2	–	–	–	–	–	–	4-Thia-1-azabicyclo[3.2.0]heptan-7-one, 6,6-dichloro-2-exo-isopropyl-2-endo-methyl-3-exo-(methylthio)-5-phenyl-		NI38803
96	114	156	110	216	128	170	86	**375**	100	56	44	39	38	33	26	16	**0.00**	16	25	9	1	–	–	–	–	–	–	–	–	–	D-Galactitol, 1,2,4,5-tetra-O-acetyl-3-acetamido-3,6-dideoxy-		NI38804
84	102	144	60	98	85	99	140	**375**	100	54	50	43	35	25	21	19	**0.00**	16	25	9	1	–	–	–	–	–	–	–	–	–	D-Galactitol, 1,3,4,5-tetra-O-acetyl-2-acetamido-2,6-dideoxy-	51268-94-1	NI38805
43	115	98	73	128	112	154	170	**375**	100	99	66	42	39	29	24	23	**0.00**	16	25	9	1	–	–	–	–	–	–	–	–	–	α-D-Glucopyranoside, methyl 3,4,6-tri-O-acetyl-2-deoxy-2-(N-methylacetamido)-	93303-54-9	NI38806
73	75	158	189	129	360	229	172	**375**	100	64	24	20	12	8	7	7	**0.00**	16	33	5	1	–	–	–	–	–	2	–	–	–	Glycine, 7-carboxyheptanoyl-, bis(trimethylsilyl) ester	18416-01-8	NI38807
376	378	380	377	379	374	343	342	**375**	100	87	32	25	24	15	10	10	**9.00**	18	12	–	3	–	–	3	–	–	–	–	–	–	2-Amino-5-chlorobenzoquinonedi-4-chloroanil		LI08159
312	343	313	344	282	254	185	141	**375**	100	50	19	11	8	8	8	5	**3.00**	18	17	8	1	–	–	–	–	–	–	–	–	–	Tetramethyl 1-(1,2-dihydro-2-pyridinylidene)cyclopenta-2,4-diene-2,3,4,5-tetracarboxylate		NI38808
213	174	149	187	117	273	231	100	**375**	100	81	37	27	27	25	19	19	**8.61**	18	21	6	3	–	–	–	–	–	–	–	–	–	1,2,3-Butanetriol, 1-(2-phenyl-2H-1,2,3-triazol-4-yl)-, triacetate (ester), [1S-(1R*,2S*,3S*)]-	34297-77-3	NI38809
340	316	375	332	167	280	342	132	**375**	100	64	48	35	30	26	22	22		20	22	4	1	–	–	1	–	–	–	–	–	–	9α-Acetoxy-6β-chloro-6-desoxyindolocodeine		LI08160
340	316	375	43	332	167	341	280	**375**	100	63	60	39	37	30	27	27		20	22	4	1	–	–	1	–	–	–	–	–	–	7a,9c-(Iminoethano)phenanthro[4,5-bcd]furan-8-ol, 5α-chloro-4aα,5,8α,9-tetrahydro-3-methoxy-12-methyl-, acetate (ester)	24735-19-1	NI38810
85	91	260	106	190	155	40	65	**375**	100	97	48	46	40	37	37	24	**0.00**	20	25	4	1	1	–	–	–	–	–	–	–	–	Benzenesulphonamide, 4-methyl-N-[1-phenyl-2-[(tetrahydro-2H-pyran-2-yl)oxy]ethyl]-, (±)-		DD01488
91	81	375	65	194	316	140	272	**375**	100	38	36	21	18	17	14	13		20	25	4	1	1	–	–	–	–	–	–	–	–	Spiro[3,4-dihydro-2H-pyrrole-2,1′-cyclohexane]-3,4-dicarboxylic acid, 5-(benzylthio)-, dimethyl ester		DD01489
229	228	227	226	170	230	169	198	**375**	100	82	47	33	27	20	10	10	**5.60**	20	25	6	1	–	–	–	–	–	–	–	–	–	N,O-Dimethylstephine	53111-18-5	NI38811
43	42	44	70	45	62	313	115	**375**	100	82	81	46	28	27	25	23	**0.00**	20	25	6	1	–	–	–	–	–	–	–	–	–	Morphinan-6-one, 4,5-epoxy-14-hydroxy-8-(2-hydroxyethoxy)-3-methoxy-17-methyl-, (5α,8β)-	55400-10-7	NI38812
138	94	73	108	42	154	137	136	**375**	100	69	47	34	33	31	27	23	**9.51**	20	29	4	1	–	–	–	–	–	1	–	–	–	Benzeneacetic acid, α-[[(trimethylsilyl)oxy]methyl]-, 9-methyl-3-oxa-9-azatricyclo[3.3.1.0^{2,4}]non-7-yl ester, [7(S)-(1α,2β,4β,5α,7β)]-	55373-83-6	NI38813
225	238	227	240	124	207	226	166	**375**	100	70	37	32	31	14	12	12	**0.70**	21	23	2	1	–	1	1	–	–	–	–	–	–	1-Butanone, 4-[4-(4-chlorophenyl)-4-hydroxy-1-piperidinyl]-1-(4-fluorophenyl)-	52-86-8	NI38814
221	235	223	237	123	165	222	95	**375**	100	69	37	32	32	13	11	11	**0.80**	21	23	2	1	–	1	1	–	–	–	–	–	–	1-Butanone, 4-[4-(4-chlorophenyl)-4-hydroxy-1-piperidinyl]-1-(4-fluorophenyl)-	52-86-8	LI08161
224	42	237	123	95	69	56	51	**375**	100	92	80	61	41	37	34	32	**1.97**	21	23	2	1	–	1	1	–	–	–	–	–	–	1-Butanone, 4-[4-(4-chlorophenyl)-4-hydroxy-1-piperidinyl]-1-(4-fluorophenyl)-	52-86-8	NI38815
251	326	325	327	340	341	200	339	**375**	100	61	39	34	32	20	20	18	**0.00**	21	28	2	2	–	–	1	–	–	–	–	–	–	Corynantheol, 10-methoxy-, α-methochloride		NI38816
104	153	238	76	125	127	105	221	**375**	100	84	75	74	29	28	28	26	**3.20**	22	14	3	1	–	–	1	–	–	–	–	–	–	Indano[2,1-d]oxazolidine-2,4-dione, 3-(4-chlorophenyl)-3a-phenyl-	98294-02-1	NI38817
330	331	375	73	59	97	109	95	**375**	100	75	71	43	40	24	23	22		22	17	5	1	–	–	–	–	–	–	–	–	–	1-Hydroxy-4-[4-(2-hydroxyethoxy)anilino]anthraquinone		IC04191
105	123	253	122	165	164	94	131	**375**	100	91	73	73	68	63	50	42	**15.00**	22	21	3	3	–	–	–	–	–	–	–	–	–	Cynodine, N[1]-demethyl-		NI38818
333	375	241	332	334	316	376	318	**375**	100	93	70	43	33	26	23	23		22	21	3	3	–	–	–	–	–	–	–	–	–	1-(4-Methoxyphenyl)-3-methyl-4,5-dihydro-4-(N-acetylanilinovinylmethylene)pyrazol-5-one		IC04192
58	59	57	49	29	91	105	42	**375**	100	5	5	4	4	3	2	2	**0.00**	22	30	2	1	–	–	1	–	–	–	–	–	–	Benzeneethanol, α-[2-(dimethylamino)-1-isopropyl]-α-phenyl-, propanoate (ester), hydrochloride, [S-(R*,S*)]-	1639-60-7	NI38819
58	57	91	59	105	77	115	42	**375**	100	8	7	7	4	4	3	3	**0.00**	22	30	2	1	–	–	1	–	–	–	–	–	–	Benzeneethanol, α-[2-(dimethylamino)-1-isopropyl]-α-phenyl-, propanoate (ester), hydrochloride, [S-(R*,S*)]-	1639-60-7	NI38820
266	209	131	340	267	208	341	210	**375**	100	80	75	67	26	21	17	14	**0.00**	22	30	2	1	–	–	1	–	–	–	–	–	–	Benzeneethanol, α-[2-(dimethylamino)-1-isopropyl]-α-phenyl-, propanoate (ester), hydrochloride, [S-(R*,S*)]-	1639-60-7	NI38821
87	58	289	72	95	42	271	171	**375**	100	26	20	20	19	11	10	10	**9.00**	22	33	4	1	–	–	–	–	–	–	–	–	–	Icaceine	74991-71-2	NI38822
129	254	375	139	137	148	344	285	**375**	100	66	60	51	50	43	37	35		22	37	2	1	–	–	–	–	–	1	–	–	–	Estra-4-en-3-one, 17-(trimethylsiloxy)-, O-methyloxime, (17β)-		NI38823
256	105	258	77	139	257	120	141	**375**	100	55	35	33	30	18	18	9	**0.10**	23	18	2	1	–	–	1	–	–	–	–	–	–	Acetophenone, 2-[5-(4-chlorophenyl)-3-phenyl-2-isoxazolin-5-yl]-	21326-94-3	NI38824
43	375	81	156	114	41	55	86	**375**	100	50	50	42	40	40	33	32		23	37	1	1	1	–	–	–	–	–	–	–	–	Spiro[5α-androstane-3,2′-thiazolidine], 3′-acetyl-	4678-96-0	NI38825
342	357	91	105	41	310	79	42	**375**	100	75	39	32	30	28	28	27	**5.00**	23	37	3	1	–	–	–	–	–	–	–	–	–	3β,17α-Dihydroxy-16β-methylpregn-5-en-20-one, methyl oxime		LI08162
312	357	91	105	41	130	342	42	**375**	100	85	75	60	60	56	55	53	**22.00**	23	37	3	1	–	–	–	–	–	–	–	–	–	3β,17α-Dihydroxypregn-5-ene-20-one, ethyl oxime		LI08163
142	70	288	84	360	143	116	57	**375**	100	27	16	15	14	13	10	9	**7.36**	23	37	3	1	–	–	–	–	–	–	–	–	–	3-Hydroxy-17-[N-methyl(ethoxycarbonyl)amino]androst-5-ene		MS08806
167	294	83	165	168	295	375	376	**375**	100	68	30	25	15	12	3	2		24	25	3	1	–	–	–	–	–	–	–	–	–	2H-1,3-Oxazine-2,4(3H)-dione, 3-cyclohexyl-6-(diphenylmethyl)-5-methyl		MS08807
71	58	316	70	72				**375**	100	68	34	25	21				**0.00**	24	41	2	1	–	–	–	–	–	–	–	–	–	Pregnane-3,20-diol, 18-isopropylideneamino-, (3β,5α,20β)-		LI08164
71	58	316	70	29	72	43	41	**375**	100	67	31	21	13	12	10	10	**4.00**	24	41	2	1	–	–	–	–	–	–	–	–	–	Pregnane-3,20-diol, 19-(isopropylideneamino)-, (3β)-	55493-80-6	NI38826
157	375	158	376	156	55	144	81	**375**	100	49	15	14	6	4	3	3		26	33	1	1	–	–	–	–	–	–	–	–	–	17-Oxo-5α-androstano[3,2-b]-N-methylindole		MS08808
41	91	284	171	57	44	42	105	**375**	100	97	92	57	52	48	43	34	**0.00**	26	33	1	1	–	–	–	–	–	–	–	–	–	Morphinan, 3-methoxy-6-methyl-17-(2-phenylethyl)-, (6α)-	55282-43-4	NI38827

MASS TO CHARGE RATIOS								M.W.	INTENSITIES								Parent	C	H	O	N	S	F	Cl	Br	I	Si	P	B	X	COMPOUND NAME	CAS Reg No	No
127	89	188	190	129	91	270	195	376	100	25	15	6	5	3	2	2	0.00	2	–	–	2	2	10	2	–	–	–	–	–	–	Dichloroethanediylidenebis(nitrilo)bis(sulphur pentafluoride)	2375-46-4	NI38828
66	129	245	243	378	380	128	131	376	100	42	34	33	26	21	18	16	13.00	5	6	1	–	–	–	6	–	–	3	–	–	–	2,4,8-Trisila-3-oxabicyclo[3.2.1]oct-6-ene, 1-methyl-2,2,4,4,8,8-hexachloro	77181-40-9	NI38829
359	361	363	357	246	289	244	210	376	100	95	51	46	46	35	35	31	0.00	8	–	–	–	–	–	8	–	–	–	–	–	–	Benzene, pentachloro(trichlorovinyl)-	29082-74-4	NI38830
310	343	345	380	308	382	378	306	376	100	99	98	97	97	90	84	74	55.46	8	–	–	–	–	–	8	–	–	–	–	–	–	Benzene, pentachloro(trichlorovinyl)-	29082-74-4	NI38831
308	310	380	378	343	345	154	382	376	100	81	79	72	71	68	60	52	27.51	8	–	–	–	–	–	8	–	–	–	–	–	–	Benzene, pentachioro(trichlorovinyl)-	29082-74-4	NI38832
207	165	163	249	205	28	247	161	376	100	92	88	78	75	63	58	56	0.00	9	21	–	–	–	–	–	–	1	–	–	–	1	Tripropyltin iodide		MS08809
313	299	378	234	287	208	248	220	376	100	85	57	30	21	21	3	3	0.00	10	8	–	–	–	–	–	2	–	–	–	–	1	Zirconium bis(π-cyclopentadienyl) dibromide		LI08165
112	56	208	152	180	236	320	292	376	100	99	38	38	31	26	22	10	0.00	11	4	8	–	–	–	–	–	–	–	–	–	2	Heptacarbonyl-μ-[(1-η:2,2α,3-η²)-2-methylene-1-oxo-1,3-propanediyl]diiron	122624-29-7	DD01490
45	29	97	121	65	153	125	27	376	100	86	80	79	52	50	40	40	8.00	11	15	2	–	3	–	2	–	–	–	1	–	–	Phosphorodithioic acid, S-[[(2,5-dichlorophenyl)thio]methyl] O,O-diethyl ester	2275-14-1	NI38833
45	186	121	97	65	214	93	153	376	100	36	36	30	20	19	19	16	5.94	11	15	2	–	3	–	2	–	–	–	1	–	–	Phosphorodithioic acid, S-[[(2,5-dichlorophenyl)thio]methyl] O,O-diethyl ester	2275-14-1	NI38834
177	185	179	157	155	181	213	159	376	100	99	75	17	15	14	10	10	0.00	11	15	2	–	3	–	2	–	–	–	1	–	–	Phosphorodithioic acid, S-[[(2,5-dichlorophenyl)thio]methyl] O,O-diethyl ester	2275-14-1	NI38835
97	125	341	285	109	153	209	65	376	100	44	35	28	23	22	19	19	0.00	11	15	4	–	2	–	2	–	–	–	1	–	–	Chlorothiophos sulphoxide (2,5-isomer)		NI38836
376	374	296	372	378	373	377	265	376	100	51	39	18	17	17	13	11		12	–	–	–	–	8	–	–	–	–	–	–	1	Dibenzoselenophene, octafluoro-	16012-85-4	NI38837
376	375	374	378	372	373	377	188	376	100	60	49	18	18	17	12	12		12	–	–	–	–	8	–	–	–	–	–	–	1	Dibenzoselenophene, octafluoro-	16012-85-4	NI38838
376	296	377	357	265	227	246	345	376	100	19	13	11	6	5	3	2		12	–	–	–	–	8	–	–	–	–	–	–	1	Dibenzoselenophene, octafluoro-	16012-85-4	LI08166
376	374	378	107	91	218	345	343	376	100	92	22	20	18	16	13	12		13	11	2	–	1	–	1	1	–	–	1	–	–	3-Bromo-8-chloro-10-methoxy-10-oxophenothiaphosphine		LI08167
89	289	287	209	286	285	288	101	376	100	33	31	27	25	23	22	21	14.66	13	12	3	–	–	–	–	–	–	–	–	–	2	Acetic acid, [(3-formylbenzo[b]selenophene-2-yl)seleno]-, ethyl ester	58751-61-4	NI38839
225	152	165	224	183	192	193	226	376	100	60	51	46	29	28	24	21	2.00	13	16	9	2	1	–	–	–	–	–	–	–	–	3-Cephem-4-carboxylic acid, 3-(acetoxymethyl)-7-(methoxycarbonylamino)-, 1,1-dioxide, methyl ester	97609-97-7	NI38840
29	303	69	78	90	275	169	28	376	100	43	30	26	24	13	12	12	6.28	14	11	4	–	–	7	–	–	–	–	–	–	–	Benzeneacetic acid, 4-[(heptafluorobutyryl)oxy]-, ethyl ester		NI38841
59	135	91	118	134	226	194	65	376	100	31	25	9	7	6	6	6	2.28	14	16	–	–	6	–	–	–	–	–	–	–	–	2,4,6,8,9,10-Hexathiaadamantane, 1-benzyl-3,5,7-trimethyl-	21404-64-8	NI38842
116	175	203	132	31	217	103	160	376	100	99	77	47	44	35	32	28	4.00	14	16	8	–	2	–	–	–	–	–	–	–	–	1,3-Dithiane-4,6-dicarboxylic acid, 5-phenyl-, 1,1,3,3-tetraoxide, dimethyl ester		NI38843
162	164	70	63	98	43	155	44	376	100	71	40	38	31	29	28	24	3.00	14	18	4	4	–	–	2	–	–	–	–	–	–	L-Arginine, N²-[(2,4-dichlorophenoxy)acetyl]-	50649-04-2	NI38844
115	98	99	103	186	228	73	140	376	100	89	83	35	31	30	22	21	0.00	15	20	11	–	–	–	–	–	–	–	–	–	–	D-Glucopyranose, 1,2,3,4-tetra-O-acetyl-6-O-formyl-	20402-36-2	NI38845
98	103	168	143	158	115	101	145	376	100	63	57	53	43	36	36	30	0.00	15	20	11	–	–	–	–	–	–	–	–	–	–	D-Glucopyranose, 1,2,3,6-tetra-O-acetyl-4-O-formyl-	82008-90-0	NI38846
69	115	86	73	98	157	109	103	376	100	93	92	82	74	59	31	30	0.00	15	20	11	–	–	–	–	–	–	–	–	–	–	D-Glucopyranose, 1,2,4,6-tetra-O-acetyl-3-O-formyl-	28219-86-5	NI38847
98	115	143	103	101	169	99	73	376	100	68	59	46	40	36	33	32	0.00	15	20	11	–	–	–	–	–	–	–	–	–	–	D-Glucopyranose, 1,3,4,6-tetra-O-acetyl-2-O-formyl-	77350-25-5	NI38848
73	169	147	361	231	286	125	243	376	100	36	34	18	14	10	8	7	3.61	15	32	5	–	–	–	–	–	–	3	–	–	–	Hexanedioic acid, 3-oxo-, tris(trimethylsilyl)-	72378-92-8	NI38849
73	169	147	361	75	231	45	286	376	100	44	43	25	24	18	15	13	4.29	15	32	5	–	–	–	–	–	–	3	–	–	–	Hexanedioic acid, 3-oxo-, tris(trimethylsilyl)-	72378-92-8	NI38850
73	169	147	361	75	231	45	286	376	100	44	43	25	24	18	15	13	4.29	15	32	5	–	–	–	–	–	–	3	–	–	–	Hexanedioic acid, [(trimethylsilyl)oxy]-, bis(trimethylsilyl) ester		NI38851
141	231	73	147	55	45	75	258	376	100	95	49	22	15	12	10	6	0.40	15	32	5	–	–	–	–	–	–	3	–	–	–	2-Hexenedioic acid, 2-[(trimethylsilyl)oxy]-, bis(trimethylsilyl) ester	55334-16-2	NI38852
277	73	262	278	45	263	279	28	376	100	79	51	23	16	11	9	6	0.00	15	32	5	–	–	–	–	–	–	3	–	–	–	2-Hexenedioic acid, 2-[(trimethylsilyl)oxy]-, bis(trimethylsilyl) ester	55334-16-2	NI38853
147	73	259	332	117	361	81	133	376	100	72	52	40	34	25	23	18	0.00	15	32	5	–	–	–	–	–	–	3	–	–	–	Propanedioic acid, propenyl[(trimethylsilyl)oxy]-, bis(trimethylsilyl) ester		MS08810
376	243	273	172	361	286	217	197	376	100	81	60	50	45	43	35	32		15	36	3	2	–	–	–	–	–	3	–	–	–	D-Xylo-4,5,6-tris(trimethylsiloxy)-1-methyl-4,5,6,7-tetrahydro-1H-1,2-diazepine		LI08168
95	121	108	81	56	67	139	237	376	100	91	86	74	67	57	22	16	0.00	16	22	2	–	–	–	–	–	–	–	–	–	1	1-Oxaspiro[4.5]decane, 3-[(4-methoxyphenyl)telluro]-		DD01491
43	115	159	103	99	129	116	112	376	100	8	6	5	5	4	4	4	0.00	16	24	10	–	–	–	–	–	–	–	–	–	–	Galactitol, 2-deoxy-, pentaacetate		MS08811
79	52	75	207	71	73	57	44	376	100	74	51	50	27	26	25	24	0.00	16	24	10	–	–	–	–	–	–	–	–	–	–	Galactitol, 1-doexy-, pentaacetate	7226-60-0	NI38854
43	81	157	115	103	98	102	44	376	100	10	6	6	6	6	5	5	0.00	16	24	10	–	–	–	–	–	–	–	–	–	–	Glucose ethylglycoside tetraacetate		MS08812
43	81	98	141	115	102	157	103	376	100	11	7	6	6	6	5	5	0.00	16	24	10	–	–	–	–	–	–	–	–	–	–	Glucose ethylglycoside tetraacetate		MS08813
170	43	128	115	129	157	103	99	376	100	99	79	76	52	39	39	39	0.00	16	24	10	–	–	–	–	–	–	–	–	–	–	L-Mannitol, 6-deoxy-, pentaacetate	7208-41-5	NI38855
115	170	128	129	103	157	99	145	376	100	99	85	63	60	45	44	40	0.00	16	24	10	–	–	–	–	–	–	–	–	–	–	L-Rhamnitol pentaacetate		LI08169
43	115	159	129	103	99	117	112	376	100	8	5	5	5	5	4	4	0.00	16	24	10	–	–	–	–	–	–	–	–	–	–	Sorbitol, 2-deoxy-, pentaacetate		MS08814
43	115	159	129	103	99	44	117	376	100	9	6	6	6	6	5	4	0.00	16	24	10	–	–	–	–	–	–	–	–	–	–	Sorbitol, 6-deoxy-, pentaacetate		MS08815
231	303	116	217	145	345	202	159	376	100	59	54	48	48	43	36	33	0.00	16	24	10	–	–	–	–	–	–	–	–	–	–	4,5,6,7-Tetraacetoxyheptanoic acid, methyl ester		LI08170
116	57	115	99	58	114	174	86	376	100	94	81	81	81	78	59	56	0.00	16	28	8	2	–	–	–	–	–	–	–	–	–	1,4,7,13,16-Pentaoxa-10,19-diazacycloheneicosane-11,14,18-trione, 10,19-dimethyl-	87989-11-5	NI38856
251	56	132	100	71	189	114	128	376	100	97	86	81	75	72	69	62	17.18	16	32	6	4	–	–	–	–	–	–	–	–	–	Bisglycyldiaza-18-crown-6	93767-94-3	NI38857
251	56	132	100	189	114	128	97	376	100	97	86	81	72	69	63	55	17.20	16	32	6	4	–	–	–	–	–	–	–	–	–	Bisglycyldiaza-18-crown-6	93767-94-3	MS08816
103	101	261	259	233	175	177	231	376	100	84	67	63	61	61	60	59	2.97	16	36	–	–	–	–	–	–	–	–	–	–	2	1,4-Digermacyclohexane, 1,1-dibutyl-4,4-diethyl-	56437-95-7	NI38858

MASS TO CHARGE RATIOS								M.W.	INTENSITIES								Parent	C	H	O	N	S	F	Cl	Br	I	Si	P	B	X	COMPOUND NAME	CAS Reg No	No
103	131	101	129	161	133	159	105	376	100	87	86	70	64	59	54	54	4.00	16	36	–	–	–	–	–	–	–	–	–	–	2	Germacyclopentane, 1-butyl-1-[2-(triethylgermyl)ethyl]-	56438-27-8	NI38859
174	73	175	176	128	317	156	59	376	100	40	19	9	9	8	7	7	4.17	16	40	2	2	–	–	–	–	–	3	–	–	–	L-Lysine, tris(trimethylsilyl)-, methyl ester	56273-13-3	NI38860
174	73	317	128	156	59	200	86	376	100	40	10	10	9	9	8	8	5.00	16	40	2	2	–	–	–	–	–	3	–	–	–	L-Lysine, tris(trimethylsilyl)-, methyl ester	56273-13-3	MS08817
174	73	175	317	176	128	156	59	376	100	39	17	9	9	8	6	6	0.00	16	40	2	2	–	–	–	–	–	3	–	–	–	L-Lysine, tris(trimethylsilyl)-, methyl ester	56273-13-3	LI08171
174	73	84	75	214	43	175	142	376	100	91	43	31	18	18	17	17	1.20	16	40	2	2	–	–	–	–	–	3	–	–	–	N,N′-Tris[(trimethylsilyl)oxy]putreanine		MS08818
73	84	75	147	246	102	77	42	376	100	50	42	36	32	28	24	24	2.40	16	40	2	2	–	–	–	–	–	3	–	–	–	N,N′-Tris[(trimethylsilyl)oxy]putreanine		MS08819
174	73	175	59	86	214	172	46	376	100	78	17	15	9	8	7	7	0.00	16	44	–	2	–	–	–	–	–	4	–	–	–	Tetrakis[(trimethylsilyl)oxy]putrescine		MS08820
350	377	378	349	376	348	322	335	376	100	96	88	76	72	72	72	64		17	14	1	2	–	–	1	1	–	–	–	–	–	1,5-Benzodiazocin-2-one, 8-bromo-6-(4-chlorophenyl)-1,2,3,4-tetrahydro-1 methyl-	63563-60-0	NI38861
376	303	269	223	285	194	267		376	100	99	20	5	2	2	1			17	20	6	4	–	–	–	–	–	–	–	–	–	4-Ethoxycarbonyl-3,4-dimethylcyclohex-2-enone, 2,4-dinitrophenylhydrazone		LI08172
376	269	267	303	194	330	285	223	376	100	40	30	25	25	20	20	20		17	20	6	4	–	–	–	–	–	–	–	–	–	6-Ethoxycarbonyl-3,6-dimethylcyclohex-2-enone, 2,4-dinitrophenylhydrazone		LI08173
376	303	269	267	223	194	205		376	100	82	10	5	4	4	2			17	20	6	4	–	–	–	–	–	–	–	–	–	4-Ethoxycarbonyl-3,4,5-trimethylcyclopent-2-enone, 2,4-dinitrophenylhydrazone		LI08174
242	171	43	243	256	44	156	31	376	100	68	60	56	51	43	38	37	2.75	17	20	6	4	–	–	–	–	–	–	–	–	–	Riboflavin	83-88-5	NI38862
242	243	213	170	156	169	43	256	376	100	50	50	50	44	34	33	27	1.08	17	20	6	4	–	–	–	–	–	–	–	–	–	Riboflavin	83-88-5	NI38863
361	376	344	137	121	316	180	153	376	100	99	50	25	12	10	10	10		18	16	9	–	–	–	–	–	–	–	–	–	–	4H-1-Benzopyran-4-one, 2-(3,4-dihydroxyphenyl)-5,7-dihydroxy-3,6,8-trimethoxy-	61451-85-2	NI38864
273	129	73	103	274	361	131	77	376	100	63	46	32	20	10	8	8	3.50	18	24	5	2	–	–	–	–	–	1	–	–	–	2,4,6(1H,3H,5H)-Pyrimidinetrione, 1,3-dimethyl-5-[2-oxo-3-[(trimethylsilyl)oxy]propyl]-5-phenyl-	57346-74-4	NI38865
146	73	361	100	147	75	117	261	376	100	41	28	19	17	16	13	12	2.74	18	28	3	2	–	–	–	–	–	2	–	–	–	2,4,6(1H,3H,5H)-Pyrimidinetrione, 5-ethyl-5-phenyl-1,3-bis(trimethylsilyl)-	52937-73-2	NI38866
146	204	73	100	117	289	147	75	376	100	53	34	22	20	14	14	12	0.50	18	28	3	2	–	–	–	–	–	2	–	–	–	2,4,6(1H,3H,5H)-Pyrimidinetrione, 5-ethyl-5-phenyl-1,3-bis(trimethylsilyl)-	52937-73-2	NI38867
114	144	100	56	301	70	315	132	376	100	73	60	53	33	30	28	23	18.00	18	36	6	2	–	–	–	–	–	–	–	–	–	4,7,13,16,21,24-Hexaoxa-1,10-diazabicyclo[8.8.8]hexacosane		MS08821
273	73	141	147	57	71	43	274	376	100	79	72	40	39	29	28	21	0.00	18	40	4	–	–	–	–	–	–	2	–	–	–	Monononanoin, bis(trimethylsilyl)-		NI38868
28	118	77	292	56	51	320	236	376	100	99	71	48	26	24	21	16	1.59	19	16	3	2	–	–	–	–	–	–	–	–	1	Iron, tricarbonyl[N,N′-(1,2-dimethyl-1,2-ethanediylidene)bis[benzenamine]-N,N′]-	54515-24-1	NI38869
118	28	292	77	320	264	56	236	376	100	90	61	56	28	20	20	18	2.30	19	16	3	2	–	–	–	–	–	–	–	–	1	Iron, tricarbonyl[N,N′-(1,2-dimethyl-1,2-ethanediylidene)bis[benzenamine]-N,N′]-	54515-24-1	NI38870
195	376	194	163	196	165	361	183	376	100	55	23	18	15	13	12	12		19	20	8	–	–	–	–	–	–	–	–	–	–	(3S)-6-Hydroxy-7,8,2′,3′-tetramethoxyisoflavanquinone		NI38871
152	223	124	376	316	180	317	348	376	100	68	53	52	48	47	46	37		19	20	8	–	–	–	–	–	–	–	–	–	–	2-trans-Propyl-3-hydroxy-5-(2-methoxycarbonyl-4,6-dimethoxybenzyl)-4-pyrone		LI08175
58	86	289	291	85	43	288	248	376	100	28	10	7	7	6	5	5	4.69	19	21	2	2	1	–	1	–	–	–	–	–	–	10H-Phenothiazin-2-ol, 8-chloro-10-[3-(dimethylamino)propyl]-, acetate (ester)	20333-81-7	NI38872
58	86	289	291	376	288	85	43	376	100	27	12	9	8	7	6	6		19	21	2	2	1	–	1	–	–	–	–	–	–	10H-Phenothiazin-2-ol, 8-chloro-10-[3-(dimethylamino)propyl]-, acetate (ester)	20333-81-7	NI38873
58	86	289	291	85	376	248	43	376	100	27	10	7	6	5	4	4		19	21	2	2	1	–	1	–	–	–	–	–	–	10H-Phenothiazin-2-ol, 8-chloro-10-[3-(dimethylamino)propyl]-, acetate (ester)	20333-81-7	LI08176
58	86	376	378	43	377	87	42	376	100	84	33	14	14	9	9	9		19	21	2	2	1	–	1	–	–	–	–	–	–	10H-Phenothiazin-3-ol, 2-chloro-10-[3-(dimethylamino)propyl]-, acetate (ester)	20333-80-6	NI38874
58	86	376	43	378	87	341	377	376	100	63	31	15	12	9	7	6		19	21	2	2	1	–	1	–	–	–	–	–	–	10H-Phenothiazin-3-ol, 2-chloro-10-[3-(dimethylamino)propyl]-, acetate (ester)	20333-80-6	LI08177
58	86	376	43	378	42	87	59	376	100	84	31	16	12	10	9	7		19	21	2	2	1	–	1	–	–	–	–	–	–	10H-Phenothiazin-3-ol, 2-chloro-10-[3-(dimethylamino)propyl]-, acetate (ester)	20333-80-6	NI38875
58	86	376	85	378	289	291	43	376	100	47	45	21	17	12	10	10		19	21	2	2	1	–	1	–	–	–	–	–	–	10H-Phenothiazin-4-ol, 8-chloro-10-[3-(dimethylamino)propyl]-, acetate (ester)	56438-24-5	NI38876
58	86	376	378	43	289	42	60	376	100	39	30	10	8	5	5	4		19	21	2	2	1	–	1	–	–	–	–	–	–	Phenothiazin-3-ol, 8-chloro-10-[3-(dimethylamino)propyl]-, acetate (ester)	14734-77-1	NI38877
43	163	70	42	292	131	113	112	376	100	93	91	77	76	74	68	66	3.86	19	24	6	2	–	–	–	–	–	–	–	–	–	N′,O,O′-Triacetyl-N-caffeoylputrescine		NI38878
376	347	319	348	377	361	73	192	376	100	85	45	32	25	23	21	15		19	28	4	2	–	–	–	–	–	1	–	–	–	2,4,6(1H,3H,5H)-Pyrimidinetrione, 1,3,5-triethyl-5-[4-[(trimethylsilyl)oxy]phenyl]-	55282-38-7	NI38879
232	73	233	376	45	144	102	75	376	100	64	21	10	8	6	6	6		19	32	2	2	–	–	–	–	–	2	–	–	–	1-Trimethylsilyl-3-[N,N-acetyl(trimethylsilyl)aminoethyl]-5-methoxyindole		NI38880
174	73	175	176	86	59	45	361	376	100	69	29	13	12	7	7	6	1.70	19	36	–	2	–	–	–	–	–	3	–	–	–	1H-Indole-3-ethanamine, N,N,1-tris(trimethylsilyl)-	55334-17-3	NI38881
130	189	131	115	103	77	143	144	376	100	25	11	7	7	7	6	4	0.00	19	36	–	2	–	–	–	–	–	3	–	–	–	1H-Indole-3-ethanamine, N,N,1-tris(trimethylsilyl)-	55334-17-3	NI38882

MASS TO CHARGE RATIOS	M.W.	INTENSITIES	Parent	C	H	O	N	S	F	Cl	Br	I	Si	P	B	X	COMPOUND NAME	CAS Reg No	No
174 73 130 86 175 59 176 45	376	100 62 25 19 17 14 9 9	0.00	19	36	–	2	–	–	–	–	–	3	–	–	–	1H-Indole-3-ethanamine, N,N,1-tris(trimethylsilyl)-	55334-17-3	MS08822
376 151 248 135 247 106 185 112	376	100 92 58 31 27 27 25 25		20	24	7	–	–	–	–	–	–	–	–	–	–	Ailanthone		LI08178
167 109 121 195 210 137 181 168	376	100 43 36 31 28 23 17 17	16.00	20	24	7	–	–	–	–	–	–	–	–	–	–	2H-1-Benzopyran-3-ol, 3,4-dihydro-5,7-dimethoxy-2-(3,4,5-trimethoxyphenyl)-, (2R-cis)-	3143-37-1	NI38883
137 138 376 163 193 149 57 55	376	100 41 30 15 13 13 13 11		20	24	7	–	–	–	–	–	–	–	–	–	–	2,3-Furandiol, tetrahydro-3,4-bis[(4-hydroxy-3-methoxyphenyl)methyl]-	87402-76-4	NI38884
195 196 255 181 121 182 180 166	376	100 90 80 43 43 32 31 19	1.00	20	24	7	–	–	–	–	–	–	–	–	–	–	1-(2-Hydroxy-4,6-dimethoxyphenyl)-2,2-dimethoxy-3-(4-methoxyphenyl)propan-1-one		NI38885
161 57 43 85 217 143 189 133	376	100 40 40 22 10 8 4 4	1.00	20	24	7	–	–	–	–	–	–	–	–	–	–	Lobatin B	84754-02-9	NI38886
138 137 153 376 151 123 43 163	376	100 70 38 18 17 9 9 8		20	24	7	–	–	–	–	–	–	–	–	–	–	Olivil, (–)-	2955-23-9	NI38887
236 57 41 73 55 43 235 69	376	100 98 50 44 41 41 38 36	0.00	20	28	5	2	–	–	–	–	–	–	–	–	–	Butanoic acid, 4-cyano-2-nitro-, 2,6-di-tert-butyl-4-methoxyphenyl ester		DD01492
376 172 130 187 229 131 43 77	376	100 84 79 69 68 59 43 25		21	16	5	2	–	–	–	–	–	–	–	–	–	N-Phthaloyl-1-acetyltryptophan		NI38888
43 376 222 235 316 378 209 207	376	100 64 41 40 36 22 22 21		21	25	4	–	–	–	1	–	–	–	–	–	–	3-Chloro-17β-acetoxy-3,5-androstadiene-2,7-dione		MS08823
59 205 43 300 360 220 344 328	376	100 99 95 90 35 35 30 30	30.00	21	28	6	–	–	–	–	–	–	–	–	–	–	17,19-Bis(abeo)-6β,7α-dihydroxy-16ξ-methoxyroyleanone		NI38889
300 302 285 284 269 205 268 257	376	100 50 40 40 40 40 35 30	0.00	21	28	6	–	–	–	–	–	–	–	–	–	–	17,19-Bis(abeo)-6β,16ξ-dihydroxy-7α-methoxyroyleanone		NI38890
99 344 100 256 258 376 150 137	376	100 38 28 18 13 12 12 12		21	28	6	–	–	–	–	–	–	–	–	–	–	1H-Indene-3a-carboxylic acid, 1,1-ethylenedioxy-5-(2-hydroxy-4-methoxyphenyl)-7aβ-methyl-2,3,3aα,4,5,6,7,7a-octahydro-, methyl ester		MS08824
330 376 312 43 137 57 28 91	376	100 75 47 46 38 31 31 29		21	28	6	–	–	–	–	–	–	–	–	–	–	Vincetogenin		LI08179
86 128 120 178 43 91 44 129	376	100 84 48 43 37 28 26 25	6.77	21	32	4	2	–	–	–	–	–	–	–	–	–	N-(N-Acetylleucyl)phenylalanine butyl ester		AI02284
358 330 106 77 134 135 302 163	376	100 35 16 14 11 10 9 3	0.00	22	16	6	–	–	–	–	–	–	–	–	–	–	5,12-Methano-5H-benzo[4,5]cyclohepta[1,2-b]naphthalene-6,11,13(12H)-trione, 1,5,7-trihydroxy-3,9-dimethyl-, (±)-	60548-82-5	NI38891
78 262 77 183 79 108 114 69	376	100 20 14 9 8 6 3 3	0.00	22	21	2	2	–	–	–	–	–	–	1	–	–	Acetic acid, [(triphenylphosphoranylidene)hydrazono]-, ethyl ester	22610-15-7	NI38892
91 41 148 93 81 107 135 105	376	100 80 72 67 62 61 59 58	1.00	22	32	3	–	1	–	–	–	–	–	–	–	–	1H-Cyclopenta[a]pentalen-7-ol, decahydro-3,3,4,7a-tetramethyl-, 4-methylbenzenesulphonate	58003-47-7	NI38893
361 43 362 343 32 149 41 55	376	100 47 21 14 14 11 11 10	4.20	22	32	5	–	–	–	–	–	–	–	–	–	–	Androst-7-en-6-one, 2,3,14,17-tetrahydroxy-, (2β,3β,5β,14α,17β)-		MS08825
43 275 121 303 159 152 105 359	376	100 66 62 61 58 53 53 52	27.20	22	32	5	–	–	–	–	–	–	–	–	–	–	Araucarolone monoacetate		NI38894
91 41 283 344 79 55 43 105	376	100 87 77 69 66 62 56 53	2.00	22	32	5	–	–	–	–	–	–	–	–	–	–	Gibberellin A_{14} methyl ester		MS08826
344 91 41 109 43 55 79 107	376	100 85 84 74 68 67 61 60	11.30	22	32	5	–	–	–	–	–	–	–	–	–	–	1-Phenanthrenecarboxylic acid, tetradecahydro-7-(2-methoxy-2-oxoethylidene)-1,4a,8-trimethyl-9-oxo-, methyl ester, [1S-(1α,4aα,4bβ,8β,8aα,10aβ)]-	57289-51-7	NI38895
129 361 359 376 320 362 130 286	376	100 43 24 23 21 14 14 11		22	36	3	–	–	–	–	–	–	1	–	–	–	Androstane-3,6-dione, 17-[(trimethylsilyl)oxy]-, (5α,17β)-		NI38896
129 75 73 376 41 79 361 286	376	100 84 84 42 33 31 27 27		22	36	3	–	–	–	–	–	–	1	–	–	–	Androstane-11,17-dione, 3-[(trimethylsilyl)oxy]-, (3α,5α)-	25876-74-8	NI38897
271 232 73 75 286 41 79 55	376	100 83 77 70 54 34 28 28	18.02	22	36	3	–	–	–	–	–	–	1	–	–	–	Androstane-11,17-dione, 3-[(trimethylsilyl)oxy]-, (3α,5β)-	5042-90-0	NI38898
271 232 73 286 75 93 272 79	376	100 76 55 54 49 23 22 22	17.00	22	36	3	–	–	–	–	–	–	1	–	–	–	Androstane-11,17-dione, 3-[(trimethylsilyl)oxy]-, (3α,5β)-	5042-90-0	NI38899
271 232 286 272 376 233 191 287	376	100 77 60 20 14 14 14 13		22	36	3	–	–	–	–	–	–	1	–	–	–	Androstane-11,17-dione, 3-[(trimethylsilyl)oxy]-, (3α,5β)-	5042-90-0	NI38900
346 103 145 73 244 304 233 169	376	100 96 60 47 43 35 30 30	10.00	22	36	3	–	–	–	–	–	–	1	–	–	–	Androstanedione, 18-[(trimethylsilyl)oxy]-, (5α)-		LI08180
209 376 179 277 223 210 361 346	376	100 80 22 20 20 18 15 8		22	36	3	–	–	–	–	–	–	1	–	–	–	4-Dodecen-3-one, 1-[3-methoxy-4-[(trimethylsilyl)oxy]phenyl]-	56700-95-9	MS08827
358 107 105 340 343 147 133 119	376	100 94 79 77 73 61 61 60	6.00	22	37	4	–	–	–	–	–	–	–	–	1	–	5α-Pregnane-3α,11β,20α,21-tetrol methyl boronate		LI08181
43 55 41 81 85 79 67 358	376	100 94 89 88 87 79 78 70	4.00	22	37	4	–	–	–	–	–	–	–	–	1	–	5α-Pregnane-3α,11β,20α,21-tetrol methyl boronate		MS08828
107 105 358 119 147 325 133 340	376	100 95 74 65 59 58 55 53	2.00	22	37	4	–	–	–	–	–	–	–	–	1	–	5α-Pregnane-3α,11β,20β,21-tetrol methyl boronate	30882-72-5	NI38901
124 43 55 41 79 81 125 111	376	100 97 81 75 65 61 60 60	1.00	22	37	4	–	–	–	–	–	–	–	–	1	–	5β-Pregnane-3α,11β,17α,20β-tetrol methyl boronate		MS08829
125 124 111 105 107 178 213 358	376	100 60 60 58 53 52 50 48	1.00	22	37	4	–	–	–	–	–	–	–	–	1	–	5β-Pregnane-3α,11β,17α,20β-tetrol methyl boronate		LI08182
376 135 253 161 361 343 107 267	376	100 71 55 42 38 34 31 30		23	20	5	–	–	–	–	–	–	–	–	–	–	4-Methyl-2,6-bis(2-hydroxy-5-methylbenzoyl)phenol		LI08183
181 271 105 358 77 182 91 272	376	100 48 17 14 14 11 10 9	6.60	23	20	5	–	–	–	–	–	–	–	–	–	–	2-Phenyl-3-hydroxy-3-phenyl-4-oxo-5,7-dimethoxy-2,3,4-trihydrobenzopyrylium		MS08830
181 271 105 358 77 182 193 91	376	100 48 18 15 14 12 10 10	6.60	23	20	5	–	–	–	–	–	–	–	–	–	–	2-Phenyl-3-hydroxy-3-phenyl-4-oxo-5,7-dimethoxy-2,3,4-trihydrobenzopyrylium		IC04193
57 55 319 69 71 83 68 376	376	100 78 68 62 58 46 44 41		23	24	3	2	–	–	–	–	–	–	–	–	–	3-Pyrrolin-2-one, 4-ethyl-3-methyl-5-[5-(4-acetoxyphenyl-2-methylene)-3,4-dimethyl-5H-pyrrolyl-2-methylene]-, (Z,Z)-		MS08831
288 55 306 41 81 67 43 79	376	100 71 50 45 34 34 32 31	8.00	23	36	4	–	–	–	–	–	–	–	–	–	–	5,17-Dioxo-6α-pentyl-4-nor-3,5-secoandrostan-3-oic acid		MS08832
221 376 305 43 287 222 55 41	376	100 54 51 37 34 33 33 30		23	36	4	–	–	–	–	–	–	–	–	–	–	6β-Hydroxy-6α-pentyl-4-oxa-5β-androstane-3,17-dione		MS08833
135 301 107 55 203 114 93 81	376	100 40 40 38 30 30 30 30	4.00	23	36	4	–	–	–	–	–	–	–	–	–	–	Labda-8(17),13-dien-15-oic acid, 3β-acetoxy-, methyl ester, (13E)-		MS08834
376 343 301 361 358	376	100 30 15 8 5		23	36	4	–	–	–	–	–	–	–	–	–	–	9-Phenanthrenol, 4b,5,6,7,8,8a,9,10-octahydro-1,3,4-trimethoxy-4b,8,8-trimethyl-2-isopropyl-, [4bS-(4bα,8aβ,9α)]-	60371-73-5	NI38902
215 330 230 284 269 161 315 175	376	100 63 49 46 43 29 23 16	0.10	23	36	4	–	–	–	–	–	–	–	–	–	–	Pregnane, 3,20-diformyl-, (3α,5β,20α)-		NI38903
255 316 298 333 283 249 231 273	376	100 80 63 37 27 23 20 17	3.00	23	36	4	–	–	–	–	–	–	–	–	–	–	Pregnan-20-one, 12-(acetoxy)-3-hydroxy-, (3α,5β,12α)-		NI38904

MASS TO CHARGE RATIOS								M.W.	INTENSITIES								Parent	C	H	O	N	S	F	Cl	Br	I	Si	P	B	X	COMPOUND NAME	CAS Reg No	No
43	97	290	93	98	95	181	55	**376**	100	99	81	73	69	69	65	65	**4.50**	23	36	4	–	–	–	–	–	–	–	–	–	–	Pregnan-20-one, 3-(acetyloxy)-14-hydroxy-, (3β,5β,14β)-	10005-75-1	NI38905
91	79	93	55	107	298	81	67	**376**	100	91	88	83	80	76	72	72	**10.61**	23	36	4	–	–	–	–	–	–	–	–	–	–	Pregnan-20-one, 3-(acetyloxy)-17-hydroxy-, (3β,5α)-	5456-44-0	WI01493
257	303	43	107	81	95	55	93	**376**	100	89	60	38	37	29	29	27	**2.53**	23	36	4	–	–	–	–	–	–	–	–	–	–	Pregnan-20-one, 21-(acetyloxy)-3-hydroxy-, (5α)-	56193-58-9	NI38906
129	271	286	229	213	253	119	268	**376**	100	60	56	46	46	35	34	31	**7.00**	23	40	2	–	–	–	–	–	–	1	–	–	–	Androst-5-en-17-ol, 17-methyl-3-[(trimethylsilyl)oxy]-, (3β,17β)-	57397-40-7	NI38907
137	95	109	108	87	136	81	121	**376**	100	97	72	66	53	48	42	36	**11.81**	23	40	2	–	–	–	–	–	–	1	–	–	–	Silane, vinylmethylbis[(1,7,7-trimethylbicyclo[2.2.1]hept-2-yl)oxy]-	74806-98-7	NI38908
44	87	86	56	260	82	346	70	**376**	100	70	62	48	47	32	20	20	**1.00**	23	40	2	2	–	–	–	–	–	–	–	–	–	Acetamide, N-[(3β)-3-amino-18-hydroxypregnan-20-yl]-	55515-16-7	NI38909
287	376	57	286	375	377	288	231	**376**	100	94	94	25	24	23	20	20		23	41	3	–	–	–	–	–	–	–	–	1	–	Boronic acid, 2,6-di-tert-butyl-4-methylphenyl-, O,O-dibutyl-		IC04194
129	188	128	84	344	376	242	377	**376**	100	37	27	19	8	7	4	2		24	24	4	–	–	–	–	–	–	–	–	–	–	1-Cyclohexene-1,4-dicarboxylic acid, 3-phenyl-4-(2-phenylethenyl)-, dimethyl ester		MS08835
91	376	361	253	226	377	240	347	**376**	100	99	46	29	29	27	24	23		24	28	2	2	–	–	–	–	–	–	–	–	–	(3-Ethyl-4-methyl-5-benzyloxycarbonyl-2-pyrrolyl)(3,5-dimethyl-4-ethyl-2 pyrrolylidene)methane		LI08184
246	43	130	42	247	186	144	91	**376**	100	48	21	19	16	15	13	11	**1.00**	24	28	2	2	–	–	–	–	–	–	–	–	–	4-(α-Hydroxybenzyl)-1-[2-(indol-3-yl)ethyl]piperidine		LI08185
146	133	376	132	106	55	41	120	**376**	100	64	42	27	14	14	14	13		24	28	2	2	–	–	–	–	–	–	–	–	–	1,8-Bis(2-oxo-3-indolinyl)octane		LI08186
84	85	57	56	375	44	292	42	**376**	100	52	34	34	31	31	25	24	**18.00**	24	28	2	2	–	–	–	–	–	–	–	–	–	Piperidine, 1,1′-([1,1′-biphenyl]-4,4′-diyldicarbonyl)bis-	52882-87-8	NI38910
91	79	93	55	107	298	81	67	**376**	100	91	88	83	80	76	72	72	**10.60**	24	40	3	–	–	–	–	–	–	–	–	–	–	3β-Acetyloxy-17α-hydroxy-5α-pregnan-20-one		AM00152
358	215	55	81	41	67	95	93	**376**	100	97	94	72	62	61	58	55	**16.00**	24	40	3	–	–	–	–	–	–	–	–	–	–	Cholan-24-oic acid, 3-hydroxy-, (3α,5β)-	434-13-9	NI38911
253	271	358	226	55	254	41	81	**376**	100	68	38	31	23	22	22	17	**0.00**	24	40	3	–	–	–	–	–	–	–	–	–	–	Chol-7-ene-3,12,24-triol, (3α,5β,12α)-	54411-71-1	NI38912
305	43	306	55	41	269	81	93	**376**	100	23	22	19	16	13	12	11	**0.00**	24	40	3	–	–	–	–	–	–	–	–	–	–	6β,17β-Dihydroxy-6α-pentyl-5α-androstan-3-one		MS08836
289	271	290	291	175	376	273	55	**376**	100	34	30	19	18	14	14	14		24	40	3	–	–	–	–	–	–	–	–	–	–	17β-Hydroxy-6-pentyl-4-nor-3,5-seco-5-androsten-3-oic acid methyl ester		MS08837
57	83	97	123	115	149	165	318	**376**	100	58	38	17	14	11	4	3	**1.00**	24	40	3	–	–	–	–	–	–	–	–	–	–	13,17-Octadecadien-15-yne, 18-[2,3-(isopropylidenedioxy)propoxy]-		MS08838
289	271	290	55	41	81	272	67	**376**	100	74	43	42	35	34	31	30	**19.00**	24	40	3	–	–	–	–	–	–	–	–	–	–	17-Oxo-6α-pentyl-4-nor-3,5-secoandrostan-3-oic acid, methyl ester		MS08839
289	271	55	81	41	67	95	93	**376**	100	64	51	45	39	38	33	29	**15.00**	24	40	3	–	–	–	–	–	–	–	–	–	–	17-Oxo-6β-pentyl-4-nor-3,5-secoandrostan-3-oic acid methyl ester		MS08840
141	265	113	283	149	264	139	235	**376**	100	29	29	20	14	12	10	8	**0.00**	24	40	3	–	–	–	–	–	–	–	–	–	–	Stearic acid, 8-hydroxy-, phenyl ester	88095-48-1	NI38913
155	127	283	265	156	249	284	264	**376**	100	31	24	24	12	9	8	8	**0.00**	24	40	3	–	–	–	–	–	–	–	–	–	–	Stearic acid, 9-hydroxy-, phenyl ester	88095-49-2	NI38914
169	265	141	283	264	149	263	170	**376**	100	42	37	24	16	13	11	10	**0.00**	24	40	3	–	–	–	–	–	–	–	–	–	–	Stearic acid, 10-hydroxy-, phenyl ester	88095-50-5	NI38915
183	265	155	283	264	149	165	277	**376**	100	80	50	42	28	24	18	15	**0.00**	24	40	3	–	–	–	–	–	–	–	–	–	–	Stearic acid, 11-hydroxy-, phenyl ester		NI38916
197	169	265	283	198	161	264	179	**376**	100	75	68	45	16	16	14	14	**0.00**	24	40	3	–	–	–	–	–	–	–	–	–	–	Stearic acid, 12-hydroxy-, phenyl ester	88095-53-8	NI38917
211	283	183	265	225	175	193	264	**376**	100	74	73	63	21	20	18	16	**0.00**	24	40	3	–	–	–	–	–	–	–	–	–	–	Stearic acid, 13-hydroxy-, phenyl ester	88095-52-7	NI38918
225	283	265	197	226	264	319	284	**376**	100	65	51	50	23	18	15	12	**0.00**	24	40	3	–	–	–	–	–	–	–	–	–	–	Stearic acid, 14-hydroxy-, phenyl ester	88095-51-6	NI38919
247	95	75	246	81	109	73	137	**376**	100	58	36	35	33	30	30	28	**2.00**	24	44	1	–	–	–	–	–	–	1	–	–	–	Androstane, 4,4-dimethyl-3-[(trimethylsilyl)oxy]-, (3β,5α)-	54412-05-4	NI38920
77	165	271	166	376	105	51	194	**376**	100	56	50	44	40	27	20	17		25	20	–	4	–	–	–	–	–	–	–	–	–	4,4′-Bis(benzeneazo)diphenylmethane		MS08841
105	376	77	107	93	91	79	67	**376**	100	18	16	2	1	1	1	1		25	28	3	–	–	–	–	–	–	–	–	–	–	Estra-1,3,5(10)-triene-3,17-diol (17β)-, 3-benzoate	50-50-0	NI38921
376	138	276	174	188	377	333	277	**376**	100	57	51	42	35	32	15	13		26	16	3	–	–	–	–	–	–	–	–	–	–	7,12-Dioxo-9-methoxy-10-methyl-7,12-dihydronaphtho[2,1,8-qra]naphthacene		LI08187
376	116	91	185	117	213	105	115	**376**	100	95	61	44	37	36	29	28		26	32	2	–	–	–	–	–	–	–	–	–	–	Androst-5-en-17-one, 3-hydroxy-16-benzylene-	74708-33-1	NI38922
69	225	376	81	239	187	121	93	**376**	100	65	42	39	36	21	19	15		26	32	2	–	–	–	–	–	–	–	–	–	–	1,4-Naphthalenedione, 2-methyl-3-(3,7,11-trimethyl-2,6,10-dodecatrienyl)-, (E,E)-	860-25-3	NI38923
249	132	105	250	133	117	79	77	**376**	100	57	21	19	14	13	9	6	**0.00**	26	36	–	2	–	–	–	–	–	–	–	–	–	Aniline, N,N′-1,2-ethanediylidenebis[2,6-diisopropyl-	74663-75-5	NI38924
191	192	141	189	178	179	165	115	**376**	100	29	26	19	16	15	14	12	**10.00**	27	20	2	–	–	–	–	–	–	–	–	–	–	9-Phenanthrenemethyl naphthalene-2-acetate	93091-04-4	NI38925
376	95	377	201	67	296	41	295	**376**	100	63	29	26	23	13	9	8		27	36	1	–	–	–	–	–	–	–	–	–	–	2,4,6-Trinorborn-2-ylphenol		MS08842
140	300	113	63	302	298	50	141	**377**	100	48	32	32	25	23	18	18	**0.00**	10	6	–	1	–	–	–	3	–	–	–	–	–	Quinoline, 2-(tribromomethyl)-	613-53-6	NI38926
151	164	377	165	69	152	149	28	**377**	100	53	23	12	12	10	8	8		14	14	3	1	–	7	–	–	–	–	–	–	–	2-(3,4-Dimethoxyphenyl)-N-(heptafluorobutyryl)ethylamine		MS08843
248	43	202	230	85	276	41	231	**377**	100	90	80	71	61	52	31	18	**0.00**	14	20	5	1	–	5	–	–	–	–	–	–	–	D-Glutamic acid, N-(pentafluoropropionyl)-, isopropyl ester		MS08844
248	43	202	230	85	276	41	249	**377**	100	89	79	71	62	55	32	18	**0.00**	14	20	5	1	–	5	–	–	–	–	–	–	–	L-Glutamic acid, N-(pentafluoropropionyl)-, isopropyl ester		MS08845
73	147	362	377	74	260	170	149	**377**	100	36	13	10	9	7	6	6		14	31	5	1	–	–	–	–	–	3	–	–	–	Gluconic acid, 2-(hydroxyimino)-, tris(trimethylsilyl)-		NI38927
89	351	191	379	63	95.5	349	190	**377**	100	96	75	70	65	49	46	44	**0.00**	15	9	1	1	–	–	–	2	–	–	–	–	–	3,1-Benzoxazepine, 7-bromo-2-(4-bromophenyl)-	19062-92-1	LI08188
89	351	191	379	63	349	353	190	**377**	100	97	76	71	67	46	44	44	**34.72**	15	9	1	1	–	–	–	2	–	–	–	–	–	3,1-Benzoxazepine, 7-bromo-2-(4-bromophenyl)-	19062-92-1	NI38928
108	377	78	43	55	42	41	39	**377**	100	50	21	19	17	17	17	17		15	15	5	5	1	–	–	–	–	–	–	–	–	3-Cyano-2,6-dihydroxy-5-(3-dimethylsulphamoyloxyphenylazo)-4-methylpyridine		IC04195
343	184	345	258	260	344	318	320	**377**	100	85	45	35	24	22	22	14	**7.30**	15	17	6	1	–	–	2	–	–	–	–	–	–	L-Glutamic acid, N-[(2,4-dichlorophenoxy)acetyl]-, dimethyl ester	56805-15-3	MS08846
116	143	175	101	84	56	59	144	**377**	100	87	53	35	35	28	26	20	**0.00**	15	23	10	1	–	–	–	–	–	–	–	–	–	β-Citryl-L-glutamic acid, tetramethyl ester		MS08847
73	260	75	45	147	98	74	261	**377**	100	50	21	19	11	11	11	10	**0.00**	15	35	4	1	–	–	–	–	–	3	–	–	–	Bis(trimethylsilyl) 2-methyl-N-(trimethylsilyl)glutamate		MS08848
77	75	73	260	170	98	147	47	**377**	100	99	95	54	44	35	34	31	**2.10**	15	35	4	1	–	–	–	–	–	3	–	–	–	Bis(trimethylsilyl) 2-(trimethylsilylamino)adipate		MS08849
73	260	55	75	128	45	217	147	**377**	100	52	30	28	19	19	15	15	**2.70**	15	35	4	1	–	–	–	–	–	3	–	–	–	Bis(trimethylsilyl) 2-(trimethylsilylamino)adipate		MS08850

MASS TO CHARGE RATIOS								M.W.	INTENSITIES								Parent	C	H	O	N	S	F	Cl	Br	I	Si	P	B	X	COMPOUND NAME	CAS Reg No	No
304	43	87	377	147	305	220	118	**377**	100	52	50	34	27	17	13	12		16	19	4	5	1	–	–	–	–	–	–	–	–	N-(β-Acetoxyethyl)-N-ethyl-3-methyl-4-(5-nitrothiazol-2-ylazo)aniline		IC04196
73	103	233	75	117	247	157	159	**377**	100	35	27	16	14	13	12	7	**0.00**	16	35	5	1	–	–	–	–	–	2	–	–	–	β-Alanine, N-[3,3-dimethyl-1-oxo-2,4-bis[(trimethylsilyl)oxy]butyl]-, methyl ester, (R)-	55591-03-2	NI38929
73	103	233	75	117	247	157	159	**377**	100	99	77	47	39	37	33	21	**0.09**	16	35	5	1	–	–	–	–	–	2	–	–	–	β-Alanine, N-[3,3-dimethyl-1-oxo-2,4-bis[(trimethylsilyl)oxy]butyl]-, methyl ester, (R)-	55591-03-2	NI38930
96	95	84	70	235	41	82	81	**377**	100	93	92	83	58	55	53	36	**4.50**	17	23	5	5	–	–	–	–	–	–	–	–	–	L-Proline, 1-[N-(5-oxo-L-prolyl)-L-histidyl]-, methyl ester	22365-00-0	NI38931
377	201	348	230	229	245	216	184	**377**	100	48	30	23	16	12	12	12		18	19	6	1	1	–	–	–	–	–	–	–	–	5H-Furo[3,2-g][1]benzopyran-5-one, 4-methoxy-7-methyl-9-[(piperidin-1-yl)sulphonyl]-		MS08851
313	230	28	41	228	377	115	284	**377**	100	80	66	64	61	42	37	31		18	19	6	1	1	–	–	–	–	–	–	–	–	2-Pyrrolidinone, 1-(5,6-dihydro-8,9-dimethoxynaphth[2,1-c][1,2]oxathiin-2-yl)-, S,S-dioxide	40535-15-7	MS08852
93	121	377	122	119	152	76	109	**377**	100	83	53	35	31	24	22	20		19	15	4	5	–	–	–	–	–	–	–	–	–	2-(4-Hydroxyphenylazo)-5-(4-nitrophenylazo)anisole		IC04197
43	163	290	166	230	208	164	174	**377**	100	83	76	76	57	50	45	29	**10.00**	20	27	6	1	–	–	–	–	–	–	–	–	–	8,13-Diacetyl-9-oxoserratinine		LI08189
245	213	244	246	230	377	243	196	**377**	100	25	21	19	17	13	9	7		20	27	6	1	–	–	–	–	–	–	–	–	–	Hasubanan-6,8-diol, 8,10-epoxy-3,4,7-trimethoxy-17-methyl-, (6β,7β,8β,10β)-	52309-77-0	NI38932
73	58	98	96	268	84	83	298	**377**	100	37	8	8	7	7	7	4	**0.00**	20	44	–	1	–	–	–	1	–	–	–	–	–	Dimethylethylcetylammonium bromide	124-03-8	NI38933
307	250	118	77	377	149	121	91	**377**	100	77	61	49	29	16	14	10		21	19	2	3	1	–	–	–	–	–	–	–	–	Pyrrolo[2,3-d][1,3]thiazin-5,6-dione, 1,2,7,7a-tetrahydro-1,7-dimethyl-2-(methylimino)-4,7a-diphenyl-	90140-43-5	NI38934
104	77	258	212	184	250	119	278	**377**	100	99	84	81	65	27	20	15	**1.00**	21	19	4	3	–	–	–	–	–	–	–	–	–	2,6,7-Triazabicyclo[2.2.2]oct-2-ene-3-carboxylic acid, 4-methyl-5,8-dioxo-6,7-diphenyl-, ethyl ester		NI38935
209	73	210	179	377	224	211	41	**377**	100	54	37	36	33	15	14	13		21	35	3	1	–	–	–	–	–	1	–	–	–	6-Nonenamide, N-[[3-methoxy-4-[(trimethylsilyl)oxy]phenyl]methyl]-8-methyl-, (E)-	56784-17-9	MS08853
335	334	377	171	376	349	276	246	**377**	100	81	48	43	31	21	19	19		22	23	3	3	–	–	–	–	–	–	–	–	–	Acetyladhatodine		LI08190
203	234	377	219	178	128	218	204	**377**	100	71	41	40	39	33	20	18		22	35	4	1	–	–	–	–	–	–	–	–	–	2,6-Di-tert-butyl-4-[[[5-(methoxycarbonyl)pentyl]amino]methyl]phenol		IC04198
128	170	115	129	197	183	127	198	**377**	100	89	89	80	78	52	50	48	**24.00**	24	19	–	5	–	–	–	–	–	–	–	–	–	11-Azatricyclo[12.2.2.2^{5,8}]eicosa-5,7,11,14,16,17,19-heptaene-2,2,13,13-tetracarbonitrile, 12-methyl-	66389-00-2	NI38936
94	91	230	42	103	95	82	77	**377**	100	24	23	18	14	11	10	9	**8.33**	24	27	3	1	–	–	–	–	–	–	–	–	–	2-Propenoic acid, 3-phenyl-, 6-hydroxy-8-methyl-2-benzyl-8-azabicyclo[3.2.1]oct-3-yl ester	55925-24-1	NI38937
377	376	266	265	198	378	158	267	**377**	100	65	50	32	30	23	20	19		24	42	–	1	–	–	–	–	–	–	–	3	–	9-Borabicyclo[3.3.1]non-9-amine, N,N-bis(9-borabicyclo[3.3.1]non-9-yl)-	65953-33-5	NI38938
105	103	230	229	228	77	152	76	**377**	100	55	51	44	29	24	13	11	**0.99**	26	19	2	1	–	–	–	–	–	–	–	–	–	1,3-Oxazin-6-one, 4,5-dihydro-4-(1-naphthalenyl)-2,4-diphenyl-	108679-32-9	NI38939
377	362	180	181	271	105	378	363	**377**	100	81	53	50	47	36	35	27		26	23	–	1	–	–	–	–	–	1	–	–	–	Phenazasiline, 5-ethyl-5,10-dihydro-10,10-diphenyl-	76-51-7	NI38940
362	377	363	378	144	182	196	183	**377**	100	76	27	22	22	10	6	5		26	35	1	1	–	–	–	–	–	–	–	–	–	3β-Hydroxy-5α-androstano[17,16-b]-N-methylindole		MS08854
109	79	145	185	80	82	47	187	**378**	100	28	23	21	17	15	8	7	**0.00**	4	7	4	–	–	–	2	2	–	–	1	–	–	Phosphoric acid, 1,2-dibromo-2,2-dichloroethyl dimethyl ester	300-76-5	NI38941
109	145	79	47	147	95	83	60	**378**	100	70	49	39	25	15	11	9	**0.10**	4	7	4	–	–	–	2	2	–	–	1	–	–	Phosphoric acid, 1,2-dibromo-2,2-dichloroethyl dimethyl ester	300-76-5	NI38942
109	145	79	185	147	35	187	47	**378**	100	44	25	19	13	7	6	6	**0.00**	4	7	4	–	–	–	2	2	–	–	1	–	–	Phosphoric acid, 1,2-dibromo-2,2-dichloroethyl dimethyl ester	300-76-5	NI38943
343	308	238	378	203	154	273	261	**378**	100	44	20	12	6	5	3	2		8	2	–	–	–	–	8	–	–	–	–	–	–	Benzene, 1,4-dichloro-2,5-bis(trichloromethyl)-	2142-29-2	NI38944
155	153	108	157	185	151	135	154	**378**	100	76	62	42	32	32	29	28	**1.00**	9	10	–	–	–	–	4	–	–	–	–	–	1	Stannane, trimethyl(2,3,4,5-tetrachlorophenyl)-	15725-06-1	NI38945
315	380	289	345	250	254	280		**378**	100	52	28	8	6	4	2		**0.00**	10	8	–	–	–	–	2	–	–	–	–	–	1	Hafnium, bis(π-cyclopentadienyl)dichloro-		LI08191
259	56	42	70	202	119	231	69	**378**	100	81	65	41	36	29	24	24	**5.20**	10	8	2	2	–	10	–	–	–	–	–	–	–	N,N'-Bis(pentafluoropropionyl)piperazine		MS08855
220	319	347	245	203	217	101	161	**378**	100	84	62	11	8	3	2	1	**0.00**	11	18	12	–	–	–	–	–	–	–	–	–	4	Beryllium, pentakis(acetato)methoxyoxotetra-	39335-38-1	NI38946
79	81	39	82	77	53	78	51	**378**	100	47	40	37	31	23	18	18	**0.28**	12	8	1	–	–	–	6	–	–	–	–	–	–	2,7:3,6-Dimethanonaphth[2,3-b]oxirene, 3,4,5,6,9,9-hexachloro-1a,2,2a,3,6,6a,7,7a-octahydro-, (1aα,2β,2aα,3β,6β,6aα,7β,7aα)-	60-57-1	NI38947
79	82	263	81	277	279	108	265	**378**	100	32	28	27	25	22	21	19	**5.05**	12	8	1	–	–	–	6	–	–	–	–	–	–	2,7:3,6-Dimethanonaphth[2,3-b]oxirene, 3,4,5,6,9,9-hexachloro-1a,2,2a,3,6,6a,7,7a-octahydro-, (1aα,2β,2aα,3β,6β,6aα,7β,7aα)-	60-57-1	NI38948
279	345	381	281	245	243	383	347	**378**	100	97	87	86	71	71	70	61	**6.74**	12	8	1	–	–	–	6	–	–	–	–	–	–	2,7:3,6-Dimethanonaphth[2,3-b]oxirene, 3,4,5,6,9,9-hexachloro-1a,2,2a,3,6,6a,7,7a-octahydro-, (1aα,2β,2aα,3β,6β,6aα,7β,7aα)-	60-57-1	NI38949
81	263	265	261	82	79	67	279	**378**	100	98	65	65	65	55	50	38	**0.00**	12	8	1	–	–	–	6	–	–	–	–	–	–	2,7:3,6-Dimethanonaphth[2,3-b]oxirene, 3,4,5,6,9,9-hexachloro-1a,2,2a,3,6,6a,7,7a-octahydro-, (1aα,2β,2aβ,3α,6α,6aβ,7β,7aα)-	72-20-8	NI38950
263	265	261	81	245	147	149	209	**378**	100	66	64	64	53	49	45	42	**1.04**	12	8	1	–	–	–	6	–	–	–	–	–	–	2,7:3,6-Dimethanonaphth[2,3-b]oxirene, 3,4,5,6,9,9-hexachloro-1a,2,2a,3,6,6a,7,7a-octahydro-, (1aα,2β,2aβ,3α,6α,6aβ,7β,7aα)-	72-20-8	NI38951
81	82	263	39	79	67	27	265	**378**	100	61	59	56	55	50	48	46	**1.30**	12	8	1	–	–	–	6	–	–	–	–	–	–	2,7:3,6-Dimethanonaphth[2,3-b]oxirene, 3,4,5,6,9,9-hexachloro-1a,2,2a,3,6,6a,7,7a-octahydro-, (1aα,2β,2aβ,3α,6α,6aβ,7β,7aα)-	72-20-8	NI38952
67	66	250	29	65	95	39	248	**378**	100	23	19	18	14	14	14	13	**0.00**	12	8	1	–	–	–	6	–	–	–	–	–	–	1,2,4-Methenocyclopenta[cd]pentalene-5-carboxaldehyde, 2,2a,3,3,4,7-hexachlorodecahydro-, (1α,2β,2aβ,4β,4aβ,5β,6aβ,6bβ,7R*)-	7421-93-4	NI38953

MASS TO CHARGE RATIOS								M.W.	INTENSITIES								Parent	C	H	O	N	S	F	Cl	Br	I	Si	P	B	X	COMPOUND NAME	CAS Reg No	No
67	27	39	66	87	79	77	113	**378**	100	28	24	20	18	17	17	16	**0.00**	12	8	1	–	–	–	6	–	–	–	–	–	–	2,5,7-Metheno-3H-cyclopenta[a]pentalen-3-one, 3b,4,5,6,6,6a-hexachlorodecahydro-, (2α,3aβ,3bβ,4β,5β,6aβ,7α,7aβ,8R*)-	53494-70-5	NI38954
345	113	343	347	317	281	315	319	**378**	100	77	66	65	44	30	28	27	**0.00**	12	8	1	–	–	–	6	–	–	–	–	–	–	2,5,7-Metheno-3H-cyclopenta[a]pentalen-3-one, 3b,4,5,6,6,6a-hexachlorodecahydro-, (2α,3aβ,3bβ,4β,5β,6aβ,7α,7aβ,8R*)-	53494-70-5	NI38955
81	79	29	39	77	27	51	73	**378**	100	23	22	12	11	11	10	8	**0.30**	12	8	1	–	–	–	6	–	–	–	–	–	–	2,4,6-Metheno-2H-cyclopenta[4,5]pentaleno[1,2-b]oxirene, 2a,3,3,4,5,5a-hexachlorodecahydro-, (1aα,1bβ,2α,2aβ,4β,5β,5aβ,5bβ,6α,6aα)-	13366-73-9	NI38956
59	98	189	157	58	248	378	150	**378**	100	28	17	14	14	12	9	9		12	14	–	2	6	–	–	–	–	–	–	–	–	2,4,6,8,9,10-Hexathiatricyclo[$3.3.1.1^{3,7}$]decane-1,3-dipropanenitrile, 5,7-dimethyl-	57274-49-4	NI38957
211	43	41	210	115	253	65	193	**378**	100	97	59	51	44	30	21	19	**5.00**	12	28	5	–	2	–	–	–	–	–	2	–	–	Thiodiphosphoric acid, tetrapropyl ester	3244-90-4	NI38958
183	380	378	344	382	343	342	345	**378**	100	93	61	59	50	50	48	46		13	6	1	2	1	–	4	–	–	–	–	–	–	Thiocyanic acid, 4-amino-2-chloro-5-(2,4,5-trichlorophenoxy)phenyl ester	13997-29-0	MS08856
44	146	148	75	111	50	28	74	**378**	100	40	26	16	14	12	12	10	**0.00**	14	6	4	–	–	–	4	–	–	–	–	–	–	Peroxide, bis(dichlorobenzoyl)	28604-90-2	WI01494
135	137	55	110	27	43	29	380	**378**	100	99	90	43	23	22	19	11	**5.50**	14	20	2	–	–	–	–	2	–	–	–	–	–	1,4-Bis(4-bromobutoxy)benzene		IC04199
57	378	80	94	120	72	321	135	**378**	100	60	38	37	33	33	27	24		14	22	8	2	1	–	–	–	–	–	–	–	–	1-(tert-Butylamino)-3-(3-methyl-2-nitro-6-sulphoxyphenoxy)-2-propanol		MS08857
115	271	113	127	143	116	111	99	**378**	100	79	32	18	14	10	8	6	**0.00**	14	26	–	4	4	–	–	–	–	–	–	–	–	1H-1,4-Diazepine, 1,1′-(dithiodicarbonothioyl)bis[hexahydro-4-methyl-	26087-98-9	NI38959
267	41	96	94	67	97	95	42	**378**	100	45	40	37	25	20	19	18	**0.00**	14	26	4	4	2	–	–	–	–	–	–	–	–	9,14-Diazatricyclo[$6.3.2.1^{2,7}$]tetradec-10-ene-9,14-disulphonamide, N,N′-dimethyl-	56666-58-1	NI38960
189	117	89	190	18	244	29	28	**378**	100	14	9	8	6	5	5	4	**0.00**	14	26	8	4	–	–	–	–	–	–	–	–	–	1,1,2,2-Tetra(ethoxycarbonylamino)ethane		MS08858
93	356	73	363	95	367	380	45	**378**	100	55	47	42	38	29	19	18	**15.15**	15	14	1	–	–	–	4	–	–	1	–	–	–	Silane, trimethyl[(2,2′,5,5′-tetrachloro[1,1′-biphenyl]-3-yl)oxy]-	55683-17-5	NI38961
93	73	365	95	363	43	380	367	**378**	100	50	35	35	26	21	19	17	**14.78**	15	14	1	–	–	–	4	–	–	1	–	–	–	Silane, trimethyl[(2,2′,5,5′-tetrachloro[1,1′-biphenyl]-4-yl)oxy]-	55683-18-6	NI38962
93	95	73	365	363	367	207	43	**378**	100	39	37	23	16	15	12	11	**7.66**	15	14	1	–	–	–	4	–	–	1	–	–	–	Silane, trimethyl[(2′,3,5′,6-tetrachloro[1,1′-biphenyl]-2-yl)oxy]-	55683-19-7	NI38963
147	146	18	378	119	44	91	148	**378**	100	47	26	25	21	13	10	7		15	14	2	–	–	8	–	–	–	–	–	–	–	1H,1H,5H-Octafluoropentyl 2,4,6-trimethylbenzoate		MS08859
147	146	378	120	45	92	148	119	**378**	100	47	25	21	14	10	8	6		15	14	2	–	–	8	–	–	–	–	–	–	–	1H,1H,5H-Octafluoropentyl 2,4,6-trimethylbenzoate		LI08192
73	147	75	217	275	143	117	45	**378**	100	29	28	24	14	14	13	13	**0.00**	15	34	5	–	–	–	–	–	–	3	–	–	–	Arabino-hexonic acid, 2-deoxy-3,5,6-tris-O-(trimethylsilyl)-, γ-lactone	74778-25-9	NI38964
73	129	147	103	75	205	246	74	**378**	100	39	23	18	14	11	9	9	**0.08**	15	34	5	–	–	–	–	–	–	3	–	–	–	D-Arabino-hexonic acid, 3-deoxy-2,5,6-tris-O-(trimethylsilyl)-, γ-lactone	33648-66-7	NI38965
73	129	261	171	147	75	363	55	**378**	100	71	57	57	40	23	19	18	**0.00**	15	34	5	–	–	–	–	–	–	3	–	–	–	Hexanedioic acid, 2-[(trimethylsilyl)oxy]-, bis(trimethylsilyl) ester		MS08860
261	73	129	171	363	147	75	262	**378**	100	95	66	63	49	47	27	23	**0.44**	15	34	5	–	–	–	–	–	–	3	–	–	–	Hexanedioic acid, 2-[(trimethylsilyl)oxy]-, bis(trimethylsilyl) ester		NI38966
73	147	75	363	129	247	45	149	**378**	100	35	26	18	18	14	12	9	**0.14**	15	34	5	–	–	–	–	–	–	3	–	–	–	Hexanedioic acid, 3-[(trimethylsilyl)oxy]-, bis(trimethylsilyl) ester		NI38967
73	147	75	247	115	45	199	231	**378**	100	34	23	20	16	12	10	9	**0.00**	15	34	5	–	–	–	–	–	–	3	–	–	–	Pentanedioic acid, 3-methyl-3-[(trimethylsilyl)oxy]-, bis(trimethylsilyl) ester	55590-95-9	NI38968
73	147	75	271	247	45	273	115	**378**	100	54	29	25	22	13	12	12	**0.00**	15	34	5	–	–	–	–	–	–	3	–	–	–	Pentanedioic acid, 3-methyl-3-[(trimethylsilyl)oxy]-, bis(trimethylsilyl) ester	55590-95-9	NI38969
73	147	247	75	115	231	363	273	**378**	100	34	25	21	14	12	11	11	**0.00**	15	34	5	–	–	–	–	–	–	3	–	–	–	Pentanedioic acid, 3-methyl-3-[(trimethylsilyl)oxy]-, bis(trimethylsilyl) ester	55590-95-9	NI38970
73	348	147	117	129	103	349	75	**378**	100	62	37	31	30	28	20	19	**3.50**	15	34	5	–	–	–	–	–	–	3	–	–	–	D-erythro-Pentonic acid, 3-deoxy-2,5-bis-O-(trimethylsilyl)-2-C-[[(trimethylsilyl)oxy]methyl]-, γ-lactone	34641-57-1	NI38971
73	348	103	147	129	117	75	349	**378**	100	54	42	33	26	26	20	17	**2.70**	15	34	5	–	–	–	–	–	–	3	–	–	–	D-threo-Pentonic acid, 3-deoxy-2,5-bis-O-(trimethylsilyl)-2-C-[[(trimethylsilyl)oxy]methyl]-, lactone	55570-79-1	NI38972
73	129	75	273	103	246	147	205	**378**	100	31	22	19	19	18	18	15	**7.60**	15	34	5	–	–	–	–	–	–	3	–	–	–	D-Ribo-hexonic acid, 3-deoxy-2,5,6-tris-O-(trimethylsilyl)-, lactone	55570-78-0	NI38973
73	217	147	117	378	218	87	75	**378**	100	54	41	36	26	23	19	17		15	34	5	–	–	–	–	–	–	3	–	–	–	D-Ribonic acid, 2-C-methyl-2,3,5-tris-O-(trimethylsilyl)-, γ-lactone	55570-77-9	NI38974
220	270	298	82	80	221	77	205	**378**	100	36	31	19	19	18	14	13	**0.00**	16	16	–	4	–	–	1	1	–	–	–	–	–	1H-1,3,4-Benzotriazepin-2(3H)-imine, 7-chloro-2-imino-1,3-dimethyl-5-phenyl-, hydrobromide		MS08861
117	127	43	261	101	99	85	161	**378**	100	23	23	19	19	19	12	8	**0.00**	16	26	10	–	–	–	–	–	–	–	–	–	–	Glucitol, 2,3-di-O-methyl-, tetraacetate	20250-47-9	NI38975
43	87	45	129	89	139	75	97	**378**	100	99	83	63	54	41	39	37	**0.00**	16	26	10	–	–	–	–	–	–	–	–	–	–	β-glycero-D-Glucoheptopyranoside, methyl 2,3,4-tri-O-acetyl-6,7-di-O-methyl-	84564-01-2	NI38976
43	129	87	45	115	74	125	142	**378**	100	99	99	93	30	28	23	22	**0.00**	16	26	10	–	–	–	–	–	–	–	–	–	–	β-glycero-D-Glucoheptopyranoside, methyl 2,3,6-tri-O-acetyl-4,7-di-O-methyl-	84563-99-5	NI38977
43	101	117	87	75	74	102	45	**378**	100	99	76	40	33	31	18	18	**0.00**	16	26	10	–	–	–	–	–	–	–	–	–	–	β-glycero-D-Glucoheptopyranoside, methyl 2,3,7-tri-O-acetyl-4,6-di-O-methyl-	84582-65-0	NI38978
43	129	161	162	101	45	127	85	**378**	100	99	63	47	44	35	30	28	**0.00**	16	26	10	–	–	–	–	–	–	–	–	–	–	β-glycero-D-Glucoheptopyranoside, methyl 2,3,7-tri-O-acetyl-4,6-di-O-methyl-	84582-65-0	NI38979
43	75	74	116	45	88	87	129	**378**	100	99	71	43	36	31	20	16	**0.00**	16	26	10	–	–	–	–	–	–	–	–	–	–	β-glycero-D-Glucoheptopyranoside, methyl 2,4,6-tri-O-acetyl-3,7-di-O-methyl-	84563-98-4	NI38980
43	75	117	74	116	87	129	69	**378**	100	99	92	67	32	23	21	15	**0.00**	16	26	10	–	–	–	–	–	–	–	–	–	–	β-glycero-D-Glucoheptopyranoside, methyl 2,4,7-tri-O-acetyl-3,6-di-O-methyl-	84564-00-1	NI38981
43	75	87	74	88	142	129	101	**378**	100	99	55	55	52	29	26	23	**0.00**	16	26	10	–	–	–	–	–	–	–	–	–	–	β-glycero-D-Glucoheptopyranoside, methyl 2,6,7-tri-O-acetyl-3,4-di-O-methyl-	84564-04-5	NI38982

MASS TO CHARGE RATIOS								M.W.	INTENSITIES								Parent	C	H	O	N	S	F	Cl	Br	I	Si	P	B	X	COMPOUND NAME	CAS Reg No	No
43	74	116	129	125	87	45	157	**378**	100	99	70	51	40	34	33	32	**0.00**	16	26	10	–	–	–	–	–	–	–	–	–	–	β-glycero-D-Glucoheptopyranoside, methyl 3,4,6-tri-O-acetyl-2,7-di-O-methyl-	84563-97-3	NI38983
43	74	117	116	87	129	75	185	**378**	100	99	68	54	49	37	34	27	**0.00**	16	26	10	–	–	–	–	–	–	–	–	–	–	β-glycero-D-Glucoheptopyranoside, methyl 3,4,7-tri-O-acetyl-2,6-di-O-methyl-	84582-64-9	NI38984
74	43	75	116	101	87	69	143	**378**	100	98	72	63	63	37	26	23	**0.00**	16	26	10	–	–	–	–	–	–	–	–	–	–	β-glycero-D-Glucoheptopyranoside, methyl 3,6,7-tri-O-acetyl-2,4-di-O-methyl-	84564-03-4	NI38985
75	88	43	74	129	87	45	73	**378**	100	83	62	12	11	11	9	8	**0.00**	16	26	10	–	–	–	–	–	–	–	–	–	–	β-glycero-D-Glucoheptopyranoside, methyl 4,6,7-tri-O-acetyl-2,3-di-O-methyl-	84564-02-3	NI38986
43	129	117	87	189	99	101	45	**378**	100	44	33	24	14	9	6	6	**0.00**	16	26	10	–	–	–	–	–	–	–	–	–	–	D-Mannitol, 2,4-di-O-methyl-, tetraacetate	24406-90-4	NI38987
43	129	87	45	113	189	99	71	**378**	100	69	60	36	26	20	19	11	**0.00**	16	26	10	–	–	–	–	–	–	–	–	–	–	Mannitol, 1,4-di-O-methyl-, tetraacetate	56270-97-4	NI38988
43	117	129	87	189	101	99	45	**378**	100	99	86	49	28	23	16	16	**0.00**	16	26	10	–	–	–	–	–	–	–	–	–	–	Mannitol, 2,4-di-O-methyl-, tetraacetate	19318-51-5	NI38989
129	189	87	99	173	190	43	113	**378**	100	58	19	16	7	6	6	5	**0.00**	16	26	10	–	–	–	–	–	–	–	–	–	–	Mannitol, 3,4-di-O-methyl-, tetraacetate	72173-15-0	NI38990
158	144	91	182	126	155	140	98	**378**	100	83	50	43	22	21	18	16	**0.00**	16	27	4	2	1	–	1	–	–	–	–	–	–	Hexanoic acid, 6-[(3-aminopropyl)[(4-methylphenyl)sulphonyl]amino]-, monohydrochloride	67370-69-8	NI38991
241	185	104	109	81	242	103	378	**378**	100	37	23	19	17	15	12	9		16	28	6	–	–	–	–	–	–	–	2	–	–	1,2-Bis(diethylphosphono)-1-phenylethane	2519-12-2	DD01493
324	360	322	378	362	326	380		**378**	100	94	76	70	35	28	26			17	11	8	–	–	–	1	–	–	–	–	–	–	2-Carboxy-5-chloro-1,4,6-trihydroxy-8-methoxy-3-methylanthraquinone		LI08193
84	166	128	277	41	28	29	184	**378**	100	51	32	19	18	16	15	13	**0.00**	17	25	4	2	–	3	–	–	–	–	–	–	–	N-Trifluoroacetyl-L-prolylpipecolic acid, butyl ester		MS08862
81	82	84	54	86	262	263	41	**378**	100	94	76	76	72	60	48	48	**7.00**	17	26	4	6	–	–	–	–	–	–	–	–	–	L-Norleucinamide, 5-oxo-L-prolyl-L-histidyl-	37666-93-6	NI38992
181	378	363	380	365	153	344	183	**378**	100	82	82	30	28	27	22	20		18	15	7	–	–	–	1	–	–	–	–	–	–	3′-Chloro-5,2′-dihydroxy-3,7,8-trimethoxyflavone		MS08863
44	299	91	92	333	334	300	335	**378**	100	98	89	55	49	36	29	22	**13.00**	18	15	7	–	–	–	1	–	–	–	–	–	–	11H-Dibenzo[b,e][1,4]dioxepin-7-carboxylic acid, 9-chloro-8-hydroxy-3-methoxy-1,4,6-trimethyl-11-oxo-	38629-31-1	NI38993
346	318	287	277	347	378	229	319	**378**	100	97	88	62	60	38	33	30		18	18	9	–	–	–	–	–	–	–	–	–	–	4,5-Benzofurandicarboxylic acid, 7-(1,2-dicarboxyvinyl)-6,7-dihydro-, tetramethyl ester	19665-39-5	NI38994
139	225	285	319	107	94	87	251	**378**	100	87	72	65	64	60	51	50	**20.00**	18	22	5	2	1	–	–	–	–	–	–	–	–	Dimethyl 7-phenoxymethylperillate		LI08194
167	362	157	81	168	109	77	66	**378**	100	54	22	21	18	18	18	18	**7.48**	18	22	7	2	–	–	–	–	–	–	–	–	–	Azoxybenzene, 3,3′,4,4′,5,5′-hexamethoxy-	24313-89-1	NI38995
127	252	56	124	363	378	72	82	**378**	100	48	44	40	33	27	24	18		18	30	3	6	–	–	–	–	–	–	–	–	–	Triimidazo[1,5-a:1′,5′-c:1″,5″-e][1,3,5]triazine-1,5,9(2H,6H,10H)-trione, hexahydro-2,3,3,6,7,7,10,11,11-nonamethyl-	67969-48-6	NI38996
321	320	139	83	81	210	55	126	**378**	100	65	56	34	22	21	21	20	**8.02**	18	33	6	–	–	–	–	–	–	–	–	3	–	allo-Inositol tri-n-butaneboronate		NI38997
126	125	139	70	83	263	81	127	**378**	100	24	23	21	14	13	12	8	**1.62**	18	33	6	–	–	–	–	–	–	–	–	3	–	chiro-Inositol tri-n-butaneboronate		NI38998
321	320	349	348	139	83	279	322	**378**	100	66	48	35	35	30	25	19	**6.56**	18	33	6	–	–	–	–	–	–	–	–	3	–	cis-Inositol tri-n-butaneboronate		NI38999
126	252	251	70	139	125	83	263	**378**	100	77	40	32	28	27	21	17	**0.74**	18	33	6	–	–	–	–	–	–	–	–	3	–	epi-Inositol tri-n-butaneboronate		NI39000
139	138	321	83	237	81	320	140	**378**	100	22	20	16	15	15	13	13	**3.14**	18	33	6	–	–	–	–	–	–	–	–	3	–	myo-Inositol tri-n-butaneboronate		NI39001
126	70	139	83	125	237	252	55	**378**	100	37	36	35	28	24	23	18	**2.50**	18	33	6	–	–	–	–	–	–	–	–	3	–	neo-Inositol tri-n-butaneboronate		NI39002
139	237	138	321	81	83	320	140	**378**	100	27	26	22	20	18	16	16	**7.10**	18	33	6	–	–	–	–	–	–	–	–	3	–	muco-Inositol tri-n-butaneboronate		NI39003
89	133	83	85	228	73	87	97	**378**	100	79	62	51	38	36	28	26	**2.50**	18	34	8	–	–	–	–	–	–	–	–	–	–	1,1′-Bis(3,6,9,12-tetraoxacyclotridecane)		MS08864
290	30	306	305	289	276	233	109	**378**	100	96	93	79	66	66	57	57	**32.07**	18	45	–	6	–	–	–	–	–	–	–	3	–	1,3,5-Tris(diethylamino)-2,4,6-triethylborazine		IC04200
98	91	182	99	102	100	101	245	**378**	100	82	81	80	78	77	76	74	**2.14**	19	26	6	2	–	–	–	–	–	–	–	–	–	β-L-Altropyranoside, benzyl 2,4-bis(acetylamino)-2,4,6-trideoxy-, 3-acetate	50611-14-8	NI39004
113	245	287	43	98	99	212	91	**378**	100	99	84	79	77	76	74	74	**1.23**	19	26	6	2	–	–	–	–	–	–	–	–	–	β-L-Idopyranoside, benzyl 2,4-bis(acetylamino)-2,4,6-trideoxy-, 3-acetate	50611-15-9	NI39005
113	85	72	71	57	56	137	309	**378**	100	81	46	45	45	41	40	36	**0.00**	19	26	6	2	–	–	–	–	–	–	–	–	–	Propanoic acid, 2,2-dimethyl-, [3-(2,2-dimethyl-1-oxopropoxy)-2,3,3a,8a-tetrahydro-5-oxo-5H-furo[3′,2′:3,4]azeto[1,2-a]pyrimidin-2-yl]methyl ester, [2S-(2α,3β,3aβ,8aaβ)]-	55429-21-5	NI39006
287	270	245	212	227	199	91	228	**378**	100	88	74	51	44	43	42	41	**3.05**	19	26	6	2	–	–	–	–	–	–	–	–	–	β-L-Talopyranoside, benzyl 2,4-bis(acetylamino)-2,4,6-trideoxy-, 3-acetate	50611-20-6	NI39007
131	174	57	86	58	188	99	100	**378**	100	71	55	50	43	35	31	30	**4.51**	19	26	6	2	–	–	–	–	–	–	–	–	–	1,4,10-Trioxa-7,13-diazacyclopentadecane-2,6,14-trione,15-benzyl-7,13-dimethyl-	91876-16-3	NI39008
378	322	121	120	92	43	379	63	**378**	100	55	48	28	27	26	22	22		20	10	8	–	–	–	–	–	–	–	–	–	–	2,2′-Binaphthalene-1,1′,4,4′-tetrone, 3,3′,5,5′-tetrahydroxy-	88381-81-1	NI39009
378	322	121	120	92	379	63	64	**378**	100	55	51	39	35	33	30	20		20	10	8	–	–	–	–	–	–	–	–	–	–	2,2′-Binaphthalene-1,1′,4,4′-tetrone, 3,3′,5,5′-tetrahydroxy-	88381-81-1	NI39010
63	92	346	362	75	332	76	64	**378**	100	81	73	70	61	60	56	50	**25.21**	20	10	8	–	–	–	–	–	–	–	–	–	–	2,2′-Binaphthalene-1,1′,4,4′-tetrone, 3,3′,8,8′-tetrahydroxy-	88381-80-0	NI39011
378	379	313	380	297	77	236	164	**378**	100	24	15	8	7	7	6	6		20	14	4	2	1	–	–	–	–	–	–	–	–	1-Amino-2-(2-dimethylaminoethoxy)-4-anilinoanthraquinone		IC04201
113	70	126	42	165	45	43	143	**378**	100	55	16	16	13	11	11	10	**0.20**	20	24	1	2	–	–	2	–	–	–	–	–	–	Piperazine, 1-[2-[α-(2-chlorophenyl)-3-chlorobenzyloxy]ethyl]-4-methyl-		MS08865
88	102	56	165	42	54	99	113	**378**	100	56	53	46	44	38	36	32	**2.10**	20	24	1	2	–	–	2	–	–	–	–	–	–	Propanol, 3-[4-[α-(2-chlorophenyl)-3-chlorobenzyl]-1-piperazinyl]-		MS08866
137	138	240	122	175	94	378	77	**378**	100	47	15	8	7	7	5	5		20	26	7	–	–	–	–	–	–	–	–	–	–	1,2,4-Butanetriol, 2,3-bis(4-hydroxy-3-methoxybenzyl)-	58139-12-1	NI39012
378	151	247	264	279	135	138	105	**378**	100	39	30	28	24	17	16	15		20	26	7	–	–	–	–	–	–	–	–	–	–	Chaparrinone		LI08195
43	99	71	55	69	41	39	53	**378**	100	83	67	62	60	57	54	51	**0.00**	20	26	7	–	–	–	–	–	–	–	–	–	–	Germacra-1(10),11(13)-dien-6,12-olide, 4,5-epoxy-9-hydroxy-, 8-(4-hydroxysenecioate), (4α,5β,8α,9α)-		NI39013

MASS TO CHARGE RATIOS								M.W.	INTENSITIES								Parent	C	H	O	N	S	F	Cl	Br	I	Si	P	B	X	COMPOUND NAME	CAS Reg No	No
137	44	59	43	39	215	77	55	378	100	52	42	34	34	33	28	27	12.00	20	26	7	–	–	–	–	–	–	–	–	–	–	Gibberellin A_8 methyl ester		MS08867
318	245	91	55	145	41	105	173	378	100	60	59	55	43	40	38	37	0.00	20	26	7	–	–	–	–	–	–	–	–	–	–	Icacinol	96881-05-9	NI39014
305	72	306	264	232	262	363	234	378	100	99	87	70	64	43	38	23	0.00	20	34	3	2	–	–	–	–	–	1	–	–	–	Benzamide, N,N-diethyl-2-(diethylcarbamoyl)-4-methoxy-3-(trimethylsilyl)-	85370-78-1	DD01494
378	360	317	379	345	335	361	43	378	100	52	49	25	14	14	13	13		21	14	7	–	–	–	–	–	–	–	–	–	–	5,12-Naphthacenedione, 8-acetyl-6,10,11-trihydroxy-1-methoxy-	20982-41-6	MS08868
334	278	277	319	378	249	206	306	378	100	72	41	37	32	28	27	26		21	18	5	2	–	–	–	–	–	–	–	–	–	Camptothecin, 10-methoxy-	19685-10-0	NI39015
378	278	334	277	319	332	317	305	378	100	84	82	70	65	51	47	47		21	18	5	2	–	–	–	–	–	–	–	–	–	Camptothecin, 10-methoxy-	19685-10-0	NI39016
378	334	319	278	349	277	306	305	378	100	99	78	65	52	52	48	43		21	18	5	2	–	–	–	–	–	–	–	–	–	Camptothecin, 11-methoxy-	39026-92-1	NI39017
334	378	319	278	277	349	306	305	378	100	90	69	56	46	45	43	38		21	18	5	2	–	–	–	–	–	–	–	–	–	Camptothecin, 11-methoxy-	70906-25-1	NI39018
378	336	43	134	219	199	173	56	378	100	72	42	39	37	35	35	27		21	22	3	4	–	–	–	–	–	–	–	–	–	2-Pyrazolin-5-one, 3-acetylamino-4-[4-(dimethylamino)benzylidene]-1-(3-methoxyphenyl)-		NI39019
289	290	245	39	45	77	378	259	378	100	26	13	12	10	8	7	4		21	22	3	4	–	–	–	–	–	–	–	–	–	1H-Pyrazolo[3,4-b]quinoxaline, 1-phenyl-3-(1,2,3-trimethoxypropyl)-, [S-(R*,S*)]-	56804-91-2	NI39020
126	66	252	67	127	93	220	41	378	100	30	20	20	14	14	8	5	4.30	21	30	–	–	3	–	–	–	–	–	–	–	–	Trispiro[1,3,5-trithiane-2,2′:4,2″:6,2‴-trisbicyclo[2.2.1]heptane]	51849-42-4	NI39021
246	378	132	345	202	145	203	190	378	100	98	50	41	9	8	7	7		21	30	2	–	2	–	–	–	–	–	–	–	–	19-Nortestosterone-2-spiro-2′-(1′,3′-dithian)-3-one		LI08196
57	104	133	246	87	41	190	29	378	100	61	27	26	26	26	21	16	0.00	21	30	4	–	1	–	–	–	–	–	–	–	–	1,3-Dioxolane-4-carbothioic acid, 2-tert-butyl-4-(1-hydroxy-3-phenyl-2-propenyl)-, S-tert-butyl ester, (2R,4R)-		DD01495
42	57	207	69	55	210	263	360	378	100	83	72	64	52	49	45	35	2.00	21	30	6	–	–	–	–	–	–	–	–	–	–	abeo-Isohumulone		LI08197
69	213	41	264	97	57	43	85	378	100	63	35	34	33	18	12	9	9.00	21	30	6	–	–	–	–	–	–	–	–	–	–	2-Cyclopenten-1-one, 3,4,5-trihydroxy-2-isovaleryl-5-(3-methyl-2-butenyl)-4-(4-methyl-3-pentenoyl)-, stereoisomer	981-03-3	NI39022
69	213	265	41	97	57	43	85	378	100	64	36	36	34	17	12	10	0.92	21	30	6	–	–	–	–	–	–	–	–	–	–	2-Cyclopenten-1-one, 3,4,5-trihydroxy-2-isovaleryl-5-(3-methyl-2-butenyl)-4-(4-methyl-3-pentenoyl)-, stereoisomer	26146-35-0	NI39023
41	69	57	180	152	97	96	43	378	100	97	18	12	12	12	12	12	0.24	21	30	6	–	–	–	–	–	–	–	–	–	–	2-Cyclopenten-1-one, 3,4,5-trihydroxy-2-isovaleryl-5-(3-methyl-2-butenyl)-4-(4-methyl-3-pentenoyl)-, stereoisomer	26110-47-4	NI39024
42	57	207	69	55	210	263	360	378	100	84	73	66	52	50	46	35	2.42	21	30	6	–	–	–	–	–	–	–	–	–	–	1,7-Dioxadispiro[4.0.5.3]tetradec-12-ene-11,14-dione, 12-hydroxy-2,2,8,8-tetramethyl-13-(3-methyl-1-oxobutyl)-	38574-28-6	NI39025
42	57	69	55	210	263	360	53	378	100	83	65	51	49	45	35	34	2.00	21	30	6	–	–	–	–	–	–	–	–	–	–	1,7-Dioxadispiro[4.0.5.3]tetradec-12-ene-11,14-dione, 12-hydroxy-2,2,8,8-tetramethyl-13-(3-methyl-1-oxobutyl)-	38574-28-6	NI39026
55	71	262	99	135	163	219	334	378	100	69	58	58	38	20	17	5	0.00	21	30	6	–	–	–	–	–	–	–	–	–	–	Ludovicin A, 9α-[3-(methylvaleryl)oxy]-		MS08869
167	195	378	109	135	55	107	360	378	100	75	43	38	32	24	21	20		21	30	6	–	–	–	–	–	–	–	–	–	–	1-Phenanthrenecarboxylic acid, 7-(carboxymethylene)tetradecahydro-9-hydroxy-1,4a,8-trimethyl-10-oxo-, 1-methyl ester, [1S-(1α,4aα,4bβ,7E,8β,8aα,9α,10aβ)]-	32231-43-9	NI39027
121	152	360	345	105	153	107	151	378	100	83	57	57	54	45	44	42	24.00	21	30	6	–	–	–	–	–	–	–	–	–	–	Picras-2-ene-1,16-dione, 11,12-dihydroxy-2-methoxy-, (11α,12β)-	24148-76-3	NI39028
91	200	105	77	121	51	165	65	378	100	99	60	47	15	14	13	13	2.00	22	18	2	–	2	–	–	–	–	–	–	–	–	Carbonodithioic acid, O-(2-oxo-1,2-diphenylethyl)-	54699-17-1	NI39029
318	259	362	258	319	260	230	274	378	100	62	48	41	24	19	14	10	1.30	22	18	6	–	–	–	–	–	–	–	–	–	–	8,9-Epoxy-2,15ξ-diacetoxy-11-methyl-8,9-secogona-1,3,5,7,9,11,13-heptaen-17-one		NI39030
378	106	135	134	163	360	77	180	378	100	18	17	13	11	10	10	8		22	18	6	–	–	–	–	–	–	–	–	–	–	Tetrahydrodiospyrin		NI39031
264	205	318	263	259	319	378	243	378	100	85	79	73	38	21	10	9		22	22	4	2	–	–	–	–	–	–	–	–	–	17-Methoxy-3,4,5,6-tetradehydrotalbotine		LI08198
378	295	360	363	264	292	361	293	378	100	59	49	36	36	24	20	20		22	22	4	2	–	–	–	–	–	–	–	–	–	4H-Pyrano[3′,2′:3,4]benzo[1,2-d]pyridazin-4-one, 3,8,9,10-tetrahydro-10-hydroxy-9-isopropenyl-5-methoxy-6-methyl-3-phenyl-, (9α,10α)-		NI39032
119	120	175	91	176	121	378	102	378	100	95	93	76	73	71	67	57		22	26	2	4	–	–	–	–	–	–	–	–	–	1,2,4,5-Tetrazine, 3,6-bis(4-butoxyphenyl)-	72243-01-7	NI39033
318	43	258	57	55	319	133	41	378	100	63	36	28	27	24	23	22	0.00	22	31	4	–	–	1	–	–	–	–	–	–	–	Estr-5(10)-ene-6β,12β-diol, 3β-fluoro-, diacetate	56890-09-6	NI39034
378	258	379	243	133	165	276	259	378	100	38	24	22	22	18	11	6		22	31	4	–	–	1	–	–	–	–	–	–	–	Estr-5(10)-ene-6β,17β-diol, 3β-fluoro-, diacetate	30882-12-3	NI39035
318	258	243	133	165	276	298		378	100	38	22	22	18	11	3		0.00	22	31	4	–	–	1	–	–	–	–	–	–	–	Estr-5(10)-ene-6β,17β-diol, 3β-fluoro-, diacetate	30882-12-3	LI08199
349	321	160	132	350	218	188	28	378	100	42	28	23	22	19	16	16	0.12	22	32	–	2	–	–	–	–	–	–	–	–	2	Aluminium, bis[μ-(benzenemethaniminato)]tetraethyldi-	12406-75-6	NI39036
217	199	43	161	105	300	318	218	378	100	42	30	25	21	19	15	15	2.00	22	34	5	–	–	–	–	–	–	–	–	–	–	6β-Acetoxy-17-oxo-4,5-secoandrostan-4-oic acid methyl ester		MS08870
318	43	55	28	82	67	41	94	378	100	88	44	43	36	35	35	33	1.00	22	34	5	–	–	–	–	–	–	–	–	–	–	5α-Androstane-3β,10β,17β-triol, 3,17-diacetate		LI08200
161	32	346	43	347	55	41	162	378	100	67	39	33	29	27	27	24	3.74	22	34	5	–	–	–	–	–	–	–	–	–	–	Benzenedodecanoic acid, 3-methoxy-2-(methoxycarbonyl)-, methyl ester	56554-83-7	NI39037
270	95	105	255	242	300	378	318	378	100	64	60	56	17	16	13	5		22	34	5	–	–	–	–	–	–	–	–	–	–	Kaurane-3,7,18-triol, 15,16-epoxy-, 3-acetate, (3α,4β,7β,15α)-	62655-12-3	NI39038
43	121	41	378	319	134	42	49	378	100	35	25	24	19	17	17	15		23	26	3	2	–	–	–	–	–	–	–	–	–	2,20-Cycloaspidospermidine-1-carboxaldehyde, 3-[(acetyloxy)methyl]-6,7-didehydro-, (2α,3β,5α,12β,19α,20R)-	56053-40-8	NI39039
105	176	77	180	378	334	252	251	378	100	73	66	36	33	20	14	14		23	26	3	2	–	–	–	–	–	–	–	–	–	2,4-Imidazolidinedione, 3-cyclohexyl-5-ethoxy-1,5-diphenyl-		DD01496
363	333	334	378	364	159	379	173	378	100	93	87	74	24	20	19	19		23	30	1	4	–	–	–	–	–	–	–	–	–	5-Pyrazolamine, 1-(3,5-di-tert-butyl-2-hydroxyphenyl)-3-(phenylamino)-		NI39040
136	154	43	41	55	57	378	137	378	100	56	51	42	38	27	23	23		23	38	2	–	1	–	–	–	–	–	–	–	–	Benzoic acid, 2-(hexadecylthio)-	2182-24-3	NI39041
43	57	55	41	98	69	73	29	378	100	74	62	58	48	35	34	34	0.00	23	38	4	–	–	–	–	–	–	–	–	–	–	5,8,11,14-Eicosatetraenoic acid, 2,3-dihydroxypropyl ester, (all-Z)-	35474-99-8	WI01495

MASS TO CHARGE RATIOS								M.W.	INTENSITIES								Parent	C	H	O	N	S	F	Cl	Br	I	Si	P	B	X	COMPOUND NAME	CAS Reg No	No
219	201	55	97	81	163	274	306	**378**	100	59	58	56	51	46	44	40	**0.00**	23	38	4	–	–	–	–	–	–	–	–	–	–	4-Hydroxy-17-oxo-4-propyl-3,4-seco-5α-androstan-3-oic acid, methyl ester		MS08871
363	231	378	230	217	215	255	75	**378**	100	60	47	45	42	42	38	33		23	42	2	–	–	–	–	–	–	1	–	–	–	Androstan-17-ol, 17-methyl-3-[(trimethylsilyl)oxy]-, (3β,5α,17β)-	56771-59-6	MS08872
205	378	77	179	76	379	206	178	**378**	100	97	44	34	33	31	24	19		24	18	1	4	–	–	–	–	–	–	–	–	–	4(3H)-Quinazolinone, 3-(1-phenyl-3-methylpyrazol-5-yl)-2-phenyl-		MS08873
198	319	378	212	43	197	194	199	**378**	100	52	47	43	42	23	20	19		24	30	2	2	–	–	–	–	–	–	–	–	–	2,20-Cycloaspidospermidine-3-methanol, 6,7-didehydro-1-ethyl-, acetate (ester), (2α,3β,5α,12β,19α,20R)-	56053-35-1	NI39042
108	79	91	255	107	77	136	378	**378**	100	85	84	63	59	52	48	32		24	30	2	2	–	–	–	–	–	–	–	–	–	4,3',5'-Trimethyl-3,4'-diethyl-5-benzyloxycarbonyl-2,2'-dipyrrolylmethane		LI08201
306	360	290	307	55	289	81	43	**378**	100	36	27	24	17	16	15	13	**2.00**	24	42	3	–	–	–	–	–	–	–	–	–	–	Androstane-3β,5α,17β-triol, 6-pentyl-, (6β)-		MS08874
273	291	55	81	41	95	67	43	**378**	100	95	53	45	42	40	31	31	**1.00**	24	42	3	–	–	–	–	–	–	–	–	–	–	17β-Hydroxy-6α-pentyl-4-nor-3,5-secoandrostan-3-oic acid, methyl ester		MS08875
300	301	285	377	363	362	196	286	**378**	100	50	29	23	20	19	13	12	**10.00**	25	22	–	2	–	–	–	–	–	1	–	–	–	10-Sila-2-azaanthracen-9-amine, 9,10-dihydro-3-methyl-10,10-diphenyl-		NI39043
299	378	379	301	300	377	272	195	**378**	100	98	32	30	30	10	10	8		25	22	–	2	–	–	–	–	–	1	–	–	–	5-Sila-5H,10H,11H-benzo[b]pyrido[4,3-e]azepine, 3-methyl-5,5-diphenyl-		NI39044
196	182	154	183	167	197	181	127	**378**	100	99	69	66	52	49	48	48	**26.80**	25	22	–	4	–	–	–	–	–	–	–	–	–	(1-Methyl-2-perimidinyl)(1-methyl-2,3-dihydro-3-perimidinyl)methane		NI39045
378	31	43	41	57	45	55	363	**378**	100	44	40	30	28	25	23	20		25	30	3	–	–	–	–	–	–	–	–	–	–	12,14-Dioxo-4,4a,7c,8-tetramethyl-2,3,4,4a,5,7,7a,8,9,10,12,14-dodecahydro-13H-dibenzo[bi]xanthene		LI08202
316	315	197	44	313	181	43	41	**378**	100	77	62	49	42	27	27	24	**3.50**	26	18	3	–	–	–	–	–	–	–	–	–	–	9-[2-(2-Carboxyphenyl)phenyl]-9-hydroxyfluorene		IC04202
378	257	336	77	335	43	118	167	**378**	100	67	53	42	35	22	21	13		26	22	1	2	–	–	–	–	–	–	–	–	–	Acetanilide, 2'-(diphenylamino)-N-phenyl-	29325-53-9	LI08203
378	257	336	77	335	379	258	334	**378**	100	65	52	40	33	30	30	21		26	22	1	2	–	–	–	–	–	–	–	–	–	Acetanilide, 2'-(diphenylamino)-N-phenyl-	29325-53-9	NI39046
378	336	167	335	261	77	43	118	**378**	100	55	42	34	29	19	19	15		26	22	1	2	–	–	–	–	–	–	–	–	–	Acetanilide, 4'-(diphenylamino)-N-phenyl-	29344-72-7	LI08204
378	336	167	335	379	261	168	43	**378**	100	55	42	35	33	30	20	18		26	22	1	2	–	–	–	–	–	–	–	–	–	Acetanilide, 4'-(diphenylamino)-N-phenyl-	29344-72-7	NI39047
301	378	165	180	167	379	152	183	**378**	100	99	59	46	41	31	31	28		26	22	1	2	–	–	–	–	–	–	–	–	–	Diazene, bis[2-benzylphenyl]-, 1-oxide	55823-09-1	NI39048
301	378	377	165	379	180	182	302	**378**	100	93	47	37	30	25	19	18		26	22	1	2	–	–	–	–	–	–	–	–	–	Phenol, 3-benzyl-2-[[2-benzylphenyl]azo]-	51284-75-4	NI39049
378	301	165	180	379	167	183	152	**378**	100	84	43	35	29	28	27	24		26	22	1	2	–	–	–	–	–	–	–	–	–	Phenol, 3-benzyl-4-[[2-benzylphenyl]azo]-	51284-77-6	NI39050
301	378	165	183	302	182	377	152	**378**	100	61	49	34	24	24	23	22		26	22	1	2	–	–	–	–	–	–	–	–	–	Phenol, 5-benzyl-2-[[2-benzylphenyl]azo]-	51284-76-5	NI39051
227	310	378	134	295	161	244	257	**378**	100	87	73	53	52	30	25	20		26	34	2	–	–	–	–	–	–	–	–	–	–	2,4-Bis(3,3,4-trans-p-menthadien-(1,8)-yl)resorcinol		LI08205
69	225	109	95	81	128	186	378	**378**	100	69	69	55	53	44	34	23		26	34	2	–	–	–	–	–	–	–	–	–	–	1,4-Naphthalenedione, 2-methyl-3-(3,7,11-trimethyl-2,10-dodecadienyl)-	56324-32-4	NI39052
43	41	55	121	57	71	58	69	**378**	100	91	79	62	58	56	54	52	**0.31**	26	50	1	–	–	–	–	–	–	–	–	–	–	1-Cyclopentyl-4-heneicosanone		AI02285
68	94	81	111	57	43	135	153	**378**	100	53	49	39	32	27	9	4	**0.00**	26	50	1	–	–	–	–	–	–	–	–	–	–	5-Hexacosyn-1-ol		DD01497
83	82	71	97	69	85	67	111	**378**	100	82	68	53	50	47	30	26	**0.08**	27	54	–	–	–	–	–	–	–	–	–	–	–	11-Cyclohexylheneicosane		AI02286
83	43	57	55	82	41	71	69	**378**	100	96	94	93	69	66	45	40	**0.25**	27	54	–	–	–	–	–	–	–	–	–	–	–	11-Cyclohexylheneicosane		AI02287
55	43	83	57	41	82	71	69	**378**	100	98	87	80	73	70	45	39	**0.00**	27	54	–	–	–	–	–	–	–	–	–	–	–	11-Cyclohexylheneicosane		AI02288
57	43	41	55	83	69	97	71	**378**	100	98	82	77	74	68	52	50	**0.78**	27	54	–	–	–	–	–	–	–	–	–	–	–	1-Cyclopentyl-2-decyldodecane		AI02289
125	69	43	55	41	57	83	111	**378**	100	77	45	43	32	31	27	17	**0.57**	27	54	–	–	–	–	–	–	–	–	–	–	–	1,3,5-Trimethyl-4-octadecylcyclohexane		AI02290
382	380	384	269	346	381	271	296	**379**	100	94	41	40	22	20	19	16	**4.60**	8	8	4	3	2	–	3	–	–	–	–	–	–	2H-1,2,4-Benzothiadiazine-7-sulphonamide, 6-chloro-3-(dichloromethyl)-3,4-dihydro-, 1,1-dioxide	133-67-5	NI39053
60	81	166	94	78	379	109	69	**379**	100	90	82	69	33	29	25	24		12	12	3	3	–	7	–	–	–	–	–	–	–	N-α-Heptafluorobutyryl-N-tele-ethoxycarbonylhistamine		MS08876
380	381	284	164	382	324	268	267	**379**	100	15	5	3	2	2	2	2	**0.00**	13	15	5	1	–	6	–	–	–	–	–	–	–	L-Proline, 1-(trifluoroacetyl)-4-[(trifluoroacetyl)oxy]-, butyl ester, trans-	5567-00-0	NI39054
164	57	41	69	67	165	279	278	**379**	100	80	48	28	17	16	15	12	**0.00**	13	15	5	1	–	6	–	–	–	–	–	–	–	L-Proline, 1-(trifluoroacetyl)-4-[(trifluoroacetyl)oxy]-, sec-butyl ester, trans-	57983-26-3	NI39055
73	75	147	234	116	45	43	276	**379**	100	29	26	25	22	20	19	13	**0.00**	13	29	6	1	–	–	–	–	–	3	–	–	–	1,3-Propanedioic acid, 2-acetamido-2-[(trimethylsilyl)oxy]-		NI39056
125	77	51	97	69	104	78	126	**379**	100	37	32	21	21	19	15	14	**12.71**	14	12	4	1	2	3	–	–	–	–	–	–	–	Methanesulphonamide, 1,1,1-trifluoro-N-[2-methyl-4-(phenylsulphonyl)phenyl]-	37924-13-3	NI39057
236	77	51	118	76	218	363	109	**379**	100	88	76	59	58	53	44	38	**0.00**	15	10	3	1	–	–	–	–	1	–	–	–	–	Iodonium, (1,2-dihydro-1,4-dihydroxy-2-oxo-3-quinolinyl)phenyl-, hydroxide, inner salt	75332-30-8	NI39058
73	147	205	215	117	103	142	75	**379**	100	38	21	14	13	13	12	9	**0.00**	15	37	4	1	–	–	–	–	–	3	–	–	–	D-erythro-Pentose, 2-deoxy-3,4,5-tris-O-(trimethylsilyl)-, O-methyloxime	56196-35-1	NI39059
309	362	310	288	363	364	57	246	**379**	100	68	57	38	35	33	33	32	**12.70**	16	14	4	3	1	–	1	–	–	–	–	–	–	10H-Phenothiazin-7-amine, N-acetyl-10-ethyl-3-chloro-8-nitro-, 5-oxide		NI39060
154	253	73	41	254	226	69	296	**379**	100	53	30	28	23	22	22	18	**0.00**	16	38	3	1	–	–	–	–	–	2	1	–	–	Phosphonic acid, [1-[(2,2-dimethylpropylidene)amino]pentyl]-, bis(trimethylsilyl) ester	55108-76-4	NI39061
379	201	350	230	245	229	158	200	**379**	100	32	23	20	14	14	8	6		17	17	7	1	1	–	–	–	–	–	–	–	–	5H-Furo[3,2-g][1]benzopyran-5-one, 4-methoxy-7-methyl-9-[(4-morpholinyl)sulphonyl]-		MS08877
224	225	192	226	164	42	182	149	**379**	100	14	8	4	4	4	3	3	**1.60**	17	18	4	1	–	–	–	1	–	–	–	–	–	3,5-Pyridinedicarboxylic acid, 4-(2-bromophenyl)-1,4-dihydro-2,6-dimethyl-, dimethyl ester		NI39062
264	91	84	106	379	43	315	274	**379**	100	74	47	38	25	25	20	13		17	21	1	3	3	–	–	–	–	–	–	–	–	Spiro[furo[2,3-d]thiazole-2(3H),4'-imidazolidine]-2',5'-dithione, tetrahydro-1',3',3a-trimethyl-3-benzyl-	23971-99-5	LI08206

MASS TO CHARGE RATIOS								M.W.	INTENSITIES								Parent	C	H	O	N	S	F	Cl	Br	I	Si	P	B	X	COMPOUND NAME	CAS Reg No	No
91	264	84	106	204	43	83	315	**379**	100	68	60	50	44	44	17	13	**11.26**	17	21	1	3	3	–	–	–	–	–	–	–	–	**Spiro[furo[2,3-d]thiazole-2(3H),4'-imidazolidine]-2',5'-dithione, tetrahydro-1',3',3a-trimethyl-3-benzyl-**	23971-99-5	**NI39063**
93	165	180	66	45	39	77	56	**379**	100	80	63	37	32	25	24	19	**1.30**	19	17	2	5	1	–	–	–	–	–	–	–	–	**1-Phenyl-2,3-dimethyl-4-(5-methoxybenzthiaz-2-ylazo)pyrazol-5-one**		**IC04203**
336	379	188	189	150	187	202	201	**379**	100	26	18	12	12	9	4	4		20	17	5	3	–	–	–	–	–	–	–	–	–	**Benzoic acid, 4-[N-acetyl-N-(2-methyl-6,7-methylenedioxyquinazolin-4-ylmethyl)amino]-**		**NI39064**
379	236	380	321	45	308	307	320	**379**	100	27	23	21	21	13	12	11		20	17	5	3	–	–	–	–	–	–	–	–	–	**4,11-Diamino-1,3,5,10-tetraoxo-2-(3-methoxypropyl)-2H-naphth[2,3-f]isoindole**		**IC04204**
94	95	122	83	138	222	96	82	**379**	100	99	95	64	54	49	44	43	**7.00**	20	29	6	1	–	–	–	–	–	–	–	–	–	**Schizanthin A**		**NI39065**
337	322	338	379	294	290	262	380	**379**	100	28	24	19	11	7	7	6		21	17	6	1	–	–	–	–	–	–	–	–	–	**9-Acetoxy-1,2,10-trimethoxy-7-oxodibenzo[de,g]quinoline**		**MS08878**
121	224	77	106	91	78	225	92	**379**	100	68	57	23	20	20	18	18	**6.00**	21	21	2	3	1	–	–	–	–	–	–	–	–	**N[2]-Methyl-N[2]-phenyl-N[1]-tosylformohydrazide phenylimide**		**MS08879**
209	379	73	179	210	380	224	43	**379**	100	73	61	38	30	24	22	22		21	37	3	1	–	–	–	–	–	1	–	–	–	**Nonanamide, N-[[3-methoxy-4-[(trimethylsilyl)oxy]phenyl]methyl]-8-methyl-**	56784-16-8	**MS08880**
214	216	77	111	75	215	51	76	**379**	100	33	30	24	24	13	12	10	**1.00**	22	18	3	1	–	–	1	–	–	–	–	–	–	**2-Methoxycarbonyl-6-methylphenyl-N-(2-chlorophenyl)benzimidate**		**NI39066**
119	91	65	243	89	90	136	230	**379**	100	98	36	18	11	10	6	5	**0.00**	22	21	5	1	–	–	–	–	–	–	–	–	–	**α-erythro-L-Pentofuranose, 1-cyano-2-deoxy-3,5-di-O-(4-toluoyl)-**		**NI39067**
119	91	243	65	230	137			**379**	100	18	6	6	3	1				22	21	5	1	–	–	–	–	–	–	–	–	–	**β-erythro-L-Pentofuranose, 1-cyano-2-deoxy-3,5-di-O-(4-toluoyl)-**		**NI39068**
246	165	123	199	109	247	213	95	**379**	100	47	36	21	19	18	14	10	**1.50**	22	22	2	3	–	1	–	–	–	–	–	–	–	**2H-Benzimidazol-2-one, 1-[1-[4-(4-fluorophenyl)-4-oxobutyl]-1,2,3,6-tetrahydro-4-pyridinyl]-1,3-dihydro-**	548-73-2	**LI08207**
247	155	124	200	109	248	213	96	**379**	100	47	35	20	18	17	14	10	**1.50**	22	22	2	3	–	1	–	–	–	–	–	–	–	**2H-Benzimidazol-2-one, 1-[1-[4-(4-fluorophenyl)-4-oxobutyl]-1,2,3,6-tetrahydro-4-pyridinyl]-1,3-dihydro-**	548-73-2	**NI39069**
163	176	217	188	175	379	191	189	**379**	100	83	70	67	56	50	50	44		22	25	1	3	1	–	–	–	–	–	–	–	–	**1-[(Benzyl-4-thiazolyl)methyl]-4-(4-methoxyphenyl)piperazine**		**MS08881**
91	200	337	65	92	379	291	274	**379**	100	25	23	15	14	13	11	6		23	25	4	1	–	–	–	–	–	–	–	–	–	**7-Benzyloxy-6-butyl-3-ethoxycarbonylquinol-4-one**		**IC04205**
310	379	70	165	152	224	281	295	**379**	100	83	46	27	27	24	21	19		23	25	4	1	–	–	–	–	–	–	–	–	–	**Dibenzo[f,h]pyrrolo[1,2-b]isoquinolin-14-ol, 9,11,12,13,13a,14-hexahydro-3,6,7-trimethoxy-, (13aS-cis)-**	62181-70-8	**NI39070**
310	379	70	165	152	224	281	163	**379**	100	80	40	25	24	20	18	15		23	25	4	1	–	–	–	–	–	–	–	–	–	**Dibenzo[f,h]pyrrolo[1,2-b]isoquinolin-14-ol, 9,11,12,13,13a,14-hexahydro-3,6,7-trimethoxy-, (13aS-trans)-**	75331-71-4	**NI39071**
123	379	110	98	55	96	70	97	**379**	100	96	67	64	60	56	56	53		23	45	1	3	–	–	–	–	–	–	–	–	–	**Pyrido[1,2-e][1,5]diazacycloheptadecin-5(6H)-one, 4-(4-aminobutyl)octadecahydro-, (R)-**	21008-79-7	**NI39072**
55	84	123	98	96	70	56	110	**379**	100	70	66	65	58	53	53	47	**27.51**	23	45	1	3	–	–	–	–	–	–	–	–	–	**2H-Pyrido[1,2-f][1,6]diazacyclooctadecin-11(6H)-one, 10-(3-aminopropyl)octadecahydro-, (R)-**	53602-28-1	**NI39073**
98	84	124	55	255	379	125	56	**379**	100	80	55	53	47	40	36	36		23	45	1	3	–	–	–	–	–	–	–	–	–	**1H-Pyrido[1,2-j][1,5,10]triazacyclodocosin-10(11H)-one, docosahydro-, (R)-**	21008-80-0	**NI39074**
205	379	77	179	89	338	90	76	**379**	100	51	32	26	26	25	24	22		24	17	2	3	–	–	–	–	–	–	–	–	–	**4(3H)-Quinazolinone, 2-phenyl-3-(3-methyl-4-phenyl-5-isoxazolyl)-**	94654-47-4	**NI39075**
117	379	116	247	90	77	132	65	**379**	100	80	61	48	30	15	14	13		24	21	–	5	–	–	–	–	–	–	–	–	–	**3H-Imidazo[1,2-b][1,2,4]triazol-2-amine, N-(4-methylphenyl)-3-(4-methylphenyl)-5-phenyl-**		**MS08882**
98	314	343	99	70	138	315	67	**379**	100	29	12	12	12	10	7	6	**0.00**	24	42	–	1	–	–	1	–	–	–	–	–	–	**Androst-2-en-17-amine, 4,4-dimethyl-N-propyl-, hydrochloride, (5α)-**	53286-43-4	**NI39076**
336	378	350	91	199	377	351	198	**379**	100	82	81	77	73	54	41	41	**6.37**	25	33	2	1	–	–	–	–	–	–	–	–	–	**6-Azaandrost-4-en-7-one, 6-benzyl-17β-hydroxy-**	16373-61-8	**NI39077**
219	232	191	91	217	203	160	76	**379**	100	94	80	52	46	46	46	46	**44.00**	26	21	2	1	–	–	–	–	–	–	–	–	–	**1H-Isoindole-1,3(2H)-dione, 2-[3-(10,11-dihydro-5H-dibenzo[a,d]cyclohepten-5-ylidene)propyl]-**	74810-79-0	**NI39078**
113	163	69	182	63	311	380	198	**380**	100	35	28	27	12	9	8	2		6	–	1	–	2	12	–	–	–	–	–	–	–	**2,2,4,4-Tetrakis(trifluoromethyl)-1,3-dithietane-1-oxide**		**DD01498**
51	69	169	119	285	95			**380**	100	51	27	9	3	2				7	1	3	–	–	13	–	–	–	–	–	–	–	**Propanoic acid, 2,3,3,3-tetrafluoro-(3-heptafluoropropoxy)-, difluoromethyl ester**		**NI39079**
59	115	28	240	112	352	87	118	**380**	100	99	99	98	98	67	48	40	**6.00**	9	1	6	–	–	3	–	–	–	–	–	–	2	**Cobalt, hexacarbonyl[μ-(3,3,3-trifluoropropyne)]di-**		**LI08208**
69	43	153	247	57	45	157	59	**380**	100	90	61	56	45	45	44	42	**0.00**	9	5	6	–	–	9	–	–	–	–	–	–	–	**Acetic acid, trifluoro-, 1,2,3-propanetriyl ester**	26158-83-8	**NI39080**
153	69	253	140	97	125	65	99	**380**	100	72	67	30	29	27	27	24	**0.00**	9	5	6	–	–	9	–	–	–	–	–	–	–	**Acetic acid, trifluoro-, 1,2,3-propanetriyl ester**	26158-83-8	**NI39081**
55	128	380	253	127	125			**380**	100	34	24	22	18	17				9	18	–	–	–	–	–	–	2	–	–	–	–	**α,ω-Diiodononane**		**LI08209**
184	212	144	326	183	268	352	296	**380**	100	86	76	45	32	27	26	25	**7.00**	11	8	6	–	1	–	–	–	–	–	–	–	2	**[μ-η[1],η[2]-propadienyl](μ-ethanethiolato)hexacarbonyldiiron**		**MS08883**
382	266	380	384	268	347	345	383	**380**	100	82	73	62	55	39	39	28		12	8	2	–	–	–	4	–	–	2	–	–	–	**1,5-Dioxa-2,6-disila-3,4:7,8-dibenzocyclooctane, 2,2,6,6-tetrachloro-**	63503-14-0	**NI39082**
190	282	338	364	134	284	236	336	**380**	100	50	49	41	37	34	33	27	**8.00**	12	14	4	–	–	–	–	2	–	–	–	–	–	**Benzoic acid, 2,4-dihydroxy-3,5-dibromo-6-pentyl-**	86791-41-5	**NI39083**
223	208	253	221	251	207	222	206	**380**	100	98	63	54	49	49	46	42	**0.00**	12	28	–	–	–	–	–	–	–	–	–	–	1	**Plumbane, tetrapropyl-**		**AI02291**
159	380	111	76	253	75	382	63	**380**	100	98	97	95	88	86	72	68		13	10	5	–	2	–	2	–	–	–	–	–	–	**4-Chloro-3'-(chlorosulphonyl)-4'-methoxydiphenyl sulphide**		**IC04206**
166	57	41	167	126	71	139	56	**380**	100	86	84	71	31	28	27	22	**0.00**	13	18	4	2	–	6	–	–	–	–	–	–	–	**L-Ornithine, N[2],N[5]-bis(trifluoroacetyl)-, butyl ester**	2804-68-4	**NI39084**
166	41	57	29	167	69	28		**380**	100	27	26	23	17	17	15		**1.00**	13	18	4	2	–	6	–	–	–	–	–	–	–	**L-Ornithine, N[2],N[5]-bis(trifluoroacetyl)-, butyl ester**	2804-68-4	**LI08210**
166	57	167	41	69	44	43	126	**380**	100	63	56	49	24	19	16	12	**0.00**	13	18	4	2	–	6	–	–	–	–	–	–	–	**L-Ornithine, N[2],N[5]-bis(trifluoroacetyl)-, sec-butyl ester**	57983-25-2	**NI39085**
43	83	55	41	177	339	175	337	**380**	100	82	74	72	61	60	46	41	**1.00**	13	25	–	–	–	–	–	1	–	–	–	–	1	**Butylisopropylcyclohexyltin bromide**		**MS08884**

MASS TO CHARGE RATIOS								M.W.	INTENSITIES								Parent	C	H	O	N	S	F	Cl	Br	I	Si	P	B	X	COMPOUND NAME	CAS Reg No	No
91	65	197	156	139	116	77	107	380	100	61	52	48	29	24	20	19	0.00	14	12	2	4	1	–	–	–	–	–	–	–	1	2,1,3-Benzoselenadiazole-4-carboxaldehyde, [(4-methylphenyl)sulphonylhydrazone]		MS08885
91	156	65	196	224	139	116	107	380	100	65	63	33	28	27	21	16	1.00	14	12	2	4	1	–	–	–	–	–	–	–	1	2,1,3-Benzoselenadiazole-5-carboxaldehyde, [(4-methylphenyl)sulphonylhydrazone]		MS08886
70	28	41	43	45	59	40	44	380	100	82	67	62	48	35	33	27	3.00	14	15	7	2	–	3	–	–	–	–	–	–	–	Uridine, 2′,3′-O-isopropylidene-, 5′-(trifluoroacetate)	1598-49-8	NI39086
73	380	117	45	275	307	189	74	380	100	24	24	13	10	9	9	9		14	28	–	–	4	–	–	–	–	2	–	–	–	2,4,6,8-Tetrathiatricyclo[$3.3.1.1^{3,7}$]decane, 9,10-dimethyl-, bis(trimethylsilyl)-	72088-23-4	NI39087
102	86	103	128	87	60	73	69	380	100	73	36	28	26	19	18	18	11.81	14	28	4	4	2	–	–	–	–	–	–	–	–	1,4,7,10-Tetraoxa-13,15,18,20-tetraazacyclodocosane-14,19-dithione		NI39088
102	86	128	103	87	60	73	69	380	100	73	39	36	26	19	18	18	11.80	14	28	4	4	2	–	–	–	–	–	–	–	–	1,4,7,10-Tetraoxa-13,15,18,20-tetraazacyclodocosane-14,19-dithione		MS08887
86	380	190	69	129	103	130	60	380	100	43	32	32	31	31	29	29		14	28	4	4	2	–	–	–	–	–	–	–	–	1,4,7,15-Tetraoxa-10,12,18,20-tetraazacyclodocosane-11,19-dithione	74804-40-3	NI39089
86	60	85	59	69	103	147	130	380	100	37	33	28	26	21	20	17	4.07	14	28	4	4	2	–	–	–	–	–	–	–	–	1,4,7,15-Tetraoxa-10,12,18,20-tetraazacyclodocosane-11,19-dithione	74804-40-3	NI39090
217	73	145	129	218	75	132	117	380	100	62	35	26	21	19	16	16	0.00	15	36	5	–	–	–	–	–	–	3	–	–	–	α-DL-Arabinofuranoside, methyl 2,3,5-tris-O-(trimethylsilyl)-	56271-62-6	NI39091
217	73	204	133	147	218	219	205	380	100	88	60	56	30	28	14	12	0.00	15	36	5	–	–	–	–	–	–	3	–	–	–	β-DL-Arabinofuranoside, methyl 2,3,5-tris-O-(trimethylsilyl)-	56271-63-7	NI39092
217	73	218	204	147	219	133	103	380	100	39	26	19	15	14	14	7	0.00	15	36	5	–	–	–	–	–	–	3	–	–	–	α-DL-Lyxofuranoside, methyl 2,3,5-tris-O-(trimethylsilyl)-	56390-03-5	NI39093
204	217	73	133	147	205	218	219	380	100	88	68	36	28	22	20	12	0.00	15	36	5	–	–	–	–	–	–	3	–	–	–	Lyxopyranoside, methyl 2,3,4-tris-O-(trimethylsilyl)-	56271-61-5	NI39094
204	73	133	147	205	218	75	45	380	100	72	31	28	20	18	11	10	0.00	15	36	5	–	–	–	–	–	–	3	–	–	–	α-D-Pyranoside, methyl 2,3,4-tris-O-(trimethylsilyl)-	18082-35-4	NI39095
217	73	133	204	218	147	219	75	380	100	83	38	36	21	21	10	9	0.00	15	36	5	–	–	–	–	–	–	3	–	–	–	α-D-Ribofuranoside, methyl 2,3,5-tris-O-(trimethylsilyl)-	56227-32-8	NI39096
146	73	191	159	147	131	204	75	380	100	83	48	33	29	19	12	12	0.00	15	36	5	–	–	–	–	–	–	3	–	–	–	D-Xylopyranose, 2-O-methyl-1,3,4-tris-O-(trimethylsilyl)-	20561-78-8	NI39097
217	73	146	133	218	147	89	131	380	100	78	67	41	22	21	15	13	0.00	15	36	5	–	–	–	–	–	–	3	–	–	–	Xylopyranose, 3-O-methyl-1,2,4-tris-O-(trimethylsilyl)-	20561-79-9	NI39098
73	204	217	147	133	205	218	129	380	100	99	65	40	35	19	13	12	0.00	15	36	5	–	–	–	–	–	–	3	–	–	–	α-D-Xylopyranoside, methyl 2,3,4-tris-O-(trimethylsilyl)-	18082-35-4	NI39099
39	345	148	132	169	99	175	325	380	100	50	50	50	46	43	31	27	0.00	16	10	–	–	–	5	1	–	–	–	–	–	1	Titanium, chlorobis(η^5-2,4-cyclopentadien-1-yl)(pentafluorophenyl)-	50648-18-5	NI39100
226	227	133	157	186	269	187	156	380	100	80	23	20	17	16	16	15	1.03	16	14	2	2	–	6	–	–	–	–	–	–	–	N,N′-1,2-Phenylenebis(5,5,5-trifluoro-4-oxopentan-2-imine)	87961-38-4	NI39101
57	195	56	197	35	185	307	177	380	100	17	16	6	6	5	4	4	0.00	16	33	4	–	–	–	1	–	–	2	–	–	–	Disiloxane, 1,3-divinyl-1,1,3-tributoxy-3-chloro-		NI39102
380	75	121	92	56	381	167	39	380	100	67	30	29	24	20	20	17		17	9	1	–	–	5	–	–	–	–	–	–	1	Ferrocene, (pentafluorobenzoyl)-	31903-78-3	NI39103
57	71	43	55	41	196	198	69	380	100	66	63	36	33	30	27	22	3.37	17	23	3	–	–	–	3	–	–	–	–	–	–	Propanoic acid, 2-(2,4,5-trichlorophenoxy)-, isooctyl ester	32534-95-5	AI02292
57	43	71	55	41	196	69	29	380	100	57	43	26	25	19	17	16	2.26	17	23	3	–	–	–	3	–	–	–	–	–	–	Propanoic acid, 2-(2,4,5-trichlorophenoxy)-, isooctyl ester	32534-95-5	AI02293
57	71	43	41	55	196	198	69	380	100	78	71	56	49	30	29	27	3.00	17	23	3	–	–	–	3	–	–	–	–	–	–	Propanoic acid, 2-(2,4,5-trichlorophenoxy)-, isooctyl ester	32534-95-5	NI39104
166	279	41	86	29	28	167	69	380	100	42	37	36	30	23	17	17	1.90	17	27	4	2	–	3	–	–	–	–	–	–	–	N-Trifluoroacetyl-L-prolylisoleucine butyl ester		MS08888
166	167	86	279	41	28	29	43	380	100	69	50	37	31	19	18	15	1.10	17	27	4	2	–	3	–	–	–	–	–	–	–	N-Trifluoroacetyl-L-prolylleucine butyl ester		MS08889
166	250	41	167	28	86	29	279	380	100	41	39	37	35	30	27	25	0.00	17	27	4	2	–	3	–	–	–	–	–	–	–	N-Trifluoroacetyl-L-prolylleucine tert-butyl ester		MS08890
166	86	167	279	41	42	70	139	380	100	72	62	42	35	26	24	17	0.00	17	27	4	2	–	3	–	–	–	–	–	–	–	N-Trifluoroacetyl-L-prolyl-α-methylvaline butyl ester		MS08891
166	167	86	279	41	29	139	28	380	100	76	47	34	28	24	15	14	3.40	17	27	4	2	–	3	–	–	–	–	–	–	–	N-Trifluoroacetyl-L-prolylnorleucine butyl ester		MS08892
380	268	296	324	213	241	78	352	380	100	39	35	27	25	21	21	16		18	12	6	4	–	–	–	–	–	–	–	–	–	N,N′-Bis(ethoxymethylene)-2,6-diamino-4,8-dihydro-4,8-dioxo-benzo[1,2-b:4,5-b′]difuran-3,7-dicarbonitrile		MS08893
380	132	382	381	352	337	249	207	380	100	80	52	20	19	10	9	9		18	14	5	–	–	–	2	–	–	–	–	–	–	4H-1-Benzopyran-4-one, 6,8-dichloro-5,7-dimethoxy-2-(4-methoxyphenyl)	85785-49-5	NI39105
77	380	188	105	78	206	218		380	100	48	44	28	26	18	11			18	16	2	6	1	–	–	–	–	–	–	–	–	Anhydro-4-hydroxy-1-methyl-3-phenyl-1,2,3-triazolium hydroxide		LI08211
192	234	193	208	163	170	57	191	380	100	79	79	78	78	76	74	73	0.00	18	24	7	2	–	–	–	–	–	–	–	–	–	3-Methyl-4-[(N-acetylalanyl)amino]-5-O-acetyl-8-methoxy-3,4,6,7-tetrahydroisocoumarin		LI08212
174	73	175	264	307	176	86	147	380	100	45	17	13	11	10	10	9	2.57	18	36	1	2	–	–	–	–	–	3	–	–	–	1-Propanone, 3-amino-1-(2-aminophenyl)-, tris(trimethylsilyl)-	56145-12-1	NI39106
57	71	43	85	55	41	69	99	380	100	70	49	48	30	27	19	15	0.00	18	37	–	–	–	–	–	–	1	–	–	–	–	Octadecane, 1-iodo-	629-93-6	LI08213
57	253	71	85	254	99	295	113	380	100	88	58	38	19	16	13	12	2.40	18	37	–	–	–	–	–	–	1	–	–	–	–	Octadecane, 1-iodo-	629-93-6	NI39107
380	381	91	306	352	106	350	75	380	100	22	20	16	13	12	10	9		19	12	7	2	–	–	–	–	–	–	–	–	–	Ethyl 2-(3-nitrophenyl)-5-oxopyrano[3,2-f]benzoxazole-7-carboxylate		LI08214
380	381	352	306	350	75	251	281	380	100	24	11	11	8	8	6	5		19	12	7	2	–	–	–	–	–	–	–	–	–	Ethyl 2-(3-nitrophenyl)-7-oxopyrano[2,3-e]benzoxazole-5-carboxylate		LI08215
380	381	306	352	75	251	205	334	380	100	22	13	8	8	5	5	4		19	12	7	2	–	–	–	–	–	–	–	–	–	Ethyl 2-(4-nitrophenyl)-5-oxopyrano[3,2-f]benzoxazole-7-carboxylate		LI08216
380	381	91	306	92	75	352	350	380	100	22	14	10	10	10	7	6		19	12	7	2	–	–	–	–	–	–	–	–	–	Ethyl 2-(4-nitrophenyl)-7-oxopyrano[2,3-e]benzoxazole-5-carboxylate		LI08217
43	111	153	108	83	273	171	107	380	100	99	66	43	41	30	19	15	0.00	19	24	8	–	–	–	–	–	–	–	–	–	–	β-L-Rhamnopyranoside, 4-methylphenyl 2,3,4-tri-O-acetyl-		NI39108
43	163	41	98	79	181	107	135	380	100	24	19	18	14	13	13	12	1.80	19	24	8	–	–	–	–	–	–	–	–	–	–	Trichothec-9-en-8-one, 3,15-diacetoxy-12,13-epoxy-7-hydroxy-, (3α,7α)-		NI39109
109	121	91	124	55	125	173	95	380	100	98	64	63	58	55	52	51	18.50	19	24	8	–	–	–	–	–	–	–	–	–	–	Trichothec-9-en-8-one, 4,15-diacetoxy-12,13-epoxy-3-hydroxy-, (3α,4β)-		NI39110
98	136	79	137	55	108	77	69	380	100	91	90	87	86	81	74	74	3.40	19	24	8	–	–	–	–	–	–	–	–	–	–	Trichothec-9-en-8-one, 7,15-diacetoxy-12,13-epoxy-3-hydroxy-, (3α,7α)-		NI39111
150	151	105	135	130	107	121	123	380	100	99	98	97	95	92	82	81	7.00	20	16	6	2	–	–	–	–	–	–	–	–	–	1-Methyl-1-(4-nitrobenzoyl)cyclopropane-2,3-(N-phenyl)dicarboximide		LI08218
135	77	352	107	353	134	76	68	380	100	9	6	4	1	1	1	1	0.80	20	20	4	4	–	–	–	–	–	–	–	–	–	4,5-Dimethyl-1-[α-(4-methoxybenzoyloxy)-4-methoxybenzylideneimino]-1,2,3-triazole		MS08894
58	322	135	133	77	59	149	103	380	100	7	4	4	4	4	3	3	1.00	20	24	–	6	1	–	–	–	–	–	–	–	–	N-Ethyl-N-(2-dimethylaminoethyl)-4-(3-phenyl-1,2,4-thiadiaz-5-ylazo)aniline		IC04207
324	245	260	306	263	246	228	289	380	100	53	49	37	31	31	30	21	3.00	20	28	7	–	–	–	–	–	–	–	–	–	–	Tricyclo[$6.4.0.0^{2,6}$]-dodec-10-ene-7,11-dicarboxylic acid, 8,10-dihydroxy-7 methyl-4-oxo-, tert-butyl methyl ester, (1α,2α,6α,7β,8α)-		MS08895

MASS TO CHARGE RATIOS								M.W.	INTENSITIES								Parent	C	H	O	N	S	F	Cl	Br	I	Si	P	B	X	COMPOUND NAME	CAS Reg No	No
163	73	131	171	164	75	55	77	**380**	100	36	35	18	11	10	8	7	**0.00**	20	32	5	–	–	–	–	–	–	1	–	–	–	Methyl 2-ethyl-4-(trimethylsilyloxy)hexylphthalate		NI39112
117	163	73	55	75	69	164	118	**380**	100	97	73	22	21	17	12	12	**0.00**	20	32	5	–	–	–	–	–	–	1	–	–	–	Methyl 2-ethyl-5-(trimethylsilyloxy)hexylphthalate		NI39113
323	73	68	41	324	75	207	40	**380**	100	50	47	30	29	29	20	13	**0.00**	20	36	3	–	–	–	–	–	–	2	–	–	–	Benzeneacetic acid, 4-[(4-tert-butyldimethylsilyl)oxy]-, tertbutyldimethylsilyl ester		MS08896
73	147	221	295	323	75	41	68	**380**	100	55	48	36	25	12	11	9	**0.00**	20	36	3	–	–	–	–	–	–	2	–	–	–	Benzeneglycolic acid, α-[(4-tert-butyldimethylsilyl)oxy]-, tert-butyldimethylsilyl ester		MS08897
179	73	380	180	75	365	381	181	**380**	100	44	32	17	15	12	10	6		20	36	3	–	–	–	–	–	–	2	–	–	–	Benzeneoctanoic acid, 4-[(trimethylsilyl)oxy]-, trimethylsilyl ester		NI39114
380	381	293	321	175	163	382	335	**380**	100	25	20	10	10	7	6	6		21	16	7	–	–	–	–	–	–	–	–	–	–	Naphtho[2,3-c]furan-1(3H)-one, 4-hydroxy-6,7-dimethoxy-9-(1,3-benzodioxol-5-yl)-	22055-22-7	NI39115
380	305	129	337	365	277	163	279	**380**	100	26	14	8	6	6	6	5		21	16	7	–	–	–	–	–	–	–	–	–	–	Naphtho[2,3-c]furan-1(3H)-one, 7-hydroxy-4,8-dimethoxy-9-(1,3-benzodioxol-5-yl)-	82012-44-0	NI39116
58	204	149	333	207	335	334	380	**380**	100	8	5	3	3	2	2	1		21	20	5	2	–	–	–	–	–	–	–	–	–	8H-1,3-Dioxolo[4,5-e]isoindol-8-one, 6-[[6-[2-(dimethylamino)ethyl]-1,3-benzodioxol-5-yl]methylene]-6,7-dihydro-, (E)-		NI39117
362	321	91	55	79	41	93	43	**380**	100	73	73	60	57	54	51	49	**23.39**	21	29	4	–	–	–	1	–	–	–	–	–	–	Pregn-4-ene-3,20-dione, 19-chloro-17,21-dihydroxy-	4050-05-9	NI39118
380	352	245	278	132	227	308	145	**380**	100	92	58	49	39	27	21	21		21	32	2	–	2	–	–	–	–	–	–	–	–	17β-Hydroxy-19-nor-5α-androstan-2-spiro-2′-[1′,3′-dithian]-3-one		LI08219
365	323	83	67	95	43	55	109	**380**	100	99	82	67	66	53	43	42	**38.00**	21	32	6	–	–	–	–	–	–	–	–	–	–	Isotetrahydrovincetogenin		LI08220
152	121	153	344	362	165	154	329	**380**	100	97	80	78	74	63	59	54	**24.00**	21	32	6	–	–	–	–	–	–	–	–	–	–	Picras-2-en-1-one, 11,12,16-trihydroxy-2-methoxy-, (11α,12β)-	30760-22-6	NI39119
362	344	43	326	91	120	105	380	**380**	100	87	58	30	29	27	24	23		21	32	6	–	–	–	–	–	–	–	–	–	–	Pregn-5-en-20-one, 3,8,11,12,14-pentahydroxy-, (3β,11α,12β,14β,17α)-	22149-67-3	NI39120
83	362	305	67	43	91	365	334	**380**	100	96	92	83	81	79	77	64	**56.00**	21	32	6	–	–	–	–	–	–	–	–	–	–	Tetrahydrovincetogenin		LI08221
286	380	313	250	382	125	320	349	**380**	100	91	91	83	44	37	34	20		22	14	2	–	–	–	2	–	–	–	–	–	–	13-Triptycenecarboxylic acid, 1,8-dichloro-, methyl ester		NI39121
286	380	313	250	125	382	320	349	**380**	100	96	89	63	61	51	33	22		22	14	2	–	–	–	2	–	–	–	–	–	–	16-Triptycenecarboxylic acid, 1,8-dichloro-, methyl ester		NI39122
380	217	365	233	148	161	337	77	**380**	100	93	90	30	23	20	18	17		22	20	6	–	–	–	–	–	–	–	–	–	–	2H,8H-Benzo[1,2-b:5,4-b′]dipyran-2-one, 4-hydroxy-5-methoxy-3-(4-methoxyphenyl)-8,8-dimethyl-	5307-59-5	NI39123
105	204	205	380	227	189	243	175	**380**	100	69	34	31	20	18	14	11		22	20	6	–	–	–	–	–	–	–	–	–	–	(+)-cis-Benzoylkhellactone methyl ether		LI08222
243	105	204	205	244	380	189	175	**380**	100	57	22	20	18	17	11	7		22	20	6	–	–	–	–	–	–	–	–	–	–	(+)-trans-Benzoylkhellactone methyl ether		LI08223
119	380	281	309	76	147	337	381	**380**	100	99	78	40	40	35	28	27		22	20	6	–	–	–	–	–	–	–	–	–	–	2,5-Cyclohexadiene-1,4-dione, 2,5-dimethoxy-3,6-bis(4-methoxyphenyl)-	51860-94-7	NI39124
380	365	334	349	335	312	295	323	**380**	100	35	23	21	19	19	15	13		22	20	6	–	–	–	–	–	–	–	–	–	–	5-O-Methylglycyrol		LI08224
197	198	105	77	104	268	165	211	**380**	100	15	15	7	5	3	3	2	**0.00**	22	21	1	–	–	–	–	1	–	–	–	–	–	1-Bromo-3-methoxy-2,3,3-triphenylpropane		MS08898
279	195	196	349	380	317	221		**380**	100	82	76	47	37	24	22			22	24	4	2	–	–	–	–	–	–	–	–	–	Alstonidine		LI08225
299	57	56	300	380	174	81	82	**380**	100	70	70	60	57	42	42	30		22	24	4	2	–	–	–	–	–	–	–	–	–	Benzo[1,2-c:4,5-c′]dipyrrole-1,3,5,7(2H,6H)-tetrone, 2,6-dicyclohexyl-	31663-85-1	MS08899
299	56	300	380	174	81	82	67	**380**	100	70	60	57	42	42	30	30		22	24	4	2	–	–	–	–	–	–	–	–	–	Benzo[1,2-c:4,5-c′]dipyrrole-1,3,5,7(2H,6H)-tetrone, 2,6-dicyclohexyl-	31663-85-1	NI39125
43	121	380	44	321	84	247	206	**380**	100	99	86	57	50	29	27	23		22	24	4	2	–	–	–	–	–	–	–	–	–	Curan-17-oic acid, 14-(acetyloxy)-2,16,19,20-tetradehydro-, methyl ester, (19E)-	56053-20-4	NI39126
227	137	229	195	170	168	214	215	**380**	100	58	53	52	52	34	33	31	**29.00**	22	24	4	2	–	–	–	–	–	–	–	–	–	2,20-Cycloaspidospermidine-3-carboxylic acid, 1-formyl-8-oxo-, methyl ester, (2α,3β,5α,12β,19α,20R)-	56086-67-0	NI39127
43	352	78	124	110	41	201	77	**380**	100	68	57	36	33	30	27	24	**12.00**	22	24	4	2	–	–	–	–	–	–	–	–	–	2,20-Cycloaspidospermidine-3-carboxylic acid, 1-formyl-10-oxo-, methyl ester, (2α,3α,5α,12β,19α,20R)-	56145-38-1	NI39128
186	185	380	199	323	144	309	171	**380**	100	31	24	21	17	9	8	7		22	24	4	2	–	–	–	–	–	–	–	–	–	Isoerinin		LI08226
380	186	185	149	381	187	309	171	**380**	100	85	28	28	25	18	17	8		22	24	4	2	–	–	–	–	–	–	–	–	–	Isoerinin		LI08227
320	279	266	380	265	261	321	207	**380**	100	67	67	48	47	26	25	25		22	24	4	2	–	–	–	–	–	–	–	–	–	17-Methoxy-3,4-didehydrotalbotine		LI08228
380	381	210	184	143	130	129	172	**380**	100	26	26	25	20	18	15	13		22	24	4	2	–	–	–	–	–	–	–	–	–	Pagicerine	99831-97-7	NI39129
380	225	130	183	210	143	144	168	**380**	100	54	42	34	31	29	28	25		22	24	4	2	–	–	–	–	–	–	–	–	–	16,19-Secostrychnidine-10,16-dione, 14-hydroxy-19-methyl-	22029-96-5	NI39130
321	381	242	174	147	236	159	365	**380**	100	48	40	37	34	30	29	25	**19.91**	22	24	4	2	–	–	–	–	–	–	–	–	–	Voachalotine oxindole, 17-deoxy-6,17-epoxy-, (6β)-	32303-66-5	NI39131
380	321	381	242	174	146	236	159	**380**	100	50	24	19	19	17	15	15		22	24	4	2	–	–	–	–	–	–	–	–	–	Voachalotine oxindole, 17-deoxy-6,17-epoxy-, (6β)-	32303-66-5	LI08229
277	350	198	278	211	380	199	363	**380**	100	36	34	28	21	17	16	15		22	24	4	2	–	–	–	–	–	–	–	–	–	Voachalotine, 6-oxo-	32303-65-4	NI39132
380	381	146	321	242	159	254	253	**380**	100	26	22	18	16	16	14	14		22	24	4	2	–	–	–	–	–	–	–	–	–	Voachalotine pseudoindoxyl, 17-deoxy-6,17-epoxy-, (6β)-	32487-54-0	NI39133
380	321	44	381	322	43	42	57	**380**	100	79	38	22	16	15	14	13		22	24	4	2	–	–	–	–	–	–	–	–	–	Vomicine	125-15-5	LI08230
380	321	251	159	146	306	160	199	**380**	100	92	18	18	18	14	13	12		22	24	4	2	–	–	–	–	–	–	–	–	–	Vomicine	125-15-5	NI39134
73	77	246	45	218	57	380	183	**380**	100	47	34	20	17	14	13	13		22	28	2	2	–	–	–	–	–	1	–	–	–	3H-Pyrazol-3-one, 4-butyl-1,2-dihydro-1,2-diphenyl-5-[(trimethylsilyl)oxy]-	74810-87-0	NI39135
265	58	252	266	323	380	251	72	**380**	100	98	60	51	48	47	26	19		22	28	2	4	–	–	–	–	–	–	–	–	–	1,4-Bis-(2-ethylaminoethylamino)anthraquinone		IC04208
380	345	382	381	223	238	28	190	**380**	100	39	35	25	21	21	19	17		23	21	3	–	–	–	1	–	–	–	–	–	–	Benzene, 1,1′,1″-(1-chloro-1-ethenyl-2-ylidene)tris[4-methoxy-	569-57-3	NI39136
380	338	337	381	339	282	281	165	**380**	100	92	72	26	20	20	13	12		23	24	5	–	–	–	–	–	–	–	–	–	–	Rotiorin		LI08231
380	213	174	321	212	190	337		**380**	100	13	10	9	8	7	5			23	28	3	2	–	–	–	–	–	–	–	–	–	O-Acetylseredamine		LI08232
380	381	43	379	213	174	321	212	**380**	100	25	23	18	17	12	10	10		23	28	3	2	–	–	–	–	–	–	–	–	–	Ajmalan-17-ol, 19,20-didehydro-10-methoxy-, acetate (ester), (17R,19E)-	6519-30-8	LI08233
380	381	379	213	174	321	212	187	**380**	100	24	18	14	11	10	10	9		23	28	3	2	–	–	–	–	–	–	–	–	–	Ajmalan-17-ol, 19,20-didehydro-10-methoxy-, acetate (ester), (17R,19E)-	6519-30-8	NI39137

MASS TO CHARGE RATIOS								M.W.	INTENSITIES								Parent	C	H	O	N	S	F	Cl	Br	I	Si	P	B	X	COMPOUND NAME	CAS Reg No	No
109	43	380	124	352	337	338	310	380	100	90	46	35	22	21	19	17		23	28	3	2	–	–	–	–	–	–	–	–	–	Aspidofractinin-17-ol, 1-acetyl-, acetate (ester)	55724-55-5	NI39138
143	144	237	164	130	55	190	238	380	100	27	20	17	12	11	8	6	2.00	23	28	3	2	–	–	–	–	–	–	–	–	–	Propanoic acid, 3-(7-oxo-2,3,4,4a,5,6,7,9,10,15-decahydro-1H-quinolino[8a,1-a]-β-carbolin-1-yl)-, methyl ester		MS08900
55	41	43	29	67	79	307	81	380	100	97	90	90	61	46	45	43	1.00	23	40	4	–	–	–	–	–	–	–	–	–	–	Methyl 3-ethoxycarbonyl-2-octylcyclopropeneoctanoate		NI39139
73	75	67	55	81	41	95	54	380	100	96	68	55	53	45	36	29	4.30	23	44	2	–	–	–	–	–	–	1	–	–	–	11,14-Eicosadienoic acid, trimethylsilyl ester		NI39140
295	87	81	138	137	95	77	241	380	100	88	66	63	51	33	31	30	21.59	23	44	2	–	–	–	–	–	–	1	–	–	–	Silane, vinylmethylbis[[5-methyl-2-isopropylcyclohexyl]oxy]-, [1α(1R*,2S*,5R*),2β,5α]-	59632-82-5	NI39141
380	181	364	381	182	168	167	199	380	100	33	29	28	28	28	20	16		24	20	1	4	–	–	–	–	–	–	–	–	–	4,4'-Bis(anilino)azoxybenzene		IC04209
91	380	261	106	129	381	107	262	380	100	99	79	67	49	27	21	19		24	20	1	4	–	–	–	–	–	–	–	–	–	1-Naphthalenol, 2,4-bis(4-tolylazo)-		IC04210
91	143	115	261	65	380	207	89	380	100	53	47	28	20	15	12	9		24	20	1	4	–	–	–	–	–	–	–	–	–	2-Naphthalenol, 1-[[2-methyl-4-[(2-methylphenyl)azo]phenyl]azo]-	85-83-6	NI39142
380	273	349	381	83	217	55	41	380	100	41	33	26	14	12	11	10		24	28	4	–	–	–	–	–	–	–	–	–	–	2-Methoxybenzaldehyde anhydrodimedone		LI08234
268	57	239	107	324	145	269	121	380	100	72	29	28	26	19	18	9	8.56	24	28	4	–	–	–	–	–	–	–	–	–	–	Phenol, 4,4'-(1,2-diethyl-1,2-ethenediyl)bis-, dipropanoate, (E)-	130-80-3	NI39143
365	366	380	339	367	381	340	173	380	100	31	24	12	9	7	4	4		24	32	2	–	–	–	–	–	–	1	–	–	–	Propylcannabinol allyldimethylsilyl ether		NI39144
105	77	275	246	135	134	106	380	380	100	37	13	10	10	8	8	6		24	32	2	2	–	–	–	–	–	–	–	–	–	N,N'-Dibenzoyl-1,10-diaminodecane		LI08235
57	223	239	279	251	83	95	131	380	100	53	22	17	15	15	14	13	1.60	24	44	3	–	–	–	–	–	–	–	–	–	–	4-Hexadecynoic acid, 2-butoxy-, butyl ester	40924-20-7	NI39145
114	115	43	57	55	41	86	69	380	100	44	35	32	22	22	20	20	9.01	25	48	2	–	–	–	–	–	–	–	–	–	–	D-Heneicosanoic acid, 4-methyl-2-methylene-, ethyl ester	18607-50-6	NI39146
114	115	43	57	55	41	69	102	380	100	44	35	32	22	22	20	19	9.00	25	48	2	–	–	–	–	–	–	–	–	–	–	D-Heneicosanoic acid, 4-methyl-2-methylene-, ethyl ester	18607-50-6	LI08236
102	83	43	69	380	55	57	41	380	100	70	69	58	53	44	42	42		25	48	2	–	–	–	–	–	–	–	–	–	–	2-Heneicosenoic acid, 2,4-dimethyl-, ethyl ester, [R-(E)]-	55429-11-3	NI39147
102	43	83	380	69	55	41	57	380	100	84	72	62	62	52	52	50		25	48	2	–	–	–	–	–	–	–	–	–	–	2-Heneicosenoic acid, 2,4-dimethyl-, ethyl ester, D-(Z)-	18607-49-3	NI39148
55	43	69	57	41	40			380	100	84	76	74	67	61			5.64	25	48	2	–	–	–	–	–	–	–	–	–	–	15-Tetracosenoic acid, methyl ester	56554-33-7	NI39149
55	41	43	69	74	83	57	28	380	100	65	60	57	50	41	35	30	4.84	25	48	2	–	–	–	–	–	–	–	–	–	–	15-Tetracosenoic acid, methyl ester	56554-33-7	NI39150
55	43	41	69	83	57	97	74	380	100	80	80	64	50	50	41	33	0.00	25	48	2	–	–	–	–	–	–	–	–	–	–	15-Tetracosenoic acid, methyl ester, (Z)-	2733-88-2	NI39151
380	379	303	202	381	183	277	304	380	100	97	83	32	27	27	22	20		26	21	1	–	–	–	–	–	–	–	1	–	–	(Benzoylmethylene)triphenylphosphorane	859-65-4	LI08237
303	380	379	183	202	277	381	201	380	100	99	88	51	47	32	25	22		26	21	1	–	–	–	–	–	–	–	1	–	–	(Benzoylmethylene)triphenylphosphorane	859-65-4	NI39152
263	380	248	264	381	379	247	91	380	100	54	47	25	17	17	15	12		26	24	1	2	–	–	–	–	–	–	–	–	–	4H-1,2-Diazepine, 5-(4-methoxyphenyl)-3,7-bis(4-methylphenyl)-	25649-77-8	NI39153
43	57	82	55	41	96	83	69	380	100	87	70	59	53	48	45	44	0.00	26	52	1	–	–	–	–	–	–	–	–	–	–	1-Hexacosanal		NI39154
380	225	172	365	171	226	198	240	380	100	90	68	60	60	46	16	12		27	40	1	–	–	–	–	–	–	–	–	–	–	19-Norcholesta-1,3,5(10)-trien-6-one, 1-methyl-	71472-80-5	NI39155
57	43	71	85	41	55	69	309	380	100	97	64	49	47	43	28	27	0.05	27	56	–	–	–	–	–	–	–	–	–	–	–	8,8-Dipentylheptadecane		AI02294
43	57	71	85	41	55	69	70	380	100	99	65	48	43	41	26	17	0.00	27	56	–	–	–	–	–	–	–	–	–	–	–	9-Ethyl-9-heptyloctadecane		AI02295
57	43	71	85	41	55	69	253	380	100	98	68	50	49	44	27	22	0.00	27	56	–	–	–	–	–	–	–	–	–	–	–	9-Ethyl-9-heptyloctadecane		AI02296
43	57	71	41	85	55	69	211	380	100	95	62	48	47	45	28	23	0.00	27	56	–	–	–	–	–	–	–	–	–	–	–	10-Ethyl-10-propyldocosane		AI02297
43	57	71	41	85	55	69	211	380	100	95	61	47	46	44	32	28	0.00	27	56	–	–	–	–	–	–	–	–	–	–	–	10-Ethyl-10-propyldocosane		LI08238
43	57	71	41	85	55	69	56	380	100	97	56	40	39	30	16	15	1.23	27	56	–	–	–	–	–	–	–	–	–	–	–	Heptacosane	593-49-7	AI02298
57	43	71	85	55	41	69	99	380	100	80	62	42	28	28	18	14	1.90	27	56	–	–	–	–	–	–	–	–	–	–	–	Heptacosane	593-49-7	NI39156
57	43	71	113	41	55	85	70	380	100	82	68	59	41	38	27	21	0.00	27	56	–	–	–	–	–	–	–	–	–	–	–	5-Methyl-5-ethyltetracosane		AI02299
57	43	71	85	41	55	69	56	380	100	94	62	45	44	41	24	24	0.00	27	56	–	–	–	–	–	–	–	–	–	–	–	10-Methyl-10-hexyleicosane		AI02300
71	85	69	267	281	295	83	309	380	100	76	46	34	30	26	25	24	0.00	27	56	–	–	–	–	–	–	–	–	–	–	–	8-Pentyl-8-hexylhexadecane		AI02301
57	43	71	41	85	55	69	239	380	100	95	65	50	48	45	27	20	0.00	27	56	–	–	–	–	–	–	–	–	–	–	–	10-Propyl-10-butyleicosane		AI02302
380	69	255	81	55	145	105	159	380	100	82	68	64	61	42	39	35		28	44	–	–	–	–	–	–	–	–	–	–	–	Ergosta-4,6,22-triene	62942-75-0	NI39157
155	43	141	55	41	57	29	225	380	100	52	27	26	21	17	14	13	7.73	28	44	–	–	–	–	–	–	–	–	–	–	–	Naphthalene, trihexyl-	74646-45-0	NI39158
368	366	370	383	381	91	65	385	381	100	78	48	23	18	13	13	11		8	10	–	3	–	–	4	–	–	–	3	–	–	1-Methyl-1-(4-methylphenyl)tetrachlorocyclotriphosphazene	87048-78-0	NI39159
325	326	270	61	327	269	271	39	381	100	98	85	83	79	77	72	66	16.79	11	23	–	–	–	–	–	–	–	–	1	–	1	Platinum, bis[(1,2,3-η)-2-methyl-2-propenyl](trimethylphosphine)-	74811-08-8	NI39160
166	41	57	71	139	167	280	53	381	100	85	83	57	43	27	16	15	0.00	13	17	5	1	–	6	–	–	–	–	–	–	–	Pentanoic acid, 2-[(trifluoroacetyl)amino]-5-[(trifluoroacetyl)oxy]-, butyl ester	74367-29-6	NI39161
266	43	69	169	71	381	312	119	381	100	14	10	7	6	2	1	1		14	18	3	1	–	7	–	–	–	–	–	–	–	Proline, N-(heptafluorobutyryl)-, 3-methylbutyl ester		NI39162
171	77	107	75	76	123	50	92	381	100	57	56	45	33	25	25	23	7.13	15	12	4	1	1	–	–	1	–	–	–	–	–	3H-Indol-3-one, 5-bromo-1,2-dihydro-1-[(4-methoxybenzene)sulphonyl]-	83372-84-3	NI39163
321	346	323	348	324	325	326	209	381	100	95	93	63	37	32	28	24	6.00	15	18	4	1	–	–	3	–	–	–	–	–	–	L-Isoleucine, N-[(2,4,5-trichlorophenoxy)acetyl]-, methyl ester	66789-92-2	NI39164
69	88	322	346	324	196	198	86	381	100	71	55	53	53	44	42	41	3.00	15	18	4	1	–	–	3	–	–	–	–	–	–	L-Leucine, N-[(2,4,5-trichlorophenoxy)acetyl]-, methyl ester	66789-91-1	NI39165
149	208	125	130	140	126	105	177	381	100	80	70	64	62	60	57	48	10.00	15	19	7	5	–	–	–	–	–	–	–	–	–	7H-Pyrrolo[2,3-]pyrimidine-5-carboxylic acid, 2-amino-3,4-dihydro-4-oxo 7-(2-acetylamino-2-deoxy-β-arabinofuranosyl)-, methyl ester	84873-19-8	NI39166
234	28	206	249	181	135	381	41	381	100	83	73	65	44	29	28	28		16	23	4	5	1	–	–	–	–	–	–	–	–	Adenosine, N-(3-methyl-2-butenyl)-2-(methylthio)-	20859-00-1	NI39167
234	206	249	367	248	181	250	292	381	100	65	59	46	39	33	26	21	0.00	16	23	4	5	1	–	–	–	–	–	–	–	–	Adenosine, N-(3-methyl-2-butenyl)-2-(methylthio)-	20859-00-1	NI39168
234	381	249	278	366	248	250	382	381	100	75	58	50	46	39	26	25		16	23	4	5	1	–	–	–	–	–	–	–	–	Adenosine, N-(3-methyl-2-butenyl)-2-(methylthio)-	20859-00-1	NI39169
227	212	43	55	69	41	57	268	381	100	91	77	61	56	53	48	45	3.20	16	26	1	1	–	7	–	–	–	–	–	–	–	1-(Heptafluorobutanoylamino)dodecane		MS08901

MASS TO CHARGE RATIOS	M.W.	INTENSITIES	Parent	C	H	O	N	S	F	Cl	Br	I	Si	P	B	X	COMPOUND NAME	CAS Reg No	No
213 121 56 129 185 211 381 73	**381**	100 84 73 67 64 43 42 37		17	15	4	3	–	–	–	–	–	–	–	–	1	1-(α-Methyl-5-nitrofurfurylidenehydrazinocarbonyl)ferrocene		LI08239
135 351 363 77 136 105 31 204	**381**	100 21 18 17 15 9 9 8	**0.00**	17	23	7	3	–	–	–	–	–	–	–	–	–	L-Serine, [N-(2-methoxybenzoyl)-β-alanylglycyl]-, methyl ester		DD01499
226 57 152 80 43 81 69 71	**381**	100 57 49 46 43 42 39 37	**1.00**	18	23	6	1	1	–	–	–	–	–	–	–	–	2-Oxa-7-azatricyclo[4.4.0.0^{3,8}]decane-7-carboxylic acid, 4-[[(4-methylphenyl)sulphonyl]oxy]-, ethyl ester	55905-56-1	NI39170
226 227 91 152 154 80 108 81	**381**	100 77 71 65 55 48 28 28	**3.00**	18	23	6	1	1	–	–	–	–	–	–	–	–	2,5-Methanofuro[3,2-b]pyridine-4(2H)-carboxylic acid, hexahydro-8-[[(4-methylphenyl)sulphonyl]oxy]-, ethyl ester, (2α,3aβ,5α,7aβ,8R*)-	49656-65-7	NI39171
264 265 366 382 263 234 148 248	**381**	100 22 10 8 8 8 8 7	**0.00**	18	31	4	1	–	–	–	–	–	2	–	–	–	1-[(Trimethylsilyl)oxycarbonyl]-1-methyl-6-[(trimethylsilyl)oxy]-7-methoxytetrahydroisoquinoline		NI39172
127 126 253 252 43 71 128 125	**381**	100 42 14 9 9 8 7 7	**0.19**	18	36	6	–	–	–	–	–	–	–	–	3	–	Mannitol butyl boronate		NI39173
91 381 290 231 232 134 65 92	**381**	100 35 34 28 20 11 11 9		19	15	2	3	2	–	–	–	–	–	–	–	–	6-Benzyl-2-(benzylthio)-7-hydroxy-5H-1,3,4-thiadiazolo[3,2-a]pyrimidin-5 one		MS08902
132 147 268 120 146 148 91 121	**381**	100 79 73 54 38 23 23 22	**0.00**	20	22	3	1	–	3	–	–	–	–	–	–	–	4,7-Dimethyl-2-(4-hydroxyphenylethyl)-6,7-dihydro-5H-2-pyrindinium trifluoroacetate		LI08240
194 165 151 166 195 43 36 28	**381**	100 89 15 12 11 10 10 10	**0.00**	20	28	4	1	–	–	1	–	–	–	–	–	–	Diphenethylamine, 3,3′,4,4′-tetramethoxy-, hydrochloride	19713-28-1	NI39174
219 43 60 161 220 203 41 57	**381**	100 38 25 23 21 21 18 16	**5.00**	20	31	4	1	1	–	–	–	–	–	–	–	–	4β-Acetamido-[(β-ethoxycarbonyl)thiomethyl]-2,6-di-tert-butylphenol		IC04211
381 366 367 352 336 351 338 183	**381**	100 36 8 8 8 6 5 5		21	19	6	1	–	–	–	–	–	–	–	–	–	10,11-Methylenedioxy-3,4,12-trimethoxy-7,8-dihydro-5H-isoindolo[1,2-b][3]benzazepin-5-one		NI39175
366 381 367 351 350 322 352 183	**381**	100 35 34 22 8 6 5 5		21	19	6	1	–	–	–	–	–	–	–	–	–	7-Methyl-9,10-methylenedioxy-3,4,11-trimethoxy-6,7-dihydro-5H-isoindolo[1,2-b]isoquinol-5-one		NI39176
91 381 92 202 352 41 306 276	**381**	100 19 19 9 7 7 6 6		22	23	5	1	–	–	–	–	–	–	–	–	–	7-Benzyloxy-3-ethoxycarbonyl-6-propoxy-1,4-dihydroquinol-4-one		IC04212
229 109 363 243 123 187 96 110	**381**	100 52 42 42 32 30 24 22	**0.70**	22	24	2	3	–	1	–	–	–	–	–	–	–	2H-Benzimidazol-2-one, 1-[1-[4-(4-fluorophenyl)-4-oxobutyl]-4-piperidinyl]-1,3-dihydro-	2062-84-2	LI08241
229 109 363 243 188 124 96 110	**381**	100 50 40 40 30 30 25 20	**0.60**	22	24	2	3	–	1	–	–	–	–	–	–	–	2H-Benzimidazol-2-one, 1-[1-[4-(4-fluorophenyl)-4-oxobutyl]-4-piperidinyl]-1,3-dihydro-	2062-84-2	NI39177
33 43 44 59 50 85 87 52	**381**	100 74 70 56 47 24 20 15	**0.00**	22	27	3	3	–	–	–	–	–	–	–	–	–	Acetamide, N-[(8α,9R)-9-hydroxy-6′-methoxycinchonan-5′-yl]-	56847-12-2	NI39178
255 256 212 199 254 184 227 339	**381**	100 75 31 21 9 9 7 5	**4.50**	23	27	4	1	–	–	–	–	–	–	–	–	–	Monascamine		LI08242
337 206 280 308 353 324 381 250	**381**	100 24 15 10 6 6 4 3		24	15	4	1	–	–	–	–	–	–	–	–	–	Furo[2,3-b]furan-2,3-dione, 5,6a-dihydro-5-(phenylimino)-4,6a-diphenyl-	91361-17-0	NI39179
259 105 77 194 260 230 195 106	**381**	100 80 44 38 20 8 8 8	**6.00**	25	19	3	1	–	–	–	–	–	–	–	–	–	1-(N-Benzoyl-2-toluidino)-2-naphthoic acid		NI39180
230 127 231 105 77 381 128 126	**381**	100 34 21 9 8 6 6 6		25	19	3	1	–	–	–	–	–	–	–	–	–	2-(Methoxycarbonyl)phenyl N-(1-naphthyl)benzimidate		NI39181
71 69 276 189 381 104 113 120	**381**	100 71 64 49 45 45 43 33		25	19	3	1	–	–	–	–	–	–	–	–	–	2-(N-Methylanilino)-4-phenyl-5-hydroxy-6,7-benzocoumar-3-one		MS08903
105 238 77 239 217 106 381 216	**381**	100 51 29 10 9 9 7 7		25	19	3	1	–	–	–	–	–	–	–	–	–	Methyl 2-(N-benzoyl-1-naphthylamino)benzoate		NI39182
379 382 91 380 290 107 97 351	**381**	100 99 80 53 40 35 33 31	**23.27**	25	35	2	1	–	–	–	–	–	–	–	–	–	6-Azaandrostan-7-one, 6-benzyl-17β-hydroxy-	16373-57-2	NI39183
55 41 296 91 352 83 270 105	**381**	100 93 90 77 47 45 42 40	**24.52**	25	40	1	1	–	–	–	–	–	–	–	1	–	Borinic acid, diethyl-, [2-(cyclohexylimino)cyclooctyl]benzyl ester	74810-35-8	NI39184
226 43 73 58 55 227 41 70	**381**	100 36 27 24 21 20 18 13	**3.70**	25	51	1	1	–	–	–	–	–	–	–	–	–	N,N-Dilaurylformamide		MS08904
180 346 181 201 269 203 381	**381**	100 98 22 10 8 3 1		26	20	–	1	–	–	1	–	–	–	–	–	–	Diphenylketimimo(diphenyl)methyl chloride		LI08243
381 365 141 156 364 366 226 172	**381**	100 58 43 40 25 18 17 9		26	39	1	1	–	–	–	–	–	–	–	–	–	19-Norcholesta-1,3,5(10)-trien-6-one oxime		NI39185
381 57 268 307 325 292 213 308	**381**	100 82 63 45 42 31 26 20		26	39	1	1	–	–	–	–	–	–	–	–	–	Pregna-5,17(20)-diene-20-carbonitrile, 3-tert-butoxy-, (3β)-	60727-73-3	NI39186
212 44 43 213 238 210 30 57	**381**	100 28 26 17 15 15 15 14	**0.00**	26	55	–	1	–	–	–	–	–	–	–	–	–	1-Tridecanamine, N-tridecyl-	5910-75-8	IC04213
212 44 210 43 57 213 55 41	**381**	100 56 40 28 20 15 15 12	**0.00**	26	55	–	1	–	–	–	–	–	–	–	–	–	1-Tridecanamine, N-tridecyl-	5910-75-8	NI39187
145 180 152 130 103 151 77 115	**381**	100 58 47 45 40 21 19 18	**0.80**	27	27	1	1	–	–	–	–	–	–	–	–	–	3-Phenyl-3-ethyl-4-sec-butylimino-2-spiro[9-fluorenyl]oxetane		MS08905
222 220 224 218 298 296 294 221	**382**	100 88 83 60 53 53 50 49	**0.64**	–	12	–	–	–	–	–	–	–	–	–	–	5	Pentagermane	15587-39-0	MS08906
255 77 127 141 51 177 155 27	**382**	100 88 63 44 31 28 22 22	**7.00**	4	4	–	–	–	4	–	–	2	–	–	–	–	Butane, 1,1,2,2-tetrafluoro-1,4-diiodo-	755-95-3	MS08907
69 97 119 363 313 181 50 31	**382**	100 29 7 4 4 4 4 3	**0.00**	7	–	2	–	–	14	–	–	–	–	–	–	–	Acetic acid, trifluoro-, undecafluoropentyl ester	42133-36-8	NI39188
69 163 181 113 131 144 119 51	**382**	100 41 39 21 20 15 15 13	**0.00**	8	1	–	–	–	15	–	–	–	–	–	–	–	Pentadecafluorocyclooctane		IC04214
382 384 383 386 385 350 162 306	**382**	100 61 19 16 13 9 8 6		10	16	2	2	4	–	–	–	–	–	–	–	1	Nickel, bis(4-morpholinecarbodithioato-S,S′)-	14024-77-2	NI39189
272 382 95 242 301 79 336 64	**382**	100 58 57 49 47 46 44 44		12	6	9	4	1	–	–	–	–	–	–	–	–	2,2′,4,4′-Tetranitrodiphenyl sulphoxide		NI39190
382 272 336 184 301 30 242 271	**382**	100 69 53 47 44 44 43 40		12	6	9	4	1	–	–	–	–	–	–	–	–	3,3′,4,4′-Tetranitrodiphenyl sulphoxide		NI39191
384 386 382 213 388 385 151 349	**382**	100 72 60 33 26 24 24 23		12	7	–	–	–	–	5	–	–	2	–	–	–	9H-9-Silafluorene, 4-(trichlorosilyl)-9,9-dichloro-		NI39192
41 55 39 56 43 240 214 161	**382**	100 71 57 40 40 35 26 26	**1.05**	12	13	3	–	–	–	–	1	–	–	–	–	1	(π-Cyclopentadienyl)(4-bromobutyl)tricarbonylmolybdenum		MS08908
382 336 383 290 278 221 384 246	**382**	100 19 17 15 7 7 5 5		14	5	4	4	1	3	–	–	–	–	–	–	–	3,7-Dinitro-9-trifluoromethylbenzimidazo[2,1-b]benzimidazole		IC04215
182 352 266 265 264 161 95 382	**382**	100 96 82 57 40 40 40 34		14	22	2	4	–	–	–	–	–	–	–	–	2	Di-μ-dimethylamino-bis(nitrosylcyclopentadienylchromium)		LI08244
353 57 309 354 71 215 242 213	**382**	100 50 31 26 24 23 21 21	**0.00**	14	34	6	–	–	–	–	–	–	3	–	–	–	Cyclotrisiloxane, 1,3,5-triethyl-1,3-dibutoxy-5-hydroxy-		NI39193
162 91 88 120 323 203 348 303	**382**	100 45 30 25 2 2 1 1	**0.00**	15	15	4	2	–	5	–	–	–	–	–	–	–	Phenylalanine, N-(pentafluoropropionyl)glycyl-, methyl ester		NI39194
121 213 211 56 185 129 138 229	**382**	100 88 82 64 58 54 40 39	**26.00**	16	14	4	4	–	–	–	–	–	–	–	–	1	1-(α-Imino-5-nitrofurfurylidenehydrazinocarbonyl)ferrocene		LI08245

MASS TO CHARGE RATIOS								M.W.	INTENSITIES								Parent	C	H	O	N	S	F	Cl	Br	I	Si	P	B	X	COMPOUND NAME	CAS Reg No	No
58	28	41	114	43	55	45	26	**382**	100	61	52	37	27	22	21	20	**0.70**	16	21	4	–	–	–	3	–	–	–	–	–	–	**Propanoic acid, 2-(2,4,5-trichlorophenoxy)-, ester with butoxypropanol**	28903-26-6	**NI39195**
350	218	200	192	191	162	97	323	**382**	100	60	39	38	23	21	18	17	**0.00**	16	22	7	4	–	–	–	–	–	–	–	–	–	**[15-Crown-5][g]quinoxaline, 2,3-bis(hydroxyimino)-1,2,3,4-tetrahydro-**		**MS08909**
382	192	326	270	354	272	103	226	**382**	100	84	75	68	48	48	32	27		17	11	4	–	–	–	–	–	–	–	–	–	1	**Rhodium, dicarbonyl(1,3-diphenyl-1,3-propanedionato-O,O')-**	24151-56-2	**NI39196**
311	312	283	382	384	284	347	255	**382**	100	61	53	49	33	25	23	22		17	12	6	–	–	–	2	–	–	–	–	–	–	**Aflatoxin B_1 2,3-dichloride**	58209-98-6	**NI39197**
235	367	73	191	137	103	251	382	**382**	100	82	71	38	37	37	26	24		17	26	6	2	–	–	–	–	–	1	–	–	–	**3'-O-Pivaloyl-5'-O-trimethylsilyl-2,2'-anhydrouridine**	40732-69-2	**LI08246**
239	137	57	169	73	297	209	145	**382**	100	98	85	71	65	61	46	46	**26.00**	17	26	6	2	–	–	–	–	–	1	–	–	–	**3'-O-Pivaloyl-5'-O-trimethylsilyl-2,2'-anhydrouridine**	40732-69-2	**NI39198**
239	137	169	73	297	367	209	382	**382**	100	98	71	65	61	55	46	26		17	26	6	2	–	–	–	–	–	1	–	–	–	**5'-O-Pivaloyl-3'-O-trimethylsilyl-2,2'-anhydrouridine**		**LI08247**
91	98	84	106	248	261	219	156	**382**	100	94	80	62	47	25	24	21	**0.40**	18	26	7	2	–	–	–	–	–	–	–	–	–	**Agropinic acid, benzylamide derivative**		**MS08910**
73	382	293	205	45	383	75	74	**382**	100	29	16	16	15	11	9	9		18	34	3	–	–	–	–	–	–	3	–	–	–	**1-Propene, 3-[(trimethylsilyl)oxy]-1-[3,4-bis[(trimethylsilyl)oxy]phenyl]-**		**NI39199**
73	297	312	367	75	311	100	298	**382**	100	99	88	80	48	37	37	28	**4.69**	18	34	3	2	–	–	–	–	–	2	–	–	–	**2,4,6(1H,3H,5H)-Pyrimidinetrione, 5-(1-methylbutyl)-5-allyl-1,3-bis(trimethylsilyl)-**	52937-71-0	**NI39200**
183	367	107	83	382	85	184	77	**382**	100	66	35	35	32	24	17	15		19	17	–	–	–	–	–	–	–	–	2	–	1	**1H-1,3,2-Benzodiphospharsole, 2,3-dihydro-2-methyl-1,3-diphenyl-, (1α,2β,3α)-**	38234-82-1	**MS08911**
198	267	156	159	199	173	382	114	**382**	100	31	28	25	23	23	20	20		19	21	3	2	–	3	–	–	–	–	–	–	–	**L-Tryptophan, 4-(3,3-dimethylallyl)-N-(trifluoroacetyl)-, methyl ester**		**LI08248**
198	267	154	156	199	164	382	114	**382**	100	32	28	24	23	23	20	20		19	21	3	2	–	3	–	–	–	–	–	–	–	**L-Tryptophan, 4-(3-methyl-2-butenyl)-N-(trifluoroacetyl)-, methyl ester**	55334-65-1	**NI39201**
82	43	41	100	55	109	95	91	**382**	100	97	35	33	21	19	16	16	**0.00**	19	26	8	–	–	–	–	–	–	–	–	–	–	**Calonectrin, 7,8-dihydroxy-**		**NI39202**
43	121	105	91	41	111	109	79	**382**	100	35	23	21	19	17	17	17	**0.00**	19	26	8	–	–	–	–	–	–	–	–	–	–	**7-Hydroxydiacetoxyscirpenol**		**NI39203**
121	159	137	91	161	138	122	185	**382**	100	36	33	33	27	27	25	24	**0.00**	19	26	8	–	–	–	–	–	–	–	–	–	–	**Neosolaniol**		**NI39204**
223	219	224	249	209	196	260	165	**382**	100	13	12	8	6	6	4	4	**0.20**	19	26	8	–	–	–	–	–	–	–	–	–	–	**Octahydrofunicone**		**LI08249**
103	353	75	105	77	43	55	69	**382**	100	96	85	72	63	59	58	43	**0.00**	19	37	1	–	–	2	1	–	–	1	–	–	–	**1-Tridecene, 2-chloro-1,1-difluoro-3-[(triethylsilyl)oxy]-**		**DD01500**
382	383	339	40	368	311	325	283	**382**	100	25	18	14	10	8	6	5		20	14	8	–	–	–	–	–	–	–	–	–	–	**10H-Benzo[b]xanthene-7,10,12-trione, 6,11-dihydroxy-3,8-dimethoxy-1-methyl-**	33390-21-5	**NI39205**
367	382	368	352	191	163	383	337	**382**	100	87	21	21	21	21	19	17		20	14	8	–	–	–	–	–	–	–	–	–	–	**4H-Furo[2,3-H]-1-benzopyran-4-one, 2-(1,3-benzodioxol-5-yl)-5-hydroxy-3,6-dimethoxy-**	77970-16-2	**NI39206**
309	291	281	382	139	74	322	252	**382**	100	57	36	26	8	8	7	6		20	14	8	–	–	–	–	–	–	–	–	–	–	**Methyl ekatetronate**		**MS08912**
338	282	323	281	253	309	310	382	**382**	100	98	97	89	85	65	64	62		20	15	4	2	–	–	1	–	–	–	–	–	–	**Camptothecin, 12-chloro-**	58546-25-1	**NI39207**
152	150	338	153	282	323	281	253	**382**	100	18	12	11	10	9	9	9	**6.00**	20	15	4	2	–	–	1	–	–	–	–	–	–	**Camptothecin, 12-chloro-**	58546-25-1	**NI39208**
91	106	192	191	382	84	276	158	**382**	100	52	47	34	18	18	14	14		20	18	2	2	2	–	–	–	–	–	–	–	–	**2,4-Bis[(N-benzylcarbamoyl)methylene]-1,3-dithiethane**		**LI08250**
149	177	134	150	178	106	382	135	**382**	100	82	67	42	36	31	30	27		20	22	4	4	–	–	–	–	–	–	–	–	–	**1,2,4,5-Tetrazine, 3,6-bis(3-ethoxy-4-methoxyphenyl)-**	77355-36-3	**NI39209**
149	177	134	150	106	178	382	166	**382**	100	89	83	41	33	30	28	17		20	22	4	4	–	–	–	–	–	–	–	–	–	**1,2,4,5-Tetrazine, 3,6-bis(3-methoxy-4-ethoxyphenyl)-**	77355-33-0	**NI39210**
183	109	155	184	111	80	55	110	**382**	100	18	13	11	9	9	8	6	**0.92**	20	30	7	–	–	–	–	–	–	–	–	–	–	**2,4-Dioxaspiro[5.5]nonane-6-carboxylic acid, 3-[2-[1-(ethoxycarbonyl)-2-oxocyclopentyl]ethyl]-, ethyl ester**		**NI39211**
203	121	263	204	159	105	175	173	**382**	100	51	50	39	37	35	33	33	**0.00**	20	30	7	–	–	–	–	–	–	–	–	–	–	**Hydroxy HT-2 toxin**	34114-98-2	**NI39212**
350	77	382	290	275	219	351	51	**382**	100	61	60	55	40	30	29	19		21	13	2	2	–	3	–	–	–	–	–	–	–	**9-Acridinecarboxylic acid, N-(phenylimino)-1-hydroxy-2,3,4-trifluoro-, methyl ester**	91035-86-8	**NI39213**
109	137	149	110	108	138	151	136	**382**	100	19	9	8	5	4	3	3	**0.00**	21	18	5	–	1	–	–	–	–	–	–	–	–	**Phenol, 4,4'-(3H-2,1-benzoxathiol-3-ylidene)bis[3-methyl-, S,S-dioxide**	2303-01-7	**NI39214**
105	77	259	263	247	91	219	277	**382**	100	82	81	63	31	31	25	23	**7.00**	21	18	7	–	–	–	–	–	–	–	–	–	–	**4H-1-Benzopyran-4-one, 6-formyl-5,8-dimethoxy-7-(benzoylmethoxy)-2-methyl-**		**NI39215**
382	275	91	92	383	165	271	227	**382**	100	72	64	41	26	22	21	21		21	19	3	–	1	–	–	–	–	–	1	–	–	**Phenothiaphosphine, 2,8-dimethyl-10-(4-tolyl)-, 5,5,10-trioxide**	23861-42-9	**NI39216**
382	77	277	105	386	113	383	209	**382**	100	83	58	54	31	28	24	19		21	22	5	2	–	–	–	–	–	–	–	–	–	**Benzoic acid, 4-(phenylazo)-, 2-[2-(tetrahydro-2-oxo-3-furanyl)ethoxy]ethyl ester**	36763-89-0	**MS08913**
254	382	381	255	353	325	326	269	**382**	100	37	18	15	8	8	5	5		21	26	3	4	–	–	–	–	–	–	–	–	–	**N-Propyl-N-pentyl-4'-nitroazobenzenecarboxamide**		**MS08914**
43	285	286	55	174	346	41	303	**382**	100	79	59	57	53	47	39	38	**0.00**	21	34	6	–	–	–	–	–	–	–	–	–	–	**Pregnan-20-one, 3,8,12,14,17-pentahydroxy-, (3β,5α,12β,14β,17α)-**	6875-17-8	**NI39217**
120	138	97	209	121	43	105	125	**382**	100	79	72	57	50	50	49	38	**2.19**	21	34	6	–	–	–	–	–	–	–	–	–	–	**Pregn-5-ene-3,8,11,12,14,20-hexol, (3β,11α,12β,14β)-**	27642-30-4	**NI39218**
351	149	382	367					**382**	100	46	38	20						22	22	6	–	–	–	–	–	–	–	–	–	–	**5-Allyl-2,3,7-trimethoxy-2-(1,3-benzodioxol-5-yl)benzofuran**		**LI08251**
367	77	115	65	91	368	51	41	**382**	100	73	28	25	25	24	18	17	**15.50**	22	23	4	–	–	–	–	–	–	–	1	–	–	**Phosphoric acid, (1,1-dimethylethyl)phenyl diphenyl ester**	56803-37-3	**NI39219**
170	291	234	382	171	168	91	292	**382**	100	40	22	17	16	13	9	7		22	26	2	2	1	–	–	–	–	–	–	–	–	**1-Naphthalenesulphonamide, 5-(dimethylamino)-N-(1,1-dimethyl-2-phenylethyl)-**	51581-20-5	**NI39220**
170	291	234	58	382	171	168	91	**382**	100	39	21	19	17	16	13	9		22	26	2	2	1	–	–	–	–	–	–	–	–	**1-Naphthalenesulphonamide, 5-(dimethylamino)-N-(1,1-dimethyl-2-phenylethyl)-**	51581-20-5	**NI39221**
171	382	170	383	172	131	168	154	**382**	100	68	27	19	13	13	12	7		22	26	2	2	1	–	–	–	–	–	–	–	–	**1-Naphthalenesulphonamide, 5-(dimethylamino)-N-(4-phenylbutyl)-**	55837-07-5	**NI39222**
171	382	170	383	172	168	91	384	**382**	100	60	28	21	14	11	7	6		22	26	2	2	1	–	–	–	–	–	–	–	–	**1-Naphthalenesulphonamide, 5-(dimethylamino)-N-(4-phenylbutyl)-**	55837-07-5	**NI39223**
105	192	57	163	77	136	177	118	**382**	100	85	68	50	24	16	12	12	**1.00**	22	26	4	2	–	–	–	–	–	–	–	–	–	**Benzeneacetic acid, α-(1-oxopropoxy)-, 2-benzoyl-1-tert-butyl-hydrazide**	67491-63-8	**NI39224**
382	383	241	162	384	354	335	325	**382**	100	25	11	10	5	5	5	5		22	26	4	2	–	–	–	–	–	–	–	–	–	**1,4-Bis(sec-butylamino)-5,8-dihydroxyanthraquinone**		**IC04216**
382	171	383	261	310	144	158	262	**382**	100	59	26	25	22	21	18	15		22	26	4	2	–	–	–	–	–	–	–	–	–	**Corymine**	6472-42-0	**NI39225**
43	283	44	45	180	382	167	121	**382**	100	85	50	31	25	23	23	23		22	26	4	2	–	–	–	–	–	–	–	–	–	**Curan-17-oic acid, 19-(acetyloxy)-2,16-didehydro-, methyl ester, (20ξ)-**	56114-74-0	**NI39226**

MASS TO CHARGE RATIOS	M.W.	INTENSITIES	Parent	C	H	O	N	S	F	Cl	Br	I	Si	P	B	X	COMPOUND NAME	CAS Reg No	No
108 382 150 125 172 156 144 130	**382**	100 53 43 30 29 24 23 22		22	26	4	2	–	–	–	–	–	–	–	–	–	Deformyl-Nb-acetyl-aspidodasycarpine		LI08252
108 382 150 172 351 144 130 323	**382**	100 69 46 22 16 16 16 15		22	26	4	2	–	–	–	–	–	–	–	–	–	Deformyl-Nb-acetyl-aspidodasycarpine		LI08253
108 382 150 172 130 159 144 156	**382**	100 64 63 19 15 14 14 11		22	26	4	2	–	–	–	–	–	–	–	–	–	Deformyl-Nb-acetyl-aspidodasycarpine		LI08254
186 382 185 133 187 199 172 383	**382**	100 53 44 24 20 18 15 14		22	26	4	2	–	–	–	–	–	–	–	–	–	Erinine, 21-deoxo-23-hydroxy-	16843-68-8	LI08255
186 382 185 144 187 199 383 171	**382**	100 55 46 25 21 18 15 15		22	26	4	2	–	–	–	–	–	–	–	–	–	Erinine, 21-deoxo-23-hydroxy-	16843-68-8	NI39227
43 57 41 45 382 55 42 69	**382**	100 88 48 42 40 40 34 33		22	26	4	2	–	–	–	–	–	–	–	–	–	2,5-Ethano-2H-azocino[4,3-b]indole-6-carboxylic acid, 6-[(acetyloxy)methyl]-4-ethylidene-1,3,4,5,6,7-hexahydro-, methyl ester, [5S-[4E,5α,6β]]-	56293-11-9	NI39228
138 382 108 244 383 139 352 257	**382**	100 69 65 40 28 25 16 16		22	26	4	2	–	–	–	–	–	–	–	–	–	Hazuntine		LI08256
186 382 199 214 381 184 156 169	**382**	100 70 62 60 47 43 37 36		22	26	4	2	–	–	–	–	–	–	–	–	–	Isomitrajavine		LI08257
167 168 169 165 166 153 152 212	**382**	100 99 98 98 96 96 96 91	**0.00**	22	26	4	2	–	–	–	–	–	–	–	–	–	Leucine, diphenylacetylglycyl-	66856-29-9	NI39229
382 138 31 108 244 32 43 383	**382**	100 73 51 47 39 39 29 24		22	26	4	2	–	–	–	–	–	–	–	–	–	Lochnerinine		LI08258
138 43 139 244 382 44 58 41	**382**	100 48 33 30 23 22 16 16		22	26	4	2	–	–	–	–	–	–	–	–	–	Minovincine, 16-methoxy-	22341-30-6	NI39230
186 382 381 214 199 156 184 169	**382**	100 83 54 52 40 30 27 25		22	26	4	2	–	–	–	–	–	–	–	–	–	Mitrajavine		LI08259
382 198 199 279 143 144 351 383	**382**	100 93 73 58 35 30 28 25		22	26	4	2	–	–	–	–	–	–	–	–	–	Sarpagan-16-carboxylic acid, 3,17-dihydroxy-1-methyl-, methyl ester, (16R)-	53632-75-0	NI39231
382 144 130 108 152 110 143 96	**382**	100 83 39 38 37 34 33 28		22	26	4	2	–	–	–	–	–	–	–	–	–	Strychnobrasiline, 14β-hydroxy-	92202-53-4	NI39232
184 280 339 198 180 382 323 336	**382**	100 47 44 40 38 36 13 11		22	26	4	2	–	–	–	–	–	–	–	–	–	Talbotine, 4-N-methyl-		LI08260
382 293 322 266 238 321 267 234	**382**	100 80 57 50 34 33 29 27		22	26	4	2	–	–	–	–	–	–	–	–	–	Talbotine, O-methyl-	30809-16-6	NI39233
382 293 322 238 184 180 294 325	**382**	100 80 57 34 24 20 14 10		22	26	4	2	–	–	–	–	–	–	–	–	–	Talbotine, O-methyl-	30809-16-6	LI08261
279 172 144 280 382 130 160 365	**382**	100 62 39 33 31 24 21 20		22	26	4	2	–	–	–	–	–	–	–	–	–	Voachalotine oxindole	26126-84-1	NI39234
311 89 135 45 57 161 113 91	**382**	100 52 41 39 33 21 20 20	**3.17**	22	38	5	–	–	–	–	–	–	–	–	–	–	Ethanol, 2-[2-[2-[2-[4-(1,1,3,3-tetramethylbutyl)phenoxy]ethoxy]ethoxy]ethoxy]-	2315-63-1	NI39235
159 57 43 41 271 55 100 70	**382**	100 17 13 12 10 9 7 7	**0.35**	22	39	3	–	–	–	–	–	–	–	1	–	–	Bis(2-ethylhexyl)phenyl phosphate		IC04217
149 292 239 173 271 284 189 221	**382**	100 55 50 48 19 14 11 9	**0.00**	22	42	3	–	–	–	–	–	–	1	–	–	–	9,12-Octadecadienoic acid, 8-(trimethylsiloxy)-, methyl ester, (Z,Z)-		NI39236
325 185 292 164 207 149 235 198	**382**	100 60 59 40 38 37 19 12	**0.00**	22	42	3	–	–	–	–	–	–	1	–	–	–	9,12-Octadecadienoic acid, 14-(trimethylsiloxy)-, methyl ester, (Z,Z)-		NI39237
73 67 55 81 80 75 79 41	**382**	100 58 46 45 44 42 38 38	**2.00**	22	42	3	–	–	–	–	–	–	1	–	–	–	9,12-Octadecadienoic acid, 18-(trimethylsiloxy)-, methyl ester	22074-68-6	NI39238
73 67 55 82 81 75 95 101	**382**	100 56 46 44 43 42 38 36	**1.50**	22	42	3	–	–	–	–	–	–	1	–	–	–	9,12-Octadecadienoic acid, 18-(trimethylsiloxy)-, methyl ester	22074-68-6	LI08262
149 56 41 57 271 205 69 55	**382**	100 98 61 44 41 41 41 37	**1.22**	23	26	5	–	–	–	–	–	–	–	–	–	–	4,4'-Bis(butoxycarbonyl)benzophenone		IC04218
298 256 215 213 255 283 200 299	**382**	100 58 40 40 36 20 20 19	**10.00**	23	26	5	–	–	–	–	–	–	–	–	–	–	Monascorubrin		LI08263
154 324 109 296 139 172 170 57	**382**	100 85 59 49 49 41 28 28	**0.00**	23	30	3	2	–	–	–	–	–	–	–	–	–	Aspidofractinine-3-carboxylic acid, 6-methoxy-1-methyl-, methyl ester, (2α,3β,5α,6β)-	55724-61-3	NI39239
109 295 382 265 43 200 352 130	**382**	100 89 56 54 51 33 32 32		23	30	3	2	–	–	–	–	–	–	–	–	–	Aspidofractinine-3-methanol, 17-methoxy-, acetate, (2α,3α,5α)-	55103-45-2	NI39240
182 382 183 122 43 354 381 210	**382**	100 26 14 13 12 10 8 8		23	30	3	2	–	–	–	–	–	–	–	–	–	Aspidospermidin-21-ol, 1-acetyl-, acetate	55208-84-9	NI39241
44 323 144 210 324 43 241 42	**382**	100 75 50 46 19 17 15 15	**14.00**	23	30	3	2	–	–	–	–	–	–	–	–	–	Curan-17-ol, 1-acetyl-, acetate, (16α)-	56053-24-8	NI39242
131 73 101 367 189 157 237 205	**382**	100 45 32 8 6 4 2 1	**0.40**	23	42	4	–	–	–	–	–	–	–	–	–	–	1-O-(2-Methoxy-4-hexynyl)-2,3-O-isopropylideneglycerol		NI39243
74 185 183 171 283 382 364 351	**382**	100 47 30 25 24 11 10 7		23	42	4	–	–	–	–	–	–	–	–	–	–	[6-(5-Methylhexyl)oxan-2-yl]-[6-(3-methoxycarbonylpropyl)oxan-2-yl]methane		LI08264
73 75 55 41 117 129 43 28	**382**	100 90 64 46 43 41 39 31	**4.62**	23	46	2	–	–	–	–	–	–	1	–	–	–	11-Eicosenoic acid, trimethylsilyl ester		NI39244
105 77 231 173 106 51 41 81	**382**	100 96 51 34 32 25 17 16	**0.02**	23	46	2	2	–	–	–	–	–	–	–	–	–	Octadecanamide, N-nitroso-N-pentyl-	74420-95-4	NI39245
105 156 77 115 261 220 192 165	**382**	100 50 31 29 26 9 6 3	**2.00**	24	18	3	2	–	–	–	–	–	–	–	–	–	N-Benzoyloxy-2-cyano-4-methyl-2,5-diphenyl-2,3-dihydrooxazole		LI08265
227 305 303 228 301 226 304 224	**382**	100 32 22 20 17 16 11 11	**0.72**	24	20	–	–	–	–	–	–	–	–	–	–	1	Tetraphenyl germane		MS08915
305 228 151 78 382	**382**	100 88 30 16 3		24	20	–	–	–	–	–	–	–	–	–	–	1	Tetraphenyl germane		LI08266
305 208 303 300 204 202 151 150	**382**	100 88 76 56 56 40 32 26	**6.00**	24	20	–	–	–	–	–	–	–	–	–	–	1	Tetraphenyl germane		LI08267
149 41 55 54 39 150 109 57	**382**	100 51 32 24 23 21 20 14	**0.06**	24	30	4	–	–	–	–	–	–	–	–	–	–	Bis(octa-2,7-dienyl) phthalate		IC04219
367 227 196 174 242 368 382 173	**382**	100 51 46 43 38 32 27 22		24	34	2	–	–	–	–	–	–	1	–	–	–	19-Nor-pregna-1,3,5(10)-trien-20-yne, 3-methoxy-17-[(trimethylsilyl)oxy]-	28426-34-8	NI39246
74 85 269 197 251 382 364 325	**382**	100 90 34 20 12 11 10 9		24	46	3	–	–	–	–	–	–	–	–	–	–	2-(3,5-Dimethylhexyl)-6-(9-methoxycarbonylnonyl)oxane		LI08268
194 91 367 77 326 382 368 195	**382**	100 96 60 40 22 20 18 18		25	26	–	4	–	–	–	–	–	–	–	–	–	3H-[1,2,4]Triazolo[4,3-a][1,5]benzodiazepine, 3a,4,5,6-tetrahydro-3a,5,5-trimethyl-1,3-diphenyl-		MS08916
382 302 221 181 287 208 364 234	**382**	100 42 27 21 20 19 10 7		25	34	3	–	–	–	–	–	–	–	–	–	–	12'-apo-β-Caroten-12'-al, 5,6-epoxy-5,6-dihydro-3-hydroxy-		NI39247
213 214 137 382 43 105 41 57	**382**	100 61 31 29 27 24 19 17		25	34	3	–	–	–	–	–	–	–	–	–	–	4-Dodecyloxy-2-hydroxybenzophenone		IC04220
105 77 43 41 106 57 55 111	**382**	100 16 9 9 7 6 6 5	**0.79**	25	34	3	–	–	–	–	–	–	–	–	–	–	3-Dodecyloxyphenyl benzoate		IC04221
121 41 107 81 55 43 122 91	**382**	100 47 41 41 39 38 36 34	**20.00**	25	34	3	–	–	–	–	–	–	–	–	–	–	D-Homo-5α-androstan-17a-one, 17-furfurylidene-3β-hydroxy-	17429-43-5	NI39248
101 102 309 74 57 87 69 71	**382**	100 35 29 25 16 15 15 13	**2.90**	25	50	2	–	–	–	–	–	–	–	–	–	–	Docosanoic acid, 3,4-dimethyl-, methyl ester		MS08917
101 102 74 309 57 71 69 87	**382**	100 27 24 23 16 14 14 13	**3.50**	25	50	2	–	–	–	–	–	–	–	–	–	–	Docosanoic acid, 3,4-dimethyl-, methyl ester		MS08918
129 87 97 69 311 148 57 18	**382**	100 80 49 30 25 24 22 20	**3.00**	25	50	2	–	–	–	–	–	–	–	–	–	–	Docosanoic acid, 4,4-dimethyl-, methyl ester	55334-81-1	NI39249

MASS TO CHARGE RATIOS								M.W.	INTENSITIES								Parent	C	H	O	N	S	F	Cl	Br	I	Si	P	B	X	COMPOUND NAME	CAS Reg No	No
129	87	97	69	57	88	71	43	**382**	100	46	32	17	13	11	9	9	**0.50**	25	50	2	–	–	–	–	–	–	–	–	–	–	Docosanoic acid, 4,4-dimethyl-, methyl ester	55334-81-1	MS08919
75	57	97	83	111	69	74	43	**382**	100	54	52	45	33	32	28	26	**1.50**	25	50	2	–	–	–	–	–	–	–	–	–	–	1-Docosanol, propanoate	55334-75-3	WI01496
75	57	97	83	111	69	74	71	**382**	100	54	52	45	33	32	28	26	**1.50**	25	50	2	–	–	–	–	–	–	–	–	–	–	1-Docosanol, propanoate	55334-75-3	MS08920
43	57	55	41	74	297	71	69	**382**	100	76	59	54	45	37	36	35	**22.00**	25	50	2	–	–	–	–	–	–	–	–	–	–	Methyl 12-hexyloctadecanoate		MS08921
43	74	57	55	69	97	41	309	**382**	100	96	81	78	76	74	67	66	**3.80**	25	50	2	–	–	–	–	–	–	–	–	–	–	Methyl 3-methyl-3-ethylheneicosanoate		MS08922
57	74	43	55	382	41	87	69	**382**	100	76	72	62	56	48	46	40		25	50	2	–	–	–	–	–	–	–	–	–	–	Methyl 18-propylheneicosanoate		MS08923
101	57	74	326	327	75	382	55	**382**	100	80	76	39	34	31	27	21		25	50	2	–	–	–	–	–	–	–	–	–	–	Methyl 3,20,20-trimethylheneicosanoate		MS08924
74	43	87	41	55	57	75	69	**382**	100	75	74	37	33	32	30	20	**4.00**	25	50	2	–	–	–	–	–	–	–	–	–	–	Tetracosanoic acid, methyl ester	2442-49-1	NI39250
74	87	43	75	57	55	69	143	**382**	100	99	99	88	80	63	45	43	**3.85**	25	50	2	–	–	–	–	–	–	–	–	–	–	Tetracosanoic acid, methyl ester	2442-49-1	NI39251
74	87	382	75	55	57	43	69	**382**	100	80	38	34	30	29	26	23		25	50	2	–	–	–	–	–	–	–	–	–	–	Tetracosanoic acid, methyl ester	2442-49-1	WI01497
339	340	57	382	41	148	177	121	**382**	100	26	16	10	9	5	4	4		26	38	2	–	–	–	–	–	–	–	–	–	–	1,1-Bis(5-tert-butyl-4-hydroxy-2-methylphenyl)butane		IC04222
339	340	382	57	176	383	341	323	**382**	100	27	14	11	6	4	4	4		26	38	2	–	–	–	–	–	–	–	–	–	–	1,1-Bis(5-tert-butyl-4-hydroxy-2-methylphenyl)butane		IC04223
382	213	367	231	191	229	349	299	**382**	100	22	20	18	17	11	8	8		26	38	2	–	–	–	–	–	–	–	–	–	–	3β-Hydroxy-16-cyclohexylmethylene-5-androsten-7-one		LI08269
339	340	41	29	148	323	162	57	**382**	100	25	14	7	5	4	4	4	**1.41**	26	38	2	–	–	–	–	–	–	–	–	–	–	1,1-Bis(2-methyl-4-hydroxy-5-iso-butylbenzene)-2-methylpropane		IC04224
43	57	55	41	83	69	71	29	**382**	100	79	70	62	45	44	35	34	**0.00**	26	54	1	–	–	–	–	–	–	–	–	–	–	1-Hexacosanol		MS08925
122	121	123	55	43	382	135	41	**382**	100	18	11	10	10	9	9	8		27	42	1	–	–	–	–	–	–	–	–	–	–	Cholesta-1,4-dien-3-one	566-91-6	NI39252
174	382	161	187	159	383	175	41	**382**	100	60	32	25	23	19	19	16		27	42	1	–	–	–	–	–	–	–	–	–	–	Cholesta-3,5-dien-7-one	567-72-6	NI39253
174	382	161	187	55	383	91	159	**382**	100	77	38	29	25	24	22	21		27	42	1	–	–	–	–	–	–	–	–	–	–	Cholesta-3,5-dien-7-one	567-72-6	NI39254
209	364	43	197	195	349	196	183	**382**	100	90	86	82	77	61	56	49	**33.00**	27	42	1	–	–	–	–	–	–	–	–	–	–	Cholesta-5,7,9(11)-trien-3β-ol		NI39255
295	313	95	69	41	382	109	81	**382**	100	95	88	69	69	57	55	52		27	42	1	–	–	–	–	–	–	–	–	–	–	Cholesta-5,17(20),24-trien-3-ol, (3β)-	56362-83-5	NI39256
271	93	69	109	81	41	382	107	**382**	100	78	76	73	68	67	66	58		27	42	1	–	–	–	–	–	–	–	–	–	–	Cholesta-5,20,24-trien-3-ol, (3β)-	41083-97-0	NI39257
382	383	191	380	190	384	191.5	381	**382**	100	31	21	16	12	9	9	6		28	14	–	–	1	–	–	–	–	–	–	–	–	Dibenzo[2,3:10,11]perylo[1,12-bcd]thiophene		MS08926
382	383	381	278	277	191	305	190	**382**	100	35	35	13	10	9	8	6		28	18	–	2	–	–	–	–	–	–	–	–	–	1,1-Bis(3-phenylisoindolylidene)		MS08927
190	95	191	67	109	41	81	55	**382**	100	97	92	64	61	60	51	46	**2.80**	28	46	–	–	–	–	–	–	–	–	–	–	–	9,9′-Biphenanthrene, octacosahydro-	55334-18-4	WI01498
95	190	191	67	109	41	81	55	**382**	100	84	74	68	67	60	55	46	**3.33**	28	46	–	–	–	–	–	–	–	–	–	–	–	9,9′-Biphenanthrene, octacosahydro-	55334-18-4	AI02303
190	95	191	109	81	79	67	41	**382**	100	96	85	60	51	26	23	19	**6.80**	28	46	–	–	–	–	–	–	–	–	–	–	–	9,9′-Biphenanthrene, octacosahydro-	55334-18-4	AI02304
69	55	382	81	43	257	109	255	**382**	100	94	79	63	63	54	53	44		28	46	–	–	–	–	–	–	–	–	–	–	–	Ergosta-4,22-diene	76866-88-1	NI39258
382	354	383	352	355	350	175	119	**382**	100	51	33	30	25	22	22	22		29	18	1	–	–	–	–	–	–	–	–	–	–	2H-Cyclopenta[b]phenanthren-2-one, 1,3-diphenyl-		IC04225
382	383	191	384	306	226	152	77	**382**	100	31	11	5	4	3	3	3		30	22	–	–	–	–	–	–	–	–	–	–	–	1,1′:3′,1″:3″,1‴:3‴:1⁗-Quinquephenyl	16716-13-5	WI01499
382	383	191	384	152	77	51	39	**382**	100	32	19	5	4	4	4	3		30	22	–	–	–	–	–	–	–	–	–	–	–	1,1′:3′:1″-Terphenyl, 2,2″-diphenyl-	5660-37-7	WI01500
306	304	308	278	310	276	280	146	**383**	100	79	47	13	10	10	7	5	**0.50**	2	5	–	3	–	–	4	1	–	–	3	–	–	1-Ethyl-1-bromotetrachlorocyclotriphosphazene	83844-15-9	NI39259
73	153	254	225	121	226	75	179	**383**	100	57	50	43	22	16	16	14	**0.00**	11	15	5	1	–	6	–	–	–	1	–	–	–	L-Threonine, N-(trifluoroacetyl)-, trimethylsilyl ester, trifluoroacetate	52558-90-4	NI39260
362	364	114	196	113	198	102	86	**383**	100	67	47	40	40	37	33	33	**0.00**	13	12	6	1	–	–	3	–	–	–	–	–	–	L-Aspartic acid, N-[(2,4,5-trichlorophenoxy)acetyl]-, 1-methyl ester	66789-93-3	NI39261
268	43	55	71	69	214	253	169	**383**	100	68	64	37	10	8	7	4	**0.00**	14	20	3	1	–	7	–	–	–	–	–	–	–	Valine, (N-heptafluorobutyryl)-, 3-methylbutyl ester		NI39262
158	73	298	159	147	299	86	45	**383**	100	52	28	15	8	7	6	6	**1.50**	14	38	3	1	–	–	–	–	–	3	1	–	–	Phosphonic acid, [1-[(trimethylsilyl)amino]pentyl]-, bis(trimethylsilyl) ester	53044-40-9	NI39263
158	73	298	147	211	207	100	75	**383**	100	52	28	8	4	4	4	4	**1.50**	14	38	3	1	–	–	–	–	–	3	1	–	–	Phosphonic acid, [1-[(trimethylsilyl)amino]pentyl]-, bis(trimethylsilyl) ester	53044-40-9	NI39264
364	163	147	365	383	148	36	118	**383**	100	26	19	17	13	13	12	11		15	18	3	5	1	–	1	–	–	–	–	–	–	N-Ethyl-N-(3-chloro-2-hydroxypropyl)-3-methyl-4-(5-nitrothiazol-2-ylazo)aniline		IC04226
195	280	181	194	208	252	383	251	**383**	100	99	91	78	72	62	49	44		16	25	4	5	1	–	–	–	–	–	–	–	–	2-Methylthio-6-(3-methylbutylamino)-9-(2-hydroxymethyl-3,4-dihydroxytetrahydrofuran-5-yl)purine		MS08928
164	107	175	108	145	147	111	43	**383**	100	99	36	35	33	29	26	26	**5.00**	17	15	5	1	–	–	2	–	–	–	–	–	–	L-Tyrosine, N-[(2,4-dichlorophenoxy)acetyl]-	50649-05-3	NI39265
181	154	182	213	95	142	123	198	**383**	100	99	95	65	63	45	41	35	**0.00**	17	21	9	1	–	–	–	–	–	–	–	–	–	3,4-Bis(methoxycarbonyl)-2,5-dihydrofuran		LI08270
43	198	41	71	226	180	152	45	**383**	100	44	37	36	35	25	25	25	**0.00**	17	28	5	1	–	3	–	–	–	–	–	–	–	L-Glutamic acid, N-(trifluoroacetyl)-, bis(1-methylbutyl) ester	57983-12-7	NI39266
73	205	293	368	308	220	307	206	**383**	100	59	34	12	11	11	10	10	**10.40**	17	29	5	1	–	–	–	–	–	2	–	–	–	2-Propenoic acid, 3-[4-methoxy-3-[(trimethylsilyl)oxy]phenyl]-α-[(trimethylsilyl)oxy]-, methoxime		MS08929
43	83	97	109	55	125	69	121	**383**	100	82	64	57	51	35	27	25	**0.00**	18	25	8	1	–	–	–	–	–	–	–	–	–	Camphanoic acid, 3-acetoxy-2-nitrocyclohexyl ester, (1R,2R,3S)-		DD01501
115	368	222	294	369	195	384	218	**383**	100	27	15	10	8	8	7	7	**0.00**	18	37	2	1	–	–	–	–	–	3	–	–	–	Synephrine, tris(trimethylsilyl)-		NI39267
174	73	175	86	368	176	147	100	**383**	100	49	29	18	14	13	9	8	**0.00**	18	37	2	1	–	–	–	–	–	3	–	–	–	Tyramine, 3-methoxytris(trimethylsilyl)-	55530-69-3	MS08930
174	73	175	86	45	59	179	176	**383**	100	82	19	13	12	10	7	7	**0.00**	18	37	2	1	–	–	–	–	–	3	–	–	–	Tyramine, 3-methoxytris(trimethylsilyl)-	55530-69-3	MS08931
174	73	175	86	176	59	368	45	**383**	100	50	19	13	9	8	6	6	**0.00**	18	37	2	1	–	–	–	–	–	3	–	–	–	Tyramine, 3-methoxytris(trimethylsilyl)-	55530-69-3	NI39268
70	84	98	112	198	196	322	154	**383**	100	75	52	48	30	28	25	25	**0.00**	18	39	1	3	–	–	2	–	–	–	–	–	–	1,5,9-Triazacycloheneicosan-10-one, dihydrochloride	67171-84-0	NI39269

MASS TO CHARGE RATIOS								M.W.	INTENSITIES								Parent	C	H	O	N	S	F	Cl	Br	I	Si	P	B	X	COMPOUND NAME	CAS Reg No	No
57	55	187	41	186	69	43	105	**383**	100	72	71	71	70	70	70	64	**1.90**	19	14	3	1	–	–	–	1	–	–	–	–	–	N-Methyl-N-(4-bromophenyl)-2-methyl-1,4-naphthoquinone-3-carboxamide		MS08932
233	383	259	245	95	56	45	57	**383**	100	41	31	30	29	24	21	19		19	21	4	5	–	–	–	–	–	–	–	–	–	Piperazine, 1-(4-amino-6,7-dimethoxy-2-quinazolinyl)-4-(2-furanylcarbonyl)-	19216-56-9	NI39270
180	43	93	120	136	94	181	119	**383**	100	64	48	46	21	21	20	17	**1.92**	19	29	7	1	–	–	–	–	–	–	–	–	–	Lycopsamine, 3′,7-diacetyl-		NI39271
180	152	83	77	181	106	153	105	**383**	100	75	20	19	11	9	6	6	**4.00**	20	21	5	3	–	–	–	–	–	–	–	–	–	3,5-Pyrazolidinedione, 4-(α-ethoxy-4-nitrobenzyl)-1-phenyl-2,4-dimethyl-		NI39272
156	91	128	383	247	155	384	183	**383**	100	88	74	61	61	57	36	34		21	21	4	1	1	–	–	–	–	–	–	–	–	2H-Pyrano[4,3-b]indol-1-one, 1,4-dihydro-3-propyl-4-(4-tolylsulphonyl)-		DD01502
206	190	162	145	132	177	91	383	**383**	100	41	14	10	9	6	5	4		21	21	6	1	–	–	–	–	–	–	–	–	–	D-Adlumine	524-46-9	NI39273
148	193	383	149	78	91	104	77	**383**	100	89	84	83	80	42	39	35		21	21	6	1	–	–	–	–	–	–	–	–	–	Bis[1,3]benzodioxolo[4,5-c:5′,6′-g]azecin-13(5H)-one, 4,6,7,14-tetrahydro-12-methoxy-5-methyl-	6014-62-6	NI39274
190	65	178	132	91	117	77	160	**383**	100	32	28	23	20	15	15	12	**11.00**	21	21	6	1	–	–	–	–	–	–	–	–	–	Hydrastine		NI39275
190	31	59	49	74	84	149	175	**383**	100	63	43	29	19	10	3	2	**0.00**	21	21	6	1	–	–	–	–	–	–	–	–	–	Hydrastine, (±)-β-		DD01503
383	352	384	353	310	294	337	368	**383**	100	53	25	14	13	13	12	10		21	21	6	1	–	–	–	–	–	–	–	–	–	Isoquinoline, 1-(2-methoxycarbonyl-4-methoxyphenyl)-5,6,7-trimethoxy-		DD01504
205	221	187	206	222	383	162	324	**383**	100	66	18	12	8	7	5	4		21	21	6	1	–	–	–	–	–	–	–	–	–	Ochratoxin B methyl ester		NI39276
177	383	368	190	206	384	369	352	**383**	100	69	67	19	17	16	16	16		21	21	6	1	–	–	–	–	–	–	–	–	–	Rheadan, 8-methoxy-16-methyl-2,3:10,11-bis[methylenebis(oxy)]-, (8β)-	2718-25-4	NI39277
383	368	384	298	340	193	192	191	**383**	100	41	25	25	22	22	22	16		21	21	6	1	–	–	–	–	–	–	–	–	–	Spiro[7H-indeno[4,5-d]-1,3-dioxole-7,1′(2′H)-isoquinolin]-8(6H)-one, 3′,4′dihydro-6-hydroxy-6′,7′-dimethoxy-2′-methyl-, cis-(+)-	34084-34-9	NI39278
191	383	192	368	365	190	161	133	**383**	100	99	52	40	36	23	22	20		21	21	6	1	–	–	–	–	–	–	–	–	–	3,4,12-Trimethoxy-10,11-methylenedioxy-7,8,13,13a-tetrahydro-5H-isoindolo[1,2-b][3]benzazepin-5-one		NI39279
217	188	180	189	179	195	167	243	**383**	100	90	79	78	76	74	66	41	**17.00**	21	22	–	3	1	–	1	–	–	–	–	–	–	1-[(2-Benzyl-4-thiazolyl)methyl]-4-(4-chlorophenyl)piperazine		MS08933
135	79	69	43	83	149	97	52	**383**	100	59	37	37	30	28	28	26	**0.00**	21	38	–	1	–	–	–	1	–	–	–	–	–	N-Hexadecylpyridinium bromide	140-72-7	NI39280
77	383	278	91	57	92	55	195	**383**	100	76	56	52	40	38	36	32		22	17	2	5	–	–	–	–	–	–	–	–	–	1-Phenylazo-4-(1,2-dihydro-1-methyl-2-oxo-4-hydroxyquinol-3-yl)benzene		IC04227
339	364	324	383	340	338	188	139	**383**	100	54	51	35	24	21	18	18		22	25	5	1	–	–	–	–	–	–	–	–	–	Isoquinoline, 3-(α-ethoxyveratryl)-6,7-dimethoxy-	24176-13-4	NI39281
91	57	79	77	326	108	107	142	**383**	100	27	22	22	20	19	18	16	**0.00**	22	25	5	1	–	–	–	–	–	–	–	–	–	3-Oxazolidinecarboxylic acid, 2-tert-butyl-4-[α-hydroxybenzyl]-5-oxo-, benzyl ester, (2S,4R,4′S)-		DD01505
244	194	383	245	261	234	175	209	**383**	100	38	33	19	17	17	17	10		22	26	2	3	–	1	–	–	–	–	–	–	–	Spiramide	510-74-7	NI39282
57	276	192	277	176	41	77	193	**383**	100	91	47	20	10	10	7	6	**0.00**	23	29	4	1	–	–	–	–	–	–	–	–	–	1-Isoquinolinemethanol, u-2-tert-butoxy-1,2,3,4-tetrahydro-6,7-dimethoxy α-phenyl-		DD01506
86	141	153	167	165	152	168	154	**383**	100	16	8	5	4	4	3	3	**0.00**	24	33	3	1	–	–	–	–	–	–	–	–	–	2-Furanpropanoic acid, tetrahydro-α-(1-naphthalenylmethyl)-, 2-(diethylamino)ethyl ester, ethanedioate (1:1)	3200-06-4	NI39283
131	103	132	91	77	383	51	292	**383**	100	36	19	19	18	5	5	2		25	21	3	1	–	–	–	–	–	–	–	–	–	O-Benzyl-N-(α-cinnamylcinnamylidene)hydroxylamine		LI08271
201	202	173	383	105	130	77	158	**383**	100	26	24	19	18	17	15	12		25	21	3	1	–	–	–	–	–	–	–	–	–	2H-Pyrano[4,3-b]indol-1-one, 3,3-diphenyl-8-methoxy-5-methyl-1,4-dihydro-		DD01507
383	183	382	108	262	185	107	152	**383**	100	38	17	15	8	7	7	3		25	22	1	1	–	–	–	–	–	–	1	–	–	N-(4-Methoxyphenyl)iminotriphenylphosphorane	14796-89-5	LI08272
383	183	384	382	108	262	185	107	**383**	100	38	25	17	15	8	7	7		25	22	1	1	–	–	–	–	–	–	1	–	–	N-(4-Methoxyphenyl)iminotriphenylphosphorane	14796-89-5	NI39284
121	252	120	132	106	383	122	105	**383**	100	29	15	14	14	10	10	10		26	29	–	3	–	–	–	–	–	–	–	–	–	Methanimidamide, N,N′-bis(2,4-dimethylphenyl)-N-[[(2,4-dimethylphenyl)imino]methyl]-	69687-74-7	NI39285
57	270	113	271	121	309	327	294	**383**	100	16	16	11	11	10	9	9	**8.00**	26	41	1	1	–	–	–	–	–	–	–	–	–	Pregn-5-ene-20-carbonitrile, 3-tert-butoxy-, (3β)-	62623-47-6	NI39286
145	105	201	284	182	77	130	103	**383**	100	75	59	45	37	35	32	20	**1.00**	27	29	1	1	–	–	–	–	–	–	–	–	–	2-sec-Butylimino-3-methyl-3,4,4-triphenyloxetane		MS08934
181	146	152	57	326	153	106	104	**383**	100	78	45	42	33	30	25	19	**8.30**	27	29	1	1	–	–	–	–	–	–	–	–	–	cis-1-tert-Butyl-2-phenyl-3-methyl-3-(4-phenylbenzoyl)azetidine	22592-76-3	NI39287
181	57	146	152	104	160	153	147	**383**	100	67	52	47	45	39	33	33	**13.70**	27	29	1	1	–	–	–	–	–	–	–	–	–	trans-1-tert-Butyl-2-phenyl-3-methyl-3-(4-phenylbenzoyl)azetidine	22592-77-4	NI39288
85	151	87	217	219	153	101	163	**384**	100	35	33	30	29	23	21	14	**0.00**	5	–	–	–	–	6	6	–	–	–	–	–	–	Pentane, 1,1,2,3,4,5-hexachloro-1,2,3,4,5,5-hexafluoro-	55975-93-4	NI39289
307	309	305	311	227	147	145	213	**384**	100	98	36	30	18	16	15	12	**2.00**	5	8	–	–	–	–	–	4	–	–	–	–	–	Pentaerythritol tetrabromide	3229-00-3	LI08273
307	309	311	305	227	147	64	145	**384**	100	99	30	30	17	16	16	14	**1.74**	5	8	–	–	–	–	–	4	–	–	–	–	–	Pentaerythritol tetrabromide	3229-00-3	NI39290
309	307	39	65	27	53	388	41	**384**	100	99	80	60	57	50	42	37	**7.30**	5	8	–	–	–	–	–	4	–	–	–	–	–	Pentaerythritol tetrabromide	3229-00-3	IC04228
67	41	39	147	149	68	65	53	**384**	100	91	88	62	56	35	35	33	**0.00**	5	8	–	–	–	–	–	4	–	–	–	–	–	1,2,3,4-Tetrabromo-2-methylbutane		IC04229
309	337	289	357	109	364	212	269	**384**	100	87	75	62	25	23	23	18	**1.00**	9	13	5	–	–	8	–	–	–	–	1	–	–	Phosphoric acid, [2,2-difluoro-2-(2,2,2-trifluoroethoxy)-1-(trifluoromethyl)ethyl] diethyl ester		MS08935
155	199	127	153	121	201	187	125	**384**	100	90	59	33	19	17	17	14	**0.00**	9	22	4	–	4	–	–	–	–	–	2	–	–	Phosphorodithioic acid, S,S′-methylene O,O,O′,O′-tetraethyl ester	563-12-2	NI39291
231	97	29	65	153	125	121	27	**384**	100	94	89	84	79	71	63	50	**42.26**	9	22	4	–	4	–	–	–	–	–	2	–	–	Phosphorodithioic acid, S,S′-methylene O,O,O′,O′-tetraethyl ester	563-12-2	NI39292
180	207	242	270	314	305	384	349	**384**	100	23	11	7	2	2	1	1		10	6	3	–	–	–	6	–	–	–	–	–	–	1,4-Epoxy-5,7-methano-1H-2,3-benzodioxepin, 1,4,6,8,9,9a-hexachloro-4,5,5a,6,7,9a-hexahydro-	75420-31-4	NI39293
181	180	207	242	235	305	270	314	**384**	100	89	84	46	46	29	20	13	**0.10**	10	6	3	–	–	–	6	–	–	–	–	–	–	1,4-Epoxy-5,7-methano-1H-2,3-benzodioxepin, 1,4,8,9,9a,10-hexachloro-4,5,5a,6,7,9a-hexahydro-	59473-46-0	NI39294

MASS TO CHARGE RATIOS								M.W.	INTENSITIES								Parent	C	H	O	N	S	F	Cl	Br	I	Si	P	B	X	COMPOUND NAME	CAS Reg No	No
384	119	315	159	63	39	245	91	384	100	88	54	17	16	9	2	2		10	7	1	–	–	6	–	–	1	–	–	–	–	2-(2-Iodo-5-methylphenyl)-1,1,1,3,3,3-hexafluoro-2-propanol		MS08936
355	356	357	59	327	43	341	197	384	100	35	23	18	16	16	9	8	0.00	10	28	6	–	–	–	–	–	–	5	–	–	–	Bicyclo[5.3.1]pentasiloxane, 1,3,5,7,9-pentaethyl-	73420-26-5	NI39295
352	324	384	354	356	41	322	120	384	100	76	73	61	52	41	36	33		12	15	2	–	–	–	–	–	–	–	–	–	1	Iridium, dicarbonyl(pentamethyl-π-cyclopentadienyl)-	32660-96-1	MS08937
73	369	147	211	299	217	133	370	384	100	88	51	50	33	33	29	26	15.91	12	29	6	–	–	–	–	–	–	3	1	–	–	2-Propenoic acid, 2-[(trimethylsilyl)oxy]-, anhydride with bis(trimethylsilyl) hydrogen phosphate	55334-82-2	NI39296
281	147	73	282	369	283	148	207	384	100	98	97	28	27	18	15	14	0.00	12	36	4	–	–	–	–	–	–	5	–	–	–	Pentasiloxane, dodecamethyl-	141-63-9	NI39297
147	73	281	369	207	266	248	191	384	100	85	80	26	15	9	5	4	0.01	12	36	4	–	–	–	–	–	–	5	–	–	–	Pentasiloxane, dodecamethyl-	141-63-9	LI08274
73	281	147	369	282	207	283	370	384	100	58	45	20	19	13	11	9	0.00	12	36	4	–	–	–	–	–	–	5	–	–	–	Tetrasiloxane, 1,1,1,3,5,5,7,7,7-nonamethyl-3-(trimethylsiloxy)-	38146-99-5	NI39298
281	147	73	369	282	283	370	267	384	100	97	97	70	51	30	28	28	0.00	12	36	4	–	–	–	–	–	–	5	–	–	–	Trisiloxane, 1,1,1,5,5,5-hexamethyl-3,3-bis[(trimethylsilyl)oxy]-	3555-47-3	NI39299
73	147	281	282	369	148	283	45	384	100	95	64	16	15	14	10	9	0.00	12	36	4	–	–	–	–	–	–	5	–	–	–	Trisiloxane, 1,1,1,5,5,5-hexamethyl-3,3-bis[(trimethylsilyl)oxy]-	3555-47-3	NI39300
73	147	281	369	45	148	282	74	384	100	72	30	15	14	10	8	8	0.00	12	36	4	–	–	–	–	–	–	5	–	–	–	Trisiloxane, 1,1,1,5,5,5-hexamethyl-3,3-bis[(trimethylsilyl)oxy]-	3555-47-3	MS08938
56	84	160	68	65	81	78	112	384	100	66	52	48	36	33	31	29	0.50	14	8	6	–	–	–	–	–	–	–	–	–	2	Iron, [μ-(η^4:η^4-7-methylene-1,3,5-cycloheptatriene)]hexacarbonyldi-	56943-70-5	MS08939
91	65	182	92	384	89	39	63	384	100	88	7	7	6	4	4	3		14	14	–	–	–	–	–	–	–	–	–	–	1	Mercury, dibenzyl-		LI08275
91	384	43	65	92	293	63	167	384	100	92	33	25	20	18	6	3		14	14	–	–	–	–	–	–	–	–	–	–	1	Mercury, di-π-tolyl-		LI08276
73	75	369	117	257	55	83	69	384	100	81	43	42	40	37	35	33	3.70	14	29	2	–	–	–	–	–	1	1	–	–	–	Undecanoic acid, 11-iodo-, trimethylsilyl ester	55334-59-3	NI39301
44	340	342	83	85	47	69	306	384	100	79	53	48	33	23	19	14	1.00	16	10	7	–	–	–	2	–	–	–	–	–	–	Spiro[benzofuran-2(3H),1′-[2,5]cyclohexadiene]-2′-carboxylic acid, 5,7-dichloro-4-hydroxy-6′-methoxy-6-methyl-3,4′-dioxo-	26891-81-6	NI39302
103	259	217	151	189	260	223	104	384	100	49	34	26	19	12	12	12	11.00	16	28	5	2	–	–	–	–	–	2	–	–	–	3′,5′-Di-O-trimethylsilyl-2,2′-anhydrothymidine		NI39303
103	259	217	151	189	260	223	384	384	100	50	34	25	18	11	11	10		16	28	5	2	–	–	–	–	–	2	–	–	–	3′,5′-Di-O-trimethylsilyl-2,2′-anhydrothymidine		LI08277
57	187	56	247	41	189	199	43	384	100	68	40	39	31	26	24	24	0.00	16	37	4	–	–	–	1	–	–	2	–	–	–	Disiloxane, 1,3-diethyl-1,1,3-tributoxy-3-chloro-		NI39304
120	200	104	140	128	146	228	168	384	100	99	71	51	41	35	27	27	0.00	17	12	3	4	2	–	–	–	–	–	–	–	–	Benzenesulphonic acid, 4-methyl-, (9-oxonaphtho[2,3-c][1,2,5]thiadiazol-4(9H)-ylidene)hydrazide	83274-13-9	NI39305
43	138	195	140	386	384	197	111	384	100	94	68	31	29	29	25	25		17	15	2	2	–	–	3	–	–	–	–	–	–	2′,5′-Dichloro-2-(4-chloroanilino)-3-methoxycrotonanilide		MS08940
384	286	386	288	252	98	68	247	384	100	90	44	38	38	35	30	29		17	17	4	–	2	–	1	–	–	–	–	–	–	Spiro[benzofuran-2(3H),1′-[3]cyclohexene]-2′,3-dione, 7-chloro-6-methoxy-6′-methyl-4,4′-bis(methylthio)-	55530-37-5	NI39306
147	237	384	145	119	91	222	77	384	100	66	32	22	20	16	12	12		17	18	2	–	–	–	–	–	–	–	–	–	1	Benzofuran, 2,3-dihydro-2-[[(4-methoxyphenyl)telluro]methyl]-7-methyl-		DD01508
73	384	45	147	341	75	252	74	384	100	21	16	14	12	11	9	9		17	32	4	–	–	–	–	–	–	3	–	–	–	Benzeneacetic acid, 2,5-bis[(trimethylsilyl)oxy]-, trimethylsilyl ester	55334-62-8	NI39307
73	384	341	385	147	252	386	75	384	100	96	42	33	24	19	16	13		17	32	4	–	–	–	–	–	–	3	–	–	–	Benzeneacetic acid, 2,5-bis[(trimethylsilyl)oxy]-, trimethylsilyl ester	55334-62-8	NI39308
384	75	73	341	385	252	386	147	384	100	90	68	41	24	22	17	15		17	32	4	–	–	–	–	–	–	3	–	–	–	Benzeneacetic acid, 2,5-bis[(trimethylsilyl)oxy]-, trimethylsilyl ester	55334-62-8	NI39309
73	384	179	267	385	75	45	369	384	100	46	38	30	16	16	12	9		17	32	4	–	–	–	–	–	–	3	–	–	–	Benzeneacetic acid, 3,4-bis[(trimethylsilyl)oxy]-, trimethylsilyl ester	37148-62-2	MS08941
73	384	77	179	267	385	75	237	384	100	59	46	43	32	21	17	13		17	32	4	–	–	–	–	–	–	3	–	–	–	Benzeneacetic acid, 3,4-bis[(trimethylsilyl)oxy]-, trimethylsilyl ester	37148-62-2	NI39310
73	384	179	267	45	75	385	74	384	100	32	32	23	14	13	11	9		17	32	4	–	–	–	–	–	–	3	–	–	–	Benzeneacetic acid, 3,4-bis[(trimethylsilyl)oxy]-, trimethylsilyl ester	37148-62-2	NI39311
267	73	268	147	269	45	341	74	384	100	99	65	37	27	21	17	15	2.49	17	32	4	–	–	–	–	–	–	3	–	–	–	Benzeneacetic acid, α,3-bis[(trimethylsilyl)oxy]-, trimethylsilyl ester	68595-69-7	NI39312
267	268	269	73	341	147	369	342	384	100	21	10	10	9	6	3	3	0.00	17	32	4	–	–	–	–	–	–	3	–	–	–	Benzeneacetic acid, α,4-bis[(trimethylsilyl)oxy]-, trimethylsilyl ester	37148-64-4	NI39313
267	73	268	269	147	341	75	77	384	100	49	25	10	10	9	9	8	0.00	17	32	4	–	–	–	–	–	–	3	–	–	–	Benzeneacetic acid, α,4-bis[(trimethylsilyl)oxy]-, trimethylsilyl ester	37148-64-4	MS08942
267	73	268	269	40	41	147	75	384	100	36	24	10	10	9	8	6	0.00	17	32	4	–	–	–	–	–	–	3	–	–	–	Benzeneacetic acid, α,4-bis[(trimethylsilyl)oxy]-, trimethylsilyl ester	37148-64-4	NI39314
117	75	73	103	55	44	129	43	384	100	89	84	51	36	35	33	29	0.00	17	34	6	–	–	–	–	–	–	1	–	2	–	α-D-Galactopyranose, 6-O-(trimethylsilyl)-, cyclic 1,2:3,4-bis(butylboronate)	72347-47-8	NI39315
117	73	75	103	129	139	43	44	384	100	93	78	74	50	41	32	27	0.00	17	34	6	–	–	–	–	–	–	1	–	2	–	α-D-Glucofuranose, 6-O-(trimethylsilyl)-, cyclic 1,2:3,5-bis(butylboronate)	72347-48-9	NI39316
139	75	73	129	103	44	213	117	384	100	79	67	50	42	41	34	29	0.00	17	34	6	–	–	–	–	–	–	1	–	2	–	Mannopyranose, 1-O-(trimethylsilyl)-, 2,3:4,6-dibutaneboronate	55712-55-5	NI39317
384	227	111	180	181	92	76	385	384	100	58	44	34	30	30	28	24		18	16	4	4	1	–	–	–	–	–	–	–	–	4-(4-Aminophenylamino)-3-nitro-benzenesulphonanilide		IC04230
369	370	55	371	77	384	73	127	384	100	32	14	13	12	8	6	5		18	20	1	–	–	4	–	–	–	2	–	–	–	Cyclo(oxydimethylsilylene-1,3-phenylene-tetrafluorodimethylene-1,3-phenylenedimethylsilylene)		LI08278
176	148	133	384	222	105	352		384	100	83	58	48	23	13	10			18	24	9	–	–	–	–	–	–	–	–	–	–	Ethyl 2-glucosyloxy-4-methoxycinnamate, trans-		LI08279
384	153	185	41	137	55	43	38.5	384	100	67	55	48	37	34	20	20		18	29	1	2	2	–	–	–	–	–	1	–	–	1,3-Bis(isopropylthio)-2-cyclohexyl-2-oxo-1,3-dihydro-1,3,2-benzodiazophosphole		IC04231
209	73	75	210	43	44	45	41	384	100	86	42	18	18	17	13	10	6.80	18	32	5	–	–	–	–	–	–	2	–	–	–	Benzenepropanoic acid, 3-methoxy-α,4-bis[(trimethylsilyl)oxy]-, ethyl ester		MS08943
105	77	303	181	384	81	113	54	384	100	53	52	36	18	17	14	14		19	20	3	4	1	–	–	–	–	–	–	–	–	1,3-Dioxolane, 2-(1-imidazolylmethyl)-2-phenyl-4-[2-(methylthio)-5-[(pyrimidinyloxy)methyl]-		MS08944
97	100	84	69	98	168	55	86	384	100	97	76	60	59	56	52	51	5.38	19	32	6	2	–	–	–	–	–	–	–	–	–	1,5-Dioxa-10,17-diazacycloheneicosane-6,9,18,21-tetrone, 3,3-dimethyl-	79688-14-5	NI39318
119	91	124	65	28	120	85	384	384	100	30	20	14	12	10	10	6		20	16	2	–	3	–	–	–	–	–	–	–	–	3,5-Bis(4-methylbenzoylmethylidene)-1,2,4-trithiolane		LI08280
384	328	356	135	271	243	299		384	100	24	17	15	11	7	6			20	16	4	–	2	–	–	–	–	–	–	–	–	6,6′-Diethoxythioindigotin		LI08281
43	323	31	296	240	29			384	100	67	61	47	44	38			7.00	20	16	8	–	–	–	–	–	–	–	–	–	–	Arianciamycinone	56437-92-4	NI39319
149	220	205	194	384	353	65	177	384	100	62	54	37	35	33	33	31		20	16	8	–	–	–	–	–	–	–	–	–	–	[2-Hydroxy-5,6-dimethoxy[4,3-furano]benzoyl]piperonylmethane	77970-13-9	NI39320

MASS TO CHARGE RATIOS								M.W.	INTENSITIES								Parent	C	H	O	N	S	F	Cl	Br	I	Si	P	B	X	COMPOUND NAME	CAS Reg No	No
288	333	286	315	313	55	314	385	**384**	100	92	92	88	80	76	63	61	**52.00**	20	21	1	2	–	–	–	1	–	–	–	–	–	1,5-Benzodiazocin-2-one, 8-bromo-1-butyl-1,2,3,4-tetrahydro-6-phenyl-	63594-50-3	NI39321
369	370	73	371	372	384	383	380	**384**	100	32	27	12	2	1	1	1		20	24	4	–	–	–	–	–	–	2	–	–	–	9,10-Anthracenedione, 1,2-bis[(trimethylsilyl)oxy]-	7336-61-0	NI39322
369	370	73	371	177	372	341	339	**384**	100	28	16	10	4	2	2	2	**1.00**	20	24	4	–	–	–	–	–	–	2	–	–	–	9,10-Anthracenedione, 1,3-bis[(trimethylsilyl)oxy]-	7336-62-1	NI39323
354	369	355	324	356	370	325	294	**384**	100	31	29	18	13	9	5	5	**2.00**	20	24	4	–	–	–	–	–	–	2	–	–	–	9,10-Anthracenedione, 1,4-bis[(trimethylsilyl)oxy]-	7336-65-4	NI39324
354	369	355	324	370	356	371	325	**384**	100	60	29	20	19	14	7	6	**0.00**	20	24	4	–	–	–	–	–	–	2	–	–	–	9,10-Anthracenedione, 1,5-bis[(trimethylsilyl)oxy]-	7336-66-5	NI39325
369	370	73	371	372	297	311	383	**384**	100	30	14	11	3	3	2	1	**0.00**	20	24	4	–	–	–	–	–	–	2	–	–	–	9,10-Anthracenedione, 1,8-bis[(trimethylsilyl)oxy]-	7336-68-7	NI39326
176	109	91	159	148	146	129	147	**384**	100	85	45	40	22	20	20	17	**2.00**	20	32	3	–	2	–	–	–	–	–	–	–	–	D-glycero-D-talo-Heptose, 2,4,6-trideoxy-2,4,6-trimethyl-7-O-benzyl-, cyclic 1,3-propanediyl mercaptal		NI39327
91	213	197	212	229	155	41	55	**384**	100	77	66	55	51	51	48	45	**2.02**	20	32	5	–	1	–	–	–	–	–	–	–	–	Dodecanoic acid, 5-[[(4-methylphenyl)sulphonyl]oxy]-, methyl ester, (R)-	42550-38-9	NI39328
71	43	95	110	167	220	264	81	**384**	100	99	68	40	34	22	18	12	**0.00**	20	32	7	–	–	–	–	–	–	–	–	–	–	5-(4-Acetoxy-2-methylbutyl)-4-(2-acetyl-1-isobutyroyloxy-ethyl)-3-methyl 2-oxo-tetrahydrofuran		LI08282
43	149	55	41	235	71	89	217	**384**	100	16	16	16	10	10	8	7	**0.00**	20	32	7	–	–	–	–	–	–	–	–	–	–	Eudesman-8-one, 6,7-dehydro-3-(2,3-dihydroxy-2-methylbutyryloxy)-4,11-dihydroxy-, (3α,4α)-		NI39329
136	137	300	342	258	343	384	259	**384**	100	28	27	25	19	11	9	9		21	20	7	–	–	–	–	–	–	–	–	–	–	1,3-Benzenediol, 4-[7-(acetyloxy)-3,4-dihydro-2H-1-benzopyran-3-yl]-, diacetate	75112-86-6	NI39330
384	137	351	297	369	299	311	281	**384**	100	96	30	26	18	11	9	9		21	20	7	–	–	–	–	–	–	–	–	–	–	4H-1-Benzopyran-4-one, 7-hydroxy-3-[2,4-dihydroxy-5-methoxy-6-(2,3-epoxy-3-methylbutyl)phenyl]-		MS08945
217	187	232	215	384	153	369	199	**384**	100	18	16	15	12	12	9	5		21	20	7	–	–	–	–	–	–	–	–	–	–	[2,6′-Bi-2H-1-benzopyran]-4(3H)-one, 3,5,7-trihydroxy-5′-methoxy-2′,2′-dimethyl-, (2S-trans)-	62498-98-0	NI39331
383	384	183	261	185	262	275	165	**384**	100	84	56	47	35	25	24	16		21	22	3	–	–	–	–	–	–	–	2	–	–	Phosphonic acid, [(triphenylphosphoranylidene)methyl]-, dimethyl ester	34204-53-0	NI39332
170	121	171	384	168	264	251	235	**384**	100	68	60	34	16	14	14	12		21	24	3	2	1	–	–	–	–	–	–	–	–	1-Naphthalenesulphonamide, 5-(dimethylamino)-N-[2-(4-methoxyphenyl)ethyl]-	5282-83-7	NI39333
170	121	171	384	168	263	250	234	**384**	100	72	60	31	17	15	13	13		21	24	3	2	1	–	–	–	–	–	–	–	–	1-Naphthalenesulphonamide, 5-(dimethylamino)-N-[2-(4-methoxyphenyl)ethyl]-	5282-83-7	NI39334
91	84	192	79	101	146	108	71	**384**	100	35	28	27	25	21	21	20	**0.00**	21	24	5	2	–	–	–	–	–	–	–	–	–	Benzoic acid, N-(benzyloxycarbonyl)-2-methylalanyl-4-amino-, ethyl ester		DD01509
45	48	69	44	91	51	105	133	**384**	100	98	80	48	37	35	30	28	**6.06**	21	27	3	–	–	3	–	–	–	–	–	–	–	Testosterone, trifluoroacetate	10589-83-0	NI39335
128	256	95	129	71	86	224	55	**384**	100	31	26	24	12	10	9	8	**2.20**	21	36	–	–	3	–	–	–	–	–	–	–	–	7,14,21-Trithiatrispiro[5.1.5.1.5.1]heneicosane, 3,11,18-trimethyl-		AI02305
304	322	43	321	44	339	55	40	**384**	100	82	79	76	75	71	60	55	**0.21**	21	36	6	–	–	–	–	–	–	–	–	–	–	Pregnane-3,8,12,14,17,20-hexol, (3β,5α,12β,14β,17α,20S)-	3080-22-6	NI39336
55	41	43	99	111	95	79	81	**384**	100	78	76	72	64	54	43	43	**0.00**	21	36	6	–	–	–	–	–	–	–	–	–	–	Prost-13-en-1-oic acid, 9,11,15-trihydroxy-6-oxo-, methyl ester, (9α,11α,13E,15S)-	63557-55-1	NI39337
151	233	328	177	384	194	136	108	**384**	100	25	10	10	6	5	5	5		22	24	2	–	2	–	–	–	–	–	–	–	–	6H-1-Benzothiocin-6-one, 2,3,4,5-tetrahydro-3,5,5-trimethyl-3-[2-(methylthio)benzoyl]-	72360-98-6	MS08946
214	384	215	311	385	143	353	187	**384**	100	42	13	11	10	7	5	5		22	24	6	–	–	–	–	–	–	–	–	–	–	Cyclopentaneacetic acid, 2-(4,6-dimethoxy-1-naphthoyl)-1-methyl-5-oxo-, methyl ester	22045-39-2	NI39338
214	384	385	215	213	283	187	353	**384**	100	71	17	11	8	7	7	6		22	24	6	–	–	–	–	–	–	–	–	–	–	8,14-seco-13ξ-Estra-1,3,5,7,9-pentaene-17-carboxylic acid, 3,6-dimethoxy-11,14-dioxo-, methyl ester	22045-38-1	NI39339
342	282	300	161	343	61	283	384	**384**	100	37	29	29	22	14	10	9		22	24	6	–	–	–	–	–	–	–	–	–	–	Estra-1,3,5(10)-trien-6,17-dione, 3,11-bis(acetyloxy)-		NI39340
384	269	385	151	270	299	295	287	**384**	100	44	24	12	10	8	8	7		22	24	6	–	–	–	–	–	–	–	–	–	–	Naphtho[2,3-c]furan-1(3H)-one, 4-(3,4-dimethoxyphenyl)-3a,4,9,9a-tetrahydro-6,7-dimethoxy-, [3aR-(3aα,4α,9aα)]-	25253-30-9	LI08283
384	269	385	28	270	299	43	55	**384**	100	60	23	23	11	8	8	7		22	24	6	–	–	–	–	–	–	–	–	–	–	Naphtho[2,3-c]furan-1(3H)-one, 4-(3,4-dimethoxyphenyl)-3a,4,9,9a-tetrahydro-6,7-dimethoxy-, [3aR-(3aα,4α,9aα)]-	25253-30-9	NI39341
384	269	385	151	238	44	299	189	**384**	100	64	26	10	9	9	8	8		22	24	6	–	–	–	–	–	–	–	–	–	–	Naphtho[2,3-c]furan-1(3H)-one, 4-(3,4-dimethoxyphenyl)-3a,4,9,9a-tetrahydro-6,7-dimethoxy-, [3aS-(3aα,4α,9aβ)]-	25925-39-7	NI39342
120	105	106	121	77	349	104	79	**384**	100	66	13	11	10	9	8	8	**4.00**	22	25	2	2	–	–	1	–	–	–	–	–	–	1,1-Cyclobutanedicarboxamide, 3-chloro-N,N′-bis(1-phenylethyl)-		NI39343
135	220	384	119	92	136	90	78	**384**	100	25	15	13	13	11	11	11		22	28	4	2	–	–	–	–	–	–	–	–	–	L-Alanine, N-[N-(1-adamantylcarbonyl)glycyl]-3-phenyl-	28415-48-7	NI39344
122	130	123	87	108	384	120	94	**384**	100	40	13	13	12	10	10	8		22	28	4	2	–	–	–	–	–	–	–	–	–	Aspidodasycarpine, N-methyl-	3909-55-5	NI39345
324	281	325	266	160	323	122	43	**384**	100	32	27	24	21	14	13	12	**1.00**	22	28	4	2	–	–	–	–	–	–	–	–	–	Aspidodispermine, O-methyl-, acetate	21451-71-8	NI39346
340	138	384	43	341	356	78	339	**384**	100	82	42	30	29	24	23	18		22	28	4	2	–	–	–	–	–	–	–	–	–	Aspidospermidin-17-ol, 1-acetyl-19,21-epoxy-16-methoxy-	2494-58-8	NI39347
239	384	130	69	145	75	117	144	**384**	100	97	76	71	65	55	47	43		22	28	4	2	–	–	–	–	–	–	–	–	–	Corynoxan-16-carboxylic acid, 16,17-didehydro-17-methoxy-2-oxo-, methyl ester, (7β,16E,20α)-	76-66-4	NI39348
57	55	105	69	384	60	149	83	**384**	100	79	61	60	58	57	48	40		22	28	4	2	–	–	–	–	–	–	–	–	–	2,4(1H)-Cyclo-3,4-secoakuammilan-16-carboxylic acid, 17-hydroxy-10-methoxy-, methyl ester, (16R)-	56259-11-1	NI39349
168	310	311	385	169	196	42	41	**384**	100	58	21	10	10	5	5	4	**3.00**	22	28	4	2	–	–	–	–	–	–	–	–	–	Cylindrocarine, 1-formyl-	22222-81-7	NI39350
168	310	311	384	169	353	356	325	**384**	100	83	31	22	11	8	7	7		22	28	4	2	–	–	–	–	–	–	–	–	–	Cylindrocarine, 1-formyl-	22222-81-7	NI39351

MASS TO CHARGE RATIOS								M.W.	INTENSITIES								Parent	C	H	O	N	S	F	Cl	Br	I	Si	P	B	X	COMPOUND NAME	CAS Reg No	No
337	296	324	323	384	366	297	295	**384**	100	92	80	68	47	36	35	32		22	28	4	2	–	–	–	–	–	–	–	–	–	Eburnamenine-14-carboxylic acid, 14,15-dihydro-14-hydroxy-11-methoxy , methyl ester, (3α,14β,16α)-	4752-37-8	NI39352
384	367	385	369	368	122	190	176	**384**	100	64	27	22	18	18	17	16		22	28	4	2	–	–	–	–	–	–	–	–	–	Ibogamine-18-carboxylic acid, 16,17-didehydro-9,17-dihydro-9-hydroxy-12-methoxy-, methyl ester, (9α)-	3464-63-9	NI39353
152	43	232	75	202	44	109	45	**384**	100	72	47	47	40	37	33	32	**8.00**	22	28	4	2	–	–	–	–	–	–	–	–	–	Indole-2-acetic acid, α-(1-acetyl-3-ethylidene-4-piperidyl)-3-(methoxymethyl)-, methyl ester	1850-26-6	NI39354
281	383	384	199	200	353	186	201	**384**	100	95	78	74	69	43	39	32		22	28	4	2	–	–	–	–	–	–	–	–	–	2β-(α-Methoxycarbonyl-β-hydroxyethyl)-3-ethylidene-10β-methoxy-1,2α,4,6,7,12,12bα-heptahydroindolo[2,3-a]quinolizine		LI08284
140	384	385	43	58	45	57	44	**384**	100	46	11	8	7	7	4	4		22	28	4	2	–	–	–	–	–	–	–	–	–	16-Methoxyminovincinine	22341-28-2	NI39355
140	384	385	43	58	45	198	184	**384**	100	46	11	8	7	7	3	3		22	28	4	2	–	–	–	–	–	–	–	–	–	16-Methoxyminovincinine	22341-28-2	LI08285
282	341	180	384	198	197	281	256	**384**	100	85	72	58	53	51	42	32		22	28	4	2	–	–	–	–	–	–	–	–	–	4-Methyl-19,20-dihydrotalbotine	30867-56-2	NI39356
384	383	385	214	199	382	198	186	**384**	100	60	23	13	11	6	6	6		22	28	4	2	–	–	–	–	–	–	–	–	–	Oxayohimban-16-carboxylic acid, 11-methoxy-19-methyl-, methyl ester	56143-19-2	NI39357
281	172	367	253	159	160	144	384	**384**	100	97	66	61	56	55	55	53		22	28	4	2	–	–	–	–	–	–	–	–	–	Voachalotine oxindole, 19,20-dihydro-	27123-64-4	NI39358
384	109	138	122	209	108	325	124	**384**	100	93	57	55	47	30	27	25		22	28	4	2	–	–	–	–	–	–	–	–	–	Voaluteine	3306-58-9	NI39359
384	383	385	200	199	214	382	353	**384**	100	98	21	9	9	8	7	6		22	28	4	2	–	–	–	–	–	–	–	–	–	Yohimban-16-carboxylic acid, 17-hydroxy-10-methoxy-, methyl ester, (16β,17α)-	15218-17-4	NI39360
384	383	199	200	214	186	174	353	**384**	100	88	15	14	12	8	7	5		22	28	4	2	–	–	–	–	–	–	–	–	–	Yohimban-16-carboxylic acid, 17-hydroxy-11-methoxy-, methyl ester, (16α,17α)-	65025-21-0	NI39361
384	383	385	83	200	199	85	186	**384**	100	93	23	18	15	14	12	11		22	28	4	2	–	–	–	–	–	–	–	–	–	17-epi-α-Yohimbine, 11-methoxy-	84710-89-4	NI39362
384	83	383	85	385	47	200	87	**384**	100	91	86	59	23	20	12	10		22	28	4	2	–	–	–	–	–	–	–	–	–	α-Yohimbine, 11-methoxy-	84710-88-3	NI39363
258	384	257	135	91	79	55	122	**384**	100	73	59	44	38	28	28	27		22	29	5	–	–	–	–	–	–	–	–	1	–	17α,21-Dihydroxypregn-4-ene-3,11,20-trione methyl boronate		MS08947
258	384	257	135	122	121	197	123	**384**	100	73	59	44	27	27	24	24		22	29	5	–	–	–	–	–	–	–	–	1	–	17α,21-Dihydroxypregn-4-ene-3,11,20-trione methyl boronate		LI08286
199	200	242	183	108	58	91	185	**384**	100	87	86	65	59	54	35	30	**0.00**	22	30	–	2	–	–	–	–	–	–	2	–	–	Phosphonous diamide, P-[(diphenylphosphino)vinyl]-N,N,N',N'-tetraethyl	33730-54-0	NI39364
384	244	385	383	218	231	386	245	**384**	100	29	28	28	12	11	8	8		22	33	3	–	–	–	–	–	–	1	–	1	–	Estra-1,3,5(10)-triene, 3-[(trimethylsilyl)oxy]-16,17-methylboronate, (16-epi)-		NI39365
281	144	143	128	129	328	356	340	**384**	100	35	31	16	14	12	8	8	**2.40**	22	42	4	1	–	–	–	–	–	–	–	–	–	2-(14-Carboxytetradecyl)-2-ethyl-4,4-dimethyl-1,3-oxazolidine N-oxide		MS08948
73	75	55	81	67	95	41	82	**384**	100	90	83	64	52	46	46	40	**15.01**	22	44	3	–	–	–	–	–	–	1	–	–	–	9-Octadecenoic acid, 18-(trimethylsiloxy)-, methyl ester	21987-17-7	NI39366
281	282	55	58	140	57	267	69	**384**	100	50	50	46	37	33	32	31	**1.00**	22	44	3	2	–	–	–	–	–	–	–	–	–	3,7-Dihydroxy-5-oxa-1,9-diazabicyclo[8.8.7]pentacosane		MS08949
300	258	215	257	217	356	202	285	**384**	100	70	40	35	35	20	20	18	**13.00**	23	28	5	–	–	–	–	–	–	–	–	–	–	Monascorubrin, dihydro-		LI08287
124	43	384	125	383	152	41	355	**384**	100	27	22	10	8	7	7	6		23	32	3	2	–	–	–	–	–	–	–	–	–	Aspidospermidine, 1-acetyl-16,17-dimethoxy-	639-26-9	NI39367
124	384	383	125	356	385	152	190	**384**	100	23	10	10	7	6	6	4		23	32	3	2	–	–	–	–	–	–	–	–	–	Aspidospermidine, 1-acetyl-16,17-dimethoxy-	639-26-9	NI39368
124	57	125	384	41	55	43	42	**384**	100	24	15	14	14	12	11	9		23	32	3	2	–	–	–	–	–	–	–	–	–	Aspidospermidin-17-ol, 16-methoxy-1-(1-oxopropyl)-	5516-66-5	NI39369
263	121	111	42	264	70	97	77	**384**	100	99	19	18	17	16	15	15	**0.00**	23	32	3	2	–	–	–	–	–	–	–	–	–	Piperazine, 1-(2-methoxy-2-phenylethyl)-4-(2-hydroxy-3-methoxy-3-phenylpropyl)-		NI39370
259	98	124	200	384	110	210	126	**384**	100	99	33	16	15	14	13	13		23	32	3	2	–	–	–	–	–	–	–	–	–	Vincaminoridine		LI08288
146	162	369	57	208	271	263	285	**384**	100	75	62	42	35	28	26	17	**12.00**	23	42	1	2	–	–	–	–	–	–	–	2	–	1,3,5,2,4-Oxadiazadiborine, 2,4-dibutyl-3,5-di-tert-butyl-6-phenylperhydro-		NI39371
43	101	28	57	41	55	71	29	**384**	100	95	77	52	40	31	25	22	**0.44**	23	44	4	–	–	–	–	–	–	–	–	–	–	2-Heptadecanone, 1-[(2,2-dimethyl-1,3-dioxolan-4-yl)methoxy]-	39033-40-4	NI39372
131	71	43	73	101	57	41	55	**384**	100	46	40	30	26	22	20	19	**0.00**	23	44	4	–	–	–	–	–	–	–	–	–	–	2,3-O-Isopropylidene-1-O-(2-methoxy-4-hexadecenyl)glycerol	16725-40-9	NI39373
131	73	101	352	157	369	189	239	**384**	100	41	12	4	4	3	3	2	**0.00**	23	44	4	–	–	–	–	–	–	–	–	–	–	2,3-O-Isopropylidene-1-O-(2-methoxy-4-hexadecenyl)glycerol	16725-40-9	NI39374
109	83	108	55	138	41	69	95	**384**	100	34	25	25	21	15	14	13	**0.71**	23	45	2	–	–	–	–	–	–	–	1	–	–	Phosphonous acid, isopropyl-, bis[5-methyl-2-isopropylcyclohexyl] ester	74630-15-2	NI39375
147	73	296	194	148	297	149	266	**384**	100	72	33	19	14	8	8	7	**0.00**	23	48	2	–	–	–	–	–	–	1	–	–	–	Eicosanoic acid, trimethylsilyl ester	55530-70-6	NI39376
73	117	75	132	43	369	145	55	**384**	100	77	67	56	48	39	38	33	**9.95**	23	48	2	–	–	–	–	–	–	1	–	–	–	Eicosanoic acid, trimethylsilyl ester	55530-70-6	NI39377
73	217	147	218	292	191	305	219	**384**	100	87	27	16	14	11	9	8	**0.00**	23	48	2	–	–	–	–	–	–	1	–	–	–	Hexadecanoic acid, 3,7,11,15-tetramethyl-, trimethylsilyl ester	55517-59-4	NI39378
159	73	117	75	43	132	69	57	**384**	100	99	78	68	61	51	43	43	**3.40**	23	48	2	–	–	–	–	–	–	1	–	–	–	Hexadecanoic acid, 3,7,11,15-tetramethyl-, trimethylsilyl ester	55517-59-4	NI39379
159	117	132	369	143	43	75	370	**384**	100	62	57	51	27	23	22	16	**12.16**	23	48	2	–	–	–	–	–	–	1	–	–	–	Hexadecanoic acid, 3,7,11,15-tetramethyl-, trimethylsilyl ester	55517-59-4	NI39380
91	128	27	64	383	358	41	32	**384**	100	94	84	62	59	58	52	49	**38.38**	24	12	–	6	–	–	–	–	–	–	–	–	–	s-Triazine, 2,4,6-tris(2-cyanophenyl)-		IC04232
384	339	77	180	385	340	51	52	**384**	100	88	71	60	30	26	21	13		24	20	3	2	–	–	–	–	–	–	–	–	–	1H-Pyrazole-4-acetic acid, 3-(4-methoxyphenyl)-1,5-diphenyl-	51431-04-0	NI39381
339	384	340	77	169	385	69	210	**384**	100	76	42	34	24	23	19	15		24	20	3	2	–	–	–	–	–	–	–	–	–	1H-Pyrazole-4-acetic acid, 5-(4-methoxyphenyl)-1,3-diphenyl-	51431-05-1	NI39382
384	104	383	367	385	209	194	77	**384**	100	37	29	27	21	19	18	16		24	20	3	2	–	–	–	–	–	–	–	–	–	2(1H)-Pyrimidinone, 4-(3-methoxyphenyl)-5-(4-methylphenoxy)-6-phenyl-	42919-59-5	NI39383
384	194	134	385	383	104	224	77	**384**	100	43	31	30	28	27	26	16		24	20	3	2	–	–	–	–	–	–	–	–	–	2(1H)-Pyrimidinone, 4-(4-methoxyphenyl)-5-(4-methylphenoxy)-6-phenyl-	42919-58-4	NI39384
215	79	107	91	233	95	366	93	**384**	100	86	74	68	64	64	61	59	**7.62**	24	32	4	–	–	–	–	–	–	–	–	–	–	Bufa-20,22-dienolide, 14,15-epoxy-3-hydroxy-, (3β,5β,15β)-	465-39-4	NI39385
215	79	366	107	91	233	95	93	**384**	100	87	86	74	68	65	65	60	**8.00**	24	32	4	–	–	–	–	–	–	–	–	–	–	Bufa-20,22-dienolide, 14,15-epoxy-3-hydroxy-, (3β,5β,15β)-	465-39-4	LI08289
123	122	108	107	216	124	105	79	**384**	100	90	90	72	69	62	62	61	**49.99**	24	32	4	–	–	–	–	–	–	–	–	–	–	Bufa-20,22-dienolide, 3-hydroxy-15-oxo-, (3β,5β,14α)-	468-86-0	NI39386
123	108	122	107	216	215	92	79	**384**	100	92	91	73	70	61	61	61	**50.00**	24	32	4	–	–	–	–	–	–	–	–	–	–	Bufa-20,22-dienolide, 3-hydroxy-15-oxo-, (3β,5β,14α)-	468-86-0	LI08290
231	91	366	136	230	79	105	93	**384**	100	92	82	81	75	72	69	54	**14.04**	24	32	4	–	–	–	–	–	–	–	–	–	–	Bufa-20,22-dienolide, 14-hydroxy-3-oxo-, (5β)-	4029-65-6	NI39387

MASS TO CHARGE RATIOS	M.W.	INTENSITIES	Parent	C	H	O	N	S	F	Cl	Br	I	Si	P	B	X	COMPOUND NAME	CAS Reg No	No
231 297 366 230 348 333 341 188	**384**	100 99 82 75 48 40 38 34	**14.00**	24	32	4	–	–	–	–	–	–	–	–	–	–	Bufa-20,22-dienolide, 14-hydroxy-3-oxo-, (5β)-	4029-65-6	NI39388
180 163 236 69 81 135 137 123	**384**	100 55 21 20 19 17 16 16	**5.00**	24	32	4	–	–	–	–	–	–	–	–	–	–	2-Propenoic acid, 3-(3,4-dihydroxyphenyl)-, albicanyl ester		NI39389
342 343 327 341 301 384 299 369	**384**	100 90 88 52 41 39 34 31		24	36	2	–	–	–	–	–	–	1	–	–	–	Propyl-Δ^1-tetrahydrocannabinol allyldimethylsilyl ether		NI39390
192 193 121 240 56 164 42 28	**384**	100 34 14 9 8 6 6 6	**0.00**	24	36	2	2	–	–	–	–	–	–	–	–	–	Ethylenediamine, N,N,N',N'-tetraethyl-1,2-bis(4-methoxyphenyl)-, (±)-		DD01510
192 193 121 28 56 164 134 29	**384**	100 14 5 4 3 2 2 2	**0.00**	24	36	2	2	–	–	–	–	–	–	–	–	–	Ethylenediamine, N,N,N',N'-tetraethyl-1,2-bis(4-methoxyphenyl)-, meso-		DD01511
101 369 43 57 370 55 41 71	**384**	100 64 25 21 11 11 11 9	**0.00**	24	48	3	–	–	–	–	–	–	–	–	–	–	Glycerol, 2,3-O-isopropylidene-1-O-octadecyl-	16725-43-2	NI39391
101 57 369 71 83 85 149 125	**384**	100 66 62 51 38 35 6 6	**1.00**	24	48	3	–	–	–	–	–	–	–	–	–	–	Glycerol, 2,3-O-isopropylidene-1-O-octadecyl-	16725-43-2	MS08950
283 97 57 83 227 111 284 125	**384**	100 34 28 26 24 24 22 14	**1.13**	24	48	3	–	–	–	–	–	–	–	–	–	–	Hexadecanoic acid, 2-butoxy-, butyl ester	40924-26-3	NI39392
57 55 384 325 83 69 90 59	**384**	100 84 80 62 49 49 31 27		24	48	3	–	–	–	–	–	–	–	–	–	–	Tricosanoic acid, 2-hydroxy-, methyl ester		MS08951
348 242 125 347 384 349 57 97	**384**	100 57 45 43 41 35 29 28		25	20	2	–	1	–	–	–	–	–	–	–	–	2-(Diphenylmethyl)diphenyl sulphone		NI39393
384 292 352 265 263 324 250 132	**384**	100 91 59 59 57 37 29 24		25	20	4	–	–	–	–	–	–	–	–	–	–	Triptycene-1,8-dicarboxylic acid, 13-methyl-, dimethyl ester		NI39394
293 384 263 265 250 132 324 353	**384**	100 71 29 25 19 19 16 14		25	20	4	–	–	–	–	–	–	–	–	–	–	Triptycene-1,8-dicarboxylic acid, 16-methyl-, dimethyl ester		NI39395
262 121 384 199 263 183 108 28	**384**	100 91 32 32 19 19 10 9		25	22	–	–	–	–	–	–	–	–	2	–	–	Phosphine, methylenebis[diphenyl-	2071-20-7	NI39396
121 262 384 199 183 77 263 91	**384**	100 81 66 62 28 26 20 20		25	22	–	–	–	–	–	–	–	–	2	–	–	Phosphine, methylenebis[diphenyl-	2071-20-7	NI39397
304 384 221 181 208 247 366 299	**384**	100 60 23 21 13 8 7 6		25	36	3	–	–	–	–	–	–	–	–	–	–	12'-apo-β-Carotene-3,12'-diol, 5,6-diepoxy-5,6-dihydro-		NI39398
304 384 221 181 208 273 366 299	**384**	100 23 15 13 11 6 5 5		25	36	3	–	–	–	–	–	–	–	–	–	–	12'-apo-β-Carotene-3,12'-diol, 5,8-epoxy-5,8-dihydro-		NI39399
286 43 57 28 327 41 259 384	**384**	100 89 84 44 41 39 37 34		25	36	3	–	–	–	–	–	–	–	–	–	–	6H-Dibenzo[b,d]pyran-1-ol, 3-(1,1-dimethylpentyl)-6a,7,10,10a-tetrahydro 6,6,9-trimethyl-, acetate, (6aR-trans)-	52493-14-8	NI39400
205 151 69 41 193 121 81 43	**384**	100 95 61 47 33 23 19 14	**8.00**	25	36	3	–	–	–	–	–	–	–	–	–	–	Phenol, 3-methoxy-5-methyl-2-(3,7,11-trimethyl-2,6,10-dodecatrienyl)-, acetate, (E,E)-	75113-21-2	NI39401
384 383 77 192 382 385 191 381	**384**	100 84 82 60 25 24 21 15		26	16	–	4	–	–	–	–	–	–	–	–	–	5,10,15,18-Tetraaza-16,17-dibenz[b,n]picene		MS08952
136 118 43 384 41 91 55 81	**384**	100 77 60 56 43 37 34 30		27	44	1	–	–	–	–	–	–	–	–	–	–	Cholecalciferol		LI08291
43 143 135 41 91 81 129 141	**384**	100 69 63 54 52 46 44 43	**0.00**	27	44	1	–	–	–	–	–	–	–	–	–	–	Cholesta-4,6-dien-3-ol, (3β)-	14214-69-8	NI39402
351 384 43 143 145 55 325 41	**384**	100 68 58 54 50 47 44 43		27	44	1	–	–	–	–	–	–	–	–	–	–	Cholesta-5,7-dien-3-ol, (3β)-	434-16-2	NI39403
351 384 43 55 325 145 143 41	**384**	100 91 62 53 47 45 44 42		27	44	1	–	–	–	–	–	–	–	–	–	–	Cholesta-5,7-dien-3-ol, (3β)-	434-16-2	NI39404
351 143 145 253 43 55 157 57	**384**	100 88 75 64 64 56 49 46	**39.00**	27	44	1	–	–	–	–	–	–	–	–	–	–	Cholesta-5,7-dien-3-ol, (3β)-	434-16-2	NI39405
384 95 385 81 82 55 69 57	**384**	100 34 30 23 21 20 18 17		27	44	1	–	–	–	–	–	–	–	–	–	–	Cholesta-5,20(22)-dien-3-ol, (3β)-	21903-21-9	NI39406
384 43 58 55 95 57 385 41	**384**	100 68 40 36 30 30 29 29		27	44	1	–	–	–	–	–	–	–	–	–	–	Cholesta-5,20-dien-3-ol, (3β)-	41083-90-3	NI39407
55 69 111 81 95 255 159 145	**384**	100 75 71 69 61 60 52 52	**28.00**	27	44	1	–	–	–	–	–	–	–	–	–	–	Cholesta-5,22-dien-3-ol, (3β)-	566-89-2	NI39408
55 69 81 255 111 95 107 105	**384**	100 80 75 63 56 50 44 44	**35.00**	27	44	1	–	–	–	–	–	–	–	–	–	–	Cholesta-5,22-dien-3-ol, (3β,22Z)-		LI08292
271 41 69 55 384 81 43 95	**384**	100 91 88 82 62 57 51 47		27	44	1	–	–	–	–	–	–	–	–	–	–	Cholesta-5,24-dien-3-ol, (3β)-	313-04-2	NI39409
271 384 369 272 296 213 273 253	**384**	100 40 40 40 29 27 20 20		27	44	1	–	–	–	–	–	–	–	–	–	–	Cholesta-5,24-dien-3-ol, (3β)-	313-04-2	NI39410
271 369 272 145 384 300 253 159	**384**	100 38 31 28 26 26 26 25		27	44	1	–	–	–	–	–	–	–	–	–	–	Cholesta-5,24-dien-3-ol, (3β)-	313-04-2	NI39411
384 43 55 41 385 57 369 69	**384**	100 43 41 33 31 28 27 26		27	44	1	–	–	–	–	–	–	–	–	–	–	Cholesta-7,9(11)-dien-3-ol, (3β,5α)-	566-96-1	NI39412
384 43 55 271 41 145 211 57	**384**	100 96 79 61 61 57 56 56		27	44	1	–	–	–	–	–	–	–	–	–	–	Cholesta-7,9(11)-dien-3-ol, (3β,5α)-	566-96-1	NI39413
369 384 43 351 55 41 238 159	**384**	100 78 53 46 34 33 32 29		27	44	1	–	–	–	–	–	–	–	–	–	–	Cholesta-7,14-dien-3-ol, (3β,5α)-		NI39414
271 109 107 105 111 121 123 119	**384**	100 52 47 47 46 41 40 38	**23.00**	27	44	1	–	–	–	–	–	–	–	–	–	–	Cholesta-7,24-dien-3-ol, (3β,5α)-		NI39415
369 384 43 351 55 41 57 238	**384**	100 82 68 51 39 36 32 31		27	44	1	–	–	–	–	–	–	–	–	–	–	Cholesta-8,14-dien-3-ol, (3β,5α)-	19431-20-0	NI39417
384 369 43 351 385 41 370 55	**384**	100 87 43 38 30 28 27 26		27	44	1	–	–	–	–	–	–	–	–	–	–	Cholesta-8,14-dien-3-ol, (3β,5α)-	19431-20-0	NI39416
96 384 41 55 81 107 369 95	**384**	100 98 83 78 58 54 52 51		27	44	1	–	–	–	–	–	–	–	–	–	–	Cholesta-8,24-dien-3-ol, (3β,5α)-	128-33-6	NI39418
384 369 385 229 135 109 95 382	**384**	100 50 26 26 25 25 23 21		27	44	1	–	–	–	–	–	–	–	–	–	–	Cholesta-8,24-dien-3-ol, (3β,5α)-	128-33-6	MS08953
384 229 385 123 108 106 110 121	**384**	100 57 32 30 29 29 28 25		27	44	1	–	–	–	–	–	–	–	–	–	–	Cholestan-5-en-3-one	601-54-7	LI08293
259 369 384 166 167 109 370 217	**384**	100 63 46 27 26 23 22 22		27	44	1	–	–	–	–	–	–	–	–	–	–	5α-Cholest-22-en-16-one, (Z)-	14949-13-4	NI39419
41 69 55 81 384 67 82 109	**384**	100 97 88 77 66 60 56 55		27	44	1	–	–	–	–	–	–	–	–	–	–	5β-Cholest-24-en-12-one	14949-29-2	NI39420
122 109 123 43 384 41 134 107	**384**	100 39 33 33 32 31 30 29		27	44	1	–	–	–	–	–	–	–	–	–	–	Cholest-1-en-3-one, (5α)-	601-55-8	NI39421
43 41 122 55 109 369 81 57	**384**	100 58 56 47 39 33 27 27	**24.44**	27	44	1	–	–	–	–	–	–	–	–	–	–	Cholest-2-en-1-one, (5α)-	601-11-6	NI39422
124 384 229 55 43 261 41 95	**384**	100 47 36 32 31 28 27 26		27	44	1	–	–	–	–	–	–	–	–	–	–	Cholest-4-en-3-one	601-57-0	NI39425
124 55 384 95 81 107 57 135	**384**	100 63 50 45 40 38 38 34		27	44	1	–	–	–	–	–	–	–	–	–	–	Cholest-4-en-3-one	601-57-0	NI39424
124 55 384 41 229 95 69 44	**384**	100 52 46 45 39 39 36 35		27	44	1	–	–	–	–	–	–	–	–	–	–	Cholest-4-en-3-one	601-57-0	NI39423
124 229 384 79 95 81 91 107	**384**	100 58 53 48 47 46 43 42		27	44	1	–	–	–	–	–	–	–	–	–	–	Cholest-5-en-3-one	601-54-7	WI01501
384 271 55 43 229 41 95 105	**384**	100 93 85 79 75 74 70 50		27	44	1	–	–	–	–	–	–	–	–	–	–	Cholest-7-en-3-one, (5α)-	15459-85-5	NI39426
384 55 369 385 271 229 57 95	**384**	100 48 39 32 32 27 27 26		27	44	1	–	–	–	–	–	–	–	–	–	–	Cholest-8(14)-en-3-one, (5α)-	15477-87-9	NI39427
384 218 385 55 271 147 43 41	**384**	100 75 30 28 25 25 25 23		27	44	1	–	–	–	–	–	–	–	–	–	–	Cholest-8(14)-en-7-one, (5α)-	20853-61-6	NI39428
384 253 385 109 369 271 136 261	**384**	100 36 35 30 28 28 28 26		27	44	1	–	–	–	–	–	–	–	–	–	–	Cholest-8(14)-en-15-one, (5α)-	40071-66-7	NI39429
384 369 232 355 356 177 161 219	**384**	100 37 16 15 13 12 10 7		27	44	1	–	–	–	–	–	–	–	–	–	–	Cholest-8-en-11-one, (14β)-		MS08954
272 274 95 109 96 69 67 81	**384**	100 24 16 15 15 14 13 12	**2.00**	27	44	1	–	–	–	–	–	–	–	–	–	–	Cholest-14-en-16-one, (5α)-	54725-43-8	NI39430

MASS TO CHARGE RATIOS								M.W.	INTENSITIES								Parent	C	H	O	N	S	F	Cl	Br	I	Si	P	B	X	COMPOUND NAME	CAS Reg No	No
369	370	55	41	69	109	384	67	**384**	100	27	22	21	20	19	18	16		27	44	1	–	–	–	–	–	–	–	–	–	–	Cholest-24-en-16-one, (5α,20ξ)-	54482-50-7	NI39431
55	43	384	41	95	81	57	93	**384**	100	88	80	76	68	64	64	50		27	44	1	–	–	–	–	–	–	–	–	–	–	4,5-Cyclo-A-homo-B-norcholestan-3-one, (4α,5α)-		NI39432
55	43	57	41	384	95	81	69	**384**	100	98	88	75	66	61	60	58		27	44	1	–	–	–	–	–	–	–	–	–	–	4,5-Cyclo-A-homo-B-norcholestan-3-one, (4β,5β)-		NI39433
55	43	57	41	95	79	69	77	**384**	100	92	77	73	69	69	62	54	**39.00**	27	44	1	–	–	–	–	–	–	–	–	–	–	3,5-Cyclocholestan-2-one, (3α,5α)-	26753-93-5	NI39434
384	136	385	366	43	369	121	95	**384**	100	36	32	32	27	24	18	15		27	44	1	–	–	–	–	–	–	–	–	–	–	3,5-Cyclocholestan-6-one, (3α,5α)-	3839-09-6	NI39435
247	137	384	271	136	135	43	369	**384**	100	99	88	78	66	66	53	47		27	44	1	–	–	–	–	–	–	–	–	–	–	5,7-Cyclocholestan-4-one, (5α,7α)-	14893-72-2	NI39436
111	55	43	137	41	384	247	135	**384**	100	90	90	80	75	65	65	65		27	44	1	–	–	–	–	–	–	–	–	–	–	5,7-Cyclocholestan-4-one, (5β,7β)-	19318-00-4	NI39437
271	384	69	369	255	213	231	300	**384**	100	95	63	37	34	27	23	12		27	44	1	–	–	–	–	–	–	–	–	–	–	26,27-Dinorcholesta-7,24-dien-3-ol, 23,23-dimethyl-, (3β)-		DD01512
384	383	341	43	385	81	41	55	**384**	100	34	33	31	30	25	23	22		27	44	1	–	–	–	–	–	–	–	–	–	–	A-Homo-B,19-dinorcholest-1(10)-ene, 3,5-epoxy-4a-methyl-, (3β,4aα,5β)-	92095-09-5	NI39438
255	111	384	95	93	68	105	145	**384**	100	85	79	75	73	72	70	65		27	44	1	–	–	–	–	–	–	–	–	–	–	27-Norergosta-5,22-dien-3-ol, (3β,22Z)-	58072-43-8	NI39439
136	118	43	384	41	91	55	158	**384**	100	77	60	56	43	37	34	31		27	44	1	–	–	–	–	–	–	–	–	–	–	9,10-Secocholesta-5,7,10(19)-trien-3-ol, (3β,5Z,7E)-	67-97-0	NI39440
41	136	118	43	384	91	55	158	**384**	100	98	77	60	55	34	34	30		27	44	1	–	–	–	–	–	–	–	–	–	–	9,10-Secocholesta-5,7,10(19)-trien-3-ol, (3β,5Z,7E)-	67-97-0	NI39441
384	385	355	326	324	162	383	356	**384**	100	32	30	17	14	13	12	10		28	16	2	–	–	–	–	–	–	–	–	–	–	9,9'-Dianthraquinone		IC04233
119	117	121	145	147	84	82	163	**384**	100	94	34	15	14	14	13	9	**0.00**	28	20	–	2	–	–	–	–	–	–	–	–	–	Pyrazine, tetraphenyl-	642-04-6	NI39442
177	109	123	135	122	192	178	111	**384**	100	62	45	42	35	28	27	25	**21.00**	28	48	–	–	–	–	–	–	–	–	–	–	–	23,28-Bisnor-17(H)-hopane, (17α)-		NI39443
177	109	122	135	178	123	192	384	**384**	100	42	34	30	27	26	22	18		28	48	–	–	–	–	–	–	–	–	–	–	–	23,28-Bisnor-17(H)-hopane, (17β)-		NI39444
384	385	369	229	230	161	122	121	**384**	100	30	26	25	20	18	14	14		28	48	–	–	–	–	–	–	–	–	–	–	–	Cholestane, 4-methylene-, (5α)-	28113-76-0	NI39445
384	316	105	229	95	385	369	121	**384**	100	44	43	35	34	33	30	30		28	48	–	–	–	–	–	–	–	–	–	–	–	Cholest-2-ene, 2-methyl-, (5α)-	22599-90-2	NI39446
384	83	85	47	385	43	369	57	**384**	100	51	44	34	32	29	26	25		28	48	–	–	–	–	–	–	–	–	–	–	–	Cholest-2-ene, 4-methyl-, (4α,5α)-	55724-44-2	NI39447
384	69	43	385	81	57	55	41	**384**	100	29	27	25	22	22	20	20		28	48	–	–	–	–	–	–	–	–	–	–	–	Cholest-3-ene, 2-methyl-, (2α,5α)-	20997-53-9	NI39448
384	369	385	355	370	356	95	81	**384**	100	47	31	30	14	9	8	6		28	48	–	–	–	–	–	–	–	–	–	–	–	Cholest-3-ene, 4-methyl-, (5α)-	6785-18-8	NI39449
300	83	55	257	109	81	95	217	**384**	100	89	85	61	58	52	50	40	**20.00**	28	48	–	–	–	–	–	–	–	–	–	–	–	Cholest-24-ene, 24-methyl-, (5β)-		LI08294
43	55	41	109	67	81	105	95	**384**	100	71	62	51	36	35	34	33	**28.00**	28	48	–	–	–	–	–	–	–	–	–	–	–	5α-Ergost-8(14)-ene	6673-69-4	NI39450
300	83	55	257	109	81	95	217	**384**	100	88	85	60	58	52	50	41	**20.79**	28	48	–	–	–	–	–	–	–	–	–	–	–	5β-Ergost-24-ene	14949-17-8	NI39451
257	94	93	81	258	95	147	67	**384**	100	38	36	32	30	29	28	27	**3.96**	28	48	–	–	–	–	–	–	–	–	–	–	–	Ergost-14-ene, (5α)-	40446-05-7	NI39452
178	384	356	385	179	357	176	177	**384**	100	83	36	26	15	11	11	7		29	20	1	–	–	–	–	–	–	–	–	–	–	2,4-Cyclopentadien-1-one, 2,3,4,5-tetraphenyl-	479-33-4	MS08955
178	384	356	385	179	357	278	279	**384**	100	87	37	27	15	11	7	6		29	20	1	–	–	–	–	–	–	–	–	–	–	2,4-Cyclopentadien-1-one, 2,3,4,5-tetraphenyl-	479-33-4	NI39453
191	167	115	384	91	293	102	192	**384**	100	92	47	44	44	36	27	21		30	24	–	–	–	–	–	–	–	–	–	–	–	1-(1,3-Diphenyl-cyclopropen-3-yl)-2,3-diphenylcyclopropane		LI08295
372	370	374	140	387	292	385	352	**385**	100	78	48	21	14	14	11	11		7	7	–	3	–	1	4	–	–	–	3	–	–	1-Methyl-1-(4-fluorophenyl)tetrachlorocyclotriphosphazene	87048-77-9	NI39454
73	212	240	370	75	213	74	117	**385**	100	66	22	18	11	9	8	7	**0.00**	10	13	4	1	1	6	–	–	–	1	–	–	–	L-Cysteine, N-(trifluoroacetyl)-, trimethylsilyl ester, trifluoroacetate	57207-35-9	NI39455
111	73	75	147	138	129	370	230	**385**	100	91	70	68	61	37	36	34	**6.00**	15	27	5	3	–	–	–	–	–	2	–	–	–	3,6-Epoxy-2H,8H-pyrimido[6,1-b][1,3]oxazocin-8-one, 3,4,5,6,9,10-hexahydro-10-imino-4,5-bis[(trimethylsilyl)oxy]-, [3R-(3α,4β,5β,6α)]-	41547-77-7	NI39456
160	294	73	161	91	295	211	147	**385**	100	29	14	12	8	6	5	5	**1.00**	17	32	3	1	–	–	–	–	–	2	1	–	–	Phosphonic acid, [1-[isopropylideneamino]-2-phenylethyl]-, bis(trimethylsilyl) ester	55108-69-5	NI39457
73	280	147	295	281	75	45	74	**385**	100	71	31	26	22	14	14	11	**0.00**	17	35	3	1	–	–	–	–	–	3	–	–	–	Pyridoxime, tris(trimethylsilyl)-		LI08296
280	73	295	281	147	296	370	282	**385**	100	83	71	26	22	21	13	12	**0.00**	17	35	3	1	–	–	–	–	–	3	–	–	–	Vitamin B_6, tris(trimethylsilyl)-		NI39458
181	120	164	121	149	313	208	92	**385**	100	60	53	34	33	32	26	25	**9.00**	18	19	5	5	–	–	–	–	–	–	–	–	–	N-Ethyl-N-(2-carboxyethyl)-4-(4-nitro-2-carbamylphenylazo)aniline		IC04234
260	258	76	157	155	77	259	75	**385**	100	99	26	18	18	18	16	16	**2.00**	19	13	1	1	–	–	1	1	–	–	–	–	–	4-Chloro-2-methoxycarbonylphenyl-N-(4-bromophenyl)benzimidate		NI39459
385	228	201	245	229	159	173	216	**385**	100	66	44	34	26	17	15	12		19	15	6	1	1	–	–	–	–	–	–	–	–	5H-Furo[3,2-g][1]benzopyran-9-sulphonamide, 4-methoxy-7-methyl-5-oxo-N-phenyl-		MS08956
221	263	342	298	222	326	67	59	**385**	100	43	20	13	13	9	9	8	**2.88**	19	19	6	3	–	–	–	–	–	–	–	–	–	3,5-Pyridinedicarboxylic acid, 2-cyano-1,4-dihydro-6-methyl-4-(3-nitrophenyl)-, 5-isopropyl 3-methyl ester		NI39460
85	59	143	135	112	93	55	107	**385**	100	78	49	32	28	25	21	17	**0.00**	19	31	5	1	1	–	–	–	–	–	–	–	–	Bornane-10,2-sultam, N-[(4R,5R)-5-ethyl-2,2,5-trimethyl-1,3-dioxolane-4-carbonyl]-		DD01513
85	59	143	112	135	55	93	57	**385**	100	87	35	27	25	25	20	19	**0.00**	19	31	5	1	1	–	–	–	–	–	–	–	–	Bornane-10,2-sultam, N-[(4S,5S)-5-ethyl-2,2,5-trimethyl-1,3-dioxolane-4-carbonyl]-		DD01514
105	77	263	265	228	51	122	106	**385**	100	50	34	20	12	10	8	8	**1.00**	20	13	3	1	–	–	2	–	–	–	–	–	–	N-Benzoyl-N-(2,4-dichlorophenyl)-N-phenylamine-2-carboxylic acid		NI39461
170	385	171	168	386	154	155	127	**385**	100	43	16	12	10	10	9	8		20	19	5	1	1	–	–	–	–	–	–	–	–	1-Dimethylaminonaphthalene-5-sulphonyloxy-2-methoxy-4-formylbenzene		LI08297
226	239	385	227	210	240	211	386	**385**	100	62	37	16	13	11	11	9		20	23	5	3	–	–	–	–	–	–	–	–	–	N-(4-Carboxybutylcarbonyloxyethyl)-2-methoxy-4-(2,2-dicyanovinyl)-5-methylaniline		IC04235
125	55	81	134	223	163	135	162	**385**	100	43	21	20	16	12	11	10	**1.00**	20	27	3	5	–	–	–	–	–	–	–	–	–	2(1H)-Quinolinone, 3,4-dihydro-6-[4-[(trans-4-hydroxycyclohexyl)-1H-5-tetrazolyl]butoxy]-		NI39462
301	302	55	167	168	153	223	180	**385**	100	23	19	14	13	12	10	9	**8.00**	21	16	3	1	–	–	–	–	–	–	–	–	1	Cymantrene, (diphenylaminomethyl)-	87122-50-7	NI39463

MASS TO CHARGE RATIOS								M.W.	INTENSITIES								Parent	C	H	O	N	S	F	Cl	Br	I	Si	P	B	X	COMPOUND NAME	CAS Reg No	No
357	385	314	342	326	283	115	298	**385**	100	93	52	47	38	36	34	33		21	23	6	1	–	–	–	–	–	–	–	–	–	**Acetamide, N-(5,6,7,9-tetrahydro-10-hydroxy-1,2,3-trimethoxy-9-oxobenzo[a]heptalen-7-yl)-, (S)-**	477-27-0	**NI39464**
206	163	385	162	354	164	207	193	**385**	100	57	23	18	16	16	12	11		21	23	6	1	–	–	–	–	–	–	–	–	–	**Rheadan-8-ol, 2,3-dimethoxy-16-methyl-10,11-[methylenebis(oxy)]-, (6α,8α)-**	2255-44-9	**NI39465**
276	109	259	126	277	110	98	385	**385**	100	44	39	37	27	18	15	14		21	24	2	1	1	–	–	–	–	–	1	–	–	**Quinuclidine, 3-[(2,8-dimethylphenothiaphosphin-10-yl)oxy]-, 10-oxide**	23855-76-7	**NI39466**
103	77	137	51	118	385	150	91	**385**	100	24	22	16	13	9	9	9		22	16	–	5	–	–	1	–	–	–	–	–	–	**3H-Imidazo[1,2-b][1,2,4]triazol-2-amine, N,3-diphenyl-5-(4-chlorophenyl)**		**MS08957**
164	179	149	150	283	206	121	77	**385**	100	23	23	15	13	13	13	13	**6.59**	22	27	5	1	–	–	–	–	–	–	–	–	–	**Dibenz[c,g]azecin-13(6H)-one, 5,7,8,14-tetrahydro-3,4,10,11-tetramethoxy 6-methyl-**	2292-20-8	**NI39467**
384	385	203	312	205	191	329	190	**385**	100	91	46	42	38	38	24	24		22	27	5	1	–	–	–	–	–	–	–	–	–	**(±)-5,6,8,8a,9,10,13,13a-Octahydro-11-oxo-11H-dibenzo[a,g]quinolizine-8a-carboxylic acid ethyl ester**		**MS08958**
385	384	370	342	311	280	354	327	**385**	100	74	50	33	33	26	21	17		22	27	5	1	–	–	–	–	–	–	–	–	–	**Thalicsmidine**		**LI08298**
385	232	354	355	386	326	233	370	**385**	100	52	47	39	37	25	18	17		23	35	2	1	–	–	–	–	–	1	–	–	–	**Estra-1,3,5(10)-trien-17-one, 1-methyl-3-[(trimethylsilyl)oxy]-, O-methyloxime**		**NI39468**
262	385	200	183	122	78	263	108	**385**	100	57	27	26	22	19	18	18		24	21	–	1	–	–	–	–	–	–	2	–	–	**Bis(diphenylphosphino)amine**		**NI39469**
224	385	338	311	193	308	239	370	**385**	100	57	29	15	14	12	12	6		24	23	–	3	1	–	–	–	–	–	–	–	–	**2-Imidazoline, 1-(2,6-dimethylphenyl)-4,4-diphenyl-5-imino-2-(methylthio)**	118514-98-0	**DD01515**
385	105	180	340	341	132	120	224	**385**	100	50	46	30	22	20	20	12		24	23	2	3	–	–	–	–	–	–	–	–	–	**4-Imidazolidinone, 5-ethoxy-3-methyl-1,5-diphenyl-2-(phenylimino)-**	116927-99-2	**DD01516**
325	294	310	326	43	91	295	105	**385**	100	44	41	25	24	16	12	12	**1.00**	24	35	3	1	–	–	–	–	–	–	–	–	–	**3β-Acetoxy-pregna-5,16-dien-20-one, methyl oxime**		**LI08299**
266	134	119	91	250	267	65	133	**385**	100	85	77	43	29	21	11	10	**0.00**	25	23	3	1	–	–	–	–	–	–	–	–	–	**4,5-Dihydro-5-(4-methoxyphenyl)-3-(4-methylphenyl)-5-benzoylmethylisoxazole**	56335-89-8	**NI39470**
385	384	274	386	67	275	273	383	**385**	100	70	33	26	20	18	18	16		25	38	–	1	–	–	–	–	–	–	–	3	–	**9-Borabicyclo[3.3.1]non-9-amine, N-9-borabicyclo[3.3.1]non-9-yl-N-(2,3-dihydro-3-methyl-1H-1-benzoborol-1-yl)-**	65953-34-6	**NI39471**
105	122	77	148	43	134	135	55	**385**	100	32	21	15	14	10	9	8	**5.00**	25	39	2	1	–	–	–	–	–	–	–	–	–	**8-Hexadecenamine, 3-acetyl-N-benzoyl-, (Z)-**		**NI39472**
370	387	384	109	272	135	110	108	**385**	100	93	87	87	80	61	60	52	**5.00**	26	43	1	1	–	–	–	–	–	–	–	–	–	**6-Azacholest-4-en-7-one**	5226-43-7	**LI08300**
370	387	384	109	272	135	110	164	**385**	100	92	88	86	80	61	60	54	**4.80**	26	43	1	1	–	–	–	–	–	–	–	–	–	**6-Azacholest-4-en-7-one**	5226-43-7	**NI39473**
205	164	163	206	165	180	152	162	**385**	100	58	19	17	9	5	4	3	**1.50**	28	19	1	1	–	–	–	–	–	–	–	–	–	**Dispiro[fluorene-9,4-Δ^2-2-methyloxazoline-5,9-fluorene]**		**MS08959**
160	90	161	111	170	71	56	155	**385**	100	56	39	36	33	25	22	17	**16.66**	28	35	–	1	–	–	–	–	–	–	–	–	–	**Cyclohexylamine, 2-allyl-N,N-dimethyl-4-(2,4-pentadienyl)-3,5-diphenyl-**	14028-83-2	**NI39474**
263	265	261	237	36	191	63	272	**386**	100	64	64	29	29	27	27	26	**1.43**	9	4	4	–	–	–	6	–	–	–	–	–	–	**Bicyclo[2.2.1]hept-5-ene-2,3-dicarboxylic acid, 1,4,5,6,7,7-hexachloro-**	115-28-6	**IC04236**
263	261	265	36	237	272	274	267	**386**	100	64	61	33	32	26	21	21	**1.00**	9	4	4	–	–	–	6	–	–	–	–	–	–	**Bicyclo[2.2.1]hept-5-ene-2,3-dicarboxylic acid, 1,4,5,6,7,7-hexachloro-**	115-28-6	**IC04237**
263	265	261	26	84	96	191	193	**386**	100	64	63	39	38	31	30	29	**0.00**	9	4	4	–	–	–	6	–	–	–	–	–	–	**Bicyclo[2.2.1]hept-5-ene-2,3-dicarboxylic acid, 1,4,5,6,7,7-hexachloro-**	115-28-6	**NI39475**
263	265	261	335	237	272	307	370	**386**	100	63	62	34	33	29	26	25	**0.00**	9	4	4	–	–	–	6	–	–	–	–	–	–	**Bicyclo[2.2.1]hept-5-ene-2,3-dicarboxylic acid, 1,4,5,6,7,7-hexachloro-, (endo,endo)-**	21678-54-6	**NI39476**
217	248	286	117	43	198	179	93	**386**	100	28	13	13	10	9	9	8	**6.80**	10	–	–	–	–	14	–	–	–	–	–	–	–	**Perfluoro-1,4-diethylbenzene**		**IC04238**
353	325	355	327	289	109	73	145	**386**	100	97	79	77	77	71	63	61	**4.30**	10	5	1	–	–	–	7	–	–	–	–	–	–	**1,3,7,8,9,10,10-Heptachlorotetracyclo[5.2.1.0^{2,6}.0^{4,8}]decan-5-one**		**NI39477**
353	325	355	41	327	57	73	289	**386**	100	97	81	80	78	75	74	69	**3.50**	10	5	1	–	–	–	7	–	–	–	–	–	–	**1,5,7,8,9,10,10-Heptachlorotetracyclo[5.2.1.0^{2,6}.0^{4,8}]decan-3-one**		**NI39478**
353	355	73	43	217	149	41	219	**386**	100	85	80	70	61	60	60	59	**3.40**	10	5	1	–	–	–	7	–	–	–	–	–	–	**1,5,7,8,9,10,10-Heptachlorotricyclo[5.2.1.0^{2,6}]dec-4-en-3-one**		**NI39479**
81	353	355	27	351	29	39	53	**386**	100	82	70	54	45	41	35	33	**5.00**	10	5	1	–	–	–	7	–	–	–	–	–	–	**2,5-Methano-2H-indeno[1,2-b]oxirene, 2,3,4,5,6,7,7-heptachloro-1a,1b,5,5a,6,6a-hexahydro-, (1aα,1bβ,2α,5α,5aβ,6β,6aα)-**	1024-57-3	**NI39480**
117	289	317	353	119	287	253	115	**386**	100	41	38	30	30	29	27	27	**0.00**	10	5	1	–	–	–	7	–	–	–	–	–	–	**2,5-Methano-2H-indeno[1,2-b]oxirene, 2,3,4,5,6,7,7-heptachloro-1a,1b,5,5a,6,6a-hexahydro-, (1aα,1bβ,2α,5α,5aβ,6β,6aα)-**	1024-57-3	**NI39481**
183	135	81	185	272	100	149	237	**386**	100	99	93	92	80	78	72	70	**1.65**	10	5	1	–	–	–	7	–	–	–	–	–	–	**2,5-Methano-2H-indeno[1,2-b]oxirene, 2,3,4,5,6,7,7-heptachloro-1a,1b,5,5a,6,6a-hexahydro-, (1aα,1bβ,2α,5α,5aβ,6β,6aα)-**	1024-57-3	**NI39482**
382	384	380	386	378	381	379	376	**386**	100	92	67	50	36	30	26	17		10	6	–	–	–	–	–	–	–	–	–	–	2	**Naphtho[1,8-cd]-1,2-ditellurole**	52875-49-7	**NI39483**
64	129	77	131	116	160	69	51	**386**	100	86	30	29	16	15	9	8	**0.01**	10	6	–	–	–	10	2	–	–	–	–	–	–	**Cyclobutane, 1,1′-(1,1,2,2-tetrafluoro-1,2-ethanediyl)bis[2-chloro-2,3,3-trifluoro-**	35208-02-7	**MS08960**
73	155	45	139	137	127	43	74	**386**	100	22	22	19	19	17	13	8	**1.14**	10	20	2	–	–	–	–	2	–	2	–	–	–	**1,3-Butene, 1,4-dibromo-2,3-bis[(trimethylsilyl)oxy]-**		**NI39484**
357	358	359	218	246	149	98	95	**386**	100	31	23	15	12	12	12	12	**0.39**	10	30	6	–	–	–	–	–	–	5	–	–	–	**Cyclopentasiloxane, 1-hydroxy-1,3,5,7,9-pentaethyl-**	75221-57-7	**NI39485**
220	330	236	171	156	140	201	155	**386**	100	84	73	60	50	48	47	46	**40.04**	12	5	2	–	–	6	1	–	–	–	–	–	1	**Iron, dicarbonyl(2-chloro-3,3,4,4,5,5-hexafluoro-1-cyclopenten-1-yl)-π-cyclopentadienyl-**	12109-43-2	**NI39486**
171	173	155	157	28	386	145	143	**386**	100	96	22	21	15	11	11	11		12	7	6	2	1	–	–	1	–	–	–	–	–	**2,4-Dinitro-4′-bromodiphenyl sulphone**	75317-05-4	**NI39487**
91	78	106	79	105	134	39	51	**386**	100	99	95	80	55	50	50	39	**0.50**	14	10	6	–	–	–	–	–	–	–	–	–	2	**Iron, hexacarbonyl-π-cycloocta-1,3,5-trienyldi-**		**MS08961**
43	274	287	57	41	155	55	71	**386**	100	62	54	36	33	31	31	27	**14.60**	15	22	2	–	–	8	–	–	–	–	–	–	–	**1,1,5-Trihydroperfluoropentyl decanoate**		**MS08962**
43	274	287	57	155	55	71	69	**386**	100	61	54	36	32	30	26	24	**14.00**	15	22	2	–	–	8	–	–	–	–	–	–	–	**1,1,5-Trihydroperfluoropentyl decanoate**		**LI08301**
73	147	129	75	371	245	69	243	**386**	100	81	74	61	57	44	44	29	**5.00**	15	26	6	2	–	–	–	–	–	2	–	–	–	**3,6-Epoxy-2H,8H-pyrimido[6,1-b][1,3]oxazocine-8,10(9H)-dione, 3,4,5,6-tetrahydro-4,5-bis[(trimethylsilyl)oxy]-, [3R-(3α,4β,5β,6α)]-**	41547-72-2	**NI39488**

MASS TO CHARGE RATIOS								M.W.	INTENSITIES								Parent	C	H	O	N	S	F	Cl	Br	I	Si	P	B	X	COMPOUND NAME	CAS Reg No	No
85	125	103	89	147	119	97	75	386	100	70	48	32	30	27	24	22	0.00	16	19	6	–	–	–	–	1	–	–	–	–	–	α-D-Glucopyranoside, methyl 4,6-O-benzylidene-2-bromo-2-deoxy-, acetate	20853-47-8	NI39489
232	386	234	233	198	246	388	154	386	100	58	42	41	34	31	26	24		17	14	1	2	1	3	1	–	–	–	–	–	–	1-(2-Chlorophenothiazin-9-yl)-3-(trifluoroacetylamino)propane		LI08302
299	301	386	388	91	193	165	219	386	100	96	63	61	54	38	33	24		17	15	2	4	–	–	–	1	–	–	–	–	–	1-(Hydrazinocarbonyl)methyl-7-bromo-5-phenyl-1,2-dihydro-3H-1,4-benzdiazepin-2-one		MS08963
138	386	388	69	344	346	152	249	386	100	73	49	33	29	20	20	19		17	16	6	–	–	–	2	–	–	–	–	–	–	Spiro[benzofuran-2(3H),1′-[2]cyclohexene]-3,4′-dione, 5,7-dichloro-2′,4,6-trimethoxy-6′-methyl-, (2S-trans)-	55555-63-0	NI39490
269	73	270	343	271	147	75	45	386	100	28	19	6	5	4	3	3	0.00	17	30	6	–	–	–	–	–	–	2	–	–	–	Benzeneacetic acid, 2,3,4-trimethoxy-α-[(trimethylsilyl)oxy]-, trimethylsilyl ester		MS08964
269	73	270	343	271	195	75	45	386	100	27	21	7	5	4	4	4	2.00	17	30	6	–	–	–	–	–	–	2	–	–	–	Benzeneacetic acid, 2,3,5-trimethoxy-α-[(trimethylsilyl)oxy]-, trimethylsilyl ester		MS08965
269	73	270	255	75	271	239	45	386	100	40	22	9	8	6	6	6	0.00	17	30	6	–	–	–	–	–	–	2	–	–	–	Benzeneacetic acid, 2,4,5-trimethoxy-α-[(trimethylsilyl)oxy]-, trimethylsilyl ester		MS08966
269	73	270	147	343	271	75	45	386	100	48	20	8	6	5	5	4	2.40	17	30	6	–	–	–	–	–	–	2	–	–	–	Benzeneacetic acid, 3,4,5-trimethoxy-α-[(trimethylsilyl)oxy]-, trimethylsilyl ester		MS08967
126	114	258	84	86	158	187	300	386	100	63	53	49	40	39	34	23	4.63	17	30	6	4	–	–	–	–	–	–	–	–	–	L-Lysine, N[6]-acetyl-N[2]-[N-(N-acetyl-L-alanyl)-L-alanyl]-, methyl ester	17353-67-2	NI39491
270	234	272	271	269	117	235	202	386	100	63	40	30	28	27	25	23	0.00	18	16	–	–	1	4	1	–	–	–	–	1	–	5-Chloro-1-ethyl-trans-2-styrylbenzo[b]thiophenium tetrafluoroborate		LI08303
28	191	43	107	149	103	77	41	386	100	49	45	38	32	29	28	25	8.33	18	18	6	4	–	–	–	–	–	–	–	–	–	Isopropyl 3-[4-(2,4-dinitrophenylazo)phenyl]propionate		NI39492
272	270	56	315	317	85	274	71	386	100	96	61	46	42	35	34	31	0.00	18	34	1	–	–	–	–	–	–	–	–	–	1	Stannane, tricyclohexylhydroxy-	13121-70-5	NI39493
83	44	39	193	386	51	77	53	386	100	92	76	75	68	63	56	55		19	14	9	–	–	–	–	–	–	–	–	–	–	7H-Isobenzofuro[4,5-b][1,4]benzodioxepin-11-carboxaldehyde, 1,3-dihydro-1,4-dihydroxy-10-methoxy-5,8-dimethyl-3,7-dioxo-	20426-13-5	NI39494
136	139	164	77	105	386	352	230	386	100	66	52	47	40	27	26	22		19	18	3	2	2	–	–	–	–	–	–	–	–	4(3H)-Pyrimidinone, 2-(ethylthio)-3-phenyl-6-tosyl-		NI39495
43	386	44	57	55	45	41	69	386	100	68	68	67	54	46	43	40		19	18	7	2	–	–	–	–	–	–	–	–	–	[2]Benzopyrano[7,6-f]indazole-8-acetic acid, 2,4,6,8,9,11-hexahydro-5,10-dihydroxy-2,6-dimethyl-4,11-dioxo-, methyl ester	15298-42-7	NI39496
283	386	133	75	284	120	104	76	386	100	31	31	25	18	15	12	11		19	22	3	4	1	–	–	–	–	–	–	–	–	N-Ethyl-N-[2-(ethylthiocarbonyl)ethyl]-4-(nitrophenylazo)aniline		IC04239
81	124	271	125	79	272	157	41	386	100	52	48	48	40	32	31	29	7.30	19	30	–	–	4	–	–	–	–	–	–	–	–	3,6,10,13-Tetrathia(4,5,11,12-trans-dicyclohexano)[6.6.1]propellane		MS08968
234	252	206	235	202	203	224	124	386	100	70	60	55	48	40	32	31	0.00	19	30	8	–	–	–	–	–	–	–	–	–	–	1,4-Diacetoxy-2-methoxycarbonyl-3-(6-methoxycarbonylhexyl)cyclopentane		LI08304
105	77	43	194	386	162	51	69	386	100	59	48	28	25	18	18	15		20	18	4	–	2	–	–	–	–	–	–	–	–	Bis(α-benzoylacetonyl) disulphide		IC04240
386	233	260	302	275	217	71	234	386	100	71	65	57	55	30	30	27		20	18	8	–	–	–	–	–	–	–	–	–	–	1,3(2H,9bH)-Dibenzofurandione, 2,6-diacetyl-9-(acetyloxy)-7-hydroxy-8,9b-dimethyl-	52941-92-1	NI39497
163	164	104	135	355	76	103	162	386	100	29	25	20	19	18	17	16	0.86	20	18	8	–	–	–	–	–	–	–	–	–	–	1,2-Bis(4-methoxycarbonylbenzoyl)ethane		IC04241
311	312	368	296	314	334	294	313	386	100	55	54	49	39	31	15	13	7.50	20	18	8	–	–	–	–	–	–	–	–	–	–	α-2-Rhodomycinone	18118-77-9	NI39498
314	296	311	350	312	286	368	297	386	100	81	38	34	33	31	24	22	9.01	20	18	8	–	–	–	–	–	–	–	–	–	–	α-Rhodomycinone	17514-43-1	NI39499
314	296	57	43	311	350	312	286	386	100	80	43	40	39	33	32	32	6.67	20	18	8	–	–	–	–	–	–	–	–	–	–	α-Rhodomycinone	17514-43-1	NI39500
309	241	270	386	242	285	311	271	386	100	44	43	38	35	34	30	25		20	19	4	2	–	–	1	–	–	–	–	–	–	5H-Isoxazolo[2,3-d][1,4]benzodiazepin-6-one, 10-chloro-1,2,7,11b-tetrahydro-2R-(methoxycarbonyl)-trans-2-methyl-cis-11b-phenyl-	89839-34-9	NI39501
193	386	195	192	388	371	194	387	386	100	88	41	38	35	35	29	25		20	19	4	2	–	–	1	–	–	–	–	–	–	Spiro[2H-1-benzopyran-2,2′-[2H]indole], 5′-chloro-1′,3′-dihydro-8-methoxy-1′,3′,3′-trimethyl-6-nitro-	14994-04-8	NI39502
371	73	372	297	386	327	178	152	386	100	40	33	31	28	18	14	12		20	26	4	–	–	–	–	–	–	2	–	–	–	4,4-Biphenyldicarboxylic acid, bis(trimethylsilyl) ester		LI08305
371	73	372	297	386	327	178	181	386	100	40	32	30	27	17	13	12		20	26	4	–	–	–	–	–	–	2	–	–	–	4,4′-Biphenyldicarboxylic acid, bis(trimethylsilyl) ester	31396-31-3	NI39503
145	69	41	143	97	132	55	226	386	100	58	57	43	41	35	32	26	16.00	20	34	3	–	2	–	–	–	–	–	–	–	–	Hexanoic acid, 6-[(2-allyl-2-methyl-6,10-dithiaspiro[4.5]decan-1-yl)oxy]-, ethyl ester	92640-65-8	NI39504
191	69	81	41	71	55	95	43	386	100	64	62	56	46	43	41	39	0.00	20	35	2	–	–	–	–	1	–	–	–	–	–	2-Naphthalenol, 6-bromodecahydro-1-(5-hydroxy-3-methyl-3-pentenyl)-2,5,5,8a-tetramethyl-, [1S-[1α(E),2α,4aβ,6α,8aα]]-	17941-22-9	NI39505
83	43	229	227	244	386	243	187	386	100	22	21	15	14	10	9	5		21	22	7	–	–	–	–	–	–	–	–	–	–	9-Acetoxy-O-isovaleryl-dihydrooroselol		LI08306
219	217	234	179	153	97	386	149	386	100	89	77	52	40	40	35	29		21	22	7	–	–	–	–	–	–	–	–	–	–	[2,6′-Bi-2H-1-benzopyran]-4(3H)-one, 3′,4′-dihydro-3,5,7-trihydroxy-5′-methoxy-2′,2′-dimethyl-, (2S-trans)-	62499-03-0	NI39506
137	368	297	386	312	232	203	243	386	100	27	17	15	11	11	11	10		21	22	7	–	–	–	–	–	–	–	–	–	–	4H-1-Benzopyran-4-one, 7-hydroxy-3-[2,4-dihydroxy-5-methoxy-6-(3-hydroxy-3-methylbutyl)phenyl]-		MS08969
386	358	180	311	343	355	91	327	386	100	78	62	55	47	42	40	38		21	22	7	–	–	–	–	–	–	–	–	–	–	11H-Dibenzo[b,e][1,4]dioxepin-7-carboxylic acid, 3,8-dimethoxy-1,4,6,9-tetramethyl-11-oxo-, methyl ester	4723-32-4	NI39507
267	150	121	137	284	151	326	153	386	100	88	53	26	24	19	18	15	6.00	21	22	7	–	–	–	–	–	–	–	–	–	–	4′,7-Dimethoxy-2,3-cis-flavan-3,4-cis-diacetate		LI08307
267	150	121	137	151	153	256	284	386	100	99	47	32	23	19	18	16	11.00	21	22	7	–	–	–	–	–	–	–	–	–	–	4′,7-Dimethoxy-2,3-cis-flavan-3,4-trans-diacetate		LI08308
267	150	121	137	151	153	386	256	386	100	99	50	46	23	18	16	15		21	22	7	–	–	–	–	–	–	–	–	–	–	4′,7-Dimethoxy-2,3-trans-flavan-3,4-cis-diacetate		LI08309
267	150	121	137	151	284	153	326	386	100	90	57	34	24	21	17	15	8.00	21	22	7	–	–	–	–	–	–	–	–	–	–	4′,7-Dimethoxy-2,3-trans-flavan-3,4-trans-diacetate		LI08310

MASS TO CHARGE RATIOS	M.W.	INTENSITIES	Parent	C	H	O	N	S	F	Cl	Br	I	Si	P	B	X	COMPOUND NAME	CAS Reg No	No
192 326 298 386 151 164 191 177	**386**	100 38 25 23 20 18 15 14		21	22	7	–	–	–	–	–	–	–	–	–	–	Peltogynol trimethyl ether acetate	14184-26-0	NI39508
192 326 386 298 327 164 151 193	**386**	100 38 25 24 13 13 13 12		21	22	7	–	–	–	–	–	–	–	–	–	–	Peltogynol trimethyl ether acetate	14184-26-0	MS08970
336 307 337 174 335 279 187 308	**386**	100 25 24 20 18 15 12 11	**0.00**	21	23	3	2	–	–	1	–	–	–	–	–	–	Schizozyginium, 4-methyl-, chloride	2474-79-5	NI39509
336 307 337 174 335 279 308 291	**386**	100 26 24 20 18 16 12 12	**0.00**	21	23	3	2	–	–	1	–	–	–	–	–	–	Schizozyginium, 4-methyl-, chloride	2474-79-5	LI08311
98 77 386 99 94 158 126 70	**386**	100 9 8 8 8 7 7 7		21	26	1	2	2	–	–	–	–	–	–	–	–	10H-Phenothiazine, 10-[2-(1-methyl-2-piperidinyl)ethyl]-2-(methylsulphinyl)-	5588-33-0	NI39510
313 314 256 257 92 87 204 91	**386**	100 19 14 8 6 6 5 5	**0.00**	21	26	5	2	–	–	–	–	–	–	–	–	–	Estrone, 2-nitro-, 3-acetate-17-methyloxime		NI39511
313 314 92 91 256 344 87 315	**386**	100 24 13 11 7 6 6 4	**0.49**	21	26	5	2	–	–	–	–	–	–	–	–	–	Estrone, 4-nitro-, 3-acetate-17-methyloxime		NI39512
55 45 69 169 54 51 71 41	**386**	100 75 72 61 58 53 47 40	**26.26**	21	29	3	–	–	3	–	–	–	–	–	–	–	Androstan-3-one, 17-[(trifluoroacetyl)oxy]-, (5β,17β)-	2600-38-6	NI39513
270 121 91 271 107 105 41 79	**386**	100 44 31 29 27 27 26 25	**0.00**	21	29	3	–	–	3	–	–	–	–	–	–	–	Androstan-17-one, 3-[(trifluoroacetyl)oxy]-, (3β,5α)-	3959-78-2	NI39514
88 43 73 71 87 41 55 29	**386**	100 93 73 46 36 30 24 22	**0.00**	21	38	6	–	–	–	–	–	–	–	–	–	–	Butanoic acid, 2-ethyl-, 1,2,3-propanetriyl ester	56554-54-2	NI39515
43 159 211 98 57 158 55 41	**386**	100 50 43 32 28 25 24 24	**0.00**	21	38	6	–	–	–	–	–	–	–	–	–	–	1,3-Diaceto-2-myristin	14290-23-4	NI39516
159 211 158 326 177 313 218 210	**386**	100 94 65 52 46 37 21 18	**0.00**	21	38	6	–	–	–	–	–	–	–	–	–	–	1,3-Diaceto-2-myristin	14290-23-4	WI01502
43 159 211 41 55 98 57 158	**386**	100 34 21 21 17 15 14 12	**0.00**	21	38	6	–	–	–	–	–	–	–	–	–	–	2,3-Diaceto-1-myristin	14473-55-3	NI39517
43 159 211 98 57 158 55 41	**386**	100 50 43 32 28 25 24 24	**0.00**	21	38	6	–	–	–	–	–	–	–	–	–	–	2,3-Diaceto-1-myristin	14473-55-3	WI01503
271 99 272 387 117 289 388 100	**386**	100 26 21 19 9 7 5 2	**0.00**	21	38	6	–	–	–	–	–	–	–	–	–	–	Hexanoic acid, 1,2,3-propanetriyl ester	621-70-5	NI39518
99 43 41 32 71 29 27 60	**386**	100 77 48 40 39 28 28 25	**0.00**	21	38	6	–	–	–	–	–	–	–	–	–	–	Hexanoic acid, 1,2,3-propanetriyl ester	621-70-5	NI39519
99 60 71 43 271 73 257 41	**386**	100 32 22 21 18 18 15 14	**0.14**	21	38	6	–	–	–	–	–	–	–	–	–	–	Hexanoic acid, 1,2,3-propanetriyl ester	621-70-5	NI39520
99 71 43 74 28 41 271 55	**386**	100 96 45 33 18 14 11 10	**0.00**	21	38	6	–	–	–	–	–	–	–	–	–	–	Pentanoic acid, 2-methyl-, 1,2,3-propanetriyl ester	56554-55-3	NI39521
236 281 386	**386**	100 50 28		22	14	5	2	–	–	–	–	–	–	–	–	–	1,3-Dioxolo[4,5-c]quinolin-4(5H)-one, 2-(4-nitrophenyl)-2-phenyl-	54616-41-0	NI39522
207 43 386 206 208 311 205 310	**386**	100 71 50 42 32 30 21 20		22	18	3	4	–	–	–	–	–	–	–	–	–	Acetamide, N-[2-[2-[2-[(2-nitrophenyl)azo]phenyl]vinyl]phenyl]-	69395-33-1	NI39523
43 65 210 327 107 208 120 77	**386**	100 80 78 73 65 61 44 43	**9.00**	22	18	3	4	–	–	–	–	–	–	–	–	–	Acetamide, N-[2-[[2-[2-(2-nitrophenyl)vinyl]phenyl]azo]phenyl]-	69395-31-9	NI39524
208 386 149 139 193 221 168 165	**386**	100 93 80 53 47 40 40 40		22	26	6	–	–	–	–	–	–	–	–	–	–	6(2H)-Benzofuranone, 3,3a-dihydro-3a-methoxy-3-methyl-5-(2-propenyl)-2-(3,4,5-trimethoxyphenyl)-, [2R-(2α,3β,3aβ)]-	75363-19-8	NI39525
208 386 165 193 221 166 317 178	**386**	100 95 67 50 45 40 27 25		22	26	6	–	–	–	–	–	–	–	–	–	–	6(2H)-Benzofuranone, 3,3a-dihydro-3a-methoxy-3-methyl-5-(2-propenyl)-2-(3,4,5-trimethoxyphenyl)-, [2S-(2α,3α,3aα)]-	75363-17-6	NI39526
208 386 193 222 165 204 178 135	**386**	100 84 31 28 26 25 23 22		22	26	6	–	–	–	–	–	–	–	–	–	–	6(2H)-Benzofuranone, 3,3a-dihydro-3a-methoxy-3-methyl-5-(2-propenyl)-2-(3,4,5-trimethoxyphenyl)-, [2S-(2α,3α,3aβ)]-	75363-18-7	NI39527
386 165 151 219 387 166 194 164	**386**	100 60 48 29 25 25 16 12		22	26	6	–	–	–	–	–	–	–	–	–	–	Diaeudesmin	16499-02-8	NI39528
163 181 182 165 180 91 164 43	**386**	100 99 66 52 38 27 26 23	**6.00**	22	26	6	–	–	–	–	–	–	–	–	–	–	1-(3,5-Dimethoxy-4-acetoxyphenyl)ethyl 2-methoxy-4-(1-propenyl)phenyl ether		NI39529
386 165 177 151 387 166 219 194	**386**	100 53 42 33 25 20 14 14		22	26	6	–	–	–	–	–	–	–	–	–	–	Epieudesmin	4375-03-5	NI39530
386 165 151 166 387 164 219 138	**386**	100 68 44 29 24 21 19 15		22	26	6	–	–	–	–	–	–	–	–	–	–	Eudesmin	526-06-7	NI39531
386 177 165 151 166 387 135 219	**386**	100 65 60 50 30 23 19 16		22	26	6	–	–	–	–	–	–	–	–	–	–	Di-O-methylpinoresinol		LI08312
203 173 77 121 155 55 111 294	**386**	100 90 42 29 25 25 21 20	**2.52**	22	26	6	–	–	–	–	–	–	–	–	–	–	Bis(2-phenoxyethyl) adipate		IC04242
43 242 302 284 227 243 174 136	**386**	100 64 43 21 16 13 13 13	**4.00**	22	26	6	–	–	–	–	–	–	–	–	–	–	5,6,7,8-Tetrahydro-7-(1-hydroxymethylvinyl)-6-methyl-6-vinylnaphthalene 1,4-diol triacetate		NI39532
194 386 83 136 138 85 206 125	**386**	100 76 65 60 42 42 36 36		22	30	–	2	2	–	–	–	–	–	–	–	–	2,3,11,12-Dibenzo-1,13-dithia-4,10-diazacyclooctadeca-2,11-diene		MS08971
41 91 43 353 70 56 65 98	**386**	100 98 60 50 45 30 23 20	**10.00**	22	30	–	2	2	–	–	–	–	–	–	–	–	Pyrimidine, 1-benzylhexahydro-3-isopropyl-2-(4,4-dimethyl-2,6-dithioxocyclohexylidene)-		MS08972
327 156 386 325 328 184 43 174	**386**	100 87 48 29 22 15 14 11		22	30	4	2	–	–	–	–	–	–	–	–	–	Aspidospermidine-20,21-diol, 1-acetyl-17-methoxy-	36459-00-4	NI39533
199 200 58 341 169 143 238 168	**386**	100 20 5 4 4 4 3 3	**1.00**	22	30	4	2	–	–	–	–	–	–	–	–	–	β-Carboline-1-butanoic acid, 2-ethyl-1,2,3,4-tetrahydro-α-(ethoxycarbonyl)-, ethyl ester		NI39534
196 386 180 190 204 371 194 224	**386**	100 30 21 8 7 5 5 4		22	30	4	2	–	–	–	–	–	–	–	–	–	2-Deoxy-2-dihydromajdinol		LI08313
218 43 57 55 41 69 386 71	**386**	100 83 79 67 66 64 59 44		22	30	4	2	–	–	–	–	–	–	–	–	–	3,9-Methano-10H-furo[3,2-d]azonine-10,11-dione, 9-[2-(dimethylamino)-3 methoxyphenyl]decahydro-2,6-dimethyl-, [2R-(2R*,3R*,3aS*,9R*,10aR*)]-	55283-40-4	NI39535
242 260 386 368 227 353 243 119	**386**	100 52 46 38 38 23 22 20		22	31	5	–	–	–	–	–	–	–	–	1	–	Cortisol methyl boronate	31012-57-4	MS08973
242 227 119 105 260 386 123 107	**386**	100 98 82 72 68 67 53 46		22	31	5	–	–	–	–	–	–	–	–	1	–	Cortisol methyl boronate	31012-57-4	NI39536
242 43 55 227 41 119 91 79	**386**	100 92 70 69 65 64 61 59	**40.00**	22	31	5	–	–	–	–	–	–	–	–	1	–	Cortisol methyl boronate	31012-57-4	MS08974
260 259 121 386 135 109 122 136	**386**	100 88 37 35 35 33 31 29		22	31	5	–	–	–	–	–	–	–	–	1	–	17α,21-Dihydroxy-5β-pregnane-3,11,20-trione methyl boronate		LI08314
260 259 55 79 41 67 43 91	**386**	100 88 76 56 54 46 45 43	**35.00**	22	31	5	–	–	–	–	–	–	–	–	1	–	17α,21-Dihydroxy-5β-pregnane-3,11,20-trione methyl boronate		MS08975
260 242 119 386 105 124 123 121	**386**	100 87 75 67 67 57 52 52		22	31	5	–	–	–	–	–	–	–	–	1	–	11α,17α,21-Trihydroxypregn-4-ene-3,20-dione methyl boronate	35314-78-4	NI39537
386 73 387 388 45 314 371 75	**386**	100 49 36 14 8 7 6 5		22	34	2	–	–	–	–	–	–	2	–	–	–	1,1′-Biphenyl, 3,5,3′,5′-tetramethyl-4,4′-bis[(trimethylsilyl)oxy]-		NI39538
185 127 171 100 57 55 59 186	**386**	100 73 56 53 50 24 21 20	**0.05**	22	42	5	–	–	–	–	–	–	–	–	–	–	Bis[2-(octanoyloxy)propyl] ether		IC04243
149 75 57 73 43 263 41 239	**386**	100 48 42 34 29 25 25 24	**0.34**	22	46	3	–	–	–	–	–	–	1	–	–	–	Hexadecanoic acid, 3-[(trimethylsilyl)oxy]propyl ester	56630-48-9	NI39539

MASS TO CHARGE RATIOS	M.W.	INTENSITIES	Parent	C	H	O	N	S	F	Cl	Br	I	Si	P	B	X	COMPOUND NAME	CAS Reg No	No
57 43 239 73 71 130 55 41	386	100 87 76 69 60 53 51 48	0.97	22	46	3	–	–	–	–	–	–	1	–	–	–	Hexadecanoic acid, 3-[(trimethylsilyl)oxy]propyl ester	56630-48-9	NI39540
73 175 89 43 41 371 55 75	386	100 90 70 60 34 32 30 28	0.00	22	46	3	–	–	–	–	–	–	1	–	–	–	Octadecanoic acid, 3-(trimethylsiloxy)-, methyl ester	21987-13-3	NI39541
73 327 43 89 57 41 371 55	386	100 67 58 44 32 31 25 23	0.63	22	46	3	–	–	–	–	–	–	1	–	–	–	Octadecanoic acid, 2-[(trimethylsilyl)oxy]-, methyl ester	56196-58-8	NI39542
73 89 327 43 57 103 75 55	386	100 41 38 31 24 23 22 21	1.00	22	46	3	–	–	–	–	–	–	1	–	–	–	Octadecanoic acid, 2-[(trimethylsilyl)oxy]-, methyl ester	56196-58-8	MS08976
73 327 89 43 371 328 103 57	386	100 99 41 35 34 26 25 24	0.89	22	46	3	–	–	–	–	–	–	1	–	–	–	Octadecanoic acid, 2-[(trimethylsilyl)oxy]-, methyl ester	56196-58-8	NI39543
187 73 301 75 51 43 41 69	386	100 78 58 45 44 35 33 28	0.00	22	46	3	–	–	–	–	–	–	1	–	–	–	Octadecanoic acid, 12-[(trimethylsilyl)oxy]-, methyl ester	15075-70-4	LI08315
187 73 301 75 54 43 41 69	386	100 79 56 45 44 34 31 30	0.00	22	46	3	–	–	–	–	–	–	1	–	–	–	Octadecanoic acid, 12-[(trimethylsilyl)oxy]-, methyl ester	15075-70-4	LI08316
187 73 301 75 55 43 41 69	386	100 80 57 47 43 34 31 30	0.04	22	46	3	–	–	–	–	–	–	1	–	–	–	Octadecanoic acid, 12-[(trimethylsilyl)oxy]-, methyl ester	15075-70-4	NI39544
386 323 279 250 340 239 295 124	386	100 50 44 29 25 15 14 14		23	14	6	–	–	–	–	–	–	–	–	–	–	Triptycene-1,8,16-tricarboxylic acid		NI39545
267 105 120 77 78 268 150 104	386	100 55 40 40 22 20 15 10	0.00	23	18	4	2	–	–	–	–	–	–	–	–	–	Acetophenone, 2-[5-(3-nitrophenyl)-3-phenyl-2-isoxazolin-5-yl]-	29972-21-2	NI39546
267 105 120 77 40 268 150 78	386	100 53 41 39 28 19 15 10	0.00	23	18	4	2	–	–	–	–	–	–	–	–	–	Acetophenone, 2-[5-(3-nitrophenyl)-3-phenyl-2-isoxazolin-5-yl]-	29972-21-2	NI39547
267 105 120 77 268 150 78 104	386	100 54 41 40 19 15 10 9	0.00	23	18	4	2	–	–	–	–	–	–	–	–	–	4,5-dihydro-3-(4-nitrophenyl)-5-phenyl-5-benzoylmethylisoxazole	56804-93-4	NI39548
327 149 328 386 113 243 279 167	386	100 84 29 25 21 16 14 13		23	30	5	–	–	–	–	–	–	–	–	–	–	Abieta-5,8,11,13-tetraen-20-oic acid, methyl ester, 11,12-dimethoxy-7-oxo-		MS08977
344 345 386 284 346 387 342 137	386	100 25 11 5 4 3 2 2		23	30	5	–	–	–	–	–	–	–	–	–	–	2-Hydroxy-3-methoxyestradiol-2,17β-diacetate		NI39549
314 315 133 243 105 316 356 326	386	100 22 17 12 11 5 3 3	1.00	23	30	5	–	–	–	–	–	–	–	–	–	–	19-Norisopimara-4(18),8(14),15-trien-7-one, 2,17-diacetoxy-, (2α)-		MS08978
386 387 244 326 297 218 388 327	386	100 31 25 11 11 10 9 7		23	34	3	–	–	–	–	–	–	1	–	–	–	Estradiol, 17β-acetate-3-(trimethylsilyl)-		NI39550
355 356 386 104 57 43 78 41	386	100 25 16 6 6 6 5 5		23	34	3	2	–	–	–	–	–	–	–	–	–	Aspidospermidine-1-ethanol, 17-hydroxy-16-methoxy-α-methyl-	54725-07-4	NI39551
141 173 124 43 95 55 57 87	386	100 79 60 38 35 35 33 31	0.00	23	46	4	–	–	–	–	–	–	–	–	–	–	Docosanoic acid, 8,9-dihydroxy-, methyl ester	56555-06-7	NI39552
43 57 55 98 41 73 69 71	386	100 85 69 62 59 47 43 41	1.00	23	46	4	–	–	–	–	–	–	–	–	–	–	Eicosanoic acid, 2-hydroxy-1-(hydroxymethyl)ethyl ester	55334-78-6	WI01504
312 97 133 313 352 355 380 42	386	100 42 32 25 18 15 14 12	9.01	23	46	4	–	–	–	–	–	–	–	–	–	–	Eicosanoic acid, 2-hydroxy-1-(hydroxymethyl)ethyl ester	55334-78-6	WI01505
241 101 371 131 73 208 209 189	386	100 53 18 13 7 6 1 1	0.00	23	46	4	–	–	–	–	–	–	–	–	–	–	2,3-O-Isopropylidene-1-O-(2-methoxyhexadecyl)glycerol	16725-39-6	NI39553
241 101 83 43 97 71 57 69	386	100 48 48 48 45 42 36 34	0.00	23	46	4	–	–	–	–	–	–	–	–	–	–	2,3-O-Isopropylidene-1-O-(2-methoxyhexadecyl)glycerol	16725-39-6	NI39554
342 343 270 326 297 386 327 344	386	100 83 47 44 40 29 20 18		24	22	3	2	–	–	–	–	–	–	–	–	–	3-Oxo-2′-amino-6′-diethylamino-spiro[phthalan-1,9′-xanthene]		IC04244
198 189 77 118 91 180 182 104	386	100 49 34 32 29 21 8 8	8.00	24	22	3	2	–	–	–	–	–	–	–	–	–	Oxazolo[5,4-d]isoxazol-5(2H)-one, tetrahydro-3a,6a-dimethyl-2,3,6-triphenyl-	55780-96-6	NI39555
172 325 93 350 91 123 79 107	386	100 71 58 56 54 53 53 51	12.96	24	34	4	–	–	–	–	–	–	–	–	–	–	Bufa-20,22-dienolide, 3,14-dihydroxy-, (3β,5β)-	465-21-4	NI39556
325 93 350 123 91 79 107 95	386	100 84 82 77 77 77 73 58	19.00	24	34	4	–	–	–	–	–	–	–	–	–	–	Bufalin		LI08317
344 270 43 345 271 147 175 227	386	100 80 43 39 25 18 15 11	9.90	24	34	4	–	–	–	–	–	–	–	–	–	–	Hinokiol diacetate	18326-15-3	NI39557
343 269 43 344 270 147 175 227	386	100 80 42 36 34 17 13 10	9.51	24	34	4	–	–	–	–	–	–	–	–	–	–	Hinokiol diacetate	18326-15-3	NI39558
318 272 386 257 342 265 273 219	386	100 95 78 64 44 36 22 22		24	34	4	–	–	–	–	–	–	–	–	–	–	2-(3,3,4-trans-p-Mentha-1,8-dienyl)-4-(carboxyethyl)-5-pentyl-resorcinol		LI08318
283 281 284 301 145 137 344 187	386	100 26 20 17 14 11 10 9	1.84	24	34	4	–	–	–	–	–	–	–	–	–	–	Pregn-4-ene-3,20-dione, 17-(acetyloxy)-6-methyl-, (6α)-	71-58-9	NI39559
43 283 301 284 244 229 91 55	386	100 94 27 22 13 13 12 12	1.30	24	34	4	–	–	–	–	–	–	–	–	–	–	Pregn-4-ene-3,20-dione, 17-(acetyloxy)-16-methyl-, (16α)-	2504-36-1	NI39560
326 251 91 327 283 105 55 265	386	100 32 29 25 23 22 22 19	0.20	24	34	4	–	–	–	–	–	–	–	–	–	–	Pregn-5-en-20-one, 3-(acetyloxy)-16,17-epoxy-6-methyl-, (3β,16α)-	55349-94-5	WI01506
386 303 330 387 73 343 331 265	386	100 53 31 30 30 20 16 16		24	38	2	–	–	–	–	–	–	1	–	–	–	Δ[1]-Tetrahydrocannabinol trimethylsilyl ether	55449-68-8	NI39562
386 371 315 387 303 372 343 330	386	100 77 49 33 27 23 21 16		24	38	2	–	–	–	–	–	–	1	–	–	–	Δ[1]-Tetrahydrocannabinol trimethylsilyl ether	55449-68-8	NI39561
386 73 371 315 303 387 343 372	386	100 87 69 42 37 29 24 23		24	38	2	–	–	–	–	–	–	1	–	–	–	Δ[1]-Tetrahydrocannabinol trimethylsilyl ether	55449-68-8	NI39563
386 371 73 303 315 367 387 343	386	100 80 77 53 45 34 32 24		24	38	2	–	–	–	–	–	–	1	–	–	–	Δ[9]-Tetrahydrocannabinol trimethylsilyl ether		MS08979
140 357 386 153 358 155 125 154	386	100 91 60 50 35 32 31 28		24	38	2	–	–	–	–	–	–	1	–	–	–	Levo-5β-dihydronorgestrel, (trimethylsilyl)-		NI39564
129 257 296 386 74 330 36 32	386	100 46 34 31 29 28 25 22		24	38	2	–	–	–	–	–	–	1	–	–	–	Pregna-5,16-dien-20-one, 3-[(trimethylsilyl)oxy]-, (3β)-		LI08319
105 386 95 288 197 93 107 199	386	100 97 92 87 81 81 79 77		24	38	2	2	–	–	–	–	–	–	–	–	–	2-Pyrrolidinecarboxamide, N-androst-5-en-17-yl-, (S)-	55448-54-9	NI39565
103 69 85 84 81 97 71 95	386	100 93 63 40 32 30 29 27	0.00	24	50	3	–	–	–	–	–	–	–	–	–	–	2,4,22-Docosanetriol, 3,5-dimethyl-	56324-80-2	NI39566
43 41 29 92 78 176 55 69	386	100 76 75 56 54 35 28 26	2.30	24	51	1	–	–	–	–	–	–	–	1	–	–	Phosphine oxide, trioctyl-	78-50-2	NI39567
371 386 105 372 251 387 343 329	386	100 67 29 26 21 19 12 10		25	22	4	–	–	–	–	–	–	–	–	–	–	Fulvinervin B	99877-72-2	NI39568
386 18 124 229 28 55 387 81	386	100 70 62 42 32 30 29 26		25	38	3	–	–	–	–	–	–	–	–	–	–	Chol-4-en-24-oic acid, 3-oxo-, methyl ester	1452-33-1	NI39569
326 327 311 121 159 91 55 93	386	100 26 23 21 19 18 17 13	0.30	25	38	3	–	–	–	–	–	–	–	–	–	–	Pregn-5-en-20-one, 3-(acetyloxy)-6,16-dimethyl-, (3β,16α)-	55335-04-1	WI01507
326 327 311 121 159 91 55 119	386	100 26 23 21 19 18 17 13	0.30	25	38	3	–	–	–	–	–	–	–	–	–	–	Pregn-5-en-20-one, 3-(acetyloxy)-6,16-dimethyl-, (3β,16α)-	55335-04-1	AM00153
111 109 256 113 167 112 149 155	386	100 60 55 51 50 45 37 29	20.00	25	42	1	2	–	–	–	–	–	–	–	–	–	20,25-Diaza-dehydrocholesterol		LI08320
388 300 389 387 297 199 194 161	386	100 99 31 28 15 15 14 12	0.00	26	30	1	2	–	–	–	–	–	–	–	–	–	1,21-Cycloaspidospermidine, 17-methoxy-21-phenyl-, (21S)-	36528-88-8	NI39571
386 387 385 193 295 282 198 43	386	100 31 21 17 16 10 10 10		26	30	1	2	–	–	–	–	–	–	–	–	–	1,21-Cycloaspidospermidine, 17-methoxy-21-phenyl-, (21S)-	36528-88-8	NI39570
343 248 83 55 69 317 81 386	386	100 80 70 61 54 48 36 33		26	43	–	–	–	–	–	–	–	–	1	–	–	Phosphine, bis[methylisopropylcyclohexyl]phenyl-	74792-91-9	NI39572
72 73 315 58 55 68 67 56	386	100 8 4 4 3 2 2 2	0.80	26	46	–	2	–	–	–	–	–	–	–	–	–	5α-Pregnane-3β,20α-diamine, N[3]-isopropylidene-N[20],N[20]-dimethyl-	27802-32-0	NI39573
72 98 73 99 386 315 82 59	386	100 50 8 5 2 2 2 2		26	46	–	2	–	–	–	–	–	–	–	–	–	Pregn-5-ene-3α,20α-diamine, N[3]-ethyl-N[3],N[20],N[20]-trimethyl-	27769-11-5	NI39574
186 200 201 187 91 96 110 97	386	100 85 38 15 12 10 9 9	4.10	27	34	–	2	–	–	–	–	–	–	–	–	–	Quinoline, decahydro-1-methyl-5,7-diphenyl-6-(3,4,5,6-tetrahydro-2-pyridinyl)-	6887-36-1	NI39575
386 43 232 55 231 81 41 95	386	100 49 47 47 45 40 36 35		27	46	1	–	–	–	–	–	–	–	–	–	–	Cholestane, 2,3-epoxy-, (2α,3α,5α)-	1753-61-3	NI39576

MASS TO CHARGE RATIOS								M.W.	INTENSITIES								Parent	C	H	O	N	S	F	Cl	Br	I	Si	P	B	X	COMPOUND NAME	CAS Reg No	No
386	43	231	387	55	232	41	81	**386**	100	34	31	30	29	28	25	22		27	46	1	–	–	–	–	–	–	–	–	–	–	**Cholestane, 3,4-epoxy-, (3α,4α,5α)-**	1249-56-5	**NI39577**
43	55	41	81	95	57	67	69	**386**	100	76	65	57	43	43	41	40	**21.00**	27	46	1	–	–	–	–	–	–	–	–	–	–	**Cholestane, 4,5-epoxy-, (4α,5α)-**	6079-19-2	**NI39578**
386	43	55	387	41	81	95	57	**386**	100	49	44	33	33	32	29	26		27	46	1	–	–	–	–	–	–	–	–	–	–	**Cholestane, 4,5-epoxy-, (4β,5β)-**	23044-74-8	**NI39579**
386	368	331	43	387	371	231	81	**386**	100	66	43	31	30	29	25	23		27	46	1	–	–	–	–	–	–	–	–	–	–	**Cholestane, 5,6-epoxy-, (5α,6α)-**	20230-22-2	**NI39580**
386	385	125	124	122	136	368	149	**386**	100	99	79	54	54	40	39	36		27	46	1	–	–	–	–	–	–	–	–	–	–	**Cholestane, 5,6-epoxy-, (5α,6α)-**	20230-22-2	**NI39581**
217	83	85	286	257	109	55	95	**386**	100	90	57	56	54	50	48	46	**21.00**	27	46	1	–	–	–	–	–	–	–	–	–	–	**Cholestane, 23,24-epoxy-, (5β)-**	54411-90-4	**NI39582**
386	43	232	124	231	56	387	41	**386**	100	53	49	42	40	33	32	32		27	46	1	–	–	–	–	–	–	–	–	–	–	**Cholestan-1-one**	55700-37-3	**NI39583**
386	54	210	387	371	96	246	56	**386**	100	42	36	33	28	26	22	21		27	46	1	–	–	–	–	–	–	–	–	–	–	**Cholestan-3-one**	15600-08-5	**NI39584**
231	123	95	93	122	136	232	135	**386**	100	92	63	42	41	38	34	29	**25.84**	27	46	1	–	–	–	–	–	–	–	–	–	–	**Cholestan-6-one**	22033-82-5	**NI39585**
386	230	123	43	109	55	387	107	**386**	100	53	51	49	43	35	31	31		27	46	1	–	–	–	–	–	–	–	–	–	–	**Cholestan-6-one**	22033-82-5	**NI39586**
231	43	123	44	109	55	41	95	**386**	100	96	93	90	68	66	65	64	**26.26**	27	46	1	–	–	–	–	–	–	–	–	–	–	**Cholestan-6-one, (5α)-**	570-46-7	**NI39587**
386	123	231	55	43	95	57	41	**386**	100	49	46	44	38	34	33	33		27	46	1	–	–	–	–	–	–	–	–	–	–	**Cholestan-6-one, (5α)-**	570-46-7	**NI39588**
386	387	231	123	273	109	232	107	**386**	100	29	29	26	21	16	12	12		27	46	1	–	–	–	–	–	–	–	–	–	–	**Cholestan-6-one, (5α)-**	570-46-7	**NI39589**
330	386	43	370	231	123	55	41	**386**	100	64	48	47	37	36	35	31		27	46	1	–	–	–	–	–	–	–	–	–	–	**Cholestan-6-one, (5β)-**	13713-79-6	**NI39590**
386	178	95	55	232	43	255	81	**386**	100	68	50	38	35	34	32	29		27	46	1	–	–	–	–	–	–	–	–	–	–	**Cholestan-7-one, (5α)-**	567-71-5	**NI39591**
386	178	43	232	231	387	41	55	**386**	100	62	48	40	33	31	31	30		27	46	1	–	–	–	–	–	–	–	–	–	–	**Cholestan-7-one, (5α)-**	567-71-5	**NI39592**
191	386	178	95	232	81	192	55	**386**	100	73	53	37	29	26	25	25		27	46	1	–	–	–	–	–	–	–	–	–	–	**Cholestan-7-one, (5α,14β)-**	40072-53-5	**NI39593**
386	289	81	95	193	109	122	276	**386**	100	76	54	43	38	37	36	32		27	46	1	–	–	–	–	–	–	–	–	–	–	**Cholestan-11-one, (5α,14β)-**		**MS08980**
218	386	209	245	55	109	95	69	**386**	100	97	80	73	65	64	53	51		27	46	1	–	–	–	–	–	–	–	–	–	–	**Cholestan-15-one, (5α)-**	40071-71-4	**NI39594**
371	259	217	372	301	121	260	135	**386**	100	68	31	29	21	14	12	11	**4.03**	27	46	1	–	–	–	–	–	–	–	–	–	–	**Cholestan-16-one**	56272-02-7	**NI39595**
371	259	217	41	372	43	55	301	**386**	100	75	33	30	29	27	26	24	**6.02**	27	46	1	–	–	–	–	–	–	–	–	–	–	**Cholestan-16-one, (5α)-**	54593-98-5	**NI39596**
371	217	372	41	55	43	301	69	**386**	100	31	28	28	25	25	22	18	**6.36**	27	46	1	–	–	–	–	–	–	–	–	–	–	**Cholestan-16-one, (5α)-**	54593-98-5	**NI39597**
355	356	386	371	387	368	372		**386**	100	25	13	5	4	2	1			27	46	1	–	–	–	–	–	–	–	–	–	–	**Cholest-1-en-19-ol, (5α)-**	30950-89-1	**NI39598**
355	386	368	356	387	369			**386**	100	40	25	25	11	7				27	46	1	–	–	–	–	–	–	–	–	–	–	**Cholest-2-en-19-ol, (5α)-**	28809-51-0	**NI39599**
368	55	105	57	81	91	106	95	**386**	100	52	48	47	45	41	40	40	**33.00**	27	46	1	–	–	–	–	–	–	–	–	–	–	**Cholest-4-en-3-ol**		**LI08321**
43	55	95	57	81	41	105	69	**386**	100	95	87	85	83	75	63	62	**57.94**	27	46	1	–	–	–	–	–	–	–	–	–	–	**Cholest-4-en-3-ol, (3α)-**	566-90-5	**NI39600**
43	44	95	55	105	81	41	91	**386**	100	99	80	80	71	71	70	67	**55.84**	27	46	1	–	–	–	–	–	–	–	–	–	–	**Cholest-4-en-3-ol, (3β)-**	517-10-2	**NI39601**
386	275	55	57	368	69	81	106	**386**	100	46	46	45	37	37	35	34		27	46	1	–	–	–	–	–	–	–	–	–	–	**Cholest-5-en-3-ol**		**LI08322**
369	151	385	370	367	386	133	127	**386**	100	51	42	29	18	14	13	13		27	46	1	–	–	–	–	–	–	–	–	–	–	**Cholest-5-en-3-ol, (3α)-**	474-77-1	**NI39602**
386	255	213	231	371	387	229	256	**386**	100	87	34	31	28	26	26	20		27	46	1	–	–	–	–	–	–	–	–	–	–	**Cholest-7-en-3-ol, (3β)-**	6036-58-4	**NI39603**
386	255	55	43	78	107	95	41	**386**	100	70	67	66	60	57	53	51		27	46	1	–	–	–	–	–	–	–	–	–	–	**Cholest-7-en-3-ol, (3β)-**	6036-58-4	**NI39604**
255	386	43	107	55	95	81	105	**386**	100	95	66	62	61	57	52	51		27	46	1	–	–	–	–	–	–	–	–	–	–	**Cholest-7-en-3-ol, (3β,5α)-**	80-99-9	**NI39605**
386	255	55	43	107	387	95	41	**386**	100	54	34	33	32	31	29	27		27	46	1	–	–	–	–	–	–	–	–	–	–	**Cholest-7-en-3-ol, (3β,5α)-**	80-99-9	**NI39606**
255	386	213	371	231	229	147	107	**386**	100	99	52	50	48	46	42	40		27	46	1	–	–	–	–	–	–	–	–	–	–	**Cholest-7-en-3-ol, (3β,5α)-**	80-99-9	**NI39607**
368	255	43	369	55	353	41	203	**386**	100	88	33	31	31	28	24	23	**10.00**	27	46	1	–	–	–	–	–	–	–	–	–	–	**Cholest-7-en-14-ol, (5α)-**	54423-70-0	**NI39608**
386	107	43	55	147	105	95	81	**386**	100	71	61	52	47	45	44	44		27	46	1	–	–	–	–	–	–	–	–	–	–	**Cholest-8(14)-en-3-ol, (3β,5α)-**		**NI39609**
386	107	43	55	387	147	41	81	**386**	100	37	37	34	31	25	25	24		27	46	1	–	–	–	–	–	–	–	–	–	–	**Cholest-8(14)-en-3α-ol**	26758-20-3	**NI39610**
386	371	107	387	229	147	135	105	**386**	100	49	49	42	40	37	34	30		27	46	1	–	–	–	–	–	–	–	–	–	–	**Cholest-8(14)-en-3β-ol**		**NI39611**
255	368	353	55	241	145	95	256	**386**	100	62	30	27	26	26	26	25	**24.24**	27	46	1	–	–	–	–	–	–	–	–	–	–	**Cholest-8(14)-en-7-ol, (5α)-**	54482-42-7	**NI39612**
386	43	107	55	95	41	105	81	**386**	100	65	58	55	53	50	49	41		27	46	1	–	–	–	–	–	–	–	–	–	–	**Cholest-8-en-3-ol, (3β)-**	7199-91-9	**NI39613**
273	255	161	94	107	93	43	274	**386**	100	49	45	43	42	38	36	32	**3.00**	27	46	1	–	–	–	–	–	–	–	–	–	–	**Cholest-14-en-3-ol, (3β,5α)-**		**NI39614**
255	273	256	193	105	95	57	55	**386**	100	63	46	32	21	19	19	19	**10.00**	27	46	1	–	–	–	–	–	–	–	–	–	–	**Cholest-14-en-7-ol, (5α,7β)-**	54411-69-7	**NI39615**
258	55	95	69	41	311	81	43	**386**	100	54	49	47	39	38	37	37	**8.08**	27	46	1	–	–	–	–	–	–	–	–	–	–	**Cholest-22-en-16-ol, (5α,16β,22E)-**	54411-86-8	**NI39616**
258	368	95	109	55	311	259	257	**386**	100	33	31	25	25	24	22	22	**9.09**	27	46	1	–	–	–	–	–	–	–	–	–	–	**Cholest-22-en-16-ol, (5α,16β,22Z)-**	14949-11-2	**NI39617**
386	43	57	387	55	275	81	69	**386**	100	43	32	29	27	22	22	22		27	46	1	–	–	–	–	–	–	–	–	–	–	**Cholesterol**	57-88-5	**NI39619**
43	81	145	95	368	105	91	107	**386**	100	98	85	85	83	80	76	74	**34.60**	27	46	1	–	–	–	–	–	–	–	–	–	–	**Cholesterol**	57-88-5	**NI39618**
43	105	91	386	145	107	57	119	**386**	100	77	66	64	57	56	51	47		27	46	1	–	–	–	–	–	–	–	–	–	–	**Cholesterol**	57-88-5	**NI39620**
95	386	121	81	273	255	93	247	**386**	100	90	87	83	77	73	67	65		27	46	1	–	–	–	–	–	–	–	–	–	–	**Dihydrotachysterol**		**LI08323**
315	257	271	316	258	386	93	41	**386**	100	75	50	23	15	6	4	4		27	46	1	–	–	–	–	–	–	–	–	–	–	**Furostan, (5α)-**	54515-00-3	**NI39621**
43	55	41	81	67	95	109	57	**386**	100	94	80	62	54	51	47	46	**38.00**	27	46	1	–	–	–	–	–	–	–	–	–	–	**B-Homo-A-norcholestan-6-one, (5α)-**	19548-94-8	**NI39622**
91	71	92	85	69	83	117	97	**386**	100	39	28	27	27	24	18	18	**1.38**	28	50	–	–	–	–	–	–	–	–	–	–	–	**11-Benzylheneicosane**		**AI02306**
91	57	43	71	85	55	92	41	**386**	100	65	49	43	31	27	25	24	**2.50**	28	50	–	–	–	–	–	–	–	–	–	–	–	**11-Benzylheneicosane**		**AI02307**
231	162	246	95	109	232	203	55	**386**	100	70	44	37	33	31	30	29	**21.07**	28	50	–	–	–	–	–	–	–	–	–	–	–	**Cholestane, 14-methyl-**	52474-84-7	**NI39623**
231	162	246	95	109	232	203	55	**386**	100	70	43	36	32	29	29	27	**21.21**	28	50	–	–	–	–	–	–	–	–	–	–	–	**Cholestane, 14-methyl-, (5α)-**	54482-34-7	**NI39624**
217	218	149	386	109	371	121	123	**386**	100	60	56	53	46	23	20	19		28	50	–	–	–	–	–	–	–	–	–	–	–	**Ergostane**		**LI08324**
123	81	67	55	122	41	95	121	**386**	100	83	52	33	32	32	29	17	**13.40**	28	50	–	–	–	–	–	–	–	–	–	–	–	**1,10-Bis(hexahydroindan-5-yl)decane**		**AI02308**

MASS TO CHARGE RATIOS								M.W.	INTENSITIES								Parent	C	H	O	N	S	F	Cl	Br	I	Si	P	B	X	COMPOUND NAME	CAS Reg No	No
385	386	167	178	77	179	207	154	**386**	100	88	73	61	49	27	24	23		29	22	1	–	–	–	–	–	–	–	–	–	–	1,1,5,5-Tetraphenylpenta-1,4-dien-3-one		LI08325
354	356	352	43	41	358	108	27	**387**	100	65	60	44	35	24	22	19	**3.07**	8	7	–	3	1	–	6	–	–	–	–	–	–	2-(Isopropylthio)-4,6-bis(trichloromethyl)-s-triazine		MS08981
354	41	356	352	43	27	108	347	**387**	100	76	73	66	48	45	43	39	**4.10**	8	7	–	3	1	–	6	–	–	–	–	–	–	2-(Propylthio)-4,6-bis(trichloromethyl)-s-triazine		MS08982
354	356	352	108	358	117	119	110	**387**	100	73	66	44	32	26	23	23	**4.00**	8	7	–	3	1	–	6	–	–	–	–	–	–	2-(Propylthio)-4,6-bis(trichloromethyl)-s-triazine		LI08326
388	386	77	87	389	390	387	51	**387**	100	74	63	60	56	54	40	31		12	5	–	7	–	–	4	–	–	–	–	–	–	N,N-Bis(4,6-dichlorotriazin-2-yl)aniline		IC04245
61	43	253	313	75	41	271	62	**387**	100	42	38	25	25	21	13	12	**8.24**	12	16	3	1	1	7	–	–	–	–	–	–	–	N-(Heptafluorobutyryl)methionine propyl ester		MS08983
387	344	236	241	198	156	92	172	**387**	100	52	33	32	32	30	28	26		12	16	4	3	2	3	–	–	–	–	–	–	–	6-Dimethylsulphamoyl-4,4-dioxo-1,3-dimethyl-7-trifluoromethyl-1,3-diaza 4-thia-1,2,3,4-tetrahydronaphthalene		MS08984
155	154	30	99	157	63	156	127	**387**	100	76	74	41	35	35	34	28	**0.00**	15	15	5	3	–	–	2	–	–	–	–	–	–	Benzamide, 5-chloro-N-(2-chloro-4-nitrophenyl)-2-hydroxy-, compd. with 2-aminoethanol (1:1)	1420-04-8	NI39625
329	73	225	313	330	91	226	75	**387**	100	60	36	30	25	23	21	21	**2.50**	15	26	3	1	1	–	–	–	–	2	1	–	–	Phosphonic acid, (1-isothiocyanato-2-phenylethyl)-, bis(trimethylsilyl) ester	55108-94-6	NI39626
134	352	389	387	102	193	192	135	**387**	100	71	20	20	12	10	10	10		16	10	–	3	2	–	–	1	–	–	–	–	–	Thiazolo[2,3-b]-1,3,4-thiadiazol-4-ium, 5-phenyl-2-[(4-bromophenyl)aminide]		MS08985
138	280	107	84	387	120	69	96	**387**	100	91	80	55	30	30	23	20		16	21	6	1	2	–	–	–	–	–	–	–	–	2-Butenoic acid, 2-hydroxy-4-oxo-4-[4-[2,2-bis(methylthio)ethylamino]-6-methyl-2-oxo-2H-pyran-3-yl]-, ethyl ester		NI39627
145	44	103	115	45	187	212	60	**387**	100	77	65	59	52	32	28	28	**0.00**	16	21	10	1	–	–	–	–	–	–	–	–	–	D-Gluconitrile, 2,3,4,5,6-penta-O-acetyl-	83830-74-4	NI39628
120	254	161	73	211	226	43	372	**387**	100	93	82	81	76	69	45	36	**26.02**	16	30	4	1	–	–	–	–	–	2	1	–	–	Phosphonic acid, [1-(acetylamino)-2-phenylethyl]-, bis(trimethylsilyl) ester	55108-86-6	NI39629
387	158	235	282	328	236	91	281	**387**	100	81	60	58	57	44	37	35		17	13	6	3	1	–	–	–	–	–	–	–	–	2-(2,4-Dinitrophenylthio)-3-indolepropionic acid		LI08327
73	388	387	389	386	390	374	372	**387**	100	70	67	66	57	17	13	13		17	18	1	3	–	–	–	1	–	1	–	–	–	Bromoazepam, trimethylsilyl-		NI39630
168	148	252	210	135	110	387	152	**387**	100	99	51	39	32	21	13	3		17	26	7	1	–	–	–	–	–	–	1	–	–	Glycine, N-[2-[(3,4-methylenedioxy)phenyl]ethyl]-N-(diisopropylphosphoryl)-	87212-49-5	NI39631
43	28	86	44	45	128	299	42	**387**	100	58	50	36	29	28	19	19	**2.00**	17	29	7	3	–	–	–	–	–	–	–	–	–	N-Acetylleucyl-(O-acetyl)threonylalanine		LI08328
73	179	75	180	45	77	74	181	**387**	100	73	14	13	12	10	9	7	**0.00**	18	18	1	1	–	5	–	–	–	1	–	–	–	Benzeneethanamine, N-[(pentafluorophenyl)methylene]-2-[(trimethylsilyl)oxy]-		MS08986
179	73	180	387	181	208	45	177	**387**	100	48	18	9	9	8	8	5		18	18	1	1	–	5	–	–	–	1	–	–	–	Benzeneethanamine, N-[(pentafluorophenyl)methylene]-4-[(trimethylsilyl)oxy]-	55334-64-0	MS08987
73	179	180	45	75	74	41	181	**387**	100	99	17	16	8	8	8	7	**2.39**	18	18	1	1	–	5	–	–	–	1	–	–	–	Benzeneethanamine, N-[(pentafluorophenyl)methylene]-4-[(trimethylsilyl)oxy]-	55334-64-0	NI39632
43	174	101	41	145	185	57	71	**387**	100	38	36	31	29	28	26	24	**18.00**	18	29	8	1	–	–	–	–	–	–	–	–	–	5-Isoxazolidinecarboxylic acid, 2-[2,3:5,6-bis-O-(isopropylidene)-α-D-mannofuranosyl]-5-methyl-, methyl ester, (R)-	64018-57-1	NI39633
43	174	101	185	145	59	372	85	**387**	100	36	30	24	23	22	20	19	**16.00**	18	29	8	1	–	–	–	–	–	–	–	–	–	5-Isoxazolidinecarboxylic acid, 5-methyl-2-[2,3:5,6-bis-O-(isopropylidene) α-D-mannofuranosyl]-, methyl ester, (S)-	69494-19-5	NI39634
84	56	100	85	41	74	94	55	**387**	100	86	79	42	42	36	28	28	**0.01**	18	33	6	3	–	–	–	–	–	–	–	–	–	1,3,5-Triazine-2,4,6-tripropanoic acid, hexahydro-, triethyl ester		NI39635
171	115	387	159	342	343	105	274	**387**	100	96	59	57	55	45	39	38		20	12	2	1	1	3	–	–	–	–	–	–	–	1,3-Oxazol-5(2H)-one, 4-(2-naphthylthio)-2-phenyl-2-(trifluoromethyl)-		DD01517
225	163	162	149	70	56	182	164	**387**	100	32	29	25	24	21	19	16	**12.70**	20	29	3	5	–	–	–	–	–	–	–	–	–	Uracil, 6-[3-[4-(2-methoxyphenyl)piperazinyl]propylamino]-1,3-dimethyl-	34661-75-1	NI39636
86	99	87	387	84	315	183	42	**387**	100	7	6	5	3	2	2	2		21	23	1	3	–	1	1	–	–	–	–	–	–	2H-1,4-Benzodiazepin-2-one, 7-chloro-1-[2-(diethylamino)ethyl]-5-(2-fluorophenyl)-1,3-dihydro-	17617-23-1	NI39637
86	32	30	87	99	58	56	42	**387**	100	25	16	10	8	8	4	4	**0.00**	21	23	1	3	–	1	1	–	–	–	–	–	–	2H-1,4-Benzodiazepin-2-one, 7-chloro-1-[2-(diethylamino)ethyl]-5-(2-fluorophenyl)-1,3-dihydro-	17617-23-1	NI39638
219	387	167	181	204	151	40	388	**387**	100	99	60	45	40	34	26	18		21	25	6	1	–	–	–	–	–	–	–	–	–	Propanenitrile, 2,3-bis(3,4,5-trimethoxyphenyl)-		NI39639
89	158	387	118	219	276	344	330	**387**	100	95	80	72	57	48	17	15		21	29	2	3	1	–	–	–	–	–	–	–	–	2,6-Diazabicyclo[2.2.2]oct-7-ene-3,5-dione, 8-(diethylamino)-1-(ethylthio)-2,6,7-trimethyl-4-phenyl-		DD01518
143	257	203	146	111	158	259	222	**387**	100	75	57	53	41	35	26	22	**0.00**	21	38	3	1	–	–	1	–	–	–	–	–	–	4-Tetradecenamide, N-(2-chloro-4-oxopentyl)-7-methoxy-N-methyl-	75332-35-3	NI39640
340	369	341	370	338	284	232	283	**387**	100	85	31	25	24	24	24	21	**2.11**	22	33	3	1	–	–	–	–	–	1	–	–	–	Estra-1,3,5(10)-trien-17-one, 14-hydroxy-3-[(trimethylsilyl)oxy]-, O-methyloxime	74299-40-4	NI39641
387	338	218	356	231	388	339	232	**387**	100	60	52	43	34	29	25	24		22	33	3	1	–	–	–	–	–	1	–	–	–	Estra-1,3,5(10)-trien-17-one, 15-hydroxy-3-[(trimethylsilyl)oxy]-, O-methyloxime, (15β)-	74299-35-7	NI39642
356	79	357	77	297	117	387	118	**387**	100	27	26	21	18	17	16	13		22	33	3	3	–	–	–	–	–	–	–	–	–	Androst-4-en-19-al, 3,17-bis(methoxyimino)-, O-methyloxime	55836-44-7	NI39643
356	91	79	357	105	77	297	132	**387**	100	30	27	26	21	21	18	18	**15.74**	22	33	3	3	–	–	–	–	–	–	–	–	–	Androst-4-en-19-al, 3,17-bis(methoxyimino)-, O-methyloxime	55836-44-7	NI39644
57	41	108	58	84	248	69	53	**387**	100	53	45	45	24	23	23	20	**0.33**	22	33	3	3	–	–	–	–	–	–	–	–	–	2,4-Bis(tert-butylcyanomethylene)-7,7,8,8-tetramethyl-6-methoxy-3,5-dioxa-1-azabicyclo[4.2.0]octane		MS08988
192	55	179	387	69	95	79	218	**387**	100	12	10	8	8	6	6	3		22	36	1	1	–	3	–	–	–	–	–	–	–	Pyrrole, 4,5-dihydro-2-methyl-1-(trifluoroacetyl)-3-[(7Z)-pentadecenyl]-		NI39645

MASS TO CHARGE RATIOS								M.W.	INTENSITIES								Parent	C	H	O	N	S	F	Cl	Br	I	Si	P	B	X	COMPOUND NAME	CAS Reg No	No
121	120	146	280	134	119	299	170	**387**	100	48	17	16	16	16	15	15	**15.13**	23	33	4	1	–	–	–	–	–	–	–	–	–	Acetic acid, [[[(17β)-17-hydroxy-17-methylandrosta-1,4-dien-3-ylidene]amino]oxy]-, methyl ester	74299-11-9	NI39646
387	386	183	389	388	185	262	108	**387**	100	70	63	33	24	23	14	12		24	19	–	1	–	–	1	–	–	–	1	–	–	N-(3-Chlorophenyl)iminotriphenylphosphorane	14796-87-3	NI39647
387	386	183	185	262	152	108	107	**387**	100	70	63	23	14	12	12	11		24	19	–	1	–	–	1	–	–	–	1	–	–	N-(3-Chlorophenyl)iminotriphenylphosphorane	14796-87-3	LI08329
105	148	122	77	43	135	134	55	**387**	100	32	20	13	10	9	8	5	**2.00**	25	41	2	1	–	–	–	–	–	–	–	–	–	Hexadecanamine, 3-acetyl-N-benzoyl-		NI39648
192	206	274	234	166	81	292	95	**387**	100	97	93	75	69	65	57	57	**10.12**	26	45	1	1	–	–	–	–	–	–	–	–	–	6-Azacholestan-7-one	5226-42-6	NI39649
256	387	257	372	243	269	255	130	**387**	100	99	99	30	25	20	20	15		27	21	–	3	–	–	–	–	–	–	–	–	–	5,6,11,12,17,18-Hexahydrocyclonona[1,2-b:4,5-b′:7,8-b″]triindole		NI39650
387	267	268	388	165	265	386	189	**387**	100	58	34	28	24	22	20	11		28	21	1	1	–	–	–	–	–	–	–	–	–	2,4,4-Triphenyl-2,3-butadienanilide		MS08989
28	388	360	276	103	390	36	332	**388**	100	7	6	6	6	5	5	4		4	–	4	–	–	–	2	–	–	–	–	–	2	Rhodium, tetracarbonyldi-μ-chlorodi-	14523-22-9	NI39651
388	103	360	304	276	332	241	288	**388**	100	96	74	63	60	50	42	11		4	–	4	–	–	–	2	–	–	–	–	–	2	Rhodium, tetracarbonyldi-μ-chlorodi-	14523-22-9	MS08990
309	311	230	63	390	392	388	69	**388**	100	99	70	54	45	26	25	19		6	–	–	–	1	5	–	2	–	–	1	–	–	Phosphonothioic dibromide, (pentafluorophenyl)-	53327-33-6	NI39652
69	119	169	131	219	100	31	181	**388**	100	27	21	13	8	7	6	3	**0.00**	7	–	–	–	–	16	–	–	–	–	–	–	–	Heptane, hexadecafluoro-		AI02309
77	92	107	388	79	63	64	202	**388**	100	34	33	27	25	22	16	15		7	7	1	–	–	–	–	1	–	–	–	–	1	(2-Methoxyphenyl)mercuric bromide		LI08330
388	107	77	92	309	65	281	202	**388**	100	29	26	21	12	12	8	7		7	7	1	–	–	–	–	1	–	–	–	–	1	(4-Methoxyphenyl)mercuric bromide		LI08331
28	304	211	93	181	63	109	124	**388**	100	98	77	55	22	22	20	17	**9.27**	9	18	9	–	–	–	–	–	–	–	2	–	1	Iron, tricarbonylbis(trimethyl phosphite-P)-	14949-85-0	NI39653
369	319	69	55	350	203	261	281	**388**	100	99	99	57	53	53	39	37	**5.20**	10	8	2	–	–	12	–	–	–	–	–	–	–	2,6-Bis(trifluoromethyl)-2,6-dihydro-1,1,1,7,7,7-hexafluoro-4-methylideneheptane		LI08332
43	155	153	69	42	27	157	15	**388**	100	89	88	88	88	87	86	86	**10.26**	10	14	4	–	–	–	2	–	–	–	–	–	1	Tin, dichlorobis(2,4-pentanedionato-O,O′)-,	16919-65-6	MS08991
229	147	125	230	73	148	31	315	**388**	100	78	42	15	13	12	12	10	**3.00**	10	22	4	–	–	5	–	–	–	2	1	–	–	Bis(trimethylsilyl) [2,2-difluoro-2-methoxy-1-(trifluoromethyl)ethyl]phosphonate		MS08992
91	92	276	66	274	304	39	65	**388**	100	48	47	36	27	22	20	18	**17.14**	11	8	4	–	–	–	–	–	–	–	–	–	1	Tungsten, π-bicyclo[2.2.1]hepta-2,5-diene-tetracarbonyl-		MS08993
276	304	66	249	360	222	55	305	**388**	100	51	51	46	30	28	28	26	**0.18**	11	8	4	–	–	–	–	–	–	–	–	–	1	Tungsten, π-cyclopentadienyl-acryloyltricarbonyl-		MS08994
57	55	60	52	56	73	69	53	**388**	100	31	21	17	16	12	12	11	**0.00**	11	18	10	1	2	–	–	–	–	–	–	–	–	Progoitrin		MS08995
390	392	264	327	132	262	388	195	**388**	100	79	73	56	54	52	50	50		12	2	2	–	–	–	6	–	–	–	–	–	–	Dibenzo[b,e][1,4]dioxin, 1,2,3,4,7,8-hexachloro-	39227-28-6	NI39654
390	392	388	327	394	329	325	264	**388**	100	81	52	35	34	23	22	22		12	2	2	–	–	–	6	–	–	–	–	–	–	Dibenzo[b,e][1,4]dioxin, 1,2,3,6,7,8-hexachloro-	57653-85-7	NI39655
390	392	388	394	327	329	325	264	**388**	100	80	52	35	29	19	18	18		12	2	2	–	–	–	6	–	–	–	–	–	–	Dibenzo[b,e][1,4]dioxin, 1,2,3,7,8,9-hexachloro-	19408-74-3	NI39656
75	230	232	150	151	311	74	309	**388**	100	89	85	84	81	70	46	35	**9.19**	12	7	–	–	–	–	–	3	–	–	–	–	–	1,1′-Biphenyl, 2,2′,5-tribromo-	59080-34-1	NI39657
75	230	390	392	232	151	150	74	**388**	100	83	82	80	79	78	78	48	**27.00**	12	7	–	–	–	–	–	3	–	–	–	–	–	1,1′-Biphenyl, 2,3′,5-tribromo-	59080-35-2	NI39658
390	392	150	230	232	151	75	388	**388**	100	99	63	60	59	56	36	29		12	7	–	–	–	–	–	3	–	–	–	–	–	1,1′-Biphenyl, 2,4,5-tribromo-		NI39659
75	150	390	392	230	232	151	74	**388**	100	99	93	91	91	88	86	44	**31.00**	12	7	–	–	–	–	–	3	–	–	–	–	–	1,1′-Biphenyl, 2,4,6-tribromo-	59080-33-0	NI39660
75	390	392	151	150	230	232	74	**388**	100	85	82	80	79	78	75	49	**28.00**	12	7	–	–	–	–	–	3	–	–	–	–	–	1,1′-Biphenyl, 2,4′,5-tribromo-	59080-36-3	NI39661
390	392	150	151	230	232	388	75	**388**	100	99	59	55	53	51	29	28		12	7	–	–	–	–	–	3	–	–	–	–	–	1,1′-Biphenyl, 3,4,5-tribromo-		NI39662
96	223	388	166	225	98	42	179	**388**	100	80	54	36	31	26	26	25		12	12	2	2	–	6	–	–	–	–	–	–	1	Nickel(II), N,N′-ethylenebis(1,1,1-trifluoro-4-iminopentan-2-onato)-	40792-92-5	NI39663
223	388	99	369	319	341	291	154	**388**	100	73	6	3	3	2	2	2		12	12	2	2	–	6	–	–	–	–	–	–	1	Nickel(II), N,N′-ethylenebis(1,1,1-trifluoro-4-iminopentan-2-onato)-	40792-92-5	NI39664
105	120	332	304	388	300	360	302	**388**	100	61	44	40	25	23	18	17		12	12	3	–	–	–	–	–	–	–	–	–	1	Tungsten, π-1,3,5-trimethylbenzene-tricarbonyl-		MS08996
298	313	52	358	388	233	299	117	**388**	100	89	84	63	59	46	36	34		12	16	2	2	2	–	–	–	–	–	–	–	2	Chromium, bis(η^5-2,4-cyclopentadien-1-yl)bis(methanthioato)dinitrosyldi-	12215-16-6	NI39665
288	237	238	219	260	218	187	205	**388**	100	87	73	73	49	30	27	24	**11.50**	13	4	–	–	–	12	–	–	–	–	–	–	–	4H,4H,5H,5H-Dodecafluorotetracyclo[6.2.2.1^{3,6}.0^{2,7}]trideca-2(7),9-diene		NI39666
237	288	219	238	260	218	187	241	**388**	100	94	88	65	62	30	27	25	**4.40**	13	4	–	–	–	12	–	–	–	–	–	–	–	4H,4H,5H,5H-Dodecafluorotetracyclo[6.2.2.1^{3,6}.0^{2,7}]trideca-2(7),9-diene		NI39667
193	195	197	157	123	194	159	196	**388**	100	64	34	9	9	8	7	6	**0.30**	13	6	1	–	–	–	6	–	–	–	–	–	–	Benzene, trichloro[(trichlorophenoxy)methyl]-	74421-52-6	NI39668
127	308	171	75	173	92	309	65	**388**	100	75	70	67	65	54	49	47	**18.00**	13	10	–	2	–	–	1	1	–	–	–	–	1	N-(4-Bromophenyl)-N′-(4-chlorophenyl)selenourea		MS08997
388	369	360	291	341	339			**388**	100	47	19	12	7	4				14	2	–	2	–	10	–	–	–	–	–	–	–	Pentafluorobenzaldazine		LI08333
107	91	63	75	90	77	64	52	**388**	100	85	71	60	53	53	47	47	**0.00**	14	8	8	6	–	–	–	–	–	–	–	–	–	Benzaldehyde, 2,4-dinitro-, azine	22014-28-4	NI39669
112	55	276	110	248	220	137	162	**388**	100	90	67	60	55	47	33	31	**12.00**	14	12	6	–	–	–	–	–	–	–	–	–	2	Iron, 1,1,1-tricarbonylferratetramethylcyclopentadiene-π-tricarbonyl-	12245-44-2	NI39670
104	282	78	284	281	388	386	103	**388**	100	96	92	88	72	71	62	58		16	16	–	–	–	–	–	–	–	–	–	–	1	Hafnium, bis(1,3,5,7-cyclooctatetraene)-	39292-59-6	NI39671
106	388	134	133	266	105	107	135	**388**	100	80	74	54	48	41	29	26		17	24	2	–	4	–	–	–	–	–	–	–	–	1,4,8,11-Tetrathiacyclotetradecane, 6-(benzoyloxy)-	91375-65-4	NI39672
388	311	105	77	106	389	126	113	**388**	100	62	35	29	28	21	17	15		18	13	2	–	–	–	–	–	1	–	–	–	–	6-Carboxyphenyl-1-iodo-2-methoxynaphthalene		MS08998
149	160	203	172	161	202	187	299	**388**	100	71	68	35	29	28	21	13	**6.00**	18	16	6	2	1	–	–	–	–	–	–	–	–	3-Cephem-4-carboxylic acid, 3-(methoxymethyl)-7-phthalimido-, methyl ester		NI39673
315	346	329	388	345	303	153	357	**388**	100	94	49	45	34	27	15	13		19	16	9	–	–	–	–	–	–	–	–	–	–	4H-1-Benzopyran-4-one, 2-(5-acetoxy-2-hydroxy-4-methoxyphenyl)-5,7-dihydroxy-3-methoxy-		MS08999
328	269	132	329	77	297	315	117	**388**	100	57	54	53	40	21	14	10	**0.00**	19	17	5	2	–	–	1	–	–	–	–	–	–	2-Quinazolineacetic acid, 6-chloro-1,2,3,4-tetrahydro-2-(methoxycarbonyl)-4-oxo-3-phenyl-, methyl ester	34804-51-8	NI39674
43	41	27	79	39	133	30	107	**388**	100	66	48	31	29	27	22	20	**3.10**	19	20	7	2	–	–	–	–	–	–	–	–	–	Benzofuran, 7-(2,4-dinitrophenoxy)-2,3-dihydro-2,2-dimethyl-3-isopropoxy-	62059-48-7	NI39675

MASS TO CHARGE RATIOS	M.W.	INTENSITIES	Parent	C	H	O	N	S	F	Cl	Br	I	Si	P	B	X	COMPOUND NAME	CAS Reg No	No
43 41 27 133 79 39 29 30	388	100 83 71 39 38 35 31 30	2.20	19	20	7	2	–	–	–	–	–	–	–	–	–	Benzofuran, 7-(2,4-dinitrophenoxy)-2,3-dihydro-2,2-dimethyl-3-propoxy-	62059-47-6	NI39676
91 84 86 169 41 68 92 56	388	100 85 74 39 29 20 15 12	5.00	19	24	5	4	–	–	–	–	–	–	–	–	–	L-Prolinamide, 5-oxo-L-prolyl-L-phenylalanyl-4-hydroxy-	34783-36-3	NI39677
91 84 86 169 41 260 120 68	388	100 84 73 39 29 27 25 20	0.60	19	24	5	4	–	–	–	–	–	–	–	–	–	Pyroglutamyl-1-phenylalanyl-1-hydroxyprolinamide		LI08334
299 153 311 300 59 55 57 312	388	100 72 56 51 38 26 23 19	5.12	20	20	8	–	–	–	–	–	–	–	–	–	–	4H-1-Benzopyran-4-one, 5,7-dihydroxy-3-[2,4-dihydroxy-3-(2,3-dihydroxy-3-methylbutyl)phenyl]-		NI39678
299 329 165 300 388 330 44 167	388	100 97 39 25 16 14 7 5		20	20	8	–	–	–	–	–	–	–	–	–	–	4H-1-Benzopyran-4-one, 5,7-dihydroxy-6-(2,3-dihydroxy-3-methylbutyl)-3 (2,4-dihydroxyphenyl)-	91681-63-9	NI39679
374 359 373 375 355 360 331 151	388	100 58 26 24 14 13 9 9	0.00	20	20	8	–	–	–	–	–	–	–	–	–	–	4H-1-Benzopyran-4-one, 2-(3,4-dimethoxyphenyl)-5-hydroxy-3,6,7-trimethoxy-	479-90-3	NI39681
87 130 374 43 115 69 359 151	388	100 99 43 41 39 38 28 22	0.00	20	20	8	–	–	–	–	–	–	–	–	–	–	4H-1-Benzopyran-4-one, 2-(3,4-dimethoxyphenyl)-5-hydroxy-3,6,7-trimethoxy-	479-90-3	NI39680
388 373 165 69 389 178 374 369	388	100 85 39 26 23 21 19 19		20	20	8	–	–	–	–	–	–	–	–	–	–	4H-1-Benzopyran-4-one, 2-(3,4-dimethoxyphenyl)-5-hydroxy-3,6,7-trimethoxy-	479-90-3	NI39682
238 58 270 136 330 245 105 181	388	100 99 78 43 35 31 31 29	1.80	20	24	4	2	1	–	–	–	–	–	–	–	–	Benzamide, 2,2′-sulphonylbis[N-isopropyl-	65838-72-4	NI39683
91 67 80 92 39 132 77 75	388	100 10 9 8 8 7 7 6	0.00	20	24	4	2	1	–	–	–	–	–	–	–	–	Carbamic acid, (thiodiethylene)di-, dibenzyl ester	26630-73-9	NI39684
162 163 119 120 91 357 326 132	388	100 10 7 5 5 4 4 4	3.00	20	24	6	2	–	–	–	–	–	–	–	–	–	2,2′-Diformyloxanilidebis(dimethylacetyl)		LI08335
315 266 388 271 287 210 343 238	388	100 32 21 19 16 16 12 11		20	24	6	2	–	–	–	–	–	–	–	–	–	3,5-Pyridinedicarboxylic acid, 1,4-dihydro-1,2,6-trimethyl-4-(4-nitrophenyl)-, diethyl ester	69796-14-1	NI39685
211 209 207 297 213 210 295 167	388	100 76 42 41 35 33 31 30	0.30	20	28	–	–	–	–	–	–	–	–	–	–	1	Dibenzyldiisopropyltin		NI39686
145 69 228 41 55 143 97 132	388	100 38 36 32 30 28 27 21	7.00	20	36	3	–	2	–	–	–	–	–	–	–	–	Hexanoic acid, 6-[(2-propyl-2-methyl-6,10-dithiaspiro[4.5]dec-1-yl)oxy]-, ethyl ester	92640-61-4	NI39687
98 112 197 70 55 181 96 71	388	100 22 20 18 18 17 17 17	1.00	20	40	5	2	–	–	–	–	–	–	–	–	–	3,13-Dihydroxy-5,8,11-trioxa-1,15-diazabicyclo[13.5.5]pentacosane		MS09000
73 233 147 257 373 75 43 28	388	100 56 54 27 26 17 12 12	0.00	20	44	3	–	–	–	–	–	–	2	–	–	–	Tetradecanoic acid, 3-[(trimethylsilyl)oxy]-, trimethylsilyl ester		NI39688
233 373 257 147 73 234 374 331	388	100 52 42 34 30 21 17 15	0.00	20	44	3	–	–	–	–	–	–	2	–	–	–	Tetradecanoic acid, 3-[(trimethylsilyl)oxy]-, trimethylsilyl ester		NI39689
41 149 55 43 57 85 71 69	388	100 92 85 84 76 54 47 47	8.00	21	24	7	–	–	–	–	–	–	–	–	–	–	(1S,5R,6S,7R,8R)-1-Allyl-3,8-dihydroxy-5-methoxy-6-(3-methoxy-4,5-methylenedioxyphenyl)-7-methyl-4-oxobicyclo[3.2.1]oct-2-ene		NI39690
180 328 151 388 346 329 222 167	388	100 76 73 59 59 56 56 50		21	24	7	–	–	–	–	–	–	–	–	–	–	2H-1-Benzopyran, 3-acetoxy-3,4-dihydro-7,8-dimethoxy-2-(3,4-dimethoxyphenyl)-, (2R,3S)-		NI39691
180 222 297 388 181 167 151 137	388	100 97 59 30 29 21 21 13		21	24	7	–	–	–	–	–	–	–	–	–	–	2H-1-Benzopyran, 3,4-dihydro-3-acetoxy-4,7-dimethoxy-2-(3,4-dimethoxyphenyl)-, (+)-(3β,4α)-, (3trans,4trans)-		MS09001
94 147 41 55 206 77 121 81	388	100 69 66 65 63 47 44 41	16.00	21	24	7	–	–	–	–	–	–	–	–	–	–	5,10b-(Epoxymethano)-10bH-naphtho[2,1-c]pyran-7-carboxylic acid, 2-(3-furanyl)dodecahydro-7-methyl-4,11-dioxo-, methyl ester, (2α,4aα,5α,6aβ,7α,10aβ,10bα)-(+)-	70447-99-3	NI39692
178 259 206 296 165 315 179 205	388	100 63 51 34 34 30 30 29	21.18	21	25	5	–	–	–	–	–	–	–	1	–	–	3-(4-Phenylphenyl)propenoic acid, 2-(diethoxyphosphinyl)-, ethyl ester	107846-64-0	NI39693
220 136 150 206 176 138 194 94	388	100 99 98 53 47 37 35 35	30.00	21	28	1	2	2	–	–	–	–	–	–	–	–	8,9,17,18-Dibenzo-13-oxa-1,7-dithia-10,16-diazacyclooctadeca-8,17-diene		MS09002
100 91 85 43 106 101 257 96	388	100 38 12 12 10 8 6 6	3.10	21	28	5	2	–	–	–	–	–	–	–	–	–	3-Benzylamino-3-C-cyanomethyl-3-desoxy-1,2,5,6-di-O-isopropylidene-α-D-glucofuranose		NI39694
58 71 72 42 167 56 78 28	388	100 61 8 7 7 6 4 4	0.00	21	28	5	2	–	–	–	–	–	–	–	–	–	Butanedioic acid, compd. with N,N-dimethyl-2-[1-phenyl-1-(2-pyridinyl)ethoxy]ethanamine (1:1)	562-10-7	NI39695
319 321 390 388 253 251 276 278	388	100 99 26 26 14 14 7 5		21	29	–	2	–	–	–	1	–	–	–	–	–	Flustramine D		NI39696
319 321 390 388 265 263 241 278	388	100 99 39 39 22 22 9 5		21	29	–	2	–	–	–	1	–	–	–	–	–	Isoflustramine D		NI39697
138 81 95 303 83 139 69 82	388	100 93 58 34 33 31 31 28	3.05	21	41	2	–	–	–	1	–	–	1	–	–	–	Silane, chlorobis(p-menth-3-yloxy)methyl-	17202-01-6	NI39698
388 76 360 75 187 343 145 189	388	100 81 64 47 37 36 33 32		22	12	7	–	–	–	–	–	–	–	–	–	–	Dinaphtho[2,3-b:2′,3′-d]furan-5,7,12,13-tetrone, 4,8-dimethoxy-		NI39699
274 245 275 58 301 77 329 388	388	100 79 68 67 60 50 40 39		22	20	3	4	–	–	–	–	–	–	–	–	–	3-(2,3-O-Isopropylidene-β-D-erythrofuranosyl)-1-phenylpyrazolo[3,4-b]quinoxaline		MS09003
331 330 329 273 243 373 240 241	388	100 98 68 15 15 13 13 11	0.00	22	20	3	4	–	–	–	–	–	–	–	–	–	1H-Pyrazolo[4,3-c]pyridazin-3-ol, 4-acetyl-2,4,5,6-tetrahydro-6-methylene-2,5-diphenyl-, acetate	35426-81-4	NI39700
194 196 195 125 55 197 127 178	388	100 32 13 11 5 4 4 3	0.00	22	26	–	2	–	–	2	–	–	–	–	–	–	1,2-Bis(2-chlorophenyl)-1,2-bis(1-pyrrolidinyl)ethane, (±)-		DD01519
194 196 195 125 55 197 130 127	388	100 32 13 10 5 4 3 3	0.00	22	26	–	2	–	–	2	–	–	–	–	–	–	1,2-Bis(3-chlorophenyl)-1,2-bis(1-pyrrolidinyl)ethane, (±)-		DD01520
194 196 195 125 197 55 127 28	388	100 33 13 10 4 4 3 3	0.00	22	26	–	2	–	–	2	–	–	–	–	–	–	1,2-Bis(4-chlorophenyl)-1,2-bis(1-pyrrolidinyl)ethane, (±)-		DD01521
194 196 195 125 250 55 197 127	388	100 32 13 11 7 5 4 4	0.00	22	26	–	2	–	–	2	–	–	–	–	–	–	1,2-Bis(2-chlorophenyl)-1,2-bis(1-pyrrolidinyl)ethane, meso-		DD01522
194 196 195 125 250 197 55 130	388	100 31 12 9 5 4 4 3	0.00	22	26	–	2	–	–	2	–	–	–	–	–	–	1,2-Bis(3-chlorophenyl)-1,2-bis(1-pyrrolidinyl)ethane, meso-		DD01523
194 196 195 125 55 127 28 197	388	100 33 13 13 5 4 4 3	0.00	22	26	–	2	–	–	2	–	–	–	–	–	–	1,2-Bis(4-chlorophenyl)-1,2-bis(1-pyrrolidinyl)ethane, meso-		DD01524
162 153 147 161 235 135 138 83	388	100 57 45 37 23 21 16 16	2.00	22	28	2	–	2	–	–	–	–	–	–	–	–	2H-1-Benzothiocin-3-methanol, 3,4,5,6-tetrahydro-6-hydroxy-3,5,5-trimethyl-α-[2-(methylthio)phenyl]-	72360-99-7	MS09004

MASS TO CHARGE RATIOS								M.W.	INTENSITIES								Parent	C	H	O	N	S	F	Cl	Br	I	Si	P	B	X	COMPOUND NAME	CAS Reg No	No
121	136	150	137	109	80	110	388	388	100	55	45	45	27	18	16	16		22	28	6	–	–	–	–	–	–	–	–	–	–	6,15-Dimethyl-2,3,11,12-dibenzo-1,4,7,10,13,16-hexaoxacyclooctadeca-2,11-diene		MS09005
213	286	214	172	328	270	61	287	388	100	86	59	30	27	21	20	18	0.00	22	28	6	–	–	–	–	–	–	–	–	–	–	Estra-1,3,5(10)-triene-3,16,17,17-tetrol, 3,17-diacetate, (16α,17β)-	69833-77-8	NI39701
136	137	121	388	80	55	110	109	388	100	97	93	57	53	53	50	50		22	28	6	–	–	–	–	–	–	–	–	–	–	6-Ethyl-2,3,11,12-dibenzo-1,4,7,10,13,16-hexaoxacyclooctadeca-2,11-diene		MS09006
43	121	161	176	260	244	201	218	388	100	42	32	31	27	25	23	19	0.10	22	28	6	–	–	–	–	–	–	–	–	–	–	2-(8-Hydroxy-3,7-dimethylocta-2,6-dienyl)hydroquinone triacetate		NI39702
388	339	269	370	389	340	151	189	388	100	68	32	31	23	14	13	12		22	28	6	–	–	–	–	–	–	–	–	–	–	Isotaxiresonoltetramethyl		NI39703
289	99	388	100	139	247	290	165	388	100	65	50	32	20	19	18	14		22	28	6	–	–	–	–	–	–	–	–	–	–	6-Oxaestra-1,3,5(10)-triene, 7,7:17,17-bis(ethylenedioxy)-3-methoxy-, (±)		MS09007
99	289	388	100	139	247	290	165	388	100	93	41	41	22	20	17	16		22	28	6	–	–	–	–	–	–	–	–	–	–	6-Oxaestra-1,3,5(10)-triene, 7,7:17,17-bis(ethylenedioxy)-3-methoxy-, (±) (8α)-		MS09008
260	310	147	131	111	224	242	214	388	100	36	31	29	26	25	21	14	0.00	22	28	6	–	–	–	–	–	–	–	–	–	–	Pseudolaric acid A	82508-32-5	NI39704
243	105	55	79	43	41	91	370	388	100	66	62	55	55	54	50	48	10.00	22	33	5	–	–	–	–	–	–	–	–	1	–	3α,17α,21-Trihydroxy-5β-pregnane-11,20-dione methyl boronate		MS09009
243	370	105	107	121	147	244	261	388	100	48	46	44	40	38	33	28	10.00	22	33	5	–	–	–	–	–	–	–	–	1	–	3α,17α,21-Trihydroxy-5β-pregnane-11-20-dione methyl boronate		LI08336
129	201	271	97	55	71	45	91	388	100	34	32	27	25	23	18	16	0.00	22	44	5	–	–	–	–	–	–	–	–	–	–	Octadecanoic acid, 9,10,12-trimethoxy-, methyl ester	55255-75-9	LI08337
129	201	97	55	71	45	91	41	388	100	35	27	26	23	18	16	12	0.00	22	44	5	–	–	–	–	–	–	–	–	–	–	Octadecanoic acid, 9,10,12-trimethoxy-, methyl ester	55255-75-9	NI39705
241	71	209	177	273	242	115	159	388	100	50	40	37	16	12	12	11	0.00	22	44	5	–	–	–	–	–	–	–	–	–	–	Octadecanoic acid, 9,12,13-trimethoxy-, methyl ester	55255-76-0	NI39706
388	387	121	389	355	311	390	210	388	100	60	47	35	32	29	17	16		23	16	–	–	3	–	–	–	–	–	–	–	–	[1,2]Dithiolo[1,5-b][1,2]dithiole-7-SIV, 2,3,4-triphenyl-	34283-92-6	NI39707
388	121	387	389	311	355	77	390	388	100	88	61	35	30	26	21	19		23	16	–	–	3	–	–	–	–	–	–	–	–	[1,2]Dithiolo[1,5-b][1,2]dithiole-7-SIV, 2,3,5-triphenyl-	16094-76-1	NI39708
309	310	105	77	203	202	102	204	388	100	29	23	20	15	15	10	7	0.00	23	17	1	–	–	–	–	1	–	–	–	–	–	Pyrylium, 2,4,6-triphenyl-, bromide	13179-84-5	NI39709
309	310	105	77	203	202	155	102	388	100	29	23	21	15	15	11	10	0.00	23	17	1	–	–	–	–	1	–	–	–	–	–	Pyrylium, 2,4,6-triphenyl-, bromide	13179-84-5	LI08338
388	194	104	390	389	138	91	228	388	100	60	50	35	32	28	28	26		23	17	2	2	–	–	1	–	–	–	–	–	–	2(1H)-Pyrimidinone, 4-(4-chlorophenyl)-5-(4-methylphenoxy)-6-phenyl-	42919-56-2	NI39710
388	118	194	387	389	390	138	180	388	100	54	44	39	36	32	29	23		23	17	2	2	–	–	1	–	–	–	–	–	–	2(1H)-Pyrimidinone, 4-(4-chlorophenyl)-6-(4-methylphenyl)-5-phenoxy-	42919-61-9	NI39711
388	118	194	387	389	391	138	180	388	100	54	44	40	37	34	29	22		23	17	2	2	–	–	1	–	–	–	–	–	–	2(1H)-Pyrimidinone, 4-(4-chlorophenyl)-6-(4-methylphenyl)-5-phenoxy-	42919-61-9	LI08339
56	177	77	201	269	388	268	241	388	100	40	24	17	15	11	9	7		23	24	2	4	–	–	–	–	–	–	–	–	–	3H-Pyrazol-3-one, 4,4′-methylenebis[1,2-dihydro-1,5-dimethyl-2-phenyl-	1251-85-0	NI39712
81	43	91	41	95	55	53	31	388	100	87	74	70	65	61	57	57	5.22	23	32	5	–	–	–	–	–	–	–	–	–	–	1α-(Acetoxymethyl)-7α,8α-dimethyl-7-(2-(3-furyl)ethyl)bicyclo[4.4.0]dec-2-ene-2-carboxylic acid methyl ester		NI39713
356	388	269	328	315	283	285	300	388	100	93	80	76	48	48	42	40		23	32	5	–	–	–	–	–	–	–	–	–	–	O-Acetylsordaricin methyl ester		LI08340
328	121	329	268	214	199	120	253	388	100	34	23	19	18	18	18	16	0.00	23	32	5	–	–	–	–	–	–	–	–	–	–	Androst-5-en-17-one, 3,16-bis(acetyloxy)-, (3β,16β)-	16597-57-2	NI39714
137	111	232	388	181	277	344	370	388	100	72	69	53	50	45	36	22		23	32	5	–	–	–	–	–	–	–	–	–	–	Anhydroatrogenin		NI39715
352	91	41	79	31	268	28	55	388	100	67	46	43	43	40	40	38	0.00	23	32	5	–	–	–	–	–	–	–	–	–	–	Corotoxigenine	468-20-2	NI39716
122	91	43	121	41	123	93	77	388	100	65	62	56	53	50	50	50	1.14	23	32	5	–	–	–	–	–	–	–	–	–	–	Ingenol-3,4-acetonide		NI39717
41	121	43	91	77	93	59	55	388	100	99	99	92	78	69	59	59	5.29	23	32	5	–	–	–	–	–	–	–	–	–	–	Ingenol-5,20-acetonide		NI39718
265	233	136	220	252	388	137	124	388	100	91	73	58	54	49	49	48		23	32	5	–	–	–	–	–	–	–	–	–	–	Methyl hydroxy-3,12-diketo-Δ-bisnorcholenate		LI08341
284	285	199	282	177	55	159	133	388	100	51	44	41	31	31	25	25	11.00	23	32	5	–	–	–	–	–	–	–	–	–	–	B-Norpregnane-6-carboxylic acid, 3-(acetyloxy)-5-hydroxy-20-oxo-, β-lactone, (3β)-	56760-81-7	NI39719
91	340	105	131	29	79	117	77	388	100	99	93	88	83	81	76	76	0.00	23	32	5	–	–	–	–	–	–	–	–	–	–	Pachygenol	546-03-2	NI39720
328	91	55	79	77	105	93	53	388	100	73	58	55	51	50	47	41	1.80	23	32	5	–	–	–	–	–	–	–	–	–	–	Pregnan-20-one, 3-(acetyloxy)-5,6:16,17-diepoxy-, (3β,5α,6α,16α)-	14231-06-2	WI01508
265	233	136	220	252	388	137	124	388	100	92	75	59	55	50	50	44		23	32	5	–	–	–	–	–	–	–	–	–	–	Pregn-4-ene-20-carboxylic acid, 9-hydroxy-3,12-dioxo-, methyl ester, (20S)-	24341-27-3	NI39721
43	159	107	55	388	105	101	91	388	100	38	28	28	26	25	24	24		23	32	5	–	–	–	–	–	–	–	–	–	–	Pregn-5-en-20-one, 21-(acetyloxy)-16,17-epoxy-3-hydroxy-, (3β,16α)-	28444-97-5	NI39722
388	287	71	389	302	329	288	232	388	100	38	36	31	18	12	11	10		23	36	3	–	–	–	–	–	–	1	–	–	–	Estra-1,3,5(10)-triene, 16,17-dimethoxy-3-[(trimethylsilyl)oxy]-	74298-79-6	NI39723
73	43	75	91	388	149	183	69	388	100	79	32	26	25	25	23	23		23	36	3	–	–	–	–	–	–	1	–	–	–	Retinoic acid, 5,6-epoxy-5,6-dihydro-, trimethylsilyl ester	50876-23-8	NI39724
73	42	40	149	271	164	388	75	388	100	88	60	56	47	38	37	37		23	36	3	–	–	–	–	–	–	1	–	–	–	Retinoic acid, 5,8-epoxy-5,8-dihydro-, trimethylsilyl ester	50876-24-9	NI39725
370	339	70	206	165	371	100	357	388	100	69	49	40	38	33	29	26	13.60	23	36	3	2	–	–	–	–	–	–	–	–	–	6α-Hydroxyprogesterone, di-methyloxime		NI39726
370	339	206	165	70	371	340	100	388	100	73	37	37	33	28	23	21	4.20	23	36	3	2	–	–	–	–	–	–	–	–	–	6β-Hydroxyprogesterone, di-methyloxime		NI39727
301	302	373	388	357	187	175	123	388	100	17	7	5	4	4	4	4		23	36	3	2	–	–	–	–	–	–	–	–	–	17β-Hydroxy-4-propyl-3,4-seco-5-androsteno[4,5,6-cd]pyrazol-3-oic acid methyl ester		MS09010
299	344	315	287	343	271	258	202	388	100	85	82	74	59	35	27	26	6.00	24	20	5	–	–	–	–	–	–	–	–	–	–	Fluorescein, diethoxy-	21934-70-3	NI39728
388	259	287	202	243	207	260	200	388	100	78	36	34	29	28	26	25		24	20	5	–	–	–	–	–	–	–	–	–	–	Fluorescein, ethoxy-, ethyl ester	87569-96-8	NI39729
125	334	335	30	336	127	36	317	388	100	68	59	39	37	36	32	29	23.00	24	21	1	2	–	–	1	–	–	–	–	–	–	1,5-Benzodiazocin-2-one, 1-(4-chlorobenzyl)-1,2,3,4-tetrahydro-8-methyl-6 phenyl-		NI39730
328	329	213	147	145	269	161	107	388	100	25	25	14	14	12	12	11	0.00	24	36	4	–	–	–	–	–	–	–	–	–	–	Androst-5-ene-17β-carboxylic acid, 3β-hydroxy-17-methyl-, methyl ester, acetate	5230-57-9	NI39731
317	43	229	318	123	93	79	145	388	100	46	34	22	22	15	15	14	3.00	24	36	4	–	–	–	–	–	–	–	–	–	–	Androst-5-en-3-one, 17-(acetyloxy)-4-hydroxy-4-propyl-		MS09011
287	388	370	278	150	269	192	245	388	100	54	40	40	37	36	33	28		24	36	4	–	–	–	–	–	–	–	–	–	–	Cholan-24-oic acid, 3,7-dioxo-, (5β)-	859-97-2	NI39732
388	247	389	248	287	370	273	245	388	100	62	27	11	9	6	5	5		24	36	4	–	–	–	–	–	–	–	–	–	–	Cholan-24-oic acid, 3,12-dioxo-, (5β)-	2958-05-6	NI39733
388	247	389	248	287	284	273	149	388	100	88	29	18	16	16	10	10		24	36	4	–	–	–	–	–	–	–	–	–	–	Cholan-24-oic acid, 3,12-dioxo-, (5β)-	2958-05-6	NI39734

MASS TO CHARGE RATIOS								M.W.	INTENSITIES								Parent	C	H	O	N	S	F	Cl	Br	I	Si	P	B	X	COMPOUND NAME	CAS Reg No	No
388	247	389	248	287	370	329	273	388	100	63	27	11	9	6	5	5		24	36	4	–	–	–	–	–	–	–	–	–	–	Cholan-24-oic acid, 3,12-dioxo-, (5β)-	2958-05-6	LI08342
247	388	370	248	389	287	245	269	388	100	65	35	18	17	17	15	12		24	36	4	–	–	–	–	–	–	–	–	–	–	Cholan-24-oic acid, 7,12-dioxo-, (5β)-	21059-35-8	NI39735
69	41	43	45	31	28	44	18	388	100	95	82	75	72	69	53	52	9.00	24	36	4	–	–	–	–	–	–	–	–	–	–	Ethyl 2,4-dihydroxy-5-geranyl-6-pentylbenzoate		MS09012
69	219	41	43	44	231	28	18	388	100	97	91	89	75	69	58	51	19.00	24	36	4	–	–	–	–	–	–	–	–	–	–	Ethyl 2,4-dihydroxy-5-geranyl-6-pentylbenzoate		MS09013
285	303	286	267	260	346	161	328	388	100	33	24	16	15	13	12	11	2.23	24	36	4	–	–	–	–	–	–	–	–	–	–	Pregnane-3,20-dione, 17-(acetyloxy)-6-methyl-, (5β,6α)-	69688-15-9	NI39736
296	91	328	101	114	105	79	71	388	100	90	88	83	81	79	69	66	0.00	24	36	4	–	–	–	–	–	–	–	–	–	–	Pregn-5-en-20-one, 3-(acetyloxy)-16-methoxy-, (3β,16α)-	55320-51-9	NI39737
271	347	75	272	81	348	253	161	388	100	77	76	38	26	22	22	18	16.64	24	40	2	–	–	–	–	–	–	1	–	–	–	3α-Allyldimethylsilyloxy-5α-androstan-17-one		NI39738
255	347	75	147	145	161	107	256	388	100	79	62	32	31	29	25	24	17.03	24	40	2	–	–	–	–	–	–	1	–	–	–	3α-Allyldimethylsilyloxy-5β-androstan-17-one		NI39739
271	347	75	81	253	272	388	348	388	100	68	35	28	23	21	19	19		24	40	2	–	–	–	–	–	–	1	–	–	–	3β-Allyldimethylsilyloxy-5β-androstan-17-one		NI39740
129	298	85	259	241	388	283	332	388	100	52	33	30	28	25	23	22		24	40	2	–	–	–	–	–	–	1	–	–	–	Pregn-5-en-20-one, 3-[(trimethylsilyl)oxy]-, (3β)-	18919-58-9	NI39741
85	298	241	259	388	130	332	283	388	100	95	66	60	59	42	41	40		24	40	2	–	–	–	–	–	–	1	–	–	–	Pregn-5-en-20-one, 3-[(trimethylsilyl)oxy]-, (3β)-	18919-58-9	NI39742
129	43	73	85	75	298	259	388	388	100	82	61	36	36	33	23	20		24	40	2	–	–	–	–	–	–	1	–	–	–	Pregn-5-en-20-one, 3-[(trimethylsilyl)oxy]-, (3β)-	18919-58-9	NI39743
129	298	388	213	259	85	283	258	388	100	74	54	51	49	43	39	31		24	40	2	–	–	–	–	–	–	1	–	–	–	Pregn-5-en-20-one, 3-[(trimethylsilyl)oxy]-, (3β,17α)-		LI08343
388	255	283	345	389	298	161	297	388	100	79	58	47	33	22	21	19		24	40	2	–	–	–	–	–	–	1	–	–	–	Pregn-16-en-20-one, 3-[(trimethylsilyl)oxy]-, (3α,5β)-		LI08344
255	298	283	388	345	373	256	299	388	100	63	63	50	38	35	28	21		24	40	2	–	–	–	–	–	–	1	–	–	–	Pregn-16-en-20-one, 3-[(trimethylsilyl)oxy]-, (3α,5β)-		LI08345
231	232	107	109	388	123	121	124	388	100	46	46	43	38	36	32	28		25	40	3	–	–	–	–	–	–	–	–	–	–	Cholan-24-oic acid, 3-oxo-, methyl ester, (5α)-	15074-03-0	NI39744
107	109	273	161	105	122	108	213	388	100	94	81	81	81	78	68	66	36.00	25	40	3	–	–	–	–	–	–	–	–	–	–	Cholan-24-oic acid, 3-oxo-, methyl ester, (5β)-	1173-32-6	NI39745
233	388	55	110	81	41	68	96	388	100	58	35	34	30	25	19	16		25	40	3	–	–	–	–	–	–	–	–	–	–	Cholan-24-oic acid, 12-oxo-, methyl ester	55870-39-8	NI39746
233	388	111	54	81	41	69	43	388	100	74	45	45	39	32	25	20		25	40	3	–	–	–	–	–	–	–	–	–	–	Cholan-24-oic acid, 12-oxo-, methyl ester, (5β)-	1173-30-4	NI39747
43	215	107	81	147	95	93	55	388	100	92	59	57	50	46	46	46	22.00	25	40	3	–	–	–	–	–	–	–	–	–	–	24-Nor-5α-cholan-22-one, 3β-hydroxy-, acetate, (20S)-	26654-77-3	NI39749
43	215	107	81	93	55	147	95	388	100	67	46	46	41	39	38	38	9.00	25	40	3	–	–	–	–	–	–	–	–	–	–	24-Nor-5α-cholan-22-one, 3β-hydroxy-, acetate, (20S)-	26654-77-3	NI39748
255	256	370	55	228	81	159	132	388	100	23	18	9	8	8	7	6	1.01	25	40	3	–	–	–	–	–	–	–	–	–	–	Chol-7-en-24-oic acid, 12-hydroxy-, methyl ester, (5β,12α)-	54411-72-2	NI39750
328	91	55	79	77	105	93	53	388	100	73	58	55	51	50	47	41	1.80	25	40	3	–	–	–	–	–	–	–	–	–	–	5α,6α,16α,17α-Diepoxypregnan-3β-ol-20-one, 3-acetate		AM00154
328	302	388	316	155	343	373	314	388	100	80	60	60	50	46	32	32		25	44	1	2	–	–	–	–	–	–	–	–	–	20,25-Diazacholesterol		LI08346
71	303	233	137	135	373	69	125	388	100	97	97	70	70	67	62	46	2.95	25	44	1	2	–	–	–	–	–	–	–	–	–	1-(Dicyanomethylene)-1-methoxy-4,8,12-trimethyloctadecane		LI08347
388	84	389	44	387	110	386	302	388	100	41	28	22	11	10	9	9		25	44	1	2	–	–	–	–	–	–	–	–	–	Pregnane, 20-acetamido-3-(dimethylamino)-, (3β,5α,20α)-	15112-55-7	NI39751
388	389	138	137	194	274	180	390	388	100	29	23	14	10	9	8	6		26	12	4	–	–	–	–	–	–	–	–	–	–	Dinaphtho[1,2-d:1′,2′-d′]benzo[1,2-b:4,5-b′]difuran-8,16-dione	3604-49-7	NI39752
388	243	244	389	387	270	77	373	388	100	98	38	30	30	29	25	21		26	20	–	4	–	–	–	–	–	–	–	–	–	1,5-Benzodiazepine, 2,3-bis(indol-3-yl)-4-methyl-	105283-36-1	NI39753
387	388	77	389	296	194	169	386	388	100	96	35	24	20	12	7	6		26	20	–	4	–	–	–	–	–	–	–	–	–	Quinazoline, 3-phenyl-2-phenylamino-4-phenylimino-		MS09014
105	388	77	389	95	298	91	78	388	100	15	12	3	2	1	1	1		26	28	3	–	–	–	–	–	–	–	–	–	–	Estra-1,3,5(10)-trien-17-one, 3-(benzoyloxy)-2-methyl-	51209-86-0	NI39754
388	387	389	297	184	203	205	156	388	100	75	67	65	56	54	50	38		26	32	1	2	–	–	–	–	–	–	–	–	–	1H-Indolizino[8,7-b]indole, 1-[2-[(benzyloxy)methyl]butyl]-2,3,5,6,11,11b-hexahydro-	26503-40-2	LI08348
388	387	389	297	223	184	225	91	388	100	75	68	65	55	55	50	40		26	32	1	2	–	–	–	–	–	–	–	–	–	1H-Indolizino[8,7-b]indole, 1-[2-[(benzyloxy)methyl]butyl]-2,3,5,6,11,11b-hexahydro-	26503-40-2	NI39755
260	91	144	82	130	168	154	96	388	100	95	35	33	25	23	18	14	13.01	26	32	1	2	–	–	–	–	–	–	–	–	–	1H-Indolizino[8,7-b]indole, 2-[2-[(benzyloxy)methyl]butyl]-2,3,5,6,11,11b-hexahydro-	14051-14-0	NI39757
184	388	149	297	83	387	91	97	388	100	81	62	60	60	58	52	47		26	32	1	2	–	–	–	–	–	–	–	–	–	1H-Indolizino[8,7-b]indole, 2-[2-[(benzyloxy)methyl]butyl]-2,3,5,6,11,11b-hexahydro-	14051-14-0	NI39756
107	81	145	55	105	95	314	271	388	100	91	87	87	85	83	81	78	24.00	26	44	2	–	–	–	–	–	–	–	–	–	–	26,27-Dinorergost-5-ene-3,24-diol, (3β)-	35882-85-0	NI39758
370	371	388	197	239	257	215	355	388	100	29	25	17	11	10	10	7		26	44	2	–	–	–	–	–	–	–	–	–	–	6-Hydroxy-19-nor-androst-5(10)-en-3-ol		LI08349
84	305	56	91	97	112	201	70	388	100	16	16	15	13	12	11	11	7.89	27	36	–	2	–	–	–	–	–	–	–	–	–	Quinoline, decahydro-1-methyl-5,6-diphenyl-6-(2-piperidyl)-	14028-82-1	NI39759
388	233	215	234	55	165	43	95	388	100	79	69	66	65	62	62	59		27	48	1	–	–	–	–	–	–	–	–	–	–	Cholestan-3-ol, (3α)-	18769-46-5	NI39760
233	215	388	234	217	373	165	216	388	100	84	78	67	47	42	34	32		27	48	1	–	–	–	–	–	–	–	–	–	–	Cholestan-3-ol, (3α,5α)-	516-95-0	NI39761
388	215	389	216	233	370	234	217	388	100	37	32	28	21	17	17	16		27	48	1	–	–	–	–	–	–	–	–	–	–	Cholestan-3-ol, (3α,5β)-	516-92-7	NI39762
43	55	41	388	81	57	95	233	388	100	84	77	62	57	54	51	49		27	48	1	–	–	–	–	–	–	–	–	–	–	Cholestan-3-ol, (3β,5α)-	80-97-7	NI39763
233	215	388	234	107	373	216	108	388	100	95	62	58	43	42	40	39		27	48	1	–	–	–	–	–	–	–	–	–	–	Cholestan-3-ol, (3β,5α)-	80-97-7	NI39764
388	233	215	234	44	370	231	95	388	100	61	53	46	38	37	37	30		27	48	1	–	–	–	–	–	–	–	–	–	–	Cholestan-3-ol, (3β,5α)-	80-97-7	NI39765
369	371	370	355	372	245	127	149	388	100	56	38	22	16	9	8	7	2.50	27	48	1	–	–	–	–	–	–	–	–	–	–	Cholestan-5-ol, (5α)-	910-19-0	NI39766
370	371	388	386	216	389	215	355	388	100	82	79	69	69	66	65	63		27	48	1	–	–	–	–	–	–	–	–	–	–	Cholestan-6-ol, (6β)-	32477-81-9	LI08350
370	371	388	216	386	389	217	147	388	100	81	80	68	67	66	63	58		27	48	1	–	–	–	–	–	–	–	–	–	–	Cholestan-6-ol, (6β)-	32477-81-9	NI39767
370	191	388	138	95	257	81	234	388	100	96	66	64	60	57	54	52		27	48	1	–	–	–	–	–	–	–	–	–	–	Cholestan-7-ol, (5α,7β,14β)-	40071-69-0	NI39768
216	217	81	109	370	95	121	355	388	100	93	86	79	76	71	50	48	23.00	27	48	1	–	–	–	–	–	–	–	–	–	–	Cholestan-11-ol, (5α,11β,14β)-		MS09015
388	389	390	165	121	310	77	308	388	100	31	9	8	7	6	6	5		28	20	–	–	1	–	–	–	–	–	–	–	–	2,3,4,5-Tetraphenylthiophene		AI02310
105	77	388	106	283	389	255	178	388	100	13	9	8	4	3	3	3		28	20	2	–	–	–	–	–	–	–	–	–	–	2-Butene-1,4-dione, 1,2,3,4-tetraphenyl-	7510-34-1	NI39769
373	152	388	374	153	179	235	194	388	100	67	65	35	31	28	26	26		28	24	–	2	–	–	–	–	–	–	–	–	–	4-Acetylbiphenyl azine		LI08351
91	297	388	298	193	194	389	65	388	100	88	29	22	22	17	9	9		28	24	–	2	–	–	–	–	–	–	–	–	–	Deoxybenzoin azine		TR00305

MASS TO CHARGE RATIOS								M.W.	INTENSITIES								Parent	C	H	O	N	S	F	Cl	Br	I	Si	P	B	X	COMPOUND NAME	CAS Reg No	No
388	282	280	389	367	281			**388**	100	52	37	31	30	27				28	24	–	2	–	–	–	–	–	–	–	–	–	**9,10-Bis(N-methylanilino)phenanthrene**		**LI08352**
43	67	41	57	121	55	135	81	**388**	100	77	75	66	49	49	48	40	**13.55**	28	52	–	–	–	–	–	–	–	–	–	–	–	**9-(5-exo-Hexahydro-4,7-methanoindanylmethyl)heptadecane**		**AI02311**
83	97	69	236	235	96	67	111	**388**	100	50	33	30	24	23	23	21	**0.06**	28	52	–	–	–	–	–	–	–	–	–	–	–	**Tritriacontane, 17-phenyl-**		**AI02312**
310	240	146	354	275	111	389	76	**389**	100	13	11	8	8	4	3	3		–	–	–	3	–	–	5	1	–	–	3	–	–	**1,3,5,2,4,6-Triazatriphosphorine, 2-bromo-2,4,4,6,6-pentachloro-2,2,4,4,6,6-hexahydro-**	14740-93-3	**NI39770**
256	155	91	57	43	41	55	83	**389**	100	91	62	33	27	25	24	20	**0.00**	15	20	4	1	1	–	–	1	–	–	–	–	–	**4-Pentenoic acid, 3-(bromomethyl)-2-(tosylamino)-, ethyl ester, (2SR,3RS)-**		**DD01525**
134	135	309	77	118	136	193	192	**389**	100	12	11	10	7	6	5	5	**0.00**	16	12	–	3	2	–	–	1	–	–	–	–	–	**2-Phenylamino-5-phenylthiazolo[2,3-b]-1,3,4-thiadiazol-4-ium bromide**		**MS09016**
91	123	105	228	125	151	149	272	**389**	100	35	27	17	14	9	9	7	**0.01**	16	14	–	1	2	–	3	–	–	–	–	–	–	**Carbamodithioic acid, N,N-bis(phenylmethyl)-, trichloromethyl ester**	91631-92-4	**NI39771**
43	114	241	96	156	139	84	72	**389**	100	30	15	11	10	10	7	6	**0.00**	16	23	10	1	–	–	–	–	–	–	–	–	–	**2-Acetamido-2-desoxy-1,3,4,6-tetra-O-acetyl-α-D-galactopyranose**	10385-50-9	**NI39772**
114	125	156	80	96	139	84	72	**389**	100	49	34	30	28	23	22	21	**0.00**	16	23	10	1	–	–	–	–	–	–	–	–	–	**α-D-Glucopyranose, 2-(acetylamino)-2-deoxy-, 1,3,4,6-tetraacetate**	7784-54-5	**NI39774**
114	156	96	139	84	241	72	126	**389**	100	34	28	23	22	21	21	13	**0.00**	16	23	10	1	–	–	–	–	–	–	–	–	–	**α-D-Glucopyranose, 2-(acetylamino)-2-deoxy-, 1,3,4,6-tetraacetate**	7784-54-5	**NI39773**
114	244	156	139	241	96	84	72	**389**	100	45	45	43	42	39	27	23	**0.00**	16	23	10	1	–	–	–	–	–	–	–	–	–	**D-Glucopyranose, 2-(acetylamino)-2-deoxy-, 1,3,4,6-tetraacetate**	14086-90-9	**NI39775**
60	84	102	115	130	112	172	114	**389**	100	91	88	81	57	57	48	43	**0.00**	16	23	10	1	–	–	–	–	–	–	–	–	–	**D-Glucopyranose, 3-(acetylamino)-3-deoxy-, 1,2,4,6-tetraacetate**	100759-98-6	**NI39776**
139	72	114	73	97	115	85	156	**389**	100	91	85	83	65	64	60	57	**0.00**	16	23	10	1	–	–	–	–	–	–	–	–	–	**D-Glucopyranose, 6-(acetylamino)-6-deoxy-, 1,2,3,4-tetraacetate**	55443-23-7	**NI39777**
142	128	43	230	244	86	100	184	**389**	100	98	61	52	50	29	26	13	**0.00**	17	27	9	1	–	–	–	–	–	–	–	–	–	**L-Glucitol, 3,6-dideoxy-3-(N-methylacetamido)-, 1,2,4,5-tetraacetate**	82486-51-9	**NI39778**
84	98	77	42	105	56	55	97	**389**	100	35	35	30	20	20	16	13	**7.00**	18	19	1	3	3	–	–	–	–	–	–	–	–	**Thieno[2,3-c]isothiazole, 5-benzoyl-4-amino-3-[2-(N-pyrrolidinyl)ethylthio]-**		**NI39779**
84	97	77	42	105	55	56	41	**389**	100	40	35	32	24	15	12	11	**2.00**	18	19	1	3	3	–	–	–	–	–	–	–	–	**Thieno[3,2-d]isothiazole, 5-benzoyl-4-amino-3-[2-(N-pyrrolidinyl)ethylthio]-**	97090-81-8	**NI39780**
272	43	110	170	212	60	124	111	**389**	100	74	55	46	27	25	20	20	**0.00**	18	31	6	1	1	–	–	–	–	–	–	–	–	**2,3-Di-O-acetyl-6-acetylamino-4,6-dideoxy-1-hexylthio-α-DL-lyxo-hexopyranoside**		**NI39781**
362	363	184	105	389				**389**	100	19	9	9	3					19	19	8	1	–	–	–	–	–	–	–	–	–	**Tetramethyl 9a-vinyl-9aH-quinolizine-1,2,3,4-tetracarboxylate**		**LI08353**
113	43	160	272	59	129	114	302	**389**	100	36	28	22	22	14	13	12	**0.80**	20	23	7	1	–	–	–	–	–	–	–	–	–	**α-D-Glucofuranose, 6-deoxy-1,2:3,5-di-O-isopropylidene-6-phthalimido-**	34213-75-7	**NI39782**
242	241	243	389	330	216	244	214	**389**	100	55	32	17	6	6	5	5		20	23	7	1	–	–	–	–	–	–	–	–	–	**Oxostephine, N,O-dimethyl-**	99964-47-3	**NI39783**
243	244	213	196	228	245	258	229	**389**	100	50	48	30	20	11	9	8	**5.00**	21	27	6	1	–	–	–	–	–	–	–	–	–	**Hasubanan-6-one, 8,10-epoxy-3,4,7,8-tetramethoxy-17-methyl-, (7α,8β,10β)-**	52309-76-9	**NI39784**
243	244	213	228	389	245	258	214	**389**	100	47	34	16	14	9	8	7		21	27	6	1	–	–	–	–	–	–	–	–	–	**Hasubanan-6-one, 8,10-epoxy-3,4,7,8-tetramethoxy-17-methyl-, (7β,8β,10β)-**	52389-15-8	**NI39785**
320	130	157	389	41	108	321	192	**389**	100	79	44	26	26	25	21	15		22	23	2	5	–	–	–	–	–	–	–	–	–	**Roquefortine**		**NI39786**
284	285	268	389	286	269	390	232	**389**	100	20	17	14	6	4	3	2		23	23	1	3	1	–	–	–	–	–	–	–	–	**3,7-Bis(dimethylamino)-10-benzoylphenothiazine**		**IC04246**
137	105	135	136	133	122	268	210	**389**	100	79	77	70	68	59	56	54	**41.00**	23	23	3	3	–	–	–	–	–	–	–	–	–	**2-Pyrrolidinone, 3-[(benzoyloxy)phenylmethyl]-1-methyl-4-(1-methyl-1H-imidazol-4-yl)-, [3R-[3α(S*),4β]]-**	50656-84-3	**NI39787**
389	153	125	390	268	137	358	129	**389**	100	53	48	30	30	30	26	24		23	39	2	1	–	–	–	–	–	1	–	–	–	**Androst-4-en-3-one, 17-[(trimethylsilyl)oxy]-, O-methyloxime, (17β)-**	56210-83-4	**NI39788**
73	253	75	105	91	41	79	374	**389**	100	99	99	68	61	50	46	43	**26.00**	23	39	2	1	–	–	–	–	–	1	–	–	–	**Androst-9(11)-en-17-one, 3-[(trimethylsilyl)oxy]-, O-methyloxime, (3α,5α)-**	57305-05-2	**NI39789**
73	389	129	268	79	91	139	358	**389**	100	82	49	30	24	23	19	17		23	39	2	1	–	–	–	–	–	1	–	–	–	**Gon-4-ene, 13-ethyl-3-(methoximino)-17-[(trimethylsilyl)oxy]-, (13β,17β)-**		**NI39790**
241	255	270	77	119	104	199	91	**389**	100	69	60	26	20	18	15	13	**0.00**	24	27	2	3	–	–	–	–	–	–	–	–	–	**2,6-Diazabicyclo[2.2.2]oct-7-ene-3,5-dione, 8-(dimethylamino)-4,7-dimethyl-2,6-diphenyl-**		**DD01526**
356	371	91	310	41	326	105	42	**389**	100	80	36	31	31	29	27	26	**2.00**	24	39	3	1	–	–	–	–	–	–	–	–	–	**3b,17a-Dihydroxy-16b-methylpregn-5-en-20-one, ethyloxime**		**LI08354**
145	158	300	75	67	55	130	81	**389**	100	57	27	25	23	23	21	21	**7.00**	24	43	1	1	–	–	–	–	–	1	–	–	–	**Pregnan-20-one, O-trimethylsilyloxime, (5α)-**		**LI08355**
324	43	339	325	105	91	42	79	**389**	100	70	66	37	32	31	27	25	**25.00**	25	27	3	1	–	–	–	–	–	–	–	–	–	**3β-Acetoxy-16-methyl-pregna-5,16-dien-20-one, methyloxime**		**LI08356**
360	389	77	316	297	226	104	318	**389**	100	96	36	35	35	29	25	24		25	31	1	3	–	–	–	–	–	–	–	–	–	**2(1H)-Pyridinone, 4,5-bis(diethylamino)-1,3-diphenyl-**		**DD01527**
72	58	91	271	59	42	44	70	**389**	100	96	18	10	9	8	7	6	**0.00**	26	31	2	1	–	–	–	–	–	–	–	–	–	**1,2-Diphenyl-1-hydroxy-1-[2-[2-(dimethylamino)ethoxy]phenyl]butane**		**IC04247**
91	153	92	161	111	41	128	115	**389**	100	28	15	14	11	11	10	10	**7.01**	26	31	2	1	–	–	–	–	–	–	–	–	–	**Morphinan, 3-methoxy-6-methyl-17-(phenylacetyl)-, (6α)-**	55281-49-7	**NI39791**
334	364	274	56	304	394	137	193	**390**	100	95	90	63	54	45	45	23	**45.00**	–	–	4	4	–	–	–	2	–	–	–	–	2	**Iron, μ-bromotetramitrosodi-**		**MS09017**
205	390	203	186	121	388	214	97	**390**	100	47	41	29	24	18	18	17		4	10	2	–	2	–	–	–	–	–	1	–	1	**Thallium(I) diethyldithiophosphate**	25502-67-4	**NI39792**
394	74	392	396	73	234	313	315	**390**	100	99	69	64	43	35	31	30	**17.50**	6	2	–	–	–	–	–	4	–	–	–	–	–	**Benzene, 1,2,4,5-tetrabromo-**	636-28-2	**NI39793**
372	390	62	189	127	245	63	186	**390**	100	58	30	30	24	20	19	17		7	4	3	–	–	–	–	–	2	–	–	–	–	**Benzoic acid, 2-hydroxy-3,5-diiodo-**	133-91-5	**NI39794**
69	389	93	79	62	136	91	63	**390**	100	7	6	4	4	3	3	3	**0.89**	7	4	3	–	–	–	–	–	2	–	–	–	–	**Benzoic acid, 4-hydroxy-3,5-diiodo-**	618-76-8	**NI39795**
321	307	113	252	70	390	127	370	**390**	100	23	21	20	16	15	11	10		8	2	–	–	2	12	–	–	–	–	–	–	–	**2-(2,2,2-Trifluoroethyl)-2,4,5-tris(trifluoromethyl)-1,3-dithiole**		**LI08357**
359	357	361	237	355	272	251	239	**390**	100	95	52	43	41	38	34	32	**5.67**	8	5	–	–	–	–	6	1	–	–	–	–	–	**Bicyclo[2.2.1]hept-2-ene, 5-(bromomethyl)-1,2,3,4,7,7-hexachloro-**	1715-40-8	**NI39796**

MASS TO CHARGE RATIOS								M.W.	INTENSITIES								Parent	C	H	O	N	S	F	Cl	Br	I	Si	P	B	X	COMPOUND NAME	CAS Reg No	No
314	90	316	312	63	250	170	168	390	100	75	50	49	47	45	37	35	14.00	8	9	2	2	1	–	1	2	–	–	–	–	–	2,6-Dibromo-4-[N-(2-chloroethyl)sulphonamido]aniline		IC04248
392	394	390	97	266	194	167	62	390	100	78	57	53	42	27	17	17		10	4	–	–	–	–	6	–	–	–	–	–	1	Ferrocene, 1,1′,2,2′,3,3′-hexachloro-		LI08358
73	116	147	195	197	101	43	45	390	100	64	28	24	21	20	20	15	0.00	10	24	2	–	–	–	–	2	–	2	–	–	–	Butane, 1,4-bibromo-2,3-bis(trimethylsiloxy)-		NI39797
73	116	147	197	195	101	43	45	390	100	60	30	25	24	17	16	14	0.00	10	24	2	–	–	–	–	2	–	2	–	–	–	Butane, 1,4-Dibromo-2,3-bis(trimethylsiloxy)-		NI39798
162	147	268	161	141	175	43	163	390	100	84	60	51	28	20	17	15	0.00	11	12	–	–	1	4	–	–	1	–	–	1	–	1-Iodomethyl-2,3-dimethylbenzo[b]thiophenium tetrafluoroborate		LI08359
59	43	56	44	42	131	41	60	390	100	55	37	32	22	19	18	16	0.00	11	20	10	1	2	–	–	–	–	–	–	–	–	Glucoconringiin		MS09018
392	394	390	364	393	263	336	74	390	100	53	53	15	13	12	10	9		12	8	5	–	–	–	–	2	–	–	–	–	–	4H-1-Benzopyran-2-carboxylic acid, 6,8-dibromo-7-hydroxy-4-oxo-, ethyl ester	30113-88-3	NI39799
392	394	390	364	393	263	336	80	390	100	53	53	15	13	12	10	9		12	8	5	–	–	–	–	2	–	–	–	–	–	4H-1-Benzopyran-2-carboxylic acid, 6,8-dibromo-7-hydroxy-4-oxo-, ethyl ester	30113-88-3	LI08360
115	59	41	57	55	43	45	116	390	100	50	27	16	11	11	8	7	2.17	12	22	4	–	–	–	–	–	–	–	–	–	2	Butanoic acid, 4,4′-diselenobis[2,2-dimethyl-	3709-10-2	NI39800
115	59	41	57	55	43	45	29	390	100	51	27	16	11	11	8	7	2.00	12	22	4	–	–	–	–	–	–	–	–	–	2	Butanoic acid, 4,4′-diselenobis[2,2-dimethyl-	3709-10-2	LI08361
73	115	69	41	55	97	43	29	390	100	95	83	52	49	35	26	21	8.67	12	22	4	–	–	–	–	–	–	–	–	–	2	Butanoic acid, 4,4′-diselenobis[3,3-dimethyl-	3709-12-4	NI39801
69	97	41	55	115	73	45	43	390	100	62	48	44	39	32	19	16	11.26	12	22	4	–	–	–	–	–	–	–	–	–	2	Hexanoic acid, 6,6′-diselenodi-	22676-35-3	NI39802
69	97	115	41	55	73	45	29	390	100	63	57	48	45	32	19	15	12.00	12	22	4	–	–	–	–	–	–	–	–	–	2	Hexanoic acid, 6,6′-diselenodi-	22676-35-3	LI08362
178	392	390	240	394	355	353	356	390	100	53	38	31	28	26	23	19		14	10	1	–	–	–	4	–	–	2	–	–	–	1,3-Disila-2-oxacyclopent-4-ene, 1,1,3,3-tetrachloro-4,5-diphenyl-		NI39803
55	41	64	87	45	67	69	81	390	100	91	57	53	52	45	39	35	0.00	14	30	4	–	4	–	–	–	–	–	–	–	–	Methanesulphonothioic acid 1,12-dodecanediyl ester		MS09019
165	247	245	147	119	107	177	179	390	100	89	89	89	78	74	70	69	4.00	15	20	2	–	–	–	–	2	–	–	–	–	–	3,8-Dioxabicyclo[5.1.1]nonane, 6-bromo-4-(1-bromopropyl)-2-(2-penten-4-ynyl)-	18762-30-6	LI08363
245	247	165	147	179	177	119	107	390	100	99	90	85	64	64	63	58	8.00	15	20	2	–	–	–	–	2	–	–	–	–	–	3,9-Dioxabicyclo[4.2.1]nonane, 7-bromo-4-(1-bromopropyl)-2-(2-penten-4-ynyl)-	19897-64-4	LI08364
245	247	165	147	177	179	119	107	390	100	99	87	83	62	61	61	56	7.82	15	20	2	–	–	–	–	2	–	–	–	–	–	3,9-Dioxabicyclo[4.2.1]nonane, 7-bromo-4-(1-bromopropyl)-2-(2-penten-4-ynyl)-	19897-64-4	NI39804
65	55	81	79	107	121	133	161	390	100	80	70	62	48	47	26	20	0.10	15	20	2	–	–	–	–	2	–	–	–	–	–	3-Pentadecen-1-yne, 6,9:7,13-bisepoxy-10,12-dibromo-, [3E-(6S,7S,9S,10S,12S,13R)]-		MS09020
209	263	261	289	317	29	291	319	390	100	91	87	61	60	57	55	49	34.48	15	20	5	–	–	–	–	1	–	–	1	–	–	3-(3-Bromophenyl)propenoic acid, 2-(diethoxyphosphinyl)-, ethyl ester		NI39805
263	261	209	29	102	81	101	182	390	100	99	97	92	86	78	66	61	15.51	15	20	5	–	–	–	–	1	–	–	1	–	–	3-(4-Bromophenyl)propenoic acid, 2-(diethoxyphosphinyl)-, ethyl ester	37176-94-6	NI39806
73	147	75	229	45	375	74	67	390	100	35	16	13	13	11	8	7	0.00	15	30	6	–	–	–	–	–	–	3	–	–	–	1-Propene-1,2,3-tricarboxylic acid, tris(trimethylsilyl) ester, (E)-	55530-72-8	NI39807
73	147	75	229	45	67	375	74	390	100	51	19	19	15	10	10	9	0.00	15	30	6	–	–	–	–	–	–	3	–	–	–	1-Propene-1,2,3-tricarboxylic acid, tris(trimethylsilyl) ester, (E)-	55530-72-8	NI39808
73	147	229	75	45	375	67	74	390	100	52	19	18	15	11	10	9	0.00	15	30	6	–	–	–	–	–	–	3	–	–	–	1-Propene-1,2,3-tricarboxylic acid, tris(trimethylsilyl) ester, (Z)-	55530-71-7	NI39809
73	147	229	375	75	45	285	74	390	100	46	23	22	16	12	10	9	0.00	15	30	6	–	–	–	–	–	–	3	–	–	–	1-Propene-1,2,3-tricarboxylic acid, tris(trimethylsilyl) ester, (Z)-	55530-71-7	NI39810
152	195	163	390	138	392	238	111	390	100	88	87	69	51	48	48	35		16	12	–	6	1	–	2	–	–	–	–	–	–	[1,2,4]Triazolo[4,3-b][1,2,4]triazole-6-amine, N,7-(4-chlorophenyl)-3-(methylthio)-		MS09021
183	91	107	77	63	52	135	29	390	100	87	73	73	67	67	57	53	12.00	16	14	8	4	–	–	–	–	–	–	–	–	–	Methyl (±)-2-hydroxy-3-[4-(2,4-dinitro-benzolazo)-5-hydroxyphenyl]propionate		NI39811
43	126	128	170	115	45	97	69	390	100	99	86	71	66	59	53	52	0.00	16	22	11	–	–	–	–	–	–	–	–	–	–	β-D-Fructopyranose, pentaacetate	20764-61-8	NI39812
126	128	170	115	97	69	101	73	390	100	86	70	65	51	50	45	42	0.00	16	22	11	–	–	–	–	–	–	–	–	–	–	β-D-Fructopyranose, pentaacetate	20764-61-8	NI39813
73	85	44	101	45	127	187	128	390	100	87	86	82	74	50	46	39	0.00	16	22	11	–	–	–	–	–	–	–	–	–	–	D-Fructose, 1,3,4,5,6-pentaacetate	6341-07-7	NI39814
43	103	143	115	145	98	245	126	390	100	99	97	91	89	70	65	59	0.00	16	22	11	–	–	–	–	–	–	–	–	–	–	β-D-Galactofuranose, pentaacetate	5531-53-3	NI39815
103	43	143	145	115	98	245	127	390	100	99	97	89	89	73	65	59	0.00	16	22	11	–	–	–	–	–	–	–	–	–	–	D-Galactofuranose, pentaacetate	62181-82-2	NI39816
331	332	153	333	211	109	169	110	390	100	13	4	3	3	3	2	2	0.00	16	22	11	–	–	–	–	–	–	–	–	–	–	β-D-Galactopyranose, pentaacetate	4163-60-4	NI39818
115	98	157	73	97	103	81	69	390	100	97	58	38	36	35	23	23	0.00	16	22	11	–	–	–	–	–	–	–	–	–	–	β-D-Galactopyranose, pentaacetate	4163-60-4	NI39817
43	347	331	330	44	389	103	287	390	100	5	4	3	3	2	2	1	0.50	16	22	11	–	–	–	–	–	–	–	–	–	–	β-D-Galactopyranose, pentaacetate	4163-60-4	LI08365
43	103	115	145	98	143	85	168	390	100	96	93	87	79	73	53	48	0.00	16	22	11	–	–	–	–	–	–	–	–	–	–	D-Glucofuranose, pentaacetate	55123-31-4	NI39819
98	115	157	103	168	73	140	109	390	100	83	78	48	39	39	35	33	0.00	16	22	11	–	–	–	–	–	–	–	–	–	–	α-D-Glucopyranose, pentaacetate	604-68-2	NI39820
43	115	98	103	109	73	145	169	390	100	96	63	31	24	23	22	20	0.00	16	22	11	–	–	–	–	–	–	–	–	–	–	α-D-Glucopyranose, pentaacetate	604-68-2	NI39821
43	115	98	157	103	169	140	109	390	100	99	71	53	34	27	24	24	0.00	16	22	11	–	–	–	–	–	–	–	–	–	–	α-D-Glucopyranose, pentaacetate	604-68-2	NI39822
115	98	157	103	140	200	145	73	390	100	71	59	34	29	24	23	21	0.00	16	22	11	–	–	–	–	–	–	–	–	–	–	β-D-Glucopyranose, pentaacetate	604-69-3	NI39823
43	115	242	98	157	103	140	200	390	100	99	79	78	68	33	28	26	0.00	16	22	11	–	–	–	–	–	–	–	–	–	–	β-D-Glucopyranose, pentaacetate	604-69-3	NI39824
115	98	157	73	103	140	200	145	390	100	76	54	35	33	23	21	20	0.00	16	22	11	–	–	–	–	–	–	–	–	–	–	β-D-Glucopyranose, pentaacetate	604-69-3	NI39825
97	45	102	60	44	126	144	103	390	100	93	31	27	26	25	24	21	0.00	16	22	11	–	–	–	–	–	–	–	–	–	–	D-Glucose, 2,3,4,5,6-pentaacetate	3891-59-6	NI39826
97	43	115	98	157	169	103	140	390	100	55	46	40	34	23	21	19	0.00	16	22	11	–	–	–	–	–	–	–	–	–	–	D-Glucose, 2,3,4,5,6-pentaacetate	3891-59-6	NI39827
43	115	98	157	103	140	200	242	390	100	20	14	11	6	5	5	4	0.00	16	22	11	–	–	–	–	–	–	–	–	–	–	D-Glucose, 2,3,4,5,6-pentaacetate	3891-59-6	NI39828
115	98	157	103	169	145	140	73	390	100	62	49	33	24	23	23	23	0.00	16	22	11	–	–	–	–	–	–	–	–	–	–	α-D-Mannopyranose, pentaacetate	4163-65-9	NI39829
115	98	157	103	73	140	85	145	390	100	80	46	36	33	24	22	20	0.00	16	22	11	–	–	–	–	–	–	–	–	–	–	β-D-Mannopyranose, pentaacetate	4026-35-1	NI39830
28	121	79	95	135	55	67	390	390	100	38	38	36	34	29	25	21		17	24	2	–	–	–	–	–	–	–	–	–	1	2-Oxabicyclo[5.4.0]undecane, 4-[(4-methoxyphenyl)telluro]-		DD01528

MASS TO CHARGE RATIOS								M.W.	INTENSITIES								Parent	C	H	O	N	S	F	Cl	Br	I	Si	P	B	X	COMPOUND NAME	CAS Reg No	No
28	95	135	43	390	121	67	55	**390**	100	68	67	57	50	41	41	32		17	24	2	–	–	–	–	–	–	–	–	–	1	2-Oxabicylco[4.4.0]decane, 3-[[(4-methoxyphenyl)telluro]methyl]-		DD01529
135	95	55	121	237	390	108	67	**390**	100	95	86	79	51	49	49	49		17	24	2	–	–	–	–	–	–	–	–	–	1	1-Oxaspiro[4.5]decane, 2-[[(4-methoxyphenyl)telluro]methyl]-		DD01530
179	79	77	85	139	73	182	210	**390**	100	85	85	61	57	48	45	43	**0.00**	17	26	10	–	–	–	–	–	–	–	–	–	–	Cyclopenta[c]pyran-4-carboxylic acid, 1-(β-D-Glucopyranosyloxy)-1,4a,5,6,7,7a-hexahydro-6-hydroxy-7-methyl-, methyl ester, [1S-(1α,4aα,6α,7α,7aα)]-	18524-94-2	NI39831
101	153	72	83	170	375	111	85	**390**	100	50	28	27	24	21	21	21	**0.00**	17	26	10	–	–	–	–	–	–	–	–	–	–	D-Glucitol, 1,2-mono-O-isopropylidene-, tetraacetate		NI39832
153	85	375	127	111	115	187	72	**390**	100	77	53	41	40	35	29	27	**0.00**	17	26	10	–	–	–	–	–	–	–	–	–	–	D-Glucitol, 2,3-mono-O-isopropylidene-, tetraacetate		NI39833
153	85	127	110	115	187	375	101	**390**	100	93	49	49	29	28	27	27	**0.00**	17	26	10	–	–	–	–	–	–	–	–	–	–	D-Glucitol, 3,4-mono-O-isopropylidene-, tetraacetate	104527-45-9	NI39834
101	153	111	375	115	171	170	129	**390**	100	64	22	18	18	17	15	15	**0.00**	17	26	10	–	–	–	–	–	–	–	–	–	–	D-Glucitol, 5,6-mono-O-isopropylidene-, tetraacetate		NI39835
43	81	115	102	98	157	112	103	**390**	100	10	8	7	6	5	5	5	**0.00**	17	26	10	–	–	–	–	–	–	–	–	–	–	Glucose propylglycoside tetraacetate		MS09022
43	81	115	157	102	98	112	103	**390**	100	12	11	10	10	8	6	6	**0.00**	17	26	10	–	–	–	–	–	–	–	–	–	–	Glucose propylglycoside tetraacetate		MS09023
101	153	111	72	115	170	375	85	**390**	100	60	25	24	20	13	11	9	**0.00**	17	26	10	–	–	–	–	–	–	–	–	–	–	D-Mannitol, 1,2-mono-O-isopropylidene-, tetraacetate	76867-27-1	NI39836
85	153	127	111	115	187	59	101	**390**	100	65	52	44	26	25	20	17	**0.00**	17	26	10	–	–	–	–	–	–	–	–	–	–	D-Mannitol, 3,4-mono-O-isopropylidene-, tetraacetate	92206-22-9	NI39837
115	153	139	375	111	85	59	101	**390**	100	65	27	15	15	15	15	11	**0.00**	17	26	10	–	–	–	–	–	–	–	–	–	–	D-Mannitol, 4,5-mono-O-isopropylidene-, tetraacetate		NI39838
305	73	341	307	271	43	284	375	**390**	100	67	52	35	33	30	24	23	**0.00**	17	31	4	2	–	–	1	–	–	1	–	–	–	2,4,6(1H,3H,5H)-Pyrimidinetrione, 5-[3-chloro-2-[(trimethylsilyl)oxy]propyl]-1,3-dimethyl-5-(1-methylbutyl)-	57397-44-1	NI39839
145	230	157	141	131	129	203	213	**390**	100	36	29	13	10	10	7	5	**0.00**	17	46	–	–	–	–	–	–	–	5	–	–	–	2,2,4,4,7,7-Hexamethyl-6,6-bis(trimethylsilyl)-2,4,7-trisilaoctane		LI08366
174	73	175	86	59	176	375	74	**390**	100	60	17	10	10	8	6	5	**0.00**	17	46	–	2	–	–	–	–	–	4	–	–	–	Cadaverine, tetrakis(trimethylsilyl)-		MS09024
150	151	162	180	196	152	122	51	**390**	100	98	95	79	48	29	22	21	**0.00**	18	14	10	–	–	–	–	–	–	–	–	–	–	Benzoic acid, 3-formyl-2,4-dihydroxy-6-methyl-, (2-carboxy-4-formyl-3,5-dihydroxyphenyl)methyl ester	529-50-0	NI39840
390	162	375	161	176	391	133	298	**390**	100	58	52	32	26	23	17	16		18	22	4	4	1	–	–	–	–	–	–	–	–	N,N-Diethyl-3-methyl-4-(3-carboethoxy-5-nitrothien-2-ylazo)aniline		IC04249
375	390	345	376	391	43	360	169	**390**	100	76	21	20	17	12	11	11		19	18	9	–	–	–	–	–	–	–	–	–	–	4H-1-Benzopyran-4-one, 5,7-dihydroxy-2-(3-hydroxy-4,5-dimethoxyphenyl)-6,8-dimethoxy-	18398-74-8	NI39841
228	213	60	185	73	199	157	98	**390**	100	62	31	13	13	3	3	3	**0.20**	19	22	7	2	–	–	–	–	–	–	–	–	–	Ruine		LI08367
168	83	334	55	41	306	28	95	**390**	100	54	44	43	39	31	30	26	**0.64**	19	26	5	–	–	–	–	–	–	–	–	–	1	Iron, tricarbonyl[[(2,3,4,5-η)-5-methyl-2-isopropylcyclohexyl]-2,4-hexadienoate]-	61249-68-1	NI39842
73	75	228	93	202	300	227	129	**390**	100	49	23	19	17	16	15	12	**0.20**	19	34	1	2	–	–	–	–	–	3	–	–	–	Indole-3-acetamide, 1-trimethylsilyl-N,N-bis(trimethylsilyl)-		NI39843
73	287	155	147	75	43	286	57	**390**	100	99	63	51	43	38	36	26	**0.00**	19	42	4	–	–	–	–	–	–	2	–	–	–	Decanoic acid, 2,3-bis(trimethylsiloxy)propyl ester	1116-64-9	WI01509
73	129	218	147	103	75	43	229	**390**	100	74	53	52	38	38	33	26	**0.00**	19	42	4	–	–	–	–	–	–	2	–	–	–	Decanoic acid, 2-[(trimethylsilyl)oxy]-1-[[(trimethylsilyl)oxy]methyl]ethyl ester	55268-48-9	WI01510
147	130	145	119	148	146	91	49	**390**	100	21	15	14	11	9	9	9	**0.00**	20	22	2	2	–	3	–	–	–	–	–	1	–	1,2,4-Oxadiazole, 3,5-dimesityl-, 4-oxide, compd. with boron fluoride (BF3) (1:1)	20434-89-3	NI39844
233	234	390	260	215	344	157	76	**390**	100	93	51	40	30	25	24	24		20	22	8	–	–	–	–	–	–	–	–	–	–	2-Benzofuranbutanoic acid, 7-acetyl-4,6-dihydroxy-α-(1-hydroxyethylidene)-3,5-dimethyl-β-oxo-, ethyl ester	55570-93-9	NI39845
390	209	359	327	316	301	300	195	**390**	100	30	15	2	2	2	2	2		20	22	8	–	–	–	–	–	–	–	–	–	–	Dimethyl 4,6,2′,6′-tetramethoxybiphenyl-2,4′-dicarboxylate		LI08368
390	209	149	359	300	329	228	285	**390**	100	9	7	6	5	4	4	2		20	22	8	–	–	–	–	–	–	–	–	–	–	Dimethyl 4,6,4′,6′-tetramethoxybiphenyl-2,2′-dicarboxylate		LI08369
157	215	314	155	216	358	390	59	**390**	100	80	61	57	49	47	42	42		20	22	8	–	–	–	–	–	–	–	–	–	–	1H,6H-8a,5:8b,4-Bis(epoxymethano)-3a,5a-ethano-as-indacene-13,14-dicarboxylic acid, hexahydro-10,12-dioxo-, dimethyl ester	32237-59-5	MS09025
358	359	157	59	210	55	129	91	**390**	100	28	20	16	14	14	13	12	**9.00**	20	22	8	–	–	–	–	–	–	–	–	–	–	8a,5-(Epoxymethano)-3,5a-methano-3a,8b-propano-6H-indeno[4,5-b]furan 4,14-dicarboxylic acid, hexahydro-2,13-dioxo-, dimethyl ester	32237-60-8	MS09026
105	77	123	106	103	165	86	73	**390**	100	18	9	7	7	6	6	6	**0.00**	20	22	8	–	–	–	–	–	–	–	–	–	–	D-Glucitol, 1,6-dibenzoate	5346-88-3	NI39846
390	359	209	164	327	301	195	179	**390**	100	14	10	5	3	2	2	2		20	22	8	–	–	–	–	–	–	–	–	–	–	4,4′-Bis(methoxycarbonyl)-2,2′,6,6′-tetramethoxybiphenyl		LI08370
136	164	192	390	138	124	125	190	**390**	100	50	45	41	37	37	33	28		20	26	2	2	2	–	–	–	–	–	–	–	–	5,6,14,15-Dibenzo-4,16-dioxa-1,10-dithia-7,13-diazacyclooctadeca-5,14-diene		MS09027
151	390	73	95	124	136	122	391	**390**	100	89	67	53	37	29	24	23		20	26	6	2	–	–	–	–	–	–	–	–	–	Dibenzo[b,k]-1,4,7,10,13,16-hexaoxacyclooctadeca-2,11-diene-4′,5″-diamine	31406-52-7	NI39847
151	95	390	124	136	73	152	195	**390**	100	49	44	35	29	24	17	16		20	26	6	2	–	–	–	–	–	–	–	–	–	Dibenzo[b,k]-1,4,7,10,13,16-hexaoxacyclooctadeca-2,11-diene-5′,5″-diamine	31352-45-1	NI39848
390	91	208	119	314	117	178	391	**390**	100	50	35	32	28	25	19	13		21	18	4	4	–	–	–	–	–	–	–	–	–	4,4′-Dimethylbenzophenone 2,4-dinitrophenylhydrazone		IC04250
288	275	289	260	245	220	43	259	**390**	100	27	23	20	15	14	11	10	**9.00**	21	18	4	4	–	–	–	–	–	–	–	–	–	1,2-Ethanediol, 1-(1-phenyl-1H-pyrazolo[3,4-b]quinoxalin-3-yl)-, diacetate, (S)-	55591-20-3	NI39849
183	163	165	184	91	77	127	181	**390**	100	16	12	9	4	4	3	2	**0.30**	21	20	3	–	–	–	2	–	–	–	–	–	–	Cyclopropanedicarboxylic acid, 3-(2,2-dichlorovinyl)-2,2-dimethyl-, (3-phenoxyphenyl)methyl ester	52645-53-1	NI39850
43	317	390	149	41	59	57	84	**390**	100	94	60	56	48	40	36	29		21	26	7	–	–	–	–	–	–	–	–	–	–	(1R,5R,6S,7R,8R)-1-Allyl-4,8-dihydroxy-5-methoxy-6-(3′-methoxy-4′,5′-methylenedioxyphenyl)-6-methyl-3-oxobicyclo[3.2.1]octane		NI39851

MASS TO CHARGE RATIOS								M.W.	INTENSITIES								Parent	C	H	O	N	S	F	Cl	Br	I	Si	P	B	X	COMPOUND NAME	CAS Reg No	No
390	224	212	211	197	179	209	181	390	100	50	30	20	20	15	14	10		21	26	7	–	–	–	–	–	–	–	–	–	–	2H-1-Benzopyran, 3,4-dihydro-6,7-dimethoxy-3-(2,3,4,5-tetramethoxyphenyl)-, (S)-	71594-01-9	NI39852
137	138	193	150	55	390	124	122	390	100	41	32	15	14	10	10	10		21	26	7	–	–	–	–	–	–	–	–	–	–	Carissanol methyl acetal (isomer a)		NI39853
137	138	193	163	150	55	390	122	390	100	41	30	16	14	13	11	11		21	26	7	–	–	–	–	–	–	–	–	–	–	Carissanol methyl acetal (isomer b)		NI39854
195	121	137	149	180	152	344	181	390	100	88	55	49	46	46	44	14	1.00	21	26	7	–	–	–	–	–	–	–	–	–	–	2,2-Dimethoxy-3-(4-methoxyphenyl)-1-(2,4,6-trimethoxyphenyl)propan-1-one		NI39855
358	91	59	41	286	55	77	79	390	100	72	69	68	60	57	56	48	4.00	21	26	7	–	–	–	–	–	–	–	–	–	–	Gibberellin A_{21} methyl ester		MS09028
136	121	109	360	110	65	57	81	390	100	86	50	46	46	43	39	29	26.80	21	26	7	–	–	–	–	–	–	–	–	–	–	6-Hydroxy-2,3,12,13-dibenzo-1,4,8,11,14,17-hexaoxacyclononadeca-2,12-diene		MS09029
390	165	262	135	214	184	391	104	390	100	95	82	40	30	27	25	22		21	26	7	–	–	–	–	–	–	–	–	–	–	O-Methyl-1-ailanthone		LI08371
43	41	283	298	311	55	217	91	390	100	94	68	54	48	44	42	39	0.00	21	26	7	–	–	–	–	–	–	–	–	–	–	1,4-Phenanthrenedione, 7-formyloxy-4b,5,6,7,8,8a,9,10-octahydro-3,9,10-trihydroxy-4b,8,8-trimethyl-2-(2-propenyl)-, (4bS,7R,8aR,9S,10S)-		DD01531
133	160	178	328	105	145	117	147	390	100	81	77	49	39	32	24	23	0.00	21	26	7	–	–	–	–	–	–	–	–	–	–	Pseudolaric acid C	82601-41-0	NI39856
80	125	124	108	83	82	81	123	390	100	99	99	99	99	99	99	90	0.10	21	28	6	1	–	–	–	–	–	–	–	–	–	(+)-7-Hydroxy-6-tigloyloxynortropan-3-yl 2-hydroxy-3-phenylpropionate		MS09030
390	76	119	44	391	75	77	63	390	100	60	27	24	23	22	21	18		22	14	7	–	–	–	–	–	–	–	–	–	–	2,2′-Binaphthalene-1,1′,4,4′-tetrone, 3-hydroxy-8,8′-dimethoxy-		NI39857
390	106	391	77	334	51	134	135	390	100	72	65	64	63	54	52	51		22	14	7	–	–	–	–	–	–	–	–	–	–	2,2′-Binaphthalene-1,1′,4,4′-tetrone, 3,8,8′-trihydroxy-6,6′-dimethyl-	88381-84-4	NI39858
390	44	334	391	69	362	290	106	390	100	42	28	25	20	17	12	11		22	14	7	–	–	–	–	–	–	–	–	–	–	2,2′-Binaphthalene-1,1′,4,4′-tetrone, 3,8,8′-trihydroxy-6,6′-dimethyl-	88381-84-4	NI39859
375	390	376	372	108	91	347	391	390	100	44	23	16	16	16	14	11		22	14	7	–	–	–	–	–	–	–	–	–	–	2,2′-Binaphthalene-1′,4′,5,8-tetrone, 1,5′,8′-trihydroxy-3,3′-dimethyl-	104531-46-6	NI39860
390	373	44	333	334	391	84	345	390	100	53	39	35	29	28	26	19		22	14	7	–	–	–	–	–	–	–	–	–	–	2,2′-Binaphthalene-1′,4′,5,8-tetrone, 4,5′,8′-trihydroxy-6′,7-dimethyl-		NI39861
390	391	150	334	392	306	135	134	390	100	16	15	13	11	8	5	4		22	14	7	–	–	–	–	–	–	–	–	–	–	Euclanone	62996-97-8	NI39862
390	196	181	195	391	162	226	153	390	100	92	74	60	24	20	19	18		22	30	2	–	2	–	–	–	–	–	–	–	–	Phenol, 2,2′-dithiobis[6-tert-butyl-4-methyl-	1620-66-2	NI39863
390	196	181	195	226	162	153	121	390	100	92	74	60	20	20	18	17		22	30	2	–	2	–	–	–	–	–	–	–	–	Phenol, 2,2′-dithiobis[6-tert-butyl-4-methyl-	1620-66-2	LI08372
135	249	93	189	109	107	121	81	390	100	74	60	54	38	38	30	27	2.00	22	30	4	–	1	–	–	–	–	–	–	–	–	2,6,10-Dodecatrienoic acid, 3,7,11-trimethyl-9-(phenylsulphonyl)-, methyl ester, (E,E)-	57683-66-6	NI39864
59	300	358	285	302	295	360	328	390	100	99	30	25	20	20	15	15	2.00	22	30	6	–	–	–	–	–	–	–	–	–	–	17,19-Bis(abeo)-6β-hydroxy-7α,16ξ-dimethoxyroyleanone		NI39865
216	312	199	96	91	294	147	83	390	100	94	69	68	68	63	58	58	0.13	22	30	6	–	–	–	–	–	–	–	–	–	–	4-Desoxy-4α-phorbol-12-monoacetate		NI39866
199	200	217	83	216	43	294	312	390	100	83	74	67	62	56	54	53	1.60	22	30	6	–	–	–	–	–	–	–	–	–	–	4-Desoxy-4α-phorbol-13-monoacetate		NI39867
41	55	43	69	135	57	241	95	390	100	99	94	82	76	72	62	57	3.00	22	30	6	–	–	–	–	–	–	–	–	–	–	Gibberellin A_{19} methyl ester		MS09031
99	390	113	157	265	100	86	55	390	100	50	30	28	26	26	26	25		22	30	6	–	–	–	–	–	–	–	–	–	–	6-Oxaestra-2,5(10)-diene, 7,7:17,17-bis(ethylenedioxy)-3-methoxy-, (±)-		MS09032
99	390	86	157	113	265	100	55	390	100	38	33	30	30	23	23	23		22	30	6	–	–	–	–	–	–	–	–	–	–	6-Oxaestra-2,5(10)-diene, 7,7:17,17-bis(ethylenedioxy)-3-methoxy-, (±)-, (8α)-		MS09033
135	245	75	73	258	333	43	390	390	100	50	39	37	20	19	17	9		22	34	2	–	1	–	–	–	–	1	–	–	–	7-Oxabicyclo[2.2.1]heptane, 2-exo-[(tert-butyldimethylsiloxy)methyl]-3-exo-vinyl-4-methyl-5-endo-(phenylthio)-, (±)-	117626-58-1	DD01532
246	228	55	43	41	79	91	174	390	100	99	81	76	74	64	60	59	1.00	22	35	5	–	–	–	–	–	–	–	–	1	–	3α,11β,17α,21-Tetrahydroxy-5β-pregnan-20-one methyl boronate		MS09034
228	246	174	213	105	107	119	227	390	100	99	59	56	55	46	39	37	1.00	22	35	5	–	–	–	–	–	–	–	–	1	–	3α,11β,17α,21-Tetrahydroxy-5β-pregnan-20-one methyl boronate		LI08373
79	43	91	93	41	55	105	67	390	100	99	97	95	94	88	82	81	25.00	22	35	5	–	–	–	–	–	–	–	–	1	–	3α,17α,20α,21-Tetrahydroxy-5β-pregnan-11-one methyl boronate		MS09035
105	318	107	121	372	119	109	269	390	100	91	89	71	63	52	52	51	30.00	22	35	5	–	–	–	–	–	–	–	–	1	–	3α,17α,20α,21-Tetrahydroxy-5β-pregnan-11-one methyl boronate		LI08374
348	333	350	390	349	335	43	49	390	100	74	39	36	26	25	24	22		23	15	4	–	–	–	1	–	–	–	–	–	–	Dinaphtho[1,2-b:1′,2′-d]furan-5-ol, 7-chloro-8-methoxy-, acetate	91913-14-3	NI39868
357	105	77	358	311	327	178	165	390	100	99	99	83	43	26	25	24	1.00	23	22	4	2	–	–	–	–	–	–	–	–	–	3-(4-Nitrophenylamino)-1,2-diphenyl-1,1-dimethoxypropene		MS09036
390	375	141	244	348	330	297	315	390	100	28	21	14	12	10	10	8		23	31	4	–	–	1	–	–	–	–	–	–	–	3β-Fluoro-5α-hydroxy-6-oxo-17β-acetoxy-17α-vinyl-androstane		LI08375
390	375	391	141	244	348	330	298	390	100	28	26	21	14	12	10	10		23	31	4	–	–	1	–	–	–	–	–	–	–	Pregn-20-yn-6-one, 17-(acetyloxy)-3-fluoro-5-hydroxy-, (3β,5β,17α)-	55268-50-3	NI39869
390	43	375	391	141	244	105	91	390	100	51	28	27	21	15	15	15		23	31	4	–	–	1	–	–	–	–	–	–	–	Pregn-20-yn-6-one, 17-(acetyloxy)-3-fluoro-5-hydroxy-, (3β,17α)-	57194-88-4	NI39870
274	67	55	109	81	79	108	95	390	100	95	95	75	63	53	48	47	0.00	23	34	3	–	1	–	–	–	–	–	–	–	–	Androst-5-en-17-one, 3-(acetyloxy)-, cyclic 17-(1,2-ethanediyl monothioacetal), (3β)-	54498-67-8	NI39871
330	43	96	93	120	312	133	109	390	100	49	42	39	38	36	32	28	2.89	23	34	5	–	–	–	–	–	–	–	–	–	–	Androstane-17-carboxylic acid, 3-(acetyloxy)-5,6-epoxy-, methyl ester, (3β,5α,6α,17β)-	56009-15-5	NI39872
201	111	108	41	149	93	219	81	390	100	62	62	54	52	52	51	50	2.08	23	34	5	–	–	–	–	–	–	–	–	–	–	Card-20(22)-enolide, 1,3,14-trihydroxy-, (1β,3β,5β)-	639-15-6	NI39873
318	124	43	55	111	201	57	219	390	100	71	71	65	45	43	41	39	0.62	23	34	5	–	–	–	–	–	–	–	–	–	–	Card-20(22)-enolide, 3,5,14-trihydroxy-, (3β,5β)-	514-39-6	NI39874
201	160	43	111	261	57	55	44	390	100	58	55	31	28	25	25	25	0.00	23	34	5	–	–	–	–	–	–	–	–	–	–	Card-20(22)-enolide, 3,14,19-trihydroxy-, (3β,5α)-	468-19-9	MS09037
143	145	57	184	158	129	91	169	390	100	89	57	43	37	37	28	15	0.00	23	34	5	–	–	–	–	–	–	–	–	–	–	Compactin, (±)-		NI39875
216	330	390	201	270	276	259	241	390	100	99	91	72	55	23	16	13		23	34	5	–	–	–	–	–	–	–	–	–	–	3α,16α-Diacetoxyandrostan-17-one		NI39876
216	201	270	330	276	390	259	315	390	100	77	38	35	22	17	15	13		23	34	5	–	–	–	–	–	–	–	–	–	–	3α,16β-Diacetoxyandrostan-17-one		NI39877
330	348	215	288	216	270	315	276	390	100	67	54	48	48	34	18	15	7.00	23	34	5	–	–	–	–	–	–	–	–	–	–	3α,17β-Diacetoxyandrostan-16-one		NI39878
201	219	262	354	372	244	202	43	390	100	79	45	35	24	24	22	15	2.50	23	34	5	–	–	–	–	–	–	–	–	–	–	Digitoxigenin, 7-hydroxy-, (7β)-	1173-21-3	NI39879
336	91	79	92	78	105	41	67	390	100	86	71	50	49	48	44	42	0.00	23	34	5	–	–	–	–	–	–	–	–	–	–	Digoxigenin		LI08376
28	85	111	201	105	55	41	219	390	100	75	66	50	50	50	50	40	1.00	23	34	5	–	–	–	–	–	–	–	–	–	–	Gomphogenin	24211-63-0	LI08377

MASS TO CHARGE RATIOS	M.W.	INTENSITIES	Parent	C	H	O	N	S	F	Cl	Br	I	Si	P	B	X	COMPOUND NAME	CAS Reg No	No
28 85 111 201 55 41 105 81	**390**	100 75 67 53 53 51 50 45	**2.36**	23	34	5	–	–	–	–	–	–	–	–	–	–	**Gomphogenin**	24211-63-0	**NI39880**
255 43 28 272 56 41 84 271	**390**	100 95 80 76 73 63 61 60	**3.00**	23	34	5	–	–	–	–	–	–	–	–	–	–	**Methyl 7β-acetoxy-15,16α-epoxy-16β-(–)-kauran-18-oate**		**MS09038**
124 267 266 334 81 69 123 125	**390**	100 76 52 40 36 36 33 29	**6.00**	23	34	5	–	–	–	–	–	–	–	–	–	–	**Methyl 7α,12α-dihydroxy-3-oxo-Δ^4-bisnorcholenate**		**LI08378**
292 305 43 362 306 293 55 41	**390**	100 95 83 46 22 22 20 20	**2.00**	23	34	5	–	–	–	–	–	–	–	–	–	–	**Pregnane-15,20-dione, 3,14-dihydroxy-, (3β,14β)-**		**LI08379**
292 293 305 43 315 81 108 109	**390**	100 22 20 15 5 5 4 3	**1.00**	23	34	5	–	–	–	–	–	–	–	–	–	–	**Pregnane-15,20-dione, 3,14-dihydroxy-, (3β,14β,17β)-**		**LI08380**
124 267 266 334 70 82 123 95	**390**	100 79 52 40 37 36 33 28	**6.67**	23	34	5	–	–	–	–	–	–	–	–	–	–	**Pregn-4-ene-20-carboxylic acid, 7,12-dihydroxy-3-oxo-, methyl ester, (7α,12α,20S)-**	55741-11-2	**NI39881**
201 244 354 107 95 372 111 93	**390**	100 60 36 33 32 30 29 25	**2.00**	23	34	5	–	–	–	–	–	–	–	–	–	–	**Sarmentogenin**		**LI08381**
354 28 372 91 43 79 55 41	**390**	100 64 59 47 47 42 42 42	**19.00**	23	34	5	–	–	–	–	–	–	–	–	–	–	**Syriogenin**		**LI08382**
122 286 123 124 105 119 91 54	**390**	100 98 95 90 84 82 82 70	**0.00**	23	34	5	–	–	–	–	–	–	–	–	–	–	**Testosterone-11α-hemisuccinate**		**NI39882**
361 362 291 93 55 363 81 211	**390**	100 16 6 6 6 5 5 4	**0.60**	23	38	3	2	–	–	–	–	–	–	–	–	–	**Heptabarbital, 1,3-dipentyl-**		**NI39883**
349 287 257 227 91 45 211 31	**390**	100 49 44 44 41 40 36 36	**0.00**	24	26	3	–	–	–	–	–	–	1	–	–	–	**Silane, allyltris(2-methoxyphenyl)-**		**DD01533**
202 390 187 186 146 347 57 375	**390**	100 60 60 54 46 45 40 26		24	26	3	2	–	–	–	–	–	–	–	–	–	**2,3a-(Epoxyethano)-2H-indole-9,10-dicarboximide, 6-tert-butyl-3,3a-dihydro-3,3-dimethyl-N-phenyl-**		**DD01534**
57 132 390 186 173 202 78 217	**390**	100 95 66 63 45 38 38 33		24	26	3	2	–	–	–	–	–	–	–	–	–	**Isoxazolo[2,3-a]indole-2,3-dicarboximide, 7-tert-butyl-2,3,3a,4-tetrahydro 4,4-dimethyl-N-phenyl-, rel-(2R,3S,3aR)-**		**DD01535**
57 173 132 217 161 390 202 200	**390**	100 90 81 71 60 47 45 25		24	26	3	2	–	–	–	–	–	–	–	–	–	**Isoxazolo[2,3-a]indole-2,3-dicarboximide, 7-tert-butyl-2,3,3a,4-tetrahydro 4,4-dimethyl-N-phenyl-, rel-(2S,3R,3aR)-**		**DD01536**
41 330 43 55 161 42 44 255	**390**	100 67 67 57 52 43 41 37	**0.00**	24	38	4	–	–	–	–	–	–	–	–	–	–	**Androstane-3,17-diol, 7-methyl-, diacetate, (3α,5α,17β)-**	54725-67-6	**NI39884**
43 29 41 55 71 81 79 67	**390**	100 62 60 38 36 34 32 32	**16.42**	24	38	4	–	–	–	–	–	–	–	–	–	–	**5α-Androstan-3-one, 1α-(ethoxycarbonylmethyl)-17β-hydroxy-17α-methyl-**	71017-11-3	**NI39885**
253 271 372 226 55 81 41 105	**390**	100 91 58 45 43 32 30 28	**2.02**	24	38	4	–	–	–	–	–	–	–	–	–	–	**Chol-7-en-24-oic acid, 3,12-dihydroxy-, (3α,5β,12α)-**	21940-27-2	**NI39886**
149 57 167 43 71 70 41 55	**390**	100 37 27 27 23 23 22 17	**0.15**	24	38	4	–	–	–	–	–	–	–	–	–	–	**Diisooctyl phthalate**	27554-26-3	**IC04251**
149 57 43 71 41 55 279 167	**390**	100 48 28 24 20 18 17 16	**0.40**	24	38	4	–	–	–	–	–	–	–	–	–	–	**Diisooctyl phthalate**	27554-26-3	**IC04252**
149 43 57 41 279 55 29 150	**390**	100 39 29 26 18 18 16 10	**1.09**	24	38	4	–	–	–	–	–	–	–	–	–	–	**Dioctyl phthalate**	117-84-0	**IC04253**
149 57 167 71 43 70 41 55	**390**	100 36 34 23 23 21 20 15	**1.00**	24	38	4	–	–	–	–	–	–	–	–	–	–	**Dioctyl phthalate**	117-84-0	**IC04254**
149 57 43 41 71 70 167 28	**390**	100 47 30 30 28 28 26 23	**0.00**	24	38	4	–	–	–	–	–	–	–	–	–	–	**Dioctyl phthalate**	117-84-0	**WI01511**
288 320 321 304 289 358 303 55	**390**	100 99 25 24 24 19 19 18	**10.00**	24	38	4	–	–	–	–	–	–	–	–	–	–	**5,17-Dioxo-6α-pentyl-4-nor-3,5-secoandrostan-3-oic acid methyl ester**		**MS09039**
303 304 55 41 83 43 285 85	**390**	100 19 17 17 16 16 15 11	**5.00**	24	38	4	–	–	–	–	–	–	–	–	–	–	**5α,6α-Epoxy-17-oxo-6β-pentyl-4-nor-3,5-secoandrostan-3-oic acid methyl ester**		**MS09040**
225 147 55 43 41 79 93 67	**390**	100 55 54 52 43 27 26 26	**7.00**	24	38	4	–	–	–	–	–	–	–	–	–	–	**5β,6β-Epoxy-17-oxo-6α-pentyl-4-nor-3,5-secoandrostan-3-oic acid methyl ester**		**MS09041**
57 44 55 41 70 43 72 56	**390**	100 99 79 65 60 46 45 28	**0.00**	24	38	4	–	–	–	–	–	–	–	–	–	–	**Bis(2-ethylhexyl) phthalate**	117-81-7	**IC04255**
149 57 169 71 70 43 55 113	**390**	100 54 33 27 27 20 19 14	**0.14**	24	38	4	–	–	–	–	–	–	–	–	–	–	**Bis(2-ethylhexyl) phthalate**	117-81-7	**IC04256**
149 57 44 42 55 167 71 56	**390**	100 71 47 42 39 37 30 28	**0.00**	24	38	4	–	–	–	–	–	–	–	–	–	–	**Bis(2-ethylhexyl) phthalate**	117-81-7	**NI39887**
319 43 301 41 55 247 67 320	**390**	100 49 44 41 40 25 24 22	**0.00**	24	38	4	–	–	–	–	–	–	–	–	–	–	**6β-Hydroxy-6α-pentyl-3-oxa-A-homo-5α-androstane-4,17-dione**		**MS09042**
96 126 93 81 191 109 135 175	**390**	100 67 27 27 24 24 21 10	**0.00**	24	38	4	–	–	–	–	–	–	–	–	–	–	**Isolinaridiol diacetate**		**NI39888**
149 57 71 41 113 69 84 43	**390**	100 49 29 19 16 15 13 13	**0.00**	24	38	4	–	–	–	–	–	–	–	–	–	–	**Di-isooctyl phthalate**	27554-26-3	**NI39889**
149 167 41 43 55 57 150 56	**390**	100 42 35 29 25 18 13 13	**0.25**	24	38	4	–	–	–	–	–	–	–	–	–	–	**Di-sec-octyl phthalate**	131-15-7	**IC04257**
149 307 43 150 57 55 41 69	**390**	100 12 11 10 8 7 7 4	**0.00**	24	38	4	–	–	–	–	–	–	–	–	–	–	**Di-sec-octyl phthalate**	131-15-7	**NI39890**
149 167 41 43 55 150 57 56	**390**	100 45 17 15 14 13 12 8	**0.14**	24	38	4	–	–	–	–	–	–	–	–	–	–	**Bis(oct-3-yl) phthalate**		**IC04258**
87 88 67 103 99 91 79 375	**390**	100 14 12 8 8 8 8 7	**0.07**	24	38	4	–	–	–	–	–	–	–	–	–	–	**Pregnan-18-oic acid, 20,20-[1,2-ethanediylbis(oxy)]-, methyl ester, (5α)-**	56053-08-8	**NI39891**
105 43 57 211 41 55 389 29	**390**	100 84 76 63 57 54 42 37	**26.04**	24	38	4	–	–	–	–	–	–	–	–	–	–	**Tetradecanoic acid, 2-phenyl-1,3-dioxan-5-yl ester**	56599-86-1	**NI39892**
43 105 57 41 55 29 91 211	**390**	100 72 70 70 64 46 34 31	**9.40**	24	38	4	–	–	–	–	–	–	–	–	–	–	**Tetradecanoic acid, 2-phenyl-1,3-dioxan-5-yl ester**	56599-86-1	**NI39893**
105 43 57 41 55 29 211 71	**390**	100 76 75 54 49 36 30 28	**9.84**	24	38	4	–	–	–	–	–	–	–	–	–	–	**Tetradecanoic acid, 2-phenyl-1,3-dioxan-5-yl ester**	56599-86-1	**NI39894**
271 347 272 75 348 81 253 161	**390**	100 68 27 26 19 13 10 8	**1.66**	24	42	2	–	–	–	–	–	–	1	–	–	–	**Androstan-17-one, 3-[(propyldimethylsilyl)oxy]-, (3α,5α)-**		**NI39895**
300 215 299 285 301 108 75 375	**390**	100 44 36 36 29 23 22 21	**13.53**	24	42	2	–	–	–	–	–	–	1	–	–	–	**Pregnan-20-one, 3-[(trimethylsilyl)oxy]-, (3α,5α)-**		**NI39896**
300 215 285 301 299 375 75 390	**390**	100 28 27 23 23 17 14 13		24	42	2	–	–	–	–	–	–	1	–	–	–	**Pregnan-20-one, 3-[(trimethylsilyl)oxy]-, (3α,5α)-**		**LI08383**
300 215 285 390 301 216 230 107	**390**	100 88 51 43 22 22 18 17		24	42	2	–	–	–	–	–	–	1	–	–	–	**Pregnan-20-one, 3-[(trimethylsilyl)oxy]-, (3α,5α,17α)-**		**LI08384**
300 285 301 215 230 375 204 108	**390**	100 47 45 43 35 33 32 25	**4.42**	24	42	2	–	–	–	–	–	–	1	–	–	–	**Pregnan-20-one, 3-[(trimethylsilyl)oxy]-, (3α,5β)-**		**NI39897**
300 155 147 107 375 145 119 215	**390**	100 38 35 33 32 30 28 27	**3.00**	24	42	2	–	–	–	–	–	–	1	–	–	–	**Pregnan-20-one, 3-[(trimethylsilyl)oxy]-, (3α,5β)-**		**LI08385**
300 215 285 75 108 145 85 230	**390**	100 46 39 32 28 26 26 23	**1.00**	24	42	2	–	–	–	–	–	–	1	–	–	–	**Pregnan-20-one, 3-[(trimethylsilyl)oxy]-, (3α,5β,17β)-**		**LI08386**
375 390 300 376 215 75 145 285	**390**	100 36 36 34 30 28 23 20		24	42	2	–	–	–	–	–	–	1	–	–	–	**Pregnan-20-one, 3-[(trimethylsilyl)oxy]-, (3β,5α)-**		**LI08387**
390 215 285 391 375 192 107 145	**390**	100 80 36 34 32 31 30 26		24	42	2	–	–	–	–	–	–	1	–	–	–	**Pregnan-20-one, 3-[(trimethylsilyl)oxy]-, (3β,5α,17α)-**		**LI08388**
215 375 390 271 376 319 77 243	**390**	100 87 36 28 24 21 21 18		25	26	4	–	–	–	–	–	–	–	–	–	–	**Fulvinervin A**		**NI39898**
390 307 257 347 203 375 313 243	**390**	100 35 30 25 20 10 10 10		25	26	4	–	–	–	–	–	–	–	–	–	–	**Isorubranine**		**LI08389**

MASS TO CHARGE RATIOS								M.W.	INTENSITIES								Parent	C	H	O	N	S	F	Cl	Br	I	Si	P	B	X	COMPOUND NAME	CAS Reg No	No
307	203	390	347	305	375			**390**	100	60	40	32	25	20				25	26	4	–	–	–	–	–	–	–	–	–	–	Rubranine		LI08390
189	105	165	201	390	285	166	190	**390**	100	84	21	20	17	17	17	14		25	27	–	2	–	–	1	–	–	–	–	–	–	Piperazine, 1-[(4-chlorophenyl)benzyl]-4-[(3-methylphenyl)methyl]-	569-65-3	NI39899
55	81	67	44	41	79	93	95	**390**	100	67	67	61	60	51	44	43	**1.95**	25	42	3	–	–	–	–	–	–	–	–	–	–	Cholan-24-oic acid, 3-hydroxy-, methyl ester, (3α,5β)-	1249-75-8	NI39900
372	215	81	216	95	93	373	230	**390**	100	97	46	42	39	38	35	33	**2.36**	25	42	3	–	–	–	–	–	–	–	–	–	–	Cholan-24-oic acid, 3-hydroxy-, methyl ester, (3α,5β)-	1249-75-8	NI39901
55	372	215	93	107	91	82	66	**390**	100	71	60	59	57	50	50	47	**0.00**	25	42	3	–	–	–	–	–	–	–	–	–	–	Cholan-24-oic acid, 3-hydroxy-, methyl ester, (3α,5β)-	1249-75-8	LI08391
257	81	258	55	154	109	95	67	**390**	100	23	22	20	16	16	16	13	**0.00**	25	42	3	–	–	–	–	–	–	–	–	–	–	Cholan-24-oic acid, 12-hydroxy-, methyl ester, (5β,12α)-	1249-70-3	NI39902
257	55	41	44	67	81	29	28	**390**	100	65	50	44	40	37	34	25	**0.24**	25	42	3	–	–	–	–	–	–	–	–	–	–	Cholan-24-oic acid, 12-hydroxy-, methyl ester, (5β,12α)-	1249-70-3	NI39903
181	153	152	77	105	390	345	51	**390**	100	94	91	77	40	31	23	14		26	18	2	2	–	–	–	–	–	–	–	–	–	2,2′-Dibenzoylazobenzene	70593-68-9	NI39904
390	389	183	182	271	208	272	195	**390**	100	31	23	16	15	14	13	12		26	22	–	4	–	–	–	–	–	–	–	–	–	18,19,20,22-Tetrahydrotetrabenzo[c,f,i,m]-1,2,5,8-tetraazacyclotetradecine		NI39905
105	360	255	269	218	268	376	374	**390**	100	30	10	6	5	4	2	2	**1.00**	26	30	3	–	–	–	–	–	–	–	–	–	–	19-Norisopimara-4(18),8(14),15-trien-7-one, 17-(benzoyloxy)-		MS09043
144	168	130	258	63	149	143	154	**390**	100	89	87	71	60	51	44	38	**29.83**	26	34	1	2	–	–	–	–	–	–	–	–	–	Indole, 3-[2-[3-[3-(benzyloxy)propyl]-3-ethyl-1-pyrrolidinyl]ethyl]-	7700-57-4	NI39906
260	91	144	82	130	168	154	390	**390**	100	95	37	33	24	23	18	15		26	34	1	2	–	–	–	–	–	–	–	–	–	N-[β-(3-Indolyl)ethyl]-3-(2-benzyloxymethylbutyl)pyrrolidine		LI08392
91	390	375	391	285	300	243	199	**390**	100	81	29	23	20	10	9	8		27	34	2	–	–	–	–	–	–	–	–	–	–	2(1H)-Phenanthrenone, 3,4,4a,9,10,10a-hexahydro-1,1,4a-trimethyl-8-isopropyl-7-(phenylmethoxy)-, (4aS-trans)-	57289-44-8	NI39907
390	391	157	195	239	138	151	144	**390**	100	29	27	16	15	13	13	10		28	22	2	–	–	–	–	–	–	–	–	–	–	Anthracene, 9,10-bis(2-methoxyphenyl)-	43217-35-2	NI39908
91	92	390	299	65	208	152	300	**390**	100	5	4	4	3	2	2	1		28	22	2	–	–	–	–	–	–	–	–	–	–	Anthracene, 1,4-bis(phenylmethoxy)-		MS09044
105	77	298	44	106	104	209	51	**390**	100	50	27	16	15	13	10	8	**0.00**	28	22	2	–	–	–	–	–	–	–	–	–	–	1,4-Butanedione, 1,2,3,4-tetraphenyl-	10516-92-4	NI39909
195	180	196	152	327	328	326	324	**390**	100	23	18	13	4	3	3	3	**0.50**	28	22	2	–	–	–	–	–	–	–	–	–	–	9,9′-Bi-9H-fluorene, 9,9′-dimethoxy-	50616-99-4	NI39910
55	83	97	41	43	69	111	57	**390**	100	81	63	54	50	39	37	37	**0.07**	28	54	–	–	–	–	–	–	–	–	–	–	–	1-Cyclohexyl-2-(cyclohexylmethyl)pentadecane		AI02313
167	91	103	223	165	77	209	168	**390**	100	58	36	35	27	26	25	19	**8.00**	29	26	1	–	–	–	–	–	–	–	–	–	–	1,1,5,5-Tetraphenylpentan-3-one		LI08393
390	375	391	180	376	180.5	360	187.5	**390**	100	36	33	33	12	10	8	7		30	30	–	–	–	–	–	–	–	–	–	–	–	1,3,5-Tris(4-ethylphenyl)benzene	55255-72-6	AM00155
390	375	391	180	376	360	195	392	**390**	100	36	33	33	12	8	6	5		30	30	–	–	–	–	–	–	–	–	–	–	–	1,3,5-Tris(4-ethylphenyl)benzene	55255-72-6	WI01512
167	168	165	91	152	390	166	115	**390**	100	15	12	9	7	5	5	2		30	30	–	–	–	–	–	–	–	–	–	–	–	1,1,6,6-Tetraphenylhexane		AI02315
91	77	115	103	39	51	41	65	**390**	100	25	24	22	22	18	14	13	**6.30**	30	30	–	–	–	–	–	–	–	–	–	–	–	1,1,6,6-Tetraphenylhexane		AI02314
57	253	255	102	237	235	115	265	**391**	100	55	50	36	22	20	16	8	**0.00**	16	23	3	1	–	–	1	1	–	–	–	–	–	5-Bromo-7-ethylbenzofuran-2-yl(hydroxy-tert-butylaminomethyl)methanol hydrochloride		NI39911
236	238	391	393	155	210	208	248	**391**	100	98	50	49	33	8	8	3		17	14	3	1	1	–	–	1	–	–	–	–	–	5H-1-Benzazepin-5-one, 4-bromo-1,2-dihydro-1-(4-tolylsulphonyl)-	20426-62-4	NI39912
173	133	351	104	353	391	105	77	**391**	100	99	37	31	24	22	21	21		17	15	2	5	–	–	2	–	–	–	–	–	–	N-Ethyl-N-(2-cyanoethyl)-4-(3,5-dichloro-4-nitrophenylazo)aniline		IC04259
43	116	158	98	99	87	74	45	**391**	100	73	38	37	26	25	25	25	**0.00**	17	29	9	1	–	–	–	–	–	–	–	–	–	D-Glucitol, 2-(acetylmethylamino)-2-deoxy-3,6-di-O-methyl-, 1,4,5-triacetate	52959-68-9	NI39913
105	146	145	331	391	318	332	106	**391**	100	26	18	16	13	13	11	9		19	21	8	1	–	–	–	–	–	–	–	–	–	5H-Pyrano[3,2-d]oxazole-6,7-diol, 5-[(acetyloxy)methyl]-3a,6,7,7a-tetrahydro-2-phenyl-, diacetate (3aα,5α,6β,7α,7aα)-	10380-87-7	NI39914
391	361	392	328	327	344	346	300	**391**	100	24	23	17	14	12	7	7		20	13	6	3	–	–	–	–	–	–	–	–	–	1-(4-Aminoanilino)-4,5-dihydroxy-8-nitroanthraquinone		IC04260
391	209	107	106	77	208	79	392	**391**	100	77	45	42	39	35	27	23		20	17	4	5	–	–	–	–	–	–	–	–	–	anti-2,4-Dinitrophenylhydrazono-5-methyl-2-aminobenzophenone		MS09045
360	391	362	376	361	393	69	252	**391**	100	53	39	26	24	20	17	14		20	22	5	1	–	–	1	–	–	–	–	–	–	Spiro[benzofuran-2(3H),1′-[2]cyclohexene]-3,4′-dione, 7-chloro-2′,6-dimethoxy-6′-methyl-4-(1-pyrrolidinyl)-	26881-67-4	NI39915
391	70	393	376	41	392	55	69	**391**	100	57	36	36	32	25	21	20		20	22	5	1	–	–	1	–	–	–	–	–	–	Spiro[benzofuran-2(3H),1′-[2]cyclohexene]-3,4′-dione, 7-chloro-4,6-dimethoxy-6′-methyl-2′-(1-pyrrolidinyl)-	26881-66-3	NI39916
137	109	363	68	391	108	138	70	**391**	100	57	19	16	15	14	12	8		20	22	5	1	–	–	1	–	–	–	–	–	–	Spiro[benzofuran-2(3H),1′-[3]cyclohexene]-2′,3-dione, 7-chloro-4,6-dimethoxy-6′-methyl-4′-(1-pyrrolidinyl)-	26942-69-8	NI39917
391	292	293	376	226	393	294	151	**391**	100	66	57	52	43	36	31	31		20	22	5	1	–	–	1	–	–	–	–	–	–	Spiro[benzofuran-2(3H),1′-[3]cyclohexene]-2′,3-dione, 7-chloro-4′,6-dimethoxy-6′-methyl-4-(1-pyrrolidinyl)-	26891-71-4	NI39918
101	232	43	210	125	376	109	102	**391**	100	14	13	11	10	9	9	6	**0.80**	20	25	5	1	1	–	–	–	–	–	–	–	–	3-C-Cyanomethyl-3-desoxy-1,2,5,6-di-O-isopropylidene-3-S-phenyl-3-thio-α-D-galactofuranose		NI39919
101	109	110	43	125	102	376	59	**391**	100	28	17	14	7	7	6	5	**1.20**	20	25	5	1	1	–	–	–	–	–	–	–	–	3-C-Cyanomethyl-3-desoxy-1,2,5,6-di-O-isopropylidene-3-S-phenyl-3-thio-α-D-glucofuranose		NI39920
258	259	257	391	227	242	243	228	**391**	100	54	44	40	36	33	13	13		20	25	7	1	–	–	–	–	–	–	–	–	–	Oxostephasunoline	91897-38-0	NI39921
86	43	128	227	28	42	218	91	**391**	100	80	54	46	44	41	40	36	**15.00**	20	29	5	3	–	–	–	–	–	–	–	–	–	L-Alanine, N-[N-(N-acetyl-L-leucyl)-L-alanyl]-3-phenyl-	20310-12-7	LI08394
77	391	105	242	91	92	270	347	**391**	100	57	43	34	28	25	19	16		21	14	3	3	–	–	1	–	–	–	–	–	–	1-Carboxy-7-chloro-2-oxo-3,5-diphenyl-2,3-dihydro-1H-benzo[e]-1,3,4-triazepine		LI08395
363	319	304	263	248	234	348	262	**391**	100	43	38	33	26	26	24	24	**22.00**	21	17	5	3	–	–	–	–	–	–	–	–	–	Camptothecin, 12-formamido-		NI39922
152	363	150	319	304	251	153	263	**391**	100	27	16	12	10	10	10	9	**6.00**	21	17	5	3	–	–	–	–	–	–	–	–	–	Camptothecin, 12-formamido-		NI39923

MASS TO CHARGE RATIOS								M.W.	INTENSITIES								Parent	C	H	O	N	S	F	Cl	Br	I	Si	P	B	X	COMPOUND NAME	CAS Reg No	No
55	256	348	81	109	105	391	322	**391**	100	23	21	20	19	19	16	15		21	20	1	1	1	3	–	–	–	–	–	–	–	Spiro[2,5-dihydro-1,3-oxazole-2,1′-cyclohexane], 5-phenyl-4-(phenylthio)-5-(trifluoromethyl)-		DD01537
245	213	230	244	246	196	243	229	**391**	100	27	23	21	18	13	8	5	**1.32**	21	29	6	1	–	–	–	–	–	–	–	–	–	Epistephamiersine, dihydro-	33111-02-3	NI39924
160	145	70	391	229	228	214		**391**	100	31	27	12	10	7	7			21	29	6	1	–	–	–	–	–	–	–	–	–	Ipalbine		LI08396
245	213	244	230	246	243	360	196	**391**	100	30	27	23	16	11	9	8	**3.20**	21	29	6	1	–	–	–	–	–	–	–	–	–	Stephamiersine, dihydro-	56423-18-8	NI39925
91	282	210	254	92	15	110	65	**391**	100	20	20	16	10	10	7	7	**0.00**	23	21	3	1	1	–	–	–	–	–	–	–	–	(Benzyloxycarbonyl)phenylalanine phenylthio ester		LI08397
349	185	316	391	262	290	288		**391**	100	28	27	14	13	11	11			23	25	3	3	–	–	–	–	–	–	–	–	–	Acetylanisotine		LI08398
391	69	43	57	322	44	41	392	**391**	100	50	47	43	41	41	40	30		23	25	3	3	–	–	–	–	–	–	–	–	–	Neoechinuline		MS09046
270	360	75	73	271	87	79	81	**391**	100	79	61	55	25	25	25	23	**1.50**	23	41	2	1	–	–	–	–	–	1	–	–	–	5α-Androstan-17-one, 3α-(trimethylsiloxy)-, O-methyloxime	13443-47-5	NI39926
360	270	361	271	376	362	87	346	**391**	100	59	32	12	11	8	7	6	**3.41**	23	41	2	1	–	–	–	–	–	1	–	–	–	5α-Androstan-17-one, 3β-(trimethylsiloxy)-, O-methyloxime	32206-63-6	NI39927
270	360	75	73	87	271	81	79	**391**	100	61	53	50	28	23	23	21	**1.12**	23	41	2	1	–	–	–	–	–	1	–	–	–	5β-Androstan-17-one, 3α-(trimethylsiloxy)-, O-methyloxime	32206-62-5	NI39928
270	360	75	73	271	361	87	81	**391**	100	58	35	31	22	19	19	15	**1.52**	23	41	2	1	–	–	–	–	–	1	–	–	–	5β-Androstan-17-one, 3α-(trimethylsiloxy)-, O-methyloxime	32206-62-5	NI39929
129	391	286	126	130	301	376	270	**391**	100	48	30	25	24	23	19	17		23	41	2	1	–	–	–	–	–	1	–	–	–	Androstan-3-one, 17-[(trimethylsilyl)oxy]-, O-methyloxime, (5β,17β)-	69833-83-6	NI39930
376	75	215	391	73	377	155	107	**391**	100	88	80	68	50	29	27	22		23	41	2	1	–	–	–	–	–	1	–	–	–	Androstan-16-one, 3-[(trimethylsilyl)oxy]-, O-methyloxime, (3β,5α)-	57305-08-5	NI39931
324	391	310	70	340	294	28	379	**391**	100	39	32	20	16	13	10	8		24	25	4	1	–	–	–	–	–	–	–	–	–	Dibenzo[f,h]pyrrolo[1,2-b]isoquinoline, 9,11,12,14-tetrahydro-3,4,6,7-tetramethoxy-	95066-38-9	NI39932
376	391	348	119	322	392	377	167	**391**	100	95	67	35	34	31	30	25		24	25	4	1	–	–	–	–	–	–	–	–	–	7H-Pyrano[2,3-c]acridin-7-one, 3,12-dihydro-6,11-dihydroxy-3,3,12-trimethyl-5-(3-methyl-2-butenyl)-	49620-08-8	MS09047
376	377	346	348	320	274	378	220	**391**	100	53	16	11	10	10	7	7	**2.61**	24	25	4	1	–	–	–	–	–	–	–	–	–	3,5-Pyridinedicarboxylic acid, 1,4-dihydro-4-methyl-2,6-diphenyl-, diethyl ester	20970-65-4	NI39934
376	377	346	348	320	274	378	220	**391**	100	26	7	5	5	5	3	3	**1.30**	24	25	4	1	–	–	–	–	–	–	–	–	–	3,5-Pyridinedicarboxylic acid, 1,4-dihydro-4-methyl-2,6-diphenyl-, diethyl ester	20970-65-4	NI39933
376	377	346	348	320	274	105	29	**391**	100	26	11	7	7	7	7	7	**1.39**	24	25	4	1	–	–	–	–	–	–	–	–	–	3,5-Pyridinedicarboxylic acid, 1,4-dihydro-4-methyl-2,6-diphenyl-, diethyl ester	20970-65-4	NI39935
105	391	77	374	239	344	313	314	**391**	100	80	47	27	24	20	16	13		26	17	3	1	–	–	–	–	–	–	–	–	–	2-Nitro-9,10-epoxy-9,10-diphenyl-9,10-dihydroanthracene		LI08399
391	392	245	217	175	92	218	393	**391**	100	41	18	18	16	14	11	10		26	33	2	1	–	–	–	–	–	–	–	–	–	1H-Dicyclooct[e,g]isoindole-1,3(2H)-dione, 3a,3b,4,5,6,7,8,9,10,11,12,13,14,15,15a,15b-hexadecahydro-2-phenyl-, (3α,3bα,15aα,15bα)-	51624-51-2	NI39936
300	42	44	91	105	56	43	41	**391**	100	66	55	52	45	31	28	28	**0.00**	26	33	2	1	–	–	–	–	–	–	–	–	–	Morphinan-14-ol, 3-methoxy-6-methyl-17-(2-phenylethyl)-, (6α)-	55268-49-0	NI39937
109	135	69	107	119	95	121	105	**391**	100	55	51	43	41	36	32	32	**3.59**	26	33	2	1	–	–	–	–	–	–	–	–	–	Retinamide, N-(4-hydroxyphenyl)-	65646-68-6	NI39938
143	204	123	115	135	189	205	187	**391**	100	91	58	53	33	26	22	19	**0.40**	26	33	2	1	–	–	–	–	–	–	–	–	–	Rosifoliol α-naphthylurethane		MS09048
113	126	55	70	43	392	69	41	**391**	100	46	38	34	28	22	20	19	**15.04**	26	49	1	1	–	–	–	–	–	–	–	–	–	Pyrrolidine, 1-(1-oxo-13-docosenyl)-	56600-11-4	NI39939
360	391	209	282	361	152	225	105	**391**	100	62	44	31	29	27	22	22		27	21	2	1	–	–	–	–	–	–	–	–	–	2-(1,1′-Biphenyl-2-yl)-3-methoxy-3-phenylphthalimidine		NI39940
280	391	279	203	281	202	392	278	**391**	100	70	47	30	27	27	24	22		27	21	2	1	–	–	–	–	–	–	–	–	–	4,9-Methano-1H-benz[f]isoindole-1,3(2H)-dione, 10-(diphenylmethylene)-3a,4,9,9a-tetrahydro-2-methyl-, (3aα,4α,9α,9aα)-	42589-27-5	NI39941
391	392	330	332	331	358	254	315	**391**	100	31	25	9	9	6	6	5		27	21	2	1	–	–	–	–	–	–	–	–	–	1-(2-methoxycarbonylphenyl)-9-methyl-4-phenylcarbazole		LI08400
121	93	391	106	181	77	283	359	**391**	100	30	28	19	17	12	11	9		27	37	1	1	–	–	–	–	–	–	–	–	–	Aniline, N-[[2-(2-methoxy-2-phenylethyl)-3,3,4,4,5,5-hexamethylcyclopent 1-en-1-yl]methyl]-	123187-17-7	DD01538
267	392	357	161	302	247	133	322	**392**	100	60	58	49	38	24	12	9		8	14	–	–	–	–	2	–	–	–	–	–	2	Palladium, bis(chloro-π-2-butenyl)di-		LI08401
161	392	302	133	357	247	337	106	**392**	100	39	27	13	11	9	5	5		8	14	–	–	–	–	2	–	–	–	–	–	2	Palladium, bis(chloro-π-methallyl)di-		LI08402
393	292	181.5	161	146	131	196.5		**392**	100	48	44	44	36	32	12		**0.00**	8	20	–	4	–	–	2	–	–	–	–	–	2	1,3,6,8-Tetramethyl-2,7-dichloro-1,3,6,8-tetra-aza-2,7-di-arsacyclodecane		LI08403
261	73	257	273	263	45	43	74	**392**	100	99	73	63	50	50	48	43	**0.00**	8	24	–	–	–	–	–	–	–	4	–	–	2	1,4-Diselenacyclohexasilane, 2,2,3,3,5,5,6,6-octamethyl-	85263-61-2	NI39942
59	165	240	135	197	167	121	91	**392**	100	60	55	46	44	29	21	19	**0.08**	9	28	9	–	–	–	–	–	–	4	–	–	–	Trisilane, 1,1,1,3,3,3-hexamethoxy-2-(trimethoxysilyl)-	53518-40-4	NI39943
197	177	195	127	113	69	51	77	**392**	100	85	49	35	31	26	26	23	**0.05**	10	6	–	–	–	14	–	–	–	–	–	–	–	3-Decene, 1,1,1,2,2,5,5,6,6,9,9,10,10,10-tetradecafluoro-	35208-08-3	NI39944
97	359	242	303	357	28	331	240	**392**	100	74	65	59	58	56	53	43	**0.00**	10	12	3	–	1	–	2	1	–	–	1	–	–	Phosphorothioic acid, O-(4-bromo-2,5-dichlorophenyl) O,O-diethyl ester	4824-78-6	NI39945
97	65	109	125	359	302	47	81	**392**	100	25	24	22	12	12	12	10	**0.00**	10	12	3	–	1	–	2	1	–	–	1	–	–	Phosphorothioic acid, O-(4-bromo-2,5-dichlorophenyl) O,O-diethyl ester	4824-78-6	NI39946
97	301	357	63	125	65	303	109	**392**	100	44	27	26	25	22	18	13	**0.00**	11	15	5	–	2	–	2	–	–	–	1	–	–	Chlorthiophos sulphone, 2,5-isomer		NI39947
394	396	324	162	163	398	161	326	**392**	100	97	82	76	52	48	48	47	**40.20**	12	3	–	–	–	–	7	–	–	–	–	–	–	1,1′-Biphenyl, 2,2′,3,3′,4,4′,5-heptachloro-	35065-30-6	NI39948
394	396	162	324	398	163	326	322	**392**	100	88	69	68	56	56	48	48	**46.80**	12	3	–	–	–	–	7	–	–	–	–	–	–	1,1′-Biphenyl, 2,2′,3,3′,4,4′,6-heptachloro-	52663-71-5	NI39949
394	324	162	396	126	36	161	108	**392**	100	99	97	86	78	66	66	64	**41.94**	12	3	–	–	–	–	7	–	–	–	–	–	–	1,1′-Biphenyl, 2,2′,3,3′,4,4′,6-heptachloro-	52663-71-5	NI39950
394	396	162	324	126	326	163	161	**392**	100	99	99	96	88	60	59	58	**46.80**	12	3	–	–	–	–	7	–	–	–	–	–	–	1,1′-Biphenyl, 2,2′,3,3′,4,5,5′-heptachloro-	52663-74-8	NI39951
394	396	162	324	163	398	326	161	**392**	100	98	83	82	54	53	50	50	**45.10**	12	3	–	–	–	–	7	–	–	–	–	–	–	1,1′-Biphenyl, 2,2′,3,3′,4,5,6′-heptachloro-	38411-25-5	NI39952
394	396	162	324	161	322	163	398	**392**	100	94	90	78	59	57	57	55	**46.50**	12	3	–	–	–	–	7	–	–	–	–	–	–	1,1′-Biphenyl, 2,2′,3,3′,4,5′,6′-heptachloro-	52663-70-4	NI39953
162	394	396	163	324	161	126	127	**392**	100	70	67	66	65	64	59	48	**31.21**	12	3	–	–	–	–	7	–	–	–	–	–	–	1,1′-Biphenyl, 2,2′,3,3′,4,5′,6′-heptachloro-	52663-70-4	NI39954

MASS TO CHARGE RATIOS	M.W.	INTENSITIES	Parent	C	H	O	N	S	F	Cl	Br	I	Si	P	B	X	COMPOUND NAME	CAS Reg No	No
394 396 162 324 398 392 161 163	**392**	100 94 66 59 51 46 41 40		12	3	–	–	–	–	7	–	–	–	–	–	–	**1,1′-Biphenyl, 2,2′,3,3′,4,6,6′-heptachloro-**	52663-65-7	**NI39955**
396 394 398 324 392 359 326 361	**392**	100 96 48 44 39 37 34 32		12	3	–	–	–	–	7	–	–	–	–	–	–	**1,1′-Biphenyl, 2,2′,3,3′,5,5′,6-heptachloro-**	52663-67-9	**NI39956**
393 396 162 324 398 161 392 163	**392**	100 97 67 66 52 46 43 43		12	3	–	–	–	–	7	–	–	–	–	–	–	**1,1′-Biphenyl, 2,2′,3,3′,5,6,6′-heptachloro-**	52663-64-6	**NI39957**
394 396 324 162 326 322 398 163	**392**	100 96 84 81 53 51 50 50	**47.50**	12	3	–	–	–	–	7	–	–	–	–	–	–	**1,1′-Biphenyl, 2,2′,3,4,4′,5,5′-heptachloro-**	35065-29-3	**NI39958**
394 396 324 162 398 326 392 322	**392**	100 97 69 64 49 45 44 43		12	3	–	–	–	–	7	–	–	–	–	–	–	**1,1′-Biphenyl, 2,2′,3,4,4′,5,6-heptachloro-**	74472-47-2	**NI39959**
324 394 396 162 326 322 398 392	**392**	100 96 90 68 63 63 48 44		12	3	–	–	–	–	7	–	–	–	–	–	–	**1,1′-Biphenyl, 2,2′,3,4,4′,5,6′-heptachloro-**	60145-23-5	**NI39960**
394 396 162 324 163 322 161 398	**392**	100 94 79 74 52 48 47 46	**42.40**	12	3	–	–	–	–	7	–	–	–	–	–	–	**1,1′-Biphenyl, 2,2′,3,4,4′,5′,6-heptachloro-**	52663-69-1	**NI39961**
394 396 162 324 163 322 326 161	**392**	100 91 86 83 59 57 56 55	**42.68**	12	3	–	–	–	–	7	–	–	–	–	–	–	**1,1′-Biphenyl, 2,2′,3,4,4′,5′,6-heptachloro-**	52663-69-1	**NI39962**
394 396 324 162 36 326 322 163	**392**	100 96 87 73 58 56 56 50	**45.58**	12	3	–	–	–	–	7	–	–	–	–	–	–	**1,1′-Biphenyl, 2,2′,3,4,5,5′,6-heptachloro-**	52712-05-7	**NI39963**
394 396 398 392 126 127 324 359	**392**	100 99 54 51 33 30 29 27		12	3	–	–	–	–	7	–	–	–	–	–	–	**1,1′-Biphenyl, 2,2′,3,4,5,5′,6-heptachloro-**	52712-05-7	**NI39964**
394 396 324 398 392 326 322 162	**392**	100 93 70 55 54 52 48 37		12	3	–	–	–	–	7	–	–	–	–	–	–	**1,1′-Biphenyl, 2,2′,3,4,5,5′,6-heptachloro-**	52712-05-7	**NI39965**
394 396 324 162 322 326 398 252	**392**	100 93 84 65 55 53 49 48	**46.00**	12	3	–	–	–	–	7	–	–	–	–	–	–	**1,1′-Biphenyl, 2,2′,3,4,5,6,6′-heptachloro-**	74472-49-4	**NI39966**
394 396 162 324 398 163 161 326	**392**	100 95 77 73 52 47 46 45	**42.60**	12	3	–	–	–	–	7	–	–	–	–	–	–	**1,1′-Biphenyl, 2,2′,3,4′,5,5′,6-heptachloro-**	52663-68-0	**NI39967**
394 396 324 162 398 392 163 322	**392**	100 92 55 54 50 43 36 34		12	3	–	–	–	–	7	–	–	–	–	–	–	**1,1′-Biphenyl, 2,3,3′,4,4′,5,5′-heptachloro-**	39635-31-9	**NI39968**
394 396 162 324 398 392 161 126	**392**	100 93 67 62 51 47 43 41		12	3	–	–	–	–	7	–	–	–	–	–	–	**1,1′-Biphenyl, 2,3,3′,4,4′,5,6-heptachloro-**	41411-64-7	**NI39969**
394 396 324 162 326 322 398 392	**392**	100 95 90 65 59 57 51 46		12	3	–	–	–	–	7	–	–	–	–	–	–	**1,1′-Biphenyl, 2,3,3′,4,5,5′,6-heptachloro-**	74472-51-8	**NI39970**
394 396 398 162 392 324 161 163	**392**	100 92 54 49 43 39 31 28		12	3	–	–	–	–	7	–	–	–	–	–	–	**1,1′-Biphenyl, 2,3,3′,4′,5,5′,6-heptachloro-**	69782-91-8	**NI39971**
394 396 324 326 398 392 162 322	**392**	100 91 82 74 51 47 44 41		12	3	–	–	–	–	7	–	–	–	–	–	–	**1,1′-Biphenyl, heptachloro-**	28655-71-2	**NI39972**
394 396 324 326 398 392 162 254	**392**	100 97 82 66 53 47 43 40		12	3	–	–	–	–	7	–	–	–	–	–	–	**1,1′-Biphenyl, heptachloro-**	28655-71-2	**NI39973**
359 361 289 394 396 357 324 287	**392**	100 82 69 61 58 55 53 53	**27.00**	12	3	–	–	–	–	7	–	–	–	–	–	–	**1,1′-Biphenyl, heptachloro-**	28655-71-2	**NI39974**
392 394 113 100 377 379 351 349	**392**	100 99 41 36 31 30 23 23		12	10	2	–	–	–	–	1	1	–	–	–	–	**Naphthalene, 1-bromo-5-iodo-2,6-dimethoxy-**	25315-13-3	**LI08404**
392 394 113 100 377 379 74 349	**392**	100 98 40 33 32 31 26 23		12	10	2	–	–	–	–	1	1	–	–	–	–	**Naphthalene, 1-bromo-5-iodo-2,6-dimethoxy-**	25315-13-3	**NI39975**
224 252 184 112 168 336 280 126	**392**	100 37 30 27 22 16 16 15	**11.82**	12	12	6	2	–	–	–	–	–	–	–	–	2	**Iron, hexacarbonylbis(μ-(2-propaniminato))di-**		**MS09049**
220 333 347 203 245 161 143 101	**392**	100 93 62 18 16 7 7 6	**0.00**	12	20	12	–	–	–	–	–	–	–	–	–	4	**Beryllium, pentakis(acetato)ethoxyoxotetra-**	12562-30-0	**NI39976**
80 64 48 392 209 183 98 375	**392**	100 48 41 12 12 11 8 5		13	8	4	6	–	–	–	–	–	–	–	–	1	**2,1,3-Benzoselenadiazole-4-carboxaldehyde 2,4-dinitrophenylhydrazone**		**DD01539**
64 392 90 344 103 298 183 130	**392**	100 40 39 30 28 26 17 17		13	8	4	6	–	–	–	–	–	–	–	–	1	**2,1,3-Benzoselenadiazole-5-carboxaldehyde 2,4-dinitrophenylhydrazone**		**DD01540**
149 99 91 105 77 259 41 39	**392**	100 81 77 45 43 42 30 28	**0.00**	13	14	4	–	–	–	–	2	–	–	–	–	–	**α-D-Galactopyranose, 2,3-dibromo-2,3-dideoxy-4,6-O-benzylene-**	56701-11-2	**NI39977**
277 81 392 377 273 129 77 125	**392**	100 99 80 40 40 40 31 29		14	6	–	–	–	10	–	–	–	1	–	–	–	**Bis(pentafluorophenyl)dimethylsilane**	801-79-6	**NI39978**
56 166 139 153 138 195 69 168	**392**	100 38 34 27 23 19 19 14	**13.00**	14	12	1	2	–	7	1	–	–	–	–	–	–	**Piperazine, 1-(2-chlorophenyl)-4-(heptafluorobutyryl)-**		**NI39979**
56 392 195 139 166 394 111 138	**392**	100 76 60 50 44 29 27 25		14	12	1	2	–	7	1	–	–	–	–	–	–	**Piperazine, 1-(4-chlorophenyl)-4-(heptafluorobutyryl)-**		**NI39980**
121 56 94 186 47 45 39 66	**392**	100 56 41 27 26 22 22 21	**0.31**	14	16	2	–	2	–	–	–	–	–	–	–	2	**Iron, di-π-cyclopentadienyl-di-μ-(π-methylthio)dicarbonyldi-**		**MS09050**
191 132 219 148 233 249 204 176	**392**	100 95 83 80 55 52 44 42	**9.00**	14	16	9	–	2	–	–	–	–	–	–	–	–	**1,3-Dithiane-4,6-dicarboxylic acid, 5-(4-hydroxyphenyl)-, 1,1,3,3-tetraoxide, dimethyl ester**		**NI39981**
147 103 77 259 148 135 257 105	**392**	100 70 64 26 26 24 22 18	**4.45**	15	8	2	–	1	–	4	–	–	–	–	–	–	**2H-Thiacyclobuta[b][1,4]benzodioxin, 4,5,6,7-tetrachloro-2a,8a-dihydro-2a-phenyl-**		**MS09051**
394 366 393 367 392 365 364 396	**392**	100 98 57 53 49 48 48 42		15	10	1	2	–	–	–	2	–	–	–	–	–	**1H-1,4-Benzodiazepin-2-one, 7-bromo-5-(3-bromophenyl)-2,3-dihydro-**	65247-11-2	**NI39982**
219 218 106 175 132 133 176 120	**392**	100 43 18 15 13 12 9 9	**0.00**	15	12	5	4	2	–	–	–	–	–	–	–	–	**Salicylidene-5,5′-bis(2-thiobarbituric acid)**		**NI39983**
392 232 234 236 103 239 208 231	**392**	100 63 62 45 29 24 24 19		15	20	2	–	–	3	–	–	–	–	–	–	1	**Rhodium, (η^4-1,2-divinylcyclohexane)(1,1,1-trifluoro-2,4-pentanedionato-O,O′)-**	74811-04-4	**NI39984**
73 259 231 69 147 287 377 75	**392**	100 78 73 65 44 32 30 28	**0.53**	15	32	6	–	–	–	–	–	–	3	–	–	–	**1,1,2-Ethanetricarboxylic acid, 2-methyl-, tris(trimethylsilyl) ester**	61713-71-1	**NI39985**
73 230 147 75 217 287 133 129	**392**	100 73 25 19 11 9 6 6	**0.50**	15	32	6	–	–	–	–	–	–	3	–	–	–	**α-D-Glucofuranurono-6,3-lactone, tri-O-(trimethylsilyl)-**	52842-26-9	**MS09052**
230 73 231 147 217 232 133 75	**392**	100 65 38 30 19 16 13 12	**0.30**	15	32	6	–	–	–	–	–	–	3	–	–	–	**α-D-Glucofuranurono-6,3-lactone, tri-O-(trimethylsilyl)-**	52842-26-9	**NI39986**
73 230 147 217 75 377 245 103	**392**	100 86 17 14 13 9 8 6	**0.50**	15	32	6	–	–	–	–	–	–	3	–	–	–	**β-D-Glucofuranurono-6,3-lactone, tri-O-(trimethylsilyl)-**		**MS09053**
73 230 147 75 217 287 133 259	**392**	100 95 27 21 14 9 7 6	**0.00**	15	32	6	–	–	–	–	–	–	3	–	–	–	**D-Glucuronolactone, 2,4,5-tri-O-(trimethylsilyl)-**	55556-97-3	**LI08405**
230 73 231 147 75 232 287 245	**392**	100 57 23 23 18 9 6 5	**0.22**	15	32	6	–	–	–	–	–	–	3	–	–	–	**D-Glucuronolactone, 2,4,5-tri-O-(trimethylsilyl)-**	55556-97-3	**NI39987**
73 230 147 75 231 287 217 74	**392**	100 67 28 25 14 11 10 8	**0.00**	15	32	6	–	–	–	–	–	–	3	–	–	–	**α-Glucuronolactone, tris-O-(trimethylsilyl)-**		**NI39988**
73 230 147 231 75 217 74 245	**392**	100 73 20 18 18 15 9 8	**0.00**	15	32	6	–	–	–	–	–	–	3	–	–	–	**β-Glucuronolactone, tris-O-(trimethylsilyl)-**		**NI39989**
73 230 231 147 75 45 217 232	**392**	100 89 22 17 16 13 10 9	**0.00**	15	32	6	–	–	–	–	–	–	3	–	–	–	**Glucuronolactone, tris-O-(trimethylsilyl)-**		**NI39990**
73 230 147 75 217 45 231 74	**392**	100 41 19 15 13 13 10 9	**0.66**	15	32	6	–	–	–	–	–	–	3	–	–	–	**Glucuronolactone, tris-O-(trimethylsilyl)-**		**NI39991**
230 73 217 75 147 129 103 158	**392**	100 83 15 15 14 11 7 5	**0.00**	15	32	6	–	–	–	–	–	–	3	–	–	–	**Gluronolactone, tris(trimethylsilyl)-**		**LI08406**
230 73 217 147 75 245 129 103	**392**	100 47 16 7 7 5 4 3	**0.50**	15	32	6	–	–	–	–	–	–	3	–	–	–	**α,β-L-Idofuranurono-6,3-lactone, tri-O-(trimethylsilyl)-**		**MS09054**
230 73 217 147 75 245 129 133	**392**	100 71 15 11 10 6 6 5	**0.00**	15	32	6	–	–	–	–	–	–	3	–	–	–	**Iduronolactone, tris(trimethylsilyl)-**		**LI08407**
73 230 147 75 217 93 129 245	**392**	100 91 75 69 21 11 10 9	**1.00**	15	32	6	–	–	–	–	–	–	3	–	–	–	**α-D-Mannofuranurono-6,3-lactone, tri-O-(trimethylsilyl)-**		**MS09055**
73 75 147 230 93 217 77 129	**392**	100 66 57 55 10 9 8 6	**0.50**	15	32	6	–	–	–	–	–	–	3	–	–	–	**β-D-Mannofuranurono-6,3-lactone, tri-O-(trimethylsilyl)-**		**MS09056**
73 230 147 75 217 259 129 189	**392**	100 90 26 17 16 11 10 9	**0.00**	15	32	6	–	–	–	–	–	–	3	–	–	–	**Mannuronolactone, tris(trimethylsilyl)-**		**LI08408**
73 69 231 259 147 75 287 377	**392**	100 67 61 59 32 25 24 20	**0.29**	15	32	6	–	–	–	–	–	–	3	–	–	–	**1,1,3-Propanetricarboxylic acid, tris(trimethylsilyl) ester**		**NI39992**

MASS TO CHARGE RATIOS								M.W.	INTENSITIES								Parent	C	H	O	N	S	F	Cl	Br	I	Si	P	B	X	COMPOUND NAME	CAS Reg No	No
73	147	75	377	185	149	217	55	392	100	52	30	26	18	15	14	12	0.00	15	32	6	–	–	–	–	–	–	3	–	–	–	1,2,3-Propanetricarboxylic acid, tris(trimethylsilyl) ester		NI39993
73	147	377	75	185	217	149	45	392	100	99	63	50	33	29	23	20	0.00	15	32	6	–	–	–	–	–	–	3	–	–	–	1,2,3-Propanetricarboxylic acid, tris(trimethylsilyl) ester		MS09057
174	73	175	102	86	291	100	59	392	100	58	18	12	12	10	7	7	2.00	15	40	2	2	–	–	–	–	–	4	–	–	–	Propanoic acid, 2,3-diamino-, N,N,N′-tris(trimethylsilyl)-, trimethylsilyl ester		MS09058
232	117	90	89	231	233	312	116	392	100	71	25	25	24	18	17	16	0.75	16	12	–	2	–	–	–	–	–	–	–	–	2	Indole, 3,3′-diselenodi-	21914-06-7	NI39994
196	194	232	392	390	117	388	116	392	100	53	44	43	40	37	24	24		16	12	–	2	–	–	–	–	–	–	–	–	2	Indole, 5,5′-diselenodi-	22129-92-6	NI39995
229	73	75	77	303	275	147	115	392	100	70	49	43	26	26	25	19	0.00	16	32	7	–	–	–	–	–	–	2	–	–	–	Citric acid, O-(trimethylsilyl)-, diethyl trimethylsilyl ester		MS09059
73	75	147	247	115	77	129	74	392	100	57	30	25	16	15	8	8	0.00	16	36	5	–	–	–	–	–	–	3	–	–	–	Pentanedioic acid, 2,3-dimethyl-3-[(trimethylsilyl)oxy]-, bis(trimethylsilyl) ester	87764-50-9	NI39996
73	147	273	129	261	74	245	133	392	100	29	28	24	20	10	8	8	0.00	16	36	5	–	–	–	–	–	–	3	–	–	–	Pentanedioic acid, 3-ethyl-3-[(trimethylsilyl)oxy]-, bis(trimethylsilyl) ester	87764-49-6	NI39997
112	82	243	392	345				392	100	3	2	1	1					17	32	6	2	1	–	–	–	–	–	–	–	–	D-erythro-α-D-Galacto-octopyranoside, methyl 6,8-dideoxy-6-[[(4-ethyl-1 methyl-2-pyrrolidinyl)carbonyl]amino]-1-thio-, (2S-trans)-	2520-24-3	NI39998
112	68	243	392	345				392	100	4	2	1	1					17	32	6	2	1	–	–	–	–	–	–	–	–	D-erythro-α-D-Galacto-octopyranoside, methyl 6,8-dideoxy-6-[[(4-propyl-2-pyrrolidinyl)carbonyl]amino]-1-thio-, (2S-trans)-	2256-16-8	NI39999
333	301	392	269	215	334	361	273	392	100	39	27	22	20	19	18	13		18	20	8	2	–	–	–	–	–	–	–	–	–	Tetramethyl-3-methyl-6,7-dihydro-pyridazino[2,3-a]azepine-6,7,8,9-tetracarboxylate		LI08409
306	275	307	203	301	217	392	243	392	100	41	18	14	10	8	7	7		18	20	8	2	–	–	–	–	–	–	–	–	–	Tetramethyl-2-methyl-6,7-dihydro-pyrimido[3,4-a]azepine-7,8,9,10-tetracarboxylate		LI08410
59	57	260	116	115	261	58	129	392	100	74	41	38	38	30	22	20	0.00	18	24	6	4	–	–	–	–	–	–	–	–	–	4a,8a-Methanophthalazine-1,4-dicarboxylic acid, 1-[N′-(tert-butoxycarbonyl)hydrazino]-1,2-dihydro-, dimethyl ester		DD01541
377	392	379	73	378	393	391	394	392	100	57	42	35	27	25	24	23		18	29	–	4	–	–	1	–	–	2	–	–	–	2,4-Pyrimidinediamine, 5-(4-chlorophenyl)-6-ethyl-N,N′-bis(trimethylsilyl)-	36972-89-1	NI40000
100	56	114	57	72	58	214	85	392	100	81	77	77	73	73	69	69	16.20	18	36	7	2	–	–	–	–	–	–	–	–	–	3,10-Dihydroxy-5,8,15,23-pentaoxa-1,12-diazabicyclo[10.5.8]pentacosane		MS09060
100	86	84	88	58	147	118	102	392	100	95	93	79	74	74	68	68	57.80	18	36	7	2	–	–	–	–	–	–	–	–	–	3,13-Dihydroxy-5,8,11,18,23-pentaoxa-1,15-diazabicyclo[13.5.5]pentacosane		MS09061
289	56	149	144	100	83	57	114	392	100	85	84	82	74	74	74	59	7.40	18	36	7	2	–	–	–	–	–	–	–	–	–	20,24-Dihydroxy-4,7,13,16,22-pentaoxa-1,10-diazabicyclo[8.8.7]pentacosane		MS09062
43	57	236	235	73	179	69	221	392	100	93	73	55	47	46	46	41	0.00	19	28	5	4	–	–	–	–	–	–	–	–	–	Butanoic acid, 4-azido-2-nitro-, 2,6-di-tert-butyl-4-methoxyphenyl ester		DD01542
73	290	291	292	102	45	75	74	392	100	90	73	24	22	15	10	9	2.70	19	36	1	2	–	–	–	–	–	3	–	–	–	1H-Indole-3-ethanamine, N,1-bis(trimethylsilyl)-5-[(trimethylsilyl)oxy]-	69937-46-8	NI40001
290	291	73	102	292	293	320	74	392	100	95	84	56	28	9	8	7	4.28	19	36	1	2	–	–	–	–	–	3	–	–	–	1H-Indole-3-ethanamine, N,1-bis(trimethylsilyl)-5-[(trimethylsilyl)oxy]-	56145-11-0	NI40002
89	133	59	55	73	81	87	93	392	100	55	46	46	45	37	33	32	0.00	19	36	8	–	–	–	–	–	–	–	–	–	–	3,3′-Methylenebis(1,5,11-tetraoxacyclotridecane)		MS09063
308	54	309	158	392	45	104	352	392	100	82	20	20	18	18	14	13		20	20	3	6	–	–	–	–	–	–	–	–	–	N-(2-Cyanoethyl)-N-[2-(2-cyanoethoxyethyl)-4-(4-nitrophenyl)azo]aniline		IC04261
164	392	168	163	196	167	227	195	392	100	12	10	3	2	2	1	1		20	24	4	–	2	–	–	–	–	–	–	–	–	2,5-Bis(4-hydroxy-3-methoxyphenyl)-3,6-dimethyl-1,4-dithiane		NI40003
233	199	357	235	197	146	168	148	392	100	99	70	70	60	35	25	25	1.00	20	25	4	2	–	–	1	–	–	–	–	–	–	1H-Pyrido[3,4-b]indole-1-butanoate, 4a-chloro-2-ethyl-2,3,4,4a-tetrahydro-α-(methoxycarbonyl)-, methyl ester		NI40004
392	135	166	57	151	377	181	393	392	100	79	77	77	34	28	24	23		20	29	2	2	1	–	–	–	–	–	1	–	–	Bis(N-tert-butylanilino)thiophosphite		IC04262
392	393	360	361	121	321	293	92	392	100	21	14	12	9	8	8	8		21	12	8	–	–	–	–	–	–	–	–	–	–	2,2′-Binaphthalene-1,1′,4,4′-tetrone, 3,5,5′-trihydroxy-3′-methoxy-	88381-87-7	NI40005
351	349	347	197	350	348	195	120	392	100	75	43	39	38	30	29	27	0.00	21	20	–	–	–	–	–	–	–	–	–	–	1	Tin, allyltriphenyl-		MS09064
392	278	165	393	135	111	229	149	392	100	48	28	24	18	16	14	14		21	28	7	–	–	–	–	–	–	–	–	–	–	Chaparrinone, 1-O-methyl-		LI08411
329	392	298	40	360	209	181	330	392	100	45	45	31	23	20	19	18		21	28	7	–	–	–	–	–	–	–	–	–	–	1-Methoxy-1,2-bis(3,4,5-trimethoxyphenyl)ethane		NI40006
181	182	211	151	167	121	165	183	392	100	75	48	40	35	32	31	29	15.00	21	28	7	–	–	–	–	–	–	–	–	–	–	2-Propanol, 1-(3,4-dimethoxyphenyl)-1-methoxy-3-(2,4,6-trimethoxyphenyl)-		NI40007
43	109	53	41	272	91	99	77	392	100	17	17	12	11	10	7	7	0.13	21	28	7	–	–	–	–	–	–	–	–	–	–	15,16,17-Trinor-13-desoxo-13α-hydroxy-13,20-diacetoxy-crotophorbolone		NI40008
43	272	109	254	332	273	99	53	392	100	98	79	71	67	37	30	30	0.65	21	28	7	–	–	–	–	–	–	–	–	–	–	15,16,17-Trinor-13-desoxo-13β-hydroxy-13,20-diacetoxy-crotophorbolone		NI40009
86	156	128	173	107	136	44	164	392	100	72	69	47	47	28	27	24	0.34	21	32	5	2	–	–	–	–	–	–	–	–	–	L-Tyrosine, N-(N-acetyl-L-leucyl)-, butyl ester	55723-91-6	NI40010
392	394	393	242	357	151	152	363	392	100	67	31	26	22	19	18	17		22	14	1	2	–	–	2	–	–	–	–	–	–	1H-1,4-Benzodiazepin-2-one, 7-chloro-3-(4-chlorobenzylidene)-2,3-dihydro-5-phenyl-	55056-34-3	NI40011
350	351	392	43	272	257	179	178	392	100	23	20	18	13	13	11	11		22	29	5	–	–	1	–	–	–	–	–	–	–	Estr-5-en-7-one, 3β-fluoro-10,17β-dihydroxy-, diacetate	33547-20-5	NI40012
350	392	272	257	179	178	332	166	392	100	19	13	13	11	11	10	8		22	29	5	–	–	1	–	–	–	–	–	–	–	Estr-5-en-7-one, 3β-fluoro-10,17β-dihydroxy-, diacetate	33547-20-5	LI08412
260	392	359	245	148	147	218	159	392	100	58	38	38	25	23	14	11		22	32	2	–	2	–	–	–	–	–	–	–	–	Testosterone-2-spiro-2′-[1′,3′-dithian]-3-one		LI08413
83	216	215	315	84	317	316	233	392	100	18	18	14	11	8	6	5	0.00	22	32	6	–	–	–	–	–	–	–	–	–	–	Eudesman-8-one, 4α-acetoxy-3α-(angeloyloxy)-11-hydroxy-6,7-dehydro-		NI40013
121	217	153	111	329	152	128	105	392	100	71	60	53	49	37	37	37	0.00	22	32	6	–	–	–	–	–	–	–	–	–	–	Picras-2-ene-1,16-dione, 11-hydroxy-2,12-dimethoxy-, (11α,12β)-	24148-77-4	NI40014
275	158	146	262	29	276	57	41	392	100	80	51	44	38	30	30	30	6.94	22	36	4	2	–	–	–	–	–	–	–	–	–	Benzene, 1,4-bis(methylcarbamic acid), dibutyl ester		IC04263
57	41	29	263	280	377	58	56	392	100	20	14	11	7	6	5	5	0.22	22	36	4	2	–	–	–	–	–	–	–	–	–	Benzene, 1,3,5-tris(2,2-dimethylpropyl)-2-methyl-4,6-dinitro-	40572-21-2	MS09065
271	105	253	107	159	119	145	131	392	100	52	50	47	38	34	32	28	1.00	22	37	5	–	–	–	–	–	–	–	–	1	–	5α-Pregnane-3β,11β,17α,20β,21-pentol methyl boronate		LI08414

MASS TO CHARGE RATIOS								M.W.	INTENSITIES								Parent	C	H	O	N	S	F	Cl	Br	I	Si	P	B	X	COMPOUND NAME	CAS Reg No	No
271	253	91	43	105	55	159	41	392	100	49	44	44	43	40	37	37	1.00	22	37	5	–	–	–	–	–	–	–	–	1	–	5α-Pregnane-3β,11β,17α,20β,21-pentol methyl boronate		MS09066
392	391	361	333	183	393	185	200	392	100	80	52	33	28	23	19	14		23	21	4	–	–	–	–	–	–	–	1	–	–	Propanedioic acid, (triphenylphosphoranylidene)-, dimethyl ester	19491-23-7	NI40015
392	391	361	333	183	393	185	201	392	100	80	52	33	28	23	20	15		23	21	4	–	–	–	–	–	–	–	1	–	–	Propanedioic acid, (triphenylphosphoranylidene)-, dimethyl ester	19491-23-7	LI08415
43	332	259	163	57	41	55	69	392	100	99	86	86	75	74	72	59	2.00	23	33	4	–	–	1	–	–	–	–	–	–	–	Androst-5-ene-17,19-diol, 3-fluoro-, diacetate, (3α,17β)-	28344-48-1	NI40016
332	259	163	319	333	260	320	164	392	100	80	62	42	24	12	8	6	2.00	23	33	4	–	–	1	–	–	–	–	–	–	–	Androst-5-ene-17,19-diol, 3-fluoro-, diacetate, (3α,17β)-	28344-48-1	NI40017
332	259	163	319	392				392	100	80	62	42	2					23	33	4	–	–	1	–	–	–	–	–	–	–	Androst-5-ene-17,19-diol, 3-fluoro-, diacetate, (3α,17β,19β)-		LI08416
43	259	163	312	332	277	260	91	392	100	99	66	53	37	28	21	20	0.00	23	33	4	–	–	1	–	–	–	–	–	–	–	Androst-5-ene-17,19-diol, 3-fluoro-, diacetate, (3β,17β)-	28344-66-3	NI40018
259	163	312	332	319	299	260	313	392	100	70	60	50	20	20	18	11	5.00	23	33	4	–	–	1	–	–	–	–	–	–	–	Androst-5-ene-17,19-diol, 3-fluoro-, diacetate, (3β,17β)-	28344-66-3	NI40019
259	163	312	332	319	299	372	239	392	100	70	60	50	20	20	10	10	5.00	23	33	4	–	–	1	–	–	–	–	–	–	–	Androst-5-ene-17,19-diol, 3-fluoro-, diacetate, (3β,17β,19β)-		LI08417
332	333	163	290	272	260	257	177	392	100	20	13	12	12	12	12	11	0.00	23	33	4	–	–	1	–	–	–	–	–	–	–	Estr-5(10)-ene-6-methanol, 17-(acetyloxy)-3-fluoro-, acetate, (3α,6β,17β)	28344-53-8	NI40020
332	43	333	163	290	272	257	177	392	100	55	24	17	13	13	12	12	0.00	23	33	4	–	–	1	–	–	–	–	–	–	–	Estr-5(10)-ene-6-methanol, 17-(acetyloxy)-3-fluoro-, acetate, (3α,6β,17β)	28344-53-8	NI40021
332	163	290	272	260	257	177	131	392	100	13	12	12	12	12	11	8	0.00	23	33	4	–	–	1	–	–	–	–	–	–	–	Estr-5(10)-ene-6-methanol, 17-(acetyloxy)-3-fluoro-, acetate, (3α,6β,17β)	28344-53-8	LI08418
332	333	43	163	320	312	272	177	392	100	27	23	13	10	10	10	9	0.00	23	33	4	–	–	1	–	–	–	–	–	–	–	Estr-5(10)-ene-6-methanol, 17-(acetyloxy)-3-fluoro-, acetate, (3β,6β,17β)	28344-73-2	NI40022
332	163	320	312	257	177	260	131	392	100	13	10	9	9	9	8	8	0.00	23	33	4	–	–	1	–	–	–	–	–	–	–	Estr-5(10)-ene-6-methanol, 17-(acetyloxy)-3-fluoro-, acetate, (3β,6β,17β)	28344-73-2	LI08419
332	333	163	320	312	257	177	131	392	100	20	13	10	9	9	9	8	0.00	23	33	4	–	–	1	–	–	–	–	–	–	–	Estr-5(10)-ene-6-methanol, 17-(acetyloxy)-3-fluoro-, acetate, (3β,6β,17β)	28344-73-2	NI40023
392	43	299	255	119	81	55	393	392	100	78	46	45	38	36	28	27		23	36	1	–	2	–	–	–	–	–	–	–	–	Pregnane-12,20-dione, cyclic 12-(ethylenemercaptole), (5α)-	7718-52-7	NI40024
278	314	42	81	55	107	95	79	392	100	63	60	36	32	31	31	26	1.44	23	36	5	–	–	–	–	–	–	–	–	–	–	Androstane-17-carboxylic acid, 3-(acetyloxy)-5-hydroxy-, methyl ester, (3β,5β,17β)-	55956-03-1	NI40025
291	97	43	95	93	156	81	94	392	100	99	97	83	81	77	72	71	31.25	23	36	5	–	–	–	–	–	–	–	–	–	–	Androstane-17-carboxylic acid, 3-(acetyloxy)-14-hydroxy-, methyl ester, (3β,5β,14β,17β)-	2900-98-3	NI40026
360	43	84	392	101	361	123	55	392	100	87	86	42	39	32	12	11		23	36	5	–	–	–	–	–	–	–	–	–	–	Androstan-6-one, 17-acetoxy-3,3-dimethoxy-, (5α,17β)-		MS09067
97	182	81	107	85	55	67	41	392	100	99	70	60	60	60	58	58	54.00	23	36	5	–	–	–	–	–	–	–	–	–	–	Cardanolide, 2,3,14-trihydroxy-, (2α,3β,5α)-	24211-69-6	LI08420
97	182	95	93	81	107	55	67	392	100	99	71	69	69	61	61	59	53.46	23	36	5	–	–	–	–	–	–	–	–	–	–	Cardanolide, 2,3,14-trihydroxy-, (2α,3β,5α)-	24211-69-6	NI40027
162	97	161	182	81	95			392	100	99	80	72	66	63			34.32	23	36	5	–	–	–	–	–	–	–	–	–	–	Cardanolide, 3,14,19-trihydroxy-, (3β,5α)-	24211-70-9	NI40028
374	356	99	303	275	71	325	343	392	100	79	60	49	34	32	25	22	5.00	23	36	5	–	–	–	–	–	–	–	–	–	–	5-Heptenoic acid, 7-[3-hydroxy-2-(5-hydroxy-1,3-decadienyl)-5-oxocyclopentyl]-, methyl ester, [1α(Z),2β(1E,3E,5S),3α]-		NI40029
374	303	99	356	275	71	343	325	392	100	52	52	43	28	27	15	12	4.00	23	36	5	–	–	–	–	–	–	–	–	–	–	5-Heptenoic acid, 7-[3-hydroxy-2-(5-hydroxy-1,3-decadienyl)-5-oxocyclopentyl]-, methyl ester, [1α(Z),2β(1E,3E,5R),3α]-		NI40030
231	335	161	291	290	160	221	275	392	100	80	77	72	67	58	52	50	0.00	23	36	5	–	–	–	–	–	–	–	–	–	–	Podocarpa-7,13-dien-15-oic acid, 18-acetoxy-13-isopropyl-11-oxo-, methyl ester		NI40031
190	119	224	192	133	127	107	91	392	100	63	59	49	47	44	37	34	0.63	23	36	5	–	–	–	–	–	–	–	–	–	–	Prosta-10,13-dien-1-oic acid, 15-(acetyloxy)-9-oxo-, (13E,15S)-	53122-03-5	NI40032
332	350	219	132	249	90	104	78	392	100	99	98	82	59	54	49	44	15.19	23	36	5	–	–	–	–	–	–	–	–	–	–	Prosta-8,13-dien-1-oic acid, 15-(acetyloxy)-9-oxo-, methyl ester, (13E,15S)-	21174-88-9	NI40033
392	248	124	348	393	320	122	246	392	100	42	42	38	27	20	15	14		24	8	6	–	–	–	–	–	–	–	–	–	–	1,3,8,10-Tetraoxaperylo[3,4-cd:9,10-c′d′]dipyran		IC04264
319	315	43	105	392	77	191	57	392	100	74	61	52	51	45	40	31		24	24	5	–	–	–	–	–	–	–	–	–	–	4H-Pyran-3,5-dicarboxylic acid, 2-methyl-4,6-diphenyl-, diethyl ester		LI08421
91	41	255	359	77	45	105	56	392	100	60	55	35	35	35	25	20	8.00	24	28	1	2	1	–	–	–	–	–	–	–	–	Pyrimidine, 1-benzylhexahydro-3-isopropyl-2-(1-phenyl-1-oxo-3-thioxobutan-2-ylidene)-		MS09068
393	168	244	183	121	169	334	243	392	100	27	25	24	23	21	20	20	0.00	24	28	3	2	–	–	–	–	–	–	–	–	–	2,20-Cycloaspidospermidine-3-methanol, 1-acetyl-6,7-didehydro-, acetate (ester), (2α,3β,5α,12β,19α,20R)-	56053-27-1	NI40034
91	269	392	393	377	270	123	255	392	100	95	85	28	21	20	18	15		24	28	3	2	–	–	–	–	–	–	–	–	–	2-Pyrrolecarboxylic acid, 4-ethyl-3-methyl-(4-ethyl-3,5-dimethylpyrrolyl), benzyl ester		LI08422
357	375	358	373	376	391	374	355	392	100	52	23	20	14	11	10	9	2.10	24	40	4	–	–	–	–	–	–	–	–	–	–	Cholan-24-oic acid, 3,6-dihydroxy-, (3α,5β,6α)-	83-49-8	NI40035
55	95	41	81	67	93	79	374	392	100	78	62	58	55	42	42	38	0.81	24	40	4	–	–	–	–	–	–	–	–	–	–	Cholan-24-oic acid, 3,6-dihydroxy-, (3α,5β,6α)-	83-49-8	NI40036
55	356	81	41	93	67	95	79	392	100	82	65	65	58	57	55	47	8.69	24	40	4	–	–	–	–	–	–	–	–	–	–	Cholan-24-oic acid, 3,7-dihydroxy-, (3α,5β,7α)-	474-25-9	NI40038
356	228	341	246	213	255	374	264	392	100	80	59	53	45	39	38	35	4.49	24	40	4	–	–	–	–	–	–	–	–	–	–	Cholan-24-oic acid, 3,7-dihydroxy-, (3α,5β,7α)-	474-25-9	NI40037
356	255	246	264	374	215	258	228	392	100	67	67	65	61	57	43	43	32.65	24	40	4	–	–	–	–	–	–	–	–	–	–	Cholan-24-oic acid, 3,7-dihydroxy-, (3α,5β,7β)-	128-13-2	NI40039
55	255	81	273	41	93	67	356	392	100	91	79	75	64	52	51	50	0.00	24	40	4	–	–	–	–	–	–	–	–	–	–	Cholan-24-oic acid, 3,12-dihydroxy-, (3α,5β,12α)-	83-44-3	NI40040
55	81	255	41	273	67	93	43	392	100	73	69	67	56	50	47	46	0.08	24	40	4	–	–	–	–	–	–	–	–	–	–	Cholan-24-oic acid, 3,12-dihydroxy-, (3α,5β,12α)-	83-44-3	NI40041
255	273	356	256	374	274	357	341	392	100	94	55	29	23	23	20	20	4.08	24	40	4	–	–	–	–	–	–	–	–	–	–	Cholan-24-oic acid, 3,12-dihydroxy-, (3α,5β,12α)-	83-44-3	NI40042
321	332	303	249	285	55	43	81	392	100	21	11	10	8	6	6	4	1.00	24	40	4	–	–	–	–	–	–	–	–	–	–	17-Homo-androstan-4-one, 6,17-dihydroxy-3-oxo-6-pentyl-, (5α,6β,17β)-		MS09069
138	95	81	83	139	55	82	99	392	100	78	69	51	45	39	28	27	0.00	24	40	4	–	–	–	–	–	–	–	–	–	–	Menthol, fumarate (2:1), (1R,3R,4S)-	34675-24-6	NI40043
57	85	115	97	117	131	145	392	392	100	46	36	29	28	6	3	1		24	40	4	–	–	–	–	–	–	–	–	–	–	13,17-Octadecadien-15-yn-1-ol, 18-[(2,3-(isoproylidenedioxy)propyl)oxy]-		MS09070
43	379	67	336	55	101	41	81	392	100	92	77	76	75	67	64	55	0.00	24	40	4	–	–	–	–	–	–	–	–	–	–	9,12,15-Octadecatrienoic acid, 2,2-dimethyl-1,3-dioxolan-4-ylmethyl ester, (Z,Z,Z)-	57156-93-1	NI40044
303	321	43	55	41	81	93	107	392	100	91	60	39	31	29	27	26	0.00	24	40	4	–	–	–	–	–	–	–	–	–	–	4,5-Secoandrostan-4-oic acid, 6-hydroxy-17-oxo-6-pentyl-, (6α,6β)-		MS09071
303	321	43	41	285	55	81	67	392	100	54	53	39	36	36	25	25	0.00	24	40	4	–	–	–	–	–	–	–	–	–	–	4,5-Secoandrostan-4-oic acid, 6-hydroxy-17-oxo-6-pentyl-, (6β,6α)-		MS09072

MASS TO CHARGE RATIOS								M.W.	INTENSITIES								Parent	C	H	O	N	S	F	Cl	Br	I	Si	P	B	X	COMPOUND NAME	CAS Reg No	No
289	321	43	55	41	81	233	290	**392**	100	95	49	32	30	24	22	21	**0.00**	24	40	4	–	–	–	–	–	–	–	–	–	–	4-Nor-3,5-secoandrostan-3-oic acid, 6-hydroxy-17-oxo-6-pentyl-, (6α,6β)-		MS09073
290	322	55	41	81	93	67	79	**392**	100	91	56	44	41	30	30	28	**5.00**	24	40	4	–	–	–	–	–	–	–	–	–	–	4-Nor-3,5-secoandrostan-3-oic acid, 17-hydroxy-5-oxo-6-pentyl-, (6α,17β)-		MS09074
321	41	322	43	91	55	275	83	**392**	100	28	22	22	21	19	18	17	**0.00**	24	40	4	–	–	–	–	–	–	–	–	–	–	2,3-Secoandrostan-3-oic acid, 2,6,17-trihydroxy-6-pentyl-, γ-lactone, (5α,6α,6β,17β)-		MS09075
392	281	394	377	393	181	282	258	**392**	100	72	36	34	33	20	15	13		25	29	–	2	–	–	1	–	–	–	–	–	–	2-Chlorophenylbis(3-methyl-4-ethylamino)methane		IC04265
100	56	55	41	98	42	101	28	**392**	100	18	12	9	8	8	7	7	**0.00**	25	32	2	2	–	–	–	–	–	–	–	–	–	Pyrrolidine, 1-[3-methyl-4-(4-morpholinyl)-1-oxo-2,2-diphenylbutyl]-, (S)-	357-56-2	NI40045
180	91	392	194	103	257	181	165	**392**	100	92	42	42	27	21	20	17		26	20	–	2	1	–	–	–	–	–	–	–	–	1,3,4-Thiadiazole, 2,3-dihydro-2,2,3,5-tetraphenyl-	36358-10-8	NI40047
180	91	194	392	103	257	165	283	**392**	100	93	42	41	28	20	17	13		26	20	–	2	1	–	–	–	–	–	–	–	–	1,3,4-Thiadiazole, 2,3-dihydro-2,2,3,5-tetraphenyl-	36358-10-8	NI40046
167	152	180	165	168	151	63	51	**392**	100	99	60	43	37	32	25	21	**0.00**	26	20	2	2	–	–	–	–	–	–	–	–	–	9-(Diphenylcarboxymethyl)-9,10,dihydro-9,10-diazaphenanthrene		MS09076
392	393	77	253	105	190	323	103	**392**	100	29	16	11	9	8	7	7		26	20	2	2	–	–	–	–	–	–	–	–	–	Oxazole, 2,2′-(1,4-phenylene)bis[4-methyl-5-phenyl-	3073-87-8	NI40048
91	92	65	196	105	195	51	63	**392**	100	25	18	11	9	8	6	4	**0.00**	26	24	–	4	–	–	–	–	–	–	–	–	–	4,9-Diazatricyclo[6.2.2.0^{2,7}]dodeca-5,11-diene-7,11-dicarbonitrile, 4,9-dibenzyl-, endo-		NI40049
105	77	147	124	106	228	55	41	**392**	100	17	12	11	9	5	5	5	**3.00**	26	32	3	–	–	–	–	–	–	–	–	–	–	Androst-4-en-3-one, 17-benzoyloxy-		LI08423
392	394	393	237	395	356			**392**	100	50	40	20	10	10				26	45	–	–	–	–	1	–	–	–	–	–	–	19-Norcholestane, 2-chloro-, (2α,5α)-		LI08424
392	237	394	393	252	395	356		**392**	100	99	50	40	25	10	10			26	45	–	–	–	–	1	–	–	–	–	–	–	19-Norcholestane, 2-chloro-, (2β,5α)-		LI08425
55	349	173	172	83	81	41	254	**392**	100	82	75	69	61	60	55	52	**17.38**	26	49	–	–	–	–	–	–	–	–	1	–	–	Phosphine, cyclohexylbis[5-methyl-2-isopropylcyclohexyl]-	61142-16-3	NI40050
182	149	225	392	167	104	106	57	**392**	100	48	43	34	33	30	28	28		27	24	1	2	–	–	–	–	–	–	–	–	–	Urea, N,N′-bis(diphenylmethyl)-		IC04266
43	45	139	61	73	111	72	56	**392**	100	79	37	32	30	18	17	16	**3.00**	28	28	–	2	–	–	–	–	–	–	–	–	–	2H-Azirin-3-amine, N,N-diethyl-2-methyl-2-(1,2,3-triphenyl-2-cyclopropen-1-yl)-	65659-99-6	NI40051
196	197	77	180	181	91	179	104	**392**	100	16	11	9	3	3	2	2	**0.00**	28	28	–	2	–	–	–	–	–	–	–	–	–	Ethylenediamine, N,N′-dimethyl-N,N′,1,2-tetraphenyl-, (±)-		DD01543
196	197	77	180	181	104	91	179	**392**	100	16	12	9	3	3	3	2	**0.00**	28	28	–	2	–	–	–	–	–	–	–	–	–	Ethylenediamine, N,N′-dimethyl-N,N′,1,2-tetraphenyl-, meso-		DD01544
363	349	392	320	72	377	304	215	**392**	100	39	32	19	12	10	9	9		28	28	–	2	–	–	–	–	–	–	–	–	–	2-Pyridinamine, N,N-diethyl-3-methyl-4,5,6-triphenyl-	65660-26-6	NI40052
57	43	55	83	69	97	71	41	**392**	100	96	81	73	66	65	53	49	**9.00**	28	56	–	–	–	–	–	–	–	–	–	–	–	Cyclooctacosane		MS09077
43	55	57	41	83	97	71	69	**392**	100	94	79	72	71	51	43	39	**0.23**	28	56	–	–	–	–	–	–	–	–	–	–	–	Dodecane, 1-cyclohexyl-2-decyl-		AI02316
55	57	83	97	71	69	251	85	**392**	100	89	85	55	47	43	37	33	**0.10**	28	56	–	–	–	–	–	–	–	–	–	–	–	Dodecane, 1-cyclohexyl-2-decyl-		AI02317
57	41	112	43	97	55	71	69	**392**	100	16	14	14	9	9	8	8	**0.80**	28	56	–	–	–	–	–	–	–	–	–	–	–	6-Tridecene, 2,2,4,10,12,12-hexamethyl-7-(3,5,5-trimethylhexyl)-		AI02318
392	91	133	115	172	220	233	273	**392**	100	76	45	42	19	17	16	15		29	28	1	–	–	–	–	–	–	–	–	–	–	Spiro[2.4]heptan-4-one, 1-phenyl-5-[(E)-benzylidene]-2-(2,4,6-trimethylphenyl)-		MS09078
392	393	265	391	389	174	390	188	**392**	100	34	26	19	13	12	10	9		31	20	–	–	–	–	–	–	–	–	–	–	–	Dibenzo[a,i]fluorene, 13-(1-naphthyl)-		AI02319
228	230	96	393	117	92	42	395	**393**	100	76	63	54	48	28	26	24		12	12	2	2	–	6	–	–	–	–	–	–	1	Copper(II), N,N′-ethylenebis(1,1,1-trifluoro-4-iminopentan-2-onato)-	40820-17-5	NI40053
228	393	63	324	199	187	131	374	**393**	100	48	8	6	6	5	3	2		12	12	2	2	–	6	–	–	–	–	–	–	1	Copper(II), N,N′-ethylenebis(1,1,1-trifluoro-4-iminopentan-2-onato)-	40820-17-5	NI40054
115	61	45	169	60	163	161	70	**393**	100	18	18	17	17	16	16	16	**0.00**	12	16	7	3	–	–	–	1	–	–	–	–	–	β-D-Glucopyranose, 1,2-dideoxy-3,4,6-tri-O-acetyl-1-azido-2-bromo-	50603-58-2	NI40055
31	30	29	15	91	44	43	92	**393**	100	94	74	63	59	55	55	46	**0.00**	12	21	6	3	–	3	–	–	–	–	–	3	–	Borazine, trifluoro-		LI08426
208	210	40	94	77	44	76	209	**393**	100	42	21	19	18	16	14	13	**0.00**	14	17	4	1	2	–	1	–	–	–	1	–	–	Phosphorodithioic acid, S-[2-chloro-1-(1,3-dihydro-1,3-dioxo-2H-isoindol-2-yl)ethyl] O,O-diethyl ester	10311-84-9	NI40057
208	210	29	65	76	27	129	97	**393**	100	42	33	22	21	19	18	16	**0.00**	14	17	4	1	2	–	1	–	–	–	1	–	–	Phosphorodithioic acid, S-[2-chloro-1-(1,3-dihydro-1,3-dioxo-2H-isoindol-2-yl)ethyl] O,O-diethyl ester	10311-84-9	NI40056
185	157	161	173	159	207	186	177	**393**	100	39	30	21	10	7	7	7	**0.00**	14	17	4	1	2	–	1	–	–	–	1	–	–	Phosphorodithioic acid, S-[2-chloro-1-(1,3-dihydro-1,3-dioxo-2H-isoindol-2-yl)ethyl] O,O-diethyl ester	10311-84-9	NI40058
77	260	51	125	119	118	69	104	**393**	100	87	84	83	76	49	38	30	**7.60**	15	14	4	1	2	3	–	–	–	–	–	–	–	Methanesulphonamide, 1,1,1-trifluoro-N-methyl-N-[2-methyl-4-(phenylsulphonyl)phenyl]-	62059-53-4	NI40059
73	267	393	193	280	179	45	268	**393**	100	58	35	23	21	21	18	14		16	26	3	1	–	3	–	–	–	2	–	–	–	Dopamine, N-(trifluoroacetyl)-O,O′-bis(trimethylsilyl)-		MS09079
73	75	306	57	321	147	55	71	**393**	100	47	31	25	19	18	16	13	**4.21**	16	31	1	5	–	–	–	–	–	3	–	–	–	2-Pteridinamine, 6-methyl-N,N-bis(trimethylsilyl)-4-[(trimethylsilyl)oxy]-	55649-43-9	NI40060
115	73	186	75	173	218	147	116	**393**	100	84	44	31	24	17	15	14	**0.00**	16	35	6	1	–	–	–	–	–	2	–	–	–	α-D-Galactopyranoside, methyl 2-(acetylamino)-2-deoxy-3-O-methyl-4,6-bis-O-(trimethylsilyl)-	56196-91-9	NI40061
115	73	186	75	69	173	218	147	**393**	100	85	44	32	26	25	17	16	**0.00**	16	35	6	1	–	–	–	–	–	2	–	–	–	α-D-Galactopyranoside, methyl 2-(acetylamino)-2-deoxy-3-O-methyl-4,6-bis-O-(trimethylsilyl)-	56196-91-9	NI40062
173	73	131	75	117	174	146	43	**393**	100	72	36	25	22	21	20	20	**0.00**	16	35	6	1	–	–	–	–	–	2	–	–	–	α-D-Galactopyranoside, methyl 2-(acetylamino)-2-deoxy-4-O-methyl-3,6-bis-O-(trimethylsilyl)-	56196-94-2	NI40063
173	73	131	45	43	175	147	75	**393**	100	72	28	28	20	18	18	18	**0.00**	16	35	6	1	–	–	–	–	–	2	–	–	–	α-D-Galactopyranoside, methyl 2-(acetylamino)-2-deoxy-6-O-methyl-3,4-bis-O-(trimethylsilyl)-	56196-93-1	NI40064
73	115	75	147	186	43	117	146	**393**	100	85	50	23	22	20	19	18	**0.00**	16	35	6	1	–	–	–	–	–	2	–	–	–	α-D-Glucopyranoside, methyl 2-(acetylamino)-2-deoxy-3-O-methyl-4,6-bis-O-(trimethylsilyl)-	56196-89-5	NI40065

MASS TO CHARGE RATIOS								M.W.	INTENSITIES								Parent	C	H	O	N	S	F	Cl	Br	I	Si	P	B	X	COMPOUND NAME	CAS Reg No	No
173	73	131	75	89	146	43	174	393	100	62	31	23	18	17	17	16	0.00	16	35	6	1	–	–	–	–	–	2	–	–	–	α-D-Glucopyranoside, methyl 2-(acetylamino)-2-deoxy-4-O-methyl-3,6-bis-O-(trimethylsilyl)-	56196-92-0	NI40066
173	73	131	69	174	175	146	226	393	100	36	29	17	16	14	14	13	0.00	16	35	6	1	–	–	–	–	–	2	–	–	–	α-D-Glucopyranoside, methyl 2-(acetylamino)-2-deoxy-6-O-methyl-3,4-bis-O-(trimethylsilyl)-	56196-90-8	NI40067
216	188	130	202	306	230	334	393	393	100	90	70	50	28	23	21	20		18	19	9	1	–	–	–	–	–	–	–	–	–	Aniline, N-[1,2-bis(methoxycarbonyl)vinyl]-2-[methoxyallyl(methoxycarbonyl)]methyl-		LI08427
362	363	318	364	393				393	100	19	5	4	3					18	19	9	1	–	–	–	–	–	–	–	–	–	Pyridine-2,3,4,5-tetracarboxylic acid, 6-(4-methoxybuta-1,3-dienyl)-, tetramethyl ester		LI08428
334	362	335	291	363	393			393	100	42	16	10	8	7				18	19	9	1	–	–	–	–	–	–	–	–	–	4H-Quinolizine-1,2,3,4-tetracarboxylic acid, 4-methoxy-, tetramethyl ester		LI08429
73	117	75	299	132	129	145	55	393	100	80	68	58	42	30	28	27	0.00	18	31	3	1	–	–	–	–	–	3	–	–	–	1H-Indole-2-carboxylic acid, 1-(trimethylsilyl)-5-[(trimethylsilyl)oxy]-, trimethylsilyl ester	74367-47-8	NI40068
73	378	379	45	393	380	231	147	393	100	93	27	18	15	13	11	10		18	31	3	1	–	–	–	–	–	3	–	–	–	1H-Indole-2-carboxylic acid, 1-(trimethylsilyl)-5-[(trimethylsilyl)oxy]-, trimethylsilyl ester	74367-47-8	NI40069
173	175	393	177	145	395	174	147	393	100	63	14	11	10	9	8	7		19	17	4	1	–	–	2	–	–	–	–	–	–	2-Propenoic acid, 3-[1-(2,6-dichlorobenzoyl)-4-acetyl-3,5-dimethylpyrrol-2-yl]-, methyl ester	95883-19-5	NI40070
296	111	209	262	261	248	310	234	393	100	21	20	14	13	10	9	8	4.00	19	23	8	1	–	–	–	–	–	–	–	–	–	Furan-3,4-dicarboxylic acid, 2,5-dihydro-, dimethyl ester		LI08430
43	57	214	164	85	215	148	135	393	100	50	22	10	5	4	4	4	0.40	19	23	8	1	–	–	–	–	–	–	–	–	–	4H-Indol-4-one, 2-(2,3,5-tri-O-acetyl-α-D-ribofuranosyl)-1,5,6,7-tetrahydro		NI40071
214	43	164	57	215	69	135	85	393	100	81	38	16	14	14	11	10	0.90	19	23	8	1	–	–	–	–	–	–	–	–	–	4H-Indol-4-one, 2-(2,3,5-tri-O-acetyl-β-D-ribofuranosyl)-1,5,6,7-tetrahydro-		NI40072
43	57	164	69	214	85	273	231	393	100	69	20	19	14	12	5	4	1.30	19	23	8	1	–	–	–	–	–	–	–	–	–	4H-Indol-4-one, 2-(2,3,4-tri-O-acetyl-α-D-ribopyranosyl)-1,5,6,7-tetrahydro-		NI40073
43	164	214	69	85	135	231	148	393	100	62	39	31	18	11	8	8	0.30	19	23	8	1	–	–	–	–	–	–	–	–	–	4H-Indol-4-one, 2-(2,3,4-tri-O-acetyl-β-D-ribopyranosyl)-1,5,6,7-tetrahydro-		NI40074
73	75	331	215	43	55	129	117	393	100	72	54	32	31	30	24	23	0.09	19	31	4	1	–	–	–	–	–	2	–	–	–	1H-Indole-3-carboxylic acid, 2-ethoxy-1-(trimethylsilyl)-5-[(trimethylsilyl)oxy]-, ethyl ester	74367-46-7	NI40075
73	319	393	246	320	364	304	394	393	100	91	52	29	25	19	15	14		19	31	4	1	–	–	–	–	–	2	–	–	–	1H-Indole-3-carboxylic acid, 2-ethoxy-1-(trimethylsilyl)-5-[(trimethylsilyl)oxy]-, ethyl ester	74367-46-7	NI40076
290	73	291	393	292	394	378	218	393	100	46	25	21	10	8	6	4		19	35	2	1	–	–	–	–	–	3	–	–	–	1H-Indole, 1-(trimethylsilyl)-5-[(trimethylsilyl)oxy]-3-[2-[(trimethylsilyl)oxy]ethyl]-	56114-64-8	NI40077
393	79	394	52	363	51	28		393	100	26	23	14	11	11	10			20	11	6	1	1	–	–	–	–	–	–	–	–	Anthraquinone, 1,5-dihydroxy-2-(phenylthio)-8-nitro-		IC04267
393	45	244	59	349	394	186	276	393	100	53	39	37	26	25	24	22		20	23	2	7	–	–	–	–	–	–	–	–	–	2,6-Bis(2-methoxyethylamino)-3-cyano-4-methyl-5-(2-cyanophenylazo)pyridine		IC04268
202	188	365	364	189	178	336	320	393	100	61	60	45	28	25	10	10	7.00	21	19	5	3	–	–	–	–	–	–	–	–	–	Benzoic acid, 4-[N-formyl-N-(2-methyl-6,7-methylenedioxyquinazolin-4-ylmethyl)amino]-, ethyl ester	85591-05-5	NI40078
176	175	44	252	45	69	95	68	393	100	64	62	46	40	32	31	28	0.00	21	27	1	7	–	–	–	–	–	–	–	–	–	Butanamide, N-[3-[3-[(aminoiminomethyl)amino]propyl]-5-(1H-indol-3-yl)pyrazinyl]-2-methyl-	17297-78-8	NI40079
238	393	237	209	329	328	314	300	393	100	36	32	22	7	4	4	1		22	19	4	1	1	–	–	–	–	–	–	–	–	11-Morphanthridinone, 5,6-dihydro-8-methoxy-5-(4-tolylsulphonyl)-	23145-78-0	NI40080
393	77	272	300	361	51	141	362	393	100	82	61	29	27	26	23	21		22	19	6	1	–	–	–	–	–	–	–	–	–	4,5-Pyridinedicarboxylic acid, 1,2-dihydro-3-methyl-2-oxo-6-phenoxy-1-phenyl-, dimethyl ester		DD01545
201	202	376	77	393	51	377	203	393	100	69	59	25	18	16	15	9		22	20	4	1	–	–	–	–	–	–	1	–	–	Styrene, β,β-dimethyl-3-nitro-α-(diphenylphosphinyl)-		MS09080
130	235	171	170	393	131	168	77	393	100	48	31	31	25	15	11	10		22	23	2	3	1	–	–	–	–	–	–	–	–	1-Naphthalenesulphonamide, 5-(dimethylamino)-N-[2-(1H-indol-3-yl)ethyl]-	13285-17-1	NI40081
130	235	171	170	393	131	168	299	393	100	50	30	29	25	15	10	8		22	23	2	3	1	–	–	–	–	–	–	–	–	1-Naphthalenesulphonamide, 5-(dimethylamino)-N-[2-(1H-indol-3-yl)ethyl]-	13285-17-1	NI40082
393	41	147	55	91	105	79	71	393	100	97	76	72	69	61	61	54		22	35	3	1	1	–	–	–	–	–	–	–	–	3′H,4′H,5′H-Mutilano[13,14-d]oxazol-2′-thione, 3-methoxy-11-oxo-		NI40083
236	88	130	45	43	56	57	221	393	100	79	68	42	35	28	23	22	0.00	22	35	5	1	–	–	–	–	–	–	–	–	–	Butanoic acid, 2-acetamido-4-methoxy-, 2,6-di-tert-butyl-4-methoxyphenyl ester		DD01546
209	73	393	179	43	210	75	41	393	100	78	56	31	26	20	20	20		22	39	3	1	–	–	–	–	–	1	–	–	–	Decanamide, N-[[3-methoxy-4-[(trimethylsilyl)oxy]phenyl]methyl]-9-methyl-	69796-06-1	MS09081
324	70	326	393	325	311	18	310	393	100	72	56	31	26	26	24	23		23	23	5	1	–	–	–	–	–	–	–	–	–	Dibenzo[f,h]pyrrolo[1,2-b]isoquinoline-4,14-diol, 9,11,12,14-tetrahydro-3,6,7-trimethoxy-	95066-41-4	NI40084
91	196	81	166	92	65	302	112	393	100	19	13	11	11	11	8	7	0.00	23	27	3	3	–	–	–	–	–	–	–	–	–	2(1H)-Pyrimidinone, 4-amino-1-[4-(phenylmethoxy)-3-[(phenylmethoxy)methyl]butyl]-	39809-23-9	NI40085

MASS TO CHARGE RATIOS								M.W.	INTENSITIES								Parent	C	H	O	N	S	F	Cl	Br	I	Si	P	B	X	COMPOUND NAME	CAS Reg No	No
57	55	69	71	95	67	94	83	**393**	100	85	59	57	53	45	40	39	**7.00**	23	27	3	3	–	–	–	–	–	–	–	–	–	**3-Pyrrolin-2-one, 4-ethyl-3-methyl-5-[5-(3,4-dimethyl-1H-pyrrolyl-2-methylene)-4-(2-carboxyethyl)-3-methyl-5H-pyrrolyl-2-methylene]-, (Z,Z)-**		**MS09082**
130	42	55	43	129	344	107	81	**393**	100	64	63	58	56	50	48	46	**21.00**	23	39	4	1	–	–	–	–	–	–	–	–	–	**Pregnan-20-one, 3b,11a,17a-trihydroxy-16a-methyl-, O-methyl oxime, (5α)-**		**LI08431**
262	393	130	105	132	131	76	77	**393**	100	55	27	19	15	12	12	8		24	15	3	3	–	–	–	–	–	–	–	–	–	**Isoindol-1-one, trimer**		**LI08432**
29	350	41	393	27	55	91	67	**393**	100	46	44	31	30	24	23	23		24	27	4	1	–	–	–	–	–	–	–	–	–	**3′H-Cycloprop[1,2]androsta-1,4,6-triene-3,17-dione, 1′-(ethoxycarbonyl)-1′-cyano-1β,2β-dihydro-**	75857-76-0	**NI40086**
324	393	325	162	309	181	152	189	**393**	100	73	60	29	20	17	15	12		24	27	4	1	–	–	–	–	–	–	–	–	–	**Dibenzo[f,h]pyrrolo[1,2-b]isoquinoline, 9,11,12,13,13a,14-hexahydro-2,3,6,7-tetramethoxy-, (S)-**	482-20-2	**NI40087**
143	165	131	67	115	123	265	66	**393**	100	58	55	52	51	26	23	20	**3.00**	24	27	4	1	–	–	–	–	–	–	–	–	–	**2-Propenamide, N-[2-acetoxy-2-[4-[(3-methyl-2-butenyl)oxy]phenyl]ethyl] 3-phenyl-, (E)-**		**NI40088**
393	252	91	43	394	253	238	41	**393**	100	70	67	40	28	27	25	23		25	35	1	3	–	–	–	–	–	–	–	–	–	**2,2′-Bipyrrole, 4-methoxy-5-[(5-undecyl-2H-pyrrol-2-ylidene)methyl]-**	14960-80-6	**MS09084**
393	91	252	43	36	253	41	394	**393**	100	87	71	43	32	31	31	29		25	35	1	3	–	–	–	–	–	–	–	–	–	**2,2′-Bipyrrole, 4-methoxy-5-[(5-undecyl-2H-pyrrol-2-ylidene)methyl]-**	14960-80-6	**MS09083**
393	91	252	43	36	253	92	41	**393**	100	87	71	43	32	31	31	30		25	35	1	3	–	–	–	–	–	–	–	–	–	**2,2′-Bipyrrole, 4-methoxy-5-[(5-undecyl-2H-pyrrol-2-ylidene)methyl]-**	14960-80-6	**NI40089**
86	128	156	393	184	57	55	41	**393**	100	39	30	27	25	24	24	20		26	51	1	1	–	–	–	–	–	–	–	–	–	**9-Octadecenamide, N,N-dibutyl-**	56630-41-2	**NI40090**
393	316	395	394	318	317	396	280	**393**	100	97	37	36	27	22	18	18		27	20	–	1	–	–	1	–	–	–	–	–	–	**Quinoline, 1-(4-chlorophenyl)-1,4-dihydro-2,3-diphenyl-**	36843-26-2	**NI40091**
118	119	211	93	105	77	90	300	**393**	100	85	77	37	23	20	17	13	**1.10**	27	23	2	1	–	–	–	–	–	–	–	–	–	**Benzamide, N-phenyl-2-(β,β-diphenyl-β-hydroxy)ethyl-**		**MS09085**
236	158	235	237	143	219	221	159	**393**	100	46	19	18	14	8	7	7	**1.00**	27	27	–	3	–	–	–	–	–	–	–	–	–	**1H-Indole, 3,3′-[1-(4-pyridinyl)-1,2-ethanediyl]bis[2-ethyl-**	37707-66-7	**NI40092**
273	57	274	91	121	393	120	41	**393**	100	24	21	15	8	7	4	3		27	39	1	1	–	–	–	–	–	–	–	–	–	**Benzamide, N-benzyl-N-methyl-2,4,6-tri-tert-butyl-**		**MS09087**
273	393	57	274	91	394	121	120	**393**	100	27	25	21	16	8	8	5		27	39	1	1	–	–	–	–	–	–	–	–	–	**Benzamide, N-benzyl-N-methyl-2,4,6-tri-tert-butyl-**		**MS09086**
55	322	43	57	41	58	29	27	**393**	100	75	75	58	54	42	42	36	**11.85**	28	43	–	1	–	–	–	–	–	–	–	–	–	**Diphenylamine, 4,4′-di-tert-octyl-**		**IC04269**
322	329	393	250	57	251	394	41	**393**	100	26	19	11	9	8	6	6		28	43	–	1	–	–	–	–	–	–	–	–	–	**4,4′-Bis(2,4,4-trimethyl-2-pentyl)phenylamine**		**IC04270**
127	267	140	394	254	139	13	12	**394**	100	55	27	11	10	10	5	3		1	1	–	–	–	–	–	–	3	–	–	–	–	**Methane, triiodo-**	75-47-8	**NI40093**
60	31	153	273	46	121	45	29	**394**	100	38	25	22	20	15	15	13	**12.00**	6	18	–	6	3	–	–	–	–	–	4	–	–	**2,4,6,8,9,10-Hexaaza-1,3,5,7-tetraphosphatricyclo[3.3.1.1^{3,7}]decane, 2,4,6,8,9,10-hexamethyl-, 1,3,5-trisulphide**	38448-55-4	**NI40094**
229	227	272	237	274	49	239	231	**394**	100	75	74	69	56	45	43	39	**0.00**	9	6	–	–	–	–	8	–	–	–	–	–	–	**Bicyclo[2.2.1]hept-2-ene, 1,2,3,4,7,7-hexachloro-5,6-bis(chloromethyl)-**	2550-75-6	**NI40095**
89	94	96	214	212	49	133	66	**394**	100	88	80	52	50	46	45	32	**0.00**	9	9	–	–	1	4	–	2	–	–	–	1	–	**2,3-Dibromo-1-methyl-2,3-dihydrobenzo[b]thiophenium tetrafluoroborate**		**LI08433**
197	177	127	77	159	113	51	95	**394**	100	76	28	25	21	20	20	15	**0.05**	10	8	–	–	–	14	–	–	–	–	–	–	–	**Decane, 1,1,1,2,2,5,5,6,6,9,9,10,10,10-tetradecafluoro-**	35278-79-6	**NI40096**
197	177	127	77	113	51	95	147	**394**	100	76	28	25	20	20	15	14	**0.05**	10	8	–	–	–	14	–	–	–	–	–	–	–	**Decane, 1,1,1,2,2,5,5,6,6,9,9,10,10,10-tetradecafluoro-**	35278-79-6	**MS09088**
72	15	18	44	42	28	73	232	**394**	100	62	36	25	24	17	13	12	**0.00**	11	11	3	2	–	–	5	–	–	–	–	–	–	**Acetic acid, trichloro-, compd. with N′-(3,4-dichlorophenyl)-N,N-dimethylurea (1:1)**	15500-95-5	**NI40097**
176	68	69	119	41	55	177	43	**394**	100	70	62	38	32	31	30	26	**6.40**	11	12	2	2	–	10	–	–	–	–	–	–	–	**1,5-Pentanediamine, N,N′-bis(pentafluoropropionyl)-**		**MS09089**
379	275	380	107	105	227	381	103	**394**	100	17	14	12	12	11	10	10	**1.18**	11	20	–	2	2	–	–	–	–	–	–	–	2	**Arsinothious acid, dimethyl-, 6-propyl-2,4-pyrimidinediyl ester**	51678-02-5	**NI40098**
78	79	95	237	272	96	77	239	**394**	100	64	63	43	39	36	32	27	**0.00**	12	8	2	–	–	–	6	–	–	–	–	–	–	**9-(syn-Epoxy)-hydroxy-1,2,3,4,10,10-hexachloro-6,7-epoxy-1,4,4a,5,6,7,8,8a,-octahydro-1,4-endo-5,8-exo-dimethano-naphthalene**		**LI08434**
43	55	27	39	53	91	70	324	**394**	100	77	64	32	25	24	24	16	**6.93**	12	18	3	–	–	–	–	–	–	–	–	–	1	**Tungsten, tris(π-2-oxobut-3-enyl)-**		**MS09090**
155	77	361	359	242	97	55	91	**394**	100	80	43	33	23	22	19	18	**0.00**	13	10	3	–	–	–	2	1	–	–	1	–	–	**Phosphonic acid, phenyl-, 4-bromo-2,5-dichlorophenyl methyl ester**	25006-32-0	**NI40099**
77	51	155	91	15	50	47	65	**394**	100	48	44	37	37	34	25	18	**0.10**	13	10	3	–	–	–	2	1	–	–	1	–	–	**Phosphonic acid, phenyl-, 4-bromo-2,5-dichlorophenyl methyl ester**	25006-32-0	**NI40100**
186	244	218	140	121	318	366	338	**394**	100	80	57	50	40	39	37	36	**23.26**	14	5	2	–	–	7	–	–	–	–	–	–	1	**Iron, dicarbonyl-π-cyclopentadienyl-[4-(heptafluoro)tolyl]-**		**MS09091**
395	396	191	180	173	397	321	284	**394**	100	19	5	3	3	2	1	1	**0.00**	14	20	4	2	–	6	–	–	–	–	–	–	–	**L-Lysine, N^2,N^2-bis(trifluoroacetyl)-, butyl ester**	2926-74-1	**NI40101**
180	41	30	57	67	126	69	29	**394**	100	65	58	49	42	30	28	25	**0.30**	14	20	4	2	–	6	–	–	–	–	–	–	–	**L-Lysine, N^2,N^6-bis(trifluoroacetyl)-, butyl ester**	55429-20-4	**NI40102**
114	395	180	321	293	339	208		**394**	100	98	46	35	35	32	15		**1.00**	14	20	4	2	–	6	–	–	–	–	–	–	–	**L-Lysine, N^2,N^6-bis(trifluoroacetyl)-, butyl ester**	55429-20-4	**NI40103**
140	67	180	334	307	297	394	335	**394**	100	69	46	12	12	5	3	3		14	20	4	2	–	6	–	–	–	–	–	–	–	**Lysine, N^2,N^6-bis(trifluoroacetyl)-N^6-methyl-, propyl ester**	91795-78-7	**NI40104**
180	57	41	44	67	69	43	55	**394**	100	63	54	41	26	23	23	18	**0.00**	14	20	4	2	–	6	–	–	–	–	–	–	–	**L-Lysine, N^2,N^6-bis(trifluoroacetyl)-, sec-butyl ester**	58072-47-2	**NI40105**
73	75	153	242	244	155	103	225	**394**	100	34	32	31	21	19	17	12	**0.00**	14	20	5	2	–	–	2	–	–	1	–	–	–	**Chloramphenicol, 3-trimethylsilyl-**		**NI40106**
73	225	44	75	208	45	74	147	**394**	100	64	39	28	25	15	14	10	**0.00**	14	20	5	2	–	–	2	–	–	1	–	–	–	**Chloramphenicol, 3-trimethylsilyl-**		**LI08435**
172	126	173	125	145	133	394		**394**	100	21	8	4	2	2	1			15	8	4	2	–	6	–	–	–	–	–	–	–	**Propane, 1,3-bis(3-nitrophenyl)hexafluoro-**		**LI08436**
45	105	61	44	59	147	60	58	**394**	100	99	34	28	13	11	10	10	**0.00**	15	22	8	–	2	–	–	–	–	–	–	–	–	**D-Arabinose, cyclic 1,2-ethanediyl mercaptal, tetraacetate**	40733-16-2	**NI40107**
73	233	105	89	234	59	215	379	**394**	100	42	41	41	38	30	22	18	**0.00**	15	30	8	–	–	–	–	–	–	2	–	–	–	**1,2,3-Propanetricarboxylic acid, 1,2-bis[(trimethylsilyl)oxy]-, trimethyl ester**	55591-00-9	**NI40108**
73	147	75	45	248	189	133	219	**394**	100	19	14	14	8	7	7	6	**0.21**	15	34	6	–	–	–	–	–	–	3	–	–	–	**Butanedioic acid, 2,3-bis[(trimethylsilyl)oxy]-, ethyl trimethylsilyl ester**		**NI40109**
197	199	118	90	89	119	200	198	**394**	100	99	93	37	19	15	10	10	**0.00**	16	12	2	–	–	–	–	2	–	–	–	–	–	**Bibenzoyl, α,α-dibromo-2,2′-dimethyl-**		**LI08437**

MASS TO CHARGE RATIOS								M.W.	INTENSITIES								Parent	C	H	O	N	S	F	Cl	Br	I	Si	P	B	X	COMPOUND NAME	CAS Reg No	No
145	321	219	217	43	115	130	232	**394**	100	88	68	68	60	50	40	28	**0.00**	16	23	10	–	–	1	–	–	–	–	–	–	–	**D-Glucitol, 2-deoxy-2-fluoro-, pentaacetate**	34339-84-9	**NI40110**
145	321	219	217	43	115	130	249	**394**	100	91	69	69	62	53	42	28	**0.00**	16	23	10	–	–	1	–	–	–	–	–	–	–	**D-Glucitol, 2-deoxy-2-fluoro-, pentaacetate**	34339-84-9	**LI08438**
145	43	219	249	128	103	147	170	**394**	100	43	27	21	18	18	16	15	**0.00**	16	23	10	–	–	1	–	–	–	–	–	–	–	**D-Glucitol, 3-deoxy-3-fluoro-, pentaacetate**		**LI08439**
145	43	219	249	103	128	147	170	**394**	100	42	28	20	18	17	16	15	**0.00**	16	23	10	–	–	1	–	–	–	–	–	–	–	**D-Glucitol, 4-deoxy-4-fluoro-, pentaacetate**	34401-81-5	**NI40111**
145	43	219	249	321	147	103	128	**394**	100	70	67	54	45	40	23	20	**0.00**	16	23	10	–	–	1	–	–	–	–	–	–	–	**D-Glucitol, 4-deoxy-4-fluoro-, pentaacetate**	34401-81-5	**NI40112**
145	219	43	249	321	147	103	228	**394**	100	70	70	55	45	40	24	22	**0.00**	16	23	10	–	–	1	–	–	–	–	–	–	–	**D-Glucitol, 4-deoxy-4-fluoro-, pentaacetate**	34401-81-5	**LI08440**
43	145	249	217	115	128	103	147	**394**	100	65	35	30	30	27	25	20	**0.00**	16	23	10	–	–	1	–	–	–	–	–	–	–	**D-Glucitol, 6-deoxy-6-fluoro-, pentaacetate**	34339-89-4	**NI40113**
43	145	249	217	115	128	103	170	**394**	100	67	35	30	30	28	25	20	**0.00**	16	23	10	–	–	1	–	–	–	–	–	–	–	**D-Glucitol, 6-deoxy-6-fluoro-, pentaacetate**	34339-89-4	**LI08441**
217	73	117	218	133	189	219	147	**394**	100	48	24	22	12	9	8	8	**0.00**	16	38	5	–	–	–	–	–	–	3	–	–	–	**α-L-Galactofuranoside, methyl 6-deoxy-2,3,5-tris-O-(trimethylsilyl)-**	56227-31-7	**NI40114**
204	73	133	205	217	147	206	75	**394**	100	51	32	23	19	17	10	7	**0.00**	16	38	5	–	–	–	–	–	–	3	–	–	–	**α-L-Galactopyranoside, methyl 6-deoxy-2,3,4-tris-O-(trimethylsilyl)-**	56271-58-0	**NI40115**
204	73	133	205	217	147	206	75	**394**	100	80	39	23	21	21	10	7	**0.00**	16	38	5	–	–	–	–	–	–	3	–	–	–	**β-L-Galactopyranoside, methyl 6-deoxy-2,3,4-tris-O-(trimethylsilyl)-**	56271-59-1	**NI40116**
204	73	133	205	217	147	206	191	**394**	100	55	22	20	17	16	9	7	**0.00**	16	38	5	–	–	–	–	–	–	3	–	–	–	**α-L-Mannopyranoside, methyl 6-deoxy-2,3,4-tris-O-(trimethylsilyl)-**	56271-60-4	**NI40117**
168	169	77	167	394	30	395	51	**394**	100	72	67	59	49	31	28	24		18	12	6	5	–	–	–	–	–	–	–	–	–	**Hydrazyl, 2,2-diphenyl-1-(2,4,6-trinitrophenyl)-**	1898-66-4	**NI40119**
396	198	229	184	252	348	394	197	**394**	100	74	62	37	36	34	30	24		18	12	6	5	–	–	–	–	–	–	–	–	–	**Hydrazyl, 2,2-diphenyl-1-(2,4,6-trinitrophenyl)-**	1898-66-4	**NI40118**
213	394	73	181	45	79	76	123	**394**	100	83	77	71	56	23	17	16		18	18	8	–	1	–	–	–	–	–	–	–	–	**Benzene, 1-(phenylsulphonyl)-4,4′-bis(acetoxymethoxy)-**		**IC04271**
105	272	77	230	335	288	243	211	**394**	100	38	10	8	7	4	4	4	**0.20**	18	18	10	–	–	–	–	–	–	–	–	–	–	**α-D-Xylofuranose, 3-C-[(carboxyoxy)methyl]-, intramol. 3,3-ester, 1,2-diacetate 5-benzoate**	70723-13-6	**NI40120**
315	317	104	118	91	105	117	131	**394**	100	99	85	45	38	33	20	18	**5.00**	18	20	–	–	–	–	–	2	–	–	–	–	–	**Butane, 4,4-bis[(4-bromomethyl)phenyl]-**		**IC04272**
364	328	303	111	365	110	83	274	**394**	100	71	71	65	55	53	35	32	**0.00**	18	20	4	4	–	2	–	–	–	–	–	–	–	**D-Arabino-hexos-2-ulose, bis[(2-fluorophenyl)hydrazone]**	51306-40-2	**NI40121**
239	91	155	83	85	240	65	56	**394**	100	71	54	21	17	13	12	12	**5.00**	18	22	4	2	2	–	–	–	–	–	–	–	–	**Piperazine, 1,4-bis[(4-methylphenyl)sulphonyl]-**	17046-84-3	**NI40122**
167	166	100	293	41	29	42	27	**394**	100	86	67	41	37	30	23	22	**0.00**	18	29	4	2	–	3	–	–	–	–	–	–	–	**L-Prolylisoleucine, α-methyl-N-(trifluoroacetyl)-, butyl ester**		**MS09092**
167	166	100	293	241	57	29	139	**394**	100	86	86	41	40	21	21	20	**0.00**	18	29	4	2	–	3	–	–	–	–	–	–	–	**L-Prolylleucine, α-methyl-N-(trifluoroacetyl)-, butyl ester**		**MS09093**
55	41	69	139	43	121	67	29	**394**	100	72	54	48	34	30	22	22	**16.66**	18	34	4	–	–	–	–	–	–	–	–	–	1	**Nonanoic acid, 9,9′-selenodi-**	22676-19-3	**NI40123**
55	41	69	139	43	121	97	29	**394**	100	73	55	48	35	30	22	22	**17.00**	18	34	4	–	–	–	–	–	–	–	–	–	1	**Nonanoic acid, 9,9′-selenodi-**	22676-19-3	**LI08442**
99	111	98	197	112	196	167	110	**394**	100	50	45	23	18	13	11	11	**0.00**	18	38	6	–	–	–	–	–	–	–	–	4	–	**Galactitol, cyclic 2,3:4,5-bis(ethylboronate) 1,6-bis(diethylborinate)**	58881-47-3	**NI40124**
43	274	173	335	275	187	215	246	**394**	100	91	83	53	52	50	49	44	**0.00**	19	22	9	–	–	–	–	–	–	–	–	–	–	**β-Allamcidin acetate**		**NI40125**
131	160	222	174	159	91	70	173	**394**	100	59	39	35	33	22	22	21	**1.56**	19	26	7	2	–	–	–	–	–	–	–	–	–	**1,4,10,13-Tetraoxa-7,16-diazacyclooctadecane-2,6,17-trione, 18-benzyl-**	74229-39-3	**NI40126**
113	85	72	71	57	137	56	309	**394**	100	80	46	45	45	40	40	36	**14.00**	19	26	7	2	–	–	–	–	–	–	–	–	–	**Uridine, 2,2′-anhydro-3′,5′-di-O-pivaloyl-**		**NI40127**
73	81	135	114	93	121	248	189	**394**	100	84	34	32	16	10	7	7	**2.00**	19	29	5	–	–	3	–	–	–	–	–	–	–	**Methyl 11-methoxy-3,7,11-trimethyl-10-(trifluoroacetoxy)-2,6-dodecadienoate**		**MS09094**
251	250	77	94	252	170	144	394	**394**	100	35	28	20	14	10	7	6		20	21	4	–	–	2	–	–	–	–	1	–	–	**Diphenyl 1-cyclohexyl-2,2-difluoroethenyl phosphate**		**DD01547**
321	224	252	196	142	394	349	304	**394**	100	67	57	24	22	17	14	14		20	30	6	2	–	–	–	–	–	–	–	–	–	**1,8-Naphthyridine-3,4a,6-tricarboxylic acid, 2,7,8a-trimethyl-1,4,4a,5,8,8a hexahydro-, triethyl ester**		**MS09095**
267	268	265	189	263	165	266	178	**394**	100	24	15	7	6	6	5	5	**0.00**	21	15	–	–	–	–	–	–	1	–	–	–	–	**Cyclopropenylium, triphenyl-, iodide**	58090-79-2	**NI40128**
394	166	117	167	395	123	104	91	**394**	100	47	40	33	25	18	18	18		21	22	4	4	–	–	–	–	–	–	–	–	–	**1,14:5,7-Dimethano-2H-[1,4]diazepino[2′,3′:3,4]cyclobuta[1,2-d][2,7]benzodiazecine-6,15,16,17-tetrone, 3,4,5a,5b,8,13,15a,15b-octahydro-5b,15a-dimethyl-**	25878-31-3	**NI40129**
394	166	117	167	123	104	91	146	**394**	100	47	38	32	18	18	18	15		21	22	4	4	–	–	–	–	–	–	–	–	–	**Thymine, 1,1′-trimethylenebis-, O-xylylene**		**LI08443**
394	273	179	219	191	178	288	260	**394**	100	17	17	7	7	7	5	5		21	24	2	2	–	–	–	–	–	–	–	–	1	**Nickel(II), N,N′-bis(salicylidene)heptanediamino-**		**NI40130**
58	135	59	336	103	177	77	42	**394**	100	54	4	3	3	2	2	2	**0.50**	21	26	–	6	1	–	–	–	–	–	–	–	–	**N-Ethyl-N-(2-dimethylaminoethyl)-3-methyl-4-(3-phenyl-1,2,4-thiadiaz-5-ylazo)aniline**		**IC04273**
144	91	105	277	131	119	205	394	**394**	100	48	42	41	33	31	23	16		21	30	7	–	–	–	–	–	–	–	–	–	–	**Hexahydrodiasin**		**NI40131**
179	394	73	379	395	180	75	207	**394**	100	56	39	20	19	17	10	8		21	38	3	–	–	–	–	–	–	2	–	–	–	**Nonanoic acid, 4-[[(trimethylsilyl)oxy]phenyl]-, trimethylsilyl ester**		**NI40132**
270	361	286	394	148	254	379	288	**394**	100	20	12	10	10	6	5	5		22	18	7	–	–	–	–	–	–	–	–	–	–	**9,10-Anthracenedione, 1,5-dihydroxy-3-methyl-8-[(2,6,7,7a-tetrahydro-4H furo[3,2-c]pyran-4-yl)oxy]-**	93513-59-8	**NI40133**
394	395	293	321	351	182	294	163	**394**	100	25	25	11	8	8	5	5		22	18	7	–	–	–	–	–	–	–	–	–	–	**Naphtho[2,3-c]furan-1(3H)-one, 4,6,7-trimethoxy-9-[3,4-(methylenedioxy)phenyl]-**	25001-57-4	**NI40134**
315	396	394	239	165	229	270	257	**394**	100	43	42	9	8	5	4	4		22	19	2	–	–	–	–	1	–	–	–	–	–	**Vinyl bromide, 1,2-dianisyl-2-phenyl-, cis-**		**LI08444**
315	394	396	239	300	270	229	107	**394**	100	45	5	4	3	2	2	2		22	19	2	–	–	–	–	1	–	–	–	–	–	**Vinyl bromide, 1,2-dianisyl-2-phenyl-, trans-**		**LI08445**
265	323	308	324	352	206	204	234	**394**	100	70	67	66	46	41	33	31	**16.00**	22	22	5	2	–	–	–	–	–	–	–	–	–	**6,21-Cyclo-4,5-secoakuammilan-17-oic acid, 5-(acetyloxy)-4,5-epoxy-, methyl ester, (5S,6α)-**	63944-62-7	**NI40135**
394	351	159	366	322	132	292	201	**394**	100	98	49	44	42	25	20	20		22	34	2	–	2	–	–	–	–	–	–	–	–	**Androstan-2-spiro-2′-(1′,3′-dithian)-3-one, 17-hydroxy-, (5α,17β)-**		**LI08446**
318	350	205	258	192	177	303	189	**394**	100	97	83	74	61	37	34	27	**9.00**	22	34	6	–	–	–	–	–	–	–	–	–	–	**15,20-Bisnorlabdane, 6,14-diacetoxy-8,13;8,14-diepoxy-**		**MS09096**
298	105	316	109	376	135	255	358	**394**	100	61	57	57	49	48	38	37	**14.00**	22	34	6	–	–	–	–	–	–	–	–	–	–	**Δ^{10}(18)-Andromedenol, acetyl-**		**LI08447**
107	396	394	41	93	283	105	145	**394**	100	88	88	84	79	78	78	73		22	35	1	–	–	–	–	1	–	–	–	–	–	**Pregn-5-en-3-ol, 21-bromo-20-methyl-, (3β)-**	55103-80-5	**NI40136**

MASS TO CHARGE RATIOS								M.W.	INTENSITIES								Parent	C	H	O	N	S	F	Cl	Br	I	Si	P	B	X	COMPOUND NAME	CAS Reg No	No
309	279	251	167	351	352	297	175	394	100	86	83	66	46	45	24	20	19.00	23	22	6	–	–	–	–	–	–	–	–	–	–	9,10-Anthracenedione, 1,8-dihydroxy-3-methoxy-6-methyl-2-(3-methyl-1-butenyl)-, monoacetate		NI40137
379	394	161	119	91	217	162	77	394	100	80	31	31	21	19	18	18		23	22	6	–	–	–	–	–	–	–	–	–	–	2H,6H-Benzo[1,2-b:5,4-b′]dipyran-6-one, 7-(2,4-dimethoxyphenyl)-5-methoxy-2,2-dimethyl-	49776-78-5	MS09097
379	394	217	363	161	119	321	202	394	100	67	17	12	7	7	5	5		23	22	6	–	–	–	–	–	–	–	–	–	–	4H,8H-Benzo[1,2-b:3,4-b′]dipyran-4-one, 3-(2,4-dimethoxyphenyl)-5-methoxy-8,8-dimethyl-	49776-77-4	MS09098
379	394	149	380	395	363	217	190	394	100	25	12	8	5	5	5	5		23	22	6	–	–	–	–	–	–	–	–	–	–	4H,8H-Benzo[1,2-b:3,4-b′]dipyran-4-one, 3-(2,4-dimethoxyphenyl)-5-methoxy-8,8-dimethyl-	49776-77-4	NI40138
192	394	191	392	177	193	393	395	394	100	31	30	28	19	16	15	10		23	22	6	–	–	–	–	–	–	–	–	–	–	[1]Benzopyran[3,4-b]furo[2,3-h][1]benzopyran-6(6aH)-one, 1,2,12,12a-tetrahydro-8,9-dimethoxy-2-isopropenyl-, [2R-(2α,6aα,12aα)]-	83-79-4	NI40139
192	191	177	394	193	208	200	202	394	100	24	17	15	14	11	7	6		23	22	6	–	–	–	–	–	–	–	–	–	–	[1]Benzopyrano[3,4-b]furo[2,3-h][1]benzopyran-6(6aH)-one, 1,2,12,12a-tetrahydro-8,9-dimethoxy-2-isopropenyl-, [2R-(2α,6aα,12aα)]-	83-79-4	NI40140
192	191	394	177	193	77	65	39	394	100	31	20	18	15	6	6	6		23	22	6	–	–	–	–	–	–	–	–	–	–	[1]Benzopyrano[3,4-b]furo[2,3-h][1]benzopyran-6(6aH)-one, 1,2,12,12a-tetrahydro-8,9-dimethoxy-2-isopropenyl-, [2R-(2α,6aα,12aα)]-	83-79-4	NI40141
192	394	392	191	393	193	177	395	394	100	35	23	23	15	15	15	12		23	22	6	–	–	–	–	–	–	–	–	–	–	3H-Bis[1]benzopyrano[3,4-b:6′,5′-e]pyran-7(7aH)-one, 13,13a-dihydro-9,10-dimethoxy-3,3-dimethyl-, (7aS-cis)-	522-17-8	NI40142
379	394	380	190	395	353	363	364	394	100	33	25	13	8	8	7	6		23	22	6	–	–	–	–	–	–	–	–	–	–	α-Isoluteone, dehydrotrimethyl-		NI40143
197	105	198	77	118	117	165	103	394	100	20	16	14	6	4	3	3	0.00	23	23	1	–	–	–	–	1	–	–	–	–	–	1-Bromo-3-methoxy-2-methyl-2,3,3-triphenylpropane		MS09099
77	121	79	321	394	91	65	78	394	100	67	67	56	51	48	47	45		23	26	4	2	–	–	–	–	–	–	–	–	–	Akuammilan-16-carboxylic acid, 17-(acetyloxy)-, methyl ester, (16R)-	1897-26-3	NI40144
394	321	335	121	168	249	180		394	100	99	43	24	20	18	17			23	26	4	2	–	–	–	–	–	–	–	–	–	Akuammilan-16-carboxylic acid, 17-(acetyloxy)-, methyl ester, (16R)-	1897-26-3	LI08448
394	379	214	395	106	120	107	117	394	100	33	33	26	21	16	15	13		23	26	4	2	–	–	–	–	–	–	–	–	–	Obscurinervan-21-one, 6,7-didehydro-16-methoxy-22-methyl-, (22α)-	6872-64-6	NI40145
335	249	169	69	394	168	107	71	394	100	47	44	42	40	33	32	27		23	26	4	2	–	–	–	–	–	–	–	–	–	Sarpagan-16-carboxylic acid, 17-(acetyloxy)-, methyl ester	14478-58-1	NI40146
169	335	168	107	249	394	121	115	394	100	90	82	70	67	47	44	44		23	26	4	2	–	–	–	–	–	–	–	–	–	Sarpagan-16-carboxylic acid, 17-(acetyloxy)-, methyl ester, (16R)-	2520-48-1	NI40147
91	197	92	113	125	379	303	153	394	100	18	10	5	4	3	3	3	0.80	23	26	4	2	–	–	–	–	–	–	–	–	–	Uracil, 1-[4-(benzyloxy)-3-[(benzyloxy)methyl]butyl]-	33498-86-1	LI08449
91	197	82	92	65	79	77	41	394	100	18	12	10	8	5	5	5	0.00	23	26	4	2	–	–	–	–	–	–	–	–	–	Uracil, 1-[4-(benzyloxy)-3-[(benzyloxy)methyl]butyl]-	33498-86-1	NI40148
136	164	196	91	165	93	107	97	394	100	85	30	30	28	27	26	25	6.93	23	26	4	2	–	–	–	–	–	–	–	–	–	Yohimban-16-carboxylic acid, 17-(acetyloxy)-19,20-didehydro-, methyl ester, (16α,17α)-	54725-62-1	NI40149
304	358	99	376	71	151	275	340	394	100	75	65	62	45	25	14	10	5.00	23	38	5	–	–	–	–	–	–	–	–	–	–	5-Heptenoic acid, 7-[3,5-bihydroxy-2-(5-hydroxy-1,3-decadienyl)cyclopentyl]-, methyl ester, [1α(Z),2β(1E,3E,5S),3α,5α]	79769-26-9	NI40150
275	99	71	304	376	358	151	340	394	100	75	50	49	20	19	15	6	1.00	23	38	5	–	–	–	–	–	–	–	–	–	–	5-Heptenoic acid, 7-[3,5-dihydroxy-2-(5-hydroxy-1,3-decadienyl)cyclopentyl]-, methyl ester, [1α(Z),2β(1E,3E,5S),3α,5α]	79769-26-9	NI40151
220	210	117	119	43	189	109	133	394	100	85	62	51	51	38	36	22	2.00	23	38	5	–	–	–	–	–	–	–	–	–	–	1-Naphthalenepentanoic acid, 3-(acetyloxy)-3,4,4a,5,6,7,8,8a-octahydro-β hydroxy-β,2,5,5,8a-pentamethyl-, methyl ester, [3S-[1(R*),3α,4aβ,8aα]]-		NI40152
105	104	76	77	291	219	119	247	394	100	55	46	45	42	27	21	20	9.40	24	14	4	2	–	–	–	–	–	–	–	–	–	Spiro[1,4,2-dioxazole-5,5′(4′H)-naphth[2,1-d]isoxazol]-4′-one, 3,3′-diphenyl-	23767-17-1	NI40153
121	43	124	144	122	293	135	93	394	100	44	26	24	24	15	14	8	2.02	24	30	3	2	–	–	–	–	–	–	–	–	–	Aspidospermidine-3-methanol, 1-acetyl-6,7-didehydro-, acetate (ester), (2β,3β,5α,12β,19α)-	56053-14-6	NI40154
43	79	377	41	67	101	55	95	394	100	83	75	70	68	66	65	46	0.00	24	42	4	–	–	–	–	–	–	–	–	–	–	9,12-Octadecadienoic acid (Z,Z)-, 2,2-dimethyl-1,3-dioxolan-4-ylmethyl ester	57156-94-2	NI40155
57	43	41	71	124	79	55	152	394	100	66	63	53	46	38	34	31	0.63	24	42	4	–	–	–	–	–	–	–	–	–	–	Phthalic acid, tetrahydro-, di(2-ethylhexyl) ester		IC04274
275	257	43	55	276	95	81	85	394	100	42	33	22	21	17	16	15	1.00	24	42	4	–	–	–	–	–	–	–	–	–	–	3,4-Secoandrostane-3,17-diol, 6-acetoxy-4-propyl-, (5α,6α,17β)-		MS09100
137	77	353	81	215	101	354	95	394	100	93	50	48	35	17	14	13	0.21	24	46	2	–	–	–	–	–	–	1	–	–	–	Silane, methylbis[[5-methyl-2-isopropylcyclohexyl]oxy]-2-propenyl-, [1α(1R*,2S*,5R*),2β,5α]-	74841-55-7	NI40156
175	394	362	396	176	166	317	352	394	100	81	64	61	49	34	33	31		25	15	1	2	–	–	1	–	–	–	–	–	–	Benzo[b]naphtho[1,2-i]phenazin-5(8H)-one, 6-chloro-8-methyl-	52736-88-6	NI40157
119	117	91	104	105	297	118	77	394	100	83	74	47	40	31	28	28	0.00	25	22	1	4	–	–	–	–	–	–	–	–	–	11-Oxatricyclo[12.2.2.2^{5,8}]eicosa-5,7,14,16,17,19-hexaene-2,2,13,13-tetracarbonitrile, 12,12-dimethyl-	66389-01-3	NI40158
209	394	185	183	395	210	207	51	394	100	84	62	55	20	15	13	10		26	20	–	–	–	–	–	–	–	–	2	–	–	Phosphine, 1,2-ethynediylbis[diphenyl-	5112-95-8	NI40159
393	183	167	392	139	394	138	211	394	100	98	55	41	30	28	28	21		26	22	2	2	–	–	–	–	–	–	–	–	–	Azobenzene, 2,2′-bis(2-methoxyphenyl)-	56701-35-0	NI40160
199	211	197	168	394	182	183	154	394	100	90	70	70	68	60	58	58		26	22	2	2	–	–	–	–	–	–	–	–	–	Azobenzene, 2,2′-bis(3-methoxyphenyl)-		LI08450
168	363	44	364	152	139	199	167	394	100	80	31	24	22	22	21	18	1.00	26	22	2	2	–	–	–	–	–	–	–	–	–	Azobenzene, 2,2′-bis(4-methoxyphenyl)-	29512-01-4	NI40161
394	183	168	393	140	139	211	198	394	100	96	58	43	32	31	22	20		26	22	2	2	–	–	–	–	–	–	–	–	–	Azobenzene, 2,2′-bis(4-methoxyphenyl)-	29512-01-4	LI08451
200	43	212	198	183	182	41	57	394	100	90	87	73	67	67	67	60	52.05	26	22	2	2	–	–	–	–	–	–	–	–	–	Azobenzene, 2,3′-bis(2-methoxyphenyl)-		NI40162
262	183	165	166	108	263	184	109	394	100	70	60	40	40	15	7	4	0.00	26	23	–	2	–	–	–	–	–	–	1	–	–	Phosphorane, [(α-methylbenzylidene)hydrazono]triphenyl-	2734-98-7	NI40163
197	198	394	395	259	199	195	181	394	100	20	18	8	7	5	5	5		26	26	–	–	–	–	–	–	–	2	–	–	–	Disilane, 1,2-dimethyl-1,1,2,2-tetraphenyl-	1172-76-5	NI40164
196	167	168	197	140	181	198	127	394	100	78	73	68	61	26	20	13	0.00	26	26	–	4	–	–	–	–	–	–	–	–	–	2,2′-Diperimidine, 1,1′-diethyl-2,2′,3,3′-tetrahydro-	93767-87-4	NI40165

MASS TO CHARGE RATIOS								M.W.	INTENSITIES								Parent	C	H	O	N	S	F	Cl	Br	I	Si	P	B	X	COMPOUND NAME	CAS Reg No	No
182	168	169	197	394	196	167	183	**394**	100	95	88	84	56	49	40	39		26	26	–	4	–	–	–	–	–	–	–	–	–	1,8-Naphthalenediamine, N[1]-ethyl-N[8]-[(1-ethyl-2-perimidinyl)methyl]-	93767-83-0	NI40166
55	339	161	394	57	177	376	340	**394**	100	50	48	47	43	34	18	16		26	34	3	–	–	–	–	–	–	–	–	–	–	Methacrylic acid, bis(2-hydroxy-3-tert-butyl-5-methylphenyl)methanyl ester		NI40167
325	69	57	394	189	121	395	309	**394**	100	63	41	19	11	10	6	5		26	34	3	–	–	–	–	–	–	–	–	–	–	Methacrylic acid, 2-(3,5-di-tert-butyl-4-hydroxybenzyl)-4-methylphenyl ester		MS09101
73	45	43	41	55	57	44	129	**394**	100	82	71	20	19	18	17	14	**1.00**	26	50	2	–	–	–	–	–	–	–	–	–	–	Cyclopropanetetradecanoic acid, 2-octyl-, methyl ester	52355-42-7	NI40168
93	75	125	124	362	330	315	55	**394**	100	94	67	53	47	38	37	29	**0.10**	26	50	2	–	–	–	–	–	–	–	–	–	–	4,8-Docosadiene, 1,1-dimethoxy-4,8-dimethyl-, [4Z,8Z(E)]-		NI40169
88	83	69	57	43	142	101	55	**394**	100	85	84	84	80	71	67	63	**57.42**	26	50	2	–	–	–	–	–	–	–	–	–	–	2-Docosenoic acid, 2,4,21-trimethyl-, methyl ester, (E)-	56630-80-9	NI40170
88	83	69	43	394	101	142	55	**394**	100	98	92	87	76	76	74	63		26	50	2	–	–	–	–	–	–	–	–	–	–	2-Docosenoic acid, 2,4,21-trimethyl-, methyl ester, (Z)-	56630-81-0	NI40171
91	41	67	55	79	65	81	39	**394**	100	18	17	16	15	14	11	11	**0.00**	27	38	2	–	–	–	–	–	–	–	–	–	–	Arachidonic acid, benzyl ester		NI40172
174	137	43	69	41	95	93	81	**394**	100	68	61	57	45	43	43	43	**18.00**	27	38	2	–	–	–	–	–	–	–	–	–	–	2,5-Cyclohexadiene-1,4-dione, 2-methyl-6-(3,7,11,15-tetramethyl-2,6,10,14-hexadecatetraenyl)-	57576-82-6	NI40173
109	149	205	394	127	239	368	267	**394**	100	59	46	30	30	10	8	8		27	38	2	–	–	–	–	–	–	–	–	–	–	8,10,12,14,16,18,20-Docosaheptaene-2,7-dione, 6,6,10,14,19-pentamethyl-, (all-E)-	66556-68-1	NI40174
225	253	83	43	95	69	57	55	**394**	100	70	24	24	22	22	20	18	**11.20**	27	54	1	–	–	–	–	–	–	–	–	–	–	2H-Pyran, 2-decyl-6-dodecyl-tetrahydro-		LI08452
394	337	57	69	119	55	338	395	**394**	100	85	47	46	38	36	35	34		27	54	1	–	–	–	–	–	–	–	–	–	–	1,10-Secocholesta-5,7,9,22-tetraene-3-one, 24-methyl-, (17β)-		MS09102
121	394	255	135	247	273	147	77	**394**	100	93	39	36	32	20	20	20		28	26	2	–	–	–	–	–	–	–	–	–	–	Cyclopentane, 1,3-bis(4-methoxybenzylidene)-2-benzylidene-, (E,E)-		MS09103
270	69	55	269	394	43	175	95	**394**	100	96	69	59	57	38	35	35		28	42	1	–	–	–	–	–	–	–	–	–	–	Ergosta-4,6,22-trien-3-one	516-77-8	NI40175
55	69	43	394	267	269	41	81	**394**	100	97	69	67	55	46	43	39		28	42	1	–	–	–	–	–	–	–	–	–	–	Ergosta-4,7,22-trien-3-one	17398-57-1	NI40176
57	43	71	85	41	55	69	56	**394**	100	89	62	42	34	31	18	14	**0.05**	28	58	–	–	–	–	–	–	–	–	–	–	–	Docosane, 7-hexyl-		AI02320
43	57	71	41	85	55	69	182	**394**	100	93	56	41	39	36	21	17	**1.55**	28	58	–	–	–	–	–	–	–	–	–	–	–	Docosane, 7-hexyl-		AI02321
43	57	71	41	85	55	281	69	**394**	100	97	58	42	41	35	20	20	**0.06**	28	58	–	–	–	–	–	–	–	–	–	–	–	Eicosane, 9-octyl-		AI02322
43	57	71	85	41	55	69	56	**394**	100	96	60	43	39	34	20	16	**0.02**	28	58	–	–	–	–	–	–	–	–	–	–	–	Eicosane, 9-octyl-		AI02323
43	57	71	85	41	55	69	56	**394**	100	95	55	38	35	30	16	15	**1.28**	28	58	–	–	–	–	–	–	–	–	–	–	–	Octacosane		AI02324
57	43	71	41	85	55	29	69	**394**	100	94	58	33	28	28	20	15	**1.57**	28	58	–	–	–	–	–	–	–	–	–	–	–	Octacosane		AI02325
57	56	41	71	55	85	58	68	**394**	100	15	9	7	5	4	4	3	**0.00**	28	58	–	–	–	–	–	–	–	–	–	–	–	Tridecane, 2,2,4,10,12,12-hexamethyl-7-(3,5,5-trimethylhexyl)-		AI02326
141	394	142	395	115	167	154	41	**394**	100	35	24	12	12	6	6	6		30	34	–	–	–	–	–	–	–	–	–	–	–	Decane, 1,10-di-(1-naphthyl)-		AI02327
141	394	142	395	28	115	154	167	**394**	100	59	21	19	14	10	7	5		30	34	–	–	–	–	–	–	–	–	–	–	–	Decane, 1,10-di-(1-naphthyl)-		AI02328
141	394	142	395	115	154	155	153	**394**	100	43	21	14	11	6	5	5		30	34	–	–	–	–	–	–	–	–	–	–	–	Decane, 1,10-di-(1-naphthyl)-		AI02329
362	364	366	108	327	360	329	110	**395**	100	96	55	55	52	46	42	38	**1.20**	6	1	–	3	–	–	8	–	–	–	–	–	–	s-Triazine, 2-(dichloromethyl)-4,6-bis(trichloromethyl)-	30362-31-3	NI40177
73	380	351	164	145	165	282	381	**395**	100	59	36	17	17	14	11	10	**0.00**	12	15	5	1	–	6	–	–	–	1	–	–	–	L-Proline, 1-(trifluoroacetyl)-4-[(trifluoroacetyl)oxy]-, trimethylsilyl ester, trans-	52558-88-0	NI40178
73	380	395	75	147	93	45	381	**395**	100	44	40	37	32	20	16	15		15	29	2	5	–	–	–	–	–	3	–	–	–	4,6-Pteridinedione, 2-amino-1,5-dihydro-, tris(trimethylsilyl)-	56272-95-8	NI40179
380	73	381	395	147	75	382	99	**395**	100	73	37	30	25	16	15	8		15	29	2	5	–	–	–	–	–	3	–	–	–	4,7(1H,8H)-Pteridinedione, 2-amino-, tris(trimethylsilyl)-	56273-03-1	NI40180
177	118	44	395	72	132	91	106	**395**	100	44	40	38	30	24	20	12		16	15	3	5	–	–	2	–	–	–	–	–	–	Aniline, N-(2-carbamylethyl)-4-(2,6-dichloro-4-nitrophenylazo)-		IC04275
93	81	66	44	39	65	55	41	**395**	100	50	34	33	23	22	16	15	**0.70**	16	18	5	3	1	–	–	–	–	–	1	–	–	Thymidine cyclic 3′,5′-phosphoranilidothioate, (R_P)-	73098-25-6	NI40181
93	81	241	55	171	39	65	66	**395**	100	84	34	32	31	26	25	24	**20.00**	16	18	5	3	1	–	–	–	–	–	1	–	–	Thymidine cyclic 3′,5′-phosphoranilidothioate, (S_P)-	73098-26-7	NI40182
73	102	147	103	89	45	203	159	**395**	100	91	90	44	32	31	30	22	**0.00**	16	37	6	1	–	–	–	–	–	2	–	–	–	D-Glucose, 2,3,6-tri-O-methyl-4,5-bis-O-(trimethylsilyl)-, O-methyloxime	56196-12-4	NI40183
169	396	395	168	229	170	394	378	**395**	100	89	86	59	40	39	25	19		18	13	6	5	–	–	–	–	–	–	–	–	–	Diphenylpicrylhydrazil		NI40184
77	43	278	395	94	141	51	41	**395**	100	66	56	43	37	36	30	28		18	21	7	1	1	–	–	–	–	–	–	–	–	2-Oxa-6-azatricyclo[3.3.1.1^{3,7}]decane-4,8-diol, 6-(phenylsulphonyl)-, diacetate (ester), (1α,3β,4β,5α,7β,8α)-	50267-35-1	NI40185
278	236	43	77	395	141	212	57	**395**	100	71	55	34	29	18	15	12		18	21	7	1	1	–	–	–	–	–	–	–	–	2,5-Methanofuro[3,2-b]pyridine-7,8-diol, octahydro-4-(phenylsulphonyl)-, diacetate (ester), (2α,3aβ,5α,7β,7aβ,8R*)-	50267-37-3	NI40186
336	274	395	244	342	168	326	318	**395**	100	64	50	39	36	32	15	15		18	21	9	1	–	–	–	–	–	–	–	–	–	Pyrrolidin-2-ol, 2,3,4,5-tetrakis(methoxycarbonyl)-1-phenyl-		LI08453
91	73	75	45	92	77	55	147	**395**	100	30	15	9	8	6	6	5	**0.40**	18	29	5	1	–	–	–	–	–	2	–	–	–	Pentanedioic acid, 2-[(phenylmethoxy)imino]-, bis(trimethylsilyl) ester	55520-98-4	NI40187
121	395	122	41	396	207	77	91	**395**	100	16	9	5	3	3	3	2		19	26	2	1	–	5	–	–	–	–	–	–	–	Benzenenonanamine, 4-methoxy-, N-(pentafluoropropionyl)-		NI40188
395	248	247	219	274	179	178	377	**395**	100	21	21	19	15	9	9	8		20	23	2	3	–	–	–	–	–	–	–	–	1	Nickel(II), N,N′-bis(salicylidene)-3,3′-bis(aminopropyl)amino-		NI40189
395	260	274	179	180	233	165	246	**395**	100	88	26	19	13	10	10	9		21	24	2	2	–	–	–	–	–	–	–	–	1	Cobalt(II), N,N′-bis(salicylidene)heptanediamino-		NI40190
364	365	274	73	324	174	173	76	**395**	100	29	17	14	13	10	9	8	**6.56**	21	37	4	1	–	–	–	–	–	1	–	–	–	1-Cyclopentene-1-propanoic acid, 5-(methoxyimino)-2-[3-[(trimethylsilyl)oxy]-1-octenyl]-, methyl ester, [S-(,E)]-	72101-40-7	NI40191
395	323	281	293	282	396	324	337	**395**	100	99	89	38	33	24	24	15		22	21	6	1	–	–	–	–	–	–	–	–	–	Actinodaphnine, 1,9-diacetyl-		LI08454
220	123	146	221	82	189	177	95	**395**	100	26	16	10	8	7	4	4	**2.00**	22	22	3	3	–	1	–	–	–	–	–	–	–	2,4(1H,3H)-Quinazolinedione, 3-[2-[4-(4-fluorobenzoyl)-1-piperidinyl]ethyl]-, (+)-	74050-98-9	NI40192
100	155	70	143	84	224	129	182	**395**	100	83	66	36	35	34	33	23	**5.00**	22	41	3	3	–	–	–	–	–	–	–	–	–	13-(7-Acetamido-4-acetyl-4-azaheptyl)-12-dodecane lactam	75701-25-6	NI40193

MASS TO CHARGE RATIOS								M.W.	INTENSITIES								Parent	C	H	O	N	S	F	Cl	Br	I	Si	P	B	X	COMPOUND NAME	CAS Reg No	No
352	70	100	98	84	112	83	297	**395**	100	60	55	46	37	36	33	30	**7.00**	22	41	3	3	–	–	–	–	–	–	–	–	–	13,17-Diacetyl-13,17-diaza-20-eicosane lactam	75701-26-7	NI40194
353	230	336	395	294	241	85	81	**395**	100	66	61	60	50	45	41	37		23	25	5	1	–	–	–	–	–	–	–	–	–	Morphinan, 3,6-bis(acetyloxy)-7,8-didehydro-4,5-diepoxy-17-(2-propenyl)-, (5α,6α)-		NI40195
244	221	118	187	130	175	273	246	**395**	100	13	13	12	11	7	5	4	**3.00**	23	26	2	3	–	1	–	–	–	–	–	–	–	1,3,8-Triazaspiro[4,5]decan-4-one, 8-[4-(4-fluorophenyl)-4-oxobutyl]-1-phenyl-	749-02-0	MS09104
244	123	165	256	98	221	95	206	**395**	100	64	57	52	34	31	24	23	**6.00**	23	26	2	3	–	1	–	–	–	–	–	–	–	1,3,8-Triazaspiro[4.5]decan-4-one, 8-[4-(4-fluorophenyl)-4-oxobutyl]-1-phenyl-	749-02-0	LI08455
244	124	166	257	98	221	95	207	**395**	100	65	56	50	35	30	25	23	**6.01**	23	26	2	3	–	1	–	–	–	–	–	–	–	1,3,8-Triazaspiro[4.5]decan-4-one, 8-[4-(4-fluorophenyl)-4-oxobutyl]-1-phenyl-	749-02-0	NI40196
120	395	204	91	176	70	339	175	**395**	100	90	53	40	37	33	28	28		23	29	3	3	–	–	–	–	–	–	–	–	–	2-Oxa-6,9-diazabicyclo[10.2.2]hexadeca-12,14,15-triene-5,8-dione, 4-amino-7-benzyl-3-isopropyl-	14642-97-8	NI40197
395	253	396	294	268	295	293	74	**395**	100	39	31	21	19	17	17	14		23	33	1	3	–	–	–	–	–	1	–	–	–	Ergoline-8-carboxamide, 9,10-didehydro-N,N-diethyl-6-methyl-1-(trimethylsilyl)-, (8β)-	55760-26-4	NI40198
86	224	225	395	239	366	209	226	**395**	100	61	33	23	22	17	14	11		23	33	1	3	–	–	–	–	–	1	–	–	–	10-Sila-2-azaanthracene, 9-[5-(diethylamino)pentanamido]-10,10-dimethyl 9,10-dihydro-	92973-67-6	NI40199
395	91	396	105	77	304	121	133	**395**	100	63	36	21	20	17	17	12		24	21	1	1	–	–	–	–	–	–	–	–	1	Ferrocene, [benzoylbenzylamino]-	12156-49-9	MS09105
29	352	41	395	57	55	27	91	**395**	100	36	36	29	28	25	24	23		24	29	4	1	–	–	–	–	–	–	–	–	–	3H-Cycloprop[1,2]androsta-1,4,6-trien-3-one, 1′-(ethoxycarbonyl)-1′-cyano-1β,2β-dihydro-17β-hydroxy-	75857-75-9	NI40200
278	395	396	279	232	231	91	115	**395**	100	84	25	23	21	16	15	14		25	21	2	3	–	–	–	–	–	–	–	–	–	4H-1,2-Diazepine, 5-(3-nitrophenyl)-3,7-bis(4-tolyl)-	25649-78-9	NI40201
105	194	77	195	106	230	336	78	**395**	100	85	30	16	10	8	7	4	**4.00**	26	21	3	1	–	–	–	–	–	–	–	–	–	2-Naphthoic acid, 1-(N-benzoyl-2-toluidino)-, methyl ester		NI40203
194	91	195	65	105	92	89	77	**395**	100	28	17	10	5	3	3	3	**2.00**	26	21	3	1	–	–	–	–	–	–	–	–	–	2-Naphthoic acid, 1-(N-benzoyl-2-toluidino)-, methyl ester		NI40202
232	233	290	289	263	48	231	38	**395**	100	20	8	8	8	6	5	3	**0.60**	26	37	2	1	–	–	–	–	–	–	–	–	–	Benzamide, N-(2-hydroxy-1-methyl-2-phenylethyl)-N-methyl-2,4,6-triisopropyl-, (R*,R*)-	66415-35-8	NI40204
198	240	43	44	380	199	226	41	**395**	100	54	34	33	32	16	15	13	**3.00**	26	53	1	1	–	–	–	–	–	–	–	–	–	Acetamide, N,N-didodecyl-		MS09106
105	286	77	395	287	122	111	67	**395**	100	75	24	20	18	17	11	11		27	25	2	1	–	–	–	–	–	–	–	–	–	2-Azetidinone, 1-benzoyl-3,3-pentamethylene-4,4-diphenyl-		NI40205
69	318	395	179	180	220	326	248	**395**	100	35	18	18	17	14	13	12		27	25	2	1	–	–	–	–	–	–	–	–	–	9,10-Dihydro-10-(1-methacryloyloxy-2-methylpropenyl)-9-phenylacridine		LI08456
395	394	211	304	122	213	121	180	**395**	100	81	22	16	14	10	5	3		27	26	–	1	–	–	–	–	–	–	1	–	–	Phosphine imide, N-phenyl-P,P,P-tris(3-tolyl)-	14796-90-8	LI08457
395	394	396	211	304	122	213	121	**395**	100	81	30	22	20	18	10	5		27	26	–	1	–	–	–	–	–	–	1	–	–	Phosphine imide, N-phenyl-P,P,P-tris(3-tolyl)-	14796-90-8	NI40206
105	196	185	183	209	214	262	108	**395**	100	83	65	63	55	41	36	26	**9.00**	27	26	–	1	–	–	–	–	–	–	1	–	–	Phosphinous amide, P,P-diphenyl-N-(1-phenylethyl)-N-benzyl-, (S)-	74952-81-1	NI40207
379	380	394	155	170	187	378	240	**395**	100	96	35	26	24	17	12	11	**7.00**	27	41	1	1	–	–	–	–	–	–	–	–	–	19-Norcholesta-1,3,5(10)-trien-6-one, 1-methyl-, oxime		NI40208
226	227	44	58	43	57	41	55	**395**	100	17	14	13	11	7	5	4	**0.00**	27	57	–	1	–	–	–	–	–	–	–	–	–	Ditridecylamine, N-methyl-		IC04276
57	41	43	44	55	56	154	83	**395**	100	77	41	38	33	25	16	15	**0.28**	27	57	–	1	–	–	–	–	–	–	–	–	–	Tris(3,5,5-trimethylhexyl)amine		IC04277
181	208	395	152	187	158	144	104	**395**	100	50	38	38	28	26	26	24		28	29	1	1	–	–	–	–	–	–	–	–	–	Azetidine, 1-cyclohexyl-2-phenyl-3-(4-phenylbenzoyl)-, cis-	13970-36-0	NI40209
181	104	214	152	378	153	395	55	**395**	100	55	48	40	38	36	34	31		28	29	1	1	–	–	–	–	–	–	–	–	–	Azetidine, 1-cyclohexyl-2-phenyl-3-(4-phenylbenzoyl)-, trans-	18599-89-8	NI40210
361	101	396	246	163	114	262	213	**396**	100	76	51	13	12	12	10	8		–	–	–	4	–	4	4	–	–	–	4	–	–	Tetrachlorotetrafluorocyclotetraphosphazene		MS09107
59	198	168	138	278	368	338	398	**396**	100	90	60	38	4	2	2	1		–	–	4	4	–	–	–	2	–	–	–	–	2	Cobalt, di-μ-bromotetranitrosodi-		MS09108
396	289	91	32	182	44	45	75	**396**	100	72	57	21	8	4	3	2		–	–	6	–	–	–	–	–	–	–	–	–	4	Arsenious oxide		IC04278
400	398	402	80	240	161	159	319	**396**	100	67	66	60	43	36	36	31	**15.00**	4	–	–	–	1	–	–	4	–	–	–	–	–	Thiophene, tetrabromo-	3958-03-0	NI40211
101	103	163	229	231	165	297	85	**396**	100	65	41	38	36	27	24	22	**3.92**	6	–	–	–	–	6	6	–	–	–	–	–	–	1-Hexene, 1,3,4,5,6,6-hexachloro-1,2,3,4,5,6-hexafluoro-		DO01116
69	48	113	67	131	179			**396**	100	88	67	58	55	38			**0.00**	6	–	2	–	2	12	–	–	–	–	–	–	–	2,2,4,4-Tetrakis(trifluoromethyl)-1,3-dithietane-1,3-dioxide		DD01548
79	80	77	240	319	28	39	321	**396**	100	86	66	61	59	59	58	57	**0.58**	6	8	–	–	–	–	–	4	–	–	–	–	–	3-Hexene, 2,3,4,5-tetrabromo-	49677-07-8	NI40212
77	279	51	396	361	50	314	47	**396**	100	53	31	23	15	14	4	3		7	5	–	–	–	–	3	–	–	–	–	–	1	Mercury, phenyl(trichloromethyl)-		LI08458
121	93	109	351	127	397	323	69	**396**	100	73	47	20	19	13	10	10	**0.00**	9	16	6	–	–	6	–	–	–	–	2	–	–	[1-(Diethoxyphosphino)-2,2,2-trifluoro-1-(trifluoromethyl)ethyl] dimethyl phosphate	111557-96-1	DD01549
160	396	227	161	229	69	398	152	**396**	100	22	20	20	12	12	11	11		10	8	2	–	2	6	–	–	–	–	–	–	1	Nickel, bis(1,1,1-trifluoro-4-thioxo-2-pentanonato-O,S)-	15744-66-8	LI08459
169	396	227	69	398	170	149	150	**396**	100	21	18	14	11	11	11	10		10	8	2	–	2	6	–	–	–	–	–	–	1	Nickel, bis(1,1,1-trifluoro-4-thioxo-2-pentanonato-O,S)-	15744-66-8	MS09109
115	116	230	128	89	229	63	57.5	**396**	100	67	25	14	13	12	12	9	**0.00**	11	7	2	–	–	–	–	–	1	–	–	–	1	Molybdenum, dicarbonyl-π-indenyliodo-		MS09110
288	100	225	150	107	121	224	193	**396**	100	89	63	59	48	44	39	34	**0.00**	11	13	7	2	1	–	–	1	–	–	–	–	–	3-Cephem-4-carboxylic acid, 3-bromomethyl-7-(methoxycarbonylamino)-, 1,1-dioxide, methyl ester	97609-90-0	NI40213
83	69	57	81	302	107	71	67	**396**	100	96	94	88	81	80	68	67	**0.70**	12	10	2	–	–	–	6	–	–	–	–	–	–	1,2,3,4,10,10-Hexachloro-trans-6,7-dihydroxy-1,4,4a,5,6,7,8,8a-octahydro-1,4-endo-5,8-exo-dimethanonaphthalene		MS09111
340	57	284	282	339	226	283	225	**396**	100	52	44	31	30	26	22	16	**13.00**	12	27	–	–	–	–	–	–	–	–	–	–	3	Tri-tert-butyl-cyclotriarsane	76173-66-5	NI40214
299	193	237	149	301	275	277	109	**396**	100	68	67	63	58	53	51	51	**10.00**	15	26	2	–	–	–	–	2	–	–	–	–	–	3,8-Dioxabicyclo[5.1.1]nonane, 6-bromo-4-(1-bromopropyl)-2-(2-penten-4-ynyl)-hexahydro-, (1S,2S)-(+)-		LI08460

MASS TO CHARGE RATIOS								M.W.	INTENSITIES								Parent	C	H	O	N	S	F	Cl	Br	I	Si	P	B	X	COMPOUND NAME	CAS Reg No	No
205	207	125	93	107	135	212	147	396	100	99	70	70	65	56	25	17	0.00	15	26	2	–	–	–	–	2	–	–	–	–	–	2H-Pyran-2-butanol, δ,5-dibromo-α-ethenyltetrahydro-α,2,6,6-tetramethyl-	79406-07-8	NI40215
73	396	381	147	397	100	75	71	396	100	81	39	30	26	21	16	13		15	28	3	4	–	–	–	–	–	3	–	–	–	Pteridine, 2,4,7-tris[(trimethylsilyl)oxy]-	31053-49-3	NI40216
73	396	381	147	397	100	75	382	396	100	84	51	35	30	25	18	15		15	28	3	4	–	–	–	–	–	3	–	–	–	Pteridine, 2,4,7-tris[(trimethylsilyl)oxy]-	31053-49-3	LI08461
149	217	231	218	147	191	150	59	396	100	36	35	21	17	12	12	9	0.00	15	36	4	–	1	–	–	–	–	3	–	–	–	D-Ribofuranose, 5-S-methyl-5-thio-1,2,3-tris-O-(trimethylsilyl)-	56700-79-9	MS09112
120	92	65	195	197	121	223	221	396	100	37	18	16	15	9	7	7		16	11	2	4	–	–	3	–	–	–	–	–	–	4H-Pyrazol-5-one, 3-[(3-aminobenzoyl)amino]-1-(2,4,6-trichlorophenyl)-		IC04279
117	198	196	118	90	194	195	89	396	100	98	53	28	22	20	19	19	13.62	16	16	–	2	–	–	–	–	–	–	–	–	2	Indoline, 5,5′-diselenodi-	22129-91-5	NI40217
396	302	397	256	45	112	228	334	396	100	27	19	18	18	17	16	10		16	16	8	2	1	–	–	–	–	–	–	–	–	Bis(4,5-dimethoxy-2-nitrophenyl) sulphide		IC04280
84	82	81	69	47	41	48	45	396	100	43	33	30	29	26	22	16	0.60	16	24	4	6	1	–	–	–	–	–	–	–	–	L-Methioninamide, 5-oxo-L-prolyl-L-histidyl-	33217-49-1	NI40218
57	41	29	45	42	43	59	27	396	100	87	85	71	37	36	26	24	0.00	17	23	4	–	–	–	3	–	–	–	–	–	–	Acetic acid, (2,4,5-trichlorophenoxy)-, 2-ethyl-2-(hydroxymethyl)hexyl ester	25417-35-0	NI40219
154	152	51	77	155	396	151	227	396	100	34	21	20	14	13	13	11		18	10	–	–	–	5	–	–	–	–	–	–	1	Arsine, diphenylpentafluorophenyl-		MS09113
398	396	295	293	268	266	397	399	396	100	95	51	51	27	27	23	21		18	13	–	4	1	–	–	1	–	–	–	–	–	Imidazole, 2-amino-4-phenyl-1-[4-(4-bromophenyl)-2-thiazolyl]-		LI08462
73	149	381	129	103	93	151	325	396	100	93	69	69	65	42	34	33	0.22	18	25	4	2	–	–	1	–	–	1	–	–	–	2,4,6(1H,3H,5H)-Pyrimidinetrione, 5-[3-chloro-2-[(trimethylsilyl)oxy]propyl]-1,3-dimethyl-5-phenyl-	57305-01-8	NI40220
73	396	219	397	75	381	45	398	396	100	99	42	25	25	20	13	12		18	32	4	–	–	–	–	–	–	3	–	–	–	Cinnamic acid, 3,4-bis(trimethylsilyloxy)-, trimethylsilyl ester	10586-03-5	NI40223
396	219	397	73	381	398	220	307	396	100	32	10	10	6	5	5	3		18	32	4	–	–	–	–	–	–	3	–	–	–	Cinnamic acid, 3,4-bis(trimethylsilyloxy)-, trimethylsilyl ester	10586-03-5	NI40222
219	396	73	397	381	191	220	398	396	100	78	48	30	27	16	15	14		18	32	4	–	–	–	–	–	–	3	–	–	–	Cinnamic acid, 3,4-bis(trimethylsilyloxy)-, trimethylsilyl ester	10586-03-5	NI40221
147	397	381	325	396	382	148	73	396	100	54	41	40	15	14	13	12		18	32	4	–	–	–	–	–	–	3	–	–	–	Cinnamic acid, α,4-bis(trimethylsilyloxy)-, trimethylsilyl ester	27750-74-9	NI40224
147	73	325	381	148	45	382	326	396	100	82	34	31	17	15	12	11	9.00	18	32	4	–	–	–	–	–	–	3	–	–	–	Cinnamic acid, α,4-bis(trimethylsilyloxy)-, trimethylsilyl ester	27750-74-9	MS09114
147	73	381	325	148	45	396	326	396	100	86	32	30	17	17	13	10		18	32	4	–	–	–	–	–	–	3	–	–	–	Cinnamic acid, α,4-bis(trimethylsilyloxy)-, trimethylsilyl ester	27750-74-9	NI40225
73	222	294	55	75	312	165	223	396	100	64	41	32	23	20	14	11	4.49	19	21	4	–	–	–	–	–	–	1	–	–	1	Cymantrene, [1-hydroxy-2-(trimethylsilyl)-1-phenylethyl]-	83617-89-4	NI40226
131	86	101	91	160	73	161	104	396	100	48	47	45	40	39	31	30	1.19	19	24	9	–	–	–	–	–	–	–	–	–	–	1,4,7,10,13,16-Hexaoxacyclooctadecane-2,5,9-trione, 3-benzyl-	79687-37-9	NI40227
43	111	153	124	83	273	171	109	396	100	99	69	58	44	33	19	14	0.00	19	24	9	–	–	–	–	–	–	–	–	–	–	β-L-Rhamnopyranoside, 2-methoxyphenyl 2,3,4-tri-O-acetyl-	18918-32-6	NI40228
44	179	42	79	95	105	91	77	396	100	44	20	18	16	14	14	14	5.24	19	24	9	–	–	–	–	–	–	–	–	–	–	Trichothec-9-en-8-one, diacetyl-12,13-epoxy-3,4,7,15-tetrahydroxy-, (3α,4β,7α)-		NI40229
247	171	91	313	172	149	217	201	396	100	13	13	7	7	7	6	6	1.00	20	20	5	4	–	–	–	–	–	–	–	–	–	7-[N-(2-Phenylmethoxycarbonylmethyl)pyridyl]-4-nitrobenzo-2-oxa-1,3-diazole		LI08463
396	219	236	235	179	178	177	237	396	100	17	14	11	11	10	7	6		20	22	3	2	–	–	–	–	–	–	–	–	1	Nickel(II), [N,N′-bis(salicylidene)-3,3′-bis(aminopropyl)oxy]-		NI40230
396	248	289	275	262	249	179	180	396	100	36	27	27	26	20	16	13		20	23	2	3	–	–	–	–	–	–	–	–	1	Cobalt(II), N,N′-bis(salicylidene)-3,3′-bis(aminopropyl)amino-		NI40231
396	181	360	329	361	298	299	398	396	100	62	56	52	40	35	31	30		20	25	6	–	–	–	1	–	–	–	–	–	–	1,2-Bis(3,4,5-trimethoxyphenyl)chloroethane		NI40232
305	364	134	304	273	79	198	231	396	100	60	56	53	31	31	25	22	5.00	20	28	8	–	–	–	–	–	–	–	–	–	–	1,4-Cyclohexene-1,4-dicarboxylic acid, 4-[4-(methoxycarbonyl)-1-butenyl]3-[(2-methoxycarbonyl)ethyl]-, dimethyl ester		MS09115
98	103	298	184	270	266	83	315	396	100	87	70	68	55	54	52	51	25.00	20	32	6	2	–	–	–	–	–	–	–	–	–	Ethanetetracarboxylic acid, 1,2-bis(cyclohexylamido), 1,2-dimethyl ester		LI08464
259	191	327	121	189	120	396	123	396	100	69	46	27	15	14	1	1		20	36	–	–	–	–	–	–	–	–	–	–	1	Tin, tetracyclopentyl-		LI08465
259	67	257	191	189	327	255	187	396	100	99	74	72	63	43	43	42	1.10	20	36	–	–	–	–	–	–	–	–	–	–	1	Tin, tetracyclopentyl-		MS09116
209	179	222	210	193	304	303	192	396	100	33	30	26	17	14	14	14	8.00	20	36	4	–	–	–	–	–	–	2	–	–	–	3-Heptanone, 1-[3-methoxy-4-[(trimethylsilyl)oxy]phenyl]-5-[(trimethylsilyl)oxy]-	56700-93-7	MS09117
173	251	105	396	77	78	319	223	396	100	75	49	23	23	20	18	16		21	14	1	–	–	6	–	–	–	–	–	–	–	Methanol, phenylbis(α,α,α-trifluoro-4-tolyl)-	21856-97-3	NI40233
270	354	355	241	118	123	153	124	396	100	80	75	38	28	25	23	23	4.00	21	16	8	–	–	–	–	–	–	–	–	–	–	Apigenin triacetate		NI40234
396	381	177	149	205	377	133	367	396	100	80	40	30	21	20	20	18		21	16	8	–	–	–	–	–	–	–	–	–	–	6H-Furo[3,2-h][1]benzopyran-6-one, 8-(1,3-benzodioxol-5-yl)-4,5,7-trimethoxy-		NI40235
190	236	149	396	78	70	118	134	396	100	54	51	49	44	33	27	26		21	24	4	4	–	–	–	–	–	–	–	–	–	Benzoic acid, 2-hydroxy-6-pentyl-4-[(1-phenyl-5-tetrazolyl)oxy]-, ethyl ester	86791-42-6	NI40236
73	176	165	166	135	45	281	100	396	100	68	56	36	36	27	24	19	0.80	21	28	2	2	–	–	–	–	–	2	–	–	–	2,4-Imidazolidinedione, 5,5-diphenyl-1,3-bis(trimethylsilyl)-	63435-72-3	NI40237
43	248	71	235	308				396	100	60	52	3	1				0.00	21	32	7	–	–	–	–	–	–	–	–	–	–	Cyclodeca[b]furan-2-one, 6-(acetoxymethyl)-6,7-epoxy-4-isobutanoyloxy-3,10-dimethyl-		LI08466
85	43	216	234	203	312	252	365	396	100	37	23	15	15	14	8	1	0.00	21	32	7	–	–	–	–	–	–	–	–	–	–	5-Heptenoic acid, 7-[5-(acetyloxy)-2-formyl-3-[(tetrahydro-2H-pyran-2-yl)oxy]cyclopentyl]-, methyl ester, [1α(Z),2β,3α,5α]-		NI40238
365	350	396	77	366	245	51	351	396	100	63	53	53	34	20	20	17		22	15	2	2	–	3	–	–	–	–	–	–	–	9-Acridinecarboxylic acid, N-(phenylimino)-1-methoxy-2,3,4-trifluoro-, methyl ester	91035-87-9	NI40239
319	279	307	340	378	318	280	308	396	100	84	83	50	48	34	33	31	21.02	22	20	7	–	–	–	–	–	–	–	–	–	–	1-Naphthacenecarboxylic acid, 2-ethyl-1,2,3,4,6,11-hexahydro-2,5,7-trihydroxy-6,11-dioxo-, methyl ester, (1R-trans)-	21179-19-1	NI40240
138	396	123	259	139	152	165	151	396	100	21	16	13	13	11	10	10		22	20	7	–	–	–	–	–	–	–	–	–	–	Succinic anhydride, bis(3,4-dimethoxybenzylidene)- trans,trans-		LI08467
396	353	397	275	225	41	238	55	396	100	33	25	15	13	11	9	9		22	24	3	2	1	–	–	–	–	–	–	–	–	Anthraquinone, 1-amino-2-(2-hydroxyethylthio)-4-cyclohexylamino-		IC04281
354	396	355	174	160	56	43	175	396	100	69	25	18	11	11	9	8		22	24	5	2	–	–	–	–	–	–	–	–	–	Aspidospermidine-10,21-dione, 1-acetyl-19,21-epoxy-17-methoxy-	36459-09-3	NI40241

MASS TO CHARGE RATIOS								M.W.	INTENSITIES								Parent	C	H	O	N	S	F	Cl	Br	I	Si	P	B	X	COMPOUND NAME	CAS Reg No	No
396	325	270	144	154	210	143	225	**396**	100	22	15	15	12	11	10	9		22	24	5	2	–	–	–	–	–	–	–	–	–	**16,19-Secostrychnidine-10,16-dione, 21,22-epoxy-21,22-dihydro-14-hydroxy-19-methyl-, (21α,22α)-**	22029-99-8	**NI40242**
367	323	396	296	352	252	73	368	**396**	100	78	66	58	57	45	44	32		22	28	3	2	–	–	–	–	–	1	–	–	–	**2,4-Imidazolidinedione, 1,3-diethyl-5-phenyl-5-[3-[(trimethylsilyl)oxy]phenyl]-**	57326-25-7	**NI40243**
396	367	73	296	323	352	268	224	**396**	100	67	52	41	39	38	38	30		22	28	3	2	–	–	–	–	–	1	–	–	–	**2,4-Imidazolidinedione, 1,3-diethyl-5-phenyl-5-[3-[(trimethylsilyl)oxy]phenyl]-**	57326-25-7	**NI40244**
396	319	324	73	296	367	352	247	**396**	100	80	53	48	45	42	40	34		22	28	3	2	–	–	–	–	–	1	–	–	–	**2,4-Imidazolidinedione, 1,3-diethyl-5-phenyl-5-[4-[(trimethylsilyl)oxy]phenyl]-**	57326-26-8	**NI40245**
323	352	224	70	252	353	324	295	**396**	100	82	63	56	24	23	23	23	**5.51**	22	28	3	2	–	–	–	–	–	1	–	–	–	**2,4-Imidazolidinedione, 1,3-diethyl-5-phenyl-5-[4-[(trimethylsilyl)oxy]phenyl]-**	57326-26-8	**NI40246**
397	120	398	208	399	396	114	108	**396**	100	59	29	16	14	10	10	10		22	28	3	2	–	–	–	–	–	1	–	–	–	**3,5-Pyrazolidinedione, 4-butyl-1-phenyl-2-[4-[(trimethylsilyl)oxy]phenyl]-**	75365-77-4	**NI40247**
254	325	396	255	326	397	312	283	**396**	100	36	18	16	12	8	7	6		22	28	3	4	–	–	–	–	–	–	–	–	–	**Benzamide, N-(2-ethylbutyl)-N-propyl-4-(4-nitrophenylazo)-**		**MS09118**
254	396	255	325	397	269	395	367	**396**	100	23	13	9	6	5	4	4		22	28	3	4	–	–	–	–	–	–	–	–	–	**Benzamide, N-hexyl-N-propyl-4-(4-nitrophenylazo)-**		**MS09119**
86	107	87	58	135	30	29	56	**396**	100	6	5	4	3	3	2	1	**0.84**	22	28	3	4	–	–	–	–	–	–	–	–	–	**1H-Benzimidazole, 3-N-[2-(diethylamino)ethyl]-2-(4-ethylbenzyl)-5-nitro-**		**NI40248**
335	353	296	43	41	223	55	69	**396**	100	68	56	55	38	35	33	28	**0.00**	22	30	4	–	–	2	–	–	–	–	–	–	–	**Pregn-4-ene-3,20-dione, 6,9-difluoro-17,21-dihydroxy-16-methyl-, (16α)-**	56145-31-4	**NI40249**
177	75	270	79	107	179	345	81	**396**	100	54	49	48	39	36	31	22	**5.00**	22	37	2	–	–	–	1	–	–	1	–	–	–	**19-Nortestosterone, 17-methyl-chlorodimethylsilyl-, (17α)-**		**NI40250**
73	339	185	209	97	99	127	74	**396**	100	21	18	17	15	10	9	9	**6.00**	22	48	–	–	–	–	–	–	–	3	–	–	–	**Silane, (1-methyl-1,2-propadien-1-yl-3-ylidene)tri-tert-butyldimethyl-**	61255-24-1	**NI40251**
396	121	186	212	238	304	213	56	**396**	100	72	62	60	35	27	20	20		23	20	1	2	–	–	–	–	–	–	–	–	1	**Ferrocene, 1-[3-(fur-2-yl)-1-phenyl-2-pyrazolin-5-yl]-**		**LI08468**
396	121	211	238	185	129	56	212	**396**	100	52	40	20	20	18	16	10		23	20	1	2	–	–	–	–	–	–	–	–	1	**Ferrocene, 1-[5-(fur-2-yl)-1-phenyl-2-pyrazolin-2-yl]-**		**LI08469**
396	365	309	179	397	161	395	311	**396**	100	49	36	31	26	26	20	20		23	24	6	–	–	–	–	–	–	–	–	–	–	**4H,6H-Benzo[1,2-b:5,4-b′]dipyran-4-one, 7,8-dihydro-3-(2,4-dimethoxyphenyl)-5-methoxy-8,8-dimethyl-**		**NI40252**
192	396	191	177	149	205	147	219	**396**	100	36	35	17	10	5	3	2		23	24	6	–	–	–	–	–	–	–	–	–	–	**[1]Benzopyrano[3,4-b]furo[2,3-h][1]benzopyran-6-one, 1,2,6,6a,12,12a-hexahydro-2-isopropyl-8,9-dimethoxy-**		**LI08470**
396	365	397	179	309	170	161	395	**396**	100	41	27	27	26	22	21	20		23	24	6	–	–	–	–	–	–	–	–	–	–	**Cyclobarpisoflavone B, dimethyl-**		**NI40253**
43	125	78	56	41	84	338	42	**396**	100	71	68	64	56	35	33	33	**0.00**	23	28	4	2	–	–	–	–	–	–	–	–	–	**Aspidofractinine-3-carboxylic acid, 1-acetyl-6-hydroxy-, methyl ester, (2α,3β,5α,6β)-**	55724-60-2	**NI40254**
338	139	307	109	42	154	41	279	**396**	100	32	22	20	18	17	17	12	**0.00**	23	28	4	2	–	–	–	–	–	–	–	–	–	**Aspidofractinine-3-carboxylic acid, 1-formyl-6-methoxy-, methyl ester, (2α,3β,5α,6β)-**	6883-09-6	**NI40255**
109	396	124	395	365	363	154	368	**396**	100	87	86	32	11	11	11	10		23	28	4	2	–	–	–	–	–	–	–	–	–	**Aspidofractinine-3-carboxylic acid, 1-formyl-17-methoxy-, methyl ester, (2α,3β,5α,21R)-**	2517-52-4	**NI40256**
109	124	340	154	396	136	110	108	**396**	100	29	21	19	17	15	10	7		23	28	4	2	–	–	–	–	–	–	–	–	–	**Aspidofractinine-3-carboxylic acid, 1-formyl-17-methoxy-, methyl ester, (2α,3β,5α,21S)-**	7013-66-3	**NI40257**
116	396	109	96	239	265	368	143	**396**	100	99	80	57	36	34	25	19		23	28	4	2	–	–	–	–	–	–	–	–	–	**Aspidofractinine-1,11-dicarboxylic acid, dimethyl ester**		**LI08471**
43	44	300	182	122	149	150	45	**396**	100	55	51	29	29	28	25	24	**2.50**	23	28	4	2	–	–	–	–	–	–	–	–	–	**Aspidospermidine-3-carboxylic acid, 20-(acetyloxy)-2,3-didehydro-, methyl ester, (5α,12β,19α,20R)-**	56143-41-0	**NI40258**
78	368	59	45	74	43	40	369	**396**	100	79	75	58	50	28	22	20	**11.00**	23	28	4	2	–	–	–	–	–	–	–	–	–	**2,20-Cyclo-8,9-secoaspidospermidine-3-carboxylic acid, 1,9-diformyl-, methyl ester, (2α,3α,5α,12β,19α,20R)-**	56053-34-0	**NI40259**
365	378	396	323	156	69	366	71	**396**	100	64	55	47	46	39	36	31		23	28	4	2	–	–	–	–	–	–	–	–	–	**4-Piperidineacetic acid, 1-acetyl-5-ethylidene-2-[3-(2-hydroxyethyl)indol-2-yl]-α-methylene-, methyl ester (ester)**	3909-65-7	**NI40260**
293	396	238	353	180	336	294	198	**396**	100	73	44	37	30	26	26	23		23	28	4	2	–	–	–	–	–	–	–	–	–	**Talbotine, 4-(N-methyl)-17-methoxy-**		**LI08472**
136	396	261	260	137	397	81		**396**	100	21	15	11	11	7	2			23	32	2	2	–	–	–	–	–	1	–	–	–	**Quinidine, 9-[(trimethylsilyl)oxy]-**		**LI08473**
273	87	274	55	288	81	67	41	**396**	100	81	63	50	47	39	31	31	**16.00**	23	37	3	–	–	–	1	–	–	–	–	–	–	**3,4-Secoandrostan-3-oic acid, 4-chloro-17-oxo-4-propyl, methyl ester, (5α)-**		**MS09120**
123	109	223	193	135	265	107	266	**396**	100	22	19	18	15	12	12	10	**0.70**	23	40	5	–	–	–	–	–	–	–	–	–	–	**1-Naphthalenepentanoic acid, 6-(acetyloxy)decahydro-2-hydroxy-β,2,5,5,8a-pentamethyl-, methyl ester, [1S-[1α(S*),2β,4aβ,6α,8aα]]**	52567-62-1	**NI40261**
197	55	43	69	81	41	95	67	**396**	100	96	72	66	64	58	54	50	**0.00**	23	45	4	–	–	–	–	–	–	–	–	1	–	**Octadecanoic acid, 9,10-dihydroxy-, methyl ester, erythro-**		**MS09121**
396	334	397	308	335	91	352	90	**396**	100	29	28	19	13	13	11	11		24	16	4	2	–	–	–	–	–	–	–	–	–	**1,2,4,5-Benzenetetracarboxylic acid, 1,2:4,5-diimide, N,N′-bis(2-tolyl)-**	31663-73-7	**NI40262**
396	91	397	65	90	74	89	191	**396**	100	40	33	24	21	21	13	11		24	16	4	2	–	–	–	–	–	–	–	–	–	**1,2,4,5-Benzenetetracarboxylic acid, 1,2:4,5-diimide, N,N′-bis(3-tolyl)-**	7143-61-5	**NI40263**
396	91	397	74	65	90	191	77	**396**	100	31	26	20	19	18	11	11		24	16	4	2	–	–	–	–	–	–	–	–	–	**1,2,4,5-Benzenetetracarboxylic acid, 1,2:4,5-diimide, N,N′-bis(4-tolyl)-**	31663-75-9	**NI40264**
396	91	397	104	378	207	368	65	**396**	100	28	27	20	18	17	14	10		24	16	4	2	–	–	–	–	–	–	–	–	–	**1,2,4,5-Benzenetetracarboxylic acid, 1,2:4,5-diimide, N,N′-dibenzyl-**	31663-79-3	**NI40265**
396	91	397	104	378	207	368	77	**396**	100	28	27	20	18	17	14	10		24	16	4	2	–	–	–	–	–	–	–	–	–	**1,2,4,5-Benzenetetracarboxylic acid, 1,2:4,5-diimide, N,N′-dibenzyl-**	31663-79-3	**MS09122**
104	77	76	172	51	121	50	247	**396**	100	85	56	50	39	34	27	15	**0.24**	24	16	4	2	–	–	–	–	–	–	–	–	–	**Spiro[1,4,2-dioxazole-5,5′(4′H)-naphth[2,1-d]isoxazol]-4′-one, 3′a,9′b-dihydro-3,3′-diphenyl-**	23767-16-0	**NI40266**
281	43	354	159	133	355	282	121	**396**	100	93	84	35	28	21	21	20	**15.00**	24	28	5	–	–	–	–	–	–	–	–	–	–	**19-Norpregna-1,3,5(10),16-tetraen-20-one, 3,21-bis(acetyloxy)-**	42224-72-6	**NI40267**
245	123	258	165	186	246	95	91	**396**	100	61	38	36	30	19	16	15	**3.40**	24	29	2	2	–	1	–	–	–	–	–	–	–	**Acetamide, N-[[1-[4-(4-fluorophenyl)-4-oxobutyl]-4-phenyl-4-piperidinyl]methyl]-**	807-31-8	**NI40268**

MASS TO CHARGE RATIOS								M.W.	INTENSITIES								Parent	C	H	O	N	S	F	Cl	Br	I	Si	P	B	X	COMPOUND NAME	CAS Reg No	No
245	123	258	165	186	246	95	158	396	100	60	37	34	29	19	15	13	3.00	24	29	2	2	–	1	–	–	–	–	–	–	–	Acetamide, N-[[1-[4-(4-fluorophenyl)-4-oxobutyl]-4-phenyl-4-piperidinyl]methyl]-	807-31-8	LI08474
43	213	181	321	396	41	241	337	396	100	32	30	25	24	24	22	18		24	32	3	2	–	–	–	–	–	–	–	–	–	2,20-Cyclo-8,9-secoaspidospermidine-3-methanol, 9-acetyl-, acetate (ester), (2α,3β,5α,12β,19α,20R)-	56145-37-0	NI40269
145	101	87	203	157	381	251	204	396	100	27	23	8	7	5	5	1	0.30	24	44	4	–	–	–	–	–	–	–	–	–	–	Glycerol, 1-O-(2-ethoxy-4-hexadecynyl)-2,3-O-isopropylidene-		NI40270
129	43	55	381	41	101	338	57	396	100	90	74	63	57	56	43	42	1.04	24	44	4	–	–	–	–	–	–	–	–	–	–	Oleic acid, (2,2-dimethyl-1,3-dioxolan-4-yl)methyl ester	33001-45-5	NI40271
136	396	137	55	91	105	145	109	396	100	50	11	9	6	5	4	4		25	32	2	–	1	–	–	–	–	–	–	–	–	Androst-5-en-17-one, 3-hydroxy-16-(phenylthio)-, (3β)-	63608-49-1	NI40272
83	149	296	227	228	167	314	229	396	100	75	50	43	38	28	14	14	6.00	25	32	4	–	–	–	–	–	–	–	–	–	–	11,19-Dihydroxyabieta-5,7,9(11),13-tetraen-12-one, 19-O-senecioate		NI40273
396	136	108	107	192	243	177	189	396	100	81	77	77	65	40	40	38		25	32	4	–	–	–	–	–	–	–	–	–	–	Δ-Octalin-1,3-diol, 4a,5-dimethyl-2,2-methylenebis-		LI08475
293	185	294	295	187	159	171	91	396	100	30	24	14	11	11	10	9	3.30	25	32	4	–	–	–	–	–	–	–	–	–	–	Pregna-1,4,6-triene-3,20-dione, 17-(acetyloxy)-6,16-dimethyl-, (16α)-	56630-83-2	MS09123
367	197	368	211	199	229	181	183	396	100	39	30	25	25	24	22	14	0.79	25	36	2	–	–	–	–	–	–	1	–	–	–	Silane, ethyl(2-methoxyphenyl)[5-methyl-2-isopropylcyclohexyloxy]phenyl-, (1α,2β,5α)-	74842-23-2	NI40274
152	339	110	396	91	41	79	105	396	100	91	61	37	32	30	26	23		25	37	3	–	–	–	–	–	–	–	–	1	–	Pregn-4-ene-3,20-dione, 21-hydroxy-, tert-butyl boronate		MS09124
152	339	110	396	105	338	151	340	396	100	91	61	37	23	22	22	21		25	37	3	–	–	–	–	–	–	–	–	1	–	Pregn-4-ene-3,20-dione, 21-hydroxy-, tert-butyl boronate		LI08476
116	117	43	57	129	55	41	69	396	100	42	33	27	24	23	16	14	4.70	25	48	3	–	–	–	–	–	–	–	–	–	–	Tetracosanoic acid, 3-oxo-, methyl ester	14531-37-4	WI01513
116	117	43	129	396	57	55	41	396	100	41	26	23	22	20	18	12		25	48	3	–	–	–	–	–	–	–	–	–	–	Tetracosanoic acid, 3-oxo-, methyl ester	14531-37-4	WI01514
43	55	41	58	57	69	74	71	396	100	61	50	44	38	37	34	27	2.14	25	48	3	–	–	–	–	–	–	–	–	–	–	Tetracosanoic acid, 23-oxo-, methyl ester	56196-16-8	NI40275
105	149	135	138	106	122	121	202	396	100	27	20	18	18	15	9	7	1.80	26	20	4	–	–	–	–	–	–	–	–	–	–	Cycloprop[a]indene-1-carboxylic acid, 6-benzoyl-1-formyl-1,1a,6,6a-tetrahydro-6a-phenyl-, methyl ester	40897-29-8	NI40276
103	77	181	51	106	50	78	75	396	100	82	45	30	9	8	7	7	0.00	26	20	4	–	–	–	–	–	–	–	–	–	–	1,2,4,5-Tetroxane, 3,3,6,6-tetraphenyl-	16204-36-7	LI08477
105	77	182	51	107	78	50	75	396	100	83	46	30	9	9	9	7	0.00	26	20	4	–	–	–	–	–	–	–	–	–	–	1,2,4,5-Tetroxane, 3,3,6,6-tetraphenyl-	16204-36-7	NI40278
105	77	181	51	107	50	78	182	396	100	83	45	31	9	9	8	7	0.00	26	20	4	–	–	–	–	–	–	–	–	–	–	1,2,4,5-Tetroxane, 3,3,6,6-tetraphenyl-	16204-36-7	NI40277
89	97	71	83	57	43	111	88	396	100	32	30	26	26	23	21	19	2.40	26	52	2	–	–	–	–	–	–	–	–	–	–	Butanoic acid, docosyl ester	55373-87-0	WI01515
201	57	43	83	55	71	69	97	396	100	58	49	38	38	34	34	33	4.24	26	52	2	–	–	–	–	–	–	–	–	–	–	Dodecanoic acid, tetradecyl ester	22412-97-1	NI40279
283	396	57	158	43	74	55	251	396	100	77	56	48	47	44	44	35		26	52	2	–	–	–	–	–	–	–	–	–	–	Heptadecanoic acid, 9-octyl-, methyl ester		MS09125
112	57	70	71	129	43	113	83	396	100	72	63	55	39	36	31	26	0.02	26	52	2	–	–	–	–	–	–	–	–	–	–	Octadecanoic acid, 2-ethylhexyl ester		IC04282
57	112	43	285	71	61	55	83	396	100	99	97	93	65	53	52	47	29.94	26	52	2	–	–	–	–	–	–	–	–	–	–	Octadecanoic acid, octyl ester	109-36-4	NI40280
74	43	87	55	41	57	75	73	396	100	83	76	47	47	46	31	31	4.00	26	52	2	–	–	–	–	–	–	–	–	–	–	Pentacosanoic acid, methyl ester	55373-89-2	NI40281
74	43	57	87	55	41	69	75	396	100	75	74	73	57	48	44	30	30.40	26	52	2	–	–	–	–	–	–	–	–	–	–	Pentacosanoic acid, methyl ester	55373-89-2	MS09126
74	87	44	396	43	75	55	57	396	100	74	43	41	37	35	33	31		26	52	2	–	–	–	–	–	–	–	–	–	–	Pentacosanoic acid, methyl ester	55373-89-2	WI01516
74	396	143	283	256	353	171	213	396	100	50	32	11	8	6	6	4		26	52	2	–	–	–	–	–	–	–	–	–	–	Tetracosanoic acid, 16-methyl-, methyl ester		LI08478
97	111	57	83	87	125	69	71	396	100	69	69	66	53	48	46	43	0.00	26	52	2	–	–	–	–	–	–	–	–	–	–	2-Tetracosanol, acetate	57289-49-3	NI40282
101	74	75	43	69	57	55	83	396	100	88	32	28	27	25	24	21	7.10	26	52	2	–	–	–	–	–	–	–	–	–	–	Tricosanoic acid, 3,5-dimethyl-, methyl ester		MS09127
149	78	396	167	248	283	95	397	396	100	83	66	53	48	27	27	20		27	40	2	–	–	–	–	–	–	–	–	–	–	Cholesta-4,6-dien-3-one, 1,2-epoxy-, (1α,2α)-	28893-44-9	NI40283
139	121	133	119	147	145	159	143	396	100	69	63	60	57	43	39	39	6.00	27	40	2	–	–	–	–	–	–	–	–	–	–	Δ(3,5)-Diosgenin	1672-65-7	NI40284
141	158	43	140	57	71	115	129	396	100	61	21	16	15	8	6	6	4.30	27	40	2	–	–	–	–	–	–	–	–	–	–	1-Naphthalenemethyl palmitate	71844-15-0	NI40285
43	57	55	41	97	71	18	83	396	100	94	66	54	52	48	46	45	0.00	27	56	1	–	–	–	–	–	–	–	–	–	–	Heptacosanol		MS09128
129	149	112	113	241	111	147	101	396	100	54	33	31	23	21	20	14	3.99	28	44	1	–	–	–	–	–	–	–	–	–	–	Cholesta-8,24-dien-3-one, 4-methyl-, (4β)-	55449-04-2	NI40286
271	81	79	297	55	67	133	159	396	100	99	51	50	41	39	31	28	0.00	28	44	1	–	–	–	–	–	–	–	–	–	–	Cholesta-5-en-3-ol, 23,23-diethenyl-, (3β)-		DD01550
69	298	124	95	271	81	43	91	396	100	73	71	62	59	59	59	54	49.41	28	44	1	–	–	–	–	–	–	–	–	–	–	Ergosta-4,22-dien-3-one	4030-92-6	NI40287
55	69	396	271	83	81	298	67	396	100	87	50	48	45	44	43	42		28	44	1	–	–	–	–	–	–	–	–	–	–	Ergosta-4,22-dien-3-one	4030-92-6	MS09129
312	313	55	269	396	41	297	69	396	100	60	50	48	41	41	36	32		28	44	1	–	–	–	–	–	–	–	–	–	–	Ergosta-4,24(28)-dien-3-one	55688-44-3	NI40288
396	363	337	211	253	381	378	229	396	100	72	28	17	13	6	6	5		28	44	1	–	–	–	–	–	–	–	–	–	–	Ergosta-5,7,24(28)-3-ol, (3β)-		LI08479
125	379	397	396	380	395	381	363	396	100	77	36	36	25	21	18	12		28	44	1	–	–	–	–	–	–	–	–	–	–	Ergosta-5,7,22-triene, 3-hydroxy-, (3β)-		NI40289
396	69	55	43	57	397	41	378	396	100	80	69	56	44	31	30	28		28	44	1	–	–	–	–	–	–	–	–	–	–	Ergosta-4,6,22-trien-3-ol, (3α)-	96790-32-8	NI40290
396	69	55	397	43	81	95	41	396	100	65	49	31	26	23	18	18		28	44	1	–	–	–	–	–	–	–	–	–	–	Ergosta-4,6,22-trien-3-ol, (3β)-	34026-93-2	NI40291
69	378	55	43	81	253	396	41	396	100	88	85	54	51	50	43	35		28	44	1	–	–	–	–	–	–	–	–	–	–	Ergosta-4,7,22-trien-3-ol, (3α)-	6538-05-2	NI40292
69	55	396	378	81	253	43	157	396	100	84	81	71	54	52	51	33		28	44	1	–	–	–	–	–	–	–	–	–	–	Ergosta-4,7,22-trien-3-ol, (3β)-	97583-19-2	NI40293
363	396	253	337	251	364	397	271	396	100	62	51	44	37	31	28	28		28	44	1	–	–	–	–	–	–	–	–	–	–	Ergosta-5,7,22-trien-3-ol, (3β,22E)-	57-87-4	NI40294
69	55	363	253	396	81	143	43	396	100	82	77	73	61	53	48	41		28	44	1	–	–	–	–	–	–	–	–	–	–	Ergosta-5,7,22-trien-3-ol, (3β,22E)-	57-87-4	NI40295
69	363	81	143	253	67	157	91	396	100	77	61	60	58	45	44	43	39.90	28	44	1	–	–	–	–	–	–	–	–	–	–	Ergosta-5,7,22-trien-3-ol, (3β,22E)-	57-87-4	NI40296
363	69	55	396	143	364	43	41	396	100	70	55	45	37	33	33	30		28	44	1	–	–	–	–	–	–	–	–	–	–	Ergosta-5,8,22-trien-3-ol, (3β,22E)-	50657-31-3	NI40297
271	253	269	145	55	69	159	107	396	100	86	80	35	32	25	24	24	5.00	28	44	1	–	–	–	–	–	–	–	–	–	–	Ergosta-8(14),15,22-trien-3-ol, (3β,5α,22E)-	56630-88-7	NI40298
81	79	55	67	95	91	105	93	396	100	62	62	44	38	35	34	28	0.00	28	44	1	–	–	–	–	–	–	–	–	–	–	27-Norcholesta-5,24(Z)-dien-3-ol, 25-ethenyl-, (3β,24Z)-		DD01551
136	118	54	69	396	119	81	91	396	100	74	57	49	46	42	36	32		28	44	1	–	–	–	–	–	–	–	–	–	–	9,10-Secoergosta-5,7,10(19),22-tetraen-3-ol, (3β,5Z,7E,22E)-	50-14-6	NI40299
136	118	56	69	396	119	81	41	396	100	75	55	50	45	40	35	35		28	44	1	–	–	–	–	–	–	–	–	–	–	9,10-Secoergosta-5,7,10(19),22-tetraen-3-ol, (3β,5Z,7E,22E)-	50-14-6	NI40300

MASS TO CHARGE RATIOS								M.W.	INTENSITIES								Parent	C	H	O	N	S	F	Cl	Br	I	Si	P	B	X	COMPOUND NAME	CAS Reg No	No
81	396	55	83	257	69	107	255	396	100	99	99	87	75	71	66	61		29	48	–	–	–	–	–	–	–	–	–	–	–	Stigmasta-4,22-diene	76866-92-7	NI40301
320	318	43	322	278	276	280	371	397	100	83	50	49	39	33	17	16	5.40	3	7	–	3	–	–	4	1	–	–	3	–	–	1-Propyl-1-bromotetrachlorocyclotriphosphazene	77589-28-7	NI40302
320	318	278	43	276	322	280	240	397	100	79	63	63	51	47	39	11	1.20	3	7	–	3	–	–	4	1	–	–	3	–	–	1-Propyl-1-bromotetrachlorocyclotriphosphazene	77589-28-7	NI40303
384	382	386	399	397	388	401	385	397	100	78	48	18	14	10	9	8		8	10	1	3	–	–	4	–	–	–	3	–	–	1-Methyl-1-(4-methoxyphenyl)tetrachlorocyclotriphosphazene	87048-79-1	NI40304
28	36	128	114	280	278	129	38	397	100	70	36	34	29	28	28	23	3.18	8	13	–	1	2	–	6	–	–	–	–	–	–	Ethylamine, N,N-diethyl-2,2-bis[(trichloromethyl)thio]-	56051-72-0	NI40305
397	278	172	399	280	69	174	149	397	100	87	80	67	57	54	50	40		13	8	1	1	–	7	2	–	–	–	–	–	–	Isoquinoline, 7,8-dichlorotetrahydro-, N-(heptafluorobutyryl)-		NI40306
397	259	30	75	398	101	378	247	397	100	34	28	23	17	16	14	14		14	6	6	5	–	3	–	–	–	–	–	–	–	Diphenylamine, 2-cyano-4′-(trifluoromethyl)-2′,4,6′-trinitro-		IC04283
43	77	41	27	30	51	131	141	397	100	70	45	41	24	20	19	18	0.00	14	24	4	1	3	–	–	–	–	–	1	–	–	Phosphorodithioic acid, O,O-diisopropyl S-[2-[(phenylsulphonyl)amino]ethyl] ester	741-58-2	NI40307
73	265	93	75	77	188	95	224	397	100	94	82	52	35	31	30	28	0.00	14	36	4	1	–	–	–	–	–	3	1	–	–	Butanoic acid, 2-[(trimethylsilyl)amino]-4-[[[(trimethylsilyl)oxy]methyl]phosphinyl]-, trimethylsilyl ester, (S)-	56772-22-6	MS09130
282	69	71	43	253	341	226	271	397	100	96	75	74	15	11	10	7	0.00	15	22	3	1	–	7	–	–	–	–	–	–	–	Isoleucine, N-(heptafluorobutyryl)-, 3-methylbutyl ester		NI40308
43	69	282	240	71	341	226	169	397	100	68	55	38	35	5	5	5	0.00	15	22	3	1	–	7	–	–	–	–	–	–	–	Leucine, N-(heptafluorobutyryl)-, 3-methylbutyl ester		NI40309
282	69	71	226	240	169	341	253	397	100	54	30	23	7	5	3	2	0.00	15	22	3	1	–	7	–	–	–	–	–	–	–	Norleucine, N-(heptafluorobutyryl)-, 3-methylbutyl ester		NI40310
229	231	397	168	147	399	153	215	397	100	59	48	40	28	24	24	21		16	14	8	–	–	–	–	–	–	–	–	–	1	Copper, bis(dehydroacetato)-		NI40311
248	250	232	181	247	182	234	246	397	100	99	88	76	72	72	62	60	7.29	16	23	5	5	1	–	–	–	–	–	–	–	–	Adenosine, N-(4-hydroxy-3-methyl-2-butenyl)-2-(methylthio)-	26190-61-4	NI40312
250	248	232	181	247	182	234	135	397	100	99	88	78	72	72	63	55	8.00	16	23	5	5	1	–	–	–	–	–	–	–	–	Adenosine, N-(4-hydroxy-3-methyl-2-butenyl)-2-(methylthio)-	26190-61-4	LI08480
43	218	57	85	69	168	278	219	397	100	37	17	3	3	2	1	1	0.00	18	23	9	1	–	–	–	–	–	–	–	–	–	Pyrrole, 3-(methoxycarbonyl)-2-methyl-5-(2,3,5-tri-O-acetyl-α-D-ribofuranosyl)-		NI40313
43	218	168	69	85	136	219	278	397	100	85	25	19	13	12	10	6	0.70	18	23	9	1	–	–	–	–	–	–	–	–	–	Pyrrole, 3-(methoxycarbonyl)-2-methyl-5-(2,3,5-tri-O-acetyl-β-D-ribofuranosyl)-		NI40314
43	168	69	218	136	167	166	57	397	100	55	46	18	11	6	6	6	3.00	18	23	9	1	–	–	–	–	–	–	–	–	–	Pyrrole, 3-(methoxycarbonyl)-2-methyl-5-(2,3,4-tri-O-acetyl-α-D-ribopyranosyl)-		NI40315
43	168	69	218	85	136	57	277	397	100	48	42	23	17	10	10	5	1.90	18	23	9	1	–	–	–	–	–	–	–	–	–	Pyrrole, 3-(methoxycarbonyl)-2-methyl-5-(2,3,4-tri-O-acetyl-β-D-ribopyranosyl)-		NI40316
151	397	152	398	347	137	107	121	397	100	43	14	8	6	6	4	3		18	24	3	1	–	5	–	–	–	–	–	–	–	Benzeneheptanamine, 3,4-dimethoxy-N-(pentafluoropropionyl)-		NI40317
73	218	45	219	75	179	91	74	397	100	63	16	13	10	9	9	9	0.00	18	35	3	1	–	–	–	–	–	3	–	–	–	L-Tyrosine, N,O-bis(trimethylsilyl)-, trimethylsilyl ester	51220-73-6	MS09131
73	218	77	47	45	75	100	147	397	100	44	37	19	14	11	10	7	0.00	18	35	3	1	–	–	–	–	–	3	–	–	–	L-Tyrosine, N,O-bis(trimethylsilyl)-, trimethylsilyl ester	51220-73-6	NI40318
218	73	147	219	281	179	220	75	397	100	48	25	21	13	12	10	7	0.13	18	35	3	1	–	–	–	–	–	3	–	–	–	L-Tyrosine, N,O-bis(trimethylsilyl)-, trimethylsilyl ester	51220-73-6	NI40319
111	83	43	109	136	153	125	55	397	100	97	88	87	61	47	45	43	0.00	19	27	8	1	–	–	–	–	–	–	–	–	–	Camphanoic acid, 3-acetoxy-2-methyl-2-nitrocyclohexyl ester, (1R,2R,3S)-		DD01552
397	188	110	285	148	176	256	166	397	100	59	58	50	50	38	36	30		19	28	6	1	–	–	–	–	–	–	1	–	–	3-Benzazepin-1-one, 2,3,4,5-tetrahydro-7,8-(methylenedioxy)-N-(dibutylphosphoryl)-	92241-55-9	NI40320
58	71	84	267	255	238	198	101	397	100	57	42	34	34	34	23	22	0.00	19	41	1	3	–	–	2	–	–	–	–	–	–	Azacyclotridecan-2-one, 1-[3-[[3-(methylamino)propyl]amino]propyl]-, dihydrochloride	67370-89-2	NI40321
184	186	201	105	199	41	43	55	397	100	97	51	43	41	40	33	30	2.90	20	16	3	1	–	–	–	1	–	–	–	–	–	1,4-Naphthoquinone-3-carboxamide, N-ethyl-N-(4-bromophenyl)-2-methyl		MS09132
397	236	220	237	180	179	291	250	397	100	21	12	11	9	9	8	7		20	22	3	2	–	–	–	–	–	–	–	–	1	Cobalt(II), [N,N′-bis(salicylidene)-3,3′-bis(aminopropyl)oxy]-		NI40322
220	120	136	93	55	43	83	297	397	100	49	41	34	34	34	31	17	2.00	20	31	7	1	–	–	–	–	–	–	–	–	–	Heliosupine	32728-78-2	LI08481
220	120	55	136	93	43	83	44	397	100	50	43	40	35	35	30	30	1.00	20	31	7	1	–	–	–	–	–	–	–	–	–	Heliosupine	32728-78-2	NI40323
220	120	119	121	136	93	221	55	397	100	73	64	46	38	35	33	32	1.20	20	31	7	1	–	–	–	–	–	–	–	–	–	Heliosupine	32728-78-2	NI40324
213	155	212	187	215	169	186	171	397	100	97	95	71	51	43	36	36	15.00	20	35	5	3	–	–	–	–	–	–	–	–	–	L-Proline, 1-[N-[N-(1-oxodecyl)glycyl]glycyl]-, methyl ester	31944-57-7	NI40325
139	91	77	125	51	119	65	39	397	100	99	42	33	25	22	22	14	0.00	21	19	3	1	2	–	–	–	–	–	–	–	–	S-Phenyl-S-[(E)-styryl]-N-(4-tolylsulphonyl)sulphoximide, (S)-		DD01553
304	379	135	302	149	380	257	77	397	100	80	55	51	51	32	28	26	18.00	22	23	4	1	1	–	–	–	–	–	–	–	–	1-(1-Adamantyl)-4-phenylsulphonyl-3-nitrobenzene		NI40326
162	397	382	366	204	236	234	188	397	100	98	89	81	50	27	27	23		22	23	6	1	–	–	–	–	–	–	–	–	–	Corydalic acid, methyl ester		NI40327
58	71	397	43	85	204	149	97	397	100	19	13	13	12	10	6	6		22	23	6	1	–	–	–	–	–	–	–	–	–	6H-Indeno[4,5-d]-1,3-dioxole-6,8(7H)-dione, 7-[2-[2-(dimethylamino)ethyl]-4,5-dimethoxyphenyl]-	94607-90-6	NI40328
162	397	192	204	163	193	205	398	397	100	79	49	34	21	17	15	14		22	23	6	1	–	–	–	–	–	–	–	–	–	Orientalidine	23943-90-0	NI40329
162	397	192	204	163	193	205	133	397	100	79	49	35	22	18	16	13		22	23	6	1	–	–	–	–	–	–	–	–	–	Orientalidine	23943-90-0	LI08482
205	221	187	206	107	160	77	57	397	100	52	23	15	14	10	9	9	5.29	22	23	6	1	–	–	–	–	–	–	–	–	–	L-Phenylalanine, N-[(3,4-dihydro-8-hydroxy-3-methyl-1-oxo-1H-2-benzopyran-7-yl)carbonyl]-, ethyl ester, (R)-	18420-71-8	NI40330
354	337	397	355	322	324	192	338	397	100	38	20	19	19	13	13	10		22	23	6	1	–	–	–	–	–	–	–	–	–	Spiro[7H-indeno[4,5-d]-1,3-dioxole-7,1′(2′H)-isoquinoline]-7′,8-diol, 3′,4′,6,8-tetrahydro-6′-methoxy-2′-methyl-, 8-acetate, trans-	30833-08-0	NI40331
268	341	218	267	397	324	215	162	397	100	91	55	52	40	33	31	29		23	27	5	1	–	–	–	–	–	–	–	–	–	Morphine, dipropanoyl-	10589-79-4	NI40332

MASS TO CHARGE RATIOS								M.W.	INTENSITIES								Parent	C	H	O	N	S	F	Cl	Br	I	Si	P	B	X	COMPOUND NAME	CAS Reg No	No
339	121	297	142	160	241	213	397	**397**	100	95	60	40	22	15	12	8		23	27	5	1	–	–	–	–	–	–	–	–	–	5′H-Pregna-1,4-dieno[17,16-d]oxazolo-3,11,20-trione, 21-hydroxy-2′-methyl-, (11β,16β)-	13649-58-6	NI40333
351	382	352	160	199	198	381	279	**397**	100	96	87	87	67	67	59	48	**15.00**	23	29	4	2	–	–	–	–	–	–	–	–	–	Macusine A, 11-methoxy-	87340-28-1	NI40334
77	275	123	397	65	114	115	113	**397**	100	80	60	48	40	36	35	35		24	19	3	3	–	–	–	–	–	–	–	–	–	Naphthalene, 2-hydroxy-3-(4-methoxyanilinocarbonyl)-1-phenylazo-		IC04284
397	194	77	175	144	398	176	51	**397**	100	57	51	41	32	26	17	16		24	19	3	3	–	–	–	–	–	–	–	–	–	4-Pyrimidinecarboxylic acid, 1,2,3,6-tetrahydro-6-oxo-1,3-diphenyl-2-(phenylimino)-, methyl ester	16135-24-3	NI40335
271	173	312	147	337	160	199	227	**397**	100	58	39	37	36	28	22	21	**13.00**	24	31	4	1	–	–	–	–	–	–	–	–	–	19-Norpregna-1,3,5(10)-triene-20-carbonitrile, 20-(acetyldioxy)-3-methoxy-	62623-55-6	NI40336
397	73	247	382	366	187	157	75	**397**	100	80	74	63	31	31	29	29		24	35	2	1	–	–	–	–	–	1	–	–	–	Estra-4,6-dien-3-one, 17-ethynyl-17-(trimethylsilyl)-, O-methyl oxime, (17α,17β)-		NI40337
208	397	91	77	209	398	382	341	**397**	100	67	52	26	22	18	18	18		26	27	1	3	–	–	–	–	–	–	–	–	–	4-Imidazolidinone, 1-tert-butyl-2-imino-5,5-diphenyl-3-benzyl-	56954-67-7	NI40338
208	397	91	77	209	398	341	382	**397**	100	67	52	26	21	20	20	18		26	27	1	3	–	–	–	–	–	–	–	–	–	4-Imidazolidinone, 1-tert-butyl-2-imino-5,5-diphenyl-3-benzyl-	56954-67-7	NI40339
250	91	251	397	194	165	341	104	**397**	100	27	23	13	13	13	12	11		26	27	1	3	–	–	–	–	–	–	–	–	–	4-Imidazolidinone, 3-tert-butyl-2-imino-5,5-diphenyl-1-benzyl-	54508-09-7	NI40340
262	246	380	264	397	232	399	381	**397**	100	76	45	38	34	26	25	25		27	31	–	3	–	–	–	–	–	–	–	–	–	2-Phenazinamine, 1,3,4,6,8,9-hexamethyl-N-(2,4,5-trimethylphenyl)-	54387-11-0	NI40341
369	174	222	397	89	106	130	370	**397**	100	90	70	55	38	36	34	30		27	43	1	1	–	–	–	–	–	–	–	–	–	7a-Aza-B-homocholesta-3,5-dien-7-one	19682-30-5	LI08483
369	174	222	397	382	188	354	154	**397**	100	89	72	55	33	22	17	17		27	43	1	1	–	–	–	–	–	–	–	–	–	7a-Aza-B-homocholesta-3,5-dien-7-one	19682-30-5	NI40342
397	319	199	305					**397**	100	20	15	10						28	19	–	3	–	–	–	–	–	–	–	–	–	Phenylrosindulin		LI08484
159	30	85	150	127	128	56	89	**398**	100	58	52	48	43	45	45	39	**0.00**	4	8	2	4	2	10	–	–	–	–	–	–	–	(Methylenediureylene)bis(sulphur pentafluoride)		NI40343
241	81	239	243	79	323	321	53	**398**	100	58	51	50	36	32	32	30	**0.03**	6	10	–	–	–	–	–	4	–	–	–	–	–	Hexane, 1,2,5,6-tetrabromo-		IC04285
241	39	41	81	27	28	79	53	**398**	100	97	95	89	84	82	56	56	**0.00**	6	10	–	–	–	–	–	4	–	–	–	–	–	Hexane, 3,3,4,4-tetrabromo-	49677-03-4	NI40344
29	69	131	100	51	79	31	93	**398**	100	57	30	20	16	10	8	6	**0.00**	8	1	1	–	–	15	–	–	–	–	–	–	–	Octanal, pentadecafluoro-	335-60-4	WI01517
163	79	85	57	45	69	43	97	**398**	100	90	55	17	17	13	11	8	**0.32**	9	18	11	–	3	–	–	–	–	–	–	–	–	α-D-Xylopyranoside, methyl, trimethanesulphonate	29709-78-2	NI40345
135	44	42	92	136	43	90	76	**398**	100	38	17	6	5	5	3	3	**0.00**	10	12	2	2	–	–	5	–	–	–	1	–	–	Phosphorodiamidic acid, tetramethyl-, pentachlorophenyl ester	1440-97-7	NI40346
199	398	155	200	181	366	379	117	**398**	100	35	35	22	19	18	10	9		12	–	–	–	2	10	–	–	–	–	–	–	–	Bis(pentafluorophenyl) disulphide	1494-06-0	NI40347
168	200	149	99	117	199	55	93	**398**	100	87	74	73	68	58	58	55	**0.00**	12	–	–	–	2	10	–	–	–	–	–	–	–	Bis(pentafluorophenyl) disulphide	1494-06-0	MS09133
182	97	103	225	86	153	95	111	**398**	100	87	68	40	23	22	15	14	**0.00**	12	12	8	–	–	6	–	–	–	–	–	–	–	β-D-Xylopyranoside, methyl 2-O-acetyl-3,4-di-O-(trifluoroacetyl)-		MS09134
278	97	279	193	165	103	95	140	**398**	100	26	15	14	10	10	10	9	**0.00**	12	12	8	–	–	6	–	–	–	–	–	–	–	β-D-Xylopyranoside, methyl 3-O-acetyl-2,4-di-O-(trifluoroacetyl)-		MS09135
225	182	97	170	114	95	338	153	**398**	100	30	19	11	11	11	9	9	**0.00**	12	12	8	–	–	6	–	–	–	–	–	–	–	β-D-Xylopyranoside, methyl 4-O-acetyl-2,3-di-O-(trifluoroacetyl)-		MS09136
370	254	284	342	226	168	398	312	**398**	100	86	77	75	69	62	47	45		13	11	8	–	–	–	–	–	–	–	–	–	1	Rhodium(I), η^5-[1,2,3-tris(methoxycarbonyl)cyclopentadienyl]bis(carbonyl)-		NI40348
104	50	398	51	76	400	126	48	**398**	100	86	85	78	65	64	60	56		14	4	4	2	2	–	2	–	–	–	–	–	–	1,4-Bis(thionylamino)-2,3-dichloroanthraquinone		IC04286
245	33	243	247	244	32	31	246	**398**	100	73	55	53	38	33	33	32	**0.00**	14	24	–	6	2	–	–	–	–	–	–	–	1	Nickel, [6,14-diisothiocyanato-1,5,9,13-tetraazacyclohexadecanato(2-)-N^1,N^5,N^9,N^{13}]-	69815-44-7	NI40349
312	104	397	69	398	77	165	130	**398**	100	60	50	50	45	35	25	15		16	10	2	2	–	6	–	–	–	–	–	2	–	2,4-Diphenyl-3-(trifluoroacetyl)-6-(trifluoromethyl)-1-oxa-3,5-diaza-2,4-dibora-5-cyclohexene		NI40350
367	369	45	398	368	183	139	400	**398**	100	60	42	20	20	20	18	12		16	16	4	4	–	–	2	–	–	–	–	–	–	N,N-Bis(2-hydroxyethyl)-3-chloro-4-(2-chloro-4-nitrophenylazo)aniline		IC04287
226	212	321	319	305	307	211	197	**398**	100	78	50	50	48	45	30	27	**0.00**	16	20	–	2	–	–	–	2	–	–	–	–	–	Pyrrole, 2-bromo-5-[[5-(bromomethyl)-4-ethyl-3-methyl-2H-pyrrol-2-ylidene]methyl]-4-ethyl-3-methyl-		LI08485
226	212	321	319	305	307	197	211	**398**	100	78	51	51	48	45	37	29	**0.00**	16	20	–	2	–	–	–	2	–	–	–	–	–	Pyrrole, 2-bromo-5-[[5-(bromomethyl)-4-ethyl-3-methyl-2H-pyrrol-2-ylidene]methyl]-4-ethyl-3-methyl-	15770-14-6	NI40351
166	28	250	61	41	324	29	167	**398**	100	50	30	29	27	23	22	20	**5.00**	16	25	4	2	1	3	–	–	–	–	–	–	–	N-(Trifluoroacetyl)-L-prolylmethionine, butyl ester		MS09137
166	250	29	41	28	167	324	70	**398**	100	42	35	33	29	24	17	15	**0.00**	16	25	4	2	1	3	–	–	–	–	–	–	–	N-(Trifluoroacetyl)-L-prolylpenicillamine, butyl ester		MS09138
73	252	129	295	147	308	398	383	**398**	100	67	57	55	55	54	25	24		16	30	4	4	–	–	–	–	–	2	–	–	–	Xanthine, 1,3-dimethyl-7-[2,3-bis-O-(trimethylsilyl)propyl]-		NI40352
57	327	58	328	258	369	56	55	**398**	100	20	8	5	5	2	2	2	**0.00**	17	11	–	–	–	8	–	–	–	–	1	–	–	5H-Dibenzophosphole, 5-(2,2-dimethylpropyl)-1,2,3,4,6,7,8,9-octafluoro-	69688-62-6	NI40353
189	247	398	370	306	399	124	208	**398**	100	94	93	82	33	26	25	19		17	17	–	–	–	–	–	–	–	–	–	–	3	Tricobalt, tri-η(5)-cyclopentadienyl-di-μ3-methylidyne-triangulo-	74127-99-4	NI40354
179	180	41	93	162	135	149	79	**398**	100	99	63	56	51	50	48	48	**5.00**	17	23	2	2	1	–	–	1	–	–	–	–	–	4(1H)-Azulenone, octahydro-8a-methyl-, [(4-bromophenyl)sulphonyl]hydrazone		NI40355
334	193	366	161	275	274	59	335	**398**	100	84	39	39	35	30	28	23	**5.50**	18	22	10	–	–	–	–	–	–	–	–	–	–	Bicyclo[3.3.1]nonane-2,4,6,8-tetracarboxylic acid, 1-methyl-3,7-dioxo, tetramethyl ester	86668-68-0	NI40356
73	179	398	45	75	267	74	180	**398**	100	44	20	18	15	14	9	8		18	34	4	–	–	–	–	–	–	3	–	–	–	Benzenepropanoic acid, 3,4-bis(trimethylsiloxy)-, trimethylsilyl ester	27750-68-1	NI40357
73	267	147	280	398	268	45	75	**398**	100	63	27	17	16	16	13	11		18	34	4	–	–	–	–	–	–	3	–	–	–	Benzenepropanoic acid, α,3-bis[(trimethylsilyl)oxy]-, trimethylsilyl ester		NI40358
73	179	147	308	180	45	75	74	**398**	100	84	24	16	14	14	9	9	**0.20**	18	34	4	–	–	–	–	–	–	3	–	–	–	Benzenepropanoic acid, α,4-bis[(trimethylsilyl)oxy]-, trimethylsilyl ester	27750-67-0	NI40360
179	308	73	147	180	75	79	77	**398**	100	71	69	54	50	35	26	25	**2.38**	18	34	4	–	–	–	–	–	–	3	–	–	–	Benzenepropanoic acid, α,4-bis[(trimethylsilyl)oxy]-, trimethylsilyl ester	27750-67-0	NI40359
179	308	180	147	309	191	281	73	**398**	100	35	16	12	7	7	6	6	**1.10**	18	34	4	–	–	–	–	–	–	3	–	–	–	Benzenepropanoic acid, α,4-bis[(trimethylsilyl)oxy]-, trimethylsilyl ester	27750-67-0	NI40361

MASS TO CHARGE RATIOS								M.W.	INTENSITIES								Parent	C	H	O	N	S	F	Cl	Br	I	Si	P	B	X	COMPOUND NAME	CAS Reg No	No
73	267	75	147	268	280	398	77	**398**	100	94	53	41	22	21	18	16		18	34	4	–	–	–	–	–	–	3	–	–	–	Benzenepropanoic acid, β,3-bis[(trimethylsilyl)oxy]-, trimethylsilyl ester		MS09139
57	45	85	56	101	125	41	199	**398**	100	96	90	90	73	64	60	46	**0.00**	18	39	7	–	–	–	–	–	–	–	1	–	–	Ethanol, 2-butoxy-, phosphate (3:1)	78-51-3	NI40362
57	251	45	56	41	85	125	94	**398**	100	87	81	70	46	45	30	27	**0.00**	18	39	7	–	–	–	–	–	–	–	1	–	–	Ethanol, 2-butoxy-, phosphate (3:1)	78-51-3	NI40363
370	170	398	70	241	100	203	101	**398**	100	25	20	20	16	13	12	9		18	44	–	6	–	–	–	–	–	–	–	–	2	Aluminium, tetraethylbis[μ-(1,1,3,3-tetramethylguanidinato)]di-	26025-00-3	NI40364
47	106	91	222	107	45	398	77	**398**	100	84	72	67	67	64	40	36		19	22	–	6	2	–	–	–	–	–	–	–	–	1,2,4-Triazole, 4-[N,N'-bis(4-methylphenyl)guanidino]-3,5-bis(methylthio)		MS09140
82	44	100	137	338	121	109	83	**398**	100	89	54	46	44	44	44	43	**2.94**	19	26	9	–	–	–	–	–	–	–	–	–	–	7,8-Dihydroxydiacetoxyscirpenol		NI40365
79	41	94	95	66	139	69	67	**398**	100	62	60	48	48	47	46	46	**1.00**	19	27	4	–	–	–	–	1	–	–	–	–	–	Propanedioic acid, (bromomethyl)methyl-, di-5-heptynyl ester	60004-36-6	NI40366
398	399	352	130	126	125	146	75	**398**	100	24	16	12	10	10	8	8		20	10	4	6	–	–	–	–	–	–	–	–	–	7,7a,12,12a-Tetraazanaphtho[4,5-a]naphthalene, dinitro-		IC04288
135	77	51	93	263	50	205	136	**398**	100	85	43	38	18	16	15	14	**0.00**	20	22	1	4	2	–	–	–	–	–	–	–	–	4-Imidazolinone, 3-phenyl-5-[(4-N-phenylthioureido)butyl]-2-thioxo-		LI08486
57	71	43	85	41	55	99	29	**398**	100	65	64	40	21	17	13	12	**0.00**	21	14	3	2	–	–	–	–	–	–	–	–	1	Iron, tricarbonyl[N-(phenyl-2-pyridinylmethylene)benzenamine-N,N']-	74764-11-7	NI40367
314	356	298	83	315	85	285	43	**398**	100	42	30	30	19	18	12	10	**0.00**	21	18	8	–	–	–	–	–	–	–	–	–	–	4H-1-Benzopyran-4-one, 5-(acetyloxy)-2-[3-(acetyloxy)-4-methoxyphenyl] 7-methoxy-	32174-63-3	NI40368
264	364	227	236	287	265	235	228	**398**	100	74	70	67	66	65	65	61	**0.00**	21	19	4	2	–	–	1	–	–	–	–	–	–	5H-Pyrrolo[1,2-d]benzodiazepine-2,6-dione, 2-acetoxy-10-chloro-1,2,7,11b-tetrahydro-trans-2-methyl-trans-11b-phenyl-, (R)-	89839-36-1	NI40369
383	384	73	385	311	184	386	368	**398**	100	30	15	12	4	4	3	2	**0.00**	21	26	4	–	–	–	–	–	–	2	–	–	–	Anthraquinone, 2-methyl-1,6-bis[(trimethylsilyl)oxy]-	91701-18-7	NI40370
383	384	385	73	311	184	386	378	**398**	100	30	13	12	4	4	3	3	**0.00**	21	26	4	–	–	–	–	–	–	2	–	–	–	Anthraquinone, 3-methyl-1,6-bis[(trimethylsilyl)oxy]-	91701-19-8	NI40371
383	384	385	311	73	386	353	397	**398**	100	30	13	4	4	3	2	1	**0.00**	21	26	4	–	–	–	–	–	–	2	–	–	–	Anthraquinone, 3-methyl-1,8-bis[(trimethylsilyl)oxy]-	7336-72-3	NI40372
383	384	385	73	367	294	176	337	**398**	100	32	14	10	6	6	5	4	**0.00**	21	26	4	–	–	–	–	–	–	2	–	–	–	Anthraquinone, 1-[(trimethylsilyl)oxy]-3-[[(trimethylsilyl)oxy]methyl]-	91701-17-6	NI40373
179	278	99	222	151	246	285	207	**398**	100	71	66	56	40	34	29	24	**1.00**	21	34	7	–	–	–	–	–	–	–	–	–	–	Cyclopentanepropanoic acid, 3,5-bis(acetyloxy)-2-(3-oxooctyl)-, methyl ester	55493-77-1	NI40374
77	293	398	399	217	93	294	65	**398**	100	98	96	27	26	26	20	20		22	18	2	6	–	–	–	–	–	–	–	–	–	1-(3-Hydroxyphenyl)-3-hydroxy-5-methyl-4-(4-phenylazophenylazo)pyrazole		IC04289
212	398	77	91	51	213	399	78	**398**	100	50	20	18	15	14	13	11		22	18	2	6	–	–	–	–	–	–	–	–	–	5-Pyrazolol, 1-phenyl-4-[3-(phenylazo)-6-hydroxyphenylazo]-3-methyl-		IC04290
398	325	380	310	365	59	295	155	**398**	100	97	89	73	61	45	35	31		22	22	7	–	–	–	–	–	–	–	–	–	–	6H-Benzofuro[3,2-c][1]benzopyran, 9-hydroxy-2-(3-hydroxy-3-methylbutyl)-1,3-dimethoxy-6-oxo-		LI08487
398	399	181	173	185	230	400	168	**398**	100	26	17	14	10	8	6	6		22	22	7	–	–	–	–	–	–	–	–	–	–	Podophyllotoxin, deoxy-	19186-35-7	NI40375
398	399	181	173	185	230	168	186	**398**	100	26	17	14	9	7	6	5		22	22	7	–	–	–	–	–	–	–	–	–	–	Podophyllotoxin, deoxy-	19186-35-7	LI08488
43	44	45	60	41	57	55	42	**398**	100	95	45	41	37	35	25	25	**3.69**	22	26	5	2	–	–	–	–	–	–	–	–	–	Curan-17-oic acid, 19-(acetyloxy)-2,16-didehydro-20-hydroxy-, methyl ester, (19S)-	2111-86-6	NI40376
186	380	185	323	199	309	171	212	**398**	100	29	24	16	10	9	5	2	**2.00**	22	26	5	2	–	–	–	–	–	–	–	–	–	Eripinic acid		LI08489
222	69	223	77	130	368	68	80	**398**	100	59	47	28	24	22	22	16	**0.00**	22	26	5	2	–	–	–	–	–	–	–	–	–	Formosanan-16-carboxylic acid, 10-methoxy-19-methyl-2-oxo-, methyl ester, (7ξ,19α,20α)-	6871-19-8	NI40377
398	180	43	399	353	190	339	356	**398**	100	50	25	24	23	23	18	15		22	26	5	2	–	–	–	–	–	–	–	–	–	Henningsoline		LI08490
398	222	236	206	237	192	160	146	**398**	100	35	32	30	27	25	25	25		22	26	5	2	–	–	–	–	–	–	–	–	–	Isorotundifoline	6884-20-4	NI40378
398	146	108	162	160	161	222	175	**398**	100	79	69	45	45	31	28	24		22	26	5	2	–	–	–	–	–	–	–	–	–	Rotundifoline	6883-25-6	NI40379
299	398	339	136	300	262	301	108	**398**	100	83	25	24	23	18	15	15		22	26	5	2	–	–	–	–	–	–	–	–	–	Tabernulosine	74627-72-8	NI40380
223	398	69	248	55	399	224	41	**398**	100	64	40	27	18	17	15	15		22	26	5	2	–	–	–	–	–	–	–	–	–	Tetraphylline oxindole B	62358-35-4	NI40381
222	223	398	381	399	176	224	208	**398**	100	74	46	21	12	12	11	7		22	26	5	2	–	–	–	–	–	–	–	–	–	Tetraphylline pseudoindoxyl	92471-49-3	NI40382
222	223	398	381	69	399	176	224	**398**	100	74	46	21	19	12	12	11		22	26	5	2	–	–	–	–	–	–	–	–	–	Tetraphylline pseudoindoxyl A		NI40383
58	128	72	98	84	56	185	112	**398**	100	66	37	34	24	20	17	17	**2.29**	22	46	2	4	–	–	–	–	–	–	–	–	–	Acetamide, N,N'-[1,4-butanediylbis[(ethylimino)-3,1-propanediyl]]bis[N-ethyl-	40563-87-9	NI40384
354	138	43	355	160	370	110	398	**398**	100	89	45	32	25	23	21	17		23	30	4	2	–	–	–	–	–	–	–	–	–	Aspidospermidine, 1-acetyl-19,21-epoxy-16,17-dimethoxy-	54725-27-8	NI40385
168	324	398	325	169	124	370	355	**398**	100	71	32	29	11	10	8	8		23	30	4	2	–	–	–	–	–	–	–	–	–	Aspidospermidin-21-oic acid, 1-acetyl-17-methoxy-, methyl ester	6858-95-3	NI40386
398	383	214	397	200	186	384	255	**398**	100	97	82	75	29	27	26	25		23	30	4	2	–	–	–	–	–	–	–	–	–	Corynan-16-carboxylic acid, 16,17-didehydro-9,17-dimethoxy-, methyl ester, (3β,16E)-	14509-92-3	NI40387
214	383	398	397	269	200	186	199	**398**	100	63	47	42	27	26	23	21		23	30	4	2	–	–	–	–	–	–	–	–	–	Corynan-16-carboxylic acid, 16,17-didehydro-9,17-dimethoxy-, methyl ester, (3β,16E,20β)-	14382-79-7	NI40388
398	397	214	383	200	399	186	199	**398**	100	90	71	50	32	29	23	21		23	30	4	2	–	–	–	–	–	–	–	–	–	Corynan-16-carboxylic acid, 16,17-didehydro-9,17-dimethoxy-, methyl ester, (16E)-	4697-67-0	NI40389
214	398	397	383	200	186	399	269	**398**	100	93	92	43	26	26	24	23		23	30	4	2	–	–	–	–	–	–	–	–	–	Corynan-16-carboxylic acid, 16,17-didehydro-9,17-dimethoxy-, methyl ester, (16E,20β)-	4098-40-2	NI40390
114	100	42	56	115	70	398	101	**398**	100	83	14	9	8	6	6	6		23	30	4	2	–	–	–	–	–	–	–	–	–	Morphinan-6-ol, 7,8-didehydro-4,5-epoxy-17-methyl-3-[2-(4-morpholinyl)ethoxy]-, (5α,6α)-	509-67-1	NI40391
398	327	44	43	58	355	42	57	**398**	100	89	73	41	34	27	25	24		23	30	4	2	–	–	–	–	–	–	–	–	–	3,4-Secocondyfolan-3-one, 1-acetyl-14,19-didehydro-10,11-dimethoxy-4-methyl-, (2β,7β,14E,15α)-	36954-68-4	NI40392
138	370	398	204	341	109	149	43	**398**	100	19	17	16	9	8	6	6		23	30	4	2	–	–	–	–	–	–	–	–	–	Spegazzinidine, 3-deoxy-O,O-dimethyl-3-oxo-	19649-33-3	NI40393

MASS TO CHARGE RATIOS								M.W.	INTENSITIES								Parent	C	H	O	N	S	F	Cl	Br	I	Si	P	B	X	COMPOUND NAME	CAS Reg No	No
398	183	397	399	108	262	185	107	398	100	76	51	25	25	24	21	13		24	19	2	2	–	–	–	–	–	–	1	–	–	Phosphine imide, N-(3-nitrophenyl)-P,P,P-triphenyl-	14796-86-2	NI40394
398	183	397	108	262	185	107	152	398	100	76	51	25	24	21	13	9		24	19	2	2	–	–	–	–	–	–	1	–	–	Phosphine imide, N-(3-nitrophenyl)-P,P,P-triphenyl-	14796-86-2	LI08491
398	183	397	108	262	185	152	107	398	100	71	59	24	23	20	12	12		24	19	2	2	–	–	–	–	–	–	1	–	–	Phosphine imide, N-(4-nitrophenyl)-P,P,P-triphenyl-	14562-02-8	LI08492
398	399	183	397	108	262	185	107	398	100	75	71	59	24	23	20	12		24	19	2	2	–	–	–	–	–	–	1	–	–	Phosphine imide, N-(4-nitrophenyl)-P,P,P-triphenyl-	14562-02-8	NI40395
398	208	43	207	206	107	380	220	398	100	84	76	59	37	37	33	32		24	22	2	4	–	–	–	–	–	–	–	–	–	Acetamide, N-[2-[2-[2-[[2-(acetylamino)phenyl]vinyl]phenyl]vinyl]phenyl]	69395-27-3	NI40396
105	104	76	77	132	50	147	28	398	100	35	26	25	14	8	5	5	0.10	24	22	2	4	–	–	–	–	–	–	–	–	–	Azimine, 2,3-di-(α-phenylethyl)-1-phthalimido-, trans-		NI40397
91	201	219	380	105	79	81	351	398	100	98	90	86	71	67	60	57	19.05	24	30	5	–	–	–	–	–	–	–	–	–	–	Bufa-20,22-dienolide, 14,15-epoxy-3-hydroxy-19-oxo-, (3β,5β,15β)-	20987-24-0	NI40398
398	383	341	356	57	124	355	43	398	100	87	49	48	44	40	36	33		24	35	1	–	–	–	–	–	–	–	–	–	1	Cobalt, [(1,2,4,5-η)-2,4-diisopropyl-6,6,7,7-tetramethylbicyclo[3.2.0]hepta-1,4-dien-3-one](η[5]-2,4-cyclopentadien-1-yl)-	37584-03-5	MS09141
98	74	55	87	69	84	44	83	398	100	88	53	50	44	40	38	34	2.90	24	46	4	–	–	–	–	–	–	–	–	–	–	Docosanedioic acid, dimethyl ester	22399-98-0	MS09142
145	101	87	157	253	203	383	352	398	100	23	20	8	7	5	4	3	0.10	24	46	4	–	–	–	–	–	–	–	–	–	–	Glycerol, 1-O-(2-ethoxy-4-hexadecenyl)-2,3-O-isopropylidene-		NI40399
57	28	69	56	41	55	118	70	398	100	43	30	26	26	24	21	20	0.00	24	46	4	–	–	–	–	–	–	–	–	–	–	Hexanedioic acid, bis(3,5,5-trimethylhexyl) ester		DO01117
57	129	71	70	255	69	55	115	398	100	51	33	28	23	21	21	19	1.40	24	46	4	–	–	–	–	–	–	–	–	–	–	Hexanedioic acid, dinonyl ester		IC04291
57	149	129	71	70	55	41	69	398	100	57	54	35	27	26	21	19	0.00	24	46	4	–	–	–	–	–	–	–	–	–	–	Hexanedioic acid, dinonyl ester		IC04292
383	101	43	57	41	55	116	129	398	100	88	86	47	43	40	39	38	0.00	24	46	4	–	–	–	–	–	–	–	–	–	–	Octadecanoic acid, (2,2-dimethyl-1,3-dioxolan-4-yl)methyl ester	32852-69-0	NI40400
167	168	165	166	152	195	194	169	398	100	95	93	86	83	66	62	62	0.00	25	22	3	2	–	–	–	–	–	–	–	–	–	Tryptophan, diphenylacetyl-	78490-38-7	NI40401
315	398	219	287	302	383	259		398	100	50	35	25	12	5	5			25	34	4	–	–	–	–	–	–	–	–	–	–	Octohydrorubranine		LI08493
135	295	136	91	296	161	187	121	398	100	85	41	37	35	35	26	25	5.80	25	34	4	–	–	–	–	–	–	–	–	–	–	Pregna-1,4-diene-3,20-dione, 17-(acetyloxy)-6,16-dimethyl-, (6α,16α)	18669-88-0	MS09143
295	296	187	55	338	313	91	145	398	100	25	13	10	9	9	9	6	2.40	25	34	4	–	–	–	–	–	–	–	–	–	–	Pregna-4,6-diene-3,20-dione, 17-(acetyloxy)-6,16-dimethyl-, (16α)-	2497-80-5	MS09144
398	69	275	71	56	329	259	95	398	100	94	85	85	84	78	65	62		25	34	4	–	–	–	–	–	–	–	–	–	–	Tricyclo[5.3.1.0[1,5]]undecane-10,11-dione, 8-hydroxy-4,4-dimethyl-7-(3-methylbut-2-enyl)-3-(1-methylvinyl)-9-(2-methylpropanoyl)-		NI40402
152	398	105	229	107	165	145	119	398	100	66	46	43	34	31	30	28		25	39	3	–	–	–	–	–	–	–	–	1	–	Pregna-5,17(20)-diene-3,20,21-triol, cyclic 20,21-(butylboronate), (3β)-	30881-81-3	NI40403
124	107	127	105	173	356	125	109	398	100	14	13	13	11	10	10	10	6.00	25	39	3	–	–	–	–	–	–	–	–	1	–	Pregn-4-en-3-one, 20,21-[(butylborylene)bis(oxy)]-, (20β)-		LI08494
124	55	41	79	43	91	107	95	398	100	20	19	17	17	16	14	14	6.00	25	39	3	–	–	–	–	–	–	–	–	1	–	Pregn-4-en-3-one, 20,21-[(butylborylene)bis(oxy)]-, (20β)-		MS09145
152	398	151	153	383	399	166	397	398	100	27	27	18	9	8	8	7		25	39	3	–	–	–	–	–	–	–	–	1	–	Pregn-17(20)-en-3-one, 20,21-[(butylborylene)bis(oxy)]-, (5α)-	35315-14-1	LI08495
152	398	151	153	383	399	166	105	398	100	27	26	18	9	8	8	7		25	39	3	–	–	–	–	–	–	–	–	1	–	Pregn-17(20)-en-3-one, 20,21-[(butylborylene)bis(oxy)]-, (5α)-	35315-14-1	NI40404
124	356	313	173	91	79	398	95	398	100	21	13	13	13	13	12	12		25	39	3	–	–	–	–	–	–	–	–	1	–	Pregn-4-en-3-one, 20,21-[[(tert-butylborylene)]bis(oxy)]-, (20β)-		MS09146
124	356	313	398	173	91	79	127	398	100	22	14	13	13	13	13	11		25	39	3	–	–	–	–	–	–	–	–	1	–	Pregn-4-en-3-one, 20,21-[[(tert-butylborylene)]bis(oxy)]-, (20β)-		LI08496
124	356	313	173	91	79	95	93	398	100	21	13	13	13	13	12	12	11.01	25	39	3	–	–	–	–	–	–	–	–	1	–	Pregn-4-en-3-one, 20,21-[[(tert-butylborylene)]bis(oxy)]-, (20R)-	30882-66-7	NI40405
152	110	151	398	153	81	55	109	398	100	36	26	24	22	18	18	14		25	39	3	–	–	–	–	–	–	–	–	1	–	Pregn-17(20)-en-3-one, 20,21-[[(tert-butylborylene)]bis(oxy)]-, (5α)-	30881-78-8	NI40406
152	110	151	398	153	81	55	166	398	100	36	26	24	22	18	18	14		25	39	3	–	–	–	–	–	–	–	–	1	–	Pregn-17(20)-en-3-one, 20,21-[[(tert-butylborylene)]bis(oxy)]-, (5α)-	30881-78-8	MS09147
339	57	71	43	97	55	83	41	398	100	89	84	74	63	61	57	47	7.80	25	50	3	–	–	–	–	–	–	–	–	–	–	Tricosanoic acid, 2-methoxy-, methyl ester		MS09148
105	215	231	136	263	202	203	232	398	100	62	54	36	24	18	14	11	0.60	26	22	4	–	–	–	–	–	–	–	–	–	–	Benzo[a]cyclopropa[cd]pentalene-2a(2H)-carboxylic acid, 1,2b,6b,6c-tetrahydro-1,2-dihydroxy-1,6c-diphenyl-, methyl ester, (1α,2α,2aα,2bα,6bα,6cα)-	40897-28-7	NI40407
280	91	105	191	119	202	189	215	398	100	85	56	54	54	51	46	42	0.40	26	22	4	–	–	–	–	–	–	–	–	–	–	Benzo[a]cyclopropa[cd]pentalene-1-carboxylic acid, 1,2,2a,2b,6b,6c-hexahydro-1,2-dihydroxy-2a,6c-diphenyl-, methyl ester, (1α,2α,2aα,2bα,6bα,6cα)-	40897-27-6	NI40408
183	398	289	185	262	370	275	108	398	100	74	64	63	51	35	30	23		26	24	–	–	–	–	–	–	–	–	2	–	–	Phosphine, 1,2-ethanediylbis[diphenyl-	1663-45-2	NI40409
183	398	185	289	262	108	275	370	398	100	51	51	47	43	35	25	21		26	24	–	–	–	–	–	–	–	–	2	–	–	Phosphine, 1,2-ethanediylbis[diphenyl-	1663-45-2	NI40410
183	398	185	289	262	370	275	399	398	100	76	65	60	46	29	28	20		26	24	–	–	–	–	–	–	–	–	2	–	–	Phosphine, 1,2-ethanediylbis[diphenyl-	1663-45-2	NI40411
208	383	91	209	77	398	57	342	398	100	32	21	20	18	16	16	14		26	26	2	2	–	–	–	–	–	–	–	–	–	2,4-Imidazolidinedione, 1-tert-butyl-5,5-diphenyl-3-benzyl-	56954-71-3	NI40412
208	383	91	77	398	209	57	342	398	100	32	21	18	16	16	16	14		26	26	2	2	–	–	–	–	–	–	–	–	–	2,4-Imidazolidinedione, 1-tert-butyl-5,5-diphenyl-3-benzyl-	56954-71-3	NI40413
398	315	231	55	43	399	41	271	398	100	46	43	36	36	30	26	21		26	38	3	–	–	–	–	–	–	–	–	–	–	6H-Dibenzo[b,d]pyran, 6a,7,10,10a-tetrahydro-6,6,9-trimethyl-3-pentyl-1-[(tetrahydro-2H-pyran-2-yl)oxy]-	41969-56-6	NI40414
43	71	57	82	41	55	70	29	398	100	93	45	41	40	38	34	27	0.00	26	54	2	–	–	–	–	–	–	–	–	–	–	Hexadecanal, diisopentyl acetal	18302-71-1	NI40415
398	383	399	91	384	277	83	41	398	100	57	30	24	16	12	9	7		27	26	3	–	–	–	–	–	–	–	–	–	–	1-Naphthalenol, 2,4-dibenzyl-5,8-dimethoxy-6-methyl-	86649-80-1	NI40416
267	252	265	354	280	325	398		398	100	37	27	25	23	5	1			27	26	3	–	–	–	–	–	–	–	–	–	–	Propaneperoxoic acid, β-[9-(10-phenylanthryl)]-, tert-butyl ester		MS09149
311	95	69	41	109	55	81	121	398	100	93	76	71	68	58	57	56	53.00	27	42	2	–	–	–	–	–	–	–	–	–	–	Cholesta-9(11),17(20),24-triene-3,6-diol, (3β,5α,6α)-	37926-43-5	NI40418
95	311	70	69	41	55	43	81	398	100	87	66	62	58	51	48	47	41.00	27	42	2	–	–	–	–	–	–	–	–	–	–	Cholesta-9(11),17(20),24-triene-3,6-diol, (3β,5α,6α)-	37926-43-5	NI40417
137	398	43	55	136	41	370	57	398	100	80	72	62	51	50	40	37		27	42	2	–	–	–	–	–	–	–	–	–	–	Cholest-4-ene-3,6-dione	984-84-9	NI40419
43	55	41	57	137	81	95	79	398	100	78	77	59	55	40	31	30	23.07	27	42	2	–	–	–	–	–	–	–	–	–	–	Cholest-4-ene-3,6-dione	984-84-9	NI40420
95	398	94	285	259	161	43	55	398	100	49	49	47	32	26	26	23		27	42	2	–	–	–	–	–	–	–	–	–	–	Cholest-7-ene-3,6-dione, (5α)-	2550-89-2	NI40421
95	398	138	81	82	69	55	272	398	100	84	66	66	63	61	60	54		27	42	2	–	–	–	–	–	–	–	–	–	–	Cholest-20(22)-ene-3,6-dione, (5α)-	41084-10-0	NI40422
398	243	55	43	57	41	244	285	398	100	99	73	64	59	59	50	46		27	42	2	–	–	–	–	–	–	–	–	–	–	3,5-Cyclocholestane-2,6-dione, (3α,5α)-	26642-11-5	NI40423
398	284	283	255	55	270	269	120	398	100	74	55	40	40	37	31	31		27	42	2	–	–	–	–	–	–	–	–	–	–	Spirost-14-ene, (5α,25R)-	24742-84-5	NI40424

MASS TO CHARGE RATIOS								M.W.	INTENSITIES								Parent	C	H	O	N	S	F	Cl	Br	I	Si	P	B	X	COMPOUND NAME	CAS Reg No	No
398	162	94	149	55	166	109	150	**398**	100	95	80	68	51	43	41	40		27	42	2	–	–	–	–	–	–	–	–	–	–	**Spirost-20(21)-ene, (5α,25R)-**	24742-74-3	**NI40425**
398	162	368	399	149	166	369	180	**398**	100	62	35	32	17	12	10	10		27	42	2	–	–	–	–	–	–	–	–	–	–	**Spirost-20(21)-ene, (5α,25R)-**	24742-74-3	**MS09150**
398	162	94	149	55	166	368	109	**398**	100	93	79	67	50	43	41	41		27	42	2	–	–	–	–	–	–	–	–	–	–	**Spirost-20(21)-ene, (5α,25R)-**	24742-74-3	**NI40426**
137	124	286	138	257	271	217	122	**398**	100	40	37	33	28	25	22	21	**15.00**	27	42	2	–	–	–	–	–	–	–	–	–	–	**Spirost-23-ene, (5α,25R)-**	24744-29-4	**NI40427**
286	398	257	287	261	258	137	399	**398**	100	61	40	23	20	20	19	18		27	42	2	–	–	–	–	–	–	–	–	–	–	**Spirost-23-ene, (25R)-**		**MS09151**
137	124	286	138	257	271	217	122	**398**	100	40	37	32	28	25	22	20	**15.00**	27	42	2	–	–	–	–	–	–	–	–	–	–	**Spirost-23-ene, (25R)-**		**MS09152**
257	68	331	137	69	55	67	41	**398**	100	88	87	87	62	51	44	37	**3.00**	27	42	2	–	–	–	–	–	–	–	–	–	–	**Spirost-24-ene, (5α)-**	24744-30-7	**NI40428**
331	257	55	137	69	67	95	41	**398**	100	83	37	35	34	31	30	30	**11.11**	27	42	2	–	–	–	–	–	–	–	–	–	–	**Spirost-24-ene, (5α)-**	24744-30-7	**NI40429**
331	257	332	286	68	328	137	271	**398**	100	35	25	17	17	9	9	8	**3.00**	27	42	2	–	–	–	–	–	–	–	–	–	–	**Spirost-24-ene, (25R)-**		**MS09153**
257	68	331	137	69	55	67	41	**398**	100	89	87	87	61	51	45	38	**3.00**	27	42	2	–	–	–	–	–	–	–	–	–	–	**Spirost-24-ene, (25R)-**		**MS09154**
398	399	400	199	367	364	363	397	**398**	100	74	22	21	17	17	13	12		28	16	–	1	1	–	–	–	–	–	–	–	–	**8H-Dinaphtho[2,3-c:2′,3′-h]phenothiazinyl (radical)**		**MS09155**
177	382	176	178	193	165	194	383	**398**	100	44	25	21	20	18	14	13	**7.20**	28	18	1	2	–	–	–	–	–	–	–	–	–	**2,2′-Azoxyphenanthrene**		**NI40430**
69	55	81	255	398	83	133	109	**398**	100	88	56	55	45	45	41	41		28	46	1	–	–	–	–	–	–	–	–	–	–	**Cholesta-5,22-dien-3-ol, 24-methyl-, (3β,24S)-**		**LI08497**
41	69	55	43	57	29	105	91	**398**	100	66	58	58	31	28	20	20	**9.44**	28	46	1	–	–	–	–	–	–	–	–	–	–	**Cholesta-8,24-dien-3-ol, 4-methyl-, (3β,4α)-**	7199-92-0	**NI40431**
398	109	121	105	107	119	383	135	**398**	100	63	60	59	56	52	48	48		28	46	1	–	–	–	–	–	–	–	–	–	–	**Cholesta-8,24-dien-3-ol, 4-methyl-, (3β,4β)-**	15737-15-2	**NI40432**
55	314	69	81	95	107	93	271	**398**	100	79	79	66	56	53	45	44	**12.50**	28	46	1	–	–	–	–	–	–	–	–	–	–	**Cholest-5-en-3-ol, 24-methylene-, (3β)-**		**LI08498**
140	398	82	139	96	261	137	57	**398**	100	26	23	22	16	11	10	10		28	46	1	–	–	–	–	–	–	–	–	–	–	**Cholest-4-en-3-one, 6-methyl-**		**LI08499**
55	57	398	124	69	95	83	81	**398**	100	99	85	76	75	71	48	48		28	46	1	–	–	–	–	–	–	–	–	–	–	**Cholest-4-en-3-one, 14-methyl-**	21857-92-1	**NI40433**
398	285	243	383	259	258	213	201	**398**	100	75	36	34	21	12	9	8		28	46	1	–	–	–	–	–	–	–	–	–	–	**Cholest-7-en-3-one, 4-methyl-, (4α)-**	13490-57-8	**NI40434**
398	243	383	285	259	258	231	201	**398**	100	40	39	35	31	22	7	7		28	46	1	–	–	–	–	–	–	–	–	–	–	**Cholest-8(14)-en-3-one, 4-methyl-, (4α)-**	82038-94-6	**NI40435**
43	55	57	41	383	69	71	83	**398**	100	73	67	63	51	48	38	36	**7.92**	28	46	1	–	–	–	–	–	–	–	–	–	–	**Cholest-8-en-3-one, 14-methyl, (5α)-**	5234-53-7	**NI40436**
285	161	398	383	213	220	206	200	**398**	100	28	5	5	4	2	2	2		28	46	1	–	–	–	–	–	–	–	–	–	–	**Cholest-14-en-3-one, 4-methyl-, (4α)-**	82038-95-7	**NI40437**
55	398	43	81	41	95	123	57	**398**	100	82	82	67	67	65	59	57		28	46	1	–	–	–	–	–	–	–	–	–	–	**4,5-Cyclo-A-homo-cholestan-3-one, (4α,5α)-**		**NI40438**
398	43	399	54	41	93	135	81	**398**	100	34	30	29	25	22	21	20		28	46	1	–	–	–	–	–	–	–	–	–	–	**4,5-Cyclo-A-homo-cholestan-6-one, (4α,5α)-**		**NI40439**
398	317	95	55	43	107	93	399	**398**	100	62	36	36	36	34	32	30		28	46	1	–	–	–	–	–	–	–	–	–	–	**4,5-Cyclo-A-homo-cholestan-6-one, (4β,5β)-**		**NI40440**
398	135	150	151	43	247	41	55	**398**	100	72	68	65	46	42	39	34		28	46	1	–	–	–	–	–	–	–	–	–	–	**5,6-Cyclo-B-homo-cholestan-4-one, (5α,6α)-**		**NI40441**
43	55	41	135	398	137	95	81	**398**	100	92	78	69	60	56	56	56		28	46	1	–	–	–	–	–	–	–	–	–	–	**5,6-Cyclo-B-homo-cholestan-4-one, (5β,6β)-**		**NI40442**
69	55	95	41	81	383	107	109	**398**	100	77	76	67	63	52	52	51	**36.00**	28	46	1	–	–	–	–	–	–	–	–	–	–	**9,19-Cyclocholestan-3-ol, 14-methyldehydro-, (3β,5α,9β)-**	75659-56-2	**NI40443**
69	343	283	398	366	107	383	121	**398**	100	27	18	17	17	17	16	13		28	46	1	–	–	–	–	–	–	–	–	–	–	**3,5-Cyclocholest-23-ene, 6-methoxy-, (3α,5α,6β,23E)-**		**MS09156**
81	95	55	247	93	43	135	107	**398**	100	84	69	65	59	59	58	57	**38.00**	28	46	1	–	–	–	–	–	–	–	–	–	–	**Cyclopropa[7,8]cholestan-3-one, 3′,7-dihydro-, (5α,7β,8α)-**	53755-19-4	**NI40444**
55	69	398	81	300	95	67	109	**398**	100	78	62	58	52	46	46	43		28	46	1	–	–	–	–	–	–	–	–	–	–	**Ergosta-3,22-dien-3-ol, (5β)-**		**MS09157**
69	255	55	398	81	95	107	105	**398**	100	75	74	70	68	58	45	44		28	46	1	–	–	–	–	–	–	–	–	–	–	**Ergosta-4,22-dien-3-ol, (3β)-**	62332-14-3	**NI40445**
43	55	41	69	57	81	71	95	**398**	100	72	63	56	50	44	38	34	**25.00**	28	46	1	–	–	–	–	–	–	–	–	–	–	**Ergosta-5,7-dien-3-ol, (3β)-**	516-79-0	**NI40446**
365	143	43	145	41	55	57	398	**398**	100	70	60	55	49	48	42	41		28	46	1	–	–	–	–	–	–	–	–	–	–	**Ergosta-5,8-dien-3-ol, (3β)-**	51724-53-9	**NI40447**
43	398	55	109	229	271	213	299	**398**	100	81	51	47	26	24	24	20		28	46	1	–	–	–	–	–	–	–	–	–	–	**Ergosta-5,20(22)-dien-3-ol, (3β,22E,24S)-**		**DD01554**
69	55	81	255	105	67	83	41	**398**	100	80	65	52	36	36	35	35	**28.00**	28	46	1	–	–	–	–	–	–	–	–	–	–	**Ergosta-5,22-dien-3-ol, (3β,22E)-**	474-67-9	**NI40448**
69	380	55	81	255	105	67	41	**398**	100	79	67	65	52	35	34	33	**15.00**	28	46	1	–	–	–	–	–	–	–	–	–	–	**Ergosta-5,22-dien-3-ol, (3β,22E,24S)-**	17472-78-5	**NI40449**
69	55	81	255	83	398	41	43	**398**	100	81	49	41	37	36	35	33		28	46	1	–	–	–	–	–	–	–	–	–	–	**Ergosta-5,22-dien-3-ol, (3β,22E,24S)-**	17472-78-5	**NI40450**
314	55	57	69	43	59	41	44	**398**	100	92	86	85	66	62	61	59	**20.20**	28	46	1	–	–	–	–	–	–	–	–	–	–	**Ergosta-5,24(28)-dien-3-ol, (3β)-**	474-63-5	**NI40452**
55	398	69	107	41	81	95	43	**398**	100	80	78	72	70	67	61	53		28	46	1	–	–	–	–	–	–	–	–	–	–	**Ergosta-5,24(28)-dien-3-ol, (3β)-**	474-63-5	**NI40451**
314	55	315	69	41	229	81	43	**398**	100	51	25	25	24	20	20	19	**2.00**	28	46	1	–	–	–	–	–	–	–	–	–	–	**Ergosta-5,24(28)-dien-3-ol, (3β)-**	474-63-5	**NI40453**
314	299	271	315	398	281	300	383	**398**	100	33	32	26	23	22	14	13		28	46	1	–	–	–	–	–	–	–	–	–	–	**Ergosta-5,24-dien-3-ol, (3β)-**	20780-41-0	**NI40454**
69	55	81	271	107	95	105	43	**398**	100	93	86	75	62	53	51	51	**32.00**	28	46	1	–	–	–	–	–	–	–	–	–	–	**Ergosta-7,22-dien-3-ol, (3β,5α,22E)-**	2465-11-4	**NI40455**
398	271	69	55	399	273	81	43	**398**	100	52	34	33	30	27	25	24		28	46	1	–	–	–	–	–	–	–	–	–	–	**Ergosta-7,22-dien-3-ol, (3β,5α,22Z)-**	54482-53-0	**NI40456**
381	125	397	271	398	382	399	383	**398**	100	93	85	38	29	28	26	16		28	46	1	–	–	–	–	–	–	–	–	–	–	**Ergosta-7,22-dien-3-ol, (3β,22E)-**	17608-76-3	**NI40457**
55	69	28	81	105	43	41	107	**398**	100	97	93	90	65	58	57	55	**14.00**	28	46	1	–	–	–	–	–	–	–	–	–	–	**Ergosta-7,22-dien-3-ol, (3β,22E)-**	17608-76-3	**NI40458**
43	398	383	55	41	365	399	57	**398**	100	95	80	46	45	34	32	30		28	46	1	–	–	–	–	–	–	–	–	–	–	**Ergosta-8,14-dien-3-ol**	55902-81-3	**NI40459**
69	55	81	272	147	107	93	91	**398**	100	94	77	64	52	52	52	46	**23.93**	28	46	1	–	–	–	–	–	–	–	–	–	–	**Ergosta-14,22-dien-3-ol, (5α)-**		**MS09158**
55	145	105	107	133	119	271	213	**398**	100	84	56	43	39	39	33	11	**0.00**	28	46	1	–	–	–	–	–	–	–	–	–	–	**27-Norcholesta-5,24-dien-3-ol, 25-ethyl-, (3β,24E)-**		**DD01555**
105	107	145	119	133	55	271	213	**398**	100	90	84	79	72	72	46	16	**0.00**	28	46	1	–	–	–	–	–	–	–	–	–	–	**27-Norcholesta-5,24-dien-3-ol, 25-ethyl-, (3β,24Z)-**		**DD01556**
123	398	95	57	81	43	105	69	**398**	100	96	86	84	64	60	56	56		28	46	1	–	–	–	–	–	–	–	–	–	–	**A-Norcholestan-3-one, 5-vinyl-, (5α)-**		**MS09159**
43	55	123	41	81	57	95	93	**398**	100	90	87	67	62	61	58	53	**39.00**	28	46	1	–	–	–	–	–	–	–	–	–	–	**A-Norcholestan-3-one, 5-vinyl-, (5β)-**		**MS09160**
145	314	105	229	161	271	213	299	**398**	100	77	63	57	46	42	40	32	**0.00**	28	46	1	–	–	–	–	–	–	–	–	–	–	**27-Norstigmasta-5,24-dien-3-ol, (3β,24E)-**		**DD01557**
55	95	386	121	81	273	255	93	**398**	100	94	86	82	80	73	69	64	**0.00**	28	46	1	–	–	–	–	–	–	–	–	–	–	**9,10-Secoergosta-5,7,22-trien-3-ol, (3β,5E,7E,10α,22E)-**	67-96-9	**NI40460**
121	55	69	135	398	133	95	81	**398**	100	50	41	37	36	32	32	32		28	46	1	–	–	–	–	–	–	–	–	–	–	**9,10-Secoergosta-5,7,22-trien-3-ol, (3β,5E,7E,10α,22E)-**	67-96-9	**NI40461**
397	398	399	191	77	396	321	199	**398**	100	61	15	8	6	4	4	3		29	22	–	2	–	–	–	–	–	–	–	–	–	**Aniline, N-(1,3,4-triphenyl-2(1H)-pyridinylidene)-**	74792-94-2	**NI40462**

MASS TO CHARGE RATIOS								M.W.	INTENSITIES								Parent	C	H	O	N	S	F	Cl	Br	I	Si	P	B	X	COMPOUND NAME	CAS Reg No	No
119	133	145	146	105	69	157	143	398	100	46	26	24	18	17	14	13	3.40	29	50	–	–	–	–	–	–	–	–	–	–	–	Benzene, 2-(1-decyl-1-undecenyl)-1,4-dimethyl-	55373-90-5	WI01518
146	119	145	43	257	41	29	55	398	100	84	63	51	45	33	29	17	13.30	29	50	–	–	–	–	–	–	–	–	–	–	–	Benzene, 2-(1-decyl-1-undecenyl)-1,4-dimethyl-	55373-90-5	AI02330
191	109	177	123	137	135	192	149	398	100	37	36	36	28	28	23	23	11.00	29	50	–	–	–	–	–	–	–	–	–	–	–	28-Nor-17(H)-hopane, (17α)-	53584-60-4	NI40463
191	109	192	123	135	177	137	122	398	100	33	28	27	26	21	21	20	17.00	29	50	–	–	–	–	–	–	–	–	–	–	–	28-Nor-17(H)-hopane, (17β)-	36728-72-0	NI40464
398	55	43	69	57	81	83	95	398	100	57	51	49	48	46	43	42		29	50	–	–	–	–	–	–	–	–	–	–	–	Stigmast-22-ene, (5α,22E)-	54412-02-1	NI40465
300	55	217	81	218	95	69	285	398	100	58	34	32	31	30	29	28	7.92	29	50	–	–	–	–	–	–	–	–	–	–	–	Stigmast-24(28)-ene, (5α)-	56143-36-3	NI40466
398	199	399	396	375	397	198	57	398	100	49	39	18	13	11	11	9		32	14	–	–	–	–	–	–	–	–	–	–	–	Ovalene	190-26-1	NI40467
69	169	92	47	100	280	119	192	399	100	95	32	18	15	14	13	10	0.00	7	–	1	1	–	15	–	–	–	–	–	–	–	N,N-Bis(heptafluoropropyl)carbamoyl fluoride		DD01558
69	92	114	180	47	119	100	131	399	100	35	18	17	16	10	8	5	0.00	7	–	1	1	–	15	–	–	–	–	–	–	–	N-(Trifluoromethyl)-N-(undecafluoropentyl)carbamoyl fluoride		DD01559
364	111	168	296	172	401	403	294	399	100	82	81	69	48	47	43	42	16.46	10	4	1	1	–	–	7	–	–	–	–	–	–	2-Pyrrolidinone, 3,3,4,4,5,5-hexachloro-1-(3-chlorophenyl)-	41910-51-4	NI40468
111	77	366	368	172	174	364	157	399	100	81	72	70	60	43	42	36	2.50	10	4	1	1	–	–	7	–	–	–	–	–	–	2-Pyrrolidinone, 3,3,4,4,5,5-hexachloro-1-(4-chlorophenyl)-	41910-50-3	NI40469
231	203	145	144	315	173	259	174	399	100	97	78	74	65	58	49	35	15.00	10	9	7	1	1	–	–	–	–	–	–	–	2	Iron, (μ-methylisocyanato)(μ-ethanethiolato)hexacarbonyldi-		MS09161
30	284	113	173	226	227	71	90	399	100	70	33	32	31	24	20	19	3.00	11	12	5	3	–	7	–	–	–	–	–	–	–	Glycine, N-[N-[N-(heptafluorobutyryl)glycyl]glycyl]-, methyl ester	19444-87-2	NI40470
163	216	120	136	52	259	135	80	399	100	28	14	13	12	11	8	8	1.75	11	18	5	3	–	–	–	–	–	–	–	–	2	Chromium, pentacarbonyltris(dimethylamino)arsinyl-		LI08500
55	67	307	42	151	243	399	63	399	100	96	60	50	46	44	42	34		14	14	5	5	1	–	1	–	–	–	–	–	–	1-(1,1-Dioxotetrahydrothien-3-yl)-3-methyl-5-hydroxy-4-(2-chloro-4-nitrophenylazo)pyrazole		IC04293
93	138	94	137	80	155	136	111	399	100	99	73	55	55	42	35	28	5.00	15	20	5	1	–	–	3	–	–	–	–	–	–	9-O-[4-(2,2,2-Trichloroethoxycarbonyl)butanoyl]retronecine		NI40471
73	309	147	384	75	266	222	296	399	100	74	51	43	32	30	29	27	12.70	17	33	4	1	–	–	–	–	–	3	–	–	–	4-Pyridinecarboxylic acid, 2-methyl-5-[(trimethylsilyl)oxy]-3-[[(trimethylsilyl)oxy]methyl]-, trimethylsilyl ester	55517-52-7	NI40472
251	284	57	340	252	41	285	237	399	100	77	30	15	12	11	9	5	3.00	18	25	9	1	–	–	–	–	–	–	–	–	–	2,3,4,5-Pyridinetetracarboxylic acid, 1-tert-butyl-1,6-dihydro-6-methoxy-, tetramethyl ester	86488-05-3	NI40473
73	175	298	89	238	269	43	208	399	100	99	60	52	39	34	33	32	8.50	18	33	5	3	–	–	–	–	–	1	–	–	–	2,4,6(1H,3H,5H)-Pyrimidinetrione, 5-[2-(methoxyimino)-3-[(trimethylsilyl)oxy]propyl]-1,3-dimethyl-5-(1-methylbutyl)-	57397-43-0	NI40474
297	73	102	75	298	45	309	77	399	100	96	51	47	25	19	14	10	0.00	18	37	3	1	–	–	–	–	–	3	–	–	–	Benzeneethylamine, N-methyl-β,3,4-tris[(trimethylsilyl)oxy]-		MS09162
73	355	75	44	45	356	74	42	399	100	41	30	20	18	12	8	6	0.00	18	37	3	1	–	–	–	–	–	3	–	–	–	Benzeneethylamine, N-methyl-β,3,4-tris[(trimethylsilyl)oxy]-		MS09163
73	102	297	45	298	74	147	75	399	100	54	48	12	10	8	5	5	0.00	18	37	3	1	–	–	–	–	–	3	–	–	–	Benzeneethylamine, N-methyl-β,3,4-tris[(trimethylsilyl)oxy]-		MS09164
310	384	355	400					399	100	90	80	20					0.00	18	37	3	1	–	–	–	–	–	3	–	–	–	Benzeneethylamine, N-methyl-β,3,4-tris[(trimethylsilyl)oxy]-, (R)-	56114-65-9	NI40475
242	182	95	181	155	153	123	368	399	100	28	25	16	15	11	11	7	4.00	19	29	8	1	–	–	–	–	–	–	–	–	–	Furan-3,4-carboxylic acid, 5-cyclohexylamino-2,5-dihydro-2-methoxy-2-[(2-methoxycarbonyl)ethyl]-, dimethyl ester		LI08501
315	399	206	178	286	218	357	110	399	100	65	49	34	32	31	28	22		19	30	6	1	–	–	–	–	–	–	1	–	–	3-Benzazepin-1-one, 2,3,4,5-tetrahydro-7-methoxy-8-ethoxy-N-(diisopropylphosphoryl)-	87212-53-1	NI40476
399	228	201	245	229	320	173	159	399	100	68	46	40	30	12	12	11		20	17	6	1	1	–	–	–	–	–	–	–	–	5H-Furo[3,2-g][1]benzopyran-9-sulphonamide, 4-methoxy-7-methyl-N-(2-methylphenyl)-5-oxo-		MS09165
399	228	201	245	229	320	216	159	399	100	77	54	38	30	20	20	20		20	17	6	1	1	–	–	–	–	–	–	–	–	5H-Furo[3,2-g][1]benzopyran-9-sulphonamide, 4-methoxy-7-methyl-N-(3-methylphenyl)-5-oxo-		MS09166
399	228	201	245	229	216	320	159	399	100	68	40	30	28	15	14	13		20	17	6	1	1	–	–	–	–	–	–	–	–	5H-Furo[3,2-g][1]benzopyran-9-sulphonamide, 4-methoxy-7-methyl-N-(4-methylphenyl)-5-oxo-		MS09167
231	230	242	273	105	298	357	315	399	100	50	43	34	24	20	14	7	3.00	20	18	4	3	–	–	1	–	–	–	–	–	–	2′-Benzoyl-4′-chloro-N-(2,2-diacetylhydrazono)methylacetanilide		LI08502
231	242	230	273	298	105	315	357	399	100	86	80	76	57	55	29	22	18.00	20	18	4	3	–	–	1	–	–	–	–	–	–	5-Chloro-2-(1,2,2-triacetyl-1-hydrazinylmethyleneamino)benzophenone		LI08503
72	121	56	44	355	399	354	143	399	100	75	55	34	27	21	20	7		20	22	–	1	1	–	1	–	–	–	–	–	1	Ferrocene, 1-[1-(dimethylamino)ethyl]-2-[(4-chlorophenyl)thio]-		MS09168
124	279	91	77	65	214	155	260	399	100	68	56	28	27	26	11	4	0.00	21	21	3	1	2	–	–	–	–	–	–	–	–	Benzenesulphonamide, N-(2-hydroxy-2-phenylethyl)-4-methyl-N-(phenylsulphenyl)-		DD01560
91	399	231	335	230	244	18	203	399	100	95	94	91	53	34	32	25		21	21	5	1	1	–	–	–	–	–	–	–	–	3H-Naphtho[1,2-e][1,2]thiazin-2(4aH)-one, 5,6-dihydro-8,9-dimethoxy-3-benzyl-, 4,4-dioxide	40535-18-0	MS09169
399	293	280	184	182	278	183	63	399	100	25	10	7	5	4	4	4		21	24	2	2	–	–	–	–	–	–	–	–	1	Copper(II), N,N′-bis(salicylidene)heptanediamino-		NI40477
312	371	43	297	399	281	298	313	399	100	53	51	45	36	34	25	23		22	25	6	1	–	–	–	–	–	–	–	–	–	Acetamide, N-(5,6,7,9-tetrahydro-1,2,3,10-tetramethoxy-9-oxobenzo[a]heptalen-7-yl)-, (S)-	64-86-8	NI40478
400	401	386	57	60	370	402	45	399	100	26	17	15	13	8	6	5	3.14	22	25	6	1	–	–	–	–	–	–	–	–	–	Acetamide, N-(5,6,7,9-tetrahydro-1,2,3,10-tetramethoxy-9-oxobenzo[a]heptalen-7-yl)-, (S)-	64-86-8	NI40479
399	179	194	204	206	195	178	400	399	100	69	65	61	60	42	31	25		22	25	6	1	–	–	–	–	–	–	–	–	–	6H-Benzo[g]-1,3-benzodioxolo[5,6-a]quinolizine-12-methanol, 5,8,13,13a-tetrahydro-10,11,14-trimethoxy-, (S)-	31098-60-9	NI40480
178	150	206	192	179	165	164	163	399	100	55	32	20	13	13	13	12	0.97	22	25	6	1	–	–	–	–	–	–	–	–	–	Dibenz[c,g]azecine-13,14-dione, 5,6,7,8-tetrahydro-3,4,10,11-tetramethoxy-6-methyl-	13061-83-1	NI40481

MASS TO CHARGE RATIOS								M.W.	INTENSITIES								Parent	C	H	O	N	S	F	Cl	Br	I	Si	P	B	X	COMPOUND NAME	CAS Reg No	No
399	340	356	400	280	357	341	298	**399**	100	90	35	26	26	24	22	18		22	25	6	1	–	–	–	–	–	–	–	–	–	Hasubanan-6,9-diol, 7,8-didehydro-4,5-epoxy-3-methoxy-17-methyl-, diacetate (ester), (5α,6α,9α,13β,14β)-	38691-63-3	NI40482
399	340	256	400	280	357	341	298	**399**	100	79	35	27	25	24	21	17		22	25	6	1	–	–	–	–	–	–	–	–	–	Hasubanan-6,9-diol, 7,8-didehydro-4,5-epoxy-3-methoxy-17-methyl-, diacetate (ester), (5α,6α,9α,13β,14β)-	38691-63-3	NI40483
399	340	356	400	280	357	342	298	**399**	100	80	32	25	24	22	20	16		22	25	6	1	–	–	–	–	–	–	–	–	–	Indolinocodeine, 6,9-diacetate, (9α)-		LI08504
356	357	399	326	340	297	358	115	**399**	100	71	26	16	14	14	12	10		22	25	6	1	–	–	–	–	–	–	–	–	–	Lumicolchicine	490-24-4	NI40484
384	399	206	385	400	368	194	192	**399**	100	93	41	27	24	23	16	15		22	25	6	1	–	–	–	–	–	–	–	–	–	Rheadan, 2,3,8-trimethoxy-16-methyl-10,11-[methylenebis(oxy)]-, (8β)-	6516-48-9	NI40485
113	399	70	141	400	259	111	71	**399**	100	88	87	71	23	22	22	17		22	29	–	3	2	–	–	–	–	–	–	–	–	10H-Phenothiazine, 2-(ethylthio)-10-[3-(4-methyl-1-piperazinyl)propyl]-	1420-55-9	NI40486
399	113	70	141	43	400	72	71	**399**	100	97	80	73	35	29	29	23		22	29	–	3	2	–	–	–	–	–	–	–	–	10H-Phenothiazine, 2-(ethylthio)-10-[3-(4-methyl-1-piperazinyl)propyl]-	1420-55-9	NI40487
169	399	170	114	401	115	400	141	**399**	100	94	49	40	37	31	30	30		23	14	2	3	–	–	1	–	–	–	–	–	–	s-Triazine, 6-chloro-2,4-bis(2-hydroxynaphth-1-yl)-		IC04294
399	400	194	382	104	398	91	77	**399**	100	29	26	24	24	15	15	15		23	17	4	3	–	–	–	–	–	–	–	–	–	2(1H)-Pyrimidinone, 5-(4-methylphenoxy)-4-(4-nitrophenyl)-6-phenyl-	42919-60-8	NI40488
351	233	103	195	352	190	281	399	**399**	100	81	71	27	26	25	14	13		23	21	–	5	1	–	–	–	–	–	–	–	–	Imidazole, 1-[(N,N′-diphenyl)guanidino]-4-phenyl-2-(methylthio)-		MS09170
357	214	213	284	283	342	358	314	**399**	100	73	52	39	23	22	21	20	0.00	23	29	5	1	–	–	–	–	–	–	–	–	–	Estra-1,3,5(10)-trien-16-ene, 3,17-bis(acetyloxy)-, 16-(O-methyloxime), (17β)-		NI40489
357	91	92	358	328	399	163	172	**399**	100	58	44	25	16	12	11	9		23	29	5	1	–	–	–	–	–	–	–	–	–	Estra-1,3,5(10)-trien-6-one, 3,17-bis(acetyloxy)-, 6-(O-methyloxime), (17β)-	69833-93-8	NI40490
357	266	267	95	358	213	249	108	**399**	100	74	31	29	26	15	14	13	0.00	23	29	5	1	–	–	–	–	–	–	–	–	–	Estra-1,3,5(10)-trien-15-one, 3,17-bis(acetyloxy)-, 15-(O-methyloxime), (17β)-		NI40491
266	357	251	211	267	265	209	146	**399**	100	39	35	31	27	25	23	16	0.00	23	29	5	1	–	–	–	–	–	–	–	–	–	Estra-1,3,5(10)-trien-17-one, 3,11-bis(acetyloxy)-, 17-(O-methyloxime), (11β)-		NI40492
266	172	267	357	297	159	211	265	**399**	100	92	61	56	27	27	23	22	0.00	23	29	5	1	–	–	–	–	–	–	–	–	–	Estra-1,3,5(10)-trien-17-one, 3,15-bis(acetyloxy)-, 17-(O-methyloxime), (15α)-		NI40493
284	266	91	357	92	285	326	159	**399**	100	73	67	56	54	24	22	21	0.00	23	29	5	1	–	–	–	–	–	–	–	–	–	Estra-1,3,5(10)-trien-17-one, 3,16-bis(acetyloxy)-, 17-(O-methyloxime), (16β)-	69833-92-7	NI40494
341	121	142	225	311	322	399	381	**399**	100	58	39	22	18	8	5	3		23	29	5	1	–	–	–	–	–	–	–	–	–	5′H-Pregna-1,4-dieno[17,16-d]oxazole-3,20-dione, 11,21-dihydroxy-2′-methyl-, (11β,16β)-	13649-57-5	NI40495
86	126	259	58	112	399	99	74	**399**	100	31	15	14	12	11	11	7		23	30	1	3	–	–	1	–	–	–	–	–	–	Acridine, 6-chloro-9-[4-(diethylamino)-1-methylbutyl]amino-2-methoxy-	83-89-6	NI40496
400	402	401	399	403	398	372	126	**399**	100	33	29	11	9	2	1	1		23	30	1	3	–	–	1	–	–	–	–	–	–	Acridine, 6-chloro-9-[4-(diethylamino)-1-methylbutyl]amino-2-methoxy-	83-89-6	NI40497
75	208	399	384	97	107	382	121	**399**	100	90	46	37	22	16	14	13		23	33	3	1	–	–	–	–	–	1	–	–	–	2-Androstanecarbonitrile, 4,5-epoxy-17-oxo-3-[(trimethylsilyl)oxy]-		NI40498
270	43	381	352	170	184	228	41	**399**	100	79	54	50	43	40	25	25	3.00	23	33	3	3	–	–	–	–	–	–	–	–	–	Hydronesechinuline		MS09171
91	399	92	400	65	40	366	165	**399**	100	23	7	6	4	4	3	3		24	17	3	1	1	–	–	–	–	–	–	–	–	6H-Pyrano[3,2-f]benzoxazol-6-one, 2-(benzylthio)-8-methyl-7-phenyl-		NI40499
399	104	77	83	85	149	194	398	**399**	100	79	77	60	41	38	35	28		24	21	3	3	–	–	–	–	–	–	–	–	–	4-Pyrimidinecarboxylic acid, hexahydro-6-oxo-1,3-diphenyl-2-(phenylimino)-, methyl ester	56630-79-6	NI40500
399	384	57	376	340	339	352	329	**399**	100	81	81	73	45	36	18	18		24	33	4	1	–	–	–	–	–	–	–	–	–	1,2-Ethenedicarboxylic acid, (5,7-di-tert-butyl-3,3-dimethyl-3H-indol-2-yl)-, dimethyl ester		DD01561
399	343	400	340	382	356	342	384	**399**	100	44	27	24	23	22	15	12		24	33	4	1	–	–	–	–	–	–	–	–	–	Lucidusculine, dehydro-		MS09172
340	343	341	149	122	167	399	342	**399**	100	41	40	35	30	18	15	13		24	33	4	1	–	–	–	–	–	–	–	–	–	Subdesculine		MS09173
384	385	259	399	209	153	137	135	**399**	100	32	29	22	22	18	14	9		24	37	2	1	–	–	–	–	–	1	–	–	–	19-Norpregn-4-en-20-yn-3-one, 17-[(trimethylsilyl)oxy]-, O-methyloxime, (17α)-	69834-05-5	NI40501
339	294	324	43	340	91	105	295	**399**	100	50	43	30	26	20	14	12	1.00	25	37	3	1	–	–	–	–	–	–	–	–	–	Pregna-5,16-dien-20-one, 3-acetyloxy-, O-ethyloxime, (3β)-		LI08505
266	134	119	91	265	250	267	65	**399**	100	85	78	42	30	30	20	13	0.50	26	25	3	1	–	–	–	–	–	–	–	–	–	Acetophenone, 2-[5-(4-methoxyphenyl)-3-(4-tolyl)-2-isoxazolin-5-yl]-4′-methyl-	21326-96-5	NI40502
354	355	340	310	283	296	399	356	**399**	100	41	23	16	14	13	11	8		26	25	3	1	–	–	–	–	–	–	–	–	–	Spiro[phthalan-1,9′-xanthene]-3-one, 6′-(diethylamino)-1′,3′dimethyl-		IC04295
57	372	260	316	298	283	213	315	**399**	100	84	56	47	42	35	31	25	4.00	26	41	2	1	–	–	–	–	–	–	–	–	–	Pregn-5-ene-20-carbonitrile, 3-tert-butoxy-20-hydroxy-, (3β)-	62623-51-2	NI40503
98	99	81	55	96	56	79	70	**399**	100	14	5	5	4	4	3	3	0.10	27	45	1	1	–	–	–	–	–	–	–	–	–	Cholest-5-en-3-ol, 22,26-epimino-, (3β,5α,22R,25S)-		MS09174
399	139	99	57	55	125	400	95	**399**	100	97	50	49	49	40	35	35		27	45	1	1	–	–	–	–	–	–	–	–	–	3-Aza-A-homocholest-4a-en-4-one	10062-39-2	NI40504
399	400	137	371	111	150	384	55	**399**	100	31	24	23	23	18	15	15		27	45	1	1	–	–	–	–	–	–	–	–	–	4-Aza-A-homocholest-4a-en-3-one	21002-96-0	NI40505
399	55	137	124	400	370	81	57	**399**	100	40	39	35	29	26	22	22		27	45	1	1	–	–	–	–	–	–	–	–	–	4-Aza-A-homocholest-1-en-3-one, (5α)-	7485-18-9	NI40506
399	398	400	367	199	364	368	366	**399**	100	37	33	32	13	12	11	11		28	17	–	1	1	–	–	–	–	–	–	–	–	8H-Dinaphtho[2,3-c:2′,3′-h]phenothiazine		MS09175
105	195	178	221	193	206	77	294	**399**	100	48	27	17	13	11	11	8	3.00	29	21	1	1	–	–	–	–	–	–	–	–	–	Pyrrole, 3-benzoyl-2,4,5-triphenyl-		NI40507
150	127	45	44	89	31	188	201	**400**	100	60	33	31	24	18	16	13	0.00	4	6	4	2	2	10	–	–	–	–	–	–	–	Ethylenebis(oxycarbonylamino)bis(sulphur pentafluoride)	89573-63-7	NI40508
69	181	131	100	31	231	93	119	**400**	100	19	13	8	6	5	5	4	0.00	8	–	–	–	–	16	–	–	–	–	–	–	–	Cyclohexane, decafluorobis(trifluoromethyl)-	26637-68-3	WI01519
69	181	131	231	100	381	119	93	**400**	100	47	24	15	9	7	7	6	0.04	8	–	–	–	–	16	–	–	–	–	–	–	–	Cyclohexane, decafluorobis(trifluoromethyl)-	26637-68-3	DO01118
69	243	381	331	293	119	143	93	**400**	100	23	11	8	8	7	5	5	0.20	8	–	–	–	–	16	–	–	–	–	–	–	–	Cyclohexane, decafluorobis(trifluoromethyl)-	26637-68-3	IC04296

MASS TO CHARGE RATIOS								M.W.	INTENSITIES								Parent	C	H	O	N	S	F	Cl	Br	I	Si	P	B	X	COMPOUND NAME	CAS Reg No	No
64	56	84	112	168	196	140	76	400	100	82	74	12	11	7	6	4	0.05	8	–	10	–	1	–	–	–	–	–	–	–	2	Iron, octacarbonyl-μ-S-sulphinatodi-		LI08506
31	69	29	131	49	32	100	61	400	100	17	14	7	7	7	5	4	0.10	8	3	1	–	–	15	–	–	–	–	–	–	–	1-Octanol, 2,2,3,3,4,4,5,5,6,6,7,7,8,8,8-pentadecafluoro-	307-30-2	WI01520
31	69	131	100	49	119	29	61	400	100	36	30	14	14	11	11	9	0.00	8	3	1	–	–	15	–	–	–	–	–	–	–	1-Octanol, 2,2,3,3,4,4,5,5,6,6,7,7,8,8,8-pentadecafluoro-	307-30-2	NI40509
77	51	400	253	95	209	69	27	400	100	38	33	15	15	12	12	10		8	6	–	–	–	9	–	–	1	–	–	–	–	1-Octene, 3,3,4,4,7,7,8,8,8-nonafluoro-5-iodo-	35208-09-4	NI40510
387	389	139	137	391	385	109	402	400	100	87	60	58	50	47	38	28	11.30	9	11	1	–	–	–	–	3	–	1	–	–	–	Silane, trimethyl(2,4,6-tribromophenoxy)-	55454-60-9	NI40511
404	402	332	406	334	330	166	400	400	100	92	90	68	66	50	48	38		10	–	–	–	–	–	8	–	–	–	–	–	–	1,3-Cyclopentadiene, 1,2,3,4-tetrachloro-5-(2,3,4,5-tetrachloro-2,4-cyclopentadien-1-ylidene)-	6298-65-3	NI40512
404	402	406	332	400	334	408	330	400	100	85	61	41	33	30	24	21		10	–	–	–	–	–	8	–	–	–	–	–	–	Naphthalene, octachloro-	2234-13-1	NI40513
404	402	406	332	334	262	330	400	400	100	92	65	60	56	42	38	36		10	–	–	–	–	–	8	–	–	–	–	–	–	Naphthalene, octachloro-	2234-13-1	IC04297
199	300	200	157	149	181	100	400	400	100	28	27	23	22	21	20	17		11	2	–	–	–	14	–	–	–	–	–	–	–	2H,6H-Tetradecafluorotricyclo[5.2.2.0^{2,6}]undec-8-ene		NI40514
91	145	28	400	146	55	92	65	400	100	40	15	11	10	10	9	7		11	14	–	–	–	–	–	–	2	–	–	–	–	Benzene, [3-iodo-2-(iodomethyl)-2-isobutyl]-	40548-52-5	NI40515
371	59	372	373	89	87	61	117	400	100	39	35	23	16	14	13	12	0.79	11	32	6	–	–	–	–	–	–	5	–	–	–	1,3,5,7,9-Pentaethyl-1-methoxycyclopentasiloxane	75221-58-8	NI40516
217	69	400	117	129	164	58	93	400	100	71	53	42	40	32	21	16		12	–	1	–	–	11	–	–	–	–	1	–	–	Phosphinic fluoride, bis(pentafluorophenyl)-	22474-69-7	NI40517
197	141	199	161	113	41	79	231	400	100	69	33	32	28	28	26	24	0.00	12	17	5	–	1	3	2	–	–	–	–	–	–	Cyclopropanecarboxylic acid, 3-[2,2-dichloro-3,3,3-trifluoro-1-(methylsulphonyloxy)propyl]-2,2-dimethyl-, ethyl ester		DD01562
250	249	251	44	350	231	331	400	400	100	21	14	12	11	10	6	5		14	4	–	–	–	12	–	–	–	–	–	–	–	4H-Dodecafluoro-5-methyltetracyclo[6.2.2.1^{3,6}.0^{2,7}]trideca-2(7),4,9-triene		NI40518
250	249	251	350	231	44	100	81	400	100	23	13	10	10	8	7	6	3.50	14	4	–	–	–	12	–	–	–	–	–	–	–	4H-Dodecafluoro-5-methyltetracyclo[6.2.2.1^{3,6}.0^{2,7}]trideca-2(7),4,9-triene		NI40519
78	79	77	80	340	52	51	39	400	100	54	35	28	27	19	17	17	4.59	14	16	2	–	–	–	–	–	–	–	–	–	1	Tungsten, dicarbonyl-di-π-cyclohexa-1,3-dienyl-		MS09176
181	73	130	103	116	101	75	161	400	100	72	38	36	30	12	11	6	0.00	14	17	4	2	–	5	–	–	–	1	–	–	–	Nitrosamine, N-[(1,2,3,4,5-pentafluorobenzyloxy)carbonylmethyl]-N-[(2-trimethylsilyloxy)ethyl]-		NI40520
232	260	288	344	174	316	112	56	400	100	85	59	49	35	24	23	20	19.00	15	12	6	–	–	–	–	–	–	–	–	–	2	Iron, hexacarbonyl allenyldi-		LI08507
260	232	288	344	174	176	112	316	400	100	83	61	56	44	33	32	27	24.00	15	12	6	–	–	–	–	–	–	–	–	–	2	Iron, hexacarbonyl allenyldi-		LI08508
232	260	176	372	288	56	112	316	400	100	78	45	41	37	31	24	23	14.00	15	12	6	–	–	–	–	–	–	–	–	–	2	Iron, hexacarbonyl allenyldi-		LI08509
267	111	280	153	81	243	213	400	400	100	70	57	20	20	8	5	1		15	14	5	4	–	–	2	–	–	–	–	–	–	9-(4,6-Di-O-acetyl-1,2,3-trideoxy-D-ribo-hex-1-enopyranos-3-yl)-2,6-dichloropurine		LI08510
153	111	267	280	81	243	340	213	400	100	30	20	10	10	8	6	5	0.10	15	14	5	4	–	–	2	–	–	–	–	–	–	9-(4,6-Di-O-acetyl-1,2,3-trideoxy-D-ribo-hex-1-enopyranos-3-yl)-2,6-dichloropurine		LI08511
301	400	217	203	201	173	401	302	400	100	65	27	16	15	13	11	11		15	21	6	–	–	–	–	–	–	–	–	–	1	Rhodium, tris(2,4-pentanedionato-O,O′)-	14284-92-5	NI40521
327	299	124	328	271	107	77	151	400	100	55	41	18	13	12	12	10	1.20	16	17	6	–	–	5	–	–	–	–	–	–	–	Phenylacetic acid, α-ethoxy-3-methoxy-4-pentafluoropropanoyl-, ethyl ester		MS09177
73	385	245	147	129	243	75	315	400	100	87	65	64	64	54	38	32	28.00	16	28	6	2	–	–	–	–	–	2	–	–	–	3,6-Epoxy-2H,8H-pyrimido[6,1-b][1,3]oxazocine-8,10(9H)-dione, 3,4,5,6-tetrahydro-9-methyl-4,5-bis[(trimethylsilyl)oxy]-, [3R-(3α,4β,5β,6α)]-	41547-74-4	NI40522
218	220	198	151	183	219	182	221	400	100	69	58	58	54	52	37	29	15.00	17	14	7	–	–	–	2	–	–	–	–	–	–	Benzoic acid, 2-(3,5-dichloro-2,6-dihydroxy-4-methylbenzoyl)-5-hydroxy-3-methoxy-, methyl ester	2151-16-8	LI08512
218	220	209	151	183	219	182	221	400	100	69	58	58	54	52	32	29	15.01	17	14	7	–	–	–	2	–	–	–	–	–	–	Benzoic acid, 2-(3,5-dichloro-2,6-dihydroxy-4-methylbenzoyl)-5-hydroxy-3-methoxy-, methyl ester	2151-16-8	NI40523
122	150	219	151	182	94	66	221	400	100	56	35	31	27	27	27	23	1.98	17	14	7	–	–	–	2	–	–	–	–	–	–	Benzoic acid, 3,5-dichloro-2,4-dihydroxy-6-methyl-, 3-hydroxy-4-(methoxycarbonyl)-5-methylphenyl ester	4382-39-2	NI40524
122	150	400	219	151	182	221	121	400	100	61	40	37	35	30	26	21		17	14	7	–	–	–	2	–	–	–	–	–	–	Benzoic acid, 3,5-dichloro-2,4-dihydroxy-6-methyl-, 3-hydroxy-4-(methoxycarbonyl)-5-methylphenyl ester	4382-39-2	LI08513
359	357	402	361	400	316	404	371	400	100	80	68	52	50	35	34	34		18	12	2	–	–	–	4	–	–	–	–	–	–	9,10-Ethenoanthracene, 1,2,3,4-tetrachloro-9,10-dihydro-11,12-dimethoxy	25283-01-6	NI40525
359	357	318	361	402	316	400	371	400	100	81	67	52	48	41	33	33		18	12	2	–	–	–	4	–	–	–	–	–	–	2,3-(Tetrachlorobenzo)-5,6-benzo-7,8-dimethoxybicyclo[2.2.2]octa-2,5,7-triene		MS09178
400	120	132	160	267	145	385	225	400	100	45	37	31	30	22	19	19		18	18	2	2	–	–	–	–	–	–	–	–	1	Palladium, [[2,2′-[(1,1-dimethyl-1,2-ethanediyl)bis(nitrilomethylidyne)]bis[phenolato]](2-)-N,N′,O,O′]-	32593-89-8	LI08514
400	402	120	404	132	398	160	146	400	100	80	45	40	40	37	32	30		18	18	2	2	–	–	–	–	–	–	–	–	1	Palladium, [[2,2′-[(1,1-dimethyl-1,2-ethanediyl)bis(nitrilomethylidyne)]bis[phenolato]](2-)-N,N′,O,O′]-	32593-89-8	NI40526
178	400	360	67	69	125	41	54	400	100	60	56	48	40	28	24	16		18	20	5	6	–	–	–	–	–	–	–	–	–	1-Cyanoethyl-3-methyl-5-hydroxy-4-(4-carbobutoxynitrophenylazo)pyrazole		IC04298
297	73	298	147	299	45	75	74	400	100	64	27	12	10	9	6	5	0.90	18	36	4	–	–	–	–	–	–	3	–	–	–	Benzeneethane-1,2-diol, 3-hydroxy-4-methoxy-, tris[(trimethylsilyl)oxy]-		MS09179
297	73	298	75	280	147	310	45	400	100	59	24	18	12	11	10	10	0.00	18	36	4	–	–	–	–	–	–	3	–	–	–	Benzeneethane-1,2-diol, 4-hydroxy-3-methoxy-, tris[(trimethylsilyl)oxy]-	68595-81-3	MS09180
297	73	298	299	147	45	75	74	400	100	59	27	10	10	7	4	4	1.10	18	36	4	–	–	–	–	–	–	3	–	–	–	Benzeneethane-1,2-diol, 4-hydroxy-3-methoxy-, tris[(trimethylsilyl)oxy]-	68595-81-3	MS09181
297	73	298	299	147	75	45	267	400	100	42	25	9	8	4	4	3	0.00	18	36	4	–	–	–	–	–	–	3	–	–	–	Benzeneethane-1,2-diol, 4-hydroxy-3-methoxy-, tris[(trimethylsilyl)oxy]-	68595-81-3	MS09182
41	271	73	69	57	43	55	272	400	100	99	99	80	65	64	55	41	0.68	18	36	4	2	–	–	–	–	–	2	–	–	–	2,4,6(1H,3H,5H)-Pyrimidinetrione, 5-ethyl-1-methyl-5-[1-methyl-3-[(trimethylsilyl)oxy]butyl]-3-(trimethylsilyl)-	57397-42-9	NI40527

MASS TO CHARGE RATIOS								M.W.	INTENSITIES								Parent	C	H	O	N	S	F	Cl	Br	I	Si	P	B	X	COMPOUND NAME	CAS Reg No	No
166	106	299	28	167	41	29	69	**400**	100	61	59	39	34	21	20	14	**0.00**	19	23	4	2	–	3	–	–	–	–	–	–	–	L-Prolylphenylglycine, N-(trifluoroacetyl)-, butyl ester		MS09183
136	105	400	164	139	77	141	244	**400**	100	83	65	62	52	41	33	29		20	20	3	2	2	–	–	–	–	–	–	–	–	4(3H)-Pyrimidinone, 2-(ethylthio)-5-methyl-3-phenyl-6-tosyl-		NI40528
398	186	212	211	267	304	147	331	**400**	100	50	28	17	16	13	11	5	**0.00**	20	21	1	–	–	–	–	–	–	–	–	1	2	Diferrocenylborinic acid		LI08515
400	253	279	252	196	224	241	210	**400**	100	65	25	22	21	14	13	13		20	23	2	3	–	–	–	–	–	–	–	–	1	Copper(II), N,N′-bis(salicylidene)-3,3′-bis(aminopropyl)amino-		NI40529
72	84	43	41	57	56	158	186	**400**	100	79	48	35	33	33	29	26	**0.98**	20	36	6	2	–	–	–	–	–	–	–	–	–	Valine, N-(N-acetyl-O-butyl-α-glutamyl)-, butyl ester		AI02331
343	73	147	75	89	344	149	133	**400**	100	99	85	73	48	27	20	20	**0.00**	20	40	4	–	–	–	–	–	–	2	–	–	–	1H-Cyclopenta[c]furan-1-one, 5-[(tert-butyldimethylsilyl)oxy]-4-[[(tert-butyldimethylsilyl)oxy]methyl]hexahydro-, [3aR-(3aα,4α,5β,6aα)]-		NI40530
149	220	131	150	103	163	161	164	**400**	100	57	54	45	45	42	33	28	**15.60**	21	20	8	–	–	–	–	–	–	–	–	–	–	Arboreol		NI40531
202	78	52	43	41	203	77	155	**400**	100	75	27	25	23	18	18	15	**12.00**	21	20	8	–	–	–	–	–	–	–	–	–	–	Oct-2-enonic acid, 2,3,5-trideoxy-2-methyl-6-C-methyl-8-O-(7-oxo-7H-furo[3,2-g][1]benzopyran-9-yl)-, γ-lactone	68729-09-9	NI40532
400	401	245	201	189	167	154	341	**400**	100	35	19	15	13	12	10	8		21	20	8	–	–	–	–	–	–	–	–	–	–	α-Peltatin		LI08516
316	400	358	301	147	315	285		**400**	100	70	56	32	21	17	13			21	20	8	–	–	–	–	–	–	–	–	–	–	Philenopteran acetate		LI08517
400	385	401	298	357	69	42	186	**400**	100	34	27	19	17	16	15	14		21	24	6	2	–	–	–	–	–	–	–	–	–	3-Nor-3,7-secodichotine, 2,7-didehydro-2-deoxy-14-hydroxy-25-oxo-, (14α)-	55320-19-9	NI40533
312	400	193	313	74	401	102	64	**400**	100	86	26	25	24	22	12	12		22	12	6	2	–	–	–	–	–	–	–	–	–	1,2,4,5-Benzenetetracarboxylic acid, 1,2:4,5-diimide, N,N′-bis(2-hydroxyphenyl)-		MS09184
400	65	401	92	74	39	102	93	**400**	100	31	30	30	27	16	15	15		22	12	6	2	–	–	–	–	–	–	–	–	–	1,2,4,5-Benzenetetracarboxylic acid, 1,2:4,5-diimide, N,N′-bis(3-hydroxyphenyl)-		MS09185
400	401	65	92	74	193	52	39	**400**	100	27	21	20	17	16	15	15		22	12	6	2	–	–	–	–	–	–	–	–	–	1,2,4,5-Benzenetetracarboxylic acid, 1,2:4,5-diimide, N,N′-bis(4-hydroxyphenyl)-		MS09186
400	250	130	401	222	102	91	65	**400**	100	30	29	25	23	20	15	10		22	16	4	4	–	–	–	–	–	–	–	–	–	1,2-Dihydro-1-oxo-2-(4-tolyl)-3-hydroxy-4-(2-nitrophenylazo)isoquinoline		IC04299
105	77	267	51	50	52	39	122	**400**	100	88	68	57	52	29	21	17	**0.30**	22	16	4	4	–	–	–	–	–	–	–	–	–	1-(4-Nitrophenyl)-4-methylenebenzoyl-5-phenyl-Δ^2-1,2,3-triazoline		MS09187
400	147	161	162	195	196	133	160	**400**	100	85	75	70	50	45	36	32		22	24	7	–	–	–	–	–	–	–	–	–	–	8-Acetoxy-1-allyl-3-hydroxy-5-methoxy-7-methyl-4-oxo-6-piperonylbicyclo[3.2.1]oct-2-ene, (1S,5S,6S,7R,8R)-		NI40534
369	355	400	343	367	339	354	385	**400**	100	86	75	62	58	48	40	37		22	24	7	–	–	–	–	–	–	–	–	–	–	1H-5-Oxaaceanthrylene-4,6-dione, 8-(2-hydroxy-1-methylethyl)-2,4,6,10b-tetrahydro-7-hydroxy-9,10-dimethoxy-3,10b-dimethyl-, (10bS,1′S)-		NI40535
31	246	45	218	29	27	248	139	**400**	100	60	37	30	27	25	23	23	**0.00**	22	25	1	2	1	–	1	–	–	–	–	–	–	1-Piperazineethanol, 4-[3-(2-chloro-9H-thioxanthen-9-ylidene)propyl]-	982-24-1	NI40536
311	184	310	313	400	312	212	122	**400**	100	98	94	54	29	27	13	12		22	28	5	2	–	–	–	–	–	–	–	–	–	Aspidospermidin-21-oic acid, 1-formyl-20-hydroxy-17-methoxy-, methyl ester	56143-42-1	NI40537
400	383	218	401	382	190	355	384	**400**	100	65	45	34	29	27	23	20		22	28	5	2	–	–	–	–	–	–	–	–	–	Ibogamine-18-carboxylic acid, 16,17-didehydro-9,17-dihydro-9,20-dihydroxy-12-methoxy-, methyl ester, (20S)-	15215-86-8	LI08518
400	383	218	190	401	355	162	108	**400**	100	67	59	36	28	28	21	21		22	28	5	2	–	–	–	–	–	–	–	–	–	Ibogamine-18-carboxylic acid, 16,17-didehydro-9,17-dihydro-9,20-dihydroxy-12-methoxy-, methyl ester, (20S)-	15215-86-8	LI08519
400	383	218	190	401	355	382	108	**400**	100	69	60	37	30	29	28	22		22	28	5	2	–	–	–	–	–	–	–	–	–	Ibogamine-18-carboxylic acid, 16,17-didehydro-9,17-dihydro-9,20-dihydroxy-12-methoxy-, methyl ester, (20S)-	15215-86-8	NI40538
400	239	224	238	208	210	240	384	**400**	100	75	33	22	20	15	14	10		22	28	5	2	–	–	–	–	–	–	–	–	–	Isometrafoline	55903-80-5	NI40539
400	239	224	208	369	210	110	146	**400**	100	68	19	12	10	10	10	7		22	28	5	2	–	–	–	–	–	–	–	–	–	3-Epi-isorotundifoline	55903-81-6	NI40540
400	239	224	240	208	238	210	384	**400**	100	65	29	20	19	18	13	12		22	28	5	2	–	–	–	–	–	–	–	–	–	Isospeciofoline	56687-49-1	NI40541
400	239	224	238	208	146	110	210	**400**	100	74	17	15	10	10	10	5		22	28	5	2	–	–	–	–	–	–	–	–	–	Mitrafoline	55903-79-2	NI40542
400	239	69	224	238	146	208	110	**400**	100	68	64	37	23	21	18	18		22	28	5	2	–	–	–	–	–	–	–	–	–	Speciofoline	5171-42-6	NI40543
57	226	201	400	227	401	256	212	**400**	100	65	57	48	41	16	16	13		22	28	5	2	–	–	–	–	–	–	–	–	–	Spiro[1-(tert-butoxycarbonyl)piperidine-4,3′-(8-methoxy-5-methyl-1,4-dihydro-1-oxo-2H-pyrano[4,3-b]indole)]		DD01563
400	108	154	180	182	181	125	247	**400**	100	78	63	62	51	51	45	38		22	28	5	2	–	–	–	–	–	–	–	–	–	Voaluteine, 20-hydroxy-, (20S)-	18646-15-6	NI40544
155	73	281	154	156	197	91	283	**400**	100	58	57	56	41	33	31	25	**14.00**	22	29	1	2	–	–	1	–	–	1	–	–	–	Benzeneacetamide, N-(4-chlorophenyl)-N-(4-piperidinyl)-, N′-(trimethylsilyl)-		MS09188
357	371	400	166	358	235	208	192	**400**	100	53	48	44	42	22	20	19		22	36	1	6	–	–	–	–	–	–	–	–	–	Bis[2-(ethylamino)-4-methyl-5-butylpyrimid-6-yl] ether		IC04300
366	336	367	352	394	250	92	350	**400**	100	36	26	24	22	12	12	6	**0.00**	22	41	–	2	1	–	1	–	–	–	–	–	–	Hydrazine, [3-(hexadecylthio)phenyl]-, monohydrochloride	69688-64-8	NI40545
73	75	117	55	98	74	69	146	**400**	100	94	55	55	45	32	27	25	**1.00**	22	44	4	–	–	–	–	–	–	1	–	–	–	Octadecanedioic acid, methyl (trimethylsilyl) ester	22396-28-7	NI40546
75	55	73	67	81	41	69	43	**400**	100	85	82	53	52	50	43	34	**0.00**	22	44	4	–	–	–	–	–	–	1	–	–	–	Octadecanoic acid, 9,10-epoxy-18-(trimethylsiloxy)-, methyl ester, cis-	22032-78-6	NI40547
297	299	402	400	401	77	115	217	**400**	100	98	64	64	41	37	34	31		23	17	–	2	–	–	–	1	–	–	–	–	–	4H-1,2-Diazepine, 5-(4-bromophenyl)-3,7-diphenyl-	25649-74-5	LI08520
297	299	400	402	77	115	217	218	**400**	100	98	65	64	37	35	31	30		23	17	–	2	–	–	–	1	–	–	–	–	–	4H-1,2-Diazepine, 5-(4-bromophenyl)-3,7-diphenyl-	25649-74-5	NI40548
91	36	129	41	65	83	39	43	**400**	100	24	21	17	14	12	12	9	**0.00**	23	28	4	–	1	–	–	–	–	–	–	–	–	5-Cyclohexanone, 4,4-dibenzyl-1,1-dimethyl-3E-mesyl-		LI08521
400	382	364	383	180	121	365	265	**400**	100	98	52	27	18	17	15	15		23	28	6	–	–	–	–	–	–	–	–	–	–	Digoxigenin, 3,7,12-tridehydro-7β-hydroxy-		NI40549
358	359	360	61	400	356	298	70	**400**	100	25	5	5	4	4	4	3		23	28	6	–	–	–	–	–	–	–	–	–	–	Estra-1,3,5(10)-trien-6-one, 3,17-bis(acetyloxy)-2-methoxy-, (17β)-		NI40550
358	83	97	85	81	95	84	111	**400**	100	78	63	62	55	41	36	35	**10.00**	23	28	6	–	–	–	–	–	–	–	–	–	–	Estra-1,3,5(10)-trien-17-one, 3,16-bis(acetyloxy)-2-methoxy-, (16α)-	51477-86-2	MS09189
358	111	137	109	359	244	125	107	**400**	100	35	31	26	23	21	21	19	**9.95**	23	28	6	–	–	–	–	–	–	–	–	–	–	Estra-1,3,5(10)-trien-17-one, 3,16-bis(acetyloxy)-2-methoxy-, (16α)-	51477-86-2	NI40551

MASS TO CHARGE RATIOS	M.W.	INTENSITIES	Parent	C	H	O	N	S	F	Cl	Br	I	Si	P	B	X	COMPOUND NAME	CAS Reg No	No
268 269 310 211 212 240 213 172	400	100 19 18 15 13 10 5 5	0.00	23	32	4	–	–	–	–	–	–	1	–	–	–	Estra-1,3,5(10)-trien-17-one, 3-(acetyloxy)- 14-[(trimethylsilyl)oxy]-, (14α)-		NI40552
124 400 168 154 125 356 401 204	400	100 46 36 20 20 18 14 12		23	32	4	2	–	–	–	–	–	–	–	–	–	Aspidospermidin-3-ol, 1-acetyl-16,17-dimethoxy-, (3α)-	54725-61-0	NI40553
69 71 60 110 73 83 81 67	400	100 87 74 71 70 55 52 51	22.70	23	32	4	2	–	–	–	–	–	–	–	–	–	4-Piperidineacetic acid, 1-acetyl-5-ethyl-2-[3-(2-hydroxyethyl)-1H-indol-2 yl]-α-methyl-, methyl ester	55724-47-5	NI40554
83 85 140 400 87 57 73 55	400	100 63 36 14 14 13 12 11		23	32	4	2	–	–	–	–	–	–	–	–	–	4,25-Secoobscurinervan-4-ol, 16-methoxy-22-methyl-, (4β,22α)-	54658-07-0	NI40555
43 57 71 55 117 85 41 69	400	100 46 29 26 21 21 21 20	0.00	23	44	5	–	–	–	–	–	–	–	–	–	–	1,2-Propanediol, 3-(hexadecyloxy)-, diacetate	21994-82-1	NI40556
125 95 83 81 99 141 139 138	400	100 15 13 13 12 11 9 9	0.00	23	45	3	–	–	–	–	–	–	–	1	–	–	Phosphonic acid, isopropyl-, bis[5-methyl-2-isopropylcyclohexyl] ester	74630-11-8	NI40557
145 73 75 116 239 103 43 71	400	100 63 56 49 40 39 39 38	0.14	23	48	3	–	–	–	–	–	–	1	–	–	–	Hexadecanoic acid, 4-[(trimethylsilyl)oxy]butyl ester	56630-47-8	NI40558
400 229 230 401 228 215 114 101	400	100 88 37 26 21 15 13 10		24	16	6	–	–	–	–	–	–	–	–	–	–	Bicyclo[2.2.2]oct-7-ene-2,3,5,6-tetracarboxylic acid, 7,8-diphenyl-, dianhydride		IC04301
145 89 180 77 136 400 164 343	400	100 92 90 82 67 66 52 47		24	20	2	2	1	–	–	–	–	–	–	–	–	1-Pyrimidinium-4-olate, 2-(ethylthio)-3,6-dihydro-6-oxo-1,3,5-triphenyl-		DD01564
400 401 213 289 77 402 214 45	400	100 35 30 16 11 10 6 6		24	20	4	–	–	–	–	–	–	1	–	–	–	Silicic acid (H4SiO4), tetraphenyl ester	1174-72-7	NI40559
77 105 119 176 118 295 91 102	400	100 65 33 30 29 26 24 21	5.00	24	20	4	2	–	–	–	–	–	–	–	–	–	1,8-Dioxa-2,6-diazaspiro[4.4]nonan-7-one, 3-benzoyl-2,6-diphenyl-	52512-33-1	NI40560
400 265 91 104 145 401 90 43	400	100 80 80 44 38 32 28 28		24	20	4	2	–	–	–	–	–	–	–	–	–	2-Oxazolidinone, 3,3′-(1,4-phenylene)bis[4-phenyl-	69974-33-0	NI40561
400 221 265 91 355 103 104 401	400	100 87 67 64 46 41 35 30		24	20	4	2	–	–	–	–	–	–	–	–	–	2-Oxazolidinone, 3,3′-(1,4-phenylene)bis[5-phenyl-	69974-32-9	NI40562
104 355 77 400 117 194 356 91	400	100 99 97 76 75 70 61 57		24	20	4	2	–	–	–	–	–	–	–	–	–	2-Oxazolidinone, 5,5′-(1,4-phenylene)bis[3-phenyl-	69974-31-8	NI40563
77 105 119 176 118 285 91 93	400	100 65 33 30 29 26 24 21	5.00	24	20	4	2	–	–	–	–	–	–	–	–	–	Oxazolo[5,4-d]isoxazol-5(2H)-one, 3-benzoyltetrahydro-6a-methyl-2,6-diphenyl-	55780-95-5	NI40564
197 69 81 205 31 137 121 95	400	100 90 61 44 37 34 29 22	0.00	24	32	5	–	–	–	–	–	–	–	–	–	–	1,2-Benzenedicarboxaldehyde, 3-hydroxy-5-methoxy-4-[(3,7,11-trimethyl-2,6,10-dodecatrienyl)oxy]-, (E,E)-	2102-72-9	NI40565
231 213 79 91 81 105 382 121	400	100 87 86 85 75 73 69 67	8.57	24	32	5	–	–	–	–	–	–	–	–	–	–	Bufa-20,22-dienolide, 14,15-epoxy-3,5-dihydroxy-, (3β,5β,15β)-	470-42-8	NI40566
107 79 91 105 67 77 41 81	400	100 84 76 70 63 58 54 53	7.62	24	32	5	–	–	–	–	–	–	–	–	–	–	Bufa-20,22-dienolide, 14,15-epoxy-3,11-dihydroxy-, (3β,5β,11α,15β)-	39005-15-7	NI40567
107 91 105 81 93 215 79 67	400	100 87 81 81 80 78 78 63	26.66	24	32	5	–	–	–	–	–	–	–	–	–	–	Bufa-20,22-dienolide, 14,15-epoxy-3,16-dihydroxy-, (3β,5β,15β,16β)-	4026-95-3	NI40568
91 79 105 131 77 313 300 41	400	100 85 67 57 57 56 54 52	26.19	24	32	5	–	–	–	–	–	–	–	–	–	–	Bufa-20,22-dienolide, 14,15-epoxy-3,19-dihydroxy-, (3β,5β,15β)-	20987-26-2	NI40569
230 170 231 371 281 156 372 229	400	100 98 71 55 55 49 40 40	18.05	24	36	3	–	–	–	–	–	–	1	–	–	–	18,19-Dinorpregn-4-en-20-yn-3-one, 16-hydroxy-17-[(trimethylsilyl)oxy]-, (16β,17β)-		NI40570
400 57 325 229 71 56 401 231	400	100 65 44 40 35 34 28 23		24	36	3	–	–	–	–	–	–	1	–	–	–	Estra-1,3,5(10)-triene-16,17-diol, 3-[(trimethylsilyl)oxy]-, cyclic isopropylidene acetal, (16α,17α)-	69688-14-8	NI40571
400 325 386 401 286 285 229 231	400	100 51 48 30 22 22 20 16		24	36	3	–	–	–	–	–	–	1	–	–	–	Estra-1,3,5(10)-triene-16,17-diol, 3-[(trimethylsilyl)oxy]-, cyclic isopropylidene acetal, (16β,17β)-	69688-05-7	NI40572
255 101 385 145 208 209 157 87	400	100 12 6 5 2 1 1 1	0.00	24	48	4	–	–	–	–	–	–	–	–	–	–	Glycerol, 1-O-(2-ethoxyhexadecyl)-2,3-O-isopropylidene-		NI40573
86 267 285 57 45 43 55 69	400	100 92 83 67 58 58 50 44	6.67	24	48	4	–	–	–	–	–	–	–	–	–	–	Propane, 1,2,3-tris(hydromethyl)-, monostearate		IC04302
219 296 323 268 400 240 309 280	400	100 60 55 55 30 17 8 5		25	20	5	–	–	–	–	–	–	–	–	–	–	5,7-Dihydroxy-6-methyl-8-(3-phenylacryloyl)flavanone		NI40574
219 268 323 296 399 398 400 372	400	100 90 85 85 80 30 20 15		25	20	5	–	–	–	–	–	–	–	–	–	–	5,7-Dihydroxy-8-methyl-6-(3-phenylacryloyl)flavanone		NI40575
400 219 268 296 192 323 240 372	400	100 90 80 45 40 30 20 3		25	20	5	–	–	–	–	–	–	–	–	–	–	3,11-Dioxo-5,11-diphenyl-9-hydroxy-8-methyl-6,14-dioxatricyclo[8.4.0.0^{2,7}]tetradeca-1(9),2(7),8-triene		NI40576
400 357 402 401 385 359 358 242	400	100 61 32 29 22 17 15 15		25	21	1	2	–	–	1	–	–	–	–	–	–	1H-1,4-Benzodiazepin-2-one, 7-chloro-2,3-dihydro-3-(4-isopropylbenzylidene)-5-phenyl-	550056-36-5	NI40577
400 121 77 275 199 185 183 91	400	100 83 60 45 44 42 35 28		25	22	1	–	–	–	–	–	–	–	2	–	–	Tetraphenyl methylene diphosphine monoxide		IC04303
358 359 400 344 147 360 343	400	100 35 8 6 6 5 4		25	36	4	–	–	–	–	–	–	–	–	–	–	Androstan-3,5-dione, 3,17-bis(acetyloxy)-2-ethyl-, (2α,17β)-		LI08522
42 400 69 71 275 55 261 332	400	100 98 94 66 65 47 42 41		25	36	4	–	–	–	–	–	–	–	–	–	–	8-Hydroxy-3-isopropyl-4,4-dimethyl-7-(3-methylbut-2-enyl)-9-(2-methylpropionyl)tricyclo[5.3.1.0^{1,5}]undecane-10,11-dione		NI40578
297 315 298 258 55 91 137 359	400	100 37 24 21 20 18 17 14	3.90	25	36	4	–	–	–	–	–	–	–	–	–	–	Pregn-4-ene-3,20-dione, 17-hydroxy-6,16-dimethyl-, acetate, (6α,16α)	14334-92-0	MS09190
152 151 400 167 153 107 166 385	400	100 25 24 20 13 13 9 8		25	41	3	–	–	–	–	–	–	–	–	1	–	Pregn-17(20)-ene-3,20,21-triol, cyclic 17,20-(butylboronate), (3α,5α)-	55521-26-1	NI40579
335 327 77 249 105 299 356 323	400	100 84 64 56 48 34 28 27	8.00	26	24	4	–	–	–	–	–	–	–	–	–	–	Ethyl 2-ethoxy-2,4-diphenyl-2H-1-benzopyran-3-carboxylate		LI08523
355 327 77 249 105 299 323 356	400	100 81 65 58 50 35 28 27	8.30	26	24	4	–	–	–	–	–	–	–	–	–	–	Ethyl 2-ethoxy-2,4-diphenyl-2H-1-benzopyran-3-carboxylate		MS09191
117 69 39 185 116 41 53 89	400	100 70 64 48 43 37 33 30	0.00	26	28	–	2	1	–	–	–	–	–	–	–	–	Bis[1-(4-cyanophenyl)-4-methylenepent-3-yl] sulphide		LI08524
368 345 400 342 385 107 121 81	400	100 91 65 54 49 47 41 37		26	40	3	–	–	–	–	–	–	–	–	–	–	3,5-Cyclochol-22-en-24-oic acid, 6-methoxy-, methyl ester, (3β,5α,6β,22E)-	56259-12-2	NI40580
271 133 122 340 123 147 135 134	400	100 74 72 70 64 63 62 62	4.00	26	40	3	–	–	–	–	–	–	–	–	–	–	9,19-Cyclopregnan-20-one, 3-acetyloxy-4,4,14-trimethyl-, (3β,5α,9β,14α)		LI08525
400 399 323 205 207 401 200 194	400	100 99 53 48 47 38 29 24		27	20	–	4	–	–	–	–	–	–	–	–	–	1H-Pyrrolo[2,3-b]pyridine, 3,3′-methylenebis[2-phenyl-	23616-61-7	NI40581
136 118 158 400 69 95 81 45	400	100 90 49 44 31 29 28 25		27	44	2	–	–	–	–	–	–	–	–	–	–	Cholecalciferol, 25-hydroxy-	19356-17-3	NI40582
136 118 158 69 81 95 400 119	400	100 90 49 31 29 28 27 26		27	44	2	–	–	–	–	–	–	–	–	–	–	Cholecalciferol, 25-hydroxy-	19356-17-3	LI08526
136 118 56 70 43 41 81 68	400	100 91 78 61 61 56 52 52	0.00	27	44	2	–	–	–	–	–	–	–	–	–	–	Cholecalciferol, 25-hydroxy-	19356-17-3	NI40583
109 271 43 69 41 82 110 55	400	100 90 68 63 47 44 36 35	0.00	27	44	2	–	–	–	–	–	–	–	–	–	–	Cholesta-5,24-diene-3,20-diol, (3β,20R)-	56362-84-6	NI40584

MASS TO CHARGE RATIOS								M.W.	INTENSITIES								Parent	C	H	O	N	S	F	Cl	Br	I	Si	P	B	X	COMPOUND NAME	CAS Reg No	No
313	95	69	41	109	55	81	314	**400**	100	72	62	42	35	32	27	24	**23.00**	27	44	2	–	–	–	–	–	–	–	–	–	–	**Cholesta-17(20),24-diene-3,6-diol, (3β,5α,6α)-**	55924-02-2	**NI40585**
313	95	69	41	109	55	81	331	**400**	100	73	63	43	36	34	28	25	**23.09**	27	44	2	–	–	–	–	–	–	–	–	–	–	**Cholesta-17(20),24-diene-3,6-diol, (3β,5α,6α)-**	55924-02-2	**NI40586**
93	69	95	289	109	81	41	107	**400**	100	89	76	71	65	60	57	41	**30.00**	27	44	2	–	–	–	–	–	–	–	–	–	–	**Cholesta-20,24-diene-3,6-diol, (3β,5α,6α)-**	41084-19-9	**NI40588**
93	69	95	289	109	81	41	133	**400**	100	89	76	71	65	60	57	43	**30.39**	27	44	2	–	–	–	–	–	–	–	–	–	–	**Cholesta-20,24-diene-3,6-diol, (3β,5α,6α)-**	41084-19-9	**NI40587**
400	245	43	55	287	401	81	95	**400**	100	58	44	40	33	32	32	31		27	44	2	–	–	–	–	–	–	–	–	–	–	**Cholestane-3,6-dione, (5α)-**	2243-09-6	**NI40589**
400	245	55	401	57	43	41	69	**400**	100	40	35	30	29	29	25	24		27	44	2	–	–	–	–	–	–	–	–	–	–	**Cholestane-3,6-dione, (5α,17α,20S)-**	41083-78-7	**NI40590**
400	124	81	95	245	137	259	135	**400**	100	36	28	25	19	16	14	13		27	44	2	–	–	–	–	–	–	–	–	–	–	**Cholestane-3,11-dione, (5α)-**		**MS09192**
400	193	289	229	263	219	246	342	**400**	100	90	53	45	39	35	30	21		27	44	2	–	–	–	–	–	–	–	–	–	–	**Cholestane-3,11-dione, (5α,8α,14β)-**		**MS09193**
289	193	400	246	382	385	219	263	**400**	100	76	70	32	16	12	10	9		27	44	2	–	–	–	–	–	–	–	–	–	–	**Cholestane-3,11-dione, (5α,14β)-**		**MS09194**
107	269	95	287	93	55	43	207	**400**	100	93	83	56	52	50	48	41	**21.00**	27	44	2	–	–	–	–	–	–	–	–	–	–	**Cholestan-7-one, 8,14-epoxy-, (5α,8α)-**	54411-78-8	**NI40591**
355	400	354	401	356	385	382	339	**400**	100	76	75	31	26	24	14	10		27	44	2	–	–	–	–	–	–	–	–	–	–	**Cholest-2-en-19-oic acid, (5α)-**	28809-50-9	**NI40592**
400	355	354	385	382	339			**400**	100	99	75	24	14	10				27	44	2	–	–	–	–	–	–	–	–	–	–	**Cholest-2-en-19-oic acid, (5α)-**	28809-50-9	**LI08527**
43	41	400	57	55	44	367	95	**400**	100	34	22	22	22	20	12	10		27	44	2	–	–	–	–	–	–	–	–	–	–	**Cholest-7-en-3-ol, 5,6-epoxy-, (3β,5α,6α)-**		**NI40593**
43	55	41	57	95	81	91	93	**400**	100	60	54	44	40	36	34	33	**28.00**	27	44	2	–	–	–	–	–	–	–	–	–	–	**Cholest-7-en-3-ol, 5,6-epoxy-, (3β,5β,6β)-**		**NI40594**
370	43	400	55	371	57	41	257	**400**	100	50	40	40	38	36	36	15		27	44	2	–	–	–	–	–	–	–	–	–	–	**Cholest-4-en-3-one, 19-hydroxy-**		**LI08528**
55	400	124	57	69	229	81	79	**400**	100	74	63	63	52	37	37	36		27	44	2	–	–	–	–	–	–	–	–	–	–	**Cholest-4-en-3-one, 26-hydroxy-**	19257-21-7	**NI40595**
124	400	229	277	358	271	243	244	**400**	100	83	40	20	17	9	6	5		27	44	2	–	–	–	–	–	–	–	–	–	–	**Cholest-4-en-3-one, 26-hydroxy-**	19257-21-7	**LI08529**
55	400	124	57	69	81	229	79	**400**	100	73	63	63	52	37	36	36		27	44	2	–	–	–	–	–	–	–	–	–	–	**Cholest-4-en-3-one, 27-hydroxy-**		**LI08530**
399	381							**400**	100	15							**0.00**	27	44	2	–	–	–	–	–	–	–	–	–	–	**Cholest-5-en-24-one, 3-hydroxy-, (3β)-**	17752-16-8	**NI40596**
400	43	71	314	382	213	289	315	**400**	100	84	40	30	28	20	18	16		27	44	2	–	–	–	–	–	–	–	–	–	–	**Cholest-5-en-24-one, 3-hydroxy-, (3β)-**	17752-16-8	**LI08531**
400	255	271	385	273	314	246	367	**400**	100	30	19	14	13	7	6	4		27	44	2	–	–	–	–	–	–	–	–	–	–	**Cholest-7-en-24-one, 3-hydroxy-**		**LI08532**
400	234	401	287	55	43	41	93	**400**	100	41	32	27	26	26	24	18		27	44	2	–	–	–	–	–	–	–	–	–	–	**Cholest-8(14)-en-7-one, 3-hydroxy-, (3β,5α)-**	566-29-0	**NI40597**
400	269	107	43	367	401	55	41	**400**	100	58	34	33	32	31	23	19		27	44	2	–	–	–	–	–	–	–	–	–	–	**Cholest-8(14)-en-15-one, 3-hydroxy-, (3β,5α)-**	50673-97-7	**NI40598**
73	159	205	319	147	217	320	103	**400**	100	99	80	79	41	30	25	22	**0.00**	27	44	2	–	–	–	–	–	–	–	–	–	–	**Cholest-8-en-7-one, 3-hydroxy-, (3β,5α)-**	69140-15-4	**NI40599**
354	353	247	355	371	241	372	199	**400**	100	33	30	28	25	22	19	15	**0.00**	27	44	2	–	–	–	–	–	–	–	–	–	–	**Cholesterol, 10-formyl-**		**LI08533**
43	345	385	105	368	400	346	386	**400**	100	63	49	47	35	25	16	14		27	44	2	–	–	–	–	–	–	–	–	–	–	**3,5-Cyclocholan-24-ol, 23ξ-ethyl-6-methoxy-, (3α,6β)-**		**MS09195**
354	247	355	241	353	199	382	371	**400**	100	40	30	27	24	17	15	13	**0.00**	27	44	2	–	–	–	–	–	–	–	–	–	–	**3,5-Cyclocholestan-19-al, 6-hydroxy-, (6β)-**	56942-37-1	**NI40600**
91	55	69	83	41	97	43	57	**400**	100	27	23	20	20	16	16	14	**0.00**	27	44	2	–	–	–	–	–	–	–	–	–	–	**Eicosenoic acid, benzyl ester**		**NI40601**
329	330	194	55	41	81	67	109	**400**	100	27	21	19	15	13	13	12	**9.00**	27	44	2	–	–	–	–	–	–	–	–	–	–	**Furostan-12-one, (5α)-**	54411-92-6	**NI40602**
400	401	181	182	302	287	257	402	**400**	100	32	23	12	10	10	8	5		27	44	2	–	–	–	–	–	–	–	–	–	–	**Furost-20(22)-en-26-ol, (5α,25R)-**	24744-54-5	**MS09196**
95	181	55	67	400	41	109	81	**400**	100	96	93	83	80	72	71	70		27	44	2	–	–	–	–	–	–	–	–	–	–	**Furost-20(22)-en-26-ol, (5α,25R)-**	24744-54-5	**NI40603**
286	400	328	401	331	287	271	330	**400**	100	83	37	26	25	23	19	17		27	44	2	–	–	–	–	–	–	–	–	–	–	**Spirostan, (5α,14β,25R)-**	24742-85-6	**MS09197**
286	328	400	257	331	109	55	271	**400**	100	91	78	55	49	48	47	46		27	44	2	–	–	–	–	–	–	–	–	–	–	**Spirostan, (5α,14β,25R)-**	24742-85-6	**NI40605**
286	328	400	257	331	109	271	55	**400**	100	92	79	54	49	48	46	45		27	44	2	–	–	–	–	–	–	–	–	–	–	**Spirostan, (5α,14β,25R)-**	24742-85-6	**NI40604**
400	181	139	328	257	109	95	55	**400**	100	64	52	40	40	37	33	30		27	44	2	–	–	–	–	–	–	–	–	–	–	**Spirostan, (5α,20R,25R)-**	24799-49-3	**NI40606**
400	181	139	257	328	109	95	401	**400**	100	62	51	41	39	35	33	32		27	44	2	–	–	–	–	–	–	–	–	–	–	**Spirostan, (5α,20R,25R)-**	24799-49-3	**NI40607**
139	257	286	271	55	122	69	81	**400**	100	40	26	22	16	15	15	14	**8.00**	27	44	2	–	–	–	–	–	–	–	–	–	–	**Spirostan, (5α,25R)-**	5012-14-6	**NI40608**
139	257	286	271	55	140	122	81	**400**	100	38	26	21	18	13	13	13	**8.01**	27	44	2	–	–	–	–	–	–	–	–	–	–	**Spirostan, (5α,25R)-**	5012-14-6	**NI40609**
139	55	41	69	43	67	81	257	**400**	100	69	55	48	35	31	28	26	**0.00**	27	44	2	–	–	–	–	–	–	–	–	–	–	**Spirostan, (5α,25R)-**	5012-14-6	**NI40610**
262	357	83	263	55	91	331	69	**400**	100	87	54	51	47	43	42	38	**32.57**	27	45	–	–	–	–	–	–	–	–	1	–	–	**Phosphine, bis[methylisopropylcyclohexyl]benzyl-**	74779-93-4	**NI40611**
400	202	210	200	401	79	91	93	**400**	100	46	28	28	25	23	22	16		28	36	–	2	–	–	–	–	–	–	–	–	–	**Diamantanone azine**		**MS09198**
400	385	401	245	246	386	316	382	**400**	100	32	31	20	15	13	11	9		28	48	1	–	–	–	–	–	–	–	–	–	–	**Cholestane, 2,3-epoxy-2-methyl-, (2α,3α,5α)-**	55402-28-3	**NI40612**
44	43	55	400	57	41	69	83	**400**	100	73	44	36	36	36	28	27		28	48	1	–	–	–	–	–	–	–	–	–	–	**Cholestane, 3,4-epoxy-2-methyl-, (2α,3α,4α,5α)-**	55724-43-1	**NI40613**
69	81	71	95	67	83	79	93	**400**	100	74	69	67	63	53	41	39	**4.59**	28	48	1	–	–	–	–	–	–	–	–	–	–	**Cholestan-3-ol, 2-methylene-, (3β,5α)-**	22599-96-8	**NI40614**
400	245	246	55	43	95	81	401	**400**	100	87	60	46	40	39	37	33		28	48	1	–	–	–	–	–	–	–	–	–	–	**5α-Cholestan-4-one, 3β-methyl-**	2634-54-0	**NI40615**
43	55	41	245	246	57	95	81	**400**	100	60	58	52	33	30	26	26	**26.00**	28	48	1	–	–	–	–	–	–	–	–	–	–	**Cholestan-3-one, 4-methyl-, (4α,5α)-**	984-87-2	**NI40616**
245	55	176	83	57	260	82	41	**400**	100	53	49	47	46	45	42	42	**28.28**	28	48	1	–	–	–	–	–	–	–	–	–	–	**Cholestan-3-one, 14-methyl-, (5α)-**	54411-55-1	**NI40617**
400	232	109	55	83	95	43	217	**400**	100	97	80	70	63	61	57	54		28	48	1	–	–	–	–	–	–	–	–	–	–	**Cholestan-15-one, 14-methyl-, (5α)-**	54515-29-6	**NI40618**
259	371	217	109	301	372	260	121	**400**	100	97	66	33	29	28	19	19	**0.00**	28	48	1	–	–	–	–	–	–	–	–	–	–	**Cholestan-16-one, 8-methyl-**	55334-69-5	**NI40619**
81	95	69	67	79	93	91	71	**400**	100	96	90	81	73	64	60	54	**50.74**	28	48	1	–	–	–	–	–	–	–	–	–	–	**Cholest-2-ene-2-methanol, (5α)-**	53287-18-6	**NI40620**
400	368	401	369	316	353	315	385	**400**	100	99	83	82	81	77	72	70		28	48	1	–	–	–	–	–	–	–	–	–	–	**Cholest-4-ene, 3-methoxy-, (3β)-**	1981-91-5	**NI40621**
368	400	353	107	81	145	55	43	**400**	100	60	57	50	46	42	41	40		28	48	1	–	–	–	–	–	–	–	–	–	–	**Cholest-5-ene, 3-methoxy-, (3β)-**	1174-92-1	**NI40622**
400	368	55	43	107	81	41	57	**400**	100	82	64	62	54	51	51	49		28	48	1	–	–	–	–	–	–	–	–	–	–	**Cholest-5-ene, 3-methoxy-, (3β)-**	1174-92-1	**NI40623**
41	43	107	145	55	105	119	57	**400**	100	92	80	77	76	73	72	71	**30.00**	28	48	1	–	–	–	–	–	–	–	–	–	–	**Cholest-5-ene, 3-methoxy-, (3β)-**	1174-92-1	**NI40624**
57	43	55	385	71	41	69	95	**400**	100	80	70	55	53	52	50	37	**9.00**	28	48	1	–	–	–	–	–	–	–	–	–	–	**Cholest-7-en-3-ol, 14-methyl-, (3β)-**	56687-74-2	**NI40625**
43	55	41	57	69	95	81	44	**400**	100	65	62	45	40	30	30	30	**4.00**	28	48	1	–	–	–	–	–	–	–	–	–	–	**Cholest-8-en-3-ol, 14-methyl-, (3β,5α)-**	6062-47-1	**NI40626**

MASS TO CHARGE RATIOS								M.W.	INTENSITIES								Parent	C	H	O	N	S	F	Cl	Br	I	Si	P	B	X	COMPOUND NAME	CAS Reg No	No
95	107	55	43	81	385	93	109	**400**	100	75	70	68	66	59	59	58	**25.00**	28	48	1	–	–	–	–	–	–	–	–	–	–	9,19-Cyclocholestan-3-ol, 14-methyl-, (3β,5α)-	1912-66-9	NI40627
400	382	95	227	93	301	107	302	**400**	100	57	54	47	43	40	40	36		28	48	1	–	–	–	–	–	–	–	–	–	–	Cyclopropa[5,6]cholestan-3-ol, 3′,6-dihydro-, (3α,5α,6β)-	35339-45-8	NI40628
329	95	107	301	119	400	93	121	**400**	100	57	47	43	43	42	42	37		28	48	1	–	–	–	–	–	–	–	–	–	–	Cyclopropa[5,6]cholestan-3-ol, 3′,6-dihydro-, (3β,5α,6β)-	35339-47-0	NI40629
400	55	121	95	81	135	107	57	**400**	100	97	93	92	91	87	86	86		28	48	1	–	–	–	–	–	–	–	–	–	–	Cyclopropa[5,6]cholestan-3-ol, 3′,6-dihydro-, (3β,5β,6α)-	38037-44-4	NI40630
385	84	135	400	81	107	126	55	**400**	100	92	86	68	67	62	60	60		28	48	1	–	–	–	–	–	–	–	–	–	–	Cyclopropa[7,8]cholestan-3-ol, 3′,7-dihydro-, (3β,5α,7β,8α)-	53755-39-8	NI40631
218	245	400	55	109	43	108	69	**400**	100	71	61	47	44	38	33	28		28	48	1	–	–	–	–	–	–	–	–	–	–	5α-Ergostan-15-one	14111-74-1	NI40632
400	257	273	302	339	255	284	382	**400**	100	86	67	65	22	14	12	8		28	48	1	–	–	–	–	–	–	–	–	–	–	5α-Ergost-22,23-ene-3β-ol		MS09199
55	69	81	109	95	67	107	83	**400**	100	89	68	64	59	52	46	43	**14.71**	28	48	1	–	–	–	–	–	–	–	–	–	–	5α-Ergost-22,23-ene-3β-ol		MS09200
55	69	81	95	109	257	67	107	**400**	100	96	77	65	59	58	51	49	**35.21**	28	48	1	–	–	–	–	–	–	–	–	–	–	5β-Ergost-22,23-ene-3α-ol		MS09202
400	302	257	273	339	255	382	284	**400**	100	88	86	85	26	8	4	4		28	48	1	–	–	–	–	–	–	–	–	–	–	5β-Ergost-22,23-ene-3α-ol		MS09201
383	399	384	400	385	401	151	382	**400**	100	51	31	22	17	11	10	5		28	48	1	–	–	–	–	–	–	–	–	–	–	Ergost-5-en-3-ol, (3β)-	4651-51-8	NI40633
43	55	57	41	81	95	107	69	**400**	100	72	58	53	52	47	45	40	**37.00**	28	48	1	–	–	–	–	–	–	–	–	–	–	Ergost-5-en-3-ol, (3β)-	4651-51-8	NI40634
400	43	107	55	382	289	315	145	**400**	100	63	49	49	44	43	37	36		28	48	1	–	–	–	–	–	–	–	–	–	–	Ergost-5-en-3-ol, (3β,24R)-	474-62-4	NI40636
213	400	255	289	382	315	385	367	**400**	100	76	61	49	47	44	37	37		28	48	1	–	–	–	–	–	–	–	–	–	–	Ergost-5-en-3-ol, (3β,24R)-	474-62-4	NI40635
43	55	255	400	41	105	107	81	**400**	100	72	64	60	56	54	50	49		28	48	1	–	–	–	–	–	–	–	–	–	–	Ergost-7-en-3-ol, (3β)-	26047-31-4	NI40637
400	255	43	55	107	95	41	81	**400**	100	71	60	47	42	35	34	33		28	48	1	–	–	–	–	–	–	–	–	–	–	Ergost-7-en-3-ol, (3β,5α)-	516-78-9	NI40638
400	43	55	107	57	41	81	69	**400**	100	59	52	43	39	33	32	32		28	48	1	–	–	–	–	–	–	–	–	–	–	Ergost-8(14)-en-3-ol, (3β)-	30365-65-2	NI40639
400	401	385	43	69	41	57	55	**400**	100	30	15	14	13	11	10	10		28	48	1	–	–	–	–	–	–	–	–	–	–	Ergost-8(14)-en-3-ol, (3β,5α)-	632-32-6	NI40640
400	269	43	95	55	105	119	81	**400**	100	95	88	80	77	63	61	60		28	48	1	–	–	–	–	–	–	–	–	–	–	4α-Methyl-5α-cholest-7-en-3β-ol		NI40641
400	43	55	121	147	105	95	41	**400**	100	82	71	61	56	56	56	52		28	48	1	–	–	–	–	–	–	–	–	–	–	4α-Methyl-5α-cholest-8(14)-en-3β-ol		NI40642
287	269	94	161	93	107	95	43	**400**	100	57	51	39	39	35	31	31	**3.00**	28	48	1	–	–	–	–	–	–	–	–	–	–	4α-Methyl-5α-cholest-14-en-3β-ol		NI40643
385	367	43	159	55	95	57	81	**400**	100	77	67	64	58	56	52	50	**11.00**	28	48	1	–	–	–	–	–	–	–	–	–	–	6-Methylcholest-5-en-3β-ol		NI40644
372	267	105	400	77	373	268	265	**400**	100	86	48	43	43	28	18	18		29	20	2	–	–	–	–	–	–	–	–	–	–	2H-Pyran-2-one, 3,4,5,6-tetraphenyl-	33524-67-3	MS09203
119	259	120	43	41	55	29	57	**400**	100	25	11	11	8	7	7	6	**5.87**	29	52	–	–	–	–	–	–	–	–	–	–	–	11-(2,5-Dimethylphenyl)heneicosane		AI02332
119	258	120	43	259	55	41	71	**400**	100	13	9	4	3	3	3	2	**0.00**	29	52	–	–	–	–	–	–	–	–	–	–	–	11-(2,5-Dimethylphenyl)heneicosane		AI02333
119	43	259	41	120	55	57	69	**400**	100	18	17	12	11	8	5	4	**4.62**	29	52	–	–	–	–	–	–	–	–	–	–	–	11-(2,5-Dimethylphenyl)heneicosane		AI02334
217	43	44	218	55	149	41	109	**400**	100	79	69	58	57	54	48	47	**34.03**	29	52	–	–	–	–	–	–	–	–	–	–	–	Stigmastane	601-58-1	WI01521
217	400	218	149	109	385	123	232	**400**	100	82	71	53	52	28	23	18		29	52	–	–	–	–	–	–	–	–	–	–	–	Stigmastane	601-58-1	LI08534
295	105	91	294	296	77	400	220	**400**	100	76	42	31	27	26	25	14		30	24	1	–	–	–	–	–	–	–	–	–	–	Cyclopentene, 2,4,5-triphenyl-3-benzoyl-	84709-76-2	NI40645
388	390	386	222	152	224	403	392	**401**	100	64	63	38	30	24	21	20	**13.00**	7	7	–	3	–	–	5	–	–	–	3	–	–	1-Methyl-1-(4-chlorophenyl)tetrachlorocyclotriphosphazene	87048-76-8	NI40646
41	56	55	368	370	108	366	29	**401**	100	88	81	70	45	45	43	38	**1.49**	9	9	–	3	1	–	6	–	–	–	–	–	–	2-(Butylthio)-4,6-bis(trichloromethyl)-s-triazine		MS09204
41	368	55	370	108	366	347	29	**401**	100	99	82	65	65	64	61	56	**1.19**	9	9	–	3	1	–	6	–	–	–	–	–	–	2-(Isobutylthio)-4,6-bis(trichloromethyl)-s-triazine		MS09205
259	213	211	371	96	399	169	60	**401**	100	30	21	19	19	16	14	13	**0.00**	11	18	5	3	–	–	–	–	–	–	1	–	1	Molybdenum, pentacarbonyl(hexamethylphosphorous triamide-P)-	14971-43-8	NI40647
76	119	261	215	401	60	163	373	**401**	100	99	50	24	16	13	10	8		11	18	5	3	–	–	–	–	–	–	1	–	1	Molybdenum, pentacarbonyl(hexamethylphosphorous triamide-P)-	14971-43-8	NI40648
119	76	259	60	163	213	371	169	**401**	100	82	34	17	15	14	9	5	**3.79**	11	18	5	3	–	–	–	–	–	–	1	–	1	Molybdenum, pentacarbonyltris(dimethylamino)phosphinyl-		MS09206
119	159	76	213	211	371	96	399	**401**	100	88	67	26	18	17	17	14		11	18	5	3	–	–	–	–	–	–	1	–	1	Molybdenum, pentacarbonyltris(dimethylamino)phosphinyl-		LI08535
277	88	275	279	62	61	53	168	**401**	100	74	69	68	27	24	21	20	**0.00**	15	17	2	1	–	–	–	2	–	–	–	–	–	Octanoic acid, 2,6-dibromo-4-cyanophenyl ester	1689-99-2	NI40650
305	277	307	303	323	317	279	221	**401**	100	59	53	49	42	36	32	31	**0.00**	15	17	2	1	–	–	–	2	–	–	–	–	–	Octanoic acid, 2,6-dibromo-4-cyanophenyl ester	1689-99-2	NI40649
81	99	149	135	127	215	189	187	**401**	100	47	27	20	19	18	17	17	**1.00**	15	24	6	5	–	–	–	–	–	–	1	–	–	D-erythro-Pentofuranoside, methyl 5-(6-amino-3H-purin-3-yl)-2,5-dideoxy-, 3-(diethyl phosphate)	62246-44-0	NI40651
401	328	158	200	235	146	282	236	**401**	100	88	70	57	55	49	45	33		18	15	6	3	1	–	–	–	–	–	–	–	–	2-(2,4-Dinitrophenylthio)-3-indolebutyric acid		LI08536
328	401	235	146	236	117	145	281	**401**	100	91	83	78	40	36	35	32		18	15	6	3	1	–	–	–	–	–	–	–	–	Ethyl 2-(2,4-dinitrophenylthio)-3-indoleacetate		LI08537
113	43	169	155	110	112	70	83	**401**	100	80	53	39	39	37	19	16	**3.98**	18	27	9	1	–	–	–	–	–	–	–	–	–	Glucopyranose, 2-deoxy-2-(1-pyrrolidinyl)-, 1,3,4,6-tetraacetate, α-D-	20744-54-1	NI40652
73	298	103	45	257	75	237	299	**401**	100	70	36	24	20	14	13	12	**1.74**	19	20	1	1	–	5	–	–	–	1	–	–	–	2-Propanamine, N-(pentafluorobenzylidene)-2-phenyl-1-(trimethylsiloxy)-		NI40653
43	188	101	159	185	100	300	59	**401**	100	98	68	50	45	45	43	40	**33.00**	19	31	8	1	–	–	–	–	–	–	–	–	–	5-Isoxazolidinecarboxylic acid, 2-[2,3:5,6-bis(O-isopropylidene)-α-D-mannofuranosyl]-3,5-dimethyl-, methyl ester, (3R-trans)-	64044-15-1	NI40654
43	101	59	41	159	300	85	69	**401**	100	30	26	25	24	22	19	18	**13.00**	19	31	8	1	–	–	–	–	–	–	–	–	–	5-Isoxazolidinecarboxylic acid, 2-[2,3:5,6-bis(O-isopropylidene)α-D-mannofuranosyl]-3,5-dimethyl-, methyl ester, (3S-cis)-	64018-58-2	NI40655
188	73	145	44	298	189	75	208	**401**	100	73	43	31	16	16	15	12	**0.09**	19	36	2	1	–	–	1	–	–	2	–	–	–	Benzeneethanamine, α-methyl-β-[(trimethylsilyl)oxy]-N-[1,1-dimethyl-2-[(trimethylsilyl)oxy]ethyl]-		NI40656
366	383	44	214	256	241	43	367	**401**	100	98	53	43	38	38	37	33	**11.11**	20	19	8	1	–	–	–	–	–	–	–	–	–	2-Naphthacenecarboxamide, 1,4,4a,5,5a,6,11,12a-octahydro-3,6,10,12,12a-pentahydroxy-6-methyl-1,11-dioxo-, [4aS-(4aα,5aα,6β,12aα)]	2444-65-7	NI40657
401	224	196	242	183	182	63	295	**401**	100	29	22	14	8	7	7	6		20	22	3	2	–	–	–	–	–	–	–	–	1	Copper(II), [N,N′-bis(salicylidene)-3,3′-bis(aminopropyl)oxy]-		NI40658
401	403	301	299	72	300	302	259	**401**	100	99	74	53	53	40	38	32		20	24	1	3	–	–	–	1	–	–	–	–	–	2-Bromolysergic acid diethylamide		MS09207

MASS TO CHARGE RATIOS								M.W.	INTENSITIES								Parent	C	H	O	N	S	F	Cl	Br	I	Si	P	B	X	COMPOUND NAME	CAS Reg No	No
43	57	59	55	41	71	56	69	**401**	100	57	50	49	43	34	28	26	**0.00**	21	23	7	1	–	–	–	–	–	–	–	–	–	**Colchifoleine**	78517-64-3	**NI40659**
136	190	162	206	175	234	178	401	**401**	100	85	44	40	38	30	22	15		21	23	7	1	–	–	–	–	–	–	–	–	–	**Nitrocalopiptin**		**LI08538**
141	123	42	41	151	169	55	96	**401**	100	94	58	49	36	33	33	30	**3.00**	21	24	1	3	1	–	1	–	–	–	–	–	–	**4-Piperidinecarboxamide, 1-[3-(2-chloro-10H-phenothiazin-10-yl)propyl]-**	84-04-8	**NI40660**
132	57	216	85	217	77	130	41	**401**	100	98	95	44	43	38	30	27	**0.00**	21	24	2	1	–	–	–	1	–	–	–	–	–	**1-Isoquinolinemethanol, u-α-(2-bromophenyl)-2-tert-butoxy-1,2,3,4-tetrahydro-**		**DD01565**
342	401	371	358	386	343	402	284	**401**	100	56	28	28	21	18	17	11		22	27	6	1	–	–	–	–	–	–	–	–	–	**Hasubanan-9-ol, 7,8-didehydro-4,5-epoxy-3,6,6-trimethoxy-17-methyl-, acetate (5α,9α,13β,14β)-**	29944-26-1	**NI40661**
222	179	208	223	206	401	178	180	**401**	100	57	18	14	10	9	9	8		22	27	6	1	–	–	–	–	–	–	–	–	–	**Rheadan-8-ol, 2,3,10,11-tetramethoxy-16-methyl-, (6α,8α)-**	14028-91-2	**NI40662**
192	193	177	206	387	194	181	182	**401**	100	13	7	4	3	3	3	2	**0.00**	22	32	5	1	–	–	–	–	–	–	–	1	–	**7-Isoquinolinol, 1,2,3,4-tetrahydro-6-methoxy-2-methyl-1-[2-(3,4,5-trimethoxyphenyl)ethyl]-, compd. with borane (1:1)**	56772-01-1	**NI40663**
264	324	173	187	159	383	174	160	**401**	100	42	14	13	11	10	10	9	**7.00**	23	31	5	1	–	–	–	–	–	–	–	–	–	**Estra-1,3,5(10)-trien-14-ol, 15-acetylamino-17-acetoxy-3-methoxy-, (14α,15β,17β)-**	85853-13-0	**NI40664**
264	324	187	383	173	159	174	249	**401**	100	21	20	15	13	10	9	7	**0.80**	23	31	5	1	–	–	–	–	–	–	–	–	–	**Estra-1,3,5(10)-trien-14-ol, 15-acetylamino-17-acetoxy-3-methoxy-, (14β,15α,17β)-**	85921-52-4	**NI40665**
342	187	264	186	282	161	254	225	**401**	100	60	44	39	37	24	23	23	**9.80**	23	31	5	1	–	–	–	–	–	–	–	–	–	**Estra-1,3,5(10)-trien-15-ol, 14-acetylamino-17-acetoxy-3-methoxy-, (14β,15α,17β)-**	85853-11-8	**NI40666**
342	401	147	282	186	212	174	173	**401**	100	48	41	40	39	33	32	32		23	31	5	1	–	–	–	–	–	–	–	–	–	**Estra-1,3,5(10)trien-15-ol, 16-acetylamino-17-acetoxy-3-methoxy-, (15β,16α,17β)-**	85853-20-9	**NI40667**
121	371	340	225	280	142	322	401	**401**	100	60	36	30	18	15	14	7		23	31	5	1	–	–	–	–	–	–	–	–	–	**5′H-Pregna-1,4-dieno[17,16-d]oxazol-3-one, 11,20,21-trihydroxy-2′-methyl-, (11β,16β,20β)-**	87539-47-7	**NI40668**
370	401	280	158	386	174	228	173	**401**	100	68	43	43	42	41	38	38		23	35	3	1	–	–	–	–	–	1	–	–	–	**Estra-1,3,5(10)-trien-16-one, 3-methoxy-17-[(trimethylsilyl)oxy]-, O-methyloxime, (17β)-**	69833-96-1	**NI40669**
401	402	371	370	403	386	372	261	**401**	100	31	17	11	9	6	6	5		23	35	3	1	–	–	–	–	–	1	–	–	–	**Estra-1,3,5(10)-trien-17-one, 2-methoxy-3-[(trimethylsilyl)oxy]-, O-methyloxime**	69833-51-8	**NI40670**
401	371	402	370	387	369	372	329	**401**	100	34	31	23	18	12	10	10		23	35	3	1	–	–	–	–	–	1	–	–	–	**Estra-1,3,5(10)-trien-17-one, 3-methoxy-4-[(trimethylsilyl)oxy]-, O-methyloxime**	74299-32-4	**NI40671**
369	229	370	96	338	339	139	126	**401**	100	33	32	26	24	17	17	13	**4.62**	23	35	3	1	–	–	–	–	–	1	–	–	–	**Estra-1,3,5(10)-trien-17-one, 6-methoxy-3-[(trimethylsilyl)oxy]-, O-methyloxime, (6α)-**	74299-37-9	**NI40672**
401	338	281	229	369	371	96	282	**401**	100	56	55	50	43	42	40	35		23	35	3	1	–	–	–	–	–	1	–	–	–	**Estra-1,3,5(10)-trien-17-one, 6-methoxy-3-[(trimethylsilyl)oxy]-, O-methyloxime, (6β)-**	74299-34-6	**NI40673**
401	338	218	370	402	339	230	231	**401**	100	76	34	32	31	27	22	18		23	35	3	1	–	–	–	–	–	1	–	–	–	**Estra-1,3,5(10)-trien-17-one, 15-methoxy-3-[(trimethylsilyl)oxy]-, O-methyloxime, (15β)-**	74299-33-5	**NI40674**
401	403	402	159	134	404	400	358	**401**	100	33	28	24	12	7	6	4		24	20	1	3	–	–	1	–	–	–	–	–	–	**1H-1,4-Benzodiazepin-2-one, 7-chloro-2,3-dihydro-3-[4-(dimethylamino)benzylidene]-5-phenyl-**	55056-35-4	**NI40675**
72	401	330	58	73	402	356	331	**401**	100	15	14	12	5	4	4	4		24	23	3	3	–	–	–	–	–	–	–	–	–	**Anthraquinone, 1-amino-2-(2-dimethylaminoethoxy)-4-anilino-**		**IC04304**
107	149	77	251	105	60	59	79	**401**	100	84	76	53	47	47	42	40	**0.00**	24	23	3	3	–	–	–	–	–	–	–	–	–	**Quinoxaline, 1-benzyl-1,2-dihydro-3-(4-nitrophenyl)-2-propoxy-**	91736-70-8	**NI40676**
206	59	60	76	179	103	77	207	**401**	100	58	56	47	43	21	18	17	**0.00**	24	23	3	3	–	–	–	–	–	–	–	–	–	**Quinoxaline, 1,2-dihydro-1-(4-nitrobenzyl)-3-phenyl-2-propoxy-**	91736-67-3	**NI40677**
73	143	149	120	75	134	161	121	**401**	100	63	46	42	33	25	24	24	**10.78**	24	39	2	1	–	–	–	–	–	1	–	–	–	**Androsta-1,4-dien-3-one, 17-methyl-17-[(trimethylsilyl)oxy]-, O-methyloxime, (17β)-**	57397-03-2	**NI40678**
386	387	140	401	153	154	372	155	**401**	100	33	28	26	20	15	12	12		24	39	2	1	–	–	–	–	–	1	–	–	–	**Estr-3-one, 17β-[(trimethylsilyl)oxy]-17α-vinyl-, methyloxime, (5α)-**		**NI40679**
386	387	140	370	401	153	196	209	**401**	100	36	28	20	18	17	16	13		24	39	2	1	–	–	–	–	–	1	–	–	–	**Estr-3-one, 17β-[(trimethylsilyl)oxy]-17α-vinyl-, methyloxime, (5β)-**		**NI40680**
91	267	106	268	134	296	92	65	**401**	100	85	21	17	12	10	8	5	**0.00**	25	27	2	3	–	–	–	–	–	–	–	–	–	**N,N′-Dibenzyl-2-(benzylamino)succinamide**		**IC04305**
300	43	359	96	155	401	358	314	**401**	100	97	93	92	89	56	50	46		25	39	3	1	–	–	–	–	–	–	–	–	–	**L-Alanine, N-acetyl-N-5α-androst-2-en-3-yl-, methyl ester**	4642-58-4	**NI40681**
401	165	192	77	400	220	191	402	**401**	100	83	77	50	45	37	30	29		26	19	–	5	–	–	–	–	–	–	–	–	–	**2-Phenyl-3-[4-(phenylazo)-phenylazo]indole**		**IC04306**
91	401	310	400	190	211	307	279	**401**	100	42	34	24	20	18	17	13		26	27	3	1	–	–	–	–	–	–	–	–	–	**(±)-11-Benzyloxy-2,3-dimethoxyberbine**		**MS09208**
357	43	55	95	246	124	401	385	**401**	100	96	81	53	31	31	30	28		26	43	2	1	–	–	–	–	–	–	–	–	–	**A-Nor-5α-cholestane-2,3-dione 3-oxime**	123239-82-7	**DD01566**
98	99	55	81	67	56	96	79	**401**	100	9	4	3	3	3	2	2	**0.10**	27	47	1	1	–	–	–	–	–	–	–	–	–	**(22R,25S)-22,26-epi-5α-Cholestan-3β-ol**		**MS09209**
43	401	55	57	81	69	95	67	**401**	100	87	80	59	47	43	42	38		27	47	1	1	–	–	–	–	–	–	–	–	–	**5α-Cholestan-2-one, oxime**	14614-13-2	**LI08539**
43	401	55	41	57	81	69	95	**401**	100	88	81	70	59	47	44	42		27	47	1	1	–	–	–	–	–	–	–	–	–	**5α-Cholestan-2-one, oxime**	14614-13-2	**NI40682**
401	384	55	402	43	95	81	41	**401**	100	48	32	30	26	25	22	21		27	47	1	1	–	–	–	–	–	–	–	–	–	**5α-Cholestan-6-one, oxime**	2735-20-8	**NI40683**
246	401	247	69	55	43	112	81	**401**	100	81	64	48	47	42	35	34		27	47	1	1	–	–	–	–	–	–	–	–	–	**Cholestan-3-one, oxime, (5α)-**	2735-21-9	**NI40684**
401	246	55	402	139	57	247	69	**401**	100	32	31	27	20	19	18	16		27	47	1	1	–	–	–	–	–	–	–	–	–	**3-Aza-A-homocholestan-4-one, (5α)-**	20283-99-2	**NI40685**
401	55	402	246	57	81	67	69	**401**	100	37	29	25	25	21	19	18		27	47	1	1	–	–	–	–	–	–	–	–	–	**4-Aza-A-homocholestan-3-one, (5α)-**	21002-92-6	**NI40686**
318	300	206	246	401	180	220	232	**401**	100	52	51	40	38	32	28	25		28	35	1	1	–	–	–	–	–	–	–	–	–	**Tubingensin A**		**DD01567**
255	254	256	146	116	77	117	91	**401**	100	26	18	10	7	7	6	6	**3.58**	29	23	1	1	–	–	–	–	–	–	–	–	–	**Spiro[azetidine-2,9′-fluoren]-4-one, 3-ethyl-1,3-diphenyl-**	15183-53-6	**NI40687**
255	254	256	146	117	77	164	163	**401**	100	25	17	9	7	5	3	3	**0.40**	29	23	1	1	–	–	–	–	–	–	–	–	–	**Spiro[azetidine-2,9′-fluoren]-4-one, 3-ethyl-1,3-diphenyl-**	15183-53-6	**MS09210**

MASS TO CHARGE RATIOS								M.W.	INTENSITIES								Parent	C	H	O	N	S	F	Cl	Br	I	Si	P	B	X	COMPOUND NAME	CAS Reg No	No
121	65	56	156	184	185	149	177	402	100	79	59	53	33	27	20	19	0.22	7	5	2	–	–	–	3	–	–	–	–	–	2	Dicarbonyl-π-cyclopentadienyliron-tin(IV)-trichloride		LI08540
43	243	328	57	30	326	81	79	402	100	65	50	50	50	37	35	33	0.00	9	22	–	4	–	–	–	2	–	–	–	–	1	Nickel, [1-bromo-N[3]-[3-[(1-bromopropyl)amino]propyl]-1,1,3-propanetriamine]	56793-01-2	NI40688
31	189	28	119	383	314	245	402	402	100	28	25	12	9	6	3	1		10	6	1	–	–	7	–	–	1	–	–	–	–	1-Fluoro-1,3-dihydro-5-methyl-3,3-bis(trifluoromethyl)-1,2-benziodoxole		MS09211
178	176	234	206	233	144	290	318	402	100	98	83	65	45	38	30	26	16.60	10	10	6	–	2	–	–	–	–	–	–	–	2	Di-μ-(ethanethiolato)hexacarbonyldiiron		MS09212
234	262	174	290	218	112	113	117	402	100	66	66	50	19	17	16	15	2.00	10	12	6	–	–	–	–	–	–	–	2	–	2	Di-μ-(dimethylphosphino)bis(pentacarbonyliron)		MS09213
387	145	42	114	346	330	57	186	402	100	73	60	54	22	19	14	11	0.00	10	24	–	2	–	–	–	–	–	–	–	–	2	(Isopropylideneamino)dimethylindium dimer		NI40689
83	305	58	82	249	147	73	361	402	100	68	42	31	30	22	12	11	0.00	12	21	4	–	–	6	–	–	–	1	1	–	–	1,3,2-Dioxaphospholane, 2,2-dihydro-2-[2,2,2-trifluoro-1-(trifluoromethyl)1-(trimethylsiloxy)ethyl]-4,4,5,5-tetramethyl-, 2-oxide		MS09214
402	379	404	381	183	76	403	217	402	100	56	44	33	28	25	19	16		13	8	1	2	1	–	1	–	1	–	–	–	–	Thiocyanic acid, 4-amino-2-chloro-5-(4-iodophenoxy)phenyl ester	13997-32-5	MS09215
195	402	167	117	327	196	93	31	402	100	40	33	20	11	11	10	10		14	–	1	2	–	10	–	–	–	–	–	–	–	1,3,4-Oxadiazole, 2,5-bis(pentafluorophenyl)-	16184-59-1	NI40690
195	402	167	117	327	196	31	346	402	100	40	33	20	11	11	10	9		14	–	1	2	–	10	–	–	–	–	–	–	–	1,3,4-Oxadiazole, 2,5-bis(pentafluorophenyl)-	16184-59-1	LI08541
52	117	145	91	173	146	80	201	402	100	45	11	11	10	7	7	5	0.45	16	10	6	–	–	–	–	–	–	–	–	–	2	Bis(π-cyclopentadienyltricarbonylchromium)		MS09216
121	122	322	402	400	320			402	100	10	2	1	1	1				16	18	2	–	–	–	–	–	–	–	–	–	2	4,4′-Bis(4-methoxybenzyl) diselenide		MS09217
41	55	69	39	83	56	70	43	402	100	94	53	49	41	35	32	32	0.00	16	36	7	–	–	–	–	–	–	–	2	–	–	Diphosphoric acid, diisooctyl ester	72101-07-6	WI01522
68	155	91	43	69	128	97	45	402	100	99	99	99	65	39	37	30	0.18	17	22	9	–	1	–	–	–	–	–	–	–	–	β-L-Arabinopyranoside, methyl, 2,3-diacetate 4-(4-methylphenyl)sulphonate	29919-81-1	NI40691
402	387	385	373	354	271	295	298	402	100	30	15	13	12	10	8	7		18	14	9	2	–	–	–	–	–	–	–	–	–	5-Hydroxy-6,7-dimethoxy-2-methyl-2′,4′-dinitroisoflavone		NI40692
293	183	107	110	109	184	402	186	402	100	78	37	24	24	21	20	20		18	15	–	–	3	–	–	–	–	–	–	–	1	Arsenotrithious acid, triphenyl ester	1776-70-1	NI40693
293	183	107	110	109	184	402	186	402	100	78	35	23	23	20	19	19		18	15	–	–	3	–	–	–	–	–	–	–	1	Arsenotrithious acid, triphenyl ester	1776-70-1	LI08542
75	345	73	215	187	343	347	213	402	100	82	80	62	53	40	38	31	0.00	18	27	1	–	–	–	1	–	–	1	–	–	1	(3β,4β)-1-Chloro-3-[[(1,1-dimethylethyl)dimethylsilyl]oxy]-4-(phenylseleno)cyclohex-1-ene		DD01568
216	104	160	187	132	172	170	76	402	100	74	73	64	62	59	38	38	17.00	19	18	6	2	1	–	–	–	–	–	–	–	–	3-Cephem-4-carboxylic acid, 3-ethoxymethyl-7-phthalimido-, methyl ester		NI40694
195	210	223	209	266	236	294	296	402	100	69	55	55	25	24	19	17	2.00	19	19	3	2	–	–	–	1	–	–	–	–	–	12bβ-Bromo-1β-ethyl-4-oxo-1,2,3,4,6,7,12,12b-octahydro-1,3-cycloindolo[2,3-a]-quinolizine-3β-carbonic acid, methyl ester		NI40695
183	167	196	217	222	106	191	103	402	100	67	48	39	30	24	20	19	1.00	19	42	3	–	–	–	–	–	–	3	–	–	–	Cyclopentane, 2-ethyl-3,5-bis[(trimethylsilyl)oxy]-1-[3-[(trimethylsilyl)oxy]-1-propenyl]-		NI40696
387	402	388	403	149	357	372	211	402	100	84	23	21	12	10	9	9		20	18	9	–	–	–	–	–	–	–	–	–	–	5-Hydroxy-3,6,7,8-tetramethoxy-2-(3,4-methylenedioxyphenyl)-1-benzopyran-4-one		NI40697
402	314	310	268	286	356	360		402	100	73	67	25	18	16	8			20	22	7	2	–	–	–	–	–	–	–	–	–	Diethyl 4a-hydro-9-hydroxy-1,3-dimethyl-6,7-methylenedioxy-β-carboline 4,9a-dicarboxylate		LI08543
345	343	341	344	342	289	349	346	402	100	74	44	38	33	25	20	18	0.00	20	26	1	–	–	–	–	–	–	–	–	–	1	10,10-Dibutylphenoxastannin		LI08544
149	107	71	109	91	235	217	233	402	100	63	59	43	35	30	17	13	0.00	20	31	6	–	–	–	1	–	–	–	–	–	–	Eudesman-8-one, 6,7-didehydro-3-(2-hydroxy-2-methyl-3-chlorobutyryloxy)-4,11-dihydroxy-, (3α,4α)-		NI40698
107	71	91	109	149	235	175	217	402	100	92	71	64	58	11	10	9	0.00	20	31	6	–	–	–	1	–	–	–	–	–	–	Eudesman-8-one, 6,7-didehydro-3-(2-hydroxy-2-methyl-3-chlorobutyryloxy)-4,11-dihyroxy-, (3α,4α)-		NI40699
185	129	259	43	57	29	41	157	402	100	57	54	54	38	27	24	22	0.00	20	34	8	–	–	–	–	–	–	–	–	–	–	Citric acid, 2-(acetyloxy)-, tributyl ester	77-90-7	NI40700
185	43	129	41	259	57	29	157	402	100	76	62	59	48	48	48	15	0.11	20	34	8	–	–	–	–	–	–	–	–	–	–	Citric acid, 3-acetyl-, tributyl ester		IC04307
185	259	129	43	57	157	41	329	402	100	79	25	21	17	14	14	13	2.80	20	34	8	–	–	–	–	–	–	–	–	–	–	Citric acid, 3-acetyl-, tributyl ester		MS09218
387	73	75	217	204	271	388	117	402	100	98	71	41	38	36	31	28	1.20	20	42	4	–	–	–	–	–	–	2	–	–	–	1,14-Tetradecanedioic acid, bis(trimethylsilyl) ester		MS09219
253	254	98	81	402	174	155	156	402	100	87	76	41	32	29	24	22		21	17	4	–	–	2	–	–	–	–	1	–	–	Diphenyl 1-benzyl-2,2-difluoroethenyl phosphate		DD01569
180	222	181	342	402	151	178	165	402	100	25	24	16	12	10	7	7		21	22	8	–	–	–	–	–	–	–	–	–	–	4H-1-Benzopyran-4-one, 3-(acetyloxy)-2,3-dimethoxy-2-(3,4-dimethoxyphenyl)-2,3-dihydro-, trans-	22469-40-5	NI40701
387	402	401	388	187	167	403	165	402	100	66	35	24	17	16	15	15		21	22	8	–	–	–	–	–	–	–	–	–	–	4H-1-Benzopyran-4-one, 3,5,6,7-tetramethoxy-2-(3,4-dimethoxyphenyl)-		MS09220
387	402	388	344	197	83	182	193	402	100	22	22	19	19	17	15	13		21	22	8	–	–	–	–	–	–	–	–	–	–	4H-1-Benzopyran-4-one, 5,6,7,8-tetramethoxy-2-(3,4-dimethoxyphenyl)-	478-01-3	NI40702
402	387	401	371	403	388	187	372	402	100	95	38	22	20	20	16	11		21	22	8	–	–	–	–	–	–	–	–	–	–	4H-1-Benzopyran-4-one, 3,5,7-trimethoxy-2-(3,4,5-trimethoxyphenyl)-	14813-27-5	NI40703
387	402	167	195	139	388	181	403	402	100	70	46	27	23	21	17	16		21	22	8	–	–	–	–	–	–	–	–	–	–	4H-1-Benzopyran-4-one, 5,7,8-trimethoxy-2-(2,3,4-trimethoxyphenyl)-	21315-69-5	NI40704
387	402	167	195	139	69	388	66	402	100	62	52	40	26	25	20	17		21	22	8	–	–	–	–	–	–	–	–	–	–	4H-1-Benzopyran-4-one, 5,7,8-trimethoxy-2-(2,4,6-trimethoxyphenyl)-	89121-54-0	NI40705
387	402	167	358	388	139	329	403	402	100	75	32	29	21	21	18	17		21	22	8	–	–	–	–	–	–	–	–	–	–	4H-1-Benzopyran-4-one, 5,7,8-trimethoxy-2-(3,4,5-trimethoxyphenyl)-	80324-51-2	NI40706
180	181	222	151	121	342	178	137	402	100	26	22	21	12	11	9	8	0.80	21	22	8	–	–	–	–	–	–	–	–	–	–	Flavanone, 3-hydroxy-2′,4′,5,7-tetramethoxy-, acetate, trans-	22425-51-0	NI40707
105	77	43	177	106	165	123	115	402	100	11	9	7	7	4	3	3	0.00	21	22	8	–	–	–	–	–	–	–	–	–	–	Xylitol, 1-acetate 2,5-dibenzoate	74793-09-2	NI40708
43	113	99	223	57	71	125	69	402	100	81	71	54	51	42	41	39	0.00	21	38	7	–	–	–	–	–	–	–	–	–	–	Erythronolide A, 12-deoxy-	3225-82-9	NI40709
43	113	99	57	223	125	85	155	402	100	73	62	44	40	35	32	25	0.00	21	38	7	–	–	–	–	–	–	–	–	–	–	Erythronolide B		LI08545
44	149	57	43	41	364	55	71	402	100	74	74	65	57	47	40	34	0.00	22	10	8	–	–	–	–	–	–	–	–	–	–	1H,3H-Naphthaceno[1,12-cd]pyran-1,3,8,13-tetrone, 5-acetyl-7,12-dihydroxy-	56784-01-1	MS09221

MASS TO CHARGE RATIOS								M.W.	INTENSITIES								Parent	C	H	O	N	S	F	Cl	Br	I	Si	P	B	X	COMPOUND NAME	CAS Reg No	No
402	404	403	405	374	372	128	91	**402**	100	99	25	24	20	20	17	14		22	15	1	2	–	–	–	1	–	–	–	–	–	1H-1,4-Benzodiazepin-2-one, 3-benzylidene-7-bromo-2,3-dihydro-5-phenyl	62642-82-4	NI40710
402	401	373	77	173	403	345	328	**402**	100	38	34	33	30	26	21	21		22	22	2	6	–	–	–	–	–	–	–	–	–	Pyrimido[5,4-d]pyrimidine, 4,8-dianilino-2,6-diethoxy-	18710-93-5	MS09222
43	57	41	55	83	71	77	181	**402**	100	89	75	72	35	17	15	13	**3.00**	22	26	7	–	–	–	–	–	–	–	–	–	–	(1S,5R,6S,7R,8R)-1-Allyl-8-hydroxy-3,5-dimethoxy-6-(3′-methoxy-4′,5′-methylenedioxyphenyl)-7-methyl-4-oxobicyclo[3.2.1]oct-2-ene		NI40711
43	324	296	281	69	28	263	278	**402**	100	99	71	66	66	53	45	44	**30.00**	22	26	7	–	–	–	–	–	–	–	–	–	–	Bisdehydrophorbol-12-acetate		NI40712
360	43	342	83	361	324	230	128	**402**	100	39	38	30	19	19	19	19	**4.11**	22	26	7	–	–	–	–	–	–	–	–	–	–	12,20-Didesoxy-12,20-dioxophorbol-13-acetate		NI40713
180	297	151	222	181	402	298	342	**402**	100	68	34	32	19	17	17	15		22	26	7	–	–	–	–	–	–	–	–	–	–	3-Flavanol, 4-ethoxy-3′,4′,7-trimethoxy-, acetate, trans-2,3,trans-3,4-	16821-17-3	NI40714
402	151	179	177	166	165	167	236	**402**	100	65	59	55	54	51	39	33		22	26	7	–	–	–	–	–	–	–	–	–	–	Gmelinol		LI08546
242	243	83	55	43	402	135	29	**402**	100	93	79	71	69	64	42	37		22	26	7	–	–	–	–	–	–	–	–	–	–	Guaia-3,10(14)-dien-12-oic acid, 6α,8,11-trihydroxy-2-oxo-, 12,6-lactone, 8-acetate 11-(2-methylcrotonate), (Z)-(8S,11R)-	33439-67-7	NI40715
242	243	83	55	43	402	199	171	**402**	100	90	65	60	53	51	24	23		22	26	7	–	–	–	–	–	–	–	–	–	–	Isoprutenin-10-one, 4-acetoxy-		LI08547
402	151	177	179	166	221	165	167	**402**	100	44	41	34	34	33	32	26		22	26	7	–	–	–	–	–	–	–	–	–	–	Neogmelinol		LI08548
151	152	107	91	402	153	135	108	**402**	100	12	3	2	1	1	1	1		22	26	7	–	–	–	–	–	–	–	–	–	–	Nortrachelogenin dimethyl ether, (–)-		NI40716
402	242	243	214	83	199	302	342	**402**	100	88	83	24	24	22	11	10		22	26	7	–	–	–	–	–	–	–	–	–	–	Prutenin-10-one, 4-acetoxy-		LI08549
237	262	189	216	233	171	187	249	**402**	100	43	32	30	29	27	25	23	**0.00**	22	26	7	–	–	–	–	–	–	–	–	–	–	Subacaulin, acetyl-		NI40717
125	136	127	137	67	402	126	122	**402**	100	60	32	17	14	11	9	7		22	27	3	2	–	–	1	–	–	–	–	–	–	Urea, 1-(4-chlorobenzyl)-1-(3-methoxycyclopentyl)-3-methyl-3-(4-methoxyphenyl)-		NI40718
328	160	402	26	154	300	269		**402**	100	34	30	18	9	8	6			22	30	5	2	–	–	–	–	–	–	–	–	–	Deoxytetrahydrodichotine		LI08550
105	79	77	100	83	82	81	78	**402**	100	45	41	30	28	26	23	20	**2.00**	23	15	2	–	–	–	–	1	–	–	–	–	–	2H-Pyran-2-one, 6-(4-bromophenyl)-3,4-diphenyl-	14966-77-9	NI40719
402	238	370	265	165				**402**	100	25	5	5	5					23	18	5	2	–	–	–	–	–	–	–	–	–	1-(2,4-Dinitrostyryl)-4-(4-methoxystyryl)benzene		LI08551
205	221	44	43	120	187	54	126	**402**	100	65	50	40	21	20	19	18	**0.00**	23	30	6	–	–	–	–	–	–	–	–	–	–	2H-Pyran-2-one, 4-methoxy-5-methyl-6-[7-methyl-8-(tetrahydro-3,4-dihydroxy-2,4,5-trimethyl-2-furanyl)-1,3,5,7-octatetraenyl]-, [2R-[2α(1E,3E,5E,7E),3β,4α,5α]]-	25425-12-1	NI40720
141	55	95	173	43	57	69	41	**402**	100	90	84	77	77	71	64	61	**0.34**	23	46	5	–	–	–	–	–	–	–	–	–	–	Docosanoic acid, 8,9,13-trihydroxy-, methyl ester	56554-25-7	NI40721
402	135	401	403	121	369	404	77	**402**	100	78	59	36	36	23	21	21		24	18	–	–	3	–	–	–	–	–	–	–	–	[1,2]Dithiolo[1,5-b][1,2]dithiole-7-SIV, 4-(4-methylphenyl)-2,3-diphenyl-	56114-48-8	NI40722
402	387	385	91	106	403	388	384	**402**	100	96	63	37	32	25	25	25		24	18	6	–	–	–	–	–	–	–	–	–	–	2,2′-Binaphthalene-1,1′,4,4′-tetrone, 5,5′-dihydroxy-3,3′,7,7′-tetramethyl-	85485-18-3	NI40723
402	387	44	385	403	388	43	106	**402**	100	79	50	42	32	22	15	14		24	18	6	–	–	–	–	–	–	–	–	–	–	2,2′-Binaphthalene-1,1′,4,4′-tetrone, 8,8′-dihydroxy-3,3′,6,6′-tetramethyl-	85485-19-4	NI40724
387	402	76	388	371	44	372	403	**402**	100	83	44	26	26	24	23	22		24	18	6	–	–	–	–	–	–	–	–	–	–	2,2′-Binaphthalene-1,1′,4,4′-tetrone, 5,5′-dimethoxy-3,3′-dimethyl-	54215-49-5	NI40725
387	402	76	44	388	371	403	104	**402**	100	88	57	45	27	25	24	19		24	18	6	–	–	–	–	–	–	–	–	–	–	2,2′-Binaphthalene-1,1′,4,4′-tetrone, 5,5′-dimethoxy-3,3′-dimethyl-	54215-49-5	NI40726
90	402	387	91	89	118	149	190	**402**	100	50	35	35	35	30	24	23		24	18	6	–	–	–	–	–	–	–	–	–	–	2,2′-Binaphthalene-1,4,5′,8′-tetrone, 1′,5-dimethoxy-3′,7-dimethyl-		MS09223
402	76	135	77	115	104	83	227	**402**	100	95	40	35	33	32	32	29		24	18	6	–	–	–	–	–	–	–	–	–	–	1a,1a′-Bi-1aH-cyclopropa[b]naphthalene-2,2′,7,7′-tetrone, 1,1′,7a,7a′-tetrahydro-6,6′-dimethoxy-		NI40727
402	309	359	291	277	387	352	334	**402**	100	30	25	25	25	20	20	7		24	28	3	–	–	2	–	–	–	–	–	–	–	6α,7α-Difluoromethylene-6β,16β-dimethyl-16α,17α-oxido-1,4-pregnadiene-3,20-dione		MS09224
239	299	257	342					**402**	100	88	43	11					**0.00**	24	34	5	–	–	–	–	–	–	–	–	–	–	3β,17β-Acetoxy-19a-methylandrost-5-en-19-one		LI08552
384	44	323	366	43	287	91	79	**402**	100	99	97	82	73	63	63	63	**21.00**	24	34	5	–	–	–	–	–	–	–	–	–	–	Bufa-20,22-dienolide, 3,5,14-trihydroxy-, (3β,5β)-	472-26-4	LI08553
384	44	323	366	43	341	287	79	**402**	100	98	96	81	71	64	63	62	**20.21**	24	34	5	–	–	–	–	–	–	–	–	–	–	Bufa-20,22-dienolide, 3,5,14-trihydroxy-, (3β,5β)-	472-26-4	NI40728
191	91	79	366	43	323	107	93	**402**	100	95	90	82	72	71	71	69	**60.21**	24	34	5	–	–	–	–	–	–	–	–	–	–	Bufa-20,22-dienolide, 3,11,14-trihydroxy-, (3β,5β,11α)-	465-11-2	NI40729
55	41	81	43	261	67	79	384	**402**	100	85	65	61	54	46	44	41	**20.00**	24	34	5	–	–	–	–	–	–	–	–	–	–	Cholan-24-oic acid, 3,7,12-trioxo-, (5β)-	81-23-2	NI40730
282	43	132	133	197	108	283	402	**402**	100	67	57	54	49	32	30	28		24	34	5	–	–	–	–	–	–	–	–	–	–	6α,7β-Diacetoxyvouacapane		NI40731
123	366	107	95	79	108	91	67	**402**	100	65	56	51	51	50	49	48	**3.25**	24	34	5	–	–	–	–	–	–	–	–	–	–	15α-Hydroxybufalin	31444-11-8	NI40732
343	301	328	162	234	204	402	354	**402**	100	99	33	28	19	19	13	13		24	34	5	–	–	–	–	–	–	–	–	–	–	Pisiferdiol 5,7-diacetate		DD01570
101	43	121	122	59	55	41	57	**402**	100	94	89	78	66	56	55	50	**5.95**	24	34	5	–	–	–	–	–	–	–	–	–	–	Pregna-1,4-dien-3-one, 11β,17,20α-21-tetrahydroxy-, cyclic 20,21-acetal with acetone	18089-32-2	NI40733
237	75	191	73	197	253	195	359	**402**	100	64	58	56	48	32	32	28	**3.00**	24	38	3	–	–	–	–	–	–	1	–	–	–	Abieta-8,11,13-trien-18-oic acid, 7α-[(trimethylsilyl)oxy]-, methyl ester		NI40734
191	234	237	73	192	359	75	155	**402**	100	33	23	19	17	16	14	12	**8.00**	24	38	3	–	–	–	–	–	–	1	–	–	–	Abieta-8,11,13-trien-18-oic acid, 7β-[(trimethylsilyl)oxy]-, methyl ester		NI40735
327	402	328	387	73	403	221	285	**402**	100	67	30	26	26	21	18	15		24	38	3	–	–	–	–	–	–	1	–	–	–	Abieta-8,11,13-trien-18-oic acid, 15-[(trimethylsilyl)oxy]-, methyl ester		NI40736
343	402	387	344	384	403	265	359	**402**	100	61	41	31	22	21	18	17		24	38	3	–	–	–	–	–	–	1	–	–	–	1β,6β-Epoxyhexahydrocannabinol trimethylsilyl ether		NI40737
387	346	402	388	347	403	331	185	**402**	100	90	72	30	29	21	16	14		24	38	3	–	–	–	–	–	–	1	–	–	–	Progesterone, 6-hydroxy-, tris(trimethylsilyl)-, (6α)-		NI40738
387	346	402	388	403	347	400	374	**402**	100	96	91	33	30	29	24	13		24	38	3	–	–	–	–	–	–	1	–	–	–	Progesterone, 6-hydroxy-, tris(trimethylsilyl)-, (6α)-		LI08554
346	387	402	388	347	403	75	331	**402**	100	96	60	30	27	19	19	15		24	38	3	–	–	–	–	–	–	1	–	–	–	Progesterone, 6-hydroxy-, tris(trimethylsilyl)-, (6β)-		NI40739
346	387	402	347	388	403	374	156	**402**	100	74	60	30	24	20	14	10		24	38	3	–	–	–	–	–	–	1	–	–	–	Progesterone, 6-hydroxy-, tris(trimethylsilyl)-, (6β)-		LI08555
371	372	402	339	151	86	384	340	**402**	100	31	17	12	11	9	7	7		24	38	3	2	–	–	–	–	–	–	–	–	–	Pregn-4-ene-3,20-dione, 17-hydroxy-6-methyl-, bis(O-methyloxime), (6α)-	69833-68-7	NI40740
191	195	189	165	402	212	179	177	**402**	100	47	33	24	16	11	11	11		25	22	5	–	–	–	–	–	–	–	–	–	–	9-Phenanthrenemethyl 3,4,5-trimethoxybenzoate		NI40741
105	252	224	77	91	120	131	293	**402**	100	39	29	24	14	11	6	4	**0.00**	25	26	3	2	–	–	–	–	–	–	–	–	–	Benzenepropanol, β-[(N-benzoyl-S-phenylalaninyl)amino]-, (S)-	58115-31-4	NI40742

MASS TO CHARGE RATIOS	M.W.	INTENSITIES	Parent	C	H	O	N	S	F	Cl	Br	I	Si	P	B	X	COMPOUND NAME	CAS Reg No	No
81 187 95 107 145 282 205 188	402	100 97 84 60 55 48 47 47	6.00	25	38	4	–	–	–	–	–	–	–	–	–	–	Butanoic acid, 3-methyl-, [5-[2-(3-furanyl)ethyl]-1,2,3,4,4a,5,8,8a-octahydro-2-hydroxy-5,6,8a-trimethyl-1-naphthalenyl]methyl ester, (1α,2α,4aβ,5β,8aα)-	51460-72-1	NI40743
41 192 55 43 402 57 69 44	402	100 88 86 71 55 54 45 44		25	38	4	–	–	–	–	–	–	–	–	–	–	5α-Cholan-24-oic acid, 3,7-dioxo-, methyl ester		LI08556
137 400 136 109 107 105 108 133	402	100 70 56 55 50 36 35 32	17.40	25	38	4	–	–	–	–	–	–	–	–	–	–	Cholan-24-oic acid, 3,6-dioxo-, methyl ester, (5β)-	1175-04-8	NI40744
247 121 105 402 135 123 119 107	402	100 42 40 39 35 35 35 32		25	38	4	–	–	–	–	–	–	–	–	–	–	Cholan-24-oic acid, 3,12-dioxo-, methyl ester, (5α)-	1174-86-3	NI40745
271 272 386 253 404 154 81 55	402	100 22 15 14 12 12 11 11	1.95	25	38	4	–	–	–	–	–	–	–	–	–	–	Chol-4-en-24-oic acid, 12-hydroxy-3-oxo-, methyl ester, (12α)-	19684-72-1	NI40746
271 272 154 386 247 270 161 402	402	100 22 14 12 11 9 9 7		25	38	4	–	–	–	–	–	–	–	–	–	–	Chol-5-en-24-oic acid, 12-hydroxy-3-oxo-, methyl ester, (12α)-	69633-06-3	NI40747
316 301 55 108 107 139 109 69	402	100 78 73 71 70 62 61 60	2.02	25	38	4	–	–	–	–	–	–	–	–	–	–	26,27-Dinorfurostan-11-one, 22,25-epoxy-3-hydroxy-, (3β,8α,14β,20β)-	57983-86-5	NI40748
43 402 41 263 69 55 71 305	402	100 87 87 85 73 68 65 56		25	38	4	–	–	–	–	–	–	–	–	–	–	8-Hydroxy-3-isopropyl-4,4-dimethyl-7-(3-methylbutyl)-9-(2-methylpropionyl)tricyclo[5.3.1.0[1,5]]undecane-10,11-dione		NI40749
342 161 133 159 131 160 221 105	402	100 60 50 36 35 32 28 27	0.00	25	38	4	–	–	–	–	–	–	–	–	–	–	Methyl 3β-acetoxy-24,23-dinor-5β-chol-5-enoate		NI40750
87 298 91 105 107 79 55 93	402	100 14 9 8 6 6 6 5	0.00	25	38	4	–	–	–	–	–	–	–	–	–	–	Pregn-5-en-20-one, 3-(acetyloxy)-, cyclic 20-(1,2-ethanediyl acetal), (3β)-	40148-10-5	WI01523
384 55 215 81 93 67 79 217	402	100 84 78 75 69 68 65 63	18.00	25	43	3	–	–	–	–	–	–	–	–	1	–	Pregnane-3,17,20-(tert-butylboronate), (3α,5β,17α,20α)-		LI08557
384 215 107 217 166 167 105 121	402	100 60 50 47 45 44 39 32	12.00	25	43	3	–	–	–	–	–	–	–	–	1	–	Pregnane-3,17,20-(tert-butylboronate), (3α,5β,17α,20β)-		LI08558
384 43 55 81 67 215 93 41	402	100 78 71 65 63 60 57 57	17.00	25	43	3	–	–	–	–	–	–	–	–	1	–	Pregnane-3,17,20-(tert-butylboronate), (3α,5β,17α,20β)-		MS09225
384 55 215 81 93 67 79 107	402	100 81 76 73 68 67 64 63	18.02	25	43	3	–	–	–	–	–	–	–	–	1	–	Pregnane-3,17,20-(tert-butylboronate), (3α,5β,17α,20S)-	55521-03-4	NI40751
107 215 167 105 217 166 384 109	402	100 99 86 83 80 77 75 72	11.01	25	43	3	–	–	–	–	–	–	–	–	1	–	Pregnane-3,17,20-(tert-butylboronate), (3α,5β,20R)-	30882-59-8	NI40752
166 215 107 167 216 105 384 135	402	100 95 93 78 75 72 70 57	8.01	25	43	3	–	–	–	–	–	–	–	–	1	–	Pregnane-3,17,20-triol, cyclic 17,20-(butylboronate), (3α,5β,20R)-	30882-60-1	NI40753
139 140 142 156 214 327 293 308	402	100 97 48 32 10 4 4 2	1.00	26	20	–	–	–	–	2	–	–	–	–	–	–	Cyclopentane, 1,3-bis(4-chlorobenzylidene)-2-benzylidene-, (E,E)-		MS09226
243 202 165 105 244 242 84 260	402	100 94 78 59 38 26 25 8	0.00	26	26	4	–	–	–	–	–	–	–	–	–	–	α-D-3-allo-Glucopyranoside, methyl 2-deoxy-3,4-epoxy-6-(triphenylmethyl)-		NI40754
161 55 41 180 121 402 175 43	402	100 72 59 42 36 33 31 30		26	42	3	–	–	–	–	–	–	–	–	–	–	Benzoic acid, 2-(12-heptadecenyl)-6-methoxy-, methyl ester	56554-65-5	NI40755
402 43 81 55 231 121 95 41	402	100 92 67 64 56 40 40 40		26	42	3	–	–	–	–	–	–	–	–	–	–	Cholan-12-one, 3-(acetyloxy)-, (3α,5β)-	42921-46-0	NI40756
402 231 81 55 110 121 403 95	402	100 47 44 39 32 31 30 26		26	42	3	–	–	–	–	–	–	–	–	–	–	Cholan-12-one, 3-(acetyloxy)-, (3α,5β)-	42921-46-0	NI40757
110 384 369 356 317 299 275 247	402	100 3 1 1 1 1 1 1	0.34	26	42	3	–	–	–	–	–	–	–	–	–	–	B-Nor-5,6-secocholest-2-en-6-oic acid, 1-oxo-, (10α)-	3949-50-6	NI40758
110 111 43 41 81 109 107 79	402	100 10 10 8 7 6 4 4	0.42	26	42	3	–	–	–	–	–	–	–	–	–	–	B-Nor-5,6-secocholest-2-en-6-oic acid, 1-oxo-, (10α)-	3949-50-6	NI40759
367 400 382 368 364 105 225 135	402	100 34 30 27 23 20 17 16	12.40	27	43	1	–	–	1	–	–	–	–	–	–	–	Cholest-4-en-6-one, 3β-fluoro-	31756-69-1	NI40760
367 382 364 225 135 338 402 119	402	100 30 23 17 16 14 12 12		27	43	1	–	–	1	–	–	–	–	–	–	–	Cholest-4-en-6-one, 3β-fluoro-	31756-69-1	LI08559
367 382 364 225 135 338 247 119	402	100 30 23 17 16 14 12 12	12.00	27	43	1	–	–	1	–	–	–	–	–	–	–	Cholest-4-en-6-one, 3β-fluoro-	31756-69-1	NI40761
402 289 135 247 374 387 155 119	402	100 44 42 38 35 28 24 24		27	43	1	–	–	1	–	–	–	–	–	–	–	Cholest-5-en-4-one, 3β-fluoro-	31756-68-0	LI08560
402 159 289 135 247 43 374 55	402	100 74 44 42 38 36 35 35		27	43	1	–	–	1	–	–	–	–	–	–	–	Cholest-5-en-4-one, 3β-fluoro-	31756-68-0	NI40762
402 43 403 194 55 91 154 57	402	100 50 37 35 23 19 18 17		27	43	1	–	–	1	–	–	–	–	–	–	–	Cholest-5-en-7-one, 3-fluoro-, (3β)-	31772-90-4	NI40763
402 194 154 161 289 382 207 179	402	100 35 18 13 12 10 10 10		27	43	1	–	–	1	–	–	–	–	–	–	–	Cholest-5-en-7-one, 3-fluoro-, (3β)-	31772-90-4	LI08561
91 108 57 43 92 71 41 55	402	100 47 37 35 21 21 19 16	0.00	27	46	2	–	–	–	–	–	–	–	–	–	–	Benzyl arachidate		NI40764
401 383 367	402	100 36 7	0.00	27	46	2	–	–	–	–	–	–	–	–	–	–	Cholestan-3-ol, 5,6-epoxy-, (3β,5α,6α)-	1250-95-9	NI40765
401 383 367	402	100 75 13	0.00	27	46	2	–	–	–	–	–	–	–	–	–	–	Cholestan-3-ol, 5,6-epoxy-, (3β,5β,6β)-	4025-59-6	NI40766
384 95 293 55 366 81 69 41	402	100 82 79 78 67 66 58 53	21.00	27	46	2	–	–	–	–	–	–	–	–	–	–	Cholestan-7-ol, 8,9-epoxy-, (5α,7α,8α)-	38819-61-3	NI40767
97 55 107 271 95 69 81 178	402	100 97 93 84 81 73 66 65	48.00	27	46	2	–	–	–	–	–	–	–	–	–	–	Cholestan-7-ol, 8,14-epoxy-, (5α,7α,8α)-	38819-60-2	NI40768
402 403 371 386 372 387	402	100 30 25 13 7 4		27	46	2	–	–	–	–	–	–	–	–	–	–	Cholestan-1-one, 19-hydroxy-, (5α)-	28809-71-4	NI40769
384 124 43 332 55 229 41 57	402	100 97 59 56 54 48 42 37	15.83	27	46	2	–	–	–	–	–	–	–	–	–	–	Cholestan-3-one, 5-hydroxy-, (5α)-	2515-01-7	NI40770
384 124 43 332 56 229 41 95	402	100 98 62 58 56 50 44 37	18.00	27	46	2	–	–	–	–	–	–	–	–	–	–	Cholestan-3-one, 5-hydroxy-, (5α)-	2515-01-7	LI08562
402 231 403 332 232 56 81 95	402	100 49 36 30 26 24 20 18		27	46	2	–	–	–	–	–	–	–	–	–	–	Cholestan-3-one, 26-hydroxy-, (5β)-	55530-50-2	NI40771
332 356 43 56 202 41 95 81	402	100 51 44 39 30 30 28 28	24.00	27	46	2	–	–	–	–	–	–	–	–	–	–	Cholestan-4-one, 5-hydroxy-, (5α)-	19043-66-4	LI08563
332 43 55 356 41 95 57 81	402	100 60 55 50 42 37 37 36	22.94	27	46	2	–	–	–	–	–	–	–	–	–	–	Cholestan-4-one, 5-hydroxy-, (5α)-	19043-66-4	NI40772
306 307 95 193 81 271 55 43	402	100 62 36 28 28 27 26 20	4.00	27	46	2	–	–	–	–	–	–	–	–	–	–	Cholestan-7-one, 9-hydroxy-, (5α)-	39031-13-5	NI40773
402 81 107 95 193 147 124 289	402	100 43 40 40 34 34 33 22		27	46	2	–	–	–	–	–	–	–	–	–	–	Cholestan-11-one, 3-hydroxy-, (3α,5α)-		MS09227
108 289 193 81 402 95 219 263	402	100 94 83 80 75 60 55 50		27	46	2	–	–	–	–	–	–	–	–	–	–	Cholestan-11-one, 3-hydroxy-, (3β,5α,8α,9β,14β)-		MS09228
108 289 81 193 95 402 219 263	402	100 99 72 70 57 56 56 48		27	46	2	–	–	–	–	–	–	–	–	–	–	Cholestan-11-one, 3-hydroxy-, (3β,5α,8α,14β)-		MS09229
43 55 41 209 330 384 108 91	402	100 75 66 62 52 40 40 39	22.59	27	46	2	–	–	–	–	–	–	–	–	–	–	Cholestan-15-one, 3-hydroxy-, (3β,5α)-	55823-04-6	NI40774
383 381 401 399 397	402	100 16 7 7 7	0.00	27	46	2	–	–	–	–	–	–	–	–	–	–	Cholest-5-ene-3,4-diol, (3β,4β)-	17320-10-4	NI40775
365 383 401 381 399	402	100 59 41 18 9	0.00	27	46	2	–	–	–	–	–	–	–	–	–	–	Cholest-5-ene-3,7-diol, (3β,7α)-	566-26-7	NI40776
365 383 401 381	402	100 79 56 21	0.00	27	46	2	–	–	–	–	–	–	–	–	–	–	Cholest-5-ene-3,7-diol, (3β,7β)-	566-27-8	NI40777
354 353 355 384 247 372 241 371	402	100 33 30 28 24 22 17 16	1.00	27	46	2	–	–	–	–	–	–	–	–	–	–	Cholest-5-ene-3,19-diol, (3β)-	561-63-7	NI40778
401 399 383 365 381 397	402	100 19 19 8 6 5	0.00	27	46	2	–	–	–	–	–	–	–	–	–	–	Cholest-5-ene-3,24-diol, (3β,24S)-	474-73-7	NI40779
401 383 381 365 399 397	402	100 57 24 20 9 9	0.00	27	46	2	–	–	–	–	–	–	–	–	–	–	Cholest-5-ene-3,25-diol, (3β)-	2140-46-7	NI40780

MASS TO CHARGE RATIOS								M.W.	INTENSITIES								Parent	C	H	O	N	S	F	Cl	Br	I	Si	P	B	X	COMPOUND NAME	CAS Reg No	No
370	108	95	55	215	355	81	43	**402**	100	38	38	34	33	32	32	32	**0.00**	27	46	2	–	–	–	–	–	–	–	–	–	–	Cholest-8(14)-ene-7,15-diol, (5α)-	54515-30-9	NI40781
271	384	95	191	81	105	91	55	**402**	100	89	46	40	27	25	25	25	**4.00**	27	46	2	–	–	–	–	–	–	–	–	–	–	Cholest-8-ene-7,11-diol, (5α)-	54514-97-5	NI40782
317	95	402	57	69	81	55	43	**402**	100	73	51	44	41	40	38	38		27	46	2	–	–	–	–	–	–	–	–	–	–	Cholest-17(20)-ene-3,6-diol, (3β,5α,6α)-	55924-01-1	NI40784
317	95	402	57	71	43	55	73	**402**	100	73	51	45	40	39	38	35		27	46	2	–	–	–	–	–	–	–	–	–	–	Cholest-17(20)-ene-3,6-diol, (3β,5α,6α)-	55924-01-1	NI40783
402	95	82	138	81	69	55	68	**402**	100	91	67	61	57	47	42	36		27	46	2	–	–	–	–	–	–	–	–	–	–	Cholest-20(22)-ene-3,6-diol, (3β,5α,6α)-	41084-08-6	NI40785
402	95	55	403	81	247	69	41	**402**	100	59	58	32	29	28	27	25		27	46	2	–	–	–	–	–	–	–	–	–	–	Cholest-20-ene-3,6-diol, (3β,5α,6α)-	41084-02-0	NI40786
402	95	55	403	247	81	69	41	**402**	100	58	33	31	28	28	27	25		27	46	2	–	–	–	–	–	–	–	–	–	–	Cholest-20-ene-3,6-diol, (3β,5α,6α)-	41084-02-0	NI40787
55	347	79	105	387	370	402	348	**402**	100	46	42	39	34	28	20	11		27	46	2	–	–	–	–	–	–	–	–	–	–	3,5-Cyclocholan-24-ol, 23ξ-ethyl-6-methoxy-, (3α,6β)-		MS09230
257	144	315	258	55	81	95	109	**402**	100	54	30	22	22	20	18	17	**2.02**	27	46	2	–	–	–	–	–	–	–	–	–	–	Furostan-26-ol, (5α,20β,25R)-	55028-77-8	NI40788
257	144	315	258	55	81	95	109	**402**	100	54	31	23	22	21	19	18	**3.00**	27	46	2	–	–	–	–	–	–	–	–	–	–	Furostan-26-ol, (5α,20S,22ξ,25R)-		MS09231
144	257	315	258	316	402	271	328	**402**	100	99	85	23	20	11	11	10		27	46	2	–	–	–	–	–	–	–	–	–	–	Furostan-26-ol, (5α,20S,22ξ,25R)-		MS09232
120	55	121	95	107	81	59	69	**402**	100	50	48	47	46	46	41	40	**17.91**	27	46	2	–	–	–	–	–	–	–	–	–	–	9,10-Secocholesta-5,7-diene-3,25-diol, (3β,5E,7E)-	25631-39-4	NI40789
120	55	121	107	95	81	135	59	**402**	100	50	47	46	46	45	40	39	**18.00**	27	46	2	–	–	–	–	–	–	–	–	–	–	9,10-Secocholesta-5,7-diene-3,25-diol, (3β,5E,7E)-	25631-39-4	LI08564
98	306	99	70	304	97	200	55	**402**	100	8	8	8	7	5	4	3	**2.58**	28	38	–	2	–	–	–	–	–	–	–	–	–	Quinoline, decahydro-1-methyl-6-(1-methyl-2-piperidyl)-5,7-diphenyl-	14028-76-3	NI40790
402	215	248	247	95	403	355	107	**402**	100	75	47	40	31	30	29	29		28	50	1	–	–	–	–	–	–	–	–	–	–	Cholestane, 3-methoxy-, (3α)-	58072-56-3	NI40791
215	402	248	247	107	217	81	95	**402**	100	69	52	52	48	46	46	45		28	50	1	–	–	–	–	–	–	–	–	–	–	Cholestane, 3-methoxy-, (3α,5α)-	2569-20-2	NI40792
402	215	95	107	248	81	108	55	**402**	100	63	57	49	43	42	39	37		28	50	1	–	–	–	–	–	–	–	–	–	–	Cholestane, 3-methoxy-, (3β)-	53109-81-2	NI40793
402	215	95	107	218	81	217	108	**402**	100	65	57	50	44	44	37	37		28	50	1	–	–	–	–	–	–	–	–	–	–	Cholestane, 3-methoxy-, (3β)-	53109-81-2	LI08565
215	402	107	248	79	95	108	247	**402**	100	92	64	61	61	60	55	54		28	50	1	–	–	–	–	–	–	–	–	–	–	Cholestane, 3-methoxy-, (3β,5α)-	1981-90-4	NI40794
370	402	371	216	230	215	217	262	**402**	100	77	75	67	64	64	61	58		28	50	1	–	–	–	–	–	–	–	–	–	–	Cholestane, 6-methoxy-, (6β)-	55528-12-6	NI40795
370	402	371	216	230	215	282	355	**402**	100	80	77	67	65	65	59	57		28	50	1	–	–	–	–	–	–	–	–	–	–	Cholestane, 6-methoxy-, (6β)-	55528-12-6	LI08566
247	178	43	55	41	95	262	57	**402**	100	89	82	65	59	49	48	48	**11.11**	28	50	1	–	–	–	–	–	–	–	–	–	–	5α-Cholestan-3β-ol, 14-methyl-	5259-27-8	NI40796
402	403	248	105	95	43	57	55	**402**	100	31	29	27	27	27	26	25		28	50	1	–	–	–	–	–	–	–	–	–	–	Cholestan-3-ol, 2-methyl-, (2β,3α,5α)-	20997-57-3	NI40797
43	58	402	55	41	95	57	69	**402**	100	42	29	17	17	15	14	13		28	50	1	–	–	–	–	–	–	–	–	–	–	Cholestan-3-ol, 4-methyl-, (3α,4α,5α)-	20997-67-5	NI40798
215	233	234	165	216	107	108	217	**402**	100	91	62	53	42	42	39	37	**34.07**	28	50	1	–	–	–	–	–	–	–	–	–	–	Ergostanol	6538-02-9	NI40799
229	95	55	43	81	121	247	179	**402**	100	86	84	84	78	74	71	60	**34.00**	28	50	1	–	–	–	–	–	–	–	–	–	–	4α-Methyl-5α-cholestan-3β-ol		NI40800
95	81	55	121	43	109	229	69	**402**	100	90	86	83	74	72	66	60	**33.00**	28	50	1	–	–	–	–	–	–	–	–	–	–	4β-Methyl-5α-cholestan-3β-ol		NI40801
121	229	95	81	43	55	247	107	**402**	100	98	70	68	67	65	59	56	**29.00**	28	50	1	–	–	–	–	–	–	–	–	–	–	6α-Methyl-5α-cholestan-3β-ol		NI40802
121	229	107	55	95	43	81	247	**402**	100	87	86	73	70	70	67	61	**30.00**	28	50	1	–	–	–	–	–	–	–	–	–	–	6β-Methyl-5α-cholestan-3β-ol		NI40803
163	81	162	43	67	41	57	55	**402**	100	61	56	41	38	29	25	23	**0.40**	29	54	–	–	–	–	–	–	–	–	–	–	–	9-(4-as-Dodecahydroindacenyl)heptadecane		AI02335
163	81	162	43	67	41	57	55	**402**	100	60	51	46	40	32	26	25	**0.33**	29	54	–	–	–	–	–	–	–	–	–	–	–	9-(4-as-Dodecahydroindacenyl)heptadecane		AI02336
402	173	213	187	229	403	172	188	**402**	100	89	88	48	37	32	25	21		30	42	–	–	–	–	–	–	–	–	–	–	–	2,2′-Butylenebis(1,3,4-trimethyl-5,6,7,8-tetrahydronaphthalene)	64094-26-4	NI40804
69	326	324	238	236	407	405	403	**403**	100	31	31	10	10	7	7	7		5	–	–	1	–	9	–	2	–	–	–	–	–	trans-1,2-Dibromo-3,3,3-trifluoro-N,N-bis(trifluoromethyl)-prop-1-enylamine		LI08567
294	187	202	186	295	188	296	250	**403**	100	75	50	43	30	25	22	18	**14.00**	13	20	4	3	2	3	–	–	–	–	–	–	–	2,4-Bis(dimethylsulphamoyl)-5-(trifluoromethyl)dimethylaniline		MS09233
91	148	190	103	43	119	131	316	**403**	100	67	39	23	20	16	14	10	**0.66**	16	16	3	1	–	7	–	–	–	–	–	–	–	N-(Heptafluorobutyryl)phenylalanine propyl ester		MS09234
178	73	106	179	298	45	135	104	**403**	100	32	24	21	15	7	6	6	**3.00**	16	34	3	1	–	–	–	–	–	3	1	–	–	Phosphonic acid, [phenyl[(trimethylsilyl)amino]methyl]-, bis(trimethylsilyl) ester	53044-42-1	NI40805
178	73	298	147	211	135	133	75	**403**	100	32	15	9	6	6	4	4	**3.00**	16	34	3	1	–	–	–	–	–	3	1	–	–	Phosphonic acid, [phenyl[(trimethylsilyl)amino]methyl]-, bis(trimethylsilyl) ester	53044-42-1	NI40806
178	73	106	179	298	147	45	104	**403**	100	32	24	21	15	9	7	6	**3.00**	16	34	3	1	–	–	–	–	–	3	1	–	–	Phosphonic acid, [phenyl[(trimethylsilyl)amino]methyl]-, bis(trimethylsilyl) ester	53044-42-1	NI40807
134	323	135	102	91	132	131	193	**403**	100	10	10	10	10	6	6	5	**0.00**	17	14	–	3	2	–	–	1	–	–	–	–	–	Thiazolo[2,3-b]-1,3,4-thiadiazol-4-ium bromide, 2-[(4-methylphenyl)amino]-5-phenyl-		MS09235
312	227	344	77	345	328	228	386	**403**	100	85	75	28	15	14	12	10	**1.70**	18	17	8	3	–	–	–	–	–	–	–	–	–	Trimethyl 4a,5,6,11-tetrahydro-4a-hydroxy-6-methyl-5-oxopyridazino[2,3 a]-quinoxaline-2,3,4-tricarboxylate		NI40808
98	112	41	77	42	55	105	91	**403**	100	40	30	27	25	20	18	13	**7.00**	19	21	1	3	3	–	–	–	–	–	–	–	–	Thieno[2,3-c]isothiazol-4-amine, 5-benzoyl-3-[2-(N-piperidinyl)ethylthio]-	97090-72-7	NI40809
98	42	41	72	305	55	105	111	**403**	100	76	73	70	54	50	38	27	**15.00**	19	21	1	3	3	–	–	–	–	–	–	–	–	Thieno[3,2-d]isothiazol-4-amine, 5-benzoyl-3-[2-(N-piperidinyl)ethylthio]-	97090-80-7	NI40810
43	283	83	174	243	186	242	91	**403**	100	41	14	12	11	10	9	9	**2.00**	19	21	7	3	–	–	–	–	–	–	–	–	–	2-Phenyl-4-(2,3,5-tri-O-acetyl-β-D-lyxofuranosyl)-1,2,3-osotriazole		MS09236
240	256	242	258	120	221	91	241	**403**	100	99	46	44	40	31	27	23	**5.83**	20	18	6	1	–	–	1	–	–	–	–	–	–	L-Phenylalanine, N-[(5-chloro-3,4-dihydro-8-hydroxy-3-methyl-1-oxo-1H-2-benzopyran-7-yl)carbonyl]-, (R)-	303-47-9	NI40811
403	77	262	172	144	234	217	188	**403**	100	73	63	41	38	36	29	28		20	21	6	1	1	–	–	–	–	–	–	–	–	Diethyl 3,3′-[1-(benzenesulphonyl)pyrrole-2,5-diyl]diprop-2-enoate	95883-04-8	NI40812
376	377	403	388	344				**403**	100	19	9	7	6					20	21	8	1	–	–	–	–	–	–	–	–	–	6-Methyl-9a-vinyl-9aH-quinolizine-1,2,3,4-tetracarboxylate		LI08568
312	104	403	344	77	283	252	313	**403**	100	43	42	40	40	27	24	20		20	21	8	1	–	–	–	–	–	–	–	–	–	Tetramethyl-2,3-dihydro-1-phenylazepine-2,3,4,5-tetracarboxylate		LI08569

MASS TO CHARGE RATIOS	M.W.	INTENSITIES	Parent	C	H	O	N	S	F	Cl	Br	I	Si	P	B	X	COMPOUND NAME	CAS Reg No	No
225 178 179 70 403 165 152 56	**403**	100 44 38 36 34 26 26 26		20	29	4	5	–	–	–	–	–	–	–	–	–	**Uracil, 6-[3-[4-(2-methoxy-4-hydroxyphenyl)piperazinyl]propylamino]-1,3 dimethyl-**	88733-12-4	**NI40813**
225 160 136 163 70 149 56 150	**403**	100 47 35 34 32 30 30 29	**0.00**	20	29	4	5	–	–	–	–	–	–	–	–	–	**Uracil, 6-[3-[4-(2-methoxyphenyl)piperazinyl-1-oxide]propylamino]-1,3-dimethyl-**	62845-30-1	**NI40814**
257 242 403 258 243 199 331 344	**403**	100 73 33 32 15 11 11 9		21	25	7	1	–	–	–	–	–	–	–	–	–	**Oxoepistephamiersine**	51804-68-3	**NI40815**
257 242 258 403 227 243 228 331	**403**	100 65 37 26 22 14 5 5		21	25	7	1	–	–	–	–	–	–	–	–	–	**Oxostephamiersine**	52466-83-8	**NI40816**
403 123 225 57 250 192 281 161	**403**	100 95 80 40 26 22 20 18		21	26	1	1	2	–	–	–	–	–	1	–	–	**1,2-Thiaphosphol-3-ene 2-sulphide, 5-tert-butyl-2-[(4-methoxyphenyl)amino]-3-(4-methylphenyl)-**	105867-42-3	**NI40817**
42 246 70 143 56 403 248 232	**403**	100 93 69 67 54 45 39 34		21	26	1	3	1	–	1	–	–	–	–	–	–	**1-Piperazineethanol, 4-[3-(2-chloro-10H-phenothiazin-10-yl)propyl]-**	58-39-9	**NI40818**
246 143 42 70 403 248 56 232	**403**	100 52 52 41 40 39 25 24		21	26	1	3	1	–	1	–	–	–	–	–	–	**1-Piperazineethanol, 4-[3-(2-chloro-10H-phenothiazin-10-yl)propyl]-**	58-39-9	**NI40819**
42 246 232 70 143 56 214 43	**403**	100 67 60 50 37 37 29 21	**10.01**	21	26	1	3	1	–	1	–	–	–	–	–	–	**1-Piperazineethanol, 4-[3-(2-chloro-10H-phenothiazin-10-yl)propyl]-**	58-39-9	**NI40820**
73 241 403 211 372 282 404 242	**403**	100 90 78 78 59 56 26 25		21	29	5	1	–	–	–	–	–	1	–	–	–	**Crinan, 1,2-didehydro-3,7-dimethoxy-11-[(trimethylsilyl)oxy]-, (3α,11S)-**	55760-25-3	**NI40821**
255 242 386 239 241 257 330 256	**403**	100 54 49 47 42 40 36 32	**6.69**	22	14	3	3	–	–	1	–	–	–	–	–	–	**1H-1,4-Benzodiazepin-2-one, 7-chloro-2,3-dihydro-3-(2-nitrobenzylidene)-5-phenyl-**	55056-32-1	**NI40822**
403 405 404 329 242 402 373 328	**403**	100 35 26 10 10 9 9 7		22	14	3	3	–	–	1	–	–	–	–	–	–	**1H-1,4-Benzodiazepin-2-one, 7-chloro-2,3-dihydro-3-(4-nitrobenzylidene)-5-phenyl-**	55056-33-2	**NI40823**
304 403 402 191 374 248 312 205	**403**	100 62 52 47 44 30 29 27		22	29	6	1	–	–	–	–	–	–	–	–	–	**11H-Dibenzo[a,g]quinolizine-8a-carboxylic acid, (±)-5,6,8,8a,9,10,12,12a,13,13a-decahydro-2,3-dimethoxy-11-oxo-, ethyl ester**		**MS09237**
216 58 217 128 186 158 230 284	**403**	100 26 15 11 4 4 3 2	**0.00**	22	45	5	1	–	–	–	–	–	–	–	–	–	**1,4,7-Trioxa-11-azacyclotetradecane, 11-[2-[2-(octyloxy)ethoxy]ethyl]-**	120547-44-6	**DD01571**
191 91 105 77 122 282 373 281	**403**	100 96 94 80 17 4 2 2	**1.40**	24	21	5	1	–	–	–	–	–	–	–	–	–	**Propanal, 2,3-bis(benzoyloxy)-, 1-[O-benzyloxime], (±)-**	71641-33-3	**NI40824**
206 197 199 207 200 198 234	**403**	100 55 53 18 5 5 4	**0.00**	24	22	–	1	–	–	–	1	–	–	–	–	–	**3-Cyclohexen-1-amine, 6-(4-bromophenyl)-2,5-diphenyl-, (1α,2β,5β,6β)-**	42715-20-8	**NI40825**
229 302 105 307 248 403 276 374	**403**	100 94 85 70 49 47 45 19		24	25	3	3	–	–	–	–	–	–	–	–	–	**Furo[2,3-d]pyrimidine-5,6-dione, 1,3-diethyl-2-(ethylimino)-1,2,3,7a-tetrahydro-4,7a-diphenyl-**	90140-45-7	**NI40826**
234 345 360 235 403 247 300 193	**403**	100 85 82 82 77 71 70 70		24	37	4	1	–	–	–	–	–	–	–	–	–	**11-Azaandrostan-6-ol, 11,14-diacetyl-, acetate**	55470-96-7	**NI40827**
43 145 91 158 128 105 79 41	**403**	100 46 41 40 38 38 36 34	**6.00**	24	37	4	1	–	–	–	–	–	–	–	–	–	**3β,21-Dihydroxy-pregn-5-en-20-one, 21-acetate, methyl oxime**		**LI08570**
157 313 284 259 158 314 403 73	**403**	100 49 37 26 20 12 11 10		24	41	2	1	–	–	–	–	–	1	–	–	–	**19-Norpregn-4-en-3-one, 17-[(trimethylsilyl)oxy]-, O-methyloxime, (17α)-**	69834-06-6	**NI40828**
374 403 91 330 77 303 104 115	**403**	100 65 41 29 27 23 23 22		26	33	1	3	–	–	–	–	–	–	–	–	–	**2(1H)-Pyridinone, 3-benzyl-4,5-bis(diethylamino)-1-phenyl-**		**DD01572**
118 105 286 403 77 90 287 165	**403**	100 72 56 37 37 28 20 16		28	21	2	1	–	–	–	–	–	–	–	–	–	**2-Azetidinone, 1-benzoyl-3,4,4-triphenyl-**	5904-47-2	**NI40829**
358 359 403 165 166 167 91 404	**403**	100 35 27 27 9 8 8 7		28	21	2	1	–	–	–	–	–	–	–	–	–	**Benzeneacetic acid, α-[4-(cyanodiphenylmethyl)phenyl]-**		**NI40830**
77 91 105 403 388 283 149 402	**403**	100 70 38 33 20 18 11 8		28	25	–	3	–	–	–	–	–	–	–	–	–	**1H-Indole, 6,7-dihydro-6,6-dimethyl-1,2-diphenyl-4-phenylazo-**		**NI40831**
257 403 272 271 273 130 402 388	**403**	100 54 36 16 14 10 4 4		28	25	–	3	–	–	–	–	–	–	–	–	–	**Tris(3-methylindol-2-yl)methane**		**NI40832**
228 316 143 297 124 142 209 146	**404**	100 96 94 69 56 42 32 32	**21.00**	–	4	–	2	–	10	–	–	–	–	4	–	1	**Bis(trifluorophosphino)bis(aminodifluorophosphino)nickel**		**NI40833**
404 258 277 127 69 150	**404**	100 59 56 34 11 10		3	–	–	–	–	6	–	–	2	–	–	–	–	**Propane, 1,1,1,3,3,3-hexafluoro-2,2-diiodo-**	38568-22-8	**NI40834**
28 266 264 294 262 235 80 263	**404**	100 14 13 12 12 11 11 10	**3.01**	9	4	5	2	–	–	–	–	–	–	–	–	1	**Tungsten, pentacarbonyl(pyridazine-N[1])-**	65761-20-8	**NI40835**
201 203 243 273 271 213 283 245	**404**	100 86 56 55 54 52 48 48	**5.82**	9	6	3	–	1	–	6	–	–	–	–	–	–	**6,9-Methano-2,4,3-benzodioxathiepin, 6,7,8,9,10,10-hexachloro-1,5,5a,6,9,9a-hexahydro-, 3-oxide**	115-29-7	**NI40836**
195 197 241 237 239 207 170 265	**404**	100 88 86 73 68 68 63 60	**2.84**	9	6	3	–	1	–	6	–	–	–	–	–	–	**6,9-Methano-2,4,3-benzodioxathiepin, 6,7,8,9,10,10-hexachloro-1,5,5a,6,9,9a-hexahydro-, 3-oxide**	115-29-7	**NI40837**
29 39 27 31 48 41 30 15	**404**	100 31 27 26 22 20 20 17	**0.00**	9	6	3	–	1	–	6	–	–	–	–	–	–	**6,9-Methano-2,4,3-benzodioxathiepin, 6,7,8,9,10,10-hexachloro-1,5,5a,6,9,9a-hexahydro-, 3-oxide, (3α,5aα,6β,9β,9aα)-**	33213-65-9	**NI40838**
29 39 195 241 27 197 48 170	**404**	100 33 29 25 22 21 21 20	**0.00**	9	6	3	–	1	–	6	–	–	–	–	–	–	**6,9-Methano-2,4,3-benzodioxathiepin, 6,7,8,9,10,10-hexachloro-1,5,5a,6,9,9a-hexahydro-, 3-oxide, (3α,5aβ,6α,9α,9aβ)-**	959-98-8	**NI40840**
195 241 197 239 207 237 277 75	**404**	100 84 77 76 71 69 68 65	**0.00**	9	6	3	–	1	–	6	–	–	–	–	–	–	**6,9-Methano-2,4,3-benzodioxathiepin, 6,7,8,9,10,10-hexachloro-1,5,5a,6,9,9a-hexahydro-, 3-oxide, (3α,5aβ,6α,9α,9aβ)-**	959-98-8	**NI40839**
57 43 121 123 41 42 235 233	**404**	100 33 31 29 26 18 14 14	**0.00**	9	11	3	–	–	–	–	3	–	–	–	–	–	**2H,4H-1,3-Dioxin-4-one, 5-bromo-2-tert-butyl-6-(dibromomethyl)-, (2R)-**		**DD01573**
375 377 376 281 378 315 253 287	**404**	100 57 39 20 19 17 17 13	**0.69**	10	29	5	–	–	–	1	–	–	5	–	–	–	**1,3,5,7,9-Pentaethyl-1-chlorocyclopentasiloxane**	73420-28-7	**NI40841**
43 67 39 227 41 225 211 65	**404**	100 31 25 23 22 20 20 20	**0.00**	11	16	1	–	–	–	–	–	–	–	–	–	3	**2,4,6-Triselenatricyclo[3.3.1.1[3,7]]decan-8-one, 1,3,5,7-tetramethyl-, (±)-**	57289-43-7	**NI40842**
54 67 79 80 39 41 93 66	**404**	100 94 50 47 46 27 22 22	**4.28**	12	12	4	–	–	–	–	–	–	–	–	–	1	**π-1,5-Cyclooctadienetetracarbonyltungsten**		**MS09238**
196 198 209 211 210 208 406 200	**404**	100 98 58 56 43 43 36 32	**18.63**	13	6	2	–	–	–	6	–	–	–	–	–	–	**Phenol, 2,2′-methylenebis[3,4,6-trichloro-**	70-30-4	**NI40843**
196 198 209 211 208 210 406 200	**404**	100 94 62 51 45 40 37 30	**20.00**	13	6	2	–	–	–	6	–	–	–	–	–	–	**Phenol, 2,2′-methylenebis[3,4,6-trichloro-**	70-30-4	**NI40844**
372 294 404 350 370 344 314 266	**404**	100 91 67 65 60 59 51 48		13	13	3	–	–	–	–	–	–	–	–	–	1	**Tricarbonyl-tetramethylene-1,2-methyl-3-cyclopentadienylrhenium**		**IC04308**
404 402 256 364 362 284 282 334	**404**	100 80 56 53 48 21 18 17		14	6	–	–	–	–	2	–	–	–	–	–	2	**3,8-Dichlorobenzoselenopheno[2,3-b]benzoselenophene**		**LI08571**
69 70 276 198 404 110 237 168	**404**	100 60 50 30 29 26 22 19		15	3	–	–	–	10	–	–	–	–	1	–	–	**Phosphine, bis(pentafluorophenyl)-1-propynyl-**	33730-52-8	**NI40845**

MASS TO CHARGE RATIOS								M.W.	INTENSITIES								Parent	C	H	O	N	S	F	Cl	Br	I	Si	P	B	X	COMPOUND NAME	CAS Reg No	No
175	72	157	56	88	86	87	71	404	100	88	86	84	81	81	72	72	49.11	16	28	8	4	–	–	–	–	–	–	–	–	–	1,4,13,16-Tetraoxa-7,10,19,22-tetraazacyclotetracosane-6,9,20,23-tetrone	79788-45-7	NI40846
73	156	203	147	45	215	74	116	404	100	24	14	14	14	12	12	10	0.00	16	36	4	2	–	–	–	–	–	3	–	–	–	Glutamine, N-acetyl-, bis(trimethylsilyl)-, trimethylsilyl ester		NI40847
127	128	77	404	307				404	100	8	4	2	2					17	10	–	–	–	10	–	–	–	–	–	–	–	1,5-Diphenyldecafluoropentane		LI08572
172	140	155	73	196	123	69	60	404	100	94	91	85	74	74	74	68	2.30	17	24	11	–	–	–	–	–	–	–	–	–	–	Hastatoside		MS09239
224	211	182	97	139	69	145	81	404	100	90	90	90	72	55	54	40	0.00	17	24	11	–	–	–	–	–	–	–	–	–	–	1,3,4,5,7-Penta-O-acetyl-2,6-anhydro-D-glycero-D-gulo-heptitol		MS09240
83	81	82	70	84	98	73	111	404	100	99	75	58	53	46	45	43	0.00	17	24	11	–	–	–	–	–	–	–	–	–	–	1,2,3′,4,6-Penta-O-acetyl-3-deoxy-3-C-(hydroxymethyl)-α-D-glucopyranose		MS09241
73	133	204	147	45	116	75	189	404	100	99	98	84	68	65	64	57	1.80	17	37	6	–	–	–	–	–	–	2	–	1	–	α-D-Galactopyranoside, methyl 2,3-bis-O-(trimethylsilyl)-, cyclic butylboronate	56211-11-1	NI40848
73	133	204	147	75	55	45	116	404	100	97	89	49	41	35	26	25	0.00	17	37	6	–	–	–	–	–	–	2	–	1	–	α-D-Galactopyranoside, methyl 2,3-bis-O-(trimethylsilyl)-, cyclic butylboronate	56211-11-1	NI40849
75	44	45	43	59	47	73	77	404	100	50	18	18	17	17	15	10	0.00	17	37	6	–	–	–	–	–	–	2	–	1	–	α-D-Galactopyranoside, methyl 2,3-bis-O-(trimethylsilyl)-, cyclic butylboronate	56211-11-1	NI40850
73	157	204	218	75	191	55	147	404	100	71	49	48	47	46	39	38	0.00	17	37	6	–	–	–	–	–	–	2	–	1	–	α-D-Galactopyranoside, methyl 2,6-bis-O-(trimethylsilyl)-, cyclic butylboronate	56211-12-2	NI40851
44	75	139	73	55	45	41	42	404	100	91	24	22	22	22	22	20	0.00	17	37	6	–	–	–	–	–	–	2	–	1	–	β-D-Galactopyranoside, methyl 2,3-bis-O-(trimethylsilyl)-, cyclic butylboronate	56211-10-0	NI40852
73	133	204	147	75	45	116	189	404	100	99	93	68	60	60	49	43	0.00	17	37	6	–	–	–	–	–	–	2	–	1	–	β-D-Galactopyranoside, methyl 2,3-bis-O-(trimethylsilyl)-, cyclic butylboronate	56211-10-0	NI40853
73	157	204	218	75	191	147	231	404	100	68	48	43	42	41	39	30	0.00	17	37	6	–	–	–	–	–	–	2	–	1	–	β-D-Galactopyranoside, methyl 2,6-bis-O-(trimethylsilyl)-, cyclic butylboronate	56211-13-3	NI40854
75	44	73	43	117	45	103	47	404	100	51	39	28	22	21	18	17	0.00	17	37	6	–	–	–	–	–	–	2	–	1	–	β-D-Galactopyranoside, methyl 2,6-bis-O-(trimethylsilyl)-, cyclic butylboronate	56211-13-3	NI40855
204	73	147	75	205	129	133	43	404	100	92	40	34	26	26	23	20	0.00	17	37	6	–	–	–	–	–	–	2	–	1	–	α-D-Glucopyranoside, methyl 2,3-bis-O-(trimethylsilyl)-, cyclic butylboronate	56211-14-4	NI40856
75	44	45	43	47	73	41	77	404	100	75	20	17	16	13	13	12	0.00	17	37	6	–	–	–	–	–	–	2	–	1	–	α-D-Glucopyranoside, methyl 2,3-bis-O-(trimethylsilyl)-, cyclic butylboronate	56211-14-4	NI40857
117	75	73	84	103	97	129	43	404	100	84	82	56	50	47	40	35	0.00	17	37	6	–	–	–	–	–	–	2	–	1	–	β-D-Glucopyranoside, methyl 2,3-bis-O-(trimethylsilyl)-, cyclic butylboronate	56211-15-5	NI40858
73	204	75	147	44	45	133	43	404	100	76	46	35	25	19	18	18	0.00	17	37	6	–	–	–	–	–	–	2	–	1	–	α-D-Mannopyranoside, methyl 2,3-bis-O-(trimethylsilyl)-, cyclic butylboronate	56211-09-7	NI40859
404	77	327	51	405	258	403	385	404	100	91	54	41	19	11	10	10		18	5	–	–	–	8	–	–	–	–	1	–	–	5H-Dibenzophosphole, 1,2,3,4,6,7,8,9-octafluoro-5-phenyl-	36284-12-5	NI40860
41	67	135	149	55	237	108	404	404	100	83	77	70	70	51	43	42		18	26	2	–	–	–	–	–	–	–	–	–	1	1-Oxaspiro[5.5]undecane, 2-[[(4-methoxyphenyl)telluro]methyl]-		DD01574
57	43	41	98	115	157	242	169	404	100	99	44	42	35	32	30	25	0.00	18	28	10	–	–	–	–	–	–	–	–	–	–	β-D-Glucopyranoside, tert-butyl 2,3,4,6-tetra-O-acetyl-	13035-50-2	NI40861
202	331	404	59	203	216	130	42	404	100	23	12	12	11	10	10	10		18	32	8	2	–	–	–	–	–	–	–	–	–	Glycine, N,N′-1,2-ethanediylbis[N-(2-ethoxy-2-oxoethyl)-, diethyl ester	3626-00-4	MS09242
265	142	70	132	56	263	114	194	404	100	66	59	53	38	37	35	33	1.27	18	36	6	4	–	–	–	–	–	–	–	–	–	Diaza-18-crown-6, bis(α-alanyl)-, (R,R)-	93767-95-4	NI40862
191	235	192	137	144	176	127	128	404	100	46	45	45	28	22	20	18	3.00	19	33	1	–	–	–	–	–	1	–	–	–	–	15-Norlabdane, 8,12-epoxy-14-iodo-, (12RS,13S)-		MS09243
123	149	151	75	119	404	124	91	404	100	33	31	16	14	12	12	12		20	24	7	2	–	–	–	–	–	–	–	–	–	1,4-Cyclohexadiene-1-butanol, 2-methoxy-δ,4-dimethyl-, 3,5-dinitrobenzoate	63646-87-7	NI40863
143	161	171	111	95	55	103	43	404	100	61	60	48	43	37	36	35	0.00	20	36	8	–	–	–	–	–	–	–	–	–	–	Pentadecanedioic acid, 3,7-anhydro-3,5,7,9,13-pentahydroxy-4,8,12-trimethyl-, dimethyl ester	85888-07-9	NI40864
103	198	361	43	273	75	71	245	404	100	97	44	43	39	30	26	23	0.00	20	36	8	–	–	–	–	–	–	–	–	–	–	Propanedioic acid, 2-ethoxy-2-(5-ethoxy-2,2-diisopropyl-1,3-dioxolan-4-yl)-, diethyl ester		DD01575
114	128	100	72	84	70	158	101	404	100	84	84	79	74	74	63	63	44.20	20	40	6	2	–	–	–	–	–	–	–	–	–	5,8,11,18,21,26-Hexaoxa-1,15-diazabicyclo[13.8.5]octacosane		MS09244
84	86	101	58	70	114	56	57	404	100	81	64	45	37	36	36	32	10.00	20	40	6	2	–	–	–	–	–	–	–	–	–	5,8,15,18,23,26-hexaoxa-1,12-diazabicyclo[10.8.8]octacosane		MS09245
28	262	183	108	154	263	184	261	404	100	75	51	26	18	15	11	10	0.15	21	15	3	–	–	–	–	–	–	–	1	–	1	Nickel, tricarbonyl(triphenylphosphine)-	14917-13-6	NI40865
137	368	232	169	138	369	297	243	404	100	32	14	10	9	8	8	8	0.00	21	21	6	–	–	–	1	–	–	–	–	–	–	4H-1-Benzopyran-4-one, 7-hydroxy-3-[6-(3-chloro-3-methylbutyl)-2,4-dihydroxy-5-methoxyphenyl]-		MS09246
404	230	229	83	248	247	246	211	404	100	77	73	70	54	50	47	40		21	24	3	–	–	–	–	–	–	–	–	–	1	Guaia-4(15),9-dieno-12,6-lactone, 3-hydroxy-11-(phenylseleno)-, (3α,6α,11β)-	119273-12-0	DD01576
312	404	340	372	285	83	258	313	404	100	90	63	57	39	38	33	32		21	24	8	–	–	–	–	–	–	–	–	–	–	Benzoic acid, 2,4-dihydroxy-5-[3-methoxy-6-(methoxycarbonyl)-2,5-dimethylphenoxy]-3,6-dimethyl-, methyl ester	38629-33-3	NI40866
193	192	28	194	135	210	164	165	404	100	70	56	47	30	21	20	19	0.00	21	24	8	–	–	–	–	–	–	–	–	–	–	Benzoic acid, 3-[(2,4-dihydroxy-6-propylbenzoyl)oxy]-2-hydroxy-4-methoxy-6-propyl-	73694-32-3	NI40867
224	404	209	225	178	211	181	166	404	100	36	19	14	10	8	5	5		21	24	8	–	–	–	–	–	–	–	–	–	–	4H-1-Benzopyran-4-one, 2,3-dihydro-6,7-dimethoxy-3-(2,3,4,5-tetramethoxyphenyl)-, (S)-	71594-03-1	NI40868

MASS TO CHARGE RATIOS								M.W.	INTENSITIES								Parent	C	H	O	N	S	F	Cl	Br	I	Si	P	B	X	COMPOUND NAME	CAS Reg No	No
317	404	285	345	238	180	120	405	404	100	84	67	42	28	28	27	24		21	28	6	2	–	–	–	–	–	–	–	–	–	4,4′-Dimethyl-3,3′-bis(methoxycarbonylethyl)-5-methoxycarbonyl-2,2′-dipyrrolylmethane		LI08573
319	321	390	388	279	277	347	345	404	100	99	26	26	15	15	10	10	2.00	21	29	1	2	–	–	–	1	–	–	–	–	–	Flustramine D N-oxide		NI40869
307	213	77	73	250	389	94	308	404	100	62	27	23	22	19	16	15	0.00	21	29	4	–	–	–	–	–	–	1	1	–	–	1-tert-Butyl-2-(trimethylsilyl)vinyl diphenyl phosphate, (Z)-	118297-83-9	DD01577
320	43	334	42	321	322	45	60	404	100	55	38	24	19	16	15	12	4.00	22	12	8	–	–	–	–	–	–	–	–	–	–	Benzo[1,2-b:4,5-b′]bisbenzofuran-6,12-dione, 3,9-bis(acetyloxy)-	51860-97-0	NI40870
302	275	247	276	301	344	43	303	404	100	76	37	33	32	26	26	23	8.00	22	20	4	4	–	–	–	–	–	–	–	–	–	1,2-Propanediol, 1-(1-phenyl-1H-pyrazolo[3,4-b]quinoxalin-3-yl)-, diacetate, [S-(R*,S*)]-	55591-19-0	NI40871
404	231	406	405	202	233	58	28	404	100	52	41	27	22	20	17	17		22	22	2	2	–	–	–	–	–	–	–	–	1	Nickel, [[3,3′-(1,2-ethanediyldinitrilo)bis[1-phenyl-1-butanonato]](2-)-N,N′,O,O′]-	15277-29-9	NI40872
344	181	404	313	151	176	345	121	404	100	87	41	36	31	26	22	17		22	28	7	–	–	–	–	–	–	–	–	–	–	2-Acetoxy-1-(3,4-dimethoxy-phenyl)-3-(2,4,6-trimethoxyphenyl)propane		LI08574
404	165	195	179	83	405	192	237	404	100	55	28	28	28	26	26	20		22	28	7	–	–	–	–	–	–	–	–	–	–	(1R,5R,6S,7R,8R)-1-Allyl-8-hydroxy-4,5-dimethoxy-6-(3-methoxy-4,5-methylenedioxyphenyl)-7-methyl-3-oxobicyclo[3.2.1]octane		NI40873
83	232	43	214	109	69	213	55	404	100	70	66	43	39	31	29	27	0.58	22	28	7	–	–	–	–	–	–	–	–	–	–	20-Desoxy-20-oxophorbol-13-acetate		NI40874
151	152	167	404	165	177	137	139	404	100	92	37	33	30	27	24	20		22	28	7	–	–	–	–	–	–	–	–	–	–	Dihydrogmelinol		LI08575
151	152	137	138	107	153	106	167	404	100	78	17	14	9	8	7	6	3.00	22	28	7	–	–	–	–	–	–	–	–	–	–	Dihydroneogmelinol		LI08576
167	137	168	166	330	151	390	344	404	100	63	59	42	39	28	14	7	1.00	22	28	7	–	–	–	–	–	–	–	–	–	–	3-(2,4-Dimethoxyphenyl)-1-(3,4-dimethoxyphenyl)-2-acetoxy-1-methoxypropane		NI40875
404	179	135	276	360	405	214	185	404	100	50	50	47	34	25	23	23		22	28	7	–	–	–	–	–	–	–	–	–	–	O-Ethyl-1-ailanthone		LI08577
151	152	404	177	207	137	107	138	404	100	36	21	16	10	10	10	9		22	28	7	–	–	–	–	–	–	–	–	–	–	2,3-Furandiol, tetrahydro-3,4-bis(3,4-dimethoxybenzyl)-		NI40876
151	404	166	138	163	152	405	233	404	100	75	33	30	26	24	19	14		22	28	7	–	–	–	–	–	–	–	–	–	–	3-Furanmethanol, 2-(3,4-dimethoxyphenyl)-4-[(3,4-dimethoxyphenyl)methyl]tetrahydro-3-hydroxy-, [2S-(2α,3β,4α)]-	10569-15-0	NI40877
151	152	137	138	107	153	106	167	404	100	78	17	14	9	8	7	6	3.00	22	28	7	–	–	–	–	–	–	–	–	–	–	3-Furanmethanol, α-(3,4-dimethoxyphenyl)tetrahydro-3-hydroxy-4-veratryl-	10569-16-1	NI40878
234	344	301	274	329	275	247	245	404	100	90	90	90	80	55	38	30	10.00	22	28	7	–	–	–	–	–	–	–	–	–	–	9,10-Phenanthrenedione, 7-[2-(acetyloxy)isopropyl]-1,2,3,4,4a,10a-hexahydro-5,6,8-trihydroxy-1,1,4a-trimethyl-, [4aS-[4aα,7(S*),10aβ]]-	66584-97-2	NI40879
329	274	344	83	330	275	261	91	404	100	90	45	43	20	20	18	17	10.00	22	28	7	–	–	–	–	–	–	–	–	–	–	9(1H)-Phenanthrenone, 7-[2-(acetyloxy)isopropyl]-2,3,4,4a-tetrahydro-5,6,8,10-tetrahydroxy-1,1,4a-trimethyl-, [R-(R*,R*)]-	66584-96-1	NI40880
247	275	151	265	262	248	293	276	404	100	95	52	40	30	30	28	26	20.90	22	32	5	–	–	–	–	–	–	1	–	–	–	1H-2-Benzoxacyclotetradecin-1,7(8H)-dione, 3,4,5,6,9,10-hexahydro-14-methoxy-3-methyl-16-[(trimethylsilyl)oxy]-, [S-(E)]-	72060-11-8	NI40881
403	405	404	242	178	152	177	151	404	100	33	29	12	11	11	10	10		23	17	3	2	–	–	1	–	–	–	–	–	–	1H-1,4-Benzodiazepin-2-one, 7-chloro-2,3-dihydro-3-(3-hydroxy-4-methoxybenzylidene)-5-phenyl-		NI40882
77	278	51	78	279	105	39	130	404	100	55	28	21	20	19	16	13	2.00	23	20	3	2	1	–	–	–	–	–	–	–	–	3,5-Pyrazolidinone, 1,2-diphenyl-4-[2-(phenylsulphinyl)ethyl]-		MS09247
327	149	328	386	243	113	313	167	404	100	79	36	23	19	17	14	13	0.00	23	32	6	–	–	–	–	–	–	–	–	–	–	Abieta-5,8,11,13-tetraen-20-oic acid, 7-hydroperoxy-11,12-dimethoxy-, methyl ester, (7α)-		MS09248
160	161	187	340	178	164	179	158	404	100	37	36	35	27	23	22	22	0.00	23	32	6	–	–	–	–	–	–	–	–	–	–	Card-20(22)-enolide, 3,5,14-trihydroxy-19-oxo-, (3β,5β)-	66-28-4	NI40883
91	28	79	55	368	67	41	111	404	100	97	93	83	80	80	80	69	0.50	23	32	6	–	–	–	–	–	–	–	–	–	–	Card-20(22)-enolide, 3,14,15-trihydroxy-19-oxo-, (3β,5α,15β)-	14155-65-8	NI40884
284	324	269	227	389	242	251	344	404	100	60	57	35	33	30	25	15	14.00	23	32	6	–	–	–	–	–	–	–	–	–	–	7α,12α-Diacetoxy-androstane-3,17-dione		NI40885
312	178	344	131	123	119	109	326	404	100	81	57	42	24	24	23	20	13.00	23	32	6	–	–	–	–	–	–	–	–	–	–	Methyl 6α-acetoxy-7β-hydroxyvouacapan-17β-oate		NI40886
312	178	43	344	69	131	55	119	404	100	81	71	56	48	42	27	24	14.00	23	32	6	–	–	–	–	–	–	–	–	–	–	Methyl 7α-acetoxy-6β-hydroxyvouacapan-17β-oate		NI40887
55	57	43	69	28	41	81	404	404	100	93	90	72	70	65	64	61		23	36	4	2	–	–	–	–	–	–	–	–	–	5α-Androstane-3,6-dione, bis(O-methyloxime)		LI08578
43	41	55	91	44	125	105	373	404	100	99	85	67	67	58	54	50	2.76	23	36	4	2	–	–	–	–	–	–	–	–	–	Pregn-4-ene-3,20-dione, 17,21-dihydroxy-, bis(O-methyloxime)	55557-09-0	NI40888
139	193	404	141	206	248	111	165	404	100	48	38	37	34	24	17	16		24	17	4	–	–	–	1	–	–	–	–	–	–	β-(9-Anthryl)propanoyl-3-chlorobenzoyl peroxide		MS09249
404	373	83	286	312	336	249	265	404	100	80	78	74	67	66	58	43		24	17	4	–	–	–	1	–	–	–	–	–	–	Triptycene-1,8-dicarboxylic acid, 13-chloro-, dimethyl ester		NI40889
404	312	83	249	373	286	344	337	404	100	83	78	67	61	57	54	45		24	17	4	–	–	–	1	–	–	–	–	–	–	Triptycene-1,8-dicarboxylic acid, 16-chloro-, dimethyl ester		NI40890
389	151	404	373	388	58	406	374	404	100	90	86	83	80	70	39	36		24	20	6	–	–	–	–	–	–	–	–	–	–	Diosindigo B	60548-89-2	NI40891
281	213	282	404	123	124	77	185	404	100	23	22	13	11	9	7	6		24	20	6	–	–	–	–	–	–	–	–	–	–	Bis(4-methoxyphenyl) benzylidenemalonate		LI08579
105	77	282	106	283	269	177	51	404	100	15	9	7	2	1	1	1	0.00	24	20	6	–	–	–	–	–	–	–	–	–	–	1,2,3-Propanetriol, tribenzoate	614-33-5	NI40892
253	170	404	131	130	254	202	241	404	100	28	22	20	18	17	16	13		24	24	4	2	–	–	–	–	–	–	–	–	–	1H-1-Benzazepine-3,4-dicarboxylic acid, 2,3-dihydro-2-(3-methylindol-3-yl)-5-methyl-, dimethyl ester, (3α)-		NI40893
404	192	273	170	131	214	405	130	404	100	57	56	51	47	32	26	24		24	24	4	2	–	–	–	–	–	–	–	–	–	1H-1-Benzazepine-3,4-dicarboxylic acid, 2,3-dihydro-2-(3-methylindol-3-yl)-5-methyl-, dimethyl ester, (3β)-		NI40894
202	160	203	404	161	143	144	125	404	100	80	26	23	12	10	9	9		24	24	4	2	–	–	–	–	–	–	–	–	–	Butanedioic acid, 2,3-bis(3-methylindol-2-yl)-, dimethyl ester		NI40895
311	404	293	279	354	336	389	361	404	100	95	81	58	40	25	23	14		24	30	3	–	–	2	–	–	–	–	–	–	–	6α,7α-Difluoromethylene-6β,16β-dimethyl-16α,17α-oxido-4-pregnen-3,20 dione		MS09250
43	255	333	273	71	133	123	93	404	100	51	34	34	28	23	20	18	5.00	24	36	5	–	–	–	–	–	–	–	–	–	–	Androstan-3-one, 17β-acetoxy-4-hydroxy-4-propyl-5,6-epoxy-		MS09251

MASS TO CHARGE RATIOS								M.W.	INTENSITIES								Parent	C	H	O	N	S	F	Cl	Br	I	Si	P	B	X	COMPOUND NAME	CAS Reg No	No
99	299	361	342	404	222	125		404	100	30	29	27	25	20	14			24	36	5	–	–	–	–	–	–	–	–	–	–	Androstan-19-one, 19a-methyl-, ethylene acetal, (5α)-		LI08580
95	79	91	121	107	191	135	112	404	100	31	25	18	18	13	12	10	0.00	24	36	5	–	–	–	–	–	–	–	–	–	–	ent-Cleroda-4(18),12-diene, 15,15-diacetoxy-15,16-epoxy-, (12Z)-		NI40896
333	273	334	305	43	55	41	287	404	100	26	21	16	12	11	10	9	0.00	24	36	5	–	–	–	–	–	–	–	–	–	–	6β-Hydroxy-17-oxo-6α-pentyl-2,3-seco-5α-androstane-2,3-dioic acid γ-lactone		MS09252
55	91	95	81	79	93	105	53	404	100	69	67	66	63	62	54	49	32.90	24	36	5	–	–	–	–	–	–	–	–	–	–	Pregnan-20-one, 3-(acetyloxy)-5,6-epoxy-17-hydroxy-16-methyl-, (3β,5α,6α,16α)-	2857-83-2	MS09253
124	58	125	41	140	121	55	43	404	100	13	8	8	7	6	6	6	0.00	24	38	4	1	–	–	–	–	–	–	–	–	–	1-Piperidinyloxy-, 2,2,6,6-tetramethyl-4-[4-(octyloxy)benzoyloxy]-		DD01578
344	239	254	345	213	240	346	255	404	100	51	39	32	16	11	9	9	0.00	24	40	3	–	–	–	–	–	–	1	–	–	–	Androst-5-ene-3β,17β-diol 3-acetate 17-trimethylsilyl ether		LI08581
129	215	314	404	216	275	274	133	404	100	44	32	21	21	16	12	12		24	40	3	–	–	–	–	–	–	1	–	–	–	Androst-5-ene-3β,17β-diol 17-acetate 3-trimethylsilyl ether		LI08582
209	404	179	277	210	223	389	251	404	100	70	23	20	20	16	10	6		24	40	3	–	–	–	–	–	–	1	–	–	–	4-Tetradecen-3-one, 1-[3-methoxy-4-[(trimethylsilyl)oxy]phenyl]-	56700-98-2	MS09254
371	386	345	372	387	251	105	343	404	100	88	55	28	22	17	16	11	6.00	25	24	5	–	–	–	–	–	–	–	–	–	–	4H,8H-Benzo[1,2-b:3,4-b′]dipyran-4-one, 5-hydroxy-6-(3-hydroxy-3-methyl-1-butenyl)-8,8-dimethyl-2-phenyl-, (E)-	104363-11-3	NI40897
389	404	390	405	361	349	333	187	404	100	50	28	14	4	4	4	4		25	24	5	–	–	–	–	–	–	–	–	–	–	2H,6H-Benzo[1,2-b:5,4-b′]dipyran-6-one, 5-hydroxy-7-(4-hydroxyphenyl)-2,2-dimethyl-10-(3-methyl-2-butenyl)-	4449-55-2	NI40898
271	386	229	262	244	247	387	154	404	100	86	39	36	33	29	26	24	9.44	25	40	4	–	–	–	–	–	–	–	–	–	–	Cholan-24-oic acid, 7-hydroxy-3-oxo-, methyl ester, (5α,7α)-	14772-96-4	NI40899
271	386	272	229	354	211	371	244	404	100	38	22	19	16	16	13	11	8.33	25	40	4	–	–	–	–	–	–	–	–	–	–	Cholan-24-oic acid, 7-hydroxy-3-oxo-, methyl ester, (5β,7α)-	14773-00-3	NI40900
271	55	386	353	81	95	105	93	404	100	75	70	49	47	43	39	37	10.10	25	40	4	–	–	–	–	–	–	–	–	–	–	Cholan-24-oic acid, 7-hydroxy-3-oxo-, methyl ester, (5β,7α)-	14773-00-3	MS09255
43	284	93	269	215	107	81	79	404	100	47	34	31	30	29	29	29	0.00	25	40	4	–	–	–	–	–	–	–	–	–	–	Pregnane-3,20-diol, diacetate, (3α,5β,20S)-	1174-69-2	NI40901
326	105	91	311	121	327	55	119	404	100	35	30	27	27	26	23	22	0.00	25	40	4	–	–	–	–	–	–	–	–	–	–	Pregnan-20-one, 3-(acetyloxy)-5-hydroxy-6,16-dimethyl-, (3β,5α,6β,16α)-	55530-52-4	WI01524
271	347	75	272	348	68	81	253	404	100	64	32	23	18	17	15	12	0.00	25	44	2	–	–	–	–	–	–	1	–	–	–	3α-tert-Butyldimethylsilyloxy-5α-androstan-17-one		NI40902
255	347	75	348	147	145	256	161	404	100	97	42	28	27	22	21	21	0.00	25	44	2	–	–	–	–	–	–	1	–	–	–	3α-tert-Butyldimethylsilyloxy-5β-androstan-17-one		NI40903
347	75	348	271	81	349	161	95	404	100	45	28	13	8	7	6	5	0.00	25	44	2	–	–	–	–	–	–	1	–	–	–	3β-tert-Butyldimethylsilyloxy-5α-androstan-17-one		NI40904
271	347	75	81	348	272	253	68	404	100	98	35	30	29	27	20	12	0.00	25	44	2	–	–	–	–	–	–	1	–	–	–	3β-tert-Butyldimethylsilyloxy-5β-androstan-17-one		NI40905
375	271	376	103	272	377	253	81	404	100	66	31	24	14	8	8	8	0.00	25	44	2	–	–	–	–	–	–	1	–	–	–	3α-Triethylsilyloxy-5α-androstan-17-one		NI40906
375	255	376	103	271	161	147	145	404	100	45	30	20	9	9	9	9	0.00	25	44	2	–	–	–	–	–	–	1	–	–	–	3α-Triethylsilyloxy-5β-androstan-17-one		NI40907
375	103	376	271	377	75	161	81	404	100	32	30	13	8	7	6	5	0.00	25	44	2	–	–	–	–	–	–	1	–	–	–	3β-Triethylsilyloxy-5α-androstan-17-one		NI40908
375	271	376	103	272	81	253	377	404	100	57	35	26	18	14	12	9	0.00	25	44	2	–	–	–	–	–	–	1	–	–	–	3β-Triethylsilyloxy-5β-androstan-17-one		NI40909
180	152	194	193	181	388	151	153	404	100	43	31	28	20	19	17	15	13.20	26	20	1	4	–	–	–	–	–	–	–	–	–	9,9′-Dimethyl-3,3′-azoxybiscarbazole	104563-30-6	NI40910
243	165	105	244	241	242	77	183	404	100	98	61	41	37	36	35	33	1.00	26	28	4	–	–	–	–	–	–	–	–	–	–	α-D-Allopyranoside, methyl 2,4-dideoxy-6-(triphenylmethyl)-		NI40911
112	143	117	115	278	372	312	253	404	100	25	20	18	12	11	8	8	6.00	26	28	4	–	–	–	–	–	–	–	–	–	–	1-Cyclohexene-1,4-dicarboxylic acid, 3-benzyl-4-(3-phenyl-1-propenyl)-, dimethyl ester		MS09256
183	105	77	165	43	273	256	243	404	100	43	17	7	7	4	2	2	0.00	26	28	4	–	–	–	–	–	–	–	–	–	–	Hexanoic acid, 3-hydroxy-, 2-hydroxy-1,2,2-triphenylethyl-, (1′R,3S)-	110744-11-1	DD01579
183	105	77	273	43	165	343	404	404	100	46	16	6	6	5	2	1		26	28	4	–	–	–	–	–	–	–	–	–	–	Pentanoic acid, 3-hydroxy-4-methyl-, 2-hydroxy-1,2,2-triphenylethyl ester, (1′R,3R)-	95061-49-7	DD01580
269	404	174	57	56	270	41	405	404	100	36	27	23	23	20	12	11		26	32	2	2	–	–	–	–	–	–	–	–	–	Aspidospermidin-21-ol, 17-methoxy-21-phenyl-	36459-04-8	NI40912
112	43	81	386	41	275	113	83	404	100	17	14	13	13	12	10	10	0.98	26	44	3	–	–	–	–	–	–	–	–	–	–	B-Nor-5,6-secocholestan-6-oic acid, 1-oxo-, (10β)-	50364-98-2	NI40913
112	386	275	247	360	358	371	319	404	100	15	14	5	3	2	1	1	1.20	26	44	3	–	–	–	–	–	–	–	–	–	–	B-Nor-5,6-secocholestan-6-oic acid, 1-oxo-, (10β)-	50364-98-2	NI40914
355	373	356	386	247	374	404	387	404	100	25	25	13	12	7	5	4		27	48	2	–	–	–	–	–	–	–	–	–	–	5α-Cholestane-2α,19-diol	28809-54-3	NI40915
355	373	386	247	404				404	100	25	13	12	5					27	48	2	–	–	–	–	–	–	–	–	–	–	5α-Cholestane-2α,19-diol	28809-54-3	LI08583
319	301	276	258	386	371	283	368	404	100	75	50	40	25	10	7	5	5.00	27	48	2	–	–	–	–	–	–	–	–	–	–	20-Iso-cholestane-3,20-diol, (3β,5α,20β)-		LI08584
355	373	233	404	317	386			404	100	60	14	9	8	4				27	48	2	–	–	–	–	–	–	–	–	–	–	Cholestane-1,19-diol, (1α,5α)-	28809-55-4	LI08585
355	373	356	374	233	404	317	386	404	100	60	25	15	14	9	8	4		27	48	2	–	–	–	–	–	–	–	–	–	–	Cholestane-1,19-diol, (1α,5α)-	28809-55-4	NI40916
386	231	232	55	43	81	371	95	404	100	53	45	43	43	38	32	32	1.20	27	48	2	–	–	–	–	–	–	–	–	–	–	Cholestane-3,5-diol, (3β,5α)-	3347-60-2	NI40917
403	385	367	383					404	100	28	22	5					0.00	27	48	2	–	–	–	–	–	–	–	–	–	–	Cholestane-3,5-diol, (3β,5α)-	3347-60-2	NI40918
403	385	367						404	100	28	28						0.00	27	48	2	–	–	–	–	–	–	–	–	–	–	Cholestane-3,5-diol, (3β,5β)-	570-97-8	NI40919
231	232	404	386	141	95	213	43	404	100	64	54	53	44	44	38	37		27	48	2	–	–	–	–	–	–	–	–	–	–	Cholestane-3,6-diol, (3β,5α,6α)-	41083-73-2	NI40920
404	95	386	43	231	55	81	57	404	100	96	76	75	65	56	47	46		27	48	2	–	–	–	–	–	–	–	–	–	–	Cholestane-3,6-diol, (3β,5α,6α,17α,20S)-	41083-77-6	NI40921
403	385	367						404	100	39	31						0.00	27	48	2	–	–	–	–	–	–	–	–	–	–	Cholestane-3,6-diol, (3β,5α,6β)-	570-85-4	NI40922
193	386	206	220	371	302	273	255	404	100	77	64	46	35	32	19	14	0.00	27	48	2	–	–	–	–	–	–	–	–	–	–	Cholestane-3,11-diol, (3β,5α,11α,14β)-		MS09257
193	386	273	371	206	220	353	255	404	100	72	62	57	57	32	17	14	0.00	27	48	2	–	–	–	–	–	–	–	–	–	–	Cholestane-3,11-diol, (3β,5α,11β,14β)-		MS09258
386	332	231	95	81	55	42	368	404	100	83	54	53	53	49	47	39	17.00	27	48	2	–	–	–	–	–	–	–	–	–	–	Cholestane-4,5-diol, (4β,5α)-	20233-47-0	LI08586
386	331	231	81	95	55	43	368	404	100	83	53	53	52	49	45	39	16.00	27	48	2	–	–	–	–	–	–	–	–	–	–	Cholestane-4,5-diol, (4β,5α)-	20233-47-0	NI40923
311	91	404	312	194	310	193	89	404	100	40	28	25	17	15	13	12		28	24	1	2	–	–	–	–	–	–	–	–	–	Benzeneacetamide, N-phenyl-α-[2-phenyl-1-(phenylamino)ethylidene]-	50297-58-0	MS09259
235	404	234	236	206	405	220	194	404	100	40	22	19	16	14	7	7		28	24	1	2	–	–	–	–	–	–	–	–	–	Spiro[3H-benz[g]indole-3,1′-cyclohexane]-2-carboxamide, N-1-naphthyl-	20944-74-5	NI40924
202	361	203	160	362	43	144	91	404	100	47	19	17	13	5	3	3	0.24	28	40	–	2	–	–	–	–	–	–	–	–	–	Aniline, N,N′-(1,2-dimethyl-1,2-ethanediylidene)bis[2,6-diisopropyl-	74663-77-7	NI40925
57	91	29	346	171	347	41	331	404	100	48	43	40	16	12	10	6	0.00	29	40	1	–	–	–	–	–	–	–	–	–	–	1,4-Epoxynaphthalene, 1,5,7-tris(tert-butyl)-1,2,3,4-tetrahydro-4-benzyl-	37754-73-7	MS09260

MASS TO CHARGE RATIOS	M.W.	INTENSITIES	Parent	C	H	O	N	S	F	Cl	Br	I	Si	P	B	X	COMPOUND NAME	CAS Reg No	No
69 328 326 96 257 255 253 247	405	100 45 45 17 11 11 11 11	0.00	5	2	–	1	–	9	–	2	–	–	–	–	–	3-[Bis(trifluoromethyl)amino]-2,3-dibromo-1,1,1-trifluoropropane		LI08587
121 278 248 65 254 183 56 186	405	100 63 45 43 37 34 33 29	1.00	5	5	1	1	–	–	–	–	2	–	–	–	1	Cyclopentadienylnitrosyldiiodoiron		LI08588
66 69 366 339 405 177 127 316	405	100 70 55 50 25 17 15 13		12	6	2	1	–	11	–	–	–	–	–	–	–	Bicyclo[2.2.1]hept-5-ene, 2,2-difluoro-3-(trifluoromethyl)-3-[5,5-bis(trifluoromethyl)-1,4,2-dioxazolin-3-yl]-	94637-19-1	NI40926
73 75 156 45 147 43 244 84	405	100 33 21 19 16 15 14 10	0.70	16	35	5	1	–	–	–	–	–	3	–	–	–	L-Glutamic acid, N-acetyl-N-(trimethylsilyl)-, bis(trimethylsilyl) ester	55517-49-2	NI40927
179 195 178 405 167 181 150 151	405	100 70 45 40 32 30 30 20		17	15	9	3	–	–	–	–	–	–	–	–	–	2,4,6-Trimethoxy-2′,4′,6′-trinitrostilbene		LI08589
187 365 147 367 118 405 91 407	405	100 48 35 31 30 27 21 18		18	17	2	5	–	–	2	–	–	–	–	–	–	N-Ethyl-N-(2-cyanoethyl)-3-methyl-4-(2,6-dichloro-4-nitrophenylazo)aniline		IC04309
100 77 114 56 105 58 55 42	405	100 60 40 40 35 35 25 25	11.00	18	19	2	3	3	–	–	–	–	–	–	–	–	Thieno[2,3-c]isothiazol-4-amine, 5-benzoyl-3-[2-(N-morpholinyl)ethylthio]	97090-74-9	NI40928
100 113 114 56 105 42 259 55	405	100 94 70 58 50 36 30 27	10.00	18	19	2	3	3	–	–	–	–	–	–	–	–	Thieno[3,2-d]isothiazol-4-amine, 5-benzoyl-3-[2-(N-morpholinyl)ethylthio]	97090-82-9	NI40929
326 405 407 327 406 296 265 390	405	100 44 41 22 21 15 15 13		19	20	4	1	–	–	–	1	–	–	–	–	–	Isoquinoline, 1-(2-bromo-4-methoxyphenyl)-5,6,7-trimethoxy-3,4-dihydro-		DD01581
258 216 174 87 129 345 147 300	405	100 55 48 26 22 19 17 15	4.70	19	23	7	3	–	–	–	–	–	–	–	–	–	1,2,3-Butanetriol, 4-methoxy-1-(2-phenyl-2H-1,2,3-triazol-4-yl)-, triacetate, [1R-(1R*,2R*,3R*)]-	35405-81-3	NI40930
188 285 230 217 212 272 242 189	405	100 15 6 6 5 4 3 2	1.50	19	23	7	3	–	–	–	–	–	–	–	–	–	1,2,3-Butanetriol, 4-methoxy-4-(2-phenyl-2H-1,2,3-triazol-4-yl)-, triacetate, [2R-(2R*,3S*,4R*)]-	35405-80-2	NI40931
258 216 174 87 345 129 147 300	405	100 55 48 26 22 22 16 15	4.70	19	23	7	3	–	–	–	–	–	–	–	–	–	N-Phenylosotriazole-6-O-methyl-D-galactose, acetate		LI08590
188 285 230 217 212 272 242 271	405	100 15 6 6 5 4 3 2	1.50	19	23	7	3	–	–	–	–	–	–	–	–	–	N-Phenylosotriazole-3-O-methyl-D-glucose, acetate		LI08591
73 284 374 89 218 103 170 405	405	100 99 61 55 41 40 38 36		19	27	5	3	–	–	–	–	–	1	–	–	–	2,4,6(1H,3H,5H)-Pyrimidinetrione, 5-[2-(methoxyimino)-3-[(trimethylsilyl)oxy]propyl]-1,3-dimethyl-5-phenyl-	57305-03-0	NI40932
73 315 202 405 74 201 273 243	405	100 32 22 11 10 9 8 7		19	35	1	3	–	–	–	–	–	3	–	–	–	1H-Indole-3-acetic acid, bis(trimethylsilyl)hydrazide, trimethylsilyl-	74367-08-1	NI40933
117 73 132 75 355 145 129 55	405	100 96 83 80 62 56 41 34	0.00	19	35	1	3	–	–	–	–	–	3	–	–	–	1H-Indole-3-acetic acid, bis(trimethylsilyl)hydrazide, trimethylsilyl-	74367-08-1	NI40934
58 84 56 100 86 289 144 275	405	100 41 40 32 27 22 22 21	1.60	19	39	6	3	–	–	–	–	–	–	–	–	–	22-Methyl-20,24-dihydroxy-4,7,13,16-tetraoxa-1,10,22-triazabicyclo[8.8.7]pentacosane		MS09261
390 405 205 391 161 43 406 133	405	100 52 30 23 20 14 13 9		20	19	3	7	–	–	–	–	–	–	–	–	–	N,N-Diethyl-3-acetamido-4-(3,5-dicyano-4-nitrophenylazo)aniline		IC04310
97 95 109 125 111 99 96 149	405	100 84 74 58 56 43 41 37	0.00	21	24	5	1	–	–	1	–	–	–	–	–	–	B-Homomorphinan-7-one, 5-chloro-5,6,8,14-tetradehydro-3-hydroxy-2,4,6 trimethoxy-17-methyl-, (9α,13α)-	56772-02-2	NI40935
258 257 405 259 346 242 227 243	405	100 76 63 63 50 48 41 15		21	27	7	1	–	–	–	–	–	–	–	–	–	Oxostephamiersine, dihydro-	51805-62-0	NI40936
105 77 302 300 51 76 78 106	405	100 71 59 59 23 18 15 13	9.00	22	16	2	1	–	–	–	1	–	–	–	–	–	1-(4-Bromanilino)-1,2-dibenzoylethylene		MS09262
363 405 361 334 319 262 263 304	405	100 86 54 54 52 39 38 37		22	19	5	3	–	–	–	–	–	–	–	–	–	Camptothecin, 12-acetamido-	58546-24-0	NI40937
91 405 81 55 362 65 77 296	405	100 25 18 17 13 13 12 11		22	22	1	1	1	3	–	–	–	–	–	–	–	Spiro[2,5-dihydro-1,3-oxazole-2,1′-cyclohexane], 4-(benzylthio)-5-phenyl-5-(trifluoromethyl)-		DD01582
405 232 59 77 406 105	405	100 25 23 20 19 16		22	22	2	2	–	–	–	–	–	–	–	–	1	Cobalt, [[3,3′-(1,2-ethanediyldinitrilo)bis[1-phenyl-1-butanonato]](2-)-N,N′,O,O′]-	36466-12-3	NI40938
87 91 72 59 105 157 171 185	405	100 65 65 65 46 34 24 18	16.00	22	31	6	1	–	–	–	–	–	–	–	–	–	Icacine	64999-57-1	NI40939
326 214 228 270 407 405 364 362	405	100 33 27 18 14 14 12 12		22	32	1	1	–	–	–	1	–	–	–	–	–	Spiro[cyclohexane-1,3′(2′H)-isoquinolin]-1′-one, 4′-bromo-1′,4′-dihydro-2′ butyl-4-tert-butyl-	120637-33-4	DD01583
73 374 75 41 79 91 67 55	405	100 99 99 80 62 60 60 60	14.96	23	39	3	1	–	–	–	–	–	1	–	–	–	Androstane-11,17-dione, 3-[(trimethylsilyl)oxy]-, 17-(O-methyloxime), (3α,5α)-	57305-07-4	NI40940
73 300 261 75 315 284 79 269	405	100 99 95 92 80 47 42 40	34.96	23	39	3	1	–	–	–	–	–	1	–	–	–	Androstane-11,17-dione, 3-[(trimethylsilyl)oxy]-, 17-(O-methyloxime), (3α,5β)-	57305-06-3	NI40941
261 315 300 284 269 405 254 316	405	100 92 91 42 42 25 23 20		23	39	3	1	–	–	–	–	–	1	–	–	–	Androstane-11,17-dione, 3-[(trimethylsilyl)oxy]-, 17-(O-methyloxime), (3α,5β)-	57305-06-3	NI40942
348 349 290 405 333 318 304 406	405	100 25 16 13 8 8 7 6		24	23	5	1	–	–	–	–	–	–	–	–	–	Acetone, 1-(12,13-dihydro-1,2-dimethoxy-12-methyl[1,3]benzodioxolo[5,6-c]phenanthridin-11-yl)-	56051-44-6	NI40943
58 71 72 121 107 93 91 79	405	100 71 8 1 1 1 1 1	0.12	24	39	4	1	–	–	–	–	–	–	–	–	–	Acetic acid, (dodecahydro-7-hydroxy-1,4b,8,8-tetramethyl-10-oxo-2(1H)-phenanthrenylidene)-, 2-(dimethylamino)ethyl ester, [1R-(1α,2E,4aα,4bβ,7β,8aα,10aβ)]-	468-76-8	NI40944
103 58 274 131 79 146 160 260	405	100 96 71 58 35 21 19 14	11.00	25	31	2	3	–	–	–	–	–	–	–	–	–	1,5,9-Triazacyclotridecan-4-one, 9-(1-oxo-3-phenyl-2-propenyl)-2-phenyl-, (Z)-(–)-	53990-48-0	NI40945
405 375 359 178 78 41 43 283	405	100 32 27 26 25 23 20 19		26	19	2	3	–	–	–	–	–	–	–	–	–	Indolizine, 1,2-pentamethylene-3-(nitrophenylamino)-		MS09263
100 85 406 83 86 72 408 407	405	100 38 27 20 18 13 9 8	3.00	26	28	1	1	–	–	1	–	–	–	–	–	–	Ethanamine, 2-[4-(2-chloro-1,2-diphenylethenyl)phenoxy]-N,N-diethyl-	911-45-5	NI40946
105 77 300 44 41 106 51 55	405	100 47 15 11 11 10 10 8	1.70	26	31	3	1	–	–	–	–	–	–	–	–	–	2,N-Dibenzoyl-12-dodecane lactam	102222-12-8	NI40947
405 390 406 391 130 182 168 144	405	100 80 30 22 17 11 10 7		27	35	2	1	–	–	–	–	–	–	–	–	–	1′H-Androst-16-eno[17,16-b]indol-3-ol, acetate, (3β,5α)-	39987-70-7	MS09264
405 370 406 371 130 168 182 159	405	100 77 30 21 16 10 9 7		27	35	2	1	–	–	–	–	–	–	–	–	–	1′H-Androst-16-eno[17,16-g]indol-3-ol, acetate, (3β,5α)-	56196-20-4	NI40948
348 334 405 349 321 322 335 406	405	100 91 54 43 43 41 26 19		30	31	–	1	–	–	–	–	–	–	–	–	–	Benzo[b]triphenylene, 9,14-imino-15-tert-butyl-9,14-dihydro-10,11,12,13-tetramethyl-	110028-23-4	DD01584

MASS TO CHARGE RATIOS								M.W.	INTENSITIES								Parent	C	H	O	N	S	F	Cl	Br	I	Si	P	B	X	COMPOUND NAME	CAS Reg No	No
230	406	202	118	87	174	123	151	**406**	100	72	63	61	61	59	52	42		8	–	8	–	–	–	–	–	–	–	–	–	3	**Bis(tetracarbonylcobalt)zinc**		**MS09265**
373	375	377	371	272	379	274	237	**406**	100	96	51	44	21	17	17	16	**0.27**	10	6	–	–	–	–	8	–	–	–	–	–	–	**γ-Chlordane**	5566-34-7	**NI40949**
373	375	377	371	101	89	68	339	**406**	100	92	52	43	42	41	39	23	**0.00**	10	6	–	–	–	–	8	–	–	–	–	–	–	**γ-Chlordane**	5566-34-7	**NI40950**
100	373	375	272	237	274	238	339	**406**	100	56	50	45	37	35	35	34	**11.97**	10	6	–	–	–	–	8	–	–	–	–	–	–	**γ-Chlordane**	5566-34-7	**AI02337**
39	65	75	27	66	49	375	373	**406**	100	64	45	45	36	31	28	28	**0.40**	10	6	–	–	–	–	8	–	–	–	–	–	–	**4,7-Methano-1H-indene, 1,2,4,5,6,7,8,8-octachloro-2,3,3a,4,7,7a-hexahydro-, (1α,2α,3aα,4β,7β,7aα)-**	5103-71-9	**NI40952**
39	65	75	373	27	375	49	66	**406**	100	63	45	45	43	41	36	32	**0.80**	10	6	–	–	–	–	8	–	–	–	–	–	–	**4,7-Methano-1H-indene, 1,2,4,5,6,7,8,8-octachloro-2,3,3a,4,7,7a-hexahydro-, (1α,2α,3aα,4β,7β,7aα)-**	5103-71-9	**NI40951**
373	375	377	237	371	272	239	274	**406**	100	99	53	50	45	42	36	32	**0.00**	10	6	–	–	–	–	8	–	–	–	–	–	–	**4,7-Methano-1H-indene, 1,2,4,5,6,7,8,8-octachloro-2,3,3a,4,7,7a-hexahydro-, (1α,2α,3aα,4β,7β,7aα)-**	5103-71-9	**NI40953**
373	375	377	371	65	75	66	237	**406**	100	96	50	45	43	34	29	26	**0.00**	10	6	–	–	–	–	8	–	–	–	–	–	–	**4,7-Methano-1H-indene, 1,2,4,5,6,7,8,8-octachloro-2,3,3a,4,7,7a-hexahydro-, (1α,2β,3aα,4β,7β,7aα)-**	5103-74-2	**NI40954**
373	375	377	371	272	237	379	274	**406**	100	94	52	45	19	18	16	14	**2.11**	10	6	–	–	–	–	8	–	–	–	–	–	–	**4,7-Methano-1H-indene, 1,2,4,5,6,7,8,8-octachloro-2,3,3a,4,7,7a-hexahydro-, (1α,2β,3aα,4β,7β,7aα)-**	5103-74-2	**NI40955**
373	375	237	109	272	119	377	75	**406**	100	96	64	53	50	48	47	45	**0.00**	10	6	–	–	–	–	8	–	–	–	–	–	–	**4,7-Methano-1H-indene, 1,2,4,5,6,7,8,8-octachloro-2,3,3a,4,7,7a-hexahydro-, (1α,2β,3aα,4β,7β,7aα)-**	5103-74-2	**NI40956**
294	296	292	266	406	322	268	320	**406**	100	90	84	70	68	66	65	60		10	6	6	–	–	–	–	–	–	–	–	–	1	**Tungsten, (methoxy-1-propynylmethylene)pentacarbonyl-**		**NI40957**
391	392	147	73	259	240	167	43	**406**	100	21	10	10	7	4	4	4	**1.00**	10	21	4	–	–	6	–	–	–	2	1	–	–	**Methyl (trimethylsilyl) [2,2,2-trifluoro-1-(trifluoromethyl)-1-(trimethylsiloxy)ethyl]phosphonate**		**MS09266**
410	408	205	275	204	203	174	172	**406**	100	94	76	71	64	61	61	59	**46.00**	12	1	1	–	–	–	7	–	–	–	–	–	–	**Dibenzofuran, 1,2,3,4,6,7,8-heptachloro-**	67562-39-4	**NI40958**
389	391	387	393	390	355	392	353	**406**	100	75	55	40	13	12	10	9	**1.10**	12	1	1	–	–	–	7	–	–	–	–	–	–	**Dibenzofuran, 1,2,3,4,6,8,9-heptachloro-**	69698-58-4	**NI40959**
64	183	91	38	48	67	43	30	**406**	100	84	53	48	32	29	29	27	**0.00**	12	11	8	4	1	–	1	–	–	–	–	–	–	**Cyclopenta[e]-1,2,3-thiadiazine, 2-(2,4-dinitrophenyl)-2,5,6,7-tetrahydro-, monoperchlorate**	57954-46-8	**NI40960**
347	220	245	203	161	178	143	101	**406**	100	84	17	13	6	4	4	3	**0.00**	12	18	13	–	–	–	–	–	–	–	–	–	4	**Beryllium, hexakis[μ-(acetato-O:O')]-μ⁴-oxotetra-**	19049-40-2	**NI40961**
347	220	245	178	161	118	101	278	**406**	100	47	13	5	5	3	3	1	**0.00**	12	18	13	–	–	–	–	–	–	–	–	–	4	**Beryllium, hexakis[μ-(acetato-O:O')]-μ⁴-oxotetra-**	19049-40-2	**LI08592**
347	222	248	203	145	182	164	144	**406**	100	99	54	36	20	18	16	14	**0.00**	12	18	13	–	–	–	–	–	–	–	–	–	4	**Beryllium, hexakis[μ-(acetato-O:O')]-μ⁴-oxotetra-**	19049-40-2	**MS09267**
347	220	245	203	178	161	143	101	**406**	100	76	14	13	6	6	5	3	**0.00**	13	22	12	–	–	–	–	–	–	–	–	–	4	**Beryllium, pentakis(acetato)oxopropoxytetra-**	39335-45-0	**NI40962**
157	216	63	107	60	77	51	64	**406**	100	95	70	44	44	40	34	27	**1.00**	14	16	2	–	4	–	–	–	–	–	2	–	–	**O,O'-Ethylenebis(phenyldithiophosphonate)**	94286-60-9	**NI40963**
205	162	264	232	219	206	188	247	**406**	100	74	73	59	50	42	40	39	**36.00**	15	18	9	–	2	–	–	–	–	–	–	–	–	**1,3-Dithiane-4,6-dicarboxylic acid, 5-(4-methoxyphenyl)-, 1,1,3,3-tetraoxide, dimethyl ester**		**NI40964**
138	51	43	180	390	92	45	358	**406**	100	99	45	41	36	32	31	28	**10.66**	16	14	7	4	1	–	–	–	–	–	–	–	–	**2,2'-Dinitro-4,4'-diacetamidodiphenyl sulphoxide**		**NI40965**
57	193	179	69	321	85	391	81	**406**	100	71	71	71	49	41	31	26	**15.71**	16	17	7	2	–	3	–	–	–	–	–	–	–	**Propanoic acid, 2,2-dimethyl-, [2,3,3a,9a-tetrahydro-6-oxo-3-[(trifluoroacetyl)oxy]-6H-furo[2',3':4,5]oxazolo[3,2-a]pyrimidin-2-yl]methyl ester, [2R-(2α,3β,3aβ,9aβ)]-**	40732-71-6	**NI40966**
57	96	406	69	137	85	177	81	**406**	100	96	76	68	67	64	51	36		16	17	7	2	–	3	–	–	–	–	–	–	–	**Propanoic acid, 2,2-dimethyl-, 2,3,3a,9a-tetrahydro-6-oxo-2-[[(trifluoroacetyl)oxy]methyl]-6H-furo[2',3':4,5]oxazolo[3,2-a]pyrimidin-3-yl ester, [2R-(2α,3β,3aβ,9aβ)]-**	40732-72-7	**NI40967**
142	226	143	185	186	184	144	244	**406**	100	81	72	59	50	47	32	10	**0.00**	16	22	10	–	1	–	–	–	–	–	–	–	–	**α-D-Galactofuranose, 1,2,3,5,6-penta-O-acetyl-4-thio-**	118918-79-9	**DD01585**
142	186	226	144	143	159	244	184	**406**	100	76	74	69	65	53	52	52	**0.00**	16	22	10	–	1	–	–	–	–	–	–	–	–	**β-D-Galactopyranose, 1,2,3,5,6-penta-O-acetyl-4-thio-**	118918-80-2	**DD01586**
43	184	243	185	244	245	226	345	**406**	100	68	66	46	44	10	7	4	**0.00**	16	22	10	–	1	–	–	–	–	–	–	–	–	**α-D-Galactopyranose, 1,2,3,6-tetra-O-acetyl-4-S-acetyl-4-thio-**	118918-81-3	**DD01587**
43	169	28	109	127	44	97	42	**406**	100	23	23	19	7	6	5	5	**0.00**	16	22	10	–	1	–	–	–	–	–	–	–	–	**β-Glucose, 1-sulphinyl-, pentacetate**	19879-84-6	**NI40968**
406	56	83	86	129	128	85	70	**406**	100	54	48	44	37	37	35	35		16	30	4	4	2	–	–	–	–	–	–	–	–	**4,7,13,16-Tetraoxa-1,10,20,23-tetraazabicyclo[8.8.6]tetracosa-19,24-dithione**		**MS09268**
55	111	86	56	57	70	60	69	**406**	100	67	59	59	48	41	33	26	**18.20**	16	30	4	4	2	–	–	–	–	–	–	–	–	**6,14,17,22-Tetraoxa-1,3,9,11-tetraazabicyclo[9.8.5]tetracosane-2,10-dithione**	110228-75-6	**MS09269**
55	86	56	57	70	60	69	58	**406**	100	59	59	48	41	33	26	26	**18.14**	16	30	4	4	2	–	–	–	–	–	–	–	–	**6,14,17,22-Tetraoxa-1,3,9,11-tetraazabicyclo[9.8.5]tetracosane-2,10-dithione**	110228-75-6	**NI40969**
73	274	147	205	117	275	89	45	**406**	100	59	38	30	25	15	10	10	**1.50**	16	34	6	–	–	–	–	–	–	3	–	–	–	**L-Ascorbic acid, 2-O-methyl-3,5,6-tris-O-(trimethylsilyl)-**	57397-19-0	**NI40970**
174	73	219	188	200	147	75	100	**406**	100	48	21	21	13	12	12	10	**6.00**	16	42	2	2	–	–	–	–	–	4	–	–	–	**Butanoic acid, 2,4-diamino-, N,N',N'-tris(trimethylsilyl)-, trimethylsilyl ester**		**MS09270**
43	129	87	189	99	127	85	57	**406**	100	66	35	28	25	22	19	15	**0.00**	17	26	11	–	–	–	–	–	–	–	–	–	–	**Galactose, 3-O-methyl-, pentaacetate**	56193-01-2	**NI40971**
43	129	87	99	85	189	127	100	**406**	100	56	26	19	19	18	16	12	**0.00**	17	26	11	–	–	–	–	–	–	–	–	–	–	**D-Glucitol, 4-O-methyl-, pentaacetate**	56083-46-6	**NI40972**
43	129	87	99	85	189	100	159	**406**	100	58	25	20	19	18	11	7	**0.01**	17	26	11	–	–	–	–	–	–	–	–	–	–	**D-Glucitol, 4-O-methyl-, pentaacetate**	56083-46-6	**LI08593**
43	45	115	125	157	87	113	102	**406**	100	99	74	72	55	40	38	38	**0.00**	17	26	11	–	–	–	–	–	–	–	–	–	–	**β-glycero-D-Glucoheptopyranoside, methyl 2,3,4,6-tetra-O-acetyl-7-O-methyl-**	84563-96-2	**NI40973**

MASS TO CHARGE RATIOS								M.W.	INTENSITIES								Parent	C	H	O	N	S	F	Cl	Br	I	Si	P	B	X	COMPOUND NAME	CAS Reg No	No
43	117	87	129	169	171	70	69	**406**	100	99	50	48	36	29	29	29	**0.00**	17	26	11	–	–	–	–	–	–	–	–	–	–	β-glycero-D-Glucoheptopyranoside, methyl 2,3,4,7-tetra-O-acetyl-6-O-methyl-	84563-95-1	NI40974
43	87	129	142	100	69	74	125	**406**	100	99	96	80	52	38	35	34	**0.00**	17	26	11	–	–	–	–	–	–	–	–	–	–	β-glycero-D-Glucoheptopyranoside, methyl 2,3,6,7-tetra-O-acetyl-4-O-methyl-	84563-94-0	NI40975
43	75	74	116	129	113	69	45	**406**	100	99	74	36	18	17	15	14	**0.00**	17	26	11	–	–	–	–	–	–	–	–	–	–	β-glycero-D-Glucoheptopyranoside, methyl 2,4,6,7-tetra-O-acetyl-3-O-methyl-	84563-93-9	NI40976
43	74	125	116	129	142	184	87	**406**	100	99	72	59	52	48	43	35	**0.00**	17	26	11	–	–	–	–	–	–	–	–	–	–	β-glycero-D-Glucoheptopyranoside, methyl 3,4,6,7-tetra-O-acetyl-2-O-methyl-	84563-92-8	NI40977
73	75	147	261	77	129	287	74	**406**	100	66	31	17	16	11	10	9	**0.00**	17	38	5	–	–	–	–	–	–	3	–	–	–	Pentanedioic acid, 3-ethyl-2-methyl-3-[(trimethylsilyl)oxy]-, bis(trimethylsilyl) ester		NI40978
73	75	147	261	129	77	74	262	**406**	100	54	32	30	23	11	10	6	**0.00**	17	38	5	–	–	–	–	–	–	3	–	–	–	Pentanedioic acid, 2,3,4-trimethyl-3-[(trimethylsilyl)oxy]-, bis(trimethylsilyl) ester		NI40979
69	181	225	227	183	139	75	89	**406**	100	98	64	53	37	34	33	29	**1.30**	18	12	4	–	–	–	1	1	–	–	–	–	–	1,3,4,6-Hexanetetrone, 1-(4-bromophenyl)-6-(4-chlorophenyl)-	58330-15-7	NI40980
163	243	241	406	103	404	164	134	**406**	100	51	25	21	13	11	11	11		18	18	4	2	–	–	–	–	–	–	–	–	1	3,5-Bis(3,5-dimethoxyphenyl)-1,2,4-selenadiazole		MS09271
361	133	362	119	104	105	77	406	**406**	100	45	21	18	13	12	11	10		18	22	5	4	1	–	–	–	–	–	–	–	–	N-Ethyl-N-methoxyethyl-4-(2-methylsulphonyl-4-nitrophenylazo)aniline		IC04311
73	160	75	246	218	45	147	289	**406**	100	33	17	15	13	13	12	9	**7.00**	18	26	5	2	–	–	–	–	–	2	–	–	–	Phthalylglycylglycine, bis(trimethylsilyl)-		NI40981
126	127	257	82	359	55	43	42	**406**	100	8	7	3	2	2	2	2	**1.60**	18	34	6	2	1	–	–	–	–	–	–	–	–	Lincomycin		LI08594
406	292	346	306	374	408	294	319	**406**	100	95	73	65	55	35	32	31		19	15	8	–	–	–	1	–	–	–	–	–	–	Methyl 2-chloro-4-formyl-3,8-dihydroxy-1,6,9-trimethyl-11-oxo-11H-dibenzo[b,e][1,4]-dioxepin-7-carboxylate		NI40982
77	187	131	130	103	129	188	102	**406**	100	99	99	99	81	58	46	40	**2.00**	19	16	4	2	–	–	2	–	–	–	–	–	–	L-Tryptophan, N-[(2,4-dichlorophenoxy)acetyl]-	50649-06-4	NI40983
77	187	130	103	129	188	162	102	**406**	100	99	99	81	58	46	40	40	**2.00**	19	16	4	2	–	–	2	–	–	–	–	–	–	L-Tryptophan, N-[(2,4-dichlorophenoxy)acetyl]-	50649-06-4	NI40984
69	48	45	46	74	43	116	47	**406**	100	43	35	22	20	19	18	18	**5.00**	19	18	1	8	1	–	–	–	–	–	–	–	–	2H-1,2,4-Triazol-5-amine, 2,3-dihydro-3-[6-methyl-3-(methylthio)-5-oxo-4,5-dihydro-1,2,4-triazin-4-yl]imino-4-phenyl-N-phenyl-		MS09272
167	166	41	139	95	29	70	94	**406**	100	74	23	18	17	14	10	9	**1.40**	19	29	4	2	–	3	–	–	–	–	–	–	–	Cyclohexanecarboxylic acid, N-(trifluoroacetyl)-L-prolyl-trans-4-(aminomethyl)-, butyl ester		MS09273
44	121	406	105	257	244	117	149	**406**	100	28	19	17	9	8	8	7		20	14	2	4	2	–	–	–	–	–	–	–	–	Anhydro-1-hydroxy-2-phenyl-4-methyl-5-oxo-7-anilinothiazolo[3,2-a]thiazolo[5,4-d]pyrimidinium hydroxide	86999-03-3	NI40985
77	371	57	41	43	55	44	406	**406**	100	70	67	63	60	52	43	41		20	15	2	6	–	–	1	–	–	–	–	–	–	3(2H)-Pyridazinone, 4-chloro-2-phenyl-5-[(1-phenyl-4-amino-6-oxo-(5H)-pyridazin-5-yl)amino]-		NI40986
55	45	366	59	54	104	77	87	**406**	100	81	78	70	59	53	45	43	**31.00**	20	18	4	6	–	–	–	–	–	–	–	–	–	N-Cyanoethyl-N-(carbomethoxyethyl)-4-(2-cyano-4-nitrophenylazo)aniline		IC04312
195	197	212	406	152	137	181	167	**406**	100	62	40	21	12	10	8	8		20	22	9	–	–	–	–	–	–	–	–	–	–	Methyl-6-(O-trimethylgalloyl)-2,4-dimethoxybenzoate		NI40987
225	151	165	255	223	209	224	181	**406**	100	46	40	39	23	21	18	18	**16.40**	21	26	8	–	–	–	–	–	–	–	–	–	–	3-(3,4-Dimethoxyphenyl)-1-(2-hydroxy-3,4-dimethoxyphenyl)-2,2-dimethoxypropan-1-one		NI40988
43	135	302	405	53	110	45	40	**406**	100	65	35	34	28	26	21	21	**7.29**	21	28	7	1	–	–	–	–	–	–	–	–	–	Senecionanium, 12-(acetyloxy)-14,15,20,21-tetradehydro-15,20-dihydro-8-hydroxy-4-methyl-11,16-dioxo-, (8ξ,12β,14Z)-	33979-15-6	NI40989
206	72	190	44	191	162	277	161	**406**	100	84	62	39	30	25	23	21	**2.19**	21	34	4	4	–	–	–	–	–	–	–	–	–	Glycinamide, L-valyl-N-[1-(1,2,3,4-tetrahydro-6,7-dimethoxy-2-methyl-1-isoquinolinyl)ethyl]-	53098-67-2	NI40990
43	56	406	41	76	104	44	135	**406**	100	71	61	51	48	41	30	28		22	14	8	–	–	–	–	–	–	–	–	–	–	2,2′-Binaphthalene-1,1′,4,4′-tetrone, 3,3′-dihydroxy-5,5′-dimethoxy-	88381-88-8	NI40991
135	406	77	134	44	350	51	106	**406**	100	96	58	48	40	37	35	29		22	14	8	–	–	–	–	–	–	–	–	–	–	2,2′-Binaphthalene-1,1′,4,4′-tetrone, 3,3′,5,5′-tetrahydroxy-7,7′-dimethyl-	104505-80-8	NI40992
406	350	135	407	134	84	77	390	**406**	100	49	37	31	25	25	24	16		22	14	8	–	–	–	–	–	–	–	–	–	–	2,2′-Binaphthalene-1,1′,4,4′-tetrone, 3,3′,8,8′-tetrahydroxy-6,6′-dimethyl-	88381-85-5	NI40993
56	406	41	84	119	54	117	350	**406**	100	91	85	81	72	67	66	61		22	14	8	–	–	–	–	–	–	–	–	–	–	2,2′-Binaphthalene-1,1′,4,4′-tetrone, 3,3′,8,8′-tetrahydroxy-6,6′-dimethyl-	88381-85-5	NI40994
57	43	406	71	391	83	41	85	**406**	100	88	83	81	75	70	53	51		22	14	8	–	–	–	–	–	–	–	–	–	–	2,2′-Binaphthalene-1,1′,4,4′-tetrone, 5,5′,8,8′-tetrahydroxy-3,3′-dimethyl-	104505-75-1	NI40995
150	57	43	71	44	122	85	169	**406**	100	97	67	60	53	48	43	40	**0.00**	22	14	8	–	–	–	–	–	–	–	–	–	–	2,2′-Binaphthalene-1,1′,4,4′-tetrone, 5,5′,8,8′-tetrahydroxy-6,6′-dimethyl-	104505-76-2	NI40996
93	77	406	65	107	121	122	109	**406**	100	55	31	26	21	19	16	13		22	22	4	4	–	–	–	–	–	–	–	–	–	2-(2,5-Dimethoxyphenylazo)-4-methoxy-5-(4-hydroxyphenyl)toluene		IC04313
164	149	406	41	186	105	91	57	**406**	100	80	74	20	12	12	8	8		22	30	2	–	–	–	–	–	–	–	–	–	1	2,2′-Selenobis(4-methyl-6-tert-butylphenol)		LI08595
43	41	55	209	388	91	44	123	**406**	100	83	68	61	49	48	48	35	**1.40**	22	30	7	–	–	–	–	–	–	–	–	–	–	Acetylcastelanolide		LI08596
135	216	136	147	346	262	151	127	**406**	100	88	67	58	33	36	33	24	**3.27**	22	30	7	–	–	–	–	–	–	–	–	–	–	Acetyl tetrahydrosubacaulin		NI40997
151	152	254	189	137	153	237	219	**406**	100	68	13	9	8	7	4	4	**1.00**	22	30	7	–	–	–	–	–	–	–	–	–	–	Carinol dimethyl ether		NI40998
300	346	241	328	356	239	374	342	**406**	100	78	78	74	66	57	56	56	**6.00**	22	30	7	–	–	–	–	–	–	–	–	–	–	Gibberellin A_{23} methyl ester		MS09274
179	278	99	222	152	207	246	225	**406**	100	71	58	55	37	33	32	27	**0.50**	22	30	7	–	–	–	–	–	–	–	–	–	–	Methyl 5β,7α-dihydroxy-11-ketotetranor-prostanoate diacetate		LI08597
179	83	121	328	310	208	43	60	**406**	100	64	61	60	47	46	41	40	**0.46**	22	30	7	–	–	–	–	–	–	–	–	–	–	Phorbol-12-monoacetate		NI40999
215	216	109	83	91	69	53	233	**406**	100	89	85	66	63	63	60	58	**0.06**	22	30	7	–	–	–	–	–	–	–	–	–	–	Phorbol-13-monoacetate		NI41000
167	406	121	165	166	153	107	105	**406**	100	53	48	47	42	27	27	27		22	30	7	–	–	–	–	–	–	–	–	–	–	Picras-2-ene-1,16-dione, 13-hydroxy-2-methoxy-11,12-[methylenebis(oxy)]-, (11α,12β)-	35334-40-8	NI41001

MASS TO CHARGE RATIOS								M.W.	INTENSITIES								Parent	C	H	O	N	S	F	Cl	Br	I	Si	P	B	X	COMPOUND NAME	CAS Reg No	No
359	121	152	165	153	331	151	105	**406**	100	84	73	70	59	50	48	34	**4.00**	22	30	7	–	–	–	–	–	–	–	–	–	–	**Picras-2-ene-1,16-dione, 13-hydroxy-2-methoxy-11,12-[methylenebis(oxy)]-, (11α,12β)-**	35334-40-8	**NI41002**
135	85	147	262	136	83	237	304	**406**	100	75	54	39	38	29	26	25	**0.00**	22	30	7	–	–	–	–	–	–	–	–	–	–	**Tetrahydroberlandin**		**NI41003**
313	334	272	349	346	291	133	254	**406**	100	61	28	23	12	12	12	11	**6.00**	23	31	5	–	–	1	–	–	–	–	–	–	–	**Androst-4-en-6-one, 17,19-bis(acetyloxy)-3-fluoro-, (3β,17β)-**	33547-24-9	**NI41004**
297	318	272	349	346	275	133	254	**406**	100	61	28	23	12	12	12	11	**0.00**	23	31	5	–	–	1	–	–	–	–	–	–	–	**Androst-4-en-6-one, 17,19-bis(acetyloxy)-3-fluoro-, (3β,17β)-**	33547-24-9	**LI08598**
43	334	406	333	335	273	386	179	**406**	100	99	40	27	25	24	23	21		23	31	5	–	–	1	–	–	–	–	–	–	–	**Androst-5-en-7-one, 17,19-bis(acetyloxy)-3-fluoro-, (3β,17β)-**	33708-70-2	**NI41005**
334	406	333	273	386	179	180	159	**406**	100	40	27	24	23	21	11	9		23	31	5	–	–	1	–	–	–	–	–	–	–	**Androst-5-en-7-one, 17,19-bis(acetyloxy)-3-fluoro-, (3β,17β)-**	33708-70-2	**LI08599**
334	406	386	179	180	159	192	147	**406**	100	40	23	21	11	9	8	7		23	31	5	–	–	1	–	–	–	–	–	–	–	**Androst-5-en-7-one, 17,19-bis(acetyloxy)-3-fluoro-, (3β,17β)-**	33708-70-2	**NI41006**
271	406	378	133	132	303	145	187	**406**	100	15	9	9	8	7	6	3		23	34	2	–	2	–	–	–	–	–	–	–	–	**19-Norandrost-5(6)-ene-2-spiro-2′-(1′,3′-dithian)-3-one, 17β-hydroxy-4,4-dimethyl-**		**LI08600**
69	57	41	81	55	110	77	252	**406**	100	80	67	63	62	28	27	8	**0.00**	23	34	4	–	1	–	–	–	–	–	–	–	–	**2-Cyclopenten-1-ol, 4-tert-butoxy-1-[1-(phenylsulphonyl)-2-octenyl]-, (1RS,1′RS,2′E,4SR)-**	119068-49-4	**DD01588**
69	57	81	41	55	110	53	77	**406**	100	90	67	64	61	35	33	26	**0.00**	23	34	4	–	1	–	–	–	–	–	–	–	–	**2-Cyclopenten-1-ol, 4-tert-butoxy-1-[1-(phenylsulphonyl)-2-octenyl]-, (1RS,1′SR,2′E,4SR)-**	119011-06-2	**DD01589**
43	95	328	268	240	81	329	388	**406**	100	93	69	34	25	17	16	15	**3.00**	23	34	6	–	–	–	–	–	–	–	–	–	–	**Androstan-17-one, 3,14-bis(acetyloxy)-5-hydroxy-, (3β,5α,14β)-**	41853-31-0	**NI41007**
207	239	221	85	208	83	375	47	**406**	100	35	18	17	15	15	14	10	**2.00**	23	34	6	–	–	–	–	–	–	–	–	–	–	**1,2,4-Benzenetricarboxylic acid, 4-dodecyl dimethyl ester**	33975-29-0	**NI41008**
318	43	96	275	248	319	55	81	**406**	100	69	45	32	30	26	23	20	**3.00**	23	34	6	–	–	–	–	–	–	–	–	–	–	**Digacetigenin, dihydro-**		**MS09275**
370	217	352	147	199	145	260	281	**406**	100	95	60	44	43	43	38	36	**3.00**	23	34	6	–	–	–	–	–	–	–	–	–	–	**Digoxigenin, 7β-hydroxy-**	72908-80-6	**NI41009**
43	55	287	41	269	229	81	95	**406**	100	33	24	21	20	20	19	18	**3.83**	23	34	6	–	–	–	–	–	–	–	–	–	–	**Pregnane-3,20-dione, 21-(acetyloxy)-11,17-dihydroxy-**	56193-59-0	**NI41010**
208	43	120	138	105	97	91	79	**406**	100	89	62	30	26	26	26	26	**14.67**	23	34	6	–	–	–	–	–	–	–	–	–	–	**Pregn-5-en-20-one, 12-(acetyloxy)-3,8,14-trihydroxy-, (3β,12β,14β)-**	55955-56-1	**NI41011**
307	74	247	132	58	308	56	90	**406**	100	45	37	33	31	22	21	18	**9.11**	23	38	4	–	–	–	–	–	–	1	–	–	–	**Prosta-5,8(12),13-trien-1-oic acid, 15-[(dimethylsilyl)oxy]-9-oxo-, methyl ester, (5Z,13E,15S)-**	53044-53-4	**NI41012**
190	74	58	56	132	118	335	70	**406**	100	82	64	64	55	55	48	45	**3.56**	23	38	4	–	–	–	–	–	–	1	–	–	–	**Prosta-5,10,13-trien-1-oic acid, 15-[(dimethylsilyl)oxy]-9-oxo-, methyl ester, (5Z,13E,15S)-**	53122-02-4	**NI41013**
406	99	125	302	111.5	78	177	198	**406**	100	27	14	13	9	9	8	6		24	6	–	8	–	–	–	–	–	–	–	–	–	**2,3,9,10-Tetracyanodibenzo[5,5a,6:11,11a,12]-1,4,7,10-tetraazanaphthacene**		**MS09276**
184	406	405	186	212	185	407	139	**406**	100	78	67	30	27	20	19	18		24	16	1	–	2	–	–	–	–	–	–	2	–	**Bis(10,9-borathiaro-10-phenanthryl) ether**		**LI08601**
106	159	145	117	105	107	91	77	**406**	100	96	92	85	81	77	77	77	**3.00**	24	22	4	–	1	–	–	–	–	–	–	–	–	**cis-2,4-Diphenyl-5-methyl-6-phenylsulphonylmethyl-1,3-dioxin**		**LI08602**
301	181	388	193	302	105	121	77	**406**	100	39	33	23	21	20	17	17	**1.50**	24	22	6	–	–	–	–	–	–	–	–	–	–	**2-(4-Anisyl)-3-hydroxy-3-phenyl-4-oxo-5,7-dimethoxy-2,3,4-trihydrobenzopyrylium**		**MS09277**
221	313	177	148	269	193	121	94	**406**	100	39	30	24	15	10	9	8	**0.24**	24	22	6	–	–	–	–	–	–	–	–	–	–	**1,4-Bis(phenoxyethoxycarbonyl)benzene**		**IC04314**
312	77	121	193	313	149	93	65	**406**	100	68	66	27	23	14	13	12	**4.71**	24	22	6	–	–	–	–	–	–	–	–	–	–	**Bis(2-phenoxyethyl) phthalate**		**IC04315**
347	262	348	406	375	263	287	277	**406**	100	38	25	13	8	7	6	5		24	23	4	–	–	–	–	–	–	–	1	–	–	**Butanedioic acid, (triphenylphosphoranylidene)-, dimethyl ester**	1104-78-5	**NI41014**
310	82	43	145	91	105	346	81	**406**	100	92	85	54	51	47	46	44	**0.00**	24	35	3	–	–	–	1	–	–	–	–	–	–	**Pregn-5-ene-20-carbonyl chloride, 3-(acetyloxy)-, (3β)-**	55103-88-3	**NI41015**
260	201	43	57	275	333	261	105	**406**	100	75	61	60	48	42	32	27	**25.00**	24	38	5	–	–	–	–	–	–	–	–	–	–	**17β-Acetoxy-3-ethyl-3-oxo-2,3-seco-5α-androstan-2-oic acid methyl ester**		**MS09278**
119	286	121	134	268	228	107	255	**406**	100	88	74	72	71	58	54	47	**11.00**	24	38	5	–	–	–	–	–	–	–	–	–	–	**3α,19-Diacetoxy-(–)-kauran-16α-ol**		**LI08603**
317	360	362	256	318	406			**406**	100	78	43	43	37	2				24	38	5	–	–	–	–	–	–	–	–	–	–	**19S-Hydroxy-19a-methyl-5α-androstane-3,17-ethylene dioxide**		**LI08604**
301	99	344	363	406	259	43	282	**406**	100	35	30	26	23	18	9	5		24	38	5	–	–	–	–	–	–	–	–	–	–	**3α-(2-Hydroxyethoxy)-19-methyl-5α-androstane-19-one 17,17-ethylene dioxide**		**LI08605**
119	286	121	134	268	228	107	93	**406**	100	87	73	71	69	57	53	47	**10.01**	24	38	5	–	–	–	–	–	–	–	–	–	–	**Kaurane-3,16,18-triol, 3,18-diacetate, (3α,4β)-**	55557-07-8	**NI41016**
406	153	374	167	168	55	137	41	**406**	100	83	78	63	58	58	42	42		24	38	5	–	–	–	–	–	–	–	–	–	–	**Pentadecanoic acid, 15-(3-methoxy-2,5-dioxobicyclo[4.1.0]hept-3-en-1-yl), methyl ester**		**MS09279**
406	374	167	168	376	105	166	122	**406**	100	82	56	55	40	36	32	32		24	38	5	–	–	–	–	–	–	–	–	–	–	**Pentadecanoic acid, 15-(6-methoxy-3-methyl-1,4-benzoquinon-2-yl)-, methyl ester**		**MS09280**
129	256	241	275	173	257	215	316	**406**	100	67	46	39	36	32	29	27	**18.00**	24	42	3	–	–	–	–	–	–	1	–	–	–	**5α-Androstane-3α,17β-diol 3-acetate 3-trimethylsilyl ether**		**LI08606**
316	256	217	255	391	315	107	155	**406**	100	49	49	46	39	36	29	28	**39.00**	24	42	3	–	–	–	–	–	–	1	–	–	–	**5α-Androstane-3α,17β-diol 3-acetate 3-trimethylsilyl ether**		**LI08607**
230	231	215	255	270	229	217	135	**406**	100	94	66	58	49	48	46	45	**27.79**	24	42	3	–	–	–	–	–	–	1	–	–	–	**Pregnane-20-one, 3,17-dihydroxy-, trimethylsilyl-, (3α,5α,17α)-**		**NI41017**
230	231	215	255	270	207	135	288	**406**	100	85	66	63	55	50	43	42	**19.00**	24	42	3	–	–	–	–	–	–	1	–	–	–	**Pregnane-20-one, 3,17-dihydroxy-, trimethylsilyl-, (3α,5α,17α)-**		**LI08608**
230	231	215	255	229	135	217	147	**406**	100	76	62	55	38	38	32	29	**17.74**	24	42	3	–	–	–	–	–	–	1	–	–	–	**Pregnane-20-one, 3,17-dihydroxy-, trimethylsilyl-, (3α,5β,17α)–**		**NI41018**
406	190	408	407	127	217	191	140	**406**	100	72	36	29	25	16	14	14		25	15	–	4	–	–	1	–	–	–	–	–	–	**2-[4-(4-Chlorostyryl)-3-cyanophenyl]-2H-naphtho[1,2-d]triazole**		**IC04316**
406	351	293	349	294	363	391	266	**406**	100	88	74	64	21	17	12	6		25	26	5	–	–	–	–	–	–	–	–	–	–	**2H-1-Benzopyran-2-one, 5,7-dihydroxy-8-(3-methyl-2-butenyl)-6-(3-methyl-1-oxobutyl)-4-phenyl-**	18483-64-2	**NI41019**
295	337	338	406	373	310	388	242	**406**	100	86	84	48	44	43	36	24		25	26	5	–	–	–	–	–	–	–	–	–	–	**Pyrano[3,2-a]xanthen-12(1H)-one, 2,3-dihydro-1,11-dihydroxy-5-methyl-8 (3-methyl-2-butenyl)-2-isopropenyl-, (1R-trans)-**	35660-46-9	**NI41020**
91	152	149	133	101	107	391	261	**406**	100	99	99	23	12	7	2	2	**0.10**	25	42	4	–	–	–	–	–	–	–	–	–	–	**1-O-(2-Benzyloxydodecyl)-2,3-O-isopropylideneglycerol**		**NI41021**
388	370	273	55	105	81	95	389	**406**	100	48	44	43	38	34	32	30	**8.90**	25	42	4	–	–	–	–	–	–	–	–	–	–	**5α-Cholan-24-oic acid, 3α,7α-dihydroxy-, methyl ester**	14772-98-6	**NI41022**

MASS TO CHARGE RATIOS								M.W.	INTENSITIES								Parent	C	H	O	N	S	F	Cl	Br	I	Si	P	B	X	COMPOUND NAME	CAS Reg No	No
370	355	255	55	388	371	213	81	**406**	100	43	39	38	34	32	28	28	**5.56**	25	42	4	–	–	–	–	–	–	–	–	–	–	Cholan-24-oic acid, 3,7-dihydroxy-, methyl ester, (3α,5β,7α)-	3057-04-3	NI41024
370	273	355	255	371	213	228	388	**406**	100	38	36	36	30	26	23	22	**7.00**	25	42	4	–	–	–	–	–	–	–	–	–	–	Cholan-24-oic acid, 3,7-dihydroxy-, methyl ester, (3α,5β,7α)-	3057-04-3	NI41023
388	273	107	81	93	109	108	120	**406**	100	67	67	67	64	56	56	55	**5.97**	25	42	4	–	–	–	–	–	–	–	–	–	–	Cholan-24-oic acid, 3,7-dihydroxy-, methyl ester, (3β,5α,7α)-	14772-97-5	NI41025
388	273	106	81	93	107	108	120	**406**	100	68	68	68	64	57	56	55	**6.00**	25	42	4	–	–	–	–	–	–	–	–	–	–	Cholan-24-oic acid, 3,7-dihydroxy-, methyl ester, (3β,5α,7α)-	14772-97-5	LI08609
57	66	71	41	43	115	70	55	**406**	100	78	56	56	54	41	35	30	**0.02**	25	42	4	–	–	–	–	–	–	–	–	–	–	Bis(2-ethylhexyl)endomethylene tetrahydrophthalate		IC04317
335	303	43	336	55	69	81	41	**406**	100	39	32	22	19	14	13	13	**0.00**	25	42	4	–	–	–	–	–	–	–	–	–	–	6α-Hydroxy-17-oxo-6β-pentyl-4,5-secoandrostan-4-oic acid methyl ester		MS09281
335	336	111	69	388	71	163	164	**406**	100	90	40	36	24	24	19	18	**7.00**	25	46	2	2	–	–	–	–	–	–	–	–	–	1H-Imidazole-2-methanol, α-heptadecyl-α-(tetrahydro-2-furanyl)-	75601-38-6	NI41026
271	301	406						**406**	100	65	25							26	18	3	2	–	–	–	–	–	–	–	–	–	1,3-Dioxolo[4,5-f][4,7]phenanthroline, 2-(4-methoxyphenyl)-2-phenyl-	54616-38-5	NI41027
323	338	69	91	233	305	191	105	**406**	100	46	36	20	9	6	5	5	**3.00**	26	30	4	–	–	–	–	–	–	–	–	–	–	2H-1-Benzopyran-6-one, 5,6-dihydro-8-hydroxy-5-(3-methyl-2-butenyl)-2,2,5-trimethyl-7-(2-phenylpropionyl)-		NI41028
323	174	324	137	187	406	337	150	**406**	100	60	42	22	20	18	12	10		26	30	4	–	–	–	–	–	–	–	–	–	–	Heminitidulan		NI41029
271	361	406						**406**	100	64	16							27	18	4	–	–	–	–	–	–	–	–	–	–	7H-Phenaleno[1,2-d][1,3]dioxol-7-one, 9-(4-methoxyphenyl)-9-phenyl-	54616-42-1	NI41030
210	196	77	120	105	406	92	132	**406**	100	71	44	32	25	23	20	16		27	22	2	2	–	–	–	–	–	–	–	–	–	Benzophenone, 2-amino-5-[(2-benzoylanilino)methyl]-	86989-01-7	NI41031
91	196	210	315	118	119	65	127	**406**	100	63	62	37	25	10	10	8	**1.00**	27	22	2	2	–	–	–	–	–	–	–	–	–	2,3-Naphthalenedicarboximide, N-[(dibenzylamino)methyl]-	31996-26-6	NI41032
55	81	67	109	95	150	136	164	**406**	100	98	89	69	35	25	22	10	**0.00**	27	50	2	–	–	–	–	–	–	–	–	–	–	5,9-Hexacosadienoic acid, methyl ester, (Z,Z)-		DD01590
140	43	80	81	55	67	57	79	**406**	100	72	69	35	34	25	25	22	**0.00**	27	50	2	–	–	–	–	–	–	–	–	–	–	5-Hexacosynoic acid, methyl ester		DD01591
363	268	172	337	83	55	69	95	**406**	100	44	36	35	35	35	29	27	**23.10**	27	51	–	–	–	–	–	–	–	–	1	–	–	Phosphine, (cyclohexylmethyl)bis[methyl(isopropyl)cyclohexyl]-	74792-88-4	NI41033
111	69	110	55	43	57	41	71	**406**	100	76	70	59	56	46	33	20	**0.17**	29	58	–	–	–	–	–	–	–	–	–	–	–	11-(2,5-Dimethylcyclohexyl)heneicosane		AI02338
111	110	69	43	55	57	83	41	**406**	100	76	33	32	31	27	17	17	**0.11**	29	58	–	–	–	–	–	–	–	–	–	–	–	11-(2,5-Dimethylcyclohexyl)heneicosane		AI02339
43	55	57	69	41	97	71	111	**406**	100	97	80	62	62	52	43	39	**3.21**	29	58	–	–	–	–	–	–	–	–	–	–	–	1,3-Dimethyl-4-(2-decyldodecyl)cyclopentane		AI02340
69	97	71	111	83	85	96	95	**406**	100	86	72	66	60	48	46	34	**0.93**	29	58	–	–	–	–	–	–	–	–	–	–	–	1,3-Dimethyl-4-(2-decyldodecyl)cyclopentane		AI02341
111	110	69	55	43	57	41	42	**406**	100	89	73	66	62	46	41	33	**0.50**	29	58	–	–	–	–	–	–	–	–	–	–	–	1,4-Dimethyl-2-(1-decylundecyl)cyclohexane		AI02342
111	69	110	71	97	83	95	85	**406**	100	82	67	26	20	19	18	17	**0.05**	29	58	–	–	–	–	–	–	–	–	–	–	–	1,4-Dimethyl-2-(1-decylundecyl)cyclohexane		AI02343
105	287	391	302	406	103	91	77	**406**	100	42	39	38	30	18	17	16		30	30	1	–	–	–	–	–	–	–	–	–	–	2,4,6-Tris-(1-phenylethyl)phenol		IC04318
147	148	119	133	107	41	105	91	**406**	100	13	12	5	5	5	2	2	**0.20**	30	46	–	–	–	–	–	–	–	–	–	–	–	Benzene, 1,1′-(1,1,10,10-tetramethyl-1,10-decanediyl)bis[3,4-dimethyl-	63934-83-8	NI41034
405	141	165	279	43	41	293	407	**406**	100	98	67	57	42	33	28	27	**0.00**	31	34	–	–	–	–	–	–	–	–	–	–	–	1,1-Bis(1-naphthyl)-1-undecene		AI02344
329	328	406	327	326	405	252		**406**	100	99	67	33	29	17	17			32	22	–	–	–	–	–	–	–	–	–	–	–	9,10-Diphenylanthracene benzyne centre ring adduct		LI08610
406	329	405	328	327	326	252	380	**406**	100	53	26	21	16	16	5	2		32	22	–	–	–	–	–	–	–	–	–	–	–	9,10-Diphenylanthracene benzyne end ring adduct		LI08611
68	407	242	228	42	254	96	41	**407**	100	48	48	42	38	37	36	33		13	14	2	2	–	6	–	–	–	–	–	–	1	Copper(II), N,N′-propylenebis(5,5,5-trifluoro-4-oxopentan-2-iminato)-	56132-47-9	NI41035
187	111	407	125	208	409	326	64	**407**	100	76	63	30	28	24	23	21		16	14	4	5	1	–	1	–	–	–	–	–	–	1-(4-Chlorophenyl)-3-methyl-5-hydroxy-4-(2-hydroxy-4-sulphamylphenylazo)pyrazole		IC04319
209	73	210	179	407	45	77	74	**407**	100	83	17	13	12	10	9	6		17	28	3	1	–	3	–	–	–	2	–	–	–	3-Methoxy-N-(trifluoroacetyl)-N,O-bis(trimethylsilyl)tyramine		MS09282
173	175	406	177	145	408	174	147	**407**	100	65	12	11	11	8	8	7	**3.00**	19	15	5	1	–	–	2	–	–	–	–	–	–	3,3′-[1-(2,6-Dichlorobenzoyl)pyrrole-2,5-diyl]diprop-2-enoic acid, dimethyl ester	96126-74-8	NI41036
392	360	348	407	274	332	376	316	**407**	100	96	53	41	38	37	31	25		19	21	9	1	–	–	–	–	–	–	–	–	–	Tetramethyl 1-methoxy-4,5-dihydropyrido[1,2-a]azepine-2,3,4,5-tetracarboxylate		NI41037
73	290	291	45	407	75	74	292	**407**	100	56	16	16	14	9	9	6		19	33	3	1	–	–	–	–	–	3	–	–	–	1H-Indole-3-acetic acid, 1-(trimethylsilyl)-5-[(trimethylsilyl)oxy]-, trimethylsilyl ester	55268-67-2	MS09283
73	290	406	291	45	407	292	75	**407**	100	79	31	23	16	12	10	9		19	33	3	1	–	–	–	–	–	3	–	–	–	1H-Indole-3-acetic acid, 1-(trimethylsilyl)-5-[(trimethylsilyl)oxy]-, trimethylsilyl ester	55268-67-2	NI41038
173	175	407	177	145	409	174	147	**407**	100	63	13	10	9	8	8	6		20	19	4	1	–	–	2	–	–	–	–	–	–	2-Propenoic acid, 3-[1-(2,6-dichlorobenzoyl)-4-acetyl-3,5-dimethylpyrrol-2-yl]-, ethyl ester	95883-20-8	NI41039
86	162	201	120	91	101	169	202	**407**	100	45	21	20	15	12	8	7	**2.60**	20	29	6	3	–	–	–	–	–	–	–	–	–	L-Alanine, N-[N-(N-carboxyglycyl)-L-Leucyl]-3-phenyl-, dimethyl ester	25159-31-3	NI41040
408	407	388	406					**407**	100	67	52	39						21	24	–	3	1	3	–	–	–	–	–	–	–	10H-Phenothiazine, 10-[3-(4-methyl-1-piperazinyl)propyl]-2-(trifluoromethyl)-	117-89-5	NI41041
113	70	43	127	248	141	266	42	**407**	100	82	57	46	45	45	43	39	**18.02**	21	24	–	3	1	3	–	–	–	–	–	–	–	10H-Phenothiazine, 10-[3-(4-methyl-1-piperazinyl)propyl]-2-(trifluoromethyl)-	117-89-5	NI41042
113	70	43	407	141	71	42	127	**407**	100	79	47	34	32	28	27	26		21	24	–	3	1	3	–	–	–	–	–	–	–	10H-Phenothiazine, 10-[3-(4-methyl-1-piperazinyl)propyl]-2-(trifluoromethyl)-	117-89-5	NI41043
43	56	31	41	27	42	39	29	**407**	100	88	85	80	64	33	33	27	**0.00**	22	21	3	3	1	–	–	–	–	–	–	–	–	Benzenesulphonic acid, 4-methyl-, [2-(methoxyimino)-1,2-diphenylethylidene]hydrazide	56667-03-9	NI41044
364	407	365	188	189	187	336	178	**407**	100	70	27	9	6	6	5	5		22	21	5	3	–	–	–	–	–	–	–	–	–	Benzoic acid, 4-[N-acetyl-N-(2-methyl-6,7-methylenedioxyquinazolin-4-ylmethyl)amino]-, ethyl ester	85591-06-6	NI41045

MASS TO CHARGE RATIOS								M.W.	INTENSITIES								Parent	C	H	O	N	S	F	Cl	Br	I	Si	P	B	X	COMPOUND NAME	CAS Reg No	No
284	160	199	407	173	186	285	147	**407**	100	46	41	32	26	25	22	22		22	24	3	1	–	3	–	–	–	–	–	–	–	Estra-1,3,5(10)-triene-17-carbonitrile, 3-methoxy-17-[(trifluoroacetyl)oxy], (17β)-	56438-19-8	NI41046
59	72	55	43	60	41	69	57	**407**	100	27	14	13	12	11	10	6	**0.38**	22	43	1	1	–	–	2	–	–	–	–	–	–	13,14-Dichlorobehenamide		IC04320
43	41	56	58	57	42	84	69	**407**	100	90	68	55	46	38	32	31	**0.10**	23	29	2	5	–	–	–	–	–	–	–	–	–	1,3,4-Oxadiazole-2,5-diamine, N²-phenyl-N⁵-[3-[3-(piperidinomethyl)phenoxy]propyl]-		MS09284
153	195	59	42	60	166	138	208	**407**	100	99	86	83	60	53	42	40	**6.00**	23	37	5	1	–	–	–	–	–	–	–	–	–	2-Pyridineundecanol, 4-(acetyloxy)-5-methoxy-α,6-dimethyl-, acetate	70001-18-2	NI41047
58	72	407	71	408	234	44	42	**407**	100	78	26	16	9	8	8	8		24	29	3	3	–	–	–	–	–	–	–	–	–	β-Carboline-2-carboxylic acid, 1,2,3,4-tetrahydro-1-[2-[2-(dimethylamino)ethoxy]phenyl]-, ethyl ester		MS09285
407	406	271	408	285	287	120	392	**407**	100	83	40	27	21	15	11	9		24	30	1	3	–	–	–	–	–	–	1	–	–	Aniline, 4,4′,4′′-phosphinylidynetris[N,N-dimethyl-	807-20-5	NI41048
144	41	42	55	43	143	98	81	**407**	100	63	60	57	56	50	45	43	**17.00**	24	41	4	1	–	–	–	–	–	–	–	–	–	3b,11a,17a-Trihydroxy-16a-methyl-5α-pregnan-20-one, ethyloxime		LI08612
323	392	324	339	309	308	70	293	**407**	100	43	28	19	19	19	18	9	**4.65**	25	29	4	1	–	–	–	–	–	–	–	–	–	Tylohirsutine, 13a-methyl-		NI41049
181	180	362	77	407	182	378	152	**407**	100	99	82	67	58	25	8	4		26	21	2	1	–	–	–	–	–	1	–	–	–	1-Oxa-3-aza-2-silacyclopentan-5-one, 2,2,3,4-tetraphenyl-	57954-42-4	NI41050
407	77	409	408	332	267	105	406	**407**	100	52	37	37	31	25	25	20		27	18	1	1	–	–	1	–	–	–	–	–	–	4(1H)-Quinolinone, 1-(4-chlorophenyl)-2,3-diphenyl-	36843-27-3	NI41051
392	43	44	393	55	30	87	41	**407**	100	50	48	33	33	29	27	25	**23.00**	27	53	1	1	–	–	–	–	–	–	–	–	–	1-Acetylazacyclohexacosane		MS09286
69	329	410	408	96	391	131	129	**408**	100	65	19	19	15	10	10	10		6	1	–	2	–	12	–	1	–	–	–	–	–	1,2-Bis[bis(trifluoromethyl)amino]-1-bromoethylene		LI08613
327	269	299	268	187	243	266	408	**408**	100	97	80	50	48	42	40	37		7	–	5	–	–	3	–	–	–	–	–	–	1	Rhenium, pentacarbonyl(trifluorovinyl)-	14882-07-6	NI41052
52	144	91	28	232	145	66	53	**408**	100	79	21	16	15	14	13	11	**6.81**	7	8	–	–	–	9	–	–	–	–	3	–	1	Chromium, [(1,2,3,4,5,6-η)-1,3,5-cycloheptatriene]tris(phosphorous trifluoride)-	59464-87-8	NI41053
103	40	311	313	105	309	94	75	**408**	100	64	51	45	33	31	28	26	**2.49**	9	4	1	–	–	–	8	–	–	–	–	–	–	4,7-Methanoisobenzofuran, 1,3,4,5,6,7,8,8-octachloro-1,3,3a,4,7,7a-hexahydro-	297-78-9	NI41054
103	311	313	275	239	237	204	206	**408**	100	74	65	61	59	59	44	42	**3.60**	9	4	1	–	–	–	8	–	–	–	–	–	–	4,7-Methanoisobenzofuran, 1,3,4,5,6,7,8,8-octachloro-1,3,3a,4,7,7a-hexahydro-	297-78-9	NI41055
333	43	369	371	290	337	335	250	**408**	100	99	70	68	68	60	42	29	**0.00**	9	21	–	–	–	–	–	2	–	–	–	–	1	Triisopropylantimony dibromide	58037-61-9	NI41056
240	52	111	296	59	242	139	380	**408**	100	93	83	80	80	67	55	54	**15.00**	10	4	6	–	–	–	2	–	–	–	–	–	2	Dicobalt, hexacarbonyl(1,4-dichlorobutyne)-		LI08614
55	128	267	394	127	139			**408**	100	37	29	27	19	10			**0.00**	11	22	–	–	–	–	–	–	2	–	–	–	–	1,11-Diiodoundecane		LI08615
328	408	409	376	117	111	87	296	**408**	100	71	5	5	5	5	5	4		12	–	–	–	1	8	–	–	–	–	–	–	1	Phenothiaselenin, octafluoro-	19638-34-7	LI08616
328	408	406	329	410	404	297	87	**408**	100	35	17	14	8	6	6	6		12	–	–	–	1	8	–	–	–	–	–	–	1	Phenothiaselenin, octafluoro-	19638-34-7	NI41057
328	408	329	404	410	405	297	111	**408**	100	34	17	8	7	7	6	5		12	–	–	–	1	8	–	–	–	–	–	–	1	Phenothiaselenin, octafluoro-	19638-34-7	NI41058
117	91	90	116	89	149	65	63	**408**	100	76	45	36	26	18	18	16	**0.00**	14	18	9	1	2	–	–	–	–	–	–	–	–	Glucotropaeolin		MS09287
361	362	363	115	348	259	319	257	**408**	100	37	36	36	35	34	31	30	**0.00**	15	8	3	–	1	–	4	–	–	–	–	–	–	2H-Thiacyclobuta[b][1,4]benzodioxin-1-oxide, 4,5,6,7-tetrachloro-2a,8a-dihydro-2a-phenyl-		MS09288
140	180	67	57	182	69	307	55	**408**	100	53	37	27	21	20	17	12	**5.75**	15	22	4	2	–	6	–	–	–	–	–	–	–	ε-N-Monomethyllysine N-butyl-N-trifluoroacetamide		NI41059
73	147	393	45	221	394	74	190	**408**	100	47	23	17	10	9	9	8	**4.45**	15	36	5	–	–	–	–	–	–	4	–	–	–	Propanedioic acid, (trimethylsilyl)[(trimethylsilyl)oxy]-, bis(trimethylsilyl) ester	55530-65-9	NI41060
73	147	393	45	394	221	74	190	**408**	100	44	28	15	10	9	9	8	**6.00**	15	36	5	–	–	–	–	–	–	4	–	–	–	Propanedioic acid, (trimethylsilyl)[(trimethylsilyl)oxy]-, bis(trimethylsilyl) ester	55530-65-9	NI41061
362	212	255	332	363	408	286	200	**408**	100	30	25	18	17	13	13	11		16	4	8	6	–	–	–	–	–	–	–	–	–	Propanedinitrile, (2,4,5,7-tetranitro-9H-fluoren-9-ylidene)-	15517-55-2	NI41062
247	97	98	248	45	48	47	94	**408**	100	60	34	20	20	18	14	12	**0.00**	16	21	4	–	3	–	1	–	–	–	–	–	–	1-Methyl-2-[2-(5-methylthenyl)]-4-[2-(5-methylthienyl)]tetrahydrothiophenium perchlorate		LI08617
105	61	60	59	58	118	53	51	**408**	100	72	64	55	41	40	40	33	**0.00**	16	24	8	–	2	–	–	–	–	–	–	–	–	D-arabino-Hexose, 2-deoxy-, cyclic 1,2-ethanediyl mercaptal, tetraacetate	55955-82-3	NI41063
105	45	61	44	147	59	107	58	**408**	100	57	33	21	11	11	9	7	**0.00**	16	24	8	–	2	–	–	–	–	–	–	–	–	L-Galactose, 6-deoxy-, cyclic 1,2-ethanediyl mercaptal, tetraacetate	55955-81-2	NI41064
51	256	106	78	150	39	64	92	**408**	100	99	78	74	70	69	67	61	**0.45**	18	12	6	6	–	–	–	–	–	–	–	–	–	1,2-Bis[(2-nitrophenyl)-ONN-azoxy]benzene, (Z,Z)-	80337-05-9	NI41065
28	67	97	95	41	111	125	29	**408**	100	76	67	57	48	43	39	39	**0.21**	18	24	7	–	–	–	–	–	–	–	–	–	1	Iron, carbonyl[(2,3,4,5-η)-diethyl 2,4-hexadienedioate][(2,3,4,5-η)-methyl 2,4-hexadienoate]-	69632-92-4	NI41066
172	116	57	43	173	204	58	55	**408**	100	52	52	38	35	34	34	33	**33.23**	18	36	–	2	4	–	–	–	–	–	–	–	–	Thioperoxydicarbonic diamide, tetrabutyl-	1634-02-2	NI41067
57	55	41	172	204	116	89	42	**408**	100	39	34	31	29	28	14	14	**4.50**	18	36	–	2	4	–	–	–	–	–	–	–	–	Thioperoxydicarbonic diamide, tetrakis(isobutyl)-	3064-73-1	NI41068
56	58	289	114	144	100	70	57	**408**	100	38	33	33	30	25	25	25	**1.10**	18	36	6	2	1	–	–	–	–	–	–	–	–	20,24-Dihydroxy-4,7,13,16-tetraoxa-22-thia-1,10-diazabicyclo[8.8.7]pentacosane		MS09289
28	324	352	268	267	253	134	120	**408**	100	49	19	11	11	11	11	10	**1.65**	19	16	5	2	–	–	–	–	–	–	–	–	1	Iron, tricarbonyl[N,N′-1,2-ethanediylidenebis[4-methoxyaniline]-N,N′]-	54446-64-9	NI41069
164	136	196	213	165	39	53	77	**408**	100	61	58	26	23	18	16	14	**2.00**	19	17	8	–	–	–	1	–	–	–	–	–	–	Benzoic acid, 3-chloro-5-formyl-4,6-dihydroxy-2-methyl-, 3-hydroxy-4-(methoxycarbonyl)-2,5-dimethylphenyl ester	479-16-3	NI41070
121	227	287	199	165	198	197	259	**408**	100	90	73	57	42	17	13	1	**0.00**	19	20	2	–	4	–	–	–	–	–	–	–	–	Carbonodithioic acid, S,S′-dibenzyl O,O′-diethyl ester	17435-01-7	NI41071

MASS TO CHARGE RATIOS								M.W.	INTENSITIES								Parent	C	H	O	N	S	F	Cl	Br	I	Si	P	B	X	COMPOUND NAME	CAS Reg No	No
221	264	236	306	231	193	235	192	408	100	65	49	31	30	28	20	19	1.00	19	20	10	–	–	–	–	–	–	–	–	–	–	4H-1-Benzopyran-4-one, 6-(diacetoxymethyl)-5,8-dimethoxy-7-acetoxy-2-methyl-		NI41072
228	229	201	245	185	275	186	408	408	100	17	15	12	12	10	10	8		19	20	10	–	–	–	–	–	–	–	–	–	–	Khellol glucoside		LI08618
213	228	161	163	214	187	185	229	408	100	40	23	22	16	9	9	8	2.00	19	21	5	–	–	–	–	1	–	–	–	–	–	Columbianetin 3-bromoangelate		LI08619
366	135	286	94	79	93	67	81	408	100	36	34	29	23	22	13	10	3.00	19	24	6	2	1	–	–	–	–	–	–	–	–	1-(1-Adamantyl)-4-isopropylsulphonyl-3,5-dinitrobenzene		NI41073
166	167	114	41	307	29	43	139	408	100	67	39	30	28	21	19	13	1.30	19	31	4	2	–	3	–	–	–	–	–	–	–	Octanoic acid, N-(trifluoroacetyl)-L-prolyl-α-amino-, butyl ester		MS09290
91	65	202	119	220	408	232		408	100	21	17	13	9	8	6			20	20	2	6	1	–	–	–	–	–	–	–	–	5,5′-Bis[1-methyl-3-(4-tolyl)-4-hydroxy-1,2,3-triazolium] sulphide		LI08620
233	275	259	408	247	232	217	289	408	100	92	53	20	17	16	16	8		20	24	9	–	–	–	–	–	–	–	–	–	–	4H-1-Benzopyran-4-one, 8-C-β-D-glucopyranosyl-2-(2-oxopropyl)-7-methoxy-5-methyl-		NI41074
187	188	229	213	228	246	189	131	408	100	81	49	23	13	12	12	12	2.39	20	24	9	–	–	–	–	–	–	–	–	–	–	2H-Furo[2,3-h]-1-benzopyran-2-one, 8-[1-(β-D-glucopyranosyloxy)isopropyl]-8,9-dihydro-, (S)-	55836-35-6	NI41075
57	290	98	58	364	224	408	308	408	100	37	32	28	19	18	15	14		20	24	9	–	–	–	–	–	–	–	–	–	–	Ginkgolide A	15291-75-5	NI41076
105	393	291	258	211	274	230	169	408	100	36	36	20	20	14	12	11	1.00	20	24	9	–	–	–	–	–	–	–	–	–	–	α-D-Xylofuranose, 3-C-[(acetyloxy)methyl]-1,2-O-(isopropylidene)-, 3-acetate 5-benzoate	70723-03-4	NI41077
77	408	91	149	259	119	105	409	408	100	86	35	29	26	24	23	22		21	20	5	4	–	–	–	–	–	–	–	–	–	Benzoic acid, 4-[(4,5-dihydro-1-phenyl-3-carboethoxy-5-oxopyrazol-4-ylidene)hydrazino]-		IC04321
294	122	295	293	251	160	43	45	408	100	40	36	25	25	19	16	13	0.00	21	23	3	2	–	3	–	–	–	–	–	–	–	20,21-Dinoraspidospermidin-5-ol, 1-acetyl-, trifluoroacetate	55162-79-3	NI41078
178	175	86	57	121	107	120	41	408	100	64	47	40	30	30	15	15	3.50	21	32	6	2	–	–	–	–	–	–	–	–	–	L-Tyrosine, N-(N-carboxy-L-Leucyl)-, N-tert-butyl methyl ester	18658-65-6	NI41079
71	127	323	41	183	131	295	224	408	100	87	46	39	36	29	23	20	11.00	21	36	4	–	–	–	–	–	–	2	–	–	–	Abscisic acid, (O-trimethylsilyl)-, trimethylsilyl ester		NI41080
59	44	87	36	220	57	336	234	408	100	98	78	38	37	35	31	31	0.10	21	44	–	–	–	–	–	–	–	4	–	–	–	3,4-Pentadien-1-yne, 1,3,5,5-tetrakis(ethyldimethylsilyl)-	41898-94-6	NI41081
408	409	83	85	365	204	337	161	408	100	26	17	11	10	9	8	7		22	16	8	–	–	–	–	–	–	–	–	–	–	Benzo[1,2-b:4,5-b′]bisbenzofuran-6,12-dione, 2,3,8,9-tetramethoxy-	3534-73-4	NI41082
408	191	365	366	190	309	337	363	408	100	92	87	50	47	37	30	8		22	20	6	2	–	–	–	–	–	–	–	–	–	Benzoic acid, 4-[N-acetyl-N-(2-methyl-6,7-methylenedioxy-4H-3,1-benzoxazin-4-ylidenemethyl)amino]-, ethyl ester		NI41083
331	105	77	182	183	28	165	179	408	100	99	80	70	60	50	29	25	0.70	22	20	6	2	–	–	–	–	–	–	–	–	–	2′,3′-Di-O-diphenylmethyleneuridine		NI41084
77	105	29	226	182	170	408	28	408	100	79	58	50	46	22	18	16		22	20	6	2	–	–	–	–	–	–	–	–	–	Propanedioic acid, (3,5-dioxo-1,2-diphenyl-4-pyrazolidinylidene)-, diethyl ester	57488-08-1	NI41085
283	284	372	408	167	373	371	295	408	100	27	12	7	7	5	5	5		22	21	2	4	–	–	1	–	–	–	–	–	–	1H-Purine-2,6-dione, 8-(2-chloro-1,2-diphenylethyl)-3,7-dihydro-1,3,7-trimethyl-, erythro-		NI41086
283	372	284	167	373	408	371	295	408	100	44	40	21	16	14	14	12		22	21	2	4	–	–	1	–	–	–	–	–	–	1H-Purine-2,6-dione, 8-(2-chloro-1,2-diphenylethyl)-3,7-dihydro-1,3,7-trimethyl-, threo-		NI41087
137	69	44	43	59	77	215	78	408	100	88	81	64	60	42	35	30	13.13	22	23	4	–	–	3	–	–	–	–	–	–	–	Estra-1,3,5(10),6-tetraene-3,17-diol, 17-acetate 3-(trifluoroacetate), (17β)-	56588-11-5	NI41088
390	376	361	315	121	165	152	111	408	100	80	77	60	60	58	49	49	24.00	22	32	7	–	–	–	–	–	–	–	–	–	–	Picras-2-ene-1,16-dione, 11,13-dihydroxy-2,12-dimethoxy-, (11α,12β)-	28387-43-1	NI41089
91	198	210	226	253	241	155	408	408	100	83	50	35	32	20	20	10		22	36	3	2	1	–	–	–	–	–	–	–	–	Benzenesulphonamide, 4-methyl-N-[3-(2-oxoazacyclotridec-1-yl)propyl]-	67370-84-7	NI41090
137	294	69	44	59	77	173	186	408	100	69	69	58	57	50	45	40	4.00	23	27	3	–	–	3	–	–	–	–	–	–	–	19-Norpregna-1,3,5(10),20-tetraen-17-ol, 3-methoxy-, trifluoroacetate	56588-07-9	NI41091
43	348	162	349	163	187	145	134	408	100	63	54	37	25	23	23	23	4.30	23	33	5	–	–	1	–	–	–	–	–	–	–	Androstane-4,19-diol, 5,6-epoxy-3-fluoro-, diacetate, (3β,4β,5β,6β)-	57207-28-0	NI41092
43	348	162	349	163	187	145	134	408	100	63	54	37	25	23	23	23	4.30	23	33	5	–	–	1	–	–	–	–	–	–	–	Androstane-17,19-diol, 5,6-epoxy-3-fluoro-, diacetate, (3β,4β,5β,6β)-	40242-92-0	MS09291
43	257	348	91	162	93	141	79	408	100	27	24	16	15	15	14	13	1.90	23	33	5	–	–	1	–	–	–	–	–	–	–	Androstane-17,19-diol, 5,6-epoxy-3-fluoro-, diacetate, (3β,5α,6α,17β)-	40242-91-9	MS09292
226	237	255	211	270	252	288	330	408	100	22	20	17	11	8	7	3	3.00	23	36	6	–	–	–	–	–	–	–	–	–	–	7α,12α-Diacetoxy-3α,17β-dihydroxy-5β-androstane		NI41093
259	305	157	337	67	93	260	81	408	100	79	78	73	69	68	67	66	0.00	23	36	6	–	–	–	–	–	–	–	–	–	–	Methyl-12-(methoxycarbonyl)-13β-hydroxy-14-oxyabietanoate		NI41094
43	55	97	113	71	165	123	85	408	100	84	41	37	31	25	25	25	0.00	23	36	6	–	–	–	–	–	–	–	–	–	–	2,6,10,14,18,22-Tricosanehexone	55110-18-4	NI41095
215	149	115	241	408	332	170	217	408	100	88	86	82	79	79	69	59		23	44	2	–	–	–	–	–	–	2	–	–	–	3β,17β-Bis(dimethylsilyloxy)-5α-androstane		NI41096
408	105	77	73	365	391	409	314	408	100	46	46	37	36	35	28	20		24	16	3	4	–	–	–	–	–	–	–	–	–	[2,2′-Bipyridine]-5-carboxamide, 5′-cyano-1,1′,6,6′-tetrahydro-6,6′-dioxo-4,4′-diphenyl-	18096-77-0	NI41097
105	408	77	269	410	139			408	100	22	16	10	9	5				24	21	4	–	–	–	1	–	–	–	–	–	–	1-Benzoyl-2-(3,4-dimethoxyphenyl)-2-(2-chlorobenzoyl)ethane		LI08621
152	104	78	103	77	50	39	91	408	100	88	82	55	36	36	29	26	10.36	24	24	–	–	–	–	–	–	–	–	–	–	2	Dititanium, tris(1,3,5,7-cyclooctatetraene)-	12149-66-5	NI41098
136	121	272	240	103	137	77	225	408	100	56	21	15	12	10	9	7	0.03	24	24	–	–	3	–	–	–	–	–	–	–	–	1,3,5-Trithiane, 2,4,6-trimethyl-2,4,6-triphenyl-		AI02345
91	180	271	151	362	92	165	149	408	100	65	31	27	11	11	10	10	3.00	24	24	6	–	–	–	–	–	–	–	–	–	–	3,4-Flavandiol, 7-(benzyloxy)-3′,4′-dimethoxy-, trans-2,3,cis-3,4-	16821-22-0	NI41099
408	204	365	334	393	350			408	100	19	6	6	5	4				24	24	6	–	–	–	–	–	–	–	–	–	–	2,3,6,7,10,11-Hexamethoxytriphenylene		LI08622
393	408	335	323	204	203	377	361	408	100	50	15	10	8	8	5	5		24	24	6	–	–	–	–	–	–	–	–	–	–	2H,6H-Pyrano[3,2-b]xanthen-6-one, 5,8-dihydroxy-9-methoxy-2,2-dimethyl-10-(3-methyl-2-butenyl)-	62326-62-9	NI41100
202	201	256	215	155	183	336	297	408	100	75	20	16	16	15	12	10	0.00	24	25	4	–	–	–	–	–	–	–	1	–	–	Cyclopentanecarboxylic acid, 2-[3-(diphenylphosphinoyl)-1-methylprop-2-enyl]-5-oxo-, methyl ester, (1RS,1′RS,2RS,2′E)-		MS09293
91	41	58	43	98	65	121	408	408	100	90	45	42	20	20	18	16		24	28	–	2	2	–	–	–	–	–	–	–	–	Pyrimidine, 1-benzyl-3-isopropyl-2-(1-phenyl-1,3-dithioxobutan-2-ylidene)hexahydro-		MS09294
335	408	275	246	180	349	247		408	100	93	42	40	35	31	29			24	28	4	2	–	–	–	–	–	–	–	–	–	Di-O-acetylaruammilinol		LI08623

MASS TO CHARGE RATIOS								M.W.	INTENSITIES								Parent	C	H	O	N	S	F	Cl	Br	I	Si	P	B	X	COMPOUND NAME	CAS Reg No	No
408	144	157	409	349	190	145	158	**408**	100	61	45	29	24	22	22	15		24	28	4	2	–	–	–	–	–	–	–	–	–	Ajmalan-16-carboxylic acid, 17-(acetyloxy)-19,20-didehydro-, methyl ester, (2α,16R,17S,19E)-	912-27-6	NI41101
43	408	144	157	190	409	145	349	**408**	100	66	52	46	20	18	17	16		24	28	4	2	–	–	–	–	–	–	–	–	–	Ajmalan-16-carboxylic acid, 17-(acetyloxy)-19,20-didehydro-, methyl ester, (2α,16R,17S,19E)-	912-27-6	NI41102
91	137	270	122	110	108	136	139	**408**	100	41	34	25	23	23	21	18	**5.00**	24	28	4	2	–	–	–	–	–	–	–	–	–	Benzyl 4-ethyl-5-(4-ethyl-3,5-dimethylpyrrolyl-2-oxocarbonyl)-3-methylpyrrole-2-carboxylate		LI08624
91	211	317	408	139	209	302	92	**408**	100	72	25	19	16	15	13	13		24	28	4	2	–	–	–	–	–	–	–	–	–	1-[4-(Benzyloxy-3-benzyloxymethyl)butyl]-5-methyl-2,4-dioxo-1,2,3,4-tetrahydropyrimidine		LI08625
408	407	409	143	169	156	170	223	**408**	100	76	24	17	13	12	10	8		24	28	4	2	–	–	–	–	–	–	–	–	–	3β,20α,19β-(2-Methylacetoacetyl)yohimb-17-one		LI08626
408	349	365	366	305	198	306	199	**408**	100	86	85	66	61	46	44	39		24	28	4	2	–	–	–	–	–	–	–	–	–	Na,O-Diacetyldemethylseredamine		LI08627
349	263	408	183	350	182	335	107	**408**	100	62	44	27	25	21	20	20		24	28	4	2	–	–	–	–	–	–	–	–	–	Sarpagan-16-carboxylic acid, 17-(acetyloxy)-1-methyl-, methyl ester, (16ξ)-	54725-24-5	NI41103
290	289	182	183	349	408	263	168	**408**	100	87	62	49	45	36	24	24		24	28	4	2	–	–	–	–	–	–	–	–	–	Sarpagan-16-carboxylic acid, 17-(acetyloxy)-1-methyl-, methyl ester, (16R)-	2912-12-1	NI41104
43	40	289	208	168	45	44	169	**408**	100	42	39	37	34	30	30	29	**3.03**	24	28	4	2	–	–	–	–	–	–	–	–	–	Sarpagan-17-ol, 16-[(acetyloxy)methyl]-, acetate	54789-46-7	NI41105
91	211	105	317	77	41	39	65	**408**	100	72	36	25	25	21	20	19	**19.00**	24	28	4	2	–	–	–	–	–	–	–	–	–	Thymine, 1-[4-(benzyloxy)-3-[(benzyloxy)methyl]butyl]-	34075-73-5	NI41106
91	71	43	203	204	99	44	175	**408**	100	74	74	64	56	50	38	22	**0.20**	24	32	2	4	–	–	–	–	–	–	–	–	–	1,2,4,5-Tetrazine, 1,4-diacetylhexahydro-2,5-bis(3-phenylpropyl)-	35029-01-7	LI08628
91	71	43	203	204	99	44	261	**408**	100	75	75	63	58	50	38	37	**3.00**	24	32	2	4	–	–	–	–	–	–	–	–	–	1,2,4,5-Tetrazine, 1,4-diacetylhexahydro-2,5-bis(3-phenylpropyl)-	35029-01-7	NI41107
271	372	253	354	343	226	247	373	**408**	100	85	82	45	35	34	28	24	**2.40**	24	40	5	–	–	–	–	–	–	–	–	–	–	Cholan-24-oic acid, 3,7,12-trihydroxy-, (3α,5β,7α,12α)-	81-25-4	NI41108
91	105	79	77	93	81	67	145	**408**	100	85	83	64	61	59	58	55	**0.00**	24	40	5	–	–	–	–	–	–	–	–	–	–	Cholan-24-oic acid, 3,7,12-trihydroxy-, (3α,5β,7α,12α)-	81-25-4	WI01525
55	41	271	81	372	43	253	93	**408**	100	67	63	63	60	55	53	50	**0.83**	24	40	5	–	–	–	–	–	–	–	–	–	–	Cholan-24-oic acid, 3,7,12-trihydroxy-, (3α,5β,7α,12α)-	81-25-4	NI41109
408	376	168	167	409	377	55	348	**408**	100	35	29	29	23	15	11	6		24	40	5	–	–	–	–	–	–	–	–	–	–	Pentadecanoic acid, 15-(2-hydroxy-3,5-dimethoxyphenyl)-, methyl ester		MS09295
84	124	109	392	269	267	268	123	**408**	100	55	55	49	42	40	36	28	**4.50**	25	28	5	–	–	–	–	–	–	–	–	–	–	4H-1-Benzopyran-4-one, 2,3-dihydro-7-hydroxy-2-(2,4-dihydroxyphenyl)-8-[5-methyl-2-(1-methylethenyl)-4-hexenyl]-		NI41110
262	183	99	108	89	105	77	146	**408**	100	52	44	41	39	35	33	32	**0.00**	26	21	1	2	–	–	–	–	–	–	1	–	–	Benzeneacetaldehyde, α-oxo-, aldehydo-[(triphenylphosphoranylidene)hydrazone]	22610-14-6	NI41111
119	209	208	83	210	97	205	85	**408**	100	97	92	92	81	68	57	54	**5.00**	26	24	1	4	–	–	–	–	–	–	–	–	–	2-Cyclohexen-1-one, 2-[[2-[2-[2-[(2-aminophenyl)azo]phenyl]vinyl]phenyl]amino]-	69395-35-3	NI41112
408	44	43	79	91	55	41	93	**408**	100	85	63	35	34	32	32	28		26	32	4	–	–	–	–	–	–	–	–	–	–	Bufa-14,16,20,22-tetraenolide, 3-(acetyloxy)-, (3β,5β)-	31444-09-4	NI41113
119	91	288	69	247	219	170	272	**408**	100	30	11	10	9	7	6	4	**0.00**	26	32	4	–	–	–	–	–	–	–	–	–	–	1,4-Dioxaspiro[4.5]decan-2-one, 3-[(1S)-1-hydroxy-2-phenylethyl]-9-methyl-6-(1-methyl-1-phenylethyl)-, (3S,5R,6S,9R)-	121445-49-6	DD01592
202	201	365	47	28	77	203	41	**408**	100	89	76	53	51	42	39	36	**22.00**	26	33	2	–	–	–	–	–	–	–	1	–	–	Cyclopentanone, 3-[3-(diphenylphosphinoyl)-1,5-dimethylhex-2-enyl]-2-methyl-, (1′RS,2SR,2′E,3RS)-		MS09296
105	77	303	274	135	106	134	148	**408**	100	31	12	10	10	9	8	6	**5.00**	26	36	2	2	–	–	–	–	–	–	–	–	–	N,N′-Dibenzoyl-1,12-diaminododecane		LI08629
105	77	303	106	390	286	51	78	**408**	100	35	13	8	7	5	5	3	**1.00**	27	20	4	–	–	–	–	–	–	–	–	–	–	2-(2-Benzoxybenzyl)phenyl benzoate		MS09297
104	221	111	54	105	222	77	241	**408**	100	80	35	33	31	24	24	23	**0.00**	27	24	2	2	–	–	–	–	–	–	–	–	–	2,3-Naphthalenedicarboximide, 1,4-imino-1,2,3,4-tetrahydro-N-(4-methylphenyl)-9-(2-phenylethyl)-, endo-	123542-30-3	DD01593
353	55	57	354	408	149	203	41	**408**	100	87	63	31	19	19	18	15		27	36	3	–	–	–	–	–	–	–	–	–	–	2-(3,5-Di-tert-butyl-4-hydroxybenzyl)-4-isopropylphenyl acrylate		MS09298
375	69	408	339	41	376	161	57	**408**	100	74	53	50	50	31	30	30		27	36	3	–	–	–	–	–	–	–	–	–	–	Bis(2-hydroxy-3-tert-butyl-5-methylphenyl)methane, mono-2-methyl-2-propenoate		NI41114
57	88	408	83	69	142	101	43	**408**	100	48	43	43	41	40	38	30		27	52	2	–	–	–	–	–	–	–	–	–	–	2-Docosenoic acid, 2,4,21,21-tetramethyl-, methyl ester, (E)-	56687-71-9	LI08630
57	408	83	69	142	101	43	41	**408**	100	44	44	41	40	38	30	30		27	52	2	–	–	–	–	–	–	–	–	–	–	2-Docosenoic acid, 2,4,21,21-tetramethyl-, methyl ester, (E)-	56687-71-9	NI41115
57	88	83	69	408	142	41	55	**408**	100	46	46	43	38	34	34	33		27	52	2	–	–	–	–	–	–	–	–	–	–	2-Docosenoic acid, 2,4,21,21-tetramethyl-, methyl ester, (Z)-	56687-70-8	NI41116
97	68	53	124	83	75	59	137	**408**	100	98	93	92	85	82	82	78	**0.00**	27	52	2	–	–	–	–	–	–	–	–	–	–	5-Hexacosenoic acid, methyl ester, (Z)-		DD01594
408	409	380	176	352	175	351	190	**408**	100	32	19	17	12	7	6	6		28	12	2	2	–	–	–	–	–	–	–	–	–	Benzo[h]benz[5,6]acridino[2,1-9,8-klmna]acridine-8,16-dione		IC04322
280	279	281	202	203	278	276	277	**408**	100	30	29	29	20	14	12	10	**9.00**	28	16	–	4	–	–	–	–	–	–	–	–	–	1,4-Methanonaphthalene-2,2,3,3-tetracarbonitrile, 9-(diphenylmethylene)-1,4-dihydro-	55723-86-9	NI41117
408	409	303	285	105	123	183	77	**408**	100	30	18	16	10	8	7	6		28	18	1	–	–	2	–	–	–	–	–	–	–	Furan, 2,4-bis(4-fluorophenyl)-3,5-diphenyl-	18741-99-6	MS09299
408	285	409	123	286	183	95	283	**408**	100	37	30	16	8	8	8	7		28	18	1	–	–	2	–	–	–	–	–	–	–	Furan, 2,5-bis(4-fluorophenyl)-3,4-diphenyl-	18749-93-4	MS09300
191	269	91	192	270	268	190	178	**408**	100	53	43	21	19	19	10	10	**0.10**	28	24	1	–	1	–	–	–	–	–	–	–	–	trans-1,2,3-Triphenylcyclopropyl-(4-tolyl) sulphoxide		MS09301
410	57	44	41	411	393	408	190	**408**	100	99	52	39	32	29	16	16		28	40	2	–	–	–	–	–	–	–	–	–	–	(2-Oxo-3,5-di-tert-butylcyclohexadienylidene)-3,5-di-tert-butylcyclohexadien-4-one		MS09302
44	57	408	41	351	56	55	409	**408**	100	66	45	40	18	17	16	15		28	40	2	–	–	–	–	–	–	–	–	–	–	2,2′,6,6′-Tetra-tert-butyldiphenoquinone		MS09303
351	352	57	408	410	409	393	309	**408**	100	72	64	60	21	19	16	15		28	40	2	–	–	–	–	–	–	–	–	–	–	2,2′,6,6′-Tetra-tert-butyldiphenoquinone		IC04323
408	177	149	91	260	69	167	232	**408**	100	99	87	64	63	55	45	42		29	28	2	–	–	–	–	–	–	–	–	–	–	Tetracyclo[$6.6.0.0^{2,6}.0^{9,13}$]tetradec-4-ene-3,14-dione, 11-(diphenylmethylidene)-1,4-dimethyl-, (1R,2R,6S,8S,9S,13S)-		DD01595

MASS TO CHARGE RATIOS								M.W.	INTENSITIES								Parent	C	H	O	N	S	F	Cl	Br	I	Si	P	B	X	COMPOUND NAME	CAS Reg No	No
57	43	309	71	41	55	69	56	**408**	100	93	68	63	43	41	27	19	**0.00**	29	60	–	–	–	–	–	–	–	–	–	–	–	8,8-Diheptylpentadecane		AI02346
43	85	71	56	54	69	83	41	**408**	100	99	99	99	88	77	73	67	**7.00**	29	60	–	–	–	–	–	–	–	–	–	–	–	Nonacosane		LI08631
43	57	71	41	85	55	69	56	**408**	100	94	54	40	38	30	16	15	**1.01**	29	60	–	–	–	–	–	–	–	–	–	–	–	Nonacosane		AI02347
365	393	366	364	408	379	308	394	**408**	100	31	29	29	14	14	13	11		29	60	–	–	–	–	–	–	–	–	–	–	–	Octacosane, 2-methyl-	1560-98-1	NI41118
365	175	121	203	162	107	135	119	**408**	100	52	48	38	37	37	35	35	**30.00**	30	48	–	–	–	–	–	–	–	–	–	–	–	A-Neooleana-3(5),12-diene	22586-84-1	NI41119
218	122	203	339	408	175	189	147	**408**	100	86	65	62	58	48	42	35		30	48	–	–	–	–	–	–	–	–	–	–	–	A-Neooleana-3,12-diene	22586-86-3	NI41120
408	255	409	119	393	109	257	256	**408**	100	40	34	33	27	20	18	18		30	48	–	–	–	–	–	–	–	–	–	–	–	A:D-Neooleana-12,14-diene, (3ξ,5α)-	55568-86-0	NI41121
217	203	218	189	105	107	119	109	**408**	100	38	25	16	15	14	13	12	**12.00**	30	48	–	–	–	–	–	–	–	–	–	–	–	24-Noroleana-4(23),12-diene, 3-methyl-, (3α)-	55331-48-1	NI41122
408	69	409	55	255	95	81	41	**408**	100	36	34	30	26	26	22	22		30	48	–	–	–	–	–	–	–	–	–	–	–	Oleana-11,13(18)-diene	54411-26-6	NI41123
69	81	40	215	97	52	119	120	**408**	100	53	36	17	16	14	12	11	**3.10**	30	48	–	–	–	–	–	–	–	–	–	–	–	Squalene, dehydro-		LI08632
271	202	69	149	229	189	272	245	**408**	100	63	62	57	54	47	46	44	**16.30**	30	48	–	–	–	–	–	–	–	–	–	–	–	Squalene, dehydro-		NI41124
69	81	271	137	119	95	408	272	**408**	100	53	24	19	12	10	9	5		30	48	–	–	–	–	–	–	–	–	–	–	–	Squalene, 12,13-didehydro-		DD01596
330	331	253	252	408	332	254	165	**408**	100	72	29	26	22	17	17	12		32	24	–	–	–	–	–	–	–	–	–	–	–	Anthracene, 4a,10-dihydro-9,10,10-triphenyl-	6549-73-1	NI41125
330	331	253	252	332	254	165	91	**408**	100	75	29	22	19	18	15	11	**4.23**	32	24	–	–	–	–	–	–	–	–	–	–	–	Anthracene, 9,10-dihydro-9,9,10-triphenyl-	809-40-5	NI41126
330	331	253	252	332	254	165	329	**408**	100	74	28	22	18	18	14	11	**4.00**	32	24	–	–	–	–	–	–	–	–	–	–	–	Anthracene, 9,10-dihydro-9,9,10-triphenyl-	809-40-5	LI08633
408	397	409	281	407	154	280	115	**409**	100	37	30	22	20	20	10	10		9	6	1	1	–	–	–	–	2	–	–	1	–	(5,7-Diiodo-8-quinolinato-O,N)borane	83210-20-2	NI41127
46	90	119	127	291	64	118	164	**409**	100	60	52	34	33	32	30	27	**0.00**	14	8	2	3	2	5	–	–	–	–	–	–	–	S,S-Bis(4-cyanophenoxy)-N-(pentafluorosulphanyl)sulphilimine		NI41128
192	148	89	146	123	160	147	132	**409**	100	43	20	15	15	14	14	14	**0.00**	16	18	5	1	–	–	3	–	–	–	–	–	–	1H-Indole-1,3-dicarboxylic acid, 2,3-dihydro-6-methoxy-, 1-(2,2,2-trichloro-1,1-dimethylethyl) 3-methyl ester		DD01597
322	240	109	83	55	253	43	41	**409**	100	86	74	60	59	40	39	38	**0.39**	16	22	3	1	–	7	–	–	–	–	–	–	–	N-(Heptafluorobutyryl)cyclohexylalanine propyl ester		MS09304
136	137	95	350	108	293	196	94	**409**	100	52	32	24	23	19	19	19	**1.50**	16	22	3	1	–	7	–	–	–	–	–	–	–	Propyl 2-((heptafluorobutyryl)amino)cyclohexanepropionate		MS09305
131	73	75	144	174	147	132	187	**409**	100	68	17	13	12	11	11	10	**1.57**	16	39	5	1	–	–	–	–	–	3	–	–	–	D-Galactopyranoside, methyl 2-amino-2-deoxy-3,4,6-tris-O-(trimethylsilyl)-	56272-07-2	NI41129
144	73	131	133	204	145	147	217	**409**	100	53	20	19	16	16	13	12	**0.00**	16	39	5	1	–	–	–	–	–	3	–	–	–	β-D-Glucopyranoside, methyl 2-amino-2-deoxy-3,4,6-tris-O-(trimethylsilyl)-	56227-30-6	NI41130
73	103	147	146	205	217	117	45	**409**	100	45	33	26	15	13	13	12	**0.00**	16	39	5	1	–	–	–	–	–	3	–	–	–	D-Ribose, 2-O-methyl-3,4,5-tris-O-(trimethylsilyl)-, O-methyloxime	56196-10-2	NI41131
73	103	147	249	146	89	45	219	**409**	100	52	24	23	19	13	13	11	**0.00**	16	39	5	1	–	–	–	–	–	3	–	–	–	D-Ribose, 3-O-methyl-2,4,5-tris-O-(trimethylsilyl)-, O-methyloxime	56196-11-3	NI41132
262	73	75	263	306	44	29	45	**409**	100	72	49	25	18	18	16	15	**0.00**	17	43	4	1	–	–	–	–	–	3	–	–	–	N,N-Bis(2-hydroxyethyl)-2-(2-hydroxyethoxy)ethylamine, tris(trimethylsilyl)-		IC04324
240	227	43	55	69	57	41	268	**409**	100	73	55	52	47	47	42	39	**4.40**	18	30	1	1	–	7	–	–	–	–	–	–	–	1-Aminotetradecane, N-(heptafluorobutyryl)-		MS09306
330	331	409	139	410	332	105	77	**409**	100	22	15	5	4	4	2	2		21	15	6	1	1	–	–	–	–	–	–	–	–	1-Hydroxy-4-(4-methylsulphonyloxyanilino)anthraquinone		IC04325
105	77	170	43	292	122	56	41	**409**	100	23	18	16	14	12	11	11	**0.00**	21	31	5	1	1	–	–	–	–	–	–	–	–	3-O-Benzoyl-6-acetylamino-4,6-dideoxy-1-N-hexylthio-α-DL-lyxo-hexopyranoside		NI41133
236	409	238	28	105	411	77	172	**409**	100	80	45	44	38	37	26	24		22	22	2	2	–	–	–	–	–	–	–	–	1	Copper, [[3,3'-(1,2-ethanediyldinitrilo)bis[1-phenyl-1-butanonato]](2-)-N,N',O,O']-	15277-26-6	NI41134
281	268	123	250	282	95	269	165	**409**	100	99	37	20	18	17	16	13	**0.80**	22	23	2	1	–	4	–	–	–	–	–	–	–	Trifluperidol		LI08634
91	105	77	79	65	92	274	318	**409**	100	57	34	16	12	10	9	6	**0.43**	23	23	6	1	–	–	–	–	–	–	–	–	–	7-Azabicyclo[2.2.1]heptane-2,7-dicarboxylic acid, 3-hydroxy-, 7-benzyl methyl ester, endo-2,exo-3-	17037-55-7	NI41135
70	326	311	395	28	283	255	165	**409**	100	91	46	21	14	12	9	7	**0.75**	24	27	5	1	–	–	–	–	–	–	–	–	–	Tylohirsutinidine, 13a-methyl-		NI41136
258	189	221	201	118	130	187	273	**409**	100	19	13	13	13	12	7	6	**1.00**	24	28	2	3	–	1	–	–	–	–	–	–	–	1,3,8-Triazabicyclo[2.2.2]oct-2-ene-5,8-dione, 8-[4-(4-fluorophenyl)-4-oxobutyl]-3-methyl-1-phenyl-		MS09307
77	104	289	105	278	290	119	131	**409**	100	56	48	40	38	25	15	8	**0.00**	25	19	3	3	–	–	–	–	–	–	–	–	–	2,6,7-Triazabicyclo[2.2.2]oct-2-ene-5,8-dione, 3-benzoyl-4-methyl-6,7-diphenyl-		NI41137
180	43	115	294	179	234	152	220	**409**	100	68	41	37	12	5	4	2	**0.10**	25	31	4	1	–	–	–	–	–	–	–	–	–	6-(2,4-Diacetoxy-1,1,3,3-tetramethylbutyl)-5,6-dihydrophenanthridine		LI08635
91	324	55	41	234	117	409	242	**409**	100	90	70	59	40	39	36	28		27	44	1	1	–	–	–	–	–	–	–	1	–	Borinic acid, diethyl-, [2-(cyclohexylimino)cyclodecyl]benzyl ester	74810-37-0	NI41138
318	44	319	91	95	81	55	41	**409**	100	71	24	17	10	10	10	9	**0.00**	28	43	1	1	–	–	–	–	–	–	–	–	–	3-(1-Benzylethylamino)-androstan-17-ol		LI08636
212	240	44	43	41	241	70	55	**409**	100	97	44	41	19	17	16	16	**0.00**	28	59	–	1	–	–	–	–	–	–	–	–	–	N-Tridecylpentadecylamine		IC04326
258	69	125	229	166	330	317	96	**410**	100	79	14	10	9	8	8	8	**0.00**	6	3	–	2	–	12	–	1	–	–	–	–	–	1,1-Bis[bis(trifluoromethyl)amino]-2-bromoethane		LI08637
69	330	166	258	125	110	78	96	**410**	100	39	25	22	20	17	16	11	**0.00**	6	3	–	2	–	12	–	1	–	–	–	–	–	1,2-Bis[bis(trifluoromethyl)amino]-1-bromoethane		LI08638
69	31	131	100	119	181	179	93	**410**	100	91	55	44	36	32	30	21	**0.00**	7	–	–	–	–	13	–	1	–	–	–	–	–	Perfluoro-trans-2-bromohept-2-ene		LI08639
380	225	69	341	391	311	410	397	**410**	100	99	99	89	70	31	30	30		7	3	4	–	–	12	–	–	–	–	1	–	–	2-Methoxy-4,4,5,5-tetrakis(trifluoromethyl)-1,3,2λ^5-dioxaphospholan-2-one	111557-90-5	DD01598

MASS TO CHARGE RATIOS								M.W.	INTENSITIES								Parent	C	H	O	N	S	F	Cl	Br	I	Si	P	B	X	COMPOUND NAME	CAS Reg No	No
213	163	169	19	113	150	119	182	410	100	40	23	13	7	4	4	3	0.00	8	–	3	–	–	14	–	–	–	–	–	–	–	Perfluoro-butyric anhydride (negative ion spectrum)		LI08640
169	69	197	119	100	31	28	147	410	100	75	22	19	11	6	5	4	0.00	8	–	3	–	–	14	–	–	–	–	–	–	–	Perfluoro-butyric anhydride (positive ion spectrum)		LI08641
294	177	178	390	410	208	238	264	410	100	99	95	84	83	80	57	46		8	18	4	4	2	–	–	–	–	–	–	–	2	Di-μ-(butanethiolato)tetranitrosodiiron		MS09308
397	395	412	399	410	414	401	377	410	100	78	58	48	45	28	10	10		9	13	–	4	–	–	4	–	–	–	3	–	–	1-Methyl-1-[4-(dimethylamino)phenyl]tetrachlorocyclotriphosphazene	87048-81-5	NI41139
241	310	260	291	210	341	272	117	410	100	27	22	19	11	10	10	9	1.20	12	–	–	–	–	14	–	–	–	–	–	–	–	Perfluorotetracyclo[6.2.2.$0^{2,7}$.$0^{5,7}$]dodeca-2,9-diene		NI41140
241	310	260	272	291	210	341	117	410	100	29	20	17	16	10	8	8	2.70	12	–	–	–	–	14	–	–	–	–	–	–	–	Perfluorotetracyclo[6.2.2.$0^{2,7}$.$0^{5,7}$]dodeca-2,9-diene		NI41141
241	310	291	260	272	410	210	117	410	100	45	19	11	10	9	6	5		12	–	–	–	–	14	–	–	–	–	–	–	–	Perfluorotricyclo[6.2.2.$0^{2,7}$]dodeca-2,6,9-triene		NI41142
171	297	77	63	299	51	109	155	410	100	86	68	45	33	33	30	22	0.00	13	10	2	–	1	–	2	1	–	–	1	–	–	Phosphonothioic acid, phenyl-, O-(4-bromo-2,4-dichlorophenyl) O-methyl ester	21609-90-5	NI41143
171	377	375	77	379	109	378	63	410	100	64	48	23	19	12	11	9	0.00	13	10	2	–	1	–	2	1	–	–	1	–	–	Phosphonothioic acid, phenyl-, O-(4-bromo-2,5-dichlorophenyl) O-methyl ester	21609-90-5	AI02348
171	377	28	375	77	155	60	379	410	100	62	60	45	37	26	21	17	0.00	13	10	2	–	1	–	2	1	–	–	1	–	–	Phosphonothioic acid, phenyl-, O-(4-bromo-2,5-dichlorophenyl) O-methyl ester	21609-90-5	NI41144
91	92	66	65	39	47	51	48	410	100	46	37	18	18	10	7	7	0.27	13	14	4	–	2	–	–	–	–	–	–	–	2	π-Bicyclo[2.2.1]hepta-2,5-dienyl-di-μ-(π-methylthio)tetracarbonyldiiron		MS09309
326	324	328	325	300	298	302	299	410	100	86	83	55	46	44	39	34	15.19	14	10	3	–	–	–	–	–	–	–	–	–	1	Tungsten, phenyltricarbonyl-π-cyclopentadienyl-	12131-09-8	NI41145
204	73	147	205	206	75	45	255	410	100	78	23	19	8	8	7	5	0.00	14	19	9	–	–	–	–	1	–	–	–	–	–	D-Galactopyranosyl bromide, tetraacetate	21698-13-5	NI41146
183	115	157	141	103	184	99	189	410	100	74	69	39	24	12	11	8	0.00	14	19	9	–	–	–	–	1	–	–	–	–	–	α-D-Glucopyranose, 6-bromo-6-deoxy-, tetraacetate	7404-34-4	NI41147
169	116	109	127	115	97	43	331	410	100	73	61	26	23	19	18	14	0.00	14	19	9	–	–	–	–	1	–	–	–	–	–	β-D-Glucopyranose, 1-deoxy-1-bromo-, tetraacetate	6919-96-6	NI41148
197	278	276	161	55	91	199	53	410	100	76	60	58	33	31	28	27	0.06	15	21	1	–	–	–	1	2	–	–	–	–	–	Chamigran-7-en-9-one, 2,10-dibromo-3-chloro-		MS09310
246	105	248	75	43	176	318	44	410	100	73	68	58	58	57	50	48	0.00	16	11	2	–	–	–	5	–	–	–	–	–	–	Benzenemethanol, 4-chloro-α-(4-chlorophenyl)-α-(trichloroethyl)-, acetate	3567-16-6	NI41149
57	41	29	197	51	353	199	195	410	100	48	35	28	22	21	19	18	0.00	16	19	–	–	–	–	–	1	–	–	–	–	1	Tin, bromodiphenyl-tert-butyl-	30191-73-2	NI41150
122	84	137	43	410	295	274	346	410	100	87	52	42	30	27	9	4		16	22	1	6	3	–	–	–	–	–	–	–	–	Spiro[3-(2-methyl-4-aminopyrimidin-5-yl) methyl-3a-methylperhydrofuro[2,3-d]thiazole-2,4′-(1′,3′-dimethylimidazolidine-2′,5′-dithione)]		LI08642
73	147	410	395	75	45	116	411	410	100	46	40	33	22	18	17	14		16	30	3	4	–	–	–	–	–	3	–	–	–	Pteridine, 6-methyl-2,4,7-tris(trimethylsiloxy)-	31053-50-6	NI41151
75	147	410	395	77	45	116	411	410	100	49	42	37	24	20	18	15		16	30	3	4	–	–	–	–	–	3	–	–	–	Pteridine, 6-methyl-2,4,7-tris(trimethylsiloxy)-	31053-50-6	LI08643
73	205	147	103	207	305	320		410	100	36	36	36	34	10	2		0.00	16	42	4	–	–	–	–	–	–	4	–	–	–	3,8-Dioxa-2,9-disiladecane, 2,2,9,9-tetramethyl-5,6-bis[(trimethylsilyl)oxy]-, (R*,S*)-	25258-02-0	LI08644
73	205	217	147	103	117	204	189	410	100	39	37	37	36	19	13	12	0.00	16	42	4	–	–	–	–	–	–	4	–	–	–	3,8-Dioxa-2,9-disiladecane, 2,2,9,9-tetramethyl-5,6-bis[(trimethylsilyl)oxy]-, (R*,S*)-	25258-02-0	NI41153
73	103	147	217	205	117	45	74	410	100	25	23	13	13	13	9	8	0.00	16	42	4	–	–	–	–	–	–	4	–	–	–	3,8-Dioxa-2,9-disiladecane, 2,2,9,9-tetramethyl-5,6-bis[(trimethylsilyl)oxy]-, (R*,S*)-	25258-02-0	NI41152
73	103	217	205	147	117	204	189	410	100	39	38	38	36	18	13	12	0.00	16	42	4	–	–	–	–	–	–	4	–	–	–	D-Threitol, 1,2,3,4-tetrakis-O-(trimethylsilyl)-	32381-52-5	NI41154
112	261	410	412	363	263	68	365	410	100	7	6	3	3	2	2	1		17	31	5	2	1	–	1	–	–	–	–	–	–	L-threo-α-D-Galacto-octopyranoside, methyl 7-chloro-6,7,8-trideoxy-6-[[(4-propyl-2-pyrrolidinyl)carbonyl]amino]-1-thio-, (2S-trans)-	22431-45-4	NI41155
45	63	107	65	136	151	27	109	410	100	83	45	28	25	24	24	18	8.58	18	28	6	–	–	–	2	–	–	–	–	–	–	Ethanol, 2,2′-[1,2-phenylenebis[(2-chloro-2,1-ethanediyl)oxy-2,1-ethanediyloxy]]bis-	74810-54-1	NI41156
192	73	75	193	45	293	176	147	410	100	53	19	16	14	12	7	7	5.10	18	34	3	2	–	–	–	–	–	3	–	–	–	Glycine, N-(4-aminobenzoyl)-, tris(trimethylsilyl)-	56272-89-0	NI41157
410	411	124	45	59	101	75	61	410	100	20	17	17	14	10	10	9		18	34	10	–	–	–	–	–	–	–	–	–	–	Hepta-O-methylprimverose		LI08645
283	55	74	69	57	83	43	97	410	100	87	78	71	62	57	57	55	4.50	18	35	2	–	–	–	–	–	1	–	–	–	–	Heptadecanoic acid, 7-iodo-, methyl ester	57289-61-9	NI41158
105	410							410	100	98								19	10	2	–	–	–	4	–	–	–	–	–	–	1,3-Benzodioxole, 4,5,6,7-tetrachloro-2,2-diphenyl-	31701-11-8	NI41159
43	97	139	157	110	259	69	199	410	100	99	78	36	36	35	16	10	0.00	19	22	10	–	–	–	–	–	–	–	–	–	–	β-D-Arabinopyranoside, 4-acetoxyphenyl 2,3,4-tri-O-acetyl-	84364-56-7	NI41160
73	81	135	114	93	189	248	410	410	100	72	27	22	15	6	4	1		19	29	5	–	–	2	1	–	–	–	–	–	–	Methyl 10-(chlorodifluoroacetoxy)-11-methoxy-3,7,11-trimethyl-2,6-dodecadienoate		MS09311
333	331	329	197	179	332	177	330	410	100	85	44	42	41	40	32	31	12.78	20	18	2	–	–	–	–	–	–	–	–	–	1	Stannane, (acetyloxy)triphenyl-	900-95-8	NI41161
333	331	51	77	179	329	120	197	410	100	85	75	67	52	41	41	40	4.20	20	18	2	–	–	–	–	–	–	–	–	–	1	Stannane, (acetyloxy)triphenyl-	900-95-8	NI41162
259	233	277	261	217	410	246	392	410	100	83	45	35	28	26	19	15		20	26	9	–	–	–	–	–	–	–	–	–	–	4H-1-Benzopyran-4-one, 8-C-β-D-glucopyranosyl-2-[(R)-2-hydroxypropyl]-7-methoxy-5-methyl-		NI41163
209	145	410	222	179	210	320	290	410	100	22	16	14	14	11	8	5		21	38	4	–	–	–	–	–	–	2	–	–	–	3-Octanone, 1-[3-methoxy-4-[(trimethylsilyl)oxy]phenyl]-5-[(trimethylsilyl)oxy]-	56700-94-8	MS09312
77	337	410	183	105	154	338	93	410	100	76	48	21	21	20	17	17		22	22	6	2	–	–	–	–	–	–	–	–	–	Propanedioic acid, (3,5-dioxo-1,2-diphenyl-4-pyrazolidinyl)-, diethyl ester	6139-70-4	NI41164
135	191	163	88	136	192	134	410	410	100	72	66	54	47	35	35	30		22	26	4	4	–	–	–	–	–	–	–	–	–	1,2,4,5-Tetrazine, 3,6-bis(3,4-diethoxyphenyl)-	77355-37-4	NI41165
108	121	152	109	165	153	137	111	410	100	76	57	57	50	46	46	46	4.00	22	34	7	–	–	–	–	–	–	–	–	–	–	Picras-2-en-1-one, 11,13,16-trihydroxy-2,12-dimethoxy-, (11α,12β)-	30248-05-6	NI41166
192	191	410	177	219	149	147		410	100	29	20	13	3	3	3			23	22	7	–	–	–	–	–	–	–	–	–	–	Amorphogenin		LI08646
197	198	105	77	134	298	119	91	410	100	16	14	10	9	6	4	4	0.00	23	23	2	–	–	–	–	1	–	–	–	–	–	1-Bromo-3-methoxy-2-(4-methoxyphenyl)-3,3-diphenylpropane		MS09313

MASS TO CHARGE RATIOS								M.W.	INTENSITIES								Parent	C	H	O	N	S	F	Cl	Br	I	Si	P	B	X	COMPOUND NAME	CAS Reg No	No
410	367	411	326	341	368	353	340	**410**	100	42	29	26	23	21	19	13		23	26	5	2	–	–	–	–	–	–	–	–	–	3H-2,3-Benzodiazepine, 3-acetyl-1-(3,4-dimethoxyphenyl)-5-ethyl-7,8-dimethoxy-	90140-65-1	NI41167
91	351	410	319	65	242	106	310	**410**	100	30	16	13	7	5	5	4		23	26	5	2	–	–	–	–	–	–	–	–	–	2,4-Cyclohexadiene-2,6-dicarboxylic acid, 3-(benzylamino)-5-methyl-4-(3,5-dimethylisoxazolyl)-, dimethyl ester		NI41168
378	410	379	215	350	122	109	201	**410**	100	47	37	19	16	15	14	10		23	26	5	2	–	–	–	–	–	–	–	–	–	4,9-Cyclo-9,19-secoaspidofractinine-1,3-dicarboxylic acid, 15-hydroxy-19-oxo-, dimethyl ester, (3β)-		MS09314
410	180	116	267	282	168	239	194	**410**	100	65	54	51	43	32	28	26		23	26	5	2	–	–	–	–	–	–	–	–	–	Na,11α-Bis(methoxycarbonyl)-3-oxoaspidofractinine		LI08647
180	410	267	116	378	168	110	282	**410**	100	77	56	56	46	37	37	33		23	26	5	2	–	–	–	–	–	–	–	–	–	Na,11β-Bis(methoxycarbonyl)-3-oxoaspidofractinine		LI08648
180	410	267	116	378	168	282	195	**410**	100	77	56	56	46	37	33	32		23	26	5	2	–	–	–	–	–	–	–	–	–	Na,11β-Bis(methoxycarbonyl)-3-oxoaspidofractinine		LI08649
410	257	411	353	412	409	395	354	**410**	100	47	44	33	30	30	30	27		23	26	5	2	–	–	–	–	–	–	–	–	–	Oxayohimban-16-carboxylic acid, 16,17,19,20-tetradehydro-10,11-dimethoxy-19-methyl-, methyl ester, (3β)-	70509-80-7	NI41169
410	70	61	69	180	167	411	73	**410**	100	72	65	52	48	46	41	41		23	26	5	2	–	–	–	–	–	–	–	–	–	Pyrido[3',4':5,6]cyclohept[1,2-b]indole-5-carboxylic acid, 2-acetyl-1,2,3,4,4a,5,6,11,12,12a-decahydro-12a-hydroxy-5-(hydroxymethyl)-12-methyl-, γ-lactone, acetate	3368-89-6	NI41170
410	214	339	58	71	57	278	240	**410**	100	36	34	32	30	27	21	21		23	26	5	2	–	–	–	–	–	–	–	–	–	16,19-Secostrychnidine-10,16-dione, 21,22-epoxy-21,22-dihydro-4-methoxy-19-methyl-, (21α,22α)-	62421-68-5	NI41171
238	350	266	291	292	212	167	294	**410**	100	68	62	47	29	28	27	26	**6.00**	23	26	5	2	–	–	–	–	–	–	–	–	–	Talbotine, 4-formyl-O-methyl-	30809-19-9	NI41172
195	243	45	99	55	410	43	41	**410**	100	59	33	28	28	25	23	21		23	38	6	–	–	–	–	–	–	–	–	–	–	6-Ethylenedioxy-17β-hydroxy-4,5-secoandrostan-4-oic acid (2-hydroxyethyl) ester		MS09315
267	268	354	326	265	165	189	87	**410**	100	24	18	8	7	7	6	5	**0.10**	24	15	3	–	–	–	–	–	–	–	–	–	1	η-(Triphenylcyclopropenyl)tricarbonylcobalt		MS09316
205	410	77	220	411	165	102	409	**410**	100	95	31	27	26	25	24	23		24	18	3	4	–	–	–	–	–	–	–	–	–	3-[2-(2-Hydroxyethyl)-1,3-dioxo-2H-benzo[c]pyrrylazo]-2-phenylindole		IC04327
410	379	395	181	380	341	355	365	**410**	100	95	28	27	25	25	17	9		24	26	6	–	–	–	–	–	–	–	–	–	–	4H-1-Benzopyran-4-one, 3-[2,4-dimethoxy-3-(3-methyl-2-butenyl)phenyl]-5,7-dimethoxy-		MS09317
410	379	411	409	381	380	377	205	**410**	100	40	26	18	13	11	9	9		24	26	6	–	–	–	–	–	–	–	–	–	–	4H-1-Benzopyran-4-one, 3-(2,4-dimethoxyphenyl)-5,7-dimethoxy-8-(3-methyl-2-butenyl)-		NI41173
151	410	260	190	259	395	152	409	**410**	100	97	20	20	17	13	11	9		24	26	6	–	–	–	–	–	–	–	–	–	–	4H-1-Benzopyran-4-one, 3-[2,4,5-trimethoxy-6-(3-methyl-2-butenyl)phenyl]-7-methoxy-		MS09318
341	299	410	342	288	41	311	69	**410**	100	27	25	22	22	16	15	12		24	26	6	–	–	–	–	–	–	–	–	–	–	1,3,6-Trihydroxy-7-methoxy-8-(3,7-dimethyl-2,6-octadienyl)xanthone		NI41174
77	91	105	103	79	104	78	121	**410**	100	99	64	55	55	41	41	36	**32.00**	24	27	4	–	–	–	–	–	–	–	1	–	–	Trixylyl phosphate		IC04328
410	28	411	121	409	122	90.5	44	**410**	100	62	27	26	21	20	19	19		24	27	4	–	–	–	–	–	–	–	1	–	–	Trixylyl phosphate		MS09319
209	410	121	122	193	107	305	106	**410**	100	85	85	82	78	58	54	42		24	27	4	–	–	–	–	–	–	–	1	–	–	Tris(3-xylyl) phosphate		IC04329
182	410	224	186	367	395	352	242	**410**	100	52	52	38	30	25	20	20		24	30	4	2	–	–	–	–	–	–	–	–	–	Ajmalinimine		MS09320
43	410	167	182	350	188	367	411	**410**	100	99	48	41	32	28	25	24		24	30	4	2	–	–	–	–	–	–	–	–	–	Aspidofractinin-6-ol, 1-acetyl-17-methoxy-, acetate (ester), (6β)-	54965-84-3	NI41175
410	323	174	180	351	208	256	173	**410**	100	60	57	46	37	25	20	18		24	30	4	2	–	–	–	–	–	–	–	–	–	Curan-17-ol, 1-acetyl-19,20-didehydro-10-methoxy-, acetate (ester), (19E)	59630-34-1	NI41176
73	410	395	411	193	396	45	74	**410**	100	55	30	20	14	11	10	7		24	34	2	–	–	–	–	–	–	2	–	–	–	Dienoestrol, bis(trimethylsilyl) ether		MS09321
135	134	201	136	410	309	103	192	**410**	100	18	15	12	10	10	10	5		25	18	2	2	1	–	–	–	–	–	–	–	–	Pyrazolo[5,1-b]thiazole, 7-benzoyl-6-(4-methoxyphenyl)-3-phenyl-		MS09322
69	313	382	410	91	105	314	93	**410**	100	91	55	28	28	25	23	23		25	30	5	–	–	–	–	–	–	–	–	–	–	1-Anthracenecarboxaldehyde, 5,8,8a,9,10,10a-hexahydro-2,3-dihydroxy-10a-methyl-4-isopropyl-6-(4-methyl-3-pentenyl)-9,10-dioxo-, cis-	64960-69-6	NI41177
410	395	354	325	367	149	396	151	**410**	100	48	44	31	19	18	14	14		25	30	5	–	–	–	–	–	–	–	–	–	–	6H-Benzofuro[3,2-c][1]benzopyran, 3,9-diethoxy-6a,11a-dihydro-7-(3,3-dimethylallyl)-8-methoxy-		MS09323
86	55	57	100	82	69	83	71	**410**	100	23	22	15	15	12	10	10	**1.80**	25	34	3	2	–	–	–	–	–	–	–	–	–	2,7-Bis[2-(diethylamino)-ethoxy]-fluoren-9-one		MS09324
159	101	157	217	395	223	43	59	**410**	100	42	10	8	5	5	5	3	**0.40**	25	46	4	–	–	–	–	–	–	–	–	–	–	1-O-(2-Propoxy-4-hexadecenyl)-2,3-O-isopropylideneglycerol		NI41178
236	77	104	130	115	91	410	133	**410**	100	63	50	35	22	21	14	11		26	22	3	2	–	–	–	–	–	–	–	–	–	Pyridazino[1,2-b]phthalazine-6,11-dione, 1-ethoxy-1,4,6,11-tetrahydro-1,4 diphenyl-		NI41179
395	396	410	397	411	412	295	40	**410**	100	33	17	9	6	2	2	2		26	38	2	–	–	–	–	–	–	1	–	–	–	Cannabinol propyldimethylsilyl ether		NI41180
337	58	308	30	338	148	264	73	**410**	100	46	33	31	25	21	14	14	**0.00**	26	42	–	4	–	–	–	–	–	–	–	–	–	Ethylenediamine, 1,2-bis[4-(N,N-diethylamino)phenyl]-N,N,N',N'-tetraethyl-, (±)-		DD01599
337	308	338	58	264	148	30	253	**410**	100	29	25	15	11	11	10	7	**0.00**	26	42	–	4	–	–	–	–	–	–	–	–	–	Ethylenediamine, 1,2-bis[4-(N,N-diethylamino)phenyl]-N,N,N',N'-tetraethyl-, meso-		DD01600
410	411	104	105	160	77	133	76	**410**	100	37	34	30	29	28	26	24		27	22	4	–	–	–	–	–	–	–	–	–	–	Bis(1,2,3,4-tetrahydro-1,4-dioxonaphtho)[3,4-b:5,6-b]tricyclo[6.2.1.0^{2,7}]undecane		NI41181
410	183	262	165	108	185	201	152	**410**	100	48	43	18	14	11	8	6		27	23	2	–	–	–	–	–	–	–	1	–	–	[(Methoxycarbonyl)benzylene]triphenylphosphorane		LI08650
43	127	109	69	41	55	145	410	**410**	100	77	77	57	43	29	28	27		27	38	3	–	–	–	–	–	–	–	–	–	–	10'-Apo-β,4-carotenal, 5,6-dihydro-5,6-dihydroxy-, (5R,6R)-	57951-40-3	NI41182
43	177	161	392	57	71	340	164	**410**	100	69	63	55	45	41	39	34	**34.00**	27	38	3	–	–	–	–	–	–	–	–	–	–	Bis(2-hydroxy-3-tert-butyl-5-methylphenyl)methane, mono-2-methylpropanoate		NI41183
69	83	143	97	95	81	85	109	**410**	100	61	58	56	42	39	38	33	**0.31**	27	54	2	–	–	–	–	–	–	–	–	–	–	1,3-Dioxane, 2,2,4,5-tetramethyl-6-(1-methyloctadecyl)-	56324-82-4	NI41184

MASS TO CHARGE RATIOS	M.W.	INTENSITIES	Parent	C	H	O	N	S	F	Cl	Br	I	Si	P	B	X	COMPOUND NAME	CAS Reg No	No
74 87 410 57 75 55 43 69	**410**	100 76 47 36 34 30 29 28		27	54	2	–	–	–	–	–	–	–	–	–	–	**Hexacosanoic acid, methyl ester**	5802-82-4	**WI01526**
74 87 43 41 55 75 57 143	**410**	100 74 73 35 32 30 25 23	**5.00**	27	54	2	–	–	–	–	–	–	–	–	–	–	**Hexacosanoic acid, methyl ester**	5802-82-4	**NI41185**
87 74 43 57 55 41 71 69	**410**	100 40 39 32 29 20 18 17	**8.20**	27	54	2	–	–	–	–	–	–	–	–	–	–	**Pentacosanoic acid, 4-methyl-, methyl ester**		**MS09325**
88 410 44 101 74 334 57 87	**410**	100 77 63 57 46 38 32 26		27	54	2	–	–	–	–	–	–	–	–	–	–	**Pentacosanoic acid, 6-methyl-, methyl ester, (S)-**	57289-52-8	**NI41186**
88 57 43 55 71 41 101 44	**410**	100 74 64 58 30 30 28 24	**19.00**	27	54	2	–	–	–	–	–	–	–	–	–	–	**Tetracosanoic acid, 2,9-dimethyl-, methyl ester**		**MS09326**
88 57 44 43 410 69 55 101	**410**	100 43 32 30 29 29 29 22		27	54	2	–	–	–	–	–	–	–	–	–	–	**Tetracosanoic acid, 2,9-dimethyl-, methyl ester, DL-(+)-**		**MS09327**
101 74 75 334 57 70 55 69	**410**	100 89 46 38 29 25 25 24	**2.70**	27	54	2	–	–	–	–	–	–	–	–	–	–	**Tetracosanoic acid, 3,6-dimethyl-, methyl ester, [S-(R*,S*)]-**	57274-47-2	**NI41187**
410 57 411 41 395 29 55 190	**410**	100 96 32 29 18 18 8 6		28	42	2	–	–	–	–	–	–	–	–	–	–	**4,4′-Dihydroxy-3,3′,5,5′-tetrabutylbiphenyl**		**IC04330**
29 56 30 46 45 57 191 410	**410**	100 38 34 29 22 13 10 9		28	42	2	–	–	–	–	–	–	–	–	–	–	**4,4′-Dihydroxy-3,3′,5,5′-tetra-tert-butylbiphenyl**		**IC04331**
96 410 94 52 95 97 411 53	**410**	100 96 75 71 47 44 42 40		28	46	–	2	–	–	–	–	–	–	–	–	–	**2′H-Cholest-2-eno[3,2-c]pyrazole, (5α)-**	28895-41-2	**NI41188**
100 56 55 54 42 41 29 28	**410**	100 97 97 97 97 97 97 97	**0.00**	28	58	1	–	–	–	–	–	–	–	–	–	–	**1-Octacosanol**	557-61-9	**NI41189**
410 55 149 395 411 97 367 81	**410**	100 40 37 35 34 33 29 29		29	46	1	–	–	–	–	–	–	–	–	–	–	**Cholesta-2,22-dien-6-one, 24-ethyl-, (5α,22E,24S)-**		**NI41190**
410 271 367 269 298 213 283 229	**410**	100 90 77 58 55 23 22 15		29	46	1	–	–	–	–	–	–	–	–	–	–	**Cholesta-4,22-dien-3-one, 24-ethyl-, (24R)-**		**NI41191**
55 271 69 410 83 123 367 97	**410**	100 56 50 47 47 39 38 36		29	46	1	–	–	–	–	–	–	–	–	–	–	**Cholesta-4,22-dien-6-one, 24-ethyl-, (22E,24S)-**		**NI41192**
43 55 155 57 169 41 211 69	**410**	100 60 59 55 52 52 49 47	**11.00**	29	46	1	–	–	–	–	–	–	–	–	–	–	**Cholesta-5,7,9(11)-trien-3-ol, 4,4-dimethyl-, (3β)-**		**NI41193**
410 395 325 246 298 283 377 328	**410**	100 56 55 41 31 31 24 21		29	46	1	–	–	–	–	–	–	–	–	–	–	**Cholesta-8,14,24-trien-3-ol, 4,4-dimethyl-, (3β,5α)-**		**MS09328**
410 55 271 81 69 367 121 79	**410**	100 90 48 45 45 43 43 42		29	46	1	–	–	–	–	–	–	–	–	–	–	**3,5-Cyclocholest-22-en-6-one, 24-ethyl-, (3α,5α,22E,24S)-**		**NI41194**
69 56 95 81 125 83 67 58	**410**	100 57 44 44 37 37 28 27	**9.90**	29	46	1	–	–	–	–	–	–	–	–	–	–	**Cycloprop[7,8]ergost-22-en-3-one, 3′,7-dihydro-, (5α,7β,8α,22E)-**	53755-18-3	**NI41195**
271 96 81 272 55 95 41 43	**410**	100 35 35 32 29 27 21 19	**5.00**	29	46	1	–	–	–	–	–	–	–	–	–	–	**33-Norgorgosta-5,24(28)-dien-3-ol, (3β)-**	55064-62-5	**NI41196**
163 191 190 410 164 192 121 119	**410**	100 56 20 15 14 11 9 9		29	46	1	–	–	–	–	–	–	–	–	–	–	**28-Norolean-17-en-3-one**	5912-72-1	**NI41197**
269 271 367 123 147 229 410 106	**410**	100 34 24 24 22 21 20 19		29	46	1	–	–	–	–	–	–	–	–	–	–	**Spinasterone**	23455-44-9	**NI41198**
43 174 410 41 44 55 57 69	**410**	100 77 66 62 60 50 42 39		29	46	1	–	–	–	–	–	–	–	–	–	–	**Stigmasta-3,5-dien-7-one**	2034-72-2	**WI01527**
312 313 69 95 81 67 93 79	**410**	100 78 73 55 55 46 42 42	**17.91**	29	46	1	–	–	–	–	–	–	–	–	–	–	**Stigmasta-4,24(28)-dien-3-one**	56143-35-2	**NI41199**
312 313 55 297 69 410 95 81	**410**	100 79 55 31 31 26 24 23		29	46	1	–	–	–	–	–	–	–	–	–	–	**Stigmasta-4,24(28)-dien-3-one**	56143-35-2	**NI41200**
312 55 313 69 41 95 81 231	**410**	100 85 80 42 37 32 32 27	**17.17**	29	46	1	–	–	–	–	–	–	–	–	–	–	**Stigmasta-4,24(28)-dien-3-one, (24E)-**	53755-09-2	**NI41201**
410 55 83 69 411 43 81 95	**410**	100 58 43 41 33 33 28 20		29	46	1	–	–	–	–	–	–	–	–	–	–	**Stigmasta-4,6,22-trien-3-ol, (3α)-**		**NI41202**
410 55 83 69 411 43 81 95	**410**	100 69 48 47 36 35 27 20		29	46	1	–	–	–	–	–	–	–	–	–	–	**Stigmasta-4,6,22-trien-3-ol, (3β)-**	96737-58-5	**NI41203**
55 392 83 43 81 69 253 41	**410**	100 72 65 57 56 56 48 42	**38.00**	29	46	1	–	–	–	–	–	–	–	–	–	–	**Stigmasta-4,7,22-trien-3-ol, (3α)-**		**NI41204**
392 55 83 69 81 410 253 43	**410**	100 89 60 55 54 51 50 50		29	46	1	–	–	–	–	–	–	–	–	–	–	**Stigmasta-4,7,22-trien-3-ol, (3β)-**	96737-57-4	**NI41205**
377 410 253 351 378 411 271 352	**410**	100 90 43 37 32 28 18 13		29	46	1	–	–	–	–	–	–	–	–	–	–	**Stigmasta-5,7,22-trien-3-ol, (3β)-**	481-19-6	**NI41206**
109 300 213 137 255 145 159 371	**410**	100 76 76 64 50 36 35 34	**10.00**	29	46	1	–	–	–	–	–	–	–	–	–	–	**Stigmasta-5,22,25-trien-3-ol, (3β,24S)-**	26315-07-1	**MS09329**
271 410 81 55 107 95 255 273	**410**	100 59 45 44 42 40 35 32		29	46	1	–	–	–	–	–	–	–	–	–	–	**Stigmasta-7,16,25-trien-3-ol, (3β,5α)-**	17808-69-4	**MS09330**
231 243 232 109 123 205 410 137	**410**	100 95 95 91 90 55 48 44		30	50	–	–	–	–	–	–	–	–	–	–	–	**Bauerene**		**NI41207**
109 205 63 41 123 110 108 77	**410**	100 22 17 12 11 10 10 10	**0.00**	30	50	–	–	–	–	–	–	–	–	–	–	–	**1,2-Dicyclohexyl-1,2-bis(2-endo-methyl-2-exo-norbornyl)ethane, meso-**		**NI41208**
204 410 286 408 218 284 205 271	**410**	100 50 45 38 37 30 30 25		30	50	–	–	–	–	–	–	–	–	–	–	–	**D-Friedoolean-14-ene**	546-14-5	**LI08651**
204 410 286 408 218 205 284 69	**410**	100 49 45 36 33 28 27 22		30	50	–	–	–	–	–	–	–	–	–	–	–	**D-Friedoolean-14-ene**	546-14-5	**NI41209**
218 95 109 69 55 121 137 81	**410**	100 77 75 58 56 55 52 46	**16.00**	30	50	–	–	–	–	–	–	–	–	–	–	–	**D:A-Friedoolean-6-ene**	56588-25-1	**NI41210**
286 271 287 122 121 133 123 55	**410**	100 33 21 9 9 8 8 8	**5.10**	30	50	–	–	–	–	–	–	–	–	–	–	–	**D:A-Friedoolean-7-ene**	18671-56-2	**NI41211**
203 177 149 123 121 395 217 135	**410**	100 93 72 71 68 47 44 43	**8.37**	30	50	–	–	–	–	–	–	–	–	–	–	–	**D:A-Friedoolean-18-ene**	3391-15-9	**NI41212**
410 191 204 68 411 95 81 395	**410**	100 77 51 48 43 40 37 28		30	50	–	–	–	–	–	–	–	–	–	–	–	**C(14a)-Homo-27-norgammacer-14-ene**	18046-86-1	**LI08652**
410 191 204 70 411 95 81 395	**410**	100 76 49 45 42 38 33 25		30	50	–	–	–	–	–	–	–	–	–	–	–	**C(14a)-Homo-27-norgammacer-14-ene**	18046-86-1	**NI41213**
410 231 395 243 411 206 205 232	**410**	100 99 50 44 34 25 23 22		30	50	–	–	–	–	–	–	–	–	–	–	–	**Isobauerene**		**LI08653**
191 410 205 218 69 55 95 81	**410**	100 63 58 55 54 53 42 39		30	50	–	–	–	–	–	–	–	–	–	–	–	**A′-Neogammacer-22(29)-ene**	1615-91-4	**NI41214**
218 203 219 118 122 120 204 135	**410**	100 25 19 10 9 9 8 8	**5.08**	30	50	–	–	–	–	–	–	–	–	–	–	–	**Olean-12-ene**	471-68-1	**NI41216**
218 203 41 55 219 191 69 43	**410**	100 43 22 21 20 19 18 18	**4.00**	30	50	–	–	–	–	–	–	–	–	–	–	–	**Olean-12-ene**	471-68-1	**NI41215**
205 109 191 218 206 189 204 107	**410**	100 84 62 57 47 38 33 29	**28.03**	30	50	–	–	–	–	–	–	–	–	–	–	–	**Olean-13(18)-ene**	3399-27-7	**NI41217**
204 189 177 191 205 120 410 395	**410**	100 73 70 49 41 32 26 25		30	50	–	–	–	–	–	–	–	–	–	–	–	**Olean-18-ene**	432-11-1	**NI41218**
204 189 177 95 69 55 41 81	**410**	100 73 69 62 56 54 54 51	**24.45**	30	50	–	–	–	–	–	–	–	–	–	–	–	**Olean-18-ene**	432-11-1	**NI41219**
69 81 95 41 137 93 67 68	**410**	100 66 26 23 18 18 18 17	**1.86**	30	50	–	–	–	–	–	–	–	–	–	–	–	**Squalene**	7683-64-9	**NI41220**
81 137 95 93 70 121 136 109	**410**	100 33 25 21 19 16 15 14	**1.10**	30	50	–	–	–	–	–	–	–	–	–	–	–	**Squalene**	7683-64-9	**WI01528**
137 123 109 151 149 191 135 111	**410**	100 38 33 24 19 18 16 15	**5.00**	30	50	–	–	–	–	–	–	–	–	–	–	–	**Squalene**	7683-64-9	**NI41221**
81 69 137 136 95 121 123 149	**410**	100 91 38 29 27 21 19 17	**2.97**	30	50	–	–	–	–	–	–	–	–	–	–	–	**Squalene, (all-E)-**	111-02-4	**NI41222**
218 41 55 69 81 95 191 43	**410**	100 37 35 32 25 24 21 20	**10.89**	30	50	–	–	–	–	–	–	–	–	–	–	–	**12-Ursene**	464-97-1	**NI41223**
333 410 334 411 409 302 105 289	**410**	100 76 31 25 11 8 8 7		31	22	1	–	–	–	–	–	–	–	–	–	–	**2,4,6-Triphenylbenzophenone**		**IC04332**
105 131 118 106 173 119 81 44	**410**	100 25 16 12 10 10 10 8	**2.25**	31	38	–	–	–	–	–	–	–	–	–	–	–	**Benzene, 1,1′,1″-[1-(3-methyl-3-butenyl)-1,3,5-pentanetriyl]tris[4-methyl-**	74685-55-5	**NI41224**

MASS TO CHARGE RATIOS	M.W.	INTENSITIES	Parent	C	H	O	N	S	F	Cl	Br	I	Si	P	B	X	COMPOUND NAME	CAS Reg No	No
334 278 332 276 336 280 57 240	411	100 95 79 70 47 45 43 39	9.00	4	9	–	3	–	–	4	1	–	–	3	–	–	1-Butyl-1-bromotetrachlorocyclotriphosphazene	77611-13-3	NI41225
57 278 276 334 280 332 336 240	411	100 67 51 35 33 28 18 14	1.00	4	9	–	3	–	–	4	1	–	–	3	–	–	1-tert-Butyl-1-bromotetrachlorocyclotriphosphazene	77589-29-8	NI41226
248 412 230 249 229 413 247 250	411	100 80 10 9 9 8 4 1	0.00	10	7	5	1	–	10	–	–	–	–	–	–	–	Serine, methyl ester, N,O-bis(pentafluoropropionyl)-		NI41227
248 412 165 216 249 413 352 380	411	100 97 29 28 20 15 12 10	0.00	10	7	5	1	–	10	–	–	–	–	–	–	–	Serine, methyl ester, N,O-bis(pentafluoropropionyl)-		NI41228
57 334 331 329 336 85 332 327	411	100 22 15 15 13 13 12 5	0.00	11	12	1	1	–	–	–	3	–	–	–	–	–	Propanamide, 2,2-dimethyl-N-(2,4,6-tribromophenyl)-	56619-92-2	NI41229
184 84 233 211 59 209 68 56	411	100 55 16 16 16 15 14 12	0.00	15	16	6	1	–	–	3	–	–	–	–	–	–	L-Glutamic acid, N-[(2,4,5-trichlorophenoxy)acetyl]-, dimethyl ester	75627-01-9	NI41230
54 67 39 80 79 27 41 66	411	100 89 57 48 46 32 26 19	0.56	16	24	–	–	–	–	–	–	–	–	–	–	1	Platinum, bis[(1,2,5,6-η)-1,5-cyclooctadiene]-	12130-66-4	NI41231
139 97 193 259 112 154 69 85	411	100 84 54 44 44 43 37 26	0.30	17	21	9	3	–	–	–	–	–	–	–	–	–	Cytidine, N-acetyl-, 2′,3′,5′-triacetate	5040-18-6	MS09331
411 413 412 137 410 36 274 414	411	100 85 72 65 60 48 45 35		18	15	–	3	–	–	3	–	–	–	–	3	–	Borazine, 2,4,6-trichloro-1,3,5-triphenyl-	981-87-3	NI41232
411 413 412 274 410 137 276 414	411	100 88 74 67 63 60 41 37		18	15	–	3	–	–	3	–	–	–	–	3	–	Borazine, 2,4,6-trichloro-1,3,5-triphenyl-	981-87-3	IC04333
153 111 273 43 213 83 274 154	411	100 91 76 55 25 20 10 9	0.09	18	21	10	1	–	–	–	–	–	–	–	–	–	β-L-Rhamnopyranoside, 2-nitrophenyl 2,3,4-tri-O-acetyl-		NI41233
267 73 179 268 45 411 269 41	411	100 66 27 24 10 9 9 8		18	29	4	1	1	–	–	–	–	2	–	–	–	Propionic acid, 2-isothiocyanato-3-[3,4-bis(trimethylsilyloxy)phenyl]-, ethyl ester		MS09332
73 277 75 74 45 190 179 147	411	100 11 10 9 8 7 5 4	0.00	18	33	4	1	–	–	–	–	–	3	–	–	–	Benzenepropanoic acid, 4-[(trimethylsilyl)oxy]-α-[[(trimethylsilyl)oxy]imino]-, trimethylsilyl ester	55530-68-2	NI41234
73 276 75 74 45 189 179 147	411	100 11 10 9 8 7 5 4	0.00	18	33	4	1	–	–	–	–	–	3	–	–	–	Benzenepropanoic acid, 4-[(trimethylsilyl)oxy]-α-[[(trimethylsilyl)oxy]imino]-, trimethylsilyl ester	55530-68-2	NI41235
193 73 396 75 324 206 45 194	411	100 80 36 30 29 28 24 18	17.00	18	33	4	1	–	–	–	–	–	3	–	–	–	Glycine, N-(trimethylsilyl)-N-[2-[(trimethylsilyl)oxy]benzoyl]-, trimethylsilyl ester	55517-53-8	NI41236
73 193 179 294 45 75 147 74	411	100 51 26 17 17 16 9 9	8.40	18	33	4	1	–	–	–	–	–	3	–	–	–	Glycine, N-(trimethylsilyl)-N-[3-[(trimethylsilyl)oxy]benzoyl]-, trimethylsilyl ester	55517-54-9	NI41237
193 73 294 75 194 396 411 295	411	100 58 32 27 16 9 8 8		18	33	4	1	–	–	–	–	–	3	–	–	–	Glycine, N-(trimethylsilyl)-N-[4-[(trimethylsilyl)oxy]benzoyl]-, trimethylsilyl ester	55517-55-0	NI41238
193 73 396 45 194 75 411 294	411	100 91 24 20 16 12 11 9		18	33	4	1	–	–	–	–	–	3	–	–	–	Salicyluric acid, tris(trimethylsilyl)-		NI41239
411 229 77 228 413 230 127 231	411	100 63 44 41 35 33 31 30		19	14	4	5	–	–	1	–	–	–	–	–	–	anti-2,4-Dinitrophenylhydrazono-5-chloro-2-aminobenzophenone		MS09333
232 73 160 147 233 174 75 114	411	100 68 33 27 22 21 17 13	0.00	19	37	3	1	–	–	–	–	–	3	–	–	–	Tyrosine, α-methyl-N,O-bis(trimethylsilyl)-, trimethylsilyl ester	56292-86-5	NI41240
220 120 119 59 221 136 93 55	411	100 63 55 48 44 34 29 29	2.00	21	33	7	1	–	–	–	–	–	–	–	–	–	2-Butenoic acid, 2-methyl-, 7-[[2,3-dihydroxy-2-(1-methoxyethyl)-3-methyl-1-oxobutoxy]methyl]-2,3,5,7a-tetrahydro-1H-pyrrolizin-1-yl ester, [1S-[1α(Z),7(2S*,3R*),7aα]]-	303-34-4	LI08654
220 120 119 59 221 136 58 55	411	100 62 57 48 45 35 33 31	2.00	21	33	7	1	–	–	–	–	–	–	–	–	–	2-Butenoic acid, 2-methyl-, 7-[[2,3-dihydroxy-2-(1-methoxyethyl)-3-methyl-1-oxobutoxy]methyl]-2,3,5,7a-tetrahydro-1H-pyrrolizin-1-yl ester, [1S-[1α(Z),7(2S*,3R*),7aα]]-	303-34-4	NI41241
120 59 118 121 220 117 83 94	411	100 35 19 15 10 9 7 5	0.07	21	33	7	1	–	–	–	–	–	–	–	–	–	2-Butenoic acid, 2-methyl-, 7-[[2,3-dihydroxy-2-(1-methoxyethyl)-3-methyl-1-oxobutoxy]methyl]-2,3,5,7a-tetrahydro-1H-pyrrolizin-1-yl ester, [1S-[1α(Z),7(2S*,3R*),7aα]]-	303-34-4	NI41242
57 41 282 284 29 357 355 43	411	100 24 21 20 18 8 8 6	0.40	21	34	2	1	–	–	–	1	–	–	–	–	–	Benzene, 2-bromo-1,3,5-tris(2,2-dimethylpropyl)-4-nitro-	40572-23-4	MS09334
176 148 147 411 218 396 380 379	411	100 76 32 25 24 13 5 5		22	21	7	1	–	–	–	–	–	–	–	–	–	1,3-Benzodioxole-5-acetic acid, 6-[N,4-dimethyl-7,8-(methylenedioxy)-1-oxo-3,4-dihydro-2H-isoquinolin-3-yl]-, methyl ester, (±)-trans-	88610-30-4	DD01601
175 119 382 139 91 77 72 411	411	100 64 43 41 41 38 31 15		22	25	3	3	1	–	–	–	–	–	–	–	–	4(3H)-Pyrimidinone, 2-(diethylamino)-5-methyl-3-phenyl-6-tosyl-		NI41243
179 106 43 28 58 29 233 180	411	100 63 58 30 24 18 17 14	2.90	22	25	5	3	–	–	–	–	–	–	–	–	–	2-(Ethoxycarbonyl)-2-[1-pyrid-3-yl-2-(ethoxycarbonyl)ethyl]-3-pyrid-3-yl-5-pyrrolidone		MS09335
91 79 93 67 135 139 107 115	411	100 75 56 38 30 29 25 20	3.00	23	25	4	1	1	–	–	–	–	–	–	–	–	1-(1-Adamantyl)-4-(3-tolylsulphonyl)-3-nitrobenzene		NI41244
162 232 218 380 411 202 142 44	411	100 63 50 48 46 40 38 33		23	25	6	1	–	–	–	–	–	–	–	–	–	1H-[1,3]Dioxino[5,4-f]isoquinoline, 9-(6-vinyl-4-methoxy-1,3-benzodioxol 5-yl)-7,8,9,10-tetrahydro-5-methoxy-8-methyl-, (S)-	22325-07-1	LI08655
162 232 218 380 411 202 142 381	411	100 63 51 48 46 41 38 34		23	25	6	1	–	–	–	–	–	–	–	–	–	1H-[1,3]Dioxino[5,4-f]isoquinoline, 9-(6-vinyl-4-methoxy-1,3-benzodioxol 5-yl)-7,8,9,10-tetrahydro-5-methoxy-8-methyl-, (S)-	22325-07-1	NI41245
254 42 143 70 56 100 113 84	411	100 87 74 63 36 32 30 30	21.02	23	29	2	3	1	–	–	–	–	–	–	–	–	2-Acetyl-10-[3-[4-(2-hydroxyethyl)piperazin-1-yl]propyl]phenothiazine	2751-68-0	NI41246
254 143 411 70 255 42 380 157	411	100 56 52 38 36 30 27 25		23	29	2	3	1	–	–	–	–	–	–	–	–	2-Acetyl-10-[3-[4-(2-hydroxyethyl)piperazin-1-yl]propyl]phenothiazine	2751-68-0	NI41247
73 410 411 322 45 44 412 368	411	100 63 40 35 24 21 17 15		23	33	2	1	–	–	–	–	–	2	–	–	–	4H-Dibenzo[de,g]quinoline, 5,6,6a,7-tetrahydro-6-methyl-10,11-bis[(trimethylsilyl)oxy]-, (R)-	74841-68-2	NI41248
271 255 57 43 97 177 354 135	411	100 92 85 58 35 34 31 28	0.00	23	33	2	1	–	–	–	–	–	2	–	–	–	Propanenitrile, 3-[(tert-butyldiphenylsilyl)oxy]-2-methyl-2-[(trimethylsilyl)oxy]-		DD01602
236 190 326 57 91 230 327 237	411	100 99 98 97 83 78 36 16	0.00	24	29	5	1	–	–	–	–	–	–	–	–	–	3-Isoquinolinecarboxylic acid, 1,2,3,4-tetrahydro-6,7-dimethoxy-2-pivaloyl-, benzyl ester, (3S)-		DD01603
324 393 325 310 309 70 340 294	411	100 27 23 14 11 10 9 9	1.30	24	29	5	1	–	–	–	–	–	–	–	–	–	Septicine, 13a-hydroxy-	95066-47-0	NI41249
289 411 123 91 288 106 412 290	411	100 71 61 37 26 26 21 21		25	21	3	3	–	–	–	–	–	–	–	–	–	1-(Tolylazo)-3-(4-methoxyanilinocarbonyl)-2-hydroxynaphthalene		IC04334

MASS TO CHARGE RATIOS								M.W.	INTENSITIES								Parent	C	H	O	N	S	F	Cl	Br	I	Si	P	B	X	COMPOUND NAME	CAS Reg No	No
43	44	120	105	91	41	77	59	**411**	100	94	84	60	50	46	34	33	**0.00**	25	33	4	1	–	–	–	–	–	–	–	–	–	**Severine**		MS09336
136	43	122	123	159	81	55	157	**411**	100	73	64	61	56	55	52	49	**28.42**	27	41	2	1	–	–	–	–	–	–	–	–	–	**Androsta-5,16-dieno[16,17-b]indolizin-3β-ol, 1′,5′,6′,7′,8′,8′aα,16α,17α-octahydro-, acetate (ester)**	28105-16-0	NI41250
125	124	110	126	396	114	145	397	**411**	100	99	95	36	27	23	10	8	**6.60**	27	41	2	1	–	–	–	–	–	–	–	–	–	**Cyclopamine**		LI08656
114	125	145	411	143	368	181	298	**411**	100	29	20	19	18	16	14	13		27	41	2	1	–	–	–	–	–	–	–	–	–	**Veramine**		LI08657
256	411	257	43	55	41	69	68	**411**	100	55	55	31	27	25	22	22		28	45	1	1	–	–	–	–	–	–	–	–	–	**Cholestan-3-one, 5-cyano-, (5α)-**		MS09337
147	59	235	148	236	69	324	78	**412**	100	93	28	23	22	18	15	15	**0.00**	–	1	–	–	–	12	–	–	–	–	4	–	1	**Tetrakis(trifluorophosphine)carbonylcobalt hydride**		MS09338
187	328	272	271	85	412	47	356	**412**	100	93	86	76	71	70	65	55		5	–	5	–	–	3	–	–	–	1	–	–	1	**(Trifluorosilyl)pentacarbonylrhenium**		LI08658
69	195	67	48	214	113	179	32	**412**	100	61	47	39	37	36	29	26	**0.00**	6	–	3	–	2	12	–	–	–	–	–	–	–	**2,2,4,4-Tetrakis(trifluoromethyl)-1,3-dithietane-1,1,3-trioxide**		DD01604
377	43	313	379	315	381	331	412	**412**	100	84	40	33	27	17	10	9		7	13	4	–	–	–	2	–	–	–	–	–	1	**Tantalum, dichlorodimethoxy(2,4-pentanedionato-O,O′)-**	19158-14-6	LI08659
377	43	313	379	315	381	39	67	**412**	100	84	40	33	27	17	14	13	**8.90**	7	13	4	–	–	–	2	–	–	–	–	–	1	**Tantalum, dichlorodimethoxy(2,4-pentanedionato-O,O′)-**	19158-14-6	NI41251
377	379	43	317	315	313	383	381	**412**	100	99	63	54	54	54	23	23	**11.90**	7	13	4	–	–	–	2	–	–	–	–	–	1	**Tantalum, dichlorodimethoxy(2,4-pentanedionato-O,O′)-**	19158-14-6	NI41252
328	330	302	304	332	326	300	386	**412**	100	97	66	62	56	55	47	44	**16.66**	8	5	3	–	–	–	–	1	–	–	–	–	1	**Tungsten, bromotricarbonylcyclopentadienyl-**	37131-50-3	NI41253
77	113	64	335	333	51	69	27	**412**	100	98	64	35	34	29	13	13	**0.30**	8	6	–	–	–	8	–	2	–	–	–	–	–	**Cyclobutane, 3-(3,4-dibromo-1,1,2,2-tetrafluorobutyl)-1,1,2,2-tetrafluoro-**	35208-00-5	MS09339
377	125	379	42	109	92	78	28	**412**	100	44	37	22	20	20	19	17	**1.21**	8	8	3	–	1	–	2	–	1	–	1	–	–	**Phosphorothioic acid, O-(2,5-dichloro-4-iodophenyl) O,O-dimethyl ester**	18181-70-9	NI41254
412	312	106	393	292	88	124	262	**412**	100	31	29	25	15	14	13	10		10	–	–	–	2	12	–	–	–	–	–	–	–	**Dodecafluoro-2,3,6,7-tetrahydro-1H,5H-dicyclopenta-1,4-dithiin**		LI08660
79	77	147	65	47	51	412	59	**412**	100	76	46	38	33	29	28	28		10	13	–	–	–	8	–	–	1	–	–	–	–	**Decane, 3,3,4,4,7,7,8,8-octafluoro-1-iodo-**	40723-82-8	NI41255
57	41	29	197	353	199	51	351	**412**	100	48	31	28	23	20	19	18	**0.00**	10	14	–	–	–	–	–	2	–	–	–	–	1	**Tin, dibromophenyl-tert-butyl-**	75042-60-3	NI41256
116	326	328	354	356	324	298	296	**412**	100	68	63	62	59	59	43	42	**41.00**	10	16	4	2	–	–	–	–	–	–	–	–	1	**(π-N,N,N′,N′-Tetramethylethylenediamine)tetracarbonyltungsten**		LI08661
58	42	116	72	71	70	354	326	**412**	100	10	6	2	2	2	1	1	**0.00**	10	16	4	2	–	–	–	–	–	–	–	–	1	**(π-N,N,N′,N′-Tetramethylethylenediamine)tetracarbonyltungsten**		MS09340
300	176	207	412	269	381	52	131	**412**	100	25	17	16	12	11	4	3		10	18	10	–	–	–	–	–	–	–	2	–	1	**Chromium, tetracarbonylbis(phosphorous acid)-, hexamethyl ester**	22614-53-5	MS09341
300	176	28	177	301	93	412	207	**412**	100	82	22	19	18	18	17	16		10	18	10	–	–	–	–	–	–	–	2	–	1	**Chromium, tetracarbonylbis(phosphorous acid)-, hexamethyl ester**	22614-53-5	NI41257
307	394	292	57	164	55	43	69	**412**	100	99	94	87	78	74	72	57	**30.00**	11	14	3	–	–	–	–	–	–	–	–	–	2	**(σ-Trimethyltin)(π-cyclopentadienyl)tricarbonylmolybdenum**		LI08662
161	107	126	75	125	163	50	52	**412**	100	60	43	41	32	28	19	15	**1.90**	14	8	–	–	–	6	2	–	–	–	–	–	1	**Chromium, bis[2-chloro-1-(trifluoromethyl)benzene]-**	53966-08-8	NI41258
174	206	146	412	175	160	102	148	**412**	100	80	63	38	23	18	18	17		14	24	8	2	2	–	–	–	–	–	–	–	–	**L-Cystine, N,N′-dicarboxy-N,N′-diethyl-, dimethyl ester**	21026-91-5	NI41259
83	119	41	91	55	105	164	162	**412**	100	23	6	5	5	4	3	3	**0.00**	15	23	1	–	–	–	1	2	–	–	–	–	–	**Chamigran-7-en-9-ol, 2,10-dibromo-3-chloro-**		MS09342
83	119	41	55	53	39	121	93	**412**	100	55	36	30	19	16	14	13	**0.00**	15	23	1	–	–	–	1	2	–	–	–	–	–	**Chamigran-9-one, 2,10-dibromo-3-chloro-**		MS09343
191	217	205	307	204	103	147	192	**412**	100	72	71	33	29	28	24	17	**0.00**	15	36	5	–	1	–	–	–	–	3	–	–	–	**D-Ribofuranose, 5-deoxy-5-(methylsulphinyl)-1,2,3-tris-O-(trimethylsilyl)**	56700-80-2	MS09344
254	256	255	136	91	257	102	149	**412**	100	64	41	32	32	27	27	23	**0.00**	16	20	2	2	–	–	4	–	–	–	–	–	–	**2-Butanone, 4,4′-[1,4-phenylenebis(methyleneimino)]bis[1,1-dichloro-**	56667-15-3	NI41260
166	250	75	29	41	324	167	263	**412**	100	34	33	27	26	24	20	17	**2.70**	17	27	4	2	1	3	–	–	–	–	–	–	–	**N-(Trifluoroacetyl)-L-prolylmethionine butyl ester**		MS09345
166	264	144	61	41	338	167	70	**412**	100	74	63	54	44	40	40	39	**0.00**	17	27	4	2	1	3	–	–	–	–	–	–	–	**N-(Trifluoroacetyl)-L-prolyl-α-methylmethionine butyl ester**		MS09346
73	280	412	235	267	75	413	281	**412**	100	72	58	28	24	22	20	16		19	36	4	–	–	–	–	–	–	3	–	–	–	**Benzenebutanoic acid, 3,4-bis[(trimethylsilyl)oxy]-, trimethylsilyl ester**		NI41261
280	412	235	267	413	281	397	179	**412**	100	87	39	37	33	26	22	16		19	36	4	–	–	–	–	–	–	3	–	–	–	**Benzenebutanoic acid, 3,4-bis[(trimethylsilyl)oxy]-, trimethylsilyl ester**		NI41262
412	251	219	179	178	218	191	177	**412**	100	38	29	15	15	14	13	13		20	22	2	2	1	–	–	–	–	–	–	–	1	**Nickel(II) sulphide, [N,N′-bis(salicylidene)-3,3′-bis(aminopropyl)]-**		NI41263
84	175	91	266	41	267	132	44	**412**	100	80	65	27	25	18	16	11	**0.10**	20	24	4	6	–	–	–	–	–	–	–	–	–	**L-Histidinamide, 5-oxo-L-prolyl-L-phenylalanyl-**	35936-95-9	NI41264
84	91	82	81	120	272	262	100	**412**	100	50	48	45	37	32	25	21	**9.10**	20	24	4	6	–	–	–	–	–	–	–	–	–	**L-Phenylalaninamide, 5-oxo-L-propyl-L-histidyl-**	27058-75-9	NI41265
313	412	286	179	395	379	394	245	**412**	100	96	68	65	50	48	46	34		21	16	9	–	–	–	–	–	–	–	–	–	–	**Arthraxin**		LI08663
77	386	384	51	414	117	412	242	**412**	100	34	34	33	29	29	28	22		21	21	–	2	1	–	–	1	–	–	–	–	–	**1H-Benzo[e][1,4]thiazepine, 2,3,5,6,7,8,9-heptahydro-1-phenyl-5-(4-bromophenylimino)-**		NI41266
397	398	399	280	73	325	191	400	**412**	100	35	11	5	5	3	3	2	**0.00**	21	24	5	–	–	–	–	–	–	2	–	–	–	**Anthraquinone-2-carboxylic acid, 4-[(trimethylsilyl)oxy]-, trimethylsilyl ester**		NI41267
393	43	139	55	41	59	109	121	**412**	100	47	46	37	29	27	19	18	**0.00**	21	26	6	–	–	2	–	–	–	–	–	–	–	**Pregna-1,4-diene-3,20-dione, 6,9-difluoro-11,16,17,21-tetrahydroxy-, (6α,11β,16α)-**	807-38-5	NI41268
43	105	324	71	341				**412**	100	52	48	43	9				**0.38**	21	32	4	–	2	–	–	–	–	–	–	–	–	**3,10-Dimethyl-6-(1,3-dithiolan-2-yl)-4-isobutyroyloxy-2-oxo-2,3,3a,4,5,8,9,10,11,11a-decahydrocyclodeca[b]furan**		LI08664
44	57	43	49	412	347	84	71	**412**	100	65	51	49	46	43	41	35		22	14	4	–	–	–	2	–	–	–	–	–	–	**2,2′-Binaphthalene-1,4-dione, 3,3′-dichloro-1′,4′-dimethoxy-**		NI41269
339	43	354	340	45	60	355	352	**412**	100	43	29	24	22	15	11	3	**0.00**	22	20	8	–	–	–	–	–	–	–	–	–	–	**Edulon A, 11,14-di-O-dehydro-**		NI41270
43	352	135	175	217	412	28	178	**412**	100	94	85	63	61	48	39	30		22	20	8	–	–	–	–	–	–	–	–	–	–	**2(3H)-Furanone, dihydro-3-(α-acetoxypiperonyl)-4-piperonyl-**		NI41271
202	412	149	135	161	219	121	220	**412**	100	37	35	27	20	10	5	3		22	20	8	–	–	–	–	–	–	–	–	–	–	**Isopaulownin acetate**		LI08665
337	309	43	362	217	339	321	319	**412**	100	42	38	30	28	21	17	17	**8.00**	22	20	8	–	–	–	–	–	–	–	–	–	–	**5,12-Naphthacenequinone, 9-acetyl-4,7-dimethoxy-7,8,9,10-tetrahydro-6,9,11-trihydroxy-, (7R)-**		NI41272
337	362	43	309	217	412	344	319	**412**	100	83	75	72	42	34	30	23		22	20	8	–	–	–	–	–	–	–	–	–	–	**5,12-Naphthacenequinone, 9-acetyl-4,7-dimethoxy-7,8,9,10-tetrahydro-6,9,11-trihydroxy-, (7S)-**		NI41273
202	135	412	149	161	219	121	220	**412**	100	44	36	27	21	7	4	3		22	20	8	–	–	–	–	–	–	–	–	–	–	**Paulownin acetate**		LI08666

MASS TO CHARGE RATIOS	M.W.	INTENSITIES	Parent	C	H	O	N	S	F	Cl	Br	I	Si	P	B	X	COMPOUND NAME	CAS Reg No	No
328 286 244 100 370 412	412	100 50 50 50 30 20		22	20	8	–	–	–	–	–	–	–	–	–	–	Piceatannol tetraacetate		LI08667
71 70 302 83 412 42 413 303	412	100 57 43 39 36 16 11 9		22	24	6	2	–	–	–	–	–	–	–	–	–	Dichotine, 2-deoxy-2,21-epoxy-, (21α)-	29474-91-7	NI41274
43 298 109 316 135 358 376 255	412	100 68 57 48 36 35 30 23	0.00	22	36	7	–	–	–	–	–	–	–	–	–	–	Acetylandromedol		LI08668
380 412 325 365 310 267 349 149	412	100 98 60 32 22 18 15 13		23	24	7	–	–	–	–	–	–	–	–	–	–	6H-Benzofuro[3,2-c][1]benzofuran-6-one, 1,3-dimethoxy-9-hydroxy-2-(3-methoxy-3-methylbutyl)-		LI08669
119 81 140 91 276 263 65 353	412	100 61 29 22 7 7 5 2	0.00	23	24	7	–	–	–	–	–	–	–	–	–	–	α-L-erythro-Pentofuranose, 1-(methoxycarbonyl)-2-deoxy-3,5-di-O-(4-toluoyl)-		NI41275
119 81 91 140 263 65 276 353	412	100 62 38 35 13 10 7 1	0.00	23	24	7	–	–	–	–	–	–	–	–	–	–	β-L-erythro-Pentofuranose, 1-(methoxycarbonyl)-2-deoxy-3,5-di-O-(4-toluoyl)-		NI41276
309 351 352 394 297 337 69 149	412	100 63 51 33 25 14 12 8	0.00	23	24	7	–	–	–	–	–	–	–	–	–	–	Vismiaquinone diacetate		NI41277
370 412 120 371 43 160 413 296	412	100 76 29 26 26 24 23 19		23	28	5	2	–	–	–	–	–	–	–	–	–	Aspidospermidin-21-oic acid, 1-acetyl-17-methoxy-10-oxo-, methyl ester	16531-05-8	NI41278
186 412 229 199 212 171 170 380	412	100 58 17 12 10 8 8 6		23	28	5	2	–	–	–	–	–	–	–	–	–	Eripine		LI08670
138 274 108 412 139 275 289 286	412	100 77 77 58 33 27 26 25		23	28	5	2	–	–	–	–	–	–	–	–	–	Hazuntinine		LI08671
91 321 108 106 325 217 132 412	412	100 12 7 7 6 6 6 3		23	28	5	2	–	–	–	–	–	–	–	–	–	Isoxazolidine-3-acetic acid, N-benzyl-5-[[(benzyloxycarbonyl)amino]methyl]-, ethyl ester, (3R*,5R*)-		DD01605
108 412 150 156 130 381 144 159	412	100 58 46 30 19 17 15 12		23	28	5	2	–	–	–	–	–	–	–	–	–	Nb-Acetyllonicerine		LI08672
412 265 239 369 116 282 143 109	412	100 69 68 35 32 25 24 22		23	28	5	2	–	–	–	–	–	–	–	–	–	Na,11α-Bis(methoxycarbonyl)-3β-hydroxyaspidofractinine		LI08673
300 412 411 397 283 413 216 311	412	100 99 68 38 25 24 21 20		23	28	5	2	–	–	–	–	–	–	–	–	–	Oxayohimban-16-carboxylic acid, 16,17-didehydro-10,11-dimethoxy-19-methyl-, methyl ester, (3β,19α,20α)-	131-02-2	NI41279
243 412 168 319 244 413 139 320	412	100 64 62 46 25 16 13 10		24	17	2	–	–	–	–	–	–	–	–	–	1	Phenoxarsine, 10-(2-phenoxyphenyl)-	27796-62-9	NI41280
412 91 201 413 77 305 120 92	412	100 47 32 28 19 17 13 11		24	20	3	4	–	–	–	–	–	–	–	–	–	1-Phenyl-3-methyl-5-hydroxy-4-[4-(benzoyloxycarbonyl)phenylazo]pyrazole		IC04335
182 412 122 183 43 384 168 210	412	100 30 17 13 13 11 10 8		24	32	4	2	–	–	–	–	–	–	–	–	–	Aspidospermidin-21-ol, 1-acetyl-17-methoxy-, acetate (ester)	36459-08-2	NI41281
182 412 324 122 138 411 325 183	412	100 33 25 14 13 12 12 12		24	32	4	2	–	–	–	–	–	–	–	–	–	Aspidospermidin-21-ol, 1-acetyl-17-methoxy-, acetate (ester)	36459-08-2	NI41282
182 43 412 122 183 59 384 45	412	100 33 23 20 13 11 10 9		24	32	4	2	–	–	–	–	–	–	–	–	–	Aspidospermidin-21-ol, 1-acetyl-17-methoxy-, acetate (ester)	36459-08-2	NI41283
397 191 398 395 36 412 338 368	412	100 38 32 23 23 18 13 10		24	32	4	2	–	–	–	–	–	–	–	–	–	DL-1,1′,2,2′-Tetramethyl-6,6′-dimethoxy-7,7′-dihydroxy-1,1′,2,2′,3′,4,4′-octahydro-8,8′-bisisoquinoline		MS09347
307 322 308 412 323 73 57 309	412	100 44 25 17 13 9 9 8		24	36	2	–	–	–	–	–	–	2	–	–	–	Estra-1,3,5,7,9-pentaene, 3,17-bis[(trimethylsilyl)oxy]-, (17α)-	69833-45-0	NI41284
412 413 281 296 307 414 322 282	412	100 37 34 24 22 13 13 11		24	36	2	–	–	–	–	–	–	2	–	–	–	Estra-1,3,5,7,9-pentaene, 3,17-bis[(trimethylsilyl)oxy]-, (17β)-	75268-15-4	NI41285
105 145 291 248 77 249 43 412	412	100 68 46 43 43 21 21 6		25	20	4	2	–	–	–	–	–	–	–	–	–	2-Anisyl-N-benzoyloxy-2-cyano-5-methyl-4-phenyl-2,3-dihydrooxazole		LI08674
244 412 229 124 105 413 243 121	412	100 74 39 29 26 23 22 22		25	37	4	–	–	–	–	–	–	–	–	1	–	Pregn-4-ene-3,20-dione, 17,21-[(butylborylene)bis(oxy)]-	30888-54-1	NI41286
244 412 91 229 55 79 95 105	412	100 73 46 44 44 42 32 31		25	37	4	–	–	–	–	–	–	–	–	1	–	Pregn-4-ene-3,20-dione, 17,21-[(butylborylene)bis(oxy)]-	30888-54-1	LI08675
244 412 91 229 55 79 124 95	412	100 70 45 44 44 41 33 30		25	37	4	–	–	–	–	–	–	–	–	1	–	Pregn-4-ene-3,20-dione, 17,21-[[tert-butylborylene]bis(oxy)]-	30888-53-0	NI41287
244 412 229 124 105 413 243 121	412	100 74 39 29 26 23 22 22		25	37	4	–	–	–	–	–	–	–	–	1	–	Pregn-4-ene-3,20-dione, 17,21-[[tert-butylborylene]bis(oxy)]-	30888-53-0	LI08676
57 267 101 41 397 133 355 99	412	100 99 75 34 28 27 24 19	2.00	25	48	4	–	–	–	–	–	–	–	–	–	–	1,3-Dioxolane, 4-[[(2-allyloxyhexadecyl)oxy]methyl]-2,2-dimethyl-		NI41288
131 71 43 101 73 57 55 41	412	100 37 34 30 25 21 17 17	1.00	25	48	4	–	–	–	–	–	–	–	–	–	–	1,3-Dioxolane, 4-[[(2-methoxy-4-octadecenyl)oxy]methyl]-2,2-dimethyl-	16725-41-0	LI08677
71 43 101 73 57 55 41 97	412	100 92 81 66 58 47 46 39	0.00	25	48	4	–	–	–	–	–	–	–	–	–	–	1,3-Dioxolane, 4-[[(2-methoxy-4-octadecenyl)oxy]methyl]-2,2-dimethyl-	16725-41-0	NI41289
159 101 157 267 217 43 352 59	412	100 41 10 6 6 4 3 3	0.10	25	48	4	–	–	–	–	–	–	–	–	–	–	1,3-Dioxolane, 4-[[(2-propyloxy-4-hexadecenyl)oxy]methyl]-2,2-dimethyl-		NI41290
57 43 41 129 55 70 56 83	412	100 33 31 28 25 23 22 21	0.00	25	48	4	–	–	–	–	–	–	–	–	–	–	Nonanedioic acid, bis(2-ethylhexyl) ester	103-24-2	NI41291
171 57 70 71 55 172 112 83	412	100 31 28 26 23 20 15 15	0.00	25	48	4	–	–	–	–	–	–	–	–	–	–	Nonanedioic acid, bis(2-ethylhexyl) ester	103-24-2	NI41292
73 147 293 308 161 148 219 190	412	100 97 29 22 14 12 8 7	0.00	25	52	2	–	–	–	–	–	–	1	–	–	–	Docosanoic acid, trimethylsilyl ester	74367-36-5	NI41293
73 117 75 132 43 397 145 28	412	100 70 67 55 55 43 42 38	13.50	25	52	2	–	–	–	–	–	–	1	–	–	–	Docosanoic acid, trimethylsilyl ester	74367-36-5	NI41294
268 269 134 394 239 412	412	100 23 17 16 11 9		26	20	5	–	–	–	–	–	–	–	–	–	–	6,7-Bis(methoxycarbonyl)-5,8-epoxy-5,6,7,8-tetrahydrodibenz[a,c]anthracene		LI08678
268 269 44 134 394 239 412	412	100 23 19 17 16 11 9		26	20	5	–	–	–	–	–	–	–	–	–	–	6,7-Bis(methoxycarbonyl)-5,8-epoxy-5,6,7,8-tetrahydrodibenz[a,c]anthracene		MS09348
69 137 191 136 43 41 81 179	412	100 84 79 77 63 61 52 46	16.00	26	36	4	–	–	–	–	–	–	–	–	–	–	1,3-Benzenediol, 5-methyl-2-(3,7,11-trimethyl-2,6,10-dodecatrienyl)-, diacetate, (E,E)-	52104-23-1	NI41295
412 123 246 107 410 229 124 91	412	100 25 21 12 11 11 11 9		26	36	4	–	–	–	–	–	–	–	–	–	–	Robustol		NI41296
412 329 413 303 327 356 371 414	412	100 43 34 23 22 18 15 11		26	40	2	–	–	–	–	–	–	1	–	–	–	$\Delta^{1(6)}$-Tetrahydrocannabinol allyldimethylsilyl ether		NI41297
412 397 413 371 357 398 356 249	412	100 56 36 27 26 22 19 16		26	40	2	–	–	–	–	–	–	1	–	–	–	$\Delta^{1(7)}$-Tetrahydrocannabinol allyldimethylsilyl ether		NI41298
371 370 355 341 412 314 397 372	412	100 89 71 71 52 41 37 29		26	40	2	–	–	–	–	–	–	1	–	–	–	Δ^{1}-Tetrahydrocannabinol allyldimethylsilyl ether		NI41299
91 289 165 121 198 197	412	100 45 25 25 14 6	0.00	27	24	–	–	2	–	–	–	–	–	–	–	–	Benzene, 1,1′-[(dibenzylthio)methylene]bis-	29055-89-8	NI41300
71 285 57 85 283 43 41 267	412	100 81 79 76 57 56 42 41	10.00	27	40	3	–	–	–	–	–	–	–	–	–	–	Cholesta-9(11),20(22)-diene-3,23-dione, 6-hydroxy-, (5α,6α)-	50676-99-8	NI41301
394 213 395 215 91 164 151 105	412	100 38 31 29 22 21 21 21	12.00	27	40	3	–	–	–	–	–	–	–	–	–	–	16,23-Cyclocholesta-5,16(23)-dien-22-one, 3,26-dihydroxy-, (3β)-	561-98-8	NI41302
394 283 269 87 79 412 397 244	412	100 50 40 40 32 25 20 20		27	40	3	–	–	–	–	–	–	–	–	–	–	4,20(22)-Furastadiene-3-one, 26-hydroxy-		NI41303

MASS TO CHARGE RATIOS								M.W.	INTENSITIES								Parent	C	H	O	N	S	F	Cl	Br	I	Si	P	B	X	COMPOUND NAME	CAS Reg No	No
339	340	412	59	57	323	365	413	412	100	26	17	16	12	11	10	5		27	40	3	–	–	–	–	–	–	–	–	–	–	1,1-Bis(2-methyl-4-tert-butyl-5-hydroxyphenyl)-3-methoxybutane		IC04336
139	298	260	412	115	283	343	342	412	100	77	39	20	20	19	16	16		27	40	3	–	–	–	–	–	–	–	–	–	–	Spirost-5-en-3-one, (3β,25R)-	6870-79-7	NI41304
357	126	380	412	397	71	43	55	412	100	92	80	57	55	54	51	47		28	44	2	–	–	–	–	–	–	–	–	–	–	3,5-Cyclocholest-22-en-24-one, 6-methoxy-, (3β,5α,6β,22E)-	55064-55-6	NI41305
352	97	55	81	255	353	145	107	412	100	78	67	59	42	29	29	27	0.00	28	44	2	–	–	–	–	–	–	–	–	–	–	26,27-Dinorergosta-5,22-dien-3-ol, acetate, (3β,22E)-	34013-73-5	NI41306
352	81	69	55	283	353	41	43	412	100	47	33	32	31	28	22	19	0.00	28	44	2	–	–	–	–	–	–	–	–	–	–	26,27-Dinorergosta-5,23-dien-3-ol, acetate, (3β)-	35882-87-2	NI41307
43	271	272	71	55	95	81	107	412	100	76	73	58	53	33	33	29	29.00	28	44	2	–	–	–	–	–	–	–	–	–	–	28,33-Dinorgorgost-5-en-24-one, 3-hydroxy-, (3β)-	55064-61-4	NI41308
261	412	285	369	314	397	379		412	100	82	75	31	30	13	12			28	44	2	–	–	–	–	–	–	–	–	–	–	Ergosta-7,22-dien-6-one, 3-hydroxy-, (3β,5α)-		LI08679
412	261	287	397	369	379			412	100	94	31	22	12	11				28	44	2	–	–	–	–	–	–	–	–	–	–	Ergosta-8(14),22-dien-6-one, 3-hydroxy-, (3β,5α)-		LI08680
136	59	118	57	55	69	83	81	412	100	97	83	82	75	59	58	54	26.11	28	44	2	–	–	–	–	–	–	–	–	–	–	9,10-Secoergosta-5,7,10(19),22-tetraene-3,25-diol, (3β,5Z,7E,22E)-	21343-40-8	NI41310
43	136	59	118	57	55	45	68	412	100	99	99	82	82	76	76	59	23.53	28	44	2	–	–	–	–	–	–	–	–	–	–	9,10-Secoergosta-5,7,10(19),22-tetraene-3,25-diol, (3β,5Z,7E,22E)-	21343-40-8	NI41309
386	387	77	149	266				412	100	60	19	17	16				0.00	29	20	1	2	–	–	–	–	–	–	–	–	–	4,5-(9,10-Phenanthreno)-1,3-diphenyl-4-imidazolin-2-one		LI08681
379	43	171	55	394	57	145	41	412	100	81	75	60	57	52	51	51	11.00	29	48	1	–	–	–	–	–	–	–	–	–	–	Cholesta-5,7-dien-3-ol, 4,4-dimethyl-, (3β)-		NI41311
412	43	55	69	41	57	379	105	412	100	73	46	41	41	40	39	37		29	48	1	–	–	–	–	–	–	–	–	–	–	Cholesta-7,9(11)-dien-3-ol, 4,4-dimethyl-, (3β,5α)-		NI41312
43	412	299	57	41	55	379	397	412	100	84	52	52	52	50	46	42		29	48	1	–	–	–	–	–	–	–	–	–	–	Cholesta-7,14-dien-3-ol, 4,4-dimethyl-, (3β,5α)-		NI41313
412	43	379	41	397	57	55	69	412	100	99	71	57	51	50	50	47		29	48	1	–	–	–	–	–	–	–	–	–	–	Cholesta-8,14-dien-3-ol, 4,4-dimethyl-, (3β,5α)-		NI41314
271	255	272	412	256	213	355	253	412	100	44	20	16	11	10	7	7		29	48	1	–	–	–	–	–	–	–	–	–	–	Cholesta-5,20(22)-dien-3-ol, 23-ethyl-, (3β,23R)-		MS09349
55	82	84	69	412	255	67	91	412	100	58	55	49	41	41	35	31		29	48	1	–	–	–	–	–	–	–	–	–	–	Cholesta-5,21-dien-3-ol, 23-ethyl-		LI08682
55	255	221	412	300	272	213	351	412	100	44	44	41	30	23	20	19		29	48	1	–	–	–	–	–	–	–	–	–	–	Cholesta-5,22-dien-3-ol, 23-ethyl-, (3β,E)-		MS09350
55	271	412	300	272	213	351	273	412	100	47	34	30	25	21	18	12		29	48	1	–	–	–	–	–	–	–	–	–	–	Cholesta-5,22-dien-3-ol, 24-ethyl-, (3β,22E,24S)-		MS09351
97	51	300	271	412	299	314	301	412	100	99	50	32	22	17	16	13		29	48	1	–	–	–	–	–	–	–	–	–	–	Cholesta-5,24-dien-3-ol, 23-ethyl-, (3β,23R)-		MS09352
97	300	412	271	314	213	255	285	412	100	75	49	46	22	16	13	12		29	48	1	–	–	–	–	–	–	–	–	–	–	Cholesta-5,24-dien-3-ol, 23-ethyl-, (3β,23R)-		MS09353
55	97	300	271	299	301	412	285	412	100	97	85	32	22	18	17	16		29	48	1	–	–	–	–	–	–	–	–	–	–	Cholesta-5,24-dien-3-ol, 23-ethyl-, (3β,23S)-		MS09354
412	271	55	299	105	145	213	69	412	100	87	82	52	51	50	47	46		29	48	1	–	–	–	–	–	–	–	–	–	–	Cholesta-5,25-dien-3-ol, 26-ethyl-, (3β,25Z)-		DD01606
314	229	315	299	213	281	296	271	412	100	31	27	26	24	22	18	16	4.00	29	48	1	–	–	–	–	–	–	–	–	–	–	Cholest-5-en-3-ol, 24-ethylidene-, (3β,24E)-		NI41315
314	229	281	299	315	213	296	211	412	100	36	35	27	24	23	21	15	3.70	29	48	1	–	–	–	–	–	–	–	–	–	–	Cholest-5-en-3-ol, 24-ethylidene-, (3β,24Z)-		NI41316
285	286	328	412	269	397	267	241	412	100	24	21	20	20	15	14	11		29	48	1	–	–	–	–	–	–	–	–	–	–	Cholest-7-en-3-ol, 4-methyl-24-methylene-, (3β,4α)-		NI41317
43	412	55	41	57	95	69	299	412	100	84	80	77	69	54	52	45		29	48	1	–	–	–	–	–	–	–	–	–	–	Cholest-7-en-3-one, 4,4-dimethyl-, (5α)-	2604-90-2	NI41318
300	271	213	283	412	255	285	215	412	100	81	38	33	26	26	21	21		29	48	1	–	–	–	–	–	–	–	–	–	–	24,26-Cyclocholest-5-en-3-ol, 23-ethyl-, (3β,23R,24R,25R)-		MS09355
271	300	412	213	255	301	299	283	412	100	38	31	23	19	16	15	14		29	48	1	–	–	–	–	–	–	–	–	–	–	24,26-Cyclocholest-5-en-3-ol, 23-ethyl-, (3β,23R,24S,25S)-		MS09356
271	300	213	412	267	283	253	314	412	100	54	24	23	19	17	17	16		29	48	1	–	–	–	–	–	–	–	–	–	–	24,26-Cyclocholest-5-en-3-ol, 23-ethyl-, (3β,23S,24R,25R)-		MS09357
300	271	283	213	412	255	231	215	412	100	84	37	35	31	23	22	18		29	48	1	–	–	–	–	–	–	–	–	–	–	24,26-Cyclocholest-5-en-3-ol, 23-ethyl-, (3β,23S,24S,25S)-		MS09358
55	357	397	145	253	380	412	358	412	100	49	33	24	21	19	16	13		29	48	1	–	–	–	–	–	–	–	–	–	–	3,5-Cyclo-27-norcholest-24-ene, 23-ethyl-6-methoxy-, (3α,5α,6β,23ξ,24Z)		MS09359
55	357	397	253	380	358	412	282	412	100	41	19	17	15	11	10	10		29	48	1	–	–	–	–	–	–	–	–	–	–	3,5-Cyclo-27-norcholest-24-ene, 23-ethyl-6-methoxy-, (3α,5α,6β,23R,24E)-		MS09360
357	314	397	380	412	205	282	213	412	100	52	51	45	40	37	33	31		29	48	1	–	–	–	–	–	–	–	–	–	–	3,5-Cyclo-27-Norcholest-24-ene, 23-ethyl-6-methoxy-, (3α,5α,6β,23S,24E)-		MS09361
69	55	135	81	269	95	83	107	412	100	57	51	46	39	38	37	35	15.15	29	48	1	–	–	–	–	–	–	–	–	–	–	Cycloprop[7,8]ergost-22-en-3-ol, 3',7-dihydro-, (3β,5α,7β,8α,22E)-	53866-87-8	NI41319
314	296	412	299	213	397	271	357	412	100	29	24	18	15	14	13	12		29	48	1	–	–	–	–	–	–	–	–	–	–	3,5-Cyclostigmast-24(28)-en-6-ol, (3α,5α,6β,24E)-		NI41320
314	297	299	412	397	213	271	356	412	100	36	22	17	12	12	11	9		29	48	1	–	–	–	–	–	–	–	–	–	–	3,5-Cyclostigmast-24(28)-en-6-ol, (3α,5α,6β,24Z)-		NI41321
271	300	213	314	283	328	281	255	412	100	46	26	25	24	21	19	19	17.00	29	48	1	–	–	–	–	–	–	–	–	–	–	23,28-Cyclostigmast-5-en-3-ol, (3β,23R,24R,28S)-		MS09362
271	213	314	215	229	217	253	231	412	100	30	24	18	15	14	13	13	12.00	29	48	1	–	–	–	–	–	–	–	–	–	–	23,28-Cyclostigmast-5-en-3-ol, (3β,23S,24S,28R)-		MS09363
314	412	69	55	54	41	83	45	412	100	73	67	66	65	43	40	37		29	48	1	–	–	–	–	–	–	–	–	–	–	23-Demethylgorgosterol		MS09364
69	97	369	351	55	81	83	255	412	100	95	86	61	54	49	44	33	7.00	29	48	1	–	–	–	–	–	–	–	–	–	–	Ergosta-5,22-dien-3-ol, 24-methyl-, (3β,22E)-	53755-22-9	NI41322
55	300	69	81	105	93	267	282	412	100	69	62	57	56	46	43	29	0.00	29	48	1	–	–	–	–	–	–	–	–	–	–	Ergosta-5,23(29)-dien-3-ol, 23-methyl-, (3β,24R)-		DD01607
412	328	55	84	271	69	413	299	412	100	86	49	39	34	34	32	25		29	48	1	–	–	–	–	–	–	–	–	–	–	Ergosta-5,25-dien-3-ol, 24-methyl-, (3β)-		DD01608
397	69	55	97	412	81	398	43	412	100	85	77	64	38	35	34	32		29	48	1	–	–	–	–	–	–	–	–	–	–	Ergosta-8,23-dien-3-ol, 14-methyl-, (3β,5α,14α,E)-		NI41323
397	55	69	95	41	412	398	231	412	100	58	54	39	37	33	33	33		29	48	1	–	–	–	–	–	–	–	–	–	–	Ergosta-8,24(28)-dien-3-ol, 14-methyl-, (3β,5α)-	33886-74-7	NI41324
397	55	69	412	41	81	95	231	412	100	94	83	56	42	40	37	33		29	48	1	–	–	–	–	–	–	–	–	–	–	Ergosta-8,24(28)-dien-3-ol, 14-methyl-, (3β,5α,14α)-		NI41325
191	69	81	95	55	395	109	41	412	100	85	78	74	64	58	45	44	3.03	29	48	1	–	–	–	–	–	–	–	–	–	–	Lup-20(29)en-28-ol, (19ξ)-	55401-95-1	NI41326
55	97	300	412	271	299	301	272	412	100	89	26	20	17	10	7	7		29	48	1	–	–	–	–	–	–	–	–	–	–	27-Norcholesta-5,25-dien-3-ol, 23-ethyl-24-methyl-, (3β,23E,24S)-		MS09365
41	55	271	213	255	339	314	281	412	100	68	27	15	14	11	11	9	7.00	29	48	1	–	–	–	–	–	–	–	–	–	–	27-Norcholesta-5,25-dien-3-ol, 23-ethyl-24-methyl-, (3β,23R,24R)-		MS09366
271	314	339	255	412	213	272	379	412	100	68	53	53	38	35	33	23		29	48	1	–	–	–	–	–	–	–	–	–	–	27-Norcholesta-5,25-dien-3-ol, 23-ethyl-24-methyl-, (3β,23S,24R)-		MS09367
271	412	300	55	397	413	69	379	412	100	81	48	41	31	26	17	16		29	48	1	–	–	–	–	–	–	–	–	–	–	27-Norcholesta-5,24-dien-3-ol, 25-isopropyl-, (3β,24E)-		DD01609
412	271	397	413	55	300	69	81	412	100	50	37	32	26	24	20	18		29	48	1	–	–	–	–	–	–	–	–	–	–	27-Norcholesta-5,24-dien-3-ol, 25-isopropyl-, (3β,24Z)-		DD01610
55	83	412	145	413	271	213	397	412	100	79	57	20	18	18	10	5		29	48	1	–	–	–	–	–	–	–	–	–	–	27-Norcholest-5-en-3-ol, 25-(2-propylidene)-, (3β)-		DD01611
412	255	300	351	273	369	397	394	412	100	99	93	80	77	69	48	36		29	48	1	–	–	–	–	–	–	–	–	–	–	32-Norlanosta-6,20(22)-dien-3-ol, (3β)-		NI41327

MASS TO CHARGE RATIOS	M.W.	INTENSITIES	Parent	C	H	O	N	S	F	Cl	Br	I	Si	P	B	X	COMPOUND NAME	CAS Reg No	No
412 55 83 255 81 69 95 97	**412**	100 89 77 67 65 60 53 43		29	48	1	–	–	–	–	–	–	–	–	–	–	**Stigmasta-4,22-dien-3-ol, (3β)-**	57815-94-8	**NI41328**
43 412 55 123 271 299 213 229	**412**	100 83 41 33 25 19 19 18		29	48	1	–	–	–	–	–	–	–	–	–	–	**Stigmasta-5,20(22)-dien-3-ol, (3β,20E,24R)-**		**DD01612**
139 395 140 411 255 413 271 396	**412**	100 15 10 8 8 6 6 5	**4.00**	29	48	1	–	–	–	–	–	–	–	–	–	–	**Stigmasta-5,22-dien-3-ol, (3β)-**		**NI41329**
394 255 412 43 57 55 41 145	**412**	100 87 78 56 49 44 40 39		29	48	1	–	–	–	–	–	–	–	–	–	–	**Stigmasta-5,22-dien-3-ol, (3β,22E)-**	83-48-7	**NI41330**
255 412 271 299 272 213 351 273	**412**	100 85 73 54 46 40 32 27		29	48	1	–	–	–	–	–	–	–	–	–	–	**Stigmasta-5,22-dien-3-ol, (3β,22E)-**	83-48-7	**NI41331**
258 55 83 412 81 69 255 97	**412**	100 16 12 11 9 9 7 7		29	48	1	–	–	–	–	–	–	–	–	–	–	**Stigmasta-5,22-dien-3-ol, (3β,22E)-**	83-48-7	**NI41332**
314 55 281 299 69 315 412 81	**412**	100 91 60 59 56 54 53 51		29	48	1	–	–	–	–	–	–	–	–	–	–	**Stigmasta-5,24(28)-dien-3-ol, (3β)-**	18472-36-1	**NI41333**
314 55 69 81 95 107 296 281	**412**	100 85 45 38 32 29 28 28	**10.00**	29	48	1	–	–	–	–	–	–	–	–	–	–	**Stigmasta-5,24(28)-dien-3-ol, (3β,24E)-**	17605-67-3	**NI41334**
314 55 69 315 81 287 95 41	**412**	100 49 28 26 23 21 20 20	**10.00**	29	48	1	–	–	–	–	–	–	–	–	–	–	**Stigmasta-5,24(28)-dien-3-ol, (3β,24E)-**	17605-67-3	**NI41336**
314 296 281 315 229 299 213 297	**412**	100 61 45 39 35 34 24 20	**13.64**	29	48	1	–	–	–	–	–	–	–	–	–	–	**Stigmasta-5,24(28)-dien-3-ol, (3β,24E)-**	17605-67-3	**NI41335**
271 412 255 273 369 246 300 272	**412**	100 56 44 41 31 25 22 22		29	48	1	–	–	–	–	–	–	–	–	–	–	**Stigmasta-7,22-dien-3-ol, (3β,5α)-**		**LI08683**
55 271 81 83 43 69 41 93	**412**	100 75 75 65 64 58 58 50	**29.00**	29	48	1	–	–	–	–	–	–	–	–	–	–	**Stigmasta-7,22-dien-3-ol, (3β,5α,22E,24R)-**	481-17-4	**NI41337**
412 271 43 69 413 255 85 83	**412**	100 47 44 37 32 28 27 25		29	48	1	–	–	–	–	–	–	–	–	–	–	**Stigmasta-7,22-dien-3-ol, (3β,5α,22Z)-**	54482-55-2	**NI41338**
271 314 246 255 412 299 397	**412**	100 83 12 11 10 8 6		29	48	1	–	–	–	–	–	–	–	–	–	–	**Stigmasta-7,24(28)-dien-3-ol**		**LI08684**
271 55 412 81 107 41 255 95	**412**	100 68 44 41 33 33 31 31		29	48	1	–	–	–	–	–	–	–	–	–	–	**Stigmasta-7,25-dien-3-ol, (3β,5α)-**	6785-58-6	**NI41339**
124 43 55 412 41 28 229 95	**412**	100 57 51 39 37 35 34 31		29	48	1	–	–	–	–	–	–	–	–	–	–	**Stigmast-4-en-3-one**		**MS09368**
412 271 229 397 119 105 244 135	**412**	100 85 53 30 27 26 22 20		29	48	1	–	–	–	–	–	–	–	–	–	–	**Stigmast-7-en-3-one**		**NI41340**
314 82 69 81 315 299 231 83	**412**	100 45 30 28 27 21 20 19	**12.00**	29	48	1	–	–	–	–	–	–	–	–	–	–	**Stigmast-24(28)-en-3-one, (5α)-**	55123-75-6	**NI41341**
314 82 55 69 81 315 95 41	**412**	100 62 53 30 28 27 27 22	**11.11**	29	48	1	–	–	–	–	–	–	–	–	–	–	**Stigmast-24(28)-en-3-one, (5α)-**	55123-75-6	**NI41342**
191 177 176 178 189 165 151 412	**412**	100 46 44 30 26 26 17 14		30	20	2	–	–	–	–	–	–	–	–	–	–	**9-Anthracenecarboxylic acid, 9-phenanthrenemethyl ester**	92174-53-3	**NI41343**
412 283 252 413 284 411 253 158	**412**	100 56 35 32 32 28 21 18		30	21	–	–	–	–	–	–	–	–	1	–	–	**Phosphine, trinaphthalenyl-**	28653-22-7	**NI41344**
149 123 412 125 121 137 135 163	**412**	100 99 81 80 60 58 56 40		30	52	–	–	–	–	–	–	–	–	–	–	–	**D:A-Friedooleanane**	559-73-9	**MS09369**
412 149 69 95 109 43 204 81	**412**	100 73 60 58 57 56 55 55		30	52	–	–	–	–	–	–	–	–	–	–	–	**D:A-Friedooleanane**	559-73-9	**NI41345**
55 109 81 95 257 300 69 83	**412**	100 91 88 80 74 73 65 61	**22.77**	30	52	–	–	–	–	–	–	–	–	–	–	–	**Gorgostane, (5α)-**	56143-37-4	**NI41346**
412 95 191 69 43 55 81 220	**412**	100 93 87 78 70 66 59 48		30	52	–	–	–	–	–	–	–	–	–	–	–	**anti-Isotirucallene**		**LI08685**
397 69 55 43 95 398 41 109	**412**	100 54 50 40 37 36 36 29	**9.90**	30	52	–	–	–	–	–	–	–	–	–	–	–	**Lanost-8-ene**	6593-21-1	**NI41347**
191 123 109 137 107 121 192 150	**412**	100 70 55 40 40 37 26 26	**12.00**	30	52	–	–	–	–	–	–	–	–	–	–	–	**Lupane**		**LI08686**
69 55 95 109 257 81 217 412	**412**	100 97 89 87 80 77 69 59		30	52	–	–	–	–	–	–	–	–	–	–	–	**24-Norcholane, 23-[2-methyl-1-isopropylcyclopropyl]-, (5α)-**	53755-13-8	**NI41348**
69 41 81 43 149 55 83 40	**412**	100 75 60 60 51 51 43 42	**29.47**	30	52	–	–	–	–	–	–	–	–	–	–	–	**Tetracosapentaene, 2,6,10,15,19,23-hexamethyl-**	26266-08-0	**NI41349**
412 308 295 206 91 413 384	**412**	100 72 61 42 42 35 26		32	28	–	–	–	–	–	–	–	–	–	–	–	**1,5-Cyclooctadiene, 1,2,5,6-tetraphenyl-**	96581-96-3	**DD01613**
412 335 413 165 215 167 297 291	**412**	100 40 37 24 19 19 14 14		32	28	–	–	–	–	–	–	–	–	–	–	–	**1,2-Dibenzhydrylidenecyclohexane**		**LI08687**
412 308 295 413 384 91 206	**412**	100 61 50 35 22 20 16		32	28	–	–	–	–	–	–	–	–	–	–	–	**Tricyclo[3.3.0.0^{2,6}]octane, 1,2,5,6-tetraphenyl-**	119391-87-6	**DD01614**
117 79 119 296 45 59 298 233	**413**	100 99 95 75 62 59 55 48	**0.00**	10	8	2	1	2	–	5	–	–	–	–	–	–	**S-[Dichloro[(trichloromethyl)thio]methyl]tetrahydrophthalimidyl sulphide**		**NI41350**
43 326 284 240 41 239 266 61	**413**	100 52 29 28 22 21 20 15	**0.00**	14	18	5	1	–	7	–	–	–	–	–	–	–	**Dipropyl N-(heptafluorobutyryl)aspartate**		**MS09370**
73 398 174 172 298 130 86 75	**413**	100 75 27 14 9 9 9 8	**0.22**	14	40	3	1	–	–	–	–	–	4	1	–	–	**Phosphonic acid, [2-[bis(trimethylsilyl)amino]ethyl]-, bis(trimethylsilyl) ester**	53044-44-3	**NI41351**
73 398 174 399 400 45 172 59	**413**	100 75 27 26 15 15 14 12	**0.00**	14	40	3	1	–	–	–	–	–	4	1	–	–	**Phosphonic acid, [2-[bis(trimethylsilyl)amino]ethyl]-, bis(trimethylsilyl) ester**	53044-44-3	**NI41352**
91 64 194 93 39 51 71 77	**413**	100 45 43 32 29 28 24 23	**0.00**	18	16	–	5	1	–	–	1	–	–	–	–	–	**2H-Tetrazolium, 2-(4,5-dimethyl-2-thiazolyl)-3,5-diphenyl-, bromide**	298-93-1	**NI41353**
298 266 238 57 354 41 206 267	**413**	100 99 83 83 47 42 33 32	**10.00**	19	27	9	1	–	–	–	–	–	–	–	–	–	**2,3,4,5-Pyridinetetracarboxylic acid, 1-tert-butyl-6-ethoxy-1,6-dihydro-, tetramethyl ester**		**NI41354**
73 58 355 75 77 45 356 207	**413**	100 84 42 33 22 19 17 15	**0.00**	19	39	3	1	–	–	–	–	–	3	–	–	–	**Adrenaline, N-methyl-O,O′,O″-tris(trimethylsilyl)-**		**MS09371**
116 73 117 45 297 74 75 118	**413**	100 74 10 9 8 6 4 3	**0.00**	19	39	3	1	–	–	–	–	–	3	–	–	–	**Silanamine, N-[2-[3-methoxy-4-[(trimethylsilyl)oxy]phenyl]-N,1,1,1-tetramethyl-2-[(trimethylsilyl)oxy]ethyl]-**	56114-63-7	**MS09372**
116 73 297 117 398 298 118 75	**413**	100 32 10 10 3 3 3 3	**0.00**	19	39	3	1	–	–	–	–	–	3	–	–	–	**Silanamine, N-[2-[3-methoxy-4-[(trimethylsilyl)oxy]phenyl]-N,1,1,1-tetramethyl-2-[(trimethylsilyl)oxy]ethyl]-**	56114-63-7	**NI41355**
413 252 266 220 179 379 218 219	**413**	100 18 10 8 8 7 7 6		20	22	2	2	1	–	–	–	–	–	–	–	1	**Cobalt(II) sulphide, [N,N′-bis(salicylidene)-3,3′-bis(aminopropyl)]-**		**NI41356**
354 355 179 296 178 382 237 193	**413**	100 25 14 7 7 5 5 4	**3.00**	21	19	8	1	–	–	–	–	–	–	–	–	–	**Tetramethyl 4H-benzo[a]quinolizine-1,2,3,4-tetracarboxylate**		**LI08688**
220 205 221 28 147 77 178 42	**413**	100 25 18 16 10 9 8 8	**0.20**	22	23	7	1	–	–	–	–	–	–	–	–	–	**1(3H)-Isobenzofuranone, 6,7-dimethoxy-3-(5,6,7,8-tetrahydro-4-methoxy-6-methyl-1,3-dioxolo[4,5-g]isoquinolin-5-yl)-, [S-(R*,S*)]-**	128-62-1	**NI41357**
220 205 221 147 42 118 77 28	**413**	100 17 15 6 5 4 4 4	**0.00**	22	23	7	1	–	–	–	–	–	–	–	–	–	**1(3H)-Isobenzofuranone, 6,7-dimethoxy-3-(5,6,7,8-tetrahydro-4-methoxy-6-methyl-1,3-dioxolo[4,5-g]isoquinolin-5-yl)-, [S-(R*,S*)]-**	128-62-1	**NI41358**
220 221 414 412 204 218 205 194	**413**	100 11 9 3 3 1 1 1	**0.50**	22	23	7	1	–	–	–	–	–	–	–	–	–	**1(3H)-Isobenzofuranone, 6,7-dimethoxy-3-(5,6,7,8-tetrahydro-4-methoxy-6-methyl-1,3-dioxolo[4,5-g]isoquinolin-5-yl)-, [S-(R*,S*)]-**	128-62-1	**NI41359**

MASS TO CHARGE RATIOS								M.W.	INTENSITIES								Parent	C	H	O	N	S	F	Cl	Br	I	Si	P	B	X	COMPOUND NAME	CAS Reg No	No
58	413	192	149	120	219	205	208	**413**	100	21	20	20	20	19	15	11		22	23	7	1	–	–	–	–	–	–	–	–	–	1(3H)-Isobenzofuranone, 3-[[6-[2-(dimethylamino)ethyl]-4-hydroxy-1,3-benzodioxol-5-yl]methylene]-6,7-dimethoxy-, (Z)-	93552-71-7	NI41360
206	178	207	150	179	208	177	220	**413**	100	66	56	31	29	22	22	20	**11.00**	22	23	7	1	–	–	–	–	–	–	–	–	–	1-Methoxy-13-oxoallocryptopine		LI08689
84	98	55	140	41	413	330	191	**413**	100	86	63	55	51	48	47	45		23	35	2	5	–	–	–	–	–	–	–	–	–	1,3,4-Oxadiazole-2,5-diamine, N^2-cyclohexyl-N^5-[3-[3-(piperidinomethyl)phenoxy]propyl]-		MS09373
413	117	415	91	132	65	77	69	**413**	100	56	38	31	19	18	15	14		24	20	–	5	–	–	1	–	–	–	–	–	–	3H-Imidazo[1,2-b][1,2,4]triazol-2-amine, N,3-bis(4-methylphenyl)-5-(4-chlorophenyl)-		MS09374
231	142	413	230	215	273	368	385	**413**	100	54	42	29	27	18	7	6		24	31	5	1	–	–	–	–	–	–	–	–	–	17β-Hydroxy-17α-(3-ethoxycarbonylisoxazol-5-yl)androst-4-en-5-one		LI08690
354	59	73	114	355	174	130	100	**413**	100	40	35	28	21	21	20	15	**0.00**	24	51	2	1	–	–	–	–	–	1	–	–	–	1-Tridecene, 1-[[[bis(2-ethoxyethyl)amino]methyl]dimethylsilyl]-, (E)-		DD01615
354	59	174	130	355	73	232	102	**413**	100	39	25	24	21	19	17	11	**0.00**	24	51	2	1	–	–	–	–	–	1	–	–	–	1-Tridecene, 3-[[[bis(2-ethoxyethyl)amino]methyl]dimethylsilyl]-		DD01616
254	354	255	267	325	253	355	266	**413**	100	88	50	30	29	24	23	17	**7.00**	25	19	5	1	–	–	–	–	–	–	–	–	–	5-(1,2-Dimethoxycarbonyl-2-oxidovinyl)-6-phenylphenanthridinium		LI08691
353	84	228	43	213	126	105	320	**413**	100	38	32	29	27	15	14	13	**0.20**	25	35	4	1	–	–	–	–	–	–	–	–	–	3β-Acetoxy-17α-(3-methylisoxazol-5-yl)androst-5-en-17β-ol		LI08692
73	273	413	75	91	79	77	137	**413**	100	37	33	32	26	26	24	22		25	39	2	1	–	–	–	–	–	1	–	–	–	Gon-4-ene, 13-ethyl-17-ethynyl-3-(methoximino)-17β-[(trimethylsilyl)oxy], (13β,17α)-		NI41361
184	262	110	384	78	109	413	186	**413**	100	91	78	69	44	43	36	35		26	25	–	1	–	–	–	–	–	–	2	–	–	N,N-Bis(diphenylphosphino)ethylamine		NI41362
413	207	104	208	250	414	161	381	**413**	100	70	63	41	31	29	21	7		26	27	–	3	1	–	–	–	–	–	–	–	–	2-Imidazoline-5-thione, 1-(2,6-dimethylphenyl)-4,4-diphenyl-2-(isopropylamino)-	118515-04-1	DD01617
207	413	208	235	165	356	264	357	**413**	100	50	36	34	34	23	16	14		26	27	–	3	1	–	–	–	–	–	–	–	–	5-Isoimidazole, 2-(tert-butylamino)-5,5-diphenyl-4-[(4-methylphenyl)thio]-	118515-01-8	DD01618
368	413	369	128	58	27	42	142	**413**	100	38	33	26	22	20	19	15		26	27	2	3	–	–	–	–	–	–	–	–	–	1-Methylamino-4-[4-(dimethylamino)methyldimethylanilino]anthraquinone		IC04337
58	342	413	59	328	72	368	343	**413**	100	12	8	5	4	4	3	3		26	27	2	3	–	–	–	–	–	–	–	–	–	1-(4-Toluidino)-4-[3-(dimethylamino)propylamino]anthraquinone		IC04338
43	338	91	353	105	79	413	41	**413**	100	36	32	26	20	20	19	18		26	39	3	1	–	–	–	–	–	–	–	–	–	3β-Acetoxy-16-methylpregna-5,16-dien-20-one, ethyloxime		LI08693
286	91	36	160	287	70	92	121	**413**	100	44	32	25	24	22	17	14	**0.00**	27	40	–	1	–	–	1	–	–	–	–	–	–	Androst-2-en-17-amine, N-(2-phenylethyl)-, hydrochloride, (5α)-	53286-42-3	NI41363
111	413	112	412	414	356	164	357	**413**	100	87	47	40	27	22	20	16		27	43	2	1	–	–	–	–	–	–	–	–	–	Korseveridine		LI08694
125	98	111	413	165	112	151	164	**413**	100	99	84	54	18	17	16	12		27	43	2	1	–	–	–	–	–	–	–	–	–	Petiline		LI08695
125	111	413	414	398	356	165	124	**413**	100	83	54	33	18	18	17	16		27	43	2	1	–	–	–	–	–	–	–	–	–	Petiline		LI08696
114	138	113	413	385	125	396	414	**413**	100	82	43	42	25	19	17	15		27	43	2	1	–	–	–	–	–	–	–	–	–	Spirosol-5-en-3-ol, (3β,22α,25R)-	126-17-0	NI41364
98	99	56	55	125	41	91	70	**413**	100	8	7	7	6	5	4	4	**1.30**	27	43	2	1	–	–	–	–	–	–	–	–	–	Veralkamine		MS09375
98	125	413	126	99	395			**413**	100	57	39	27	20	18				27	43	2	1	–	–	–	–	–	–	–	–	–	Veralkamine		LI08697
131	69	119	181	51	31	169	162	**414**	100	42	34	22	22	18	16	8	**0.00**	8	1	2	–	–	15	–	–	–	–	–	–	–	Octanoic acid, pentadecafluoro-	335-67-1	IC04339
131	69	44	181	119	231	162	31	**414**	100	42	20	14	10	5	4	4	**0.00**	8	1	2	–	–	15	–	–	–	–	–	–	–	Octanoic acid, pentadecafluoro-	335-67-1	IC04340
131	69	31	44	93	181	119	100	**414**	100	83	30	27	19	11	10	9	**0.00**	8	1	2	–	–	15	–	–	–	–	–	–	–	Octanoic acid, pentadecafluoro-	335-67-1	WI01529
238	236	414	326	119	103	190	205	**414**	100	25	19	16	14	9	4	3		10	15	–	–	–	6	–	–	–	–	2	–	1	Rhodium, [(1,2,3,4,5-η)-1,2,3,4,5-pentamethyl-2,4-cyclopentadien-1-yl]bis(phosphorous trifluoride)-	34822-29-2	MS09376
113	259	261	381	383	263	237	257	**414**	100	50	47	46	42	42	40	38	**8.48**	11	8	4	–	–	–	6	–	–	–	–	–	–	Bicyclo[2.2.1]hept-5-ene-2,3-dicarboxylic acid, 1,4,5,6,7,7-hexachloro-, dimethyl ester	54002-92-5	NI41365
76	50	379	223	75	315	64	36	**414**	100	85	69	54	54	46	39	39	**4.62**	12	8	6	–	3	–	2	–	–	–	–	–	–	Bis[4-(chlorosulphonyl)phenyl] sulphone		IC04341
66	107	79	77	143	83	65	55	**414**	100	86	75	38	35	26	24	24	**0.09**	12	9	1	–	–	–	7	–	–	–	–	–	–	Dieldrin chlorohydrin		NI41366
66	29	39	79	27	107	31	77	**414**	100	90	74	69	58	46	43	39	**0.00**	12	9	1	–	–	–	7	–	–	–	–	–	–	1,4:5,8-Dimethanonaphthalen-2-ol, 3,5,6,7,8,9,9-heptachloro-1,2,3,4,4a,5,8,8a-octahydro-, (1α,2α,3β,4α,4aβ,5α,8α,8β)-	62059-42-1	NI41367
154	412	207	77	414	205	284	410	**414**	100	71	71	70	68	67	56	48		12	10	–	–	–	–	–	–	–	–	–	–	2	Ditelluride, diphenyl-	32294-60-3	NI41368
414	264	115	69	234	236	105	149	**414**	100	52	20	20	18	12	11	9		15	6	–	–	–	12	–	–	–	–	–	–	–	Dodecafluoro-4,5-dimethyltetracyclo[6.2.2.1^{3,6}.0^{2,7}]trideca-2(7),4,9-triene		NI41369
264	249	345	265	414	241	245	242	**414**	100	54	14	13	11	10	9	9		15	6	–	–	–	12	–	–	–	–	–	–	–	Dodecafluoro-4,5-dimethyltetracyclo[6.2.2.1^{3,6}.0^{2,7}]trideca-2(7),4,9-triene		NI41370
85	41	55	93	91	43	57	319	**414**	100	40	25	9	9	9	8	7	**0.00**	15	25	1	–	–	–	1	2	–	–	–	–	–	Chamigran-9-ol, 2,10-dibromo-3-chloro-		MS09377
88	75	101	45	266	235	76	77	**414**	100	75	30	30	23	13	8	7	**6.67**	16	22	9	4	–	–	–	–	–	–	–	–	–	β-D-gluco-Hexodialdo-1,5-pyranoside, methyl 2,3,4-tri-O-methyl-, (2,4-dinitrophenyl)hydrazone	41545-31-7	NI41371
277	45	75	101	97	261	172	70	**414**	100	81	77	30	23	19	17	11	**6.23**	16	22	9	4	–	–	–	–	–	–	–	–	–	β-D-ribo-Hexopyranosid-3-ulose, methyl 2,4,6-tri-O-methyl-, (2,4-dinitrophenyl)hydrazone	41545-27-1	NI41372
45	281	75	88	149	101	369	282	**414**	100	63	40	31	20	20	19	13	**0.00**	16	22	9	4	–	–	–	–	–	–	–	–	–	β-D-xylo-Hexopyranosid-4-ulose, methyl 2,3,6-tri-O-methyl-, (2,4-dinitrophenyl)hydrazone	41545-25-9	NI41373
166	41	61	250	29	324	167	263	**414**	100	30	27	25	20	19	17	15	**0.00**	16	25	5	2	1	3	–	–	–	–	–	–	–	N-Trifluoroacetyl-L-prolylmethionine sulphoxide butyl ester		MS09378
73	399	400	309	45	401	147	74	**414**	100	64	25	15	14	13	11	10	**2.75**	16	38	1	4	–	–	–	–	–	4	–	–	–	1H-Imidazole-4-carboxamide, 5-amino-, tetrakis(trimethylsilyl)-	56272-85-6	NI41374

MASS TO CHARGE RATIOS	M.W.	INTENSITIES	Parent	C	H	O	N	S	F	Cl	Br	I	Si	P	B	X	COMPOUND NAME	CAS Reg No	No
218 197 220 164 69 53 165 59	**414**	100 73 65 42 42 40 29 28	**2.00**	17	12	8	–	–	–	2	–	–	–	–	–	–	Spiro[4H-1,3-benzodioxin-2,1′-[2,5]cyclohexadiene]-2′-carboxylic acid, 6,8-dichloro-5-hydroxy-6′-methoxy-7-methyl-4,4′-dioxo-, methyl ester	428-21-7	NI41375
159 161 81 163 54 89 27 335	**414**	100 62 42 10 9 8 7 7	**1.10**	18	14	1	2	–	–	4	–	–	–	–	–	–	1H-Imidazole, 1-[2-(2,4-dichlorophenyl)-2-[(2,4-dichlorophenyl)methoxy]ethyl]-	22916-47-8	NI41376
297 73 298 299 296 294 295 147	**414**	100 48 27 11 7 7 6 6	**1.34**	18	34	5	–	–	–	–	–	–	3	–	–	–	Benzeneacetic acid, 3-methoxy-α,4-bis[(trimethylsilyl)oxy]-, trimethylsilyl ester	55268-66-1	NI41379
73 297 298 45 74 147 299 75	**414**	100 66 17 17 9 8 7 7	**0.00**	18	34	5	–	–	–	–	–	–	3	–	–	–	Benzeneacetic acid, 3-methoxy-α,4-bis[(trimethylsilyl)oxy]-, trimethylsilyl ester	55268-66-1	NI41378
297 298 299 73 371 399 267 147	**414**	100 24 10 7 5 4 3 3	**1.23**	18	34	5	–	–	–	–	–	–	3	–	–	–	Benzeneacetic acid, 3-methoxy-α,4-bis[(trimethylsilyl)oxy]-, trimethylsilyl ester	55268-66-1	NI41377
355 356 357 414 415 399 358 237	**414**	100 33 15 10 4 4 4 4		18	34	5	–	–	–	–	–	–	3	–	–	–	Benzeneacetic acid, α,3,4-tris[(trimethylsilyl)oxy]-, methyl ester	55268-65-0	NI41380
355 73 356 357 354 45 353 89	**414**	100 65 36 18 9 8 7 7	**5.60**	18	34	5	–	–	–	–	–	–	3	–	–	–	Benzeneacetic acid, α,3,4-tris[(trimethylsilyl)oxy]-, methyl ester	55268-65-0	NI41381
73 355 45 356 74 357 89 75	**414**	100 53 20 17 9 8 8 6	**2.30**	18	34	5	–	–	–	–	–	–	3	–	–	–	Benzeneacetic acid, α,3,4-tris[(trimethylsilyl)oxy]-, methyl ester	55268-65-0	MS09379
70 112 203 213 100 154 142 144	**414**	100 99 78 65 52 51 50 47	**33.20**	19	34	6	4	–	–	–	–	–	–	–	–	–	Hexanoic acid, 3-(acetylamino)-6-[[3,6-bis(acetylamino)-1-oxohexyl]amino]-, methyl ester, [S-(R*,R*)]-	33861-49-3	NI41382
112 101 113 44 84 70 126 72	**414**	100 81 80 75 56 55 38 36	**20.00**	19	34	6	4	–	–	–	–	–	–	–	–	–	Hexanoic acid, 6-(acetylamino)-3-[[3,6-bis(acetylamino)-1-oxohexyl]amino]-, methyl ester, [S-(R*,R*)]-	33861-54-0	NI41383
112 113 101 45 55 70 54 53	**414**	100 69 69 64 55 47 42 37	**17.17**	19	34	6	4	–	–	–	–	–	–	–	–	–	Hexanoic acid, 6-(acetylamino)-3-[[3,6-bis(acetylamino)-1-oxohexyl]amino]-, methyl ester, [S-(R*,R*)]-	33861-54-0	NI41384
72 86 43 114 270 145 185 213	**414**	100 67 50 42 32 31 28 24	**2.65**	19	34	6	4	–	–	–	–	–	–	–	–	–	L-Leucine, N-[N-[N-(N-acetyl-L-alanyl)-L-valyl]glycyl]-, methyl ester	4249-30-3	NI41385
178 177 176 152 414 151 413 179	**414**	100 50 36 28 21 21 18 17		20	14	–	–	–	–	–	–	–	–	–	–	2	1,3-Benzodiselenole, 2-(diphenylmethylidene)-		DD01619
328 128 298 191 166 329 190 129	**414**	100 60 35 32 20 18 16 15	**13.00**	20	18	8	2	–	–	–	–	–	–	–	–	–	Trimethyl 7-nitro-10,11-dihydroazepino[1,2-a]quinoline-8,9,10-tricarboxylate		LI08698
193 178 135 179 150 194 118 414	**414**	100 89 68 67 61 59 53 45		20	22	6	4	–	–	–	–	–	–	–	–	–	1,2,4,5-Tetrazine, 3,6-bis(3,4,5-trimethoxyphenyl)-	81258-52-8	NI41386
166 148 28 204 167 91 41 149	**414**	100 72 65 43 22 21 18 15	**0.00**	20	25	4	2	–	3	–	–	–	–	–	–	–	N-Trifluoroacetyl-L-prolylphenylalanine butyl ester		MS09380
330 246 288 245 372 152 153 121	**414**	100 89 87 78 45 32 26 24	**1.00**	21	18	9	–	–	–	–	–	–	–	–	–	–	Benzophenone, 2,4,4′,6-tetraacetoxy-		NI41387
84 58 46 44 414 42 85 187	**414**	100 70 53 44 41 19 15 12		21	22	7	2	–	–	–	–	–	–	–	–	–	2-Naphthacenecarboxamide, 4-(dimethylamino)-1,4,4a,5,5a,6,11,12a-octahydro-3,10,12,12a-tetrahydroxy-1,11-dioxo-, [4S-(4α,4aα,5aα,12aα)]-	808-26-4	NI41388
57 302 29 246 414 41 43 358	**414**	100 31 29 27 23 22 21 14		21	35	–	–	–	–	–	–	1	–	–	–	–	Benzene, 1,3,5-tris(2,2-dimethylpropyl)-2-iodo-	25347-04-0	MS09381
322 302 414 180 193 321 93 252	**414**	100 46 45 24 20 19 19 12		22	17	1	2	1	3	–	–	–	–	–	–	–	Cinnamanilide, α-[(trifluoromethyl)thio]-β-(phenylamino)-		DD01620
137 368 297 203 312 352 313 281	**414**	100 50 28 22 21 15 14 14	**0.00**	22	22	8	–	–	–	–	–	–	–	–	–	–	4H-1-Benzopyran-4-one, 3-[2,4-dihydroxy-5-methoxy-6-(3-formyloxy-3-methylbutyl)]-7-hydroxy-		MS09382
105 43 106 77 232 207 190 177	**414**	100 22 7 7 5 4 4 4	**0.00**	22	22	8	–	–	–	–	–	–	–	–	–	–	1,2,3,4-Butanetetrol, 2,3-diacetate 1,4-dibenzoate, (R*,R*)-	74792-93-1	NI41389
105 43 77 106 232 207 190 177	**414**	100 34 11 7 4 4 4 4	**0.00**	22	22	8	–	–	–	–	–	–	–	–	–	–	1,2,3,4-Butanetetrol, 2,3-diacetate 1,4-dibenzoate, (R*,S*)-	74793-38-7	NI41390
105 77 43 106 207 232 51 78	**414**	100 14 11 7 6 3 2 1	**0.00**	22	22	8	–	–	–	–	–	–	–	–	–	–	1,2,3,4-Butanetetrol, 1,4-diacetate 2,3-dibenzoate, [S-(R*,R*)]-	74793-35-4	NI41391
312 396 297 313 397 168 156 254	**414**	100 69 29 19 16 14 11 9	**4.29**	22	22	8	–	–	–	–	–	–	–	–	–	–	Furo[3′,4′:6,7]naphtho[2,3-d]-1,3-dioxol-6(5aH)-one, 5,8,8a,9-tetrahydro-9 hydroxy-5-(3,4,5-trimethoxyphenyl)-, [5R-(5α,5aα,8aα,9α)]-	477-47-4	NI41392
414 415 168 169 181 189 201 153	**414**	100 26 13 10 7 6 5 5		22	22	8	–	–	–	–	–	–	–	–	–	–	Furo[3′,4′:6,7]naphtho[2,3-d]-1,3-dioxol-6(5aH)-one, 5,8,8a,9-tetrahydro-9 hydroxy-5-(3,4,5-trimethoxyphenyl)-, [5R-(5α,5aβ,8aα,9α)]-	518-28-5	NI41393
414 415 181 189 246 201 168 190	**414**	100 19 15 14 13 12 6 5		22	22	8	–	–	–	–	–	–	–	–	–	–	Furo[3′,4′:6,7]naphtho[2,3-d]-1,3-dioxol-6(5aH)-one, 5,8,8a,9-tetrahydro-10-hydroxy-5-(3,4,5-trimethoxyphenyl)-, [5R-(5α,5aβ,8aα)]-	518-29-6	NI41394
43 208 165 166 250 209 134 133	**414**	100 91 45 36 14 11 9 9	**0.00**	22	22	8	–	–	–	–	–	–	–	–	–	–	4,4′-Bis(methoxycarbonyl)hydrobenzoin diacetate		IC04342
285 43 354 313 326 257 414 91	**414**	100 79 70 64 45 45 20 18		22	22	8	–	–	–	–	–	–	–	–	–	–	Naphtho[2,3-b]furan-2(3H),5,8-trione, 7-(2-acetoxyisopropyl)-6,9-dihydroxy-3,4-dimethyl-3-(2-propenyl)-, (2′ζ,3R)-		DD01621
339 43 354 45 60 311 283 321	**414**	100 50 20 19 15 12 10 8	**5.00**	22	22	8	–	–	–	–	–	–	–	–	–	–	2H-Phenaleno[1,9-bc]pyran-2,8(4H)-dione, 10-[2-(acetyloxy)-1-methylethyl]-5,5a-dihydro-6,7,9-trihydroxy-3,5a-dimethyl-	68421-12-5	NI41395
169 151 414 194 180 167 152 168	**414**	100 80 68 66 31 30 29 23		22	26	4	2	1	–	–	–	–	–	–	–	–	1-Naphthalenesulphonamide, N-[2-(3,4-dimethoxyphenyl)ethyl]-5-(dimethylamino)-	5282-85-9	NI41396
170 151 171 414 250 263 168 152	**414**	100 84 59 57 19 16 16 16		22	26	4	2	1	–	–	–	–	–	–	–	–	1-Naphthalenesulphonamide, N-[2-(3,4-dimethoxyphenyl)ethyl]-5-(dimethylamino)-	5282-85-9	NI41397
170 151 171 414 250 168 263 152	**414**	100 85 61 58 20 18 16 16		22	26	4	2	1	–	–	–	–	–	–	–	–	1-Naphthalenesulphonamide, N-[2-(3,4-dimethoxyphenyl)ethyl]-5-(dimethylamino)-	5282-85-9	NI41398
414 357 43 57 69 314 55 44	**414**	100 95 67 46 40 37 36 35		22	26	6	2	–	–	–	–	–	–	–	–	–	Dichotine (neutral)	27530-76-3	NI41399
70 57 44 414 357 58 42 314	**414**	100 49 49 48 47 39 31 1		22	26	6	2	–	–	–	–	–	–	–	–	–	Dichotine (neutral)	27530-76-3	LI08699

MASS TO CHARGE RATIOS	M.W.	INTENSITIES	Parent	C	H	O	N	S	F	Cl	Br	I	Si	P	B	X	COMPOUND NAME	CAS Reg No	No
70 57 44 414 357 58 42 71	**414**	100 50 50 48 47 40 31 27		22	26	6	2	–	–	–	–	–	–	–	–	–	Dichotine (neutral)	27530-76-3	NI41400
91 108 79 107 77 43 28 42	**414**	100 93 58 50 34 31 31 24	**0.00**	22	26	6	2	–	–	–	–	–	–	–	–	–	DL-Lysine, N[2],N[6]-bis[benzyloxycarbonyl]-	55592-85-3	NI41401
414 138 94 239 226 269 268 283	**414**	100 99 75 45 36 31 30 24		22	26	6	2	–	–	–	–	–	–	–	–	–	Spiro[indoline-3,7′-1′-azabicyclo[4.3.0]nonan]-2-one, 3′-acetyl-4′-(bismethoxycarbonylmethyl) cis-		LI08700
414 138 94 239 226 269 268 383	**414**	100 99 73 32 31 25 24 18		22	26	6	2	–	–	–	–	–	–	–	–	–	Spiro[indoline-3,7′-1′-azabicyclo[4.3.0]nonan]-2-one, 3′-acetyl-4′-(bismethoxycarbonylmethyl)-, trans-		LI08701
72 326 327 381 341 342 223 328	**414**	100 75 63 34 28 22 22 20	**0.00**	22	30	2	2	1	–	–	–	–	1	–	–	–	Benzamide, N,N-diethyl-6-(phenylthiocarbamoyl)-3-methoxy-2-(trimethylsilyl)-	85370-77-0	DD01622
162 414 135 322 150 149 151 384	**414**	100 95 70 64 58 56 52 51		23	26	7	–	–	–	–	–	–	–	–	–	–	rel-(1S,5S,6S,7R,8R)-8-Acetoxy-3,5-dimethoxy-7-methyl-4-oxo-6-piperonylbicyclo[3.2.1]oct-2-ene		NI41402
412 344 343 329 301 315 413 165	**414**	100 96 75 62 62 29 25 23	**0.00**	23	26	7	–	–	–	–	–	–	–	–	–	–	4H-1-Benzopyran-4-one, 2-(3,4-dimethoxyphenyl)-2,3-dihydro-5-hydroxy-3-methoxy-7-[(3-methyl-1-butenyl)oxy]-	56771-65-4	MS09383
192 191 414 177 149 219 147	**414**	100 50 23 11 8 2 2		23	26	7	–	–	–	–	–	–	–	–	–	–	2,3-Dimethoxy-9-hydroxy-8-(4-hydroxy-3-methylbutyl)-12-oxo-6,6a,12,12a-tetrahydro[1]benzopyrano[3,4-b][1]benzopyran		LI08702
105 91 259 96 112 323 281 200	**414**	100 65 52 45 38 30 30 30	**0.00**	23	30	3	2	1	–	–	–	–	–	–	–	–	cis-1-Acetamido-2-(N-phenethyl-4-toluenesulphonamido)cyclohexane		MS09384
281 323 60 140 91 98 81 110	**414**	100 80 60 45 41 40 31 17	**0.00**	23	30	3	2	1	–	–	–	–	–	–	–	–	cis-1-Acetamido-3-(N-phenethyl-4-toluenesulphonamido)cyclohexane		MS09385
281 323 140 91 60 81 98 105	**414**	100 85 75 70 58 39 30 24	**0.00**	23	30	3	2	1	–	–	–	–	–	–	–	–	cis-1-Acetamido-4-(N-phenethyl-4-toluenesulphonamido)cyclohexane		MS09386
140 81 259 91 98 96 105 112	**414**	100 74 70 54 51 50 49 42	**0.00**	23	30	3	2	1	–	–	–	–	–	–	–	–	trans-1-Acetamido-2-(N-phenethyl-4-toluenesulphonamido)cyclohexane		MS09387
140 60 200 91 81 98 323 105	**414**	100 35 20 20 15 13 10 8	**0.00**	23	30	3	2	1	–	–	–	–	–	–	–	–	trans-1-Acetamido-3-(N-phenethyl-4-toluenesulphonamido)cyclohexane		MS09388
140 323 91 60 81 98 259 155	**414**	100 33 21 20 13 10 8 8	**0.00**	23	30	3	2	1	–	–	–	–	–	–	–	–	trans-1-Acetamido-4-(N-phenethyl-4-toluenesulphonamido)cyclohexane		MS09389
138 44 414 160 386 370 41 43	**414**	100 49 33 26 25 19 16 15		23	30	5	2	–	–	–	–	–	–	–	–	–	Aspidospermidine, 19,21-epoxy-1-formyl-15,16,17-trimethoxy-	54751-73-4	NI41403
184 325 324 327 414 326 212 174	**414**	100 84 69 48 39 30 18 14		23	30	5	2	–	–	–	–	–	–	–	–	–	Aspidospermidin-21-oic acid, 1-acetyl-20-hydroxy-17-methoxy-, methyl ester	55123-71-2	NI41404
184 325 324 327 326 414 43 186	**414**	100 76 60 46 30 25 19 16		23	30	5	2	–	–	–	–	–	–	–	–	–	Aspidospermidin-21-oic acid, 1-acetyl-20-hydroxy-17-methoxy-, methyl ester	55123-71-2	NI41405
414 415 209 355 399 45 138 44	**414**	100 71 59 55 48 43 42 41		23	30	5	2	–	–	–	–	–	–	–	–	–	Conopharyngine oxindole, (all-ξ)-	16790-92-4	NI41406
414 415 209 355 399 44 138 43	**414**	100 70 59 55 48 44 41 41		23	30	5	2	–	–	–	–	–	–	–	–	–	Conopharyngine oxindole, (all-ξ)-		LI08703
186 185 382 325 414 199 144 171	**414**	100 47 41 16 12 12 12 10		23	30	5	2	–	–	–	–	–	–	–	–	–	Dihydroeripine A		LI08704
186 382 325 414 199 144 171 212	**414**	100 30 21 13 10 10 8 7		23	30	5	2	–	–	–	–	–	–	–	–	–	Dihydroeripine B		LI08705
69 71 138 83 81 67 85 97	**414**	100 83 62 62 48 46 43 42	**0.92**	23	30	5	2	–	–	–	–	–	–	–	–	–	13a,3a-(Epoxyethano)-1H-indolizino[8,1-cd]carbazol-7-ol, 6-acetyl-2,3,4,5,5a,6,11,12-octahydro-8,9-dimethoxy-	2122-26-1	NI41407
43 45 44 74 60 42 152 352	**414**	100 95 83 50 49 44 40 25	**2.78**	23	30	5	2	–	–	–	–	–	–	–	–	–	Indole-2-acetic acid, α-(1-acetyl-3-ethylidene-4-piperidyl)-α-(hydroxymethyl)-3-(methoxymethyl)-, methyl ester	2202-18-8	NI41408
414 72 70 58 204 44 415 285	**414**	100 47 44 44 38 30 28 28		23	30	5	2	–	–	–	–	–	–	–	–	–	4H-3,8a-Methanofuro[2′,3′:6,7]azonino[5,4-b]indol-14-one, 13-acetyl-2,3,3a,5,6,7,8,13,13a,13b-decahydro-11,12-dimethoxy-2,6-dimethyl-	29484-55-7	NI41409
414 413 415 384 244 230 383 229	**414**	100 47 20 13 13 12 10 7		23	30	5	2	–	–	–	–	–	–	–	–	–	Oxayohimban-16-carboxylic acid, 10,11-dimethoxy-19-methyl-, methyl ester, (16β,19α)-	5308-79-2	NI41410
414 383 415 384 351 416 399 352	**414**	100 33 32 15 11 10 7 6		23	34	3	2	–	–	–	–	–	1	–	–	–	Estra-1,3,5(10)-triene-6,17-dione, 3-[(trimethylsilyl)oxy]-, bis(O-methyloxime)	69833-98-3	NI41411
414 415 383 352 92 91 384 218	**414**	100 35 31 29 20 20 18 15		23	34	3	2	–	–	–	–	–	1	–	–	–	Estra-1,3,5(10)-triene-16,17-dione, 3-[(trimethylsilyl)oxy]-, bis(O-methyloxime)	75113-02-9	NI41412
159 98 239 158 92 43 218 84	**414**	100 62 55 52 49 43 35 28	**0.00**	23	42	6	–	–	–	–	–	–	–	–	–	–	Hexadecanoic acid, 2-(acetyloxy)-1-[(acetyloxy)methyl]ethyl ester	55268-69-4	WI01531
43 159 57 41 55 98 239 84	**414**	100 24 24 20 19 18 14 13	**0.00**	23	42	6	–	–	–	–	–	–	–	–	–	–	Hexadecanoic acid, 2-(acetyloxy)-1-[(acetyloxy)methyl]ethyl ester	55268-69-4	WI01530
159 43 98 57 41 55 84 29	**414**	100 81 59 58 56 50 43 35	**0.00**	23	42	6	–	–	–	–	–	–	–	–	–	–	Hexadecanoic acid, 2,3-bis(acetyloxy)propyl ester	55268-70-7	WI01533
159 239 158 177 98 145 355 84	**414**	100 28 24 20 17 12 10 9	**0.00**	23	42	6	–	–	–	–	–	–	–	–	–	–	Hexadecanoic acid, 2,3-bis(acetyloxy)propyl ester	55268-70-7	WI01532
173 217 73 103 83 210 299 178	**414**	100 35 34 17 11 9 6 6	**0.20**	23	46	4	–	–	–	–	–	–	1	–	–	–	9-Octadecenoic acid, 12-methoxy-13-[(trimethylsilyl)oxy]-, methyl ester	40707-93-5	NI41413
312 354 158 313 252 170 157 237	**414**	100 30 29 23 19 17 17 14	**0.00**	24	30	6	–	–	–	–	–	–	–	–	–	–	Estra-1,3,5(10)-triene-3,6,7-triol, triacetate, (6α,17β)-	6626-42-2	NI41414
312 354 158 330 313 252 157 170	**414**	100 31 27 25 25 18 17 15	**0.61**	24	30	6	–	–	–	–	–	–	–	–	–	–	Estra-1,3,5(10)-triene-3,6,17-triol, triacetate, (6β,17β)-	6944-48-5	NI41415
310 235 352 250 311 209 197 157	**414**	100 15 14 14 12 7 6 6	**0.00**	24	30	6	–	–	–	–	–	–	–	–	–	–	Estra-1,3,5(10)-triene-3,11,17-triol, triacetate, (11β,17β)-	74299-24-4	NI41416
372 373 252 237 158 160 224 414	**414**	100 24 13 12 10 8 7 6		24	30	6	–	–	–	–	–	–	–	–	–	–	Estra-1,3,5(10)-triene-3,15,17-triol, triacetate, (15β,17β)-	74299-23-3	NI41417
372 373 160 252 414 270 159 172	**414**	100 24 22 17 8 7 7 6		24	30	6	–	–	–	–	–	–	–	–	–	–	Estra-1,3,5(10)-triene-3,16,17-triol, triacetate, (16α,17β)-	2284-32-4	NI41418
372 160 252 159 414 133 172 146	**414**	100 27 17 17 14 13 11 10		24	30	6	–	–	–	–	–	–	–	–	–	–	Estra-1,3,5(10)-triene-3,16,17-triol, triacetate, (16α,17β)-	2284-32-4	NI41419
124 414 43 355 83 356 47 85	**414**	100 89 75 69 48 42 38 31		24	34	4	2	–	–	–	–	–	–	–	–	–	Aspidospermidine-1-acetic acid, 17-hydroxy-16-methoxy-α-methyl-, methyl ester	54725-05-2	NI41420
140 414 141 125 109 399 123 91	**414**	100 13 11 8 7 6 6 6		24	34	4	2	–	–	–	–	–	–	–	–	–	4,25-Secoobscurinervan, 15,16-dimethoxy-22-methyl-, (22α)-	54658-14-9	NI41421
309 324 414 307 310 325 415 323	**414**	100 93 45 37 27 26 17 15		24	38	2	–	–	–	–	–	–	2	–	–	–	Estra-1,3,5(10),7-tetraene, 3,17-[(trimethylsilyl)oxy]-, (17α)-	69688-20-6	NI41422

MASS TO CHARGE RATIOS								M.W.	INTENSITIES								Parent	C	H	O	N	S	F	Cl	Br	I	Si	P	B	X	COMPOUND NAME	CAS Reg No	No
244	75	169	324	414	232	183	73	414	100	82	79	62	59	50	44	40		24	38	2	–	–	–	–	–	–	2	–	–	–	Estra-1,3,5(10),11-tetraene, 3,17-[(trimethylsilyl)oxy]-, (17α)-	5150-60-7	NI41423
101	43	400	57	55	41	83	69	414	100	57	18	14	9	9	8	8	0.00	24	46	5	–	–	–	–	–	–	–	–	–	–	2-Hexadecanol, 1-[(2,2-dimethyl-1,3-dioxolan-4-yl)methoxy]-, acetate	56599-80-5	NI41424
73	355	89	43	399	356	57	103	414	100	93	40	40	33	30	28	25	1.43	24	50	3	–	–	–	–	–	–	1	–	–	–	Eicosanoic acid, 2-[(trimethylsilyl)oxy]-, methyl ester		NI41425
180	56	57	214	152	55	181	165	414	100	68	58	40	24	23	19	18	2.00	25	16	–	2	–	–	2	–	–	–	–	–	–	9,9'-Bis(3-chlorocarbazyl)methane	82434-76-2	NI41426
243	56	39	28	165	244	99	115	414	100	65	65	65	53	45	37	34	0.10	25	22	4	2	–	–	–	–	–	–	–	–	–	2,5-Pyrrolidinedione, 1-[[[(triphenylmethyl)amino]acetyl]oxy]-	3397-28-2	NI41427
414	221	245	220	415	399	222	206	414	100	98	40	38	29	29	29	24		25	22	4	2	–	–	–	–	–	–	–	–	–	Spiro[2H-1-benzopyran-2,2'-[2H]indole], 1',3'-dihydro-8-methoxy-3',3'-dimethyl-6-nitro-1-phenyl-	2067-32-5	NI41428
297	149	312	43	55	57	339	354	414	100	91	86	77	74	70	56	55	45.00	25	34	5	–	–	–	–	–	–	–	–	–	–	6H-Dibenzo[b,d]pyran-1,8-diol, 6a,7,8,9,10,10a-hexahydro-6,6-dimethyl-9 methylene-3-pentyl-, diacetate, [6aR-(6aα,8α,10aβ)]-	52522-57-3	NI41429
149	312	270	354	43	297	269	167	414	100	84	83	74	63	56	56	44	11.00	25	34	5	–	–	–	–	–	–	–	–	–	–	6H-Dibenzo[b,d]pyran-1,8-diol, 6a,7,8,9,10,10a-hexahydro-6,6-dimethyl-9 methylene-3-pentyl-, diacetate, [6aR-(6aα,8β,10aβ)]-	27262-27-7	NI41430
57	43	149	312	83	129	372	414	414	100	99	78	63	48	40	39	37		25	34	5	–	–	–	–	–	–	–	–	–	–	6H-Dibenzo[b,d]pyran-9-methanol, 1-(acetyloxy)-6a,7,10,10a-tetrahydro-6,6-dimethyl-3-pentyl-, acetate, (6aR-trans)-	41969-62-4	NI41431
354	43	281	355	91	105	121	81	414	100	76	70	25	17	16	15	15	0.00	25	34	5	–	–	–	–	–	–	–	–	–	–	Pregna-5,16-dien-20-one, 3,21-bis(acetyloxy)-, (3β)-	37413-93-7	NI41432
354	281	355	282	145	121	107	339	414	100	84	28	20	13	13	9	8	0.00	25	34	5	–	–	–	–	–	–	–	–	–	–	Pregna-5,16-dien-20-one, 3,21-bis(acetyloxy)-, (3β)-	37413-93-7	MS09390
246	414	105	107	231	121	119	109	414	100	46	43	38	36	31	28	28		25	39	4	–	–	–	–	–	–	–	–	1	–	5β-Pregnane-3,20-dione, 17,21-dihydroxy-, cyclic 1-butaneboronate	31012-68-7	NI41433
246	357	107	105	121	231	119	215	414	100	64	43	43	40	36	29	28	25.00	25	39	4	–	–	–	–	–	–	–	–	1	–	5α-Pregn-4-ene-3,20-dione, 17α,21-dihydroxy-, tert-butyl boronate		LI08706
246	55	41	357	81	43	79	67	414	100	96	69	64	63	60	58	58	25.00	25	39	4	–	–	–	–	–	–	–	–	1	–	5α-Pregn-4-ene-3,20-dione, 17α,21-dihydroxy-, tert-butyl boronate		MS09391
244	412	229	91	55	79	124	105	414	100	71	44	44	43	41	33	30	4.00	25	39	4	–	–	–	–	–	–	–	–	1	–	Pregn-4-ene-3,20-dione, 17α,21-dihydroxy-, tert-butyl boronate		MS09392
414	287	269	415	147	413	124	267	414	100	46	39	33	30	29	29	26		25	39	4	–	–	–	–	–	–	–	–	1	–	Pregn-4-en-3-one, 17,21-[(butylborylene)bis(oxy)]-20-hydroxy-, (20R)-	26611-26-7	NI41434
287	269	414	288	124	229	267	105	414	100	46	35	29	27	25	23	21		25	39	4	–	–	–	–	–	–	–	–	1	–	Pregn-4-en-3-one, 17,21-[(butylborylene)bis(oxy)]-20-hydroxy-, (20S)-	24376-83-8	NI41435
269	287	91	79	93	55	105	67	414	100	72	36	30	28	28	26	26	22.02	25	39	4	–	–	–	–	–	–	–	–	1	–	Pregn-4-en-3-one, 17,21-[[tert-butylborylene]bis(oxy)]-20-hydroxy-, (20S)	30888-39-2	NI41436
287	269	414	288	124	229	267	270	414	100	46	35	29	27	25	23	21		25	39	4	–	–	–	–	–	–	–	–	1	–	Pregn-4-en-3-one, 17α,20α,21-trihydroxy-, butyl boronate		LI08707
269	287	91	79	55	93	105	145	414	100	73	37	32	29	28	27	25	23.00	25	39	4	–	–	–	–	–	–	–	–	1	–	Pregn-4-en-3-one, 17α,20α,21-trihydroxy-, tert-butyl boronate		LI08708
287	269	105	107	124	145	229	121	414	100	54	40	34	32	28	26	26	25.00	25	39	4	–	–	–	–	–	–	–	–	1	–	Pregn-4-en-3-one, 17α,20β,21-trihydroxy-, tert-butyl boronate		LI08709
287	43	91	269	79	41	55	105	414	100	60	55	54	53	47	46	40	25.00	25	39	4	–	–	–	–	–	–	–	–	1	–	Pregn-4-en-3-one, 17α,20β,21-trihydroxy-, tert-butyl boronate		MS09393
269	399	101	97	57	83	71	111	414	100	59	38	31	30	29	22	21	1.00	25	50	4	–	–	–	–	–	–	–	–	–	–	1,3-Dioxolane, 4-[[(2-methoxyoctadecyl)oxy]methyl]-2,2-dimethyl-	16725-42-1	NI41437
269	227	101	399	43	159	208	59	414	100	14	14	8	8	5	3	2	0.10	25	50	4	–	–	–	–	–	–	–	–	–	–	1,3-Dioxolane, 4-[[(2-propyloxyhexadecyl)oxy]methyl]-2,2-dimethyl-		NI41438
270	268	113	269	241	93	272	271	414	100	63	48	35	27	27	23	23	0.00	26	22	5	–	–	–	–	–	–	–	–	–	–	23-Oxapentacyclo$[18.2.1.0^{2,19}.0^{5,10}.0^{11,16}]$tricosa-2(19),3,5(10),6,8,11(16),12,14,17-nonaene-21,22-dicarboxylic acid, dimethyl ester		LI08710
306	307	118	224	77	222	93	91	414	100	24	19	17	15	13	7	7	4.00	26	26	3	2	–	–	–	–	–	–	–	–	–	Oxazolo[5,4-d]isoxazol-5(2H)-one, tetrahydro-6a-methyl-2,6-diphenyl-3-(2,4,6-trimethylphenyl)-	52512-32-0	NI41439
87	355	43	57	370	356	369	170	414	100	89	39	37	30	30	29	25	8.00	26	26	3	2	–	–	–	–	–	–	–	–	–	Spiro[isobenzofuran-1(3H),9'-[9H]xanthen]-3-one, 3',6'-bis(ethylamino)-2',7'-dimethyl-	41382-37-0	IC04343
355	369	326	370	356	353	327	311	414	100	54	45	42	35	28	23	16	2.30	26	26	3	2	–	–	–	–	–	–	–	–	–	Spiro[isobenzofuran-1(3H),9'-[9H]xanthen]-3-one, 3',6'-bis(ethylamino)-2',7'-dimethyl-	41382-37-0	NI41440
69	289	41	277	345	275	290	346	414	100	76	74	50	41	31	22	21	17.02	26	38	4	–	–	–	–	–	–	–	–	–	–	2,4-Cyclohexadien-1-one, 3,5-dihydroxy-2,6,6-tris(3-methyl-2-butenyl)-4-(3-methyl-1-oxobutyl)-	468-28-0	NI41441
69	41	39	55	53	67	27	43	414	100	87	43	27	27	25	24	23	0.70	26	38	4	–	–	–	–	–	–	–	–	–	–	2,4-Cyclohexadien-1-one, 3,5-dihydroxy-2,6,6-tris(3-methyl-2-butenyl)-4-(3-methyl-1-oxobutyl)-	468-28-0	NI41442
124	414	137	166	291	231			414	100	48	16	3	2	2				26	38	4											1,14-Bis(3,5-dihydroxyphenyl)tetradecane		LI08711
414	399	343	415	400	331	344	358	414	100	63	50	34	21	19	15	14		26	42	2	–	–	–	–	–	–	1	–	–	–	Δ^1-Tetrahydrocannabinol propyldimethylsilyl ether		NI41443
373	75	119	374	145	155	281	159	414	100	54	34	31	29	23	15	11	8.99	26	42	2	–	–	–	–	–	–	1	–	–	–	Pregn-5-en-20-one, 3-[(allyldimethylsilyl)oxy]-, (3β)-		NI41444
100	58	355	282	74	101	60	56	414	100	70	22	17	17	7	7	7	0.00	26	42	2	2	–	–	–	–	–	–	–	–	–	Pregn-5-ene, 3-(acetylamino)-20-(N-methylacetamido)-		LI08712
100	58	355	282	74	43	60	56	414	100	70	20	18	17	10	8	8	0.00	26	42	2	2	–	–	–	–	–	–	–	–	–	Pregn-5-ene, 3-(acetylamino)-20-(N-methylacetamido)-, (3α)-	55568-85-9	NI41445
86	341	44	74	70	342	282	43	414	100	98	95	75	60	26	25	20	3.00	26	42	2	2	–	–	–	–	–	–	–	–	–	Pregn-5-ene, 3-(acetylamino)-20-(N-methylacetamido)-, (3β)-	55555-61-8	NI41446
86	341	44	74	70	342	282	43	414	100	97	97	74	56	24	24	19	0.00	26	42	2	2	–	–	–	–	–	–	–	–	–	Pregn-5-ene, 20-(acetylamino)-3-(N-methylacetamido)-		LI08713
43	57	95	269	85	69	41	55	414	100	86	80	75	74	72	65	63	47.00	27	42	3	–	–	–	–	–	–	–	–	–	–	Cholesta-9(11),20(22)-dien-23-one, 3,6-dihydroxy-, (3β,5α,6α)-	37717-05-8	NI41447
124	229	55	414	95	81	79	69	414	100	60	52	47	45	43	40	38		27	42	3	–	–	–	–	–	–	–	–	–	–	Cholest-4-en-26-oic acid, 3-oxo-	23017-97-2	NI41448
359	382	399	255	414	356	253	213	414	100	71	52	36	28	16	13	12		27	42	3	–	–	–	–	–	–	–	–	–	–	3,5-Cyclochol-22-enoic acid, 6-methoxy-, ethyl ester, (3α,6β,22E)-		MS09394
286	122	271	55	257	81	95	41	414	100	87	75	69	60	55	47	45	0.00	27	42	3	–	–	–	–	–	–	–	–	–	–	Spirostan, 23,24-epoxy-, (5α,25S)-	55028-74-5	NI41449
286	257	41	271	43	57	69	55	414	100	40	38	37	35	32	30	30	3.03	27	42	3	–	–	–	–	–	–	–	–	–	–	Spirostan, 24,25-epoxy-, (5α)-	55028-75-6	NI41450
153	286	257	271	122	69	95	84	414	100	61	60	52	41	39	28	27	1.00	27	42	3	–	–	–	–	–	–	–	–	–	–	Spirostan, 24,25-epoxy-, (5α,24R,25S)-	24744-34-1	MS09395
153	286	257	271	122	69	95	81	414	100	61	59	52	41	39	28	27	1.00	27	42	3	–	–	–	–	–	–	–	–	–	–	Spirostan, 24,25-epoxy-, (5α,24R,25S)-	24744-34-1	NI41451
153	286	69	271	257	84	122	43	414	100	45	41	37	36	29	28	26	1.00	27	42	3	–	–	–	–	–	–	–	–	–	–	Spirostan, 24,25-epoxy-, (5α,24S,25R)-	24744-33-0	NI41452

MASS TO CHARGE RATIOS								M.W.	INTENSITIES								Parent	C	H	O	N	S	F	Cl	Br	I	Si	P	B	X	COMPOUND NAME	CAS Reg No	No
286	271	257	122	55	81	287	109	**414**	100	70	57	56	31	30	24	24	**0.40**	27	42	3	–	–	–	–	–	–	–	–	–	–	**Spirostan, 23,24-epoxy-, (5α,23R,24R,25S)-**		**MS09396**
286	287	257	271	122	272	258	331	**414**	100	25	21	19	7	6	5	4	**0.40**	27	42	3	–	–	–	–	–	–	–	–	–	–	**Spirostan, 23,24-epoxy-, (5α,23R,24R,25S)-**		**MS09397**
286	271	122	384	257	55	81	109	**414**	100	70	57	46	46	30	29	24	**12.12**	27	42	3	–	–	–	–	–	–	–	–	–	–	**Spirostan, 23,24-epoxy-, (5α,22S,23R,24R,25S)-**	24744-31-8	**NI41453**
286	257	271	344	414	287	356	258	**414**	100	89	46	44	24	24	20	20		27	42	3	–	–	–	–	–	–	–	–	–	–	**Spirostan, 23,24-epoxy-, (5α,23S,24S,25S)-**		**MS09399**
153	257	55	286	271	81	41	67	**414**	100	89	75	58	51	50	48	46	**0.53**	27	42	3	–	–	–	–	–	–	–	–	–	–	**Spirostan, 23,24-epoxy-, (5α,23S,24S,25S)-**		**MS09398**
153	257	55	286	271	81	41	67	**414**	100	89	76	58	51	50	48	46	**0.00**	27	42	3	–	–	–	–	–	–	–	–	–	–	**Spirostan, 23,24-epoxy-, (5α,22S,23S,24S,25S)-**	24744-32-9	**NI41454**
181	414	139	109	95	55	271	69	**414**	100	64	49	41	40	40	37	32		27	42	3	–	–	–	–	–	–	–	–	–	–	**Spirostan-3-one, (5α,20β,25R)-**	55028-78-9	**NI41455**
181	414	139	109	95	55	271	69	**414**	100	65	47	41	41	40	36	31		27	42	3	–	–	–	–	–	–	–	–	–	–	**Spirostan-3-one, (5α,20R,25R)-**		**MS09400**
414	181	415	342	182	271	344	115	**414**	100	40	32	18	12	11	10	10		27	42	3	–	–	–	–	–	–	–	–	–	–	**Spirostan-3-one, (5α,20R,25R)-**		**MS09401**
139	55	69	271	41	43	115	81	**414**	100	69	57	51	44	32	31	29	**6.25**	27	42	3	–	–	–	–	–	–	–	–	–	–	**Spirostan-3-one, (5α,25R)-**	470-07-5	**NI41456**
139	271	115	69	55	342	41	81	**414**	100	31	21	17	15	13	13	12	**4.00**	27	42	3	–	–	–	–	–	–	–	–	–	–	**Spirostan-3-one, (5α,25R)-**	470-07-5	**NI41457**
139	271	115	69	55	342	140	81	**414**	100	30	20	17	16	13	13	13	**4.00**	27	42	3	–	–	–	–	–	–	–	–	–	–	**Spirostan-3-one, (5α,25R)-**	470-07-5	**NI41458**
271	139	272	355	285	181	414	161	**414**	100	34	22	21	14	12	11	11		27	42	3	–	–	–	–	–	–	–	–	–	–	**Spirostan-3-one, (5β,25S)-**	639-96-3	**NI41459**
139	126	300	257	414	140	55	342	**414**	100	32	17	13	12	12	11	9		27	42	3	–	–	–	–	–	–	–	–	–	–	**Spirostan-12-one, (5α,25R)-**	54965-88-7	**NI41460**
257	55	56	67	331	81	95	258	**414**	100	41	36	28	27	27	25	24	**1.00**	27	42	3	–	–	–	–	–	–	–	–	–	–	**Spirostan-23-one, (5α,22S,25R)-**	24744-41-0	**NI41462**
257	331	258	386	109	55	41	67	**414**	100	23	22	14	13	12	11	10	**0.10**	27	42	3	–	–	–	–	–	–	–	–	–	–	**Spirostan-23-one, (5α,22S,25R)-**	24744-41-0	**NI41461**
257	331	258	386	109	55	41	95	**414**	100	23	23	14	13	13	12	11	**0.13**	27	42	3	–	–	–	–	–	–	–	–	–	–	**Spirostan-23-one, (5α,25R)-**		**MS09402**
331	257	386	332	286	387	258	330	**414**	100	86	65	27	23	21	20	16	**0.33**	27	42	3	–	–	–	–	–	–	–	–	–	–	**Spirostan-23-one, (5α,25R)-**		**MS09403**
286	271	287	122	257	272	384	218	**414**	100	27	25	11	8	7	5	5	**3.00**	27	42	3	–	–	–	–	–	–	–	–	–	–	**Spirostan-24-one, (5α,25R)-**	24744-42-1	**MS09404**
286	153	271	122	257	95	81	287	**414**	100	71	70	59	30	27	27	26	**3.00**	27	42	3	–	–	–	–	–	–	–	–	–	–	**Spirostan-24-one, (5α,25R)-**	24744-42-1	**NI41464**
286	153	270	122	257	81	95	109	**414**	100	97	96	85	48	48	47	43	**0.00**	27	42	3	–	–	–	–	–	–	–	–	–	–	**Spirostan-24-one, (5α,25R)-**	24744-42-1	**NI41463**
139	55	68	66	79	57	91	77	**414**	100	79	69	27	25	25	22	22	**1.63**	27	42	3	–	–	–	–	–	–	–	–	–	–	**Spirost-5-en-3-ol, (3β,25R)-**	512-04-9	**NI41465**
139	282	300	69	55	271	115	41	**414**	100	38	21	17	17	16	14	14	**7.00**	27	42	3	–	–	–	–	–	–	–	–	–	–	**Spirost-5-en-3-ol, (3β,25R)-**	512-04-9	**NI41466**
139	55	69	45	43	41	42	67	**414**	100	96	83	78	77	75	33	32	**1.00**	27	42	3	–	–	–	–	–	–	–	–	–	–	**Spirost-5-en-3-ol, (3β,25R)-**	512-04-9	**NI41467**
85	300	58	43	84	301	56	40	**414**	100	61	37	21	16	15	10	10	**3.00**	27	46	1	2	–	–	–	–	–	–	–	–	–	**Pregnan-18-ol, 3,20-bis(isopropylideneamino)-**		**LI08714**
85	300	58	44	399	84	301	41	**414**	100	60	35	20	18	15	14	10	**3.00**	27	46	1	2	–	–	–	–	–	–	–	–	–	**Pregnan-18-ol, 3,20-bis(isopropylideneamino)-, (3β)-**	55331-87-8	**NI41468**
58	71	355	70	300	72	414	391	**414**	100	98	40	35	24	17	13	13		27	46	1	2	–	–	–	–	–	–	–	–	–	**Pregnan-20-ol, 3,18-bis(isopropylideneamino)-**		**LI08715**
58	71	355	70	300	72	55	41	**414**	100	95	37	31	22	16	15	15	**12.00**	27	46	1	2	–	–	–	–	–	–	–	–	–	**Pregnan-20-ol, 3,18-bis(isopropylideneamino)-, (3β)-**	55331-85-6	**NI41469**
315	285	81	107	69	55	95	93	**414**	100	92	80	67	55	50	47	35	**10.00**	28	46	2	–	–	–	–	–	–	–	–	–	–	**26,27-Dinorergost-23-en-3-ol, acetate, (3β,5α)-**	35882-92-9	**NI41470**
269	109	43	55	122	81	121	41	**414**	100	82	79	57	56	50	48	42	**18.81**	28	46	2	–	–	–	–	–	–	–	–	–	–	**Ergost-1-en-3-one, 12-hydroxy-, (5β,12α)-**	56052-07-4	**NI41471**
43	124	269	55	354	396	177	41	**414**	100	87	84	66	64	48	47	47	**28.00**	28	46	2	–	–	–	–	–	–	–	–	–	–	**Ergost-4-en-3-one, 12-hydroxy-, (12α)-**	56052-06-3	**NI41472**
43	414	256	399	273	257	271	381	**414**	100	73	56	20	20	13	11	8		28	46	2	–	–	–	–	–	–	–	–	–	–	**Stigmast-7-en-25-one, 3-hydroxy-, (3β)-**		**LI08716**
273	316	301	285	302	257	215	255	**414**	100	30	19	19	16	15	15	11	**6.00**	29	50	1	–	–	–	–	–	–	–	–	–	–	**Calystanol, dihydro-, (5α,3S,24S,28R)-**		**MS09405**
43	55	414	41	95	57	81	69	**414**	100	82	70	63	57	56	55	53		29	50	1	–	–	–	–	–	–	–	–	–	–	**Cholestan-3-one, 4,4-dimethyl-, (5α)-**	2097-85-0	**NI41473**
368	329	107	95	81	353	119	105	**414**	100	89	78	78	78	67	67	67	**67.00**	29	50	1	–	–	–	–	–	–	–	–	–	–	**Cholest-5-ene, 3-ethoxy-, (3β)-**	986-19-6	**NI41474**
396	57	381	55	414	95	105	82	**414**	100	90	88	82	64	48	42	41		29	50	1	–	–	–	–	–	–	–	–	–	–	**Cholest-5-en-3-ol, 4,4-dimethyl-**		**LI08717**
414	415	396	381	331	399	397	382	**414**	100	87	87	87	76	75	74	74		29	50	1	–	–	–	–	–	–	–	–	–	–	**Cholest-5-en-3-ol, 4,4-dimethyl-, (3β)-**	1253-88-9	**NI41476**
381	43	396	149	95	55	57	81	**414**	100	88	79	79	71	67	57	56	**26.00**	29	50	1	–	–	–	–	–	–	–	–	–	–	**Cholest-5-en-3-ol, 4,4-dimethyl-, (3β)-**	1253-88-9	**NI41475**
135	43	283	95	55	119	105	414	**414**	100	98	78	78	74	69	69	67		29	50	1	–	–	–	–	–	–	–	–	–	–	**Cholest-7-en-3-ol, 2,2-dimethyl-, (3β,5α)-**		**NI41477**
43	135	105	55	414	69	57	147	**414**	100	92	80	74	70	68	67	66		29	50	1	–	–	–	–	–	–	–	–	–	–	**Cholest-7-en-3-ol, 4,4-dimethyl-, (3β,5α)-**		**NI41478**
135	43	414	147	55	95	109	69	**414**	100	67	61	49	47	44	39	39		29	50	1	–	–	–	–	–	–	–	–	–	–	**Cholest-8(14)-en-3-ol, 2,2-dimethyl-, (3β,5α)-**		**NI41479**
135	43	414	55	147	95	57	69	**414**	100	84	66	65	64	61	56	55		29	50	1	–	–	–	–	–	–	–	–	–	–	**Cholest-8(14)-en-3-ol, 4,4-dimethyl-, (3β,5α)-**		**NI41480**
301	283	146	43	94	93	107	95	**414**	100	87	79	67	58	51	49	48	**11.00**	29	50	1	–	–	–	–	–	–	–	–	–	–	**Cholest-14-en-3-ol, 4,4-dimethyl-, (3β,5α)-**		**NI41481**
414	164	160	162	213	255	396	400	**414**	100	81	80	76	70	64	60	59		29	50	1	–	–	–	–	–	–	–	–	–	–	**Cholest-5-en-3-ol, 23-ethyl-**		**LI08718**
55	81	95	57	69	107	145	414	**414**	100	66	63	61	59	53	45	44		29	50	1	–	–	–	–	–	–	–	–	–	–	**Cholest-5-en-3-ol, 24-ethyl-, (3β,24R)-**		**LI08719**
255	414	213	231	399	229	415	273	**414**	100	86	40	39	29	29	27	26		29	50	1	–	–	–	–	–	–	–	–	–	–	**Cholest-7-en-3-ol, 24-ethyl-, (3β,5α,24R)-**		**NI41482**
399	43	55	69	41	57	95	97	**414**	100	90	67	49	49	48	38	36	**18.00**	29	50	1	–	–	–	–	–	–	–	–	–	–	**Ergost-8-en-3-ol, 14-methyl-, (3β,5α)-**	33860-48-9	**NI41483**
273	215	274	414	272	255	233	316	**414**	100	42	30	26	23	23	19	18		29	50	1	–	–	–	–	–	–	–	–	–	–	**Petrostanol, (5α,24R,25R,26R)-**		**MS09406**
55	81	109	95	67	41	149	43	**414**	100	82	76	76	70	69	61	61	**4.00**	29	50	1	–	–	–	–	–	–	–	–	–	–	**Stigmastane, 23,24-epoxy-, (5α)-**	54411-91-5	**NI41484**
43	55	414	178	95	41	81	232	**414**	100	69	63	63	60	58	41	37		29	50	1	–	–	–	–	–	–	–	–	–	–	**Stigmastan-7-one**	55331-88-9	**WI01534**
91	55	95	108	119	396	414	145	**414**	100	80	60	50	35	25	20	20		29	50	1	–	–	–	–	–	–	–	–	–	–	**Stigmast-5-en-3-ol, (3β)-**	83-46-5	**NI41486**
81	91	95	107	67	105	93	69	**414**	100	95	88	87	84	83	82	79	**12.81**	29	50	1	–	–	–	–	–	–	–	–	–	–	**Stigmast-5-en-3-ol, (3β)-**	83-46-5	**WI01535**
414	213	303	255	231	396	273	329	**414**	100	82	71	59	56	53	53	47		29	50	1	–	–	–	–	–	–	–	–	–	–	**Stigmast-5-en-3-ol, (3β)-**	83-46-5	**NI41485**
91	81	69	105	67	95	79	107	**414**	100	96	96	84	81	78	77	73	**5.91**	29	50	1	–	–	–	–	–	–	–	–	–	–	**Stigmast-5-en-3-ol, (3β,24S)-**	83-47-6	**WI01536**
43	55	81	41	69	95	107	57	**414**	100	90	68	60	55	53	50	50	**24.00**	29	50	1	–	–	–	–	–	–	–	–	–	–	**Stigmast-5-en-3-ol, (3β,24S)-**	83-47-6	**NI41487**
414	43	107	55	415	399	147	105	**414**	100	56	43	40	33	30	29	27		29	50	1	–	–	–	–	–	–	–	–	–	–	**Stigmast-7-en-3-ol, (3β)-**	18069-99-3	**NI41488**

MASS TO CHARGE RATIOS								M.W.	INTENSITIES								Parent	C	H	O	N	S	F	Cl	Br	I	Si	P	B	X	COMPOUND NAME	CAS Reg No	No
414	255	43	55	107	41	95	81	**414**	100	74	61	47	38	37	35	35		29	50	1	–	–	–	–	–	–	–	–	–	–	Stigmast-7-en-3-ol, (3β,5α,24S)-	18525-35-4	NI41489
95	55	396	381	399	57	288	414	**414**	100	41	39	34	27	26	24	16		29	50	1	–	–	–	–	–	–	–	–	–	–	Stigmast-9(11)-en-3-ol, (3β,5α)-	77794-81-1	NI41490
191	178	179	192	189	165	176	414	**414**	100	44	30	29	18	17	13	12		30	22	2	–	–	–	–	–	–	–	–	–	–	9-Phenanthrenemethyl 2-phenylcinnamate		NI41491
207	344	43	165	69	345	267	265	**414**	100	74	44	43	35	30	29	21	**15.00**	30	26	–	2	–	–	–	–	–	–	–	–	–	Methanamine, N-[4-(diphenylmethylene)-1-methyl-3,3-diphenyl-2-azetidinylidene]-	39129-62-9	MS09407
191	69	95	81	55	123	109	43	**414**	100	72	57	57	57	43	43	43	**43.22**	30	54	–	–	–	–	–	–	–	–	–	–	–	Baccharane		NI41492
191	69	95	81	55	414	125	109	**414**	100	72	57	57	57	44	43	43		30	54	–	–	–	–	–	–	–	–	–	–	–	Baccharane		LI08720
137	81	95	67	55	136	41	69	**414**	100	63	53	36	33	30	28	25	**8.30**	30	54	–	–	–	–	–	–	–	–	–	–	–	1,10-Di(decahydronaphth-1-yl)decane		AI02349
259	190	274	69	109	95	81	43	**414**	100	92	50	49	47	45	43	38	**12.84**	30	54	–	–	–	–	–	–	–	–	–	–	–	Lanostane	474-20-4	NI41493
69	83	41	55	81	57	43	95	**414**	100	94	89	76	73	49	41	35	**22.36**	30	54	–	–	–	–	–	–	–	–	–	–	–	Tetracosatetraene, 2,6,10,15,19,23-hexamethyl-	26982-04-7	NI41494
83	97	67	81	69	95	109	82	**414**	100	42	32	25	19	18	17	15	**0.04**	30	54	–	–	–	–	–	–	–	–	–	–	–	1,1,6,6-Tetracyclohexylhexane		AI02351
55	83	41	67	81	69	109	95	**414**	100	85	49	29	22	19	16	16	**0.09**	30	54	–	–	–	–	–	–	–	–	–	–	–	1,1,6,6-Tetracyclohexylhexane		AI02350
127	131	128	170	226	100	69	104	**415**	100	29	27	23	20	16	11	10	**0.00**	6	2	3	1	1	13	–	–	–	–	–	–	–	(5-Carboxy-2,2,3,3,4,4,5,5-octafluoropentanamido)sulphur pentafluoride		NI41495
41	382	43	42	69	384	380	29	**415**	100	98	83	74	71	70	61	45	**1.20**	10	11	–	3	1	–	6	–	–	–	–	–	–	2-(Amylthio)-4,6-bis(trichloromethyl)-s-triazine		MS09408
119	78	268	253	77	69	107	106	**415**	100	83	82	67	66	55	54	46	**32.00**	13	7	3	1	–	10	–	–	–	–	–	–	–	Benzylamine, 2-hydroxy-, bis(pentafluoropropionyl)-		NI41496
253	415	119	252	251	225	69	77	**415**	100	96	63	55	35	27	24	21		13	7	3	1	–	10	–	–	–	–	–	–	–	Benzylamine, 4-hydroxy-, bis(pentafluoropropionyl)-		NI41497
73	315	299	400	173	89	45	142	**415**	100	77	61	57	39	38	25	23	**0.50**	13	34	6	1	–	–	–	–	–	3	1	–	–	Phosphoric acid, 2-(methoxyimino)-3-[(trimethylsilyl)oxy]propyl bis(trimethylsilyl) ester	55319-85-2	NI41498
73	299	328	160	211	315	174	158	**415**	100	91	50	40	38	29	26	25	**0.90**	13	34	6	1	–	–	–	–	–	3	1	–	–	Phosphoric acid, 3-(methoxyimino)-2-[(trimethylsilyl)oxy]propyl bis(trimethylsilyl) ester, (±)-	55319-84-1	NI41499
71	43	61	75	341	271	252	415	**415**	100	99	89	31	16	12	10	7		14	20	3	1	1	7	–	–	–	–	–	–	–	Methionine, N-(heptafluorobutyryl)-, 3-methylbutyl ester		NI41500
91	162	117	218	254	256	207	131	**415**	100	99	74	65	60	56	47	44	**5.00**	18	16	4	1	–	–	3	–	–	–	–	–	–	L-Phenylalanine, N-[(2,4,5-trichlorophenoxy)acetyl]-, methyl ester	66789-96-6	NI41501
87	342	158	43	54	415	344	171	**415**	100	77	51	49	44	27	26	17		19	18	4	5	–	–	1	–	–	–	–	–	–	N-(2-Cyanoethyl)-N-(2-acetoxyethyl)-4-(2-chloro-4-nitrophenylazo)aniline		IC04344
45	191	375	54	59	231	415	104	**415**	100	71	66	65	64	55	48	45		19	18	4	5	–	–	1	–	–	–	–	–	–	N-Cyanoethyl-N-(2-carbomethoxyethyl)-4-(2-chloro-4-nitrophenylazo)aniline		IC04345
43	114	81	127	141	169	156	115	**415**	100	67	33	19	18	17	14	13	**2.00**	19	29	9	1	–	–	–	–	–	–	–	–	–	D-Glucosylamine, N-cyclopentyl-2,3,4,6-tetra-O-acetyl-		NI41502
148	280	135	168	224	110	415	222	**415**	100	46	40	32	16	15	7	4		19	30	7	1	–	–	–	–	–	–	1	–	–	Glycine, N-[2-[(3,4-methylenedioxy)phenyl]ethyl]-N-(dibutylphosphoryl)-	92241-53-7	NI41503
415	228	229	201	245	159	216	173	**415**	100	47	21	19	14	7	6	6		20	17	7	1	1	–	–	–	–	–	–	–	–	5H-Furo[3,2-g][1]benzopyran-9-sulphonamide, 4-methoxy-N-(2-methoxyphenyl)-7-methyl-5-oxo-		MS09409
320	336	201	351	245	415	321	229	**415**	100	70	53	38	38	34	30	25		20	17	7	1	1	–	–	–	–	–	–	–	–	5H-Furo[3,2-g][1]benzopyran-9-sulphonamide, 4-methoxy-N-(3-methoxyphenyl)-7-methyl-5-oxo-		MS09410
174	144	43	227	114	186	77	175	**415**	100	56	40	21	17	15	10	9	**1.00**	20	21	5	3	1	–	–	–	–	–	–	–	–	Oxacillin methyl ester		NI41504
380	162	133	105	119	147	76	218	**415**	100	54	47	38	17	16	11	8	**5.00**	20	24	2	2	–	–	1	–	–	–	–	–	1	Chlorobis(N-propylsalicylaldiminato)iron(III)		MS09411
43	202	101	173	314	59	141	99	**415**	100	42	37	33	28	25	23	20	**10.00**	20	33	8	1	–	–	–	–	–	–	–	–	–	5-Isoxazolidinecarboxylic acid, 2-[2,3:5,6-bis-O-isopropylidene-α-D-mannofuranosyl]-3,3,5-trimethyl-, methyl ester, (S)-	64018-59-3	NI41505
86	158	130	144	57	40	42	56	**415**	100	90	85	45	43	39	31	26	**0.10**	20	37	6	3	–	–	–	–	–	–	–	–	–	Glycine, N-[N-[(N-tert-butoxycarbonyl)-L-leucyl]-L-leucyl]-, methyl ester	27545-11-5	NI41506
356	415	297	179	384	266	180	59	**415**	100	63	46	45	38	36	25	24		21	21	8	1	–	–	–	–	–	–	–	–	–	Tetramethyl 4,5-dihydro-3-methylcyclopenta[a]quinolizine-1,2,4,5-tetracarboxylate		NI41507
201	94	91	202	65	95	183	308	**415**	100	99	98	79	56	55	45	44	**0.00**	21	25	4	3	1	–	–	–	–	–	–	–	–	Imidazole, 1-[(dimethylamino)sulphonyl]-2,4-bis[(benzyloxy)methyl]-	118599-83-0	DD01623
98	99	133	55	96	42	112	104	**415**	100	7	3	3	2	2	1	1	**0.70**	21	26	2	5	–	–	1	–	–	–	–	–	–	N-Ethyl-N-[(2-piperid-1-yl)ethyl]-4-(2-chloro-4-nitrophenylazo)aniline		IC04346
57	168	326	43	210	120	136	41	**415**	100	95	93	58	54	52	50	45	**0.00**	21	37	3	1	2	–	–	–	–	–	–	–	–	1-Acetyl-2-acetoxy-3,6-di-tert-butylthio-4-tert-butyl-1,2,3,6-tetrahydropyridine		LI08721
312	415	387	31	297	340	281	313	**415**	100	53	49	47	44	35	35	24		22	25	7	1	–	–	–	–	–	–	–	–	–	Colchifoline		NI41508
415	324	43	356	313	416	325	383	**415**	100	82	74	58	56	38	27	26		22	25	7	1	–	–	–	–	–	–	–	–	–	Colchinol, 10-(methoxycarbonyl)-N-acetyl-		NI41509
342	356	343	43	296	415	355	214	**415**	100	30	19	16	11	10	9	9		22	25	7	1	–	–	–	–	–	–	–	–	–	3,11-Diacetyloxycephalotaxine		NI41510
58	338	397	352	193	383	179	204	**415**	100	50	47	40	34	30	30	14	**12.00**	22	25	7	1	–	–	–	–	–	–	–	–	–	Hydrastine, N-methyl-		NI41511
312	31	297	415	387	281	313	282	**415**	100	52	48	47	44	35	26	23		22	25	7	1	–	–	–	–	–	–	–	–	–	Isocolchifoline		NI41512
343	150	73	415	75	222	44	43	**415**	100	95	94	58	54	43	40	29		22	33	3	1	–	–	–	–	–	2	–	–	–	Morphinan, 7,8-didehydro-4,5-epoxy-3,6-bis[(trimethylsilyl)oxy]-, (5α,6α)	55319-88-5	NI41513
116	43	57	41	55	85	69	415	**415**	100	78	70	63	62	50	44	2		22	41	6	1	–	–	–	–	–	–	–	–	–	Dimethyl N-(3-hydroxy-13-methyltetradecanoyl)glutamate		LI08722
357	142	260	121	256	338	397	415	**415**	100	80	28	19	14	7	6	5		23	29	6	1	–	–	–	–	–	–	–	–	–	5′H-Pregna-1,4-dieno[17,16-d]oxazolo-3,20-dione, 6,11,21-trihydroxy-2′-methyl-, (6α,11β,16β)-	87539-45-5	NI41514
357	142	256	121	274	338	415	397	**415**	100	60	32	23	16	10	3	2		23	29	6	1	–	–	–	–	–	–	–	–	–	5′H-Pregna-1,4-dieno[17,16-D]oxazolo-3,20-dione, 6,11,21-trihydroxy-2′-methyl-, (6β,11β,16β)-	72099-45-7	NI41515

MASS TO CHARGE RATIOS								M.W.	INTENSITIES								Parent	C	H	O	N	S	F	Cl	Br	I	Si	P	B	X	COMPOUND NAME	CAS Reg No	No
193	206	222	400	415	194	311	401	**415**	100	37	35	32	27	15	14	11		23	29	6	1	–	–	–	–	–	–	–	–	–	Rheadan, 2,3,8,10,11-pentamethoxy-16-methyl-, (6α,8β)-	14028-90-1	NI41516
130	243	131	86	43	415	44	159	**415**	100	23	11	11	9	6	5	4		23	33	4	3	–	–	–	–	–	–	–	–	–	N^2-(N-Acetylleucyl)tryptophan butyl ester		AI02352
415	251	237	148	149	134	165	164	**415**	100	90	82	75	50	40	25	25		24	21	4	3	–	–	–	–	–	–	–	–	–	1-(2,4-Dinitrostyryl)-4-(4-dimethylaminostyryl)benzene		LI08723
415	168	385	416	386	354	329	100	**415**	100	32	29	27	16	12	7	7		24	37	3	3	–	–	–	–	–	–	–	–	–	Pregn-4-ene-3,6,20-trione, tris(methyloxime)		NI41517
386	140	155	415	384	153	154	387	**415**	100	90	65	61	56	53	47	39		25	41	2	1	–	–	–	–	–	1	–	–	–	levo-5β-Dihydronorgestrel, trimethylsilyl-, methyloxime		NI41518
73	286	129	75	415	91	310	105	**415**	100	99	99	93	75	66	49	49		25	41	2	1	–	–	–	–	–	1	–	–	–	Pregna-5,16-dien-20-one, 3-[(trimethylsilyl)oxy]-, O-methyloxime, (3β)-	56210-84-5	NI41519
148	415	91	414	149	324	416	266	**415**	100	60	48	25	23	21	17	12		26	25	4	1	–	–	–	–	–	–	–	–	–	Cheilanthifoline, O-benzyl-, (±)-		NI41520
148	91	149	415	324	69	414	41	**415**	100	40	36	32	23	14	12	11		26	25	4	1	–	–	–	–	–	–	–	–	–	Groenlandicine, O-benzyltetrahydro-, (±)-		NI41521
218	217	274	273	233	57	288	111	**415**	100	99	83	71	69	58	56	55	**0.00**	26	25	4	1	–	–	–	–	–	–	–	–	–	2,3-Triphenylenedicarboxylic acid, 1,4-imino-13-tert-butyl-1,4-dihydro-, dimethyl ester	110028-21-2	DD01624
327	371	370	328	415	311	163	354	**415**	100	36	31	27	23	13	12	11		26	29	2	3	–	–	–	–	–	–	–	–	–	Isobenzofuran-3-one, 1,1-bis[4-(dimethylamino)phenyl]-5-(dimethylamino)		IC04347
143	397	88	87	55	129	398	81	**415**	100	52	24	23	22	21	18	18	**18.00**	26	41	3	1	–	–	–	–	–	–	–	–	–	Morpholine, 4-[(3β)-3-hydroxy-20-methyl-21-oxopregn-5-en-21-yl]-	55103-89-4	NI41522
105	77	103	104	122	130	258	106	**415**	100	95	55	41	14	12	7	7	**0.00**	26	41	3	1	–	–	–	–	–	–	–	–	–	4-Oxazoleoctanoic acid, 4,5-dihydro-5-octyl-2-phenyl-, methyl ester, cis-	55530-33-1	NI41523
105	103	77	104	122	302	106	130	**415**	100	59	43	39	13	7	7	6	**0.00**	26	41	3	1	–	–	–	–	–	–	–	–	–	5-Oxazoleoctanoic acid, 4,5-dihydro-4-octyl-2-phenyl-, methyl ester, cis-	55530-31-9	NI41524
105	103	77	104	258	130	122	259	**415**	100	55	50	41	40	17	15	11	**0.00**	26	41	3	1	–	–	–	–	–	–	–	–	–	4-Oxazoleoctanoic acid, 4,5-dihydro-5-octyl-2-phenyl-, methyl ester, trans-	55530-34-2	NI41525
105	103	104	77	302	130	122	303	**415**	100	55	48	42	30	17	16	8	**0.00**	26	41	3	1	–	–	–	–	–	–	–	–	–	5-Oxazoleoctanoic acid, 4,5-dihydro-4-octyl-2-phenyl-, methyl ester, trans-	55530-32-0	NI41526
222	415	174	150	400	192	397	387	**415**	100	33	25	25	20	20	12	4		27	45	2	1	–	–	–	–	–	–	–	–	–	7a-Aza-B-homocholest-5-en-7-one, 3-hydroxy-, (3β)-	17398-64-0	NI41527
222	55	43	53	415	41	150	174	**415**	100	38	37	35	33	28	26	25		27	45	2	1	–	–	–	–	–	–	–	–	–	7a-Aza-B-homocholest-5-en-7-one, 3-hydroxy-, (3β)-	17398-64-0	LI08724
98	99	140	150	97	56	55	96	**415**	100	7	5	2	2	2	2	1	**0.17**	27	45	2	1	–	–	–	–	–	–	–	–	–	16,28-Secosolanid-5-ene-3,16-diol, (3β,16α,22α)-		NI41528
301	139	43	69	56	55	57	82	**415**	100	99	99	87	87	87	76	69	**39.00**	27	45	2	1	–	–	–	–	–	–	–	–	–	Spiro[(8H)-naphtho[2′,1′:4,5]indeno[2,1-b]furan[8,2′][2H]pyran]-4b-ol, 2-amino-4a,6a,7,5′-tetramethyl-, (2β,4aβ,4bα,5′β,6aβ,7α,12bα)-		LI08725
114	138	113	415	387	139	388	386	**415**	100	90	51	43	30	22	12	12		27	45	2	1	–	–	–	–	–	–	–	–	–	Spirosolan-3-ol, (3β,5α,22β,25S)-	77-59-8	NI41530
114	138	113	125	415	387	139	111	**415**	100	87	49	14	13	13	13	9		27	45	2	1	–	–	–	–	–	–	–	–	–	Spirosolan-3-ol, (3β,5α,22β,25S)-	77-59-8	NI41529
114	113	138	415	387	139	125	98	**415**	100	85	68	31	22	13	12	10		27	45	2	1	–	–	–	–	–	–	–	–	–	Spirosolan-3-ol, (3β,5α,22β,25S)-	77-59-8	MS09412
43	301	139	69	56	55	57	82	**415**	100	95	95	84	83	83	73	67	**39.68**	27	45	2	1	–	–	–	–	–	–	–	–	–	Spirostan-3-amine, (3β,5α,25S)-	6084-44-2	NI41531
260	126	415	400	230	153	416	104	**415**	100	81	75	27	27	27	25	24		28	49	1	1	–	–	–	–	–	–	–	–	–	Cholestan-3-one, O-methyloxime	55555-44-7	NI41532
84	415	45	75	105	135	416	107	**415**	100	72	45	41	33	26	24	14		29	53	–	1	–	–	–	–	–	–	–	–	–	Cholestan-2-amine, N,N-dimethyl-	55331-90-3	NI41533
84	415	416	98	99	71	85	43	**415**	100	65	22	10	9	9	7	4		29	53	–	1	–	–	–	–	–	–	–	–	–	Cholestan-2-amine, N,N-dimethyl-, (2α)-	62057-64-1	NI41534
84	415	45	75	105	416	135	107	**415**	100	81	50	45	36	28	28	15		29	53	–	1	–	–	–	–	–	–	–	–	–	Cholestan-2-amine, N,N-dimethyl-, (5α)-	54548-14-0	NI41535
84	43	69	44	71	57	55	41	**415**	100	84	81	78	74	73	57	49	**13.64**	29	53	–	1	–	–	–	–	–	–	–	–	–	Cholestan-6-amine, N,N-dimethyl-, (5α,6β)-	2222-54-0	NI41536
415	110	84	58	43	71	416	55	**415**	100	86	45	31	31	29	26	19		29	53	–	1	–	–	–	–	–	–	–	–	–	Cholestan-7-amine, N,N-dimethyl-	55331-89-0	NI41537
415	110	84	58	43	71	416	55	**415**	100	87	44	30	30	29	26	19		29	53	–	1	–	–	–	–	–	–	–	–	–	Cholestan-7-amine, N,N-dimethyl-, (5α,7β)-	1254-01-9	NI41538
356	182	299	386	184	244	214	183	**416**	100	99	81	44	38	33	30	26	**22.00**	8	18	4	4	2	–	–	–	–	–	–	–	2	Di(μ-butanethiolato)tetranitrosodicobalt		MS09413
190	276	248	220	115	388	205	75	**416**	100	99	99	94	71	70	61	55	**29.00**	9	6	7	–	–	–	–	–	–	–	–	–	3	Tricarbonyl-μ-(dimethylarsino)(tetracarbonyliron)cobalt(Co-Fe)		NI41539
88	44	43	296	58	42	73	64	**416**	100	34	31	26	26	20	14	13	**4.00**	9	18	–	3	6	–	–	–	–	–	–	–	1	Iron, tris(dimethylcarbamodithioato-S,S′)-	14484-64-1	NI41540
88	296	177	76	176	44	73	42	**416**	100	44	27	18	15	15	14	13	**0.00**	9	18	–	3	6	–	–	–	–	–	–	–	1	Iron, tris(dimethylcarbamodithioato-S,S′)-	14484-64-1	AI02353
127	177	199	65	39	77	330	63	**416**	100	42	29	17	17	11	10	6	**2.37**	11	5	3	–	–	7	–	–	–	–	–	–	1	(π-Cyclopentadienyl)(heptafluoropropyl)tricarbonylmolybdenum		MS09414
416	292	168	385	262	103	247	230	**416**	100	33	26	10	8	3	2	2		11	23	6	–	–	–	–	–	–	–	2	–	1	Rhodium, cyclopentadienylbis(trimethylphosphite)		NI41541
217	416	129	249	397	69	218	417	**416**	100	41	17	15	14	13	8	7		12	–	–	–	1	11	–	–	–	–	1	–	–	Phosphinothioic fluoride, bis(pentafluorophenyl)-	53327-22-3	NI41542
43	249	85	55	235	69	67	65	**416**	100	64	49	39	28	19	19	19	**9.00**	12	12	2	–	–	12	–	–	–	–	–	–	–	2,6-Bis(trifluoromethyl)-2,6-dihydroxy-1,1,1,7,7,7-hexafluoro-4-isopropylhept-3-ene		LI08726
133	45	134	135	416	313	164	39	**416**	100	8	6	5	3	3	3	3		13	6	–	–	2	10	–	–	–	–	–	–	–	Thiophene, 2,2′-(1,1,2,2,3,3,4,4,5,5-decafluoro-1,5-pentanediyl)bis-	24014-44-6	NI41543
77	120	138	356	354	92	353	78	**416**	100	47	29	24	21	18	14	13	**4.21**	13	10	3	–	–	–	–	–	–	–	–	–	1	Mercury, (2-hydroxybenzoato-O^1,O^2)phenyl-	28086-13-7	NI41544
56	192	96	220	112	248	136	304	**416**	100	95	75	55	55	44	36	26	**0.00**	14	8	8	–	–	–	–	–	–	–	–	–	2	Iron, heptacarbonyl-μ-[(1,4,4α,5-η^4:3,3α-η^2)-3,4-bis(methylene)-1-oxo-1,5 pentanediyl]di-	122624-32-2	DD01625
206	360	358	416	357	354	276	388	**416**	100	10	7	6	6	6	5	4		14	22	4	–	2	–	–	–	–	–	–	–	1	(2,2,7,7-Tetramethyl-3,6-dithiaoctane)molybdenum tetracarbonyl		LI08727
81	53	96	82	416	73	358	75	**416**	100	21	13	7	6	5	4	4		16	21	5	2	1	–	1	–	–	1	–	–	–	Benzoic acid, 4-chloro-2-[(2-furanylmethyl)amino]-5-[[(trimethylsilyl)amino]sulphonyl]-, methyl ester	74793-84-3	NI41545
81	89	61	417	73	114	419	82	**416**	100	63	50	31	25	16	13	13	**3.12**	16	21	5	2	1	–	1	–	–	1	–	–	–	Benzoic acid, 4-chloro-2-[(2-furanylmethyl)amino]-5-[[(trimethylsilyl)amino]sulphonyl]-, methyl ester	74793-84-3	NI41546
138	295	69	416	338	296	374	279	**416**	100	86	67	62	49	47	43	39		17	17	8	–	1	–	1	–	–	–	–	–	–	Spiro[benzofuran-2(3H),1′-[2]cyclohexene]-3,4′-dione, 7-chloro-4-hydroxy 2′,6-dimethoxy-6′-methyl-, methanesulphonate	26881-65-2	NI41547

MASS TO CHARGE RATIOS								M.W.	INTENSITIES								Parent	C	H	O	N	S	F	Cl	Br	I	Si	P	B	X	COMPOUND NAME	CAS Reg No	No
129	258	101	75	202	51	105	74	**416**	100	33	25	22	16	15	10	9	**0.00**	18	10	2	–	–	–	–	2	–	–	–	–	–	2,5-Cyclohexadiene-1,4-dione, 2,5-dibromo-3,6-diphenyl-	28293-39-2	NI41548
216	416	418	137	111	73	217	250	**416**	100	72	50	40	22	16	14	12		18	14	–	6	1	–	2	–	–	–	–	–	–	[1,2,4]Triazolo[1,5-d][1,2,4]triazylium-2-(3-chlorophenyl)aminide, 1-(3-chlorophenyl)-6,8-dimethyl-5-thioxo-		MS09415
216	416	418	137	111	217	139	250	**416**	100	72	50	42	22	16	14	12		18	14	–	6	1	–	2	–	–	–	–	–	–	[1,2,4]Triazolo[1,5-d][1,2,4]triazylium-2-(4-chlorophenyl)aminide, 1-(4-chlorophenyl)-6,8-dimethyl-5-thioxo-		MS09416
43	107	132	133	176	77	78	59	**416**	100	87	59	26	25	23	20	20	**13.11**	18	16	8	4	–	–	–	–	–	–	–	–	–	4-Acetoxyphenyl pyruvate 2,4-dinitrophenylhydrazone		NI41549
179	73	310	166	206	237	96	117	**416**	100	53	32	26	6	5	4	3	**0.00**	19	27	3	2	–	3	–	–	–	1	–	–	–	Norephedrine, N-[(trifluoroacetyl)prolyl]-O-(trimethylsilyl)-		NI41550
126	55	69	100	56	113	70	98	**416**	100	46	32	29	29	28	24	23	**0.00**	19	32	8	2	–	–	–	–	–	–	–	–	–	1,5,13,14-Tetraoxa-10,19-diazacyclotricosane-6,9,20,23-tetrone, 3,3-dimethyl-	79688-16-7	NI41551
44	73	82	218	300	388	330	356	**416**	100	40	16	14	9	7	6	4	**0.00**	20	16	10	–	–	–	–	–	–	–	–	–	–	Dispiro[furan-2(5H),4′(9′H)-benzo[1,2-b:4,5-b′]dipyran-9′,2″(5″H)-furan]-5,5′,5″,10′-tetrone, 2′,3′,7′,8′-tetrahydro-2′,7′-dimethoxy-	80234-46-4	NI41552
83	45	149	160	230	104	132	187	**416**	100	92	56	15	13	13	11	10	**2.00**	20	20	6	2	1	–	–	–	–	–	–	–	–	3-Cephem-4-carboxylic acid, 3-(isopropoxymethyl)-3-phthalimido-, methyl ester		NI41553
73	75	359	89	159	133	145	360	**416**	100	64	31	28	24	15	12	9	**0.00**	20	40	5	–	–	–	–	–	–	2	–	–	–	1H-Cyclopenta[c]furan-1-one, 5-[(1,1-dimethylethyl)dimethylsilyloxy]-4-[[(1,1-dimethylethyl)dimethylsilyloxy]methyl]hexahydro-6a-hydroxy , [3aS-(3aα,4α,5β,6aα)]-		NI41554
167	165	52	192	276	180	168	115	**416**	100	52	45	31	30	30	27	27	**1.10**	21	16	6	–	–	–	–	–	–	–	–	–	1	Chromium, (1-methoxy-3,3-diphenylpropylidene)pentacarbonyl-	54873-10-8	NI41555
150	178	374	69	43	223	327	163	**416**	100	98	95	81	74	52	51	42	**0.00**	21	20	9	–	–	–	–	–	–	–	–	–	–	Benzoic acid, 2-[[5-acetoxy-4-oxo-6-(1-propenyl)-4H-pyran-3-yl]carbonyl] 3,5-dimethoxy-, methyl ester		LI08728
178	207	150	210	179	166	75	77	**416**	100	75	70	60	25	15	15	12	**5.00**	21	20	9	–	–	–	–	–	–	–	–	–	–	Benzoic acid, 3,5-diformyl-2,4-dihydroxy-6-methyl-, 3-hydroxy-4-(methoxycarbonyl)-2,5,6-trimethylphenyl ester	27839-39-0	MS09417
401	73	402	193	403	358	75	344	**416**	100	37	30	23	12	11	7	5	**2.60**	21	28	5	–	–	–	–	–	–	2	–	–	–	6H-Dibenzo[b,d]pyran-6-one, 3,7-bis[(trimethylsilyl)oxy]-9-methoxy-1-methyl-		NI41556
356	167	314	166	179	137	208	416	**416**	100	56	51	47	30	27	23	21		22	24	8	–	–	–	–	–	–	–	–	–	–	(+)-Catechin, 5,7,3′-trimethyl-3,4′-diacetyl-		LI08729
387	416	69	401	388	164	135	166	**416**	100	54	36	31	28	19	19	17		22	24	8	–	–	–	–	–	–	–	–	–	–	3,6′-O-Diethyl-3,7,4′-O-trimethylquercetagetin		NI41557
297	180	298	151	416	167	165	222	**416**	100	95	37	36	27	23	20	15		22	24	8	–	–	–	–	–	–	–	–	–	–	3,4-Flavandiol, 3′,4′,7-trimethoxy-, diacetate, cis,cis-	19456-09-8	NI41558
297	180	151	416	165	222	194	286	**416**	100	96	37	27	20	15	15	13		22	24	8	–	–	–	–	–	–	–	–	–	–	3,4-Flavandiol, 3′,4′,7-trimethoxy-, diacetate, cis,cis-	19456-09-8	LI08730
182	150	297	121	416	286	314	183	**416**	100	55	50	44	29	28	24	19		22	24	8	–	–	–	–	–	–	–	–	–	–	3,4-Flavandiol, 4′,7,8-trimethoxy-, diacetate, cis,cis-		LI08731
297	180	151	222	165	194	137	152	**416**	100	75	40	15	15	13	12	10	**7.00**	22	24	8	–	–	–	–	–	–	–	–	–	–	3,4-Flavandiol, 3′,4′,7-trimethoxy-, diacetate, cis-2,3,trans-3,4-	19456-10-1	LI08732
297	180	151	298	115	167	222	165	**416**	100	76	40	34	18	16	15	14	**6.30**	22	24	8	–	–	–	–	–	–	–	–	–	–	3,4-Flavandiol, 3′,4′,7-trimethoxy-, diacetate, cis-2,3,trans-3,4-	19456-10-1	NI41559
182	297	150	121	314	416	286	183	**416**	100	93	70	45	40	37	32	20		22	24	8	–	–	–	–	–	–	–	–	–	–	3,4-Flavandiol, 4′,7,8-trimethoxy-, diacetate, cis-2,3,trans-3,4-		LI08733
297	180	298	151	167	416	222	165	**416**	100	87	28	25	23	17	15	12		22	24	8	–	–	–	–	–	–	–	–	–	–	3,4-Flavandiol, 3′,4′,7-trimethoxy-, diacetate, trans-2,3,cis-3,4-	20184-42-3	NI41560
297	180	151	416	222	165	286	153	**416**	100	88	25	17	14	13	8	8		22	24	8	–	–	–	–	–	–	–	–	–	–	3,4-Flavandiol, 3′,4′,7-trimethoxy-, diacetate, trans-2,3,cis-3,4-	20184-42-3	LI08734
297	180	298	151	167	416	314	165	**416**	100	60	31	26	14	11	10	10		22	24	8	–	–	–	–	–	–	–	–	–	–	3,4-Flavandiol, 3′,4′,7-trimethoxy-, diacetate, trans,trans-	19456-08-7	NI41561
297	180	151	416	314	165	356	286	**416**	100	60	25	11	10	10	7	6		22	24	8	–	–	–	–	–	–	–	–	–	–	3,4-Flavandiol, 3′,4′,7-trimethoxy-, diacetate, trans,trans-	19456-08-7	LI08735
197	210	374	416	332				**416**	100	82	55	30	15					22	24	8	–	–	–	–	–	–	–	–	–	–	Lonchocarpan diacetate		LI08736
248	57	332	85	41	416	140	108	**416**	100	90	70	56	33	30	25	23		22	28	4	2	1	–	–	–	–	–	–	–	–	Bis[4-(valerylamino)phenyl] sulphone		IC04348
73	233	147	75	401	117	55	285	**416**	100	53	45	41	24	24	23	22	**0.00**	22	48	3	–	–	–	–	–	–	2	–	–	–	Hexadecanoic acid, 3-[(trimethylsilyl)oxy]-, trimethylsilyl ester		NI41562
416	417	418	208	148	240	213	110	**416**	100	26	5	5	5	4	4	4		23	12	8	–	–	–	–	–	–	–	–	–	–	5,7,12,14-Tetrahydroxy-6,13-dioxopentacene-3-carboxylic acid		IC04349
262	416	155	108	259	121	274	261	**416**	100	93	50	37	32	29	26	24		23	22	3	–	–	–	2	–	–	–	–	–	–	4-Methyl-2,6-bis(2-hydroxy-3-chloro-5-methylbenzyl)phenol		LI08737
119	47	57	43	41	370	149	77	**416**	100	60	50	50	50	40	30	30	**20.00**	23	28	7	–	–	–	–	–	–	–	–	–	–	6(2H)-Benzofuranone, 3,5-dihydro-5,7-dimethoxy-3-methyl-5-(2-propenyl) 2-(3,4,5-trimethoxyphenyl)-, [2S-(2α,3β,5β)]-	64343-02-8	NI41563
416	314	417	345	315	342	332	299	**416**	100	50	28	17	17	16	14	12		23	28	7	–	–	–	–	–	–	–	–	–	–	Gomisin A	58546-54-6	NI41564
416	360	137	417	151	55	398	194	**416**	100	54	39	27	21	21	15	15		23	28	7	–	–	–	–	–	–	–	–	–	–	Gomisin O	72960-22-6	NI41565
416	398	208	360	417	182	209	399	**416**	100	64	37	31	29	26	25	21		23	28	7	–	–	–	–	–	–	–	–	–	–	Isogomisin O	83916-76-1	NI41566
248	416	42	372	234	58	57	44	**416**	100	41	32	31	30	30	30	26		23	32	5	2	–	–	–	–	–	–	–	–	–	3,9-Methano-10H-furo[3,2-d]azonine-10,11-dione, 9-[2-(dimethylamino)3,4-dimethoxyphenyl]decahydro-1,5-dimethyl-	29686-32-6	NI41567
416	174	384	385	188	186	200	187	**416**	100	54	27	17	13	11	9	9		23	32	5	2	–	–	–	–	–	–	–	–	–	Methyl 3,4-secoreserpate		LI08738
65	239	224	400	160	175	398	208	**416**	100	99	97	63	47	45	43	38	**14.00**	23	32	5	2	–	–	–	–	–	–	–	–	–	anti-Rotundifoline N-oxide		NI41568
69	224	400	208	210	146	239	160	**416**	100	89	85	59	46	38	28	26	**9.00**	23	32	5	2	–	–	–	–	–	–	–	–	–	syn-Rotundifoline N-oxide		NI41569
105	172	159	147	77	146	294	160	**416**	100	57	46	41	31	27	25	23	**0.07**	24	21	6	–	–	–	–	–	–	–	–	1	–	1,3,2-Dioxaborinane-4-methanol, 5-(benzoyloxy)-2-phenyl-, benzoate, cis-	74810-26-7	NI41570
105	172	159	77	171	173	158	106	**416**	100	67	30	30	16	8	7	7	**0.00**	24	21	6	–	–	–	–	–	–	–	–	1	–	1,3,2-Dioxaborolane-4,5-dimethanol, 2-phenyl-, dibenzoate, trans-	74793-48-9	NI41571
340	43	339	374	55	341	41	267	**416**	100	65	45	40	26	24	24	20	**2.60**	24	32	4	–	1	–	–	–	–	–	–	–	–	Pregn-4-ene-21-carboxylic acid, 7-(acetylthio)-17-hydroxy-3-oxo-, γ-lactone, (7α,17α)-	52-01-7	NI41572
341	43	340	374	342	54	41	105	**416**	100	64	45	39	25	25	23	19	**2.00**	24	32	4	–	1	–	–	–	–	–	–	–	–	Pregn-4-ene-21-carboxylic acid, 7-(acetylthio)-17-hydroxy-3-oxo-, γ-lactone, (7α,17α)-	52-01-7	LI08739

MASS TO CHARGE RATIOS	M.W.	INTENSITIES	Parent	C	H	O	N	S	F	Cl	Br	I	Si	P	B	X	COMPOUND NAME	CAS Reg No	No
107 43 231 91 213 81 79 105	416	100 57 50 45 38 38 38 35	7.62	24	32	6	–	–	–	–	–	–	–	–	–	–	Bufa-20,22-dienolide, 14,15-epoxy-3,5,16-trihydroxy-, (3β,5β,15β,16β)-	4099-30-3	NI41573
107 43 231 91 213 81 79 121	416	100 58 52 47 38 38 38 35	9.00	24	32	6	–	–	–	–	–	–	–	–	–	–	Bufa-20,22-dienolide, 14,15-epoxy-3,5,16-trihydroxy-, (3β,5β,15β,16β)-	4099-30-3	LI08740
204 355 175 191 416 398 370 43	416	100 64 62 55 49 48 35 32		24	32	6	–	–	–	–	–	–	–	–	–	–	Bufa-20,22-dienolide, 3,11,14-trihydroxy-12-oxo-, (3β,5β,11α)-	464-74-4	NI41574
136 121 137 55 416 109 80 181	416	100 91 89 73 59 59 59 46		24	32	6	–	–	–	–	–	–	–	–	–	–	6-Butyl-2,3,11,12-dibenzo-1,4,7,10,13,16-hexaoxacyclooctadeca-2,11-diene		MS09418
286 328 230 287 218 243 215 204	416	100 66 40 39 25 21 17 17	0.80	24	32	6	–	–	–	–	–	–	–	–	–	–	Carnosic acid diacetate		MS09419
43 71 138 91 298 109 107 55	416	100 70 53 45 37 37 37 35	4.66	24	32	6	–	–	–	–	–	–	–	–	–	–	3-Deoxy-3-oxoingen-5-yl butyrate		NI41575
43 296 131 356 183 314 285 268	416	100 53 43 33 22 19 17 16	7.00	24	32	6	–	–	–	–	–	–	–	–	–	–	6α,7β-Diacetoxyvouacapan-14β-al		NI41576
416 417 418 387 419 401 388 274	416	100 66 51 36 15 11 11 7		24	36	4	–	–	–	–	–	–	1	–	–	–	Estrane, 17-acetyloxy-3-[(trimethylsilyl)oxy]-2-methoxy-, (17β)-		NI41577
140 416 401 141 41 248 43 44	416	100 27 11 10 10 8 7 6		24	36	4	2	–	–	–	–	–	–	–	–	–	Aspidospermidin-21-ol, 1-ethyl-15,16,17-trimethoxy-	54751-75-6	NI41578
416 417 256 418 345 326 346 257	416	100 38 18 13 11 11 7 7		24	40	2	–	–	–	–	–	–	2	–	–	–	Estra-1,3,5(10)-triene, 3,16-bis[(trimethylsilyl)oxy]-, (16β)-	74299-36-8	NI41579
416 285 417 129 326 286 232 418	416	100 62 38 24 23 18 17 14		24	40	2	–	–	–	–	–	–	2	–	–	–	Estra-1,3,5(10)-triene, 3,17-bis[(trimethylsilyl)oxy]-, (17α)-	33283-14-6	NI41580
73 129 75 231 285 205 74 45	416	100 99 99 72 70 63 60 60	40.00	24	40	2	–	–	–	–	–	–	2	–	–	–	Estra-1,3,5(10)-triene, 3,17-bis[(trimethylsilyl)oxy]-, (17α)-	33283-14-6	NI41581
285 416 202 129 28 326 232 73	416	100 96 83 46 44 34 30 30		24	40	2	–	–	–	–	–	–	2	–	–	–	Estra-1,3,5(10)-triene, 3,17-bis[(trimethylsilyl)oxy]-, (17β)-	5150-62-9	NI41582
416 285 417 129 326 286 232 418	416	100 67 37 27 21 19 17 12		24	40	2	–	–	–	–	–	–	2	–	–	–	Estra-1,3,5(10)-triene, 3,17-bis[(trimethylsilyl)oxy]-, (17β)-	5150-62-9	NI41584
416 285 73 417 129 418 286 232	416	100 40 38 24 20 15 15 14		24	40	2	–	–	–	–	–	–	2	–	–	–	Estra-1,3,5(10)-triene, 3,17-bis[(trimethylsilyl)oxy]-, (17β)-	5150-62-9	NI41583
398 261 416 283 301 399 417 343	416	100 95 89 34 30 28 24 19		25	36	5	–	–	–	–	–	–	–	–	–	–	Cholan-24-oic acid, 3,7,12-trioxo-, methyl ester	55555-42-5	NI41585
99 87 43 354 55 269 129 147	416	100 58 47 24 18 16 15 14	1.10	25	36	5	–	–	–	–	–	–	–	–	–	–	11β,19-Cyclopregn-5-ene-3,20-dione, 11-hydroxy-, cyclic bis(ethylene acetal)	33585-92-1	NI41586
41 43 347 69 348 416 305 329	416	100 86 79 69 53 40 34 21		25	36	5	–	–	–	–	–	–	–	–	–	–	2-(1-Hydroxycyclopropyl)-6-hydroxy-7-isobutryl-5,5-di-(isopent-2-enyl)-2,3-dihydrobenzo[b]furan		LI08741
243 398 213 161 261 115 85 147	416	100 63 49 43 41 41 40 39	6.00	25	36	5	–	–	–	–	–	–	–	–	–	–	Methyl 7α-3,12-dioxo-4-cholenate		LI08742
247 211 248 387 212 301 416 283	416	100 90 22 15 15 12 11 10		25	36	5	–	–	–	–	–	–	–	–	–	–	Methyl 3,7,12-trioxo-5β-cholanate		LI08743
99 87 43 55 100 416 44 91	416	100 42 11 8 7 5 4 3		25	36	5	–	–	–	–	–	–	–	–	–	–	Pregn-5-ene-3,11,20-trione, cyclic 3,20-bis(1,2-ethanediyl acetal)	2302-12-7	NI41587
281 296 356 211 297 357 43 416	416	100 95 81 73 66 54 48 47		25	36	5	–	–	–	–	–	–	–	–	–	–	Pregn-9(11)-en-20-one, 3,6-bis(acetyloxy)-, (3β,5α,6α,17α)-	56362-35-7	NI41588
81 98 55 69 93 109 121 147	416	100 71 71 60 52 41 37 26	4.00	25	36	5	–	–	–	–	–	–	–	–	–	–	Salvileucolide-6,23-lactone		NI41589
283 373 145 374 284 143 241 137	416	100 88 64 28 23 13 12 11	3.35	25	40	3	–	–	–	–	–	–	1	–	–	–	Pregn-4-ene-3,20-dione, 6-methyl-17-[(trimethylsilyl)oxy]-, (6α)-	69833-55-2	NI41590
44 327 86 87 60 49 268 43	416	100 85 75 62 33 30 25 23	0.00	25	40	3	2	–	–	–	–	–	–	–	–	–	Acetamide, N,N′-[(3β)-18-hydroxypregn-5-ene-3,20-diyl]bis-	55555-60-7	NI41591
44 327 86 87 60 49 268 357	416	100 83 73 63 34 29 27 25	0.00	25	40	3	2	–	–	–	–	–	–	–	–	–	Acetamide, N,N′-[(3β)-18-hydroxypregn-5-ene-3,20-diyl]bis-	55555-60-7	LI08744
229 230 107 215 105 147 398 121	416	100 47 35 34 33 28 25 25	4.00	25	41	4	–	–	–	–	–	–	–	–	1	–	Pregnan-20-one, 17,21-[(butylborylene)bis(oxy)]-3-hydroxy-, (3α,5β)-	24376-80-5	NI41592
229 230 107 215 105 147 395 121	416	100 47 35 34 33 28 25 25	4.00	25	41	4	–	–	–	–	–	–	–	–	1	–	Pregnan-20-one, 17,21-[(butylborylene)bis(oxy)]-3-hydroxy-, (3α,5β)-	24376-80-5	LI08745
229 230 398 215 107 105 147 109	416	100 45 39 37 35 32 30 25	8.01	25	41	4	–	–	–	–	–	–	–	–	1	–	Pregnan-20-one, 17,21-[[tert-butylborylene]bis(oxy)]-3-hydroxy-, (3α,5β)-	30888-61-0	NI41593
229 230 398 215 107 105 147 182	416	100 45 39 37 35 32 30 25	8.00	25	41	4	–	–	–	–	–	–	–	–	1	–	Pregnan-20-one, 17,21-[[tert-butylborylene]bis(oxy)]-3-hydroxy-, (3α,5β)-	30888-61-0	LI08746
209 208 416 181 77 210 398 221	416	100 91 24 19 18 16 9 8		26	20	–	6	–	–	–	–	–	–	–	–	–	1,2-Bis(2-amino-1-phenazyl)ethane		LI08747
209 208 416 181 77 210 398 417	416	100 91 24 19 18 16 9 8		26	20	–	6	–	–	–	–	–	–	–	–	–	1,2-Bis(2-amino-1-phenazyl)ethane		MS09420
221 220 195 77 78 222 208 51	416	100 99 95 80 75 63 56 54	13.40	26	20	–	6	–	–	–	–	–	–	–	–	–	1,2-Bis[2-(pyrid-2-yl)benzimidazol-1-yl]ethane		IC04350
296 120 416 297 121 174 77 160	416	100 81 24 22 22 18 17 14		26	28	3	2	–	–	–	–	–	–	–	–	–	Aspidospermidin-10-one, 17-methoxy-21-oxo-21-phenyl-	54966-54-0	NI41594
87 312 91 105 107 121 313 55	416	100 56 19 18 17 15 14 14	0.00	26	40	4	–	–	–	–	–	–	–	–	–	–	Pregn-5-en-20-one, 3-(acetyloxy)-16-methyl-, cyclic 20-(1,2-ethanediyl acetal), (3β,16α)-	56792-49-5	MS09421
416 135 265 79 93 385 67 359	416	100 38 17 17 14 12 9 5		27	32	2	2	–	–	–	–	–	–	–	–	–	1,3-Bis(1-adamantyl)-4-cyano-5-nitrobenzene		NI41595
117 229 144 187 228 230 146 84	416	100 89 81 71 68 50 46 46	0.00	27	32	2	2	–	–	–	–	–	–	–	–	–	1,7-Dioxa-2,8-diazadispiro[4.0.4.3]trideca-2,8-diene,11,11,12,12,13,13-hexamethyl-3,9-dimethyl-		DD01626
399 187 146 229 77 188 315 296	416	100 97 71 57 40 39 38 37	0.00	27	32	2	2	–	–	–	–	–	–	–	–	–	1,2-Oxazaspiro[4.4]non-2-ene, 6-[2-(hydroxyimino)-2-phenylethylidene]-7,7,8,8,9,9-hexamethyl-3-phenyl-		DD01627
95 109 69 82 93 41 81 55	416	100 81 77 73 61 54 53 37	0.00	27	44	3	–	–	–	–	–	–	–	–	–	–	Cholesta-9(11),24-diene-3,6,20-triol, (3β,5α,6α,20R)-	37926-46-8	NI41596
331 129 313 138 69 288 111 287	416	100 68 60 56 50 47 47 35	2.94	27	44	3	–	–	–	–	–	–	–	–	–	–	Cholestane-3,6-dione, 20-hydroxy-, (5α,20R)-	56362-85-7	NI41597
299 99 81 124 71 69 70 177	416	100 98 67 54 47 42 34 32	0.00	27	44	3	–	–	–	–	–	–	–	–	–	–	Δ^4-Cholestene-3-one, 22,25-dihydroxy-		LI08748
300 99 81 43 124 71 44 69	416	100 95 67 65 53 47 47 43	0.00	27	44	3	–	–	–	–	–	–	–	–	–	–	Δ^4-Cholestene-3-one, 22,25-dihydroxy-		MS09422
43 152 41 55 44 57 95 81	416	100 84 45 39 28 24 15 15	0.00	27	44	3	–	–	–	–	–	–	–	–	–	–	5,6:7,8-Diepoxycholestan-3-ol, (3β)-		NI41598
149 383 398 355 279 401 311 365	416	100 69 44 44 35 31 19 17	15.00	27	44	3	–	–	–	–	–	–	–	–	–	–	24-Epoxy-25-norcycloartane-3,24-diol, (3β,16β,24ξ)-		NI41599
55 361 401 384 416 213 255 362	416	100 40 30 26 16 13 12 11		27	44	3	–	–	–	–	–	–	–	–	–	–	Ethyl 6β-methoxy-3α,5-cyclocholanate		MS09423
85 314 315 86 67 55 43 41	416	100 36 11 6 6 5 4 4	0.12	27	44	3	–	–	–	–	–	–	–	–	–	–	Pregn-5-ene-20-methanol, 3-[(tetrahydro-2H-pyran-2-yl)oxy]-, (3β)-	55123-76-7	NI41600
134 55 59 135 81 69 152 105	416	100 76 72 59 53 52 43 40	13.40	27	44	3	–	–	–	–	–	–	–	–	–	–	9,10-Secocholesta-5,7,10(19)-triene-1,3,25-triol, (3β,5Z,7E)-	32511-63-0	NI41601
136 118 119 91 55 81 95 145	416	100 85 39 36 36 34 33 31	24.00	27	44	3	–	–	–	–	–	–	–	–	–	–	9,10-Secocholesta-5,7,10(19)-triene-3,21,25-triol, (3β,5Z,7E)-	26369-40-4	NI41602
43 136 55 44 118 59 57 69	416	100 99 99 99 92 84 71 67	12.07	27	44	3	–	–	–	–	–	–	–	–	–	–	9,10-Secocholesta-5,7,10(19)-triene-3,24,25-triol, (3β,5Z,7E)-	40013-87-4	NI41603
55 118 67 136 91 105 95 79	416	100 96 93 88 88 87 83 78	8.85	27	44	3	–	–	–	–	–	–	–	–	–	–	9,10-Secocholesta-5,7,10(19)-triene-3,25,26-triol, (3β,5Z,7E)-	29261-12-9	NI41604
302 287 398 115 84 303 259 399	416	100 91 49 38 28 25 18 16	0.23	27	44	3	–	–	–	–	–	–	–	–	–	–	5α-Spirostan-20-ol, (20S,25R)-	24742-73-2	MS09424

MASS TO CHARGE RATIOS								M.W.	INTENSITIES								Parent	C	H	O	N	S	F	Cl	Br	I	Si	P	B	X	COMPOUND NAME	CAS Reg No	No
287	84	115	302	43	288	109	55	**416**	100	63	57	42	29	23	23	22	**2.00**	27	44	3	–	–	–	–	–	–	–	–	–	–	5α-Spirostan-20-ol, (20S,25R)-	24742-73-2	NI41605
257	331	383	258	109	55	95	81	**416**	100	36	22	22	12	12	10	10	**1.20**	27	44	3	–	–	–	–	–	–	–	–	–	–	5α-Spirostan-23-ol, (23R,25R)-		MS09426
331	257	332	258	416	333	330	286	**416**	100	40	24	9	5	4	3	3		27	44	3	–	–	–	–	–	–	–	–	–	–	5α-Spirostan-23-ol, (23R,25R)-		MS09425
331	257	332	258	416	333	286	417	**416**	100	37	25	9	7	3	3	2		27	44	3	–	–	–	–	–	–	–	–	–	–	5α-Spirostan-23-ol, (23S,25R)-		MS09427
257	331	258	109	55	95	332	81	**416**	100	40	22	12	12	11	10	10	**1.83**	27	44	3	–	–	–	–	–	–	–	–	–	–	5α-Spirostan-23-ol, (23S,25R)-		MS09428
155	286	257	271	156	122	81	55	**416**	100	29	29	21	14	14	12	12	**4.00**	27	44	3	–	–	–	–	–	–	–	–	–	–	5α-Spirostan-24-ol, (24R,25S)-	24744-38-5	NI41606
286	257	155	271	287	131	416	331	**416**	100	50	34	29	24	20	18	12		27	44	3	–	–	–	–	–	–	–	–	–	–	5α-Spirostan-24-ol, (24R,25S)-	24744-38-5	MS09429
155	286	257	271	55	41	81	67	**416**	100	47	39	26	23	23	21	21	**9.00**	27	44	3	–	–	–	–	–	–	–	–	–	–	5α-Spirostan-24-ol, (24R,25S)-	24744-38-5	NI41607
155	257	286	271	122	81	95	67	**416**	100	43	39	35	20	17	15	15	**1.80**	27	44	3	–	–	–	–	–	–	–	–	–	–	5α-Spirostan-24-ol, (24S,25S)-		MS09430
328	155	257	55	385	81	67	329	**416**	100	79	59	30	29	27	27	26	**4.04**	27	44	3	–	–	–	–	–	–	–	–	–	–	5α-Spirostan-25-ol, (25R)-	24744-40-9	NI41608
328	155	257	385	329	286	271	109	**416**	100	84	57	31	27	15	15	15	**2.00**	27	44	3	–	–	–	–	–	–	–	–	–	–	5α-Spirostan-25-ol, (25R)-	24744-40-9	MS09431
328	155	257	385	329	286	55	67	**416**	100	84	56	31	27	15	15	14	**2.00**	27	44	3	–	–	–	–	–	–	–	–	–	–	5α-Spirostan-25-ol, (25R)-	24744-40-9	NI41609
139	273	302	115	287	344	122	55	**416**	100	35	26	18	16	14	13	13	**10.00**	27	44	3	–	–	–	–	–	–	–	–	–	–	Spirostan-3-ol, (3β,5α,25R)-	77-60-1	NI41610
139	273	302	115	287	140	122	69	**416**	100	35	26	18	16	13	13	12	**10.30**	27	44	3	–	–	–	–	–	–	–	–	–	–	Spirostan-3-ol, (3β,5α,25R)-	77-60-1	MS09432
139	273	416	181	55	357	107	95	**416**	100	66	45	37	34	29	29	29		27	44	3	–	–	–	–	–	–	–	–	–	–	Spirostan-3-ol, (3β,5β,25S)-	126-19-2	NI41611
155	328	257	385	286	386	271	109	**416**	100	61	59	47	18	17	17	17	**0.00**	27	44	3	–	–	–	–	–	–	–	–	–	–	Spirostan-25-ol, (5α,25S)-	24744-39-6	MS09433
155	328	257	385	286	95	81	55	**416**	100	61	59	47	18	17	17	17	**1.00**	27	44	3	–	–	–	–	–	–	–	–	–	–	Spirostan-25-ol, (5α,25S)-	24744-39-6	NI41612
328	257	155	67	55	43	385	81	**416**	100	89	78	31	30	29	28	28	**1.01**	27	44	3	–	–	–	–	–	–	–	–	–	–	Spirostan-25-ol, (5α,25S)-	24744-39-6	NI41613
257	331	258	109	55	95	332	81	**416**	100	40	22	12	12	11	10	10	**2.00**	27	44	3	–	–	–	–	–	–	–	–	–	–	5α-Spirostan-23-ol, (22S,23R,25R)-	24744-35-2	NI41614
257	331	258	109	55	95	81	41	**416**	100	36	22	12	12	11	10	9	**2.00**	27	44	3	–	–	–	–	–	–	–	–	–	–	5α-Spirostan-23-ol, (22S,23S,25R)-	24744-36-3	NI41615
105	77	311	106	78	416	312	107	**416**	100	35	15	8	4	2	2	1		28	20	2	2	–	–	–	–	–	–	–	–	–	Ethanedione, diphenyl-, mono[(2-oxo-1,2-diphenylethylidene)hydrazone]	3893-33-2	NI41616
362	338	297	363	284	283	339	53	**416**	100	40	29	28	20	16	13	11	**0.30**	28	24	–	2	–	–	–	–	–	1	–	–	–	3-Methyl-10,10-diphenyl-9-(β-cyanoethyl)-9,10-dihydro-10-sila-2-azaanthracene	84854-43-3	NI41617
194	179	195	417	78	18	210	208	**416**	100	38	32	24	18	14	13	10	**0.00**	28	24	–	4	–	–	–	–	–	–	–	–	–	9,9′-Diethyl-3,3′-azobiscarbazole	95600-24-1	NI41618
240	144	77	416	268				**416**	100	43	30	25	25					28	36	1	2	–	–	–	–	–	–	–	–	–	Cyclohexen-1-yl-glyoxylic acid-1-(2,6-diethyl-anilide)-2-(2,6-diethyl-anil)		LI08749
416	151	417	150	43	191	55	57	**416**	100	60	29	28	23	16	14	13		28	48	2	–	–	–	–	–	–	–	–	–	–	2H-1-Benzopyran-6-ol, 3,4-dihydro-2,5,8-trimethyl-2-(4,8,12-trimethyltridecyl)-	148-03-8	NI41619
151	416	150	191	152	417	43	57	**416**	100	49	22	20	17	16	16	10		28	48	2	–	–	–	–	–	–	–	–	–	–	2H-1-Benzopyran-6-ol, 3,4-dihydro-2,7,8-trimethyl-2-(4,8,12-trimethyltridecyl)-	7616-22-0	NI41620
416	369	55	95	331	384	57	109	**416**	100	96	82	79	71	67	57	56		28	48	2	–	–	–	–	–	–	–	–	–	–	5α-Cholestan-6-one, 3α-methoxy-	21513-82-6	NI41621
64	48	43	41	44	55	45	57	**416**	100	52	45	33	31	29	21	20	**3.00**	28	48	2	–	–	–	–	–	–	–	–	–	–	5α-Cholestan-6-one, 3β-methoxy-	5837-39-8	NI41622
416	43	55	95	41	387	93	57	**416**	100	57	50	49	44	38	36	35		28	48	2	–	–	–	–	–	–	–	–	–	–	5α-Cholestan-6-one, 3β-methoxy-	5837-39-8	NI41623
416	95	387	93	417	153	384	91	**416**	100	50	39	37	30	19	18	17		28	48	2	–	–	–	–	–	–	–	–	–	–	Cholestan-6-one, 3-methoxy-, (3β)-	55700-38-4	NI41624
44	43	41	55	95	57	69	383	**416**	100	75	43	40	30	30	29	21	**3.00**	28	48	2	–	–	–	–	–	–	–	–	–	–	Cholest-8-ene-3,6-diol, 14-methyl-, (3β,5α,6α)-	2126-69-4	NI41625
354	371	355	353	372	398	384	370	**416**	100	38	33	23	12	10	8	7	**0.00**	28	48	2	–	–	–	–	–	–	–	–	–	–	Cholest-5-en-3-ol, 19-methoxy-, (3β)-	1106-13-4	NI41626
271	43	55	81	398	272	299	286	**416**	100	49	38	31	27	27	26	26	**2.97**	28	48	2	–	–	–	–	–	–	–	–	–	–	Ergostan-3-one, 12-hydroxy-, (5β,12α)-	56052-05-2	NI41627
43	55	271	57	398	41	81	71	**416**	100	70	65	61	55	50	47	42	**20.00**	28	48	2	–	–	–	–	–	–	–	–	–	–	Ergost-5-ene-3,12-diol, (3β,12α)-	56052-10-9	NI41628
398	59	81	55	273	95	107	69	**416**	100	99	94	88	87	87	85	72	**43.56**	28	48	2	–	–	–	–	–	–	–	–	–	–	Ergost-5-ene-3,25-diol, (3β)-	56362-42-6	NI41629
124	122	134	55	81	314	70	41	**416**	100	69	68	67	54	52	46	42	**27.00**	28	48	2	–	–	–	–	–	–	–	–	–	–	Ergost-24(28)-ene-3,12-diol, (3α,5β,12α)-	56052-69-8	NI41630
354	353	386	247	355	241	199	387	**416**	100	45	34	34	30	25	12	11	**0.00**	28	48	2	–	–	–	–	–	–	–	–	–	–	19-Hydroxy-3-methoxycholest-5-ene		LI08750
99	86	100	43	125	55	41	317	**416**	100	24	22	15	14	14	11	10	**9.33**	28	48	2	–	–	–	–	–	–	–	–	–	–	B-Norcholestan-3-one, cyclic 1,2-ethanediyl acetal, (5α)-	54498-53-2	NI41631
99	317	416	91	100	43	55	41	**416**	100	31	29	26	21	16	14	12		28	48	2	–	–	–	–	–	–	–	–	–	–	B-Norcholestan-3-one, cyclic 1,2-ethanediyl acetal, (5β)-	54498-58-7	NI41632
99	317	416	91	100	43	55	41	**416**	100	33	32	29	23	17	16	13		28	48	2	–	–	–	–	–	–	–	–	–	–	B-Norcholestan-3-one, cyclic 1,2-ethanediyl acetal, (5β)-	54498-58-7	NI41633
194	165	166	416	195	417	167	164	**416**	100	89	69	29	13	9	8	7		29	20	3	–	–	–	–	–	–	–	–	–	–	2,4(3H,5H)-Furandione, 5-(diphenylmethylene)-3,3-diphenyl-	67073-68-1	NI41634
243	105	183	165	244	55	41	259	**416**	100	56	43	43	36	18	17	16	**2.34**	29	36	2	–	–	–	–	–	–	–	–	–	–	1-Decanol, 10-(triphenylmethoxy)-	38204-88-5	NI41635
401	283	416	402	284	417	133	267	**416**	100	80	50	31	17	16	8	5		29	37	–	–	–	–	–	–	–	–	1	–	–	Phosphine, [2-[bis(2,4,6-trimethylphenyl)methyl]-4,6-dimethylphenyl]dimethyl-	74793-19-4	NI41636
58	188	97	84	186	91	70	59	**416**	100	11	11	11	9	7	5	5	**4.21**	29	40	–	2	–	–	–	–	–	–	–	–	–	Piperidine, 2-[3-allyl-4-(dimethylamino)-2,6-diphenylcyclohexyl]-1-methyl	14028-79-6	NI41637
58	188	142	96	84	186	91	59	**416**	100	11	11	11	11	10	8	5	**4.21**	29	40	–	2	–	–	–	–	–	–	–	–	–	Quinoline, 6-[5-(dimethylamino)-1-pentenyl]decahydro-1-methyl-5,6-diphenyl-	14028-84-3	NI41638
215	95	81	107	109	108	55	416	**416**	100	96	88	85	76	74	73	65		29	52	1	–	–	–	–	–	–	–	–	–	–	Cholestane, 3-ethoxy-, (3β,5α)-	2089-02-3	NI41639
43	135	95	109	81	55	69	57	**416**	100	97	91	79	79	78	75	63	**29.00**	29	52	1	–	–	–	–	–	–	–	–	–	–	4,4-Dimethyl-5α-cholestan-3β-ol		NI41640
57	233	215	234	416	216	401	217	**416**	100	82	76	64	50	32	30	23		29	52	1	–	–	–	–	–	–	–	–	–	–	(23R)-23-ethylcholestan-3β-ol		MS09434
43	233	215	234	416	216	401	217	**416**	100	85	77	58	48	36	29	24		29	52	1	–	–	–	–	–	–	–	–	–	–	(23S)-23-ethylcholestan-3β-ol		MS09435
345	360	416	346	401	43	361	41	**416**	100	56	40	26	22	22	16	14		29	52	1	–	–	–	–	–	–	–	–	–	–	3-Pentadecyl-4,6-di-tert-butylphenol		IC04351
215	233	416	401	290	201	219	383	**416**	100	98	43	25	13	13	12	11		29	52	1	–	–	–	–	–	–	–	–	–	–	Stigmast-3β-ol		MS09436

MASS TO CHARGE RATIOS								M.W.	INTENSITIES								Parent	C	H	O	N	S	F	Cl	Br	I	Si	P	B	X	COMPOUND NAME	CAS Reg No	No
178	165	179	416	43	166	191	417	**416**	100	88	35	24	23	15	11	8		30	24	2	–	–	–	–	–	–	–	–	–	–	9H-Fluorene-9-methanol, 9-(9H-fluoren-9-ylmethyl)-, acetate	63839-87-2	NI41641
416	69	55	43	105	91	81	95	**416**	100	75	67	65	46	44	43	42		30	40	1	–	–	–	–	–	–	–	–	–	–	all-trans-β-Apo-8'-carotenal		MS09437
137	136	81	55	95	41	83	69	**416**	100	83	68	67	59	57	40	38	**0.22**	30	56	–	–	–	–	–	–	–	–	–	–	–	1-Cyclohexyl-4-(1'-decahydronaphthyl)-tetradecane		AI02355
137	136	55	81	95	83	41	69	**416**	100	89	73	68	61	42	40	39	**0.82**	30	56	–	–	–	–	–	–	–	–	–	–	–	1-Cyclohexyl-4-(1'-decahydronaphthyl)-tetradecane		AI02354
137	136	81	55	95	41	84	69	**416**	100	88	80	77	65	60	44	43	**0.23**	30	56	–	–	–	–	–	–	–	–	–	–	–	1-Cyclohexyl-4-(1'-decahydronaphthyl)-tetradecane		AI02356
104	29	207	91	57	78	208	416	**416**	100	74	30	14	10	9	5	2		32	32	–	–	–	–	–	–	–	–	–	–	–	[2_4]Paracyclophane	283-81-8	DD01628
292	290	294	296	146	220	240	242	**417**	100	79	49	9	4	4	4	3	**0.01**	1	3	–	3	–	–	4	–	1	–	3	–	–	1-Methyl-1-iodotetrachlorocyclotriphosphazene	77589-30-1	NI41642
69	117	63	67	131	100	82	31	**417**	100	99	83	69	60	50	46	44	**0.00**	4	–	1	1	1	9	3	–	–	–	1	–	–	Phosphorimidic trichloride, [(nonafluorobutyl)sulphinyl]-	51735-88-7	NI41643
127	69	170	100	131	43	128	119	**417**	100	83	35	29	25	25	21	20	**0.00**	6	1	2	1	1	14	–	–	–	–	–	–	–	[2,2,3,3,4,4,5,5-Octafluoro-5-(fluoroformyl)pentanamido]sulphur pentafluoride	91940-22-6	NI41644
333	152	55	153	334	417	72	196	**417**	100	34	34	26	17	13	13	12		17	16	6	1	1	–	–	–	–	–	–	–	1	Tricarbonyl[η-1-(N,N-dimethylsulphamoyl)-2-(α-hydroxybenzyl)-cyclopentadienyl]manganese		NI41645
192	73	227	298	326	211	193	120	**417**	100	92	30	22	20	19	18	15	**2.40**	17	36	3	1	–	–	–	–	–	3	1	–	–	Phosphonic acid, [2-phenyl-1-[(trimethylsilyl)amino]ethyl]-, bis(trimethylsilyl) ester	53122-00-2	NI41646
192	73	227	298	326	211	147	91	**417**	100	92	30	22	20	19	14	12	**2.40**	17	36	3	1	–	–	–	–	–	3	1	–	–	Phosphonic acid, [2-phenyl-1-[(trimethylsilyl)amino]ethyl]-, bis(trimethylsilyl) ester	53122-00-2	NI41647
386	147	387	328	268	358	156	162	**417**	100	32	21	20	12	11	11	9	**3.00**	19	19	8	3	–	–	–	–	–	–	–	–	–	Trimethyl 4a,5,6,11-tetrahydro-4a-methoxy-6-methyl-5-oxopyridazino[2,3-a]quinoxaline-2,3,4-tricarboxylate		NI41648
209	73	417	210	179	418	208	402	**417**	100	45	26	18	17	7	7	6		19	20	2	1	–	5	–	–	–	1	–	–	–	2-(3'-Methoxy-4'-trimethylsiloxyphenyl)-N-pentafluorophenylmethylene-ethylamine	55556-69-9	MS09438
73	179	209	208	45	207	74	59	**417**	100	21	18	18	15	9	8	8	**6.59**	19	20	2	1	–	5	–	–	–	1	–	–	–	2-(3'-Methoxy-4'-trimethylsiloxyphenyl)-N-pentafluorophenylmethylene-ethylamine	55556-69-9	NI41649
209	179	417	210	73	418	208	211	**417**	100	27	26	18	17	7	7	5		19	20	2	1	–	5	–	–	–	1	–	–	–	2-(4'-Methoxy-3'-trimethylsiloxyphenyl)-N-pentafluorophenylmethylene-ethylamine		MS09439
178	168	252	137	210	417	165	110	**417**	100	70	43	42	36	34	29	7		19	32	7	1	–	–	–	–	–	–	1	–	–	N-[2-[(3-Methoxy-4-ethoxy)phenyl]ethyl]-N-(diisopropylphosphoryl)glycine	87212-48-4	NI41650
417	258	284	222	196	210	257	311	**417**	100	73	11	11	10	10	8	8		20	22	2	2	1	–	–	–	–	–	–	–	1	[N,N'-bis(salicylidene)-3,3'-bis(aminopropyl) sulphide]copper(II)		NI41651
73	99	158	144	59	187	44	74	**417**	100	58	34	29	25	23	22	20	**0.00**	20	51	–	5	–	–	–	–	–	2	–	–	–	N(1),N(5)-bis(dimethyl-t-butylsilyl)tetraethylenepentamine		NI41652
239	255	241	221	257	240	57	193	**417**	100	77	35	27	25	17	17	13	**4.95**	21	20	6	1	–	–	1	–	–	–	–	–	–	Ochratoxin A methyl ester		NI41653
104	113	173	84	175	114	141	77	**417**	100	88	80	57	54	49	42	42	**1.00**	21	21	2	3	–	–	2	–	–	–	–	–	–	1-(2,4-Dichlorobenzoyl)-2-phenyl-4-piperidino-5-hydroxy-4,5-dihydroimidazole	80365-99-7	NI41654
132	100	146	245	232	117	231	83	**417**	100	96	71	69	51	41	38	30	**0.00**	21	47	1	3	1	–	–	–	–	1	–	–	–	2-Dodecanethione, 7,10-diamino-8-butyl-5-(ethylamino)-12-[(trimethylsilyl)oxy]-	56196-56-6	NI41655
179	399	151	400	180	222	340	150	**417**	100	86	24	23	22	21	20	16	**1.00**	22	27	7	1	–	–	–	–	–	–	–	–	–	Alkaloid E N-oxide		LI08751
359	142	223	236	239	258	417	399	**417**	100	90	35	29	26	23	5	3		23	31	6	1	–	–	–	–	–	–	–	–	–	(6α,11β,16β)-6,11,25-Trihydroxy-2'-methyl-5'H-pregna-5α-1-eno[17,16-d]oxazolo-3,20-dione		NI41656
160	203	257	313	273	301	340	417	**417**	100	37	12	11	5	4	3	1		23	35	4	3	–	–	–	–	–	–	–	–	–	L-Alanine, N-[N-(N-benzylidene-L-valyl)-L-isoleucyl]-, ethyl ester	37580-78-2	NI41657
58	72	372	346	417	360	77	59	**417**	100	52	14	10	4	4	4	4		24	23	2	3	1	–	–	–	–	–	–	–	–	1-Amino-2-(2-dimethylaminoethylthio)-4-anilinoanthraquinone		IC04352
91	148	162	43	326	136	135	92	**417**	100	44	28	14	13	13	13	13	**0.00**	24	27	2	5	–	–	–	–	–	–	–	–	–	Adenine, N-[4-(benzyloxy)-3-[(benzyloxy)methyl]butyl]-	33498-85-0	NI41658
91	148	162	326	136	135	92	205	**417**	100	44	28	13	13	13	13	9	**3.00**	24	27	2	5	–	–	–	–	–	–	–	–	–	Adenine, N-[4-(benzyloxy)-3-[(benzyloxy)methyl]butyl]-	33498-85-0	LI08752
91	326	190	28	327	311	92	310	**417**	100	56	16	14	13	13	13	11	**1.00**	24	27	2	5	–	–	–	–	–	–	–	–	–	9H-Purin-6-amine, 9-[4-benzyloxy-3-(benzyloxymethyl)butyl]-	33498-84-9	NI41659
91	326	162	311	92	410	135	148	**417**	100	56	17	13	13	11	10	6	**1.00**	24	27	2	5	–	–	–	–	–	–	–	–	–	9H-Purin-6-amine, 9-[4-benzyloxy-3-(benzyloxymethyl)butyl]-	33498-84-9	LI08753
266	357	284	91	92	326	267	358	**417**	100	95	94	87	70	33	28	25	**0.00**	24	35	5	1	–	–	–	–	–	–	–	–	–	Androst-5-en-17-one, 3,16-bis(acetyloxy)-, 17-(O-methyloxime), (3β,16β)-	69688-32-0	NI41660
282	329	367	399	314	340	339	326	**417**	100	64	43	39	39	36	36	36	**0.00**	24	35	5	1	–	–	–	–	–	–	–	–	–	Dimethyl (5,7-di-tert-butyl-3,3-dimethyl-2-indolinyl)oxalacetate		DD01629
258	200	57	256	242	417	186	402	**417**	100	91	87	82	67	65	62	60		24	35	5	1	–	–	–	–	–	–	–	–	–	Dimethyl rel-(2R,3aS,9S,10S)-5,7-di-tert-butyl-3,3a-dihydro-3,3-dimethyl-3a,2-(epoxyethano)-2H-indole-9,10-dicarboxylate		DD01630
288	258	241	329	186	57	314	200	**417**	100	96	96	78	74	52	43	39	**0.00**	24	35	5	1	–	–	–	–	–	–	–	–	–	Dimethyl rel-(2S,3S,3aR)-6,8-di-tert-butyl-2,3,3a,4-tetrahydro-4,4-dimethylisoxazolo[2,3-a]indole-2,3-dicarboxylate		DD01631
388	389	416	417	212	285	104	361	**417**	100	25	17	5	5	4	4	3		24	35	5	1	–	–	–	–	–	–	–	–	–	N-Ethoxycarbonyl-17β-acetoxy-4α,5α-imino-3-oxo-androstane		LI08754
231	203	43	371	232	41	55	29	**417**	100	34	24	22	18	18	15	14	**14.40**	24	35	5	1	–	–	–	–	–	–	–	–	–	Ethyl 6-decyloxy-7-ethoxy-4-hydroxyquinoline-3-carboxylate		MS09440
417	371	386	100	42	418	41	87	**417**	100	60	49	41	29	28	25	21		24	39	3	3	–	–	–	–	–	–	–	–	–	5α-Pregnane-3,6,20-trione-tris-O-methyloxime		LI08755
138	386	41	100	85	55	417	79	**417**	100	86	80	67	45	45	42	41		24	39	3	3	–	–	–	–	–	–	–	–	–	5β-Pregnane-3,6,20-trione-tris-O-methyloxime		LI08756
43	159	91	128	105	81	79	41	**417**	100	44	42	40	39	33	32	30	**4.00**	25	39	4	1	–	–	–	–	–	–	–	–	–	3β,21-Dihydroxy-pregn-5-en-20-one, 21-acetate, ethyloxime		LI08757

MASS TO CHARGE RATIOS								M.W.	INTENSITIES								Parent	C	H	O	N	S	F	Cl	Br	I	Si	P	B	X	COMPOUND NAME	CAS Reg No	No
314	289	282	332	315	258	91	300	**417**	100	56	36	27	25	19	18	15	**7.36**	25	39	4	1	–	–	–	–	–	–	–	–	–	Pregnane-3,20-dione, 17-(acetyloxy)-6-methyl-, 3-(O-methyloxime), (5β,6α)-	69833-69-8	NI41661
100	87	386	70	402	312	129	239	**417**	100	80	79	55	51	51	44	39	**16.32**	25	43	2	1	–	–	–	–	–	1	–	–	–	Pregn-5-en-20-one, 3β-(trimethylsiloxy)-, O-methyloxime	25830-38-0	NI41662
73	100	75	87	129	70	91	42	**417**	100	99	78	71	60	55	38	37	**12.00**	25	43	2	1	–	–	–	–	–	1	–	–	–	Pregn-5-en-20-one, 3β-(trimethylsiloxy)-, O-methyloxime	25830-38-0	NI41663
73	100	87	75	129	70	386	402	**417**	100	99	66	61	56	48	44	37	**13.92**	25	43	2	1	–	–	–	–	–	1	–	–	–	Pregn-5-en-20-one, 3β-(trimethylsiloxy)-, O-methyloxime	25830-38-0	NI41664
208	120	253	121	327	417	296	252	**417**	100	89	66	65	64	62	60	46		26	31	2	3	–	–	–	–	–	–	–	–	–	Bis(4-dimethylaminophenyl)(2-carboxy-4-dimethylaminophenyl)methane		IC04353
176	164	139	126	121	291	303	122	**417**	100	48	22	22	20	18	16	15	**13.14**	26	43	3	1	–	–	–	–	–	–	–	–	–	11-Azaspirostan-3-ol, (3β,5β,25S)-	55399-48-9	NI41665
262	417	43	55	69	81	95	248	**417**	100	66	61	60	39	33	30	24		26	43	3	1	–	–	–	–	–	–	–	–	–	3-Hydroxy-3-aza-5α-cholestane-2,4-dione	123263-93-4	DD01632
98	140	99	150	399	397	151	141	**417**	100	10	7	5	4	3	3	3	**1.26**	27	47	2	1	–	–	–	–	–	–	–	–	–	16,28-Secosolanidane-3,16-diol, (3β,5α,16β,22ξ,25ξ)-	55902-91-5	NI41666
319	417	320	28	221	104	209	222	**417**	100	48	27	27	24	22	21	19		28	35	2	1	–	–	–	–	–	–	–	–	–	4-Cycloocten-1-ol, 8,8′-(iminodi-2,1-phenylene)bis-	61142-10-7	NI41667
205	203	418	343	416	301	131	341	**418**	100	44	41	31	19	19	17	14		6	14	2	–	2	–	–	–	–	–	1	–	1	Thallium(I) di-N-propyldithiophosphate	30903-33-4	NI41668
263	276	418	306	278	334	248	362	**418**	100	90	84	77	74	58	58	52		8	10	4	–	2	–	–	–	–	–	–	–	1	π-2,5-Dithiahexane-tetracarbonyl tungsten		LI08758
61	122	75	74	79	102	229	228	**418**	100	68	37	19	3	2	1	1	**0.44**	8	10	4	–	2	–	–	–	–	–	–	–	1	π-2,5-Dithiahexane-tetracarbonyl tungsten		MS09441
238	240	236	203	201	242	205	182	**418**	100	70	64	29	26	21	20	15	**0.00**	10	2	1	–	–	–	8	–	–	–	–	–	–	1,3,4-Metheno-2H-cyclobuta[cd]pentalen-2-one, 1a,3,3a,4,5,5,5a,6-octachlorooctahydro-	55570-85-9	NI41669
238	240	236	203	201	109	242	287	**418**	100	68	62	43	33	25	24	23	**0.08**	10	2	1	–	–	–	8	–	–	–	–	–	–	1,3,4-Metheno-2H-cyclobuta[cd]pentalen-2-one, 1a,3,3a,4,5,5,5a,6-octachlorooctahydro-		MS09442
349	245	90	69	314	28	383	217	**418**	100	92	79	39	28	26	24	23	**4.00**	10	6	1	–	–	6	1	–	1	–	–	–	–	1-Chloro-1,3-dihydro-5-methyl-3,3-bis(trifluoromethyl)-1,2-benziodoxole		MS09443
59	15	131	274	374	69	124	100	**418**	100	54	13	11	9	7	5	5	**0.00**	10	6	4	–	–	12	–	–	–	–	–	–	–	Dimethyl dodecafluorooctanoate		LI08759
73	147	245	75	45	297	295	74	**418**	100	33	20	19	12	11	10	9	**0.00**	10	20	4	–	–	–	–	2	–	2	–	–	–	Succinic acid, 2,3-dibromo-bis(trimethylsilyl)-		NI41670
57	69	93	119	43	74	45	99	**418**	100	64	33	26	16	11	11	10	**0.00**	11	7	2	–	–	13	–	–	–	–	–	–	–	1,2-Epoxy-3-(1-methylenetridecafluoroheptyloxy)propane		DD01633
361	359	357	305	247	360	303	358	**418**	100	74	42	40	37	31	30	28	**0.79**	12	27	–	–	–	–	–	–	1	–	–	–	1	Tributyltin iodide		MS09444
57	361	305	303	359	247	245	301	**418**	100	88	84	66	64	54	41	39	**0.00**	12	27	–	–	–	–	–	–	1	–	–	–	1	Triisobutyltin iodide		MS09445
96	95	70	418	97	177	53	81	**418**	100	90	55	50	48	43	41	39		13	8	7	2	–	6	–	–	–	–	–	–	–	Acetic acid, trifluoro-, [2,3,3a,9a-tetrahydro-6-oxo-3-[(trifluoroacetyl)oxy]-6H-furo[2′,3′:4,5]oxazolo[3,2-c]pyrimidin-2-yl]methyl ester, [2R-(2α,3β,3aβ,9aβ)]-	55319-77-2	NI41672
69	96	95	70	418	97	177	53	**418**	100	37	26	16	15	14	13	13		13	8	7	2	–	6	–	–	–	–	–	–	–	Acetic acid, trifluoro-, [2,3,3a,9a-tetrahydro-6-oxo-3-[(trifluoroacetyl)oxy]-6H-furo[2′,3′:4,5]oxazolo[3,2-c]pyrimidin-2-yl]methyl ester, [2R-(2α,3β,3aβ,9aβ)]-	55319-77-2	NI41671
69	96	95	70	418	97	177	53	**418**	100	30	27	17	15	15	13	12		13	8	7	2	–	6	–	–	–	–	–	–	–	3′,5′-Di-O-trifluoroacetyl-2,2′-anhydro-1-(β-D-arabinofuranosyl)-uracil		NI41673
28	250	54	151	41	56	43	278	**418**	100	30	28	21	20	14	8	7	**0.39**	13	10	7	2	–	–	–	–	–	–	–	–	2	Iron, heptacarbonyl[1,2-diazaspiro[2.5]octanato(2-)]di-	61178-33-4	NI41674
418	211	210	157	207	208	209	181	**418**	100	98	43	37	34	33	27	27		13	13	2	–	–	6	–	–	–	–	–	–	1	Rhodium, [(1,2,5,6-η)-1,5-cyclooctadiene](1,1,1,5,5,5-hexafluoro-2,4-pentanedionato-O,O′)-	32610-47-2	NI41675
211	157	418	103	155	210	209	181	**418**	100	86	37	34	23	11	11	11		13	13	2	–	–	6	–	–	–	–	–	–	1	Rhodium, (η⁴-1,2-diethenylcyclobutane)(1,1,1,5,5,5-hexafluoro-2,4-pentanedionato-O,O′)-	74810-95-0	NI41676
211	225	418	193	117	149	419	63	**418**	100	89	71	46	17	15	14	13		14	–	–	2	1	10	–	–	–	–	–	–	–	1,3,4-Thiadiazole, 2,5-bis(pentafluorophenyl)-	16065-71-7	NI41677
211	225	418	193	117	149	419	93	**418**	100	89	71	46	17	15	14	13		14	–	–	2	1	10	–	–	–	–	–	–	–	1,3,4-Thiadiazole, 2,5-bis(pentafluorophenyl)-	16065-71-7	LI08760
56	112	222	250	192	134	190	278	**418**	100	85	65	50	50	45	40	35	**0.00**	14	10	8	–	–	–	–	–	–	–	–	–	2	Hexacarbonyl-μ-[(1-3-η:4,4α,5-η)-2-(methoxymethyl)-4-methylene-1-oxo-2-pentene-1,5-diyl]diiron	122624-38-8	DD01634
418	236	63	183	75	77	64	51	**418**	100	68	65	64	57	50	37	37		14	10	8	8	–	–	–	–	–	–	–	–	–	Ethanedial, bis[(2,4-dinitrophenyl)hydrazone]	1177-16-8	NI41678
129	69	83	41	55	59	87	111	**418**	100	60	37	31	28	27	23	22	**3.00**	14	26	4	–	–	–	–	–	–	–	–	–	2	5,6-Diseleno-1,10-dicarboxy-2,2,9,9-tetramethyl-decane		LI08761
83	55	69	111	41	95	67	45	**418**	100	93	77	68	65	30	23	23	**13.00**	14	26	4	–	–	–	–	–	–	–	–	–	2	Heptanoic acid, 7,7′-diselenodi-		LI08762
83	55	69	111	41	43	95	45	**418**	100	95	76	69	66	31	30	23	**12.54**	14	26	4	–	–	–	–	–	–	–	–	–	2	Heptanoic acid, 7,7′-diselenodi-	22676-36-4	NI41679
69	43	41	129	87	150	70	67	**418**	100	58	24	18	10	6	6	6	**2.50**	14	26	4	–	–	–	–	–	–	–	–	–	2	1-Pentanol, 4,4′-diselenobis-, diacetate	23243-53-0	NI41680
69	43	41	149	87	67	129	61	**418**	100	59	22	11	9	9	5	5	**4.61**	14	26	4	–	–	–	–	–	–	–	–	–	2	1-Pentanol, 5,5′-diselenobis-, diacetate	23243-52-9	LI08763
69	43	41	149	87	67	147	61	**418**	100	46	23	11	10	8	6	5	**4.83**	14	26	4	–	–	–	–	–	–	–	–	–	2	1-Pentanol, 5,5′-diselenobis-, diacetate	23243-52-9	NI41681
129	69	83	219	41	55	59	87	**418**	100	60	37	34	31	28	27	23	**2.88**	14	26	4	–	–	–	–	–	–	–	–	–	2	Valeric acid, 5,5′-diselenobis[3,3-dimethyl-	3709-14-6	NI41682
107	140	55	41	81	222	83	141	**418**	100	79	67	48	37	33	29	25	**0.70**	14	28	2	–	4	–	–	–	–	–	2	–	–	O,O′-Ethylenebis(cyclohexyldithiophosphonate)	94286-63-2	NI41683
75	229	345	76	43	346	286	227	**418**	100	73	65	34	20	18	15	13	**2.22**	17	34	6	–	–	–	–	–	–	3	–	–	–	5-Carboxy-4,6-diketoheptanoate, tris(trimethylsilyl)-		NI41684
126	144	99	127	128	100	55	188	**418**	100	78	61	55	43	43	29	28	**0.00**	18	30	9	2	–	–	–	–	–	–	–	–	–	1,4,7,10,18-Pentaoxa-15,21-diazacyclopentacosane-11,14,22,25-tetrone		MS09446
126	144	55	99	100	142	127	98	**418**	100	44	40	37	28	23	23	22	**0.55**	18	30	9	2	–	–	–	–	–	–	–	–	–	1,4,7,15,18-Pentaoxa-12,21-diazacyclopentacosane-8,11,22,25-tetrone	89964-89-6	NI41685
418	417	190	419	420	416	157	384	**418**	100	58	46	34	21	17	14	10		20	12	1	–	4	–	–	–	–	–	–	2	–	Bis(4-dithieno[3,2-b:2′,3′-f]borepinyl) ether		LI08764
223	105	418	133	205	401	77	384	**418**	100	35	30	27	15	12	12	8		21	14	6	4	–	–	–	–	–	–	–	–	–	3-Benzoyl-phthalide-2,4-dinitrophenylhydrazone		LI08765
195	346	194	388	331	179	418	165	**418**	100	95	85	75	28	19	14	13		21	22	9	–	–	–	–	–	–	–	–	–	–	(3S)-6-Acetyl-7,8,2′,3′-tetramethoxyisoflavanquinone		NI41686

MASS TO CHARGE RATIOS								M.W.	INTENSITIES								Parent	C	H	O	N	S	F	Cl	Br	I	Si	P	B	X	COMPOUND NAME	CAS Reg No	No
292	163	131	70	113	112	134	114	**418**	100	81	69	59	51	50	35	24	**0.59**	21	26	7	2	–	–	–	–	–	–	–	–	–	**2-Propenamide, N-acetyl-N-[4-(acetylamino)butyl]-3-[3,4-bis(acetyloxy)phenyl]-**	56818-05-4	**MS09447**
75	331	68	73	55	41	196	44	**418**	100	57	49	48	40	40	30	26	**1.00**	21	30	5	–	–	–	–	–	–	2	–	–	–	**Benzophenone, 4,4′-dimethoxy-2,2′-di-trimethylsilyloxy-**	55724-92-0	**NI41687**
73	129	218	147	102	43	75	57	**418**	100	98	82	58	47	42	40	32	**0.00**	21	46	4	–	–	–	–	–	–	2	–	–	–	**Dodecanoic acid, 2-[(trimethylsilyl)oxy]-1-[[(trimethylsilyl)oxy]methyl]ethyl ester**	1188-53-0	**WI01538**
218	129	219	403	315	220	257	404	**418**	100	26	21	18	12	9	8	6	**0.00**	21	46	4	–	–	–	–	–	–	2	–	–	–	**Dodecanoic acid, 2-[(trimethylsilyl)oxy]-1-[[(trimethylsilyl)oxy]methyl]ethyl ester**	1188-53-0	**WI01537**
315	73	75	147	43	183	57	41	**418**	100	76	68	40	39	38	32	24	**0.24**	21	46	4	–	–	–	–	–	–	2	–	–	–	**Lauric acid, 2,3-bis(trimethylsiloxy)propyl ester**	1116-65-0	**NI41689**
315	73	75	147	43	183	57	41	**418**	100	98	68	40	39	38	32	24	**0.00**	21	46	4	–	–	–	–	–	–	2	–	–	–	**Lauric acid, 2,3-bis(trimethylsiloxy)propyl ester**	1116-65-0	**WI01539**
43	41	27	39	42	29	57	28	**418**	100	98	98	97	94	81	44	40	**0.00**	21	46	4	–	–	–	–	–	–	2	–	–	–	**Lauric acid, 2,3-bis(trimethylsiloxy)propyl ester**	1116-65-0	**NI41688**
418	279	170	183	280	144	182	181	**418**	100	53	40	29	26	17	15	14		22	21	3	2	–	3	–	–	–	–	–	–	–	**Alstophyllan-19-one, 4-demethyl-4-(trifluoroacetyl)-**	67510-47-8	**NI41690**
363	56	307	169	41	345	29	76	**418**	100	68	43	38	36	32	32	23	**0.53**	22	26	6	–	1	–	–	–	–	–	–	–	–	**4,4′-Bis(butoxycarbonyl)-diphenyl sulphone**		**IC04354**
138	208	193	177	44	162	166	151	**418**	100	97	75	73	68	60	41	40	**0.00**	22	26	8	–	–	–	–	–	–	–	–	–	–	**Benzoic acid, 2-hydroxy-3-[(4-hydroxy-2-methoxy-6-propylbenzoyl)oxy]-4-methoxy-6-propyl-**	69563-43-5	**NI41691**
260	177	131	224	111	242	214	400	**418**	100	42	34	33	27	25	11	8	**0.00**	22	26	8	–	–	–	–	–	–	–	–	–	–	**Demethylpseudolaric acid b**		**NI41692**
58	361	57	175	244	287	69	418	**418**	100	73	63	62	45	44	44	42		22	30	6	2	–	–	–	–	–	–	–	–	–	**Tetrahydrodichotine**	29474-90-6	**LI08766**
40	43	57	55	69	44	58	42	**418**	100	99	68	63	61	55	51	51	**32.00**	22	30	6	2	–	–	–	–	–	–	–	–	–	**Tetrahydrodichotine**	29474-90-6	**NI41693**
307	251	213	73	57	77	94	323	**418**	100	91	86	66	39	37	20	14	**0.00**	22	31	4	–	–	–	–	–	–	1	1	–	–	**1-(2,2-Dimethylpropyl)-2-(trimethylsilyl)vinyl diphenyl phosphate**		**DD01635**
202	179	177	125	234	291	58	232	**418**	100	99	93	88	85	76	70	65	**0.00**	22	34	–	–	–	–	–	–	–	–	–	–	1	**2-Naphthyltributyltin**		**MS09448**
257	183	161	204	118	101	184	149	**418**	100	92	79	64	64	50	48	48	**10.00**	22	34	4	4	–	–	–	–	–	–	–	–	–	**L-Alanine, N-[N-[N-(2-pyridinylmethylene)-L-valyl]-L-isoleucyl]-, ethyl ester**	37580-31-7	**NI41694**
204	161	274	313	418	302	276	373	**418**	100	79	12	9	6	5	4	2		22	34	4	4	–	–	–	–	–	–	–	–	–	**L-Alanine, N-[N-[N-(3-pyridinylmethylene)-L-valyl]-L-isoleucyl]-, ethyl ester**	37580-32-8	**NI41695**
161	274	204	258	314	302	373	340	**418**	100	98	59	28	12	11	5	2	**1.40**	22	34	4	4	–	–	–	–	–	–	–	–	–	**L-Alanine, N-[N-[N-(4-pyridinylmethylene)-L-valyl]-L-isoleucyl]-, ethyl ester**	37580-33-9	**NI41696**
217	191	204	239	313				**418**	100	72	30	18	10				**0.00**	22	42	7	–	–	–	–	–	–	–	–	–	–	**α-D-Glucopyranose, 1-hexadecanoate**	59473-41-5	**NI41697**
204	217	191	313	239				**418**	100	54	44	12	8				**0.00**	22	42	7	–	–	–	–	–	–	–	–	–	–	**α-D-Glucopyranose, 6-hexadecanoate**	17651-09-1	**NI41698**
217	204	191	239	313				**418**	100	35	25	11	6				**0.00**	22	42	7	–	–	–	–	–	–	–	–	–	–	**β-D-Glucopyranose, 1-hexadecanoate**	39848-71-0	**NI41699**
359	375	418	356	328	329	327	300	**418**	100	46	37	29	17	15	13	12		23	18	6	2	–	–	–	–	–	–	–	–	–	**Alstonilidine**		**LI08767**
418	419	99	121	122	401	296	148	**418**	100	88	28	15	12	7	7	7		23	22	4	4	–	–	–	–	–	–	–	–	–	**1-(4-Methoxyphenyl)-3-methyl-4-(1-(4-methoxyphenyl)-3-methyl-5-oxo-pyrazol-4-methylidyl)-pyrazole**		**IC04355**
354	41	91	79	55	93	43	400	**418**	100	98	86	86	84	79	79	74	**24.00**	23	30	7	–	–	–	–	–	–	–	–	–	–	**Sarverogenine**		**LI08768**
418	73	375	419	207	75	403	359	**418**	100	73	37	30	26	17	14	11		23	34	5	–	–	–	–	–	–	1	–	–	–	**Gibbane-1,10-dicarboxylic acid, 4a-hydroxy-1-methyl-8-methylene-7-[(trimethylsilyl)oxy]-, 1,4a-lactone, 10-methyl ester, (1α,4aα,4bβ,10β)-**		**NI41700**
418	73	156	75	358	403	225	207	**418**	100	99	90	51	24	19	18	17		23	34	5	–	–	–	–	–	–	1	–	–	–	**Gibbane-1,10-dicarboxylic acid, 4a-hydroxy-1-methyl-8-methylene-9-[(trimethylsilyl)oxy]-, 1,4a-lactone, 10-methyl ester, (1α,4aα,4bβ,9β,10β)-**	75311-70-5	**NI41701**
57	105	183	41	122	77	84	305	**418**	100	62	35	31	20	20	19	16	**0.00**	23	34	5	2	–	–	–	–	–	–	–	–	–	**(2R,4S)-3-Benzoyl-2-tert-butyl-4-[4-[(tert-butyloxycarbonyl)amino]butyl]-5-oxazolidinone**		**DD01636**
185	271	159	71	45	213	75	55	**418**	100	41	26	26	25	22	17	17	**0.00**	23	46	6	–	–	–	–	–	–	–	–	–	–	**Octadecanoic acid, 9,10,12,13-tetramethoxy-, methyl ester**	55319-80-7	**NI41702**
185	303	159	71	45	213	311	323	**418**	100	30	26	26	25	21	18	17	**0.00**	23	46	6	–	–	–	–	–	–	–	–	–	–	**Octadecanoic acid, 9,10,12,13-tetramethoxy-, methyl ester**	55319-80-7	**LI08769**
403	375	404	135	387	106	77	388	**418**	100	32	25	18	17	17	15	12	**0.00**	24	18	7	–	–	–	–	–	–	–	–	–	–	**8,8′-Dihydroxy-3-methoxy-3′,6,6′-trimethyl-2,2′-binaphthalene-1,1′,4,4′-tetrone**		**NI41703**
418	403	44	419	83	43	373	76	**418**	100	35	31	28	17	14	14	14		24	18	7	–	–	–	–	–	–	–	–	–	–	**3,3′-Dimethyl-1′-hydroxy-5,8-dimethoxy-2,2′-binaphthalene-1,4,5′,8′-tetrone**	104506-03-8	**NI41704**
91	312	327	221	191	92	137	136	**418**	100	19	16	14	11	10	9	8	**0.60**	24	26	3	4	–	–	–	–	–	–	–	–	–	**Hypoxanthine, 9-[4-(benzyloxy)-3-[(benzyloxy)methyl]butyl]-**	34075-72-4	**LI08770**
91	18	28	312	327	221	191	92	**418**	100	40	20	19	16	14	11	10	**0.00**	24	26	3	4	–	–	–	–	–	–	–	–	–	**Hypoxanthine, 9-[4-(benzyloxy)-3-[(benzyloxy)methyl]butyl]-**	34075-72-4	**NI41705**
43	45	41	55	111	60	109	69	**418**	100	53	42	33	31	26	24	24	**0.84**	24	34	6	–	–	–	–	–	–	–	–	–	–	**4,10-Desoxido-4-hydroxy-9-x-desoxo-9-hydroxy-ingol-diacetate**		**NI41706**
43	122	121	41	91	123	93	71	**418**	100	97	74	61	55	50	48	45	**1.16**	24	34	6	–	–	–	–	–	–	–	–	–	–	**Ingenol-3-butyrate**		**NI41707**
315	43	316	213	121	108	93	105	**418**	100	69	25	23	23	23	23	20	**0.00**	24	34	6	–	–	–	–	–	–	–	–	–	–	**3(2H)-Phenanthrenone, 2-(acetyloxy)-7-[(acetyloxy)acetyl]-1,4,4a,4b,5,6,7,8,10,10a-decahydro-1,1,4a,7-tetramethyl-, [2S-(2α,4aα,4bβ,7β,10aβ)]-**	57345-29-6	**NI41708**
418	419	287	297	129	420	181	313	**418**	100	32	25	11	10	9	9	6		24	38	4	–	–	–	–	–	–	1	–	–	–	**Estra-1,3,5(10)-triene, 2,3,4-trimethoxy-17-[trimethylsiloxy]-, (17β)-**	74298-87-6	**NI41709**
83	57	71	113	43	41	55	70	**418**	100	39	32	28	22	15	10	8	**0.00**	24	51	3	–	–	–	–	–	–	–	1	–	–	**Phosphorous acid, tris(2-ethylhexyl) ester**	301-13-3	**NI41710**
245	161	329	83	105	43	41	55	**418**	100	79	47	15	14	14	11	10	**0.00**	24	56	–	–	–	–	–	–	–	–	–	–	1	**Tetrahexylgermane**		**MS09449**

MASS TO CHARGE RATIOS	M.W.	INTENSITIES	Parent	C	H	O	N	S	F	Cl	Br	I	Si	P	B	X	COMPOUND NAME	CAS Reg No	No
277 278 361 418 301 362 303 279	**418**	100 50 33 17 10 7 7 7		25	23	4	–	–	–	–	–	–	–	1	–	–	**Hexanoic acid, 3,5-dioxo-2-(triphenylphosphoranylidene)-, methyl ester**	26535-32-0	**NI41711**
43 41 331 202 93 143 55 107	**418**	100 26 24 24 22 22 21 21	**10.04**	25	38	5	–	–	–	–	–	–	–	–	–	–	**17β-Acetoxy-1α-carboethoxymethyl-5α-androstan-3-one**		**NI41712**
43 418 71 255 375 315 316 345	**418**	100 70 70 50 46 31 30 28		25	38	5	–	–	–	–	–	–	–	–	–	–	**17β-Acetoxy-4-oxo-4-propyl-3,4-seco-5-androsten-3-oic acid methyl ester**		**MS09450**
285 400 55 286 43 95 47 41	**418**	100 27 23 17 15 12 12 12	**5.00**	25	38	5	–	–	–	–	–	–	–	–	–	–	**5α-Cholan-24-oic acid, 12α-hydroxy-3,7-dioxo-, methyl ester**	16656-66-9	**LI08771**
285 400 55 286 43 58 95 41	**418**	100 27 23 16 15 13 12 12	**4.46**	25	38	5	–	–	–	–	–	–	–	–	–	–	**5α-Cholan-24-oic acid, 12α-hydroxy-3,7-dioxo-, methyl ester**	16656-66-9	**NI41713**
124 267 122 123 382 133 109 95	**418**	100 82 27 25 20 17 17 17	**1.68**	25	38	5	–	–	–	–	–	–	–	–	–	–	**Chol-4-en-24-oic acid, 7,12-dihydroxy-3-oxo-, methyl ester (7α,12α)-**	55319-79-4	**NI41714**
124 267 122 123 382 81 109 95	**418**	100 82 28 24 20 20 17 17	**2.00**	25	38	5	–	–	–	–	–	–	–	–	–	–	**Chol-4-en-24-oic acid, 7,12-dihydroxy-3-oxo-, methyl ester (7α,12α)-**	55319-79-4	**LI08772**
347 273 348 55 43 274 41 287	**418**	100 75 40 20 18 15 14 13	**0.00**	25	38	5	–	–	–	–	–	–	–	–	–	–	**6β-Hydroxy-17-oxo-6α-pentyl-2,3-seco-5α-androstane-2,3-dioic acid methyl ester γ-lactone**		**MS09451**
358 276 343 301 273 233 300 275	**418**	100 83 70 70 70 70 68 65	**51.00**	25	38	5	–	–	–	–	–	–	–	–	–	–	**9-Phenanthrenol, 4b,5,6,7,8,8a,9,10-octahydro-1,3,4-trimethoxy-4b,8,8-trimethyl-2-isopropyl-, acetate, [4bS-(4bα,8aβ,9α)]-**	60371-76-8	**NI41715**
358 343 359 298 122 283 231 176	**418**	100 41 26 23 19 16 14 12	**5.66**	25	38	5	–	–	–	–	–	–	–	–	–	–	**6-Oxo-5α-pregnane-3β,20β-diol-diacetate**		**NI41716**
87 91 88 79 105 93 77 67	**418**	100 6 5 5 4 4 4 4	**0.05**	25	38	5	–	–	–	–	–	–	–	–	–	–	**Pregnan-20-one, 3-(acetyloxy)-5,6-epoxy-, cyclic 20-(1,2-ethanediyl acetal), (3β,5α,6α)-**	56784-22-6	**MS09452**
298 43 358 213 299 95 107 81	**418**	100 55 46 38 35 34 31 31	**0.00**	25	38	5	–	–	–	–	–	–	–	–	–	–	**Pregnan-20-one, 3,6-bis(acetyloxy)-, (3β,5α,6α)-**	37772-22-8	**NI41717**
400 167 166 105 107 246 215 213	**418**	100 48 48 44 42 37 33 31	**2.00**	25	43	4	–	–	–	–	–	–	–	–	1	–	**5β-Pregnane-3α,11β,17α,20β-tetrol butyl boronate**		**LI08773**
79 52 51 44 43 50 55 41	**418**	100 74 35 28 28 24 22 20	**1.00**	25	43	4	–	–	–	–	–	–	–	–	1	–	**5β-Pregnane-3α,11β,20α,21-tetrol butyl boronate**		**MS09453**
166 400 167 105 107 246 215 109	**418**	100 48 48 44 42 37 33 31	**2.00**	25	43	4	–	–	–	–	–	–	–	–	1	–	**Pregnane-3,11,17,20-tetrol, cyclic 17,20-(butylboronate), (3α,5β,11β,20R)-**	30882-64-5	**NI41718**
54 43 80 41 66 94 92 107	**418**	100 98 95 93 92 91 87 86	**5.00**	25	43	4	–	–	–	–	–	–	–	–	1	–	**5α-Pregnane-3α,11β,20α,21-tetrol tert-butyl boronate**		**MS09454**
107 400 105 382 147 385 133 119	**418**	100 84 76 70 70 64 49 47	**6.00**	25	43	4	–	–	–	–	–	–	–	–	1	–	**5α-Pregnane-3α,11β,20α,21-tetrol tert-butyl boronate**		**LI08774**
81 93 41 107 55 91 79 67	**418**	100 89 88 87 87 86 85 83	**1.00**	25	43	4	–	–	–	–	–	–	–	–	1	–	**5α-Pregnane-3α,11β,20β,21-tetrol tert-butyl boronate**		**MS09455**
107 105 400 147 367 119 382 133	**418**	100 92 72 63 61 61 60 52	**1.00**	25	43	4	–	–	–	–	–	–	–	–	1	–	**5α-Pregnane-3α,11β,20β,21-tetrol tert-butyl boronate**		**LI08775**
166 43 55 41 67 81 105 167	**418**	100 99 87 80 76 66 65 64	**1.00**	25	43	4	–	–	–	–	–	–	–	–	1	–	**5β-Pregnane-3α,11β,17α,20β-tetrol tert-butyl boronate**		**MS09456**
167 166 105 107 215 400 213 119	**418**	100 64 64 61 49 46 42 42	**1.00**	25	43	4	–	–	–	–	–	–	–	–	1	–	**5β-Pregnane-3α,11β,17α,20β-tetrol tert-butyl boronate**		**LI08776**
418 270 272 44 76 419 271 209	**418**	100 75 69 51 38 29 17 16		26	14	4	2	–	–	–	–	–	–	–	–	–	**1,8-Bis(1,3-dihydro-1-oxo-isobenzfur-3-ylideneamino)naphthalene**		**IC04356**
418 419 390 124 125 420 248 126	**418**	100 30 16 16 10 6 6 6		26	14	4	2	–	–	–	–	–	–	–	–	–	**1,3,8,10-Tetradoxo-2,9-dimethylanthra[2,1,9-def:6,5,10-d'e'f']diisoquinoline**		**IC04357**
403 349 418 404 387 350 419 361	**418**	100 48 24 24 11 10 9 6		26	26	5	–	–	–	–	–	–	–	–	–	–	**4H,8H-Benzo[1,2-b:3,4-b']dipyran-4-one, 3-(4-hydroxyphenyl)-5-methoxy 8,8-dimethyl-6-(3-methyl-2-butenyl)-**	5233-97-6	**NI41719**
403 418 363 375 404 419 347 364	**418**	100 60 39 32 23 15 11 9		26	26	5	–	–	–	–	–	–	–	–	–	–	**2H,6H-Benzo[1,2-b:5,4-b']dipyran-6-one, 5-hydroxy-7-(p-methoxyphenyl) 2,2-dimethyl-10-(3-methyl-2-butenyl)-**	4225-28-9	**NI41720**
143 130 91 144 418 131 419 92	**418**	100 92 83 60 52 25 18 18		26	30	3	2	–	–	–	–	–	–	–	–	–	**Succinimide, 2-[2-[(benzyloxy)methyl]butyl]-N-(2-indol-3-ylethyl)-**	14058-66-3	**NI41721**
137 130 144 167 123 119 264 125	**418**	100 86 85 73 70 67 65 64	**0.00**	26	30	3	2	–	–	–	–	–	–	–	–	–	**Succinimide, 2-[3-(benzyloxy)propyl]-2-ethyl-N-(2-indol-3-ylethyl)-**	7691-04-5	**NI41722**
57 149 71 111 127 70 293 55	**418**	100 56 47 30 26 21 18 18	**0.08**	26	42	4	–	–	–	–	–	–	–	–	–	–	**1,2-Benzenedicarboxylic acid, bis(3,5,5-trimethylhexyl) ester**		**IC04358**
149 43 41 150 57 55 29 279	**418**	100 17 15 10 9 8 7 5	**0.00**	26	42	4	–	–	–	–	–	–	–	–	–	–	**1,2-Benzenedicarboxylic acid, decyl octyl ester**	119-07-3	**NI41723**
149 71 57 43 55 41 85 69	**418**	100 38 38 36 22 22 21 17	**0.00**	26	42	4	–	–	–	–	–	–	–	–	–	–	**1,2-Benzenedicarboxylic acid, diisononyl ester**	28553-12-0	**NI41724**
57 149 71 43 31 70 41 108	**418**	100 94 48 45 45 34 34 29	**0.00**	26	42	4	–	–	–	–	–	–	–	–	–	–	**1,2-Benzenedicarboxylic acid, dinonyl ester**	84-76-4	**LI08777**
57 149 41 71 70 69 127 111	**418**	100 98 50 27 15 15 9 8	**1.00**	26	42	4	–	–	–	–	–	–	–	–	–	–	**1,2-Benzenedicarboxylic acid, dinonyl ester**	84-76-4	**MS09457**
57 149 71 43 31 70 41 55	**418**	100 94 48 45 45 34 34 31	**0.00**	26	42	4	–	–	–	–	–	–	–	–	–	–	**1,2-Benzenedicarboxylic acid, dinonyl ester**	84-76-4	**WI01540**
105 43 57 41 55 239 29 71	**418**	100 97 89 60 59 44 38 36	**24.04**	26	42	4	–	–	–	–	–	–	–	–	–	–	**Hexadecanoic acid, 2-phenyl-1,3-dioxan-5-yl ester**	56599-87-2	**NI41726**
43 105 57 41 55 29 91 69	**418**	100 99 78 64 56 42 30 27	**9.34**	26	42	4	–	–	–	–	–	–	–	–	–	–	**Hexadecanoic acid, 2-phenyl-1,3-dioxan-5-yl ester**	56599-87-2	**NI41725**
43 105 57 41 55 29 77 69	**418**	100 88 75 69 59 44 29 27	**0.00**	26	42	4	–	–	–	–	–	–	–	–	–	–	**Hexadecanoic acid, 2-phenyl-1,3-dioxan-5-yl ester, cis-**	10588-87-1	**NI41727**
43 105 41 55 57 29 69 87	**418**	100 67 67 60 59 45 29 25	**4.04**	26	42	4	–	–	–	–	–	–	–	–	–	–	**Hexadecanoic acid, 2-phenyl-1,3-dioxolan-4-yl methyl ester, cis-**	42495-31-8	**NI41728**
141 113 277 136 123 248 281 265	**418**	100 84 65 35 32 26 9 8	**0.00**	26	42	4	–	–	–	–	–	–	–	–	–	–	**8-Hydroxystearic acid phenacyl ester**		**NI41729**
155 127 291 136 137 262 263 281	**418**	100 68 52 29 22 14 10 7	**0.00**	26	42	4	–	–	–	–	–	–	–	–	–	–	**9-Hydroxystearic acid phenacyl ester**		**NI41730**
169 141 305 142 136 276 281 265	**418**	100 60 49 40 39 13 10 7	**0.00**	26	42	4	–	–	–	–	–	–	–	–	–	–	**10-Hydroxystearic acid phenacyl ester**		**NI41731**
183 155 136 319 137 265 290 281	**418**	100 70 50 43 42 14 12 9	**0.00**	26	42	4	–	–	–	–	–	–	–	–	–	–	**11-Hydroxystearic acid phenacyl ester**		**NI41732**
197 169 136 333 135 281 265 198	**418**	100 61 56 44 38 21 21 20	**0.00**	26	42	4	–	–	–	–	–	–	–	–	–	–	**12-Hydroxystearic acid phenacyl ester**		**NI41733**
211 347 136 183 212 135 265 281	**418**	100 64 64 54 32 28 21 19	**0.00**	26	42	4	–	–	–	–	–	–	–	–	–	–	**13-Hydroxystearic acid phenacyl ester**		**NI41734**
225 197 361 136 135 226 281 265	**418**	100 74 62 58 42 21 19 18	**0.00**	26	42	4	–	–	–	–	–	–	–	–	–	–	**14-Hydroxystearic acid phenacyl ester**		**NI41735**
386 55 28 81 41 95 105 93	**418**	100 75 48 44 38 35 33 33	**26.90**	26	42	4	–	–	–	–	–	–	–	–	–	–	**Methyl 7α-methoxy-3-oxo-5β-cholanoate**		**MS09458**
143 95 142 138 83 81 55 69	**418**	100 22 21 17 17 15 14 9	**1.01**	26	43	2	–	–	–	–	–	–	–	1	–	–	**Phosphonous acid, phenyl-, bis(5-methyl-2-isopropylcyclohexyl) ester, [1R-[1α(1R*,2S*,5R*),2β,5α]]-**	58359-50-5	**NI41736**
179 418 43 163 41 180 57 55	**418**	100 22 21 13 12 11 9 8		26	46	2	2	–	–	–	–	–	–	–	–	–	**N-Methyl-N-octadecyl-2-amino-4-carboxy-aniline**		**IC04359**
172 105 171 173 145 107 143 119	**418**	100 36 31 25 18 18 16 16	**10.01**	27	35	3	–	–	–	–	–	–	–	–	1	–	**Pregna-5,17(20)-diene-3,20,21-triol, cyclic 20,21-(phenylboronate), (3β)-**	30881-82-4	**NI41737**

MASS TO CHARGE RATIOS	M.W.	INTENSITIES	Parent	C	H	O	N	S	F	Cl	Br	I	Si	P	B	X	COMPOUND NAME	CAS Reg No	No
172 105 171 173 145 107 229 143	**418**	100 36 31 25 18 18 16 16	**10.00**	27	35	3	–	–	–	–	–	–	–	–	1	–	**Pregna-5,17(20)-diene-3,20,21-triol, cyclic 20,21-(phenylboronate), (3β)-**	30881-82-4	**LI08778**
124 187 105 107 173 376 149 133	**418**	100 33 24 15 12 10 10 10	**10.00**	27	35	3	–	–	–	–	–	–	–	–	1	–	**Pregn-4-en-3-one, 20,21-[(phenylborylene)bis(oxy)]-, (20R)-**	30882-69-0	**LI08779**
124 147 105 107 173 133 125 119	**418**	100 33 24 15 12 10 10 10	**10.01**	27	35	3	–	–	–	–	–	–	–	–	1	–	**Pregn-4-en-3-one, 20,21-[(phenylborylene)bis(oxy)]-, (20R)-**	30882-69-0	**NI41738**
400 356 401 388 418 357 389 419	**418**	100 30 28 18 16 8 6 5		27	46	3	–	–	–	–	–	–	–	–	–	–	**5α-Cholestan-19-oic acid, 2β-hydroxy-**	24637-59-0	**NI41739**
43 55 418 81 318 57 41 95	**418**	100 96 83 74 69 67 65 64		27	46	3	–	–	–	–	–	–	–	–	–	–	**Cholestan-6-one, 3,5-dihydroxy-, (3β,5α)-**	13027-33-3	**NI41740**
43 255 273 81 55 71 41 93	**418**	100 69 65 59 58 57 43 42	**8.00**	27	46	3	–	–	–	–	–	–	–	–	–	–	**Cholestan-24-one, 3,12-dihydroxy-, (3α,5β,12α)-**	56052-68-7	**NI41741**
387 369 333 315 351 400	**418**	100 45 16 10 8 6	**0.00**	27	46	3	–	–	–	–	–	–	–	–	–	–	**Cholest-5-ene-3,20,21-triol, (3β)-**	26273-31-4	**LI08780**
399 417 381 397	**418**	100 48 28 9	**0.00**	27	46	3	–	–	–	–	–	–	–	–	–	–	**Cholest-5-ene-3,20,21-triol, (3β)-**	26273-31-4	**NI41742**
43 55 400 161 57 124 125 81	**418**	100 62 59 57 56 53 51 47	**10.14**	27	46	3	–	–	–	–	–	–	–	–	–	–	**Cholest-6-ene-3β,5β,8β-triol**		**NI41743**
44 43 41 55 57 60 371 69	**418**	100 86 50 48 42 40 31 29	**0.00**	27	46	3	–	–	–	–	–	–	–	–	–	–	**Cholest-7-ene-3β,5α,6β-triol**		**NI41744**
43 55 41 400 57 95 139 69	**418**	100 78 66 57 57 55 53 46	**37.96**	27	46	3	–	–	–	–	–	–	–	–	–	–	**Cholest-7-ene-3β,5β,6β-triol**		**NI41745**
400 287 43 269 55 385 367 41	**418**	100 88 78 74 59 54 44 43	**1.01**	27	46	3	–	–	–	–	–	–	–	–	–	–	**Cholest-8(14)-ene-3,7,15-triol, (3β,5α)-**	54411-51-7	**NI41746**
381 399 415 397 365 401 417	**418**	100 27 22 22 22 16 13	**0.00**	27	46	3	–	–	–	–	–	–	–	–	–	–	**Cholest-6-en-3-ol, 5-hydroperoxy-, (3β,5α)-**	3328-25-4	**NI41747**
123 95 55 43 41 81 93 57	**418**	100 86 80 78 73 71 59 59	**6.50**	27	46	3	–	–	–	–	–	–	–	–	–	–	**3α,7β-Dihydroxy-5β,6β-epoxycholestane**		**NI41748**
418 142 143 55 57 331 419 400	**418**	100 98 83 49 48 45 29 26		27	46	3	–	–	–	–	–	–	–	–	–	–	**3β,5-Epoxy-4aα-methyl-a-homo-b,19-dinor-5β,10α-cholestane-1α,10-diol**		**NI41749**
43 41 332 81 83 79 121 119	**418**	100 84 80 72 44 44 33 24	**3.31**	27	46	3	–	–	–	–	–	–	–	–	–	–	**4,5-Secocholestan-4-oic acid, 5-oxo-**	53512-58-6	**NI41751**
247 317 334 406 413 400	**418**	100 60 29 24 22 22	**0.00**	27	46	3	–	–	–	–	–	–	–	–	–	–	**4,5-Secocholestan-4-oic acid, 5-oxo-**	53512-58-6	**NI41750**
112 56 43 58 41 113 81 247	**418**	100 12 12 11 11 9 9 7	**0.38**	27	46	3	–	–	–	–	–	–	–	–	–	–	**5,6-Secocholestan-6-oic acid, 1-oxo-, (10α)-**	53512-63-3	**NI41752**
112 113 43 81 107 83 41 105	**418**	100 13 7 6 5 5 5 4	**1.11**	27	46	3	–	–	–	–	–	–	–	–	–	–	**B-Nor-5,6-secocholestan-6-oic acid, 5-oxo-, methyl ester, (10α)-**	1253-62-9	**NI41753**
209 418 165 166 178 419 210 44	**418**	100 67 52 24 23 22 17 16		28	22	2	2	–	–	–	–	–	–	–	–	–	**Methanone, [4-[[4-[2-(4-methoxyphenyl)ethenyl]phenyl]azo]phenyl]phenyl**	55937-87-6	**NI41754**
355 386 387 400 368 313 372 418	**418**	100 99 26 25 25 25 21 20		28	50	2	–	–	–	–	–	–	–	–	–	–	**5α-Cholestan-19-ol, 2α-methoxy-**	28809-60-1	**NI41755**
355 386 400 368 372 418	**418**	100 99 25 25 21 20		28	50	2	–	–	–	–	–	–	–	–	–	–	**5α-Cholestan-19-ol, 2α-methoxy-**	28809-60-1	**LI08781**
43 255 55 41 81 273 95 57	**418**	100 62 61 46 40 36 28 28	**0.00**	28	50	2	–	–	–	–	–	–	–	–	–	–	**Ergostane-3,12-diol, (3α,5β,12α)-**	56052-70-1	**NI41756**
386 101 43 58 57 55 413 41	**418**	100 65 64 46 44 42 40 37	**0.00**	28	50	2	–	–	–	–	–	–	–	–	–	–	**B-Norcholestane, 3,3-dimethoxy-**	56324-21-1	**NI41757**
360 163 169 362 419 128 165 69	**419**	100 66 35 34 32 27 22 22		15	12	4	1	–	6	1	–	–	–	–	–	–	**N-Heptafluorobutyryl 4-chloroindolylacetate methyl ester**		**NI41758**
206 160 419 147 220 175 277 159	**419**	100 89 62 61 60 56 46 41		16	21	8	1	2	–	–	–	–	–	–	–	–	**1,3-Dithiane-4,6-dicarboxylic acid, 5-[4-(dimethylamino)phenyl]-,1,1,3,3-tetraoxide, dimethyl ester**		**NI41759**
131 89 187 127 85 115 145 173	**419**	100 95 92 50 47 34 31 30	**2.19**	17	25	11	1	–	–	–	–	–	–	–	–	–	**2,3,4,5,6-Penta-O-acetyl-D-glucose-O-methyloxime**	91869-00-0	**NI41760**
116 43 158 98 142 74 261 201	**419**	100 70 52 45 25 22 20 10	**0.00**	18	29	10	1	–	–	–	–	–	–	–	–	–	**3-O-Methyl-1,4,5,6-tetra-O-acetyl-2-deoxy-2-(N-methylacetamido)-glucitol**		**NI41761**
178 172 242 275 345 201 179 199	**419**	100 94 68 50 44 24 17 15	**1.00**	18	33	6	3	1	–	–	–	–	–	–	–	–	**Isobutyl permethylglycylmethionylglycine**		**LI08782**
419 228 201 245 421 229 159 216	**419**	100 82 68 52 38 36 25 24		19	14	6	1	1	–	1	–	–	–	–	–	–	**N-(3-Chlorophenyl)-4-methoxy-7-methyl-5-oxo-5H-furo[3,2-g][1]benzopyran-9-sulphonamide**		**MS09459**
419 228 201 245 421 229 173 159	**419**	100 90 74 64 46 45 20 19		19	14	6	1	1	–	1	–	–	–	–	–	–	**N-(4-Chlorophenyl)-4-methoxy-7-methyl-5-oxo-5H-furo[3,2-g][1]benzopyran-9-sulphonamide**		**MS09460**
73 147 45 348 419 300 74 75	**419**	100 34 21 13 10 9 8 6		20	33	3	1	–	–	–	–	–	3	–	–	–	**2-Propenoic acid, 2-[(trimethylsilyl)oxy]-3-[1-(trimethylsilyl)-1H-indol-3-yl]-, trimethylsilyl ester**		**MS09461**
181 86 72 182 85 152 108 99	**419**	100 67 62 25 12 11 10 9	**0.00**	21	26	3	1	–	–	–	1	–	–	–	–	–	**Ethanaminium, N,N-diethyl-N-methyl-2-[(9H-xanthen-9-ylcarbonyl)oxy]-, bromide**	53-46-3	**NI41762**
137 232 139 419 152 102 421 116	**419**	100 62 36 22 17 16 14 13		22	15	–	5	–	–	2	–	–	–	–	–	–	**2-(4-Chlorophenyl)amino-3-(4-chlorophenyl)-5-phenyl-3H-imidazo[1,2-b][1,2,4]triazole**		**MS09462**
43 44 42 45 115 70 230 55	**419**	100 82 58 34 32 32 25 23	**12.01**	22	29	7	1	–	–	–	–	–	–	–	–	–	**4,15:5,10-Dimethanobenzofuro[3′,2′:7,8][1,4]dioxonino[6,7-c]pyridin-4aβ(5H)-ol, 1,2,3,4,7,8,10,10aβ-octahydro-10-(2-hydroxyethoxy)-12 methoxy-3-methyl-**	24945-93-5	**NI41763**
300 302 105 77 120 185 183 301	**419**	100 98 81 48 27 24 24 21	**0.10**	23	18	2	1	–	–	–	1	–	–	–	–	–	**Acetophenone, 2-[5-(p-bromophenyl)-3-phenyl-2-isoxazolin-5-yl]-**	29809-07-2	**NI41764**
296 332 278 88 419 404 297 314	**419**	100 70 50 50 47 31 25 20		24	37	5	1	–	–	–	–	–	–	–	–	–	**Batrachotoxinin A, 7,8-dihydro-, (8β)-**	38930-41-5	**NI41765**
419 344 345 420 285 55 44 58	**419**	100 62 38 26 21 21 21 20		24	37	5	1	–	–	–	–	–	–	–	–	–	**1-Phenanthrenecarboxylic acid, tetradecahydro-1,4a,8-trimethyl-7-[2-[2-(methylamino)ethoxy]-2-oxoethylidene]-9-oxo-, methyl ester, [1S-(1α,4aα,4bβ,7E,8β,8aα,10aβ)]-**	36150-72-8	**NI41766**
374 340 375 303 195 376 43 330	**419**	100 90 88 70 62 52 52 42	**30.00**	25	22	3	1	–	–	1	–	–	–	–	–	–	**2′-Chloro-3′-methyl-6′-diethylamino-spiro(phthalan-1,9′-xanthene)**		**IC04360**
336 419 418 420 337 376 320 278	**419**	100 92 34 25 24 19 13 13		25	25	5	1	–	–	–	–	–	–	–	–	–	**(±)-Valachine**	95307-59-8	**NI41767**
143 73 144 130 75 145 131 115	**419**	100 27 17 15 11 6 6 5	**0.60**	25	45	2	1	–	–	–	–	–	1	–	–	–	**Androstan-3-one, 1,17-dimethyl-17-[(trimethylsilyl)oxy]-, O-methyloxime, (1α,5α,17β)-**	57397-02-1	**NI41768**
100 388 87 70 389 241 298 113	**419**	100 70 55 27 24 14 11 9	**5.20**	25	45	2	1	–	–	–	–	–	1	–	–	–	**3α-Hydroxy-5α-pregnan-20-one-methyloxime-TMS**		**NI41769**

MASS TO CHARGE RATIOS								M.W.	INTENSITIES								Parent	C	H	O	N	S	F	Cl	Br	I	Si	P	B	X	COMPOUND NAME	CAS Reg No	No
100	388	87	298	70	389	256	241	**419**	100	53	50	28	27	19	13	13	**9.70**	25	45	2	1	–	–	–	–	–	1	–	–	–	3α-Hydroxy-5β-pregnan-20-one-methyloxime-TMS		NI41770
404	419	144	405	420	55	182	57	**419**	100	74	35	31	25	18	15	15		28	37	2	1	–	–	–	–	–	–	–	–	–	3β-Acetoxy-5α-androstano[17,16-b]N-methylindole		MS09463
113	43	44	70	49	84	55	86	**419**	100	91	89	80	78	65	51	43	**41.14**	28	53	1	1	–	–	–	–	–	–	–	–	–	Pyrrolidine, 1-(1-oxo-15-tetracosenyl)-	56600-17-0	NI41771
419	330	420	331	346	345	315	421	**419**	100	47	32	20	18	11	9	7		29	25	2	1	–	–	–	–	–	–	–	–	–	1-[(2-Ethoxycarbonyl)phenyl]-9-ethyl-4-phenylcarbazole		LI08783
293	295	263	265	151	248	153	250	**420**	100	75	60	42	40	30	28	21	**0.00**	3	9	–	–	–	–	–	–	2	–	–	–	1	Trimethylantimony diiodide	13077-53-7	NI41772
293	420	55	89	137	69	166	152	**420**	100	43	38	33	30	29	26	19		8	6	4	–	2	10	–	–	–	–	–	–	–	6-Oxo-2,4-bis(pentafluorosulphurmethyl)-6H-pyran-3-carboxylic acid		NI41773
29	39	229	30	272	18	27	274	**420**	100	46	43	41	40	38	36	34	**0.80**	9	6	4	–	1	–	6	–	–	–	–	–	–	6,9-Methano-2,4,3-benzodioxathiepin, 6,7,8,9,10,10-hexachloro-1,5,5a,6,9,9a-hexahydro-, 3,3-dioxide	1031-07-8	NI41774
272	274	229	387	270	227	239	237	**420**	100	92	78	67	58	58	56	53	**11.29**	9	6	4	–	1	–	6	–	–	–	–	–	–	6,9-Methano-2,4,3-benzodioxathiepin, 6,7,8,9,10,10-hexachloro-1,5,5a,6,9,9a-hexahydro-, 3,3-dioxide	1031-07-8	NI41775
115	51	29	117	149	185	87	187	**420**	100	85	62	37	35	32	32	29	**0.00**	10	4	1	–	–	–	8	–	–	–	–	–	–	2,5-Methano-2H-indeno[1,2-b]oxirene, 2,3,4,5,6,6a,7,7-octachloro-1a,1b,5,5a,6,6a-hexahydro-, (1aα,1bβ,2α,5α,5aβ,6β,6aα)-	27304-13-8	NI41776
51	115	29	87	117	85	63	73	**420**	100	75	55	48	45	41	38	34	**0.00**	10	4	1	–	–	–	8	–	–	–	–	–	–	2,5-Methano-2H-indeno[1,2-b]oxirene, 2,3,4,5,6,6a,7,7-octachloro-1a,1b,5,5a,6,6a-hexahydro-, (1aα,1bβ,2α,5α,5aβ,6β,6aα)-	27304-13-8	NI41777
115	185	187	149	51	387	117	389	**420**	100	57	52	46	27	25	25	24	**0.00**	10	4	1	–	–	–	8	–	–	–	–	–	–	2,5-Methano-2H-indeno[1,2-b]oxirene, 2,3,4,5,6,6a,7,7-octachloro-1a,1b,5,5a,6,6a-hexahydro-, (1aα,1bβ,2α,5α,5aβ,6β,6aα)-	27304-13-8	NI41778
115	185	187	149	51	117	237	151	**420**	100	50	40	39	32	28	22	20	**0.00**	10	4	1	–	–	–	8	–	–	–	–	–	–	2,5-Methano-2H-indeno[1,2-b]oxirene, 2,3,4,5,6,6a,7,7-octachloro-1a,1b,5,5a,6,6a-hexahydro-, (1aα,1bβ,2α,5α,5aβ,6β,6aα)-	27304-13-8	NI41779
77	305	73	333	405	147	377	285	**420**	100	79	76	68	55	48	34	24	**1.00**	11	23	4	–	–	6	–	–	–	2	1	–	–	Ethyl(trimethylsilyl) [2,2,2-trifluoro-1-(trifluoromethyl)-1-(trimethylsiloxy)ethyl]phosphonate		MS09464
61	82	129	55	44	100	87	60	**420**	100	45	26	24	16	15	14	13	**0.00**	12	22	9	1	3	–	–	–	–	–	–	–	–	Glucoerucian		MS09465
39	270	140	75	65	155	55	117	**420**	100	99	99	96	87	67	40	37	**25.02**	13	5	2	–	–	9	–	–	–	–	–	–	1	Iron, dicarbonyl(η^5-2,4-cyclopentadien-1-yl)(2,3,3,4,4,5,5,6,6-nonafluoro-1 cyclohexen-1-yl)-	12215-33-7	NI41780
183	137	135	136	110	153	168	107	**420**	100	45	42	33	18	17	15	15		13	13	8	4	1	–	1	–	–	–	–	–	–	2H-1,2,3-Benzothiadiazine, 2-(2,4-dinitrophenyl)-5,6,7,8-tetrahydro-, monoperchlorate		NI41781
259	391	111	278	119	95	129	277	**420**	100	52	50	30	30	30	27	24	**11.40**	16	10	–	–	–	10	–	–	–	1	–	–	–	Silane, diethylbis(pentafluorophenyl)-	35369-97-2	NI41782
111	402	45	261	154	39	295	252	**420**	100	36	29	27	24	23	17	13	**3.00**	16	12	4	4	3	–	–	–	–	–	–	–	–	1,4-Bis(thien-2-oylhydrazinocarbonyl)benzene		LI08784
56	420	86	144	55	115	114	100	**420**	100	91	40	36	33	31	27	27		16	28	5	4	2	–	–	–	–	–	–	–	–	3,8,11,14,19-Pentaoxa-1,5,17,21-tetraazatricyclo[19.2.1.1[17,21]]pentacosa-24,25-dithione		MS09466
56	420	86	115	144	188	145	72	**420**	100	52	42	25	23	21	21	21		16	28	5	4	2	–	–	–	–	–	–	–	–	3,8,11,16,21-Pentaoxa-1,5,14,18-tetraazatricyclo[16.5.1.1[14,18]]pentacosa-24,25-dithione		MS09467
363	127	211	153	139	279	109	193	**420**	100	53	44	37	35	33	30	27	**1.00**	16	31	9	–	–	–	–	–	–	–	1	2	–	β-D-Fructopyranose, cyclic 2,3:4,5-bis(butylboronate) 1-(dimethyl phosphate)	58634-46-1	NI41783
127	43	363	362	139	193	195	109	**420**	100	76	60	30	25	22	21	21	**4.26**	16	31	9	–	–	–	–	–	–	–	1	2	–	α-D-Glucofuranose, cyclic 1,2:3,5-bis(butylboronate) 6-(dimethyl phosphate)	58634-44-9	MS09468
127	363	195	43	193	153	139	181	**420**	100	61	25	25	23	23	22	19	**1.00**	16	31	9	–	–	–	–	–	–	–	1	2	–	α-D-Glucofuranose, cyclic 1,2:3,5-bis(butylboronate) 6-(dimethyl phosphate)	58634-44-9	NI41784
73	75	191	147	117	204	115	45	**420**	100	95	91	84	81	72	66	53	**0.00**	16	37	6	–	–	–	–	–	–	3	–	1	–	α-D-Galactopyranose, 1,2,3-tris-O-(trimethylsilyl)-, cyclic methylboronate	56196-95-3	NI41785
174	73	175	86	405	147	176	75	**420**	100	59	19	11	10	10	10	7	**0.00**	16	40	3	2	–	–	–	–	–	4	–	–	–	Glycylglycine, tetrakis(trimethylsilyl)-		NI41786
387	389	385	391	388	315	317	158	**420**	100	65	60	21	19	19	18	12	**3.85**	17	6	–	–	–	–	6	–	–	–	–	–	–	7,8,9,10,11,11-Hexachloro-7,10-dihydro-7,10-methanofluoranthene		MS09469
81	44	141	73	103	98	243	140	**420**	100	99	63	62	46	43	40	26	**0.00**	17	28	10	–	–	–	–	–	–	1	–	–	–	1-O-Trimethylsilyl-2,3,4,6-tetra-O-acetyl-α-D-glucopyranoside	27641-18-5	NI41787
420	373	327	404	372	50	47	400	**420**	100	99	90	67	64	63	63	61		18	5	1	–	–	8	–	–	–	–	1	–	–	5H-Dibenzophosphole, 1,2,3,4,6,7,8,9-octafluoro-5-phenyl-, 5-oxide	36126-97-3	NI41788
420	199	217	280	216	327	262	187	**420**	100	85	72	35	34	32	31	26		18	15	6	–	–	–	–	–	–	–	3	–	–	1,3,5,2,4,6-Trioxatriphosphorinane, 2,4,6-triphenyl-, 2,4,6-trioxide	57156-84-0	NI41789
130	131	260	77	210	103	340	208	**420**	100	44	29	21	20	15	13	11	**3.41**	18	16	–	2	–	–	–	–	–	–	–	–	2	1H-Indole, 3,3′-diselenobis[2-methyl-	1233-38-1	NI41790
260	131	130	261	245	340	210	129	**420**	100	47	45	22	20	16	16	10	**0.95**	18	16	–	2	–	–	–	–	–	–	–	–	2	Indole, 3,3′-diselenobis[1-methyl-	1160-39-0	NI41791
129	130	102	76	51	128	103	131	**420**	100	92	50	20	18	17	15	14	**0.00**	18	16	–	2	–	–	–	–	–	–	–	–	2	Indole, 3,3′-(diselenodimethylene)di-	21903-69-5	NI41792
129	130	102	76	51	128	103	160	**420**	100	92	50	20	19	18	15	14	**0.00**	18	16	–	2	–	–	–	–	–	–	–	–	2	Indole, 3,3′-(diselenodimethylene)di-		LI08785
163	164	134	258	192	420	148	216	**420**	100	76	59	29	14	10	10	8		18	24	6	6	–	–	–	–	–	–	–	–	–	Acetamide, N-[4-(acetyloxy)-2-[(acetyloxy)methyl]-5-[6-(dimethylamino)-9H-purin-9-yl]tetrahydro-3-furanyl]-	56051-71-9	NI41793
163	164	134	258	192	148	420	216	**420**	100	76	59	29	14	10	9	8		18	24	6	6	–	–	–	–	–	–	–	–	–	Puromycin nucleoside triacetate		MS09470
43	143	336	378	337	379	144	246	**420**	100	99	97	44	16	8	7	5	**0.00**	19	14	4	–	–	6	–	–	–	–	–	–	–	1,3-Bis(3-acetoxyphenyl)hexafluoropropane		LI08786
420	223	277	366	124	258	183	61	**420**	100	99	93	36	36	17	17	17		19	19	–	–	–	6	–	–	–	–	–	–	1	π-(2,3-Bis(trifluoromethyl)-5,6,7,8-tetramethylbicyclo[2.2.2]octa-2,5,7-trienyl)-π-cyclopentadienyl cobalt		MS09471

MASS TO CHARGE RATIOS	M.W.	INTENSITIES	Parent	C	H	O	N	S	F	Cl	Br	I	Si	P	B	X	COMPOUND NAME	CAS Reg No	No
203 77 246 227 189 245 185 59	**420**	100 67 46 25 25 21 21 20	**2.00**	19	22	–	–	–	6	–	–	–	2	–	–	–	1,3-Di-[3-(dimethylsilyl)-phenyl]-hexafluoro-propane	37601-87-9	NI41794
420 108 43 182 378 181 270 192	**420**	100 94 62 60 50 48 40 38		20	16	5	6	–	–	–	–	–	–	–	–	–	N-(4-Acetamidophenyl)-3-nitro-4-(4-nitrophenylazo)aniline		IC04361
132 77 420 131 118 305 91 104	**420**	100 93 76 36 32 25 22 21		20	20	1	8	1	–	–	–	–	–	–	–	–	1-Methyl-4-phenyl-5-(phenylimino)-3-[6-methyl-3-(methylthio)-5-oxo-4,5-dihydro-1,2,4-triazin-4-yl]imino-2H-2,3-dihydro-1,2,4-triazole		MS09472
256 107 86 420 306 204 291 198	**420**	100 23 23 15 15 15 7 7		20	24	6	2	1	–	–	–	–	–	–	–	–	tert-Butyl 3-(4-hydroxybenzyl)-7-carbomethoxy amino-2-cephem-4-carboxylate		NI41795
58 357 72 73 269 105 59 89	**420**	100 96 95 93 83 80 71 63	**7.99**	20	28	6	4	–	–	–	–	–	–	–	–	–	4',4'',5',5''-tetraaminodibenzo[b,k]-1,4,7,10,13,16-hexaoxacyclooctadeca-2,11-diene		NI41796
202 73 203 45 291 74 204 75	**420**	100 85 20 12 7 7 6 6	**0.00**	20	36	2	2	–	–	–	–	–	3	–	–	–	L-Tryptophan, N,1-bis(trimethylsilyl)-, trimethylsilyl ester	55429-28-2	NI41797
202 73 203 291 45 218 204 74	**420**	100 72 18 9 7 6 6 6	**0.00**	20	36	2	2	–	–	–	–	–	3	–	–	–	L-Tryptophan, N,1-bis(trimethylsilyl)-, trimethylsilyl ester	55429-28-2	MS09473
202 73 203 291 45 204 74 218	**420**	100 62 20 8 7 6 6 5	**0.00**	20	36	2	2	–	–	–	–	–	3	–	–	–	L-Tryptophan, N,1-bis(trimethylsilyl)-, trimethylsilyl ester	55429-28-2	NI41798
55 41 74 155 67 43 69 87	**420**	100 99 99 91 84 75 66 48	**0.00**	20	36	7	–	1	–	–	–	–	–	–	–	–	Dimethyl threo-9,10-sulphinyldioxyoctadecanedioate	88357-50-0	NI41799
141 195 124 140 125 280 139 405	**420**	100 56 35 32 25 19 16 12	**5.00**	21	36	3	6	–	–	–	–	–	–	–	–	–	Triimidazo[1,5-a:1',5'-c:1'',5''-e][1,3,5]triazine-1,5,9(2H,6H,10H)-trione, 2,6,10-triethylhexahydro-3,3,7,7,11,11-hexamethyl-	67969-49-7	NI41800
89 133 73 69 59 87 91 58	**420**	100 72 56 56 44 43 37 30	**0.00**	21	40	8	–	–	–	–	–	–	–	–	–	–	3,3'-Isopropylidenebis(1,5,8,11-tetraoxacyclotridecane)		MS09474
89 133 120 73 87 67 59 55	**420**	100 59 41 37 31 30 29 29	**1.10**	21	40	8	–	–	–	–	–	–	–	–	–	–	3,3'-Trimethylenebis(1,5,8,11-tetraoxacyclotridecane)		MS09475
224 152 363 197 57 55 225 69	**420**	100 52 41 39 37 32 27 27	**3.00**	21	41	6	–	–	–	–	–	–	–	1	–	–	Diethyl [1-(ethoxycarbonyl)-2-(2-oxopropyl)-4,8-dimethylnonyl]phosphonate		NI41801
135 209 210 106 78 420 73 134	**420**	100 41 15 14 14 10 10 5		22	28	–	–	4	–	–	–	–	–	–	–	–	7,8:16,17-Bisbenzo-1,5,10,14-tetrathiacyclooctadecane	22887-75-8	NI41802
106 105 104 91 210 420 103 212	**420**	100 80 65 50 45 36 27 25		22	28	–	–	4	–	–	–	–	–	–	–	–	7,10:18,21-Bisbenz-1,5,12,16-tetrathiacyclodocosane		MS09476
420 104 208 105 180 106 207 103	**420**	100 91 67 61 56 43 40 38		22	28	–	–	4	–	–	–	–	–	–	–	–	7,9:17,19-Bisbenz-1,5,11,15-tetrathiacycloeicosane		MS09477
327 345 360 273 274 299 261 312	**420**	100 34 29 28 26 20 19 15	**10.00**	22	28	8	–	–	–	–	–	–	–	–	–	–	(2S,2'S,4aR)-2-Acetoxy-2,3,4,4a-tetrahydro-5,6,8,10-tetrahydroxy-7-(2-hydroxy-1-methylethyl)-1,1,4a-trimethylphenanthren-9(1H)-one		NI41803
327 345 273 299 275 313 360 312	**420**	100 43 29 21 17 15 14 14	**10.00**	22	28	8	–	–	–	–	–	–	–	–	–	–	(2'S,3R,4aR)-3-Acetoxy-2,3,4,4a-tetrahydro-5,6,8,10-tetrahydroxy-7-(2-hydroxy-1-methylethyl)-1,1,4a-trimethylphenanthren-9(1H)-one		NI41804
420 224 212 211 209 197 181 178	**420**	100 45 31 30 25 20 12 11		22	28	8	–	–	–	–	–	–	–	–	–	–	2H-1-Benzopyran, 3,4-dihydro-6,7,8-trimethoxy-3-(2,3,4,5-tetramethoxyphenyl)-, (S)-	71594-02-0	NI41805
241 388 181 209 300 328 360 269	**420**	100 80 76 64 46 38 33 20	**15.00**	22	28	8	–	–	–	–	–	–	–	–	–	–	Dicyclobuta[b,g]naphthalene-2a,4a,6a,8a(3H,7H)-tetracarboxylic acid, 1,2,4,5,6,8-hexahydro-, tetramethyl ester	68244-81-5	NI41806
269 77 301 183 104 154 155 211	**420**	100 86 71 29 28 27 24 23	**0.00**	23	20	6	2	–	–	–	–	–	–	–	–	–	Dimethyl 4-methyl-3,5-dioxo-2,6-diphenyl-2,6-diazabicyclo[2.2.2]oct-7-ene-7,8-dicarboxylate		DD01637
123 135 81 79 136 107 55 93	**420**	100 99 99 98 97 89 81 76	**0.60**	23	32	7	–	–	–	–	–	–	–	–	–	–	Allethrin hemisuccinate		NI41807
120 113 300 285 360 43 105 28	**420**	100 99 91 78 73 56 43 43	**0.23**	23	32	7	–	–	–	–	–	–	–	–	–	–	14β-Androst-5-en-17-one, 3β,8,12β,14-tetrahydroxy-, 3,12-diacetate	18607-75-5	NI41808
264 338 91 82 131 160 105 161	**420**	100 89 77 76 75 74 71 68	**0.00**	23	32	7	–	–	–	–	–	–	–	–	–	–	Antiarigenin		LI08787
31 179 91 44 29 338	**420**	100 94 89 86 86 79	**0.00**	23	32	7	–	–	–	–	–	–	–	–	–	–	Card-20(22)-enolide, 3,5,11,14-tetrahydroxy-19-oxo-, (3β,5β,11α)-	6785-70-2	NI41809
328 300 329 282 268 28 283 310	**420**	100 34 33 33 31 23 22 21	**2.00**	23	32	7	–	–	–	–	–	–	–	–	–	–	Gibberellin A13 methyl ester		MS09478
300 328 241 388 41 55 301 329	**420**	100 92 52 45 45 40 37 29	**1.00**	23	32	7	–	–	–	–	–	–	–	–	–	–	Gibberellin A17 methyl ester		MS09479
360 345 361 121 165 358 152 151	**420**	100 30 25 24 20 18 18 17	**1.00**	23	32	7	–	–	–	–	–	–	–	–	–	–	Picras-2-ene-1,16-dione, 12-(acetyloxy)-11-hydroxy-2-methoxy-, (11α,12β)-	24148-79-6	NI41810
377 420 240 285 318 303 286 147	**420**	100 32 14 13 11 7 7 6		24	36	2	–	2	–	–	–	–	–	–	–	–	17β-Hydroxy-4,4-dimethylandrost-5-en-2-spiro-2'-(1',3'-dithian)-3-one		LI08788
420 405 349 422 421 407 406 337	**420**	100 63 42 36 34 27 20 19		24	37	2	–	–	–	1	–	–	1	–	–	–	Δ[1]-Tetrahydrocannabinol chloromethyldimethylsilyl ether		NI41811
420 125 159 117 107 255 145 143	**420**	100 62 60 44 44 40 34 34		24	40	4	–	–	–	–	–	–	1	–	–	–	Methyl all-trans-5,6-erythro-6-hydroxy-5-trimethylsilyloxy-5,6-dihydroretinoate		NI41812
420 109 161 143 133 186 156 159	**420**	100 35 28 28 26 24 20 16		24	40	4	–	–	–	–	–	–	1	–	–	–	Methyl all-trans-5,6-threo-6-hydroxy-5-trimethylsilyloxy-5,6-dihydroretinoate		NI41813
321 75 322 247 420 349 173 146	**420**	100 34 26 26 24 22 17 17		24	40	4	–	–	–	–	–	–	1	–	–	–	Prosta-5,8(12),13-trien-1-oic acid, 9-oxo-15-[(trimethylsilyl)oxy]-, methyl ester, (5Z,13E,15S)-	53044-52-3	NI41814
349 190 199 74 118 132 128 80	**420**	100 90 68 55 43 29 28 28	**5.05**	24	40	4	–	–	–	–	–	–	1	–	–	–	Prosta-5,10,13-trien-1-oic acid, 9-oxo-15-[(trimethylsilyl)oxy]-, methyl ester, (5Z,13E,15S)-	53122-01-3	NI41815
80 139 122 420 181 360 258 220	**420**	100 61 54 49 48 15 15 15		24	40	4	2	–	–	–	–	–	–	–	–	–	2R-(3E,5E-tridecadienyl)-4R-acetoxy-1-acetyl-5S-(acetylamino)piperidine		MS09480
80 139 122 181 420 360 258 220	**420**	100 65 56 50 38 14 14 14		24	40	4	2	–	–	–	–	–	–	–	–	–	2R-(3E,5Z-tridecadienyl)-4R-acetoxy-1-acetyl-5S-(acetylamino)piperidine		MS09481
330 240 331 199 225 129 146 144	**420**	100 33 32 18 14 14 13 12	**1.88**	24	44	2	–	–	–	–	–	–	2	–	–	–	Estra-5(10)-ene, 3,17-bis[(trimethylsilyl)oxy]-, (3α,17β)-	75113-13-2	NI41816
233 97 155 234 209 420 151 179	**420**	100 95 94 86 80 72 70 36		24	48	–	–	–	–	–	–	–	3	–	–	–	1-Methyl-1,2-pentadien-4-yne, 1,3,5-tris(tert-butyldimethylsilyl)-	61227-85-8	NI41817
275 276 347 183 375 185 348 303	**420**	100 80 66 56 52 50 42 38	**33.00**	25	25	4	–	–	–	–	–	–	–	1	–	–	Dicarboethoxy methylene triphenyl phosphorane		LI08789
273 43 360 274 55 81 93 95	**420**	100 55 41 28 20 16 14 12	**1.00**	25	40	5	–	–	–	–	–	–	–	–	–	–	6-Acetoxy-17-oxo-4-propyl-3,4-seco-5β-androstan-3-oic acid methyl ester		MS09482

MASS TO CHARGE RATIOS								M.W.	INTENSITIES								Parent	C	H	O	N	S	F	Cl	Br	I	Si	P	B	X	COMPOUND NAME	CAS Reg No	No
273	274	43	360	333	55	217	81	420	100	40	31	20	17	17	14	13	3.00	25	40	5	–	–	–	–	–	–	–	–	–	–	6β-Acetoxy-17-oxo-4-propyl-3,4-seco-5α-androstan-3-oic acid methyl ester		MS09483
269	384	287	270	260	154	385	159	420	100	60	29	23	19	19	17	15	0.00	25	40	5	–	–	–	–	–	–	–	–	–	–	Cholan-24-oic acid, 7,12-dihydroxy-3-oxo-, methyl ester, (5α,7α,12α)-	14772-92-0	NI41818
269	384	402	270	420	287	262	385	420	100	49	25	23	21	18	18	17		25	40	5	–	–	–	–	–	–	–	–	–	–	Cholan-24-oic acid, 7,12-dihydroxy-3-oxo-, methyl ester, (5β,7α,12α)-	14772-99-7	NI41819
295	171	127	170	114	296	57	99	420	100	75	55	55	29	24	21	19	0.00	25	44	3	–	–	–	–	–	–	1	–	–	–	Hete methyl ester, trimethylsilyl-		NI41820
79	77	52	51	105	183	243	165	420	100	86	68	31	13	10	9	9	0.00	26	28	5	–	–	–	–	–	–	–	–	–	–	Methyl 2-deoxy-6-(triphenylmethyl)-α-D-glucopyranoside		NI41821
296	420	297	135	149	295	162	123	420	100	96	38	25	13	17	17	15		26	28	5	–	–	–	–	–	–	–	–	–	–	Nitiducol		NI41822
337	338	173	420	187	169	151	405	420	100	30	15	14	8	6	6	5		26	28	5	–	–	–	–	–	–	–	–	–	–	Nitidulan		NI41823
370	55	81	355	93	95	105	107	420	100	70	51	44	44	42	41	40	0.20	26	44	4	–	–	–	–	–	–	–	–	–	–	Cholan-24-oic acid, 3-hydroxy-7-methoxy-, methyl ester, (3α,5β,7α)-		MS09484
370	55	81	93	95	67	41	107	420	100	95	78	70	69	59	56	54	1.20	26	44	4	–	–	–	–	–	–	–	–	–	–	Cholan-24-oic acid, 7-hydroxy-3-methoxy-, methyl ester, (3α,5β,7α)-		MS09485
83	40	85	41	55	57	369	69	420	100	97	65	46	43	32	27	24	0.00	27	45	1	–	–	–	1	–	–	–	–	–	–	Cholestan-6-one, 3-chloro-, (3α,5α)-	21072-86-6	NI41824
420	422	421	307	95	265	93	384	420	100	36	29	26	24	23	23	22		27	45	1	–	–	–	1	–	–	–	–	–	–	Cholestan-6-one, 3-chloro-, (3β)-	22033-80-3	NI41825
369	384	43	41	55	44	356	95	420	100	74	57	44	42	35	32	30	0.00	27	45	1	–	–	–	1	–	–	–	–	–	–	Cholestan-6-one, 3-chloro-, (3β,5α)-	1056-93-5	NI41826
420	422	421	55	43	307	57	41	420	100	36	30	29	29	25	25	24		27	45	1	–	–	–	1	–	–	–	–	–	–	Cholestan-6-one, 3-chloro-, (3β,5α)-	1056-93-5	NI41827
43	41	55	163	57	93	91	79	420	100	63	58	46	46	32	32	31	14.70	27	45	2	–	–	1	–	–	–	–	–	–	–	Cholestan-19-ol, 5,6-epoxy-3-fluoro-, (3β,5α,6α)-	23924-39-2	NI41828
390	372	402	391	371	373	234	217	420	100	94	58	44	43	37	35	26	12.00	27	45	2	–	–	1	–	–	–	–	–	–	–	Cholestan-19-ol, 5,6-epoxy-3-fluoro-, (3β,5β,6β)-	23924-38-1	MS09486
402	43	55	81	384	95	57	41	420	100	87	81	69	59	58	58	56	6.19	27	48	3	–	–	–	–	–	–	–	–	–	–	Cholestane-3,5,6-triol, (3β,5α,6β)-	1253-84-5	NI41829
419	417	401	383	399	365	381		420	100	71	50	46	43	36	11		0.00	27	48	3	–	–	–	–	–	–	–	–	–	–	Cholestane-3,5,6-triol, (3β,5α,6β)-	1253-84-5	NI41830
420	402	95	247	81	93	265	43	420	100	86	66	52	48	45	41	40		27	48	3	–	–	–	–	–	–	–	–	–	–	Cholestane-3,6,7-triol, (3β,5α,6α,7α)-	56588-31-9	NI41831
402	420	141	108	95	109	276	43	420	100	89	82	78	78	75	73	67		27	48	3	–	–	–	–	–	–	–	–	–	–	Cholestane-3,6,7-triol, (3β,5α,6β,7α)-	56588-32-0	NI41832
417	43	95	81	55	93	57	69	420	100	72	71	55	53	47	47	45	19.39	27	48	3	–	–	–	–	–	–	–	–	–	–	Cholestane-3,6,7-triol, (3β,5α,6β,7β)-	56588-33-1	NI41833
69	335	95	55	271	94	129	43	420	100	90	90	76	63	62	54	50	0.00	27	48	3	–	–	–	–	–	–	–	–	–	–	Cholestane-3,6,20-triol, (3β,5α,6α,20R)-	56362-86-8	NI41834
402	249	246	250	403	264	273	213	420	100	49	31	29	27	20	19	18	15.00	27	48	3	–	–	–	–	–	–	–	–	–	–	16-Deoxymyxinol		LI08790
420	419	421	210	196	197	194	422	420	100	49	29	10	10	6	6	4		28	18	–	2	–	2	–	–	–	–	–	–	–	Pyrazine, 2,5-bis(4-fluorophenyl)-3,6-diphenyl-	22158-34-5	MS09487
420	388	421	197	389	194	178	152	420	100	63	34	26	21	19	18	15		28	20	–	–	2	–	–	–	–	–	–	–	–	1,4-Dithiin, 2,5-bis([1,1′-biphenyl]-4-yl)-	37989-52-9	NI41835
178	420	64	356	372	179	421	105	420	100	40	22	15	13	13	11	8		28	20	2	–	1	–	–	–	–	–	–	–	–	Thiophene, tetraphenyl-, 1,1-dioxide	1059-75-2	LI08791
178	420	64	356	179	372	421	176	420	100	40	21	16	16	14	13	9		28	20	2	–	1	–	–	–	–	–	–	–	–	Thiophene, tetraphenyl-, 1,1-dioxide	1059-75-2	NI41836
105	77	106	420	51	78	421	299	420	100	59	23	14	7	5	4	1		28	20	4	–	–	–	–	–	–	–	–	–	–	2-Dibenzoyloxy-1,2-diphenyl ethylene		IC04362
378	243	244	420	379	43	336	421	420	100	93	43	38	30	20	15	13		28	24	2	2	–	–	–	–	–	–	–	–	–	Acetamide, N,N′-[1,1′-biphenyl]-2,2′-diylbis[N-phenyl-	29325-51-7	NI41837
224	77	420	105	91	196	146	405	420	100	23	22	10	9	6	5	4		28	24	2	2	–	–	–	–	–	–	–	–	–	5-[(O-Benzoylanilino)methyl]-2-(methylamino)benzophenone		NI41838
196	420	392	421	178	197	393	194	420	100	99	37	32	16	15	11	9		29	18	1	–	–	2	–	–	–	–	–	–	–	2,4-Cyclopentadien-1-one, 2,4-bis(4-fluorophenyl)-3,5-diphenyl-	22818-67-3	MS09488
420	196	392	178	214	98	117	81	420	100	80	35	13	7	5	1	1		29	18	1	–	–	2	–	–	–	–	–	–	–	2,4-Cyclopentadien-1-one, 2,4-bis(4-fluorophenyl)-3,5-diphenyl-		LI08792
420	196	392	178	421	214	197	393	420	100	98	36	34	31	17	15	11		29	18	1	–	–	2	–	–	–	–	–	–	–	2,4-Cyclopentadien-1-one, 2,5-bis(4-fluorophenyl)-3,4-diphenyl-	23101-20-4	MS09489
420	196	392	421	178	214	197	393	420	100	88	32	31	28	15	13	10		29	18	1	–	–	2	–	–	–	–	–	–	–	2,4-Cyclopentadien-1-one, 3,4-bis(4-fluorophenyl)-2,5-diphenyl-	56805-29-9	MS09490
402	420	403	180	208	421	270	194	420	100	66	33	32	31	22	15	14		29	28	1	2	–	–	–	–	–	–	–	–	–	Acetamide, N-(4-methylphenyl)-N-[4-methyl-2-[[2-(phenylamino)phenyl]methyl]phenyl]-	52812-78-9	NI41839
420	217	204	201	421	55	121	324	420	100	95	54	36	33	26	23	19		29	40	2	–	–	–	–	–	–	–	–	–	–	Bis[2-hydroxy-5-methyl-3-(1′-methylcyclohexyl)phenyl]methane		IC04364
420	217	204	55	201	421	121	147	420	100	99	57	41	39	33	31	22		29	40	2	–	–	–	–	–	–	–	–	–	–	Bis[2-hydroxy-5-methyl-3-(1′-methylcyclohexyl)phenyl]methane		IC04363
57	97	83	71	420	69	43	111	420	100	69	65	55	54	50	46	39		30	60	–	–	–	–	–	–	–	–	–	–	–	Cyclotriacontane	297-35-8	NI41840
43	167	41	141	153	420	57	165	420	100	80	60	55	44	37	34	32		31	48	–	–	–	–	–	–	–	–	–	–	–	11-(1′-Naphthyl)-10-heneicosene		AI02357
328	327	329	251	326	325	32	43	420	100	42	28	14	11	10	9	7	1.50	33	24	–	–	–	–	–	–	–	–	–	–	–	9,10[1′,2′]-Benzenoanthracene, 9,10-dihydro-9-phenyl-10-benzyl-	56701-34-9	NI41841
306	304	77	308	109	423	108	65	421	100	99	53	45	43	41	38	38	20.77	11	5	–	3	1	–	6	–	–	–	–	–	–	2-Phenylthio-4,6-bis(trichloromethyl)-s-triazine		MS09491
242	421	178	199	406	374	352	391	421	100	35	4	3	2	2	2	1		14	16	2	2	–	6	–	–	–	–	–	–	1	N,N,2,3-Butylenebis(trifluoroacetylacetoniminato)copper(II)		NI41842
206	138	421	207	110	423	69	96	421	100	91	91	72	48	42	39	36		14	16	2	2	–	6	–	–	–	–	–	–	1	N,N′-Butylenebis(5,5,5-trifluoro-4-oxopentan-2-iminato)copper(II)		NI41843
235	260	277	218	193	192	234	319	421	100	52	47	46	37	36	25	24	1.75	17	19	8	5	–	–	–	–	–	–	–	–	–	2-Acetylamino-7-(1,2,3-triacetoxypropyl)-3,4-dihydropteridin-4-one		LI08793
179	180	308	181	421	406	309	293	421	100	17	10	5	3	3	3	2		17	26	4	1	–	3	–	–	–	2	–	–	–	L-Tyrosine, N-(trifluoroacetyl)-O-(trimethylsilyl)-, trimethylsilyl ester	52558-10-8	NI41844
173	73	130	131	174	259	362	204	421	100	35	20	17	13	11	8	7	0.00	17	39	5	1	–	–	–	–	–	3	–	–	–	2-Acetamido-2,6-dideoxy-1,3,4-tri-O-trimethylsilyl glucose		MS09492
106	232	406	421	216				421	100	75	20	10	8					19	27	4	3	1	–	–	–	–	1	–	–	–	4-Thia-1-azabicyclo[3.2.0]heptane-2-carboxylic acid, 6-[(aminophenylacetyl)amino]-3,3-dimethyl-7-oxo-, trimethylsilyl ester, [2S-[2α,5α,6β(S*)]]-	50617-61-3	NI41845
406	407	73	408	420	318	45	75	421	100	38	27	16	5	5	5	4	3.70	19	31	4	1	–	–	–	–	–	3	–	–	–	2-Quinolinecarboxylic acid, 4,8-bis[(trimethylsilyl)oxy]-, trimethyl silyl ester		MS09493
160	234	291	376	188	161	104	130	421	100	61	21	18	18	14	12	10	0.00	20	27	7	3	–	–	–	–	–	–	–	–	–	Ethyl α-(O-ethoxycarbonylmethylcarbamoylbenzamido) iso-butyrate		LI08794
216	390	96	45	391	174	281	43	421	100	70	50	32	21	20	20	15	8.90	20	31	5	5	–	–	–	–	–	–	–	–	–	Pentamethylzeatin riboside		NI41846

MASS TO CHARGE RATIOS								M.W.	INTENSITIES								Parent	C	H	O	N	S	F	Cl	Br	I	Si	P	B	X	COMPOUND NAME	CAS Reg No	No
75	290	73	44	421	77	45	291	**421**	100	70	50	30	29	23	21	17		20	35	3	1	–	–	–	–	–	3	–	–	–	1H-Indole-3-propanoic acid, 1-(trimethylsilyl)-5-[(trimethylsilyl)oxy]-, trimethylsilyl ester		MS09494
202	203	147	204	292	129	130	102	**421**	100	43	30	12	10	8	7	6	**1.30**	20	35	3	1	–	–	–	–	–	3	–	–	–	1H-Indole-3-propanoic acid, 1-(trimethylsilyl)-α-[(trimethylsilyl)oxy]-, trimethylsilyl ester	55319-91-0	NI41847
202	73	203	421	204	147	422	406	**421**	100	24	19	11	6	6	4	3		20	35	3	1	–	–	–	–	–	3	–	–	–	1H-Indole-3-propanoic acid, 1-(trimethylsilyl)-α-[(trimethylsilyl)oxy]-, trimethylsilyl ester	55319-91-0	MS09495
266	421	220	422	376	335	357		**421**	100	68	59	16	10	2	1			21	27	6	1	1	–	–	–	–	–	–	–	–	Butanoic acid, 4-[N-(3,4-dimethoxyphenyl)-4-toluenesulphonamido]-, ethyl ester	24310-38-1	NI41848
84	91	98	106	77	133	107	105	**421**	100	74	35	26	26	24	24	24	**0.10**	24	31	2	5	–	–	–	–	–	–	–	–	–	2-(Benzylamino)-5-[3-[3-(piperidinomethyl)phenoxy]propylamino]-1,3,4-oxadiazole		MS09496
84	157	231	98	100	421	43	190	**421**	100	76	67	55	45	39	39	33		24	31	2	5	–	–	–	–	–	–	–	–	–	4,5-Dihydro-4-methyl-2-(phenylamino)-5-[3-[3-(piperidinomethyl)phenoxy]propylimino]-1,3,4-oxadiazole		MS09497
421	58	44	422	55	43	329	83	**421**	100	31	27	26	26	26	24	23		24	39	5	1	–	–	–	–	–	–	–	–	–	1-Phenanthrenecarboxylic acid, tetradecahydro-9-hydroxy-1,4a,8-trimethyl-7-[2-[2-(methylamino)ethoxy]-2-oxoethylidene]-, methyl ester	36150-73-9	NI41849
271	316	421						**421**	100	45	40							25	15	4	3	–	–	–	–	–	–	–	–	–	1,3-Dioxolo[4,5-f][4,7]phenanthroline, 2-(4-nitrophenyl)-2-phenyl-	54616-39-6	NI41850
361	360	310	324	333	352	309	294	**421**	100	83	75	53	52	42	37	20	**5.00**	25	27	5	1	–	–	–	–	–	–	–	–	–	Dibenzo[f,h]pyrrolo[1,2-b]isoquinolin-14-ol, 9,11,12,13,13a,14-hexahydro-3,6,7-trimethoxy-, acetate, (13aS-cis)-	62057-93-6	NI41851
336	28	422	421	190	337	176	423	**421**	100	67	36	30	25	24	19	18		25	27	5	1	–	–	–	–	–	–	–	–	–	Dihydrovalachine	95307-60-1	NI41852
106	70	69	91	77	160	146	243	**421**	100	81	76	66	55	53	52	48	**39.60**	25	31	3	3	–	–	–	–	–	–	–	–	–	13-(α-Hydroxybenzyl)-7-phenyl-1,6,10-triazabicyclo[10.2.1]pentadecan-9,14-dione		NI41853
105	122	164	177	77	194	147	206	**421**	100	42	35	18	17	10	9	1	**0.00**	25	43	4	1	–	–	–	–	–	–	–	–	–	DL-lyxo-2-Benzamido-1,3,4-trihydroxy-octadecane		LI08795
105	147	164	122	194	77	177	280	**421**	100	43	38	30	22	18	13	1	**0.00**	25	43	4	1	–	–	–	–	–	–	–	–	–	DL-ribo-2-Benzamido-1,3,4-trihydroxy-octadecane		LI08796
271	316	421						**421**	100	40	20							26	15	5	1	–	–	–	–	–	–	–	–	–	7H-Phenaleno[1,2-d][1,3]dioxol-7-one, 9-(4-nitrophenyl)-9-phenyl-	54616-43-2	NI41854
59	421	73	150	364	422	197	315	**421**	100	91	65	50	34	25	19	18		26	35	2	1	–	–	–	–	–	1	–	–	–	3-[4′-(Trimethylsilyl)oxyphenoxy]-N-methylmorphinan		NI41855
405	55	87	69	378	421	364	56	**421**	100	98	46	45	42	40	38	35		27	51	2	1	–	–	–	–	–	–	–	–	–	Azacyclohexacosan-14-one, 1-acetyl-	15939-31-8	NI41856
421	91	57	422	40	41	420	44	**421**	100	94	79	75	69	47	44	37		28	39	2	1	–	–	–	–	–	–	–	–	–	1H-Phenoxazin-1-one, 2,4,6,8-tetra-tert-butyl-		MS09498
197	195	150	101	146	152	144	103	**422**	100	76	67	49	47	43	35	32	**0.00**	4	–	–	–	–	4	4	2	–	–	–	–	–	1,1,2,4-Tetrachloro-1,2,3,4-tetrafluoro-3,4-dibromobutane		DO01119
69	254	127	219	100	131	119	31	**422**	100	48	48	35	32	29	22	22	**0.00**	4	–	–	–	–	13	–	–	1	–	–	–	–	Tetrafluoro(perfluorobutyl)iodine		MS09499
422	183	366	310	56	338	127	394	**422**	100	75	43	42	40	29	20	15		4	–	4	–	–	–	–	–	2	–	–	–	1	Iron tetracarbonyl di-iodide		IC04365
422	183	366	56	310	84	338	239	**422**	100	72	53	53	42	42	40	25		4	–	4	–	–	–	–	–	2	–	–	–	1	Iron tetracarbonyl di-iodide		MS09500
282	310	262	185	420	255	271	211	**422**	100	65	58	50	42	40	28	28	**0.00**	8	2	5	–	–	3	–	–	–	–	–	–	1	Rhenium, pentacarbonyl(3,3,3-trifluoropropenyl)-, (Z)-	18285-39-7	NI41857
73	281	355	407	267	282	379	356	**422**	100	94	53	51	32	27	21	19	**0.00**	8	23	4	–	–	–	–	–	1	4	–	–	–	Cyclotetrasiloxane, (iodomethyl)heptamethyl-	17882-88-1	NI41858
59	165	227	121	211	91	197	255	**422**	100	81	63	45	44	42	40	29	**0.15**	10	30	10	–	–	–	–	–	–	4	–	–	–	Tris(trimethoxysilyl)methoxysilane	56120-91-3	NI41859
69	41	53	153	29	13	181	166	**422**	100	38	35	30	30	28	22	22	**0.00**	11	7	7	–	–	9	–	–	–	–	–	–	–	3-Desoxy-1,2,5-tris-O-(trifluoracetyl)-α-DL-erythro-pentofuranose		MS09501
69	53	41	181	29	153	81	59	**422**	100	35	32	25	25	22	21	19	**0.00**	11	7	7	–	–	9	–	–	–	–	–	–	–	3-Desoxy-1,2,5-tris-O-(trifluoracetyl)-β-DL-erythro-pentofuranose		MS09502
69	53	41	181	153	166	81	29	**422**	100	30	30	27	26	21	20	20	**0.00**	11	7	7	–	–	9	–	–	–	–	–	–	–	3-Desoxy-1,2,5-tris-O-(trifluoracetyl)-β-DL-threo-pentofuranose		MS09503
69	81	195	53	166	45	79	97	**422**	100	65	26	25	21	17	15	14	**0.00**	11	7	7	–	–	9	–	–	–	–	–	–	–	Tris-trifluoroacetyl-desoxyribofuranose		MS09504
69	81	45	153	181	53	79	51	**422**	100	79	24	23	22	19	18	17	**0.00**	11	7	7	–	–	9	–	–	–	–	–	–	–	Tris-trifluoroacetyl-desoxyribopyranose		MS09505
83	93	41	55	306	147	69	125	**422**	100	84	32	25	22	20	20	13	**0.00**	11	18	6	–	–	6	–	–	–	–	2	–	–	2-[1-(Dimethoxyphosphinyl)-2,2,2-trifluoro-1-(trifluoromethyl)ethoxy]-4,4,5,5-tetramethyl-1,3,2-dioxaphospholane	111557-98-3	DD01638
64	41	61	68	67	63	60	47	**422**	100	79	60	58	31	30	28	26	**0.00**	11	20	10	1	3	–	–	–	–	–	–	–	–	Glucoiberin		MS09506
426	424	149	108	142	212	428	213	**422**	100	77	70	64	55	54	48	48	**32.37**	12	1	2	–	–	–	7	–	–	–	–	–	–	1,2,3,4,6,7,8-Heptachlorodibenzodioxin	35822-46-9	NI41860
43	81	304	302	223	201	127	126	**422**	100	51	5	5	3	2	2	2	**0.01**	14	19	8	2	–	–	–	1	–	–	–	–	–	Bromo-5-hydroxy-6-deoxy-2′-β-ribofuranosyl-1-thymine-diacetyl		LI08797
43	81	304	302	141	223	201	127	**422**	100	23	9	9	4	2	2	2	**0.00**	14	19	8	2	–	–	–	1	–	–	–	–	–	Bromo-5-hydroxy-6-deoxy-2′-α-ribopyranosyl-1-thymine-diacetyl		LI08798
43	81	141	304	302	201	127	126	**422**	100	38	9	8	8	5	2	2	**0.00**	14	19	8	2	–	–	–	1	–	–	–	–	–	Bromo-5-hydroxy-6-deoxy-2′-β-ribopyranosyl-1-thymine-diacetyl		LI08799
32	91	131	65	44	92	77	51	**422**	100	68	25	9	6	5	5	4	**0.00**	15	20	9	1	2	–	–	–	–	–	–	–	–	Gluconasturtiin		MS09507
387	389	391	424	422	359	361		**422**	100	93	35	26	21	20	16			16	10	5	–	–	–	4	–	–	–	–	–	–	11H-Dibenzo[b,e][1,4]dioxepin-11-one, 2,4,7,9-tetrachloro-3-hydroxy-8-methoxy-1,6-dimethyl-		LI08800
387	389	391	424	359	422	388	361	**422**	100	93	34	25	22	20	20	20		16	10	5	–	–	–	4	–	–	–	–	–	–	11H-Dibenzo[b,e][1,4]dioxepin-11-one, 2,4,7,9-tetrachloro-3-hydroxy-8-methoxy-1,6-dimethyl-		LI08801
387	389	391	424	422	361	359	388	**422**	100	98	33	25	20	20	20	19		16	10	5	–	–	–	4	–	–	–	–	–	–	11H-Dibenzo[b,e][1,4]dioxepin-11-one, 2,4,7,9-tetrachloro-3-hydroxy-8-methoxy-1,6-dimethyl-	527-93-5	NI41861
73	262	231	103	28	133	147	45	**422**	100	26	15	12	12	10	9	9	**0.00**	16	38	5	2	–	–	–	–	–	3	–	–	–	xylo-Pentodialdose, 2,3,4-tris-O-(trimethylsilyl)-, bis(O-methyloxime)	62108-15-0	NI41862

MASS TO CHARGE RATIOS								M.W.	INTENSITIES								Parent	C	H	O	N	S	F	Cl	Br	I	Si	P	B	X	COMPOUND NAME	CAS Reg No	No
44	43	80	69	299	45	82	342	**422**	100	54	41	41	40	40	37	26	**0.00**	18	19	5	2	–	–	–	1	–	–	–	–	–	**Cyanamide, [2-(1-bromo-6a,7,7a,10a,11,11a-hexahydro-4-hydroxy-6-oxo-6H-[1,3]benzodioxolo[5,6-c][1]benzopyran-2-yl)ethyl]methyl-**	55401-69-9	**NI41863**
117	43	44	89	87	59	377	45	**422**	100	88	22	21	12	11	11	11	**0.00**	18	30	11	–	–	–	–	–	–	–	–	–	–	**1,3,4,5-Tetra-O-acetyl-2,6,7-tri-O-methyl-L-glycero-D-mannoheptitol**		**NI41864**
117	43	233	101	127	99	85	159	**422**	100	94	19	18	12	8	7	7	**0.00**	18	30	11	–	–	–	–	–	–	–	–	–	–	**1,3,5,7-Tetra-O-acetyl-2,4,6-tri-O-methyl-L-glycero-D-mannoheptitol**		**NI41865**
127	69	84	354	393	315	265	421	**422**	100	63	56	18	12	9	4	1	**0.00**	18	40	–	2	–	–	–	–	–	–	–	–	2	**Gallium, tetraethyldi-μ-1-piperidinyldi-**	42777-03-7	**MS09508**
422	405	358	91	194	77	423	102	**422**	100	54	38	38	35	31	29	29		19	14	6	6	–	–	–	–	–	–	–	–	–	**5-Nitro-2-aminobenzophenone,2,4-dinitrophenylhydrazone**		**MS09509**
273	260	274	369	404	285	272	287	**422**	100	64	49	34	20	20	20	18	**1.00**	19	18	11	–	–	–	–	–	–	–	–	–	–	**Mangiferin**		**LI08802**
91	155	213	171	229	197	380	320	**422**	100	23	19	13	11	11	9	9	**0.00**	19	22	9	2	–	–	–	–	–	–	–	–	–	**Methyl 2-(benzyloxycarbonyl)-7,8-diacetoxy-3-oxa-2,5-diazabicyclo[2.2.2]octane-5-carboxylate**		**DD01639**
205	422	161	407	206	424	409	162	**422**	100	35	35	34	17	14	13	13		19	23	3	4	1	–	1	–	–	–	–	–	–	**N,N-Diethyl-3-acetamido-4-(4-methylsulphonyl-2-chlorophenylazo)-aniline**		**IC04366**
73	43	117	218	217	75	145	129	**422**	100	84	41	29	29	29	27	23	**0.00**	19	26	7	–	–	–	–	–	–	2	–	–	–	**Methyl β-galactopyranoside-2,3-diacetate-di-TMS**		**NI41866**
73	132	43	217	117	75	44	243	**422**	100	89	73	51	48	43	36	33	**0.00**	19	26	7	–	–	–	–	–	–	2	–	–	–	**Methyl β-galactopyranoside-3,6-diacetate-di-TMS**		**NI41867**
204	73	43	133	75	117	205	142	**422**	100	67	44	21	19	17	16	11	**0.00**	19	26	7	–	–	–	–	–	–	2	–	–	–	**Methyl β-galactopyranoside-4,6-diacetate-di-TMS**		**NI41868**
422	202	421	201	45	203	423	157	**422**	100	53	50	47	43	38	31	30		20	16	1	–	4	–	–	–	–	–	–	2	–	**4H-Borepino[3,2-b:6,7-b′]dithiophene, 4,4′-oxybis[8,9-dihydro-**	51081-19-7	**NI41869**
44	57	171	139	127	362	209	239	**422**	100	85	27	26	26	24	23	18	**0.00**	20	30	6	4	–	–	–	–	–	–	–	–	–	**3,3′-(1,6-Hexanediyl)bis[1-(2-hydroxyethyl)-6-methyl-2,4(1H,3H)-pyrimidinedione]**		**MS09510**
55	69	41	43	135	153	422	83	**422**	100	75	71	48	37	35	32	30		20	38	4	–	–	–	–	–	–	–	–	–	1	**Decanoic acid, 10,10′-selenodi-**	22676-20-6	**NI41870**
89	133	87	73	228	97	59	96	**422**	100	57	33	32	25	25	18	17	**0.00**	20	38	9	–	–	–	–	–	–	–	–	–	–	**3-(3,6,9,12-Tetraoxacyclotridecyl)-1,5,8,11,14-pentaoxacyclohexadecane**		**MS09511**
57	71	56	393	349	55	69	29	**422**	100	95	79	63	57	54	45	38	**0.00**	20	46	5	–	–	–	–	–	–	2	–	–	–	**1,3-Diethyl-1,1,3,3-tetrabutoxydisiloxane**	94048-03-0	**NI41871**
55	134	118	337	259	366	188	322	**422**	100	64	58	48	42	40	40	28	**0.00**	21	23	4	2	–	–	–	–	–	–	–	–	1	**Dicarbonyl(5-ethoxy-1,3-dimethyl-2-oxo-5-phenyl-4-imidazolidinylidene)(methylcyclopentadienyl)manganese**	116928-08-6	**DD01640**
135	190	199	242	260	243	362	303	**422**	100	62	48	30	28	21	20	20	**3.50**	21	26	9	–	–	–	–	–	–	–	–	–	–	**2α-Acetoxy-8-desacylglaucolide-E-acetate**		**MS09512**
43	163	98	181	137	290	41	136	**422**	100	23	18	12	11	9	9	8	**1.10**	21	26	9	–	–	–	–	–	–	–	–	–	–	**Triacetyldeoxynivalenol**		**NI41872**
422	407	162	161	176	423	147	408	**422**	100	57	56	39	35	27	16	15		22	22	3	4	1	–	–	–	–	–	–	–	–	**N,N-Diethyl-3-methyl-4-(3-benzyl-5-nitrothien-2-ylazo)aniline**		**IC04367**
197	422	225	198	109	423	169	199	**422**	100	41	33	19	19	14	11	9		22	30	4	–	2	–	–	–	–	–	–	–	–	**1,2-Bis(2,5-diethoxyphenylthio)ethane**	93885-72-4	**NI41873**
93	135	79	109	163	191	168	195	**422**	100	75	73	46	40	33	30	23	**0.00**	22	30	6	–	1	–	–	–	–	–	–	–	–	**Bis[methyl (1RS,2RS,4RS,5SR)-4-methyl-8-oxobicyclo[3.3.0]octan-2-yl-1 carboxylate] sulphide**		**MS09513**
57	85	43	242	260	320	393	391	**422**	100	76	70	35	25	6	2	2	**2.00**	22	30	8	–	–	–	–	–	–	–	–	–	–	**9α-Acetoxy-14,15-dihydroxy-8β-(2-methylbutyryloxy)-14-oxoacanthospermolide**		**NI41874**
43	83	69	109	41	53	45	28	**422**	100	31	18	14	14	11	8	8	**0.17**	22	30	8	–	–	–	–	–	–	–	–	–	–	**12β-Acetoxy-6,7α-epoxy-4,9,13,20-tetrahydroxy-tigli-1-en-3-one**		**NI41875**
43	83	109	41	69	53	45	55	**422**	100	99	86	73	63	57	35	29	**0.06**	22	30	8	–	–	–	–	–	–	–	–	–	–	**12β-Acetoxy-6,7β-epoxy-4,9,13,20-tetrahydroxy-tigli-1-en-3-one**		**NI41876**
85	43	250	59	321	190	181	189	**422**	100	94	87	85	30	13	12	11	**2.00**	22	30	8	–	–	–	–	–	–	–	–	–	–	**Lobatin A**	84754-03-0	**NI41877**
43	57	85	71	69	148	41	176	**422**	100	97	65	42	39	37	37	26	**2.20**	22	30	8	–	–	–	–	–	–	–	–	–	–	**Valtrate**	18296-44-1	**MS09514**
422	59	87	73	292	264	199	60	**422**	100	90	40	20	16	9	9	7		22	46	–	–	–	–	–	–	–	4	–	–	–	**3,4-Hexadien-1-yne, 1,3,5,6-tetrakis(ethyldimethylsilyl)-**	61227-84-7	**NI41878**
43	422	267	280	137	69	44	254	**422**	100	51	50	41	37	33	33	31		23	25	4	–	–	3	–	–	–	–	–	–	–	**Estra-1,3,5(10),6-tetraene-3,17-diol, 1-methyl-, 17-acetate 3-(trifluoroacetate), (17β)-**	56588-08-0	**NI41879**
211	331	209	91	329	207	327	210	**422**	100	80	79	67	61	47	35	34	**0.90**	23	26	–	–	–	–	–	–	–	–	–	–	1	**Tribenzylethyltin**		**NI41880**
244	248	187	175	130	118	300	146	**422**	100	37	18	18	14	12	8	7	**1.10**	23	26	4	4	–	–	–	–	–	–	–	–	–	**8-[4-(4-Nitrophenyl)-4-oxobutyl]-1-phenyl-1,3,8-triazaspiro[4,5]decan-4-one**		**MS09515**
85	278	43	260	338	229	391	364	**422**	100	77	50	39	36	13	7	5	**0.20**	23	34	7	–	–	–	–	–	–	–	–	–	–	**7-[5-Acetyloxy-2-(2-formylvinyl)-3-[(tetrahydro-2H-pyran-2-yl)oxy]cyclopentyl]-5-heptenoic acid, methyl ester, [1α(Z),2β(1E),3α,5α]-**	60894-66-8	**NI41881**
41	55	43	105	91	29	113	57	**422**	100	99	99	83	83	76	73	68	**0.52**	23	34	7	–	–	–	–	–	–	–	–	–	–	**Pregn-5-en-20-one, 12-(acetyloxy)-3,8,14,17-tetrahydroxy-, (3β,12β,14β,17α)-**	3513-02-8	**NI41882**
91	155	198	116	70	210	267	188	**422**	100	83	47	34	32	17	16	9	**0.00**	23	38	3	2	1	–	–	–	–	–	–	–	–	**Benzenesulphonamide, N,4-dimethyl-N-[3-(2-oxoazacyclotridec-1-yl)propyl]-**	67370-85-8	**NI41883**
212	166	139	167	164	165	140	422	**422**	100	80	50	26	25	22	19	18		24	14	4	4	–	–	–	–	–	–	–	–	–	**1,9′-Bis(3-nitrocarbazole)**		**NI41884**
256	179	422	255	152	57	83	165	**422**	100	79	72	67	50	46	39	35		24	22	7	–	–	–	–	–	–	–	–	–	–	**Obtusifolin**		**LI08803**
319	260	362	303	245	232	422	278	**422**	100	30	29	22	16	10	8	8		24	22	7	–	–	–	–	–	–	–	–	–	–	**1α,2β,15ξ-Triacetoxy-1,2,15,16-tetrahydro-11-methylcyclopenta[a]phenanthren-17-one**		**NI41885**
183	422	55	41	181	43	182	390	**422**	100	46	46	41	40	27	25	23		24	38	6	–	–	–	–	–	–	–	–	–	–	**Methyl 15-(5,6-dimethoxy-1,4-benzoquinon-2-yl)pentadecanoate**		**MS09516**
212	211	379	96	97	422	53	68	**422**	100	38	26	24	23	20	20	16		24	42	4	2	–	–	–	–	–	–	–	–	–	**Azimine**		**LI08804**
310	366	422	282	311	367	178	423	**422**	100	67	45	23	22	18	14	12		25	26	6	–	–	–	–	–	–	–	–	–	–	**2H,6H,10H-Benzo[1,2-b:3,4-b′:5,6-b″]tripyran-2-one, 7,8,11,12-tetrahydro-4-hydroxy-3-(4-hydroxyphenyl)-7,7,11,11-tetramethyl-**	56211-22-4	**NI41886**
333	354	353	283	255	422	351	404	**422**	100	51	44	40	32	28	17	10		25	26	6	–	–	–	–	–	–	–	–	–	–	**Tajixanthone**	35660-48-1	**NI41887**

MASS TO CHARGE RATIOS	M.W.	INTENSITIES	Parent	C	H	O	N	S	F	Cl	Br	I	Si	P	B	X	COMPOUND NAME	CAS Reg No	No
422 157 363 144 158 176 170	**422**	100 30 25 25 22 18 12		25	30	4	2	–	–	–	–	–	–	–	–	–	**O,O-Diacetylvincamajinol**		**LI08805**
379 422 31 118 29 380 423 218	**422**	100 74 38 35 35 25 21 14		25	30	4	2	–	–	–	–	–	–	–	–	–	**Tetraglycidyldiaminodiphenylmethane**		**NI41888**
369 387 370 388 385 386 367 368	**422**	100 73 27 18 18 12 8 7	**0.20**	25	42	5	–	–	–	–	–	–	–	–	–	–	**3α,7α,12α-Trihydroxy-5β-cholan-24-oic acid, methyl ester**		**NI41889**
253 271 386 368 254 272 269 357	**422**	100 91 44 34 30 27 20 15	**0.00**	25	42	5	–	–	–	–	–	–	–	–	–	–	**3α,7α,12α-Trihydroxy-5β-cholan-24-oic acid, methyl ester**		**LI08806**
212 154 183 422 68 155 213 127	**422**	100 55 38 23 22 19 16 16		26	18	4	2	–	–	–	–	–	–	–	–	–	**Dibenz[b,b][1,4]oxazepin-6(11H)-one dimer**		**MS09517**
366 365 421 189 367 216 214 197	**422**	100 70 54 43 27 27 21 21	**0.00**	26	22	2	4	–	–	–	–	–	–	–	–	–	**2-tert-Butyl-1-(4-nitrophenyldiazenyl)-2H-dibenz[e,g]isoindole**	110028-28-9	**DD01641**
233 335 69 41 43 57 71 179	**422**	100 81 73 71 68 66 43 41	**28.00**	26	30	5	–	–	–	–	–	–	–	–	–	–	**Desoxyhomoflemingin**		**MS09518**
339 69 354 91 331 107 123 109	**422**	100 66 50 32 25 11 5 5	**4.00**	26	30	5	–	–	–	–	–	–	–	–	–	–	**5,6-Dihydro-8-hydroxy-5-(3-methylbut-2-en-1-yl)-2,2,5-trimethyl-7-[2-(4-hydroxyphenyl)propionyl]-2H-1-benzopyran-6-one**		**NI41890**
147 407 91 351 83 205 196 55	**422**	100 40 34 30 29 23 21 21	**20.00**	26	30	5	–	–	–	–	–	–	–	–	–	–	**Meso-1,1′-bis-[2-senecionyl-4-(1-ethyl)-phenoxy]ether**		**NI41891**
339 173 422 187 153 407 166 145	**422**	100 74 70 52 49 31 11 11		26	30	5	–	–	–	–	–	–	–	–	–	–	**Nitidulin**	66446-87-5	**NI41892**
147 187 55 83 203 91 379	**422**	100 96 78 50 40 33 10	**0.00**	26	30	5	–	–	–	–	–	–	–	–	–	–	**2-Senecionyl-4-(1-acetoxyethyl)-phenol**		**NI41893**
147 203 407 204 196 91 422 83	**422**	100 98 75 48 48 48 38 37		26	30	5	–	–	–	–	–	–	–	–	–	–	**1,1′-Bis-[2-senecionyl-4-(1-ethyl)-phenoxy]ether**		**NI41894**
127 124 250 153 111 152 278 57	**422**	100 62 54 54 38 27 17 16	**6.92**	26	46	4	–	–	–	–	–	–	–	–	–	–	**Dinonyl 3,4,5,6-tetrahydrophthalate**		**IC04368**
339 340 257 83 55 115 81 41	**422**	100 22 12 5 5 4 4 3	**1.38**	26	48	–	–	–	–	–	–	–	–	2	–	–	**Phosphine, 1,2-ethanediylbis[dicyclohexyl-**	23743-26-2	**NI41895**
404 105 215 217 405 187 186 422	**422**	100 54 47 37 30 30 28 26		27	39	3	–	–	–	–	–	–	–	–	1	–	**Pregnane-3,17,20-triol, 17,20-phenylboronate, (3α,5β,20S)-**	18929-38-9	**NI41896**
391 422 386 355 251 268 424 393	**422**	100 97 93 50 50 38 33 33		27	47	1	–	–	–	1	–	–	–	–	–	–	**5α-Cholestan-19-ol, 1β-chloro-**	28809-52-1	**NI41897**
105 77 317 422 106 423 318 78	**422**	100 21 13 11 8 3 3 2		28	22	4	–	–	–	–	–	–	–	–	–	–	**2-(3-Benzoxybenzyl)-4-methylphenyl benzoate**		**MS09519**
105 77 317 106 422 318 78 51	**422**	100 26 10 8 3 3 2 2		28	22	4	–	–	–	–	–	–	–	–	–	–	**2-(4-Benzoxybenzyl)-4-methylphenyl benzoate**		**MS09520**
183 185 422 149 152 57 107 108	**422**	100 59 54 21 14 14 13 12		28	24	–	–	–	–	–	–	–	–	2	–	–	**1,4-Bis(diphenylphosphinyl)but-2-yne**		**NI41898**
244 229 422 179 245 212 226 178	**422**	100 37 32 26 18 18 14 14		28	26	2	2	–	–	–	–	–	–	–	–	–	**3,6-Dimethoxy-9,10,19,20-tetrahydrotetrabenzo[b,d,l,j][1,6]diazacyclododecine**		**LI08807**
44 140 169 168 167 127 181 170	**422**	100 54 42 42 42 40 38 36	**5.60**	28	30	–	4	–	–	–	–	–	–	–	–	–	**1-Propylamino-8-[(1-propyl-2-perimidinyl)methylamino]naphthalene**	93767-84-1	**NI41899**
69 41 353 57 354 422 133 149	**422**	100 92 72 64 22 15 15 11		28	38	3	–	–	–	–	–	–	–	–	–	–	**2-(3,5-Di-tert-butyl-4-hydroxybenzyl)-4-isopropylphenyl methacrylate**		**MS09521**
117 88 352 422 294 409 322 169	**422**	100 56 4 2 2 1 1 1		28	54	2	–	–	–	–	–	–	–	–	–	–	**Methyl mycolipenate**		**LI08808**
295 171 155 170 310 422 311	**422**	100 79 79 42 26 21 16		29	58	1	–	–	–	–	–	–	–	–	–	–	**Ginnol**		**LI08809**
225 241 240 226 109 180 111 110	**422**	100 32 20 18 13 12 10 8	**4.00**	29	58	1	–	–	–	–	–	–	–	–	–	–	**15-Nonacosanone**	2764-73-0	**MS09522**
422 365 308 424 423 366 421 309	**422**	100 55 37 33 33 17 14 11		30	30	2	–	–	–	–	–	–	–	–	–	–	**1,7-Diamylperixanthenoxanthene**		**IC04369**
422 365 423 308 366 420 309 321	**422**	100 70 32 28 15 13 11 8		30	30	2	–	–	–	–	–	–	–	–	–	–	**1,8-Diamylperixanthenoxanthene**		**IC04370**
56 44 30 422 365 308 309 64	**422**	100 66 61 31 28 24 22 20		30	30	2	–	–	–	–	–	–	–	–	–	–	**5,11-Diamylperixanthenoxanthene**		**MS09523**
155 127 170 156 55 41 422 141	**422**	100 57 40 13 6 6 5 5		30	30	2	–	–	–	–	–	–	–	–	–	–	**1,10-Di-(1′-naphthyl)-1,10-decanedione**		**AI02358**
136 121 203 119 187 145 137 131	**422**	100 60 29 28 27 27 17 17	**6.52**	30	46	1	–	–	–	–	–	–	–	–	–	–	**Lupa-12,20(29)-dien-3-one**	55887-95-1	**NI41900**
422 218 255 133 269 203 407 392	**422**	100 72 33 27 18 18 12 6		30	46	1	–	–	–	–	–	–	–	–	–	–	**Ursan-9(11),12-dien-3-one**		**NI41901**
57 43 71 85 41 55 56 69	**422**	100 83 67 32 32 30 22 19	**0.00**	30	62	–	–	–	–	–	–	–	–	–	–	–	**2,6,10,15,19,23-Hexamethyltetracosane**	111-01-3	**WI01541**
57 43 71 85 41 55 56 69	**422**	100 80 65 30 30 29 22 19	**0.01**	30	62	–	–	–	–	–	–	–	–	–	–	–	**2,6,10,15,19,23-Hexamethyltetracosane**	111-01-3	**AI02359**
57 71 43 85 55 113 41 99	**422**	100 99 97 91 45 44 43 41	**0.00**	30	62	–	–	–	–	–	–	–	–	–	–	–	**2,6,10,15,19,23-Hexamethyltetracosane**	111-01-3	**MS09524**
393 394 392 365 364 422 407 379	**422**	100 62 55 22 12 11 11 10		30	62	–	–	–	–	–	–	–	–	–	–	–	**Nonacosane, 3-methyl-**	14167-67-0	**NI41902**
43 57 71 41 55 85 69 56	**422**	100 91 50 46 41 34 26 21	**0.59**	30	62	–	–	–	–	–	–	–	–	–	–	–	**9-Octyldocosane**		**AI02360**
57 71 422 43 85 55 69 423	**422**	100 82 81 79 72 34 30 27		30	62	–	–	–	–	–	–	–	–	–	–	–	**Triacontane**	638-68-6	**NI41903**
43 57 71 85 41 55 69 56	**422**	100 99 58 40 37 32 17 15	**0.85**	30	62	–	–	–	–	–	–	–	–	–	–	–	**Triacontane**	638-68-6	**AI02361**
67 81 135 95 284 285 79 69	**422**	100 96 90 88 77 71 46 40	**5.17**	31	50	–	–	–	–	–	–	–	–	–	–	–	**13-(1′-Decahydronaphthyl)-perhydrodibenzo[a,i]fluorene**		**AI02362**
67 135 81 95 284 41 285 55	**422**	100 99 96 85 82 82 75 67	**6.88**	31	50	–	–	–	–	–	–	–	–	–	–	–	**13-(1′-Decahydronaphthyl)-perhydrodibenzo[a,i]fluorene**		**AI02363**
141 422 142 115 155 154 153 128	**422**	100 71 15 6 4 4 4 3		32	38	–	–	–	–	–	–	–	–	–	–	–	**Naphthalene, 1,1′-(1,12-dodecanediyl)bis-**	38412-20-3	**NI41904**
410 408 65 412 292 290 368 146	**423**	100 78 56 48 27 21 21 19	**12.30**	11	16	–	3	–	–	4	–	–	–	3	–	–	**1-Methyl-1-(p-tert-butylphenyl)tetrachlorocyclotriphosphazene**	87048-84-8	**NI41905**
134 125 152 137 127 169 192 113	**423**	100 42 22 20 20 18 17 13	**0.00**	16	11	–	3	2	–	1	1	–	–	–	–	–	**2-(4-Chlorophenyl)amino-5-phenylthiazolo[2,3-b]-1,3,4-thiadiazol-4-iumbromide**		**MS09525**
192 148 160 89 350 352 146 147	**423**	100 43 24 20 16 15 15 14	**0.00**	17	20	5	1	–	–	3	–	–	–	–	–	–	**2,2,2-Trichloro-1,1-dimethylethyl 6-methoxy-3-[(methoxycarbonyl)methyl]-2,3-dihydro-1H-indole-1-carboxylate**		**DD01642**
297 73 298 299 75 45 112 74	**423**	100 88 23 10 10 9 8 7	**2.60**	17	28	4	1	–	3	–	–	–	2	–	–	–	**N-(Trifluoroacetyl)-O,O′-bis(trimethylsilyl)normetanephrine**		**MS09526**
217 139 183 63 107 185 152 423	**423**	100 37 11 9 7 3 3 1		18	12	3	1	1	–	2	–	–	–	1	–	–	**O-(2,6-Dichloro-4-nitrophenyl)-diphenylphosphinothionate**		**MS09527**
136 335 246 135 245 177 97 180	**423**	100 92 53 40 35 32 30 26	**4.00**	18	25	5	5	1	–	–	–	–	–	–	–	–	**D-Xylitol, 1-C-(6-amino-9H-purin-9-yl)-2,5-anhydro-1-S-isobutyl-1-thio-, 3,4-diacetate**	40031-70-7	**NI41906**
187 423 189 425 172 140 43 141	**423**	100 42 32 25 24 15 15 11		19	19	4	3	–	–	2	–	–	–	–	–	–	**4′-Chloro-2-(4-chloro-2-methylphenylazo)-2′,5′-dimethoxyacetoacetanilide**		**IC04371**
194 195 179 181 215 213 425 423	**423**	100 23 21 16 14 14 12 12		19	22	5	1	–	–	–	1	–	–	–	–	–	**N-(2′,3′,4′-Trimethoxyphenethyl)-2-bromo-4-methoxybenzamide**		**DD01643**

MASS TO CHARGE RATIOS								M.W.	INTENSITIES								Parent	C	H	O	N	S	F	Cl	Br	I	Si	P	B	X	COMPOUND NAME	CAS Reg No	No
324	266	264	169	364	237	423	292	**423**	100	40	36	36	35	22	18	18		23	25	5	3	–	–	–	–	–	–	–	–	–	α-Cyano-1α-ethyl-2-methoxycarbonyl-3-oxo-2,3,5,6,11,11bβ-hexahydro-1H-indolizino[8,7-b]indol-1β-propionic acid, methyl ester		NI41907
222	268	238	379	378	423	394	358	**423**	100	24	13	11	4	2	1	1		24	25	4	1	1	–	–	–	–	–	–	–	–	Morphanthridine, 11-ethoxy-5,6-dihydro-8-methoxy-5-(p-tolylsulphonyl)-	23145-79-1	NI41908
425	423	344	426	424	315	345	314	**423**	100	99	45	30	30	28	20	20		25	14	1	1	–	–	–	1	–	–	–	–	–	6-Bromo-7-phenylindolo[3,2,1-de]phenanthridin-13-one		LI08810
77	423	204	128	390	118	116	422	**423**	100	81	50	40	21	20	15	11		25	21	–	5	1	–	–	–	–	–	–	–	–	7-(Methylthio)-5-phenyl-4-(p-tolylamino)-2-(p-tolyl)imidazo[5,1-f][1,2,4]triazine		MS09528
259	258	84	218	120	177	147	150	**423**	100	94	67	61	50	33	22	18	**0.00**	25	29	5	1	–	–	–	–	–	–	–	–	–	2-Propenoic acid, 3-(4-hydroxyphenyl)-, octahydro-4-(3-hydroxy-4-methoxyphenyl)-2H-quinolizin-2-yl ester, [2α(E),4β,9aβ]-, (±)-	66512-91-2	NI41909
423	132	106	93	119	380			**423**	100	70	64	51	43	37				27	37	3	1	–	–	–	–	–	–	–	–	–	16β-Anilino-17aβ-methyl-D-homoandrost-5-en-3β,17α-diol-17-one		LI08811
124	58	140	125	41	55	43	57	**423**	100	21	19	8	8	7	7	6	**0.00**	27	53	2	1	–	–	–	–	–	–	–	–	–	2,2,6,6-Tetramethyl-4-[(1-oxooctadecyl)oxy]piperidine		DD01644
226	254	58	44	43	255	227	57	**423**	100	94	27	26	20	19	17	16	**0.00**	29	61	–	1	–	–	–	–	–	–	–	–	–	N-Methyl-N-tridecylpentadecylamine		IC04372
117	119	309	277	190	79	311	192	**424**	100	95	88	86	79	69	63	61	**0.00**	4	–	–	–	3	–	8	–	–	–	–	–	–	trans-2,3-Bis[(trichloromethyl)thio]-2,3-dichlorothiirane	91631-90-2	NI41910
230	231	69	212	260	259	114	167	**424**	100	65	35	29	27	19	19	13	**0.00**	4	12	–	4	–	5	–	–	–	–	3	–	1	1,3,5,2,4,6-Triazatriphosphorine, 2,2,4,4,6-pentafluoro-2,2,4,4,6,6-hexahydro-6-[methyl(trimethylstannyl)amino]-	41006-31-9	NI41911
105	79	267	53	39	77	41	91	**424**	100	57	48	48	43	32	31	28	**0.00**	8	12	–	–	–	–	–	4	–	–	–	–	–	Cyclohexane, 1,2-dibromo-4-(1,2-dibromoethyl)-	3322-93-8	NI41912
105	79	185	187	41	107	106	53	**424**	100	96	46	45	36	33	32	31	**0.00**	8	12	–	–	–	–	–	4	–	–	–	–	–	Cyclooctane, 1,2,5,6-tetrabromo-	3194-57-8	NI41913
294	292	296	426	238	279	267	216	**424**	100	49	45	35	29	25	23	20	**3.70**	14	18	5	–	–	–	–	2	–	–	–	–	–	17,18-Dibromo-2,3-benzo-1,4,7,10,13-pentaoxacyclopentadecane		MS09529
43	78	133	106	42	58	77	45	**424**	100	80	66	61	60	56	53	53	**0.00**	14	18	10	1	2	–	–	–	–	–	–	–	–	Sinalbin		MS09530
115	105	103	77	116	180	104	89	**424**	100	48	46	41	21	18	18	13	**0.74**	15	8	4	–	1	–	4	–	–	–	–	–	–	4,5,6,7-Tetrachloro-2a,8a-dihydro-2a-phenyl-2H-thiacyclobuta(b)(1,4)benzodioxin 1,1-dioxide		MS09531
340	91	338	342	249	339	247	65	**424**	100	99	93	68	52	49	48	41	**0.65**	15	12	3	–	–	–	–	–	–	–	–	–	1	Tungsten, benzyltricarbonyl-π-cyclopentadienyl-	38547-50-1	NI41914
340	338	91	342	92	339	336	337	**424**	100	98	77	60	51	47	42	27	**7.91**	15	12	3	–	–	–	–	–	–	–	–	–	1	Tungsten, (p-tolyl)tricarbonyl-π-cyclopentadienyl	82615-26-7	NI41915
73	307	147	409	308	75	74	45	**424**	100	57	46	20	16	14	9	9	**0.00**	15	36	6	–	–	–	–	–	–	4	–	–	–	Propanedioic acid, bis[(trimethylsilyl)oxy]-, bis(trimethylsilyl) ester	42827-95-2	NI41916
56	228	115	172	284	256	312	368	**424**	100	44	42	36	18	16	6	5	**0.00**	16	8	7	–	–	–	–	–	–	–	–	–	2	Heptacarbonyl-μ-[(1-3)-(η^3:2-η)-(1E)-1-phenyl-1-propene-2,3-diyl]diiron	105335-16-8	DD01645
426	266	428	424	120	101	133	75	**424**	100	71	58	50	43	37	32	30		16	10	–	–	2	–	–	2	–	–	–	–	–	1,4-Dithiin, 2,5-bis(4-bromophenyl)-	37989-50-7	NI41917
73	173	247	235	131	121	115	179	**424**	100	34	24	10	5	5	5	3	**1.46**	16	32	2	–	–	–	–	–	–	4	–	–	1	Iron, dicarbonyl(η^5-2,4-cyclopentadien-1-yl)(nonamethyltetrasilanyl)-	56784-38-4	MS09532
86	147	60	146	55	69	130	103	**424**	100	44	40	38	37	35	33	33	**23.07**	16	32	5	4	2	–	–	–	–	–	–	–	–	1,4,7,10,18-Pentaoxa-13,15,21,23-tetraazacyclopentacosane-14,22-dithione		NI41918
73	75	55	129	117	217	45	74	**424**	100	52	24	16	14	14	10	10	**0.00**	16	33	4	–	–	–	–	1	–	2	–	–	–	2-Bromosebacic acid, bis(trimethylsilyl) ester		NI41919
73	147	75	41	292	205	45	117	**424**	100	32	19	19	17	11	10	8	**0.00**	16	40	5	–	–	–	–	–	–	4	–	–	–	2,3,4-Trihydroxybutanoic acid tetra-TMS		MS09533
73	147	292	205	220	117	103	217	**424**	100	39	32	27	20	14	13	10	**0.00**	16	40	5	–	–	–	–	–	–	4	–	–	–	2,3,4-Trihydroxybutanoic acid, tetra-TMS		LI08812
43	135	45	159	75	139	44	47	**424**	100	99	99	58	47	43	43	38	**3.00**	17	28	8	–	2	–	–	–	–	–	–	–	–	D-Arabinose, diethyl mercaptal, tetraacetate	5329-45-3	NI41920
43	45	135	159	75	139	44	47	**424**	100	90	38	25	20	19	18	16	**1.00**	17	28	8	–	2	–	–	–	–	–	–	–	–	D-Arabinose, diethyl mercaptal, tetraacetate	5329-45-3	NI41921
73	424	147	409	75	116	425	100	**424**	100	65	65	42	35	27	25	16		17	32	3	4	–	–	–	–	–	3	–	–	–	Pteridine, 6-ethyl-2,4,7-tris(trimethylsiloxy)-	31053-51-7	LI08813
73	147	424	409	75	116	425	100	**424**	100	59	58	37	32	24	21	15		17	32	3	4	–	–	–	–	–	3	–	–	–	Pteridine, 6-ethyl-2,4,7-tris(trimethylsiloxy)-	31053-51-7	NI41922
159	73	191	147	160	89	217	45	**424**	100	52	17	17	13	9	8	8	**0.00**	17	40	6	–	–	–	–	–	–	3	–	–	–	Galactofuranose, 2,6-di-O-methyl-1,3,5-tris-O-(trimethylsilyl)-	55400-18-5	NI41923
146	73	191	147	89	159	75	101	**424**	100	46	35	15	14	11	11	10	**0.00**	17	40	6	–	–	–	–	–	–	3	–	–	–	Galactopyranose, 2,4-di-O-methyl-1,3,6-tris-O-(trimethylsilyl)-	55449-69-9	NI41924
146	73	191	159	147	45	131	89	**424**	100	54	46	30	24	14	12	12	**0.00**	17	40	6	–	–	–	–	–	–	3	–	–	–	Galactopyranose, 2,6-di-O-methyl-1,3,4-tris-O-(trimethylsilyl)-	55400-22-1	NI41925
133	146	73	159	147	89	131	71	**424**	100	80	69	60	19	18	13	10	**0.00**	17	40	6	–	–	–	–	–	–	3	–	–	–	Galactopyranose, 3,4-di-O-methyl-1,2,6-tris-O-(trimethylsilyl)-	55400-20-9	NI41926
133	88	73	159	89	147	117	134	**424**	100	83	69	40	17	14	11	10	**0.00**	17	40	6	–	–	–	–	–	–	3	–	–	–	Glucopyranose, 2,3-di-O-methyl-1,4,6-tris-O-(trimethylsilyl)-	55400-17-4	NI41927
146	73	191	147	89	131	75	159	**424**	100	41	34	17	13	11	10	9	**0.00**	17	40	6	–	–	–	–	–	–	3	–	–	–	Glucopyranose, 2,4-di-O-methyl-1,3,6-tris-O-(trimethylsilyl)-	55400-19-6	NI41928
146	217	73	133	147	89	218	45	**424**	100	80	71	64	21	19	16	16	**0.00**	17	40	6	–	–	–	–	–	–	3	–	–	–	Glucopyranose, 3,6-di-O-methyl-1,2,4-tris-O-(trimethylsilyl)-	55400-21-0	NI41929
204	73	191	205	147	159	146	45	**424**	100	58	35	22	14	12	12	12	**0.00**	17	40	6	–	–	–	–	–	–	3	–	–	–	Glucopyranose, 4,6-di-O-methyl-1,2,3-tris-O-(trimethylsilyl)-	55400-23-2	NI41930
73	204	133	89	146	147	205	117	**424**	100	65	34	23	18	15	14	14	**0.00**	17	40	6	–	–	–	–	–	–	3	–	–	–	α-D-Glucopyranoside, methyl 4-O-methyl-2,3,6-tris-O-(trimethylsilyl)-	52430-40-7	NI41931
204	73	133	205	159	146	89	147	**424**	100	68	42	23	21	17	16	13	**0.00**	17	40	6	–	–	–	–	–	–	3	–	–	–	β-D-Glucopyranoside, methyl 4-O-methyl-2,3,6-tris-O-(trimethylsilyl)-	32388-33-3	NI41932
103	73	219	147	129	104	220	116	**424**	100	89	40	18	10	10	9	8	**0.00**	17	44	4	–	–	–	–	–	–	4	–	–	–	D-erythro-Pentitol, 2-deoxy-1,3,4,5-tetrakis-O-(trimethylsilyl)-	33648-72-5	NI41933
73	147	101	116	75	217	133	103	**424**	100	62	60	59	39	33	30	23	**0.00**	17	44	4	–	–	–	–	–	–	4	–	–	–	erythro-Pentitol, 2-deoxy-1,3,4,5-tetrakis-O-(trimethylsilyl)-	56271-71-7	NI41934
73	103	219	147	205	307	321	319	**424**	100	70	60	53	49	47	20	13	**0.00**	17	44	4	–	–	–	–	–	–	4	–	–	–	erythro-Pentitol, 2-deoxy-1,3,4,5-tetrakis-O-(trimethylsilyl)-	56271-71-7	MS09534
60	44	170	319	214	63	126	246	**424**	100	99	79	31	28	26	25	21	**0.00**	18	16	4	–	4	–	–	–	–	–	–	–	–	DL-Bis(1-carboxyethyl) 2,6-naphthylenebis(dithioic acid)	110699-23-5	DD01646
73	55	57	75	81	95	71	127	**424**	100	47	35	29	28	24	21	16	**0.00**	18	27	3	–	–	–	3	–	–	1	–	–	–	(E)-3-[1-Methyl-4-[1-oxo-3-(trimethylsilyl)-2-propenyl]-3-cyclohexenyl]propyl trichloroacetate		DD01647
126	82	275	426	424	379	377	277	**424**	100	4	3	1	1	1	1	1		18	33	5	2	1	–	1	–	–	–	–	–	–	L-threo-α-D-Galacto-octopyranoside, methyl 7-chloro-6,7,8-trideoxy-6-[[(1-methyl-4-propyl-2-pyrrolidinyl)carbonyl]amino]-1-thio-, (2S-trans)-	18323-44-9	NI41935

MASS TO CHARGE RATIOS								M.W.	INTENSITIES								Parent	C	H	O	N	S	F	Cl	Br	I	Si	P	B	X	COMPOUND NAME	CAS Reg No	No
73	204	133	147	116	221	45	75	**424**	100	99	97	88	67	62	61	60	**0.00**	19	33	6	–	–	–	–	–	–	2	–	1	–	α-D-Galactopyranoside, methyl 2,3-bis-O-(trimethylsilyl)-, cyclic phenylboronate	56211-20-2	NI41936
84	83	43	128	42	75	85	39	**424**	100	75	66	59	48	31	28	25	**0.00**	19	33	6	–	–	–	–	–	–	2	–	1	–	α-D-Galactopyranoside, methyl 2,3-bis-O-(trimethylsilyl)-, cyclic phenylboronate	56211-20-2	NI41937
75	44	59	45	43	47	73	77	**424**	100	54	19	19	19	18	17	13	**0.00**	19	33	6	–	–	–	–	–	–	2	–	1	–	α-D-Galactopyranoside, methyl 2,6-bis-O-(trimethylsilyl)-, cyclic phenylboronate	56211-16-6	NI41938
75	44	73	143	110	45	47	147	**424**	100	71	63	48	32	24	22	17	**0.00**	19	33	6	–	–	–	–	–	–	2	–	1	–	α-D-Galactopyranoside, methyl 2,6-bis-O-(trimethylsilyl)-, cyclic phenylboronate	56211-16-6	NI41939
204	73	147	205	245	75	45	133	**424**	100	92	54	43	39	35	33	31	**0.00**	19	33	6	–	–	–	–	–	–	2	–	1	–	α-D-Glucopyranoside, methyl 2,3-bis-O-(trimethylsilyl)-, cyclic phenylboronate	56211-21-3	NI41940
202	218	146	320	92	192	147	207	**424**	100	81	46	44	44	27	26	17	**0.59**	19	36	3	2	–	–	–	–	–	3	–	–	–	Benzenebutanoic acid, γ-oxo-α,2-bis[(trimethylsilyl)amino]-, trimethylsilyl ester	72088-03-0	NI41941
73	192	307	75	218	45	425	147	**424**	100	45	33	24	20	18	11	11	**0.00**	19	36	3	2	–	–	–	–	–	3	–	–	–	Benzenebutanoic acid, γ-oxo-α,2-bis[(trimethylsilyl)amino]-, trimethylsilyl ester	72088-03-0	MS09535
424	77	78	283	141	425	51	266	**424**	100	75	28	26	26	24	20	17		20	16	5	4	1	–	–	–	–	–	–	–	–	O-Phenylsulphonyl-4-((1,2,3,6-tetrahydro-2,6-dioxo-4-methyl-5-cyanopyridin-3-ylidene)hydrazino)phenol		IC04373
424	241	66	359	213	128	65		**424**	100	72	24	16	14	10	6			20	18	–	–	–	–	2	–	–	–	–	–	2	Chlorotitanocene		LI08814
91	43	220	65	39	119	41	92	**424**	100	46	38	30	30	28	28	16	**0.00**	20	20	3	6	1	–	–	–	–	–	–	–	–	5,5'-Bis(1-methyl-3-p-tolyl-4-hydroxy-1,2,3-triazolium)sulphoxide		LI08815
169	43	109	331	127	94	81	170	**424**	100	99	68	43	38	25	14	11	**0.00**	20	24	10	–	–	–	–	–	–	–	–	–	–	Phenyl 2,3,4,6-tetra-O-acetyl-β-D-glucopyranoside	4468-72-8	NI41942
43	109	181	41	137	138	97	98	**424**	100	71	59	48	44	38	31	30	**0.93**	22	32	8	–	–	–	–	–	–	–	–	–	–	12-O-Acetyl-5,6-dihydro-ingol-5,6-epoxide		NI41943
165	246	205	159	121	153	152	248	**424**	100	91	85	83	75	73	67	62	**2.00**	22	32	8	–	–	–	–	–	–	–	–	–	–	Picras-2-ene-1,16-dione, 11,13,14-trihydroxy-2,12-dimethoxy-, (11α,12β)-	30248-06-7	NI41944
44	57	85	121	41	203	105	122	**424**	100	89	58	48	41	40	35	28	**1.00**	22	32	8	–	–	–	–	–	–	–	–	–	–	Trichothec-9-ene-3,4,8,15-tetrol, 12,13-epoxy-, 15-acetate 8-isopropanoate, (3α,4β,8α)-	26934-87-2	NI41945
56	114	315	70	100	144	58	57	**424**	100	90	66	39	37	33	23	23	**22.00**	22	36	6	2	–	–	–	–	–	–	–	–	–	5,6-Benzo-4,7,13,16,21,24-hexaoxa-1,10-diazabicyclo[8.8.8]hexacosane		MS09536
73	367	351	131	75	368	352	339	**424**	100	76	74	25	25	22	22	22	**0.00**	22	40	4	–	–	–	–	–	–	2	–	–	–	4-Hydroxymandelic acid ethyl ester bis(tert-butyldimethylsilyl)-		MS09537
209	334	222	210	179	424	409		**424**	100	16	15	11	7	4	3			22	40	4	–	–	–	–	–	–	2	–	–	–	3-Nonanone, 1-[3-methoxy-4-[(trimethylsilyl)oxy]phenyl]-5-[(trimethylsilyl)oxy]-	56700-96-0	MS09538
91	196	118	210	228	65	209	333	**424**	100	45	35	18	12	12	10	9	**4.00**	23	18	2	2	–	–	2	–	–	–	–	–	–	4,5-Dichloro-N-(dibenzylaminomethyl)phthalimide	96919-15-2	NI41946
337	43	424	336	322	338	323	306	**424**	100	86	42	31	27	26	21	18		23	24	6	2	–	–	–	–	–	–	–	–	–	Cyanocolchicine		NI41947
324	365	263	245	169	424	144	337	**424**	100	91	90	83	64	51	40	38		23	24	6	2	–	–	–	–	–	–	–	–	–	1α-Ethyl-3α-methoxycarbonyl-β,4-dioxo-1,2,3,4,6,7,12,12bβ-octahydro-1,3-cycloindolo[2,3-a]quinolizin-12-propionic acid, methyl ester		NI41948
324	365	263	245	169	424	144	249	**424**	100	95	90	79	64	53	50	33		23	24	6	2	–	–	–	–	–	–	–	–	–	1β-Ethyl-3β-methoxycarbonyl-β,4-dioxo-1,2,3,4,6,7,12,12bβ-octahydro-1,3-cycloindolo[2,3-a]quinolizin-12-propionic acid, methyl ester		NI41949
69	44	137	43	59	364	77	424	**424**	100	95	92	80	55	41	41	39		23	27	4	–	–	3	–	–	–	–	–	–	–	Estra-1,3,5(10)-triene-3,17-diol, 1-methyl-, 17-acetate 3-(trifluoroacetate), (17β)-	56588-09-1	NI41950
336	276	262	131	159	168	258	147	**424**	100	90	31	26	24	18	17	16	**2.00**	23	33	6	–	–	1	–	–	–	–	–	–	–	Androstan-6-one, 17,19-bis(acetyloxy)-3-fluoro-5-hydroxy-, (3β,5α,17α)-	55401-67-7	NI41951
43	424	91	93	79	81	133	41	**424**	100	99	33	29	29	28	26	26		23	33	6	–	–	1	–	–	–	–	–	–	–	Androstan-6-one, 17,19-bis(acetyloxy)-3-fluoro-5-hydroxy-, (3β,5α,17β)-	33547-31-8	NI41952
424	133	336	276	131	157	129	273	**424**	100	26	21	20	18	16	15	13		23	33	6	–	–	1	–	–	–	–	–	–	–	Androstan-6-one, 17,19-bis(acetyloxy)-3-fluoro-5-hydroxy-, (3β,5α,17β)-		LI08816
424	425	133	334	274	131	157	129	**424**	100	26	26	21	20	18	16	15		23	33	6	–	–	1	–	–	–	–	–	–	–	Androstan-6-one, 17,19-bis(acetyloxy)-3-fluoro-5-hydroxy-, (3β,5β,17α)-	55401-66-6	NI41953
43	346	276	93	133	262	79	81	**424**	100	99	90	40	35	31	31	30	**2.30**	23	33	6	–	–	1	–	–	–	–	–	–	–	Androstan-6-one, 17,19-bis(acetyloxy)-3-fluoro-5-hydroxy-, (3β,5β,17β)-	33547-32-9	NI41954
336	276	262	131	159	168	258	326	**424**	100	90	31	26	24	18	17	16	**2.00**	23	33	6	–	–	1	–	–	–	–	–	–	–	Androstan-6-one, 17,19-bis(acetyloxy)-3-fluoro-5-hydroxy-, (3β,5β,17β)-		LI08817
146	348	259	119	91	175	366	424	**424**	100	70	69	60	55	37	32	25		23	36	3	–	2	–	–	–	–	–	–	–	–	(±)-2,4,6-Trideoxy-2,4,6-trimethyl-3,5-O-(1-methylethylidene)-7-O-(phenylmethyl)-D-glycero-D-talo-heptose cyclic 1,3-propanediyl mercaptal		NI41955
214	114	311	91	128	269	155	140	**424**	100	68	48	32	28	18	15	12	**4.00**	23	40	3	2	1	–	–	–	–	–	–	–	–	Heptanamide, N-hexyl-N-[3-[[(4-methylphenyl)sulphonyl]amino]propyl]-	67138-92-5	NI41956
85	280	282	366	281	249	221	75	**424**	100	60	28	24	22	12	9	8	**1.20**	23	40	5	–	–	–	–	–	–	1	–	–	–	1H-Indene-5-propanoic acid, 2,3,3a,6,7,7a-hexahydro-β,7a-dimethyl-1-[(tetrahydro-2H-pyran-2-yl)oxy]-3a-[(trimethylsilyl)oxy]-, methyl ester	67884-43-9	NI41957
91	78	104	28	103	51	106	77	**424**	100	84	71	61	42	40	36	35	**14.48**	24	24	–	–	–	–	–	–	–	–	–	–	2	Iron, tris(1,3,5,7-cyclootatetraene)di-	61279-69-4	NI41958
424	425	382	354	383	426	326	367	**424**	100	28	22	6	5	5	4	2		24	28	5	2	–	–	–	–	–	–	–	–	–	3-Acetyl-1-(3,4-dimethoxyphenyl)-5-ethyl-4,5-dihydro-7,8-dimethoxy-4-methylene-3H-2,3-benzodiazepine	54080-15-8	NI41959
424	341	326	340	425	342	381	409	**424**	100	86	40	38	27	19	17	14		24	28	5	2	–	–	–	–	–	–	–	–	–	3-Acetyl-1-(3,4-dimethoxyphenyl)-5-ethyl-7,8-dimethoxy-4-methyl-3H-2,3 benzodiazepine	90140-50-4	NI41960
238	392	364	305	226	180	424	167	**424**	100	65	51	44	36	32	27	26		24	28	5	2	–	–	–	–	–	–	–	–	–	4-N-Acetyl-17-methoxytalbotine		LI08818
424	78	351	365	121	77	425	79	**424**	100	68	63	43	42	38	29	29		24	28	5	2	–	–	–	–	–	–	–	–	–	Akuammilan-16-carboxylic acid, 17-(acetyloxy)-10-methoxy-, methyl ester, (16R)-	38734-62-2	NI41961

MASS TO CHARGE RATIOS								M.W.	INTENSITIES								Parent	C	H	O	N	S	F	Cl	Br	I	Si	P	B	X	COMPOUND NAME	CAS Reg No	No
424	365	351	121	69	198	122	60	**424**	100	66	62	56	52	44	38	38		24	28	5	2	–	–	–	–	–	–	–	–	–	**1,16-Cyclocorynan-16-carboxylic acid, 17-(acetyloxy)-19,20-didehydro-10 methoxy-, methyl ester, (16ξ,19E)-**	55724-49-7	**NI41962**
305	364	424	180	183	181	169	263	**424**	100	69	44	33	29	19	18	17		24	28	5	2	–	–	–	–	–	–	–	–	–	**1H-Indolo[3,2,1-de][1,5]naphthyridine-6-carboxylic acid, 3-acetyl-5-[1-[(acetyloxy)methyl]-1-propenyl]-2,3,3a,4,5,6-hexahydro-, methyl ester, [3aS-[3aα,5β(E),6α]]-**	30809-34-8	**NI41963**
424	174	425	222	162	321	365	132	**424**	100	35	27	23	20	19	18	15		24	28	5	2	–	–	–	–	–	–	–	–	–	**11-Methoxyhenningsamin**		**LI08819**
409	424	107	410	198	425	244	111	**424**	100	88	28	27	26	25	24	24		24	28	5	2	–	–	–	–	–	–	–	–	–	**Obscurinervan-21-one, 6,7-didehydro-15,16-dimethoxy-22-methyl-, (22α)-**	7168-67-4	**NI41965**
69	71	409	424	85	83	81	60	**424**	100	71	54	47	42	37	35	28		24	28	5	2	–	–	–	–	–	–	–	–	–	**Obscurinervan-21-one, 6,7-didehydro-15,16-dimethoxy-22-methyl-, (22α)-**	7168-67-4	**NI41964**
238	392	364	305	266	226	292	180	**424**	100	65	51	44	36	36	34	32	**27.00**	24	28	5	2	–	–	–	–	–	–	–	–	–	**Talbotine, 4-acetyl-O-methyl-**	30809-18-8	**NI41966**
365	172	279	366	424	160	159	142	**424**	100	70	50	30	22	20	18	15		24	28	5	2	–	–	–	–	–	–	–	–	–	**Voachalotine oxindole, acetate (ester)**	26144-10-5	**NI41967**
86	100	87	101	72	58	396	424	**424**	100	92	14	13	9	9	6	4		25	36	2	4	–	–	–	–	–	–	–	–	–	**9-Hydrazono-2,7-bis[2-(diethylamino)ethoxy]fluorene**		**MS09539**
212	77	424	360	109	51	213	425	**424**	100	57	56	36	33	20	17	16		26	20	–	2	2	–	–	–	–	–	–	–	–	**Tetraphenyldithiooxamide**		**MS09540**
395	424	396	425	194	353	382	224	**424**	100	52	23	22	19	17	16	14		26	28	2	2	–	–	–	–	–	–	–	–	1	**Magnesium, bis(2-butyl-8-quinolinolato)-**	30049-13-9	**NI41968**
44	328	137	43	161	150	41	148	**424**	100	41	38	29	23	23	17	16	**0.00**	26	32	5	–	–	–	–	–	–	–	–	–	–	**Deacetyldeoxygedunin**		**MS09541**
173	101	115	57	157	231	223	409	**424**	100	27	24	20	11	7	6	5	**0.40**	26	48	4	–	–	–	–	–	–	–	–	–	–	**1-O-(2-Butoxy-4-hexadecynyl)-2,3-O-isopropylideneglycerol**		**NI41969**
332	165	333	254	77	255	166	126	**424**	100	33	28	20	14	13	12	11	**0.00**	27	20	3	–	1	–	–	–	–	–	–	–	–	**3,3,4,4-Tetraphenyl-1,2-oxathiolan-5-one 2-oxide**		**MS09542**
104	221	203	105	188	222	220	77	**424**	100	82	75	29	28	25	24	24	**0.00**	27	24	3	2	–	–	–	–	–	–	–	–	–	**endo-1,2,3,4-Tetrahydro-N-(4-methoxyphenyl)-9-(β-phenylethyl)-1,4-iminonaphthalene-2,3-dicarboximide**	123542-31-4	**DD01648**
219	58	175	150	84	204	91	424	**424**	100	77	67	60	60	57	57	53		27	36	4	–	–	–	–	–	–	–	–	–	–	**α-β-Tocopherol spiro dimer**		**LI08820**
409	410	424	411	425	412	395	309	**424**	100	33	20	9	7	2	2	2		27	40	2	–	–	–	–	–	–	1	–	–	–	**Cannabinol trimethylsilyl ether**		**NI41970**
346	271	409	137	347	331	272	75	**424**	100	63	46	45	29	23	23	21	**0.00**	27	40	2	–	–	–	–	–	–	1	–	–	–	**3α-Phenyldimethylsilyloxy-5α-androstan-17-one**		**NI41971**
124	382	424	341	107	383	381	173	**424**	100	91	56	35	28	25	25	23		27	41	3	–	–	–	–	–	–	–	–	1	–	**Pregn-4-en-3-one, 20,21-[(cyclohexylborylene)bis(oxy)]-, (20R)-**	30882-68-9	**NI41972**
200	143	55	43	57	83	71	185	**424**	100	83	78	62	61	59	58	57	**2.61**	27	52	3	–	–	–	–	–	–	–	–	–	–	**Hexacosanoic acid, 9-oxo-, methyl ester**	55955-49-2	**NI41973**
144	55	43	57	185	31	74	41	**424**	100	53	51	40	38	33	32	32	**7.50**	27	52	3	–	–	–	–	–	–	–	–	–	–	**2H-Pyran-2-butanoic acid, 6-heptadecyltetrahydro-, methyl ester**	55955-48-1	**NI41974**
105	319	424	216	200	248	232	353	**424**	100	40	20	20	20	10	10	8		28	28	2	2	–	–	–	–	–	–	–	–	–	**1-(Dibenzoylamino)-3-(diethylamino)-2-methylindene**	123775-95-1	**DD01649**
105	324	202	334	306	246	152	151	**424**	100	60	22	20	12	10	10	10	**0.00**	28	28	2	2	–	–	–	–	–	–	–	–	–	**endo-6-(Diethylcarbamoyl)-exo-6-methyl-1,3,5-triphenyl-2-oxa-4-azabicyclo[3.1.0]hex-3-ene**	123775-96-2	**DD01650**
173	43	57	55	83	69	97	71	**424**	100	46	45	32	29	28	27	27	**12.54**	28	56	2	–	–	–	–	–	–	–	–	–	–	**Decanoic acid, octadecyl ester**	34689-06-0	**NI41975**
201	57	43	83	55	97	69	71	**424**	100	52	44	36	36	32	32	30	**6.84**	28	56	2	–	–	–	–	–	–	–	–	–	–	**Dodecanoic acid, hexadecyl ester**	20834-06-4	**NI41976**
74	43	87	57	41	75	55	69	**424**	100	87	76	56	40	38	36	27	**7.00**	28	56	2	–	–	–	–	–	–	–	–	–	–	**Heptacosanoic acid, methyl ester**	55682-91-2	**NI41977**
88	43	57	55	41	101	69	424	**424**	100	68	47	42	40	31	24	20		28	56	2	–	–	–	–	–	–	–	–	–	–	**Methyl 2-methylhexacosanoate**		**MS09543**
74	43	55	57	41	75	87	69	**424**	100	69	48	46	43	35	34	33	**23.00**	28	56	2	–	–	–	–	–	–	–	–	–	–	**Methyl 5-methylhexacosanoate**		**MS09544**
101	88	57	55	69	424	83	71	**424**	100	70	18	12	11	10	10	8		28	56	2	–	–	–	–	–	–	–	–	–	–	**Methyl 2,4,6-trimethyltetracosanoate**		**MS09545**
285	57	43	140	55	71	83	69	**424**	100	93	89	87	56	52	44	44	**35.24**	28	56	2	–	–	–	–	–	–	–	–	–	–	**Octadecanoic acid, decyl ester**	32509-55-0	**NI41978**
88	18	57	43	55	69	71	41	**424**	100	82	74	64	58	43	30	30	**0.00**	28	56	2	–	–	–	–	–	–	–	–	–	–	**Pentacosanoic acid, 2,10-dimethyl-, methyl ester, [S-(R*,S*)]-**	57289-53-9	**NI41979**
229	57	43	55	71	83	69	97	**424**	100	85	75	53	49	48	47	42	**5.80**	28	56	2	–	–	–	–	–	–	–	–	–	–	**Tetradecanoic acid, tetradecyl ester**	3234-85-3	**NI41980**
57	43	74	424	18	71	55	87	**424**	100	80	72	67	65	61	52	49		28	56	2	–	–	–	–	–	–	–	–	–	–	**Tricosanoic acid, 10,14,18,22-tetramethyl-, methyl ester**	55429-72-6	**MS09546**
57	43	74	424	71	55	87	69	**424**	100	80	72	67	61	52	49	42		28	56	2	–	–	–	–	–	–	–	–	–	–	**Tricosanoic acid, 10,14,18,22-tetramethyl-, methyl ester**	55429-72-6	**NI41981**
424	409	327	325	213	301	231	314	**424**	100	21	19	17	17	15	15	10		29	44	2	–	–	–	–	–	–	–	–	–	–	**Androstan-5-one, 3-hydroxy-16-(2,2,4,4-tetramethylcyclopentyl)methylene, (3β)-**		**LI08821**
424	409	57	425	410	219	197	367	**424**	100	91	48	31	28	25	24	20		29	44	2	–	–	–	–	–	–	–	–	–	–	**Bis(3,5-di-tert-butyl-4-hydroxyphenyl)methane**	10077-20-0	**NI41982**
69	40	42	55	143	364	158	145	**424**	100	90	68	57	53	52	44	33	**3.47**	29	44	2	–	–	–	–	–	–	–	–	–	–	**Cholesta-5,7,24-trien-3-ol, acetate, (3β)-**	17137-77-8	**NI41983**
295	95	69	43	81	41	296	133	**424**	100	31	25	24	23	23	21	21	**7.00**	29	44	2	–	–	–	–	–	–	–	–	–	–	**Cholesta-5,17(20),24-trien-3-ol, acetate, (3β)-**	56362-41-5	**NI41984**
364	282	93	43	109	253	121	107	**424**	100	66	55	51	45	42	37	37	**4.00**	29	44	2	–	–	–	–	–	–	–	–	–	–	**Cholesta-5,20(22),24-trien-3-ol, acetate, (3β)-**	41083-99-2	**NI41985**
364	253	69	109	93	81	43	41	**424**	100	62	55	54	48	48	45	44	**1.00**	29	44	2	–	–	–	–	–	–	–	–	–	–	**Cholesta-5,20,24-trien-3-ol, acetate, (3β)-**	41083-98-1	**NI41986**
141	158	43	57	140	71	115	129	**424**	100	53	25	21	12	8	6	4	**3.00**	29	44	2	–	–	–	–	–	–	–	–	–	–	**1-Naphthalenemethyl stearate**	71843-68-0	**NI41987**
424	409	425	410	367	219	197	129	**424**	100	97	32	31	22	19	17	6		29	44	2	–	–	–	–	–	–	–	–	–	–	**Phenol, 4,4′-methylenebis[2,6-di-tert-butyl-**	118-82-1	**WI01542**
424	409	57	425	197	219	43	410	**424**	100	41	34	31	17	14	13	12		29	44	2	–	–	–	–	–	–	–	–	–	–	**Phenol, 4,4′-methylenebis[2,6-di-tert-butyl-**		**MS09547**
424	108	110	109	149	425	107	97	**424**	100	79	69	30	24	23	23	18		29	48	–	2	–	–	–	–	–	–	–	–	–	**2′H-Cholest-2-eno[3,2-c]pyrazole, 4-methyl-, (5α)-**	56247-65-5	**NI41988**
83	157	97	69	297	111	125	139	**424**	100	69	65	65	36	24	14	12	**1.00**	29	60	1	–	–	–	–	–	–	–	–	–	–	**Nonacosan-10-ol**	504-55-2	**NI41989**
83	97	111	157	297	125	139	406	**424**	100	72	38	30	26	22	14	12	**0.00**	29	60	1	–	–	–	–	–	–	–	–	–	–	**Nonacosan-10-ol**	504-55-2	**MS09548**
43	57	55	97	41	83	69	71	**424**	100	89	69	55	53	48	46	41	**0.00**	29	60	1	–	–	–	–	–	–	–	–	–	–	**Nonacosanol**		**MS09549**
212	141	149	57	213	43	56	41	**424**	100	82	36	25	23	19	18	18	**0.00**	30	36	–	2	–	–	–	–	–	–	–	–	–	**meso-N,N,N′,N′-Tetraethyl-1,2-bis(1-naphthalenyl)ethylenediamine**		**DD01651**
212	213	28	56	351	280	141	58	**424**	100	47	19	16	15	15	14	12	**0.00**	30	36	–	2	–	–	–	–	–	–	–	–	–	**meso-N,N,N′,N′-Tetraethyl-1,2-bis(2-naphthalenyl)ethylenediamine**		**DD01652**
141	212	182	28	196	213	56	115	**424**	100	94	27	26	23	21	21	19	**0.00**	30	36	–	2	–	–	–	–	–	–	–	–	–	**(±)-N,N,N′,N′-Tetraethyl-1,2-bis(1-naphthalenyl)ethylenediamine**		**DD01653**
212	213	28	58	141	129	57	71	**424**	100	17	11	8	6	4	4	3	**0.00**	30	36	–	2	–	–	–	–	–	–	–	–	–	**(±)-N,N,N′,N′-Tetraethyl-1,2-bis(2-naphthalenyl)ethylenediamine**		**DD01654**

MASS TO CHARGE RATIOS								M.W.	INTENSITIES								Parent	C	H	O	N	S	F	Cl	Br	I	Si	P	B	X	COMPOUND NAME	CAS Reg No	No
257	409	424	125	109	137	119	107	**424**	100	93	60	57	43	41	40	38		30	48	1	–	–	–	–	–	–	–	–	–	–	**Aborenone**		**NI41990**
177	204	424	189	133	121	109	136	**424**	100	77	74	71	52	51	47	46		30	48	1	–	–	–	–	–	–	–	–	–	–	**Allobetul-1-ene**		**MS09550**
134	189	424	121	109	205	177	163	**424**	100	99	88	88	76	64	60	53		30	48	1	–	–	–	–	–	–	–	–	–	–	**Allobetul-2-ene**		**MS09551**
150	135	381	149	191	151	424	189	**424**	100	84	83	58	56	32	30	26		30	48	1	–	–	–	–	–	–	–	–	–	–	**16-Oxo-19αH-ψ-taraxene**		**NI41991**
245	205	119	257	424	135	409	218	**424**	100	22	18	17	13	13	11	10		30	48	1	–	–	–	–	–	–	–	–	–	–	**Bauerenone**		**LI08822**
253	81	96	55	145	95	296	41	**424**	100	58	46	44	41	37	36	33	**6.86**	30	48	1	–	–	–	–	–	–	–	–	–	–	**3,5-Cyclo-33-norgorgost-24(28)-ene, 6-methoxy-, (3β,5α,6β)-**	55064-59-0	**NI41992**
55	83	69	81	95	97	43	41	**424**	100	84	65	54	53	51	37	34	**29.00**	30	48	1	–	–	–	–	–	–	–	–	–	–	**Cyclopropa[5,6]stigmast-22-en-3-one, 3′,6-dihydro-, (5β,6α,22E)-**	53755-40-1	**NI41993**
253	369	287	227	213	409	392	285	**424**	100	8	6	6	6	5	5	5	**2.00**	30	48	1	–	–	–	–	–	–	–	–	–	–	**(23R,24R,28S)-25-dehydro-6β-methoxy-3α,5:23,28-dicyclostigmastane**		**MS09552**
253	392	213	227	377	369	287	285	**424**	100	11	10	8	6	6	6	6	**2.00**	30	48	1	–	–	–	–	–	–	–	–	–	–	**(23S,24S,28R)-25-dehydro-6β-methoxy-3α,5:23,28-dicyclostigmastane**		**MS09553**
299	424	203	219	409	216	201	341	**424**	100	85	66	64	49	48	48	47		30	48	1	–	–	–	–	–	–	–	–	–	–	**4α,14α-Dimethyl-24-methylene-9,19-cyclocholestan-3-one**		**LI08823**
218	205	245	257	219	203	123	121	**424**	100	52	34	32	22	22	17	17	**6.42**	30	48	1	–	–	–	–	–	–	–	–	–	–	**D:C-Friedooleanan-3-one**	5945-53-9	**NI41994**
41	55	204	300	69	43	133	285	**424**	100	95	82	67	62	60	56	50	**0.00**	30	48	1	–	–	–	–	–	–	–	–	–	–	**D-Friedoolean-14-en-3-one**	514-07-8	**NI41995**
204	300	133	285	205	121	119	149	**424**	100	83	70	60	55	40	35	29	**0.00**	30	48	1	–	–	–	–	–	–	–	–	–	–	**D-Friedoolean-14-en-3-one**	514-07-8	**NI41996**
218	205	245	257	119	135	424	409	**424**	100	53	36	33	14	12	7	7		30	48	1	–	–	–	–	–	–	–	–	–	–	**D:C-Friedoolean-7-en-one**	5945-53-9	**LI08824**
205	257	245	424	119	135	218	409	**424**	100	60	46	18	18	16	11	7		30	48	1	–	–	–	–	–	–	–	–	–	–	**D:C-Friedoolean-8-en-3-one**	22611-26-3	**LI08825**
205	257	245	206	121	424	135	119	**424**	100	60	45	22	21	16	16	16		30	48	1	–	–	–	–	–	–	–	–	–	–	**D:C-Friedoolean-8-en-3-one**	22611-26-3	**NI41997**
245	109	123	205	107	257	246	121	**424**	100	29	28	22	21	19	19	19	**12.01**	30	48	1	–	–	–	–	–	–	–	–	–	–	**D:C-Friedours-7-en-3-one**	6895-55-2	**NI41998**
245	424	205	257	425	246			**424**	100	80	60	32	28	24				30	48	1	–	–	–	–	–	–	–	–	–	–	**Isobauerenone**		**LI08826**
424	229	409	189	203	205	285	218	**424**	100	83	70	57	46	37	32	28		30	48	1	–	–	–	–	–	–	–	–	–	–	**Lupa-1,20(29)-dien-3β-ol**		**NI41999**
150	95	55	93	149	151	424	203	**424**	100	34	25	19	14	12	11	11		30	48	1	–	–	–	–	–	–	–	–	–	–	**Lup-1-en-3-one, (+)-**	6155-08-4	**NI42000**
205	124	109	424	121	107	123	135	**424**	100	97	81	62	60	60	52	41		30	48	1	–	–	–	–	–	–	–	–	–	–	**Lup-20(29)-en-3-one**	1617-70-5	**NI42001**
189	81	95	424	55	93	205	107	**424**	100	42	36	34	33	28	27	27		30	48	1	–	–	–	–	–	–	–	–	–	–	**A′-Neogammacer-22(29)-en-3-one**	25615-11-6	**NI42002**
189	424	205	105	119	190	409	204	**424**	100	46	33	33	30	26	20	20		30	48	1	–	–	–	–	–	–	–	–	–	–	**A′-Neogammacer-22(29)-en-3-one, (21β)-**	1812-63-1	**NI42003**
299	424	300	203	218	409	328	301	**424**	100	64	49	32	29	27	26	25		30	48	1	–	–	–	–	–	–	–	–	–	–	**31-Norcyclolaudenone**		**LI08827**
203	41	69	55	81	43	95	191	**424**	100	96	79	76	64	60	48	44	**2.10**	30	48	1	–	–	–	–	–	–	–	–	–	–	**Olean-12-en-28-al**	10070-76-5	**NI42004**
424	273	232	425	177	43	95	55	**424**	100	80	53	32	13	12	11	11		30	48	1	–	–	–	–	–	–	–	–	–	–	**Olean-12-en-11-one**	10070-81-2	**NI42005**
274	55	69	41	95	43	259	81	**424**	100	86	78	75	67	61	54	51	**15.70**	30	48	1	–	–	–	–	–	–	–	–	–	–	**Simiarenone**		**NI42006**
424	123	381	97	283	55	43	138	**424**	100	82	69	67	65	65	64	60		30	48	1	–	–	–	–	–	–	–	–	–	–	**Stigmasta-4,22-dien-3-one, 4-methyl-, (22E,24R)-**	56312-54-0	**NI42007**
150	135	149	191	381	424	231	189	**424**	100	99	68	59	50	25	23	22		30	48	1	–	–	–	–	–	–	–	–	–	–	**16-Oxo-ψ-taraxene**		**NI42008**
203	191	133	69	55	192	204	81	**424**	100	45	33	30	29	24	23	23	**1.60**	30	48	1	–	–	–	–	–	–	–	–	–	–	**Urs-12-en-28-al**	13250-38-9	**NI42009**
227	228	157	183	142	199	226	198	**424**	100	26	20	13	13	12	11	9	**4.36**	31	36	1	–	–	–	–	–	–	–	–	–	–	**1,4-Dihydro-5-isopropyl-4-[(7-isopropyl-1-methyl-4-azulenyl)methyl]-3,8-dimethyl-2-azulenecarboxaldehyde**		**MS09554**
227	228	149	157	142	43	199	424	**424**	100	19	18	15	10	10	9	7		31	36	1	–	–	–	–	–	–	–	–	–	–	**3,6-Dihydro-7-isopropyl-6-[(7-isopropyl-1-methyl-4-azulenyl)methyl]-1,4-dimethyl-2-azulenecarboxaldehyde**		**MS09555**
141	406	29	165	279	43	41	267	**424**	100	89	80	75	74	68	63	46	**0.00**	31	36	1	–	–	–	–	–	–	–	–	–	–	**1,1-Di-(1′-naphthyl)-1-undecanol**		**AI02364**
81	95	41	67	137	55	136	135	**424**	100	84	75	74	68	59	46	45	**3.72**	31	52	–	–	–	–	–	–	–	–	–	–	–	**Tri-(1′-decahydronaphthyl)methane**		**AI02365**
347	330	348	270	331	269	252	241	**424**	100	34	29	23	17	10	10	9	**6.51**	32	24	1	–	–	–	–	–	–	–	–	–	–	**9-Anthrol, 9,10-dihydro-9,10,10-triphenyl-**	6636-11-9	**NI42010**
424	105	77	425	165	347	243	319	**424**	100	83	50	26	19	17	17	16		32	24	1	–	–	–	–	–	–	–	–	–	–	**1-(4-Benzoylbenzyl)-4-(4-biphenylyl)benzene**		**IC04374**
347	330	348	270	331	269	252	241	**424**	100	34	30	24	18	10	10	10	**5.00**	32	24	1	–	–	–	–	–	–	–	–	–	–	**10-Hydroxy-9,9,10-triphenyldihydroanthracene**		**LI08828**
160	158	162	81	79	80	82	113	**425**	100	53	50	11	11	10	7	6	**0.00**	–	–	–	1	2	6	–	2	–	–	–	–	1	**Dibromodithionitronium hexafluoroarsenate(V)**	88446-27-9	**NI42011**
425	179	43	28	57	55	41	69	**425**	100	97	85	76	55	40	37	33		10	5	2	1	–	–	–	–	2	–	–	–	–	**N-Phenyldiiodomaleimide**		**MS09556**
262	426	89	230	71	61	85	263	**425**	100	57	37	33	20	19	16	9	**0.00**	11	9	5	1	–	10	–	–	–	–	–	–	–	**Threonine methyl ester, N,O-dipentafluorpropionate**		**NI42012**
93	66	65	39	341	397	92	314	**425**	100	63	59	59	45	30	30	24	**0.27**	14	11	3	1	–	–	–	–	–	–	–	–	1	**π-Cyclopentadienyl-π-2-methylpyridyltricarbonyltungsten**		**MS09557**
139	259	97	152	69	174	115	126	**425**	100	69	65	32	28	23	20	19	**0.00**	18	23	9	3	–	–	–	–	–	–	–	–	–	**Cytidine, N-acetyl-N-methyl-, 2′,3′,5′-triacetate**	55760-29-7	**NI42013**
105	244	43	77	260	218	202	122	**425**	100	50	40	38	22	18	18	12	**0.00**	20	19	6	5	–	–	–	–	–	–	–	–	–	**3-(2-Acetamido-3,4-dihydro-4-oxopteridin-6-yl)-2-acetoxypropyl benzoate**		**DD01655**
170	114	128	171	127	71	227	98	**425**	100	28	23	16	15	8	6	5	**0.10**	20	44	–	1	–	–	–	–	1	–	–	–	–	**Tetra-iso-amylammonium iodide**	5424-26-0	**NI42014**
170	171	71	114	298	227	198	100	**425**	100	18	17	13	12	11	11	3	**0.03**	20	44	–	1	–	–	–	–	1	–	–	–	–	**Tetra-N-amylammonium iodide**	15959-61-2	**NI42015**
297	425	298	56	147	118	275	100	**425**	100	30	20	16	12	10	9	9		22	27	4	5	–	–	–	–	–	–	–	–	–	**N-Ethyl-N-(2-oxooxazolidin-3-ylbutyl)-3-methyl-4-(4-nitrophenylazo)aniline**		**IC04375**
268	143	425	55	70	41	40	269	**425**	100	62	59	55	51	49	48	42		24	31	2	3	1	–	–	–	–	–	–	–	–	**1-Propanone, 1-[10-[3-[4-(2-hydroxyethyl)-1-piperazinyl]propyl]-10H-phenothiazin-2-yl]-**	2622-30-2	**NI42016**
43	84	85	41	57	102	60	42	**425**	100	65	56	34	19	17	17	17	**0.00**	24	43	5	1	–	–	–	–	–	–	–	–	–	**Acetamide, N-[2-(acetyloxy)-1-[(acetyloxy)methyl]-3-heptadecenyl]-, [R-[R*,S*(E)]]-**	2482-37-3	**NI42017**
338	397	276	339	194	216	152	105	**425**	100	47	35	26	26	24	24	22	**2.10**	25	31	5	1	–	–	–	–	–	–	–	–	–	**14-Acetyl-8-benzoylserratinine**	5876-35-7	**NI42018**

MASS TO CHARGE RATIOS								M.W.	INTENSITIES								Parent	C	H	O	N	S	F	Cl	Br	I	Si	P	B	X	COMPOUND NAME	CAS Reg No	No
355	268	100	267	232	162	81	215	**425**	100	79	74	47	46	37	35	33	**24.90**	25	31	5	1	–	–	–	–	–	–	–	–	–	Dibutanoylmorphine	66641-03-0	NI42019
425	177	84	260	258	136	107	259	**425**	100	72	71	70	67	65	63	50		25	31	5	1	–	–	–	–	–	–	–	–	–	trans-2-(p-Hydroxylhydrocinnamoyloxy)-4-(3-hydroxy-4-methoxyphenyl)quinolizidine		LI08829
86	100	87	58	72	56	88	84	**425**	100	55	13	11	7	6	6	5	**3.40**	25	35	3	3	–	–	–	–	–	–	–	–	–	9-Oximo-2,7-bis[2-(diethylamino)ethoxy]fluorene		MS09558
81	41	55	95	302	69	30	67	**425**	100	48	46	35	26	26	25	20	**3.79**	30	51	–	1	–	–	–	–	–	–	–	–	–	Tris(2,7-dimethyl-2,7-octadien-1-yl)amine	88399-36-4	NI42020
243	242	256	244	241	229	228	270	**425**	100	93	44	25	20	10	9	8	**1.30**	31	39	–	1	–	–	–	–	–	–	–	–	–	5-Tetradecylbenzo[i]phenanthridine		NI42021
245	217	219	255	256	243	230	189	**425**	100	26	23	20	7	7	6	6	**1.40**	31	39	–	1	–	–	–	–	–	–	–	–	–	6-Tetradecylbenzo[c]phenanthridine		NI42022
175	69	113	194	163	94	144	219	**426**	100	96	89	41	35	35	33	26	**15.38**	4	–	–	–	–	12	–	–	–	–	–	–	2	Tetrakis(trifluoromethyl)diarsine		MS09559
60	153	121	273	89	350	149	31	**426**	100	87	62	48	40	35	35	31	**23.00**	6	18	–	6	4	–	–	–	–	–	4	–	–	2,4,6,8,9,10-Hexaaza-1,3,5,7-tetraphosphatricyclo[3.3.1.1(3,7)]decane, 2,4,6,8,9,10-hexamethyl-, 1,3,5,7-tetrasulphide	37747-07-2	NI42023
296	230	382	363	197	194	338	275	**426**	100	60	57	21	19	19	15	9	**0.00**	6	18	–	7	–	6	–	–	–	–	4	–	–	Tris(dimethylamino)hexafluorocyclotetraphosphazene		MS09560
41	67	39	55	27	29	43	109	**426**	100	53	52	45	38	30	29	27	**0.83**	8	14	–	–	–	–	–	4	–	–	–	–	–	1,2,7,8-Tetrabromooctane		IC04376
165	59	274	227	211	121	259	135	**426**	100	99	39	30	30	30	25	22	**0.00**	9	27	9	–	–	–	1	–	–	4	–	–	–	Tris(trimethoxysilyl)chlorosilane	55580-30-8	NI42024
130	312	314	340	342	338	368	370	**426**	100	26	24	23	21	19	18	16	**15.00**	11	18	4	2	–	–	–	–	–	–	–	–	1	1-(N,N,N′,N′-Tetramethyl-1,3-diaminopropane)-tetracarbonyltungsten		LI08830
296	426	424	422	227	421	134	265	**426**	100	91	86	52	22	19	16	13		12	–	–	–	–	8	–	–	–	–	–	–	1	Dibenzotellurophene, octafluoro-	16012-86-5	NI42025
426	424	296	422	227	421	420	265	**426**	100	98	85	59	24	20	13	12		12	–	–	–	–	8	–	–	–	–	–	–	1	Dibenzotellurophene, octafluoro-	16012-86-5	NI42026
430	428	179	180	358	432	144	360	**426**	100	87	74	67	63	60	51	47	**31.70**	12	2	–	–	–	–	8	–	–	–	–	–	–	1,1′-Biphenyl, 2,2′,3,3′,4,4′,5,5′-octachloro-	35694-08-7	NI42027
430	428	432	358	360	426	179	356	**426**	100	97	68	65	59	43	39	37		12	2	–	–	–	–	8	–	–	–	–	–	–	1,1′-Biphenyl, 2,2′,3,3′,4,4′,5,5′-octachloro-	35694-08-7	NI42028
430	428	179	358	180	144	432	143	**426**	100	86	81	77	76	71	67	61	**36.06**	12	2	–	–	–	–	8	–	–	–	–	–	–	1,1′-Biphenyl, 2,2′,3,3′,4,4′,5,5′-octachloro-	35694-08-7	NI42029
430	428	358	432	360	179	180	144	**426**	100	86	78	63	56	55	51	44	**31.17**	12	2	–	–	–	–	8	–	–	–	–	–	–	1,1′-Biphenyl, 2,2′,3,3′,4,4′,5,6-octachloro-	52663-78-2	NI42030
430	428	179	180	432	358	144	143	**426**	100	88	75	63	61	56	53	43	**35.70**	12	2	–	–	–	–	8	–	–	–	–	–	–	1,1′-Biphenyl, 2,2′,3,3′,4,4′,5,6′-octachloro-	42740-50-1	NI42031
430	428	432	179	180	358	426	144	**426**	100	92	64	55	51	47	35	34		12	2	–	–	–	–	8	–	–	–	–	–	–	1,1′-Biphenyl, 2,2′,3,3′,4,4′,6,6′-octachloro-	33091-17-7	NI42032
430	428	432	426	434	358	360	431	**426**	100	87	65	34	27	27	20	14		12	2	–	–	–	–	8	–	–	–	–	–	–	1,1′-Biphenyl, 2,2′,3,3′,4,4′,6,6′-octachloro-	33091-17-7	NI42033
430	428	179	358	180	432	144	360	**426**	100	95	73	73	68	66	61	59	**37.15**	12	2	–	–	–	–	8	–	–	–	–	–	–	1,1′-Biphenyl, 2,2′,3,3′,4,5,5′,6-octachloro-	68194-17-2	NI42034
430	428	179	180	432	358	144	143	**426**	100	89	72	68	64	56	44	39	**31.30**	12	2	–	–	–	–	8	–	–	–	–	–	–	1,1′-Biphenyl, 2,2′,3,3′,4,5,5′,6′-octachloro-	52663-75-9	NI42035
430	428	432	179	180	358	360	144	**426**	100	96	66	60	49	46	39	36	**34.20**	12	2	–	–	–	–	8	–	–	–	–	–	–	1,1′-Biphenyl, 2,2′,3,3′,4,5,6,6′-octachloro-	52663-73-7	NI42036
430	428	432	179	180	358	144	426	**426**	100	90	68	62	55	44	37	35		12	2	–	–	–	–	8	–	–	–	–	–	–	1,1′-Biphenyl, 2,2′,3,3′,4,5′,6,6′-octachloro-	40186-71-8	NI42037
430	428	358	179	180	432	360	144	**426**	100	90	80	73	70	63	62	58	**36.73**	12	2	–	–	–	–	8	–	–	–	–	–	–	1,1′-Biphenyl, 2,2′,3,3′,4,5′,6,6′-octachloro-	40186-71-8	NI42038
430	428	179	432	180	358	144	143	**426**	100	93	59	58	51	48	42	34	**31.90**	12	2	–	–	–	–	8	–	–	–	–	–	–	1,1′-Biphenyl, 2,2′,3,3′,5,5′,6,6′-octachloro-	2136-99-4	NI42039
430	428	432	358	426	360	356	179	**426**	100	90	65	51	42	41	30	29		12	2	–	–	–	–	8	–	–	–	–	–	–	1,1′-Biphenyl, 2,2′,3,3′,5,5′,6,6′-octachloro-	2136-99-4	NI42040
430	428	179	144	180	358	432	143	**426**	100	89	79	74	71	65	62	60	**36.44**	12	2	–	–	–	–	8	–	–	–	–	–	–	1,1′-Biphenyl, 2,2′,3,3′,5,5′,6,6′-octachloro-	2136-99-4	NI42041
430	428	432	358	179	144	180	360	**426**	100	94	63	55	51	46	45	43	**35.41**	12	2	–	–	–	–	8	–	–	–	–	–	–	1,1′-Biphenyl, 2,2′,3,4,4′,5,6,6′-octachloro-	74472-52-9	NI42042
347	278	309	297	348	327	247	328	**426**	100	33	14	12	10	10	7	5	**0.40**	13	1	–	–	–	10	–	1	–	–	–	–	–	Benzene, 1,1′-(bromomethylene)bis[2,3,4,5,6-pentafluoro-	5736-49-2	NI42043
347	278	309	297	327	247	328	117	**426**	100	33	14	12	10	7	5	4	**0.35**	13	1	–	–	–	10	–	1	–	–	–	–	–	Decafluorobenzhydryl bromide		LI08831
258	144	200	286	342	314	230	175	**426**	100	60	54	43	40	32	31	20	**11.00**	13	14	7	–	1	–	–	–	–	–	–	–	2	[μ-η(2)-tert-butylcarbonyl](μ-ethanethiolato)hexacarbonyldiiron		MS09561
199	227	155	426	213	117	200	228	**426**	100	88	46	17	16	12	11	9		14	4	–	–	2	10	–	–	–	–	–	–	–	Benzene, 1,1′-[1,2-ethanediylbis(thio)]bis[2,3,4,5,6-pentafluoro-	31597-89-4	NI42044
213	215	214	212	426	185	199	142	**426**	100	40	30	19	15	11	10	10		16	26	2	8	2	–	–	–	–	–	–	–	–	Bis(6-diethylamino-4-methoxy-S-triazin-2-yl) disulphide		IC04377
284	57	281	29	267	283	43	55	**426**	100	99	79	70	50	50	44	39	**0.00**	16	42	5	–	–	–	–	–	–	4	–	–	–	1,3,5,7-Tetraethyl-1,7-dibutoxytetrasiloxane	73420-33-4	NI42045
341	183	185	343	339	43	170	155	**426**	100	97	92	51	49	42	41	32	**0.00**	17	16	3	–	–	–	–	2	–	–	–	–	–	Benzeneacetic acid, 4-bromo-α-(4-bromophenyl)-α-hydroxy-, isopropyl ester	18181-80-1	NI42046
43	75	41	27	185	183	104	155	**426**	100	98	59	58	54	50	47	36	**0.00**	17	16	3	–	–	–	–	2	–	–	–	–	–	Benzeneacetic acid, 4-bromo-α-(4-bromophenyl)-α-hydroxy-, isopropyl ester	18181-80-1	NI42047
411	185	183	231	341	229	413	409	**426**	100	91	90	72	69	67	52	51	**0.00**	17	16	3	–	–	–	–	2	–	–	–	–	–	Benzeneacetic acid, 4-bromo-α-(4-bromophenyl)-α-hydroxy-, isopropyl ester	18181-80-1	NI42048
395	396	162	397	293	220	28	364	**426**	100	19	5	4	4	4	4	3	**2.34**	18	18	12	–	–	–	–	–	–	–	–	–	–	Benzenehexacarboxylic acid, hexamethyl ester	6237-59-8	NI42049
73	45	44	40	59	294	411	75	**426**	100	26	19	13	12	11	10	9	**8.40**	19	34	5	–	–	–	–	–	–	3	–	–	–	Methyl catecholpyruvate tris(trimethylsilyl) ether		MS09562
73	147	411	45	148	426	355	75	**426**	100	91	25	23	15	14	11	9		19	34	5	–	–	–	–	–	–	3	–	–	–	Trimethylsilyl O,O′-bis(trimethylsilyl)vanilpyruvate	68595-71-1	MS09563
73	147	148	411	426	74	355	75	**426**	100	67	12	11	9	9	8	8		19	34	5	–	–	–	–	–	–	3	–	–	–	Trimethylsilyl O,O′-bis(trimethylsilyl)vanilpyruvate	68595-71-1	NI42050
73	147	45	411	426	148	355	74	**426**	100	80	21	17	14	13	10	9		19	34	5	–	–	–	–	–	–	3	–	–	–	Vanillylpyruvic acid tri-TMS		NI42051
426	73	267	427	205	75	268	411	**426**	100	94	78	37	32	23	21	21		20	38	4	–	–	–	–	–	–	3	–	–	–	(3,4-Dihydroxyphenyl)pentanoic acid, tris(trimethylsilyl)-		NI42052
150	92	104	110	66	76	134	120	**426**	100	64	52	37	35	30	20	17	**0.00**	21	18	8	2	–	–	–	–	–	–	–	–	–	Bicyclo[2.2.1]heptane-2,7-diol, bis(4-nitrobenzoate)	74367-28-5	NI42053
112	154	142	141	125	155	179	153	**426**	100	99	99	99	98	42	33	27	**2.50**	21	38	5	4	–	–	–	–	–	–	–	–	–	L-Valine, N-[N-(1-alloisoleucyl-L-prolyl)-N-methylglycyl]-N-methyl-, methyl ester	55555-43-6	NI42054
43	300	342	282	384	281	283	367	**426**	100	91	76	60	43	36	26	25	**14.46**	22	18	9	–	–	–	–	–	–	–	–	–	–	Barpisoflavone a triacetate		NI42055

MASS TO CHARGE RATIOS								M.W.	INTENSITIES								Parent	C	H	O	N	S	F	Cl	Br	I	Si	P	B	X	COMPOUND NAME	CAS Reg No	No
58	426	170	313	84	98	44	427	**426**	100	57	25	20	20	19	18	14		22	22	7	2	–	–	–	–	–	–	–	–	–	2-Naphthacenecarboxamide, 4-(dimethylamino)-1,4,4a,5,12,12a-hexahydro-3,10,11,12a-tetrahydroxy-6-methyl-1,12-dioxo-, [4S-(4α,4aα,12aα)]-	1665-56-1	NI42056
339	77	341	130	428	398	426	57	**426**	100	99	93	51	47	47	47	46		22	23	–	2	1	–	–	1	–	–	–	–	–	2,3,5,6,7,8,9,10-Octahydro-1-phenyl-5-(p-bromophenylimino)-1H-cyclohepta[e][1,4]thiazepine		NI42057
86	69	382	197	182	282	169	167	**426**	100	88	80	61	53	40	37	37	**1.03**	22	38	6	2	–	–	–	–	–	–	–	–	–	Cyclic (D-alloisoleucyl-L-α-hydroxyisovaleryl-L-isoleucyl-L-α-hydroxyisovaleryl)	2441-03-4	NI42058
337	43	309	339	321	217	362	319	**426**	100	57	43	30	30	29	28	23	**6.00**	23	22	8	–	–	–	–	–	–	–	–	–	–	(7R)-9-acetyl-7-ethoxy-4-methoxy-7,8,9,10-tetrahydro-6,9,11-trihydroxy-5,12-naphthacenequinone		NI42059
337	309	43	362	217	426	339	321	**426**	100	63	56	36	35	28	24	24		23	22	8	–	–	–	–	–	–	–	–	–	–	(7S)-9-acetyl-7-ethoxy-4-methoxy-7,8,9,10-tetrahydro-6,9,11-trihydroxy-5,12-naphthacenequinone		NI42060
49	43	84	115	142	285	170	150	**426**	100	65	45	15	10	3	3	3	**0.00**	23	26	4	2	1	–	–	–	–	–	–	–	–	trans-5-Ethyl-2-[1-(phenylsulphonyl)-2-indolyl]-4-piperidone ethylene acetal		DD01656
426	57	58	71	230	174	72	70	**426**	100	60	59	38	24	24	21	21		23	26	6	2	–	–	–	–	–	–	–	–	–	16,19-Secostrychnidine-10,16-dione, 21,22-epoxy-21,22-dihydro-14-hydroxy-4-methoxy-19-methyl-, (21α,22α)-	62421-69-6	NI42061
109	123	69	43	95	141	167	227	**426**	100	85	83	65	55	28	25	21	**0.00**	23	38	7	–	–	–	–	–	–	–	–	–	–	Methyl [1S-[1α(R*),2β(R*)]]-2-[2-(acetyloxy)-3-oxobutyl]-β-hydroxy-β,1,3,3-tetramethyl-ε-oxocyclohexanehexanoic acid		NI42062
229	426	230	231	30	391	228	243	**426**	100	61	61	50	46	40	40	26		24	18	4	4	–	–	–	–	–	–	–	–	–	2,5-Cyclohexadien-1-one, 4,4-diphenyl-, (2,4-dinitrophenyl)hydrazone	74779-64-9	NI42063
311	382	312	228	325	383	339	309	**426**	100	32	23	18	16	7	5	5	**0.00**	24	26	7	–	–	–	–	–	–	–	–	–	–	Siphulin		LI08832
411	426	71	69	85	427	412	83	**426**	100	93	55	51	34	25	25	25		24	30	5	2	–	–	–	–	–	–	–	–	–	Obscurinervidine, 6,7-dihydro-	6887-30-5	NI42064
426	44	124	427	112	43	57	42	**426**	100	40	30	27	18	18	16	16		24	30	5	2	–	–	–	–	–	–	–	–	–	Rindline		LI08833
426	425	396	395	200	199	186	174	**426**	100	73	10	10	7	6	6	6		24	30	5	2	–	–	–	–	–	–	–	–	–	Yohimban-16-carboxylic acid, 17-(acetyloxy)-11-methoxy-, methyl ester, (3β,16α,17α)-	65025-23-2	NI42065
59	31	85	55	41	29	67	54	**426**	100	89	69	53	51	46	31	22	**0.00**	24	42	6	–	–	–	–	–	–	–	–	–	–	Propanedioic acid, [12-[(tetrahydro-2H-pyran-2-yl)oxy]-3-dodecenyl]-, diethyl ester, (Z)-	69494-17-3	NI42066
89	355	57	45	113	59	135	161	**426**	100	87	81	73	72	63	62	56	**1.61**	24	42	6	–	–	–	–	–	–	–	–	–	–	3,6,9,12-Tetraoxatetradecan-1-ol, 14-[4-(1,1,3,3-tetramethylbutyl)phenoxy]-	2315-64-2	NI42067
73	58	127	99	85	129	113	74	**426**	100	99	98	85	83	69	60	59	**19.31**	24	54	–	–	–	–	–	–	–	3	–	–	–	Hexa-tert-butylcyclotrisilane	89463-49-0	NI42068
284	410	28	285	18	165	43	123	**426**	100	88	81	58	52	50	45	42	**36.00**	25	30	6	–	–	–	–	–	–	–	–	–	–	2-Methyl-2-isohexyl-5,8-dihydroxy-7-[1-oxo-3-(p-quinon-2-yl)propyl]chroman		MS09564
258	426	257						**426**	100	77	54							25	35	5	–	–	–	–	–	–	–	–	1	–	Cortisone butyl boronate		LI08834
258	257	426	91	55	135	122	43	**426**	100	53	47	32	32	27	26	26		25	35	5	–	–	–	–	–	–	–	–	1	–	17α,21-Dihydroxypregn-4-ene-3,11,20-trione butyl boronate		MS09565
258	257	426	135	122	121	123	105	**426**	100	53	47	27	26	24	22	19		25	35	5	–	–	–	–	–	–	–	–	1	–	17α,21-Dihydroxypregn-4-ene-3,11,20-trione butyl boronate		LI08835
258	426	257	41	91	55	43	122	**426**	100	74	55	34	29	28	26	25		25	35	5	–	–	–	–	–	–	–	–	1	–	17α,21-Dihydroxypregn-4-ene-3,11,20-trione tert-butyl boronate		MS09566
258	426	257	122	121	425	123	105	**426**	100	74	55	25	24	20	20	20		25	35	5	–	–	–	–	–	–	–	–	1	–	17α,21-Dihydroxypregn-4-ene-3,11,20-trione tert-butyl boronate		LI08836
426	427	411	412	428	393	205.5	18	**426**	100	26	23	8	6	6	6	6		26	18	6	–	–	–	–	–	–	–	–	–	–	6,11-Dioxo-4,5,7,12-tetrahydroxy-2-(α-methylbenzyl)naphthacene		MS09567
105	167	152	183	91	227	165	180	**426**	100	36	36	32	31	22	19	17	**0.00**	26	23	1	–	–	–	–	–	–	–	–	–	1	(2-Hydroxy-2,2-diphenylethyl)diphenylarsane		DD01657
426	69	427	57	151	231	301	149	**426**	100	61	30	21	15	13	11	10		26	34	3	–	1	–	–	–	–	–	–	–	–	2-tert-Butyl-6-(3-tert-butyl-2-hydroxy-5-methylphenylthio)-4-methylphenyl methacrylate		MS09568
55	227	101	133	71	355	411	113	**426**	100	68	29	24	21	6	5	3	**1.00**	26	50	4	–	–	–	–	–	–	–	–	–	–	1-O-(2-Butenyloxyhexadecyl)-2,3-O-isopropylideneglycerol		NI42069
173	101	57	115	157	231	281	411	**426**	100	27	21	17	12	6	5	4	**0.10**	26	50	4	–	–	–	–	–	–	–	–	–	–	1-O-(2-Butoxy-4-hexadecenyl)-2,3-O-isopropylideneglycerol		NI42070
129	57	43	71	85	55	287	69	**426**	100	99	84	82	66	64	54	51	**2.41**	26	50	4	–	–	–	–	–	–	–	–	–	–	Diisodecyl trimethyladipate		IC04378
185	57	112	70	71	43	55	113	**426**	100	74	54	53	50	38	30	28	**0.39**	26	50	4	–	–	–	–	–	–	–	–	–	–	Dioctylsebacate		IC04379
411	43	101	57	129	55	41	116	**426**	100	93	83	56	42	39	37	32	**0.00**	26	50	4	–	–	–	–	–	–	–	–	–	–	Eicosanoic acid, 2,2-dimethyl-1,3-dioxolan-4-ylmethyl ester	57156-96-4	NI42071
185	57	71	70	112	43	55	113	**426**	100	49	36	35	32	24	21	19	**0.00**	26	50	4	–	–	–	–	–	–	–	–	–	–	Bis(2-ethylhexyl)sebacate		DO01120
109	43	127	69	145	91	105	155	**426**	100	65	55	42	35	35	33	27	**23.90**	27	38	4	–	–	–	–	–	–	–	–	–	–	Azafrin	507-61-9	MS09569
109	43	69	91	59	426	133	408	**426**	100	65	42	35	32	24	18	14		27	38	4	–	–	–	–	–	–	–	–	–	–	Azafrin	507-61-9	LI08837
95	177	69	426	167	137	41	175	**426**	100	84	56	30	28	28	28	22		27	38	4	–	–	–	–	–	–	–	–	–	–	6-(2,5-Dihydroxy-3-methylphenyl)-1-[1,2-dimethyl-2-(4-methyl-1-oxo-3-pentenyl)cyclopentyl]-4-methyl-4-hexen-2-one	97743-94-7	NI42072
312	126	426	297	313	127	55	427	**426**	100	51	44	30	27	20	15	14		27	38	4	–	–	–	–	–	–	–	–	–	–	5α-Spirost-14-ene-3,12-dione, (25R)-	24742-83-4	NI42073
137	95	108	109	136	121	81	135	**426**	100	54	51	49	42	31	23	20	**9.58**	27	42	2	–	–	–	–	–	–	1	–	–	–	Silane, methylphenylbis[(1,7,7-trimethylbicyclo[2.2.1]hept-2-yl)oxy]-	74806-99-8	NI42074
157	230	213	155	185	426	237	235	**426**	100	26	14	12	7	4	4	4		28	26	4	–	–	–	–	–	–	–	–	–	–	DL-diethyl 2,3-di-1-naphthylsuccinate		NI42075
213	426	141	185	380	157	352	306	**426**	100	37	35	25	24	12	6	4		28	26	4	–	–	–	–	–	–	–	–	–	–	DL-diethyl 2,3-di-2-naphthylsuccinate		NI42076
155	157	185	258	214	168	228	426	**426**	100	64	23	19	13	10	5	1		28	26	4	–	–	–	–	–	–	–	–	–	–	Meso-diethyl 2,3-di-1-naphthylsuccinate		NI42077
155	213	185	256	229	300	279	426	**426**	100	36	20	18	18	8	7	4		28	26	4	–	–	–	–	–	–	–	–	–	–	Meso-diethyl 2,3-di-2-naphthylsuccinate		NI42078
105	120	426	106	104	77	188	121	**426**	100	82	15	13	12	10	9	9		28	30	2	2	–	–	–	–	–	–	–	–	–	3-Phenyl-1,1-bis[(1-phenylethyl)carbamoyl]cyclobutane		NI42079

MASS TO CHARGE RATIOS								M.W.	INTENSITIES								Parent	C	H	O	N	S	F	Cl	Br	I	Si	P	B	X	COMPOUND NAME	CAS Reg No	No
81	283	426	343	255	314	313		**426**	100	54	50	48	26	24	23			28	42	3	–	–	–	–	–	–	–	–	–	–	**3β-Acetoxy-27-nor-5α-cholest-7-en-23-yn-22-ol**		**LI08838**
105	426	314	351	384	383	366		**426**	100	78	76	70	56	50	46			29	46	2	–	–	–	–	–	–	–	–	–	–	**3-Acetoxy-5β-cholesta-2,7-diene**		**LI08839**
426	313	366	253	213	351	411	342	**426**	100	85	63	51	42	31	22	20		29	46	2	–	–	–	–	–	–	–	–	–	–	**3β-Acetoxycholest-5,22-diene**	26033-11-4	**NI42080**
384	426	385	427	368	57	55	41	**426**	100	53	28	13	5	5	5	4		29	46	2	–	–	–	–	–	–	–	–	–	–	**Cholesta-3,5-dien-3-ol, acetate**	2309-32-2	**NI42081**
143	42	366	56	158	40	81	58	**426**	100	97	80	75	72	72	62	62	**5.15**	29	46	2	–	–	–	–	–	–	–	–	–	–	**Cholesta-5,7-dien-3-ol, acetate, (3β)-**	1059-86-5	**NI42082**
281	366	43	95	157	351	212	55	**426**	100	58	29	28	27	26	25	25	**6.06**	29	46	2	–	–	–	–	–	–	–	–	–	–	**Cholesta-5,17(20)-dien-3-ol, acetate, (3β,17E)-**	56312-72-2	**NI42083**
366	281	95	81	43	367	57	55	**426**	100	43	36	35	34	30	29	27	**0.00**	29	46	2	–	–	–	–	–	–	–	–	–	–	**Cholesta-5,17(20)-dien-3-ol, acetate, (3β,17Z)-**	56312-71-1	**NI42084**
384	385	426	218	271	55	410	95	**426**	100	30	27	27	17	16	14	12		29	46	2	–	–	–	–	–	–	–	–	–	–	**Cholesta-7,14-dien-7-ol, acetate, (5α)-**	54411-68-6	**NI42085**
71	371	43	426	394	411	253	95	**426**	100	94	87	65	57	51	49	41		29	46	2	–	–	–	–	–	–	–	–	–	–	**3,5-Cyclo-28,33-dinorgorgostan-24-one, 6-methoxy-, (3β,5α,6β)-**	55064-56-7	**NI42086**
366	367	253	351	145	81	245	147	**426**	100	32	31	18	15	15	13	13	**0.00**	29	46	2	–	–	–	–	–	–	–	–	–	–	**Desmosterol acetate**		**MS09570**
43	57	41	55	85	189	83	81	**426**	100	93	60	55	49	45	35	35	**3.00**	29	46	2	–	–	–	–	–	–	–	–	–	–	**A′-Neo-30-norgammacerane-3,22-dione**	1177-74-8	**NI42087**
426	411	213	81	427	229	109	366	**426**	100	43	38	35	34	34	23	22		29	46	2	–	–	–	–	–	–	–	–	–	–	**Zymosterol acetate**		**MS09571**
191	426	109	136	177	123	203	149	**426**	100	82	66	65	62	61	51	50		30	50	1	–	–	–	–	–	–	–	–	–	–	**Allobetulane**		**MS09572**
426	427	300	314	271	57	81	281	**426**	100	33	26	24	24	23	22	14		30	50	1	–	–	–	–	–	–	–	–	–	–	**25-tert-Butyl-27-norcholesta-5,24(E)-dien-3β-ol**		**DD01658**
314	69	55	315	426	41	299	281	**426**	100	32	25	24	20	19	18	17		30	50	1	–	–	–	–	–	–	–	–	–	–	**Cholest-5-en-3-ol, 24-propylidene-, (3β)-**	56362-45-9	**NI42088**
314	55	69	41	81	281	43	95	**426**	100	99	83	49	41	36	36	36	**5.60**	30	50	1	–	–	–	–	–	–	–	–	–	–	**Cholest-5-en-3-ol, 24-propylidene-, (3β)-**	56362-45-9	**NI42089**
55	69	41	95	43	57	68	79	**426**	100	80	65	56	50	41	39	33	**7.00**	30	50	1	–	–	–	–	–	–	–	–	–	–	**9,19-Cycloergost-24(28)-en-3-ol, 4,14-dimethyl-, (3β,4α,5α)-**	469-39-6	**NI42090**
69	41	55	109	81	95	107	93	**426**	100	58	55	44	44	43	40	37	**13.00**	30	50	1	–	–	–	–	–	–	–	–	–	–	**9,19-Cyclolanost-24-en-3-ol, (3β)-**	469-38-5	**NI42091**
408	394	426	286	411	409	339	203	**426**	100	82	77	69	66	35	34	27		30	50	1	–	–	–	–	–	–	–	–	–	–	**9,19-Cyclolanost-24-en-3-ol, (3β)-**	469-38-5	**NI42092**
411	393	408	300	201	301	283	299	**426**	100	89	77	65	53	49	46	43	**37.00**	30	50	1	–	–	–	–	–	–	–	–	–	–	**Cyclomusalenol**		**LI08840**
83	43	69	55	81	95	107	93	**426**	100	83	82	80	60	50	47	40	**14.00**	30	50	1	–	–	–	–	–	–	–	–	–	–	**Cyclopropa[5,6]-33-norgorgostan-3-ol, 3′,6-dihydro-, (3β,5β,6α,22ξ,23ξ)-**	53761-94-7	**NI42093**
83	55	69	81	107	95	97	93	**426**	100	65	57	47	36	34	28	28	**28.00**	30	50	1	–	–	–	–	–	–	–	–	–	–	**Cyclopropa[5,6]stigmast-22-en-3-ol, 3′,6-dihydro-, (3β,5β,6α,22E)-**	53755-20-7	**NI42094**
84	126	43	189	232	41	57	161	**426**	100	55	46	39	32	30	29	25	**3.00**	30	50	1	–	–	–	–	–	–	–	–	–	–	**Cyclopropa[5,6]stigmast-22-en-3-ol, 3′,6-dihydro-, (3β,5β,6α,22E)-**	53755-20-7	**NI42095**
55	83	69	371	255	426	95	41	**426**	100	96	63	59	51	48	39	37		30	50	1	–	–	–	–	–	–	–	–	–	–	**3,5-Cyclostigmast-22-ene, 6-methoxy-, (3β,5α,6β,22E)-**	53603-94-4	**NI42096**
411	426	55	69	81	412	95	427	**426**	100	71	56	55	38	36	33	27		30	50	1	–	–	–	–	–	–	–	–	–	–	**4α,14-Dimethyl-5α-ergosta-8,24(28)-dien-3β-ol**		**NI42097**
69	55	81	95	271	91	93	83	**426**	100	88	84	63	50	50	46	42	**9.70**	30	50	1	–	–	–	–	–	–	–	–	–	–	**26,26-Dimethyl-5,23-ergostadien-3β-ol**		**NI42098**
314	69	55	81	67	271	43	79	**426**	100	47	47	33	33	33	33	30	**16.70**	30	50	1	–	–	–	–	–	–	–	–	–	–	**26,26-Dimethyl-5,24(28)-ergostadien-3β-ol**		**NI42099**
411	245	412	426	233	201	227	215	**426**	100	60	53	39	39	39	37	34		30	50	1	–	–	–	–	–	–	–	–	–	–	**4α,14α-Dimethyl-24-methylenecholest-8-en-3β-ol**		**NI42100**
408	201	393	411	217	300	216	215	**426**	100	89	81	77	67	66	55	55	**22.00**	30	50	1	–	–	–	–	–	–	–	–	–	–	**4α,14α-Dimethyl-24-methylene-9,19-cyclocholestan-3β-ol**		**NI42101**
393	408	202	411	300	301	217	394	**426**	100	91	66	63	46	43	43	42	**10.00**	30	50	1	–	–	–	–	–	–	–	–	–	–	**4α,14α-Dimethyl-24-methylene-9,19-cyclocholestan-3β-ol**		**LI08841**
411	55	69	95	41	426	81	109	**426**	100	75	63	52	45	40	40	37		30	50	1	–	–	–	–	–	–	–	–	–	–	**Ergosta-8,24(28)-dien-3-ol, 4,14-dimethyl-, (3β,4α,5α)-**	16910-32-0	**NI42102**
55	371	97	411	426	394	253	372	**426**	100	30	23	19	14	12	9	8		30	50	1	–	–	–	–	–	–	–	–	–	–	**(23R,24R,25R)-23-Ethyl-6β-methoxy-3α,5:24,26-dicyclocholestane**		**MS09573**
55	371	253	411	97	394	372	285	**426**	100	64	61	36	35	20	18	18	**16.00**	30	50	1	–	–	–	–	–	–	–	–	–	–	**(23R,24S,25S)-23-Ethyl-6β-methoxy-3α,5:24,26-dicyclocholestane**		**MS09574**
55	371	411	253	394	426	372	285	**426**	100	32	20	16	12	11	9	7		30	50	1	–	–	–	–	–	–	–	–	–	–	**(23S,24R,25R)-23-Ethyl-6β-methoxy-3α,5:24,26-dicyclocholestane**		**MS09575**
55	371	411	97	426	394	372	412	**426**	100	73	45	35	33	32	19	14		30	50	1	–	–	–	–	–	–	–	–	–	–	**(23S,24S,25S)-23-Ethyl-6β-methoxy-3α,5:24,26-dicyclocholestane**		**MS09576**
43	55	41	57	69	95	81	426	**426**	100	65	65	56	45	40	36	35		30	50	1	–	–	–	–	–	–	–	–	–	–	**9β-Euph-7-en-3-one**		**MS09577**
259	411	241	137	412	426	121	393	**426**	100	86	50	36	30	29	28	24		30	50	1	–	–	–	–	–	–	–	–	–	–	**Fernenol**	4966-00-1	**LI08842**
259	95	55	81	411	69	107	109	**426**	100	93	80	77	72	71	62	58	**14.00**	30	50	1	–	–	–	–	–	–	–	–	–	–	**Fernenol**	4966-00-1	**NI42103**
411	259	426	241	189	393	273	247	**426**	100	96	48	48	29	27	17	16		30	50	1	–	–	–	–	–	–	–	–	–	–	**D:C-Friedo-B′:A′-neogammacer-8-en-3-ol, (3β)-**	4575-73-9	**NI42104**
43	81	95	379	243	257	229	235	**426**	100	90	78	32	26	25	25	23	**21.00**	30	50	1	–	–	–	–	–	–	–	–	–	–	**D:A-Friedooleanane, 25,26-epoxy-**	52647-78-6	**MS09578**
273	125	232	205	247	426	302	179	**426**	100	99	88	85	70	60	60	60		30	50	1	–	–	–	–	–	–	–	–	–	–	**D:A-Friedooleanan-1-one**	43230-72-4	**MS09579**
43	69	55	57	41	426	95	81	**426**	100	89	87	86	74	72	68	62		30	50	1	–	–	–	–	–	–	–	–	–	–	**D:A-Friedooleanan-2-one**	17947-04-5	**NI42105**
426	69	95	55	125	123	273	205	**426**	100	98	91	81	63	63	45	45		30	50	1	–	–	–	–	–	–	–	–	–	–	**D:A-Friedooleanan-3-one**	559-74-0	**NI42107**
426	55	43	69	41	57	81	95	**426**	100	98	90	88	78	76	68	67		30	50	1	–	–	–	–	–	–	–	–	–	–	**D:A-Friedooleanan-3-one**	559-74-0	**NI42106**
109	125	123	107	121	205	137	163	**426**	100	94	93	56	48	45	42	39	**25.32**	30	50	1	–	–	–	–	–	–	–	–	–	–	**D:A-Friedooleanan-3-one**	559-74-0	**NI42108**
426	193	205	274	427	287	302	273	**426**	100	58	40	37	32	21	20	15		30	50	1	–	–	–	–	–	–	–	–	–	–	**D:A-Friedooleanan-7-one**	18671-54-0	**NI42109**
426	193	69	55	123	205	95	43	**426**	100	56	43	42	39	38	38	37		30	50	1	–	–	–	–	–	–	–	–	–	–	**D:A-Friedooleanan-7-one**	18671-54-0	**NI42110**
204	69	109	95	135	81	121	107	**426**	100	69	61	60	59	59	57	55	**22.00**	30	50	1	–	–	–	–	–	–	–	–	–	–	**D-Friedoolean-14-en-3-ol, (3β)-**	127-22-0	**NI42111**
204	302	426	218	287	411			**426**	100	55	40	24	20	10				30	50	1	–	–	–	–	–	–	–	–	–	–	**D-Friedoolean-14-en-3-ol, (3β)-**	127-22-0	**LI08843**
205	426	259	135	119	411	247	229	**426**	100	89	87	52	52	40	39	37		30	50	1	–	–	–	–	–	–	–	–	–	–	**D:C-Friedoolean-8-en-3-ol, (3β)-**	24462-48-4	**LI08844**
69	137	411	41	95	55	426	81	**426**	100	92	78	71	64	64	56	50		30	50	1	–	–	–	–	–	–	–	–	–	–	**D:B-Friedo-18,19-secolup-19-ene, 3,10-epoxy-, (3β,10β)-**	35060-26-5	**NI42112**
123	426	109	121	341	107	137	205	**426**	100	45	42	34	32	32	23	22		30	50	1	–	–	–	–	–	–	–	–	–	–	**D:A-Friedoursan-3-one**		**MS09580**
247	229	426	205	411	259	248	206	**426**	100	44	38	32	28	24	20	17		30	50	1	–	–	–	–	–	–	–	–	–	–	**D:C-Friedours-7-en-3-ol, (3β)-**	6466-94-0	**MS09581**
314	55	271	69	83	81	108	107	**426**	100	99	85	84	77	71	55	55	**44.55**	30	50	1	–	–	–	–	–	–	–	–	–	–	**Gorgost-5-en-3-ol, (3β)-**	29782-65-8	**NI42114**
55	43	69	83	41	81	67	105	**426**	100	85	72	68	60	48	45	42	**16.00**	30	50	1	–	–	–	–	–	–	–	–	–	–	**Gorgost-5-en-3-ol, (3β)-**	29782-65-8	**NI42113**

MASS TO CHARGE RATIOS								M.W.	INTENSITIES								Parent	C	H	O	N	S	F	Cl	Br	I	Si	P	B	X	COMPOUND NAME	CAS Reg No	No
426	247	205	259	119	229	411	135	**426**	100	85	58	53	40	38	34	34		30	50	1	-	-	-	-	-	-	-	-	-	-	10-[3-[4-[2-(2-Hydroxyethoxy)ethyl]piperazine-1-yl]-isobutyl]phenothiazine		LI08845
134	133	191	135	231	408	187	232	**426**	100	92	68	62	58	35	24	13	**0.00**	30	50	1	-	-	-	-	-	-	-	-	-	-	16α-Hydroxy-ψ-taraxene		NI42115
191	426	192	187	205	408	189	201	**426**	100	20	20	11	10	7	7	5		30	50	1	-	-	-	-	-	-	-	-	-	-	16β-Hydroxy-ψ-taraxene		NI42116
247	426	229	411	119	259	135	205	**426**	100	68	35	25	25	24	22	20		30	50	1	-	-	-	-	-	-	-	-	-	-	Isobauerenol		LI08846
426	247	411	427	259	229			**426**	100	68	57	52	46	31				30	50	1	-	-	-	-	-	-	-	-	-	-	Isobauerenol		LI08847
69	55	272	271	81	43	133	67	**426**	100	93	49	44	40	39	35	32	**17.70**	30	50	1	-	-	-	-	-	-	-	-	-	-	24-Isopropyl-5,23-cholestadien-3β-ol		NI42117
314	69	55	97	426	81	315	67	**426**	100	74	59	35	30	30	26	26		30	50	1	-	-	-	-	-	-	-	-	-	-	24-Isopropyl-5,24-cholestadien-3β-ol		NI42118
411	69	55	109	95	91	33	81	**426**	100	85	47	38	36	34	34	27	**17.00**	30	50	1	-	-	-	-	-	-	-	-	-	-	Lanosta-8,24-dien-3-ol, (3β)-	79-63-0	NI42119
409	425	426	410	427	192	205	125	**426**	100	52	38	30	18	16	15	15		30	50	1	-	-	-	-	-	-	-	-	-	-	Lanosta-8,24-dien-3-ol, (3β)-	79-63-0	NI42120
411	43	55	41	70	57	95	81	**426**	100	81	77	69	56	53	43	38	**22.80**	30	50	1	-	-	-	-	-	-	-	-	-	-	Lanost-7-en-3-one		MS09582
125	427	245	425	426	126	428	411	**426**	100	25	18	17	9	9	8	4		30	50	1	-	-	-	-	-	-	-	-	-	-	Lanost-8-en-3-one		NI42121
411	426	55	43	69	109	427	41	**426**	100	36	20	18	17	15	13	13		30	50	1	-	-	-	-	-	-	-	-	-	-	Lanost-8-en-7-one	52474-71-2	NI42122
426	219	274	427	69	55	43	205	**426**	100	44	41	34	22	20	19	17		30	50	1	-	-	-	-	-	-	-	-	-	-	Lanost-8-en-11-one	52474-68-7	NI42123
135	273	95	426	149	69	81	55	**426**	100	40	17	14	14	14	12	12		30	50	1	-	-	-	-	-	-	-	-	-	-	Lanost-9(11)-en-12-one	52474-76-7	NI42124
205	123	426	163	109	191	149	107	**426**	100	95	87	72	58	56	51	48		30	50	1	-	-	-	-	-	-	-	-	-	-	Lupan-3-one	3186-72-9	NI42125
189	207	426	218	203	408			**426**	100	86	81	56	55	16				30	50	1	-	-	-	-	-	-	-	-	-	-	Lup-20(29)-en-3-ol, (3β)-		LI08848
43	69	55	57	83	97	218	111	**426**	100	88	88	87	86	58	30	29	**14.00**	30	50	1	-	-	-	-	-	-	-	-	-	-	Lup-20(29)-en-3-ol, (3β)-	545-47-1	NI42126
285	328	286	267	329	310	260	269	**426**	100	53	26	15	14	11	9	8	**4.00**	30	50	1	-	-	-	-	-	-	-	-	-	-	4α-Methyl-(24Z)-24-ethylidenecholest-7-en-3β-ol		NI42127
55	69	81	79	67	426	109	105	**426**	100	85	80	78	74	72	71	67		30	50	1	-	-	-	-	-	-	-	-	-	-	24-Methyl-5,28-stigmastadien-3β-ol		NI42128
189	207	190	81	135	95	69	56	**426**	100	34	34	30	29	28	22	22	**21.00**	30	50	1	-	-	-	-	-	-	-	-	-	-	A′-Neogammacer-22(29)-en-3-ol, (3β,21β)-	1678-31-5	NI42129
314	55	281	229	299	213	271	255	**426**	100	50	25	22	21	12	10	5	**4.00**	30	50	1	-	-	-	-	-	-	-	-	-	-	27-Norstigmasta-5,24(28)-dien-3-ol, 25-ethyl-, (3β,24Z,25S)-	69081-88-5	NI42130
189	95	81	55	93	107	69	135	**426**	100	68	61	58	51	46	46	45	**6.00**	30	50	1	-	-	-	-	-	-	-	-	-	-	Hop-22(29)-en-3β-ol	58801-23-3	NI42131
191	81	95	69	55	395	109	67	**426**	100	81	79	67	65	51	46	45	**39.39**	30	50	1	-	-	-	-	-	-	-	-	-	-	Lup-20(29)-en-28-ol	13952-76-6	NI42132
55	43	41	203	69	57	81	95	**426**	100	76	76	74	69	53	40	33	**0.00**	30	50	1	-	-	-	-	-	-	-	-	-	-	Urs-12-en-28-ol	10153-88-5	NI42133
218	203	219	189	95	109	69	135	**426**	100	26	18	10	9	8	8	7	**5.98**	30	50	1	-	-	-	-	-	-	-	-	-	-	Olean-12-en-3-ol, (3β)-	559-70-6	NI42134
426	69	427	123	81	95	232	109	**426**	100	48	41	40	32	28	25	24		30	50	1	-	-	-	-	-	-	-	-	-	-	β-Serratan-3-one	24739-07-9	NI42135
426	69	427	123	81	95	232	109	**426**	100	50	42	42	35	32	30	28		30	50	1	-	-	-	-	-	-	-	-	-	-	β-Serratan-3-one	24739-07-9	LI08849
274	55	95	69	81	260	134	109	**426**	100	97	96	92	69	60	60	56	**9.20**	30	50	1	-	-	-	-	-	-	-	-	-	-	Simiarenol		NI42136
384	43	229	95	81	55	93	107	**426**	100	63	62	58	56	55	54	52	**4.00**	30	50	1	-	-	-	-	-	-	-	-	-	-	Spiro[cholestane-3,1′-cyclobutan]-2′-one, (3α)-	74345-22-5	NI42137
230	122	55	95	81	43	57	229	**426**	100	90	82	76	76	75	64	62	**7.00**	30	50	1	-	-	-	-	-	-	-	-	-	-	Spiro[cholestane-3,1′-cyclobutan]-2′-one, (3β)-	74310-24-0	NI42138
55	83	255	81	426	69	133	159	**426**	100	98	75	74	73	67	47	45		30	50	1	-	-	-	-	-	-	-	-	-	-	Stigmasta-5,22-diene, 3-methoxy-, (3β,22E)-	10453-25-5	NI42139
207	189	135	190	191	107	205	426	**426**	100	87	48	45	40	37	34	33		30	50	1	-	-	-	-	-	-	-	-	-	-	Taraxasterol	1059-14-9	NI42140
218	203	189	426	207	408			**426**	100	16	13	12	11	2				30	50	1	-	-	-	-	-	-	-	-	-	-	Urs-12-en-3-ol, (3β)-		LI08850
218	219	203	189	135	122	95	426	**426**	100	18	16	11	10	10	10	9		30	50	1	-	-	-	-	-	-	-	-	-	-	Urs-12-en-3-ol, (3β)-	638-95-9	NI42141
189	121	122	190	123	119	204	203	**426**	100	60	36	33	31	30	27	26	**0.00**	30	50	1	-	-	-	-	-	-	-	-	-	-	Urs-20-en-3-ol, (3β,18α,19α)-	464-98-2	NI42142
426	262	183	427	165	63	241	98	**426**	100	48	42	29	29	29	26	25		31	23	-	-	-	-	-	-	-	-	1	-	-	Phosphorane, 9H-fluoren-9-ylidenetriphenyl-	4756-25-6	LI08851
426	262	183	57	43	427	165	83	**426**	100	48	43	39	35	29	29	29		31	23	-	-	-	-	-	-	-	-	1	-	-	Phosphorane, 9H-fluoren-9-ylidenetriphenyl-	4756-25-6	NI42143
145	285	43	41	55	29	146	143	**426**	100	29	26	17	14	14	13	12	**5.02**	31	54	-	-	-	-	-	-	-	-	-	-	-	11-(5,6,7,8-Tetrahydro-1-naphthyl)heneicosane		AI02367
145	285	43	41	29	146	143	55	**426**	100	36	33	20	18	15	13	13	**9.48**	31	54	-	-	-	-	-	-	-	-	-	-	-	11-(5,6,7,8-Tetrahydro-1-naphthyl)heneicosane		AI02366
321	105	426	77	91	306			**426**	100	99	58	45	12	8				32	26	1	-	-	-	-	-	-	-	-	-	-	β-Dypnopicalone		LI08852
426	105	322	321	77	411			**426**	100	78	33	23	15	10				32	26	1	-	-	-	-	-	-	-	-	-	-	Iso-dypnopinacolone		LI08853
105	321	322	426	77	306	60	228	**426**	100	99	77	61	54	30	23	20		32	26	1	-	-	-	-	-	-	-	-	-	-	6-Methyl-2,4,6-triphenylcyclohexa-2,4-dienyl phenyl ketone		LI08854
252	43	280	340	85	298	41	253	**427**	100	71	60	51	51	40	34	13	**0.63**	15	20	5	1	-	7	-	-	-	-	-	-	-	Dipropyl N-(heptafluorobutyryl)glutamate		MS09583
73	412	174	413	354	45	59	414	**427**	100	50	37	18	16	15	14	10	**0.00**	15	42	3	1	-	-	-	-	-	4	1	-	-	Phosphonic acid, [3-[bis(trimethylsilyl)amino]propyl]-, bis(trimethylsilyl) ester	29836-73-5	NI42144
73	412	174	354	86	147	298	135	**427**	100	50	37	16	10	9	7	7	**0.25**	15	42	3	1	-	-	-	-	-	4	1	-	-	Phosphonic acid, [3-[bis(trimethylsilyl)amino]propyl]-, bis(trimethylsilyl) ester	29836-73-5	NI42145
218	73	219	209	310	220	147	77	**427**	100	41	19	17	9	8	6	5	**2.00**	19	37	4	1	-	-	-	-	-	3	-	-	-	N,O-Bis(trimethylsilyl)-3-methoxytyrosine trimethylsilyl ester		MS09584
163	187	228	134	149	147	203	63	**427**	100	88	80	79	55	40	39	39	**23.56**	20	21	2	5	2	-	-	-	-	-	-	-	-	N-(β-Cyanoethyl)-N-ethyl-3-methyl-4(6-methylsulphonylbenzthiazol-2-ylazo)aniline		IC04380
217	188	189	225	226	224	223	243	**427**	100	88	48	25	24	24	23	20	**6.00**	21	22	-	3	1	-	-	1	-	-	-	-	-	1-[(2-Benzyl-4-thiazolyl)methyl]-4-(3-bromophenyl)piperazine		MS09585
154	182	411	295	366	338			**427**	100	12	6	5	3	1			**0.00**	21	37	6	3	-	-	-	-	-	-	-	-	-	L-Alanine, N-[N-[N-(2,4-dioxopentyl)-L-valyl]-L-isoleucyl]-, ethyl ester	37580-34-0	NI42146

MASS TO CHARGE RATIOS								M.W.	INTENSITIES								Parent	C	H	O	N	S	F	Cl	Br	I	Si	P	B	X	COMPOUND NAME	CAS Reg No	No
341	310	342	427	283	311	281	252	427	100	35	21	14	11	7	7	6		22	21	8	1	–	–	–	–	–	–	–	–	–	Tetramethyl 4,5-dihydroazepino[2,1-a]isoquinoline-1,2,3,4-tetracarboxylate		LI08855
341	310	342	427	336	238	149	252	427	100	25	21	18	14	11	11	9		22	21	8	1	–	–	–	–	–	–	–	–	–	Tetramethyl 10,11-dihydroazepino[1,2-a]quinoline-7,8,9,10-tetracarboxylate		LI08856
354	355	179	336	178	427	191	396	427	100	22	9	8	6	5	5	4		22	21	8	1	–	–	–	–	–	–	–	–	–	Trimethyl 4-methoxycarbonylmethyl-4H-benzo[a]quinolizine-1,2,3-tricarboxylate		LI08857
220	156	276	251	128	205	87	221	427	100	37	36	28	28	22	19	17	6.56	22	37	7	1	–	–	–	–	–	–	–	–	–	Cyclopentanepropanoic acid, 3,5-bis(acetyloxy)-2-[3-(methoxyimino)octyl]-, methyl ester, [1R-(1α,2β,3α,5β)]-	72150-75-5	NI42147
255	91	256	79	77	168	152	167	427	100	63	59	39	28	26	26	25	3.00	23	25	5	1	1	–	–	–	–	–	–	–	–	Tricyclo[3.3.1.1(^{3,7})]decan-2-ol, 1-(4-nitrophenyl)-, 4-methylbenzenesulphonate (ester)	55937-88-7	NI42148
340	43	325	427	328	341	368	326	427	100	77	63	62	48	23	22	22		23	25	7	1	–	–	–	–	–	–	–	–	–	Formyl colchicine		NI42149
427	56	384	78	398	326	356	399	427	100	99	80	70	51	50	36	25		24	21	3	5	–	–	–	–	–	–	–	–	–	Ethyl 1,6-diethyl-2-oxo-3-(3,6-di-2-pyridyl-4-pyridazinyl)-1,2-dihydropyridine-5-carboxylate		MS09586
91	363	228	155	280	188	344	272	427	100	71	21	20	19	19	16	15	14.00	24	29	4	1	1	–	–	–	–	–	–	–	–	Benzenesulphonamide, N,4-dimethyl-N-(6a,7,10,10a-tetrahydro-1-hydroxy-6,6,9-trimethyl-6H-dibenzo[b,d]pyran-3-yl)-, (6aR-trans)-	40248-05-3	NI42150
91	82	69	272	105	188	216	133	427	100	79	45	42	33	30	28	28	18.00	24	29	4	1	1	–	–	–	–	–	–	–	–	Benzenesulphonamide, N,4-dimethyl-N-(6a,7,10,10a-tetrahydro-3-hydroxy-6,6,9-trimethyl-6H-dibenzo[b,d]pyran-1-yl)-, (6aR-trans)-	40248-04-2	NI42151
212	42	187	45	70	180	56	98	427	100	76	69	65	62	59	55	47	15.01	24	33	2	3	1	–	–	–	–	–	–	–	–	10-[3-[4-[2-(2-Hydroxyethoxy)ethyl]piperazin-1-yl]-2-methylpropyl]phenothiazine	2470-73-7	NI42152
107	380	77	427	379	240	117	190	427	100	29	29	26	23	23	21	18		25	25	–	5	1	–	–	–	–	–	–	–	–	1-[N,N'-bis(4-methylphenyl)]guanidino-4-phenyl-2-methylthioimidazole		MS09587
368	369	325	267	280	412	281	268	427	100	25	25	16	15	13	13	10	3.00	26	21	5	1	–	–	–	–	–	–	–	–	–	5-(1,2-Dimethoxycarbonyl-2-oxidovinyl)-6-o-tolylphenanthridinium		LI08858
183	384	427	185	108	241	109	210	427	100	97	60	50	29	23	23	20		27	27	–	1	–	–	–	–	–	–	2	–	–	N,N-Bis(diphenylphosphino)propylamine		NI42153
207	104	208	427	339	264	165	395	427	100	50	47	34	17	16	16	12		27	29	–	3	1	–	–	–	–	–	–	–	–	2-(tert-Butylamino)-1-(2,6-dimethylphenyl)-4,4-diphenyl-2-imidazoline-5-thione	118515-06-3	DD01659
114	111	110	98	139	124	427		427	100	81	27	13	10	3	1			27	41	3	1	–	–	–	–	–	–	–	–	–	Korsevinine		LI08859
105	103	77	104	122	302	130		427	100	58	43	38	14	8	7		0.00	27	41	3	1	–	–	–	–	–	–	–	–	–	cis-2-Phenyl-4-octyl-5-(7-methoxycarbonylheptyl)oxazoline		LI08860
105	103	104	77	302	130	122		427	100	58	50	44	33	18	17		0.00	27	41	3	1	–	–	–	–	–	–	–	–	–	trans-2-Phenyl-4-octyl-5-(7-methoxycarbonylheptyl)oxazoline		LI08861
427	202	380	397	379	203	426	213.5	427	100	14	8	7	7	7	6	5		30	21	2	1	–	–	–	–	–	–	–	–	–	1-(4-Nitrostyryl)-4-(2-anthr-9-ylvinyl)benzene		LI08862
385	42	55	44	70	59	69	386	427	100	51	50	43	39	37	30	29	0.00	30	53	–	1	–	–	–	–	–	–	–	–	–	Cholestan-2-amine, N-isopropylidene-, (2β,5α)-	56052-60-9	NI42154
214	195	69	67	48	100	97	113	428	100	99	90	64	42	29	22	21	0.00	6	–	4	–	2	12	–	–	–	–	–	–	–	2,2,4,4-Tetrakis(trifluoromethyl)-1,3-dithietane-1,1,3,3-tetraoxide		DD01660
288	286	217	314	426	398	185	245	428	100	63	61	45	44	44	42	34	0.00	6	–	5	–	1	3	–	–	–	–	–	–	1	Trifluoromethylthio(pentacarbonyl)rhenium		LI08863
29	86	28	202	41	342	200	43	428	100	87	71	15	14	11	11	11	8.00	8	10	4	4	–	–	–	–	–	–	–	–	1	Mercury, bis(diazoacetic ester)		LI08864
428	430	316	187	372	344	121	400	428	100	60	25	24	23	23	21	19		8	20	4	–	4	–	–	–	–	–	2	–	1	Nickel[II] bis(diethyldithiophosphate)	15701-66-3	NI42155
75	99	191	77	49	39	193	209	428	100	82	35	34	32	26	22	18	0.00	9	15	4	–	–	–	6	–	–	–	1	–	–	Tris(1,3-dichloro-2-propyl)phosphate	13674-87-8	NI42156
75	99	191	77	49	193	39	209	428	100	72	41	31	28	26	25	22	0.00	9	15	4	–	–	–	6	–	–	–	1	–	–	Tris(1,3-dichloro-2-propyl)phosphate	13674-87-8	NI42157
75	99	191	381	209	193	383	379	428	100	87	81	60	51	49	40	38	10.00	9	15	4	–	–	–	6	–	–	–	1	–	–	Tris(2,3-dichloropropyl)phosphate		IC04381
232	315	288	400	174	143	372	260	428	100	67	55	41	38	34	26	26	2.00	12	6	10	–	–	–	–	–	–	–	–	–	2	Cobalt, (acetylenedicarboxylic acid)hexacarbonyldi-, dimethyl ester	37685-64-6	NI42158
232	344	316	288	400	174	143	372	428	100	74	67	55	41	38	34	26	1.00	12	6	10	–	–	–	–	–	–	–	–	–	2	Cobalt, (acetylenedicarboxylic acid)hexacarbonyldi-, dimethyl ester		LI08865
395	335	397	237	223	399	272	183	428	100	93	82	79	72	66	64	62	8.88	12	10	4	–	–	–	6	–	–	–	–	–	–	Methyl [1,4,5,6,7,7-hexachloro-2-(methoxycarbonyl)bicyclo[2.2.1]hept-5-en-3-yl]acetate	104472-55-1	NI42159
184	202	315	44	130	69	170	59	428	100	83	42	35	32	30	26	26	4.54	12	14	6	2	1	6	–	–	–	–	–	–	–	L-Alanine, 3,3'-thiobis[N-(trifluoroacetyl)-, dimethyl ester	26527-25-3	NI42160
39	40	132	314	51	38	91	78	428	100	80	73	66	20	19	18	18	14.66	14	12	4	–	–	–	–	–	–	–	–	–	1	π-Tricyclo[5.2.1.0(^{2,6})]deca-3,8-diene(tetracarbonyl) tungsten		MS09588
327	73	399	298	102	147	203	328	428	100	99	70	52	47	36	34	26	1.00	14	41	3	2	–	–	–	–	–	4	1	–	–	Phosphonic acid, [1,2-bis[(trimethylsilyl)amino]ethyl]-, bis(trimethylsilyl) ester	56196-71-5	NI42161
428	430	128	429	427	432	127	99	428	100	96	56	53	35	33	33	29		18	12	4	–	–	–	3	–	–	–	1	–	–	Tris(p-chlorophenyl)phosphate		IC04382
93	141	131	170	148	99	95	120	428	100	82	57	47	47	45	45	43	4.00	18	32	6	2	–	–	–	–	–	2	–	–	–	2(1H)-Pyrimidinone, 1-[2,3-O-isopropylidene-5-O-(trimethylsilyl)-β-D-ribofuranosyl]-4-(trimethylsiloxy)-	29015-21-2	NI42162
297	73	75	147	45	299	44	40	428	100	80	19	14	13	10	10	10	0.00	18	36	4	2	–	–	–	–	–	3	–	–	–	Bis(4-hydroxy-3-methoxyphenylglycol)piperazine salt, trimethylsilyl-		NI42163
350	428	351	304	429	380	336	183	428	100	58	38	31	28	20	15	15		19	16	8	4	–	–	–	–	–	–	–	–	–	4-(2'-Dinitrophenylhydrazono)-6-hydroxy-3-hydroxymethyl-7-methoxy-1,2,3,4-tetrahydro-2-naphthoic acid lactone		MS09589
355	73	356	357	75	45	77	74	428	100	63	36	16	13	8	7	5	0.00	19	36	5	–	–	–	–	–	–	3	–	–	–	Benzeneacetic acid, α,3,4-tris[(trimethylsilyl)oxy]-, ethyl ester		MS09590
355	73	356	357	75	45	267	251	428	100	49	32	16	9	6	5	5	2.40	19	36	5	–	–	–	–	–	–	3	–	–	–	Benzeneacetic acid, α,3,4-tris[(trimethylsilyl)oxy]-, ethyl ester		MS09591
209	73	75	210	338	147	77	179	428	100	59	17	16	14	14	13	10	0.00	19	36	5	–	–	–	–	–	–	3	–	–	–	Benzenepropanoic acid, 3-methoxy-α,4-bis[(trimethylsilyl)oxy]-, trimethylsilyl ester		MS09592

MASS TO CHARGE RATIOS								M.W.	INTENSITIES								Parent	C	H	O	N	S	F	Cl	Br	I	Si	P	B	X	COMPOUND NAME	CAS Reg No	No
209	73	210	338	147	45	179	75	**428**	100	75	16	12	11	10	9	6	**4.70**	19	36	5	–	–	–	–	–	–	3	–	–	–	Benzenepropanoic acid, 3-methoxy-α,4-bis[(trimethylsilyl)oxy]-, trimethylsilyl ester		MS09593
209	73	210	147	338	179	46	74	**428**	100	97	18	18	13	11	11	9	**4.84**	19	36	5	–	–	–	–	–	–	3	–	–	–	Benzenepropanoic acid, 3-methoxy-α,4-bis[(trimethylsilyl)oxy]-, trimethylsilyl ester	55887-61-1	NI42164
297	73	298	75	147	299	45	74	**428**	100	71	25	15	11	10	8	7	**5.60**	19	36	5	–	–	–	–	–	–	3	–	–	–	3-Hydroxy-3-(4′-hydroxy-3′-methoxyphenyl)propionic acid, tris(trimethylsilyl)-		NI42165
73	267	268	179	45	269	74	75	**428**	100	90	21	21	19	8	8	7	**6.60**	19	36	5	–	–	–	–	–	–	3	–	–	–	Methyl catechollactate tris(trimethylsilyl) ether		MS09594
209	73	210	147	338	45	179	74	**428**	100	97	18	18	13	12	11	9	**0.00**	19	36	5	–	–	–	–	–	–	3	–	–	–	Vanillyllactic acid-tri-TMS		NI42166
413	377	428	217	392	117	133	90	**428**	100	82	77	76	64	52	46	45		20	21	3	6	–	–	1	–	–	–	–	–	–	N,N-Diethyl-3-(3-chloropropionamido)-4-(2-cyano-4-nitrophenylazo)aniline		IC04383
312	328	310	226	256	313	282	227	**428**	100	76	65	38	37	36	36	36	**13.00**	21	16	10	–	–	–	–	–	–	–	–	–	–	Aflatoxin Q1 hemisuccinate		NI42167
243	244	165	192	270	115	91	268	**428**	100	22	20	10	9	8	8	6	**4.00**	21	18	–	–	–	–	–	2	–	–	–	–	–	1,2-Dibromo-3,3,3-triphenylpropane		MS09595
342	311	343	428	239	279	312	253	**428**	100	35	22	15	10	8	7	6		21	20	8	2	–	–	–	–	–	–	–	–	–	Tetramethyl 8,9-dihydroazepino[3,4-a]quinazoline-9,10,11,12-tetracarboxylate		LI08866
342	311	428	343	194	253	239	337	**428**	100	44	29	22	14	10	10	9		21	20	8	2	–	–	–	–	–	–	–	–	–	Tetramethyl 10,11-dihydroazepino[1,2-a]quinoxaline-7,8,9,10-tetracarboxylate		LI08867
342	311	337	343	428	194	239	253	**428**	100	44	25	24	17	13	12	10		21	20	8	2	–	–	–	–	–	–	–	–	–	Tetramethyl 10,11-dihydroazepino[1,2-a]quinoxaline-7,8,9,11-tetracarboxylate		LI08868
155	153	211	209	336	151	157	299	**428**	100	74	60	46	40	37	36	33	**11.50**	21	21	–	–	–	–	1	–	–	–	–	–	1	Stannane, chlorotribenzyl-	3151-41-5	NI42168
309	326	59	119	45	91	90	58	**428**	100	74	57	46	42	35	29	24	**22.00**	21	24	6	4	–	–	–	–	–	–	–	–	–	1-Ethyl-4-methyl-5-cyano-6-hydroxy-3-[4-[2-(2-methoxyethoxy)ethoxycarbonyl]phenylazo]pyrid-2-one		IC04384
321	411	196	268	265	325	235	286	**428**	100	28	15	10	10	5	5	3	**0.00**	21	40	5	2	–	–	–	–	–	1	–	–	–	Cyclopentanepropanoic acid, 5-(methoxyimino)-2-[3-(methoxyimino)octyl]-3-[(trimethylsilyl)oxy]-, [1R-(1α,2β,3α)]-	55401-47-3	NI42169
233	260	386	275	428	302	217	344	**428**	100	99	92	92	80	80	29	28		22	20	9	–	–	–	–	–	–	–	–	–	–	1,3(2H,9bH)-Dibenzofurandione, 6-acetyl-7,9-bis(acetyloxy)-2-(1-hydroxyethylidene)-8,9b-dimethyl-	55570-94-0	NI42170
392	43	57	78	55	41	69	44	**428**	100	72	68	59	48	48	41	38	**21.00**	22	20	9	–	–	–	–	–	–	–	–	–	–	1-Naphthacenecarboxylic acid, 2-ethyl-1,2,3,4,6,11-hexahydro-2,4,5,7,10-pentahydroxy-6,11-dioxo-, methyl ester	54773-77-2	NI42171
392	333	351	393	428	410	322	294	**428**	100	36	29	28	21	14	14	14		22	20	9	–	–	–	–	–	–	–	–	–	–	1-Naphthacenecarboxylic acid, 2-ethyl-1,2,3,4,6,11-hexahydro-2,4,5,7,10-pentahydroxy-6,11-dioxo-, methyl ester, [1R-(1α,2β,4β)]-	21288-61-9	NI42172
428	339	311	351	368	312	340	378	**428**	100	79	78	57	42	42	39	35		22	20	9	–	–	–	–	–	–	–	–	–	–	1-Naphthacenecarboxylic acid, 2-ethyl-1,2,3,4,6,11-hexahydro-2,5,7,10,12 pentahydroxy-6,11-dioxo-, methyl ester	3569-04-8	NI42173
428	44	274	83	69	429	41	245	**428**	100	88	38	31	29	26	22	18		22	24	7	2	–	–	–	–	–	–	–	–	–	Dichotine, 21-oxo-	29484-60-4	NI42174
70	100	71	85	69	83	97	84	**428**	100	99	96	81	63	50	42	38	**0.00**	22	24	7	2	–	–	–	–	–	–	–	–	–	2-Naphthacenecarboxamide, 4-(dimethylamino)-1,4,4a,5,5a,6,11,12a-octahydro-3,6,10,12-tetrahydroxy-6-methyl-1,11-dioxo-, [4S-(4α,4aα,5aα,6β,12aα)]-	4199-36-4	NI42175
413	414	415	73	416	383	398	427	**428**	100	38	14	9	2	2	1	1	**0.50**	22	28	5	–	–	–	–	–	–	2	–	–	–	1,6-Dihydroxy-8-methoxy-3-methylanthraquinone, bis(trimethylsilyl)-	91701-23-4	NI42176
413	414	415	73	370	355	416	397	**428**	100	38	15	7	4	3	3	2	**0.60**	22	28	5	–	–	–	–	–	–	2	–	–	–	1,7-Dihydroxy-3-methoxy-6-methylanthraquinone, bis(trimethylsilyl)-	91701-24-5	NI42177
413	414	415	73	399	416	400	417	**428**	100	38	14	8	4	3	1	1	**0.20**	22	28	5	–	–	–	–	–	–	2	–	–	–	1,8-Dihydroxy-6-methoxy-3-methylanthraquinone, bis(trimethylsilyl)-	7336-75-6	NI42178
428	413	399	411	383	414	427	410	**428**	100	30	24	13	12	11	11	8		22	28	5	–	–	–	–	–	–	2	–	–	–	2,7-Dihydroxy-5-methoxy-3-methylanthraquinone, bis(trimethylsilyl)-	91701-32-5	NI42179
332	257	163	272	333	312	319	273	**428**	100	30	30	25	22	20	18	5	**1.00**	22	33	5	–	1	1	–	–	–	–	–	–	–	Androst-5-ene-17,19-diol, 3-fluoro-, 17-acetate 3-methanesulphonate, (3β,17β)-	28344-71-0	NI42181
43	332	91	129	41	177	117	77	**428**	100	99	48	33	33	32	29	27	**0.00**	22	33	5	–	1	1	–	–	–	–	–	–	–	Androst-5-ene-17,19-diol, 3-fluoro-, 17-acetate 3-methanesulphonate, (3β,17β)-	28344-71-0	NI42180
332	257	272	163	312	320	319	428	**428**	100	30	23	15	10	6	6	2		22	33	5	–	1	1	–	–	–	–	–	–	–	Androst-5-ene-17β,19-diol, 3α-fluoro-, 17-acetate methanesulphonate	28344-52-7	LI08869
332	257	272	333	163	312	320	319	**428**	100	30	23	22	15	10	6	6	**2.00**	22	33	5	–	1	1	–	–	–	–	–	–	–	Androst-5-ene-17β,19-diol, 3α-fluoro-, 17-acetate methanesulphonate	28344-52-7	NI42182
332	43	96	79	257	333	272	81	**428**	100	60	56	45	29	25	23	23	**2.30**	22	33	5	–	1	1	–	–	–	–	–	–	–	Androst-5-ene-17β,19-diol, 3α-fluoro-, 17-acetate methanesulphonate	28344-52-7	NI42183
332	272	257	177	131	312	145	260	**428**	100	27	27	23	14	12	11	10	**1.00**	22	33	5	–	1	1	–	–	–	–	–	–	–	Estr-5(10)-ene-6β-methanol, 3α-fluoro-17β-hydroxy-, 17-acetate methanesulphonate	28526-85-4	LI08870
332	272	257	177	333	131	312	145	**428**	100	27	27	23	20	14	12	11	**1.00**	22	33	5	–	1	1	–	–	–	–	–	–	–	Estr-5(10)-ene-6β-methanol, 3α-fluoro-17β-hydroxy-, 17-acetate methanesulphonate	28526-85-4	NI42184
332	43	252	272	333	165	177	163	**428**	100	33	29	28	27	25	24	19	**0.81**	22	33	5	–	1	1	–	–	–	–	–	–	–	Estr-5(10)-ene-6β-methanol, 3α-fluoro-17β-hydroxy-, 17-acetate methanesulphonate	28526-85-4	NI42185
312	96	43	332	91	79	237	129	**428**	100	63	56	49	31	29	28	28	**0.17**	22	33	5	–	1	1	–	–	–	–	–	–	–	Estr-5(10)-ene-6β-methanol, 3β-fluoro-17β-hydroxy-, 17-acetate methanesulphonate	28344-76-5	NI42186
312	332	131	237	211	145	272	252	**428**	100	50	31	30	28	22	20	17	**1.00**	22	33	5	–	1	1	–	–	–	–	–	–	–	Estr-5(10)-ene-6β-methanol, 3β-fluoro-17β-hydroxy-, 17-acetate methanesulphonate	28344-76-5	LI08871

MASS TO CHARGE RATIOS								M.W.	INTENSITIES								Parent	C	H	O	N	S	F	Cl	Br	I	Si	P	B	X	COMPOUND NAME	CAS Reg No	No
312	332	131	237	211	145	313	272	428	100	50	31	30	28	22	20	20	1.00	22	33	5	–	1	1	–	–	–	–	–	–	–	Estr-5(10)-ene-6β-methanol, 3β-fluoro-17β-hydroxy-, 17-acetate methanesulphonate	28344-76-5	NI42187
319	332	257	163	272	312	408	368	428	100	99	30	30	25	20	1	1	1.00	22	33	5	–	1	1	–	–	–	–	–	–	–	3β-Fluoro-19-methylsulphonyloxy-17-acetoxy-5-androstene		LI08872
57	43	316	41	260	428	71	29	428	100	43	41	41	39	28	28	28		22	37	–	–	–	–	–	–	1	–	–	–	–	Benzene, 1,3,5-tris(2,2-dimethylpropyl)-2-iodo-4-methyl-	41080-91-5	MS09596
43	69	41	81	95	55	191	71	428	100	68	53	50	41	39	37	37	0.40	22	37	3	–	–	–	–	1	–	–	–	–	–	2-Naphthalenol, 6-bromodecahydro-1-(5-hydroxy-3-methyl-3-pentenyl)-2,5,5,8a-tetramethyl-, [1S-(1α,2α,4aβ,6α,8aα)]-	56009-17-7	NI42188
326	43	308	284	428	366	295	283	428	100	82	57	39	34	31	29	28		23	24	8	–	–	–	–	–	–	–	–	–	–	Athrotaxindimethyletherdiacetate		NI42189
137	368	297	312	313	203	281	428	428	100	42	21	18	16	15	14	1		23	24	8	–	–	–	–	–	–	–	–	–	–	7,2′,4′-Trihydroxy-5′-methoxy-6′-(3-acetoxy-3-methylbutyl)isoflavone		MS09597
428	223	59	208	180	204	205	206	428	100	46	35	13	9	6	5	4		23	28	6	2	–	–	–	–	–	–	–	–	–	Isomajdine		LI08873
428	223	59	208	180	204	205	206	428	100	46	35	13	9	6	5	4		23	28	6	2	–	–	–	–	–	–	–	–	–	Majdine	20497-42-1	LI08874
105	222	77	181	209	208	195	194	428	100	66	31	18	17	16	14	12	7.10	23	32	4	4	–	–	–	–	–	–	–	–	–	N-α-Benzoyldeguanino-5-(4′-spirocyclopentyl-1′-methyl-5′-oxo-2′-imidazolinyl-aminomethyl)arginine		LI08875
302	303	304	300	428	301	305		428	100	62	55	53	38	36	27			24	13	–	–	–	–	–	–	1	–	–	–	–	1-Ethynyl-8-(1′-iodo-8′-naphthylethynyl)naphthalene		LI08876
215	186	297	428	133	268	104	51	428	100	90	62	54	52	44	37	29		24	16	2	2	2	–	–	–	–	–	–	–	–	Leuco-indophenine		LI08877
351	120	197	274	77	275	199	135	428	100	71	57	42	7	2	1	1	0.20	24	20	–	–	–	–	–	–	–	–	–	–	1	Tetraphenyltin		LI08878
55	327	193	428	326	248	77	72	428	100	70	50	40	30	30	20	20		24	21	4	–	–	–	–	–	–	–	–	–	1	4,4-Diphenyl-4-hydroxybutylcyclopentadienylmanganese tricarbonyl		LI08879
286	344	132	287	229	231	215	228	428	100	39	21	20	16	14	13	11	8.99	24	21	4	–	–	–	–	–	–	–	–	–	1	α-Propoxybenzhydrylcymantrene	82441-29-0	NI42190
325	43	327	231	343	326	428	233	428	100	97	35	30	24	21	16	12		24	25	5	–	–	–	1	–	–	–	–	–	–	19-Nor-3-chloro-17α-acetoxy-2,4-diformyl-1,3,5(10),6-pregnatetraen-20-one		MS09598
386	284	387	266	173	186	269	285	428	100	40	26	21	15	14	12	9	5.96	24	28	7	–	–	–	–	–	–	–	–	–	–	6-Oxo-estriol triacetate		NI42191
379	410	355	193	181	341	397	380	428	100	69	68	30	30	29	28	27	1.00	24	28	7	–	–	–	–	–	–	–	–	–	–	5,7,2′,4′-Tetramethoxy-3′-(3-hydroxy-3-methylbutyl)isoflavone		MS09599
138	384	400	385	382	160	43	161	428	100	57	35	27	24	21	21	15	0.00	24	32	5	2	–	–	–	–	–	–	–	–	–	Aspidospermidine, 1-acetyl-19,21-epoxy-15,16,17-trimethoxy-	13467-46-4	NI42192
138	384	428	110	57	370	160	109	428	100	50	47	35	25	24	22	22		24	32	5	2	–	–	–	–	–	–	–	–	–	Aspidospermidin-17-ol, 19,21-epoxy-15,16-dimethoxy-1-(1-oxopropyl)-	2122-25-0	NI42193
69	71	428	83	81	85	137	67	428	100	73	51	42	39	36	32	32		24	32	5	2	–	–	–	–	–	–	–	–	–	4,25-Secoobscurinervan-4-ol, 6,7-didehydro-15,16-dimethoxy-22-methyl-, (4β,22α)-	7236-84-2	NI42194
356	341	357	428	342	57	299	258	428	100	44	17	13	11	11	10	10		24	36	3	–	–	–	–	–	–	2	–	–	–	Estra-1,3,5(10),6-tetraen-17-one, 3,6-bis[(trimethylsilyl)oxy]-	69833-76-7	NI42195
43	85	88	74	41	57	73	27	428	100	85	43	36	25	19	18	12	0.00	24	44	6	–	–	–	–	–	–	–	–	–	–	Pentanoic acid, 2,2-dimethyl-, 1,2,3-propanetriyl ester	57346-62-0	NI42196
41	69	305	43	57	247			428	100	99	55	46	37	29			5.15	25	37	5	–	–	–	–	–	–	–	–	1	–	1,3,2-Benzodioxaborol-4(3aH)-one, 2-butyl-6-hydroxy-3a,7-bis(3-methyl-2 butenyl)-5-(3-methyl-1-oxobutyl)-	55429-03-3	NI42197
259	260	246	428	121	107	135	109	428	100	29	26	21	14	14	11	11		25	37	5	–	–	–	–	–	–	–	–	1	–	17α,21-Dihydroxy-5β-pregnane-3,11,20-trione, butyl boronate		LI08880
259	260	246	428	121	107	429	105	428	100	29	26	21	14	14	12	11		25	37	5	–	–	–	–	–	–	–	–	1	–	17α,21-Dihydroxy-5β-pregnane-3,11,20-trione, butyl boronate	30888-25-6	NI42198
259	371	260	41	43	79	67	93	428	100	84	70	65	56	50	49	43	18.00	25	37	5	–	–	–	–	–	–	–	–	1	–	17α,21-Dihydroxy-5β-pregnane-3,11,20-trione, tert-butyl boronate		MS09600
259	371	260	121	135	109	107	122	428	100	84	70	38	35	35	33	32	18.00	25	37	5	–	–	–	–	–	–	–	–	1	–	17α,21-Dihydroxy-5β-pregnane-3,11,20-trione, tert-butyl boronate		LI08881
242	227	260	119	410	428	121	105	428	100	57	55	48	47	43	35	35		25	37	5	–	–	–	–	–	–	–	–	1	–	11β,17α,21-Trihydroxypregn-4-ene-3,20-dione, butyl boronate		LI08882
242	43	55	227	41	260	119	91	428	100	80	67	52	50	47	45	44	30.00	25	37	5	–	–	–	–	–	–	–	–	1	–	11β,17α,21-Trihydroxypregn-4-ene-3,20-dione, butyl boronate		MS09601
242	260	428	410	227	243	395	124	428	100	45	39	38	30	20	18	17		25	37	5	–	–	–	–	–	–	–	–	1	–	11β,17α,21-Trihydroxypregn-4-ene-3,20-dione, butyl boronate		MS09602
260	242	119	124	105	123	121	227	428	100	95	80	78	73	62	62	58	42.00	25	37	5	–	–	–	–	–	–	–	–	1	–	11α,17α,21-Trihydroxypregn-4-ene-3,20-dione, tert-butyl boronate		LI08883
242	43	55	41	227	119	260	91	428	100	85	66	64	58	52	49	48	38.00	25	37	5	–	–	–	–	–	–	–	–	1	–	11β,17α,21-Trihydroxypregn-4-ene-3,20-dione, tert-butyl boronate		MS09603
260	242	119	124	105	123	121	107	428	100	95	80	78	73	62	62	58	42.04	25	37	5	–	–	–	–	–	–	–	–	1	–	11β,17α,21-Trihydroxypregn-4-ene-3,20-dione, tert-butyl boronate	35314-79-5	NI42199
242	260	428	410	227	243	124	395	428	100	51	43	41	32	21	21	20		25	37	5	–	–	–	–	–	–	–	–	1	–	11β,17α,21-Trihydroxypregn-4-ene-3,20-dione, tert-butyl boronate		MS09604
242	227	119	260	105	121	123	124	428	100	60	60	53	52	44	42	40	29.03	25	37	5	–	–	–	–	–	–	–	–	1	–	11β,17α,21-Trihydroxypregn-4-ene-3,20-dione, tert-butyl boronate	30888-31-4	NI42200
399	57	55	41	414	83	69	28	428	100	88	68	65	58	56	56	56	0.00	25	48	5	–	–	–	–	–	–	–	–	–	–	2-Heptadecanol, 1-[(2,2-dimethyl-1,3-dioxolan-4-yl)methoxy]-, acetate	57346-64-2	NI42201
428	397	337	125	368	250	310	293	428	100	31	27	26	19	19	13	13		26	20	6	–	–	–	–	–	–	–	–	–	–	Trimethyl triptycene-1,8,13-tricarboxylate		NI42202
428	337	397	125	250	368	183	278	428	100	85	36	33	29	22	22	19		26	20	6	–	–	–	–	–	–	–	–	–	–	Trimethyl triptycene-1,8,16-tricarboxylate		NI42203
358	43	428	71	164	359	149	57	428	100	53	46	31	27	24	24	18		26	36	3	–	1	–	–	–	–	–	–	–	–	2-tert-Butyl-6-(3-tert-butyl-2-hydroxy-5-methylphenylthio)-4-methylphenyl 2-methylpropanoate		MS09605
214	43	325	91	79	93	107	105	428	100	86	75	71	71	69	66	66	2.88	26	36	5	–	–	–	–	–	–	–	–	–	–	Bufa-20,22-dienolide, 3-(acetyloxy)-14-hydroxy-, (3β,5β)-	4029-66-7	NI42204
214	43	325	91	79	93	109	107	428	100	83	74	69	69	67	65	65	3.00	26	36	5	–	–	–	–	–	–	–	–	–	–	Bufa-20,22-dienolide, 3-(acetyloxy)-14-hydroxy-, (3β,5β)-	4029-66-7	LI08884
214	325	350	147	368	215	188	145	428	100	74	54	41	37	37	36	36	3.00	26	36	5	–	–	–	–	–	–	–	–	–	–	Bufa-20,22-dienolide, 3-(acetyloxy)-14-hydroxy-, (3β,5β)-	4029-66-7	NI42205
43	91	41	77	122	93	79	39	428	100	58	55	49	42	40	32	32	12.26	26	36	5	–	–	–	–	–	–	–	–	–	–	Ingenol-3,4,5,20-diacetonide		NI42206
283	57	227	101	413	173	208	115	428	100	37	27	14	9	6	3	3	0.20	26	52	4	–	–	–	–	–	–	–	–	–	–	1-O-(2-Butoxyhexadecyl)-2,3-O-isopropylideneglycerol		NI42207
57	283	227	101	413	173	115	73	428	100	15	12	7	3	1	1	1	0.40	26	52	4	–	–	–	–	–	–	–	–	–	–	1-O-(2-Isobutoxyhexadecyl)-2,3-O-isopropylideneglycerol		NI42208
428	413	427	429	248	250	414	249	428	100	58	40	33	30	22	20	15		27	20	–	6	–	–	–	–	–	–	–	–	–	1,2-Benzenediamine, N-(1-methyl-3-phenylpyrimido[5,4-a]phenazin-5-yl)-	61416-97-5	NI42209
113	428	85	245	286	285			428	100	98	53	32	19	17				27	40	4	–	–	–	–	–	–	–	–	–	–	22,26-Epoxy-3,6,26-trioxocholestane		LI08885
314	43	41	69	57	44	77	71	428	100	78	77	75	70	68	46	43	15.84	27	40	4	–	–	–	–	–	–	–	–	–	–	Spirost-8-en-11-one, 3-hydroxy-, (3β,5α,14β,20β,22β,25R)-	58072-54-1	NI42210
314	299	315	428	55	41	43	281	428	100	31	28	25	25	22	21	19		27	40	4	–	–	–	–	–	–	–	–	–	–	Spirost-8-en-11-one, 3-hydroxy-, (3β,5α,14β,25R)-	54965-96-7	NI42211

MASS TO CHARGE RATIOS	M.W.	INTENSITIES	Parent	C	H	O	N	S	F	Cl	Br	I	Si	P	B	X	COMPOUND NAME	CAS Reg No	No
428 345 372 303 429 371 373 346	**428**	100 55 52 38 36 28 23 16		27	44	2	–	–	–	–	–	–	1	–	–	–	**Δ¹(⁶)-Tetrahydrocannabinol tert-butyldimethylsilyl ether**		**NI42212**
371 428 372 429 413 357 373 345	**428**	100 66 34 23 23 23 11 10		27	44	2	–	–	–	–	–	–	1	–	–	–	**Δ¹-Tetrahydrocannabinol tert-butyldimethylsilyl ether**		**NI42213**
428 345 372 429 373 346 115 386	**428**	100 53 41 36 22 18 17 12		27	44	2	–	–	–	–	–	–	1	–	–	–	**Δ¹(⁶)-Tetrahydrocannabinol triethylsilyl ether**		**NI42214**
428 413 357 429 414 358 345 372	**428**	100 59 54 34 19 17 16 12		27	44	2	–	–	–	–	–	–	1	–	–	–	**Δ¹-Tetrahydrocannabinol triethylsilyl ether**		**NI42215**
410 215 81 55 93 67 217 193	**428**	100 51 51 50 40 40 38 38	**0.00**	27	45	3	–	–	–	–	–	–	–	–	1	–	**5β-Pregnane-3α,17α,20α-triol, cyclohexyl boronate**		**MS09606**
410 215 217 193 107 411 105 192	**428**	100 51 38 38 38 32 30 28	**11.00**	27	45	3	–	–	–	–	–	–	–	–	1	–	**5β-Pregnane-3α,17α,20α-triol, cyclohexyl boronate**		**LI08886**
43 171 214 199 385 301 410 343	**428**	100 70 20 15 14 10 8 8	**1.00**	28	28	4	–	–	–	–	–	–	–	–	–	–	**(1α,2α,3β,4β)-1,3-Diacetyl-2,4-bis[(E)-3-(2-acetylethenyl)phenyl]cyclobutane**		**MS09607**
171 172 209 214 215 315 385 267	**428**	100 75 24 16 13 8 7 6	**1.00**	28	28	4	–	–	–	–	–	–	–	–	–	–	**(1α,2β,3α,4β)-1,2-Diacetyl-3,4-bis[(E)-2-(2-acetylethenyl)phenyl]cyclobutane**		**MS09608**
43 385 386 129 171 271 229 215	**428**	100 42 20 18 17 10 10 10	**2.00**	28	28	4	–	–	–	–	–	–	–	–	–	–	**(1α,4α,5α,6α,7α,10α,11α,12α)-5,6,11,12-tetraacetyldibenzo[b,h]tricyclo[8.2.0.0⁴·⁷]dodecane**		**MS09609**
42 171 172 129 215 157 229 315	**428**	100 37 15 14 10 9 8 5	**2.00**	28	28	4	–	–	–	–	–	–	–	–	–	–	**(1α,4α,5α,6β,7β,10β,11β,12α)-5,6,11,12-Tetraacetyldibenzo[b,h]tricyclo[8.2.0.0⁴·⁷]dodecane**		**MS09610**
324 428 198 43 429 56 325 41	**428**	100 84 80 40 31 28 25 25		28	32	2	2	–	–	–	–	–	–	–	–	–	**Aspidospermidine, 1-acetyl-20,21-didehydro-17-methoxy-21-phenyl-**	55103-31-6	**NI42216**
324 428 198 429 281 325 141 91	**428**	100 93 79 29 25 24 23 22		28	32	2	2	–	–	–	–	–	–	–	–	–	**Aspidospermidine, 1-acetyl-20,21-didehydro-17-methoxy-21-phenyl-**	55103-31-6	**NI42217**
43 344 343 283 284 428 313 255	**428**	100 31 22 22 10 7 3 3		28	44	3	–	–	–	–	–	–	–	–	–	–	**3β-Acetoxy-27-nor-5α-chlolest-7,23-dien-22-ol**		**LI08887**
124 428 57 229 69 55 71 95	**428**	100 99 95 82 71 70 66 64		28	44	3	–	–	–	–	–	–	–	–	–	–	**Cholest-4-en-26-oic acid, 3-oxo-, methyl ester**	53481-64-4	**NI42218**
69 55 81 41 83 57 95 93	**428**	100 94 47 43 36 33 31 28	**4.40**	28	44	3	–	–	–	–	–	–	–	–	–	–	**Ergosterol peroxide**	2061-64-5	**IC04385**
249 251 266 267 392 376 253 248	**428**	100 98 69 63 48 37 36 35	**16.00**	28	44	3	–	–	–	–	–	–	–	–	–	–	**Ergosterol peroxide**	2061-64-5	**LI08888**
396 84 69 55 41 70 397 57	**428**	100 60 40 37 35 32 30 28	**0.00**	28	44	3	–	–	–	–	–	–	–	–	–	–	**Ergosterol peroxide**	2061-64-5	**NI42219**
290 428 396 177 311 381 219 413	**428**	100 75 61 56 47 28 25 19		28	44	3	–	–	–	–	–	–	–	–	–	–	**24-Methoxy-16β,24ξ-epoxy-25-norcycloartan-3-one**		**MS09611**
428 124 57 229 69 55 95 71	**428**	100 99 98 83 74 73 67 67		28	44	3	–	–	–	–	–	–	–	–	–	–	**Methyl 3-oxo-cholest-4-en-27-oate**		**LI08889**
368 81 105 107 91 147 145 95	**428**	100 51 37 35 35 34 34 33	**0.00**	29	48	2	–	–	–	–	–	–	–	–	–	–	**Cholest-5-en-3-ol, (3β)-, acetate**	604-35-3	**WI01543**
368 369 147 247 81 260 145 353	**428**	100 29 29 19 16 15 15 14	**0.00**	29	48	2	–	–	–	–	–	–	–	–	–	–	**Cholest-5-en-3-ol, (3β)-, acetate**	604-35-3	**MS09612**
427 59 367 365	**428**	100 70 9 9	**0.00**	29	48	2	–	–	–	–	–	–	–	–	–	–	**Cholest-5-en-3-ol, (3β)-, acetate**	604-35-3	**NI42220**
42 255 428 55 81 105 95 40	**428**	100 60 54 50 44 39 38 38		29	48	2	–	–	–	–	–	–	–	–	–	–	**Cholest-7-en-3-ol, (3β)-, acetate**	17137-70-1	**NI42221**
428 107 429 55 81 105 95 57	**428**	100 35 34 30 29 26 22 22		29	48	2	–	–	–	–	–	–	–	–	–	–	**Cholest-8(14)-en-3-ol, (3β)-, acetate**	55569-52-3	**NI42222**
42 55 40 81 105 57 107 69	**428**	100 45 45 38 34 34 33 29	**24.14**	29	48	2	–	–	–	–	–	–	–	–	–	–	**Cholest-8-en-3β-ol, acetate**	17137-74-5	**NI42223**
315 94 255 107 93 316 81 161	**428**	100 58 54 39 39 33 31 29	**10.00**	29	48	2	–	–	–	–	–	–	–	–	–	–	**Cholest-14-en-3-ol, acetate, (3β)-**	41853-29-6	**NI42224**
315 413 255 107 93 81 57 55	**428**	100 89 61 59 54 42 42 40	**13.00**	29	48	2	–	–	–	–	–	–	–	–	–	–	**Cholest-14-en-3-ol, acetate, (3β,17α)-**	58072-58-5	**NI42225**
258 368 257 311 259 95 369 353	**428**	100 36 24 23 20 13 12 12	**1.00**	29	48	2	–	–	–	–	–	–	–	–	–	–	**Cholest-22-en-16-ol, acetate, (5α,16β,22Z)-**	54411-85-7	**NI42226**
428 99 291 165 413	**428**	100 35 27 16 4		29	48	2	–	–	–	–	–	–	–	–	–	–	**Cholest-4-en-6-one, cyclic 1,2-ethanediyl acetal**	38404-90-9	**NI42227**
366 367 253 211 428 351 199 197	**428**	100 30 27 20 17 15 15 8		29	48	2	–	–	–	–	–	–	–	–	–	–	**5,19-Cyclo-5β-cholestan-3β-ol, acetate**	21072-75-3	**NI42228**
291 373 87 428 165 99 86 413	**428**	100 99 71 66 66 46 38 25		29	48	2	–	–	–	–	–	–	–	–	–	–	**3,5-Cyclocholestan-6-one, cyclic 1,2-ethanediyl acetal, (3β,5α)-**	35868-79-2	**NI42229**
71 253 353 373 81 55 43 95	**428**	100 99 91 83 74 71 70 62	**58.00**	29	48	2	–	–	–	–	–	–	–	–	–	–	**3,5-Cyclo-28,33-dinorgorgostan-24-ol, 6-methoxy-, (3β,5α,6β)-**	56298-03-4	**NI42230**
373 41 55 43 413 95 81 107	**428**	100 64 62 62 60 60 51 50	**50.00**	29	48	2	–	–	–	–	–	–	–	–	–	–	**3,5-Cycloergostane, 22,23-epoxy-6-methoxy-, (3β,5α,6β)-**	55103-86-1	**NI42231**
55 373 396 255 213 413 428	**428**	100 35 22 22 20 14 11		29	48	2	–	–	–	–	–	–	–	–	–	–	**3,5-Cyclo-27-norstigmastan-25-one, 6-methoxy-, (3β,5α,6β,24ξ)-**	69081-93-2	**NI42232**
410 428 377 253 395 157 392 87	**428**	100 72 69 64 54 42 35 32		29	48	2	–	–	–	–	–	–	–	–	–	–	**26,27-Dimethylcholesta-5,7-diene-3β,25-diol**		**MS09613**
428 57 43 429 99 41 73 81	**428**	100 49 49 40 36 35 33 31		29	48	2	–	–	–	–	–	–	–	–	–	–	**6,6-Ethylenedioxycholest-4-ene**		**LI08890**
291 373 87 428 165 43 167 99	**428**	100 97 68 67 65 52 48 44		29	48	2	–	–	–	–	–	–	–	–	–	–	**6,6-Ethylenedioxy-3α,5-cyclo-5α-cholestane**		**LI08891**
383 43 398 55 384 410 57 41	**428**	100 91 53 49 47 35 33 33	**23.00**	29	48	2	–	–	–	–	–	–	–	–	–	–	**19-Hydroxy-β-sitost-4-en-3-one**		**LI08892**
71 69 55 373 428 396 282 413	**428**	100 80 64 54 47 39 36 34		29	48	2	–	–	–	–	–	–	–	–	–	–	**(24R)-6β-Methoxy-3α,5-cyclo-5α-ergostan-23-one**		**DD01661**
271 385 255 410 213 428 231 367	**428**	100 51 41 32 30 27 26 21		29	48	2	–	–	–	–	–	–	–	–	–	–	**5α-Stigmasta-7,28-diene-3β,24-diol**		**LI08893**
44 55 43 28 57 69 98 41	**428**	100 33 30 29 26 24 20 20	**13.50**	29	48	2	–	–	–	–	–	–	–	–	–	–	**5α-Stigmastan-3,6-dione**		**MS09614**
428 429 69 44 430 70 45 230	**428**	100 39 12 12 9 7 7 6		29	52	–	2	–	–	–	–	–	–	–	–	–	**Cholestan-3-one, dimethylhydrazone, (5α)-**	54966-63-1	**NI42233**
428 91 43 69 133 109 83 322	**428**	100 73 34 24 23 18 13 10		30	36	2	–	–	–	–	–	–	–	–	–	–	**2,6,10,15,19,25-Hexamethyltetracosa-2,4,6,8,10,12,14,16,18,20,22-undecan-1,24-dial**		**LI08894**
43 428 55 41 69 81 57 273	**428**	100 69 52 51 40 35 35 34		30	52	1	–	–	–	–	–	–	–	–	–	–	**Cholestan-3-one, 2-isopropyl-, (2α,5α)-**	55162-60-2	**NI42234**
357 428 396 358 381 429 275 122	**428**	100 55 36 29 20 19 14 14		30	52	1	–	–	–	–	–	–	–	–	–	–	**Cholest-5-ene, 3-methoxy-4,4-dimethyl-, (3β)-**	35490-53-0	**NI42235**
95 93 105 107 382 91 119 121	**428**	100 94 74 71 65 64 59 56	**11.06**	30	52	1	–	–	–	–	–	–	–	–	–	–	**Cyclopropa[5,6]cholestane, 3-ethoxy-3′,6-dihydro-, (3α,5α,6β)-**	35339-43-6	**NI42236**
99 95 110 106 120 93 81 104	**428**	100 92 80 70 65 62 62 59	**17.86**	30	52	1	–	–	–	–	–	–	–	–	–	–	**Cyclopropa[5,6]cholestane, 3-ethoxy-3′,6-dihydro-, (3β,5α,6β)-**	51414-68-7	**NI42237**
397 95 109 81 69 259 123 137	**428**	100 56 51 51 51 38 36 34	**2.06**	30	52	1	–	–	–	–	–	–	–	–	–	–	**Desoxocanophyllol**		**MS09615**
109 123 107 111 105 121 135 108	**428**	100 90 68 62 61 57 47 47	**7.92**	30	52	1	–	–	–	–	–	–	–	–	–	–	**Ergost-5-en-3-ol, 22,23-dimethyl-, (3β)-**	55724-23-7	**NI42238**
428 43 57 55 109 275 69 165	**428**	100 65 52 49 48 46 42 41		30	52	1	–	–	–	–	–	–	–	–	–	–	**D:A-Friedooleanan-2-ol**	56588-26-2	**NI42239**
43 428 57 55 231 109 123 69	**428**	100 99 87 82 74 73 70 68		30	52	1	–	–	–	–	–	–	–	–	–	–	**D:A-Friedooleanan-3-ol, (3α)-**	5085-72-3	**NI42240**

MASS TO CHARGE RATIOS								M.W.	INTENSITIES								Parent	C	H	O	N	S	F	Cl	Br	I	Si	P	B	X	COMPOUND NAME	CAS Reg No	No
428	57	43	55	165	69	107	109	**428**	100	76	74	71	65	63	62	61		30	52	1	–	–	–	–	–	–	–	–	–	–	**D:A-Friedooleanan-3-ol, (3β)-**	16844-71-6	**NI42241**
275	428	205	413	207	189	304	341	**428**	100	88	75	55	38	30	13	3		30	52	1	–	–	–	–	–	–	–	–	–	–	**D:A-Friedooleanan-3-ol, (3β)-**	16844-71-6	**LI08895**
95	109	205	193	410	395	191	231	**428**	100	67	32	30	28	19	19	12	**2.00**	30	52	1	–	–	–	–	–	–	–	–	–	–	**D:A-Friedooleanan-7-ol, (7α)-**	18671-57-3	**MS09616**
177	95	69	109	55	81	123	205	**428**	100	88	78	73	59	58	54	45	**2.00**	30	52	1	–	–	–	–	–	–	–	–	–	–	**D:A-Friedooleanan-7-ol, (7α)-**	18671-57-3	**NI42242**
109	205	123	397	231	149	95	137	**428**	100	95	80	65	50	45	45	36	**1.00**	30	52	1	–	–	–	–	–	–	–	–	–	–	**D:A-Friedooleanan-24-ol**	2130-12-3	**MS09617**
259	204	428	141	233	217	218	413	**428**	100	91	90	80	73	63	59	50		30	52	1	–	–	–	–	–	–	–	–	–	–	**D:A-Friedooleanan-29-ol, (20α)-**	39903-18-9	**MS09618**
69	81	121	95	136	123	68	67	**428**	100	85	74	71	61	59	56	53	**15.92**	30	52	1	–	–	–	–	–	–	–	–	–	–	**C(14a)-Homo-27-nor-14β-gammaceran-3α-ol**	24739-08-0	**NI42243**
69	81	121	95	136	123	67	109	**428**	100	85	74	72	63	60	54	51	**19.00**	30	52	1	–	–	–	–	–	–	–	–	–	–	**C(14a)-Homo-27-nor-14β-gammaceran-3α-ol**	24739-08-0	**LI08896**
428	69	95	81	55	413	43	109	**428**	100	72	57	55	55	54	53	42		30	52	1	–	–	–	–	–	–	–	–	–	–	**Lanostane, 11,18-epoxy-, (11β)-**	52474-94-9	**NI42244**
53	56	69	95	288	81	428	175	**428**	100	92	71	60	51	48	47	45		30	52	1	–	–	–	–	–	–	–	–	–	–	**Lanostan-1-one**	16196-36-4	**NI42245**
273	288	204	43	245	55	83	69	**428**	100	63	62	58	56	46	42	41	**8.00**	30	52	1	–	–	–	–	–	–	–	–	–	–	**Lanostan-3-one**	4639-29-6	**NI42246**
206	69	428	55	164	95	123	109	**428**	100	52	51	45	39	37	33	33		30	52	1	–	–	–	–	–	–	–	–	–	–	**Lanostan-7-one**	52475-04-4	**NI42247**
206	428	55	164	95	109	123	57	**428**	100	50	45	38	37	34	33	31		30	52	1	–	–	–	–	–	–	–	–	–	–	**Lanostan-7-one**	52475-04-4	**NI42248**
205	428	303	95	69	429	81	55	**428**	100	98	78	43	41	32	31	27		30	52	1	–	–	–	–	–	–	–	–	–	–	**Lanostan-11-one**	50633-28-8	**NI42250**
205	303	95	69	429	81	290	55	**428**	100	78	44	41	33	32	30	28	**9.64**	30	52	1	–	–	–	–	–	–	–	–	–	–	**Lanostan-11-one**	50633-28-8	**NI42249**
275	276	69	95	55	81	44	41	**428**	100	22	12	9	9	7	7	6	**4.04**	30	52	1	–	–	–	–	–	–	–	–	–	–	**Lanostan-12-one**	52474-77-8	**NI42251**
413	69	55	395	95	57	83	119	**428**	100	68	66	63	55	48	38	35	**22.00**	30	52	1	–	–	–	–	–	–	–	–	–	–	**Lanost-8-en-3-ol, (3β)-**	79-62-9	**NI42253**
411	427	428	247	412	413	429	125	**428**	100	54	38	36	30	25	15	14		30	52	1	–	–	–	–	–	–	–	–	–	–	**Lanost-8-en-3-ol, (3β)-**	79-62-9	**NI42252**
191	207	428	192	189	208	190	206	**428**	100	52	41	22	20	11	10	9		30	52	1	–	–	–	–	–	–	–	–	–	–	**Wallichiniol**		**LI08897**
28	428	274	29	55	95	205	403	**428**	100	87	81	71	55	51	50	48		31	56	–	–	–	–	–	–	–	–	–	–	–	**3-Butylcholestane**		**MS09619**
370	315	371	215	355	216	257	217	**428**	100	36	31	31	26	23	15	10	**5.00**	31	56	–	–	–	–	–	–	–	–	–	–	–	**3-Butylcholestane**		**MS09620**
95	81	136	137	67	55	69	41	**428**	100	97	95	90	58	51	46	45	**0.14**	31	56	–	–	–	–	–	–	–	–	–	–	–	**1,1-Di-(1′-decahydronaphthyl)undecane**		**AI02368**
91	259	43	105	104	57	92	41	**428**	100	35	18	16	13	11	9	9	**7.51**	31	56	–	–	–	–	–	–	–	–	–	–	–	**13-Phenylpentacosane**		**AI02370**
91	43	105	57	41	119	104	55	**428**	100	33	22	20	18	12	12	12	**0.00**	31	56	–	–	–	–	–	–	–	–	–	–	–	**13-Phenylpentacosane**		**AI02369**
91	105	57	43	259	119	133	104	**428**	100	22	17	17	16	13	11	10	**1.80**	31	56	–	–	–	–	–	–	–	–	–	–	–	**13-Phenylpentacosane**		**AI02371**
428	429	427	426	430	425	424	423	**428**	100	36	15	14	7	6	4	2		32	16	–	2	–	–	–	–	–	–	–	–	–	**5,10-Diazabenzo[rst]phenaleno[1,2,3-de]pentaphene**		**NI42254**
428	429	427	426	430	425	424	399	**428**	100	36	13	13	7	4	4	4		32	16	–	2	–	–	–	–	–	–	–	–	–	**5,17-Diazadibenzo[a,rst]naphtho[8,12-cde]pentaphene**		**NI42255**
428	429	426	430	427	425	399	424	**428**	100	36	8	7	7	3	3	2		32	16	–	2	–	–	–	–	–	–	–	–	–	**5,14-Diazadinaphtho[1,2,3-cd;1′,2′,3′-lm]perylene**	188-85-2	**NI42256**
428	429	427	426	425	430	424	399	**428**	100	36	21	11	7	6	6	3		32	16	–	2	–	–	–	–	–	–	–	–	–	**3,12-Diazatetrabenzo[a,cd,j,lm]perylene**		**NI42257**
428	427	429	426	425	424	430	423	**428**	100	77	31	22	9	8	6	3		32	16	–	2	–	–	–	–	–	–	–	–	–	**5,14-Diazatetrabenzo[a,cd,lm,o]perylene**		**NI42258**
398	396	108	400	110	394	117	119	**429**	100	89	69	64	43	35	34	33	**2.00**	6	–	–	3	–	–	9	–	–	–	–	–	–	**1,3,5-Triazine, 2,4,6-tris(trichloromethyl)-**	6542-67-2	**NI42259**
429	121	125	317	345	153	373	188	**429**	100	55	46	41	32	30	24	22		8	20	4	–	4	–	–	–	–	–	2	–	1	**Cobalt[II] bis(diethyldithiophosphate)**	37511-99-2	**NI42260**
55	41	396	43	398	394	56	83	**429**	100	93	77	77	54	47	45	36	**1.60**	11	13	–	3	1	–	6	–	–	–	–	–	–	**2-(Hexylthio)-4,6-bis(trichloromethyl)-s-triazine**		**MS09621**
56	69	85	170	57	311	198	113	**429**	100	48	38	28	20	17	17	17	**0.00**	13	13	5	3	2	–	1	–	–	–	–	–	1	**Potassium 4-[(5-chloro-2-oxo-3(2H)-benzothiazolyl)acetyl]-1-piperazine sulphonate**		**NI42261**
159	265	253	119	266	90	118	91	**429**	100	87	80	77	75	62	55	52	**2.20**	14	9	3	1	–	10	–	–	–	–	–	–	–	**Propanamide, 2,2,3,3,3-pentafluoro-N-(1-hydroxy-2-phenylethyl)-, mono(2,2,3,3,3-pentafluoro-1-oxopropyl)-**	72347-72-9	**NI42262**
253	266	119	225	105	159	77	91	**429**	100	73	72	53	36	33	22	19	**1.20**	14	9	3	1	–	10	–	–	–	–	–	–	–	**Propanamide, 2,2,3,3,3-pentafluoro-N-(2-hydroxy-2-phenylethyl)-, mono(2,2,3,3,3-pentafluoro-1-oxopropyl)-**		**MS09622**
266	176	147	253	119	267	91	103	**429**	100	36	33	21	19	18	7	4	**0.00**	14	9	3	1	–	10	–	–	–	–	–	–	–	**Propanamide, 2,2,3,3,3-pentafluoro-N-[2-(2-hydroxyphenyl)ethyl]-, mono(2,2,3,3,3-pentafluoro-1-oxopropyl)-**		**NI42263**
266	176	147	253	119	267	148	103	**429**	100	36	33	21	19	18	4	4	**0.00**	14	9	3	1	–	10	–	–	–	–	–	–	–	**Propanamide, 2,2,3,3,3-pentafluoro-N-[2-(2-hydroxyphenyl)ethyl]-, mono(2,2,3,3,3-pentafluoro-1-oxopropyl)-**	72347-70-7	**NI42264**
266	176	119	147	253	267	91	78	**429**	100	46	39	26	24	18	14	14	**0.00**	14	9	3	1	–	10	–	–	–	–	–	–	–	**Propanamide, 2,2,3,3,3-pentafluoro-N-[2-(2-hydroxyphenyl)ethyl]-, mono(2,2,3,3,3-pentafluoro-1-oxopropyl)-**		**MS09623**
266	176	267	253	119	103	91	147	**429**	100	23	16	9	6	6	4	3	**1.30**	14	9	3	1	–	10	–	–	–	–	–	–	–	**Propanamide, 2,2,3,3,3-pentafluoro-N-[2-(3-hydroxyphenyl)ethyl]-, mono(2,2,3,3,3-pentafluoro-1-oxopropyl)-**		**NI42265**
266	176	267	253	119	103	147		**429**	100	23	16	9	6	6	3		**0.00**	14	9	3	1	–	10	–	–	–	–	–	–	–	**Propanamide, 2,2,3,3,3-pentafluoro-N-[2-(3-hydroxyphenyl)ethyl]-, mono(2,2,3,3,3-pentafluoro-1-oxopropyl)-**	72347-71-8	**NI42266**
266	176	119	267	253	103	78	69	**429**	100	33	30	16	15	12	12	10	**1.20**	14	9	3	1	–	10	–	–	–	–	–	–	–	**Propanamide, 2,2,3,3,3-pentafluoro-N-[2-(3-hydroxyphenyl)ethyl]-, mono(2,2,3,3,3-pentafluoro-1-oxopropyl)-**		**MS09624**
266	119	253	267	78	225	69	91	**429**	100	45	25	19	13	12	10	8	**0.00**	14	9	3	1	–	10	–	–	–	–	–	–	–	**Propanamide, 2,2,3,3,3-pentafluoro-N-[2-(4-hydroxyphenyl)ethyl]-, mono(2,2,3,3,3-pentafluoro-1-oxopropyl)-**		**MS09625**

MASS TO CHARGE RATIOS								M.W.	INTENSITIES								Parent	C	H	O	N	S	F	Cl	Br	I	Si	P	B	X	COMPOUND NAME	CAS Reg No	No
266	119	253	267	176	225	78	91	**429**	100	27	20	13	12	10	5	4	**0.00**	14	9	3	1	–	10	–	–	–	–	–	–	–	Propanamide, 2,2,3,3,3-pentafluoro-N-[2-(4-hydroxyphenyl)ethyl]-, mono(2,2,3,3,3-pentafluoro-1-oxopropyl)-		NI42267
266	119	253	267	176	225	109	91	**429**	100	27	20	13	12	10	4	4	**0.00**	14	9	3	1	–	10	–	–	–	–	–	–	–	Propanamide, 2,2,3,3,3-pentafluoro-N-[2-(4-hydroxyphenyl)ethyl]-, mono(2,2,3,3,3-pentafluoro-1-oxopropyl)-	72347-73-0	NI42268
73	315	414	103	299	187	211	75	**429**	100	63	39	33	32	29	18	17	**0.60**	14	36	6	1	–	–	–	–	–	3	1	–	–	Phosphoric acid, 2-(ethoxyimino)-3-[(trimethylsilyl)oxy]propyl bis(trimethylsilyl) ester, (E)-	55334-93-5	NI42269
73	315	299	414	187	225	103	316	**429**	100	65	45	40	40	33	33	22	**0.60**	14	36	6	1	–	–	–	–	–	3	1	–	–	Phosphoric acid, 2-(ethoxyimino)-3-[(trimethylsilyl)oxy]propyl bis(trimethylsilyl) ester, (Z)-	55334-94-6	NI42270
73	299	227	384	211	315	328	174	**429**	100	71	65	61	56	38	36	26	**1.80**	14	36	6	1	–	–	–	–	–	3	1	–	–	Phosphoric acid, 3-(ethoxyimino)-2-[(trimethylsilyl)oxy]propyl bis(trimethylsilyl) ester, (E)-(±)-	55334-95-7	NI42271
73	299	328	174	211	384	315	142	**429**	100	93	78	77	65	51	40	32	**0.90**	14	36	6	1	–	–	–	–	–	3	1	–	–	Phosphoric acid, 3-(ethoxyimino)-2-[(trimethylsilyl)oxy]propyl bis(trimethylsilyl) ester, (Z)-(±)-	55334-96-8	NI42272
73	299	188	174	172	300	315	328	**429**	100	99	99	99	88	58	50	45	**0.00**	14	40	4	1	–	–	–	–	–	4	1	–	–	Phosphoric acid, 2-[bis(trimethylsilyl)amino]ethyl bis(trimethylsilyl) ester	55334-92-4	NI42273
73	299	188	174	172	315	300	114	**429**	100	69	47	42	31	22	22	20	**0.30**	14	40	4	1	–	–	–	–	–	4	1	–	–	Phosphoric acid, 2-[bis(trimethylsilyl)amino]ethyl bis(trimethylsilyl) ester	55334-92-4	NI42274
299	73	188	174	172	315	300	114	**429**	100	99	67	61	44	31	31	29	**0.00**	14	40	4	1	–	–	–	–	–	4	1	–	–	Phosphoric acid, 2-[bis(trimethylsilyl)amino]ethyl bis(trimethylsilyl) ester	55334-92-4	NI42275
114	115	430	129	374	328	101	141	**429**	100	53	24	24	16	13	13	11	**0.00**	17	17	5	1	–	6	–	–	–	–	–	–	–	L-Tyrosine, N-(trifluoroacetyl)-, butyl ester, trifluoroacetate (ester)	5282-98-4	NI42276
203	260	316	69	57	261	41	175	**429**	100	95	43	24	21	18	17	15	**0.00**	17	17	5	1	–	6	–	–	–	–	–	–	–	L-Tyrosine, N-(trifluoroacetyl)-, butyl ester, trifluoroacetate (ester)	5282-98-4	NI42277
400	398	104	77	399	401	431	429	**429**	100	97	53	26	22	21	20	18		18	16	3	5	–	–	–	1	–	–	–	–	–	N-Hydroxyethyl-N-(2-bromoallyl)-4-(2-cyano-4-nitrophenylazo)aniline		IC04386
240	312	122	73	75	313	241	80	**429**	100	77	73	70	27	22	19	15	**1.79**	19	39	4	1	–	–	–	–	–	3	–	–	–	Kainic acid, tris(trimethylsilyl)-		NI42278
184	156	43	287	115	429	56	185	**429**	100	24	21	17	14	13	11	9		22	23	8	1	–	–	–	–	–	–	–	–	–	Dimethyl 2′-[6,7-bis(methoxycarbonyl)-3,4-benzo-2-azabicyclo[3.2.0]hepta 3,6-dienyl]fumarate		NI42279
184	156	287	56	429	115	185	71	**429**	100	24	20	17	15	13	10	8		22	23	8	1	–	–	–	–	–	–	–	–	–	Dimethyl 2′-[6,7-bis(methoxycarbonyl)-3,4-benzo-2-azabicyclo[3.2.0]hepta 3,6-dienyl]maleate		NI42280
184	156	115	56	398	185	429	287	**429**	100	40	22	14	9	9	7	5		22	23	8	1	–	–	–	–	–	–	–	–	–	Dimethyl 2′-[2,5-dimethyl-3,4-bis(methoxycarbonyl)-6,7-benzo-1H-azepin 1-yl]maleate		NI42281
311	310	312	252	253	343	429	338	**429**	100	40	35	30	22	20	15	15		22	23	8	1	–	–	–	–	–	–	–	–	–	Tetramethyl 5,6,10,11-tetrahydroazepino[1,2-a]quinoline-7,8,9,10-tetracarboxylate		LI08898
311	312	310	343	252	429	253	338	**429**	100	30	30	29	24	20	20	10		22	23	8	1	–	–	–	–	–	–	–	–	–	Tetramethyl 5,6,10,11-tetrahydroazepino[1,2-a]quinoline-7,8,9,11-tetracarboxylate		LI08899
139	141	111	73	312	429	140	75	**429**	100	32	19	19	17	15	9	8		22	24	4	1	–	–	1	–	–	1	–	–	–	1H-Indole-3-acetic acid, 1-(4-chlorobenzoyl)-5-methoxy-2-methyl-, trimethylsilyl ester	55334-98-0	NI42282
139	141	73	111	140	75	312	113	**429**	100	36	30	29	14	14	12	12	**10.70**	22	24	4	1	–	–	1	–	–	1	–	–	–	1H-Indole-3-acetic acid, 1-(4-chlorobenzoyl)-5-methoxy-2-methyl-, trimethylsilyl ester	55334-98-0	NI42283
139	141	429	111	73	140	431	113	**429**	100	32	18	17	12	8	7	6		22	24	4	1	–	–	1	–	–	1	–	–	–	1H-Indole-3-acetic acid, 1-(4-chlorobenzoyl)-2-methyl-5-[(trimethylsilyl)oxy]-, methyl ester	55334-99-1	NI42284
324	43	429	309	326	325	311	310	**429**	100	73	62	43	35	29	19	18		23	27	7	1	–	–	–	–	–	–	–	–	–	(Hydroxymethyl)colchicine		NI42285
414	430	340	371					**429**	100	74	52	18					**0.00**	23	35	3	1	–	–	–	–	–	2	–	–	–	Morphinan, 7,8-didehydro-4,5-epoxy-17-methyl-3,6-bis[(trimethylsilyl)oxy]-, (5α,6α)-	55449-66-6	NI42286
429	73	430	414	236	431	196	401	**429**	100	54	45	37	36	17	17	14		23	35	3	1	–	–	–	–	–	2	–	–	–	Morphinan, 7,8-didehydro-4,5-epoxy-17-methyl-3,6-bis[(trimethylsilyl)oxy]-, (5α,6α)-	55449-66-6	NI42287
98	84	55	58	41	43	36	60	**429**	100	78	72	36	36	33	26	21	**0.00**	23	43	6	1	–	–	–	–	–	–	–	–	–	1-Propanaminium, 3-carboxy-2-[(15-carboxy-1-oxopentadecyl)oxy]-N,N,N-trimethyl-	55570-92-8	NI42288
105	205	77	197	104	429	282	119	**429**	100	96	96	72	65	55	45	45		24	19	5	3	–	–	–	–	–	–	–	–	–	1-Benzoyl-5-(methoxycarbonylamino)-3,5-diphenylimidazolidine-2,4-dione		MS09626
387	388	61	386	385	389	357	70	**429**	100	23	7	6	6	5	5	5	**0.00**	24	31	6	1	–	–	–	–	–	–	–	–	–	Estradiol-6-one, 3,17-bis(acetyloxy)-2-methoxy-, O-(methyloxime), (17β)-		NI42289
369	370	338	284	339	73	371	283	**429**	100	33	28	13	12	11	10	10	**0.00**	24	35	4	1	–	–	–	–	–	1	–	–	–	14α-Hydroxy-estrone-3-acetate-methyloxime-TMS		NI42290
429	430	358	431	428	73	340	427	**429**	100	35	19	12	7	5	4	3		24	39	2	1	–	–	–	–	–	2	–	–	–	2-Aminoestrone DITMS		NI42291
357	358	342	359	233	343	272	234	**429**	100	31	14	8	5	4	4	3	**0.58**	24	39	2	1	–	–	–	–	–	2	–	–	–	4-Aminoestrone DITMS		NI42292
91	204	232	92	119	104	65	160	**429**	100	99	82	79	77	68	63	62	**0.80**	25	19	6	1	–	–	–	–	–	–	–	–	–	Dibenzyl phthalimidomalonate	38068-70-1	NI42293
260	339	170	274	429	259	156	400	**429**	100	67	55	54	50	43	39	37		25	39	3	1	–	–	–	–	–	1	–	–	–	18,19-Dinorpregn-4-en-20-yn-3-one, 13-ethyl-16β-hydroxy-17β-[(trimethylsilyl)oxy]-, O-(methyloxime)		NI42294
330	86	30	100	58	44	331	42	**429**	100	78	44	43	37	27	23	21	**3.00**	26	27	3	3	–	–	–	–	–	–	–	–	–	1-Amino-2-(4-hydroxyphenyl)-4-[(diethylamino)ethylamino]anthraquinone		IC04387
364	301	149	429	244	245	230	402	**429**	100	99	84	78	60	55	44	38		27	19	1	5	–	–	–	–	–	–	–	–	–	5-(1,1,2,2-Tetracyanoethyl)-1-(2-methylphenyl)-2-phenyl-4-oxo-4,5,6,7-tetrahydroindole	106012-23-1	NI42295

MASS TO CHARGE RATIOS								M.W.	INTENSITIES								Parent	C	H	O	N	S	F	Cl	Br	I	Si	P	B	X	COMPOUND NAME	CAS Reg No	No
402	364	375	400	301	244	245	273	**429**	100	99	85	50	50	42	40	37	**25.00**	27	19	1	5	–	–	–	–	–	–	–	–	–	**5-(1,1,2,2-Tetracyanoethyl)-1-(3-methylphenyl)-2-phenyl-4-oxo-4,5,6,7-tetrahydroindole**	106012-24-2	**NI42296**
91	105	107	115	135	149	131	299	**429**	100	50	30	20	19	18	18	16	**0.00**	27	19	1	5	–	–	–	–	–	–	–	–	–	**5-(1,1,2,2-Tetracyanoethyl)-1-(4-methylphenyl)-2-phenyl-4-oxo-4,5,6,7-tetrahydroindole**	106012-25-3	**NI42297**
254	132	268	146	307	204	170	359	**429**	100	90	50	45	15	11	10	9	**0.00**	27	27	4	1	–	–	–	–	–	–	–	–	–	**1-(2-Benzoyloxyethylphenylamino)-2-benzoyloxybut-3-ene**		**LI08900**
429	128	430	428	127	57	180	114	**429**	100	97	43	29	23	19	16	14		27	43	3	1	–	–	–	–	–	–	–	–	–	**Korsine**		**LI08901**
429	430	400	28	372	428	315	431	**429**	100	67	35	30	20	18	17	14		28	15	4	1	–	–	–	–	–	–	–	–	–	**Dianthramide**		**IC04388**
43	55	81	216	95	41	215	67	**429**	100	88	72	69	69	69	66	62	**34.00**	28	47	–	1	1	–	–	–	–	–	–	–	–	**Cholestane, 3-thiocyanato-, (3α,5α)-**	20997-49-3	**NI42298**
44	58	43	429	70	41	274	55	**429**	100	77	69	61	53	53	52	50		28	47	–	1	1	–	–	–	–	–	–	–	–	**Cholestane, 3-thiocyanato-, (3β,5α)-**	20997-50-6	**NI42299**
138	110	43	55	412	81	95	69	**429**	100	69	62	61	52	45	44	44	**0.00**	28	47	2	1	–	–	–	–	–	–	–	–	–	**A-Homo-5α-cholestane-4,4a-dione 4a-oxime**	123239-92-9	**DD01662**
412	43	81	57	274	95	69	429	**429**	100	92	80	71	69	66	66	51		28	47	2	1	–	–	–	–	–	–	–	–	–	**A-Homo-5α-cholestane-3,4-dione 3-oxime**	123239-93-0	**DD01663**
60	43	149	370	41	57	55	107	**429**	100	77	57	40	37	29	28	21	**21.00**	29	51	1	1	–	–	–	–	–	–	–	–	–	**Cholestane, 3-acetamido-, (3β,5α)-**	1912-64-7	**NI42300**
429	370	430	215	355	135	371	147	**429**	100	67	39	28	25	24	20	20		29	51	1	1	–	–	–	–	–	–	–	–	–	**Cholestane, 4-acetamido-, (4α,5α)-**	13944-35-9	**NI42301**
110	98	401	71	36	28	429	58	**429**	100	48	45	32	30	25	21	21		29	51	1	1	–	–	–	–	–	–	–	–	–	**Cholestan-3-one, 2-(dimethylamino)-, (2β,5α)-**	54725-09-6	**NI42302**
99	42	55	58	140	57	69	81	**429**	100	34	30	29	28	23	21	19	**13.33**	29	51	1	1	–	–	–	–	–	–	–	–	–	**Cholestan-3-one, 4,4-dimethyl-, oxime, (5α)-**	14732-99-1	**NI42303**
427	105	350	428	180	248	194	77	**429**	100	73	45	32	19	11	9	8	**2.00**	30	23	2	1	–	–	–	–	–	–	–	–	–	**2,3-Dibenzoyl-cis-4,5-diphenyl-δ(2)-pyrroline**		**NI42304**
427	105	350	428	180	194	248	77	**429**	100	71	39	31	17	13	11	8	**2.00**	30	23	2	1	–	–	–	–	–	–	–	–	–	**2,3-Dibenzoyl-trans-4,5-diphenyl-δ(2)-pyrroline**		**NI42305**
193	249	429	194	155	180	156	250	**429**	100	17	10	9	8	7	4	3		31	27	1	1	–	–	–	–	–	–	–	–	–	**3,3-Diphenyl-4-butylimino-2-spiro-(9-fluorenyl)-oxetane**		**MS09627**
309	414	352	294	429	249	119	129	**429**	100	85	51	40	33	9	9	8		31	32	–	1	–	–	–	–	–	–	–	1	–	**Boranamine, N-(diphenylmethylene)-1,1-bis(2,4,6-trimethylphenyl)-**	23115-37-9	**NI42306**
213	215	211	337	335	151	149	135	**430**	100	52	52	47	39	25	25	24	**0.00**	6	10	2	–	–	–	–	4	–	–	–	–	–	**2,5-Hexanediol, 1,3,4,6-tetrabromo-**	35972-79-3	**NI42307**
77	318	89	402	249	298	305	223	**430**	100	78	71	60	56	53	39	35	**2.34**	10	5	4	–	–	3	–	–	–	–	–	–	1	**Tungsten, (π-cyclopentadienyl)trifluoroacetyltricarbonyl-**		**MS09628**
51	69	145	101	279	209	95	77	**430**	100	36	27	20	19	19	18	17	**0.05**	10	6	–	–	–	16	–	–	–	–	–	–	–	**Decane, 1,1,2,2,3,3,4,4,7,7,8,8,9,9,10,10-hexadecafluoro-**	40723-66-8	**MS09629**
253	266	105	225	103	119	267	91	**430**	100	97	63	49	47	38	21	15	**0.00**	14	8	4	–	–	10	–	–	–	–	–	–	–	**Propanoic acid, pentafluoro-, 1-phenyl-1,2-ethanediyl ester**	55683-13-1	**NI42308**
43	269	44	42	271	267	287	148	**430**	100	58	51	38	36	33	29	24	**0.00**	14	38	7	–	–	–	–	–	–	4	–	–	–	**3,3,5,5-Tetraethoxy-1,1,1,7,7,7-hexamethyltetrasiloxane**		**MS09630**
73	131	227	211	75	130	285	299	**430**	100	27	20	20	19	18	16	15	**0.00**	15	39	6	–	–	–	–	–	–	3	1	–	–	**Bis-1,2-propanediol phosphate, tris(trimethylsilyl)-**		**NI42309**
142	430	127	216	114	143	141	128	**430**	100	31	17	13	12	10	9	8		16	10	–	–	–	12	–	–	–	–	–	–	–	**4,5-Dimethyl-1,3,6,8,11,11,12,12,13,13,14,14-dodecafluorotetracyclo[6.2.2.2^{3,6}.0^{2,7}]tetradeca-4,9-diene**		**NI42310**
172	69	430	215	173	77	104	103	**430**	100	45	40	38	9	6	6	6		16	16	–	6	–	4	–	–	–	–	2	–	–	**5,5,10,10-Tetrafluoro-1,6-dimethyl-3,8-diphenyl[1,3,2,4]diazadiphospheto[2.1-c:4.3-c']bis[1,2,4,3λ(5)]triazaphosphole**		**NI42311**
166	167	41	29	56	57	69	139	**430**	100	63	33	32	28	15	14	12	**2.80**	16	25	6	2	1	3	–	–	–	–	–	–	–	**N-TFA-L-Prolylmethionine sulphone butyl ester**		**MS09631**
77	107	136	350	430	428	348		**430**	100	13	12	2	1	1	1			18	22	2	–	–	–	–	–	–	–	–	–	2	**4,4'-Bis(4-ethoxybenzyl) diselenide**		**MS09632**
73	268	75	133	74	269	59	239	**430**	100	37	35	11	11	10	10	9	**3.00**	18	34	6	2	–	–	–	–	–	2	–	–	–	**Fumarylacetoacetate diethoxime, bis(trimethylsilyl)-**		**NI42312**
417	415	401	419	28	399	403	331	**430**	100	77	49	48	47	38	24	23	**8.40**	19	14	3	–	–	–	4	–	–	–	–	–	–	**1,2,3,4-Tetrachloro-9,10-dihydro-9,11,12-trimethoxy-9,10-ethenoanthracene**	57187-60-7	**NI42313**
91	92	312	188	358	16	371	328	**430**	100	9	8	8	5	5	4	4	**1.00**	19	18	6	4	1	–	–	–	–	–	–	–	–	**2-(2',4'-Dinitrophenylthio)-melatonin**		**LI08902**
351	353	43	55	41	271	69	57	**430**	100	95	36	35	30	26	26	26	**2.00**	19	28	1	–	–	–	–	2	–	–	–	–	–	**Androstan-15-one, 14,16-dibromo-, (5α,14ξ)-**	54411-96-0	**NI42314**
58	166	73	251	179	324	194	96	**430**	100	52	43	36	24	9	8	5	**0.00**	20	29	3	2	–	3	–	–	–	1	–	–	–	**N-(trifluoroacetyl)prolyl-o-(trimethylsilyl)ephedrine**		**NI42315**
228	301	317	344	270	373	357	400	**430**	100	42	33	26	9	7	6	1	**0.00**	20	34	8	2	–	–	–	–	–	–	–	–	–	**3,10-Bis(tert-butoxymethyl)-2,5,9,12-tetraoxo-1,8-dioxa-4,11-diazocyclotetradecane**		**LI08903**
57	149	172	104	160	85	83	132	**430**	100	70	63	50	40	35	35	30	**4.70**	21	22	6	2	1	–	–	–	–	–	–	–	–	**Methyl 3-butoxymethyl-7-phthalimido-3-cephem-4-carboxylate**		**NI42316**
81	63	47	175	256	430	57	239	**430**	100	18	8	5	4	3	3	2		21	22	8	2	–	–	–	–	–	–	–	–	–	**Demethyldoxycycline**		**NI42317**
44	84	58	46	73	42	387	98	**430**	100	42	39	25	24	17	11	10	**7.91**	21	22	8	2	–	–	–	–	–	–	–	–	–	**2-Naphthacenecarboxamide, 4-(dimethylamino)-1,4,4a,5,5a,6,11,12a-octahydro-3,6,10,12,12a-pentahydroxy-1,11-dioxo-, [4S-(4α,4aα,5aα,6β,12aα)]-**	987-02-0	**NI42318**
371	430	372	312	399	339	253	182	**430**	100	22	20	20	10	9	8	8		21	22	8	2	–	–	–	–	–	–	–	–	–	**Tetramethyl 5-methyl-9,10-dihydro-5H-azepino[1,2-a]benzimidazole-7,8,9,10-tetracarboxylate**		**LI08904**
73	430	429	45	431	147	432	75	**430**	100	32	32	21	20	14	12	11		21	27	2	2	–	–	1	–	–	2	–	–	–	**2H-1,4-Benzodiazepin-2-one, 7-chloro-1,3-dihydro-5-phenyl-1-(trimethylsilyl)-3-[(trimethylsilyl)oxy]-**	55319-93-2	**NI42319**
117	310	167	221	311	189	220	182	**430**	100	67	48	41	20	19	14	14	**0.00**	21	38	7	–	–	–	–	–	–	1	–	–	–	**Cyclopentaneheptanoic acid, 3,5-dihydroxy-2-[(trimethylsiloxy)methyl]-, methyl ester, diacetate**	19236-94-3	**NI42320**
318	56	183	28	262	319	164	108	**430**	100	96	41	40	30	22	15	15	**2.44**	22	15	4	–	–	–	–	–	–	–	1	–	1	**Iron, tetracarbonyl(triphenylphosphine)-**	14649-69-5	**NI42321**
339	269	311	59	398	430	340	310	**430**	100	73	49	29	22	20	20	16		22	22	9	–	–	–	–	–	–	–	–	–	–	**Dimethyl 3,11-(1',2'-dicarbomethoxyethano)-1,2,3,11-tetrahydrodibenzofuran-1,2-dicarboxylate**		**LI08905**

MASS TO CHARGE RATIOS								M.W.	INTENSITIES								Parent	C	H	O	N	S	F	Cl	Br	I	Si	P	B	X	COMPOUND NAME	CAS Reg No	No
430	42	196	43	216	415	431	41	**430**	100	30	29	29	28	27	26	24		22	26	7	2	–	–	–	–	–	–	–	–	–	3-Nor-3,7-secodichotine, 2,7-didehydro-2-deoxy-14-hydroxy-11-methoxy-25-oxo-, (14α)-	55283-42-6	NI42322
130	44	300	43	299	117	42	28	**430**	100	55	33	24	23	22	21	21	**5.66**	22	30	5	4	–	–	–	–	–	–	–	–	–	L-Leucine, N-[N-(N-acetyl-L-alanyl)-L-tryptophyl]-	56272-48-1	NI42323
271	347	75	272	207	348	81	253	**430**	100	65	28	22	20	18	13	9	**0.00**	22	36	2	–	–	–	2	–	–	1	–	–	–	5α-[(Dichloromethyl)dimethylsilyloxy]-5α-androstan-17-one		NI42324
57	141	43	73	127	145	41	71	**430**	100	82	61	44	43	39	38	37	**0.00**	23	46	5	–	–	–	–	–	–	1	–	–	–	Glycerol, 1-nonanoyl-3-octanoyl diester, 2-(trimethylsilyl)-		NI42325
346	43	388	347	430	329	289	92	**430**	100	99	33	33	22	19	11	11		24	14	8	–	–	–	–	–	–	–	–	–	–	5,5′-Diacetoxy-2,2′-binaphthalene-1,1′,4,4′-tetrone		NI42326
279	123	412	165	281	206	91	292	**430**	100	72	65	50	35	35	31	26	**3.00**	24	28	2	2	–	1	1	–	–	–	–	–	–	Amiperone	1580-71-8	LI08906
279	123	412	165	281	206	197	292	**430**	100	72	64	51	36	36	32	27	**3.83**	24	28	2	2	–	1	1	–	–	–	–	–	–	Amiperone	1580-71-8	NI42327
128	206	180	224	168	150	138	178	**430**	100	56	56	54	52	52	50	48	**0.00**	24	30	7	–	–	–	–	–	–	–	–	–	–	Benzoic acid, 2,4-dihydroxy-6-pentyl-, 4-carboxy-3-hydroxy-5-pentylphenyl ester	641-68-9	NI42328
140	141	109	67	110	96	81	69	**430**	100	10	10	9	7	7	7	7	**6.00**	24	34	5	2	–	–	–	–	–	–	–	–	–	4,25-Secoobscurinervan-4-ol, 15,16-dimethoxy-22-methyl-, (4β,22α)-	54678-26-1	NI42329
430	129	340	415	299	431	73	258	**430**	100	78	74	65	61	40	30	25		24	38	3	–	–	–	–	–	–	2	–	–	–	Estra-1,3,5(10)-trien-6-one, 3,17-bis[(trimethylsilyl)oxy]-, (17β)-	69688-40-0	NI42330
430	129	285	431	245	143	312	258	**430**	100	69	45	38	31	29	27	26		24	38	3	–	–	–	–	–	–	2	–	–	–	Estra-1,3,5(10)-trien-16-one, 3,17β-bis[(trimethylsilyl)oxy]-	25876-84-0	NI42331
430	431	73	432	429	428	341	345	**430**	100	38	19	15	10	10	9	6		24	38	3	–	–	–	–	–	–	2	–	–	–	Estra-1,3,5(10)-trien-17-one, 2,3-bis[(trimethylsilyl)oxy]-	51497-38-2	NI42332
430	431	432	341	342	345	325	306	**430**	100	39	14	11	9	6	5	5		24	38	3	–	–	–	–	–	–	2	–	–	–	Estra-1,3,5(10)-trien-17-one, 3,4-bis[(trimethylsilyl)oxy]-	51497-39-3	NI42333
340	430	341	431	280	75	429	218	**430**	100	86	35	31	19	14	13	13		24	38	3	–	–	–	–	–	–	2	–	–	–	Estra-1,3,5(10)-trien-17-one, 3,6-bis[(trimethylsilyl)oxy]-, (6β)-	69688-25-1	NI42334
430	340	242	218	431	243	313	341	**430**	100	70	56	55	37	34	28	23		24	38	3	–	–	–	–	–	–	2	–	–	–	Estra-1,3,5(10)-trien-17-one, 3,11-bis[(trimethylsilyl)oxy]-, (11β)-	69688-38-6	NI42335
340	341	283	148	339	312	192	342	**430**	100	27	10	10	9	9	9	8	**7.14**	24	38	3	–	–	–	–	–	–	2	–	–	–	Estra-1,3,5(10)-trien-17-one, 3,12-bis[(trimethylsilyl)oxy]-, (12β)-	69688-39-7	NI42336
415	323	430	321	218	416	431	324	**430**	100	76	67	47	43	35	24	21		24	38	3	–	–	–	–	–	–	2	–	–	–	Estra-1,3,5(10)-trien-17-one, 3,12-bis[(trimethylsilyl)oxy]-, (12β)-	69688-39-7	NI42337
340	341	312	283	73	284	270	244	**430**	100	30	16	14	14	12	12	12	**4.93**	24	38	3	–	–	–	–	–	–	2	–	–	–	Estra-1,3,5(10)-trien-17-one, 3,14-bis[(trimethylsilyl)oxy]-	74299-19-7	NI42338
431	286	244	432	218	245	231	430	**430**	100	83	42	41	40	39	33	28		24	38	3	–	–	–	–	–	–	2	–	–	–	Estra-1,3,5(10)-trien-17-one, 3,15-bis[(trimethylsilyl)oxy]-, (15α)-	69688-01-3	NI42339
430	431	286	218	245	244	432	258	**430**	100	38	32	32	15	15	14	13		24	38	3	–	–	–	–	–	–	2	–	–	–	Estra-1,3,5(10)-trien-17-one, 3,15-bis[(trimethylsilyl)oxy]-, (15β)-	74231-49-5	NI42340
286	287	430	231	218	129	431	288	**430**	100	26	19	8	8	8	7	7		24	38	3	–	–	–	–	–	–	2	–	–	–	Estra-1,3,5(10)-trien-17-one, 3,16-bis[(trimethylsilyl)oxy]-, (16β)-		NI42341
286	287	430	231	218	244	431	180	**430**	100	24	17	12	12	10	8	8		24	38	3	–	–	–	–	–	–	2	–	–	–	Estra-1,3,5(10)-trien-17-one, 3,16α-bis[(trimethylsilyl)oxy]-	28416-72-0	NI42342
286	73	430	287	231	218	77	75	**430**	100	36	28	26	11	10	10	8		24	38	3	–	–	–	–	–	–	2	–	–	–	Estra-1,3,5(10)-trien-17-one, 3,16α-bis[(trimethylsilyl)oxy]-	28416-72-0	LI08907
340	341	339	312	430	342	283	148	**430**	100	34	11	10	9	9	9	9		24	38	3	–	–	–	–	–	–	2	–	–	–	Estra-1,3,5(10)-trien-17-one, 3,17-bis[(trimethylsilyl)oxy]-, (17α)-		NI42343
399	400	415	96	87	401	383	82	**430**	100	32	17	10	10	9	8	8	**1.24**	24	38	3	2	–	–	–	–	–	1	–	–	–	Androstane-2-carbonitrile, 4,5-epoxy-3-(methoxyimino)-17-[(trimethylsilyl)oxy]-, (2α,4α,5α,17β)-	74299-06-2	NI42344
187	73	75	345	97	95	103	149	**430**	100	48	28	24	23	22	20	20	**0.00**	24	54	2	–	–	–	–	–	–	2	–	–	–	Octadecane-1,12-diol, bis(trimethylsilyl)-		NI42345
229	303	215	317	304	230	83	97	**430**	100	84	38	28	27	24	12	10	**1.00**	24	54	2	–	–	–	–	–	–	2	–	–	–	1,9-Bis[(trimethylsilyl)oxy]octadecane		NI42346
215	317	229	303	216	318	83	97	**430**	100	66	31	23	21	20	11	10	**1.00**	24	54	2	–	–	–	–	–	–	2	–	–	–	1,10-Bis[(trimethylsilyl)oxy]octadecane		NI42347
254	430	339	255	431	359	429	340	**430**	100	25	18	16	7	5	4	4		25	26	3	4	–	–	–	–	–	–	–	–	–	N-Benzyl-N-pentyl-4′-nitroazobenzenecarboxamide		MS09633
122	43	430	95	370	91	149	107	**430**	100	50	41	41	40	39	31	31		25	34	6	–	–	–	–	–	–	–	–	–	–	Di-O-acetylisodihydroanhydrohirundigenin		LI08908
43	41	91	122	312	121	77	39	**430**	100	25	17	16	15	14	14	14	**1.30**	25	34	6	–	–	–	–	–	–	–	–	–	–	Ingenol-5,20-monoacetonide-3-acetate		NI42348
243	147	412	107	105	121	135	109	**430**	100	45	41	39	37	36	35	35	**13.01**	25	39	5	–	–	–	–	–	–	–	–	1	–	Pregnane-11,20-dione, 17,21-[(butylborylene)bis(oxy)]-3-hydroxy-, (3α,5β)-	30888-30-3	NI42349
243	55	43	41	412	79	93	67	**430**	100	59	57	57	49	49	47	44	**11.00**	25	39	5	–	–	–	–	–	–	–	–	1	–	Pregnane-11,20-dione, 17,21-[(tert-butylborylene)bis(oxy)]-3-hydroxy-, (3α,5β)-		MS09634
415	57	43	41	45	55	90	71	**430**	100	68	56	56	53	41	39	36	**33.82**	26	22	6	–	–	–	–	–	–	–	–	–	–	5,5′-Dimethoxy-3,3′,7,7′-tetramethyl-2,2′-binaphthalene-1,1′,4,4′-tetrone	85485-13-8	NI42350
200	105	415	172	430	77	133	186	**430**	100	80	73	72	63	60	57	45		26	26	2	2	1	–	–	–	–	–	–	–	–	Benzoxazole, 2,2′-(2,5-thiophenediyl)bis[5-(1,1-dimethylethyl)-	7128-64-5	NI42351
93	75	81	109	55	107	135	95	**430**	100	75	65	60	48	41	35	31	**0.00**	26	38	3	–	1	–	–	–	–	–	–	–	–	(E,E,E)-(5-Phenylsulphonylgeranyl)geraniol		NI42352
123	43	122	135	253	91	151	28	**430**	100	95	78	68	66	60	53	51	**0.00**	26	38	5	–	–	–	–	–	–	–	–	–	–	9-Desoxo-9(R)-hydroxy-ingenol-3,4,5,20-diacetonide		NI42353
43	41	59	91	77	55	79	53	**430**	100	71	46	39	32	32	29	26	**0.00**	26	38	5	–	–	–	–	–	–	–	–	–	–	9-Desoxo-9(S)-hydroxy-ingenol-3,4,5,20-diacetonide		NI42354
355	430	271	431	356	55	41	253	**430**	100	99	63	31	27	19	19	18		27	42	2	–	1	–	–	–	–	–	–	–	–	Furost-5-en-3-ol, 22,26-epithio-, (3β,22α,25R)-	55028-76-7	NI42355
255	316	107	105	147	159	161	119	**430**	100	59	50	50	43	36	34	32	**0.00**	27	42	4	–	–	–	–	–	–	–	–	–	–	Chol-3-en-24-oic acid, 12-(acetyloxy)-, methyl ester, (5β,12α)-	60354-34-9	NI42356
252	109	105	107	147	111	145	370	**430**	100	99	94	91	90	85	81	71	**0.00**	27	42	4	–	–	–	–	–	–	–	–	–	–	Chol-5-en-24-oic acid, 3-(acetyloxy)-, methyl ester, (3β)-	31823-53-7	NI42357
105	119	228	131	107	145	255	213	**430**	100	93	88	88	85	73	71	68	**0.50**	27	42	4	–	–	–	–	–	–	–	–	–	–	Chol-11-en-24-oic acid, 3-(acetyloxy)-, methyl ester, (3α,5β)-	2242-12-8	NI42358
413	395	414	396	377	415	397	379	**430**	100	40	28	14	9	6	6	5	**0.00**	27	42	4	–	–	–	–	–	–	–	–	–	–	Cholest-5-ene-16,22-dione, 3,26-dihydroxy-	55925-29-6	NI42359
115	69	55	316	41	139	97	43	**430**	100	80	50	48	48	43	38	37	**0.00**	27	42	4	–	–	–	–	–	–	–	–	–	–	Cholest-5-ene-16,22-dione, 3,26-dihydroxy-	55925-29-6	NI42360
412	180	181	413	166	124	125	43	**430**	100	88	52	27	20	9	8	7	**0.00**	27	42	4	–	–	–	–	–	–	–	–	–	–	Kryptogenin	468-99-5	NI42361
180	181	124	412	166	125	55	44	**430**	100	46	45	38	28	22	17	16	**0.00**	27	42	4	–	–	–	–	–	–	–	–	–	–	Kryptogenin	468-99-5	MS09635
412	180	181	413	166	124	125	165	**430**	100	88	52	27	20	10	9	8	**0.00**	27	42	4	–	–	–	–	–	–	–	–	–	–	Kryptogenin	468-99-5	MS09636
105	107	145	119	131	106	159	201	**430**	100	48	45	39	38	38	33	31	**0.00**	27	42	4	–	–	–	–	–	–	–	–	–	–	24-Norchol-22-ene-3,7-diol, diacetate, (3α,5β,7α)-	57071-90-6	NI42362
255	105	107	146	161	145	119	256	**430**	100	98	78	77	56	54	47	46	**0.00**	27	42	4	–	–	–	–	–	–	–	–	–	–	24-Norchol-22-ene-3,12-diol, diacetate, (3α,5β,12α)-	21152-87-4	NI42363
126	153	149	127	139	105	180	155	**430**	100	62	46	42	34	22	21	20	**15.30**	27	42	4	–	–	–	–	–	–	–	–	–	–	Pennogenin	507-89-1	NI42364
273	221	45	97	46	71	274	83	**430**	100	53	38	35	31	29	27	27	**1.40**	27	42	4	–	–	–	–	–	–	–	–	–	–	5α-Spirostan-23-one, 3β-hydroxy-, (22S,25R)-	28404-66-2	NI42365

MASS TO CHARGE RATIOS								M.W.	INTENSITIES								Parent	C	H	O	N	S	F	Cl	Br	I	Si	P	B	X	COMPOUND NAME	CAS Reg No	No
168	139	126	169	140	82	69	55	**430**	100	88	21	12	11	9	9	8	**0.00**	27	42	4	–	–	–	–	–	–	–	–	–	–	**Spirostan-15-one, 3-hydroxy-, (3β,5α,14β,25R)-**	54965-91-2	**NI42366**
168	139	126	430	169	140	55	82	**430**	100	79	25	23	13	10	10	9		27	42	4	–	–	–	–	–	–	–	–	–	–	**Spirostan-15-one, 3-hydroxy-, (3β,5α,14β,25R)-**	54965-91-2	**MS09637**
126	316	298	299	430	412	55	107	**430**	100	90	54	50	44	44	43	42		27	42	4	–	–	–	–	–	–	–	–	–	–	**5α-Spirost-14-ene-3β,12α-diol, (25R)-**	24742-82-3	**NI42367**
155	282	131	300	156	271	55	283	**430**	100	44	22	21	18	17	14	12	**0.00**	27	42	4	–	–	–	–	–	–	–	–	–	–	**Spirost-5-ene-3,27-diol, (3β,25S)-**	7050-40-0	**NI42368**
155	156	282	131	300	271	267	54	**430**	100	43	41	22	18	17	12	12	**5.00**	27	42	4	–	–	–	–	–	–	–	–	–	–	**Spirost-5-ene-3,27-diol, (3β,25S)-**	7050-40-0	**LI08909**
155	156	282	131	300	271	283	56	**430**	100	42	41	22	19	17	12	12	**4.91**	27	42	4	–	–	–	–	–	–	–	–	–	–	**Spirost-5-ene-3,27-diol, (3β,25S)-**	7050-40-0	**NI42369**
137	352	345	81	139	138	55	95	**430**	100	27	23	21	20	19	16	13	**3.32**	27	46	2	–	–	–	–	–	–	1	–	–	–	**Silane, methylbis[[4-methyl-2-isopropylcyclohexyl]oxy]phenyl-, [1α(1R*,2R*,4R*),2α,4β]-**	74764-48-0	**NI42370**
137	352	345	139	138	81	55	41	**430**	100	30	25	23	19	19	14	12	**1.80**	27	46	2	–	–	–	–	–	–	1	–	–	–	**Silane, methylbis[[5-methyl-2-isopropylcyclohexyl]oxy]phenyl-, [1α(1R*,2S*,5R*),2β,5α]-**	68269-74-9	**NI42371**
401	171	402	103	173	147	281	297	**430**	100	34	32	25	15	14	12	10	**1.33**	27	46	2	–	–	–	–	–	–	1	–	–	–	**3β-Triethylsiloxypregn-5-en-20-one**		**NI42372**
254	241	43	228	430	415	91	119	**430**	100	92	71	69	57	48	25	21		28	34	2	2	–	–	–	–	–	–	–	–	–	**5,5-Dimethyl-3-(2,4,6-trimethylphenyl)-4-[[5-methyl-3-(2,4,6-trimethylphenyl)-2-isoxazolin-5-yl]methylene]-2-isoxazoline**		**DD01664**
430	289	145	285	431	146	43	131	**430**	100	86	41	39	35	34	31	27		28	38	–	4	–	–	–	–	–	–	–	–	–	**1-Dodecylbenzimidazol-2yl-(1-methylbenzimidazol-2-yl)methane**		**IC04389**
43	384	55	81	95	93	57	41	**430**	100	94	86	67	64	64	63	63	**0.00**	28	46	3	–	–	–	–	–	–	–	–	–	–	**26,27-Dinorergostan-3-ol, 6,19-epoxy-, acetate, (3β,6β)-**	57983-94-5	**NI42373**
384	43	55	330	147	41	81	57	**430**	100	84	66	64	64	56	53	46	**0.00**	28	46	3	–	–	–	–	–	–	–	–	–	–	**26,27-Dinorergostan-11-one, 3-(acetyloxy)-, (3α)-**	57983-99-0	**NI42374**
370	371	81	352	55	43	59	107	**430**	100	30	30	28	26	25	21	20	**0.00**	28	46	3	–	–	–	–	–	–	–	–	–	–	**26,27-Dinorergost-5-ene-3,24-diol, 3-acetate, (3β)-**	35882-86-1	**NI42375**
251	376	394	252	249	377	253	361	**430**	100	48	34	25	25	19	19	15	**0.83**	28	46	3	–	–	–	–	–	–	–	–	–	–	**Ergosta-7,22-diene-3,5,6-triol, (3β,5α,6β,22E)-**	516-37-0	**NI42376**
302	55	284	430	18	287	43	303	**430**	100	77	42	34	27	24	24	23		28	46	3	–	–	–	–	–	–	–	–	–	–	**Itesmol**	51771-53-0	**LI08910**
55	319	107	95	81	145	412	213	**430**	100	94	91	86	74	71	69	69	**68.50**	28	46	3	–	–	–	–	–	–	–	–	–	–	**Methyl (25RS)-3β-hydroxy-5-cholesten-26-oate**	56845-86-4	**NI42377**
110	111	81	109	247	43	107	41	**430**	100	15	15	12	11	11	10	10	**3.11**	28	46	3	–	–	–	–	–	–	–	–	–	–	**5,6-Secocholest-2-en-6-oic acid, 1-oxo-, methyl ester, (10α)-**	14772-40-8	**NI42378**
247	357	430	321	370	289	399	398	**430**	100	46	29	24	15	15	14	12		28	46	3	–	–	–	–	–	–	–	–	–	–	**5,6-Secocholest-2-en-6-oic acid, 1-oxo-, methyl ester, (10α)-**	14772-40-8	**NI42379**
154	55	136	135	125	57	41	69	**430**	100	46	41	28	27	23	23	18	**0.00**	28	46	3	–	–	–	–	–	–	–	–	–	–	**9,10-Secoergosta-5,10(19),22-triene-3,7,8-triol, (3β,5Z,8ξ,22E)-**	54661-20-0	**NI42380**
69	125	55	303	128	110	81	83	**430**	100	99	98	95	95	67	67	52	**0.39**	28	46	3	–	–	–	–	–	–	–	–	–	–	**9,10-Secoergosta-7,10(19),22-triene-3,5,6-triol, (3β)-**	56282-28-1	**NI42381**
370	215	316	355	204	43	371	216	**430**	100	57	48	41	34	34	29	29	**0.00**	29	50	2	–	–	–	–	–	–	–	–	–	–	**Cholestan-3-ol, acetate, (3β)-**	42995-53-9	**NI42382**
97	43	257	55	81	107	344	95	**430**	100	91	73	71	69	57	47	45	**0.00**	29	50	2	–	–	–	–	–	–	–	–	–	–	**Cholestan-3-ol, acetate, (3β,5α)-**	1255-88-5	**MS09638**
370	215	316	355	43	216	371	81	**430**	100	63	54	47	39	33	32	26	**0.00**	29	50	2	–	–	–	–	–	–	–	–	–	–	**Cholestan-3-ol, acetate, (3β,5α)-**	1255-88-5	**NI42383**
215	43	216	370	107	55	81	93	**430**	100	74	49	46	44	42	41	39	**15.90**	29	50	2	–	–	–	–	–	–	–	–	–	–	**Cholestan-3-ol, acetate, (3β,5α)-**	1255-88-5	**NI42384**
370	355	215	371	316	257	93	95	**430**	100	42	36	31	25	23	23	21	**0.00**	29	50	2	–	–	–	–	–	–	–	–	–	–	**Cholestan-3-ol, acetate, (3β,5β)-**	4947-63-1	**NI42385**
99	125	100	430	112	81	55	43	**430**	100	36	14	13	10	5	5	5		29	50	2	–	–	–	–	–	–	–	–	–	–	**Cholestan-3-one, cyclic 1,2-ethanediyl acetal, (5α)-**	1858-14-6	**NI42386**
430	99	125	431	43	57	55	100	**430**	100	99	34	29	17	15	13	12		29	50	2	–	–	–	–	–	–	–	–	–	–	**Cholestan-3-one, cyclic 1,2-ethanediyl acetal, (5α)-**	1858-14-6	**NI42387**
99	125	43	55	57	41	69	430	**430**	100	45	32	31	28	28	20	19		29	50	2	–	–	–	–	–	–	–	–	–	–	**Cholestan-3-one, cyclic 1,2-ethanediyl acetal, (5α)-**	1858-14-6	**NI42388**
99	125	55	43	57	41	69	71	**430**	100	50	27	27	23	22	15	13	**9.00**	29	50	2	–	–	–	–	–	–	–	–	–	–	**Cholestan-3-one, cyclic 1,2-ethanediyl acetal, (5β)-**	25328-53-4	**NI42389**
43	55	41	125	99	57	73	81	**430**	100	74	67	62	51	43	40	35	**20.98**	29	50	2	–	–	–	–	–	–	–	–	–	–	**Cholestan-7-one, cyclic 1,2-ethanediyl acetal, (5α)-**	54498-52-1	**NI42390**
368	45	369	81	43	95	55	57	**430**	100	60	52	32	32	30	29	24	**0.00**	29	50	2	–	–	–	–	–	–	–	–	–	–	**Cholest-5-ene, 3-(methoxymethoxy)-, (3β)-**	2626-17-7	**NI42391**
87	412	255	273	397	379	394	402	**430**	100	57	27	23	22	12	9	7	**4.98**	29	50	2	–	–	–	–	–	–	–	–	–	–	**26,27-Dimethylcholest-5-ene-3β,25-diol**		**MS09639**
43	430	389	57	398	375	55	415	**430**	100	79	77	72	70	69	59	57		29	50	2	–	–	–	–	–	–	–	–	–	–	**(23S,24R)-6β-Methoxy-3α,5-cyclo-5α-ergostan-23-ol**		**DD01665**
59	55	105	375	255	213	398	430	**430**	100	67	59	14	14	12	9	8		29	50	2	–	–	–	–	–	–	–	–	–	–	**(24S)-6β-Methoxy-3α,5-cyclo-5α-ergostan-25-ol**		**DD01666**
41	43	71	55	430	398	375	415	**430**	100	89	66	59	50	48	46	38		29	50	2	–	–	–	–	–	–	–	–	–	–	**(23R,24R)-6β-Methoxy-3α,5-ergostan-23-ol**		**DD01667**
165	430	163	43	56	203	54	41	**430**	100	42	39	27	20	18	18	16		29	50	2	–	–	–	–	–	–	–	–	–	–	**α-Tocopherol**	59-02-9	**NI42392**
430	165	164	431	205	166	432	69	**430**	100	95	34	31	10	10	6	4		29	50	2	–	–	–	–	–	–	–	–	–	–	**α-Tocopherol**	59-02-9	**MS09640**
165	430	164	428	203	205	429	176	**430**	100	57	46	40	27	25	20	16		29	50	2	–	–	–	–	–	–	–	–	–	–	**α-Tocopherol**	59-02-9	**LI08911**
160	83	71	85	91	161	84	58	**430**	100	32	26	20	16	15	12	7	**5.26**	30	42	–	2	–	–	–	–	–	–	–	–	–	**Cyclohexylamine, 2-allyl-4-[5-(dimethylamino)-1-pentenyl]-N,N-dimethyl-3,5-diphenyl-**	14028-80-9	**NI42393**
206	275	43	95	69	290	55	134	**430**	100	71	49	48	45	42	41	40	**22.77**	30	54	1	–	–	–	–	–	–	–	–	–	–	**Lanostan-3-ol, (3β)-**	4581-87-7	**NI42394**
84	69	95	60	397	81	62	109	**430**	100	39	28	28	20	20	20	18	**0.00**	30	54	1	–	–	–	–	–	–	–	–	–	–	**Lanostan-11-ol, (11β)-**	54411-50-6	**NI42395**
398	95	81	69	109	399	137	55	**430**	100	48	42	42	32	30	30	30	**0.00**	30	54	1	–	–	–	–	–	–	–	–	–	–	**Lanostan-18-ol**	54411-54-0	**NI42396**
207	430	135	95	189	123	109	69	**430**	100	92	86	57	50	47	46	43		30	54	1	–	–	–	–	–	–	–	–	–	–	**18,19-Secolupan-3-ol, (3β,17ξ)-**	30211-96-2	**NI42397**
95	121	81	43	55	107	369	41	**430**	100	88	81	79	76	74	72	67	**0.00**	30	54	1	–	–	–	–	–	–	–	–	–	–	**Tetrahydrodammaradienol**		**NI42398**
50	105	77	222	52	209	221	58	**430**	100	50	44	37	32	29	20	19	**3.00**	31	26	2	–	–	–	–	–	–	–	–	–	–	**2-Hydroxy-2,4,6-triphenyl-cyclohex-3-enyl phenyl ketone**		**LI08912**
183	248	105	77	91	247	184	117	**430**	100	56	56	32	24	17	16	14	**0.00**	32	30	1	–	–	–	–	–	–	–	–	–	–	**1-(Hydroxydiphenylmethyl)-2-benzhydrylidenecyclohexane**		**LI08913**
306	304	308	240	242	278	276	310	**431**	100	77	48	26	26	21	16	11	**0.01**	2	5	–	3	–	–	4	–	1	–	3	–	–	**1-Ethyl-1-iodotetrachlorocyclotriphosphazene**	77589-31-2	**NI42399**
131	69	45	100	181	119	93	51	**431**	100	67	40	23	22	21	12	12	**0.00**	8	4	2	1	–	15	–	–	–	–	–	–	–	**Ammonium perfluoro-octanoate**		**IC04390**

MASS TO CHARGE RATIOS								M.W.	INTENSITIES								Parent	C	H	O	N	S	F	Cl	Br	I	Si	P	B	X	COMPOUND NAME	CAS Reg No	No
166	55	210	165	347	137	123	41	431	100	91	83	69	67	43	43	43	5.60	18	18	6	1	1	–	–	–	–	–	–	–	1	Tricarbonyl[η-1-(N,N-dimethylsulphamoyl)-2-(1-hydroxy-1-phenylethyl)cyclopentadienyl]manganese		NI42400
91	148	70	71	103	218	131	119	431	100	52	43	34	18	17	14	14	0.20	18	20	3	1	–	7	–	–	–	–	–	–	–	3-Methylbutyl n-(heptafluorobutyryl)phenylalaninate		NI42401
43	371	97	157	28	198	85	172	431	100	38	34	25	25	25	24	22	0.70	19	29	8	1	1	–	–	–	–	–	–	–	–	4,5,6,7-Tetraacetoxydecyl isothiocyanate	57103-44-3	NI42402
174	144	77	243	114	75	175	116	431	100	37	17	14	13	13	10	10	0.40	20	21	6	3	1	–	–	–	–	–	–	–	–	Hydroxyoxacillin methyl ester		NI42403
372	373	312	400	374	227	401	167	431	100	22	15	13	5	4	3	2	1.00	20	21	8	3	–	–	–	–	–	–	–	–	–	Tetramethyl 5-ethyl-5H-pyridazino[1,2-a]benzotriazole-7,8,9,10-tetracarboxylate		NI42404
239	255	241	221	257	240	176	193	431	100	67	35	25	24	15	13	11	3.35	22	22	6	1	–	–	1	–	–	–	–	–	–	L-Phenylalanine, N-[(5-chloro-3,4-dihydro-8-hydroxy-3-methyl-1-oxo-1H-2-benzopyran-7-yl)carbonyl]-, ethyl ester, (R)-	4865-85-4	NI42405
105	77	96	183	51	42	182	28	431	100	65	56	32	29	26	25	25	0.00	22	26	3	1	–	–	–	1	–	–	–	–	–	1-Azoniabicyclo[2.2.2]octane, 3-[(hydroxydiphenylacetyl)oxy]-1-methyl-, bromide	3485-62-9	NI42406
431	433	183	430	262	432	434	185	431	100	99	69	54	35	30	25	24		24	19	–	1	–	–	–	1	–	–	1	–	–	Aniline, 4-bromo-N-(triphenylphosphoranylidene)-	14987-96-3	NI42407
431	432	433	416	300	434	430	417	431	100	37	15	9	6	3	3	3		24	41	2	1	–	–	–	–	–	2	–	–	–	4-Aminoestradiol-17β-DITMS		NI42408
276	165	277	210	80	431	197	91	431	100	25	23	19	19	18	16	14		25	21	2	1	2	–	–	–	–	–	–	–	–	2-Tosyl-2-aza-3-thiabicyclo[3.3.0]oct-7-ene-4-spiro-9′-fluorene		MS09641
388	310	389	328	311	296	413	254	431	100	64	28	24	15	14	11	11	7.52	25	37	5	1	–	–	–	–	–	–	–	–	–	Pregn-4-ene-3,20-dione, 17-(acetyloxy)-21-hydroxy-6-methyl-, 3-(O-methyloxime), (6β)-	74299-08-4	NI42409
387	386	304	352	270	388	289	389	431	100	86	84	81	61	56	46	35	7.00	26	22	3	1	–	–	1	–	–	–	–	–	–	Cyclohexylamino-chloro-3-oxo-spiro(phthalan-1,9′-xanthene)		IC04391
114	73	129	87	75	400	84	77	431	100	96	42	42	42	40	26	25	14.00	26	45	2	1	–	–	–	–	–	1	–	–	–	Pregn-5-en-20-one, 16-methyl-3-[(trimethylsilyl)oxy]-, O-(methyloxime), (3β,16α)-		LI08914
114	73	101	75	129	400	84	81	431	100	66	31	28	27	22	20	19	10.00	26	45	2	1	–	–	–	–	–	1	–	–	–	Pregn-5-en-20-one, 17-methyl-3-[(trimethylsilyl)oxy]-, O-(methyloxime), (3β,17α)-		LI08915
139	43	431	56	55	69	41	317	431	100	98	84	83	78	70	63	62		27	45	3	1	–	–	–	–	–	–	–	–	–	Spirostan-6-ol, 3-amino-, (3β,5α,6α,25R)-	23656-00-0	MS09642
139	56	69	55	57	71	299	70	431	100	82	79	79	72	59	44	43	39.34	27	45	3	1	–	–	–	–	–	–	–	–	–	Spirostan-9-ol, 3-amino-, (3β,5α,25R)-	16577-35-8	NI42410
372	267	105	354	326	373	266	294	431	100	85	75	46	44	30	26	25	8.00	29	21	3	1	–	–	–	–	–	–	–	–	–	5-(1-Methoxycarbonyl-2-oxido-2-phenylvinyl)-6-phenylphenanthridinium		LI08916
230	105	127	231	77	288	170	114	431	100	41	26	24	12	9	9	7	4.00	29	21	3	1	–	–	–	–	–	–	–	–	–	Methyl 1-(N-benzoyl-1-naphthylamino)-2-naphthoate		NI42411
105	230	288	77	431	267	106	231	431	100	53	37	30	12	12	12	11		29	21	3	1	–	–	–	–	–	–	–	–	–	Methyl 1-(N-benzoyl-1-naphthylamino)-2-naphthoate		NI42412
259	430	340	181	431	260	354	105	431	100	72	69	40	36	32	30	26		29	25	1	1	–	–	–	–	–	1	–	–	–	3-(N-Methylanilino)-2-(triphenylsilyl)-2-cyclobuten-1-one	79139-22-3	NI42413
199	276	122	77	149	200	107	198	431	100	35	32	28	23	20	17	16	3.00	29	25	1	1	–	–	–	–	–	1	–	–	–	N-Methyl-2-(triphenylsilyl)-2,3-butadienanilide	79139-19-8	NI42414
98	431	115	344	58	71	432	84	431	100	80	47	44	36	28	26	22		29	53	1	1	–	–	–	–	–	–	–	–	–	Cholestan-3-ol, 2-(dimethylamino)-, (2β,3α,5α)-	2454-45-7	NI42415
193	105	85	77	165	332	182	208	431	100	99	83	75	65	42	42	30	0.00	31	29	1	1	–	–	–	–	–	–	–	–	–	2-sec-Butylimino-3,3,4,4-tetraphenyl-oxetane		MS09643
101	103	85	151	167	169	153	117	432	100	64	40	30	25	21	20	20	0.00	6	1	–	–	–	6	7	–	–	–	–	–	–	Hexafluoroheptachlorohexane		DO01121
43	428	426	430	429	432	431	427	432	100	51	47	44	44	35	33	31		8	12	8	–	–	–	–	–	–	–	–	–	2	Molybdenum, tetrakis[μ-(acetato-O:O′)]di-	14221-06-8	NI42416
432	346	389	303	345	302	194	237	432	100	81	74	56	35	30	21	19		8	24	–	8	–	4	–	–	–	–	4	–	–	Tetrakis(dimethylamino)tetrafluorocyclotetraphosphazene		MS09644
389	431	391	372	390	253	371	77	432	100	26	12	12	10	7	5	5	0.00	9	6	4	–	–	–	–	–	2	–	–	–	–	Benzoic acid, 4-(acetyloxy)-3,5-diiodo-	41072-98-4	NI42417
203	55	201	303	83	301	199	41	432	100	83	77	71	61	54	40	34	0.00	10	14	2	4	–	–	6	–	–	–	–	–	–	Biformylchlorazin	26644-46-2	NI42418
233	69	432	397	235	145	63	85	432	100	57	55	41	39	37	31	26		12	–	–	–	1	10	1	–	–	–	1	–	–	Phosphinothioic chloride, bis(pentafluorophenyl)-	53327-28-9	NI42419
171	173	92	261	352	75	65	263	432	100	95	60	46	43	42	42	36	7.00	13	10	–	2	–	–	–	2	–	–	–	–	1	N,N′-Bis(4-bromophenyl)selenourea		MS09645
135	153	107	154	384	202	136	382	432	100	38	24	15	11	10	10	9	0.00	13	10	2	–	1	–	–	–	–	–	–	–	1	Mercury, (2-mercaptobenzoato-S)phenyl-	56247-85-9	NI42420
178	305	176	179	432	152	177	151	432	100	16	15	14	9	9	7	7		14	10	–	–	–	–	–	–	2	–	–	–	–	1,1′-(1,2-Diiodo-1,2-ethanediyl)bisbenzene	74752-96-8	NI42421
69	432	193	57	81	125	97	71	432	100	63	37	36	32	28	28	25		14	10	7	2	–	6	–	–	–	–	–	–	–	Acetic acid, trifluoro-, [2,3,3a,9a-tetrahydro-7-methyl-6-oxo-3-[(trifluoroacetyl)oxy]-6H-furo[2′,3′:4,5]oxazolo[3,2-a]pyrimidin-2-yl]methyl ester, [2R-(2α,3β,3aβ,9aβ)]-	55319-76-1	NI42422
329	73	177	299	45	159	89	101	432	100	69	47	41	30	25	20	19	0.00	14	37	7	–	–	–	–	–	–	3	1	–	–	Phosphoric acid, 2,2-dimethoxy-3-[(trimethylsilyl)oxy]propyl bis(trimethylsilyl) ester	55319-96-5	NI42423
189	191	225	223	209	211	193	227	432	100	76	53	52	35	33	23	16	9.32	15	10	2	–	–	–	6	–	–	–	–	–	–	2,2′-Methylenebis[3,4,6-trichloroanisole]	100965-64-8	NI42424
250	432	77	63	75	415	79	249	432	100	89	68	60	59	50	46	45		15	12	8	8	–	–	–	–	–	–	–	–	–	Methylglyoxal 2,4-dinitrophenylosazone	1107-69-3	NI42425
119	117	296	294	115	298	121	292	432	100	75	60	58	55	46	28	27	11.00	15	30	–	–	–	–	–	–	–	–	–	–	3	Germane, (1-Methyl-1,2-pentadien-4-yne-1,3,5-triyl)tris(trimethyl)-	61227-86-9	NI42426
189	124	432	404	150	209	274	339	432	100	13	12	6	5	4	2	1		16	15	1	–	1	–	–	–	–	–	–	–	3	Tris(π-cyclopentadienylcobalt)carbonylsulphide		LI08917
139	45	401	403	104	124	75	138	432	100	95	84	81	45	41	41	40	15.00	16	15	4	4	–	–	3	–	–	–	–	–	–	N,N-Bis(2-hydroxyethyl)-3-chloro-4-(2,6-dichloro-4-nitrophenylazo)aniline		IC04392
43	145	372	205	177	89	77	63	432	100	52	29	26	25	21	19	17	0.00	18	16	9	4	–	–	–	–	–	–	–	–	–	Methyl (±)-2-acetoxy-3-[4-(2,4-dinitro-phenylazo)-5-hydroxyphenyl] propionate		NI42427
168	126	210	115	157	103	109	97	432	100	65	57	45	27	21	17	12	0.00	18	24	12	–	–	–	–	–	–	–	–	–	–	allo-Inositol, hexaacetate	29267-04-7	NI42428
168	210	126	115	157	109	103	199	432	100	63	61	50	34	19	19	13	0.00	18	24	12	–	–	–	–	–	–	–	–	–	–	D-chiro-Inositol, hexaacetate	29267-03-6	NI42429

MASS TO CHARGE RATIOS								M.W.	INTENSITIES								Parent	C	H	O	N	S	F	Cl	Br	I	Si	P	B	X	COMPOUND NAME	CAS Reg No	No
168	126	210	115	157	109	103	97	**432**	100	70	63	63	41	25	24	17	**0.00**	18	24	12	–	–	–	–	–	–	–	–	–	–	epi-Inositol, hexaacetate	20108-71-8	NI42430
43	168	210	115	126	157	44	103	**432**	100	26	16	15	14	10	6	5	**0.00**	18	24	12	–	–	–	–	–	–	–	–	–	–	Inositol, hexaacetate		MS09646
168	210	115	126	157	109	103	199	**432**	100	66	65	58	50	26	24	21	**0.00**	18	24	12	–	–	–	–	–	–	–	–	–	–	myo-Inositol, hexaacetate	1254-38-2	NI42431
168	126	115	210	157	103	109	73	**432**	100	60	60	56	40	23	22	16	**0.00**	18	24	12	–	–	–	–	–	–	–	–	–	–	neo-Inositol, hexaacetate	20097-40-9	NI42432
168	115	126	210	157	109	103	73	**432**	100	68	63	62	48	26	25	21	**0.00**	18	24	12	–	–	–	–	–	–	–	–	–	–	Inositol, hexaacetate, cis-	29307-62-8	NI42433
168	126	210	115	157	103	109	97	**432**	100	66	47	37	22	16	14	13	**0.00**	18	24	12	–	–	–	–	–	–	–	–	–	–	muco-Inositol, hexaacetate	18779-57-2	NI42434
43	81	173	131	217	141	103	115	**432**	100	27	9	7	6	6	6	5	**0.00**	18	24	12	–	–	–	–	–	–	–	–	–	–	1,2,4,5,6-Penta-O-acetyl-3,8-anhydro-β-D-altro-L-glycerooctulopyranose	78174-74-0	NI42435
168	115	210	126	157	109	103	199	**432**	100	78	59	59	56	31	29	23	**0.00**	18	24	12	–	–	–	–	–	–	–	–	–	–	scyllo-Inositol, hexaacetate	20108-52-5	NI42436
57	389	237	219	41	56	317	277	**432**	100	48	43	38	37	31	26	24	**5.49**	18	36	6	–	–	–	–	–	–	3	–	–	–	1,3,5-Trivinyl-1,3,5-tributoxycyclotrisiloxane	94048-06-3	NI42437
73	75	107	103	55	432	225	74	**432**	100	40	18	15	12	11	10	9		18	36	6	2	–	–	–	–	–	2	–	–	–	Succinylacetoacetate diethoxime, bis(trimethylsilyl)-		NI42438
279	280	266	292	165	278	297	69	**432**	100	52	49	41	27	26	25	25	**0.00**	21	20	10	–	–	–	–	–	–	–	–	–	–	4H-1-Benzopyran-4-one, 6-β-D-glucopyranosyl-5,7-dihydroxy-2-(4-hydroxyphenyl)-		LI08918
43	270	55	57	121	29	69	118	**432**	100	97	68	63	57	52	48	47	**0.00**	21	20	10	–	–	–	–	–	–	–	–	–	–	4H-1-Benzopyran-4-one, 8β-D-glucopyranosyl-5,7-dihydroxy-2-(4-hydroxyphenyl)-	3681-93-4	NI42439
283	31	43	165	29	121	284	44	**432**	100	51	50	46	35	28	17	16	**0.55**	21	20	10	–	–	–	–	–	–	–	–	–	–	4H-1-Benzopyran-4-one, 8β-D-glucopyranosyl-5,7-dihydroxy-2-(4-hydroxyphenyl)-	3681-93-4	MS09647
270	168	271	69	140	73	77	70	**432**	100	23	18	9	6	5	3	3	**0.00**	21	20	10	–	–	–	–	–	–	–	–	–	–	4H-1-Benzopyran-4-one, 7-(β-D-glucopyranosyloxy)-5,6-dihydroxy-2-phenyl-	57396-78-8	NI42440
130	199	139	157	259	432	97	174	**432**	100	96	66	60	45	44	30	21		21	24	8	2	–	–	–	–	–	–	–	–	–	Diacetylneosidomycin		MS09648
417	432	418	431	401	433	97	111	**432**	100	60	24	17	17	15	10	6		22	24	9	–	–	–	–	–	–	–	–	–	–	3,3′,4′,5,5′,6,7-Heptamethoxyflavone		LI08919
313	43	314	115	275	372	77	245	**432**	100	64	26	22	16	13	13	9	**8.00**	23	20	5	4	–	–	–	–	–	–	–	–	–	3-(2,3-Di-O-acetyl-β-D-erythrofuranosyl)-1-phenylpyrazolo[3,4-b]quinoxaline		MS09649
348	146	376	28	349	105	277	307	**432**	100	47	33	32	25	24	16	10	**1.54**	23	24	3	2	–	–	–	–	–	–	–	–	1	Iron, tricarbonyl[N,N′-(1,2-dimethyl-1,2-ethanediylidene)bis[2,6-dimethylaniline]-N,N′]-	74764-10-6	NI42441
135	105	91	432	191	137	265	75	**432**	100	46	21	18	17	17	15	15		23	28	4	–	2	–	–	–	–	–	–	–	–	2,4:3,5-Di-O-benzylidene-D-arabinose diethyl dithioacetal		MS09650
236	432	343	342	165	193	312	360	**432**	100	91	91	67	64	56	52	44		23	28	8	–	–	–	–	–	–	–	–	–	–	Deangeloylgomisin B	58546-59-1	NI42442
181	137	432	166	165	151	167	149	**432**	100	24	23	18	12	9	8	5		23	28	8	–	–	–	–	–	–	–	–	–	–	1-(3,4-Dimethoxyphenyl)-3-(2-acetoxy-4-methoxyphenyl)-2-acetoxy-1-methoxypropane		NI42443
262	191	166	152	280	251	322	269	**432**	100	40	20	17	15	8	7	7	**0.00**	23	28	8	–	–	–	–	–	–	–	–	–	–	Methyl 5α,7α,11-trihydroxytetranorprostanoate triacetate		LI08920
260	224	131	105	242	191	261	214	**432**	100	37	27	27	26	24	21	21	**0.00**	23	28	8	–	–	–	–	–	–	–	–	–	–	Pseudolaric acid B	82508-31-4	NI42444
193	162	179	208	151	134	77	65	**432**	100	48	25	24	21	20	16	16	**1.00**	23	28	8	–	–	–	–	–	–	–	–	–	–	o-Pyrocatechuic acid, 4-methoxy-6-propyl-, methyl ester, 3-(2-hydroxy-6-propyl-p-anisate)	6161-77-9	NI42445
358	432	190	58	44	359	126	43	**432**	100	33	32	29	27	24	18	16		23	32	6	2	–	–	–	–	–	–	–	–	–	Dichotine, 2-deoxytetrahydro-11-methoxy-	29537-52-8	NI42446
358	432	190	126	330	154	299		**432**	100	33	32	18	11	8	6			23	32	6	2	–	–	–	–	–	–	–	–	–	Dichotine, 2-deoxytetrahydro-11-methoxy-		LI08921
83	41	85	415	69	43	57	55	**432**	100	93	72	71	67	64	57	53	**46.00**	23	32	6	2	–	–	–	–	–	–	–	–	–	3,4-Secocondyfolan-3-one, 1-acetyl-14,19-dihydroxy-10,11-dimethoxy-4-methyl-, (2β,7β,14ξ,15α)-	36950-31-9	NI42447
223	205	311	432	279	342	402	327	**432**	100	95	56	54	38	37	26	26		23	36	4	2	–	–	–	–	–	1	–	–	–	7α-Hydroxyandrostenedione-O-(methyloxime) 7-[(trimethylsilyl)oxy]-		LI08922
73	351	335	352	263	338	293	353	**432**	100	84	33	31	30	20	20	16	**0.05**	23	44	–	–	–	–	–	–	–	4	–	–	–	8,9,10,11-Tetrakis(trimethylsilyl)tricyclo[5.2.2.0^{2,6}]undeca-3,8,10-triene		NI42448
215	217	216	181	304	341	128	397	**432**	100	44	28	25	13	12	11	9	**2.00**	24	18	–	4	–	–	2	–	–	–	–	–	–	(1α,2α,3β,4β)-1,3-Bis(6-chloro-3-pyradazinyl)-2,4-diphenylcyclobutane		MS09651
432	108	433	80	65	44	184	45	**432**	100	78	25	22	19	19	11	11		24	20	4	2	1	–	–	–	–	–	–	–	–	4,4′-Bis(4-aminophenoxy)diphenyl sulphone		IC04393
263	221	169	171	377	404	222	406	**432**	100	76	61	55	51	48	47	45	**40.09**	24	21	1	2	–	–	–	1	–	–	–	–	–	1-(3-Bromobenzyl)-1,2,3,4-tetrahydro-8-methyl-6-phenyl-1,5-benzodiazocin-2-one	76061-61-5	NI42449
294	43	312	216	69	125	41	91	**432**	100	83	65	54	54	49	46	43	**1.88**	24	32	7	–	–	–	–	–	–	–	–	–	–	4-Desoxy-4α-phorbol-12,13-diacetate		NI42450
199	83	43	200	217	312	294	91	**432**	100	96	88	64	58	51	45	26	**0.21**	24	32	7	–	–	–	–	–	–	–	–	–	–	4-Desoxy-4α-phorbol-13,20-diacetate		NI42451
432	330	433	361	358	315	331	348	**432**	100	38	28	15	12	12	11	10		24	32	7	–	–	–	–	–	–	–	–	–	–	Schizandrin	7432-28-2	NI42452
105	197	57	77	142	106	319	41	**432**	100	99	51	48	25	24	20	20	**0.00**	24	36	5	2	–	–	–	–	–	–	–	–	–	(2S,4S)-3-Benzoyl-2-tert-butyl-4-[4-[(tert-butyloxycarbonyl)amino]butyl]-4-methyl-5-oxazolidinone		DD01668
432	218	85	433	342	244	324	283	**432**	100	45	42	35	35	35	33	25		24	40	3	–	–	–	–	–	–	2	–	–	–	Estra-1,3,5(10)-trien-3-ol, 11,17-bis[(trimethylsilyl)oxy]-, (11β,17β)-	75113-11-0	NI42453
414	309	324	415	283	230	298	310	**432**	100	63	45	40	32	24	21	18	**13.20**	24	40	3	–	–	–	–	–	–	2	–	–	–	Estra-1,3,5(10)-trien-7-ol, 3,17-bis[(trimethylsilyl)oxy]-, (7α,17β)-	75113-12-1	NI42454
414	324	415	270	283	309	73	285	**432**	100	49	38	31	21	19	16	15	**0.00**	24	40	3	–	–	–	–	–	–	2	–	–	–	Estra-1,3,5(10)-trien-14-ol, 3,17-bis[(trimethylsilyl)oxy]-, (17β)-	69688-16-0	NI42455
73	401	129	75	343	147	402	145	**432**	100	47	40	27	26	20	17	17	**11.86**	24	40	3	2	–	–	–	–	–	1	–	–	–	Androst-4-en-19-al, 3-(methoxyimino)-17-[(trimethylsilyl)oxy]-, O-methyloxime, (17β)-		NI42456
401	129	75	343	147	145	402	130	**432**	100	86	57	56	43	37	36	29	**25.94**	24	40	3	2	–	–	–	–	–	1	–	–	–	Androst-4-en-19-al, 3-(methoxyimino)-17-[(trimethylsilyl)oxy]-, O-methyloxime, (17β)-	55836-45-8	NI42457
311	432	125	342	327	279	312	75	**432**	100	72	40	38	36	31	27	27		24	40	3	2	–	–	–	–	–	1	–	–	–	Androst-4-ene-3,17-dione, 7-[(trimethylsilyl)oxy]-, bis(O-methyloxime), (7α)-	69833-73-4	NI42458

MASS TO CHARGE RATIOS								M.W.	INTENSITIES								Parent	C	H	O	N	S	F	Cl	Br	I	Si	P	B	X	COMPOUND NAME	CAS Reg No	No
432	417	401	311	268	433	418	402	**432**	100	99	74	49	35	32	32	30		24	40	3	2	–	–	–	–	–	1	–	–	–	**Androst-4-ene-3,17-dione, 12-[(trimethylsilyl)oxy]-, bis(O-methyloxime), (12β)-**	69688-35-3	**NI42459**
75	103	105	77	79	158	226	143	**432**	100	89	42	41	38	34	33	32	**3.30**	24	40	3	2	–	–	–	–	–	1	–	–	–	**Androst-4-ene-3,17-dione, 19-[(trimethylsilyl)oxy]-, bis(O-methyloxime)**	55836-46-9	**NI42460**
85	83	43	84	111	193	79	151	**432**	100	84	51	37	36	34	33	28	**0.00**	24	52	4	–	–	–	–	–	–	1	–	–	–	**Silicic acid, tetrakis(2-ethylbutyl) ester**	78-13-7	**NI42461**
403	431	404	432	417	384	201.5	194	**432**	100	84	30	27	17	12	12	9		25	24	5	2	–	–	–	–	–	–	–	–	–	**1-(But-2-ylamino)-4,8-dihydroxy-5-(4-methoxyanilino)-anthraquinone**		**IC04394**
251	329	269	389	252	312	213	214	**432**	100	44	29	24	20	16	14	8	**0.00**	25	36	6	–	–	–	–	–	–	–	–	–	–	**Pregn-5-en-20-one, 3,16-bis(acetyloxy)-17-hydroxy-, (3β,16α)-**	23357-25-7	**NI42462**
251	329	269	372	295	312	213	226	**432**	100	82	76	61	61	51	51	33	**0.00**	25	36	6	–	–	–	–	–	–	–	–	–	–	**Pregn-5-en-20-one, 3,16-bis(acetyloxy)-17-hydroxy-, (3β,16β)-**	23357-24-6	**NI42463**
371	143	130	386	388	387	372	402	**432**	100	64	42	37	34	32	30	17	**5.24**	25	40	4	–	–	–	–	–	–	1	–	–	–	**Pregn-4-ene-3,20-dione, 17-hydroxy-6-methyl-21-[(trimethylsilyl)oxy]-, (6α)-**	74298-96-7	**NI42464**
414	432	360	287	107	269	121	108	**432**	100	95	79	57	54	51	44	37		25	41	5	–	–	–	–	–	–	–	–	1	–	**Pregnan-11-one, 17,21-[(butylborylene)bis(oxy)]-3,20-dihydroxy-, (3α,5β,20S)-**	30888-51-8	**NI42465**
246	228	244	174	107	105	119	213	**432**	100	60	54	41	38	35	30	29	**5.00**	25	41	5	–	–	–	–	–	–	–	–	1	–	**Pregnan-20-one, 17,21-[(butylborylene)bis(oxy)]-3,11-dihydroxy-, (3α,5β,11β)-**	30888-35-8	**NI42466**
360	93	43	41	91	105	79	414	**432**	100	92	90	90	87	84	81	80	**0.00**	25	41	5	–	–	–	–	–	–	–	–	1	–	**Pregnan-11-one, 17,21-[(tert-butylborylene)bis(oxy)]-3,20-dihydroxy-, (3α,5β,17α,20α)-**		**MS09652**
246	228	43	55	41	79	57	81	**432**	100	84	63	62	54	44	43	42	**1.00**	25	41	5	–	–	–	–	–	–	–	–	1	–	**Pregnan-20-one, 17,21-[(tert-butylborylene)bis(oxy)]-3,11-dihydroxy-, (3α,5β,11β,17α)-**		**MS09653**
73	156	75	168	270	432	327	169	**432**	100	99	95	91	68	65	65	57		25	44	2	–	–	–	–	–	–	2	–	–	–	**Androsta-2,16-diene, 11,17-bis[(trimethylsilyl)oxy]-, (5α,11β)-**	57397-13-4	**NI42467**
432	433	129	270	434	360	209	417	**432**	100	38	23	16	14	14	11	10		25	44	2	–	–	–	–	–	–	2	–	–	–	**Androsta-3,5-diene, 3,17-bis[(trimethylsilyl)oxy]-, (17β)-**	25495-26-5	**NI42468**
75	43	73	215	342	217	81	107	**432**	100	73	70	68	65	51	45	41	**10.00**	25	45	3	–	–	–	–	–	–	1	–	1	–	**Pregnane-3,17,20-triol, 3-[(trimethylsilyl)oxy]-, methylboronate, (3α,5β,17α,20α)-**		**MS09654**
75	73	215	342	217	81	107	93	**432**	100	66	54	46	43	42	40	39	**6.00**	25	45	3	–	–	–	–	–	–	1	–	1	–	**Pregnane-3,17,20-triol, 3-[(trimethylsilyl)oxy]-, methylboronate, (3α,5β,17α,20β)-**		**MS09655**
43	49	44	84	86	51	45	57	**432**	100	81	75	62	51	30	29	26	**17.42**	26	24	6	–	–	–	–	–	–	–	–	–	–	**6,6′-Dimethyl-5,5′,8,8′-tetramethoxy-2,2′-binaphthylidene-1,1′-dione**	101539-81-5	**NI42469**
87	91	81	93	88	79	77	68	**432**	100	7	6	5	5	5	5	5	**0.00**	26	40	5	–	–	–	–	–	–	–	–	–	–	**Pregnan-20-one, 3-(acetyloxy)-5,6-epoxy-6-methyl-, cyclic 20-(1,2-ethanediyl acetal), (3β,5α,6α)-**	56784-23-7	**MS09656**
87	91	93	88	79	328	99	81	**432**	100	6	5	5	5	4	4	4	**0.22**	26	40	5	–	–	–	–	–	–	–	–	–	–	**Pregnan-20-one, 3-(acetyloxy)-5,6-epoxy-6-methyl-, cyclic 20-(1,2-ethanediyl acetal), (3β,5β,6β)-**	56784-25-9	**MS09657**
87	328	91	79	99	77	93	67	**432**	100	30	24	23	22	18	17	16	**0.00**	26	40	5	–	–	–	–	–	–	–	–	–	–	**Pregnan-20-one, 3-(acetyloxy)-5,6-epoxy-16-methyl-, cyclic 20-(1,2-ethanediyl acetal), (3β,5α,6α,16α)-**	56784-24-8	**MS09658**
129	243	342	244	432	119	277	161	**432**	100	91	48	39	32	29	25	23		26	44	3	–	–	–	–	–	–	1	–	–	–	**Pregn-5-ene-3,20-diol, 20-(acetyloxy)-3-[(trimethylsilyl)oxy]-, (3β,20α)-**		**LI08923**
244	105	124	432	229	119	121	107	**432**	100	57	45	41	38	35	30	28		27	33	4	–	–	–	–	–	–	–	–	1	–	**Pregn-4-ene-3,20-dione, 17,21-[(phenylborylene)bis(oxy)]-**	30888-55-2	**NI42470**
372	215	257	216	230	357	373	217	**432**	100	62	32	27	26	23	22	19	**0.40**	27	44	4	–	–	–	–	–	–	–	–	–	–	**Cholan-24-oic acid, 3-(acetyloxy)-, methyl ester, (3α)-**	56085-36-0	**NI42471**
215	107	105	146	121	161	109	108	**432**	100	81	73	64	53	49	49	49	**0.00**	27	44	4	–	–	–	–	–	–	–	–	–	–	**Cholan-24-oic acid, 3-(acetyloxy)-, methyl ester, (3α,5β)-**	3253-69-8	**NI42472**
139	55	404	249	343	57	69	137	**432**	100	87	79	64	63	63	59	56	**4.00**	27	44	4	–	–	–	–	–	–	–	–	–	–	**Cholest-7-en-6-one, 3,5,14-trihydroxy-, (3β,5β,14α)-**		**LI08924**
139	115	140	289	70	318	56	271	**432**	100	49	34	29	21	20	16	15	**9.23**	27	44	4	–	–	–	–	–	–	–	–	–	–	**Spirostan-2,3-diol, (2β,3α,5β,25R)-**	2460-96-0	**NI42473**
139	55	69	44	95	42	115	57	**432**	100	42	36	35	32	28	23	22	**17.39**	27	44	4	–	–	–	–	–	–	–	–	–	–	**Spirostan-3,6-diol, (3β,5α,6α,25S)-**	511-91-1	**NI42474**
168	139	126	300	317	55	432	169	**432**	100	95	76	44	42	32	30	26		27	44	4	–	–	–	–	–	–	–	–	–	–	**Spirostan-3,15-diol, (3β,5α,15β,25R)-**	6877-35-6	**NI42475**
155	43	54	41	69	81	67	95	**432**	100	78	53	43	42	33	33	29	**1.86**	27	44	4	–	–	–	–	–	–	–	–	–	–	**Spirostan-3,27-diol, (3α)-**	55759-97-2	**NI42476**
147	165	201	167	166	105	203	117	**432**	100	76	75	62	49	48	37	36	**4.90**	28	33	–	2	–	–	1	–	–	–	–	–	–	**Piperazine, 1-[(4-chlorophenyl)phenylmethyl]-4-[[4-tert-butylphenyl]methyl]-**	82-95-1	**NI42477**
152	153	135	154	107	109	77	136	**432**	100	46	23	8	6	5	5	4	**0.00**	28	48	3	–	–	–	–	–	–	–	–	–	–	**Benzoic acid, 3-methoxy-, eicosyl ester**	56954-77-9	**NI42478**
152	153	135	154	136	77	107	92	**432**	100	38	19	4	2	2	1	1	**0.00**	28	48	3	–	–	–	–	–	–	–	–	–	–	**Benzoic acid, 4-methoxy-, eicosyl ester**	56954-79-1	**NI42479**
372	373	355	432	374	400	356	414	**432**	100	48	28	22	13	12	10	6		28	48	3	–	–	–	–	–	–	–	–	–	–	**5α-Cholestan-19-oic acid, 2α-hydroxy-, methyl ester**	24637-67-0	**NI42480**
400	401	355	372	432	414	433	373	**432**	100	30	25	24	22	16	7	7		28	48	3	–	–	–	–	–	–	–	–	–	–	**5α-Cholestan-19-oic acid, 2β-hydroxy-, methyl ester**	24637-61-4	**NI42481**
432	400	356	386	433	388	372	387	**432**	100	50	50	45	30	30	25	18		28	48	3	–	–	–	–	–	–	–	–	–	–	**5α-Cholestan-19-oic acid, 2α-methoxy-**	28809-61-2	**NI42482**
356	200	432	357	414	417	433	201	**432**	100	40	25	25	17	10	8	8		28	48	3	–	–	–	–	–	–	–	–	–	–	**5α-Cholestan-19-oic acid, 2β-methoxy-**	24637-69-2	**NI42483**
304	286	289	271	414	381	399	396	**432**	100	78	60	38	22	15	6	3	**3.00**	28	48	3	–	–	–	–	–	–	–	–	–	–	**Dihydroitesmol**		**LI08925**
55	69	81	95	41	109	70	93	**432**	100	91	81	72	66	59	51	45	**7.00**	28	48	3	–	–	–	–	–	–	–	–	–	–	**Ergost-25-ene-3,5,6-triol, (3β,5α,6β)-**	56143-28-3	**NI42484**
215	414	95	55	81	216	107	93	**432**	100	74	55	54	54	50	45	37	**0.80**	28	48	3	–	–	–	–	–	–	–	–	–	–	**Methyl (25RS)-3α-hydroxy-5β-cholestan-26-oate**		**NI42485**
112	43	247	113	81	41	107	83	**432**	100	15	12	12	12	11	10	8	**0.56**	28	48	3	–	–	–	–	–	–	–	–	–	–	**5,6-Secocholestan-6-oic acid, 1-oxo-, ethyl ester, (10α)-**	53512-64-4	**NI42486**
182	247	249	251	154	232	129	115	**432**	100	82	62	55	17	13	11	10	**3.00**	29	28	–	4	–	–	–	–	–	–	–	–	–	**17-Norcorynan, 19,20-didehydro-16-(9H-pyrido[3,4-b]indol-1-yl)-, (19E)-**	36150-14-8	**NI42487**
43	55	370	81	41	95	57	108	**432**	100	65	59	55	53	48	48	40	**6.47**	29	52	–	–	1	–	–	–	–	–	–	–	–	**Cholestane, 3-(ethylthio)-, (3β,5α)-**	54498-89-4	**NI42488**
101	399	432	386	43	57	55	41	**432**	100	77	75	56	56	53	49	46		29	52	2	–	–	–	–	–	–	–	–	–	–	**Cholestane, 2,2-dimethoxy-, (5α)-**	18003-84-4	**NI42489**
101	400	127	401	85	316	301	55	**432**	100	57	26	22	22	15	15	12	**11.00**	29	52	2	–	–	–	–	–	–	–	–	–	–	**Cholestane, 3,3-dimethoxy-, (5α)-**		**NI42490**
432	43	400	101	105	401	57	69	**432**	100	64	63	62	39	38	37	36		29	52	2	–	–	–	–	–	–	–	–	–	–	**Cholestane, 3,3-dimethoxy-, (5α)-**	16159-03-8	**NI42491**

MASS TO CHARGE RATIOS								M.W.	INTENSITIES								Parent	C	H	O	N	S	F	Cl	Br	I	Si	P	B	X	COMPOUND NAME	CAS Reg No	No
45	215	432	216	95	81	107	55	432	100	76	73	60	60	58	49	47		29	52	2	–	–	–	–	–	–	–	–	–	–	Cholestane, 3-(methoxymethoxy)-, (3β,5α)-	4707-81-7	NI42492
327	387	105	77	299	388	328	355	432	100	72	56	42	35	26	25	14	4.60	30	24	3	–	–	–	–	–	–	–	–	–	–	2-Ethoxy-3-benzoyl-2,4-diphenyl-2H-1-benzopyran		MS09659
327	387	105	299	77	328	388	355	432	100	74	56	36	34	26	25	14	5.00	30	24	3	–	–	–	–	–	–	–	–	–	–	2-Ethoxy-3-benzoyl-2,4-diphenyl-2H-1-benzopyran		LI08926
328	155	127	43	265	77	372	343	432	100	90	42	13	10	4	3	3	0.00	30	24	3	–	–	–	–	–	–	–	–	–	–	(R)-1,1-Di-2-naphthyl-2-phenyl-1,2-ethanediol-2-acetate	110743-90-3	DD01669
137	136	43	81	95	41	57	69	432	100	94	74	64	54	52	51	37	0.92	31	60	–	–	–	–	–	–	–	–	–	–	–	11-(2′-Bicyclo[4.4.0]decyl)heneicosane		AI02372
137	136	43	81	95	57	55	41	432	100	84	60	51	47	43	40	37	0.32	31	60	–	–	–	–	–	–	–	–	–	–	–	11-(2′-Bicyclo[4.4.0]decyl)heneicosane		AI02373
432	433	77	177	78	194	176	353	432	100	36	22	15	14	13	11	11		34	24	–	–	–	–	–	–	–	–	–	–	–	1,2,3,4-Tetraphenylnaphthalene	751-38-2	NI42493
354	240	146	275	398	111	170	76	433	100	13	8	7	5	4	3	3	3.00	–	–	–	3	–	–	4	2	–	–	3	–	–	1,3,5,2,4,6-Triazatriphosphorine, 2,4-dibromo-2,4,6,6-tetrachloro-2,2,4,4,6,6-hexahydro-	15964-99-5	NI42494
254	73	208	211	255	207	75	147	433	100	49	34	30	18	14	14	10	0.00	17	36	4	1	–	–	–	–	–	3	1	–	–	Phosphonic acid, [1-amino-2-(4-hydroxyphenyl)ethyl]-, tris(trimethylsilyl)	56211-23-5	NI42495
254	73	208	211	75	147	326	195	433	100	49	34	30	14	10	9	7	0.50	17	36	4	1	–	–	–	–	–	3	1	–	–	Phosphonic acid, [1-amino-2-[4-[(trimethylsilyl)oxy]phenyl]ethyl]-, bis(trimethylsilyl) ester	53044-35-2	NI42496
360	87	433	88	300	361	77	199	433	100	97	78	68	46	42	42	33		18	19	6	5	1	–	–	–	–	–	–	–	–	2-[4-[Bis(acetoxyethyl)amino]phenyl]-5-nitro-2H-thieno[2,3-d]-v-triazole		IC04395
43	84	126	85	151	168	60	56	433	100	97	43	42	23	19	16	15	0.00	18	27	11	1	–	–	–	–	–	–	–	–	–	D-Glucitol, 2-(acetylamino)-2-deoxy-, 1,3,4,5,6-pentaacetate	56211-19-9	NI42497
84	144	102	60	139	156	86	85	433	100	45	45	36	31	24	19	18	0.00	18	27	11	1	–	–	–	–	–	–	–	–	–	1,3,4,5,6-Penta-O-acetyl-2-acetamido-2-deoxy-D-altrositol		NI42498
96	114	84	126	156	186	216	168	433	100	58	55	44	41	38	32	32	0.00	18	27	11	1	–	–	–	–	–	–	–	–	–	1,2,4,5,6-Penta-O-acetyl-3-acetamido-3-deoxy-D-glucitol		NI42499
342	374	433	256	197	185	402	315	433	100	90	73	39	28	28	23	21		20	19	8	1	1	–	–	–	–	–	–	–	–	Tetramethyl 9,10-dihydroazepino[2,1-b]benzthiazole-7,8,9,10-tetracarboxylate		LI08927
271	313	229	174	115	212	433	216	433	100	60	54	42	30	27	26	21		20	23	8	3	–	–	–	–	–	–	–	–	–	1,2,3,4-Butanetetrol, 1-(2-phenyl-2H-1,2,3-triazol-4-yl)-, tetraacetate, [1R (1R*,2S*,3R*)]-	7770-63-0	NI42500
374	318	316	286	360	317	344	375	433	100	61	52	38	37	33	26	24	8.00	21	23	9	1	–	–	–	–	–	–	–	–	–	Trimethyl 1-propionyl-2-(methyloxycarbonylmethyl)-2H-quinolizine-2,3,4 tricarboxylate		NI42501
73	232	202	45	114	74	233	75	433	100	31	30	16	10	8	7	7	0.00	21	37	2	2	–	–	–	–	–	3	–	–	–	α-Methyltryptamine, tris(trimethylsilyl)-		MS09660
105	77	146	51	28	251	157	155	433	100	54	36	14	13	11	11	11	0.50	22	16	2	3	–	–	–	1	–	–	–	–	–	1-(4-Bromophenyl)-4-benzoylmethylene-5-hydroxy-5-phenyl-Δ^2-1,2,3-triazoline		MS09661
194	77	55	93	139	111	104	119	433	100	71	45	38	28	23	19	16	15.00	23	20	–	5	1	–	1	–	–	–	–	–	–	1-(N,N′-diphenyl)guanidino-4-(4-chlorophenyl)-2-methylthioimidazole		MS09662
247	91	262	171	155	248	229	65	433	100	69	53	30	26	20	19	17	0.00	25	23	2	1	2	–	–	–	–	–	–	–	–	3′,6′-Dihydro-5′-methyl-1′-tosyliminospiro(fluorene-9,2′-thiopyran)		MS09663
29	43	41	27	69	58	45	56	433	100	97	53	51	47	44	39	37	6.08	25	39	5	1	–	–	–	–	–	–	–	–	–	Batrachotoxinin A, 7,8-dihydro-O[3]-methyl-, (8β,20ξ)-	35596-59-9	NI42502
336	433	337	434	432	49	168	189	433	100	76	23	22	19	17	17	14		26	27	5	1	–	–	–	–	–	–	–	–	–	(±)-Karachine	80908-02-7	NI42503
433	298	91	434	77	269	268	299	433	100	65	35	29	27	18	17	15		27	19	3	3	–	–	–	–	–	–	–	–	–	1-(1′,3′-Diphenyl-2′-oxo-propylidene)-2-nitro-1,5-dihydrophenazine		MS09664
433	105	312	434	432	311	162	313	433	100	82	65	30	30	26	23	17		27	31	4	1	–	–	–	–	–	–	–	–	–	15-Benzoylpseudokobusine		MS09665
433	91	167	286	141	356	312	313	433	100	67	54	47	39	30	24	12		28	23	2	3	–	–	–	–	–	–	–	–	–	1,3,4,5,6,7-Hexahydro-N,4-diphenyl-7-[(E)-phenylmethylene]-2H-cyclopenta[d]pyradazine-2,3-dicarboximide		MS09666
433	183	248	262	434	432	172	77	433	100	59	48	38	31	28	28	25		28	24	–	3	–	–	–	–	–	–	1	–	–	Phosphorane, [(2-phenyl-5-methyl-1,2,3-triazol-4-yl)methylene]triphenyl-		NI42504
85	151	153	87	101	285	283	163	434	100	57	36	32	19	18	14	13	0.00	6	–	–	–	–	8	6	–	–	–	–	–	–	1,2,3,4,5,6-Hexachlorooctafluorohexane		DO01122
434	436	357	359	438	361	193	373	434	100	76	72	48	47	37	31	24		8	20	4	–	4	–	–	–	–	–	2	–	1	Zinc[II] bis(diethyldithiophosphate)	16569-75-8	NI42505
399	401	315	313	77	51	355	353	434	100	99	42	42	37	33	21	21	16.50	9	11	4	–	–	–	2	–	–	–	–	–	1	Tantalum, dichloro(2-hydroxybenzaldehydato-O,O′)dimethoxy-	52690-27-4	NI42506
399	77	51	401	313	36	29	15	434	100	49	44	33	32	31	26	26	12.00	9	11	4	–	–	–	2	–	–	–	–	–	1	Tantalum, dichloro(2-hydroxybenzaldehydato-O,O′)dimethoxy-	52690-27-4	NI42507
167	100	342	344	193	166	165	140	434	100	44	32	30	30	22	22	21	0.10	12	13	7	2	1	–	3	–	–	–	–	–	–	2,2,2-Trichloroethyl 7-methoxycarbonylamino-3-methyl-3-cephem-4-carboxylate 1,1-dioxide	97609-89-7	NI42508
266	294	244	406	378	212	176	322	434	100	39	30	26	26	25	23	22	2.00	13	6	6	–	2	–	–	–	–	–	–	–	2	Diiron, μ-(π-toluene-3,4-dithio)hexacarbonyl-		MS09667
434	415	385	296	346	327	315	269	434	100	56	47	17	13	13	12	9		14	–	–	–	–	14	–	–	–	–	–	–	–	4,4′-Bis(trifluoromethyl)octafluorobiphenyl		LI08928
226	224	223	225	210	212	211	209	434	100	96	80	79	74	65	62	60	0.00	14	8	3	–	–	–	6	–	–	–	–	–	–	Tris(2,3,5-trichloro-6-hydroxy-benzyl) ether		IC04396
126	81	69	127	55	53	82	83	434	100	95	25	21	11	11	9	8	6.01	14	12	7	2	–	6	–	–	–	–	–	–	–	Thymidine, 3′,5′-bis(trifluoroacetate)	18354-10-4	NI42509
142	434	435	143	217	322	173	65	434	100	60	11	9	3	2	2	2		17	12	–	2	–	10	–	–	–	–	–	–	–	1,5-Bis(m-aminophenyl)decafluoropentane		LI08929
207	75	79	44	149	52	208	315	434	100	72	71	53	46	31	26	24	0.00	18	26	12	–	–	–	–	–	–	–	–	–	–	D-Galactitol, hexaacetate	14330-96-2	NI42510
43	115	187	139	157	145	127	217	434	100	18	13	10	8	8	8	7	0.00	18	26	12	–	–	–	–	–	–	–	–	–	–	Galactitol hexaacetate		MS09668
43	115	145	187	103	157	139	128	434	100	32	24	18	16	15	15	15	0.00	18	26	12	–	–	–	–	–	–	–	–	–	–	D-Glucitol, hexaacetate	7208-47-1	NI42511
115	43	145	187	128	70	139	170	434	100	99	63	61	45	45	44	43	0.00	18	26	12	–	–	–	–	–	–	–	–	–	–	D-Glucitol, hexaacetate	7208-47-1	NI42512
43	115	145	103	187	128	170	139	434	100	13	12	7	7	7	6	6	0.00	18	26	12	–	–	–	–	–	–	–	–	–	–	D-Glucitol, hexaacetate	7208-47-1	NI42513
43	115	103	187	145	44	157	128	434	100	14	7	6	6	6	5	5	0.00	18	26	12	–	–	–	–	–	–	–	–	–	–	Iditol hexaacetate		MS09669
79	75	207	73	52	44	39	96	434	100	99	92	59	52	51	43	34	0.00	18	26	12	–	–	–	–	–	–	–	–	–	–	D-Mannitol, hexaacetate	642-00-2	NI42514

MASS TO CHARGE RATIOS								M.W.	INTENSITIES								Parent	C	H	O	N	S	F	Cl	Br	I	Si	P	B	X	COMPOUND NAME	CAS Reg No	No
43	115	145	139	217	157	103	187	**434**	100	14	11	10	7	7	6	6	**0.00**	18	26	12	–	–	–	–	–	–	–	–	–	–	**D-Mannitol, hexaacetate**	642-00-2	**NI42515**
43	115	187	139	145	157	128	103	**434**	100	16	12	12	11	8	8	8	**0.00**	18	26	12	–	–	–	–	–	–	–	–	–	–	**Mannitol hexaacetate**		**MS09670**
153	43	115	111	102	103	157	110	**434**	100	99	71	71	50	49	48	48	**0.29**	18	26	12	–	–	–	–	–	–	–	–	–	–	**Methyl 2,3,4,6,7-penta-O-acetyl-β-glycero-D-glucoheptopyranoside**	104012-36-4	**NI42516**
43	115	187	145	139	103	128	127	**434**	100	16	12	10	9	8	7	7	**0.00**	18	26	12	–	–	–	–	–	–	–	–	–	–	**Sorbitol hexaacetate**		**MS09671**
73	204	419	147	75	361	420	97	**434**	100	48	36	35	19	13	12	12	**0.01**	18	38	6	–	–	–	–	–	–	3	–	–	–	**2-Methyl-3,3-dicarboxy-hexanoate, tris(trimethylsilyl)-**		**NI42517**
73	273	147	419	111	275	75	55	**434**	100	44	42	41	25	16	16	15	**0.10**	18	38	6	–	–	–	–	–	–	3	–	–	–	**2-Methyl-3,3-dicarboxy-hexanoate, tris(trimethylsilyl)-**		**NI42518**
73	174	317	318	156	92	128	230	**434**	100	84	65	21	18	18	17	15	**3.50**	18	46	2	2	–	–	–	–	–	4	–	–	–	**L-Lysine, N^2,N^6,N^6-tris(trimethylsilyl)-, trimethylsilyl ester**	55429-07-7	**NI42519**
174	73	312	313	230	175	156	128	**434**	100	81	78	28	21	20	17	16	**8.00**	18	46	2	2	–	–	–	–	–	4	–	–	–	**L-Lysine, N^2,N^6,N^6-tris(trimethylsilyl)-, trimethylsilyl ester**	55429-07-7	**NI42520**
73	218	174	317	219	45	100	74	**434**	100	59	22	13	12	11	11	10	**0.63**	18	46	2	2	–	–	–	–	–	4	–	–	–	**L-Lysine, N^2,N^6,N^6-tris(trimethylsilyl)-,trimethylsilyl ester**	55429-07-7	**NI42521**
317	73	174	318	156	93	84	128	**434**	100	92	83	35	32	31	25	22	**10.00**	18	46	2	2	–	–	–	–	–	4	–	–	–	**L-Lysine, tetrakis(trimethylsilyl)-**	25737-20-6	**NI42522**
73	174	317	156	318	128	175	147	**434**	100	70	53	27	19	17	16	11	**0.00**	18	46	2	2	–	–	–	–	–	4	–	–	–	**L-Lysine, tetrakis(trimethylsilyl)-**		**MS09672**
73	156	84	45	245	74	102	59	**434**	100	58	31	17	15	15	9	8	**0.00**	18	46	2	2	–	–	–	–	–	4	–	–	–	**α-Methylornithine tetra-TMS**		**MS09673**
87	43	169	109	55	115	97	157	**434**	100	94	27	22	14	13	8	6	**0.00**	19	30	11	–	–	–	–	–	–	–	–	–	–	**(1RS)-1-Methoxybutyl 2,3,4,6-tetra-O-acetyl-β-D-glucopyranoside**		**MS09674**
73	317	75	147	109	129	318	55	**434**	100	47	35	33	15	15	13	12	**0.00**	19	42	5	–	–	–	–	–	–	3	–	–	–	**2-Trimethylsiloxysebacic acid, bis(trimethylsilyl) ester**		**NI42523**
73	419	233	147	303	217	420	75	**434**	100	46	42	41	25	21	16	14	**0.00**	19	42	5	–	–	–	–	–	–	3	–	–	–	**3-Trimethylsiloxysebacic acid, bis(trimethylsilyl) ester**	76735-13-2	**NI42524**
140	271	96	373	434				**434**	100	10	6	4	3					20	38	6	2	1	–	–	–	–	–	–	–	–	**D-erythro-α-D-Galacto-octopyranoside, ethyl 6,8-dideoxy-6-[[(1-ethyl-4-propyl-2-pyrrolidinyl)carbonyl]amino]-1-thio-, (2S-trans)-**	21085-65-4	**NI42525**
154	285	96	387	434				**434**	100	8	4	2	1					20	38	6	2	1	–	–	–	–	–	–	–	–	**D-erythro-α-D-Galacto-octopyranoside, methyl 6-[[(4-butyl-1-ethyl-2-pyrrolidinyl)carbonyl]amino]-6,8-dideoxy-1-thio-, (2S-trans)-**	17057-68-0	**NI42526**
120	165	416	285	398	272	69	55	**434**	100	99	44	40	36	31	31	28	**0.00**	21	22	10	–	–	–	–	–	–	–	–	–	–	**4H-1-Benzopyran-4-one, 6β-D-glucopyranosyl-2,3-dihydro-5,7-dihydroxy 2-(4-hydroxyphenyl)-, (S)-**	3682-03-9	**NI42527**
264	434	191	320	218	121	69	158	**434**	100	26	23	18	18	18	18	15		21	26	6	2	1	–	–	–	–	–	–	–	–	**tert-Butyl 3-(2-hydroxy-5-methylbenzyl)-7-carbomethoxyamino-2-cephem 4-carboxylate**		**NI42528**
105	168	77	140	167	99	84	66	**434**	100	47	28	22	18	18	18	18	**0.10**	21	26	8	2	–	–	–	–	–	–	–	–	–	**8-[2-(Benzoyloxy)-1-hydroxy-1-isopropyl]-6-hydroxy-N,9-dimethyl-5-methylene-10-oxo-2,7-dioxa-9-azabicyclo[4.2.2]decane-1-carboxamide**	89291-84-9	**NI42529**
259	180	346	120	108	197	185	106	**434**	100	47	36	35	25	22	22	22	**0.00**	21	26	8	2	–	–	–	–	–	–	–	–	–	**4,4′-Dimethyl-3,3′-bis(methoxycarbonylethyl)-5,5′-carbonyl-2,2′-dipyrrolylmethane**		**LI08930**
290	73	291	303	304	292	434	74	**434**	100	59	27	23	10	10	9	5		21	38	2	2	–	–	–	–	–	3	–	–	–	**1H-Indole-3-ethanamine, 1-acetyl-N,N-bis(trimethylsilyl)-5-[(trimethylsilyl)oxy]-**	56114-61-5	**NI42530**
333	43	434	115	54	334	346	306	**434**	100	49	30	28	22	21	19	18		22	22	4	6	–	–	–	–	–	–	–	–	–	**N-Cyanoethyl-N-(2-isopropionyloxyethyl)-4-(2-cyano-4-nitrophenylazo)aniline**		**IC04397**
224	434	211	225	209	210	404	178	**434**	100	45	30	20	20	15	10	10		22	26	9	–	–	–	–	–	–	–	–	–	–	**4H-1-Benzopyran-4-one, 2,3-dihydro-6,7,8-trimethoxy-3-(2,3,4,5-tetramethoxyphenyl)-, (S)-**	71594-04-2	**NI42531**
113	229	171	230	434	257	59	145	**434**	100	36	26	24	21	19	14	13		22	26	9	–	–	–	–	–	–	–	–	–	–	**3a,5a-Ethano-as-indacene-4,5,9,10-tetracarboxylic acid, 1,2,3,4,5,6,7,8-octahydro-1-oxo-, tetramethyl ester**	32251-43-7	**MS09675**
69	41	228	128	199	256	43	161	**434**	100	15	14	13	10	10	10	8	**0.00**	22	26	9	–	–	–	–	–	–	–	–	–	–	**Melcanthin D**	79405-87-1	**NI42532**
28	44	91	43	41	29	27	77	**434**	100	96	91	83	71	60	54	50	**43.05**	23	30	8	–	–	–	–	–	–	–	–	–	–	**24-Nor-5β,14β-chol-20(22)-ene-19,23-dioic acid, 1β,3β,5,11α,14,21-hexahydroxy-, di-γ-lactone**	15718-20-4	**NI42533**
181	182	374	343	165	151	121	167	**434**	100	45	42	39	36	35	32	25	**24.00**	23	30	8	–	–	–	–	–	–	–	–	–	–	**1-(3,4-Dimethoxyphenyl)-1-methoxy-3-(2,4,6-trimethoxyphenyl)-2-acetoxypropane**		**NI42534**
144	120	148	204	186	214	91	43	**434**	100	77	74	53	48	46	41	41	**13.99**	23	34	6	2	–	–	–	–	–	–	–	–	–	**N-(N-Acetyl-O-butyl-α-aspartyl)phenylalanine butyl ester**		**AI02374**
434	44	45	73	435	43	378	90	**434**	100	61	36	35	29	27	26	24		24	18	8	–	–	–	–	–	–	–	–	–	–	**3,3′-Dihydroxy-5,5′-dimethoxy-7,7′-dimethyl-2,2′-binaphthalene-1,1′,4,4′-tetrone**	88381-91-3	**NI42535**
44	90	418	148	434	416	192	50	**434**	100	30	23	19	17	16	15	14		24	18	8	–	–	–	–	–	–	–	–	–	–	**3,3′-Dihydroxy-8,8′-dimethoxy-6,6′-dimethyl-2,2′-binaphthalene-1,1′,4,4′-tetrone**	88381-86-6	**NI42536**
419	434	44	403	404	420	92	388	**434**	100	90	59	52	52	43	39	38		24	18	8	–	–	–	–	–	–	–	–	–	–	**7,7′-Dihydroxy-8,8′-dimethoxy-3,3′-dimethyl-2,2′-binaphthalene-1,1′,4,4′-tetrone**		**NI42537**
44	434	403	83	435	85	404	373	**434**	100	53	33	14	14	10	10	7		24	18	8	–	–	–	–	–	–	–	–	–	–	**1,1′,4,4′-Tetramethoxy-2,2′-binaphthalene-5,5′,8,8′-tetrone**	104505-91-1	**NI42538**
76	104	75	63	389	134	91	77	**434**	100	37	35	31	31	25	24	24	**0.00**	24	18	8	–	–	–	–	–	–	–	–	–	–	**3,3′,8,8′-Tetramethoxy-2,2′-binaphthalene-1,1′,4,4′-tetrone**		**NI42539**
93	112	163	114	106	94	92	77	**434**	100	60	24	17	12	10	10	9	**2.12**	24	18	8	–	–	–	–	–	–	–	–	–	–	**5,5′,8,8′-Tetramethoxy-2,2′-binaphthalene-1,1′,4,4′-tetrone**	104505-93-3	**NI42540**
434	419	435	436	420	43	76	389	**434**	100	57	33	17	16	9	9	8		24	18	8	–	–	–	–	–	–	–	–	–	–	**5,5′,8,8′-Tetramethoxy-2,2′-binaphthalene-1,1′,4,4′-tetrone**	104505-93-3	**NI42541**
217	262	139	183	434	77	63	107	**434**	100	96	92	48	44	40	40	23		24	20	–	–	2	–	–	–	–	–	2	–	–	**Tetraphenyldiphosphane disulphide**		**MS09676**
43	69	93	91	83	131	107	95	**434**	100	32	26	26	26	25	25	23	**1.36**	24	34	7	–	–	–	–	–	–	–	–	–	–	**3-Desoxo-3β-hydroxy-4-desoxy-phorbol-12,13-diacetate**		**NI42542**
43	41	122	69	55	121	109	67	**434**	100	36	29	26	26	23	23	22	**0.00**	24	34	7	–	–	–	–	–	–	–	–	–	–	**6,7-Dihydro-16-hydroxy-20-desoxy-ingenol-diacetate**		**NI42543**
359	360	217	121	105	127	111	374	**434**	100	84	73	48	37	34	33	32	**2.00**	24	34	7	–	–	–	–	–	–	–	–	–	–	**Picras-2-ene-1,16-dione, 11-(acetyloxy)-2,12-dimethoxy-, (11α,12β)-**	24148-78-5	**NI42544**

MASS TO CHARGE RATIOS								M.W.	INTENSITIES								Parent	C	H	O	N	S	F	Cl	Br	I	Si	P	B	X	COMPOUND NAME	CAS Reg No	No
434	435	388	403	129	389	298	404	434	100	36	34	29	17	13	13	12		24	42	3	2	–	–	–	–	–	1	–	–	–	Androstane-3,6-dione, 17-[(trimethylsilyl)oxy]-, bis(O-methyloxime), (5α,17β)-		NI42545
99	55	41	70	57	113	112	69	434	100	62	49	41	27	26	24	20	0.00	24	51	4	–	–	–	–	–	–	–	1	–	–	Tris(2-ethylhexyl) phosphate		IC04398
99	113	57	71	55	323	43	41	434	100	30	23	16	16	13	13	12	0.25	24	51	4	–	–	–	–	–	–	–	1	–	–	Trioctyl phosphate		IC04399
99	113	57	71	55	43	41	69	434	100	24	24	15	15	14	14	11	0.00	24	51	4	–	–	–	–	–	–	–	1	–	–	Trioctyl phosphate		IC04400
57	41	43	99	55	71	113	70	434	100	64	61	58	55	52	46	30	0.00	24	51	4	–	–	–	–	–	–	–	1	–	–	Trioctyl phosphate		MS09677
43	71	332	434	55	271	133	93	434	100	72	60	31	30	28	28	18		25	38	6	–	–	–	–	–	–	–	–	–	–	17β-Acetoxy-4-oxo-4-propyl-5,6-epoxy-3,4-secoandrostan-3-oic acid methyl ester		MS09678
43	356	55	41	81	107	93	79	434	100	99	83	77	76	74	74	74	3.00	25	38	6	–	–	–	–	–	–	–	–	–	–	Di-O-Acetyltetrahydrostapelogenin		NI42546
374	229	253	314	281	289	296	299	434	100	90	48	42	25	20	15	11	1.00	25	38	6	–	–	–	–	–	–	–	–	–	–	3α,7α-Diacetoxy-12α-hydroxy-5β-pregn-20-one		NI42547
314	296	281	253	271	299	200	211	434	100	84	80	80	58	40	40	38	1.00	25	38	6	–	–	–	–	–	–	–	–	–	–	7α,12α-Diacetoxy-3α-hydroxy-5β-pregn-20-one		NI42548
85	55	74	97	69	138	167	96	434	100	58	57	32	30	29	16	15	0.20	25	42	4	–	–	–	–	–	–	1	–	–	–	Methyl 2,4,6-trideoxy-(3R)-(tert-butyldimethylsiloxy)-(5R)-6-(4-oxo-2,7,8,10-dodecatetraenyl)-α-D-allopyranoside		NI42549
271	105	253	107	159	119	145	131	434	100	52	50	47	38	34	32	28	0.00	25	43	5	–	–	–	–	–	–	–	–	1	–	Pregnane-3,11,17,20,21-pentol, cyclic 17,21-tert-butylboronate, (3β,5α,11β,20R)-	30888-48-3	NI42550
73	344	75	169	105	129	143	146	434	100	99	99	94	61	57	55	47	17.00	25	46	2	–	–	–	–	–	–	2	–	–	–	Androst-2-ene, 11,17-bis[(trimethylsilyl)oxy]-, (5α,11β)-	33283-02-2	NI42551
143	142	434	202	435	144	75	344	434	100	70	68	20	19	16	16	10		25	46	2	–	–	–	–	–	–	2	–	–	–	Androst-4-ene, 3,17-bis[(trimethylsilyl)oxy]-, (3α,17β)-		LI08931
73	434	143	142	75	81	105	91	434	100	99	99	99	99	37	36	35		25	46	2	–	–	–	–	–	–	2	–	–	–	Androst-4-ene, 3,17-bis[(trimethylsilyl)oxy]-, (3β,17β)-	33283-03-3	NI42552
73	75	143	142	434	91	105	129	434	100	71	69	47	28	28	25	20		25	46	2	–	–	–	–	–	–	2	–	–	–	Androst-4-ene, 3,17-bis[(trimethylsilyl)oxy]-, (3β,17β)-	33283-03-3	NI42553
143	142	434	129	127	202	435	405	434	100	71	53	33	32	20	19	18		25	46	2	–	–	–	–	–	–	2	–	–	–	Androst-4-ene, 3,17-bis[(trimethylsilyl)oxy]-, (5β,17β)-	69833-82-5	NI42554
73	169	75	143	105	228	129	107	434	100	99	99	63	57	56	54	52	13.00	25	46	2	–	–	–	–	–	–	2	–	–	–	Androst-4-ene, 11,17-bis[(trimethylsilyl)oxy]-, (11β,17β)-	33283-04-4	NI42555
129	215	145	239	344	254	119	107	434	100	86	40	39	30	30	28	19	16.00	25	46	2	–	–	–	–	–	–	2	–	–	–	Androst-5-ene, 3,16-bis[(trimethylsilyl)oxy]-, (3β,16α)-	49774-92-7	NI42556
129	73	215	344	75	434	145	239	434	100	86	84	58	56	51	51	45		25	46	2	–	–	–	–	–	–	2	–	–	–	Androst-5-ene, 3,16-bis[(trimethylsilyl)oxy]-, (3β,16β)-	33283-05-5	NI42557
129	73	215	344	75	434	145	239	434	100	85	82	56	53	50	50	44		25	46	2	–	–	–	–	–	–	2	–	–	–	Androst-5-ene, 3,16-bis[(trimethylsilyl)oxy]-, (3β,16β)-	33283-05-5	LI08932
129	215	239	254	213	344	119	305	434	100	72	53	49	40	35	28	23	11.00	25	46	2	–	–	–	–	–	–	2	–	–	–	Androst-5-ene, 3,17-bis[(trimethylsilyl)oxy]-, (3β,17α)-	13111-27-8	NI42558
215	239	254	129	344	213	329	305	434	100	86	83	82	75	48	45	38	29.72	25	46	2	–	–	–	–	–	–	2	–	–	–	Androst-5-ene, 3,17-bis[(trimethylsilyl)oxy]-, (3β,17α)-	13111-27-8	NI42559
255	419	345	129	433	420	256	343	434	100	66	52	26	24	22	21	18	12.50	25	46	2	–	–	–	–	–	–	2	–	–	–	Androst-5-ene, 3,17-bis[(trimethylsilyl)oxy]-, (3β,17β)-		NI42560
73	129	75	215	344	239	254	213	434	100	66	56	42	40	39	31	28	17.00	25	46	2	–	–	–	–	–	–	2	–	–	–	Androst-5-ene, 3,17-bis[(trimethylsilyl)oxy]-, (3β,17β)-	13110-76-4	NI42561
239	215	344	254	129	213	329	305	434	100	93	89	88	88	54	52	52	37.06	25	46	2	–	–	–	–	–	–	2	–	–	–	Androst-5-ene, 3,17-bis[(trimethylsilyl)oxy]-, (3β,17β)-	13110-76-4	NI42562
73	419	75	434	169	329	45	74	434	100	99	99	95	72	70	50	48		25	46	2	–	–	–	–	–	–	2	–	–	–	Androst-16-ene, 3,17-bis[(trimethylsilyl)oxy]-, (3α,5α)-	57305-20-1	NI42563
73	434	75	419	329	169	45	74	434	100	99	99	86	82	58	42	41		25	46	2	–	–	–	–	–	–	2	–	–	–	Androst-16-ene, 3,17-bis[(trimethylsilyl)oxy]-, (3α,5β)-	57305-19-8	NI42564
434	285	419	435	420	245	121	286	434	100	71	65	29	24	15	14	13		26	26	6	–	–	–	–	–	–	–	–	–	–	2H,8H-Benzo[1,2-b:3,4-b′]dipyran-2-one, 4-hydroxy-3-(4-hydroxyphenyl)-5-methoxy-8,8-dimethyl-6-(3-methyl-2-butenyl)-	5084-00-4	NI42565
434	419	285	435	420	245	121	257	434	100	86	37	29	25	24	11	9		26	26	6	–	–	–	–	–	–	–	–	–	–	2H,8H-Benzo[1,2-b:5,4-b′]dipyran-2-one, 4-hydroxy-3-(4-hydroxyphenyl)-5-methoxy-8,8-dimethyl-10-(3-methyl-2-butenyl)-	5490-47-1	NI42566
189	246	190	175	161	247	162	115	434	100	39	23	21	18	8	8	8	5.00	26	26	6	–	–	–	–	–	–	–	–	–	–	1,3-Isobenzofurandione, 5,5′-(1,1,6,6-tetramethyl-1,6-hexanediyl)bis-	63317-78-2	NI42567
119	203	120	434	85	84	204	435	434	100	99	97	95	95	95	94	90		26	34	2	4	–	–	–	–	–	–	–	–	–	3,6-Bis(p-N-hexoxyphenyl)-1,2,4,5-tetrazine	72243-03-9	NI42568
227	296	121	138	228	240	93	211	434	100	64	48	23	20	16	9	7	2.00	27	30	5	–	–	–	–	–	–	–	–	–	–	2α,11-Dihydroxyabieta-5,7,9(11),13-tetraen-12-one, 2α-O-(4-hydroxybenzoate)		NI42569
121	296	228	242	227	434	281	229	434	100	24	22	18	16	12	10	7		27	30	5	–	–	–	–	–	–	–	–	–	–	11,19-Dihydroxyabieta-5,7,9(11),13-tetraen-12-one, 19-O-(4-hydroxybenzoate)		NI42570
246	434	105	231	107	121	119	161	434	100	62	48	35	33	30	28	24		27	35	4	–	–	–	–	–	–	–	–	1	–	Pregnane-3,20-dione, 17,21-[(phenylborylene)bis(oxy)]-, (5β)-	30888-59-6	NI42571
105	147	124	159	107	434	121	119	434	100	74	57	53	50	46	43	42		27	35	4	–	–	–	–	–	–	–	–	1	–	Pregn-4-en-3-one, 20-hydroxy-17,21-[(phenylborylene)bis(oxy)]-, (20R)-	30888-46-1	NI42572
57	41	55	263	29	291	81	43	434	100	52	32	30	30	27	20	17	2.00	27	46	4	–	–	–	–	–	–	–	–	–	–	17β-tert-Butoxy-6-oxo-4-propyl-3,4-seco-5α-androstan-3-oic acid dimethyl ester		MS09679
57	41	55	378	81	291	207	93	434	100	43	40	27	27	24	24	23	3.00	27	46	4	–	–	–	–	–	–	–	–	–	–	17β-tert-Butoxy-6-oxo-4-propyl-3,4-seco-5β-androstan-3-oic acid dimethyl ester		MS09680
99	55	81	43	41	105	107	57	434	100	41	40	38	30	23	22	22	1.00	27	46	4	–	–	–	–	–	–	–	–	–	–	Cholest-5-ene-3,16,22,26-tetrol	55925-28-5	NI42573
99	399	283	271	417	381	255	400	434	100	84	65	60	42	35	30	24	0.00	27	46	4	–	–	–	–	–	–	–	–	–	–	Cholest-5-ene-3,16,22,26-tetrol	55925-28-5	NI42574
151	161	401	416	398	201	233	383	434	100	85	77	69	69	54	50	46	4.30	27	46	4	–	–	–	–	–	–	–	–	–	–	25-Norcycloartane-3β,6α,16β,24-tetraol		MS09681
194	195	158	77	90	114	76	54	434	100	20	14	10	7	3	3	3	0.00	28	34	4	–	–	–	–	–	–	–	–	–	–	[2,2′-Bifuran]-5,5′(2H,2′H)-dione, tetrahydro-3,3′-bis(2,4a,5,6,7,8-hexahydro-1-naphthalenyl)-	56701-33-8	NI42575
125	199	377	197	139	160	323	183	434	100	77	38	32	32	28	27	27	0.00	28	38	2	–	–	–	–	–	–	1	–	–	–	4(R)-[(tert-Butyldiphenylsilyl)oxy]dodeca-2(E),6(Z)-dienal	100311-75-9	DD01670
434	331	302	387	359	332	303	59	434	100	33	20	15	15	11	10	8		29	22	4	–	–	–	–	–	–	–	–	–	–	2-Methoxy-5,6,7-triphenyl-3-methoxycarbonyl-benzofuran		LI08933
180	374	43	416	256	181	375	254	434	100	56	26	19	19	17	11	10	4.80	29	26	2	2	–	–	–	–	–	–	–	–	–	Acetamide, N,N′-(methylenedi-2,1-phenylene)bis[N-phenyl-	52812-77-8	NI42576

MASS TO CHARGE RATIOS								M.W.	INTENSITIES								Parent	C	H	O	N	S	F	Cl	Br	I	Si	P	B	X	COMPOUND NAME	CAS Reg No	No
247	250	249	183	184	248	251	235	**434**	100	95	95	95	75	50	30	23	**3.00**	29	30	–	4	–	–	–	–	–	–	–	–	–	17-Norcorynan, 19,20-didehydro-16-(4,9-dihydro-3H-pyrido[3,4-b]indol-1-yl)-, (19E)-	36150-15-9	NI42577
70	265	434	266	71	148	121	44	**434**	100	75	25	20	12	10	10	9		29	42	1	2	–	–	–	–	–	–	–	–	–	Pregn-5-ene, 20-[(2-hydroxybenzyl)imino]-3-(methylamino)-, (3β,20S)-	27769-08-0	NI42578
434	436	435	437	44	203	119	57	**434**	100	38	34	12	5	4	4	4		30	42	2	–	–	–	–	–	–	–	–	–	–	1,2-Bis(4′-oxo-3′,5′-di-tert-butylcyclohexadienylidene)ethane		MS09682
434	57	435	436	219	41	419	55	**434**	100	54	36	18	18	16	5	5		30	42	2	–	–	–	–	–	–	–	–	–	–	3,3′,5,5′-Tetra-tert-butylstilbene 4,4′-quinone		IC04401
57	217	161	41	434	202	218	203	**434**	100	50	16	15	14	10	9	9		30	46	–	2	–	–	–	–	–	–	–	–	–	2,2′-Dimethyl-4,4′,6,6′-tetra-tert-butylazobenzene		MS09683
83	57	43	55	82	41	71	69	**434**	100	77	76	69	59	45	41	33	**0.08**	31	62	–	–	–	–	–	–	–	–	–	–	–	13-Cyclohexylpentacosane		AI02375
83	57	82	43	55	71	97	69	**434**	100	98	87	86	73	55	47	45	**0.08**	31	62	–	–	–	–	–	–	–	–	–	–	–	13-Cyclohexylpentacosane		AI02376
434	435	177	279	355	265	139	44	**434**	100	36	13	12	8	8	8	8		34	26	–	–	–	–	–	–	–	–	–	–	–	Dibenz[a,h]anthracene, 5,6,12,13-tetrahydro-5,12-diphenyl-	14474-64-7	NI42579
434	435	279	265	278	436	355	178	**434**	100	38	12	9	8	7	7	7		34	26	–	–	–	–	–	–	–	–	–	–	–	Dibenz[a,h]anthracene, 5,6,12,13-tetrahydro-5,12-diphenyl-	14474-65-8	NI42580
434	435	265	139	341	436	343	279	**434**	100	39	19	17	13	10	10	10		34	26	–	–	–	–	–	–	–	–	–	–	–	p-Terphenyl, 2′,5′-distyryl-	14474-63-6	NI42581
422	420	424	437	435	400	402	190	**435**	100	78	48	22	17	14	13	12		8	7	–	3	–	3	4	–	–	–	3	–	–	1-Methyl-1-(4-trifluoromethylphenyl)tetrachlorocyclotriphosphazene	87048-80-4	NI42582
69	76	119	366	416	435	171	126	**435**	100	73	41	38	20	15	13	13		9	–	–	3	–	15	–	–	–	–	–	–	–	1,3,5-Triazine, 2,4,6-tris(pentafluoroethyl)-	858-46-8	NI42583
91	65	402	404	36	437	400	92	**435**	100	15	12	8	8	7	7	6	**3.67**	12	7	–	3	1	–	6	–	–	–	–	–	–	2-Benzylthio-4,6-bis(trichloromethyl)-s-triazine		MS09684
320	318	91	437	439	322	45	36	**435**	100	99	54	46	38	37	34	31	**23.07**	12	7	–	3	1	–	6	–	–	–	–	–	–	2-(4′-Tolylthio)-4,6-bis(trichloromethyl)-s-triazine		MS09685
81	306	43	41	82	29			**435**	100	82	77	60	11	7			**4.83**	15	16	4	3	–	7	–	–	–	–	–	–	–	3-Acetyl-N-(heptafluorobutyryl)histidine propyl ester		MS09686
100	56	377	437	375	435	378	379	**435**	100	77	65	60	50	48	38	35		17	13	4	1	–	–	4	–	–	–	–	–	–	Dimethyl 6,7,8,9-tetrachloro-3a,9b-dihydro-1-methylbenz[g]indole-2,3-dicarboxylate		NI42584
45	75	43	356	88	102	58	44	**435**	100	56	49	43	40	31	31	30	**1.30**	18	15	2	5	1	–	2	–	–	–	–	–	–	8-Chloro-6-(2-chlorophenyl-1-[(methanesulphonamido)methyl])-4H-1,2,4-triazolo[4,3-a]-1,4-benzodiazepine		MS09687
312	302	343	313	274	345	330	361	**435**	100	55	47	40	33	31	30	28	**6.00**	20	21	10	1	–	–	–	–	–	–	–	–	–	Tetramethyl 1-acetoxy-4,5-dihydropyrido[1,2-a]azepine-2,3,4,5-tetracarboxylate		NI42585
124	125	165	436	435	272	149	97	**435**	100	10	6	1	1	1	1	1		20	22	4	1	–	5	–	–	–	–	–	–	–	Hyoscyamine pentafluoropropionate		NI42586
105	264	77	437	435	51	231	229	**435**	100	76	57	24	23	10	8	8		21	14	1	3	1	–	–	1	–	–	–	–	–	Benzamide, N-[4-(p-bromophenyl)-3-phenyl-Δ^2-1,2,4-thiadiazolin-5-ylidene]-	33863-65-9	NI42587
173	175	177	145	174	435	147	176	**435**	100	63	11	10	8	7	6	5		21	19	5	1	–	–	2	–	–	–	–	–	–	Diethyl 3,3′-[1-(2,6-dichlorobenzoyl)pyrrole-2,5-diyl]diprop-2-enoate	95883-18-4	NI42588
57	187	155	186	188	156	43	41	**435**	100	63	52	46	43	43	42	40	**0.00**	23	33	3	1	2	–	–	–	–	–	–	–	–	Pyridine, 1-acetyl-2-acetoxy-4-phenyl-3,6-di-tert-butylmercapto-1,2,3,6-tetrahydro-		MS09688
57	187	155	186	188	156	43	346	**435**	100	63	52	46	43	43	42	38	**0.00**	23	33	3	1	2	–	–	–	–	–	–	–	–	Pyridine, 1-acetyl-2-acetoxy-4-phenyl-3,6-di-tert-butylmercapto-1,2,3,6-tetrahydro-		LI08934
435	360	109	361	436	44	58	55	**435**	100	43	40	35	27	24	23	23		24	37	6	1	–	–	–	–	–	–	–	–	–	1-Phenanthrenecarboxylic acid, tetradecahydro-9-hydroxy-1,4a,8-trimethyl-7-[2-[2-(methylamino)ethoxy]-2-oxoethylidene]-10-oxo-, methyl ester, [1S-(1α,4aα,4bβ,7e,8β,8aα,9α)]-	36150-74-0	NI42589
418	146	163	165	160	176	330	435	**435**	100	95	70	68	65	64	59	41		25	29	4	3	–	–	–	–	–	–	–	–	–	13-Benzoyl-6-hydroxy-7-phenyl-1,6,10-triazabicyclo[10.2.1]pentadecane-9,14-dione		NI42590
404	116	296	314	386	86	272	271	**435**	100	70	64	55	49	38	35	32	**23.17**	25	45	3	1	–	–	–	–	–	1	–	–	–	5α-Pregnan-20-one, 3α,17α-dihydroxy-, methyloxime, TMS		NI42591
296	404	314	116	86	297	272	279	**435**	100	77	56	55	36	31	29	26	**22.05**	25	45	3	1	–	–	–	–	–	1	–	–	–	5β-Pregnan-20-one, 3α,17α-dihydroxy-, methyloxime, TMS		NI42592
420	435	344	362	182	130	44	436	**435**	100	98	82	78	76	72	72	57		27	33	4	1	–	–	–	–	–	–	–	–	–	2H-1-Benzopyrano[5′,6′:6,7]indeno[1,2-b]indol-3(4bH)-one, 5,6,6a,7,12,12b,12c,13,14,14a-decahydro-4b-hydroxy-2-(1-hydroxyisopropyl)-12b,12c-dimethyl-	57186-25-1	NI42593
280	435	175	107	436	103	43	281	**435**	100	74	61	28	22	22	21	20		27	33	4	1	–	–	–	–	–	–	–	–	–	4-(4-Dodecylphenylcarbamoyl)phthalic anhydride		IC04402
420	419	421	435	328	218	418	434	**435**	100	67	26	20	18	18	17	13		27	36	–	3	–	–	–	–	–	–	–	3	–	Borazine, 1,3,5-triethyl-2,4,6-tris(2-methylphenyl)-	52176-10-0	NI42594
143	399	129	435	253	400	291	255	**435**	100	74	37	24	23	22	22	22		27	49	3	1	–	–	–	–	–	–	–	–	–	Cholestane-3,16,22-triol, 26-amino-, (3β,5α,16β)-	50686-94-7	NI42595
93	234	407	145	288	235	173	144	**435**	100	54	30	21	18	18	15	15	**1.00**	28	25	2	3	–	–	–	–	–	–	–	–	–	1-Anilino-2,7-diphenyl-2,7-diazatetracyclo[4,4,2,$0^{3,12}$.$0^{8,11}$]dodecane-4,10-dione		MS09689
91	105	257	147	135	79	133	107	**435**	100	96	94	47	26	26	25	25	**0.00**	28	37	3	1	–	–	–	–	–	–	–	–	–	Estr-4-en-3-one, 17-(1-oxo-3-phenylpropoxy)-, 3-(O-methyloxime), (17β)-	57397-27-0	NI42597
105	91	139	133	137	79	138	106	**435**	100	80	44	35	28	19	15	15	**4.71**	28	37	3	1	–	–	–	–	–	–	–	–	–	Estr-4-en-3-one, 17-(1-oxo-3-phenylpropoxy)-, 3-(O-methyloxime), (17β)-	57397-27-0	NI42596
172	348	69	260	143	146	329	88	**436**	100	84	81	80	78	75	66	44	**23.00**	2	5	1	–	–	11	–	–	–	–	4	–	1	Tris(trifluorophosphine)(difluoroethoxyphosphine)nickel(0)		LI08935
260	172	348	69	88	234	329	124	**436**	100	85	67	59	48	34	28	26	**22.00**	2	5	1	–	–	11	–	–	–	–	4	–	1	Tris(trifluorophosphine)(difluoroethoxyphosphine)nickel(0)		LI08936
93	95	58	218	183	186	151	220	**436**	100	35	30	28	27	24	23	21	**8.21**	8	8	–	–	6	–	4	–	–	–	–	–	–	2,4,6,8,9,10-Hexathiaadamantane, 1,3,5,7-tetrakis(chloromethyl)-	13639-09-3	NI42598
28	42	55	57	112	56	58	350	**436**	100	34	28	21	19	13	9	8	**3.71**	11	12	5	2	–	–	–	–	–	–	–	–	1	Tungsten, pentacarbonyl(1,4-diazabicyclo[2.2.2]octane)-	35419-59-1	NI42599
436	271	106	147	174	202	417	367	**436**	100	75	24	19	9	8	6	6		12	12	2	2	–	6	–	–	–	–	–	–	1	N,N′-Ethylenebis(1,1,1-trifluoro-4-iminopentan-2-onato)palladium(II)	40792-96-9	NI42600

MASS TO CHARGE RATIOS								M.W.	INTENSITIES								Parent	C	H	O	N	S	F	Cl	Br	I	Si	P	B	X	COMPOUND NAME	CAS Reg No	No
79	81	80	39	77	53	27	41	**436**	100	71	64	36	35	31	30	28	**0.64**	12	18	–	–	–	–	–	2	–	–	–	–	2	Nickel, di-μ-bromobis[(1,2,3-η)-2-cyclohexen-1-yl]di-	32877-98-8	NI42601
43	238	240	237	263	236	261	272	**436**	100	15	11	11	10	10	9	8	**0.15**	14	10	3	–	–	–	6	–	–	–	–	–	–	9-(syn-Epoxy)acetoxy-1,2,3,4,10,10-hexachloro-6,7-epoxy-1,4,4a,5,6,7,8,8a,-octahydro-1,4-endo-5,8-exo-dimethanonaphthalene		MS09690
124	189	436	241	91	371	150	306	**436**	100	76	13	8	4	3	3	2		15	15	–	–	2	–	–	–	–	–	–	–	3	Tris(π-cyclopentadienyl-cobalt)disulphide		LI08937
43	338	436	100	85	59	42	41	**436**	100	82	32	31	23	11	11	11		15	21	6	–	–	–	–	–	–	–	–	–	1	Lanthanum, tris(2,4-pentanedionato)-	14284-88-9	NI42602
226	227	31	43	45	187	157	29	**436**	100	83	28	19	12	12	12	11	**3.50**	16	12	2	2	–	6	–	–	–	–	–	–	1	N,N'-O-Phenylenebis(5,5,5-trifluoro-4-oxopentan-2-iminato)nickel(II)	87962-10-5	NI42603
223	208	221	207	253	237	222	206	**436**	100	99	55	51	50	47	46	44	**0.00**	16	36	–	–	–	–	–	–	–	–	–	–	1	Tetraisobutyllead		AI02377
73	147	421	45	74	148	75	221	**436**	100	47	19	16	9	8	6	5	**2.69**	16	36	6	–	–	–	–	–	–	4	–	–	–	2-Butenedioic acid, 2,3-bis[(trimethylsilyl)oxy]-, bis(trimethylsilyl) ester, (Z)-	55887-89-3	NI42604
73	147	421	45	148	74	422	436	**436**	100	46	19	15	9	9	8	5		16	36	6	–	–	–	–	–	–	4	–	–	–	2-Butenedioic acid, 2,3-bis[(trimethylsilyl)oxy]-, bis(trimethylsilyl) ester, (Z)-	55887-89-3	NI42605
401	403	405	438	373	375	436	402	**436**	100	98	35	26	23	21	20	20		17	12	5	–	–	–	4	–	–	–	–	–	–	Diploicin methyl ether	19314-80-8	NI42606
401	403	405	438	373	436	375	402	**436**	100	95	34	25	23	20	20	19		17	12	5	–	–	–	4	–	–	–	–	–	–	Diploicin methyl ether	19314-80-8	LI08938
94	172	160	187	207	329	357	269	**436**	100	39	17	15	10	9	5	4	**0.36**	17	13	5	2	1	–	–	1	–	–	–	–	–	Methyl 3-bromomethyl-7-phthalimido-3-cephem-4-carboxylate		NI42607
295	281	227	204	173	350	243	54	**436**	100	99	99	92	92	59	53	53	**3.30**	17	14	3	–	–	6	–	–	–	–	–	–	1	π-[2,3-Bis(trifluoromethyl)-5,6,7,8-tetramethylbicyclo[2.2.2]octa-2,5,7-trienyl]tricarbonyliron		MS09691
73	273	147	229	75	319	303	77	**436**	100	56	55	53	36	27	26	23	**0.00**	17	36	7	–	–	–	–	–	–	3	–	–	–	Citric acid ethyl ester tri-TMS		MS09692
56	268	212	141	156	115	112	296	**436**	100	80	60	60	50	50	50	45	**0.00**	18	12	6	–	–	–	–	–	–	–	–	–	2	Hexacarbonyl-μ-[(1,2,2α-η^3:3,3α,4-η^3)-3-(E)-benzylidene-2-methylene-1,4 butanediyl]diiron	122624-36-6	DD01671
436	434	419	417	421	406	407	430	**436**	100	68	34	19	18	12	11	9		18	27	–	–	–	–	–	–	–	–	–	–	1	1,5-Cyclooctadiene(pentamethylcyclopentadienyl)iridium		LI08939
73	217	205	147	103	129	131	189	**436**	100	50	32	22	17	17	16	12	**0.03**	18	40	6	–	–	–	–	–	–	3	–	–	–	1,2-O-Isopropylidene-3,5,6-tri-O-trimethylsilyl-D-glucofuranose		NI42608
42	91	155	281	65	225			**436**	100	37	13	12	12	10			**0.50**	19	24	4	4	2	–	–	–	–	–	–	–	–	3,7-Bis(p-tolylsulphonyl)-1,3,5,7-tetra-azabicyclo[3,3,1]nonane		LI08940
55	43	57	44	105	69	45	71	**436**	100	98	95	85	73	69	64	56	**7.80**	20	8	3	–	–	–	4	–	–	–	–	–	–	2-Carboxytetrachlorophenyl biphenylen-2-yl ketone		IC04403
43	214	159	256	231	189	103	71	**436**	100	10	10	9	9	5	5	3	**0.10**	20	25	10	–	–	–	–	–	–	–	–	1	–	D-Glucitol, cyclic 3,4-(phenylboronate) 1,2,5,6-tetraacetate	74810-61-0	NI42609
232	204	73	147	117	219	103	304	**436**	100	67	62	60	41	11	11	9	**7.00**	20	36	3	2	–	–	–	–	–	3	–	–	–	(1'R,2'R)-2-[1',2',3'-tris(trimethylsilyloxy)propyl]quinoxaline		NI42610
290	73	291	75	292	406	45	147	**436**	100	63	32	25	12	10	10	7	**0.00**	20	36	3	2	–	–	–	–	–	3	–	–	–	L-Tryptophan, 5-hydroxy-, tris(trimethylsilyl) -	56272-77-6	NI42611
114	100	147	144	84	56	86	72	**436**	100	92	80	80	80	80	76	76	**22.80**	20	40	8	2	–	–	–	–	–	–	–	–	–	3,3-Dihydroxy-5,8,11,18,21,26-hexaoxa-1,15-diazabicyclo[13.8.5]octacosane		MS09693
289	56	144	100	114	58	57	70	**436**	100	98	87	87	80	55	55	51	**6.90**	20	40	8	2	–	–	–	–	–	–	–	–	–	3,10-Dihydroxy-5,8,15,18,23,26-hexaoxa-1,12-diazabicyclo[10.8.8]octaosane		MS09694
28	148	352	77	380	92	311	56	**436**	100	66	45	20	17	12	11	11	**0.95**	21	20	5	2	–	–	–	–	–	–	–	–	1	Iron, tricarbonyl[N,N'-(1,2-dimethyl-1,2-ethanediylidene)bis[4-methoxybenzenamine]-N,N']-	65016-01-5	NI42612
121	199	165	198	301	197	241	259	**436**	100	80	65	35	20	15	11	1	**0.00**	21	24	2	–	4	–	–	–	–	–	–	–	–	Carbonodithioic acid, S,S'-(diphenylmethylene) O,O'-diisopropyl ester	26564-33-0	NI42613
436	438	74	211	437	111	102	128	**436**	100	69	40	29	25	22	20	18		22	10	4	2	–	–	2	–	–	–	–	–	–	N,N'-Di-(m-chlorophenyl)pyromellitimide		MS09695
401	403	74	402	220	183	164	404	**436**	100	35	28	24	22	22	10	9	**6.00**	22	10	4	2	–	–	2	–	–	–	–	–	–	N,N'-Di-(o-chlorophenyl)pyromellitimide		MS09696
436	438	211	437	439	111	102	392	**436**	100	67	28	27	17	16	16	15		22	10	4	2	–	–	2	–	–	–	–	–	–	N,N'-Di-(o-chlorophenyl)pyromellitimide		MS09697
91	207	436	92	65	208	51	44	**436**	100	15	13	7	5	3	3	3		22	17	2	2	–	5	–	–	–	–	–	–	–	N-Acetyl-5-benzyloxytryptamine PFP		MS09698
98	55	352	365	281	99	436		**436**	100	32	28	10	8	8	7			22	24	4	6	–	–	–	–	–	–	–	–	–	N,N-Bis(2-acrylylaminoethyl)-4-(4-nitrophenylazo)aniline		IC04404
203	189	73	204	190	436	258	347	**436**	100	99	36	21	18	17	17	13		22	28	2	4	–	–	–	–	–	2	–	–	–	1H-Benzimidazole-2-carboxaldehyde, 1-(trimethylsilyl)-	56195-99-4	NI42614
71	43	128	228	83	256	91	77	**436**	100	75	26	24	24	20	16	15	**0.00**	22	28	9	–	–	–	–	–	–	–	–	–	–	Melcanthin E	79405-86-0	NI42615
232	56	233	100	174	144	114	130	**436**	100	15	12	10	8	6	6	6	**1.00**	22	32	7	2	–	–	–	–	–	–	–	–	–	13-(5-Phthalimido-3-oxapentyl)-1,4,7,10-tetraoxa-13-azacyclopentadecane		MS09699
309	310	105	77	254	203	202	102	**436**	100	25	17	15	13	10	10	8	**0.00**	23	17	1	–	–	–	–	–	1	–	–	–	–	Pyrylium, 2,4,6-triphenyl-, iodide	3495-60-1	NI42616
309	310	105	77	254	155	203	202	**436**	100	27	18	16	13	13	10	10	**0.00**	23	17	1	–	–	–	–	–	1	–	–	–	–	Pyrylium, 2,4,6-triphenyl-, iodide	3495-60-1	LI08941
371	336	401	227	335	292	208	262	**436**	100	70	58	22	20	15	14	7	**0.00**	23	17	2	2	–	–	1	–	–	–	–	–	1	Titanium, chloro(η-5-2,4-cyclopentadien-1-yl)bis(8-quinolinolato-N[1],O[8])-	35256-08-7	NI42617
138	73	55	436	299	41	139	28	**436**	100	33	25	24	18	14	13	12		23	31	1	2	–	3	–	–	–	1	–	–	–	7'-Trifluoromethyl-dihydrocinchonidine, trimethylsilyl-		NI42618
400	191	372	113	137	357	256	151	**436**	100	21	18	18	17	16	16	16	**0.00**	24	17	6	–	–	–	1	–	–	–	–	–	–	Sphagnorubin B	84018-30-4	NI42619
124	137	180	258	243	162	299	151	**436**	100	99	86	78	42	37	23	15	**3.00**	24	20	8	–	–	–	–	–	–	–	–	–	–	7H-1,4-Dioxino[2,3-c]xanthen-7-one, 2,3-dihydro-3-(4-hydroxy-3-methoxyphenyl)-2-(hydroxymethyl)-5-methoxy-, trans-(±)-	64280-48-4	NI42620
219	131	436	102	175	201	78	130	**436**	100	65	58	55	44	41	30	28		24	20	8	–	–	–	–	–	–	–	–	–	–	Malonic acid, benzylidene-, bimol. cyclic ethylene ester	21620-33-7	NI42621
91	92	239	330	155	329	167	154	**436**	100	9	7	5	3	2	2	2	**0.10**	24	25	2	4	–	–	1	–	–	–	–	–	–	9-(4'-Benzyloxy-3'-benzyloxymethylbutyl)-6-chloropurine		LI08942
91	92	239	65	330	55	41	39	**436**	100	9	7	7	5	4	4	3	**0.00**	24	25	2	4	–	–	1	–	–	–	–	–	–	9H-Purine, 6-chloro-9-[4-(phenylmethoxy)-3-[(phenylmethoxy)methyl]butyl]-	33498-83-8	NI42622
211	209	345	343	207	91	210	341	**436**	100	79	71	52	46	43	36	31	**1.50**	24	28	–	–	–	–	–	–	–	–	–	–	1	Tribenzylisopropyltin		NI42623
258	189	248	118	130	201	187	146	**436**	100	37	32	29	25	22	16	9	**1.10**	24	28	4	4	–	–	–	–	–	–	–	–	–	3-Methyl-8-[4-(4-nitrophenyl)-4-oxobutyl]-1-phenyl-1,3,8-triazaspiro[4,5]decan-4-one		MS09700

MASS TO CHARGE RATIOS								M.W.	INTENSITIES								Parent	C	H	O	N	S	F	Cl	Br	I	Si	P	B	X	COMPOUND NAME	CAS Reg No	No
69	97	143	115	145	41	185	96	**436**	100	90	51	51	50	50	29	29	**7.00**	24	36	3	–	2	–	–	–	–	–	–	–	–	Ethyl 7-oxa-7-(2-benzyl-2-methyl-6,10-dithiaspiro[4.5]decan-1-yl)heptanoate	92640-63-6	NI42624
298	209	240	358	316	300	330	280	**436**	100	99	57	47	44	44	42	36	**3.20**	24	36	7	–	–	–	–	–	–	–	–	–	–	Grayanotoxin II 3,6-diacetate		NI42625
77	236	67	103	200	105	51	129	**436**	100	97	96	57	45	39	36	33	**0.00**	25	20	2	6	–	–	–	–	–	–	–	–	–	3H-Pyrazol-3-one, 4-[(4,5-dihydro-3-methyl-5-oxo-1-phenyl-1H-pyrazol-4 yl)azo]-2,4-dihydro-2,5-diphenyl-	56666-63-8	MS09701
436	352	394	283	338	351	295	267	**436**	100	95	90	64	46	40	40	30		25	24	7	–	–	–	–	–	–	–	–	–	–	5-O-methyllupiwighteone diacetate		NI42626
376	119	333	122	144	130	317	120	**436**	100	46	43	28	26	16	15	15	**11.00**	25	28	5	2	–	–	–	–	–	–	–	–	–	12α,23-Diacetoxy-12β,13α-dihydroisostrychnine		NI42627
436	377	130	190	143	437	264	185	**436**	100	57	42	37	34	28	25	22		25	28	5	2	–	–	–	–	–	–	–	–	–	Quebrachidine, N-acetyl-, acetate (ester)	4914-01-6	NI42628
257	316	258	317	147	121	259	109	**436**	100	32	24	12	9	9	8	7	**0.00**	25	40	6	–	–	–	–	–	–	–	–	–	–	Methyl 1,11-diacetyloxyabietanoate		NI42629
197	436	196	404	55	41	438	195	**436**	100	63	38	36	29	23	23	21		25	40	6	–	–	–	–	–	–	–	–	–	–	Methyl 15-(5,6-dimethoxy-3-methyl-1,4-benzoquinon-2-yl)pentadecanoate		MS09702
436	183	197	167	55	196	181	437	**436**	100	51	38	36	31	30	30	28		25	40	6	–	–	–	–	–	–	–	–	–	–	Methyl 15-(3,4-dimethoxy-6-methyl-2,5-dioxobicyclo[4.1.0]hept-3-en-1-yl)pentadecanoate		MS09703
43	55	41	67	81	57	83	29	**436**	100	33	27	23	21	21	14	14	**0.00**	25	40	6	–	–	–	–	–	–	–	–	–	–	9,12,15-Octadecatrienoic acid, 2-(acetyloxy)-1-[(acetyloxy)methyl]ethyl ester, (Z,Z,Z)-	55320-01-9	WI01545
262	159	263	81	155	352	171	98	**436**	100	95	23	22	17	15	13	13	**4.00**	25	40	6	–	–	–	–	–	–	–	–	–	–	9,12,15-Octadecatrienoic acid, 2-(acetyloxy)-1-[(acetyloxy)methyl]ethyl ester, (Z,Z,Z)-	55320-01-9	WI01544
43	55	41	57	81	79	67	83	**436**	100	23	16	12	11	11	11	10	**0.00**	25	40	6	–	–	–	–	–	–	–	–	–	–	9,12,15-Octadecatrienoic acid, 2,3-bis(acetyloxy)propyl ester, (Z,Z,Z)-	55320-02-0	WI01546
159	352	74	79	380	219	161	92	**436**	100	23	21	20	18	14	10	10	**0.00**	25	40	6	–	–	–	–	–	–	–	–	–	–	9,12,15-Octadecatrienoic acid, 2,3-bis(acetyloxy)propyl ester, (Z,Z,Z)-	55320-02-0	WI01547
129	241	215	256	148	107	217	130	**436**	100	61	43	36	34	34	33	32	**24.41**	25	48	2	–	–	–	–	–	–	2	–	–	–	5α-Androstane-3α,17α-diol, bis(trimethylsilyl)-	10426-35-4	NI42630
241	346	148	130	421	149	256	331	**436**	100	96	91	84	83	81	80	64	**47.79**	25	48	2	–	–	–	–	–	–	2	–	–	–	5α-Androstane-3α,17β-diol, bis(trimethylsilyl)-	13260-01-0	NI42631
129	241	256	215	346	148	130	217	**436**	100	90	85	57	53	46	42	40	**27.21**	25	48	2	–	–	–	–	–	–	2	–	–	–	5α-Androstane-3α,17β-diol, bis(trimethylsilyl)-		NI42632
129	75	73	241	130	215	148	107	**436**	100	89	87	61	47	43	40	40	**31.00**	25	48	2	–	–	–	–	–	–	2	–	–	–	5α-Androstane-3α,17β-diol, bis(trimethylsilyl)-	10426-36-5	NI42633
73	156	75	95	107	81	256	67	**436**	100	99	99	85	80	80	70	70	**25.00**	25	48	2	–	–	–	–	–	–	2	–	–	–	5α-Androstane-3β,11β-diol, bis(trimethylsilyl)-	33283-00-0	NI42634
73	144	75	421	255	93	81	107	**436**	100	99	99	50	50	50	50	45	**13.50**	25	48	2	–	–	–	–	–	–	2	–	–	–	5α-Androstane-3β,16α-diol, bis(trimethylsilyl)-	33283-01-1	NI42635
75	73	144	421	255	81	93	107	**436**	100	62	45	29	21	21	20	19	**7.74**	25	48	2	–	–	–	–	–	–	2	–	–	–	5α-Androstane-3β,16β-diol, bis(trimethylsilyl)-	37977-24-5	NI42636
129	241	346	148	149	256	421	215	**436**	100	62	59	46	41	40	39	32	**20.39**	25	48	2	–	–	–	–	–	–	2	–	–	–	5α-Androstane-3β,17β-diol, bis(trimethylsilyl)-		NI42637
256	129	241	346	215	257	217	199	**436**	100	51	45	40	34	26	19	19	**4.29**	25	48	2	–	–	–	–	–	–	2	–	–	–	5β-Androstane-3α,17β-diol, bis(trimethylsilyl)-		NI42639
256	241	129	346	215	257	199	201	**436**	100	51	40	37	34	23	22	16	**4.28**	25	48	2	–	–	–	–	–	–	2	–	–	–	5β-Androstane-3α,17β-diol, bis(trimethylsilyl)-		NI42640
73	256	129	75	241	215	346	81	**436**	100	99	99	99	51	48	43	42	**9.50**	25	48	2	–	–	–	–	–	–	2	–	–	–	5β-Androstane-3α,17β-diol, bis(trimethylsilyl)-	13111-26-7	NI42638
129	75	73	241	256	215	199	130	**436**	100	99	88	81	74	59	43	43	**22.00**	25	48	2	–	–	–	–	–	–	2	–	–	–	5β-Androstane-3α,17β-diol, bis(trimethylsilyl)-	13111-26-7	NI42641
129	75	73	241	256	215	130	199	**436**	100	94	85	71	62	52	42	35	**13.00**	25	48	2	–	–	–	–	–	–	2	–	–	–	5β-Androstane-3α,17β-diol, bis(trimethylsilyl)-	13111-26-7	NI42642
129	75	73	130	107	256	241	93	**436**	100	90	87	38	36	33	33	30	**7.00**	25	48	2	–	–	–	–	–	–	2	–	–	–	5β-Androstane-3β,17β-diol, bis(trimethylsilyl)-	18880-48-3	NI42643
436	421	203	219	217	403	437	422	**436**	100	97	48	37	35	32	29	25		26	28	6	–	–	–	–	–	–	–	–	–	–	2′-O-Methylcajanane		MS09704
436	201	267	268	437	185	294	391	**436**	100	83	43	38	37	31	20	17		26	29	4	–	–	–	–	–	–	–	1	–	–	Ethyl (1RS,1′RS,2E,2RS)-[2-[3′-(diphenylphosphinoyl)cyclopent-2′-enyl]-5-oxocyclopentyl]acetate		MS09705
57	43	302	201	41	320	107	81	**436**	100	48	41	40	27	26	25	23	**0.00**	26	44	5	–	–	–	–	–	–	–	–	–	–	6β-Acetoxy-17β-tert-butoxy-4,5-secoandrostan-4-oic acid methyl ester		MS09706
43	55	41	57	69	81	44	29	**436**	100	91	87	80	61	55	49	48	**0.00**	26	44	5	–	–	–	–	–	–	–	–	–	–	Ethyl iso-allocholate		NI42644
57	305	362	257	55	275	344	41	**436**	100	96	56	41	38	37	35	33	**0.00**	27	48	4	–	–	–	–	–	–	–	–	–	–	17β-tert-Butoxy-6-hydroxy-4-propyl-3,4-seco-5β-androstan-3-oic acid methyl ester		MS09707
57	305	257	362	275	55	41	43	**436**	100	53	30	27	26	22	21	16	**2.00**	27	48	4	–	–	–	–	–	–	–	–	–	–	17β-tert-Butoxy-6β-hydroxy-4-propyl-3,4-seco-5α-androstan-3-oic acid methyl ester		MS09708
402	266	95	107	420	384	81	43	**436**	100	97	79	52	51	51	51	46	**0.00**	27	48	4	–	–	–	–	–	–	–	–	–	–	Cholestane-3,5,6,7-tetrol, (3β,5α,6α,7β)-	56588-30-8	NI42645
105	77	436	106	51	44	437	79	**436**	100	28	11	8	5	5	4	3		29	24	4	–	–	–	–	–	–	–	–	–	–	2,2-Bis(4-benzoyloxyphenyl)propane		IC04405
105	331	77	78	314	436	332	106	**436**	100	30	27	15	13	10	8	8		29	24	4	–	–	–	–	–	–	–	–	–	–	2,2′-Methylenebis(4-methylphenol) dibenzoate		NI42646
403	436	367	69	404	368	175	191	**436**	100	47	45	35	31	18	18	16		29	40	3	–	–	–	–	–	–	–	–	–	–	Bis(2-hydroxy-3-tert-butyl-5-ethylphenyl)methane, mono-2-methylprop-2-enoate		NI42647
100	239	281	253	296	278	220	309	**436**	100	43	41	34	25	21	19	18	**8.00**	29	56	2	–	–	–	–	–	–	–	–	–	–	12,14-Nonacosanedione	58141-94-9	NI42648
436	57	437	41	219	29	203	438	**436**	100	88	38	24	11	11	9	7		30	44	2	–	–	–	–	–	–	–	–	–	–	1,2-Bis(3,5-di-tert-butyl-4-hydroxyphenyl)ethylene		IC04406
436	437	57	203	421	218	189	41	**436**	100	33	25	7	5	4	4	3		30	44	2	–	–	–	–	–	–	–	–	–	–	1,2-Bis(3,5-di-tert-butyl-4-hydroxyphenyl)ethylene		IC04407
436	200	109	187	69	108	95	119	**436**	100	75	59	53	53	43	43	36		30	44	2	–	–	–	–	–	–	–	–	–	–	3-Isopropylidene-a-nor-5β-methyl-δ9(10)-oleanenolactone		NI42649
57	43	71	85	55	295	41	69	**436**	100	80	59	42	28	24	24	22	**0.03**	31	64	–	–	–	–	–	–	–	–	–	–	–	11-Decylheneicosane		AI02378
436	437	309	323	338	351	364	379	**436**	100	37	30	29	28	26	24	22		31	64	–	–	–	–	–	–	–	–	–	–	–	Hentriacontane	630-04-6	NI42650
203	379	323	161	436	218	267	147	**436**	100	42	26	18	17	16	14	13		32	52	–	–	–	–	–	–	–	–	–	–	–	4′(1,1-Dimethylethyl)-1′,2,5-tris[(E)-2,2-dimethylpropylidene]-2′,3′,6′,7′-tetrahydrospiro[cyclopentan-1,5′(4′H)-indene]		MS09709
408	331	330	409	253	332	252	254	**436**	100	87	69	35	28	21	20	15	**2.26**	33	24	1	–	–	–	–	–	–	–	–	–	–	2-Indanone, 1,1,3,3-tetraphenyl-	20396-40-1	NI42651
408	330	329	409	253	331	252	254	**436**	100	90	70	36	30	20	20	16	**2.00**	33	24	1	–	–	–	–	–	–	–	–	–	–	2-Indanone, 1,1,3,3-tetraphenyl-	20396-40-1	LI08943

MASS TO CHARGE RATIOS								M.W.	INTENSITIES								Parent	C	H	O	N	S	F	Cl	Br	I	Si	P	B	X	COMPOUND NAME	CAS Reg No	No
218	73	291	219	220	292	147	100	437	100	51	34	21	12	9	9	8	0.10	15	39	4	3	–	–	–	–	–	4	–	–	–	Tetrakis(trimethylsilyl)-L-alanosine		MS09710
297	73	298	45	299	74	75	77	437	100	95	23	12	9	7	6	5	1.30	18	30	4	1	–	3	–	–	–	2	–	–	–	N-Trifluoroacetyl-O,O′-bis(trimethylsilyl)metanephrine		MS09711
274	378	318	437	286	244	324	344	437	100	92	50	42	32	30	25	16		20	23	10	1	–	–	–	–	–	–	–	–	–	1-Phenyl-2-acetoxy-2,3,4,5-tetrakis(methoxycarbonyl)pyrrolidine		LI08944
268	227	43	55	57	69	41	83	437	100	49	44	36	35	32	32	20	4.40	20	34	1	1	–	7	–	–	–	–	–	–	–	1-Aminohexadecane HFB		MS09712
280	70	42	143	113	281	57	406	437	100	43	43	41	26	21	21	20	13.72	22	26	1	3	1	3	–	–	–	–	–	–	–	Fluphenazine	69-23-8	NI42652
280	42	143	70	113	57	406	98	437	100	44	42	42	26	22	20	20	14.00	22	26	1	3	1	3	–	–	–	–	–	–	–	Fluphenazine		LI08945
280	42	43	70	143	45	56	41	437	100	92	75	75	51	48	45	40	34.80	22	26	1	3	1	3	–	–	–	–	–	–	–	Fluphenazine	69-23-8	NI42653
204	178	282	235	210	91	203	179	437	100	41	31	31	29	24	19	19	17.00	24	23	3	1	2	–	–	–	–	–	–	–	–	3′-Ethoxy-2′-tosylspiro[fluorene-9,5′-(1,3)thiazolidine]		MS09713
58	45	44	43	170	59	42	41	437	100	21	11	6	4	4	4	4	2.90	24	27	3	3	1	–	–	–	–	–	–	–	–	O-Dansylbufotenine	20289-22-9	NI42654
58	59	170	437	171	168	169	235	437	100	5	4	3	3	3	2	1		24	27	3	3	1	–	–	–	–	–	–	–	–	O-Dansylbufotenine	20289-22-9	NI42655
170	172	262	292	202	277	186	437	437	100	93	70	69	44	26	16	14		24	27	3	3	1	–	–	–	–	–	–	–	–	O-Dansylbufotenine	20289-22-9	LI08946
437	91	218	128	404	118	116	436	437	100	82	43	38	31	19	15	13		26	23	–	5	1	–	–	–	–	–	–	–	–	7-(Methylthio)-5-phenyl-4-[(m-tolylmethyl)amino]-2-(p-tolyl)imidazo[5,1-f][1,2,4]triazine		MS09714
105	77	41	335	106	55	51	42	437	100	40	12	11	10	10	10	10	3.50	26	31	5	1	–	–	–	–	–	–	–	–	–	Ethyl 3-(1′-benzoyl-3′-(2-benzoyloxyethyl)-piperid-4-yl)-propionate		NI42656
91	84	82	240	110	300	364	437	437	100	48	27	19	15	13	12	8		26	31	5	1	–	–	–	–	–	–	–	–	–	1-Quinolizidinecarboxylic acid, 4-(3-benzyloxy-4-methoxyphenyl)-2-oxo-, ethyl ester		LI08947
43	71	41	57	56	55	29	44	437	100	71	16	11	9	9	9	8	2.00	26	47	4	1	–	–	–	–	–	–	–	–	–	1-N-Isobutyl-N-octadecylaminosuccinic anhydride		IC04408
70	149	105	310	194	367	339	180	437	100	22	17	13	12	10	9	8	0.00	27	23	3	3	–	–	–	–	–	–	–	–	–	1,2,7,7a-Tetrahydro-1-methyl-2-(methylimino)-7-(4-methylphenyl)-4,7a-diphenylpyrrolo[2,3-D][1,3]oxazin-5,6-dione	90140-42-4	NI42657
91	166	79	255	77	271	51	254	437	100	65	27	25	25	23	23	22	0.00	29	27	1	1	1	–	–	–	–	–	–	–	–	Dispiro[9H-fluorene-9,3′-[1,4,2]oxathiazolidine-5′,2″-tricyclo[3.3.1.1^{3,7}]decane], 2′-phenyl-	50455-54-4	NI42658
360	344	105	373	256	361	345	343	437	100	98	37	33	27	25	25	24	12.00	32	23	1	1	–	–	–	–	–	–	–	–	–	10,14b-Dihydro-10,14b-diphenylisoindolo[2,1-f]phenanthridin-10-ol		NI42659
254	437	360	255	256	105	438	77	437	100	58	52	48	25	23	20	20		32	23	1	1	–	–	–	–	–	–	–	–	–	2-(6-Phenanthridyl)phenyl(diphenyl)methanol		NI42660
344	342	346	265	263	184	267	261	438	100	68	66	59	56	30	23	19	0.00	2	–	–	–	–	1	–	5	–	–	–	–	–	Pentabromofluoroethane		DO01123
409	407	411	405	413	237	272	109	438	100	92	66	39	28	22	19	19	0.00	10	3	–	–	–	–	9	–	–	–	–	–	–	4,7-Methano-1H-indene, 1,2,3,4,5,6,7,8,8-nonachloro-3a,4,7,7a-tetrahydro	21641-70-3	NI42661
43	444	442	440	438	401	399	397	438	100	49	49	48	48	47	47	45		10	5	3	2	–	–	–	3	–	–	–	–	–	1,2,3,4-Tetrahydropyrimidine, 1-(3′,5′-dibromo-4′-hydroxyphenyl)-5-bromo-2,4-dioxo-	83809-84-1	NI42662
64	183	44	48	91	52	63	153	438	100	48	32	30	24	21	19	14	0.00	10	8	4	4	1	–	–	2	–	–	–	–	–	2H-1,2,3-Thiadiazine, 2-(2,4-dinitrophenyl)-6-methyl-, compd. with bromine (1:1)	57954-51-5	NI42663
173	113	265	175	174	177	205	145	438	100	34	31	9	6	5	3	3	0.00	10	12	–	–	6	6	–	–	–	–	–	–	–	S,S′-Bis(2-trifluoromethyl-1,3-dithia-cyclopent-2-yl)dithioglycol		MS09715
240	438	298	205	145	335	354	224	438	100	99	99	59	57	45	43	40		11	–	5	–	–	9	–	–	–	–	–	–	1	Manganese, pentacarbonyl(2,3,3,4,4,5,5,6,6-nonafluoro-1-cyclohexen-1-yl)	14837-18-4	NI42664
438	298	241	28	205	143	335	326	438	100	99	99	99	59	58	46	40		11	–	5	–	–	9	–	–	–	–	–	–	1	Manganese, pentacarbonyl(2,3,3,4,4,5,5,6,6-nonafluoro-1-cyclohexen-1-yl)	14837-18-4	LI08948
425	423	427	440	438	292	442	290	438	100	78	48	32	25	17	15	14		11	17	–	4	–	–	4	–	–	–	3	–	–	1-Methyl-1-(4-N,N-diethylaminophenyl)tetrachlorocyclotriphosphazene	87048-82-6	NI42665
68	41	44	81	64	80	67	79	438	100	68	60	28	28	24	24	21	0.00	11	20	11	1	3	–	–	–	–	–	–	–	–	Glucocheirolin		MS09716
352	326	354	105	324	328	77	350	438	100	98	92	84	84	75	64	57	0.00	15	10	4	–	–	–	–	–	–	–	–	–	1	Tungsten, benzoyltricarbonyl-π-cyclopentadienyl-	41308-20-7	NI42666
28	134	215	79	55	52	256	200	438	100	72	53	39	39	32	29	24	0.07	15	11	7	2	–	–	–	–	–	–	–	–	2	Chromium, pentacarbonyl[dicarbonyl(η^5-2,4-cyclopentadien-1-yl)manganese][μ-(3,3-dimethyl-3H-diazirine-N[1]:N[2])]-	74779-85-4	NI42667
73	147	292	219	293	423	74	45	438	100	37	32	21	9	8	8	7	0.49	16	38	6	–	–	–	–	–	–	4	–	–	–	meso-Tartaric acid tetra-TMS	38165-94-5	NI42668
73	147	292	45	189	74	219	423	438	100	41	19	10	9	9	8	7	0.20	16	38	6	–	–	–	–	–	–	4	–	–	–	Tartaric acid tetra-TMS	18602-86-3	NI42669
73	147	292	219	423	293	189	74	438	100	70	51	31	15	15	15	14	0.00	16	38	6	–	–	–	–	–	–	4	–	–	–	Tartaric acid tetra-TMS	18602-86-3	MS09717
73	147	292	219	423	293	74	305	438	100	37	33	22	8	8	8	4	0.50	16	38	6	–	–	–	–	–	–	4	–	–	–	Bis(trimethylsilyl)-2,3-bis(trimethylsilylol)succinate		LI08949
45	47	127	262	161	111	73	152	438	100	59	54	49	49	49	48	47	11.00	17	16	–	6	2	–	2	–	–	–	–	–	–	4-[N,N′-bis(4-chlorophenyl)]guanidino-3,5-bis(methylthio)[1,2,4]triazole		MS09718
217	73	133	218	219	277	159	147	438	100	83	33	32	17	16	14	14	0.00	17	38	7	–	–	–	–	–	–	3	–	–	–	α-D-Galactofuranosiduronic acid, methyl 2,3,5-tris-O-(trimethylsilyl)-, methyl ester	56211-18-8	NI42670
217	73	218	133	204	147	219	75	438	100	77	20	17	16	12	9	9	0.00	17	38	7	–	–	–	–	–	–	3	–	–	–	β-D-Galactofuranosiduronic acid, methyl 2,3,5-tris-O-(trimethylsilyl)-, methyl ester	56211-17-7	NI42671
204	217	73	205	147	129	218	133	438	100	59	56	20	16	14	13	12	0.00	17	38	7	–	–	–	–	–	–	3	–	–	–	α-D-Galactopyranosiduronic acid, methyl 2,3,4-tris-O-(trimethylsilyl)-, methyl ester	56271-13-7	NI42672
217	73	218	147	219	103	75	74	438	100	65	19	15	9	7	7	5	0.00	17	42	5	–	–	–	–	–	–	4	–	–	–	Arabinofuranose TMS	55399-49-0	LI08950
217	73	218	147	219	103	75	191	438	100	66	19	15	10	7	7	6	0.00	17	42	5	–	–	–	–	–	–	4	–	–	–	Arabinofuranose TMS	55399-49-0	NI42673
73	217	204	191	147	218	205	75	438	100	71	68	48	27	14	13	10	0.00	17	42	5	–	–	–	–	–	–	4	–	–	–	α-D-Arabinopyranose TMS	20585-61-9	NI42674
217	73	204	191	147	218	189	205	438	100	90	74	48	40	32	18	17	0.00	17	42	5	–	–	–	–	–	–	4	–	–	–	α-DL-Arabinopyranose TMS	56271-65-9	NI42675
73	217	204	191	147	205	218	75	438	100	58	57	42	25	12	11	10	0.00	17	42	5	–	–	–	–	–	–	4	–	–	–	β-D-Arabinopyranose TMS		LI08951
217	73	204	147	191	218	75	189	438	100	98	43	37	31	25	22	15	0.00	17	42	5	–	–	–	–	–	–	4	–	–	–	β-DL-Arabinopyranose TMS	56271-64-8	NI42676

MASS TO CHARGE RATIOS								M.W.	INTENSITIES								Parent	C	H	O	N	S	F	Cl	Br	I	Si	P	B	X	COMPOUND NAME	CAS Reg No	No
73	217	204	191	147	205	218	75	**438**	100	58	58	42	25	12	11	10	**0.00**	17	42	5	–	–	–	–	–	–	4	–	–	–	β-L-Arabinopyranose TMS	32166-73-7	NI42677
73	217	204	191	147	218	205	74	**438**	100	49	47	34	22	11	10	9	**0.00**	17	42	5	–	–	–	–	–	–	4	–	–	–	D-Arabinose, tetrakis(trimethylsilyl)-	18622-97-4	NI42678
73	103	423	147	231	305	424	321	**438**	100	41	38	33	23	20	19	15	**1.47**	17	42	5	–	–	–	–	–	–	4	–	–	–	Butanoic acid, 2,4-bis[(trimethylsilyl)oxy]-2-[[(trimethylsilyl)oxy]methyl], trimethylsilyl ester	55521-19-2	NI42679
73	103	147	231	305	321	115	335	**438**	100	39	32	21	20	14	10	8	**0.10**	17	42	5	–	–	–	–	–	–	4	–	–	–	Per-TMS-3-deoxy-2-C-(hydroxymethyl)-tetronic acid		LI08952
73	147	75	205	117	74	217	129	**438**	100	21	17	8	8	8	7	7	**0.00**	17	42	5	–	–	–	–	–	–	4	–	–	–	Per-TMS-2-deoxy-erythro-pentonic acid	74742-30-6	NI42680
73	147	204	217	335	233	306	203	**438**	100	30	21	18	12	12	11	11	**0.00**	17	42	5	–	–	–	–	–	–	4	–	–	–	Per-TMS-2-deoxy-erythro-pentonic acid	74742-30-6	LI08953
73	245	147	231	205	75	74	69	**438**	100	58	37	22	11	10	9	9	**0.00**	17	42	5	–	–	–	–	–	–	4	–	–	–	Per-TMS-3-deoxy-erythro-pentonic acid		LI08954
306	307	234	205	103	204	308	217	**438**	100	76	62	48	38	31	27	27	**0.40**	17	42	5	–	–	–	–	–	–	4	–	–	–	3,9-Dioxa-2,10-disilaundecan-5-one, 2,2,10,10-tetramethyl-6,7-bis[(trimethylsilyl)oxy]-, (R*,R*)-	38321-21-0	NI42681
204	73	217	191	147	205	189	75	**438**	100	72	40	30	27	20	17	9	**0.00**	17	42	5	–	–	–	–	–	–	4	–	–	–	α-DL-Lyxopyranose TMS	56271-67-1	NI42682
73	217	204	147	191	75	45	205	**438**	100	96	96	47	41	29	27	24	**0.00**	17	42	5	–	–	–	–	–	–	4	–	–	–	β-DL-Lyxopyranose TMS	56271-66-0	NI42683
73	348	147	423	205	217	349	335	**438**	100	46	30	28	21	17	15	12	**0.00**	17	42	5	–	–	–	–	–	–	4	–	–	–	Pentonic acid, 2-deoxy-3,4,5-tris-O-(trimethylsilyl)-, trimethylsilyl ester	55521-17-0	NI42684
73	147	75	205	217	117	129	74	**438**	100	28	15	14	13	10	9	8	**0.00**	17	42	5	–	–	–	–	–	–	4	–	–	–	Pentonic acid, 2-deoxy-3,4,5-tris-O-(trimethylsilyl)-, trimethylsilyl ester	55521-17-0	NI42685
73	245	335	147	333	231	423	336	**438**	100	58	54	35	32	24	22	16	**0.00**	17	42	5	–	–	–	–	–	–	4	–	–	–	Pentonic acid, 3-deoxy-2,4,5-tris-O-(trimethylsilyl)-, trimethylsilyl ester	55521-18-1	NI42686
217	73	218	147	219	75	103	191	**438**	100	79	27	23	13	11	9	8	**0.00**	17	42	5	–	–	–	–	–	–	4	–	–	–	D-Ribofuranose TMS	56271-69-3	NI42687
73	116	101	147	75	103	217	117	**438**	100	79	45	39	38	28	18	17	**0.00**	17	42	5	–	–	–	–	–	–	4	–	–	–	D-Ribopyranose TMS	56271-70-6	NI42688
73	217	204	191	147	75	45	218	**438**	100	98	93	67	52	31	27	26	**0.00**	17	42	5	–	–	–	–	–	–	4	–	–	–	D-Ribopyranose TMS	56271-70-6	NI42689
73	217	204	147	191	205	218	103	**438**	100	72	67	30	22	18	15	10	**0.00**	17	42	5	–	–	–	–	–	–	4	–	–	–	D-Ribose, 2,3,4,5-tetrakis(trimethylsilyl)-	33648-69-0	NI42690
73	217	204	191	147	205	218	75	**438**	100	55	51	39	24	11	11	9	**0.00**	17	42	5	–	–	–	–	–	–	4	–	–	–	D-Ribose, 2,3,4,5-tetrakis(trimethylsilyl)-	33648-69-0	NI42691
75	217	93	95	73	204	103	191	**438**	100	67	65	24	23	18	12	8	**2.00**	17	42	5	–	–	–	–	–	–	4	–	–	–	Ribose TMS		LI08955
306	307	234	205	103	204	308	217	**438**	100	77	63	48	38	32	29	28	**0.40**	17	42	5	–	–	–	–	–	–	4	–	–	–	1,2,3-Tris(trimethylsilylol)propyl trimethylsilylol-methyl ketone		LI08956
217	73	218	147	219	103	191	75	**438**	100	68	25	22	12	11	9	9	**0.00**	17	42	5	–	–	–	–	–	–	4	–	–	–	D-Xylofuranose TMS	56271-68-2	NI42692
204	73	217	147	191	205	189	75	**438**	100	89	50	35	32	23	14	12	**0.00**	17	42	5	–	–	–	–	–	–	4	–	–	–	α-D-Xylopyranose, 1,2,3,4-tetrakis(trimethylsilyl)-	14251-20-8	NI42693
204	73	217	147	191	205	75	189	**438**	100	99	49	43	37	26	22	19	**0.00**	17	42	5	–	–	–	–	–	–	4	–	–	–	β-D-Xylopyranose, 1,2,3,4-tetrakis(trimethylsilyl)-	18623-27-3	NI42694
73	204	217	147	191	205	45	75	**438**	100	72	38	25	22	17	10	9	**0.00**	17	42	5	–	–	–	–	–	–	4	–	–	–	D-Xylopyranose TMS	55555-45-8	NI42695
73	204	217	191	147	205	75	74	**438**	100	80	30	27	26	16	9	9	**0.00**	17	42	5	–	–	–	–	–	–	4	–	–	–	D-Xylose, tetrakis(trimethylsilyl)-	18623-22-8	NI42696
73	204	217	191	147	205	74	75	**438**	100	83	35	33	22	17	9	8	**0.00**	17	42	5	–	–	–	–	–	–	4	–	–	–	D-Xylose, tetrakis(trimethylsilyl)-	18623-22-8	NI42697
212	91	44	39	211	51	78	77	**438**	100	99	99	75	70	60	55	55	**10.00**	18	16	4	6	–	–	–	–	–	–	–	–	1	7,8,15,16,17,18-Hexahydro-3,12-dinitrodibenzo[e,m][1,4,8,11]tetraazacyclotetradecinatonickel(II)		LI08957
122	438	43	84	137	347	326	398	**438**	100	34	32	24	15	11	3	1		18	26	1	6	3	–	–	–	–	–	–	–	–	Spiro(3-(2-methyl-4-aminopyrimidin-5-yl)methyl-3a-methylperhydrofuro[2,3-d]thiazole-2,4′-(1′,3′-diallylimidazolidine-2′,5′-dithione))		LI08958
122	323	84	438	43	137	302	374	**438**	100	47	41	34	30	27	9	7		18	26	1	6	3	–	–	–	–	–	–	–	–	Spiro(3-(2-methyl-4-aminopyrimidin-5-yl)methyl-3a-methylperhydrofuro(2,3-d)thiazole-2,4′-(1′,3′-diethylimidazolidine-2′,5′-dithione))		LI08959
43	155	45	61	44	60	42	59	**438**	100	38	34	21	16	13	13	12	**0.60**	18	30	8	–	2	–	–	–	–	–	–	–	–	D-Arabino-hexose, 2-dioxy-, diethyl mercaptal, tetraacetate	16885-34-0	NI42698
43	45	135	44	173	47	42	41	**438**	100	39	25	16	14	13	11	11	**0.40**	18	30	8	–	2	–	–	–	–	–	–	–	–	L-Rhamnose, diethyl mercaptal, tetraacetate	24807-89-4	NI42699
73	75	438	147	423	116	439	424	**438**	100	48	44	33	30	30	15	13		18	34	3	4	–	–	–	–	–	3	–	–	–	Pteridine, 6-propyl-2,4,7-tris(trimethylsiloxy)-	31083-63-3	LI08960
73	75	438	147	423	116	439	410	**438**	100	44	40	30	28	27	15	15		18	34	3	4	–	–	–	–	–	3	–	–	–	Pteridine, 6-propyl-2,4,7-tris(trimethylsiloxy)-	31083-63-3	NI42700
409	57	241	410	41	365	56	43	**438**	100	46	39	32	29	22	21	20	**0.00**	18	42	6	–	–	–	–	–	–	3	–	–	–	1,3,5-Triethyl-1,3,5-tributoxycyclotrisiloxane	94048-05-2	NI42701
73	155	168	147	103	75	142	45	**438**	100	21	18	17	15	13	10	10	**0.00**	18	46	4	–	–	–	–	–	–	4	–	–	–	3,8-Dioxa-2,9-disiladecane, 2,2,9,9-tetramethyl-5,6-bis[[(trimethylsilyl)oxy]methyl]-	74779-61-6	NI42702
86	131	219	91	103	266	438	406	**438**	100	80	60	60	40	13	4	3		19	23	4	2	–	5	–	–	–	–	–	–	–	N-(pentafluoropropionyl)phenylalanylleucine, methyl ester		NI42703
87	43	438	365	440	367	206	91	**438**	100	57	18	13	7	7	5	5		19	23	4	4	1	–	1	–	–	–	–	–	–	N,N-Bis(2-acetoxyethyl)-3-methyl-4-(3-methyl-4-chloro-isothiaz-5-ylazo)aniline		IC04409
166	167	29	41	57	28	194	144	**438**	100	70	28	25	22	14	12	12	**5.60**	19	29	6	2	–	3	–	–	–	–	–	–	–	N-(Trifluoroacetyl)-L-prolylaspartic acid dibutyl ester		MS09719
187	438	220	90	251	373	406	405	**438**	100	44	28	16	11	8	3	2		20	14	4	4	2	–	–	–	–	–	–	–	–	Bis(2-methyl-6,7-methylenedioxyquinazolin-4-yl) disulphide		NI42704
43	87	105	51	178	53	52	118	**438**	100	78	58	50	48	38	33	32		20	18	4	6	1	–	–	–	–	–	–	–	–	N-(2-Cyanoethyl)-N-(2-acetoxyethyl)-4-(nitrobenzthiaz-2-ylazo)aniline		IC04410
107	209	135	158	238	134	438	181	**438**	100	33	25	24	17	13	11	11		20	22	6	–	–	–	–	–	–	–	–	–	1	3-Ethoxy-1,8-dioxo-9-phenylseleno-2-oxaspiro[4.5]dec-6-en-3-carboxylic acid, ethyl ester		NI42705
107	438	158	134	294	135	263	208	**438**	100	58	31	25	22	22	21	18		20	22	6	–	–	–	–	–	–	–	–	–	1	3-Ethoxy-1,8-dioxo-9-phenylseleno-2-oxaspiro[4.5]dec-6-en-3-carboxylic acid, ethyl ester		NI42706
297	73	298	299	147	45	428	74	**438**	100	72	26	10	10	7	7	6	**0.00**	20	34	5	–	–	–	–	–	–	3	–	–	–	Vanillylhydroxyacrylic acid, tris(trimethylsilyl)-		NI42707
73	89	87	58	59	86	133	99	**438**	100	89	89	48	46	43	42	39	**4.43**	20	38	10	–	–	–	–	–	–	–	–	–	–	(1S,17S)-3,6,9,12,15,18,21,24,27,30-Decaoxabicyclo[15.13.0]triacontane	73346-56-2	NI42708
73	87	89	59	99	86	58	133	**438**	100	96	88	63	42	42	42	40	**3.10**	20	38	10	–	–	–	–	–	–	–	–	–	–	(1S,17S)-3,6,9,12,15,18,21,24,27,30-Decaoxabicyclo[15.13.0]triacontane	73346-56-2	MS09720

MASS TO CHARGE RATIOS	M.W.	INTENSITIES	Parent	C	H	O	N	S	F	Cl	Br	I	Si	P	B	X	COMPOUND NAME	CAS Reg No	No
87 73 89 59 58 88 75 57	438	100 90 71 68 48 28 26 25	0.00	20	38	10	–	–	–	–	–	–	–	–	–	–	(2S,2'S)-2,2'-Bis(1,4,7,10,13-pentaoxacyclopentadecane)	95721-96-3	MS09721
73 87 89 99 59 86 133 71	438	100 94 91 36 36 35 35 33	1.44	20	38	10	–	–	–	–	–	–	–	–	–	–	(2S,2'S)-2,2'-Bis[1,4,7,10,13-pentaoxacyclopentadecane]	95721-96-3	NI42709
135 137 361 359 83 69 97 85	438	100 98 96 94 91 85 84 84	0.00	20	40	–	–	–	–	–	2	–	–	–	–	–	Eicosane, 1,20-dibromo-	14296-16-3	NI42710
365 438 55 189 161 366 439 423	438	100 58 56 26 25 22 15 15		21	22	5	6	–	–	–	–	–	–	–	–	–	N-Ethyl-N-carboxyethoxyethyl-3-acetamido-4-(2-cyano-4-nitrophenylazo)aniline		IC04411
43 169 109 127 331 108 97 81	438	100 38 38 17 11 10 6 6	0.01	21	26	10	–	–	–	–	–	–	–	–	–	–	β-D-Glucopyranoside, 2-methylphenyl, tetraacetate	5346-66-7	MS09722
131 70 72 56 160 174 91 85	438	100 70 59 56 54 44 44 40	2.75	21	30	8	2	–	–	–	–	–	–	–	–	–	12-Benzyl-1,4,7,13,16-pentaoxa-10,19-diazacycloheneicosane-11,14,18-trione	79688-24-7	NI42711
154 307 349 381 438 423	438	100 11 10 6 4 2		21	30	8	2	–	–	–	–	–	–	–	–	–	5-tert-Butylimino-2-methoxy-2-(1-pyrrolidino-2-methoxy-carbonyl-vinyl)-3,4-bis(methoxycarbonyl)-2,5-dihydrofuran		LI08961
43 159 243 201 147 160 105 91	438	100 31 29 24 17 12 11 8	4.81	22	24	8	–	–	–	–	–	–	–	–	2	–	D-Glucitol, cyclic 1,3:2,4-bis(phenylboronate) 5,6-diacetate	74825-22-2	NI42712
191 171 208 139 438 75 222 192	438	100 69 49 42 34 29 13 13		22	30	9	–	–	–	–	–	–	–	–	–	–	Cyclohexanecarboxylic acid, 3-[[3-(3,4-dimethoxyphenyl)-1-oxo-2-propenyl]oxy]-1,4,5-trimethoxy-, methyl ester, [1S-(1α,3β,4α,5α)]-	55837-04-2	NI42713
378 350 290 379 124 106 175 197	438	100 45 37 31 29 27 18 17	0.00	22	34	7	–	–	–	–	–	–	1	–	–	–	Diacetoxyscirpenol TMS		NI42714
209 222 438 179 210 348 223 193	438	100 29 27 27 21 15 12 12		23	42	4	–	–	–	–	–	–	2	–	–	–	3-Decanone, 1-[3-methoxy-4-[(trimethylsilyl)oxy]phenyl]-5-[(trimethylsilyl)oxy]-	56700-97-1	MS09723
438 146 437 159 312 160 105 439	438	100 72 62 59 32 29 25 23		24	21	6	–	–	–	–	–	–	–	–	3	–	allo-Inositol tri-benzeneboronate		NI42715
438 437 159 146 105 160 439 436	438	100 75 72 69 31 24 23 18		24	21	6	–	–	–	–	–	–	–	–	3	–	cis-Inositol tri-benzeneboronate		NI42716
159 438 437 158 146 160 105 439	438	100 35 23 22 16 15 11 9		24	21	6	–	–	–	–	–	–	–	–	3	–	myo-Inositol tri-benzeneboronate		NI42717
312 159 438 311 437 104 146 105	438	100 83 69 63 55 39 38 31		24	21	6	–	–	–	–	–	–	–	–	3	–	muco-Inositol tri-benzeneboronate		NI42718
138 396 43 258 139 438 165	438	100 18 15 14 14 13 12		24	22	8	–	–	–	–	–	–	–	–	–	–	4-Acetoxy-3-(α-hydroxy-3,4-dimethoxy-benzyl)-6,7-dimethoxy-2-naphthoic acid lactone		LI08962
270 271 85 438 41 29 354 55	438	100 32 31 22 18 18 15 15		24	26	6	2	–	–	–	–	–	–	–	–	–	Anthraquinone, 1,5-bis(2-amino-2,3,5,6-tetrahydropyranyl)-4,8-dihydroxy-		IC04412
307 306 396 107 246 295 380 247	438	100 80 40 34 27 22 16 16	13.00	24	26	6	2	–	–	–	–	–	–	–	–	–	6,21-Cyclo-4,5-secoakuammilan-17-oic acid, 4,5-bis(acetyloxy)-, methyl ester, (6α)-	63944-65-0	NI42719
323 307 213 188 324 151 231 251	438	100 89 50 40 18 17 10 6	0.00	24	27	4	–	–	–	–	–	–	1	1	–	–	(Z)-2-Methyl-1-phenyl-2-(trimethylsilyl)vinyl diphenyl phosphate	118297-87-3	DD01672
70 283 214 114 91 128 155 198	438	100 53 32 27 15 14 8 6	0.50	24	42	3	2	1	–	–	–	–	–	–	–	–	N-[3-(p-Methyltosylamino)propyl]-N-hexyl-heptanamide		NI42720
73 225 295 75 79 91 131 43	438	100 57 31 27 24 18 17 16	0.00	24	42	5	–	–	–	–	–	–	1	–	–	–	Prosta-5,13-dien-1-oic acid, 9,11-epidioxy-15-[[(4-methylphenyl)sulphonyl]oxy]-, methyl ester, (5Z,9α,11α,13E,15S)-	65147-48-0	NI42721
267 326 354 268 265 148 382 189	438	100 46 38 27 23 15 14 12	0.20	25	15	4	–	–	–	–	–	–	–	–	–	1	2,4-η-(2,3,4-Triphenylcyclobut-2-en-1-on-4-yl)tricarbonylcobalt		MS09724
382 120 183 55 383 262 185 108	438	100 45 39 29 25 10 9 9	4.99	25	20	2	–	–	–	–	–	–	–	1	–	1	Cyclopentadienylmanganesedicarbonyltriphenylphosphine	12100-41-3	NI42722
43 167 320 380 182 107 321 42	438	100 31 28 27 18 18 14 14	0.00	25	30	5	2	–	–	–	–	–	–	–	–	–	Aspidofractinine-3-carboxylic acid, 1-acetyl-6-(acetyloxy)-, methyl ester, (2α,3β,5α,6β)-	55724-59-9	NI42723
378 149 144 379 291 438 365 119	438	100 80 80 61 25 24 24 21		25	30	5	2	–	–	–	–	–	–	–	–	–	Curan-17,18-diol, 1-acetyl-19,20-didehydro-, diacetate (ester), (19E)-	24182-67-0	NI42724
319 144 320 180 130 378 61 143	438	100 47 40 25 24 22 18 17	0.23	25	30	5	2	–	–	–	–	–	–	–	–	–	Curan-19,20-diol, 1-acetyl-16,17-didehydro-, diacetate (ester), (19S)-	56143-40-9	NI42725
438 122 379 121 124 365 77 86	438	100 68 59 56 51 46 41 38		25	30	5	2	–	–	–	–	–	–	–	–	–	1,16-Cyclocorynan-16-methanol, 17-(acetyloxy)-19,20-didehydro-10-methoxy-, acetate (ester), (19E)-	55724-52-2	NI42726
382 438 369 354 383 326 439 370	438	100 75 63 29 27 27 25 17		25	30	5	2	–	–	–	–	–	–	–	–	–	1-(3,4-Dimethoxyphenyl)-5-ethyl-4,5-dihydro-7,8-dimethoxy-4-methylene-3-propionyl-3H-2,3-benzodiazepine	90140-51-5	NI42727
438 341 439 326 340 342 381 424	438	100 80 30 30 25 18 12 10		25	30	5	2	–	–	–	–	–	–	–	–	–	1-(3,4-Dimethoxyphenyl)-5-ethyl-7,8-dimethoxy-4-methyl-3-propionyl-3H-2,3-benzodiazepine	90140-53-7	NI42728
438 423 409 439 244 424 440 410	438	100 57 47 29 19 17 12 12		25	30	5	2	–	–	–	–	–	–	–	–	–	Obscurinervine	7097-00-9	NI42729
438 423 409 439 244 424 410 57	438	100 55 47 30 18 17 13 12		25	30	5	2	–	–	–	–	–	–	–	–	–	Obscurinervine	7097-00-9	NI42730
438 197 439 406 121 41 69 407	438	100 99 35 34 30 28 26 23		25	42	6	–	–	–	–	–	–	–	–	–	–	Methyl 15-(2-hydroxy-3,4,5-trimethoxyphenyl)pentadecanoate		MS09725
160 354 109 383 99 172 261 93	438	100 44 29 24 23 22 19 16	0.00	25	42	6	–	–	–	–	–	–	–	–	–	–	9,12-Octadecadienoic acid (Z,Z)-, 2-(acetyloxy)-1-[(acetyloxy)methyl]ethyl ester	55320-03-1	WI01549
43 55 41 67 57 29 83 81	438	100 30 23 20 19 16 13 13	0.00	25	42	6	–	–	–	–	–	–	–	–	–	–	9,12-Octadecadienoic acid (Z,Z)-, 2-(acetyloxy)-1-[(acetyloxy)methyl]ethyl ester	55320-03-1	WI01548
43 55 41 155 29 57 159 27	438	100 24 23 19 19 17 16 13	0.00	25	42	6	–	–	–	–	–	–	–	–	–	–	9,12-Octadecadienoic acid (Z,Z)-, 2,3-bis(acetyloxy)propyl ester	55320-04-2	WI01551
159 155 56 138 82 98 158 145	438	100 24 16 15 14 13 12 12	0.00	25	42	6	–	–	–	–	–	–	–	–	–	–	9,12-Octadecadienoic acid (Z,Z)-, 2,3-bis(acetyloxy)propyl ester	55320-04-2	WI01550
43 55 41 29 57 159 155 27	438	100 24 23 19 17 16 14 13	0.00	25	42	6	–	–	–	–	–	–	–	–	–	–	9,12-Octadecadienoic acid (Z,Z)-, 2,3-bis(acetyloxy)propyl ester	55320-04-2	NI42731
75 72 69 163 55 145 95 137	438	100 86 66 65 64 56 56 55	0.00	25	46	4	–	–	–	–	–	–	1	–	–	–	(2S,4S,6S,8R,9S)-4-(dimethyl-tert-butylsiloxy)-2-[(5R,2E)-5-formyl-3-methylhex-2-enyl]-8,9-dimethyl-1,7-dioxaspiro[5.5]undecane		MS09726
438 183 439 185 209 252 178 207	438	100 43 38 38 32 25 20 14		26	20	–	–	–	–	–	–	–	–	1	–	1	Phosphine, [(diphenylarsino)ethynyl]diphenyl-	33730-55-1	NI42732

MASS TO CHARGE RATIOS								M.W.	INTENSITIES								Parent	C	H	O	N	S	F	Cl	Br	I	Si	P	B	X	COMPOUND NAME	CAS Reg No	No
222	217	149	95	135	91	223	105	**438**	100	24	19	19	18	16	14	14	**9.00**	26	30	6	–	–	–	–	–	–	–	–	–	–	(1′S,4a′S,8a′R)-6,8-dimethoxy-7-[(1,2,3,4,4a,5,6,8a-octahydro-5,5,8a-trimethyl-2-methylene-6-oxo-1-naphthalenyl)methoxy]-2H-1-benzopyran-2-one		NI42733
222	95	91	135	121	223	149	105	**438**	100	17	16	15	15	14	13	13	**6.50**	26	30	6	–	–	–	–	–	–	–	–	–	–	(1′S,4a′S,8a′R)-7-[(1,4,4a,6,6,8a-hexahydro-2,5,5,8a-tetramethyl-6-oxo-1-naphthalenyl)methoxy]-6,8-dimethoxy-2H-1-benzopyran-2-one		NI42734
233	179	41	351	298	69	297	180	**438**	100	80	56	48	39	38	35	22	**20.00**	26	30	6	–	–	–	–	–	–	–	–	–	–	Homoflemingin		MS09727
315	44	41	95	316	91	43	28	**438**	100	56	28	24	22	19	17	17	**0.40**	26	30	6	–	–	–	–	–	–	–	–	–	–	7-Oxogedunin		MS09728
438	366	439	29	44	365	336	337	**438**	100	55	30	18	17	15	10	9		27	22	4	2	–	–	–	–	–	–	–	–	–	1-Carbethoxy-3-methyl-6-(3-toluidino)-2,3-dihydro-2,7-dioxo-7H-dibenz[f,ij]isoquinoline		IC04413
438	392	364	439	393	365	321	355	**438**	100	73	40	30	21	11	10	7		27	22	4	2	–	–	–	–	–	–	–	–	–	1-Carbethoxy-4-methyl-6-toluidino-2,3-dihydro-2,7-dioxo-7H-dibenz[f,ij]isoquinoline		IC04414
319	77	119	91	64	304	51	115	**438**	100	32	28	25	13	11	11	10	**0.00**	27	22	4	2	–	–	–	–	–	–	–	–	–	Methyl 4-benzyl-3,5-dioxo-2,6-diphenyl-2,6-diazabicyclo[2.2.2]oct-7-ene-7 carboxylate		DD01673
216	113	180	77	293	63	181	155	**438**	100	95	50	50	48	28	23	22	**2.00**	27	23	–	2	–	–	1	–	–	1	–	–	–	Silanediamine, 1-chloro-N,N′-bis(diphenylmethylene)-1-methyl-	38033-06-6	NI42735
244	124	229	438	105	121	123	107	**438**	100	47	40	34	33	28	26	26		27	39	4	–	–	–	–	–	–	–	–	1	–	Pregn-4-ene-3,20-dione, 17,21-dihydroxy-, cyclohexylboronate	31012-67-6	NI42736
211	210	178	227	165	197	212	209	**438**	100	68	68	49	47	43	40	40	**38.88**	28	22	1	–	2	–	–	–	–	–	–	–	–	Dibenzo[b,f]thiepin, 10,10′-oxybis[10,11-dihydro-	16661-24-8	NI42737
210	178	227	211	91	179	194	165	**438**	100	83	68	54	40	31	30	19	**0.00**	28	22	1	–	2	–	–	–	–	–	–	–	–	Bis-(6,11-dihydrodibenzo[b,e]thiepin-11-yl) ether		MS09729
150	149	57	223	288	41	191	438	**438**	100	93	60	38	33	27	22	20		28	38	4	–	–	–	–	–	–	–	–	–	–	[1α,3(E)4αβ,7αβ]-(+)-4-methoxy-2-methyl-6-[3-methyl-4-(5,6,7,7a-tetrahydro-4a,5′,5′,7a-tetramethylspiro[cyclopenta[c]pyran-1(4aH),2′(5′H)-furan]-3-yl)-2-butenyl]phenol	96253-60-0	NI42738
271	347	75	272	348	81	253	223	**438**	100	68	35	22	20	14	11	8	**4.42**	28	42	2	–	–	–	–	–	–	1	–	–	–	3α-Benzyldimethylsilyloxy-5α-androstan-17-one		NI42739
209	183	105	238	77	239	210	296	**438**	100	99	77	64	39	13	13	4	**0.00**	29	26	4	–	–	–	–	–	–	–	–	–	–	(4R,5R)-α,α,α′,α′-Tetraphenyl-1,3-dioxolane-4,5-dimethanol		DD01674
124	183	251	184	223	171	247	130	**438**	100	90	85	85	75	75	70	70	**55.00**	29	34	–	4	–	–	–	–	–	–	–	–	–	20-Epiochrolifuanine A	51820-26-9	NI42740
438	183	251	247	184	130	223	171	**438**	100	80	75	75	70	70	65	63		29	34	–	4	–	–	–	–	–	–	–	–	–	20-Epiochrolifuanine B	51820-25-8	NI42741
131	44	17	130	83	56	43	69	**438**	100	36	35	34	28	28	24	22	**7.00**	29	58	2	–	–	–	–	–	–	–	–	–	–	Docosyl heptanoate		LI08963
131	44	18	130	97	83	57	43	**438**	100	36	35	34	32	28	28	24	**8.10**	29	58	2	–	–	–	–	–	–	–	–	–	–	Docosyl heptanoate		MS09730
43	438	57	28	61	55	102	41	**438**	100	69	64	58	50	40	35	32		29	58	2	–	–	–	–	–	–	–	–	–	–	Hexacosanoic acid, propyl ester	40710-38-1	MS09731
74	87	43	55	75	41	57	69	**438**	100	77	74	38	33	33	29	24	**9.00**	29	58	2	–	–	–	–	–	–	–	–	–	–	Methyl octacosanoate	55682-92-3	NI42742
437	439	438	440	143	131	129	117	**438**	100	85	43	35	20	18	18	16		29	58	2	–	–	–	–	–	–	–	–	–	–	Methyl octacosanoate	55682-92-3	NI42743
74	87	75	43	57	55	18	69	**438**	100	75	61	53	37	37	31	28	**17.70**	29	58	2	–	–	–	–	–	–	–	–	–	–	Methyl octacosanoate	55682-92-3	NI42744
69	396	378	55	125	81	143	253	**438**	100	99	67	59	57	41	39	36	**27.50**	30	46	2	–	–	–	–	–	–	–	–	–	–	3β-Acetoxyergosta-4,6,22-triene	96737-73-4	NI42745
219	220	57	438	204	41	203	43	**438**	100	19	15	10	7	5	4	4		30	46	2	–	–	–	–	–	–	–	–	–	–	1,2-Bis(3,5-di-tert-butyl-4-hydroxyphenyl)ethane		IC04415
438	233	95	189	107	299	121	207	**438**	100	57	54	45	39	36	36	24		30	46	2	–	–	–	–	–	–	–	–	–	–	3,11-Diketo-lup-20(29)-ene	4924-77-0	NI42746
395	135	150	423	438	149	189	205	**438**	100	47	40	27	18	18	13	8		30	46	2	–	–	–	–	–	–	–	–	–	–	3,16-Dioxo-19αH-ψ-taraxene		NI42747
395	135	150	438	149	175	189	219	**438**	100	87	80	58	32	28	19	17		30	46	2	–	–	–	–	–	–	–	–	–	–	3,16-Dioxo-ψ-taraxene		NI42748
108	109	135	150	395	438	149	175	**438**	100	69	45	42	37	31	22	16		30	46	2	–	–	–	–	–	–	–	–	–	–	3,16-Dioxotaraxene		NI42749
423	217	438	393	391	255	205	189	**438**	100	65	50	42	40	35	30	30		30	46	2	–	–	–	–	–	–	–	–	–	–	D:A-Friedoolean-1-en-3-one, 25,26-epoxy-	56792-48-4	MS09732
78	378	43	379	77	55	51	52	**438**	100	63	35	20	18	16	16	15	**1.00**	30	46	2	–	–	–	–	–	–	–	–	–	–	1,10-Seco-17β-24-methyl-3-acetoxy-5,7,9,22-cholestatetraene		MS09733
394	323	379	284	409	395	393	322	**438**	100	78	76	56	41	35	35	32	**27.02**	30	46	2	–	–	–	–	–	–	–	–	–	–	Olean-12-en-28-al, 3-oxo-	33608-08-1	NI42750
137	150	134	122	438	121	107	216	**438**	100	98	59	57	56	56	56	52		30	46	2	–	–	–	–	–	–	–	–	–	–	1-Oxoallobetul-2-ene		MS09734
137	150	438	215	203	135	121	109	**438**	100	99	85	82	50	45	45	44		30	46	2	–	–	–	–	–	–	–	–	–	–	3-Oxoallobetul-1-ene		MS09735
203	247	438	229	205	382	219	395	**438**	100	92	62	48	28	12	8	6		30	46	2	–	–	–	–	–	–	–	–	–	–	Thurberin-3-12-dione		LI08964
122	311	326	312	55	96	121	123	**438**	100	96	51	27	26	25	21	20	**0.00**	30	50	–	2	–	–	–	–	–	–	–	–	–	2′H-Cholest-2-eno[3,2-c]pyrazole, 4,4-dimethyl-, (5α)-	56247-66-6	NI42751
438	410	333	302	289	202			**438**	100	26	20	10	7	6				32	22	2	–	–	–	–	–	–	–	–	–	–	4,5,6,7-Tetraphenyl-coumaran-2-one		LI08965
105	91	255	256	253	254	252	79	**438**	100	57	25	16	16	11	11	11	**9.46**	34	30	–	–	–	–	–	–	–	–	–	–	–	p-Terphenyl, 2′,5′-diphenethyl-	14474-61-4	NI42752
69	319	273	80	291	439	245	227	**439**	100	69	40	30	28	16	16	16		12	5	12	7	–	–	–	–	–	–	–	–	–	Dipicrylamine	131-73-7	NI42753
43	189	188	41	119	70	139	69	**439**	100	66	42	21	20	13	11	8	**0.00**	12	11	5	1	–	10	–	–	–	–	–	–	–	N,O-bis(pentafluoropropionyl)-D-serine isopropyl ester		MS09736
43	189	188	119	41	70	27	69	**439**	100	74	48	22	21	14	9	8	**0.00**	12	11	5	1	–	10	–	–	–	–	–	–	–	N,O-bis(pentafluoropropionyl)-L-serine isopropyl ester		MS09737
78	92	81	68	94	82	80	104	**439**	100	93	93	87	82	74	74	69	**0.00**	17	21	9	5	–	–	–	–	–	–	–	–	–	Dnp-Gly-Ser-Pro-OMe		NI42754
103	211	321	319	75	367	131	365	**439**	100	75	60	58	47	37	37	36	**2.00**	19	22	6	1	–	–	–	1	–	–	–	–	–	Propanedioic acid, (4-bromobutyl)(1,3-dihydro-1,3-dioxo-2H-isoindol-2-yl)-, diethyl ester	32723-82-3	NI42755
191	338	379	339	192	147	121	115	**439**	100	64	40	18	18	14	14	14	**1.00**	22	17	3	1	3	–	–	–	–	–	–	–	–	1-Allylthio-3-oxo-4,N-diphenyl-2,7-dithiabicyclo[2.2.1]heptan-5-endo,6-endo-dicarboximide		NI42756

MASS TO CHARGE RATIOS								M.W.	INTENSITIES								Parent	C	H	O	N	S	F	Cl	Br	I	Si	P	B	X	COMPOUND NAME	CAS Reg No	No
300	164	121	318	301	134	190	163	**439**	100	77	39	34	23	23	18	18	**0.98**	22	29	3	7	–	–	–	–	–	–	–	–	–	Benzenepropanamide, α-amino-N-[3-[6-(dimethylamino)-9H-purin-9-yl]-2-hydroxycyclopentyl]-4-methoxy-, [1R-[1α(S*),2α,3β]]-	34597-43-8	MS09738
408	409	318	324	410	439	174	72	**439**	100	24	10	9	7	6	6	6		22	37	6	1	–	–	–	–	–	1	–	–	–	1-Cyclopentene-1-propanoic acid, 5-(methoxyimino)-2-[8-methoxy-8-oxo-3-[(trimethylsilyl)oxy]-1-octenyl]-, methyl ester	69502-89-2	NI42757
183	262	185	238	383	317	55	108	**439**	100	99	99	73	67	67	67	53	**16.67**	24	19	2	1	–	–	–	–	–	–	1	–	1	Manganese, dicarbonyl-π-pyrrolyl(triphenylphosphine)-	32826-10-1	NI42758
55	262	185	183	238	383	317	108	**439**	100	15	15	15	11	10	10	8	**2.50**	24	19	2	1	–	–	–	–	–	–	1	–	1	Manganese, dicarbonyl-π-pyrrolyl(triphenylphosphine)-	32826-10-1	LI08966
396	439	397	379	440	366	354	380	**439**	100	44	40	19	16	11	11	8		24	25	7	1	–	–	–	–	–	–	–	–	–	Furmaritine diacetate	30833-09-1	NI42759
241	242	302	390	421	301	439		**439**	100	80	42	28	22	22	4			24	38	4	1	–	–	1	–	–	–	–	–	–	5β-Hydroxy-17β-acetoxy-19-(N-methyl-N-chloroacetylamino)androstane		LI08967
439	440	314	65	77	298	241	151	**439**	100	30	25	23	21	17	16	14		26	17	4	1	1	–	–	–	–	–	–	–	–	1-Anilino-2-phenysulphonylanthraquinone		IC04416
91	190	55	120	81	107	95	121	**439**	100	73	60	39	39	33	31	30	**5.40**	27	37	4	1	–	–	–	–	–	–	–	–	–	1-Benzyl-3a,5-dihydroxy-7-methyl-4-(4-methyl-8-oxo-non-1-enyl)-6-methylene-3-oxo-3a,4,5,6,7,7a-hexahydroisoindoline		IC04417
144	115	89	116	280	145	63	377	**439**	100	52	42	34	21	12	12	8	**4.00**	28	25	4	1	–	–	–	–	–	–	–	–	–	1-(4-Hydroxynaphthyl)-1-(4-diethylamino-2-hydroxyphenyl)-3-oxo-isobenzofuran		IC04418
276	198	122	110	153	166	181	199	**439**	100	90	72	63	57	56	55	43	**7.00**	28	29	2	1	–	–	–	–	–	1	–	–	–	3-Diethylamino-2-[1-(triphenylsiloxy)vinyl]-2-cyclobuten-1-one		NI42760
58	199	44	43	77	122	115	181	**439**	100	81	41	41	32	29	27	26	**5.00**	28	29	2	1	–	–	–	–	–	1	–	–	–	N,N-diethyl-4-(triphenylsiloxy)-1,2,4-pentatrien-3-carboxamide		NI42761
368	57	43	72	81	55	147	56	**439**	100	35	33	32	27	24	21	21	**0.22**	30	49	1	1	–	–	–	–	–	–	–	–	–	2-Propenamide, N-[(3β)-cholest-5-en-3-yl]-	38759-52-3	MS09739
348	91	321	322	349	439	65	440	**439**	100	99	85	44	37	35	25	20		33	29	–	1	–	–	–	–	–	–	–	–	–	15-Benzyl-9,14-dihydro-10,11,12,13-tetramethyl-9,14-iminobenzo[b]triphenylene	110028-25-6	DD01675
176	97	264	135	96	144	185	273	**440**	100	92	89	79	64	53	41	36	**11.40**	3	5	–	–	–	9	–	1	–	–	3	–	1	π-Allyl-tris(trifluorophosphine)iron bromide		NI42762
127	170	270	100	150	109	297	89	**440**	100	59	37	26	21	20	17	13	**0.30**	4	2	2	2	2	14	–	–	–	–	–	–	–	2,2,3,3-Tetrafluoro-1,4-dioxo-1,4-butylenebis(iminosulphur pentafluoride)	80409-48-9	NI42763
77	50	279	440	361	82	405	165	**440**	100	35	25	5	3	2	1	1		7	5	–	–	–	–	2	1	–	–	–	–	1	Phenyl(bromodichloromethyl)mercury		LI08968
43	31	405	331	27	29	36	285	**440**	100	83	75	53	45	39	30	29	**9.50**	9	17	4	–	–	–	2	–	–	–	–	–	1	Tantalum, dichlorodiethoxy(2,4-pentanedionato-O,O′)-	17035-44-8	NI42764
43	407	405	333	331	289	287	285	**440**	100	99	99	70	70	51	51	51	**16.90**	9	17	4	–	–	–	2	–	–	–	–	–	1	Tantalum, dichlorodiethoxy(2,4-pentanedionato-O,O′)-	17035-44-8	NI42765
43	405	331	285	395	407	341	287	**440**	100	75	53	30	27	25	23	19	**9.50**	9	17	4	–	–	–	2	–	–	–	–	–	1	Tantalum, dichlorodiethoxy(2,4-pentanedionato-O,O′)-	17035-44-8	LI08969
39	109	73	85	99	83	65	111	**440**	100	91	83	66	65	54	52	50	**0.00**	10	5	–	–	–	–	9	–	–	–	–	–	–	cis-Nonachlor	5103-73-1	NI42766
409	407	411	405	413	237	109	272	**440**	100	91	65	37	27	21	19	18	**2.50**	10	5	–	–	–	–	9	–	–	–	–	–	–	Nonachlor	3734-49-4	LI08970
55	57	43	109	71	97	83	69	**440**	100	80	80	60	60	53	46	40	**0.00**	10	5	–	–	–	–	9	–	–	–	–	–	–	Nonachlor	3734-49-4	NI42767
409	407	411	405	413	237	272	109	**440**	100	86	62	33	25	20	18	18	**0.00**	10	5	–	–	–	–	9	–	–	–	–	–	–	Nonachlor	3734-49-4	NI42768
135	408	410	137	406	407	412	405	**440**	100	99	63	60	57	38	23	22	**0.00**	10	5	–	–	–	–	9	–	–	–	–	–	–	trans-Nonachlor	39765-80-5	NI42769
135	409	137	407	411	405	413	408	**440**	100	67	67	60	43	23	16	15	**0.00**	10	5	–	–	–	–	9	–	–	–	–	–	–	trans-Nonachlor	39765-80-5	NI42770
442	444	293	291	279	295	280	440	**440**	100	90	79	46	46	42	41	39		10	7	3	2	–	–	–	3	–	–	–	–	–	Pyrimidine, 1-(3′,5′-dibromo-4′-hydroxyphenyl)-5-bromo-2,4-dioxohexahydro-	83809-78-3	NI42771
423	425	427	421	389	429	391	424	**440**	100	95	55	48	27	20	20	18	**2.00**	12	–	1	–	–	–	8	–	–	–	–	–	–	Dibenzofuran, octachloro-	39001-02-0	NI42773
444	442	446	440	448	377	309	379	**440**	100	89	73	39	21	14	12	11		12	–	1	–	–	–	8	–	–	–	–	–	–	Dibenzofuran, octachloro-	39001-02-0	NI42772
444	442	446	309	440	379	381	307	**440**	100	92	66	45	35	31	29	28		12	–	1	–	–	–	8	–	–	–	–	–	–	Dibenzofuran, octachloro-	39001-02-0	NI42774
440	298	425	100	283	113	441	397	**440**	100	26	21	18	17	16	13	10		12	10	2	–	–	–	–	–	–	–	–	–	2	Naphthalene, 1,5-diiodo- 2,6-dimethoxy-	25315-12-2	NI42775
39	41	144	184	142	103	185	40	**440**	100	54	37	31	23	23	22	22	**2.44**	12	20	–	–	–	–	2	–	–	–	–	–	2	Di-μ-chlorotetraallyldirhodium	12090-11-8	NI42776
185	364	440	322	254	183	442	144	**440**	100	43	26	25	25	21	17	17		12	20	–	–	–	–	2	–	–	–	–	–	2	Di-μ-chlorotetraallyldirhodium	12090-11-8	LI08971
55	41	83	67	45	278	69	225	**440**	100	86	79	63	54	48	34	32	**3.27**	12	22	–	–	–	–	–	2	–	–	–	–	2	Nickel, di-μ-bromobis[(1,2,3-η)-1,1-dimethyl-2-butenyl]di-	62265-44-5	NI42777
209	286	78	51	77	154	50	52	**440**	100	76	34	18	14	13	8	7	**0.00**	18	15	–	–	–	–	–	–	–	–	–	–	1	Triphenylbismuth	603-34-9	MS09740
245	51	77	167	244	166	246	115	**440**	100	83	72	44	41	31	19	19		18	15	–	–	–	–	–	–	–	–	–	–	1	Triphenylbismuth	603-34-9	NI42778
286	209	287	154	363	153	152	39	**440**	100	97	6	6	3	3	3	3	**0.04**	18	15	–	–	–	–	–	–	–	–	–	–	1	Triphenylbismuth	603-34-9	IC04419
210	214	216	56	300	244	242	134	**440**	100	77	65	63	56	56	52	50	**7.29**	18	16	6	–	–	–	–	–	–	–	–	–	2	Iron, hexacarbonyl[μ-[(4a,5,10,10a-η:5,10-η)-1,2,3,4,5,6,7,8,9-octahydro-5,10-benzocyclooctenediyl]]di-	38999-65-4	MS09741
117	166	73	41	396	29	75	167	**440**	100	98	88	34	28	28	25	15	**0.00**	18	31	5	2	–	3	–	–	–	1	–	–	–	N-Trifluoroacetyl-L-prolylthreonine butyl ester, trimethylsilyl-		MS09742
126	377	379	82	275	442	440	277	**440**	100	13	4	3	2	1	1	1		18	33	6	2	1	–	1	–	–	–	–	–	–	L-Threo-α-D-galacto-octopyranose, 7-chloro-1,6,7,8-tetradeoxy-6-[[(1-methyl-4-propyl-2-pyrrolidinyl)carbonyl]amino]-1-(methylsulphinyl), (2S-trans)-	22431-46-5	NI42779
73	208	323	218	209	75	280	74	**440**	100	87	43	15	15	15	14	13	**5.70**	19	36	4	2	–	–	–	–	–	3	–	–	–	TMS 3-(2-amino-3-TMSoxybenzoyl)-2-TMS-aminopropionate		MS09743
73	208	323	218	209	75	280	74	**440**	100	87	43	15	15	15	14	13	**5.70**	19	36	4	2	–	–	–	–	–	3	–	–	–	Benzenebutanoic acid, γ-oxo-α,2-bis[(trimethylsilyl)amino]-3-[(trimethylsilyl)oxy]-	55401-61-1	NI42780
199	241	157	283	381	367	325	242	**440**	100	93	85	46	44	30	25	15	**0.00**	19	44	7	–	–	–	–	–	–	2	–	–	–	Silicic acid, butyl pentapropyl ester	55683-31-3	NI42781
361	333	362	92	63	75	119	64	**440**	100	60	58	32	28	18	18	17	**2.04**	20	9	7	–	–	–	–	1	–	–	–	–	–	3-Bromo-3,8,8′-trihydroxy-2,2′-binaphthalene-1,1′,4,4′-tetrone	88381-92-4	NI42782
361	80	82	362	333	92	79	63	**440**	100	35	32	30	29	20	17	17	**0.00**	20	9	7	–	–	–	–	1	–	–	–	–	–	3-Bromo-3′,5,5′-trihydroxy-2,2′-binaphthalene-1,1′,4,4′-tetrone	88381-89-9	NI42783

MASS TO CHARGE RATIOS								M.W.	INTENSITIES								Parent	C	H	O	N	S	F	Cl	Br	I	Si	P	B	X	COMPOUND NAME	CAS Reg No	No
135	105	98	440					**440**	100	40	20	14						20	12	3	–	–	–	4	–	–	–	–	–	–	1,3-Benzodioxole, 4,5,6,7-tetrachloro-2-(3-methoxyphenyl)-2-phenyl-	61233-45-2	NI42784
105	240	239	196	224	222	133	225	**440**	100	77	26	21	20	16	15	14	**0.00**	20	16	6	4	1	–	–	–	–	–	–	–	–	Benzoic acid, 2-[1-(2-hydroxy-5-sulphophenyl)-3-phenyl-5-formazano]-	135-52-4	NI42785
165	240	221	121	222	65	77	39	**440**	100	98	83	48	43	38	30	27	**0.00**	20	16	6	4	1	–	–	–	–	–	–	–	–	Benzoic acid, 2-[1-(2-hydroxy-5-sulphophenyl)-3-phenyl-5-formazano]-	135-52-4	NI42786
147	73	425	440	426	251	368	279	**440**	100	98	75	38	28	26	21	16		20	36	5	–	–	–	–	–	–	3	–	–	–	4-Hydroxy-3-ethoxyphenylpyruvic acid tri-TMS		MS09744
73	440	267	179	441	75	425	268	**440**	100	76	60	47	29	29	17	17		21	40	4	–	–	–	–	–	–	3	–	–	–	(3,4-Dihydroxyphenyl)hexanoic acid, tris(trimethylsilyl)-		NI42787
150	104	291	120	92	274	440	290	**440**	100	10	3	3	3	2	1	1		23	12	6	4	–	–	–	–	–	–	–	–	–	α,α-Di(4-nitrobenzoyl)-O-cyanobenzyl cyanide		LI08972
137	381	272	339	314	356	297	255	**440**	100	99	86	68	67	60	46	42	**24.00**	23	20	9	–	–	–	–	–	–	–	–	–	–	2,4,3′,4′-Tetraacetoxychalcone	90826-62-3	NI42788
279	191	369	280	253	129	254	57	**440**	100	33	26	26	13	11	9	8	**0.00**	23	44	4	–	–	–	–	–	–	2	–	–	–	Cyclopentanepropanoic acid, 2-(1,5-octadienyl)-3,5-bis[(trimethylsilyl)oxy]-, methyl ester	56248-21-6	NI42789
147	159	91	146	105	440	104	158	**440**	100	57	45	43	25	17	17	14		24	23	6	–	–	–	–	–	–	–	–	3	–	D-Glucitol, cyclic 1,3:2,4:5,6-tris(phenylboronate)	6638-74-0	NI42790
337	43	339	321	309	362	217	319	**440**	100	94	53	53	44	38	37	35	**4.00**	24	24	8	–	–	–	–	–	–	–	–	–	–	(7R)-9-acetyl-7-isopropoxy-4-methoxy-7,8,9,10-tetrahydro-6,9,11-trihydroxy-5,12-naphthacenequinone		NI42791
337	309	43	362	339	321	217	319	**440**	100	63	60	46	46	35	32	29	**13.00**	24	24	8	–	–	–	–	–	–	–	–	–	–	(7S)-9-Acetyl-7-isopropoxy-4-methoxy-7,8,9,10-tetrahydro-6,9,11-trihydroxy-5,12-naphthacenequinone	90363-03-4	NI42792
202	203	97	239	440	149	201	189	**440**	100	31	18	17	16	15	14	14		24	24	8	–	–	–	–	–	–	–	–	–	–	7H-Furo[3,2-g][1]benzopyran-7-one, 9-[[5-[(2,5-dihydro-4-methyl-5-oxo-2-furanyl)methyl]-2,2,5-trimethyl-1,3-dioxolan-4-yl]methoxy]-	68725-64-4	NI42793
394	299	77	283	115	142	185	130	**440**	100	64	57	51	45	44	43	42	**0.00**	24	28	4	2	1	–	–	–	–	–	–	–	–	trans-5-Ethyl-1-methyl-2-[1-(phenylsulphonyl)-2-indolyl]-4-piperidone ethylene acetal		DD01676
43	45	60	320	293	380	44	42	**440**	100	76	25	24	18	14	14	14	**0.95**	24	28	6	2	–	–	–	–	–	–	–	–	–	Compactinervine, diacetate (ester)	2111-85-5	NI42794
177	149	145	70	85	101	45	44	**440**	100	78	60	41	32	31	25	25	**0.00**	24	28	6	2	–	–	–	–	–	–	–	–	–	trans-N,N′-Diferuloylputrescine	42369-86-8	NI42795
380	43	440	83	381	190	85	57	**440**	100	57	55	30	26	26	20	20		24	28	6	2	–	–	–	–	–	–	–	–	–	Henningsoline, acetate (ester)	18797-84-7	NI42796
440	58	381	369	71	57	244	72	**440**	100	28	26	20	20	20	14	14		24	28	6	2	–	–	–	–	–	–	–	–	–	16,19-Secostrychnidine-10,16-dione, 21,22-epoxy-21,22-dihydro-3,4-dimethoxy-19-methyl-, (21α,22α)-	62421-67-4	NI42797
383	440	338	240	297	221	280	255	**440**	100	90	53	40	33	33	29	26		24	40	–	4	–	–	–	–	–	2	–	–	–	Cyclodisilazane-1,3-diamine, 2,4-di-tert-butyl-N,N,N′,N′-tetramethyl-2,4-diphenyl-	66436-36-0	NI42798
73	75	440	369	130	225	79	67	**440**	100	74	55	53	45	34	24	21		24	48	3	–	–	–	–	–	–	2	–	–	–	Octadecadienoic acid, [(trimethylsilyl)oxy]-, trimethylsilyl ester	72361-09-2	MS09745
144	380	381	293	121	321	70	208	**440**	100	89	64	63	47	38	27	25	**21.00**	25	32	5	2	–	–	–	–	–	–	–	–	–	17,18-seco-NA,18-Diacetyl-17-acetoxy-3-desoxyisostrychnosplendine		LI08973
440	411	425	55	441	244	56	57	**440**	100	52	49	35	29	22	19	18		25	32	5	2	–	–	–	–	–	–	–	–	–	Obscurinervine, dihydro-	7097-01-0	NI42799
380	33	154	139	109	339	381	354	**440**	100	83	46	33	30	28	27	25	**0.00**	25	32	5	2	–	–	–	–	–	–	–	–	–	Pyrifoline	6878-72-4	NI42800
382	43	154	139	109	339	383	354	**440**	100	83	47	33	30	27	26	24	**0.00**	25	32	5	2	–	–	–	–	–	–	–	–	–	Pyrifoline	6878-72-4	NI42801
105	179	230	216	193	120	69	93	**440**	100	76	57	52	50	45	43	39	**28.00**	25	32	5	2	–	–	–	–	–	–	–	–	–	4,25-Secoobscurinervan-4-ol, 6,7-didehydro-16-methoxy-22-methyl-, 21-acetate, (4β,22α)-	54658-05-8	NI42802
182	43	45	41	44	55	122	57	**440**	100	88	78	49	44	43	34	31	**10.00**	25	32	5	2	–	–	–	–	–	–	–	–	–	4,25-Secoobscurinervan-4-one, O-acetyl-16-methoxy-22-methyl-, (22α)-	54678-25-0	NI42803
43	55	41	67	69	57	29	159	**440**	100	36	35	18	17	17	17	16	**0.00**	25	44	6	–	–	–	–	–	–	–	–	–	–	9-Octadecenoic acid (Z)-, 2-(acetyloxy)-1-[(acetyloxy)methyl]ethyl ester	55401-63-3	WI01552
43	55	159	41	69	81	57	29	**440**	100	28	24	24	15	12	12	10	**0.00**	25	44	6	–	–	–	–	–	–	–	–	–	–	9-Octadecenoic acid (Z)-, 2,3-bis(acetyloxy)propyl ester	55401-64-4	WI01553
295	75	73	147	69	95	99	55	**440**	100	81	75	49	49	43	40	40	**0.00**	25	48	4	–	–	–	–	–	–	1	–	–	–	(2S,4S,6S,8R,9S)-4-(dimethyl-tert-butylsiloxy)-2-[(5R,2E)-6-hydroxy-3,5-dimethylhex-2-enyl]-8,9-dimethyl-1,7-dioxaspiro[5.5]undecane		MS09746
165	136	43	91	105	79	93	107	**440**	100	90	80	50	49	47	45	44	**13.57**	26	32	6	–	–	–	–	–	–	–	–	–	–	Bufa-20,22-dienolide, 3-(acetyloxy)-14,15-epoxy-16-oxo-, (3β,5β,15β)-	36615-16-4	NI42804
380	231	151	43	161	91	107	81	**440**	100	86	86	57	54	52	43	43	**4.76**	26	32	6	–	–	–	–	–	–	–	–	–	–	Bufa-20,22-dienolide, 16-(acetyloxy)-14,15-epoxy-3-oxo-, (5β,15β,16β)-	6869-66-5	NI42805
299	102	137	41	95	149	91	93	**440**	100	81	62	62	56	53	43	41	**4.80**	26	32	6	–	–	–	–	–	–	–	–	–	–	Deacetylgedunin		MS09747
222	223	121	93	107	91	440	95	**440**	100	15	14	9	8	8	8	7		26	32	6	–	–	–	–	–	–	–	–	–	–	(1′S,4′aS,8′aR)-7-[(decahydro-5,5,8a-trimethyl-2-methylene-6-oxo-1-naphthalenyl)methoxy]-6,8-dimethoxy-2H-1-benzopyran-2-one		NI42806
29	348	394	333	280	223	305	41	**440**	100	72	55	44	37	34	33	27	**16.68**	26	32	6	–	–	–	–	–	–	–	–	–	–	1′,1′-Dicarboethoxy-1β,2β-dihydro-3′H-cycloprop[1,2]androsta-1,4,6-trien-3,17-dione		NI42807
380	231	151	43	161	91	107	81	**440**	100	88	85	57	55	53	43	43	**5.00**	26	32	6	–	–	–	–	–	–	–	–	–	–	3-Keto-cinobufagin		LI08974
347	44	84	43	127	70	348	397	**440**	100	59	44	42	29	26	22	20	**1.00**	26	36	4	2	–	–	–	–	–	–	–	–	–	Putrescine, N,N′-diacetyl-N,N′-di(γ-phenoxypropyl)-		LI08975
425	285	300	440	196	426	232	208	**440**	100	46	43	42	41	37	31	20		26	40	2	–	–	–	–	–	–	2	–	–	–	19-Norpregna-1,3,5(10)-trien-20-yne, 3,17-bis[(trimethylsilyl)oxy]-, (17α)-	28426-35-9	NI42808
143	225	144	130	226	297	224	131	**440**	100	46	35	26	9	8	5	5	**3.00**	27	28	2	4	–	–	–	–	–	–	–	–	–	3-(3-Oxo-2,3,5,6,11,11b-hexahydro-1H-pyrrolo[2,1-a]-β-carbolin-11b-yl)propionic acid N-2-(3-indolyl)ethylamide		MS09748
246	55	81	67	79	231	93	95	**440**	100	57	40	36	32	30	30	28	**16.00**	27	41	4	–	–	–	–	–	–	–	–	1	–	17α,21-Dihydroxy-5β-pregnane-3,20-dione cyclohexyl boronate		MS09749
246	231	107	105	121	357	215	213	**440**	100	30	25	25	24	23	22	21	**16.00**	27	41	4	–	–	–	–	–	–	–	–	1	–	17α,21-Dihydroxy-5β-pregnane-3,20-dione cyclohexyl boronate		LI08976
105	124	107	440	287	147	269	121	**440**	100	99	90	87	83	77	75	62		27	41	4	–	–	–	–	–	–	–	–	1	–	17α,20β,21-Trihydroxypregn-4-en-3-one cyclohexyl boronate		LI08977
73	84	216	75	158	156	318	43	**440**	100	90	76	63	58	48	42	29	**0.00**	27	56	2	–	–	–	–	–	–	1	–	–	–	Tetracosanoic acid, trimethylsilyl ester	74367-37-6	NI42809
109	440	43	69	91	133	422	253	**440**	100	62	59	39	24	19	15	11		28	40	4	–	–	–	–	–	–	–	–	–	–	Azafrin methyl ester		LI08978
339	340	57	94	440	41	55	43	**440**	100	26	19	15	10	8	5	5		28	40	4	–	–	–	–	–	–	–	–	–	–	6,6-Bis-(2-methyl-4-hydroxy-5-tert-butyl-phenyl)-hexanoic acid		IC04420
79	334	41	105	67	55	91	95	**440**	100	96	74	73	71	66	57	53	**8.34**	28	40	4	–	–	–	–	–	–	–	–	–	–	9,12,15-Octadecatrienoic acid, 2-phenyl-1,3-dioxan-2-yl ester, (Z,Z)-	56554-38-2	NI42810

MASS TO CHARGE RATIOS								M.W.	INTENSITIES								Parent	C	H	O	N	S	F	Cl	Br	I	Si	P	B	X	COMPOUND NAME	CAS Reg No	No
105	28	79	41	55	67	43	57	440	100	73	64	56	54	50	41	40	2.54	28	40	4	–	–	–	–	–	–	–	–	–	–	9,12,15-Octadecatrienoic acid, 2-phenyl-1,3-dioxan-5-yl ester	56700-76-6	NI42811
44	105	43	41	55	77	57	91	440	100	90	76	64	62	53	42	37	1.04	28	40	4	–	–	–	–	–	–	–	–	–	–	9,12,15-Octadecatrienoic acid, (2-phenyl-1,3-dioxolan-4-yl)methyl ester	56847-06-4	NI42813
105	79	55	67	41	95	91	43	440	100	76	64	63	61	53	53	48	4.54	28	40	4	–	–	–	–	–	–	–	–	–	–	9,12,15-Octadecatrienoic acid, (2-phenyl-1,3-dioxolan-4-yl)methyl ester	56847-06-4	NI42812
399	398	341	383	440	314	425	400	440	100	91	78	61	57	49	35	34		28	44	2	–	–	–	–	–	–	1	–	–	–	Heptyl-Δ^{1}-tetrahydrocannabinol allyldimethylsilyl ether		NI42814
137	69	83	81	97	95	155	67	440	100	90	87	81	77	77	64	55	0.00	29	60	2	–	–	–	–	–	–	–	–	–	–	Nonacosan-5,10-diol	51995-92-7	NI42815
73	440	77	57	180	58	441	71	440	100	78	52	43	33	32	27	25		30	20	2	2	–	–	–	–	–	–	–	–	–	2,2′-Diphenyl-4,4′-biisoquinoline-3,3′(2H,2′H)-dione		LI08979
81	69	55	313	255	107	105	95	440	100	99	87	69	60	57	56	50	34.31	30	48	2	–	–	–	–	–	–	–	–	–	–	3β-Acetoxy-5α-ergost-7,22-diene		MS09750
314	55	69	81	315	93	440	107	440	100	42	38	32	31	31	29	25		30	48	2	–	–	–	–	–	–	–	–	–	–	3β-Acetoxy-5α-ergost-14,22,diene		MS09751
380	356	313	213	253	440	296	273	440	100	48	47	40	28	13	12	10		30	48	2	–	–	–	–	–	–	–	–	–	–	3β-Acetoxy-24-methylenecholest-5,22-diene		NI42816
57	383	105	327	384	163	221	165	440	100	59	38	24	18	15	10	10	3.00	30	48	2	–	–	–	–	–	–	–	–	–	–	11,12-Di-tert-butyl-3,8-bis-(neopentanoyl)tricyclo[4.3.2$^{(6,9)}$.1]dodeca-2,7-diene		LI08980
137	28	81	95	69	109	55	65	440	100	78	73	69	69	56	53	50	4.61	30	48	2	–	–	–	–	–	–	–	–	–	–	Canophyllal		MS09752
43	41	55	69	57	81	225	95	440	100	33	32	18	17	14	12	12	8.50	30	48	2	–	–	–	–	–	–	–	–	–	–	5α-Cholesta-7,9(11)-dien-3β-ol, 14-methyl-, acetate	5596-02-1	NI42817
43	143	380	57	55	41	69	311	440	100	54	36	36	33	33	30	26	0.00	30	48	2	–	–	–	–	–	–	–	–	–	–	Cholesta-2,8-dien-6-ol, 14-methyl-, acetate, (5α,6α)-	56293-01-7	NI42818
69	55	43	95	81	41	93	147	440	100	86	86	84	73	70	65	51	5.00	30	48	2	–	–	–	–	–	–	–	–	–	–	9,19-Cyclocholest-24-en-3-ol, 14-methyl-, acetate, (3β,5α)-	68654-81-9	NI42819
440	425	55	69	441	57	426	121	440	100	92	45	34	33	33	30	23		30	48	2	–	–	–	–	–	–	–	–	–	–	3,7-Diketoeuph-8-ene		MS09753
440	233	55	343	69	441	412	57	440	100	94	75	70	35	34	32	28		30	48	2	–	–	–	–	–	–	–	–	–	–	3,11-Diketoeuph-8-ene		MS09754
440	205	206	190	219	218	422	425	440	100	95	66	29	26	19	18	17		30	48	2	–	–	–	–	–	–	–	–	–	–	3,16-Dioxo-20βH-taraxane		NI42820
55	440	205	206	219	422	425	354	440	100	72	55	50	16	14	13	13		30	48	2	–	–	–	–	–	–	–	–	–	–	3,16-Dioxotaraxane		NI42821
380	55	83	81	381	296	69	145	440	100	43	40	38	32	29	29	26	0.00	30	48	2	–	–	–	–	–	–	–	–	–	–	Ergosta-5,24-dien-3-ol, acetate, (3β)-	33444-85-8	NI42822
193	440	191	153	288	259	411	205	440	100	90	85	85	80	70	60	60		30	48	2	–	–	–	–	–	–	–	–	–	–	D:A-Friedooleanan-24-al, 1-oxo-	43230-80-4	MS09755
139	273	231	425	218	440	232	234	440	100	83	41	25	25	23	23	21		30	48	2	–	–	–	–	–	–	–	–	–	–	D:A-Friedooleanan-24-al, 3-oxo-	15353-29-4	MS09756
440	316	288	287	260	205	246	261	440	100	99	99	99	60	60	55	50		30	48	2	–	–	–	–	–	–	–	–	–	–	D:A-Friedooleanane-1,3-dione	32768-97-1	MS09757
440	288	441	207	205	55	69	43	440	100	35	34	30	30	30	28	25		30	48	2	–	–	–	–	–	–	–	–	–	–	D:A-Friedooleanane-3,7-dione	18671-50-6	NI42823
207	205	440	288	316	191	301	273	440	100	61	50	40	23	23	20	18		30	48	2	–	–	–	–	–	–	–	–	–	–	D:A-Friedooleanane-3,7-dione	18671-50-6	LI08981
315	440	204	408	393	217	147	121	440	100	70	70	63	40	40	36	30		30	48	2	–	–	–	–	–	–	–	–	–	–	D:A-Friedooleanan-1-one, 25,26-epoxy-	56784-08-8	MS09758
203	422	216	189	204	205	217	175	440	100	99	85	85	77	72	70	52	38.00	30	48	2	–	–	–	–	–	–	–	–	–	–	Lup-20(29)-en-3-one, 6-hydroxy-, (6α)-	71298-27-6	NI42824
98	134	422	440	203	343	407	177	440	100	54	42	23	20	18	16	14		30	48	2	–	–	–	–	–	–	–	–	–	–	1α-Hydroxyallobetul-2-ene		MS09759
98	343	134	440	187	177	203	341	440	100	34	34	32	32	24	23	22		30	48	2	–	–	–	–	–	–	–	–	–	–	1β-Hydroxyallobetul-2-ene		MS09760
121	245	134	440	98	95	187	109	440	100	98	95	81	67	62	57	57		30	48	2	–	–	–	–	–	–	–	–	–	–	3β-Hydroxyallobetul-1-ene		MS09761
218	203	440	119	189	121	425	135	440	100	36	30	15	12	11	10	10		30	48	2	–	–	–	–	–	–	–	–	–	–	3β-Hydroxyolean-12-ene-1-one		NI42825
135	150	149	440	397	189	133	175	440	100	81	53	41	34	28	28	19		30	48	2	–	–	–	–	–	–	–	–	–	–	3β-Hydroxy-16-oxo-ψ-taraxene		NI42826
205	440	187	422	189	203	202	191	440	100	79	46	44	29	25	25	16		30	48	2	–	–	–	–	–	–	–	–	–	–	16β-Hydroxy-3-oxo-ψ-taraxene		NI42827
440	426	441	69	427	219	274	81	440	100	49	32	30	17	17	15	14		30	48	2	–	–	–	–	–	–	–	–	–	–	Lanost-8-ene-7,11-dione	54411-45-9	NI42828
380	55	69	81	57	145	95	83	440	100	23	22	19	13	8	8	8	0.00	30	48	2	–	–	–	–	–	–	–	–	–	–	(24S)-24-Methylcholesta-5,22E-dien-3β-yl acetate		NI42829
248	204	191	440	235	221	218	287	440	100	98	70	45	32	32	15	14		30	48	2	–	–	–	–	–	–	–	–	–	–	Oleanane-1,5-carbolactone		MS09762
40	248	204	43	203	69	55	105	440	100	99	53	48	39	39	39	37	16.00	30	48	2	–	–	–	–	–	–	–	–	–	–	Olean-12-en-28-oic acid	17990-43-1	NI42830
248	271	440	425	249	69	55	41	440	100	38	24	22	21	21	17	16		30	48	2	–	–	–	–	–	–	–	–	–	–	Olean-9(11)-en-12-one, 3β-hydroxy-	4354-40-9	NI42831
152	440	109	203	139	177	149	121	440	100	77	66	61	59	58	57	56		30	48	2	–	–	–	–	–	–	–	–	–	–	1-Oxoallobetulane		MS09763
440	138	151	369	135	119	109	191	440	100	92	74	60	58	57	56	54		30	48	2	–	–	–	–	–	–	–	–	–	–	2-Oxoallobetulane		MS09764
440	205	369	109	125	149	191	177	440	100	73	69	67	66	63	61	60		30	48	2	–	–	–	–	–	–	–	–	–	–	3-Oxoallobetulane		MS09765
289	271	205	191	440	411	179	221	440	100	28	22	16	13	13	12	6		30	48	2	–	–	–	–	–	–	–	–	–	–	3-Oxofriedelan-26-al		NI42832
57	41	28	105	43	39	29	27	440	100	63	50	28	26	26	26	17	0.00	30	48	2	–	–	–	–	–	–	–	–	–	–	1-Propanone, 1,1′-[11,12-bis(1,1-dimethylethyl)tricyclo[5.2.2.12,6]dodeca-3,8-diene-4,8-diyl]bis[2,2-dimethyl-	56772-03-3	NI42833
425	440	273	241	393	426	441	137	440	100	76	68	46	41	34	28	28		31	52	1	–	–	–	–	–	–	–	–	–	–	Arundoin		LI08982
273	425	241	109	393	137	440	119	440	100	99	80	52	50	50	48	44		31	52	1	–	–	–	–	–	–	–	–	–	–	Arundoin	4555-56-0	NI42834
55	95	107	81	41	69	109	93	440	100	97	70	70	68	65	60	58	14.00	31	52	1	–	–	–	–	–	–	–	–	–	–	9,19-Cyclolanostan-3β-ol, 24-methylene-	1449-09-8	NI42835
216	203	407	201	425	422	217	353	440	100	91	87	74	56	51	51	39	25.71	31	52	1	–	–	–	–	–	–	–	–	–	–	9,19-Cyclolanostan-3β-ol, 24-methylene-	1449-09-8	NI42836
69	95	55	41	81	107	109	93	440	100	96	85	66	62	58	57	51	14.00	31	52	1	–	–	–	–	–	–	–	–	–	–	9,19-Cyclolanost-25-en-3-ol, 24-methyl-, (3β,24S)-	511-61-5	NI42837
440	425	273	393	441	426	241	121	440	100	65	45	44	34	26	26	26		31	52	1	–	–	–	–	–	–	–	–	–	–	Cylindrin		LI08983
204	440	316	218	205	441	69	121	440	100	73	52	47	35	30	26	25		31	52	1	–	–	–	–	–	–	–	–	–	–	D-Friedoolean-14-ene, 3β-methoxy-		LI08984
204	440	316	218	205	441	135	69	440	100	72	51	46	33	27	23	23		31	52	1	–	–	–	–	–	–	–	–	–	–	D-Friedoolean-14-ene, 3β-methoxy-	14021-23-9	NI42838
204	135	121	133	316	301	109	107	440	100	48	40	36	34	30	30	30	14.00	31	52	1	–	–	–	–	–	–	–	–	–	–	D-Friedoolean-14-ene, 3β-methoxy-	14021-23-9	NI42839
261	229	440	109	123	425	134	107	440	100	76	46	40	35	32	30	30		31	52	1	–	–	–	–	–	–	–	–	–	–	D:C-Friedours-7-ene, 3β-methoxy-	14021-29-5	NI42840
440	204	441	95	81	221	69	109	440	100	62	45	42	39	38	36	35		31	52	1	–	–	–	–	–	–	–	–	–	–	C-Homo-27-norgammacer-14-ene, 3β-methoxy-	24433-22-5	NI42841
425	55	440	407	426	69	41	81	440	100	50	43	39	37	37	33	32		31	52	1	–	–	–	–	–	–	–	–	–	–	Lanost-8-en-3-ol, 24-methylene-, (3β)-	6890-88-6	NI42842

MASS TO CHARGE RATIOS								M.W.	INTENSITIES								Parent	C	H	O	N	S	F	Cl	Br	I	Si	P	B	X	COMPOUND NAME	CAS Reg No	No
218	219	136	106	134	203	189	105	**440**	100	20	20	19	18	17	17	17	**12.50**	31	52	1	–	–	–	–	–	–	–	–	–	–	Urs-12-ene, 3-methoxy-, (3β)-	14021-28-4	NI42843
440	441	204	94	221	81	69	109	**440**	100	46	46	43	40	40	38	37		31	52	1	–	–	–	–	–	–	–	–	–	–	3β-Methoxy-Δ[14]-serratene		LI08985
218	203	219	189	135	109	190	107	**440**	100	26	19	15	15	15	14	14	**8.00**	31	52	1	–	–	–	–	–	–	–	–	–	–	Olean-12-ene, 3-methoxy-, (3β)-	14021-26-2	NI42844
407	346	347	269	330	239	408	331	**440**	100	86	72	63	39	37	36	21	**1.97**	32	24	2	–	–	–	–	–	–	–	–	–	–	10-Hydroperoxy-9,9,10-triphenyldihydroanthracene	6549-72-0	NI42845
440	371	77	106	306	372	307	399	**440**	100	99	80	73	35	32	27	26		32	28	–	2	–	–	–	–	–	–	–	–	–	9,10-Bis(N-allylanilino)phananthrene		LI08986
227	226	228	187	157	207	133	186	**441**	100	62	12	11	10	8	7	7	**4.49**	16	12	2	2	–	6	–	–	–	–	–	–	1	N,N′-O-Phenylenebis(5,5,5-trifluoro-4-oxopentan-2-iminato)copper(II)	88007-88-9	NI42846
372	441	175	106	78	225	69	373	**441**	100	79	35	35	35	28	26	20		18	10	4	2	–	6	–	–	–	–	–	–	1	Bis(4,4,4-trifluoro-1-(3-pyridyl)-1,3-butaned ionato)beryllium	30983-54-1	NI42847
372	441	106	78	225	69	51	328	**441**	100	79	34	34	28	27	8	7		18	10	4	2	–	6	–	–	–	–	–	–	1	Bis(4,4,4-trifluoro-1-(3-pyridyl)-1,3-butaned ionato)beryllium	30983-54-1	LI08987
44	84	120	137	28	83	41	43	**441**	100	53	27	21	19	15	14	14	**0.00**	19	19	6	7	–	–	–	–	–	–	–	–	–	L-Glutamic acid, N-[4-[[(2-amino-1,4-dihydro-4-oxo-6-pteridinyl)methyl]amino]benzoyl]-	59-30-3	NI42848
55	105	134	211	268	309	385	229	**441**	100	82	42	40	24	14	12	10	**0.00**	20	20	3	1	2	–	–	–	–	–	–	–	1	Dicarbonyl(5-ethoxy-3-methyl-5-phenyl-2-thioxo-4-thiazolidinylidene)(methylcyclopentadienyl)manganese	116928-07-5	DD01677
346	347	101	229	15	228	201	18	**441**	100	33	25	21	21	20	15	13	**6.57**	20	27	8	1	1	–	–	–	–	–	–	–	–	Butanoic acid, 4-(acetylamino)-, [3,4-dihydro-6,7-dimethoxy-2-(methoxysulphonyl)-1(2H)-naphthalenylidene]methyl ester	56701-28-1	MS09766
218	73	223	219	179	324	220	75	**441**	100	48	29	18	10	8	8	8	**0.00**	20	39	4	1	–	–	–	–	–	3	–	–	–	3-Ethoxytyrosine tri-TMS		MS09767
174	73	75	175	40	86	45	176	**441**	100	57	20	17	8	7	7	6	**0.00**	20	43	2	1	–	–	–	–	–	4	–	–	–	Dopamine, tetrakis(trimethylsilyl)-	55606-74-1	MS09768
174	73	175	176	426	86	100	74	**441**	100	31	17	7	6	6	3	3	**0.00**	20	43	2	1	–	–	–	–	–	4	–	–	–	Dopamine, tetrakis(trimethylsilyl)-	55606-74-1	NI42849
174	73	175	176	86	45	59	74	**441**	100	49	19	8	8	5	5	4	**0.00**	20	43	2	1	–	–	–	–	–	4	–	–	–	Dopamine, tetrakis(trimethylsilyl)-	55606-74-1	NI42850
174	73	175	176	267	147	86	426	**441**	100	30	16	7	6	5	5	3	**0.00**	20	43	2	1	–	–	–	–	–	4	–	–	–	Octopamine, tetrakis(trimethylsilyl)-	55556-99-5	NI42851
174	267	175	73	177	176	86	147	**441**	100	43	22	12	10	10	5	4	**0.02**	20	43	2	1	–	–	–	–	–	4	–	–	–	Octopamine, tetrakis(trimethylsilyl)-	55556-99-5	NI42852
84	98	127	107	56	77	41	191	**441**	100	54	42	35	35	27	27	23	**4.00**	23	28	2	5	–	–	1	–	–	–	–	–	–	2-(4-Chlorophenylamino)-5-[3-[3-(piperidinomethyl)phenoxy]propylamino] 1,3,4-oxadiazole		MS09769
347	66	317	413	65	55	120	382	**441**	100	66	50	33	33	33	18	10	**1.00**	24	21	2	1	–	–	–	–	–	–	1	–	1	Manganese, carbonyl[(1,2,3,4-η)-1,3-cyclopentadiene]nitrosyl(triphenylphosphine)-	43064-63-7	MS09770
91	29	170	65	121	186	107	350	**441**	100	99	61	33	32	31	18	7	**4.00**	24	27	5	1	1	–	–	–	–	–	–	–	–	Ethyl 2-(2-benzyloxy-1-benzyloxymethylethoxymethyl)thiazole-4-carboxylate		MS09771
91	43	105	174	121	77	57	84	**441**	100	27	22	11	11	11	10	8	**2.00**	24	27	7	1	–	–	–	–	–	–	–	–	–	1-Benzyloxycarbonyl-3,4-diacetoxy-2-(4-methoxybenzyl)-pyrrolidine		LI08988
174	73	158	312	326	130	370	441	**441**	100	55	37	31	14	12	7	5		24	31	5	1	–	–	–	–	–	1	–	–	–	3a(R),9b(S)-dihydro-5-methoxy-9b-[2-[N-methyl-N-(((2-(trimethylsilyl)ethyl)oxy)carbonyl)amino]ethyl]phenanthro[4,4a,4b,5 bcd]furan-3(8H)-one		NI42853
55	115	159	256	81	105	398	343	**441**	100	47	28	25	21	20	19	17	**0.00**	25	22	1	1	1	3	–	–	–	–	–	–	–	4-(2-Naphthylthio)-5-phenyl-5-(trifluoromethyl)spiro[2,5-dihydro-1,3-oxazole-2,1′-cyclohexane]		DD01678
286	377	91	230	358	243	294	218	**441**	100	80	55	37	36	19	18	16	**8.00**	25	31	4	1	1	–	–	–	–	–	–	–	–	Benzenesulphonamide, N,4-dimethyl-N-(6a,7,10,10a-tetrahydro-1-methoxy-6,6,9-trimethyl-6H-dibenzo[b,d]pyran-3-yl)-, (6aR-trans)-	40248-06-4	NI42854
382	383	385	441	384	398	370	386	**441**	100	53	39	15	15	13	13	9		26	35	5	1	–	–	–	–	–	–	–	–	–	12-Acetyldehydrolucidusculine		MS09772
261	183	384	441	185	109	215	262	**441**	100	98	94	55	52	27	24	22		28	29	–	1	–	–	–	–	–	–	2	–	–	N,N-Bis(diphenylphosphino)butylamine		NI42855
127	55	381	98	366	43	113	72	**441**	100	40	30	29	28	26	23	20	**2.10**	28	43	3	1	–	–	–	–	–	–	–	–	–	Pyrrolidine, [(3β)-3-(acetyloxy)-20-methyl-21-oxopregn-5-en-21-yl]-	56143-22-7	NI42856
441	125	137	153	442	411	268	317	**441**	100	90	45	35	32	25	22	20		28	43	3	1	–	–	–	–	–	–	–	–	–	Testosterone-17β-cypionate-3-methyloxime		NI42857
314	315	160	105	70	91	93	67	**441**	100	25	23	23	23	14	9	8	**0.00**	29	44	–	1	–	–	1	–	–	–	–	–	–	Androst-2-en-17-amine, 4,4-dimethyl-N-(2-phenylethyl)-, hydrochloride (5α)-	53286-45-6	NI42858
441	316	68	54	94	41	410	411	**441**	100	74	70	68	50	49	32	28		29	47	2	1	–	–	–	–	–	–	–	–	–	Ergosta-7,22-dien-6-one, 3-hydroxy-, O-methyloxime (3β,5α)-		NI42859
441	316	68	54	94	41	410	89	**441**	100	74	70	68	50	49	32	27		29	47	2	1	–	–	–	–	–	–	–	–	–	Ergosta-7,22-dien-6-one, 3-hydroxy-, O-methyloxime, (3β,5α,22E)-	53336-26-8	NI42860
441	316	56	41	442	43	411	58	**441**	100	74	68	46	40	34	32	28		29	47	2	1	–	–	–	–	–	–	–	–	–	Ergosta-7,22-dien-6-one, 3-hydroxy-, O-methyloxime, (3β,5α,22E)-	53336-26-8	NI42861
441	105	101	442	100	77	336	99	**441**	100	70	44	35	35	25	21	16		31	23	2	1	–	–	–	–	–	–	–	–	–	4-Benzoyl-N-(4-methylbenzylidene)-3,5-diphenyl-2-furanamine		NI42862
110	43	441	125	41	55	57	69	**441**	100	91	62	61	60	58	42	27		31	55	–	1	–	–	–	–	–	–	–	–	–	Cholest-2-en-3-amine, N,N-diethyl-, (5α)-	55320-50-8	NI42863
302	330	190	414	442	246	218	266	**442**	100	73	71	55	51	51	46	43		10	1	9	–	–	–	–	–	–	–	–	–	3	Methinyl tricobalt enneacarbonyl		LI08989
374	376	446	444	372	448	378	442	**442**	100	80	72	67	58	39	34	31		12	2	1	–	–	–	8	–	–	–	–	–	–	Octachlorodiphenyl ether	40356-57-8	NI42864
261	263	102	259	182	183	181	180	**442**	100	78	60	46	45	33	33	22	**14.00**	14	8	–	2	–	–	–	2	–	–	–	–	1	3,5-Bis(4-bromophenyl)-1,2,4-selenadiazole		MS09773
182	92	442	440	221	219	312	310	**442**	100	77	30	30	28	26	25	24		14	14	–	–	–	–	–	–	–	–	–	–	2	Ditelluride, bis(4-methylphenyl)-	32294-57-8	NI42865
83	55	84	68	41	387	56	111	**442**	100	88	78	49	47	28	28	27	**19.00**	14	14	2	–	–	12	–	–	–	–	–	–	–	1,1,7-Trihydroperfluoroheptyl cyclohexanecarboxylate		MS09774
162	175	164	63	42	99	98	54	**442**	100	98	98	98	98	96	96	96	**1.00**	14	16	6	2	2	–	2	–	–	–	–	–	–	L-Cystine, N-[(2,4-dichlorophenoxy)acetyl]-		NI42866
355	356	57	327	59	383	56	43	**442**	100	41	23	23	16	15	12	12	**0.00**	14	38	6	–	–	–	–	–	–	5	–	–	–	1,3,5,7,9-Pentaethyl-1-butoxycyclopentasiloxane	73420-31-2	NI42867

MASS TO CHARGE RATIOS	M.W.	INTENSITIES	Parent	C	H	O	N	S	F	Cl	Br	I	Si	P	B	X	COMPOUND NAME	CAS Reg No	No
355 356 57 357 58 56 383 327	**442**	100 36 35 25 21 15 12 9	**0.00**	14	38	6	–	–	–	–	–	–	5	–	–	–	1,3,5,7-Tetraethyl-1-ethylbutoxysiloxycyclotetrasiloxane	73420-30-1	NI42868
186 188 123 121 302 244 246 300	**442**	100 45 35 33 31 16 14 9	**1.50**	15	10	5	–	–	–	–	–	–	–	–	–	3	Dicyclopentadienyldinickelironpentacarbonyl		LI08990
73 221 59 147 233 45 74 75	**442**	100 28 22 14 14 11 9 7	**0.00**	16	42	2	–	2	–	–	–	–	4	–	–	–	Dithioerythritol, tetrakis(trimethylsilyl)-		NI42869
73 221 233 59 147 217 336 222	**442**	100 47 20 18 17 14 12 10	**1.77**	16	42	2	–	2	–	–	–	–	4	–	–	–	Dithiothreitol, tetrakis(trimethylsilyl)-		NI42870
198 442 77 255 51 127 69 275	**442**	100 94 54 51 49 44 43 27		18	5	–	–	–	10	–	–	–	–	1	–	–	Decafluorotriphenylphosphine	5074-71-5	NI42871
198 77 442 69 127 51 255 110	**442**	100 54 53 53 52 50 38 29		18	5	–	–	–	10	–	–	–	–	1	–	–	Decafluorotriphenylphosphine	5074-71-5	NI42872
198 51 442 127 77 69 255 110	**442**	100 43 41 41 40 40 36 22		18	5	–	–	–	10	–	–	–	–	1	–	–	Decafluorotriphenylphosphine	5074-71-5	NI42873
351 197 120 152 398 427 274 229	**442**	100 43 24 16 9 7 5 5	**1.00**	20	18	–	–	2	–	–	–	–	–	–	–	1	Methyl triphenylstannanedithiocarboxylate	73137-42-5	NI42874
247 249 248 251 250 232 234 189	**442**	100 65 12 11 8 6 4 3	**0.00**	20	20	7	–	–	–	2	–	–	–	–	–	–	4,2-Cresotic acid, 6-methoxy-, methyl ester, 3,5-dichloro-4,6-dimethoxy-o-toluate	5859-23-4	NI42875
91 213 442 155 65 41 172 92	**442**	100 57 41 40 25 23 14 14		20	26	7	–	2	–	–	–	–	–	–	–	–	3,3-Bis(toluene-p-sulphonyloxy) dipropyl ether		IC04421
267 268 179 269 442 45 352 44	**442**	100 25 12 10 9 5 4 4		20	38	5	–	–	–	–	–	–	3	–	–	–	3,4-Dihydroxyphenyllactic acid ethyl ester tri-TMS		MS09775
267 73 268 179 75 45 193 442	**442**	100 68 22 13 10 10 9 8		20	38	5	–	–	–	–	–	–	3	–	–	–	3,4-Dihydroxyphenyllactic acid ethyl ester tri-TMS		MS09776
73 217 147 75 218 43 45 41	**442**	100 99 75 36 23 20 19 17	**16.00**	20	38	5	2	–	–	–	–	–	2	–	–	–	2,4,6(1H,3H,5H)-Pyrimidinetrione, 5-[2,3-bis[(trimethylsilyl)oxy]-2-propenyl]-1,3-dimethyl-5-(1-methylbutyl)-	57397-41-8	NI42876
141 115 142 281 362 139 442 89	**442**	100 21 17 10 8 6 5 4		22	18	–	–	–	–	–	–	–	–	–	–	2	Bis(1-naphthomethyl) diselenide		MS09777
324 351 356 325 208 266 442 238	**442**	100 40 36 33 33 26 25 24		22	22	8	2	–	–	–	–	–	–	–	–	–	Tetramethyl-6-methyl-10,11-dihydro-[1,2-a]quinoxaline-7,8,9,11-tetracarboxylate		LI08991
321 411 322 196 192 268 210 298	**442**	100 66 28 16 13 12 8 6	**0.15**	22	42	5	2	–	–	–	–	–	1	–	–	–	Cyclopentanepropanoic acid, 5-(methoxyimino)-2-[3-(methoxyimino)octyl]-3-[(trimethylsilyl)oxy]-, methyl ester, [1R-(1α,2β,3α)]-	72101-42-9	NI42877
321 411 196 268 265 325 286 235	**442**	100 28 14 9 9 5 4 4	**0.01**	22	42	5	2	–	–	–	–	–	1	–	–	–	1-Methoxyimino-2-(2-methoxycarbonylethyl)-3-(3-methoxyimino-octyl)-4-trimethylsilyloxy cyclopentane		LI08992
202 121 69 241 145 97 174 203	**442**	100 77 51 34 34 29 25 17	**16.00**	23	22	9	–	–	–	–	–	–	–	–	–	–	7H-Furo[3,2-g][1]benzopyran-7-one, 9-[2-(acetyloxy)-4-(2,5-dihydro-4-methyl-5-oxo-2-furanyl)-3-hydroxy-3-methylbutoxy]-	68725-66-6	NI42878
358 123 137 400 316 274 383 341	**442**	100 80 62 48 48 39 27 18	**5.00**	23	22	9	–	–	–	–	–	–	–	–	–	–	2,4,3′,4′-Tetraacetoxydihydrochalcone	94143-70-1	NI42879
71 58 43 70 83 59 442 332	**442**	100 88 76 53 43 27 26 26		23	26	7	2	–	–	–	–	–	–	–	–	–	Dichotine, 2-deoxy-2,21-epoxy-11-methoxy-, (21α)-	55283-47-1	NI42880
442 58 57 71 72 70 190 189	**442**	100 66 52 39 19 19 12 12		23	26	7	2	–	–	–	–	–	–	–	–	–	16,19-Secostrychnidine-10,16-dione, 21,22-epoxy-21,22-dihydro-4,14-dihydroxy-3-methoxy-19-methyl-, (21α,22α)-	62421-66-3	NI42881
115 57 91 79 117 346 427 288	**442**	100 57 22 21 10 4 1 1	**0.50**	23	38	6	–	1	–	–	–	–	–	–	–	–	16-[[2,3-(Isopropylidenedioxy)propyl]oxy]-11,15-hexadecadien-13-yn-1-yl-methanesulphonate		MS09778
262 191 166 152 280 180 269 251	**442**	100 40 20 17 15 14 8 8	**0.00**	23	38	8	–	–	–	–	–	–	–	–	–	–	Cyclopentanepropanoic acid, 3,5-bis(acetyloxy)-2-[3-(acetyloxy)octyl]-, methyl ester	55401-45-1	NI42882
259 260 77 181 105 442 364 155	**442**	100 32 23 16 15 13 13 11		24	15	1	–	–	5	–	–	–	1	–	–	–	Silane, (pentafluorophenoxy)triphenyl-	22529-93-7	NI42883
365 363 361 364 362 366 369 367	**442**	100 76 43 41 32 22 17 16	**0.00**	24	18	1	–	–	–	–	–	–	–	–	–	1	10,10-Diphenylphenoxastannin		LI08993
367 43 368 442 382 339 336 311	**442**	100 34 27 24 21 17 17 13		24	26	8	–	–	–	–	–	–	–	–	–	–	(10bS,1′S)-8-(2-acetoxy-1-methylethyl)-2,4,6,10b-tetrahydro-7-hydroxy-9,10-dimethoxy-3,10b-dimethyl-1H-5-oxaaceanthrylene-4,6-dione		NI42884
43 105 149 91 107 85 106 77	**442**	100 84 41 31 16 16 15 15	**10.24**	24	26	8	–	–	–	–	–	–	–	–	–	–	D-Glucitol, 1,3:2,4-bis-O-(phenylmethylene)-, diacetate	74842-25-4	NI42885
369 370 442 341 340 383 371 443	**442**	100 26 18 10 8 6 5 5		24	30	6	2	–	–	–	–	–	–	–	–	–	N-[(3,4-Dimethoxyphenyl)[2-(1-ethyl-2-oxopropyl)-4,5-dimethoxyphenyl]methylene] acetohydrazide	90140-55-9	NI42886
138 442 94 269 226 267 411 311	**442**	100 90 60 35 32 30 20 20		24	30	6	2	–	–	–	–	–	–	–	–	–	cis-2-Ethoxy-3H-indole-3-spiro-7′-(1′-aza-3′-acetyl-4′-bismethoxycarbonylmethyl-bicyclo[4.3.0]nonane)		LI08994
138 94 442 269 226 267 268 311	**442**	100 65 39 33 27 23 17 16		24	30	6	2	–	–	–	–	–	–	–	–	–	trans-2-Ethoxy-3H-indole-3-spiro-7′-(1′aza-3′-acetyl-4′-bismethoxycarbonylmethyl-bicyclo[4.3.0]nonane)		LI08995
342 400 309 161 282 300 277 61	**442**	100 73 71 52 47 38 29 29	**19.49**	24	30	6	2	–	–	–	–	–	–	–	–	–	11β-Hydroxy-estrone-6-one-di-acetate-di-methyloxime		NI42887
349 223 173 222 77 56 130 57	**442**	100 93 67 43 30 30 28 28	**0.00**	24	30	6	2	–	–	–	–	–	–	–	–	–	1,4,7,10-Tetraoxa-13-(N-phenylphthalamil)azacyclopentadecane		MS09779
180 73 75 181 179 442 205 165	**442**	100 99 26 24 22 16 13 12		24	34	4	–	–	–	–	–	–	2	–	–	–	Enterolactone, bis(trimethylsilyl)-	77545-19-8	NI42888
298 129 326 116 228 257 229 214	**442**	100 94 47 45 44 18 18 12	**1.07**	24	34	4	4	–	–	–	–	–	–	–	–	–	L-Alanine, N-[N-[N-[(4-cyanophenyl)methylene]-L-valyl]-L-isoleucyl]-, ethyl ester	37580-26-0	NI42889
183 262 223 73 313 263 69 133	**442**	100 76 46 41 32 19 19 16	**7.80**	24	50	3	–	–	–	–	–	–	2	–	–	–	Silane, [[4-(7-methoxyheptyl)-5-(3-methyl-2-butenyl)-1,3-cyclopentanediyl]bis(oxy)]bis[trimethyl-, (1α,3α,4β,5α)-	61177-08-0	NI42890
313 314 73 442 217 69 315 147	**442**	100 32 31 24 18 16 14 12		24	50	3	–	–	–	–	–	–	2	–	–	–	Silane, [[4-(7-methoxyheptyl)-5-(3-methyl-2-butenyl)-1,3-cyclopentanediyl]bis(oxy)]bis[trimethyl-, (1α,3α,4β,5β)-	61177-07-9	NI42891
313 183 73 314 69 442 184 75	**442**	100 90 66 29 28 19 16 16		24	50	3	–	–	–	–	–	–	2	–	–	–	Silane, [[4-(7-methoxyheptyl)-5-(3-methyl-2-butenyl)-1,3-cyclopentanediyl]bis(oxy)]bis[trimethyl-, (1α,3β,4α,5β)-	61177-12-6	NI42892
73 187 75 55 103 188 328 97	**442**	100 98 37 27 18 17 14 11	**0.00**	24	50	3	–	–	–	–	–	–	2	–	–	–	12-Trimethylsiloxy-9-octadecenoic acid, trimethylsilyl ester		NI42893

MASS TO CHARGE RATIOS	M.W.	INTENSITIES	Parent	C	H	O	N	S	F	Cl	Br	I	Si	P	B	X	COMPOUND NAME	CAS Reg No	No
187 73 75 55 188 328 103 97	**442**	100 86 32 21 17 16 16 10	**0.00**	24	50	3	–	–	–	–	–	–	2	–	–	–	**12-Trimethylsiloxy-9-octadecenoic acid, trimethylsilyl ester**		**NI42894**
73 133 147 59 57 99 173 159	**442**	100 60 43 28 26 18 14 14	**5.80**	24	54	1	–	–	–	–	–	–	3	–	–	–	**2,2,3,3,4,4-Hexa-tert-butyl-1-oxa-2,3,4-trisiletane**	93194-12-8	**NI42895**
318 386 183 262 28 164 319 240	**442**	100 44 33 29 25 23 22 21	**1.90**	25	23	2	–	–	–	–	–	–	–	1	–	1	**Iron, dicarbonyl[(1,2,3,4-η)-2-methyl-1,3-butadiene](triphenylphosphine)-**	74710-06-8	**NI42896**
83 55 43 242 243 82 29 199	**442**	100 96 68 64 61 29 29 26	**4.29**	25	30	7	–	–	–	–	–	–	–	–	–	–	**4-Hydroxypruteninone angelicate**	33439-66-6	**NI42897**
242 243 199 171 214 442 342	**442**	100 94 45 40 27 10 6		25	30	7	–	–	–	–	–	–	–	–	–	–	**4-Hydroxypruteninone angelicate**		**LI08996**
138 57 43 110 160 59 44 41	**442**	100 56 49 33 31 31 25 23	**0.00**	25	34	5	2	–	–	–	–	–	–	–	–	–	**Aspidospermidine, 19,21-epoxy-15,16,17-trimethoxy-1-(1-oxopropyl)-**	54751-72-3	**NI42898**
236 106 221 57 161 77 237 442	**442**	100 99 27 23 20 20 17 6		25	34	5	2	–	–	–	–	–	–	–	–	–	**2,6-Di-tert-butyl-4-methoxyphenyl 2-nitro-4-(phenylamino)butanoate**		**DD01679**
138 137 124 151 123 442 107 109	**442**	100 84 81 57 56 55 39 38		25	34	5	2	–	–	–	–	–	–	–	–	–	**4,25-Secoobscurinervan-4-ol, 6,7-didehydro-22-ethyl-15,16-dimethoxy-, (4β,22α)-**	6887-31-6	**NI42899**
140 124 69 141 67 109 81 260	**442**	100 27 13 10 10 9 8 7	**5.46**	25	34	5	2	–	–	–	–	–	–	–	–	–	**4,25-Secoobscurinervan-4-one, 22-ethyl-15,16-dimethoxy-, (22α)-**	54658-09-2	**NI42900**
264 159 382 265 262 353 98 383	**442**	100 85 38 29 26 15 12 10	**0.00**	25	46	6	–	–	–	–	–	–	–	–	–	–	**Octadecanoic acid, 2-(acetyloxy)-1-[(acetyloxy)methyl]ethyl ester**	55401-62-2	**WI01556**
159 382 267 158 177 266 383 218	**442**	100 95 68 57 55 30 26 25	**0.00**	25	46	6	–	–	–	–	–	–	–	–	–	–	**Octadecanoic acid, 2-(acetyloxy)-1-[(acetyloxy)methyl]ethyl ester**	55401-62-2	**WI01555**
43 159 57 98 267 55 158 73	**442**	100 71 42 36 34 31 30 28	**0.00**	25	46	6	–	–	–	–	–	–	–	–	–	–	**Octadecanoic acid, 2-(acetyloxy)-1-[(acetyloxy)methyl]ethyl ester**	55401-62-2	**WI01554**
43 74 57 159 55 41 87 69	**442**	100 65 64 63 59 54 44 35	**0.00**	25	46	6	–	–	–	–	–	–	–	–	–	–	**Octadecanoic acid, 2,3-bis(acetyloxy)propyl ester**	33599-07-4	**WI01557**
159 264 380 265 160 88 107 102	**442**	100 41 15 12 8 7 5 4	**0.00**	25	46	6	–	–	–	–	–	–	–	–	–	–	**Octadecanoic acid, 2,3-bis(acetyloxy)propyl ester**	33599-07-4	**WI01558**
105 259 77 229 91 167 266 154	**442**	100 79 79 39 33 29 27 27	**0.00**	26	23	2	–	–	–	–	–	–	–	–	–	1	**(2-Hydroxy-2,2-diphenylethyl)diphenylarsane oxide**		**DD01680**
215 43 91 107 105 93 81 79	**442**	100 90 71 64 61 60 60 60	**38.09**	26	34	6	–	–	–	–	–	–	–	–	–	–	**3β-Acetoxy-16β-desacetylcinobufagin**	4026-96-4	**NI42901**
364 43 45 91 60 79 105 328	**442**	100 90 74 62 52 47 45 42	**6.67**	26	34	6	–	–	–	–	–	–	–	–	–	–	**3β-Acetoxymarinobufagin**	4029-68-9	**NI42902**
232 382 43 123 231 442 292 151	**442**	100 88 71 62 57 46 46 39		26	34	6	–	–	–	–	–	–	–	–	–	–	**3β-Acetoxy-15-oxobufalin**	31444-12-9	**NI42903**
382 43 151 215 233 107 105 79	**442**	100 88 71 67 64 62 54 50	**5.95**	26	34	6	–	–	–	–	–	–	–	–	–	–	**Cinobufagin**	470-37-1	**NI42904**
222 223 95 109 107 93 442 91	**442**	100 16 9 6 6 6 5 5		26	34	6	–	–	–	–	–	–	–	–	–	–	**(1′R,2′S,4′aS,5′R,8′aR)-7-[(decahydro-1,2,4a,5-tetramethyl-6-oxo-1-naphthalenyl)methoxy]-6,8-dimethoxy-2H-1-benzopyran-2-one**		**NI42905**
29 396 350 41 223 55 43 282	**442**	100 39 33 30 28 27 26 26	**8.78**	26	34	6	–	–	–	–	–	–	–	–	–	–	**1′,1′-Dicarboethoxy-1β,2β-dihydro-17β-hydroxy-3′H-cycloprop[1,2]androsta-1,4,6-trien-3-one**		**NI42906**
301 41 43 149 95 77 55 39	**442**	100 52 46 45 43 42 37 33	**2.90**	26	34	6	–	–	–	–	–	–	–	–	–	–	**Dihydrodeacetylgedunin**		**MS09780**
442 443 427 43 413 412 428 149	**442**	100 26 24 16 15 15 12 10		26	34	6	–	–	–	–	–	–	–	–	–	–	**3-Ethoxy-17,20,20,21-bismethylenedioxy-2-formyl-2,4,6-pregnatriene**		**MS09781**
279 185 280 281 159 186 171 187	**442**	100 29 23 7 7 6 6 5	**0.00**	26	34	6	–	–	–	–	–	–	–	–	–	–	**Pregna-4,6-diene-3,20-dione, 2α,17-dihydroxy-6-methyl-, diacetate**	5234-56-0	**NI42907**
279 185 280 91 281 159 186 171	**442**	100 45 28 11 10 10 8 8	**0.00**	26	34	6	–	–	–	–	–	–	–	–	–	–	**Pregna-4,6-diene-3,20-dione, 2β,17-dihydroxy-6-methyl-, diacetate**	6113-79-7	**NI42908**
75 55 73 43 69 41 74 395	**442**	100 77 60 60 45 44 43 34	**1.00**	26	54	3	–	–	–	–	–	–	1	–	–	–	**Docosanoic acid, 22-(trimethylsiloxy)-, methyl ester**		**LI08997**
75 55 73 43 69 41 74 57	**442**	100 75 60 60 44 44 42 35	**1.00**	26	54	3	–	–	–	–	–	–	1	–	–	–	**Docosanoic acid, 22-(trimethylsiloxy)-, methyl ester**	21987-15-5	**NI42909**
73 382 89 43 57 75 103 83	**442**	100 47 44 43 34 27 26 24	**1.00**	26	54	3	–	–	–	–	–	–	1	–	–	–	**Docosanoic acid, 2-[(trimethylsilyl)oxy]-, methyl ester**	56784-02-2	**MS09782**
73 383 43 89 57 427 384 103	**442**	100 90 45 36 34 32 31 25	**2.07**	26	54	3	–	–	–	–	–	–	1	–	–	–	**Docosanoic acid, 2-[(trimethylsilyl)oxy]-, methyl ester**	56784-02-2	**NI42910**
150 168 274 137 178 95 113 175	**442**	100 58 41 33 14 14 13 12	**11.50**	27	38	5	–	–	–	–	–	–	–	–	–	–	**Cystoseirol B**	102396-15-6	**NI42911**
150 177 95 168 137 83 274 82	**442**	100 68 39 26 24 18 18 15	**5.70**	27	38	5	–	–	–	–	–	–	–	–	–	–	**Cystoseirol C**	10291-66-7	**NI42912**
150 137 168 177 175 82 95 135	**442**	100 56 53 34 33 28 27 26	**3.30**	27	38	5	–	–	–	–	–	–	–	–	–	–	**Cystoseirol D**	102396-16-7	**NI42913**
175 150 168 137 177 82 95 135	**442**	100 90 77 51 38 23 22 17	**1.50**	27	38	5	–	–	–	–	–	–	–	–	–	–	**Cystoseirol E**	102490-77-7	**NI42914**
427 442 399 428 413 443 415 414	**442**	100 65 40 31 28 24 22 16		28	30	3	2	–	–	–	–	–	–	–	–	–	**Benzoic acid, 2-[6-(ethylamino)-3-(ethylimino)-2,7-dimethyl-3H-xanthen-9-yl-, ethyl ester**		**IC04422**
355 370 356 369 353 326 311 371	**442**	100 41 36 34 11 11 10 9	**2.40**	28	30	3	2	–	–	–	–	–	–	–	–	–	**Benzoic acid, 2-[6-(ethylamino)-3-(ethylimino)-2,7-dimethyl-3H-xanthen-9-yl]-, ethyl ester**	3373-01-1	**NI42915**
57 191 442 41 206 223 443 192	**442**	100 54 43 39 38 35 14 10		28	42	2	–	1	–	–	–	–	–	–	–	–	**Bis(2-hydroxy-3,5-di-tert-butylphenyl) sulphide**		**IC04423**
371 57 259 135 41 43 372 29	**442**	100 87 50 46 42 29 27 23	**14.42**	28	42	2	–	1	–	–	–	–	–	–	–	–	**Bis(2-hydroxy-5-(1,1,3,3-tetramethylbutyl)phenyl) sulphide**		**IC04424**
336 67 55 105 41 79 81 129	**442**	100 41 39 36 34 29 28 24	**3.74**	28	42	4	–	–	–	–	–	–	–	–	–	–	**9,12-Octadecadienoic acid, 2-phenyl-1,3-dioxan-5-yl ester**	56648-80-7	**NI42916**
336 105 67 55 41 81 57 79	**442**	100 92 81 80 63 57 56 50	**3.64**	28	42	4	–	–	–	–	–	–	–	–	–	–	**9,12-Octadecadienoic acid, 2-phenyl-1,3-dioxan-5-yl ester, cis-**	56687-50-4	**NI42917**
105 67 55 91 41 81 79 43	**442**	100 92 85 72 67 65 47 46	**14.64**	28	42	4	–	–	–	–	–	–	–	–	–	–	**9,12-Octadecadienoic acid, (2-phenyl-1,3-dioxolan-4-yl)methyl ester, cis-**	56599-47-4	**NI42918**
105 91 67 55 41 81 43 28	**442**	100 76 57 54 46 41 35 32	**9.64**	28	42	4	–	–	–	–	–	–	–	–	–	–	**9,12-Octadecadienoic acid, (2-phenyl-1,3-dioxolan-4-yl)methyl ester, trans-**	56599-48-5	**NI42919**
138 442 137 151 139 441 123 152	**442**	100 39 37 14 11 4 4 2		28	42	4	–	–	–	–	–	–	–	–	–	–	**Striatol**		**LI08998**
159 160 146 173 147 442 175 174	**442**	100 34 29 9 7 6 4 4		28	46	2	2	–	–	–	–	–	–	–	–	–	**5-Hydroxy-N-stearoyltryptamine**		**MS09783**
151 150 71 69 149 85 125 91	**442**	100 41 13 9 8 8 7 6	**0.00**	29	14	5	–	–	–	–	–	–	–	–	–	–	**9,10-Anthracenedione, 1,1′-carbonylbis-**	55401-51-9	**WI01559**
235 442 151 236 207 443 150 428	**442**	100 37 22 17 14 12 9 6		29	14	5	–	–	–	–	–	–	–	–	–	–	**9,10-Anthracenedione, 1,1′-carbonylbis-**	55401-51-9	**NI42920**
55 124 69 81 95 79 107 67	**442**	100 93 70 62 51 50 48 48	**31.00**	29	46	3	–	–	–	–	–	–	–	–	–	–	**27-Acetoxy cholest-4-en-3-one**		**LI08999**
442 163 109 95 224 43 443 108	**442**	100 63 63 46 42 34 32 32		29	46	3	–	–	–	–	–	–	–	–	–	–	**20-Acetyl-(20R,25R)-5α-spirostan**	24799-50-6	**NI42921**
442 443 224 169 223 328 163 444	**442**	100 34 23 13 11 8 8 6		29	46	3	–	–	–	–	–	–	–	–	–	–	**20-Acetyl-(20R,25R)-5α-spirostan**		**MS09784**
353 354 382 247 242 199 248 228	**442**	100 43 15 10 10 9 8 8	**0.00**	29	46	3	–	–	–	–	–	–	–	–	–	–	**Cholest-5-en-19-al, 3-(acetyloxy)-, (3β)-**		**LI09000**
353 354 382 248 246 242 199 226	**442**	100 43 15 10 10 10 10 9	**0.00**	29	46	3	–	–	–	–	–	–	–	–	–	–	**Cholest-5-en-19-al, 3-(acetyloxy)-, (3β)-**	1107-90-0	**NI42922**

MASS TO CHARGE RATIOS	M.W.	INTENSITIES	Parent	C	H	O	N	S	F	Cl	Br	I	Si	P	B	X	COMPOUND NAME	CAS Reg No	No
73 124 442 55 95 107 93 81	442	100 22 19 19 18 15 15 12		29	46	3	–	–	–	–	–	–	–	–	–	–	Cholest-4-en-26-al, 3-oxo-, cyclic 26-(ethylene acetal)	23405-51-8	NI42923
55 124 69 81 95 79 107 67	442	100 93 71 60 50 50 49 46	31.08	29	46	3	–	–	–	–	–	–	–	–	–	–	Cholest-4-en-3-one, 26-(acetyloxy)-	53481-63-3	NI42924
382 43 383 82 145 147 71 55	442	100 48 31 26 23 22 22 20	0.00	29	46	3	–	–	–	–	–	–	–	–	–	–	Cholest-5-en-24-one, 3-(acetyloxy)-, (3β)-	20981-59-3	NI42925
95 43 94 93 91 55 81 41	442	100 40 39 23 23 21 17 17	3.25	29	46	3	–	–	–	–	–	–	–	–	–	–	Cholest-7-en-6-one, 3-(acetyloxy)-, (3β,5α)-	988-19-2	NI42927
95 43 94 442 55 382 175 318	442	100 48 43 39 36 33 31 30		29	46	3	–	–	–	–	–	–	–	–	–	–	Cholest-7-en-6-one, 3-(acetyloxy)-, (3β,5α)-	988-19-2	NI42926
442 163 109 95 443 108 224 67	442	100 38 36 36 32 32 22 20		29	46	3	–	–	–	–	–	–	–	–	–	–	Furost-20(22)-en-26-ol, acetate, (5α,25R)-	24744-53-4	NI42928
442 163 109 95 108 443 224 123	442	100 38 36 36 33 32 21 20		29	46	3	–	–	–	–	–	–	–	–	–	–	Furost-20(22)-en-26-ol, acetate, (5α,25R)-		MS09785
105 57 77 43 426 71 338 44	442	100 80 68 62 60 50 40 40	3.00	30	22	2	2	–	–	–	–	–	–	–	–	–	2,5-Dibenzoyl-3,6-diphenyl-1,4-dihydropyrazine		LI09001
442 214 443 427 199 197 192 191	442	100 66 42 30 22 22 22 22		30	38	1	2	–	–	–	–	–	–	–	–	–	2,5-Cyclohexadien-1-one, 4-[(4'-amino-2,2',3,3',5,5',6,6'-octamethyl[1,1'-biphenyl]-4-yl)imino]-2,3,5,6-tetramethyl-	54245-96-4	NI42929
69 55 81 257 95 107 109 93	442	100 90 81 76 64 60 59 55	17.23	30	50	2	–	–	–	–	–	–	–	–	–	–	3α-Acetoxy-5β-ergost-22-ene		MS09786
315 442 257 344 339 255 382 284	442	100 83 78 72 24 23 16 15		30	50	2	–	–	–	–	–	–	–	–	–	–	3α-Acetoxy-5β-ergost-22-ene		MS09787
257 442 315 344 284 382 339 255	442	100 69 47 39 33 31 29 29		30	50	2	–	–	–	–	–	–	–	–	–	–	3β-Acetoxy-5α-ergost-22-ene		MS09788
69 55 81 107 109 95 257 93	442	100 93 91 68 64 64 57 57	16.39	30	50	2	–	–	–	–	–	–	–	–	–	–	3β-Acetoxy-5α-ergost-22-ene		MS09789
358 215 315 255 275 427 442 367	442	100 46 42 24 16 10 8 8		30	50	2	–	–	–	–	–	–	–	–	–	–	3β-Acetoxy-24-methylenecholest-5-ene		NI42930
441 442 440 443 439 207 410 409	442	100 48 16 15 12 6 4 4		30	50	2	–	–	–	–	–	–	–	–	–	–	Lup-20(29)-en-28-oic acid, 3-hydroxy-, (3β)-	473-98-3	LI09002
189 95 207 203 135 411 81 121	442	100 78 75 71 69 64 53 51	38.46	30	50	2	–	–	–	–	–	–	–	–	–	–	Lup-20(29)-en-28-oic acid, 3-hydroxy-, (3β)-	473-98-3	NI42931
43 55 41 57 69 367 95 81	442	100 40 40 32 30 29 28 28	6.00	30	50	2	–	–	–	–	–	–	–	–	–	–	Cholest-8-en-3-ol, 14-methyl-, acetate, (3β,5α)-	5502-23-8	NI42932
57 43 368 147 81 105 145 55	442	100 62 49 43 42 40 36 34	0.00	30	50	2	–	–	–	–	–	–	–	–	–	–	Cholest-5-en-3-ol (3β)-, propanoate	633-31-8	NI42933
387 305 101 179 442 44 100 388	442	100 92 37 33 32 30 29 27		30	50	2	–	–	–	–	–	–	–	–	–	–	3,5-Cyclocholestan-6-one, cyclic 1-methyl-1,2-ethanediyl acetal, (3α,5α)-		LI09003
387 305 101 442 179 100 113 427	442	100 93 39 36 36 31 24 17		30	50	2	–	–	–	–	–	–	–	–	–	–	3,5-Cyclocholestan-6-one, cyclic 1-methyl-1,2-ethanediyl acetal, (3β,5α)-	38404-85-2	NI42934
109 95 55 43 81 107 69 121	442	100 92 78 78 77 70 70 63	6.12	30	50	2	–	–	–	–	–	–	–	–	–	–	9,19-Cyclolanost-23-ene-3,25-diol, (3β,23E)-	54482-57-4	NI42935
205 109 95 218 442 69 81 155	442	100 84 74 44 42 36 32 24		30	50	2	–	–	–	–	–	–	–	–	–	–	2α,3α-Dihydroxyolean-13(18)-ene		MS09790
424 134 133 135 247 189 207 187	442	100 93 93 74 38 32 27 25	1.00	30	50	2	–	–	–	–	–	–	–	–	–	–	3β,16α-Dihydroxy-ψ-taraxene		NI42936
207 189 190 442 187 424 191 203	442	100 90 32 30 28 26 22 20		30	50	2	–	–	–	–	–	–	–	–	–	–	3β,16β-Dihydroxy-ψ-taraxene		NI42937
108 207 135 189 424 203 191 190	442	100 45 34 32 15 15 14 14	12.00	30	50	2	–	–	–	–	–	–	–	–	–	–	3β,16β-Dihydroxytaraxene		NI42938
207 189 424 205 442 409 222 204	442	100 92 75 75 65 61 52 51		30	50	2	–	–	–	–	–	–	–	–	–	–	29,30-Dinorgammaceran-3-one, 22-hydroxy-21,21-dimethyl-, (8α,9β,13α,14β,17α,18β,22α)-	43206-44-6	NI42939
43 442 330 427 409 194 329 193	442	100 84 24 23 23 23 14 14		30	50	2	–	–	–	–	–	–	–	–	–	–	11β,18-Epoxylanostan-3-one		LI09004
43 442 55 69 41 81 95 57	442	100 85 77 69 68 62 61 59		30	50	2	–	–	–	–	–	–	–	–	–	–	11β,18-Epoxylanostan-3-one	25116-73-8	NI42940
203 442 411 189 207 231 220 385	442	100 88 83 70 60 36 18 15		30	50	2	–	–	–	–	–	–	–	–	–	–	Erythrodiol		LI09005
203 43 41 55 57 69 204 81	442	100 70 68 60 54 45 32 28	3.03	30	50	2	–	–	–	–	–	–	–	–	–	–	Erythrodiol	545-48-2	NI42941
205 137 442 177 149 273 318 189	442	100 80 73 60 55 50 40 40		30	50	2	–	–	–	–	–	–	–	–	–	–	Friedelalactone		LI09006
259 155 204 250 217 427 218 442	442	100 90 67 65 65 55 47 45		30	50	2	–	–	–	–	–	–	–	–	–	–	D:A-Friedooleanan-29-oic acid, (20α)-	39944-58-6	MS09791
273 411 137 109 247 95 81 69	442	100 84 58 36 35 35 32 32	2.59	30	50	2	–	–	–	–	–	–	–	–	–	–	D:A-Friedooleanan-3-one, 28-hydroxy-		MS09792
141 427 411 287 236 442 218 221	442	100 83 83 47 46 42 40 29		30	50	2	–	–	–	–	–	–	–	–	–	–	D:A-Friedooleanan-3-one, 29-hydroxy-, (20α)-	39903-21-4	MS09793
424 191 43 425 219 205 69 55	442	100 95 33 32 32 29 26 26	9.90	30	50	2	–	–	–	–	–	–	–	–	–	–	D:A-Friedooleanan-7-one, 3-hydroxy-	56614-56-3	NI42942
209 205 290 303 191 442 245	442	100 58 37 30 30 26 24		30	50	2	–	–	–	–	–	–	–	–	–	–	D:A-Friedooleanan-7-one, 3-hydroxy-, (3α)-	21681-22-1	LI09007
204 424 442 341 189 177 123 139	442	100 87 81 67 62 62 51 47		30	50	2	–	–	–	–	–	–	–	–	–	–	1α-Hydroxyallobetulane		MS09794
424 442 204 177 341 136 139 189	442	100 95 82 72 64 57 54 53		30	50	2	–	–	–	–	–	–	–	–	–	–	1β-Hydroxyallobetulane		MS09795
442 424 189 123 177 205 134 149	442	100 99 98 85 84 80 80 79		30	50	2	–	–	–	–	–	–	–	–	–	–	2β-Hydroxyallobetulane		MS09796
189 121 207 109 135 149 442 203	442	100 98 97 91 87 73 71 67		30	50	2	–	–	–	–	–	–	–	–	–	–	3β-Hydroxyallobetulane		MS09797
442 384 410 427 412 424 271 253	442	100 60 38 22 18 13 12 10		30	50	2	–	–	–	–	–	–	–	–	–	–	(24R)-4β-Hydroxy-24-ethylcholesta-5,22-diene-3β-methyl ether		NI42943
291 273 205 191 162 255 424 395	442	100 82 70 55 42 25 15 15	9.00	30	50	2	–	–	–	–	–	–	–	–	–	–	3β-Hydroxyfriedelan-26-al		NI42944
43 253 285 227 387 255 213 199	442	100 32 14 14 10 9 9 7	6.00	30	50	2	–	–	–	–	–	–	–	–	–	–	(23R,24R,28S)-25-hydroxy-6β-methoxy-3α,5:23,28-dicyclostigmastane		MS09798
253 387 227 442 255 392 283 213	442	100 14 11 10 10 9 8 8		30	50	2	–	–	–	–	–	–	–	–	–	–	(23S,24S,28R)-25-hydroxy-6β-methoxy-3α,5:23,28-dicyclostigmastane		MS09799
189 207 424 206 381 190 442 409	442	100 79 67 47 43 35 32 32		30	50	2	–	–	–	–	–	–	–	–	–	–	3β-Hydroxy-16-oxotaraxane		NI42945
205 442 207 206 191 210 424 189	442	100 70 41 41 40 35 30 29		30	50	2	–	–	–	–	–	–	–	–	–	–	3-Oxo-16β-hydroxytaraxane		NI42946
219 442 95 69 303 137 81 83	442	100 61 51 45 42 27 26 24		30	50	2	–	–	–	–	–	–	–	–	–	–	Lanostane-3,11-dione	50764-37-9	NI42948
219 442 95 303 81 137 83 443	442	100 61 51 44 28 27 25 20		30	50	2	–	–	–	–	–	–	–	–	–	–	Lanostane-3,11-dione	50764-37-9	NI42947
289 290 442 75 329 109 121 107	442	100 21 8 7 4 4 3 3		30	50	2	–	–	–	–	–	–	–	–	–	–	Lanostane-3,12-dione	54411-48-2	NI42949
289 69 285 95 427 81 55 290	442	100 86 78 66 63 63 63 60	57.57	30	50	2	–	–	–	–	–	–	–	–	–	–	Lanostan-18-oic acid, 11-hydroxy-, γ-lactone, (11β)-	52474-95-0	NI42950
205 189 218 442 187 223 229 424	442	100 87 82 76 55 25 20 16		30	50	2	–	–	–	–	–	–	–	–	–	–	Lup-20(29)-ene-3,23-diol, (3α,4α)-	32451-85-7	MS09800
55 109 127 387 410 427 255 213	442	100 53 50 34 18 17 16 16	0.00	30	50	2	–	–	–	–	–	–	–	–	–	–	(24R)-6β-Methoxy-3α,5-cyclo-5α-stigmastan-22-one		DD01681
85 282 387 427 410 314 299 442	442	100 42 38 20 17 15 11 10		30	50	2	–	–	–	–	–	–	–	–	–	–	(24S)-6β-Methoxy-3α,5-cyclo-5α-stigmastan-23-one		DD01682
382 135 383 367 147 134 269 158	442	100 38 31 25 12 10 9 9	5.00	30	50	2	–	–	–	–	–	–	–	–	–	–	4β-Methyl-cholesteryl acetate		MS09801

MASS TO CHARGE RATIOS								M.W.	INTENSITIES								Parent	C	H	O	N	S	F	Cl	Br	I	Si	P	B	X	COMPOUND NAME	CAS Reg No	No
241	411	247	79	207	230	221	91	**442**	100	99	99	99	98	97	97	97	**73.11**	30	50	2	–	–	–	–	–	–	–	–	–	–	Myrtifolol		NI42951
234	219	235	208	209	442	220	441	**442**	100	25	19	16	9	6	6	3		30	50	2	–	–	–	–	–	–	–	–	–	–	Sophoradiol	6822-47-5	NI42952
43	57	41	44	55	32	400	386	**442**	100	99	67	65	63	52	48	48	**0.95**	31	54	1	–	–	–	–	–	–	–	–	–	–	Cholestan-3-one, 2-tert-butyl-, (2β,5α)-	55162-57-7	NI42953
221	204	442	252	443	441	232	351	**442**	100	30	29	25	11	7	7	4		32	30	–	2	–	–	–	–	–	–	–	–	–	N,N'-Diphenyl[2.2[4,7]]isoindolinophane	102343-35-1	NI42954
77	28	445	366	447	44	367	64	**443**	100	75	69	69	68	54	53	44	**24.06**	11	4	–	5	–	–	–	3	–	–	–	–	–	7-Bromo-4-cyano-2-dibromomethyl-1,3,6-triazacycl[3.3.3]azine		NI42955
80	82	28	64	79	81	83	366	**443**	100	98	62	42	37	36	30	27	**6.00**	11	4	–	5	–	–	–	3	–	–	–	–	–	9-Bromo-4-cyano-2-dibromomethyl-1,3,6-triazacycl[3.3.3]azine		NI42956
190	266	119	267	42	253	191	103	**443**	100	79	28	15	8	7	5	5	**0.30**	15	11	3	1	–	10	–	–	–	–	–	–	–	4-Hydroxy-N-methylphenylethylamine di-PFP		NI42957
177	133	119	262	58	280	91	117	**443**	100	48	35	31	31	26	20	19	**5.60**	15	11	3	1	–	10	–	–	–	–	–	–	–	3-(2-Hydroxyphenyl)propylamine di-PFP		MS09802
177	280	119	116	58	253	115	103	**443**	100	59	37	36	34	28	25	25	**14.80**	15	11	3	1	–	10	–	–	–	–	–	–	–	3-(3-Hydroxyphenyl)propylamine di-PFP		MS09803
280	177	119	253	133	58	443	116	**443**	100	94	50	43	39	37	26	25		15	11	3	1	–	10	–	–	–	–	–	–	–	3-(4-Hydroxyphenyl)propylamine di-PFP		MS09804
190	266	119	267	42	43	140	69	**443**	100	76	31	12	11	10	8	7	**0.00**	15	11	3	1	–	10	–	–	–	–	–	–	–	N-Methyltyramine di-PFP		MS09805
371	307	428	323	266	443	187	186	**443**	100	50	43	42	36	35	27	25		16	24	4	3	2	3	–	–	–	–	–	–	–	6-Diethylsulphamoyl-4,4-dioxo-1,3-diethyl-7-trifluoromethyl-1,3-diaza-4-thia-1,2,3,4-tetrahydrnaphthalene		MS09806
214	186	74	356	157	158	323	264	**443**	100	98	61	45	44	40	30	26	**9.50**	17	21	9	3	1	–	–	–	–	–	–	–	–	1,6-Dihydro-5-formyl-2-methylthio-4-(2,3,4-tri-O-acetyl-β-D-xylopyranosylamino)-1H-pyrimidin-6-one		MS09807
128	352	129	384	196	156	216	101	**443**	100	45	25	24	24	20	17	14	**7.00**	22	21	9	1	–	–	–	–	–	–	–	–	–	Tetramethyl 8-hydroxy-8,11-dihydro-azepino[1,2-a]quinolinetetracarboxylate		LI09008
305	292	306	123	307	294	95	293	**443**	100	79	70	50	40	29	21	16	**2.00**	22	22	2	1	–	4	1	–	–	–	–	–	–	Clofluperidol	10457-91-7	NI42958
306	201	307	202	77	47	51	28	**443**	100	55	21	16	14	5	4	4	**0.39**	23	27	4	1	–	–	–	–	–	–	2	–	–	Diethyl 1-diphenylphosphinamino-1-phenylmethanephosphonate		MS09808
105	77	205	211	443	133	104	283	**443**	100	86	85	84	83	83	83	67		25	21	5	3	–	–	–	–	–	–	–	–	–	1-Benzoyl-5-(methoxycarbonylamino)-5-phenyl-3-(p-tolyl)imidazolidine-2,4-dione		MS09809
133	77	41	69	57	39	94	107	**443**	100	55	39	37	28	24	23	22	**0.00**	25	33	6	1	–	–	–	–	–	–	–	–	–	2,6-Di-tert-butyl-4-methoxyphenyl-2-nitro-4-phenoxybutanoate		DD01683
186	299	144	314	229	215	257	116	**443**	100	40	32	24	18	16	14	9	**2.60**	25	37	4	3	–	–	–	–	–	–	–	–	–	L-Alanine, N-[N-[N-(3-phenyl-2-propenylidene)-1-valyl]-L-isoleucyl]-, ethyl ester	57174-05-7	NI42959
130	201	186	131	202	159	257	243	**443**	100	99	26	24	19	10	9	9	**3.61**	25	37	4	3	–	–	–	–	–	–	–	–	–	L-Tryptophan, N-[N-(1-oxodecyl)-β-alanyl]-, methyl ester	15939-57-8	NI42960
443	400	384	324	444	382	284	340	**443**	100	74	48	34	30	29	28	26		26	37	5	1	–	–	–	–	–	–	–	–	–	12-Acetyllucidusculine		MS09810
99	105	119	369	107	383	131	145	**443**	100	21	17	14	13	11	9	7	**2.00**	26	37	5	1	–	–	–	–	–	–	–	–	–	Pregn-5-ene-20-carbonitrile, 20-(acetyldioxy)-3,3-[1,2-ethanediylbis(oxy)]-	62623-52-3	NI42961
383	43	294	75	91	73	327	368	**443**	100	84	67	52	50	47	39	38	**4.00**	26	41	3	1	–	–	–	–	–	1	–	–	–	3-β-Acetoxy-pregna-5,16-diene-20-one, TMS oxime		LI09009
124	125	113	110	114	97	443		**443**	100	24	14	12	10	5	1			27	41	4	1	–	–	–	–	–	–	–	–	–	Edpetine		LI09010
384	443	229	444	288	230	385	43	**443**	100	99	49	42	41	39	38	32		29	49	–	1	1	–	–	–	–	–	–	–	–	Thiocyanic acid, 2α-methyl-5α-cholestan-3α-yl ester	20997-51-7	NI42962
43	55	443	57	69	81	41	45	**443**	100	97	95	92	86	76	73	68		29	49	–	1	1	–	–	–	–	–	–	–	–	Thiocyanic acid, 2α-methyl-5α-cholestan-3β-yl ester	20997-56-2	NI42963
43	104	57	91	55	41	443	103	**443**	100	83	47	46	43	42	37	35		29	49	–	1	1	–	–	–	–	–	–	–	–	Thiocyanic acid, 2α-methyl-5α-cholestan-3β-yl ester	20997-56-2	NI42964
443	384	288	95	230	229	81	55	**443**	100	73	48	47	41	40	40	39		29	49	–	1	1	–	–	–	–	–	–	–	–	Thiocyanic acid, 2β-methyl-5α-cholestan-3α-yl ester	20997-62-0	NI42965
69	41	43	55	81	57	443	95	**443**	100	95	87	76	72	66	52	50		29	49	–	1	1	–	–	–	–	–	–	–	–	Thiocyanic acid, 4α-methyl-5α-cholestan-3α-yl ester	20997-64-2	NI42966
443	288	289	43	44	81	229	69	**443**	100	80	66	47	42	41	39	36		29	49	–	1	1	–	–	–	–	–	–	–	–	Thiocyanic acid, 4α-methyl-5α-cholestan-3β-yl ester	20997-69-7	NI42967
443	410	382	414	444	411	428	383	**443**	100	85	68	35	34	27	21	21		31	25	–	1	1	–	–	–	–	–	–	–	–	Pyridine, 2-(ethylthio)-3,4,5,6-tetraphenyl-	74630-02-7	NI42968
84	85	71	443	43	69	57	55	**443**	100	6	4	3	3	2	2	2		31	57	–	1	–	–	–	–	–	–	–	–	–	Cholestan-3-amine, N,N,4,4-tetramethyl-, (3β,5α)-	54498-47-4	NI42969
88	276	76	266	176	182	56	444	**444**	100	38	27	26	21	19	19	17		8	–	6	–	2	4	–	–	–	–	–	–	2	μ-(π-Tetrafluoroethylenedithio)hexacarbonyldiiron		MS09811
169	69	101	85	170	72	444	154	**444**	100	98	97	54	51	48	40	23		10	8	2	–	2	6	–	–	–	–	–	–	1	Bis(1,1,1-trifluoro-4-thioxo-2-pentanonato)palladium	55029-41-9	NI42970
69	365	129	217	110	198	296	168	**444**	100	87	48	40	34	31	30	27	**14.60**	12	–	–	–	–	10	–	1	–	–	1	–	–	Bis(pentafluorophenyl)bromophosphine		MS09812
127	177	199	65	39	330	77	161	**444**	100	66	28	18	16	8	8	7	**0.02**	12	5	4	–	–	7	–	–	–	–	–	–	1	π-Cyclopentadienyl-heptafluorobutyryl-tricarbonyl molybdenum		MS09813
73	341	429	342	430	343	325	147	**444**	100	81	52	27	23	19	18	16	**0.00**	12	36	6	–	–	–	–	–	–	6	–	–	–	Cyclohexasiloxane, dodecamethyl-	540-97-6	NI42971
355	415	356	327	59	87	357	416	**444**	100	46	36	35	34	33	26	25	**1.09**	12	36	6	–	–	–	–	–	–	6	–	–	–	Cyclohexasiloxane, 1,3,5,7,9,11-hexaethyl-	17002-87-8	NI42972
28	261	360	259	362	358	304	302	**444**	100	22	20	19	18	17	17	17	**8.15**	13	8	6	–	–	–	–	–	–	–	–	–	1	Tungsten, pentacarbonyl(methoxyphenylmethylene)-	37823-96-4	NI42973
261	56	197	185	198	78	120	149	**444**	100	72	61	56	51	43	42	35	**0.83**	13	10	2	–	–	–	2	–	–	–	–	–	2	Dicarbonyl-π-cyclopentadienyliron-phenyltin(IV) dichloride		LI09011
73	147	267	355	429	341	59	281	**444**	100	60	46	38	35	29	28	26	**0.00**	13	40	5	–	–	–	–	–	–	6	–	–	–	1,1,1,5,7,7,7-Heptamethyl-3,3-bis(trimethylsiloxy)tetrasiloxane	38147-00-1	NI42974
360	362	334	358	332	336	359	333	**444**	100	92	85	67	67	59	49	32	**15.37**	14	9	3	–	–	–	1	–	–	–	–	–	1	Tungsten, cyclopentadienyl-(4-chlorophenyl)tricarbonyl		NI42975
43	446	444	209	140	448	44	167	**444**	100	98	75	64	55	49	45	32		17	12	2	4	–	–	4	–	–	–	–	–	–	1-(2,4,6-Trichlorophenyl)-3-(2-chloro-4-acetylamino-anilino)-4,5-dihydropyrazol-5-one		IC04425
444	202	44	174	261	146	28	445	**444**	100	75	52	45	41	31	30	23		18	12	10	4	–	–	–	–	–	–	–	–	–	6-[[(2,4-Dinitrophenyl)hydrazono]methyl]-7,8-dihydroxy-3-(methoxycarbonyl)isocoumarin		NI42976

MASS TO CHARGE RATIOS	M.W.	INTENSITIES	Parent	C	H	O	N	S	F	Cl	Br	I	Si	P	B	X	COMPOUND NAME	CAS Reg No	No
332 276 304 360 112 388 56 138	444	100 83 74 53 50 49 42 26	26.00	18	20	6	–	–	–	–	–	–	–	–	–	2	Iron, hexacarbonyl[μ-(1,2,3,4-η:1,4-η)-1,2,3,4-tetraethyl-1,3-butadiene-1,4 diyl]di-	33310-07-5	NI42977
42 294 119 307 151 67 69 169	444	100 83 75 71 63 52 46 38	3.00	19	17	5	6	–	–	1	–	–	–	–	–	–	1-(Anilinocarbonylethoxy)-3-methyl-4-(2-chloro-4-nitrophenylazo)-5-hydroxypyrazole		IC04426
87 43 163 195 90 222 133 117	444	100 66 64 62 59 57 46 40	5.00	19	20	5	6	1	–	–	–	–	–	–	–	–	N,N-Bis-(2-hydroxyethyl)-3-acetamido-4-(5-nitro-2,1-benzisothiaz-3-ylazo)-aniline		IC04427
202 203 444 204 445 446 205 73	444	100 19 7 5 2 1 1 1		19	27	3	2	–	3	–	–	–	2	–	–	–	L-Tryptophan, N-(trifluoroacetyl)-1-(trimethysilyl)-, trimethylsilyl ester	52558-11-9	NI42978
444 58 57 445 429 56 77 51	444	100 74 57 30 30 30 26 26		20	11	3	–	–	6	–	–	–	–	1	–	–	5H-Dibenzophosphole, 1,2,4,6,8,9-hexafluoro-3,7-dimethoxy-5-phenyl-, 5-oxide	36357-20-7	NI42979
73 81 135 114 93 107 121 280	444	100 80 24 24 18 12 8 5	1.00	20	29	5	–	–	5	–	–	–	–	–	–	–	Methyl 11-methoxy-3,7,11-trimethyl-10-(pentafluoropropionyloxy)-2,6-dodecadienoate		MS09814
57 278 43 41 334 29 390 276	444	100 61 48 48 41 35 32 32	7.80	21	34	–	–	–	–	–	2	–	–	–	–	–	Benzene, 2,4-dibromo-1,3,5-tris(2,2-dimethylpropyl)-	25347-06-2	MS09815
46 100 36 58 444 201 72 187	444	100 96 93 59 23 17 12 10		22	24	8	2	–	–	–	–	–	–	–	–	–	Doxycycline	564-25-0	NI42980
84 58 44 46 98 444 42 426	444	100 99 61 56 26 22 22 21		22	24	8	2	–	–	–	–	–	–	–	–	–	2-Naphthacenecarboxamide, 4-(dimethylamino)-1,4,4a,5,5a,6,11,12a-octahydro-3,6,10,12,12a-pentahydroxy-6-methyl-1,11-dioxo-, [4S-(4α,4aα,5aα,6β,12aα)]-	60-54-8	NI42981
385 444 386 326 413 353 267 169	444	100 26 23 18 10 8 6 5		22	24	8	2	–	–	–	–	–	–	–	–	–	Tetramethyl-5-ethyl-9,10-dihydro-5H-azepino[1,2-a]benzimidazole-7,8,9,10-tetracarboxylate		LI09012
343 444 344 299 326 268 354 311	444	100 34 22 13 10 8 7 7		22	24	8	2	–	–	–	–	–	–	–	–	–	Tetramethyl-6-methyl-5,6,10,11-tetrahydroazepino[1,2-a]quinoxaline-7,8,9,11-tetracarboxylate		LI09013
202 247 248 201 429 218 220 232	444	100 94 84 75 72 44 37 22	0.00	22	28	6	4	–	–	–	–	–	–	–	–	–	1-(2′,4′-Dinitroanilino)-2,4,4,7-tetramethyl-3-ethoxycarbonyl-1,4,5,6,7,8-hexahydroquinoline		LI09014
235 233 215 217 216 189 232 191	444	100 55 51 43 27 24 20 19	0.00	22	33	7	–	–	–	1	–	–	–	–	–	–	4α-Acetoxy-3α-(2′-hydroxy-2′-methyl-3′-chlorobutyryloxy)-11-hydroxy-6,7-dehydroeudesman-8-one		NI42982
369 235 43 233 371 217 215 370	444	100 86 82 37 36 30 22 20	0.00	22	33	7	–	–	–	1	–	–	–	–	–	–	4α-Acetoxy-3α-(2′-hydroxy-2′-methyl-3′-chlorobutyryloxy)-11-hydroxy-6,7-dehydroeudesman-8-one		NI42983
70 444 387 57 58 44 71 43	444	100 47 42 40 37 35 22 22		23	28	7	2	–	–	–	–	–	–	–	–	–	Dichotine, 11-methoxy-	29537-51-7	NI42984
444 384 385 325 324 412 353 283	444	100 73 39 27 19 14 11 10		24	28	8	–	–	–	–	–	–	–	–	–	–	Dimethyl rel-(1R,1′S,2R,4R,5S,8′R,9′R)-2,8′-dimethyl-3,7,11-trioxospiro[bicyclo[3.3.0]octane-4,5′-[6]oxatricyclo[7.3.0.0^{2,7}]dodec[2(7)]ene]-2,8′-dicarboxylate		MS09816
415 69 416 166 444 135 164 178	444	100 35 25 25 21 21 14 13		24	28	8	–	–	–	–	–	–	–	–	–	–	Flavone, 3′,5,6-triethoxy-3,4′,7-trimethoxy-	14481-48-2	NI42985
429 444 69 430 151 166 135 371	444	100 36 36 29 28 25 23 16		24	28	8	–	–	–	–	–	–	–	–	–	–	Flavone, 4′,5,7-triethoxy-3,3′,6-trimethoxy-	14397-67-2	NI42986
444 156 258 232 445 429 327 73	444	100 51 44 39 38 27 27 27		24	36	4	–	–	–	–	–	–	2	–	–	–	11β-Hydroxy-6-oxoestrone di-TMS		NI42987
135 82 272 93 265 121 136 79	444	100 33 31 25 24 23 22 22	12.01	24	36	4	4	–	–	–	–	–	–	–	–	–	L-Leucine, N-[N-(tricyclo[3.3.1.1$^{(3,7)}$]dec-2-ylcarbonyl)-L-histidyl]-, methyl ester	55401-52-0	NI42988
273 141 57 71 43 73 429 215	444	100 72 43 32 32 29 25 23	0.00	24	48	5	–	–	–	–	–	–	1	–	–	–	1,3-Dinonanoin, trimethylsilyl-		NI42989
217 73 71 75 41 45 115 55	444	100 87 63 56 35 33 32 29	0.00	24	48	5	–	–	–	–	–	–	1	–	–	–	Methyl (11R,12R,13R)-(Z)-12-trimethylsilyloxy-11,13-dimethoxy-9-octadecenoate	83303-85-9	NI42990
173 73 174 240 103 55 89 83	444	100 59 14 11 11 9 8 8	0.00	24	48	5	–	–	–	–	–	–	1	–	–	–	Methyl (12S,13S)-(E)-13-trimethylsilyloxy-9,12-dimethoxy-10-octadecenoate	88303-86-0	NI42991
201 73 71 45 137 169 173 95	444	100 87 57 46 37 29 28 27	0.00	24	48	5	–	–	–	–	–	–	1	–	–	–	Methyl (13S)-(E)-13-trimethylsilyloxy-9,10-dimethoxy-11-octadecenoate	88303-87-1	NI42992
239 224 73 240 28 225 45 167	444	100 64 63 20 14 12 11 7	0.00	24	52	3	–	–	–	–	–	–	2	–	–	–	Octadecanoic acid, 2-(trimethylsiloxy)-, trimethylsilyl ester	74367-66-1	NI42993
73 147 233 43 429 75 313 41	444	100 49 49 22 18 17 16 11	0.00	24	52	3	–	–	–	–	–	–	2	–	–	–	Octadecanoic acid, 3-(trimethylsiloxy)-, trimethylsilyl ester		NI42994
73 187 359 75 129 360 55 188	444	100 77 58 57 23 18 15 13	0.00	24	52	3	–	–	–	–	–	–	2	–	–	–	Octadecanoic acid, 12-(trimethylsiloxy)-, trimethylsilyl ester	22396-22-1	NI42995
73 187 75 359 55 129 69 188	444	100 61 59 24 21 16 11 10	0.00	24	52	3	–	–	–	–	–	–	2	–	–	–	Octadecanoic acid, 12-(trimethylsiloxy)-, trimethylsilyl ester	22396-22-1	NI42996
444 415 445 417 105 77 213 371	444	100 55 28 16 16 13 11 9		25	16	8	–	–	–	–	–	–	–	–	–	–	3-Carbethoxy-5,7,12,14-tetrahydroxy-6,13-dioxopentacene		IC04428
150 182 206 122 238 163 178 107	444	100 99 98 94 91 88 86 52	0.00	25	32	7	–	–	–	–	–	–	–	–	–	–	Benzoic acid, 2,4-dihydroxy-6-pentyl-, 3-hydroxy-4-(methoxycarbonyl)-5-pentylphenyl ester	64185-31-5	NI42997
138 124 66 137 41 177 182 69	444	100 48 30 27 27 25 22 21	0.00	25	32	7	–	–	–	–	–	–	–	–	–	–	Benzoic acid, 2-hydroxy-4-[(4-hydroxy-2-methoxy-6-pentylbenzoyl)oxy]-6 pentyl-	19314-66-0	NI42998
300 43 360 231 149 402 444	444	100 84 58 33 28 25 12		25	32	7	–	–	–	–	–	–	–	–	–	–	7-Oxo-carnosic acid diacetate		MS09817
402 403 252 444 145 173 253 404	444	100 53 33 17 13 12 11 10		25	36	5	–	–	–	–	–	–	1	–	–	–	15β-Hydroxyestradiol 3,17β-diacetate TMS		NI42999
78 140 111 109 429 445 110 123	444	100 30 25 20 15 14 14 13	0.00	25	36	5	2	–	–	–	–	–	–	–	–	–	4,25-Secoobscurinervan-4-ol, 22-ethyl-15,16-dimethoxy-, (4β,22α)-	54658-11-6	NI43000
384 203 135 341 107 93 79 95	444	100 85 84 77 74 70 68 60	6.42	26	36	6	–	–	–	–	–	–	–	–	–	–	Bufotalin	471-95-4	NI43002
384 203 135 341 323 178 147 118	444	100 85 84 78 60 44 43 42	6.00	26	36	6	–	–	–	–	–	–	–	–	–	–	Bufotalin	471-95-4	NI43001
384 203 135 341 107 93 79 323	444	100 84 83 78 74 71 68 60	7.00	26	36	6	–	–	–	–	–	–	–	–	–	–	Bufotalin	471-95-4	LI09015

MASS TO CHARGE RATIOS								M.W.	INTENSITIES								Parent	C	H	O	N	S	F	Cl	Br	I	Si	P	B	X	COMPOUND NAME	CAS Reg No	No
136	137	121	80	69	110	109	55	444	100	91	91	57	57	44	44	39	31.30	26	36	6	–	–	–	–	–	–	–	–	–	–	6-Hexyl-2,3,11,12-dibenzo-1,4,7,10,13,16-hexaoxacyclooctadeca-2,11-diene		MS09818
122	77	93	121	123	79	177	105	444	100	98	74	73	57	56	46	45	0.76	26	36	6	–	–	–	–	–	–	–	–	–	–	16-Hydroxy-ingenol-3,4,5,20-diacetonide		NI43003
121	122	147	43	91	225	41	135	444	100	99	39	39	35	30	27	24	0.00	26	36	6	–	–	–	–	–	–	–	–	–	–	Pregna-1,4-diene-3,20-dione, 21-(2,2-dimethyl-1-oxopropoxy)-11,17-dihydroxy-, (11β)-	1107-99-9	NI43004
429	339	196	209	214	199	354	430	444	100	54	54	49	46	39	38	34	7.49	26	44	2	–	–	–	–	–	–	2	–	–	–	17α-Ethynyl-5(10)-estrene-3α,17β-diol di-TMS		NI43005
429	339	196	430	214	209	199	354	444	100	53	46	41	38	38	34	27	8.03	26	44	2	–	–	–	–	–	–	2	–	–	–	17α-Ethynyl-5(10)-estrene-3β,17β-diol di-TMS		NI43006
105	252	224	293	384	353	323	444	444	100	42	26	4	3	3	2	2		27	28	4	2	–	–	–	–	–	–	–	–	–	Anomalamide		MS09819
105	252	77	224	91	293	384	131	444	100	32	22	17	15	14	12	6	3.00	27	28	4	2	–	–	–	–	–	–	–	–	–	Aurantiamide acetate		NI43007
312	175	283	70	314	444	84	77	444	100	54	44	38	34	30	30	28		27	29	–	4	–	–	1	–	–	–	–	–	–	3-Phenyl-5-chloro-1-[4-[4-(1-phenylpiperazin)-butyl]]-indazole		MS09820
121	229	147	173	227	384	230	161	444	100	78	30	20	18	17	17	16	14.10	27	40	5	–	–	–	–	–	–	–	–	–	–	Chol-9(11)-en-24-oic acid, 3-(acetyloxy)-12-oxo-, methyl ester, (3α,5β)-	4472-02-0	NI43009
121	229	444	43	81	55	230	147	444	100	84	31	24	21	18	16	16		27	40	5	–	–	–	–	–	–	–	–	–	–	Chol-9(11)-en-24-oic acid, 3-(acetyloxy)-12-oxo-, methyl ester, (3α,5β)-	4472-02-0	NI43008
171	359	145	444	142	105	85	99	444	100	90	73	67	60	58	55	50		27	40	5	–	–	–	–	–	–	–	–	–	–	14β,15-Dehydro-anosmagenin	40922-10-9	NI43010
105	77	192	103	416	178	51	89	444	100	28	9	7	6	6	5	2	0.50	28	20	2	4	–	–	–	–	–	–	–	–	–	4,5-Diphenyl-1-(α-benzoyloxy-benzylidene-imino)-1,2,3-triazole		MS09821
324	325	281	56	57	41	266	105	444	100	25	11	10	8	8	7	7	0.00	28	32	3	2	–	–	–	–	–	–	–	–	–	Aspidospermidin-21-one, 1-acetyl-17-methoxy-21-phenyl-	36455-19-3	NI43011
444	429	282	57	443	297	296	373	444	100	83	48	35	25	25	25	23		28	32	3	2	–	–	–	–	–	–	–	–	–	(E,Z)-4-(5,7-Di-tert-butyl-3,3-dimethyl-2-indolinylidene)-1-phenyl-2,3,5-pyrrolidinetrione		DD01684
55	41	43	69	70	135	95	81	444	100	87	85	71	63	52	46	44	3.00	28	44	4	–	–	–	–	–	–	–	–	–	–	Ergost-25-ene-3,6-dione, 5,12-dihydroxy-, (5α,12β)-	56052-99-4	NI43012
125	55	69	81	95	70	109	67	444	100	96	79	55	49	41	39	37	0.00	28	44	4	–	–	–	–	–	–	–	–	–	–	Ergost-25-ene-6,12-dione, 3,5-dihydroxy-, (3β,5α)-	56052-98-3	NI43013
55	338	129	41	43	57	105	69	444	100	99	84	77	68	62	59	52	4.24	28	44	4	–	–	–	–	–	–	–	–	–	–	9-Octadecenoic acid, 2-phenyl-1,3-dioxan-5-yl ester	16106-28-8	NI43014
73	45	43	105	28	55	57	29	444	100	84	67	42	39	30	29	29	0.00	28	44	4	–	–	–	–	–	–	–	–	–	–	9-Octadecenoic acid, (2-phenyl-1,3-dioxolan-4-yl)methyl ester, cis-	56599-45-2	NI43016
43	55	57	41	105	69	28	29	444	100	90	63	61	49	49	49	41	0.00	28	44	4	–	–	–	–	–	–	–	–	–	–	9-Octadecenoic acid, (2-phenyl-1,3-dioxolan-4-yl)methyl ester, cis-	56599-45-2	NI43015
91	43	105	73	28	45	107	29	444	100	52	40	37	31	30	26	26	0.00	28	44	4	–	–	–	–	–	–	–	–	–	–	9-Octadecenoic acid, (2-phenyl-1,3-dioxolan-4-yl)methyl ester, trans-	56599-46-3	NI43017
367	385	386	365	383	427	207	368	444	100	70	22	21	20	19	19	18	3.50	29	48	3	–	–	–	–	–	–	–	–	–	–	3β-Acetoxy-5,6-epoxycholestane		NI43018
113	87	99	86	444	125	85	81	444	100	99	98	78	61	52	45	39		29	48	3	–	–	–	–	–	–	–	–	–	–	Cholestane-1,3-dione, cyclic 3-(1,2-ethanediyl acetal), (5β)-	56247-67-7	NI43019
371	444	384	353	402				444	100	78	60	25	9					29	48	3	–	–	–	–	–	–	–	–	–	–	Cholestan-19-ol, 2,3-epoxy-, acetate, (2α,3α,5α)-	30993-72-7	LI09016
371	444	384	445	353	402			444	100	78	60	25	25	9				29	48	3	–	–	–	–	–	–	–	–	–	–	Cholestan-19-ol, 2,3-epoxy-, acetate, (2α,3α,5α)-	30993-72-7	NI43020
384	385	369	55	43	57	41	95	444	100	30	13	13	12	10	10	9	1.00	29	48	3	–	–	–	–	–	–	–	–	–	–	Cholestan-6-one, 3-(acetyloxy)-, (3α,5α)-	21513-83-7	NI43022
43	44	369	384	41	55	95	81	444	100	69	66	59	54	50	35	35	0.00	29	48	3	–	–	–	–	–	–	–	–	–	–	Cholestan-6-one, 3-(acetyloxy)-, (3α,5α)-	21513-83-7	NI43021
384	95	43	93	55	69	81	57	444	100	81	73	59	59	49	45	41	37.00	29	48	3	–	–	–	–	–	–	–	–	–	–	Cholestan-6-one, 3-(acetyloxy)-, (3β,5α)-	1256-83-3	NI43023
384	43	369	136	41	55	95	79	444	100	86	69	49	49	48	39	37	0.00	29	48	3	–	–	–	–	–	–	–	–	–	–	Cholestan-6-one, 3-(acetyloxy)-, (3β,5α)-	1256-83-3	NI43024
43	444	236	41	56	445	93	290	444	100	84	51	28	27	26	25	24		29	48	3	–	–	–	–	–	–	–	–	–	–	Cholestan-7-one, 3-(acetyloxy)-, (3β)-	52993-63-2	NI43025
43	444	236	41	55	445	93	289	444	100	85	50	29	28	26	25	24		29	48	3	–	–	–	–	–	–	–	–	–	–	Cholestan-7-one, 3-(acetyloxy)-, (3β,5α)-	6038-71-7	NI43026
384	444	371	385	445	372	402	403	444	100	50	50	28	15	14	8	2		29	48	3	–	–	–	–	–	–	–	–	–	–	5α-Cholestan-2-one, 19-hydroxy-, acetate	28809-68-9	NI43027
353	354	366	212	197	352	384	355	444	100	50	14	14	14	13	12	12	0.00	29	48	3	–	–	–	–	–	–	–	–	–	–	Cholest-5-ene-3,19-diol, 3-acetate, (3β)-	750-59-4	LI09017
353	354	366	212	197	355	352	384	444	100	50	15	15	15	14	14	13	0.00	29	48	3	–	–	–	–	–	–	–	–	–	–	Cholest-5-ene-3,19-diol, 3-acetate, (3β)-	750-59-4	NI43028
366	69	81	43	129	367	55	95	444	100	60	40	39	35	32	32	29	0.00	29	48	3	–	–	–	–	–	–	–	–	–	–	Cholest-5-ene-3,20-diol, 3-acetate, (3β,20R)-	7429-99-4	NI43029
105	390	408	134	251	87	426	152	444	100	90	79	71	48	41	33	26	3.45	29	48	3	–	–	–	–	–	–	–	–	–	–	1α,25-Dihydroxy-26,27-dimethylvitamin D3		MS09822
134	105	87	251	152	390	116	408	444	100	58	40	26	25	18	17	13	3.42	29	48	3	–	–	–	–	–	–	–	–	–	–	5,6-Trans-1α,25-dihydroxy-26,27-dimethylvitamin D3		MS09823
389	412	444	429	255	390	386	297	444	100	72	62	57	33	28	26	25		29	48	3	–	–	–	–	–	–	–	–	–	–	Ethyl 23ξ-ethyl-6β-methoxy-3α,5-cyclocholanate		MS09824
444	121	81	43	445	55	291	41	444	100	51	41	40	34	34	33	25		29	48	3	–	–	–	–	–	–	–	–	–	–	Pregnan-12-one, 3-(acetyloxy)-20-hexyl-, (3α)-	57984-01-7	NI43030
373	374	72	358	429	443	444	430	444	100	28	27	19	12	11	6	6		29	52	1	2	–	–	–	–	–	–	–	–	–	Pregnan-4-ol, 20-(dimethylamino)-3-[3-isopropyl-1-azetidinyl]-, (3β,4β)-	56687-72-0	NI43031
288	444	289	287	274	258	443	222	444	100	43	25	22	19	14	13	13		30	24	2	2	–	–	–	–	–	–	–	–	–	N,N′-Bis(O-hydroxynaphthylidene)-4,5-dimethyl-O-phenylenediamine	103110-79-8	NI43032
325	310	119	205	91	324	191	203	444	100	90	86	59	52	39	31	28	3.00	30	24	2	2	–	–	–	–	–	–	–	–	–	4-Methyl-6,6-diphenyl-2-(N-phenylcarbamoyl)-2-azatricyclo[5.2.2.0^{1,5}]undeca-4,8,10-trien-3-one		MS09825
444	443	445	222	91	338	442	65	444	100	74	30	19	13	10	8	7		30	28	–	4	–	–	–	–	–	–	–	–	–	3-p-Tolyl-2-p-tolylamino-4-p-tolylimino-quinazoline		MS09826
105	426	77	427	339	106	161	57	444	100	39	18	13	11	8	7	7	7.00	30	36	3	–	–	–	–	–	–	–	–	–	–	Bis(2-hydroxy-3-tert-butyl-5-methylphenyl)methane, monobenzoate		NI43033
229	384	55	95	43	81	369	121	444	100	97	68	66	66	61	60	58	13.00	30	52	2	–	–	–	–	–	–	–	–	–	–	5α-Cholestan-3α-ol, 2α-methyl-, acetate	20997-52-8	NI43034
384	83	85	47	48	49	43	385	444	100	83	77	73	51	43	32	31	23.00	30	52	2	–	–	–	–	–	–	–	–	–	–	5α-Cholestan-3α-ol, 4α-methyl-, acetate	20997-65-3	NI43035
444	414	371	261	412	426	386	276	444	100	34	22	11	10	8	8	8		30	52	2	–	–	–	–	–	–	–	–	–	–	Cholestan-4-one, 24-ethyl-3-methoxy-, (3β,24S)-		NI43036
99	100	43	41	55	444	95	57	444	100	16	10	7	6	5	4	4		30	52	2	–	–	–	–	–	–	–	–	–	–	Cholestan-3-one, 4-methyl-, cyclic 1,2-ethanediyl acetal, (4α,5α)-	54498-66-7	NI43037
59	369	368	57	55	95	81	43	444	100	33	32	15	15	14	13	13	0.02	30	52	2	–	–	–	–	–	–	–	–	–	–	Cholest-5-ene, 3-(2-methoxyethoxy)-, (3β)-	57156-77-1	NI43038
95	69	55	109	81	107	121	59	444	100	86	75	70	68	60	52	51	7.00	30	52	2	–	–	–	–	–	–	–	–	–	–	9,19-Cyclo-9β-lanostane-3β,25-diol	26525-84-8	NI43039
207	189	204	190	191	426	444	217	444	100	86	41	40	33	19	15	14		30	52	2	–	–	–	–	–	–	–	–	–	–	3β,16α-Dihydroxytaraxane		NI43040
207	189	190	191	426	218	444	231	444	100	66	40	30	25	19	15	14		30	52	2	–	–	–	–	–	–	–	–	–	–	3β,16β-Dihydroxytaraxane		NI43041

MASS TO CHARGE RATIOS								M.W.	INTENSITIES								Parent	C	H	O	N	S	F	Cl	Br	I	Si	P	B	X	COMPOUND NAME	CAS Reg No	No
207	189	136	191	234	204	275	426	**444**	100	87	65	45	25	24	23	22	**17.00**	30	52	2	–	–	–	–	–	–	–	–	–	–	29,30-Dinorgammacerane-3,22-diol, 21,21-dimethyl-, (3β,8α,9β,13α,14β,17α,18β,22α)-	21688-72-2	NI43042
215	384	444	216	43	276	275	68	**444**	100	99	97	76	76	69	52	49		30	52	2	–	–	–	–	–	–	–	–	–	–	Ergostan-3-ol, acetate, (3β,5α)-	4356-09-6	NI43043
413	109	95	395	205	199	179	81	**444**	100	72	62	61	60	59	44	36	**0.00**	30	52	2	–	–	–	–	–	–	–	–	–	–	Friedelane-3β,26-diol		NI43044
95	109	413	395	123	121	243	205	**444**	100	75	50	50	50	40	35	35	**1.00**	30	52	2	–	–	–	–	–	–	–	–	–	–	D:A-Friedooleanane-1,24-diol	56816-62-7	MS09827
141	444	231	233	220	236	221	217	**444**	100	77	77	35	35	31	27	27		30	52	2	–	–	–	–	–	–	–	–	–	–	D:A-Friedooleanane-3,29-diol, (3α,20α)-	56816-11-6	MS09828
43	95	55	413	69	444	273	290	**444**	100	94	77	74	70	65	63	50		30	52	2	–	–	–	–	–	–	–	–	–	–	Lanostan-3β-ol, 11β,18-epoxy-		LI09018
444	43	445	69	55	95	81	429	**444**	100	43	36	36	32	31	28	27		30	52	2	–	–	–	–	–	–	–	–	–	–	Lanostan-3β-ol, 11β,18-epoxy-	25116-67-0	NI43045
43	95	56	413	69	444	273	41	**444**	100	91	76	72	72	67	63	62		30	52	2	–	–	–	–	–	–	–	–	–	–	Lanostan-3β-ol, 11β,19-epoxy-	22417-94-3	NI43046
444	43	69	55	445	95	81	429	**444**	100	45	38	38	33	31	30	29		30	52	2	–	–	–	–	–	–	–	–	–	–	Lanostan-3β-ol, 11β,19-epoxy-		LI09019
221	444	95	69	303	81	84	109	**444**	100	57	55	55	42	33	28	26		30	52	2	–	–	–	–	–	–	–	–	–	–	Lanostan-11-one, 3-hydroxy-, (3β)-	2130-15-6	NI43047
411	444	123	304	109	412	289	445	**444**	100	60	50	40	40	30	25	20		30	52	2	–	–	–	–	–	–	–	–	–	–	5α-Lanost-7-ene-3β,11α-diol		MS09829
426	191	177	237	411	218	205	408	**444**	100	60	25	24	22	18	16	14	**10.00**	30	52	2	–	–	–	–	–	–	–	–	–	–	Lupane-3,6-diol, (3β,6α)-	71298-20-9	NI43048
69	55	389	284	429	412	444	301	**444**	100	92	77	52	40	29	25	24		30	52	2	–	–	–	–	–	–	–	–	–	–	(22R,24R)-6β-Methoxy-3α,5-cyclo-5α-stigmastan-22-ol		DD01685
69	55	283	373	253	323	394	429	**444**	100	73	46	41	33	26	9	4	**0.00**	30	52	2	–	–	–	–	–	–	–	–	–	–	(23S,24R)-23-Methyl-6β-methoxy-3α,5-cyclo-5α-ergostan-23-ol		DD01686
207	189	149	109	444	191	190	121	**444**	100	91	73	39	36	36	36	30		30	52	2	–	–	–	–	–	–	–	–	–	–	Zeorin		LI09020
311	91	246	353	220	218	310	367	**444**	100	38	28	23	21	12	10	9	**8.00**	31	28	1	2	–	–	–	–	–	–	–	–	–	7-Methyl-3,3-dibenzyl-3H-indole-2-2′-methyl-carboxanilide		MS09830
105	207	77	321	221	223	225	426	**444**	100	52	50	40	40	37	35	21	**2.00**	32	28	2	–	–	–	–	–	–	–	–	–	–	2-Hydroxy-6-methyl-2,4,6-triphenyl-cyclohex-3-enyl phenyl ketone		LI09021
105	222	77	221	325	444			**444**	100	27	23	12	10	4				32	28	2	–	–	–	–	–	–	–	–	–	–	5-Methyl-1,3,5,7-tetraphenyl-hept-2-ene-1,7-dione		LI09022
278	320	276	318	322	280	43	240	**445**	100	91	78	74	48	48	48	17	**0.20**	3	7	–	3	–	–	4	–	1	–	3	–	–	1-Isopropyl-1-iodotetrachlorocyclotriphosphazene	77589-34-5	NI43049
320	318	322	278	276	324	240	242	**445**	100	77	49	16	12	11	9	9	**0.01**	3	7	–	3	–	–	4	–	1	–	3	–	–	1-Propyl-1-iodotetrachlorocyclotriphosphazene		NI43050
345	131	69	100	176	426	376	181	**445**	100	77	75	55	53	46	22	14	**0.50**	9	–	–	1	–	17	–	–	–	–	–	–	–	Perfluoro-(N-cyclobutylpiperidine)		LI09023
100	133	69	426	119	169	445	181	**445**	100	50	36	25	11	9	6	5		9	–	–	1	–	17	–	–	–	–	–	–	–	trans-Perfluoroquinolizidine		DD01687
282	136	445	162	283	119	135	281	**445**	100	85	73	68	66	50	37	36		14	9	4	1	–	10	–	–	–	–	–	–	–	2-Hydroxy-3-methoxybenzylamine di-PFP		MS09831
445	283	119	430	298	107	282	446	**445**	100	70	28	26	26	25	19	17		14	9	4	1	–	10	–	–	–	–	–	–	–	3-Hydroxy-4-methoxybenzylamine di-PFP		MS09832
332	317	73	203	216	430	333	318	**445**	100	75	60	56	31	26	21	15	**1.00**	16	17	5	1	–	6	–	–	–	1	–	–	–	L-Tyrosine, N-(trifluoroacetyl)-, trimethylsilyl ester, trifluoroacetate (ester)	52558-85-7	NI43051
87	98	217	185	58	45	156	145	**445**	100	60	28	22	16	16	11	10	**3.33**	16	19	–	3	6	–	–	–	–	–	–	–	–	2,4,6,8,9,10-Hexathiatricyclo[3.3.1.1^{3,7}]decane-1,3,5-tripropanenitrile, 7-propyl-	57274-51-8	NI43052
177	219	211	73	43	254	178	344	**445**	100	70	62	57	45	41	38	34	**11.01**	18	32	6	1	–	–	–	–	–	2	1	–	–	Phosphonic acid, [1-(acetylamino)-2-[4-(acetyloxy)phenyl]ethyl]-, bis(trimethylsilyl) ester	56196-76-0	NI43053
158	315	258	202	342	283	343	201	**445**	100	23	19	16	14	12	10	9	**0.50**	20	35	8	3	–	–	–	–	–	–	–	–	–	1-Isobutoxy-permethyl-γ-glutamyl-alanyl-glycine		LI09024
177	176	211	195	187	178	331	237	**445**	100	50	22	22	22	20	18	18	**0.00**	21	18	3	5	–	3	–	–	–	–	–	–	–	N^5-Methyl-6,7-diphenyl-5,8-dihydropterin trifluoroacetate		MS09833
58	234	44	59	427	42	41	36	**445**	100	11	8	5	4	4	4	4	**0.20**	23	27	8	1	–	–	–	–	–	–	–	–	–	Benzoic acid, 6-[[6-[2-(dimethylamino)ethyl]-4-methoxy-1,3-benzodioxol-5-yl]acetyl]-2,3-dimethoxy-	131-28-2	NI43054
58	355	179	42	192	178	206	77	**445**	100	53	44	38	36	31	30	29	**0.00**	23	27	8	1	–	–	–	–	–	–	–	–	–	Benzoic acid, 6-[[6-[2-(dimethylamino)ethyl]-4-methoxy-1,3-benzodioxol-5-yl]acetyl]-2,3-dimethoxy-	131-28-2	NI43055
43	42	70	55	56	246	185	87	**445**	100	93	89	66	64	61	57	57	**25.02**	23	28	2	3	1	–	1	–	–	–	–	–	–	1-Piperazineethanol, 4-[3-(2-chloro-10H-phenothiazin-10-yl)propyl]-, acetate (ester)	84-06-0	NI43056
246	185	445	70	125	154	213	87	**445**	100	94	82	74	66	60	56	51		23	28	2	3	1	–	1	–	–	–	–	–	–	1-Piperazineethanol, 4-[3-(2-chloro-10H-phenothiazin-10-yl)propyl]-, acetate (ester)	84-06-0	NI43057
57	41	120	29	39	43	135	92	**445**	100	67	42	31	26	18	16	16	**0.00**	23	43	1	1	3	–	–	–	–	–	–	–	–	1-Acetyl-4-tert-butyl-2,3,6-tri-tert-butylmercapto-1,2,3,6-tetrahydropyridine		MS09834
132	79	93	52	73	95	51	342	**445**	100	82	72	55	51	26	23	20	**0.00**	24	55	2	1	–	–	–	–	–	2	–	–	–	1,3-Octadecanediol, 2-amino-, bis(trimethylsilyl)-, R-(R*,S*)-	72088-28-9	NI43058
386	370	445	428	387	429	371	446	**445**	100	80	39	27	27	20	19	10		25	35	6	1	–	–	–	–	–	–	–	–	–	Yesoxine		MS09835
69	105	77	158	291	263	417	445	**445**	100	51	48	27	14	12	5	4		27	31	3	3	–	–	–	–	–	–	–	–	–	1,2,3,7a-Tetrahydro-1,3-diisopropyl-2-(isopropylimino)-4,7a-diphenylfuro[2,3-D]pyrimidin-5,6-dione	90140-46-8	NI43059
372	373	356	185.5	186	29	340	177.5	**445**	100	30	22	12	10	9	8	7	**4.00**	28	35	2	3	–	–	–	–	–	–	–	–	–	Tris(4-dimethylaminophenyl)carbethoxymethane		IC04429
209	165	104	29	91	77			**445**	100	48	43	42	35	27			**18.49**	30	27	1	3	–	–	–	–	–	–	–	–	–	Ethanamine, N-[[4-[[4-[2-(4-methoxyphenyl)vinyl]phenyl]azo]phenyl]benzylene]-	55937-86-5	NI43060
376	411	306	341	135.5	329	446	271	**446**	100	99	71	38	22	21	20	18		8	–	–	–	–	–	10	–	–	–	–	–	–	Perchloro-p-xylene	2142-35-0	LI09025
276	411	306	341	329	446	271	188	**446**	100	98	70	38	21	20	18	16		8	–	–	–	–	–	10	–	–	–	–	–	–	Perchloro-p-xylene	2142-35-0	NI43061

MASS TO CHARGE RATIOS								M.W.	INTENSITIES								Parent	C	H	O	N	S	F	Cl	Br	I	Si	P	B	X	COMPOUND NAME	CAS Reg No	No
104	85	45	127	105	47	44	77	**446**	100	99	88	82	82	75	75	71	**0.00**	8	8	2	4	2	10	–	–	–	–	–	–	–	(1,4-Phenylenediureylene)bis(sulphur pentafluoride)	90598-05-3	NI43062
205	131	187	203	57	446	335	301	**446**	100	50	46	42	39	35	31	27		8	18	2	–	2	–	–	–	–	–	1	–	1	Thallium(I) di-n-butyldithiophosphate		NI43063
205	131	187	203	446	335	301	357	**446**	100	50	46	42	35	31	27	19		8	18	2	–	2	–	–	–	–	–	1	–	1	Thallium[I] (O,O-di-n-butyl)dithiophosphate	30903-39-0	NI43064
115	369	371	128	63	129	209	130	**446**	100	97	91	85	81	78	76	71	**0.00**	10	10	–	–	–	–	–	4	–	–	–	–	–	Benzene, 1,2,4,5-tetrakis(bromomethyl)-	15442-91-8	NI43065
66	65	334	332	39	232	167	40	**446**	100	75	64	64	64	50	35	33	**26.07**	14	10	4	–	–	–	–	–	–	–	–	–	2	π-Cyclopentadienyldicarbonylruthenium dimer		MS09836
446	193	219	180	153	333	447	69	**446**	100	31	29	25	22	21	19	19		14	12	6	4	–	6	–	–	–	–	–	–	–	7-((2′,3′-Di-O-trifluoroacetyl)-propyl)-1,3-dimethylxanthine		NI43066
329	73	299	191	330	211	147	43	**446**	100	75	38	24	23	20	20	18	**0.00**	14	35	8	–	–	–	–	–	–	3	1	–	–	Propanoic acid, 3-[[bis[(trimethylsilyl)oxy]phosphinyl]oxy]-2,2-dimethoxy-, trimethylsilyl ester	55401-57-5	NI43067
28	173	175	137	336	138	174	172	**446**	100	99	34	27	18	18	17	15	**0.00**	15	6	4	–	–	–	4	–	–	–	–	–	1	Iron, tetracarbonyl(1,2,3,4-tetrachlorospiro[4.6]undeca-1,3,6,8,10-pentaene)-	61178-34-5	NI43068
55	83	125	41	97	43	69	81	**446**	100	47	46	46	44	30	25	19	**10.91**	16	30	4	–	–	–	–	–	–	–	–	–	2	Octanoic acid, 8,8′-diselenodi-	22686-32-4	NI43069
162	88	120	240	91	131	69	86	**446**	100	28	25	21	20	18	15	14	**0.50**	17	17	4	2	–	7	–	–	–	–	–	–	–	L-Phenylalanine, N-[N-(2,2,3,3,4,4,4-heptafluoro-1-oxobutyl)-L-alanyl]-, methyl ester	55401-56-4	NI43070
73	75	283	147	211	45	356	385	**446**	100	20	20	18	12	12	11	10	**3.07**	19	38	6	–	–	–	–	–	–	3	–	–	–	5-Carbethoxysuccinylacetone, tris(trimethylsilyl)-		NI43071
44	73	42	207	58	224	72	45	**446**	100	60	45	16	12	10	10	10	**5.00**	20	22	8	4	–	–	–	–	–	–	–	–	–	N,N′-Bis(2-(1,1-ethylenedioxymethyl)-4-nitrophenyl)ethylenediamine		LI09026
58	55	99	57	158	114	127	112	**446**	100	72	68	61	56	49	44	39	**2.09**	20	34	9	2	–	–	–	–	–	–	–	–	–	12,21-Dimethyl-1,4,7,15,18-pentaoxa-12,21-diazacyclopentacosane-8,11,22,25-tetrone	91876-19-6	NI43072
158	58	99	114	57	390	141	140	**446**	100	77	72	64	50	49	42	40	**3.25**	20	34	9	2	–	–	–	–	–	–	–	–	–	15,21-Dimethyl-1,4,7,10,18-pentaoxa-15,21-diazacyclopentacosane-11,14,22,25-tetrone	91876-20-9	NI43073
271	336	206	380	141	305	381	445	**446**	100	63	49	30	8	8	7	2	**1.50**	21	23	1	1	–	–	–	–	–	–	–	–	1	Dicyclopentadienyl[(4-phenylimino)-2-pentanonato]praseodymium		NI43074
284	257	258	69	229	117	91	77	**446**	100	82	71	55	54	50	47	42	**0.00**	22	22	10	–	–	–	–	–	–	–	–	–	–	5,7-Dihydroxy-3′-methyl-4′-(O-α-L-rhamnopyranosyl)flavone		NI43075
193	31	149	385	341	211	386	104	**446**	100	48	47	42	36	29	20	20	**0.01**	22	22	10	–	–	–	–	–	–	–	–	–	–	1,2-Bis[(4-hydroxyethoxycarbonyl)benzoyloxy]ethane		IC04430
297	298	43	313	179	299	60	55	**446**	100	78	23	20	16	15	11	10	**0.56**	22	22	10	–	–	–	–	–	–	–	–	–	–	Swertisin	6991-10-2	NI43076
297	298	43	312	299	179	60	61	**446**	100	79	24	21	15	15	11	10	**0.50**	22	22	10	–	–	–	–	–	–	–	–	–	–	Swertisin	6991-10-2	LI09027
284	132	285	241	152	124	60	133	**446**	100	33	19	12	12	11	10	9	**0.00**	22	22	10	–	–	–	–	–	–	–	–	–	–	Tilianin	4291-60-5	NI43077
297	428	284	298	165	429	181	356	**446**	100	58	26	21	21	14	13	11	**3.42**	22	22	10	–	–	–	–	–	–	–	–	–	–	Trematin	2326-34-3	NI43078
297	428	284	165	298	429	356	349	**446**	100	58	26	22	20	14	11	10	**4.00**	22	22	10	–	–	–	–	–	–	–	–	–	–	Trematin	2326-34-3	LI09028
113	70	127	43	71	42	319	141	**446**	100	87	64	55	37	37	28	23	**21.74**	22	30	2	4	2	–	–	–	–	–	–	–	–	Thioperazine	316-81-4	NI43079
70	113	43	42	127	71	44	56	**446**	100	99	86	60	57	42	37	28	**22.02**	22	30	2	4	2	–	–	–	–	–	–	–	–	Thioperazine	316-81-4	NI43080
227	73	217	283	282	356	243	431	**446**	100	99	93	61	40	34	34	30	**0.00**	22	46	5	–	–	–	–	–	–	2	–	–	–	Cyclopentaneacetic acid, 2-(7-methoxyheptyl)-3,5-bis[(trimethylsilyl)oxy], methyl ester, (1α,2α,3β,5β)-	61177-14-8	NI43081
227	55	217	73	283	431	243	228	**446**	100	95	94	91	74	35	29	28	**0.00**	22	46	5	–	–	–	–	–	–	2	–	–	–	Cyclopentaneacetic acid, 2-(7-methoxyheptyl)-3,5-bis[(trimethylsilyl)oxy], methyl ester, (1α,2β,3α,5β)-	61177-13-7	NI43082
243	217	73	227	283	431	356	282	**446**	100	19	19	18	17	8	8	8	**0.00**	22	46	5	–	–	–	–	–	–	2	–	–	–	Cyclopentaneacetic acid, 2-(7-methoxyheptyl)-3,5-bis[(trimethylsilyl)oxy], methyl ester, (1α,2β,3β,5β)-	61177-11-5	NI43083
180	182	327	167	344	196	316	151	**446**	100	81	80	79	44	33	27	27	**18.00**	23	26	9	–	–	–	–	–	–	–	–	–	–	2H-1-Benzopyran-3,4-diol, 2-(3,4-dimethoxyphenyl)-3,4-dihydro-7,8-dimethoxy-, diacetate, [2R-(2α,3α,4α)]-	72150-29-9	NI43084
180	327	344	446	151	328	316	222	**446**	100	93	61	52	51	50	50	49		23	26	9	–	–	–	–	–	–	–	–	–	–	(2R,3S,4S)-2,3-trans-3,4-cis-3,4-diacetoxy-3′,4′,7,8-tetramethoxyflavan		NI43085
446	167	196	362	184	183	179	404	**446**	100	66	58	38	29	24	23	21		23	26	9	–	–	–	–	–	–	–	–	–	–	2,5-Diacetyl-3′,4′,6,7-tetramethoxyisoflavan		NI43086
237	45	238	44	191	57	55	43	**446**	100	22	15	13	9	7	6	6	**0.00**	23	26	9	–	–	–	–	–	–	–	–	–	–	Isophthalic acid, 2,4-dimethoxy-6-methyl-, 3-methyl ester, 1-carboxy-3,6-dimethyl-2-methoxyphenyl ester	19314-77-3	NI43087
327	210	195	181	446	137	252	151	**446**	100	99	34	31	27	19	18	14		23	26	9	–	–	–	–	–	–	–	–	–	–	3′,4′,5′,7-Tetramethoxy-2,3-cis flavan-3,4-cis diacetate		LI09029
180	182	167	344	446	151	222	316	**446**	100	80	77	41	37	29	24	22		23	26	9	–	–	–	–	–	–	–	–	–	–	3′,4′,7,8-Tetramethoxy-2,3-cis flavan-3,4-cis diacetate		LI09030
327	210	446	181	195	252	344	151	**446**	100	50	36	18	14	13	12	10		23	26	9	–	–	–	–	–	–	–	–	–	–	3′,4′,5′,7-Tetramethoxy-2,3-trans flavan-3,4-cis diacetate		LI09031
327	210	167	181	446	195	344	386	**446**	100	49	30	24	20	19	12	11		23	26	9	–	–	–	–	–	–	–	–	–	–	3′,4′,5′,7-Tetramethoxy-2,3-trans flavan-3,4-diacetate		LI09032
73	187	259	55	155	75	147	103	**446**	100	44	33	24	18	18	16	15	**0.00**	23	50	4	–	–	–	–	–	–	2	–	–	–	Hexadecanoic acid, 9,10-bis(trimethylsiloxy)-, methyl ester		LI09033
73	187	259	153	75	147	333		**446**	100	46	36	17	17	15	3		**0.00**	23	50	4	–	–	–	–	–	–	2	–	–	–	Hexadecanoic acid, 9,10-bis(trimethylsiloxy)-, methyl ester		LI09034
73	273	75	275	95	103	129	81	**446**	100	77	50	43	40	30	26	21	**0.00**	23	50	4	–	–	–	–	–	–	2	–	–	–	Hexadecanoic acid, 10,16-bis(trimethylsiloxy)-, methyl ester	21987-16-6	NI43088
73	273	75	275	95	55	103	129	**446**	100	77	49	45	40	35	31	27	**0.00**	23	50	4	–	–	–	–	–	–	2	–	–	–	Hexadecanoic acid, 10,16-bis(trimethylsiloxy)-, methyl ester	21987-16-6	LI09035
73	273	75	275	95	55	103	129	**446**	100	76	47	44	40	33	30	27	**0.00**	23	50	4	–	–	–	–	–	–	2	–	–	–	Hexadecanoic acid, 10,16-bis(trimethylsiloxy)-, methyl ester	21987-16-6	LI09036
343	73	147	43	211	344	57	75	**446**	100	58	32	32	28	27	26	23	**0.00**	23	50	4	–	–	–	–	–	–	2	–	–	–	Myristic acid, 2,3-bis(trimethylsiloxy)propyl ester	1188-73-4	WI01560
343	344	431	345	432	145	218	205	**446**	100	27	15	7	5	5	4	4	**0.00**	23	50	4	–	–	–	–	–	–	2	–	–	–	Myristic acid, 2,3-bis(trimethylsiloxy)propyl ester	1188-73-4	WI01561
73	218	147	103	43	75	57	41	**446**	100	81	52	50	50	44	36	31	**0.00**	23	50	4	–	–	–	–	–	–	2	–	–	–	Myristic acid, 2-(trimethylsiloxy)-1-[(trimethylsiloxy)methyl]ethyl ester	14473-56-4	WI01562
218	431	219	129	343	220	432	247	**446**	100	23	21	21	18	9	8	8	**0.00**	23	50	4	–	–	–	–	–	–	2	–	–	–	Myristic acid, 2-(trimethylsiloxy)-1-[(trimethylsiloxy)methyl]ethyl ester	14473-56-4	WI01563
251	59	60	76	77	205	178	151	**446**	100	50	48	45	36	35	28	25	**0.00**	24	22	5	4	–	–	–	–	–	–	–	–	–	1-(4-Nitrobenzyl)-2-propoxy-3-(4-nitrophenyl)-1,2-dihydroquinoxaline	91736-68-4	NI43089

MASS TO CHARGE RATIOS	M.W.	INTENSITIES	Parent	C	H	O	N	S	F	Cl	Br	I	Si	P	B	X	COMPOUND NAME	CAS Reg No	No
193 236 138 177 28 210 134 43	**446**	100 74 42 39 31 28 28 28	**0.00**	24	30	8	–	–	–	–	–	–	–	–	–	–	**Benzoic acid, 2-hydroxy-3-[(4-hydroxy-2-methoxy-6-propylbenzoyl)oxy]-4-methoxy-6-pentyl-**	69563-44-6	**NI43090**
163 123 302 283 206 193 55 41	**446**	100 66 64 49 48 47 46 44	**0.00**	24	30	8	–	–	–	–	–	–	–	–	–	–	**Cortisol triformate**		**MS09837**
446 447 181 43 195 55 207 69	**446**	100 26 24 19 18 18 16 16		24	30	8	–	–	–	–	–	–	–	–	–	–	**1H,3H-Furo[3,4-c]furan, tetrahydro-1,4-bis(3,4,5-trimethoxyphenyl)-, [1S (1α,3aα,4α,6aα)]-**	13060-14-5	**NI43091**
43 242 119 174 161 123 326 302	**446**	100 38 37 26 26 26 13 10	**0.50**	24	30	8	–	–	–	–	–	–	–	–	–	–	**2-(8-Hydroxy-3-hydroxymethyl-7-methylocta-2,6-dienyl)hydroquinone-tetraacetate**		**NI43092**
274 224 131 125 242 191 386 214	**446**	100 33 29 22 21 17 6 6	**0.00**	24	30	8	–	–	–	–	–	–	–	–	–	–	**Pseudolaric acid B methyl ester**		**NI43093**
446 447 448 450 311 152 137 119	**446**	100 45 38 32 20 10 10 10		24	36	2	–	–	–	–	–	–	–	–	–	1	**Zirconium, [2-butene-2,3-diolato(2-)-O,O']bis[(1,2,3,4,5-η)-1,2,3,4,5-pentamethyl-2,4-cyclopentadien-1-yl]-**	67108-82-1	**NI43094**
171 57 127 99 87 73 55 43	**446**	100 63 45 33 30 21 21 19	**0.00**	24	46	7	–	–	–	–	–	–	–	–	–	–	**Hexanoic acid, 2-ethyl-, oxybis(2,1-ethanediyloxy-2,1-ethanediyl) ester**	18268-70-7	**NI43095**
446 77 217 185 65 447 141 115	**446**	100 78 69 69 31 29 28 28		25	18	6	–	1	–	–	–	–	–	–	–	–	**4-Phenoxy-4'-(4-carboxy-phenoxy)diphenyl sulphone**		**IC04431**
326 312 178 131 229 386 137 123	**446**	100 46 25 25 21 19 18 18	**1.00**	25	34	7	–	–	–	–	–	–	–	–	–	–	**Methyl 6α,7β-diacetoxyvouacapan-14β-oate**		**NI43096**
99 55 43 41 57 100 45 69	**446**	100 83 70 55 53 48 48 46	**25.67**	25	34	7	–	–	–	–	–	–	–	–	–	–	**Pregn-5-ene-3,11-dione, 17,20:20,21-bis[methylenebis(oxy)]-, cyclic 3-(1,2 ethanediyl acetal)**	52248-40-5	**NI43097**
386 43 344 387 446 274 329 262	**446**	100 88 37 25 23 20 17 16		25	34	7	–	–	–	–	–	–	–	–	–	–	**2β,3β,17β-Triacetoxy-5α-androst-7-ene-6-one**		**MS09838**
387 55 121 388 109 91 79 135	**446**	100 47 34 27 24 21 20 19	**3.40**	25	34	7	–	–	–	–	–	–	–	–	–	–	**(22R)-6α,11β,21-Trihydroxy-16α,17α-propylmethylenedioxypregna-1,4-diene-3,20-dione**		**NI43098**
387 55 49 121 84 388 91 279	**446**	100 55 47 41 33 26 26 24	**4.00**	25	34	7	–	–	–	–	–	–	–	–	–	–	**(22R)-6β,11β,21-Trihydroxy-16α,17α-propylmethylenedioxypregna-1,4-diene-3,20-dione**		**NI43099**
387 55 121 91 109 173 79 279	**446**	100 86 71 40 33 37 36 36	**5.10**	25	34	7	–	–	–	–	–	–	–	–	–	–	**(22S)-6β,11β,21-Trihydroxy-16α,17α-propylmethylenedioxypregna-1,4-diene-3,20-dione**		**NI43100**
343 401 159 81 403 130 251 341	**446**	100 78 70 64 60 52 46 32	**6.00**	25	38	5	–	–	–	–	–	–	1	–	–	–	**3β-Acetoxy-16α,17α-dihydroxy-pregn-5-en-20-one dimethyl siliconide**		**LI09037**
446 359 447 267 360 143 431 57	**446**	100 77 37 24 20 19 17 17		25	42	3	–	–	–	–	–	–	2	–	–	–	**Androsta-3,5-dien-17-one, 3,12-bis[(trimethylsilyl)oxy]-, (12β)-**	69688-29-5	**NI43101**
356 225 357 266 172 251 171 75	**446**	100 36 29 29 22 20 19 12	**1.16**	25	42	3	–	–	–	–	–	–	2	–	–	–	**6α-Hydroxy-3-methoxy-17β-estradiol di-TMS**	74298-81-0	**NI43102**
356 225 446 266 75 171 357 447	**446**	100 79 73 61 38 34 30 29		25	42	3	–	–	–	–	–	–	2	–	–	–	**6β-Hydroxy-3-methoxy-17β-estradiol di-TMS**		**NI43103**
356 251 266 357 172 225 240 182	**446**	100 66 56 31 24 22 18 15	**4.87**	25	42	3	–	–	–	–	–	–	2	–	–	–	**7α-Hydroxy-3-methoxy-17β-estradiol di-TMS**		**NI43104**
356 266 186 239 160 446 157 357	**446**	100 99 99 51 49 43 43 37		25	42	3	–	–	–	–	–	–	2	–	–	–	**11β-Hydroxy-3-methoxy-17β-estradiol di-TMS**		**NI43105**
356 266 212 357 225 251 227 238	**446**	100 97 64 47 40 37 36 29	**0.00**	25	42	3	–	–	–	–	–	–	2	–	–	–	**14α-Hydroxy-3-methoxy-17β-estradiol di-TMS**		**NI43106**
446 217 447 187 169 191 218 240	**446**	100 65 39 36 26 23 21 19		25	42	3	–	–	–	–	–	–	2	–	–	–	**15α-Hydroxy-3-methoxy-17β-estradiol di-TMS**		**NI43107**
253 446 287 328 356 129 266 212	**446**	100 95 79 72 68 53 52 45		25	42	3	–	–	–	–	–	–	2	–	–	–	**16α-Hydroxy-3-methoxy-17β-estradiol di-TMS**	18880-86-9	**NI43108**
253 446 287 356 328 129 266 212	**446**	100 86 85 68 68 60 51 44		25	42	3	–	–	–	–	–	–	2	–	–	–	**16β-Hydroxy-3-methoxy-17β-estradiol di-TMS**	74298-83-2	**NI43109**
370 371 355 287 329 314 372 327	**446**	100 36 17 15 13 10 9 6	**1.07**	25	42	3	–	–	–	–	–	–	2	–	–	–	**6α-Hydroxy-Δ^1-tetrahydrocannabinol dimethylsilyl ether**		**NI43110**
357 358 431 283 446 359 249 432	**446**	100 29 19 15 11 8 8 7		25	42	3	–	–	–	–	–	–	2	–	–	–	**7-Hydroxy-Δ^1-tetrahydrocannabinol dimethylsilyl ether**		**NI43111**
431 432 159 258 232 299 433 311	**446**	100 38 38 30 26 24 14 14	**0.00**	25	42	3	–	–	–	–	–	–	2	–	–	–	**1-Methoxy-17β-estradiol di-TMS**	74299-25-5	**NI43112**
446 447 315 448 416 129 431 325	**446**	100 47 29 18 16 15 12 11		25	42	3	–	–	–	–	–	–	2	–	–	–	**2-Methoxy-17β-estradiol di-TMS**	29825-47-6	**NI43113**
414 415 283 324 416 309 230 229	**446**	100 38 23 19 17 16 14 13	**3.25**	25	42	3	–	–	–	–	–	–	2	–	–	–	**6-Methoxy-17β-estradiol di-TMS**	74298-82-1	**NI43114**
446 223 447 444 325 202 204 448	**446**	100 54 43 40 36 31 28 18		26	22	–	–	–	–	–	–	–	–	–	–	2	**Ferrocene, 1,1''-(1,4-phenylene)bis-**	56712-08-4	**NI43115**
343 386 339 399 357 446 314 368	**446**	100 89 85 57 53 52 49 29		26	38	6	–	–	–	–	–	–	–	–	–	–	**3-β-Acetoxy-14-hydroxy-14,20-oxy(1-hydroxy)methylene-5-β,14-β-card-20(22)-enolide**		**LI09038**
75 203 73 356 338 431 147 107	**446**	100 73 65 48 48 34 30 30	**13.11**	26	42	4	–	–	–	–	–	–	1	–	–	–	**Card-20(22)-enolide, 14-hydroxy-3-[(trimethylsilyl)oxy]-, (3β,5β)-**	20759-23-3	**MS09839**
75 203 73 356 338 431 111 413	**446**	100 74 65 48 48 37 25 19	**15.00**	26	42	4	–	–	–	–	–	–	1	–	–	–	**Card-20(22)-enolide, 14-hydroxy-3-[(trimethylsilyl)oxy]-, (3β,5β)-**	20759-23-3	**NI43116**
75 203 73 356 338 147 107 93	**446**	100 75 66 49 48 29 29 26	**0.00**	26	42	4	–	–	–	–	–	–	1	–	–	–	**Card-20(22)-enolide, 14-hydroxy-3-[(trimethylsilyl)oxy]-, (3β,5β)-**	20759-23-3	**NI43117**
75 203 73 356 338 431 147 111	**446**	100 75 67 49 48 33 30 28	**14.00**	26	42	4	–	–	–	–	–	–	1	–	–	–	**3-Trimethylsilyloxydigitoxigenin**		**LI09039**
91 92 355 386 371 356 387 372	**446**	100 82 49 38 34 14 11 10	**0.00**	26	42	4	2	–	–	–	–	–	–	–	–	–	**Pregnane-3,20-dione, 17-(acetyloxy)-6-methyl-, 3,20-bis(O-methyloxime), (5β,6α)-**	69833-58-5	**NI43118**
431 432 446 140 196 155 433 154	**446**	100 67 52 50 30 26 24 24		26	46	2	–	–	–	–	–	–	2	–	–	–	**17α-Ethynyl-5α-estrane-3α,17β-diol di-TMS**		**NI43119**
431 432 446 140 433 155 154 447	**446**	100 37 33 25 15 14 14 13		26	46	2	–	–	–	–	–	–	2	–	–	–	**17α-Ethynyl-5α-estrane-3β,17β-diol di-TMS**		**NI43120**
431 432 140 446 196 209 155 433	**446**	100 64 46 41 35 29 28 21		26	46	2	–	–	–	–	–	–	2	–	–	–	**17α-Ethynyl-5β-estrane-3α,17β-diol di-TMS**		**NI43121**
431 432 446 140 209 196 75 155	**446**	100 35 26 25 20 18 17 14		26	46	2	–	–	–	–	–	–	2	–	–	–	**17α-Ethynyl-5β-estrane-3β,17β-diol di-TMS**		**NI43122**
258 91 257 446 105 79 41 77	**446**	100 72 56 55 52 38 34 33		27	31	5	–	–	–	–	–	–	–	–	1	–	**17α,21-Dihydroxypregn-4-ene-3,11,20-trione phenyl boronate**		**MS09840**
258 257 446 105 121 125 120 119	**446**	100 56 55 52 30 29 29 23		27	31	5	–	–	–	–	–	–	–	–	1	–	**17α,21-Dihydroxypregn-4-ene-3,11,20-trione phenyl boronate**		**LI09040**
446 447 445 244 218 231 217 245	**446**	100 31 29 23 12 9 7 5		27	35	3	–	–	–	–	–	–	1	–	1	–	**3-Epiestriol 3-TMS phenyl boronate**		**NI43123**
361 55 105 57 91 85 93 81	**446**	100 37 33 30 28 28 27 27	**5.00**	27	42	5	–	–	–	–	–	–	–	–	–	–	**Anosmagenin**	41451-80-3	**NI43124**
361 55 327 326 105 57 91 85	**446**	100 37 35 35 33 30 28 28	**5.00**	27	42	5	–	–	–	–	–	–	–	–	–	–	**Anosmagenin**	41451-80-3	**NI43125**
107 105 109 135 121 119 253 132	**446**	100 93 92 79 79 74 73 69	**5.90**	27	42	5	–	–	–	–	–	–	–	–	–	–	**Cholan-24-oic acid, 3-(acetyloxy)-7-oxo-, methyl ester, (3α,5β)-**	10452-65-0	**NI43126**

MASS TO CHARGE RATIOS	M.W.	INTENSITIES	Parent	C	H	O	N	S	F	Cl	Br	I	Si	P	B	X	COMPOUND NAME	CAS Reg No	No
231 121 109 105 107 146 135 123	**446**	100 92 44 44 40 34 32 29	**0.00**	27	42	5	–	–	–	–	–	–	–	–	–	–	Cholan-24-oic acid, 3-(acetyloxy)-12-oxo-, methyl ester, (3α,5β)-	5143-55-5	NI43127
231 446 121 81 43 154 386 109	**446**	100 81 46 42 38 37 35 32		27	42	5	–	–	–	–	–	–	–	–	–	–	Cholan-24-oic acid, 3-(acetyloxy)-12-oxo-, methyl ester, (3α,5β)-	5143-55-5	NI43128
104 119 130 106 145 159 133 120	**446**	100 69 56 51 49 47 38 38	**0.00**	27	42	5	–	–	–	–	–	–	–	–	–	–	Cholan-24-oic acid, 7-(acetyloxy)-3-oxo-, methyl ester, (5β,7β)-	60354-41-8	NI43129
271 161 154 105 107 272 119 121	**446**	100 30 26 26 24 23 20 18	**0.00**	27	42	5	–	–	–	–	–	–	–	–	–	–	Cholan-24-oic acid, 12-(acetyloxy)-3-oxo-, methyl ester, (5α,12α)-	60354-43-0	NI43130
271 104 253 106 147 145 161 119	**446**	100 71 70 56 50 49 47 42	**0.00**	27	42	5	–	–	–	–	–	–	–	–	–	–	Cholan-24-oic acid, 12-(acetyloxy)-3-oxo-, methyl ester, (5β,12α)-	7753-75-5	NI43131
139 314 332 374 446 387 428 359	**446**	100 97 24 21 15 10 7 4		27	42	5	–	–	–	–	–	–	–	–	–	–	9(11)-Dehydroagapanthagenin		MS09841
155 428 298 131 410 142 340 280	**446**	100 76 28 14 9 9 8 6	**5.00**	27	42	5	–	–	–	–	–	–	–	–	–	–	27-Hydroxyruscogenin		MS09842
149 169 142 128 143 129 115 105	**446**	100 73 73 66 59 45 45 43	**7.30**	27	42	5	–	–	–	–	–	–	–	–	–	–	Pennogenin, 24-hydroxy-	113960-61-5	NI43132
446 239 447 240 223 139 413 39	**446**	100 55 34 12 9 8 7 5		28	14	4	–	1	–	–	–	–	–	–	–	–	1,2′-Bianthraquinonyl sulphide		IC04432
105 77 446 106 78 51 447 76	**446**	100 95 12 8 7 5 4 3		28	18	4	2	–	–	–	–	–	–	–	–	–	Bis(1,5-benzoylamino)anthraquinone		IC04433
311 446 312 174 326 445 447 109	**446**	100 34 23 23 19 17 12 10		28	34	3	2	–	–	–	–	–	–	–	–	–	Aspidospermidin-21-ol, 1-acetyl-17-methoxy-21-phenyl-	55103-35-0	NI43133
57 119 200 77 91 258 256 187	**446**	100 30 27 22 21 20 18 15	**0.00**	28	34	3	2	–	–	–	–	–	–	–	–	–	rel-(2R,3aS,9R,10S)-5,7-Di-tert-butyl-3,3a-dihydro-3,3-dimethyl-N-phenyl 3a,2-(epoxyethano)-2H-indole-9,10-dicarboximide		DD01688
57 77 258 431 446 200 242 190	**446**	100 83 47 45 41 41 32 25		28	34	3	2	–	–	–	–	–	–	–	–	–	rel-(2R,3aS,9S,10R)-5,7-Di-tert-butyl-3,3a-dihydro-3,3-dimethyl-N-phenyl 3a,2-(epoxyethano)-2H-indole-9,10-dicarboximide		DD01689
149 43 307 57 150 55 41 71	**446**	100 18 14 14 10 10 9 6	**0.67**	28	46	4	–	–	–	–	–	–	–	–	–	–	Didecyl phthalate		IC04434
149 57 71 43 55 85 69 41	**446**	100 56 44 41 32 29 24 24	**0.00**	28	46	4	–	–	–	–	–	–	–	–	–	–	Di-isodecyl phthalate	26761-40-0	IC04435
149 57 43 71 85 55 41 69	**446**	100 30 26 25 24 19 16 15	**0.00**	28	46	4	–	–	–	–	–	–	–	–	–	–	Di-isodecyl phthalate	26761-40-0	NI43134
149 57 43 71 85 55 41 69	**446**	100 44 42 36 32 27 23 21	**0.00**	28	46	4	–	–	–	–	–	–	–	–	–	–	Di-isodecyl phthalate	26761-40-0	NI43135
105 57 43 55 41 113 267 71	**446**	100 73 72 47 42 36 33 33	**17.04**	28	46	4	–	–	–	–	–	–	–	–	–	–	Octadecanoic acid, 2-phenyl-1,3-dioxan-5-yl ester	56554-39-3	NI43136
43 105 41 55 29 57 27 28	**446**	100 76 67 47 46 43 31 30	**0.00**	28	46	4	–	–	–	–	–	–	–	–	–	–	Octadecanoic acid, 2-phenyl-1,3-dioxolan-4-yl methyl ester	56599-43-0	NI43137
105 43 57 41 55 77 29 69	**446**	100 50 31 31 29 24 17 16	**2.34**	28	46	4	–	–	–	–	–	–	–	–	–	–	Octadecanoic acid, 2-phenyl-1,3-dioxolan-4-yl methyl ester, cis-	56599-88-3	NI43138
105 43 57 446 324 162 55 41	**446**	100 63 59 44 42 38 38 30		28	46	4	–	–	–	–	–	–	–	–	–	–	Stearic acid, 2-phenyl-m-dioxan-5-yl ester, cis-	10588-88-2	NI43139
43 105 57 41 55 29 69 71	**446**	100 91 81 60 56 39 28 27	**5.84**	28	46	4	–	–	–	–	–	–	–	–	–	–	Stearic acid, 2-phenyl-m-dioxan-5-yl ester, trans-	10564-35-9	NI43140
403 271 404 131 405 272 89 255	**446**	100 39 32 18 10 9 9 7	**0.00**	28	50	2	–	–	–	–	–	–	1	–	–	–	3α-Tri-propylsilyloxy-5α-androstan-17-one		NI43141
403 404 255 131 405 271 147 89	**446**	100 31 28 17 9 7 7 7	**0.00**	28	50	2	–	–	–	–	–	–	1	–	–	–	3α-Tri-propylsilyloxy-5β-androstan-17-one		NI43142
403 404 131 271 89 405 361 161	**446**	100 34 26 15 14 10 7 7	**0.00**	28	50	2	–	–	–	–	–	–	1	–	–	–	3β-Tri-propylsilyloxy-5α-androstan-17-one		NI43143
403 404 271 131 255 405 89 81	**446**	100 42 39 25 14 12 10 10	**0.00**	28	50	2	–	–	–	–	–	–	1	–	–	–	3β-Tri-propylsilyloxy-5β-androstan-17-one		NI43144
446 91 77 415 253 105 167 121	**446**	100 40 33 17 16 16 15 15		29	34	4	–	–	–	–	–	–	–	–	–	–	5,24,25-Trimethoxy-2-oxatricyclo[20.2.2.1^{3,7}]heptacosa-3,5,7(27),22,24,25-hexaene-13,15-diyne		NI43145
258 361 301 343 283 428 431 413	**446**	100 55 40 25 7 5 3 3	**1.00**	29	50	3	–	–	–	–	–	–	–	–	–	–	3-Acetoxy-20-iso-5α-cholestane-3β,20β-diol		LI09041
428 228 368 291 429 120 229 152	**446**	100 70 51 50 34 34 33 33	**20.00**	29	50	3	–	–	–	–	–	–	–	–	–	–	Cholestan-3β,7β-diol 3-acetate		MS09843
373 446 386 355 447 374 387 356	**446**	100 92 80 50 28 26 23 13		29	50	3	–	–	–	–	–	–	–	–	–	–	5α-Cholestane-1α,19-diol 19-acetate	28809-70-3	NI43146
386 373 355 368 232 428 446	**446**	100 50 50 48 30 25 10		29	50	3	–	–	–	–	–	–	–	–	–	–	5α-Cholestane-2α,19-diol 2-acetate	28809-69-0	LI09042
386 373 355 368 232 387 428 374	**446**	100 50 50 48 30 28 25 15	**10.01**	29	50	3	–	–	–	–	–	–	–	–	–	–	5α-Cholestane-2α,19-diol 2-acetate	28809-69-0	NI43147
386 373 358 355 428 368 446	**446**	100 25 25 21 18 18 13		29	50	3	–	–	–	–	–	–	–	–	–	–	5α-Cholestane-2α,19-diol 19-acetate	28809-67-8	LI09043
386 387 373 358 355 428 368 446	**446**	100 28 25 25 21 18 18 13		29	50	3	–	–	–	–	–	–	–	–	–	–	5α-Cholestane-2α,19-diol 19-acetate	28809-67-8	NI43148
356 357 242 341 386 373 314 243	**446**	100 26 25 20 10 10 8 5	**0.00**	29	50	3	–	–	–	–	–	–	–	–	–	–	5α-Cholestane-2β,19-diol 19-acetate	30950-87-9	NI43149
356 243 341 386 373 314 446	**446**	100 25 20 10 10 8 3		29	50	3	–	–	–	–	–	–	–	–	–	–	5α-Cholestane-2β,19-diol 19-acetate	30950-87-9	LI09044
386 332 43 368 55 81 95 57	**446**	100 73 70 50 41 39 36 31	**22.00**	29	50	3	–	–	–	–	–	–	–	–	–	–	Cholestane-4,5-diol, 4-acetate, (4β,5α)-	5789-00-4	NI43150
446 414 386 354 447 415 387 355	**446**	100 82 50 48 33 26 15 14		29	50	3	–	–	–	–	–	–	–	–	–	–	5α-Cholestan-19-oic acid, 2β-methoxy-, methyl ester	24649-36-3	NI43151
446 447 77 141 168 115 223 139	**446**	100 32 22 12 12 12 7 7		30	22	4	–	–	–	–	–	–	–	–	–	–	Benzene, 1,3-bis(3-phenoxyphenoxy)-	2455-71-2	NI43152
353 445 446 354 199 179 183 152	**446**	100 23 21 18 17 12 8 8		30	23	2	–	–	–	–	–	–	–	1	–	–	Phenyl-di-(o-phenoxyphenyl)-phosphine		LI09045
371 59 95 81 109 55 83 57	**446**	100 64 56 53 43 42 39 35	**0.70**	30	54	2	–	–	–	–	–	–	–	–	–	–	Cholestane, 3-(2-methoxyethoxy)-, (3β,5α)-	57156-79-3	NI43153
446 386 263 412 213 215 243 232	**446**	100 44 32 26 26 20 18 18		30	54	2	–	–	–	–	–	–	–	–	–	–	24S-Ethylcholestane-3β,4β-diol 3β-methyl ether		NI43154
81 67 109 69 95 141 180 121	**446**	100 64 61 47 35 29 23 22	**7.00**	30	54	2	–	–	–	–	–	–	–	–	–	–	Methyl 5,9,23-nonacosatrienoate		MS09844
446 447 291 448 165 367 289 267	**446**	100 37 14 7 7 5 5 5		35	26	–	–	–	–	–	–	–	–	–	–	–	Benzene, 1,1′,1″,1‴,1⁗-(1,3-cyclopentadiene-1,2,3,4,5-pentayl)pentakis-	2519-10-0	NI43155
233 273 44 104 316 78 246 56	**447**	100 96 72 61 57 33 21 20	**0.40**	15	25	7	3	–	3	–	–	–	–	1	–	–	L-Alanine, γ-(methoxymethylphosphinyl)-N-(trifluoroacetyl)-L-α-aminobutyryl-L-alanyl-, methyl ester	56890-06-3	MS09845
58 130 44 59 117 42 155 43	**447**	100 71 50 48 28 27 19 19	**0.00**	16	19	9	2	2	–	–	–	–	–	–	–	–	Glucobrassicin		MS09846
448 322 127 449 430 431 323 450	**447**	100 38 38 31 9 7 7 6	**0.00**	19	37	7	5	–	–	–	–	–	–	–	–	–	Sisomicin	32385-11-8	NI43156
160 118 145 110 84 304 127 86	**447**	100 87 45 43 41 39 38 38	**4.71**	19	37	7	5	–	–	–	–	–	–	–	–	–	Sisomicin	32385-11-8	NI43158
448 322 127 160 449 163 323 430	**447**	100 97 46 35 22 21 15 12	**0.00**	19	37	7	5	–	–	–	–	–	–	–	–	–	Sisomicin	32385-11-8	NI43157

MASS TO CHARGE RATIOS								M.W.	INTENSITIES								Parent	C	H	O	N	S	F	Cl	Br	I	Si	P	B	X	COMPOUND NAME	CAS Reg No	No
218	368	244	242	396	367	231	229	**447**	100	57	33	33	18	11	8	8	**1.00**	21	22	5	1	–	–	–	1	–	–	–	–	–	2-(β-(2-Bromo-4,5-dimethoxyphenethyl))-6,7-dimethoxy-3-methylenephthalimidine		LI09046
285	43	28	118	328	17	97	139	**447**	100	85	57	53	48	47	42	37	**0.00**	22	25	9	1	–	–	–	–	–	–	–	–	–	3-Methyl-3-phenyl-1-(2,3,5-tri-O-acetyl-β-D-ribofuranosyl)-2,5-pyrrolidinedione		NI43159
169	109	447	117	331	127	145	130	**447**	100	52	44	44	29	20	14	14		22	25	9	1	–	–	–	–	–	–	–	–	–	1-(2,3,4,6-Tetra-O-acetyl-β-D-glucopyranosyl)indole		LI09047
169	109	447	117	331	146	140	118	**447**	100	52	46	45	29	15	12	12		22	25	9	1	–	–	–	–	–	–	–	–	–	1-(2,3,4,6-Tetra-O-acetyl-β-D-glucopyranosyl)indole	5059-38-1	NI43160
93	276	275	274	273	447	445	449	**447**	100	58	47	35	29	7	4	2		23	29	1	1	1	–	–	–	–	–	–	–	1	2-[[(cis-4-tert-Butylcyclohexyl)thio]seleno]-N-phenylbenzamide	118398-41-7	DD01690
93	276	275	277	447	445	449	448	**447**	100	55	29	23	7	4	2	2		23	29	1	1	1	–	–	–	–	–	–	–	1	2-[[(trans-4-tert-Butylcyclohexyl)thio]seleno]-N-phenylbenzamide	118398-42-8	DD01691
190	233	73	303	134	402	331	257	**447**	100	83	82	67	12	11	9	2	**1.20**	24	37	5	3	–	–	–	–	–	–	–	–	–	L-Alanine, N-[N-[N-[(4-methoxyphenyl)methylene]-L-valyl]-L-isoleucyl]-, ethyl ester	37580-30-6	NI43161
73	447	75	129	448	432	221	197	**447**	100	59	59	30	18	15	12	12		25	45	2	1	–	–	–	–	–	2	–	–	–	13β-Ethyl-3-oximinogon-4-en-17β-ol 3,17, bis(trimethylsilyl)-		NI43162
73	75	357	238	129	329	210	358	**447**	100	64	52	49	39	35	22	21	**4.00**	25	45	2	1	–	–	–	–	–	2	–	–	–	13β-Ethyl-3-oximinogon-5-en-17β-ol 3,17, bis(trimethylsilyl)-		NI43163
261	91	276	65	171	262	155	107	**447**	100	82	44	30	28	24	24	21	**0.50**	26	25	2	1	2	–	–	–	–	–	–	–	–	3′,6′-Dihydro-4′,5′-dimethyl-1′-tosyliminospiro(fluorene-9,2′-thiopyran)		MS09847
81	119	91	120	82	353	96	95	**447**	100	67	19	6	6	5	5	3	**0.09**	26	25	6	1	–	–	–	–	–	–	–	–	–	1-(3′,5′-Di-O-p-toluyl-2-desoxy-β-D-ribofuranosyl)-2(1H)-pyridone		LI09048
388	125	356	387	389	372	357	447	**447**	100	78	52	28	25	25	20	18		26	41	5	1	–	–	–	–	–	–	–	–	–	6-Oxo-5α-pregnane-3β,20β-diol-diacetate-methyloxime		NI43164
91	73	447	341	340	96	77	75	**447**	100	59	19	17	17	17	14	13		28	37	2	1	–	–	–	–	–	1	–	–	–	Estra-1,3,5(10)-trien-17-one, 3-[(trimethylsilyl)oxy]-, O-benzyloxime	57305-09-6	NI43165
166	120	148	159	119	121	133	95	**447**	100	91	66	63	56	50	46	46	**30.68**	29	37	3	1	–	–	–	–	–	–	–	–	–	N-Retinoyl DL-phenylalanine (all-trans)	97885-88-6	NI43166
166	120	148	159	119	121	133	94	**447**	100	91	66	63	56	50	46	46	**30.68**	29	37	3	1	–	–	–	–	–	–	–	–	–	N-Retinoyl DL-phenylalanine (all-trans)	97885-88-6	NI43167
167	280	447	168	165	152	432	448	**447**	100	37	34	20	17	13	12	11		31	29	2	1	–	–	–	–	–	–	–	–	–	1,2-Dimethoxy-6-diphenylmethyl-6-aza-5,6,6a,7-tetrahydro-4H-benz[mn]anthracene		MS09848
280	153	392	56	127	254	364	420	**448**	100	39	36	31	29	25	22	18	**1.90**	6	2	4	–	–	–	–	–	2	–	–	–	1	Iron, tetracarbonyl(η-2-1,2-diiodoethene)-	52613-72-6	NI43168
69	131	31	119	100	181	179	93	**448**	100	79	74	51	51	45	38	21	**4.00**	7	–	–	–	–	15	–	1	–	–	–	–	–	Perfluoro-2-bromoheptane		LI09049
165	69	64	169	100	135	32	48	**448**	100	75	75	67	58	50	50	42	**0.00**	7	9	2	–	1	9	–	–	–	–	–	–	1	Stannane, trimethyl[[(nonafluorobutyl)sulphinyl]oxy]-	51735-78-5	NI43169
63	69	119	65	429	281	181	231	**448**	100	90	73	30	17	11	11	10	**0.00**	8	–	2	–	–	15	1	–	–	–	–	–	–	1-Perfluoro-3-ethylpent-3-yl chloroformate		IC04436
110	127	89	150	81	109	82	53	**448**	100	69	39	35	33	25	20	19	**0.11**	8	6	4	2	2	10	–	–	–	–	–	–	–	1,4-Phenylenebis(oxycarbonylamino)bis(sulphur pentafluoride)	90597-97-0	NI43170
308	280	420	183	59	87	184	335	**448**	100	56	44	43	39	32	31	27	**27.00**	10	–	6	–	–	6	–	–	–	–	–	–	2	Cobalt, hexacarbonyl(1,1,1,4,4,4-hexafluoro-2-butyne)di-	37685-63-5	NI43171
194	44	69	110	42	60	46	379	**448**	100	14	11	10	8	7	2	1	**0.01**	10	12	–	4	1	12	–	–	–	–	–	–	–	N,N′-Bis(hexafluoro-α-dimethylaminoisopropyl)sulphur di-imide		LI09050
124	448	59	420	298	98	69	58	**448**	100	16	14	10	10	7	7	6		18	5	1	–	–	8	–	–	–	–	–	–	1	Cobalt, carbonyl(η-5-2,4-cyclopentadien-1-yl)(3,3′,4,4′,5,5′,6,6′-octafluoro[1,1′-biphenyl]-2,2′-diyl)-	51509-31-0	NI43172
128	64	65	63	39	130	92	28	**448**	100	62	40	39	32	30	25	19	**2.99**	18	12	4	–	–	–	3	–	–	–	–	–	1	Tris(4-chlorophenoxo)oxovanadium	14735-52-5	NI43173
232	172	102	219	243	259	128	360	**448**	100	83	78	62	49	36	26	22	**0.00**	18	44	3	2	–	–	–	–	–	4	–	–	–	4-Hydroxylysine, tetrakis(trimethylsilyl)-		NI43174
73	246	174	147	84	200	156	59	**448**	100	71	55	30	26	15	15	15	**1.20**	19	48	2	2	–	–	–	–	–	4	–	–	–	Putreanine tetra-TMS		MS09849
195	238	308	280	167	140	307	413	**448**	100	34	29	19	17	11	9	7	**0.00**	20	18	2	2	–	5	1	–	–	–	–	–	–	Butanilicaine pentafluorobenzamide		NI43175
144	130	143	115	145	117	225	142	**448**	100	53	32	16	11	9	8	6	**1.03**	20	20	–	2	–	–	–	–	–	–	–	–	2	Indole, 3,3′-(diselenodiethylene)di-	1919-99-9	NI43176
144	130	143	115	145	225	117	77	**448**	100	54	32	16	12	8	8	6	**0.84**	20	20	–	2	–	–	–	–	–	–	–	–	2	Indole, 3,3′-(diselenodiethylene)di-	1919-99-9	LI09051
107	105	77	91	448	167	149	299	**448**	100	96	28	27	25	11	11	8		20	20	10	2	–	–	–	–	–	–	–	–	–	Methyl 4,6-O-benzylidene-2-O-(2,4-dinitrophenyl)-α-D-glucopyranoside	57565-81-8	NI43177
448	107	159	105	167	299	91	77	**448**	100	79	57	55	45	31	23	18		20	20	10	2	–	–	–	–	–	–	–	–	–	Methyl 4,6-O-benzylidene-3-O-(2,4-dinitrophenyl)-α-D-glucopyranoside	57565-80-7	NI43178
43	113	115	98	159	69	157	81	**448**	100	27	13	13	12	12	11	10	**0.00**	20	32	11	–	–	–	–	–	–	–	–	–	–	β-D-Glucopyranoside, 1-ethyl-3-hydroxybutyl, 2,3,4,6-tetraacetate, [S-(R*,R*)]-	56805-09-5	MS09850
126	299	82	448	401				**448**	100	14	6	2	2					20	36	7	2	1	–	–	–	–	–	–	–	–	D-erythro-D-Galacto-octopyranoside, methyl 6,8-dideoxy-6-(1-methyl-4-propyl-l-2-pyrrolidinecarboxamido)-1-thio-, 7-acetate, trans-, α-	17147-41-0	NI43179
73	81	101	224	222	135	263	261	**448**	100	6	5	4	4	3	1	1	**0.00**	20	37	4	–	–	–	–	1	–	1	–	–	–	Methyl 10-[(bromomethyl)dimethylsilyloxy]-11-methoxy-3,7,11-trimethyl 2,6-dodecadienoate		MS09851
448	431	230	136	184	312	200	159	**448**	100	23	14	12	9	8	8	7		21	16	6	6	–	–	–	–	–	–	–	–	–	4-[[5-Hydroxy-3-methyl-1-(4-nitrophenyl)-2-pyrazolyl]methylene]-3-methyl-1-(4-nitrophenyl)-2-pyrazolin-5-one		NI43180
286	60	153	287	73	43	258	57	**448**	100	43	25	19	19	15	14	13	**0.00**	21	20	11	–	–	–	–	–	–	–	–	–	–	Cynaroside	5373-11-5	NI43181
257	73	347	117	75	43	348	232	**448**	100	95	80	58	41	32	23	22	**1.00**	21	32	5	2	–	–	–	–	–	2	–	–	–	2,4,6(1H,3H,5H)-Pyrimidinetrione, 5-[2,3-bis[(trimethylsilyl)oxy]-2-propenyl]-1,3-dimethyl-5-phenyl-	57346-73-3	NI43182
448	446	433	431	403	418	416	401	**448**	100	92	67	64	9	8	8	8		22	24	–	–	–	–	–	–	–	–	–	–	2	[1]Benzoselenopheno[2,3-b][1]benzoselenophene, 3,8-di-tert-butyl-	25855-83-8	LI09052
448	446	433	431	444	447	445	429	**448**	100	92	67	63	54	53	40	37		22	24	–	–	–	–	–	–	–	–	–	–	2	[1]Benzoselenopheno[2,3-b][1]benzoselenophene, 3,8-di-tert-butyl-	25855-83-8	NI43183
101	70	56	114	58	100	72	57	**448**	100	77	77	73	73	63	57	55	**12.10**	22	44	7	2	–	–	–	–	–	–	–	–	–	5,8,11,18,21,26,29-Heptaoxa-1,15-diazabicyclo[13.8.8]hentriacontane		MS09852
317	205	448	391	58	274	44	114	**448**	100	86	78	75	74	65	63	43		23	32	7	2	–	–	–	–	–	–	–	–	–	Dichotine, tetrahydro-11-methoxy-	29474-88-2	NI43184
317	205	448	391	374	58	274	44	**448**	100	86	79	75	74	72	63	61		23	32	7	2	–	–	–	–	–	–	–	–	–	Dichotine, tetrahydro-11-methoxy-		LI09053

MASS TO CHARGE RATIOS								M.W.	INTENSITIES								Parent	C	H	O	N	S	F	Cl	Br	I	Si	P	B	X	COMPOUND NAME	CAS Reg No	No
77	141	69	94	95	150	156	127	**448**	100	25	24	21	14	11	10	10	**0.00**	24	32	6	–	1	–	–	–	–	–	–	–	–	2,6-Decadienedioic acid, 3,7-dimethyl-9-(2-methyl-1-propenyl)-9-(phenylsulphonyl)-, dimethyl ester, (E,E)-	57683-65-5	NI43185
179	328	122	121	208	123	109	105	**448**	100	83	74	72	55	39	39	36	**0.05**	24	32	8	–	–	–	–	–	–	–	–	–	–	Acetoxycrotophorbolon-20-acetate		LI09054
105	149	166	202	433	355	293	415	**448**	100	56	28	19	17	6	5	2	**0.00**	24	32	8	–	–	–	–	–	–	–	–	–	–	1α-Benzoyloxy-9α-acetoxy-6β,8β,4β-trihydroxydihydro-β-agarofuran		MS09853
136	121	448	224	137	45	80	43	**448**	100	67	52	46	35	26	21	19		24	32	8	–	–	–	–	–	–	–	–	–	–	Dibenzo-24-crown-8	14174-09-5	NI43186
136	121	137	80	109	110	108	73	**448**	100	73	57	31	24	20	16	16	**13.30**	24	32	8	–	–	–	–	–	–	–	–	–	–	2,3,14,15-Dibenzo-1,4,7,10,13,16,19,22-octaoxacyclotetracoza-2,14-diene		MS09854
43	83	41	69	28	53	39	45	**448**	100	81	71	55	55	53	45	44	**0.25**	24	32	8	–	–	–	–	–	–	–	–	–	–	Phorbol-12,20-diacetate		NI43187
268	253	240	328	373	225	286	388	**448**	100	44	37	36	32	27	21	20	**8.00**	25	36	7	–	–	–	–	–	–	–	–	–	–	3α,7α,12α-Triacetoxy-5β-androstan-17-one		NI43188
346	388	43	264	448	347	277	389	**448**	100	64	62	42	36	26	25	16		25	36	7	–	–	–	–	–	–	–	–	–	–	2β,3β,17β-Triacetoxy-5α-androst-6-one		MS09855
332	333	129	242	334	433	75	200	**448**	100	27	10	9	8	6	6	4	**3.00**	25	44	3	–	–	–	–	–	–	2	–	–	–	Androst-4-en-3-one, 1β,17β-dihydroxy-, di-TMS		LI09055
433	434	435	431	432	332	129	74	**448**	100	39	14	10	6	6	5	4	**1.00**	25	44	3	–	–	–	–	–	–	2	–	–	–	Androst-4-en-3-one, 2β,17β-dihydroxy-, di-TMS		LI09056
392	448	433	393	449	434	211	129	**448**	100	93	93	46	44	42	27	27		25	44	3	–	–	–	–	–	–	2	–	–	–	Androst-4-en-3-one, 6α,17β-dihydroxy-, di-TMS		LI09057
358	448	191	269	268	359	73	343	**448**	100	66	52	50	46	30	30	29		25	44	3	–	–	–	–	–	–	2	–	–	–	Androst-4-en-3-one, 16α,17β-dihydroxy-, di-TMS	23261-29-2	NI43189
358	268	129	191	317	217	448		**448**	100	84	76	34	28	28	4			25	44	3	–	–	–	–	–	–	2	–	–	–	Androst-4-en-3-one, 17β,18-dihydroxy-, di-TMS		LI09058
129	392	319	393	130	377	448	394	**448**	100	32	18	14	14	9	6	6		25	44	3	–	–	–	–	–	–	2	–	–	–	Androst-5-en-11-one, 3β,17β-dihydroxy-, di-TMS	49774-90-5	NI43190
73	129	191	75	214	433	143	117	**448**	100	99	58	51	27	25	25	25	**18.00**	25	44	3	–	–	–	–	–	–	2	–	–	–	Androst-5-en-16-one, 3,17α-dihydroxy-, di-TMS	21952-35-2	NI43191
129	448	171	143	214	213	145	130	**448**	100	21	20	16	15	15	15	13		25	44	3	–	–	–	–	–	–	2	–	–	–	Androst-5-en-16-one, 3β,17β-dihydroxy-, di-TMS	21952-35-2	NI43192
129	433	171	130	448	145	214	143	**448**	100	30	23	19	17	16	15	14		25	44	3	–	–	–	–	–	–	2	–	–	–	Androst-5-en-16-one, 3β,17β-dihydroxy-, di-TMS	21952-35-2	NI43193
214	304	131	175	199	145	196	129	**448**	100	65	43	33	30	29	26	24	**6.00**	25	44	3	–	–	–	–	–	–	2	–	–	–	Androst-5-en-17-one, 3α,16α-dihydroxy-, di-TMS		LI09059
214	304	129	199	175	117	196	88	**448**	100	59	52	40	39	35	33	22	**8.00**	25	44	3	–	–	–	–	–	–	2	–	–	–	Androst-5-en-17-one, 3β,16α-dihydroxy-, di-TMS	13111-28-9	LI09060
214	129	304	117	196	199	175	215	**448**	100	52	47	41	38	36	35	19	**2.00**	25	44	3	–	–	–	–	–	–	2	–	–	–	Androst-5-en-17-one, 3β,16α-dihydroxy-, di-TMS	13111-28-9	NI43194
214	304	196	129	199	175	117	305	**448**	100	76	33	33	32	29	20	16	**6.47**	25	44	3	–	–	–	–	–	–	2	–	–	–	Androst-5-en-17-one, 3β,16α-dihydroxy-, di-TMS	13111-28-9	NI43195
214	129	304	199	196	175	117	215	**448**	100	56	50	40	38	27	26	15	**5.27**	25	44	3	–	–	–	–	–	–	2	–	–	–	Androst-5-en-17-one, 3β,16β-dihydroxy-, di-TMS	40822-67-1	NI43196
214	304	129	196	175	117	199	433	**448**	100	76	73	36	33	31	30	23	**15.00**	25	44	3	–	–	–	–	–	–	2	–	–	–	Androst-5-en-17-one, 3β,16β-dihydroxy-, di-TMS	40822-67-1	LI09061
158	103	248	129	418	343	213	159	**448**	100	45	35	35	26	26	26	22	**10.96**	25	44	3	–	–	–	–	–	–	2	–	–	–	Androst-5-en-17-one, 3β,19-dihydroxy-, di-TMS	57397-08-7	NI43197
392	448	433	393	449	434	129	358	**448**	100	94	94	47	44	40	28	24		25	44	3	–	–	–	–	–	–	2	–	–	–	6-α-Hydroxytestosterone, TMS ether		LI09062
100	224	208	153	293	238	196	70	**448**	100	70	65	65	60	60	60	60	**3.00**	25	44	3	4	–	–	–	–	–	–	–	–	–	13-(4,8-Diacetyl-10-cyano-4,8-diazadecyl)-12-dodecane lactam	75701-27-8	NI43198
117	234	199	351	377	214	316	276	**448**	100	98	87	65	62	54	53	42	**11.70**	26	36	–	4	–	–	–	–	–	–	–	4	–	1-Tert-butyl-2,4,6,8-tetramethyl-3,5,7-triphenyloctahydro-1,3,5,7-tetraza-2,4,6,8-tetraborocine		NI43199
448	99	125	43	259	344	300	275	**448**	100	84	75	56	23	14	13	8		26	40	6	–	–	–	–	–	–	–	–	–	–	19R-Acetoxy-19-methyl-5α-androstane-3,3:17,17-bisethylene dioxide		LI09063
95	81	121	107	198	175	191	253	**448**	100	49	37	35	30	29	27	13	**0.00**	26	40	6	–	–	–	–	–	–	–	–	–	–	Isolinaritriol triacetate		NI43200
109	81	98	69	174	121	234	134	**448**	100	80	60	55	24	24	21	20	**0.70**	26	40	6	–	–	–	–	–	–	–	–	–	–	Salvileucolide methyl ester		NI43201
209	43	41	190	55	179	57	203	**448**	100	52	39	33	32	29	27	15	**3.00**	26	44	4	2	–	–	–	–	–	–	–	–	–	N-Methyl-N-octadecyl-2-nitro-4-carboxyaniline		IC04437
73	143	253	268	75	227	129	213	**448**	100	65	61	59	55	26	26	24	**3.61**	26	48	2	–	–	–	–	–	–	2	–	–	–	Androst-5-ene, 3,17-bis[(trimethylsilyl)oxy]-17-methyl-, (3β,17β)-	39780-59-1	NI43202
73	143	253	268	75	129	227	213	**448**	100	56	53	51	47	23	22	21	**3.11**	26	48	2	–	–	–	–	–	–	2	–	–	–	Androst-5-ene, 3,17-bis[(trimethylsilyl)oxy]-17-methyl-, (3β,17β)-	39780-59-1	MS09856
268	253	143	227	213	144	226	211	**448**	100	92	78	41	40	29	28	26	**2.00**	26	48	2	–	–	–	–	–	–	2	–	–	–	Androst-5-ene, 3,17-bis[(trimethylsilyl)oxy]-17-methyl-, (3β,17β)-	39780-59-1	NI43203
259	448	260	246	105	109	121	107	**448**	100	36	30	24	21	19	16	15		27	33	5	–	–	–	–	–	–	–	–	1	–	5β-Pregnane-3,11,20-trione, 17α,21-dihydroxy-, phenyl boronate	30888-27-8	NI43204
155	55	131	289	318	303	41	122	**448**	100	53	40	31	20	20	18	14	**3.50**	27	44	5	–	–	–	–	–	–	–	–	–	–	Crestagenin		MS09857
289	363	271	253	327	345	448		**448**	100	40	10	8	7	5	2			27	44	5	–	–	–	–	–	–	–	–	–	–	Isoplexigenin C		LI09064
99	55	141	105	386	353	271	73	**448**	100	95	83	66	63	59	58	56	**4.30**	27	44	5	–	–	–	–	–	–	–	–	–	–	Methyl 7α-hydroxy-3,3-ethylenedioxy-5β-cholanoate		MS09858
363	345	57	289	43	55	71	448	**448**	100	54	43	40	38	34	31	30		27	44	5	–	–	–	–	–	–	–	–	–	–	Paniculogenin	16750-37-1	NI43205
155	131	156	289	271	318	303	360	**448**	100	52	25	23	15	14	12	11	**8.05**	27	44	5	–	–	–	–	–	–	–	–	–	–	Spirostan-2,3,27-triol, (2β,3α,5β,25R)-	55401-44-0	NI43206
155	131	156	289	318	271	303	360	**448**	100	53	25	22	15	15	12	11	**8.00**	27	44	5	–	–	–	–	–	–	–	–	–	–	Spirostan-2,3,27-triol, (2β,3α,5β,25S)-	21152-89-6	LI09065
343	55	177	69	430	137	203	420	**448**	100	99	93	90	73	67	45	43	**3.00**	27	44	5	–	–	–	–	–	–	–	–	–	–	2β,3β,5β,14α-Tetrahydroxy-5β-cholest-7-en-6-one		LI09066
155	352	81	137	363	157	215	136	**448**	100	43	42	41	38	30	25	25	**1.81**	27	45	2	–	–	1	–	–	–	1	–	–	–	Silane, (4-fluorophenyl)methylbis[[5-methyl-2-isopropylcyclohexyl]oxy]-, [1α(1R*,2S*,5R*),2β,5α]-	74841-57-9	NI43207
412	430	255	413	273	228	213	397	**448**	100	56	40	36	32	30	28	27	**8.00**	28	48	4	–	–	–	–	–	–	–	–	–	–	Cholestan-26-oic acid, 3,7-dihydroxy-, methyl ester, (3α,5β,7α)-	13316-62-6	NI43208
55	69	41	81	95	109	93	83	**448**	100	99	76	68	66	47	43	43	**2.00**	28	48	4	–	–	–	–	–	–	–	–	–	–	Ergost-25-ene-3,5,6,12-tetrol, (3β,5α,6β,12β)-	56052-97-2	NI43209
105	448	77	121	449	415	371	106	**448**	100	68	50	39	21	14	9	9		29	20	1	–	2	–	–	–	–	–	–	–	–	Acetophenone, 2-(4,5-diphenyl-3H-1,2-dithiol-3-ylidene)-2-phenyl-	16074-89-8	NI43210
448	433	449	377	434	355	370	314	**448**	100	59	44	37	24	21	17	15		29	40	2	–	–	–	–	–	–	1	–	–	–	Δ^1-Tetrahydrocannabinol phenyldimethylsilyl ether		NI43211
430	412	247	244	230	397	211	290	**448**	100	81	47	45	39	23	20	16	**7.00**	29	52	3	–	–	–	–	–	–	–	–	–	–	5α-Stigmastane-3β,5,6β-triol	20835-91-0	NI43212
196	448	420	392	449	421	178	197	**448**	100	64	57	29	22	19	18	16		30	18	2	–	–	2	–	–	–	–	–	–	–	p-Benzoquinone, 2,5-bis(p-fluorophenyl)-3,6-diphenyl-	22030-92-8	MS09859
196	448	420	392	449	421	197	194	**448**	100	56	48	26	18	15	15	12		30	18	2	–	–	2	–	–	–	–	–	–	–	p-Benzoquinone, 2,6-bis(p-fluorophenyl)-3,5-diphenyl-	22030-93-9	MS09860
196	448	420	392	178	214	98	81	**448**	100	59	53	27	17	9	3	1		30	18	2	–	–	2	–	–	–	–	–	–	–	2,6-Bis(p-fluorophenyl)-3,5-diphenyl benzoquinone		LI09067
196	448	420	392	178	214	98	81	**448**	100	57	49	27	11	6	5	1		30	18	2	–	–	2	–	–	–	–	–	–	–	2,6-Bis(p-fluorophenyl)-3,5-diphenyl benzoquinone		LI09068

MASS TO CHARGE RATIOS								M.W.	INTENSITIES								Parent	C	H	O	N	S	F	Cl	Br	I	Si	P	B	X	COMPOUND NAME	CAS Reg No	No
119	206	191	189	312	295	392	220	**448**	100	65	22	14	13	10	5	5	**2.00**	30	32	–	4	–	–	–	–	–	–	–	–	–	2H-1,2,3-Triazol-4-amine, 2-[1,2-bis(4-methylphenyl)-3-phenyl-2-cyclopropen-1-yl]-N,N-diethyl-5-methyl-	65660-06-2	NI43213
448	405	449	406	366	340	109	353	**448**	100	41	35	14	12	12	9	8		30	40	3	–	–	–	–	–	–	–	–	–	–	Chamaecydin	86746-82-9	NI43214
91	357	92	251	105	358	97	83	**448**	100	23	11	8	8	6	6	6	**0.35**	30	40	3	–	–	–	–	–	–	–	–	–	–	4-Hexadecynoic acid, 2-benzyloxy-, benzyl ester	41011-99-8	NI43215
448	405	449	406	340	109	366	352	**448**	100	69	33	22	9	8	7	7		30	40	3	–	–	–	–	–	–	–	–	–	–	Isochamaecydin	86699-53-8	NI43216
131	69	41	146	70	159	255	388	**448**	100	31	26	25	25	22	18	15	**14.46**	30	40	3	–	–	–	–	–	–	–	–	–	–	2-Propen-1-one, 1-[(3β,5β,17β)-3-(acetyloxy)androstan-17-yl]-3-phenyl-	74464-42-9	NI43217
85	167	55	97	153	181	139	377	**448**	100	80	60	41	9	8	7	6	**0.00**	30	56	2	–	–	–	–	–	–	–	–	–	–	2-(4-Pentacosynyloxy)tetrahydro-2H-pyran		DD01692
365	366	297	255	448	433	367		**448**	100	23	8	6	5	5	5			31	44	2	–	–	–	–	–	–	–	–	–	–	Biscannabichromen		LI09069
365	366	448	285	433	367	243	297	**448**	100	25	8	6	5	5	5	3		31	44	2	–	–	–	–	–	–	–	–	–	–	Biscannabicyclol		LI09070
297	365	380	327	448	298	271	241	**448**	100	57	48	43	39	27	21	21		31	44	2	–	–	–	–	–	–	–	–	–	–	2,4-Bis(3,3,4-trans-p-menthadien-(1,8)yl)-5-pentyl resorcinol		LI09071
69	169	92	47	131	100	364	119	**449**	100	35	25	14	13	12	10	10	**0.00**	8	–	1	1	–	17	–	–	–	–	–	–	–	N-(Nonafluorobutyl)-N-(heptafluoropropyl)carbamoyl fluoride		DD01693
69	169	92	47	131	100	364	119	**449**	100	35	25	14	13	12	10	10	**0.00**	8	–	1	1	–	17	–	–	–	–	–	–	–	Perfluoro(n-propyl-n-butyl-carbamoyl) fluoride		MS09861
46	311	63	64	89	139	184	127	**449**	100	55	55	46	40	33	26	16	**0.00**	12	8	6	3	2	5	–	–	–	–	–	–	–	S,S-Bis(4-nitrophenoxy)-n-pentafluorosulphanylsulphilimine		NI43218
451	453	449	32	379	455	381	386	**449**	100	64	63	40	27	21	21	20		17	8	3	1	–	–	5	–	–	–	–	–	–	Naphthalene, 3-methoxy-1-nitro-2-(pentachlorophenyl)-	57187-61-8	NI43219
87	43	449	376	450	159	419	294	**449**	100	42	28	27	7	7	6	6		19	23	6	5	1	–	–	–	–	–	–	–	–	N,N-Bis(2-acetoxyethyl)-3-methyl-4-(3-methyl-4-nitro-isothiaz-5-ylazo)aniline		IC04438
129	160	56	54	120	70	80	78	**449**	100	90	59	56	51	38	35	33	**0.56**	19	39	7	5	–	–	–	–	–	–	–	–	–	Gentamicin C1α	26098-04-4	NI43220
450	451	110	142	449	111	128	126	**449**	100	24	9	5	3	3	2	2		19	39	7	5	–	–	–	–	–	–	–	–	–	Gentamicin C1α	26098-04-4	NI43221
138	139	450	165	149	286	137	122	**449**	100	17	9	6	4	3	3	3	**0.50**	20	20	5	1	–	5	–	–	–	–	–	–	–	Scopolamine pentafluoropropionate		NI43222
174	43	114	178	180	261	220	263	**449**	100	73	46	43	16	13	10	5	**0.01**	20	20	5	3	1	–	1	–	–	–	–	–	–	Cloxacillin methyl ester		MS09862
148	449	81	109	169	119	161	147	**449**	100	96	74	49	48	42	29	29		22	27	9	1	–	–	–	–	–	–	–	–	–	1-(2,3,4,6-Tetra-O-acetyl-β-D-glucopyranosyl)indoline		LI09072
343	449	345	451	289	344	169	113	**449**	100	72	66	50	48	42	40	23		24	17	2	3	–	–	2	–	–	–	–	–	–	1-(2,4-Dichlorophenylazo)-2-hydroxy-3-(2-methylphenylcarbamoyl)-naphthalene		IC04439
278	89	221	279	43	178	207	249	**449**	100	44	29	20	20	13	12	11	**0.00**	26	27	6	1	–	–	–	–	–	–	–	–	–	Epanorin methyl ester	22149-30-0	NI43223
255	39	81	117	55	273	107	41	**449**	100	93	53	52	52	36	36	35	**1.13**	26	43	5	1	–	–	–	–	–	–	–	–	–	Glycine, N-[(3α,5β,12α)-3,12-dihydroxy-24-oxocholan-24-yl]-	360-65-6	NI43224
380	294	140	43	381	57	55	69	**449**	100	75	65	44	31	30	27	23	**0.80**	26	50	1	1	–	3	–	–	–	–	–	–	–	N,N-Didodecyltrifluoroacetamide		MS09863
307	91	308	43	449	306	216	300	**449**	100	86	26	25	24	17	15	12		29	23	4	1	–	–	–	–	–	–	–	–	–	Dimethyl 13-benzyl-1,4-dihydro-1,4-iminotriphenylene-2,3-dicarboxylate	110028-22-3	DD01694
377	378	78	449	72	332	379	333	**449**	100	31	11	8	5	5	4	4		30	27	3	1	–	–	–	–	–	–	–	–	–	4-[(N,N-Dimethylcarbamoyl)diphenylmethyl]diphenylacetic acid		NI43225
360	44	361	43	91	81	95	93	**449**	100	85	27	25	22	13	10	10	**0.00**	30	43	2	1	–	–	–	–	–	–	–	–	–	Androst-4-en-17-ol, 3-[(1-methyl-2-phenylethyl)amino]-, acetate (ester), (3β,17β)-	55401-53-1	NI43226
331	269	256	253	330	332	165	254	**449**	100	49	18	18	16	14	11	10	**5.00**	33	23	1	1	–	–	–	–	–	–	–	–	–	3,3-Diphenyl-4-phenylimino-2-spiro-(9-fluorenyl)-oxetane		MS09864
255	112	91	256	254	99	105	86	**449**	100	26	25	21	19	16	14	9	**8.00**	33	23	1	1	–	–	–	–	–	–	–	–	–	1,3,3-Triphenyl-4-oxo-2-spiro-(9-fluorenyl)-azetidine		MS09865
217	215	15	214	216	213	202	14	**450**	100	93	88	74	63	49	49	46	**14.00**	2	6	1	–	–	–	–	–	–	–	–	–	2	Bis(methylmercury)oxide		LI09073
105	270	255	92	225	89	165	91	**450**	100	96	90	90	80	70	60	60	**38.00**	5	15	–	–	–	–	–	–	–	–	–	–	5	Pentamethylcyclopentaarsine		LI09075
255	270	105	225	450	165	195	435	**450**	100	99	97	53	42	33	28	25		5	15	–	–	–	–	–	–	–	–	–	–	5	Pentamethylcyclopentaarsine		LI09074
31	51	131	69	29	49	61	101	**450**	100	20	13	11	9	8	5	4	**0.01**	9	3	1	–	–	17	–	–	–	–	–	–	–	Perfluoro-octyl methanol		MS09866
373	375	214	212	377	371	216	294	**450**	100	99	77	55	33	33	25	23	**0.09**	10	14	–	–	–	–	–	4	–	–	–	–	–	Cyclohexene, 4,5-dibromo-1-(1,2-dibromoisopropyl)-4-methyl-, [1(R*),4α,5β]-(±)-	70168-60-4	NI43227
199	350	299	207	150	181	100	200	**450**	100	65	26	21	11	10	10	8	**7.40**	12	2	–	–	–	16	–	–	–	–	–	–	–	2H,7H-Hexadecafluorotricyclo[6.2.2.0^{2,7}]dodec-9-ene		NI43228
330	176	420	390	253	360	450	165	**450**	100	99	79	60	58	46	23	23		12	10	4	4	2	–	–	–	–	–	–	–	2	Di-μ-benzenethiolato-tetranitroso-di-iron		MS09867
226	338	310	282	254	52	114	394	**450**	100	84	84	78	38	37	36	32	**30.00**	12	12	8	–	–	–	–	–	–	–	2	–	2	Di-μ-dimethylphosphinobis(tetracarbonylchromium)		MS09868
127	281	250	350	231	219	450	249	**450**	100	69	57	38	22	21	19	17		15	4	–	–	–	14	–	–	–	–	–	–	–	4H-Tetradecafluoro-6-methylhexacyclo[7.2.2.1^{3,7}.0^{2,4}.0^{2,8}.0^{6,8}]tetradec-10-ene		NI43229
350	281	335	250	285	231	113	69	**450**	100	81	52	45	33	33	31	21	**11.90**	15	4	–	–	–	14	–	–	–	–	–	–	–	4H-Tetradecafluoro-6-methylpentacyclo[7.2.2.1^{3,7}.0^{4,6}.0^{2,8}]tetradeca-2(8),10-diene		NI43230
363	69	450	115	196	117	197	97	**450**	100	72	54	53	51	44	41	28		15	10	2	–	–	12	–	–	–	–	–	–	–	2,6-Bis(trifluoromethyl)-2,6-dihydroxy-1,1,1,7,7,7-hexafluoro-4-phenylhept-3-ene		LI09076
127	139	156	181	265	140	153	109	**450**	100	37	26	22	21	18	17	17	**1.00**	17	33	10	–	–	–	–	–	–	–	1	2	–	D-Gluconic acid, methyl ester, cyclic 2,3:4,5-bis(butylboronate) 6-(dimethyl phosphate)	58634-47-2	NI43231
56	84	149	112	282	226	97	254	**450**	100	30	25	25	20	20	20	15	**0.00**	18	10	7	–	–	–	–	–	–	–	–	–	2	Hexacarbonyl-μ-[(1-3-η^3:4,4α,5-η^3)-4-methylene-2-phenyl-1-oxo-2-pentene-1,5-diyl]diiron	122624-34-4	DD01695
374	376	450	452	375	298	377	196	**450**	100	51	45	27	24	16	13	13		18	20	–	2	4	–	–	–	–	–	–	–	1	Nickel, bis(ethylphenylcarbamodithioato-S,S′)-, (SP-4-1)-	30145-37-0	NI43232

MASS TO CHARGE RATIOS								M.W.	INTENSITIES								Parent	C	H	O	N	S	F	Cl	Br	I	Si	P	B	X	COMPOUND NAME	CAS Reg No	No
86	83	85	56	57	132	60	87	**450**	100	83	77	38	36	27	26	23	**5.44**	18	34	5	4	2	–	–	–	–	–	–	–	–	**6,9,17,20,25-Pentaoxa-1,3,12,14-tetraazabicyclo[12.8.5]heptacosane-2,13-dithione**	110228-76-7	**NI43233**
86	61	85	132	56	70	100	74	**450**	100	88	70	58	54	42	36	32	**10.99**	18	34	5	4	2	–	–	–	–	–	–	–	–	**6,14,17,22,25-Pentaoxa-1,3,9,11-tetraazabicyclo[9.8.8]heptacosane-2,10-dithione**	89663-10-5	**NI43234**
73	217	147	131	220	129	231	75	**450**	100	62	30	25	23	23	22	18	**0.50**	18	38	7	–	–	–	–	–	–	3	–	–	–	**Trimethylsilyl 1,2-O-isopropylidene-3,6-di-O-trimethylsilyl-D-glucofuranuronate**		**NI43235**
73	102	77	75	232	172	147	74	**450**	100	90	47	46	38	28	28	18	**1.60**	18	46	3	2	–	–	–	–	–	4	–	–	–	**Trimethylsilyl 2,6-bis(trimethylsilylamino)-5-(trimethylsilyloxy)hexanoate**	72361-02-5	**NI43236**
196	225	345	152	240	272	121	61	**450**	100	67	48	48	43	33	33	33	**0.00**	20	18	4	–	4	–	–	–	–	–	–	–	–	**DL-Bis(1-carboxyethyl) 4,4′-biphenylylenebis(dithiocarboxylate)**	110699-22-4	**DD01696**
166	181	73	182	120	450	79	151	**450**	100	95	55	51	38	35	31	26		20	22	10	2	–	–	–	–	–	–	–	–	–	**4′,5″-Dinitrodibenzo[b,k]-1,4,7,10,13,16-hexaoxacyclooctadeca-2,11-diene**	32082-45-4	**NI43237**
166	181	182	73	120	79	450	107	**450**	100	80	57	43	40	33	29	27		20	22	10	2	–	–	–	–	–	–	–	–	–	**4′,4″-Dinitrodibenzo[b,k]-1,4,7,10,13,16-hexaoxacyclooctadecan-2,11-diene**	32082-46-5	**NI43238**
65	284	120	78	164	268	39	92	**450**	100	75	67	54	48	37	36	34	**2.15**	21	18	6	6	–	–	–	–	–	–	–	–	–	**1,2-Bis[(Z)-(4′-methyl-2′-nitrophenyl)-ONN-azoxy]-4-methylbenzene**		**NI43239**
280	234	267	207	395	219	137	450	**450**	100	21	21	20	18	14	14	10		21	26	7	2	1	–	–	–	–	–	–	–	–	**tert-Butyl 3-(2-hydroxy-5-methoxybenzyl)-7-carbomethoxyamino-2-cephem-4-carboxylate**		**NI43240**
280	207	450	281	234	137	336	156	**450**	100	24	22	22	22	22	19	19		21	26	7	2	1	–	–	–	–	–	–	–	–	**tert-Butyl 3-(4-hydroxy-3-methoxybenzyl)-7-carbomethoxyamino-2-cephem-4-carboxylate**		**NI43241**
246	217	73	205	361	348	147	271	**450**	100	33	26	24	23	23	16	13	**3.00**	21	38	3	2	–	–	–	–	–	3	–	–	–	**(2′S,3′R)-2-[2′,3′,4′-Tris(trimethylsilyloxy)butyl]quinoxaline**	86204-48-0	**NI43242**
246	73	361	348	271	272	206	218	**450**	100	92	55	41	40	37	33	32	**4.00**	21	38	3	2	–	–	–	–	–	3	–	–	–	**(2′S)-2-Trimethylsilyloxymethyl-3-[2′,3′-bis(trimethylsilyloxy)propyl]quinoxaline**	86204-47-9	**NI43243**
73	147	233	75	305	217	191	74	**450**	100	45	24	22	13	11	11	11	**0.00**	21	50	4	–	–	–	–	–	–	3	–	–	–	**3,10-Bis(trimethylsiloxy)dodecanoic acid, bis(trimethylsilyl) ester**		**NI43244**
450	379	346	333	451	392	331	418	**450**	100	45	29	28	27	14	13	12		23	18	6	2	1	–	–	–	–	–	–	–	–	**1,3-Dioxo-5-methoxy-2-(3-methoxypropyl)-11-nitro-1,3-dihydrothioxantheno[3,1,9-def]isoquinoline**		**IC04440**
83	149	85	167	278	113	160	112	**450**	100	92	56	36	20	18	14	12	**0.90**	23	18	6	2	1	–	–	–	–	–	–	–	–	**Methyl 3-(4-hydroxybenzyl)-7-phthalimido-3-cephem-4-carboxylate**		**NI43245**
85	57	228	128	43	256	161	200	**450**	100	84	28	28	27	21	17	14	**0.00**	23	30	9	–	–	–	–	–	–	–	–	–	–	**Melcanthin F**	79405-85-9	**NI43246**
331	391	330	178	179	312	390	258	**450**	100	99	95	86	85	69	66	63	**0.30**	23	30	9	–	–	–	–	–	–	–	–	–	–	**4,7β,9-Trihydroxy-12β,13-diacetoxy-20-nor-tigli-1-en-3,6-dione**		**LI09077**
77	450	325	357	451	51	76	168	**450**	100	95	61	39	29	29	26	24		24	18	5	–	2	–	–	–	–	–	–	–	–	**Bis(phenylsulphonyl)diphenyl ether**		**IC04441**
43	324	134	366	296	408	133	106	**450**	100	78	41	36	30	20	17	16	**5.00**	24	18	9	–	–	–	–	–	–	–	–	–	–	**Tri-O-acetylgrevillin A**		**NI43247**
450	217	139	201	294	183	185	107	**450**	100	56	56	47	27	16	15	13		24	20	1	–	2	–	–	–	–	–	2	–	–	**Diphenylphosphinothioic anhydride**		**MS09869**
450	217	139	201	294	183	185	63	**450**	100	56	56	47	27	16	15	13		24	20	1	–	2	–	–	–	–	–	2	–	–	**Diphenylphosphinothioic anhydride**		**LI09078**
270	159	157	133	146	145	77	43	**450**	100	52	52	48	45	32	27	27	**15.16**	24	26	5	4	–	–	–	–	–	–	–	–	–	**Estra-1,3,5(10)-trien-17-one, 3-(2,4-dinitrophenylazo)-**		**NI43248**
255	256	128	115	141	41	450	143	**450**	100	22	19	18	14	14	13	13		24	26	5	4	–	–	–	–	–	–	–	–	–	**Estra-1,3,5(10)-trien-17-one, 3-(2,4-dinitrophenylazo)-**		**NI43249**
84	98	107	77	191	108	202	56	**450**	100	66	26	26	22	22	20	20	**0.10**	24	30	3	6	–	–	–	–	–	–	–	–	–	**5-Amino-3-phenoxy-1-[3-[3-(piperidinomethyl)phenoxy]propyl]carbamoyl-1H-1,2,4-triazole**		**MS09870**
43	83	69	111	312	91	55	95	**450**	100	58	37	23	21	21	21	15	**0.82**	24	34	8	–	–	–	–	–	–	–	–	–	–	**3-Desoxo-3β-hydroxy-phorbol-12,13-diacetate**		**NI43250**
343	390	165	121	418	375	358	313	**450**	100	75	50	45	40	40	37	37	**16.00**	24	34	8	–	–	–	–	–	–	–	–	–	–	**Picras-2-ene-1,16-dione, 11-(acetyloxy)-13-hydroxy-2,12-dimethoxy-, (11α,12β)-**	28360-79-4	**NI43251**
414	400	191	151	113	385	357	200	**450**	100	78	34	27	25	22	19	19	**0.00**	25	19	6	–	–	–	1	–	–	–	–	–	–	**Sphagnorubin C**	84018-31-5	**NI43252**
217	393	450	391	408	435	394	351	**450**	100	98	58	53	40	36	27	15		25	22	8	–	–	–	–	–	–	–	–	–	–	**Barpisoflavone C diacetate**		**NI43253**
366	311	408	450	296				**450**	100	44	29	26	10					25	22	8	–	–	–	–	–	–	–	–	–	–	**Glycyrol diacetate**		**LI09079**
138	194	151	165	176	450	258	299	**450**	100	99	99	45	30	29	26	20		25	22	8	–	–	–	–	–	–	–	–	–	–	**O-Methylkielcorin**		**NI43254**
435	213	128	436	77	73	291	307	**450**	100	39	33	27	25	19	12	11	**0.00**	25	27	4	–	–	–	–	–	–	1	1	–	–	**Diphenyl 2-(trimethylsilyl)-3,4-dihydro-1-naphthalenyl phosphate**	118297-84-0	**DD01697**
211	209	91	207	210	208	120	215	**450**	100	79	73	46	33	29	19	18	**0.30**	25	30	–	–	–	–	–	–	–	–	–	–	1	**Tribenzyl-tert-butyltin**		**NI43255**
73	202	216	45	450	75	203	74	**450**	100	37	25	13	10	10	9	9		25	34	2	2	–	–	–	–	–	2	–	–	–	**1-TMS-3-(p-TMS-Oxyphenylethylcarbonylaminoethyl)indole**		**NI43256**
43	58	330	29	42	27	26	41	**450**	100	99	88	88	84	76	74	72	**16.00**	25	38	7	–	–	–	–	–	–	–	–	–	–	**Pregn-5-ene-3,11,12,14,20-pentol, 11,12-diacetate, (3β,11α,12β,14β)-**	20230-39-1	**NI43257**
226	255	237	211	270	252	330	286	**450**	100	24	24	16	12	10	8	8	**1.00**	25	38	7	–	–	–	–	–	–	–	–	–	–	**3α,7α,12β-Triacetoxy-17α-hydroxy-5β-androstane**		**NI43258**
407	226	237	252	211	255	269	270	**450**	100	69	54	45	43	38	32	28	**1.00**	25	38	7	–	–	–	–	–	–	–	–	–	–	**7α,12α,17β-Triacetoxy-3α-hydroxy-5β-androstane**		**NI43259**
73	319	75	320	270	129	105	93	**450**	100	77	41	20	19	14	14	12	**10.76**	25	46	3	–	–	–	–	–	–	2	–	–	–	**5α-Androstan-3-one, 7α,17β-dihydroxy-, di-TMS**	61233-49-6	**NI43260**
360	129	270	217	191	319	147	271	**450**	100	92	88	60	48	42	33	32	**2.00**	25	46	3	–	–	–	–	–	–	2	–	–	–	**5α-Androstan-3-one, 17β,18-dihydroxy-, di-TMS**		**LI09080**
129	435	450	215	216	117	436	217	**450**	100	58	40	38	37	26	25	18		25	46	3	–	–	–	–	–	–	2	–	–	–	**5α-Androstan-16-one, 3β,17β-dihydroxy-, di-TMS**		**LI09081**
270	360	271	255	213	361	243	129	**450**	100	85	76	40	28	26	25	21	**3.00**	25	46	3	–	–	–	–	–	–	2	–	–	–	**5α-Androstan-17-one, 3α,7α-dihydroxy-, di-TMS**		**LI09082**
156	73	75	199	184	157	450	186	**450**	100	93	58	40	30	19	18	16		25	46	3	–	–	–	–	–	–	2	–	–	–	**5α-Androstan-17-one, 3α,11β-dihydroxy-, di-TMS**	17562-92-4	**NI43261**
73	156	75	199	184	79	450	157	**450**	100	99	79	36	29	27	20	19		25	46	3	–	–	–	–	–	–	2	–	–	–	**5α-Androstan-17-one, 3α,11β-dihydroxy-, di-TMS**	17562-92-4	**NI43262**
156	199	184	450	394	186	157	360	**450**	100	39	35	28	23	19	16	10		25	46	3	–	–	–	–	–	–	2	–	–	–	**5α-Androstan-17-one, 3α,11β-dihydroxy-, di-TMS**	17562-92-4	**LI09083**
216	117	106	306	201	107	108	191	**450**	100	78	46	35	32	25	24	12	**3.00**	25	46	3	–	–	–	–	–	–	2	–	–	–	**5α-Androstan-17-one, 3α,16α-dihydroxy-, di-TMS**		**LI09084**
360	270	435	269	243	129	361	255	**450**	100	82	64	35	34	34	31	31	**6.00**	25	46	3	–	–	–	–	–	–	2	–	–	–	**5α-Androstan-17-one, 3β,7α-dihydroxy-, di-TMS**		**LI09085**

MASS TO CHARGE RATIOS								M.W.	INTENSITIES								Parent	C	H	O	N	S	F	Cl	Br	I	Si	P	B	X	COMPOUND NAME	CAS Reg No	No
435	436	450	243	129	332	437	360	**450**	100	40	36	35	28	16	13	12		25	46	3	–	–	–	–	–	–	2	–	–	–	5α-Androstan-17-one, 3β,7β-dihydroxy-, di-TMS	56210-89-0	NI43263
143	450	435	216	106	74	116	145	**450**	100	52	35	30	29	24	20	17		25	46	3	–	–	–	–	–	–	2	–	–	–	5α-Androstan-17-one, 3β,15α-dihydroxy-, di-TMS		LI09086
216	117	217	106	306	107	201	108	**450**	100	91	64	60	34	23	21	20	**3.00**	25	46	3	–	–	–	–	–	–	2	–	–	–	5α-Androstan-17-one, 3β,16α-dihydroxy-, di-TMS		LI09088
117	216	106	107	306	119	215	160	**450**	100	75	60	36	29	27	24	24	**3.00**	25	46	3	–	–	–	–	–	–	2	–	–	–	5α-Androstan-17-one, 3β,16α-dihydroxy-, di-TMS		LI09087
117	216	106	306	107	217	190	201	**450**	100	89	52	31	17	14	14	11	**1.00**	25	46	3	–	–	–	–	–	–	2	–	–	–	5α-Androstan-17-one, 3β,16α-dihydroxy-, di-TMS		LI09089
435	436	450	243	74	360	143	253	**450**	100	40	35	31	26	21	21	20		25	46	3	–	–	–	–	–	–	2	–	–	–	5α-Androstan-17-one, 3β,17β-dihydroxy-, di-TMS		LI09090
270	257	271	360	256	239	106	242	**450**	100	30	24	16	13	11	8	7	**1.00**	25	46	3	–	–	–	–	–	–	2	–	–	–	5α-Androstan-17-one, 3β,19-dihydroxy-, di-TMS		LI09091
73	75	270	319	360	211	450	129	**450**	100	45	43	34	32	32	21	21		25	46	3	–	–	–	–	–	–	2	–	–	–	5β-Androstan-3-one, 7α,17β-dihydroxy-, di-TMS	55801-52-0	NI43264
270	271	147	223	360	129	305	253	**450**	100	56	40	30	23	23	22	21	**13.00**	25	46	3	–	–	–	–	–	–	2	–	–	–	5β-Androstan-17-one, 3α,6α-dihydroxy-, di-TMS		NI43265
73	156	75	108	184	157	270	79	**450**	100	99	64	32	30	17	15	14	**8.50**	25	46	3	–	–	–	–	–	–	2	–	–	–	5β-Androstan-17-one, 3α,11β-dihydroxy-, di-TMS	17562-89-9	NI43267
73	156	75	79	52	43	108	184	**450**	100	99	99	67	41	38	37	31	**10.00**	25	46	3	–	–	–	–	–	–	2	–	–	–	5β-Androstan-17-one, 3α,11β-dihydroxy-, di-TMS	17562-89-9	NI43266
216	117	201	162	120	217	174	107	**450**	100	69	38	19	10	9	8	7	**2.00**	25	46	3	–	–	–	–	–	–	2	–	–	–	5β-Androstan-17-one, 3α,16α-dihydroxy-, di-TMS	56210-91-4	NI43268
419	450	434	420	435	203	232	451	**450**	100	48	37	31	23	23	17	14		26	26	7	–	–	–	–	–	–	–	–	–	–	Cajaisoflavone		MS09871
91	269	450	108	134	451	270	178	**450**	100	48	29	15	12	10	10	10		26	30	5	2	–	–	–	–	–	–	–	–	–	(3-Ethyl-4-methyl-5-benzyl-oxycarbonyl-2-pyrrolyl)-(3,5-dimethyl-4-(2-methoxy-carbonyl-ethyl)-2-pyrrolyl)-ketone		LI09092
212	240	42	96	114	52	211	97	**450**	100	81	76	54	50	43	39	38	**29.00**	26	46	4	2	–	–	–	–	–	–	–	–	–	Azcarpine		LI09093
143	130	144	435	145	131	436	360	**450**	100	20	18	13	5	5	4	4	**3.00**	26	50	2	–	–	–	–	–	–	2	–	–	–	Androstane, 3,17-bis[(trimethylsilyl)oxy]-17-methyl-, (3β,5α,17β)-	39780-71-7	MS09872
84	57	85	337	393	251	195	351	**450**	100	96	65	15	9	3	3	2	**0.00**	26	56	–	2	–	–	–	–	–	–	–	–	2	Aluminium, tetrakis(2-methylpropyl)di-μ-1-piperidinyldi-	42859-27-8	MS09873
450	419	299	451	435	420	281	151	**450**	100	72	65	30	23	21	20	17		27	30	6	–	–	–	–	–	–	–	–	–	–	Cyclotriveratrylene	1180-60-5	NI43269
300	450	420	281	268	451	435		**450**	100	88	69	31	28	25	19			27	30	6	–	–	–	–	–	–	–	–	–	–	Cyclotriveratrylene	1180-60-5	LI09094
299	451	420	151	149	281	268	452	**450**	100	89	89	57	48	34	33	32	**0.00**	27	30	6	–	–	–	–	–	–	–	–	–	–	Cyclotriveratrylene	1180-60-5	MS09874
227	228	296	177	213	137	229	110	**450**	100	96	63	37	25	24	23	15	**9.00**	27	30	6	–	–	–	–	–	–	–	–	–	–	2α,11-Dihydroxyabieta-5,7,9(11),13-tetraen-12-one, 2α-O-3,4-dihydroxybenzoate		NI43270
57	43	137	100	229	71	227	149	**450**	100	90	69	63	62	54	42	42	**3.00**	27	30	6	–	–	–	–	–	–	–	–	–	–	11,19-Dihydroxyabieta-5,7,9(11),13-tetraen-12-one, 19-O-3′,4′-dihydroxybenzoate		NI43271
151	299	450	419	281	268	300	312	**450**	100	83	60	60	43	43	28	27		27	30	6	–	–	–	–	–	–	–	–	–	–	2,3-Dimethyl-5-isobutenyl-8-phenyl-1,6,8-triazabicyclo[4.3.0]-non-3-ene7,9-dione		LI09095
120	121	92	137	65	39	93	64	**450**	100	84	82	64	64	40	32	28	**14.00**	27	34	4	2	–	–	–	–	–	–	–	–	–	17-(2-Hydroxy-1-oxoethyl)-3-salicyloylhydrazono-4-estrene		MS09875
271	253	414	289	226	273	254	201	**450**	100	82	36	32	30	27	22	19	**2.00**	27	46	5	–	–	–	–	–	–	–	–	–	–	Cholestan-26-oic acid, 3,7,12-trihydroxy-, (3α,5β,7α,12α)-	547-98-8	NI43272
379	43	319	259	372	271	331	380	**450**	100	73	67	44	36	36	31	25	**3.00**	27	46	5	–	–	–	–	–	–	–	–	–	–	6-Pentyl-4-nor-3,5-secoandrostane 3,6,17β-triol 3,17-diacetate		MS09876
43	135	450	204	95	81	271	55	**450**	100	45	40	33	33	33	32	31		27	46	5	–	–	–	–	–	–	–	–	–	–	6-Pentyl-4-nor-3,5-secoandrostane-3,5β,17β-triol 3,17-diacetate		MS09877
184	43	268	347	226	183	348	450	**450**	100	58	48	42	37	31	23	19		28	18	6	–	–	–	–	–	–	–	–	–	–	meso-1,1′-Bis(1-acetoxy-2-oxoacenaphthene)	120417-89-2	DD01698
327	163	450	217	286	164	409	141	**450**	100	75	60	55	35	35	19	15		28	34	5	–	–	–	–	–	–	–	–	–	–	4H-1-Benzopyran-4-one, 2,3-dihydro-7-methoxy-2-(2,4-dimethoxyphenyl)-8-[5-methyl-2-(1-methylethenyl)-4-hexenyl]-		NI43273
43	269	55	81	57	95	41	69	**450**	100	83	70	60	43	42	42	40	**0.00**	28	50	4	–	–	–	–	–	–	–	–	–	–	Ergostane-3,5,6,12-tetrol, (3β,5α,6β,12α)-	56052-11-0	NI43274
59	55	81	289	271	95	93	69	**450**	100	59	56	50	45	45	41	39	**0.00**	28	50	4	–	–	–	–	–	–	–	–	–	–	Ergostane-3,5,6,25-tetrol, (3β,5α,6β)-	56143-31-8	NI43275
105	77	345	106	450	346	51	78	**450**	100	34	18	8	7	5	4	3		30	26	4	–	–	–	–	–	–	–	–	–	–	1,1-Bis(2-benzoxy-5-methylphenyl)ethane		MS09878
43	55	41	57	69	83	29	56	**450**	100	81	67	59	54	33	27	25	**2.74**	30	58	2	–	–	–	–	–	–	–	–	–	–	9-Hexadecenoic acid, tetradecyl ester, (Z)-	22393-82-4	NI43276
43	55	41	69	57	82	83	29	**450**	100	92	76	55	50	35	33	33	**0.64**	30	58	2	–	–	–	–	–	–	–	–	–	–	Myristic acid, 9-hexadecenyl ester, (Z)-	22393-89-1	NI43277
373	450	374	128	451	270	204	165	**450**	100	35	29	16	12	12	7	7		31	22	–	4	–	–	–	–	–	–	–	–	–	[1,2,4]Triazolo[4,3-a]quinoline, 1,2-dihydro-1,1-diphenyl-2-(2-quinolinyl)-	51659-06-4	NI43278
43	255	450	315	435	451	313	213	**450**	100	38	34	18	12	11	10	9		31	46	2	–	–	–	–	–	–	–	–	–	–	(3β,5α)-3-Acetoxy-24,24-dimethylcholesta-7,25-dien-22-yne	118112-71-3	DD01699
57	141	71	58	157	55	59	337	**450**	100	78	70	53	51	47	39	31	**15.60**	31	62	1	–	–	–	–	–	–	–	–	–	–	Hentriacontan-9-one		LI09096
239	255	240	71	43	58	450	149	**450**	100	65	36	34	34	27	26	26		31	62	1	–	–	–	–	–	–	–	–	–	–	Hentriacontan-16-one	502-73-8	MS09879
450	77	422	105	345	217	434	373	**450**	100	36	30	27	21	12	6	4		32	22	1	2	–	–	–	–	–	–	–	–	–	Benzoyl phenyl diquinol-2-yl methane		LI09097
77	205	233	217	101	229	156	105	**450**	100	72	43	29	22	14	14	14	**0.00**	32	22	1	2	–	–	–	–	–	–	–	–	–	Benzoyl phenyl diquinol-8-yl methane		LI09098
43	57	71	41	85	55	69	56	**450**	100	93	53	39	37	36	22	17	**0.06**	32	66	–	–	–	–	–	–	–	–	–	–	–	11-Decyldocosane		AI02380
57	43	71	85	55	41	69	83	**450**	100	91	61	41	32	32	20	15	**0.00**	32	66	–	–	–	–	–	–	–	–	–	–	–	11-Decyldocosane		AI02379
57	43	71	85	41	55	69	56	**450**	100	93	60	42	32	30	19	16	**1.62**	32	66	–	–	–	–	–	–	–	–	–	–	–	Dotriacontane		AI02381
57	43	71	41	85	55	69	29	**450**	100	94	45	30	29	29	18	18	**1.64**	32	66	–	–	–	–	–	–	–	–	–	–	–	Dotriacontane		AI02382
43	57	71	85	41	55	69	56	**450**	100	99	54	38	32	31	18	16	**0.59**	32	66	–	–	–	–	–	–	–	–	–	–	–	Dotriacontane		AI02383
421	422	420	393	395	423	435	407	**450**	100	86	71	29	18	14	13	11	**7.22**	32	66	–	–	–	–	–	–	–	–	–	–	–	Hentriacontane, 3-methyl-	4981-99-1	NI43279
43	57	71	85	41	55	69	56	**450**	100	99	58	41	38	36	22	17	**0.00**	32	66	–	–	–	–	–	–	–	–	–	–	–	9-Octyltetracosane		AI02384
58	450	232	218	191	178	451	57	**450**	100	86	48	41	41	40	32	13		33	26	–	2	–	–	–	–	–	–	–	–	–	Bis(9-(2-cyanoethyl)-fluoren-9-yl)methane		IC04442
450	451	448	225	449	452	446	447	**450**	100	40	23	11	10	8	4	3		36	18	–	–	–	–	–	–	–	–	–	–	–	Diacenaphtho[1,2-j:1′,2′-l]fluoranthene	191-48-0	NI43280
318	319	302	189	317	303	289	165	**450**	100	45	40	36	35	26	25	21	**0.00**	36	18	–	–	–	–	–	–	–	–	–	–	–	11,12:13,14-Dibenzo-3,8-methanopentahendecafulvalene		LI09099

MASS TO CHARGE RATIOS								M.W.	INTENSITIES								Parent	C	H	O	N	S	F	Cl	Br	I	Si	P	B	X	COMPOUND NAME	CAS Reg No	No
334	336	137	453	455	338	138	65	**451**	100	99	61	46	37	37	37	24	**23.37**	12	7	1	3	1	–	6	–	–	–	–	–	–	2-(2′-Methoxyphenylthio)-4,6-bis(trichloromethyl)-s-triazine		MS09880
57	255	253	111	311	257	406	438	**451**	100	89	72	54	44	43	42	41	**9.00**	18	17	4	1	–	–	4	–	–	–	–	–	–	5,6,7,8-Tetrachloro-2,3-dicarbomethoxy-1,4-dihydro-1,4-tert-butyliminonaphthalene		NI43281
209	73	210	179	451	75	77	45	**451**	100	50	16	11	8	8	6	6		18	28	5	1	–	3	–	–	–	2	–	–	–	3-Methoxy-N-(trifluoroacetyl)-O-(trimethylsilyl)tyrosine trimethylsilyl ester		MS09881
143	189	115	89	144	28	116	63	**451**	100	48	35	23	22	17	9	8	**0.00**	18	41	6	1	–	–	–	–	–	3	–	–	–	D-Galactopyranoside, methyl 2-(acetylamino)-2-deoxy-3,4,6-tris-O-(trimethylsilyl)-	74410-43-8	NI43282
173	73	75	131	174	147	204	226	**451**	100	59	21	14	13	13	11	10	**0.00**	18	41	6	1	–	–	–	–	–	3	–	–	–	α-D-Glucopyranoside, methyl 2-acetylamino-2-deoxy-3,4,6-tris-O-(trimethylsilyl)-	18434-96-3	NI43283
173	73	217	75	174	131	147	116	**451**	100	93	40	25	23	19	16	16	**0.00**	18	41	6	1	–	–	–	–	–	3	–	–	–	α-D-Glucopyranoside, methyl 3-(acetylamino)-3-deoxy-2,4,6-tris-O-(trimethylsilyl)-	55282-23-0	NI43284
171	73	75	131	172	147	204	226	**451**	100	81	29	18	17	16	12	10	**0.00**	18	41	6	1	–	–	–	–	–	3	–	–	–	D-Glucopyranoside, methyl 2-(acetylamino)-2-deoxy-3,4,6-tris-O-(trimethylsilyl)-	56272-06-1	NI43285
279	142	128	127	281	280	158	251	**451**	100	39	39	39	25	25	22	19	**0.00**	21	28	1	2	–	–	–	–	1	–	–	–	–	ϵ^2-Dihydromavacurin-methyl iodide		LI09100
217	188	189	203	208	263	247	231	**451**	100	88	51	40	38	35	28	24	**3.00**	22	21	–	3	1	3	1	–	–	–	–	–	–	1-[(2-Benzyl-4-thiazolyl)methyl]-4-(4-chloro-3-trifluoromethyl)phenyl]piperazine		MS09882
105	334	110	43	170	152	77	124	**451**	100	70	34	31	18	18	18	16	**0.00**	23	33	6	1	1	–	–	–	–	–	–	–	–	2-O-Acetyl-3-O-benzoyl-6-acetylamino-4,6-dideoxy-1-n-hexylthio-α-DL-lyxo-hexopyranoside		NI43286
105	43	292	170	77	41	56	55	**451**	100	32	23	21	21	17	16	16	**0.00**	23	33	6	1	1	–	–	–	–	–	–	–	–	3-O-Acetyl-2-O-benzoyl-6-acetylamino-4,6-dideoxy-1-n-hexylthio-α-DL-lyxo-hexopyranoside		NI43287
73	420	380	421	308	419	381	75	**451**	100	84	55	28	21	18	15	15	**10.00**	25	45	4	1	–	–	–	–	–	1	–	–	–	Prosta-8(12),13-dien-1-oic acid, 9-(methoxyimino)-15-[(trimethylsilyl)oxy], methyl ester, (13E,15S)-	53044-56-7	NI43288
73	129	380	148	199	75	420	173	**451**	100	64	31	31	29	29	21	18	**3.00**	25	45	4	1	–	–	–	–	–	1	–	–	–	Prosta-10,13-dien-1-oic acid, 9-(methoxyimino)-15-[(trimethylsilyl)oxy]-, methyl ester, (13E,15S)-	55821-11-9	NI43289
105	286	394	77	108	106	91	79	**451**	100	98	61	54	49	48	45	45	**0.00**	26	33	4	3	–	–	–	–	–	–	–	–	–	(2S,5S)-1-Benzoyl-5-[4-[(benzyloxycarbonyl)amino]propyl]-2-tert-butyl-3-methyl-4-imidazolidinone		DD01700
145	175	44	416	69	174	176	55	**451**	100	70	70	60	50	40	38	35	**3.00**	26	42	3	1	–	–	1	–	–	–	–	–	–	Tetradecanamide, N-[2-chloro-3-(2-hydroxy-3-methylphenyl)-2-propenyl]-8-methoxy-N-methyl-	75332-38-6	NI43290
57	29	27	133	451	408	395	41	**451**	100	68	11	11	10	9	9	8		27	33	5	1	–	–	–	–	–	–	–	–	–	1′-Carboethoxy-1′-cyano-1β,2β-dihydro-17β-propionoxy-3′H-cycloprop[1,2]androsta-1,4,6-trien-3-one		NI43291
435	18	28	281	132	437	418	43	**451**	100	70	39	38	38	37	37	36	**0.00**	27	46	2	1	–	–	1	–	–	–	–	–	–	Cholestan-6-one, 3-chloro-5-hydroxy-, oxime, (3β,5α)-		MS09883
58	100	84	98	55	436	71	97	**451**	100	99	81	52	51	48	43	39	**34.78**	28	57	1	3	–	–	–	–	–	–	–	–	–	2-Hexadecenamide, N,N-bis[4-(dimethylamino)butyl]-, (E)-	17232-86-9	NI43292
326	205	78	299	325	327	451	97	**451**	100	55	41	40	35	24	18	18		29	29	2	3	–	–	–	–	–	–	–	–	–	(1R*,7R*)-2-(n-cyclohexylcarbamoyl)-4-methyl-6,6-diphenyl-2,8-dicyclo[5.2.2.0(1,5)]undeca-4,8,10-trien-3-one		MS09884
119	303	77	91	332	317	104	64	**451**	100	94	59	55	50	38	31	30	**0.00**	29	29	2	3	–	–	–	–	–	–	–	–	–	8-(Diethylamino)-7-methyl-2,4,6-triphenyl-2,6-diazabicyclo[2.2.2]oct-7-ene-3,5-dione		DD01701
451	292	293	422	217	158	170		**451**	100	65	59	43	31	31	25			29	29	2	3	–	–	–	–	–	–	–	–	–	2,2′-(p-Nitrobenzylidene)-bis(3-propylindole)		LI09101
451	41	436	43	452	140	56	44	**451**	100	43	39	39	33	28	27	17		29	41	3	1	–	–	–	–	–	–	–	–	–	3α-(3′-Oxo-5′,5′-dimethylcyclohex-1-enylamino)-20β-hydroxy-pregn-5-en-18-carboxylic acid lactone		MS09885
360	44	361	43	91	81	95	93	**451**	100	85	26	24	20	12	10	9	**0.00**	30	45	2	1	–	–	–	–	–	–	–	–	–	Androstane, 17-(acetyloxy)-3-(1-benzylethylamino)-		LI09102
451	450	452	183	266	264	262	225.5	**451**	100	40	32	31	19	10	7	7		32	22	–	1	–	–	–	–	–	–	1	–	–	Fluorene-2-carbonitrile, 9-(triphenylphosphoranylidene)-	7293-78-9	LI09103
451	450	452	183	266	264	190	185	**451**	100	42	32	31	19	10	8	8		32	22	–	1	–	–	–	–	–	–	1	–	–	Fluorene-2-carbonitrile, 9-(triphenylphosphoranylidene)-	7293-78-9	NI43293
451	91	233	452	232	234	218	453	**451**	100	73	56	36	36	11	7	6		32	37	1	1	–	–	–	–	–	–	–	–	–	Androstano[3,2-b]indole, N-benzyl-17-oxo-, (5α)-		MS09886
452	439	373	294	249	230	197	152	**452**	100	88	65	33	32	32	32	32		–	–	–	4	–	6	–	2	–	–	4	–	–	Dibromohexafluorocyclotetraphosphazene		MS09887
147	373	452	291	253	332	106	411	**452**	100	32	25	25	25	14	7	3		6	10	–	–	–	–	–	2	–	–	–	–	2	Palladium, di-π-allyldibromodi-		LI09104
272	274	270	276	236	238	237	234	**452**	100	86	53	40	22	18	17	16	**0.00**	10	1	1	–	–	–	9	–	–	–	–	–	–	1,3,4-Metheno-2H-cyclobuta[cd]pentalen-2-one, 1,1a,3,4,5,5,5a,5b,6-nonachlorooctahydro-	53308-47-7	NI43294
272	274	270	276	237	321	149	235	**452**	100	83	51	38	29	22	21	20	**0.23**	10	1	1	–	–	–	9	–	–	–	–	–	–	Monohydrokepone		MS09888
249	425	277	377	397	432	151	45	**452**	100	95	89	62	31	8	8	8	**7.00**	10	12	5	–	–	11	–	–	–	–	1	–	–	[2,2-Difluoro-2-(1′,1′,1′,3′,3′,3′-hexafluoroisopropoxy)-1-(trifluoromethyl)ethyl]diethylphosphate		MS09889
375	377	373	379	458	456	454	452	**452**	100	35	34	33	1	1	1	1		10	16	–	–	–	–	–	4	–	–	–	–	–	Cyclohexane, 1,2,4,5-tetrabromo-1-methyl-4-isopropyl-, (1α,2β,4α,5β)-(±)-	70223-04-0	NI43295
296	294	298	377	375	379	373	454	**452**	100	52	48	14	14	4	4	1	**0.01**	10	16	–	–	–	–	–	4	–	–	–	–	–	Cyclohexane, 1,2,5-tribromo-4-(1-bromo-1-methylethyl)-1-methyl-, (1α,2β,4α,5α)-(±)-	70168-62-6	NI43296

MASS TO CHARGE RATIOS								M.W.	INTENSITIES								Parent	C	H	O	N	S	F	Cl	Br	I	Si	P	B	X	COMPOUND NAME	CAS Reg No	No
143	124	305	123	65	59	39	333	**452**	100	34	31	30	23	17	17	16	**1.00**	12	5	1	–	–	12	–	–	–	–	–	–	1	Cyclopentadienyl-heptafluorobut-1-enyl-pentafluoroethyl-cobalt carbonyl		NI43297
279	278	97	165	193	265	129	157	**452**	100	98	84	72	61	50	44	41	**0.00**	12	9	8	–	–	9	–	–	–	–	–	–	–	Methyl 2,3,4-tri-O-trifluoroacetyl-β-D-xylopyranoside		MS09890
44	55	82	43	45	46	41	54	**452**	100	66	44	29	27	22	21	20	**0.00**	12	22	11	1	3	–	–	–	–	–	–	–	–	Glucoerysolin		MS09891
453	454	435	230	144	455	434	212	**452**	100	12	2	2	2	1	1	1	**0.00**	13	14	4	2	–	10	–	–	–	–	–	–	–	Lysine methyl ester, di-N-pentafluoropropionate		NI43298
226	76	208	75	30	120	178	210	**452**	100	32	30	28	26	23	21	16	**0.00**	14	8	12	6	–	–	–	–	–	–	–	–	–	Hexanitrobibenzyl	5180-53-0	NI43299
197	452	221	255	212	433	409	288	**452**	100	56	28	22	8	4	3	3		14	16	2	–	2	6	–	–	–	–	–	–	1	Bis(1,1,1-trifluoro-4-mercapto-5-methylhex-3-en-2-onato)nickel(II)		NI43300
368	370	366	424	367	316	426	422	**452**	100	91	88	55	52	49	46	43	**6.93**	16	9	3	–	–	1	–	–	–	–	–	–	1	Tungsten, cyclopentadienyl-[(4-fluorophenyl)ethynyl]tricarbonyl		NI43301
148	204	164	120	200	220	176	124	**452**	100	90	59	53	30	28	25	5	**0.20**	16	19	2	–	2	–	–	–	–	–	1	–	2	μ-(Carbon disulphide)(η^5-cyclopentadienyl)dicarbonyl(η^5-cyclopentadienylmanganese)(triphenylphosphine)cobalt		MS09892
228	284	172	256	56	396	112	368	**452**	100	98	70	56	53	47	42	35	**0.00**	17	8	8	–	–	–	–	–	–	–	–	–	2	Heptacarbonyl-μ-[(1-η:2,2α,3-η^3)-2-(E)-benzylidene-1-oxo-1,3-propanediyl]diiron	122624-33-3	DD01702
73	117	275	274	452	189	74	45	**452**	100	24	17	16	14	9	9	9		17	36	–	–	4	–	–	–	–	3	–	–	–	Silane, (9,10-dimethyl-2,4,6,8-tetrathiatricyclo[3.3.1.1^{3,7}]decane-1,3,5-triyl)tris[trimethyl-	57289-38-0	NI43302
73	147	292	233	220	75	74	189	**452**	100	28	25	15	8	8	8	7	**0.00**	17	40	6	–	–	–	–	–	–	4	–	–	–	Pentanedioic acid, 2,3-bis[(trimethylsilyl)oxy]-, bis(trimethylsilyl) ester		LI09105
73	147	292	233	75	74	220	45	**452**	100	26	22	13	9	9	8	7	**0.00**	17	40	6	–	–	–	–	–	–	4	–	–	–	Pentanedioic acid, 2,3-bis[(trimethylsilyl)oxy]-, bis(trimethylsilyl) ester, (R*,S*)-	55282-53-6	NI43303
73	245	147	246	437	319	217	335	**452**	100	84	48	18	14	12	10	9	**0.00**	17	40	6	–	–	–	–	–	–	4	–	–	–	Pentanedioic acid, 2,4-bis[(trimethylsilyl)oxy]-, bis(trimethylsilyl) ester		LI09106
73	245	147	246	437	320	219	75	**452**	100	83	49	17	12	12	10	10	**0.00**	17	40	6	–	–	–	–	–	–	4	–	–	–	Pentanedioic acid, 2,4-bis[(trimethylsilyl)oxy]-, bis(trimethylsilyl) ester, (R*,S*)-	38166-12-0	NI43304
421	363	422	336	394	452	364	362	**452**	100	46	44	43	36	35	35	35		18	16	10	2	1	–	–	–	–	–	–	–	–	2,4,5-Pyridinetricarboxylic acid, 6-(4,5-dicarboxy-2-thiazolyl)-, pentamethyl ester	23850-47-7	NI43305
73	204	147	217	218	191	117	205	**452**	100	53	32	28	21	20	17	13	**0.00**	18	44	5	–	–	–	–	–	–	4	–	–	–	2-Deoxy-D-glucose, tetrakis(trimethylsilyl)-		NI43306
73	204	147	217	103	205	191	75	**452**	100	50	24	23	14	13	11	10	**0.00**	18	44	5	–	–	–	–	–	–	4	–	–	–	2-Deoxy-galactose, tetrakis(trimethylsilyl)-		NI43307
73	204	147	217	205	103	75	218	**452**	100	67	24	16	15	14	11	11	**0.00**	18	44	5	–	–	–	–	–	–	4	–	–	–	2-Deoxy-galactose, tetrakis(trimethylsilyl)-		NI43308
217	218	191	259	219	147	129	204	**452**	100	70	55	28	21	19	18	17	**2.20**	18	44	5	–	–	–	–	–	–	4	–	–	–	1-Deoxyglucose, tetrakis(trimethylsilyl)-		NI43309
217	73	204	147	191	218	117	75	**452**	100	99	74	37	33	24	24	19	**0.00**	18	44	5	–	–	–	–	–	–	4	–	–	–	α-L-Galactofuranose, 6-deoxy-1,2,3,5-tetrakis-O-(trimethylsilyl)-	56227-36-2	NI43310
204	73	217	191	147	205	189	75	**452**	100	87	57	31	27	23	13	13	**0.00**	18	44	5	–	–	–	–	–	–	4	–	–	–	α-D-Galactopyranose, 6-deoxy-1,2,3,4-tetrakis-O-(trimethylsilyl)-	32727-30-3	NI43311
204	73	191	217	147	57	205	75	**452**	100	88	80	33	25	23	20	17	**0.00**	18	44	5	–	–	–	–	–	–	4	–	–	–	β-L-Galactopyranose, 6-deoxy-1,2,3,4-tetrakis-O-(trimethylsilyl)-	32727-31-4	NI43312
217	73	204	191	147	117	218	75	**452**	100	96	56	41	40	39	37	24	**0.00**	18	44	5	–	–	–	–	–	–	4	–	–	–	α-L-Mannofuranose, 6-deoxy-1,2,3,5-tetrakis-O-(trimethylsilyl)-	56227-38-4	NI43313
217	73	204	117	147	191	218	75	**452**	100	98	36	30	29	23	22	21	**0.00**	18	44	5	–	–	–	–	–	–	4	–	–	–	β-L-Mannofuranose, 6-deoxy-1,2,3,5-tetrakis-O-(trimethylsilyl)-	56227-37-3	NI43314
204	73	205	147	191	189	217	206	**452**	100	54	24	19	16	14	10	9	**0.00**	18	44	5	–	–	–	–	–	–	4	–	–	–	α-L-Mannopyranose, 6-deoxy-1,2,3,4-tetrakis-O-(trimethylsilyl)-	55057-21-1	NI43315
204	73	147	205	191	189	217	75	**452**	100	90	34	32	27	23	16	15	**0.00**	18	44	5	–	–	–	–	–	–	4	–	–	–	β-L-Mannopyranose, 6-deoxy-1,2,3,4-tetrakis-O-(trimethylsilyl)-	55057-22-2	NI43316
204	73	191	147	205	217	189	206	**452**	100	59	25	20	19	12	9	8	**0.00**	18	44	5	–	–	–	–	–	–	4	–	–	–	L-Mannose, 6-deoxy-2,3,4,5-tetrakis-O-(trimethylsilyl)-	19127-15-2	NI43317
73	204	205	147	191	206	74	75	**452**	100	97	19	18	16	9	8	8	**0.00**	18	44	5	–	–	–	–	–	–	4	–	–	–	Mannose, 6-deoxy-2,3,4,5-tetrakis-O-(trimethylsilyl)-, L-	19127-15-2	NI43318
226	339	452	227	283	130	129	340	**452**	100	30	17	14	12	7	7	6		19	18	4	2	–	6	–	–	–	–	–	–	–	L-Tryptophan, N,1-bis(trifluoroacetyl)-, butyl ester	55282-24-1	NI43319
166	158	68	41	29	73	258	57	**452**	100	71	39	37	34	33	28	25	**0.00**	19	31	5	2	–	3	–	–	–	1	–	–	–	L-Prolylhydroxyproline, N-(trifluoroacetyl)-, butyl ester, trimethylsilyl-		MS09893
135	454	452	90	77	76	63	103	**452**	100	50	48	32	22	19	15	14		20	13	2	4	1	–	–	1	–	–	–	–	–	Δ^2-1,2,4-Thiadiazoline, 4-(4-bromophenyl)-5-[(4-nitrophenyl)imino]-3-phenyl-	33863-70-6	NI43320
127	452	185	393	96	392	110	335	**452**	100	79	55	40	38	37	32	31		20	28	4	4	2	–	–	–	–	–	–	–	–	12,22-Dimethyl-3,4-dithia[6.6](1.3)-1,2,3,4-tetrahydropyrimidinophane		MS09894
166	167	29	41	28	84	57	194	**452**	100	58	31	29	27	23	19	14	**0.00**	20	31	6	2	–	3	–	–	–	–	–	–	–	L-Prolylglutamic acid, N-(trifluoroacetyl)-, dibutyl ester		MS09895
73	132	217	147	218	133	75	43	**452**	100	43	39	32	31	19	18	16	**0.00**	20	32	6	–	–	–	–	–	–	3	–	–	–	Methyl β-galactopyranoside-2-acetate-tri-TMS		NI43321
73	132	217	129	174	147	117	75	**452**	100	69	61	34	28	27	26	24	**0.00**	20	32	6	–	–	–	–	–	–	3	–	–	–	Methyl β-galactopyranoside-3-acetate-tri-TMS		NI43322
73	204	117	75	205	133	43	147	**452**	100	99	28	28	26	22	21	16	**0.00**	20	32	6	–	–	–	–	–	–	3	–	–	–	Methyl β-galactopyranoside-4-acetate-tri-TMS		NI43323
204	73	217	147	133	205	43	75	**452**	100	73	43	23	23	22	16	15	**0.00**	20	32	6	–	–	–	–	–	–	3	–	–	–	Methyl β-galactopyranoside-6-acetate-tri-TMS		NI43324
105	168	393	317	154	229	209	215	**452**	100	18	15	15	14	13	10	9	**0.00**	21	24	11	–	–	–	–	–	–	–	–	–	–	α-D-Xylofuranose, 3-C-[(acetyloxy)methyl]-, 1,2,3-triacetate 5-benzoate	70723-04-5	NI43325
452	454	383	381	409	178	192	179	**452**	100	99	90	83	61	52	51	49		22	13	6	–	–	–	–	1	–	–	–	–	–	3-Bromo-7,7′-dimethyl-4,4′-dihydroxy-1,1′-binaphthalene-5,5′,8,8′-tetrone	101459-16-9	NI43326
283	372	226	295	373	167	371	454	**452**	100	10	9	7	5	5	4	2	**2.00**	22	21	2	4	–	–	–	1	–	–	–	–	–	8-(2-Bromo-1,2-diphenylethyl)-3,7-dihydro-1,3,7-trimethyl-1H-purine-2,6-dione		NI43327
451	452	153	227	186	453	77	258	**452**	100	76	54	41	39	32	19	17		24	16	–	6	2	–	–	–	–	–	–	–	–	Bis(5-cyano-2-methyl-6-phenyl-4-pyrimidinyl) disulphide	82141-06-8	NI43328
452	454	83	437	85	453	439	407	**452**	100	86	56	40	37	34	28	24		24	17	7	–	–	–	1	–	–	–	–	–	–	4′-Chloro-1′-hydroxy-5,8-dimethoxy-3,3′-dimethyl-2,2′-binaphthalene-1,4,5′,8′-tetrone	104506-02-7	NI43329
274	137	180	452	259	124	420	231	**452**	100	71	68	39	36	32	24	15		24	20	9	–	–	–	–	–	–	–	–	–	–	7H-1,4-Dioxino[2,3-c]xanthen-7-one, 2,3-dihydro-8-hydroxy(3-hydroxy-4-methoxyphenyl)(hydroxymethyl)-5-methoxy-, (2S-trans)-	64280-46-2	NI43330
205	287	369	121	120	452	203	123	**452**	100	92	75	33	17	13	13	13		24	44	–	–	–	–	–	–	–	–	–	–	1	Tetracyclohexyltin		LI09107
205	287	203	81	369	285	367	201	**452**	100	90	83	75	71	67	59	53	**1.20**	24	44	–	–	–	–	–	–	–	–	–	–	1	Tetracyclohexyltin		MS09896

MASS TO CHARGE RATIOS								M.W.	INTENSITIES								Parent	C	H	O	N	S	F	Cl	Br	I	Si	P	B	X	COMPOUND NAME	CAS Reg No	No
209	376	179	222	210	223	201		452	100	23	22	20	19	15	12		0.00	24	44	4	–	–	–	–	–	–	2	–	–	–	3-Undecanone, 1-[3-methoxy-4-[(trimethylsilyl)oxy]phenyl]-5-[(trimethylsilyl)oxy]-		MS09897
55	30	452	98	100	97	41	227	452	100	99	97	63	58	48	41	32		24	44	4	4	–	–	–	–	–	–	–	–	–	1,8,15,22-Tetraaza-9,14,23,28-tetraoxo-cyclooctacosane		IC04443
410	320	452	350	43	319	349	335	452	100	58	57	56	50	42	35	30		25	24	8	–	–	–	–	–	–	–	–	–	–	1-Acetoxy-2-methyl-3-acetoxymethyl-4-[3,4-(methylenedioxy)phenyl]-6,7-dimethoxynaphthalene		NI43331
167	392	238	305	393	333	225	180	452	100	77	39	33	31	29	23	23	5.00	25	28	6	2	–	–	–	–	–	–	–	–	–	Talbotine, 4-acetyl-, acetate (ester)	30809-23-5	LI09108
167	392	166	238	305	393	333	107	452	100	76	59	38	33	31	29	26	5.09	25	28	6	2	–	–	–	–	–	–	–	–	–	Talbotine, 4-acetyl-, acetate (ester)	30809-23-5	NI43332
297	43	298	296	57	55	452	85	452	100	90	42	40	29	27	26	24		25	40	3	–	2	–	–	–	–	–	–	–	–	Pregn-9(11)-en-20-one, 3,6-bis[(methylthio)methoxy]-, (3β,5α,6α)-	56312-74-4	NI43333
84	91	297	155	214	115	128	254	452	100	60	43	25	21	19	13	7	0.00	25	44	3	2	1	–	–	–	–	–	–	–	–	N-[3-(p-Ethyltosylamino)propyl]-N-hexyl-heptanamide		NI43334
344	209	434	291	345	347	129	435	452	100	81	72	42	39	32	32	27	0.73	25	48	3	–	–	–	–	–	–	2	–	–	–	Androstan-5-ol, 3,6-bis[(trimethylsilyl)oxy]-, (3β,5α)-	56248-35-2	NI43335
77	105	143	452	115	169	144	106	452	100	74	63	15	10	8	7	6		26	20	4	4	–	–	–	–	–	–	–	–	–	Benzoyl-(4-methyl-2-nitrophenylazo)-naphth-1-ylaminocarbonylmethane		IC04444
313	183	452	73	314	453	69	75	452	100	53	35	32	29	13	13	12		26	52	2	–	–	–	–	–	–	2	–	–	–	Silane, [[4-(3-methyl-2-butenyl)-5-(8-methyl-7-nonenyl)-1,3-cyclopentanediyl]bis(oxy)]bis[trimethyl-	56306-71-9	NI43336
83	157	91	294	154	314	312	450	452	100	16	15	11	11	10	9	1	0.30	27	20	–	2	–	–	–	–	–	–	–	–	1	Propanedinitrile, [5,6-diphenyl-3-(phenylseleno)bicyclo[3.1.0]hex-2-ylidene]-	70238-99-2	NI43337
452	165	423	137	288	424	287	285	452	100	95	50	29	18	15	15	15		27	32	6	–	–	–	–	–	–	–	–	–	–	7,2',4'-Triethoxy-5'-methoxy-6'-(3,3-dimethylallyl)isoflavone		MS09898
118	77	91	251	160	103	145	117	452	100	31	30	29	27	24	23	22	8.00	27	33	4	–	–	–	–	–	–	–	1	–	–	Phosphoric acid, O,O,O-tricumyl-		IC04445
91	103	77	121	211	107	79	65	452	100	65	63	46	24	24	22	18	2.45	27	33	4	–	–	–	–	–	–	–	1	–	–	Phosphoric acid, O,O,O-tricumyl-	26967-76-0	NI43338
258	257	452	55	91	41	121	122	452	100	57	50	37	32	30	27	26		27	37	5	–	–	–	–	–	–	–	–	1	–	17α,21-Dihydroxypregn-4-ene-3,11,20-trione cyclohexyl boronate		MS09899
258	257	452	121	122	105	123	163	452	100	57	50	27	26	22	21	19		27	37	5	–	–	–	–	–	–	–	–	1	–	17α,21-Dihydroxypregn-4-ene-3,11,20-trione cyclohexyl boronate		LI09109
105	178	179	452	64	315	226	180	452	100	78	14	4	4	2	1	1		28	20	4	–	1	–	–	–	–	–	–	–	–	cis-Dibenzoylstilbene episulphone		LI09110
163	77	105	257	102	258	392	155	452	100	36	26	21	21	16	15	15	11.00	28	24	4	2	–	–	–	–	–	–	–	–	–	5,8-Ethenobenzo[3,4]cyclobuta[1,2-d]pyridazine-6,7-dicarboxylic acid, 4,4b,5,8,8a,8b-hexahydro-1,4-diphenyl-, dimethyl ester	56083-43-3	NI43339
395	423	191	452	261	410	169		452	100	23	18	17	11	8	8			28	36	5	–	–	–	–	–	–	–	–	–	–	2,6-Dipentanoyl-4-(3-pentanoyl-4-hydroxybenzyl)phenol		LI09111
395	423	191	452	261	410	169	424	452	100	23	18	17	11	8	8	6		28	36	5	–	–	–	–	–	–	–	–	–	–	Valerophenone, 5'-valeryl-3',3'''-methylenebis[6'-hydroxy-	33342-94-8	NI43340
395	57	423	396	41	452	133	191	452	100	29	24	24	22	18	18	17		28	36	5	–	–	–	–	–	–	–	–	–	–	Valerophenone, 5'-valeryl-3',3'''-methylenebis[6'-hydroxy-	33342-94-8	NI43341
452	355	368	453	43	55	356	95	452	100	70	46	32	31	24	20	20		29	48	–	4	–	–	–	–	–	–	–	–	–	Cyclocholestane, 6-(5-methyltetrazol-2-yl)-, (3α,5α,6β)-		MS09900
246	42	247	377	91	172	47	28	452	100	18	17	12	9	6	6	6	3.80	30	32	2	2	–	–	–	–	–	–	–	–	–	4-Piperidinecarboxylic acid, 1-(3-cyano-3,3-diphenylpropyl)-4-phenyl-, ethyl ester	915-30-0	NI43342
107	205	105	119	125	121	109	135	452	100	99	91	83	75	75	72	59	0.00	30	44	3	–	–	–	–	–	–	–	–	–	–	Lup-20(30)-en-28-oic acid, 19β-hydroxy-3-oxo-, γ-lactone	510-78-1	NI43343
452	409	217	453	269	437	121	105	452	100	40	32	31	21	20	12	12		30	44	3	–	–	–	–	–	–	–	–	–	–	25-Noroleana-9,12-dien-29-oic acid, 5-methyl-11-oxo-, (18α)-	55282-47-8	NI43344
201	57	43	55	83	97	71	69	452	100	52	48	34	32	29	29	29	15.74	30	60	2	–	–	–	–	–	–	–	–	–	–	Dodecanoic acid, octadecyl ester	3234-84-2	NI43345
101	88	57	43	69	55	83	41	452	100	87	35	32	30	30	22	21	9.51	30	60	2	–	–	–	–	–	–	–	–	–	–	Hexacosanoic acid, 2,4,6-trimethyl-, 1-methyl ester, (2R,4S,6R)-(–)-	27829-53-4	NI43346
57	257	43	71	55	83	69	97	452	100	97	87	59	59	54	52	46	7.74	30	60	2	–	–	–	–	–	–	–	–	–	–	Hexadecanoic acid, tetradecyl ester	4536-26-9	NI43347
451	453	452	454	129	143	131	159	452	100	90	46	38	33	28	25	18		30	60	2	–	–	–	–	–	–	–	–	–	–	Methyl nonacosanoate		NI43348
285	57	43	168	71	55	83	69	452	100	83	72	67	46	44	39	39	29.34	30	60	2	–	–	–	–	–	–	–	–	–	–	Octadecanoic acid, dodecyl ester	5303-25-3	NI43349
229	57	43	55	83	71	69	97	452	100	66	56	41	40	38	37	35	6.54	30	60	2	–	–	–	–	–	–	–	–	–	–	Tetradecanoic acid, hexadecyl ester	2599-01-1	NI43350
396	340	397	341	452	381	41	191	452	100	90	47	45	44	43	39	33		31	32	3	–	–	–	–	–	–	–	–	–	–	2H,5H-Benzo[1,2-b:5,4-b']dipyran-5-one, 10-(3,3-diphenyl-2-propenylidene)-3,4,6,7,8,10-hexahydro-2,2,8,8-tetramethyl-	56336-13-1	NI43351
43	313	452	315	255	409	229	437	452	100	42	41	30	27	26	23	18		31	48	2	–	–	–	–	–	–	–	–	–	–	(3β,5α)-3-Acetoxy-24,24-dimethylcholest-7-en-22-yne	118112-73-5	DD01703
55	83	43	69	410	81	392	143	452	100	92	85	75	60	48	31	29	18.10	31	48	2	–	–	–	–	–	–	–	–	–	–	3β-Acetoxystigmasta-4,6,22-triene	96737-71-2	NI43352
43	255	452	409	229	437	453	213	452	100	59	41	26	23	18	14	13		31	48	2	–	–	–	–	–	–	–	–	–	–	(24R)-(3β,5α)-3-Acetoxystigmast-7-en-22-yne	118112-68-8	DD01704
247	231	246	57	211	191	248	232	452	100	85	71	70	29	20	18	17	15.40	31	48	2	–	–	–	–	–	–	–	–	–	–	2,2-Bis(3,5-di-tert-butyl-2-hydroxyphenyl)propane		IC04446
57	438	437	211	452	41	247	212	452	100	99	99	99	98	71	55	41		31	48	2	–	–	–	–	–	–	–	–	–	–	Bis(3,5-di-tert-butyl-4-hydroxyphenyl)propane		IC04447
43	313	255	452	342	423	392	377	452	100	68	37	21	15	7	3	3		31	48	2	–	–	–	–	–	–	–	–	–	–	Elasterol acetate		LI09112
313	81	255	95	55	356	314	452	452	100	31	28	28	28	27	27	25		31	48	2	–	–	–	–	–	–	–	–	–	–	Elasterol acetate		MS09901
57	28	41	55	161	43	149	12	452	100	82	55	52	44	35	24	24	2.35	31	48	2	–	–	–	–	–	–	–	–	–	–	2,2'-Methylene-bis(4-methyl-6-tert-octylphenol)		IC04448
55	452	339	69	83	437	409	312	452	100	88	57	53	48	30	29	23		31	48	2	–	–	–	–	–	–	–	–	–	–	Stigmasta-8,14(Z),24(28)-trienyl acetate		LI09113
83	55	271	69	314	299	255	452	452	100	22	16	10	8	5	5	2		32	52	1	–	–	–	–	–	–	–	–	–	–	26-(1,1,2-Trimethyl-2-propenyl)-27-norcholesta-5,24-dien-3β-ol		DD01705
43	203	202	57	41	119	84	153	453	100	90	58	28	26	21	14	10	0.00	13	13	5	1	–	10	–	–	–	–	–	–	–	N,O-Bis(pentafluoropropionyl)-D-threonine isopropyl ester		MS09902
203	43	202	57	41	119	84	204	453	100	95	64	28	24	21	15	10	0.00	13	13	5	1	–	10	–	–	–	–	–	–	–	N,O-Bis(pentafluoropropionyl)-L-threonine isopropyl ester		MS09903
201	175	176	186	259	84	158	70	453	100	28	24	16	15	14	14	13	0.00	15	25	9	3	–	–	–	–	–	–	2	–	–	1,4-Pentanediamine, N4-(6-methoxy-8-quinolinyl)-, phosphate (1:2)	63-45-6	NI43353
289	129	276	119	115	162	290	453	453	100	93	68	62	61	55	30	12		16	9	3	1	–	10	–	–	–	–	–	–	–	Propanoic acid, pentafluoro-, 2-[1-(2,2,3,3,3-pentafluoro-1-oxopropyl)-1H indol-3-yl]ethyl ester	56761-69-4	MS09904

MASS TO CHARGE RATIOS								M.W.	INTENSITIES								Parent	C	H	O	N	S	F	Cl	Br	I	Si	P	B	X	COMPOUND NAME	CAS Reg No	No
453	362	291	329	233	407	422	319	453	100	94	43	28	28	26	19	17		19	19	10	1	1	–	–	–	–	–	–	–	–	Benzo[b]thiophene-3,4,5,6,7-pentacarboxylic acid, 2-amino-, 3-ethyl 4,5,6,7-tetramethyl ester	66385-68-0	NI43354
309	294	266	310	154.5	311	280	250	453	100	73	37	35	20	17	9	5	0.00	19	20	4	1	–	–	–	–	1	–	–	–	–	Cryptowoline iodide		LI09114
137	102	139	152	266	111	455	453	453	100	32	29	21	17	13	9	9		22	14	–	5	–	–	3	–	–	–	–	–	–	2-(4-Chlorophenyl)amino-3,5-bis(4-chlorophenyl)-3H-imidazo[1,2-b][1,2,4]triazole		MS09905
438	420	422	436	264	58	421	439	453	100	97	66	47	37	32	30	27	11.00	24	39	7	1	–	–	–	–	–	–	–	–	–	Aconitane-1,7,8,14-tetrol, 20-ethyl-6,16-dimethoxy-4-(methoxymethyl)-, (1α,6β,14α,16β)-	545-56-2	NI43355
422	58	71	41	45	43	420	438	453	100	82	54	45	37	37	36	35	5.00	24	39	7	1	–	–	–	–	–	–	–	–	–	N-ethyl-4-(hydroxymethyl)-1-α,14α,16β-trimethoxyaconitane-6,7,8-triol		NI43356
91	241	58	155	198	181	298	268	453	100	78	52	38	35	25	21	19	11.00	25	27	5	1	1	–	–	–	–	–	–	–	–	3a(R),6(S),8,9b(S)-Tetrahydro-5-methoxy-6-hydroxy-9b-[2-[N-methyl-N-((4-methylphenyl)sulphonyl)amino]ethyl]phenanthro[4,4a,4b,5-bcd]furan		NI43357
155	149	127	453	167	57	100	85	453	100	50	46	26	19	13	12	11		25	27	7	1	–	–	–	–	–	–	–	–	–	4-(1,2:3,4-Di-O-isopropylidene-α-D-galactopyranuronamide)benzophenone		NI43358
453	77	128	234	92	420	118	107	453	100	48	28	24	22	21	20	17		26	23	1	5	1	–	–	–	–	–	–	–	–	4-(4-Methoxyphenylamino)-7-(methylthio)-5-phenyl-2-(p-tolyl)-imidazo[5,1-f][1,2,4]triazine		MS09906
91	128	218	453	134	103	77	133	453	100	47	46	43	28	25	24	23		26	23	1	5	1	–	–	–	–	–	–	–	–	2-(4-Methoxyphenyl)-7-(methylthio)-5-phenyl-4-(p-tolylamino)imidazo[5,1 f][1,2,4]triazine		MS09907
58	438	100	84	43	395	98	41	453	100	64	64	60	60	46	42	38	38.29	28	59	1	3	–	–	–	–	–	–	–	–	–	Hexadecanamide, N,N-bis[4-(dimethylamino)butyl]-	17232-85-8	NI43359
453	158	121	147	257	185	454	225	453	100	55	45	41	35	35	32	25		29	43	3	1	–	–	–	–	–	–	–	–	–	Acetamide, N-(3-oxofurosta-1,4-dien-26-yl)-	50686-93-6	NI43360
86	368	43	81	56	57	55	41	453	100	98	26	23	20	19	17	16	7.00	31	51	1	1	–	–	–	–	–	–	–	–	–	Cholest-5-ene, 3-(pyrrolidin-2-one-1-yl)-, (3β)-	38759-55-6	MS09908
327	127	177	100	131	69	454	181	454	100	44	41	37	31	30	26	26		4	–	–	–	–	8	–	–	2	–	–	–	–	Butane, 1,1,2,2,3,3,4,4-octafluoro-1,4-diiodo-	375-50-8	NI43361
314	398	370	199	426	342	454	326	454	100	24	21	20	18	18	12	12		5	–	5	–	–	–	–	–	1	–	–	–	1	Rhenium, pentacarbonyliodo-		MS09909
69	299	155	454	129	305	230	144	454	100	75	64	49	40	29	27	25		5	9	–	–	–	–	–	–	–	–	–	–	5	Tricyclo[3.3.1.0^{3,7}]nonane, 2,8-diselena-5-methyl-1,3,7-triarsa-	77609-87-1	NI43362
456	458	293	41	294	39	291	292	454	100	99	68	64	52	46	46	38	38.00	11	9	3	2	–	–	–	3	–	–	–	–	–	1-(3′,5′-Dibromo-4′-hydroxyphenyl)-5-bromo-6-methyl-2,4-dioxohexahydropyrimidine		NI43363
296	227	148	456	228	246	454	117	454	100	68	67	52	42	37	35	26		12	–	–	–	–	8	–	2	–	–	–	–	–	4,4′-Dibromooctafluorobiphenyl	10386-84-2	NI43364
387	454	413	388	435	400	367	425	454	100	20	19	12	8	8	8	7		15	14	2	–	–	12	–	–	–	–	–	–	–	1,1,7-Trihydroperfluoroheptyl norbornane-2-carboxylate		LI09115
296	297	295	237	456	377	375	281	454	100	19	17	15	7	6	6	6	3.00	16	16	2	4	–	–	–	2	–	–	–	–	–	Erythro-8-(1,2-dibromo-2-phenylethyl)-3,7-dihydro-1,3,7-trimethyl-1H-purine-2,6-dione		NI43365
57	56	41	29	184	69	342	55	454	100	75	38	36	29	15	11	11	0.00	16	20	8	–	–	6	–	–	–	–	–	–	–	Butanedioic acid, 2,3-bis[(trifluoroacetyl)oxy]-, dibutyl ester, [R-(R*,R*)]-	56114-67-1	NI43366
334	43	83	332	119	41	91	336	454	100	94	91	62	25	24	23	15	0.00	17	25	2	–	–	–	1	2	–	–	–	–	–	9-Acetoxy-2,10-dibromo-3-chlorochamigran-7-ene		MS09910
225	196	197	97	275	69	454	335	454	100	64	48	46	45	35	34	34		18	22	8	4	1	–	–	–	–	–	–	–	–	2-Methyl-6-(methylthio)-4-[N-(2,3,4-tri-O-acetyl-β-D-xylopyranosyl)amino]oxazolo[5,4-d]pyrimidine		NI43367
146	73	75	114	89	147	59	131	454	100	70	41	29	20	16	15	13	0.00	18	42	7	–	–	–	–	–	–	3	–	–	–	Hexopyranosid-3-ulose, methyl 2,4,6-tris-O-(trimethylsilyl)-, dimethyl acetal	74685-73-7	NI43368
150	120	454	59	456	118	119	45	454	100	52	40	35	26	25	20	18		19	20	5	4	–	–	2	–	–	–	–	–	–	N-[(2-Methoxyethoxy)carbonylethyl]-4-(2,6-dichloro-4-nitrophenylazo)-2-methylaniline		IC04449
101	88	45	201	75	85	71	59	454	100	51	49	46	37	22	16	14	0.00	19	34	12	–	–	–	–	–	–	–	–	–	–	Methyl 3,4-di-O-methyl-2-O-(methyl 2,3,4-tri-O-methyl-α-D-glucopyanosyl uronate)-α-D-xylopyranoside		LI09116
41	55	74	43	69	263	295	67	454	100	98	77	72	60	54	51	51	0.30	19	36	2	–	–	–	–	2	–	–	–	–	–	Octadecanoic acid, 9,10-dibromo-, methyl ester	25456-04-6	NI43369
375	377	293	295	456	376	454	104	454	100	91	58	39	26	22	22	20		20	10	4	–	–	–	4	–	–	–	–	–	–	3,3-Bis(2,6-dichloro-1-hydroxyphenyl)phthalide		NI43370
235	454	234	75	453	219	455	422	454	100	94	74	33	28	28	26	21		20	38	11	–	–	–	–	–	–	–	–	–	–	Cellobiose octa-O-methyl ether	56247-29-1	LI09117
88	279	45	101	75	187	219	71	454	100	54	39	35	34	22	16	16	0.00	20	38	11	–	–	–	–	–	–	–	–	–	–	Cellobiose octa-O-methyl ether	56247-29-1	NI43371
88	101	45	75	59	71	129	187	454	100	68	67	59	57	50	24	20	0.00	20	38	11	–	–	–	–	–	–	–	–	–	–	D-Fructofuranoside, methyl 1,3,4-tri-O-methyl-6-O-(2,3,4,6-tetra-O-methyl-α-D-glucopyranosyl)-	25531-74-2	NI43372
115	88	101	75	45	279	409	71	454	100	77	76	53	46	36	32	25	0.00	20	38	11	–	–	–	–	–	–	–	–	–	–	D-Fructopyranoside, methyl 1,3,4-tri-O-methyl-5-O-(2,3,4,6-tetra-O-methyl-β-D-galactopyranosyl)-	55298-60-7	NI43373
235	454	234	75	453	219	455	222	454	100	95	74	33	28	28	26	22		20	38	11	–	–	–	–	–	–	–	–	–	–	β-D-Glucopyranoside, methyl 2,3,6-tri-O-methyl-4-O-(tetra-O-methyl-β-D-glucopyranosyl)-	10225-71-5	NI43374
88	101	45	129	71	75	187	59	454	100	80	77	72	64	57	28	28	0.00	20	38	11	–	–	–	–	–	–	–	–	–	–	Leucrose permethyl ether	60618-00-0	NI43375
187	45	219	101	59	71	111	75	454	100	85	75	75	24	23	20	19	0.00	20	38	11	–	–	–	–	–	–	–	–	–	–	Sucrose permethyl ether	5346-73-6	NI43376
101	45	187	111	75	71	155	88	454	100	51	32	19	13	13	8	7	0.00	20	38	11	–	–	–	–	–	–	–	–	–	–	Trehalose permethyl ether	25018-29-5	NI43377
409	101	88	187	45	75	111	71	454	100	85	67	63	45	34	32	31	0.00	20	38	11	–	–	–	–	–	–	–	–	–	–	Turanose permethyl ether	57174-13-7	NI43378
199	241	157	381	283	395	255	339	454	100	94	88	52	43	33	31	25	0.00	20	46	7	–	–	–	–	–	–	2	–	–	–	Silicic acid, dibutyl tetrapropyl ester	72347-78-5	NI43379

MASS TO CHARGE RATIOS								M.W.	INTENSITIES								Parent	C	H	O	N	S	F	Cl	Br	I	Si	P	B	X	COMPOUND NAME	CAS Reg No	No
454	455	439	226	227	453	423	42	**454**	100	30	19	17	11	9	7	7		22	17	–	2	–	6	–	–	–	–	1	–	–	5H-Dibenzophosphole-3,7-diamine, 1,2,4,6,8,9-hexafluoro-N,N,N′,N′-tetramethyl-5-phenyl-	36126-96-2	NI43380
93	79	135	107	163	109	195	191	**454**	100	75	74	72	50	40	33	32	**0.00**	22	30	6	–	2	–	–	–	–	–	–	–	–	Bis[methyl (1RS,2RS,4RS,5SR)-4-methyl-8-oxobicyclo[3.3.0]octan-2-yl-1 carboxylate] disulphide		MS09911
114	243	238	86	454	239	310	126	**454**	100	99	95	90	74	62	55	52		22	38	6	4	–	–	–	–	–	–	–	–	–	L-Valine, N-[N-[1-(N-formyl-L-isoleucyl)-L-prolyl]-N-methylglycyl]-N-methyl-, methyl ester	55821-22-2	NI43381
73	454	267	179	455	75	439	268	**454**	100	78	61	43	32	19	18	18		22	42	4	–	–	–	–	–	–	3	–	–	–	(3,4-Dihydroxyphenyl)heptanoic acid, tris(trimethyl)-		NI43382
328	286	370	329	285	287	257	412	**454**	100	73	56	18	15	13	13	8	**2.00**	23	18	10	–	–	–	–	–	–	–	–	–	–	Scutellarein tetraacetate		LI09118
454	295	73	131	455	117	217	143	**454**	100	71	42	41	40	39	38	38		23	42	5	–	–	–	–	–	–	2	–	–	–	1-Cyclopentene-1-propanoic acid, 2-[3,7-bis[(trimethylsilyl)oxy]-1-octenyl]-5-oxo-, methyl ester	69502-88-1	NI43383
454	295	422	455	235	423	221	73	**454**	100	51	49	36	23	20	17	17		23	42	5	–	–	–	–	–	–	2	–	–	–	1-Cyclopentene-1-propanoic acid, 2-[3,8-bis[(trimethylsilyl)oxy]-1-octenyl]-5-oxo-, methyl ester	69502-87-0	NI43384
213	166	307	322	250	135	231	151	**454**	100	84	75	64	57	35	30	28	**0.00**	24	27	5	–	–	–	–	–	–	1	1	–	–	(Z)-1-(4-Methoxyphenyl)-2-(trimethylsilyl)vinyl diphenyl phosphate	118297-93-1	DD01706
454	456	439	441	383	371	455	385	**454**	100	72	66	45	43	37	33	30		24	36	2	–	–	–	2	–	–	1	–	–	–	Δ^1-Tetrahydrocannabinol (dichloromethyl)dimethylsilyl ether		NI43385
337	321	43	362	339	344	309	319	**454**	100	76	72	64	60	41	41	35	**5.00**	25	26	8	–	–	–	–	–	–	–	–	–	–	(7R)-9-Acetyl-7-isobutoxy-4-methoxy-7,8,9,10-tetrahydro-6,9,11-trihydroxy-5,12-naphthacenequinone		NI43386
337	309	362	43	321	339	217	319	**454**	100	70	69	67	64	60	43	40	**16.00**	25	26	8	–	–	–	–	–	–	–	–	–	–	(7S)-9-Acetyl-7-isobutoxy-4-methoxy-7,8,9,10-tetrahydro-6,9,11-trihydroxy-5,12-naphthacenequinone		NI43387
186	185	454	144	171	199	197	212	**454**	100	51	49	26	18	15	11	10		25	30	6	2	–	–	–	–	–	–	–	–	–	Eripine, O-acetyl-		LI09119
108	150	412	156	187	454	130	144	**454**	100	45	25	25	20	18	16	13		25	30	6	2	–	–	–	–	–	–	–	–	–	Lonicerine, N,O-diacetyl-		LI09120
225	226	351	238	394	169	184	180	**454**	100	46	31	20	16	16	12	11	**2.00**	25	30	6	2	–	–	–	–	–	–	–	–	–	Talbotine, O,4-diacetyl-19,20-dihydro-		LI09121
225	226	351	238	394	169	212	184	**454**	100	46	31	20	16	16	14	12	**2.00**	25	30	6	2	–	–	–	–	–	–	–	–	–	Talbotine, O,4-diacetyl-19,20-dihydro-, (20α)-	30809-31-5	NI43388
291	255	191	127	102	293	383	454	**454**	100	80	75	69	47	36	33	30		26	16	–	4	–	–	2	–	–	–	–	–	–	1,2-Bis[(E)-2-(3-chloro-2-quinoxalinyl)ethenyl]benzene		MS09912
163	395	259	260	78	77	205	156	**454**	100	35	34	23	21	19	15	14	**8.00**	26	22	4	4	–	–	–	–	–	–	–	–	–	5,8-Etheno[3,4]cyclobuta[1,2-d]pyridazine-6,7-dicarboxylic acid, 4a,4b,5,8,8a,8b-hexahydro-1,4-di-2-pyridinyl-, dimethyl ester	56083-42-2	NI43389
336	296	155	321	337	454	297	411	**454**	100	81	39	34	30	24	24	21		26	38	3	–	–	–	–	–	–	2	–	–	–	19-Norpregna-1,3,5,7,9-pentaen-21-al, 3,17-bis[(trimethylsilyl)oxy]-, (17α)	74299-05-1	NI43390
289	371	241	290	372	242	207	160	**454**	100	53	40	15	12	7	6	5	**3.48**	26	48	2	–	–	–	–	–	–	–	2	–	–	Phosphinous acid, dicyclohexyl-, 1,2-ethanediyl ester	74685-86-2	NI43391
105	77	106	305	122	79	91	348	**454**	100	19	8	6	5	4	3	2	**0.20**	27	22	5	2	–	–	–	–	–	–	–	–	–	(2R,4S,5R)-5-Benzoyloxy-4-(1-benzoylpyrazol-3-yl)-2-phenyl-1,3-dioxan		NI43392
259	260	135	122	121	109	123	107	**454**	100	90	43	43	43	41	37	35	**10.01**	27	39	5	–	–	–	–	–	–	–	–	1	–	17,21-Dihydroxy-5β-pregnane-3,11,20-trione cyclohexyl boronate	30888-26-7	NI43393
259	260	135	122	121	109	123	136	**454**	100	90	43	43	43	41	37	35	**10.00**	27	39	5	–	–	–	–	–	–	–	–	1	–	17,21-Dihydroxy-5β-pregnane-3,11,20-trione cyclohexyl boronate	30888-26-7	LI09122
384	229	382	383	299	247	385	454	**454**	100	57	42	37	37	32	30	29		27	44	1	–	–	–	2	–	–	–	–	–	–	5α,6β-Dichloro-3β,4β-epoxycholestane		LI09123
73	69	395	89	43	57	439	83	**454**	100	42	40	40	40	38	36	35	**4.00**	27	54	3	–	–	–	–	–	–	1	–	–	–	Tricosenoic acid, 2-[(trimethylsilyl)oxy]-, methyl ester	56847-15-5	MS09913
253	183	201	47	185	77	128	51	**454**	100	24	18	17	16	16	15	10	**2.19**	28	24	2	–	–	–	–	–	–	–	2	–	–	1,4-Bis(diphenyl-oxophosphinyl)but-2-yne		NI43394
454	360	77	224	455	439	51	28	**454**	100	79	75	39	36	36	35	35		28	26	–	2	2	–	–	–	–	–	–	–	–	1,2-Bis(diphenylamino)-1,2-bis(methylthio)ethylene		MS09914
91	92	65	39	31	285	63	51	**454**	100	66	15	11	10	8	7	7	**0.00**	28	28	–	–	–	–	–	–	–	–	–	–	1	Tetrabenzylzirconium	24356-01-2	NI43395
409	367	98	88	55	101	69	410	**454**	100	77	63	40	34	28	28	27	**14.00**	28	54	4	–	–	–	–	–	–	–	–	–	–	Tetracosanedioic acid, diethyl ester	1472-91-9	NI43396
245	139	316	192	161	454	436		**454**	100	45	41	33	30	28	23			29	42	4	–	–	–	–	–	–	–	–	–	–	3β,22R-Dihydroxy-24-carboxymethylenecholest-5-en-7-one 24′,22-lactone		LI09124
47	139	316	245	436	454	192	161	**454**	100	49	43	30	28	20	18	17		29	42	4	–	–	–	–	–	–	–	–	–	–	3β,22S-Dihydroxy-24-carboxymethylenecholest-5-en-7-one 24′,22-lactone		LI09125
95	69	41	97	175	135	205	165	**454**	100	90	42	39	38	28	26	26	**20.00**	29	42	4	–	–	–	–	–	–	–	–	–	–	6-(2,5-Dimethoxy-3-methylphenyl)-1-[1,2-dimethyl-2-(4-methyl-1-oxo-3-pentenyl)cyclopentyl]-4-methyl-4-hexen-2-one	97730-92-2	NI43397
165	396	135	205	436	231	151	365	**454**	100	83	76	50	46	43	24	13	**12.00**	29	42	4	–	–	–	–	–	–	–	–	–	–	[3aα,6(E),7aα]-6-[4-(2,5-dimethoxy-3-methylphenyl)-2-methyl-2-butenyl]-1,2,3,3a,7,7a-hexahydro-5-(2-hydroxy-2-methylpropyl)-3a,7a-dimethyl-4H-inden-4-one	91893-80-0	NI43398
454	151	135	152	91	288	121	136	**454**	100	14	14	13	12	9	9	7		29	42	4	–	–	–	–	–	–	–	–	–	–	5,24,25-Trimethoxy-2-oxatricyclo[20.2.2.1^{3,7}]heptacosa-3,5,7(27),22,24,25-hexaene		NI43399
243	454	327	269	255	303	394	329	**454**	100	58	55	43	39	36	27	24		30	46	3	–	–	–	–	–	–	–	–	–	–	3β-Acetoxy-5α-ergosta-7,22-dien-6-one		LI09126
243	454	255	269	303	394	329	379	**454**	100	63	53	27	25	24	19	13		30	46	3	–	–	–	–	–	–	–	–	–	–	3β-Acetoxy-5α-ergosta-8(14),22-dien-6-one		LI09127
454	243	394	379	269	255	327	329	**454**	100	86	80	60	52	36	28	24		30	46	3	–	–	–	–	–	–	–	–	–	–	3β-Acetoxy-5β-ergosta-7,22-dien-6-one		LI09128
125	395	455	396	393	453	271	126	**454**	100	67	32	20	18	15	15	8	**8.00**	30	46	3	–	–	–	–	–	–	–	–	–	–	3β-Acetoxyergosta-7,22-dien-6-one		NI43400
245	257	235	95	121	119	189	109	**454**	100	90	90	90	60	45	44	34	**2.00**	30	46	3	–	–	–	–	–	–	–	–	–	–	Bauer-7-en-3-on-28-oic acid		NI43401
81	205	69	55	95	107	41	93	**454**	100	95	92	86	83	80	76	74	**65.00**	30	46	3	–	–	–	–	–	–	–	–	–	–	Lup-20(29)-ene-3,21-dione, 28-hydroxy-	13952-73-3	NI43402
153	454	107	383	121	149	166	168	**454**	100	80	61	57	57	56	54	52		30	46	3	–	–	–	–	–	–	–	–	–	–	1,3-Dioxoallobetulane		MS09915
153	105	149	193	119	454	111	191	**454**	100	87	75	73	73	70	70	30		30	46	3	–	–	–	–	–	–	–	–	–	–	D:A-Friedooleanan-24-al, 1,3-dioxo-	35162-67-5	MS09916
454	329	407	301	163	139	439	235	**454**	100	40	27	27	27	27	25	17		30	46	3	–	–	–	–	–	–	–	–	–	–	D:A-Friedooleanane-1,3-dione, 25,26-epoxy-	52647-80-0	MS09917
454	367	43	395	69	421	55	57	**454**	100	88	79	63	54	49	49	43		30	46	3	–	–	–	–	–	–	–	–	–	–	Lanosta-7,9(11)-dien-18-oic acid, 3,20-dihydroxy-, γ-lactone, (3β)-	24041-68-7	NI43403
57	71	43	259	85	83	63	97	**454**	100	68	64	46	46	41	38	36	**0.51**	30	62	2	–	–	–	–	–	–	–	–	–	–	1,2-Ditetradecycloxyethane		IC04450

MASS TO CHARGE RATIOS	M.W.	INTENSITIES	Parent	C	H	O	N	S	F	Cl	Br	I	Si	P	B	X	COMPOUND NAME	CAS Reg No	No
163 262 183 108 165 164 166 263	**454**	100 88 54 38 20 20 18 13	**1.80**	31	23	–	2	–	–	–	–	–	–	1	–	–	9H-Fluoren-9-one, (triphenylphosphoranylidene)hydrazone	751-35-9	NI43404
81 313 255 454 229 315 288 213	**454**	100 75 67 54 37 30 20 20		31	50	2	–	–	–	–	–	–	–	–	–	–	(22Z)-(3β,5α)-3-Acetoxy-24,24-dimethylcholesta-7,22-diene	118112-72-4	DD01707
296 313 356 213 394 253 273 379	**454**	100 98 56 31 26 18 8 7	**6.00**	31	50	2	–	–	–	–	–	–	–	–	–	–	3β-Acetoxy-24-ethylidenecholest-5-ene		NI43405
439 379 454 227 440 442 287 268	**454**	100 58 56 53 46 41 40 37		31	50	2	–	–	–	–	–	–	–	–	–	–	Cholesta-8,24-dien-3-ol, 4,14-dimethyl-, acetate, (3β,4α)-	55331-92-5	LI09129
439 379 454 227 440 442 287 269	**454**	100 57 56 55 46 41 41 38		31	50	2	–	–	–	–	–	–	–	–	–	–	Cholesta-8,24-dien-3-ol, 4,14-dimethyl-, acetate, (3β,4α)-	55331-92-5	NI43406
313 356 43 255 454 439 329 384	**454**	100 63 29 25 13 12 9 7		31	50	2	–	–	–	–	–	–	–	–	–	–	25,26-Dihydroelasterol acetate		LI09130
313 454 255 314 81 55 43 356	**454**	100 40 29 28 28 25 21 20		31	50	2	–	–	–	–	–	–	–	–	–	–	Dihydroelasterol acetate		MS09918
43 379 454 56 54 41 69 394	**454**	100 60 40 40 40 40 35 26		31	50	2	–	–	–	–	–	–	–	–	–	–	4,4-Dimethyl-5α-cholesta-8,14-dien-3β-ol, acetate		LI09131
327 370 328 228 269 454 439 241	**454**	100 39 29 29 28 21 18 14		31	50	2	–	–	–	–	–	–	–	–	–	–	Ergosta-7,24(28)-dien-3-ol, 4-methyl-, acetate, (3β,4α)-	55399-28-5	NI43407
169 249 262 264 257 215 203 231	**454**	100 56 40 36 20 17 16 13	**8.80**	31	50	2	–	–	–	–	–	–	–	–	–	–	D:A-Friedoolean-3-en-29-oic acid, methyl ester, (20α)-	39903-16-7	MS09919
153 439 454 301 261 221 205 179	**454**	100 65 25 17 15 12 10 10		31	50	2	–	–	–	–	–	–	–	–	–	–	D:A-Friedoolean-2-en-1-one, 3-methoxy-	43230-67-7	MS09920
203 69 95 71 135 81 67 107	**454**	100 76 72 68 62 55 48 45	**17.82**	31	50	2	–	–	–	–	–	–	–	–	–	–	C(14a)-Homo-27-norgammacer-13-en-21-one, 3-methoxy-, (3α)-	24759-09-9	NI43408
203 69 95 135 81 71 67 107	**454**	100 78 74 63 57 51 50 47	**20.00**	31	50	2	–	–	–	–	–	–	–	–	–	–	3α-Methoxy-21-oxo-Δ^{13}-serratene		LI09132
373 367 454 439 411	**454**	100 70 35 9 5		31	50	2	–	–	–	–	–	–	–	–	–	–	3,4-Secolupa-4(23),20(29)-dien-3-oic acid, methyl ester	1179-91-5	NI43409
218 454 203 372 411 367	**454**	100 75 48 20 19 7		31	50	2	–	–	–	–	–	–	–	–	–	–	3,4-Secooleana-4(23),12-dien-3-oic acid, methyl ester	59076-77-6	NI43410
218 454 203 372 411 367	**454**	100 75 45 20 16 7		31	50	2	–	–	–	–	–	–	–	–	–	–	3,4-Secoursa-4(23),12-dien-3-oic acid, methyl ester	40286-37-1	NI43411
394 55 81 83 69 255 43 41	**454**	100 91 73 66 51 40 37 32	**0.00**	31	50	2	–	–	–	–	–	–	–	–	–	–	Stigmasta-5,22-dien-3-ol, acetate, (3β,22Z)-	54482-54-1	NI43412
80 394 296 145 395 95 297 107	**454**	100 98 98 37 32 28 26 25	**0.00**	31	50	2	–	–	–	–	–	–	–	–	–	–	Stigmasta-5,24(28)-dien-3-ol, acetate, (3β,24E)-	51297-12-2	NI43413
55 57 81 313 43 83 69 71	**454**	100 96 92 87 87 79 70 61	**32.40**	31	50	2	–	–	–	–	–	–	–	–	–	–	Stigmasta-7,22-dienyl acetate		MS09921
255 394 395 213 253 351 229 256	**454**	100 33 31 24 23 19 19 15	**0.00**	31	50	2	–	–	–	–	–	–	–	–	–	–	Stigmasterol acetate	4651-48-3	NI43414
394 43 149 395 255 57 55 41	**454**	100 43 38 32 28 28 28 27	**6.00**	31	50	2	–	–	–	–	–	–	–	–	–	–	Stigmasterol acetate	4651-48-3	NI43415
394 55 81 83 255 395 69 43	**454**	100 54 51 42 36 34 34 19	**0.00**	31	50	2	–	–	–	–	–	–	–	–	–	–	Stigmasterol acetate	4651-48-3	NI43416
95 436 314 175 421 454 203 439	**454**	100 64 58 56 46 42 38 35		32	54	1	–	–	–	–	–	–	–	–	–	–	Cycloneolitsol		LI09133
327 439 407 454 440 328 408 215	**454**	100 84 81 47 28 28 27 18		32	54	1	–	–	–	–	–	–	–	–	–	–	3β-Methoxy-24S-methyllanosta-9(11),25-diene		MS09922
454 262 183 78 455 193 269 108	**454**	100 65 47 44 37 27 26 21		33	27	–	–	–	–	–	–	–	–	1	–	–	Phosphorane, (2,7-dimethylfluoren-9-ylidene)triphenyl-	18916-67-1	LI09134
454 262 183 78 455 193 269 453	**454**	100 65 47 43 37 27 24 23		33	27	–	–	–	–	–	–	–	–	1	–	–	Phosphorane, (2,7-dimethylfluoren-9-ylidene)triphenyl-	18916-67-1	NI43417
340 338 342 108 459 457 110 461	**455**	100 72 53 49 46 46 30 25	**20.27**	11	4	–	3	1	–	7	–	–	–	–	–	–	2-(4′-Chlorophenylthio)-4,6-bis(trichloromethyl)-s-triazine		MS09923
43 190 368 103 119 41 220 188	**455**	100 73 24 18 17 15 10 8	**0.00**	12	11	4	1	1	10	–	–	–	–	–	–	–	N,S-Bis(pentafluoropropionyl)-L-cysteine isopropyl ester		MS09924
119 133 455 100 308 107 237 280	**455**	100 72 57 51 40 24 23 21		15	7	4	1	–	10	–	–	–	–	–	–	–	5-Hydroxy-6-methoxyindole, bis(pentafluoropropionyl)-		NI43418
119 100 455 308 133 280 107 147	**455**	100 90 43 34 28 14 14 13		15	7	4	1	–	10	–	–	–	–	–	–	–	6-Hydroxy-5-methoxyindole, bis(pentafluoropropionyl)-		NI43419
115 223 221 302 301 457 459 455	**455**	100 75 75 38 38 26 14 13		17	15	2	1	1	–	–	2	–	–	–	–	–	1H-Cyclopropa[c]quinoline, 1,1-dibromo-1a,2,3,7b-tetrahydro-3-(p-tolylsulphonyl)-	24310-26-7	NI43420
455 457 194 77 274 275 273 105	**455**	100 98 80 60 58 52 44 44		19	14	4	5	–	–	–	1	–	–	–	–	–	2,4-Dinitrophenylhydrazono-5-bromo-2-aminobenzophenone		MS09925
364 260 350 455 306 318 337 292	**455**	100 41 39 30 21 20 18 18		19	21	10	1	1	–	–	–	–	–	–	–	–	Benzo[b]thiophene-3,4,5,6,7-pentacarboxylic acid, 2-amino-4,5-dihydro-, 3-ethyl 4,5,6,7-tetramethyl ester	75314-23-7	NI43421
192 58 142 177 193 121 190 127	**455**	100 37 19 17 13 9 5 5	**0.00**	20	26	3	1	–	–	–	–	1	–	–	–	–	(R)-(–)-1-(4-Methoxybenzyl)-1,2,3,4-tetrahydro-8-hydroxy-7-methoxy-2,2-dimethylisoquinolinium iodide		MS09926
455 457 376 456 426 424 458 377	**455**	100 62 60 34 32 30 26 22		22	18	5	1	–	–	–	1	–	–	–	–	–	1H-Inden-1-one, 2-bromo-3-(6,7-dimethoxy-1-isoquinolinyl)-5,6-dimethoxy-	23434-75-5	NI43422
84 87 260 44 45 43 218 192	**455**	100 85 74 74 63 60 59 58	**0.50**	22	25	6	5	–	–	–	–	–	–	–	–	–	1-Pyrrolidinecarboxylic acid, 2-[[[5-[[(3-amino-3-oxopropyl)amino]carbonyl]-1-methyl-1H-pyrrol-3-yl]amino]carbonyl]-5-oxo-, benzyl ester	55356-25-7	NI43423
455 457 456 454 458 459 253 122	**455**	100 80 79 54 43 35 17 15		24	17	–	1	–	–	3	–	–	–	1	–	–	Phosphine imide, P,P,P-tris(p-chlorophenyl)-N-phenyl-	14796-92-0	NI43424
455 454 253 122 251 142 344 364	**455**	100 54 17 15 12 9 7 6		24	17	–	1	–	–	3	–	–	–	1	–	–	Phosphine imide, P,P,P-tris(p-chlorophenyl)-N-phenyl-	14796-92-0	LI09135
169 141 455 115 91 39 36 43	**455**	100 72 59 46 26 20 19 18		24	17	5	5	–	–	–	–	–	–	–	–	–	2-Hydroxy-4′-(2,4-dinitro-phenylamino)-4-phenyl-azobenzene		IC04451
151 300 268 284 305 455 165 155	**455**	100 71 26 9 8 6 3 3		24	25	6	1	1	–	–	–	–	–	–	–	–	Anthranilic acid, N-(p-tolylsulphonyl)-N-veratryl-, methyl ester	23145-61-1	NI43425
212 155 211 185 299 255 298 273	**455**	100 71 44 39 32 25 17 15	**1.99**	24	45	5	3	–	–	–	–	–	–	–	–	–	N-Decanoylglycyl-DL-leucyl-L-valine methyl ester	56272-50-5	NI43426
86 72 212 132 456 71 245 85	**455**	100 39 38 37 34 32 27 22	**1.00**	24	45	5	3	–	–	–	–	–	–	–	–	–	N-Decanoylglycyl-L-leucyl-L-valine methyl ester	31944-59-9	NI43427
212 155 211 194 298 255 213 196	**455**	100 77 46 40 37 27 20 16	**3.60**	24	45	5	3	–	–	–	–	–	–	–	–	–	N-Decanoylglycyl-L-leucyl-L-valine methyl ester	31944-59-9	NI43429
212 155 211 184 298 213 186 297	**455**	100 75 49 36 33 16 13 12	**4.67**	24	45	5	3	–	–	–	–	–	–	–	–	–	N-Decanoylglycyl-L-leucyl-L-valine methyl ester	31944-59-9	NI43428
73 309 174 157 129 310 116 247	**455**	100 87 65 42 33 23 23 21	**0.00**	24	49	3	1	–	–	–	–	–	2	–	–	–	Acetamide, N-[2-[(trimethylsilyl)oxy]-1-[[(trimethylsilyl)oxy]methyl]-3,11 pentadecadienyl]-	55429-52-2	NI43430
363 77 91 455 364 119 146 65	**455**	100 60 30 24 22 21 20 18		25	21	4	5	–	–	–	–	–	–	–	–	–	4-((1-Phenyl-3-carboethoxy-4,5-dihydropyrazol-4-ylidene)hydrazino)-N-phenylbenzamide		IC04452

MASS TO CHARGE RATIOS								M.W.	INTENSITIES								Parent	C	H	O	N	S	F	Cl	Br	I	Si	P	B	X	COMPOUND NAME	CAS Reg No	No
120	43	55	165	135	65	66	382	**455**	100	72	50	38	34	34	26	13	**2.00**	25	23	2	1	–	–	–	–	–	–	1	–	1	**Manganese, acetyl(η^5-2,4-cyclopentadien-1-yl)nitrosyl(triphenylphosphine)-**	43064-64-8	**MS09927**
347	317	79	55	427	80	381	65	**455**	100	56	53	38	21	10	5	5	**1.00**	25	23	2	1	–	–	–	–	–	–	1	–	1	**Manganese, carbonyl[(1,2,3,4-η)-5-methyl-1,3-cyclopentadiene]nitrosyl(triphenylphosphine)-, stereoisomer**	43135-67-7	**MS09928**
426	455	225	255	456	427	253	384	**455**	100	99	42	33	30	28	23	20		26	28	2	2	–	–	–	–	–	–	–	–	1	**Manganese, bis(2-butyl-8-quinolinolato)-**	30049-12-8	**NI43431**
123	304	98	165	206	221	130	118	**455**	100	60	60	50	39	29	28	24	**0.50**	26	31	2	3	–	2	–	–	–	–	–	–	–	**8-[4-(4-Fluorophenyl)-4-oxobutyl]-3-(3-fluoropropyl)-1-phenyl-11,3,8-triazaspiro[4,5]decan-4-one**		**MS09929**
455	456	57	315	457	194	440	75	**455**	100	36	27	16	14	11	10	10		26	41	2	1	–	–	–	–	–	2	–	–	–	**2-Amino-17α-ethynyl-estradiol-di-O-TMS**		**NI43432**
455	456	315	457	300	440	75	247	**455**	100	40	26	18	17	14	13	9		26	41	2	1	–	–	–	–	–	2	–	–	–	**4-Amino-17α-ethynyl-estradiol-di-O-TMS**		**NI43433**
396	397	308	296	455	309	364	322	**455**	100	67	26	16	8	7	6	6		28	25	5	1	–	–	–	–	–	–	–	–	–	**5-(1,2-Dimethoxycarbonyl-2-oxidovinyl)-6-mesityl-phenanthridinium**		**LI09136**
199	378	166	98	259	276	122	181	**455**	100	83	74	50	43	41	33	28	**7.00**	28	29	3	1	–	–	–	–	–	1	–	–	–	**3-Diethylamino-2-(triphenylsiloxyacetyl)-2-cyclobuten-1-one**		**NI43434**
384	183	455	185	109	214	110	184	**455**	100	89	54	52	25	20	15	13		29	31	–	1	–	–	–	–	–	–	2	–	–	**N,N-Bis(diphenylphosphino)pentylamine**		**NI43435**
455	232	454	456	378	105	77	233	**455**	100	65	33	31	28	23	21	11		30	22	4	–	–	–	–	–	–	–	–	–	1	**Beryllium, bis(1,3-diphenyl-1,3-propanedionato-O,O')-**	19368-60-6	**NI43436**
455	232	456	378	454	105	77	233	**455**	100	70	33	32	31	24	24	12		30	22	4	–	–	–	–	–	–	–	–	–	1	**Beryllium, bis(1,3-diphenyl-1,3-propanedionato-O,O')-**	19368-60-6	**NI43437**
455	232	454	378	105	77	426	154	**455**	100	70	31	31	24	24	11	6		30	22	4	–	–	–	–	–	–	–	–	–	1	**Beryllium, bis(1,3-diphenyl-1,3-propanedionato-O,O')-**	19368-60-6	**LI09137**
455	440	456	441	180	218	457	217	**455**	100	93	37	30	14	10	7	6		31	37	2	1	–	–	–	–	–	–	–	–	–	**Benzo[e]naphth[2',1':4,5]indeno[1,2-b]indol-2-ol, 1,2,3,4,4a,4b,5,6,6a,7,14,14a,14b,15,16,16a-hexadecahydro-4a,6a-dimethyl-, acetate (ester), [2S-(2α,4aα,4bβ,6aα,14aβ,14bα)]-**	40039-62-1	**MS09930**
440	455	441	180	218	456	217	194	**455**	100	50	33	27	18	14	11	11		31	37	2	1	–	–	–	–	–	–	–	–	–	**Benzo[g]naphth[2',1':4,5]indeno[1,2-b]indol-2-ol, 1,2,3,4,4a,4b,5,6,6a,7,14,14a,14b,15,16,16a-hexadecahydro-4a,6a-dimethyl-, acetate (ester), [2S-(2α,4aα,4bβ,6aα,14aβ,14bα)]-**	39987-75-2	**MS09931**
88	369	368	43	147	81	145	57	**455**	100	99	99	69	40	40	35	30	**8.70**	31	53	1	1	–	–	–	–	–	–	–	–	–	**Propanamide, N-[(3β)-cholest-5-en-3-yl]-2-methyl-**	38759-53-4	**MS09932**
456	331	428	275	302	221	400	233	**456**	100	65	19	16	10	9	6	5		7	1	4	–	–	6	–	–	–	–	–	–	1	**Iridium, dicarbonyl(1,1,1,5,5,5-hexafluoro-2,4-pentanedionato-O,O')-**	14049-69-5	**NI43438**
230	456	201	114	173	202	87	115	**456**	100	90	69	69	66	55	53	46		8	–	8	–	–	–	–	–	–	–	–	–	3	**Cadmium, bis(tetracarbonylcobalt)-**		**LI09138**
107	67	41	39	227	225	105	213	**456**	100	98	94	79	71	64	49	43	**31.43**	10	16	–	–	–	–	–	–	–	–	–	–	4	**2,4,6,8-Tetraselenaadamantane, 1,3,5,7-tetramethyl-**	25133-47-5	**NI43439**
316	428	288	260	344	204	202	232	**456**	100	64	48	46	45	42	36	32	**22.00**	11	3	9	–	–	–	–	–	–	–	–	–	3	**Cobalt, nonacarbonyl-μ^3-ethylidenetri-, triangulo-**		**MS09933**
192	112	164	220	304	56	52	332	**456**	100	75	50	49	44	34	28	26	**0.00**	11	3	9	–	–	–	–	–	–	–	–	–	3	**Cobalt, nonacarbonyl-μ^3-ethylidynetri-, triangulo-**	13682-04-7	**NI43440**
376	456	160	296	159	135	117	265	**456**	100	53	34	25	8	7	6	5		12	–	–	–	–	8	–	–	–	–	–	–	2	**Selenanthrene, octafluoro-**	16259-99-7	**LI09139**
376	296	374	456	454	160	378	158	**456**	100	52	48	33	30	21	18	18		12	–	–	–	–	8	–	–	–	–	–	–	2	**Selenanthrene, octafluoro-**	16259-99-7	**NI43441**
376	374	296	456	160	452	378	372	**456**	100	47	44	33	18	17	17	17		12	–	–	–	–	8	–	–	–	–	–	–	2	**Selenanthrene, octafluoro-**	16259-99-7	**NI43442**
452	450	454	448	456	397	394	453	**456**	100	95	69	39	29	22	18	17		12	–	2	–	–	–	8	–	–	–	–	–	–	**Dibenzo[b,e][1,4]dioxin, octachloro-**	3268-87-9	**NI43443**
460	458	462	456	464	397	395	461	**456**	100	90	65	36	27	16	16	15		12	–	2	–	–	–	8	–	–	–	–	–	–	**Dibenzo[b,e][1,4]dioxin, octachloro-**	3268-87-9	**NI43444**
460	458	462	456	464	395	397	322	**456**	100	93	60	39	25	17	14	9		12	–	2	–	–	–	8	–	–	–	–	–	–	**Dibenzo[b,e][1,4]dioxin, octachloro-**	3268-87-9	**NI43445**
59	336	426	456	396	182	366	123	**456**	100	58	50	25	25	23	20	20		12	10	4	4	2	–	–	–	–	–	–	–	2	**Cobalt, di-μ-benzenethiolato-tetranitrosodi-**		**MS09934**
232	260	175	344	288	203	174	204	**456**	100	78	68	65	61	60	60	44	**27.00**	12	10	8	–	2	–	–	–	–	–	–	–	2	**Ethanethiolato-tetracarbonylmanganese-dimer**		**MS09935**
346	232	55	428	316	147	116	70	**456**	100	99	86	84	79	63	63	52	**5.00**	12	12	8	–	–	–	–	–	–	–	2	–	2	**Di-μ-dimethyl-phosphino-bis(tetracarbonylmanganese)**		**MS09936**
131	199	187	103	171	153	185	169	**456**	100	79	70	35	33	26	21	14	**0.00**	12	26	6	–	4	–	–	–	–	–	2	–	–	**Phosphorodithioic acid, S,S'-1,4-dioxane-2,3-diyl O,O,O',O'-tetraethyl ester**	78-34-2	**NI43447**
97	125	65	153	93	45	121	73	**456**	100	67	47	38	34	34	32	32	**0.00**	12	26	6	–	4	–	–	–	–	–	2	–	–	**Phosphorodithioic acid, S,S'-1,4-dioxane-2,3-diyl O,O,O',O'-tetraethyl ester**	78-34-2	**NI43446**
97	135	29	270	73	65	153	93	**456**	100	67	66	47	43	40	31	30	**0.00**	12	26	6	–	4	–	–	–	–	–	2	–	–	**Phosphorodithioic acid, S,S'-1,4-dioxane-2,3-diyl O,O,O',O'-tetraethyl ester**	78-34-2	**NI43448**
43	44	83	55	374	31	82	18	**456**	100	99	96	75	67	59	58	45	**1.74**	15	16	2	–	–	12	–	–	–	–	–	–	–	**1,1,7-Trihydroperfluoroheptyl cyclohexylacetate**		**MS09937**
147	202	148	41	39	187	27	91	**456**	100	18	16	15	13	12	12	10	**0.00**	15	22	–	–	–	–	–	–	2	–	–	–	–	**1,2,4,5-Tetramethyl-3-(2,2-diiodomethylpropyl)benzene**		**AI02385**
242	43	44	171	256	243	213	156	**456**	100	62	50	35	32	26	18	13	**0.00**	17	21	9	4	–	–	–	–	–	–	1	–	–	**Riboflavin 5'-(dihydrogen phosphate)**	146-17-8	**NI43449**
290	456	124	166	189	217	411	246	**456**	100	38	14	12	11	10	9	8		17	35	6	–	–	–	–	–	–	–	2	–	1	**Cyclopentadienyl-cobalt-bis(triethylphosphite)**		**LI09140**
73	441	456	457	442	147	382	45	**456**	100	51	49	21	20	17	14	12		17	36	3	4	–	–	–	–	–	4	–	–	–	**Uric acid tetra-TMS**	55530-45-5	**MS09938**
73	456	75	441	147	77	442	45	**456**	100	32	32	31	29	28	18	17		17	36	3	4	–	–	–	–	–	4	–	–	–	**Uric acid tetra-TMS**	55530-45-5	**NI43450**
136	77	196	120	168	109	135	104	**456**	100	98	18	14	12	12	11	11	**5.50**	18	20	–	2	4	–	–	–	–	–	–	–	1	**Zinc bis-(N-ethyl-N-phenyldithiocarbamate)**		**IC04453**
73	147	139	231	441	129	157	148	**456**	100	57	38	19	18	18	16	11	**10.86**	20	42	6	–	–	–	–	–	–	2	–	2	–	**Scyllo-Inositol di-butaneboronate di-trimethylsilyl ether**		**NI43451**
223	102	121	186	279	56	251	39	**456**	100	87	74	72	43	32	30	30	**0.00**	22	16	4	–	–	–	–	–	–	–	–	–	2	**7,8-Bis(cyclopentadienylirondicarbonyl)bicyclo[4.2.0]octatriene**		**NI43452**
365	366	425	265	210	115	196	456	**456**	100	25	24	23	20	20	15	1		22	36	8	2	–	–	–	–	–	–	–	–	–	**Cyclopentaneoctanoic acid, 5-(acetyloxy)-ϵ,3-bis(methoxyimino)-2-(3-methoxy-3-oxopropyl)-, methyl ester**		**LI09141**

MASS TO CHARGE RATIOS								M.W.	INTENSITIES								Parent	C	H	O	N	S	F	Cl	Br	I	Si	P	B	X	COMPOUND NAME	CAS Reg No	No
365	366	425	265	210	115	196	197	**456**	100	25	23	23	20	20	15	7	**0.00**	22	36	8	2	–	–	–	–	–	–	–	–	–	Cyclopentaneoctanoic acid, 5-(acetyloxy)-ε,3-bis(methoxyimino)-2-(3-methoxy-3-oxopropyl)-, methyl ester, [1R-(1α,2β,5β)]-	55669-97-1	NI43453
365	366	115	210	425	192	76	265	**456**	100	22	22	21	17	16	16	14	**0.46**	22	36	8	2	–	–	–	–	–	–	–	–	–	Cyclopentaneoctanoic acid, 5-(acetyloxy)-ε,3-bis(methoxyimino)-2-(3-methoxy-3-oxopropyl)-, methyl ester, [1R-(1α,2β,5β)]-	55669-97-1	NI43454
253	43	295	91	414	92	120	252	**456**	100	66	62	62	44	38	26	23	**20.00**	23	16	5	6	–	–	–	–	–	–	–	–	–	N-(4-Acetamidophenyl)-5-(4-methyl-5-cyano-2,6-dioxo-1,2,3,6-tetrahydropyrid-3-ylidenehydrazino)-phthalimide		IC04454
179	387	175	76	254	207	217	147	**456**	100	97	68	59	55	54	52	48	**2.70**	23	44	5	–	–	–	–	–	–	2	–	–	–	Cyclopentanepropanoic acid, 2-(3-oxo-1-octenyl)-3,5-bis[(trimethylsilyl)oxy]-, methyl ester	56273-16-6	NI43455
105	232	274	77	170	397	191	43	**456**	100	17	16	12	11	9	8	8	**0.00**	24	24	9	–	–	–	–	–	–	–	–	–	–	β-D-Xylofuranose, 3-C-methyl-, 1,2-diacetate 3,5-dibenzoate	70774-27-5	NI43456
396	43	456	57	41	58	55	42	**456**	100	72	47	38	38	32	29	25		24	28	7	2	–	–	–	–	–	–	–	–	–	Dichotine, acetate (ester)	29474-87-1	NI43457
351	197	120	309	77	274	105	154	**456**	100	54	45	31	28	23	22	21	**0.70**	25	20	1	–	–	–	–	–	–	–	–	–	1	Benzoyltriphenyltin		LI09142
105	76	121	106	73	75	122	74	**456**	100	95	64	29	25	23	16	14	**0.08**	25	20	5	4	–	–	–	–	–	–	–	–	–	4-Benzoyl-2,6-dibenzoyl-oxyiminopiperazine		MS09939
105	321	159	77	320	173	104	160	**456**	100	59	38	38	31	28	24	17	**1.37**	25	22	7	–	–	–	–	–	–	–	–	2	–	1-O-Benzoyl-β-D-fructopyranose 2,3:4,5-bis(benzeneboronate)	53829-59-7	NI43458
105	159	146	334	77	188	147	160	**456**	100	27	27	16	15	13	12	11	**1.80**	25	22	7	–	–	–	–	–	–	–	–	2	–	6-O-Benzoyl-α-D-glucofuranose 1,2:3,5-bis(benzeneboronate)	53829-57-5	NI43459
161	162	180	160	149	135	150	179	**456**	100	99	80	80	80	80	75	70	**0.00**	25	28	8	–	–	–	–	–	–	–	–	–	–	Bicyclo[3.2.1]oct-3-en-2-one, 3,8-bis(acetyloxy)-7-(1,3-benzodioxol-5-ylmethyl)-1-methoxy-6-methyl-5-(2-propenyl)-	75332-12-6	NI43460
44	412	123	178	413	456	235	41	**456**	100	97	43	31	26	22	20	19		25	28	8	–	–	–	–	–	–	–	–	–	–	11H-Dibenzo[b,e][1,4]dioxepin-7-carboxylic acid, 8-hydroxy-3-methoxy-11-oxo-1-(1-oxopentyl)-6-pentyl-	522-53-2	NI43461
379	410	355	193	380	325	181	327	**456**	100	62	60	29	27	22	22	20	**3.00**	25	28	8	–	–	–	–	–	–	–	–	–	–	5,7,2′,4′-Tetramethoxy-3′-(3-formyloxy-3-methylbutyl)isoflavone		MS09940
226	325	456	43	160	324	369	227	**456**	100	58	28	25	23	22	18	16		25	32	6	2	–	–	–	–	–	–	–	–	–	Aspidospermidin-21-oic acid, 1-acetyl-20-(acetyloxy)-17-methoxy-, methyl ester	55123-70-1	NI43462
160	456	161	412	383	457	159	174	**456**	100	98	68	42	34	30	20	19		25	32	6	2	–	–	–	–	–	–	–	–	–	Aspidospermidin-21-oic acid, 19-hydroxy-15,16,17-trimethoxy-1-(1-oxopropyl)-, γ-lactone	13467-47-5	NI43463
160	161	456	138	136	412	383	83	**456**	100	58	55	51	28	27	24	19		25	32	6	2	–	–	–	–	–	–	–	–	–	Aspidospermidin-21-oic acid, 19-hydroxy-15,16,17-trimethoxy-1-(1-oxopropyl)-, γ-lactone	13467-47-5	NI43464
369	370	341	340	383	384	371	456	**456**	100	24	11	10	5	5	4	4		25	32	6	2	–	–	–	–	–	–	–	–	–	N-Æ(3,4-Dimethoxyphenyl)[2-(1-ethyl-2-oxopropyl)-4,5-dimethoxyphenyl]methyleneŒ propionohydrazine	90140-56-0	NI43465
108	130	150	144	106	77	156	60	**456**	100	56	41	25	21	21	20	17	**5.05**	25	32	6	2	–	–	–	–	–	–	–	–	–	4,5-Secoakuammilan-17-oic acid, 4-acetyl-5-(acetyloxy)-1,2-dihydro-16-(hydroxymethyl)-, methyl ester, (2α)-	14514-02-4	NI43466
161	173	188	122	281	296	381	308	**456**	100	90	89	89	40	31	17	10	**7.00**	25	32	6	2	–	–	–	–	–	–	–	–	–	Vindoline	2182-14-1	NI43467
199	314	312	242	116	144	73	143	**456**	100	80	64	37	29	19	17	10	**3.20**	25	36	4	4	–	–	–	–	–	–	–	–	–	L-Alanine, N-[N-[N-(1H-indol-3-ylmethylene)-L-valyl]-L-isoleucyl]-, ethyl ester	37580-24-8	NI43468
105	77	228	229	196	456	106	230	**456**	100	64	60	18	18	12	11	5		26	20	2	2	2	–	–	–	–	–	–	–	–	2,2′-Dithiobenzanilide	2527-63-1	NI43469
318	400	240	319	183	164	401	262	**456**	100	52	23	22	22	22	15	14	**2.61**	26	25	2	–	–	–	–	–	–	–	1	–	1	Iron, dicarbonyl[(1,2,3,4-η)-2,3-dimethyl-1,3-butadiene](triphenylphosphine)-	74753-27-8	NI43470
456	427	457	227	254	229	428	240	**456**	100	38	30	23	17	14	12	10		26	28	2	2	–	–	–	–	–	–	–	–	1	Iron, bis(2-butyl-8-quinolinolato)-	30106-81-1	NI43471
315	43	41	95	55	91	316	29	**456**	100	77	61	36	32	26	22	22	**0.00**	26	32	7	–	–	–	–	–	–	–	–	–	–	Epoxydeacetylgedunin		MS09941
43	41	55	69	309	456	115	57	**456**	100	95	70	59	47	38	37	31		26	32	7	–	–	–	–	–	–	–	–	–	–	Spiro[benzofuran-3(2H),1′-[2,5]cyclohexadiene]-5-carboxylic acid, 6-hydroxy-2′-methoxy-2,4′-dioxo-4-pentyl-, methyl ester	19314-88-6	NI43472
456	397	355	124	457	177	414	413	**456**	100	69	54	37	27	23	20	20		26	36	5	2	–	–	–	–	–	–	–	–	–	Aspidospermidine-1-acetic acid, 17-(acetyloxy)-16-methoxy-α-methyl-, methyl ester	54725-06-3	NI43473
456	457	91	274	241	239	215	213	**456**	100	38	34	16	12	12	12	12		27	24	5	2	–	–	–	–	–	–	–	–	–	Bis(4-methyl-5-benzyloxy-carbonyl-3-pyrrolyl) ketone		LI09143
166	165	396	43	79	91	55	93	**456**	100	82	81	75	73	71	66	60	**36.00**	27	41	5	–	–	–	–	–	–	–	–	1	–	17α,20β,21-Trihydroxypregn-4-en-3-one 20β-acetate n-butyl boronate		MS09942
73	43	89	397	57	129	103	75	**456**	100	42	39	38	37	33	33	33	**2.00**	27	56	3	–	–	–	–	–	–	1	–	–	–	Tricosanoic acid, 2-[(trimethylsilyl)oxy]-, methyl ester	56784-03-3	MS09943
228	200	180	157	125	211	185	103	**456**	100	79	58	32	32	15	14	13	**0.40**	28	24	6	–	–	–	–	–	–	–	–	–	–	cis-1R,2-bis[6-(4-methoxy-2-pyronyl)]-trans-3,4-diphenylcyclobutane		NI43474
255	201	325	262	77	256	183	456	**456**	100	54	34	30	25	20	10	8		28	26	2	–	–	–	–	–	–	–	2	–	–	1,4-Bis(diphenylphosphino)-but-2-ene dioxide		IC04455
150	88	58	420	43	137	189	288	**456**	100	78	56	44	32	29	28	27	**0.60**	28	40	5	–	–	–	–	–	–	–	–	–	–	[1α,5α,6β(R*),7α(E)]-(+)-1-(Dihydro-4′-hydroxy-1,5,5′,5′-tetramethylspiro[bicyclo[3.2.0]heptane-6,2′(3′H)-furan]-7-yl)-5-(2-hydroxy-5-methoxy-3-methylphenyl)-3-methyl-3-penten-1-one	92675-09-7	NI43475
227	91	228	319	230	189	291	263	**456**	100	29	26	15	14	14	9	9	**0.00**	29	20	2	4	–	–	–	–	–	–	–	–	–	2-Benzyl-1-(4-nitrophenyldiazenyl)-2H-dibenz[e,g]isoindole	110028-29-0	DD01708
315	99	316	295	71	297	358	313	**456**	100	72	25	20	19	18	13	12	**0.00**	29	44	4	–	–	–	–	–	–	–	–	–	–	Pregn-4-ene-3,20-dione, 6,16-dimethyl-17-[(1-oxohexyl)oxy]-, (6α,16α)-	56784-28-2	MS09944
456	99	342	457	125	341	112	55	**456**	100	83	40	33	28	17	13	12		29	44	4	–	–	–	–	–	–	–	–	–	–	5α-Spirost-14-en-3-one, cyclic ethylene acetal, (25R)-	24799-51-7	NI43476
456	99	342	457	125	341	112	328	**456**	100	83	41	32	28	18	13	12		29	44	4	–	–	–	–	–	–	–	–	–	–	5α-Spirost-14-en-3-one, cyclic ethylene acetal, (25R)-	24799-51-7	MS09945
456	457	342	328	343	458	99	384	**456**	100	32	28	8	7	6	6	4		29	44	4	–	–	–	–	–	–	–	–	–	–	5α-Spirost-14-en-3-one, cyclic ethylene acetal, (25R)-	24799-51-7	MS09946
456	357	441	457	442	373	372	358	**456**	100	46	43	42	17	16	16	14		29	48	2	–	–	–	–	–	–	1	–	–	–	Heptyl-Δ^{1}-tetrahydrocannabinol triethylsilyl ether		NI43477

MASS TO CHARGE RATIOS								M.W.	INTENSITIES								Parent	C	H	O	N	S	F	Cl	Br	I	Si	P	B	X	COMPOUND NAME	CAS Reg No	No
441	456	442	72	455	439	371	73	**456**	100	43	35	32	20	20	18	16		29	48	2	2	–	–	–	–	–	–	–	–	–	**2-Azetidinone, 1-[(3β,5α,20S)-20-(dimethylamino)-4-oxopregnan-3-yl]-3-isopropyl-, (3R)-**	6156-99-6	**NI43478**
441	442	456	73	373	344	72	56	**456**	100	45	38	37	30	25	25	20		29	48	2	2	–	–	–	–	–	–	–	–	–	**Pregn-3-eno[4,3-b][1,4]oxazepin-7′(4′H)-one, 20-(dimethylamino)-3,4,5′,6′ tetrahydro-6′-isopropylidene-, (3α,4α,5α,20S)-**	15027-63-1	**NI43479**
325	353	180	105	296	456	457	428	**456**	100	65	30	25	24	11	4	2		30	20	3	2	–	–	–	–	–	–	–	–	–	**1,4,6a-Triphenyl-6-phenylimino-2,3,6,6a-tetrahydro-1H-furo[3,4-b]pyrrole 2,3-dione**	76678-12-1	**NI43480**
170	141	176	142	171	121	165	263	**456**	100	41	29	25	20	20	19	17	**0.00**	30	32	4	–	–	–	–	–	–	–	–	–	–	**(2R,3R)-1,7,7-Trimethyl-2-(1-naphthyl)bicyclo[2.2.1]hept-3-yl 2-(4-methoxyphenyl)-3-oxopropanoate**		**DD01709**
375	376	55	167	83	293	125	81	**456**	100	30	22	20	20	10	10	9	**7.00**	30	36	2	2	–	–	–	–	–	–	–	–	–	**3-Cyclohexyl-2-(cyclohexylimino)-6-(diphenylmethyl)-2,3-dihydro-5-methyl-4H-1,3-oxazin-4-one**		**MS09947**
375	206	194	166	55	376	125	83	**456**	100	85	82	30	26	22	21	17	**7.00**	30	36	2	2	–	–	–	–	–	–	–	–	–	**3-Cyclohexyl-2-(cyclohexylimino)-6-(diphenylmethylidene)-2,3,5,6-tetrahydro-5-methyl-4H-1,3-oxazin-4-one**		**MS09948**
206	205	234	55	331	83	191	456	**456**	100	81	80	70	65	40	33	25		30	36	2	2	–	–	–	–	–	–	–	–	–	**N,N′-dicyclohexyl-N-(4,4-diphenyl-2-methyl-2,3-butadienoyl)urea**		**MS09949**
248	203	133	249	189	59	207	81	**456**	100	60	32	20	20	20	16	16	**4.00**	30	48	3	–	–	–	–	–	–	–	–	–	–	**Urs-12-en-28-oic acid, 3-hydroxy-, (3β)-**	77-52-1	**NI43481**
189	135	121	108	119	106	109	207	**456**	100	57	55	55	50	45	40	37	**11.00**	30	48	3	–	–	–	–	–	–	–	–	–	–	**Betulinic acid**	472-15-1	**MS09950**
189	456	207	190	175	248	203	187	**456**	100	70	50	40	40	35	35	35		30	48	3	–	–	–	–	–	–	–	–	–	–	**Betulinic acid**	472-15-1	**NI43482**
189	207	456	248	220	219			**456**	100	69	59	50	22	19				30	48	3	–	–	–	–	–	–	–	–	–	–	**Betulinic acid**	472-15-1	**MS09951**
43	396	41	159	381	55	57	44	**456**	100	51	33	31	30	28	21	20	**0.00**	30	48	3	–	–	–	–	–	–	–	–	–	–	**5α-Cholest-8-en-3-one, 6α-hydroxy-14-methyl-, acetate**	5259-17-6	**NI43483**
95	233	81	189	438	207	299	398	**456**	100	77	77	63	62	38	34	28	**6.00**	30	48	3	–	–	–	–	–	–	–	–	–	–	**3,11-Diketo-lupan-20-ol**	79081-62-2	**NI43484**
396	43	269	55	57	397	41	71	**456**	100	69	57	44	41	33	33	27	**6.86**	30	48	3	–	–	–	–	–	–	–	–	–	–	**Ergost-4-en-3-one, 12-(acetyloxy)-, (12α)-**	56052-08-5	**NI43485**
243	456	269	396	381	329	441	327	**456**	100	52	46	45	39	37	25	13		30	48	3	–	–	–	–	–	–	–	–	–	–	**Ergost-7-en-6-one, 3-(acetyloxy)-, (3β,5α)-**		**LI09144**
107	71	69	105	251	456	91	81	**456**	100	85	84	63	61	58	54	53		30	48	3	–	–	–	–	–	–	–	–	–	–	**Ergost-8(14)-en-15-one, 3-(acetyloxy)-, (3β,5α)-**	40446-09-1	**NI43486**
401	255	441	424	213	456	253	229	**456**	100	56	50	50	39	36	29	27		30	48	3	–	–	–	–	–	–	–	–	–	–	**Ethyl 6-methoxy-3,5:23,28-dicyclo-26,27-dinorstigmastan-25-oate**		**MS09952**
305	123	259	137	395	287	109	325	**456**	100	98	90	67	40	36	33	30	**5.00**	30	48	3	–	–	–	–	–	–	–	–	–	–	**Friedelan-26-oic acid, 3-oxo-**		**NI43487**
205	456	163	179	191	233	221	441	**456**	100	64	57	43	35	31	26	25		30	48	3	–	–	–	–	–	–	–	–	–	–	**D:A-Friedooleanane-1,3-dione, 7-hydroxy-, (7α)-**	35162-66-4	**MS09953**
273	259	425	179	261	191	151	205	**456**	100	86	80	70	55	50	40	38	**0.00**	30	48	3	–	–	–	–	–	–	–	–	–	–	**D:A-Friedooleanane-1,3-dione, 24-hydroxy-**	43230-65-5	**MS09954**
155	273	163	250	231	161	456	235	**456**	100	83	58	42	42	38	35	29		30	48	3	–	–	–	–	–	–	–	–	–	–	**D:A-Friedooleanan-29-oic acid, 3-oxo-, (20α)-**	33600-93-0	**MS09955**
81	119	69	55	121	109	107	95	**456**	100	99	99	99	98	98	98	95	**8.00**	30	48	3	–	–	–	–	–	–	–	–	–	–	**3α-Hydroxybauer-7-en-28-oic acid**		**NI43488**
456	43	277	69	457	55	95	57	**456**	100	61	36	36	34	34	33	32		30	48	3	–	–	–	–	–	–	–	–	–	–	**Lanostane-3,7,11-trione**	6593-13-1	**NI43489**
248	203	57	73	69	60	256	249	**456**	100	93	84	81	78	76	40	37	**0.00**	30	48	3	–	–	–	–	–	–	–	–	–	–	**Oleanolic acid**	508-02-1	**NI43490**
351	325	456	364	352	326	131	144	**456**	100	80	37	29	29	24	22	20		30	52	1	–	–	–	–	–	–	1	–	–	–	**7-Dehydrocholesterol trimethylsilyl ether**		**MS09956**
129	343	327	366	456	351	253	441	**456**	100	58	55	43	32	26	26	22		30	52	1	–	–	–	–	–	–	1	–	–	–	**Desmosterol trimethylsilyl ether**		**MS09957**
55	69	111	129	456	73	366	81	**456**	100	93	79	68	59	59	56	50		30	52	1	–	–	–	–	–	–	1	–	–	–	**24-Nor-22,23-methylenecholest-5-en-3-ol trimethylsilyl ether, (3β)-**		**MS09958**
253	456	199	81	457	366	351	254	**456**	100	79	49	31	29	29	24	22		30	52	1	–	–	–	–	–	–	1	–	–	–	**9,10-Secocholesta-5(10),6,8-triene, 3-[(trimethylsilyl)oxy]-, (3β,E)-**	27281-76-1	**NI43491**
69	75	73	107	81	456	95	105	**456**	100	70	58	52	39	36	36	33		30	52	1	–	–	–	–	–	–	1	–	–	–	**Zymosterol trimethylsilyl ether**	55515-19-0	**NI43492**
456	457	441	229	213	107	351	147	**456**	100	41	37	24	24	24	21	16		30	52	1	–	–	–	–	–	–	1	–	–	–	**Zymosterol trimethylsilyl ether**	55515-19-0	**MS09959**
259	197	260	456	457	181	198	261	**456**	100	41	24	22	10	10	8	6		31	28	–	–	–	–	–	–	–	2	–	–	–	**Disilane, methylpentaphenyl-**	1450-22-2	**NI43493**
456	255	396	457	213	229	441	381	**456**	100	67	59	33	29	20	18	12		31	52	2	–	–	–	–	–	–	–	–	–	–	**Cholest-7-ene, 24,24-dimethyl-, 3-acetoxy-, (3β,5α)-**	105097-84-5	**DD01710**
456	396	255	213	441	275	381	315	**456**	100	67	52	25	19	19	17	12		31	52	2	–	–	–	–	–	–	–	–	–	–	**Cholest-5-ene, 24-ethyl-, acetoxy-, (3β)-**		**NI43494**
42	55	40	69	95	57	105	135	**456**	100	42	37	33	32	32	30	29	**28.37**	31	52	2	–	–	–	–	–	–	–	–	–	–	**Cholest-7-en-3-ol, 4,4-dimethyl-, acetate, (3β)-**	17137-72-3	**NI43495**
99	100	456	147	87	57	356	55	**456**	100	98	95	91	90	90	89	89		31	52	2	–	–	–	–	–	–	–	–	–	–	**Cholest-5-en-3-one, 4,4-dimethyl-, cyclic 1,2-ethanediyl acetal**	35490-52-9	**NI43496**
139	413	456	441	423	395	317	299	**456**	100	19	1	1	1	1	1	1		31	52	2	–	–	–	–	–	–	–	–	–	–	**α-Dehydroalnincanol**		**NI43497**
139	413	441	456	423	395	317	299	**456**	100	17	2	1	1	1	1	1		31	52	2	–	–	–	–	–	–	–	–	–	–	**β-Dehydroalnincanol**		**NI43498**
139	413	456	441	423	395	317	299	**456**	100	8	1	1	1	1	1	1		31	52	2	–	–	–	–	–	–	–	–	–	–	**δ-Dehydroalnincanol**		**NI43499**
139	413	456	441	423	395	317	299	**456**	100	15	1	1	1	1	1	1		31	52	2	–	–	–	–	–	–	–	–	–	–	**γ-Dehydroalnincanol**		**NI43500**
169	259	264	205	217	441	218	233	**456**	100	63	53	48	43	38	29	19	**17.50**	31	52	2	–	–	–	–	–	–	–	–	–	–	**D:A-Friedooleanan-29-oic acid, methyl ester, (20α)-**	39903-17-8	**MS09960**
69	95	135	203	189	81	71	107	**456**	100	91	81	67	64	60	57	52	**38.88**	31	52	2	–	–	–	–	–	–	–	–	–	–	**C(14a)-Homo-27-norgammacer-13-en-21-ol, 3-methoxy-, (3α,21β)-**	24433-34-9	**NI43501**
69	95	135	203	189	81	71	107	**456**	100	90	82	68	65	61	59	53	**0.00**	31	52	2	–	–	–	–	–	–	–	–	–	–	**3α-Methoxy-21β-hydroxy-Δ^{13}-serratene**		**LI09145**
375	456	369	413	441				**456**	100	58	32	21	13					31	52	2	–	–	–	–	–	–	–	–	–	–	**3,4-Secolup-4(23)-en-3-oic acid, methyl ester**	40286-43-9	**NI43502**
396	163	397	147	381	81	288	145	**456**	100	45	36	20	12	11	10	10	**1.00**	31	52	2	–	–	–	–	–	–	–	–	–	–	**Sitosterol acetate**		**MS09961**
456	255	43	81	213	95	55	53	**456**	100	90	50	46	43	37	35	35		31	52	2	–	–	–	–	–	–	–	–	–	–	**Stigmast-7-en-1-ol, acetate**		**MS09962**
257	456	344	315	353	115	316	457	**456**	100	93	64	59	55	40	39	34		31	52	2	–	–	–	–	–	–	–	–	–	–	**Stigmast-22-en-3β-ol, acetate**		**LI09146**
456	125	457	200	28	400	398	199	**456**	100	45	39	24	18	15	12	11		34	16	2	–	–	–	–	–	–	–	–	–	–	**Dibenzanthrone**		**IC04456**
241	456	457	228	439	239	242	455	**456**	100	77	31	30	27	25	21	17		36	24	–	–	–	–	–	–	–	–	–	–	–	**Hexa-o-phenylene centrosymmetric**		**MS09963**
456	440	457	228	455	454	441	426	**456**	100	56	40	37	33	24	24	24		36	24	–	–	–	–	–	–	–	–	–	–	–	**Hexa-o-phenylene screw form**		**MS09964**
456	457	455	228	458	303	143	139	**456**	100	40	15	10	9	6	5	5		36	24	–	–	–	–	–	–	–	–	–	–	–	**1-(2-o-Tetraphenyl)-triphenylene**		**MS09965**

MASS TO CHARGE RATIOS								M.W.	INTENSITIES								Parent	C	H	O	N	S	F	Cl	Br	I	Si	P	B	X	COMPOUND NAME	CAS Reg No	No
69	131	236	181	186	155	245	405	457	100	34	23	23	15	11	10	9	6.50	10	–	–	1	–	17	–	–	–	–	–	–	–	Tricyclo[3.3.1.1^{3,7}]decan-1-amine, N,N,2,2,3,4,4,5,6,6,7,8,8,9,9,10,10-heptadecafluoro-	67700-18-9	NI43503
233	289	177	317	56	373	204	401	457	100	72	47	45	44	34	29	21	8.00	13	15	6	1	2	–	–	–	–	–	–	–	2	(μ-Ethyl isothiocyanato)(μ-tert-butylthiolato)hexacarbonyldiiron		MS09966
268	73	370	269	75	147	45	236	457	100	75	30	21	20	17	16	15	0.00	15	40	5	1	–	–	–	–	–	4	1	–	–	Alanine, 3-[bis[(trimethylsilyl)oxy]phosphinyl]-N-(trimethylsilyl)-, trimethylsilyl ester	56196-74-8	NI43504
253	276	177	119	176	147	131	130	457	100	48	40	38	36	28	26	25	16.20	16	13	3	1	–	10	–	–	–	–	–	–	–	(2-Hydroxyphenyl)butylamine, bis(pentafluoropropionyl)-		NI43505
253	41	119	457	225	177	338	178	457	100	26	24	19	15	14	14	14		16	13	3	1	–	10	–	–	–	–	–	–	–	4-(3′-Hydroxyphenyl)butylamine, bis(pentafluoropropionyl)-		NI43506
134	87	86	89	101	145	330	104	457	100	68	64	60	56	42	41	41	0.10	16	15	2	1	2	–	4	–	–	–	–	–	–	8,8-Dichloro-3,3-dimethyl-5-(dichloromethylidene)-2-endo-(methylthio)-8a phenyl-1,3-thiazolo[3,2-c]-1,3-oxazin-7-one		NI43507
88	228	200	338	370	172	278	268	457	100	52	37	33	30	19	17	15	7.40	18	23	9	3	1	–	–	–	–	–	–	–	–	1,6-Dihydro-5-formyl-1-methyl-2-methylthio-4-(2,3,4-tri-O-acetyl-β-D-xylopyranosylamino)-1H-pyrimidin-6-one		MS09967
73	183	147	442	169	129	230	193	457	100	94	85	40	36	30	28	28	18.00	18	35	5	3	–	–	–	–	–	3	–	–	–	3,6-Epoxy-2H,8H-pyrimido[6,1-b][1,3]oxazocin-8-one, 3,4,5,6,9,10-hexahydro-10-[(trimethylsilyl)imino]-4,5-bis[(trimethylsilyl)oxy]-, [3R-(3α,4β,5β,6α)]-	41547-78-8	NI43508
73	355	102	356	45	74	357	147	457	100	43	40	13	11	8	6	5	0.00	20	43	3	1	–	–	–	–	–	4	–	–	–	Noradrenaline tetra-TMS	56145-09-6	MS09968
355	73	356	102	357	147	75	74	457	100	54	33	28	15	5	5	5	0.00	20	43	3	1	–	–	–	–	–	4	–	–	–	Noradrenaline tetra-TMS	56145-09-6	NI43509
154	457	219	69	41	218	458	153	457	100	73	67	66	42	23	22	20		22	27	6	5	–	–	–	–	–	–	–	–	–	3-Cyano-4-methyl-6-hydroxy-5-[4-nitro-2-(5-hydroxypentyloxy)-phenylazo]-pyrid-2-one		IC04457
208	396	91	59	397	56	209	92	457	100	74	35	26	19	14	13	12	2.40	23	39	8	1	–	–	–	–	–	–	–	–	–	(1S,2S)-1,2-Dihydroxy-1,3-benzyl-4,7,10,16,19,22-hexaoxa-13-azacyclotricosane		MS09969
32	105	77	353	367	121	60	425	457	100	80	75	58	25	15	15	11	0.00	25	19	2	3	2	–	–	–	–	–	–	–	–	O-Methyl 3-[α-(3-benzoyl-5-phenyl-4-isothiazolyl)benzylidene]thiocarbazate	110315-37-2	DD01711
457	128	238	111	77	118	459	424	457	100	74	63	59	49	45	41	30		25	20	–	5	1	–	1	–	–	–	–	–	–	4-(4-Chlorophenylamino)-7-(methylthio)-5-phenyl-2-(p-tolyl)imidazo[5,1-f][1,2,4]triazine		MS09970
415	416	264	298	282	201	172	163	457	100	27	12	11	10	10	9	9	0.00	25	31	7	1	–	–	–	–	–	–	–	–	–	Estra-1,3,5(10)-trien-6-one, 3,16,17-tris(acetyloxy)-, 6-(O-methyloxime), (16α,17β)-	74299-47-1	NI43510
385	356	354	57	96	386	355	341	457	100	50	49	48	44	31	22	22	11.03	25	39	3	1	–	–	–	–	–	2	–	–	–	Estra-1,3,5(10),6-tetraen-17-one, 3,6-bis[(trimethylsilyl)oxy]-, O-methyloxime	69833-97-2	NI43511
120	106	134	27	135	16	51	39	457	100	66	65	48	46	46	39	38	0.00	26	33	–	3	–	–	2	–	–	–	–	–	–	Methyl green	82-94-0	AM00156
120	121	27	16	50	77	51	39	457	100	63	48	46	43	41	39	38	0.00	26	33	–	3	–	–	2	–	–	–	–	–	–	Methyl green	82-94-0	WI01564
236	235	391	165	237	203	91	204	457	100	73	27	27	22	22	18	14	10.00	27	23	2	1	2	–	–	–	–	–	–	–	–	3′-Tosylspiro[fluorene-9,5′-4-thia-3-azatricyclo[5,2,1,0^{2,6}]dec(8)ene]		MS09971
43	382	457	308	75	397	91	292	457	100	74	67	62	54	52	50	44		27	43	3	1	–	–	–	–	–	1	–	–	–	3β-Acetoxy-16-methyl-pregna-5,16-dien-20-one, TMS oxime		LI09147
98	99	43	55	67	150	56	41	457	100	19	17	10	10	9	8	8	0.34	29	47	3	1	–	–	–	–	–	–	–	–	–	(22S,25S)-22,26-Epiminocholest-5-ene-3β,16α-diol 16-acetate	36069-45-1	NI43512
222	174	369	43	150	397	382	354	457	100	99	77	77	55	54	22	15	9.00	29	47	3	1	–	–	–	–	–	–	–	–	–	7a-Aza-B-homocholest-5-en-7-one, 3-(acetyloxy)-, (3β)-	17398-63-9	NI43513
222	174	369	43	107	93	56	150	457	100	94	80	74	67	60	56	55	7.00	29	47	3	1	–	–	–	–	–	–	–	–	–	7a-Aza-B-homocholest-5-en-7-one, 3-(acetyloxy)-, (3β)-	17398-63-9	LI09148
101	426	84	43	102	88	41	57	457	100	8	8	8	6	6	6	5	2.00	30	51	2	1	–	–	–	–	–	–	–	–	–	3,3-Dimethoxy-5β-cyano-cholestane		MS09972
88	43	457	55	84	57	41	81	457	100	31	25	14	13	12	12	11		31	55	1	1	–	–	–	–	–	–	–	–	–	Acetamide, N-[(3β,5α)-cholestan-3-yl]-N-ethyl-	55320-47-3	NI43514
382	414	415	457	424	383	458	416	457	100	74	52	50	35	32	18	16		32	27	–	1	1	–	–	–	–	–	–	–	–	Pyridine, 2-[isopropylthio]-3,4,5,6-tetraphenyl-	74630-10-7	NI43515
366	367	105	43	302	212	58	41	457	100	29	16	16	15	9	9	8	2.00	32	59	–	1	–	–	–	–	–	–	–	–	–	N,N-Didodecyl-2-phenylethylamine		MS09973
457	458	459	456	57	209	55	442	457	100	37	7	5	5	4	4	3		33	47	–	1	–	–	–	–	–	–	–	–	–	Cholest-5-eno[3,4-b]indole		MS09974
440	442	438	444	443	441	436	439	458	100	87	87	81	66	66	60	56	12.57	–	14	–	–	–	–	–	–	–	–	–	–	6	Hexagermane	37867-35-9	MS09975
318	345	268	458	242	243	373	429	458	100	56	40	38	25	18	17	16		8	–	5	–	–	5	–	–	–	–	–	–	1	Rhenium, pentacarbonyl(pentafluoropropenyl)-	15492-11-2	NI43516
97	69	358	92	269	154	389	339	458	100	99	52	49	46	20	15	12	1.80	8	1	3	2	–	15	–	–	–	–	–	–	–	N-trifluoromethyl-N-[5,5-bis(trifluoromethyl)-1,4,2-dioxazolin-3-yl]hydroxylamine		NI43517
78	91	65	39	52	51	402	50	458	100	55	31	24	19	19	14	14	4.02	9	7	2	–	–	–	–	–	1	–	–	–	1	π-Cycloheptatrienyl-dicarbonyl tungsten monoiodide		MS09976
131	96	462	460	133	259	464	257	458	100	82	80	75	68	58	50	48	32.03	10	2	–	–	–	–	8	–	–	–	–	–	1	Ferrocene, 1,1′,2,2′,3,3′,4,4′-octachloro-	33306-56-8	NI43518
131	462	96	460	133	257	255	61	458	100	82	82	76	68	58	49	45	0.00	10	2	–	–	–	–	8	–	–	–	–	–	1	Ferrocene, 1,1′,2,2′,3,3′,4,4′-octachloro-		LI09149
69	131	181	119	293	100	343	243	458	100	25	11	9	7	7	5	5	0.00	11	3	–	–	–	17	–	–	–	–	–	–	–	1-Methylperfluorodecalin		IC04458
254	182	208	282	110	56	209	138	458	100	82	43	41	36	36	34	26	8.05	11	14	5	–	–	6	–	–	–	–	2	–	1	Iron, carbonyl[(2,3,4,5-η)-diethyl 2,4-hexadienedioate]bis(phosphorous trifluoride)-	74792-81-7	NI43519
458	126	63	323	228	200	67	28	458	100	45	36	34	33	28	25	20		12	6	4	2	–	–	–	2	–	–	–	–	1	Nickel, bis(4-bromo-O-benzoquinone 2-oximato)-	30993-29-4	NI43520
290	234	318	174	247	178	374	430	458	100	99	68	43	35	30	29	23	0.00	12	12	8	–	–	–	–	–	–	–	2	–	2	μ-Tetramethyldiphosphine-bis(tetracarbonyliron)		MS09977
429	430	369	59	87	341	45	431	458	100	36	30	28	26	20	17	15	0.59	12	34	7	–	–	–	–	–	–	6	–	–	–	1,3,5,7,9,11-Hexaethylbicyclo[5.5.1]hexasiloxane	73113-17-4	NI43521
261	197	211	458	389	415	439	361	458	100	99	53	33	20	13	7	7		14	16	2	–	2	6	–	–	–	–	–	–	1	Bis(1,1,1-trifluoro-4-mercapto-5-methylhex-3-en-2-onato)zinc(II)		NI43522

MASS TO CHARGE RATIOS	M.W.	INTENSITIES	Parent	C	H	O	N	S	F	Cl	Br	I	Si	P	B	X	COMPOUND NAME	CAS Reg No	No
178 177 188 234 290 145 144 318	**458**	100 90 74 61 50 43 41 23	**7.80**	14	18	6	–	2	–	–	–	–	–	–	–	2	Di-μ-butanethiolato-hexacarbonyl-di-iron		MS09978
115 331 157 129 99 169 141 173	**458**	100 74 59 55 35 34 30 28	**0.00**	14	19	9	–	–	–	–	–	1	–	–	–	–	α-D-Glucopyranose, 6-deoxy-6-iodo-, tetraacetate	24871-54-3	NI43523
73 147 221 281 207 355 443 369	**458**	100 56 40 30 10 7 4 3	**0.01**	14	42	5	–	–	–	–	–	–	6	–	–	–	Hexasiloxane, tetradecamethyl-	107-52-8	LI09150
73 221 147 281 355 267 222 443	**458**	100 93 61 38 30 22 22 18	**0.00**	14	42	5	–	–	–	–	–	–	6	–	–	–	Hexasiloxane, tetradecamethyl-	107-52-8	NI43524
73 147 221 355 443 281 267 74	**458**	100 36 30 28 25 24 22 19	**0.00**	14	42	5	–	–	–	–	–	–	6	–	–	–	1,1,1,3,5,7,7,7-Octamethyl-3,5-bis(trimethylsiloxy)tetrasiloxane	2003-92-1	NI43525
43 18 374 127 387 57 41 55	**458**	100 77 54 43 41 39 29 27	**5.72**	15	18	2	–	–	12	–	–	–	–	–	–	–	1H,1H,7H-Dodecafluoroheptyl octanoate		MS09979
43 374 127 386 57 55 69 85	**458**	100 54 43 42 39 28 17 12	**6.00**	15	18	2	–	–	12	–	–	–	–	–	–	–	1H,1H,7H-Dodecafluoroheptyl octanoate		LI09151
91 155 65 106 184 74 42 60	**458**	100 44 34 21 17 17 16 12	**12.00**	18	22	4	2	4	–	–	–	–	–	–	–	–	1,5-Ditosyl-2,6-dithia-1,5-diazocine		LI09152
29 41 57 27 401 287 39 28	**458**	100 77 41 27 18 17 17 16	**0.00**	18	27	–	–	–	5	–	–	–	–	–	–	1	Stannane, tributyl(pentafluorophenyl)-	1045-56-3	NI43526
73 147 443 243 273 75 129 245	**458**	100 84 48 38 37 34 30 24	**13.00**	18	34	6	2	–	–	–	–	–	3	–	–	–	3,6-Epoxy-2H,8H-pyrimido[6,1-b][1,3]oxazocine-8,10(9H)-dione, 3,4,5,6-tetrahydro-9-(trimethylsilyl)-4,5-bis[(trimethylsilyl)oxy]-, [3R-(3α,4β,5β,6α)]-	41547-73-3	NI43527
141 51 79 59 78 260 52 63	**458**	100 92 79 69 68 64 56 47	**0.00**	19	18	8	6	–	–	–	–	–	–	–	–	–	Dimethyl 1-[N'-(2,4-dinitrophenyl)hydrazino]-1,2-dihydro-4a,8a-methanophthalazine-1,4-dicarboxylate		DD01712
443 73 77 444 281 445 355 45	**458**	100 92 57 41 24 21 16 15	**0.00**	19	38	5	–	–	–	–	–	–	4	–	–	–	2,3,4-Trihydroxybenzoic acid, tetrakis(trimethylsilyl)-		MS09980
443 444 73 77 445 355 147 446	**458**	100 39 31 24 21 9 8 6	**2.40**	19	38	5	–	–	–	–	–	–	4	–	–	–	2,4,6-Trihydroxybenzoic acid, tetrakis(trimethylsilyl)-		MS09981
281 458 73 443 459 282 460 444	**458**	100 66 52 32 27 22 13 12		19	38	5	–	–	–	–	–	–	4	–	–	–	3,4,5-Trihydroxybenzoic acid, tetrakis(trimethylsilyl)-	2078-17-3	MS09982
73 281 458 443 45 282 147 74	**458**	100 57 46 18 15 14 9 9		19	38	5	–	–	–	–	–	–	4	–	–	–	3,4,5-Trihydroxybenzoic acid, tetrakis(trimethylsilyl)-	2078-17-3	NI43528
73 281 458 74 459 282 443 179	**458**	100 30 17 8 8 7 7 6		19	38	5	–	–	–	–	–	–	4	–	–	–	3,4,5-Trihydroxybenzoic acid, tetrakis(trimethylsilyl)-	2078-17-3	NI43529
171 173 65 92 39 184 182 50	**458**	100 97 50 26 16 15 15 14	**0.00**	20	16	1	2	–	–	–	2	–	–	–	–	–	β,β-Bis(4-bromoanilino)-acetophenone		MS09983
169 43 109 127 128 331 97 170	**458**	100 99 88 35 16 10 10 10	**0.00**	20	23	10	–	–	–	1	–	–	–	–	–	–	p-Chlorophenyl-2,3,4,6-tetra-O-acetyl-β-D-glucopyranoside	5041-92-9	NI43530
70 57 314 41 199 171 44 172	**458**	100 42 40 25 20 15 15 13	**1.20**	20	34	8	4	–	–	–	–	–	–	–	–	–	L-Threonine, N-[N-[1-[N-[(1,1-dimethylethoxy)carbonyl]glycyl]-L-prolyl]L-alanyl]-, methyl ester	41863-54-1	NI43531
355 73 356 357 147 75 45 281	**458**	100 49 31 14 10 5 5 4	**0.00**	20	42	4	–	–	–	–	–	–	4	–	–	–	(3,4-Dihydroxyphenyl)ethylene glycol tetra-TMS	56114-62-6	MS09985
355 73 356 357 147 45 75 74	**458**	100 82 34 15 14 10 6 6	**0.90**	20	42	4	–	–	–	–	–	–	4	–	–	–	(3,4-Dihydroxyphenyl)ethylene glycol tetra-TMS	56114-62-6	MS09984
355 73 356 357 147 281 75 74	**458**	100 49 33 16 12 5 4 4	**0.95**	20	42	4	–	–	–	–	–	–	4	–	–	–	(3,4-Dihydroxyphenyl)ethylene glycol tetra-TMS	56114-62-6	NI43532
271 43 107 81 105 95 272 93	**458**	100 40 28 26 24 22 21 20	**0.69**	21	31	3	–	–	–	–	–	1	–	–	–	–	Androstan-11-one, 3-(acetyloxy)-17-iodo-, (17α)-	56784-27-1	MS09986
271 398 399 105 81 303 107 93	**458**	100 74 69 61 60 54 49 49	**9.37**	21	31	3	–	–	–	–	–	1	–	–	–	–	Androstan-11-one, 3-(acetyloxy)-17-iodo-, (17β)-	56784-26-0	MS09987
412 413 338 164 183 220 74 305	**458**	100 28 27 20 18 16 14 12	**6.00**	22	10	8	4	–	–	–	–	–	–	–	–	–	N,N'-Bis(2-nitrophenyl)pyromellitimide		MS09988
458 412 220 459 73 366 74 102	**458**	100 45 32 30 27 25 20 18		22	10	8	4	–	–	–	–	–	–	–	–	–	N,N'-Bis(3-nitrophenyl)pyromellitimide		LI09153
458 412 220 459 73 366 164 74	**458**	100 45 32 30 27 25 25 20		22	10	8	4	–	–	–	–	–	–	–	–	–	N,N'-Bis(3-nitrophenyl)pyromellitimide		MS09989
458 338 98 428 99 164 102 74	**458**	100 45 36 35 31 28 27 27		22	10	8	4	–	–	–	–	–	–	–	–	–	N,N'-Bis(4-nitrophenyl)pyromellitimide		MS09990
87 43 458 385 459 278 88 386	**458**	100 50 28 19 7 6 5 4		22	26	7	4	–	–	–	–	–	–	–	–	–	N,N-Bis(2-acetoxyethyl)-4-(2-methoxy-5-nitrophenylazo)aniline		IC04459
299 91 300 131 117 153 141 105	**458**	100 40 33 24 19 19 19 15	**7.10**	22	35	2	–	–	–	–	–	1	–	–	–	–	Methyl (4-iodophenyl)pentadecanoate	88337-00-2	NI43533
278 217 267 227 279 218 255 206	**458**	100 46 43 35 22 20 19 16	**2.04**	22	42	6	–	–	–	–	–	–	2	–	–	–	Cyclopentanebutanoic acid, 2-(5-methoxy-5-oxo-2-butenyl)-3,5-bis[(trimethylsilyl)oxy]-, methyl ester, [1R-(1α,2β,3β,5β)]-	55282-46-7	NI43534
458 304 44 459 414 305 83 41	**458**	100 65 61 27 19 18 18 16		23	26	8	2	–	–	–	–	–	–	–	–	–	Dichotinamide, 11-methoxy-	29484-59-1	NI43535
152 414 44 153 412 124 415 357	**458**	100 95 67 38 30 25 24 23	**9.00**	23	26	8	2	–	–	–	–	–	–	–	–	–	Dichotine, 11-methoxy-25-oxo-	55283-43-7	NI43536
254 217 247 368 255 281 353 99	**458**	100 85 50 45 31 30 25 25	**0.00**	23	46	5	–	–	–	–	–	–	2	–	–	–	Cyclopentanepropanoic acid, 2-(3-oxooctyl)-3,5-bis[(trimethylsilyl)oxy]-, methyl ester		LI09154
254 217 368 281 99 353 278 255	**458**	100 83 45 30 27 25 22 20	**1.00**	23	46	5	–	–	–	–	–	–	2	–	–	–	Cyclopentanepropanoic acid, 2-(3-oxooctyl)-3,5-bis[(trimethylsilyl)oxy]-, methyl ester, (1α,2β,3β,5α)-	55669-99-3	NI43537
259 260 200 181 261 201 154 105	**458**	100 26 16 12 9 8 7 7	**0.60**	24	15	–	–	1	5	–	–	–	1	–	–	–	Silane, [(pentafluorophenyl)thio]triphenyl-	22530-03-6	NI43538
134 185 183 303 459 457 102 157	**458**	100 86 86 73 63 63 60 56	**30.00**	24	15	1	2	1	–	–	1	–	–	–	–	–	7-Benzoyl-6-(4-bromophenyl)-3-phenylpyrazolo[5,1-b]thiazole		MS09991
458 163 323 295 291 459 426 399	**458**	100 64 48 40 36 30 29 23		24	18	6	4	–	–	–	–	–	–	–	–	–	4,4'-Bis(2-methoxycarbonyl-benzoyl)-5,5'-dipyrazolyl		MS09992
430 286 431 458 210 183 287 168	**458**	100 48 26 18 12 10 9 7		24	20	1	–	–	–	–	–	–	–	1	–	1	Rhodium, carbonyl(η-5-2,4-cyclopentadien-1-yl)(triphenylphosphine)-	12203-88-2	NI43539
422 331 304 330 440 423 329 305	**458**	100 82 76 56 39 32 23 23	**19.00**	24	26	2	–	–	–	–	–	–	–	–	–	2	Ferrocene, 1',1'''-bis(α-hydroxyethyl)di-		MS09993
73 75 204 443 217 129 117 55	**458**	100 80 62 53 34 32 31 24	**4.00**	24	50	4	–	–	–	–	–	–	2	–	–	–	Octadecanedioic acid, bis(trimethylsilyl) ester	22396-20-9	LI09155
73 75 204 443 217 129 117 327	**458**	100 80 63 50 35 32 30 23	**5.00**	24	50	4	–	–	–	–	–	–	2	–	–	–	Octadecanedioic acid, bis(trimethylsilyl) ester	22396-20-9	NI43540
73 345 401 59 346 127 56 57	**458**	100 62 35 29 22 21 19 18	**0.00**	24	54	–	–	1	–	–	–	–	3	–	–	–	Hexa-tert-butylthiatrisiletane	93194-14-0	NI43541
73 205 207 219 233 191 221 59	**458**	100 59 51 40 39 34 33 33	**0.00**	24	54	2	–	–	–	–	–	–	3	–	–	–	1,1,2,2,4,4-Hexa-tert-butyl-3,5-dioxa-1,2,4-trisilolane	93194-13-9	NI43542
443 179 82 110 251 445 73 125	**458**	100 74 72 62 49 46 45 40	**22.00**	25	35	2	2	–	–	1	–	–	1	–	–	–	N-(Chlorophenyl)-4-hydroxy-N-(1-isopropyl-4-piperidinyl)benzeneacetamide, trimethylsilyl-		MS09994
43 159 283 115 57 189 55 41	**458**	100 81 24 22 13 10 10 8	**0.00**	25	46	7	–	–	–	–	–	–	–	–	–	–	1,2-Propanediol, 3-[[2-(acetyloxy)hexadecyl]oxy]-, diacetate	56599-36-1	NI43543
374 43 375 416 356 357 78 417	**458**	100 40 24 16 6 6 5 4	**0.00**	26	18	8	–	–	–	–	–	–	–	–	–	–	1,1'-Diacetoxy-6,6'-dimethyl-2,2'-binaphthalene-5,5',8,8'-tetrone		NI43544

MASS TO CHARGE RATIOS								M.W.	INTENSITIES								Parent	C	H	O	N	S	F	Cl	Br	I	Si	P	B	X	COMPOUND NAME	CAS Reg No	No
412	458	384	414	460	459	413	386	**458**	100	79	60	33	29	24	20	20		26	19	4	2	–	–	1	–	–	–	–	–	–	**6-Chloroanilino-4-methyl-1-carbethoxy-2,7-dioxo-2,3-dihydro-7H-dibenz[f,ij]isoquinoline**		**IC04460**
257	259	458	230	228	231	255	261	**458**	100	74	40	33	33	29	27	26		26	28	2	2	–	–	–	–	–	–	–	–	1	**Nickel, bis(2-butyl-8-quinolinolato)-**	30049-17-3	**NI43545**
362	334	43	333	91	143	120	197	**458**	100	95	75	72	67	41	39	38	**2.00**	26	34	7	–	–	–	–	–	–	–	–	–	–	**3β-Acetoxyhellebrigenin**	4064-09-9	**LI09156**
362	334	43	333	91	44	143	120	**458**	100	96	77	72	67	44	41	41	**2.15**	26	34	7	–	–	–	–	–	–	–	–	–	–	**3β-Acetoxyhellebrigenin**	4064-09-9	**NI43546**
380	231	213	151	91	43	81	79	**458**	100	65	53	50	40	39	34	33	**6.19**	26	34	7	–	–	–	–	–	–	–	–	–	–	**Cinobufotalin**	1108-68-5	**NI43547**
380	231	213	151	381	398	249	362	**458**	100	42	34	32	19	18	18	16	**4.00**	26	34	7	–	–	–	–	–	–	–	–	–	–	**Cinobufotalin**	1108-68-5	**NI43548**
380	231	213	151	91	43	81	105	**458**	100	66	55	50	42	40	35	32	**7.00**	26	34	7	–	–	–	–	–	–	–	–	–	–	**Cinobufotalin**	1108-68-5	**LI09157**
159	300	119	107	161	145	121	175	**458**	100	94	87	73	72	72	71	70	**0.35**	26	34	7	–	–	–	–	–	–	–	–	–	–	**Retinoyl-β-glucuronide 6′,3′-lactone**	101470-87-5	**NI43549**
135	93	69	109	121	108	107	81	**458**	100	75	44	43	42	40	38	37	**1.00**	27	38	4	–	1	–	–	–	–	–	–	–	–	**2,6,10,14-Hexadecatetraenoic acid, 3,7,11,15-tetramethyl-9-(phenylsulphonyl)-, methyl ester, (E,E,E)-**	74367-17-2	**NI43550**
261	301	287	262	274	440	151	185	**458**	100	33	22	18	17	16	15	14	**7.67**	28	42	5	–	–	–	–	–	–	–	–	–	–	**Cholestan-26-oic acid, 3,7,12-trioxo-, methyl ester, (5β)-**	56247-73-5	**NI43551**
152	69	139	41	43	109	153	95	**458**	100	37	35	31	21	19	18	16	**0.00**	28	42	5	–	–	–	–	–	–	–	–	–	–	**15-Hydroxy-1-(5-hydroxy-2-methoxy-3-methylphenyl)-3,7,11,15-tetramethyl-2,6-hexadecadiene-5,12-dione**	99816-41-8	**NI43552**
91	101	207	133	265	149	157	313	**458**	100	15	10	10	9	6	4	3	**1.00**	29	46	4	–	–	–	–	–	–	–	–	–	–	**1-O-(2-Benzyloxy-4-hexadecynyl)-2,3-O-isopropylideneglycerol**		**NI43553**
43	304	425	55	95	41	57	69	**458**	100	64	56	50	46	44	42	34	**12.82**	29	46	4	–	–	–	–	–	–	–	–	–	–	**Cholest-7-en-6-one, 3-(acetyloxy)-9-hydroxy-, (3β,5α)-**	863-40-1	**NI43554**
139	315	149	344	115	55	386	329	**458**	100	26	23	22	20	14	13	13	**11.11**	29	46	4	–	–	–	–	–	–	–	–	–	–	**Spirostan-3-ol, acetate, (3β,5α,25R)-**	2530-07-6	**NI43555**
99	458	344	125	386	459	100	389	**458**	100	34	27	27	13	11	11	9		29	46	4	–	–	–	–	–	–	–	–	–	–	**Spirostan-3-one, cyclic 1,2-ethanediyl acetal, (5α,14β,25R)-**	24742-86-7	**MS09996**
458	344	99	459	345	386	389	100	**458**	100	70	56	32	20	17	11	9		29	46	4	–	–	–	–	–	–	–	–	–	–	**Spirostan-3-one, cyclic 1,2-ethanediyl acetal, (5α,14β,25R)-**	24742-86-7	**MS09995**
99	458	125	344	386	459	100	314	**458**	100	34	27	26	12	11	10	8		29	46	4	–	–	–	–	–	–	–	–	–	–	**Spirostan-3-one, cyclic 1,2-ethanediyl acetal, (5α,14β,25R)-**	24742-86-7	**NI43556**
458	459	99	386	344	181	182	388	**458**	100	32	22	20	16	15	11	10		29	46	4	–	–	–	–	–	–	–	–	–	–	**Spirostan-3-one, cyclic 1,2-ethanediyl acetal, (5α,20β,25R)-**	55028-81-4	**MS09997**
99	458	181	139	125	386	55	459	**458**	100	61	40	36	35	23	22	21		29	46	4	–	–	–	–	–	–	–	–	–	–	**Spirostan-3-one, cyclic 1,2-ethanediyl acetal, (5α,20β,25R)-**	55028-81-4	**NI43557**
99	139	125	344	69	315	55	140	**458**	100	96	40	38	20	18	17	13	**12.01**	29	46	4	–	–	–	–	–	–	–	–	–	–	**Spirostan-3-one, cyclic 1,2-ethanediyl acetal, (5α,20β,25R)-**	55028-81-4	**NI43558**
99	139	125	344	69	315	55	140	**458**	100	95	40	36	20	18	17	13	**11.00**	29	46	4	–	–	–	–	–	–	–	–	–	–	**Spirostan-3-one, cyclic 1,2-ethanediyl acetal, (5α,25R)-**		**MS09998**
344	99	139	345	458	315	386	389	**458**	100	66	45	36	31	29	17	14		29	46	4	–	–	–	–	–	–	–	–	–	–	**Spirostan-3-one, cyclic 1,2-ethanediyl acetal, (5α,25R)-**		**MS09999**
458	459	344	396	293	308	221	332	**458**	100	30	24	20	14	13	11	8		29	46	4	–	–	–	–	–	–	–	–	–	–	**Spirostan-12-one, cyclic 1,2-ethanediyl acetal, (5α,25R)-**	55028-80-3	**MS10000**
99	458	221	344	282	126	55	459	**458**	100	72	30	25	24	24	24	23		29	46	4	–	–	–	–	–	–	–	–	–	–	**Spirostan-12-one, cyclic 1,2-ethanediyl acetal, (5α,25R)-**	55028-80-3	**NI43559**
443	444	458	445	459	367	223	207	**458**	100	38	26	12	10	10	5	5		30	38	2	–	–	–	–	–	–	1	–	–	–	**Cannabinol benzyldimethylsilyl ether**		**NI43560**
95	81	69	382	383	120	93	67	**458**	100	93	93	74	71	66	64	62	**11.85**	30	50	1	–	1	–	–	–	–	–	–	–	–	**Cholest-2-ene-2-methanethiol, acetate, (5α)-**	55283-36-8	**NI43561**
165	109	95	121	81	107	69	123	**458**	100	77	76	72	66	64	61	58	**6.66**	30	50	3	–	–	–	–	–	–	–	–	–	–	**Canophyllic acid**		**MS10001**
398	399	400	443	384	227	401	383	**458**	100	81	67	48	41	28	26	23	**0.00**	30	50	3	–	–	–	–	–	–	–	–	–	–	**Cholest-4-en-6-ol, 3-methoxy-, acetate, (3β,6β)-**	2701-05-5	**NI43562**
353	354	352	398	366	195	213	197	**458**	100	30	25	8	7	5	4	4	**0.00**	30	50	3	–	–	–	–	–	–	–	–	–	–	**Cholest-5-en-3-ol, 19-methoxy-, acetate, (3β)-**	1108-65-2	**NI43563**
151	416	193	191	150	57	152	55	**458**	100	28	20	15	13	12	11	10	**1.90**	30	50	3	–	–	–	–	–	–	–	–	–	–	**3,4-Dihydro-2,5,8-trimethyl-2-(4,8,12-trimethyltridecyl)-2H-1-benzopyran 6-acetate**		**NI43564**
109	123	203	149	187	137	220	177	**458**	100	88	85	83	81	81	77	72	**54.00**	30	50	3	–	–	–	–	–	–	–	–	–	–	**1α,3α-Dihydroxyallobetulane**		**MS10002**
458	43	398	231	55	41	81	291	**458**	100	98	74	59	45	43	39	36		30	50	3	–	–	–	–	–	–	–	–	–	–	**Ergostan-12-one, 3-(acetyloxy)-, (3β)-**	55823-03-5	**NI43565**
43	271	55	398	41	81	105	95	**458**	100	57	56	50	41	36	35	32	**4.00**	30	50	3	–	–	–	–	–	–	–	–	–	–	**Ergost-5-ene-3,12-diol, 12-acetate, (3β,12α)-**	56052-09-6	**NI43566**
193	191	205	153	217	203	175	164	**458**	100	72	37	33	30	30	28	28	**28.00**	30	50	3	–	–	–	–	–	–	–	–	–	–	**D:A-Friedooleanan-1-one, 3,24-dihydroxy-**	56816-61-6	**MS10003**
307	152	149	123	290	175	440	165	**458**	100	78	42	42	36	36	33	33	**9.00**	30	50	3	–	–	–	–	–	–	–	–	–	–	**3α-Hydroxyfriedelan-26-oic acid**		**NI43567**
307	148	123	290	175	440	165	189	**458**	100	42	42	36	36	33	33	27	**9.00**	30	50	3	–	–	–	–	–	–	–	–	–	–	**3β-Hydroxyfriedelan-26-oic acid**		**NI43568**
81	95	109	55	69	107	93	67	**458**	100	79	70	69	65	64	51	51	**11.61**	30	50	3	–	–	–	–	–	–	–	–	–	–	**22α-Hydroxy-3,4-secostict-4(23)-en-3-oic acid**		**MS10004**
55	69	95	81	109	121	422	203	**458**	100	88	87	85	77	63	59	48	**0.00**	30	50	3	–	–	–	–	–	–	–	–	–	–	**3-Keto-lupane-11α,20-diol**		**NI43569**
71	28	422	407	98	41	112	85	**458**	100	95	86	86	84	81	72	63	**4.65**	30	50	3	–	–	–	–	–	–	–	–	–	–	**20ξ-Lanosta-7,9(11)-diene-3β,18,20-triol**	25116-58-9	**NI43570**
69	85	109	55	211	81	127	245	**458**	100	75	74	71	67	57	47	44	**8.00**	30	50	3	–	–	–	–	–	–	–	–	–	–	**Malabaricol**	18456-09-2	**LI09158**
250	235	251	208	458	236	252	237	**458**	100	80	30	20	5	5	3	3		30	50	3	–	–	–	–	–	–	–	–	–	–	**Olean-12-ene-3,21,22-triol, (3β,21β,22β)-**	83718-68-7	**NI43571**
219	234	235	216	224	217	458	440	**458**	100	99	50	20	15	5	3	2		30	50	3	–	–	–	–	–	–	–	–	–	–	**Olean-12-ene-3,22,23-triol, (3β,4β,22β)-**	595-15-3	**NI43572**
201	219	220	250	207	202	232	221	**458**	100	57	42	29	25	19	12	12	**6.04**	30	50	3	–	–	–	–	–	–	–	–	–	–	**Olean-12-ene-3,28,29-triol, (3β,20β)-**	3767-05-3	**NI43573**
203	234	216	189	133	187	458	427	**458**	100	20	10	8	8	6	4	3		30	50	3	–	–	–	–	–	–	–	–	–	–	**Olean-12-ene-3β,6β,28-triol**		**LI09159**
234	219	224	235	440	225	217	216	**458**	100	45	40	15	5	5	5	5	**3.00**	30	50	3	–	–	–	–	–	–	–	–	–	–	**Olean-12-ene3,21,23-triol, (3β,4β,21β)-**	96820-47-2	**NI43574**
109	440	371	359	415	397	443		**458**	100	42	32	18	12	6	5		**0.00**	30	50	3	–	–	–	–	–	–	–	–	–	–	**3,4-Secodammara-4(28),24-dien-3-oic acid, 20-hydroxy-**	34336-09-9	**NI43575**
458	255	459	75	107	443	213	229	**458**	100	58	40	35	24	24	23	19		30	54	1	–	–	–	–	–	–	1	–	–	–	**5α-Cholest-7-ene, 3β-(trimethylsiloxy)-**	2665-03-4	**NI43576**
368	106	105	213	255	314	353	201	**458**	100	50	34	33	32	31	22	18	**1.00**	30	54	1	–	–	–	–	–	–	1	–	–	–	**5α-Cholest-2-en-6α-ol trimethylsilyl ether**		**NI43577**
368	106	255	105	213	314	353	201	**458**	100	85	35	35	20	18	17	16	**1.00**	30	54	1	–	–	–	–	–	–	1	–	–	–	**5α-Cholest-2-en-6β-ol trimethylsilyl ether**		**NI43578**
255	458	213	229	107	459	147	353	**458**	100	80	49	46	33	30	29	21		30	54	1	–	–	–	–	–	–	1	–	–	–	**5α-Cholest-7-en-3β-ol trimethylsilyl ether**		**LI09160**
458	255	459	229	213	443	353	107	**458**	100	82	50	48	46	33	30	25		30	54	1	–	–	–	–	–	–	1	–	–	–	**5α-Cholest-7-en-3β-ol trimethylsilyl ether**		**NI43579**
458	255	459	443	229	213	256	107	**458**	100	59	39	18	17	14	13	12		30	54	1	–	–	–	–	–	–	1	–	–	–	**5α-Cholest-7-en-3β-ol trimethylsilyl ether**		**MS10005**

MASS TO CHARGE RATIOS	M.W.	INTENSITIES	Parent	C	H	O	N	S	F	Cl	Br	I	Si	P	B	X	COMPOUND NAME	CAS Reg No	No
458 459 229 107 255 213 443 147	**458**	100 36 33 26 20 20 19 16		30	54	1	–	–	–	–	–	–	1	–	–	–	5α-Cholest-8(14)-en-3β-ol trimethylsilyl ether		MS10006
129 329 368 121 95 353 81 107	**458**	100 60 41 31 28 23 22 21	**10.98**	30	54	1	–	–	–	–	–	–	1	–	–	–	Cholest-4-en-3β-ol trimethylsilyl ether	53084-65-4	NI43580
143 458 142 75 368 459 106 144	**458**	100 68 68 45 36 26 21 16		30	54	1	–	–	–	–	–	–	1	–	–	–	Cholest-4-en-3β-ol trimethylsilyl ether	53084-65-4	MS10007
57 129 73 55 95 81 75 69	**458**	100 99 99 84 73 72 67 65	**8.33**	30	54	1	–	–	–	–	–	–	1	–	–	–	Cholest-5-en-3α-ol trimethylsilyl ether	16134-40-0	NI43582
43 55 81 57 95 41 145 107	**458**	100 92 79 73 72 71 60 60	**0.00**	30	54	1	–	–	–	–	–	–	1	–	–	–	Cholest-5-en-3α-ol trimethylsilyl ether	16134-40-0	NI43581
255 213 229 458 256 443 353 214	**458**	100 40 29 25 21 16 15 10		30	54	1	–	–	–	–	–	–	1	–	–	–	Cholest-7-en-3β-ol trimethylsilyl ether	56687-73-1	NI43583
329 129 368 353 458 369 330 121	**458**	100 89 80 35 33 25 25 25		30	54	1	–	–	–	–	–	–	1	–	–	–	Cholesterol trimethylsilyl ether	1856-05-9	NI43584
129 73 43 329 75 57 368 55	**458**	100 89 88 76 61 60 59 45	**43.82**	30	54	1	–	–	–	–	–	–	1	–	–	–	Cholesterol trimethylsilyl ether	1856-05-9	NI43585
73 129 329 45 75 43 368 74	**458**	100 22 15 15 14 14 11 9	**4.78**	30	54	1	–	–	–	–	–	–	1	–	–	–	Cholesterol trimethylsilyl ether	1856-05-9	NI43586
141 397 415 458 443 425 317 299	**458**	100 4 3 1 1 1 1 1		31	54	2	–	–	–	–	–	–	–	–	–	–	α-Alnincanol		NI43587
141 415 397 458 443 425 317 299	**458**	100 6 6 1 1 1 1 1		31	54	2	–	–	–	–	–	–	–	–	–	–	β-Alnincanol		NI43588
141 397 415 458 443 425 317 299	**458**	100 6 5 1 1 1 1 1		31	54	2	–	–	–	–	–	–	–	–	–	–	δ-Alnincanol		NI43589
141 415 397 458 443 425 317 299	**458**	100 6 6 1 1 1 1 1		31	54	2	–	–	–	–	–	–	–	–	–	–	γ-Alnincanol		NI43590
99 100 43 55 57 69 41 81	**458**	100 14 9 8 7 6 6 5	**3.14**	31	54	2	–	–	–	–	–	–	–	–	–	–	Cholestan-3-one, 4,4-dimethyl-, cyclic 1,2-ethanediyl acetal, (5α)-	54498-64-5	NI43591
199 44 94 186 458 206 229 400	**458**	100 94 81 80 79 58 57 56		34	18	2	–	–	–	–	–	–	–	–	–	–	[3,3′-Bi-7H-benz[de]anthracene]-7,7′-dione	116-96-1	NI43592
458 55 57 69 81 95 459 83	**458**	100 83 70 60 45 42 36 34		34	18	2	–	–	–	–	–	–	–	–	–	–	3,3′-Dibenzanthronyl		IC04461
458 457 200 459 429 400 69 199	**458**	100 47 42 33 25 23 19 18		34	18	2	–	–	–	–	–	–	–	–	–	–	6,6′-Dibenzanthronyl		IC04462
211 119 91 248 212 440	**458**	100 48 30 21 19 16	**0.00**	34	34	1	–	–	–	–	–	–	–	–	–	–	2-Benzhydrylidene-1(hydroxydi-p-tolylmethyl)cyclohexane		LI09161
458 459 229 457 460 228 152 226	**458**	100 40 36 26 8 6 5 4		36	26	–	–	–	–	–	–	–	–	–	–	–	m-Sexiphenyl	4740-51-6	NI43593
458 229 306 289 152 441 365 276	**458**	100 21 4 4 3 2 2 2		36	26	–	–	–	–	–	–	–	–	–	–	–	m-Sexiphenyl		MS10008
334 332 336 278 276 240 280 242	**459**	100 79 52 45 35 21 21 20	**0.00**	4	9	–	3	–	–	4	–	1	–	3	–	–	1-Butyl-1-iodotetrachlorocyclotriphosphazene	77589-33-4	NI43594
57 278 276 334 280 332 336 240	**459**	100 46 38 29 19 18 12 7	**0.10**	4	9	–	3	–	–	4	–	1	–	3	–	–	1-tert-Butyl-1-iodotetrachlorocyclotriphosphazene	77589-35-6	NI43595
463 419 461 465 421 417 232 151	**459**	100 71 68 67 47 47 45 31	**0.00**	8	1	2	1	–	–	–	4	–	–	–	–	–	Tetrabromophthalimide		IC04463
277 184 88 279 275 63 278 53	**459**	100 75 71 58 51 48 42 42	**0.00**	13	7	6	3	–	–	–	2	–	–	–	–	–	Benzaldehyde, 3,5-dibromo-4-hydroxy-, O-(2,4-dinitrophenyl)oxime	13181-17-4	NI43596
430 459 269 431 267 119 284 256	**459**	100 96 60 53 41 41 38 36		15	11	4	1	–	10	–	–	–	–	–	–	–	4-Ethoxy-3-hydroxybenzylamine, bis(pentafluoropropionyl)-		MS10009
296 283 119 136 135 255 108 459	**459**	100 88 66 65 52 49 45 34		15	11	4	1	–	10	–	–	–	–	–	–	–	(3-Methoxyphenyl)ethanolamine, bis(pentafluoropropionyl)-		NI43597
296 149 283 297 119 107 459 176	**459**	100 96 56 37 37 20 18 17		15	11	4	1	–	10	–	–	–	–	–	–	–	3-Methoxytyramine, bis(pentafluoropropionyl)-		MS10010
296 149 119 283 69 297 107 77	**459**	100 62 47 29 25 19 17 16	**9.10**	15	11	4	1	–	10	–	–	–	–	–	–	–	3-Methoxytyramine, bis(pentafluoropropionyl)-		MS10011
296 324 297 188 276 336 137 298	**459**	100 14 10 10 6 4 4 3	**0.00**	15	11	4	1	–	10	–	–	–	–	–	–	–	Octopamine 4′-methyl ether, bis(pentafluoropropionyl)-		NI43598
71 72 299 357 223 314 300 73	**459**	100 90 85 76 45 44 44 42	**7.56**	16	42	6	1	–	–	–	–	–	3	1	–	–	Phosphoric acid, 2,3-bis(trimethylsiloxy)propyl 2-(dimethylamino)ethyl trimethylsilyl ester	32046-30-3	NI43599
57 386 330 41 403 29 347 444	**459**	100 40 40 25 22 20 14 7	**5.10**	21	34	2	1	–	–	–	–	1	–	–	–	–	Benzene, 1,3,5-tris(2,2-dimethylpropyl)-2-iodo-4-nitro-	41080-95-9	MS10012
282 314 283 340 459 315 250 143	**459**	100 34 20 14 10 7 6 6		22	21	10	1	–	–	–	–	–	–	–	–	–	Tetramethyl 5,7,10,11-tetrahydro-6-hydroxy-5-oxazepino[1,2-a]-quinoline 7,8,9,10-tetracarboxylate		LI09162
84 258 177 259 136 260 150 218	**459**	100 80 80 70 70 50 28 25	**8.00**	23	26	4	1	–	–	–	1	–	–	–	–	–	Rel-(2S,4S,10R)-2-(p-Bromobenzoyloxy)-4-(3-hydroxy-4-methoxyphenyl)quinolizidine	75659-55-1	NI43600
69 41 109 236 65 39 18 57	**459**	100 56 36 35 30 30 27 26	**0.00**	25	33	5	1	1	–	–	–	–	–	–	–	–	2,6-Di-tert-butyl-4-methoxyphenyl 2-nitro-4-(phenylthio)butanoate		DD01713
369 139 370 57 338 339 368 56	**459**	100 31 30 30 24 23 15 13	**6.09**	25	41	3	1	–	–	–	–	–	2	–	–	–	Estra-1,3,5(10)-trien-16-one, 3,17-bis[(trimethylsilyl)oxy]-, O-methyloxime, (17α)-	69834-01-1	NI43601
459 428 444 286 338 460 174 158	**459**	100 85 50 48 46 38 37 36		25	41	3	1	–	–	–	–	–	2	–	–	–	Estra-1,3,5(10)-trien-16-one, 3,17-bis[(trimethylsilyl)oxy]-, O-methyloxime, (17β)-	69833-48-3	NI43602
459 460 427 428 461 73 429 306	**459**	100 39 22 15 14 13 12 12		25	41	3	1	–	–	–	–	–	2	–	–	–	Estra-1,3,5(10)-trien-17-one, 2,3-bis[(trimethylsilyl)oxy]-, O-methyloxime	74299-41-5	NI43603
459 460 461 73 429 428 370 306	**459**	100 39 14 14 12 12 8 6		25	41	3	1	–	–	–	–	–	2	–	–	–	Estra-1,3,5(10)-trien-17-one, 3,4-bis[(trimethylsilyl)oxy]-, O-methyloxime	74299-16-4	NI43604
459 369 460 338 370 281 280 461	**459**	100 70 39 31 25 22 19 14		25	41	3	1	–	–	–	–	–	2	–	–	–	Estra-1,3,5(10)-trien-17-one, 3,6-bis[(trimethylsilyl)oxy]-, O-methyloxime, (6β)-	69834-02-2	NI43605
369 370 338 139 368 371 339 283	**459**	100 30 18 18 9 8 8 7	**6.39**	25	41	3	1	–	–	–	–	–	2	–	–	–	Estra-1,3,5(10)-trien-17-one, 3,7-bis[(trimethylsilyl)oxy]-, O-methyloxime, (7α)-	74299-43-7	NI43606
459 428 338 429 460 356 218 444	**459**	100 85 50 40 38 33 32 22		25	41	3	1	–	–	–	–	–	2	–	–	–	Estra-1,3,5(10)-trien-17-one, 3,11-bis[(trimethylsilyl)oxy]-, O-methyloxime, (11β)-	69834-00-0	NI43607
218 444 139 459 428 219 338 342	**459**	100 38 36 35 26 20 18 17		25	41	3	1	–	–	–	–	–	2	–	–	–	Estra-1,3,5(10)-trien-17-one, 3,12-bis[(trimethylsilyl)oxy]-, O-methyloxime, (12β)-	69833-50-7	NI43608
218 444 459 139 428 219 342 338	**459**	100 47 39 38 30 20 19 19		25	41	3	1	–	–	–	–	–	2	–	–	–	Estra-1,3,5(10)-trien-17-one, 3,12-bis[(trimethylsilyl)oxy]-, O-methyloxime, (12β)-	69833-50-7	NI43609

MASS TO CHARGE RATIOS								M.W.	INTENSITIES								Parent	C	H	O	N	S	F	Cl	Br	I	Si	P	B	X	COMPOUND NAME	CAS Reg No	No
340	369	341	284	283	370	338	312	**459**	100	52	30	22	21	16	16	16	**0.00**	25	41	3	1	–	–	–	–	–	2	–	–	–	Estra-1,3,5(10)-trien-17-one, 3,14-bis[(trimethylsilyl)oxy]-, O-methyloxime	74299-46-0	NI43610
459	338	460	428	339	232	73	75	**459**	100	91	36	26	26	23	23	22		25	41	3	1	–	–	–	–	–	2	–	–	–	Estra-1,3,5(10)-trien-17-one, 3,15-bis[(trimethylsilyl)oxy]-, O-methyloxime, (13β)-	74299-39-1	NI43611
459	338	460	73	339	400	429	75	**459**	100	71	39	33	25	18	16	16		25	41	3	1	–	–	–	–	–	2	–	–	–	Estra-1,3,5(10)-trien-17-one, 3,15-bis[(trimethylsilyl)oxy]-, O-methyloxime, (15α)-	69833-47-2	NI43612
459	184	460	285	429	338	428	323	**459**	100	56	39	39	36	31	30	22		25	41	3	1	–	–	–	–	–	2	–	–	–	Estra-1,3,5(10)-trien-17-one, 3,16-bis[(trimethylsilyl)oxy]-, O-methyloxime, (16α)-	69833-46-1	NI43613
258	459	229	430	231	257	243	259	**459**	100	50	36	33	25	22	18	17		26	28	2	2	–	–	–	–	–	–	–	–	1	Cobalt, bis(2-butyl-8-quinolinolato)-	30049-16-2	NI43614
143	121	280	459	120	146	161	73	**459**	100	57	47	40	37	33	27	23		26	41	4	1	–	–	–	–	–	1	–	–	–	Acetic acid, [[[(17β)-17-methyl-17-[(trimethylsilyl)oxy]androsta-1,4-dien-3-ylidene]amino]oxy]-, methyl ester	74299-10-8	NI43615
238	304	67	91	164	459	236	178	**459**	100	90	64	42	38	30	27	25		27	25	2	1	2	–	–	–	–	–	–	–	–	3′-Tosylspiro[fluorene-9,5′-4-thia-3-azatricyclo[5,2,1,0^{2,6}]decane]		MS10013
185	273	114	170	101	459	158	107	**459**	100	63	56	51	43	38	31	27		29	49	3	1	–	–	–	–	–	–	–	–	–	Acetamide, N-[(3β,5α)-3-hydroxyfurostan-26-yl]-	50837-85-9	NI43616
222	152	43	459	431	399	346	416	**459**	100	83	83	66	33	33	25	16		29	49	3	1	–	–	–	–	–	–	–	–	–	7a-Aza-B-homocholestan-7-one, 3-(acetyloxy)-, (3β,5α)-	21843-22-1	NI43617
222	53	43	152	459	52	41	93	**459**	100	99	90	87	67	66	61	60		29	49	3	1	–	–	–	–	–	–	–	–	–	7a-Aza-B-homocholestan-7-one, 3-(acetyloxy)-, (3β,5α)-	21843-22-1	LI09163
143	459	460	144	57	55	182	130	**459**	100	50	18	16	6	6	4	4		33	49	–	1	–	–	–	–	–	–	–	–	–	5α-Cholestano[3,2-b]indole		MS10014
459	460	182	170	130	183	144	143	**459**	100	37	17	14	11	10	10	10		33	49	–	1	–	–	–	–	–	–	–	–	–	5α-Cholestano[3,4-b]indole	34535-55-2	NI43618
459	460	182	170	143	130	142	444	**459**	100	38	18	15	13	13	11	8		33	49	–	1	–	–	–	–	–	–	–	–	–	5α-Cholestano[3,4-b]indole	34535-55-2	LI09164
459	460	182	143	130	183	170	157	**459**	100	36	23	22	21	17	16	14		33	49	–	1	–	–	–	–	–	–	–	–	–	5β-Cholestano[3,4-b]indole		LI09165
459	460	182	143	130	157	183	170	**459**	100	36	22	22	20	16	15	15		33	49	–	1	–	–	–	–	–	–	–	–	–	5β-Cholestano[3,4-b]indole		MS10015
31	45	66	427	429	101	36	35	**460**	100	59	49	45	43	38	38	35	**3.00**	–	–	–	4	–	–	8	–	–	–	4	–	–	Octachlorotetraphosphonitrile		LI09166
420	464	72	418	232	162	422	466	**460**	100	98	78	69	69	64	62	56	**20.00**	8	–	3	–	–	–	–	4	–	–	–	–	–	1,3-Isobenzofurandione, 4,5,6,7-tetrabromo-	632-79-1	IC04464
420	464	418	462	422	232	72	466	**460**	100	88	68	64	64	62	62	60	**17.00**	8	–	3	–	–	–	–	4	–	–	–	–	–	1,3-Isobenzofurandione, 4,5,6,7-tetrabromo-	632-79-1	NI43619
72	232	420	151	153	418	230	234	**460**	100	81	65	55	52	42	39	37	**4.83**	8	–	3	–	–	–	–	4	–	–	–	–	–	1,3-Isobenzofurandione, 4,5,6,7-tetrabromo-	632-79-1	NI43620
376	374	378	432	434	430	350	348	**460**	100	86	83	77	69	66	63	60	**20.99**	8	5	3	–	–	–	–	–	1	–	–	–	1	Cyclopentadienyltungstentricarbonyl iodide	31870-69-6	NI43621
335	337	333	331	43	334	339	332	**460**	100	87	79	46	39	38	33	31	**0.41**	8	12	2	–	–	–	–	–	–	–	–	–	4	Ethaneselenoic acid, Se,Se′-(2,4-dimethyl-1,3-diselenetane-2,4-diyl) ester	57289-34-6	NI43622
429	427	431	287	107	289	285	215	**460**	100	85	78	70	45	45	39	39	**0.22**	9	2	–	–	–	–	10	–	–	–	–	–	–	1-Chloro-2,4,6-tris(trichloromethyl)benzene		NI43623
464	462	466	392	394	468	196	197	**460**	100	84	80	47	44	41	40	39	**34.17**	12	1	–	–	–	–	9	–	–	–	–	–	–	1,1′-Biphenyl, 2,2′,3,3′,4,4′,5,5′,6-nonachloro-	40186-72-9	NI43624
465	463	467	464	466	462	461	469	**460**	100	82	69	57	46	43	36	31	**14.20**	12	1	–	–	–	–	9	–	–	–	–	–	–	1,1′-Biphenyl, 2,2′,3,3′,4,4′,5,5′,6-nonachloro-	40186-72-9	NI43625
464	36	462	196	394	392	197	161	**460**	100	85	80	79	79	77	75	74	**26.29**	12	1	–	–	–	–	9	–	–	–	–	–	–	1,1′-Biphenyl, 2,2′,3,3′,4,4′,5,5′,6-nonachloro-	40186-72-9	NI43626
464	462	466	392	394	468	197	460	**460**	100	78	74	51	48	36	28	28		12	1	–	–	–	–	9	–	–	–	–	–	–	1,1′-Biphenyl, 2,2′,3,3′,4,4′,5,6,6′-nonachloro-	52663-79-3	NI43627
464	196	197	462	392	466	394	161	**460**	100	82	82	80	77	74	73	68	**28.31**	12	1	–	–	–	–	9	–	–	–	–	–	–	1,1′-Biphenyl, 2,2′,3,3′,4,5,5′,6,6′-nonachloro-	52663-77-1	NI43628
198	68	58	138	118	230	86	40	**460**	100	67	47	41	39	33	24	21	**20.00**	12	14	6	2	2	6	–	–	–	–	–	–	–	N,N′-Bis(trifluoroacetyl)-L-cystine dimethyl ester	1492-83-7	NI43629
55	101	415	29	18	129	28	51	**460**	100	85	78	70	66	41	40	29	**2.37**	13	12	4	–	–	12	–	–	–	–	–	–	–	1H,1H,7H-Dodecafluoroheptyl ethyl succinate		MS10016
55	415	101	129	51	41	57	56	**460**	100	89	85	41	30	27	23	22	**3.00**	13	12	4	–	–	12	–	–	–	–	–	–	–	1H,1H,7H-Dodecafluoroheptyl ethyl succinate		LI09167
460	320	335	376	290	222	212	350	**460**	100	93	72	46	42	28	27	19		14	32	6	–	–	–	–	–	–	–	2	–	1	Bis(2-methylallyl)bis(trimethylphosphite)ruthenium		LI09168
296	283	119	297	69	77	149	121	**460**	100	67	38	32	25	24	17	16	**11.70**	15	10	5	–	–	10	–	–	–	–	–	–	–	(3-Hydroxy-4-methoxyphenyl)ethanol bis(perfluoropropionate)		MS10017
149	296	119	69	297	77	460	91	**460**	100	85	58	41	33	27	24	24		15	10	5	–	–	10	–	–	–	–	–	–	–	(4-Hydroxy-3-methoxyphenyl)ethanol bis(perfluoropropionate)		MS10018
73	357	299	147	358	445	315	300	**460**	100	99	89	30	29	26	24	22	**0.00**	15	41	6	–	–	–	–	–	–	4	1	–	–	Phosphoric acid, bis(trimethylsilyl) 2,3-bis[(trimethylsilyl)oxy]propyl ester	31038-11-6	NI43630
73	299	357	147	75	45	103	101	**460**	100	32	31	18	16	13	12	12	**0.00**	15	41	6	–	–	–	–	–	–	4	1	–	–	Phosphoric acid, bis(trimethylsilyl) 2,3-bis[(trimethylsilyl)oxy]propyl ester	31038-11-6	NI43631
73	357	299	358	300	445	315	103	**460**	100	83	75	25	20	16	16	16	**0.24**	15	41	6	–	–	–	–	–	–	4	1	–	–	Phosphoric acid, bis(trimethylsilyl) 2,3-bis[(trimethylsilyl)oxy]propyl ester	31038-11-6	NI43632
73	243	299	129	147	211	244	103	**460**	100	81	39	25	23	20	15	15	**0.00**	15	41	6	–	–	–	–	–	–	4	1	–	–	Phosphoric acid, 2-[(trimethylsilyl)oxy]-1-[[(trimethylsilyl)oxy]methyl]ethyl ester	31038-12-7	NI43633
202	187	201	185	186	203	175	188	**460**	100	70	40	40	25	23	17	13	**0.00**	16	14	–	–	–	–	–	–	2	–	–	–	–	Naphthalene, 1,8-bis(1-iodo-1-propenyl)-, (E,E)-	56701-40-7	NI43634
103	73	75	173	299	129	47	343	**460**	100	62	61	34	31	30	30	25	**0.00**	16	41	7	–	–	–	–	–	–	3	1	–	–	Phosphoric acid, 3,3-diethoxy-2-[(trimethylsilyl)oxy]propyl bis(trimethylsilyl) ester	55282-72-9	NI43635
205	207	161	63	44	460	381	206	**460**	100	33	22	20	14	11	9	8		17	8	4	–	–	6	2	–	–	–	–	–	–	1,3-Bis(m-chloroformyloxyphenyl)hexafluoropropane		LI09169
110	66	95	401	365	173	46	142	**460**	100	50	48	45	41	34	34	33	**8.44**	17	18	6	2	–	6	–	–	–	–	–	–	–	L-Lysine, N^6-[2-(2-furanyl)-2-oxoethyl]-N^2,N^6-bis(trifluoroacetyl)-, methyl ester	55282-73-0	NI43636
91	155	184	198	65	92	156	139	**460**	100	73	57	22	21	12	7	7	**4.00**	18	24	4	2	4	–	–	–	–	–	–	–	–	2-Tosylamidoethyl disulphide		LI09170
113	461	259	185	141	257	445	157	**460**	100	60	55	33	31	29	25	25		18	36	6	2	–	–	–	–	–	3	–	–	–	Uridine tri-TMS		NI43637

MASS TO CHARGE RATIOS								M.W.	INTENSITIES								Parent	C	H	O	N	S	F	Cl	Br	I	Si	P	B	X	COMPOUND NAME	CAS Reg No	No
462	460	447	431	464	429	449	445	**460**	100	88	64	63	58	39	38	30		20	16	4	–	–	–	4	–	–	–	–	–	–	Dibenzo(a,f)cyclopropa(cd)pentalene, 1,2,3,4-tetrachloro-4b,8b,8c,8d-tetrahydro-4b,8b,8c,8d-tetramethoxy-	42187-42-8	NI43638
447	445	449	411	413	80	415	91	**460**	100	79	62	23	22	22	15	15	**10.50**	20	16	4	–	–	–	4	–	–	–	–	–	–	1,2,3,4-Tetrachloro-9,10-dihydro-9,10,11,12-tetramethoxy-9,10-ethenoanthracene	42081-38-9	NI43639
113	242	169	200	340	331	280	400	**460**	100	26	21	20	11	5	3	2	**0.00**	20	28	12	–	–	–	–	–	–	–	–	–	–	2′,3′,4′,6′-Tetra-O-acetylparasorboside		LI09171
44	43	41	365	55	363	56	70	**460**	100	72	70	68	61	59	47	47	**0.20**	20	36	4	–	–	–	–	–	–	–	–	–	1	1,3,2-Dioxastannepin-4,7-dione, 2,2-dioctyl-	16091-18-2	NI43640
46	100	44	58	98	460	201	170	**460**	100	81	81	69	48	33	31	19		22	24	9	2	–	–	–	–	–	–	–	–	–	2-Naphthacenecarboxamide, 4-(dimethylamino)-1,4,4a,5,5a,6,11,12a-octahydro-3,5,6,10,12,12a-hexahydroxy-6-methyl-1,11-dioxo-, [4S-(4α,4aα,5α,5aα,6β,12aα)]-	79-57-2	NI43641
230	43	373	42	231	144	88	41	**460**	100	16	13	12	11	8	6	5	**4.30**	22	40	8	2	–	–	–	–	–	–	–	–	–	(Ethylenedinitrilo)tetraacetic acid tetrapropyl ester	22414-25-1	MS10019
41	57	91	89	131	133	55	43	**460**	100	46	45	41	40	39	32	32	**0.00**	22	48	–	–	–	–	–	–	–	–	–	–	2	1,5-Digermacyclooctane, 1,1,5,5-tetrabutyl-	56437-93-5	NI43642
43	173	162	215	163	180	460	134	**460**	100	41	36	33	32	28	20	17		23	28	8	2	–	–	–	–	–	–	–	–	–	Penta-acetyl paucine		LI09172
43	327	95	81	94	91	189	79	**460**	100	43	31	26	19	17	13	13	**0.10**	24	28	9	–	–	–	–	–	–	–	–	–	–	7β,19-Diacetoxy-4α,18,15,16-diepoxy-6-keto-neo-cleroda-13(16),14-dien-20,12S-olide		NI43643
230	231	29	28	174	159	202	127	**460**	100	23	15	13	7	7	6	5	**0.00**	24	30	–	2	–	6	–	–	–	–	–	–	–	meso-N,N,N′,N′-Tetraethyl-1,2-bis[4-(trifluoromethyl)phenyl]ethylenediamine		DD01714
230	231	29	28	202	159	174	127	**460**	100	14	10	9	7	6	4	4	**0.00**	24	30	–	2	–	6	–	–	–	–	–	–	–	(±)-N,N,N′,N′-Tetraethyl-1,2-bis[4-(trifluoromethyl)phenyl]ethylenediamine		DD01715
43	117	403	58	57	55	44	460	**460**	100	94	62	55	46	42	38	36		24	32	7	2	–	–	–	–	–	–	–	–	–	3,4-Secocondyfolan-3-one, 12-[2-(acetyloxy)-1-methoxyethoxy]-16,19-epoxy-2-hydroxy-4-methyl-, [12(S),14β,16β,19R]-	55283-38-0	NI43644
148	83	55	41	149	175	231	230	**460**	100	66	51	39	38	27	25	19	**1.41**	24	46	4	–	–	–	–	–	–	–	2	–	–	Phosphinic acid, dicyclohexyl-, dimer	74793-78-5	NI43645
259	215	260	332	216	188	214	155	**460**	100	78	24	22	16	14	13	11	**0.10**	24	52	4	–	–	–	–	–	–	2	–	–	–	Methyl 9,10-bis(trimethylsiloxy)octadecanoate		LI09173
165	142	223	166	77	51	146	50	**460**	100	72	56	56	50	30	10	10	**0.00**	25	22	–	–	–	4	–	–	–	–	2	–	–	Bis(difluorodiphenylphosphoranyl)methane		LI09174
137	460	312	109	95	83	55	294	**460**	100	94	80	44	39	39	39	36		25	32	8	–	–	–	–	–	–	–	–	–	–	Di-O-acetyl-vincetogenin		LI09175
221	139	222	138	140	166	44	165	**460**	100	40	35	29	24	19	19	15	**0.00**	25	32	8	–	–	–	–	–	–	–	–	–	–	Benzoic acid, 2,4-dihydroxy-3-[(4-hydroxy-2-methoxy-6-pentylbenzoyl)oxy]-6-pentyl-	2948-07-4	NI43646
300	360	231	194	386	285	215	344	**460**	100	52	46	40	38	32	32	30	**0.00**	25	32	8	–	–	–	–	–	–	–	–	–	–	6-Oxo-7β-hydroxycarnosic acid diacetate		MS10020
203	246	460	147	387	231	344	316	**460**	100	44	29	29	22	8	6	6		25	40	4	4	–	–	–	–	–	–	–	–	–	L-Alanine, N-[N-[N-[[4-(dimethylamino)phenyl]methylene]-L-valyl]-L-isoleucyl]-, ethyl ester	37580-35-1	NI43647
285	225	243	218	175	354	372	460	**460**	100	84	25	17	7	4	3	1		25	40	4	4	–	–	–	–	–	–	–	–	–	L-Leucine, N-[N-[N-(2-pyridinylmethylene)-L-leucyl]-L-leucyl]-, methyl ester	42547-80-8	NI43648
243	370	355	316	244	75	93	67	**460**	100	55	46	30	30	30	26	25	**8.00**	25	41	5	–	–	–	–	–	–	1	–	1	–	3α,17α,21-Trihydroxy-5β-pregnane-11,20-dione, 3-trimethylsilyl-, methyl boronate		MS10021
324	108	110	376	404	136	149	460	**460**	100	88	40	38	36	36	32	28		26	17	3	–	–	–	–	–	–	–	1	–	1	2,4,6-Triphenylphosphorine-chromium(0) tricarbonyl		LI09176
418	403	419	404	460	43	44	91	**460**	100	85	29	24	23	20	18	13		26	20	8	–	–	–	–	–	–	–	–	–	–	4-Acetoxy-6′,7-dimethyl-5′,8′-dimethoxy-1,2′-binaphthalene-1′,4′,5,8-tetrone	104505-99-9	NI43649
374	43	375	356	416	357	376	358	**460**	100	70	51	17	16	15	14	14	**0.00**	26	20	8	–	–	–	–	–	–	–	–	–	–	1,1′-Diacetoxy-6,6′-dimethyl-2,2′-binaphthalene-5,5′,8,8′-tetrone		NI43650
55	221	193	81	205	204	222	96	**460**	100	93	59	44	42	37	34	33	**4.55**	26	36	7	–	–	–	–	–	–	–	–	–	–	1H-Indene-4-propanoic acid, octahydro-7a-methyl-1,5-dioxo-, 4-(2-carboxyethyl)octahydro-7a-methyl-5-oxo-1H-inden-1-yl ester, [1α(3aR*,4R*,7aR*),3aβ,4β,7aα]-	55821-17-5	NI43651
398	429	151	167	91	399	92	430	**460**	100	69	49	34	33	30	30	19	**0.00**	26	40	5	2	–	–	–	–	–	–	–	–	–	Pregn-4-ene-3,20-dione, 17-(acetyloxy)-21-hydroxy-6-methyl-, 3,20-bis(O-methyloxime), (6α)-	74299-07-3	NI43652
151	369	400	429	167	125	338	329	**460**	100	97	81	76	50	49	39	36	**7.76**	26	40	5	2	–	–	–	–	–	–	–	–	–	Pregn-4-ene-3,20-dione, 21-(acetyloxy)-17-hydroxy-6-methyl-, 3,20-bis(O-methyloxime), (6α)-	74299-01-7	NI43653
143	281	209	144	282	208	210	460	**460**	100	44	38	18	10	9	7	6		26	44	3	–	–	–	–	–	–	2	–	–	–	17α-Methyl-6β,17β-dihydroxy-1,4-androstadien-3-one, bis(trimethylsilyl)	67896-65-5	NI43654
445	413	446	460	429	414	428	447	**460**	100	35	31	20	19	16	15	9		26	44	3	2	–	–	–	–	–	1	–	–	–	Pregn-4-ene-3,20-dione, 6-[(trimethylsilyl)oxy]-, 3,20-bis(O-methyloxime), (6α)-		NI43655
445	413	446	428	429	460	414	415	**460**	100	48	35	21	20	18	18	17		26	44	3	2	–	–	–	–	–	1	–	–	–	Pregn-4-ene-3,20-dione, 6-[(trimethylsilyl)oxy]-, 3,20-bis(O-methyloxime), (6β)-		NI43656
460	459	461	92	230	133	326	134	**460**	100	65	32	17	15	11	10	10		27	24	–	8	–	–	–	–	–	–	–	–	–	1,3,5-Triazine-2,4,6-triamine, N′,N″-bis(3-aminophenyl)-N,N-diphenyl-	33933-66-3	NI43657
245	109	107	121	135	105	133	119	**460**	100	79	47	41	37	36	31	28	**9.60**	27	40	6	–	–	–	–	–	–	–	–	–	–	Cholan-24-oic acid, 3-(acetyloxy)-7,12-dioxo-, methyl ester, (3α,5β)-	7753-73-3	NI43658
121	245	154	147	105	109	401	243	**460**	100	50	43	42	35	34	33	32	**4.30**	27	40	6	–	–	–	–	–	–	–	–	–	–	Cholan-24-oic acid, 7-(acetyloxy)-3,12-dioxo-, methyl ester, (5β,7α)-	60354-42-9	NI43659
142	143	460	127	267	357	461	215	**460**	100	92	82	52	38	35	30	30		27	44	4	–	–	–	–	–	–	1	–	–	–	Pregn-3-en-20-one, 17-(acetyloxy)-6-methyl-3-[(trimethylsilyl)oxy]-, (6α)-	69833-72-3	NI43660
460	431	140	155	154	432	461	153	**460**	100	93	90	64	47	45	37	34		27	48	2	–	–	–	–	–	–	2	–	–	–	Levo-3α,5β-tetrahydronorgestrel di-TMS		NI43661
244	73	143	147	245	255	75	91	**460**	100	55	29	26	22	19	15	10	**4.00**	27	48	2	–	–	–	–	–	–	2	–	–	–	5α-Pregna-2,20-diene-17,20-diol di-TMS	57397-24-7	NI43662
43	57	18	58	59	44	55	41	**460**	100	93	88	77	76	76	67	67	**0.00**	27	56	5	–	–	–	–	–	–	–	–	–	–	Dimethoxyglycerol docosyl ether		NI43663

MASS TO CHARGE RATIOS								M.W.	INTENSITIES								Parent	C	H	O	N	S	F	Cl	Br	I	Si	P	B	X	COMPOUND NAME	CAS Reg No	No
460	461	77	75	416	178	63	166	**460**	100	34	26	25	15	15	12	9		28	16	5	2	–	–	–	–	–	–	–	–	–	**1H-Isoindole-1,3(2H)-dione, 5,5′-oxybis[2-phenyl-**	33734-37-1	**NI43664**
168	386	460	105	387	77	461	169	**460**	100	60	43	34	25	15	14	11		28	32	4	2	–	–	–	–	–	–	–	–	–	**Cylindrocarine, 1-benzoyl-**	22222-82-8	**NI43665**
167	352	137	215	136	169	81	161	**460**	100	58	53	45	43	40	25	21	**0.76**	28	48	3	–	–	–	–	–	–	1	–	–	–	**Silane, (4-methoxyphenyl)methylbis[(5-methyl-2-isopropylcyclohexyl)oxy]-, [1α(1R*,2S*,5R*),2β,5α]-**	74841-56-8	**NI43666**
91	207	101	265	133	315	149	157	**460**	100	8	8	7	7	4	4	2	**0.40**	29	48	4	–	–	–	–	–	–	–	–	–	–	**1-O-(2-Benzyloxy-4-hexadecenyl)-2,3-O-isopropylideneglycerol**		**NI43667**
400	401	415	416	356	354	402	417	**460**	100	60	50	30	28	28	18	10	**6.01**	29	48	4	–	–	–	–	–	–	–	–	–	–	**5α-Cholestan-19-oic acid, 2α-hydroxy-, acetate**	28809-58-7	**NI43668**
400	401	415	416	356	354	460		**460**	100	60	50	30	28	28	6			29	48	4	–	–	–	–	–	–	–	–	–	–	**5α-Cholestan-19-oic acid, 2α-hydroxy-, acetate**	28809-58-7	**LI09177**
107	327	95	135	163	370	205	309	**460**	100	88	88	84	34	31	30	26	**0.00**	29	48	4	–	–	–	–	–	–	–	–	–	–	**Nepetidone**	104104-55-4	**NI43669**
314	139	428	230	101	315	429	285	**460**	100	95	70	60	60	55	35	35	**19.00**	29	48	4	–	–	–	–	–	–	–	–	–	–	**Smilagenone dimethyl acetal**		**MS10022**
101	181	428	460	139	429	109	55	**460**	100	73	61	47	46	36	33	30		29	48	4	–	–	–	–	–	–	–	–	–	–	**Spirostan, 3,3-dimethoxy-, (5α,20β,25R)-**	55028-82-5	**NI43670**
101	181	428	460	139	429	109	95	**460**	100	73	60	46	46	35	33	30		29	48	4	–	–	–	–	–	–	–	–	–	–	**Spirostan, 3,3-dimethoxy-, (5α,20β,25R)-**	55028-82-5	**MS10023**
460	428	101	429	388	461	430	181	**460**	100	96	87	49	35	34	27	27		29	48	4	–	–	–	–	–	–	–	–	–	–	**Spirostan, 3,3-dimethoxy-, (5α,20β,25R)-**	55028-82-5	**MS10024**
139	101	314	69	55	428	230	85	**460**	100	57	29	27	21	19	16	15	**7.00**	29	48	4	–	–	–	–	–	–	–	–	–	–	**Spirostan, 3,3-dimethoxy-, (5α,25R)-**	54965-94-5	**NI43671**
101	314	139	428	315	460	429	346	**460**	100	62	32	25	23	22	20	13		29	48	4	–	–	–	–	–	–	–	–	–	–	**Spirostan, 3,3-dimethoxy-, (5α,25R)-**		**MS10025**
57	257	41	347	329	348	330	275	**460**	100	36	32	23	23	19	19	19	**1.00**	30	52	3	–	–	–	–	–	–	–	–	–	–	**17β-tert-Butoxy-4-propyl-3,4-seco-5-androsten-3-oic acid tert-butyl ester**		**MS10026**
189	207	223	442	205	275	187	229	**460**	100	90	73	65	65	55	55	53	**15.00**	30	52	3	–	–	–	–	–	–	–	–	–	–	**29,30-Dinorgammacerane-2,3,22-triol, 21,21-dimethyl-, (2α,3β,8α,9β,13α,14β,17α,18β,22α)-**	43206-43-5	**NI43672**
69	95	55	81	107	121	424	135	**460**	100	90	90	85	65	62	61	61	**0.00**	30	52	3	–	–	–	–	–	–	–	–	–	–	**Lupane-3,11,20-triol, (3β,11α)-**	79081-66-6	**NI43673**
345	346	373	363	355	347	403	460	**460**	100	43	22	12	11	11	9	7		30	52	3	–	–	–	–	–	–	–	–	–	–	**Methyl 4-methyl-4-formyl-3,4-secocholestan-3-oate**		**LI09178**
216	460	215	370	145	217	230	144	**460**	100	95	95	85	70	63	60	60		30	56	1	–	–	–	–	–	–	1	–	–	–	**5α-Cholestan-2β-ol TMS ether**		**MS10027**
215	370	355	75	216	460	217	107	**460**	100	81	70	60	45	42	37	34		30	56	1	–	–	–	–	–	–	1	–	–	–	**5α-Cholestan-3α-ol TMS ether**	18880-50-7	**NI43674**
215	370	216	355	217	230	460	148	**460**	100	65	49	46	41	36	25	20		30	56	1	–	–	–	–	–	–	1	–	–	–	**5α-Cholestan-3α-ol TMS ether**	18880-50-7	**MS10028**
215	216	460	445	217	106	306	75	**460**	100	77	76	68	54	41	39	37		30	56	1	–	–	–	–	–	–	1	–	–	–	**5α-Cholestan-3β-ol TMS ether**	18880-51-8	**MS10029**
75	215	107	445	216	81	73	55	**460**	100	99	53	47	47	40	38	37	**37.00**	30	56	1	–	–	–	–	–	–	1	–	–	–	**5α-Cholestan-3β-ol TMS ether**	18880-51-8	**NI43675**
370	215	445	216	230	355	257	197	**460**	100	83	60	50	34	33	23	20	**9.00**	30	56	1	–	–	–	–	–	–	1	–	–	–	**5α-Cholestan-6α-ol TMS ether**		**NI43676**
370	215	147	216	230	148	355	445	**460**	100	58	46	45	36	36	22	18	**8.00**	30	56	1	–	–	–	–	–	–	1	–	–	–	**5α-Cholestan-6β-ol TMS ether**		**NI43677**
75	370	215	73	95	216	81	55	**460**	100	99	82	58	54	53	53	48	**0.75**	30	56	1	–	–	–	–	–	–	1	–	–	–	**5β-Cholestan-3α-ol TMS ether**	55331-93-6	**NI43678**
370	215	216	371	355	257	230	217	**460**	100	47	31	29	21	18	15	15	**0.90**	30	56	1	–	–	–	–	–	–	1	–	–	–	**5β-Cholestan-3α-ol TMS ether**	55331-93-6	**MS10030**
370	371	355	215	108	216	257	217	**460**	100	30	25	24	21	17	14	11	**1.70**	30	56	1	–	–	–	–	–	–	1	–	–	–	**5β-Cholestan-3β-ol TMS ether**	55331-94-7	**MS10031**
370	215	108	355	216	371	257	135	**460**	100	70	50	47	37	34	31	29	**0.99**	30	56	1	–	–	–	–	–	–	1	–	–	–	**5β-Cholestan-3β-ol TMS ether**	55331-94-7	**NI43679**
370	75	73	215	95	355	81	57	**460**	100	75	50	49	42	38	37	37	**1.44**	30	56	1	–	–	–	–	–	–	1	–	–	–	**5β-Cholestan-3β-ol TMS ether**	55331-94-7	**NI43680**
414	129	43	415	99	45	55	81	**460**	100	74	51	49	40	39	38	33	**32.00**	31	56	2	–	–	–	–	–	–	–	–	–	–	**5α-Cholestan-3-one, diethyl acetal**	3621-03-2	**NI43681**
81	67	109	69	95	141	180	135	**460**	100	69	57	51	36	35	28	24	**6.00**	31	56	2	–	–	–	–	–	–	–	–	–	–	**Methyl 5,9,23-tricontatrienoate**		**MS10032**
290	460	127	180	154	356	181	363	**460**	100	95	51	45	30	22	21	18		32	32	1	2	–	–	–	–	–	–	–	–	–	**Cyclohexane-1-spiro-3′-[3′H]benzo[g]indole-2′-carboxamide, 3,3,5,5-tetramethyl-N-1-naphthyl-**		**LI09179**
83	293	460	220	227	85	233	91	**460**	100	84	81	73	65	65	55	34		33	32	2	–	–	–	–	–	–	–	–	–	–	**Benzene, 1,1′-[(2,2-dimethyl-3,3-diphenylcyclopropyl)vinylidene]bis[3-methoxy-**	67437-10-9	**NI43682**
293	167	294	280	233	91	227	220	**460**	100	24	22	22	19	19	18	18	**13.00**	33	32	2	–	–	–	–	–	–	–	–	–	–	**Benzene, 1,1′-(3,3-dimethyl-5,5-diphenyl-1,4-pentadienylidene)bis[3-methoxy-**	67437-03-0	**NI43683**
293	460	167	280	294	417	91	227	**460**	100	84	51	49	32	27	24	20		33	32	2	–	–	–	–	–	–	–	–	–	–	**Benzene, 1,1′-[3-(2,2-diphenylvinyl)-2,2-dimethylcyclopropylidene]bis[3-methoxy-**	67437-11-0	**NI43684**
458	459	460	167	365	200	381	302	**460**	100	40	12	10	7	6	5	5		36	28	–	–	–	–	–	–	–	–	–	–	–	**Cyclohexadiene, pentaphenyl-**		**IC04465**
78	165	69	135	60	242	77	100	**461**	100	72	50	32	18	16	14	13	**0.00**	8	12	1	1	1	9	–	–	–	–	–	–	1	**1-Butanesulphinamide, 1,1,2,2,3,3,4,4,4-nonafluoro-N-methyl-N-(trimethylstannyl)-**	41006-33-1	**MS10033**
155	91	107	65	30	64	276	156	**461**	100	85	24	17	17	15	14	9	**1.40**	14	11	11	3	2	–	–	–	–	–	–	–	–	**Benzenesulphonic acid, 2,4,6-trinitro-, [(4-methylphenyl)sulphonyl]methyl ester**	56177-35-6	**NI43685**
73	357	299	147	103	358	101	315	**461**	100	44	44	22	15	13	13	12	**0.00**	15	42	6	–	–	–	–	–	–	4	1	–	–	**α-Glycerophosphoric acid, tetrakis(trimethylsilyl)-**		**NI43686**
463	420	41	465	150	164	461	55	**461**	100	58	51	50	50	49	48	45		20	17	2	1	–	–	–	2	–	–	–	–	–	**1-Cyclohexylamino-2,4-dibromoanthraquinone**		**IC04466**
77	461	356	91	199	462	105	64	**461**	100	73	46	25	24	20	16	15		22	19	3	7	1	–	–	–	–	–	–	–	–	**4-(4-Phenylazo-phenylazo)-5-hydroxy-1-(3-sulphamylphenyl)-pyrazole**		**IC04467**
44	285	42	43	72	41	162	55	**461**	100	98	79	56	45	40	37	33	**0.00**	23	27	9	1	–	–	–	–	–	–	–	–	–	**β-D-Glucopyranosiduronic acid, (5α,6α)-7,8-didehydro-4,5-epoxy-6-hydroxy-17-methylmorphinan-3-yl**	20290-09-9	**NI43687**
169	109	461	131	130	127	115	159	**461**	100	71	66	56	30	24	17	16		23	27	9	1	–	–	–	–	–	–	–	–	–	**Indole, 1-β-D-glucopyranosyl-2-methyl-, 2′,3′,4′,6′-tetraacetate**	29742-54-9	**LI09180**
169	109	460	131	130	127	461	160	**461**	100	71	66	57	31	24	19	16		23	27	9	1	–	–	–	–	–	–	–	–	–	**Indole, 1-β-D-glucopyranosyl-2-methyl-, 2′,3′,4′,6′-tetraacetate**	29742-54-9	**NI43688**
194	267	220	165	73	268	195	151	**461**	100	70	43	37	25	17	13	11	**0.00**	24	39	4	1	–	–	–	–	–	2	–	–	–	**Denopamine, O,O-bis(trimethylsilyl)-**		**NI43689**

MASS TO CHARGE RATIOS	M.W.	INTENSITIES	Parent	C	H	O	N	S	F	Cl	Br	I	Si	P	B	X	COMPOUND NAME	CAS Reg No	No
341 205 105 118 282 237 104 77	**461**	100 99 99 97 89 70 68 62	**2.00**	25	23	6	3	–	–	–	–	–	–	–	–	–	5-Hydroxy-1-(methoxycarbonyl)-2-(methoxycarbonylamino)-2,3,5-triphenylimidazolidin-4-one		MS10034
117 91 106 77 65 150 132 107	**461**	100 39 28 21 21 15 15 14	**12.00**	25	24	–	5	1	–	1	–	–	–	–	–	–	1-[N,N'-Bis(4-methylphenyl)]guanidino-4-(4-chlorophenyl)-2-methylthioimidazole		MS10035
83 210 97 70 153 126 112 250	**461**	100 40 27 25 21 20 20 18	**0.00**	25	39	3	3	1	–	–	–	–	–	–	–	–	1,5-Diazacycloheptadecane-1-propanenitrile, 5-[(4-methylphenyl)sulphonyl]-17-oxo-	67171-88-4	NI43690
143 105 119 77 103 342 91 64	**461**	100 65 61 42 26 23 18 12	**10.60**	27	15	5	3	–	–	–	–	–	–	–	–	–	Spiro[benzo[1,2-d:4,3-d']diisoxazole-4(5H),5'-[1,4,2]dioxazol]-5-one, 3,3',6 triphenyl-	23872-06-2	NI43691
335 121 105 264 278 279 149 292	**461**	100 36 30 26 19 17 15 13	**6.00**	27	31	2	3	1	–	–	–	–	–	–	–	–	1,2,7,7a-Tetrahydro-1,7-diisopropyl-2-(isopropylimino)-4,7a-diphenylpyrrolo[2,3-D][1,3]thiazin-5,6-dione	90140-44-6	NI43692
392 393 18 55 69 43 57 41	**461**	100 32 27 20 14 14 11 11	**7.00**	27	50	1	1	–	3	–	–	–	–	–	–	–	1-(Trifluoroacetyl)azacyclohexacosane		MS10036
109 230 461 217 187 231 96 462	**461**	100 53 28 20 17 14 12 10		28	29	1	3	–	2	–	–	–	–	–	–	–	Pimozide	2062-78-4	NI43693
230 231 42 82 109 187 96 217	**461**	100 82 76 39 36 35 31 27	**6.19**	28	29	1	3	–	2	–	–	–	–	–	–	–	Pimozide	2062-78-4	NI43694
230 461 217 187 231 96 462 109	**461**	100 54 37 33 27 23 18 18		28	29	1	3	–	2	–	–	–	–	–	–	–	Pimozide	2062-78-4	LI09181
187 258 461 201 232 199 178 230	**461**	100 86 75 59 51 44 42 41		28	35	3	3	–	–	–	–	–	–	–	–	–	(±)-9-Demethyltubulosine		DD01716
127 335 462 333 334 460 461 208	**462**	100 52 36 27 23 17 15 10		–	–	–	–	–	–	–	–	2	–	–	–	1	Lead iodide	12684-19-4	NI43695
69 131 293 243 393 119 343 100	**462**	100 76 62 49 32 17 15 12	**0.20**	10	–	–	–	–	18	–	–	–	–	–	–	–	Perfluoro-cis-decalin		MS10037
69 131 293 243 119 100 93 181	**462**	100 72 30 27 14 13 12 11	**1.00**	10	–	–	–	–	18	–	–	–	–	–	–	–	Perfluorodecalin	306-94-5	IC04468
69 131 293 243 100 93 393 181	**462**	100 95 47 28 14 12 10 10	**1.30**	10	–	–	–	–	18	–	–	–	–	–	–	–	Perfluorodecalin	306-94-5	MS10038
69 131 293 243 100 119 93 181	**462**	100 94 28 14 11 10 10 8	**1.40**	10	–	–	–	–	18	–	–	–	–	–	–	–	Perfluorodecalin	306-94-5	IC04469
131 69 293 243 119 181 443 193	**462**	100 71 38 22 9 7 6 5	**1.40**	10	–	–	–	–	18	–	–	–	–	–	–	–	Perfluoro-trans-decalin		MS10039
245 395 393 314 383 217 464 462	**462**	100 53 53 37 22 20 1 1		10	6	1	–	–	6	–	1	1	–	–	–	–	1-Bromo-1,3-dihydro-5-methyl-3,3-bis(trifluoromethyl)-1,2-benziodoxole		MS10040
466 464 468 462 276 139 274 168	**462**	100 98 39 35 32 29 20 12		14	9	3	–	–	–	–	3	–	–	–	–	–	Spiro[benzofuran-2(3H),1'-[2,5]cyclohexadien]-4'-one, 3',5',7-tribromo-5-(hydroxymethyl)-	52498-93-8	NI43696
78 104 231 233 103 77 51 39	**462**	100 84 59 55 50 39 31 24	**0.00**	16	16	–	–	–	–	–	2	–	–	–	–	2	Titanium, di-μ-bromobis(η^8-1,3,5,7-cyclooctatetraene)di-	62319-91-9	NI43697
389 262 417 231 287 259 51 175	**462**	100 48 31 11 8 8 6 4	**0.00**	17	30	12	–	–	–	–	–	–	–	–	–	4	Beryllium, μ-ethoxy-μ^4-oxopentakis[μ-(propanoato-O:O')]tetra-	70225-64-8	NI43698
115 43 143 138 139 103 245 157	**462**	100 99 55 45 44 40 38 32	**0.00**	19	26	13	–	–	–	–	–	–	–	–	–	–	Acetyl 2,3,5,6,7-penta-O-acetyl-D-glycero-D-glucoheptofuranoside	90213-26-6	NI43699
115 43 157 152 139 110 97 103	**462**	100 99 68 60 46 44 32 28	**0.00**	19	26	13	–	–	–	–	–	–	–	–	–	–	Acetyl 2,3,4,6,7-penta-O-acetyl-D-glycero-α-D-glucoheptopyranoside		NI43700
115 43 157 152 110 139 97 116	**462**	100 99 70 65 49 37 28 26	**0.00**	19	26	13	–	–	–	–	–	–	–	–	–	–	Acetyl 2,3,4,6,7-penta-O-acetyl-D-glycero-β-D-glucoheptopyranoside		NI43701
43 127 69 55 128 101 147 126	**462**	100 97 53 27 16 13 7 6	**1.00**	19	27	8	–	–	–	–	1	–	–	–	–	–	Propanedioic acid, (bromomethyl)methyl-, bis(5,6-dioxoheptyl) ester	60004-37-7	NI43702
73 204 147 205 75 74 206 45	**462**	100 88 25 17 10 9 8 7	**3.10**	19	42	5	–	–	–	–	–	–	4	–	–	–	1-Cyclohexene-1-carboxylic acid, 3,4,5-tris[(trimethylsilyl)oxy]-, trimethylsilyl ester, [3R-(3α,4α,5β)]-	55520-78-0	NI43703
204 73 147 205 206 74 45 75	**462**	100 100 38 34 16 13 12 11	**4.46**	19	42	5	–	–	–	–	–	–	4	–	–	–	1-Cyclohexene-1-carboxylic acid, 3,4,5-tris[(trimethylsilyl)oxy]-, trimethylsilyl ester, [3R-(3α,4α,5β)]-	55520-78-0	NI43704
204 372 255 357 73 462 254 447	**462**	100 80 77 66 64 54 45 36		19	42	5	–	–	–	–	–	–	4	–	–	–	3,4,5-Tris(trimethylsiloxy)-1-cyclohexene-1-carboxylic acid, trimethylsilyl ester	70244-15-4	NI43705
217 129 131 201 157 213 203 230	**462**	100 27 6 5 5 4 4 3	**0.00**	20	54	–	–	–	–	–	–	–	6	–	–	–	2,2,4,4,7,7-Hexamethyl-1,6,6-tris(trimethylsilyl)-2,4,7-trisilaoctane		LI09182
464 466 462 64 125 249 465 28	**462**	100 52 51 49 43 27 23 20		22	8	2	–	–	–	–	2	–	–	–	–	–	Dibromoanthanthrone		IC04470
444 408 313 391 300 426 340 393	**462**	100 79 58 51 46 43 37 34	**0.00**	22	22	11	–	–	–	–	–	–	–	–	–	–	4H-1-Benzopyran-4-one, 8-β-D-glucopyranosyl-5,7-dihydroxy-3-(4-hydroxy-3-methoxyphenyl)-	40522-83-6	MS10041
300 301 60 257 43 73 57 229	**462**	100 21 18 17 14 12 12 10	**0.00**	22	22	11	–	–	–	–	–	–	–	–	–	–	4H-1-Benzopyran-4-one, 7-(β-D-glucopyranosyloxy)-5-hydroxy-2-(3-hydroxy-4-methoxyphenyl)-	20126-59-4	NI43706
383 384 341 313 462 436 434 151	**462**	100 29 21 15 10 10 10 10		22	23	6	–	–	–	–	1	–	–	–	–	–	7-Bromo-3-hydroxy-8-methoxy-1,9-dimethyl-6-sec-butyldibenzo[b,e][1,4]dioxepin-11-one		NI43707
136 205 206 190 234 178 162 175	**462**	100 67 60 57 28 21 17 15	**6.00**	22	26	9	2	–	–	–	–	–	–	–	–	–	Dinitroveraguensin		LI09183
169 109 127 331 81 115 139 462	**462**	100 88 32 30 26 25 23 22		22	26	9	2	–	–	–	–	–	–	–	–	–	1-(2,3,4,6-Tetra-O-acetyl-β-D-glucopyranosyl)-5-nitroindole		LI09184
359 466 464 361 444 252 446 250	**462**	100 99 96 92 88 85 83 77	**0.00**	22	27	4	2	–	–	–	1	–	–	–	–	–	Voachalotine oxindole, 10-bromo-19,20-dihydro-	29019-57-6	NI43708
43 237 209 124 195 161 462 163	**462**	100 42 22 19 16 15 12 11		23	26	8	–	1	–	–	–	–	–	–	–	–	3-(4'-Hydroxy-5'-methoxyphenyl)-3-mercapto-2-(2'-methoxyphenoxy)-propanol triacetate		NI43709
447 462 234 165 231 417 197 432	**462**	100 75 12 12 11 10 10 7		23	26	10	–	–	–	–	–	–	–	–	–	–	Exoticin	13364-94-8	MS10042
447 431 462 197 151 195 179 225	**462**	100 41 40 27 17 15 11 9		23	26	10	–	–	–	–	–	–	–	–	–	–	3,5,6,7,8,2',4',5'-Octamethoxyflavone	79105-52-5	NI43710
229 314 313 228 286 230 201 86	**462**	100 96 50 50 48 25 25 25	**2.33**	23	30	6	2	1	–	–	–	–	–	–	–	–	2-Pyrrolidinone, 1-[[3,4-dihydro-6,7-dimethoxy-2-(1-piperidinylsulphonyl) 1-naphthalenyl]acetyl]-	56701-29-2	MS10043

MASS TO CHARGE RATIOS	M.W.	INTENSITIES	Parent	C	H	O	N	S	F	Cl	Br	I	Si	P	B	X	COMPOUND NAME	CAS Reg No	No
462 403 375 431 343 238 371 432	**462**	100 95 87 55 55 49 35 30		23	30	8	2	–	–	–	–	–	–	–	–	–	**4,4′-Dimethyl-3,3′-bis(methoxycarbonylethyl)-5,5′-bis(methoxycarbonyl)-2,2′-dipyrrolylmethane**		**LI09185**
370 343 298 371 416 324 344 325	**462**	100 90 57 35 29 27 26 25	**19.00**	23	30	8	2	–	–	–	–	–	–	–	–	–	**4,4′-Dimethyl-3,5,3′,5′-tetrakis(ethoxycarbonyl)-2,2′-dipyrrolylmethane**		**LI09186**
175 313 205 149 318 248 257 346	**462**	100 37 37 32 29 25 22 15	**0.40**	23	34	6	4	–	–	–	–	–	–	–	–	–	**L-Alanine, N-[N-[N-[(4-nitrophenyl)methylene]-L-valyl]-L-isoleucyl]-, ethyl ester**	37580-25-9	**NI43711**
275 43 360 301 318 317 276 247	**462**	100 83 78 41 37 29 28 19	**14.00**	24	22	6	4	–	–	–	–	–	–	–	–	–	**1,2,3-Propanetriol, 1-(1-phenyl-1H-pyrazolo[3,4-b]quinoxalin-3-yl)-, triacetate, [S-(R*,S*)]-**	55591-21-4	**NI43712**
360 275 43 301 318 317 359 220	**462**	100 81 57 50 44 37 28 26	**14.00**	24	22	6	4	–	–	–	–	–	–	–	–	–	**1,2,3-Propanetriol, 1-(1-phenyl-1H-pyrazolo[3,4-b]quinoxalin-3-yl)-, triacetate, [S-(R*,S*)]-**	55591-21-4	**NI43713**
462 463 461 433 203 77 107 447	**462**	100 30 30 23 16 16 14 12		24	26	4	6	–	–	–	–	–	–	–	–	–	**Pyrimido[5,4-d]pyrimidine, 4,8-bis(m-anisidino)-2,6-diethoxy-**	18710-90-2	**MS10044**
73 151 55 45 125 41 75 260	**462**	100 30 23 19 16 15 14 12	**2.00**	24	38	5	–	–	–	–	–	–	2	–	–	–	**1H-2-Benzoxacyclotetradecin-1,7(8H)-dione, 3,4,5,6,9,10-hexahydro-3-methyl-14,16-bis[(trimethylsilyl)oxy]-, (E)-**	74841-67-1	**NI43714**
73 151 333 260 462 429 305 261	**462**	100 50 42 33 26 23 23 22		24	38	5	–	–	–	–	–	–	2	–	–	–	**1H-2-Benzoxacyclotetradecin-1,7(8H)-dione, 3,4,5,6,9,10-hexahydro-3-methyl-14,16-bis[(trimethylsilyl)oxy]-, (E)-**	74841-67-1	**NI43715**
73 151 333 260 305 429 317 261	**462**	100 61 54 32 30 26 25 21	**15.20**	24	38	5	–	–	–	–	–	–	2	–	–	–	**1H-2-Benzoxacyclotetradecin-1,7(8H)-dione, 3,4,5,6,9,10-hexahydro-3-methyl-14,16-bis[(trimethylsilyl)oxy]-, (E)-**	74841-67-1	**NI43716**
73 333 151 260 305 429 317 75	**462**	100 28 25 16 15 13 13 13	**6.30**	24	38	5	–	–	–	–	–	–	2	–	–	–	**1H-2-Benzoxacyclotetradecin-1,7(8H)-dione, 3,4,5,6,9,10-hexahydro-3-methyl-14,16-bis[(trimethylsilyl)oxy]-, (Z)-**		**NI43717**
155 209 447 308 124 265 405 139	**462**	100 97 70 62 43 29 28 24	**15.00**	24	42	3	6	–	–	–	–	–	–	–	–	–	**Triimidazo[1,5-a:1′,5′-c:1″,5″-e][1,3,5]triazine-1,5,9(2H,6H,10H)-trione, hexahydro-3,3,7,7,11,11-hexamethyl-2,6,10-triisopropyl-**	67969-50-0	**NI43718**
89 133 87 73 81 59 95 55	**462**	100 68 32 30 27 23 21 21	**3.00**	24	46	8	–	–	–	–	–	–	–	–	–	–	**3,3′-Hexamethylenebis(1,5,8,11-tetraoxacyclotridecane)**		**MS10045**
462 298 179 164 430 196 432 165	**462**	100 33 9 9 8 8 7 7		25	22	7	2	–	–	–	–	–	–	–	–	–	**1-(2,4-Dinitrostyryl)-4-(2,4,6-trimethoxystyryl)-benzene**		**LI09187**
43 159 124 267 327 387 341 447	**462**	100 80 48 40 31 23 17 7	**6.00**	25	34	8	–	–	–	–	–	–	–	–	–	–	**Methyl 6α,7β-diacetoxy-14-hydroxyvinhaticoate**		**NI43719**
310 384 342 324 311 370 385 343	**462**	100 83 63 43 37 30 25 24	**3.00**	25	34	8	–	–	–	–	–	–	–	–	–	–	**4-O-Methylphorbol-12,13-diacetate**		**LI09188**
162 120 102 256 284 259 287 103	**462**	100 50 50 35 33 29 28 27	**8.67**	25	38	6	2	–	–	–	–	–	–	–	–	–	**L-Phenylalanine, N-[N-(1-oxodecyl)-L-α-aspartyl]-, dimethyl ester**	15939-53-4	**NI43720**
462 447 464 448 463 449 465 417	**462**	100 78 38 35 29 23 10 9		26	22	8	–	–	–	–	–	–	–	–	–	–	**5,5′,8,8′-Tetramethoxy-6,6′-methyl-2,2′-binaphthalene-1,1′,4,4′-tetrone**	104548-70-1	**NI43721**
56 119 147 91 113 42 294 64	**462**	100 49 44 42 37 32 29 28	**0.00**	26	34	2	6	–	–	–	–	–	–	–	–	–	**3a,6b-Bis(dimethylamino)-1,3b,4,6b-tetramethyl-2,5-diphenyloctahydrocyclobuta[1,2-c′,4-c]dipyrazole-3,6-dione**		**MS10046**
73 100 341 75 70 42 143 41	**462**	100 99 60 55 47 45 38 33	**2.50**	26	46	3	2	–	–	–	–	–	1	–	–	–	**Pregnane-3,20-dione, 11-[(trimethylsilyl)oxy]-, bis(O-methyloxime), (5β,11β)-**	57305-27-8	**NI43722**
73 431 156 75 432 172 188 158	**462**	100 99 51 38 36 32 31 31	**15.00**	26	46	3	2	–	–	–	–	–	1	–	–	–	**Pregnane-3,20-dione, 17-[(trimethylsilyl)oxy]-, bis(O-methyloxime), (5α)-**	57305-28-9	**NI43723**
73 431 156 188 432 172 158 55	**462**	100 99 64 46 41 38 38 34	**22.00**	26	46	3	2	–	–	–	–	–	1	–	–	–	**Pregnane-3,20-dione, 17-[(trimethylsilyl)oxy]-, bis(O-methyloxime), (5β)-**	57305-29-0	**NI43724**
73 431 188 89 288 79 93 81	**462**	100 99 96 77 65 61 60 59	**40.00**	26	46	3	2	–	–	–	–	–	1	–	–	–	**Pregnane-3,20-dione, 21-[(trimethylsilyl)oxy]-, bis(O-methyloxime), (5α)-**	57305-26-7	**NI43725**
255 104 147 254 106 145 119 130	**462**	100 62 60 56 53 52 42 41	**0.00**	27	42	6	–	–	–	–	–	–	–	–	–	–	**Pregnane-20-carboxylic acid, 3,12-bis(acetyloxy)-, methyl ester, (3α,5β,12α,20S)-**	60354-44-1	**NI43726**
361 287 325 444 343 139 105 157	**462**	100 74 71 60 55 52 44 41	**13.00**	27	42	6	–	–	–	–	–	–	–	–	–	–	**(25S)-Spirosta-5-ene-1β,3β,23,24-tetrol**	51803-71-5	**NI43727**
361 287 325 444 343 139 105 157	**462**	100 74 71 60 55 52 44 41	**13.00**	27	42	6	–	–	–	–	–	–	–	–	–	–	**(25S)-Spirosta-5-ene-1β,3β,23,24-tetrol**	51803-71-5	**NI43728**
277 43 193 278 379 55 149 41	**462**	100 51 27 18 15 15 14 14	**1.49**	27	42	6	–	–	–	–	–	–	–	–	–	–	**Trimellitic acid, trihexyl ester**		**IC04471**
73 405 75 462 406 363 215 299	**462**	100 99 79 58 38 32 30 28		27	46	4	–	–	–	–	–	–	1	–	–	–	**Prosta-5,8(12),13-trien-1-oic acid, 15-[(isopropyldimethylsilyl)oxy]-9-oxo-, methyl ester, (5Z,13E,15S)-**	53044-55-6	**NI43729**
73 75 405 406 215 81 391 91	**462**	100 92 54 25 24 23 21 18	**3.59**	27	46	4	–	–	–	–	–	–	1	–	–	–	**Prosta-5,10,13-trien-1-oic acid, 15-[(isopropyldimethylsilyl)oxy]-9-oxo-, methyl ester, (5Z,13E,15S)-**	53122-04-6	**NI43730**
269 359 270 360 244 243 229 173	**462**	100 38 22 14 14 13 11 11	**1.71**	27	46	4	–	–	–	–	–	–	1	–	–	–	**5β,3α-Tetrahydromedroxyprogesterone acetate TMS**		**NI43731**
73 143 117 75 357 111 194 81	**462**	100 99 99 99 75 70 58 50	**30.00**	27	50	2	–	–	–	–	–	–	2	–	–	–	**Pregn-4-ene-11β,21-diol, bis(trimethylsilyl)-**	33283-09-9	**NI43732**
117 129 118 119 372 145 132 282	**462**	100 17 13 7 6 5 5 4	**3.82**	27	50	2	–	–	–	–	–	–	2	–	–	–	**Pregn-5-ene-3α,20-diol, bis(trimethylsilyl)-**	55515-25-8	**NI43733**
117 118 129 119 73 372 282 75	**462**	100 11 9 6 6 3 3 3	**2.40**	27	50	2	–	–	–	–	–	–	2	–	–	–	**Pregn-5-ene-3β,20α-diol, bis(trimethylsilyl)-**		**NI43734**
117 73 75 129 118 119 81 55	**462**	100 37 17 14 12 9 4 4	**1.56**	27	50	2	–	–	–	–	–	–	2	–	–	–	**Pregn-5-ene-3β,20α-diol, bis(trimethylsilyl)-**	13110-77-5	**NI43735**
117 73 75 118 129 119 372 143	**462**	100 31 16 11 11 10 4 3	**1.80**	27	50	2	–	–	–	–	–	–	2	–	–	–	**Pregn-5-ene-3β,20α-diol, bis(trimethylsilyl)-**	13110-77-5	**NI43736**
117 73 75 129 118 119 372 81	**462**	100 31 14 11 11 9 4 3	**2.00**	27	50	2	–	–	–	–	–	–	2	–	–	–	**Pregn-5-ene-3β,20β-diol, bis(trimethylsilyl)-**	33283-11-3	**NI43737**
117 118 372 129 119 462 73 282	**462**	100 15 12 12 11 6 6 5		27	50	2	–	–	–	–	–	–	2	–	–	–	**Pregn-5-ene-3β,20β-diol, bis(trimethylsilyl)-**		**NI43738**
117 73 75 129 118 372 149 143	**462**	100 32 16 12 11 3 3 3	**1.90**	27	50	2	–	–	–	–	–	–	2	–	–	–	**Pregn-5-ene-3β,20β-diol, bis(trimethylsilyl)-**	33283-11-3	**NI43739**
73 129 117 372 75 333 215 243	**462**	100 99 40 32 32 30 26 25	**16.00**	27	50	2	–	–	–	–	–	–	2	–	–	–	**Pregn-5-ene-3β,21-diol, bis(trimethylsilyl)-**	33283-12-4	**NI43740**
447 421 365 391 462 394 353 309	**462**	100 55 55 50 45 45 45 40		28	30	6	–	–	–	–	–	–	–	–	–	–	**5,9,11-Trihydroxy-3,3-dimethyl-6,8-bis-(3-methylbut-2-enyl)-3H,12H-pyrano[3,2-a]xanthen-12-one**		**NI43741**
120 119 98 217 99 121 83 218	**462**	100 99 99 97 93 92 90 88	**67.00**	28	38	2	4	–	–	–	–	–	–	–	–	–	**3,6-Bis(p-N-heptoxyphenyl)-1,2,4,5-tetrazine**	72243-04-0	**NI43742**
155 99 100 55 156 101 69 81	**462**	100 80 18 15 12 9 8 7	**6.20**	28	46	5	–	–	–	–	–	–	–	–	–	–	**Methyl 7α-methoxy-3,3-ethylenedioxy-5β-cholanoate**		**MS10047**

MASS TO CHARGE RATIOS								M.W.	INTENSITIES								Parent	C	H	O	N	S	F	Cl	Br	I	Si	P	B	X	COMPOUND NAME	CAS Reg No	No
215	372	216	257	357	230	373	161	**462**	100	92	44	36	29	27	24	20	**2.22**	28	50	3	–	–	–	–	–	–	1	–	–	–	Cholan-24-oic acid, 3-[(trimethylsilyl)oxy]-, methyl ester, (3α,5β)-	59953-47-8	NI43743
353	402	260	404	403	354	366	247	**462**	100	99	48	35	30	30	28	20	**0.00**	29	47	2	–	–	–	1	–	–	–	–	–	–	Cholest-5-en-3β-ol, 19-chloro-, acetate	21072-67-3	NI43744
43	163	55	41	57	69	95	81	**462**	100	65	59	54	53	46	43	42	**4.30**	29	47	3	–	–	1	–	–	–	–	–	–	–	Cholestan-19-ol, 5,6-epoxy-3-fluoro-, acetate, (3β,5α,6α)-	40242-96-4	MS10048
43	41	55	141	57	81	91	67	**462**	100	45	43	36	34	27	25	23	**5.20**	29	47	3	–	–	1	–	–	–	–	–	–	–	Cholestan-19-ol, 5,6-epoxy-3-fluoro-, acetate, (3β,5β,6β)-	40242-95-3	MS10049
462	43	131	57	55	463	69	41	**462**	100	53	45	37	35	34	31	31		29	50	–	–	2	–	–	–	–	–	–	–	–	5α-Cholestan-4-one, cyclic ethylene mercaptole	20233-42-5	NI43745
43	323	57	255	55	95	71	41	**462**	100	85	64	44	36	33	32	29	**9.26**	29	50	–	–	2	–	–	–	–	–	–	–	–	5α-Cholestan-6-one, cyclic ethylene mercaptole	17690-27-6	NI43746
91	133	101	107	447	462	317	149	**462**	100	28	11	8	2	1	1	1		29	50	4	–	–	–	–	–	–	–	–	–	–	1-O-(2-Benzyloxyhexadecyl)-2,3-O-isopropylideneglycerol		NI43747
378	186	69	420	81	225	462	326	**462**	100	49	33	21	15	12	11	11		30	38	4	–	–	–	–	–	–	–	–	–	–	1,4-Naphthalenediol, 2-methyl-3-(3,7,11-trimethyl-2,6,10-dodecatrienyl)-, diacetate	56298-82-9	NI43748
371	462	391	372	463	447	392	373	**462**	100	77	34	30	29	23	12	11		30	42	2	–	–	–	–	–	–	1	–	–	–	Δ^1-Tetrahydrocannabinol benzyldimethylsilyl ether		NI43749
464	449	465	462	216	57	218	463	**462**	100	99	30	26	21	21	19	18		30	42	2	2	–	–	–	–	–	–	–	–	–	1,4-Bis(4′-oxo-3′,5′-di-tert-butylcyclohexadienylidene)-2,3-diazabutane		MS10050
57	350	291	263	349	347	235	55	**462**	100	52	52	37	32	23	22	21	**5.00**	30	54	3	–	–	–	–	–	–	–	–	–	–	3,17β-Di-tert-butoxy-4-propyl-3,4-seco-5α-androstan-6-one		MS10051
193	301	414	411	220	283	429	444	**462**	100	88	37	26	23	21	20	18	**0.00**	30	54	3	–	–	–	–	–	–	–	–	–	–	Lanostane-3β,11β,18-triol	24041-79-0	LI09189
193	301	43	69	95	55	81	57	**462**	100	88	65	54	50	46	41	40	**0.00**	30	54	3	–	–	–	–	–	–	–	–	–	–	Lanostane-3β,11β,18-triol	24041-79-0	NI43750
414	399	426	413	396	381	411	444	**462**	100	64	55	48	32	26	20	17	**1.00**	30	54	3	–	–	–	–	–	–	–	–	–	–	Lanostane-3β,11β,19-triol	24041-79-0	LI09190
414	399	426	413	84	95	396	415	**462**	100	64	55	47	38	37	32	31	**0.00**	30	54	3	–	–	–	–	–	–	–	–	–	–	Lanostane-3β,11β,19-triol	25116-75-0	NI43751
359	373	43	360	55	81	57	41	**462**	100	42	28	26	16	14	12	12	**0.00**	30	54	3	–	–	–	–	–	–	–	–	–	–	3-Methoxy-3,6α-dipentyl-5α-androstane-6β,17β-diol		MS10052
402	462	444	208	180	403	420	43	**462**	100	70	47	43	37	34	31	28		31	30	2	2	–	–	–	–	–	–	–	–	–	Acetamide, N-[2-[[2-[acetyl(4-methylphenyl)amino]-5-methylphenyl]methyl]phenyl]-N-phenyl-	52812-79-0	NI43752
43	91	105	119	444	462	426	411	**462**	100	38	31	24	22	20	20	3		31	42	3	–	–	–	–	–	–	–	–	–	–	Paracentrone		LI09191
43	91	105	119	462	444	426	411	**462**	100	24	18	15	7	6	3	1		31	42	3	–	–	–	–	–	–	–	–	–	–	Paracentrone		LI09192
292	154	143	462	291	319	69	127	**462**	100	43	35	34	26	25	22	20		32	34	1	2	–	–	–	–	–	–	–	–	–	Cyclohexane-1-spiro-3′-benzo[g]indoline-2′-carboxamide, 3,3,5,5-tetramethyl-N-1-naphthyl-		LI09193
462	447	157	463	144	448	182	194	**462**	100	63	42	38	38	24	24	16		33	38	–	2	–	–	–	–	–	–	–	–	–	5α-Androstano[3,2-b:17,16-b]-N,N′-dimethyldiindole		LI09194
462	447	157	463	144	448	182	158	**462**	100	62	41	40	39	22	21	20		33	38	–	2	–	–	–	–	–	–	–	–	–	5α-Androstano[3,2-b:17,16-b]-N,N′-dimethyldiindole		MS10053
231	230	202	216	101	154	149	115	**462**	100	92	20	14	12	9	9	7	**0.00**	34	26	–	2	–	–	–	–	–	–	–	–	–	(1α,2α,3β,4β)-1,3-Bis(4-quinolinyl)-2,4-diphenylcyclobutane		MS10054
135	136	93	79	67	231	190	55	**462**	100	10	9	9	6	5	5	5	**1.00**	34	54	–	–	–	–	–	–	–	–	–	–	–	Meso-1,2-di-1-adamantyl-1,2-dicyclohexylethane		NI43753
462	463	178	105	385	77	356	279	**462**	100	38	25	12	9	9	8	8		35	26	1	–	–	–	–	–	–	–	–	–	–	2,4-Cyclopentadien-1-ol, 1,2,3,4,5-pentaphenyl-	2137-74-8	LI09195
462	463	178	279	356	105	385	357	**462**	100	36	12	11	10	9	8	8		35	26	1	–	–	–	–	–	–	–	–	–	–	2,4-Cyclopentadien-1-ol, 1,2,3,4,5-pentaphenyl-	2137-74-8	NI43754
44	203	201	62	64	53	145	143	**463**	100	93	93	86	82	66	60	60	**15.43**	12	6	4	2	–	–	–	2	–	–	–	–	1	Copper, bis(4-bromo-O-benzoquinone 2-oximato)-	31971-31-0	NI43755
63	44	203	201	62	64	53	145	**463**	100	31	29	29	27	25	21	19	**1.00**	12	6	4	2	–	–	–	2	–	–	–	–	1	Copper, bis(4-bromo-O-benzoquinone 2-oximato)-	31971-31-0	MS10055
43	77	336	141	41	69	80	220	**463**	100	66	54	50	48	44	39	37	**5.00**	16	18	5	1	1	–	–	–	1	–	–	–	–	2-Oxa-6-azatricyclo[3.3.1.1^{3,7}]decan-4-ol, 8-iodo-6-(phenylsulphonyl)-, acetate, (1α,3β,4β,5α,7β,8α)-	50267-29-3	NI43756
73	70	75	45	147	74	142	43	**463**	100	28	14	14	12	12	11	9	**0.00**	18	45	3	3	–	–	–	–	–	4	–	–	–	Tetra(trimethylsilyl)citrulline		NI43757
228	465	201	463	245	229	173	159	**463**	100	72	72	70	59	44	25	21		19	14	6	1	1	–	–	1	–	–	–	–	–	N-(4-Bromophenyl)-4-methoxy-7-methyl-5-oxo-5H-furo[3,2-g][1]benzopyran-9-sulphonamide		MS10056
143	160	120	70	80	54	56	71	**463**	100	81	57	51	45	45	44	40	**0.57**	20	41	7	5	–	–	–	–	–	–	–	–	–	Gentamycin C2	25876-11-3	NI43758
464	322	465	104	466	463	305	149	**463**	100	32	17	7	6	5	5	4		20	41	7	5	–	–	–	–	–	–	–	–	–	Gentamycin C2	25876-11-3	NI43759
162	463	81	169	109	161	464	175	**463**	100	85	55	38	31	28	25	22		23	29	9	1	–	–	–	–	–	–	–	–	–	Indoline, 1-β-D-glucopyranosyl-2-methyl-, 2′,3′,4′,6′-tetraacetate		LI09196
162	81	463	133	139	109	169	175	**463**	100	75	73	39	35	35	31	27		23	29	9	1	–	–	–	–	–	–	–	–	–	Quinoline, 1,2,3,4-tetrahydro-1-(2,3,4,6-tetra-O-acetyl-β-D-glucopyranosyl)-	29742-56-1	NI43760
162	81	463	169	133	109	141	175	**463**	100	75	71	48	39	36	35	27		23	29	9	1	–	–	–	–	–	–	–	–	–	Quinoline, 1,2,3,4-tetrahydro-1-(2,3,4,6-tetra-O-acetyl-β-D-glucopyranosyl)-	29742-56-1	LI09197
323	294	125	143	77	141	142	89	**463**	100	98	94	90	65	64	56	50	**40.00**	25	19	2	3	–	–	2	–	–	–	–	–	–	1-(4-Chloro-2-methylphenylazo)-2-hydroxy-3-(4-chloro-2-methylanilinocarbonyl)naphthalene		IC04472
262	264	463	263	233	465	234	221	**463**	100	92	48	43	38	35	35	30		26	28	2	2	–	–	–	–	–	–	–	–	1	Copper, bis(2-butyl-8-quinolinolato)-	30049-19-5	NI43761
432	433	418	402	448	372	463	460	**463**	100	39	14	12	7	6	4	3		26	41	6	1	–	–	–	–	–	–	–	–	–	Pacilime		MS10057
43	183	57	41	55	71	29	69	**463**	100	81	76	54	51	30	30	27	**13.41**	26	48	2	1	–	3	–	–	–	–	–	–	–	Dodecanamide, N-dodecyl-N-(trifluoroacetyl)-	33845-35-1	NI43762
225	182	181	57	168	226	196	237	**463**	100	82	52	49	20	16	14	12	**3.77**	27	45	5	1	–	–	–	–	–	–	–	–	–	4-Tridecanone, 13-(4-hydroxy-5,6-dimethoxy-3-methyl-2-pyridyl)-3,5,7,11 tetramethyl-, acetate (ester)	5024-36-2	NI43763
262	463	183	108	263	261	27	184	**463**	100	61	32	26	20	11	10	9		30	26	2	1	–	–	–	–	–	–	1	–	–	Methyl 5-phenyl-2-(triphenylphosphazeno)penta-2,4-dienoate		NI43764
463	407	464	462	408	448	387	90	**463**	100	54	39	24	24	20	20	15		33	53	–	1	–	–	–	–	–	–	–	–	–	6-Azacholestane, 6-benzyl-	16373-52-7	NI43765

MASS TO CHARGE RATIOS								M.W.	INTENSITIES								Parent	C	H	O	N	S	F	Cl	Br	I	Si	P	B	X	COMPOUND NAME	CAS Reg No	No
28	183	112	131	211	31	56	93	**464**	100	99	70	62	60	57	53	42	**0.00**	7	–	4	–	–	7	–	–	–	–	–	–	2	Iron, tetracarbonyl(heptafluoropropyl)iodo-	22925-82-2	NI43766
267	209	296	29	239	238	464	436	**464**	100	95	80	80	30	25	4	4		10	15	4	–	–	–	–	–	–	–	–	–	2	Tetracarbonyl(triethylbismuthine)iron	90188-72-0	NI43767
197	198	138	149	165	117	192	121	**464**	100	37	24	23	20	20	18	14	**0.00**	13	15	8	2	1	–	3	–	–	–	–	–	–	2,2,2-Trichloroethyl 7-methoxycarbonylamino-3-methoxymethyl-3-cephem-4-carboxylate 1,1-dioxide	97609-96-6	NI43768
449	73	464	361	147	433	133	75	**464**	100	35	24	6	4	3	3	2		13	38	6	–	–	–	–	–	–	4	2	–	–	Phosphonic acid, methylenebis-, tetrakis(trimethylsilyl) ester	53044-26-1	NI43769
449	73	450	464	451	465	361	452	**464**	100	35	30	24	20	7	6	5		13	38	6	–	–	–	–	–	–	4	2	–	–	Phosphonic acid, methylenebis-, tetrakis(trimethylsilyl) ester	53044-26-1	NI43770
403	217	69	400	129	117	164	404	**464**	100	76	42	24	20	19	13	11	**0.00**	14	5	–	–	1	12	–	–	–	–	1	–	–	Phosphorane, (ethylthio)difluorobis(pentafluorophenyl)-	65132-59-4	NI43771
104	42	76	98	105	27	162	161	**464**	100	36	16	11	10	10	8	8	**0.00**	16	18	6	–	–	–	–	2	–	–	–	–	–	β-L-Fructopyranoside, methyl 4,6-dibromo-1,4-dideoxy-, 3-acetate5-benzoate, (R)-	56701-39-4	NI43772
131	350	352	103	351	380	378	354	**464**	100	64	58	51	47	38	36	36	**0.00**	17	12	4	–	–	–	–	–	–	–	–	–	1	Tungsten, cyclopentadienyl-cinnamoyl tricarbonyl		NI43773
73	171	75	45	74	43	100	59	**464**	100	50	48	37	27	23	18	16	**0.00**	17	44	3	4	–	–	–	–	–	4	–	–	–	Tetra(trimethylsilyl)canavanine		NI43774
73	171	75	45	74	43	100	218	**464**	100	50	48	37	27	23	18	16	**0.10**	17	44	3	4	–	–	–	–	–	4	–	–	–	Tetra(trimethylsilyl)canavanine		MS10058
73	332	147	205	333	117	374	75	**464**	100	64	32	31	20	15	9	9	**2.30**	18	40	6	–	–	–	–	–	–	4	–	–	–	L-Ascorbic acid, 2,3,5,6-tetrakis-O-(trimethylsilyl)-	55517-56-1	NI43775
73	332	147	205	117	333	45	74	**464**	100	34	27	23	16	11	10	9	**1.00**	18	40	6	–	–	–	–	–	–	4	–	–	–	L-Ascorbic acid, 2,3,5,6-tetrakis-O-(trimethylsilyl)-	55517-56-1	NI43776
332	147	205	333	117	75	73	334	**464**	100	51	46	31	24	24	24	15	**1.71**	18	40	6	–	–	–	–	–	–	4	–	–	–	L-Ascorbic acid, 2,3,5,6-tetrakis-O-(trimethylsilyl)-	55517-56-1	NI43777
184	182	156	198	181	209	226	180	**464**	100	52	52	33	31	29	27	24	**0.00**	19	12	4	–	–	–	–	–	–	–	–	–	2	1-Selenocoumarin, 3,3′-methylenebis[4-hydroxy-	26452-59-5	NI43778
324	178	237	352	296	436	59	380	**464**	100	87	72	53	49	32	18	17	**6.00**	20	10	6	–	–	–	–	–	–	–	–	–	2	Cobalt, hexacarbonyl[μ-[1,1′-(η²:η²-1,2-ethynediyl)]bis[benzene]]di-,	14515-69-6	NI43779
302	303	73	69	43	137	29	31	**464**	100	50	23	23	23	22	22	21	**0.00**	21	20	12	–	–	–	–	–	–	–	–	–	–	4H-1-Benzopyran-4-one, 2-(3,4-dihydroxyphenyl)-3-(β-D-galactopyranosyloxy)-5,7-dihydroxy-	482-36-0	NI43780
84	44	58	46	421	464	42	98	**464**	100	83	68	45	25	24	24	19		21	21	8	2	–	–	1	–	–	–	–	–	–	2-Naphthacenecarboxamide, 7-chloro-4-(dimethylamino)-1,4,4a,5,5a,6,11,12a-octahydro-3,6,10,12,12a-pentahydroxy-1,11-dioxo-, [4S-(4α,4aα,5aα,6β,12aα)]-	127-33-3	NI43781
256	255	309	69	45	368	141	115	**464**	100	74	67	60	58	56	54	54	**0.00**	22	22	4	–	–	6	–	–	–	–	–	–	–	Estradiol, bis(trifluoroacetate)	1549-15-1	NI43782
174	73	175	86	45	176	59	74	**464**	100	78	18	13	9	8	8	7	**0.00**	22	44	1	2	–	–	–	–	–	4	–	–	–	Serotonin, tetra-TMS		MS10059
174	73	175	290	449	291	176	464	**464**	100	22	20	17	12	8	8	6		22	44	1	2	–	–	–	–	–	4	–	–	–	Serotonin, tetra-TMS		MS10060
57	391	167	41	392	157	42	58	**464**	100	83	58	35	23	22	22	17	**0.00**	22	48	6	–	–	–	–	–	–	2	–	–	–	1-Vinyl-1,1,3,3,3-pentabutoxydisiloxane	94048-04-1	NI43783
405	419	179	139	121	221	161	195	**464**	100	96	90	53	50	32	31	30	**0.00**	22	50	6	–	–	–	–	–	–	–	–	–	2	Aluminium, μ-tert-butoxytri-tert-butoxy-μ-isopropoxydi-	28899-45-8	NI43784
256	229	348	257	109	349	201	128	**464**	100	85	75	75	60	50	40	40	**0.00**	23	28	10	–	–	–	–	–	–	–	–	–	–	Enhydrin		MS10061
73	238	57	152	69	55	46	71	**464**	100	99	65	63	57	55	51	35	**0.00**	23	45	7	–	–	–	–	–	–	–	1	–	–	Diethyl [8-methoxy-1-(isopropoxycarbonyl)-2-(2-oxopropyl)-4,8-dimethylnonyl]phosphonate		NI43785
104	371	230	76	158	132	94	190	**464**	100	97	66	43	36	34	29	28	**1.10**	24	20	6	2	1	–	–	–	–	–	–	–	–	Methyl 3-(2-hydroxy-5-methylbenzyl)-7-phthalimido-3-cephem-4-carboxylate		NI43786
405	406	464	178	229	177	375	151	**464**	100	28	22	13	11	10	8	8		24	20	8	2	–	–	–	–	–	–	–	–	–	Tetramethyl benzo[c]pyridazino[1,2-a]cinnoline-6,7,8,9-tetracarboxylate		NI43787
93	464	121	76	122	65	103	45	**464**	100	92	85	85	58	52	35	30		24	24	6	4	–	–	–	–	–	–	–	–	–	1-(4-Hydroxyphenylazo)-2-(2-methoxyethoxy)-4-(4-carbomethoxyphenylazo)-5-methoxybenzene		IC04473
43	69	41	405	109	149	83	91	**464**	100	43	34	25	25	24	24	19	**0.28**	24	32	9	–	–	–	–	–	–	–	–	–	–	12β-13-Diacetoxy-4,7β,9,20-tetrahydroxy-1,5-tigliadien-3-one		NI43788
311	326	293	283	295	280	269	241	**464**	100	80	70	65	60	50	50	45	**2.00**	24	32	9	–	–	–	–	–	–	–	–	–	–	Spiro[cyclopropane-1,2′(1′H)-phenanthrene]-1′,4′(3′H)-dione, 7′-(acetyloxy)-8′-[(acetyloxy)methyl]-4′b,5′,6′,7′,8′,8′a,9′,10′-octahydro 3′,9′,10′-trihydroxy-2,4′b,8′-trimethyl-	66607-73-6	NI43789
344	329	108	121	165	153	152	345	**464**	100	78	74	68	64	50	46	45	**2.00**	25	36	8	–	–	–	–	–	–	–	–	–	–	Picras-2-en-1-one, 12,16-bis(acetyloxy)-11-hydroxy-2-methoxy-, (11α,12β)-	30908-30-6	NI43790
212	197	117	302	230	129	94	196	**464**	100	98	53	50	29	28	26	22	**6.00**	25	44	4	–	–	–	–	–	–	2	–	–	–	Androst-5-en-17-one, 11-hydroxy-3,16-bis[(trimethylsilyl)oxy]-, (3β,11β,16α)-	49774-88-1	NI43791
338	43	323	422	380	379	337	270	**464**	100	91	82	80	66	63	59	51	**2.20**	26	24	8	–	–	–	–	–	–	–	–	–	–	5,7,4′-Triacetoxy-8-(3,3-dimethylallyl)isoflavone		NI43792
105	77	177	106	165	122	51	178	**464**	100	18	7	7	4	3	3	2	**0.00**	26	24	8	–	–	–	–	–	–	–	–	–	–	DL-Xylitol, 1,2,5-tribenzoate	74841-64-8	NI43793
435	464	437	439	466	436	468	438	**464**	100	70	63	45	42	30	28	25		26	28	2	2	–	–	–	–	–	–	–	–	1	Zinc, bis(2-butyl-8-quinolinolato)-	15685-77-5	NI43794
159	421	464	436	174	392	132	121	**464**	100	80	35	15	15	14	13	13		26	40	3	–	2	–	–	–	–	–	–	–	–	17β-Acetoxy-4,4-dimethyl-5α-androstan-2-spiro-2′-[1′,3′-dithiane]		LI09198
99	81	55	69	330	300	250	279	**464**	100	98	76	68	56	38	26	18	**0.00**	27	44	6	–	–	–	–	–	–	–	–	–	–	Cholest-7-en-6-one, 2,3,14,22,25-pentahydroxy-, (2β,3β,5β,22R)-	3604-87-3	NI43795
329	330	55	347	331	234	285	81	**464**	100	78	61	54	51	43	35	34	**1.56**	27	44	6	–	–	–	–	–	–	–	–	–	–	Cholest-7-en-6-one, 3,14,20,22,25-pentahydroxy-, (3β,5β,22R)-	17942-08-4	NI43796
329	330	55	347	331	59	234	285	**464**	100	79	62	54	51	48	43	36	**2.00**	27	44	6	–	–	–	–	–	–	–	–	–	–	Cholest-7-en-6-one, 3,14,20,22,25-pentahydroxy-, (3β,5β,22R)-	17942-08-4	LI09199
311	329	99	269	330	312	81	55	**464**	100	83	62	55	50	50	50	44	**1.28**	27	44	6	–	–	–	–	–	–	–	–	–	–	Cholest-7-en-6-one, 3α,14,20,22,25-pentahydroxy-, (5β,22R)-	22785-88-2	NI43797
329	347	311	81	99	143	234	395	**464**	100	42	17	15	13	8	7	2	**0.07**	27	44	6	–	–	–	–	–	–	–	–	–	–	2-Deoxy-20-hydroxy-ecdysone		NI43798
73	284	75	81	95	285	147	55	**464**	100	99	99	60	55	50	50	45	**6.50**	27	52	2	–	–	–	–	–	–	2	–	–	–	3α,6α-Bis(trimethylsilyloxy)-5β-pregnane	33283-07-7	NI43799
284	239	147	161	129	269	374	229	**464**	100	63	45	37	35	32	21	20	**9.00**	27	52	2	–	–	–	–	–	–	2	–	–	–	3α,6α-Bis(trimethylsilyloxy)-5β-pregnane	33283-07-7	NI43800
117	118	284	119	116	73	269	103	**464**	100	12	5	5	5	5	4	4	**0.00**	27	52	2	–	–	–	–	–	–	2	–	–	–	3α,20β-Bis(trimethylsilyloxy)-5β-pregnane		NI43801
117	118	119	284	269	105	116	107	**464**	100	99	58	41	39	32	27	23	**0.00**	27	52	2	–	–	–	–	–	–	2	–	–	–	3α,20R-Bis(trimethylsilyloxy)-5α-pregnane	55429-65-7	NI43802

MASS TO CHARGE RATIOS								M.W.	INTENSITIES								Parent	C	H	O	N	S	F	Cl	Br	I	Si	P	B	X	COMPOUND NAME	CAS Reg No	No
73	118	117	75	119	116	81	284	**464**	100	99	99	99	46	37	26	25	**0.96**	27	52	2	–	–	–	–	–	–	2	–	–	–	3α,20S-Bis(trimethylsilyloxy)-5β-pregnane	16134-56-8	NI43803
151	296	228	227	464	229	253	242	**464**	100	28	21	14	12	9	8	7		28	32	6	–	–	–	–	–	–	–	–	–	–	11,19-Dihydroxy-abieta-5,7,9(11),13-tetraen-12-one, 19-O-vanilloate		NI43804
310	328	282	292	295	464	446	428	**464**	100	48	36	13	8	6	5	1		28	32	6	–	–	–	–	–	–	–	–	–	–	5,12b-Epoxy-3,6a-methano-1H-azuleno[5,4-d][1,3]dioxocin-10(9H)-one, 2,3,9a,12a-tetrahydro-9a-hydroxy-8-(hydroxymethyl)-1,11-dimethyl-3-isopropenyl-5-benzyl-	75659-59-5	NI43805
121	464	387	463	77	465	285	355	**464**	100	53	50	26	24	21	16	15		29	20	–	–	3	–	–	–	–	–	–	–	–	[1,2]Dithiolo[1,5-b][1,2]dithiole-7-SIV, 2,3,4,5-tetraphenyl-	16094-77-2	NI43806
391	464	404	369	466	393	465	406	**464**	100	99	90	50	33	33	30	30		29	49	2	–	–	–	1	–	–	–	–	–	–	Cholestan-19-ol, 2-chloro-, acetate, (2α,5α)-	28809-78-1	NI43807
125	464	466	465	99	43	112	55	**464**	100	68	27	24	19	19	17	14		29	49	2	–	–	–	1	–	–	–	–	–	–	Cholestan-3-one, 2-chloro-, cyclic 1,2-ethanediyl acetal, (2α,5α)-	54552-54-4	NI43808
291	99	464	428	201	178	165	466	**464**	100	29	9	7	7	5	5	4		29	49	2	–	–	–	1	–	–	–	–	–	–	Cholestan-6-one, 3-chloro-, cyclic 1,2-ethanediyl acetal, (3β,5α)-	35868-80-5	NI43809
291	99	292	43	53	464	41	127	**464**	100	29	22	12	11	9	9	8		29	49	2	–	–	–	1	–	–	–	–	–	–	Cholestan-6-one, 3-chloro-, cyclic 1,2-ethanediyl acetal, (3β,5α)-	35868-80-5	LI09200
428	271	376	316	287	253	410	385	**464**	100	94	77	68	58	53	44	18	**0.00**	29	49	2	–	–	–	1	–	–	–	–	–	–	Cholest-22-en-3,5-diol, 6-chloro-24-ethyl-, (3β,5α,6β,24R)-		NI43810
464	446	28	465	447	43	41	363	**464**	100	94	72	37	33	28	21	19		30	40	4	–	–	–	–	–	–	–	–	–	–	Chamaecydinol	86699-52-7	NI43811
380	186	69	422	464	225	105	229	**464**	100	62	46	29	18	15	10	3		30	40	4	–	–	–	–	–	–	–	–	–	–	1,4-Naphthalenediol, 2-methyl-3-(3,7,11-trimethyl-2,10-dodecadienyl)-, diacetate	56298-81-8	NI43812
43	100	57	41	71	69	55	83	**464**	100	83	76	53	47	44	44	21	**6.67**	31	60	2	–	–	–	–	–	–	–	–	–	–	14,16-Hentriacontanedione	24724-84-3	NI43813
100	211	239	281	253	268	464	296	**464**	100	45	43	41	34	28	25	25		31	60	2	–	–	–	–	–	–	–	–	–	–	14,16-Hentriacontanedione	24724-84-3	NI43814
149	167	447	85	113	97	95	464	**464**	100	34	31	25	24	24	24	23		32	20	2	2	–	–	–	–	–	–	–	–	–	2-Hydroxy-10,13,16-triphenyl-14,15-diaza-9-oxatetracyclo[9.6.0.0^{3,8}.0^{13,16}]heptadeca-1,3(8),4,6,10,12(17),14-heptaene		NI43815
155	421	197	449	253	154	422	420	**464**	100	76	47	29	26	23	20	17	**3.66**	33	68	–	–	–	–	–	–	–	–	–	–	–	Dotriacontane, 2-methyl-	1720-11-2	NI43816
155	197	154	253	196	153	351	156	**464**	100	54	37	28	21	17	13	11	**10.11**	33	68	–	–	–	–	–	–	–	–	–	–	–	Tritriacontane	630-05-7	NI43817
387	359	360	105	77	281	464	165	**464**	100	74	68	66	52	20	14	13		34	24	2	–	–	–	–	–	–	–	–	–	–	3-Benzoyl-2,2,4-triphenyl-2H-1-benzopyran		LI09201
387	359	360	105	77	388	281	361	**464**	100	76	66	66	53	30	20	15	**13.50**	34	24	2	–	–	–	–	–	–	–	–	–	–	3-Benzoyl-2,2,4-triphenyl-2H-1-benzopyran		MS10062
359	360	464	105	253	77	252	465	**464**	100	33	19	16	14	10	9	7		34	24	2	–	–	–	–	–	–	–	–	–	–	2,3-Diphenyl-5-(2-oxo-1,2-diphenylethyl)benzofuran		NI43818
464	105	465	77	232	91	104	359	**464**	100	44	37	28	26	22	18	11		34	24	2	–	–	–	–	–	–	–	–	–	–	2,2',6,6'-Tetraphenyl-4,4'-bis(pyranylidene)		NI43819
208	138	177	176	150	209	156	465	**465**	100	48	35	28	25	21	19	10		19	23	9	5	–	–	–	–	–	–	–	–	–	Methyl 2-amino-3,4-dihydro-4-oxo-7-(2-acetylamino-2-deoxy-3,4-diacetyl-β-arabinofuranosyl)-7H-pyrrolo[2,3-d]pyrimidine-5-carboxylate	84873-20-1	NI43820
73	267	268	179	465	45	193	74	**465**	100	74	19	17	16	14	9	8		19	34	3	1	–	3	–	–	–	3	–	–	–	N-(Trifluoroacetyl)-N,O,O'-tris(trimethylsilyl)dopamine		MS10063
174	178	114	277	180	236	279	238	**465**	100	28	18	11	10	7	4	2	**2.00**	20	20	6	3	1	–	1	–	–	–	–	–	–	5-Hydroxymethylcloxacillin methyl ester		MS10064
296	227	43	57	55	268	69	41	**465**	100	59	54	51	46	41	38	35	**5.20**	22	38	1	1	–	7	–	–	–	–	–	–	–	Octadecylamine, HFB		MS10065
400	428	323	429	324	352	401	399	**465**	100	83	83	48	39	35	26	26	**0.00**	25	26	–	3	–	–	1	–	–	–	2	–	–	Aminodiphenylphosphino-methylaminodiphenyl-phosphoranylidinimine chloride		LI09202
204	178	207	203	91	210	205	176	**465**	100	26	21	20	20	19	19	15	**11.00**	26	27	3	1	2	–	–	–	–	–	–	–	–	3'-Isobutoxy-2'-tosylspiro[fluorene-9,5'-(1,3)thiazolidine]		MS10066
323	325	269	143	234	69	465	430	**465**	100	35	25	20	13	10	8	8		26	40	4	1	–	–	1	–	–	–	–	–	–	4-Tetradecenamide, N-[2-chloro-3-(3-methyl-2-oxo-7-oxabicyclo[4.1.0]hept-3-en-1-yl)-2-propenyl]-7-methoxy-N-methyl-	69121-75-1	NI43821
105	77	91	302	411	79	28	197	**465**	100	44	27	25	17	14	14	13	**0.00**	27	35	4	3	–	–	–	–	–	–	–	–	–	(2S,5S)-1-Benzoyl-5-[4-[(benzyloxycarbonyl)amino]butyl]-2-tert-butyl-3-methyl-4-imidazolidinone		DD01717
105	196	91	408	286	77	106	409	**465**	100	42	37	25	15	13	8	6	**0.00**	27	35	4	3	–	–	–	–	–	–	–	–	–	(2S,5R)-1-Benzoyl-5-[3-[(benzyloxycarbonyl)amino]propyl]-2-tert-butyl-3,5-dimethyl-4-imidazolidinone		DD01718
225	183	195	168	238	182	43	42	**465**	100	81	21	21	20	17	17	17	**2.90**	27	47	5	1	–	–	–	–	–	–	–	–	–	2-Pyridinedecanol, α-sec-butyl-4-hydroxy-5,6-dimethoxy-β,δ,θ,3-tetramethyl-, 4-acetate	20238-88-4	NI43822
91	73	129	75	268	358	207	105	**465**	100	47	45	24	22	19	15	14	**2.48**	29	43	2	1	–	–	–	–	–	1	–	–	–	Androst-5-en-17-one, 3-[(trimethylsilyl)oxy]-, O-benzyloxime, (3β)-	57305-10-9	NI43823
317	119	91	77	274	198	104	64	**465**	100	49	36	32	29	17	17	16	**0.00**	30	31	2	3	–	–	–	–	–	–	–	–	–	4-Benzyl-8-(diethylamino)-7-methyl-2,6-diphenyl-2,6-diazabicyclo[2.2.2]oct-7-ene-3,5-dione		DD01719
403	468	389	377	310	298	338	324	**466**	100	65	59	32	20	8	6	3	**0.00**	10	8	–	–	–	–	–	2	–	–	–	–	1	Hafnium, dibromobis(π-cyclopentadienyl)-		LI09203
150	74	75	310	389	391	308	98	**466**	100	48	44	44	24	23	22	21	**1.51**	12	6	–	–	–	–	–	4	–	–	–	–	–	Biphenyl, 2,2',4',5-tetrabromo-	60044-24-8	NI43824
150	75	74	310	155	98	62	50	**466**	100	54	52	34	22	22	21	21	**1.04**	12	6	–	–	–	–	–	4	–	–	–	–	–	Biphenyl, 2,2',5,5'-tetrabromo-	59080-37-4	NI43825
150	74	75	310	98	62	155	389	**466**	100	49	48	35	22	20	19	19	**1.30**	12	6	–	–	–	–	–	4	–	–	–	–	–	Biphenyl, 2,2',5,6'-tetrabromo-	60044-25-9	NI43826
470	150	468	472	310	74	75	308	**466**	100	88	65	63	54	29	26	21	**7.86**	12	6	–	–	–	–	–	4	–	–	–	–	–	Biphenyl, 3,3',5,5'-tetrabromo-	16400-50-3	NI43827
451	73	452	453	466	299	207	45	**466**	100	83	35	21	17	15	10	10		12	36	7	–	–	–	–	–	–	4	2	–	–	Silanol, trimethyl-, pyrophosphate (4:1)	18395-45-4	NI43828
73	451	452	45	453	299	75	74	**466**	100	61	21	15	12	12	10	9	**8.92**	12	36	7	–	–	–	–	–	–	4	2	–	–	Silanol, trimethyl-, pyrophosphate (4:1)	18395-45-4	NI43829
186	410	166	180	466	195			**466**	100	31	29	25	18	12				13	5	2	–	–	9	–	–	–	–	–	–	1	π-Heptadienyldicarbonylnonafluorocyclohexeneruthenium		LI09204
119	382	142	219	339	466	263	176	**466**	100	28	27	15	9	8	7	7		15	36	3	6	–	–	–	–	–	–	2	–	1	Bis[tris(dimethylamino)phosphine]tricarbonyl iron		MS10067

MASS TO CHARGE RATIOS								M.W.	INTENSITIES								Parent	C	H	O	N	S	F	Cl	Br	I	Si	P	B	X	COMPOUND NAME	CAS Reg No	No
89	76	382	466	410	131	176	263	466	100	72	62	33	25	19	17	15		15	36	3	6	–	–	–	–	–	–	2	–	1	Bis[tris(dimethylamino)phosphine]tricarbonyl iron		NI43830
119	382	76	410	219	466	263	163	466	100	72	65	32	25	22	15	14		15	36	3	6	–	–	–	–	–	–	2	–	1	Bis[tris(dimethylamino)phosphine]tricarbonyl iron		MS10068
182	117	52	466	234	318	145	173	466	100	69	64	40	35	33	25	23		16	10	6	–	–	–	–	–	–	–	–	–	3	Bis(cyclopentadienyltricarbonylchromium)zinc		MS10069
451	155	159	125	91	77	81	28	466	100	68	65	56	56	56	53	51	3.00	16	12	1	–	–	10	–	–	–	2	–	–	–	Disiloxane, 1,1,3,3-tetramethyl-1,3-bis(pentafluorophenyl)-	19091-32-8	NI43831
139	246	187	202	63	107	108	203	466	100	90	80	66	55	45	27	23	0.50	16	20	4	–	4	–	–	–	–	–	2	–	–	O,O′-Ethylenebis(4-methoxyphenyldithiophosphonate)	94286-62-1	NI43832
466	125	93	203	109	79	467	47	466	100	80	66	62	60	34	22	21		16	20	6	–	3	–	–	–	–	–	2	–	–	Phosphorothioic acid, O,O′-(thiodi-4,1-phenylene) O,O,O′,O′-tetramethyl ester	3383-96-8	NI43835
466	125	93	47	109	79	15	63	466	100	51	38	35	31	28	21	18		16	20	6	–	3	–	–	–	–	–	2	–	–	Phosphorothioic acid, O,O′-(thiodi-4,1-phenylene) O,O,O′,O′-tetramethyl ester	3383-96-8	NI43833
125	109	79	93	46	203	341	63	466	100	45	36	30	24	20	18	14	0.00	16	20	6	–	3	–	–	–	–	–	2	–	–	Phosphorothioic acid, O,O′-(thiodi-4,1-phenylene) O,O,O′,O′-tetramethyl ester	3383-96-8	NI43834
184	169	69	223	185	41	183	224	466	100	76	74	42	41	25	19	12	0.00	17	21	5	2	–	7	–	–	–	–	–	–	–	Butanoic acid, heptafluoro-, 3-(5-ethylhexahydro-1,3-dimethyl-2,4,6-trioxo-5-pyrimidinyl)-1-methylbutyl ester	55429-80-6	NI43836
73	225	208	75	224	226	451	361	466	100	61	22	22	21	12	2	2	0.02	17	28	5	2	–	–	2	–	–	2	–	–	–	Chloramphenicol bis(trimethylsilyl) ether		LI09205
225	73	226	208	75	224	227	93	466	100	53	17	17	12	10	5	5	0.00	17	28	5	2	–	–	2	–	–	2	–	–	–	Chloramphenicol bis(trimethylsilyl) ether		NI43837
73	225	208	75	224	226	74	243	466	100	57	23	23	12	10	9	4	0.01	17	28	5	2	–	–	2	–	–	2	–	–	–	Chloramphenicol bis(trimethylsilyl) ether		LI09206
105	45	44	61	147	107	60	59	466	100	84	26	22	11	9	9	8	0.00	18	26	10	–	2	–	–	–	–	–	–	–	–	D-Glucose, cyclic 1,2-ethanediyl mercaptal, pentaacetate	17429-98-0	NI43838
57	466	29	116	41	468	172	128	466	100	61	61	55	55	35	32	26		18	36	–	2	4	–	–	–	–	–	–	–	1	Nickel bis(dibutylcarbamodithioato-S,S′)-		IC04474
73	147	217	205	103	75	74	189	466	100	27	26	11	10	9	9	8	3.10	18	42	6	–	–	–	–	–	–	4	–	–	–	Allonic acid, 2,3,5,6-tetrakis-O-(trimethylsilyl)-, lactone	55515-34-9	NI43839
73	217	147	103	205	75	218	74	466	100	44	23	12	11	11	9	9	3.70	18	42	6	–	–	–	–	–	–	4	–	–	–	Altronic acid, 2,3,5,6-tetrakis-O-(trimethylsilyl)-, lactone	55528-73-9	NI43840
73	305	422	147	451	361	221	133	466	100	40	33	32	28	26	18	17	0.00	18	42	6	–	–	–	–	–	–	4	–	–	–	Ditrimethylsilyl trimethylsiloxypropyl(trimethylsiloxy)dimalonate		MS10070
73	204	230	147	217	75	258	74	466	100	28	17	16	15	15	9	8	0.00	18	42	6	–	–	–	–	–	–	4	–	–	–	2-Furanacetaldehyde, tetrahydro-α,3,4,5-tetrakis[(trimethylsilyl)oxy]-	74685-70-4	NI43841
73	217	147	103	205	189	219	218	466	100	93	31	28	26	20	20	18	7.10	18	42	6	–	–	–	–	–	–	4	–	–	–	Galactonic acid 1,4-lactone, tetrakis(trimethylsilyl)-		NI43842
217	73	466	205	147	334	219	335	466	100	81	33	32	28	27	24	21		18	42	6	–	–	–	–	–	–	4	–	–	–	D-Galactonic acid, 2,3,5,6-tetrakis-O-(trimethylsilyl)-, γ-lactone	32384-64-8	NI43843
73	217	147	205	103	189	219	75	466	100	44	20	16	15	12	11	11	6.91	18	42	6	–	–	–	–	–	–	4	–	–	–	Galactonic acid, 2,3,5,6-tetrakis-O-(trimethylsilyl)-, lactone	55528-74-0	NI43844
73	319	147	129	220	75	204	103	466	100	26	22	13	11	11	9	9	2.50	18	42	6	–	–	–	–	–	–	4	–	–	–	Gluconic acid, 2,3,4,6-tetrakis-O-(trimethylsilyl)-, lactone	55515-29-2	NI43845
73	217	147	205	204	103	75	218	466	100	42	21	12	10	10	9	8	1.90	18	42	6	–	–	–	–	–	–	4	–	–	–	Gluconic acid, 2,3,5,6-tetrakis-O-(trimethylsilyl)-, lactone	55515-33-8	NI43846
73	217	147	204	103	219	205	189	466	100	44	38	20	20	18	18	17	9.81	18	42	6	–	–	–	–	–	–	4	–	–	–	L-Gluconic acid, 2,3,5,6-tetrakis-O-(trimethylsilyl)-, lactone	56298-43-2	NI43847
73	217	147	103	219	205	204	189	466	100	25	25	15	13	13	13	11	9.80	18	42	6	–	–	–	–	–	–	4	–	–	–	Gulonic acid, 2,3,5,6-tetrakis-O-(trimethylsilyl)-, lactone	55528-75-1	NI43848
204	73	147	205	129	293	363	217	466	100	82	21	19	19	14	12	11	0.71	18	42	6	–	–	–	–	–	–	4	–	–	–	threo-2,5-Hexodiulose, 1,3,4,6-tetrakis-O-(trimethylsilyl)-, lactone	56114-57-9	NI43849
73	217	147	103	205	189	244	75	466	100	43	25	19	15	11	10	10	2.30	18	42	6	–	–	–	–	–	–	4	–	–	–	Idonic acid, 2,3,5,6-tetrakis-O-(trimethylsilyl)-, lactone	55515-32-7	NI43850
73	147	319	129	220	217	103	74	466	100	24	21	13	10	9	9	9	1.10	18	42	6	–	–	–	–	–	–	4	–	–	–	Mannoonic acid, 2,3,4,6-tetrakis-O-(trimethylsilyl)-, lactone	55515-28-1	NI43851
73	217	147	218	103	75	74	219	466	100	59	25	12	10	8	8	7	1.70	18	42	6	–	–	–	–	–	–	4	–	–	–	Mannoonic acid, 2,3,5,6-tetrakis-O-(trimethylsilyl)-, lactone	55515-30-5	NI43853
73	147	217	75	275	189	143	117	466	100	30	25	23	14	14	14	12	0.00	18	42	6	–	–	–	–	–	–	4	–	–	–	Mannoonic acid, 2,3,5,6-tetrakis-O-(trimethylsilyl)-, lactone	55515-30-5	NI43852
73	259	147	233	260	75	43	45	466	100	74	40	25	17	14	12	10	0.00	18	42	6	–	–	–	–	–	–	4	–	–	–	2-Methyl-2,4-bis(trimethylsiloxy)bis(trimethylsilyl)glutarate		NI43854
73	217	147	103	205	74	361	189	466	100	29	21	12	10	8	7	7	3.10	18	42	6	–	–	–	–	–	–	4	–	–	–	Talonic acid, 2,3,5,6-tetrakis-O-(trimethylsilyl)-, lactone	55515-31-6	MS10071
73	217	147	103	205	361	74	75	466	100	29	21	12	10	8	8	7	3.10	18	42	6	–	–	–	–	–	–	4	–	–	–	Talonic acid, 2,3,5,6-tetrakis-O-(trimethylsilyl)-, lactone	55515-31-6	NI43855
299	268	151	69	230	198	108	110	466	100	70	70	27	23	23	23	20	10.00	20	5	–	–	–	10	–	–	–	–	1	–	–	Phosphine, bis(pentafluorophenyl)(phenylethynyl)-	33730-53-9	NI43856
466	77	233	118	224	234	159	144	466	100	59	57	37	36	35	34	34		20	18	2	8	2	–	–	–	–	–	–	–	–	Bis(2-anilino-4-methoxy-s-triazin-6-yl) disulphide		IC04475
466	176	271	107	467	319	465	183	466	100	30	28	26	23	20	12	12		21	12	–	–	–	9	–	–	–	–	1	–	–	Phosphine, tris[4-(trifluoromethyl)phenyl]-	13406-29-6	NI43857
69	48	157	311	116	110	310	45	466	100	46	42	41	33	28	27	27	0.00	21	22	3	8	1	–	–	–	–	–	–	–	–	4-(3-Methoxyphenyl)-5-(3-methoxyphenyl)amino-3-[6-methyl-3-(methylthio)-5-oxo-4,5-dihydro-1,2,4-triazin-4-yl]imino-2H-2,3-dihydro-1,2,4-triazole		MS10072
69	311	296	48	157	45	116	47	466	100	53	41	40	39	33	32	32	0.00	21	22	3	8	1	–	–	–	–	–	–	–	–	4-(4-Methoxyphenyl)-5-(4-methoxyphenyl)amino-3-[6-methyl-3-(methylthio)-5-oxo-4,5-dihydro-1,2,4-triazin-4-yl]imino-2H-2,3-dihydro-1,2,4-triazole		MS10073
166	167	98	41	172	29	365	70	466	100	67	54	42	39	26	22	19	1.10	21	33	6	2	–	3	–	–	–	–	–	–	–	N-Trifluoroacetyl-L-prolyl-α-methylglutamic acid dibutyl ester		MS10074
240	238	236	242	410	234	438	466	466	100	52	48	36	33	29	8	7		22	19	2	–	–	5	–	–	–	–	–	–	1	Iron, dicarbonyl[(4a,4b,9a,10,10a-η)-1,3,4,5,6,7,8,9-octahydrobenz[a]azulen-4a(2H)-yl](pentafluorophenyl)-	41684-55-3	MS10075
466	422	409	467	410	356	407	448	466	100	48	41	23	23	23	18	16		22	21	3	–	–	7	–	–	–	–	–	–	–	Estra-1,3,5(10)-trien-3-ol, 17-oxo-, heptafluorobutyrate		MS10076
229	228	331	195	151	121	271	197	466	100	50	41	31	31	31	27	18	0.00	22	26	3	–	4	–	–	–	–	–	–	–	–	Carbonodithioic acid, S,S′-[(4-methoxyphenyl)benzylene] O,O′-diisopropyl ester	26416-50-2	NI43858
109	169	127	331	139	103	97	145	466	100	98	72	54	40	38	38	34	0.00	22	26	11	–	–	–	–	–	–	–	–	–	–	4-Acetylphenyl-β-D-2,3,4,6-tetracetylglucopyranoside		MS10077
89	133	87	73	59	177	97	91	466	100	32	32	27	23	20	18	17	0.00	22	42	10	–	–	–	–	–	–	–	–	–	–	1,1′-bis(3,6,9,12,15-pentaoxacyclohexadecane)		MS10078
89	133	73	87	59	228	97	83	466	100	66	42	41	35	25	23	18	0.00	22	42	10	–	–	–	–	–	–	–	–	–	–	3-(3,6,9,12-Tetraoxacyclotridecyl)-1,5,8,11,14,17-hexaoxacyclononadecane		MS10079

MASS TO CHARGE RATIOS								M.W.	INTENSITIES								Parent	C	H	O	N	S	F	Cl	Br	I	Si	P	B	X	COMPOUND NAME	CAS Reg No	No
258	231	407	203	351	339	216	175	**466**	100	75	60	60	50	40	40	30	**1.00**	23	30	10	–	–	–	–	–	–	–	–	–	–	Dihydroenhydrin	33880-86-3	MS10080
362	43	135	465	264	110	363	151	**466**	100	67	39	35	35	35	34	26	**12.00**	23	32	9	1	–	–	–	–	–	–	–	–	–	Senecionanium, 12,14-bis(acetyloxy)-8-hydroxy-4-methyl-11,16-dioxo-, (12ξ,13ξ)-	34429-54-4	NI43859
131	174	92	187	188	57	86	100	**466**	100	78	67	44	43	41	41	36	**3.40**	23	34	8	2	–	–	–	–	–	–	–	–	–	12-Benzyl-10,19-dimethyl-1,4,7,13,16-pentaoxa-10,19-diazacycloheneicosane-11,14,18-trione		MS10081
141	466	152	326	65	185	93	467	**466**	100	94	57	42	34	30	30	26		24	18	6	–	2	–	–	–	–	–	–	–	–	4,4′-Bis(4-hydroxyphenylsulphonyl)diphenyl		IC04476
91	106	77	126	92	128	127	107	**466**	100	88	11	10	9	8	8	8	**7.00**	24	22	4	2	2	–	–	–	–	–	–	–	–	1,5-Bis(benzylaminosulphonyl)naphthalene		IC04477
105	111	222	77	110	106	315	209	**466**	100	36	13	10	9	8	7	7	**0.00**	24	28	8	–	–	–	–	–	–	–	–	2	–	D-Glucitol, cyclic 2,4:3,5-bis(ethylboronate) 1,6-dibenzoate	74793-53-6	NI43860
364	180	121	278	43	185	105	125	**466**	100	81	72	62	62	56	45	44	**2.66**	24	34	9	–	–	–	–	–	–	–	–	–	–	Trichothec-9-ene-3,4,8,15-tetrol, 12,13-epoxy-, 4,15-diacetate 8-isopentanoate, (3α,4β,8α)-	21259-20-1	NI43861
127	325	55	466	77	380	154	99	**466**	100	98	58	37	30	25	23	19		25	26	5	2	1	–	–	–	–	–	–	–	–	trans-3-Ethyl-2,2-(ethylenedioxy)-12-(phenylsulphonyl)-1,2,3,4,12,12b-hexahydro-7H-indolo[2,3-a]quinolizin-6-one		DD01720
268	120	138	269	43	28	121	105	**466**	100	77	52	50	44	44	42	40	**32.00**	25	38	8	–	–	–	–	–	–	–	–	–	–	Pregn-5-ene-3,8,11,12,14,20-hexol, 11,12-diacetate, (3β,11α,12β,14β)-	27526-89-2	LI09207
268	120	138	269	43	28	121	105	**466**	100	56	52	50	44	43	40	38	**31.04**	25	38	8	–	–	–	–	–	–	–	–	–	–	Pregn-5-ene-3,8,11,12,14,20-hexol, 11,12-diacetate, (3β,11α,12β,14β)-	27526-89-2	NI43862
91	467	65	92	468				**466**	100	26	13	11	8				**0.00**	26	26	4	–	2	–	–	–	–	–	–	–	–	Diethyl 2,5-bis(benzylthio)terephthalate		LI09208
137	59	83	85	151	152	43	125	**466**	100	76	71	48	35	29	25	23	**0.00**	26	26	8	–	–	–	–	–	–	–	–	–	–	3-Methoxy-4-O-acetylbenzyl-2′-methoxyphenyl-4′-benzyl-2″-methoxy-4″-formylphenyl ether		NI43863
466	465	184	434	433	170	169	467	**466**	100	61	54	38	34	33	30	28		26	30	6	2	–	–	–	–	–	–	–	–	–	Methyl 3β,20α-17-hydroxy-19α-(2-methylacetoacetyl)yohimb-16-ene-16-carboxylate		LI09209
434	466	433	435	169	156	467	170	**466**	100	77	62	38	34	33	23	23		26	30	6	2	–	–	–	–	–	–	–	–	–	Methyl 3β,20α-17-hydroxy-19β-(2-methylacetoacetyl)yohimb-16-ene-16-carboxylate		LI09210
91	69	139	107	343	153	419	165	**466**	100	47	25	20	14	13	12	11	**0.10**	26	42	5	–	1	–	–	–	–	–	–	–	–	Methyl (±)-2,4,6,8,10-pentadeoxy-4,6,8,10-tetramethyl-7,9-O-(1-methylethylidene)-11-O-(phenylmethyl)-1-thio-D-arabino-β-L-manno-undecopyranosid-3-ulose		NI43864
410	411	55	183	148	70	262	263	**466**	100	28	24	16	10	9	9	8	**3.99**	27	24	2	–	–	–	–	–	–	–	1	–	1	(Ethylcyclopentadienyl)manganesedicarbonyltriphenylphosphine	33336-99-1	NI43865
162	466	467	163	91	133	118	451	**466**	100	69	22	19	19	17	16	14		27	26	2	6	–	–	–	–	–	–	–	–	–	1-(4-Nitrophenylazo)-4-(4-diethylamino-2-methylphenylazo)naphthalene		IC04478
135	343	451	466	423	371	150	41	**466**	100	93	67	58	40	36	28	28		27	30	5	–	1	–	–	–	–	–	–	–	–	Thymol blue		IC04479
86	100	58	87	56	57	72	84	**466**	100	39	16	12	9	9	8	6	**1.80**	27	38	3	4	–	–	–	–	–	–	–	–	–	9-Acetylhydrazono-2,7-bis[2-(diethylamino)-ethoxy]-fluorene		MS10082
328	44	43	31	329	28	370	137	**466**	100	60	44	28	24	21	19	17	**0.10**	28	34	6	–	–	–	–	–	–	–	–	–	–	Deoxygedunin		MS10083
149	167	57	71	70	150	87	113	**466**	100	32	20	17	13	12	12	10	**1.00**	29	42	3	2	–	–	–	–	–	–	–	–	–	2,7-Bis[4-(diethylamino)-butoxy]fluorenone		MS10084
226	225	189	227	228	240	191	239	**466**	100	82	42	40	33	31	27	20	**14.00**	30	20	1	–	–	–	2	–	–	–	–	–	–	Bis(10-chloro-9-anthryl) methyl ether		LI09211
260	261	215	41	55	91	178	135	**466**	100	78	77	76	60	58	55	55	**15.43**	30	42	4	–	–	–	–	–	–	–	–	–	–	Olean-12-en-28-oic acid, 21-hydroxy-3,16-dioxo-, γ-lactone, (21β)-	30993-62-5	NI43866
451	452	466	453	467	468	423	295	**466**	100	36	22	11	8	5	2	1		30	46	2	–	–	–	–	–	–	1	–	–	–	Cannabinol tripropylsilyl ether		NI43867
466	423	95	43	159	57	241	55	**466**	100	72	62	55	48	48	44	40		31	46	3	–	–	–	–	–	–	–	–	–	–	Naphth[2,3-b]oxirene-2,7-dione, 1a,7a-dihydro-1a-methyl-7a-(3,7,11,15-tetramethyl-2-hexadecenyl)-	25486-55-9	NI43868
466	423	443	159	57	467	55	95	**466**	100	72	55	47	47	40	40	38		31	46	3	–	–	–	–	–	–	–	–	–	–	Naphth[2,3-b]oxirene-2,7-dione, 1a,7a-dihydro-1a-methyl-7a-(3,7,11,15-tetramethyl-2-hexadecenyl)-	25486-55-9	LI09212
423	135	466	451	424	317	121	108	**466**	100	60	38	30	30	28	25	23		31	46	3	–	–	–	–	–	–	–	–	–	–	A-Neooleana-3(5),12-dien-29-oic acid, 11-oxo-, methyl ester, (20β)-	5092-06-8	NI43869
135	316	121	451	175	257	105	107	**466**	100	88	43	40	34	33	32	30	**28.00**	31	46	3	–	–	–	–	–	–	–	–	–	–	A-Neooleana-3(5),12-dien-30-oic acid, 11-oxo-, methyl ester, (18α)-	5573-15-9	NI43870
135	423	466	397	120	189	175	107	**466**	100	80	75	52	48	35	35	32		31	46	3	–	–	–	–	–	–	–	–	–	–	A-Neooleana-3,12-dien-29-oic acid, 11-oxo-, methyl ester, (20β)-	5092-03-5	NI43871
135	451	466	121	119	105	107	276	**466**	100	92	90	46	40	40	37	36		31	46	3	–	–	–	–	–	–	–	–	–	–	A-Neooleana-3,12-dien-30-oic acid, 11-oxo-, methyl ester, (18α)-	6471-55-2	NI43872
135	121	451	105	119	107	466	133	**466**	100	58	55	53	52	48	38	32		31	46	3	–	–	–	–	–	–	–	–	–	–	Oleana-2,12-dien-29-oic acid, 11-oxo-, methyl ester, (20α)-	55555-62-9	NI43873
135	466	451	276	121	119	105	107	**466**	100	87	72	50	43	42	38	36		31	46	3	–	–	–	–	–	–	–	–	–	–	Oleana-2,12-dien-29-oic acid, 11-oxo-, methyl ester, (20β)-	10301-75-4	NI43874
73	75	81	55	43	41	123	69	**466**	100	78	75	73	70	64	63	53	**21.50**	31	50	1	–	–	–	–	–	–	1	–	–	–	Ergosta-5,7,22,24(28)-tetraene, 3-[(trimethylsilyl)oxy]-, (3β)-		LI09213
74	43	87	57	55	75	41	143	**466**	100	89	77	50	40	39	36	26	**4.00**	31	62	2	–	–	–	–	–	–	–	–	–	–	Triacontanoic acid, methyl ester	629-83-4	NI43875
465	467	466	468	143	145	171	131	**466**	100	85	55	50	33	30	29	28		31	62	2	–	–	–	–	–	–	–	–	–	–	Triacontanoic acid, methyl ester	629-83-4	NI43876
220	78	246	219	245	142	116	221	**466**	100	58	56	48	37	31	20	19	**7.01**	32	26	–	4	–	–	–	–	–	–	–	–	–	Pyrazolo[3″,4″:3′,4′]cyclobuta[1′,2′:3,4]cyclobuta[1,2-c]pyrazole, 1,3a,3b,3c,6,6a,6b,6c-octahydro-1,3,4,6-tetraphenyl-	34835-67-1	NI43877
231	81	466	91	260	167	109	191	**466**	100	67	40	35	32	32	32	19		32	34	3	–	–	–	–	–	–	–	–	–	–	(1R,2R,4R,5R,8S,9S,11S,12S,16S)-5-Hydroxy-14-(diphenylmethylidene)-1,4-dimethylpentacyclo[9.6.0.0^{2,9}.0^{4,8}.0^{12,16}]heptadeca-3,17-dione		DD01721
231	109	91	167	466	260	191	215	**466**	100	35	35	32	29	26	21	19		32	34	3	–	–	–	–	–	–	–	–	–	–	(1R,2R,4R,5S,8S,9S,11S,12S,16S)-5-Hydroxy-14-(diphenylmethylidene)-1,4-dimethylpentacyclo[9.6.0.0^{2,9}.0^{4,8}.0^{12,16}]heptadeca-3,17-dione		DD01722
466	255	133	313	149	391	203	295	**466**	100	76	26	20	18	16	12	10		32	50	2	–	–	–	–	–	–	–	–	–	–	3β-Acetoxyursan-9(11),12-diene	41607-46-9	NI43878
466	467	253	240	353	391	313	406	**466**	100	38	28	23	22	16	15	14		32	50	2	–	–	–	–	–	–	–	–	–	–	Lanosta-7,9(11),24-triene, 3-(acetyloxy)-, (3β)-		MS10085
407	406	229	189	203	187	285	391	**466**	100	95	86	60	57	37	35	29	**27.00**	32	50	2	–	–	–	–	–	–	–	–	–	–	Lupa-1,20(29)-diene, 3-(acetyloxy)-, (3β)-		NI43879
57	55	83	69	97	71	56	85	**466**	100	84	63	62	48	48	39	29	**0.00**	32	66	1	–	–	–	–	–	–	–	–	–	–	1-Dotriacontanol	6624-79-9	WI01565

MASS TO CHARGE RATIOS	M.W.	INTENSITIES	Parent	C	H	O	N	S	F	Cl	Br	I	Si	P	B	X	COMPOUND NAME	CAS Reg No	No
225 57 43 71 83 69 55 97	**466**	100 91 70 52 42 39 39 30	**3.03**	32	66	1	–	–	–	–	–	–	–	–	–	–	**Hexadecane, 1,1′-oxybis-**	4113-12-6	**NI43880**
57 69 43 83 71 55 126 85	**466**	100 93 84 75 58 58 48 41	**0.04**	32	66	1	–	–	–	–	–	–	–	–	–	–	**1-Octacosanol, 2,4,6,8-tetramethyl-, (all-R)-**	27829-63-6	**NI43881**
94 69 56 81 41 121 109 123	**466**	100 65 61 57 55 41 40 30	**22.00**	34	58	–	–	–	–	–	–	–	–	–	–	–	**Isobo-tryococcene**		**LI09214**
43 57 85 41 29 353 269 129	**466**	100 19 15 12 10 9 7 7	**0.00**	34	58	–	–	–	–	–	–	–	–	–	–	–	**Naphthalene, tetrahexyldihydro-**	74663-70-0	**NI43882**
230 43 67 41 68 176 231 394	**467**	100 28 28 13 13 11 10 9	**0.00**	14	17	4	2	–	10	–	–	–	–	–	–	–	**N,N′-Bis(pentafluoropropionyl)-D-lysine isopropyl ester**		**MS10086**
230 43 67 28 231 394 41 176	**467**	100 24 21 14 11 11 11 10	**0.00**	14	17	4	2	–	10	–	–	–	–	–	–	–	**N,N′-Bis(pentafluoropropionyl)-L-lysine isopropyl ester**		**MS10087**
134 135 117 389 387 192 193 213	**467**	100 11 10 8 8 7 6 5	**0.00**	16	11	–	3	2	–	–	2	–	–	–	–	–	**2-(4-Bromophenyl)amino-5-phenylthiazolo[2,3-b]-1,3,4-thiadiazol-4-ium bromide**		**MS10088**
73 103 307 217 147 160 75 74	**467**	100 64 25 19 18 13 7 7	**0.00**	18	45	5	1	–	–	–	–	–	4	–	–	–	**Butanal, 2,3,4-tris[(trimethylsilyl)oxy]-3-[[(trimethylsilyl)oxy]methyl]-, O methyloxime, (S)-**	56196-22-6	**NI43883**
131 73 132 144 75 147 74 103	**467**	100 63 12 11 10 9 8 7	**0.86**	18	45	5	1	–	–	–	–	–	4	–	–	–	**D-Galactosamine, tetrakis(trimethylsilyl)-**		**NI43884**
131 73 144 132 147 103 232 204	**467**	100 72 13 12 9 8 6 6	**1.00**	18	45	5	1	–	–	–	–	–	4	–	–	–	**D-Galactose, 2-amino-2-deoxy-3,4,5,6-tetrakis-O-(trimethylsilyl)-**	56248-47-6	**LI09215**
131 73 144 132 75 147 103 74	**467**	100 72 13 13 13 9 8 6	**1.20**	18	45	5	1	–	–	–	–	–	4	–	–	–	**D-Galactose, 2-amino-2-deoxy-3,4,5,6-tetrakis-O-(trimethylsilyl)-**	56248-47-6	**NI43885**
131 73 144 132 147 232 204 75	**467**	100 32 11 11 8 6 6 6	**0.95**	18	45	5	1	–	–	–	–	–	4	–	–	–	**Galactose, 2-amino-2-dioxy-, tetrakis(trimethylsilyl)-**	56145-07-4	**NI43886**
131 73 132 75 144 74 147 204	**467**	100 62 12 10 9 8 8 7	**0.00**	18	45	5	1	–	–	–	–	–	4	–	–	–	**D-Glucosamine, tetrakis(trimethylsilyl)-**		**NI43887**
131 73 132 204 144 147 191 133	**467**	100 30 11 8 8 7 5 5	**0.00**	18	45	5	1	–	–	–	–	–	4	–	–	–	**D-Glucose, 2-amino-2-deoxy-, tetrakis(trimethylsilyl)-**	56193-55-6	**NI43888**
73 103 307 217 147 204 189 205	**467**	100 69 32 28 22 12 12 10	**0.00**	18	45	5	1	–	–	–	–	–	4	–	–	–	**D-Ribose, 2,3,4,5-tetrakis-O-(trimethylsilyl)-, O-methyloxime**	56196-08-8	**NI43889**
73 103 307 217 147 205 204 189	**467**	100 66 32 30 24 14 14 10	**0.00**	18	45	5	1	–	–	–	–	–	4	–	–	–	**D-Xylose, 2,3,4,5-tetrakis-O-(trimethylsilyl)-, O-methyloxime**	56196-07-7	**NI43890**
288 467 256 419 289 468 77 240	**467**	100 58 20 14 14 11 9 7		19	9	2	1	1	8	–	–	–	–	–	–	–	**2-Anilino-4′-methoxyoctafluorodiphenylsulphoxide**		**NI43891**
174 43 196 114 198 279 238 281	**467**	100 80 33 27 11 8 6 3	**1.00**	20	19	5	3	1	1	1	–	–	–	–	–	–	**Flucloxacillin methyl ester**		**MS10089**
137 467 102 232 139 469 152 103	**467**	100 35 32 31 27 25 14 13		23	19	–	5	1	–	2	–	–	–	–	–	–	**1-[N,N′-Bis(4-chlorophenyl)]guanidino-4-phenyl-2-methylthioimidazole**		**MS10090**
467 468 436 450 341 452 272 437	**467**	100 27 27 17 15 12 8 7		23	24	6	1	–	3	–	–	–	–	–	–	–	**B-Homomorphinan-7-one, 5,6,8,14-tetradehydro-2,3,4,6-tetramethoxy-17-(trifluoroacetyl)-, (9α,13α)-**	56793-04-5	**NI43892**
331 45 147 133 132 136 149 104	**467**	100 76 70 58 50 48 46 40	**20.00**	24	26	3	5	–	–	1	–	–	–	–	–	–	**N-Ethyl-N-(4-dimethylaminophenoxyethyl)-4-(2-chloro-4-nitrophenylazo)aniline**		**IC04480**
452 434 450 436 453 451 58 278	**467**	100 73 70 61 28 25 23 22	**15.00**	25	41	7	1	–	–	–	–	–	–	–	–	–	**Aconitane-1,7,8-triol, 20-ethyl-6,14,16-trimethoxy-4-(methoxymethyl)-, (1α,6β,14α,16β)-**	509-18-2	**NI43893**
225 73 117 75 238 452 195 153	**467**	100 75 52 30 16 11 9 9	**4.00**	25	49	3	1	–	–	–	–	–	2	–	–	–	**Pyridine, 3-methoxy-2-methyl-4-[(trimethylsilyl)oxy]-6-[11-[(trimethylsilyl)oxy]undecyl]-**	70001-17-1	**NI43894**
210 215 253 313 238 341 323 154	**467**	100 99 96 48 48 33 31 27	**19.00**	27	37	4	3	–	–	–	–	–	–	–	–	–	**L-Alanine, N-[N-[N-(1-naphthalenylmethylene)-L-valyl]-L-isoleucyl]-, ethyl ester**	57237-92-0	**NI43895**
215 253 323 313 154 210 116 230	**467**	100 88 28 21 20 16 9 8	**3.50**	27	37	4	3	–	–	–	–	–	–	–	–	–	**L-Alanine, N-[N-[N-(2-naphthalenylmethylene)-L-valyl]-L-isoleucyl]-, ethyl ester**	57237-93-1	**NI43896**
43 55 57 414 95 81 69 41	**467**	100 92 81 71 71 68 67 57	**0.32**	27	46	3	1	–	–	1	–	–	–	–	–	–	**Cholestan-3-ol, 5-chloro-6-nitro-, (3β,5α,6β)-**		**MS10091**
91 348 55 120 110 43 358 41	**467**	100 99 94 82 82 78 63 63	**14.00**	29	41	4	1	–	–	–	–	–	–	–	–	–	**1-Benzyl-6,7-dimethyl-3a-hydroxy-4-[4-methyl-7-(5-oxotetrahydrofuran-2-yl)heptyl]-3-oxo-3a,4,5,7a-tetrahydroisoindoline**		**IC04481**
91 75 270 73 360 96 77 41	**467**	100 32 28 26 20 13 12 11	**0.96**	29	45	2	1	–	–	–	–	–	1	–	–	–	**Androstan-17-one, 3-[(trimethylsilyl)oxy]-, O-benzyloxime, (3α,5α)-**	57305-13-2	**NI43897**
91 73 75 270 360 96 147 207	**467**	100 49 40 30 22 19 16 15	**1.68**	29	45	2	1	–	–	–	–	–	1	–	–	–	**Androstan-17-one, 3-[(trimethylsilyl)oxy]-, O-benzyloxime, (3α,5β)-**	57305-12-1	**NI43898**
376 467 377 91 158 468 159 130	**467**	100 37 28 24 15 13 13 4		32	37	2	1	–	–	–	–	–	–	–	–	–	**17-Oxo-5α-androstano[3,2-b][5′]benzyloxyindole**		**MS10092**
118 116 234 215 117 29 28 69	**468**	100 69 59 30 27 19 19 9	**0.00**	4	12	–	4	–	12	–	–	–	–	4	–	–	**Phosphorimidic trifluoride, methyl-, tetramer**	69782-83-8	**NI43899**
59 15 424 324 131 69 124 100	**468**	100 51 11 11 11 10 5 5	**0.00**	11	6	4	–	–	14	–	–	–	–	–	–	–	**Nonanedioic acid, tetradecafluoro-, dimethyl ester**	22116-90-1	**NI43900**
186 300 216 328 272 244 55 356	**468**	100 92 80 66 59 54 47 35	**19.00**	11	6	9	–	–	–	–	–	–	–	–	–	3	**Tetracarbonyliron-dimethylarsenic-pentacarbonylmanganese**		**NI43901**
294 292 296 279 238 470 277 281	**468**	100 56 50 23 22 14 14 13	**5.60**	16	22	6	–	–	–	–	2	–	–	–	–	–	**20,21-Dibromo-2,3-benzo-1,4,7,10,13,16-hexaoxacyclooctadecane**		**MS10093**
226 355 468 227 356 453 340 469	**468**	100 44 18 13 10 9 7 5		18	18	4	2	–	6	–	–	–	1	–	–	–	**L-Tryptophan, N,1-bis(trifluoroacetyl)-, trimethylsilyl ester**	52558-86-8	**NI43902**
163 146 308 104 205 162 469 324	**468**	100 63 30 29 26 21 20 20	**0.00**	18	36	10	4	–	–	–	–	–	–	–	–	–	**D-Streptamine, O-2-amino-2-deoxy-α-D-glucopyranosyl-(1-4)-O-[3-deoxy 3-(methylamino)-α-D-xylopyranosyl-(1-6)]-2-deoxy-**	13291-74-2	**NI43903**
169 81 109 180 141 167 115 468	**468**	100 76 61 54 52 37 37 30		20	24	11	2	–	–	–	–	–	–	–	–	–	**β-D-Glucopyranosylamine, N-(4-nitrophenyl)-, 2,3,4,6-tetraacetate**	26302-39-6	**LI09216**
43 169 468 180 109 331 118 115	**468**	100 23 11 11 11 10 8 8		20	24	11	2	–	–	–	–	–	–	–	–	–	**β-D-Glucopyranosylamine, N-(4-nitrophenyl)-, 2,3,4,6-tetraacetate**	26302-39-6	**MS10094**
169 81 109 180 140 167 468 331	**468**	100 76 62 54 53 38 31 30		20	24	11	2	–	–	–	–	–	–	–	–	–	**β-D-Glucopyranosylamine, N-(4-nitrophenyl)-, 2,3,4,6-tetraacetate**	26302-39-6	**NI43904**
88 201 101 75 45 319 233 73	**468**	100 84 77 47 25 18 12 12	**0.00**	20	36	12	–	–	–	–	–	–	–	–	–	–	**D-Glucopyranose, 4-O-β-D-glucopyranuronosyl-, heptamethyl-**		**LI09217**
88 101 75 71 45 367 89 201	**468**	100 77 62 17 11 10 9 8	**0.02**	20	36	12	–	–	–	–	–	–	–	–	–	–	**D-Glucopyranose, 6-O-β-D-glucopyranuronosyl-, heptamethyl-**		**LI09218**

MASS TO CHARGE RATIOS								M.W.	INTENSITIES								Parent	C	H	O	N	S	F	Cl	Br	I	Si	P	B	X	COMPOUND NAME	CAS Reg No	No
88	101	293	75	45	201	187	233	468	100	35	28	26	19	15	15	14	0.00	20	36	12	–	–	–	–	–	–	–	–	–	–	D-Glucopyranosiduronic acid, methyl 2,3-di-O-methyl-4-O-(2,3,4,6-tetra-O-methyl-α-D-glucopyranosyl)-, methyl ester	55836-34-5	NI43905
177	470	468	57	396	207	398	135	468	100	85	82	68	65	60	55	40		21	17	2	4	1	–	–	1	–	–	–	–	–	Shermilamine A		DD01723
157	199	241	395	255	213	297	283	468	100	93	68	60	54	37	34	25	0.00	21	48	7	–	–	–	–	–	–	2	–	–	–	Silicic acid, tributyl tripropyl ester	72347-77-4	NI43906
169	43	121	109	331	127	77	115	468	100	99	85	84	33	30	20	14	0.00	22	28	11	–	–	–	–	–	–	–	–	–	–	(1RS)-1-Phenyloxyethyl 2,3,4,6-tetra-O-acetyl-β-D-glucopyranoside		MS10095
384	43	426	83	298	324	85	309	468	100	63	31	31	26	22	21	17	2.50	23	16	11	–	–	–	–	–	–	–	–	–	–	8,9-Diacetoxy-2,3-dihydro-2-methoxyspiro[4H-benzofuro[2,3-g]-1-benzopyran-4,2'(5'H)-furan]-5,5',11-trione	80234-48-6	NI43907
329	269	43	410	408	330	93	287	468	100	28	27	24	24	24	21	17	9.20	23	33	5	–	–	–	–	1	–	–	–	–	–	Androst-6-one, 2-bromo-3,17-bis(acetyloxy)-, (2α,3β,5α,17β)-		MS10096
135	176	394	219	84	79	93	136	468	100	50	40	16	16	16	14	13	0.80	23	36	6	2	1	–	–	–	–	–	–	–	–	L-Glutamic acid, N-[N-(1-adamantylcarbonyl)-L-methionyl]-, dimethyl ester	28417-16-5	NI43908
234	236	235	164	66	470	84	238	468	100	92	33	28	28	24	21	20	10.00	24	12	–	2	–	–	4	–	–	–	–	–	–	9,9'-Bis(3,6-dichlorocarbazole)	18628-11-0	NI43909
219	405	83	404	362	85	468	220	468	100	75	42	34	30	28	21	20		25	24	9	–	–	–	–	–	–	–	–	–	–	8-(2-Methyl-5-hydroxy-7-O-methylchromone-6-ylmethyl)-2-methyl-5-hydroxy-6-methoxyethylchromone		LI09219
327	77	127	185	115	213	309	423	468	100	37	32	25	25	15	8	7	0.00	25	28	5	2	1	–	–	–	–	–	–	–	–	3-Ethyl-2,2-(ethylenedioxy)-7-hydroxy-12-(phenylsulphonyl)-1,2,3,4,6,7,12,12b-octahydroindolo[2,3-a]quinolizine		DD01724
73	468	45	77	57	218	125	74	468	100	15	14	11	10	9	9	9		25	36	3	2	–	–	–	–	–	2	–	–	–	3H-Pyrazol-3-one, 4-butyl-1,2-dihydro-2-phenyl-5-[(trimethylsilyl)oxy]-1-[4-[(trimethylsilyl)oxy]phenyl]-	74793-81-0	NI43910
73	75	77	246	93	45	117	218	468	100	22	21	16	12	12	11	10	1.00	25	36	3	2	–	–	–	–	–	2	–	–	–	3H-Pyrazol-3-one, 1,2-dihydro-1,2-diphenyl-5-[(trimethylsilyl)oxy]-4-[3-[(trimethylsilyl)oxy]butyl]-	74793-80-9	NI43911
383	104	149	384	132	468	338	76	468	100	29	28	26	25	22	12	12		26	28	8	–	–	–	–	–	–	–	–	–	–	Neopentyl glycol, terephthalic acid ester cyclic dimer		IC04482
414	432	450	396	322	285	321	303	468	100	60	50	50	50	50	45	25	0.00	27	48	6	–	–	–	–	–	–	–	–	–	–	5α-Cholestane-3β,5,6β,15α,16β,26-hexol	93368-83-3	NI43912
95	225	207	303	285	321	450	251	468	100	63	34	27	20	16	14	14	0.40	27	48	6	–	–	–	–	–	–	–	–	–	–	5α-Cholestane-3β,6α,8β,15α,16β,26-hexol		NI43913
414	267	285	432	450	303	321	349	468	100	90	70	60	50	50	30	25	0.00	27	48	6	–	–	–	–	–	–	–	–	–	–	5α-Cholestane-3β,6β,7α,15α,16β,26-hexol	93368-84-4	NI43914
243	244	105	112	241	242	166	69	468	100	25	21	20	17	16	16	14	2.00	28	24	5	2	–	–	–	–	–	–	–	–	–	6H-Furo[2',3':4,5]oxazolo[3,2-a]pyrimidin-6-one, 2,3,3a,9a-tetrahydro-3-hydroxy-2-[(triphenylmethoxy)methyl]-, [2S-(2α,3β,3aβ,9aβ)]-	3249-94-3	NI43915
468	425	411	356	299	426	355	412	468	100	98	98	59	50	40	40	37		28	36	6	–	–	–	–	–	–	–	–	–	–	1,2-Dihydro-5,9,11-trihydroxy-3,3-dimethyl-6,8-bis-(3-methylbutyl)-3H,12H-pyrano[3,2-a]xanthen-12-one		NI43916
109	125	135	121	319	136	107	147	468	100	44	39	34	22	22	11	10	3.00	28	36	6	–	–	–	–	–	–	–	–	–	–	(+)-Jaborol		MS10097
73	57	43	55	41	89	83	71	468	100	52	48	44	35	34	33	32	2.00	28	56	3	–	–	–	–	–	–	1	–	–	–	Tetracosenoic acid, 2-[(trimethylsilyl)oxy]-, methyl ester	56847-14-4	MS10098
468	363	469	241	105	195	364	163	468	100	37	30	17	13	11	10	10		29	24	6	–	–	–	–	–	–	–	–	–	–	2,2'-Dimethoxy-4,4'-dibenzoyl-5,5'-dihydroxydiphenylmethane		IC04483
468	469	379	127	234	470	380	227	468	100	33	15	13	12	7	7	7		30	16	4	2	–	–	–	–	–	–	–	–	–	N,N'-Di-(1-naphthyl)-pyromellitimide		MS10099
468	469	127	99	126	98	42	41	468	100	37	27	25	24	20	18	16		30	16	4	2	–	–	–	–	–	–	–	–	–	N,N'-Di-(2-naphthyl)-pyromellitimide		MS10100
468	57	85	43	323	41	55	83	468	100	79	66	53	49	49	48	46		30	44	4	–	–	–	–	–	–	–	–	–	–	Lanost-9(11)-en-18-oic acid, 20-hydroxy-3,23-dioxo-, (20ξ)-	36871-81-5	NI43917
468	440	437	149	269	453	302	287	468	100	4	4	4	3	2	2	2		30	44	4	–	–	–	–	–	–	–	–	–	–	Tetramethoxyturriane		LI09220
193	192	55	128	194	175	83	41	468	100	42	15	14	11	9	8	7	0.84	30	45	2	–	–	–	–	–	–	–	1	–	–	Phosphonous acid, 1-naphthalenyl-, bis[5-methyl-2-isopropylcyclohexyl] ester	74645-99-1	NI43918
240	121	468	133	225	223	228	108	468	100	81	67	49	47	40	34	34		31	32	4	–	–	–	–	–	–	–	–	–	–	Bis[2-hydroxy-3-(2-hydroxy-5-methyl benzyl)-5-methylphenyl]methane		LI09221
240	121	468	133	241	225	223	239	468	100	80	67	59	58	47	41	38		31	32	4	–	–	–	–	–	–	–	–	–	–	Bis[2-hydroxy-3-(2-hydroxy-5-methyl benzyl)-5-methylphenyl]methane		LI09222
218	219	203	135	133	122	109	107	468	100	20	17	16	16	16	14	14	4.60	31	48	3	–	–	–	–	–	–	–	–	–	–	Urs-12-en-24-oic acid, 3-oxo-, methyl ester, (+)-	20475-86-9	NI43919
203	262	189	202	204	133	119	249	468	100	44	28	26	18	17	14	11	8.62	31	48	3	–	–	–	–	–	–	–	–	–	–	Olean-12-en-28-oic acid, 3-oxo-, methyl ester	1721-58-0	NI43920
189	249	203	119	187	190	468	133	468	100	35	30	29	26	24	17	17		31	48	3	–	–	–	–	–	–	–	–	–	–	Olean-18-en-28-oic acid, 3-oxo-, methyl ester	55887-94-0	NI43921
363	337	468	364	378	253	131	69	468	100	60	58	32	30	29	29	29		31	52	1	–	–	–	–	–	–	1	–	–	–	Ergosta-5,7,22-triene, 3-[(trimethylsilyl)oxy]-, (3β)-	2625-45-8	MS10101
69	73	363	337	81	119	131	143	468	100	97	76	55	43	37	36	34	17.02	31	52	1	–	–	–	–	–	–	1	–	–	–	Ergosta-5,7,22-triene, 3-[(trimethylsilyl)oxy]-, (3β)-	2625-45-8	NI43922
99	117	98	381	76	65	77	64	468	100	87	27	24	18	18	16	16	0.00	31	64	2	–	–	–	–	–	–	–	–	–	–	2,6-Hentriacontanediol	7796-18-1	NI43923
468	469	391	105	196	313	77	51	468	100	35	34	27	12	9	7	7		32	20	4	–	–	–	–	–	–	–	–	–	–	5,6-Dibenzoyl-8-hydroxy-3-phenyl-2-oxatricyclo[7.4.0.0(4,7)]trideca-1(9),3,5,7,10,12-hexaene		NI43924
105	91	77	351	246	377	468	280	468	100	95	83	21	9	7	3	2		32	24	2	2	–	–	–	–	–	–	–	–	–	1,4-Bis(benzylimino)-3,6-diphenyl-1H,4H-furo[3,4-C]furan	85830-81-5	NI43925
77	468	345	450	205	105	128	217	468	100	44	27	22	17	17	13	6		32	24	2	2	–	–	–	–	–	–	–	–	–	1,2-Diphenyl-1,2-diquinol-2-yl-1,2-dihydroxy-ethane		LI09223
205	156	77	234	128	105	233	101	468	100	57	57	29	29	29	20	9	0.00	32	24	2	2	–	–	–	–	–	–	–	–	–	1,2-Diphenyl-1,2-diquinol-8-yl-1,2-dihydroxy-ethane		LI09224
356	468	412	413	357	233	167	469	468	100	70	60	33	33	28	27	25		32	36	3	–	–	–	–	–	–	–	–	–	–	2H,6H-Benzo[1,2-b:5,4-b']dipyran, 10-(3,3-diphenylallyl)-3,4,7,8-tetrahydro-5-methoxy-2,2,8,8-tetramethyl-	26537-42-8	NI43926
468	412	413	469	167	189	233	357	468	100	80	49	37	37	33	29	28		32	36	3	–	–	–	–	–	–	–	–	–	–	2H,8H-Benzo[1,2-b:3,4-b']dipyran, 6-(3,3-diphenyl-2-propenyl)-3,4,9,10-tetrahydro-5-methoxy-2,2,8,8-tetramethyl-	56336-12-0	NI43927
43	189	408	218	203	365	297	229	468	100	55	35	35	30	20	10	10	10.00	32	52	2	–	–	–	–	–	–	–	–	–	–	Lup-20(29)-en-3-ol, acetate, (3β)-	1617-68-1	NI43928
189	468	203	218	204	408	205		468	100	45	42	36	36	33	27			32	52	2	–	–	–	–	–	–	–	–	–	–	Lup-20(29)-en-3-ol, acetate, (3β)-	1617-68-1	LI09225
189	468	203	218	249	276	453	357	468	100	66	43	35	24	19	14	14		32	52	2	–	–	–	–	–	–	–	–	–	–	Lup-20(29)-en-3-ol, acetate, (3β)-	1617-68-1	NI43929

MASS TO CHARGE RATIOS								M.W.	INTENSITIES								Parent	C	H	O	N	S	F	Cl	Br	I	Si	P	B	X	COMPOUND NAME	CAS Reg No	No
218	189	203	468	408	453	393		**468**	100	50	35	15	10	6	6			32	52	2	–	–	–	–	–	–	–	–	–	–	Urs-12-en-3-ol, acetate, (3β)-	863-76-3	NI43930
218	203	189	468	408				**468**	100	18	16	9	7					32	52	2	–	–	–	–	–	–	–	–	–	–	Urs-12-en-3-ol, acetate, (3β)-	863-76-3	LI09226
134	133	135	231	191	408	187	232	**468**	100	92	60	57	50	40	23	12	**0.00**	32	52	2	–	–	–	–	–	–	–	–	–	–	18α,19βH-Urs-20-en-16-ol, acetate, (16α)-		NI43931
134	408	133	191	135	231	187	205	**468**	100	93	80	67	65	44	27	20	**0.00**	32	52	2	–	–	–	–	–	–	–	–	–	–	18α,19βH-Urs-20-en-16-ol, acetate, (16β)-		NI43932
408	134	191	135	133	203	202	187	**468**	100	48	26	26	24	16	16	15	**1.00**	32	52	2	–	–	–	–	–	–	–	–	–	–	18α,19βH-Urs-20-en-16-ol, acetate, (16β)-		NI43933
408	393	468	453	353	309	285	227	**468**	100	59	19	7	7	5	5	3		32	52	2	–	–	–	–	–	–	–	–	–	–	9,19-Cyclocholest-24-en-3-ol, 4,14,24-trimethyl-		NI43934
408	393	283	300	453	468	365	343	**468**	100	51	13	10	8	7	7	5		32	52	2	–	–	–	–	–	–	–	–	–	–	9,19-Cycloergost-24(28)-en-3-ol, 4,14-dimethyl-, acetate, (3β,4α,5α)-	10376-42-8	NI43935
408	393	286	339	271	297	468	255	**468**	100	52	45	30	25	21	20	15		32	52	2	–	–	–	–	–	–	–	–	–	–	9,19-Cyclolanost-24-en-3-ol, acetate, (3β)-	1259-10-5	NI43936
453	468	393	287	227	369	301	408	**468**	100	60	24	22	20	9	9	4		32	52	2	–	–	–	–	–	–	–	–	–	–	Ergosta-8,24(28)-dien-3-ol, 4,14-dimethyl-, acetate, (3β,4α,5α)-	55866-28-9	NI43937
453	468	393	369	227	287	384	301	**468**	100	55	29	11	10	8	4	4		32	52	2	–	–	–	–	–	–	–	–	–	–	Ergosta-8,24-dien-3-ol, 4,14-dimethyl-, acetate, (3β,4α,5α)-	64543-37-9	NI43938
468	469	43	240	393	253	313	288	**468**	100	34	31	24	23	21	19	19		32	52	2	–	–	–	–	–	–	–	–	–	–	Eupha-7,9(11)-dien-3-ol, acetate		MS10102
204	69	189	269	393	218	408	468	**468**	100	90	45	40	20	20	19	2		32	52	2	–	–	–	–	–	–	–	–	–	–	D-Friedoolean-14-en-3-ol, acetate, (3β)-	2189-80-2	NI43939
43	229	408	241	69	393	218	365	**468**	100	25	20	20	20	15	15	10	**7.00**	32	52	2	–	–	–	–	–	–	–	–	–	–	D:C-Friedoolean-7-en-3-ol, acetate, (3β)-	53298-81-0	NI43940
289	205	468	229	119	301	135	241	**468**	100	93	77	69	50	49	37	32		32	52	2	–	–	–	–	–	–	–	–	–	–	D:C-Friedours-8-en-3-ol, acetate, (3β)-		LI09227
408	296	409	255	133	283	145	107	**468**	100	60	32	30	26	24	24	24	**1.00**	32	52	2	–	–	–	–	–	–	–	–	–	–	Gorgost-5-en-3-ol, acetate, (3β)-	55724-22-6	NI43941
409	410	468	467	129	469	117	411	**468**	100	30	22	22	22	20	20	11		32	52	2	–	–	–	–	–	–	–	–	–	–	Lanosta-7,9(11)-dien-3-ol, acetate, (3β)-	5600-01-1	NI43942
468	253	240	469	313	393	408	453	**468**	100	43	40	38	27	26	20	17		32	52	2	–	–	–	–	–	–	–	–	–	–	Lanosta-7,9(11)-dien-3-ol, acetate, (3β)-	5600-01-1	MS10103
468	469	470	453	253	455	240	149	**468**	100	30	15	10	8	2	1	1		32	52	2	–	–	–	–	–	–	–	–	–	–	Lanosta-7,9(11)-dien-3-ol, acetate, (3β)-	5600-01-1	MS10104
453	468	393	241	301	255	408	229	**468**	100	79	62	14	13	11	10	10		32	52	2	–	–	–	–	–	–	–	–	–	–	Lanosta-8,24-dien-3-ol, acetate, (3β)-	2671-68-3	NI43943
468	453	69	469	109	97	83	81	**468**	100	52	52	35	26	26	26	26		32	52	2	–	–	–	–	–	–	–	–	–	–	Lanosta-8,24-dien-3-ol, acetate, (3β)-	2671-68-3	NI43944
393	109	468	69	454	394	95	469	**468**	100	70	60	40	35	33	28	22		32	52	2	–	–	–	–	–	–	–	–	–	–	Lanosta-8,24-dien-3-ol, acetate, (3β)-		MS10105
189	190	202	81	43	95	93	69	**468**	100	32	31	30	29	28	24	23	**18.00**	32	52	2	–	–	–	–	–	–	–	–	–	–	A′-Neogammacer-22(29)-en-3-ol, acetate, (3β,21β)-	2085-25-8	NI43945
218	203	189	468	408	453	398		**468**	100	70	40	4	4	3	3			32	52	2	–	–	–	–	–	–	–	–	–	–	Olean-12-en-3-ol, acetate, (3β)-	1616-93-9	NI43946
436	421	314	437	468	393	175	453	**468**	100	41	33	32	28	27	25	20		33	56	1	–	–	–	–	–	–	–	–	–	–	Cycloneolitsin		LI09228
436	95	421	393	175	314	367	203	**468**	100	57	44	38	38	34	27	26	**20.00**	33	56	1	–	–	–	–	–	–	–	–	–	–	Cycloneolitsin		LI09229
354	352	356	473	471	475	469	121	**469**	100	76	50	42	42	22	18	16		12	6	–	3	1	–	7	–	–	–	–	–	–	2-(3′-Methyl-4′-chlorophenylthio)-4,6-bis(trichloromethyl)-s-triazine		MS10106
71	43	354	239	69	284	312	266	**469**	100	91	14	11	7	5	4	3	**0.00**	18	26	5	1	–	7	–	–	–	–	–	–	–	Bis(3-methylbutyl) n-heptafluorobutyrylaspartate		NI43947
169	43	109	127	331	97	139	81	**469**	100	99	70	36	24	11	10	10	**0.00**	20	23	12	1	–	–	–	–	–	–	–	–	–	p-Nitrophenyl 2,3,4,6-tetra-O-acetyl-β-D-glucopyranoside	5987-78-0	NI43948
325	310	326	162.5	266	327	282	296	**469**	100	92	27	27	20	13	13	10	**0.00**	20	24	4	1	–	–	–	–	1	–	–	–	–	Cryptaustoline iodide		LI09230
142	178	327	326	149	176	127	192	**469**	100	80	41	38	38	30	28	27	**0.00**	20	24	4	1	–	–	–	–	1	–	–	–	–	Discretamine methoiodide		NI43949
178	327	326	142	176	150	328	310	**469**	100	99	86	61	40	32	25	21	**0.00**	20	24	4	1	–	–	–	–	1	–	–	–	–	Scoulerine methoiodide		NI43950
142	178	327	326	149	341	192	340	**469**	100	90	59	56	48	47	47	44	**0.00**	20	24	4	1	–	–	–	–	1	–	–	–	–	Stepholidine methoiodide		NI43951
327	326	178	341	176	142	340	310	**469**	100	87	41	34	31	29	28	25	**0.00**	20	24	4	1	–	–	–	–	1	–	–	–	–	Steponine methoiodide		NI43952
143	157	271	228	168	451	289	180	**469**	100	79	45	39	39	34	28	25	**7.00**	22	31	10	1	–	–	–	–	–	–	–	–	–	1-[(1-Deoxylactit-1-yl)amino]naphthalene		NI43953
274	182	195	469	65	275	181		**469**	100	45	32	19	14	13	9			23	26	6	1	–	3	–	–	–	–	–	–	–	7-Isoquinolinol, 1,2,3,4-tetrahydro-6-methoxy-2-(trifluoroacetyl)-1-[2-(3,4,5-trimethoxyphenyl)ethyl]-	56771-99-4	NI43954
100	155	268	426	256	242	312	298	**469**	100	50	40	30	30	30	23	21	**3.00**	23	43	5	5	–	–	–	–	–	–	–	–	–	Hexanamide, 6-[acetyl[3-[acetyl[3-(acetylamino)propyl]amino]propyl]amino]-N-[3-(acetylamino)propyl]-	67370-66-5	NI43955
347	95	348	108	135	136	107	106	**469**	100	35	33	23	18	13	12	12	**0.00**	26	29	8	–	–	–	–	–	–	–	–	–	–	Limonin		LI09231
242	185	183	243	406	404	122	155	**469**	100	28	28	24	20	20	15	8	**1.00**	26	32	2	1	–	–	–	1	–	–	–	–	–	Benzoic acid, 4-bromo-, 2-[dodecahydro-6-(2-penten-4-ynyl)pyrrolo[1,2-a]quinolin-1-yl]ethyl ester, [1S-[1α,3aβ,5aα,6α(Z),9aα]]-	63983-57-3	NI43956
471	469	423	425	344	472	470	315	**469**	100	88	62	60	40	37	30	26		27	20	2	1	–	–	–	1	–	–	–	–	–	3-Bromo-1-(2-methoxycarbonylphenyl)-9-methyl-4-phenyl-carbazole		LI09232
145	89	88	120	290	63	91	117	**469**	100	95	88	63	54	47	43	38	**4.00**	28	23	6	1	–	–	–	–	–	–	–	–	–	Rhizocarpic acid	18463-11-1	NI43957
290	145	88	469	120	90	291	117	**469**	100	75	50	41	25	23	22	20		28	23	6	1	–	–	–	–	–	–	–	–	–	Rhizocarpic acid	18463-11-1	NI43958
396	469	180	350	351	77	368	323	**469**	100	51	50	41	34	28	17	16		29	27	5	1	–	–	–	–	–	–	–	–	–	4,5-Bis(ethoxycarbonyl)-1,3,6-triphenyl-3,4-dihydropyridin-2(1H)-one		MS10107
131	92	181	119	169	231	108	107	**469**	100	90	89	82	80	42	40	38	**0.00**	30	31	4	1	–	–	–	–	–	–	–	–	–	Benzenemethanol, α-[[[[4-methoxy-3-(phenylmethoxy)phenyl]methyl]amino]methyl]-4-(phenylmethoxy)-	56772-04-4	NI43959
378	379	18	105	28	43	55	17	**469**	100	32	31	12	10	9	8	8	**2.00**	33	59	–	1	–	–	–	–	–	–	–	–	–	1-(2-Phenylethyl)azacyclohexacosane		MS10108
469	470	222	471	208	194	57	55	**469**	100	40	15	10	7	7	6	6		34	47	–	1	–	–	–	–	–	–	–	–	–	Cholesta-4,6-dieno[3,2-b]N-methylindole		MS10109
330	385	470	358	279	441	415	299	**470**	100	99	72	72	68	46	23	20		9	–	5	–	–	5	–	–	–	–	–	–	1	Rhenium, pentacarbonyl(2,3,3,4,4-pentafluoro-1-cyclobuten-1-yl)-	15038-33-2	NI43960
237	239	235	241	404	402	332	334	**470**	100	63	62	21	16	14	13	11	**0.79**	10	–	–	–	–	–	10	–	–	–	–	–	–	Bi-2,4-cyclopentadien-1-yl, 1,1′,2,2′,3,3′,4,4′,5,5′-decachloro-	2227-17-0	NI43961

MASS TO CHARGE RATIOS								M.W.	INTENSITIES								Parent	C	H	O	N	S	F	Cl	Br	I	Si	P	B	X	COMPOUND NAME	CAS Reg No	No
237	239	235	241	332	18	334	28	**470**	100	65	62	26	20	20	17	16	**2.24**	10	-	-	-	-	-	10	-	-	-	-	-	-	Bi-2,4-cyclopentadien-1-yl, 1,1′,2,2′,3,3′,4,4′,5,5′-decachloro-	2227-17-0	NI43962
237	239	235	241	332	404	167	166	**470**	100	65	64	20	11	10	10	10	**0.00**	10	-	-	-	-	-	10	-	-	-	-	-	-	Bi-2,4-cyclopentadien-1-yl, 1,1′,2,2′,3,3′,4,4′,5,5′-decachloro-	2227-17-0	NI43963
69	198	168	400	217	110	366	365	**470**	100	66	58	51	45	39	38	35	**0.00**	12	-	-	-	-	10	3	-	-	-	1	-	-	Bis(pentafluorophenyl)trichlorophosphorane		MS10110
441	442	396	443	57	43	397	207	**470**	100	40	25	23	14	14	13	11	**0.00**	14	34	8	-	-	-	-	-	-	5	-	-	-	1-Butoxyperethylhomotetrasilsesquioxane		NI43964
393	391	200	312	228	470	284	256	**470**	100	98	74	40	28	21	13	7		20	8	4	-	-	-	-	2	-	-	-	-	-	1,2-Phthaloyl-2a,8a-dibromo-2a,3,8,8a-tetrahydro-3,8-diketonaphtho[b]cyclobutadiene		LI09233
130	470	355	158	103	71	69	257	**470**	100	5	2	2	2	1	1	1		20	21	3	2	-	7	-	-	-	-	-	-	-	3-Methylbutyl n-heptafluorobutyryltryptophanate		NI43965
235	187	145	133	219	177	381	305	**470**	100	33	18	8	7	5	4	4	**0.00**	21	42	11	-	-	-	-	-	-	-	-	-	-	Nonamethyl 2-O-(β-D-glucopyranosyl)-D-sorbitol		LI09234
187	235	219	295	133	145	425	381	**470**	100	53	22	19	15	11	3	3	**0.00**	21	42	11	-	-	-	-	-	-	-	-	-	-	Nonamethyl 4-O-(α-D-glucopyranosyl)-D-sorbitol		LI09235
235	295	187	145	219	133	177	337	**470**	100	60	52	30	23	18	13	8	**0.00**	21	42	11	-	-	-	-	-	-	-	-	-	-	Nonamethyl 6-O-(β-D-glucopyranosyl)-D-sorbitol		LI09236
203	470	236	147	267	439	438	437	**470**	100	60	49	27	23	6	4	3		22	22	4	4	2	-	-	-	-	-	-	-	-	Bis(6,7-dimethoxy-2-methylquinazolin-4-yl) disulphide	92340-02-8	NI43966
131	287	159	470	155	315	314		**470**	100	32	24	7	7	6	6			23	22	5	2	2	-	-	-	-	-	-	-	-	3H-1,5-Benzodiazepin-3-one, 1,2,4,5-tetrahydro-1,5-bis[(4-methylphenyl)sulphonyl]-	1179-18-6	NI43967
379	265	380	210	196	76	69	43	**470**	100	39	33	28	26	19	19	19	**0.39**	23	38	8	2	-	-	-	-	-	-	-	-	-	Cyclopentanepropanoic acid, 3-(acetyloxy)-2-[8-(acetyloxy)-3-(methoxyimino)octyl]-5-(methoxyimino)-, methyl ester	69502-86-9	NI43968
155	188	212	354	269	185	156	184	**470**	100	69	65	40	25	21	21	20	**3.00**	23	42	6	4	-	-	-	-	-	-	-	-	-	Glycine, N-[N-[N-[N-(1-oxodecyl)glycyl]glycyl]-L-leucyl]-, methyl ester	31944-48-6	NI43969
55	415	438	414	469	455	213	255	**470**	100	20	14	11	9	9	7	6	**6.00**	24	39	1	-	-	-	-	-	1	-	-	-	-	3a,5-Cyclonorcholane, 23-iodo-6-methoxy-, (3α,6β)-		NI43970
146	105	159	250	145	104	147	470	**470**	100	33	28	26	26	26	17	8		25	25	7	-	-	-	-	-	-	-	-	3	-	α-D-Galactopyranoside, methyl, cyclic 4,6-(phenylboronate), cyclic 2,3-ester with phenylboronic acid bimol. monoanhydride	41356-05-2	NI43971
146	250	105	159	145	249	160	104	**470**	100	28	21	19	19	14	14	14	**0.23**	25	25	7	-	-	-	-	-	-	-	-	3	-	α-D-Glucopyranoside, methyl, cyclic 4,6-(phenylboronate), cyclic ester with phenylboronic acid bimol. monoanhydride	41356-06-3	NI43972
77	115	156	127	42	55	439	170	**470**	100	79	77	74	70	41	12	8	**0.00**	25	30	5	2	1	-	-	-	-	-	-	-	-	trans-5-Ethyl-2-[1-(phenylsulphonyl)-2-indolyl]-1-(2-hydroxyethyl)-4-piperidone ethylene acetal		DD01725
169	246	291	307	156	231	183	279	**470**	100	31	28	26	18	15	13	11	**6.00**	25	30	7	2	-	-	-	-	-	-	-	-	-	21(R)-(β-D-glucopyranosyl)-hydroxysarpagan-17-al		MS10111
115	79	91	316	105	133	376	470	**470**	100	28	16	11	11	5	3	1		25	42	6	-	1	-	-	-	-	-	-	-	-	18-[[2,3-(Isopropylidenedioxy)propyl]oxy]-13,17-octadecadien-15-yn-1-yl-methanesulphonate		MS10112
235	237	165	236	200	199	239	238	**470**	100	66	44	20	17	16	12	10	**0.20**	26	18	-	-	-	-	4	-	-	-	-	-	-	1,2-Bis(2-chlorophenyl)-1,2-bis(3-chlorophenyl)ethane		MS10113
316	374	148	95	317	133	375	358	**470**	100	40	20	20	15	13	8	8	**0.00**	26	30	8	-	-	-	-	-	-	-	-	-	-	Deoxyrutaevin		LI09237
124	43	426	41	264	55	123	69	**470**	100	83	78	59	49	40	37	30	**6.00**	26	30	8	-	-	-	-	-	-	-	-	-	-	11H-Dibenzo[b,e][1,4]dioxepin-7-carboxylic acid, 3,8-dihydroxy-11-oxo-1-(2-oxoheptyl)-6-pentyl-	84-24-2	NI43973
151	135	57	152	28	217	85	105	**470**	100	16	15	10	6	3	3	2	**2.00**	26	30	8	-	-	-	-	-	-	-	-	-	-	Dihydrohelianthoidin		LI09238
147	219	234	204	203	266	470	438	**470**	100	79	47	34	32	19	3	1		26	31	-	-	3	-	-	-	-	-	1	-	-	3,8-Di-tert-butyl-5,6-diphenyl-2,9-dithia-1-phosphabicyclo[4.3.0]nona-3,7-diene 1-sulphide	82518-04-5	NI43974
470	469	409	251	199	200	186	174	**470**	100	72	41	17	14	12	11	8		26	34	6	2	-	-	-	-	-	-	-	-	-	Reserpinediol, diacetate		LI09239
470	69	78	74	71	105	246	455	**470**	100	76	62	58	53	51	48	44		26	34	6	2	-	-	-	-	-	-	-	-	-	4,25-Secoobscurinervan-4-ol, 6,7-didehydro-15,16-dimethoxy-22-methyl-, 21-acetate, (4β,22α)-	54658-06-9	NI43975
470	182	43	246	45	224	260	41	**470**	100	78	76	68	58	49	43	43		26	34	6	2	-	-	-	-	-	-	-	-	-	4,25-Secoobscurinervan-4-one, O-acetyl-15,16-dimethoxy-22-methyl-, (22α)-	54658-10-5	NI43976
84	85	42	97	41	135	82	43	**470**	100	65	29	25	25	19	19	19	**1.90**	26	38	4	4	-	-	-	-	-	-	-	-	-	Ceanothine C	18658-42-9	NI43977
43	159	98	57	55	84	71	69	**470**	100	71	54	46	35	34	29	27	**0.00**	27	50	6	-	-	-	-	-	-	-	-	-	-	Eicosanoic acid, 2-(acetyloxy)-1-[(acetyloxy)methyl]ethyl ester	55429-68-0	WI01566
159	295	158	98	177	112	84	160	**470**	100	37	37	30	16	13	13	9	**0.00**	27	50	6	-	-	-	-	-	-	-	-	-	-	Eicosanoic acid, 2-(acetyloxy)-1-[(acetyloxy)methyl]ethyl ester	55429-68-0	WI01567
159	158	98	73	177	295	117	84	**470**	100	19	18	15	12	11	10	10	**0.00**	27	50	6	-	-	-	-	-	-	-	-	-	-	Eicosanoic acid, 2,3-bis(acetyloxy)propyl ester	55429-67-9	WI01568
43	159	57	98	55	41	84	73	**470**	100	44	32	27	27	23	19	18	**0.00**	27	50	6	-	-	-	-	-	-	-	-	-	-	Eicosanoic acid, 2,3-bis(acetyloxy)propyl ester	55429-67-9	WI01569
327	471	472	328	345	443	401	299	**470**	100	91	26	20	14	6	4	4	**0.00**	27	50	6	-	-	-	-	-	-	-	-	-	-	Octanoic acid, 1,2,3-propanetriyl ester	538-23-8	NI43978
57	127	43	60	41	55	73	84	**470**	100	99	47	45	33	31	29	16	**0.00**	27	50	6	-	-	-	-	-	-	-	-	-	-	Octanoic acid, 1,2,3-propanetriyl ester	538-23-8	NI43979
327	328	329	326	325	299	242	201	**470**	100	24	4	4	3	2	1	1	**0.00**	27	50	6	-	-	-	-	-	-	-	-	-	-	Octanoic acid, 1,2,3-propanetriyl ester	538-23-8	NI43980
127	57	43	327	55	41	201	128	**470**	100	69	18	15	12	12	9	9	**0.00**	27	50	6	-	-	-	-	-	-	-	-	-	-	Octanoic acid, 1,2,3-propanetriyl ester	538-23-8	NI43981
277	77	278	201	183	199	51	43	**470**	100	79	65	56	47	44	44	42	**35.00**	28	23	3	-	1	-	-	-	-	-	1	-	-	Triphenylphosphonio(vinylsulphonyl-benzoyl-methanide)		LI09240
205	161	149	187	265	175	251	162	**470**	100	15	15	14	8	8	7	5	**0.40**	28	38	6	-	-	-	-	-	-	-	-	-	-	Furan, 2,2′-[1,2-ethanediylbis(oxy)]bis[tetrahydro-5-(2-methoxy-4-methylphenyl)-5-methyl-	63646-82-2	NI43982
124	95	141	123	131	43	67	55	**470**	100	96	81	67	57	51	48	47	**6.00**	28	38	6	-	-	-	-	-	-	-	-	-	-	Withaferin A	5119-48-2	NI43983
124	90	141	131	197	347	470		**470**	100	95	83	59	35	26	7			28	38	6	-	-	-	-	-	-	-	-	-	-	Withaferin A		LI09241
73	43	411	89	57	129	83	455	**470**	100	44	38	37	33	32	30	28	**2.00**	28	58	3	-	-	-	-	-	-	1	-	-	-	Tetracosanoic acid, 2-[(trimethylsilyl)oxy]-, methyl ester	56784-04-4	MS10114
73	411	43	57	89	412	455	75	**470**	100	93	49	38	38	32	31	26	**2.87**	28	58	3	-	-	-	-	-	-	1	-	-	-	Tetracosanoic acid, 2-[(trimethylsilyl)oxy]-, methyl ester	56784-04-4	NI43984
151	121	440	199	229	314	470	299	**470**	100	90	85	48	36	20	18	18		29	42	5	-	-	-	-	-	-	-	-	-	-	3-Acetoxy-23-hydroxy-23-methyl-12-oxo-23a,27-epoxyfurost-9(11)-ene		LI09242
83	44	470	85	47	471	472	43	**470**	100	88	75	58	27	26	24	23		30	14	6	-	-	-	-	-	-	-	-	-	-	2,2′:3′,2″-Ternaphthalene-1,1′,1″,4,4′,4″-hexone		NI43985

MASS TO CHARGE RATIOS	M.W.	INTENSITIES	Parent	C	H	O	N	S	F	Cl	Br	I	Si	P	B	X	COMPOUND NAME	CAS Reg No	No
470 471 438 437 406 472 404 235	**470**	100 36 27 20 18 16 13 13		30	18	–	2	2	–	–	–	–	–	–	–	–	7,12-Dihydro-benzo[4]phenothiazino[3,4-c]benzo[d]phenothiazine		MS10115
455 470 456 471 469 124 453 415	**470**	100 67 32 23 17 17 16 11		30	18	4	2	–	–	–	–	–	–	–	–	–	N,N′-Dicyclopropyl-3,4:9,10-perylenebis(dicarboximide)	110590-24-4	MS10116
470 247 471 105 472 101 77 248	**470**	100 60 46 28 24 22 22 18		30	22	4	–	–	–	–	–	–	–	–	–	1	Magnesium, bis(1-benzoyl-2-oxo-2-phenylethyl)-	864-19-7	NI43986
220 222 205 237 57 251 55 470	**470**	100 22 15 7 6 5 5 4		30	46	2	–	1	–	–	–	–	–	–	–	–	3,3′,5,5′-Tetra-tert-butyl-2,2′-dihydroxy-6,6′-dimethylphenyl sulphide		IC04484
152 470 165 151 194 139 137 179	**470**	100 57 18 13 4 3 3 2		30	46	4	–	–	–	–	–	–	–	–	–	–	1,14-Bis(3,5-dimethoxyphenyl)tetradecane		LI09243
125 117 453 207 223 411 152 111	**470**	100 22 10 10 9 8 8 8	**2.50**	30	46	4	–	–	–	–	–	–	–	–	–	–	Ergosta-7,22-dien-5-ol, 3-(acetyloxy)-6-oxo-, (3β,5α)-		NI43987
70 55 41 69 137 81 83 341	**470**	100 80 62 55 48 35 34 30	**0.00**	30	46	4	–	–	–	–	–	–	–	–	–	–	Ergost-4-ene-3,6-dione, 25-(acetyloxy)-	56143-32-9	NI43988
151 123 205 452 137 470 163 320	**470**	100 90 76 55 55 52 45 40		30	46	4	–	–	–	–	–	–	–	–	–	–	D:A-Friedooleanan-24-oic acid, 1,3-dioxo-	43230-66-6	MS10117
365 69 470 43 127 41 55 59	**470**	100 94 92 88 68 65 63 57		30	46	4	–	–	–	–	–	–	–	–	–	–	Lanosta-7,9(11)-dien-18-oic acid, 3,20,25-trihydroxy-, γ-lactone, (3β,20ξ)	24041-67-6	NI43989
470 43 57 85 83 97 55 71	**470**	100 92 73 68 68 61 60 57		30	46	4	–	–	–	–	–	–	–	–	–	–	Lanost-9(11)-en-18-oic acid, 20,23-dihydroxy-3-oxo-, γ-lactone, (20ξ)-	36872-79-4	NI43990
190 244 191 41 81 43 55 69	**470**	100 92 60 56 52 52 50 48	**4.00**	30	46	4	–	–	–	–	–	–	–	–	–	–	Olean-12-en-28-oic acid, 3,16,21-trihydroxy-, γ-lactone, (3β,16β,21β)-	30950-05-1	LI09244
190 244 191 41 81 55 69 95	**470**	100 92 62 59 54 53 51 49	**3.66**	30	46	4	–	–	–	–	–	–	–	–	–	–	Olean-12-en-28-oic acid, 3,16,21-trihydroxy-, γ-lactone, (3β,16β,21β)-	30950-05-1	NI43991
425 190 191 236 427 189 192 470	**470**	100 89 84 69 54 46 45 42		30	46	4	–	–	–	–	–	–	–	–	–	–	Semimoronic acid	102053-38-3	NI43992
413 57 75 414 73 133 41 357	**470**	100 85 53 40 20 15 15 14	**8.00**	30	50	2	–	–	–	–	–	–	1	–	–	–	4-[(2,4,6-Tri-tert-butylphenoxy)dimethylsiloxy]-1,2,3,5,6,7,8,8a-octahydroazulene		NI43993
470 387 471 414 415 388 428 472	**470**	100 48 39 37 15 15 14 12		30	50	2	–	–	–	–	–	–	1	–	–	–	$\Delta^{1(6)}$-Tetrahydrocannabinol tripropylsilyl ether		NI43994
470 399 471 455 428 385 400 357	**470**	100 50 43 36 30 17 16 14		30	50	2	–	–	–	–	–	–	1	–	–	–	Δ^{1}-Tetrahydrocannabinol tripropylsilyl ether		NI43995
455 456 470 457 471 73 427 87	**470**	100 38 10 10 5 2 1 1		30	50	2	–	–	–	–	–	–	1	–	–	–	Δ^{3}-Tetrahydrocannabinol tripropylsilyl ether		NI43996
159 160 146 470 471 469 312 295	**470**	100 33 27 6 2 1 1 1		30	50	2	2	–	–	–	–	–	–	–	–	–	N-Arachidoyl-5-hydroxytryptamine		MS10118
180 77 91 105 44 351 353 234	**470**	100 81 51 38 32 18 12 3	**3.00**	31	22	3	2	–	–	–	–	–	–	–	–	–	6-Benzyloxy-1-oxo-2,4-diphenyl-1H-pyrimido[1,2-a]quinolin-4-ium-3-olate		NI43997
189 207 107 119 135 175 105 121	**470**	100 53 46 44 43 42 42 38	**19.78**	31	50	3	–	–	–	–	–	–	–	–	–	–	Lup-20(29)-en-28-oic acid, 3-hydroxy-, methyl ester, (3β)-	2259-06-5	NI43998
189 207 470 262 220 203 411 452	**470**	100 60 55 55 30 30 25 10		31	50	3	–	–	–	–	–	–	–	–	–	–	Lup-20(29)-en-28-oic acid, 3-hydroxy-, methyl ester, (3β)-	2259-06-5	NI43999
189 119 135 175 207 203 262 147	**470**	100 63 51 48 42 34 29 28	**18.47**	31	50	3	–	–	–	–	–	–	–	–	–	–	Lup-20(29)-en-28-oic acid, 3-hydroxy-, methyl ester, (3β)-	2259-06-5	NI44000
262 203 43 41 55 69 119 81	**470**	100 98 94 79 75 53 45 41	**2.19**	31	50	3	–	–	–	–	–	–	–	–	–	–	Urs-12-en-28-oic acid, 3-hydroxy-, methyl ester, (3β)-	32208-45-0	NI44001
133 190 207 470 203 411 262 452	**470**	100 70 60 50 50 41 30 26		31	50	3	–	–	–	–	–	–	–	–	–	–	Urs-12-en-28-oic acid, 3-hydroxy-, methyl ester, (3β)-	32208-45-0	LI09245
203 262 207 263 204 202 249 470	**470**	100 99 23 20 10 8 6 5		31	50	3	–	–	–	–	–	–	–	–	–	–	Urs-12-en-28-oic acid, 3-hydroxy-, methyl ester, (3β)-	32208-45-0	NI44002
153 123 121 439 287 273 275 261	**470**	100 20 20 15 15 14 10 6	**1.00**	31	50	3	–	–	–	–	–	–	–	–	–	–	D:A-Friedooleanan-1-en-3-one, 24-hydroxy-1-methoxy-	43230-82-6	MS10119
319 257 205 470 438 191 181 410	**470**	100 54 54 14 12 12 12 7		31	50	3	–	–	–	–	–	–	–	–	–	–	D:A-Friedooleanan-26-oic acid, 3-oxo-, methyl ester		NI44003
169 273 470 163 223 264 246 249	**470**	100 50 43 43 38 30 23 22		31	50	3	–	–	–	–	–	–	–	–	–	–	D:A-Friedooleanan-29-oic acid, 3-oxo-, methyl ester, (20α)-	39903-11-2	MS10120
153 121 453 455 207 287 439 221	**470**	100 50 30 28 28 25 20 20	**2.00**	31	50	3	–	–	–	–	–	–	–	–	–	–	D:A-Friedoolean-2-en-1-one, 24-hydroxy-3-methoxy-	43230-81-5	MS10121
189 247 259 250 229 147 119 69	**470**	100 98 96 96 95 95 95 94	**55.00**	31	50	3	–	–	–	–	–	–	–	–	–	–	D:C-Friedours-7-en-28-oic acid, 3-hydroxy-, methyl ester, (3α)-		NI44004
203 262 202 207 263 204 470 249	**470**	100 87 27 21 17 17 8 5		31	50	3	–	–	–	–	–	–	–	–	–	–	Methyl 3β-hydroxyolean-12-en-28-oate	35933-00-7	NI44005
203 262 189 133 452 249 470 410	**470**	100 49 25 15 3 3 2 2		31	50	3	–	–	–	–	–	–	–	–	–	–	Methyl 3β-hydroxyolean-12-en-28-oate	35933-00-7	MS10122
203 262 133 189 207 249 452 410	**470**	100 95 68 35 24 9 6 3	**2.00**	31	50	3	–	–	–	–	–	–	–	–	–	–	Methyl 3β-hydroxyurs-12-en-28-oate		MS10123
69 73 75 55 121 470 95 81	**470**	100 70 59 57 50 47 41 41		31	54	1	–	–	–	–	–	–	1	–	–	–	Cholesta-8,24-diene, 4-methyl-3-[(trimethylsilyl)oxy]-, (3β,4α,5α)-	55429-69-1	NI44006
470 121 471 227 455 109 365 243	**470**	100 46 42 33 30 30 29 28		31	54	1	–	–	–	–	–	–	1	–	–	–	Cholesta-8(14),24-diene, 4-methyl-3-[(triphenylsilyl)oxy]-, (3β,4α)-	55429-70-4	NI44007
341 380 386 296 257 365 253 255	**470**	100 81 67 67 61 53 53 50	**33.33**	31	54	1	–	–	–	–	–	–	1	–	–	–	Ergosta-5,24(28)-diene, 3-[(trimethylsilyl)oxy]-, (3β)-	22042-04-2	NI44008
470 255 343 43 471 57 55 229	**470**	100 66 60 49 39 39 34 32		31	54	1	–	–	–	–	–	–	1	–	–	–	Ergosta-7,22-diene, 3-[(trimethylsilyl)oxy]-, (3β,5α)-	40272-63-7	NI44009
69 75 81 73 255 107 343 105	**470**	100 95 67 65 60 46 41 36	**20.02**	31	54	1	–	–	–	–	–	–	1	–	–	–	Ergosta-7,22-diene, 3-[(trimethylsilyl)oxy]-, (3β,22E)-	55527-93-0	NI44010
470 107 227 455 213 150 365 229	**470**	100 62 60 54 54 50 46 42		31	54	1	–	–	–	–	–	–	1	–	–	–	Ergosta-8,24(28)-diene, 3-[(trimethylsilyl)oxy]-, (3β,5α)-	55429-71-5	NI44011
255 121 120 470 147 69 159 256	**470**	100 94 53 42 32 29 26 24		31	54	1	–	–	–	–	–	–	1	–	–	–	9,10-Secoergosta-5,7,22-triene, 3-[(trimethylsilyl)oxy]-, (3β,5E,7E,22E)-	69688-07-9	NI44012
83 91 92 178 177 176 155 205	**470**	100 30 10 7 7 4 3 2	**1.00**	32	22	4	–	–	–	–	–	–	–	–	–	–	2-Anthracenecarboxylic acid, 1,2-ethanediyl ester	61549-17-5	NI44013
349 77 180 377 350 470 348 320	**470**	100 66 63 53 33 21 14 14		32	26	2	2	–	–	–	–	–	–	–	–	–	1,5-Diphenyl-3-(α,N-diphenyl-carbamoyl-propylidene)pyrrol-2-one		IC04485
125 470 87 43 427 55 401 41	**470**	100 45 40 36 26 26 23 22		32	54	2	–	–	–	–	–	–	–	–	–	–	Cholestan-3-one, 2-isopropenyl-, cyclic 1,2-ethanediyl acetal, (5a)-	54498-62-3	NI44014
43 96 97 107 41 55 45 44	**470**	100 90 77 62 53 47 47 42	**0.00**	32	54	2	–	–	–	–	–	–	–	–	–	–	Cholestan-3-one, 2-isopropylidene-, cyclic 1,2-ethanediyl acetal, (5α)-	54498-61-2	NI44015
85 368 369 57 55 41 67 43	**470**	100 25 9 9 8 8 7 7	**0.00**	32	54	2	–	–	–	–	–	–	–	–	–	–	Cholest-5-ene, 3-(tetrahydro-2H-pyran-2-yl)-	6252-45-5	NI44016
43 57 95 55 69 41 410 149	**470**	100 48 46 46 43 39 38 29	**13.00**	32	54	2	–	–	–	–	–	–	–	–	–	–	9,19-Cyclolanostan-3-ol, acetate, (3β)-	4575-74-0	NI44017
410 395 288 367 297 203 470 341	**470**	100 72 49 30 28 28 24 24		32	54	2	–	–	–	–	–	–	–	–	–	–	9,19-Cyclolanostan-3-ol, acetate, (3β)-	4575-74-0	NI44018
455 395 456 95 43 396 69 55	**470**	100 82 34 32 31 26 26 21	**19.90**	32	54	2	–	–	–	–	–	–	–	–	–	–	Dihydrobutyrospermyl acetate		MS10124
105 109 107 368 121 123 119 135	**470**	100 81 75 53 51 47 46 45	**2.00**	32	54	2	–	–	–	–	–	–	–	–	–	–	Ergost-5-en-3-ol, 22,23-dimethyl-, acetate, (3β)-	55724-21-5	NI44019
43 455 69 395 135 95 55 81	**470**	100 89 84 82 75 75 70 58	**0.00**	32	54	2	–	–	–	–	–	–	–	–	–	–	Euph-7-enyl acetate, (9β)-		MS10125
455 395 43 69 95 456 55 396	**470**	100 92 48 44 41 34 34 29	**23.00**	32	54	2	–	–	–	–	–	–	–	–	–	–	Euph-8-enyl acetate		MS10126
455 395 43 456 95 69 55 396	**470**	100 80 36 34 34 31 26 25	**19.80**	32	54	2	–	–	–	–	–	–	–	–	–	–	Euph-9(11)-enyl acetate, (8α)-		MS10127
177 95 81 69 109 205 55 123	**470**	100 62 44 44 40 38 35 31	**1.00**	32	54	2	–	–	–	–	–	–	–	–	–	–	D:A-Friedooleanan-7-ol, acetate, (7α)-	56588-24-0	NI44020
470 95 69 43 81 357 220 123	**470**	100 99 76 64 57 54 49 46		32	54	2	–	–	–	–	–	–	–	–	–	–	Isoeuphenyl acetate		MS10128
411 412 469 470 409 289 471 456	**470**	100 40 33 23 22 20 13 13		32	54	2	–	–	–	–	–	–	–	–	–	–	Lanost-8-en-3-ol, acetate, (3β)-	1724-19-2	NI44021

MASS TO CHARGE RATIOS								M.W.	INTENSITIES								Parent	C	H	O	N	S	F	Cl	Br	I	Si	P	B	X	COMPOUND NAME	CAS Reg No	No
455	395	470	296	297	301	357	315	470	100	60	40	12	9	3	1	1		32	54	2	–	–	–	–	–	–	–	–	–	–	Lanost-8-en-3-ol, acetate, (3β)-	1724-19-2	NI44022
248	316	95	69	43	190	109	55	470	100	75	67	57	57	49	49	48	0.00	32	54	2	–	–	–	–	–	–	–	–	–	–	Lanost-9(11)-en-3-ol, acetate, (3β)-	1180-88-7	NI44023
395	455	43	396	95	135	69	456	470	100	74	38	32	29	27	27	26	24.50	32	54	2	–	–	–	–	–	–	–	–	–	–	Lanost-7-enyl acetate		MS10129
455	395	43	456	69	470	396	41	470	100	88	45	34	30	29	28	27		32	54	2	–	–	–	–	–	–	–	–	–	–	Lanost-8-enyl acetate		MS10130
455	395	43	95	456	69	470	55	470	100	87	47	41	37	37	36	31		32	54	2	–	–	–	–	–	–	–	–	–	–	Lanost-9(11)-enyl acetate		MS10131
99	100	69	83	55	109	95	81	470	100	24	12	10	10	8	8	8	0.00	32	54	2	–	–	–	–	–	–	–	–	–	–	Lupan-3-one, cyclic 1,2-ethanediyl acetal	54498-65-6	NI44024
73	282	354	56	75	128	384	147	471	100	74	61	58	26	22	20	19	1.48	16	42	5	1	–	–	–	–	–	4	1	–	–	Butanoic acid, 4-[bis[(trimethylsilyl)oxy]phosphinyl]-2-[(trimethylsilyl)amino]-, trimethylsilyl ester	56196-75-9	NI44025
253	266	119	471	177	176	218	225	471	100	34	25	24	19	18	16	15		17	15	3	1	–	10	–	–	–	–	–	–	–	(4-Hydroxyphenyl)pentylamine, bis(pentafluoropropionyl)-		NI44026
58	44	133	120	104	72	59	413	471	100	68	6	3	3	3	3	2	0.40	20	27	2	5	1	–	2	–	–	–	–	–	–	N-Ethyl-N-(2-dimethylaminoethyl)-4-(4-dimethylaminosulphonyl-2,6-dichlorophenylazo)-aniline		IC04486
32	262	183	294	293	185	108	30	471	100	99	79	73	46	45	22	21	0.10	21	20	2	1	2	–	–	–	–	–	1	–	1	(Triphenylphosphino)(O-ethyldithiocarbonato)(nitrosyl)nickel		NI44027
116	73	117	355	356	118	74	45	471	100	39	12	11	4	4	3	3	0.09	21	45	3	1	–	–	–	–	–	4	–	–	–	Adrenaline tetra-TMS		MS10132
116	73	117	45	355	74	75	118	471	100	67	11	10	7	7	4	3	0.00	21	45	3	1	–	–	–	–	–	4	–	–	–	Adrenaline tetra-TMS		MS10133
174	73	175	75	176	86	297	45	471	100	53	18	9	7	7	6	6	0.00	21	45	3	1	–	–	–	–	–	4	–	–	–	Normetanephrine, tetra-TMS		MS10134
116	355	117	73	356	75	118	357	471	100	15	11	10	5	5	4	2	0.02	21	45	3	1	–	–	–	–	–	4	–	–	–	Silanamine, N-[2-[3,4-bis[(trimethylsilyl)oxy]phenyl]-2-[(trimethylsilyl)oxy]ethyl]-N,1,1,1-tetramethyl-, (R)-	55429-87-3	NI44028
116	73	355	117	356	147	118	74	471	100	35	18	10	6	3	3	3	0.00	21	45	3	1	–	–	–	–	–	4	–	–	–	Silanamine, N-[2-[3,4-bis[(trimethylsilyl)oxy]phenyl]-2-[(trimethylsilyl)oxy]ethyl]-N,1,1,1-tetramethyl-, (R)-	55429-87-3	NI44029
174	73	175	176	297	86	456	147	471	100	28	17	8	7	5	4	4	0.00	21	45	3	1	–	–	–	–	–	4	–	–	–	Silanamine, N-[1-[3-methoxy-4-[(trimethylsilyl)oxy]phenyl]-2-[(trimethylsilyl)oxy]ethyl]-1,1,1-trimethyl-N-(trimethylsilyl)-	56114-60-4	NI44030
121	164	134	163	150	350	332	192	471	100	98	58	37	30	29	28	27	9.00	22	29	5	7	–	–	–	–	–	–	–	–	–	Puromycin		MS10135
115	320	200	87	43	412	187	220	471	100	83	78	77	52	51	47	43	23.77	23	37	9	1	–	–	–	–	–	–	–	–	–	Cyclopentaneoctanoic acid, 3,5-bis(acetyloxy)-ε-(methoxyimino)-2-(3-methoxy-3-oxopropyl)-, methyl ester	55869-54-0	NI44031
115	320	200	87	43	412	187	220	471	100	85	79	77	53	51	48	43	25.00	23	37	9	1	–	–	–	–	–	–	–	–	–	Dimethyl 5,7-diacetoxy-11-methoxyimino-tetranorprosta-1,16-dioate		LI09246
105	268	211	309	415	471	310	370	471	100	50	48	44	36	16	8	4		26	26	4	1	–	–	–	–	–	–	–	–	1	Dicarbonyl(5-ethoxy-3-methyl-2,5-diphenyl-4-oxazolidinylidene)(methylcyclopentadienyl)manganese	116928-05-3	DD01726
236	43	126	221	149	56	237	57	471	100	68	51	48	46	39	34	30	0.00	27	37	4	1	1	–	–	–	–	–	–	–	–	2,6-Di-tert-butyl-4-methoxyphenyl 2-acetamido-4-(phenylthio)butanoate		DD01727
308	320	184	158	277	278	262	276	471	100	59	57	50	48	45	33	32	0.00	27	37	6	1	–	–	–	–	–	–	–	–	–	Pregna-4,6-diene-3,20-dione, 2,17-dihydroxy-6-methyl-, diacetate, oxime, (2α)-		NI44032
308	411	320	277	278	184	158	262	471	100	99	49	46	40	40	38	33	0.00	27	37	6	1	–	–	–	–	–	–	–	–	–	Pregna-4,6-diene-3,20-dione, 2,17-dihydroxy-6-methyl-, diacetate, oxime, (2β)-		NI44033
368	369	278	471	280	75	370	279	471	100	31	17	14	10	9	8	7		27	41	4	1	–	–	–	–	–	1	–	–	–	Pregna-4,6-diene-3,20-dione, 17-(acetyloxy)-6-methyl-, 3-[O-(trimethylsilyl)oxime]	75113-25-6	NI44034
73	75	331	442	153	471	91	155	471	100	42	30	25	24	20	17	16		27	45	2	1	–	–	–	–	–	2	–	–	–	13β-Ethyl-17α-ethynyl-3-oximinogon-4-en-17β-ol, 3,17-bis(trimethylsilyl)		NI44035
380	91	83	55	381	40	80	110	471	100	44	34	31	29	24	21	19	1.80	29	45	4	1	–	–	–	–	–	–	–	–	–	3a,5(4H)-Isoindolinediol, 1-benzyl-4-(8,11-dihydroxy-4-methyl-1-undecenyl)tetrahydro-7-methyl-6-methylene-	14110-73-7	NI44036
380	91	84	55	381	41	110	30	471	100	44	34	31	29	25	19	17	1.80	29	45	4	1	–	–	–	–	–	–	–	–	–	7-Methyl-6-methylene-3a,5-dihydroxy-1-benzyl-4-(4′-methyl-8′,11′-dihydroxyundecyl)-hexahydroisoindoline		LI09247
43	69	57	50	106	107	91	77	471	100	83	70	70	67	64	60	52	35.00	31	21	4	1	–	–	–	–	–	–	–	–	–	5,5a,6,8,13,14,15,16-Octahydro-5,8,13,16-tetraoxo-6-(p-tolyl)dinaphtho[2,3-c][2′,3′-g]indole		MS10136
471	456	472	180	194	57	457	55	471	100	43	38	28	26	18	16	16		33	45	1	1	–	–	–	–	–	–	–	–	–	6-Oxo-cholest-4-eno[3,2-b]indole		MS10137
471	472	473	470	223	456	57	55	471	100	38	8	6	4	3	3	3		34	49	–	1	–	–	–	–	–	–	–	–	–	Cholest-5-eno[3,2-b]N-methylindole		MS10138
127	62	61	472	218	63	91	189	472	100	50	34	32	32	30	19	17		6	3	1	–	–	–	–	–	3	–	–	–	–	Phenol, 2,4,6-triiodo-	609-23-4	NI44037
236	238	234	201	203	440	237	438	472	100	63	60	40	20	7	7	6	0.00	10	2	–	–	–	–	10	–	–	–	–	–	–	1,3,4-Metheno-1H-cyclobuta[cd]pentalene, 1a,2,2,3,3a,4,5,5,5a,6-decachlorooctahydro-	55570-84-8	NI44038
476	63	478	474	99	98	77	88	472	100	79	73	66	60	57	50	37	16.80	10	4	2	–	–	–	–	4	–	–	–	–	–	3,3-Dihydroxy-1,4,6,7-tetrabromonaphthalene		IC04487
349	478	284	413	258	220	155	387	472	100	31	31	8	8	8	6	2	0.00	10	8	–	–	–	–	–	–	2	–	–	–	1	Zirconium bis(π-cyclopentadienyl)di-iodine		LI09248
234	290	262	374	318	174	346	218	472	100	60	60	42	34	29	15	12	0.00	10	12	6	–	–	–	2	–	–	–	2	–	2	Iron, μ-tetramethyldiphosphine-bis(chlorotricarbonyl)di-		MS10139
472	77	69	65	39	159	51	140	472	100	93	42	21	19	9	9	8		12	5	4	–	–	12	–	–	–	–	1	–	–	2-Oxo-2-phenoxy-4,4,5,5-tetrakis(trifluoromethyl)1,3,2$\lambda^5\sigma^4$-dioxaphospholane		MS10140
443	444	445	207	369	446	370	355	472	100	38	22	15	14	11	8	8	0.49	12	32	8	–	–	–	–	–	–	6	–	–	–	1,3,5,7,9,11-Hexaethyltricyclo[5.5.1.1^{3,11}]hexasiloxane	80177-42-0	NI44039

MASS TO CHARGE RATIOS								M.W.	INTENSITIES								Parent	C	H	O	N	S	F	Cl	Br	I	Si	P	B	X	COMPOUND NAME	CAS Reg No	No
179	148	45	296	178	60	44	281	472	100	51	50	34	34	34	31	21	7.59	12	36	–	12	–	–	–	–	–	–	4	–	–	2,4,6,8,10-Pentakis(dimethylamino)-2,4,6,8,10-pentaaza-1,3,5,7,-tetraphosphatricyclo[3.3.1.1^{3,7}]decane		LI09249
248	220	56	192	360	304	276	332	472	100	87	74	71	67	60	47	44	5.00	14	8	8	–	–	–	–	–	–	–	–	–	3	Iron, octacarbonyl[μ^3-[(1,2-η:1,2,3,4-η:1,2,3,4-η)-1-ethyl-1,3-butadiene-1,4-diyl]]tri-	74398-33-7	NI44040
444	133	330	358	445	387	69	105	472	100	38	33	32	28	27	25	17	11.01	18	15	5	2	–	7	–	–	–	–	–	–	–	Butanoic acid, 2,2,3,3,4,4,4-heptafluoro-, 4-(5-ethylhexahydro-1,3-dimethyl-2,4,6-trioxo-5-pyrimidinyl)phenyl ester	55429-88-4	NI44041
183	91	92	52	28	63	30	152	472	100	55	51	48	42	41	38	32	29.00	18	16	8	8	–	–	–	–	–	–	–	–	–	1,2-Cyclohexanedione, bis[(2,4-dinitrophenyl)hydrazone]	1468-24-2	NI44042
278	472	371	284	77	208	209	120	472	100	91	91	83	63	52	49	48		18	18	2	2	2	6	–	–	–	–	–	–	–	2-(4-Morpholinyl)-N-phenyl-1,3-bis[(trifluoromethyl)thio]-2-cyclopentenecarboxamide		DD01728
457	77	73	94	65	51	39	233	472	100	76	31	30	20	17	16	15	1.00	18	19	4	–	–	6	–	–	–	1	1	–	–	Diphenyl [2,2,2-trifluoro-1-(trifluoromethyl)-1-(trimethylsiloxy)ethyl]phosphonate		MS10141
140	472	41	474	116	55	476	268	472	100	9	9	7	7	7	5	4		18	36	–	2	4	–	–	–	–	–	–	–	1	Zinc bis(N,N-dibutylthiocarbamate)		IC04488
313	286	314	200	258	393	229	472	472	100	66	65	62	35	32	24	1		20	10	4	–	–	–	–	2	–	–	–	–	–	5b,11a-Dibromo-5,12-dihydroxy-5b,6,11,11a-tetrahydro-6,11-dioxodibenzo[b,h]biphenylene		LI09250
474	472	459	457	219	473	475	57	472	100	97	91	89	72	26	25	25		20	21	3	6	–	–	–	1	–	–	–	–	–	N,N-Diethyl-3-propionylamino-4-(2-bromo-6-cyano-4-nitrophenylazo)-aniline		IC04489
285	198	138	258	156	139	96	157	472	100	13	13	11	11	9	9	5	1.10	20	28	11	2	–	–	–	–	–	–	–	–	–	D-erythro-2-Acetoxymethyl-L-lyxol 5,6-diacetoxy-7-acetoxymethyl-cis-anti-cis-perhydro-di-dipyrrolo[2,1-b:2',3'-d]oxazole		LI09251
355	73	356	357	382	267	147	45	472	100	55	32	14	9	9	9	8	2.40	20	40	5	–	–	–	–	–	–	4	–	–	–	Benzeneacetic acid, 2,2,5-tris[(trimethylsilyl)oxy]-, trimethylsilyl ester		MS10142
73	472	355	147	267	429	437	193	472	100	50	43	11	9	7	6	3		20	40	5	–	–	–	–	–	–	4	–	–	–	Benzeneacetic acid, 2,4,5-tris[(trimethylsilyl)oxy]-, trimethylsilyl ester	55823-12-6	NI44043
355	356	73	429	357	457	295	147	472	100	28	21	15	15	11	11	8	6.60	20	40	5	–	–	–	–	–	–	4	–	–	–	Benzeneacetic acid, α,3,4-tris[(trimethylsilyl)oxy]-, trimethylsilyl ester	37148-65-5	NI44044
355	73	356	357	147	75	429	267	472	100	35	30	16	11	9	8	6	1.20	20	40	5	–	–	–	–	–	–	4	–	–	–	Benzeneacetic acid, α,3,4-tris[(trimethylsilyl)oxy]-, trimethylsilyl ester	37148-65-5	NI44045
73	355	356	147	74	357	75	295	472	100	46	16	9	8	8	4	3	0.62	20	40	5	–	–	–	–	–	–	4	–	–	–	Benzeneacetic acid, α,3,4-tris[(trimethylsilyl)oxy]-, trimethylsilyl ester	37148-65-5	NI44046
73	472	355	473	474	356	147	45	472	100	83	49	37	20	16	12	12		20	40	5	–	–	–	–	–	–	4	–	–	–	2,5-Bis(trimethylsiloxy)-bis(trimethylsilyl)mandelate		NI44047
73	472	473	147	474	45	267	74	472	100	61	27	16	15	11	9	8		20	40	5	–	–	–	–	–	–	4	–	–	–	Tris(trimethylsiloxy)phenyl(trimethylsilyl)acetate	55823-12-6	NI44048
43	175	87	419	134	436	133	195	472	100	95	85	75	65	55	39	33	1.00	21	21	5	6	–	–	1	–	–	–	–	–	–	N-(2-Cyanoethyl)-N-(2-acetoxyethyl)-3-acetylamino-4-(2-chloro-4-nitrophenylazo)-aniline		IC04490
73	255	385	356	147	75	467	472	472	100	30	15	15	15	15	8	1		21	44	4	–	–	–	–	–	–	4	–	–	–	2,3,7,8-Tetrakis(trimethylsiloxy)spiro[4.4]nona-2,7-diene		LI09252
44	43	45	63	60	78	327	149	472	100	91	61	49	44	41	20	20	0.00	22	16	12	–	–	–	–	–	–	–	–	–	–	2',7'-Diacetoxy-2',3',7',8'-tetrahydrodispiro[furan-2(5H),4'(9'H)-benzo[1,2 b;4,5-b']dipyran-9',2''(5''H)-furan]-5,5',5'',10-tetraone	80234-45-3	NI44049
111	213	153	101	60	370	113	328	472	100	62	62	49	43	33	30	28	0.00	22	32	11	–	–	–	–	–	–	–	–	–	–	α-D-Galactopyranose, 6-O-(2,3-dideoxy-α-D-erythro-hex-2-enopyranosyl) 1,2:3,4-di-O-isopropylidene-, diacetate	19940-05-7	NI44050
442	443	457	444	458	412	370	459	472	100	37	24	15	10	6	6	5	2.60	23	32	5	–	–	–	–	–	–	3	–	–	–	1,2,4-Trihydroxyanthraquinone, tris(trimethylsilyl)-	7336-71-2	NI44051
116	98	274	270	273	176	271	117	472	100	86	73	69	62	48	46	35	2.95	23	40	8	2	–	–	–	–	–	–	–	–	–	L-Glutamic acid, N-(N-decanoyl-L-α-glutamyl)-, trimethyl ester	15939-46-5	NI44052
144	242	116	360	230	216	145	176	472	100	43	38	13	13	11	10	9	2.32	23	40	8	2	–	–	–	–	–	–	–	–	–	L-Glutamic acid, N-[N-(1-oxodecyl)-L-γ-glutamyl]-, trimethyl ester	15939-48-7	NI44053
310	325	368	222	311	353	43	44	472	100	52	51	44	30	25	25	24	2.94	24	28	8	2	–	–	–	–	–	–	–	–	–	Dichotine, 19-hydroxy-, 2-acetate	55283-44-8	NI44054
416	183	418	417	185	262	108	325	472	100	43	35	25	20	17	14	11	5.49	25	19	2	–	–	–	1	–	–	–	1	–	1	(Chlorocyclopentadienyl)manganesedicarbonyltriphenylphosphine	33309-62-5	NI44055
472	152	167	91	306	227	229	473	472	100	95	88	85	64	53	31	23		25	22	–	–	–	–	–	–	–	–	–	–	2	Methane, bis(diphenylarsino)-		LI09253
57	85	41	29	156	28	58	371	472	100	38	14	9	8	8	5	4	0.00	25	44	8	–	–	–	–	–	–	–	–	–	–	Propanoic acid, 2,2-dimethyl-, 2,2-bis[(2,2-dimethyl-1-oxopropoxy)methyl]-1,3-propanediyl ester	5178-17-6	NI44056
73	173	275	299	270	185	147	103	472	100	91	80	75	29	26	22	15	0.00	25	52	4	–	–	–	–	–	–	2	–	–	–	9-Octadecenoic acid, 12,13-bis[(trimethylsilyl)oxy]-, methyl ester	75299-46-6	NI44057
119	121	120	236	91	129	117	75	472	100	8	5	3	3	2	1	1	0.86	26	32	–	–	4	–	–	–	–	–	–	–	–	3-Nonen-5-yne, (E)-4-ethyl-	74744-59-5	NI44058
370	328	310	312	371	311	352	329	472	100	91	82	27	26	25	21	21	0.00	26	32	8	–	–	–	–	–	–	–	–	–	–	Estra-1,3,5(10)-triene-3,6,7,17-tetrol, tetraacetate, (6α,7β,17β)-	7323-92-4	NI44059
310	352	370	311	312	328	250	157	472	100	39	32	30	22	19	16	11	0.00	26	32	8	–	–	–	–	–	–	–	–	–	–	Estra-1,3,5(10)-triene-3,6,7,17-tetrol, tetraacetate, (6β,7α,17β)-	69688-24-0	NI44060
150	41	124	68	121	206	43	64	472	100	58	53	46	43	39	31	30	0.00	26	32	8	–	–	–	–	–	–	–	–	–	–	Olivetoric acid	491-47-4	LI09254
150	41	124	69	121	206	43	65	472	100	53	48	43	40	36	29	28	0.00	26	32	8	–	–	–	–	–	–	–	–	–	–	Olivetoric acid	491-47-4	NI44061
310	295	267	293	282	281	199	91	472	100	80	40	30	30	30	30	22	0.00	26	32	8	–	–	–	–	–	–	–	–	–	–	Spiro[cyclopropane-1,2'(1'H)-phenanthrene]-1',4'(3'H)-dione, 3',9',10'-tris(acetyloxy)-4'b,5',6',7',8',8'a,9',10'-octahydro-2,4'b,7'-trimethyl-8'-methylene-	66610-36-4	NI44062
122	83	78	183	41	55	45	44	472	100	70	67	53	53	45	45	45	0.00	26	36	6	2	–	–	–	–	–	–	–	–	–	Aspidospermidin-21-ol, 1-acetyl-15,16,17-trimethoxy-, acetate	54751-74-5	NI44063
69	182	71	122	83	60	81	67	472	100	77	62	58	46	43	40	38	10.76	26	36	6	2	–	–	–	–	–	–	–	–	–	4,25-Secoobscurinervan-4-ol, 15,16-dimethoxy-22-methyl-, 21-acetate, (4β,22α)-	54658-13-8	NI44064
472	429	400	387	219	473	86	58	472	100	90	60	46	34	31	29	29		26	40	4	4	–	–	–	–	–	–	–	–	–	4,7,10-Triazabicyclo[12.3.1]octadeca-1(18),2,14,16-tetraene-5,8,11-trione, 12-(dimethylamino)-15-methoxy-9-sec-butyl-6-isobutyl-	38840-27-6	NI44065
159	473	115	297	57	55	41	28	472	100	30	28	26	23	20	18	15	0.82	26	48	7	–	–	–	–	–	–	–	–	–	–	1,2-Propanediol, 3-[[2-(acetyloxy)heptadecyl]oxy]-, diacetate	57346-63-1	NI44066
57	127	259	43	73	41	55	75	472	100	78	52	34	31	23	20	20	0.00	26	52	5	–	–	–	–	–	–	1	–	–	–	Dicaprylin, trimethylsilyl		NI44067

MASS TO CHARGE RATIOS	M.W.	INTENSITIES	Parent	C	H	O	N	S	F	Cl	Br	I	Si	P	B	X	COMPOUND NAME	CAS Reg No	No
472 473 259 456 457 194 382 75	**472**	100 51 36 26 24 23 20 16		27	44	3	–	–	–	–	–	–	2	–	–	–	18,19-Dinorpregn-4-en-20-yn-3-one, 13-ethyl-16,17-bis[(trimethylsilyl)oxy]-, (16β,17α)-		NI44068
44 388 43 387 357 472 358 389	**472**	100 44 22 19 17 13 12 11		28	24	7	–	–	–	–	–	–	–	–	–	–	3,4-Diacetoxy-9,12-dimethoxy-5,10-dimethyldibenzo[c,kl]xanthene		NI44069
136 137 55 121 154 110 472 192	**472**	100 89 87 74 63 63 56 56		28	40	6	–	–	–	–	–	–	–	–	–	–	6,15-Dibutyl-2,3,11,12-dibenzo-1,4,7,10,13,16-hexaoxacyclooctadeca-2,11-diene		MS10143
298 125 97 252 299 121 174 156	**472**	100 53 50 29 29 25 20 18	**2.40**	28	40	6	–	–	–	–	–	–	–	–	–	–	5α,14α-Dihydroxy-1-oxo-6α,7α-epoxy-20R,22R-witha-2-enolide		MS10144
472 103 77 75 473 250 178 428	**472**	100 93 69 69 39 35 20 18		29	16	5	2	–	–	–	–	–	–	–	–	–	1H-Isoindole-1,3(2H)-dione, 5,5′-carbonylbis 2-phenyl-	6097-13-8	NI44070
43 115 91 55 105 41 69 376	**472**	100 94 77 72 66 52 49 46	**0.00**	29	44	5	–	–	–	–	–	–	–	–	–	–	Cholest-5-ene-16,22-dione, 3β,26-dihydroxy-, 3-acetate, (20S,25R)-	29853-25-6	NI44071
165 43 95 41 139 135 235 55	**472**	100 91 83 83 75 75 61 50	**25.00**	29	44	5	–	–	–	–	–	–	–	–	–	–	6-(2,5-Dimethoxy-3-methylphenyl)-1-[2-(4-hydroxy-4-methyl-1-oxopentyl)-1,2-dimethylcyclopentyl]-4-methyl-4-hexen-2-one	93236-29-4	NI44072
165 436 135 95 235 205 43 396	**472**	100 77 69 61 54 46 38 31	**0.80**	29	44	5	–	–	–	–	–	–	–	–	–	–	6-[1-(2,5-Dimethoxy-3-methylphenyl)-2-methyl-2-butenyl]octahydro-6-hydroxy-5-(2-hydroxy-2-methylpropyl)-3a,7a-dimethyl-4H-inden-4-one	93236-30-7	NI44073
139 126 43 69 55 140 358 107	**472**	100 38 27 19 16 14 11 10	**7.10**	29	44	5	–	–	–	–	–	–	–	–	–	–	Spirostan-12-one, 3-acetyloxy-, (3β,5α)-		IC04491
43 189 187 205 235 219 411 436	**472**	100 51 50 39 21 21 9 8	**6.00**	30	48	4	–	–	–	–	–	–	–	–	–	–	Careyagenolide	84749-88-2	NI44074
44 119 205 189 187 223 454 426	**472**	100 37 10 9 7 5 2 2	**1.80**	30	48	4	–	–	–	–	–	–	–	–	–	–	2α,3β-Dihydroxy-taraxastan-20-en-28-oic acid		NI44075
472 123 436 408 454 121 332 191	**472**	100 68 64 61 53 50 42 42		30	48	4	–	–	–	–	–	–	–	–	–	–	D:A-Friedooleanan-28-oic acid, 21-hydroxy-3-oxo-, (21α)-	64543-32-4	NI44076
248 203 223 204 175 206 133 193	**472**	100 98 75 63 59 56 55 50	**11.00**	30	48	4	–	–	–	–	–	–	–	–	–	–	Hederagenin	465-99-6	NI44077
472 69 43 55 57 353 41 473	**472**	100 57 46 44 35 34 34 33		30	48	4	–	–	–	–	–	–	–	–	–	–	Lanost-9(11)-en-18-oic acid, 3,20,23-trihydroxy-, (3β,20ξ)-	36872-80-7	NI44078
55 81 412 147 145 105 107 93	**472**	100 94 88 77 72 69 66 66	**0.00**	30	48	4	–	–	–	–	–	–	–	–	–	–	Methyl (25RS)-3β-acetoxy-5-cholesten-26-oate		NI44079
472 382 174 473 75 129 367 142	**472**	100 37 37 35 32 28 20 20		30	52	2	–	–	–	–	–	–	1	–	–	–	Cholesterol, 7-oxo-, trimethylsilyl ether		MS10145
129 382 343 472 127 367 383 344	**472**	100 81 81 35 28 26 25 22		30	52	2	–	–	–	–	–	–	1	–	–	–	Cholesterol, 24-oxo-, trimethylsilyl ether		MS10146
429 430 131 133 281 431 129 89	**472**	100 33 21 15 13 9 9 9	**0.00**	30	52	2	–	–	–	–	–	–	1	–	–	–	Pregn-5-ene-20-one, 3-[(tripropylsilyl)oxy]-, (3β)-		NI44080
381 472 382 242 239 331 291 332	**472**	100 95 80 30 30 20 20 6		31	40	2	2	–	–	–	–	–	–	–	–	–	5-Benzylamino-2-hydroxy-3-undecyl-1,4-benzoquinone-1-benzylimine		NI44081
430 165 431 164 43 58 472 56	**472**	100 89 35 31 22 14 12 12		31	52	3	–	–	–	–	–	–	–	–	–	–	2H-1-Benzopyran-6-ol, 3,4-dihydro-2,5,7,8-tetramethyl-2-(4,8,12-trimethyltridecyl)-, acetate, [2R-[2R*(4R*,8R*)]]-	58-95-7	NI44082
165 164 430 207 57 166 55 431	**472**	100 37 36 27 16 12 11 11	**4.76**	31	52	3	–	–	–	–	–	–	–	–	–	–	2H-1-Benzopyran-6-ol, 3,4-dihydro-2,5,7,8-tetramethyl-2-(4,8,12-trimethyltridecyl)-, acetate, [2R-[2R*(4R*,8R*)]]-	58-95-7	NI44083
327 83 123 440 175 189 203 303	**472**	100 70 33 25 16 15 12 10	**3.00**	31	52	3	–	–	–	–	–	–	–	–	–	–	D:A-Friedooleanan-26-oic acid, 3-hydroxy-, methyl ester, (3α,20α)-	56816-10-5	NI44084
169 264 275 233 220 223 249 231	**472**	100 62 48 33 33 31 27 26	**12.38**	31	52	3	–	–	–	–	–	–	–	–	–	–	D:A-Friedooleanan-29-oic acid, 3-hydroxy-, methyl ester, (3α,20α)-	56816-10-5	MS10147
109 454 385 373 411 457 439	**472**	100 18 16 15 5 2 2	**0.00**	31	52	3	–	–	–	–	–	–	–	–	–	–	3,4-Secodammara-4(28),25-dien-3-oic acid, 20-hydroxy-, methyl ester	56393-87-4	NI44085
343 382 255 367 344 215 342 383	**472**	100 50 30 29 29 22 21 18	**17.86**	31	56	1	–	–	–	–	–	–	1	–	–	–	Ergost-5-ene, 3-[(trimethylsilyl)oxy]-, (3β,24ξ)-	55429-60-2	NI44086
129 343 382 367 107 121 215 472	**472**	100 97 89 44 41 40 38 37		31	56	1	–	–	–	–	–	–	1	–	–	–	Ergost-5-ene, 3-[(trimethylsilyl)oxy]-, (3β,24R)-	55429-62-4	NI44087
129 343 382 472 367 344 383 342	**472**	100 94 77 32 31 26 24 20		31	56	1	–	–	–	–	–	–	1	–	–	–	Ergost-5-ene, 3-[(trimethylsilyl)oxy]-, (3β,24R)-	55429-62-4	MS10148
75 73 255 81 69 472 83 71	**472**	100 94 73 59 59 53 53 39		31	56	1	–	–	–	–	–	–	1	–	–	–	Ergost-7-ene, 3-[(trimethylsilyl)oxy]-, (3β,5α)-	40272-64-8	NI44088
75 255 73 472 55 107 81 95	**472**	100 75 57 56 49 43 40 37		31	56	1	–	–	–	–	–	–	1	–	–	–	Ergost-7-ene, 3-[(trimethylsilyl)oxy]-, (3β,5ξ)-	18880-54-1	NI44089
472 75 107 73 55 147 229 105	**472**	100 75 72 59 53 45 43 41		31	56	1	–	–	–	–	–	–	1	–	–	–	Ergost-8(14)-ene, 3-[(trimethylsilyl)oxy]-, (3β)-	55515-24-7	NI44090
269 472 147 227 95 121 161 367	**472**	100 78 50 49 42 40 39 34		31	56	1	–	–	–	–	–	–	1	–	–	–	Methosteryl trimethylsilyl ether		LI09255
472 269 473 121 135 161 147 382	**472**	100 64 41 20 18 17 17 16		31	56	1	–	–	–	–	–	–	1	–	–	–	4α-Methyl-5α-cholest-7-en-3β-ol trimethylsilyl ether		MS10149
343 129 344 73 472 382 135 95	**472**	100 48 27 27 26 25 25 20		31	56	1	–	–	–	–	–	–	1	–	–	–	4β-Methyl-5α-cholest-7-en-3β-ol trimethylsilyl ether		LI09256
125 43 141 472 112 55 41 57	**472**	100 35 33 26 22 19 19 17		32	56	2	–	–	–	–	–	–	–	–	–	–	Cholestan-3-one, 2-isopropyl-, cyclic 1,2-ethanediyl acetal, (2α,5α)-	55162-59-9	NI44091
472 471 473 210 197 43 470 57	**472**	100 66 34 15 10 10 9 8		34	48	1	–	–	–	–	–	–	–	–	–	–	Cholest-4-en-3-one, 2-benzylene-	74421-21-9	NI44092
185 155 165 205 135 120 170 150	**473**	100 48 44 37 28 23 16 16	**0.00**	3	9	2	3	1	–	4	–	–	–	2	–	1	4H-1,2,4,6,3,5-Thiatriazadiphosphorine, 3,3,5,5-tetrachloro-3,3,5,5-tetrahydro-4-(trimethylstannyl)-, 1,1-dioxide	41083-44-7	MS10150
100 150 69 323 112 238 93 404	**473**	100 70 58 48 24 16 15 14	**0.00**	10	–	1	1	–	17	–	–	–	–	–	–	–	6-Oxo-4-(trifluoromethyl)perfluoroquinolizidine		DD01729
73 299 384 357 147 328 211 458	**473**	100 61 38 35 29 27 27 25	**0.00**	15	40	6	1	–	–	–	–	–	4	1	–	–	Phosphoric acid, bis(trimethylsilyl) 2-[(trimethylsilyl)oxy]-3-[[(trimethylsilyl)oxy]imino]propyl ester, (±)-	55429-92-0	NI44093
73 299 384 227 357 211 147 75	**473**	100 69 46 34 32 29 27 18	**0.00**	15	40	6	1	–	–	–	–	–	4	1	–	–	Phosphoric acid, bis(trimethylsilyl) 2-[(trimethylsilyl)oxy]-3-[[(trimethylsilyl)oxy]imino]propyl ester, (±)-	55429-92-0	NI44094
73 299 315 147 231 316 458 225	**473**	100 77 72 44 26 23 20 17	**0.80**	15	40	6	1	–	–	–	–	–	4	1	–	–	2-Propanone, 1,3-bis[(trimethylsilyl)oxy]-, O-[bis[(trimethylsilyl)oxy]phosphinyl]oxime	55429-50-0	NI44095
269 297 119 270 298 69 267 43	**473**	100 79 19 10 9 9 6 5	**0.00**	16	13	4	1	–	10	–	–	–	–	–	–	–	β-Ethoxyoctopamine, bis(pentafluoropropionyl)-		MS10151
283 473 310 177 119 284 147 77	**473**	100 50 34 28 19 14 13 12		16	13	4	1	–	10	–	–	–	–	–	–	–	3-Hydroxy-4-methoxyphenylpropylamine, bis(pentafluoropropionyl)-		NI44096
296 190 149 119 297 140 283 42	**473**	100 81 29 23 18 14 11 9	**0.00**	16	13	4	1	–	10	–	–	–	–	–	–	–	N-Methyl-3-methoxytyramine, bis(pentafluoropropionyl)-		MS10152

MASS TO CHARGE RATIOS								M.W.	INTENSITIES								Parent	C	H	O	N	S	F	Cl	Br	I	Si	P	B	X	COMPOUND NAME	CAS Reg No	No
105	77	354	352	356	51	106	166	473	100	50	21	12	11	11	8	7	1.00	20	13	3	1	–	–	–	2	–	–	–	–	–	N-Benzoyl-N-phenyl-4,6-dibromoaniline-2-carboxylic acid		NI44097
294	248	73	416	295	211	249	179	473	100	57	47	29	22	19	13	12	0.00	20	40	4	1	–	–	–	–	–	3	1	–	–	Phosphonic acid, [1-[isopropylideneamino]-2-[4-[(trimethylsilyl)oxy]phenyl]ethyl]-, bis(trimethylsilyl) ester	56227-26-0	NI44098
57	344	361	400	41	29	417	43	473	100	48	31	30	25	18	16	8	7.90	22	36	2	1	–	–	–	–	1	–	–	–	–	Benzene, 1,3,5-tris(2,2-dimethylpropyl)-2-iodo-4-methyl-6-nitro-	41080-94-8	MS10153
255	105	420	94	286	308	77	136	473	100	70	67	60	55	42	41	40	20.00	25	23	5	5	–	–	–	–	–	–	–	–	–	N-Cyanoethyl-N-(2-(phenoxyacetoxy)ethyl)-4-(4-nitrophenylazo)aniline		IC04492
107	72	44	279	43	42	77	41	473	100	57	50	43	35	32	26	24	0.00	25	35	6	3	–	–	–	–	–	–	–	–	–	N,N-Bis[3-(4′-carbamoylmethylphenoxy)-2-hydroxypropyl]isopropylamine		NI44099
74	157	299	247	103	458	370	174	473	100	98	50	25	16	10	10	8	0.00	25	55	3	1	–	–	–	–	–	2	–	–	–	N-Acetyl-O-TMS-15-methyl-hexadecasphinganine		LI09257
473	280	370	474	225	75	209	388	473	100	46	42	38	29	25	23	16		27	43	4	1	–	–	–	–	–	1	–	–	–	Pregn-4-ene-3,20-dione, 17-(acetyloxy)-6-methyl-, 3-[O-(trimethylsilyl)oxime], (6α)-	74299-45-9	NI44100
473	384	474	417	458	129	344	385	473	100	50	43	37	34	31	23	18		27	47	2	1	–	–	–	–	–	2	–	–	–	Pregna-5,16-dien-20-one, 3-[(trimethylsilyl)oxy-, 20-[O-(trimethylsilyl)oxime]-, (3β)-		LI09258
105	77	135	136	44	151	51	121	473	100	73	57	32	25	22	18	17	2.00	28	27	6	1	–	–	–	–	–	–	–	–	–	1-Acetyl-3,4-bis(benzoyloxy)-2-(4-methoxybenzyl)pyrrolidine		LI09259
473	254	127	117	472	128	77	440	473	100	30	30	21	16	15	15	13		29	23	–	5	1	–	–	–	–	–	–	–	–	7-(Methylthio)-4-(naphthylamino)-5-phenyl-2-(p-tolyl)imidazo[5,1-f][1,2,4]triazine		MS10154
18	57	43	95	55	81	69	28	473	100	94	93	84	82	78	67	64	0.96	29	47	4	1	–	–	–	–	–	–	–	–	–	3β-Acetoxy-6-nitro-5-cholestene	1912-54-5	MS10155
43	57	55	95	81	41	69	107	473	100	77	72	63	63	54	52	50	0.00	29	47	4	1	–	–	–	–	–	–	–	–	–	3β-Acetoxy-6-nitro-5-cholestene	1912-54-5	NI44101
473	143	474	144	130	475	182	57	473	100	93	37	15	10	8	7	7		33	47	1	1	–	–	–	–	–	–	–	–	–	6-Oxo-5α-cholestano[3,2-b]indole		MS10156
157	473	474	158	55	57	69	81	473	100	64	26	20	15	14	11	7		34	51	–	1	–	–	–	–	–	–	–	–	–	Cholestano[3,2-b]5′-methylindole, (5α)-		MS10157
473	474	196	144	157	197	55	57	473	100	40	21	21	20	17	16	15		34	51	–	1	–	–	–	–	–	–	–	–	–	Cholestano[3,4-b]5′-methylindole, (5β)-		MS10158
157	473	474	158	57	55	475	144	473	100	80	30	17	7	6	5	5		34	51	–	1	–	–	–	–	–	–	–	–	–	Cholestano[3,2-b]N-methylindole, (5α)-		MS10159
473	474	472	471	196	475	144	57	473	100	37	11	9	8	7	6	6		34	51	–	1	–	–	–	–	–	–	–	–	–	Cholestano[3,4-b]N-methylindole, (5α)-		MS10160
180	292	104	312	179	77	200	151	474	100	99	95	70	50	50	45	40	0.00	6	18	–	6	–	–	–	–	–	–	–	–	4	2,4,6,8,9,10-Hexaaza-1,3,5,7-tetraarsatricyclo[3.3.1.1^{3,7}]decane, 2,4,6,8,9,10-hexamethyl-	2030-90-2	NI44102
28	306	334	194	222	168	112	56	474	100	5	4	4	3	3	3	3	1.08	12	2	10	–	–	–	–	–	–	–	–	–	3	Iron, decacarbonyl(ethyne)tri-	56902-53-5	NI44103
250	191	63	107	60	252	193	206	474	100	75	72	71	54	42	28	22	2.00	14	14	2	–	4	–	2	–	–	–	2	–	–	O,O′-Ethylenebis(4-chlorophenyldithiophosphonate)	94286-61-0	NI44104
474	252	415	59	475	387	283	253	474	100	67	62	42	18	13	9	8		15	8	6	–	–	10	–	–	–	–	–	–	–	3,4-Dihydroxyphenylacetic acid, bis(pentafluoropropionate)-, methyl ester	55683-24-4	NI44105
59	119	252	415	69	65	474	77	474	100	54	36	30	29	25	22	18		15	8	6	–	–	10	–	–	–	–	–	–	–	3,4-Dihydroxyphenylacetic acid, bis(pentafluoropropionate)-, methyl ester	55683-24-4	MS10161
311	117	312	129	165	149	103	313	474	100	20	18	17	16	7	6	5	0.00	15	8	6	–	–	10	–	–	–	–	–	–	–	α,4-Dihydroxyphenylacetic acid, bis(pentafluoropropionate)-, methyl ester		NI44106
73	299	147	357	227	315	211	387	474	100	67	64	59	58	38	33	29	0.00	15	39	7	–	–	–	–	–	–	4	1	–	–	3,5-Dioxa-4-phospha-2-silaoctan-8-oic acid, 2,2-dimethyl-4,7-bis[(trimethylsilyl)oxy]-, trimethylsilyl ester, 4-oxide	31038-13-8	NI44107
73	357	299	147	227	459	387	315	474	100	51	44	35	32	25	20	20	0.00	15	39	7	–	–	–	–	–	–	4	1	–	–	3,5-Dioxa-4-phospha-2-silaoctan-8-oic acid, 2,2-dimethyl-4,7-bis[(trimethylsilyl)oxy]-, trimethylsilyl ester, 4-oxide	31038-13-8	NI44108
73	299	459	315	369	147	217	387	474	100	36	32	28	24	20	18	17	0.40	15	39	7	–	–	–	–	–	–	4	1	–	–	Tetrakis(trimethylsilyl)-2-phosphoglyceric acid	31038-14-9	NI44109
73	327	341	328	459	343	329	74	474	100	42	19	12	11	11	9	9	0.00	15	42	7	–	–	–	–	–	–	5	–	–	–	3,3,5-Triethoxy-1,1,1,7,7,7-hexamethyl-5-(trimethylsiloxy)tetrasiloxane		MS10162
191	88	187	101	71	111	219	75	474	100	90	65	40	37	23	19	19	0.00	17	30	15	–	–	–	–	–	–	–	–	–	–	Gluco-1,4-gluco-1,5-xylose		LI09260
55	69	41	139	43	97	83	121	474	100	60	60	39	39	36	25	21	5.52	18	34	4	–	–	–	–	–	–	–	–	–	2	Nonanoic acid, 9,9′-diselenodi-	22686-33-5	NI44110
69	157	48	116	74	110	319	45	474	100	33	31	29	24	23	21	21	0.00	19	16	1	8	1	–	2	–	–	–	–	–	–	4-(3-Chlorophenyl)-5-(3-chlorophenyl)amino-3-[6-methyl-3-(methylthio)-5-oxo-4,5-dihydro-1,2,4-triazin-4-yl]imino-2H-2,3-dihydro-1,2,4-triazole		MS10163
69	45	46	110	74	319	321	127	474	100	89	65	44	42	28	21	21	0.00	19	16	1	8	1	–	2	–	–	–	–	–	–	4-(4-Chlorophenyl)-5-(4-chlorophenyl)amino-3-[6-methyl-3-(methylthio)-5-oxo-4,5-dihydro-1,2,4-triazin-4-yl]imino-2H-2,3-dihydro-1,2,4-triazole		MS10164
43	155	441	81	355	257	91	214	474	100	67	66	61	44	43	43	40	0.19	20	26	11	–	1	–	–	–	–	–	–	–	–	Methyl 2,3,4-tri-O-acetyl-6-O-tosyl-α-D-glucopyranoside		NI44111
185	140	98	212	169	290	217	41	474	100	94	64	38	37	34	33	33	8.01	21	31	6	2	–	–	–	–	–	2	–	1	–	2(1H)-Pyrimidinone, 4-(trimethylsiloxy)-1-[5-O-(trimethylsilyl)-β-D-ribofuranosyl]-, cyclic benzeneboronate	29015-23-4	NI44112
132	454	455	303	474	58	151	133	474	100	50	19	18	17	17	16	16		22	10	–	–	–	8	–	–	–	–	–	–	1	Titanium, bis(η^5-2,4-cyclopentadien-1-yl)(3,3′,4,4′,5,5′,6,6′-octafluoro[1,1′-biphenyl]-2,2′-diyl)-	12278-18-1	NI44113
105	268	224	226	154	106	182	195	474	100	98	84	70	61	60	54	39	0.00	22	26	8	4	–	–	–	–	–	–	–	–	–	L-Histidine, N-(N-benzoyl-L-α-glutamyl)-1-(ethoxycarbonyl)-, 5-methyl ester	55521-20-5	NI44114
270	335	336	333	168	233	259	334	474	100	85	69	46	40	34	23	23	12.99	23	23	1	–	–	–	–	–	–	–	–	–	2	η^4-2-[(ferrocenyl(hydroxy)methyl)norbornadiene](η^5-cyclopentadienyl)rhodium	72950-50-6	NI44115

MASS TO CHARGE RATIOS	M.W.	INTENSITIES	Parent	C	H	O	N	S	F	Cl	Br	I	Si	P	B	X	COMPOUND NAME	CAS Reg No	No
474 173 144 169 109 81 186 475	**474**	100 80 63 60 46 46 35 34		23	26	9	2	–	–	–	–	–	–	–	–	–	1-(2,3,4,6-Tetra-O-acetyl-β-D-glucopyranosyl)-6-aminoquinoline		LI09261
470 173 144 169 109 81 471 145	**474**	100 80 63 61 46 46 34 31	**0.00**	23	26	9	2	–	–	–	–	–	–	–	–	–	1-(2,3,4,6-Tetra-O-acetyl-β-D-glucopyranosyl)-6-aminoquinoline		NI44116
474 476 108 459 457 206 415 207	**474**	100 96 52 44 42 23 16 15		23	27	4	2	–	–	–	1	–	–	–	–	–	2-Hydroxy-10-bromo-O-acetylseredamine		LI09262
73 459 460 222 461 75 74 387	**474**	100 77 27 20 14 11 9 8	**1.40**	23	34	5	–	–	–	–	–	–	3	–	–	–	3,7,9-Tris[(trimethylsilyl)oxy]-1-methyl-6H-dibenzo[b,d]pyran-6-one		NI44117
318 319 320 284 321 75 456 101	**474**	100 72 48 32 24 16 13 13	**0.60**	24	16	1	6	–	–	2	–	–	–	–	–	–	7-Chloro-5-(2-chlorophenyl)-2-(cinnolinylcarbonylhydrazono)-2,3-dihydro-1H-1,4-benzodiazepine		MS10165
152 227 229 151 154 51 153 77	**474**	100 58 33 26 25 24 20 18	**14.00**	24	20	1	–	–	–	–	–	–	–	–	–	2	Bis(diphenylarsine) oxide	2215-16-9	NI44118
474 209 475 341 311 325 340 193	**474**	100 62 34 29 17 16 12 11		24	26	10	–	–	–	–	–	–	–	–	–	–	Flavone, 8-β-D-glucopyranosyl-4',5,7-trimethoxy-	18469-71-1	NI44119
169 109 144 127 331 115 145 170	**474**	100 60 40 25 23 14 11 9	**0.69**	24	26	10	–	–	–	–	–	–	–	–	–	–	α-Naphthyl-2,3,4,6-tetra-O-acetyl-β-D-glucopyranoside	14581-88-5	NI44120
151 90 59 96 58 88 78 73	**474**	100 91 50 44 44 43 43 40	**15.50**	24	30	8	2	–	–	–	–	–	–	–	–	–	20,24-Diacetamido-2,3,11,12-dibenzo-1,4,7,10,13,16-hexaoxacyclooctadeca-2,11-diene		MS10166
281 282 73 283 209 459 193 75	**474**	100 25 18 10 6 4 3 2	**0.00**	24	38	4	–	–	–	–	–	–	3	–	–	–	1-[2,4-Bis(trimethylsiloxy)phenyl]-2-[(4-trimethylsiloxy)phenyl]propan-1-one		NI44121
417 375 319 193 277 207 291 177	**474**	100 83 77 71 65 57 48 42	**0.00**	24	54	3	–	–	–	–	–	–	3	–	–	–	Hexa-tert-butylcyclotrisiloxane	78393-18-7	NI44122
325 324 474 387 244 326 61 413	**474**	100 97 28 27 27 22 22 16		25	34	5	2	1	–	–	–	–	–	–	–	–	Aspidospermidin-21-oic acid, 1-acetyl-17-methoxy-20-[(methylthio)methoxy]-, methyl ester	55103-36-1	NI44123
287 288 375 361 218 289 145 75	**474**	100 22 14 12 9 6 6 5	**0.00**	25	54	4	–	–	–	–	–	–	2	–	–	–	Hexadecanoic acid, 2,3-bis[(trimethylsilyl)oxy]propyl ester	1188-74-5	WI01570
371 73 43 57 147 372 55 41	**474**	100 86 58 43 42 32 32 30	**0.00**	25	54	4	–	–	–	–	–	–	2	–	–	–	Hexadecanoic acid, 2,3-bis[(trimethylsilyl)oxy]propyl ester	1188-74-5	NI44124
73 43 371 57 41 55 147 71	**474**	100 82 55 55 39 35 33 26	**0.00**	25	54	4	–	–	–	–	–	–	2	–	–	–	Hexadecanoic acid, 2,3-bis[(trimethylsilyl)oxy]propyl ester	1188-74-5	NI44125
218 341 219 313 371 220 387 129	**474**	100 40 20 14 8 8 7 7	**0.00**	25	54	4	–	–	–	–	–	–	2	–	–	–	Hexadecanoic acid, 2-[(trimethylsilyl)oxy]-1-[[(trimethylsilyl)oxy]methyl]ethyl ester	53212-97-8	WI01571
218 219 375 129 287 220 376 294	**474**	100 20 17 15 10 10 8 6	**0.00**	25	54	4	–	–	–	–	–	–	2	–	–	–	Hexadecanoic acid, 2-[(trimethylsilyl)oxy]-1-[[(trimethylsilyl)oxy]methyl]ethyl ester	53212-97-8	WI01572
73 129 218 43 103 147 57 41	**474**	100 92 63 57 53 47 41 33	**0.00**	25	54	4	–	–	–	–	–	–	2	–	–	–	Hexadecanoic acid, 2-[(trimethylsilyl)oxy]-1-[[(trimethylsilyl)oxy]methyl]ethyl ester	53212-97-8	WI01573
73 217 257 185 75 147 113 290	**474**	100 53 41 17 15 14 10 6	**0.00**	25	54	4	–	–	–	–	–	–	2	–	–	–	Octadecanoic acid, 6,7-bis[(trimethylsilyl)oxy]-, methyl ester		LI09263
73 259 215 75 155 147 103 332	**474**	100 37 37 18 15 15 13 4	**0.00**	25	54	4	–	–	–	–	–	–	2	–	–	–	Octadecanoic acid, 9,10-bis[(trimethylsilyl)oxy]-, methyl ester	55429-57-7	LI09264
73 259 215 69 155 147 75 44	**474**	100 79 75 23 20 19 17 17	**0.00**	25	54	4	–	–	–	–	–	–	2	–	–	–	Octadecanoic acid, 9,10-bis[(trimethylsilyl)oxy]-, methyl ester	55429-57-7	LI09265
73 215 259 69 75 55 155 147	**474**	100 37 36 18 17 16 15 15	**0.00**	25	54	4	–	–	–	–	–	–	2	–	–	–	Octadecanoic acid, 9,10-bis[(trimethylsilyl)oxy]-, methyl ester	55429-57-7	NI44126
73 259 215 69 155 147 55 44	**474**	100 80 72 22 18 18 15 15	**0.00**	25	54	4	–	–	–	–	–	–	2	–	–	–	Octadecanoic acid, 9,10-bis[(trimethylsilyl)oxy]-, methyl ester, (R*,S*)-	22032-79-7	NI44127
73 259 215 155 147 69 260 44	**474**	100 79 75 21 21 21 17 17	**0.00**	25	54	4	–	–	–	–	–	–	2	–	–	–	Octadecanoic acid, 9,10-bis[(trimethylsilyl)oxy]-, methyl ester, (R*,S*)-	22032-79-7	LI09266
73 301 287 173 187 75 147 55	**474**	100 85 73 67 57 35 31 28	**0.00**	25	54	4	–	–	–	–	–	–	2	–	–	–	Octadecanoic acid, 11,12-bis[(trimethylsilyl)oxy]-, methyl ester		LI09267
294 43 414 372 354 312 295 83	**474**	100 81 56 53 49 42 35 35	**0.87**	26	34	8	–	–	–	–	–	–	–	–	–	–	4-Desoxy-4α-lumiphorbol-12,13,20-triacetate		NI44128
43 294 83 149 69 312 121 414	**474**	100 54 52 51 35 34 33 28	**0.25**	26	34	8	–	–	–	–	–	–	–	–	–	–	4-Desoxy-phorbol-12,13,20-triacetate		NI44129
294 414 372 354 312 474	**474**	100 64 53 46 43 10		26	34	8	–	–	–	–	–	–	–	–	–	–	9-Hydroxy-12,13,20-triacetoxy-4-desoxy-1,6-tigliadien-3-one		LI09268
43 121 122 91 93 135 41 77	**474**	100 67 63 47 38 35 33 29	**1.80**	26	34	8	–	–	–	–	–	–	–	–	–	–	1H-2,8a-Methanocyclopenta[a]cyclopropa[e]cyclodecen-11-one, 5,6-bis(acetyloxy)-4-[(acetyloxy)methyl]-1a,2,5,5a,6,9,10,10a-octahydro-5a-hydroxy-1,1,7,9-tetramethyl-, [1aR-(1aα,2α,5β,5aβ)]-	30220-45-2	NI44130
283 152 127 153 359 165 121 223	**474**	100 81 80 76 72 72 72 53	**49.00**	26	34	8	–	–	–	–	–	–	–	–	–	–	24-Nor-16,17-secochol-2-ene-16,23-dioic acid, 7,11,21-trihydroxy-2-methoxy-4,8-dimethyl-1,17-dioxo-, γ-lactone δ-lactone, (4α,5α,7α,11α,13β,14α,20R)-	27368-79-2	NI44131
474 384 324 244 297 475 385 218	**474**	100 84 63 51 43 37 34 33		26	42	4	–	–	–	–	–	–	2	–	–	–	11β-Hydroxy-estradiol-17β-acetate di-TMS		NI44132
474 409 381 475 304 410 472 237	**474**	100 66 39 37 23 20 13 13		27	22	1	–	–	–	–	–	–	–	–	–	2	1'-Benzoylbiferrocene		MS10167
120 255 474 92 475 256 119 41	**474**	100 77 44 32 14 12 12 11		27	26	6	2	–	–	–	–	–	–	–	–	–	1-Amino-2-[6-(2-aminobenzoxy)hexyloxy]-4-hydroxyanthraquinone		IC04493
256 164 106 324 75 175 107 417	**474**	100 32 25 23 20 18 16 15	**0.00**	27	30	4	4	–	–	–	–	–	–	–	–	–	4,20-Dimethoxy-2,24-dioxa-10,13,16-triaza-9,10,16,17-diene[2.9.17.24](1,4)diphenyl[1,25](2,6)piridinophan		MS10168
43 203 44 32 354 55 162 41	**474**	100 87 73 70 68 59 55 55	**43.71**	27	38	7	–	–	–	–	–	–	–	–	–	–	5α-Card-20(22)-enolide, 3β,14,15β-trihydroxy-, 3,15-diacetate	14155-62-5	NI44133
221 55 193 205 204 81 236 39	**474**	100 80 49 44 39 30 29 24	**6.06**	27	38	7	–	–	–	–	–	–	–	–	–	–	7aβ-Methyl-1,5-dioxo-3aα-hexahydro-4-indanpropionyl-7aβ-methyl-5-oxo-1β-hydroxy-3aα-hexahydro-4-indanpropionic acid methyl ester	23212-01-3	NI44134
221 55 193 205 204 236 80 220	**474**	100 64 39 34 32 24 24 11	**4.00**	27	38	7	–	–	–	–	–	–	–	–	–	–	7aβ-Methyl-1,5-dioxo-3aα-hexahydro-4-indanpropionyl-7aβ-methyl-5-oxo-1β-hydroxy-3aα-hexahydro-4-indanpropionic acid methyl ester	23212-01-3	LI09269
431 279 294 251 354 253 356 296	**474**	100 80 63 62 50 44 38 35	**1.00**	27	38	7	–	–	–	–	–	–	–	–	–	–	Pregn-16-en-20-one, 3,7,12-tris(acetyloxy)-, (3α,5β,7α,12α)-		NI44135
138 330 137 372 256 123 136 175	**474**	100 68 48 39 22 20 19 18	**3.24**	27	42	5	–	–	–	–	–	–	1	–	–	–	Pregna-3,5-dien-20-one, 21-(acetyloxy)-17-hydroxy-6-methyl-3-[(trimethylsilyl)oxy]-	74299-00-6	NI44136
418 281 419 282 474 299 459 403	**474**	100 97 33 23 19 19 17 17		27	42	5	–	–	–	–	–	–	1	–	–	–	Pregn-4-ene-3,20-dione, 17-(acetyloxy)-6-methyl-6-[(trimethylsilyl)oxy]-, (6β)-	69833-71-2	NI44137

MASS TO CHARGE RATIOS								M.W.	INTENSITIES								Parent	C	H	O	N	S	F	Cl	Br	I	Si	P	B	X	COMPOUND NAME	CAS Reg No	No
303	474	384	265	304	475	385	369	**474**	100	66	48	29	28	25	21	21		27	46	3	–	–	–	–	–	–	2	–	–	–	5β-Hydroxy-$\Delta^{1(6)}$-tetrahydrocannabinol bis(trimethylsilyl) ether		NI44138
443	444	287	474	353	188	158	331	**474**	100	39	31	24	22	16	15	14		27	46	3	2	–	–	–	–	–	1	–	–	–	Pregn-4-ene-3,20-dione, 6-methyl-17-[(trimethylsilyl)oxy]-, bis(O-methyloxime), (6α)-	69833-56-3	NI44139
388	472	357	387	358	389	43	430	**474**	100	48	38	36	27	27	24	15	**3.01**	28	26	7	–	–	–	–	–	–	–	–	–	–	3,4-Diacetoxy-8,11-dimethoxy-6,10-dimethyldibenzo[c,kl]xanthene		NI44140
105	474	386	387	475	77	182	281	**474**	100	97	94	66	33	28	27	25		28	30	5	2	–	–	–	–	–	–	–	–	–	Cylindrocarine, 1-benzoyl-20-oxo-	22226-27-3	NI44141
403	404	135	57	134	133	291	41	**474**	100	27	19	19	13	13	11	11	**6.40**	28	42	4	–	1	–	–	–	–	–	–	–	–	Bis[2-hydroxy-5-(2,4,4,-trimethylpentyl)] sulphone		IC04494
122	57	43	123	41	135	121	93	**474**	100	82	60	55	50	47	47	46	**0.42**	28	42	6	–	–	–	–	–	–	–	–	–	–	Ingenol-3-octanoate		NI44142
75	215	73	217	384	81	107	43	**474**	100	70	70	58	52	44	42	42	**10.00**	28	51	3	–	–	–	–	–	–	1	–	1	–	5β-Pregnane-3α,17α,20α-triol 3-TMS tert-butyl-boronate		MS10169
105	369	237	209	325	132	77	474	**474**	100	95	30	30	20	20	20	8		30	18	6	–	–	–	–	–	–	–	–	–	–	3,3′-Bis-(3-benzoylphthalidyl)		LI09270
355	354	353	325	428	475	337	326	**474**	100	6	4	4	3	2	1	1	**0.50**	30	22	4	2	–	–	–	–	–	–	–	–	–	4-Benzoyl-1,5-diphenyl-3-hydroxypyrrol-2-on-3-carboxanilide		NI44143
180	181	77	78	459	154	16	15	**474**	100	47	46	22	14	9	9	9	**0.20**	30	32	–	2	–	–	–	–	–	–	–	–	2	Dimethyl(α-phenylbenzylideneamino)aluminium dimer		LI09271
411	69	81	76	95	71	83	67	**474**	100	82	69	68	59	53	51	49	**39.00**	30	50	2	–	1	–	–	–	–	–	–	–	–	Cholest-2-ene-2-carbothioic acid, 3-hydroxy-, O-ethyl ester, (5α)-	55283-37-9	NI44144
217	248	199	173	218	207	249	200	**474**	100	82	40	24	18	17	16	14	**2.00**	30	50	4	–	–	–	–	–	–	–	–	–	–	Camelliagenin A		MS10170
105	57	43	55	41	113	71	69	**474**	100	71	70	45	37	36	34	27	**6.04**	30	50	4	–	–	–	–	–	–	–	–	–	–	Eicosanoic acid, 2-phenyl-1,3-dioxan-5-yl ester	57156-90-8	NI44145
95	107	135	323	341	201	384	438	**474**	100	95	60	34	32	22	12	8	**0.00**	30	50	4	–	–	–	–	–	–	–	–	–	–	Nepedinol	104139-53-9	NI44146
143	459	125	387	375				**474**	100	20	15	2	2				**0.00**	30	50	4	–	–	–	–	–	–	–	–	–	–	3,4-Secodammar-4(28)-en-3-oic acid, 20,24-epoxy-25-hydroxy-, (24S)-	21671-00-1	NI44147
250	235	251	224	474	234	223	236	**474**	100	35	15	10	5	5	5	3		30	50	4	–	–	–	–	–	–	–	–	–	–	Soyasapogenol A	508-01-0	NI44148
167	149	57	55	43	321	71	41	**474**	100	82	38	17	17	14	14	13	**0.00**	30	50	4	–	–	–	–	–	–	–	–	–	–	Bis(undecyl)phthalate	3648-20-2	NI44149
149	321	57	43	55	41	167	71	**474**	100	48	35	25	18	15	13	12	**0.18**	30	50	4	–	–	–	–	–	–	–	–	–	–	Bis(undecyl)phthalate	3648-20-2	IC04495
384	230	229	244	321	459	124	147	**474**	100	64	56	52	31	27	23	22	**11.00**	30	54	2	–	–	–	–	–	–	1	–	–	–	6β-Trimethylsiloxy-5α-cholestan-3-one		NI44150
459	321	384	229	129	403	211	173	**474**	100	95	20	18	18	17	16	15	**11.00**	30	54	2	–	–	–	–	–	–	1	–	–	–	6β-Trimethylsiloxy-5β-cholestan-3-one		NI44151
397	104	383	474	134	78	357	103	**474**	100	77	54	36	33	28	26	24		31	22	5	–	–	–	–	–	–	–	–	–	–	Nordracorubin		MS10171
75	384	108	215	57	55	257	81	**474**	100	96	92	76	74	68	62	57	**0.00**	31	58	1	–	–	–	–	–	–	1	–	–	–	24-Methylcoprostanol, trimethylsilyl-		NI44152
73	417	179	418	147	177	45	195	**475**	100	59	53	24	19	16	15	14	**0.00**	18	34	4	1	1	–	–	–	–	3	1	–	–	Phosphonic acid, [1-isothiocyanato-2-[4-[(trimethylsilyl)oxy]phenyl]ethyl], bis(trimethylsilyl) ester	56196-77-1	NI44153
294	248	73	417	295	211	249	418	**475**	100	59	53	39	25	21	18	16	**0.00**	19	38	5	1	–	–	–	–	–	3	1	–	–	Phosphonic acid, [1-(acetylamino)-2-[4-[(trimethylsilyl)oxy]phenyl]ethyl]-, bis(trimethylsilyl) ester	56227-25-9	NI44154
73	388	386	360	477	475	362	75	**475**	100	30	29	17	16	15	13	11		20	26	2	3	–	–	–	1	–	2	–	–	–	3-Hydroxybromoazepam, bis(trimethylsilyl)-		NI44155
73	360	477	362	475	147	75	74	**475**	100	15	15	14	14	11	10	9		20	26	2	3	–	–	–	1	–	2	–	–	–	3-Hydroxybromoazepam, bis(trimethylsilyl)-		NI44156
267	73	475	179	45	268	208	476	**475**	100	98	30	30	18	17	12	10		21	26	2	1	–	5	–	–	–	2	–	–	–	Phenylethylamine, 3,4-bis[(trimethylsilyl)oxy]-N-pentafluorobenzyl-	50314-21-1	MS10172
73	267	179	268	45	475	74	208	**475**	100	79	24	19	17	15	8	8		21	26	2	1	–	5	–	–	–	2	–	–	–	Phenylethylamine, 3,4-bis[(trimethylsilyl)oxy]-N-pentafluorobenzyl-	50314-21-1	NI44157
267	73	268	269	45	75	74	460	**475**	100	56	25	10	8	7	6	4	**0.00**	21	26	2	1	–	5	–	–	–	2	–	–	–	Phenylethylamine, α,4-bis[(trimethylsilyl)oxy]-N-pentafluorobenzyl-		MS10173
73	267	44	75	74	268	209	77	**475**	100	30	15	9	8	8	5	4	**0.00**	21	26	2	1	–	5	–	–	–	2	–	–	–	Phenylethylamine, β,4-bis[(trimethylsilyl)oxy]-N-pentafluorobenzyl-		NI44158
477	475	479	317	315	478	399	480	**475**	100	47	46	32	32	19	15	13		24	15	–	1	–	–	–	2	–	–	–	–	–	3,6-Dibromo-1,4-diphenyl-carbazole		LI09272
246	248	372	70	42	113	98	56	**475**	100	33	31	26	23	21	17	16	**10.71**	24	34	1	3	1	–	1	–	–	1	–	–	–	Trimethylsilylperphenazine		NI44159
60	212	167	73	43	31	89	83	**475**	100	40	31	26	25	24	23	23	**0.01**	24	45	8	1	–	–	–	–	–	–	–	–	–	Dihydrodeoxypseudopederine		LI09273
458	237	105	205	118	367	91	457	**475**	100	99	98	97	97	89	84	68	**2.00**	26	25	6	3	–	–	–	–	–	–	–	–	–	3-Benzyl-2-hydroxy-1-(methoxycarbonyl)-5-(methoxycarbonylamino)-2,5-diphenylimidazolidin-4-one		MS10174
105	237	91	205	118	458	77	210	**475**	100	99	98	92	92	87	54	53	**5.00**	26	25	6	3	–	–	–	–	–	–	–	–	–	3-Benzyl-5-hydroxy-1-(methoxycarbonyl)-2-(methoxycarbonylamino)-2,5-diphenylimidazolidin-4-one		MS10175
105	296	118	104	77	205	237	268	**475**	100	96	94	79	72	67	62	38	**6.00**	26	25	6	3	–	–	–	–	–	–	–	–	–	5-Hydroxy-1-(methoxycarbonyl)-2-(methoxycarbonylamino)-2,5-diphenyl-3-(p-tolyl)imidazolidin-4-one		MS10176
256	105	78	106	104	407	257	225	**475**	100	83	73	62	55	53	50	26	**0.00**	27	29	5	3	–	–	–	–	–	–	–	–	–	4,20-Dimethoxy-2,13,24-trioxa-10,16-diaza-9,10,16,17-dieno[2.9.17.24](1.4)diphenyl[1.25](2.6)pyridinophane		MS10177
145	73	158	75	129	130	386	70	**475**	100	43	40	36	18	17	15	15	**9.00**	27	49	2	1	–	–	–	–	–	2	–	–	–	3β-Trimethylsiloxy-pregn-5-en-20-one, TMS oxime		LI09274
91	92	108	276	128	340	294	181	**475**	100	9	7	3	2	1	1	1	**0.00**	28	29	6	1	–	–	–	–	–	–	–	–	–	Benzyl 2,4-dideoxy-2-benzyloxycarbonylamino-3-O-benzyl-4,5-dihydro-α-D-glucopyranoside		LI09275
475	220	165	192	193	107	77	476	**475**	100	57	50	43	40	38	38	32		29	25	2	5	–	–	–	–	–	–	–	–	–	2-Phenyl-3-(4-tolylazo)-2,5-dimethoxyphenylazoindole		IC04496
244	109	300	245	203	175	187	138	**475**	100	28	24	20	12	12	8	8	**5.20**	29	31	1	3	–	2	–	–	–	–	–	–	–	Fluspirilene	1841-19-6	NI44160
44	91	366	55	43	190	384	92	**475**	100	54	52	21	21	18	16	15	**1.70**	29	33	5	1	–	–	–	–	–	–	–	–	–	1-Benzyl-12,19-dimethyl-18-methylene-3,5,8,17-tetraoxo-4-oxa-6,14-diene perhydro-cyclotetradeca[2,3-d]isoindole		IC04497
187	246	192	191	272	205	201	176	**475**	100	95	79	71	64	62	52	41	**33.03**	29	37	3	3	–	–	–	–	–	–	–	–	–	Tubulosan-8′-ol, 10,11-dimethoxy-	2632-29-3	NI44161
110	28	458	68	18	398	43	459	**475**	100	84	80	69	69	40	31	29	**25.27**	29	49	4	1	–	–	–	–	–	–	–	–	–	3β-Acetoxy-5-hydroxy-5α-cholestan-6-one oxime		MS10178
377	378	332	333	98	475	379	334	**475**	100	34	34	19	17	5	4	2		32	29	3	1	–	–	–	–	–	–	–	–	–	Diphenylacetic acid, 4-[(pyrrolidinocarbonyl)diphenylmethyl]-	70008-34-3	NI44162

MASS TO CHARGE RATIOS								M.W.	INTENSITIES								Parent	C	H	O	N	S	F	Cl	Br	I	Si	P	B	X	COMPOUND NAME	CAS Reg No	No
91	448	150	477	200	431	199	122	475	100	92	89	84	81	73	72	68	4.44	33	49	1	1	–	–	–	–	–	–	–	–	–	6-Azacholest-4-en-7-one, 6-benzyl-	16373-60-7	NI44163
103	28	448	182	366	450	131	478	476	100	52	19	11	9	7	7	6	3.21	4	–	4	–	–	–	–	2	–	–	–	–	2	Rhodium, di-μ-bromotetracarbonyldi-	21475-96-7	NI44164
308	224	448	336	280	252	364	392	476	100	88	54	54	49	49	23	20	16.81	10	–	9	–	–	–	1	–	–	–	–	–	3	Cobalt, μ-chloro-μ-carbononacarbonyltri-		MS10179
63	69	397	279	129	277	217	110	476	100	92	46	15	15	14	11	10	10.20	12	–	–	–	1	10	–	1	–	–	1	–	–	Phosphinothioic bromide, bis(pentafluorophenyl)-	53327-34-7	NI44165
73	147	203	75	45	74	156	43	476	100	27	19	12	11	9	9	6	0.62	19	44	4	2	–	–	–	–	–	4	–	–	–	N-acetylglutamine, tetrakis(trimethylsilyl)-		NI44166
43	129	115	69	97	81	169	98	476	100	66	28	26	20	13	11	11	0.00	21	32	12	–	–	–	–	–	–	–	–	–	–	Pentanoic acid, 4-methyl-5-[(2,3,4,6-tetra-O-acetyl-D-glucopyranosyl)oxy]-, methyl ester	56246-40-3	NI44167
101	43	159	113	461	87	85	59	476	100	92	67	64	60	58	58	50	0.00	22	36	11	–	–	–	–	–	–	–	–	–	–	α-D-Glucofuranose, 3-[2-hydroxy-2-(tetrahydro-6-methoxy-2,2-dimethylfuro[2,3-d]-1,3-dioxol-5-yl)ethoxy]-1,2:5,6-di-O-isopropylidene-	29587-06-2	NI44168
434	104	149	132	177	257	259	122	476	100	62	60	41	38	33	32	31	0.00	23	28	9	2	–	–	–	–	–	–	–	–	–	Propanedioic acid, [4-[[(acetyloxy)acetyl]amino]butyl](1,3-dihydro-1,3-dioxo-2H-isoindol-2-yl)-, diethyl ester	56247-58-6	NI44169
83	55	476	247	345	245	100	262	476	100	90	40	30	22	22	19	15		25	32	9	–	–	–	–	–	–	–	–	–	–	6α-Tigloyloxychaparrinone		MS10180
180	254	238	179	222	208	194	476	476	100	70	65	56	45	42	41	30		26	16	4	6	–	–	–	–	–	–	–	–	–	1,2-Bis(2-nitrophenazin-1-yl)ethane		MS10181
148	458	426	135	283	325	440	476	476	100	71	62	60	48	39	36	35		26	32	3	6	–	–	–	–	–	–	–	–	–	2-(2′-6N-Adenyl)-acetamidoestradiol 3-methyl ether	81560-76-1	NI44170
218	41	117	39	91	177	45	187	476	100	80	63	62	57	39	38	37	0.00	26	34	–	2	1	–	2	–	–	–	–	–	–	Bis[(6′-chloro-4′-chloro-4′-cyanocyclohex-1′-enyl)-4-methylene-pent-3-yl] sulphide		LI09276
43	41	69	83	91	79	93	55	476	100	58	40	39	35	32	29	25	0.60	26	36	8	–	–	–	–	–	–	–	–	–	–	3-Desoxo-3β-hydroxy-4-desoxy-phorbol-3,12,13-triacetate		NI44171
43	69	356	105	83	131	95	91	476	100	26	25	21	21	20	20	20	0.24	26	36	8	–	–	–	–	–	–	–	–	–	–	3-Desoxo-3β-hydroxy-4-desoxy-phorbol-3,12,13-triacetate		NI44172
43	28	296	356	149	69	83	41	476	100	43	31	29	29	28	26	24	4.28	26	36	8	–	–	–	–	–	–	–	–	–	–	3-Desoxo-3β-hydroxy-4-desoxy-phorbol-12,13,20-triacetate		NI44173
43	69	55	41	123	93	79	81	476	100	50	44	43	30	29	27	26	0.00	26	36	8	–	–	–	–	–	–	–	–	–	–	6,7-Dihydro-16-hydroxy-20-desoxy-ingenol-3,5,16-triacetate		NI44174
301	159	177	300	302	181	283	317	476	100	73	65	54	49	44	34	31	0.43	26	36	8	–	–	–	–	–	–	–	–	–	–	1-O-Retinoyl-β-D-glucopyranuronic acid	401-10-5	NI44175
476	477	478	345	446	355	239	129	476	100	40	15	10	7	6	6	5		26	44	4	–	–	–	–	–	–	2	–	–	–	Estra-2,17-diol, 3,4-dimethoxy-, di-TMS, (17β)-		NI44176
476	477	478	461	345	446	355	479	476	100	30	17	7	7	6	6	4		26	44	4	–	–	–	–	–	–	2	–	–	–	Estra-3,17-diol, 2,4-dimethoxy-, di-TMS, (17β)-		NI44177
476	477	446	461	345	239	478	129	476	100	42	10	8	8	5	4	4		26	44	4	–	–	–	–	–	–	2	–	–	–	Estra-4,17-diol, 2,3-dimethoxy-, di-TMS, (17β)-		NI44178
460	304	476	474	305	461	477	459	476	100	96	80	40	39	34	28	23		27	24	1	–	–	–	–	–	–	–	–	–	2	1′-(α-Hydroxybenzyl)biferrocene		MS10182
253	356	281	296	313	200	226	211	476	100	99	87	86	28	23	15	15	1.00	27	40	7	–	–	–	–	–	–	–	–	–	–	3α,7α,12α-Triacetoxy-5β-pregnan-20-one		NI44179
415	117	129	213	416	172	119	417	476	100	86	73	54	43	26	20	15	0.00	27	48	3	–	–	–	–	–	–	2	–	–	–	3,20-Dihydroxypregn-5-en-16-one, bis(trimethylsilyl)-		LI09277
73	129	75	43	172	157	143	159	476	100	99	90	65	55	53	51	50	9.00	27	48	3	–	–	–	–	–	–	2	–	–	–	3β,16α-Dihydroxypregn-5-en-20-one, bis(trimethylsilyl)-	40822-79-5	NI44181
129	172	386	157	159	296	257	109	476	100	82	62	62	60	53	44	43	12.93	27	48	3	–	–	–	–	–	–	2	–	–	–	3β,16α-Dihydroxypregn-5-en-20-one, bis(trimethylsilyl)-	40822-79-5	NI44180
129	172	159	386	296	157	109	257	476	100	85	57	49	45	44	40	39	15.00	27	48	3	–	–	–	–	–	–	2	–	–	–	3β,16α-Dihydroxypregn-5-en-20-one, bis(trimethylsilyl)-	40822-79-5	LI09278
296	386	172	157	159	75	109	253	476	100	60	49	49	44	37	28	26	0.00	27	48	3	–	–	–	–	–	–	2	–	–	–	3β,16β-Dihydroxypregn-5-en-20-one, bis(trimethylsilyl)-		LI09279
73	255	373	75	129	81	145	143	476	100	99	44	40	30	28	23	23	12.50	27	48	3	–	–	–	–	–	–	2	–	–	–	3β,21-Dihydroxypregn-5-en-20-one, bis(trimethylsilyl)-	40822-83-1	NI44182
184	387	105	476	386	389	388	77	476	100	84	74	70	67	59	42	25		28	32	5	2	–	–	–	–	–	–	–	–	–	Aspidospermidin-21-oic acid, 1-benzoyl-20-hydroxy-17-methoxy-, methyl ester	56143-44-3	NI44183
356	43	255	65	81	357	107	93	476	100	87	81	37	35	31	31	31	0.00	28	44	6	–	–	–	–	–	–	–	–	–	–	Cholan-24-oic acid, 3,12-bis(acetyloxy)-, (3α,5β,12α)-	33628-48-7	NI44184
104	118	106	145	130	120	159	147	476	100	72	66	63	62	58	56	56	0.00	28	44	6	–	–	–	–	–	–	–	–	–	–	24-Norcholan-23-oic acid, 3,11-bis(acetyloxy)-, methyl ester, (3α,5β,11α)	60354-48-5	NI44185
255	356	105	119	107	121	147	256	476	100	49	34	29	29	28	27	26	0.00	28	44	6	–	–	–	–	–	–	–	–	–	–	24-Norcholan-23-oic acid, 3,12-bis(acetyloxy)-, methyl ester, (3α,5β,12α)	60354-47-4	NI44186
382	354	383	353	197	398	352	199	476	100	32	29	28	23	18	15	13	0.00	29	48	1	–	2	–	–	–	–	–	–	–	–	Cholest-5-en-19-al, 3β-hydroxy-, cyclic ethylene mercaptal	1107-85-3	NI44187
110	60	45	43	111	42	109	81	476	100	19	19	19	14	10	8	8	0.00	29	48	5	–	–	–	–	–	–	–	–	–	–	5,6-Secocholestan-6-oic acid, 3-(acetyloxy)-5-oxo-, (3β)-	10473-42-4	NI44189
110	247	416	357	331	401	398	370	476	100	3	2	2	2	1	1	1	0.00	29	48	5	–	–	–	–	–	–	–	–	–	–	5,6-Secocholestan-6-oic acid, 3-(acetyloxy)-5-oxo-, (3β)-	10473-42-4	NI44188
105	476	77	32	57	41	477	44	476	100	56	44	32	27	24	19	18		30	20	2	–	2	–	–	–	–	–	–	–	–	2,4-Bis(α-benzoylbenzylidene)-1,3-dithietane	19018-13-4	NI44190
43	313	173	159	425	443	336	203	476	100	68	65	63	60	55	48	48	30.00	30	52	4	–	–	–	–	–	–	–	–	–	–	24S-cycloartane-3β,16β,24,25-tetraol		NI44191
327	476	328	236	262	477	250	385	476	100	37	27	15	14	13	8	7		31	28	3	2	–	–	–	–	–	–	–	–	–	5-Methoxy-3,3-dibenzyl-3H-indole-2-carboxylic acid 4-methoxyanilide		MS10183
327	476	328	236	262	250	385	399	476	100	37	27	15	13	8	6	5		31	28	3	2	–	–	–	–	–	–	–	–	–	5-Methoxy-3,3-dibenzyl-3H-indole-2-carboxylic acid 4-methoxyanilide		LI09280
385	476	178	386	91	105	108	326	476	100	45	35	30	28	26	22	12		31	28	3	2	–	–	–	–	–	–	–	–	–	2-(4-Methoxyphenylimino)-4-phenyl-3-benzyl-3-butenoic acid 4-methoxyanilide		LI09281
385	476	178	386	91	105	108	78	476	100	45	35	30	28	25	22	18		31	28	3	2	–	–	–	–	–	–	–	–	–	2-(4-Methoxyphenylimino)-4-phenyl-3-benzyl-3-butenoic acid 4-methoxyanilide		MS10184
296	297	57	475	263	236	238	253	476	100	95	72	62	62	62	48	45	19.00	32	28	4	–	–	–	–	–	–	–	–	–	–	1,2-Dibenzoyl-3,4-bis(4-methoxyphenyl)cyclobutane		NI44192
105	91	77	371	372	160	106	51	476	100	47	38	21	11	11	9	7	1.00	32	32	2	2	–	–	–	–	–	–	–	–	–	N,N′-Dibenzoyl-N,N′-dibenzyl-putrescine		LI09282
72	405	406	332	333	241	253	407	476	100	90	32	19	7	6	5	5	4.40	32	32	2	2	–	–	–	–	–	–	–	–	–	N,N-Dimethyl-4-[(N,N-dimethylcarbamoyl)benzyl]triphenylacetamide		NI44193
55	69	41	43	83	67	81	56	476	100	69	67	62	40	36	29	29	4.14	32	60	2	–	–	–	–	–	–	–	–	–	–	9-Hexadecenoic acid, 9-hexadecenyl ester, (Z,Z)-	22393-97-1	NI44194
274	476	477	124	448	109	105	275	476	100	99	36	35	34	28	27	24		33	36	1	2	–	–	–	–	–	–	–	–	–	Aspidospermidine, 20,21-didehydro-17-methoxy-1-methyl-21,21-diphenyl-	36459-06-0	NI44195

MASS TO CHARGE RATIOS								M.W.	INTENSITIES								Parent	C	H	O	N	S	F	Cl	Br	I	Si	P	B	X	COMPOUND NAME	CAS Reg No	No
368	105	369	81	147	43	55	77	**476**	100	43	34	27	25	24	20	18	**0.00**	33	48	2	–	–	–	–	–	–	–	–	–	–	26,27-Dinorergost-5-en-3-ol, benzoate, (3β)-	58003-48-8	NI44196
271	232	194	245	213	238	192	218	**476**	100	89	75	59	59	55	53	45	**10.00**	34	20	3	–	–	–	–	–	–	–	–	–	–	2,5-Bis(9-oxo-9,10-dihydroanthracen-10-ylidenemethyl)furan		MS10185
383	73	427	384	75	57	74	43	**476**	100	50	40	33	23	22	21	21	**0.00**	34	52	1	–	–	–	–	–	–	–	–	–	–	Cholest-7-ene, 3-benzyloxy-, (3β,5α)-	74420-83-0	NI44197
262	183	263	261	108	184	185	78	**476**	100	29	19	12	10	7	5	5	**3.80**	35	25	–	–	–	–	–	–	–	–	1	–	–	Phosphorane, 11H-benzo[a]fluoren-11-ylidenetriphenyl-	31083-20-2	NI44198
262	183	476	108	277	477	230	215	**476**	100	57	38	34	18	14	13	13		35	25	–	–	–	–	–	–	–	–	1	–	–	Phosphorane, 11H-benzo[b]fluoren-11-ylidenetriphenyl-	31083-21-3	LI09283
262	183	476	108	277	263	51	477	**476**	100	57	37	34	19	18	16	15		35	25	–	–	–	–	–	–	–	–	1	–	–	Phosphorane, 11H-benzo[b]fluoren-11-ylidenetriphenyl-	31083-21-3	NI44199
398	240	284	146	477	205	170	111	**477**	100	12	8	6	4	4	4	4		–	–	–	3	–	–	3	3	–	–	3	–	–	1,3,5,2,4,6-Triazatriphosphorine, 2,4,6-tribromo-2,4,6-trichloro-2,2,4,4,6,6 hexahydro-	16032-52-3	NI44200
375	128	311	313	115	129	434	161	**477**	100	87	82	81	80	72	62	58	**0.00**	19	29	1	1	1	–	–	2	–	–	–	–	–	S-[2,6-Bis(3-bromopropyl)-4-tert-butyl-1-phenyl] N,N-dimethylthiocarbamate		DD01730
43	143	117	101	418	199	122	116	**477**	100	22	22	20	11	7	7	6	**0.00**	20	31	12	1	–	–	–	–	–	–	–	–	–	D-Glycero-D-galacto-2-nonulopyranosidonic acid, methyl 5-(acetylamino) 3,5-dideoxy-8-O-methyl-, methyl ester, 4,7,9-triacetate	56282-35-0	NI44201
364	477	365	478	380	311	363	307	**477**	100	57	23	14	9	9	6	6		21	17	5	1	–	6	–	–	–	–	–	–	–	3,6-Di(trifluoroacetyl)morphine		NI44202
355	357	147	477	356	479	359	358	**477**	100	69	24	20	20	14	12	12		21	21	4	5	–	–	2	–	–	–	–	–	–	N-Ethyl-N-(2-cyanoethoxycarbonylethyl)-3-methyl-4-(2,6-dichloro-4-nitrophenylazo)-aniline		IC04498
157	60	119	160	82	71	94	72	**477**	100	99	70	66	55	49	47	44	**1.12**	21	43	7	5	–	–	–	–	–	–	–	–	–	Gentamicin C1	25876-10-2	NI44204
478	479	157	322	160	480	323	158	**477**	100	29	26	20	7	5	3	3	**1.00**	21	43	7	5	–	–	–	–	–	–	–	–	–	Gentamicin C1	25876-10-2	NI44203
32	43	31	41	91	117	55	441	**477**	100	49	49	37	36	32	32	30	**14.10**	22	27	7	3	1	–	–	–	–	–	–	–	–	Griseoviridin		MS10186
130	262	86	79	131	93	234	41	**477**	100	72	48	34	28	28	25	23	**0.40**	26	43	5	3	–	–	–	–	–	–	–	–	–	L-Alanine, N-[N-[N-(tricyclo[3.3.1.1^{3,7}]dec-1-ylcarbonyl)-L-valyl]-L-isoleucyl]-, methyl ester	28415-47-6	NI44205
133	134	387	477	72	445	462	135	**477**	100	10	8	7	7	6	4	4		26	47	3	1	–	–	–	–	–	2	–	–	–	1β,17β-Dihydroxyandrost-4-en-3-one methyloxime, bis(trimethylsilyl)-		LI09284
448	461	387	449	462	477	450	388	**477**	100	56	44	40	22	20	16	14		26	47	3	1	–	–	–	–	–	2	–	–	–	2β,17β-Dihydroxyandrost-4-en-3-one methyloxime, bis(trimethylsilyl)-		LI09285
73	446	266	75	129	356	174	105	**477**	100	99	84	80	70	46	37	33	**14.96**	26	47	3	1	–	–	–	–	–	2	–	–	–	3β,16α-Dihydroxyandrost-5-en-17-one methyloxime, bis(trimethylsilyl)-	55557-05-6	NI44206
446	266	356	174	129	447	431	239	**477**	100	94	57	49	44	41	25	22	**10.95**	26	47	3	1	–	–	–	–	–	2	–	–	–	3β,16α-Dihydroxyandrost-5-en-17-one methyloxime, bis(trimethylsilyl)-	55557-05-6	NI44207
129	446	266	462	356	447	106	239	**477**	100	68	42	27	26	21	21	19	**11.00**	26	47	3	1	–	–	–	–	–	2	–	–	–	3β,16α-Dihydroxyandrost-5-en-17-one methyloxime, bis(trimethylsilyl)-	55557-05-6	LI09286
446	129	447	174	57	266	356	265	**477**	100	48	44	38	38	29	27	24	**7.09**	26	47	3	1	–	–	–	–	–	2	–	–	–	3β,16β-Dihydroxyandrost-5-en-17-one methyloxime, bis(trimethylsilyl)-	69597-48-4	NI44208
73	158	75	174	129	446	133	173	**477**	100	99	99	83	73	70	65	55	**16.00**	26	47	3	1	–	–	–	–	–	2	–	–	–	3β,17β-Dihydroxyandrost-5-en-16-one methyloxime, bis(trimethylsilyl)-	56210-92-5	NI44209
462	463	430	445	464	446	477	417	**477**	100	43	31	18	17	17	16	14		26	47	3	1	–	–	–	–	–	2	–	–	–	6α,17β-Dihydroxyandrost-4-en-3-one methyloxime, bis(trimethylsilyl)-		LI09287
477	153	478	356	387	266	446	191	**477**	100	44	40	32	31	29	18	18		26	47	3	1	–	–	–	–	–	2	–	–	–	16α,17β-Dihydroxyandrost-4-en-3-one methyloxime, bis(trimethylsilyl)-	69688-34-2	NI44210
75	477	103	478	129	373	417	315	**477**	100	59	54	24	24	20	18	18		26	47	3	1	–	–	–	–	–	2	–	–	–	17β,19-Dihydroxyandrost-4-en-3-one methyloxime, bis(trimethylsilyl)-	55836-47-0	NI44211
168	43	181	183	120	106	105	83	**477**	100	62	59	53	42	42	42	38	**33.00**	28	31	6	1	–	–	–	–	–	–	–	–	–	Neoaureothin		MS10187
185	441	255	254	170	158	442	186	**477**	100	94	54	53	45	37	32	16	**6.64**	29	51	4	1	–	–	–	–	–	–	–	–	–	Acetamide, N-[(3β,5α,16β)-3,16,22-trihydroxycholestan-26-yl]-	50686-95-8	NI44212
180	420	77	93	421	329	144	104	**477**	100	99	49	35	34	30	15	15	**7.00**	30	27	3	3	–	–	–	–	–	–	–	–	–	Benzenecarboximidamide, N-[2-(2,2-dimethyl-1-oxopropyl)-3,6-dioxo-4-(phenylamino)-1,4-cyclohexadien-1-yl]-N-phenyl-	61417-00-3	NI44213
91	95	401	479	97	96	476	93	**477**	100	94	91	80	70	70	66	63	**7.73**	33	51	1	1	–	–	–	–	–	–	–	–	–	6-Azacholestan-7-one, 6-benzyl-	16373-56-1	NI44214
78	477	52	77	51	50	447		**477**	100	30	17	16	16	13	7			34	23	2	1	–	–	–	–	–	–	–	–	–	6-Nitro-1,2,3,4-tetraphenyl-naphthalene		LI09288
88	303	215	480	391	232	179	137	**478**	100	95	35	22	13	1	1	1	**0.00**	–	–	–	–	–	12	2	–	–	–	4	–	1	Tetrakis(trifluorophosphine)iron dichloride		NI44215
276	400	435	310	478	357	444	206	**478**	100	95	76	76	59	48	45	32		4	12	–	6	–	–	6	–	–	–	4	–	–	Bis(dimethylamino)hexachlorocyclotetraphosphazene		MS10188
461	459	463	460	165	457	462	163	**478**	100	95	81	60	59	46	45	41	**8.63**	12	18	–	–	–	4	–	–	–	–	–	–	2	Stannane, (2,3,5,6-tetrafluoro-4-phenylene)bis[trimethyl-	23653-80-7	NI44216
165	163	135	139	161	133	137	169	**478**	100	72	62	47	46	38	34	32	**0.00**	12	18	–	–	–	4	–	–	–	–	–	–	2	Stannane, (3,4,5,6-tetrafluoro-2-phenylene)bis[trimethyl-	23653-79-4	NI44217
389	449	391	355	361	451	450	363	**478**	100	65	50	49	40	33	27	27	**0.49**	12	35	6	–	–	–	1	–	–	6	–	–	–	1,3,5,7,9,11-Hexaethyl-1-chlorocyclohexasiloxane	73420-32-3	NI44218
73	45	198	196	171	90	155	117	**478**	100	94	51	51	43	40	31	31	**5.00**	16	12	–	6	1	–	–	2	–	–	–	–	–	6-(4-Bromophenyl)amino-7-(4-bromophenyl)-3-methylthio[1,2,4]triazolo[4,3-b][1,2,4]triazole		MS10189
195	196	197	69	41	138	169	223	**478**	100	97	32	26	24	20	14	13	**3.50**	18	21	5	2	–	7	–	–	–	–	–	–	–	Butanoic acid, 2,2,3,3,4,4,4-heptafluoro-, 3-[hexahydro-1,3-dimethyl-2,4,6 trioxo-5-(2-propenyl)-5-pyrimidinyl]-1,3-dimethylpropyl ester	55429-89-5	NI44219
91	155	243	96	160	188	178	176	**478**	100	67	42	39	36	20	20	20	**0.00**	21	23	4	2	1	–	–	1	–	–	–	–	–	Benzenesulphonamide, N-(3-bromopropyl)-N-[3-(1,3-dihydro-1,3-dioxo-2H-isoindol-2-yl)propyl]-4-methyl-	75444-62-1	NI44220
310	478	268	115	255	141	311	254	**478**	100	39	28	28	22	22	18	17		22	20	5	–	–	6	–	–	–	–	–	–	–	Estra-1,3,5(10)-trien-17-one, 3,16α-dihydroxy-, bis(trifluoroacetate)	30519-83-6	NI44221
43	313	31	354	61	60	29	58	**478**	100	95	75	46	37	35	28	27	**0.00**	22	22	12	–	–	–	–	–	–	–	–	–	–	4H-1-Benzopyran-4-one, 8-(β-D-glucopyranosyloxy)-5,7-dihydroxy-2-(4-hydroxy-3-methoxyphenyl)-	57396-70-0	NI44222

MASS TO CHARGE RATIOS								M.W.	INTENSITIES								Parent	C	H	O	N	S	F	Cl	Br	I	Si	P	B	X	COMPOUND NAME	CAS Reg No	No
84	58	44	46	98	42	478	71	**478**	100	90	79	60	25	23	22	20		22	23	8	2	–	–	1	–	–	–	–	–	–	**2-Naphthacenecarboxamide, 7-chloro-4-(dimethylamino)-1,4,4a,5,5a,6,11,12a-octahydro-3,6,10,12,12a-pentahydroxy-6-methyl 1,11-dioxo-, [4S-(4α,4aα,5aα,6β,12aα)]-**	57-62-5	**NI44223**
297	299	480	115	482	478	217	298	**478**	100	98	72	39	37	37	34	24		23	16	–	2	–	–	–	2	–	–	–	–	–	**4H-1,2-Diazepine, 3,7-bis(4-bromophenyl)-5-phenyl-**	25649-79-0	**NI44224**
478	69	137	364	129	155	157	365	**478**	100	61	45	37	36	35	34	32		23	24	4	–	–	6	–	–	–	–	–	–	–	**Estra-1,3,5(10)-triene-3,17-diol, 1-methyl-, bis(trifluoroacetate), (17β)-**	56438-18-7	**NI44225**
478	365	364	479	269	155	157	243	**478**	100	31	31	30	24	23	22	18		23	24	4	–	–	6	–	–	–	–	–	–	–	**Estra-1,3,5(10)-triene-3,17-diol, 1-methyl-, bis(trifluoroacetate), (17β)-**	56438-18-7	**NI44226**
323	168	155	91	82	198	324	184	**478**	100	61	36	36	24	22	20	16	**1.00**	24	34	4	2	2	–	–	–	–	–	–	–	–	**3,3,7,7-Tetramethyl-1,5-diaza-1,5-di(p-tosyl)cyclooctane**	87338-07-6	**NI44227**
75	185	171	159	45	229	71	101	**478**	100	94	59	52	44	38	38	33	**0.00**	25	50	8	–	–	–	–	–	–	–	–	–	–	**Octadecanoic acid, 6,7,9,10,12,13-hexamethoxy-, methyl ester**	20207-68-5	**NI44228**
75	185	171	159	319	45	331	71	**478**	100	94	59	52	46	46	37	37	**0.00**	25	50	8	–	–	–	–	–	–	–	–	–	–	**Octadecanoic acid, 6,7,9,10,12,13-hexamethoxy-, methyl ester**	20207-68-5	**LI09289**
75	143	117	213	271	45	85	71	**478**	100	72	41	35	29	29	28	25	**0.00**	25	50	8	–	–	–	–	–	–	–	–	–	–	**Octadecanoic acid, 9,10,12,13,15,16-hexamethoxy-, methyl ester**	20207-67-4	**NI44229**
75	143	117	213	271	85	73	71	**478**	100	72	40	35	30	27	26	26	**0.00**	25	50	8	–	–	–	–	–	–	–	–	–	–	**Octadecanoic acid, 9,10,12,13,15,16-hexamethoxy-, methyl ester**	20207-67-4	**LI09290**
379	305	173	478	388	380	463	73	**478**	100	78	49	35	33	33	30	30		26	46	4	–	–	–	–	–	–	2	–	–	–	**Prosta-5,8(12),13-trien-1-oic acid, 9-oxo-15-[(trimethylsilyl)oxy]-, trimethylsilyl ester, (5Z,13E,15S)-**	53044-54-5	**NI44230**
72	379	74	305	116	478	463	380	**478**	100	82	65	56	38	30	30	29		26	46	4	–	–	–	–	–	–	2	–	–	–	**Prosta-5,8(12),13-trien-1-oic acid, 9-oxo-15-[(trimethylsilyl)oxy]-, trimethylsilyl ester, (5Z,13E,15S)-**	53044-54-5	**NI44231**
190	407	317	199	119	73	388	75	**478**	100	53	50	45	37	34	33	29	**9.00**	26	46	4	–	–	–	–	–	–	2	–	–	–	**Prosta-5,10,13-trien-1-oic acid, 9-oxo-15-[(trimethylsilyl)oxy]-, trimethylsilyl ester, (5Z,13E,15S)-**	41577-91-7	**NI44232**
190	317	407	199	118	116	388	128	**478**	100	73	67	59	55	53	42	35	**6.58**	26	46	4	–	–	–	–	–	–	2	–	–	–	**Prosta-5,10,13-trien-1-oic acid, 9-oxo-15-[(trimethylsilyl)oxy]-, trimethylsilyl ester, (5Z,13E,15S)-**	41577-91-7	**NI44233**
114	115	85	42	131	91	84	58	**478**	100	11	5	4	3	3	3	3	**0.12**	27	34	4	4	–	–	–	–	–	–	–	–	–	**Aralionine, debenzoyl-**	21761-50-2	**NI44234**
478	73	479	224	463	480	225	464	**478**	100	52	43	26	18	17	16	8		27	39	–	–	–	–	–	–	–	3	1	–	–	**Phosphine, tris[4-(trimethylsilyl)phenyl]-**	18848-96-9	**NI44235**
478	73	479	224	463	480	464	225	**478**	100	52	43	26	18	17	8	6		27	39	–	–	–	–	–	–	–	3	1	–	–	**Phosphine, tris[4-(trimethylsilyl)phenyl]-**	18848-96-9	**NI44236**
43	399	55	41	57	81	69	95	**478**	100	72	67	57	46	40	36	33	**0.00**	27	43	2	–	–	–	–	1	–	–	–	–	–	**5β,6β-Epoxy-7-bromocholestan-3-one**		**NI44237**
478	480	399	286	331	257	481	479	**478**	100	98	56	48	38	38	32	32		27	43	2	–	–	–	–	1	–	–	–	–	–	**5α-Spirostan, 23-bromo-, (23R,25R)**		**MS10190**
41	55	69	67	42	81	257	95	**478**	100	98	70	67	67	59	55	46	**14.00**	27	43	2	–	–	–	–	1	–	–	–	–	–	**5α-Spirostan, 23-bromo-, (23R,25R)**		**MS10191**
286	398	478	480	257	271	331	399	**478**	100	54	46	45	45	35	25	24		27	43	2	–	–	–	–	1	–	–	–	–	–	**5α-Spirostan, 23-bromo-, (23S,25R)**		**MS10192**
257	286	55	271	67	81	41	95	**478**	100	59	57	48	44	43	43	40	**24.00**	27	43	2	–	–	–	–	1	–	–	–	–	–	**5α-Spirostan, 23-bromo-, (23S,25R)**		**MS10193**
55	41	257	69	43	139	95	81	**478**	100	93	84	79	72	70	67	67	**0.00**	27	43	2	–	–	–	–	1	–	–	–	–	–	**5α-Spirostan, 23-bromo-, (22S,23R,25R)-**	4988-84-5	**NI44238**
257	95	109	93	91	107	139	161	**478**	100	90	66	63	47	45	44	40	**1.83**	27	43	2	–	–	–	–	1	–	–	–	–	–	**5α-Spirostan, 23-bromo-, (22S,23R,25R)-**	4988-84-5	**NI44239**
43	298	73	75	147	81	299	95	**478**	100	99	99	63	45	45	44	40	**10.50**	27	50	3	–	–	–	–	–	–	2	–	–	–	**3α,6α-Dihydroxy-5β-pregnan-20-one, bis(trimethylsilyl)-**	33287-43-3	**NI44240**
298	299	251	283	161	388	243	81	**478**	100	39	31	26	25	24	21	21	**9.30**	27	50	3	–	–	–	–	–	–	2	–	–	–	**3α,6α-Dihydroxy-5β-pregnan-20-one, bis(trimethylsilyl)-**	33287-43-3	**NI44241**
298	147	251	161	129	388	283	213	**478**	100	57	50	47	37	31	28	28	**15.00**	27	50	3	–	–	–	–	–	–	2	–	–	–	**3α,6α-Dihydroxy-5β-pregnan-20-one, bis(trimethylsilyl)-**	33287-43-3	**NI44242**
298	478	255	479	213	388	143	240	**478**	100	92	56	46	46	44	43	39		27	50	3	–	–	–	–	–	–	2	–	–	–	**3α,11β-Dihydroxy-5β-pregnan-20-one, bis(trimethylsilyl)-**		**NI44243**
172	173	174	129	388	157	74	32	**478**	100	21	7	6	4	4	4	4	**2.00**	27	50	3	–	–	–	–	–	–	2	–	–	–	**3α,15α-Dihydroxy-5α-pregnan-20-one, bis(trimethylsilyl)-**		**LI09291**
172	109	388	157	159	255	96	186	**478**	100	85	68	55	49	44	35	32	**3.56**	27	50	3	–	–	–	–	–	–	2	–	–	–	**3α,16α-Dihydroxy-5α-pregnan-20-one, bis(trimethylsilyl)-**	56247-71-3	**NI44244**
255	345	435	147	256	145	436	107	**478**	100	27	23	23	22	19	17	10	**1.82**	27	50	3	–	–	–	–	–	–	2	–	–	–	**3α,17α-Dihydroxy-5α-pregnan-20-one, bis(trimethylsilyl)-**		**NI44245**
255	435	256	215	436	345	143	145	**478**	100	32	25	14	13	13	8	7	**1.17**	27	50	3	–	–	–	–	–	–	2	–	–	–	**3α,17α-Dihydroxy-5β-pregnan-20-one, bis(trimethylsilyl)-**		**NI44246**
285	298	286	284	172	267	85	117	**478**	100	28	28	28	25	24	20	18	**4.00**	27	50	3	–	–	–	–	–	–	2	–	–	–	**3α,19-Dihydroxy-5α-pregnan-20-one, bis(trimethylsilyl)-**	56247-72-4	**LI09292**
285	286	284	267	85	298	147	189	**478**	100	20	19	18	17	15	13	6	**0.22**	27	50	3	–	–	–	–	–	–	2	–	–	–	**3α,19-Dihydroxy-5α-pregnan-20-one, bis(trimethylsilyl)-**	56247-72-4	**NI44247**
117	388	283	298	282	478	244	118	**478**	100	25	19	18	15	12	12	12		27	50	3	–	–	–	–	–	–	2	–	–	–	**3α,20β-Dihydroxy-5β-pregnan-11-one, bis(trimethylsilyl)-**		**NI44248**
375	257	258	376	161	135	175	147	**478**	100	99	75	34	31	31	28	27	**2.86**	27	50	3	–	–	–	–	–	–	2	–	–	–	**3α,21-Dihydroxy-5α-pregnan-20-one, bis(trimethylsilyl)-**		**NI44249**
257	73	375	75	81	258	143	95	**478**	100	50	40	40	22	21	19	19	**0.79**	27	50	3	–	–	–	–	–	–	2	–	–	–	**3α,21-Dihydroxy-5β-pregnan-20-one, bis(trimethylsilyl)-**	33287-45-5	**NI44250**
257	258	376	375	175	161	95	81	**478**	100	25	16	16	15	12	11	11	**1.01**	27	50	3	–	–	–	–	–	–	2	–	–	–	**3α,21-Dihydroxy-5β-pregnan-20-one, bis(trimethylsilyl)-**		**NI44251**
478	479	197	392	143	110	388	240	**478**	100	42	32	26	26	22	19	19		27	50	3	–	–	–	–	–	–	2	–	–	–	**3β,11α-Dihydroxy-5α-pregnan-20-one, bis(trimethylsilyl)-**		**LI09293**
298	285	299	284	267	255	388	147	**478**	100	40	26	23	16	15	12	10	**2.00**	27	50	3	–	–	–	–	–	–	2	–	–	–	**3β,19-Dihydroxy-5α-pregnan-20-one, bis(trimethylsilyl)-**		**LI09294**
375	258	257	121	376	149	147	135	**478**	100	57	48	32	31	30	30	27	**1.82**	27	50	3	–	–	–	–	–	–	2	–	–	–	**3β,21-Dihydroxy-5β-pregnan-20-one, bis(trimethylsilyl)-**		**NI44252**
253	117	129	272	360	271	213	119	**478**	100	95	65	59	55	47	45	38	**7.81**	27	50	3	–	–	–	–	–	–	2	–	–	–	**Pregn-5-ene-3β,17α,20α-triol, bis(trimethylsilyl)-**		**NI44253**
271	73	117	75	253	129	360	119	**478**	100	85	56	42	36	28	26	24	**3.38**	27	50	3	–	–	–	–	–	–	2	–	–	–	**Pregn-4-en-17-ol, 3,20-bis[(trimethylsilyl)oxy]-, (3β,20S)-**	41259-41-0	**NI44254**
433	75	462	73	119	343	434	157	**478**	100	94	89	78	56	49	37	37	**1.94**	27	50	3	–	–	–	–	–	–	2	–	–	–	**Pregn-5-en-20-ol, 3,17-bis[(trimethylsilyl)oxy]-, (3β,20S)-**	41164-18-5	**NI44255**
183	185	410	253	108	107	154	263	**478**	100	86	63	42	35	32	30	28	**1.00**	28	24	–	4	–	–	–	–	–	–	2	–	–	**Cyanamide, [1,2-ethanediylbis(diphenylphosphoranylidyne)]bis-**	66055-14-9	**NI44256**
427	442	399	425	428	415	440	413	**478**	100	61	61	42	35	35	27	27	**0.00**	28	31	3	2	–	–	1	–	–	–	–	–	–	**Benzoic acid, 2-[6-(ethylamino)-3-(ethylimino)-2,7-dimethyl-3H-xanthen-9-yl]-, ethyl ester, monohydrochloride**	989-38-8	**NI44257**
397	398	326	399	327	353	369	184	**478**	100	99	74	26	21	20	17	17	**0.00**	28	31	3	2	–	–	1	–	–	–	–	–	–	**Ethanaminium, N-[9-(2-carboxyphenyl)-6-(diethylamino)-3H-xanthen-3-ylidene]-N-ethyl-, chloride**	81-88-9	**NI44258**
240	239	478	96	222	97	53	335	**478**	100	52	26	26	24	22	22	16		28	50	4	2	–	–	–	–	–	–	–	–	–	**Carpine**		**LI09295**
73	75	67	91	133	145	55	147	**478**	100	64	50	48	44	42	40	38	**0.30**	28	54	2	–	–	–	–	–	–	2	–	–	–	**Docosa-8,14-diyn-cis-1,22-diol, bis(trimethylsilyl)-**		**NI44259**

MASS TO CHARGE RATIOS	M.W.	INTENSITIES	Parent	C	H	O	N	S	F	Cl	Br	I	Si	P	B	X	COMPOUND NAME	CAS Reg No	No
244 205 273 206 272 190 121 149	478	100 78 60 60 59 50 50 48	16.01	29	38	4	2	–	–	–	–	–	–	–	–	–	Emetine, 1′,2′-didehydro-	70866-75-0	NI44260
418 384 383 443 478	478	100 76 72 54 12		29	47	3	–	–	–	1	–	–	–	–	–	–	Cholestane, 3-acetoxy-4-chloro-5,6-epoxy-, (3β,4α,5β,6β)-		LI09296
55 95 371 57 43 81 370 83	478	100 96 93 93 92 89 75 72	0.40	29	50	1	–	2	–	–	–	–	–	–	–	–	Cholestan-3-ol, methyl carbonodithioate, (3β,5α)-	5211-17-6	NI44261
43 55 57 41 45 44 69 95	478	100 45 44 42 41 34 30 24	0.00	29	50	3	–	1	–	–	–	–	–	–	–	–	6α-[(2-Hydroxyethyl)thio]cholest-7-ene-3β,5β-diol		NI44262
365 43 44 45 366 41 55 60	478	100 75 39 31 30 26 26 25	0.00	29	50	3	–	1	–	–	–	–	–	–	–	–	6β-[(2-Hydroxyethyl)thio]cholest-7-ene-3β,5α-diol		NI44263
78 77 141 337 309 310 388 302	478	100 92 39 38 38 32 29 28	22.00	30	19	2	–	1	–	1	–	–	–	–	–	–	2-(4-Chlorobenzylidene)-3,8-diphenyl-2H-naphtho[3,2,b]-thiete-1,1-dioxide		LI09297
300 179 178 301 478 165 180 194	478	100 45 28 18 11 10 7 5		30	26	4	2	–	–	–	–	–	–	–	–	–	Dimethyl 9,10,19,20-tetrahydrotetrabenzo[b,d,l,j][1,6]diazacyclododecine-2,7-dicarboxylate		LI09298
237 300 179 178 478 238 219 165	478	100 63 62 37 25 16 13 12		30	26	4	2	–	–	–	–	–	–	–	–	–	Dimethyl 9,10,19,20-tetrahydrotetrabenzo[b,d,l,j][1,6]diazacyclododecine-4,5-dicarboxylate		LI09299
105 149 389 148 122 419 390 268	478	100 85 60 59 25 22 20 18	0.00	30	42	3	2	–	–	–	–	–	–	–	–	–	3β-Acetamido-20α-benzamido-18-hydroxypregn-5-ene	55515-23-6	LI09300
105 149 389 148 159 122 419 390	478	100 85 63 60 55 22 20 20	0.00	30	42	3	2	–	–	–	–	–	–	–	–	–	3β-Acetamido-20α-benzamido-18-hydroxypregn-5-ene	55515-23-6	NI44264
435 436 44 43 120 55 18 30	478	100 35 20 20 14 14 14 13	6.00	30	58	2	2	–	–	–	–	–	–	–	–	–	1,15-Diacetyl-1,15-diazacyclooctacosane		MS10194
240 225 238 226 241 210 76 239	478	100 87 49 38 23 20 15 12	9.00	31	26	5	–	–	–	–	–	–	–	–	–	–	Tabebuin		MS10195
478 359 242 240 227	478	100 66 48 40 29		32	30	4	–	–	–	–	–	–	–	–	–	–	3,3′-Bis(1-4-methoxyphenylvinyl)-5,5′-dimethoxybiphenyl		LI09301
43 55 57 41 69 83 71 29	478	100 79 65 63 53 34 26 26	2.44	32	62	2	–	–	–	–	–	–	–	–	–	–	9-Hexadecenoic acid, hexadecyl ester, (Z)-	22393-83-5	NI44265
43 55 41 57 69 83 82 67	478	100 88 72 66 55 40 39 34	0.74	32	62	2	–	–	–	–	–	–	–	–	–	–	Myristic acid, 9-octadecenyl ester, (Z)-	22393-93-7	NI44266
57 55 43 264 69 44 83 97	478	100 89 82 78 78 71 68 61	3.54	32	62	2	–	–	–	–	–	–	–	–	–	–	9-Octadecenoic acid, tetradecyl ester, (Z)-	22393-85-7	NI44267
55 57 43 264 69 83 41 71	478	100 98 97 82 78 65 58 54	6.84	32	62	2	–	–	–	–	–	–	–	–	–	–	9-Octadecenoic acid, tetradecyl ester, (Z)-	22393-85-7	NI44268
43 55 57 41 69 83 71 29	478	100 77 70 65 52 37 27 27	2.64	32	62	2	–	–	–	–	–	–	–	–	–	–	9-Octadecenoic acid, tetradecyl ester, (Z)-	22393-85-7	NI44269
82 96 55 222 69 83 110 43	478	100 99 82 75 74 68 56 56	0.74	32	62	2	–	–	–	–	–	–	–	–	–	–	Palmitic acid, 9-hexadecenyl ester, (Z)-	22393-90-4	NI44270
57 43 85 29 99 478 113 449	478	100 61 21 20 17 14 14 7		33	66	1	–	–	–	–	–	–	–	–	–	–	Tritriacontan-3-one	79097-23-7	NI44271
286 285 194 193 165 45 208 44	478	100 99 99 99 97 70 64 51	23.50	34	22	3	–	–	–	–	–	–	–	–	–	–	2-(9-Oxo-9,10-dihydroanthracen-10-ylidenemethyl)-5-(9-oxo-9,10-dihydroanthracen-10-methyl)furan		MS10196
71 85 57 43 55 69 70 99	478	100 97 85 80 59 46 44 42	0.00	34	70	–	–	–	–	–	–	–	–	–	–	–	Botryococcane		LI09302
57 43 71 85 55 41 69 83	478	100 86 61 40 30 28 20 16	0.00	34	70	–	–	–	–	–	–	–	–	–	–	–	11-Decyltetracosane		AI02386
57 43 71 85 55 41 69 83	478	100 85 62 42 31 27 20 17	0.00	34	70	–	–	–	–	–	–	–	–	–	–	–	9-Octylhexacosane		AI02387
57 43 71 85 55 41 74 69	478	100 71 67 45 29 23 22 20	0.00	34	70	–	–	–	–	–	–	–	–	–	–	–	Tetratriacontane		MS10197
449 155 197 450 448 253 435 153	478	100 79 42 37 24 23 20 18	1.82	34	70	–	–	–	–	–	–	–	–	–	–	–	Tritriacontane, 3-methyl-	14167-69-2	NI44272
133 173 77 104 78 105 91 441	479	100 98 95 65 40 39 34 25	5.00	17	15	2	5	–	–	–	2	–	–	–	–	–	N-(2-Cyanoethyl)-4-(2,6-dibromo-4-nitrophenylazo)-N-ethylaniline		IC04499
171 186 43 45 44 60 160 73	479	100 92 66 50 37 36 34 29	0.00	17	23	10	2	2	–	–	–	–	–	–	–	–	4-Methoxyglucobrassicin	83327-21-3	NI44273
481 483 479 401 485 399 482 314	479	100 66 60 29 26 23 22 22		18	10	4	1	–	–	5	–	–	–	–	–	–	Naphthalene, 2,3-dimethoxy-1-nitro-4-(pentachlorophenyl)-	57187-62-9	NI44274
363 216 148 479 144 142 331 364	479	100 83 53 32 26 22 21 19		18	29	10	3	1	–	–	–	–	–	–	–	–	L-Glutamine, N^2-carboxy-N-[1-[(carboxymethyl)carbamoyl]-2-mercaptoethyl]-, N^2-ethyl dimethyl ester, ethyl carbonate (ester)	21026-92-6	NI44275
203 73 204 131 75 79 147 217	479	100 82 23 22 20 18 14 10	0.30	19	45	5	1	–	–	–	–	–	4	–	–	–	Methyl 2-deoxy-tri-O-(trimethylsilyl)-2-(trimethylsilylamino)-D-galactoside		LI09303
91 359 103 420 360 391 462 404	479	100 32 26 22 15 14 12 12	1.00	24	21	8	3	–	–	–	–	–	–	–	–	–	Trimethyl 6-benzyl-4a,5,6,11-tetrahydro-4a-hydroxy-5-oxopyridazino[2,3-a]quinoxaline-2,3,4-tricarboxylate		NI44276
479 103 149 286 480 376 255 73	479	100 80 44 42 41 38 36 31		25	45	4	1	–	–	–	–	–	2	–	–	–	3α,18-Dihydroxy-5α-androstan-17-one, oxime, di-TMS		LI09304
400 199 200 323 324 198 338 428	479	100 75 74 68 42 42 40 35	0.00	26	28	–	3	–	–	1	–	–	–	2	–	–	(Aminodiphenylphosphinoimino)-(ethylamino)diphenylphosphorane chloride		NI44277
419 328 388 360 420 359 279 141	479	100 74 45 39 37 32 26 23	2.73	26	41	7	1	–	–	–	–	–	–	–	–	–	PGE-2, methyl ester, diacetate		NI44278
328 359 70 218 57 219 141 419	479	100 65 63 59 56 56 53 52	2.35	26	41	7	1	–	–	–	–	–	–	–	–	–	PGE-2, methyl ester, diacetate		NI44279
73 448 75 213 125 358 268 449	479	100 99 99 69 50 47 47 42	7.44	26	49	3	1	–	–	–	–	–	2	–	–	–	3α,11β-Dihydroxy-5α-androstan-17-one, methoxime, bis(trimethylsilyl)-	32221-29-7	NI44280
73 448 125 75 213 268 449 93	479	100 99 99 99 92 72 40 40	10.96	26	49	3	1	–	–	–	–	–	2	–	–	–	3α,11β-Dihydroxy-5β-androstan-17-one, methoxime, bis(trimethylsilyl)-	32206-64-7	NI44281
448 147 449 215 129 464 174 107	479	100 47 39 33 32 24 21 21	3.00	26	49	3	1	–	–	–	–	–	2	–	–	–	3α,16α-Dihydroxy-5α-androstan-17-one, methoxime, bis(trimethylsilyl)-		LI09305
73 448 75 213 268 358 449 125	479	100 99 99 74 58 48 47 47	7.00	26	49	3	1	–	–	–	–	–	2	–	–	–	3β,11β-Dihydroxy-5α-androstan-17-one, methoxime, bis(trimethylsilyl)-	39780-65-9	NI44282
448 449 174 464 450 241 433 268	479	100 43 32 15 15 13 12 10	3.00	26	49	3	1	–	–	–	–	–	2	–	–	–	3β,16β-Dihydroxy-5α-androstan-17-one, methoxime, bis(trimethylsilyl)-		LI09306
448 158 449 174 464 133 173 358	479	100 58 55 55 41 35 34 31	6.00	26	49	3	1	–	–	–	–	–	2	–	–	–	3β,17β-Dihydroxy-5α-androstan-16-one, methoxime, bis(trimethylsilyl)-		LI09307
464 465 479 424 480 141 466 120	479	100 31 26 9 8 6 5 4		27	29	7	1	–	–	–	–	–	–	–	–	–	2H,8H-Benzo[1,2-b:5,4-b′]dipyran-10-propanol, 5-methoxy-2,2,8,8-tetramethyl-, 4-nitrobenzoate	56335-87-6	NI44283
311 41 479 69 268 70 312 214	479	100 85 80 63 47 40 28 27		27	33	5	3	–	–	–	–	–	–	–	–	–	Fumitremorgin B	12626-17-4	NI44284
311 42 268 479 69 70 45 200	479	100 96 78 64 63 51 51 48		27	33	5	3	–	–	–	–	–	–	–	–	–	Fumitremorgin B	12626-17-4	NI44285

MASS TO CHARGE RATIOS								M.W.	INTENSITIES								Parent	C	H	O	N	S	F	Cl	Br	I	Si	P	B	X	COMPOUND NAME	CAS Reg No	No
105	91	77	314	210	422	106	300	**479**	100	46	32	25	24	20	14	11	**0.00**	28	37	4	3	–	–	–	–	–	–	–	–	–	(2S,5R)-1-Benzoyl-5-[4-[(benzyloxycarbonyl)amino]butyl]-2-tert-butyl-3,5 dimethyl-4-imidazolidinone		DD01731
225	101	183	422	380	168	226	182	**479**	100	90	60	25	22	15	11	11	**10.00**	28	49	5	1	–	–	–	–	–	–	–	–	–	Octahydropiericidin B acetate	20238-89-5	LI09308
225	101	183	422	380	45	168	70	**479**	100	91	61	22	21	20	15	15	**10.00**	28	49	5	1	–	–	–	–	–	–	–	–	–	Octahydropiericidin B acetate	20238-89-5	NI44286
91	370	388	461	352	174	479	190	**479**	100	86	70	61	34	27	24	23		29	37	5	1	–	–	–	–	–	–	–	–	–	Phomine	14930-96-2	NI44287
129	91	73	75	130	423	105	77	**479**	100	62	36	15	13	11	11	9	**1.00**	29	41	3	1	–	–	–	–	–	1	–	–	–	Androst-5-ene-11,17-dione, 3-[(trimethylsilyl)oxy]-, 17-[O-benzyloxime], (3β)-	57325-70-9	NI44288
479	382	452	480	247	425	383	381	**479**	100	59	44	43	34	24	24	20		34	29	–	1	–	–	–	–	–	1	–	–	–	7-Silabicyclo[2.2.1]hept-5-ene-2-carbonitrile, 7-vinyl-7-methyl-1,4,5,6-tetraphenyl-	56805-08-4	MS10198
161	401	480	291	133	346	106	370	**480**	100	40	32	13	13	9	6	4		8	14	–	–	–	–	–	2	–	–	–	–	2	Palladium, dibromodi-π-crotyldi-		LI09309
480	267	401	346	161	291	322	239	**480**	100	89	83	67	67	39	17	17		8	14	–	–	–	–	–	2	–	–	–	–	2	Palladium, dibromodi-π-crotyldi-		LI09310
226	55	69	169	41	54	227	267	**480**	100	84	48	42	39	38	36	34	**4.58**	12	10	2	2	–	14	–	–	–	–	–	–	–	N,N'-Bis(heptafluorobutyryl)putrescine		MS10199
98	83	55	85	67	53	100	58	**480**	100	78	72	42	39	31	15	14	**0.17**	12	22	2	–	–	–	2	–	–	–	–	–	2	Palladium, di-μ-chloro-di-(4-methoxy-2-methylbut-2-enyl)di-		LI09311
368	480	396	424	452	312	340	310	**480**	100	85	76	50	41	36	32	30		14	12	6	–	–	–	–	–	–	–	–	–	2	Ruthenium, hexacarbonyl[μ-(1,2,3,4-tetramethyl-1,3-butadienylene)]di-, (Ru-Ru)	33310-09-7	NI44289
55	59	216	333	176	42	218	119	**480**	100	90	60	57	36	30	22	18	**1.20**	15	18	4	2	–	10	–	–	–	–	–	–	–	Putreanine ethyl ester bis(pentafluoropropionate)		MS10200
195	52	176	145	266	127	247	214	**480**	100	96	81	31	22	22	20	17	**6.50**	16	8	–	–	–	12	–	–	–	–	–	–	1	Chromium, bis(1,3-di(trifluoromethyl)benzene)-	53966-05-5	NI44290
195	145	176	480	52	126	69	75	**480**	100	73	44	38	38	33	26	14		16	8	–	–	–	12	–	–	–	–	–	–	1	Chromium, bis(1,4-di(trifluoromethyl)benzene)-	53966-06-6	NI44291
43	447	449	445	107	44	451	448	**480**	100	30	19	18	18	8	7	6	**0.60**	16	14	4	–	–	–	6	–	–	–	–	–	–	1,2,3,4,10,10-Hexachloro-trans-6,7-diacetyloxy-1,4,4a,5,6,7,8,8a-octahydro-1,4-endo-5,8-exo-dimethanonaphthalene		MS10201
73	147	75	273	201	45	74	149	**480**	100	30	26	20	18	11	9	8	**0.00**	18	40	7	–	–	–	–	–	–	4	–	–	–	Citric acid tetra-TMS	14330-97-3	NI44292
73	147	273	75	363	45	375	347	**480**	100	46	43	16	13	12	11	11	**0.00**	18	40	7	–	–	–	–	–	–	4	–	–	–	Citric acid tetra-TMS	14330-97-3	NI44293
73	147	75	245	157	246	190	77	**480**	100	44	44	30	25	14	7	5	**0.15**	18	40	7	–	–	–	–	–	–	4	–	–	–	Dehydroascorbic acid tetra-TMS		NI44294
73	147	273	75	363	375	274	45	**480**	100	42	41	16	11	10	10	10	**0.00**	18	40	7	–	–	–	–	–	–	4	–	–	–	Isocitric acid tetra-TMS	55517-57-2	NI44295
73	147	273	245	75	465	319	45	**480**	100	43	24	20	15	12	11	11	**0.00**	18	40	7	–	–	–	–	–	–	4	–	–	–	Isocitric acid tetra-TMS	55517-57-2	NI44296
73	147	273	245	75	45	375	74	**480**	100	42	33	19	14	11	9	9	**1.00**	18	40	7	–	–	–	–	–	–	4	–	–	–	Isocitric acid tetra-TMS	55517-57-2	NI44297
73	217	147	204	244	219	218	305	**480**	100	61	34	16	12	12	12	8	**7.36**	18	40	7	–	–	–	–	–	–	4	–	–	–	Saccharo-1,4-lactone tetra-TMS	56272-60-7	NI44298
204	73	217	205	147	191	206	129	**480**	100	69	39	21	18	11	9	9	**0.00**	19	44	6	–	–	–	–	–	–	4	–	–	–	D-Altro-2-heptulose, anhydrotetrakis-O-(trimethylsilyl)-	56270-89-4	NI44299
129	73	147	157	363	191	273	201	**480**	100	89	64	36	28	22	18	18	**0.00**	19	44	6	–	–	–	–	–	–	4	–	–	–	Ditrimethylsilyl 2,6-bis(trimethylsiloxy)diheptanoate		MS10202
75	43	169	109	331	115	271	132	**480**	100	24	14	8	7	3	2	2	**0.00**	20	32	13	–	–	–	–	–	–	–	–	–	–	(1RS)-1,3,3-Trimethoxypropyl 2,3,4,6-tetra-O-acetyl-β-D-glucopyranoside		MS10203
88	43	45	101	187	305	71	75	**480**	100	96	86	78	46	31	31	29	**0.00**	21	36	12	–	–	–	–	–	–	–	–	–	–	α-D-Galactopyranosiduronic acid, methyl 3,4-O-isopropylidene-2-(2,3,4,6 tetra-O-methyl-β-D-glucopyranosyl)-, methyl ester	39102-75-5	NI44300
69	310	325	157	116	110	74	326	**480**	100	36	30	30	30	25	23	10	**5.00**	22	24	3	8	1	–	–	–	–	–	–	–	–	1-Methyl-4-(4-methoxyphenyl)-5-(4-methoxyphenyl)imino-3-[6-methyl-3-(methylthio)-5-oxo-4,5-dihydro-1,2,4-triazin-4-yl]imino-2H-2,3-dihydro-1,2,4-triazole		MS10204
240	242	241	243	161	178	29	28	**480**	100	98	13	12	11	8	8	8	**0.00**	22	30	–	2	–	–	–	2	–	–	–	–	–	(±)-1,2-Bis(4-bromophenyl)-N,N,N',N'-tetraethylethylenediamine		DD01732
240	242	241	243	161	28	29	178	**480**	100	96	45	43	37	31	30	24	**0.00**	22	30	–	2	–	–	–	2	–	–	–	–	–	meso-1,2-Bis(4-bromophenyl)-N,N,N',N'-tetraethylethylenediamine		DD01733
289	144	114	100	56	70	57	58	**480**	100	90	82	81	80	77	72	69	**18.00**	22	44	9	2	–	–	–	–	–	–	–	–	–	3,13-Dihydroxy-5,8,11,18,21,26,29-heptoxa-1,15-diazabicyclo[13.8.8]hentriacontane		MS10205
336	401	415	335	208	480	318	192	**480**	100	75	43	34	15	10	10	8		23	17	2	2	–	–	–	1	–	–	–	–	1	Titanium, bromo(η^5-2,4-cyclopentadien-1-yl)bis(8-quinolinolato-N[1],O[8])-	35256-09-8	NI44301
478	316	60	304	479	43	317	345	**480**	100	76	55	49	46	46	42	26	**13.59**	24	24	7	4	–	–	–	–	–	–	–	–	–	1H-Pyrazole-4,5-dione, 1-phenyl-3-[1,2,3-tris(acetyloxy)propyl]-, 4-(phenylhydrazone), (R*,R*)-(–)-	56143-18-1	NI44302
309	295	384	358	283	281	324	239	**480**	100	70	55	40	40	40	35	30	**0.00**	24	32	10	–	–	–	–	–	–	–	–	–	–	17-Abeo-3α,18-diacetoxy-6β,7α,16ξ-trihydroxyroyleanone		NI44303
43	57	41	83	85	60	148	55	**480**	100	92	33	25	24	20	8	8	**0.15**	24	32	10	–	–	–	–	–	–	–	–	–	–	Acevaltrate		MS10206
83	104	77	71	58	55	85	160	**480**	100	87	85	80	54	53	50	46	**5.00**	26	28	7	2	–	–	–	–	–	–	–	–	–	Propanedioic acid, (1,3-dihydro-1,3-dioxo-2H-isoindol-2-yl)[4-[benzyleneamino]butyl]-, diethyl ester, N-oxide, (Z)-	38068-84-7	NI44304
215	73	143	202	216	249	130	480	**480**	100	75	33	29	22	13	12	11		26	36	3	2	–	–	–	–	–	2	–	–	–	Methoxyphenylethylcarbonylaminoethylindole, bis(trimethylsilyl)-		NI44305
155	225	165	197	323	147	123	166	**480**	100	85	85	61	35	35	35	30	**0.80**	26	40	6	–	1	–	–	–	–	–	–	–	–	Methyl 12-oxo-13-tosyloxyoctadec-9-enoate		LI09312
409	319	199	410	73	173	130	411	**480**	100	53	53	40	30	28	26	17	**6.00**	26	48	4	–	–	–	–	–	–	2	–	–	–	Prostaglandin A1, bis(trimethylsilyl)-	55429-54-4	NI44306
73	409	75	319	410	381	55	465	**480**	100	65	43	27	23	15	14	12	**9.01**	26	48	4	–	–	–	–	–	–	2	–	–	–	Prostaglandin A1, bis(trimethylsilyl)-	55429-54-4	NI44307
381	480	382	409	465	481	383	410	**480**	100	34	34	29	22	14	13	10		26	48	4	–	–	–	–	–	–	2	–	–	–	Prostaglandin B1, bis(trimethylsilyl)-	56009-41-7	NI44308
424	162	70	425	183	382	262	55	**480**	100	78	47	35	24	23	15	14	**5.49**	27	22	3	–	–	–	–	–	–	–	1	–	1	(Acetylcyclopentadienyl)manganesedicarbonyltriphenylphosphine	33309-65-8	NI44309

MASS TO CHARGE RATIOS								M.W.	INTENSITIES								Parent	C	H	O	N	S	F	Cl	Br	I	Si	P	B	X	COMPOUND NAME	CAS Reg No	No
254	287	178	281	480	113	343	423	**480**	100	30	25	24	22	19	14	12		27	41	–	2	–	–	–	–	–	1	–	1	1	3,3-Bis(cyclopentadienyl)-2-tert-butyl-1-[tert-butyl(trimethylsilyl)amino]-2-aza-1-bora-3-titanaindane		NI44310
91	107	153	433	197	343	465	480	**480**	100	26	21	19	19	18	5	0		27	44	5	–	1	–	–	–	–	–	–	–	–	Methyl (±)-2,4,6,8,10-pentadeoxy-4,6,8,10-tetramethyl-3-O-methyl-7,9-O (1-methylethylidene)-11-O-(phenylmethyl)-1-thio-D-manno-β-L-galacto-undecopyranoside		NI44311
28	43	345	363	346	327	328	364	**480**	100	72	22	20	12	8	6	4	**1.00**	27	44	7	–	–	–	–	–	–	–	–	–	–	Ecdysterone		MS10207
426	410	99	408	427	345	344	429	**480**	100	38	38	32	30	27	24	23	**0.01**	27	44	7	–	–	–	–	–	–	–	–	–	–	Ecdysterone		MS10208
345	328	327	344	300	329	346	426	**480**	100	98	96	93	78	71	70	58	**0.61**	27	44	7	–	–	–	–	–	–	–	–	–	–	Ecdysterone, (β)-	5289-74-7	NI44312
300	181	142	291	115	211	93	444	**480**	100	60	48	45	45	40	38	33	**1.00**	27	44	7	–	–	–	–	–	–	–	–	–	–	Ecdysterone, 26-Hydroxy-		NI44313
148	81	236	183	314	211	130	95	**480**	100	80	32	13	10	10	9	8	**0.00**	27	44	7	–	–	–	–	–	–	–	–	–	–	Locin	102071-99-8	NI44314
43	55	41	401	57	95	81	69	**480**	100	69	62	56	51	44	42	33	**0.00**	27	45	2	–	–	–	–	1	–	–	–	–	–	5β,6β-Epoxy-7α-bromocholestan-3β-ol		NI44315
255	117	274	273	119	256	107	105	**480**	100	58	34	14	13	12	10	8	**1.00**	27	52	3	–	–	–	–	–	–	2	–	–	–	Pregnane-3,17,20-triol, bis(trimethylsilyl) ether, (3α,5β,20S)-	56211-30-4	NI44316
255	73	75	117	273	272	119	256	**480**	100	71	55	50	48	22	21	18	**1.00**	27	52	3	–	–	–	–	–	–	2	–	–	–	Pregnan-17-ol, 3,20-bis[(trimethylsilyl)oxy]-, (3α,5β,20S)-	17562-94-6	NI44317
120	92	240	480	64	360	39	76	**480**	100	91	50	42	18	16	12	7		28	16	8	–	–	–	–	–	–	–	–	–	–	Benzyl alcohol, 4-hydroxy-		MS10209
267	480	252	280	139	156	265	325	**480**	100	37	23	21	21	18	16	14		30	21	4	–	–	–	1	–	–	–	–	–	–	β-[9-(10-Phenylanthryl)]propanoyl-3-chlorobenzoyl peroxide		MS10210
361	407	408	451	182	362	360	270	**480**	100	99	87	54	49	40	27	23	**5.00**	30	28	4	2	–	–	–	–	–	–	–	–	–	Benzoic acid, 2,2′-([1,1′-biphenyl]-4,4′-diyldiimino)bis-, diethyl ester	65591-24-4	NI44318
131	73	101	189	465	157	335	303	**480**	100	52	24	8	5	4	2	1	**0.40**	30	56	4	–	–	–	–	–	–	–	–	–	–	1-O-(2-Methoxy-4-tricosynyl)-2,3-O-isopropylideneglycerol		NI44319
43	55	100	57	69	71	211	395	**480**	100	93	88	87	58	54	51	49	**1.47**	31	60	3	–	–	–	–	–	–	–	–	–	–	14,16-Hentriacontanedione, 25-hydroxy-	52262-75-6	NI44320
394	195	480	437	197	182	393	144	**480**	100	81	68	61	56	55	52	46		32	40	–	4	–	–	–	–	–	–	–	–	–	Isoborreverine		MS10211
57	43	257	55	83	71	69	41	**480**	100	91	59	59	50	50	50	45	**7.73**	32	64	2	–	–	–	–	–	–	–	–	–	–	Hexadecanoic acid, hexadecyl ester	540-10-3	IC04500
257	57	43	71	55	83	69	97	**480**	100	82	69	48	48	47	43	41	**9.14**	32	64	2	–	–	–	–	–	–	–	–	–	–	Hexadecanoic acid, hexadecyl ester	540-10-3	NI44322
285	57	43	71	28	55	83	69	**480**	100	74	63	44	44	43	41	38	**0.00**	32	64	2	–	–	–	–	–	–	–	–	–	–	Hexadecanoic acid, hexadecyl ester	540-10-3	NI44321
285	57	43	71	83	55	69	97	**480**	100	90	71	55	49	49	46	44	**10.04**	32	64	2	–	–	–	–	–	–	–	–	–	–	Octadecanoic acid, tetradecyl ester	17661-50-6	NI44323
285	57	43	71	55	196	83	69	**480**	100	97	82	56	54	50	50	48	**23.64**	32	64	2	–	–	–	–	–	–	–	–	–	–	Octadecanoic acid, tetradecyl ester	17661-50-6	NI44324
32	229	57	40	43	55	71	69	**480**	100	46	34	34	31	21	19	19	**7.74**	32	64	2	–	–	–	–	–	–	–	–	–	–	Tetradecanoic acid, octadecyl ester	3234-81-9	NI44325
229	43	57	55	69	71	83	41	**480**	100	88	80	56	44	41	40	40	**26.04**	32	64	2	–	–	–	–	–	–	–	–	–	–	Tetradecanoic acid, octadecyl ester	3234-81-9	NI44326
480	124	302	178	240	59	237		**480**	100	64	62	21	11	9	6			33	25	–	–	–	–	–	–	–	–	–	–	1	Cobalt, (π-cyclopentadienyl)(π-tetraphenylcyclobutadienyl)-	1278-02-0	NI44327
178	165	179	215	302	191	176	462	**480**	100	23	18	17	16	16	16	12	**12.00**	34	24	3	–	–	–	–	–	–	–	–	–	–	3,4-Bis[2-(2′-formyl-2-biphenyl)vinyl]furan		LI09313
485	483	487	325	323	327	162.5	244	**481**	100	68	66	54	36	33	33	18		12	7	–	1	–	–	–	4	–	–	–	–	–	2,2′,4,4′-Tetrabromodiphenylamine		IC04501
73	204	218	147	75	233	189	74	**481**	100	62	60	60	60	45	32	32	**0.80**	18	43	6	1	–	–	–	–	–	4	–	–	–	D-Glucuronamide, tetrakis(trimethylsilyl)-		NI44328
73	355	356	45	357	75	74	77	**481**	100	64	19	11	10	9	8	4	**1.60**	19	34	4	1	–	3	–	–	–	3	–	–	–	N-(Trifluoroacetyl)-O,O′,O″-tris(trimethylsilyl)norepinephrine		MS10212
73	217	103	75	57	481	55	71	**481**	100	48	44	40	35	23	21	19		19	35	4	5	–	–	–	–	–	3	–	–	–	Tris(trimethylsilyl)-8,2′-aminoanhydroinosine		NI44329
73	481	103	482	466	260	483	169	**481**	100	99	54	41	37	29	19	19		19	35	4	5	–	–	–	–	–	3	–	–	–	Tris(trimethylsilyl)-8,2′-anhydroadenosine		NI44330
73	481	260	350	482	466	75	103	**481**	100	99	62	45	42	26	21	19		19	35	4	5	–	–	–	–	–	3	–	–	–	Tris(trimethylsilyl)-8,3′-anhydroadenosine		NI44331
162	91	253	28	134	119	163	116	**481**	100	41	39	31	25	13	10	7	**0.00**	19	47	5	1	–	–	–	–	–	4	–	–	–	D-Galactopyranoside, methyl 2-deoxy-3,4,6-tris-O-(trimethylsilyl)-2-[(trimethylsilyl)amino]-	74420-76-1	NI44332
117	73	160	147	118	219	277	75	**481**	100	57	13	11	10	7	6	5	**0.00**	19	47	5	1	–	–	–	–	–	4	–	–	–	D-Galactose, 6-deoxy-2,3,4,5-tetrakis-O-(trimethylsilyl)-, O-methyloxime	56196-05-5	NI44333
181	390	391	182	91	65			**481**	100	90	15	11	8	4			**0.11**	22	13	–	1	–	10	–	–	–	–	–	–	–	N,N-Bis(pentafluorophenylmethyl)-2-phenylethylamine		MS10213
162	43	161	176	133	119	104	65	**481**	100	98	33	21	16	9	7	7	**0.04**	24	23	8	3	–	–	–	–	–	–	–	–	–	Bis[3-(O-acetoxyphenyl)-isoxazolin-5-yl]acetoxyamine		NI44334
145	113	159	55	101	59	173	15	**481**	100	96	35	32	30	30	24	21	**0.00**	24	35	9	1	–	–	–	–	–	–	–	–	–	1,1-Dimethyl 4-(2,6-di-tert-butyl-4-methoxyphenyl) 4-nitrobutane-1,1,4-tricarboxylate		DD01734
239	270	483	481	183	121	282	373	**481**	100	45	30	30	30	30	20	5		25	24	4	1	–	–	–	1	–	–	–	–	–	Cherylline, 2-(4-bromobenzoyl)-2-demethyl-O,O-dimethyl-	23330-78-1	MS10214
281	133	165	77	200	282	79	202	**481**	100	92	39	28	22	20	15	10	**0.00**	26	24	4	1	1	–	1	–	–	–	–	–	–	(1S*,4R*,5S*,8R*,9S*,10S*)-N-Phenyl-8-methyl-9-methoxy-9-(phenylthio)-10-chloro-11-oxatricyclo[6.2.1.0^{2,7}]undec-2(7)-ene-4,5-dicarboximide	117626-68-3	DD01735
44	164	241	297	396	268	252	470	**481**	100	85	45	40	38	30	25	23	**0.00**	28	51	5	1	–	–	–	–	–	–	–	–	–	Tetradecanamide, N-[3-[4-(acetyloxy)-3-methyl-7-oxabicyclo[4.1.0]hept-1-yl]propyl]-7-methoxy-N-methyl-	69121-68-2	NI44335
55	91	81	41	67	43	190	380	**481**	100	80	47	47	40	40	36	30	**2.00**	29	39	5	1	–	–	–	–	–	–	–	–	–	1-Benzyl-3a,5-dihydroxy-7-methyl-4[4-methyl-7-(5-oxo-tetrahydrofuran-2-yl]hept-1-enyl)-6-methylene-3-oxo-3a,4,5,6,7,7a-hexahydroisoindoline		IC04502
91	55	372	190	81	120	380	95	**481**	100	64	46	39	39	34	32	28	**10.70**	29	39	5	1	–	–	–	–	–	–	–	–	–	1-Benzyl-8,17-dihydroxy-12,19-dimethyl-3,5-dioxo-8-methylene-4-oxa-14-ene-perhydrocyclotetradeca[2,3-d]isoindole		IC04503
91	107	89	92	151	188	180	232	**481**	100	67	49	40	25	23	12	9	**0.00**	29	39	5	1	–	–	–	–	–	–	–	–	–	rel-[2S,6S,2(3R)]-N-[(Benzyloxy)carbonyl]-2-[3-[[(benzyloxy)carbonyl]oxy]heptyl]-6-methylpiperidine		DD01736

MASS TO CHARGE RATIOS								M.W.	INTENSITIES								Parent	C	H	O	N	S	F	Cl	Br	I	Si	P	B	X	COMPOUND NAME	CAS Reg No	No
91	73	75	147	68	41	207	92	**481**	100	40	26	18	15	12	10	10	**0.32**	29	43	3	1	–	–	–	–	–	1	–	–	–	Androstane-11,17-dione, 3-[(trimethylsilyl)oxy]-, 17-[O-benzyloxime], (3α,5α)-	57305-11-0	NI44336
259	98	166	42	181	70	199	72	**481**	100	79	72	60	48	46	44	40	**10.00**	30	31	3	1	–	–	–	–	–	1	–	–	–	3-Diethylamino-2-[3-(triphenylsiloxy)-3-butenoyl]-2-cyclobuten-1-one		NI44337
485	487	483	489	481	486	491	488	**482**	100	83	52	42	12	9	7	6	**0.00**	7	3	–	–	–	–	–	5	–	–	–	–	–	Benzene, pentabromomethyl-	87-83-2	NI44338
486	488	407	490	406	87	405	485	**482**	100	92	65	54	54	54	48	46	**10.00**	7	3	–	–	–	–	–	5	–	–	–	–	–	Benzene, pentabromomethyl-	87-83-2	IC04504
44	440	45	28	58	30	397	31	**482**	100	99	55	31	30	24	14	13	**0.00**	10	30	–	6	–	–	4	–	–	–	–	–	4	Bis(methylimino)tetrakis(dimethylamino)tetraaluminium tetrachloride		LI09314
125	15	109	79	93	47	43	30	**482**	100	31	28	26	25	25	23	13	**1.10**	16	20	7	–	3	–	–	–	–	–	2	–	–	Phosphorothioic acid, O,O′-(sulphinyldi-4,1-phenylene) O,O,O′,O′-tetramethyl ester	17210-55-8	NI44339
73	103	93	75	57	95	77	259	**482**	100	99	99	99	94	82	73	69	**35.00**	19	34	5	4	–	–	–	–	–	3	–	–	–	Tris(trimethylsilyl)-8,2′-anhydroinosine		NI44340
73	482	351	467	258	93	379	169	**482**	100	99	92	68	61	55	51	47		19	34	5	4	–	–	–	–	–	3	–	–	–	Tris(trimethylsilyl)-8,3′-anhydroinosine		NI44341
160	322	163	483	118	324	162	142	**482**	100	95	77	70	54	34	30	30	**0.00**	19	38	10	4	–	–	–	–	–	–	–	–	–	D-Streptamine, O-2-amino-2-deoxy-α-D-glucopyranosyl-(1-4)-O-[3-deoxy 4-C-methyl-3-(methylamino)-β-L-arabinopyranosyl-(1-6)]-2-deoxy-	36889-17-5	NI44342
160	163	118	322	483	205	324	142	**482**	100	52	52	51	26	18	16	14	**0.00**	19	38	10	4	–	–	–	–	–	–	–	–	–	D-Streptamine, O-6-amino-6-deoxy-α-D-glucopyranosyl-(1-4)-O-[3-deoxy 4-C-methyl-3-(methylamino)-β-L-arabinopyranosyl-(1-6)]-2-deoxy-	36889-15-3	NI44343
217	257	218	279	379	147	103	133	**482**	100	29	22	20	19	17	17	13	**0.00**	19	46	6	–	–	–	–	–	–	4	–	–	–	α-D-Fructofuranoside, methyl 1,3,4,6-tetrakis-O-(trimethylsilyl)-	30788-71-7	NI44344
217	73	147	133	218	79	319	75	**482**	100	82	21	21	20	14	11	11	**0.20**	19	46	6	–	–	–	–	–	–	4	–	–	–	α-D-Galactofuranoside, methyl 2,3,5,6-tetrakis-O-(trimethylsilyl)-	6736-93-2	NI44345
204	73	133	217	205	147	75	129	**482**	100	72	35	34	24	22	17	13	**0.00**	19	46	6	–	–	–	–	–	–	4	–	–	–	α-D-Galactopyranoside, methyl 2,3,4,6-tetrakis-O-(trimethylsilyl)-	4133-45-3	NI44346
204	73	133	217	205	147	129	75	**482**	100	75	36	35	28	22	14	13	**0.00**	19	46	6	–	–	–	–	–	–	4	–	–	–	β-D-Galactopyranoside, methyl 2,3,4,6-tetrakis-O-(trimethylsilyl)-	2296-39-1	NI44347
73	204	133	217	147	205	129	75	**482**	100	94	27	25	22	21	14	12	**0.00**	19	46	6	–	–	–	–	–	–	4	–	–	–	α-D-Galactoside, methyl tetrakis-O-(trimethylsilyl)-	74725-78-3	NI44348
217	73	129	204	218	132	147	117	**482**	100	99	70	44	38	33	23	23	**0.00**	19	46	6	–	–	–	–	–	–	4	–	–	–	α-D-Glucofuranoside, methyl 2,3,5,6-tetrakis-O-(trimethylsilyl)-	6736-96-5	NI44349
146	73	191	147	159	133	103	75	**482**	100	68	40	27	9	9	8	8	**0.00**	19	46	6	–	–	–	–	–	–	4	–	–	–	β-D-Glucopyranose, 2-O-methyl-1,3,4,6-tetrakis-O-(trimethylsilyl)-	55515-37-2	NI44350
133	73	146	217	147	218	129	89	**482**	100	93	91	73	31	17	14	14	**0.00**	19	46	6	–	–	–	–	–	–	4	–	–	–	β-D-Glucopyranose, 3-O-methyl-1,2,4,6-tetrakis-O-(trimethylsilyl)-	55515-36-1	NI44351
204	73	191	217	205	147	206	129	**482**	100	63	29	25	19	17	9	8	**0.00**	19	46	6	–	–	–	–	–	–	4	–	–	–	β-D-Glucopyranose, 6-O-methyl-1,2,3,4-tetrakis-O-(trimethylsilyl)-	55515-35-0	NI44352
204	73	133	217	205	147	206	129	**482**	100	83	62	52	41	36	18	15	**0.00**	19	46	6	–	–	–	–	–	–	4	–	–	–	α-D-Glucopyranoside, methyl 2,3,4,6-tetrakis-O-(trimethylsilyl)-	2641-79-4	NI44353
204	73	133	217	205	147	206	129	**482**	100	80	37	27	24	24	10	10	**0.00**	19	46	6	–	–	–	–	–	–	4	–	–	–	α-D-Glucopyranoside, methyl 2,3,4,6-tetrakis-O-(trimethylsilyl)-	2641-79-4	NI44355
73	204	133	217	147	205	75	191	**482**	100	90	40	25	23	18	11	10	**0.00**	19	46	6	–	–	–	–	–	–	4	–	–	–	α-D-Glucopyranoside, methyl 2,3,4,6-tetrakis-O-(trimethylsilyl)-	2641-79-4	NI44354
204	73	133	217	205	147	75	129	**482**	100	93	51	33	30	24	18	15	**0.00**	19	46	6	–	–	–	–	–	–	4	–	–	–	β-D-Glucopyranoside, methyl 2,3,4,6-tetrakis-O-(trimethylsilyl)-	2296-40-4	NI44356
73	217	129	116	218	204	117	75	**482**	100	81	55	44	37	35	26	26	**0.00**	19	46	6	–	–	–	–	–	–	4	–	–	–	α-D-Mannofuranoside, methyl 2,3,5,6-tetrakis-O-(trimethylsilyl)-	6737-01-5	NI44357
204	73	133	217	205	147	129	206	**482**	100	87	31	25	23	21	10	9	**0.00**	19	46	6	–	–	–	–	–	–	4	–	–	–	α-D-Mannopyranoside, methyl 2,3,4,6-tetrakis-O-(trimethylsilyl)-	1769-06-8	NI44359
204	73	133	205	217	147	206	129	**482**	100	86	48	30	28	22	12	11	**0.00**	19	46	6	–	–	–	–	–	–	4	–	–	–	α-D-Mannopyranoside, methyl 2,3,4,6-tetrakis-O-(trimethylsilyl)-	1769-06-8	NI44358
204	73	133	217	205	147	117	75	**482**	100	93	40	28	24	22	21	20	**0.00**	19	46	6	–	–	–	–	–	–	4	–	–	–	β-D-Mannopyranoside, methyl 2,3,4,6-tetrakis-O-(trimethylsilyl)-	3504-69-6	NI44360
204	147	379	205	133	103	257	217	**482**	100	36	29	20	14	12	11	11	**0.00**	19	46	6	–	–	–	–	–	–	4	–	–	–	L-Sorbopyranoside, methyl 1,3,4,5-tetrakis-O-(trimethylsilyl)-	30788-70-6	NI44361
169	43	109	127	110	81	331	97	**482**	100	99	86	30	25	11	9	9	**0.00**	22	26	12	–	–	–	–	–	–	–	–	–	–	O-Acetoxyphenyl 2,3,4,6-tetra-O-acetyl-β-D-glucopyranoside	17019-75-9	NI44362
169	109	110	127	331	139	145	115	**482**	100	98	50	40	25	19	17	9	**0.00**	22	26	12	–	–	–	–	–	–	–	–	–	–	β-D-Glucopyranoside, 4-(acetyloxy)phenyl, tetraacetate	14698-56-7	NI44363
482	483	69	119	105	91	150	55	**482**	100	25	17	15	14	14	13	13		23	25	3	–	–	7	–	–	–	–	–	–	–	Androsta-3,5-dien-17-one, 3-heptafluorobutyryl-		MS10215
169	135	109	91	43	331	145	211	**482**	100	99	99	78	68	63	26	21	**0.00**	23	30	11	–	–	–	–	–	–	–	–	–	–	(1RS)-1-Methoxy-2-phenylethyl 2,3,4,6-tetra-O-acetyl-β-D-glucopyranoside		MS10216
73	482	267	179	467	75	268	484	**482**	100	88	57	45	20	20	18	17		24	46	4	–	–	–	–	–	–	3	–	–	–	(3,4-Dihydroxyphenyl)nonanoic acid, tris(trimethylsilyl)-		NI44364
274	210	167	259	180	154	315	231	**482**	100	48	42	38	29	19	17	17	**8.00**	25	22	10	–	–	–	–	–	–	–	–	–	–	7H-1,4-Dioxino[2,3-c]xanthen-7-one, 2,3-dihydro-8-hydroxy(5-hydroxy-2,4-dimethoxyphenyl)(hydroxymethyl)-5-methoxy-, (2S-trans)-	64280-47-3	NI44365
126	270	269	166	284	152	153	124	**482**	100	67	34	27	18	17	16	15	**2.70**	25	22	10	–	–	–	–	–	–	–	–	–	–	Isosilychristin	77182-66-2	NI44366
381	423	440	306	395	307	107	246	**482**	100	65	46	46	33	31	31	28	**4.00**	26	30	7	2	–	–	–	–	–	–	–	–	–	6,21-Cyclo-4,5-secoakuammilan-17-oic acid, 1-acetyl-4,5-bis(acetyloxy)-1,2-dihydro-, methyl ester, (2ξ,6α)-	63944-68-3	NI44367
73	467	75	468	482	44	173	377	**482**	100	62	24	23	20	14	10	9		26	34	5	–	–	–	–	–	–	2	–	–	–	3H,7H-Furo[3,2-d:5,4-f′]bis[1]benzopyran, 3,3-dimethyl-6b,10[12bH]-bis[(trimethylsilyl)oxy]-, cis-(–)-	55429-53-3	NI44368
438	482	265	162	134	464	465	237	**482**	100	75	53	48	35	33	24	11		27	14	9	–	–	–	–	–	–	–	–	–	–	Wikstrosin	65556-47-0	NI44369
105	283	161	317	311	189	255	284	**482**	100	82	73	55	36	36	34	33	**22.70**	27	30	8	–	–	–	–	–	–	–	–	–	–	Daphnetoxin	28164-88-7	NI44370
214	44	73	242	96	42	124	215	**482**	100	37	34	30	20	20	17	16	**13.00**	27	38	2	2	1	–	–	–	–	1	–	–	–	10H-Phenothiazine, 2-acetyl-10-[3-[4-[2-[(trimethylsilyl)oxy]ethyl]-1-piperidinyl]propyl]-	74793-82-1	NI44371
70	84	267	98	238	138	113	132	**482**	100	80	50	40	32	31	26	22	**8.00**	27	58	1	6	–	–	–	–	–	–	–	–	–	Azacyclotridecan-2-one, 1-(19-amino-4,8,12,16-tetraazanonadec-1-yl)-	75422-10-5	NI44372
70	84	98	482	113	127	310		**482**	100	70	45	30	22	20	18			27	58	1	6	–	–	–	–	–	–	–	–	–	13,17,21,25,29-Pentaaza-32-dotriacontanelactam	75422-12-7	NI44373
299	149	43	44	137	95	328	300	**482**	100	38	35	34	28	27	26	21	**2.30**	28	34	7	–	–	–	–	–	–	–	–	–	–	Gedunin		MS10217
239	464	221	225	446	303	285	428	**482**	100	62	60	45	44	40	38	30	**0.00**	28	50	6	–	–	–	–	–	–	–	–	–	–	24-ξ-methyl-5α-cholestane-3β,6α,8,15α,16β,26-hexol		NI44374
43	482	439	425	453	383	397	55	**482**	100	80	73	73	59	55	48	48		28	54	4	2	–	–	–	–	–	–	–	–	–	1,4-Dioxa-7,18-diazacycloeicosane-6,19-dione, 7,18-dihexyl-	67456-25-1	NI44375

MASS TO CHARGE RATIOS								M.W.	INTENSITIES								Parent	C	H	O	N	S	F	Cl	Br	I	Si	P	B	X	COMPOUND NAME	CAS Reg No	No
271	43	55	41	105	253	91	107	**482**	100	68	58	58	52	50	48	47	**0.00**	29	43	5	–	–	–	–	–	–	–	–	1	–	5α-Pregnane-3β,11β,17α,20β,21-pentol tert-butyl boronate		MS10218
368	369	247	213	353	255	260	229	**482**	100	37	21	21	18	18	14	9	**0.71**	29	45	2	–	–	3	–	–	–	–	–	–	–	Cholest-5-en-3-ol, trifluoroacetate, (3β)-	2665-02-3	NI44376
255	482	369	327	213	467	483	229	**482**	100	74	64	59	41	37	30	30		29	45	2	–	–	3	–	–	–	–	–	–	–	Cholest-7-en-3-ol, trifluoroacetate, (3β)-	55515-22-5	NI44377
73	57	43	55	69	423	41	83	**482**	100	48	46	44	43	40	36	34	**2.00**	29	58	3	–	–	–	–	–	–	1	–	–	–	Pentacosenoic acid, 2-[(trimethylsilyl)oxy]-, methyl ester	56784-06-6	MS10219
150	210	149	77	209	124	135	123	**482**	100	47	44	19	18	16	15	15	**0.00**	30	26	6	–	–	–	–	–	–	–	–	–	–	10-[1-(4-Hydroxy-3-methoxyphenyl)-2-(2-methoxyphenoxy)ethyl]-10-hydroxy-9-anthrone	74838-64-5	NI44378
482	308	226	468	281	439	294	240	**482**	100	47	47	37	30	27	26	20		30	30	4	2	–	–	–	–	–	–	–	–	–	Pteledimeridine		MS10220
131	73	101	157	467	450	189	337	**482**	100	42	11	4	3	3	3	2	**0.10**	30	58	4	–	–	–	–	–	–	–	–	–	–	1-O-(2-Methoxy-4-tricosenyl)-2,3-O-isopropylideneglycerol		NI44379
40	293	81	44	121	43	69	41	**482**	100	99	58	58	48	47	41	31	**15.00**	31	46	4	–	–	–	–	–	–	–	–	–	–	Cholesta-9(11),17(20),24-triene-3,6-diol, diacetate, (3β,5α,6α)-	37926-47-9	NI44380
423	216	217	424	147	107	105	276	**482**	100	59	34	33	29	29	29	26	**1.16**	31	46	4	–	–	–	–	–	–	–	–	–	–	Olean-12-en-28-oic acid, 3,16-dioxo-, methyl ester	22478-83-7	NI44381
75	467	57	43	73	103	97	83	**482**	100	76	56	55	48	42	38	34	**0.00**	31	66	1	–	–	–	–	–	–	1	–	–	–	Octacosanol trimethylsilyl ether		MS10221
422	407	69	95	119	239	81	107	**482**	100	80	67	57	47	45	45	40	**12.00**	32	50	3	–	–	–	–	–	–	–	–	–	–	3β-Acetoxy-9a-homo-19-norlanosta-9a(10),9(11)-dien-7-one		NI44382
422	440	119	261	407	105	380	69	**482**	100	77	75	64	43	41	36	34	**27.00**	32	50	3	–	–	–	–	–	–	–	–	–	–	3β-Acetoxy-9a-homo-19-norlanosta-1(10),9(11)-dien-7-one		NI44383
134	120	422	187	203	98	245	190	**482**	100	63	58	45	39	35	33	33	**5.00**	32	50	3	–	–	–	–	–	–	–	–	–	–	1α-Acetoxyallobetul-2-ene		MS10222
98	140	134	341	482	422	187	120	**482**	100	98	59	55	50	49	49	44		32	50	3	–	–	–	–	–	–	–	–	–	–	1β-Acetoxyallobetul-2-ene		MS10223
134	121	245	422	187	107	467	149	**482**	100	77	74	53	50	48	45	41	**3.00**	32	50	3	–	–	–	–	–	–	–	–	–	–	3β-Acetoxyallobetul-1-ene		MS10224
273	135	232	175	482	422	297	407	**482**	100	97	84	44	12	12	12	11		32	50	3	–	–	–	–	–	–	–	–	–	–	3β-Acetoxy-11-oxoursan-12-ene	2348-66-5	NI44384
203	189	453	191	190	204	163	187	**482**	100	96	79	72	56	37	37	36	**6.00**	32	50	3	–	–	–	–	–	–	–	–	–	–	3-Acetylmoraldehyde		NI44385
422	409	123	407	177	147	191	137	**482**	100	80	75	50	50	50	38	37	**10.00**	32	50	3	–	–	–	–	–	–	–	–	–	–	D:A-Friedoolean-2-en-1-one, 24-(acetyloxy)-	43230-77-9	MS10225
381	127	382	380	201	399	379	126	**482**	100	31	30	20	14	12	11	9	**0.10**	32	50	3	–	–	–	–	–	–	–	–	–	–	3β-Hydroxycholesta-5,22-dien-24-one, O-tetrahydropyranyl-		NI44386
467	482	407	241	301	283	439	297	**482**	100	64	57	14	13	8	5	3		33	54	2	–	–	–	–	–	–	–	–	–	–	Lanost-8-en-3-ol, 24-methylene-, acetate, (3β)-		NI44387
467	69	482	43	301	323	383	422	**482**	100	60	30	30	25	20	15	10		33	54	2	–	–	–	–	–	–	–	–	–	–	Lanost-8-en-3-ol, 24-methylene-, acetate, (3β,13α,14β,17α,20S)-	14787-39-4	NI44388
422	407	379	300	297	353	482	285	**482**	100	84	36	36	34	17	16	15		33	54	2	–	–	–	–	–	–	–	–	–	–	24-Methylenecycloartanol acetate		NI44389
64	205	95	109	81	189	424	409	**482**	100	96	82	80	46	30	22	10	**0.00**	33	54	2	–	–	–	–	–	–	–	–	–	–	Olean-13(18)-ene-2,3-acetonide		MS10226
327	435	467	328	288	482	255	287	**482**	100	58	56	31	30	29	27	25	.	34	58	1	–	–	–	–	–	–	–	–	–	–	O-Methylpertyol		MS10227
73	468	75	147	116	469	74	483	**483**	100	89	83	64	51	42	35	27		18	37	3	5	–	–	–	–	–	4	–	–	–	4,6,7(1H)-Pteridinetrione, 2-amino-5,8-dihydro-, tetrakis(trimethylsilyl)-	56272-96-9	NI44390
71	298	252	368	280	69	326	253	**483**	100	24	19	16	16	5	4	3	**0.30**	19	28	5	1	–	7	–	–	–	–	–	–	–	Bis(3-methylbutyl) n-heptafluorobutyrylglutamate		NI44391
174	43	212	214	114	295	297	254	**483**	100	37	24	15	14	10	7	6	**3.00**	20	19	5	3	1	–	2	–	–	–	–	–	–	Dicloxacillin methyl ester		MS10228
174	196	114	295	198	297	254	256	**483**	100	24	14	11	8	4	4	2	**1.00**	20	19	6	3	1	1	1	–	–	–	–	–	–	5-Hydroxymethylflucloxacillin methyl ester		MS10229
192	142	340	341	149	190	354	355	**483**	100	76	62	59	52	39	32	30	**0.00**	21	26	4	1	–	–	–	–	1	–	–	–	–	Corydalmine methoiodide		NI44392
142	149	164	341	340	354	355	192	**483**	100	63	50	41	41	35	33	20	**0.00**	21	26	4	1	–	–	–	–	1	–	–	–	–	α-Hainanine methoiodide		NI44393
142	149	164	127	355	354	150	190	**483**	100	28	26	21	19	16	12	10	**0.00**	21	26	4	1	–	–	–	–	1	–	–	–	–	Tetrahydrocolumbamine methoiodide		NI44394
152	427	227	55	229	282	361	132	**483**	100	86	78	69	33	21	17	16	**2.82**	24	19	2	1	–	–	–	–	–	–	–	–	2	Manganese, dicarbonyl-π-pyrrolyl(triphenylarsine)-	32826-11-2	NI44395
410	364	483	191	411	438	336	264	**483**	100	44	39	38	26	21	19	19		26	29	8	1	–	–	–	–	–	–	–	–	–	Tetraethyl 10,11-dihydroazepino[1,2-a]quinoline-8,9,10,11-tetracarboxylate		LI09315
73	309	174	157	129	310	116	247	**483**	100	86	64	41	33	23	22	21	**0.00**	26	53	3	1	–	–	–	–	–	2	–	–	–	N-Acetyl-1,3-bis(O-trimethylsilyl)sphinga-4,14-dienine		LI09316
73	309	174	157	75	43	129	116	**483**	100	58	55	39	33	33	29	27	**0.00**	26	53	3	1	–	–	–	–	–	2	–	–	–	N-Acetylbis(O-trimethylsilyl)sphinga-4,8-dienine		MS10230
226	116	410	144	269	229	214	254	**483**	100	15	12	11	6	5	5	4	**0.00**	27	37	5	3	–	–	–	–	–	–	–	–	–	L-Alanine, N-[N-[N-[(2-hydroxy-1-naphthalenyl)methylene]-L-valyl]-L-isoleucyl]-, ethyl ester	37580-37-3	NI44396
57	71	452	155	73	369	451	97	**483**	100	97	96	94	82	74	59	46	**17.99**	27	41	3	1	–	–	–	–	–	2	–	–	–	19-Norpregna-1,3,5,7,9-pentaen-21-al, 3,17-bis[(trimethylsilyl)oxy]-, O-methyloxime, (17α)-	74299-04-0	NI44397
375	440	214	319	56	164	403	483	**483**	100	96	91	55	37	29	19	18		28	30	1	1	–	–	–	–	–	–	1	–	1	Carbonyl(cyclopentadienyl)methyl[[(S)-N-methyl-1-phenylethylamino]diphenylphosphine]iron		NI44398
221	89	279	249	91	305	178	145	**483**	100	90	69	58	45	44	43	37	**19.13**	29	25	6	1	–	–	–	–	–	–	–	–	–	Methyl rhizocarpate	22628-23-5	NI44399
278	279	221	89	305	483	145	91	**483**	100	23	20	20	15	11	11	11		29	25	6	1	–	–	–	–	–	–	–	–	–	Methyl rhizocarpate	22628-23-5	NI44400
91	65	167	137	92	107	108	90	**483**	100	14	12	11	9	6	5	4	**0.00**	31	33	4	1	–	–	–	–	–	–	–	–	–	Benzenemethanol, α-[[[[4-methoxy-3-benzyloxyphenyl]methyl]methylamino]methyl]-4-benzyloxy-	56847-11-1	NI44401
60	79	81	141	139	61	220	488	**484**	100	93	92	85	84	53	45	44	**4.30**	6	1	1	–	–	–	–	5	–	–	–	–	–	Phenol, pentabromo-	608-71-9	NI44402
82	242	240	244	238	130	45	482	**484**	100	40	33	23	19	16	15	13	**10.00**	8	4	–	–	2	–	–	–	–	–	–	–	4	Dithieno[3,4-S][3′,4′-D]tetraselenocine	88419-19-6	NI44403
161	196	202	101	135	133	66	39	**484**	100	91	62	51	43	34	33	33	**3.00**	8	5	3	–	–	–	1	–	–	–	–	–	2	Molybdenum, (π-cyclopentadienyl)tricarbonyl(σ-chloromercury)-		LI09317
186	121	484	214	482	125	142	137	**484**	100	81	53	44	43	37	33	30		8	20	4	–	4	–	–	–	–	–	2	–	1	Cadmium[II] bis(diethyldithiophosphate)	16569-74-7	NI44404
484	316	486	318	358	400	442	187	**484**	100	71	56	40	39	35	26	24		12	28	4	–	4	–	–	–	–	–	2	–	1	Nickel[II] bis(dipropyldithiophosphate)	30111-44-5	NI44405

MASS TO CHARGE RATIOS	M.W.	INTENSITIES	Parent	C	H	O	N	S	F	Cl	Br	I	Si	P	B	X	COMPOUND NAME	CAS Reg No	No
139 137 135 321 469 319 138 136	**484**	100 76 49 45 31 30 26 25	**5.49**	14	6	–	–	–	10	–	–	–	–	–	–	1	**Stannane, dimethylbis(pentafluorophenyl)-**	801-79-6	**NI44406**
102 74 201 103 48 170 96 348	**484**	100 23 16 8 7 6 6 5	**0.00**	14	15	4	–	–	–	–	3	–	–	–	–	–	**Altropyranoside, methyl 2,3,6-tribromo-2,3,6-trideoxy-, benzoate, α-D-**	23103-18-6	**NI44407**
217 218 484 117 69 167 174 148	**484**	100 9 5 5 5 4 3 3		15	–	–	–	–	16	–	–	–	–	–	–	–	**Perfluoro-1,3-diphenylpropane**	25078-75-5	**NI44408**
217 218 484 117 69 167 198 179	**484**	100 9 5 5 5 4 3 3		15	–	–	–	–	16	–	–	–	–	–	–	–	**Perfluoro-1,3-diphenylpropane**	25078-75-5	**LI09318**
43 139 137 157 115 169 140 242	**484**	100 45 45 13 13 12 12 10	**0.00**	17	25	11	–	–	–	–	1	–	–	–	–	–	**(1RS)-2-Bromo-1-methoxyethyl 2,3,4,6-tetra-O-acetyl-β-D-glucopyranoside**		**MS10231**
73 147 469 484 325 470 251 148	**484**	100 95 42 35 29 18 16 15		21	40	5	–	–	–	–	–	–	4	–	–	–	**(3,4-Dihydroxyphenyl)pyruvic acid, tetrakis(trimethylsilyl)-**		**NI44409**
73 147 45 325 469 148 74 75	**484**	100 55 18 9 8 8 8 7	**6.30**	21	40	5	–	–	–	–	–	–	4	–	–	–	**Trimethylsilyl catecholpyruvate tris(trimethylsilyl) ether**		**MS10232**
88 101 187 249 111 75 45 71	**484**	100 54 50 38 38 37 36 26	**0.00**	21	40	12	–	–	–	–	–	–	–	–	–	–	**Methyl octa-O-methylcellobionate**		**NI44410**
88 101 187 75 249 111 73 45	**484**	100 79 59 56 53 43 38 37	**0.00**	21	40	12	–	–	–	–	–	–	–	–	–	–	**Methyl octa-O-methylisomaltonate**		**NI44411**
88 101 45 75 249 111 187 71	**484**	100 61 43 40 37 37 33 33	**0.00**	21	40	12	–	–	–	–	–	–	–	–	–	–	**Methyl octa-O-methyllactobionate**		**NI44412**
88 101 45 75 187 249 129 111	**484**	100 63 46 37 31 29 26 26	**0.00**	21	40	12	–	–	–	–	–	–	–	–	–	–	**Methyl octa-O-methyllaminaribionate**		**NI44413**
88 101 187 111 75 45 249 89	**484**	100 57 48 45 42 35 33 27	**0.00**	21	40	12	–	–	–	–	–	–	–	–	–	–	**Methyl octa-O-methylmaltonate**		**NI44414**
88 101 249 75 187 111 45 71	**484**	100 74 61 46 38 37 35 34	**0.00**	21	40	12	–	–	–	–	–	–	–	–	–	–	**Methyl octa-O-methylmelibionate**		**NI44415**
57 104 41 49 248 132 43 76	**484**	100 51 51 41 40 36 36 35	**0.13**	23	21	5	2	–	–	–	1	–	–	–	–	–	**(Z)-1-(tert-Butoxycarbonylmethyl)-4-(2-bromophenyl)-3-phthalimidoazetidin-2-one**		**NI44416**
242 167 243 484 140 151 120.5 214	**484**	100 30 25 16 7 4 3 2		24	18	–	2	–	–	–	–	–	–	–	–	2	**10,10'(5H,5'H)-Biphenarsazine**	32435-30-6	**LI09319**
242 167 243 484 166 241 140 139	**484**	100 30 25 15 13 11 5 5		24	18	–	2	–	–	–	–	–	–	–	–	2	**10,10'(5H,5'H)-Biphenarsazine**	32435-30-6	**NI44417**
484 73 485 486 381 483 469 45	**484**	100 94 50 21 20 10 10 10		24	36	3	2	–	–	–	–	–	3	–	–	–	**3,4-Dihydroxybenzyl quinoxalinol tri-TMS**		**MS10233**
149 141 283 97 202 167 113 111	**484**	100 71 60 57 50 49 45 33	**10.00**	25	24	10	–	–	–	–	–	–	–	–	–	–	**7H-Furo[3,2-g][1]benzopyran-7-one, 9-[2,3-bis(acetyloxy)-4-(2,5-dihydro-4-methyl-5-oxo-2-furanyl)-3-methylbutoxy]-**	68725-65-5	**NI44418**
384 265 365 484 385 267 221 239	**484**	100 42 31 26 25 25 25 24		25	28	8	2	–	–	–	–	–	–	–	–	–	**Tetraethyl-10,11-dihydroazepino[1,2-a]quinoxaline-7,8,9,10-tetracarboxylate**		**LI09320**
55 399 428 180 134 321 384 383	**484**	100 45 40 40 40 26 22 18	**0.00**	26	25	4	2	–	–	–	–	–	–	–	–	1	**Dicarbonyl(5-ethoxy-3-methyl-2-oxo-1,5-diphenyl-4-imidazolidinylidene)(methylcyclopentadienyl)manganese**	116928-09-7	**DD01737**
43 45 144 60 364 291 119 42	**484**	100 55 45 28 18 16 15 8	**0.00**	26	32	7	2	–	–	–	–	–	–	–	–	–	**Compactinervine, N-acetyl-2,16-dihydro-, diacetate**	2111-91-3	**NI44419**
136 137 484 81 349 95	**484**	100 19 11 8 5 4		26	40	3	2	–	–	–	–	–	2	–	–	–	**1,2-Dihydroquinidine-2-one di-TMS**		**LI09321**
329 161 330 163 484 187 189 36	**484**	100 64 44 38 31 31 19 15		27	20	5	2	1	–	–	–	–	–	–	–	–	**1-Amino-2-phenoxy-4-(p-tolysulphonamido)-anthraquinone**		**IC04505**
78 79 77 80 76 74 63 182	**484**	100 23 17 16 16 14 14 11	**8.92**	27	36	6	2	–	–	–	–	–	–	–	–	–	**4,25-Secoobscurinervan-4-one, O-acetyl-22-ethyl-15,16-dimethoxy-, (22α)**	54658-08-1	**NI44420**
314 315 299 161 145 91 121 107	**484**	100 26 18 18 16 16 15 15	**0.00**	27	39	4	–	–	3	–	–	–	–	–	–	–	**Chol-5-en-24-oic acid, 3-[(trifluoroacetyl)oxy]-, methyl ester, (3β)-**	56468-53-2	**NI44421**
225 448 207 283 241 430 319 301	**484**	100 62 60 42 42 37 32 30	**0.00**	27	48	7	–	–	–	–	–	–	–	–	–	–	**5α-Cholestane-3β,4β,6α,8,15α,16β,26-heptaol**	93368-88-8	**NI44422**
95 293 448 136 154 430 301 412	**484**	100 98 85 73 58 55 46 41	**0.40**	27	48	7	–	–	–	–	–	–	–	–	–	–	**5α-Cholestane-3β,6α,7α,8β,15α,16β,26-heptaol**		**NI44423**
242 241 225 155 185 184 116 127	**484**	100 70 39 36 26 25 15 14	**2.00**	28	28	4	4	–	–	–	–	–	–	–	–	–	**(1α,2α,3β,4β)-1,3-Bis(1,3-dimethyl-6-uracilyl)-2,4-diphe nylcyclobutane**		**MS10234**
373 313 315 355 109 151 269 111	**484**	100 78 78 53 43 37 34 22	**9.90**	28	36	7	–	–	–	–	–	–	–	–	–	–	**(23R)-5β,6β-12,22-diepoxy-12β,17β-dihydroxy-1-oxoergosta-2,24-dien-23,26-olide**		**NI44424**
44 301 43 149 55 41 95 28	**484**	100 86 76 34 32 32 28 25	**4.10**	28	36	7	–	–	–	–	–	–	–	–	–	–	**Dihydrogedunin**		**MS10235**
105 319 122 77 301 283 43 113	**484**	100 99 96 83 74 63 57 52	**0.62**	28	36	7	–	–	–	–	–	–	–	–	–	–	**14β,17α-Pregn-5-en-20-one, 3β,8,12β,14,17-pentahydroxy-, 12-benzoate**	19308-44-2	**NI44425**
105 319 122 77 103 283 43 113	**484**	100 99 92 81 73 62 57 53	**7.00**	28	36	7	–	–	–	–	–	–	–	–	–	–	**14β,17α-Pregn-5-en-20-one, 3β,8,12β,14,17-pentahydroxy-, 12-benzoate**	19308-44-2	**LI09322**
225 199 28 154 226 184 210 200	**484**	100 81 61 17 16 14 12 11	**0.00**	29	28	5	2	–	–	–	–	–	–	–	–	–	**Urea, N,N'-bis(2',4'-dimethoxy[1,1'-biphenyl]-4-yl)-**	56909-15-0	**NI44426**
442 400 443 121 401 484	**484**	100 68 31 31 19 11		29	40	6	–	–	–	–	–	–	–	–	–	–	**6-(1'-Acetoxyethylidene)-3,17β-diacetoxy-2-ethyl-androsta-2,4-dione**		**LI09323**
329 330 55 484 81 43 95 107	**484**	100 67 61 58 58 57 56 49		29	47	2	–	–	3	–	–	–	–	–	–	–	**Cholestan-3-ol, trifluoroacetate, (3β,5α)-**	2839-20-5	**NI44427**
73 43 57 425 89 55 69 41	**484**	100 62 58 41 38 38 32 32	**2.00**	29	60	3	–	–	–	–	–	–	1	–	–	–	**Pentacosanoic acid, 2-[(trimethylsilyl)oxy]-, methyl ester**	56784-05-5	**MS10236**
300 482 481 242 302 299 483 301	**484**	100 91 80 69 67 63 54 54	**25.00**	30	22	2	–	–	–	2	–	–	–	–	–	–	**1,3-Dibenzoyl-2,4-di(4-chlorophenyl)cyclobutane**		**NI44428**
242 227 243 211 199 228 241 212	**484**	100 75 59 28 16 12 9 8	**8.00**	30	28	6	–	–	–	–	–	–	–	–	–	–	**2H-1-Benzopyran-2-one, 7-methoxy-6-[2-[2-(7-methoxy-2-oxo-2H-1-benzopyran-6-yl)-1,4-dimethyl-3-cyclohexen-1-yl]vinyl]-**		**LI09324**
242 227 243 211 199 228 241 156	**484**	100 73 58 27 16 12 10 10	**8.01**	30	28	6	–	–	–	–	–	–	–	–	–	–	**2H-1-Benzopyran-2-one, 7-methoxy-6-[2-[2-(7-methoxy-2-oxo-2H-1-benzopyran-6-yl)-1,4-dimethyl-3-cyclohexen-1-yl]vinyl]-**	18458-66-7	**NI44429**
55 83 57 71 484 97 81 95	**484**	100 99 99 92 90 90 86 83		30	44	5	–	–	–	–	–	–	–	–	–	–	**Lanosta-7,9(11)-dien-18-oic acid, 22,25-epoxy-3,17,20-trihydroxy-, γ-lactone, (3β)-**	56143-25-0	**NI44430**
366 129 381 104 161 367 57 43	**484**	100 87 82 67 64 29 29 25	**3.49**	30	60	4	–	–	–	–	–	–	–	–	–	–	**Heptacosanoic acid, 2,4-dimethoxy-, methyl ester, [S-(R*,S*)]-**	35301-08-7	**NI44431**
44 43 454 45 55 41 468 60	**484**	100 40 27 17 16 15 15 15	**0.96**	31	16	6	–	–	–	–	–	–	–	–	–	–	**12-Hydroxy-17-methoxy-5,6,11,18-trinaphthylenetetrone**	96722-27-9	**NI44432**
189 203 190 191 219 221 315 234	**484**	100 64 58 49 42 41 37 27	**12.00**	31	48	4	–	–	–	–	–	–	–	–	–	–	**Cyclamigenin A2**		**LI09325**
73 69 365 43 484 55 127 41	**484**	100 60 52 45 42 41 40 40		31	48	4	–	–	–	–	–	–	–	–	–	–	**Lanosta-7,9(11)-dien-18-oic acid, 3,20-dihydroxy-25-methoxy-, γ-lactone, (3β,20ξ)-**	24041-66-5	**NI44433**
317 135 276 175 262 189 484 121	**484**	100 81 73 44 34 34 33 33		31	48	4	–	–	–	–	–	–	–	–	–	–	**Methyl glycyrrhetate**	1477-44-7	**NI44434**
317 135 276 484 175 485 121 189	**484**	100 78 66 65 37 24 24 22		31	48	4	–	–	–	–	–	–	–	–	–	–	**Methyl glycyrrhetate**		**LI09326**

MASS TO CHARGE RATIOS								M.W.	INTENSITIES								Parent	C	H	O	N	S	F	Cl	Br	I	Si	P	B	X	COMPOUND NAME	CAS Reg No	No
203	262	189	133	484	249	187	425	**484**	100	50	21	14	11	9	9	5		31	48	4	–	–	–	–	–	–	–	–	–	–	Methyl 3β-hydroxy-6-oxo-olean-12-en-28-oate		LI09327
203	262	189	133	484	249	187	466	**484**	100	53	25	14	10	10	10	6		31	48	4	–	–	–	–	–	–	–	–	–	–	Methyl 3-oxo-6β-hydroxyolean-12-en-28-oate		LI09328
189	121	119	175	133	203	147	145	**484**	100	83	78	66	59	51	36	36	**36.00**	31	48	4	–	–	–	–	–	–	–	–	–	–	A(1)-Norlup-20(29)-en-28-oic acid, 2-formyl-3-hydroxy-, methyl ester	67594-78-9	NI44435
191	207	216	190	217	424	107	199	**484**	100	89	81	74	71	65	65	59	**16.47**	31	48	4	–	–	–	–	–	–	–	–	–	–	Olean-12-en-28-oic acid, 3β-hydroxy-21-oxo-, methyl ester	33512-86-6	NI44436
311	310	206	173	282	283	234	129	**484**	100	60	30	30	16	12	11	10	**5.50**	32	24	3	2	–	–	–	–	–	–	–	–	–	Endo-1-benzyl-4,7-diphenyl[8,9]benzo-4,10-diazatricyclo[2.2.0.3]-8-undecene-3,5,11-trione		NI44437
311	310	173	206	282	283	234	129	**484**	100	57	32	28	17	11	10	10	**4.40**	32	24	3	2	–	–	–	–	–	–	–	–	–	Exo-1-benzyl-4,7-diphenyl[8,9]benzo-4,10-diazatricyclo[2.2.0.3]-8-undecene-3,5,11-trione		NI44438
204	191	424	484	177	218	149	136	**484**	100	87	79	53	52	45	44	41		32	52	3	–	–	–	–	–	–	–	–	–	–	1α-Acetoxyallobetulane		MS10237
424	177	189	203	136	204	484	123	**484**	100	76	73	70	69	68	64	63		32	52	3	–	–	–	–	–	–	–	–	–	–	1β-Acetoxyallobetulane		MS10238
424	484	189	218	134	204	205	119	**484**	100	82	79	72	71	64	62	57		32	52	3	–	–	–	–	–	–	–	–	–	–	2β-Acetoxyallobetulane		MS10239
189	136	424	149	355	203	121	109	**484**	100	91	88	79	78	78	77	72	**68.00**	32	52	3	–	–	–	–	–	–	–	–	–	–	3β-Acetoxyallobetulane		MS10240
484	469	485	424	470	288	425	409	**484**	100	68	35	28	25	20	10	10		32	52	3	–	–	–	–	–	–	–	–	–	–	3β-Acetoxy-7β,8β-epoxy-5α-lanost-9(11)-ene		MS10241
484	455	343	485	456	371	303	469	**484**	100	80	50	35	30	30	20	15		32	52	3	–	–	–	–	–	–	–	–	–	–	3β-Acetoxy-9β,11α-epoxy-5α-lanost-7-ene		MS10242
234	189	135	219	191	177	171	193	**484**	100	83	57	53	53	53	47	33	**20.00**	32	52	3	–	–	–	–	–	–	–	–	–	–	3β-Acetoxy-12α,13-epoxyoleanane		NI44439
273	333	123	121	255	205	189	395	**484**	100	80	80	80	75	65	40	32	**8.00**	32	52	3	–	–	–	–	–	–	–	–	–	–	3β-Acetoxyfriedelan-26-al		NI44440
234	466	255	468	484	191	133	235	**484**	100	96	37	36	32	28	28	20		32	52	3	–	–	–	–	–	–	–	–	–	–	3β-Acetoxy-11α-hydroxyursan-12-ene		NI44441
425	485	483	426	486	484	469		**484**	100	80	38	30	25	18	18			32	52	3	–	–	–	–	–	–	–	–	–	–	3β-Acetoxylanost-8-en-7-one		NI44442
43	55	81	69	109	95	41	107	**484**	100	99	90	85	84	83	72	63	**3.00**	32	52	3	–	–	–	–	–	–	–	–	–	–	9,19-Cyclolanost-23-ene-3,25-diol, 3-acetate, (3β,23E)-	54482-56-3	NI44443
189	424	205	191	409	203	381	187	**484**	100	90	70	65	60	48	45	45	**12.00**	32	52	3	–	–	–	–	–	–	–	–	–	–	29,30-Dinorgammaceran-3-one, 22-(acetyloxy)-21,21-dimethyl-, (8α,9β,13α,14β,17α,18β,22α)-	43206-65-1	NI44444
424	411	272	125	273	247	177	123	**484**	100	90	70	70	65	65	62	60	**6.00**	32	52	3	–	–	–	–	–	–	–	–	–	–	D:A-Friedooleanan-1-one, 24-(acetyloxy)-	43230-85-9	MS10243
251	245	205	332	191	484	345	303	**484**	100	83	79	48	43	34	24	24		32	52	3	–	–	–	–	–	–	–	–	–	–	D:A-Friedooleanan-7-one, 3α-hydroxy-, acetate		LI09329
251	43	484	205	123	69	121	55	**484**	100	50	47	42	39	38	37	37		32	52	3	–	–	–	–	–	–	–	–	–	–	D:A-Friedooleanan-7-one, 3β-hydroxy-, acetate	18671-60-8	NI44445
469	484	470	485	69	55	57	95	**484**	100	91	36	31	30	29	25	22		32	52	3	–	–	–	–	–	–	–	–	–	–	7-Ketoeuph-8-enyl acetate		MS10244
484	277	485	424	69	55	332	95	**484**	100	43	36	27	24	23	19	15		32	52	3	–	–	–	–	–	–	–	–	–	–	11-Ketoeuph-8-enyl acetate		MS10245
135	331	332	175	271	484	136	329	**484**	100	31	14	12	11	9	7	6		32	52	3	–	–	–	–	–	–	–	–	–	–	Lanost-9(11)-en-12-one, 3-(acetyloxy)-, (3β)-	55724-72-6	NI44446
234	123	469	219	121	135	470	137	**484**	100	77	66	55	37	33	21	21	**5.68**	32	52	3	–	–	–	–	–	–	–	–	–	–	Ursan-12-one, 3-(acetyloxy)-, (3β)-	55887-96-2	NI44447
69	73	135	75	95	81	109	105	**484**	100	92	77	49	39	34	32	29	**22.02**	32	56	1	–	–	–	–	–	–	1	–	–	–	Cholesta-8,24-diene, 4,4-dimethyl-3-[(trimethylsilyl)oxy]-, (3β,5α)-	55429-93-1	NI44448
55	73	75	69	121	95	81	105	**484**	100	92	81	77	75	73	64	61	**0.00**	32	56	1	–	–	–	–	–	–	1	–	–	–	Ergosta-7,24(28)-diene, 4-methyl-3-[(trimethylsilyl)oxy]-, (3β,4α)-	23648-45-5	NI44449
129	43	396	75	95	145	73	55	**484**	100	46	38	38	35	31	31	31	**0.78**	32	56	1	–	–	–	–	–	–	1	–	–	–	Stigmasta-5,22-diene, 3-[(trimethylsilyl)oxy]-, (3β,22E)-	14030-29-6	NI44451
83	129	394	255	484	69	81	139	**484**	100	43	29	29	22	22	20	19		32	56	1	–	–	–	–	–	–	1	–	–	–	Stigmasta-5,22-diene, 3-[(trimethylsilyl)oxy]-, (3β,22E)-	14030-29-6	NI44450
386	296	257	253	255	281	215	213	**484**	100	97	61	53	47	44	44	40	**6.67**	32	56	1	–	–	–	–	–	–	1	–	–	–	Stigmasta-5,24(28)-diene, 3-[(trimethylsilyl)oxy]-, (3β)-	55527-94-1	NI44452
343	386	344	213	387	255	371	296	**484**	100	61	31	22	21	17	16	15	**8.33**	32	56	1	–	–	–	–	–	–	1	–	–	–	Stigmasta-7,24(28)-diene, 3-[(trimethylsilyl)oxy]-, (3β)-	55527-95-2	NI44453
102	280	77	350	382	421	484	450	**484**	100	50	31	27	25	21	11	9		33	25	–	–	1	–	–	–	–	–	1	–	–	2,3-Benzo-7,10,11-triphenyl-8-phosphabicyclo[6.2.2.0^{1,6}]dodeca-6,10,11-triene 8-sulphide		NI44454
424	95	175	409	302	203	297	381	**484**	100	73	62	56	50	47	40	31	**12.00**	33	56	2	–	–	–	–	–	–	–	–	–	–	3β-Acetoxy-24,24-dimethyl-26,27-dinor-9,19-cyclolanostan-25-al		LI09330
57	44	41	43	486	71	55	69	**484**	100	42	33	26	25	25	25	20	**6.00**	34	44	2	–	–	–	–	–	–	–	–	–	–	1,4-Bis-(4′-oxo-3′,5′-di-tert-butyl-cyclohexadienylidene)-cyclohexadiene diylidene		MS10246
91	105	57	43	119	288	71	133	**484**	100	15	14	12	9	8	8	7	**0.70**	35	64	–	–	–	–	–	–	–	–	–	–	–	15-Phenylnonacosane		AI02388
91	287	43	105	104	288	57	92	**484**	100	43	20	16	14	13	13	10	**10.34**	35	64	–	–	–	–	–	–	–	–	–	–	–	15-Phenylnonacosane		AI02389
91	43	287	57	41	105	55	104	**484**	100	44	28	26	20	19	15	13	**4.26**	35	64	–	–	–	–	–	–	–	–	–	–	–	15-Phenylnonacosane		AI02390
317	485	139	359	131	401	107	188	**485**	100	78	59	49	43	41	36	32		12	28	4	–	4	–	–	–	–	–	2	–	1	Cobalt[II] bis(dipropyldithiophosphate)	37913-62-5	NI44455
408	406	487	489	485	77	327	407	**485**	100	98	71	36	36	27	26	25		21	13	3	1	–	–	–	2	–	–	–	–	–	1-Hydroxy-4-(2,6-dibromo-4-methylanilino)-anthraquinone		IC04506
73	218	219	267	45	74	220	100	**485**	100	95	19	16	10	9	8	7	**0.00**	21	43	4	1	–	–	–	–	–	4	–	–	–	3,4-Bis(trimethylsiloxy)-bis(trimethylsilyl)phenylalanine		NI44456
73	218	267	219	74	45	75	179	**485**	100	46	9	9	9	9	8	7	**0.00**	21	43	4	1	–	–	–	–	–	4	–	–	–	L-Tyrosine, N,O-bis(trimethylsilyl)-3-[(trimethylsilyl)oxy]-, trimethylsilyl ester	55638-45-4	NI44457
218	73	219	267	45	220	74	75	**485**	100	96	18	12	9	8	8	7	**0.60**	21	43	4	1	–	–	–	–	–	4	–	–	–	L-Tyrosine, N,O-bis(trimethylsilyl)-3-[(trimethylsilyl)oxy]-, trimethylsilyl ester	55638-45-4	MS10247
218	179	219	267	77	147	220	293	**485**	100	42	35	34	30	17	16	12	**0.25**	21	43	4	1	–	–	–	–	–	4	–	–	–	L-Tyrosine, N,O-bis(trimethylsilyl)-3-[(trimethylsilyl)oxy]-, trimethylsilyl ester	55638-45-4	NI44458
322	349	163	350	334	277	177	336	**485**	100	91	90	75	74	72	72	65	**0.00**	23	19	11	1	–	–	–	–	–	–	–	–	–	6,7-Diacetoxy-2-formyl-1,2,3,4-tetrahydroisoquinoline-1-spiro-2′-1′-hydroxy-6′,7′-methylenedioxy-3′-oxo-2′,3′-dihydroindene		NI44459

MASS TO CHARGE RATIOS								M.W.	INTENSITIES								Parent	C	H	O	N	S	F	Cl	Br	I	Si	P	B	X	COMPOUND NAME	CAS Reg No	No
322	349	163	350	334	277	177	336	**485**	100	91	90	75	74	72	72	65	**0.00**	23	19	11	1	–	–	–	–	–	–	–	–	–	**Ethaneperoxoic acid, 2′-formyl-3′,4′-dihydro-8-hydroxy-6-oxospiro[7H-indeno[4,5-d]-1,3-dioxole-7,1′(2′H)-isoquinoline]-6′,7′-diyl ester**	55649-79-1	**NI44460**
334	43	214	312	87	426	220	69	**485**	100	63	60	55	55	52	51	50	**16.73**	24	39	9	1	–	–	–	–	–	–	–	–	–	**Cyclopentanepropanoic acid, 3,5-bis(acetyloxy)-2-[8-(acetyloxy)-3-(methoxyimino)octyl]-, methyl ester**	55759-99-4	**NI44461**
334	43	214	312	87	426	220	69	**485**	100	64	60	55	55	51	49	49	**17.00**	24	39	9	1	–	–	–	–	–	–	–	–	–	**Methyl 5,7,16-triacetoxy-11-methoxyimino-tetranorprostanoate**		**LI09331**
241	312	137	324	225	133	130	75	**485**	100	85	84	75	42	42	36	30	**0.67**	24	47	5	1	–	–	–	–	–	2	–	–	–	**Cyclopentanepropanoic acid, 5-(methoxyimino)-3-[(trimethylsilyl)oxy]-2-[3-[(trimethylsilyl)oxy]-1-octenyl]-, methyl ester, [1R-[1α,2β(1E,3S*),3α]]-**	72121-38-1	**NI44462**
73	147	217	377	205	204	319	197	**485**	100	34	27	26	26	25	23	23	**0.00**	24	47	5	1	–	–	–	–	–	2	–	–	–	**5-Heptenoic acid, 7-[2-[3-(methoxyimino)butyl]-3,5-bis[(trimethylsilyl)oxy]cyclopentyl]-, methyl ester, [1R-(1α,2β,3α,5α)]-**	74367-64-9	**NI44463**
470	471	196	330	472	153	331	455	**485**	100	71	60	55	25	21	14	13	**11.45**	26	39	4	1	–	–	–	–	–	2	–	–	–	**2-Nitro-17α-ethynyl-estradiol-17β-di-TMS**		**NI44464**
470	196	471	153	472	154	75	169	**485**	100	50	42	23	18	13	13	12	**8.37**	26	39	4	1	–	–	–	–	–	2	–	–	–	**4-Nitro-17α-ethynyl-estradiol-17β-di-TMS**		**NI44465**
43	84	144	102	145	60	57	85	**485**	100	93	50	36	35	31	20	18	**0.00**	26	47	7	1	–	–	–	–	–	–	–	–	–	**Acetamide, N-[2,3-dihydroxy-1-(hydroxymethyl)heptadecyl]-, triacetate**	3613-96-5	**NI44466**
84	105	86	211	268	429	309	120	**485**	100	76	64	28	22	20	20	20	**0.00**	27	28	4	1	–	–	–	–	–	–	–	–	1	**Dicarbonyl(5-ethoxy-2,3-dimethyl-2,5-diphenyl-4-oxazolidinylidene)(methylcyclopentadienyl)manganese**	116928-06-4	**DD01738**
124	43	78	106	105	79	125	123	**485**	100	99	98	94	92	86	79	60	**0.05**	27	35	7	1	–	–	–	–	–	–	–	–	–	**Rostratamine**		**NI44467**
57	55	43	69	41	81	95	403	**485**	100	99	99	82	82	76	69	63	**2.14**	27	45	2	1	–	–	2	–	–	–	–	–	–	**3β,5-Dichloro-6β-nitro-5α-cholestane**		**MS10248**
426	341	121	285	206	313	159	135	**485**	100	72	27	25	23	21	19	18	**1.00**	28	39	6	1	–	–	–	–	–	–	–	–	–	**3-Acetoxy-22-methoxycarbonyl-27-nitrilo-12-oxo-23-oxa-5β-furost-9(11)-ene**		**LI09332**
394	91	376	358	55	190	120	69	**485**	100	95	92	71	46	42	39	38	**11.30**	29	43	5	1	–	–	–	–	–	–	–	–	–	**1-Benzyl-3a,5-dihydroxy-6,7-dimethyl-4-(4-methyl-7-(5-oxo-tetrahydrofuran-2-yl)heptyl)-3-oxo-3a,4,5,6,7,7a-hexahydro-isoindoline**		**IC04507**
286	272	485	287	245	273	470	230	**485**	100	46	31	25	17	12	9	7		30	47	4	1	–	–	–	–	–	–	–	–	–	**Desacetyl daphniphylline**		**LI09333**
157	487	488	158	196	144	197	489	**485**	100	81	30	19	18	10	9	7		34	47	1	1	–	–	–	–	–	–	–	–	–	**1′H-Cholest-2-eno[3,2-b]indol-6-one, 1′-methyl-, (5β)-**	34535-61-0	**NI44468**
485	470	486	194	208	471	195	57	**485**	100	38	37	24	22	14	11	11		34	47	1	1	–	–	–	–	–	–	–	–	–	**1′H-Cholest-4-eno[3,2-b]indol-6-one, 1′-methyl-**		**MS10249**
69	409	407	490	488	486	131	129	**486**	100	61	61	31	31	31	31	31		6	–	–	2	–	12	–	2	–	–	–	–	–	**1,2-Bis[trifluoromethylamino]-1,2-dibromoethylene**		**LI09334**
272	274	270	237	276	239	235	218	**486**	100	93	58	45	41	28	28	21	**1.59**	10	–	1	–	–	–	10	–	–	–	–	–	–	**1,3,4-Metheno-2H-cyclobuta[cd]pentalen-2-one, 1,1a,3,3a,4,5,5,5a,5b,6-decachlorooctahydro-**	143-50-0	**NI44469**
272	274	237	270	143	276	239	141	**486**	100	77	56	48	39	34	34	31	**0.00**	10	–	1	–	–	–	10	–	–	–	–	–	–	**1,3,4-Metheno-2H-cyclobuta[cd]pentalen-2-one, 1,1a,3,3a,4,5,5,5a,5b,6-decachlorooctahydro-**	143-50-0	**NI44470**
272	274	270	237	276	239	235	218	**486**	100	79	52	35	34	24	22	18	**0.65**	10	–	1	–	–	–	10	–	–	–	–	–	–	**1,3,4-Metheno-2H-cyclobuta[cd]pentalen-2-one, 1,1a,3,3a,4,5,5,5a,5b,6-decachlorooctahydro-**		**AI02391**
417	467	69	486	367	348	329	279	**486**	100	91	49	29	21	15	9	8		12	–	–	–	–	18	–	–	–	–	–	–	–	**Hexakis-(trifluoromethyl)-benzene**		**LI09335**
417	467	69	367	348	486	329	317	**486**	100	89	61	26	14	11	7	6		12	–	–	–	–	18	–	–	–	–	–	–	–	**Hexakis-(trifluoromethyl)-benzvalene**		**LI09336**
69	417	467	367	31	279	348	486	**486**	100	83	73	16	11	10	9	8		12	–	–	–	–	18	–	–	–	–	–	–	–	**Hexakis-(trifluoromethyl)-prismane**		**LI09337**
328	180	488	490	486	329	261	259	**486**	100	67	63	36	32	14	13	13		12	–	–	–	1	8	–	2	–	–	–	–	–	**Benzene, 1,1′-thiobis[2-bromo-3,4,5,6-tetrafluoro-**	17728-68-6	**NI44471**
328	180	488	164	490	486	215	329	**486**	100	80	76	52	38	38	32	17		12	–	–	–	1	8	–	2	–	–	–	–	–	**Benzene, 1,1′-thiobis[2-bromo-3,4,5,6-tetrafluoro-**	17728-68-6	**NI44472**
205	203	298	377	486	39	296	94	**486**	100	42	23	22	19	18	10	10		12	10	2	–	2	–	–	–	–	–	1	–	1	**Thallium(I) O,O′-diphenyldithiophosphate**		**MS10250**
457	458	459	486	487	460	58	488	**486**	100	41	31	23	13	9	8	6		12	30	9	–	–	–	–	–	–	6	–	–	–	**Perethylhexasilsesquioxane**	18971-66-9	**NI44473**
488	486	490	124	123	428	277	275	**486**	100	48	46	38	31	30	28	27		14	4	8	2	–	–	–	2	–	–	–	–	–	**2,6-Dibromo-1,5-dihydroxy-4,8-dinitroanthraquinone**		**IC04508**
333	283	486	149	199	233	263	180	**486**	100	12	10	9	4	2	1	1		15	12	6	–	–	9	–	–	–	–	–	–	1	**Aluminium, tris-(1,1,1-trifluoro-2,4-pentanedionato)-**		**LI09338**
212	211	275	467	486	444	225	417	**486**	100	35	32	31	15	12	12	10		16	20	2	–	2	6	–	–	–	–	–	–	1	**Bis(1,1,1-trifluoro-4-mercapto-6-methylhept-3-en-2-onato)zinc(II)**	56260-82-3	**NI44474**
57	319	301	317	299	432	302	177	**486**	100	61	52	42	36	32	29	28	**10.00**	16	36	1	–	–	–	–	–	–	–	–	–	2	**tert-Butylstibinous acid anhydride**	80536-70-5	**NI44475**
43	18	55	374	387	41	57	155	**486**	100	50	48	38	37	35	33	28	**12.70**	17	22	2	–	–	12	–	–	–	–	–	–	–	**1,1,7-Trihydroperfluoroheptyl decanoate**		**MS10251**
43	55	374	387	57	155	68	70	**486**	100	47	38	37	33	28	27	25	**12.00**	17	22	2	–	–	12	–	–	–	–	–	–	–	**1,1,7-Trihydroperfluoroheptyl decanoate**		**LI09339**
242	51	77	244	486	299	171	113	**486**	100	23	20	16	14	13	13	10		18	5	–	–	–	10	–	–	–	–	–	–	1	**Bis(pentafluorophenyl)-phenylarsine**		**MS10252**
97	139	182	157	440	160	115	214	**486**	100	50	38	38	31	19	19	15	**0.00**	18	22	10	4	1	–	–	–	–	–	–	–	–	**1,6-Dihydro-2-methylthio-5-[(E)-2-nitrovinyl]-4-(2,3,4-tri-O-acetyl-β-D-xylopyronosylamino)-1H-pyrimidin-6-one**		**MS10253**
43	269	267	427	271	303	253	41	**486**	100	61	51	34	29	28	27	25	**0.00**	18	46	7	–	–	–	–	–	–	4	–	–	–	**3,3,5,5-Tetraisopropoxy-1,1,1,7,7,7-hexamethyltetrasiloxane**		**MS10254**
197	261	185	120	78	56	353	430	**486**	100	71	66	59	53	48	38	33	**0.93**	19	15	2	–	–	–	1	–	–	–	–	–	2	**Dicarbonyl-π-cyclopentadienyliron-diphenyltin(IV)chloride**		**LI09340**
150	389	260	220	124	136	417	164	**486**	100	59	52	47	33	29	28	28	**6.84**	19	24	4	4	–	6	–	–	–	–	–	–	–	**N[5]-(4,6-Dimethylpyrimidin-2-yl)-N[2],N[5]-bis(trifluoroacetyl)ornithine butyl ester**		**MS10255**
387	43	389	388	391	84	100	205	**486**	100	62	38	38	33	27	17	12	**0.00**	20	28	8	–	–	–	–	–	–	–	–	–	1	**Zirconium, tetrakis(2,4-pentanedionato-O,O′)-**	17501-44-9	**NI44476**

MASS TO CHARGE RATIOS								M.W.	INTENSITIES								Parent	C	H	O	N	S	F	Cl	Br	I	Si	P	B	X	COMPOUND NAME	CAS Reg No	No
73	267	268	147	179	396	45	74	**486**	100	76	19	15	13	12	12	7	**4.90**	21	42	5	–	–	–	–	–	–	4	–	–	–	Trimethylsilyl catechollactate tris(trimethylsilyl) ether		MS10256
374	93	108	375	486	78	161	176	**486**	100	66	30	21	15	15	13	8		22	23	5	6	–	–	1	–	–	–	–	–	–	N-Ethyl-N-(2-(2,5-dioxopyrrolidin-1-ylethyl))-3-acetamido-4-(2-chloro-4-nitrophenylazo)aniline		IC04509
105	106	77	91	172	65	112	313	**486**	100	98	96	79	54	18	12	11	**1.00**	23	22	8	2	1	–	–	–	–	–	–	–	–	Uridine, 2′,3′-O-benzylene-, 5′-(4-methylbenzenesulphonate)	35837-30-0	NI44477
365	455	196	115	268	265	369	279	**486**	100	26	10	10	8	6	5	3	**0.30**	23	42	7	2	–	–	–	–	–	1	–	–	–	Cyclopentaneoctanoic acid, ε,3-bis(methoxyimino)-2-(3-methoxy-3-oxopropyl)-5-[(trimethylsilyl)oxy]-, methyl ester, (1α,2β,5β)-	55669-98-2	NI44478
243	168	244	486	139	169	487	113	**486**	100	30	16	12	10	5	4	3		24	16	2	–	–	–	–	–	–	–	–	–	2	10,10′-Bis(10H-phenoxarsine)	15430-08-7	NI44479
402	359	360	403	401	69	486	151	**486**	100	39	26	24	16	14	10	9		24	22	11	–	–	–	–	–	–	–	–	–	–	4H-1-Benzopyran-4-one, 5,6-bis(acetyloxy)-2-[3-(acetyloxy)-4-methoxyphenyl]-3,7-dimethoxy-	14397-72-9	NI44480
402	359	360	403	401	69	486	444	**486**	100	39	26	24	16	14	10	9		24	22	11	–	–	–	–	–	–	–	–	–	–	4H-1-Benzopyran-4-one, 5,6-bis(acetyloxy)-2-[3-(acetyloxy)-4-methoxyphenyl]-3,7-dimethoxy-		MS10257
444	360	402	345	359	445	361	342	**486**	100	94	57	44	29	28	21	18	**13.00**	24	22	11	–	–	–	–	–	–	–	–	–	–	4H-1-Benzopyran-4-one, 5,7-bis(acetyloxy)-2-[4-(acetyloxy)-3-methoxyphenyl]-3,6-dimethoxy-	14397-72-9	MS10258
471	472	73	473	474	367	455	475	**486**	100	40	17	17	5	4	3	1	**0.30**	24	34	5	–	–	–	–	–	–	3	–	–	–	1,8-Dihydroxy-3-(hydroxymethyl)anthraquinone, tris(trimethylsilyl)-	91701-26-7	NI44481
471	472	73	473	474	475	485	456	**486**	100	39	19	18	5	1	1	1	**0.40**	24	34	5	–	–	–	–	–	–	3	–	–	–	1,3,8-Trihydroxy-6-methylanthraquinone, tris(trimethylsilyl)-	7336-74-5	NI44482
456	471	457	458	472	473	426	384	**486**	100	43	39	20	15	7	6	5	**1.40**	24	34	5	–	–	–	–	–	–	3	–	–	–	1,4,5-Trihydroxy-2-methylanthraquinone, tris(trimethylsilyl)-		NI44483
456	471	457	458	472	473	426	459	**486**	100	55	43	23	19	8	7	6	**0.90**	24	34	5	–	–	–	–	–	–	3	–	–	–	1,4,5-Trihydroxy-3-methylanthraquinone, tris(trimethylsilyl)-	91701-22-3	NI44484
456	457	471	413	458	472	414	473	**486**	100	49	37	26	18	13	12	8	**0.70**	24	34	5	–	–	–	–	–	–	3	–	–	–	1,5,6-Trihydroxy-2-methylanthraquinone, tris(trimethylsilyl)-	91701-25-6	NI44485
456	471	457	472	458	473	426	459	**486**	100	49	36	19	19	8	5	5	**1.20**	24	34	5	–	–	–	–	–	–	3	–	–	–	1,5,8-Trihydroxy-3-methylanthraquinone, tris(trimethylsilyl)-	25458-54-2	NI44486
59	247	407	31	406	246	408	41	**486**	100	93	83	34	32	24	18	15	**0.00**	24	60	6	–	–	–	–	–	–	–	–	–	6	2-Propanol, 2-methyl-, lithium salt, hexamer	14322-83-9	NI44487
105	43	77	106	207	279	115	177	**486**	100	31	11	7	4	3	3	2	**0.00**	25	26	10	–	–	–	–	–	–	–	–	–	–	DL-Xylitol, 1,3,4-triacetate 2,5-dibenzoate	74810-28-9	NI44488
160	91	155	184	188	198	187	104	**486**	100	98	85	65	62	32	30	20	**0.00**	25	30	6	2	1	–	–	–	–	–	–	–	–	Hexanoic acid, 6-[[3-(1,3-dihydro-1,3-dioxo-2H-isoindol-2-yl)propyl][(4-methylphenyl)sulphonyl]amino]-, methyl ester	67370-68-7	NI44489
427	43	428	486	44	57	70	42	**486**	100	30	27	23	20	18	15	14		25	30	8	2	–	–	–	–	–	–	–	–	–	Dichotine, 11-methoxy-, acetate	29474-86-0	NI44490
217	282	281	182	396	191	99	218	**486**	100	75	65	43	40	31	28	23	**1.76**	25	50	5	–	–	–	–	–	–	2	–	–	–	Cyclopentanepentanoic acid, 2-(3-oxooctyl)-3,5-bis[(trimethylsilyl)oxy]-, methyl ester, [1R-(1α,2β,3α,5α)]-	55760-04-8	NI44491
73	75	426	354	427	201	183	147	**486**	100	82	66	33	27	26	22	20	**12.00**	25	50	5	–	–	–	–	–	–	2	–	–	–	Digoxigenin, 3,12-bis[(trimethylsilyl)oxy]-	751-49-5	LI09341
361	469	95	470	206	362	91	109	**486**	100	90	58	27	27	24	24	15	**0.00**	26	30	9	–	–	–	–	–	–	–	–	–	–	Limonin diosphenol		LI09342
140	95	141	96	363	364	486	468	**486**	100	41	36	18	9	5	2	1		26	30	9	–	–	–	–	–	–	–	–	–	–	Limonoic acid, 19-deoxy-6,19-epoxy-, 16,17-lactone	751-49-5	NI44492
140	95	141	47	32	242	215	96	**486**	100	44	38	27	25	23	23	18	**1.60**	26	30	9	–	–	–	–	–	–	–	–	–	–	Limonoic acid, 19-deoxy-6,19-epoxy-, 16,17-lactone		LI09343
95	345	105	121	119	303	135	128	**486**	100	60	46	27	27	24	24	24	**0.00**	26	30	9	–	–	–	–	–	–	–	–	–	–	Rutaevin		LI09344
150	206	182	238	69	122	121	163	**486**	100	82	80	48	46	34	32	30	**0.00**	27	34	8	–	–	–	–	–	–	–	–	–	–	Methyl olivetorate	19314-71-7	NI44493
486	441	186	187	238	160	312	161	**486**	100	35	30	20	18	17	15	15		27	38	6	2	–	–	–	–	–	–	–	–	–	4-Methyl-3,4-secoreserpine di-acetate		LI09345
486	182	487	122	471	457	426	472	**486**	100	99	87	79	73	72	35	32		27	38	6	2	–	–	–	–	–	–	–	–	–	4,25-Secoobscurinervan-4-ol, 25-ethyl-15,16-dimethoxy-, 25-acetate, (4β,22α)-	72101-38-3	NI44494
69	55	81	95	372	107	93	68	**486**	100	74	62	52	50	47	45	44	**4.50**	27	41	4	–	–	3	–	–	–	–	–	–	–	5β-Cholan-24-oic acid, 3α-hydroxy-, methyl ester, trifluoroacetate	3793-41-7	NI44495
372	215	329	108	216	257	149	357	**486**	100	84	57	53	42	36	35	32	**4.00**	27	41	4	–	–	3	–	–	–	–	–	–	–	5β-Cholan-24-oic acid, 3β-hydroxy-, methyl ester, trifluoroacetate		LI09346
486	487	488	143	245	155	453	233	**486**	100	39	14	11	9	8	8	8		27	42	4	–	–	–	–	–	–	2	–	–	–	Aldosterone, bis(trimethylsilyl)-		NI44496
229	173	73	272	441	342			**486**	100	77	41	37	9	3			**0.00**	27	42	4	4	–	–	–	–	–	–	–	–	–	L-Alanine, N-[N-[N-[3-[4-(dimethylamino)phenyl]-2-propenylidene]-L-valyl]-L-isoleucyl]-, ethyl ester	37580-28-2	NI44497
100	85	97	135	86	443	190	182	**486**	100	39	27	23	15	5	5	5	**1.00**	27	42	4	4	–	–	–	–	–	–	–	–	–	Butanamide, 2-(dimethylamino)-3-methyl-N-[3-isopropyl]-7-isobutyl-5,8-dioxo-2-oxa-6,9-diazabicyclo[10.2.2]hexadeca-10,12,14,15-tetraen-4-yl-	64408-09-9	NI44498
269	301	217	139	183	185	345	63	**486**	100	91	91	57	40	25	15	11	**5.82**	28	24	–	–	2	–	–	–	–	–	2	–	–	1,4-Bis(diphenyl-thiophosphinyl)but-2-yne		NI44499
235	236	41	151	178	135	91	43	**486**	100	15	4	3	2	2	2	1	**0.00**	28	38	7	–	–	–	–	–	–	–	–	–	–	Methyl 2,2′,4-tri-O-methylanziate	5366-08-5	NI44500
57	178	43	61	71	55	161	85	**486**	100	78	78	67	50	44	39	36	**8.00**	28	54	4	–	1	–	–	–	–	–	–	–	–	Dihendecyl thiodipropionate		IC04510
149	75	97	83	147	73	103	69	**486**	100	68	40	40	38	38	33	33	**0.60**	28	62	2	–	–	–	–	–	–	2	–	–	–	3,26-Dioxa-2,27-disilaoctacosane, 2,2,27,27-tetramethyl-	22396-30-1	NI44501
340	486	105	341	487	445	471	349	**486**	100	50	26	23	18	12	11	10		29	30	5	2	–	–	–	–	–	–	–	–	–	3-Benzoyl-1-(3,4-dimethoxyphenyl)-5-ethyl-7,8-dimethoxy-4-methyl-3H-2,3-benzodiazepine	90140-54-8	NI44502
125	181	199	43	41	55	126	67	**486**	100	34	32	24	22	20	18	17	**8.80**	29	42	6	–	–	–	–	–	–	–	–	–	–	2,7-Deoxy-2,3-dihydro-3-methoxy-withaferin A		NI44503
125	181	199	331	288	486			**486**	100	35	32	14	13	2				29	42	6	–	–	–	–	–	–	–	–	–	–	2,7-Deoxy-2,3-dihydro-3-methoxy-withaferin A		LI09347
445	253	75	446	329	155	149	369	**486**	100	58	49	39	31	24	19	17	**12.75**	29	50	2	–	–	–	–	–	–	2	–	–	–	3β,17β-Bis(allyldimethylsilyloxy)androst-5-ene		NI44504
121	175	189	133	248	147	219	203	**486**	100	65	60	54	46	43	36	33	**25.00**	30	46	5	–	–	–	–	–	–	–	–	–	–	A(1)-Norlup-20(29)-en-28-oic acid, 2-carboxy-3-hydroxy-, (2α,3β)-	21302-79-4	NI44505
159	160	146	486	147	175	384	225	**486**	100	19	17	4	4	3	2	2		30	50	3	2	–	–	–	–	–	–	–	–	–	5-Hydroxy-N-(20-hydroxyarachidoyl)tryptamine		MS10259
181	152	153	182	196	355	286	134	**486**	100	36	25	19	12	10	10	6	**5.00**	31	22	2	2	1	–	–	–	–	–	–	–	–	6-(4-Biphenyl)-7-(4-methoxybenzoyl)-3-phenylpyrazolo[5,1-b]thiazole		MS10260
189	262	175	203	119	205	249	223	**486**	100	44	42	41	41	40	39	35	**18.18**	31	50	4	–	–	–	–	–	–	–	–	–	–	Lup-20(29)-en-28-oic acid, 2,3-dihydroxy-, methyl ester, (2α,3β)-	6831-09-0	NI44506

MASS TO CHARGE RATIOS								M.W.	INTENSITIES								Parent	C	H	O	N	S	F	Cl	Br	I	Si	P	B	X	COMPOUND NAME	CAS Reg No	No
179	146	207	191	426	201	220	147	**486**	100	43	32	30	24	22	19	17	**9.27**	31	50	4	–	–	–	–	–	–	–	–	–	–	Urs-12-en-28-oic acid, 3,19-dihydroxy-, methyl ester, (3β)-	14511-72-9	NI44507
366	353	367	211	253	197	199	354	**486**	100	56	31	30	20	19	17	16	**0.00**	31	50	4	–	–	–	–	–	–	–	–	–	–	Cholest-5-ene-3,19-diol, 3,19-diacetate		LI09348
366	353	367	211	253	197	212	199	**486**	100	58	32	30	20	18	17	17	**0.00**	31	50	4	–	–	–	–	–	–	–	–	–	–	Cholest-5-ene-3,19-diol, diacetate, (3β)-	21072-68-4	NI44508
426	282	366	351	326				**486**	100	11	10	8	6				**0.00**	31	50	4	–	–	–	–	–	–	–	–	–	–	Cholest-5-ene-3,23-diol, diacetate, (3β,23α,23R)-		LI09349
43	70	138	155	142	42	28	96	**486**	100	74	70	51	50	45	28	27	**0.00**	31	50	4	–	–	–	–	–	–	–	–	–	–	Cholest-8(14)-ene-3,7-diol, diacetate, (3β,5α,7β)-	74420-84-1	NI44509
138	70	43	155	142	42	96	94	**486**	100	99	93	91	85	68	46	44	**0.00**	31	50	4	–	–	–	–	–	–	–	–	–	–	Cholest-8(14)-ene-3,15-diol, diacetate, (3β,5α,15α)-	69140-08-5	NI44510
327	73	371	328	89	103	83	97	**486**	100	64	38	28	24	21	17	16	**0.00**	31	50	4	–	–	–	–	–	–	–	–	–	–	Cholest-8(14)-ene-3,15-diol, diacetate, (3β,5α,15β)-	69140-09-6	NI44511
138	95	426	81	82	43	69	227	**486**	100	97	89	81	79	73	72	69	**57.00**	31	50	4	–	–	–	–	–	–	–	–	–	–	Cholest-20(22)-ene-3,6-diol, diacetate, (3β,5α,6α)-	41084-09-7	NI44512
426	427	43	163	51	71	53	41	**486**	100	59	55	50	42	41	38	37	**8.00**	31	50	4	–	–	–	–	–	–	–	–	–	–	Cholest-4-en-6-one, 3-(acetyloxy)-, cyclic 6-(1,2-ethanediyl acetal), (3β)-	38404-91-0	LI09350
426	43	163	91	99	291	411	444	**486**	100	56	50	28	26	22	14	9	**7.00**	31	50	4	–	–	–	–	–	–	–	–	–	–	Cholest-4-en-6-one, 3-(acetyloxy)-, cyclic 6-(1,2-ethanediyl acetal), (3β)-	38404-91-0	NI44513
135	87	134	117	195	390	251	408	**486**	100	78	60	43	35	24	24	22	**0.00**	31	50	4	–	–	–	–	–	–	–	–	–	–	5,6-Trans-1α,25-dihydroxy-26,27-dimethylvitamin D3 3-acetate		MS10261
262	203	249	189	486	223	468	426	**486**	100	90	16	16	13	10	6	6		31	50	4	–	–	–	–	–	–	–	–	–	–	Methyl 2α,3β-dihydroxyolean-12-en-28-oate	22425-82-7	MS10262
203	262	202	204	263	249	486	223	**486**	100	78	27	18	17	10	8	7		31	50	4	–	–	–	–	–	–	–	–	–	–	Methyl 2α,3β-dihydroxyolean-12-en-28-oate	22425-82-7	NI44514
262	203	223	189	249	133	486	468	**486**	100	97	12	10	7	7	4	2		31	50	4	–	–	–	–	–	–	–	–	–	–	Methyl 3β,23-dihydroxyolean-12-en-28-oate		MS10263
189	175	441	107	381	396	363	207	**486**	100	98	93	93	89	69	66	64	**7.30**	31	50	4	–	–	–	–	–	–	–	–	–	–	Methyl 3β,27-dihydroxyolean-12-en-28-oate		NI44515
262	203	133	249	189	223	233	486	**486**	100	72	19	13	12	11	6	4		31	50	4	–	–	–	–	–	–	–	–	–	–	Methyl 2α,3β-dihydroxyurs-12-en-28-oate	4518-70-1	MS10264
262	203	263	249	204	223	202	205	**486**	100	60	22	14	12	11	10	8	**6.00**	31	50	4	–	–	–	–	–	–	–	–	–	–	Methyl 2α,3β-dihydroxyurs-12-en-28-oate	4518-70-1	NI44516
262	203	223	133	189	249	486	426	**486**	100	57	15	13	10	8	3	2		31	50	4	–	–	–	–	–	–	–	–	–	–	Methyl 3β,23-dihydroxyurs-12-en-28-oate		MS10265
262	203	43	486	55	69	189	223	**486**	100	99	48	41	38	37	36	33		31	50	4	–	–	–	–	–	–	–	–	–	–	Methyl hydroxyoleanolate		LI09351
247	201	278	207	187	219	218	455	**486**	100	43	33	24	16	14	14	13	**7.50**	31	50	4	–	–	–	–	–	–	–	–	–	–	Methyl mesembryanthemoidigenate		MS10266
247	201	278	207	218	219	455	486	**486**	100	44	34	25	15	14	13	8		31	50	4	–	–	–	–	–	–	–	–	–	–	Methyl mesembryanthemoidigenate		LI09352
471	319	241	486	307	301	333	205	**486**	100	78	44	38	23	23	22	21		31	50	4	–	–	–	–	–	–	–	–	–	–	Methyl retigerate A		NI44517
119	147	455	486	426	471			**486**	100	54	29	14	12	3				31	50	4	–	–	–	–	–	–	–	–	–	–	Methyl ribifolate		NI44518
203	262	189	133	187	486	426	247	**486**	100	52	29	19	13	5	5	5		31	50	4	–	–	–	–	–	–	–	–	–	–	Methyl sumaresinolate		LI09353
186	207	232	189	412	358	262	440	**486**	100	81	72	70	26	18	17	14	**5.00**	31	50	4	–	–	–	–	–	–	–	–	–	–	12-Oleanene, 3,16-diol, 28,30-epoxy-30-methoxy-		NI44519
203	262	186	486	427	426	409	453	**486**	100	78	22	6	4	4	3	2		31	50	4	–	–	–	–	–	–	–	–	–	–	Olean-12-en-28-oic acid, 2,3-dihydroxy-, methyl ester, (2α,3aα)-	26563-64-4	MS10267
105	379	486	77	246	91	352	468	**486**	100	81	42	35	14	10	4	3		32	26	3	2	–	–	–	–	–	–	–	–	–	4-Benzoyl-N-(4-methylphenyl)-5-(4-methylphenylamino)-2-phenylfuran-3-carboxamide	85830-83-7	NI44520
303	486	247	233	487	304	105	177	**486**	100	88	44	33	32	25	23	17		32	38	4	–	–	–	–	–	–	–	–	–	–	2H,6H-Benzo[1,2-b:5,4-b']dipyran-10-propanol, 3,4,7,8-tetrahydro-5-methoxy-2,2,8,8-tetramethyl-α,α-diphenyl-	26537-41-7	NI44521
303	486	247	487	233	304	105	177	**486**	100	87	44	33	33	25	20	18		32	38	4	–	–	–	–	–	–	–	–	–	–	2H,6H-Benzo[1,2-b:5,4-b']dipyran-10-propanol, 3,4,7,8-tetrahydro-5-methoxy-2,2,8,8-tetramethyl-α,α-diphenyl-	26537-41-7	NI44522
486	303	487	289	233	191	247	105	**486**	100	82	36	35	31	23	22	22		32	38	4	–	–	–	–	–	–	–	–	–	–	2H,8H-Benzo[1,2-b:3,4-b']dipyran-6-propanol, 3,4,9,10-tetrahydro-5-methoxy-2,2,8,8-tetramethyl-α,α-diphenyl-	56335-88-7	NI44523
303	95	43	290	304	69	263	55	**486**	100	44	35	34	31	30	23	23	**15.00**	32	54	3	–	–	–	–	–	–	–	–	–	–	11β,19-Cyclolanostane-3β,11-diol, 3-acetate	23827-55-6	NI44524
189	136	207	191	220	468	471	426	**486**	100	70	63	49	36	35	33	31	**10.00**	32	54	3	–	–	–	–	–	–	–	–	–	–	29,30-Dinorgammacerane-3,22-diol, 21,21-dimethyl-, 3-acetate, (3β,8α,9β,13α,14β,17α,18β,22α)-	43206-38-8	NI44525
251	43	484	205	123	69	121	55	**486**	100	50	48	41	39	39	37	37	**3.69**	32	54	3	–	–	–	–	–	–	–	–	–	–	D:A-Friedooleanane-3,7-diol, 3-acetate	55515-38-3	NI44526
486	43	411	471	393	194	453	193	**486**	100	99	43	41	33	32	20	18		32	54	3	–	–	–	–	–	–	–	–	–	–	Lanostan-3β-ol, 11β,18-epoxy-, acetate	25116-68-1	LI09354
486	43	69	95	55	81	57	82	**486**	100	98	92	84	71	67	61	56		32	54	3	–	–	–	–	–	–	–	–	–	–	Lanostan-3β-ol, 11β,18-epoxy-, acetate	25116-68-1	NI44527
456	393	410	95	69	43	57	457	**486**	100	80	64	56	44	43	39	38	**0.00**	32	54	3	–	–	–	–	–	–	–	–	–	–	Lanostan-3β-ol, 11β,19-epoxy-, acetate	22417-93-2	NI44529
95	43	69	255	55	243	332	57	**486**	100	97	76	71	64	53	51	49	**31.00**	32	54	3	–	–	–	–	–	–	–	–	–	–	Lanostan-3β-ol, 11β,19-epoxy-, acetate	22417-93-2	NI44528
43	69	95	41	55	57	81	109	**486**	100	72	68	62	52	42	38	32	**18.30**	32	54	3	–	–	–	–	–	–	–	–	–	–	Lanostan-11-one, 3-(acetyloxy)-, (3β)-	10049-93-1	MS10268
43	69	263	95	55	303	41	57	**486**	100	79	72	68	65	58	52	51	**15.30**	32	54	3	–	–	–	–	–	–	–	–	–	–	Lanostan-11-one, 3-(acetyloxy)-, (3β)-	10049-93-1	MS10269
263	303	95	43	69	486	426	55	**486**	100	74	35	34	32	27	27	21		32	54	3	–	–	–	–	–	–	–	–	–	–	Lanostan-11-one, 3-(acetyloxy)-, (3β)-	10049-93-1	NI44530
486	487	361	123	205	413	348	136	**486**	100	35	34	19	16	13	13	11		32	54	3	–	–	–	–	–	–	–	–	–	–	Lanostan-11-one, 18-(acetyloxy)-	54411-53-9	NI44531
468	469	393	453	253	470	339	312	**486**	100	40	15	8	8	5	5	5	**1.00**	32	54	3	–	–	–	–	–	–	–	–	–	–	Lanost-8-en-7-ol, 3-(acetyloxy)-, (3β,5α,7α)-		MS10270
381	471	382	472	486				**486**	100	54	34	31	19					32	58	1	–	–	–	–	–	–	1	–	–	–	4α,14α-Dimethyl-5α-cholest-7-en-3β-ol		LI09355
381	471	382	472	486				**486**	100	56	34	31	19					32	58	1	–	–	–	–	–	–	1	–	–	–	4α,14α-Dimethyl-5α-cholest-8-en-3β-ol		LI09356
486	135	487	260	396	381	261	147	**486**	100	63	41	38	31	19	19	19		32	58	1	–	–	–	–	–	–	1	–	–	–	4,4-Dimethyl-5α-cholest-8-en-3β-ol-TMS		LI09357
357	358	129	356	443	149	210	123	**486**	100	28	25	14	13	9	7	7	**2.50**	32	58	1	–	–	–	–	–	–	1	–	–	–	4,4-Dimethylcholesterol trimethylsilyl ether		MS10271
257	55	255	486	83	75	374	69	**486**	100	88	60	59	48	48	47	44		32	58	1	–	–	–	–	–	–	1	–	–	–	24-Ethyl-δ(22)-coprostenol, trimethylsilyl-		NI44532
396	129	486	381	397	357	358	121	**486**	100	96	43	40	34	32	28	28		32	58	1	–	–	–	–	–	–	1	–	–	–	Stigmast-5-ene, 3-[(trimethylsilyl)oxy]-, (3β)-	2625-46-9	NI44533
129	357	121	396	95	71	81	57	**486**	100	46	39	37	37	34	32	26	**9.80**	32	58	1	–	–	–	–	–	–	1	–	–	–	Stigmast-5-ene, 3-[(trimethylsilyl)oxy]-, (3β)-	2625-46-9	NI44534
73	83	129	55	69	207	81	147	**486**	100	74	49	43	33	32	30	27	**1.37**	32	58	1	–	–	–	–	–	–	1	–	–	–	Stigmast-5-ene, 3-[(trimethylsilyl)oxy]-, (3β)-	2625-46-9	NI44535
73	75	255	69	81	486	83	213	**486**	100	95	78	67	60	56	56	55		32	58	1	–	–	–	–	–	–	1	–	–	–	Stigmast-7-ene, 3-[(trimethylsilyl)oxy]-, (3β,5α,24ξ)-	56248-34-1	NI44536

MASS TO CHARGE RATIOS	M.W.	INTENSITIES	Parent	C	H	O	N	S	F	Cl	Br	I	Si	P	B	X	COMPOUND NAME	CAS Reg No	No
119 347 301 487 299 359 76 443	487	100 64 31 29 16 13 13 11		11	18	5	3	–	–	–	–	–	–	1	–	1	Tungsten, pentacarbonyltris(dimethylamino)phosphinyl-		LI09358
269 297 77 119 298 270 42 190	487	100 90 40 15 12 10 8 7	0.00	17	15	4	1	–	10	–	–	–	–	–	–	–	β-Ethoxysympathol di-PFP		MS10272
229 216 255 230 455 311 396 217	487	100 75 27 27 20 20 18 18	12.00	23	41	8	3	–	–	–	–	–	–	–	–	–	Dimethyl 7-acetyl-2,12-bis(acetylmethylamino)-9-methoxy-7-aza-tridecanedioate		LI09359
91 92 487 43 267 161 488 65	487	100 8 5 4 3 2 1 1		24	21	5	7	–	–	–	–	–	–	–	–	–	N-Benzyl-N-ethyl-3-acetamido-4-(2,4-dinitro-6-cyanophenylazo)aniline		IC04511
139 73 141 487 111 370 75 140	487	100 35 25 15 15 13 9 8		24	30	4	1	–	–	1	–	–	2	–	–	–	1H-Indole-3-acetic acid, 1-(4-chlorobenzoyl)-2-methyl-5-[(trimethylsilyl)oxy]-, trimethylsilyl ester	55471-00-6	NI44537
253 366 276 456 254 217 311 179	487	100 50 50 25 20 18 15 12	1.00	24	49	5	1	–	–	–	–	–	2	–	–	–	Cyclopentanepropanoic acid, 2-[3-(methoxyimino)octyl]-3,5-bis[(trimethylsilyl)oxy]-, methyl ester, (1α,2β,3β,5α)-	55670-01-4	NI44538
253 276 366 254 87 178 217 156	487	100 44 39 23 21 16 15 15	0.36	24	49	5	1	–	–	–	–	–	2	–	–	–	Cyclopentanepropanoic acid, 2-[3-(methoxyimino)octyl]-3,5-bis[(trimethylsilyl)oxy]-, methyl ester, [1R-(1α,2β,3α,5α)]-	55821-18-6	NI44539
487 95 109 154 137 152 198 96	487	100 64 57 52 51 43 42 37		26	37	6	3	–	–	–	–	–	–	–	–	–	20-Hydroxy-7-isopropyl-4,8,16-trimethyl-6,23-dioxa-3,12,25-triazabicyclo[20.2.1]pentacosa-1(24),9,14,16,22(25)-pentaene-2,5,11-trione		MS10273
471 412 410 469 428 470 398 396	487	100 87 81 65 59 53 15 6	5.00	27	37	7	1	–	–	–	–	–	–	–	–	–	Macrodaphnine		LI09360
57 149 43 55 71 41 83 85	487	100 74 52 49 36 34 32 25	2.23	28	25	7	1	–	–	–	–	–	–	–	–	–	3′,8,8′-Trimethoxy-3-piperidyl-2,2′-binaphthalene-1,1′,4,4′-tetrone		NI44540
91 186 187 118 132 106 99 160	487	100 94 61 61 56 50 34 33	5.60	29	33	4	3	–	–	–	–	–	–	–	–	–	4,21-Dimethoxy-2,25-dioxa-10,17-diaza-9,10,17,18-dieno[2.9.18.25](1,4)diphenyl[1,26](2,6)pyridinophan		MS10274
157 487 488 196 158 197 144 459	487	100 80 29 18 18 9 9 5		34	49	1	1	–	–	–	–	–	–	–	–	–	1′H-Cholest-2-ano[3,2-b]indol-6-one, 1′-methyl-, (5β)-		LI09361
487 157 488 158 144 55 196 57	487	100 90 37 16 10 9 8 8		34	49	1	1	–	–	–	–	–	–	–	–	–	1′H-Cholestano[3,2-b]indol-6-one, 5′-methyl-, (5α)-		MS10275
157 487 488 158 144 489 196 57	487	100 95 35 15 10 7 7 7		34	49	1	1	–	–	–	–	–	–	–	–	–	1′H-Cholestano[3,4-b]indol-6-one, 1′-methyl-, (5α)-		MS10276
487 488 196 197 157 459 144 489	487	100 37 31 17 11 8 8 7		34	49	1	1	–	–	–	–	–	–	–	–	–	1′H-cholestano[3,4-b]indol-6-one, 1′-methyl-, (5β)-		MS10277
487 488 196 197 157 459 181 473	487	100 40 34 19 14 9 7 5		34	49	1	1	–	–	–	–	–	–	–	–	–	1′H-Cholestano[3,4-b]indol-6-one, 1′-methyl-, (5β)-		LI09362
312 224 183 97 273 271 185 144	488	100 99 99 85 80 65 63 63	61.50	3	5	–	–	–	9	–	–	1	–	3	–	1	π-Allyl-tris(trifluorophosphine)iron iodide		NI44541
220 218 222 182 183 184 185 237	488	100 81 48 29 29 29 28 18	0.00	10	2	1	–	–	–	10	–	–	–	–	–	–	1,3,4-Metheno-1H-cyclobuta[cd]pentalen-2-ol, 1,1a,3,3a,4,5,5,5a,5b,6-decachlorooctahydro-	1034-41-9	NI44542
473 105 103 83 227 89 85 107	488	100 71 61 53 47 42 34 32	1.58	10	19	–	2	3	–	–	–	–	–	–	–	3	Arsinothious acid, dimethyl-, 2,4,6-pyrimidinetriyl ester	51678-03-6	NI44543
473 443 195 105 263 353 229 173	488	100 40 35 15 12 10 10 10	1.00	13	32	–	–	–	–	–	–	–	–	–	–	4	Tetrakis(dimethylarsinomethyl)methane		NI44544
459 460 369 461 59 89 371 370	488	100 43 35 30 20 20 16 12	0.89	13	36	8	–	–	–	–	–	–	6	–	–	–	1,3,5,7,9,11-Hexaethyl-3-methoxybicyclo[5.5.1]hexasiloxane	73113-18-5	NI44545
404 406 378 402 408 376 405 380	488	100 96 66 52 44 39 38 37	7.48	14	9	3	–	–	–	–	1	–	–	–	–	1	Tungsten, cyclopentadienyl-(4-bromophenyl)tricarbonyl		NI44546
137 246 244 109 63 75 101 100	488	100 50 48 42 30 28 25 22	0.00	16	10	4	–	2	–	–	2	–	–	–	–	–	2-Bromothianaphthene-1,1 dioxide dimer		MS10278
324 293 325 253 283 429 265 43	488	100 36 19 16 7 4 4 2	0.00	16	10	6	–	–	10	–	–	–	–	–	–	–	Benzenepropanoic acid, α,4-bis(2,2,3,3,3-pentafluoro-1-oxopropoxy)-, methyl ester		MS10279
324 293 325 253 283 429 265 326	488	100 36 19 16 7 4 4 3	0.00	16	10	6	–	–	10	–	–	–	–	–	–	–	Benzenepropanoic acid, α,4-bis(2,2,3,3,3-pentafluoro-1-oxopropoxy)-, methyl ester	55683-25-5	NI44547
415 119 416 44 269 43 488 296	488	100 70 67 47 43 30 23 23		16	10	6	–	–	10	–	–	–	–	–	–	–	2,5-Dihydroxyphenylacetic acid ethyl ester di-PFP		MS10280
415 119 269 387 416 65 77 69	488	100 72 65 44 33 29 28 23	18.40	16	10	6	–	–	10	–	–	–	–	–	–	–	3,4-Dihydroxyphenylacetic acid ethyl ester di-PFP		MS10281
415 269 119 387 416 77 105 65	488	100 80 52 49 39 27 21 19	17.20	16	10	6	–	–	10	–	–	–	–	–	–	–	3,4-Dihydroxyphenylacetic acid ethyl ester di-PFP		MS10282
415 119 387 416 267 147 253 252	488	100 95 32 30 17 17 15 15	2.40	16	10	6	–	–	10	–	–	–	–	–	–	–	4-Hydroxymandelic acid ethyl ester di-PFP		MS10283
415 119 416 387 267 253 252 147	488	100 76 31 31 17 15 14 14	2.40	16	10	6	–	–	10	–	–	–	–	–	–	–	4-Hydroxymandelic acid ethyl ester di-PFP		MS10284
357 73 299 231 147 315 300 75	488	100 99 94 36 35 33 30 30	0.00	16	41	7	–	–	–	–	–	–	4	1	–	–	Phosphoric acid, 4-oxo-2,3-bis[(trimethylsilyl)oxy]butyl bis(trimethylsilyl) ester, [R-(R*,R*)]-	55723-94-9	NI44550
73 357 299 147 358 211 315 75	488	100 94 54 28 26 26 24 23	0.00	16	41	7	–	–	–	–	–	–	4	1	–	–	Phosphoric acid, 4-oxo-2,3-bis[(trimethylsilyl)oxy]butyl bis(trimethylsilyl) ester, [R-(R*,R*)]-	55723-94-9	NI44548
73 357 299 231 147 315 300 75	488	100 83 78 30 29 28 25 25	0.30	16	41	7	–	–	–	–	–	–	4	1	–	–	Phosphoric acid, 4-oxo-2,3-bis[(trimethylsilyl)oxy]butyl bis(trimethylsilyl) ester, [R-(R*,R*)]-	55723-94-9	NI44549
488 490 166 489 152 447 372 491	488	100 74 61 29 24 23 21 20		20	18	1	8	1	–	2	–	–	–	–	–	–	1-Methyl-4-(4-chlorophenyl)-5-(4-chlorophenyl)imino-3-[6-methyl-3-(methylthio)-5-oxo-4,5-dihydro-1,2,4-triazin-4-yl]imino-2H-2,3-dihydro-1,2,4-triazole		MS10285
46 45 333 335 69 334 332 153	488	100 69 64 50 36 31 30 24	10.00	20	18	1	8	1	–	2	–	–	–	–	–	–	2-Methyl-4-(4-chlorophenyl)-5-(4-chlorophenyl)imino-3-[6-methyl-3-(methylthio)-5-oxo-4,5-dihydro-1,2,4-triazin-4-yl]imino-2H-2,3-dihydro-1,2,4-triazole		MS10286
73 239 371 281 370 147 327 74	488	100 34 20 18 16 14 10 9	1.96	20	40	6	–	–	–	–	–	–	4	–	–	–	Fumarylacetoacetic acid, tetrakis(trimethylsilyl)-		NI44551
107 145 490 488 255 165 115 133	488	100 91 77 60 58 51 46 43		22	20	4	–	–	–	4	–	–	–	–	–	–	Bis(dichloroacetyl)diethylstilbestrol		NI44552

MASS TO CHARGE RATIOS								M.W.	INTENSITIES								Parent	C	H	O	N	S	F	Cl	Br	I	Si	P	B	X	COMPOUND NAME	CAS Reg No	No
127	229	200	69	98	81	100	85	**488**	100	99	60	50	47	45	35	35	**0.00**	22	32	12	–	–	–	–	–	–	–	–	–	–	α-D-Galactopyranose, 6-O-(2,4-di-O-acetyl-3,6-anhydro-β-D-galactopyranosyl)-1,2:3,4-bis-O-isopropylidene-	51885-42-8	MS10287
305	303	301	304	151	307	149	306	**488**	100	72	56	36	29	25	24	22	**0.00**	24	15	1	–	–	5	–	–	–	–	–	–	1	Germane, (pentafluorophenoxy)triphenyl-	22529-94-8	NI44553
403	429	488	415	167	473	373	431	**488**	100	94	75	69	65	58	54	50		24	24	11	–	–	–	–	–	–	–	–	–	–	Flavone, 2′,5′-dihydroxy-3,4′,5,6,7-pentamethoxy-, diacetate	34318-22-4	NI44554
351	188	264	263	161	104	103	233	**488**	100	52	40	30	30	30	30	28	**0.20**	24	29	7	2	–	–	–	–	–	–	1	–	–	[1,2′-Biaziridine]-2,3′-dicarboxylic acid, 3-(diethoxyphosphinyl)-2′,3-diphenyl-, dimethyl ester	33866-43-2	NI44555
351	188	264	263	161	104	103	233	**488**	100	52	40	30	30	30	30	28	**0.20**	24	29	7	2	–	–	–	–	–	–	1	–	–	Diethyl N-(2-phenyl-3-methoxycarbonylaziridin-2-yl)-2-phenyl-3-methoxy carbonylaziridin-2-yl-phosphonate		LI09363
86	307	100	132	263	70	114	56	**488**	100	41	29	27	18	17	16	16	**0.97**	24	48	6	4	–	–	–	–	–	–	–	–	–	(R,R)-Bisleucyldiaza-18-crown-6	93788-00-2	NI44556
282	281	488	309	354	326	169	325	**488**	100	55	34	25	23	17	13	9		25	28	10	–	–	–	–	–	–	–	–	–	–	Egonol glucoside	77690-83-6	NI44557
201	202	77	47	185	472	108	488	**488**	100	60	56	28	18	16	7	2		26	22	2	2	1	–	–	–	–	–	2	–	–	Bis(diphenylphosphinyliminomethyl) sulphide		LI09364
209	473	110	82	73	251	125	475	**488**	100	80	65	60	50	43	37	34	**29.00**	26	37	3	2	–	–	1	–	–	1	–	–	–	N-(4-Chlorophenyl)-4-hydroxy-3-methoxy-N-[1-isopropyl-4-piperidinyl]benzene-acetamide TMS		MS10288
73	273	215	69	75	147	55	83	**488**	100	43	37	23	20	16	16	14	**0.00**	26	56	4	–	–	–	–	–	–	2	–	–	–	Octadecanoic acid, 9,10-bis[(trimethylsilyl)oxy]-, ethyl ester	55470-99-0	NI44558
73	273	215	69	75	147	155	55	**488**	100	43	39	23	21	17	16	16	**0.00**	26	56	4	–	–	–	–	–	–	2	–	–	–	Octadecanoic acid, 9,10-bis[(trimethylsilyl)oxy]-, ethyl ester		LI09365
185	183	490	488	305	157	155	449	**488**	100	99	24	24	22	14	14	11		27	21	4	–	–	–	–	1	–	–	–	–	–	3-(4-Bromobenzoyloxy)-5,7,8-trimethyl-2-(2-propenyl)phenanthrene-1,4-dione		NI44559
387	55	121	388	109	135	297	91	**488**	100	96	85	60	55	46	44	43	**7.40**	27	36	8	–	–	–	–	–	–	–	–	–	–	(22R)-21-acetoxy-6α,11β-dihydroxy-16α,17α-propylmethylenedioxypregna-1,4-diene-3,20-dione		NI44560
387	55	121	279	91	109	388	159	**488**	100	62	50	30	29	28	26	25	**3.80**	27	36	8	–	–	–	–	–	–	–	–	–	–	(22R)-21-Acetoxy-6β,11β-dihydroxy-16α,17α-propylmethylenedioxypregna-1,4-diene-3,20-dione	93789-68-5	NI44561
55	121	387	149	91	109	122	135	**488**	100	92	84	74	58	56	53	51	**6.60**	27	36	8	–	–	–	–	–	–	–	–	–	–	(22S)-21-Acetoxy-6α,11β-dihydroxy-16α,17α-propylmethylenedioxypregna-1,4-diene-3,20-dione	93789-66-3	NI44562
121	55	387	109	91	279	147	107	**488**	100	92	72	50	43	42	41	40	**4.80**	27	36	8	–	–	–	–	–	–	–	–	–	–	(22S)-21-Acetoxy-6β,11β-dihydroxy-16α,17α-propylmethylenedioxypregna-1,4-diene-3,20-dione	93789-69-6	NI44563
320	43	28	45	111	44	60	91	**488**	100	62	48	45	43	42	41	40	**1.31**	27	36	8	–	–	–	–	–	–	–	–	–	–	5α-Card-20(22)-enolide, 3β,14,15β-trihydroxy-19-oxo-, 3,15-diacetate	14155-64-7	NI44564
488	487	473	486	174	176	188	186	**488**	100	76	63	45	45	25	20	18		27	40	6	2	–	–	–	–	–	–	–	–	–	9,12-Dihydro-3,4-secoreserpine-diol diacetate		LI09366
488	303	432	489	73	433	265	490	**488**	100	58	49	38	27	18	18	17		27	44	4	–	–	–	–	–	–	2	–	–	–	Bis(trimethylsilyl)-Δ^1-tetrahydrocannabinol		NI44565
43	313	29	41	55	93	91	81	**488**	100	50	28	22	19	18	17	17	**3.49**	28	40	7	–	–	–	–	–	–	–	–	–	–	17β-Acetoxy-1′,1′-dicarboethoxy-1β,2β-dihydrocycloprop[1,2]-5α-androst-1-en-3-one		NI44566
344	343	87	326	43	57	314	271	**488**	100	72	29	25	22	17	16	16	**12.00**	28	40	7	–	–	–	–	–	–	–	–	–	–	Hirundoside A	23284-21-1	LI09367
432	183	262	55	170	132	115	108	**488**	100	22	19	13	11	11	11	10	**4.00**	29	22	2	–	–	–	–	–	–	–	1	–	1	Manganese, dicarbonyl-π-indenyl(triphenylphosphine)-	31871-85-9	NI44567
253	313	226	143	145	105	254	314	**488**	100	63	54	40	36	33	30	29	**0.00**	29	44	6	–	–	–	–	–	–	–	–	–	–	Chol-8(14)-en-24-oic acid, 3,12-bis(acetyloxy)-, methyl ester, (3α,5β,12α)	60354-36-1	NI44568
253	226	157	105	143	211	159	199	**488**	100	76	63	62	59	52	48	46	**0.00**	29	44	6	–	–	–	–	–	–	–	–	–	–	Chol-11-en-24-oic acid, 3,7-bis(acetyloxy)-, methyl ester, (3α,5β,7α)-	2284-36-8	NI44569
253	313	368	428	226	211	353	281	**488**	100	76	38	24	24	24	9	8	**1.00**	29	44	6	–	–	–	–	–	–	–	–	–	–	Chol-12-en-24-oic acid, 3,12-bis(acetyloxy)-, methyl ester, (3α,5β,12α)-		NI44570
253	104	143	145	159	157	130	106	**488**	100	64	49	46	42	40	37	31	**0.00**	29	44	6	–	–	–	–	–	–	–	–	–	–	24-Norchol-22-ene-3,7,12-triol, triacetate, (3α,5β,7α,12α)-	60354-49-6	NI44571
255	447	371	75	448	256	331	147	**488**	100	58	46	32	24	22	20	17	**9.28**	29	52	2	–	–	–	–	–	–	2	–	–	–	3α,17β-Bis(allyldimethylsilyloxy)-5α-androstane		NI44572
255	175	75	256	447	273	176	147	**488**	100	88	30	23	22	18	17	11	**5.44**	29	52	2	–	–	–	–	–	–	2	–	–	–	3α,17β-Bis(allyldimethylsilyloxy)-5β-androstane		NI44573
447	255	75	448	371	256	449	147	**488**	100	70	48	39	37	18	16	13	**6.27**	29	52	2	–	–	–	–	–	–	2	–	–	–	3β,17β-Bis(allyldimethylsilyloxy)-5α-androstane		NI44574
70	55	43	69	137	60	45	41	**488**	100	97	80	75	74	72	70	69	**0.00**	30	48	5	–	–	–	–	–	–	–	–	–	–	Ergostane-3,6-dione, 25-(acetyloxy)-5-hydroxy-, (5α)-	56143-30-7	NI44575
55	84	27	28	29	83	56	39	**488**	100	65	43	41	40	33	31	28	**0.00**	30	48	5	–	–	–	–	–	–	–	–	–	–	Methyl 7α-(2′-tetrahydropyranyloxy)-3-oxo-5β-cholanoate		MS10289
189	207	185	214	220	232	251	279	**488**	100	85	80	45	42	36	18	15	**2.00**	30	48	5	–	–	–	–	–	–	–	–	–	–	3β,16α,22β-Trihydroxy-13,28-epoxyolean-29-al		NI44576
264	369	119	353	236	105	133	383	**488**	100	70	25	11	10	6	1	1	**0.18**	31	24	4	2	–	–	–	–	–	–	–	–	–	4-Benzoyl-5-hydroxy-1-(4-methylphenyl)-3-phenyl-5-phenylcarbamoyl-2(5H)-pyrrolone		NI44577
125	488	43	55	41	57	112	69	**488**	100	35	35	26	19	18	17	15		31	52	4	–	–	–	–	–	–	–	–	–	–	Cholestane-2-carboxylic acid, 3,3-[1,2-ethanediylbis(oxy)]-, methyl ester, (2β,5α)-	54498-63-4	NI44578
368	355	428	369	415	356	429	416	**488**	100	50	31	25	18	13	10	6	**5.00**	31	52	4	–	–	–	–	–	–	–	–	–	–	Cholestane-1,19-diol, diacetate, (1α,5α)-	28809-57-6	NI44579
368	355	428	415	488				**488**	100	50	31	18	5					31	52	4	–	–	–	–	–	–	–	–	–	–	Cholestane-1,19-diol, diacetate, (1α,5α)-	28809-57-6	LI09368
368	428	355	386	488				**488**	100	87	50	20	9					31	52	4	–	–	–	–	–	–	–	–	–	–	Cholestane-2,19-diol, diacetate, (2α,5α)-	28809-56-5	LI09369
368	428	355	429	369	386	356	488	**488**	100	87	50	26	25	20	13	9		31	52	4	–	–	–	–	–	–	–	–	–	–	Cholestane-2,19-diol, diacetate, (2α,5α)-	28809-56-5	NI44580
369	429	370	367	427	368	207	430	**488**	100	32	28	15	13	10	9	8	**0.00**	31	52	4	–	–	–	–	–	–	–	–	–	–	Cholestane-3,6-diol, diacetate, (3β,5α,6α)-	3514-28-1	NI44581
368	181	43	428	81	213	55	44	**488**	100	74	46	45	36	34	30	30	**0.00**	31	52	4	–	–	–	–	–	–	–	–	–	–	Cholestane-3,6-diol, diacetate, (3β,5α,6α)-	3514-28-1	NI44582
291	99	488	43	225	428	178	165	**488**	100	16	7	7	6	2	2	2		31	52	4	–	–	–	–	–	–	–	–	–	–	Cholestan-6-one, 3-(acetyloxy)-, cyclic 1,2-ethanediyl acetal, (3β,5α)-	19518-70-8	NI44583
291	292	99	488	43	225	51	95	**488**	100	21	16	8	7	6	6	5		31	52	4	–	–	–	–	–	–	–	–	–	–	Cholestan-6-one, 3-(acetyloxy)-, cyclic 1,2-ethanediyl acetal, (3β,5α)-	19518-70-8	LI09370
143	125	389	473	470	401			**488**	100	15	7	5	2	2			**0.00**	31	52	4	–	–	–	–	–	–	–	–	–	–	3,4-Secodammar-4(28)-en-3-oic acid, 20,24-epoxy-25-hydroxy-, methyl ester, (24S)-	21671-01-2	NI44584

MASS TO CHARGE RATIOS								M.W.	INTENSITIES								Parent	C	H	O	N	S	F	Cl	Br	I	Si	P	B	X	COMPOUND NAME	CAS Reg No	No
431	271	432	159	103	433	272	255	**488**	100	42	34	13	12	9	9	8	**0.00**	31	56	2	–	–	–	–	–	–	1	–	–	–	3α-Tributylsilyloxy-5α-androstan-17-one		NI44585
431	432	255	159	433	103	271	161	**488**	100	34	30	15	10	10	7	6	**0.00**	31	56	2	–	–	–	–	–	–	1	–	–	–	3α-Tributylsilyloxy-5β-androstan-17-one		NI44586
431	432	159	271	103	375	433	255	**488**	100	34	15	14	14	12	11	7	**0.00**	31	56	2	–	–	–	–	–	–	1	–	–	–	3β-Tributylsilyloxy-5α-androstan-17-one		NI44587
431	432	271	159	103	433	255	81	**488**	100	36	27	16	12	11	10	7	**0.00**	31	56	2	–	–	–	–	–	–	1	–	–	–	3β-Tributylsilyloxy-5β-androstan-17-one		NI44588
384	397	488	104	103	385	78	489	**488**	100	58	51	47	24	23	20	18		32	24	5	–	–	–	–	–	–	–	–	–	–	Dracorubin		MS10290
84	95	69	43	55	135	57	41	**488**	100	33	29	29	25	22	20	19	**0.40**	32	56	3	–	–	–	–	–	–	–	–	–	–	Lanostane-3β,11β-diol, 3-acetate	28328-14-5	NI44589
398	399	75	55	383	57	215	108	**488**	100	35	28	26	25	21	19	14	**0.50**	32	60	1	–	–	–	–	–	–	1	–	–	–	24-Ethylcoprostanol, trimethylsilyl-		NI44590
215	75	216	473	398	217	106	488	**488**	100	88	68	46	44	43	39	37		32	60	1	–	–	–	–	–	–	1	–	–	–	Stigmastanol trimethylsilyl ether		MS10291
488	143	456	489	457	182	472	471	**488**	100	56	53	39	31	30	27	26		33	48	1	2	–	–	–	–	–	–	–	–	–	6-Oximino-5α-cholestano[3,2-b]indole		MS10292
369	119	167	91	174	291	202	321	**488**	100	98	36	33	25	20	20	5	**5.00**	34	32	3	–	–	–	–	–	–	–	–	–	–	2H-1-Benzopyran-2-one, 3-(1,2-dimethyl-2-phenylpropyl)-6-(2,2-diphenylethyl)-7-hydroxy-	39081-74-8	MS10293
43	143	135	41	91	129	141	128	**488**	100	69	63	54	50	44	43	41	**0.00**	34	48	2	–	–	–	–	–	–	–	–	–	–	Cholesta-4,6-dien-3-ol, benzoate, (3β)-	25485-34-1	NI44591
384	366	248	253	350	212	275	229	**488**	100	33	10	6	4	4	3	3	**0.00**	34	48	2	–	–	–	–	–	–	–	–	–	–	7-Dehydrocholesteryl benzoate		MS10294
374	376	372	493	142	491	495	108	**489**	100	64	59	38	34	33	25	22	**13.20**	11	3	–	3	1	–	8	–	–	–	–	–	–	2-(3′,4′-Dichlorophenylthio)-4,6-bis(trichloromethyl)-s-triazine		MS10295
43	489	85	487	100	114	486	87	**489**	100	68	56	48	43	30	28	28		15	21	6	–	–	–	–	–	–	–	–	–	1	Osmium, tris(2,4-pentanedionato-O,O′)-	15635-86-6	NI44592
491	489	308	228	309	306	493	91	**489**	100	79	46	42	31	28	27	27		19	13	4	5	–	–	1	1	–	–	–	–	–	anti-2,4-Dinitrophenylhydrazono-5-bromo-2′-chloro-2-aminobenzophenone		MS10296
57	41	29	362	360	58	364	43	**489**	100	19	14	9	5	5	4	4	**0.10**	21	33	2	1	–	–	–	2	–	–	–	–	–	Benzene, 1,3-dibromo-2,4,6-tris(2,2-dimethylpropyl)-5-nitro-	40572-24-5	MS10297
118	145	112	86	160	110	100	84	**489**	100	80	72	63	62	60	59	53	**0.47**	21	39	8	5	–	–	–	–	–	–	–	–	–	D-Streptamine, 4-O-[3-(acetylamino)-6-(aminomethyl)-3,4-dihydro-2H-pyran-2-yl]-2-deoxy-6-O-[3-deoxy-4-C-methyl-3-(methylamino)-β-L-arabinopyranosyl]-, (2S-cis)-	55649-82-6	NI44593
96	94	97	43	89	95	71	60	**489**	100	97	93	91	83	79	77	77	**0.15**	24	43	9	1	–	–	–	–	–	–	–	–	–	Pseudopederin		LI09371
474	475	476	459	460	473	472	75	**489**	100	64	21	15	9	7	7	7	**5.36**	25	35	7	1	–	–	–	–	–	1	–	–	–	2-Nitro-estriol-3-TMS-16α,17β-diacetate		NI44594
474	489	475	490	472	473	459	457	**489**	100	94	38	36	35	25	22	15		25	35	7	1	–	–	–	–	–	1	–	–	–	4-Nitro-estriol-3-TMS-16α,17β-diacetate		NI44595
396	182	113	397	276	121	70	120	**489**	100	42	26	26	16	15	15	6	**5.30**	27	27	6	3	–	–	–	–	–	–	–	–	–	Tris(2-phenoxyethoxy)isocyanurate		NI44596
145	73	75	159	172	400	129	158	**489**	100	53	41	37	34	29	21	19	**13.00**	28	51	2	1	–	–	–	–	–	2	–	–	–	3β-Trimethylsiloxy-16α-methyl-pregn-5-en-20-one, TMS-oxime		LI09372
105	432	57	376	130	203	43	77	**489**	100	29	29	25	25	19	19	15	**0.00**	29	35	4	3	–	–	–	–	–	–	–	–	–	(2S,5S)-1-Benzoyl-2-tert-butyl-5-[[1-(tert-butoxycarbonyl)-1H-indol-3-yl]methyl]-3-methyl-4-imidazolidinone		DD01739
201	246	192	272	215	191	205	244	**489**	100	45	44	40	35	35	29	28	**17.00**	30	39	3	3	–	–	–	–	–	–	–	–	–	Tubulosan, 8′,10,11-trimethoxy-	5263-31-0	NI44597
86	453	124	43	368	36	425	438	**489**	100	70	67	35	33	33	22	17	**4.30**	31	52	1	1	–	–	1	–	–	–	–	–	–	Butanamide, 4-chloro-N-[(3β)-cholest-5-en-3-yl]-	38759-54-5	MS10298
384	184	93	186	106	78	199	123	**489**	100	29	20	17	10	8	7	7	**0.90**	32	29	–	1	–	–	–	–	–	–	2	–	–	N,N-Bis(diphenylphosphino)-1-phenylethylamine		NI44598
122	368	105	369	81	489	43	57	**489**	100	58	52	17	17	14	14	13		34	51	1	1	–	–	–	–	–	–	–	–	–	3β-Benzamido-5-cholestene		MS10299
173	489	490	174	158	491	57	55	**489**	100	81	30	19	11	5	5	4		34	51	1	1	–	–	–	–	–	–	–	–	–	5α-Cholestano[3,2-b]5′-methoxyindole		MS10300
127	170	150	128	321	100	89	301	**490**	100	32	15	14	9	9	7	7	**0.00**	5	2	2	2	2	16	–	–	–	–	–	–	–	(2,2,3,3,4,4-Hexafluoro-1,5-dioxo-1,5-pentylene)bis(iminosulphur pentafluoride)	91940-20-4	NI44599
322	462	350	266	112	84	56	294	**490**	100	91	91	91	91	91	91	52	**0.00**	8	–	9	–	–	4	–	–	–	–	2	–	2	μ-Oxo-bis(difluorophosphineiron tetracarbonyl)		NI44600
305	303	301	490	304	277	249	488	**490**	100	72	42	33	33	31	31	25		8	20	4	–	4	–	–	–	–	–	2	–	1	Tin[II] bis(diethyldithiophosphate)	58204-29-8	NI44601
322	131	324	490	364	289	366	326	**490**	100	99	75	65	65	54	50	50		12	28	4	–	4	–	–	–	–	–	2	–	1	Zinc[II] bis(dipropyldithiophosphate)	70844-72-3	NI44602
73	341	147	207	475	342	327	325	**490**	100	23	23	12	11	9	9	9	**0.00**	13	42	6	–	–	–	–	–	–	7	–	–	–	1,1,1,3,5,7,9,11,11,11-Decamethyl-5-(trimethylsiloxy)hexasiloxane	50694-26-3	NI44603
490	488	391	389	491	347	291	289	**490**	100	58	42	27	12	8	8	8		15	21	6	–	–	–	–	–	–	–	–	–	1	Iridium, tris(2,4-pentanedionato-O,O′)-	15635-87-7	NI44604
76	119	163	60	259	215	195	172	**490**	100	80	23	17	1	1	1	1	**0.44**	16	36	4	6	–	–	–	–	–	–	2	–	1	Bis[tris(dimethylamino)phosphine]tetracarbonylchromium		MS10301
119	378	215	163	95	172	259	490	**490**	100	25	24	24	14	12	7	6		16	36	4	6	–	–	–	–	–	–	2	–	1	Bis[tris(dimethylamino)phosphine]tetracarbonylchromium, trans-		MS10302
165	194	166	167	168	154	105	77	**490**	100	71	58	49	44	40	24	18	**1.80**	18	10	5	–	–	–	–	–	–	–	–	–	1	Tungsten, pentacarbonyl(diphenylmethylene)-	50276-12-5	NI44605
416	266	286	230	419	45	172	210	**490**	100	80	49	33	20	12	11	10	**0.00**	18	30	13	–	–	–	–	–	–	–	–	–	4	Beryllium, μ^4-oxohexakis[μ-(propanoato-O:O′)]tetra-	36593-61-0	MS10303
417	262	287	231	51	76	175	107	**490**	100	36	13	11	5	4	3	3	**0.00**	18	30	13	–	–	–	–	–	–	–	–	–	4	Beryllium, μ^4-oxohexakis[μ-(propanoato-O:O′)]tetra-	36593-61-0	NI44606
194	119	462	490	434	93	94	406	**490**	100	65	23	19	19	16	13	12		19	10	7	2	–	–	–	–	–	–	–	–	2	μ-N,N-Diphenylurea-bis(tricarbonyliron)		MS10304
322	378	56	350	490	161	462	121	**490**	100	78	61	47	42	36	33	29		19	10	7	2	–	–	–	–	–	–	–	–	2	μ-N,N-Diphenylurea-bis(tricarbonyliron)		MS10305
322	378	350	490	462	406	434	294	**490**	100	87	48	35	35	26	23	15		19	10	7	2	–	–	–	–	–	–	–	–	2	μ-N,N-Diphenylurea-bis(tricarbonyliron)		MS10306
73	211	75	169	147	345	45	313	**490**	100	26	26	24	21	18	15	11	**3.49**	20	42	6	–	–	–	–	–	–	4	–	–	–	Succinylacetoacetate, tetrakis(trimethylsilyl)-		NI44607
435	83	179	490	408	437	55	164	**490**	100	94	72	63	62	36	35	23		21	23	6	6	–	–	1	–	–	–	–	–	–	N-(1,3-Dimethylbut-2-enyl)-2-methoxy-5-acetamido-4-(2,4-dinitro-6-chlorophenylazo)aniline		IC04512
364	304	406	448	283	230	388	346	**490**	100	82	76	49	29	14	12	12	**8.00**	22	18	13	–	–	–	–	–	–	–	–	–	–	4,4′,5′-Triacetoxy-6-methoxybiphenyl-2,2′,3-tricarboxylic acid	87998-50-3	NI44608
73	338	247	265	264	173	323	131	**490**	100	76	41	38	34	29	21	19	**4.03**	22	35	–	–	–	–	–	1	–	4	–	–	–	9-(9-Bromofluorenyl)tris(trimethylsilyl)silane	84457-49-8	NI44609

MASS TO CHARGE RATIOS	M.W.	INTENSITIES	Parent	C	H	O	N	S	F	Cl	Br	I	Si	P	B	X	COMPOUND NAME	CAS Reg No	No
338 73 265 264 323 131 195 255	**490**	100 91 44 42 24 17 11 9	**5.60**	22	35	–	–	–	–	–	1	–	4	–	–	–	**Bromo-9-(9-trimethylsilylfluorenyl)bis(trimethylsilyl)silane**	84457-50-1	**NI44610**
58 99 158 57 55 127 114 112	**490**	100 71 66 66 53 47 42 38	**3.41**	22	38	10	2	–	–	–	–	–	–	–	–	–	**15,24-Dimethyl-1,4,7,10,18,21-hexaoxa-15,24-diazacyclooctacosane-11,14,25,28-tetrone**	91876-17-4	**NI44611**
58 99 92 158 57 55 127 114	**490**	100 71 71 66 66 53 47 42	**3.40**	22	38	10	2	–	–	–	–	–	–	–	–	–	**15,24-Dimethyl-1,4,7,10,18,21-hexaoxa-15,24-diazacyclooctacosane-11,14,25,28-tetrone**	91876-17-4	**MS10307**
56 70 58 100 57 114 102 86	**490**	100 52 50 48 45 39 39 35	**5.15**	22	42	8	4	–	–	–	–	–	–	–	–	–	**3,12-Dimethyl-6,9,17,20,25,28-hexaoxa-1,3,12,14-tetraazabicyclo[12.8.8]triacontane-2,13-dione**	89863-14-9	**NI44612**
475 230 490 73 476 99 491 231	**490**	100 87 59 45 39 26 23 17		24	30	2	6	–	–	–	–	–	2	–	–	–	**1H-1,2,4-Triazole-3-carboxaldehyde, 5-phenyl-1-(trimethylsilyl)-, dimer**	56196-01-1	**NI44613**
117 43 433 58 44 317 490 57	**490**	100 63 15 14 12 10 9 9		25	34	8	2	–	–	–	–	–	–	–	–	–	**3,4-Secocondyfolan-3-one, 12-[2-(acetyloxy)-1-methoxyethoxy]-16,19-epoxy-2-hydroxy-11-methoxy-4-methyl-, [12(S),14β,16β,19R]-**	55283-39-1	**NI44614**
243 472 310 370 328 430 388 282	**490**	100 61 61 54 52 33 31 28	**0.00**	26	34	9	–	–	–	–	–	–	–	–	–	–	**Neophorbol-3,13,20-triacetate**		**LI09373**
43 388 448 330 490	**490**	100 89 14 2 1		26	34	9	–	–	–	–	–	–	–	–	–	–	**Triacetylcastelanolide**		**LI09374**
73 400 55 91 41 75 369 120	**490**	100 84 84 74 67 66 62 62	**10.18**	26	42	5	2	–	–	–	–	–	1	–	–	–	**Pregn-4-ene-3,11,20-trione, 17,21-dihydroxy-, bis(O-methyloxime), mono(trimethylsilyl) ether**	72088-34-7	**NI44615**
139 335 121 490 111 180 141 337	**490**	100 89 43 39 39 36 34 32		27	16	5	–	–	–	2	–	–	–	–	–	–	**Bis[4-(4-chloro-benzoyl)phenyl] carbonate**		**IC04513**
83 55 43 41 29 39 53 27	**490**	100 61 54 21 15 14 13 12	**5.38**	27	38	8	–	–	–	–	–	–	–	–	–	–	**12-O-Acetyl-ingol-8-tigliate**		**NI44616**
83 43 55 41 109 69 165 29	**490**	100 73 58 23 16 13 12 12	**4.24**	27	38	8	–	–	–	–	–	–	–	–	–	–	**12-O-Acetyl-ingol-8-tigliate**		**NI44617**
43 83 312 372 294 412 95 131	**490**	100 57 50 46 46 43 43 35	**11.07**	27	38	8	–	–	–	–	–	–	–	–	–	–	**3-Desoxo-3β-hydroxy-phorbol-3,4-acetonide-12,13-diacetate**		**NI44618**
43 230 41 267 55 197 490 310	**490**	100 66 62 60 53 50 39 38		27	38	8	–	–	–	–	–	–	–	–	–	–	**Dihydro-digacetigenin-diacetate**		**MS10308**
430 281 415 397 282 327 431 73	**490**	100 75 52 40 36 34 33 33	**0.00**	27	42	6	–	–	–	–	–	–	1	–	–	–	**6β,21-Dihydroxy-medroxyprogesterone-acetate-21-TMS**		**NI44619**
490 491 492 143 133 129 147 145	**490**	100 42 17 10 8 7 7 7		27	46	4	–	–	–	–	–	–	2	–	–	–	**3β,5α-Tetrahydroaldosterone, bis(trimethylsilyl)-**		**NI44620**
490 491 341 240 269 143 492 400	**490**	100 45 22 21 21 19 16 14		27	46	4	–	–	–	–	–	–	2	–	–	–	**Tetrahydroaldosterone, bis(trimethylsilyl)-**		**NI44621**
369 400 459 151 329 114 387 460	**490**	100 84 71 44 38 34 29 28	**5.39**	27	46	4	2	–	–	–	–	–	1	–	–	–	**Pregn-4-ene-3,20-dione, 17-hydroxy-6-methyl-21-[(trimethylsilyl)oxy]-, bis(O-methyloxime), (6α)-**	74298-98-9	**NI44622**
212 113 67 43 165 197 441 243	**490**	100 57 55 44 32 27 26 25	**3.00**	28	42	7	–	–	–	–	–	–	–	–	–	–	**Colletotrichin**		**NI44623**
283 376 348 490 282 139 55 315	**490**	100 91 78 60 48 36 35 34		29	46	2	–	2	–	–	–	–	–	–	–	–	**5α-Spirostan-12-one, cyclic ethylene mercaptole, (25R)-**	24742-76-5	**NI44624**
104 106 145 147 130 119 159 120	**490**	100 68 62 59 57 49 44 44	**0.00**	29	46	6	–	–	–	–	–	–	–	–	–	–	**Cholan-24-oic acid, 3,6-bis(acetyloxy)-, methyl ester, (3α,5β,6α)-**	1181-65-3	**NI44625**
370 105 145 213 147 107 255 121	**490**	100 60 49 45 41 39 37 35	**0.00**	29	46	6	–	–	–	–	–	–	–	–	–	–	**Cholan-24-oic acid, 3,6-bis(acetyloxy)-, methyl ester, (3α,5β,6β)-**	60354-35-0	**NI44626**
370 213 255 105 355 107 145 119	**490**	100 74 73 72 53 49 46 42	**0.00**	29	46	6	–	–	–	–	–	–	–	–	–	–	**Cholan-24-oic acid, 3,7-bis(acetyloxy)-, methyl ester, (3α,5β,7α)-**	2616-71-9	**NI44627**
255 370 105 145 119 107 159 213	**490**	100 84 83 60 58 58 55 49	**0.00**	29	46	6	–	–	–	–	–	–	–	–	–	–	**Cholan-24-oic acid, 3,7-bis(acetyloxy)-, methyl ester, (3α,5β,7β)-**	60384-30-7	**NI44628**
370 79 93 43 255 355 44 52	**490**	100 58 46 46 41 38 29 28	**0.00**	29	46	6	–	–	–	–	–	–	–	–	–	–	**Cholan-24-oic acid, 3,7-bis(acetyloxy)-, methyl ester, (3α,7α)-**	56085-34-8	**NI44629**
370 255 355 371 213 201 228 316	**490**	100 67 38 30 29 20 18 15	**0.00**	29	46	6	–	–	–	–	–	–	–	–	–	–	**Cholan-24-oic acid, 3,7-bis(acetyloxy)-, methyl ester, (3α,7β)-**	56085-35-9	**NI44630**
255 105 146 107 145 161 119 159	**490**	100 49 42 36 33 29 29 27	**0.00**	29	46	6	–	–	–	–	–	–	–	–	–	–	**Cholan-24-oic acid, 3,12-bis(acetyloxy)-, methyl ester, (3α,5β,12α)-**	1181-44-8	**NI44631**
255 370 43 316 256 87 315 147	**490**	100 67 26 22 22 22 20 18	**0.00**	29	46	6	–	–	–	–	–	–	–	–	–	–	**Cholan-24-oic acid, 3,12-bis(acetyloxy)-, methyl ester, (3α,12α)-**	56085-33-7	**NI44632**
255 315 107 105 161 119 147 121	**490**	100 84 49 49 43 41 40 34	**0.00**	29	46	6	–	–	–	–	–	–	–	–	–	–	**Cholan-24-oic acid, 3,12-bis(acetyloxy)-, methyl ester, (3β,5α,12α)-**	60384-34-1	**NI44633**
305 69 91 337 323 217 77 379	**490**	100 99 70 60 47 31 31 25	**1.50**	30	34	6	–	–	–	–	–	–	–	–	–	–	**Grenoblone diacetate**		**NI44634**
475 117 119 44 83 488 91 55	**490**	100 98 95 86 50 40 40 40	**32.00**	30	34	6	–	–	–	–	–	–	–	–	–	–	**11-Hydroxy-5,9-dimethoxy-3,3-dimethyl-6,8-bis-(3-methylbut-2-enyl)-3H,12H-pyrano[3,2-a]xanthen-12-one**		**NI44635**
189 147 190 161 41 188 43 175	**490**	100 83 61 26 21 15 13 12	**1.10**	30	34	6	–	–	–	–	–	–	–	–	–	–	**1,3-Isobenzofurandione, 5,5′-(1,1,10,10-tetramethyl-1,10-decanediyl)bis-**	63322-28-1	**NI44636**
120 119 231 112 83 84 121 82	**490**	100 99 95 95 93 88 79 68	**18.00**	30	42	2	4	–	–	–	–	–	–	–	–	–	**3,6-Bis(4-N-octyloxyphenyl)-1,2,4,5-tetrazine**	72242-52-5	**NI44637**
235 266 224 236 490 267 269 237	**490**	100 60 10 7 5 5 3 1		30	50	5	–	–	–	–	–	–	–	–	–	–	**Kudzusapogenol A**	96820-46-1	**NI44638**
215 264 246 207 105 197 282 107	**490**	100 72 62 58 54 52 47 47	**3.00**	30	50	5	–	–	–	–	–	–	–	–	–	–	**Olean-12-ene-3β,16α,21β,22α,28-pentol**		**LI09375**
215 264 246 207 197 105 107 282	**490**	100 70 60 57 52 52 47 46	**3.00**	30	50	5	–	–	–	–	–	–	–	–	–	–	**Olean-12-ene-3β,16α,21β,22α,28-pentol**		**MS10309**
199 454 472 44 43 200 455 81	**490**	100 93 77 37 33 28 27 26	**0.00**	30	50	5	–	–	–	–	–	–	–	–	–	–	**Olean-12-ene-3β,16β,21β,23,28-pentol**		**LI09376**
105 159 189 203 207 233 417 441	**490**	100 64 46 45 42 21 10 9	**2.00**	30	50	5	–	–	–	–	–	–	–	–	–	–	**3β,16α,22β,30-Tetrahydroxy-13,28-epoxyoleanane**	94450-54-1	**NI44639**
137 253 490 199 81 377 215 492	**490**	100 71 43 43 34 22 21 19		30	51	1	–	–	–	1	–	–	1	–	–	–	**9,10-Secocholesta-5(10),6,8(14)triene, 3-[(chloromethyl)dimethyl]oxy-, (3β,6E)-**	69688-10-4	**NI44640**
165 167 295 194 166 296 152 489	**490**	100 81 64 55 54 22 20 15	**6.00**	31	26	2	2	1	–	–	–	–	–	–	–	–	**Benzeneacetic acid, α-phenyl-, 2,2-diphenyl-1-(2-thioxo-1-imidazolidinyl)vinyl ester**	67845-18-5	**NI44641**
394 43 57 55 71 83 41 69	**490**	100 77 56 51 47 44 44 40	**33.66**	31	51	2	–	–	–	1	–	–	–	–	–	–	**Stigmast-22-en-3-ol, 5-chloro-, acetate, (3β,5α,22E)-**	55724-18-0	**NI44642**
143 125 472 389 475	**490**	100 15 8 7 6	**0.00**	31	54	4	–	–	–	–	–	–	–	–	–	–	**3,4-Secodammaran-3-oic acid, 20,24-epoxy-25-hydroxy-, methyl ester, (24S)-**	21671-02-3	**NI44643**
490 57 491 433 462 475 284 259	**490**	100 46 38 14 12 11 9 9		33	46	3	–	–	–	–	–	–	–	–	–	–	**Bis(4-hydroxy-3,5-di-tert-butylstyryl) ketone**		**IC04514**
244 245 140 243 167 168 242 166	**490**	100 68 49 40 38 29 19 17	**14.10**	34	26	–	4	–	–	–	–	–	–	–	–	–	**1-Anilino-8-[(1-phenyl-2-perimidinyl)methylamino]naphthalene**	93767-85-2	**NI44644**
57 43 55 83 82 71 41 69	**490**	100 95 79 77 74 57 50 41	**0.08**	35	70	–	–	–	–	–	–	–	–	–	–	–	**15-Cyclohexylnonacosane**		**AI02393**
83 82 71 97 85 111 293 292	**490**	100 75 55 47 37 22 21 16	**0.00**	35	70	–	–	–	–	–	–	–	–	–	–	–	**15-Cyclohexylnonacosane**		**AI02392**

MASS TO CHARGE RATIOS								M.W.	INTENSITIES								Parent	C	H	O	N	S	F	Cl	Br	I	Si	P	B	X	COMPOUND NAME	CAS Reg No	No
43	57	55	41	83	69	97	71	**490**	100	82	61	48	47	44	38	38	**0.50**	35	70	–	–	–	–	–	–	–	–	–	–	–	17-Pentatriacontene	6971-40-0	WI01574
43	57	55	41	83	69	97	71	**490**	100	82	61	48	47	44	41	36	**0.48**	35	70	–	–	–	–	–	–	–	–	–	–	–	17-Pentatriacontene		AM00157
41	57	76	27	115	34	32	77	**491**	100	83	67	59	45	41	26	24	**0.60**	10	20	–	2	4	–	–	–	–	–	–	–	1	Bis(N-tert-butyldithiocarbamato)platinum(II)	90491-33-1	NI44645
476	477	478	491	479	492	493	480	**491**	100	71	48	24	24	17	15	9		17	49	–	3	–	–	–	–	–	7	–	–	–	1,3,5,7-Tetrakis(trimethylsilyl)-2,2,6,6-tetramethyl-2,4,6-trisila-1,3,5-triazaspiro[3.3]heptane		LI09377
149	105	164	234	246	189	279	488	**491**	100	99	27	14	11	10	6	3	**1.00**	25	31	10	–	–	–	–	–	–	–	–	–	–	1α-Benzoyloxy-8β,9α-diacetoxy-4β,6β,15-trihydroxydihy dro-β-agarofuran		MS10310
105	164	202	213	246	147	231	324	**491**	100	24	8	6	5	5	4	3	**2.00**	25	31	10	–	–	–	–	–	–	–	–	–	–	1α-Benzoyloxy-9-α,15-diacetoxy-4-β,6β,8β-trihydroxydi hydro-β-agarofuran		MS10311
194	297	165	250	298	73	195	299	**491**	100	78	38	35	28	21	14	10	**0.00**	25	41	5	1	–	–	–	–	–	2	–	–	–	Denopamine iso-M2, O,O-bis(trimethylsilyl)-		NI44646
297	194	165	250	298	73	195	75	**491**	100	97	40	37	33	28	14	13	**0.00**	25	41	5	1	–	–	–	–	–	2	–	–	–	Denopamine-M2, O,O-bis(trimethylsilyl)-		NI44647
491	153	169	339	493	184	138	492	**491**	100	90	76	69	35	31	31	28		26	22	5	3	–	–	1	–	–	–	–	–	–	2-Hydroxy-3-(2,4-dimethoxyanilinocarbonyl)-1-(4-chloro-2-methoxyphenylazo)naphthalene		IC04515
73	75	129	86	145	91	105	93	**491**	100	98	80	44	33	32	25	23	**20.00**	27	49	3	1	–	–	–	–	–	2	–	–	–	3β-Trimethylsiloxy-17α-hydroxy-pregn-5-en-20-one, TMS-oxime		LI09378
493	491	492	494	447	91	449	77	**491**	100	98	38	32	32	18	12	12		28	18	1	3	–	–	–	1	–	–	–	–	–	7-Oxo-2-phenyl-4-bromo-6-(4-toluidino)-7H-benzo[e]perimidine		IC04516
55	196	41	69	83	95	81	67	**491**	100	68	53	37	36	35	35	27	**6.40**	29	49	5	1	–	–	–	–	–	–	–	–	–	1-Cyclohexylmethyl-3a,5-dihydroxy-6,7-dimethyl-4-(4-methyl-7-(5-oxotetrahydrofuran-2-yl)heptyl)-3-oxo-3a,4-5,6,7,7a-hexahydro-isoindoline		IC04517
91	73	491	129	75	105	492	476	**491**	100	27	26	18	15	11	9	9		31	45	2	1	–	–	–	–	–	1	–	–	–	Pregna-5,16-dien-20-one, 3-[(trimethylsilyl)oxy]-, O-benzyloxime, (3β)-	57325-71-0	NI44648
43	41	55	29	491	57	448	95	**491**	100	41	36	34	32	29	20	18		32	45	3	1	–	–	–	–	–	–	–	–	–	3′H-Cycloprop(1,2)cholesta-1,4,6-trien-3-one, 1′-carboethoxy-1′-cyano-1β,2β-dihydro-	75857-78-2	NI44649
473	57	55	338	104	83	91	118	**491**	100	66	60	57	41	41	39	37	**6.50**	33	49	2	1	–	–	–	–	–	–	–	–	–	6-Azacholest-4-en-7-one, 6-benzyl-3α-hydroxy-	17373-01-2	LI09379
473	57	55	338	105	83	91	111	**491**	100	66	61	58	52	52	50	47	**1.97**	33	49	2	1	–	–	–	–	–	–	–	–	–	6-Azacholest-4-en-7-one, 6-benzyl-3α-hydroxy-	17373-01-2	NI44650
91	492	57	473	55	493	68	490	**491**	100	65	55	44	44	42	42	38	**17.00**	33	49	2	1	–	–	–	–	–	–	–	–	–	2α-Hydroxy-N-benzyl-6-aza-4-cholesten-7-one		LI09380
373	90	491	474	316	56	54	475	**491**	100	84	80	70	66	54	50	46		33	49	2	1	–	–	–	–	–	–	–	–	–	3β-Hydroxy-N-benzyl-6-aza-4-cholesten-7-one		LI09381
246	216	186	59	119	462	372	492	**492**	100	50	40	30	24	18	16	10		–	–	4	4	–	–	–	–	2	–	–	–	2	Di-μ-iodo-tetranitroso-di-cobalt		MS10312
253	352	324	238	294	163	309	56	**492**	100	56	50	24	22	22	17	17	**6.00**	13	11	6	–	–	–	–	–	–	–	–	–	3	Cyclo-tetracarbonyliron-dimethylarsenic-molybdenumdicarbonylcyclopentadiene		NI44651
145	126	95	75	125	345	57	492	**492**	100	22	16	15	12	11	10	9		14	8	–	–	–	6	–	–	–	–	–	–	1	Mercury, bis[p-(trifluoromethyl)phenyl]-		LI09382
225	165	138	192	164	107	167	183	**492**	100	41	41	38	29	25	24	18	**0.10**	14	15	9	2	1	–	3	–	–	–	–	–	–	2,2,2-Trichloroethyl 3-acetoxymethyl-7-methoxycarbonylamino-3-cephem 4-carboxylate 1,1-dioxide	97609-98-8	NI44652
492	454	416	412	310	308	414	346	**492**	100	33	32	17	16	15	10	10		16	24	–	–	–	–	2	–	–	–	–	–	2	Rhodium, di-μ-chlorobis[(1,2,5,6-η)-1,5-cyclooctadiene]di-	12092-47-6	LI09383
54	67	39	80	79	27	41	36	**492**	100	79	50	39	35	26	24	24	**10.15**	16	24	–	–	–	–	2	–	–	–	–	–	2	Rhodium, di-μ-chlorobis[(1,2,5,6-η)-1,5-cyclooctadiene]di-	12092-47-6	NI44653
54	39	79	27	67	80	53	41	**492**	100	39	24	23	16	14	12	12	**0.00**	16	24	–	–	–	–	2	–	–	–	–	–	2	Rhodium, di-μ-chlorobis(η[4]-1,2-vinylcyclobutane)di-	74825-26-6	NI44654
168	464	103	169	142	492	465	57	**492**	100	29	11	9	9	8	7	6		18	5	1	–	–	8	–	–	–	–	–	–	1	Rhodium, carbonyl(η[5]-2,4-cyclopentadien-1-yl)(3,3′,4,4′,5,5′,6,6′-octafluoro[1,1′-biphenyl]-2,2′-diyl)-	51509-32-1	NI44655
492	345	119	493	181	69	318	54	**492**	100	29	28	21	17	15	14	14		18	10	3	2	–	10	–	–	–	–	–	–	–	N-Acetylserotonin di-PFP		MS10313
43	145	169	109	81	157	115	98	**492**	100	50	14	12	12	11	11	11	**0.00**	20	28	14	–	–	–	–	–	–	–	–	–	–	Methyl 3-O-acetyl-2-O-(2′,3′,4′,6′-tetra-O-acetyl-α-D-mannopyranosyl)-D glycerate		MS10314
57	43	324	41	326	29	436	380	**492**	100	21	19	19	18	14	13	12	**10.00**	21	34	–	–	–	–	–	1	1	–	–	–	–	Benzene, 2-bromo-1,3,5-tris(2,2-dimethylpropyl)-4-iodo-	25347-09-5	MS10315
169	109	331	81	114	492	97	103	**492**	100	90	30	27	25	23	21	12		22	24	11	2	–	–	–	–	–	–	–	–	–	1H-Indole, 5-nitro-1-(2,3,4,6-tetra-O-acetyl-β-D-glucopyranosyl)-	29742-55-0	NI44656
354	492	45	246	213	344	265	104	**492**	100	96	93	74	60	46	46	44		24	12	–	–	6	–	–	–	–	–	–	–	–	Cyclododeca[1,2-b[4],3-b′[5],6-b″[8],7-b‴[9],10-b″″[1]2,11-b‴″]hexathiophene		NI44657
73	81	195	135	114	93	248	347	**492**	100	34	18	18	6	6	3	1	**0.60**	24	29	5	–	–	5	–	–	–	–	–	–	–	Methyl 11-methoxy-3,7,11-trimethyl-10-(pentabromobenzoyloxy)-2,6-dodecadienoate		MS10316
492	433	493	494	190	434	257	206	**492**	100	42	32	16	16	13	12	12		26	24	8	2	–	–	–	–	–	–	–	–	–	Tetramethyl 3,12-dimethylbenzo[c]pyridazine[1,2-a]cinnoline-6,7,8,9-tetracarboxylate		NI44658
83	69	111	55	173	95	43	79	**492**	100	76	56	47	38	38	37	35	**0.50**	26	36	9	–	–	–	–	–	–	–	–	–	–	3-Desoxo-3α-hydroxy-phorbol-12,13,20-triacetate		NI44659
43	83	69	111	55	41	173	91	**492**	100	43	31	23	16	16	15	14	**0.23**	26	36	9	–	–	–	–	–	–	–	–	–	–	3-Desoxo-3β-hydroxy-phorbol-12,13,20-triacetate		NI44660
390	330	375	372	408	432	450	492	**492**	100	38	25	23	18	13	10	4		26	36	9	–	–	–	–	–	–	–	–	–	–	Diacetylchaparrolide		LI09384
77	295	294	105	51	106	310	78	**492**	100	72	72	72	61	51	32	30	**0.00**	27	24	7	–	1	–	–	–	–	–	–	–	–	α-D-Xylopyranoside, methyl 4-thio-, tribenzoate	15076-96-7	NI44661
252	237	226	449	269	297	312	211	**492**	100	90	75	65	56	51	48	37	**1.00**	27	40	8	–	–	–	–	–	–	–	–	–	–	3α,7α,12α,17β-Tetraacetoxy-5β-androstane		NI44662
271	389	161	143	390	272	477	299	**492**	100	84	37	28	26	26	22	22	**12.00**	27	48	4	–	–	–	–	–	–	2	–	–	–	11β,21-Dihydroxy-5α-pregnane-3,20-dione-di-TMS		LI09385

MASS TO CHARGE RATIOS	M.W.	INTENSITIES	Parent	C	H	O	N	S	F	Cl	Br	I	Si	P	B	X	COMPOUND NAME	CAS Reg No	No
271 389 161 143 299 390 272 477	**492**	100 86 46 29 28 26 25 24	**12.00**	27	48	4	–	–	–	–	–	–	2	–	–	–	11β,21-Dihydroxy-5α-pregnane-3,20-dione-di-TMS		LI09386
492 117 147 159 173 285 231 197	**492**	100 13 12 8 5 4 4 4		27	48	4	–	–	–	–	–	–	2	–	–	–	Methyl all-trans-5,6-erythro-5,6-bis(trimethylsilyloxy)-5,6-dihydroretinoate		NI44663
186 187 155 59 128 492 461 143	**492**	100 94 54 37 36 28 23 21		28	28	8	–	–	–	–	–	–	–	–	–	–	(1α,4α,5α,6α,7α,10α,11α,12α)-Dibenzo[b,h]tricyclo[8.2.0.0^{4,7}]dodecane-5,6,11,12-tetracarboxylate		MS10317
186 187 155 128 143 214 146 461	**492**	100 80 53 46 23 9 9 6	**2.00**	28	28	8	–	–	–	–	–	–	–	–	–	–	(1α,2α,3β,4β)-3,4-Bis[(E)-2-(2-methoxycarbonylethenyl)phenyl]cycobutane-1,2-dicarboxylate		MS10318
30 84 44 98 41 70 56 492	**492**	100 98 90 83 83 67 63 45		28	36	4	4	–	–	–	–	–	–	–	–	–	Chaenorpin		DD01740
98 84 70 251 169 252 238 223	**492**	100 95 47 24 24 23 23 23	**23.00**	28	36	4	4	–	–	–	–	–	–	–	–	–	4H-1,16-Etheno-5,15-(propaniminoethano)furo[3,4-l][1,5,10]triazacyclohexadecine-4,21-dione, 3,3a,6,7,8,9,10,11,12,13,14,15-dodecahydro-3-(4-hydroxyphenyl)-	75363-58-5	NI44664
271 253 372 226 354 247 211 339	**492**	100 80 75 50 35 33 33 22	**1.00**	28	44	7	–	–	–	–	–	–	–	–	–	–	Methyl 7α,12α-diacetoxy-3α-hydroxy-24-nor-5β-cholan-23-oate		NI44665
312 268 145 85 105 91 294 43	**492**	100 86 83 73 68 64 60 60	**9.09**	28	44	7	–	–	–	–	–	–	–	–	–	–	Pregn-5-ene-3,11,12,14,20-pentol, 11-acetate 12-(3-methylbutanoate), (3β,11α,12β,14β)-	20230-38-0	NI44666
313 312 268 294 279 373 269 330	**492**	100 99 97 85 84 83 68 66	**2.00**	28	44	7	–	–	–	–	–	–	–	–	–	–	Pregn-5-ene-3,11,12,14,20-pentol, 11-acetate 12-(3-methylbutanoate), (3β,11α,12β,14β)-	20230-38-0	LI09387
399 492 371 493 400 343 372 199.5	**492**	100 82 65 26 25 17 16 14		29	16	8	–	–	–	–	–	–	–	–	–	–	3-Benzoxy-5,7,12,14-tetrahydroxy-6,13-dioxopentacene		IC04518
239 183 77 42 44 51 64 253	**492**	100 74 71 67 54 51 50 49	**11.00**	29	26	–	4	–	–	–	–	–	–	2	–	–	Cyanamide, [1,3-propanediylbis(diphenylphosphoranylidyne)]bis-	66055-15-0	NI44667
474 311 456 397 441 423 329 459	**492**	100 96 87 87 85 74 70 65	**4.20**	30	52	5	–	–	–	–	–	–	–	–	–	–	Cyclocanthogenin		MS10319
329 311 441 456 423 397 474 415	**492**	100 86 79 60 58 57 55 35	**6.00**	30	52	5	–	–	–	–	–	–	–	–	–	–	245-Lanost-9(11)-ene-3β,6α,16β,24,25-pentaol		MS10320
184 321 492 171 185 156 183 279	**492**	100 65 44 41 35 29 27 23		31	32	2	4	–	–	–	–	–	–	–	–	–	Roxburghine	27802-03-5	NI44668
147 95 396 255 107 105 81 69	**492**	100 87 78 76 75 75 73 65	**10.00**	31	53	2	–	–	–	1	–	–	–	–	–	–	Stigmastan-3-ol, 5-chloro-, acetate, (3β,5α)-	55724-19-1	NI44669
165 492 164 262 166 491 493 350	**492**	100 27 24 13 13 11 10 10		35	30	–	–	–	–	–	–	–	–	1	1	–	Phosphine, [1-[(diphenylboryl)benzylidene]-2-methyl-2-propenyl]diphenyl-	74646-16-5	NI44670
57 43 71 85 55 41 69 56	**492**	100 96 62 47 41 37 29 21	**0.00**	35	72	–	–	–	–	–	–	–	–	–	–	–	10-Heptyl-10-octyleicosane		AI02394
57 43 71 85 55 41 69 83	**492**	100 80 61 41 27 22 18 15	**0.42**	35	72	–	–	–	–	–	–	–	–	–	–	–	Pentatriacontane		AI02395
57 71 85 127 141 155 295 169	**492**	100 77 58 23 13 10 10 8	**0.20**	35	72	–	–	–	–	–	–	–	–	–	–	–	Tritriacontane, 15,19-dimethyl-		NI44671
267 73 268 69 45 41 43 77	**493**	100 72 25 25 20 20 19 18	**0.10**	18	26	3	1	–	7	–	–	–	2	–	–	–	1-(4′-Trimethylsiloxyphenyl)-1-trimethylsiloxy-2-heptafluorobutyramido-ethane		MS10321
407 376 408 462 304 377 318 493	**493**	100 35 21 10 8 7 7 6		22	23	12	1	–	–	–	–	–	–	–	–	–	Hexamethyl 6,7-dihydropyrido[1,2-a]azepine-2,3,7,8,9,10-hexacarboxylate		LI09388
178 232 262 478 216 493 247	**493**	100 65 25 20 10 9 7		22	35	4	3	1	–	–	–	–	2	–	–	–	4-Thia-1-azabicyclo[3.2.0]heptane-2-carboxylic acid, 3,3-dimethyl-7-oxo-6[[phenyl[(trimethylsilyl)amino]acetyl]amino]-, trimethylsilyl ester, [2S-[2α,5α,6β(S*)]]-	26508-25-8	NI44672
409 228 229 55 410 226 227 272	**493**	100 46 43 33 25 14 10 9	**6.30**	23	20	6	1	1	–	–	–	–	–	–	–	1	Tricarbonyl[η-1-(N,N-dimethylsulphamoyl)-2-(diphenylhydroxymethyl)cyclopentadienyl]manganese		NI44673
56 169 43 41 67 82 27 39	**493**	100 39 28 25 22 16 16 14	**0.00**	23	28	5	3	1	–	1	–	–	–	–	–	–	Benzamide, 5-chloro-N-[2-[4-[[[(cyclohexylamino)carbonyl]amino]sulphonyl]phenyl]ethyl]-2-methoxy-	10238-21-8	NI44674
251 493 170 494 252 250 478 171	**493**	100 52 25 20 16 15 14 10		24	31	8	1	1	–	–	–	–	–	–	–	–	1,2:3,4-Di-O-isopropylidene-6-O-(5-dimethylamino-1-naphthalenesulphonyl)-α-D-galactopyranose		NI44675
43 309 42 291 101 59 60 281	**493**	100 79 46 38 38 35 19 15	**0.04**	25	19	10	1	–	–	–	–	–	–	–	–	–	Ekatetrone triacetate		MS10322
434 435 462 92 436 312 316 100	**493**	100 30 10 8 5 4 3 3	**1.00**	25	23	8	3	–	–	–	–	–	–	–	–	–	Tetramethyl 5-benzyl-5H-pyridazino[1,2-a]benzotriazole-7,8,9,10-tetracarboxylate		NI44676
462 91 463 404 402 92 90 434	**493**	100 90 38 14 13 10 10 7	**3.00**	25	23	8	3	–	–	–	–	–	–	–	–	–	Trimethyl 6-benzyl-4a,5,6,11-tetrahydro-4a-methoxy-5-oxopyridazino[2,3-a]quinoxaline-2,3,4-tricarboxylate		NI44677
121 119 105 149 165 365 363 460	**493**	100 87 80 73 67 60 53 27	**0.00**	26	16	1	5	–	–	–	1	–	–	–	–	–	5-(1,1,2,2-Tetracyanoethyl)-1-(4-bromophenyl)-2-phenyl-4-oxo-4,5,6,7-tetrahydroindole	106012-26-4	NI44678
400 428 414 323 459 458 200 183	**493**	100 7 5 4 2 2 2 2	**0.00**	27	30	–	3	–	–	1	–	–	–	2	–	–	(Aminodiphenylphosphinoimino)-(propyl)diphenylphosphane chloride		NI44679
91 172 143 92 279 280 101 402	**493**	100 50 50 42 39 29 28 26	**0.30**	29	32	5	1	–	1	–	–	–	–	–	–	–	α-D-Glucopyranoside, benzyl 2-(acetylamino)-2,6-dideoxy-6-fluoro-3,4-bis-O-benzyl-	41341-94-0	NI44680
165 493 311 312 494 182 492 328	**493**	100 96 76 55 31 30 19 10		29	35	6	1	–	–	–	–	–	–	–	–	–	15-Veratroylpseudokobusine		MS10323
436 478 493 351 437 407 336 352	**493**	100 88 83 77 72 63 57 56		32	47	3	1	–	–	–	–	–	–	–	–	–	9-Spiro(3′,5′-di-tert-butylcyclohexa-3′,5′-dien-2-one)-4,6-di-tert-butyl-1-aza-2,12-dioxa-tricyclo[8.4.0.0^{3,8}]-tetradeca-1(10),11,13-triene		LI09389
177 493 179 495 494 178 57 55	**493**	100 53 33 20 20 17 13 12		33	48	–	1	–	–	1	–	–	–	–	–	–	5α-Cholestano[3,2-b]5′-chloroindole		MS10324

MASS TO CHARGE RATIOS								M.W.	INTENSITIES								Parent	C	H	O	N	S	F	Cl	Br	I	Si	P	B	X	COMPOUND NAME	CAS Reg No	No
169	254	59	45	83	199	74	85	**493**	100	80	65	49	42	41	38	31	**12.00**	34	27	1	1	–	–	–	–	–	1	–	–	–	**N,N-Diphenyl-2-(triphenylsilyl)-2,3-butadienamide**	79139-23-4	**NI44681**
182	388	77	104	105	389	195	183	**493**	100	57	57	24	21	16	12	12	**0.00**	34	27	1	3	–	–	–	–	–	–	–	–	–	**3-Benzoyl-1,2,4,5-tetraphenyl-1,2,3,4-tetrahydro-as-triazine**		**MS10325**
194	256	167	166	165	195	299	257	**493**	100	33	33	24	19	15	7	7	**1.91**	35	27	2	1	–	–	–	–	–	–	–	–	–	**2-Azetidinone, 1-(diphenylacetyl)-3,3,4-triphenyl-**	64187-60-6	**NI44682**
479	230	139	165	135	85	449	169	**494**	100	98	86	72	72	72	67	58	**42.00**	3	9	1	4	–	7	–	–	–	–	4	–	1	**1,3,5,2,4,6-Triazatriphosphorine, 2-[(difluorophosphinyl)(trimethylstannyl)amino]-2,4,4,6,6-pentafluoro-2,2,4,4,6,6-hexahydro-**	41006-39-7	**MS10326**
98	99	61	88	49	97	73	129	**494**	100	61	45	36	35	34	31	30	**0.10**	8	3	–	–	–	–	–	5	–	–	–	–	–	**Benzene, pentabromoethenyl-**	53097-59-9	**NI44683**
114	76	72	70	253	107	59	44	**494**	100	79	71	32	26	19	16	16	**0.00**	10	16	–	2	4	–	–	–	1	–	–	–	1	**Iodobis(pyrrolidyldithiocarbamato)arsine**	59196-52-0	**NI44684**
354	304	438	410	466	382	355	110	**494**	100	46	42	31	16	14	11	11	**6.01**	11	–	5	–	–	5	–	–	–	–	–	–	1	**Rhenium, pentacarbonyl(1,3,4,5,6-pentafluorobicyclo[2.2.0]hexa-2,5-dien-2-yl)-**	20523-94-8	**NI44685**
354	304	410	494	382	466	438	110	**494**	100	50	47	35	25	17	17	15		11	–	5	–	–	5	–	–	–	–	–	–	1	**Rhenium, pentacarbonyl(pentafluorophenyl)-**	14837-16-2	**NI44686**
73	448	451	449	491	493	494	492	**494**	100	94	92	90	87	84	82	80		11	22	3	–	–	–	–	–	–	2	–	4	1	**Tricarbonyl-2,3-bis(trimethylsilyl)-2,3-dicarba-1-osmaheptaborane(6)**	904124-67-0	**NI44687**
498	500	496	502	428	426	356	430	**494**	100	84	69	47	40	35	28	26	**21.00**	12	–	–	–	–	–	10	–	–	–	–	–	–	**Decachlorobiphenyl**		**IC04519**
107	214	142	109	178	179	498	110	**494**	100	72	69	69	53	44	38	29	**1.00**	12	–	–	–	–	–	10	–	–	–	–	–	–	**Decachlorobiphenyl**		**IC04520**
498	36	214	500	178	496	428	213	**494**	100	95	87	82	73	72	63	63	**21.63**	12	–	–	–	–	–	10	–	–	–	–	–	–	**Decachlorobiphenyl**	2051-24-3	**NI44688**
163	137	81.5	66	247	219	39	98	**494**	100	38	31	31	23	19	18	14	**0.00**	16	10	6	–	–	–	–	–	–	–	–	–	2	**Di(π-cyclopentadienyl-tricarbonyl molybdenum)**		**MS10327**
163	247	219	164	191	248	220	192	**494**	100	37	28	15	14	4	4	4	**0.00**	16	10	6	–	–	–	–	–	–	–	–	–	2	**Di(π-cyclopentadienyl-tricarbonyl molybdenum)**		**LI09390**
354	438	382	410	219	191	163	247	**494**	100	22	13	11	8	4	4	3	**0.00**	16	10	6	–	–	–	–	–	–	–	–	–	2	**Di(π-cyclopentadienyl-tricarbonyl molybdenum)**		**MS10328**
204	206	247	101	249	103	89	77	**494**	100	64	49	32	31	29	21	18	**0.60**	16	16	–	6	–	–	4	–	–	–	2	–	–	**5,5,10,10-Tetrachloro-1,6-dimethyl-3,8-diphenyl[1,3,2,4]diazadiphospheto[2,1-c:4,3-c′]bis[1,2,4,3λ^5]triazaphosphole**	63148-50-5	**NI44689**
172	126	173	127	114	125	145	142	**494**	100	24	10	5	5	4	3	2	**0.10**	17	8	4	2	–	10	–	–	–	–	–	–	–	**1,5-Bis(m-nitrophenyl)-decafluoropentane**		**LI09391**
69	198	59	166	97	85	179	91	**494**	100	92	85	64	44	44	38	33	**3.00**	17	20	6	2	1	6	–	–	–	–	–	–	–	**1H-Thieno[3,4-d]imidazole-4-pentanoic acid, hexahydro-1-(methoxycarbonyl)-2-oxo-3-[3,3,3-trifluoro-1-oxo-2-(trifluoromethyl)propyl]-, methyl ester**	67845-11-8	**NI44690**
327	494	258	328	495	475	188	69	**494**	100	59	25	18	14	9	7	7		18	–	–	–	–	13	–	–	–	–	1	–	–	**5H-Dibenzophosphole, 1,2,3,4,6,7,8,9-octafluoro-5-(pentafluorophenyl)-**	69688-63-7	**NI44691**
73	287	147	361	288	75	389	377	**494**	100	86	61	23	20	19	17	13	**0.00**	19	42	7	–	–	–	–	–	–	4	–	–	–	**Methylcitric acid, tetrakis(trimethylsilyl)-**		**MS10329**
73	287	147	479	377	361	288	389	**494**	100	95	61	29	29	27	23	22	**0.00**	19	42	7	–	–	–	–	–	–	4	–	–	–	**Methylcitric acid, tetrakis(trimethylsilyl)-**		**NI44692**
142	87	86	101	89	114	56	73	**494**	100	99	89	83	83	81	75	74	**0.00**	20	34	12	2	–	–	–	–	–	–	–	–	–	**1,4,7,10,13,16,22,25-Octaoxa-19,28-diazacyclottriacontane-3,14,18,29-tetrone**		**MS10330**
86	61	85	87	132	56	83	57	**494**	100	55	48	33	31	31	27	26	**4.50**	20	38	6	4	2	–	–	–	–	–	–	–	–	**6,9,12,20,23,28-Hexaoxa-1,3,15,17-tetraazabicyclo[15.8.5]triacontane-2,16-dithione**		**MS10331**
86	85	132	56	74	100	60	70	**494**	100	46	43	36	30	24	24	23	**11.85**	20	38	6	4	2	–	–	–	–	–	–	–	–	**6,9,17,20,25,28-Hexaoxa-1,3,12,14-tetraazabicyclo[12.8.8]triacontane-2,13-dithione**	89863-11-6	**NI44693**
99	267	266	111	69	98	113	81	**494**	100	89	65	60	37	30	25	25	**0.00**	20	39	10	–	–	–	–	–	–	–	–	5	–	**DL-Xylitol, cyclic 1,4:2,3-bis(ethylboronate) 5,5′-(ethylboronate)**	74792-96-4	**NI44694**
345	435	255	191	204	314	73	299	**494**	100	96	96	94	92	89	89	86	**6.00**	20	46	6	–	–	–	–	–	–	4	–	–	–	**Cyclohexanecarboxylic acid, 1,3,4,5-tetrakis[(trimethylsilyl)oxy]-, methyl ester, [1S-(1α,3α,4β,5β)]-**	50459-27-3	**NI44695**
73	81	135	93	107	114	121	248	**494**	100	72	35	26	24	20	18	6	**0.10**	21	29	5	–	–	7	–	–	–	–	–	–	–	**Methyl 10-(heptafluorobutyryloxy)-11-methoxy-3,7,11-trimethyldodecadienoate**		**MS10332**
169	109	81	193	494	206	331	99	**494**	100	74	57	53	27	27	21	15		22	26	11	2	–	–	–	–	–	–	–	–	–	**1H-Indole, 2,3-dihydro-5-nitro-1-(2,3,4,6-tetra-O-acetyl-β-D-glucopyranosyl)-**	26386-09-4	**NI44696**
169	109	81	193	141	206	494	115	**494**	100	74	56	52	27	26	25	24		22	26	11	2	–	–	–	–	–	–	–	–	–	**1H-Indole, 2,3-dihydro-5-nitro-1-(2,3,4,6-tetra-O-acetyl-β-D-glucopyranosyl)-**	26386-09-4	**LI09392**
351	352	142	127	156	337	223	169	**494**	100	99	99	71	68	40	28	28	**0.00**	22	27	3	2	–	–	–	–	1	–	–	–	–	**Melinonine-A iodide**		**LI09393**
380	41	381	69	186	267	137	159	**494**	100	44	26	21	15	14	14	11	**9.00**	23	24	5	–	–	6	–	–	–	–	–	–	–	**Estra-1,3,5(10)-triene-11,17-diol, 3-methoxy-, bis(trifluoroacetate), (11α,17β)-**	56588-10-4	**NI44697**
494	495	69	173	186	147	160	159	**494**	100	27	27	22	19	19	18	17		23	24	5	–	–	6	–	–	–	–	–	–	–	**Estra-1,3,5(10)-triene-16α,17β-diol, 3-methoxy-, bis(trifluoroacetate)**	34210-15-6	**NI44698**
160	93	91	110	81	79	107	173	**494**	100	60	49	48	43	42	41	39	**12.63**	24	25	3	–	–	7	–	–	–	–	–	–	–	**Ethisterone-HFB**		**NI44699**
73	494	479	495	211	227	243	480	**494**	100	97	54	44	36	26	23	18		24	39	5	–	–	–	–	–	–	2	1	–	–	**Estrone, bis(trimethylsilyl) phosphate**	33745-62-9	**NI44700**
44	77	43	41	57	55	69	51	**494**	100	72	44	44	39	39	27	27	**5.56**	25	18	7	–	2	–	–	–	–	–	–	–	–	**Bis[4-(phenylsulphonyl)phenyl] carbonate**		**IC04521**
73	55	151	494	83	247	91	95	**494**	100	53	33	31	29	28	24	22		25	34	10	–	–	–	–	–	–	–	–	–	–	**Glaucarubinone**		**LI09394**
43	57	41	97	85	60	148	42	**494**	100	92	31	30	24	18	8	7	**0.15**	25	34	10	–	–	–	–	–	–	–	–	–	–	**Isohomoacevaltrate**		**MS10333**
435	494	495	436	317	209	227	208	**494**	100	99	31	29	25	21	14	13		26	26	8	2	–	–	–	–	–	–	–	–	–	**Tetramethyl 6,9-dihydro-3,12-dimethylbenzo[c]pyridazino[1,2-a]cinnoline 6,7,8,9-tetracarboxylate**		**NI44701**

MASS TO CHARGE RATIOS								M.W.	INTENSITIES								Parent	C	H	O	N	S	F	Cl	Br	I	Si	P	B	X	COMPOUND NAME	CAS Reg No	No
323	414	413	400	459	122	399	338	**494**	100	94	94	94	72	39	38	38	**0.00**	26	29	–	4	–	–	1	–	–	–	2	–	–	(Aminodiphenylphosphinoimino)-(N′,N′-dimethylhydrazino)diphenylphosphane chloride		NI44702
233	43	452	410	435	451	44	409	**494**	100	95	86	67	54	47	46	45	**40.19**	27	26	9	–	–	–	–	–	–	–	–	–	–	Barpisoflavone B triacetate		NI44703
137	452	179	410	232	494	453	231	**494**	100	54	48	40	21	20	16	14		27	26	9	–	–	–	–	–	–	–	–	–	–	7,2′,4′-Triacetoxy-5′-methoxy-6′-(3,3-dimethylallyl)isoflavone		MS10334
262	466	436	91	134	214	183	319	**494**	100	96	69	65	47	45	35	25	**13.00**	27	27	2	2	–	–	–	–	–	–	1	–	1	Chromium, carbonyl(η^5-2,4-cyclopentadien-1-yl)[N-methyl-P,P-diphenyl-N-(1-phenylethyl)phosphinous amide]nitrosyl-	75520-17-1	NI44704
314	315	187	191	212	117	174	226	**494**	100	27	24	20	17	17	15	14	**0.00**	27	42	8	–	–	–	–	–	–	–	–	–	–	PGE-2, α-methyl ester triacetate		NI44705
73	287	117	75	119	43	55	93	**494**	100	99	99	99	49	43	40	38	**9.44**	27	50	4	–	–	–	–	–	–	2	–	–	–	Pregnan-11-one, 17-hydroxy-3,20-bis[(trimethylsilyl)oxy]-, (3α,5β,20S)-	51166-36-0	NI44706
100	70	155	238	84	224	143	112	**494**	100	54	48	46	25	23	22	21	**1.00**	27	50	4	4	–	–	–	–	–	–	–	–	–	Acetamide, N-[3-(acetylamino)propyl]-N-[3-[acetyl[3-(2-oxoazacyclotridec-1-yl)propyl]amino]propyl]-	67473-76-1	NI44707
79	100	70	451	112	297	125	139	**494**	100	75	65	60	43	25	25	20	**4.00**	27	50	4	4	–	–	–	–	–	–	–	–	–	13,17,21-Triacetyl-13,17,21-triaza-24-tetracosane lactam	75701-28-9	NI44708
368	253	386	369	458	271	343	254	**494**	100	92	60	45	33	30	24	23	**6.25**	28	46	7	–	–	–	–	–	–	–	–	–	–	Cholan-24-oic acid, 3-[(ethoxycarbonyl)oxy]-7,12-dihydroxy-, methyl ester, (3α,5β,7α,12α)-	21059-36-9	NI44709
368	253	386	369	458	277	343	254	**494**	100	82	60	46	33	30	24	23	**6.00**	28	46	7	–	–	–	–	–	–	–	–	–	–	Cholan-24-oic acid, 3-[(ethoxycarbonyl)oxy]-7,12-dihydroxy-, methyl ester, (3α,5β,7α,12α)-		LI09395
165	494	180	205	138	137	122	152	**494**	100	88	85	28	25	20	15	13		30	38	6	–	–	–	–	–	–	–	–	–	–	1,14-(1,3-Dimethoxycarbonyl)-3,9-dodecadiene		LI09396
178	494	230	94	495	359	231	179	**494**	100	40	40	24	11	10	9	9		30	42	2	2	1	–	–	–	–	–	–	–	–	Neothiobinupharidine, (7S,13S)-	30343-72-7	NI44710
439	441	479	253	462	481	440	464	**494**	100	63	48	48	42	34	31	30	**27.00**	30	48	1	–	–	–	2	–	–	–	–	–	–	(23R,24ξ,25ξ)-26,26-Dichloro-23-ethyl-6β-methoxy-3α,5;24,26-dicyclocholestane		MS10335
55	439	253	441	479	462	481	440	**494**	100	97	78	63	50	35	33	30	**22.00**	30	48	1	–	–	–	2	–	–	–	–	–	–	(23S,24R,25R)-26,26-Dichloro-23-ethyl-6β-methoxy-3α,5;24,26-dic yclocholestane		MS10336
55	439	441	479	253	462	481	255	**494**	100	87	59	46	40	37	31	31	**25.00**	30	48	1	–	–	–	2	–	–	–	–	–	–	(23S,24S,25S)-26,26-dichloro-23-ethyl-6β-methoxy-3α,5;24,26-dic yclocholestane		MS10337
153	181	245	60	129	227	95	41	**494**	100	52	36	25	18	14	14	14	**11.00**	31	42	5	–	–	–	–	–	–	–	–	–	–	(+)-Milbemycin β^3		MS10338
208	77	105	183	179	197	225	167	**494**	100	96	95	92	78	75	53	50	**0.00**	33	34	4	–	–	–	–	–	–	–	–	–	–	(4R,5R)-2-tert-Butyl-α,α,α′,α′-tetraphenyl-1,3-dioxolane-4,5-dimethanol		DD01741
283	494	188	124	495	284	105	148	**494**	100	64	46	24	23	23	9	8		33	38	2	2	–	–	–	–	–	–	–	–	–	Aspidospermidin-21-ol, 17-methoxy-1-methyl-21,21-diphenyl-	55103-34-9	NI44711
494	495	95	57	496	110	81	109	**494**	100	36	12	12	11	11	11	10		33	50	1	–	1	–	–	–	–	–	–	–	–	Cholestan-3-one, 2-(phenylthio)-, (5α)-	63608-48-0	NI44712
100	101	88	57	71	69	44	85	**494**	100	99	90	21	17	14	14	12	**5.77**	33	66	2	–	–	–	–	–	–	–	–	–	–	Nonacosanoic acid, 2,4,6-trimethyl-, methyl ester	56282-31-6	NI44713
101	88	111	494	129	57	115	139	**494**	100	90	38	30	25	22	20	18		33	66	2	–	–	–	–	–	–	–	–	–	–	Octacosanoic acid, 2,4,6,8-tetramethyl-, methyl ester		LI09397
101	88	57	43	69	41	71	56	**494**	100	97	91	73	54	47	46	46	**18.91**	33	66	2	–	–	–	–	–	–	–	–	–	–	Octacosanoic acid, 2,4,6,8-tetramethyl-, methyl ester, (all-R)-(-)-	27829-61-4	NI44714
219	220	58	276	41	83	55	221	**494**	100	30	25	14	10	6	6	4	**1.00**	34	54	2	–	–	–	–	–	–	–	–	–	–	4-(3,5-Di-tert-butyl-4-hydroxy-benzyl)-2,4,6-tri-tert-butyl-5-oxo-bicyclo[4.1.0]hept-2-ene		MS10339
197	105	180	271	77	193	81	91	**494**	100	48	39	27	23	22	20	19	**1.44**	36	46	1	–	–	–	–	–	–	–	–	–	–	2-(3β-Hydroxybisnorchol-5-enyl)-1,1-diphenylethylene		NI44715
131	69	100	476	181	169	119	114	**495**	100	57	50	12	8	8	8	6	**0.00**	10	–	–	1	–	19	–	–	–	–	–	–	–	Perfluoro-1-azabicyclo[5.4.0]undecane		DD01742
69	426	476	100	131	169	119	181	**495**	100	79	56	55	47	45	19	15	**0.00**	10	–	–	1	–	19	–	–	–	–	–	–	–	4-(Trifluoromethyl)perfluoroquinolizidine		DD01743
496	498	494	497	524	257	397	165	**495**	100	35	22	20	15	8	7	7	**6.00**	18	21	3	3	–	7	1	–	–	–	–	–	–	Metaclopramide heptafluorobutyrate derivative		NI44716
73	147	423	408	75	424	93	290	**495**	100	45	39	36	20	15	14	12	**1.05**	19	37	3	5	–	–	–	–	–	4	–	–	–	6-Pteridinecarboxylic acid, 2-[bis(trimethylsilyl)amino]-1,4-dihydro-4-oxo 1-(trimethylsilyl)-, trimethylsilyl ester	55649-41-7	NI44717
73	75	204	309	147	246	217	307	**495**	100	57	39	35	30	27	24	23	**0.20**	19	45	6	1	–	–	–	–	–	4	–	–	–	N-Methyl-D-glucuronamide tetrakis(trimethylsilyl)-		NI44718
297	73	298	299	45	74	75	77	**495**	100	80	25	10	7	6	5	4	**1.00**	20	36	4	1	–	3	–	–	–	3	–	–	–	N-(Trifluoroacetyl)-N,O,O′-tris(trimethylsilyl)normetanephrine		MS10340
73	355	356	45	357	74	75	77	**495**	100	68	20	11	10	8	6	4	**0.80**	20	36	4	1	–	3	–	–	–	3	–	–	–	N-(Trifluoroacetyl)-O,O′,O″-tris(trimethylsilyl)epinephrine		MS10341
185	257	183	166	214	72	108	242	**495**	100	90	68	66	50	30	29	15	**1.00**	21	20	5	1	–	–	–	–	–	–	1	–	1	(Diethylamino)-diphenylphosphine molybdenum pentacarbonyl		IC04522
282	495	283	281	496	229	225	266	**495**	100	66	23	20	18	8	8	7		22	20	4	1	–	7	–	–	–	–	–	–	–	Heptafluorobutyryl-codeine		NI44719
43	91	345	92	79	77	44	115	**495**	100	94	14	7	6	6	6	5	**0.00**	23	29	11	1	–	–	–	–	–	–	–	–	–	Glucose benzyloxime pentaacetate		MS10342
480	462	464	478	58	481	479	463	**495**	100	54	52	47	32	29	24	20	**15.00**	26	41	8	1	–	–	–	–	–	–	–	–	–	Aconitane-1,7,8,14-tetrol, 20-ethyl-6,16-dimethoxy-4-(methoxymethyl)-, 14-acetate, (1α,6β,14α,16β)-	50676-21-6	NI44720
161	277	203	146	118	435	263	162	**495**	100	72	57	48	37	26	24	22	**7.00**	27	33	6	3	–	–	–	–	–	–	–	–	–	Capparisinine	105694-42-6	NI44721
296	84	213	98	55	70	110	124	**495**	100	28	25	14	13	12	9	8	**2.00**	28	53	4	3	–	–	–	–	–	–	–	–	–	N,N-Diacetyl-oncinotinic acid, methyl ester		LI09398
296	297	84	100	55	110	98	53	**495**	100	21	9	7	6	5	5	5	**2.98**	28	53	4	3	–	–	–	–	–	–	–	–	–	2-Piperidineundecanoic acid, 1-[4-[acetyl[3-(acetylamino)propyl]amino]butyl]-, methyl ester, (R)-	53602-40-7	NI44722
196	126	223	81	222	95	83	71	**495**	100	62	50	44	38	38	35	32	**1.50**	29	53	5	1	–	–	–	–	–	–	–	–	–	1-Cyclohexylmethyl-3a,5-dihydroxy-4-(8,11-dihydroxy-4-methyl-undecyl)-6,7-dimethyl-3-oxo-3a,4,5,6,7-7a-hexahydro-isoindoline		IC04523
352	351	337	495	480	353	438	338	**495**	100	99	90	56	49	36	32	26		32	49	3	1	–	–	–	–	–	–	–	–	–	2-(3,5-Di-tert-butyl-2-hydroxyphenyl)-4,6-di-tert-butyl-6-morpholinocyclohexadienone		MS10343

MASS TO CHARGE RATIOS								M.W.	INTENSITIES								Parent	C	H	O	N	S	F	Cl	Br	I	Si	P	B	X	COMPOUND NAME	CAS Reg No	No
43	55	41	57	340	29	81	69	**495**	100	62	55	49	43	39	37	36	**35.66**	32	49	3	1	–	–	–	–	–	–	–	–	–	3′H-Cycloprop(1,2)-5-cholest-1-en-3-one, 1′-carboethoxy-1′-cyano-1,2-dihydro-	75857-80-6	NI44723
352	351	337	495	480	438	408	393	**495**	100	99	90	62	54	38	26	10		32	49	3	1	–	–	–	–	–	–	–	–	–	2,4-Dibutyl-2-morpholino-6-(1-hydroxy-2,4-dibutyl-phen-2-yl)-cyclohexa-3,5-dienone		LI09399
91	495	480	496	481	220	420	435	**495**	100	69	69	25	25	12	9	7		34	41	2	1	–	–	–	–	–	–	–	–	–	3β-Acetoxy-5α-androstano[17,16-b]N-benzylindole		MS10344
180	500	502	485	74	487	85	130	**496**	100	98	94	92	89	88	84	81	**10.00**	8	5	–	–	–	–	–	5	–	–	–	–	–	Benzene, pentabromoethyl-	85-22-3	NI44724
217	184	170	141	139	156	125	140	**496**	100	87	87	82	82	80	80	60	**0.00**	10	30	10	–	–	–	–	–	–	–	–	–	2	Diniobium decamethoxide		MS10345
382	354	270	210	326	240	298	225	**496**	100	86	58	48	45	41	39	35	**0.00**	12	12	8	–	–	–	–	–	–	–	2	–	2	1-Tetracarbonylchromium-2-tetracarbonylmolybdenum-tetramethyldiphosphine		NI44725
384	337	468	440	248	382	381	378	**496**	100	88	78	72	70	69	66	66	**61.00**	14	16	4	–	–	–	–	–	–	–	–	–	3	3[o-Phenylene-bis(dimethylarsino)]molybdenum-tetracarbonyl		LI09400
158	212	39	41	27	54	36	28	**496**	100	81	78	68	61	50	44	41	**0.00**	16	28	–	–	–	–	2	–	–	–	–	–	2	Rhodium, tetrakis(1,2,3-η)-2-butenyl-di-μ-chlorodi-	12308-60-0	NI44726
212	213	158	157	156	155	56	55	**496**	100	62	51	45	42	39	35	33	**1.03**	16	28	–	–	–	–	2	–	–	–	–	–	2	Rhodium, tetrakis(2-methylallyl)-di-μ-chlorodi-		LI09401
73	147	273	245	75	45	375	319	**496**	100	37	29	19	12	10	10	10	**0.00**	18	40	8	–	–	–	–	–	–	4	–	–	–	Trimethylsiloxyisocitric acid, tris(trimethylsilyl)-		NI44727
73	218	147	18	146	100	75	219	**496**	100	55	23	20	18	18	15	12	**0.00**	18	44	4	2	1	–	–	–	–	4	–	–	–	Tetra(trimethylsilyl)lanthionine		NI44728
157	97	139	259	225	335	199	496	**496**	100	96	93	74	68	54	54	51		20	24	9	4	1	–	–	–	–	–	–	–	–	4-Acetamido-2-methyl-6-(methylthio)-4-N-(2,3,4-tri-O-acetyl-β-D-xylopyranosylamino)oxazolo[5,4-d]pyrimidine		NI44729
60	47	59	61	58	46	55	53	**496**	100	86	81	58	53	47	34	34	**0.09**	20	32	10	–	2	–	–	–	–	–	–	–	–	D-Fructose, diethyl mercaptal, pentaacetate	7621-93-4	NI44730
135	47	60	75	59	73	61	46	**496**	100	99	75	66	64	63	61	45	**1.18**	20	32	10	–	2	–	–	–	–	–	–	–	–	D-Galactose, diethyl mercaptal, pentaacetate	6935-10-0	NI44731
73	217	147	218	103	319	75	74	**496**	100	82	28	19	13	11	11	10	**0.00**	20	48	6	–	–	–	–	–	–	4	–	–	–	β-D-Galactofuranoside, ethyl 2,3,5,6-tetrakis-O-(trimethylsilyl)-	55493-81-7	LI09402
73	217	147	218	103	75	74	319	**496**	100	87	28	18	12	12	10	9	**0.00**	20	48	6	–	–	–	–	–	–	4	–	–	–	β-D-Galactofuranoside, ethyl 2,3,5,6-tetrakis-O-(trimethylsilyl)-	55493-81-7	NI44732
382	107	69	91	105	93	41	383	**496**	100	41	40	38	37	32	31	30	**0.00**	23	26	5	–	–	6	–	–	–	–	–	–	–	Androst-5-en-16-one, 3,17-bis[(trifluoroacetyl)oxy]-, (3β,17β)-	56438-14-3	NI44733
151	259	225	258	301	257	361		**496**	100	42	28	25	15	12	10		**0.00**	23	28	4	–	4	–	–	–	–	–	–	–	–	Carbonodithioic acid, S,S′-[bis(4-methoxyphenyl)methylene] O,O′-bis(isopropyl) ester	26416-51-3	NI44734
331	169	43	127	97	139	115	78	**496**	100	99	97	95	87	86	84	82	**0.00**	23	28	12	–	–	–	–	–	–	–	–	–	–	Salicin, pentaacetate	16643-37-1	MS10346
169	109	331	127	139	115	145	110	**496**	100	96	40	29	21	21	17	13	**0.00**	23	28	12	–	–	–	–	–	–	–	–	–	–	Salicin, pentaacetate	16643-37-1	NI44735
43	159	109	173	296	331	374	356	**496**	100	95	69	62	51	38	33	30	**0.00**	25	36	10	–	–	–	–	–	–	–	–	–	–	Methyl 6α,7β-diacetoxy-12,16-dihydro-12,14-dihydroxy-16-oxovinhaticoate		NI44736
496	73	267	179	481	498	268	75	**496**	100	71	40	30	21	19	12	11		25	48	4	–	–	–	–	–	–	3	–	–	–	(3,4-Dihydroxyphenyl)decanoic acid, tris(trimethylsilyl)-		NI44737
164	138	136	182	121	107	44	109	**496**	100	90	89	75	73	73	73	72	**0.00**	26	24	10	–	–	–	–	–	–	–	–	–	–	4,2′-O-Dimethylgyrophoric acid		NI44738
165	150	346	122	182	151	196	315	**496**	100	82	70	58	30	30	19	12	**0.00**	26	24	10	–	–	–	–	–	–	–	–	–	–	Peltigerin		LI09403
165	122	151	182	164	94	136	332	**496**	100	93	89	59	36	36	13	4	**0.00**	26	24	10	–	–	–	–	–	–	–	–	–	–	Tenuiorin		LI09404
348	440	263	318	183	262	264	185	**496**	100	83	57	48	30	26	11	10	**7.99**	27	22	4	–	–	–	–	–	–	–	1	–	1	(Methoxycarbonylcyclopentadienyl)manganesedicarbonyltriphenylphosphine	33309-67-0	NI44739
316	226	334	317	117	173	131	187	**496**	100	41	32	23	23	17	17	14	**0.00**	27	44	8	–	–	–	–	–	–	–	–	–	–	Prost-13-en-1-oic acid, 9,11,15-tris(acetyloxy)-, methyl ester, (9α,11α,13E,15S)-	56890-05-2	NI44740
73	55	29	41	147	155	43	57	**496**	100	97	84	72	70	65	63	55	**0.00**	27	52	4	–	–	–	–	–	–	2	–	–	–	9,12,15-Octadecatrienoic acid, 2,3-bis[(trimethylsilyl)oxy]propyl ester, (Z,Z,Z)-	55521-22-7	WI01575
75	149	155	207	131	117	129	57	**496**	100	29	20	18	17	17	16	15	**0.00**	27	52	4	–	–	–	–	–	–	2	–	–	–	9,12,15-Octadecatrienoic acid, 2,3-bis[(trimethylsilyl)oxy]propyl ester, (Z,Z,Z)-	55521-22-7	WI01576
41	73	55	28	43	27	29	133	**496**	100	87	76	71	70	70	65	55	**0.00**	27	52	4	–	–	–	–	–	–	2	–	–	–	9,12,15-Octadecatrienoic acid, 2-[(trimethylsilyl)oxy]-1-[[(trimethylsilyl)oxy]methyl]ethyl ester, (Z,Z,Z)-	55521-23-8	WI01577
75	155	59	60	47	98	131	84	**496**	100	90	47	41	38	36	31	28	**0.00**	27	52	4	–	–	–	–	–	–	2	–	–	–	9,12,15-Octadecatrienoic acid, 2-[(trimethylsilyl)oxy]-1-[[(trimethylsilyl)oxy]methyl]ethyl ester, (Z,Z,Z)-	55521-23-8	WI01578
105	315	77	122	121	295	91	464	**496**	100	33	32	21	18	11	11	10	**0.00**	29	33	6	–	–	1	–	–	–	–	–	–	–	Pregna-1,4-diene-3,20-dione, 17-(benzoyloxy)-9-fluoro-11,21-dihydroxy-16-methyl-, (11β,16β)-	22298-29-9	NI44741
261	236	136	98	219	106	233	43	**496**	100	73	43	42	40	31	25	20	**0.00**	29	40	5	2	–	–	–	–	–	–	–	–	–	2,6-Di-tert-butyl-4-methoxyphenyl 2-acetamido-4-(N-phenylacetamido)butanoate		DD01744
41	60	45	44	43	42	56	55	**496**	100	70	70	70	70	70	15	10	**0.82**	30	40	6	–	–	–	–	–	–	–	–	–	–	Ergosta-2,24-dien-26-oic acid, 27-(acetyloxy)-5,6-epoxy-22-hydroxy-1-oxo, δ-lactone	55902-88-0	NI44742
139	83	95	138	81	55	69	57	**496**	100	50	33	31	28	25	18	16	**0.20**	30	57	3	–	–	–	–	–	–	–	1	–	–	Cyclohexanol, 5-methyl-2-isopropyl-, phosphite (3:1)	74793-74-1	NI44743
412	234	178	440	56	206	468	496	**496**	100	86	49	47	39	31	26	6		31	20	3	–	–	–	–	–	–	–	–	–	1	Tetraphenyl-cyclobutadiene-iron-tricarbonyl	31811-56-0	NI44744
146	43	55	57	170	68	79	74	**496**	100	97	77	69	48	47	43	43	**1.91**	31	60	4	–	–	–	–	–	–	–	–	–	–	Butanedioic acid, pentacosyl-, dimethyl ester	55712-67-9	NI44745
173	44	171	175	91	43	93	454	**496**	100	89	48	44	20	19	15	15	**1.06**	32	16	6	–	–	–	–	–	–	–	–	–	–	6-Acetoxy-11,12-epoxy-17-hydroxy-5,18-trinaphthylenedione		NI44746
230	332	201	229	202	200	164	151	**496**	100	90	75	50	45	40	25	20	**10.00**	32	20	4	2	–	–	–	–	–	–	–	–	–	1-(2,4-Dinitrostyryl)-4-(2-pyren-3-yl-vinyl)-benzene		LI09405

MASS TO CHARGE RATIOS	M.W.	INTENSITIES	Parent	C	H	O	N	S	F	Cl	Br	I	Si	P	B	X	COMPOUND NAME	CAS Reg No	No
360 227 135 252 361 240 121 239	**496**	100 96 60 36 30 30 30 28	**0.00**	32	32	5	–	–	–	–	–	–	–	–	–	–	**Furan, tetrahydro-2,3,4,5-tetrakis(4-methoxyphenyl)-**		**MS10347**
360 227 135 252 240 121 329 478	**496**	100 96 60 36 30 30 28 8	**0.00**	32	32	5	–	–	–	–	–	–	–	–	–	–	**Furan, tetrahydro-2,3,4,5-tetrakis(methoxyphenyl)-**	74367-04-7	**NI44747**
301 496 454 139 218 163 439 259	**496**	100 80 24 22 15 11 9 9		32	48	4	–	–	–	–	–	–	–	–	–	–	**D:A-Friedoolean-2-en-1-one, 3-(acetyloxy)-25,26-epoxy-**	56784-09-9	**MS10348**
496 367 43 437 421 69 497 368	**496**	100 97 93 83 74 43 38 38		32	48	4	–	–	–	–	–	–	–	–	–	–	**Lanosta-7,9(11)-dien-18-oic acid, 3-(acetyloxy)-20-hydroxy-, γ-lactone, (3β)-**	24041-70-1	**NI44748**
55 293 309 69 269 97 295 81	**496**	100 75 66 47 39 34 32 26	**7.00**	32	48	4	–	–	–	–	–	–	–	–	–	–	**Lanosta-7,9(11)-dien-21-oic acid, 16-hydroxy-24-methylene-3-oxo-, methyl ester (16α)-**		**MS10349**
43 57 55 71 41 69 83 85	**496**	100 96 67 57 55 43 41 37	**1.54**	32	64	3	–	–	–	–	–	–	–	–	–	–	**Palmitic acid, 2-(tetradecyloxy)ethyl ester**	28843-33-6	**NI44749**
57 43 71 82 55 83 85 69	**496**	100 85 61 52 51 46 43 43	**0.54**	32	64	3	–	–	–	–	–	–	–	–	–	–	**Palmitic acid, 2-(tetradecyloxy)ethyl ester**	28843-33-6	**NI44750**
57 43 71 82 239 83 85 55	**496**	100 72 69 65 56 56 52 49	**4.74**	32	64	3	–	–	–	–	–	–	–	–	–	–	**Palmitic acid, 2-(tetradecyloxy)ethyl ester**	28843-33-6	**NI44751**
69 73 75 55 253 496 95 129	**496**	100 80 42 42 38 32 27 26		33	56	1	–	–	–	–	–	–	1	–	–	–	**Lanosta-7,9(11),24-triene, 3-[(trimethylsilyl)oxy]-, (3β)-**	55538-96-0	**NI44752**
496 419 91 248 497 216 385 121	**496**	100 51 41 37 36 25 15 15		34	24	–	–	2	–	–	–	–	–	–	–	–	**2,2′,6,6′-Tetraphenyl-4,4′-di(thiopyranylidene)**		**NI44753**
220 149 105 102 248 391 348	**496**	100 99 40 39 36 7 4	**0.00**	34	24	4	–	–	–	–	–	–	–	–	–	–	**2,6-Diphenyl-4H-pyran-4-one dimer**		**LI09406**
193 249 496 219 233 204 179 74	**496**	100 61 59 57 26 18 18 10		34	24	4	–	–	–	–	–	–	–	–	–	–	**2,2′-Diphenyl[2.2](4,7)-1,3-indanedionophane**	88277-90-1	**NI44754**
436 95 175 421 203 314 496 297	**496**	100 69 43 42 37 35 28 25		34	56	2	–	–	–	–	–	–	–	–	–	–	**Cycloneolitsol acetate**		**LI09407**
73 497 498 482 103 499 217 237	**497**	100 99 40 35 35 24 21 19		19	35	3	5	1	–	–	–	–	3	–	–	–	**Tris(trimethylsilyl)-8,2′-thioanhydroadenosine**		**NI44755**
497 73 482 498 57 425 499 56	**497**	100 89 47 40 40 35 23 22		19	35	3	5	1	–	–	–	–	3	–	–	–	**Tris(trimethylsilyl)-8,3′-thioanhydroadenosine**		**NI44756**
142 149 355 164 190 127 356 165	**497**	100 57 55 39 24 23 22 22	**0.00**	22	28	4	1	–	–	–	–	1	–	–	–	–	**Tetrahydropalmatine methoiodide**		**NI44757**
300 302 105 185 183 77 301 303	**497**	100 99 60 31 31 22 20 17	**0.00**	23	17	2	1	–	–	–	2	–	–	–	–	–	**Acetophenone, 2-[4′-bromo-5-(p-bromophenyl)-3-phenyl-2-isoxazolin-5-yl]**	56804-90-1	**NI44758**
300 302 105 185 183 77 301 303	**497**	100 99 59 31 31 22 21 17	**0.00**	23	17	2	1	–	–	–	2	–	–	–	–	–	**Acetophenone, 4′-bromo-2-[3-(p-bromophenyl)-5-phenyl-2-isoxazolin-5-yl]**	21326-93-2	**NI44759**
105 122 77 51 50 78 106 74	**497**	100 81 81 40 24 13 10 10	**0.00**	25	23	10	1	–	–	–	–	–	–	–	–	–	**Tetramethyl 1-benzoyloxy-4,5-dihydropyrido[1,2-a]azepine-2,3,4,5-tetracarboxylate**		**NI44760**
263 247 149 236 123 264 150 248	**497**	100 99 99 87 84 66 24 23	**0.00**	30	43	3	1	1	–	–	–	–	–	–	–	–	**2,4,6-Tri-tert-butylphenyl 2-acetamido-4-(phenylthio)butanoate**		**DD01745**
219 203 161 220 218 41 466 175	**497**	100 18 18 17 12 10 9 8	**0.44**	32	51	3	1	–	–	–	–	–	–	–	–	–	**N,N-Bis(3,5-di-tert-butyl-4-hydroxybenzyl)ethanolamine**		**IC04524**
330 176 197 165 442 144 358 498	**498**	100 65 31 28 16 16 13 8		18	10	6	–	2	–	–	–	–	–	–	–	2	**Diiron, bis(μ-benzeneethiolato)hexacarbonyl-**		**MS10350**
43 153 151 169 115 109 157 140	**498**	100 28 28 14 11 10 8 8	**0.00**	18	27	11	–	–	–	–	1	–	–	–	–	–	**(1RS)-2-Bromo-1-ethoxyethyl 2,3,4,6-tetra-O-acetyl-β-D-glucopyranoside**		**MS10351**
259 103 217 223 169 147 243 115	**498**	100 91 77 68 48 48 40 37	**8.13**	20	38	5	4	–	–	–	–	–	3	–	–	–	**Inosine, 1-methyl-2′,3′,5′-tris-O-(trimethylsilyl)-**	55556-96-2	**NI44761**
88 101 45 75 71 57 85 55	**498**	100 78 70 54 20 20 19 19	**0.00**	21	38	13	–	–	–	–	–	–	–	–	–	–	**Methyl 2,3,6-tri-O-methyl-4-O-(methyl 2,3,4-tri-O-methyl-α-D-galactopyranosyluronate)-α-D-galactopyranoside**		**LI09408**
101 88 75 45 71 201 73 89	**498**	100 96 69 26 16 15 11 10	**0.00**	21	38	13	–	–	–	–	–	–	–	–	–	–	**Methyl 2,3,4-tri-O-methyl-6-O-(methyl 2,3,4-tri-O-methyl-β-D-glucopyranosyluronate)-β-D-galactopyranoside**		**LI09409**
101 201 88 45 75 71 85 73	**498**	100 58 50 45 41 19 14 12	**0.00**	21	38	13	–	–	–	–	–	–	–	–	–	–	**Methyl 3,4,6-tri-O-methyl-2-O-(methyl 2,3,4-tri-O-methyl-β-D-glucopyranosyluronate)-β-D-mannopyranoside**		**LI09410**
258 43 169 276 171 318 109 229	**498**	100 69 58 55 55 53 50 46	**0.59**	25	38	10	–	–	–	–	–	–	–	–	–	–	**Cyclopentanepropanoic acid, 3,5-bis(acetyloxy)-2-[3,7-bis(acetyloxy)-1-octenyl]-, methyl ester**	55760-01-5	**NI44762**
318 336 171 204 184 115 319 185	**498**	100 87 48 40 32 29 27 27	**0.00**	25	38	10	–	–	–	–	–	–	–	–	–	–	**Cyclopentanepropanoic acid, 3,5-bis(acetyloxy)-2-[3,8-bis(acetyloxy)-1-octenyl]-, methyl ester**	55760-00-4	**NI44763**
105 241 122 253 109 123 281 213	**498**	100 69 61 59 53 50 49 48	**23.00**	27	30	9	–	–	–	–	–	–	–	–	–	–	**12-Hydroxydaphnetoxin**	35302-50-2	**NI44764**
73 207 147 129 75 67 55 34	**498**	100 55 47 35 32 30 28 26	**0.00**	27	54	4	–	–	–	–	–	–	2	–	–	–	**9,12-Octadecadienoic acid (Z,Z)-, 2,3-bis[(trimethylsilyl)oxy]propyl ester**	54284-45-6	**WI01579**
129 262 131 75 203 395 103 305	**498**	100 70 41 38 36 35 33 31	**0.00**	27	54	4	–	–	–	–	–	–	2	–	–	–	**9,12-Octadecadienoic acid (Z,Z)-, 2,3-bis[(trimethylsilyl)oxy]propyl ester**	54284-45-6	**WI01580**
73 103 129 41 55 67 81 75	**498**	100 87 72 61 57 51 45 43	**0.00**	27	54	4	–	–	–	–	–	–	2	–	–	–	**9,12-Octadecadienoic acid (Z,Z)-, 2-[(trimethylsilyl)oxy]-1-[[(trimethylsilyl)oxy]methyl]ethyl ester**	54284-46-7	**WI01582**
262 129 409 81 218 263 336 131	**498**	100 71 61 30 27 25 22 21	**0.00**	27	54	4	–	–	–	–	–	–	2	–	–	–	**9,12-Octadecadienoic acid (Z,Z)-, 2-[(trimethylsilyl)oxy]-1-[[(trimethylsilyl)oxy]methyl]ethyl ester**	54284-46-7	**WI01581**
222 223 107 91 93 217 121 95	**498**	100 16 10 10 9 8 8 8	**4.70**	28	34	8	–	–	–	–	–	–	–	–	–	–	**(1′S,4′aS,7′R,8′aR)-7-[(7-acetoxydecahydro-5,5,8a-trimethyl-2-methylene-6-oxo-1-naphthalenyl)methoxy]-6,8-dimethoxy-2H-1-benzopyran-2-one**		**NI44765**
398 83 55 498 32 399 100 499	**498**	100 71 68 64 45 44 38 21		28	34	8	–	–	–	–	–	–	–	–	–	–	**Angeloylgomisin O**	83864-69-1	**NI44766**
398 83 498 399 55 100 499 181	**498**	100 99 82 57 40 32 27 20		28	34	8	–	–	–	–	–	–	–	–	–	–	**Angeloylisogomisin O**	83864-70-4	**NI44767**
315 43 44 95 41 316 28 149	**498**	100 74 45 30 29 26 25 19	**3.20**	28	34	8	–	–	–	–	–	–	–	–	–	–	**Epoxygedunin**		**MS10352**
498 247 262 483 219 499 261 263	**498**	100 67 60 54 43 32 30 27		28	34	8	–	–	–	–	–	–	–	–	–	–	**Uliginosin B**	19809-79-1	**LI09411**

MASS TO CHARGE RATIOS								M.W.	INTENSITIES								Parent	C	H	O	N	S	F	Cl	Br	I	Si	P	B	X	COMPOUND NAME	CAS Reg No	No
57	29	452	27	406	338	55	223	**498**	100	69	13	10	8	6	5	5	**4.52**	29	38	7	–	–	–	–	–	–	–	–	–	–	1′,1′-Dicarboethoxy-1β,2β-dihydro-17β-propionoxy-3′H-cycloprop[1,2]androsta-1,4,6-trien-3-one		NI44768
141	57	127	43	71	55	41	341	**498**	100	87	52	50	34	31	30	25	**0.00**	29	54	6	–	–	–	–	–	–	–	–	–	–	Dinonanoin monocaprylin		NI44769
129	236	262	77	103	91	105	237	**498**	100	75	45	43	35	30	17	16	**0.00**	30	22	2	6	–	–	–	–	–	–	–	–	–	3H-Pyrazol-3-one, 4,4′-azobis[2,4-dihydro-2,5-diphenyl-	56666-59-2	MS10353
180	165	498	151	137	193	122	470	**498**	100	86	68	19	15	11	10	7		30	42	6	–	–	–	–	–	–	–	–	–	–	1,12-Bis(3,5-dimethoxybenzoyl)dodecane		LI09412
81	43	109	211	55	378	95	93	**498**	100	91	83	67	56	53	50	47	**7.00**	31	46	5	–	–	–	–	–	–	–	–	–	–	Cholesta-9(11),20(22)-dien-23-one, 3,6-bis(acetyloxy)-, (3β,5α,6α)-	37717-06-9	NI44770
331	135	498	290	480	175	332	121	**498**	100	83	81	59	38	35	25	25		31	46	5	–	–	–	–	–	–	–	–	–	–	Methyl liquorate		LI09413
44	454	156	468	157	149	150	412	**498**	100	39	23	22	21	18	16	14	**2.01**	32	18	6	–	–	–	–	–	–	–	–	–	–	12,17-Dimethoxy-5,6,11,18-trinaphthylenetetrone		NI44771
442	441	443	470	221	498	415	220	**498**	100	35	31	14	12	8	7	6		32	22	4	2	–	–	–	–	–	–	–	–	–	N,N′-Dicyclobutyl-3,4:9,10-perylenebis(dicarboximide)	80509-46-2	MS10354
442	441	443	470	221	498	415	220	**498**	100	35	31	14	12	8	7	6		32	22	4	2	–	–	–	–	–	–	–	–	–	N,N′-Dicyclobutyl-3,4:9,10-perylenebis(dicarboximide)	80509-46-2	DD01746
453	189	203	163	191	177	190	175	**498**	100	40	22	21	17	15	14	14	**0.00**	32	50	4	–	–	–	–	–	–	–	–	–	–	3-Acetylmorolic acid		NI44772
189	438	248	498	202	249	219	262	**498**	100	48	31	21	21	19	10	7		32	50	4	–	–	–	–	–	–	–	–	–	–	Lup-20(29)-en-28-oic acid, 3-(acetyloxy)-, (3β)-		MS10355
81	43	211	109	378	363	55	71	**498**	100	89	76	72	63	58	56	53	**2.02**	32	50	4	–	–	–	–	–	–	–	–	–	–	Ergosta-9(11),20(22)-diene-3,6-diol, diacetate, (3β,5α,6α)-	56362-37-9	NI44773
153	498	287	423	439	451	331	275	**498**	100	85	85	60	33	25	25	25		32	50	4	–	–	–	–	–	–	–	–	–	–	D:A-Friedoolean-1-en-24-oic acid, 1-methoxy-3-oxo-, methyl ester	43230-21-3	MS10356
153	423	439	287	498	451	233	248	**498**	100	40	35	35	26	22	20	18		32	50	4	–	–	–	–	–	–	–	–	–	–	D:A-Friedoolean-2-en-24-oic acid, 3-methoxy-1-oxo-, methyl ester	43230-91-7	MS10357
498	499	43	69	55	57	41	302	**498**	100	37	27	18	17	14	14	13		32	50	4	–	–	–	–	–	–	–	–	–	–	Lanost-8-ene-7,11-dione, 3-(acetyloxy)-, (3β)-	2115-49-3	NI44774
55	69	295	81	95	67	105	83	**498**	100	68	61	48	46	36	34	33	**5.00**	32	50	4	–	–	–	–	–	–	–	–	–	–	Methyl 7,11-dehydrotumulosate		MS10358
248	203	249	438	190	189	423	204	**498**	100	51	24	22	17	16	12	10	**10.00**	32	50	4	–	–	–	–	–	–	–	–	–	–	Olean-12-en-28-oic acid, 3-(acetyloxy)-, (3β)-	4339-72-4	MS10359
357	417	411	498	483	455			**498**	100	80	66	60	10	6				32	50	4	–	–	–	–	–	–	–	–	–	–	3,4-Secolupa-4(23),20(29)-diene-3,28-dioic acid, dimethyl ester	35928-16-6	NI44775
498	499	399	483	456	414	413	400	**498**	100	46	43	28	26	16	14	13		32	54	2	–	–	–	–	–	–	1	–	–	–	Δ^1-Tetrahydrocannabinol, O-(tripropylsilyl)heptyl-		NI44776
159	160	146	498	147	175	340	174	**498**	100	19	15	4	4	3	2	2		32	54	2	2	–	–	–	–	–	–	–	–	–	N-Behenoyl-5-hydroxytryptamine		MS10360
368	353	43	369	69	107	89	81	**498**	100	35	35	34	30	29	29	27	**22.85**	32	55	3	–	–	–	–	–	–	–	–	1	–	1,3,2-Dioxaborinane, 2-[[(3β)-cholest-5-en-3-yl]oxy]-4,6-dimethyl-	55162-73-7	NI44777
43	109	95	55	81	69	107	121	**498**	100	92	92	92	91	90	81	72	**4.00**	33	54	3	–	–	–	–	–	–	–	–	–	–	9,19-Cyclolanost-23-en-3-ol, 25-methoxy-, acetate, (3β,23E)-	54411-94-8	NI44778
139	455	498	483	423	395	359	299	**498**	100	19	1	1	1	1	1	1		33	54	3	–	–	–	–	–	–	–	–	–	–	α-Dehydroalnincanol acetate		NI44779
139	455	483	498	423	395	359	299	**498**	100	14	2	1	1	1	1	1		33	54	3	–	–	–	–	–	–	–	–	–	–	β-Dehydroalnincanol acetate		NI44780
139	455	498	483	423	395	359	299	**498**	100	9	1	1	1	1	1	1		33	54	3	–	–	–	–	–	–	–	–	–	–	δ-Dehydroalnincanol acetate		NI44781
139	455	299	498	483	423	395	359	**498**	100	14	2	1	1	1	1	1		33	54	3	–	–	–	–	–	–	–	–	–	–	γ-Dehydroalnincanol acetate		NI44782
439	205	440	179	499	136	137	109	**498**	100	47	32	26	19	17	8	8	**2.00**	33	54	3	–	–	–	–	–	–	–	–	–	–	Lanost-7-en-11-one, 3-(acetyloxy)-7-methyl-, (3β)-		NI44783
69	73	95	55	81	109	107	93	**498**	100	70	46	44	38	36	33	31	**1.00**	33	58	1	–	–	–	–	–	–	1	–	–	–	9,19-Cyclolanost-24-ene, 3-[(trimethylsilyl)oxy]-, (3β,9β)-	17608-55-8	NI44784
498	499	253	393	240	408	500	483	**498**	100	45	41	23	23	15	14	11		33	58	1	–	–	–	–	–	–	1	–	–	–	Dihydroagnosterol trimethylsilyl ether		MS10361
393	483	498	394	95	484	97	109	**498**	100	76	37	34	32	31	30	28		33	58	1	–	–	–	–	–	–	1	–	–	–	Ergosta-8,24(28)-diene, 4,14-dimethyl-3-[(trimethylsilyl)oxy]-, [3β,4α,5α]	55622-52-1	NI44785
55	43	73	69	83	75	129	41	**498**	100	92	89	86	68	64	60	48	**13.86**	33	58	1	–	–	–	–	–	–	1	–	–	–	Gorgost-5-ene, 3-[(trimethylsilyl)oxy]-, (3β)-	55103-85-0	NI44786
69	393	75	55	95	109	81	129	**498**	100	48	42	39	37	31	26	25	**8.01**	33	58	1	–	–	–	–	–	–	1	–	–	–	Lanosta-9(11),24-diene, 3-[(trimethylsilyl)oxy]-, (3β)-	55538-95-9	NI44787
393	483	394	498	109	484	499	241	**498**	100	34	34	27	20	13	11	11		33	58	1	–	–	–	–	–	–	1	–	–	–	Lanost-8,24-diene, 3-[(trimethylsilyl)oxy]-, (3β)-	55493-84-0	NI44789
69	393	73	109	55	95	75	81	**498**	100	84	51	46	39	38	33	29	**18.02**	33	58	1	–	–	–	–	–	–	1	–	–	–	Lanost-8,24-diene, 3-[(trimethylsilyl)oxy]-, (3β)-	55493-84-0	NI44788
393	498	483	394	109	499	135	484	**498**	100	44	39	34	26	19	18	16		33	58	1	–	–	–	–	–	–	1	–	–	–	Lanost-8,24-diene, 3-[(trimethylsilyl)oxy]-, (3β)-	55493-84-0	MS10362
393	69	483	498	73	394	109	94	**498**	100	63	43	40	39	33	33	26		33	58	1	–	–	–	–	–	–	1	–	–	–	Lanosterol TMS ether		LI09414
393	69	483	498	73	109	93	55	**498**	100	63	43	40	39	33	26	25		33	58	1	–	–	–	–	–	–	1	–	–	–	Lanosterol TMS ether		LI09415
218	219	73	203	190	69	95	75	**498**	100	18	18	17	16	16	13	13	**2.50**	33	58	1	–	–	–	–	–	–	1	–	–	–	Olean-12-ene, 3-[(trimethylsilyl)oxy]-, (3β)-	1721-67-1	NI44790
91	398	363	391	272	364	399	365	**498**	100	40	32	23	18	13	11	10	**0.00**	34	30	2	2	–	–	–	–	–	–	–	–	–	3-(γ-Benzoylidene-N-benzylpropionamide)-N-benzyl-5-phenylpyrroline-2-one	104924-60-9	NI44791
498	119	91	69	41	145	55	157	**498**	100	68	57	48	41	40	38	37		35	46	2	–	–	–	–	–	–	–	–	–	–	Neurosporaxanthin		MS10363
498	91	69	133	43	406	109	83	**498**	100	57	48	31	27	17	17	14		35	46	2	–	–	–	–	–	–	–	–	–	–	Neurosporaxanthin		LI09416
69	219	131	92	100	119	47	330	**499**	100	28	24	22	11	10	9	8	**0.00**	9	–	1	1	–	19	–	–	–	–	–	–	–	N,N-Bis(nonafluorobutyl)carbamoyl fluoride		DD01747
69	169	92	119	280	100	47	131	**499**	100	47	26	21	15	13	12	10	**0.00**	9	–	1	1	–	19	–	–	–	–	–	–	–	N-(Undecafluoropentyl)-N-(heptafluoropropyl)carbamoyl fluoride		DD01748
119	320	145	117	352	499	177	118	**499**	100	87	70	38	35	24	23	22		16	7	6	1	–	10	–	–	–	–	–	–	–	Methyl 5,6-dihydroxyindole-2-carboxylate, bis(pentafluoropropionyl)-		NI44792
253	55	119	83	41	107	177	78	**499**	100	38	36	30	28	27	24	23	**10.00**	19	19	3	1	–	10	–	–	–	–	–	–	–	(4-Hydroxyphenyl)heptylamine, bis(pentafluoropropionyl)-		NI44793
174	114	212	214	311	313	270	272	**499**	100	18	17	11	7	5	4	3	**0.01**	20	19	6	3	1	–	2	–	–	–	–	–	–	5-(Hydroxymethyl) dicloxacillin methyl ester		MS10364
97	210	182	238	288	348	320	282	**499**	100	54	38	17	13	8	8	8	**4.00**	20	25	10	3	1	–	–	–	–	–	–	–	–	1,6-Dihydro-2-methylthio-5-((E)-2-methoxycarbonylvinyl)-4-(2,3,4-tri-O-acetyl-β-D-xylopyranosylamino)-1H-pyrimidin-6-one		MS10365
73	232	114	233	45	74	147	75	**499**	100	97	20	19	17	11	9	9	**0.00**	22	45	4	1	–	–	–	–	–	4	–	–	–	3,4-Dihydroxyphenyl-2-methylalanine tetra-TMS		MS10366
105	439	240	350	440	312	410	159	**499**	100	58	22	18	15	15	14	13	**5.00**	23	25	6	5	1	–	–	–	–	–	–	–	–	D-Xylitol, 2,5-anhydro-1-C-[6-(benzoylamino)-9H-purin-9-yl]-1-S-ethyl-1-thio-, 3,4-diacetate	40031-66-1	NI44794
178	378	215	164	179	163	134	162	**499**	100	21	16	14	12	8	8	5	**0.90**	24	33	5	7	–	–	–	–	–	–	–	–	–	Dimethylpuromycin		MS10367

MASS TO CHARGE RATIOS								M.W.	INTENSITIES								Parent	C	H	O	N	S	F	Cl	Br	I	Si	P	B	X	COMPOUND NAME	CAS Reg No	No
91	202	364	499	408	187	336	454	**499**	100	38	37	19	15	11	5	4		28	25	6	3	–	–	–	–	–	–	–	–	–	Ethyl 4-[N-(benzyloxycarbonyl)-N-(2-methyl-6,7-methylenedioxyquinazolin-4-ylmethyl)amino]benzoate	85591-07-7	NI44795
163	43	109	95	439	114	162	222	**499**	100	98	97	87	81	75	59	58	**36.70**	31	49	4	1	–	–	–	–	–	–	–	–	–	Spirosolan-3-ol, 28-acetyl-, acetate, (3β,5α,22β,25S)-	1181-86-8	NI44796
74	75	374	73	45	92	36	127	**500**	100	81	73	49	45	42	41	37	**0.00**	7	3	2	–	–	–	–	–	3	–	–	–	–	Benzoic acid, 2,3,5-triiodo-	88-82-4	NI44797
360	332	472	304	276	444	248	190	**500**	100	95	88	60	52	35	30	30	**5.00**	12	3	11	–	–	–	–	–	–	–	–	–	3	Tricobalt, nonacarbonyl[μ^3-(methoxyoxoethylidyne)]tri-, triangulo-	36834-85-2	NI44798
500	168	334	276	231	260	455	290	**500**	100	56	41	18	14	14	13	11		17	35	6	–	–	–	–	–	–	–	2	–	1	Cyclopentadienylrhodiumbis(triethylphosphite)	53556-33-5	NI44799
174	73	399	147	86	75	298	207	**500**	100	94	57	16	16	12	11	11	**0.00**	17	49	3	2	–	–	–	–	–	5	1	–	–	Phosphonic acid, [2-[bis(trimethylsilyl)amino]-1-[(trimethylsilyl)amino]ethyl]-, bis(trimethylsilyl) ester	53044-45-4	NI44800
174	73	399	400	175	147	86	45	**500**	100	94	57	21	20	16	16	12	**0.00**	17	49	3	2	–	–	–	–	–	5	1	–	–	Phosphonic acid, (1,2-diaminoethyl)-, pentakis(trimethylsilyl)-	56211-24-6	NI44801
57	41	355	29	341	43	27	56	**500**	100	65	65	57	50	49	45	34	**0.00**	18	48	6	–	–	–	–	–	–	5	–	–	–	1,3,5,7,9-Pentaethyl-1,9-dibutoxypentasiloxane	73420-35-6	NI44802
256	135	91	121	191	89	134	257	**500**	100	77	34	33	28	28	16	16	**10.00**	19	12	2	–	1	–	4	–	–	–	–	–	1	4,5,6,7-Tetrachloro-2a,8a-dihydro-2a-ferrocenyl-2H-thiacyclobuta(b)(1,4)benzodioxin		MS10368
196	454	139	157	88	224	412	199	**500**	100	71	42	36	30	9	9	7	**0.00**	19	24	10	4	1	–	–	–	–	–	–	–	–	1,6-Dihydro-1-methyl-2-methylthio-5-((E)-2-nitrovinyl)-4-(2,3,4-tri-O-acetyl-β-D-xylopyranosylamino)-1H-pyrimidin-6-one		MS10369
487	485	489	423	421	267	341	502	**500**	100	52	51	46	46	38	37	36	**18.00**	20	22	5	–	–	–	–	2	–	–	–	–	–	2H,8H-Benzo[1,2-b:3,4-b′]dipyran-6-propanoic acid, 3,9-dibromo-5-methoxy-2,2,8,8-tetramethyl-	56336-18-6	NI44803
150	220	403	274	431	136	232	192	**500**	100	34	27	25	24	23	11	11	**8.53**	20	26	4	4	–	6	–	–	–	–	–	–	–	N[6]-(4,6-Dimethylpyrimidin-2-yl)-N[2],N[6]-bis(trifluoroacetyl)lysine butyl ester		MS10370
229	257	256	349	348	275	377	289	**500**	100	85	85	70	65	30	25	20	**0.00**	23	29	10	–	–	–	1	–	–	–	–	–	–	Enhydrin chlorohydrin	38096-49-0	MS10371
241	166	242	167	140	240	151	139	**500**	100	57	24	23	11	9	9	8	**0.30**	24	18	1	2	–	–	–	–	–	–	–	–	2	Bis(5,10-dihydrophenarsazine)oxide	4095-45-8	NI44804
485	486	487	73	488	295	413	489	**500**	100	39	21	16	4	4	3	1	**0.00**	24	32	6	–	–	–	–	–	–	3	–	–	–	4,5-Dihydroxyanthraquinone-2-carboxylic acid, tris(trimethylsilyl)-	7336-81-4	NI44805
441	500	43	442	56	44	58	41	**500**	100	43	36	28	22	20	14	14		25	28	9	2	–	–	–	–	–	–	–	–	–	Dichotinamide, 11-methoxy-, acetate	29552-32-7	NI44806
260	191	320	152	108	178	99	153	**500**	100	84	79	60	55	53	49	44	**0.00**	25	40	10	–	–	–	–	–	–	–	–	–	–	Cyclopentanepropanoic acid, 3,5-bis(acetyloxy)-2-[3,7-bis(acetyloxy)octyl]-, methyl ester	55760-03-7	NI44807
320	191	152	166	321	155	115	43	**500**	100	49	28	25	23	23	23	22	**0.00**	25	40	10	–	–	–	–	–	–	–	–	–	–	Cyclopentanepropanoic acid, 3,5-bis(acetyloxy)-2-[3,8-bis(acetyloxy)octyl]-, methyl ester	55760-02-6	NI44808
201	77	282	43	262	51			**500**	100	71	67	51	42	36			**9.22**	26	21	3	–	–	3	–	–	–	–	2	–	–	Diphenylphosphinic acid, (1-diphenylphosphoryl-2,2,2-trifluoroethyl) ester		LI09417
201	77	282	262	423	29	499	325	**500**	100	64	60	38	32	23	21	19	**9.22**	26	21	3	–	–	3	–	–	–	–	2	–	–	Diphenylphosphinic acid, (1-diphenylphosphoryl-2,2,2-trifluoroethyl) ester		LI09418
55	134	444	180	415	400	243	320	**500**	100	40	32	30	26	18	18	16	**0.00**	26	25	3	2	1	–	–	–	–	–	–	–	1	Dicarbonyl(5-ethoxy-3-methyl-1,5-diphenyl-2-thioxo-4-imidazolidinylidene)(methylcyclopentadienyl)manganese	116928-10-0	DD01749
105	43	169	109	77	393	333	231	**500**	100	25	10	9	9	7	1	1	**0.00**	26	28	10	–	–	–	–	–	–	–	–	–	–	β-D-Glucopyranoside, 2-methylphenyl, 3,4,6-triacetate 2-benzoate	35558-60-2	MS10372
358	427	343	500	374	373	485		**500**	100	51	35	18	12	11	6			27	21	–	–	–	–	–	–	1	1	–	–	–	Naphthalene, 1-(2-bromovinyl)-8-(1-iodo-8-naphthylethynyl)-, trans-		LI09419
154	95	155	96	377	427	428	378	**500**	100	65	31	24	9	7	5	5	**1.30**	27	32	9	–	–	–	–	–	–	–	–	–	–	Limonoic acid, 19-deoxy-6,19-epoxy-, 16,17-lactone, methyl ester, (6β)-	1063-13-4	NI44809
185	183	136	300	123	502	121	500	**500**	100	99	77	46	44	43	43	42		27	33	4	–	–	–	–	1	–	–	–	–	–	Cleomeolide 4-bromobenzoate		MS10373
185	485	483	468	470	347	345	500	**500**	100	35	34	15	14	14	14	7		27	33	4	–	–	–	–	1	–	–	–	–	–	Pisiferdiol 7-(4-bromobenzoate)		DD01750
500	186	174	200	187	173	201	270	**500**	100	76	36	35	35	25	20	19		27	36	7	2	–	–	–	–	–	–	–	–	–	O,N-Diacetyl methyl 3,4-secoreserpate		LI09420
129	152	482	263	317	299	465	293	**500**	100	80	35	34	30	28	26	26	**0.00**	27	48	8	–	–	–	–	–	–	–	–	–	–	5α-Cholestane-3β,4β,6α,7α,8β,15α,16β,26-octaol		NI44810
73	55	129	41	43	103	147	69	**500**	100	54	50	41	38	34	32	31	**2.04**	27	56	4	–	–	–	–	–	–	2	–	–	–	Monoolein, bis(trimethylsilyl)-		NI44811
73	55	129	41	43	147	103	69	**500**	100	67	62	55	51	48	44	35	**1.00**	27	56	4	–	–	–	–	–	–	2	–	–	–	9-Octadecenoic acid (Z)-, 2,3-bis[(trimethylsilyl)oxy]propyl ester	54284-47-8	WI01584
129	130	203	131	117	397	221	75	**500**	100	60	55	52	50	49	48	48	**0.00**	27	56	4	–	–	–	–	–	–	2	–	–	–	9-Octadecenoic acid (Z)-, 2,3-bis[(trimethylsilyl)oxy]propyl ester	54284-47-8	WI01583
129	131	103	264	218	203	98	147	**500**	100	49	48	33	33	29	27	21	**0.00**	27	56	4	–	–	–	–	–	–	2	–	–	–	9-Octadecenoic acid (Z)-, 2-[(trimethylsilyl)oxy]-1-[[(trimethylsilyl)oxy]methyl]ethyl ester	54284-48-9	WI01585
73	103	55	41	147	43	131	69	**500**	100	74	68	62	52	52	44	39	**0.00**	27	56	4	–	–	–	–	–	–	2	–	–	–	9-Octadecenoic acid (Z)-, 2-[(trimethylsilyl)oxy]-1-[[(trimethylsilyl)oxy]methyl]ethyl ester	54284-48-9	WI01586
73	103	129	41	55	67	81	75	**500**	100	87	72	61	57	51	45	43	**0.00**	27	56	4	–	–	–	–	–	–	2	–	–	–	9-Octadecenoic acid, 2-[(trimethylsilyl)oxy]-1-[[(trimethylsilyl)oxy]methyl]ethyl ester	56554-42-8	NI44812
221	500	264	457	165	222	501	265	**500**	100	60	48	44	34	14	12	12		28	36	8	–	–	–	–	–	–	–	–	–	–	Dihydrouliginosin B		LI09421
165	221	500	264	43	41	193	40	**500**	100	86	73	72	64	50	48	38		28	36	8	–	–	–	–	–	–	–	–	–	–	Uliginosin A		LI09422
114	97	86	85	84	41	135	42	**500**	100	2	2	2	2	2	1	1	**0.03**	28	44	4	4	–	–	–	–	–	–	–	–	–	Frangulanine		MS10374
114	72	97	53	135	86	58	51	**500**	100	7	4	4	3	3	3	3	**0.10**	28	44	4	4	–	–	–	–	–	–	–	–	–	Frangulanine		LI09423
114	115	41	85	97	86	84	42	**500**	100	71	21	19	17	17	17	15	**0.30**	28	44	4	4	–	–	–	–	–	–	–	–	–	Frangulanine		LI09424

MASS TO CHARGE RATIOS								M.W.	INTENSITIES								Parent	C	H	O	N	S	F	Cl	Br	I	Si	P	B	X	COMPOUND NAME	CAS Reg No	No
91	43	60	427	121	444	458	440	**500**	100	88	83	67	50	46	31	29	**0.00**	31	20	1	2	2	–	–	–	–	–	–	–	–	**5,6,8,8b-Tetraphenyl-5H,8bH-thieno[3′,4′:3,4]pyrrolo[2,1-b]-1,3,4-thiadiazol-4-ium-2-olate**	110315-34-9	**DD01751**
109	43	82	44	81	69	93	41	**500**	100	94	91	89	76	76	61	56	**0.00**	31	48	5	–	–	–	–	–	–	–	–	–	–	**Cholesta-9(11),24-diene-3,6,20-triol, 3,6-diacetate, (3β,5α,6α,20R)-**	56312-50-6	**NI44813**
116	43	158	145	45	74	142	98	**500**	100	95	60	43	43	42	39	38	**0.00**	31	48	5	–	–	–	–	–	–	–	–	–	–	**Cholest-8(14)-en-3-one, 7,15-bis(acetyloxy)-, (5α,7α,15α)-**	69140-11-0	**NI44814**
217	371	254	396	218	129	191	185	**500**	100	48	47	41	24	22	21	18	**0.00**	31	48	5	–	–	–	–	–	–	–	–	–	–	**Cholest-8(14)-en-3-one, 7,15-bis(acetyloxy)-, (5α,7α,15α)-**	69140-11-0	**NI44815**
203	262	189	202	119	106	133	108	**500**	100	33	33	24	21	21	20	17	**0.00**	31	48	5	–	–	–	–	–	–	–	–	–	–	**A-Homo-3-oxaolean-12-en-28-oic acid, 2-hydroxy-23-oxo-, methyl ester**	56362-30-2	**NI44816**
55	43	41	69	105	191	95	81	**500**	100	95	94	85	83	45	42	39	**0.00**	32	52	–	–	2	–	–	–	–	–	–	–	–	**1,3-Dithiolane, 2-(28-norurs-12-en-17-yl)-**	10153-89-6	**NI44817**
105	69	41	55	43	95	81	191	**500**	100	71	68	59	55	50	49	47	**0.98**	32	52	–	–	2	–	–	–	–	–	–	–	–	**Olean-12-en-28-al, cyclic 1,2-ethanediyl mercaptal**	54446-81-0	**NI44818**
349	440	289	175	316	189	425	203	**500**	100	77	60	60	56	46	45	45	**3.00**	32	52	4	–	–	–	–	–	–	–	–	–	–	**3α-Acetoxyfriedelan-26-oic acid**		**NI44819**
349	440	175	289	316	189	425	203	**500**	100	77	66	60	56	46	45	45	**3.00**	32	52	4	–	–	–	–	–	–	–	–	–	–	**3β-Acetoxyfriedelan-26-oic acid**		**NI44820**
189	121	105	119	109	407	173	133	**500**	100	91	78	70	70	64	64	61	**1.00**	32	52	4	–	–	–	–	–	–	–	–	–	–	**3β-Acetoxy-11β-hydroxy-19(10-9β)abeo-lanost-1(10)-en-7-one**		**NI44821**
123	422	195	203	407	318	189	339	**500**	100	78	38	34	23	22	14	12	**0.00**	32	52	4	–	–	–	–	–	–	–	–	–	–	**11α-Acetoxy-3-keto-lupan-20-ol**	79081-61-1	**NI44822**
43	365	44	440	41	380	55	366	**500**	100	55	42	32	29	28	24	22	**0.00**	32	52	4	–	–	–	–	–	–	–	–	–	–	**5α-Cholest-8-ene-3β,6α-diol, 14-methyl-, diacetate**	5259-11-0	**NI44823**
112	454	357	207	455	246	189	292	**500**	100	31	27	17	12	12	8	3	**1.00**	32	52	4	–	–	–	–	–	–	–	–	–	–	**Cyclamigenin A**		**LI09425**
112	357	454	207	455	246	189	292	**500**	100	79	27	16	14	8	7	4	**2.00**	32	52	4	–	–	–	–	–	–	–	–	–	–	**Cyclamigenin C**		**LI09426**
43	69	55	41	57	95	500	121	**500**	100	64	59	59	51	44	39	34		32	52	4	–	–	–	–	–	–	–	–	–	–	**Lanostane-7,11-dione, (acetyloxy)-**		**MS10375**
277	43	41	69	55	136	57	95	**500**	100	85	52	48	45	43	37	35	**14.90**	32	52	4	–	–	–	–	–	–	–	–	–	–	**Lanostane-7,11-dione, (acetyloxy)-, (13α,14β,17α)-**		**MS10376**
500	43	69	95	55	277	57	41	**500**	100	77	66	56	55	48	47	45		32	52	4	–	–	–	–	–	–	–	–	–	–	**Lanostane-7,11-dione, 3-(acetyloxy)-, (3β)-**	6593-12-0	**NI44824**
43	69	343	95	347	55	81	135	**500**	100	64	57	48	47	46	41	39	**31.68**	32	52	4	–	–	–	–	–	–	–	–	–	–	**Lanostan-18-oic acid, 3β,11β-dihydroxy-, γ-lactone, acetate**	24041-80-3	**NI44825**
467	55	69	81	109	67	95	107	**500**	100	96	87	69	53	49	46	44	**29.00**	32	52	4	–	–	–	–	–	–	–	–	–	–	**Methyl polyporenate A**		**MS10377**
99	500	81	109	123	153	485	456	**500**	100	85	32	20	16	15	12	8		32	52	4	–	–	–	–	–	–	–	–	–	–	**3-Oxofriedelan-26-oic acid ethylene acetal**		**NI44826**
43	84	56	69	42	41	86	57	**500**	100	80	73	69	65	64	62	59	**2.75**	32	52	4	–	–	–	–	–	–	–	–	–	–	**Lup-20(29)-ene-3,21,28-triol, 28-acetate, (3β,21β)-**	55401-94-0	**NI44827**
115	69	215	95	217	216	107	41	**500**	100	87	67	55	54	51	51	50	**48.51**	32	57	3	–	–	–	–	–	–	–	–	1	–	**1,3,2-Dioxaborinane, 2-[[(3β,5α)-cholestan-3-yl]oxy]-4,6-dimethyl-**	55162-72-6	**NI44828**
87	43	41	55	74	57	113	69	**500**	100	76	69	65	40	40	38	35	**0.00**	33	56	3	–	–	–	–	–	–	–	–	–	–	**Cholestan-15-one, 3-(acetyloxy)-14-butyl-, (3β,5α)-**	74420-87-4	**NI44829**
69	43	28	75	41	73	55	81	**500**	100	94	93	91	85	83	83	80	**49.50**	33	60	1	–	–	–	–	–	–	1	–	–	–	**Cholest-22-en-3-ol, 4,23,24-trimethyl-3-[(trimethylsilyl)oxy]-, (3β,4α,5α)-**		**MS10378**
471	103	472	171	75	95	87	473	**500**	100	83	40	33	33	14	13	12	**3.00**	33	60	1	–	–	–	–	–	–	1	–	–	–	**Cholesterol, (triethylsilyl)-**	7604-85-5	**LI09427**
471	103	472	171	75	30	57	87	**500**	100	83	41	33	33	30	14	13	**2.00**	33	60	1	–	–	–	–	–	–	1	–	–	–	**Cholesterol, (triethylsilyl)-**	7604-85-5	**NI44830**
471	103	472	75	171	95	87	473	**500**	100	82	42	33	32	14	13	12	**3.30**	33	60	1	–	–	–	–	–	–	1	–	–	–	**Cholesterol, (triethylsilyl)-**	7604-85-5	**NI44831**
395	396	485	500	135	486	501	227	**500**	100	35	30	24	14	13	10	7		33	60	1	–	–	–	–	–	–	1	–	–	–	**Lanost-8-en-3-ol, 3-[(trimethylsilyl)oxy]-, (3β)-**	18985-29-0	**MS10379**
395	396	485	500	486	135	227	95	**500**	100	33	24	11	10	9	7	6		33	60	1	–	–	–	–	–	–	1	–	–	–	**Lanost-8-en-3-ol, 3-[(trimethylsilyl)oxy]-, (3β)-**	18985-29-0	**NI44832**
91	410	28	105	132	501	69	92	**501**	100	50	33	29	14	13	9	8		16	10	–	1	–	15	–	–	–	–	–	–	–	**2-Phenyl-N-(pentadecafluoroheptylmethylene)ethylamine**		**MS10380**
137	266	139	453	152	267	138	268	**501**	100	61	38	35	34	21	21	20	**7.00**	23	18	–	5	1	–	3	–	–	–	–	–	–	**1-[N,N′-Bis(4-chlorophenyl)]guanidino-4-4-chlorophenyl-2-methylthioimidazole**		**MS10381**
140	152	98	457	224	381	373	440	**501**	100	97	46	33	15	13	13	5	**1.00**	23	35	11	1	–	–	–	–	–	–	–	–	–	**1-(N-Cyclopentylacetamido)-1-desoxy-2,3,4,5,6-penta-O-acetyl-D-galactitol**		**NI44833**
152	140	457	224	153	381	373	458	**501**	100	78	24	14	11	10	9	7	**0.30**	23	35	11	1	–	–	–	–	–	–	–	–	–	**1-(N-Cyclopentylacetamido)-1-desoxy-2,3,4,5,6-penta-O-acetyl-D-mannitol**		**NI44834**
152	140	98	224	153	381	373	440	**501**	100	80	29	14	10	9	9	4	**0.30**	23	35	11	1	–	–	–	–	–	–	–	–	–	**1-(N-Cyclopentylacetamido)-1-desoxy-2,3,4,5,6-penta-O-acetyl-D-sorbitol**		**NI44835**
60	155	71	45	95	43	41	31	**501**	100	65	65	65	60	54	54	45	**0.00**	25	43	9	1	–	–	–	–	–	–	–	–	–	**Pederone**		**MS10382**
60	155	71	45	95	43	31	89	**501**	100	65	65	65	60	54	45	43	**0.00**	25	43	9	1	–	–	–	–	–	–	–	–	–	**Pederone**		**LI09428**
105	77	501	500	355	91	178	424	**501**	100	83	66	60	56	53	46	42		28	21	1	1	–	6	–	–	–	–	–	–	–	**5,5-Diphenyl-4-[3,3,3-trifluoro-2-(trifluoromethyl)-1-propylidene]-3-(2,4,6-trimethylphenyl)-2-isoxazoline**		**DD01752**
228	242	153	411	229	227	472	382	**501**	100	76	44	41	31	29	23	23	**17.37**	28	47	3	1	–	–	–	–	–	2	–	–	–	**DL-Norgestrel, 16,17-bis[(trimethylsilyl)oxy]-, O-methyloxime, (16β)-**		**NI44836**
320	500	321	501	103	307	91	410	**501**	100	25	20	9	6	5	5	4		31	23	4	3	–	–	–	–	–	–	–	–	–	**2-(2,3-Diphenylpropyl)-1,3,5,10-tetrahydro-4,11-diamino-1,3,5,10-tetroxo-2H-naphtha[2,3-f]isoindole**		**IC04525**
127	375	502	377	121	504	248	123	**502**	100	43	36	32	29	26	24	22		–	–	–	–	–	–	–	–	3	–	–	–	1	**Antimony triiodide**	7790-44-5	**NI44837**
143	305	124	123	333	93	59	39	**502**	100	10	10	9	4	4	4	4	**0.17**	13	5	1	–	–	14	–	–	–	–	–	–	1	**Cobalt, (π-cyclopentadienyl)(heptafluoroprop-1-enyl)(heptafluoropropyl)carbonyl**		**NI44838**
473	474	475	443	476	384	222	41	**502**	100	39	22	19	11	11	11	10	**1.89**	13	34	9	–	–	–	–	–	–	6	–	–	–	**1,3,5,7,9,11-Hexaethyl-5-methoxytricyclo [5.5.1.1^{3,11}]hexasiloxane**		**NI44839**
206	362	364	360	358	290	276	292	**502**	100	59	55	54	46	32	31	30	**21.55**	14	22	4	–	2	–	–	–	–	–	–	–	1	**Tungsten, tetracarbonyl(2,2,7,7-tetramethyl-3,6-dithiaoctane)**		**LI09429**
57	291	389	302	447	300	293	335	**502**	100	23	16	15	10	10	9	7	**1.00**	16	36	–	–	1	–	–	–	–	–	–	–	2	**tert-Butylstibinous acid thioanhydride**	80536-71-6	**NI44840**
428	429	431	119	415	457	281	265	**502**	100	23	22	18	14	12	10	8	**8.40**	17	12	6	–	–	10	–	–	–	–	–	–	–	**3,4-Dihydroxyphenylpropionic acid ethyl ester, bis(pentafluoropropionyl)**		**MS10383**

MASS TO CHARGE RATIOS	M.W.	INTENSITIES	Parent	C	H	O	N	S	F	Cl	Br	I	Si	P	B	X	COMPOUND NAME	CAS Reg No	No
73 299 81 315 147 227 217 211	**502**	100 42 40 35 35 27 27 23	**0.00**	17	43	7	–	–	–	–	–	–	4	1	–	–	**D-erythro-Pentofuranose, 2-deoxy-1,3-bis-O-(trimethylsilyl)-, bis(trimethylsilyl) phosphate**	69688-43-3	**NI44841**
296 390 446 502 474 370 418	**502**	100 96 50 34 23 23 20		20	10	4	–	–	4	–	–	–	–	–	–	2	**Diiron, tetracarbonylbis(η^5-2,4-cyclopentadien-1-yl)[μ-(2,3,5,6-tetrafluoro 1,4-phenylene)]**	37366-22-6	**NI44842**
296 426 390 474 502 398 370 418	**502**	100 27 25 19 18 15 10 3		20	10	4	–	–	4	–	–	–	–	–	–	2	**Diiron, tetracarbonylbis(η^5-2,4-cyclopentadien-1-yl)[μ-(2,4,5,6-tetrafluoro 1,3-phenylene)]**	37355-34-3	**NI44843**
55 69 41 43 135 153 83 18	**502**	100 93 77 45 40 39 38 31	**17.63**	20	38	4	–	–	–	–	–	–	–	–	–	2	**10,10′-Diselenodidecanoic acid**	22686-34-6	**NI44844**
179 73 292 166 180 41 29 293	**502**	100 54 51 39 16 16 14 12	**0.00**	23	33	5	2	–	3	–	–	–	1	–	–	–	**N-Trifluoroacetyl-L-Prolyltyrosine butyl ester, trimethylsilyl-**		**MS10384**
243 168 502 244 139 503 169 103	**502**	100 58 31 30 17 9 8 4		24	16	3	–	–	–	–	–	–	–	–	–	2	**Bis(phenoxarsin-10-yl) ether**	58-36-6	**NI44845**
502 470 351 471 438 439 45 219	**502**	100 76 47 46 37 35 34 27		24	22	8	–	2	–	–	–	–	–	–	–	–	**Tetramethyl 2,2′-(1,4-phenylene)bis(5-methyl-3,4-thiophenedicarboxylate)**	110699-33-7	**DD01753**
224 223 105 192 77 113 59 327	**502**	100 87 85 54 40 26 21 20	**0.71**	24	26	10	2	–	–	–	–	–	–	–	–	–	**4-Benzoylamino-2-(1,2,3,4-tetramethoxycarbonyl-2-butylidene)amino-2-buten-4-one**		**NI44846**
254 217 412 255 143 325 291 228	**502**	100 72 40 20 20 18 15 15	**1.00**	24	46	7	–	–	–	–	–	–	2	–	–	–	**Cyclopentaneoctanoic acid, 2-(3-methoxy-3-oxopropyl)-ϵ-oxo-3,5-bis[(trimethylsilyl)oxy]-, methyl ester, [1R-(1α,2β,3β,5α)]-**	55670-02-5	**NI44847**
127 77 142 128 115 170 55 214	**502**	100 23 10 9 8 5 4 2	**0.00**	25	27	5	2	1	–	1	–	–	–	–	–	–	**trans-1-(Chloroacetyl)-5-ethyl-2-[1-(phenylsulphonyl)-2-indolyl]-4-piperidone ethylene acetal**		**DD01754**
442 401 44 58 43 383 70 42	**502**	100 84 62 49 43 40 31 27	**0.00**	25	30	9	2	–	–	–	–	–	–	–	–	–	**Dichotine, 16,19-deepoxy-14,19-didehydro-16,19-dihydroxy-11-methoxy-, 2-acetate, (14Z,16β)-**	55283-45-9	**NI44848**
443 401 43 44 58 444 383 57	**502**	100 56 47 34 30 26 25 17	**0.00**	25	30	9	2	–	–	–	–	–	–	–	–	–	**Dichotine, 16,19-deepoxy-14,19-didehydro-16,19-dihydroxy-11-methoxy-, 2-acetate, (14Z,16β)-**	55283-45-9	**NI44849**
339 355 398 341 383 251 43 44	**502**	100 60 57 32 31 24 22 20	**4.90**	25	30	9	2	–	–	–	–	–	–	–	–	–	**Dichotine, 19-hydroxy-11-methoxy-, 2-acetate**	29484-56-8	**NI44850**
209 73 179 210 502 193 180 223	**502**	100 70 43 28 19 12 10 7		26	38	6	–	–	–	–	–	–	2	–	–	–	**(3R)-trans-Dihydro-3,4-bis[(4-hydroxy-3-methoxyphenyl)methyl]-2(3H)-furanone, bis(trimethylsilyl)-**		**NI44851**
73 275 173 185 71 75 55 276	**502**	100 45 26 17 16 11 10 9	**0.00**	26	54	5	–	–	–	–	–	–	2	–	–	–	**Methyl (Z)-12,13-bis(trimethylsilyloxy)-11-methoxy-9-octadecenoate**	88303-88-2	**NI44852**
502 43 91 93 79 94 55 442	**502**	100 86 22 21 19 17 17 14		27	34	9	–	–	–	–	–	–	–	–	–	–	**Di-O-acetylsarverogenin**		**LI09430**
105 85 108 44 81 110 80 41	**502**	100 96 52 51 49 44 43 42	**10.00**	27	34	9	–	–	–	–	–	–	–	–	–	–	**Verrucarin A**	3148-09-2	**NI44853**
412 296 413 322 281 307 242 295	**502**	100 45 36 28 28 25 22 17	**6.95**	27	46	3	–	–	–	–	–	–	3	–	–	–	**Estra-1,3,5(10),6-tetraene, 3,11,17-tris[(trimethylsilyl)oxy]-, (11β,17β)-**	74298-80-9	**NI44854**
399 73 43 57 147 400 129 55	**502**	100 76 52 42 37 30 30 28	**0.00**	27	58	4	–	–	–	–	–	–	2	–	–	–	**Octadecanoic acid, 2,3-bis[(trimethylsilyl)oxy]propyl ester**	1188-75-6	**NI44855**
73 399 43 147 57 129 55 41	**502**	100 92 68 61 59 39 39 33	**0.00**	27	58	4	–	–	–	–	–	–	2	–	–	–	**Octadecanoic acid, 2,3-bis[(trimethylsilyl)oxy]propyl ester**	1188-75-6	**NI44856**
399 400 487 383 401 473 488 218	**502**	100 31 17 15 9 8 7 4	**0.00**	27	58	4	–	–	–	–	–	–	2	–	–	–	**Octadecanoic acid, 2,3-bis[(trimethylsilyl)oxy]propyl ester**	1188-75-6	**WI01587**
218 487 219 488 397 220 347 267	**502**	100 33 20 14 10 9 7 7	**0.00**	27	58	4	–	–	–	–	–	–	2	–	–	–	**Octadecanoic acid, 2-[(trimethylsilyl)oxy]-1-[[(trimethylsilyl)oxy]methyl]ethyl ester**	53336-13-3	**WI01588**
129 73 218 103 147 43 57 55	**502**	100 82 76 52 46 45 38 27	**0.00**	27	58	4	–	–	–	–	–	–	2	–	–	–	**Octadecanoic acid, 2-[(trimethylsilyl)oxy]-1-[[(trimethylsilyl)oxy]methyl]ethyl ester**	53336-13-3	**WI01589**
73 399 57 147 55 129 400 71	**502**	100 85 54 40 32 31 28 25	**0.00**	27	58	4	–	–	–	–	–	–	2	–	–	–	**Bis(trimethylsilyl)monostearin**	68000-49-7	**NI44857**
73 129 57 218 43 103 55 147	**502**	100 91 60 58 56 52 38 37	**0.00**	27	58	4	–	–	–	–	–	–	2	–	–	–	**O,O″-Trimethylsilyl-O′-stearoylglycerol**		**IC04526**
178 91 183 77 262 113 185 108	**502**	100 70 69 54 50 47 39 34	**24.00**	28	25	–	–	–	–	–	–	–	–	3	–	1	**Titanium, bis(π-cyclopentadienyl)(1,2,3-triphenyltriphosphonato-P1,P3)-**		**MS10385**
43 267 29 327 41 57 107 161	**502**	100 49 37 32 31 31 29 29	**4.33**	29	42	7	–	–	–	–	–	–	–	–	–	–	**17β-Acetoxy-1′,1′-dicarboethoxy-1β,2β-dihydro-17α-methyl-3′H-cycloprop[1,2]-5α-androst-1-en-3-one**		**NI44858**
93 77 119 172 183 320 502 91	**502**	100 45 36 18 16 15 14 12		30	22	4	4	–	–	–	–	–	–	–	–	–	**[4,4′-Bipyrazolidine]-3,3′,5,5′-tetrone, 1,1′,2,2′-tetraphenyl-**	3474-99-5	**NI44859**
189 139 156 111 246 243 190 141	**502**	100 61 56 36 31 31 25 22	**22.02**	30	30	7	–	–	–	–	–	–	–	–	–	–	**Coumarin, 6,6′-[(3,6-dimethyl-7-oxabicyclo[4.1.0]hept-3,2-ylene)ethylene]bis[7-methoxy-**	28753-30-2	**NI44860**
184 502 413 131 415 412 414 372	**502**	100 80 65 58 39 39 38 28		30	34	5	2	–	–	–	–	–	–	–	–	–	**Aspidospermidin-21-oic acid, 20-hydroxy-17-methoxy-1-(1-oxo-3-phenyl-2 propenyl)-, methyl ester, [1(E)]-**	56143-43-2	**NI44861**
248 203 249 204 189 502 233 202	**502**	100 75 27 19 14 10 10 8		30	46	6	–	–	–	–	–	–	–	–	–	–	**Medicagenic acid**	599-07-5	**NI44862**
179 146 442 205 55 201 119 219	**502**	100 50 31 30 30 27 24 21	**9.76**	31	50	5	–	–	–	–	–	–	–	–	–	–	**Urs-12-en-28-oic acid, 2,3,19-trihydroxy-, methyl ester, (2α,3β)-**	13850-15-2	**NI44863**
425 487 407 502 453 205 335 443	**502**	100 61 60 49 49 33 31 27		31	50	5	–	–	–	–	–	–	–	–	–	–	**25-Carbomethoxy-fern-$\Delta^{9,11}$-ene-2,3,23-triol**		**NI44864**
383 201 443 203 165 257 384 503	**502**	100 78 53 49 41 40 29 28	**0.50**	31	50	5	–	–	–	–	–	–	–	–	–	–	**Cholestan-2-one, 3,6-bis(acetyloxy)-**		**NI44865**
255 297 43 71 382 105 95 81	**502**	100 80 57 56 39 35 35 35	**3.80**	31	50	5	–	–	–	–	–	–	–	–	–	–	**Cholestan-24-one, 3,12-bis(acetyloxy)-, (3α,5β,12α)-**	56052-67-6	**NI44866**
43 44 364 81 69 349 211 95	**502**	100 83 76 70 67 62 54 52	**0.00**	31	50	5	–	–	–	–	–	–	–	–	–	–	**Cholest-9(11)-ene-3,6,20-triol, 3,6-diacetate, (3β,5α,6α,20R)-**	56312-51-7	**NI44867**
262 203 263 249 204 202 250 205	**502**	100 68 23 14 11 10 4 4	**3.00**	31	50	5	–	–	–	–	–	–	–	–	–	–	**Methyl 2α,3β,23-trihydroxyurs-12-en-28-oic acid, 2,3,23-trihydroxy-, methyl ester, (2α,3β)**	20016-63-1	**NI44868**
262 203 133 249 189 502 442 466	**502**	100 71 22 14 12 3 3 2		31	50	5	–	–	–	–	–	–	–	–	–	–	**Methyl 2α,3β,23-trihydroxyurs-12-en-28-oic acid, 2,3,23-trihydroxy-, methyl ester, (2α,3β)**	20016-63-1	**MS10386**
43 55 41 69 95 81 189 114	**502**	100 55 47 45 43 42 35 34	**10.10**	31	50	5	–	–	–	–	–	–	–	–	–	–	**30-Norlupan-28-oic acid, 3-hydroxy-21-methoxy-20-oxo-, methyl ester, (3β)-**	55401-91-7	**NI44869**

MASS TO CHARGE RATIOS								M.W.	INTENSITIES								Parent	C	H	O	N	S	F	Cl	Br	I	Si	P	B	X	COMPOUND NAME	CAS Reg No	No
55	41	246	201	260	133	442	200	**502**	100	99	40	28	24	20	11	7	**2.00**	31	50	5	–	–	–	–	–	–	–	–	–	–	Olean-12-en-28-oic acid, 2,3,21-trihydroxy-, methyl ester, (2α,3β,21β)-		LI09431
262	203	202	189	133	204			**502**	100	99	37	37	37	31			**14.62**	31	50	5	–	–	–	–	–	–	–	–	–	–	Olean-12-en-28-oic acid, 2,3,23-trihydroxy-, methyl ester, (2α,3β,4α)-	22452-82-0	MS10387
203	149	262	105	150	207	189	106	**502**	100	78	59	58	47	42	35	29	**0.00**	31	50	5	–	–	–	–	–	–	–	–	–	–	Olean-12-en-28-oic acid, 2,3,23-trihydroxy-, methyl ester, (2α,3β,4α)-	22452-82-0	NI44870
203	262	189	249	133	407	442	502	**502**	100	73	17	11	7	5	4	3		31	50	5	–	–	–	–	–	–	–	–	–	–	Olean-12-en-28-oic acid, 2,3,23-trihydroxy-, methyl ester, (2α,3β,4α)-		MS10388
203	262	202	204	263	249	205	201	**502**	100	82	26	20	18	10	6	6	**5.00**	31	50	5	–	–	–	–	–	–	–	–	–	–	Olean-12-en-28-oic acid, 2,3,23-trihydroxy-, methyl ester, (2α,3β,4α)-	22452-82-0	NI44871
203	262	28	189	263	204	55	69	**502**	100	88	45	27	22	22	19	18	**5.00**	31	50	5	–	–	–	–	–	–	–	–	–	–	Olean-12-en-28-oic acid, 2,3,23-trihydroxy-, methyl ester, (2β,3β,4α)-	22452-82-0	MS10389
217	235	207	190	294	234	203	189	**502**	100	70	37	25	23	23	19	19	**10.00**	31	50	5	–	–	–	–	–	–	–	–	–	–	Olean-12-en-28-oic acid, 3,15,16-trihydroxy-, methyl ester, (3β,15α,16α)	56114-49-9	NI44872
217	235	207	294	234	203	218	208	**502**	100	70	37	24	24	19	18	15	**9.33**	31	50	5	–	–	–	–	–	–	–	–	–	–	Olean-12-en-28-oic acid, 3β,21α,22α-trihydroxy-, methyl ester	13843-95-3	NI44873
129	73	55	373	75	95	81	107	**502**	100	85	61	52	46	44	38	37	**8.20**	31	54	3	–	–	–	–	–	–	1	–	–	–	Methyl (25RS)-3β-hydroxy-5-cholesten-26-oate, trimethylsilyl ether		NI44874
263	319	289	264	233	95	177	167	**502**	100	27	27	15	14	10	8	8	**1.00**	32	38	5	–	–	–	–	–	–	–	–	–	–	1,2-Propanediol, 3-(3,4,7,8-tetrahydro-5-methoxy-2,2,8,8-tetramethyl-2H,6H-benzo[1,2-b:5,4-b']dipyran-10-yl)-1,1-diphenyl-	56336-21-1	NI44875
189	406	203	279	134	216	162	466	**502**	100	83	80	77	77	66	63	60	**0.00**	32	54	4	–	–	–	–	–	–	–	–	–	–	3-Acetoxylupane-11α,20-diol	79081-65-5	NI44876
279	219	95	43	424	69	83	55	**502**	100	83	66	65	58	52	45	45	**11.65**	32	54	4	–	–	–	–	–	–	–	–	–	–	Lanostan-11-one, 3β,19-dihydroxy-, 3-acetate	23827-56-7	NI44877
161	57	43	73	284	203	218	219	**502**	100	92	92	88	83	81	72	70	**66.50**	33	58	3	–	–	–	–	–	–	–	–	–	–	4-Hydroxy-3,5-di-tert-butylbenzyl stearate		IC04527
430	431	92	98	502	432	91	503	**502**	100	37	11	10	9	5	4	3		34	34	2	2	–	–	–	–	–	–	–	–	–	Acetamide, N,N-dimethyl-4-(pyrrolidinocarbonylbenzyl)triphenyl-	70008-33-2	NI44878
251	501	235	134	503	502	218	204	**502**	100	84	74	66	52	33	31	27		34	34	2	2	–	–	–	–	–	–	–	–	–	N,N-Bis(4-anisyl)[2.2](4,7)-isoindolinophane		NI44879
502	459	208	503	460	208.5	504	312	**502**	100	61	53	38	23	13	4	4		34	34	2	2	–	–	–	–	–	–	–	–	–	1,5-Bis(4-butylanilino)anthraquinone		IC04528
404	405	502	98	406	503	332	446	**502**	100	35	11	9	5	4	2	2		34	34	2	2	–	–	–	–	–	–	–	–	–	Pyrrolidine, 4-((N,N-dimethylcarbamoyl)benzyl)triphenylacetyl-	70008-32-1	NI44880
188	160	119	335	132	145	503	117	**503**	100	79	70	53	50	39	35	33		16	8	5	1	–	11	–	–	–	–	–	–	–	1,1,1,3,3,3-Hexafluoroisopropyl 5-hydroxy-6-methoxyindole-2-carboxylate, pentafluoropropionyl-		NI44881
335	160	119	132	145	188	100	117	**503**	100	95	94	57	50	47	42	40	**14.19**	16	8	5	1	–	11	–	–	–	–	–	–	–	1,1,1,3,3,3-Hexafluoroisopropyl 6-hydroxy-5-methoxyindole-2-carboxylate, pentafluoropropionyl-		NI44882
327	299	124	271	107	328	77	119	**503**	100	56	42	25	18	14	14	13	**0.00**	17	15	5	1	–	10	–	–	–	–	–	–	–	β-Ethoxynormetadrenaline, bis(pentafluorppropionyl)-		MS10390
327	384	328	181	340	221	277	193	**503**	100	27	15	15	14	14	8	8	**6.40**	17	15	5	1	–	10	–	–	–	–	–	–	–	2-(2,4,6-Trimethoxyphenyl)ethylamine, bis(pentafluoropropionyl)-		MS10391
181	253	251	255	174	172	209	208	**503**	100	77	32	30	22	21	18	17	**0.00**	22	19	3	1	–	–	–	2	–	–	–	–	–	δ-Methrin		NI44883
155	26	96	60	29	71	89	208	**503**	100	80	70	69	50	45	44	42	**0.05**	25	45	9	1	–	–	–	–	–	–	–	–	–	Pederin		LI09432
221	283	222	169	503	284	192	223	**503**	100	19	13	9	3	3	2	1		26	33	9	1	–	–	–	–	–	–	–	–	–	Nocamycin		MS10392
109	95	137	110	44	43	153	67	**503**	100	72	67	54	53	50	49	49	**6.00**	26	37	7	3	–	–	–	–	–	–	–	–	–	18,20-Dihydroxy-7-isopropyl-4,8,16-trimethyl-6,23-dioxa-3,12,25-triazabicyclo[20.2.1]pentacosa-1(24),9,14,16,22(25)-pentaene-2,5,11-trione		MS10393
431	432	433	503	416	194	430	75	**503**	100	66	23	13	13	12	10	9		27	49	2	1	–	–	–	–	–	3	–	–	–	2-Aminoestradiol-17β, tris(trimethylsilyl)-		NI44884
431	432	433	416	300	434	429	417	**503**	100	37	13	7	5	4	3	2	**0.00**	27	49	2	1	–	–	–	–	–	3	–	–	–	4-Aminoestradiol, tris(trimethylsilyl)-		NI44885
488	472	489	147	473	310	503	490	**503**	100	48	37	28	19	16	13	11		28	45	5	1	–	–	–	–	–	1	–	–	–	Pregn-4-ene-3,20-dione, 17-(acetyloxy)-6-methyl-6-[(trimethylsilyl)oxy]-, 3-(O-methyloxime), (6β)-	69833-70-1	NI44886
369	400	459	57	151	114	84	329	**503**	100	66	62	58	56	50	47	41	**4.79**	28	45	5	1	–	–	–	–	–	1	–	–	–	Pregn-4-ene-3,20-dione, 17-(acetyloxy)-6-methyl-21-[(trimethylsilyl)oxy]-, 3-(O-methyloxime), (6α)-	74312-90-6	NI44887
105	273	106	217	446	130	57	390	**503**	100	75	32	31	29	26	25	16	**0.00**	30	37	4	3	–	–	–	–	–	–	–	–	–	(2R,5R)-1-Benzoyl-2-tert-butyl-5-[[1-(tert-butoxycarbonyl)-1H-indol-3-yl]methyl]-3,5-dimethyl-4-imidazolidinone		DD01755
377	43	164	379	166	250	248	334	**504**	100	99	85	73	65	55	50	40	**0.00**	9	21	–	–	–	–	–	–	2	–	–	–	1	Triisopropylantimony diiodide	26591-86-6	NI44888
69	223	185	281	504	472	204	506	**504**	100	96	94	58	55	39	31	25		10	2	2	–	2	12	–	–	–	–	–	–	1	Nickel, bis(1,1,1,5,5,5-hexafluoro-4-thioxo-2-pentanonato-O,S)-	31541-97-6	LI09433
69	223	185	508	85	281	504	91	**504**	100	96	91	81	61	57	53	46		10	2	2	–	2	12	–	–	–	–	–	–	1	Nickel, bis(1,1,1,5,5,5-hexafluoro-4-thioxo-2-pentanonato-O,S)-	31541-97-6	MS10394
504	485	208	252	133	347	369		**504**	100	14	7	5	4	3	1			10	15	–	–	–	6	–	–	–	–	2	–	1	Iridium, [(1,2,3,4,5-η)-1,2,3,4,5-pentamethyl-2,4-cyclopentadien-1-yl]bis(phosphorous trifluoride)-	34822-30-5	MS10395
280	55	196	111	250	168	278	446	**504**	100	85	61	56	51	47	43	37	**2.00**	12	3	12	–	–	–	–	–	–	–	–	–	3	Manganese tetracarbonylhydrido trimer		MS10396
475	476	477	223	45	478	445	447	**504**	100	39	26	12	12	9	7	5	**1.29**	12	32	10	–	–	–	–	–	–	6	–	–	–	1,3,5,7,9,11-Hexaethyl-5,9-dioxytricyclo[5.5.1.1^{3,11}]hexasiloxane		NI44889
106	304	292	210	393	435	234	265	**504**	100	85	57	43	42	37	33	25	**18.00**	15	17	5	4	–	9	–	–	–	–	–	–	–	Tris(trifluoroacetylarginine) isopropyl ester		LI09434
445	119	69	417	59	297	151	77	**504**	100	90	30	22	21	18	17	16	**14.40**	16	10	7	–	–	10	–	–	–	–	–	–	–	Isovanillylmandelic acid, methyl ester, pentafluoropropionate		MS10397
119	445	59	504	151	417	298	297	**504**	100	99	46	39	30	22	21	19		16	10	7	–	–	10	–	–	–	–	–	–	–	Vanillylmandelic acid, methyl ester, pentafluoropionate		NI44890
237	139	238	268	504	269	119	209	**504**	100	99	64	41	34	29	25	23		16	18	4	6	1	6	–	–	–	–	–	–	–	Bis(1,3-dimethyl-2,4-dioxo-6-(2,2,2-trifluoroethylamino)-1,2,3,4-tetrahydro-5-pyrimidyl) sulphide		LI09435
327	299	124	271	77	119	107	43	**504**	100	56	33	20	19	16	16	15	**3.20**	17	14	6	–	–	10	–	–	–	–	–	–	–	2-(4-Hydroxy-3-methoxyphenyl)-2-ethoxyethanol, bis(pentafluoropropionyl)-		MS10398

MASS TO CHARGE RATIOS								M.W.	INTENSITIES								Parent	C	H	O	N	S	F	Cl	Br	I	Si	P	B	X	COMPOUND NAME	CAS Reg No	No
449	451	447	171	173	71	369	367	**504**	100	50	50	44	38	32	30	30	**8.00**	18	14	–	6	1	–	–	2	–	–	–	–	–	1-(4-Bromophenyl)-6,8-dimethyl-5-thioxo[1,2,4]thiazolo[1,5-d][1,2,4]triazylium-2-(4-bromophenyl)aminide		MS10399
489	77	43	246	231	105	261	175	**504**	100	99	99	91	90	59	54	30	**2.10**	18	15	2	–	–	6	–	–	1	–	–	–	–	1,3-Dihydro-3,3-dimethyl-1-[1-phenyl-1-(trifluoromethyl)-2,2,2-trifluoroethoxy]-1,2-benziodoxole		MS10400
235	88	101	187	115	71	111	75	**504**	100	99	75	56	52	50	42	42	**0.00**	18	32	16	–	–	–	–	–	–	–	–	–	–	Galacto-(1-4)-gluco-(1-6)-galactose		LI09436
235	88	101	187	219	145	115	71	**504**	100	46	29	27	22	10	10	10	**0.00**	18	32	16	–	–	–	–	–	–	–	–	–	–	Galacto-(1-6)-gluco-(1-6)-galactose		LI09437
88	187	75	71	101	235	111	115	**504**	100	55	51	44	43	42	38	33	**0.00**	18	32	16	–	–	–	–	–	–	–	–	–	–	Gluco-(1-4)-gluco-(1-4)-glucose		LI09438
187	88	235	101	111	75	71	219	**504**	100	95	67	29	17	17	17	14	**0.00**	18	32	16	–	–	–	–	–	–	–	–	–	–	Gluco-(1-4)-gluco-(1-6)-glucose		LI09439
325	163	343	505	181	487			**504**	100	43	14	12	8	2			**0.00**	18	32	16	–	–	–	–	–	–	–	–	–	–	α-D-Glucopyranoside, O-α-D-glucopyranosyl-(1-3)-β-D-fructofuranosyl-	597-12-6	NI44891
469	471	473	470	472	254	74	217	**504**	100	98	36	27	26	23	21	16	**5.00**	22	8	4	2	–	–	4	–	–	–	–	–	–	N,N'-Bis(2,6-dichlorophenyl)pyromellitimide		MS10401
183	294	185	293	295	77	51	107	**504**	100	92	88	58	22	19	17	16	**1.00**	22	15	4	–	1	–	–	–	–	–	1	–	1	Molybdenum, tetracarbonyl(phenylthio)(diphenylphosphine)		IC04529
305	303	301	304	307	306	151	149	**504**	100	75	54	35	21	21	16	15	**0.00**	24	15	–	–	1	5	–	–	–	–	–	–	1	Germane, [(pentafluorophenyl)thio]triphenyl-	2034-07-3	NI44892
43	169	109	127	271	331	174	105	**504**	100	66	36	18	14	13	12	12	**0.12**	24	24	12	–	–	–	–	–	–	–	–	–	–	2-(2',3',4',6'-Tetra-O-acetyl-β-D-glucopyranosyloxy)-1,4-naphthoquinone	41431-92-9	NI44893
43	109	169	174	127	118	146	105	**504**	100	63	57	35	27	21	17	9	**0.43**	24	24	12	–	–	–	–	–	–	–	–	–	–	5-(2',3',4',6'-Tetra-O-acetyl-β-D-glucopyranosyloxy)-1,4-naphthoquinone	76768-53-1	NI44894
462	504	100	183	195	196	378	184	**504**	100	70	69	66	48	43	39	32		25	28	11	–	–	–	–	–	–	–	–	–	–	2',5',6'-Triacetyl-3',4',7,8-tetramethoxyisoflavan		NI44895
504	325	369	397	174	160	324	384	**504**	100	75	54	49	48	48	45	36		25	32	5	2	2	–	–	–	–	–	–	–	–	Aspidospermidin-21-oic acid, 1-acetyl-17-methoxy-20-[(methylthio)thioxomethoxy]-, methyl ester	56192-90-6	NI44896
217	201	77	218	185	63	488	108	**504**	100	77	66	61	57	31	27	25	**3.00**	26	22	1	2	2	–	–	–	–	–	2	–	–	1,1,5,5-Tetraphenyl-2,3-diimino-1-oxo-5-thioxo-1,5-diphospha-3-thiapentane		LI09440
91	504	109	123	95	121	105	119	**504**	100	82	65	55	55	47	47	38		26	30	3	–	–	6	–	–	–	–	–	–	–	2α,3α:6α,7α-Bis(difluoromethylene)-3-(difluoromethoxy)-6β,16β-dimethyl-16α,17α-epoxypregn-4-en-20-one		MS10402
91	504	109	123	105	95	454	121	**504**	100	36	29	25	24	24	18	17		26	30	3	–	–	6	–	–	–	–	–	–	–	2β,3β:6α,7α-Bis(difluoromethylene)-3-(difluoromethoxy)-6β,16β-dimethyl-16α,17α-epoxypregn-4-en-20-one		MS10403
43	69	255	324	41	109	83	444	**504**	100	59	47	33	32	31	31	29	**0.58**	26	32	10	–	–	–	–	–	–	–	–	–	–	12β,13,20-Triacetoxy-4,9-dihydroxy-1,5-tigliadien-3,7-dione		NI44897
349	91	194	350	155	504	195	193	**504**	100	56	29	25	24	23	18	12		27	24	4	2	2	–	–	–	–	–	–	–	–	3-(N-Methyl-N-tosylamino)-9-tosylcarbazole	94127-20-5	NI44898
311	401	283	255	402	312	265	75	**504**	100	82	40	37	24	21	16	16	**0.98**	27	44	5	–	–	–	–	–	–	2	–	–	–	Pregn-4-en-18-al, 3,20-dioxo-11,21-bis[(trimethylsilyl)oxy]-, (11β)-	69833-66-5	NI44899
504	505	217	169	191	506	218	298	**504**	100	47	45	24	22	21	21	17		27	48	3	–	–	–	–	–	–	3	–	–	–	Estra-1,3,4(10)-triene, 3,15,17-tris[(trimethylsilyl)oxy]-, (15β,17β)-	69833-74-5	NI44900
504	73	505	506	75	507	267	74	**504**	100	61	48	22	8	7	6	6		27	48	3	–	–	–	–	–	–	3	–	–	–	Estra-1,3,5(10)-triene, 2,3,17-tris[(trimethylsilyl)oxy]-, (17β)-	51497-42-8	NI44902
504	505	506	73	373	503	507	325	**504**	100	48	20	13	9	8	6	6		27	48	3	–	–	–	–	–	–	3	–	–	–	Estra-1,3,5(10)-triene, 2,3,17-tris[(trimethylsilyl)oxy]-, (17β)-	51497-42-8	NI44901
73	414	229	75	415	230	45	283	**504**	100	56	56	56	41	29	29	28	**1.00**	27	48	3	–	–	–	–	–	–	3	–	–	–	Estra-1,3,5(10)-triene, 3,6,17-tris[(trimethylsilyl)oxy]-, (6α,17β)-	33287-47-7	NI44903
414	415	283	324	309	230	416	229	**504**	100	38	23	21	17	15	14	13	**1.79**	27	48	3	–	–	–	–	–	–	3	–	–	–	Estra-1,3,5(10)-triene, 3,6,17-tris[(trimethylsilyl)oxy]-, (6α,17β)-	33287-47-7	NI44904
504	414	505	283	324	415	280	506	**504**	100	82	46	45	41	34	24	20		27	48	3	–	–	–	–	–	–	3	–	–	–	Estra-1,3,5(10)-triene, 3,6,17-tris[(trimethylsilyl)oxy]-, (6β,17β)-	69688-19-3	NI44905
414	309	324	415	310	230	283	298	**504**	100	62	46	36	20	19	18	16	**6.10**	27	48	3	–	–	–	–	–	–	3	–	–	–	Estra-1,3,5(10)-triene, 3,7,17-tris[(trimethylsilyl)oxy]-, (7α,17β)-	69688-37-5	NI44906
504	415	244	505	324	157	297	218	**504**	100	65	49	47	43	33	32	29		27	48	3	–	–	–	–	–	–	3	–	–	–	Estra-1,3,5(10)-triene, 3,11,17-tris[(trimethylsilyl)oxy]-, (11β,17β)-	69688-36-4	NI44907
504	324	505	325	283	414	183	129	**504**	100	69	46	46	35	22	20	20		27	48	3	–	–	–	–	–	–	3	–	–	–	Estra-1,3,5(10)-triene, 3,12,17-tris[(trimethylsilyl)oxy]-, (12β,17β)-	69597-47-3	NI44908
414	324	415	270	309	283	325	73	**504**	100	54	38	28	22	22	18	18	**7.36**	27	48	3	–	–	–	–	–	–	3	–	–	–	Estra-1,3,5(10)-triene, 3,14,17-tris[(trimethylsilyl)oxy]-, (17β)-	69688-23-9	NI44909
504	505	217	506	245	218	324	191	**504**	100	47	35	21	20	14	10	10		27	48	3	–	–	–	–	–	–	3	–	–	–	Estra-1,3,5(10)-triene, 3,15,17-tris[(trimethylsilyl)oxy]-, (15α,17β)-	69833-61-0	NI44910
73	504	345	386	129	505	311	297	**504**	100	56	33	26	26	24	23	17		27	48	3	–	–	–	–	–	–	3	–	–	–	Estra-1,3,5(10)-triene, 3,16,17-tris[(trimethylsilyl)oxy]-		MS10404
73	504	129	345	311	386	297	505	**504**	100	27	20	18	18	15	14	13		27	48	3	–	–	–	–	–	–	3	–	–	–	Estra-1,3,5(10)-triene, 3,16,17-tris[(trimethylsilyl)oxy]-		LI09441
73	504	129	311	75	505	386	297	**504**	100	34	22	21	19	16	14	14		27	48	3	–	–	–	–	–	–	3	–	–	–	Estra-1,3,5(10)-triene, 3,16,17-tris[(trimethylsilyl)oxy]-		LI09442
73	504	129	311	147	75	345	505	**504**	100	86	84	68	59	58	56	48		27	48	3	–	–	–	–	–	–	3	–	–	–	Estra-1,3,5(10)-triene, 3,16,17-tris[(trimethylsilyl)oxy]-, (16α,17α)-	57305-23-4	NI44911
504	311	505	345	386	414	297	324	**504**	100	48	46	43	39	37	29	22		27	48	3	–	–	–	–	–	–	3	–	–	–	Estra-1,3,5(10)-triene, 3,16,17-tris[(trimethylsilyl)oxy]-, (16α,17α)-	57305-23-4	NI44912
504	311	386	505	345	414	129	297	**504**	100	55	46	45	45	41	38	34		27	48	3	–	–	–	–	–	–	3	–	–	–	Estra-1,3,5(10)-triene, 3,16,17-tris[(trimethylsilyl)oxy]-, (16α,17β)-	18888-17-0	NI44913
73	129	504	311	345	147	75	386	**504**	100	68	67	52	47	46	39	38		27	48	3	–	–	–	–	–	–	3	–	–	–	Estra-1,3,5(10)-triene, 3,16,17-tris[(trimethylsilyl)oxy]-, (16β,17α)-	57305-22-3	NI44914
229	504	311	129	345	505	386	297	**504**	100	86	53	45	42	40	40	35		27	48	3	–	–	–	–	–	–	3	–	–	–	Estra-1,3,5(10)-triene, 3,16,17-tris[(trimethylsilyl)oxy]-, (16β,17α)-	57305-22-3	NI44915
73	504	129	311	345	505	386	147	**504**	100	88	88	68	66	43	43	41		27	48	3	–	–	–	–	–	–	3	–	–	–	Estra-1,3,5(10)-triene, 3,16,17-tris[(trimethylsilyl)oxy]-, (16β,17β)-	57305-21-2	NI44916
504	311	345	505	386	414	297	129	**504**	100	56	48	46	43	40	35	35		27	48	3	–	–	–	–	–	–	3	–	–	–	Estra-1,3,5(10)-triene, 3,16,17-tris[(trimethylsilyl)oxy]-, (16β,17β)-	57305-21-2	NI44917
489	169	113	57	279	490	41	124	**504**	100	65	53	38	34	29	24	22	**6.00**	27	48	3	6	–	–	–	–	–	–	–	–	–	Triimidazo[1,5-a:1',5'-c:1'',5''-e][1,3,5]triazine-1,5,9(2H,6H,10H)-trione, 2,6,10-tri-tert-butyl-hexahydro-3,3,7,7,11,11-hexamethyl-	67969-51-1	NI44918
504	177	137	121	57	136	192	505	**504**	100	96	65	56	43	41	36	32		28	40	8	–	–	–	–	–	–	–	–	–	–	26-tert-butyl-2,3,14,15-dibenzo-1,4,7,10,13,16,19,22-octaoxacyclotetracoza-2,4-diene		MS10405
394	395	78	77	396	304	211	121	**504**	100	33	31	26	22	22	21	21	**10.00**	29	28	1	4	–	–	–	–	–	–	–	–	1	Ferrocene, 1-(3-fur-2'-yl-1-phenylhydrazino-3-phenylhydrazonopropyl)-		LI09443
100	405	135	101	68	404	120	67	**504**	100	9	9	5	5	3	3	3	**0.94**	29	36	4	4	–	–	–	–	–	–	–	–	–	5,8-Ethenopyrrolo[3,2-b][1,5,8]oxadiazacyclotetradecine-12,15(1H,11H)-dione, 2,3,3a,13,14,15a-hexahydro-1-[3-methyl-2-(methylamino)-1-oxopentyl]-13-benzyl-	52617-26-2	NI44919
444	243	269	251	384	311	289	353	**504**	100	64	43	43	39	32	32	20	**7.14**	29	44	7	–	–	–	–	–	–	–	–	–	–	Cholan-24-oic acid, 3,7-bis(acetyloxy)-12-oxo-, methyl ester, (3α,5β,7α)-	28535-81-1	NI44920

MASS TO CHARGE RATIOS								M.W.	INTENSITIES								Parent	C	H	O	N	S	F	Cl	Br	I	Si	P	B	X	COMPOUND NAME	CAS Reg No	No
105	229	121	146	131	107	119	145	**504**	100	99	94	61	61	61	56	55	**0.00**	29	44	7	–	–	–	–	–	–	–	–	–	–	Cholan-24-oic acid, 3,7-bis(acetyloxy)-12-oxo-, methyl ester, (3α,5β,7α)-	28535-81-1	NI44921
269	105	251	133	131	157	119	161	**504**	100	67	66	45	41	39	37	35	**0.00**	29	44	7	–	–	–	–	–	–	–	–	–	–	Cholan-24-oic acid, 3,12-bis(acetyloxy)-7-oxo-, methyl ester, (3α,5β,12α)	21066-20-6	NI44922
269	104	159	143	145	267	130	211	**504**	100	43	42	39	36	31	29	28	**0.00**	29	44	7	–	–	–	–	–	–	–	–	–	–	Cholan-24-oic acid, 7,12-bis(acetyloxy)-3-oxo-, methyl ester, (5β,7α,12α)	4947-65-3	NI44924
269	384	314	229	369	279	353	311	**504**	100	45	11	10	9	9	7	7	**2.00**	29	44	7	–	–	–	–	–	–	–	–	–	–	Cholan-24-oic acid, 7,12-bis(acetyloxy)-3-oxo-, methyl ester, (5β,7α,12α)	4947-65-3	NI44923
269	504	268	270	220	237	152	139	**504**	100	25	24	18	18	15	12	12		30	20	4	2	1	–	–	–	–	–	–	–	–	1,1′-Bis[4-(methylamino)anthraquinonyl] sulphide		IC04530
504	105	77	506	282	505	223	203	**504**	100	89	59	46	46	33	26	21		30	22	4	–	–	–	–	–	–	–	–	–	1	Nickel, bis(1,3-diphenyl-1,3-propanedionato-O,O′)-	14405-47-1	NI44925
504	505	210	43	462	237	290	107	**504**	100	36	36	30	27	23	12	12		30	24	4	4	–	–	–	–	–	–	–	–	–	1,8-Bis(4-acetamidoanilino)anthraquinone		IC04531
135	178	77	339	152	192	103	107	**504**	100	73	12	10	7	6	6	5	**0.20**	30	24	4	4	–	–	–	–	–	–	–	–	–	4,5-Diphenyl-1-[α-(p-methoxybenzoyloxy)-p-methoxybenzylideneimino]-1,2,3-triazole		MS10406
504	505	322	161	506	307	120	162	**504**	100	46	31	23	21	14	11	10		30	32	–	4	–	–	–	–	–	2	–	–	–	1,2,4,5-Tetraaza-3,6-disilacyclohexane, 3,6-dimethyl-1,2,4,5-tetraphenyl-3,6-divinyl-	17082-85-8	NI44926
184	372	454	285	83	365	366	212	**504**	100	49	40	36	31	29	22	22	**14.00**	30	36	5	2	–	–	–	–	–	–	–	–	–	Aspidospermidin-21-oic acid, 20-hydroxy-17-methoxy-1-(1-oxo-3-phenylpropyl)-, methyl ester	56143-46-5	NI44927
264	246	201	231	67	119	131	55	**504**	100	94	85	41	41	32	29	26	**2.00**	30	48	6	–	–	–	–	–	–	–	–	–	–	Sericic acid	55306-03-1	NI44928
253	94	109	137	80	123	159	283	**504**	100	62	56	41	39	33	30	28	**1.00**	30	49	1	–	–	–	–	1	–	–	–	–	–	(23R,24R,28R)-28-Bromo-23,28-cyclostigmast-5-en-3β-ol		MS10407
370	355	356	255	105	12	20	201	**504**	100	48	41	28	26	24	23	22	**0.00**	31	52	5	–	–	–	–	–	–	–	–	–	–	Methyl 3α-methoxy-7α-(2′-tetrahydropyranyloxy)-5β-cholanoate		MS10408
121	255	137	120	147	69	133	159	**504**	100	71	53	35	29	27	22	20	**14.32**	31	53	1	–	–	–	1	–	–	1	–	–	–	9,10-Secoergosta-5,7,22-triene, 3-[[(chloromethyl)dimethylsilyl]oxy]-, (3β,5Z,7E,22E)-	69688-08-0	NI44929
75	215	55	81	95	73	107	414	**504**	100	81	71	60	58	55	47	44	**0.60**	31	56	3	–	–	–	–	–	–	1	–	–	–	Methyl (25RS)-3α-hydroxy-5β-cholestan-26-oate, trimethylsilyl ether		NI44930
43	486	91	105	119	426	504	411	**504**	100	42	37	26	24	15	14	4		33	44	4	–	–	–	–	–	–	–	–	–	–	Paracentrone acetate		LI09444
55	69	83	97	82	96	81	67	**504**	100	90	80	63	62	58	48	47	**5.14**	34	64	2	–	–	–	–	–	–	–	–	–	–	9-Hexadecenoic acid, 9-octadecenyl ester, (Z,Z)-	22393-98-2	NI44931
55	69	83	43	41	97	82	67	**504**	100	82	62	54	48	42	41	41	**15.14**	34	64	2	–	–	–	–	–	–	–	–	–	–	9-Octadecenoic acid, 9-hexadecenyl ester, (Z,Z)-	22393-99-3	NI44932
55	69	83	97	82	96	81	67	**504**	100	96	84	63	61	58	48	47	**4.84**	34	64	2	–	–	–	–	–	–	–	–	–	–	9-Octadecenoic acid, 9-hexadecenyl ester, (Z,Z)-	22393-99-3	NI44933
103	76	281	56	337	91	177	365	**505**	100	56	34	30	23	20	13	12	**2.00**	17	15	6	1	2	–	–	–	–	–	–	–	2	(μ-Phenyl isothiocyanato)(μ-tert-butylthiolato)hexacarbonyldiiron		MS10409
75	73	227	44	43	45	47	103	**505**	100	89	48	43	39	37	34	26	**0.00**	20	40	9	1	–	–	–	–	–	2	–	1	–	B-N-Acetylneuraminic, methyl ester 2-methyl-7,9-methylboronate-3,8-di-TMS		NI44934
75	73	227	45	43	44	47	103	**505**	100	91	42	42	34	33	30	28	**0.00**	20	40	9	1	–	–	–	–	–	2	–	1	–	B-N-Acetylneuraminic, methyl ester 2-methyl-8,9-methylboronate-3,7-di-TMS		NI44935
73	254	326	227	211	280	208	147	**505**	100	56	41	37	28	23	21	17	**0.00**	20	44	4	1	–	–	–	–	–	4	1	–	–	Phosphonic acid, [1-[(trimethylsilyl)amino]-2-[4-[(trimethylsilyl)oxy]phenyl]ethyl]-, bis(trimethylsilyl) ester	53044-43-2	NI44936
73	254	326	227	211	280	490	147	**505**	100	56	41	37	28	23	17	17	**0.28**	20	44	4	1	–	–	–	–	–	4	1	–	–	Phosphonic acid, [1-[(trimethylsilyl)amino]-2-[4-[(trimethylsilyl)oxy]phenyl]ethyl]-, bis(trimethylsilyl) ester	53044-43-2	NI44937
43	143	101	54	446	55	52	42	**505**	100	34	28	16	15	11	10	9	**0.00**	21	31	13	1	–	–	–	–	–	–	–	–	–	D-Glycero-D-galacto-2-nonulopyranosidonic acid, methyl 5-(acetylamino)3,5-dideoxy-, methyl ester, 4,7,8,9-tetraacetate	56323-64-9	NI44938
419	421	388	390	316	318	330	422	**505**	100	99	43	36	31	30	21	20	**11.00**	22	20	8	1	–	–	–	1	–	–	–	–	–	Tetramethyl-3-bromo-10,11-dihydroazepino[1,2-a]quinoline-7,8,9,10-tetracarboxylate		LI09445
421	419	388	390	422	420	318	316	**505**	100	99	27	24	21	21	18	18	**12.00**	22	20	8	1	–	–	–	1	–	–	–	–	–	Tetramethyl-3-bromo-10,11-dihydroazepino[1,2-a]quinoline-7,8,9,11-tetracarboxylate		LI09446
297	73	298	299	45	490	208	74	**505**	100	61	26	11	7	5	5	5	**0.90**	22	28	3	1	–	5	–	–	–	2	–	–	–	Phenylethylamine, 3-methoxy-β,4-bis[(trimethylsilyl)oxy]-N-(pentafluorophenyl)methylene-	55517-87-8	MS10410
297	73	298	299	43	45	74	194	**505**	100	72	28	11	11	9	6	4	**0.00**	22	28	3	1	–	5	–	–	–	2	–	–	–	Phenylethylamine, 3-methoxy-β,4-bis[(trimethylsilyl)oxy]-N-(pentafluorophenyl)methylene-	55517-87-8	NI44939
60	157	83	45	31	28	71	41	**505**	100	50	45	44	40	39	38	34	**0.00**	25	47	9	1	–	–	–	–	–	–	–	–	–	Dihydropederin		LI09447
73	129	75	145	91	188	105	43	**505**	100	93	92	55	37	30	30	28	**7.00**	28	51	3	1	–	–	–	–	–	2	–	–	–	Pregn-5-en-20-one, 3,17-bis[(trimethylsilyl)oxy]-16-methyl-, oxime, (3β,16β,17α)-		LI09448
312	73	280	313	147	129	103	281	**505**	100	81	31	26	25	22	14	13	**4.50**	28	51	3	1	–	–	–	–	–	2	–	–	–	Pregn-4-en-3-one, 20,21-bis[(trimethylsilyl)oxy]-, O-methyloxime, (20R)-	57305-30-3	NI44940
73	474	75	129	475	188	156	505	**505**	100	55	42	36	22	18	18	16		28	51	3	1	–	–	–	–	–	2	–	–	–	Pregn-5-en-20-one, 3,17-bis[(trimethylsilyl)oxy]-, O-methyloxime, (3β)-	39876-70-5	NI44941
73	129	75	89	103	175	474	91	**505**	100	99	99	94	64	60	55	50	**47.00**	28	51	3	1	–	–	–	–	–	2	–	–	–	Pregn-5-en-20-one, 3,21-bis[(trimethylsilyl)oxy]-, O-methyloxime, (3β)-	56196-40-8	NI44942
402	131	403	445	312	70	43	91	**505**	100	41	38	23	21	19	14	13	**1.10**	29	35	5	3	–	–	–	–	–	–	–	–	–	13-(α-Acetoxybenzyl)-6-acetyl-7-phenyl-1,6,10-triazabicyclo[10.2.1]pentadecan-9,14-dione	85702-15-4	NI44943
505	105	506	77	283	204	282	428	**505**	100	56	33	32	23	14	12	10		30	22	4	–	–	–	–	–	–	–	–	–	1	Cobalt, bis(1,3-diphenyl-1,3-propanedionato-O,O′)-	14405-50-6	NI44944
242	91	152	151	200	452	376	469	**505**	100	58	15	13	4	1	1	0	**0.00**	30	32	4	1	–	–	1	–	–	–	–	–	–	1,2-Bis(3-benzyloxy-4-methoxyphenyl)ethylamine hydrochloride		NI44945

MASS TO CHARGE RATIOS								M.W.	INTENSITIES								Parent	C	H	O	N	S	F	Cl	Br	I	Si	P	B	X	COMPOUND NAME	CAS Reg No	No
272	236	274	238	234	201	270	235	**506**	100	95	78	67	65	62	49	46	**0.00**	10	1	–	–	–	–	11	–	–	–	–	–	–	**1,3,4-Metheno-1H-cyclobuta[cd]pentalene, 1,1a,2,2,3,3a,4,5,5,5a,5b-undecachlorooctahydro-**	39801-14-4	**NI44946**
350	378	406	431	461	476	505	491	**506**	100	63	56	25	5	4	1	1	**0.00**	10	18	8	–	–	4	–	–	–	–	2	–	1	**Ruthenium, dicarbonyltetrafluoroethylenebis(trimethoxyphosphine)-**		**LI09449**
69	253	165	265	278	139	379	279	**506**	100	35	13	10	9	8	7	6	**0.90**	12	6	8	–	–	12	–	–	–	–	–	–	–	**Erythritol, tetrakis(trifluoroacetate), meso-**	27088-70-6	**NI44947**
43	295	338	253	209	506	364	478	**506**	100	45	40	35	30	14	14	4		13	21	4	–	–	–	–	–	–	–	–	–	2	**Tetracarbonyl(tripropylbismuthine)iron**	90188-73-1	**NI44948**
226	174	166	113	210	104	135	239	**506**	100	99	46	41	20	20	19	18	**4.00**	14	12	10	–	–	–	–	–	–	–	2	–	2	**Chromium, bis(pentacarbonyl)-μ-tetramethyldiphosphinedi-**		**MS10411**
178	41	29	57	69	139	126	152	**506**	100	84	82	80	75	42	33	32	**0.00**	16	19	6	2	–	9	–	–	–	–	–	–	–	**Hexanoic acid, 2,6-bis(trifluoroacetylamino)-5-trifluoroacetoxy-, butyl ester**		**NI44949**
278	284	506	405	208	209	352	126	**506**	100	86	83	76	54	48	36	34		18	17	2	2	2	6	1	–	–	–	–	–	–	**N-(4-Chlorophenyl)-2-(4-morpholinyl)-1,3-bis[(trifluoromethyl)thio]-2-cyclopentenecarboxamide**		**DD01756**
107	79	307	91	139	93	108	77	**506**	100	47	31	28	25	17	16	15	**0.90**	20	18	8	4	2	–	–	–	–	–	–	–	–	**1-Cyclohexene, 1,2-bis[[(2,4-dinitrophenyl)thio]methyl]-**		**MS10412**
43	115	187	139	157	145	127	103	**506**	100	13	9	8	7	7	5	5	**0.00**	21	30	14	–	–	–	–	–	–	–	–	–	–	**Glucoheptitol, heptaacetate**		**MS10413**
43	115	187	139	169	157	259	145	**506**	100	12	8	8	7	7	6	6	**0.00**	21	30	14	–	–	–	–	–	–	–	–	–	–	**Perseitol, heptaacetate**		**MS10414**
506	238	304	232	440	351	257	387	**506**	100	53	48	48	32	31	30	30		23	23	3	–	–	–	–	–	–	–	–	–	2	**Rhodium, (2-ferrocenoylnorbornadiene)acetylacetonato-**		**NI44950**
474	289	286	430	183	475	262	168	**506**	100	62	47	42	30	28	26	18	**0.00**	24	20	–	–	2	–	–	–	–	–	1	–	1	**Rhodium, (carbon disulphide-S)(η^5-2,4-cyclopentadien-1-yl)(triphenylphosphine)-**	50563-05-8	**NI44951**
73	149	57	59	167	71	55	393	**506**	100	86	85	44	38	34	33	31	**0.61**	24	54	–	–	–	–	–	–	–	3	–	–	1	**Hexa-tert-butylselenatrisiletane**	93194-15-1	**NI44952**
69	258	240	300	446	387	386	506	**506**	100	6	4	3	1	1	0	0		25	30	11	–	–	–	–	–	–	–	–	–	–	**Cistiglaucolide-8-O-methacrylate**		**MS10415**
305	307	304	303	309	302	306	506	**506**	100	84	67	63	40	33	32	22		26	28	2	2	–	–	–	–	–	–	–	–	1	**Palladium, bis(2-butyl-8-quinolinolato)-**	30049-18-4	**NI44953**
43	69	447	109	83	41	149	91	**506**	100	31	23	16	16	13	12	12	**0.26**	26	34	10	–	–	–	–	–	–	–	–	–	–	**Tiglia-1,5-dien-3-one, 12,13,20-triacetyloxy-4,7,9-trihydroxy-, (7β)-**		**NI44954**
43	83	69	326	109	53	91	41	**506**	100	36	28	22	21	16	14	14	**0.32**	26	34	10	–	–	–	–	–	–	–	–	–	–	**Tiglia-1,6-dien-3-one, 12,13,20-triacetyloxy-4,5,9-trihydroxy-, (12β)-**		**NI44955**
231	478	238	506	258	430			**506**	100	87	86	23	21	9				26	39	5	–	–	1	–	–	–	2	–	–	–	**Pregna-1,4-diene-3,20-dione, 9-fluoro-11-hydroxy-16-methyl-17,21-bis[(trimethylsilyl)oxy]-, (11β,16β)-**		**LI09450**
69	157	116	48	110	74	45	47	**506**	100	40	40	24	23	20	19	16	**0.00**	27	22	1	8	1	–	–	–	–	–	–	–	–	**3-[6-Methyl-3-(methylthio)-5-oxo-4,5-dihydro-1,2,4-triazin-1-yl]imino-4-(1 naphthyl)-5-(1-naphthyl)amino-2H-2,3-dihydro-1,2,4-triazole**		**MS10416**
183	506	376	287	262	121	165	77	**506**	100	89	82	75	68	61	53	46		27	23	4	–	2	–	–	–	–	–	1	–	–	**Triphenylphosphonio-(vinylsulphonyl-phenyl-sulphonyl-methanide)**		**LI09451**
157	235	400	360	263	216	320	277	**506**	100	4	3	2	2	1	1	1	**0.30**	27	26	6	2	1	–	–	–	–	–	–	–	–	**6-Hydroxy-4-methoxy-N-(2-methylphenyl)-5-[3-[(2-methylphenyl)amino]-1-oxo-2-butenyl]-benzofuran-7-sulphonamide**		**MS10417**
43	506	111	507	137	138	99	122	**506**	100	83	25	24	16	15	15	14		27	38	9	–	–	–	–	–	–	–	–	–	–	**Ingol, 3,7,12-triacetyloxy-8-O-methyl-**		**NI44956**
435	199	345	385	475	416	398	155	**506**	100	69	53	16	13	10	8	7	**3.50**	27	50	3	2	–	–	–	–	–	2	–	–	–	**9-Oxo-11,15-dihydroxyprosta-5,13-dienenitrile, bis(trimethylsilyl)-**		**MS10418**
262	225	133	155	385	333	308	199	**506**	100	45	20	13	12	11	7	7	**1.70**	27	50	3	2	–	–	–	–	–	2	–	–	–	**9-Oxo-11,15-dihydroxyprosta-5,13-dienenitrile, bis(trimethylsilyl)-**		**MS10419**
245	244	210	247	246	165	212	261	**506**	100	44	40	38	28	20	16	12	**6.00**	28	20	1	–	2	–	2	–	–	–	–	–	–	**Bis(2-chloro-10,11-dihydrodibenzo[b,f]thiepin-10-yl) ether**		**MS10420**
43	57	328	149	55	69	71	267	**506**	100	80	77	77	76	73	71	63	**10.00**	28	42	8	–	–	–	–	–	–	–	–	–	–	**Pregn-5-en-20-one, 11-(acetyloxy)-3,14-dihydroxy-12-(2-hydroxy-3-methyl-1-oxobutoxy)-, (3β,11α,12β,14β)-**	20230-37-9	**NI44957**
84	98	70	121	237	155	100	110	**506**	100	72	65	34	32	32	32	28	**14.00**	29	38	4	4	–	–	–	–	–	–	–	–	–	**4H-1,16-Etheno-5,15-(propaniminoethano)furo[3,4-l][1,5,10]triazacyclohexadecine-4,21-dione, 3,3a,6,7,8,9,10,11,12,13,14,15-dodecahydro-3-(4-methoxyphenyl)-**	69721-60-4	**NI44958**
506	463	434	421	57	58	42	41	**506**	100	95	61	46	45	39	38	36		29	38	4	4	–	–	–	–	–	–	–	–	–	**4,7,10-Triazabicyclo[12.3.1]octadeca-1(18),2,14,16-tetraene-5,8,11-trione, 12-(dimethylamino)-15-methoxy-9-sec-butyl-6-(phenylmethyl)-**	38840-25-4	**NI44959**
253	368	313	428	261	271	226	211	**506**	100	38	38	24	21	14	14	14	**1.00**	29	46	7	–	–	–	–	–	–	–	–	–	–	**Cholan-24-oic acid, 3,7-diacetyloxy-12-hydroxy-, methyl ester, (3α,7α,12α)-**		**NI44960**
506	508	507	509	253	510	255	180	**506**	100	74	35	25	21	17	17	17		30	16	2	2	–	–	2	–	–	–	–	–	–	**Triphenodioxazine, 6,13-dichloro-**		**IC04532**
139	91	77	64	154	51	48	44	**506**	100	98	76	58	55	54	47	44	**6.00**	30	28	–	4	–	–	–	–	–	–	2	–	–	**Cyanamide, [1,4-butanediylbis(diphenylphosphoranylidyne)]bis-**	66055-16-1	**NI44961**
310	370	328	282	292	341	295	506	**506**	100	56	21	20	19	12	10	6		30	34	7	–	–	–	–	–	–	–	–	–	–	**20-O-Acetylresiniferonol-9,13,14-ortho-phenylacetate**		**NI44962**
506	446	431	353	413				**506**	100	32	20	20	14					30	50	2	–	2	–	–	–	–	–	–	–	–	**Cholestan-19-oic acid, 1-oxo-, methyl ester, cyclic 1-(ethylene mercaptole), (5α)-**	28809-63-4	**LI09452**
506	507	446	431	353	414	447	508	**506**	100	34	32	20	20	14	11	9		30	50	2	–	2	–	–	–	–	–	–	–	–	**Cholestan-19-oic acid, 1-oxo-, methyl ester, cyclic 1-(ethylene mercaptole), (5α)-**	28809-63-4	**NI44963**
447	506	448	507	353	508	449	354	**506**	100	88	33	30	14	9	9	4		30	50	2	–	2	–	–	–	–	–	–	–	–	**Cholestan-19-oic acid, 2-oxo-, methyl ester, cyclic 2-(ethylene mercaptole), (5α)-**	28809-62-3	**NI44964**
145	43	101	59	95	81	73	55	**506**	100	53	43	33	32	28	28	28	**0.20**	30	50	6	–	–	–	–	–	–	–	–	–	–	**9,19-Cycloergostan-3-one, 22,24,25,28-tetrahydroxy-4-(hydroxymethyl)-14-methyl-, (5α)-**	70237-87-5	**NI44965**
249	280	207	190	189	262	231	175	**506**	100	72	69	60	50	44	44	43	**1.00**	30	50	6	–	–	–	–	–	–	–	–	–	–	**Olean-12-ene-3,15,16,21,22,28-hexol, (3β,15α,16α,21β,22α)-**		**MS10421**
249	280	207	190	189	231	262	298	**506**	100	72	68	59	50	44	43	42	**2.00**	30	50	6	–	–	–	–	–	–	–	–	–	–	**Olean-12-ene-3,16,21,22,23,28-hexol, (3β,15α,16α,21β,22α)-**		**LI09453**
215	197	119	107	105	157	131	121	**506**	100	80	63	57	57	48	48	48	**2.00**	30	50	6	–	–	–	–	–	–	–	–	–	–	**Olean-12-ene-3,16,21,22,23,28-hexol, (3β,16α,21β,22α)-**		**LI09454**
215	197	81	55	43	264	157	31	**506**	100	95	69	57	56	50	48	48	**0.00**	30	50	6	–	–	–	–	–	–	–	–	–	–	**Olean-12-ene-3,16,21,22,23,28-hexol, (3β,16β,21β,22α)-**		**LI09455**

MASS TO CHARGE RATIOS								M.W.	INTENSITIES								Parent	C	H	O	N	S	F	Cl	Br	I	Si	P	B	X	COMPOUND NAME	CAS Reg No	No
215	175	264	246	197	105	121	119	**506**	100	70	61	59	59	40	39	38	**0.00**	30	50	6	–	–	–	–	–	–	–	–	–	–	Olean-12-ene-3,16,21,22,24,28-hexol, (3β,16α,21β,22α)-		MS10422
81	55	43	93	69	105	41	215	**506**	100	91	88	68	68	67	67	64	**2.00**	30	50	6	–	–	–	–	–	–	–	–	–	–	Olean-12-ene-3,16,21,22,24,28-hexol, (3β,16α,21β,22α)-		LI09456
215	197	119	107	105	157	131	121	**506**	100	81	61	58	58	48	48	48	**1.00**	30	50	6	–	–	–	–	–	–	–	–	–	–	Theasapogenol A		MS10423
410	349	253	367	392	470	329	385	**506**	100	84	64	52	48	42	35	32	**18.00**	31	51	3	–	–	–	1	–	–	–	–	–	–	Cholest-22-en-5-ol, 3-acetoyloxy-6-chloro-24-ethyl-, (5α,6β,24R)-		NI44966
105	91	77	148	118	252	224	415	**506**	100	43	38	25	11	8	7	6	**0.60**	32	30	4	2	–	–	–	–	–	–	–	–	–	(-)-Anabellamide		MS10424
254	310	196	195	77	208	167	222	**506**	100	70	46	44	38	32	25	24	**3.00**	32	30	4	2	–	–	–	–	–	–	–	–	–	11,22-Diphenyl-7,8,18,19-tetrahydro-6H-17H-dibenzo[h,q][1,5,10,14,6,15]tetraoxadiazacyclooctadecine	89572-40-7	NI44967
91	506	92	365	385	352	69	409	**506**	100	13	9	8	7	7	7	5		33	46	4	–	–	–	–	–	–	–	–	–	–	(E)-Cyclopentadec-2-en-1-one 1,4-di-O-benzyl-D-threitol ketal		NI44968
91	92	69	506	181	105	81	507	**506**	100	8	5	4	3	3	3	2		33	46	4	–	–	–	–	–	–	–	–	–	–	(Z)-Cyclopentadec-2-en-1-one 1,4-di-O-benzyl-D-threitol ketal		NI44969
105	99	183	77	225	167	141	106	**506**	100	49	38	20	18	8	8	8	**0.00**	34	34	4	–	–	–	–	–	–	–	–	–	–	(2R,3R)-α,α,α′,α′-Tetraphenyl-1,4-dioxaspiro[4.5]decane-2,3-dimethanol		DD01757
250	55	82	43	96	69	83	57	**506**	100	85	77	75	74	65	58	58	**2.24**	34	66	2	–	–	–	–	–	–	–	–	–	–	Hexadecanoic acid, 9-octadecenyl ester, (Z)-	2906-55-0	NI44970
82	96	55	83	250	69	57	43	**506**	100	99	78	70	69	69	56	56	**0.84**	34	66	2	–	–	–	–	–	–	–	–	–	–	Hexadecanoic acid, 9-octadecenyl ester, (Z)-	2906-55-0	NI44971
57	236	55	69	83	43	97	71	**506**	100	93	92	90	79	78	69	64	**2.84**	34	66	2	–	–	–	–	–	–	–	–	–	–	9-Hexadecenoic acid, octadecyl ester, (Z)-	22393-84-6	NI44972
96	82	222	55	69	83	43	110	**506**	100	99	79	76	71	66	55	53	**0.74**	34	66	2	–	–	–	–	–	–	–	–	–	–	Octadecanoic acid, 9-hexadecenyl ester, (Z)-	22393-91-5	NI44973
27	43	57	55	264	69	83	41	**506**	100	85	84	84	70	68	57	47	**13.54**	34	66	2	–	–	–	–	–	–	–	–	–	–	9-Octadecenoic acid, hexadecyl ester, (Z)-	22393-86-8	NI44974
264	57	55	69	83	43	97	71	**506**	100	98	85	83	78	73	70	63	**3.34**	34	66	2	–	–	–	–	–	–	–	–	–	–	9-Octadecenoic acid, hexadecyl ester, (Z)-	22393-86-8	NI44975
387	267	268	388	165	265	189	252	**506**	100	98	63	56	40	32	17	16	**0.20**	35	26	2	2	–	–	–	–	–	–	–	–	–	N,N′-Diphenyl-N-(2,4,4-triphenyl-2,3-butadienoyl)urea		MS10425
119	387	267	91	268	388	64	265	**506**	100	55	30	30	20	18	17	12	**0.00**	35	26	2	2	–	–	–	–	–	–	–	–	–	4,6,6-Triphenyl-2-(N-phenylcarbamoyl)-2-azatricyclo[5.2.2.0^{1,5}]undeca-4,8,10-trien-3-one		MS10426
267	283	71	57	239	43	85	282	**506**	100	77	68	65	58	54	48	44	**0.00**	35	70	1	–	–	–	–	–	–	–	–	–	–	18-Pentatriacontanone	504-53-0	LI09457
71	267	69	85	83	283	97	82	**506**	100	55	47	37	36	35	23	19	**1.60**	35	70	1	–	–	–	–	–	–	–	–	–	–	18-Pentatriacontanone	504-53-0	WI01590
57	43	477	85	129	29	506	185	**506**	100	83	33	33	28	17	11	11		35	70	1	–	–	–	–	–	–	–	–	–	–	Pentatriacontan-3-one	79097-22-6	NI44976
57	43	71	55	41	58	267	69	**506**	100	85	65	50	31	30	25	23	**1.00**	35	70	1	–	–	–	–	–	–	–	–	–	–	Di-stearyl ketone		IC04533
57	43	71	85	55	41	69	83	**506**	100	87	61	43	30	27	21	18	**0.00**	36	74	–	–	–	–	–	–	–	–	–	–	–	Hexatriacontane		AI02397
57	43	71	85	55	41	69	56	**506**	100	80	58	36	23	21	16	13	**0.51**	36	74	–	–	–	–	–	–	–	–	–	–	–	Hexatriacontane		AI02396
57	43	71	55	85	41	69	29	**506**	100	84	42	28	27	25	19	17	**0.96**	36	74	–	–	–	–	–	–	–	–	–	–	–	Hexatriacontane		AI02398
57	43	71	85	55	41	69	83	**506**	100	99	60	43	35	33	24	20	**0.00**	36	74	–	–	–	–	–	–	–	–	–	–	–	Pentacosane, 13-undecyl-		AI02399
44	112	69	64	42	43	41	84	**507**	100	74	53	42	34	32	26	21	**0.00**	17	25	13	5	–	–	–	–	–	–	–	–	–	β-D-Allofuranuronic acid, 5-[[2-amino-5-O-(aminocarbonyl)-2-deoxy-L-xylonoyl]amino]-1,5-dideoxy-1-[3,4-dihydro-5-(hydroxymethyl)-2,4-dioxo-1(2H)-pyrimidinyl]-	19396-06-6	NI44977
208	138	250	176	177	109	219	251	**507**	100	99	64	56	52	39	35	26	**18.00**	21	25	10	5	–	–	–	–	–	–	–	–	–	Methyl 2-acetylamino-3,4-dihydro-4-oxo-7-(2-acetylamino-2-deoxy-3,4-diacetyl-β-arabinofuranosyl)-7H-pyrrolo[2,3-d]pyrimidine-5-carboxylate	84933-22-2	NI44978
492	479	480	506	481	507	465	467	**507**	100	98	84	64	52	43	22	20		22	25	–	1	–	–	–	–	–	–	–	1	1	Iridium, [(2,3,4,5,6-η)-1-cyano-1-phenylboratabenzene][(1,2,3,4,5-η)-1,2,3,4,5-pentamethyl-2,4-cyclopentadien-1-yl]-	69910-64-1	NI44979
43	55	42	83	41	70	169	114	**507**	100	94	90	73	68	62	59	59	**1.00**	24	48	6	3	–	–	–	–	–	–	–	3	–	2-Aziridinyl-4,4,6-trimethyl-1,3,2-dioxaborinane		MS10427
476	477	478	73	507	173	162	75	**507**	100	43	18	13	11	10	10	8		27	49	4	1	–	–	–	–	–	2	–	–	–	Prosta-5,8(12),13-trien-1-oic acid, 9-(methoxyimino)-15-[(trimethylsilyl)oxy]-, trimethylsilyl ester, (5Z,13E,15S)-	56009-42-8	NI44980
400	323	428	200	112	183	161	414	**507**	100	71	41	40	31	23	19	17	**2.40**	28	32	–	3	–	–	1	–	–	–	2	–	–	(Aminodiphenylphosphinoimino)-(butyl)diphenylphosphane chloride		NI44981
100	87	73	70	75	386	476	296	**507**	100	45	45	25	22	20	19	19	**2.00**	28	53	3	1	–	–	–	–	–	2	–	–	–	Pregnan-20-one, 3,6-bis[(trimethylsilyl)oxy]-, O-methyloxime, (3α,5β,6α)	33287-44-4	NI44982
100	87	476	296	386	70	241	115	**507**	100	40	27	23	19	18	15	12	**2.39**	28	53	3	1	–	–	–	–	–	2	–	–	–	Pregnan-20-one, 3,6-bis[(trimethylsilyl)oxy]-, O-methyloxime, (3α,5β,6α)	33287-44-4	NI44983
201	476	170	87	387	477	202	100	**507**	100	82	72	72	35	34	32	28	**7.00**	28	53	3	1	–	–	–	–	–	2	–	–	–	Pregnan-20-one, 3,15-bis[(trimethylsilyl)oxy]-, O-methyloxime, (3α,5α,15α)-		LI09458
73	476	75	188	156	477	158	507	**507**	100	99	80	62	50	42	40	34		28	53	3	1	–	–	–	–	–	2	–	–	–	Pregnan-20-one, 3,17-bis[(trimethylsilyl)oxy]-, O-methyloxime, (3α,5α)-	33287-42-2	NI44984
73	476	75	188	156	477	158	507	**507**	100	99	99	50	45	44	34	30		28	53	3	1	–	–	–	–	–	2	–	–	–	Pregnan-20-one, 3,17-bis[(trimethylsilyl)oxy]-, O-methyloxime, (3α,5β)-	57305-32-5	NI44985
476	174	477	188	57	364	142	462	**507**	100	49	47	45	41	39	34	32	**14.88**	28	53	3	1	–	–	–	–	–	2	–	–	–	Pregnan-20-one, 3,17-bis[(trimethylsilyl)oxy]-, O-methyloxime, (3α,5β,17α)-		NI44986
73	476	75	188	156	477	158	81	**507**	100	99	94	57	54	41	41	30	**29.00**	28	53	3	1	–	–	–	–	–	2	–	–	–	Pregnan-20-one, 3,17-bis[(trimethylsilyl)oxy]-, O-methyloxime, (3β,5α)-	39780-68-2	NI44987
314	100	476	404	315	477	87	70	**507**	100	74	43	39	26	20	20	17	**4.00**	28	53	3	1	–	–	–	–	–	2	–	–	–	Pregnan-20-one, 3,19-bis[(trimethylsilyl)oxy]-, O-methyloxime, (3α,5α)-		LI09459
73	188	75	476	175	507	477	479	**507**	100	99	97	90	70	60	38	35		28	53	3	1	–	–	–	–	–	2	–	–	–	Pregnan-20-one, 3,21-bis[(trimethylsilyl)oxy]-, O-methyloxime, (3α,5α)-	57305-33-6	NI44988
73	188	75	175	476	507	81	477	**507**	100	99	86	73	67	59	30	29		28	53	3	1	–	–	–	–	–	2	–	–	–	Pregnan-20-one, 3,21-bis[(trimethylsilyl)oxy]-, O-methyloxime, (3α,5β)-	56196-41-9	NI44989
225	183	42	168	196	238	226	182	**507**	100	75	41	26	19	18	18	13	**4.64**	29	49	6	1	–	–	–	–	–	–	–	–	–	4-Tridecanol, 13-(4-hydroxy-5,6-dimethoxy-3-methyl-2-pyridyl)-3,5,7,11-tetramethyl-, diacetate	26469-05-6	NI44990
189	137	165	507	342	326	325	492	**507**	100	94	91	88	71	63	59	31		30	37	6	1	–	–	–	–	–	–	–	–	–	Yesoline		MS10428

MASS TO CHARGE RATIOS	M.W.	INTENSITIES	Parent	C	H	O	N	S	F	Cl	Br	I	Si	P	B	X	COMPOUND NAME	CAS Reg No	No
169 507 170 508 114 337 115 141	**507**	100 77 37 29 27 22 21 17		33	21	3	3	–	–	–	–	–	–	–	–	–	**2,4,6-Tris(2-hydroxynapth-1-yl)-s-triazine**		**IC04534**
507 177 179 509 508 178 57 55	**507**	100 92 30 28 28 15 14 14		33	46	1	1	–	–	1	–	–	–	–	–	–	**Cholestano[3,2-b]indole, 5′-chloro-6-oxo-, (5α)-**		**MS10429**
113 93 69 162 143 163 290 182	**508**	100 82 61 43 43 36 23 19	**8.41**	9	–	5	–	–	7	–	–	–	–	–	–	1	**Rhenium, pentacarbonylheptafluorobuten-1-yl-**		**NI44991**
493 135 165 313 283 147 253 255	**508**	100 40 27 13 13 12 9 8	**0.00**	10	28	–	–	–	–	–	–	–	–	–	–	3	**Stannane, methylidynetris[trimethyl-**	56177-42-5	**NI44992**
300 176 93 301 424 508 207 177	**508**	100 49 18 16 14 13 13 11		12	27	12	–	–	–	–	–	–	–	3	–	1	**Chromium, tricarbonyltris(trimethyl phosphite-P)-**	62413-59-6	**NI44993**
139 27 137 120 118 135 28 168	**508**	100 75 74 64 49 46 46 43	**0.00**	16	6	–	–	–	10	–	–	–	–	–	–	1	**Stannane, bis(pentafluorophenyl)divinyl-**	1247-12-7	**NI44994**
340 236 312 424 91 221 144 176	**508**	100 98 96 82 76 73 70 60	**7.00**	16	12	6	–	3	–	–	–	–	–	–	–	2	**(μ-benzyldithioacetato)(μ-ethanethiolato)hexacarbonyldiiron**		**MS10430**
508 360 400 252 453 305 212 480	**508**	100 99 57 25 24 19 18 11		18	32	6	–	–	–	–	–	–	–	2	–	1	**Ruthenium, bis(2-methylallyl)bis(4-methyl-2,6,7-trioxaphosphabicyclo[2.2.2]octanyl)-**		**LI09460**
495 493 510 508 235 191 43 509	**508**	100 98 92 92 61 46 36 25		19	21	6	6	–	–	–	1	–	–	–	–	–	**Aniline-5-acetamide, N,N-diethyl-2-methoxy-4-(2,4-dinitro-6-bromophenylazo)-**		**IC04535**
43 508 435 59 179 67 205 510	**508**	100 67 59 40 33 28 26 22		21	25	7	6	–	–	1	–	–	–	–	–	–	**Aniline-5-acetamide, N-(1,3-dimethyl-3-hydroxybutyl)-2-methoxy-4-(2,4-dinitro-6-chlorophenylazo)-**		**IC04536**
105 254 121 76 99 178 107 89	**508**	100 90 61 50 23 21 11 10	**0.00**	22	20	2	–	6	–	–	–	–	–	–	–	–	**3,8,9,10-Tetrathiatricyclo[4.2.1.1^{2,5}]decane-4,7-dione, 1,2-bis(ethylthio)-5,6-diphenyl-, endo-**		**NI44995**
73 81 135 128 121 147 217 262	**508**	100 74 36 32 16 12 11 1	**0.70**	22	31	5	–	–	7	–	–	–	–	–	–	–	**Ethyl 10-(heptafluorobutyryloxy)-11-methoxy-3,7,11-trimethyl-2,6-dodecadienoate**		**MS10431**
87 81 99 149 121 114 135 265	**508**	100 68 34 28 26 26 19 16	**0.30**	22	31	5	–	–	7	–	–	–	–	–	–	–	**Methyl 10-(heptafluorobutyryloxy)-11-methoxy-3,7,11-trimethyl-2,6-tridecadienoate**		**MS10433**
87 81 99 149 114 135 93 121	**508**	100 76 33 32 30 28 28 27	**0.70**	22	31	5	–	–	7	–	–	–	–	–	–	–	**Methyl 10-(heptafluorobutyryloxy)-11-methoxy-3,7,11-trimethyl-2,6-tridecadienoate**		**MS10432**
73 75 508 465 74 41 45 77	**508**	100 27 20 19 14 14 13 12		23	36	5	2	1	–	–	–	–	2	–	–	–	**Benzoic acid, 3-(aminosulphonyl)-5-(butylamino)-4-phenoxy-, N-(trimethylsilyl)-, trimethylsilyl ester**		**NI44996**
73 290 291 45 75 292 74 218	**508**	100 48 20 14 9 8 8 6	**0.00**	23	44	3	2	–	–	–	–	–	4	–	–	–	**Tryptophan, 5-hydroxy-, tetrakis(trimethylsilyl)-**	74367-80-9	**MS10434**
73 332 147 205 117 333 45 74	**508**	100 51 46 31 21 12 10 8	**0.00**	23	44	3	2	–	–	–	–	–	4	–	–	–	**Tryptophan, 5-hydroxy-, tetrakis(trimethylsilyl)-**	74367-80-9	**NI44997**
290 73 291 292 74 45 218 75	**508**	100 62 30 11 4 4 3 3	**0.00**	23	44	3	2	–	–	–	–	–	4	–	–	–	**Tryptophan, 5-hydroxy-, tetrakis(trimethylsilyl)-**	74367-80-9	**NI44998**
302 428 510 508 382 380	**508**	100 61 4 4 4 4		24	14	–	–	–	–	–	1	1	–	–	–	–	**Naphthalene, 1-(2-bromovinyl)-8-(1-iodo-8-naphthylethynyl)-, cis-**		**LI09461**
302 510 508 428 429 383 381 382	**508**	100 35 35 8 7 3 3 2		24	14	–	–	–	–	–	1	1	–	–	–	–	**Naphthalene, 1-(2-bromovinyl)-8-(1-iodo-8-naphthylethynyl)-, trans-**		**LI09462**
91 78 104 43 398 401 400 402	**508**	100 75 74 73 67 64 64 57	**9.03**	24	24	–	–	–	–	–	–	–	–	–	–	2	**Molybdenum, tris(1,3,5,7-cyclooctatetraene)di-**	12391-50-3	**NI44999**
73 377 75 392 378 74 403 147	**508**	100 62 25 20 18 10 9 7	**0.50**	24	40	6	–	–	–	–	–	–	3	–	–	–	**2,3,4,4a-Tetrahydro-2,3,7-tris[(trimethylsilyl)oxy]-9-methoxy-4a-methyl-6H-dibenzo[b,d]pyran-6-one**		**NI45000**
57 85 121 124 180 105 185 125	**508**	100 77 74 49 43 42 34 32	**0.00**	26	36	10	–	–	–	–	–	–	–	–	–	–	**Acetyl toxin T2**		**NI45001**
73 55 508 262 165 263 95 83	**508**	100 75 65 55 50 45 39 39		26	36	10	–	–	–	–	–	–	–	–	–	–	**Glaucarubinone, 1-O-methyl-**		**LI09463**
294 105 77 295 51 106 310 78	**508**	100 99 99 72 61 51 32 30	**0.00**	27	24	8	–	1	–	–	–	–	–	–	–	–	**Methyl 2,3-di-O-benzoyl-4-S-benzoyl-4-thio-α-D-xylopyranoside**		**LI09464**
458 120 29 31 43 74	**508**	100 66 46 43 37 35	**0.00**	27	40	9	–	–	–	–	–	–	–	–	–	–	**Pregn-5-ene-3,8,11,12,14,20-hexol, triacetate(3β,11α,12β,14β)-**	26855-45-8	**NI45002**
451 75 73 452 365 135 101 125	**508**	100 49 42 31 25 24 24 22	**0.00**	27	44	5	–	1	–	–	–	–	1	–	–	–	**(2S,4S,6S,8R,9S)-4-(dimethyl-tert-butylsiloxy)-8,9-dimethyl-2-[(2E)-3-phenylsulphonylbut-2-enyl]-1,7-dioxaspiro[5.5]undecane**		**MS10435**
179 180 508 195 221 220 225 178	**508**	100 90 77 56 42 42 33 33		28	24	4	6	–	–	–	–	–	–	–	–	–	**1,6-Bis(2-(2′-nitrodiphenylamino))-2,5-diazahexa-1,5-diene**		**LI09465**
377 508 361 376 432 375 390 348	**508**	100 58 46 29 17 15 14 13		28	28	9	–	–	–	–	–	–	–	–	–	–	**4-(6-Deoxy-3-C-methyl-β-gulopyranosyl)-8-vinyl-1-hydroxy-10,12-dimethoxy-6H-benzo[d]naphtho[1,2-b]pyran-6-one**	82196-88-1	**NI45003**
91 299 508 208 134 108 509 313	**508**	100 48 43 26 19 18 14 10		28	32	7	2	–	–	–	–	–	–	–	–	–	**4,3′,5′-Trimethyl-3,4′-bis-(2-methoxycarbonyl-ethyl)-5-benzyloxycarbonyl 2,2′-bis(pyrrolyl)ketone**		**LI09466**
326 508 477 150 94 332 339 327	**508**	100 89 79 56 50 36 28 24		29	36	6	2	–	–	–	–	–	–	–	–	–	**Gelsamydine**		**DD01758**
125 99 43 429 57 55 510 430	**508**	100 84 48 40 40 36 28 26	**26.13**	29	49	2	–	–	–	–	1	–	–	–	–	–	**Cholestan-3-one, 2-bromo-, cyclic 1,2-ethanediyl acetal, (2α,5α)-**	1923-43-9	**NI45004**
99 125 43 57 55 41 69 112	**508**	100 70 27 23 19 14 13 12	**12.00**	29	49	2	–	–	–	–	1	–	–	–	–	–	**Cholestan-3-one, 2-bromo-, cyclic 1,2-ethanediyl acetal, (2β,5α)-**	56052-89-2	**NI45005**
508 77 510 69 79 509 61 57	**508**	100 82 68 46 45 38 38 38		30	18	2	2	–	–	2	–	–	–	–	–	–	**6,6′-Dichloro-2,2′-diphenyl-4,4′(2H,2′H)-biisoquinoline-3,3′-dione**		**LI09467**
105 77 28 238 508 51 106 269	**508**	100 36 20 11 9 9 7 5		30	20	2	–	3	–	–	–	–	–	–	–	–	**Acetophenone, 2,2″-(1,2,4-trithiolane-3,5-diylidene)bis[2-phenyl-**	19018-17-8	**NI45006**
105 77 238 51 508 106	**508**	100 37 10 10 8 6		30	20	2	–	3	–	–	–	–	–	–	–	–	**Acetophenone, 2,2″-(1,2,4-trithiolane-3,5-diylidene)bis[2-phenyl-**	19018-17-8	**LI09468**
43 55 41 69 81 95 93 107	**508**	100 95 76 65 64 55 45 38	**0.00**	30	52	6	–	–	–	–	–	–	–	–	–	–	**Ergostane-3,5,6,12,25-pentol, 25-acetate, (3β,5α,6β,12β)-**	56053-00-0	**NI45007**
473 198 129 383 508	**508**	100 42 40 35 11		30	53	2	–	–	–	1	–	–	1	–	–	–	**Cholestane, 4-chloro-5,6-epoxy-3-[(trimethylsilyl)oxy]-, (3β,4α,5β,6β)-**		**LI09469**
352 415 77 183 507 165 65 353	**508**	100 83 81 78 48 40 32 26	**24.00**	31	26	3	–	–	–	–	–	–	–	2	–	–	**Phosphonic acid, [(triphenylphosphoranylidene)methyl]-, diphenyl ester**	22400-41-5	**NI45008**
253 394 367 351 395 396 508 213	**508**	100 81 48 46 38 31 27 27		31	47	2	–	–	3	–	–	–	–	–	–	–	**Stigmasta-5,22-dien-3-ol, trifluoroacetate, (3β,22E)-**	3870-50-6	**NI45009**
97 83 55 69 81 95 123 145	**508**	100 88 88 76 72 65 57 56	**52.48**	31	47	2	–	–	3	–	–	–	–	–	–	–	**Stigmasta-5,22-dien-3-ol, trifluoroacetate, (3β,22E)-**	3870-50-6	**NI45010**
77 138 81 423 83 139 137 369	**508**	100 65 62 54 50 46 42 37	**30.72**	31	60	3	–	–	–	–	–	–	1	–	–	–	**1-Cyclohexanol, 2-isopropyl-5-methyl-, trimethylsilyl ether**	74685-38-4	**NI45011**

MASS TO CHARGE RATIOS								M.W.	INTENSITIES								Parent	C	H	O	N	S	F	Cl	Br	I	Si	P	B	X	COMPOUND NAME	CAS Reg No	No
119	91	64	360	289	77	317	508	**508**	100	70	38	36	31	19	8	3		32	36	2	4	–	–	–	–	–	–	–	–	–	**7,8-Bis(diethylamino)-2,4,6-triphenyl-2,6-diazabicyclo[2.2.2]oct-7-ene-3,5-dione**		**DD01759**
153	181	69	259	95	129	81	241	**508**	100	68	50	42	29	22	21	20	**10.00**	32	44	5	–	–	–	–	–	–	–	–	–	–	**(+)-Milbemycin β^3 methyl ether**		**MS10436**
72	44	493	105	508	491	77	443	**508**	100	91	79	78	67	55	43	37		32	48	3	2	–	–	–	–	–	–	–	–	–	**Benzamide, N-[20-(dimethylamino)-4-hydroxy-pregnan-3-yl]-, acetate, (3β,4β,5α,20α)-**	5874-20-4	**NI45012**
91	119	51	93	50	77	52	92	**508**	100	99	80	68	34	28	19	16	**0.00**	33	24	2	4	–	–	–	–	–	–	–	–	–	**7H-Acenaphtho[1,2-c]pyrazole-7,8(9H)-dicarboxanilide, 9-phenyl-**	22531-81-3	**NI45013**
119	91	51	92	50	77	64	62	**508**	100	99	80	68	34	28	20	20	**0.00**	33	24	2	4	–	–	–	–	–	–	–	–	–	**7H-Acenaphtho[1,2-c]pyrazole-7,8(9H)-dicarboxanilide, 9-phenyl-**	22531-81-3	**LI09470**
493	494	508	509	495	510	451	452	**508**	100	41	31	13	11	4	3	1		33	52	2	–	–	–	–	–	–	1	–	–	–	**Cannabinol tributylsilyl ether**		**NI45014**
256	126	55	43	224	199	310	69	**508**	100	63	55	55	53	50	45	44	**10.00**	33	64	3	–	–	–	–	–	–	–	–	–	–	**Tricontanoic acid, 2-ethyl-11-oxo-, methyl ester**		**LI09471**
508	506	509	507	254	510	256	493	**508**	100	39	38	34	30	22	16	15		34	44	–	4	–	–	–	–	–	–	–	–	–	**21H,23H-Porphine, 2,7,12,17-tetraethyl-22,24-dihydro-3,8,13,18,21,23-hexamethyl-**	56630-99-0	**MS10437**
355	73	399	356	89	83	103	69	**508**	100	53	38	29	21	20	19	19	**0.00**	34	52	3	–	–	–	–	–	–	–	–	–	–	**Cholestan-7-ol, 8,9-epoxy-3-benzyloxy-, (3β,5α,7α,8α)-**	74420-81-8	**NI45015**
327	73	371	328	89	83	103	97	**508**	100	95	43	30	25	22	21	20	**0.00**	34	52	3	–	–	–	–	–	–	–	–	–	–	**Cholestan-7-ol, 8,14-epoxy-3-benzyloxy-, (3β,5α,7α,8α)-**	74420-82-9	**NI45016**
43	57	313	55	71	69	41	83	**508**	100	99	76	58	51	46	43	42	**31.04**	34	68	2	–	–	–	–	–	–	–	–	–	–	**Eicosanoic acid, tetradecyl ester**	22413-04-3	**NI45017**
257	57	43	55	71	69	83	41	**508**	100	80	80	52	43	42	40	35	**24.04**	34	68	2	–	–	–	–	–	–	–	–	–	–	**Hexadecanoic acid, octadecyl ester**	2598-99-4	**NI45018**
257	57	43	28	55	71	83	69	**508**	100	79	68	50	45	44	42	40	**27.04**	34	68	2	–	–	–	–	–	–	–	–	–	–	**Hexadecanoic acid, octadecyl ester**	2598-99-4	**NI45019**
285	57	43	71	83	55	69	97	**508**	100	80	63	50	47	44	43	41	**11.04**	34	68	2	–	–	–	–	–	–	–	–	–	–	**Octadecanoic acid, hexadecyl ester**	1190-63-2	**NI45020**
57	43	224	70	71	84	69	55	**508**	100	89	76	64	58	47	46	46	**1.00**	34	68	2	–	–	–	–	–	–	–	–	–	–	**Octadecanoic acid, hexadecyl ester**	1190-63-2	**IC04537**
229	43	57	55	69	83	71	41	**508**	100	95	89	61	46	44	44	43	**26.04**	34	68	2	–	–	–	–	–	–	–	–	–	–	**Tetradecanoic acid, eicosyl ester**	22413-00-9	**NI45021**
508	509	57	43	71	73	55	69	**508**	100	38	35	26	19	16	14	13		34	68	2	–	–	–	–	–	–	–	–	–	–	**Tetratriacontanoic acid**	38232-04-1	**WI01591**
507	509	508	510	129	157	100	208	**508**	100	91	48	31	24	22	17	16		34	68	2	–	–	–	–	–	–	–	–	–	–	**Tritriacontanoic acid, methyl ester**		**NI45022**
441	506	442	507	384	365	443	124	**508**	100	45	32	17	17	16	9	8	**3.00**	35	29	–	–	–	–	–	–	–	–	–	–	1	**Cobalt, [1,1′,1″,1‴-[(1,2,3,4-η)-1,3-cyclohexadiene-1,2,3,4-tetrayl]tetrakis[benzene]](η⁵-2,4-cyclopentadien-1-yl)-**	55518-03-1	**NI45023**
83	69	71	82	97	68	96	85	**508**	100	98	97	87	68	55	53	52	**0.10**	35	72	1	–	–	–	–	–	–	–	–	–	–	**1-Pentatriacontanol**	55517-90-3	**WI01592**
91	508	69	402	506	133	43	73	**508**	100	60	31	30	24	24	17	12		37	48	1	–	–	–	–	–	–	–	–	–	–	**β-Apo-2′-carotenal**		**LI09472**
197	253	180	193	105	198	254	81	**508**	100	60	30	26	24	20	16	15	**3.00**	37	48	1	–	–	–	–	–	–	–	–	–	–	**2-(3α,5-Cyclo-6β-methoxynorcholanyl)-1,1-diphenylethylene**		**NI45024**
43	374	372	36	268	266	28		**509**	100	6	6	6	5	5	5		**0.00**	16	19	9	1	–	–	4	–	–	–	–	–	–	**1,3,4,6-Tetra-O-acetyl-2-deoxy-2-tetrachloroethylideneamino-β-D-glucopyranose**		**MS10438**
267	73	179	268	509	45	269	75	**509**	100	98	27	23	13	12	10	10		20	34	5	1	–	3	–	–	–	3	–	–	–	**L-Phenylalanine, N-(trifluoroacetyl)-3,4-bis[(trimethylsilyl)oxy]-, trimethylsilyl ester**		**MS10439**
173	233	147	174	305	218	217	204	**509**	100	43	23	16	10	10	10	9	**0.00**	20	47	6	1	–	–	–	–	–	4	–	–	–	**α-D-Galactopyranose, 2-(acetylamino)-2-deoxy-1,3,4,6-tetrakis-O-(trimethylsilyl)-**	53110-72-8	**NI45025**
173	73	233	75	147	174	131	43	**509**	100	78	45	32	20	17	16	13	**0.00**	20	47	6	1	–	–	–	–	–	4	–	–	–	**α-D-Galactopyranose, 2-(acetylamino)-2-deoxy-1,3,4,6-tetrakis-O-(trimethylsilyl)-**	53110-72-8	**NI45026**
173	73	233	75	147	131	174	116	**509**	100	93	66	57	29	23	22	22	**0.00**	20	47	6	1	–	–	–	–	–	4	–	–	–	**β-D-Galactopyranose, 2-(acetylamino)-2-deoxy-1,3,4,6-tetrakis-O-(trimethylsilyl)-**	53110-73-9	**NI45028**
173	233	147	305	174	218	234	217	**509**	100	62	21	16	16	15	13	12	**0.00**	20	47	6	1	–	–	–	–	–	4	–	–	–	**β-D-Galactopyranose, 2-(acetylamino)-2-deoxy-1,3,4,6-tetrakis-O-(trimethylsilyl)-**	53110-73-9	**NI45027**
173	233	73	305	174	147	131	204	**509**	100	72	50	14	14	14	13	10	**0.00**	20	47	6	1	–	–	–	–	–	4	–	–	–	**D-Galactose, 2-(acetylamino)-2-deoxy-1,3,4,6-tetrakis-O-(trimethylsilyl)-**		**MS10440**
173	73	75	233	131	147	217	103	**509**	100	99	47	33	22	21	13	12	**0.00**	20	47	6	1	–	–	–	–	–	4	–	–	–	**D-Galactose, 2-(acetylamino)-2-deoxy-3,4,5,6-tetrakis-O-(trimethylsilyl)-**	55721-25-0	**LI09473**
73	173	75	233	131	147	174	217	**509**	100	99	46	31	21	20	16	12	**0.00**	20	47	6	1	–	–	–	–	–	4	–	–	–	**D-Galactose, 2-(acetylamino)-2-deoxy-3,4,5,6-tetrakis-O-(trimethylsilyl)-**	55721-25-0	**NI45029**
73	173	75	147	233	131	174	43	**509**	100	74	24	20	17	17	12	11	**0.00**	20	47	6	1	–	–	–	–	–	4	–	–	–	**D-Galactose, 2-(acetylamino)-2-deoxy-3,4,5,6-tetrakis-O-(trimethylsilyl)-**	55721-25-0	**MS10441**
173	73	75	131	147	174	217	204	**509**	100	90	25	21	20	18	14	13	**0.00**	20	47	6	1	–	–	–	–	–	4	–	–	–	**α-D-Glucopyranose, 2-acetamido-2-deoxy-1,3,4,6-tetrakis-O-(trimethylsilyl)-**	31980-72-0	**NI45030**
173	147	174	259	217	204	233	218	**509**	100	18	16	15	14	11	7	7	**0.00**	20	47	6	1	–	–	–	–	–	4	–	–	–	**α-D-Glucopyranose, 2-(acetylamino)-2-deoxy-1,3,4,6-tetrakis-O-(trimethylsilyl)-**	53110-68-2	**NI45031**
173	217	259	174	147	233	218	204	**509**	100	17	15	15	14	11	10	10	**0.00**	20	47	6	1	–	–	–	–	–	4	–	–	–	**β-D-Glucopyranose, 2-(acetylamino)-2-deoxy-1,3,4,6-tetrakis-O-(trimethylsilyl)-**	53110-69-3	**NI45032**
173	73	259	217	174	233	147	131	**509**	100	33	20	17	14	12	10	10	**0.00**	20	47	6	1	–	–	–	–	–	4	–	–	–	**D-Glucose, 2-(acetylamino)-2-deoxy-1,3,4,6-tetrakis-O-(trimethylsilyl)-**		**MS10443**
173	73	75	131	147	174	217	204	**509**	100	90	25	21	20	18	14	13	**0.00**	20	47	6	1	–	–	–	–	–	4	–	–	–	**D-Glucose, 2-(acetylamino)-2-deoxy-1,3,4,6-tetrakis-O-(trimethylsilyl)-**		**MS10442**
173	73	233	131	147	75	174	305	**509**	100	51	37	18	17	16	15	12	**0.00**	20	47	6	1	–	–	–	–	–	4	–	–	–	**D-Glucose, 2-(acetylamino)-2-deoxy-3,4,5,6-tetrakis-O-(trimethylsilyl)-**	55529-74-3	**NI45033**
173	147	174	259	217	204	233	218	**509**	100	19	16	15	15	11	8	7	**0.00**	20	47	6	1	–	–	–	–	–	4	–	–	–	**α-D-Mannopyranose, 2-(acetylamino)-2-deoxy-1,3,4,6-tetrakis-O-(trimethylsilyl)-**	53110-70-6	**NI45034**

MASS TO CHARGE RATIOS	M.W.	INTENSITIES	Parent	C	H	O	N	S	F	Cl	Br	I	Si	P	B	X	COMPOUND NAME	CAS Reg No	No
173 73 174 75 259 217 204 147	**509**	100 51 24 24 22 22 16 16	**0.00**	20	47	6	1	–	–	–	–	–	4	–	–	–	**α-D-Mannopyranose, 2-(acetylamino)-2-deoxy-1,3,4,6-tetrakis-O-(trimethylsilyl)-**	53110-70-6	**NI45035**
173 73 75 174 131 217 147 259	**509**	100 73 45 27 24 22 20 19	**0.00**	20	47	6	1	–	–	–	–	–	4	–	–	–	**β-D-Mannopyranose, 2-(acetylamino)-2-deoxy-1,3,4,6-tetrakis-O-(trimethylsilyl)-**	53110-71-7	**NI45036**
173 147 217 174 259 204 218 233	**509**	100 20 16 16 15 11 8 7	**0.00**	20	47	6	1	–	–	–	–	–	4	–	–	–	**β-D-Mannopyranose, 2-(acetylamino)-2-deoxy-1,3,4,6-tetrakis-O-(trimethylsilyl)-**	53110-71-7	**NI45037**
173 73 131 174 147 217 204 75	**509**	100 46 17 15 11 10 9 9	**0.00**	20	47	6	1	–	–	–	–	–	4	–	–	–	**D-Mannose, 2-(acetylamino)-2-deoxy-, tetrakis-O-(trimethylsilyl)-**	56298-44-3	**NI45038**
173 73 259 217 174 147 233 131	**509**	100 33 19 16 14 12 11 11	**0.00**	20	47	6	1	–	–	–	–	–	4	–	–	–	**D-Mannose, 2-(acetylamino)-2-deoxy-1,3,4,6-tetrakis-O-(trimethylsilyl)-**		**MS10444**
87 43 436 88 510 509 44 30	**509**	100 46 5 5 4 4 4 4		24	23	6	5	1	–	–	–	–	–	–	–	–	**Aniline, N,N-bis(2-acetoxyethyl)-3-methyl-4-(1-cyano-5-nitro-2-benzthienylazo)-**		**IC04538**
255 127 58 57 283 201 171 256	**509**	100 99 28 24 21 19 19 18	**0.00**	24	48	8	1	–	–	–	–	–	–	1	–	–	**Choline, hydroxide, dihydrogen phosphate, inner salt**	19191-91-4	**NI45039**
280 406 281 73 509 70 407 248	**509**	100 47 19 18 14 13 12 12		25	34	1	3	1	3	–	–	–	1	–	–	–	**Trimethylsilylfluphenazine**		**NI45040**
73 478 75 479 43 132 74 55	**509**	100 53 40 23 10 9 9 9	**4.50**	27	51	4	1	–	–	–	–	–	2	–	–	–	**Prosta-8(12),13-dien-1-oic acid, 9-(methoxyimino)-15-[(trimethylsilyl)oxy], trimethylsilyl ester**	72150-31-3	**NI45041**
478 479 480 494 477 75 438 388	**509**	100 42 16 12 11 9 8 8	**8.00**	27	51	4	1	–	–	–	–	–	2	–	–	–	**Prosta-8(12),13-dien-1-oic acid, 9-(methoxyimino)-15-[(trimethylsilyl)oxy], trimethylsilyl ester, (13E,15S)-**	56009-43-9	**NI45042**
73 75 438 129 478 388 199 419	**509**	100 33 21 21 17 14 14 13	**2.30**	27	51	4	1	–	–	–	–	–	2	–	–	–	**Prosta-10,13-dien-1-oic acid, 9-(methoxyimino)-15-[(trimethylsilyl)oxy]-, trimethylsilyl ester**	55530-44-4	**NI45043**
199 438 148 419 478 129 388 73	**509**	100 84 70 66 65 65 62 52	**8.00**	27	51	4	1	–	–	–	–	–	2	–	–	–	**Prosta-10,13-dien-1-oic acid, 9-(methoxyimino)-15-[(trimethylsilyl)oxy]-, trimethylsilyl ester, (9E,13E,15S)-**	56227-28-2	**NI45044**
148 129 199 478 388 438 73 419	**509**	100 59 43 39 37 34 31 29	**6.00**	27	51	4	1	–	–	–	–	–	2	–	–	–	**Prosta-10,13-dien-1-oic acid, 9-(methoxyimino)-15-[(trimethylsilyl)oxy]-, trimethylsilyl ester, (9Z,13E,15S)-**	56227-29-3	**NI45045**
57 101 73 55 85 83 69 129	**509**	100 83 75 72 57 49 29 28	**0.00**	28	35	4	3	1	–	–	–	–	–	–	–	–	**16-Tosyl-2,3,11,12-dibenzo-1,13-dioxa-4,10,16-triazacyclooctadeca-2,11-diene**		**MS10445**
105 91 77 43 45 360 196 106	**509**	100 61 32 23 15 11 10 10	**0.00**	29	39	5	3	–	–	–	–	–	–	–	–	–	**(2S,5S)-1-Benzoyl-5-[4-[(benzyloxycarbonyl)amino]butyl]-2-tert-butyl-5-(1-hydroxyethyl)-3-methyl-4-imidazolidinone**		**DD01760**
91 252 452 105 344 222 196 360	**509**	100 76 65 53 42 39 23 20	**0.00**	29	39	5	3	–	–	–	–	–	–	–	–	–	**(2R,5S)-5-[1-(Benzoyloxy)ethyl]-5-[4-[(benzyloxycarbonyl)amino]butyl]-2-tert-butyl-3-methyl-4-imidazolidinone**		**DD01761**
105 223 77 285 509 287 511 224	**509**	100 50 44 42 38 27 20 16		30	22	4	–	–	–	–	–	–	–	–	–	1	**Copper, bis(1,3-diphenyl-1,3-propanedionato-O,O′)-,**	14405-48-2	**NI45046**
91 216 180 260 466 374 314 270	**509**	100 36 33 23 17 4 3 3	**0.00**	31	43	5	1	–	–	–	–	–	–	–	–	–	**rel-[2S,6S,2(3R)]-N-[(Benzyloxy)carbonyl]-2-[3-[[(benzyloxy)carbonyl]oxy]heptyl]-6-propylpiperidine**		**DD01762**
73 86 509 168 394 424 125 451	**509**	100 29 24 20 18 16 15 14		32	35	3	3	–	–	–	–	–	–	–	–	–	**N,N′,N″-Tripropyltriptycene-1,8,16-tricarboxamide**		**NI45047**
186 171 268 270 185 266 240 212	**510**	100 62 35 33 29 28 23 23	**3.00**	16	10	6	2	–	–	–	–	–	–	–	–	1	**[5-Acetyl-4-phenylpyrazol-2-yl]pentacarbonyltungsten**		**NI45048**
391 209 510 205 115 181 81 180	**510**	100 84 63 63 55 51 50 48		21	26	11	4	–	–	–	–	–	–	–	–	–	**2-Methyl-6-methoxy-4-[(N-2,3,4,6-tetra-O-acetyl-β-D-glucopyranosyl)amino]oxazolo[5,4-d]pyrimidine**		**NI45049**
233 290 59 41 55 258 235 73	**510**	100 72 40 40 32 30 28 28	**0.00**	24	34	10	2	–	–	–	–	–	–	–	–	–	**Tetramethyl 4,10-dimethoxy-6,6,12,12-tetramethyl-1:7-diazatricyclo[8.2.0.0^{4,7}]dodeca-2,8-diene-2,3,8,9-tetracarboxylate**		**NI45050**
170 41 96 97 42 171 138 43	**510**	100 67 65 53 52 51 45 43	**15.00**	24	42	–	6	3	–	–	–	–	–	–	–	–	**Triimidazo[1,5-a:1′,5′-c:1″,5″-e][1,3,5]triazine-1,5,9(2H,6H,10H)-trithione, hexahydro-3,3,7,7,11,11-hexamethyl-2,6,10-triisopropyl-**	67969-53-3	**NI45051**
89 73 133 59 58 97 55 105	**510**	100 66 40 29 26 21 19 19	**0.00**	24	46	11	–	–	–	–	–	–	–	–	–	–	**3-(3,6,9,12,15-Pentaoxacyclohexadecyl)-1,5,8,11,14,17-hexaocyclononadecane**		**MS10446**
269 213 157 437 325 381 270 41	**510**	100 78 71 39 27 21 18 16	**0.00**	24	54	7	–	–	–	–	–	–	2	–	–	–	**Silicic acid (H6Si2O7), hexabutyl ester**	4422-63-3	**NI45052**
510 147 511 105 79 69 67 133	**510**	100 43 29 29 29 28 27 25		25	29	3	–	–	7	–	–	–	–	–	–	–	**Pregna-3,5-dien-20-one, 3-(heptafluorobutanoyl)-**		**MS10447**
68 41 73 351 40 453 425 352	**510**	100 60 45 41 26 13 12 12	**0.00**	26	50	4	–	–	–	–	–	–	3	–	–	–	**Phenylacetic, α,4-bis[(tert-butyldimethylsilyl)oxy]-, tert-butyldimethylsilyl ester**		**MS10448**
314 315 243 215 271 231 230 329	**510**	100 32 28 12 11 6 6 5	**0.00**	27	19	2	4	–	–	–	1	–	–	–	–	–	**2-Phenyl-6-oxo-8,10-diphenyl-4H,6H-pyrido[2,1-f][1,3,4]oxadiazino[2,3-c][1,2,4]triazin-11-ium bromide**		**MS10449**
225 173 337 243 195 129 111 121	**510**	100 36 30 27 23 23 22 20	**7.45**	27	50	5	–	–	–	–	–	–	2	–	–	–	**Prosta-7,13-dien-1-oic acid, 6,9-epoxy-11,15-bis[(trimethylsilyl)oxy]-, methyl ester, (13E,15S)-**	56248-51-2	**NI45053**
323 510 351 392 511 352 75	**510**	100 95 63 39 37 26 24		27	50	5	–	–	–	–	–	–	2	–	–	–	**Prostaglandin B1, 19-hydroxy-, bis(trimethylsilyl) ether**		**LI09474**
323 510 351 511 324 512 352 65	**510**	100 90 55 36 29 16 16 16		27	50	5	–	–	–	–	–	–	2	–	–	–	**Prostaglandin B1, 20-hydroxy-, bis(trimethylsilyl) ether**		**LI09475**
377 450 43 510 317 378 391 226	**510**	100 89 60 56 48 42 40 39		28	34	7	2	–	–	–	–	–	–	–	–	–	**4-N,16,16,20-Tetraacetyl-tris(hydroxymethyl)-talbotine**		**LI09476**

MASS TO CHARGE RATIOS								M.W.	INTENSITIES								Parent	C	H	O	N	S	F	Cl	Br	I	Si	P	B	X	COMPOUND NAME	CAS Reg No	No
468	43	44	42	28	469	41	55	**510**	100	59	50	40	29	28	24	22	**8.01**	29	34	8	–	–	–	–	–	–	–	–	–	–	3-Cyclohexene-1-acetic acid, 4-[1α-(3-furyl-3,7,8,8a-tetrahydro-8aα-methyl-3-oxo-1H-2-benzopyran-5-yl)methyl]-3-hydroxy-2α,6,6-trimethyl-5-oxo-, methyl ester, acetate	30576-19-3	**NI45054**
510	253	197	296	366	257	437	465	**510**	100	92	77	72	9	9	6	1		29	42	4	4	–	–	–	–	–	–	–	–	–	L-Alanine, N-[N-[N-[[4-(dimethylamino)-1-naphthalenyl]methylene]-L-valyl]-L-isoleucyl]-, ethyl ester	57237-94-2	**NI45055**
105	510	77	512	514	511	513	106	**510**	100	31	31	19	12	12	8	7		30	22	4	–	–	–	–	–	–	–	–	–	1	Zinc, bis(1,3-diphenyl-1,3-propanedionato-O,O′)-	21333-45-9	**NI45056**
510	511	447	152	89	121	420	255	**510**	100	33	26	18	16	14	12	10		30	22	4	–	2	–	–	–	–	–	–	–	–	Dimethyl 2,2′-(1,4-phenylene)bis(5-phenyl-3-thiophenecarboxylate)	110699-45-1	**DD01763**
510	511	121	224	56	84	479	255	**510**	100	33	15	12	11	8	7	7		30	22	4	–	2	–	–	–	–	–	–	–	–	Dimethyl 2,2′-(1,4-phenylene)bis(5-phenyl-4-thiophenecarboxylate)	110699-44-0	**DD01764**
510	511	479	121	255	447	419	77	**510**	100	33	16	15	10	6	6	4		30	22	4	–	2	–	–	–	–	–	–	–	–	Dimethyl 2,2′-(1,3-phenylene)bis(5-phenylthiophene)-3,4′dicarboxylate	110699-48-4	**DD01765**
510	41	57	121	511	77	105	479	**510**	100	75	37	34	33	25	16	11		30	22	4	–	2	–	–	–	–	–	–	–	–	Dimethyl 2,2′-(1,4-phenylene)bis(5-phenylthiophene)-3,4′-dicarboxylate	110699-46-2	**DD01766**
254	191	223	121	224	325	414	313	**510**	100	60	30	26	15	11	2	2	**0.00**	30	23	–	–	3	–	–	–	–	–	1	–	–	3,5,6,8-Tetraphenyl-2,9-dithia-1-phosphabicyclo[4.3.0]nona-3,7-diene 1-sulphide	82518-06-7	**NI45057**
243	165	244	193	105	267	191	137	**510**	100	36	26	16	16	12	12	10	**7.00**	30	26	6	2	–	–	–	–	–	–	–	–	–	6H-Furo[2′,3′:4,5]oxazolo[3,2-a]pyrimidin-6-one, 3-(acetyloxy)-2,3,3a,9a-tetrahydro-2-[(triphenylmethoxy)methyl]-, [2R-(2α,3β,3aβ,9aβ)]-	40732-55-6	**NI45058**
125	109	136	135	137	121	97	319	**510**	100	50	45	39	37	33	30	20	**1.00**	30	38	7	–	–	–	–	–	–	–	–	–	–	(+)-Jaborol acetate		**MS10450**
493	230	94	107	79	81	178	510	**510**	100	54	28	16	14	12	7	6		30	42	3	2	1	–	–	–	–	–	–	–	–	anti-Thiobinupharidine sulphoxide		**NI45059**
493	230	492	94	107	510	176	178	**510**	100	95	40	35	26	20	18	10		30	42	3	2	1	–	–	–	–	–	–	–	–	syn-Thiobinupharidine sulphoxide		**NI45060**
510	388	91	43	360	89	373	44	**510**	100	95	76	61	54	53	50	49		31	26	7	–	–	–	–	–	–	–	–	–	–	8′-Benzyloxy-1,1′,4,4′-tetramethoxy-2,2′-binaphthalene-7,8-dione		**NI45061**
255	396	510	369	327	213	342	229	**510**	100	68	67	56	50	32	23	23		31	49	2	–	–	3	–	–	–	–	–	–	–	Stigmast-7-en-3-ol, trifluoroacetate, (3β,24ξ)-	55517-92-5	**NI45062**
43	510	69	41	511	435	109	55	**510**	100	93	79	38	36	34	33	25		32	46	5	–	–	–	–	–	–	–	–	–	–	Lanosta-9(11),25-dien-18-oic acid, 3-(acetyloxy)-20-hydroxy-16-oxo-, γ-lactone, (3β)-	56052-65-4	**NI45063**
201	260	200	202	69	261	245	67	**510**	100	66	18	17	17	13	11	6	**3.23**	33	50	4	–	–	–	–	–	–	–	–	–	–	Oleana-12,15-dien-28-oic acid, 3-(acetyloxy)-, methyl ester, (3β)-	39701-82-1	**NI45064**
57	86	142	43	171	215	55	69	**510**	100	78	58	53	49	44	44	42	**1.27**	33	66	3	–	–	–	–	–	–	–	–	–	–	3-Dotriacontanone, 9,11-dihydroxy-4-methyl-, [4R-(4R*,9R*,11R*)]-	28808-36-8	**NI45065**
270	87	43	57	55	41	98	83	**510**	100	86	80	68	63	47	46	43	**0.00**	33	66	3	–	–	–	–	–	–	–	–	–	–	Octadecanoic acid, 3-hydroxy-2-tetradecyl-, methyl ester	17369-87-8	**NI45066**
43	57	74	87	41	75	55	82	**510**	100	91	87	75	73	72	69	61	**0.00**	33	66	3	–	–	–	–	–	–	–	–	–	–	Octadecanoic acid, 3-hydroxy-2-tetradecyl-, methyl ester, (2R,3R)-	18951-36-5	**NI45067**
57	71	43	85	83	287	55	97	**510**	100	72	67	49	45	42	40	38	**0.44**	34	70	2	–	–	–	–	–	–	–	–	–	–	Hexadecane, 1,1′-[1,2-ethanediylbis(oxy)]bis-	17367-09-8	**NI45068**
510	407	451	435	408	389	300	105	**510**	100	28	11	11	11	10	10	10		35	26	4	–	–	–	–	–	–	–	–	–	–	1H-Cyclopropabenzene-1,1-dicarboxylic acid, 2,3,4,5-tetraphenyl-, dimethyl ester		**LI09477**
510	495	121	227	149	255	280	240	**510**	100	40	27	26	19	17	15	12		35	30	2	2	–	–	–	–	–	–	–	–	–	Benzonitrile, 4,4′-[5,5-bis(4-methoxyphenyl)-3,3-dimethyl-1,4-pentadienylidene]bis-	74367-26-3	**NI45069**
510	149	280	135	227	495	121	293	**510**	100	53	47	44	42	40	39	36		35	30	2	2	–	–	–	–	–	–	–	–	–	Benzonitrile, 4,4′-[3-[2,2-bis(4-methoxyphenyl)vinyl]-2,2-dimethylcyclopropylidene]bis-	67437-14-3	**NI45070**
510	91	243	369	230	117	229	165	**510**	100	28	16	13	13	13	11	11		38	54	–	–	–	–	–	–	–	–	–	–	–	Hexacosa-11,15-dien-13-yne, 12,15-diphenyl-		**NI45071**
358	511	359	43	224	155	288	205	**511**	100	36	22	21	19	15	12	12		15	12	6	–	–	9	–	–	–	–	–	–	1	Chromium, tris(1,1,1-trifluoro-2,4-pentanedionato-O,O′)-	14592-89-3	**NI45072**
358	511	224	155	205	135	288	174	**511**	100	28	15	12	11	11	8	7		15	12	6	–	–	9	–	–	–	–	–	–	1	Chromium, tris(1,1,1-trifluoro-2,4-pentanedionato-O,O′)-	14592-89-3	**LI09478**
358	155	224	205	174	135	238	359	**511**	100	89	78	61	54	43	32	22	**8.10**	15	12	6	–	–	9	–	–	–	–	–	–	1	Chromium, tris(1,1,1-trifluoro-2,4-pentanedionato-O,O′)-	14592-89-3	**NI45073**
438	144	511	119	388	439	291	116	**511**	100	38	33	29	27	20	16	15		18	11	5	1	–	10	–	–	–	–	–	–	–	Indoleacetic acid, 5-hydroxy-, 15-bis(pentafluoropropanoyl)-, ethyl ester		**MS10451**
205	73	261	160	147	217	219	206	**511**	100	91	40	35	31	22	20	18	**0.00**	20	49	6	1	–	–	–	–	–	4	–	–	–	D-Glucose, 3-O-methyl-2,4,5,6-tetrakis-O-(trimethylsilyl)-, O-methyloxime	56196-03-3	**NI45074**
60	43	57	45	15	41	149	187	**511**	100	95	85	80	73	70	47	46	**0.00**	22	33	9	5	–	–	–	–	–	–	–	–	–	N[5]-Hydroxy-N[2]-methyl-(N-salicoyl-β-alanylglycyl-L-seryl)-L-ornithine methyl ester		**DD01767**
219	235	228	171	189	260	295	337	**511**	100	79	69	52	40	38	15	10	**0.00**	23	45	11	1	–	–	–	–	–	–	–	–	–	GlcNAc-β-(1-3)-galol, methylated		**MS10452**
93	77	105	201	107	67	61	108	**511**	100	96	45	40	38	38	37	33	**3.75**	24	25	8	5	–	–	–	–	–	–	–	–	–	3H-Pyrazol-3-one, 2,4-dihydro-4-[[5-hydroxy-1-phenyl-3-(1,2,3-trihydroxypropyl)-1H-pyrazol-4-yl]imino]-2-phenyl-5-(1,2,3-trihydroxypropyl)-	35454-88-7	**NI45075**
512	514	434	515	513	435	404	467	**511**	100	97	64	26	23	23	13	13	**0.00**	26	24	6	–	–	–	–	1	–	–	–	–	–	3-Bromo-5,5′,8,8′-tetramethoxy-2,3′-dimethyl-1,2′-binaphthalene-1′,4-diol		**NI45076**
77	105	106	104	103	28	91	51	**511**	100	65	60	55	42	40	39	39	**0.00**	30	29	5	3	–	–	–	–	–	–	–	–	–	Hydroxylamine, N,N-bis[1-phenyl-3-hydroxyimino-3-(2-hydroxyphenyl)propyl]-		**MS10453**
89	77	136	51	91	220	180	511	**511**	100	78	41	41	37	34	34	27		31	33	2	3	1	–	–	–	–	–	–	–	–	8-(Diethylamino)-1-(ethylthio)-7-methyl-2,4,6-triphenyl-2,6-diazabicyclo[2.2.2]oct-7-ene-3,5-dione		**DD01768**
169	511	375	217	334	512	141	300	**511**	100	43	24	17	15	14	12	11		31	33	4	3	–	–	–	–	–	–	–	–	–	2-Naphthalenol, 3-(4-ethoxyanilinocarbonyl)-1-(4-hexyloxyphenylazo)-		**IC04539**
69	131	181	100	293	119	93	243	**512**	100	33	13	13	12	10	10	6	**0.17**	11	–	–	–	–	20	–	–	–	–	–	–	–	Decalin, 1-methyl-perfluoro-		**AI02400**

MASS TO CHARGE RATIOS								M.W.	INTENSITIES								Parent	C	H	O	N	S	F	Cl	Br	I	Si	P	B	X	COMPOUND NAME	CAS Reg No	No
69	131	181	100	293	343	119	93	**512**	100	30	12	12	8	7	7	7	**0.29**	11	–	–	–	–	20	–	–	–	–	–	–	–	Decalin, 1-methyl-perfluoro-		IC04540
164	107	136	132	407	405	167	166	**512**	100	87	76	51	48	48	48	48	**0.00**	12	12	7	2	1	–	3	1	–	–	–	–	–	2,2,2-Trichloroethyl 7-methoxycarbonylamino-3-bromomethyl-3-cephem-4 carboxylate 1,1-dioxide	97609-91-1	NI45077
360	59	208	255	240	165	121	225	**512**	100	90	32	15	15	15	13	12	**0.00**	12	36	12	–	–	–	–	–	–	5	–	–	–	Tetrakis(trimethoxysilyl)silane	20138-20-9	NI45078
225	206	400	187	156	512			**512**	100	99	61	55	50	32				13	3	4	–	–	9	–	–	–	–	–	–	2	Cobalt, tetracarbonyl-π-3,3,3-trifluoropropynyldi-		LI09479
112	45	161	85	163	27	113	39	**512**	100	22	20	20	11	11	8	8	**0.00**	14	22	6	2	4	–	2	–	–	–	–	–	–	Chlormethiazole ethanedisulphonate	1867-58-9	NI45079
177	288	232	400	316	178	176	344	**512**	100	74	71	66	45	43	33	29	**21.00**	16	18	8	–	2	–	–	–	–	–	–	–	2	Manganese, tetracarbonylbutanethiolato-, dimer		MS10454
95	204	109	81	67	91	205	149	**512**	100	88	87	84	82	71	70	70	**0.00**	19	30	–	–	–	–	–	–	2	–	–	–	–	Androstane, 17,18-diiodo-, (5α,17β)-	56053-11-3	NI45080
73	103	217	205	147	307	319	422	**512**	100	45	40	30	30	20	10	1	**0.00**	20	52	5	–	–	–	–	–	–	5	–	–	–	Arabinitol, pentakis-O-(trimethylsilyl)-		LI09480
73	103	217	205	147	307	129	319	**512**	100	46	41	32	30	20	14	12	**0.00**	20	52	5	–	–	–	–	–	–	5	–	–	–	Arabinitol, pentakis-O-(trimethylsilyl)-	25138-28-7	NI45081
73	217	103	147	205	307	74	129	**512**	100	29	28	21	17	10	9	9	**0.00**	20	52	5	–	–	–	–	–	–	5	–	–	–	Arabinitol, pentakis-O-(trimethylsilyl)-	14199-73-6	NI45082
73	103	217	205	147	319	307	129	**512**	100	44	39	36	30	19	19	16	**0.00**	20	52	5	–	–	–	–	–	–	5	–	–	–	Ribitol, 1,2,3,4,5-pentakis-O-(trimethylsilyl)-	32381-53-6	NI45083
73	103	147	217	205	129	74	45	**512**	100	26	15	10	8	8	8	8	**0.00**	20	52	5	–	–	–	–	–	–	5	–	–	–	Ribitol, 1,2,3,4,5-pentakis-O-(trimethylsilyl)-	32381-53-6	NI45084
73	103	217	147	205	319	129	307	**512**	100	28	27	26	23	13	11	11	**0.00**	20	52	5	–	–	–	–	–	–	5	–	–	–	Ribitol, 1,2,3,4,5-pentakis-O-(trimethylsilyl)-	32381-53-6	NI45085
73	103	217	204	147	307	129	319	**512**	100	53	47	36	32	27	16	15	**0.00**	20	52	5	–	–	–	–	–	–	5	–	–	–	Xylitol, 1,2,3,4,5-pentakis-O-(trimethylsilyl)-	14199-72-5	NI45086
73	217	103	147	205	307	319	129	**512**	100	62	47	38	34	28	18	16	**0.00**	20	52	5	–	–	–	–	–	–	5	–	–	–	Xylitol, 1,2,3,4,5-pentakis-O-(trimethylsilyl)-	14199-72-5	NI45087
73	217	103	147	205	74	307	117	**512**	100	26	25	25	10	8	8	8	**0.00**	20	52	5	–	–	–	–	–	–	5	–	–	–	Xylitol, 1,2,3,4,5-pentakis-O-(trimethylsilyl)-	14199-72-5	NI45088
344	172	400	428	288	91	115	484	**512**	100	58	40	38	36	26	24	15	**0.00**	24	16	6	–	–	–	–	–	–	–	–	–	2	Hexacarbonyl-μ-[(1,2,2α-η^3:3,3α,4-η^3)-2,3-di-(E)-benzylidene-1,3-anti-butadiene]diiron	122624-37-7	DD01769
344	56	228	172	400	288	115	428	**512**	100	79	40	39	38	33	28	27	**0.00**	24	16	6	–	–	–	–	–	–	–	–	–	2	Hexacarbonyl-μ-[(1,2,2α-η^3:3,3α,4-η^3)-2,3-di-(E)-benzylidene-1,3-syn-butadiene]diiron	122675-05-2	DD01770
344	288	115	56	172	400	228	112	**512**	100	49	49	39	34	30	27	20	**0.50**	24	16	6	–	–	–	–	–	–	–	–	–	2	Iron, hexacarbonylbis(π-phenylallyl)di-		LI09481
182	238	177	274	138	276	192	123	**512**	100	84	66	48	48	36	21	15	**0.00**	25	30	7	–	–	–	2	–	–	–	–	–	–	Benzoic acid, 3,5-dichloro-2,4-dihydroxy-6-pentyl-, 4-carboxy-3-methoxy 5-pentylphenyl ester	64185-14-4	NI45089
43	329	107	81	45	93	69	55	**512**	100	66	65	63	62	56	53	50	**31.00**	25	34	4	–	–	6	–	–	–	–	–	–	–	Pregnane-3,20-diol, bis(trifluoroacetate), (3β,5α)-	56614-55-2	NI45090
380	381	293	175	321	263	43	295	**512**	100	25	17	9	8	7	7	6	**0.00**	26	24	11	–	–	–	–	–	–	–	–	–	–	4-(D-Xylosyloxy)-6,7-dimethoxy-9-[3,4-(methylenedioxy)phenyl]naphtho[2,3-c]furan-1(3H)-one		NI45091
97	115	326	325	344	316	343	300	**512**	100	60	35	27	25	25	24	23	**0.00**	27	44	9	–	–	–	–	–	–	–	–	–	–	1β,2β,3β,14α,20R,22R,24ξ,25-octahydroxy-5β-cholest-7-en-6-one	102099-20-7	NI45092
73	225	173	75	337	243	195	129	**512**	100	44	17	16	16	11	10	9	**0.63**	27	52	5	–	–	–	–	–	–	2	–	–	–	Prostacyclin, bis(trimethylsilyl)-		NI45093
82	31	137	115	43	81	105	95	**512**	100	89	82	77	72	62	61	56	**8.18**	29	36	8	–	–	–	–	–	–	–	–	–	–	Verrucarin A, 2′,3′-didehydro-7′-deoxo-2′-deoxy-7′,5′-(ethylideneoxy)-	29953-50-2	NI45094
82	31	137	43	81	105	95	247	**512**	100	90	83	73	65	61	57	55	**8.00**	29	36	8	–	–	–	–	–	–	–	–	–	–	Verrucarin A, 2′,3′-didehydro-7′-deoxo-2′-dioxy-7′,5′-(ethylideneoxy)-	29953-50-2	LI09482
497	455	241	220	184	263	199	512	**512**	100	16	10	7	6	5	3	2		29	44	4	–	–	–	–	–	–	2	–	–	–	1-Pentanone, 1,1′-[methylenebis[6-[(trimethylsilyl)oxy]-3,1-phenylene]]bis	55517-96-9	LI09483
497	73	75	147	498	425	44	205	**512**	100	61	46	44	39	38	31	30	**2.00**	29	44	4	–	–	–	–	–	–	2	–	–	–	1-Pentanone, 1,1′-[methylenebis[6-[(trimethylsilyl)oxy]-3,1-phenylene]]bis	55517-96-9	NI45095
179	110	121	77	45	256	147	370	**512**	100	16	10	7	6	5	3	0	**0.00**	30	24	–	–	4	–	–	–	–	–	–	–	–	3,6-Diphenyl-4-phenylthio-3-[(2-phenylthio)ethenyl]-3,4-dihydro-1,2-dithiin		MS10455
512	162	513	269	134	241	136	240	**512**	100	44	35	24	22	21	18	15		30	32	4	4	–	–	–	–	–	–	–	–	–	5,10,15,20(22H,24H)-Porphinetetrone, 2,7,12-triethyl-3,8,13,18-tetramethyl-	27934-21-0	NI45096
125	43	124	389	285	311	434	446	**512**	100	35	27	17	13	7	5	4	**2.20**	30	40	7	–	–	–	–	–	–	–	–	–	–	Ergosta-2,24-dien-26-oic acid, 18-(acetyloxy)-5,6-epoxy-4,22-dihydroxy-1-oxo-, δ-lactone, (4β,5β,6β,20S,22R)-		LI09484
125	43	124	131	97	41	389	126	**512**	100	35	30	18	18	18	15	15	**2.36**	30	40	7	–	–	–	–	–	–	–	–	–	–	Ergosta-2,24-dien-26-oic acid, 18-(acetyloxy)-5,6-epoxy-4,22-dihydroxy-1-oxo-, δ-lactone, (4β,5β,6β,22R)-	21902-99-8	NI45097
123	43	81	95	312	83	294	59	**512**	100	30	24	20	19	18	17	16	**4.00**	30	40	7	–	–	–	–	–	–	–	–	–	–	Phorbol, 13-acetyloxy-4-deoxy-12-O-octadienoyl-, (2Z,4E)-		MS10456
57	83	41	56	60	103	43	29	**512**	100	37	25	22	17	14	14	13	**0.00**	30	56	6	–	–	–	–	–	–	–	–	–	–	Hexanoic acid, 3,5,5-trimethyl-, 1,2,3-propanetriyl ester	56554-53-1	NI45098
141	57	71	43	355	55	215	41	**512**	100	44	31	30	25	17	15	15	**0.00**	30	56	6	–	–	–	–	–	–	–	–	–	–	Trinonanoin	126-53-4	NI45099
270	43	212	154	245	184	85	114	**512**	100	99	77	60	53	40	33	27	**15.00**	30	60	4	2	–	–	–	–	–	–	–	–	–	Acetamide, 2,2′-[1,2-ethanediylbis(oxy)]bis N,N-dihexyl-	67456-22-8	NI45100
512	513	514	440	422	515	481	385	**512**	100	41	16	12	10	6	5	4		31	48	4	–	–	–	–	–	–	1	–	–	–	10′-Apo-β,ψ-carotenoic acid, 5,6-dihydro-6-hydroxy-5-[(trimethylsilyl)oxy]-, methyl ester, (5R,6R)-	56051-70-8	NI45101
512	513	28	200	456	469	400	413	**512**	100	37	36	30	26	19	18	15		32	32	6	–	–	–	–	–	–	–	–	–	–	Diamyl binaphthalene dioxide quinone		IC04541
385	285	259	512	361	469	497		**512**	100	68	66	61	56	19	10			32	48	5	–	–	–	–	–	–	–	–	–	–	Ergosta-7,22-dien-6-one, 2,3-bis(acetyloxy)-, (2β,3β,5β)-		LI09485
393	452	55	69	512	127	295	437	**512**	100	49	48	47	41	38	35	34		32	48	5	–	–	–	–	–	–	–	–	–	–	Lanost-9(11)-en-18-oic acid, 23-(acetyloxy)-20-hydroxy-3-oxo-, γ-lactone, (20ξ)-	36872-78-3	NI45102
99	413	57	329	414	85	83	41	**512**	100	29	20	15	12	10	9	8	**8.00**	32	48	5	–	–	–	–	–	–	–	–	–	–	Lanost-9(11)-en-18-oic acid, 3,3-[1,2-ethanediylbis(oxy)]-20-hydroxy-23-oxo-, γ-lactone, (20ξ)-	36871-83-7	NI45103
182	183	169	77	150	168	104	167	**512**	100	15	14	14	12	9	8	4	**0.70**	33	28	–	4	1	–	–	–	–	–	–	–	–	Benzimidazole, 2,3-dihydro-1,3-bis[(diphenylamino)ethyl]-2-thio-		IC04542
383	439	497	271	454	313	398	258	**512**	100	92	77	46	44	18	13	13	**2.60**	33	52	4	–	–	–	–	–	–	–	–	–	–	24,25-Dihydroxy-24S-cycloartane-3,16-dione 24,25-acetonide		NI45104
262	203	189	202	263	204	190	249	**512**	100	98	24	22	20	19	19	18	**7.00**	33	52	4	–	–	–	–	–	–	–	–	–	–	Olean-12-en-28-oic acid, 3-(acetyloxy)-, methyl ester, (3β)-	1721-57-9	MS10457

MASS TO CHARGE RATIOS								M.W.	INTENSITIES								Parent	C	H	O	N	S	F	Cl	Br	I	Si	P	B	X	COMPOUND NAME	CAS Reg No	No
203	262	202	70	204	263	249	147	**512**	100	59	26	23	18	12	9	6	**1.47**	33	52	4	–	–	–	–	–	–	–	–	–	–	Olean-12-en-28-oic acid, 3-(acetyloxy)-, methyl ester, (3β)-	1721-57-9	NI45105
512	513	456	429	457	514	430	143	**512**	100	44	33	31	15	14	10	9		33	56	2	–	–	–	–	–	–	1	–	–	–	$\Delta^{1(6)}$-Tetrahydrocannabinol tri-butylsilyl ether		NI45106
512	441	513	456	497	457	442	498	**512**	100	40	36	31	24	15	15	14		33	56	2	–	–	–	–	–	–	1	–	–	–	Δ^{1}-Tetrahydrocannabinol tri-butylsilyl ether		NI45107
497	498	199	512	499	513	143	200	**512**	100	42	17	14	14	6	5	3		33	56	2	–	–	–	–	–	–	1	–	–	–	Δ^{3}-Tetrahydrocannabinol tributylsilyl ether		NI45108
512	414	382	440	105	304	221	135	**512**	100	51	39	34	23	22	22	20		34	28	3	–	–	–	–	–	–	1	–	–	–	4,7-Silanoisobenzofuran-1,3-dione, 3a,4,7,7a-tetrahydro-8,8-dimethyl-4,5,6,7-tetraphenyl-, (3aα,4α,7α,7aα)-	51528-48-4	MS10458
407	408	497	512	241	95	123	109	**512**	100	34	33	25	21	18	17	17		34	60	1	–	–	–	–	–	–	1	–	–	–	Lanost-8-ene, 24-methylene-3-[(trimethylsilyl)oxy]-, (3β)-	55518-07-5	NI45109
512	91	69	133	420	510	406	109	**512**	100	23	17	14	13	12	10	7		36	48	2	–	–	–	–	–	–	–	–	–	–	Neurosporaxanthin, methyl ester		LI09486
512	119	105	91	157	145	69	93	**512**	100	28	24	23	18	17	17	15		36	48	2	–	–	–	–	–	–	–	–	–	–	Neurosporaxanthin, methyl ester		MS10459
207	165	249	120	285	240	513		**513**	100	85	76	58	24	24	2			18	43	–	1	–	–	–	–	–	–	–	–	2	Bis(tripropylstannyl)amine		LI09487
73	236	75	188	175	148	61	192	**513**	100	47	42	29	27	20	20	16	**0.30**	20	39	3	5	1	–	–	–	–	3	–	–	–	Adenosine, 5′-S-methyl-5′-thio-N-(trimethylsilyl)-2′,3′-bis-O-(trimethylsilyl)-	54623-28-8	NI45110
175	188	164	169	236	176	189	135	**513**	100	85	58	34	26	17	13	12	**1.00**	20	39	3	5	1	–	–	–	–	3	–	–	–	Adenosine, 5′-S-methyl-5′-thio-N-(trimethylsilyl)-2′,3′-bis-O-(trimethylsilyl)-	54623-28-8	MS10460
167	273	107	79	91	183	240		**513**	100	56	50	36	23	17	6		**0.00**	21	15	11	5	–	–	–	–	–	–	–	–	–	Tyrosine, bis(2,4-dinitrophenyl)-		LI09488
119	136	436	434	414	416	438	296	**513**	100	47	20	17	15	14	10	8	**0.00**	21	27	3	1	1	–	4	–	–	–	–	–	–	(1RS,4SR,5SR,6SR,7SR,8SR)-5,6,7,8-Tetrakis(chloromethyl)bicyclo[2.2.2]octan-2-one O-(2,4,6-trimethylbenzenesulphonyl)oxime		DD01771
97	196	224	210	252	302	296	208	**513**	100	51	38	15	13	10	9	9	**2.60**	21	27	10	3	1	–	–	–	–	–	–	–	–	1,6-Dihydro-1-methyl-2-methylthio-5-((E)-2-methoxycarbonylvinyl)-4-(2,3,4-tri-O-acetyl-β-D-xylopyranosylamino)-1H-pyrimidin-6-one		MS10461
97	210	182	238	194	288	320	296	**513**	100	41	37	12	10	8	8	7	**6.30**	21	27	10	3	1	–	–	–	–	–	–	–	–	1,6-Dihydro-2-methylthio-5-((E)-2-ethoxycarbonylvinyl)-4-(2,3,4-tri-O-acetyl-β-D-xylopyranosylamino)-1H-pyrimidin-6-one		MS10462
171	170	57	172	135	121	173	87	**513**	100	87	77	60	60	58	54	54	**9.00**	24	12	–	3	–	9	–	–	–	–	–	–	–	s-Triazine, 2,4,6-tris(p-trifluoromethylphenyl)-		MS10463
208	209	73	117	207	118	119	75	**513**	100	57	31	24	20	16	11	10	**0.51**	24	47	5	1	–	–	–	–	–	3	–	–	–	Indicine, 5,7a-didehydro-tris(trimethylsilyl)-		MS10464
513	105	514	77	290	212	168	144	**513**	100	74	34	34	18	8	7	7		30	22	5	–	–	–	–	–	–	–	–	–	1	Vanadium, bis(1,3-diphenyl-1,3-propanedionato-O,O′)oxo-	15022-46-5	NI45111
73	513	75	514	483	388	484	166	**513**	100	99	77	53	42	42	32	29		32	55	2	1	–	–	–	–	–	1	–	–	–	Ergosta-7,22-dien-6-one, 3-[(trimethylsilyl)oxy]-, O-methyloxime, (3β,5α,22E)-	53286-61-6	NI45112
73	513	75	54	69	514	483	388	**513**	100	99	75	65	52	50	45	45		32	55	2	1	–	–	–	–	–	1	–	–	–	Ergosta-7,22-dien-6-one, 3-[(trimethylsilyl)oxy]-, O-methyloxime, (3β,5α,22E)-	53286-61-6	NI45113
512	513	514	515	416	256.5	294	247.5	**513**	100	88	27	5	4	3	2	2		35	22	–	1	–	3	–	–	–	–	–	–	–	Pyridine, 2,4,6-tris(p-fluorophenyl)-3,5-diphenyl-		LI09489
512	513	514	515	416	60	57	55	**513**	100	88	27	5	4	2	2	2		35	22	–	1	–	3	–	–	–	–	–	–	–	Pyridine, 2,4,6-tris(p-fluorophenyl)-3,5-diphenyl-	22158-33-4	MS10465
422	105	44	423	91	77	81	55	**513**	100	94	94	32	23	13	12	11	**0.00**	35	47	2	1	–	–	–	–	–	–	–	–	–	Androstan-17-ol, 3-(1-benzylethylamino)-, benzoate		LI09490
513	353	314	498	146	145	368	223	**513**	100	30	30	20	20	15	10	10		36	39	–	3	–	–	–	–	–	–	–	–	–	1,4-Bis(4,7-dimethylindol-2-yl)-1,4,5,8-tetramethyl-1,2,3,4-tetrahydrocarbazole		LI09491
483	485	481	487	224	455	225	479	**514**	100	79	64	32	31	26	22	20	**1.52**	8	–	2	–	–	–	2	4	–	–	–	–	–	Terephthaloyl chloride, tetrabromo-	54120-56-8	NI45114
346	374	190	318	486	290	189	262	**514**	100	78	70	54	52	50	48	42	**2.00**	13	5	11	–	–	–	–	–	–	–	–	–	3	Cobalt, nonacarbonyl-μ^3-(carboxyethylidyne)tri-, methyl ester, triangulo-	18433-90-4	NI45115
205	203	312	514	391	310	108	77	**514**	100	42	30	22	17	13	13	10		14	14	2	–	2	–	–	–	–	–	1	–	1	Thallium(I) O,O′-bis(2-methylphenyl)dithiophosphate		MS10466
205	203	312	170	514	77	391	39	**514**	100	42	25	20	17	15	14	13		14	14	2	–	2	–	–	–	–	–	1	–	1	Thallium(I) O,O′-bis(3-methylphenyl)dithiophosphate		MS10467
205	203	312	391	514	310	108	107	**514**	100	43	37	21	16	16	13	13		14	14	2	–	2	–	–	–	–	–	1	–	1	Thallium(I) O,O′-bis(4-methylphenyl)dithiophosphate		MS10468
84	485	115	486	402	173	397	198	**514**	100	27	13	5	5	5	4	4	**0.00**	18	40	–	2	–	–	–	–	–	–	–	–	2	Indium, tetraethylbis[μ-(piperidinyl)]di-	42745-49-3	MS10469
174	73	327	298	100	147	86	121	**514**	100	92	41	24	21	18	16	15	**0.10**	18	51	3	2	–	–	–	–	–	5	1	–	–	Phosphonic acid, [3-[bis(trimethylsilyl)amino]-1-[(trimethylsilyl)amino]propyl]-, bis(trimethylsilyl) ester	53044-46-5	NI45116
174	73	327	298	175	100	147	86	**514**	100	92	41	24	21	21	18	16	**0.15**	18	51	3	2	–	–	–	–	–	5	1	–	–	Phosphonic acid, (1,3-diaminopropyl)-, pentakis(trimethylsilyl)-	56211-25-7	NI45117
43	386	374	58	56	183	69	71	**514**	100	46	45	41	35	27	22	17	**14.00**	19	26	2	–	–	12	–	–	–	–	–	–	–	1H,1H,7H-Dodecafluoroheptyl dodecanoate		LI09492
43	387	374	57	55	31	41		**514**	100	46	45	41	35	33	31		**14.60**	19	26	2	–	–	12	–	–	–	–	–	–	–	1H,1H,7H-Dodecafluoroheptyl dodecanoate		MS10470
127	257	255	512	514	253	254	510	**514**	100	54	49	41	39	38	33	32		20	14	–	–	–	–	–	–	–	–	–	–	2	Ditelluride, di-1-naphthalenyl	32294-58-9	NI45118
512	514	127	510	516	508	511	257	**514**	100	90	85	68	53	37	30	29		20	14	–	–	–	–	–	–	–	–	–	–	2	Ditelluride, di-2-naphthalenyl	1666-12-2	NI45119
103	59	87	425	44	427	104	207	**514**	100	68	53	18	14	12	12	9	**2.00**	20	20	8	4	–	–	2	–	–	–	–	–	–	Aniline, N,N-bis(2-carbomethoxyethoxy)-4-(2,6-dichloro-4-nitrophenylazo)-		IC04543
97	69	243	139	183	157	172	225	**514**	100	70	63	55	42	39	36	33	**5.00**	20	26	10	4	1	–	–	–	–	–	–	–	–	5-(N,N-Diacetylamino)-2-(methylthio)-4-(2,3,4-tri-O-acetyl-β-D-xylopyranosylamino)-1H-pyrimidin-6-one		NI45120

MASS TO CHARGE RATIOS								M.W.	INTENSITIES								Parent	C	H	O	N	S	F	Cl	Br	I	Si	P	B	X	COMPOUND NAME	CAS Reg No	No
501	503	499	516	437	435	518	267	**514**	100	55	55	49	31	31	27	26	**25.00**	21	24	5	–	–	–	–	2	–	–	–	–	–	2H,8H-Benzo[1,2-b:5,4-b']dipyran-10-propanoic acid, 3,7-dibromo-5-methoxy-2,2,8,8-tetramethyl-, methyl ester	56336-17-5	NI45121
129	147	145	158	103	101	219	105	**514**	100	57	48	41	35	32	30	28	**0.61**	27	54	5	–	–	–	–	–	–	2	–	–	–	Prostan-1-oic acid, 6,9-epoxy-11,15-bis[(trimethylsilyl)oxy]-, methyl ester, (15S)-	56248-52-3	NI45122
258	514	274	515	243	516	69	259	**514**	100	91	67	35	23	21	16	15		28	18	10	–	–	–	–	–	–	–	–	–	–	Ustilaginoidin A		NI45123
83	55	514	269	300	343	312	301	**514**	100	98	56	55	54	34	19	19		28	34	9	–	–	–	–	–	–	–	–	–	–	Gomisin B	58546-55-7	NI45124
55	43	343	83	342	300	514	414	**514**	100	99	96	92	81	66	62	48		28	34	9	–	–	–	–	–	–	–	–	–	–	Gomisin F	62956-47-2	NI45125
196	177	137	221	164	252	195	41	**514**	100	70	45	40	32	27	23	23	**0.00**	29	38	8	–	–	–	–	–	–	–	–	–	–	o-Anisic acid, 4-hydroxy-6-pentyl-, methyl ester, ester with 2-hydroxy-6-(2-oxoheptyl)-p-anisic acid	19314-73-9	NI45126
196	177	150	221	252	137	206	182	**514**	100	55	42	40	37	35	34	33	**0.00**	29	38	8	–	–	–	–	–	–	–	–	–	–	o-Anisic acid, 4-hydroxy-6-pentyl-, methyl ester, ester with 2-hydroxy-6-(2-oxoheptyl)-p-anisic acid	19314-73-9	NI45127
514	169	168	155	154	153	515	167	**514**	100	81	71	49	46	41	31	30		29	38	8	–	–	–	–	–	–	–	–	–	–	p-Benzoquinone, 5-methyl-2,2'-(8-hexadecenylene)bis[3,6-dihydroxy-, (Z)-	21596-60-1	NI45128
257	300	514	370	214	469	144	229	**514**	100	68	40	18	9	7	5	4		29	46	4	4	–	–	–	–	–	–	–	–	–	L-Alanine, N-[N-[N-[3-[4-(diethylamino)phenyl]-2-propenylidene]-L-valyl]-L-isoleucyl]-, ethyl ester	37580-29-3	NI45129
514	516	515	517	120	518	121	77	**514**	100	70	35	23	17	15	11	7		30	24	2	2	–	–	2	–	–	–	–	–	–	Anthraquinone, 6,7-dichloro-1,4-bis(2,6-dimethylanilino)-		IC04544
514	115	144	360	515	131	116	104	**514**	100	46	43	37	36	36	28	18		30	34	4	4	–	–	–	–	–	–	–	–	–	Pyrrolo[3,4-c]pyrrolo[3',4':4,5]pyrazolo[1,2-a]pyrazole-1,3,6,8(2H,3aH,5H,7H)-tetrone, 2,7-dibutyltetrahydro-5,10-diphenyl-		LI09493
514	115	144	360	515	131	116	77	**514**	100	45	42	37	36	35	27	18		30	34	4	4	–	–	–	–	–	–	–	–	–	Pyrrolo[3,4-c]pyrrolo[3',4':4,5]pyrazolo[1,2-a]pyrazole-1,3,6,8(2H,3aH,5H,7H)-tetrone, 2,7-dibutyltetrahydro-5,10-diphenyl-	42244-31-5	NI45130
43	57	55	41	69	83	56	71	**514**	100	92	87	74	58	45	45	42	**4.50**	30	58	4	–	1	–	–	–	–	–	–	–	–	Dilaurylthio dipropionate		IC04545
43	57	55	69	83	71	41	97	**514**	100	98	69	56	46	45	45	32	**29.00**	30	58	4	–	1	–	–	–	–	–	–	–	–	Dilaurylthio dipropionate		IC04546
43	57	55	143	71	69	178	132	**514**	100	96	85	56	48	48	41	37	**7.40**	30	58	4	–	1	–	–	–	–	–	–	–	–	Dilaurylthio dipropionate		MS10471
117	73	75	149	118	69	103	147	**514**	100	32	27	19	10	9	8	6	**0.08**	30	66	2	–	–	–	–	–	–	2	–	–	–	3,27-Dioxa-2,28-disilanonacosane, 2,2,4,28,28-pentamethyl-	56196-19-1	NI45131
117	73	75	118	149	119	76	74	**514**	100	19	15	11	6	6	5	3	**0.12**	30	66	2	–	–	–	–	–	–	2	–	–	–	3,26-Dioxa-2,27-disilaoctacosane, 2,2,4,25,27,27-hexamethyl-	56196-18-0	NI45132
91	55	482	459	213	228	255	499	**514**	100	70	10	10	10	9	7	5	**5.00**	31	46	4	–	1	–	–	–	–	–	–	–	–	3,5-Cyclonorcholestan-23-ol, 6-methoxy-, tolysylate, (3α,6β)-		NI45133
115	394	43	97	297	298	325	69	**514**	100	34	22	20	19	16	14	14	**0.00**	31	46	6	–	–	–	–	–	–	–	–	–	–	Cholest-5-ene-16,22-dione, 3,26-bis(acetyloxy)-, (3β,25R)-	7554-95-2	NI45134
394	454	325	297	298	395	280	455	**514**	100	46	41	41	37	29	16	15	**0.30**	31	46	6	–	–	–	–	–	–	–	–	–	–	Cholest-5-ene-16,22-dione, 3,26-bis(acetyloxy)-, (3β,25R)-		MS10472
185	197	184	282	43	441	69	454	**514**	100	59	43	36	29	20	20	18	**0.21**	31	46	6	–	–	–	–	–	–	–	–	–	–	Furost-5-ene-3,26-diol, 22,25-epoxy-, diacetate, (3β,22α,25S)-	56196-24-8	NI45135
347	480	512	434	407	494			**514**	100	45	40	25	20	15				31	46	6	–	–	–	–	–	–	–	–	–	–	Methyl 9α,11α-epoxy-16,23-methyleneanhydrotigogenoate acetate		LI09494
446	447	393	514	380	448	407	394	**514**	100	40	30	18	18	15	15	11		31	54	2	–	–	–	–	–	–	2	–	–	–	Cannabidiol, bis(propyldimethylsilyl)-		NI45136
514	515	103	130	129	257	102		**514**	100	38	32	27	25	19	9			32	18	–	8	–	–	–	–	–	–	–	–	–	Phthalocyanine	574-93-6	IC04547
514	130	515	129	516	131	128	257	**514**	100	50	40	30	9	7	7	5		32	18	–	8	–	–	–	–	–	–	–	–	–	Phthalocyanine	574-93-6	NI45137
43	69	395	514	81	55	41	57	**514**	100	27	26	25	24	24	20	17		32	50	5	–	–	–	–	–	–	–	–	–	–	Lanost-9(11)-en-18-oic acid, 23-(acetyloxy)-3,20-dihydroxy-, γ-lactone, (3β,20ξ)-	36872-76-1	NI45138
329	415	43	99	55	397	57	69	**514**	100	65	60	56	45	43	43	35	**31.37**	32	50	5	–	–	–	–	–	–	–	–	–	–	Lanost-9(11)-en-18-oic acid, 3,3-[1,2-ethanediylbis(oxy)]-20,23-dihydroxy-, γ-lactone, (20ξ)-	36871-82-6	NI45139
121	133	175	189	147	203	159	251	**514**	100	74	71	64	43	42	32	30	**20.77**	32	50	5	–	–	–	–	–	–	–	–	–	–	A(1)-Norup-20(29)-ene-2,28-dioic acid, 3-hydroxy-, dimethyl ester, (2α,3β)-		NI45140
217	246	207	199	436	190	208	200	**514**	100	89	83	67	63	60	45	45	**9.33**	32	50	5	–	–	–	–	–	–	–	–	–	–	Olean-12-en-28-al, 16-(acetyloxy)-3,22-dihydroxy-, (3β,16α,22α)-	26339-85-5	NI45141
414	154	167	499	183	143	361	343	**514**	100	99	40	30	30	20	15	12	**4.00**	32	50	5	–	–	–	–	–	–	–	–	–	–	Spirostan-3-ol, 9,11-epoxy-16,26,26-trimethyl-, acetate, (5α,9α,11α,16α,25R)-		LI09495
159	160	146	514	147	412	338	175	**514**	100	23	19	3	3	2	2	2		32	54	3	2	–	–	–	–	–	–	–	–	–	Tryptamine, 5-hydroxy-N-(22-hydroxybehenoyl)-		MS10473
43	41	55	327	69	81	99	95	**514**	100	34	33	29	27	24	23	21	**8.00**	33	54	4	–	–	–	–	–	–	–	–	–	–	(22R,23S)-22-Acetoxy-24-methylene-5α-lanost-8-ene-3β-22-diol		MS10474
363	69	81	152	482	119	105	175	**514**	100	52	46	22	18	16	14	13	**10.00**	33	54	4	–	–	–	–	–	–	–	–	–	–	D:A-Friedooleanan-26-oic acid, 3-(acetyloxy)-, methyl ester, (3α)-		NI45142
363	69	81	152	482	119	175	189	**514**	100	52	46	22	18	16	13	12	**10.00**	33	54	4	–	–	–	–	–	–	–	–	–	–	D:A-Friedooleanan-26-oic acid, 3-(acetyloxy)-, methyl ester, (3β)-		NI45143
514	317	318	153	499	485	169	99	**514**	100	76	66	26	22	22	21	20		33	54	4	–	–	–	–	–	–	–	–	–	–	D:A-Friedooleanan-29-oic acid, 3,3-[1,2-ethanediylbis(oxy)]-, methyl ester, (20α)-	39903-14-5	MS10475
443	424	189	514	187	204	315	381	**514**	100	95	60	47	45	40	39	37		33	58	2	–	–	–	–	–	–	1	–	–	–	29,30-Dinorgammaceran-3-one, 21,21-dimethyl-22-[(trimethylsilyl)oxy]-, (8α,9β,13α,14β,17α,18β,22α)-	69774-02-3	NI45144
457	458	159	147	103	281	459	297	**514**	100	41	16	16	11	10	9	9	**0.00**	33	58	2	–	–	–	–	–	–	1	–	–	–	Pregn-5-ene-20-one, 3-[(tributylsilyl)oxy]-, (3β)-		NI45145
256	257	283	284	514	229	398	230	**514**	100	72	28	23	17	16	14	11		34	22	–	6	–	–	–	–	–	–	–	–	–	(1α,2α,3β,4β)-1,2-Bis(2-quinoxalinyl)-3,4-bis(4-cyanophe nyl)cyclobutane		MS10476
95	454	175	203	297	436	439	332	**514**	100	71	64	44	41	31	27	25	**7.00**	34	58	3	–	–	–	–	–	–	–	–	–	–	9,19-Cyclolanostan-25-ol, 3-(acetyloxy)-24,24-dimethyl-, (3β)-		LI09496
75	454	439	332	297	411	385	514	**514**	100	96	54	46	36	20	20	18		34	58	3	–	–	–	–	–	–	–	–	–	–	9,19-Cyclolanostan-25-ol, 3-(acetyloxy)-24,24-dimethyl-, (3β)-		LI09497
105	91	183	77	285	208	207	296	**514**	100	89	37	37	22	16	15	5	**0.00**	35	30	4	–	–	–	–	–	–	–	–	–	–	(4R,5R)-α,α,α',α',2-Pentaphenyl-1,3-dioxolane-4,5-dimethanol		DD01772

MASS TO CHARGE RATIOS								M.W.	INTENSITIES								Parent	C	H	O	N	S	F	Cl	Br	I	Si	P	B	X	COMPOUND NAME	CAS Reg No	No
514	382	414	135	105	441	221	305	**514**	100	35	26	26	15	9	9	8		35	34	2	–	–	–	–	–	–	1	–	–	–	7-Silabicyclo[2.2.1]hept-5-ene-2-carboxylic acid, 7,7-dimethyl-1,4,5,6-tetraphenyl-, ethyl ester	56805-07-3	MS10477
55	57	143	178	69	56	71	132	**514**	100	56	30	28	25	23	19	17	**8.00**	35	62	2	–	–	–	–	–	–	–	–	–	–	Methyl tetratricontatetraenoate		MS10478
205	234	78	63	206	45	43	57	**514**	100	94	63	60	40	36	36	33	**0.00**	36	26	–	4	–	–	–	–	–	–	–	–	–	5,10-Methanobenzo[g]phthalazine, 11-(diphenylmethylene)-4a,5,10,10a-tetrahydro-1,4-di-2-pyridinyl-, (4aα,5β,10β,10aα)-	42589-26-4	NI45146
164	276	246	134	41	291	261	169	**515**	100	88	36	33	30	21	21	18	**0.27**	9	9	5	1	–	–	–	–	–	–	–	–	2	Tungsten, pentacarbonylcyanotrimethylstannyl-		LI09498
164	134	41	149	119	40	175	402	**515**	100	33	30	15	15	13	10	1	**0.27**	9	9	5	1	–	–	–	–	–	–	–	–	2	Tungsten, pentacarbonylisocyanotrimethylstannyl-		MS10479
85	171	173	346	344	103	219	217	**515**	100	68	66	51	49	34	20	20	**2.75**	12	8	2	1	2	5	–	2	–	–	–	–	–	S,S-Bis(4-bromophenoxy)-N-pentafluorosulphanylsulphilimine		NI45147
208	81	69	138	95	51	53	446	**515**	100	96	96	66	51	41	39	35	**5.97**	15	10	7	3	–	9	–	–	–	–	–	–	–	Cytidine, 2′-deoxy-N-(trifluoroacetyl)-, 3′,5′-bis(trifluoroacetate)	35221-99-9	NI45148
362	159	515	312	293	209	473	228	**515**	100	46	17	17	12	9	3	3		15	12	6	–	–	9	–	–	–	–	–	–	1	Iron, tris(1,1,1-trifluoro-2,4-pentanedionato-O,O′)-		LI09499
80	82	79	81	217	96	55	94	**515**	100	83	49	37	11	6	5	5	**0.00**	15	24	2	3	–	–	–	3	–	–	–	–	–	5-Hydroxy-6-methoxy-8-[(4-amino-1-methylbutyl)amino]quinoline trihydrobromide	53111-25-4	NI45149
253	310	352	42	428	119	266	353	**515**	100	91	87	42	27	22	20	14	**0.00**	18	15	5	1	–	10	–	–	–	–	–	–	–	N,O-Bis(pentafluoropropionyl)-D-tyrosine isopropyl ester		MS10480
253	310	352	42	428	119	266	225	**515**	100	92	88	45	27	23	20	15	**0.00**	18	15	5	1	–	10	–	–	–	–	–	–	–	N,O-Bis(pentafluoropropionyl)-L-tyrosine isopropyl ester		MS10481
186	214	74	322	158	157	184	280	**515**	100	68	36	33	31	26	25	24	**1.60**	20	25	11	3	1	–	–	–	–	–	–	–	–	1,6-Dihydro-5-formyl-2-methylthio-4-(2,3,4,6-tetra-O-acetyl-β-D-glucopyranosylamino)-1H-pyrimidin-6-one		MS10482
210	211	94	73	208	117	93	120	**515**	100	55	44	33	30	30	30	19	**0.46**	24	49	5	1	–	–	–	–	–	3	–	–	–	Indicine, tris[(trimethylsilyl)oxy]-		MS10483
109	77	258	83	91	125	93	65	**515**	100	84	74	72	67	52	52	47	**0.00**	26	26	6	1	1	–	1	–	–	–	–	–	–	(1R*,2S*,4S*,5R*,7R*,8S*,9R*,10R*)-N-Phenyl-2-hydroxy-8-methyl-9-methoxy-9-(phenylthio)-10-chloro-11,12-dioxatricyclo[6.2.1.1^{2,7}]dodecane-4,5-dicarboximide	117626-70-7	DD01773
286	258	245	215	287	229	243	228	**515**	100	27	27	25	24	23	19	17	**0.00**	28	25	7	3	–	–	–	–	–	–	–	–	–	11-(But-3-enyl)-1,2,3,4-tetrahydrochrysene picrate		MS10484
298	299	43	150	28	515	266	99	**515**	100	38	36	31	31	27	27	19		28	37	8	1	–	–	–	–	–	–	–	–	–	Deoxyharringtonine		NI45150
500	501	369	73	515	375	75	153	**515**	100	39	35	32	25	25	25	19		28	45	4	1	–	–	–	–	–	2	–	–	–	Norethisterone, 3-(O-carboxymethyl)oxime, bis[(trimethylsilyl)oxy]-		NI45151
105	146	77	119	104	133	103	118	**515**	100	99	66	46	46	40	31	30	**4.00**	31	21	5	3	–	–	–	–	–	–	–	–	–	Dispiro[1,4,2-dioxazole-5,4′(5′H)-naphth[2,1-d]isoxazole-5′,5″-[1,4,2]dioxazole], 3′a,9′b-dihydro-3,3′,3″-triphenyl-	23767-18-2	NI45152
105	146	77	119	133	205	219	51	**515**	100	97	66	46	40	25	15	15	**0.40**	31	21	5	3	–	–	–	–	–	–	–	–	–	Dispiro[1,4,2-dioxazole-5,4′(5′H)-naphth[2,1-d]isoxazole-5′,5″-[1,4,2]dioxazole], 3′a,9′b-dihydro-3,3′,3″-triphenyl-	23767-18-2	LI09500
91	160	119	56	132	55	57	106	**515**	100	99	91	68	64	57	55	48	**10.90**	31	37	4	3	–	–	–	–	–	–	–	–	–	4,23-Dimethoxy-2,27-dioxa-10,19-diaza-9,10,19,20-dieno[2.9.20.27](1,4)diphenyl[1.28](2,6)pyridinophane		MS10485
469	470	468	409	393	304	190	485	**515**	100	70	55	35	32	30	28	20	**1.00**	32	53	4	1	–	–	–	–	–	–	–	–	–	Lanost-8-en-7-yl, 3-acetyloxy-, nitrite ester, (3β,5α,7α)-		MS10486
174	106	107	77	43	56	82	51	**515**	100	59	52	41	34	28	17	17	**0.00**	33	29	3	1	–	–	–	–	–	1	–	–	–	3-(N-Methylanilino)-2-[3-(triphenylsiloxy)-3-butenoyl]-2-cyclobuten-1-one	79139-21-2	NI45153
515	227	517	516	226	228	257	514	**515**	100	89	40	40	39	28	17	15		35	18	–	3	–	–	1	–	–	–	–	–	–	s-Triazine, 6-chloro-2,4-bis(pyren-1-yl)-		IC04548
472	473	86	57	43	55	41	81	**515**	100	37	21	20	19	18	16	14	**12.00**	35	65	1	1	–	–	–	–	–	–	–	–	–	Cholestan-3-ol, 2-(dibutylamino)-, (2β,3α,5α)-	54725-10-9	NI45154
340	199	141	428	287	268	409	481	**516**	100	79	39	21	11	11	8	7	**0.00**	8	20	–	2	–	10	–	–	–	–	4	–	1	Nickel, bis(trifluorophosphino)bis(diethylaminodifluorophosphino)-		NI45155
220	516	518	514	216	218	214	488	**516**	100	95	89	81	75	70	64	50		15	24	4	–	2	–	–	–	–	–	–	–	1	Tungsten, tetracarbonyl(2,2,8,8-tetramethyl-3,7-dithianonanyl)-		LI09501
117	52	145	173	114	201	182	234	**516**	100	95	24	17	13	12	6	2	**1.70**	16	10	6	–	–	–	–	–	–	–	–	–	3	Cadmium, bis(cyclopentadienyltricarbonylchromium)-		MS10487
163	119	59	310	516	69	313	343	**516**	100	79	69	66	58	52	40	33		17	10	7	–	–	10	–	–	–	–	–	–	–	Phenylpyranic acid, α,4-bis(pentafluoropropanoyl)-, methyl ester		MS10488
73	43	327	343	341	457	41	375	**516**	100	45	20	16	15	13	10	9	**0.00**	18	48	7	–	–	–	–	–	–	5	–	–	–	Tetrasiloxane, 3,5,5-triisopropoxy-1,1,1,7,7,7-hexamethyl-5-[(trimethylsilyl)oxy]-		MS10489
87	43	518	516	445	443	520	447	**516**	100	39	30	30	27	27	10	8		20	19	6	4	–	–	3	–	–	–	–	–	–	Aniline, N,N-bis(2-acetoxyethyl)-3-chloro-4-(2,6-dichloro-4-nitrophenylazo)-		IC04549
43	87	443	516	444	88	30	517	**516**	100	88	15	8	4	4	4	2		22	24	9	6	–	–	–	–	–	–	–	–	–	Aniline, N,N-bis(2-acetoxyethyl)-3-methyl-4-(2,6-dinitro-3-carbamylphenylazo)-		IC04550
41	39	307	173	149	28	174	69	**516**	100	83	41	41	40	38	28	26	**0.00**	23	21	–	2	2	–	–	–	1	–	–	–	–	Benzothiazolium, 3-(2-propenyl)-2-[3-[3-(2-propenyl)-2(3H)-benzothiazolylidene]-1-propenyl]-, iodide	34263-37-1	NI45156
208	439	285	206	438	207	437	283	**516**	100	90	67	47	42	42	40	31	**0.05**	24	20	–	–	–	–	–	–	–	–	–	–	1	Plumbane, tetraphenyl-	595-89-1	NI45157
208	206	207	51	78	52	50	39	**516**	100	45	43	8	6	5	5	5	**0.00**	24	20	–	–	–	–	–	–	–	–	–	–	1	Plumbane, tetraphenyl-	595-89-1	NI45158
208	439	285	205	438	283	51	154	**516**	100	88	76	45	39	28	12	11	**0.00**	24	20	–	–	–	–	–	–	–	–	–	–	1	Plumbane, tetraphenyl-	595-89-1	NI45159
179	166	73	306	180	41	29	308	**516**	100	48	44	30	18	15	13	8	**0.00**	24	35	5	2	–	3	–	–	–	1	–	–	–	L-Propyltyrosine, α-methyl-N-(trifluoroacetyl)-4-[(trimethylsilyl)oxy]-, butyl ester		MS10490
501	502	503	73	504	369	485	486	**516**	100	40	20	15	5	5	3	2	**0.60**	25	36	6	–	–	–	–	–	–	3	–	–	–	1,6-Dihydroxy-3-(hydroxymethyl)-8-methoxyanthraquinone, tris(trimethylsilyl)-	91701-33-6	NI45160

MASS TO CHARGE RATIOS								M.W.	INTENSITIES								Parent	C	H	O	N	S	F	Cl	Br	I	Si	P	B	X	COMPOUND NAME	CAS Reg No	No
501	502	503	413	471	504	472	414	**516**	100	39	22	9	9	6	4	4	**0.40**	25	36	6	–	–	–	–	–	–	3	–	–	–	**1,2,8-Trihydroxy-3-methoxy-6-methylanthraquinone tritms**	25466-65-3	**NI45161**
501	502	73	503	413	504	370	486	**516**	100	40	22	21	7	5	4	2	**0.00**	25	36	6	–	–	–	–	–	–	3	–	–	–	**1,2,8-Trihydroxy-6-methoxy-3-methylanthraquinone tritms**	91701-31-4	**NI45162**
486	501	487	443	502	488	503	489	**516**	100	55	45	24	22	20	11	5	**4.30**	25	36	6	–	–	–	–	–	–	3	–	–	–	**1,4,5-Trihydroxy-2-methoxy-7-methylanthraquinone tritms**	25458-56-4	**NI45163**
486	487	501	488	502	443	503	444	**516**	100	45	39	20	16	15	8	6	**1.10**	25	36	6	–	–	–	–	–	–	3	–	–	–	**1,4,5-Trihydroxy-7-methoxy-2-methylanthraquinone tritms**		**NI45164**
73	191	75	74	45	104	167	147	**516**	100	29	17	9	9	8	7	7	**2.90**	25	40	4	2	–	–	–	–	–	3	–	–	–	**2,4-Imidazolidinedione, 5-[3,4-bis[(trimethylsilyl)oxy]phenyl]-3-methyl-5-phenyl-1-(trimethylsilyl)-**	55517-85-6	**NI45165**
73	191	75	255	444	516	354	45	**516**	100	29	17	15	11	9	9	9		25	40	4	2	–	–	–	–	–	3	–	–	–	**2,4-Imidazolidinedione, 5-(3,4-dihydroxy-1,5-cyclohexadien-1-yl)-5-phenyl-, methyl, tris(trimethylsilyl)-**	72088-26-7	**NI45166**
81	250	267	516	293	435	346	345	**516**	100	47	42	23	10	9	8	8		26	26	1	2	2	–	2	–	–	–	–	–	–	**Bis(2-chloro-5a,6,7,8,9,9a-hexahydro-1,4-dibenzo[b,f]thiazepin-5a,9a-trans-11-yl)oxide, trans-**		**LI09502**
457	383	43	44	45	58	57	70	**516**	100	91	90	67	65	44	40	35	**28.28**	26	32	9	2	–	–	–	–	–	–	–	–	–	**Dichotine, 11,19-dimethoxy-, acetate (ester)**	55283-46-0	**NI45167**
258	56	41	44	43	40	57	42	**516**	100	63	56	53	33	31	29	29	**2.70**	26	48	8	2	–	–	–	–	–	–	–	–	–	**Glycine, N,N'-1,2-ethanediylbis[N-(2-butoxy-2-oxoethyl)-, dibutyl ester**	14531-12-5	**MS10491**
168	354	130	180	383	339	143	196	**516**	100	71	45	43	32	25	21	20	**12.00**	27	36	8	2	–	–	–	–	–	–	–	–	–	**Acetylnorrauglucine 17-O-acetyl-21-O-β-D-glucopyranosylnorajmaline**		**MS10492**
117	43	400	58	118	44	70	57	**516**	100	41	16	7	6	6	5	5	**3.00**	27	36	8	2	–	–	–	–	–	–	–	–	–	**Dichotine, 1-acetyl-2-deoxytetrahydro-11-methoxy-, acetate (ester)**	31592-09-3	**NI45168**
487	485	347	486	345	483	484	343	**516**	100	78	70	52	52	48	43	35	**0.00**	28	28	2	–	–	–	–	–	–	–	–	–	1	**Tin, diethylbis(2-phenoxyphenyl)-**		**LI09503**
409	516	381	517	410	304	347	78	**516**	100	91	46	33	30	30	27	20		29	24	2	–	–	–	–	–	–	–	–	–	2	**Ferrocene, 1'-acetyl-1'''-benzoyldi-**		**MS10493**
107	121	93	77	79	133	81	95	**516**	100	99	93	86	84	83	83	72	**0.00**	29	40	6	–	1	–	–	–	–	–	–	–	–	**2,6,10-Tetradecatrienedioic acid, 3,7,11-trimethyl-13-(2-methyl-1-propenyl)-13-(phenylsulphonyl)-, dimethyl ester, (E,E,E)-**	57683-69-9	**NI45169**
83	43	55	71	310	121	122	179	**516**	100	44	41	33	19	17	16	15	**0.23**	29	40	8	–	–	–	–	–	–	–	–	–	–	**Isobutyryl-12-desoxy-16-hydroxy-phorbol-angelate**		**NI45170**
207	179	43	288	260	41	57	270	**516**	100	44	39	33	30	27	24	22	**2.44**	31	48	6	–	–	–	–	–	–	–	–	–	–	**15,16,17-Trinorcrotophorbolone-20-myristate**		**NI45171**
103	283	475	75	143	129	476	104	**516**	100	38	27	24	17	13	12	10	**0.00**	31	56	2	–	–	–	–	–	–	2	–	–	–	**Pregnane, 3,20-bis[(allyldimethylsilyl)oxy]-, (3α,5β,20β)-**		**NI45172**
475	75	476	103	143	477	283	149	**516**	100	57	42	33	32	15	14	14	**1.09**	31	56	2	–	–	–	–	–	–	2	–	–	–	**Pregnane, 3,20-bis[(allyldimethylsilyl)oxy]-, (3β,5α,20α)-**		**NI45173**
103	475	75	476	143	477	283	104	**516**	100	80	38	34	30	13	13	9	**0.00**	31	56	2	–	–	–	–	–	–	2	–	–	–	**Pregnane, 3,20-bis[(allyldimethylsilyl)oxy]-, (3β,5α,20β)-**		**NI45174**
44	396	381	55	42	378	95	57	**516**	100	59	36	33	33	29	27	27	**0.00**	32	52	5	–	–	–	–	–	–	–	–	–	–	**Cholestane, 3,6-bis(acetyloxy)-8,9-epoxy-14-methyl-, (3β,5α,6α,8α)-**	5259-15-4	**NI45175**
43	55	69	81	95	83	123	109	**516**	100	78	75	68	60	47	44	39	**1.00**	32	52	5	–	–	–	–	–	–	–	–	–	–	**Cholestan-5-ol, 3,6-bis(acetyloxy)-24-methylene-, (3β,5α,6β)-**		**MS10494**
43	81	44	69	55	211	121	83	**516**	100	69	64	57	54	46	42	42	**0.00**	32	52	5	–	–	–	–	–	–	–	–	–	–	**Ergost-9(11)-en-20-ol, 3,6-bis(acetyloxy)-, (3β,5α,6α,20R)-**	56362-38-0	**NI45176**
410	456	95	43	69	55	57	41	**516**	100	79	69	52	50	45	35	35	**4.00**	32	52	5	–	–	–	–	–	–	–	–	–	–	**Lanostan-19-oic acid, 3-(acetyloxy)-11-oxo-, (3β)-**	25116-71-6	**NI45177**
410	456	95	438	393	395	498	377	**516**	100	80	68	23	16	10	7	7	**5.00**	32	52	5	–	–	–	–	–	–	–	–	–	–	**Lanostan-19-oic acid, 3-(acetyloxy)-11-oxo-, (3β)-**	25116-71-6	**LI09504**
230	383	212	213	159	268	105	96	**516**	100	65	34	31	23	22	20	20	**3.00**	33	44	3	2	–	–	–	–	–	–	–	–	–	**Acetone, 1,3-bis(octahydro-6-hydroxy-7-methyl-6-phenyl-5-indolizinyl)-, (5α,6β,7β,8aα)-[5'R*(5'α,6'β,7'β,8'aα)]-(±)-**	51020-39-4	**NI45178**
383	230	70	77	96	140	105	212	**516**	100	92	44	40	36	34	32	22	**4.00**	33	44	3	2	–	–	–	–	–	–	–	–	–	**Furo[3,2-e]indolizin-2-ol, decahydro-4-methyl-2-[(octahydro-6-hydroxy-7-methyl-6-phenyl-5-indolizinyl)methyl]-3a-phenyl-, [2α,2(5S*,6S*,7R*,8aS*),3aβ,4α,5aβ,9β]-(±)-**	50906-94-0	**NI45179**
149	425	313	501	173	407	440	203	**516**	100	38	32	24	24	18	16	15	**4.10**	33	56	4	–	–	–	–	–	–	–	–	–	–	**Cyclofoetigenin A 24,25-acetonide**		**NI45180**
434	205	435	179	499	136	500	432	**516**	100	33	23	20	19	10	8	7	**1.40**	33	56	4	–	–	–	–	–	–	–	–	–	–	**Lanostan-11-one, 3-acetyloxy-7-hydroxy-7-methyl-, (3β,7α,7β)-**		**NI45181**
131	368	81	147	103	43	55	105	**516**	100	76	60	59	56	44	43	42	**0.00**	36	52	2	–	–	–	–	–	–	–	–	–	–	**Cholest-5-en-3-ol (3β)-, 3-phenyl-2-propenoate**	1990-11-0	**NI45182**
412	394	380	253	366	275	266	271	**516**	100	80	15	15	8	5	5	3	**0.00**	36	52	2	–	–	–	–	–	–	–	–	–	–	**Sitosterol, 7-dehydro-, benzoate**		**MS10495**
167	258	516	336	259	179	180	425	**516**	100	97	38	26	23	20	15	9		40	36	–	–	–	–	–	–	–	–	–	–	–	**2',3',6',7'-Tetrahydro-4'-phenyl-1',2,5-tris[(E)-phenylmethylene]spiro[cyclopentan-1,5'(4'H)-indene]**		**MS10496**
104	91	105	69	65	131	119	92	**517**	100	35	26	12	7	4	4	4	**0.15**	16	10	1	1	–	15	–	–	–	–	–	–	–	**Phenylethylamine, N-(pentadecafluorooctanoyl)-**		**MS10497**
354	283	119	207	355	517	107	284	**517**	100	82	25	23	16	12	12	11		17	13	6	1	–	10	–	–	–	–	–	–	–	**Tyrosine, 3-methyoxy-N,O-bis(pentafluoropropanoyl)-, methyl ester**		**MS10498**
357	73	299	358	147	359	68	41	**517**	100	98	38	35	21	18	17	14	**0.00**	17	44	7	1	–	–	–	–	–	4	1	–	–	**Phosphorimidic acid, methoxy-, 4-oxo-2,3-bis[(trimethylsilyl)oxy]butyl bis(trimethylsilyl) ester, [R-(R*,R*)]-**	55517-61-8	**NI45183**
327	299	124	328	271	190	119	77	**517**	100	46	26	14	10	8	8	8	**0.00**	18	17	5	1	–	10	–	–	–	–	–	–	–	**Metadrenaline, β-ethoxy-N,O-bis(pentafluoropropanoyl)-**		**MS10499**
346	155	159	216	302	273	281	224	**517**	100	87	84	77	72	72	64	53	**1.87**	24	43	7	3	1	–	–	–	–	–	–	–	–	**L-Valine, N-[N-[3-[(carboxymethyl)thio]-N-decanoyl-L-alanyl]glycyl]-, dimethyl ester**	19729-27-2	**NI45184**
115	159	318	160	105	57	121	97	**517**	100	81	43	26	24	24	23	15	**0.00**	27	17	1	1	1	6	–	–	–	–	–	–	–	**4,5-Bis(trifluoromethyl)-4,5-dihydro-2-(2-naphthylthio)-4,5-diphenyl-1,3-oxazole**		**DD01774**
105	115	160	84	159	86	77	232	**517**	100	78	77	48	39	30	27	25	**0.00**	27	17	1	1	1	6	–	–	–	–	–	–	–	**trans-2,5-Bis(trifluoromethyl)-2,5-dihydro-4-(2-naphthylthio)-2,5-diphenyl 1,3-oxazole**		**DD01775**
98	82	57	149	84	85	58	71	**517**	100	91	78	73	60	53	46	36	**0.00**	27	33	3	3	–	–	2	–	–	–	–	–	–	**6-Oximo-3,6-dichloro-2,7-bis(2-piperidinoethoxy)-fluorene**		**MS10500**
162	160	440	517	448	502	119	130	**517**	100	97	28	22	19	13	11	10		29	25	1	1	–	6	–	–	–	–	–	–	–	**2-Methyl-5,5-diphenyl-4-[3,3,3-trifluoro-2-(trifluoromethyl)-1-propenylidene]-3-(2,4,6-trimethylphenyl)isoxazolidine**		**DD01776**

MASS TO CHARGE RATIOS	M.W.	INTENSITIES	Parent	C	H	O	N	S	F	Cl	Br	I	Si	P	B	X	COMPOUND NAME	CAS Reg No	No
517 133 370 207 265 169 383 327	**517**	100 50 30 21 20 17 15 10		34	35	2	3	–	–	–	–	–	–	–	–	–	**1,3,4,5,6,7-Hexahydro-N-phenyl-4-(2,4,6-trimethylphenyl)-7-[(E)-(2,4,6-trimethylphenyl)methylene]-2H-cyclopenta[d]pyridazine-2,3-dicarboxamide**		**MS10501**
522 293 524 520 442 291 280 440	**518**	100 66 65 64 55 50 44 44	**17.70**	10	6	3	2	–	–	–	4	–	–	–	–	–	**1-(3′,5′-Dibromo-4′-hydroxyphenyl)-5,5-dibromo-2,4-dioxohexahydropyrimidine**	83809-80-7	**NI45185**
59 15 131 474 374 69 274 100	**518**	100 49 12 11 11 7 5 4	**0.00**	12	6	4	–	–	16	–	–	–	–	–	–	–	**Decanedioic acid, hexadecafluoro-, dimethyl ester**		**LI09505**
57 69 43 93 119 131 169 99	**518**	100 46 28 22 20 10 8 6	**0.00**	13	7	2	–	–	17	–	–	–	–	–	–	–	**1,2-Epoxy-3-(1-methyleneheptadecafluorononyloxy)propane**		**DD01777**
59 87 341 369 429 355 489 356	**518**	100 80 42 40 39 38 38 30	**0.49**	14	42	7	–	–	–	–	–	–	7	–	–	–	**1,3,5,7,9,11,13-Heptaethylcycloheptasiloxane**	17909-36-3	**NI45186**
365 296 162 212 518 350 476 254	**518**	100 70 39 31 27 14 11 2		15	12	6	–	–	9	–	–	–	–	–	–	1	**Cobalt, tris(1,1,1-trifluoro-2,4-pentanedionato-O,O′)-**		**LI09506**
41 69 57 166 70 292 304 449	**518**	100 93 76 74 60 59 55 50	**13.30**	16	19	5	4	–	9	–	–	–	–	–	–	–	**Arginine, N-(trifluoroacetyl)-, butyl ester**		**NI45187**
73 327 503 74 75 45 328 59	**518**	100 14 8 8 5 5 4 4	**0.00**	16	46	7	–	–	–	–	–	–	6	–	–	–	**Tetrasiloxane, 3,5-diethoxy-1,1,1,7,7,7-hexamethyl-3,5-bis[(trimethylsilyl)oxy]-**		**MS10502**
445 119 518 446 151 417 298 283	**518**	100 80 31 26 26 24 19 16		17	12	7	–	–	10	–	–	–	–	–	–	–	**4-Hydroxy-3-methoxymandelic acid, 3-methoxy-α,4-bis(pentafluoropropanoyl)-, ethyl ester**		**MS10503**
354 119 207 283 59 69 323 107	**518**	100 80 60 57 42 38 27 21	**19.20**	17	12	7	–	–	10	–	–	–	–	–	–	–	**Phenyllactic acid, 3-methoxy-2,4-bis(pentafluoropropanoyl)-, methyl ester**		**MS10504**
149 281 128 171 143 518 202 370	**518**	100 80 44 40 30 22 22 12		19	21	2	–	–	–	1	–	–	–	–	–	1	**Mercury, chloro(3,17-dioxoandrosta-1,4,6-trien-2-yl)-**	36794-29-3	**NI45188**
149 281 282 171 283 134 128 143	**518**	100 71 68 55 54 37 37 30	**10.00**	19	21	2	–	–	–	1	–	–	–	–	–	1	**Mercury, chloro(3,17-dioxoandrosta-1,4,6-trien-2-yl)-**	36794-29-3	**NI45189**
518 193 519 287 211 187 173 39	**518**	100 48 25 7 5 5 5 4		22	8	–	–	–	10	–	–	–	–	–	–	1	**Ferrocene, 1,1′-bis(pentafluorophenyl)-**	55517-82-3	**NI45190**
50 309 77 307 197 51 195 311	**518**	100 85 74 63 56 44 37 33	**0.74**	24	15	–	–	–	5	–	–	–	–	–	–	1	**Stannane, (pentafluorophenyl)triphenyl-**	1058-08-8	**NI45191**
518 111 391 168 359 76 75 520	**518**	100 74 63 58 47 47 37 7		24	16	5	–	2	–	2	–	–	–	–	–	–	**Bis[(4-chlorophenylsulphonyl)phenyl] ether**		**IC04551**
43 236 417 135 255 315 176 296	**518**	100 38 34 30 27 18 17 16	**0.00**	24	38	12	–	–	–	–	–	–	–	–	–	–	**1,3,5,7,9,11-Hexaacetoxy-2-methylundecane**		**NI45192**
154 39 78 54 153 27 53 51	**518**	100 51 46 39 30 28 26 26	**0.12**	26	27	–	–	–	–	–	–	–	–	–	–	2	**Cobalt, bis(η^4-1,3-butadiene)(triphenylstannyl)-**	74811-07-7	**NI45193**
43 44 113 41 85 144 55 109	**518**	100 73 47 37 22 21 20 17	**0.00**	26	30	11	–	–	–	–	–	–	–	–	–	–	**Rubratoxin B**	21794-01-4	**MS10505**
113 43 44 109 41 97 55 85	**518**	100 95 52 38 38 27 27 24	**0.00**	26	30	11	–	–	–	–	–	–	–	–	–	–	**Rubratoxin B**	21794-01-4	**NI45194**
259 73 155 260 109 55 75 129	**518**	100 90 30 20 18 18 16 14	**0.00**	26	54	6	–	–	–	–	–	–	2	–	–	–	**Octadecanedioic acid, 9,10-bis(trimethylsilyloxy)-, dimethyl ester**		**LI09507**
259 73 155 260 109 75 129 487	**518**	100 88 30 20 17 15 14 13	**0.00**	26	54	6	–	–	–	–	–	–	2	–	–	–	**Octadecanedioic acid, 9,10-bis(trimethylsilyloxy)-, dimethyl ester, erythro-**		**LI09508**
259 73 155 260 109 75 67 129	**518**	100 90 32 22 17 15 15 14	**0.00**	26	54	6	–	–	–	–	–	–	2	–	–	–	**Octadecanedioic acid, 9,10-bis(trimethylsilyloxy)-, dimethyl ester, threo-**	22032-80-0	**NI45195**
367 253 27 368 105 55 159 96	**518**	100 60 56 26 14 14 13 12	**0.00**	27	41	6	–	–	3	–	–	–	–	–	–	–	**Chol-7-en-24-oic acid, 3,12-dihydroxy-, trifluoroacetate, methyl ester, (3α,5β,12α)-**		**LI09509**
367 368 340 253 482 27 369 145	**518**	100 29 26 18 16 12 8 6	**0.00**	27	41	6	–	–	3	–	–	–	–	–	–	–	**Chol-8(14)-en-24-oic acid, 3,12-dihydroxy-, trifluoroacetate, methyl ester, (3α,5β,12α)-**		**LI09510**
416 72 74 519 217 83 104 225	**518**	100 94 85 78 73 78 73 62	**1.00**	27	42	6	4	–	–	–	–	–	–	–	–	–	**L-Alanine, N-[N-[N-(3-phenyl-N-propionyl-L-alanyl)-L-valyl]-L-leucyl]-, methyl ester**	19526-15-9	**NI45196**
72 120 303 86 204 176 104 131	**518**	100 41 34 30 17 12 9 8	**0.30**	27	42	6	4	–	–	–	–	–	–	–	–	–	**L-Alanine, N-[N-[N-(3-phenyl-N-propionyl-L-alanyl)-L-valyl]-L-leucyl]-, methyl ester**	19526-15-9	**NI45197**
100 107 84 56 129 70 72 55	**518**	100 63 56 48 46 45 33 32	**5.34**	27	42	6	4	–	–	–	–	–	–	–	–	–	**Benzenepropanamide, N-[4-[acetyl[3-(acetylamino)propyl]amino]butyl]-N[3-(acetylamino)propyl]-4-(acetyloxy)-**	42920-02-5	**NI45198**
518 519 325 428 338 520 299 245	**518**	100 47 30 24 24 21 21 19		27	46	4	–	–	–	–	–	–	3	–	–	–	**Estra-1,3,5(10)-trien-6-one, 3,16,17-tris[(trimethylsilyl)oxy]-, (16α,17β)-**	69833-52-9	**NI45199**
518 519 520 521 503 428 517 504	**518**	100 47 23 7 5 5 4 3		27	46	4	–	–	–	–	–	–	3	–	–	–	**Estra-1,3,5(10)-trien-17-one, 2,3,4-tris[(trimethylsilyl)oxy]-**		**NI45200**
43 143 131 69 105 278 96 83	**518**	100 18 18 18 16 15 15 15	**0.28**	28	38	9	–	–	–	–	–	–	–	–	–	–	**Phorbol, 3,12,13,20-tetrabis(acetyloxy)-3-hydroxy-3-desoxo-4-desoxy-, (3β)-**		**NI45201**
354 145 105 91 81 43 107 201	**518**	100 99 99 81 63 61 58 54	**1.24**	29	42	8	–	–	–	–	–	–	–	–	–	–	**Carda-5,20(22)-dienolide, 3-[(6-deoxy-α-L-mannopyranosyl)oxy]-14-hydroxy-, (3β)-**	35536-76-6	**NI45202**
73 170 75 518 183 519 147 500	**518**	100 73 43 38 18 16 16 14		29	50	4	–	–	–	–	–	–	2	–	–	–	**Card-20(22)-enolide, 12,14-dihydroxy-3,23-bis[(trimethylsilyl)oxy]-, (5β)-**		**LI09511**
73 170 75 518 183 147 500 338	**518**	100 72 43 38 20 18 15 8		29	50	4	–	–	–	–	–	–	2	–	–	–	**24-Norchola-20,22-dien-14-ol, 21,23-epoxy-3,23-bis[(trimethylsilyl)oxy]-, (3β,5β,14β)-**	40837-83-0	**NI45203**
73 170 75 518 183 147 519 500	**518**	100 72 43 38 18 16 15 14		29	50	4	–	–	–	–	–	–	2	–	–	–	**24-Norchola-20,22-dien-14-ol, 21,23-epoxy-3,23-bis[(trimethylsilyl)oxy]-, (3β,5β,14β)-**	40837-83-0	**NI45204**
73 170 75 518 183 147 500 107	**518**	100 72 43 37 19 15 14 10		29	50	4	–	–	–	–	–	–	2	–	–	–	**24-Norchola-20,22-dien-14-ol, 21,23-epoxy-3,23-bis[(trimethylsilyl)oxy]-, (3β,5β,14β)-**	40837-83-0	**MS10506**
518 434 476 419 519 75 389 435	**518**	100 81 68 59 33 26 24 23		30	30	8	–	–	–	–	–	–	–	–	–	–	**4,4′-Diacetoxy-6,6′-dimethyl-5,5′,8,8′-tetramethoxy-1,1′-binaphthalene**	101459-22-7	**NI45205**
518 503 504 519 476 489 473 501	**518**	100 86 41 35 22 19 19 18		30	30	8	–	–	–	–	–	–	–	–	–	–	**6,6′,7,7′-Tetraethoxy-3,3′-dimethyl-2,2′-binaphthalene-1,1′,4,4′-tetrone**		**NI45206**

MASS TO CHARGE RATIOS								M.W.	INTENSITIES								Parent	C	H	O	N	S	F	Cl	Br	I	Si	P	B	X	COMPOUND NAME	CAS Reg No	No
434	476	518	435	477	217	404	419	**518**	100	89	32	28	27	24	17	15		30	30	8	–	–	–	–	–	–	–	–	–	–	**5,5′,8,8′-Tetramethoxy-3,3′-dimethyl-1′,4-diacetoxy-2,2′-binaphthalene**	89475-27-4	**NI45207**
189	242	363	108	227	159	131	139	**518**	100	76	47	40	38	23	23	21	**0.00**	30	30	8	–	–	–	–	–	–	–	–	–	–	**Thamnosindiol**	18458-68-9	**NI45208**
189	242	363	227	108	159	131	140	**518**	100	78	46	40	40	24	24	22	**0.00**	30	30	8	–	–	–	–	–	–	–	–	–	–	**Thamnosindiol**	18458-68-9	**LI09512**
226	105	518	387	77	160	519	254	**518**	100	88	62	47	25	24	22	20		30	34	6	2	–	–	–	–	–	–	–	–	–	**Aspidospermidin-21-oic acid, 20-(acetyloxy)-1-benzoyl-17-methoxy-, methyl ester**	56143-45-4	**NI45209**
85	374	434	99	330	43	151	356	**518**	100	72	64	44	37	29	21	6	**0.10**	30	46	7	–	–	–	–	–	–	–	–	–	–	**7-[5-(Acetyloxy)-3-[(tetrahydro-2H-pyran-2-yl)oxy]-2-(5-oxo-1,3-decadienyl)-cyclopentyl]-5-heptenoic acid, methyl ester, [1α(Z),2β(1E,3E)3α,5α]-**	79743-18-3	**NI45210**
87	59	71	43	41	58	57	203	**518**	100	90	89	88	88	84	82	79	**0.05**	30	46	7	–	–	–	–	–	–	–	–	–	–	**Card-20(22)-enolide, 3,14-dihydroxy-, mono(2,6-dideoxy-3-O-methyl-D-ribo-hexosyl) ether, (3β,5β)-**	12708-27-9	**NI45211**
429	519	430	201	428	520	339	229	**518**	100	48	29	25	18	15	15	13	**2.50**	30	46	7	–	–	–	–	–	–	–	–	–	–	**Chol-5-ene-12-ene, 3,24-bis(ethoxyformyloxy)-**		**NI45212**
105	353	354	352	485	412	241	197	**518**	100	65	28	28	7	3	3	3	**0.00**	31	50	2	–	2	–	–	–	–	–	–	–	–	**Cholest-5-en-19-al, 3β-hydroxy-, cyclic ethylene mercaptal, acetate**	1109-44-0	**LI09513**
353	354	352	458	247	241	227	197	**518**	100	40	40	10	5	5	5	5	**0.00**	31	50	2	–	2	–	–	–	–	–	–	–	–	**Cholest-5-en-19-al, 3β-hydroxy-, cyclic ethylene mercaptal, acetate**	1109-44-0	**NI45213**
43	209	41	181	191	57	55	290	**518**	100	70	59	54	53	41	38	34	**0.68**	31	50	6	–	–	–	–	–	–	–	–	–	–	**13-Deoxo-13α-hydroxy-15,16,17-trinorcrotophorbolone-20-myristate**		**NI45214**
209	43	191	41	272	181	57	55	**518**	100	71	50	45	42	42	41	38	**0.82**	31	50	6	–	–	–	–	–	–	–	–	–	–	**13-Deoxo-13β-hydroxy-15,16,17-trinorcrotophorbolone-20-myristate**		**NI45215**
294	206	224	266	235	234	295	225	**518**	100	50	25	20	20	18	5	5	**3.00**	31	50	6	–	–	–	–	–	–	–	–	–	–	**Kudzusapogenol B methyl ester**		**NI45216**
201	518	200	500	458	260	440	133	**518**	100	48	47	38	30	26	18	18		31	50	6	–	–	–	–	–	–	–	–	–	–	**Olean-12-en-28-oic acid, 2,3,21,23-tetrahydroxy-, methyl ester, (2α,3β,21β)-**		**LI09514**
273	290	272	254	244	109	81	69	**518**	100	99	99	98	98	98	96	96	**2.47**	31	50	6	–	–	–	–	–	–	–	–	–	–	**15,16,17-Trinor-12-desoxy-13β-hydroxyphorbol-13-tetradecanoate**		**NI45217**
43	28	41	57	272	55	44		**518**	100	76	67	57	55	52	42		**0.28**	31	50	6	–	–	–	–	–	–	–	–	–	–	**15,16,17-Trinor-12,13-didesoxy-13β-hydroxyphorbol-13-tetradecanoate**		**NI45218**
461	253	462	75	329	463	68	254	**518**	100	50	40	35	33	15	12	11	**0.00**	31	58	2	–	–	–	–	–	–	2	–	–	–	**Androst-5-ene, 3,17-bis[(tert-butyldimethylsilyl)oxy]-, (3β,17β)-**		**NI45219**
120	119	85	126	97	98	84	121	**518**	100	99	97	96	96	92	92	91	**23.00**	32	46	2	4	–	–	–	–	–	–	–	–	–	**1,2,4,5-Tetrazine, 3,6-bis(4-nonyloxyphenyl)-**	72242-53-6	**NI45220**
43	55	69	81	123	99	57	93	**518**	100	65	63	60	55	55	48	38	**1.00**	32	54	5	–	–	–	–	–	–	–	–	–	–	**Cholestan-3,5,6-triol, 24-methyl-, 3,6-diacetate, (3β,5α,6β,24ξ)-**		**MS10507**
145	95	43	93	91	358	109	69	**518**	100	99	99	93	81	79	79	73	**0.00**	32	54	5	–	–	–	–	–	–	–	–	–	–	**Lanost-9(11)-ene-3,18,20,23-tetrol, 23-acetate, (3β,20ξ)-**	56259-25-7	**NI45221**
305	113	460	255	443	192	518	179	**518**	100	29	8	8	5	5	4	3		33	58	4	–	–	–	–	–	–	–	–	–	–	**Cholestan-6-one, 3-(2-hydroxypropoxy)-, cyclic 1,2-propanediyl acetal, (3β,5α)-**	38404-88-5	**NI45222**
305	113	306	43	93	41	90	195	**518**	100	29	24	22	19	19	11	9	**5.00**	33	58	4	–	–	–	–	–	–	–	–	–	–	**Cholestan-6-one, 3-(2-hydroxypropoxy)-, cyclic 1,2-propanediyl acetal, (3β,5α)-**	38404-88-5	**LI09515**
167	437	55	129	145	438	268	351	**518**	100	76	50	39	30	28	12	7	**2.00**	35	38	2	2	–	–	–	–	–	–	–	–	–	**3-Cyclohexyl-2-(cyclohexylimino)-6-(diphenylmethyl)-2,3-dihydro-5-phenyl-4H-1,3-oxazin-4-one**		**MS10508**
268	67	97	267	82	393	69	83	**518**	100	93	71	60	57	52	30	22	**2.00**	35	38	2	2	–	–	–	–	–	–	–	–	–	**N,N′-Dicyclohexyl-N-(2,4,4-triphenyl-2,3-butadienoyl)urea**		**MS10509**
259	260	181	518	105	519	261	180	**518**	100	23	14	12	8	6	6	4		36	30	–	–	–	–	–	–	–	2	–	–	–	**Disilane, hexaphenyl-**	1450-23-3	**NI45223**
518	273	519	152	217	269	274	202	**518**	100	72	40	32	16	14	12	10		37	26	3	–	–	–	–	–	–	–	–	–	–	**Benzophenone, 4,4′-bis(4-phenylphenoxy)-**		**IC04552**
87	43	446	519	57	521	448	520	**519**	100	47	23	19	11	8	8	5		23	26	7	5	–	–	1	–	–	–	–	–	–	**Aniline-3-propanamide, N,N-bis(2-acetoxyethyl)-4-(3-chloro-4-nitrophenylazo)-**		**IC04553**
252	267	220	223	73	253	268	75	**519**	100	53	43	29	26	20	13	12	**0.00**	26	45	4	1	–	–	–	–	–	3	–	–	–	**Denopamine iso-M1, O,O,O-tris(trimethylsilyl)-**		**NI45224**
252	267	220	223	73	253	268	209	**519**	100	60	51	30	28	20	15	13	**0.00**	26	45	4	1	–	–	–	–	–	3	–	–	–	**Denopamine M1, O,O,O-tris(trimethylsilyl)-**		**NI45225**
519	520	75	518	521	312	360	129	**519**	100	48	10	9	6	6	5	5		27	49	3	1	–	–	–	–	–	3	–	–	–	**Estra-1,3,5(10)-triene, 4-amino-3,16,17-tris[(trimethylsilyl)oxy]-, (16α,17β)-**		**NI45226**
131	73	75	144	132	90	207	55	**519**	100	63	50	37	34	33	25	23	**5.56**	30	53	4	1	–	–	–	–	–	1	–	–	–	**Glycine, N-[(3α,5β)-24-oxo-3-[(trimethylsilyl)oxy]cholan-24-yl]-, methyl ester**	57326-15-5	**NI45227**
105	283	441	518	248	91	427	284	**519**	100	11	9	8	7	7	5	5	**2.00**	37	29	2	1	–	–	–	–	–	–	–	–	–	**5-Benzyl-2,3-dibenzoyl-cis-4,5-diphenyl-δ²-pyrroline**		**NI45228**
167	165	310	519	205	194	152	77	**519**	100	50	19	18	18	15	12	10		37	29	2	1	–	–	–	–	–	–	–	–	–	**2-(Diphenylacetyl)-4-methyl-6,6-diphenyl-2-azatricyclo[5.2.2.0^{1,5}]undeca-4,8,10-trien-3-one**		**MS10510**
127	139	393	266	520	174	47	301	**520**	100	31	22	5	5	2	1	1		1	–	–	–	–	–	–	–	4	–	–	–	–	**Methane, tetraiodo-**	507-25-5	**NI45229**
225	491	283	312	301	520	341	105	**520**	100	75	62	55	53	50	41	37		10	25	–	–	–	–	–	–	–	–	–	–	5	**Cyclopentaarsine, pentaethyl-**		**LI09516**
69	57	41	152	140	198	56	168	**520**	100	76	64	52	41	35	25	22	**2.40**	15	17	6	4	–	9	–	–	–	–	–	–	–	**Butanoic acid, 2-(trifluoroacetyl)amino-4-[1′,3′-bis(trifluoroacetyl)guanidioxy]-, butyl ester**		**MS10511**
203	201	202	188	69	71	101	86	**520**	100	97	90	72	70	67	45	44	**0.00**	16	12	–	–	–	–	–	4	–	–	–	–	–	**Naphthalene, 1,8-bis(1,2-dibromo-1-propenyl)-, (Z,Z)-**	56701-41-8	**NI45230**
122	159	121	283	202	284	358	43	**520**	100	70	70	45	45	30	27	16	**15.00**	19	23	2	–	–	–	1	–	–	–	–	–	1	**Mercury, chloro(3,17-dioxoandrosta-1,4-dien-2-yl)-**	36794-35-1	**NI45231**
43	122	121	283	159	358	202	520	**520**	100	54	37	34	33	15	10	5		19	23	2	–	–	–	1	–	–	–	–	–	1	**Mercury, chloro(3,17-dioxoandrosta-1,4-dien-2-yl)-**	36794-35-1	**NI45232**
107	151	133	171	134	283	284	128	**520**	100	81	81	79	59	49	48	46	**7.00**	19	23	2	–	–	–	1	–	–	–	–	–	1	**Mercury, chloro[(17β)-17-hydroxy-3-oxoandrosta-1,4,6-trien-2-yl]-**	36794-30-6	**NI45233**

MASS TO CHARGE RATIOS	M.W.	INTENSITIES	Parent	C	H	O	N	S	F	Cl	Br	I	Si	P	B	X	COMPOUND NAME	CAS Reg No	No
324 352 296 408 183 142 55 520	**520**	100 84 80 60 56 44 36 26		20	10	8	–	–	–	–	–	–	–	1	–	2	**Manganese, tetracarbonyliron-μ-diphenylphosphido-**		**IC04554**
352 324 296 408 380 520 436 464	**520**	100 92 81 60 40 36 33 15		20	10	8	–	–	–	–	–	–	–	1	–	2	**Manganese, tetracarbonyliron-μ-diphenylphosphido-**		**LI09517**
43 352 42 44 353 64 41 394	**520**	100 78 57 25 22 22 22 17	**4.00**	26	16	12	–	–	–	–	–	–	–	–	–	–	**Benzo[1,2-b:4,5-b′]bisbenzofuran-6,12-dione, 2,3,8,9-tetrakis(acetyloxy)-**	51860-95-8	**NI45234**
367 365 348 350 210 239 209 181	**520**	100 99 56 52 42 28 28 28	**3.54**	26	18	2	–	–	–	–	2	–	–	–	–	–	**Anthracene-9,10-diol, 9,10-dihydro-9,10-bis(4-bromophenyl)-, cis-**		**MS10512**
217 218 185 146 108 77 63 488	**520**	100 86 73 54 50 45 45 14	**0.08**	26	22	–	2	3	–	–	–	–	–	2	–	–	**Bis(diphenylthiophosphinyliminomethyl) sulphide**		**LI09518**
320 318 316 319 317 277 275 355	**520**	100 82 50 38 37 34 24 20	**12.01**	26	28	2	2	–	–	–	–	–	–	–	–	1	**Tin, bis(2-butyl-8-quinolinolato)-**	30106-82-2	**NI45235**
43 83 69 41 109 125 53 79	**520**	100 97 83 73 53 44 44 41	**1.66**	27	36	10	–	–	–	–	–	–	–	–	–	–	**4-O-Methyl-6,7-dihydro-6β,7β-epoxy-phorbol-12,13,20-triacetate**		**NI45236**
43 83 69 124 125 109 149 165	**520**	100 52 44 42 31 27 20 18	**0.85**	27	36	10	–	–	–	–	–	–	–	–	–	–	**12β,13,20-Triacetoxy-9-hydroxy-4-methoxy-6α,7α-epoxy-4αH-1,6-tigliadien-3-one**		**NI45237**
43 83 28 69 41 125 124 109	**520**	100 35 29 28 20 17 16 15	**2.76**	27	36	10	–	–	–	–	–	–	–	–	–	–	**12β,13,20-Triacetoxy-9-hydroxy-4-methoxy-6β,7β-epoxy-4αH-1,6-tigliadien-3-one**		**NI45238**
369 105 119 213 255 81 107 95	**520**	100 86 72 67 64 64 63 60	**0.00**	27	43	6	–	–	3	–	–	–	–	–	–	–	**Cholan-24-oic acid, 3,7-dihydroxy-, trifluoroacetate, methyl ester, (3α,5β,7α)-**		**LI09519**
369 255 370 107 81 55 105 28	**520**	100 64 54 40 38 38 34 34	**0.00**	27	43	6	–	–	3	–	–	–	–	–	–	–	**Cholan-24-oic acid, 3,7-dihydroxy-, trifluoroacetate, methyl ester, (3α,5β,7α)-**		**LI09520**
502 295 412 242 503 322 296 75	**520**	100 59 58 58 42 42 39 36	**0.00**	27	48	4	–	–	–	–	–	–	3	–	–	–	**Estra-1,3,5(10)-trien-11-ol, 3,6,17-tris[(trimethylsilyl)oxy]-, (6α,11β,17β)-**	74298-85-4	**NI45239**
136 520 119 120 521 42 168 77	**520**	100 80 33 26 25 23 22 20		28	28	4	2	2	–	–	–	–	–	–	–	–	**Biphenyl, bis(N,N-dimethylaminophenylsulphonyl)-**		**IC04555**
358 269 448 191 268 147 73 169	**520**	100 55 54 53 52 42 42 38	**11.36**	28	52	3	–	–	–	–	–	–	3	–	–	–	**Androsta-3,5-diene, 3,16,17-tris[(trimethylsilyl)oxy]-, (16α,17β)-**	69705-66-4	**NI45240**
73 505 520 75 506 147 521 74	**520**	100 99 82 55 45 45 38 26		28	52	3	–	–	–	–	–	–	3	–	–	–	**Androsta-5,15-diene, 3,16,17-tris[(trimethylsilyl)oxy]-, (3β,17β)-**	57397-22-5	**NI45241**
304 330 502 486 331 257 153 305	**520**	100 70 57 49 45 34 28 25	**10.50**	29	28	2	–	–	–	–	–	–	–	–	–	2	**Biferrocene, 1′-(2-hydroxybenzyl)-1‴-(1-hydroxyethyl)-**		**MS10513**
365 366 91 92 199 198 520 160	**520**	100 85 60 35 19 15 14 12		29	32	5	2	1	–	–	–	–	–	–	–	–	**Seredamine, O′-acetyl-O′-demethyl-, tosylate**		**LI09521**
253 340 105 313 254 143 145 107	**520**	100 43 36 33 28 27 25 22	**0.00**	29	44	8	–	–	–	–	–	–	–	–	–	–	**Pregnane-20-carboxylic acid, 3,7,12-tris(acetyloxy)-, methyl ester, (3α,5β,7α,12α,20S)-**	60354-46-3	**NI45242**
157 300 344 327 301 183 299 345	**520**	100 47 32 29 29 29 24 20	**1.00**	29	44	8	–	–	–	–	–	–	–	–	–	–	**Stigmast-7-en-26-oic acid, 2,3,14,20,22,28-hexahydroxy-6-oxo-, γ-lactone, (2β,3β,5β,22R,24S,25S,28R)-**	17086-76-9	**MS10514**
327 345 300 328 484 302 301 466	**520**	100 94 57 56 53 50 48 43	**1.32**	29	44	8	–	–	–	–	–	–	–	–	–	–	**Stigmast-7-en-26-oic acid, 2,3,14,20,22,28-hexahydroxy-6-oxo-, γ-lactone, (2β,3β,5β,22R,24S,25S,28R)-**	17086-76-9	**NI45243**
73 59 137 39 78 215 77 237	**520**	100 83 65 60 50 47 47 40	**5.00**	29	60	–	–	–	–	–	–	–	4	–	–	–	**3,4-Pentadien-1-yne, 1,3,5,5-tetrakis(trimethylsilyl)-**	41898-92-4	**NI45244**
84 463 70 111 98 107 520 223	**520**	100 63 56 51 47 27 26 24		30	40	4	4	–	–	–	–	–	–	–	–	–	**4H-1,16-Etheno-5,15-(propaniminoethano)furo[3,4-l][1,5,10]triazacyclohexadecine-4,21-dione, 3,3a,6,7,8,9,10,11,12,13,14,15-dodecahydro-3-(4-hydroxyphenyl)-10,14-dimethyl-**	69721-88-6	**NI45245**
43 357 339 254 255 81 55 107	**520**	100 56 53 48 46 43 42 37	**1.00**	30	48	5	–	1	–	–	–	–	–	–	–	–	**Cholane-24-thioic acid, 3,12-bis(acetyloxy)-, S-ethyl ester, (3β,5β,12α)-**	56052-66-5	**NI45246**
253 285 105 226 143 145 211 157	**520**	100 90 41 37 36 30 27 26	**0.00**	30	48	7	–	–	–	–	–	–	–	–	–	–	**Cholan-24-oic acid, 7,12-bis(acetyloxy)-3-methoxy-, methyl ester, (3α,5β)-**	69779-06-2	**NI45247**
363 345 301	**520**	100 23 8	**0.00**	30	48	7	–	–	–	–	–	–	–	–	–	–	**Cholest-7-en-6-one, 2,3,14,20,22,25-hexahydroxy-, 6-acetonide, (2β,3β,5β,22R)-**		**LI09522**
75 73 257 215 81 55 430 95	**520**	100 79 44 31 29 28 28 25	**2.52**	30	56	3	–	–	–	–	–	–	2	–	–	–	**3-α-Trimethylsiloxycholanic acid, trimethylsilyl ester**		**NI45248**
262 203 189 263 105 202 133 119	**520**	100 50 49 47 47 46 46 46	**0.00**	31	52	6	–	–	–	–	–	–	–	–	–	–	**Ursan-23-oic acid, 2,3,6,24-tetrahydroxy-, methyl ester, (3β,4β)-**	55517-91-4	**NI45249**
255 463 387 464 331 75 256 388	**520**	100 84 43 35 26 26 24 15	**0.00**	31	60	2	–	–	–	–	–	–	2	–	–	–	**Androstane, 3,17-bis[(tert-butyldimethylsilyl)oxy]-, (3α,5α,17β)-**		**NI45250**
255 463 464 75 256 387 331 68	**520**	100 95 39 33 31 26 19 15	**0.00**	31	60	2	–	–	–	–	–	–	2	–	–	–	**Androstane, 3,17-bis[(tert-butyldimethylsilyl)oxy]-, (3α,5β,17β)-**		**NI45251**
463 255 464 75 387 68 465 73	**520**	100 56 40 40 24 18 15 12	**0.00**	31	60	2	–	–	–	–	–	–	2	–	–	–	**Androstane, 3,17-bis[(tert-butyldimethylsilyl)oxy]-, (3β,5α,17β)-**		**NI45252**
255 491 103 359 492 387 256 388	**520**	100 73 37 35 32 32 27 23	**6.07**	31	60	2	–	–	–	–	–	–	2	–	–	–	**Androstane, 3,17-bis[(triethylsilyl)oxy]-, (3α,5α,17β)-**		**NI45253**
255 491 256 359 103 492 257 388	**520**	100 69 54 34 33 31 27 17	**2.14**	31	60	2	–	–	–	–	–	–	2	–	–	–	**Androstane, 3,17-bis[(triethylsilyl)oxy]-, (3α,5β,17β)-**		**NI45254**
491 492 255 103 493 387 388 359	**520**	100 43 39 33 18 13 11 11	**2.83**	31	60	2	–	–	–	–	–	–	2	–	–	–	**Androstane, 3,17-bis[(triethylsilyl)oxy]-, (3β,5α,17β)-**		**NI45255**
103 145 477 146 75 104 478 147	**520**	100 81 18 11 11 9 8 5	**0.00**	31	60	2	–	–	–	–	–	–	2	–	–	–	**Pregnane, 3,20-bis[(propyldimethylsilyl)oxy]-, (3β,5α,20β)-**		**NI45256**
520 503 521 253 475 504 252 237	**520**	100 41 30 25 17 16 15 11		34	20	4	2	–	–	–	–	–	–	–	–	–	**1,2,4,5-Benzenetetracarboxylic acid, 1,2:4,5-diimide, N,N′-bis(2-biphenylyl)-**	31664-78-5	**NI45257**
520 503 521 253 475 504 252 237	**520**	100 41 40 25 17 16 15 11		34	20	4	2	–	–	–	–	–	–	–	–	–	**1,2,4,5-Benzenetetracarboxylic acid, 1,2:4,5-diimide, N,N′-bis(2-biphenylyl)-**	31664-78-5	**MS10515**
520 152 521 102 74 153 324 128	**520**	100 91 43 29 24 22 17 16		34	20	4	2	–	–	–	–	–	–	–	–	–	**1,2,4,5-Benzenetetracarboxylic acid, 1,2:4,5-diimide, N,N′-bis(4-biphenylyl)-**	33529-33-8	**NI45258**
520 152 521 102 74 153 324 260	**520**	100 91 43 29 24 22 17 16		34	20	4	2	–	–	–	–	–	–	–	–	–	**1,2,4,5-Benzenetetracarboxylic acid, 1,2:4,5-diimide, N,N′-bis(4-biphenylyl)-**	33529-33-8	**MS10516**
91 92 339 105 69 181 175 520	**520**	100 9 7 6 4 3 3 2		34	48	4	–	–	–	–	–	–	–	–	–	–	**(1S,15S)-Bicyclo[13.1.0]hexadecan-2-one 1,4-di-O-benzyl-D-threitolketal**		**NI45259**

MASS TO CHARGE RATIOS								M.W.	INTENSITIES								Parent	C	H	O	N	S	F	Cl	Br	I	Si	P	B	X	COMPOUND NAME	CAS Reg No	No
520	280	227	477	121	265	281	228	**520**	100	87	82	43	39	37	23	23		35	36	4	–	–	–	–	–	–	–	–	–	–	Benzene, 1,1′-[[2,2-bis(3-methoxyphenyl)-3,3-dimethylcyclopropyl]ethenylidene]bis[3-methoxy-	67437-09-6	NI45260
105	183	77	208	207	197	373	225	**520**	100	67	29	26	25	19	17	15	**0.00**	35	36	4	–	–	–	–	–	–	–	–	–	–	(4R,5R)-2-Cyclohexyl-α,α,α′,α′-tetraphenyl-1,3-dioxolane-4,5-dimethanol		DD01778
520	91	521	522	430	383	355	397	**520**	100	50	44	15	11	11	8	6		36	56	–	–	1	–	–	–	–	–	–	–	–	Cholest-2-ene, 4,4-dimethyl-3-benzylthio-, (5α)-	69841-47-0	NI45261
442	284	170	190	76	146	111	205	**521**	100	24	8	7	7	6	6	5	**2.20**	–	–	–	3	–	–	2	4	–	–	3	–	–	1,3,5,2,4,6-Triazatriphosphorine, 2,2,4,6-tetrabromo-4,6-dichloro-2,2,4,4,6,6-hexahydro-	15965-00-1	NI45262
169	69	214	314	402	119	164	114	**521**	100	31	23	5	4	2	1	1	**0.00**	9	–	–	1	–	21	–	–	–	–	–	–	–	Tripropylamine, perfluoro-		MS10517
57	41	44	28	446	525	444	445	**521**	100	99	75	75	65	64	61	59	**12.00**	11	3	–	5	–	–	–	4	–	–	–	–	–	1,3,6-Triazacyclo[3.3.3]azine, 7,9-dibromo-4-cyano-2-(dibromomethyl)-		NI45263
91	174	430	152	341	180	90	176	**521**	100	32	21	20	19	18	16	13	**5.00**	21	26	7	3	1	3	–	–	–	–	–	–	–	TFA-γ-Gly-Cys-(Bzl)-Gly-(OMe)2		LI09523
165	123	221	370	98	383	206	313	**521**	100	84	82	60	29	23	18	5	**1.10**	23	25	2	3	–	1	–	–	1	–	–	–	–	8-[4-(4-Fluorophenyl)-4-oxobutyl]-1-(4-iodophenyl)-1,3,8-triazaspiro[4,5]decan-4-one		MS10518
464	442	444	462	466	446	443	465	**521**	100	99	97	76	50	35	28	26	**17.00**	25	19	3	1	–	–	4	–	–	–	–	–	–	Spiro[phthalan-1,9′-xanthene]-, 4,5,6,7-tetrachloro-6′-diethylamino-2′-methyl-3-oxo-		IC04556
188	73	490	75	175	521	491	103	**521**	100	97	50	49	43	25	24	21		28	51	4	1	–	–	–	–	–	2	–	–	–	Pregnane-11,20-dione, 3,21-bis[(trimethylsilyl)oxy]-, 20-(O-methyloxime), (3α,5α)-	57305-34-7	NI45264
73	188	75	521	175	431	400	103	**521**	100	99	51	37	37	32	28	28		28	51	4	1	–	–	–	–	–	2	–	–	–	Pregnane-11,20-dione, 3,21-bis[(trimethylsilyl)oxy]-, 20-(O-methyloxime), (3α,5β)-	57305-35-8	NI45265
73	188	75	175	521	431	103	400	**521**	100	66	41	27	26	26	24	23		28	51	4	1	–	–	–	–	–	2	–	–	–	Pregnane-11,20-dione, 3,21-bis[(trimethylsilyl)oxy]-, 20-(O-methyloxime), (3α,5β)-	56211-33-7	NI45266
476	477	75	478	73	521	93	173	**521**	100	42	25	17	16	14	13	11		28	51	4	1	–	–	–	–	–	2	–	–	–	Prosta-5,8(12),13-trien-1-oic acid, 9-(ethoxyimino)-15-[(trimethylsilyl)oxy]-, trimethylsilyl ester, (5Z,13E,15S)-	56009-44-0	NI45267
400	324	428	183	185	162	414	108	**521**	100	70	32	28	15	15	14	12	**1.50**	29	34	–	3	–	–	1	–	–	–	2	–	–	(Aminodiphenylphosphinoimino)-(pentyl)diphenylphosphane chloride		NI45268
258	170	522	346	56	434	141	327	**522**	100	97	92	56	53	47	27	16		2	5	1	–	–	14	–	–	–	–	5	–	1	Iron, difluoroethoxyphosphinetetrakis(trifluorophosphine)-		LI09524
522	170	258	346	503	327	415	434	**522**	100	77	72	60	58	58	56	54		2	5	1	–	–	14	–	–	–	–	5	–	1	Iron, difluoroethoxyphosphinetetrakis(trifluorophosphine)-		LI09525
354	298	382	326	270	242	522	187	**522**	100	82	75	50	45	32	25	13		10	–	10	–	–	–	–	–	–	–	–	–	2	Manganese, pentacarbonyl-pentacarbonylrhenium		MS10519
203	205	168	103	204	142	271	115	**522**	100	39	30	8	7	6	5	4	**0.19**	14	9	2	–	–	–	6	–	–	–	–	–	1	(η^4-1,2,3,4,7,7,-Hexachloro-5-methoxycarbonylnorbornadiene)(η^5-cyclopentadienyl)rhodium	79234-84-7	NI45269
403	522	404	472	119	255	523	69	**522**	100	25	22	9	8	7	6	5		19	12	4	2	–	10	–	–	–	–	–	–	–	6-Hydroxymelatonin, bis(pentafluoropropionyl)-	98451-71-9	NI45270
285	202	122	286	244	121	522		**522**	100	80	75	60	50	42	2			19	25	2	–	–	–	1	–	–	–	–	–	1	Mercury, chloro-(3,17-dioxoandrost-1-en-2-yl)-, (5α)-	36794-33-9	NI45272
285	122	121	202	159	286	244	358	**522**	100	63	58	45	36	35	19	16	**1.00**	19	25	2	–	–	–	1	–	–	–	–	–	1	Mercury, chloro-(3,17-dioxoandrost-1-en-2-yl)-, (5α)-	36794-33-9	NI45271
41	139	55	69	58	42	158	43	**522**	100	86	80	75	68	66	65	60	**8.50**	19	31	8	4	–	–	–	1	–	–	–	–	–	Propanedioic acid, (bromomethyl)methyl-, bis[5,6-bis(hydroxyimino)heptyl] ester	60004-38-8	NI45273
87	85	99	107	163	114	135	121	**522**	100	60	35	32	25	23	17	16	**0.20**	23	33	5	–	–	7	–	–	–	–	–	–	–	2,6-Tridecanoic acid, 7-ethyl-10-hydroxy-11-methoxy-3,11-dimethyl-, (heptafluorobutanoate)		MS10520
74	55	43	41	69	81	67	75	**522**	100	40	30	27	25	20	20	18	**0.50**	23	36	6	–	–	6	–	–	–	–	–	–	–	Octadecanoic acid, 9,10-dihydroxy-, methyl ester, bis(trifluoroacetate)	21987-19-9	NI45274
74	137	123	294	54	124	138	263	**522**	100	47	46	40	40	34	33	29	**7.00**	23	36	6	–	–	6	–	–	–	–	–	–	–	Octadecanoic acid, 9,10-dihydroxy-, methyl ester, bis(trifluoroacetate)	21987-19-9	LI09526
354	312	269	396	268	438	340	270	**522**	100	57	57	24	24	21	15	8	**0.80**	24	18	10	4	–	–	–	–	–	–	–	–	–	Benzo[1,2-b:4,5-b′]difuran-3,7-dicarbonitrile, N,N,N′,N′-tetraacetatyl-4,8-bis(acetyloxy)-		MS10521
73	290	232	291	45	114	74	75	**522**	100	48	23	14	12	10	7	5	**0.00**	24	46	3	2	–	–	–	–	–	4	–	–	–	Tryptamine, 5-hydroxy-α-methyl-, tetrakis(trimethylsilyl)-		MS10522
73	81	135	93	248	490	377	280	**522**	100	34	18	6	3	1	1	1	**0.30**	25	31	6	–	–	5	–	–	–	–	–	–	–	2,6-Dodecanoic acid, 11-methoxy-3,7,11-trimethyl-10-(pentafluorophenoxyacetoxy)-, methyl ester		MS10523
274	250	273	522	191	77	69	245	**522**	100	62	57	36	28	25	24	22		26	18	12	–	–	–	–	–	–	–	–	–	–	γ-Rubromycin		NI45275
180	77	123	226	184	487	261	165	**522**	100	64	51	46	44	42	42	21	**0.00**	26	20	–	2	–	–	4	–	–	–	–	2	–	Boron, bis[dichlorodiphenylketimino]di-		LI09527
137	175	360	311	187	241	342	180	**522**	100	94	89	88	79	78	71	67	**27.00**	26	34	11	–	–	–	–	–	–	–	–	–	–	Isolariciresinol, mono-β-D-glucoside, (+)-		LI09528
43	28	83	69	41	44	53	109	**522**	100	44	40	33	32	30	26	25	**0.75**	26	34	11	–	–	–	–	–	–	–	–	–	–	1,6-Tigliadien-3-one, 12,13,20-tris(acetyloxy)-4,5,9-trihydroxy-6,7-epoxy-, (5ξ,6β,7β,12β)-		NI45276
402	313	384	296	295	356	314	297	**522**	100	84	78	71	67	64	59	53	**0.80**	26	34	11	–	–	–	–	–	–	–	–	–	–	6,7-seco-Tiglia-1-en-3,6-dione, 12,13,20-tris(acetyloxy)-7,9-dihydroxy-4,7-epoxy-, (7β,12β)-		LI09529
396	438	395	437	353	480	341	354	**522**	100	83	83	78	72	61	57	33	**4.00**	28	26	10	–	–	–	–	–	–	–	–	–	–	5,7,2′,4′-Tetraacetoxy-3′-(3,3-dimethylallyl)isoflavone		NI45277
73	215	202	219	216	320	74	45	**522**	100	67	33	30	14	11	10	10	**3.20**	28	42	2	2	–	–	–	–	–	3	–	–	–	Indole, 1-trimethylsilyl-3-[4-[(trimethylsilyl)oxyphenylethylcarbonyl-(trimethylsilylaminoethyl)]-		NI45278
73	147	75	342	252	129	103	393	**522**	100	99	71	70	44	44	38	35	**21.00**	28	54	3	–	–	–	–	–	–	3	–	–	–	Androst-5-ene, 3,4,17-tris[(trimethylsilyl)oxy]-, (3β,4α,17β)-	33287-28-4	NI45279

MASS TO CHARGE RATIOS								M.W.	INTENSITIES								Parent	C	H	O	N	S	F	Cl	Br	I	Si	P	B	X	COMPOUND NAME	CAS Reg No	No
73	129	75	432	237	143	169	213	**522**	100	99	99	91	90	85	76	74	**33.00**	28	54	3	–	–	–	–	–	–	3	–	–	–	Androst-5-ene, 3,11,17-tris[(trimethylsilyl)oxy]-, (3β,11β,17β)-	33287-29-5	NI45280
73	75	129	432	237	143	169	213	**522**	100	27	21	19	19	18	16	15	**6.71**	28	54	3	–	–	–	–	–	–	3	–	–	–	Androst-5-ene, 3,11,17-tris[(trimethylsilyl)oxy]-, (3β,11β,17β)-	33287-29-5	NI45281
237	327	252	342	213	169	432	360	**522**	100	85	72	69	67	65	59	52	**12.49**	28	54	3	–	–	–	–	–	–	3	–	–	–	Androst-5-ene, 3,11,17-tris[(trimethylsilyl)oxy]-, (3β,11β,17β)-	33287-29-5	NI45282
432	239	147	129	329	213	252	237	**522**	100	96	75	75	50	43	39	39	**35.00**	28	54	3	–	–	–	–	–	–	3	–	–	–	Androst-5-ene, 3,16,17-tris[(trimethylsilyl)oxy]-, (3α,16α,17β)-		LI09530
239	129	432	147	75	329	213	252	**522**	100	88	80	78	77	70	65	53	**27.00**	28	54	3	–	–	–	–	–	–	3	–	–	–	Androst-5-ene, 3,16,17-tris[(trimethylsilyl)oxy]-, (3β,16α,17β)-	33287-32-0	NI45283
73	239	129	432	147	75	329	213	**522**	100	30	27	24	24	24	21	20	**8.41**	28	54	3	–	–	–	–	–	–	3	–	–	–	Androst-5-ene, 3,16,17-tris[(trimethylsilyl)oxy]-, (3β,16α,17β)-	33287-32-0	NI45284
432	239	329	129	213	342	254	327	**522**	100	78	68	57	50	46	46	43	**36.00**	28	54	3	–	–	–	–	–	–	3	–	–	–	Androst-5-ene, 3,16,17-tris[(trimethylsilyl)oxy]-, (3β,16β,17α)-		LI09531
239	73	129	329	213	432	252	237	**522**	100	82	69	61	60	54	53	51	**10.00**	28	54	3	–	–	–	–	–	–	3	–	–	–	Androst-5-ene, 3,16,17-tris[(trimethylsilyl)oxy]-, (3β,16β,17β)-	33287-31-9	NI45286
239	329	432	252	213	237	129	342	**522**	100	65	63	61	56	51	48	45	**13.00**	28	54	3	–	–	–	–	–	–	3	–	–	–	Androst-5-ene, 3,16,17-tris[(trimethylsilyl)oxy]-, (3β,16β,17β)-	25876-85-1	NI45285
73	147	75	129	213	239	105	74	**522**	100	99	99	72	47	46	46	44	**6.96**	28	54	3	–	–	–	–	–	–	3	–	–	–	Androst-5-ene, 3,16,17-tris[(trimethylsilyl)oxy]-, (3β,16β,17β)-	33287-31-9	NI45287
388	387	405	343					**522**	100	86	58	13					**0.00**	29	46	8	–	–	–	–	–	–	–	–	–	–	Cholest-7-en-6-one, 2-(acetyloxy)-3,14,20,22,25-pentahydroxy-, (2β,3β,5β)-		LI09532
388	387	405	343					**522**	100	99	63	18					**0.00**	29	46	8	–	–	–	–	–	–	–	–	–	–	Cholest-7-en-6-one, 3-(acetyloxy)-2,14,20,22,25-pentahydroxy-, (2β,3β,5β)-		LI09533
43	99	60	426	365	81	344	327	**522**	100	77	46	36	32	31	30	20	**0.01**	29	46	8	–	–	–	–	–	–	–	–	–	–	Cholest-7-en-6-one, 25-(acetyloxy)-2,3,14,20,22-pentahydroxy-, (2β,3β,5β,22R)-	22033-96-1	MS10524
383	384	43	55	121	57	229	95	**522**	100	38	36	30	28	27	24	24	**1.00**	29	47	3	–	–	–	–	1	–	–	–	–	–	Cholestan-3-ol, 5-bromo-6,19-epoxy-, acetate, (3β,5α,6β)-	1258-07-7	NI45288
398	43	399	41	44	57	55	81	**522**	100	87	32	28	24	22	21	21	**0.00**	29	47	3	–	–	–	–	1	–	–	–	–	–	5β,6β-Epoxy-7α-bromocholestane-3β-acetate		NI45289
43	44	57	41	42	77	55	69	**522**	100	67	30	26	24	20	19	18	**0.00**	30	20	–	6	–	–	–	–	–	–	–	–	1	Nickel, [5,6,17,18-tetrahydrotetrabenzo[b,f,j,n][1,5,9,13]tetraazacyclohexadecine-6,18-dicarbonitrilato(2-)-N5,N11,N17,N23]-, (sp-4-1)-	72101-34-9	NI45290
103	57	248	104	76	274	77	55	**522**	100	81	33	33	31	28	28	17	**0.00**	31	22	1	8	–	–	–	–	–	–	–	–	–	Urea, 1,3-bis(4,6-diphenyl-s-triazin-2-yl)-	29366-70-9	NI45291
44	153	55	43	45	41	388	57	**522**	100	99	88	85	82	78	71	68	**1.00**	31	59	2	2	–	–	–	–	–	–	1	–	–	Cholestan-3-ol, tetramethylphosphorodiamidate, (3β,5α)-	54423-69-7	NI45292
262	522	183	261	521	77	185	108	**522**	100	70	64	50	43	37	35	32		32	28	3	–	–	–	–	–	–	–	2	–	–	Phosphonic acid, [1-(triphenylphosphoranylidene)ethyl]-, diphenyl ester	57356-83-9	NI45293
387	373	126	79	107	147	94	110	**522**	100	27	26	25	22	19	15	15	**9.42**	32	46	2	2	1	–	–	–	–	–	–	–	–	2,4-Dispirobis[6-[4-(3-furyl)-1-methyl-3-butenyl]-1-methylpiperidin-3-yl]tetrahydrothiophene		NI45294
119	91	64	374	52	403	331	522	**522**	100	94	49	25	18	16	4	2		33	38	2	4	–	–	–	–	–	–	–	–	–	4-Benzyl-7,8-bis(diethylamino)-2,6-diphenyl-2,6-diazabicyclo[2.2.2]oct-7-ene-3,5-dione		DD01779
197	199	367	336	139	211	136	229	**522**	100	72	69	66	65	51	42	37	**0.85**	33	50	3	–	–	–	–	–	–	1	–	–	–	Silane, (2-methoxyphenyl)bis[[5-methyl-2-isopropylcyclohexyl]oxy]phenyl, [1α(1R*,2R*,5R*),2β,5α]-	74793-60-5	NI45295
105	43	77	122	44	51	41	55	**522**	100	79	61	57	45	39	35	31	**0.00**	34	50	4	–	–	–	–	–	–	–	–	–	–	Cholest-7-en-3β,5α-diol-6α-benzoate		NI45296
105	43	77	486	122	41	55	471	**522**	100	50	28	21	20	19	18	16	**0.00**	34	50	4	–	–	–	–	–	–	–	–	–	–	Cholest-7-en-3β,5β-diol-6β-benzoate		NI45297
255	43	256	57	55	99	41	69	**522**	100	21	20	14	11	10	10	7	**0.24**	34	66	3	–	–	–	–	–	–	–	–	–	–	Myristic acid, 2-(1-octadecenyloxy)ethyl ester, (E)-	30760-02-2	NI45298
255	256	43	57	55	99	41	69	**522**	100	20	19	14	11	10	9	7	**0.14**	34	66	3	–	–	–	–	–	–	–	–	–	–	Myristic acid, 2-(1-octadecenyloxy)ethyl ester, (Z)-	30760-01-1	NI45299
521	523	522	524	157	143	111	507	**522**	100	93	48	29	28	24	22	18		35	70	2	–	–	–	–	–	–	–	–	–	–	Tetratriacontanoic acid, methyl ester		NI45300
523	359	79	524	195	69	290	360	**523**	100	48	33	21	14	14	9	8		18	6	–	3	–	12	–	–	–	–	1	–	–	Phosphine, tris(aminotetrafluorophenyl)-		MS10525
43	44	86	69	70	340	41	63	**523**	100	86	55	55	54	42	38	35	**4.90**	20	21	8	5	2	–	–	–	–	–	–	–	–	2-Propanamine, N-[2,2-bis[(2,4-dinitrophenyl)thio]vinyl]-N-isopropyl-	42362-45-8	NI45301
355	168	357	126	359	508	466	418	**523**	100	78	38	35	4	3	3	3	**0.45**	22	40	4	1	–	3	–	–	–	3	–	–	–	Acetamide, N-[2-[3,5-bis[(trimethylsilyl)oxy]phenyl]-2-[(trimethylsilyl)oxy]ethyl]-2,2,2-trifluoro-N-isopropyl-	40629-68-3	NI45302
344	342	525	217	216	102	527	523	**523**	100	99	68	59	44	38	35	35		23	15	2	3	–	–	–	2	–	–	–	–	–	4H-1,2-Diazepine, 3,7-bis(4-bromophenyl)-5-(4-nitrophenyl)-	25649-80-3	NI45303
105	122	120	77	210	197	281	106	**523**	100	92	42	33	21	15	12	10	**4.00**	25	33	11	1	–	–	–	–	–	–	–	–	–	4-[(1-Deoxylactit-1-yl)amino]benzophenone		NI45304
478	479	480	477	508	523	452	75	**523**	100	40	15	14	10	9	8	8		28	53	4	1	–	–	–	–	–	2	–	–	–	Prosta-8(12),13-dien-1-oic acid, 9-(ethoxyimino)-15-[(trimethylsilyl)oxy]-, trimethylsilyl ester, (13E,15S)-	56009-45-1	NI45305
162	75	433	478	199	452	388	129	**523**	100	86	78	68	68	62	56	46	**13.00**	28	53	4	1	–	–	–	–	–	2	–	–	–	Prosta-10,13-dien-1-oic acid, 9-(ethoxyimino)-15-[(trimethylsilyl)oxy]-, trimethylsilyl ester, (9E,13E,15S)-	56009-46-2	NI45306
162	75	478	77	129	199	173	73	**523**	100	83	41	41	40	36	35	34	**9.00**	28	53	4	1	–	–	–	–	–	2	–	–	–	Prosta-10,13-dien-1-oic acid, 9-(ethoxyimino)-15-[(trimethylsilyl)oxy]-, trimethylsilyl ester, (9Z,13E,15S)-	56009-47-3	NI45307
41	278	39	103	139	141	75	44	**524**	100	37	34	28	26	26	25	23	**0.20**	9	5	1	–	–	–	–	5	–	–	–	–	–	Benzene, pentabromo(2-propenyloxy)-	3555-11-1	NI45308
69	181	131	169	31	100	231	150	**524**	100	14	14	10	10	5	3	3	**0.07**	12	–	–	–	–	20	–	–	–	–	–	–	–	Tricyclo[3.3.1.1^{3,7}]decane, 1,2,2,3,4,4,6,6,8,8,9,9,10,10-tetradecafluoro-5,7-bis(trifluoromethyl)-	36481-20-6	NI45309
144	56	200	244	216	300	328	356	**524**	100	87	56	33	30	28	25	20	**0.00**	13	8	10	–	1	–	–	–	–	–	–	–	3	[μ^3-Ethanethiolato][μ-eta^2-methylcarbonyl]nonacarbonyltriiron		MS10526

MASS TO CHARGE RATIOS								M.W.	INTENSITIES								Parent	C	H	O	N	S	F	Cl	Br	I	Si	P	B	X	COMPOUND NAME	CAS Reg No	No
160	73	28	161	147	75	74	45	**524**	100	77	15	12	8	8	6	6	**0.00**	20	48	6	2	–	–	–	–	–	4	–	–	–	Galacto-hexodialdose, 2,3,4,5-tetrakis-O-(trimethylsilyl)-, bis(O-methyloxime)	62181-81-1	NI45310
73	103	147	217	307	75	205	74	**524**	100	62	18	17	12	9	8	8	**0.00**	20	48	6	2	–	–	–	–	–	4	–	–	–	D-Gluco-hexodialdose, 2,3,4,5-tetrakis-O-(trimethylsilyl)-, bis(O-methyloxime)	62108-39-8	NI45311
73	205	147	117	230	103	75	28	**524**	100	30	26	14	13	9	9	8	**0.00**	20	48	6	2	–	–	–	–	–	4	–	–	–	Ribo-hexos-3-ulose, 2,4,5,6-tetrakis-O-(trimethylsilyl)-, bis(O-methyloxime)	62108-14-9	NI45312
73	147	75	262	103	217	89	231	**524**	100	16	12	11	11	9	9	8	**0.00**	20	48	6	2	–	–	–	–	–	4	–	–	–	Xylo-hexos-5-ulose, 2,3,4,6-tetrakis-O-(trimethylsilyl)-, bis(O-methyloxime)	62108-13-8	NI45313
423	525	139	55	140	303	141	260	**524**	100	88	70	50	36	30	30	28		21	20	6	1	1	–	–	–	–	–	–	–	2	Manganese, tricarbonyl[η-1-(N,N-dimethylsulphamoyl)-2-(hydroxymethylferrocenyl)cyclopentadienyl]-		NI45314
57	255	35	245	56	247	283	257	**524**	100	17	8	7	6	4	4	4	**0.00**	22	45	6	–	–	–	1	–	–	3	–	–	–	1,3,5-Trivinyl-1,1,3,5-tetrabutoxy-5-chlorotrisiloxane		NI45315
524	522	523	504	525	502	526	411	**524**	100	78	42	35	29	27	26	24		24	10	–	–	–	8	–	–	–	–	–	–	1	5H-Dibenzogermole, 1,2,3,4,6,7,8,9-octafluoro-5,5-diphenyl-	10380-43-5	NI45316
105	133	134	90	193	103	91	135	**524**	100	45	42	24	20	18	15	14	**0.00**	25	21	2	2	2	–	–	1	–	–	–	–	–	2-Benzoyl-3-(4-methoxyphenyl)-6-phenyl-2,3-dihydro-4H-thiazolo[2,3-b][1,3,4]thiadiazin-5-ium bromide		MS10527
135	134	103	193	104	107	102	276	**524**	100	21	17	14	13	8	8	5	**0.00**	25	21	2	2	2	–	–	1	–	–	–	–	–	3,6-Diphenyl-2-(4-methoxybenzolyl)-2,3-dihydro-4H-thiazolo[2,3-b][1,3,4]thiadiazin-5-ium bromide		MS10528
43	83	109	69	125	41	114	53	**524**	100	85	52	48	36	31	27	22	**0.07**	26	36	11	–	–	–	–	–	–	–	–	–	–	Tigli-1-en-3-one, 7,12,13-tris(acetyloxy)-4,6,9,20-tetrahydroxy-, (6β,7α,12β)-		NI45317
313	372	312	390	330	446	377	464	**524**	100	75	75	60	53	45	45	43	**0.00**	26	36	11	–	–	–	–	–	–	–	–	–	–	Tigli-1-en-3-one, 12,13,20-tris(acetyloxy)-4,6,7,9-tetrahydroxy-, (6β,7β,12β)-		LI09534
43	83	69	109	41	28	125	53	**524**	100	40	26	21	20	20	15	12	**0.00**	26	36	11	–	–	–	–	–	–	–	–	–	–	Tigli-1-en-3-one, 12,13,20-tris(acetyloxy)-4,6,7-trihydroxy-7βH-, (7α,12β)-		NI45318
107	242	225	106	152	330	419	226	**524**	100	71	71	58	57	48	45	45	**3.00**	26	40	7	2	1	–	–	–	–	–	–	–	–	Dec-Cys(CH2COOMe)-Tyr-O-Methyl		LI09535
509	510	43	524	479	42	41	57	**524**	100	29	21	19	16	13	10	7		27	28	9	2	–	–	–	–	–	–	–	–	–	2H,8H-Benzo[1,2-b:3,4-b']dipyran-6-propanol, 5-methoxy-2,2,8,8-tetramethyl-, 3,5-dinitrobenzoate	56335-86-5	NI45319
509	510	524	525	511	141	297	127	**524**	100	30	24	7	6	6	5	4		27	28	9	2	–	–	–	–	–	–	–	–	–	2H,8H-Benzo[1,2-b:5,4-b']dipyran-10-propanol, 5-methoxy-2,2,8,8-tetramethyl-, 3,5-dinitrobenzoate	26537-43-9	NI45320
509	510	524	525	141	511	479	297	**524**	100	24	19	6	6	5	4	4		27	28	9	2	–	–	–	–	–	–	–	–	–	2H,8H-Benzo[1,2-b:5,4-b']dipyran-10-propanol, 5-methoxy-2,2,8,8-tetramethyl-, 3,5-dinitrobenzoate	26537-43-9	NI45321
155	252	323	324	200	156	253	199	**524**	100	28	15	10	10	9	8	5	**0.74**	27	48	6	4	–	–	–	–	–	–	–	–	–	L-Valine, N-[N-[N-[1-(1-oxodecyl)-L-prolyl]-L-alanyl]-L-alanyl]-, methyl ester	56087-13-9	NI45322
389	285	180	345	393	480	313	346	**524**	100	75	75	55	50	40	35	30	**0.00**	28	32	8	2	–	–	–	–	–	–	–	–	–	4,4'-Dimethyl-3,3'-bis(methoxycarbonylethyl)-5'-benzyloxycarbonyl-5-carboxy-2,2'dipyrrolyl-methane		LI09536
100	226	128	127	71	99	85	83	**524**	100	99	71	35	29	18	18	17	**0.30**	28	52	5	4	–	–	–	–	–	–	–	–	–	DL-Valinamide, N-methyl-N-(1-oxodecyl)glycyl-N-methyl-DL-leucyl-N-formyl-N,N2-dimethyl-	55822-83-8	NI45323
509	255	434	138	143	345	142	344	**524**	100	96	84	80	65	61	56	55	**53.00**	28	56	3	–	–	–	–	–	–	3	–	–	–	Androstane, 2,3,17-tris[(trimethylsilyl)oxy]-, (2β,3α,5α,17β)-		LI09537
509	255	129	434	345	510	142	143	**524**	100	80	66	63	52	46	43	40	**20.00**	28	56	3	–	–	–	–	–	–	3	–	–	–	Androstane, 2,3,17-tris[(trimethylsilyl)oxy]-, (2β,3α,5α,17β)-		LI09538
509	255	345	434	129	142	143	344	**524**	100	86	60	56	56	50	48	38	**26.00**	28	56	3	–	–	–	–	–	–	3	–	–	–	Androstane, 2,3,17-tris[(trimethylsilyl)oxy]-, (2β,3β,5α,17β)-		LI09539
344	434	254	223	129	215	345	228	**524**	100	54	50	50	43	33	32	30	**4.00**	28	56	3	–	–	–	–	–	–	3	–	–	–	Androstane, 3,6,17-tris[(trimethylsilyl)oxy]-, (3α,5α,6β,17β)-		LI09540
344	434	254	223	129	345	215	228	**524**	100	64	49	43	40	32	30	24	**3.00**	28	56	3	–	–	–	–	–	–	3	–	–	–	Androstane, 3,6,17-tris[(trimethylsilyl)oxy]-, (3α,5α,6β,17β)-		LI09541
129	344	434	254	223	130	255	345	**524**	100	90	76	60	60	40	31	30	**12.00**	28	56	3	–	–	–	–	–	–	3	–	–	–	Androstane, 3,6,17-tris[(trimethylsilyl)oxy]-, (3β,5α,6β,17β)-		LI09542
393	73	75	394	434	524	344	129	**524**	100	95	60	40	38	34	34	33		28	56	3	–	–	–	–	–	–	3	–	–	–	Androstane, 3,7,17-tris[(trimethylsilyl)oxy]-, (3α,5α,7α,17β)-	61233-48-5	NI45324
393	344	394	434	435	255	254	524	**524**	100	46	45	40	30	30	30	28		28	56	3	–	–	–	–	–	–	3	–	–	–	Androstane, 3,7,17-tris[(trimethylsilyl)oxy]-, (3α,5α,7α,17β)-		LI09543
73	434	75	129	435	344	107	93	**524**	100	67	60	34	27	17	17	17	**12.00**	28	56	3	–	–	–	–	–	–	3	–	–	–	Androstane, 3,7,17-tris[(trimethylsilyl)oxy]-, (3α,5α,7β,17β)-	55801-53-1	NI45325
73	344	239	75	254	129	345	211	**524**	100	83	69	58	38	36	33	32	**5.00**	28	56	3	–	–	–	–	–	–	3	–	–	–	Androstane, 3,7,17-tris[(trimethylsilyl)oxy]-, (3α,5β,7α,17β)-	55801-54-2	NI45326
393	73	75	394	524	434	129	344	**524**	100	80	50	39	37	28	28	18		28	56	3	–	–	–	–	–	–	3	–	–	–	Androstane, 3,7,17-tris[(trimethylsilyl)oxy]-, (3β,5α,7α,17β)-	55801-51-9	NI45327
73	434	75	129	435	393	147	93	**524**	100	67	65	40	28	25	17	16	**16.00**	28	56	3	–	–	–	–	–	–	3	–	–	–	Androstane, 3,7,17-tris[(trimethylsilyl)oxy]-, (3β,5α,7β,17β)-	61233-47-4	NI45328
73	75	344	129	211	345	254	147	**524**	100	65	59	34	26	21	21	21	**4.11**	28	56	3	–	–	–	–	–	–	3	–	–	–	Androstane, 3,7,17-tris[(trimethylsilyl)oxy]-, (3β,5β,7α,17β)-	55786-08-8	NI45329
217	434	129	344	191	254	393	255	**524**	100	68	60	47	36	24	16	16	**1.00**	28	56	3	–	–	–	–	–	–	3	–	–	–	Androstane, 3,7,18-tris[(trimethylsilyl)oxy]-, (3α,5α,17β)-		LI09544
217	434	191	129	344	254	255	393	**524**	100	62	48	48	41	26	25	10	**3.00**	28	56	3	–	–	–	–	–	–	3	–	–	–	Androstane, 3,7,18-tris[(trimethylsilyl)oxy]-, (3β,5α,17β)-		LI09545
73	434	344	75	254	169	147	143	**524**	100	99	97	88	79	73	58	47	**27.50**	28	56	3	–	–	–	–	–	–	3	–	–	–	Androstane, 3,11,17-tris[(trimethylsilyl)oxy]-, (3α,5β,11α,17β)-	33287-30-8	NI45330
217	191	218	219	434	332	192	74	**524**	100	50	28	17	15	12	11	10	**1.00**	28	56	3	–	–	–	–	–	–	3	–	–	–	Androstane, 3,15,17-tris[(trimethylsilyl)oxy]-, (3α,5α,15α,17β)-		LI09546
191	434	524	147	129	344	419	255	**524**	100	79	64	64	36	35	34	34		28	56	3	–	–	–	–	–	–	3	–	–	–	Androstane, 3,16,17-tris[(trimethylsilyl)oxy]-, (3α,5α,16α,17β)-		LI09547
191	434	524	525	435	344	205	169	**524**	100	61	60	27	25	25	20	20		28	56	3	–	–	–	–	–	–	3	–	–	–	Androstane, 3,16,17-tris[(trimethylsilyl)oxy]-, (3α,5α,16α,17β)-		LI09548
191	434	524	344	205	525	435	255	**524**	100	47	38	22	22	20	18	16		28	56	3	–	–	–	–	–	–	3	–	–	–	Androstane, 3,16,17-tris[(trimethylsilyl)oxy]-, (3β,5α,16α,17β)-		LI09549
191	434	524	344	205	168	73	255	**524**	100	50	46	21	21	18	18	17		28	56	3	–	–	–	–	–	–	3	–	–	–	Androstane, 3,16,17-tris[(trimethylsilyl)oxy]-, (3β,5α,16β,17β)-		LI09550

MASS TO CHARGE RATIOS								M.W.	INTENSITIES								Parent	C	H	O	N	S	F	Cl	Br	I	Si	P	B	X	COMPOUND NAME	CAS Reg No	No
253	199	81	524	137	411	526	413	**524**	100	64	59	54	53	41	39	28		30	50	1	–	–	–	2	–	–	1	–	–	–	9,10-Secocholesta-5(10),6,8(14)-triene, 3-[(dichloromethyl)dimethyloxy]-, (3β,6E)-	69688-11-5	NI45331
270	87	43	299	57	55	98	41	**524**	100	92	77	65	62	62	45	45	**0.00**	33	64	4	–	–	–	–	–	–	–	–	–	–	Octadecanoic acid, 3-hydroxy-2-(1-oxotetradecyl)-, methyl ester	56554-23-5	NI45332
130	83	81	75	73	71	69	57	**524**	100	99	99	99	99	99	99	99	**6.50**	33	68	2	–	–	–	–	–	–	1	–	–	–	Tricosan-15-on-1-ol, trimethylsilyl-		NI45333
524	255	480	525	225	300	435	317	**524**	100	54	46	40	25	19	15	15		34	20	6	–	–	–	–	–	–	–	–	–	–	Phthalide, 3-(2′-fluoranyl)-3-(4-hydroxyphenyl)-		LI09551
524	255	480	525	225	479	300	181	**524**	100	55	45	38	25	23	20	17		34	20	6	–	–	–	–	–	–	–	–	–	–	Spiro[isobenzofuran-1(3H),9′-[9H]xanthen]-3-one, 2′-[1,3-dihydro-1-(4-hydroxyphenyl)-3-oxo-1-isobenzofuranyl]-	55530-30-8	NI45334
91	92	300	284	374	524	328	262	**524**	100	64	26	26	19	17	14	14		34	28	2	4	–	–	–	–	–	–	–	–	–	2,3,4,4a,10a,11,12,13-Octahydro-3,3,12,12-tetramethyl-1,4a,10a,14-tetraazaviolanthrone	87710-93-8	NI45335
43	69	313	451	315	55	95	81	**524**	100	47	42	40	34	34	31	29	**12.05**	34	52	4	–	–	–	–	–	–	–	–	–	–	Lanosta-7,9(11),20(22)-triene-3β,18-diol, diacetate	25116-61-4	NI45336
43	69	451	55	449	464	313	97	**524**	100	41	37	29	28	25	25	25	**14.05**	34	52	4	–	–	–	–	–	–	–	–	–	–	Lanosta-7,9(11),20-triene-3β,18-diol, diacetate	24041-73-4	NI45337
57	43	71	82	83	55	85	69	**524**	100	82	61	57	50	49	44	43	**5.04**	34	68	3	–	–	–	–	–	–	–	–	–	–	Hexadecanoic acid, 2-(hexadecyloxy)ethyl ester	28843-32-5	NI45338
202	73	166	130	203	41	29	69	**525**	100	55	36	35	19	10	9	6	**0.00**	25	34	4	3	–	3	–	–	–	1	–	–	–	L-Prolyltryptopham, N-(trifluoracetyl)-N-(trimethylsilyl)-, butyl ester		MS10529
383	294	143	323	385	69	325	111	**525**	100	50	50	40	33	25	15	15	**5.00**	28	44	6	1	–	–	1	–	–	–	–	–	–	4-Tetradecenamide, N-[3-[4-(acetyloxy)-3-methyl-2-oxo-7-oxabicyclo[4.1.0]hept-1-yl]-2-chloro-2-propenyl]-7-methoxy-N-methyl-	69121-74-0	NI45339
330	332	62	119	121	201	141	143	**526**	100	97	86	68	64	58	39	38	**0.00**	9	7	1	–	–	–	–	5	–	–	–	–	–	Propane, 1-(2,4,6-tribromophenoxy)-2,3-dibromo-	35109-60-5	NI45340
530	532	528	293	295	130	167	165	**526**	100	80	70	70	53	48	45	45	**26.02**	10	–	–	–	–	–	10	–	–	–	–	–	1	Ferrocene, decachloro-	11121-63-4	NI45341
47	171	73	90	102	161	196	198	**526**	100	70	50	41	40	33	28	26	**5.00**	17	16	–	6	2	–	–	2	–	–	–	–	–	4-[N,N′-Bis(4-bromophenyl)]guanidino-3,5-bis(methylthio)[1,2,4]triazole		MS10530
184	167	73	74	168	127	311	325	**526**	100	85	83	55	35	15	14	11	**0.00**	19	18	14	4	–	–	–	–	–	–	–	–	–	Methyl 2,3-bis-O-(2,4-dinitrophenyl)-α-D-glucopyranoside		NI45342
41	29	40	57	44	20	152	198	**526**	100	99	97	80	58	42	35	34	**0.00**	19	28	6	2	1	6	–	–	–	–	–	–	–	L-Homocysteine, S-[3-butoxy-3-oxo-2-[(trifluoroacetyl)amino]propyl]-, N (trifluoroacetyl)-, butyl ester, (R)-	55518-00-8	NI45343
73	292	103	147	217	205	307	293	**526**	100	32	30	27	15	11	9	9	**0.00**	20	50	6	–	–	–	–	–	–	5	–	–	–	Arabinonic acid, tetrakis-O-(trimethylsilyl)-, trimethylsilyl ester		LI09552
292	307	293	217	333	294	103	205	**526**	100	30	28	20	15	15	15	10	**0.00**	20	50	6	–	–	–	–	–	–	5	–	–	–	Pentanoic acid, 2,3,4,5-tetrakis-O-(trimethylsilyl)-, trimethylsilyl ester		LI09553
292	307	293	217	333	294	103	205	**526**	100	29	28	18	14	14	12	9	**0.00**	20	50	6	–	–	–	–	–	–	5	–	–	–	Ribonic acid, 2,3,4,5-tetrakis-O-(trimethylsilyl)-, trimethylsilyl ester	57197-35-0	NI45344
73	147	103	292	217	205	75	74	**526**	100	29	29	24	17	10	10	8	**0.00**	20	50	6	–	–	–	–	–	–	5	–	–	–	Ribonic acid, 2,3,4,5-tetrakis-O-(trimethylsilyl)-, trimethylsilyl ester	57197-35-0	NI45345
73	292	147	103	217	205	307	75	**526**	100	39	33	31	23	12	12	12	**0.00**	20	50	6	–	–	–	–	–	–	5	–	–	–	Ribonic acid, 2,3,4,5-tetrakis-O-(trimethylsilyl)-, trimethylsilyl ester	57197-35-0	NI45346
526	407	225	196	197	81	169	333	**526**	100	52	46	34	27	25	24	21		21	26	10	4	1	–	–	–	–	–	–	–	–	2-Methyl-6-(methylthio)-4-[N-(2,3,4,6-tetra-O-acetyl-β-D-glucopyranosyl)amino]oxazolo[5,4-d]pyrimidine		NI45347
103	73	219	147	231	217	74	104	**526**	100	99	34	20	11	11	10	9	**0.00**	21	54	5	–	–	–	–	–	–	5	–	–	–	D-Arabino-hexitol, 2-deoxy-1,3,4,5,6-pentakis-O-(trimethylsilyl)-	33648-64-5	NI45348
73	117	205	147	319	103	75	217	**526**	100	64	23	19	16	13	11	9	**0.00**	21	54	5	–	–	–	–	–	–	5	–	–	–	D-Glucitol, 6-deoxy-1,2,3,4,5-pentakis-O-(trimethylsilyl)-	53537-99-8	NI45349
73	231	129	147	205	232	103	69	**526**	100	68	33	27	26	15	13	11	**0.00**	21	54	5	–	–	–	–	–	–	5	–	–	–	D-Ribo-hexitol, 3-deoxy-1,2,4,5,6-pentakis-O-(trimethylsilyl)-	34665-31-1	NI45350
42	111	152	153	194	195	364	95	**526**	100	83	76	52	42	41	36	29	**0.00**	22	26	13	2	–	–	–	–	–	–	–	–	–	D-Gluco-octitol, 3,7-anhydro-1,4-dideoxy-2-C-methyl-, 5,6,8-triacetate 2-(3,5-dinitrobenzoate)	50692-73-4	NI45351
219	221	150	43	261	151	220	122	**526**	100	63	39	32	29	26	25	22	**0.00**	23	20	10	–	–	–	2	–	–	–	–	–	–	Benzoic acid, 2,4-bis(acetyloxy)-3,5-dichloro-6-methyl-, 3-(acetyloxy)-4-(methoxycarbonyl)-5-methylphenyl ester	5859-24-5	NI45352
143	175	88	101	75	235	115	71	**526**	100	98	69	52	35	30	28	21	**0.00**	23	42	13	–	–	–	–	–	–	–	–	–	–	Methyl-2,3-di-O-methyl-4-O-[2,3-di-O-methyl-4-O-(2,3,4-tri-O-methyl-β-D-xylopyranosyl)-β-D-xylopyranosyl]-β-D-xylopyranoside		NI45353
175	115	101	143	395	111	75	88	**526**	100	88	66	62	56	44	36	29	**0.00**	23	42	13	–	–	–	–	–	–	–	–	–	–	Methyl-2,3-di-O-methyl-4-O-[2,4-di-O-methyl-3-O-(2,3,4-tri-O-methyl-α-D-xylopyranosyl)-β-D-xylopyranosyl]-β-D-xylopyranoside	73654-68-9	NI45354
175	115	101	395	143	75	88	114	**526**	100	81	77	51	51	43	34	30	**0.00**	23	42	13	–	–	–	–	–	–	–	–	–	–	Methyl-2,3-di-O-methyl-4-O-[2,4-di-O-methyl-3-O-(2,3,4-tri-O-methyl-β-D-xylopyranosyl)-β-D-xylopyranosyl]-β-D-xylopyranoside	73654-68-9	NI45355
175	143	88	303	101	75	115	99	**526**	100	48	40	25	25	19	14	14	**0.00**	23	42	13	–	–	–	–	–	–	–	–	–	–	Methyl-2,3-di-O-methyl-4-O-[3,4-di-O-methyl-2-O-(2,3,4-tri-O-methyl-β-D-xylopyranosyl)-β-D-xylopyranosyl]-β-D-xylopyranoside	72521-38-1	NI45356
175	88	143	101	395	129	75	84	**526**	100	46	38	32	24	15	12	11	**0.00**	23	42	13	–	–	–	–	–	–	–	–	–	–	Methyl-2-O-methyl-3-O-(2,3,4-tri-O-methyl-β-D-xylopyranosyl)-4-O-(2,3,4-tri-O-methyl-β-D-xylopyranosyl)-β-D-xylopyranoside	72521-33-6	NI45357
175	143	101	88	75	99	115	145	**526**	100	93	69	61	21	19	16	14	**0.00**	23	42	13	–	–	–	–	–	–	–	–	–	–	Methyl-3-O-methyl-2-O-(2,3,4-tri-O-methyl-β-D-xylopyranosyl)-4-O-(2,3,4-tri-O-methyl-β-D-xylopyranosyl)-β-D-xylopyranoside	71072-45-2	NI45358
101	115	75	88	143	71	175	45	**526**	100	73	68	65	63	46	43	43	**0.00**	23	42	13	–	–	–	–	–	–	–	–	–	–	D-Xylose, O-2,3,4-tri-O-methyl-β-D-xylopyranosyl-(1-3)-O-2,4-di-O-methyl-β-D-xylopyranosyl-(1-4)-2,3,5-tri-O-methyl-	56248-58-9	NI45359
56	330	358	414	386	470	442	274	**526**	100	80	65	44	40	30	30	30	**0.00**	24	14	7	–	–	–	–	–	–	–	–	–	2	Hexacarbonyl-μ-[(1-3-η^3:4,4α,5-η^3)-4-methylene-2,3-diphenyl-1-oxo-2-pentene-1,5-diyl]diiron	122624-30-0	DD01780

MASS TO CHARGE RATIOS								M.W.	INTENSITIES								Parent	C	H	O	N	S	F	Cl	Br	I	Si	P	B	X	COMPOUND NAME	CAS Reg No	No
73	89	87	103	59	57	99	75	**526**	100	90	88	54	51	37	32	32	**0.00**	24	46	12	–	–	–	–	–	–	–	–	–	–	(1S,20S)-Bicyclo[18.16.0]-3,6,9,12,15,18,21,24,27,30,33,36-dodecaoxahexatriacontane		MS10531
87	89	73	59	58	133	131	86	**526**	100	64	58	49	24	24	22	22	**1.70**	24	46	12	–	–	–	–	–	–	–	–	–	–	(1S,1'S)-1,1'-Bis(2,5,8,11,14,17-hexaoxacyclooctadecane)	95637-58-4	MS10532
73	89	87	63	78	59	149	71	**526**	100	75	67	62	46	40	31	27	**1.41**	24	46	12	–	–	–	–	–	–	–	–	–	–	(1S,1'S)-1,1'-Bis(2,5,8,11,14,17-hexaoxacyclooctadecane)	95637-58-4	NI45360
526	147	149	135	104	93	105	121	**526**	100	92	76	58	53	46	42	39		25	29	4	–	–	7	–	–	–	–	–	–	–	Androsta-2,4-diene-3,17-diol, 17-acetate 3-(heptafluorobutanoate), (17β)-	49566-69-0	MS10533
146	526	133	147	105	91	148	93	**526**	100	95	82	64	41	39	30	28		25	29	4	–	–	7	–	–	–	–	–	–	–	Androsta-3,5-diene-3,17-diol, 17-acetate 3-(heptafluorobutanoate), (17β)-	18072-24-7	MS10534
185	186	273	245	303	466	304	103	**526**	100	86	84	69	62	60	59	59	**0.00**	25	42	8	–	–	–	–	–	–	2	–	–	–	Trichothec-9-ene-4,8-diol, 12,13-epoxy-3,15-bis[(trimethylsilyl)oxy]-, diacetate, (3α,4β,8α)-	74299-12-0	NI45361
44	69	41	528	526	43	94	55	**526**	100	96	95	74	72	43	26	22		26	23	7	–	–	–	–	1	–	–	–	–	–	3-Bromo-4-hydroxy-2,3'-dimethyl-5,5',8,8'-tetramethoxy-1,2'-binaphthalene-1',4'-dione	101459-34-1	NI45362
498	393	379	468	526	183	105	185	**526**	100	55	44	22	21	11	11	4		26	25	2	2	–	–	–	–	–	–	1	–	1	Molybdenum, carbonyl(η^5-2,4-cyclopentadien-1-yl)[P,P-diphenyl-N-(1-phenylethyl)phosphinous amide-P]nitrosyl-, stereoisomer	75520-16-0	NI45363
77	105	68	91	106	232	69	51	**526**	100	99	99	98	95	79	54	43	**0.00**	27	26	9	–	1	–	–	–	–	–	–	–	–	β-L-Arabinopyranoside, methyl, 2,3-dibenzoate 4-p-toluenesulphonate	13143-92-5	NI45364
85	425	55	424	67	526	91	95	**526**	100	28	23	22	22	16	14	12		27	43	2	–	–	–	–	–	1	–	–	–	–	(20S)-21-Iodo-20-methylpregn-7-en-3β-ol tetrahydropyranyl ether		DD01781
458	460	335	459	405	461	407	419	**526**	100	76	42	37	32	31	27	19	**10.05**	27	44	2	–	–	–	2	–	–	2	–	–	–	Cannabidiol bis[(chloromethyldimethylsilyl)oxy]- ether		NI45365
72	226	44	43	41	30	87	28	**526**	100	71	66	25	20	20	19	15	**0.32**	27	50	6	4	–	–	–	–	–	–	–	–	–	L-Leucine, N-[N-[N-(N-decanoyl-L-alanyl)-L-valyl]glycyl]-, methyl ester	4560-72-9	NI45366
284	295	456	285	240	296	242	469	**526**	100	87	79	72	64	63	52	50	**39.00**	28	35	2	–	–	5	–	–	–	1	–	–	–	Androst-4-en-3-one, 17-[[dimethyl(pentafluorophenyl)silyl]oxy]-17-methyl, (17β)-	57691-91-5	NI45367
284	295	456	285	296	240	242	469	**526**	100	86	79	72	70	65	52	50	**41.00**	28	35	2	–	–	5	–	–	–	1	–	–	–	Androst-4-en-3-one, 17-[[dimethyl(pentafluorophenyl)silyl]oxy]-17-methyl, (17β)-		MS10535
230	94	228	79	491	107	81	508	**526**	100	41	28	28	26	22	19	18	**0.50**	30	42	4	2	1	–	–	–	–	–	–	–	–	Syn-6-hydroxythiobinupharidine sulphoxide		NI45368
230	492	94	228	107	79	81	509	**526**	100	65	20	19	16	14	11	10	**0.00**	30	42	4	2	1	–	–	–	–	–	–	–	–	Syn-6'-hydroxythiobinupharidine sulphoxide		NI45369
94	107	230	176	262	245	447	446	**526**	100	55	47	43	11	6	2	2	**0.30**	30	42	4	2	1	–	–	–	–	–	–	–	–	Thionuphlutine-A, 6,6'-dihydroxy-		LI09554
176	230	107	94	262	445	245	490	**526**	100	66	22	22	6	3	3	1	**0.40**	30	42	4	2	1	–	–	–	–	–	–	–	–	Thionuphlutine-B, 6,6'-dihydroxy-		LI09555
467	183	262	121	108	201	185	77	**526**	100	99	66	60	59	58	52	46	**15.00**	31	28	4	–	–	–	–	–	–	–	2	–	–	1,3-Diphosphacyclopenta-1,3-diene, 4,5-bis(methoxycarbonyl)-1,1,3,3-tetraphenyl-		MS10536
467	183	262	201	185	199	468	370	**526**	100	99	66	58	52	41	30	20	**15.00**	31	28	4	–	–	–	–	–	–	–	2	–	–	1,3-Diphosphacyclopenta-1,3-diene, 4,5-bis(methoxycarbonyl)-1,1,3,3-tetraphenyl-		LI09556
150	177	137	210	95	135	83	93	**526**	100	45	43	38	16	15	12	11	**6.60**	31	42	7	–	–	–	–	–	–	–	–	–	–	Cystoseirol B acetate		NI45370
150	219	137	177	210	135	95	175	**526**	100	60	36	34	27	14	11	11	**5.40**	31	42	7	–	–	–	–	–	–	–	–	–	–	Cystoseirol C acetate		NI45371
261	525	301	526	262	165	302	260	**526**	100	57	30	21	16	6	5	5		31	49	3	–	–	3	–	–	–	–	–	–	–	α-Tocophenol, trifluoroacetate		LI09557
261	525	301	526	262	302	165	527	**526**	100	48	26	22	13	5	5	3		31	49	3	–	–	3	–	–	–	–	–	–	–	α-Tocophenol, trifluoroacetate	56272-54-9	NI45372
261	525	301	526	262	165	260	302	**526**	100	51	30	18	15	7	5	4		31	49	3	–	–	3	–	–	–	–	–	–	–	α-Tocophenol, trifluoroacetate		LI09558
526	527	485	443	363	133	369	329	**526**	100	45	24	18	15	12	11	11		31	50	3	–	–	–	–	–	–	2	–	–	–	$\Delta^{1(6)}$-Tetrahydrocannabinol, 7-hydroxy-, bis(allyldimethylsilyl)-		NI45373
410	411	369	395	412	327	370	353	**526**	100	33	26	12	10	9	8	6	**5.31**	31	50	3	–	–	–	–	–	–	2	–	–	–	Δ^1-Tetrahydrocannabinol, 6-hydroxy-, bis(allyldimethylsilyl)-, (6α)-		NI45374
397	355	398	356	526	369	399	368	**526**	100	56	34	18	16	11	9	9		31	50	3	–	–	–	–	–	–	2	–	–	–	Δ^1-Tetrahydrocannabinol, 7-hydroxy-, bis(allyldimethylsilyl)-		NI45375
526	194	179	207	165	152	151	508	**526**	100	60	50	10	8	6	6	4		32	46	6	–	–	–	–	–	–	–	–	–	–	Dodecane, 1,12-bis(4-methyl-3,5-dimethoxybenzoyl)-		LI09559
526	99	397	451	527	43	81	127	**526**	100	49	48	41	35	34	31	23		32	46	6	–	–	–	–	–	–	–	–	–	–	Lanosta-7,9(11)-dien-18-oic acid, 3-(acetyloxy)-22,25-epoxy-17,20-dihydroxy-, γ-lactone, (3β)-	56143-26-1	NI45376
43	469	45	60	484	44	42	470	**526**	100	18	17	13	11	11	10	8	**0.79**	33	18	7	–	–	–	–	–	–	–	–	–	–	17-Acetoxy-12-methoxy-5,6,11,18-trinaphthylenetetrone		NI45377
189	175	121	119	203	133	149	161	**526**	100	75	71	61	49	49	34	29	**6.77**	33	50	5	–	–	–	–	–	–	–	–	–	–	A(1)-Norlup-20(29)-en-28-oic acid, 3-acetyl-2-formyl-, methyl ester	75714-78-2	NI45378
203	262	189	133	187	466	407	249	**526**	100	45	22	13	11	9	9	8	**4.00**	33	50	5	–	–	–	–	–	–	–	–	–	–	Olean-12-en-28-oic acid, 3-(acetyloxy)-6-oxo-, methyl ester		LI09560
482	43	55	119	44	95	105	81	**526**	100	65	52	43	41	40	38	37	**32.00**	33	50	5	–	–	–	–	–	–	–	–	–	–	Ursa-11-en-13,28-olide, 2,3,23-trihydroxy-3,23-O-isopropylidene-, (2α,3β,13β)-		MS10537
526	435	91	234	262	448	179	132	**526**	100	88	70	32	14	12	12	11		34	26	4	2	–	–	–	–	–	–	–	–	–	N,N'-Dibenzyl[2.2]paracyclophane-4,5,12,13-tetracarboxylic acid diimide		NI45379
390	391	526	458	459	527	392	346	**526**	100	46	35	23	15	13	10	10		34	26	4	2	–	–	–	–	–	–	–	–	–	N,N'-Dicyclopentyl-3,4:9,10-perylenebis(dicarboximide)	80509-52-0	MS10538
205	109	95	189	69	84	526	119	**526**	100	78	56	38	38	30	20	18		34	54	4	–	–	–	–	–	–	–	–	–	–	2α,3α-Diacetoxyolean-13(18)-ene		MS10539
189	107	135	466	216	249	276	423	**526**	100	61	60	35	19	10	8	7	**3.00**	34	54	4	–	–	–	–	–	–	–	–	–	–	Lup-20(29)-ene-3,12-diol, diacetate, (3β,12β)-		LI09561
187	189	218	526	307	406	247	466	**526**	100	94	49	48	22	20	15	11		34	54	4	–	–	–	–	–	–	–	–	–	–	Lup-20(29)-ene-3,23-diol, diacetate, (3α)-		LI09562
187	189	218	526	307	406	247	229	**526**	100	94	49	48	22	20	15	11		34	54	4	–	–	–	–	–	–	–	–	–	–	Lup-20(29)-ene-3,23-diol, diacetate, (3α,4α)-	56830-88-7	MS10540
511	391	239	359	406	526	205	253	**526**	100	94	88	84	82	79	36	32		34	54	4	–	–	–	–	–	–	–	–	–	–	D:C-Friedo-B':A'-neogammacer-9(11)-ene-3,12-diol, diacetate		NI45380
189	202	108	135	133	106	203	119	**526**	100	69	61	52	51	49	46	45	**1.16**	34	54	4	–	–	–	–	–	–	–	–	–	–	D-Friedoolean-14-ene-3β,28-diol, diacetate	17884-89-8	NI45381
43	392	257	69	55	57	466	145	**526**	100	69	45	44	40	39	32	28	**8.97**	34	54	4	–	–	–	–	–	–	–	–	–	–	Lanosta-7,9(11)-diene-3,18-diol, diacetate, (3β,20ξ)-	56298-04-5	NI45382
203	216	189	204	190	201	466	187	**526**	100	66	30	23	23	21	20	20	**1.73**	34	54	4	–	–	–	–	–	–	–	–	–	–	Olean-12-ene-3,28-diol, diacetate, (3β)-	1896-77-1	NI45383
276	148	201	203	107	277	109	119	**526**	100	29	24	21	21	19	19	17	**1.00**	34	54	4	–	–	–	–	–	–	–	–	–	–	Olean-12-ene-3,28-diol, diacetate, (3β)-	1896-77-1	NI45384

MASS TO CHARGE RATIOS								M.W.	INTENSITIES								Parent	C	H	O	N	S	F	Cl	Br	I	Si	P	B	X	COMPOUND NAME	CAS Reg No	No
203	189	204	190	466	187	119	105	**526**	100	28	21	21	17	17	17	17	**1.00**	34	54	4	–	–	–	–	–	–	–	–	–	–	**Olean-12-ene-3,28-diol, diacetate, (3β)-**	1896-77-1	**NI45385**
262	203	186	526	468	453	409	511	**526**	100	78	16	8	5	5	3	2		34	54	4	–	–	–	–	–	–	–	–	–	–	**Olean-12-en-28-oic acid, 2,3-[isopropylidenebis(oxy)]-, methyl ester, (2β,3β)-**	56868-92-9	**MS10541**
133	466	134	135	189	187	406	289	**526**	100	91	78	55	37	30	23	22	**0.00**	34	54	4	–	–	–	–	–	–	–	–	–	–	**ψ-Taraxene-3,16-diol, diacetate, (3β,16α)-**		**NI45386**
43	466	135	133	189	216	187	190	**526**	100	79	40	40	21	20	20	9	**3.00**	34	54	4	–	–	–	–	–	–	–	–	–	–	**ψ-Taraxene-3,16-diol, diacetate, (3β,16β,19H)-**		**NI45387**
466	134	135	203	133	189	406	187	**526**	100	70	50	40	40	26	25	25	**0.50**	34	54	4	–	–	–	–	–	–	–	–	–	–	**ψ-Taraxene-3,16-diol, diacetate, (3β,16β,19H)-**		**NI45388**
107	466	108	189	202	190	203	467	**526**	100	99	99	83	47	45	43	41	**17.00**	34	54	4	–	–	–	–	–	–	–	–	–	–	**Taraxene-3,16-diol, diacetate, (3β,16β)-**		**NI45389**
159	146	160	526	147	368	175	174	**526**	100	23	21	4	4	3	3	2		34	58	2	2	–	–	–	–	–	–	–	–	–	**Tryptamine, N-lignoceroyl-5-hydroxy-**		**MS10542**
481	375	376	332	526	410	482	130	**526**	100	98	27	24	23	23	21	18		35	30	3	2	–	–	–	–	–	–	–	–	–	**Spiro[phthalan-1,9′-xanthene], 3-oxo-2-(p-toluidino)-**		**IC04557**
368	147	145	43	57	81	105	95	**526**	100	57	51	51	43	41	39	36	**0.00**	36	62	2	–	–	–	–	–	–	–	–	–	–	**Cholest-5-en-3-ol (3β)-, nonanoate**	1182-66-7	**NI45390**
105	43	57	526	41	104	55	527	**526**	100	70	41	39	27	19	19	17		38	70	–	–	–	–	–	–	–	–	–	–	–	**Benzene, 1,4-dihexadecyl-**	55517-88-9	**WI01593**
262	183	108	263	107	184	261	185	**526**	100	60	37	18	18	16	13	11	**4.50**	39	27	–	–	–	–	–	–	–	–	1	–	–	**Phosphorane, 13H-dibenzo[a,c]fluoren-13-ylidenetriphenyl-**		**LI09563**
262	183	108	263	107	261	184	185	**526**	100	60	34	21	15	12	12	8	**2.00**	39	27	–	–	–	–	–	–	–	–	1	–	–	**Phosphorane, 13H-dibenzo[a,i]fluoren-13-ylidenetriphenyl-**	31083-22-4	**NI45391**
262	183	108	263	107	184	261	51	**526**	100	69	37	19	19	14	13	13	**4.70**	39	27	–	–	–	–	–	–	–	–	1	–	–	**Phosphorane, 13H-indeno[1,2-l]phenanthren-13-ylidenetriphenyl-**	31083-23-5	**NI45392**
253	69	55	107	527	408	119	177	**527**	100	25	22	20	19	19	19	19		21	23	3	1	–	10	–	–	–	–	–	–	–	**(4-Hydroxyphenyl)nonylamine, bis(pentafluoropropionyl)-**		**NI45393**
73	75	236	189	147	202	169	192	**527**	100	67	51	36	21	19	17	16	**0.30**	21	41	3	5	1	–	–	–	–	3	–	–	–	**Adenosine, 5′-S-ethyl-5′-thio-N-(trimethylsilyl)-2′,3′-bis-O-(trimethylsilyl)**	54623-29-9	**NI45394**
97	196	454	310	302	252	208	362	**527**	100	51	17	13	13	13	11	9	**6.50**	22	29	10	3	1	–	–	–	–	–	–	–	–	**1,6-Dihydro-1-methyl-2-methylthio-5-((E)-2-ethoxycarbonylvinyl)-4-(2,3,4 tri-O-acetyl-β-D-xylopyranosylamino)-1H-pyrimidin-6-one**		**MS10543**
91	392	420	312	320	276	220	200	**527**	100	24	7	7	5	5	5	5	**5.00**	25	31	5	1	1	–	2	–	–	–	–	–	–	**1-Piperidinecarboxylic acid, 4-[3,3-dichloro-2-[[(4-methylphenyl)sulphonyl]oxy]propyl]-3-ethyl-, phenylmethyl ester, [3α,4α(S*)]-(±)-**	66087-85-2	**NI45395**
135	86	220	131	350	136	107	333	**527**	100	67	30	26	18	11	11	10	**1.20**	29	41	6	3	–	–	–	–	–	–	–	–	–	**L-Tyrosine, N-[N-[N-(1-adamantylcarbonyl)glycyl]-L-leucyl]-, methyl ester**	28415-42-1	**NI45396**
484	289	466	442	527				**527**	100	99	30	20	5					32	49	5	1	–	–	–	–	–	–	–	–	–	**27-Azaspirostan, 3-(acetyloxy)-9,11-epoxy-16-isopropyl-, (3β,9α,11α,16α)-**		**LI09564**
286	272	442	527	287	512	443	230	**527**	100	38	34	33	22	18	10	10		32	49	5	1	–	–	–	–	–	–	–	–	–	**Daphniphylline hydrochloride**		**LI09565**
286	272	245	527	230	426	512	350	**527**	100	23	18	14	13	4	3	3		32	49	5	1	–	–	–	–	–	–	–	–	–	**Isodaphniphylline hydrochloride**		**LI09566**
100	131	509	69	309	114	528	397	**528**	100	47	45	43	15	12	10	10		10	–	–	2	–	20	–	–	–	–	–	–	–	**Bipiperidyl, 1,1′-perfluoro-**		**LI09567**
184	39	41	144	142	103	143	185	**528**	100	68	60	57	46	46	36	35	**0.53**	12	20	–	–	–	–	–	2	–	–	–	–	2	**Rhodium, di-μ-bromotetrakis(η^3-2-propenyl)di-**	12307-73-2	**NI45397**
73	218	75	146	45	147	100	74	**528**	100	26	24	23	19	18	17	15	**0.00**	18	44	4	2	2	–	–	–	–	4	–	–	–	**L-Cystine, N,N′-bis(trimethylsilyl)-, bis(trimethylsilyl) ester**	69688-44-4	**NI45398**
73	218	146	147	100	45	75	219	**528**	100	45	32	16	16	12	11	9	**0.00**	18	44	4	2	2	–	–	–	–	4	–	–	–	**L-Cystine, N,N′-bis(trimethylsilyl)-, bis(trimethylsilyl) ester**	69688-44-4	**MS10544**
350	148	73	365	130	45	336	203	**528**	100	53	53	44	26	20	13	13	**4.00**	18	54	–	4	–	–	–	–	–	5	2	–	–	**1,3,2,4-Diazadiphosphetidine, 4-[bis(trimethylsilyl)amino]-2,2,4,4-tetrahydro-2,2,2-trimethyl-1,3-bis(trimethylsilyl)-4-[(trimethylsilyl)imino]-**	67277-83-2	**NI45399**
146	364	365	333	161	301	188	342	**528**	100	61	25	13	13	8	7	6	**0.00**	18	57	–	3	–	–	–	–	–	6	–	–	1	**Silanamine, 1,1,1-trimethyl-N-(trimethylsilyl)-, scandium(3+) salt**	37512-28-0	**NI45400**
142	73	298	174	172	147	287	207	**528**	100	99	48	37	20	18	11	11	**8.00**	19	53	3	2	–	–	–	–	–	5	1	–	–	**Phosphonic acid, [4-[bis(trimethylsilyl)amino]-1-[(trimethylsilyl)amino]butyl]-, bis(trimethylsilyl) ester**	53044-47-6	**NI45401**
73	142	298	174	172	147	299	143	**528**	100	95	46	35	19	17	16	14	**7.61**	19	53	3	2	–	–	–	–	–	5	1	–	–	**Phosphonic acid, (1,4-diaminobutyl)-, pentakis(trimethylsilyl)-**	56211-26-8	**NI45402**
486	257	528	367	197	97	228	239	**528**	100	63	43	30	30	24	18	16		21	28	10	4	1	–	–	–	–	–	–	–	–	**5-(N,N-Diacetylamino)-1-methyl-2-(methylthio)-4-(2,3,4-tri-O-acetyl-β-D-xylopyranosylamino)-1H-pyrimidin-6-one**		**NI45403**
57	85	41	78	362	360	29	528	**528**	100	19	13	9	8	8	7	6		24	23	7	–	–	–	3	–	–	–	–	–	–	**Xanthen-9-one, 2,4,7-trichloro-1,3,6-trihydroxy-8-methyl-, bis(2,2-dimethylpropanoate) ester**		**LI09568**
84	120	59	138	121	92	229	331	**528**	100	9	5	4	4	4	3	2	**0.50**	24	36	9	2	1	–	–	–	–	–	–	–	–	**D-Erythro-α-D-Galacto-octopyranoside, 2-[(2-hydroxybenzoyl)oxy]ethyl 6,8-dideoxy-7-O-methyl-6-[[(1-methyl-2-pyrrolidinyl)carbonyl]amino]-1-thio-, (S)-**	2520-21-0	**NI45404**
351	78	185	197	120	149	56	395	**528**	100	53	44	38	34	29	28	25	**1.40**	25	20	2	–	–	–	–	–	–	–	–	–	2	**Iron, dicarbonyl-π-cyclopentadienyl-triphenyltin(IV)**		**LI09569**
91	288	105	160	77	70	122	51	**528**	100	91	53	32	30	20	18	15	**0.00**	27	32	7	2	1	–	–	–	–	–	–	–	–	**α-DL-lyxo-Hexopyranoside, 3-O-benzoyl-6-(N-carbobenzoxyprolineamide)-1-methylthio-**		**NI45405**
528	473	529	511	474	289	233	261	**528**	100	76	37	23	21	17	15	12		27	32	9	2	–	–	–	–	–	–	–	–	–	**2H,6H-Benzo[1,2-b:5,4-b′]dipyran-10-propanol, 3,4,7,8-tetrahydro-5-methoxy-2,2,8,8-tetramethyl-, 3,5-dinitrobenzoate**	26537-44-0	**NI45406**
528	473	529	511	474	289	233	44	**528**	100	65	32	19	18	15	13	10		27	32	9	2	–	–	–	–	–	–	–	–	–	**2H,6H-Benzo[1,2-b:5,4-b′]dipyran-10-propanol, 3,4,7,8-tetrahydro-5-methoxy-2,2,8,8-tetramethyl-, 3,5-dinitrobenzoate**	26537-44-0	**NI45407**

MASS TO CHARGE RATIOS								M.W.	INTENSITIES								Parent	C	H	O	N	S	F	Cl	Br	I	Si	P	B	X	COMPOUND NAME	CAS Reg No	No
106	327	291	104	132	263	105	392	**528**	100	43	31	27	27	25	22	18	**2.66**	28	40	6	4	–	–	–	–	–	–	–	–	–	(R,R)-Bis(phenylglycyl)diaza-18-crown-6	93788-01-3	NI45408
43	28	29	207	91	79	77	36	**528**	100	68	16	14	14	10	10	10	**5.39**	29	36	9	–	–	–	–	–	–	–	–	–	–	Satratoxin H		NI45409
69	71	528	122	182	81	408	60	**528**	100	60	54	51	49	45	44	41		29	40	7	2	–	–	–	–	–	–	–	–	–	4,25-Secoobscurinervan-4-ol, 22-ethyl-15,16-dimethoxy-, diacetate (ester), (4β,22α)-	54658-12-7	NI45410
169	528	168	530	139	529	153	167	**528**	100	96	81	48	37	34	32	31		30	40	8	–	–	–	–	–	–	–	–	–	–	2,5-Cyclohexadiene-1,4-dione, 2,5-dihydroxy-3-[16-(2-hydroxy-5-methoxy 3,6-dioxo-1,4-cyclohexadien-1-yl)-8-hexadecenyl]-6-methyl-, (Z)-	18761-04-1	NI45411
169	168	528	139	530	153	167	529	**528**	100	79	51	45	32	31	30	27		30	40	8	–	–	–	–	–	–	–	–	–	–	2,5-Cyclohexadiene-1,4-dione, 2,2′-(8-hexadecene-1,16-diyl)bis[3-hydroxy-6-methoxy-, (Z)-	18799-05-8	NI45412
151	283	528	255	301	313	342	329	**528**	100	93	42	38	24	23	22	19		30	40	8	–	–	–	–	–	–	–	–	–	–	Excoecariatoxin	92219-48-2	NI45413
83	55	328	310	429	29	53	39	**528**	100	87	17	17	15	13	11	10	**0.98**	30	40	8	–	–	–	–	–	–	–	–	–	–	Phorbol, 12,13-bis[(E)-2-methylcrotonate]		NI45414
123	81	83	95	43	69	310	53	**528**	100	82	77	70	70	51	46	42	**1.49**	30	40	8	–	–	–	–	–	–	–	–	–	–	Phorbol, 13-O-acetyl-12-O-octa-2,6-dienoyl-, (2′Z,6′E)-		NI45415
105	43	29	83	31	328			**528**	100	79	75	72	68	67			**1.00**	30	40	8	–	–	–	–	–	–	–	–	–	–	Pregn-5-ene-3,8-tetrol, 12-O-benzoyl-20-O-acetyl-, (3β,12β,14β,17α,20α)		LI09570
528	496	529	530	83	85	494	264	**528**	100	87	42	27	24	16	12	6		32	16	–	–	4	–	–	–	–	–	–	–	–	Cycloocta[1,2-b^{4},3-b'^{5},6-b''^{8},7-b‴]tetrathionaphthene		NI45416
43	57	55	41	138	71	69	298	**528**	100	99	64	56	53	51	50	48	**5.55**	32	48	6	–	–	–	–	–	–	–	–	–	–	Ingenol, 3-desoxy-3-oxo-5-dodecanoate		NI45417
43	69	109	510	41	55	59	435	**528**	100	62	49	42	36	31	24	21	**2.00**	32	48	6	–	–	–	–	–	–	–	–	–	–	Lanost-9(11)-en-18-oic acid, 3-(acetyloxy)-20,25-dihydroxy-16-oxo-, γ-lactone, (3β)-	53534-45-5	NI45418
280	121	293	135	81	528	91	165	**528**	100	43	38	35	32	28	25	23		33	30	2	–	–	–	2	–	–	–	–	–	–	Benzene, 1,1′-[5,5-bis(4-chlorophenyl)-3,3-dimethyl-1,4-pentadienylidene]bis[4-methoxy-	67437-04-1	NI45419
148	352	176	161	528	105	77	133	**528**	100	85	67	33	18	14	14	7		33	36	6	–	–	–	–	–	–	–	–	–	–	Tri-O-thymotide		MS10545
203	262	189	133	187	249	247	468	**528**	100	67	24	14	10	6	5	4	**2.00**	33	52	5	–	–	–	–	–	–	–	–	–	–	Olean-12-en-28-oic acid, 3-(acetyloxy)-6-hydroxy-, methyl ester, (3β,6β)-		LI09571
203	262	189	187	133	468	249	247	**528**	100	63	18	18	16	8	7	5	**3.00**	33	52	5	–	–	–	–	–	–	–	–	–	–	Olean-12-en-28-oic acid, 6-(acetyloxy)-3-hydroxy-, methyl ester, (3β,6β)-		LI09572
276	217	207	334	216	208	277	199	**528**	100	70	70	66	54	30	29	26	**1.01**	33	52	5	–	–	–	–	–	–	–	–	–	–	Olean-12-en-28-oic acid, 3-hydroxy-15,16-[isopropylidenebis(oxy)]-, (3β,15α,16α)-	55759-93-8	NI45420
189	136	468	202	203	204	453	191	**528**	100	51	45	40	39	38	37	37	**9.70**	34	56	4	–	–	–	–	–	–	–	–	–	–	29,30-Dinorgammacerane-3,22-diol, 21,21-dimethyl-, diacetate, (3β,8α,9β,13α,14β,17α,18β,22α)-	43206-35-5	NI45421
317	278	231	468	183	275	257	217	**528**	100	99	99	76	55	52	51	51	**31.00**	34	56	4	–	–	–	–	–	–	–	–	–	–	D:A-Friedooleanane-3,29-diol, diacetate, (3α,20α)-	56816-09-2	MS10546
205	264	464	496	167	497	329	528	**528**	100	25	22	20	18	12	11	10		36	32	4	–	–	–	–	–	–	–	–	–	–	Dimethyl 3,3-diphenyl-4-(2,2-diphenylvinyl)cyclohex-1-ene-1,4-dicarboxylate		DD01782
98	55	431	56	165	99	528	432	**528**	100	56	28	14	5	5	5	4		36	36	2	2	–	–	–	–	–	–	–	–	–	Pyrrolidine, 4-(pyrrolidinocarbonylbenzyl)triphenylacetyl-	70008-31-0	NI45422
91	120	160	364	365	186	409	92	**528**	100	65	54	38	18	18	13	12	**4.00**	36	52	1	2	–	–	–	–	–	–	–	–	–	Pregnan-20-ol, 18-(benzylamino)-3-(benzylmethylamino)-, (5α,20β)-	32095-46-8	LI09573
91	120	160	264	265	186	409	92	**528**	100	70	55	35	18	18	13	12	**4.00**	36	52	1	2	–	–	–	–	–	–	–	–	–	Pregnan-20-ol, 18-(benzylamino)-3-(benzylmethylamino)-, (5α,20β)-	32095-46-8	NI45423
316	69	226	169	103	317	91	78	**529**	100	41	34	31	29	22	20	17	**0.00**	16	9	3	1	–	14	–	–	–	–	–	–	–	(2-Hydroxyphenyl)ethylamine, bis(heptafluorobutanoyl)-		NI45424
316	226	169	317	69	303	119	147	**529**	100	38	24	22	22	20	20	17	**0.70**	16	9	3	1	–	14	–	–	–	–	–	–	–	(2-Hydroxyphenyl)ethylamine, bis(heptafluorobutanoyl)-		NI45425
316	119	69	317	303	169	275	226	**529**	100	28	22	22	21	19	12	11	**0.00**	16	9	3	1	–	14	–	–	–	–	–	–	–	(3-Hydroxyphenyl)ethylamine, bis(heptafluorobutanoyl)-		NI45426
69	169	303	316	275	105	77	103	**529**	100	92	89	70	55	55	44	34	**1.50**	16	9	3	1	–	14	–	–	–	–	–	–	–	Phenylethylamine, β-hydroxy-N,N-bis(heptafluorobutanoyl)-		MS10547
316	69	119	317	303	169	28	275	**529**	100	27	23	22	20	17	16	10	**0.30**	16	9	3	1	–	14	–	–	–	–	–	–	–	Tyramine, N,O-bis(heptafluorobutanoyl)-		MS10548
152	81	69	222	221	301	53	153	**529**	100	82	39	23	19	15	13	9	**2.90**	16	12	7	3	–	9	–	–	–	–	–	–	–	Cytidine, 2′-deoxy-5-methyl-N-(trifluoroacetyl)-, 3′,5′-bis(trifluoroacetate)	35170-13-9	NI45427
88	200	228	410	172	198	171	336	**529**	100	50	33	25	17	15	13	12	**1.00**	21	27	11	3	1	–	–	–	–	–	–	–	–	1,6-Dihydro-5-formyl-1-methyl-2-methylthio-4-(2,3,4,6-tetra-O-acetyl-β-D-glucopyranosylamino)-1H-pyrimidin-6-one		MS10549
450	452	531	453	451	533	529	227	**529**	100	98	49	29	28	25	24	15		22	13	5	1	–	–	–	2	–	–	–	–	–	Anthraquinone, 1-(dibromo-4-carbomethoxyanilino)-4-hydroxy-, (3α,20α)		IC04558
40	75	44	68	73	45	77	41	**529**	100	87	82	41	40	39	36	27	**0.00**	23	51	3	1	–	–	–	–	–	5	–	–	–	Norepinephrine, N,N,O,O′,O″-pentakis(trimethylsilyl)-	56114-59-1	MS10550
174	73	175	355	176	147	86	356	**529**	100	35	18	10	8	4	4	3	**0.00**	23	51	3	1	–	–	–	–	–	5	–	–	–	Norepinephrine, N,N,O,O′,O″-pentakis(trimethylsilyl)-	56114-59-1	NI45428
198	154	77	41	67	39	153	152	**529**	100	57	31	28	23	19	17	13	**0.51**	24	19	2	1	–	–	–	–	–	–	–	–	2	Manganese, dicarbonyl-π-pyrrolyl(triphenylstibine)-	32826-12-3	NI45429
470	471	353	219.5	498	472	529		**529**	100	25	8	6	5	5	3			25	23	12	1	–	–	–	–	–	–	–	–	–	1H-Benzo[c]quinolizine-1,2,3,4,7,8-hexacarboxylic acid, hexamethyl ester		LI09574
107	91	108	79	178	86	77	170	**529**	100	97	82	71	49	43	42	20	**1.00**	27	35	8	3	–	–	–	–	–	–	–	–	–	L-Tyrosine, N-[N-[N-[(phenylmethoxy)carbonyl]-L-seryl]-L-leucyl]-, methyl ester	55517-95-8	NI45430
124	106	123	105	78	120	373	55	**529**	100	55	38	36	31	28	27	24	**4.60**	29	39	8	1	–	–	–	–	–	–	–	–	–	Rostratine		NI45431
98	296	310	256	84	100	110	70	**529**	100	70	38	30	28	21	12	12	**0.00**	29	56	4	3	–	1	–	–	–	–	–	–	–	Piperidinium, 1-[4-[acetyl[3-(acetylamino)propyl]amino]butyl]-2-(11-methoxy-11-oxoundecyl)-1-methyl-, fluoride, (R)-	55823-08-0	NI45432
91	103	190	55	120	408	57	426	**529**	100	68	64	52	48	46	44	40	**3.00**	31	47	6	1	–	–	–	–	–	–	–	–	–	1-Benzyl-3a,5-dihydroxy-4-(8,11-dihydroxy-9-ethoxy-4-methyl-undec-1-enyl)-7-methyl-6-methylene-3-oxo-3a,4,5,6,7,7a-hexahydro-isoindoline		IC04559

MASS TO CHARGE RATIOS								M.W.	INTENSITIES								Parent	C	H	O	N	S	F	Cl	Br	I	Si	P	B	X	COMPOUND NAME	CAS Reg No	No
529	356	486	355	341	313	514	528	**529**	100	84	59	35	28	16	6	4		37	43	–	3	–	–	–	–	–	–	–	–	–	Tris(3-methyl-1-propylindol-2-yl)methane		NI45433
324	69	155	117	205	112	167	362	**530**	100	50	47	42	41	33	28	26	**19.00**	10	–	4	–	–	12	–	–	–	–	2	–	1	Iron, tetracarbonyl-1,2,3,4-tetrakis(trifluoromethyl)-1,2-diphosphacyclobut-3-enyl-		NI45434
139	119	158	127	65	169	502	138	**530**	100	92	79	75	70	58	52	52	**0.62**	12	5	4	–	–	7	–	–	–	–	–	–	1	Tungsten, tricarbonyl π-cyclopentadienyl heptafluorobutanoyl-		MS10551
305	264	257	497	80	59	495	255	**530**	100	77	74	66	63	62	61	56	**0.10**	15	18	8	–	2	–	4	–	–	–	–	–	–	1,8,9,10-Tetrachloro-3,6-bis(mesyloxy)-11,11-dimethoxy-endo-tricyclo[6.2.1.0^{2,7}]undeca-4,9-diene		MS10552
93	254	378	211	304	181	182	63	**530**	100	64	63	57	56	32	22	19	**0.00**	17	32	11	–	–	–	–	–	–	–	2	–	1	Iron, carbonyl[(2,3,4,5-η)-diethyl 2,4-hexadienedioate]bis(trimethylphosphite-P)-	74792-82-8	NI45435
73	299	75	315	243	227	147	515	**530**	100	53	29	27	25	18	18	15	**0.00**	18	43	8	–	–	–	–	–	–	4	1	–	–	D-Arabino-hexonic acid, 2-deoxy-3,5-bis-O-(trimethylsilyl)-, γ-lactone, bis(trimethylsilyl) phosphate	55517-62-9	NI45436
299	73	75	315	243	227	147	515	**530**	100	99	55	51	47	33	33	28	**0.00**	18	43	8	–	–	–	–	–	–	4	1	–	–	2-Deoxy-6-phosphogluconolactone, tetrakis(trimethylsilyl)-		NI45437
195	168	167	117	362	99	93	31	**530**	100	42	36	29	22	22	11	11	**0.60**	19	1	1	–	–	15	–	–	–	–	–	–	–	Tris(pentafluorophenyl)methanol		MS10553
44	532	534	43	530	49	84	461	**530**	100	80	45	44	42	37	33	26		22	12	6	–	–	–	–	2	–	–	–	–	–	3,3′-Dibromo-4,4′-dihydroxy-7,7′-dimethyl-1,1′-binaphthalene-5,5′,8,8′-tetrone		NI45438
362	390	284	418	360	446	474	282	**530**	100	66	40	38	38	31	24	23	**8.00**	23	15	6	–	–	–	–	–	–	–	1	–	2	Iron, tetracarbonyl (dicarbonyl-π-cyclopentadienyliron)-μ-(diphenylphosphido)-		LI09575
295	296	516	514	120	253	294	77	**530**	100	27	12	12	10	5	3	3	**0.00**	24	23	5	2	1	–	–	1	–	–	–	–	–	20,21-Dinoraspidospermidin-10-one, 1-[[(4-bromophenyl)sulphonyl]oxy]-5,19-didehydro-17-methoxy-	54966-53-9	NI45439
515	516	517	73	518	325	443	250	**530**	100	44	21	16	6	5	5	4	**0.50**	25	34	7	–	–	–	–	–	–	3	–	–	–	4,5-Dihydroxy-7-methoxyanthraquinone-2-carboxylic acid, tris(trimethylsilyl)-		NI45440
530	297	532	531	515	534	299	233	**530**	100	67	63	46	46	43	38	36		26	34	4	4	–	–	–	–	–	–	–	–	1	Zinc, bis[[5,5′-methylenebis[3,4-dihydro-4,4-dimethyl-2H-pyrrol-2-onato]](1-)-N1,N1′]-	69782-63-4	NI45441
279	254	191	350	369	280	253	217	**530**	100	79	55	30	24	24	24	20	**0.00**	26	54	5	–	–	–	–	–	–	3	–	–	–	Cyclopentanepropanoic acid, 3,5-bis[(trimethylsilyl)oxy]-2-[3-[(trimethylsilyl)oxy]-1-octenyl]-, methyl ester, [1R-[1α,2β(1E,3S*),3α,5α]]-		LI09576
279	254	191	350	280	253	368	217	**530**	100	81	56	31	25	25	24	21	**0.19**	26	54	5	–	–	–	–	–	–	3	–	–	–	Cyclopentanepropanoic acid, 3,5-bis[(trimethylsilyl)oxy]-2-[3-[(trimethylsilyl)oxy]-1-octenyl]-, methyl ester, [1R-[1α,2β(1E,3S*),3α,5α]]-	55759-98-3	NI45442
167	311	324	217	440	350	191	243	**530**	100	97	85	81	47	44	40	39	**1.00**	26	54	5	–	–	–	–	–	–	3	–	–	–	Methyl 8-[3,5-bis(trimethylsilyloxy)-2-[(3-trimethylsilyloxy)-1-propenyl]cyclopent-1-yl]octanoate		NI45443
329	275	344	488	386	317	428	404	**530**	100	95	60	55	55	55	45	40	**30.00**	28	34	10	–	–	–	–	–	–	–	–	–	–	9(1H)-Phenanthrenone, 5,6,8-tris(acetyloxy)-7-[2-(acetyloxy)-isopropyl]-2,3,4,4a-tetrahydro-10-hydroxy-1,1,4a-trimethyl-, [R-(R*,R*)]-	75379-00-9	NI45444
182	144	112	109	397	169	158	157	**530**	100	62	54	33	26	25	25	23	**12.00**	28	38	8	2	–	–	–	–	–	–	–	–	–	17-O-acetyl-21-O-β-D-glucopyranosylajmaline acetylrauglucine		MS10554
387	91	69	415	111	79	67	81	**530**	100	95	94	89	86	85	79	77	**18.00**	29	38	9	–	–	–	–	–	–	–	–	–	–	Uscharidin		LI09577
427	428	117	515	130	73	429	116	**530**	100	34	19	12	11	11	10	9	**0.00**	29	62	4	–	–	–	–	–	–	2	–	–	–	Eicosanoic acid, 2,3-bis[(trimethylsilyl)oxy]propyl ester	55517-94-7	WI01594
43	73	57	41	55	117	147	75	**530**	100	74	70	50	49	48	37	34	**0.00**	29	62	4	–	–	–	–	–	–	2	–	–	–	Eicosanoic acid, 2,3-bis[(trimethylsilyl)oxy]propyl ester	55517-94-7	WI01595
129	73	43	218	147	103	57	41	**530**	100	73	65	63	57	57	48	39	**0.00**	29	62	4	–	–	–	–	–	–	2	–	–	–	Eicosanoic acid, 2-[(trimethylsilyl)oxy]-1-[[(trimethylsilyl)oxy]methyl]ethyl ester	55517-93-6	WI01596
218	129	219	130	427	117	203	131	**530**	100	43	26	23	18	18	13	13	**0.00**	29	62	4	–	–	–	–	–	–	2	–	–	–	Eicosanoic acid, 2-[(trimethylsilyl)oxy]-1-[[(trimethylsilyl)oxy]methyl]ethyl ester	55517-93-6	WI01597
151	268	275	380	167	182	530		**530**	100	81	53	9	5	3	1			30	30	5	2	1	–	–	–	–	–	–	–	–	p-Benzotoluidide, 2-(N-veratryl-p-toluenesulphonamido)-	23145-64-4	NI45445
399	400	401	530	531	515	59	40	**530**	100	33	9	8	4	4	4	3		31	54	3	–	–	–	–	–	–	2	–	–	–	Δ[1]-Tetrahydrocannabinol, 7-hydroxy-, bis[(propyldimethylsilyl)oxy]-		NI45446
207	214	220	203	232	442	452	470	**530**	100	72	65	60	30	24	19	18	**7.00**	32	50	6	–	–	–	–	–	–	–	–	–	–	22β-Acetoxy-3β,16α-dihydroxy-13,28-epoxyolean-29-al	94450-45-0	NI45447
189	119	220	175	203	262	273	187	**530**	100	96	90	86	84	77	75	73	**39.00**	32	50	6	–	–	–	–	–	–	–	–	–	–	Dimethyl granulosate		NI45448
453	471	515	205	231	363	435	393	**530**	100	79	76	76	57	50	40	40	**40.00**	32	50	6	–	–	–	–	–	–	–	–	–	–	Dimethyl retigerate B		NI45449
122	43	121	57	55	123	41	93	**530**	100	98	93	70	65	60	60	58	**0.65**	32	50	6	–	–	–	–	–	–	–	–	–	–	Ingenol-3-dodecanoate		NI45450
248	203	484	249	187	189	233	281	**530**	100	84	30	20	20	16	15	13	**1.00**	32	50	6	–	–	–	–	–	–	–	–	–	–	2,23-Di-O-methylmedicagenic acid		NI45451
530	502	364	487	69	405	215	531	**530**	100	85	60	48	41	40	38	31		33	38	6	–	–	–	–	–	–	–	–	–	–	1,5-Methano-1H,7H,11H-furo[3,4-g]pyrano[3,2-b]xanthene-7,15-dione, 3,3a,4,5-tetrahydro-8-hydroxy-3,3,11,11-tetramethyl-1,13-bis(3-methyl-2-butenyl)-, [1R-(1α,3aβ,5α,14aS*)]-	1064-34-2	NI45452
178	150	177	91	122	179	163	149	**530**	100	55	44	44	18	16	16	16	**10.00**	33	38	6	–	–	–	–	–	–	–	–	–	–	β-Tocophenol, trimer		LI09578
410	470	397	377	339	530			**530**	100	75	74	50	20	6				33	54	5	–	–	–	–	–	–	–	–	–	–	Cholestane, 3-acetyl-2,19-diacetoxy-, (2α,3α,5α)-		LI09579
390	530	391	460	531	272	129	461	**530**	100	67	43	32	25	25	22	18		34	30	4	2	–	–	–	–	–	–	–	–	–	N,N′-Bis(1-ethylpropyl)-3,4:9,10-perylenebis(dicarboximide)	110590-81-3	MS10555
149	349	57	71	43	167	85	69	**530**	100	85	72	52	45	39	36	35	**5.88**	34	58	4	–	–	–	–	–	–	–	–	–	–	Adipic acid, diisodecyl ester		IC04560

MASS TO CHARGE RATIOS								M.W.	INTENSITIES								Parent	C	H	O	N	S	F	Cl	Br	I	Si	P	B	X	COMPOUND NAME	CAS Reg No	No
57	149	71	43	55	167	69	85	**530**	100	97	66	52	37	36	35	33	**0.00**	34	58	4	–	–	–	–	–	–	–	–	–	–	Phthalic acid, ditridecyl ester	119-06-2	IC04561
57	149	73	55	71	69	85	167	**530**	100	71	57	50	48	43	26	17	**0.00**	34	58	4	–	–	–	–	–	–	–	–	–	–	Phthalic acid, ditridecyl ester	119-06-2	NI45453
149	57	43	71	55	69	85	167	**530**	100	67	45	44	34	32	29	28	**0.00**	34	58	4	–	–	–	–	–	–	–	–	–	–	Phthalic acid, ditridecyl ester	119-06-2	NI45454
530	57	531	515	43	55	41	219	**530**	100	54	39	37	24	22	22	20		35	62	3	–	–	–	–	–	–	–	–	–	–	Phenol, 2,6-di-tert-butyl-4-[(2-octadecyloxycarbonyl)ethyl]-		IC04562
530	320	399	487	501	180			**530**	100	52	47	40	39	34				36	26	1	4	–	–	–	–	–	–	–	–	–	2-[4-(5-Oxo-4,4-diphenyl-4,5-dihydro-1H-imidazol-2-yl)phenyl]-4,5-diphenyl-1H-imidazole	87019-89-4	NI45455
530	174	352	237	178	115	59	265	**530**	100	49	33	27	27	18	16	7		37	27	–	–	–	–	–	–	–	–	–	–	1	Cobalt, (π-indenyl-π-tetraphenylcyclobutadienyl)-		LI09581
199	154	77	41	67	39	153	152	**530**	100	57	31	28	23	19	17	13	**0.51**	37	27	–	–	–	–	–	–	–	–	–	–	1	Cobalt, (π-indenyl-π-tetraphenylcyclobutadienyl)-		LI09580
91	530	181	439	271	167	272	258	**530**	100	54	53	52	45	45	39	29		41	38	–	–	–	–	–	–	–	–	–	–	–	(E)-1,3-Bis(2-phenylmethyl-5-phenylmethylene-1-cyclopentenyl)propene		MS10556
79	36	42	142	114	179	177	81	**531**	100	75	55	45	41	37	36	33	**0.02**	6	6	–	3	3	–	9	–	–	–	–	–	–	1,3,5-Triazine, hexahydro-1,3,5-tris[(trichloromethyl)thio]-	37424-91-2	NI45456
163	120	375	403	343	301	345	329	**531**	100	10	8	7	5	5	3	3	**0.32**	11	18	5	3	–	–	–	–	–	–	–	–	2	Tungsten, pentacarbonyltris(dimethylamino)arsine-		LI09582
404	77	141	136	220	41	80	67	**531**	100	91	38	30	22	22	21	16	**2.00**	14	15	3	1	1	–	–	–	2	–	–	–	–	2-Oxa-6-azatricyclo[3.3.1.1^{3,7}]decane, 4,8-diiodo-6-(phenylsulphonyl)-, (1α,3β,4α,5α,7β,8β)-	50267-21-5	NI45457
404	77	141	79	220	80	531	136	**531**	100	60	30	25	23	20	19	14		14	15	3	1	1	–	–	–	2	–	–	–	–	2-Oxa-6-azatricyclo[3.3.1.1^{3,7}]decane, 4,8-diiodo-6-(phenylsulphonyl)-, (1α,3β,4β,5α,7β,8α)-	50267-19-1	NI45458
77	404	79	141	80	41	220	81	**531**	100	97	44	35	33	26	22	22	**3.00**	14	15	3	1	1	–	–	–	2	–	–	–	–	2-Oxa-6-azatricyclo[3.3.1.1^{3,7}]decane, 4,8-diiodo-6-(phenylsulphonyl)-, (1α,3β,4β,5α,7β,8β)-	50267-20-4	NI45459
93	128	92	254	127	66	39	65	**531**	100	60	55	47	35	34	33	27	**0.00**	18	19	–	3	–	–	–	–	2	–	–	–	–	4-[4-(4-Methylpyridinium-1-ylmethyl)pyridinium-1-ylmethyl]pyridine diiodide		LI09583
73	315	357	299	101	147	75	68	**531**	100	52	29	29	18	17	17	15	**0.30**	18	46	7	1	–	–	–	–	–	4	1	–	–	D-erythro-Pentose, 2-deoxy-3,4-bis-O-(trimethylsilyl)-, 5-[bis(trimethylsilyl) methoxyphosphorimidate]	55517-63-0	NI45460
315	73	357	299	101	147	75	68	**531**	100	99	56	55	34	33	32	29	**0.00**	18	46	7	1	–	–	–	–	–	4	1	–	–	D-erythro-Pentose, 2-deoxy-3,4-bis-O-(trimethylsilyl)-, O-methyloxime, 5-[bis(trimethylsilyl) phosphate]	55520-88-2	NI45461
146	161	528	367	351	440	335	303	**531**	100	11	3	3	2	1	1	1	**0.00**	18	57	–	3	–	–	–	–	–	6	–	–	1	Silanamine, 1,1,1-trimethyl-N-(trimethylsilyl)-, titanium(3+) salt	37512-29-1	NI45462
73	223	217	245	147	195	184	348	**531**	100	67	48	20	15	14	14	13	**0.42**	21	45	5	3	–	–	–	–	–	4	–	–	–	Cytidine, tetrakis(trimethylsilyl)-	56666-34-3	MS10557
253	410	320	500	217	254	115	87	**531**	100	58	53	50	22	20	16	15	**0.00**	25	49	7	1	–	–	–	–	–	2	–	–	–	Cyclopentaneoctanoic acid, ϵ-(methoxyimino)-2-(3-methoxy-3-oxopropyl)-3,5-bis[(trimethylsilyl)oxy]-, methyl ester, [1R-(1α,2β,3β,5α)]-	55670-00-3	NI45463
253	410	320	500	254	217	411	179	**531**	100	60	55	48	20	20	18	17	**0.00**	25	49	7	1	–	–	–	–	–	2	–	–	–	Cyclopropaneheptanoic acid, γ-(methylimino)-2-[(methoxycarbonyl)ethyl] 3,5-bis[(trimethylsilyl)oxy]-, methyl ester		LI09584
77	43	531	114	351	105	246	141	**531**	100	34	22	20	19	14	13	13		27	25	5	5	1	–	–	–	–	–	–	–	–	Naphthalen-2-ol, 1-(4-phenylazo)-6-[N-methyl-2-(acetoxyethylsulphonamido)]-		IC04563
298	299	531	43	266	150	116	41	**531**	100	30	25	25	19	18	18	13		28	37	9	1	–	–	–	–	–	–	–	–	–	Harringtonine		MS10558
298	299	43	150	28	531	266	41	**531**	100	23	21	17	15	14	13	9		28	37	9	1	–	–	–	–	–	–	–	–	–	Isoharringtonine		MS10559
278	532	405	127	151	254	279	128	**532**	100	78	60	51	28	13	2	2		2	–	–	–	–	–	–	–	4	–	–	–	–	Ethylene, tetraiodo-	513-92-8	NI45464
127	278	151	405	532	254	139	279	**532**	100	60	32	24	20	17	9	1		2	–	–	–	–	–	–	–	4	–	–	–	–	Ethylene, tetraiodo-		LI09585
127	278	151	404	531	254	12	24	**532**	100	64	34	28	18	16	16	12	**0.30**	2	–	–	–	–	–	–	–	4	–	–	–	–	Ethylene, tetraiodo-	513-92-8	NI45465
294	279	504	476	448	420	392	323	**532**	100	65	1	1	1	1	1	1	**1.00**	10	9	5	–	–	–	–	–	–	–	–	–	4	Molybdenum, pentacarbonyl 4-methyl-1,2,6-triarsatricyclo[2.2.1.0^{2,6}]heptanyl-		MS10560
392	394	73	420	422	532	534	448	**532**	100	85	60	50	38	30	22	12		14	27	5	–	–	–	–	–	–	3	–	–	2	Pentacarbonyltris(trimethylsilyl)stibinechromium	80753-45-3	NI45466
503	504	505	31	29	237	506	489	**532**	100	42	28	24	16	10	10	8	**0.99**	14	36	10	–	–	–	–	–	–	6	–	–	–	1,3,5,7,9,11-Hexaethyl-5,9-dimethoxytricyclo[5.5.1.1^{3,11}]hexasiloxane		NI45467
443	444	87	59	115	503	415	445	**532**	100	49	41	34	28	20	17	15	**1.29**	14	40	8	–	–	–	–	–	–	7	–	–	–	1,3,5,7,9,11,13-Heptaethylbicyclo[7.5.1]heptasiloxane	73113-19-6	NI45468
73	221	147	341	517	355	325	207	**532**	100	27	27	26	24	17	15	15	**0.00**	16	48	6	–	–	–	–	–	–	7	–	–	–	1,1,1,5,7,7,7-Heptamethyl-3,3,5-tris(trimethylsiloxy)tetrasiloxane	34915-06-5	NI45469
73	221	147	295	207	222	355	341	**532**	100	86	38	27	22	21	20	17	**0.00**	16	48	6	–	–	–	–	–	–	7	–	–	–	Heptasiloxane, hexadecamethyl-	541-01-5	NI45470
532	365	69	296	217	129	533	227	**532**	100	61	58	38	33	31	20	14		18	–	–	–	–	15	–	–	–	–	1	–	–	Phosphine, tris(pentafluorophenyl)-		IC04564
532	365	69	533	296	217	129	79	**532**	100	43	21	20	17	14	14	11		18	–	–	–	–	15	–	–	–	–	1	–	–	Phosphine, tris(pentafluorophenyl)-		MS10561
288	51	77	290	159	161	217	140	**532**	100	85	78	75	31	23	22	17	**4.20**	18	5	–	–	–	10	–	–	–	–	–	–	1	Stibine, bis(pentafluorophenyl)phenyl-		MS10562
368	323	283	221	119	340	296	44	**532**	100	52	42	28	24	19	14	13	**0.00**	18	14	7	–	–	10	–	–	–	–	–	–	–	Phenyllactic acid, 3-methoxy-α,4-bis(pentafluoropanoyl)-, ethyl ester		MS10563
517	519	43	229	518	520	131	214	**532**	100	99	48	35	25	24	24	16	**7.00**	21	21	6	6	–	–	–	1	–	–	–	–	–	1,2-Dihydro-2,2,4-trimethyl-5-acetamido-8-methoxy-6-(2,4-dinitro-6-bromophenylazo)quinoline		IC04565
294	293	183	185	295	108	107	110	**532**	100	65	65	63	25	25	23	20	**1.00**	23	15	5	–	1	–	–	–	–	–	1	–	1	Molybdenum, pentacarbonyldiphenyl(phenylthio)phosphine-		IC04566
118	158	266	121	129	130	122	115	**532**	100	76	66	66	63	46	33	33	**0.00**	24	20	2	–	6	–	–	–	–	–	–	–	–	3,8,9,10-Tetrathiatricyclo[4.2.1.1^{2,5}]decane-4,7-dione, 1,2-bis(allylthio)-5,6 diphenyl-		NI45471

MASS TO CHARGE RATIOS	M.W.	INTENSITIES	Parent	C	H	O	N	S	F	Cl	Br	I	Si	P	B	X	COMPOUND NAME	CAS Reg No	No
101 289 43 273 230 143 111 169	**532**	100 95 74 48 43 41 41 35	**0.29**	25	40	12	–	–	–	–	–	–	–	–	–	–	**D-Mannitol, 1,2:4,5-di-O-isopropylidene-, cyclic orthocarbonate (2:1)**	7727-02-8	**NI45472**
91 65 230 93 79 77 108 109	**532**	100 10 9 6 6 6 5 4	**0.16**	26	32	8	2	1	–	–	–	–	–	–	–	–	**L-Alanine, 3,3′-thiobis N-carboxy-, N,N′-dibenzyl diethyl ester**	26527-27-5	**NI45473**
213 212 121 56 214 41 147 28	**532**	100 93 48 33 17 14 13 12	**11.00**	27	32	–	–	2	–	–	–	–	–	–	–	2	**Propane, 1,2-bis[(ferrocenyl)ethylthio]-**		**NI45474**
212 213 121 56 214 106 41 147	**532**	100 99 51 36 21 17 17 13	**9.00**	27	32	–	–	2	–	–	–	–	–	–	–	2	**Propane, 1,3-bis[(1-ferrocenyl)ethylthio]-**		**NI45475**
354 355 356 178 390 278 279 357	**532**	100 65 59 41 34 34 29 26	**0.41**	28	20	–	–	–	–	2	–	–	–	–	–	1	**Palladium, dichlorotetraphenylcyclobutadienyl-**		**MS10564**
105 248 202 127 267 159 412 490	**532**	100 6 6 5 4 4 3 2	**0.00**	28	36	10	–	–	–	–	–	–	–	–	–	–	**1α-Benzoyloxy-6β,9β,15-triacetoxy-4β-hydroxydihydro-β- agarofuran**		**MS10565**
281 279 295 310 225 296 238 214	**532**	100 40 32 30 30 28 26 25	**1.00**	28	36	10	–	–	–	–	–	–	–	–	–	–	**Spiro[cyclopropane-1,2′(1′H)-phenanthrene]-1′,4′(3′H)-dione, 3′,7′,9′,10′-tetrakis(acetyloxy)-4′b,5′,6′,7′,8′,8′a,9′,10′-octahydro-2,4′b,8′,8′-tetramethyl-, [2′S-[2′α(R*),3′α,4′bβ,7′α]]-**	66584-89-2	**NI45476**
476 55 152 115 132 227 306 116	**532**	100 86 82 35 31 27 25 20	**5.00**	29	22	2	–	–	–	–	–	–	–	–	–	2	**Manganese, dicarbonyl-π-indenyl(triphenylarsine)-**	31973-99-6	**NI45477**
43 69 412 83 95 94 294 199	**532**	100 46 44 40 24 24 21 21	**1.59**	29	40	9	–	–	–	–	–	–	–	–	–	–	**3-Desoxo-3β-hydroxy-phorbol-3,4-acetonide-12,13,20-triacetate**		**NI45478**
85 44 81 42 79 46 105 91	**532**	100 74 49 46 37 37 36 31	**3.20**	29	40	9	–	–	–	–	–	–	–	–	–	–	**Roridin A**		**NI45479**
398 429 151 167 399 125 430 397	**532**	100 65 57 36 30 23 20 17	**0.00**	29	48	5	2	–	–	–	–	–	1	–	–	–	**Pregn-4-ene-3,20-dione, 17-(acetyloxy)-6-methyl-21-[(trimethylsilyl)oxy]-, 3,20-bis(O-methyloxime), (6α)-**	74298-99-0	**NI45480**
502 441 287 532 411 442 57 168	**532**	100 94 60 47 45 40 40 38		29	48	5	2	–	–	–	–	–	1	–	–	–	**Pregn-4-ene-3,20-dione, 21-(acetyloxy)-6-methyl-17-[(trimethylsilyl)oxy]-, 3,20-bis(O-methyloxime), (6α)-**	74299-02-8	**NI45481**
74 27 105 31 91 145 354 29	**532**	100 98 96 83 76 73 72 48	**0.00**	30	44	8	–	–	–	–	–	–	–	–	–	–	**Carda-4,20(22)-dienolide, 3-[(6-deoxy-3-O-methyl-α-L-mannopyranosyl)oxy]-14-hydroxy-, (3β)-**	20248-01-5	**NI45482**
212 113 67 43 165 197 439 385	**532**	100 46 37 37 22 19 13 11	**9.00**	30	44	8	–	–	–	–	–	–	–	–	–	–	**Colletotrichin, 3-acetate**		**NI45483**
283 155 329 342 255 313 324 301	**532**	100 61 40 32 31 30 28 28	**9.00**	30	44	8	–	–	–	–	–	–	–	–	–	–	**Simplexin**	1404-62-2	**NI45484**
266 231 265 267 77 268 103 128	**532**	100 96 84 45 42 33 24 13	**2.00**	32	22	–	4	–	–	2	–	–	–	–	–	–	**(1α,2α,3β,4β)-1,3-Bis(3-chloro-2-quinoxalinyl)-2,4-diphe nylcyclobutane**		**MS10566**
105 119 201 145 412 107 213 255	**532**	100 65 56 50 49 48 46 43	**0.00**	32	52	6	–	–	–	–	–	–	–	–	–	–	**Cholestan-26-oic acid, 3,7-bis(acetyloxy)-, methyl ester, (3α,5β,7α)-**	60354-52-1	**NI45485**
43 55 69 81 95 41 83 109	**532**	100 62 51 42 38 38 36 31	**0.00**	32	52	6	–	–	–	–	–	–	–	–	–	–	**Ergostan-6-one, 3,25-bis(acetyloxy)-5-hydroxy-, (3β,5α)-**	56143-29-4	**NI45486**
445 385 233 473 413 501 221 251	**532**	100 75 46 33 33 25 25 20	**4.00**	32	52	6	–	–	–	–	–	–	–	–	–	–	**D:A-Friedo-A(1)-nor-2,3-secooleanane-2,3,24-trioic acid, trimethyl ester, (4R)-**	43230-22-4	**MS10567**
165 194 167 337 166 295 338 152	**532**	100 74 68 64 55 50 25 20	**0.00**	33	28	3	2	1	–	–	–	–	–	–	–	–	**Benzeneacetic acid, α-phenyl-, 1-(3-acetyl-2-thioxo-1-imidazolidinyl)-2,2-diphenylethenyl ester**	67845-16-3	**NI45487**
517 514 311 201 456 423 405 496	**532**	100 58 58 58 50 50 42 42	**3.30**	33	56	5	–	–	–	–	–	–	–	–	–	–	**Cyclocanthogenium 24,25-acetonide**		**MS10568**
149 532 213 427 256 485 395 500	**532**	100 18 18 16 14 10 10 9		33	56	5	–	–	–	–	–	–	–	–	–	–	**3β-Hydroxy-16β,24,25-trimethoxy-24R-cycloartan-6-one**		**MS10569**
397 326 532 369 398 533 353 282	**532**	100 95 90 47 36 34 31 26		34	36	2	4	–	–	–	–	–	–	–	–	–	**Spiro[1H-isoindole-1,9′-xanthene], 3′,6′-bis(diethylamino)-2-4-anilino-2,3-dihydro-3-oxo-**		**IC04567**
55 69 83 97 82 96 81 57	**532**	100 96 93 74 66 64 52 50	**6.64**	36	68	2	–	–	–	–	–	–	–	–	–	–	**9-Octadecenoic acid (Z)-, 9-octadecenyl ester, (Z)-**	3687-45-4	**NI45488**
91 69 109 532 133 426 83 440	**532**	100 21 16 13 11 8 4 3		40	52	–	–	–	–	–	–	–	–	–	–	–	**Bisdehydrolycopene**		**LI09586**
133 532 69 91 440 426 43 371	**532**	100 73 50 38 11 11 10 7		40	52	–	–	–	–	–	–	–	–	–	–	–	**Chlorobactin**		**LI09587**
429 259 430 146 428 518 260 130	**533**	100 69 58 58 54 50 50 47	**42.90**	18	57	–	5	–	–	–	–	–	6	–	2	–	**Iminobis[[bis(trimethylsilyl)amino][(trimethylsilyl)amino]borane]**		**MS10570**
105 164 151 202 488 246 530 324	**533**	100 25 13 9 7 7 3 3	**2.00**	27	33	11	–	–	–	–	–	–	–	–	–	–	**1α-Benzoyloxy-6β,9α,15-triacetoxy-4β,8β-dihydroxydihydro-β-agarofuran**		**MS10571**
266 267 165 102 73 268 220 179	**533**	100 25 17 13 12 7 5 4	**0.00**	27	47	4	1	–	–	–	–	–	3	–	–	–	**Denopamine N,O,O-tris(trimethylsilyl)-**		**NI45489**
132 73 204 129 116 299 312 75	**533**	100 89 43 32 24 20 16 16	**0.00**	27	63	3	1	–	–	–	–	–	3	–	–	–	**3,8-Dioxa-2,9-disiladecan-5-amine, 2,2,9,9-tetramethyl-7-tetradecyl-6-[(trimethylsilyl)oxy]-**	55517-84-5	**NI45490**
132 73 204 129 116 299 75 340	**533**	100 86 42 30 23 19 16 15	**0.00**	27	63	3	1	–	–	–	–	–	3	–	–	–	**Octadecane, 2-amino-1,3,4-tris[(trimethylsilyl)oxy]-**		**LI09588**
203 73 75 43 129 216 117 91	**533**	100 72 56 39 33 25 21 19	**10.00**	29	51	4	1	–	–	–	–	–	2	–	–	–	**Pregn-5-en-20-one, 21-acetyloxy-3-[(trimethylsilyl)oxy]-, O-trimethylsilyl oxime**		**LI09589**
243 244 165 55 105 43 228 59	**533**	100 57 36 10 9 9 6 6	**0.00**	31	35	7	1	–	–	–	–	–	–	–	–	–	**β-Alanine, N-hydroxy-N-[2,3-O-isopropylidene-5-O-(triphenylmethyl)-β-D-ribofuranosyl]-, methyl ester**	64018-66-2	**NI45491**
327 273 394 422 342 534 309 478	**534**	100 22 18 14 7 5 5 4		10	9	8	–	–	–	–	–	–	–	–	–	4	**7-Methyl-1,3,5-triarsa-2,4,9-trioxa-adamantan-as-pentacarbonylchromium**	80950-87-4	**NI45492**
69 193 265 97 83 45 79 51	**534**	100 26 19 19 15 15 13 11	**0.00**	13	6	9	–	–	12	–	–	–	–	–	–	–	**D-Ribofuranose, tetrakis(trifluoroacetate)**	56942-82-6	**MS10572**
69 278 193 265 97 41 165 43	**534**	100 34 32 27 25 16 13 11	**0.00**	13	6	9	–	–	12	–	–	–	–	–	–	–	**D-Ribopyranose, tetrakis(trifluoroacetate)**	56942-83-7	**MS10573**
69 193 278 265 97 165 153 279	**534**	100 97 82 53 35 23 21 20	**0.00**	13	6	9	–	–	12	–	–	–	–	–	–	–	**D-Xylose, 2,3,4,5-tetrakis(trifluoroacetate)**	38424-94-1	**NI45493**
259 260 73 261 332 258 130 519	**534**	100 52 48 43 39 38 34 31	**29.40**	18	56	1	4	–	–	–	–	–	6	–	2	–	**Oxybis[[bis(trimethylsilyl)amino][(trimethylsilyl)amino]borane]**		**MS10574**
73 147 171 70 244 187 74 142	**534**	100 55 48 48 42 34 18 15	**4.20**	21	54	2	4	–	–	–	–	–	5	–	–	–	**L-Ornithine, N^5-[[bis(trimethylsilyl)amino]iminomethyl]-N^2,N^2-bis(trimethylsilyl)-, trimethylsilyl ester**	55517-81-2	**NI45494**

MASS TO CHARGE RATIOS								M.W.	INTENSITIES								Parent	C	H	O	N	S	F	Cl	Br	I	Si	P	B	X	COMPOUND NAME	CAS Reg No	No
43	287	103	173	145	131	115	347	**534**	100	30	10	9	7	7	7	6	**0.00**	22	30	15	–	–	–	–	–	–	–	–	–	–	1,2,3,4,5,7,8-Hepta-O-acetyl-β-D-altro-L-glycero-3-octulofuranose	78174-70-6	NI45495
43	97	46	173	157	98	131	115	**534**	100	13	8	7	7	7	6	5	**0.00**	22	30	15	–	–	–	–	–	–	–	–	–	–	1,2,3,4,5,6,8-Hepta-O-acetyl-α-D-gluco-L-glycero-3-octulopyranose	78174-71-7	NI45496
187	127	169	331	242	200	503	461	**534**	100	34	7	6	3	3	1	1	**0.00**	23	34	14	–	–	–	–	–	–	–	–	–	–	Hexanoic acid, 5-acetyloxy-3-(2,3,4,6-tetra-O-acetyl-β-D-glucopyranosyloxy)-, methyl ester, (3S,5S)-		LI09590
187	43	127	85	188	169	109	98	**534**	100	63	36	20	11	9	7	6	**0.00**	23	34	14	–	–	–	–	–	–	–	–	–	–	L-erythro-Hexonic acid, 2,4,6-trideoxy-3-O-β-D-glucopyranosyl-, methyl ester, pentaacetate	33275-52-4	MS10575
351	349	184	154	347	136	350	78	**534**	100	72	63	58	44	44	41	38	**0.00**	24	15	1	–	–	5	–	–	–	–	–	–	1	Stannane, (pentafluorophenoxy)triphenyl-	22529-95-9	NI45497
243	426	168	244	139	534	169	128	**534**	100	17	15	7	5	4	3	3		24	16	5	–	–	–	–	–	–	–	–	–	2	10H-Phenoxarsine, 10,10′-oxybis-, 10,10′-dioxide	55517-80-1	NI45498
69	183	84	466	325	127	154	210	**534**	100	97	91	40	30	26	21	19	**0.00**	26	56	–	2	–	–	–	–	–	–	–	–	2	Gallium, tetraisobutyldi-μ-1-piperidinyldi-	42777-04-8	MS10576
43	83	69	95	91	105	79	55	**534**	100	34	31	19	19	15	15	15	**0.15**	28	38	10	–	–	–	–	–	–	–	–	–	–	3-Desoxo-3β-hydroxy-phorbol-3,12,13,20-tetraacetate		NI45499
43	534	122	109	69	139	138	121	**534**	100	78	56	53	49	48	45	39		28	38	10	–	–	–	–	–	–	–	–	–	–	Ingol-3,7,8,12-tetraacetate		NI45500
534	535	536	504	519	532	537	445	**534**	100	48	21	12	9	8	7	7		28	50	4	–	–	–	–	–	–	3	–	–	–	Estra-1,3,5(10)-triene, 2-methoxy-3,4,17-tris[(trimethylsilyl)oxy]-, (17β)-		NI45501
534	535	536	129	375	341	327	390	**534**	100	48	22	12	11	9	8	7		28	50	4	–	–	–	–	–	–	3	–	–	–	Estra-1,3,5(10)-triene, 2-methoxy-3,16,17-tris[(trimethylsilyl)oxy]-, (16α,17β)-	51497-48-4	NI45502
534	535	504	536	505	519	533	532	**534**	100	54	25	24	12	8	6	6		28	50	4	–	–	–	–	–	–	3	–	–	–	Estra-1,3,5(10)-triene, 3-methoxy-2,4,17-tris[(trimethylsilyl)oxy]-, (17β)-		NI45503
444	184	445	264	354	237	238	75	**534**	100	45	43	43	41	41	37	29	**5.14**	28	50	4	–	–	–	–	–	–	3	–	–	–	Estra-1,3,5(10)-triene, 3-methoxy-6,11,17-tris[(trimethylsilyl)oxy]-, (6α,11β,17β)-		NI45504
534	535	536	537	519	445	532	413	**534**	100	54	26	7	7	6	5	5		28	50	4	–	–	–	–	–	–	3	–	–	–	Estra-1,3,5(10)-triene, 4-methoxy-2,3,17-tris[(trimethylsilyl)oxy]-, (17β)-		NI45505
358	131	359	340	386	488	534	403	**534**	100	44	30	19	18	14	12	11		29	42	9	–	–	–	–	–	–	–	–	–	–	Helveticoside		LI09591
73	75	426	354	201	183	147	336	**534**	100	82	65	33	27	21	20	18	**13.01**	29	50	5	–	–	–	–	–	–	2	–	–	–	Card-20(22)-enolide, 12,14-dihydroxy-3-[(trimethylsilyl)oxy]-, (3β,5β,12β)-	28371-18-8	MS10577
73	75	426	354	201	183	147	336	**534**	100	81	63	32	27	21	19	18	**12.00**	29	50	5	–	–	–	–	–	–	2	–	–	–	Card-20(22)-enolide, 12,14-dihydroxy-3-[(trimethylsilyl)oxy]-, (3β,5β,12β)-	28371-18-8	NI45507
73	75	426	354	427	201	183	147	**534**	100	67	53	27	22	22	18	16	**10.38**	29	50	5	–	–	–	–	–	–	2	–	–	–	Card-20(22)-enolide, 12,14-dihydroxy-3-[(trimethylsilyl)oxy]-, (3β,5β,12β)-	28371-18-8	NI45506
339	74	357	85	340	203	356	105	**534**	100	67	48	45	41	37	34	28	**0.20**	30	46	8	–	–	–	–	–	–	–	–	–	–	Neriifolin		LI09592
253	354	444	313	226	339	247	211	**534**	100	99	45	40	31	25	23	22	**1.00**	30	46	8	–	–	–	–	–	–	–	–	–	–	24-Norcholan-23-oic acid, 3,7,12-triacetyloxy-, methyl ester, (3α,7α,12α)		NI45508
227	43	296	137	154	240	228	165	**534**	100	69	51	50	32	21	20	11	**0.50**	31	34	8	–	–	–	–	–	–	–	–	–	–	2α,11-Dihydroxyabieta-5,7,9(11),13-tetraen-12-one, 2α-O-3′,4′-diacetoxybenzoate		NI45509
296	179	227	137	228	229	534	281	**534**	100	92	81	77	53	21	17	14		31	34	8	–	–	–	–	–	–	–	–	–	–	11,19-Dihydroxyabieta-5,7,9(11),13-tetraen-12-one, 19-O-3′,4′-diacetoxybenzoate		NI45510
84	477	534	70	98	111	237	250	**534**	100	73	52	51	49	48	25	23		31	42	4	4	–	–	–	–	–	–	–	–	–	4H-1,16-Etheno-5,15-(propaniminoethano)furo[3,4-l][1,5,10]triazacyclohexadecine-4,21-dione, 3,3a,6,7,8,9,10,11,12,13,14,15-dodecahydro-3-(4-methoxyphenyl)-10,14-dimethyl-	69721-61-5	NI45511
148	133	134	86	443	135	91	42	**534**	100	6	5	5	4	4	4	4	**0.10**	31	42	4	4	–	–	–	–	–	–	–	–	–	Frangufolin		LI09593
114	135	131	85	44	43	42	41	**534**	100	2	2	2	2	2	2	2	**0.06**	31	42	4	4	–	–	–	–	–	–	–	–	–	Integerrenin		MS10578
253	299	105	226	145	143	119	157	**534**	100	63	46	40	35	33	30	27	**0.00**	31	50	7	–	–	–	–	–	–	–	–	–	–	Cholan-24-oic acid, 7,12-bis(acetyloxy)-3-ethoxy-, methyl ester, (3α,5β)-	69779-07-3	NI45512
268	370	165	166	164	149	152	151	**534**	100	70	50	45	20	20	15	15	**0.00**	32	26	6	2	–	–	–	–	–	–	–	–	–	Stilbene, 4-(2,4-dinitrostyryl)-4′-(2,4-dimethoxystyryl)-		LI09594
534	536	535	519	537	520	521	504	**534**	100	45	38	22	17	12	11	11		32	36	–	4	–	–	–	–	–	–	–	–	1	Nickel, [2,8,12,18-tetraethyl-3,7,13,17-tetramethylporphinato(2-)]-	22925-21-9	NI45513
368	384	360	396	414	444	456	426	**534**	100	99	86	61	60	49	48	40	**0.00**	32	54	6	–	–	–	–	–	–	–	–	–	–	Acetyl 3β-acetoxy-5-hydroxy-6α-cholestanyl formal		LI09595
534	173	535	174	57	536	158	55	**534**	100	53	38	16	8	7	7	7		34	50	3	2	–	–	–	–	–	–	–	–	–	1′H-Cholest-2-eno[3,2-b]indole, 5′-methoxy-6-nitro-, (5α,6α)-	34534-85-5	MS10579
534	173	535	174	536	158	160	210	**534**	100	51	38	16	8	8	7	5		34	50	3	2	–	–	–	–	–	–	–	–	–	1′H-Cholest-2-eno[3,2-b]indole, 5′-methoxy-6-nitro-, (5α,6α)-	34534-85-5	NI45514
534	173	535	174	536	158	487	210	**534**	100	53	38	17	8	8	5	5		34	50	3	2	–	–	–	–	–	–	–	–	–	1′H-Cholest-2-eno[3,2-b]indole, 5′-methoxy-6-nitro-, (5α,6α)-	34534-85-5	LI09596
96	82	222	55	69	83	57	43	**534**	100	98	87	76	75	71	60	57	**0.74**	36	70	2	–	–	–	–	–	–	–	–	–	–	Eicosanoic acid, 9-hexadecenyl ester, (Z)-	22393-92-6	NI45515
57	55	69	236	83	43	97	71	**534**	100	91	88	87	79	76	68	64	**2.74**	36	70	2	–	–	–	–	–	–	–	–	–	–	9-Hexadecenoic acid, eicosyl ester, (Z)-	22522-34-5	NI45516
82	96	250	55	83	69	57	43	**534**	100	99	78	76	73	69	60	59	**0.94**	36	70	2	–	–	–	–	–	–	–	–	–	–	Octadecanoic acid, 9-octadecenyl ester, (Z)-	17673-50-6	NI45517
57	264	55	69	83	43	97	71	**534**	100	86	84	81	77	76	67	63	**2.64**	36	70	2	–	–	–	–	–	–	–	–	–	–	9-Octadecenoic acid (Z)-, octadecyl ester	17673-49-3	NI45518
43	57	71	85	55	41	69	83	**534**	100	99	57	38	34	33	23	18	**0.00**	38	78	–	–	–	–	–	–	–	–	–	–	–	Hexacosane, 13-dodecyl-		AI02401
267	534	268	535	178	105	457	265	**534**	100	43	24	22	12	12	6	6		42	30	–	–	–	–	–	–	–	–	–	–	–	Benzene, 1,1′,1″,1‴,1⁗,1′′′′′-[bi-2-cyclopropen-1-yl]-1,1′,2,2′,3,3′-hexaylhexakis-	4997-62-0	NI45519
534	535	536	363	225	219	213	178	**534**	100	46	12	3	3	3	3	3		42	30	–	–	–	–	–	–	–	–	–	–	–	Benzene, hexaphenyl-	992-04-1	NI45520
534	535	536	458	459	457	365	207	**534**	100	43	9	4	3	3	3	3		42	30	–	–	–	–	–	–	–	–	–	–	–	Benzene, hexaphenyl-	992-04-1	NI45521
534	535	45	12	27	536	46	41	**534**	100	43	17	13	10	9	6	6		42	30	–	–	–	–	–	–	–	–	–	–	–	Benzene, hexaphenyl-	992-04-1	IC04568

MASS TO CHARGE RATIOS								M.W.	INTENSITIES								Parent	C	H	O	N	S	F	Cl	Br	I	Si	P	B	X	COMPOUND NAME	CAS Reg No	No
371	355	295	340	517	309	389	325	**535**	100	62	61	24	8	8	6	6	**0.00**	18	57	–	3	–	–	–	–	–	6	–	–	1	Chromium, tris(1,1,1,3,3,3-hexamethyldisilazanato-N)-	37512-31-5	NI45522
73	217	317	147	75	204	187	159	**535**	100	25	21	15	13	12	12	11	**0.00**	19	37	7	3	1	–	–	–	–	3	–	–	–	1,2,3-Thiadiazole-5-carboxaldoxime, glucuronic acid conjugate, tris(trimethylsilyl)-, methyl ester		NI45523
230	377	149	148	98	273	158	331	**535**	100	75	52	49	40	32	29	21	**18.20**	22	37	10	3	1	–	–	–	–	–	–	–	–	N,S-Bis(ethoxycarbonyl)-(L-α-aminoadipyl)-L-cysteinyl-D-valine, methyl ester		MS10580
204	186	73	217	205	114	199	147	**535**	100	79	36	24	21	18	13	13	**2.01**	23	53	5	1	–	–	–	–	–	4	–	–	–	Piperidine, 1-[2,3,4,6-tetrakis-O-(trimethylsilyl)-β-D-glucopyranosyl]-	72021-24-0	NI45524
144	288	185	172	347	208	147	216	**535**	100	78	56	49	46	46	40	38	**21.02**	26	37	9	3	–	–	–	–	–	–	–	–	–	L-Proline, 1-[N-(N-carboxy-L-valyl)-L-tyrosyl]-, N-ethyl methyl ester, ethyl carbonate (ester)	21026-93-7	NI45525
43	147	85	157	41	45	71	55	**535**	100	28	20	20	18	18	17	15	**8.13**	26	41	5	5	1	–	–	–	–	–	–	–	–	14-[(3-Amino-1,2,4-triazol-5-yl)thioacetoxy]-13-formamido-3-methoxy-mutilan-11-one		NI45526
43	504	505	58	71	476	446	535	**535**	100	72	21	18	17	7	3	1		28	41	9	1	–	–	–	–	–	–	–	–	–	14-Acetyldictyocarpine	75659-26-6	NI45527
446	447	75	265	129	266	356	191	**535**	100	44	43	39	33	32	28	28	**16.06**	28	53	3	1	–	–	–	–	–	3	–	–	–	Androst-5-en-17-one, 3,16-bis[(trimethylsilyl)oxy]-, O-(trimethylsilyl)oxime, (3β,16β)-	69833-81-4	NI45528
112	113	111	98	110	125	99	96	**535**	100	9	4	4	3	2	2	2	**0.60**	29	45	8	1	–	–	–	–	–	–	–	–	–	Cevane-3,4,14,15,16,20-hexol, 4,9-epoxy-, 3-acetate, (3β,4α,15α,16β)-	2777-79-9	NI45529
194	166	165	493	256	299	43	167	**535**	100	68	64	39	35	30	20	18	**0.00**	37	29	3	1	–	–	–	–	–	–	–	–	–	2-Azetidinone, 1-(1,3-dioxo-2,2-diphenylbutyl)-3,3,4-triphenyl-	74764-45-7	NI45530
521	522	253	523	505	524	238	245	**536**	100	42	38	33	12	12	9	9	**1.79**	8	24	12	–	–	–	–	–	–	8	–	–	–	Permethyloctasilsesquioxane	17865-85-9	NI45531
148	82	118	167	230	452	261	63	**536**	100	99	80	49	35	26	25	23	**5.00**	10	–	8	–	2	6	–	–	–	–	–	–	2	Manganese, bis[(trifluoromethylthio)tetracarbonyl]di-		LI09597
41	409	43	67	389	81	71	39	**536**	100	45	22	12	11	11	10	10	**2.52**	10	10	–	–	–	8	–	–	2	–	–	–	–	1-Decene, 7,7,8,8,9,9,10,10-octafluoro-5,10-diiodo-	74779-66-1	NI45532
167	117	536	369	98	93	148	79	**536**	100	38	21	17	9	9	6	5		12	–	–	–	–	10	–	–	–	–	–	–	1	Mercury, bis(pentafluorophenyl)-		LI09598
149	77	105	311	150	89	90	51	**536**	100	59	55	46	42	38	32	32	**0.00**	15	14	2	4	2	10	–	–	–	–	–	–	–	Methylenebis(1,4-phenylenediureylene)bis(sulphur pentafluoride)	90598-06-4	NI45533
119	76	163	60	195	371	259	213	**536**	100	90	23	21	13	1	1	1	**0.12**	16	36	4	6	–	–	–	–	–	–	2	–	1	Molybdenum, tetracarbonylbis[tris(dimethylamino)phosphine]-		MS10581
119	117	115	398	400	396	121	118	**536**	100	74	54	45	40	36	26	24	**6.00**	17	36	–	–	–	–	–	–	–	–	–	–	4	Germane, 3,4-pentadien-1-yne-1,3-diyl-5-ylidenetetrakis[trimethyl-	41898-95-7	NI45534
119	117	398	400	396	115	402	394	**536**	100	75	67	59	55	55	35	32	**0.00**	17	36	–	–	–	–	–	–	–	–	–	–	4	Germane, 1,3-pentadiyn-1-yl-5-ylidynetetrakis[trimethyl-	41898-96-8	NI45535
536	374	212	481	521	319	266	495	**536**	100	98	25	20	16	13	12	10		20	36	6	–	–	–	–	–	–	–	2	–	1	Ruthenium, bis(2-methylallyl)bis(4-ethyl-2,6,7-trioxaphosphabicyclo[2.2.1]octanoyl)-		LI09599
43	229	345	85	187	29	289	171	**536**	100	99	75	75	73	70	60	58	**0.00**	22	32	15	–	–	–	–	–	–	–	–	–	–	Methyl 3,4-di-O-acetyl-2-O-(methyl 2,3-di-O-acetyl-4-O-methyl-α-D-glucopyranosyluronate)-α-D-xylopyranoside		LI09600
147	169	109	189	331	164	494	536	**536**	100	99	80	77	63	57	20	14		25	28	13	–	–	–	–	–	–	–	–	–	–	β-D-Glucopyranose, 2,3,4,6-tetraacetate 1-[3-[4-(acetyloxy)phenyl]-2-propenoate]	18449-70-2	NI45536
43	147	169	44	57	189	109	55	**536**	100	72	64	64	62	48	48	42	**2.20**	25	28	13	–	–	–	–	–	–	–	–	–	–	β-D-Glucose, 1-O-(p-coumaroyl)-2,3,4,5,6-pentaacetate		MS10582
43	452	454	44	60	42	49	84	**536**	100	64	61	58	56	50	49	40	**1.68**	26	17	8	–	–	–	–	1	–	–	–	–	–	4,4′-Diacetoxy-3-bromo-7,7′-dimethyl-1,1′-binaphthalene-5,5′,8,8′-tetrone		NI45537
73	197	446	305	183	307	169	431	**536**	100	19	17	16	15	14	14	11	**7.30**	27	48	5	–	–	–	–	–	–	3	–	–	–	cis-Zearalen-α-ol, tris(trimethylsilyl)-		NI45538
73	446	333	305	307	260	197	151	**536**	100	20	20	18	13	13	12	11	**10.70**	27	48	5	–	–	–	–	–	–	3	–	–	–	cis-Zearalen-β-ol, tris(trimethylsilyl)-		NI45539
73	197	305	446	183	307	169	431	**536**	100	23	22	21	18	17	16	15	**10.50**	27	48	5	–	–	–	–	–	–	3	–	–	–	trans-Zearalen-α-ol, tris(trimethylsilyl)-		NI45540
73	446	305	536	333	197	307	260	**536**	100	28	22	22	16	14	14	12		27	48	5	–	–	–	–	–	–	3	–	–	–	trans-Zearalen-β-ol, tris(trimethylsilyl)-		NI45541
43	69	109	122	55	95	121	91	**536**	100	40	30	27	23	22	19	19	**0.80**	28	40	10	–	–	–	–	–	–	–	–	–	–	9-Desoxo-9-hydroxy-ingol-3,7,8,12-tetraacetate		NI45542
43	55	41	343	79	153	149	57	**536**	100	29	29	25	22	21	19	19	**0.00**	29	44	9	–	–	–	–	–	–	–	–	–	–	Stigmast-7-en-26-oic acid, 2,3,5,14,20,22,28-heptahydroxy-6-oxo-, γ-lactone, (2β,3β,5β,20ξ)-		LI09601
267	130	268	536	157	245	156	384	**536**	100	70	56	21	10	6	6	4		30	28	4	6	–	–	–	–	–	–	–	–	–	2,5,16-Triazatetracyclo[7.7.0.0^{2,7}.0^{10,15}]hexadeca-10,12,14-triene, 2-methyl 4-methylene-3,6-dioxo-9,9′-bis-		LI09602
343	105	342	414	59	300	536	77	**536**	100	90	77	66	66	64	58	49		30	32	9	–	–	–	–	–	–	–	–	–	–	Gomisin C	58546-56-8	NI45543
100	85	135	107	97	136	77	190	**536**	100	45	30	28	27	20	8	6	**1.70**	30	40	5	4	–	–	–	–	–	–	–	–	–	Butanamide, 2-(dimethylamino)-N-[7-[(4-hydroxyphenyl)methyl]-3-isopropyl-5,8-dioxo-2-oxa-6,9-diazabicyclo[10.2.2]hexadeca-10,12,14,15-tetraen-4-yl]-3-methyl-	64408-10-2	NI45544
215	141	154	83	321	197	221	249	**536**	100	30	20	19	13	12	8	6	**0.00**	30	48	8	–	–	–	–	–	–	–	–	–	–	Spathulasin	122899-03-0	DD01783
240	274	536	275	155	121	239	133	**536**	100	91	85	73	65	59	57	57		31	30	4	–	–	–	2	–	–	–	–	–	–	Methane, bis[3-(3′-chloro-2′-hydroxy-5′-methylbenzyl)-2-hydroxy-5-methylphenyl]-		LI09603
114	115	131	85	100	44	43	42	**536**	100	8	2	2	1	1	1	1	**0.04**	31	44	4	4	–	–	–	–	–	–	–	–	–	Integerrenin, dihydro-		MS10583
134	268	106	536	402	521	387	253	**536**	100	94	91	64	50	1	1	1		32	24	8	–	–	–	–	–	–	–	–	–	–	Tetra-O-cresotide		MS10584
478	296	72	58	210	340	282	297	**536**	100	89	77	75	64	62	48	41	**18.06**	34	40	2	4	–	–	–	–	–	–	–	–	–	1,2-Ethanediamine, N-[2-(dibenz[b,e][1,4]oxazepin-5(11H)-ylmethyl)phenyl]-N-[2-[2-(dimethylamino)ethoxy]phenyl]-N′,N′-dimethyl-	16882-84-1	NI45545
536	463	268	521	448	505	477	433	**536**	100	22	11	9	8	5	5	4		34	40	2	4	–	–	–	–	–	–	–	–	–	3,6,8-Triethyl-1-[(2″-methoxycarbonyl)ethyl]-2,4,5,7-tetramethylporphin		LI09604
73	43	353	325	57	41	55	75	**536**	100	63	37	36	33	28	25	18	**2.00**	34	68	2	–	–	–	–	–	–	1	–	–	–	Hentriacontane-14,16-dione, trimethylsilyl-		MS10585

MASS TO CHARGE RATIOS								M.W.	INTENSITIES								Parent	C	H	O	N	S	F	Cl	Br	I	Si	P	B	X	COMPOUND NAME	CAS Reg No	No
57	197	86	43	55	28	69	83	**536**	100	90	50	50	44	37	36	30	**2.58**	35	68	3	–	–	–	–	–	–	–	–	–	–	3-Octanone, 8-(6-heneicosyl-2-methyl-1,3-dioxan-4-yl)-4-methyl-, [4R-[4α(R*),6β]]-	56247-57-5	NI45546
43	57	313	55	69	41	83	97	**536**	100	98	69	59	45	44	42	34	**26.04**	36	72	2	–	–	–	–	–	–	–	–	–	–	Eicosanoic acid, hexadecyl ester	22413-05-4	NI45547
257	57	43	55	71	69	83	41	**536**	100	95	95	58	48	47	46	40	**30.04**	36	72	2	–	–	–	–	–	–	–	–	–	–	Hexadecanoic acid, eicosyl ester	22413-01-0	NI45548
285	57	43	71	55	83	69	97	**536**	100	81	69	48	46	44	40	38	**29.14**	36	72	2	–	–	–	–	–	–	–	–	–	–	Octadecanoic acid, octadecyl ester	2778-96-3	NI45549
43	57	55	41	69	71	83	285	**536**	100	77	58	55	38	35	34	28	**2.30**	36	72	2	–	–	–	–	–	–	–	–	–	–	Octadecanoic acid, octadecyl ester	2778-96-3	WI01598
536	537	444	538	430	445	534	378	**536**	100	43	18	12	8	6	5	3		40	56	–	–	–	–	–	–	–	–	–	–	–	α-Carotene		LI09605
536	537	444	538	445	430	380	341	**536**	100	43	26	11	10	5	5	5		40	56	–	–	–	–	–	–	–	–	–	–	–	β,β-Carotene	7235-40-7	NI45550
69	119	105	55	95	41	91	81	**536**	100	84	75	66	63	62	56	56	**54.04**	40	56	–	–	–	–	–	–	–	–	–	–	–	β,β-Carotene	7235-40-7	NI45551
28	145	43	41	69	55	105	536	**536**	100	70	67	56	51	51	37	35		40	56	–	–	–	–	–	–	–	–	–	–	–	β,β-Carotene	7235-40-7	NI45552
536	537	444	538	430	445	534	378	**536**	100	43	18	13	9	6	5	4		40	56	–	–	–	–	–	–	–	–	–	–	–	β,ε-Carotene	432-70-2	NI45553
43	536	69	55	105	41	91	119	**536**	100	86	66	60	58	58	47	41		40	56	–	–	–	–	–	–	–	–	–	–	–	β-Carotene		NI45554
536	537	444	538	445	430	380	346	**536**	100	43	26	12	10	5	5	5		40	56	–	–	–	–	–	–	–	–	–	–	–	β-Carotene		LI09606
69	106	91	105	536	119	107	92	**536**	100	87	75	73	53	38	38	37		40	56	–	–	–	–	–	–	–	–	–	–	–	ψ,ψ-Carotene	502-65-8	NI45555
69	43	41	55	44	95	91	81	**536**	100	96	82	79	60	44	42	41	**39.30**	40	56	–	–	–	–	–	–	–	–	–	–	–	all-trans-β-Carotene		MS10586
69	41	536	105	55	119	91	81	**536**	100	93	90	90	89	85	82	76		40	56	–	–	–	–	–	–	–	–	–	–	–	neo-α-Carotene		NI45556
536	43	69	195	57	41	537	55	**536**	100	99	64	58	57	57	49	48		40	56	–	–	–	–	–	–	–	–	–	–	–	neo-β-Carotene		NI45557
69	41	105	91	145	93	81	119	**536**	100	57	47	47	38	36	36	34	**21.70**	40	56	–	–	–	–	–	–	–	–	–	–	–	Lycopene		MS10587
69	91	536	133	109	43	83		**536**	100	47	22	22	19	16	8			40	56	–	–	–	–	–	–	–	–	–	–	–	Lycopene		LI09607
167	536	170	369	191	291	178	165	**536**	100	32	15	14	11	10	10	9		42	32	–	–	–	–	–	–	–	–	–	–	–	Tricyclohexane, hexaphenyl-		LI09608
57	41	29	44	410	408	56	43	**537**	100	28	18	9	7	7	7	7	**1.00**	21	33	2	1	–	–	–	1	1	–	–	–	–	Benzene, 1-bromo-2,4,6-tris(2,2-dimethylpropyl)-3-iodo-5-nitro-	40572-26-7	MS10588
73	204	147	75	217	133	288	230	**537**	100	90	84	58	53	36	26	22	**0.50**	22	51	6	1	–	–	–	–	–	4	–	–	–	N-Butyl-D-glucuronamide tetrakis(trimethylsilyl)-		NI45558
127	302	304	509	207	511	111	129	**537**	100	83	57	44	42	40	35	32	**2.00**	28	22	2	3	–	–	3	–	–	–	–	–	–	1-(m-Chloroanilino)-2,7-bis(m-chlorophenyl)-2,7-diazatetracyclo[4,4,2,$0^{3,12}$,$0^{8,11}$]dodecane-4,10-dione		MS10589
326	528	526	253	44	296	28	73	**537**	100	47	46	30	29	23	23	22	**13.00**	28	42	5	–	–	–	–	1	–	–	–	–	–	14,15-Secopregna-5,8(14)-dien-3-ol, (4-bromobenzoyl)-14,16:15,20:18,20-triepoxy-, (3β,16α,20α)-		LI09609
326	241	271	232	114	506	286	308	**537**	100	96	90	73	62	59	54	53	**0.00**	28	51	5	1	–	–	–	–	–	2	–	–	–	Pregnane-11,20-dione, 17-hydroxy-3,21-bis[(trimethylsilyl)oxy]-, 20-(O-methyloxime), (3α,5β)-	75113-26-7	NI45559
73	326	75	241	506	55	447	107	**537**	100	99	99	89	59	59	58	56	**6.00**	28	51	5	1	–	–	–	–	–	2	–	–	–	Pregnane-11,20-dione, 3,17,21-trihydroxy-, mono(O-methyloxime), bis(trimethylsilyl) ether, (3α,5β)-	72088-35-8	NI45560
105	106	77	124	149	451	95	81	**537**	100	74	27	25	18	14	7	7	**1.00**	30	35	8	1	–	–	–	–	–	–	–	–	–	1α-Benzoyloxy-6β-nicotinoyloxy-9β-acetoxy-4β-hydroxydihyd ro-β-agarofuran		MS10590
537	536	179	358	538	458	357	457	**537**	100	63	60	41	36	30	20	17		36	30	–	3	–	–	–	–	–	–	–	3	–	Borazine, hexaphenyl-	16672-48-3	NI45561
30	267	286	426	217	158	130	32	**538**	100	99	97	84	83	50	48	44	**24.00**	12	–	4	–	–	12	–	–	–	–	–	–	1	Ruthenium, tricarbonyltetrakis(trifluoromethyl)cyclopentadienonyl-		LI09610
81	73	75	323	299	227	211	82	**538**	100	33	12	8	7	7	7	6	**0.40**	19	39	8	2	–	–	–	–	–	3	1	–	–	Thymidine, 3′-O-(trimethylsilyl)-, 5′-[bis(trimethylsilyl) phosphate]	32645-60-6	NI45562
73	297	314	316	225	315	298	75	**538**	100	87	78	55	39	22	22	19	**0.00**	20	36	5	2	–	–	2	–	–	3	–	–	–	Chloramphenicol, tris[(trimethylsilyl)oxy]-		NI45563
380	381	379	540	542	382	77	80	**538**	100	48	48	38	22	21	21	20	**18.87**	21	16	5	–	1	–	–	2	–	–	–	–	–	Phenol, 4,4′-(3H-2,1-benzoxathiol-3-ylidene)bis[2-bromo-6-methyl-, S,S-dioxide	115-40-2	NI45564
73	305	204	217	306	191	307	75	**538**	100	84	50	37	33	25	23	20	**3.72**	21	50	6	–	–	–	–	–	–	5	–	–	–	myo-Inosose-2, 1,3,4,5,6-pentakis-O-(trimethylsilyl)-	14251-19-5	MS10591
73	305	204	217	306	191	307	75	**538**	100	46	28	20	18	14	13	11	**4.00**	21	50	6	–	–	–	–	–	–	5	–	–	–	myo-Inosose-2, 1,3,4,5,6-pentakis-O-(trimethylsilyl)-	14251-19-5	LI09611
73	204	217	306	191	307	75	233	**538**	100	27	20	18	14	13	11	8	**1.80**	21	50	6	–	–	–	–	–	–	5	–	–	–	myo-Inosose-2, 1,3,4,5,6-pentakis-O-(trimethylsilyl)-	14251-19-5	NI45565
107	241	134	195	132	121	194	122	**538**	100	83	41	22	18	18	17	17	**1.20**	22	28	2	2	4	–	–	–	–	–	–	–	1	Bis(2,2-dimethylthio-N-salicylideneethylaminato)nickel		NI45566
86	56	57	74	132	85	70	58	**538**	100	77	57	43	40	37	37	37	**3.70**	22	42	7	4	2	–	–	–	–	–	–	–	–	6,9,12,20,23,28,31-Heptaoxa-1,3,15,17-tetraazabicyclo[15.8.8]tritriaconta-2,16-dithione		MS10592
318	370	259	371	452	424	538	368	**538**	100	52	37	23	14	14	12	6		24	18	8	–	–	–	–	–	–	–	–	–	2	Chromium, hexacarbonyl[μ-[(1,11,12,13,14,15-η:4,5,6,7,8,16-η)-methyl tricyclo[9.3.1.$1^{4,8}$]hexadeca-1(15),4,6,8(16),11,13-hexaene-5-carboxylate]]di-, stereoisomer	41424-18-4	MS10593
274	236	520	69	286	250	77	228	**538**	100	86	62	43	41	37	36	33	**0.00**	26	18	13	–	–	–	–	–	–	–	–	–	–	Purpuromycin	53969-01-0	NI45567
157	235	416	376	253	225	368	417	**538**	100	6	3	2	2	2	1	1	**0.50**	27	26	8	2	1	–	–	–	–	–	–	–	–	6-Hydroxy-4-methoxy-N-(4-methoxyphenyl)-5-[3-[(4-methoxyphenyl)amino]-1-oxo-2-butenyl]-benzofuran-7-sulphonamide		MS10594

MASS TO CHARGE RATIOS								M.W.	INTENSITIES								Parent	C	H	O	N	S	F	Cl	Br	I	Si	P	B	X	COMPOUND NAME	CAS Reg No	No
150	344	120	104	538	345			**538**	100	60	35	32	15	14				27	26	10	2	–	–	–	–	–	–	–	–	–	Bis-p-nitrobenzoyloxy-1,3-dimethoxy-4-methyl-6-(1-methyl-3-hydroxypropyl)benzene		LI09612
73	43	57	122	85	350	105	117	**538**	100	99	88	79	68	53	47	44	**0.00**	27	42	9	–	–	–	–	–	–	1	–	–	–	T-2 Toxin, trimethylsilyl ether		NI45568
179	180	165	121	44	41	55	43	**538**	100	12	11	10	9	9	8	8	**0.00**	29	30	10	–	–	–	–	–	–	–	–	–	–	4,2-Cresotic acid, 6-methoxy-, bimol. ester, methyl ester, 4,6-dimethoxy-o-toluate	19314-74-0	NI45569
520	45	538	521	79	52	46	539	**538**	100	94	86	62	40	38	38	23		30	18	10	–	–	–	–	–	–	–	–	–	–	[1,1′-Bianthracene]-9,9′,10,10′-tetrone, 2,2′,4,4′,5,5′-hexahydroxy-7,7′-dimethyl-	602-06-2	NI45570
538	521	539	269	493	522	360	55	**538**	100	31	30	25	23	15	15	11		30	18	10	–	–	–	–	–	–	–	–	–	–	[1,1′-Bianthracene]-9,9′,10,10′-tetrone, 4,4′,5,5′,8,8′-hexahydroxy-7,7′-dimethyl-	568-42-3	NI45571
538	539	445	234	269	386	385	387	**538**	100	35	27	20	14	12	12	11		30	18	10	–	–	–	–	–	–	–	–	–	–	3,3″-Biflavone, 4′,4‴,5,5″,7,7″-hexahydroxy-	27090-20-6	NI45572
121	255	69	147	137	120	133	95	**538**	100	56	26	25	23	21	20	19	**10.17**	31	52	1	–	–	–	2	–	–	1	–	–	–	9,10-Secoergosta-5,7,22-triene, 3-[(dichloromethyl)dimethyloxy]-, (3β,5Z,7E,22E)-	69688-09-1	NI45573
514	257	499	471	528	512	264	542	**538**	100	15	12	12	5	4	2	1	**0.00**	32	26	8	–	–	–	–	–	–	–	–	–	–	Dibenzo[fg,op]naphthacene-1,8-quinone, 2,5,6,9,12,13-hexamethoxy-		LI09613
514	528	499	471	257	512	264	542	**538**	100	20	10	10	10	5	4	3	**0.00**	32	26	8	–	–	–	–	–	–	–	–	–	–	Dibenzo[fg,op]naphthacene-1,10-quinone, 2,5,6,9,12,13-hexamethoxy-		LI09614
269	270	241	271	242	227	255	198	**538**	100	58	36	12	7	6	5	5	**1.30**	32	26	8	–	–	–	–	–	–	–	–	–	–	Physcion-10,10′-bianthrone	21871-90-9	NI45574
335	113	396	183	201	49	336	185	**538**	100	70	60	33	26	25	24	20	**5.00**	32	28	4	–	–	–	–	–	–	–	2	–	–	2,3-Bis(methoxycarbonyl)-1,1,4,4-tetraphenyl-1,4-diphosphabenzene		MS10595
454	412	496	229	123	163	246	107	**538**	100	52	36	32	23	16	10	9	**8.00**	32	42	7	–	–	–	–	–	–	–	–	–	–	2-Oxatricyclo[20.2.2.1^{3,7}]heptacosa-3,5,7(27),22,24,25-hexaene, 5,24,25-triacethyloxy-		NI45575
131	55	77	330	103	71	147	83	**538**	100	42	36	34	30	28	27	27	**0.00**	32	42	7	–	–	–	–	–	–	–	–	–	–	Pregnan-20-one, 12-acetyloxy-11-cinnamoyloxy-3,14-dihydroxy-		MS10596
408	296	409	255	406	283	253	159	**538**	100	42	32	21	17	16	15	11	**1.98**	32	52	2	–	–	–	2	–	–	–	–	–	–	Gorgostan-3-ol, 5,6-dichloro-, acetate, (3β,5α)-	55724-20-4	NI45576
108	262	482	165	183	166	83	55	**538**	100	94	80	69	61	57	12	11	**0.47**	33	24	2	–	–	–	–	–	–	–	1	–	1	Manganese, dicarbonyl-π-fluorenyl(triphenylphosphine)-	31760-88-0	NI45577
538	177	540	539	179	178	95	541	**538**	100	92	39	37	32	25	16	13		33	47	2	2	–	–	1	–	–	–	–	–	–	Cholestano[3,2-b]indole, 5′-chloro-6-nitro-, (5α,6α)-		MS10597
454	455	453	397	44	398	426	311	**538**	100	68	48	36	30	27	27	22	**0.00**	34	18	7	–	–	–	–	–	–	–	–	–	–	6,17-Diacetoxy-11,12-epoxy-5,18-trinaphthylenedione	96722-08-6	NI45578
43	29	57	41	55	492	71	69	**538**	100	42	40	35	28	23	15	14	**5.71**	34	50	5	–	–	–	–	–	–	–	–	–	–	1′,1′-Dicarboethoxy-1β,2β-dihydro-3′H-cycloprop[1,2]cholesta-1,4,6-trien 3-one		NI45579
283	43	239	99	282	98	55	28	**538**	100	42	41	39	29	29	27	27	**0.94**	34	66	4	–	–	–	–	–	–	–	–	–	–	Hexadecanoic acid, 1,2-ethanediyl ester	624-03-3	NI45580
255	294	538	495	256	239	539	449	**538**	100	72	45	28	24	24	17	10		35	22	6	–	–	–	–	–	–	–	–	–	–	Phthalide, 3-(4-methoxyphenyl)-3-(2′-fluoranyl)-		LI09615
255	494	538	495	256	239	493	539	**538**	100	72	45	30	24	23	22	15		35	22	6	–	–	–	–	–	–	–	–	–	–	Spiro[isobenzofuran-1(3H),9′-[9H]xanthen]-3-one, 2′-[1,3-dihydro-1-(4-methoxyphenyl)-3-oxo-1-isobenzofuranyl]-	55517-71-0	NI45581
327	57	71	328	43	85	55	69	**538**	100	40	25	24	20	18	15	14	**0.54**	35	70	3	–	–	–	–	–	–	–	–	–	–	1,3-Dioxane, 4-(hexadecyloxy)-2-pentadecyl-	56599-40-7	NI45582
327	57	43	28	71	328	55	83	**538**	100	49	33	33	27	24	23	21	**0.94**	35	70	3	–	–	–	–	–	–	–	–	–	–	1,3-Dioxane, 5-(hexadecyloxy)-2-pentadecyl-, cis-	34298-21-0	NI45583
327	57	43	71	328	55	83	69	**538**	100	48	33	28	25	23	20	20	**0.00**	35	70	3	–	–	–	–	–	–	–	–	–	–	1,3-Dioxane, 5-(hexadecyloxy)-2-pentadecyl-, trans-	34315-34-9	NI45584
327	57	328	71	43	83	55	69	**538**	100	39	24	23	22	19	18	17	**0.44**	35	70	3	–	–	–	–	–	–	–	–	–	–	1,3-Dioxane, 5-(hexadecyloxy)-2-pentadecyl-, trans-	34315-34-9	NI45585
199	538	537	170	200	152	369	169	**538**	100	24	15	15	12	11	9	9		36	27	3	–	–	–	–	–	–	–	1	–	–	[1,1′-Biphenyl]-2-ol, phosphite (3:1)	2752-19-4	NI45586
57	43	71	55	85	83	69	28	**538**	100	81	68	48	47	44	41	39	**0.14**	36	74	2	–	–	–	–	–	–	–	–	–	–	Octadecane, 1-[2-(hexadecyloxy)ethoxy]-	17367-10-1	NI45587
57	71	85	43	83	97	69	55	**538**	100	83	62	59	58	52	46	43	**1.04**	36	74	2	–	–	–	–	–	–	–	–	–	–	Octadecane, 1-[2-(hexadecyloxy)ethoxy]-	17367-10-1	NI45588
180	538	461	77	181	539	462	537	**538**	100	28	25	17	15	12	10	9		40	30	–	2	–	–	–	–	–	–	–	–	–	Pyridazine, 1,2-dihydro-1,2,3,4,5,6-hexaphenyl-	74810-58-5	NI45589
81	119	105	538	95	123	93	91	**538**	100	92	87	81	80	71	65	59		40	58	–	–	–	–	–	–	–	–	–	–	–	γ-Carotene, 7′,8′-dihydro-		LI09616
207	76	165	206	104	179	208	105	**539**	100	52	46	43	36	28	17	13	**0.00**	14	10	1	1	–	–	6	–	–	–	–	–	1	3-Phenyl-1-oxoisoindolium hexachloroantimonate		LI09617
43	239	238	41	69	70	169	210	**539**	100	49	37	26	23	17	16	8	**0.00**	14	11	5	1	–	14	–	–	–	–	–	–	–	L-Serine, N,O-bis(heptafluorobutanoyl)-, propyl ester		MS10598
162	81	231	119	69	161	232	53	**539**	100	95	83	71	56	41	40	36	**4.07**	16	10	6	5	–	9	–	–	–	–	–	–	–	Adenosine, 2′-deoxy-N-(trifluoroacetyl)-, 3′,5′-bis(trifluoroacetate)	35170-10-6	NI45590
161	375	289	536	521	506	361	346	**539**	100	9	7	1	1	1	1	1	**0.00**	18	57	–	3	–	–	–	–	–	6	–	–	1	Iron, tris(1,1,1,3,3,3-hexamethyldisilazanato-N)-	22999-67-3	NI45591
203	73	204	147	304	205	217	75	**539**	100	67	26	13	12	12	10	10	**2.60**	21	53	5	1	–	–	–	–	–	5	–	–	–	D-Galactose, 2-deoxy-3,4,5,6-tetrakis-O-(trimethylsilyl)-2-[(trimethylsilyl)amino]-	56248-46-5	NI45592
241	134	46	132	122	121	59	45	**539**	100	92	83	67	67	58	54	54	**9.20**	22	28	2	2	4	–	–	–	–	–	–	–	1	Bis(2,2-dimethylthio-N-salicylideneethylaminato)cobalt	90078-40-3	NI45593
202	73	42	203	29	130	41	45	**539**	100	88	29	24	22	18	14	13	**0.00**	26	36	4	3	–	3	–	–	–	1	–	–	–	L-Prolyl-α-methyltryptophan, N-(trifluoroacetyl)-O-(trimethylsilyl)-, butyl ester		MS10599
227	112	157	213	199	185	113	539	**539**	100	87	67	57	33	26	17	11		26	45	7	5	–	–	–	–	–	–	–	–	–	L-Prolyl-α-methyltryptophan, N-(trifluoroacetyl)-O-(trimethylsilyl)-, butyl ester		LI09618
224	294	132	179	225	172	365	295	**539**	100	52	24	22	20	14	13	12	**0.00**	28	53	5	1	–	–	–	–	–	2	–	–	–	PGE, methyl ester, bis(trimethylsilyl)-		NI45594
73	310	75	132	55	107	105	43	**539**	100	99	99	77	72	71	71	70	**1.00**	28	53	5	1	–	–	–	–	–	2	–	–	–	Pregnan-20-one, 3,11-dihydroxy-17,21-bis[(trimethylsilyl)oxy]-, O-methyloxime, (3α,5β,11β)-	57305-36-9	NI45595
225	295	133	180	226	173	366	296	**539**	100	52	24	22	20	14	13	12	**1.53**	28	53	5	1	–	–	–	–	–	2	–	–	–	Prosta-5,13-dien-1-oic acid, 9-(methoxyimino)-11,15-bis[(trimethylsilyl)oxy]-, methyl ester, (5Z,11α,13E,15S)-	24524-81-0	NI45596

MASS TO CHARGE RATIOS								M.W.	INTENSITIES								Parent	C	H	O	N	S	F	Cl	Br	I	Si	P	B	X	COMPOUND NAME	CAS Reg No	No
378	217	539	105	379	77	133	43	**539**	100	31	29	25	23	13	8	8		30	27	6	–	–	–	–	–	–	–	–	–	1	Iron, tris(1-phenyl-1,3-butanedionato-O,O')-	14323-17-2	NI45597
83	85	149	56	71	55	99	167	**539**	100	97	96	85	42	36	29	28	0.00	31	41	7	1	–	–	–	–	–	–	–	–	–	Allethrin tyramine derivative		NI45598
539	541	540	524	542	526	525	509	**539**	100	52	39	21	18	12	11	10		32	36	–	4	–	–	–	–	–	–	–	–	1	Copper, [2,8,12,18-tetraethyl-3,7,13,17-tetramethyl-21H,23H-porphinato(2-)-N^{21},N^{22},N^{23},N^{24}]-	14055-18-6	NI45599
539	541	540	524	269.5	542	526	270.5	**539**	100	52	39	21	20	18	12	11		32	36	–	4	–	–	–	–	–	–	–	–	1	Copper, [2,8,12,18-tetraethyl-3,7,13,17-tetramethyl-21H,23H-porphinato(2-)-N^{21},N^{22},N^{23},N^{24}]-	14055-18-6	AM00158
127	170	128	351	131	100	150	69	**540**	100	33	22	10	9	8	8	7	0.00	6	2	2	2	2	18	–	–	–	–	–	–	–	2,2,3,3,4,4,5,5-Octafluoro-1,6-dioxo-1,6-hexylenebis(iminosulphur)pentafluoride	91940-21-5	NI45600
272	274	237	239	235	270	276	241	**540**	100	94	80	53	50	49	30	18	0.00	10	–	–	–	–	–	12	–	–	–	–	–	–	1,3,4-Metheno-1H-cyclobuta[cd]pentalene, 1,1a,2,2,3,3a,4,5,5,5a,5b,6-dodecachlorooctahydro-	2385-85-5	NI45601
272	274	270	237	276	239	235	332	**540**	100	75	54	46	34	29	28	14	0.00	10	–	–	–	–	–	12	–	–	–	–	–	–	1,3,4-Metheno-1H-cyclobuta[cd]pentalene, 1,1a,2,2,3,3a,4,5,5,5a,5b,6-dodecachlorooctahydro-	2385-85-5	NI45602
237	239	235	236	241	238	509	234	**540**	100	64	64	27	21	21	17	16	0.00	10	–	–	–	–	–	12	–	–	–	–	–	–	1,3,4-Metheno-1H-cyclobuta[cd]pentalene, 1,1a,2,2,3,3a,4,5,5,5a,5b,6-dodecachlorooctahydro-	2385-85-5	NI45603
114	76	72	394	44	70	59	147	**540**	100	97	75	44	41	39	24	22	0.00	10	16	–	2	4	–	–	–	1	–	–	–	1	Iodobis(pyrrolidyldithiocarbamato)stibine	59196-58-6	NI45604
131	173	214	130	172	107	139	256	**540**	100	39	34	29	26	17	14	12	6.00	12	28	4	–	4	–	–	–	–	–	2	–	1	Cadmium[II] bis(dipropyldithiophosphate)	43159-08-6	NI45605
43	100	73	87	129	142	99	141	**540**	100	32	23	19	16	12	11	9	0.00	13	19	8	–	–	–	1	–	–	–	–	–	1	β-D-Glucopyranoside, methyl 2-(chloromercurio)-2-deoxy-, triacetate	6641-37-8	NI45606
529	527	531	544	293	525	63	530	**540**	100	66	64	25	25	20	20	18	4.00	15	12	2	–	–	–	–	4	–	–	–	–	–	Phenol, [4,4'-isopropylidenebis]2,6-dibromo-	79-94-7	NI45607
540	316	542	131	318	372	428	484	**540**	100	76	56	48	42	34	28	27		16	36	4	–	4	–	–	–	–	–	2	–	1	Nickel[II] bis(dibutyldithiophosphate)	37913-22-7	NI45608
211	541	73	227	119	165	226	69	**540**	100	100	95	70	62	54	49	44	36.21	20	20	14	4	–	–	–	–	–	–	–	–	–	4',4'',5',5''-Tetranitrodibenzo[b,k]-1,4,7,10,13,16-hexaoxacyclooctadecane-2,11-diene	94270-85-6	NI45609
73	292	147	293	277	189	74	321	**540**	100	38	29	12	12	10	9	8	0.00	20	48	7	–	–	–	–	–	–	5	–	–	–	Arabinaric acid, 2,3,4-tris-O-(trimethylsilyl)-, bis(trimethylsilyl) ester	38165-95-6	NI45610
292	293	321	277	525	294	333	73	**540**	100	32	24	22	16	16	9	9	0.00	20	48	7	–	–	–	–	–	–	5	–	–	–	Glutanic acid, 2,3,4-tris-O-(trimethylsilyl)-, bis(trimethylsilyl) ester		LI09619
73	292	147	277	293	189	321	407	**540**	100	38	28	14	10	10	8	6	0.00	20	48	7	–	–	–	–	–	–	5	–	–	–	Glutaric acid, 2,3,4-tris-O-(trimethylsilyl)-, bis(trimethylsilyl) ester		LI09620
292	293	321	277	525	294	333	73	**540**	100	31	25	21	15	15	9	8	0.00	20	48	7	–	–	–	–	–	–	5	–	–	–	Ribaric acid, 2,3,4-tris-O-(trimethylsilyl)-, bis(trimethylsilyl) ester	57197-34-9	NI45611
57	372	428	484	41	43	540	29	**540**	100	78	46	35	26	22	21	17		21	34	–	–	–	–	–	–	2	–	–	–	–	Benzene, 1,3,5-tris(2,2-dimethylpropyl)-2,4-diiodo-	53173-11-8	NI45612
204	73	191	217	205	147	218	192	**540**	100	60	45	43	19	19	11	8	0.00	21	52	6	–	–	–	–	–	–	5	–	–	–	D-Altrose, 2,3,4,5,6-pentakis-O-(trimethylsilyl)-	56114-56-8	NI45613
217	73	191	218	147	319	219	205	**540**	100	53	24	21	18	13	9	7	0.00	21	52	6	–	–	–	–	–	–	5	–	–	–	L-Altrose, 2,3,4,5,6-pentakis-O-(trimethylsilyl)-	56114-55-7	NI45614
73	217	335	147	103	306	307	205	**540**	100	33	30	30	24	20	19	15	0.00	21	52	6	–	–	–	–	–	–	5	–	–	–	arabino-Hexonic acid, 2-deoxy-3,4,5,6-tetrakis-O-(trimethylsilyl)-, trimethylsilyl ester		LI09621
73	245	147	246	129	75	335	103	**540**	100	45	33	17	17	12	11	10	0.00	21	52	6	–	–	–	–	–	–	5	–	–	–	arabino-Hexonic acid, 3-deoxy-2,4,5,6-tetrakis-O-(trimethylsilyl)-, trimethylsilyl ester	55517-66-3	LI09622
73	245	147	246	129	335	75	525	**540**	100	76	33	17	17	12	12	11	0.00	21	52	6	–	–	–	–	–	–	5	–	–	–	arabino-Hexonic acid, 3-deoxy-2,4,5,6-tetrakis-O-(trimethylsilyl)-, trimethylsilyl ester	55517-66-3	NI45615
73	204	191	379	217	147	133	75	**540**	100	72	66	64	15	15	8	8	0.00	21	52	6	–	–	–	–	–	–	5	–	–	–	α-D-Fructopyranose, 1,2,3,4,5-pentakis-O-(trimethylsilyl)-	53538-02-6	NI45616
73	204	217	147	437	205	103	438	**540**	100	44	31	24	22	16	9	9	0.00	21	52	6	–	–	–	–	–	–	5	–	–	–	D-Fructose, 1,3,4,5,6-pentakis-O-(trimethylsilyl)-	19126-98-8	NI45617
217	73	147	79	218	191	319	75	**540**	100	63	39	33	21	18	17	11	0.20	21	52	6	–	–	–	–	–	–	5	–	–	–	α-D-Galactofuranose, 1,2,3,5,6-pentakis-O-(trimethylsilyl)-	55529-72-1	NI45618
217	73	218	191	147	319	219	74	**540**	100	82	21	21	20	13	9	7	0.00	21	52	6	–	–	–	–	–	–	5	–	–	–	β-D-Galactofuranose, 1,2,3,5,6-pentakis-O-(trimethylsilyl)-	7045-52-5	NI45619
204	191	217	147	205	192	218	206	**540**	100	95	63	32	19	16	13	8	0.00	21	52	6	–	–	–	–	–	–	5	–	–	–	α-D-Galactopyranose, 1,2,3,4,6-pentakis-O-(trimethylsilyl)-	32166-80-6	NI45621
204	73	191	217	147	205	129	206	**540**	100	78	48	27	22	21	11	9	0.00	21	52	6	–	–	–	–	–	–	5	–	–	–	α-D-Galactopyranose, 1,2,3,4,6-pentakis-O-(trimethylsilyl)-	32166-80-6	NI45620
204	191	217	147	205	192	218	206	**540**	100	93	49	33	22	18	11	9	0.00	21	52	6	–	–	–	–	–	–	5	–	–	–	β-D-Galactopyranose, 1,2,3,4,6-pentakis-O-(trimethylsilyl)-	32166-97-5	NI45622
73	204	191	147	217	205	129	75	**540**	100	99	41	32	31	22	13	13	0.00	21	52	6	–	–	–	–	–	–	5	–	–	–	D-Galactose, 2,3,4,5,6-pentakis-O-(trimethylsilyl)-	6736-94-3	NI45623
73	204	191	217	147	205	129	206	**540**	100	96	44	25	24	19	11	9	0.00	21	52	6	–	–	–	–	–	–	5	–	–	–	D-Galactose, 2,3,4,5,6-pentakis-O-(trimethylsilyl)-	6736-94-3	NI45624
73	204	191	147	217	205	192	75	**540**	100	86	63	25	21	17	12	12	0.10	21	52	6	–	–	–	–	–	–	5	–	–	–	α-D-Glucopyranose, 1,2,3,4,6-pentakis-O-(trimethylsilyl)-	3327-61-5	NI45626
204	73	191	147	205	217	206	129	**540**	100	89	40	24	21	17	9	8	0.00	21	52	6	–	–	–	–	–	–	5	–	–	–	α-D-Glucopyranose, 1,2,3,4,6-pentakis-O-(trimethylsilyl)-	3327-61-5	NI45625
73	204	191	147	217	205	192	75	**540**	100	88	65	26	22	18	12	12	0.00	21	52	6	–	–	–	–	–	–	5	–	–	–	α-D-Glucopyranose, 1,2,3,4,6-pentakis-O-(trimethylsilyl)-	3327-61-5	NI45627
204	73	191	205	147	217	206	192	**540**	100	62	35	19	19	17	8	7	0.00	21	52	6	–	–	–	–	–	–	5	–	–	–	β-D-Glucopyranose, 1,2,3,4,6-pentakis-O-(trimethylsilyl)-	2775-90-8	NI45628
204	73	191	147	205	217	206	74	**540**	100	77	38	21	20	17	9	7	0.00	21	52	6	–	–	–	–	–	–	5	–	–	–	β-D-Glucopyranose, 1,2,3,4,6-pentakis-O-(trimethylsilyl)-	2775-90-8	NI45629
204	73	191	147	205	217	206	74	**540**	100	100	47	23	21	18	9	9	0.00	21	52	6	–	–	–	–	–	–	5	–	–	–	D-Glucose, 2,3,4,5,6-pentakis-O-(trimethylsilyl)-	6736-97-6	NI45630
204	73	205	191	147	217	206	218	**540**	100	45	19	16	13	12	8	5	0.00	21	52	6	–	–	–	–	–	–	5	–	–	–	Gulose, 2,3,4,5,6-pentakis-O-(trimethylsilyl)-	56192-86-0	NI45631
73	204	147	217	191	75	361	205	**540**	100	87	40	38	38	32	24	21	0.00	21	52	6	–	–	–	–	–	–	5	–	–	–	Hexopyranose, 1,2,3,4,6-pentakis-O-(trimethylsilyl)-	55521-15-8	NI45632
204	73	191	205	147	217	206	74	**540**	100	78	39	20	19	16	9	7	0.00	21	52	6	–	–	–	–	–	–	5	–	–	–	α-D-Mannopyranose, 1,2,3,4,6-pentakis-O-(trimethylsilyl)-	24707-99-1	NI45633
204	73	191	147	205	217	206	75	**540**	100	78	37	23	20	18	9	8	0.00	21	52	6	–	–	–	–	–	–	5	–	–	–	D-Mannopyranose, 1,2,3,4,6-pentakis-O-(trimethylsilyl)-	55529-69-6	NI45634
204	73	191	147	205	217	206	74	**540**	100	98	40	22	20	17	9	9	0.00	21	52	6	–	–	–	–	–	–	5	–	–	–	D-Mannopyranose, 1,2,3,4,6-pentakis-O-(trimethylsilyl)-	55529-69-6	NI45635

MASS TO CHARGE RATIOS								M.W.	INTENSITIES								Parent	C	H	O	N	S	F	Cl	Br	I	Si	P	B	X	COMPOUND NAME	CAS Reg No	No
73	437	147	435	243	438	347	205	**540**	100	46	34	24	24	18	18	18	**0.00**	21	52	6	–	–	–	–	–	–	5	–	–	–	erythro-Pentonic acid, 3-deoxy-2,4,5-tris-O-(trimethylsilyl)-2-C-[[(trimethylsilyl)oxy]methyl]-, trimethylsilyl ester	55517-79-8	NI45636
73	147	243	347	205	129	74	75	**540**	100	33	23	17	17	10	8	7	**0.00**	21	52	6	–	–	–	–	–	–	5	–	–	–	erythro-Pentonic acid, 3-deoxy-2,4,5-tris-O-(trimethylsilyl)-2-C-[[(trimethylsilyl)oxy]methyl]-, trimethylsilyl ester	55517-79-8	LI09623
73	103	306	307	147	217	205	305	**540**	100	51	43	41	33	27	21	14	**0.00**	21	52	6	–	–	–	–	–	–	5	–	–	–	Ribonic acid, 2-C-methyl-2,3,4,5-tetrakis-O-(trimethylsilyl)-, trimethylsilyl ester	38166-01-7	LI09624
73	103	306	307	147	217	205	308	**540**	100	51	44	43	35	29	22	15	**0.00**	21	52	6	–	–	–	–	–	–	5	–	–	–	Ribonic acid, 2-C-methyl-2,3,4,5-tetrakis-O-(trimethylsilyl)-, trimethylsilyl ester	38166-01-7	NI45637
204	147	437	205	217	438	206	129	**540**	100	30	27	26	14	12	11	7	**0.00**	21	52	6	–	–	–	–	–	–	5	–	–	–	L-Sorbopyranose, 1,2,3,4,5-pentakis-O-(trimethylsilyl)-	30645-02-4	NI45638
73	204	437	147	205	217	74	438	**540**	100	58	28	25	13	11	9	8	**0.00**	21	52	6	–	–	–	–	–	–	5	–	–	–	Sorbose, pentakis(trimethylsilyl)-		NI45639
204	73	191	217	205	147	206	192	**540**	100	55	39	27	20	17	8	7	**0.00**	21	52	6	–	–	–	–	–	–	5	–	–	–	Talose, 2,3,4,5,6-pentakis-O-(trimethylsilyl)-	56192-85-9	NI45640
150	498	456	341	299	145	540	383	**540**	100	18	18	13	11	10	8	5		24	32	10	2	1	–	–	–	–	–	–	–	–	Tetra-acetyl-5-S-cysteinyl-DOPA diethyl ester (isomer a)		LI09625
150	145	341	456	299	383	498	540	**540**	100	80	34	15	15	13	10	7		24	32	10	2	1	–	–	–	–	–	–	–	–	Tetra-acetyl-5-S-cysteinyl-DOPA diethyl ester (isomer b)		LI09626
152	303	151	77	153	271	108	304	**540**	100	63	50	48	37	32	32	28	**0.00**	25	16	–	–	2	–	–	–	–	–	–	–	2	2-[5-(1,3-Benzodiselenol-2-ylidene)bicyclo[4.4.1]undeca-3,6,8,10-tetraen-2-ylidene]-1,3-benzodithiole		DD01784
43	91	169	115	73	193	157	97	**540**	100	33	11	11	10	8	7	7	**0.00**	26	36	12	–	–	–	–	–	–	–	–	–	–	(1RS)-3-Benzyloxy-1-ethoxypropyl 2,3,4,6-tetra-O-acetyl-β-D-glucopyranoside		MS10600
105	348	512	183	109	185	377	379	**540**	100	62	58	43	42	28	27	23	**18.00**	27	27	2	2	–	–	–	–	–	–	1	–	1	Molybdenum, carbonyl(η^5-2,4-cyclopentadien-1-yl)[N-methyl-P,P-diphenyl-N-(1-phenylethyl)phosphinous amide-P]nitrosyl-, stereoisomer	59657-12-4	NI45641
314	315	243	215	271	329	231	230	**540**	100	32	27	11	10	4	4	4	**0.00**	28	21	3	4	–	–	–	1	–	–	–	–	–	2-(4-Methoxyphenyl)-6-oxo-8,10-diphenyl-4H,6H-pyrido[2,1-f][1,3,4]oxadiazino[2,3-c][1,2,4]triazin-11-ium bromide		MS10601
105	281	253	271	317	540	311	299	**540**	100	71	54	36	33	28	27	26		29	32	10	–	–	–	–	–	–	–	–	–	–	12-Acetoxydaphnetoxin	82870-43-7	NI45642
216	186	62	78	110	108	217	187	**540**	100	74	32	24	18	17	12	10	**0.00**	30	25	–	–	–	–	–	–	–	–	5	–	–	Pentaphospholane, pentaphenyl-	3376-52-1	LI09627
262	183	185	108	324	109	263	139	**540**	100	72	60	45	42	35	23	21	**12.05**	30	25	–	–	–	–	–	–	–	–	5	–	–	Pentaphospholane, pentaphenyl-	3376-52-1	NI45643
270	384	271	156	213	540	128	385	**540**	100	27	22	22	20	19	19	18		30	28	6	4	–	–	–	–	–	–	–	–	–	(1α,2α,3β,4β)-1,3-Bis(1,2,3,4,5,6-hexahydro-1,3-dimethyl -2,4,6-trioxo-5-pyrimidinylidenemethyl)-2,4-diphenylcyclobutane		MS10602
540	359	181	509	358	328	91	541	**540**	100	88	86	70	50	43	22	19		30	36	9	–	–	–	–	–	–	–	–	–	–	1,2,3-Tris(3,4,5-trimethoxyphenyl)propene		NI45644
179	124	293	121	123	180	249	181	**540**	100	35	30	26	23	12	11	11	**0.00**	32	28	–	–	4	–	–	–	–	–	–	–	–	4-(4-methylphenylthio)-3,6-diphenyl-3-[2-(4-methylphenylthio)ethenyl]-3,4 dihydro-1,2-dithiin		MS10603
540	541	162	118	148	255	283	104	**540**	100	37	26	20	16	15	13	8		32	36	4	4	–	–	–	–	–	–	–	–	–	5,10,15,20(22H,24H)-Porphinetetrone, 2,3,7,17-tetraethyl-8,12,13,18-tetramethyl-	27800-02-8	NI45645
162	111	105	432	125	127	113	109	**540**	100	77	75	54	40	38	38	35	**0.00**	32	36	4	4	–	–	–	–	–	–	–	–	–	5,10,15,20(22H,24H)-Porphinetetrone, 2,7,12,17-tetraethyl-3,8,13,18-tetramethyl-	27800-00-6	NI45646
540	162	541	269	136	242	404	109	**540**	100	50	47	23	13	10	9	8		32	36	4	4	–	–	–	–	–	–	–	–	–	5,10,15,20(22H,24H)-Porphinetetrone, 2,7,12,17-tetraethyl-3,8,13,18-tetramethyl-	27800-00-6	NI45647
149	57	43	71	167	41	55	368	**540**	100	85	84	81	80	61	58	51	**7.00**	32	55	1	–	–	3	–	–	–	1	–	–	–	Cholesterol [(3,3,3-trifluoropropyl)dimethylsilyloxy]-		NI45648
270	135	227	271	121	492	77	28	**540**	100	52	38	19	11	9	9	8	**1.73**	33	32	7	–	–	–	–	–	–	–	–	–	–	2,3-Dihydro-2-methoxy-2,3,5,6-tetrakis(4-methoxyphenyl)-1,4-dioxin		NI45649
243	241	206	354	269	244	242	339	**540**	100	64	60	35	27	20	18	14	**0.00**	33	48	4	–	1	–	–	–	–	–	–	–	–	Cholestene, 10-formyl-3-tosyloxy-		LI09628
121	136	177	415	137	357	80	91	**540**	100	80	65	48	44	28	28	26	**6.50**	34	36	6	–	–	–	–	–	–	–	–	–	–	20,24-Dibenzyl-2,3:11,12-dibenzo-1,4,7,10,13,16-hexaoxacyclooctadeca-2,11-diene		MS10604
247	205	181	335	481	317	275	540	**540**	100	98	83	70	67	50	50	33		34	52	5	–	–	–	–	–	–	–	–	–	–	D:A-Friedooleanane-1,3-dione, 2-acetyl-7-(acetyloxy)-, (7α)-	43230-93-9	MS10605
290	329	525	135	540	291	121	189	**540**	100	39	31	24	23	21	20	19		34	52	5	–	–	–	–	–	–	–	–	–	–	Olean-9(11)-en-12-one, 3,28-bis(acetyloxy)-, (3β)-	4409-10-3	NI45650
291	231	121	147	480	292	119	207	**540**	100	33	27	23	20	20	20	19	**5.23**	34	52	5	–	–	–	–	–	–	–	–	–	–	Olean-12-en-15-one, 3,28-bis(acetyloxy)-, (3β)-	39701-80-9	NI45651
91	105	71	119	133	147	104	85	**540**	100	17	11	10	9	7	7	7	**0.25**	39	72	–	–	–	–	–	–	–	–	–	–	–	Tritriacontane, 17-phenyl-		AI02402
91	43	57	315	105	41	55	71	**540**	100	41	31	22	21	20	18	16	**3.97**	39	72	–	–	–	–	–	–	–	–	–	–	–	Tritriacontane, 17-phenyl-		AI02403
91	43	57	315	105	41	55	104	**540**	100	47	32	24	20	20	16	15	**3.45**	39	72	–	–	–	–	–	–	–	–	–	–	–	Tritriacontane, 17-phenyl-		AI02404
540	69	541	81	403	542	404	446	**540**	100	59	42	27	21	11	7	3		40	60	–	–	–	–	–	–	–	–	–	–	–	ζ-Carotene		NI45652
443	540	457	270	281	350	268	227	**540**	100	91	85	84	81	67	60	56		40	60	–	–	–	–	–	–	–	–	–	–	–	4′-Cyclohexyl-1′,2,5-tris[(E)-cyclohexylmethylene]-2′,3′,6′,7′-tetrahydrospiro[cyclopentan-1,5′(4′H)-indene]		MS10606
261	471	541	331	401	506			**541**	100	76	21	17	14	13				13	–	–	–	–	–	11	–	–	–	–	–	–	Methyl, chlorobis(pentachlorophenyl)-	3225-61-4	NI45653
261	471	235.5	200.5	541	331	401	506	**541**	100	76	53	34	21	17	14	13		13	–	–	–	–	–	11	–	–	–	–	–	–	Methyl, chlorobis(pentachlorophenyl)-	3225-61-4	LI09629
317	541	131	373	115	429	187	188	**541**	100	74	54	45	39	36	31	30		16	36	4	–	4	–	–	–	–	–	2	–	1	Cobalt[II] bis(dibutyldithiophosphate)	38625-34-2	NI45654
69	372	57	29	41	428	56	77	**541**	100	93	92	72	71	43	35	29	**0.00**	19	16	7	1	–	9	–	–	–	–	–	–	–	L-Tyrosine, N,O-bis(trifluoroacetyl)-3-[(trifluoroacetyl)oxy]-, butyl ester	55493-76-0	NI45655

MASS TO CHARGE RATIOS								M.W.	INTENSITIES								Parent	C	H	O	N	S	F	Cl	Br	I	Si	P	B	X	COMPOUND NAME	CAS Reg No	No
87	43	458	88	541	460	543	45	**541**	100	20	6	5	4	3	2	2		21	24	8	5	1	–	1	–	–	–	–	–	–	N,N-Bis(2-acetoxyethyl)-3-methylsulphonamido-4-(4-nitro-2-chlorophenylazo)-aniline		IC04569
75	73	217	147	132	204	103	218	**541**	100	94	59	20	18	17	17	14	**0.00**	21	55	5	1	–	–	–	–	–	5	–	–	–	D-Glucitol, 2-amino-2-deoxy-1,3,4,5,6-pentakis-O-(trimethylsilyl)-	56196-64-6	NI45656
162	253	91	28	134	119	163	116	**541**	100	39	31	29	24	19	13	7	**0.00**	21	55	5	1	–	–	–	–	–	5	–	–	–	D-Glucitol, 2-amino-2-deoxy-1,3,4,5,6-pentakis-O-(trimethylsilyl)-	56196-64-6	NI45657
75	73	217	147	132	103	204	117	**541**	100	94	57	21	20	20	19	10	**0.00**	21	55	5	1	–	–	–	–	–	5	–	–	–	D-Glucitol, 2-amino-2-deoxy-1,3,4,5,6-pentakis-O-(trimethylsilyl)-	56196-64-6	LI09630
73	236	75	203	147	216	169	90	**541**	100	60	60	42	33	31	25	20	**0.20**	22	43	3	5	1	–	–	–	–	3	–	–	–	Adenosine, 5′-S-propyl-5′-thio-N-(trimethylsilyl)-2′,3′-bis-O-(trimethylsilyl)-	54623-30-2	NI45658
183	185	237	283	300	209	200	198	**541**	100	99	98	71	66	64	63	63	**7.00**	27	26	7	–	–	–	–	1	–	–	–	–	–	Gibberellin A3 p-bromophenacyl ester		NI45659
192	135	385	137	204	190	124	123	**541**	100	85	77	55	54	46	42	39	**3.80**	28	35	2	3	3	–	–	–	–	–	–	–	–	13-Tosyl-8,9,17,18-dibenzo-1,7-dithia-10,13,16-triazacyclooctadeca-8,17-diene		MS10607
73	75	470	129	199	380	471	420	**541**	100	79	50	40	35	27	19	16	**0.85**	28	55	5	1	–	–	–	–	–	2	–	–	–	PGE 1, methyl ester, methoxyoxime, bis(trimethylsilyl)-		NI45660
73	297	368	75	130	133	298	69	**541**	100	94	57	52	31	28	23	22	**0.46**	28	55	5	1	–	–	–	–	–	2	–	–	–	PGE 1, methyl ester, methoxyoxime, bis(trimethylsilyl)-		NI45661
265	340	510	420	470	173	199	75	**541**	100	77	51	38	35	30	22	21	**6.00**	28	55	5	1	–	–	–	–	–	2	–	–	–	Prost-5-enoic acid, 9-keto-11,15-bis[(trimethylsilyl)oxy]-, O-methyloxime, methyl ester, (11α)-		LI09631
400	323	504	401	200	505	399	324	**541**	100	64	34	27	23	21	21	17	**0.00**	31	30	–	3	–	–	1	–	–	–	2	–	–	Aminodiphenylphosphino-benzylaminodiphenyl-phosphoranylidinimine chloride		LI09632
140	98	141	44	43	99	139	81	**541**	100	63	24	10	9	5	5	4	**0.00**	33	51	5	1	–	–	–	–	–	–	–	–	–	N-Acetyl-(22S,25S)-22,26-epiminocholest-5-ene-3β,16α-diol 16-acetate (O,N-diacetylmuldamine)		NI45662
243	298	244	165	83	85	43	41	**541**	100	34	34	33	21	13	13	11	**4.00**	34	39	5	1	–	–	–	–	–	–	–	–	–	1,2-Benzisoxazole, octahydro-2-[2,3-O-isopropylidene-5-O-(triphenylmethyl)-β-D-ribofuranosyl]-, (3aR-cis)-	64018-63-9	NI45663
402	430	109	346	374	218	272	458	**542**	100	84	70	63	58	58	49	48	**35.00**	12	12	8	–	–	–	–	–	–	–	2	–	2	Molybdenum, [bis(tetracarbonyl)di-μ-dimethylphosphino]di-		MS10608
155	153	527	525	165	529	185	163	**542**	100	81	77	72	71	61	60	56	**0.00**	12	18	–	–	–	–	4	–	–	–	–	–	2	Stannane, (2,3,5,6-tetrachloro-4-phenylene)bis[trimethyl-	18689-05-9	NI45664
165	163	155	153	161	185	527	135	**542**	100	77	73	51	47	36	33	32	**0.00**	12	18	–	–	–	–	4	–	–	–	–	–	2	Stannane, (3,4,5,6-tetrachloro-2-phenylene)bis[trimethyl-	15725-05-0	NI45665
63	91	265	203	201	44	465	263	**542**	100	55	33	30	30	30	26	25	**0.00**	17	11	4	3	–	–	–	2	–	–	–	–	1	Copper(II), bis(4-bromo-1-quinone-2-oximato)pyridine (1:1 adduct)		MS10609
407	405	341	109	136	343	100	193	**542**	100	99	76	75	73	72	72	63	**2.00**	18	17	7	2	2	–	3	–	–	–	–	–	–	2,2,2-Trichloroethyl 7-methoxycarbonylamino-3-phenylthiomethyl-3-cephem-4-carboxylate 1,1-dioxide	97609-92-2	NI45666
176	359	177	155	183	125	69	156	**542**	100	34	22	21	10	9	9	8	**1.00**	20	4	2	–	–	14	–	–	–	–	–	–	–	Benzene, 1,2,4,5-tetrafluoro-3,6-bis[(pentafluorophenoxy)methyl]-	38795-55-0	LI09633
176	359	177	155	183	125	69	150	**542**	100	34	22	21	10	9	9	8	**1.00**	20	4	2	–	–	14	–	–	–	–	–	–	–	Benzene, 1,2,4,5-tetrafluoro-3,6-bis[(pentafluorophenoxy)methyl]-	38795-55-0	NI45667
73	41	57	55	269	271	81	267	**542**	100	80	78	49	41	39	32	28	**0.00**	22	54	7	–	–	–	–	–	–	4	–	–	–	Tetrasiloxane, 3,5,5,5-tetrabutoxy-1,1,1,7,7,7-hexamethyl-		MS10610
542	193	175	162	543	527	287	39	**542**	100	36	35	35	26	18	13	11		24	14	2	–	–	8	–	–	–	–	–	–	1	Ferrocene, 1,1′-bis(2,3,4,5-tetrafluoro-6-methoxyphenyl)-	55494-18-3	NI45668
155	173	527	252	115	113	139	127	**542**	100	17	14	10	9	9	7	7	**0.00**	24	30	10	–	2	–	–	–	–	–	–	–	–	Methyl 2,3-O-isopropylidene-5,6-di-O-toluene-p-sulphonyl-β-D-allofuranoside		LI09634
45	71	101	191	75	143	115	103	**542**	100	92	88	66	57	55	41	39	**0.00**	24	46	13	–	–	–	–	–	–	–	–	–	–	L-Xylitol, O-2,3,4-tri-O-methyl-β-D-xylopyranosyl-(1-3)-O-2,4-di-O-methyl-β-D-xylopyranosyl-(1-3)-1,2,4,5-tetra-O-methyl-	32581-27-4	NI45669
43	83	109	69	125	41	53	91	**542**	100	99	65	58	41	41	39	26	**0.00**	26	35	10	–	–	–	1	–	–	–	–	–	–	Tigli-1-en-3-one, 12,13,20-tris(acetyloxy)-7-chloro-4,6,9-trihydroxy-, (7α,12β)-		NI45670
544	542	43	41	44	57	56	42	**542**	100	97	94	86	70	55	49	41		27	27	7	–	–	–	–	1	–	–	–	–	–	3-Bromo-1′,4-dihydroxy-2,3′-dimethyl-4′,5,5′,8,8′-pentamethoxy-1,2′-binaphthalene	101459-33-0	NI45671
91	108	92	301	299	107	285	228	**542**	100	13	9	6	6	6	5	5	**0.00**	28	31	7	2	–	–	1	–	–	–	–	–	–	4,3′-Dimethyl-5′-chloromethyl-3,4′-bis[(2-methoxycarbonyl)ethyl]-5-benzyloxycarbonyl-2,2′-bis-pyrrolyl-ketone		LI09635
126	57	58	69	120	91	153	147	**542**	100	92	76	73	60	58	54	43	**10.00**	30	22	10	–	–	–	–	–	–	–	–	–	–	4H-1-Benzopyran-4-one, 8-[5-(3,4-dihydro-5,7-dihydroxy-4-oxo-2H-1-benzopyran-2-yl)-2-hydroxyphenyl]-2,3-dihydro-5,7-dihydroxy-2-(4-hydroxyphenyl)-, [S-(R*,R*)]-	72149-91-8	NI45672
120	147	161	44	91	258	119	94	**542**	100	43	39	35	34	33	27	23	**6.90**	30	22	10	–	–	–	–	–	–	–	–	–	–	Rhusflavanone		MS10611
256	272	271	270	542	257	273	43	**542**	100	21	19	19	17	12	11	10		30	22	10	–	–	–	–	–	–	–	–	–	–	Rugulosin	23537-16-8	NI45673
504	505	252	508	506	449	509	464	**542**	100	45	45	38	36	27	22	22	**12.69**	30	22	10	–	–	–	–	–	–	–	–	–	–	6H,14H,2,4,9,11-Tetraoxadibenzo[bc,kl]coronene-6,14-dione, 1,3,3a,8,10,10a-hexahydro-3a,7,10a,13-tetrahydroxy-1,3,8,10-tetramethyl-, [1R-(1α,3β,3aα,8α,10β,10aα)]-	13861-94-4	NI45674
239	238	240	262	134	106	146	133	**542**	100	78	50	30	25	24	23	21	**0.00**	30	26	–	2	4	–	–	–	–	–	–	–	–	1,4,11,14-Tetrathia-7,18-diaza-5,9,15,19-tetrabenzo-cyclo-5,7,9,15,17,19-eico-hexene		NI45675
380	162	218	133	105	119	147	189	**542**	100	28	23	23	20	12	10	9	**4.00**	30	36	3	3	–	–	–	–	–	–	–	–	1	Iron(III), tris(propylsalicylaldiminato)-		MS10612
196	164	248	150	43	165	192	99	**542**	100	76	74	70	68	41	38	35	**0.00**	30	38	9	–	–	–	–	–	–	–	–	–	–	Methyl arthoniate		LI09636
230	94	229	446	79	81	216	107	**542**	100	38	34	22	22	21	20	20	**0.10**	30	42	5	2	1	–	–	–	–	–	–	–	–	cis-6,6′-Dihydroxythiobinupharidine sulphoxide		NI45676
224	486	132	262	487	183	55	225	**542**	100	87	61	29	28	24	21	13	**0.00**	32	24	3	–	–	–	–	–	–	–	1	–	1	(Benzoylcyclopentadienyl)manganesedicarbonyltriphenylphosphine	102777-72-0	NI45677

MASS TO CHARGE RATIOS	M.W.	INTENSITIES	Parent	C	H	O	N	S	F	Cl	Br	I	Si	P	B	X	COMPOUND NAME	CAS Reg No	No
527 542 47 528 41 57 87 149	**542**	100 83 67 37 36 34 33 32		32	30	8	–	–	–	–	–	–	–	–	–	–	5,5′,6,6′,7,7′,8,8′-Octahydro-9,9′,10,10′-tetramethoxy-2,2′-bianthracene-1,1′,4,4′-tetrone	99968-11-3	NI45678
87 57 43 55 189 170 374 134	**542**	100 33 29 15 9 8 7 7	**5.20**	32	62	4	–	1	–	–	–	–	–	–	–	–	Butanoic acid, 4,4′-thiobis-, didocecyl ester		MS10613
87 57 43 71 55 129 61 41	**542**	100 39 35 19 19 18 18 18	**10.00**	32	62	4	–	1	–	–	–	–	–	–	–	–	Butanoic acid, 4,4′-thiobis-, didocecyl ester		IC04570
57 71 43 69 55 85 178 83	**542**	100 64 62 51 47 40 34 32	**2.00**	32	62	4	–	1	–	–	–	–	–	–	–	–	Propanoic acid, 3,3′-thiobis-, ditridecyl ester		IC04571
73 510 43 44 511 109 84 69	**542**	100 97 61 56 38 36 35 33	**1.00**	33	50	6	–	–	–	–	–	–	–	–	–	–	Lanost-9(11)-en-18-oic acid, 3-(acetyloxy)-20-hydroxy-25-methoxy-16-oxo-,γ-lactone, (3β)-	53534-46-6	NI45679
474 475 421 407 542 476 435 436	**542**	100 43 30 25 23 15 14 10		33	58	2	–	–	–	–	–	–	2	–	–	–	Cannabidiol, bis[(triethylsilyl)oxy]-		NI45680
542 318 273 525 497 274 225 197	**542**	100 58 39 1 1 1 1 1		34	22	7	–	–	–	–	–	–	–	–	–	–	2,4-Bis(3-p-hydroxyphenylphthalid-3-yl)phenol		LI09637
542 318 273 274 225 541 543 317	**542**	100 60 40 35 32 30 29 25		34	22	7	–	–	–	–	–	–	–	–	–	–	1(3H)-Isobenzofuranone, 3,3′-(4-hydroxy-1,3-phenylene)bis[3-(4-hydroxyphenyl)-	55517-70-9	NI45681
151 152 135 240 227 136 225 121	**542**	100 43 22 10 10 10 9 8	**0.30**	34	38	6	–	–	–	–	–	–	–	–	–	–	Benzene, 1,1′,1″,1‴-(1,4-dimethoxy-1,2,3,4-butanetetrayl)tetrakis[methoxy-		MS10614
367 43 57 55 29 41 71 69	**542**	100 81 64 53 52 44 32 31	**10.69**	34	54	5	–	–	–	–	–	–	–	–	–	–	3′H-Cycloprop(1,2)-5α-cholest-1-en-3-one, 1′,1′-dicarboethoxy-1β,2β-dihydro-	75857-79-3	NI45682
43 69 451 55 57 449 41 95	**542**	100 99 70 58 54 48 39 36	**2.00**	34	54	5	–	–	–	–	–	–	–	–	–	–	Lanosta-7,9(11)-diene-3β,18,20-triol, 3,18-diacetate, (20R)-	24041-72-3	NI45683
55 69 80 95 83 109 67 79	**542**	100 84 71 56 55 54 46 43	**4.00**	34	54	5	–	–	–	–	–	–	–	–	–	–	Methyl 3-acetylpolyporenate A		MS10615
262 45 43 203 60 55 69 41	**542**	100 99 82 51 39 36 35 34	**18.00**	34	54	5	–	–	–	–	–	–	–	–	–	–	Methyl-3,23-O-isopropylideneasiatate		MS10616
276 217 207 334 216 190 208 277	**542**	100 70 69 66 54 35 29 28	**5.00**	34	54	5	–	–	–	–	–	–	–	–	–	–	Olean-12-en-28-oic acid, 3-hydroxy-15,16-[isopropylidenebis(oxy)]-, methyl ester, (3β,15α,16α)-	56114-50-2	NI45684
203 262 202 189 133 105 119 107	**542**	100 80 26 25 21 20 19 19	**4.74**	34	54	5	–	–	–	–	–	–	–	–	–	–	Olean-12-en-28-oic acid, 2α,3β,23-trihydroxy-, methyl ester, cyclic 3,23-acetal with acetone	10258-29-4	NI45685
69 43 41 81 44 55 95 57	**542**	100 65 65 62 56 54 32 28	**22.00**	40	62	–	–	–	–	–	–	–	–	–	–	–	ψ,ψ-Carotene, 7,7′,8,8′,11,12-hexahydro-	540-05-6	NI45686
227 330 133 69 303 58 169 116	**543**	100 97 49 47 37 32 32 32	**17.20**	17	11	3	1	–	14	–	–	–	–	–	–	–	(4-Hydroxyphenyl)propylamine, bis(heptafluorobutanoyl)-		NI45687
240 316 169 317 241 103 119 212	**543**	100 81 26 17 7 6 6 6	**0.00**	17	11	3	1	–	14	–	–	–	–	–	–	–	N-Methyltyramine, bis(heptafluorobutanoyl)-		NI45688
304 306 255 107 303 240 257 241	**543**	100 88 88 72 52 52 48 40	**23.10**	22	28	2	2	4	–	–	–	–	–	–	–	1	Bis(2,2-dimethylthio-N-salicylideneethylaminato)copper	90078-39-0	NI45689
543 544 528 529 513 271.5 545 498	**543**	100 37 32 13 13 11 9 8		32	36	1	4	–	–	–	–	–	–	–	–	1	Vanadylaetioporphyrin I		MS10617
371 174 172 145 372 370 389 175	**543**	100 42 42 31 26 17 14 9	**7.00**	33	30	3	3	–	–	–	–	–	–	–	–	1	Aluminium, tris(2,5-dimethyl-8-quinolinato)-		LI09638
207 230 202 115 87 174 59 118	**544**	100 13 13 11 10 8 8 7	**6.70**	8	–	8	–	–	–	–	–	–	–	–	–	3	Bis(tetracarbonylcobalt)mercury		LI09639
202 101 59 115 87 66 143 69	**544**	100 25 5 4 4 3 2 2	**0.30**	8	–	8	–	–	–	–	–	–	–	–	–	3	Bis(tetracarbonylcobalt)mercury		MS10618
230 115 87 174 59 118 262 143	**544**	100 85 77 63 60 58 49 29	**6.70**	8	–	8	–	–	–	–	–	–	–	–	–	3	Bis(tetracarbonylcolbalt)mercury		MS10619
74 75 149 228 230 98 115 114	**544**	100 51 50 50 49 48 44 42	**1.05**	12	5	–	–	–	–	–	5	–	–	–	–	–	Biphenyl, 2,2′,4,5,5′-pentabromo-	67888-96-4	NI45690
74 228 230 149 388 98 390 99	**544**	100 95 91 66 59 58 57 41	**0.72**	12	5	–	–	–	–	–	5	–	–	–	–	–	Biphenyl, 2,2′,4,5′,6-pentabromo-	59080-39-6	NI45691
293 253 119 308 88 380 69 428	**544**	100 75 65 60 25 20 20 1	**1.00**	18	14	6	2	–	10	–	–	–	–	–	–	–	O,N-Bis(pentafluoropropionyl)tyrosylglycine, methyl ester		NI45692
531 546 529 544 181 451 545 466	**544**	100 98 97 95 41 29 23 22		18	21	7	6	1	–	–	1	–	–	–	–	–	N,N-Diethyl-2-methoxy-5-methylsulphonylamino-4-(2,4-dinitro-6-bromophenylazo)aniline		IC04572
57 41 184 240 69 165 55 104	**544**	100 82 38 36 22 19 18 17	**4.00**	18	26	6	2	2	6	–	–	–	–	–	–	–	L-Cystine, N,N′-bis(trifluoroacetyl)-, dibutyl ester	5283-00-1	NI45693
544 277 406 375 129 81 545 227	**544**	100 75 43 42 42 38 33 25		19	3	–	–	–	15	–	–	–	1	–	–	–	Silane, methyltris(pentafluorophenyl)-	18920-98-4	NI45694
212 544 332 264 434 378	**544**	100 21 20 13 3 1		20	44	6	–	–	–	–	–	–	–	2	–	1	Bis(2-methylallyl)bis(triethylphosphite)ruthenium		LI09640
79 208 64 285 206 154 207 283	**544**	100 91 87 50 46 46 41 27	**0.00**	21	20	2	–	1	–	–	–	–	–	–	–	1	Plumbane, triphenyl[(2-propenylsulphinyl)oxy]-	47415-74-7	LI09641
78 208 285 206 154 207 283 48	**544**	100 92 55 48 46 41 31 29	**0.00**	21	20	2	–	1	–	–	–	–	–	–	–	1	Plumbane, triphenyl[(2-propenylsulphinyl)oxy]-	47415-74-7	MS10620
242 256 241 258 132 243 107 260	**544**	100 72 67 56 51 44 44 39	**13.30**	22	28	2	2	4	–	–	–	–	–	–	–	1	Bis(2,2-dimethylthio-N-salicylideneethylaminato)zinc		NI45695
43 242 243 485 171 256 501 383	**544**	100 92 75 31 18 16 15 14	**0.00**	25	28	10	4	–	–	–	–	–	–	–	–	–	Riboflavin, 2′,3′,4′,5′-tetraacetate	752-13-6	NI45696
242 243 486 485 43 383 256 244	**544**	100 99 34 33 28 26 21 19	**5.38**	25	28	10	4	–	–	–	–	–	–	–	–	–	Riboflavin, 2′,3′,4′,5′-tetraacetate	752-13-6	NI45697
460 376 461 375 387 105 418 345	**544**	100 26 25 25 21 19 17 17	**8.00**	26	24	13	–	–	–	–	–	–	–	–	–	–	Flavone, 2′,5,5′,6-tetrahydroxy-3,4′,7-trimethoxy-, tetraacetate	34318-39-3	NI45698
314 243 315 230 231 77 215 139	**544**	100 32 30 20 18 10 7 6	**0.00**	27	18	2	4	–	–	1	1	–	–	–	–	–	2-(4-Chlorophenyl)-6-oxo-8,10-diphenyl-4H,6H-pyrido[2,1-f][1,3,4]oxadiazino[2,3-c][1,2,4]triazin-11-ium bromide		MS10621
77 105 78 51 44 94 118 174	**544**	100 99 57 43 42 27 11 10	**0.00**	27	28	8	–	2	–	–	–	–	–	–	–	–	1,3-Dioxolane-4,5-dimethanol, 2-(2-phenylethenyl)-, bis(4-methylbenzenesulphonate)	75332-40-0	NI45699
460 461 28 458 76 62 41 404	**544**	100 29 18 12 11 7 7 6	**5.05**	27	54	3	–	–	–	–	–	–	–	2	–	1	Iron, tricarbonylbis(tributylphosphine)-, (TB-5-11)-	49655-14-3	NI45700
210 181 151 179 327 167 195 91	**544**	100 50 18 17 15 11 8 8	**0.00**	28	32	9	–	1	–	–	–	–	–	–	–	–	O-Ethylisomelacacidin, tetramethyl ether, 3-(p-toluenesulphonate)		NI45701
227 432 149 375 57 433 105 228	**544**	100 73 49 35 29 23 21 17	**0.00**	28	32	11	–	–	–	–	–	–	–	–	–	–	2-Hydroxy-4-[(3,5-di-tert-butyl-4-hydroxyphenyl)methoxy]benzophenone		IC04573

MASS TO CHARGE RATIOS	M.W.	INTENSITIES	Parent	C	H	O	N	S	F	Cl	Br	I	Si	P	B	X	COMPOUND NAME	CAS Reg No	No
91 92 65 544 240 123 106 148	**544**	100 21 21 20 18 15 15 9		30	28	–	2	4	–	–	–	–	–	–	–	–	Bis(dibenzylthiocarbamoyl) disulphide	10591-85-2	NI45702
110 136 39 162 69 137 55 163	**544**	100 63 27 24 20 16 16 15	**2.80**	32	32	8	–	–	–	–	–	–	–	–	–	–	4,6,10,12,16,18,22,24-Octahydroxy-2,8,14,20-tetramethylpentacyclo-[19.3.1.1^{3,7}.1^{9,13}.1^{15,19}]octacosa-1(25),3,5,7(28),9,11,13,(27),15,17,19(26),2,23-dodecaene		IC04574
117 43 129 45 101 161 87 71	**544**	100 96 83 66 51 47 21 21	**0.00**	33	52	6	–	–	–	–	–	–	–	–	–	–	Cholest-8(14)-ene-3,7,15-triol, triacetate, (3β,5α,7α,15α)-	69140-10-9	NI45703
203 262 484 189 261 453 281 249	**544**	100 63 24 19 18 13 10 10	**2.00**	33	52	6	–	–	–	–	–	–	–	–	–	–	Medicagenic acid, 3,23,28-tri-O-methyl-		NI45704
327 107 135 95 163 205 309 231	**544**	100 99 84 68 34 30 26 14	**0.00**	33	52	6	–	–	–	–	–	–	–	–	–	–	Nepetidone 3,11-diacetate		NI45705
466 203 216 189 135 451 255 526	**544**	100 40 22 22 19 17 16 15	**0.00**	34	56	5	–	–	–	–	–	–	–	–	–	–	3β,11α-Diacetoxy-lupan-20-ol	79081-63-3	NI45706
43 385 29 55 57 161 384 228	**544**	100 66 60 52 52 50 47 46	**5.79**	34	56	5	–	–	–	–	–	–	–	–	–	–	1α-Dicarboethoxymethyl-5α-cholestan-3-one		NI45707
361 43 484 135 263 471 544 193	**544**	100 69 48 33 29 20 18 17		34	56	5	–	–	–	–	–	–	–	–	–	–	20-Epilanostan-11-one, 3β,18-dihydroxy-, diacetate		LI09642
361 43 484 69 263 135 95 362	**544**	100 52 49 42 35 30 29 28	**24.75**	34	56	5	–	–	–	–	–	–	–	–	–	–	Lanostan-11-one, 3β,18-dihydroxy-, diacetate	24041-78-9	NI45708
361 43 484 263 135 544 471 175	**544**	100 53 40 35 30 25 20 14		34	56	5	–	–	–	–	–	–	–	–	–	–	Lanostan-11-one, 3β,18-dihydroxy-, diacetate	24041-78-9	LI09643
361 43 69 484 95 81 55 93	**544**	100 68 57 48 43 41 36 33	**16.83**	34	56	5	–	–	–	–	–	–	–	–	–	–	Lanostan-11-one, 3β,18-dihydroxy-, diacetate, (20S)-	24041-77-8	NI45709
232 257 163 233 258 393 452 180	**544**	100 43 42 31 23 21 13 13	**0.00**	35	32	4	2	–	–	–	–	–	–	–	–	–	5,10-Azo-1,4-etheno-6,9-methanobenzo[b]biphenylene-2,3-dicarboxylic acid, 1,4,4a,4b,5,5a,6,9,9a,10,10a,10b-dodecahydro-5,10-diphenyl-, dimethyl ester	56083-45-5	NI45710
339 340 205 191 544 353 323 206	**544**	100 27 19 11 10 5 5 4		37	52	3	–	–	–	–	–	–	–	–	–	–	1,1,3-Tris(5-tert-butyl-4-hydroxy-2-methylphenyl)butane	89552-71-6	IC04575
339 340 57 544 205 191 545 338	**544**	100 29 22 20 19 14 8 7		37	52	3	–	–	–	–	–	–	–	–	–	–	1,1,3-Tris(5-tert-butyl-4-hydroxy-2-methylphenyl)butane	89552-71-6	NI45711
375 169 205 376 374 141 140 360	**544**	100 93 40 28 24 24 20 10	**6.00**	38	28	2	2	–	–	–	–	–	–	–	–	–	(1R*,7R*)-4-Methyl-2-[N-(1-naphthyl)carbamoyl]-6,6-diphenyl-2-az[8.9]tricyclo[5.2.2.0^{1,5}]undeca-4,8,10-trien-3-one		MS10622
69 544 81 41 55 57 141 545	**544**	100 69 60 58 47 42 40 36		40	64	–	–	–	–	–	–	–	–	–	–	–	ψ,ψ-Carotene, 7,7′,8,8′,11,11′,12,12′-octahydro-	540-04-5	NI45712
181 91 453 272 142 364 166 115	**544**	100 67 42 41 41 39 27 18	**12.00**	42	40	–	–	–	–	–	–	–	–	–	–	–	3′,4′,5′,6′,7′,8′-Hexahydro-1′-phenyl-2,5′,6-tris[(E)-phenylmethylene]spiro[cyclohexane-1,2′(1′H)-naphthalene]		MS10623
69 349 377 169 150 545	**545**	100 36 16 15 12 1		8	1	4	3	–	18	–	–	–	–	–	–	–	Acetamide, N,N-bis(trifluoromethyl)-2,2-bis[bis(trifluoromethyl)aminooxy]-		LI09644
236 73 530 208 310 192 337 75	**545**	100 85 35 35 30 30 23 23	**4.00**	19	36	6	5	–	–	–	–	–	3	1	–	–	Adenosine, N-(trimethylsilyl)-2′-O-(trimethylsilyl)-, cyclic 3′,5′-[trimethylsilyl phosphate]	32645-64-0	NI45713
412 36 293 411 413 283 294 38	**545**	100 75 72 59 29 28 25 25	**0.00**	28	20	–	4	–	–	2	–	–	–	–	–	1	Copper, [6,18-dichloro-5,6,17,18-tetrahydrotetrabenzo[b,f,j,n][1,5,9,13]tetraazacyclohexadecinato(2-) N^5,N^{11},N^{17},N^{23}]-, (SP-4-1)-	69815-52-7	NI45714
298 299 545 266 205 150 315 314	**545**	100 23 18 11 11 9 6 6		29	39	9	1	–	–	–	–	–	–	–	–	–	Homoharringtonine		MS10624
243 244 167 105 302 183 59 43	**545**	100 30 20 19 8 7 6 6	**0.10**	32	35	7	1	–	–	–	–	–	–	–	–	–	5-Isoxazolidinecarboxylic acid, 2-[2,3-O-isopropylidene-5-O-(triphenylmethyl)-β-D-ribofuranosyl]-, methyl ester, (S)-	64018-64-0	NI45715
552 550 554 548 232 556 72 44	**546**	100 77 73 32 31 30 27 24	**5.00**	6	–	–	–	–	–	–	6	–	–	–	–	–	Benzene, hexabromo-	87-82-1	NI45716
347 345 349 346 527 79 529 343	**546**	100 99 98 93 85 84 81 78	**0.00**	6	18	–	–	3	–	–	–	–	–	–	–	3	2,2,4,4,6,6-Hexamethyl-1,3,5-trithia-2,4,6-tristannacyclohexane		NI45717
28 253 223 251 208 252 506 221	**546**	100 74 51 36 36 31 30 28	**0.00**	7	18	–	2	–	–	–	–	–	–	–	–	2	Bis(trimethyllead)diazomethane		LI09645
333 331 249 247 329 291 131 546	**546**	100 75 56 44 41 41 34 33		12	28	4	–	4	–	–	–	–	–	2	–	1	Tin[II] bis(dipropyldithiophosphate)	88478-79-9	NI45718
517 518 519 59 55 57 520 56	**546**	100 54 39 27 17 16 14 12	**0.79**	14	38	9	–	–	–	–	–	–	7	–	–	–	1,3,5,7,9,11,13-Heptaethyl-5,9,13-trihydrotricyclo[5.5.1.3^{3,11}]heptasiloxane	80177-43-1	NI45719
131 435 187 437 379 323 289 381	**546**	100 64 62 49 44 39 36 35	**22.00**	16	36	4	–	4	–	–	–	–	–	2	–	1	Zinc[II] bis(dibutyldithiophosphate)	24645-45-2	NI45720
73 43 327 74 487 341 147 41	**546**	100 27 15 8 5 5 5 5	**0.00**	18	50	7	–	–	–	–	–	–	6	–	–	–	3,5-Diisopropyloxy-1,1,1,7,7,7-hexamethyl-3,5-bis(trimethylsiloxy)tetrasiloxane		MS10625
235 233 303 301 248 246 548 550	**546**	100 90 78 75 27 26 12 6	**6.00**	20	28	4	4	–	–	–	2	–	–	–	–	–	3,3′-(1,6-Hexanediyl)bis[1-(2-bromoethyl)-6-methyl-2,4(1H,3H)-pyrimidinedione]		MS10626
43 171 159 202 546	**546**	100 19 15 11 4		21	25	2	–	–	–	1	–	–	–	–	–	1	Mercury, chloro(3,20-dioxopregna-1,4,6-trien-2-yl)-	36794-32-8	NI45721
43 202 171 310 159 546	**546**	100 71 27 26 21 5		21	25	2	–	–	–	1	–	–	–	–	–	1	Mercury, chloro(3,20-dioxopregna-1,4,6-trien-2-yl)-	36794-32-8	NI45722
527 546 206 366 528 526 121 547	**546**	100 95 68 63 49 40 37 25		24	12	–	–	–	10	–	–	–	–	–	–	1	Ferrocene, 1,1′-bis[(pentafluorophenyl)methyl]-	55494-16-1	NI45723
105 149 218 205 281 444 343 504	**546**	100 98 42 33 18 10 9 6	**2.00**	28	34	11	–	–	–	–	–	–	–	–	–	–	1α-Benzoyloxy-6-β,9α,15-triacetoxy-8-oxo-4β-hydroxydihyd ro-β-agarofuran		MS10627
371 161 207 133 341 223 134 370	**546**	100 27 13 13 12 11 7 6	**4.50**	29	38	10	–	–	–	–	–	–	–	–	–	–	2,4,4′,6-Tetramethyl-3-(tetramethyl-6-C-glucosyl)chalcone		NI45724
281 531 358 57 546 75 532 456	**546**	100 83 56 53 51 44 38 33		30	50	5	–	–	–	–	–	–	2	–	–	–	Pregna-2,4-dien-20-one, 17-(acetyloxy)-6-methyl-3,6-bis[(trimethylsilyl)oxy]-, (6β)-	69833-57-4	NI45725

MASS TO CHARGE RATIOS								M.W.	INTENSITIES								Parent	C	H	O	N	S	F	Cl	Br	I	Si	P	B	X	COMPOUND NAME	CAS Reg No	No
283	414	399	284	486	73	371	357	**546**	100	39	39	26	24	23	18	18	**6.07**	30	50	5	–	–	–	–	–	–	2	–	–	–	Pregna-3,5-dien-20-one, 17-(acetyloxy)-6-methyl-3,21-bis[(trimethylsilyl)oxy]-	74298-95-6	NI45726
283	373	145	374	256	284	137	138	**546**	100	95	64	29	28	24	20	18	**3.66**	30	50	5	–	–	–	–	–	–	2	–	–	–	Pregna-3,5-dien-20-one, 21-(acetyloxy)-6-methyl-3,17-bis[(trimethylsilyl)oxy]-	74299-09-5	NI45727
433	432	43	434	195	57	431	69	**546**	100	73	35	28	23	21	18	17	**8.62**	30	57	6	–	–	–	–	–	–	–	–	3	–	allo-Inositol tri-octaneboronate		NI45728
433	432	43	434	335	334	57	41	**546**	100	65	38	26	26	20	20	18	**12.28**	30	57	6	–	–	–	–	–	–	–	–	3	–	cis-Inositol tri-octaneboronate		NI45729
195	433	432	43	194	81	293	83	**546**	100	55	35	28	23	23	20	16	**5.62**	30	57	6	–	–	–	–	–	–	–	–	3	–	myo-Inositol tri-octaneboronate		NI45730
195	433	293	43	432	81	194	546	**546**	100	41	34	31	28	24	23	22		30	57	6	–	–	–	–	–	–	–	–	3	–	muco-Inositol tri-octaneboronate		NI45731
195	546	351	335	167	130	143	416	**546**	100	68	30	22	13	10	8	3		31	34	7	2	–	–	–	–	–	–	–	–	–	O-(3,4,5-Trimethoxybenzoyl)quebrachidine		LI09646
43	426	380	531	546	457	105	77	**546**	100	36	31	23	20	19	19	17		33	26	2	2	2	–	–	–	–	–	–	–	–	2,5-Dimethoxy-5,6,8,8b-tetraphenyl-5H,8bH-thieno[3′,4′:3,4]pyrrolo[2,1-b]-1,3,4-thiadiazole	110315-31-6	DD01785
163	257	232	258	180	77	194	78	**546**	100	56	41	30	24	21	14	14	**0.50**	33	30	4	4	–	–	–	–	–	–	–	–	–	5,10-Azo-1,4-etheno-6,9-methanobenzo[b]biphenylene-2,3-dicarboxylic acid, 1,4,4a,4b,5,5a,6,9,9a,10,10a,10b-dodecahydro-5,10-di-2-pyridinyl-, dimethyl ester	56083-44-4	NI45732
197	375	180	195	259	531	181	376	**546**	100	98	78	66	51	48	48	36	**0.60**	33	34	2	–	–	–	–	–	–	3	–	–	–	Trisiloxane, 1,3,5-trimethyl-1,1,3,5,5-pentaphenyl-		NI45733
410	470	397	377	411	471	398	339	**546**	100	75	74	50	32	26	24	20	**0.00**	33	54	6	–	–	–	–	–	–	–	–	–	–	Cholestane-2,3,19-triol, triacetate, (2α,3β,5α)-	55570-88-2	NI45734
305	57	71	70	332	193	113	435	**546**	100	76	52	42	34	32	31	28	**0.00**	33	54	6	–	–	–	–	–	–	–	–	–	–	Tri(2-ethylhexyl) trimellitate		IC04576
57	305	71	113	43	435	55	69	**546**	100	78	67	44	43	32	32	28	**0.00**	33	54	6	–	–	–	–	–	–	–	–	–	–	Triisooctyl trimellitate		IC04577
456	403	441	143	142	531	546	194	**546**	100	85	72	36	34	32	22	19		33	62	2	–	–	–	–	–	–	2	–	–	–	Cholest-4-ene, 3,6-bis[(trimethylsilyl)oxy]-, (3α,6β)-		NI45735
456	403	441	194	546	143	142	531	**546**	100	54	43	20	19	17	14	9		33	62	2	–	–	–	–	–	–	2	–	–	–	Cholest-4-ene, 3,6-bis[(trimethylsilyl)oxy]-, (3β,6α)-		NI45736
73	456	403	75	57	43	147	55	**546**	100	99	99	99	85	80	68	60	**34.00**	33	62	2	–	–	–	–	–	–	2	–	–	–	Cholest-4-ene, 3,6-bis[(trimethylsilyl)oxy]-, (3β,6β)-	33403-39-3	NI45737
403	456	441	531	143	546	194	142	**546**	100	78	35	22	21	19	19	19		33	62	2	–	–	–	–	–	–	2	–	–	–	Cholest-4-ene, 3,6-bis[(trimethylsilyl)oxy]-, (3β,6β)-	33403-39-3	NI45738
73	147	43	57	129	55	366	71	**546**	100	99	89	69	54	42	39	38	**13.00**	33	62	2	–	–	–	–	–	–	2	–	–	–	Cholest-5-ene, 3,4-bis[(trimethylsilyl)oxy]-, (3β,4β)-	33287-25-1	NI45739
456	73	457	75	43	95	81	57	**546**	100	63	38	25	15	13	13	13	**3.00**	33	62	2	–	–	–	–	–	–	2	–	–	–	Cholest-5-ene, 3,7-bis[(trimethylsilyl)oxy]-, (3β,7β)-	33287-26-2	NI45740
456	457	73	75	81	55	95	43	**546**	100	32	32	18	10	10	9	9	**2.50**	33	62	2	–	–	–	–	–	–	2	–	–	–	Cholest-5-ene, 3,7-bis[(trimethylsilyl)oxy]-, (3β,7β)-	33287-26-2	NI45741
129	73	149	75	71	69	109	83	**546**	100	99	92	90	90	90	80	80	**5.00**	33	62	2	–	–	–	–	–	–	2	–	–	–	Cholest-5-ene, 3,25-bis[(trimethylsilyl)oxy]-, (3β)-	33287-27-3	MS10628
131	73	75	129	456	271	132	327	**546**	100	38	29	20	14	14	13	11	**7.01**	33	62	2	–	–	–	–	–	–	2	–	–	–	Cholest-5-ene, 3,25-bis[(trimethylsilyl)oxy]-, (3β)-	33287-27-3	NI45742
131	75	129	132	271	456	327	133	**546**	100	17	16	11	10	7	7	6	**2.30**	33	62	2	–	–	–	–	–	–	2	–	–	–	Cholest-5-ene, 3β,24α-bis(trimethylsilyloxy)-		MS10629
145	413	159	129	414	503	327	323	**546**	100	80	48	32	29	24	19	16	**6.50**	33	62	2	–	–	–	–	–	–	2	–	–	–	Cholest-5-ene, 3β,24α-bis(trimethylsilyloxy)-		MS10630
159	517	103	131	518	160	171	519	**546**	100	38	38	30	17	15	12	8	**1.46**	33	62	2	–	–	–	–	–	–	2	–	–	–	Pregn-5-en, 3,20-bis[(trimethylsilyl)oxy]-, (3β,20α)-		NI45743
43	41	29	57	461	135	107	241	**546**	100	54	42	38	36	31	31	29	**8.00**	36	51	2	–	–	–	–	–	–	–	1	–	–	P,P-Bis(4-nonylphenoxy)phenylphosphine		IC04578
273	274	546	197	469	271	55	57	**546**	100	92	86	65	53	52	45	43		38	26	–	–	2	–	–	–	–	–	–	–	–	10,10′-Bis(10-phenyl-9,10-dihydro-9-thiaanthryl)		LI09647
273	274	77	197	78	198	165		**546**	100	95	84	68	52	44	35		**0.00**	38	26	–	–	2	–	–	–	–	–	–	–	–	9,9′-Bis(10-phenyl-9-thiaanthryl)		LI09648
57	43	83	82	55	71	41	69	**546**	100	94	76	73	72	56	44	40	**0.06**	39	78	–	–	–	–	–	–	–	–	–	–	–	17-Cyclohexyltritriacontane		AI02405
83	82	71	69	97	85	67	111	**546**	100	69	48	41	40	32	22	17	**0.00**	39	78	–	–	–	–	–	–	–	–	–	–	–	17-Cyclohexyltritriacontane		AI02406
69	109	83	43	546	133	91	409	**546**	100	12	5	4	3	3	3	2		40	66	–	–	–	–	–	–	–	–	–	–	–	ψ,ψ-Carotene, 7,7′,8,8′,11,11′,12,12′,15,15′-decahydro-	502-62-5	LI09649
69	81	137	95	121	41	109	93	**546**	100	58	19	18	15	15	12	12	**3.10**	40	66	–	–	–	–	–	–	–	–	–	–	–	ψ,ψ-Carotene, 7,7′,8,8′,11,11′,12,12′,15,15′-decahydro-	502-62-5	MS10631
69	41	43	546	83	55	85	68	**546**	100	60	44	42	32	30	26	21		40	66	–	–	–	–	–	–	–	–	–	–	–	ψ,ψ-Carotene, 7,7′,8,8′,11,11′,12,12′,15,15′-decahydro-	502-62-5	NI45744
349	350	348	276	277	163	248	190	**547**	100	30	29	28	27	22	20	20	**0.00**	23	18	10	1	–	–	–	1	–	–	–	–	–	6,7-Diacetoxy-2-formyl-1,2,3,4-tetrahydroisoquinoline-1-spiro-2′-1′-bromo 6′,7′-methylenedioxy-3′-oxo-2′,3′-dihydroindene		NI45745
105	122	77	133	239	211	177	164	**547**	100	62	38	19	10	5	2	2	**0.00**	27	25	8	5	–	–	–	–	–	–	–	–	–	Benzamide, N-[6-[2-O-acetyl-5-O-benzoyl-3-C-(hydroxymethyl)-β-D-xylofuranosyl]-9H-purin-9-yl]-	75314-21-5	NI45746
400	310	548	336	426	549	129	516	**547**	100	97	90	86	50	44	36	35	**0.00**	28	49	4	1	–	–	–	–	–	3	–	–	–	Estra-1,3,5(10)-trien-6-one, 3,16,17-tris[(trimethylsilyl)oxy]-, O-methyloxime, (16α,17β)-	69855-54-5	NI45747
91	236	237	392	65	238	208	92	**547**	100	74	54	42	36	23	22	18	**5.30**	29	29	4	3	2	–	–	–	–	–	–	–	–	3,6-Bis(N-methyl-N-tosylamino)-9-methylcarbazole	98785-96-7	NI45748
180	77	119	371	427	428	399	547	**547**	100	80	40	11	10	6	4	1		36	25	3	3	–	–	–	–	–	–	–	–	–	7,7a-Dihydro-1,3,4,7,7a-pentaphenyl-1H-pyrrolo[2,3-D]pyrimidin-2,5,6(3H)trione	90140-26-4	NI45749
547	548	91	549	546	57	55	532	**547**	100	44	20	10	5	4	3	2		40	53	–	1	–	–	–	–	–	–	–	–	–	N-Benzylcholest-5-eno[3,2-b]indole		MS10632
533	534	535	253	521	258	536	254	**548**	100	47	37	27	21	14	14	10	**0.99**	9	24	12	–	–	–	–	–	–	8	–	–	–	Heptamethylvinyloctasilsesquioxane		NI45750
69	141	154	265	97	252	45	57	**548**	100	57	25	21	17	16	13	12	**0.00**	14	8	9	–	–	12	–	–	–	–	–	–	–	α-L-Mannopyranose, 6-deoxy-, tetrakis(trifluoroacetate)	49561-09-3	MS10633
323	380	260	548	520	406	209	462	**548**	100	80	60	24	12	12	10	4		16	27	4	–	–	–	–	–	–	–	–	–	2	Tetracarbonyl(tributylbismuthine)iron	90188-74-2	NI45751
117	365	296	69	265	548	167	183	**548**	100	92	46	45	37	36	35	34		18	–	1	–	–	15	–	–	–	–	1	–	–	Tris(pentafluorophenyl)phosphine oxide		MS10634

MASS TO CHARGE RATIOS								M.W.	INTENSITIES								Parent	C	H	O	N	S	F	Cl	Br	I	Si	P	B	X	COMPOUND NAME	CAS Reg No	No
439	481	453	440	497	394	86	69	**548**	100	20	17	14	9	8	7	6	**2.00**	18	15	2	–	–	15	–	–	–	–	–	–	–	3-Perfluorooctanoylcamphor		NI45752
201	43	141	202	169	113	99	115	**548**	100	70	42	12	11	11	10	7	**0.00**	24	36	14	–	–	–	–	–	–	–	–	–	–	L-erythro-Hexonic acid, 2,4,6-trideoxy-3-O-β-D-glucopyranosyl-, ethyl ester, pentaacetate	33275-51-3	NI45753
70	91	57	44	405	88	213	41	**548**	100	86	60	51	43	35	33	33	**13.00**	27	40	8	4	–	–	–	–	–	–	–	–	–	Glycine, N-[N-[2-[1-[2-[(butoxycarbonyl)amino]-1-oxopropyl]-2-pyrrolidinyl]-2-oxo-1-phenylethyl]-L-seryl]-, ethyl ester	55781-94-7	NI45754
105	164	149	106	189	246	488	306	**548**	100	61	35	22	20	6	5	2	**0.00**	28	36	11	–	–	–	–	–	–	–	–	–	–	1α-Benzoyloxy-8β,9α,15-triacetoxy-4β,6β-dihydroxydihydro-β-agarofuran		MS10635
43	83	69	91	109	79	77	105	**548**	100	63	42	31	28	17	17	15	**0.15**	28	36	11	–	–	–	–	–	–	–	–	–	–	5,12β,13,20-Tetraacetoxy-4,9-dihydroxy-1,6-tigliadien-3-one		NI45755
43	69	83	28	45	41	91	15	**548**	100	23	22	22	16	15	14	13	**0.14**	28	36	11	–	–	–	–	–	–	–	–	–	–	7β,12β,13,20-Tetraacetoxy-4,9-dihydroxy-1,5-tigliadien-3-one		NI45756
151	111	123	236	487	472	193	147	**548**	100	81	75	72	68	59	55	48	**1.20**	28	37	7	2	–	–	1	–	–	–	–	–	–	4,5-Deoxymaytansinol		NI45757
113	128	129	29	114	43	41	39	**548**	100	53	8	8	7	7	7	7	**0.00**	29	40	10	–	–	–	–	–	–	–	–	–	–	Proceroside	25323-74-4	NI45758
73	103	89	75	188	91	427	74	**548**	100	99	90	80	70	51	48	46	**43.00**	29	52	4	2	–	–	–	–	–	2	–	–	–	Pregn-4-ene-3,20-dione, 11β,21-bis(trimethylsiloxy)-, bis(O-methyloxime)	32206-60-3	NI45759
73	517	361	518	548	75	188	158	**548**	100	99	65	62	60	41	40	40		29	52	4	2	–	–	–	–	–	2	–	–	–	Pregn-4-ene-3,20-dione, 11,17-bis[(trimethylsilyl)oxy]-, bis(O-methyloxime), (11β)-	57305-37-0	NI45760
105	43	77	106	269	366	177	51	**548**	100	16	12	7	3	2	2	2	**0.00**	30	28	10	–	–	–	–	–	–	–	–	–	–	DL-Xylitol, 2,3-diacetate 1,4,5-tribenzoate	74793-67-2	NI45761
43	91	81	120	87	107	67	44	**548**	100	76	74	67	65	51	47	46	**0.00**	30	44	9	–	–	–	–	–	–	–	–	–	–	Carda-4,20(22)-dienolide, 3-[(6-deoxy-3-O-methyl-α-D-allopyranosyl)oxy]1,14-dihydroxy-, (1β,3β)-	56701-08-7	NI45762
340	87	160	91	43	131	95	79	**548**	100	88	80	61	60	43	43	43	**0.10**	30	44	9	–	–	–	–	–	–	–	–	–	–	Card-20(22)-enolide, 3-[(2,6-dideoxy-3-O-methyl-β-D-ribo-hexopyranosyl)oxy]-5,14-dihydroxy-19-oxo-, (3β,5β)-	508-77-0	NI45763
340	87	160	91	43	145	131	79	**548**	100	88	81	62	62	49	47	45	**0.10**	30	44	9	–	–	–	–	–	–	–	–	–	–	Cymarin		LI09650
43	84	70	279	42	287	44	56	**548**	100	47	36	32	32	30	30	29	**11.21**	31	40	5	4	–	–	–	–	–	–	–	–	–	7-Oxa-15,20,24,27-tetraazatetracyclo[13.9.6.2^{8,11}.1^{2,6}]tritriaconta-2,4,6(33),8,10,12,31-heptaene-14,26-dione, 20-acetyl-5-methoxy-, [S-(Z)]-	40135-65-7	NI45764
168	548	126	105	549	43	520	160	**548**	100	79	57	55	29	24	22	22		31	48	4	–	2	–	–	–	–	–	–	–	–	5α-Spirostan-15-one, 3β-hydroxy-, cyclic ethylene mercaptole, acetate, (25R)-	24742-80-1	NI45765
386	159	253	105	107	211	157	145	**548**	100	71	52	44	40	34	34	33	**0.00**	31	48	8	–	–	–	–	–	–	–	–	–	–	Cholan-24-oic acid, 3,6,7-tris(acetyloxy)-, methyl ester, (3α,5β,6α,7α)-	2616-70-8	NI45766
386	159	105	253	107	368	121	119	**548**	100	36	32	30	28	24	18	18	**0.00**	31	48	8	–	–	–	–	–	–	–	–	–	–	Cholan-24-oic acid, 3,6,7-tris(acetyloxy)-, methyl ester, (3α,5β,6β,7α)-	60354-37-2	NI45767
386	159	105	253	107	119	146	115	**548**	100	79	37	30	30	25	23	22	**0.00**	31	48	8	–	–	–	–	–	–	–	–	–	–	Cholan-24-oic acid, 3,6,7-tris(acetyloxy)-, methyl ester, (3α,5β,6β,7β)-	60354-38-3	NI45768
253	313	105	145	107	368	131	428	**548**	100	63	41	35	32	31	29	28	**0.00**	31	48	8	–	–	–	–	–	–	–	–	–	–	Cholan-24-oic acid, 3,7,12-tris(acetyloxy)-, methyl ester, (3α,5α,7α,12α)-	60384-35-2	NI45769
253	105	159	143	157	145	131	254	**548**	100	38	35	32	28	27	25	24	**0.00**	31	48	8	–	–	–	–	–	–	–	–	–	–	Cholan-24-oic acid, 3,7,12-tris(acetyloxy)-, methyl ester, (3α,5β,7α,12α)-	6818-44-6	NI45770
253	368	313	428	353	226	211	281	**548**	100	48	42	26	21	20	18	9	**1.00**	31	48	8	–	–	–	–	–	–	–	–	–	–	Cholan-24-oic acid, 3,7,12-tris(acetyloxy)-, methyl ester, (3α,5β,7α,12α)-	6818-44-6	NI45771
253	105	313	145	368	254	159	119	**548**	100	40	37	34	27	27	27	25	**0.00**	31	48	8	–	–	–	–	–	–	–	–	–	–	Cholan-24-oic acid, 3,7,12-tris(acetyloxy)-, methyl ester, (3α,5β,7β,12α)-	60384-32-9	NI45772
253	368	313	314	428	254	226	353	**548**	100	47	40	26	22	22	19	18	**0.00**	31	48	8	–	–	–	–	–	–	–	–	–	–	Cholan-24-oic acid, 3,7,12-tris(acetyloxy)-, methyl ester, (3α,7α,12α)-	56085-37-1	NI45773
253	104	159	254	157	143	145	130	**548**	100	42	36	30	30	29	27	26	**0.00**	31	48	8	–	–	–	–	–	–	–	–	–	–	Cholan-24-oic acid, 3,7,12-tris(acetyloxy)-, methyl ester, (3β,5β,7α,12α)-	60384-31-8	NI45774
428	418	429	255	201	213	374	228	**548**	100	43	37	35	31	28	16	15	**0.00**	31	48	8	–	–	–	–	–	–	–	–	–	–	Cholan-24-oic acid, 3,7,23-tris(acetyloxy)-, methyl ester, (3α,7α)-	56087-14-0	NI45775
228	302	246	242	260	548	488	530	**548**	100	70	60	50	20	2	2	1		32	36	8	–	–	–	–	–	–	–	–	–	–	Artanomaloide		MS10636
199	530	248	498	217	280	467	548	**548**	100	55	52	51	38	11	10	1		33	56	6	–	–	–	–	–	–	–	–	–	–	3β,21β,28-Trihydroxy-2β,16β,23-trimethoxyolean-12-ene		MS10637
458	403	368	321	129	228	213	255	**548**	100	84	73	49	31	27	26	25	**2.00**	33	64	2	–	–	–	–	–	–	2	–	–	–	Cholestane, 3,6-bis[(trimethylsilyl)oxy]-, (3α,5β,6β)-		NI45776
204	369	368	458	191	329	129	161	**548**	100	94	78	72	70	65	45	42	**28.00**	33	64	2	–	–	–	–	–	–	2	–	–	–	Cholestane, 3,6-bis[(trimethylsilyl)oxy]-, (3β,5α,6α)-		NI45777
368	458	403	228	321	353	213	215	**548**	100	66	37	37	24	21	21	18	**3.00**	33	64	2	–	–	–	–	–	–	2	–	–	–	Cholestane, 3,6-bis[(trimethylsilyl)oxy]-, (3β,5α,6β)-		NI45778
458	368	369	228	304	321	318	305	**548**	100	67	64	40	31	20	20	20	**11.00**	33	64	2	–	–	–	–	–	–	2	–	–	–	Cholestane, 3,6-bis[(trimethylsilyl)oxy]-, (3β,5α,6β)-		NI45779
458	368	73	75	369	459	95	321	**548**	100	84	67	54	43	42	41	32	**5.50**	33	64	2	–	–	–	–	–	–	2	–	–	–	Cholestane, 3,6-bis[(trimethylsilyl)oxy]-, (3β,5β,6β)-	33403-38-2	NI45780
458	368	321	129	353	213	329	143	**548**	100	96	59	52	41	39	32	24	**7.00**	33	64	2	–	–	–	–	–	–	2	–	–	–	Cholestane, 3,6-bis[(trimethylsilyl)oxy]-, (3β,5β,6β)-	33403-38-2	NI45781
129	73	75	329	95	368	81	43	**548**	100	75	70	65	50	42	42	40	**0.00**	33	64	2	–	–	–	–	–	–	2	–	–	–	Cholestane, 3β,5α-bis(trimethylsilyloxy)-	33283-15-7	NI45782
129	329	111	73	368	75	57	43	**548**	100	76	71	67	60	51	37	36	**0.00**	33	64	2	–	–	–	–	–	–	2	–	–	–	Cholestane, 3β,5α-bis(trimethylsilyloxy)-	33283-15-7	NI45783
103	145	491	283	159	75	492	117	**548**	100	67	66	59	39	31	29	25	**0.00**	33	64	2	–	–	–	–	–	–	2	–	–	–	Pregnane, 3,20-bis[(tert-butyldimethylsilyl)oxy]-, (3α,5β,20β)-		NI45784
491	75	492	103	159	493	283	117	**548**	100	55	45	41	37	16	15	11	**0.00**	33	64	2	–	–	–	–	–	–	2	–	–	–	Pregnane, 3,20-bis[(tert-butyldimethylsilyl)oxy]-, (3β,5α,20α)-		NI45785
103	491	75	159	492	283	493	104	**548**	100	60	36	29	27	12	10	9	**0.00**	33	64	2	–	–	–	–	–	–	2	–	–	–	Pregnane, 3,20-bis[(tert-butyldimethylsilyl)oxy]-, (3β,5α,20β)-		NI45786
131	519	159	520	283	103	521	132	**548**	100	80	64	35	34	26	13	12	**0.00**	33	64	2	–	–	–	–	–	–	2	–	–	–	Pregnane, 3,20-bis[(triethylsilyl)oxy]-, (3α,5β,20β)-		NI45787
519	159	103	520	131	521	160	283	**548**	100	76	65	40	30	15	9	8	**0.00**	33	64	2	–	–	–	–	–	–	2	–	–	–	Pregnane, 3,20-bis[(triethylsilyl)oxy]-, (3β,5α,20α)-		NI45788
131	519	159	103	520	521	132	283	**548**	100	80	52	37	31	12	12	10	**0.00**	33	64	2	–	–	–	–	–	–	2	–	–	–	Pregnane, 3,20-bis[(triethylsilyl)oxy]-, (3β,5α,20β)-		NI45789
548	166	165	167	549	152	530	104	**548**	100	83	73	49	40	22	16	10		36	24	4	2	–	–	–	–	–	–	–	–	–	N,N'-Bis(diphenylmethyl)pyromellitimide		LI09651
548	166	165	167	549	152	530	550	**548**	100	83	73	49	40	22	16	10		36	24	4	2	–	–	–	–	–	–	–	–	–	N,N'-Bis(diphenylmethyl)pyromellitimide		MS10638
516	517	488	489	544	518	236	213	**548**	100	67	47	25	17	17	12	12	**8.33**	37	72	2	–	–	–	–	–	–	–	–	–	–	Cyclopropanepentadecanoic acid, 2-octadecyl-, methyl ester	55517-72-1	NI45790
548	457	366	189	281				**548**	100	91	27	22	17					43	32	–	–	–	–	–	–	–	–	–	–	–	Bis(10-benzylanthracen-9-yl)methane		LI09652

MASS TO CHARGE RATIOS								M.W.	INTENSITIES								Parent	C	H	O	N	S	F	Cl	Br	I	Si	P	B	X	COMPOUND NAME	CAS Reg No	No
84	28	64	44	63	56	30	55	**549**	100	74	60	57	53	47	43	39	**0.00**	22	23	8	5	2	–	–	–	–	–	–	–	–	Cyclohexanamine, N-[2,2-bis[(2,4-dinitrophenyl)thio]ethenyl]-N-ethyl-	42362-44-7	NI45791
91	155	65	211	394	549			**549**	100	60	20	18	13	8				24	27	6	3	3	–	–	–	–	–	–	–	–	1,3,5-Tris-p-tolylsulphonylhexahydro-1,3,5-triazine		LI09653
252	297	250	223	298	73	253	209	**549**	100	72	32	29	24	22	20	9	**0.00**	27	47	5	1	–	–	–	–	–	3	–	–	–	Denopamine-M3 isomer 1, O,O,O-tris(trimethylsilyl)-		NI45792
252	297	250	223	298	73	253	251	**549**	100	66	47	32	24	24	19	12	**0.00**	27	47	5	1	–	–	–	–	–	3	–	–	–	Denopamine-M3 isomer 2, O,O,O-tris(trimethylsilyl)-		NI45793
252	297	250	223	73	298	253	75	**549**	100	55	41	31	23	21	20	12	**0.00**	27	47	5	1	–	–	–	–	–	3	–	–	–	Denopamine-M3 isomer 3, O,O,O-tris(trimethylsilyl)-		NI45794
252	297	250	223	298	73	253	209	**549**	100	88	47	35	29	29	20	14	**0.00**	27	47	5	1	–	–	–	–	–	3	–	–	–	Denopamine-M3, O,O,O-tris(trimethylsilyl)-		NI45795
459	390	356	431	549	369	460	129	**549**	100	61	49	47	45	44	40	38		27	47	5	1	–	–	–	–	–	3	–	–	–	Estriol, tris-O-trimethylsilyl-4-nitro-		NI45796
142	127	98	129	128	105	95	135	**549**	100	84	80	75	71	68	67	66	**2.00**	29	39	4	7	–	–	–	–	–	–	–	–	–	N-Methylpyrrolidin-2-on-5-yloxo-N,N′-dimethylhistidyl-N,N′-dimethyltryptophan dimethylamide		NI45797
112	214	549	260	213	261	232	157	**549**	100	23	17	17	12	11	10	9		32	31	4	5	–	–	–	–	–	–	–	–	–	4-(1,3-Dimethyl-2-oxo-4,5-diphenylcyclopent-4-en-1-yl)methyl-5-morpholino-1-(4-nitrophenyl)-1,2,3-triazole		DD01786
161	121	70	134	209	197	133	110	**549**	100	46	37	20	18	17	14	14	**2.96**	32	43	5	3	–	–	–	–	–	–	–	–	–	Benzenepropanamide, N-[4-[acetyl-1-propenylamino]butyl]-4-methoxy-N-[3-[[3-(4-methoxyphenyl)-1-oxo-2-propenyl]amino]propyl]-	55622-38-3	NI45798
370	387	371	400	119	82	453	331	**549**	100	73	62	55	35	35	34	29	**0.00**	33	59	5	1	–	–	–	–	–	–	–	–	–	N-(5α-Cholestan-3β-yl)-D-galactosylamine		NI45799
219	220	57	549	67	41	203	550	**549**	100	17	7	4	4	4	3	1		35	51	4	1	–	–	–	–	–	–	–	–	–	1,1-Bis(4-hydroxy-3,5-di-tert-butylbenzyl)-1-cyanoacetic acid, ethyl ester		IC04579
549	233	550	91	232	57	55	234	**549**	100	49	44	42	28	12	12	10		40	55	–	1	–	–	–	–	–	–	–	–	–	5α-Cholestano[3,2-b]indole, N-benzyl-		MS10639
549	91	233	550	232	551	548	218	**549**	100	53	51	46	28	7	7	6		40	55	–	1	–	–	–	–	–	–	–	–	–	5α-Cholestano[3,2-b]indole, N-benzyl-		LI09654
549	91	550	548	551	547	220	218	**549**	100	49	43	13	8	8	8	4		40	55	–	1	–	–	–	–	–	–	–	–	–	5α-Cholestano[3,4-b]indole, N-benzyl-		LI09655
549	91	550	55	67	57	548	551	**549**	100	47	44	14	13	13	12	10		40	55	–	1	–	–	–	–	–	–	–	–	–	5α-Cholestano[3,4-b]indole, N-benzyl-		MS10640
549	233	91	550	232	234	55	57	**549**	100	64	56	44	23	18	15	14		40	55	–	1	–	–	–	–	–	–	–	–	–	5β-Cholestano[3,4-b]indole, N-benzyl-		MS10641
549	253	91	550	232	234	551	218	**549**	100	65	56	44	23	17	10	7		40	55	–	1	–	–	–	–	–	–	–	–	–	5β-Cholestano[3,4-b]indole, N-benzyl-		LI09656
69	63	219	119	131	64	169	100	**550**	100	58	33	31	29	23	19	19	**0.20**	8	–	3	–	2	18	–	–	–	–	–	–	–	1-Butanesulphinic acid, 1,1,2,2,3,3,4,4,4-nonafluoro-, anhydride	51735-74-1	NI45800
535	521	522	536	537	253	523	505	**550**	100	80	48	46	41	40	36	19	**2.39**	9	26	12	–	–	–	–	–	–	8	–	–	–	Heptamethylethyloctasilsesquioxane	77626-17-6	NI45801
84	56	168	112	196	140	326	68	**550**	100	74	26	26	18	12	8	8	**0.00**	12	–	12	–	–	–	–	–	–	–	–	–	3	Ruthenium, tetracarbonyl(di-μ-carbonylhexacarbonyldiiron)-	20468-34-2	NI45802
59	15	131	474	374	69	274	100	**550**	100	49	12	11	11	7	5	4	**0.00**	12	6	6	–	–	16	–	–	–	–	–	–	–	Carbonic acid, 1,1,2,2,3,3,4,4,5,5,6,6,7,7,8,8-hexadecafluoro-1,8-octanediyl dimethyl ester	55521-14-7	NI45803
269	271	267	190	188	191	189	109	**550**	100	55	55	43	43	26	26	26	**0.60**	12	10	1	–	2	–	–	4	–	–	–	–	–	Di-1-[2-bromo-1-(5-bromothiophene)] ethyl ether		MS10642
326	438	382	410	219	324	163	191	**550**	100	23	14	12	8	5	5	4	**0.00**	18	10	8	–	–	–	–	–	–	–	–	–	2	Di-π-cyclopentadienyloctacarbonyldimolybdenum		MS10643
105	77	56	382	298	326	176	466	**550**	100	49	27	18	15	13	11	7	**0.00**	18	14	7	–	3	–	–	–	–	–	–	–	2	(μ-benzoyldithioacetato)(μ-tert-butylthiolato)hexacarbonyldiiron		MS10644
90	43	550	552	263	63	261	183	**550**	100	95	69	67	66	66	63	56		21	23	7	6	–	–	–	1	–	–	–	–	–	N-(2-Hydroxycyclohexyl)-2-methoxy-5-acetamido-4-(2,4-dinitro-6-bromophenylazo)aniline		IC04580
57	73	41	153	120	75	136	101	**550**	100	36	16	10	9	7	6	6	**0.50**	21	29	5	2	–	7	–	–	–	1	–	–	–	1-[tert-Butyl(heptafluorobutyryl)amino]-3-(3-methyl-2-nitrophenoxy)-2-propanol trimethylsilyl ether		MS10645
225	111	224	99	295	125	110	294	**550**	100	90	55	25	23	23	23	18	**0.45**	22	39	12	–	–	–	–	–	–	–	–	5	–	β-D-Fructopyranose, 1,1′-(ethylboronate) cyclic 2,3:2′,3′:4,5:4′,5′-tetrakis(ethylboronate)	64780-49-0	NI45804
351	349	78	200	154	197	347	77	**550**	100	78	72	54	51	48	44	43	**0.30**	24	15	–	–	1	5	–	–	–	–	–	–	1	Stannane, [(pentafluorophenyl)thio]triphenyl-	17314-31-7	NI45805
126	89	55	70	59	58	87	99	**550**	100	38	30	23	18	18	18	15	**0.00**	24	42	12	2	–	–	–	–	–	–	–	–	–	1,4,7,10,13,21,24,27-Octaoxa-18,30-diazacyclotetratriacontane-14,17,31,34-tetrone	79688-19-0	NI45806
69	45	51	97	70	50	95	85	**550**	100	87	62	59	56	52	49	48	**0.00**	25	24	7	–	–	6	–	–	–	–	–	–	–	Pregna-1,4-diene-3,11,20-trione, 17,21-dihydroxy-, bis(trifluoroacetate)	425-40-1	NI45807
73	333	305	260	147	307	377	334	**550**	100	99	86	47	47	46	41	37	**13.80**	27	46	6	–	–	–	–	–	–	3	–	–	–	1H-2-Benzoxacyclotetradecin-1,7(8H)-dione, 3,4,5,6,9,10-hexahydro-3-methyl-5,14,16-tris[(trimethylsilyl)oxy]-, [3S-(3R*,5R*,11E)]-	72060-12-9	NI45808
333	147	305	221	260	307	306	334	**550**	100	96	89	60	59	49	41	36	**10.00**	27	46	6	–	–	–	–	–	–	3	–	–	–	1H-2-Benzoxacyclotetradecin-1,7(8H)-dione, 3,4,5,6,9,10-hexahydro-3-methyl-5,14,16-tris[(trimethylsilyl)oxy]-, [3S-(3R*5R*11E)]-	72060-12-9	NI45809
43	41	69	55	85	15	28	95	**550**	100	56	42	38	35	33	31	29	**3.26**	28	38	11	–	–	–	–	–	–	–	–	–	–	5,6-Epoxyingol 3,7,8,12-tetraacetate		NI45810
43	84	109	111	239	126	69	257	**550**	100	29	24	21	20	20	20	18	**0.39**	28	38	11	–	–	–	–	–	–	–	–	–	–	4β,12β,13,20-Tetraacetoxy-7α,9-epoxy-4,6-dihydroxy-6βH-tigli-1-ene		NI45811
490	120	43	121	491	44			**550**	100	62	46	35	31	31			**0.00**	29	42	10	–	–	–	–	–	–	–	–	–	–	Pregn-5-ene-3,8,11,12,14,20-hexol, 3,11,12,20-tetraacetate, (3β,11α,12β,14β)-	72173-04-7	NI45812
73	188	75	363	429	103	143	519	**550**	100	99	48	43	41	40	33	31	**20.00**	29	54	4	2	–	–	–	–	–	2	–	–	–	Pregnane-3,20-dione, 11,21-bis[(trimethylsilyl)oxy]-, bis(O-methyloxime), (5α,11β)-	57305-39-2	NI45813
73	519	520	429	103	75	550	147	**550**	100	99	44	38	31	26	25	23		29	54	4	2	–	–	–	–	–	2	–	–	–	Pregnane-3,20-dione, 17,21-bis[(trimethylsilyl)oxy]-, bis(O-methyloxime), (5β)-	57305-38-1	NI45814
130	201	131	197	202	170	159	72	**550**	100	94	27	21	15	15	11	11	**10.56**	30	38	6	4	–	–	–	–	–	–	–	–	–	L-Tryptophan, N-[N-[N-[(phenylmethoxy)carbonyl]-L-valyl]-L-valyl]-, methyl ester	55517-83-4	NI45815

MASS TO CHARGE RATIOS								M.W.	INTENSITIES								Parent	C	H	O	N	S	F	Cl	Br	I	Si	P	B	X	COMPOUND NAME	CAS Reg No	No
74	87	85	355	373	41	93	43	**550**	100	99	99	98	92	91	77	77	**4.58**	30	46	9	–	–	–	–	–	–	–	–	–	–	**Card-20(22)-enolide, 3-[(6-deoxy-3-O-methyl-α-L-talopyranosyl)oxy]-1,14 dihydroxy-, (1β,3β,5β)-**	663-95-6	**NI45816**
117	73	460	75	147	157	156	141	**550**	100	99	48	38	33	30	29	28	**0.25**	30	58	3	–	–	–	–	–	–	3	–	–	–	**Pregn-5-ene, 3β,16α,20α-tris(trimethylsilyloxy)-**	33287-38-6	**NI45817**
117	460	461	445	156	157	141	462	**550**	100	90	66	46	40	38	37	26	**0.00**	30	58	3	–	–	–	–	–	–	3	–	–	–	**Pregn-5-ene, 3β,16α,20α-tris(trimethylsilyloxy)-**	33287-38-6	**NI45818**
460	117	461	445	156	157	462	141	**550**	100	66	42	28	25	21	20	14	**0.00**	30	58	3	–	–	–	–	–	–	3	–	–	–	**Pregn-5-ene, 3β,16α,20α-tris(trimethylsilyloxy)-**	33287-38-6	**LI09657**
73	253	433	117	75	147	343	254	**550**	100	81	41	32	30	25	22	17	**3.42**	30	58	3	–	–	–	–	–	–	3	–	–	–	**Pregn-5-ene, 3β,17,21α-tris(trimethylsilyloxy)-**	41164-19-6	**NI45819**
253	73	433	343	434	147	117	75	**550**	100	91	61	35	25	24	23	21	**1.00**	30	58	3	–	–	–	–	–	–	3	–	–	–	**Pregn-5-ene, 3β,17,21α-tris(trimethylsilyloxy)-**	41164-19-6	**NI45820**
271	253	129	117	272	360	75	462	**550**	100	39	33	31	26	25	20	17	**0.00**	30	58	3	–	–	–	–	–	–	3	–	–	–	**Pregn-5-ene, 3β,17α,20α-tris(trimethylsilyloxy)-**		**LI09658**
267	73	268	147	129	133	157	75	**550**	100	61	22	22	20	18	12	11	**0.06**	30	58	3	–	–	–	–	–	–	3	–	–	–	**Pregn-5-ene, 3β,20α,21-tris(trimethylsilyloxy)-**	57305-40-5	**NI45821**
215	83	115	109	168	101	197	335	**550**	100	98	73	66	56	52	50	28	**0.00**	31	50	8	–	–	–	–	–	–	–	–	–	–	**Spathulasin methyl ester**	122899-04-1	**DD01787**
370	371	355	75	255	147	95	81	**550**	100	36	19	19	16	16	14	14	**0.34**	31	58	4	–	–	–	–	–	–	2	–	–	–	**Cholan-24-oic acid, 3,7-bis[(trimethylsilyl)oxy]-, methyl ester, (3α,5β,7α)-**	52759-80-5	**NI45822**
370	255	262	355	213	249	228	243	**550**	100	34	20	17	16	15	8	4	**0.00**	31	58	4	–	–	–	–	–	–	2	–	–	–	**Cholan-24-oic acid, 3,7-bis[(trimethylsilyl)oxy]-, methyl ester, (3α,5β,7α)-**	52759-80-5	**NI45823**
255	73	75	370	208	256	81	55	**550**	100	76	52	26	23	22	21	20	**0.29**	31	58	4	–	–	–	–	–	–	2	–	–	–	**Cholan-24-oic acid, 3,12-bis[(trimethylsilyl)oxy]-, methyl ester, (3α,5β,12α)-**	6818-41-3	**NI45824**
460	370	255	213	355	243	228	249	**550**	100	67	53	25	23	18	12	12	**0.70**	31	58	4	–	–	–	–	–	–	2	–	–	–	**Methyl 3α,7β-bis(trimethylsiloxy)-5β-cholan-24-oate**	63813-73-0	**NI45825**
370	355	255	262	213	460	249	228	**550**	100	33	31	22	19	18	15	6	**0.00**	31	58	4	–	–	–	–	–	–	2	–	–	–	**Methyl 3β,7α-bis(trimethylsiloxy)-5β-cholan-24-oate**	81445-73-0	**NI45826**
370	460	255	355	213	249	262	228	**550**	100	55	48	34	25	20	13	7	**0.00**	31	58	4	–	–	–	–	–	–	2	–	–	–	**Methyl 3β,7β-bis(trimethylsiloxy)-5β-cholan-24-oate**	78964-21-3	**NI45827**
377	218	318	365	173	349	235	203	**550**	100	30	23	11	10	8	6	6	**1.00**	32	26	7	2	–	–	–	–	–	–	–	–	–	**Endo-1-methyl-4,7-diphenyl-10-[1,2-bis(methoxycarbonyl)vinyl][8,9]benzo 4,10-diazatricyclo[2.2.0.3]-8-undecene-3,5,11-trione**		**NI45828**
550	552	304	395	548	275	394	305	**550**	100	42	27	15	13	13	10	8		34	30	–	–	–	–	–	–	–	–	–	–	2	**1′,1‴-Dibenzylbiferrocene**		**MS10646**
58	107	79	77	492	98	339	384	**550**	100	31	22	11	6	2	1	1	**0.41**	34	50	2	2	1	–	–	–	–	–	–	–	–	**2,4-Bis[7-(3-furyl)-4-methylhepta-2,6-dienyl]-2,4-bis(N,N-dimethylaminomethyl)tetrahydrothiophene**	95391-71-2	**NI45829**
86	57	161	87	550	219	85	43	**550**	100	99	42	41	27	20	20	18		34	50	4	2	–	–	–	–	–	–	–	–	–	**1,4-Bis[5-hydroxy-3-(1-methylhexyl)-2,3-dihydrobenzofuryl]piperazine**		**IC04581**
283	284	43	57	99	55	41	69	**550**	100	22	18	14	11	10	9	6	**0.24**	36	70	3	–	–	–	–	–	–	–	–	–	–	**Palmitic acid, 2-(1-octadecenyloxy)ethyl ester, (E)-**	30760-04-4	**NI45830**
41	57	219	220	55	56	83	203	**550**	100	92	51	25	22	16	14	13	**0.00**	38	62	2	–	–	–	–	–	–	–	–	–	–	**Bicyclo[4.1.0]hept-2-en-5-one, 2,4,6-tri-tert-butyl-4-[(1,3,5-tri-tert-butyl-4 oxocyclohexa-2,5-dien-1-yl)methyl]-**		**MS10647**
550	91	105	119	157	551	145	131	**550**	100	78	58	55	44	43	43	35		40	54	1	–	–	–	–	–	–	–	–	–	–	**Anhydrolutein 11**	752-29-4	**NI45831**
550	458	69	548	133	91	549	109	**550**	100	13	10	9	9	9	5	4		40	54	1	–	–	–	–	–	–	–	–	–	–	**Echinenone**		**LI09659**
550	458	119	548	105	69	133	91	**550**	100	13	12	10	10	10	9	9		40	54	1	–	–	–	–	–	–	–	–	–	–	**Echinenone**		**MS10648**
69	550	119	105	55	41	133	91	**550**	100	83	80	73	73	67	57	54		40	54	1	–	–	–	–	–	–	–	–	–	–	**Echinenone**		**LI09660**
69	43	550	91	83	133	109	444	**550**	100	81	70	69	40	31	25	23		40	54	1	–	–	–	–	–	–	–	–	–	–	**Lycopen-16-al**		**LI09661**
549	550	551	447	165	166	548	267	**550**	100	92	31	14	12	9	8	8		41	30	–	2	–	–	–	–	–	–	–	–	–	**Aniline, N-(1,3,4,5,6-pentaphenyl-2(1H)-pyridinylidene)-**	71704-80-8	**NI45832**
522	550	523	551	165	367	183	524	**550**	100	47	47	22	19	17	13	11		42	30	1	–	–	–	–	–	–	–	–	–	–	**2,4-Cyclohexadien-1-one, 2,3,4,5,6,6-hexaphenyl-**	74645-97-9	**NI45833**
73	147	75	232	230	217	164	135	**551**	100	76	37	27	27	25	21	19	**19.09**	24	45	4	5	–	–	–	–	–	3	–	–	–	**Adenosine, N-(3-methyl-2-butenyl)-2′,3′,5′-tris-O-(trimethylsilyl)-**	32352-57-1	**NI45834**
73	147	75	230	232	217	164	551	**551**	100	77	37	27	26	25	21	20		24	45	4	5	–	–	–	–	–	3	–	–	–	**Adenosine, N-(3-methyl-2-butenyl)-2′,3′,5′-tris-O-(trimethylsilyl)-**	32352-57-1	**LI09662**
73	232	230	148	551	217	147	75	**551**	100	79	64	59	54	41	30	25		24	45	4	5	–	–	–	–	–	3	–	–	–	**Adenosine, N-(3-methyl-3-butenyl)-2′,3′,5′-tris-O-(trimethylsilyl)-**	32352-58-2	**NI45835**
73	232	230	148	551	217	147	552	**551**	100	79	63	59	54	40	30	25		24	45	4	5	–	–	–	–	–	3	–	–	–	**Adenosine, N-(3-methyl-3-butenyl)-2′,3′,5′-tris-O-(trimethylsilyl)-**	32352-58-2	**LI09663**
185	214	194	313	270	57	86	108	**551**	100	63	60	52	40	39	38	30	**1.50**	25	28	5	1	–	–	–	–	–	–	1	–	1	**[(Dibutylamino)diphenylphosphine]pentacarbonylmolybdenum**		**IC04582**
91	551	460	373	479	251	552	329	**551**	100	25	14	13	8	7	5	5		30	37	7	3	–	–	–	–	–	–	–	–	–	**4,4′-Dimethyl-3,3′-bis-(methoxycarbonylethyl)-5-benzyloxycarbonyl-5′-dimethylcarboxamido-2,2′-dipyrrolylmethane**		**LI09664**
105	91	77	160	339	122	429	399	**551**	100	94	58	18	9	8	2	1	**0.30**	33	29	7	1	–	–	–	–	–	–	–	–	–	**D-erythro-Pentose, 2-deoxy-, O-(phenylmethyl)oxime, 3,4,5-tribenzoate**	71646-95-2	**NI45836**
553	551	223	221	552	554	55	57	**551**	100	94	69	69	37	36	18	16		33	46	1	1	–	–	–	1	–	–	–	–	–	**5′-Bromo-6-oxo-5α-cholestano[3,2-b]indole**		**MS10649**
223	232	234	231	224	230	236	329	**552**	100	15	14	11	8	7	6	5	**3.30**	10	2	2	–	2	12	–	–	–	–	–	–	1	**Palladium, bis(1,1,1,5,5,5-hexafluoro-4-thioxo-2-pentanonato-O,S)-**	55521-28-3	**NI45837**
124	241	300	182	176	269	180	297	**552**	100	68	41	27	25	15	12	10	**0.10**	14	5	9	–	–	–	–	–	–	–	–	–	4	**Cyclopentadienylnickeltricobalt nonacarbonyl**		**LI09665**
59	179	191	43	552	554	164	481	**552**	100	90	75	55	50	49	44	41		21	25	7	6	–	–	–	1	–	–	–	–	–	**N-(1,3-Dimethyl-3-hydroxybutyl)-2-methoxy-5-acetamido-4-(2,4-dinitro-6-bromophenylazo)aniline**		**IC04583**
73	552	553	103	217	537	554	147	**552**	100	99	42	30	28	27	26	22		22	44	3	6	–	–	–	–	–	4	–	–	–	**Tetrakis(trimethylsilyl)-8,2′-aminoanhydroadenosine**		**NI45838**
107	139	138	552	243	170	275	211	**552**	100	90	50	15	12	9	7	7		24	24	–	8	4	–	–	–	–	–	–	–	–	**2,11,20,29-Tetrathia[3.3.3.3](2,5)pyrazinophane**	123624-98-6	**DD01788**
174	73	175	176	378	86	552	379	**552**	100	33	17	8	7	6	4	3		25	52	2	2	–	–	–	–	–	5	–	–	–	**1H-Indole-3-ethylamine, N,N,1-tris(trimethylsilyl)-5,6-bis[(trimethylsilyl)oxy]-**	56114-58-0	**NI45839**

MASS TO CHARGE RATIOS								M.W.	INTENSITIES								Parent	C	H	O	N	S	F	Cl	Br	I	Si	P	B	X	COMPOUND NAME	CAS Reg No	No
466	468	467	42	43	437	44	436	**552**	100	94	52	48	43	40	34	31	**16.95**	28	25	7	–	–	–	–	1	–	–	–	–	–	3,4-Diacetoxy-2-bromo-8,11-dimethoxy-6,10-dimethyl-dibenzo[c,kl]xanthene		NI45840
43	356	416	357	338	107	97	55	**552**	100	79	44	29	29	26	22	22	**0.90**	29	44	10	–	–	–	–	–	–	–	–	–	–	11-Glycoloyloxy-3,12,20-triacetoxy-14-hydroxypregnane		MS10650
462	463	117	447	156	464	157	141	**552**	100	42	39	25	24	16	16	14	**2.00**	30	60	3	–	–	–	–	–	–	3	–	–	–	5α-Pregnane, 3α,16α,20α-tris(trimethylsilyloxy)-		LI09666
255	73	435	75	147	256	436	117	**552**	100	90	78	40	38	37	32	30	**6.00**	30	60	3	–	–	–	–	–	–	3	–	–	–	5α-Pregnane, 3α,17,20α-tris(trimethylsilyloxy)-	57305-41-6	NI45841
255	73	147	435	75	117	145	107	**552**	100	97	44	39	35	25	18	18	**1.25**	30	60	3	–	–	–	–	–	–	3	–	–	–	5α-Pregnane, 3α,17,20β-tris(trimethylsilyloxy)-	33287-34-2	NI45842
269	359	449	134	360	147	450	149	**552**	100	78	55	35	24	24	20	20	**1.00**	30	60	3	–	–	–	–	–	–	3	–	–	–	5α-Pregnane, 3α,20α,21-tris(trimethylsilyloxy)-		LI09667
73	117	75	462	141	156	147	157	**552**	100	99	77	61	54	48	39	26	**0.75**	30	60	3	–	–	–	–	–	–	3	–	–	–	5α-Pregnane, 3β,16α,20α-tris(trimethylsilyloxy)-	33287-36-4	NI45843
255	435	73	436	256	147	346	75	**552**	100	88	77	48	38	38	36	36	**4.18**	30	60	3	–	–	–	–	–	–	3	–	–	–	5α-Pregnane, 3β,17,20β-tris(trimethylsilyloxy)-	57305-42-7	NI45844
269	449	134	359	450	161	147	129	**552**	100	40	22	19	17	17	16	16	**1.00**	30	60	3	–	–	–	–	–	–	3	–	–	–	5α-Pregnane, 3β,20β,21-tris(trimethylsilyloxy)-		LI09668
117	73	75	143	118	282	119	81	**552**	100	68	27	13	13	11	9	8	**1.06**	30	60	3	–	–	–	–	–	–	3	–	–	–	5β-Pregnane, 3α,11β,20β-tris(trimethylsilyloxy)-	33287-46-6	NI45845
255	73	435	75	43	147	256	117	**552**	100	88	55	36	28	26	20	20	**1.18**	30	60	3	–	–	–	–	–	–	3	–	–	–	5β-Pregnane, 3α,17,20α-tris(trimethylsilyloxy)-	17846-09-2	NI45846
73	255	435	75	147	117	256	345	**552**	100	99	59	52	40	33	27	25	**7.00**	30	60	3	–	–	–	–	–	–	3	–	–	–	5β-Pregnane, 3α,17,20α-tris(trimethylsilyloxy)-	17846-09-2	NI45847
43	327	387	137	123	285	179	161	**552**	100	83	65	57	57	51	32	30	**2.89**	31	36	9	–	–	–	–	–	–	–	–	–	–	3-(4′-Hydroxy-3′-methoxyphenyl)-3-(4′-propyl-2′-methoxyphenoxy)-2-(2′-methoxyphenoxy)propanol diacetate		NI45848
552	553	276	554	262	519	491	277	**552**	100	34	9	6	5	1	1	1		32	24	9	–	–	–	–	–	–	–	–	–	–	Anhydrophlegmacin-9,10-quinone B2	64181-94-8	NI45849
321	456	441	129	367	366	477	552	**552**	100	73	46	35	32	32	31	20		32	57	3	–	–	–	1	–	–	1	–	–	–	3β-Acetoxy-5α-chloro-6β-(trimethylsilyloxy)cholestane		LI09669
291	411	552	177	467	235	412	69	**552**	100	73	60	32	22	22	21	18		33	44	7	–	–	–	–	–	–	–	–	–	–	1,5-Methano-1H,3H,8H-furo[3,4-g]pyrano[3,2-b]xanthene-1-butanal, 3a,4,5,6,6a,7,9,10-octahydro-8-hydroxy-α,3,3,11,11-pentamethyl-13-(3-methylbutyl)-7,15-dioxo-	55452-66-9	NI45850
551	168	552	178	141	41	103	43	**552**	100	80	60	32	30	20	18	16		34	25	1	–	–	–	–	–	–	–	–	–	1	Tetraphenylcyclopentadienone-π-cyclopentadienylrhodium		MS10651
551	552	276	550	91	118	553	18	**552**	100	39	17	16	14	9	8	5		34	32	–	8	–	–	–	–	–	–	–	–	–	Pyrimido[5,4-d]pyrimidine, 2,4,6,8-tetra-m-toluidino-	18711-05-2	MS10652
551	552	446	276	403	91	106	118	**552**	100	42	37	34	23	21	17	16		34	32	–	8	–	–	–	–	–	–	–	–	–	Pyrimido[5,4-d]pyrimidine, 2,4,6,8-tetra-o-toluidino-	18711-04-1	MS10653
105	233	284	234	99	81	77	69	**552**	100	82	66	66	59	59	59	59	**4.00**	34	48	6	–	–	–	–	–	–	–	–	–	–	2-Deoxy-α-ecdysone, 22-O-benzoate		NI45851
341	313	57	342	239	43	55	71	**552**	100	24	24	23	19	19	12	11	**0.94**	35	68	4	–	–	–	–	–	–	–	–	–	–	Hexadecanoic acid, 2-pentadecyl-1,3-dioxan-5-yl ester, cis-	41563-16-0	NI45852
341	342	313	239	57	43	71	55	**552**	100	24	24	23	23	17	11	11	**0.80**	35	68	4	–	–	–	–	–	–	–	–	–	–	Hexadecanoic acid, 2-pentadecyl-1,3-dioxan-5-yl ester, trans-	41563-17-1	NI45853
341	342	57	43	313	55	71	41	**552**	100	24	15	14	8	8	6	5	**0.44**	35	68	4	–	–	–	–	–	–	–	–	–	–	Hexadecanoic acid, (2-pentadecyl-1,3-dioxolan-4-yl)methyl ester	41563-11-5	NI45854
297	296	43	57	239	100	28	113	**552**	100	36	32	31	29	28	26	25	**0.64**	35	68	4	–	–	–	–	–	–	–	–	–	–	Hexadecanoic acid, 1,3-propanediyl ester	818-21-3	NI45855
59	264	257	239	296	222	283	356	**552**	100	36	29	24	21	9	8	7	**4.00**	35	68	4	–	–	–	–	–	–	–	–	–	–	Methyl 18-palmitoyloxystearate		NI45856
57	43	71	82	83	55	97	69	**552**	100	76	68	63	60	55	51	51	**5.14**	36	72	3	–	–	–	–	–	–	–	–	–	–	Hexadecanoic acid, 2-(octadecyloxy)ethyl ester	29899-13-6	NI45858
57	43	71	55	83	85	69	82	**552**	100	88	58	52	46	42	42	38	**0.64**	36	72	3	–	–	–	–	–	–	–	–	–	–	Hexadecanoic acid, 2-(octadecyloxy)ethyl ester	29899-13-6	NI45857
57	71	43	82	83	97	85	55	**552**	100	70	68	59	57	51	51	49	**5.74**	36	72	3	–	–	–	–	–	–	–	–	–	–	Octadecanoic acid, 2-(hexadecyloxy)ethyl ester	34639-56-0	NI45859
218	219	408	203	189	524	190	409	**552**	100	76	64	50	47	27	26	16	**7.00**	38	64	2	–	–	–	–	–	–	–	–	–	–	Urs-12-en-3-ol, octanoate, (3β)-		LI09670
218	203	408	189	524	219	190	204	**552**	100	77	60	59	49	44	42	27	**4.00**	38	64	2	–	–	–	–	–	–	–	–	–	–	Olean-12-en-3-ol, octanoate, (3β)-		LI09671
105	119	285	91	69	95	93	81	**552**	100	88	78	78	78	74	67	67	**40.00**	40	56	1	–	–	–	–	–	–	–	–	–	–	β,β-Carotene, 5,6-epoxy-5,6-dihydro-	1923-89-3	LI09672
205	165	552	472	336	192	218	231	**552**	100	61	54	39	37	25	23	20		40	56	1	–	–	–	–	–	–	–	–	–	–	β,β-Carotene, 5,6-epoxy-5,6-dihydro-	1923-89-3	NI45860
205	69	105	91	119	55	95	40	**552**	100	91	80	74	72	68	65	63	**22.00**	40	56	1	–	–	–	–	–	–	–	–	–	–	β,β-Carotene, 5,8-epoxy-5,8-dihydro-		LI09673
91	41	145	119	93	81	107	55	**552**	100	66	46	46	45	43	39	39	**6.60**	40	56	1	–	–	–	–	–	–	–	–	–	–	Lycoxanthin		MS10654
91	69	43	133	109				**552**	100	82	30	29	24				**0.00**	40	56	1	–	–	–	–	–	–	–	–	–	–	Lycoxanthin		LI09674
552	69	91	133	446	109	43	460	**552**	100	65	34	26	24	17	16	15		40	56	1	–	–	–	–	–	–	–	–	–	–	Rubixanthin		LI09675
552	534	460	442	446	496			**552**	100	76	13	10	4	1				40	56	1	–	–	–	–	–	–	–	–	–	–	Zeinoxanthin		LI09676
253	252	57	84	280	169	69	54	**553**	100	82	55	16	13	11	11	11	**0.00**	15	13	5	1	–	14	–	–	–	–	–	–	–	N,O-Bis(heptafluorobutyryl)threonine propyl ester		MS10655
176	81	484	256	177	53	485	119	**553**	100	33	29	10	10	6	4	4	**0.68**	17	12	6	5	–	9	–	–	–	–	–	–	–	Acetamide, N-[9-[2-deoxy-3,5-bis-O-(trifluoroacetyl)-β-D-erythro-pentofuranosyl]-1,9-dihydro-1-methyl-6H-purin-6-ylidene]-2,2,2-trifluoro-	35170-14-0	NI45861
148	176	69	119	149	53	246	247	**553**	100	37	34	14	10	10	9	8	**4.44**	17	12	6	5	–	9	–	–	–	–	–	–	–	Adenosine, 2′-deoxy-N-methyl-N-(trifluoroacetyl)-, 3′,5′-bis(trifluoroacetate)	35170-12-8	NI45862
87	43	482	555	278	484	480	557	**553**	100	94	43	37	36	28	26	23	**22.00**	21	20	4	3	–	–	5	–	–	–	–	–	–	N,N-Bis(2-acetoxyethyl)-3-methyl-4-(pentachlorophenylazo)aniline		IC04584
73	355	356	357	45	74	75	77	**553**	100	57	17	8	8	7	5	3	**0.50**	22	42	4	1	–	3	–	–	–	4	–	–	–	N-(Trifluoroacetyl)-N,O,O′,O″-tetrakis(trimethylsilyl)norepinephrine		MS10656
255	73	75	79	103	52	256	129	**553**	100	88	42	33	30	24	21	21	**0.00**	26	47	6	3	–	–	–	–	–	2	–	–	–	Aniline, 2,4-dinitro-N-[1-[[(trimethylsilyl)oxy]methyl]-2-[(trimethylsilyl)oxy]-3-tridecenyl]-, [R-[R*,S*,(E)]]-		LI09677
255	73	75	79	103	52	256	129	**553**	100	87	40	33	30	23	21	19	**0.06**	26	47	6	3	–	–	–	–	–	2	–	–	–	Aniline, 2,4-dinitro-N-[1-[[(trimethylsilyl)oxy]methyl]-2-[(trimethylsilyl)oxy]-3-tridecenyl]-, [R-[R*,S*,(E)]]-	56272-05-0	NI45863

MASS TO CHARGE RATIOS								M.W.	INTENSITIES								Parent	C	H	O	N	S	F	Cl	Br	I	Si	P	B	X	COMPOUND NAME	CAS Reg No	No
105	124	77	271	305	95	467	259	**553**	100	39	23	12	10	10	9	9	**1.00**	30	35	9	1	–	–	–	–	–	–	–	–	–	1α-Benzoyloxy-6β-nicotinoyloxy-9β-acetoxy-8β,4β-dihydroxydihydro-β-agarofuran		MS10657
91	276	44	81	55	445	120	41	**553**	100	63	63	62	56	35	35	35	**0.70**	32	43	7	1	–	–	–	–	–	–	–	–	–	5-Acetoxy-1-benzyl-3a-hydroxy-7-methyl-4-[4-methyl-7-(5-oxotetrahydrofuran-2-yl)-1-heptenyl]-6-methylene-3-oxo-3a,4,5,6,7,7a-hexahydroisoindolin-5-yl peroxypropionate		IC04585
91	73	75	446	129	144	447	74	**553**	100	99	30	18	17	15	9	9	**1.44**	32	51	3	1	–	–	–	–	–	2	–	–	–	Androst-5-en-16-one, 3,17-bis[(trimethylsilyl)oxy]-, O-benzyloxime, (3β,17β)-	57305-16-5	NI45864
73	91	75	446	105	129	447	77	**553**	100	99	63	47	37	35	26	24	**6.50**	32	51	3	1	–	–	–	–	–	2	–	–	–	Androst-5-en-17-one, 3,11-bis[(trimethylsilyl)oxy]-, O-benzyloxime, (3β,11β)-	57325-80-1	NI45865
91	73	75	446	266	129	105	356	**553**	100	82	30	21	21	20	13	10	**1.44**	32	51	3	1	–	–	–	–	–	2	–	–	–	Androst-5-en-17-one, 3,16-bis[(trimethylsilyl)oxy]-, O-benzyloxime, (3β,16α)-	57305-14-3	NI45866
91	73	446	75	447	250	129	266	**553**	100	67	59	36	34	30	29	20	**2.00**	32	51	3	1	–	–	–	–	–	2	–	–	–	Androst-5-en-17-one, 3,16-bis[(trimethylsilyl)oxy]-, O-benzyloxime, (3β,16β)-	57305-15-4	NI45867
205	203	128	63	554	332	65	64	**554**	100	46	14	12	11	11	7	6		12	8	2	–	2	–	2	–	–	–	1	–	1	Thallium(I) O,O′-bis(3-chlorophenyl)dithiophosphate		MS10658
205	203	332	128	554	63	556	77	**554**	100	45	12	10	9	8	5	5		12	8	2	–	2	–	2	–	–	–	1	–	1	Thallium(I) O,O′-bis(4-chlorophenyl)dithiophosphate		MS10659
93	31	29	79	80	63			**554**	100	88	45	35	31	29			**6.10**	12	27	12	–	–	–	–	–	–	–	3	–	1	Molybdenum, tricarbonyltris(trimethyl phosphite-P)-	22261-14-9	NI45868
427	69	467	429	279	97	485	349	**554**	100	74	62	37	32	26	18	16	**0.19**	13	8	3	–	–	18	–	–	–	–	–	–	–	2,6-Bis(trifluoromethyl)-2,6-dihydroxy-1,1,1,7,7,7-hexafluoro-4-(2-hydroxy-2-trifluoromethyl-3,3,3-trifluoropropylidene)heptane		LI09678
135	539	389	540	541	390	451	136	**554**	100	90	48	43	27	17	15	14	**2.19**	20	34	7	–	–	–	–	–	–	6	–	–	–	3,3,5,5,9,9,11,11-Octamethyl-1,7-diphenylbicyclo[5.5.1]hexasiloxane	49538-51-4	NI45869
73	147	335	217	292	305	74	189	**554**	100	31	28	26	14	10	10	7	**0.00**	21	50	7	–	–	–	–	–	–	5	–	–	–	Adipic acid, 2,3,4-tris(trimethylsilyloxy)-, bis(trimethylsilyl) ester		LI09679
73	245	147	292	246	74	347	217	**554**	100	74	38	18	17	10	9	7	**0.00**	21	50	7	–	–	–	–	–	–	5	–	–	–	Adipic acid, 2,3,5-tris(trimethylsilyloxy)-, bis(trimethylsilyl) ester		LI09680
73	147	335	217	292	305	74	75	**554**	100	30	28	24	12	10	9	8	**0.00**	21	50	7	–	–	–	–	–	–	5	–	–	–	arabino-Hexaric acid, 2-deoxy-3,4,5-tris-O-(trimethylsilyl)-, bis(trimethylsilyl) ester	38165-98-9	NI45870
73	245	147	292	246	347	74	75	**554**	100	74	37	17	16	9	9	8	**0.00**	21	50	7	–	–	–	–	–	–	5	–	–	–	arabino-Hexaric acid, 3-deoxy-2,4,5-tris-O-(trimethylsilyl)-, bis(trimethylsilyl) ester	38165-99-0	NI45871
73	204	147	217	74	305	292	205	**554**	100	26	20	13	8	6	6	5	**0.00**	21	50	7	–	–	–	–	–	–	5	–	–	–	Galacturonic acid, pentakis(trimethylsilyl)-	65337-27-1	NI45872
204	73	217	147	205	305	292	191	**554**	100	84	52	29	19	16	15	15	**0.00**	21	50	7	–	–	–	–	–	–	5	–	–	–	Galacturonic acid, 2,3,4,5-tetrakis-O-(trimethylsilyl)-, trimethylsilyl ester	56192-87-1	NI45873
73	204	217	147	205	292	305	191	**554**	100	63	37	25	12	12	12	11	**0.00**	21	50	7	–	–	–	–	–	–	5	–	–	–	Galacturonic acid, 2,3,4,5-tetrakis-O-(trimethylsilyl)-, trimethylsilyl ester	56192-87-1	NI45874
73	217	291	218	147	191	305	75	**554**	100	49	42	30	30	23	14	14	**0.05**	21	50	7	–	–	–	–	–	–	5	–	–	–	D-Glucofuranuronic acid, 1,2,3,5-tetrakis-O-(trimethylsilyl)-, trimethylsilyl ester	52783-57-0	NI45875
217	73	204	147	191	305	292	233	**554**	100	80	67	30	28	27	27	10	**0.00**	21	50	7	–	–	–	–	–	–	5	–	–	–	α-D-Glucopyranuronic acid, 1,2,3,4-tetrakis-O-(trimethylsilyl)-, trimethylsilyl ester	52842-25-8	MS10660
73	217	204	147	75	305	290	68	**554**	100	74	49	37	27	18	18	16	**0.06**	21	50	7	–	–	–	–	–	–	5	–	–	–	α-D-Glucopyranuronic acid, 1,2,3,4-tetrakis-O-(trimethylsilyl)-, trimethylsilyl ester	52842-25-8	NI45876
217	204	73	191	75	305	218	292	**554**	100	85	35	24	23	20	19	18	**0.00**	21	50	7	–	–	–	–	–	–	5	–	–	–	α-D-Glucopyranuronic acid, 1,2,3,4-tetrakis-O-(trimethylsilyl)-,trimethylsilyl ester	52842-25-8	NI45877
217	204	73	147	191	303	292	218	**554**	100	81	80	42	29	22	21	21	**0.00**	21	50	7	–	–	–	–	–	–	5	–	–	–	β-D-Glucopyranuronic acid, 1,2,3,4-tetrakis-O-(trimethylsilyl)-, trimethylsilyl ester	52842-24-7	NI45879
73	217	204	305	293	218	205	75	**554**	100	83	64	21	20	17	14	14	**0.01**	21	50	7	–	–	–	–	–	–	5	–	–	–	β-D-Glucopyranuronic acid, 1,2,3,4-tetrakis-O-(trimethylsilyl)-, trimethylsilyl ester	52842-24-7	NI45878
73	217	204	147	191	305	292	218	**554**	100	64	54	27	17	14	13	13	**0.00**	21	50	7	–	–	–	–	–	–	5	–	–	–	D-Glucuronic acid, 2,3,4,5-tetrakis-O-(trimethylsilyl)-, trimethylsilyl ester	55530-80-8	NI45880
217	204	191	305	292	218	205	233	**554**	100	97	26	24	23	20	19	10	**0.01**	21	50	7	–	–	–	–	–	–	5	–	–	–	D-Glucuronic acid, 2,3,4,5-tetrakis-O-(trimethylsilyl)-, trimethylsilyl ester	55530-80-8	NI45881
73	204	217	147	74	191	75	45	**554**	100	24	24	19	9	7	7	6	**0.00**	21	50	7	–	–	–	–	–	–	5	–	–	–	D-Glucuronic acid, 2,3,4,5-tetrakis-O-(trimethylsilyl)-, trimethylsilyl ester	55530-80-8	NI45882
73	204	217	147	305	191	74	218	**554**	100	34	33	24	11	10	9	8	**0.00**	21	50	7	–	–	–	–	–	–	5	–	–	–	D-Glucuronic acid, 2,3,4,5-tetrakis-O-(trimethylsilyl)-, trimethylsilyl ester	55530-80-8	NI45883
73	147	217	437	74	438	75	204	**554**	100	19	16	11	8	5	5	5	**0.00**	21	50	7	–	–	–	–	–	–	5	–	–	–	2-Keto-D-gluconic acid, pentakis(trimethylsilyl)-		NI45884
73	217	147	451	333	218	74	75	**554**	100	49	25	16	12	11	9	7	**0.00**	21	50	7	–	–	–	–	–	–	5	–	–	–	5-Keto-D-gluconic acid, pentakis(trimethylsilyl)-		NI45885
333	306	292	334	423	335	359	307	**554**	100	55	35	29	28	26	25	20	**0.50**	21	50	7	–	–	–	–	–	–	5	–	–	–	ribo-5-Hexulosonic acid, 2,3,4,6-tetrakis-O-(trimethylsilyl)-, trimethylsilyl ester	38166-08-4	NI45886

MASS TO CHARGE RATIOS	M.W.	INTENSITIES	Parent	C	H	O	N	S	F	Cl	Br	I	Si	P	B	X	COMPOUND NAME	CAS Reg No	No
217 73 292 147 191 305 75 93	**554**	100 59 25 17 12 7 7 6	**0.00**	21	50	7	–	–	–	–	–	–	5	–	–	–	**Per-O-(trimethylsilyl)-α-D-galactofuranuronic acid**		**MS10661**
217 73 75 292 147 93 191 77	**554**	100 66 30 25 20 20 12 12	**0.00**	21	50	7	–	–	–	–	–	–	5	–	–	–	**Per-O-(trimethylsilyl)-β-D-galactofuranuronic acid**		**MS10662**
204 217 73 147 292 305 75 191	**554**	100 92 92 33 27 24 23 19	**0.00**	21	50	7	–	–	–	–	–	–	5	–	–	–	**Per-O-(trimethylsilyl)-α-D-galactopyranuronic acid**		**MS10663**
204 73 217 147 292 305 75 191	**554**	100 90 82 29 27 22 22 19	**0.00**	21	50	7	–	–	–	–	–	–	5	–	–	–	**Per-O-(trimethylsilyl)-β-D-galactopyranuronic acid**		**MS10664**
217 73 292 147 191 75 319 305	**554**	100 56 25 20 13 8 6 6	**0.00**	21	50	7	–	–	–	–	–	–	5	–	–	–	**Per-O-(trimethylsilyl)-α,β-D-glucofuranuronic acid**		**MS10665**
217 73 292 147 191 305 204 75	**554**	100 60 32 22 16 14 13 8	**0.00**	21	50	7	–	–	–	–	–	–	5	–	–	–	**Per-O-(trimethylsilyl)-α-D-mannofuranuronic acid**		**MS10666**
217 73 292 191 147 319 305 75	**554**	100 47 26 14 12 7 6 4	**0.00**	21	50	7	–	–	–	–	–	–	5	–	–	–	**Per-O-(trimethylsilyl)-β-D-mannofuranuronic acid**		**MS10667**
217 73 204 305 292 147 191 233	**554**	100 75 70 26 25 24 23 10	**0.00**	21	50	7	–	–	–	–	–	–	5	–	–	–	**Per-O-(trimethylsilyl)-α-D-mannopyranuronic acid**		**MS10668**
217 73 147 292 191 75 305 319	**554**	100 80 30 17 13 13 8 5	**0.00**	21	50	7	–	–	–	–	–	–	5	–	–	–	**Per-O-(trimethylsilyl)-α-L-idofuranuronic acid**		**MS10669**
217 73 147 75 292 204 191 305	**554**	100 97 33 18 15 15 13 9	**0.00**	21	50	7	–	–	–	–	–	–	5	–	–	–	**Per-O-(trimethylsilyl)-β-L-idofuranuronic acid**		**MS10670**
73 204 217 147 75 305 191 292	**554**	100 76 44 41 26 10 9 7	**0.00**	21	50	7	–	–	–	–	–	–	5	–	–	–	**Per-O-(trimethylsilyl)-α,β-L-idopyranuronic acid**		**MS10671**
57 217 76 73 103 87 88 553	**554**	100 99 99 99 87 83 73 66	**31.00**	22	44	4	5	–	–	–	–	–	4	–	–	–	**Tetrakis(trimethylsilyl)-8,2′-aminoanhydroinosine**		**NI45887**
43 127 122 169 109 331 139 97	**554**	100 58 57 53 43 33 24 23	**0.00**	25	30	14	–	–	–	–	–	–	–	–	–	–	**Salirepin hexaacetate**		**MS10672**
180 479 139 376 77 477 374 556	**554**	100 99 95 93 88 75 70 63	**47.00**	26	20	–	2	–	–	4	–	–	–	–	–	2	**Aluminium, tetrachlorobis[μ-(1,1-diphenylmethyleniminato)]di-**	31390-30-4	**NI45888**
85 43 207 169 105 331 109 77	**554**	100 62 60 28 28 15 13 8	**0.00**	26	34	13	–	–	–	–	–	–	–	–	–	–	**(1RS,3RS)-3-Benzoyloxy-1-methoxybutyl-2,3,4,6-tetra-O-acetyl-β-D-glucopyranoside**		**MS10673**
89 73 133 87 59 58 72 88	**554**	100 71 48 48 44 30 19 18	**0.00**	26	50	12	–	–	–	–	–	–	–	–	–	–	**1,1′-Bis(3,6,9,12,15,18-hexaoxacyclononadecane)**		**MS10674**
135 270 136 91 412 554 271 77	**554**	100 80 73 36 30 24 23 20		28	26	4	–	4	–	–	–	–	–	–	–	–	**Bis[4-(2-phenylsulphonylethyl)phenyl] disulphide**		**IC04586**
43 380 101 74 381 293 85 44	**554**	100 34 9 9 8 6 6 6	**0.00**	28	26	12	–	–	–	–	–	–	–	–	–	–	**4-(D-Xylosyloxy)-6,7-dimethoxy-9-[3,4-(methylenedioxy)phenyl]naphtho[2,3-c]furan-1(3H)-one monoacetate**		**NI45889**
189 276 278 277 57 554 261 263	**554**	100 69 67 60 59 30 22 16		28	34	4	4	2	–	–	–	–	–	–	–	–	**Bis[2-(tert-butyl)-6,7-dimethoxyquinazolin-4-yl] disulphide**		**NI45890**
172 170 554 240 204 303 203 556	**554**	100 86 10 8 8 6 6 5		28	34	4	4	2	–	–	–	–	–	–	–	–	**1-Naphthalenesulphonamide, N,N′-1,4-butanediylbis[5-(dimethylamino)-**	13285-10-4	**LI09681**
171 170 169 554 168 57 69 154	**554**	100 77 45 38 32 24 22 16		28	34	4	4	2	–	–	–	–	–	–	–	–	**1-Naphthalenesulphonamide, N,N′-1,4-butanediylbis[5-(dimethylamino)-**	13285-10-4	**NI45891**
303 304 305 306 307 331 269 333	**554**	100 92 91 86 81 70 64 60	**36.00**	28	36	7	–	–	–	2	–	–	–	–	–	–	**Benzoic acid, 3,5-dichloro-2,4-dimethoxy-6-pentyl-, 3-methoxy-4-(methoxycarbonyl)-5-pentylphenyl ester**	64185-28-0	**NI45892**
43 38 519 536 42 520 60 26	**554**	100 36 16 12 9 7 5 5	**4.12**	30	18	11	–	–	–	–	–	–	–	–	–	–	**Aurantioskyrin**		**NI45893**
43 71 554 73 42 119 40 26	**554**	100 8 7 7 7 6 6 6		30	18	11	–	–	–	–	–	–	–	–	–	–	**Oxyskyrin**		**NI45894**
271 270 269 272 189 191 206 558	**554**	100 92 92 87 62 32 29 20	**19.00**	30	20	1	–	–	–	–	2	–	–	–	–	–	**Bis[(10-bromo-9-anthryl)methyl] ether**		**LI09682**
494 508 495 301 151 327 180 509	**554**	100 60 55 51 49 46 46 45	**1.24**	30	34	10	–	–	–	–	–	–	–	–	–	–	**(2R,3R,4S)-3-Acetoxy-4-(2,4,6-trimethoxyphenyl)-5,7,3′,4′-tetramethoxyflavan**		**NI45895**
134 135 463 86 131 224 215 187	**554**	100 37 23 13 9 6 4 3	**0.80**	33	38	4	4	–	–	–	–	–	–	–	–	–	**Aralionine B**		**LI09683**
100 101 131 114 85 44 103 91	**554**	100 7 3 2 2 2 1 1	**0.03**	33	38	4	4	–	–	–	–	–	–	–	–	–	**Integerressin**		**MS10675**
366 215 107 81 105 79 91 93	**554**	100 98 88 78 75 73 71 62	**1.90**	33	46	7	–	–	–	–	–	–	–	–	–	–	**Bufa-20,22-dienolide, 14,15-epoxy-3-[(8-methoxy-1,8-dioxooctyl)oxy]-, (3β,5β,15β)-**	20987-32-0	**NI45896**
155 383 382 229 43 384 57 71	**554**	100 94 39 28 26 24 24 21	**0.64**	33	62	6	–	–	–	–	–	–	–	–	–	–	**Decanoic acid, 1,2,3-propanetriyl ester**	621-71-6	**NI45897**
43 470 45 60 42 44 471 469	**554**	100 37 19 15 15 12 12 10	**0.00**	34	18	8	–	–	–	–	–	–	–	–	–	–	**5,6-Diacetoxy-11,12,17,18-trinaphthylenetetrone**	96722-31-5	**NI45898**
43 42 44 45 60 41 470 83	**554**	100 77 53 50 29 21 7 4	**0.12**	34	18	8	–	–	–	–	–	–	–	–	–	–	**11,18-Diacetoxy-5,6,12,17-trinaphthylenetetrone**	96722-33-7	**NI45899**
43 470 454 468 450 45 42 44	**554**	100 42 37 27 26 21 21 18	**1.22**	34	18	8	–	–	–	–	–	–	–	–	–	–	**12,17-Diacetoxy-5,6,11,18-trinaphthylenetetrone**	96722-26-8	**NI45900**
353 364 251 366 354 212 365 198	**554**	100 98 70 58 48 47 37 35	**0.00**	34	50	4	–	1	–	–	–	–	–	–	–	–	**Cholest-5-en-19-al, 3β-hydroxy-, p-toluenesulphonate**	3865-49-4	**NI45901**
342 458 253 261 75 459 144 146	**554**	100 82 74 63 38 32 26 24	**0.00**	34	66	5	–	–	–	–	–	–	–	–	–	–	**3α,7α,12α-Trihydroxy-5α-cholanoate**		**LI09684**
390 391 554 472 473 392 346 555	**554**	100 49 23 23 17 12 12 10		36	30	4	2	–	–	–	–	–	–	–	–	–	**N,N′-Dicyclohexyl-3,4:9,10-perylenebis(dicarboximide)**	41572-86-5	**DD01789**
390 391 554 472 473 392 346 555	**554**	100 49 23 23 17 12 12 10		36	30	4	2	–	–	–	–	–	–	–	–	–	**N,N′-Dicyclohexyl-3,4:9,10-perylenebis(dicarboximide)**	41572-86-5	**MS10676**
69 41 81 55 57 105 95 119	**554**	100 58 50 48 43 34 33 31	**6.70**	40	58	1	–	–	–	–	–	–	–	–	–	–	**Rhodopin**		**MS10677**
69 91 83 109 133 536 430 448	**554**	100 30 30 26 21 13 9 8	**7.00**	40	58	1	–	–	–	–	–	–	–	–	–	–	**Rhodopin**		**LI09685**
43 240 468 145 69 41 169 266	**555**	100 36 33 33 29 25 24 22	**1.26**	14	11	4	1	1	14	–	–	–	–	–	–	–	**N,S-Bis(heptafluorobutyryl)cysteine propyl ester**		**MS10678**
81 51 69 247 50 53 97 57	**555**	100 67 63 47 47 20 18 16	**3.39**	16	10	7	5	–	9	–	–	–	–	–	–	–	**Guanosine, 2′-deoxy-N-(trifluoroacetyl)-, 3′,5′-bis(trifluoroacetate)**	35170-11-7	**NI45902**
230 236 245 217 540 231 73 243	**555**	100 93 50 39 29 23 21 18	**5.54**	22	45	4	5	–	–	–	–	–	4	–	–	–	**Adenosine, tetrakis(trimethylsilyl)-**		**NI45903**
463 319 391 509 464 68 418 555	**555**	100 65 38 31 27 27 25 24		24	33	12	3	–	–	–	–	–	–	–	–	–	**2,4,6-Tris(diethoxycarbonylmethylene)hexahydro-s-triazine**		**NI45904**
199 198 186 169 130 194 354 168	**555**	100 62 60 45 45 30 27 23	**1.00**	26	33	7	7	–	–	–	–	–	–	–	–	–	**4-Dimethylaminonaphthylidene-glycyl-glycyl-glycyl-glycyl-glycyl-glycyl methyl ester**		**LI09686**
199 184 169 130 354 84 200 154	**555**	100 61 45 44 27 17 15 15	**0.80**	26	33	7	7	–	–	–	–	–	–	–	–	–	**Glycine, N-[N-[N-[N-[N-[N-[[N-(dimethylamino)-1-naphthalenyl]methylene]glycyl]glycyl]glycyl]glycyl]glycyl]-, methyl ester**	57237-91-9	**NI45905**

MASS TO CHARGE RATIOS								M.W.	INTENSITIES								Parent	C	H	O	N	S	F	Cl	Br	I	Si	P	B	X	COMPOUND NAME	CAS Reg No	No
314	230	231	315	243	215	150	122	**555**	100	22	20	16	10	10	2	2	**0.00**	27	18	4	5	–	–	–	1	–	–	–	–	–	**2-(4-Nitrophenyl)-6-oxo-8,10-diphenyl-4H,6H-pyrido[2,1-f][1,3,4]oxadiazino[2,3-c][1,2,4]triazin-11-ium bromide**		**MS10679**
78	108	124	80	79	400	200	323	**555**	100	93	70	63	55	50	45	40	**0.00**	32	32	–	3	–	–	1	–	–	–	2	–	–	**(Aminodiphenylphosphinoimino)(1-phenylethylamino)diphenylphosphane chloride**		**NI45906**
91	73	75	448	268	147	358	105	**555**	100	94	74	49	31	30	29	23	**2.72**	32	53	3	1	–	–	–	–	–	2	–	–	–	**Androstan-17-one, 3,11-bis[(trimethylsilyl)oxy]-, O-benzyloxime, (3α,5α,11β)-**	57305-18-7	**NI45907**
73	91	75	268	448	147	105	93	**555**	100	99	60	21	20	20	15	15	**1.04**	32	53	3	1	–	–	–	–	–	2	–	–	–	**Androstan-17-one, 3,11-bis[(trimethylsilyl)oxy]-, O-benzyloxime, (3α,5β,11β)-**	57305-17-6	**NI45908**
105	91	286	376	390	77	106	287	**555**	100	48	43	29	22	20	13	8	**0.00**	34	41	4	3	–	–	–	–	–	–	–	–	–	**(2S,5S)-1-Benzoyl-5-benzyl-5-[4-[(benzyloxycarbonyl)amino]butyl]-2-tert-butyl-3-methyl-4-imidazolidinone**		**DD01790**
118	28	186	158	59	214	298	416	**556**	100	99	60	39	35	32	23	21	**11.00**	13	4	10	–	–	–	–	–	–	–	–	–	4	**Decacarbonyl-μ-propynetetracobalt**		**MS10680**
321	424	319	426	422	323	271	558	**556**	100	96	60	57	57	56	24	23	**9.29**	14	11	4	–	–	–	6	–	–	–	–	–	1	**(η^4-1,2,3,4,7,7-hexachloro-5-methoxycarbonylnorbornadiene)acetylacetonatorhodium**	79209-99-7	**NI45909**
253	556	126	557	125	254	145	127	**556**	100	50	30	8	8	7	7	5		15	8	–	–	–	6	–	–	2	–	–	–	–	**1,3-Bis(m-iodophenyl)hexafluoropropane**		**LI09687**
462	460	464	458	466	560	424	426	**556**	100	87	65	33	26	10	10	9	**3.40**	19	16	2	–	–	–	8	–	–	–	–	–	–	**Pentachlorophenyl trichloro-2-(heptyloxy)phenyl ether**		**MS10681**
73	128	56	75	45	278	160	147	**556**	100	58	25	20	17	16	15	13	**0.00**	20	48	4	2	2	–	–	–	–	4	–	–	–	**Butanoic acid, 4,4′-dithiobis[2-[(trimethylsilyl)amino]-, bis(trimethylsilyl) ester, [S-(R*,R*)]-**	69688-45-5	**NI45910**
541	557	542	209	558	281	230	237	**556**	100	66	48	48	39	30	29	27	**0.00**	22	44	5	4	–	–	–	–	–	4	–	–	–	**Inosine, tetrakis(trimethylsilyl)-**		**NI45911**
44	415	462	460	79	147	207	194	**556**	100	46	28	28	28	27	22	19	**0.00**	23	23	4	–	–	6	–	1	–	–	–	–	–	**Estra-1,3,5(10)-triene-3,17-diol, 2-bromo-1-methyl-, bis(trifluoroacetate), (17β)-**	56438-20-1	**NI45912**
169	331	109	209	556	525	497		**556**	100	57	57	9	1	1	1			25	32	14	–	–	–	–	–	–	–	–	–	–	**Dehydrologanin tetraacetate**		**LI09688**
169	331	127	109	115	139	209	557	**556**	100	25	25	25	20	15	12	1	**0.00**	25	32	14	–	–	–	–	–	–	–	–	–	–	**Secologanin tetraacetate**		**LI09689**
91	124	199	214	65	556	79	246	**556**	100	67	60	45	38	29	28	20		26	24	4	2	4	–	–	–	–	–	–	–	–	**Bis(2-tosylamidophenyl) disulphide**		**LI09690**
170	114	156	155	113	197	83	42	**556**	100	97	80	59	58	48	48	44	**0.00**	26	38	5	4	–	–	2	–	–	–	–	–	–	**4,4′-Bipyridinium, 1,1′-bis[2-(3,5-dimethyl-4-morpholinyl)-2-oxoethyl]-, dichloride, dihydrate**	72060-08-3	**NI45913**
257	258	331	558	67	81	55	95	**556**	100	24	22	21	20	18	18	16	**11.00**	27	42	2	–	–	–	–	2	–	–	–	–	–	**5α-Spirostane, 23,23-dibromo-, (22S,25R)-**	24744-26-1	**NI45914**
257	558	331	560	556	559	258	286	**556**	100	79	52	42	40	24	23	20		27	42	2	–	–	–	–	2	–	–	–	–	–	**5α-Spirostane, 23,23-dibromo-, (25R)-**		**MS10682**
556	525	360	59	343	247	209	510	**556**	100	12	8	8	6	4	4	3		28	28	12	–	–	–	–	–	–	–	–	–	–	**Benzoic acid, 3-methoxy-4-[5-(methoxycarbonyl)-2,3-dimethoxyphenoxy]-5-[4-(methoxycarbonyl)-2-methoxyphenoxy]-, methyl ester**		**LI09691**
147	120	164	259	217	177	277	392	**556**	100	34	28	21	11	10	6	2	**0.60**	29	32	11	–	–	–	–	–	–	–	–	–	–	**8-C-β-D-[2-O-(E)-p-coumaroyl]glucopyranosyl-2-[(R)-2-hydroxy]propyl-7 methoxy-5-methylchromone**		**NI45915**
125	99	428	43	112	57	55	81	**556**	100	81	41	34	27	26	25	19	**11.88**	29	49	2	–	–	–	–	–	1	–	–	–	–	**Cholestan-3-one, 2-iodo-, cyclic 1,2-ethanediyl acetal, (2α,5α)-**	56052-90-5	**NI45916**
137	342	330	327	360	44	57	45	**556**	100	69	27	18	15	14	12	12	**0.00**	30	36	10	–	–	–	–	–	–	–	–	–	–	**Leptolepisol A**		**MS10683**
120	341	448	70	198	56	100	132	**556**	100	31	28	25	20	18	16	16	**0.00**	30	44	6	4	–	–	–	–	–	–	–	–	–	**(S,S)-Bis(benzylalanyl)diaza-18-crown-6**	93767-77-2	**MS10684**
120	341	448	70	56	100	132	263	**556**	100	31	27	25	18	16	16	13	**0.74**	30	44	6	4	–	–	–	–	–	–	–	–	–	**(S,S)-Bis[benzylalanyl]diaza-18-crown-6**	93767-77-2	**NI45917**
556	121	183	165	398	273	262	201	**556**	100	75	50	32	29	21	21	18		31	25	4	–	2	–	–	–	–	–	1	–	–	**Triphenylphosphonio(bisphenylsulphonylmethanide)**		**LI09692**
270	538	539	227	271	269	255	152	**556**	100	75	27	27	18	14	13	6	**2.00**	32	28	9	–	–	–	–	–	–	–	–	–	–	**Anhydrophlegmacin B2**	40451-45-4	**NI45918**
538	556	539	507	540	270	262	255	**556**	100	44	35	11	9	5	5	5		32	28	9	–	–	–	–	–	–	–	–	–	–	**Torosanin**	84813-71-8	**NI45919**
99	105	478	43	122	291	286	77	**556**	100	84	49	34	28	27	25	25	**1.50**	32	44	8	–	–	–	–	–	–	–	–	–	–	**Esteraglycone 1-ethylene ketal**		**MS10685**
439	73	147	75	440	43	57	129	**556**	100	97	51	39	35	33	29	24	**0.00**	32	68	3	–	–	–	–	–	–	2	–	–	–	**Hexacosanoic acid, 2-[(trimethylsilyl)oxy]-, trimethylsilyl ester**	55517-64-1	**NI45920**
105	313	77	43	57	55	69	45	**556**	100	75	64	47	43	31	23	20	**1.00**	33	24	5	4	–	–	–	–	–	–	–	–	–	**3-(2,3-Di-O-benzoyl-β-D-erythrofuranosyl)-1-phenylpyrazolo[3,4-b]quinoxaline**		**MS10686**
100	101	131	120	91	85	44	42	**556**	100	7	4	3	3	2	2	2	**0.09**	33	40	4	4	–	–	–	–	–	–	–	–	–	**Dihydrointegerressin**		**MS10687**
44	43	45	60	42	41	454	76	**556**	100	92	84	68	15	6	4	3	**0.00**	34	20	8	–	–	–	–	–	–	–	–	–	–	**6,17-Diacetoxy-11,12-dihydroxy-5,18-trinaphthylenedione**	96722-32-6	**NI45921**
43	45	454	49	60	42	455	84	**556**	100	42	41	34	24	21	21	20	**5.52**	34	20	8	–	–	–	–	–	–	–	–	–	–	**11,12-Diacetoxy-6,17-dihydroxy-5,18-trinaphthylenedione**	96722-30-4	**NI45922**
105	77	51	122	41	50	39	91	**556**	100	71	24	23	16	13	13	11	**0.09**	34	36	7	–	–	–	–	–	–	–	–	–	–	**Ingenol 3,5-dibenzoate**		**NI45923**
211	366	354	241	199	197	212	253	**556**	100	99	74	55	50	45	40	35	**0.00**	34	52	4	–	1	–	–	–	–	–	–	–	–	**Cholest-5-ene-3β,19-diol, 3-p-toluenesulphonate**	21072-66-2	**NI45924**
43	59	55	481	119	69	41	189	**556**	100	71	42	31	31	28	28	27	**3.00**	34	52	6	–	–	–	–	–	–	–	–	–	–	**Urs-12-en-28-oic acid, 2-hydroxy-3,23-(isopropylidenedioxyl)-11-oxo-, methyl ester**		**MS10688**
55	436	454	187	119	234	203	421	**556**	100	77	75	65	58	39	39	24	**18.00**	34	52	6	–	–	–	–	–	–	–	–	–	–	**Careyagenolide diacetate**		**NI45925**
190	274	247	306	262	215	175	497	**556**	100	43	31	24	24	22	18	17	**8.00**	34	52	6	–	–	–	–	–	–	–	–	–	–	**Urs-12-ene-27,28-dioic acid, 3β-hydroxy-, dimethyl ester, acetate**		**MS10689**
556	496	43	557	497	69	55	57	**556**	100	93	76	40	39	38	33	32		34	52	6	–	–	–	–	–	–	–	–	–	–	**Lanost-8-ene-7,11-dione, 3,18-bis(acetyloxy)-, (3β,20ξ)-**	56298-05-6	**NI45926**
187	306	246	247	186	190	496	173	**556**	100	82	70	41	37	35	31	30	**4.26**	34	52	6	–	–	–	–	–	–	–	–	–	–	**Olean-12-ene-28,29-dioic acid, 3β-hydroxy-, dimethyl ester, acetate**	4871-89-0	**NI45927**
262	496	556	437					**556**	100	53	16	16						34	52	6	–	–	–	–	–	–	–	–	–	–	**Rubifolic acid diacetate**		**NI45928**

MASS TO CHARGE RATIOS								M.W.	INTENSITIES								Parent	C	H	O	N	S	F	Cl	Br	I	Si	P	B	X	COMPOUND NAME	CAS Reg No	No
105	136	556	135	422	77	134	378	**556**	100	40	23	23	22	20	17	13		36	48	3	2	–	–	–	–	–	–	–	–	–	Benzamide, N-(18-benzamido-20-hydroxypregnan-3β-yl)-N-methyl-		LI09693
105	136	135	422	556	77	134	60	**556**	100	40	25	23	22	20	16	10		36	48	3	2	–	–	–	–	–	–	–	–	–	Benzamide, N-(18-benzamido-20β-hydroxy-5α-pregnan-3β-yl)-N-methyl-	32095-52-6	NI45929
105	149	77	183	147	373	150	91	**556**	100	94	24	23	21	16	10	6	**0.00**	38	36	4	–	–	–	–	–	–	–	–	–	–	(4R,5R)-2-(2,4,6-Trimethylphenyl)-α,α,α',α'-tetraphenyl-1,3-dioxolane-4,5 dimethanol		DD01791
218	190	189	203	408	191	556	219	**556**	100	83	71	65	58	52	51	50		39	56	2	–	–	–	–	–	–	–	–	–	–	Olean-12-en-3-ol, 3-phenyl-2-propenoate, (3β)-	21593-81-7	NI45930
218	190	203	408	556	219	189	204	**556**	100	75	68	57	50	50	50	35		39	56	2	–	–	–	–	–	–	–	–	–	–	Urs-12-en-3-ol, 3-phenyl-2-propenoate, (3β)-	13161-35-8	LI09694
218	190	203	408	556	219	202	409	**556**	100	75	68	56	50	50	48	28		39	56	2	–	–	–	–	–	–	–	–	–	–	Urs-12-en-3-ol, 3-phenyl-2-propenoate, (3β)-	13161-35-8	NI45931
73	317	223	217	75	43	318	74	**557**	100	38	22	17	17	12	11	10	**0.20**	24	43	8	1	–	–	–	–	–	3	–	–	–	β-D-Glucopyranosiduronic acid, 4-(acetylamino)phenyl 2,3,4-tris-O-(trimethylsilyl)-, methyl ester	55591-02-1	NI45932
362	205	363	191	556	345	557	190	**557**	100	73	60	42	38	30	28	27		30	39	9	1	–	–	–	–	–	–	–	–	–	14-β-Hydroxy-13-β-methyleneoxy-7,8-dimethoxy-15-β-(3,4,5-trimethoxybenzoxy)alloberban		NI45933
187	427	428	188	429	84	68	86	**557**	100	96	47	17	13	5	4	3	**0.08**	32	39	4	5	–	–	–	–	–	–	–	–	–	5,8-Ethenopyrrolo[3,2-b][1,5,8]oxadiazacyclotetradecine-12,15(1H,11H)-dione, 1-[2-(dimethylamino)-3-(1H-indol-3-yl)-1-oxopropyl]-2,3,3a,13,14,15a-hexahydro-13-isobutyl-	52617-27-3	NI45934
255	241	269	285	201	202	239	225	**557**	100	19	18	15	15	14	12	12	**4.00**	34	43	4	3	–	–	–	–	–	–	–	–	–	Tripyrrane A		LI09695
244	279	558	174	209	418	456	488	**558**	100	97	53	39	26	24	18	16		12	–	–	–	2	–	10	–	–	–	–	–	–	Bis(pentachlorophenyl) disulphide	22441-21-0	MS10690
246	281	244	562	283	248	279	564	**558**	100	74	74	51	50	50	47	44	**10.00**	12	–	–	–	2	–	10	–	–	–	–	–	–	Bis(pentachlorophenyl) disulphide	22441-21-0	NI45935
69	375	234	193	278	166	194	165	**558**	100	95	67	45	40	40	38	30	**0.00**	14	10	6	4	–	12	–	–	–	–	–	–	–	N,N',N'',N'''-Tetrakis(trifluoroacetyl)arginine	85985-01-9	NI45936
109	182	74	210	512	229	211	331	**558**	100	52	38	19	10	10	10	5	**0.00**	21	26	12	4	1	–	–	–	–	–	–	–	–	1,6-Dihydro-2-methylthio-5-((E)-2-nitrovinyl)-4-(2,3,4,6-tetra-O-acetyl-β-D-glucopyranosylamino)-1H-pyrimidin-6-one		MS10691
73	75	204	43	147	117	129	97	**558**	100	91	89	75	73	67	61	56	**0.00**	21	41	11	–	–	–	–	–	–	2	–	3	–	D-Glucose, 6-O-α-D-galactopyranosyl-, bis-O-(trimethylsilyl)-, cyclic tris(methylboronate)	72347-76-3	NI45938
73	75	191	117	115	218	43	204	**558**	100	98	55	51	47	46	45	44	**0.00**	21	41	11	–	–	–	–	–	–	2	–	3	–	D-Glucose, 6-O-α-D-galactopyranosyl-, bis-O-(trimethylsilyl)-, cyclic tris(methylboronate)	72347-76-3	NI45937
73	41	57	327	343	341	485	55	**558**	100	62	56	29	27	19	11	10	**0.00**	21	54	7	–	–	–	–	–	–	5	–	–	–	3,3,5-Tributoxy-1,1,1,7,7,7-hexamethyl-5-(trimethylsiloxy)tetrasiloxane		MS10692
558	197	186	257	228	439	169	109	**558**	100	39	32	30	27	18	17	15		22	30	11	4	1	–	–	–	–	–	–	–	–	5-Acetamido-1-methyl-2-(methylthio)-4-(2,3,4,6-tetra-O-acetyl-β-D-glucopyranosylamino)-1H-pyrimidin-6-one		NI45939
73	147	205	234	117	103	89	45	**558**	100	39	27	24	19	17	9	9	**0.00**	22	54	8	–	–	–	–	–	–	4	–	–	–	arabino-Hexos-2-ulose, 3,4,5,6-tetrakis-O-(trimethylsilyl)-, bis(dimethyl acetal)	74685-71-5	NI45940
139	134	207	193	310	141	102	133	**558**	100	62	54	50	43	34	33	29	**0.00**	25	20	2	2	2	–	1	1	–	–	–	–	–	2-(4-Chlorobenzoyl)-3-(4-methoxyphenyl)-6-phenyl-2,3-dihydro-4H-thiazolo[2,3-b][1,3,4]thiadiazin-5-ium bromide		MS10693
331	169	211	193	179	558	527		**558**	100	16	7	2	2	1	1			25	34	14	–	–	–	–	–	–	–	–	–	–	5-Epiloganin tetra acetate		LI09696
204	73	91	133	217	205	147	191	**558**	100	73	61	32	30	30	20	16	**0.00**	25	50	6	–	–	–	–	–	–	4	–	–	–	Benzyl-2,3,4,6-tetrakis-O-trimethylsilyl-β-D-glucopyranoside		LI09697
64	44	66	43	68	474	207	369	**558**	100	76	59	46	40	15	8	5	**0.17**	26	22	14	–	–	–	–	–	–	–	–	–	–	2',5',7',10'-Tetraacetoxy-2',3',7',8'-tetrahydrodispiro[furan-2(5H),4'-(9'H)-benzo[1,2-b][4,5-b']dipyran-9',2''(5''H)-furan]-5,5''-dione		NI45941
256	499	257	199	43	105	500	171	**558**	100	99	99	67	67	61	59	56	**8.97**	26	30	10	4	–	–	–	–	–	–	–	–	–	Riboflavin, 3-methyl-, 2',3',4',5'-tetraacetate	21066-33-1	NI45942
256	499	257	199	43	105	500	171	**558**	100	99	99	66	66	60	58	54	**10.00**	26	30	10	4	–	–	–	–	–	–	–	–	–	Riboflavin, 3-methyl-, 2',3',4',5'-tetraacetate	21066-33-1	LI09698
105	43	77	106	187	279	207	115	**558**	100	56	9	7	6	5	5	5	**0.00**	28	30	12	–	–	–	–	–	–	–	–	–	–	Hexitol, 1,3,4,6-tetraacetate 2,5-dibenzoate	74741-51-8	NI45943
105	107	43	231	106	393	109	77	**558**	100	49	45	20	14	13	9	8	**0.01**	28	30	12	–	–	–	–	–	–	–	–	–	–	Populin tetraacetate		MS10694
105	43	169	106	77	393	109	107	**558**	100	35	11	11	9	8	8	2	**0.00**	28	30	12	–	–	–	–	–	–	–	–	–	–	Tremuloidin tetraacetate		MS10695
282	191	397	281	468	217	487	173	**558**	100	87	77	43	39	39	35	35	**0.83**	28	58	5	–	–	–	–	–	–	3	–	–	–	Cyclopentanepentanoic acid, 3,5-bis[(trimethylsilyl)oxy]-2-[3-[(trimethylsilyl)oxy]-1-octenyl]-, methyl ester	55759-96-1	NI45944
282	191	397	281	468	217	487	173	**558**	100	89	77	44	38	38	35	35	**0.00**	28	58	5	–	–	–	–	–	–	3	–	–	–	Methyl 7α,9α,13-tri(trimethylsilyl)dinorprost-11-enoate		LI09699
43	57	71	78	432	85	58	55	**558**	100	94	60	56	43	39	34	29	**18.00**	31	26	10	–	–	–	–	–	–	–	–	–	–	Violet quinone		NI45945
83	43	55	71	310	122	121	179	**558**	100	68	59	40	29	21	21	16	**0.13**	31	42	9	–	–	–	–	–	–	–	–	–	–	20-O-Acetylisobutyryl-12-deoxy-16-hydroxyphorbol angelate		NI45946
426	427	385	428	411	386	355	412	**558**	100	36	13	11	9	6	5	4	**1.34**	33	58	3	–	–	–	–	–	–	2	–	–	–	6α-Hydroxy-Δ^1-tetrahydrocannabinol bis(triethylsilyl) ether		NI45947
413	414	558	415	559	544	416	560	**558**	100	39	13	11	6	4	4	2		33	58	3	–	–	–	–	–	–	2	–	–	–	7-Hydroxy-Δ^1-tetrahydrocannabinol bis(triethylsilyl) ether		NI45948
413	414	558	415	501	73	68	411	**558**	100	34	11	10	7	7	7	5		33	58	3	–	–	–	–	–	–	2	–	–	–	7-Hydroxy-Δ^1-tetrahydrocannabinol tert-butyldimethylsilyl ether		NI45949
456	438	498	43	45	457	423	439	**558**	100	94	58	47	45	40	36	34	**3.00**	34	54	6	–	–	–	–	–	–	–	–	–	–	Ergost-7-ene-3,6,23-triol, triacetate, (3β,5α,6α,23R)-	56312-73-3	NI45950
558	43	456	121	423	191	483	498	**558**	100	56	25	23	15	14	11	10		34	54	6	–	–	–	–	–	–	–	–	–	–	Lanostane-7,11-dione, 3,18-bis(acetyloxy)-, (3β,20ξ)-	56298-06-7	LI09700
558	43	559	69	456	121	57	55	**558**	100	56	40	33	25	23	22	22		34	54	6	–	–	–	–	–	–	–	–	–	–	Lanostane-7,11-dione, 3,18-bis(acetyloxy)-, (3β,20ξ)-	56298-06-7	NI45951

MASS TO CHARGE RATIOS	M.W.	INTENSITIES	Parent	C	H	O	N	S	F	Cl	Br	I	Si	P	B	X	COMPOUND NAME	CAS Reg No	No
189 468 136 204 191 202 203 483	**558**	100 72 72 56 53 51 48 44	**12.00**	35	62	3	–	–	–	–	–	–	1	–	–	–	**29,30-Dinorgammaceran-3-ol, 21,21-dimethyl-22-[(trimethylsilyl)oxy]-, acetate, (3β,8α,9β,13α,14β,17α,18β,22α)-**	69774-01-2	**NI45952**
178 279 113 202 558 102	**558**	100 81 46 21 19 19		36	30	–	–	–	–	–	–	–	–	–	–	2	**(Phenylethynyl)dicyclopentadienyldititanium**		**LI09701**
91 149 76 65 106 34 117 44	**559**	100 50 30 24 16 8 7 5	**0.30**	16	16	–	2	4	–	–	–	–	–	–	–	1	**Bis(N-benzyldithiocarbamato)platinum(II)**	68229-21-0	**NI45953**
346 149 69 347 333 28 169 107	**559**	100 53 32 23 22 21 18 11	**6.11**	17	11	4	1	–	14	–	–	–	–	–	–	–	**2-(3′-Hydroxy-4′-methoxyphenyl)-N,N-bis(heptafluorobutyryl)ethylamine**		**MS10696**
346 333 347 69 28 334 169 149	**559**	100 54 23 14 14 7 7 7	**3.43**	17	11	4	1	–	14	–	–	–	–	–	–	–	**2-(4′-Hydroxy-3′-methoxyphenyl)-N,N-bis(heptafluorobutyryl)ethylamine**		**MS10697**
55 180 134 442 503 380 324 325	**559**	100 50 42 40 32 30 30 24	**0.00**	32	30	3	3	–	–	–	–	–	–	–	–	1	**Dicarbonyl[5-ethoxy-3-methyl-1,5-diphenyl-2-(phenylimino)-4-imidazolidinylidene](methylcyclopentadienyl)manganese**	116928-11-1	**DD01792**
234 290 262 374 318 174 346 402	**560**	100 70 68 55 40 20 19 13	**0.00**	10	12	6	–	–	–	–	2	–	–	2	–	2	**μ-Tetramethyldiphosphinebis(bromotricarbonyliron)**		**MS10698**
545 533 546 534 547 259 535 548	**560**	100 52 50 29 27 27 23 17	**2.69**	10	24	12	–	–	–	–	–	–	8	–	–	–	**Hexamethyldivinyloctasilsesquioxane**	92888-99-8	**NI45954**
531 532 533 251 534 59 230 87	**560**	100 49 37 15 14 14 13 9	**0.00**	14	36	10	–	–	–	–	–	–	7	–	–	–	**1-Hydroperethylhomohexasilsesquioxane**	73895-14-4	**NI45955**
475 57 476 477 531 487 445 457	**560**	100 47 41 27 19 14 14 13	**0.00**	16	40	10	–	–	–	–	–	–	6	–	–	–	**1,3,5,7,9,11-Hexaethyl-5-oxy-9-butoxytricyclo[5.5.1.1^{3,11}]hexasiloxane**		**NI45956**
91 92 65 156 102 70 139 64	**560**	100 88 86 64 38 30 29 27	**0.00**	20	16	4	8	4	–	–	–	–	–	–	–	–	**4H,8H-Benzo[1,2-C:4,5-C′]bis[1,2,5]thiadiazol-4,8-dion-di-p-tosyl-hydrazon**	83274-19-5	**NI45957**
73 560 561 562 147 563 545 45	**560**	100 96 52 31 22 11 10 8		23	48	6	–	–	–	–	–	–	5	–	–	–	**Tetrakis(trimethylsiloxy)phenyl(trimethylsilyl)acetate**		**NI45958**
503 334 278 419 333 391 335 447	**560**	100 94 85 57 54 46 46 28	**0.20**	24	34	4	–	–	–	–	–	–	–	2	–	2	**Di-μ,μ′-[1-tert-butyl-3,4-dimethylphosphole]bis[dicarbonyliron]**	77611-24-6	**NI45959**
530 545 531 532 546 547 533 544	**560**	100 54 43 26 26 13 7 5	**1.80**	26	40	6	–	–	–	–	–	–	4	–	–	–	**1,2,5,8-Tetrahydroxyanthraquinone, tetrakis(trimethylsilyl)-**	91701-27-8	**NI45960**
78 197 135 73 117 51 132 94	**560**	100 73 57 31 26 26 25 20	**0.00**	26	52	7	–	–	–	–	–	–	3	–	–	–	**Cyclopentanebutanoic acid, 2-(7-methoxy-7-oxo-2-heptenyl)-α,3,5-tris[(trimethylsilyl)oxy]-, [1R-[1α(R*),2β(Z),3β,5β]]-**	74367-63-8	**NI45961**
245 229 544 246 91 230 213 122	**560**	100 95 90 90 90 85 85 80	**2.50**	30	30	1	2	2	–	–	–	–	–	2	–	–	**Di-p-tolylphosphinyliminomethyl di-p-tolylthiophosphinyiminomethyl sulphide**		**LI09702**
262 183 108 429 505 428 431 222	**560**	100 75 12 11 7 7 6 6	**0.00**	30	36	–	–	–	–	–	–	–	–	2	–	1	**Ruthenium, (η^5-2,4-cyclopentadien-1-yl)(2-methyl-2-propenyl)(trimethylphosphine)(triphenylphosphine)-**	74558-73-9	**NI45962**
86 473 105 179 71 127 207 28	**560**	100 99 97 90 90 86 83 75	**2.51**	30	40	10	–	–	–	–	–	–	–	–	–	–	**Carda-16,20(22)-dienolide, 3-[(4,6-dideoxy-3-O-methylhexopyranos-2-ulos 1-yl)oxy]-6,11,14(or 7,11,14)-trihydroxy-12-oxo-, (3β,5β,11α)-**	29610-97-7	**NI45963**
199 43 41 73 131 97 75 139	**560**	100 90 75 57 48 40 40 39	**5.00**	31	48	7	–	–	–	–	–	–	1	–	–	–	**3-Oxocolletotrichin, monotrimethylsilyl ether**		**NI45964**
560 44 561 544 542 558 545 135	**560**	100 69 39 36 27 25 21 20		33	20	9	–	–	–	–	–	–	–	–	–	–	**Isogalpinone**	69938-73-4	**NI45965**
560 106 135 134 562 545 561 543	**560**	100 71 56 50 40 32 31 25		33	20	9	–	–	–	–	–	–	–	–	–	–	**[1,2′:7′,2″-Ternaphthalene]-1″,4″,5,5′,8,8′-hexone, 1′,4,8″-trihydroxy-2,3′,6″-trimethyl-**	61774-50-3	**NI45966**
69 287 245 215 285 405 231 363	**560**	100 39 30 27 23 22 22 21	**19.00**	33	36	8	–	–	–	–	–	–	–	–	–	–	**2-Butenoic acid, 2-methyl-4-[3a,4,5,7-tetrahydro-8-hydroxy-3,3,11,11-tetramethyl-13-(3-methyl-2-butenyl)-7,15-dioxo-1,5-methano-1H,3H,11H-furo[3,4-g]pyrano[3,2-b]xanthen-1-yl]-**	5304-71-2	**NI45967**
403 385 341	**560**	100 17 15	**0.00**	33	52	7	–	–	–	–	–	–	–	–	–	–	**Crustecdysone diacetonide**		**LI09703**
466 73 129 151 150 467 75 213	**560**	100 98 57 51 42 41 40 37	**0.00**	33	60	3	–	–	–	–	–	–	2	–	–	–	**Cholest-5-en-22-one, 3,26-bis[(trimethylsilyl)oxy]-, (3β,25R)-**	54965-95-6	**NI45968**
560 561 350 562 175 351 335 280	**560**	100 67 42 21 17 14 11 9		34	40	–	4	–	–	–	–	–	2	–	–	–	**1,2,4,5-Tetraaza-3,6-disilacyclohexane, 3,6-dimethyl-1,2,4,5-tetra-m-tolyl-3,6-divinyl-**	17082-87-0	**NI45969**
129 73 75 81 83 95 69 85	**560**	100 99 85 61 59 56 54 51	**8.00**	34	64	2	–	–	–	–	–	–	2	–	–	–	**Cholest-5-ene, 3,25-bis(trimethylsilyloxy)-24-methyl-, (3β)-**		**MS10699**
305 101 43 113 297 560 192 443	**560**	100 29 27 22 9 5 4 3		35	60	5	–	–	–	–	–	–	–	–	–	–	**Cholestan-6-one, 3-[2-(acetyloxy)propoxy]-, cyclic 1,2-propanediyl acetal, (3β,5α)-**	38404-89-6	**NI45970**
305 101 43 113 306 96 53 55	**560**	100 30 27 22 21 12 10 9	**5.00**	35	60	5	–	–	–	–	–	–	–	–	–	–	**Cholestan-6-one, 3-[2-(acetyloxy)propoxy]-, cyclic 1,2-propanediyl acetal, (3β,5α)-**	38404-89-6	**LI09704**
560 561 562 483 563 280 485 315	**560**	100 45 41 28 17 15 12 12		37	25	–	4	–	–	1	–	–	–	–	–	–	**2H-Naphtho[1,2-d]triazole, 2-[1-(4-chlorophenyl)-1,4-dihydro-2,3-diphenyl-7-quinolinyl]-**	36858-31-8	**NI45971**
561 91 562 546 284 547 563 57	**561**	100 51 48 36 16 14 10 10		40	51	1	1	–	–	–	–	–	–	–	–	–	**Cholest-4-eno[3,2-b]indole, N-benzyl-6-oxo-**		**MS10700**
336 282 169 518 223 562 317 543	**562**	100 54 46 41 37 22 21 11		7	21	–	4	–	10	–	–	–	–	5	–	1	**Tris[difluoro(dimethylamino)phosphine][methyliminobis(diphenylphosphine)]iron**		**MS10701**
62 172 91 90 63 170 171 299	**562**	100 93 88 83 82 69 68 64	**4.60**	12	6	4	–	1	–	–	4	–	–	–	–	–	**Phenol, 4,4′-sulphonylbis[2,6-dibromo-**	39635-79-5	**NI45972**
43 409 85 562 28 69 154 271	**562**	100 68 49 42 42 38 17 13		15	12	6	–	–	9	–	–	–	–	–	–	1	**Rhodium, tris(1,1,1-trifluoro-2,4-pentanedionato-O,O′)-**	14652-54-1	**NI45973**

MASS TO CHARGE RATIOS								M.W.	INTENSITIES								Parent	C	H	O	N	S	F	Cl	Br	I	Si	P	B	X	COMPOUND NAME	CAS Reg No	No
533	443	534	87	115	535	444	445	**562**	100	80	54	50	40	37	35	34	**0.49**	15	42	9	–	–	–	–	–	–	7	–	–	–	1,3,5,7,9,11,13-Heptaethyl-11-methoxybicyclo[7.5.1]heptasiloxane	73113-20-9	NI45974
73	147	547	74	281	327	221	45	**562**	100	24	9	8	7	5	5	5	**0.00**	17	50	7	–	–	–	–	–	–	7	–	–	–	3-Ethoxy-1,1,1,7,7,7-hexamethyl-3,5,5-tris(trimethylsiloxy)tetrasiloxane		MS10702
282	157	59	43	131	117	184	217	**562**	100	95	81	78	68	68	55	53	**45.10**	18	26	–	–	10	–	–	–	–	–	–	–	–	2,4,6,8-Tetrathiatricyclo[3.3.1.1^{3,7}]decane, 1,1′-dithiobis[3,5,7-trimethyl-	57274-54-1	NI45975
123	124	121	122	502	429	159	145	**562**	100	89	88	74	68	38	33	32	**0.00**	23	25	8	–	–	7	–	–	–	–	–	–	–	4,5-Diacetoxy-3-(heptafluorobutyryl)scirpenol		NI45976
45	43	44	55	59	31	241	58	**562**	100	92	70	49	46	39	38	34	**0.60**	23	26	11	6	–	–	–	–	–	–	–	–	–	N-(2-Methoxyethoxycarbonylethyl)-3-acetylamino-6-methoxy-4-(3,5-dinitro-2-methoxycarbonylphenylazo)aniline		IC04587
480	295	52	322	565	257	43	227	**562**	100	54	34	14	12	10	10	6	**0.00**	24	30	13	–	–	–	–	–	–	–	–	–	4	Oxohexakis(crotonato)tetraberyllium(II)		MS10703
479	480	321	297	42	37	67	254	**562**	100	25	22	16	15	10	8	7	**0.00**	24	30	13	–	–	–	–	–	–	–	–	–	4	Oxohexakis(vinylacetato)tetraberyllium(II)		MS10704
111	203	159	168	75	76	251	63	**562**	100	67	67	58	58	50	37	33	**20.83**	25	16	7	–	2	–	2	–	–	–	–	–	–	Bis(4-chlorophenylsulphonylphenyl) carbonate		IC04588
57	41	27	237	227	58	265	405	**562**	100	46	43	29	23	16	10	10	**0.00**	26	54	7	–	–	–	–	–	–	3	–	–	–	1,3,5-Trivinyl-1,1,3,5,5-pentabutoxytrisiloxane	110991-09-8	NI45977
213	212	121	56	214	562	147	28	**562**	100	99	48	34	18	12	12	12		28	34	1	–	2	–	–	–	–	–	–	–	2	Bis[2-(1-ferrocenylethylthio)ethyl] ether		NI45978
73	259	303	75	147	129	81	54	**562**	100	54	32	23	17	17	16	16	**0.00**	28	62	5	–	–	–	–	–	–	3	–	–	–	Octadecanoic acid, 9,10,18-tris[(trimethylsilyl)oxy]-, methyl ester	40707-76-4	LI09705
73	259	303	75	147	129	81	55	**562**	100	55	33	23	18	17	17	17	**0.00**	28	62	5	–	–	–	–	–	–	3	–	–	–	Octadecanoic acid, 9,10,18-tris[(trimethylsilyl)oxy]-, methyl ester	40707-76-4	NI45979
273	73	173	274	75	169	129	147	**562**	100	42	41	21	10	9	9	8	**0.00**	28	62	5	–	–	–	–	–	–	3	–	–	–	Octadecanoic acid, 10,12,13-tris[(trimethylsilyl)oxy]-, methyl ester	56196-69-1	NI45980
73	259	75	81	55	146	68	303	**562**	100	30	19	17	16	12	12	11	**0.00**	28	62	5	–	–	–	–	–	–	3	–	–	–	Octadecanoic acid, 9,10,18-tris[(trimethylsilyl)oxy]-, methyl ester, erythro-		LI09706
73	259	75	303	129	147	81	54	**562**	100	48	25	22	22	21	19	17	**0.00**	28	62	5	–	–	–	–	–	–	3	–	–	–	Octadecanoic acid, 9,10,18-tris[(trimethylsilyl)oxy]-, methyl ester, threo-	22032-81-1	LI09707
73	259	75	303	147	129	81	55	**562**	100	50	25	24	21	21	20	18	**0.00**	28	62	5	–	–	–	–	–	–	3	–	–	–	Octadecanoic acid, 9,10,18-tris[(trimethylsilyl)oxy]-, methyl ester, threo-	22032-81-1	NI45981
486	433	488	453	224	346	468	236	**562**	100	43	40	39	38	37	36	32	**0.00**	29	39	7	2	–	–	1	–	–	–	–	–	–	9-O-Methyl-4,5-deoxymaytansinol	75349-70-1	NI45982
474	103	471	489	73	475	75	472	**562**	100	79	77	45	44	36	31	30	**7.00**	29	50	5	2	–	–	–	–	–	2	–	–	–	Pregn-4-en-18-al, 3-(methoxyimino)-20-oxo-11,21-bis[(trimethylsilyl)oxy]-, 18-(O-methyloxime), (11β,17α)-	69855-55-6	NI45983
73	531	75	103	441	562	532	84	**562**	100	99	94	77	71	46	45	42		29	50	5	2	–	–	–	–	–	2	–	–	–	Pregn-4-ene-3,11,20-trione, 17,21-bis[(trimethylsilyl)oxy]-, 3,20-bis(O-methyloxime)	57305-43-8	NI45984
148	108	347	122	96	149	121	163	**562**	100	44	37	30	28	26	21	19	**16.21**	29	50	5	6	–	–	–	–	–	–	–	–	–	L-Alanine, N-[N-[N^2-(1-oxodecyl)-N^5-pyrimidinyl-L-ornithyl]-L-leucyl]-, methyl ester	72088-17-6	NI45985
148	347	149	123	108	562	163	122	**562**	100	94	75	64	60	56	53	41		29	50	5	6	–	–	–	–	–	–	–	–	–	L-Alanine, N-[N-[N^2-(1-oxodecyl)-N^5-pyrimidinyl-L-ornithyl]-L-leucyl]-, methyl ester	72088-17-6	NI45986
148	562	563	84	123	108	564	347	**562**	100	81	65	58	47	44	42	37		29	50	5	6	–	–	–	–	–	–	–	–	–	L-Alanine, N-[N-[N^2-(1-oxodecyl)-N^5-pyrimidinyl-L-ornithyl]-L-leucyl]-, methyl ester	72088-17-6	NI45987
283	561	149	562	111	167	284	168	**562**	100	98	82	71	29	28	26	19		30	26	4	8	–	–	–	–	–	–	–	–	–	8,8′-(9,10-Phenanthrenediyl)bis(3,7-dihydro-1,3,7-trimethyl-1H-purine-2,6 dione)		NI45988
79	402	105	91	55	41	93	43	**562**	100	99	99	99	99	99	97	94	**0.42**	30	42	10	–	–	–	–	–	–	–	–	–	–	Card-20(22)-enolide, 3-[(4,6-dideoxy-3-O-methylhexopyranos-2-ulos-1-yl)oxy]-6,11,14(or 7,11,14)-trihydroxy-12-oxo-, (3β,11α)-	38890-62-9	NI45989
281	282	75	397	502	283	399	73	**562**	100	72	34	27	25	16	15	15	**0.00**	30	50	6	–	–	–	–	–	–	2	–	–	–	Pregn-4-ene-3,20-dione, 17-(acetyloxy)-6-methyl-6,21-bis[(trimethylsilyl)oxy]-, (6β)-	69833-78-9	NI45990
531	532	441	287	533	442	459	73	**562**	100	47	39	19	18	16	15	14	**7.08**	30	54	4	2	–	–	–	–	–	2	–	–	–	Pregn-4-ene-3,20-dione, 6-methyl-17,21-bis[(trimethylsilyl)oxy]-, bis(O-methyloxime), (6α)-	74299-03-9	NI45991
227	267	382	241	502	442	327	367	**562**	100	99	93	76	72	51	35	33	**4.00**	31	46	9	–	–	–	–	–	–	–	–	–	–	Methyl 3α,7α,11α-triacetoxy-12-oxo-5β-cholan-24-oate		NI45992
382	267	442	367	487	400	427	502	**562**	100	90	75	60	30	25	20	15	**10.00**	31	46	9	–	–	–	–	–	–	–	–	–	–	Methyl 3α,7α,12β-triacetoxy-11-oxo-5β-cholan-24-oate		NI45993
562	475	547	531					**562**	100	18	2	2						34	34	4	4	–	–	–	–	–	–	–	–	–	Pyromethyl phaophorbide		LI09708
335	562	561	168	336	349	563	321	**562**	100	69	51	42	35	31	24	24		35	34	5	2	–	–	–	–	–	–	–	–	–	Rodiasine, 6′,7-didemethoxy-2′-demethyl-6′,7-epoxy-, (1α,1′α)-	27577-49-7	MS10705
562	335	561	168	336	563	349	321	**562**	100	89	72	40	36	32	26	22		35	34	5	2	–	–	–	–	–	–	–	–	–	Rodiasine, 6′,7-didemethoxy-2′-demethyl-6′,7-epoxy-, (1ξ,1′ξ)-	26426-60-8	MS10706
166	259	199	181	98	57	41	72	**562**	100	60	44	32	31	30	30	28	**10.00**	35	38	3	2	–	–	–	–	–	1	–	–	–	6-[2-(Diethylamino)-4-oxo-1-cyclobutenyl]-2,2-dimethyl-6-oxo-4-(triphenylsiloxy)-3-hexen-3-carbonitrile		NI45994
91	43	95	117	105	121	81	69	**562**	100	92	73	62	62	49	46	46	**3.00**	37	54	4	–	–	–	–	–	–	–	–	–	–	9,19-Cyclo-25,26,27-trinorlanostane, 3β,24-diacetoxy-24-phenyl-		LI09709
82	96	250	83	55	69	57	43	**562**	100	99	89	74	74	70	63	58	**1.04**	38	74	2	–	–	–	–	–	–	–	–	–	–	Eicosanoic acid, 9-octadecenyl ester, (Z)-	22393-96-0	NI45995
55	69	83	57	43	97	82	41	**562**	100	87	79	77	72	65	63	54	**5.84**	38	74	2	–	–	–	–	–	–	–	–	–	–	9-Octadecene, 1,1′-[1,2-ethanediylbis(oxy)]bis-, (Z,Z)-	17367-13-4	NI45996
57	264	55	69	83	43	97	71	**562**	100	86	84	80	79	75	68	63	**2.74**	38	74	2	–	–	–	–	–	–	–	–	–	–	Oleic acid, eicosyl ester	22393-88-0	NI45997
203	40	91	119	105	145	143	133	**562**	100	99	85	82	67	63	51	49	**37.81**	40	50	2	–	–	–	–	–	–	–	–	–	–	Allogrohiarubin		NI45998
562	91	456	133	470	414	560	109	**562**	100	89	43	28	24	21	15	15		40	50	2	–	–	–	–	–	–	–	–	–	–	Rhodoxanthin		LI09710
562	91	105	145	456	119	157	135	**562**	100	89	50	48	43	43	39	39		40	50	2	–	–	–	–	–	–	–	–	–	–	Rhodoxanthin		MS10707
69	81	55	135	93	95	41	109	**562**	100	67	32	30	28	26	23	19	**1.39**	40	66	1	–	–	–	–	–	–	–	–	–	–	2,6,10,14,18,22,26,30-Dotriacontaoctaen-1-ol, 3,7,11,15,19,23,27,31-octamethyl-, (all-E)-	33569-79-8	NI45999
57	71	43	99	85	98	70	55	**562**	100	47	42	33	28	27	23	20	**0.00**	40	82	–	–	–	–	–	–	–	–	–	–	–	Octatriacontane, 3,5-dimethyl-	13897-20-6	NI46000

MASS TO CHARGE RATIOS								M.W.	INTENSITIES								Parent	C	H	O	N	S	F	Cl	Br	I	Si	P	B	X	COMPOUND NAME	CAS Reg No	No
355	73	356	357	45	75	74	358	**563**	100	84	34	17	9	7	7	4	**1.15**	24	34	3	1	–	5	–	–	–	3	–	–	–	Benzeneethanamine, N-[(pentafluorophenyl)methylene]-β,3,4-tris[(trimethylsilyl)oxy]-		MS10708
73	355	356	357	45	75	74	44	**563**	100	83	28	13	11	10	9	5	**0.91**	24	34	3	1	–	5	–	–	–	3	–	–	–	Benzeneethanamine, N-[(pentafluorophenyl)methylene]-β,3,4-tris[(trimethylsilyl)oxy]-	55429-13-5	NI46001
519	131	130	520	477	188	502	373	**563**	100	46	35	31	20	19	8	7	**1.00**	28	37	11	1	–	–	–	–	–	–	–	–	–	1-[N-((±)-1,2,3,4-tetrahydro-1-naphthyl)acetamido]-1-desoxy-2,3,4,5,6-penta-O-acetyl-D-galactitol		NI46002
266	267	165	102	73	268	209	297	**563**	100	21	15	12	10	6	4	3	**0.00**	28	49	5	1	–	–	–	–	–	3	–	–	–	Denopamine iso-M2, N,O,O-tris(trimethylsilyl)-		NI46003
266	267	165	102	73	268	297	209	**563**	100	21	15	12	10	6	4	4	**0.00**	28	49	5	1	–	–	–	–	–	3	–	–	–	Denopamine-M2, N,O,O-tris(trimethylsilyl)-		NI46004
71	532	531	122	84	98	110	150	**563**	100	62	17	16	13	10	9	8	**2.00**	30	45	9	1	–	–	–	–	–	–	–	–	–	Glaucerine	78018-29-8	NI46005
233	563	564	129	234	460	75	147	**563**	100	84	50	44	23	22	20	19		30	57	3	1	–	–	–	–	–	3	–	–	–	Pregn-5-en-20-one, 3β,21-bis(trimethylsilyloxy)-, O-(trimethylsilyl)oxime		LI09711
286	442	563	272	467	548	468	244	**563**	100	44	43	32	23	12	12	8		31	49	6	1	1	–	–	–	–	–	–	–	–	Daphniphylline methanesulphonate		LI09712
105	77	106	122	215	403	319	51	**563**	100	97	94	21	13	13	13	3	**0.00**	33	25	8	1	–	–	–	–	–	–	–	–	–	Tetra-O-benzoyl-D-arabinononitrile	20744-60-9	NI46006
77	105	106	281	51	269	319	122	**563**	100	99	94	40	25	19	10	8	**0.00**	33	25	8	1	–	–	–	–	–	–	–	–	–	Tetra-O-benzoyl-D-xylononitrile	71439-38-8	NI46007
563	233	91	564	232	57	55	234	**563**	100	64	63	46	33	20	19	15		40	53	1	1	–	–	–	–	–	–	–	–	–	Chloestano[3,2-b]indole, N-benzyl-6-oxo-, (5α)-		MS10709
535	549	536	550	537	538	125	253	**564**	100	69	47	39	28	11	10	9	**2.19**	10	28	12	–	–	–	–	–	–	8	–	–	–	Hexamethyldiethyloctasilsesquioxane	77680-72-9	NI46008
43	100	198	141	183	129	87	142	**564**	100	51	49	39	30	30	26	24	**0.60**	15	22	10	–	–	–	–	–	–	–	–	–	1	β-D-Glucopyranoside, methyl 2-[(acetyloxy)mercurio]-2-deoxy-, triacetate	35822-13-0	NI46009
365	69	532	296	117	129	217	227	**564**	100	99	75	66	65	60	41	35	**7.50**	18	–	–	–	1	15	–	–	–	–	1	–	–	Tris(pentafluorophenyl)phosphine sulphide		MS10710
396	480	338	452	280	308	366	163	**564**	100	52	52	34	18	16	15	15	**11.00**	18	12	14	–	–	–	–	–	–	–	–	–	2	Iron, hexacarbonyl[μ-[(1,2,3,4-η:1,4-η)-1,2,3,4-tetrakis(methoxycarbonyl) 1,3-butadiene-1,4-diyl]]di-, (Fe-Fe)	41551-07-9	MS10711
508	183	509	262	185	108	55	380	**564**	100	39	26	23	20	14	13	7	**2.99**	25	19	2	–	–	–	–	–	1	–	1	–	1	(Iodocyclopentadienyl)manganesedicarbonyltriphenylphosphine	33309-64-7	NI46010
43	162	163	180	480	420	45	42	**564**	100	91	53	34	26	17	17	17	**1.30**	26	28	14	–	–	–	–	–	–	–	–	–	–	Cyclohexanecarboxylic acid, 1,3,4-tris(acetyloxy)-5-[[3-[3,4-bis(acetyloxy)phenyl]-1-oxo-2-propenyl]oxy]-, [1R-(1α,3α,4α,5β)]-	55822-94-1	NI46011
91	109	92	79	108	77	64	74	**564**	100	36	36	31	29	21	15	14	**7.63**	26	32	8	2	2	–	–	–	–	–	–	–	–	L-Cystine, N,N′-bis[(phenylmethoxy)carbonyl]-, diethyl ester	23926-58-1	NI46012
55	111	236	184	238	262	201	485	**564**	100	75	41	28	21	20	18	16	**0.00**	28	37	8	2	–	–	1	–	–	–	–	–	–	Maytansinol	57103-68-1	NI46013
73	533	75	103	443	289	534	84	**564**	100	99	81	56	45	45	44	34	**31.18**	29	52	5	2	–	–	–	–	–	2	–	–	–	Pregn-4-ene-3,20-dione, 11-hydroxy-17,21-bis[(trimethylsilyl)oxy]-, bis(O methyloxime), (11β)-	57305-44-9	NI46014
73	447	448	147	267	357	117	75	**564**	100	99	41	31	30	28	25	25	**1.50**	30	56	4	–	–	–	–	–	–	3	–	–	–	Pregn-5-en-11-one, 3,17,20-tris[(trimethylsilyl)oxy]-, (3β,20R)-	57305-25-6	NI46015
73	447	117	75	267	448	357	147	**564**	100	99	44	43	40	38	33	33	**2.50**	30	56	4	–	–	–	–	–	–	3	–	–	–	Pregn-5-en-11-one, 3,17,20-tris[(trimethylsilyl)oxy]-, (3β,20S)-	40822-87-5	NI46016
430	429	447	385					**564**	100	58	39	9					**0.00**	31	48	9	–	–	–	–	–	–	–	–	–	–	Crustecdysone 2,3-diacetate		LI09713
126	405	387	343					**564**	100	95	56	31					**0.00**	31	48	9	–	–	–	–	–	–	–	–	–	–	Crustecdysone 3,22-diacetate		LI09714
384	269	385	270	549	337	261	207	**564**	100	83	39	26	16	16	16	15	**0.19**	31	56	5	–	–	–	–	–	–	2	–	–	–	5α-Cholan-24-oic acid, 3-oxo-7α,12α-bis[(trimethylsilyl)oxy]-, methyl ester	15093-98-8	NI46017
269	207	384	281	270	549	385	208	**564**	100	68	65	29	27	17	16	15	**7.69**	31	56	5	–	–	–	–	–	–	2	–	–	–	5β-Cholan-24-oic acid, 3-oxo-7α,12α-bis[(trimethylsilyl)oxy]-, methyl ester	15093-99-9	NI46018
269	207	384	281	270	549	385	208	**564**	100	68	49	29	26	16	16	14	**7.80**	31	56	5	–	–	–	–	–	–	2	–	–	–	5β-Cholan-24-oic acid, 3-oxo-7α,12α-bis[(trimethylsilyl)oxy]-, methyl ester	15093-99-9	MS10712
459	460	73	75	549	95	55	461	**564**	100	38	38	21	16	12	11	11	**1.50**	31	56	5	–	–	–	–	–	–	2	–	–	–	Methyl 3α,7β-dihydroxy-6-oxo-5β-cholan-24-oate, bis(trimethylsilyl) ether		NI46019
564	565	282	154	52	282.5	566	281.5	**564**	100	47	33	22	19	17	14	13		32	16	–	8	–	–	–	–	–	–	–	–	1	Chromium phthalocyanine		IC04589
129	403	321	456	143	546	147	133	**564**	100	96	88	80	50	47	40	40	**2.96**	33	64	3	–	–	–	–	–	–	2	–	–	–	Cholestan-5-ol, 3,6-bis[(trimethylsilyl)oxy]-, (3β,5α,6β)-	40272-82-0	NI46020
564	409	381	565	410	304	382	562	**564**	100	79	55	42	24	19	17	13		34	28	1	–	–	–	–	–	–	–	–	–	2	1′-Benzoyl-1‴-benzylbiferrocene		MS10713
564	562	563	458	472	560	348	308	**564**	100	33	16	11	10	8	6	6		38	44	4	–	–	–	–	–	–	–	–	–	–	3-Cyclopentene-1,2-dione, 4,4′-(3,7,12,16-tetramethyl-1,3,5,7,9,11,13,15,17 octadecanonaene-1,18-diyl)bis[3,5,5-trimethyl-, (all-E)-	22453-06-1	LI09715
564	562	563	458	472	560	308	461	**564**	100	83	40	27	25	22	15	14		38	44	4	–	–	–	–	–	–	–	–	–	–	3-Cyclopentene-1,2-dione, 4,4′-(3,7,12,16-tetramethyl-1,3,5,7,9,11,13,15,17 octadecanonaene-1,18-diyl)bis[3,5,5-trimethyl-, (all-E)-	22453-06-1	NI46021
549	564	534	267	282	520	262	457	**564**	100	22	20	17	14	10	7	5		38	52	–	4	–	–	–	–	–	–	–	–	–	α,γ-Dimethyl-α,γ-dihydrooctaethylporphyrin		LI09716
564	565	562	563	566	283	282	535	**564**	100	43	19	17	15	10	10	9		38	52	–	4	–	–	–	–	–	–	–	–	–	21H,23H-Porphine, 2,3,7,8,12,13,17,18-octaethyl-22,24-dihydro-21,23-dimethyl-	56630-82-1	MS10714
257	563	97	239	111	83	69	258	**564**	100	31	29	22	19	18	18	17	**14.30**	38	76	2	–	–	–	–	–	–	–	–	–	–	Docosanyl palmitate		NI46022
313	57	43	71	83	97	69	55	**564**	100	83	60	52	46	42	42	42	**13.04**	38	76	2	–	–	–	–	–	–	–	–	–	–	Eicosanoic acid, octadecyl ester	22432-79-7	NI46023
313	57	43	71	55	83	69	97	**564**	100	89	85	52	51	45	45	40	**29.57**	38	76	2	–	–	–	–	–	–	–	–	–	–	Octadecanoic acid, eicosyl ester	22413-02-1	NI46024
285	57	43	55	83	71	69	97	**564**	100	92	69	57	56	56	52	50	**8.44**	38	76	2	–	–	–	–	–	–	–	–	–	–	Octadecanoic acid, eicosyl ester	22413-02-1	NI46025
57	250	83	55	82	69	71	96	**564**	100	85	77	77	74	71	68	67	**9.34**	38	76	2	–	–	–	–	–	–	–	–	–	–	9-Octadecene, 1-[2-(octadecyloxy)ethoxy]-	56599-41-8	NI46026

MASS TO CHARGE RATIOS								M.W.	INTENSITIES								Parent	C	H	O	N	S	F	Cl	Br	I	Si	P	B	X	COMPOUND NAME	CAS Reg No	No
57	55	43	69	83	71	250	82	**564**	100	77	71	68	67	66	61	60	**1.04**	38	76	2	–	–	–	–	–	–	–	–	–	–	9-Octadecene, 1-[2-(octadecyloxy)ethoxy]-, (Z)-	17367-12-3	NI46027
564	69	41	55	565	91	43	105	**564**	100	53	53	51	45	43	43	42		40	52	2	–	–	–	–	–	–	–	–	–	–	Canthaxanthin		LI09717
83	564	43	91	69	133	109	472	**564**	100	41	26	19	15	14	8	7		40	52	2	–	–	–	–	–	–	–	–	–	–	Canthaxanthin		LI09718
486	328	170	190	76	249	565	213	**565**	100	20	5	4	4	3	2	2		–	–	–	3	–	–	1	5	–	–	3	–	–	1,3,5,2,4,6-Triazatriphosphorine, 2,2,4,4,6-pentabromo-6-chloro-2,2,4,4,6,6-hexahydro-	15608-37-4	NI46028
264	67	43	41	69	265	169	478	**565**	100	20	15	12	11	10	7	6	**0.42**	16	13	5	1	–	14	–	–	–	–	–	–	–	N,O-Bis(heptafluorobutyryl)hydroxyproline propyl ester		MS10715
178	232	334	216	550	305	460	319	**565**	100	50	45	20	12	11	10	8	**8.00**	25	43	4	3	1	–	–	–	–	3	–	–	–	4-Thia-1-azabicyclo[3.2.0]heptane-2-carboxylic acid, 6-[[[bis(trimethylsilyl)amino]phenylacetyl]amino]-3,3-dimethyl-7-oxo-, trimethylsilyl ester, [2S-[2α,5α,6β(S*)]]-	74231-50-8	NI46029
163	43	214	273	177	324	220	206	**565**	100	80	75	2	2	1	1	1	**0.10**	27	35	12	1	–	–	–	–	–	–	–	–	–	Isoipecoside		LI09719
283	155	55	71	43	57	339	41	**565**	100	91	59	39	38	36	31	24	**0.00**	28	56	8	1	–	–	–	–	–	–	1	–	–	3,5,9-Trioxa-4-phosphanonadecan-1-aminium, 4-hydroxy-N,N,N-trimethyl 10-oxo-7-[(1-oxodecyl)oxy]-, hydroxide, inner salt, 4-oxide, (R)-	3436-44-0	NI46030
86	437	324	69	393	508	438	281	**565**	100	96	91	75	51	46	45	45	**42.04**	30	55	5	5	–	–	–	–	–	–	–	–	–	Cyclic(L-isoleucyl-L-leucyl-L-isoleucyl-L-leucyl-L-leucyl)	38184-76-8	NI46031
565	533	478	445	490	550	521		**565**	100	46	44	19	7	6	5			33	35	4	5	–	–	–	–	–	–	–	–	–	γ-Cyano-2-devinyl-2-methoxycarbonylpyrrochlorin, methyl ester		LI09720
158	43	79	141	108	107	91	80	**565**	100	38	28	23	17	17	17	14	**0.21**	33	47	5	3	–	–	–	–	–	–	–	–	–	Glycinamide, N-[(3α)-20-oxopregnan-3-yl](benzyloxycarbonyl)glycyl-	56710-71-5	NI46032
474	565	475	159	566	158	476	249	**565**	100	64	48	30	28	15	14	8		40	55	1	1	–	–	–	–	–	–	–	–	–	5α-Cholestano[3,2-b]indole, 5′-benzyloxy-		LI09721
474	565	475	159	566	91	158	476	**565**	100	66	50	31	27	24	14	11		40	55	1	1	–	–	–	–	–	–	–	–	–	5α-Cholestano[3,2-b]indole, 5′-benzyloxy-		MS10716
474	565	475	566	146	91	476	57	**565**	100	81	47	36	18	17	10	9		40	55	1	1	–	–	–	–	–	–	–	–	–	5β-Cholestano[3,4-b]indole, 5′-benzyloxy-		MS10717
474	565	475	566	146	476	567	198	**565**	100	82	48	39	19	12	10	8		40	55	1	1	–	–	–	–	–	–	–	–	–	5β-Cholestano[3,4-b]indole, 5′-benzyloxy-		LI09722
88	568	392	304	216	480	488	244	**566**	100	99	99	50	30	20	1	1	**0.00**	–	–	–	–	–	12	–	2	–	–	4	–	1	Tetrakis(trifluorophosphine)iron dibromide		NI46033
312	185	439	566	311	309	293	166	**566**	100	42	30	20	8	5	4	4		12	32	–	4	–	8	–	–	–	–	4	–	1	Tetrakis(propylaminodifluorophosphino)nickel		NI46034
266	43	41	69	267	70	169	71	**566**	100	42	16	13	12	10	8	7	**0.98**	16	16	4	2	–	14	–	–	–	–	–	–	–	N,N′-Bis(heptafluorobutyryl)ornithine propyl ester		MS10718
342	286	398	340	284	283	339	396	**566**	100	95	86	83	79	77	71	68	**13.70**	20	13	8	–	–	–	–	–	–	–	1	–	2	[[η^4-(1-Phenyl-3,4-dimethylphosphole)tricarbonyliron]-P]pentacarbonylmolybdenum	77611-26-8	NI46035
102	465	348	103	288	219	147	73	**566**	100	11	10	10	9	9	8	8	**0.24**	22	54	5	2	–	–	–	–	–	5	–	–	–	Hexanedioic acid, 5-[bis(trimethylsilyl)amino]-2-[[bis(trimethylsilyl)amino]methyl]-2-[(trimethylsilyl)oxy]-	56247-64-4	NI46036
217	147	85	132	204	129	305	220	**566**	100	51	51	45	28	16	14	14	**0.30**	23	50	8	–	–	–	–	–	–	4	–	–	–	Isovaleryl-β-D-glucuronide, tetrakis(trimethylsilyl)-		NI46037
73	58	204	272	218	84	147	85	**566**	100	98	90	85	77	67	55	55	**8.00**	23	54	6	2	–	–	–	–	–	4	–	–	–	N-[2-(N,N-Dimethylamino)propyl]-D-glucuronamide, tetrakis(trimethylsilyl)-		NI46038
566	57	71	85	43	567	98	100	**566**	100	69	53	38	38	27	19	17		24	22	8	–	4	–	–	–	–	–	–	–	–	Tetramethyl 2,2′-(1,4-phenylene)bis[5-(methylthio)-3,4-thiophenedicarboxylate]	110699-34-8	DD01794
566	57	71	85	43	98	100	503	**566**	100	69	53	38	38	19	17	15		24	22	8	–	4	–	–	–	–	–	–	–	–	Tetramethyl 2,2′-(1,4-phenylene)bis[5-(methylthio)-3,4-thiophenedicarboxylate]	110699-34-8	DD01793
116	229	169	176	114	86	74	117	**566**	100	88	61	60	50	40	38	33	**7.00**	25	34	11	4	–	–	–	–	–	–	–	–	–	N-Acetylactinobolin-5-pyrazolone hexaacetate		LI09723
114	566	100	505	56	491	144	315	**566**	100	93	75	73	70	59	58	50		25	50	8	4	1	–	–	–	–	–	–	–	–	4,7,15,18,24,27,32,35-Octaoxa-1,10,12,21-tetraazabicyclo[9.8.8]heptatriaconta-11-thione		MS10719
328	310	398	370	338	356	440	380	**566**	100	93	77	77	77	59	57	48	**4.50**	28	22	13	–	–	–	–	–	–	–	–	–	–	Penta-O-acetylgrevillin D		NI46039
75	253	185	247	319	71	287	317	**566**	100	38	37	36	30	30	28	27	**0.00**	29	58	10	–	–	–	–	–	–	–	–	–	–	Eicosanoic acid, 5,6,8,9,11,12,14,15-octamethoxy-, methyl ester		LI09724
202	269	346	535	566	507	417		**566**	100	41	16	14	13	4	4			30	31	9	–	–	–	–	–	–	–	1	–	–	2,3,4,5-Tetrakis(methoxycarbonyl)-1,1,5-triphenyl-1-phosphacyclopenta-1,3-diene		LI09725
435	436	152	493	180	263	303	165	**566**	100	31	19	17	15	10	9	8	**5.00**	30	34	9	2	–	–	–	–	–	–	–	–	–	Tetraethyl 6(or 7)-ethoxy-6,7-dihydrobenzo[c]pyridazino[1,2-a]cinnoline-6,7,8,9-tetracarboxylate		NI46040
73	449	75	147	117	359	450	55	**566**	100	99	99	90	82	56	40	40	**4.00**	30	58	4	–	–	–	–	–	–	3	–	–	–	Pregnan-11-one, 3,17,20-tris[(trimethylsilyl)oxy]-, (3α,5β,20S)-	55557-06-7	NI46041
73	449	359	450	147	269	75	117	**566**	100	89	48	32	26	24	23	18	**2.28**	30	58	4	–	–	–	–	–	–	3	–	–	–	Pregnan-11-one, 3,17,20-tris[(trimethylsilyl)oxy]-, (3α,5β,20S)-	55557-06-7	NI46042
73	373	283	463	147	374	129	75	**566**	100	99	89	50	42	33	27	27	**7.00**	30	58	4	–	–	–	–	–	–	3	–	–	–	Pregnan-11-one, 3,20,21-tris[(trimethylsilyl)oxy]-, (3α,5β,20S)-	57377-54-5	NI46043
73	391	255	75	147	143	256	81	**566**	100	99	99	99	65	50	47	46	**0.00**	30	58	4	–	–	–	–	–	–	3	–	–	–	Pregnan-20-one, 3,11,21-tris[(trimethylsilyl)oxy]-, (3α,5α,11β)-	57397-18-9	NI46044
255	463	143	159	345	162	164	373	**566**	100	35	35	22	18	18	16	14	**11.00**	30	58	4	–	–	–	–	–	–	3	–	–	–	Pregnan-20-one, 3,11,21-tris[(trimethylsilyl)oxy]-, (3α,5β,11β)-	56196-42-0	NI46045
407	476	433	408	217	477	143	434	**566**	100	65	55	42	30	26	24	23	**0.00**	30	58	4	–	–	–	–	–	–	3	–	–	–	Pregnan-20-one, 3,15,16-tris[(trimethylsilyl)oxy]-, (3β,5α,15α,16α)-		LI09726
407	476	75	433	408	477	217	71	**566**	100	78	57	42	36	32	30	30	**3.50**	30	58	4	–	–	–	–	–	–	3	–	–	–	Pregnan-20-one, 3,15,16-tris[(trimethylsilyl)oxy]-, (3ξ,5ξ,16β,16α)-		LI09727
73	435	255	147	103	75	345	231	**566**	100	99	99	99	99	99	72	70	**57.00**	30	58	4	–	–	–	–	–	–	3	–	–	–	Pregnan-20-one, 3,17,21-tris[(trimethylsilyl)oxy]-, (3α,5β)-	57397-20-3	NI46046
73	566	567	230	147	75	568	231	**566**	100	74	40	30	26	26	18	15		30	58	4	–	–	–	–	–	–	3	–	–	–	Pregn-20-en-11-ol, 3,20,21-tris[(trimethylsilyl)oxy]-, (3α,5β,11β)-	56248-33-0	NI46047

MASS TO CHARGE RATIOS								M.W.	INTENSITIES								Parent	C	H	O	N	S	F	Cl	Br	I	Si	P	B	X	COMPOUND NAME	CAS Reg No	No
348	300	347	346	349	318	301	314	**566**	100	31	25	24	21	19	15	12	**0.00**	32	22	10	–	–	–	–	–	–	–	–	–	–	4H-1-Benzopyran-4-one, 5,7-dihydroxy-8-[5-(5-hydroxy-7-methoxy-4-oxo 4H-1-benzopyran-2-yl)-2-methoxyphenyl]-2-(4-hydroxyphenyl)-	481-46-9	NI46048
566	345	191	567	369	219	551	440	**566**	100	70	53	33	17	16	11	7		32	55	–	–	–	–	–	–	1	–	–	–	–	30-(2-Iodoethyl)hopane		DD01795
125	272	443	397	127	399	445	398	**566**	100	73	66	60	41	30	22	22	**5.00**	34	28	2	2	–	–	2	–	–	–	–	–	–	3-(γ-Benzoylidene-N-p-chlorobenzylpropionamide)-N-p-chlorobenzyl-5-phenylpyrroline-2-one	104924-50-7	NI46049
566	568	567	569	570	146	91	144	**566**	100	73	38	27	15	14	13	9		34	28	2	2	–	–	2	–	–	–	–	–	–	1,4-Bis(5,6,7,8-tetrahydronaphth-2-ylamino)-6,7-dichloroanthraquinone		IC04590
550	551	395	394	304	564	149	393	**566**	100	43	33	33	31	23	23	22	**15.30**	34	30	1	–	–	–	–	–	–	–	–	–	2	1′-(α-Hydroxybenzyl)-1‴-benzylbiferrocene		MS10720
566	567	537	538	255	568	493	465	**566**	100	44	40	17	15	11	8	8		36	46	2	4	–	–	–	–	–	–	–	–	–	1,1,4,4,5,6,7,8-Octaethyl-2,3-dioxo-1,2,3,4-tetrahydroporphine		LI09728
73	55	43	57	71	98	239	84	**566**	100	99	93	83	77	68	52	43	**1.14**	36	70	4	–	–	–	–	–	–	–	–	–	–	Hexadecanoic acid, 1,4-butanediyl ester	26719-63-1	NI46050
239	311	57	43	55	28	71	127	**566**	100	57	52	48	43	41	32	29	**0.00**	36	70	4	–	–	–	–	–	–	–	–	–	–	Hexadecanoic acid, 1,1-dimethyl-1,2-ethanediyl ester	56630-27-4	NI46051
311	55	114	310	57	43	127	239	**566**	100	67	58	40	40	40	33	31	**0.20**	36	70	4	–	–	–	–	–	–	–	–	–	–	Hexadecanoic acid, 1-methyl-1,3-propanediyl ester	33599-10-9	NI46052
43	57	283	98	99	55	311	84	**566**	100	84	67	59	57	56	52	45	**0.00**	36	70	4	–	–	–	–	–	–	–	–	–	–	Octadecanoic acid, 2-[(1-oxohexadecyl)oxy]ethyl ester	26158-81-6	NI46053
43	99	283	28	57	311	55	98	**566**	100	91	89	86	79	74	64	54	**1.74**	36	70	4	–	–	–	–	–	–	–	–	–	–	Octadecanoic acid, 2-[(1-oxohexadecyl)oxy]ethyl ester	26158-81-6	NI46054
74	98	55	43	87	57	69	83	**566**	100	97	84	82	75	67	66	50	**22.52**	36	70	4	–	–	–	–	–	–	–	–	–	–	Tetratriacontanedioic acid, dimethyl ester	55429-46-4	NI46055
431	326	566	432	327	91	567	340	**566**	100	95	51	36	25	25	23	20		38	34	3	2	–	–	–	–	–	–	–	–	–	3-Oxo-6′-diethylamino-2′-dibenzylaminospiro[phthalan-1,9′-xanthene]		IC04591
83	85	564	47	43	91	41	119	**566**	100	63	41	26	26	19	18	16	**0.00**	40	54	2	–	–	–	–	–	–	–	–	–	–	Canthaxanthin		MS10721
566	564	474	549	283	456	551	472	**566**	100	45	21	19	17	15	14	7		40	54	2	–	–	–	–	–	–	–	–	–	–	β,ϵ-Caroten-3′-one, 3-hydroxy-, (3R,6′R)-	23984-55-6	NI46056
566	474	460	548	475	456	368	302	**566**	100	16	9	8	7	5	5	5		40	54	2	–	–	–	–	–	–	–	–	–	–	β,ϵ-Caroten-3′-one, 3-hydroxy-, (3R,6′R)-	23984-55-6	LI09729
91	566	43	460	69	59	73	133	**566**	100	52	48	37	37	36	26	21		40	54	2	–	–	–	–	–	–	–	–	–	–	2′-Dehydroplectaniaxanthin		LI09730
91	73	106	105	77	39	92	69	**566**	100	96	54	29	14	13	11	10	**1.30**	41	58	1	–	–	–	–	–	–	–	–	–	–	Anhydrorhodovibrin		MS10722
91	73	69	458	43	133	109	566	**566**	100	96	10	5	4	2	2	1		41	58	1	–	–	–	–	–	–	–	–	–	–	Anhydrorhodovibrin		LI09731
135	136	216	147	282	146	283	285	**566**	100	98	98	97	84	63	52	48	**4.20**	42	62	–	–	–	–	–	–	–	–	–	–	–	1,1,2,2-Tetrakis(adamant-1-yl)ethane		NI46057
280	281	567	279	203	258	307	202	**566**	100	30	25	25	20	19	15	15	**5.00**	43	34	1	–	–	–	–	–	–	–	–	–	–	1,4:9,10-Dimethanoanthracen-12-one, 11-(diphenylmethylene)-1,4,4a,9,9a,10-hexahydro-9,10-dimethyl-2,3-diphenyl-	55937-91-2	NI46058
236	78	220	130	77	208	165	202	**567**	100	53	38	28	28	27	24	18	**0.00**	16	14	1	1	–	–	6	–	–	–	–	–	1	2-Ethyl-3-phenyl-1-oxoisoindolium hexachloroantimonate		LI09732
43	71	69	238	169	283	119	226	**567**	100	69	23	14	9	2	2	1	**0.00**	16	15	5	1	–	14	–	–	–	–	–	–	–	3-Methylbutyl N,O-bis(heptafluorobutyryl)serinate		NI46059
360	374	567	181	167	117	131	568	**567**	100	67	65	29	25	25	24	17		20	–	–	3	–	15	–	–	–	–	–	–	–	3,4,5-Tris(pentafluorophenyl)-1,2,4-triazole		LI09733
225	226	422	240	423	269	168	155	**567**	100	77	69	67	65	61	60	58	**48.97**	32	61	5	3	–	–	–	–	–	–	–	–	–	Leucine, N-(N[2],N[5]-didecanoyl-L-ornithyl)-, methyl ester, L-	16859-09-9	NI46060
59	15	131	524	424	69	324	549	**568**	100	38	12	11	9	9	6	4	**0.00**	13	6	4	–	–	18	–	–	–	–	–	–	–	Undecanedioic acid, octadecafluoro-, dimethyl ester	22116-91-2	LI09734
59	15	131	524	69	424	324	100	**568**	100	38	12	11	10	9	6	4	**0.00**	13	6	4	–	–	18	–	–	–	–	–	–	–	Undecanedioic acid, octadecafluoro-, dimethyl ester	22116-91-2	NI46061
77	83	73	469	471	41	453	468	**568**	100	97	68	35	27	27	25	17	**0.00**	15	21	5	–	–	12	–	–	–	1	1	–	–	4,4,5,5-Tetrakis(trifluoromethyl)-4′,4′,5′,5′-tetramethyl-2-(trimethylsiloxy) spiro[1′,3′,2′λ[5]σ[5]]-dioxaphospholane-2,2′-[1,3,2λ[5]σ[5]]-dioxaphospholane		MS10723
523	508	568	359	210	342	538	344	**568**	100	95	93	57	55	48	45	45		18	9	–	6	–	12	–	–	–	–	1	–	–	Tris(hydrazinotetrafluorophenyl)phosphine		MS10724
231	134	568	232	149	121	163	77	**568**	100	40	17	17	9	9	6	6		20	12	2	–	2	6	–	–	–	–	–	–	1	Bis(1,1,1-trifluoro-4-phenyl-4-thioxo-2-butanonato)palladium	53776-98-0	NI46062
29	41	511	509	455	507	510	57	**568**	100	65	58	43	22	21	19	18	**7.64**	20	18	–	–	–	10	–	–	–	–	–	–	1	Stannane, dibutylbis(pentafluorophenyl)-	1059-35-4	NI46063
568	210	389	569	229	193	167	69	**568**	100	50	38	30	26	10	10	9		21	9	3	–	–	12	–	–	–	–	1	–	–	Tris(methoxytetrafluorophenyl)phosphine		MS10725
73	217	147	204	103	218	75	232	**568**	100	54	20	11	11	9	9	8	**0.60**	22	52	7	–	–	–	–	–	–	5	–	–	–	D-Glycero-D-gulo-heptanoic acid, 2,3,5,6,7-pentakis-O-(trimethylsilyl)-, γ-lactone	10589-41-0	LI09735
73	217	147	205	218	103	75	74	**568**	100	54	20	13	11	11	8	8	**1.20**	22	52	7	–	–	–	–	–	–	5	–	–	–	D-Glycero-D-gulo-heptanoic acid, 2,3,5,6,7-pentakis-O-(trimethylsilyl)-, γ-lactone	10589-41-0	NI46064
217	73	147	218	103	75	219	74	**568**	100	99	23	22	9	9	8	8	**0.10**	22	52	7	–	–	–	–	–	–	5	–	–	–	D-Glycero-L-manno-heptanoic acid, 2,3,5,6,7-pentakis-O-(trimethylsilyl)-, γ-lactone	27531-31-3	NI46065
73	217	216	202	218	45			**568**	100	92	36	27	17	16			**0.00**	22	56	5	2	–	–	–	–	–	5	–	–	–	D-Glucose, 2-deoxy-3,4,5,6-tetrakis-O-(trimethylsilyl)-2-[(trimethylsilyl)amino]-, O-methyloxime	74367-65-0	NI46066
169	109	81	197	225	196	407	239	**568**	100	70	55	50	36	36	34	26	**6.00**	23	28	11	4	1	–	–	–	–	–	–	–	–	4-Acetamido-2-methyl-6-(methylthio)-4-[N-(2,3,4,6-tetra-O-acetyl-β-D-glucopyranosyl)amino]oxazolo[5,4-d]pyrimidine		NI46067
259	57	41	241	260	271	253	261	**568**	100	78	34	33	24	22	20	15	**0.00**	26	60	7	–	–	–	–	–	–	3	–	–	–	1,3,5-Triethyl-1,1,3,5,5-pentabutoxytrisiloxane	110991-10-1	NI46068
73	568	569	553	437	129	554	211	**568**	100	91	54	44	32	27	26	26		27	49	5	–	–	–	–	–	–	3	1	–	–	Estra-1,3,5(10)-trien-3-ol, 17β-(trimethylsiloxy)-, bis(trimethylsilyl) phosphate	33745-64-1	NI46069
568	218	227	73	326	211	243	569	**568**	100	91	83	75	55	52	45	38		27	49	5	–	–	–	–	–	–	3	1	–	–	Estra-1,3,5(10)-trien-17β-(trimethylsiloxy)-, bis(trimethylsilyl) phosphate	33745-63-0	NI46070

MASS TO CHARGE RATIOS	M.W.	INTENSITIES	Parent	C	H	O	N	S	F	Cl	Br	I	Si	P	B	X	COMPOUND NAME	CAS Reg No	No
156 528 526 44 43 99 513 511	**568**	100 85 69 62 55 43 41 39	**31.70**	28	25	8	–	–	–	–	1	–	–	–	–	–	4-Acetoxy-3-bromo-2,3′-dimethyl-5,5′,8,8′-tetramethoxy-1,2′-binaphthalene-1′,4′-dione	104506-01-6	NI46071
43 96 109 41 125 123 69 55	**568**	100 17 12 12 11 10 10 9	**1.79**	28	40	12	–	–	–	–	–	–	–	–	–	–	5,6-Dihydroxyingol 3,7,8,12-tetraacetate		NI46072
171 170 169 568 168 57 569 55	**568**	100 64 57 52 29 20 18 17		29	36	4	4	2	–	–	–	–	–	–	–	–	1-Naphthalenesulphonamide, N,N′-1,5-pentanediylbis[5-(dimethylamino)-	55521-24-9	NI46073
253 81 95 75 121 107 145 159	**568**	100 59 56 56 54 51 44 41	**5.00**	29	46	1	–	–	–	–	2	–	–	–	–	–	(23R,24R)-23,24-(Dibromomethylene)-6β-methoxy-3α,5-cyclo-5α-cholestane		MS10726
225 173 395 243 226 147 129 111	**568**	100 29 27 20 20 20 19 18	**4.25**	29	56	5	–	–	–	–	–	–	3	–	–	–	Prosta-7,13-dien-1-oic acid, 6,9-epoxy-11,15-bis[(trimethylsilyl)oxy]-, trimethylsilyl ester, (13E,15S)-	56248-53-4	NI46074
225 173 75 190 353 73 199 407	**568**	100 98 70 57 44 43 42 41	**0.50**	29	56	5	–	–	–	–	–	–	3	–	–	–	Prosta-5,11-dien-1-oic acid, 9-oxo-11,15-bis[(trimethylsilyl)oxy]-, trimethylsilyl ester, (5Z,11α,13E,15S)-	39003-20-8	NI46075
568 128 569 284 156 184 284.5 56	**568**	100 47 39 37 37 27 16 15		32	16	–	8	–	–	–	–	–	–	–	–	1	Iron phthalocyanine		IC04592
481 568 451 554 538 524 509	**568**	100 92 85 69 57 21 15		33	36	5	4	–	–	–	–	–	–	–	–	–	2-Devinyl-γ-formyl-2-methoxycarbonylpyrrochlorin, methyl ester		LI09736
150 137 210 219 95 233 135 177	**568**	100 60 39 29 26 24 22 21	**1.70**	33	44	8	–	–	–	–	–	–	–	–	–	–	Cystoseirol D acetate		NI46076
148 149 477 131 91 42 147 133	**568**	100 95 78 50 40 33 30 30	**0.50**	34	40	4	4	–	–	–	–	–	–	–	–	–	Benzenepropanamide, α-(dimethylamino)-N-[7-isobutyl-5,8-dioxo-3-phenyl-2-oxa-6,9-diazabicyclo[10.2.2]hexadeca-10,12,14,15-tetraen-4-yl]-, [3R-[3R*,4S*(S*),7S*]]-	23926-98-9	NI46077
366 211 354 241 199 197 212 253	**568**	100 99 74 53 50 46 38 29	**0.00**	35	52	4	–	1	–	–	–	–	–	–	–	–	19-Hydroxy-3-tosyloxycholestene		LI09737
43 57 55 41 98 73 28 60	**568**	100 95 80 64 61 60 54 50	**0.34**	35	68	5	–	–	–	–	–	–	–	–	–	–	Hexadecanoic acid, 1-(hydroxymethyl)-1,2-ethanediyl ester	761-35-3	NI46078
331 313 551 332 552 391 314 239	**568**	100 40 31 21 13 8 8 6	**0.00**	35	68	5	–	–	–	–	–	–	–	–	–	–	Hexadecanoic acid, 1-(hydroxymethyl)-1,2-ethanediyl ester	761-35-3	NI46079
551 569 313 552 570 257 138 129	**568**	100 71 62 39 28 27 15 15	**0.00**	35	68	5	–	–	–	–	–	–	–	–	–	–	Hexadecanoic acid, 1-(hydroxymethyl)-1,2-ethanediyl ester	761-35-3	NI46080
43 57 55 41 28 73 60 29	**568**	100 78 72 72 67 59 50 43	**0.00**	35	68	5	–	–	–	–	–	–	–	–	–	–	Hexadecanoic acid, 2-hydroxy-1,3-propanediyl ester	502-52-3	NI46081
551 313 552 331 314 553 257 239	**568**	100 66 45 17 17 11 6 3	**0.00**	35	68	5	–	–	–	–	–	–	–	–	–	–	Hexadecanoic acid, 2-hydroxy-1,3-propanediyl ester	502-52-3	NI46082
551 313 552 569 314 331 257 553	**568**	100 68 38 15 15 13 11 8	**0.00**	35	68	5	–	–	–	–	–	–	–	–	–	–	Hexadecanoic acid, 2-hydroxy-1,3-propanediyl ester	502-52-3	NI46083
239 137 193 136 240 328 327 103	**568**	100 58 51 50 45 42 12 7	**2.80**	37	28	6	–	–	–	–	–	–	–	–	–	–	Phenylbis(6-hydroxyflavanon-3-yl)methane		NI46084
568 229 270 244 553 269 283 227	**568**	100 25 20 19 17 17 16 16		39	52	3	–	–	–	–	–	–	–	–	–	–	16-(3-Oxo-4-androsten-17-ylmethylene)-4-androstene-3,17-dione		LI09738
568 448 550 396 449 551 462 397	**568**	100 42 24 19 15 10 10 8		40	56	2	–	–	–	–	–	–	–	–	–	–	ψ,ψ-Caroten-20-al, 1,2-dihydro-1-hydroxy-, 13-cis-	17884-87-6	NI46085
91 119 105 145 41 55 93 107	**568**	100 69 69 51 50 46 43 41	**22.50**	40	56	2	–	–	–	–	–	–	–	–	–	–	β,ε-Carotene-3,3′-diol, (3R,3′R,6′R)-	127-40-2	NI46086
91 119 105 145 41 55 93 324	**568**	100 69 69 51 50 46 43 41	**22.50**	40	56	2	–	–	–	–	–	–	–	–	–	–	β,ε-Carotene-3,3′-diol, (3R,3′R,6′R)-	127-40-2	MS10727
568 550 569 551 457 570 548 444	**568**	100 78 43 34 17 14 14 14		40	56	2	–	–	–	–	–	–	–	–	–	–	β,ε-Carotene-3,3′-diol, (3R,3′R,6′R)-	127-40-2	LI09739
105 568 550 476 458 462 392 410	**568**	100 53 24 21 7 6 6 3		40	56	2	–	–	–	–	–	–	–	–	–	–	β,ε-Carotene-3,3′-diol, (3R,3′S,6′R)-	52842-48-5	NI46087
225 165 95 91 105 93 69 119	**568**	100 64 43 38 35 34 33 28	**24.00**	40	56	2	–	–	–	–	–	–	–	–	–	–	β-Carotene 5,8,5′,8′-dioxide		LI09740
221 568 488 181 352 569 489 287	**568**	100 78 69 55 43 34 27 19		40	56	2	–	–	–	–	–	–	–	–	–	–	Cryptoflavin	30311-63-8	NI46088
41 56 54 568 69 71 119 105	**568**	100 99 68 56 54 49 40 38		40	56	2	–	–	–	–	–	–	–	–	–	–	Lutein	62624-08-2	NI46089
43 91 568 69 462 109 476 83	**568**	100 81 35 35 25 19 8 8		40	56	2	–	–	–	–	–	–	–	–	–	–	Lycophyll		LI09741
43 91 105 119 93 145 107 81	**568**	100 81 59 46 46 45 40 39	**35.40**	40	56	2	–	–	–	–	–	–	–	–	–	–	Lycophyll		MS10728
91 462 69 43 133 568 109 463	**568**	100 39 34 30 25 17 15 13		40	56	2	–	–	–	–	–	–	–	–	–	–	Plectaniaxanthin		LI09742
105 106 119 462 69 145 43 107	**568**	100 87 72 61 53 50 47 44	**27.23**	40	56	2	–	–	–	–	–	–	–	–	–	–	Plectaniaxanthin		MS10729
568 43 91 133 69 109 476 83	**568**	100 95 76 53 45 37 34 24		40	56	2	–	–	–	–	–	–	–	–	–	–	Zeaxanthin	144-68-3	LI09743
568 43 119 105 91 145 55 41	**568**	100 95 94 78 76 63 59 59		40	56	2	–	–	–	–	–	–	–	–	–	–	Zeaxanthin	144-68-3	MS10730
568 119 569 105 91 476 93 157	**568**	100 49 42 34 34 32 30 29		40	56	2	–	–	–	–	–	–	–	–	–	–	Zeaxanthin	144-68-3	NI46090
569 73 75 93 570 554 147 571	**569**	100 84 82 55 52 35 28 26		22	43	5	5	–	–	–	–	–	4	–	–	–	Tetrakis(trimethylsilyl)-8,2′-anhydroguanosine		NI46091
73 103 307 217 147 308 74 104	**569**	100 74 45 32 20 13 8 7	**0.00**	22	55	6	1	–	–	–	–	–	5	–	–	–	D-Fructose, 1,3,4,5,6-pentakis-O-(trimethylsilyl)-, O-methyloxime	56196-14-6	NI46092
319 205 73 160 147 320 217 103	**569**	100 85 79 40 34 29 28 21	**0.00**	22	55	6	1	–	–	–	–	–	5	–	–	–	D-Glucose, 2,3,4,5,6-pentakis-O-(trimethylsilyl)-, O-methyloxime	34152-44-8	NI46094
319 205 73 160 147 320 217 364	**569**	100 84 78 41 34 30 28 23	**0.00**	22	55	6	1	–	–	–	–	–	5	–	–	–	D-Glucose, 2,3,4,5,6-pentakis-O-(trimethylsilyl)-, O-methyloxime	34152-44-8	NI46093
190 572 253 570 255 445 127 192	**570**	100 85 76 66 65 57 52 48		–	–	–	–	–	–	–	–	3	–	–	–	3	Cuprous iodide trimer		NI46095
430 117 168 187 458 570 487 280	**570**	100 70 57 48 34 30 27 25		11	–	5	–	–	9	–	–	–	–	–	–	1	Rhenium, pentacarbonyl(2,3,3,4,4,5,5,6,6-nonafluoro-1-cyclohexen-1-yl)-	14837-19-5	NI46096
429 117 167 185 457 569 485 279	**570**	100 71 56 48 33 30 29 25	**0.00**	11	–	5	–	–	9	–	–	–	–	–	–	1	Rhenium, pentacarbonyl(2,3,3,4,4,5,5,6,6-nonafluoro-1-cyclohexen-1-yl)-	14837-19-5	LI09744
430 402 374 570 84 346 56 318	**570**	100 86 71 70 70 52 47 38		12	1	12	–	–	–	–	–	–	–	–	–	4	Hydridododecacarbonylirontricobalt		LI09745
200 28 312 118 172 59 228 284	**570**	100 99 62 60 46 31 30 22	**7.00**	14	6	10	–	–	–	–	–	–	–	–	–	4	(μ-But-2-yne)decacarbonyltetracobalt		MS10731
403 69 404 117 255 217 129 167	**570**	100 21 13 8 6 6 6 5	**3.60**	18	–	–	–	–	17	–	–	–	–	1	–	–	Phosphorane, difluorotris(pentafluorophenyl)-	22474-72-2	NI46097
204 73 191 217 147 117 129 103	**570**	100 69 38 18 17 11 6 4	**0.00**	22	54	7	–	–	–	–	–	–	5	–	–	–	D-Glucose, 6-O-[2-(trimethylsilyloxy)ethyl]-tetrakis-O-(trimethylsilyl)-		LI09746

MASS TO CHARGE RATIOS	M.W.	INTENSITIES	Parent	C	H	O	N	S	F	Cl	Br	I	Si	P	B	X	COMPOUND NAME	CAS Reg No	No
530 528 572 570 529 531 43 44	**570**	100 99 52 49 35 29 26 25		28	27	8	–	–	–	–	1	–	–	–	–	–	4′-Acetoxy-3-bromo-1′,4-dihydroxy-2,3′-dimethyl-5,5′,8,8′-tetramethoxy-1,2′-binaphthalene	101476-48-6	NI46098
570 59 539 209 359 327 194 507	**570**	100 16 12 7 5 5 5 4		29	30	12	–	–	–	–	–	–	–	–	–	–	Benzene-1,2-dicarboxylic acid, 3,4-dimethoxy-5-[2-methoxy-6-[5-(methoxycarbonyl)-2,3-dimethoxyphenyl]phenoxy]-, dimethyl ester		LI09747
225 173 353 478 143 75 463 73	**570**	100 62 44 37 35 35 33 32	**0.00**	29	58	5	–	–	–	–	–	–	3	–	–	–	Prost-13-en-1-oic acid, 9-oxo-11,15-bis[(trimethylsilyl)oxy]-, trimethylsilyl ester, (8β,11α,13E,15S)-	56144-50-4	NI46099
355 499 409 500 465 313 75 480	**570**	100 86 53 40 36 36 33 31	**6.00**	29	58	5	–	–	–	–	–	–	3	–	–	–	Prost-13-en-1-oic acid, 9-oxo-11,15-bis[(trimethylsilyl)oxy]-, trimethylsilyl ester, (11α,13E,15S)-	39003-19-5	NI46100
570 572 571 573 285 574 287 286	**570**	100 50 40 20 20 10 10 10		32	16	–	8	–	–	–	–	–	–	–	–	1	Nickel phthalocyanine		IC04593
570 572 571 573 574 285 575 186	**570**	100 54 54 22 14 6 5 4		32	16	–	8	–	–	–	–	–	–	–	–	1	Nickel phthalocyanine	14055-02-8	NI46101
66 486 30 458 32 514 39 55	**570**	100 99 99 67 65 58 50 44	**28.00**	32	20	4	–	–	–	–	–	–	–	–	–	1	Tetracyclonetricarbonylruthenium		LI09748
570 552 571 553 276 572 521 262	**570**	100 57 37 18 9 8 6 5		32	26	10	–	–	–	–	–	–	–	–	–	–	Torosanin-9,10-quinone		NI46102
57 71 43 147 113 41 55 129	**570**	100 61 61 43 39 39 24 19	**0.00**	32	58	8	–	–	–	–	–	–	–	–	–	–	Tri(2-ethylhexyl) acetylcitrate		IC04594
148 479 131 120 86 107 229 201	**570**	100 22 6 6 5 4 1 1	**0.10**	34	42	4	4	–	–	–	–	–	–	–	–	–	N-Methyldihydroaralionine B		LI09749
510 43 511 269 55 452 69 57	**570**	100 65 42 42 33 30 29 29	**18.63**	34	50	7	–	–	–	–	–	–	–	–	–	–	Lanost-9(11)-en-18-oic acid, 3,23-bis(acetyloxy)-20-hydroxy-12-oxo-, γ-lactone, (3β,20ξ)-	36872-81-8	NI46103
43 57 55 41 28 71 69 178	**570**	100 81 70 56 41 37 37 33	**10.00**	34	66	4	–	1	–	–	–	–	–	–	–	–	Dimyristyl thiodipropionate		IC04595
253 366 282 211 206 353 367 199	**570**	100 90 40 35 30 28 23 23	**0.00**	35	54	4	–	1	–	–	–	–	–	–	–	–	Cholest-5-en-3β-ol, 19-methoxy-, p-toluenesulphonate	1110-53-8	NI46104
253 366 282 211 353 206 367 199	**570**	100 88 40 36 29 29 25 23	**0.00**	35	54	4	–	1	–	–	–	–	–	–	–	–	Cholest-5-en-3β-ol, 19-methoxy-, p-toluenesulphonate	1110-53-8	LI09750
203 262 189 187 249 133 510 450	**570**	100 66 21 17 13 12 8 7	**3.00**	35	54	6	–	–	–	–	–	–	–	–	–	–	Methyl 3β,6β-diacetoxyolean-12-en-28-oate		LI09751
283 403 435 205 271 450 555 297	**570**	100 55 46 38 30 25 23 19	**13.00**	35	54	6	–	–	–	–	–	–	–	–	–	–	Methyl retigerate A diacetate		NI46105
43 190 321 189 202 119 191 107	**570**	100 96 60 56 49 39 36 36	**23.11**	35	54	6	–	–	–	–	–	–	–	–	–	–	Methyl rubifolate diacetate		NI46106
570 83 57 97 85 55 43 69	**570**	100 69 69 65 60 54 52 50		36	74	–	–	2	–	–	–	–	–	–	–	–	Distearyl disulphide		IC04596
69 43 55 109 105 570 91 119	**570**	100 98 52 46 44 43 41 40		40	58	2	–	–	–	–	–	–	–	–	–	–	β,β-Carotene, 5,6-dihydro-5,6-dihydroxy-	75442-63-6	NI46107
570 448 464 449 552 376 568 396	**570**	100 82 44 39 36 35 33 30		40	58	2	–	–	–	–	–	–	–	–	–	–	ψ,ψ-Carotene, 1,2-dihydro-1,20-dihydroxy-	31589-42-1	NI46108
570 109 127 243 119 69 145 95	**570**	100 39 26 23 22 21 20 18		40	58	2	–	–	–	–	–	–	–	–	–	–	β,β-Carotene, 5,6-dihydro-5,6-dihydroxy-, 9′-cis-	75442-64-7	NI46109
109 210 182 194 282 238 332 252	**571**	100 55 45 15 14 10 7 7	**4.20**	23	29	12	3	1	–	–	–	–	–	–	–	–	1,6-Dihydro-2-methylthio-5-((E)-2-methoxycarbonylvinyl)-4-(2,3,4,6-tetra-O-acetyl-β-D-glucopyranosylamino)-1H-pyrimidin-6-one		MS10732
343 240 344 339 170 213 185 171	**571**	100 87 77 75 74 61 53 52	**26.00**	30	45	6	5	–	–	–	–	–	–	–	–	–	L-Alanine, N-[N-[N-[N-(1-oxodecyl)glycyl]-L-tryptophyl]-L-alanyl]-, methyl ester	19716-78-0	LI09752
343 344 339 213 185 159 158 155	**571**	100 77 75 60 52 50 38 28	**26.25**	30	45	6	5	–	–	–	–	–	–	–	–	–	L-Alanine, N-[N-[N-[N-(1-oxodecyl)glycyl]-L-tryptophyl]-L-alanyl]-, methyl ester	19716-78-0	NI46110
84 28 56 27 25 99 42 26	**571**	100 40 32 17 16 15 15 15	**0.00**	34	35	–	5	–	–	–	–	–	–	–	–	1	Nickel, [N,N,18-triethyl-5,6,17,18-tetrahydrotetrabenzo[b,f,j,n][1,5,9,13]tetraazacyclohexadecin-6-aminato(2-)-N[5],N[11],N[17],N[23]]-	69815-42-5	NI46111
251 104 76 90 195 105 283 78	**571**	100 99 98 70 52 49 45 40	**0.00**	34	41	5	3	–	–	–	–	–	–	–	–	–	(2S,5S)-1-Benzoyl-5-[4-[(benzyloxycarbonyl)amino]butyl]-2-tert-butyl-5-(α-hydroxybenzyl)-3-methyl-4-imidazolidinone		DD01796
91 360 252 105 77 361 111 42	**571**	100 66 56 42 21 15 14 13	**0.00**	34	41	5	3	–	–	–	–	–	–	–	–	–	(2R,5S)-5-[α-(Benzoyloxy)benzyl]-5-[4-[(benzyloxycarbonyl)amino]butyl]-2-tert-butyl-3-methyl-4-imidazolidinone		DD01797
557 545 558 559 546 547 265 560	**572**	100 75 48 37 36 28 22 13	**4.29**	11	24	12	–	–	–	–	–	–	8	–	–	–	Pentamethyltrivinyloctasilsesquioxane		NI46112
432 320 59 236 177 348 292 118	**572**	100 80 73 71 56 52 52 52	**31.20**	12	–	12	–	–	–	–	–	–	–	–	–	4	Dodecacarbonyltetracobalt		MS10733
323 380 260 572 317 460 373 261	**572**	100 80 60 26 15 13 12 11		17	27	5	–	–	–	–	–	–	–	–	–	2	Pentacarbonyl(tributylbismuthine)chromium	90188-75-3	NI46113
152 57 227 41 69 171 153 140	**572**	100 47 33 28 22 16 16 8	**0.00**	20	30	6	2	2	6	–	–	–	–	–	–	–	Butanoic acid, 4,4′-dithiobis[2-[(trifluoroacetyl)amino]-, dibutyl ester, [S-(R*,R*)]-	69779-40-4	NI46114
196 88 526 224 211 484 271 364	**572**	100 93 33 15 11 7 7 4	**0.00**	22	28	12	4	1	–	–	–	–	–	–	–	–	1,6-Dihydro-1-methyl-2-methylthio-5-((E)-2-nitrovinyl)-4-(2,3,4,6-tetra-O-acetyl-β-D-glucopyranosylamino)-1H-pyrimidin-6-one		MS10734
460 28 432 461 429 401 343 56	**572**	100 73 28 24 22 15 12 12	**0.00**	23	20	12	2	–	–	–	–	–	–	–	–	1	Iron, tetracarbonyl[dimethyl 1-[(2-η)-3-methoxy-1-(methoxycarbonyl)-3-oxo-1-propenyl]-4-phenyl-3,3-pyrazolidinedicarboxylato(2-)-N[2]]-	58080-14-1	NI46115
185 183 276 154 193 157 155 102	**572**	100 98 46 35 26 26 26 24	**0.00**	24	18	1	2	2	–	–	2	–	–	–	–	–	2-(4-Bromobenzoyl)-3,6-diphenyl-2,3-dihydro-4H-thiazolo[2,3-b][1,3,4]thiadiazin-5-ium bromide		MS10735

MASS TO CHARGE RATIOS								M.W.	INTENSITIES								Parent	C	H	O	N	S	F	Cl	Br	I	Si	P	B	X	COMPOUND NAME	CAS Reg No	No
224	117	75	309	500	237	103	147	**572**	100	83	68	57	50	49	48	42	**0.07**	25	44	9	–	–	–	–	–	–	3	–	–	–	8,9,10-Trimethoxy-2-[[(trimethylsilyl)oxy]methyl]-3,4-bis[(trimethylsilyl)oxy]-3,4,4a,10b-tetrahydropyrano[3,2-c][2]benzopyran-6(2H)-one		NI46116
41	105	60	77	42	107	69	39	**572**	100	99	87	63	58	49	49	48	**0.00**	28	38	8	–	–	–	2	–	–	–	–	–	–	Phorbol 12,13-di-4-chlorobutyrate		NI46117
171	213	392	349	529	572			**572**	100	90	45	45	10	2				29	36	10	2	–	–	–	–	–	–	–	–	–	N-Acetylisovincoside		LI09753
171	213	392	349	529	572			**572**	100	90	45	45	10	5				29	36	10	2	–	–	–	–	–	–	–	–	–	N-Acetylvincoside		LI09754
262	183	108	263	184	261	107	91	**572**	100	75	54	20	17	15	14	12	**0.05**	30	29	6	–	–	–	–	–	–	–	1	–	1	Iron, dicarbonyl[(2,3,4,5-η)-diethyl 2,4-hexadienedioate](triphenylphosphine)-	62337-82-0	NI46118
55	91	41	43	179	105	93	178	**572**	100	99	97	95	87	79	75	74	**33.33**	30	36	11	–	–	–	–	–	–	–	–	–	–	Carda-16,20(22)-dienolide, 3-[(6-deoxy-3,4-O-methylenehexopyranos-2-ulos-1-yl)oxy]-7,8-epoxy-11,14-dihydroxy-12-oxo-, (3β,5β,7β,11α)-	38945-72-1	NI46119
43	246	264	201	146	187	189	173	**572**	100	48	45	43	43	33	28	20	**7.00**	34	52	7	–	–	–	–	–	–	–	–	–	–	Urs-12-en-28-oic acid, 2,3-diacetyloxy-18-hydroxy-, (2α,3β,18α)-		LI09755
143	393	201	513	437	557	554	572	**572**	100	50	25	16	13	7	3	0		34	52	7	–	–	–	–	–	–	–	–	–	–	6α,16β-Diacetoxy-25-hydroxy-20S,24R-epoxycycloartan-3-one		MS10736
43	246	264	201	146	187	189	572	**572**	100	47	45	43	43	33	28	7		34	52	7	–	–	–	–	–	–	–	–	–	–	2,3-Diacetoxy-tormentic acid		LI09756
83	69	55	215	81	105	264	207	**572**	100	93	88	82	77	72	65	63	**1.59**	35	56	6	–	–	–	–	–	–	–	–	–	–	Olean-12-ene-3,16,21,22,28-pentol, 21-(2-methyl-2-butenoate), [3β,16α,21β(Z),22α]-	20089-98-9	NI46120
554	572	536	466	555	538	480	462	**572**	100	96	55	54	43	36	34	23		38	52	4	–	–	–	–	–	–	–	–	–	–	3-Cyclopentene-1,2-diol, 4,4′-(3,7,12,16-tetramethyl-1,3,5,7,9,11,13,15,17-octadecanonaene-1,8-diyl)bis[3,5,5-tetramethyl-		LI09757
554	572	536	466	555	538	480	556	**572**	100	96	55	53	43	34	34	27		38	52	4	–	–	–	–	–	–	–	–	–	–	3-Cyclopentene-1,2-diol, 4,4′-(3,7,12,16-tetramethyl-1,3,5,7,9,11,13,15,17-octadecanonaene-1,18-diyl)bis[3,5,5-trimethyl-	22467-23-8	NI46121
91	69	133	43	109	466	59	572	**572**	100	77	28	27	25	20	20	18		40	60	2	–	–	–	–	–	–	–	–	–	–	Dihydroxylycopene		LI09758
103	286	271	181	179	364	155	572	**572**	100	64	62	62	62	61	54	52		44	44	–	–	–	–	–	–	–	–	–	–	–	2′,3′,6′,7′-Tetrahydro-4′-(4-methylphenyl)-1′,2,5-tris[(E)-(4-methylphenyl)methylene]spiro[cyclopentan-1,5′(4′H)-indene]		MS10737
336	64	59	160	176	57	256	192	**573**	100	92	82	68	61	58	57	54	**4.00**	16	12	–	6	2	–	–	2	–	–	–	–	1	Copper, [[1,1′-[1,2-bis(p-bromophenyl)ethylene]bis[3-thiosemicarbazidato]](2-)]-	33148-03-7	NI46122
165	135	246	230	169	139	559	69	**574**	100	60	55	50	50	42	40	33	**0.00**	6	18	–	4	–	5	–	–	–	–	3	–	2	1,3,5,2,4,6-Triazatriphosphorine, 2-[bis(trimethylstannyl)amino]-2,4,4,6,6-pentafluoro-2,2,4,4,6,6-hexahydro-	41083-43-6	MS10738
73	41	57	327	74	415	328	341	**574**	100	31	30	16	8	6	5	4	**0.00**	20	54	7	–	–	–	–	–	–	6	–	–	–	3,5-Dibutoxy-1,1,1,7,7,7-hexamethyl-3,5-bis(trimethylsiloxy)tetrasiloxane		MS10739
166	70	114	263	41	167	139	71	**574**	100	75	28	25	20	19	10	10	**4.50**	23	32	6	4	–	6	–	–	–	–	–	–	–	N,N-Bis[(trifluoroacetyl)-L-prolyl]ornithine butyl ester		MS10740
574	92	167	575	75	39	193	56	**574**	100	73	55	25	20	19	15	13		24	8	2	–	–	10	–	–	–	–	–	–	1	Ferrocene, 1,1′-bis(pentafluorobenzoyl)-	55450-72-1	NI46123
305	490	45	46	331	31	70	33	**574**	100	90	66	60	48	48	43	39	**2.00**	24	42	13	–	–	–	–	–	–	–	–	–	4	Oxohexakisbutyratotetraberyllium(II)		MS10741
487	305	490	265	330	293	204	470	**574**	100	30	24	10	8	8	6	5	**0.00**	24	42	13	–	–	–	–	–	–	–	–	–	4	Oxohexakisisobutyratotetraberyllium(II)		MS10742
209	210	183	104	185	57	63	39	**574**	100	17	16	14	10	10	9	9	**6.00**	26	10	–	–	–	10	–	–	–	–	2	–	–	Phosphine, [[bis(pentafluorophenyl)phosphino]ethynyl]diphenyl-	55712-69-1	NI46124
544	545	559	546	560	561	547	472	**574**	100	43	41	23	22	11	7	5	**0.70**	27	42	6	–	–	–	–	–	–	4	–	–	–	1,4,5,7-Tetrahydroxy-2-methylanthraquinone, tetrakis(trimethylsilyl)-	7336-83-6	NI46125
544	559	545	560	546	561	471	547	**574**	100	51	47	28	24	14	10	7	**1.80**	27	42	6	–	–	–	–	–	–	4	–	–	–	1,4,5,8-Tetrahydroxy-2-methylanthraquinone, tetrakis(trimethylsilyl)-	91701-30-3	NI46126
43	108	259	97	157	68	109	159	**574**	100	99	41	28	27	24	23	21	**0.00**	28	46	12	–	–	–	–	–	–	–	–	–	–	1,3,5,7,11,13-Hexaacetoxy-2,6,10-trimethyltridecane		NI46127
83	271	574	385	286	514	474	414	**574**	100	4	0	0	0	0	0	0		29	34	12	–	–	–	–	–	–	–	–	–	–	14-O-Senecioylvernocistifolide-8-O-angelate		MS10743
126	94	69	45	270	86	110	50	**574**	100	67	35	30	25	22	19	18	**0.24**	30	22	12	–	–	–	–	–	–	–	–	–	–	[3,8′-Bi-4H-1-benzopyran]-4,4′-dione, 2′-(3,4-dihydroxyphenyl)-2,2′,3,3′-tetrahydro-3′,5,5′,7,7′-pentahydroxy-2-(4-hydroxyphenyl)-	18913-18-3	NI46128
31	29	32	44	50	83	30	78	**574**	100	89	86	70	42	27	21	20	**0.00**	30	22	12	–	–	–	–	–	–	–	–	–	–	[8,8′-Bi-1H-naphtho[2,3-c]pyran]-1,1′,6,6′,9,9′-hexone, 3,3′,4,4′-tetrahydro 10,10′-dihydroxy-7,7′-dimethoxy-3,3′-dimethyl-, [R-(R*,R*)]-	1685-91-2	NI46129
31	29	32	44	50	83	30	78	**574**	100	90	86	72	44	29	23	22	**1.39**	30	22	12	–	–	–	–	–	–	–	–	–	–	[8,8′-Bi-1H-naphtho[2,3-c]pyran]-1,1′,6,6′,9,9′-hexone, 3,3′,4,4′-tetrahydro 10,10′-dihydroxy-7,7′-dimethoxy-3,3′-dimethyl-, [R-(R*,R*)]-	1685-91-2	NI46130
270	53	286	51	271	288	50	52	**574**	100	84	37	28	26	24	20	19	**5.06**	30	22	12	–	–	–	–	–	–	–	–	–	–	(–)-Rubroskyrin		NI46131
63	270	286	257	271	287	288	121	**574**	100	96	85	67	28	24	22	18	**7.73**	30	22	12	–	–	–	–	–	–	–	–	–	–	Rugulosin, 8,8′-dihydroxy-, (1S,1′S,2R,2′R,3S,3′S,9aR,9′aR)-	21884-44-6	NI46132
574	573	77	575	170	418	65	94	**574**	100	57	40	32	21	15	12	9		30	24	8	–	–	–	–	–	–	–	2	–	–	1,3-Bis(diphenylphosphono)benzene		IC04597
55	179	41	105	43	109	81	69	**574**	100	99	74	72	69	62	62	62	**48.53**	30	38	11	–	–	–	–	–	–	–	–	–	–	Carda-16,20(22)-dienolide, 3-[(6-deoxy-3,4-O-methylenehexopyranos-2-ulos-1-yl)oxy]-5,11,14-trihydroxy-12-oxo-, (3β,5α,11α)-	29428-87-3	NI46133
137	208	182	300	138	149	164	151	**574**	100	94	16	14	13	9	8	8	**0.00**	30	38	11	–	–	–	–	–	–	–	–	–	–	Leptolepisol B		MS10744
490	532	43	448	406	58	374	203	**574**	100	74	55	39	21	16	9	9	**5.00**	32	30	10	–	–	–	–	–	–	–	–	–	–	Diosindigo B leucotetraacetate		NI46134
442	77	443	121	59	410	542	428	**574**	100	57	27	25	15	6	4	4	**0.00**	33	26	4	4	1	–	–	–	–	–	–	–	–	Methyl 3-[α-[1,5-diphenyl-3-(thiobenzoyl)-4-pyrazolyl]benzylidene]-2-(methoxycarbonyl)carbazate	110315-38-3	DD01798
310	328	311	329	473	412	309	430	**574**	100	95	38	28	22	17	16	9	**5.00**	33	50	8	–	–	–	–	–	–	–	–	–	–	12-(2-Methylbutyryl)-13-octanoylphorbol		LI09759

MASS TO CHARGE RATIOS								M.W.	INTENSITIES								Parent	C	H	O	N	S	F	Cl	Br	I	Si	P	B	X	COMPOUND NAME	CAS Reg No	No
197	128	373	402	342	100	444	387	**574**	100	38	29	27	27	26	22	18	**4.90**	33	50	8	–	–	–	–	–	–	–	–	–	–	Spirostan, 2,3,27-triacetyloxy-, (2α,3β,5α,25S)-		MS10745
426	105	70	77	121	45	394	471	**574**	100	92	87	64	59	56	29	28	**0.00**	35	30	2	2	2	–	–	–	–	–	–	–	–	2,5-Diethoxy-5,6,8,8b-tetraphenyl-5H,8bH-thieno[3′,4′:3,4]pyrrolo[2,1-b]-1,3,4-thiadiazole	110315-32-7	DD01799
573	574	288	287	575	227	289	77	**574**	100	78	71	59	34	25	18	18		36	22	–	4	2	–	–	–	–	–	–	–	–	2,2′-Bis(3-cyano-4,6-diphenylpyridyl) disulphide	78564-24-6	NI46135
308	410	202	203	279	276	176	138	**574**	100	80	45	33	23	20	20	18	**0.00**	38	26	4	2	–	–	–	–	–	–	–	–	–	4-(2,4-Dinitrostyryl)-4′-(2-anthr-9-ylvinyl)stilbene		LI09760
73	357	445	299	461	103	174	188	**575**	100	40	29	29	22	17	16	15	**0.43**	20	54	6	1	–	–	–	–	–	5	1	–	–	Phosphoric acid, 2,3-bis(trimethylsiloxy)propyl 2-[bis(trimethylsilyl)amino]ethyl trimethylsilyl ester	32046-29-0	NI46136
193	107	47	45	120	160	59	132	**575**	100	92	75	70	67	58	58	57	**0.00**	22	28	4	2	4	–	–	–	–	–	–	–	1	Bis(2-methylsulphinyl-2-methylthio-N-salicylideneethylaminato)copper		NI46137
132	73	218	187	186	116	276	103	**575**	100	92	39	29	29	21	20	20	**0.00**	29	65	4	1	–	–	–	–	–	3	–	–	–	Acetamide, N-[2,3-bis[(trimethylsilyl)oxy]-1-[[(trimethylsilyl)oxy]methyl]heptadecyl]-, [1S-(1R*,2R*,3S*)]-	15811-81-1	NI46138
170	171	168	169	203	115	342	172	**575**	100	89	30	24	14	14	13	12	**7.57**	30	29	5	3	2	–	–	–	–	–	–	–	–	1-Naphthalenesulphonic acid, 5-(dimethylamino)-, 4-[[[5-(dimethylamino)-1-naphthalenyl]sulphonyl]amino]phenyl ester	55836-97-0	NI46139
170	171	42	168	154	128	44	127	**575**	100	89	52	31	28	27	27	25	**6.50**	30	29	5	3	2	–	–	–	–	–	–	–	–	1-Naphthalenesulphonic acid, 5-(dimethylamino)-, 4-[[[5-(dimethylamino)-1-naphthalenyl]sulphonyl]amino]phenyl ester	55836-97-0	NI46140
575	577	576	287.5	128	191	288.5	288	**575**	100	50	40	37	27	23	19	15		32	16	–	8	–	–	–	–	–	–	–	–	1	Copper phthalocyanine		IC04598
377	285.5	376	375	286.5	378	286	191	**575**	100	80	77	59	39	38	31	25	**0.00**	32	16	–	8	–	–	–	–	–	–	–	–	1	Copper phthalocyanine		IC04599
575	577	576	578	191	193	579	192	**575**	100	67	47	23	17	8	4	3		32	16	–	8	–	–	–	–	–	–	–	–	1	Copper phthalocyanine		NI46141
576	506	436	366	253				**576**	100	82	16	3	2					13	–	–	–	–	–	12	–	–	–	–	–	–	Benzene, pentachloro[chloro(2,3,4,4,5,6-hexachloro-2,5-cyclohexadien-1-ylidene)methyl]-	33500-13-9	NI46142
547	548	549	550	259	41	73	74	**576**	100	50	36	14	12	11	9	8	**0.00**	14	36	11	–	–	–	–	–	–	7	–	–	–	1-Oxyperethylhomohexasilsesquioxane	73895-15-5	NI46143
547	548	549	533	531	550	231	443	**576**	100	55	38	18	15	14	12	10	**0.00**	15	40	10	–	–	–	–	–	–	7	–	–	–	1,3,5,7,9,11,13-Heptaethyl-5-methoxytricyclo[7.5.1.1^{3,11}1]heptasiloxane		NI46144
242	113	296	261	409	576	129	277	**576**	100	82	63	61	54	51	33	18		18	–	–	–	–	15	–	–	–	–	–	–	1	Tris(pentafluorophenyl)arsine		MS10746
73	147	43	281	221	74	415	327	**576**	100	35	17	12	12	8	7	7	**0.00**	18	52	7	–	–	–	–	–	–	7	–	–	–	3-Isopropoxy-1,1,1,7,7,7-hexamethyl-3,5,5-tris(trimethylsiloxy)tetrasiloxane		MS10747
290	108	302	136	147	163	291	333	**576**	100	80	80	50	44	36	32	29	**9.40**	23	23	9	–	–	7	–	–	–	–	–	–	–	3,15-Diacetyl-7-(heptafluorobutyryl)deoxynivalenol		NI46145
57	27	41	307	43	297	69	73	**576**	100	34	31	24	19	16	13	13	**0.00**	24	48	8	–	–	–	–	–	–	4	–	–	–	1,3,5,7-Tetravinyl-1,3,5,7-tetrabutoxycyclotetrasiloxane	110991-12-3	NI46146
169	187	229	109	213	228	331	127	**576**	100	71	70	47	32	29	27	17	**7.19**	28	32	13	–	–	–	–	–	–	–	–	–	–	Glucopyranoside, 1-(8,9-dihydro-2-oxo-2H-furo[2,3-H]-1-benzopyran-8-yl)-1-isopropyl-, tetraacetate, (S)-β-D-	20594-06-3	NI46147
170	171	168	42	169	154	155	127	**576**	100	27	22	22	19	16	14	13	**5.60**	30	28	6	2	2	–	–	–	–	–	–	–	–	1-Naphthalenesulphonic acid, 5-(dimethylamino)-, 1,2-phenylene ester	13285-14-8	NI46148
170	171	168	169	576	167	172	174	**576**	100	27	22	20	7	5	4	3		30	28	6	2	2	–	–	–	–	–	–	–	–	1-Naphthalenesulphonic acid, 5-(dimethylamino)-, 1,2-phenylene ester	13285-14-8	NI46149
170	168	169	171	154	155	127	576	**576**	100	29	24	23	18	16	15	14		30	28	6	2	2	–	–	–	–	–	–	–	–	1-Naphthalenesulphonic acid, 5-(dimethylamino)-, 1,3-phenylene ester	55836-98-1	NI46150
170	168	169	171	576	343	203	167	**576**	100	30	24	22	15	8	5	5		30	28	6	2	2	–	–	–	–	–	–	–	–	1-Naphthalenesulphonic acid, 5-(dimethylamino)-, 1,3-phenylene ester	55836-98-1	NI46151
170	42	171	168	169	154	155	127	**576**	100	35	30	23	21	17	16	16	**10.60**	30	28	6	2	2	–	–	–	–	–	–	–	–	1-Naphthalenesulphonic acid, 5-(dimethylamino)-, 1,4-phenylene ester	55836-99-2	NI46152
170	171	168	169	576	101	343	203	**576**	100	30	23	21	10	7	6	5		30	28	6	2	2	–	–	–	–	–	–	–	–	1-Naphthalenesulphonic acid, 5-(dimethylamino)-, 1,4-phenylene ester	55836-99-2	NI46153
55	348	109	43	105	181	402	83	**576**	100	99	99	95	87	85	75	75	**67.78**	30	40	11	–	–	–	–	–	–	–	–	–	–	Card-20(22)-enolide, 3-[(6-deoxy-3,4-O-methylenehexopyranos-2-ulos-1-yl)oxy]-5,11,14-trihydroxy-12-oxo-, (3β,5α,11α)-	29428-86-2	NI46154
149	57	86	71	167	83	70	85	**576**	100	62	56	45	42	39	29	25	**0.00**	30	42	3	4	–	–	2	–	–	–	–	–	–	9-(2′,2′-Dimethylpropanoylhydrazono)-3,6-dichloro-2,7-bis[2-(diethylamino)ethoxy]-fluorene		MS10748
576	73	331	577	147	578	332	373	**576**	100	89	40	37	35	22	15	12		30	52	5	–	–	–	–	–	–	3	–	–	–	Pregna-4,20-diene-3,11-dione, 17,20,21-tris[(trimethylsilyl)oxy]-	42599-97-3	NI46155
143	371	388	130	387	402	504	488	**576**	100	92	58	58	51	49	43	41	**13.32**	31	56	4	–	–	–	–	–	–	3	–	–	–	Pregna-3,5-dien-20-one, 6-methyl-3,17,21-tris[(trimethylsilyl)oxy]-	74298-97-8	NI46156
576	578	580	577	288	579	289	564	**576**	100	66	46	39	38	32	26	24		32	16	–	8	–	–	–	–	–	–	–	–	1	Zinc phthalocyanine		IC04600
43	533	56	84	534	70	265	576	**576**	100	78	35	34	28	24	18	7		32	40	6	4	–	–	–	–	–	–	–	–	–	2,6-Diacetylchaenorpin		DD01800
480	28	411	77	500	481	412	459	**576**	100	85	77	50	42	36	28	25	**7.37**	34	26	–	–	–	6	–	–	–	1	–	–	–	Silacyclohepta-2,4,6-triene, 1,1-dimethyl-2,3,4,5-tetraphenyl-6,7-bis(trifluoromethyl)-	56784-29-3	MS10749
185	210	131	43	69	548	95	55	**576**	100	38	36	36	28	26	25	23	**7.00**	34	56	3	–	2	–	–	–	–	–	–	–	–	Lanostane-7,11-dione, 3-(acetyloxy)-, cyclic 7-(1,2-ethanediyl mercaptole), (3β)-	56259-20-2	NI46157
575	576	545	174	561	145	146	288	**576**	100	72	29	25	20	18	16	15		35	32	6	2	–	–	–	–	–	–	–	–	–	(+)-Norstephasubine		NI46158
105	77	576	499	577	500	471	220	**576**	100	50	48	45	19	19	10	10		36	20	6	2	–	–	–	–	–	–	–	–	–	N,N′-Bis(p-benzoylphenyl)pyromellitimide	5994-98-9	MS10750
349	576	175	350	335	575	333	319	**576**	100	50	50	32	32	28	17	6		36	36	5	2	–	–	–	–	–	–	–	–	–	Dinklacorin		NI46159
197	575	576	198	577	289	274	578	**576**	100	65	36	19	14	9	8	5		38	36	–	2	–	–	–	–	–	2	–	–	–	Hydrazine, 1,2-bis(methyldiphenylsilyl)-1,2-diphenyl-		LI09761
197	576	577	198	578	289	274	579	**576**	100	65	36	19	12	10	8	5		38	36	–	2	–	–	–	–	–	2	–	–	–	Hydrazine, 1,2-bis(methyldiphenylsilyl)-1,2-diphenyl-	5994-98-9	NI46160
309	310	55	43	99	69	57	41	**576**	100	25	20	18	16	14	12	11	**0.14**	38	72	3	–	–	–	–	–	–	–	–	–	–	9-Octadecenoic acid (Z)-, 2-(9-octadecenyloxy)ethyl ester, (E)-	30760-08-8	NI46161

MASS TO CHARGE RATIOS								M.W.	INTENSITIES								Parent	C	H	O	N	S	F	Cl	Br	I	Si	P	B	X	COMPOUND NAME	CAS Reg No	No
309	28	55	69	83	43	57	44	**576**	100	91	79	70	67	52	48	45	**10.00**	38	72	3	–	–	–	–	–	–	–	–	–	–	**9-Octadecenoic acid (Z)-, 2-(9-octadecenyloxy)ethyl ester, (Z)-**	56554-29-1	**NI46163**
309	55	69	83	43	265	57	41	**576**	100	83	66	58	53	45	43	39	**10.84**	38	72	3	–	–	–	–	–	–	–	–	–	–	**9-Octadecenoic acid (Z)-, 2-(9-octadecenyloxy)ethyl ester, (Z)-**	56554-29-1	**NI46162**
576	476	121	164	197	135	105	382	**576**	100	38	37	32	26	25	24	18		40	36	2	–	–	–	–	–	–	1	–	–	–	**7-Silabicyclo[2.2.1]hept-5-ene-2-carboxylic acid, 7-methyl-1,4,5,6,7-pentaphenyl-, ethyl ester**	56805-06-2	**MS10751**
57	71	43	85	55	98	99	69	**576**	100	53	43	38	32	29	27	27	**2.10**	41	84	–	–	–	–	–	–	–	–	–	–	–	**Octatriacontane, 3,5,23-trimethyl-**	13897-16-0	**NI46164**
520	492	435	407	577	521	454	493	**577**	100	72	63	59	33	33	27	24		29	47	7	5	–	–	–	–	–	–	–	–	–	**Destruxin A**	6686-70-0	**LI09762**
520	492	435	407	577	521	493	464	**577**	100	75	65	60	37	35	28	27		29	47	7	5	–	–	–	–	–	–	–	–	–	**Destruxin A**	6686-70-0	**NI46165**
85	71	546	122	547	150	166	518	**577**	100	97	87	28	27	16	15	6	**0.50**	31	47	9	1	–	–	–	–	–	–	–	–	–	**Glaucenine**	78018-27-6	**NI46166**
86	100	57	127	83	71	85	149	**577**	100	40	14	14	14	10	9	9	**0.00**	32	40	3	5	–	–	1	–	–	–	–	–	–	**9-[4-(4′-Chlorophenyl)semicarbozono]-2,7-bis[2-(diethylamino)-ethoxy]-fluorene**		**MS10752**
393	186	391	392	121	214	337	365	**578**	100	57	46	43	41	30	24	22	**14.00**	8	20	4	–	4	–	–	–	–	–	2	–	1	**Lead[II] bis(diethyldithiophosphate)**	16569-73-6	**NI46167**
549	550	563	551	564	565	260	552	**578**	100	46	40	30	22	17	15	14	**1.59**	11	30	12	–	–	–	–	–	–	8	–	–	–	**Pentamethyltriethyloctasilsesquioxane**	77680-73-0	**NI46168**
73	147	393	578	522	421	365	494	**578**	100	70	35	30	30	30	30	25		14	27	5	–	–	–	–	–	–	3	–	–	2	**Pentacarbonyltris(trimethylsilyl)stibinemolybdenum**	80753-46-4	**NI46169**
73	147	563	475	564	565	476	74	**578**	100	52	33	26	17	14	12	11	**0.00**	14	42	9	–	–	–	–	–	–	8	–	–	–	**Bis(heptamethylcyclotetrasiloxy)siloxane**	17909-39-6	**NI46170**
73	147	475	563	74	476	401	564	**578**	100	84	46	31	19	18	17	16	**0.00**	14	42	9	–	–	–	–	–	–	8	–	–	–	**Bis(pentamethylcyclotrisiloxy)tetramethyldisiloxane**	17909-18-1	**NI46171**
69	177	157	170	81	44	61	149	**578**	100	40	35	28	27	25	22	18	**0.30**	15	10	10	–	–	12	–	–	–	–	–	–	–	**α-D-Mannopyranoside, methyl, tetrakis(trifluoroacetate)**	28034-57-3	**MS10753**
270	45	69	85	271	60	73	71	**578**	100	32	31	28	19	15	14	12	**0.00**	27	30	14	–	–	–	–	–	–	–	–	–	–	**4H-1-Benzopyran-4-one, 7-[[6-O-(6-deoxy-α-L-mannopyranosyl)-β-D-glucopyranosyl]oxy]-5-hydroxy-2-(4-hydroxyphenyl)-**	552-57-8	**NI46172**
212	213	121	56	28	214	147	61	**578**	100	92	48	32	21	17	12	10	**6.00**	28	34	–	–	3	–	–	–	–	–	–	–	2	**Bis[2-(1-ferrocenylethylthio)ethyl] sulphide**		**NI46173**
198	55	522	115	116	77	132	275	**578**	100	55	49	39	27	23	22	13	**3.37**	29	22	2	–	–	–	–	–	–	–	–	–	2	**Manganese, dicarbonyl-π-indenyl(triphenylstibine)-**	31871-86-0	**NI46174**
199	154	55	115	116	77	132	276	**578**	100	67	55	39	27	23	22	13	**3.37**	29	22	2	–	–	–	–	–	–	–	–	–	2	**Manganese, dicarbonyl-π-indenyl(triphenylstibine)-**	31871-86-0	**LI09763**
111	236	144	186	502	184	466	222	**578**	100	70	47	44	35	34	31	27	**2.00**	29	39	8	2	–	–	1	–	–	–	–	–	–	**9-O-Methylmaytansinol**	75340-68-0	**NI46175**
91	160	43	461	70	204	92	161	**578**	100	45	26	18	18	15	8	6	**0.00**	29	42	8	2	1	–	–	–	–	–	–	–	–	**2,3-Di-O-acetyl-6-(N-carbobenzoxyprolineamido)-1-N-hexylthio-α-DL-lyxo-hexopyranoside**		**NI46176**
578	546	579	460	342	257	214	531	**578**	100	41	33	28	24	23	18	16		30	26	8	–	2	–	–	–	–	–	–	–	–	**Tetramethyl 2,2′-(4,4′-biphenylylene)bis(5-methyl-3,4-thiophenedicarboxylate)**	110699-39-3	**DD01801**
163	355	164	149	356	207	104	162	**578**	100	72	18	17	15	14	13	11	**0.42**	30	26	12	–	–	–	–	–	–	–	–	–	–	**Terephthalic acid, bis(4-methoxycarbonylbenzoylethyl) ester**		**IC04601**
73	116	373	75	117	147	101	103	**578**	100	99	75	58	45	35	27	25	**3.00**	30	54	5	–	–	–	–	–	–	3	–	–	–	**Pregn-4-ene-3,11-dione, 17,20,21-tris[(trimethylsilyl)oxy]-, (20R)-**	57397-21-4	**NI46177**
578	547	577	365	367	195	366	351	**578**	100	75	66	32	25	25	20	20		32	38	8	2	–	–	–	–	–	–	–	–	–	**Deserpidine**	131-01-1	**MS10754**
365	221	195	28	366	31	29	212	**578**	100	57	57	44	41	36	35	33	**25.02**	32	38	8	2	–	–	–	–	–	–	–	–	–	**Deserpidine**	131-01-1	**NI46178**
365	578	577	195	221	333	351	211	**578**	100	90	75	65	41	39	36	21		32	38	8	2	–	–	–	–	–	–	–	–	–	**Isodeserpidine**		**MS10755**
409	578	381	149	410	579	304	382	**578**	100	59	55	34	29	26	19	16		34	26	2	–	–	–	–	–	–	–	–	–	2	**1′,1‴-Dibenzoylbiferrocene**		**MS10756**
578	103	281	577	254	280	268	219	**578**	100	86	80	22	14	13	13	13		36	18	1	8	–	–	–	–	–	–	–	–	–	**Bis[4-(5,6-dicyano-3-phenyl-2-pyrazinyl)phenyl] ether**		**MS10757**
57	263	55	43	69	83	98	71	**578**	100	96	85	83	72	63	58	57	**1.50**	38	74	3	–	–	–	–	–	–	–	–	–	–	**9-Octadecenoic acid (Z)-, 2-(octadecyloxy)ethyl ester**	56847-04-2	**NI46179**
57	264	43	55	69	83	98	71	**578**	100	94	94	83	73	62	58	58	**15.54**	38	74	3	–	–	–	–	–	–	–	–	–	–	**9-Octadecenoic acid, 2-(octadecyloxy)ethyl ester**	56599-42-9	**NI46180**
311	312	43	57	99	55	41	83	**578**	100	25	18	15	10	10	8	6	**0.00**	38	74	3	–	–	–	–	–	–	–	–	–	–	**Stearic acid, 2-(1-octadecenyloxy)ethyl ester, (E)-**	30760-06-6	**LI09764**
311	312	43	57	99	55	41	69	**578**	100	25	18	15	10	10	8	6	**0.24**	38	74	3	–	–	–	–	–	–	–	–	–	–	**Stearic acid, 2-(1-octadecenyloxy)ethyl ester, (E)-**	30760-06-6	**NI46181**
311	312	43	57	55	99	41	69	**578**	100	25	19	15	13	9	9	8	**0.24**	38	74	3	–	–	–	–	–	–	–	–	–	–	**Stearic acid, 2-(1-octadecenyloxy)ethyl ester, (Z)-**	30760-05-5	**NI46182**
311	57	43	55	41	28	69	71	**578**	100	82	78	66	54	53	50	44	**0.00**	38	74	3	–	–	–	–	–	–	–	–	–	–	**Stearic acid, 2-(9-octadecenyloxy)ethyl ester, (Z)-**	29027-97-2	**NI46183**
57	43	71	55	83	69	97	82	**578**	100	75	69	67	63	63	48	45	**29.54**	39	78	2	–	–	–	–	–	–	–	–	–	–	**9-Octadecene, 1-[3-(octadecyloxy)propoxy]-, (Z)-**	17367-39-4	**NI46184**
578	296	282	579	281	395	297	560	**578**	100	64	62	52	46	36	30	23		40	26	1	4	–	–	–	–	–	–	–	–	–	**cis-10-[N-(2-Amino-2′-biphenyl)-1-isoindol-3-onylidene]isoindolo[2,1-a]dibenzo[d,f]-1,3-diazepine**		**LI09765**
282	281	462	283	461	463	254	560	**578**	100	56	30	22	20	12	11	3	**2.00**	40	26	1	4	–	–	–	–	–	–	–	–	–	**trans-10-[N-(2-Amino-2′-biphenyl)-1-isoindol-3-onylidene]isoindolo[2,1-a]dibenzo[d,f]-1,3-diazepine**		**LI09766**
91	578	562	579	560	545	561	57	**578**	100	89	53	39	35	30	26	24		40	54	1	2	–	–	–	–	–	–	–	–	–	**5α-Cholestano[3,2-b]indole, N-benzyl-6-oximino-**		**MS10758**
193	43	91	115	55	95	69	167	**578**	100	48	43	36	33	29	26	25	**8.00**	41	54	2	–	–	–	–	–	–	–	–	–	–	**3β-Acetoxy-24,24-diphenyl-9,19-cyclo-25,26,27-trisnorianost-23-ene**		**LI09767**
133	91	69	73	546	578	472	109	**578**	100	80	51	35	22	15	8	7		41	54	2	–	–	–	–	–	–	–	–	–	–	**Okenone**		**LI09768**
522	437	494	409	579	495	523	410	**579**	100	80	75	62	40	32	30	28		29	49	7	5	–	–	–	–	–	–	–	–	–	**Cyclic(β-alanyl-D-α-hydroxy-valeryl-L-prolyl-L-isoleucyl-N-methyl-L-valyl-N-methyl-L-alanyl)**	30845-05-7	**NI46185**
476	477	280	196	195	579	519	278	**579**	100	40	19	10	10	8	8	7		31	34	4	1	–	5	–	–	–	–	–	–	–	**Megestrol acetate 3-O-(pentafluorobenzyl)oxime**		**NI46186**

MASS TO CHARGE RATIOS								M.W.	INTENSITIES								Parent	C	H	O	N	S	F	Cl	Br	I	Si	P	B	X	COMPOUND NAME	CAS Reg No	No
254	186	127	128	121	187	56	43	**580**	100	36	21	14	12	7	6	5	**0.00**	11	11	–	–	–	–	–	–	3	–	–	–	1	Cyclopentadienylironbenzene triiodide		NI46187
440	412	382	397	56	580	468	341	**580**	100	37	31	24	19	17	16	13		13	13	6	–	–	–	–	–	–	–	–	–	3	Tungsten, dicarbonylcyclopentadienyl(μ-dimethylarsenic)tetracarbonyliron-, (Fe-W)-		NI46188
79	54	67	39	80	41	27	93	**580**	100	59	49	29	25	15	15	10	**2.70**	16	24	–	–	–	–	–	2	–	–	–	–	2	Rhodium, di-μ-bromobis[(1,2,5,6-η)-1,5-cyclooctadiene]di-	12092-45-4	NI46189
580	498	390	416	310	496	308	412	**580**	100	23	17	15	14	12	12	9		16	24	–	–	–	–	–	2	–	–	–	–	2	Rhodium, di-μ-bromobis[(1,2,5,6-η)-1-5-cyclooctadiene]di-	12092-45-4	LI09769
280	67	43	41	281	226	69	55	**580**	100	29	21	13	10	8	7	6	**0.37**	17	18	4	2	–	14	–	–	–	–	–	–	–	N,N′-Bis(heptafluorobutyryl)lysine propyl ester		MS10759
411	523	333	439	467	412	524	277	**580**	100	79	47	29	26	26	23	17	**0.00**	26	30	4	–	–	–	–	–	–	–	2	–	2	μ-[1-phenyl-3,4-dimethylphosphole]-μ′-[1-tert-butyl-3,4-dimethylphosphole]bisÆdicarbonylironŒ		NI46190
173	73	156	55	174	276	75	83	**580**	100	36	15	15	13	12	10	4	**0.00**	27	32	14	–	–	–	–	–	–	–	–	–	–	4H-1-Benzopyran-4-one, 7-[[2-O-(6-deoxy-α-L-mannopyranosyl)-β-D-glucopyranosyl]oxy]-2,3-dihydro-5-hydroxy-2-(4-hydroxyphenyl)-, (S)-	10236-47-2	NI46191
272	153	271	120	179	166	273	152	**580**	100	64	52	36	24	22	19	14	**0.00**	27	32	14	–	–	–	–	–	–	–	–	–	–	4H-1-Benzopyran-4-one, 7-[[2-O-(6-deoxy-α-L-mannopyranosyl)-β-D-glucopyranosyl]oxy]-2,3-dihydro-5-hydroxy-2-(4-hydroxyphenyl)-, (S)-	10236-47-2	NI46192
176	88	57	106	204	76	317	89	**580**	100	86	71	57	56	54	51	51	**2.94**	27	40	8	4	1	–	–	–	–	–	–	–	–	L-Cysteine, S-(2-methoxy-2-oxoethyl)-N-[N-[N-[N-(1-oxopropyl)-L-phenylalanyl]-L-leucyl]glycyl]-, methyl ester	35146-63-5	NI46193
340	281	298	400	297	341	401	382	**580**	100	95	74	73	52	48	45	35	**0.00**	29	40	12	–	–	–	–	–	–	–	–	–	–	4,9-Dihydroxy-3β,6β,7β,12β,13-pentaacetoxy-20-nortigli-1-ene		LI09770
144	143	108	109	146	249	179	145	**580**	100	73	66	50	37	36	36	32	**0.00**	30	22	–	–	4	–	2	–	–	–	–	–	–	4-(p-Chlorophenylthio)-3,6-diphenyl-3-[2-(p-chlorophenylthio)ethenyl]-3,4-dihydro-1,2-dithiin		MS10760
380	69	41	43	293	448			**580**	100	55	45	35	8	2			**0.00**	30	28	12	–	–	–	–	–	–	–	–	–	–	4-(D-Xylosyloxy)-6,7-dimethoxy-9-[3,4-(methylenedioxy)phenyl]naphtho[2,3-c]furan-1(3H)-one monocrotonate		NI46194
387	565	269	388	371	566	297	260	**580**	100	58	42	31	30	28	25	25	**9.00**	30	56	5	–	–	–	–	–	–	3	–	–	–	3α,15α,21-Tris(trimethylsilyloxy)-5α-pregna-11,12-dione		LI09771
178	73	580	91	563	260	402	55	**580**	100	75	51	49	46	46	41	36		31	36	9	2	–	–	–	–	–	–	–	–	–	Crotonic acid, 2-(3-formamido-o-anisamido)-, ester with 5-benzyldihydro-4-hydroxy-3,3-dimethyl-2(3H)-furanone 2-hydroxy-3-methylbutyrate	22862-48-2	NI46195
41	83	55	45	109	43	39	367	**580**	100	85	64	55	53	50	47	40	**3.00**	32	36	10	–	–	–	–	–	–	–	–	–	–	16-O-Camphanoyl-16-O-desacetyl-11,12-O-dimethyledulon A		NI46196
454	412	496	538	426	468	482	580	**580**	100	99	92	21	16	14	13	2		34	44	8	–	–	–	–	–	–	–	–	–	–	Turriane tetraacetate		LI09772
580	507	237	434	433	521	547	419	**580**	100	22	11	9	8	7	5	5		35	40	4	4	–	–	–	–	–	–	–	–	–	2-Ethyl-6,7-bis[2-(methoxycarbonyl)ethyl]-1,3,4,5,8-pentamethylporphine		LI09773
69	43	57	28	239	55	41	71	**580**	100	71	66	62	53	43	39	34	**0.34**	37	72	4	–	–	–	–	–	–	–	–	–	–	Hexadecanoic acid, 1-isopropyl-1,2-ethanediyl ester	56599-93-0	NI46197
69	68	43	87	57	98	239	41	**580**	100	93	67	65	63	56	44	44	**3.24**	37	72	4	–	–	–	–	–	–	–	–	–	–	Hexadecanoic acid, 1,5-pentanediyl ester	26933-79-9	NI46198
239	57	86	43	69	55	71	41	**580**	100	62	61	60	52	51	40	39	**0.00**	37	72	4	–	–	–	–	–	–	–	–	–	–	Hexadecanoic acid, 1,1,2-trimethyl-1,2-ethanediyl ester	56599-99-6	NI46199
59	264	285	296	267	222	356	580	**580**	100	37	29	26	24	11	9	8		37	72	4	–	–	–	–	–	–	–	–	–	–	Methyl 18-stearoyloxystearate		NI46200
223	324	254	201	121	356	580	548	**580**	100	76	43	36	33	31	1	1		38	29	–	–	2	–	–	–	–	–	1	–	–	3,5,7,8,10-Pentaphenyl-1-phospha-2-thiabicyclo[4.4.0]deca-3,7,9-triene-1-sulphide	108672-35-1	NI46201
580	581	161	275	565	407	582	41	**580**	100	44	35	21	17	15	10	10		38	36	2	4	–	–	–	–	–	–	–	–	–	1,2-Bis[4-[2-(4-tert-butylphenyl)-1,3,4-oxadiazol-5-yl]phenyl]ethylene		IC04602
57	43	71	55	83	85	69	97	**580**	100	84	62	55	47	45	44	39	**1.44**	38	76	3	–	–	–	–	–	–	–	–	–	–	Octadecanoic acid, 2-(octadecyloxy)ethyl ester	28843-25-6	NI46202
57	71	43	85	55	41	253	69	**580**	100	72	72	52	41	33	30	30	**0.24**	39	80	2	–	–	–	–	–	–	–	–	–	–	Octadecane, 1,1′-[(1-methyl-1,2-ethanediyl)bis(oxy)]bis-	35545-51-8	NI46203
281	57	71	43	85	83	58	97	**580**	100	84	72	51	50	45	45	40	**1.34**	39	80	2	–	–	–	–	–	–	–	–	–	–	Octadecane, 1,1′-[1,3-propanediylbis(oxy)]bis-	17367-38-3	NI46204
119	91	145	203	105	157	133	121	**580**	100	83	76	72	72	61	50	46	**0.00**	40	52	3	–	–	–	–	–	–	–	–	–	–	β-Doradecin	31460-54-5	NI46205
126	96	176	372	374	55	110	204	**581**	100	77	73	69	57	56	55	49	**0.00**	14	18	7	3	–	–	6	–	–	–	1	–	–	N-(5′-Deoxythymidinyl) di-O-(β,β,β-trichloroethyl) phosphoramidate		MS10761
43	71	252	69	169	241	213	119	**581**	100	85	22	15	12	3	2	1	**0.00**	17	17	5	1	–	14	–	–	–	–	–	–	–	3-Methylbutyl N,O-bis(heptafluorobutyryl)threoninate		NI46206
474	91	73	475	75	264	156	129	**581**	100	53	40	28	19	17	16	15	**1.00**	34	55	3	1	–	–	–	–	–	2	–	–	–	Pregn-5-en-20-one, 3,16-bis[(trimethylsilyl)oxy]-, O-benzyloxime, (3β,16α)-	57325-87-8	NI46207
490	73	91	474	491	75	475	129	**581**	100	94	80	67	45	34	31	31	**6.00**	34	55	3	1	–	–	–	–	–	2	–	–	–	Pregn-5-en-20-one, 3,17-bis[(trimethylsilyl)oxy]-, O-benzyloxime, (3β)-	57326-12-2	NI46208
91	73	251	474	129	75	581	103	**581**	100	61	39	26	24	23	22	20		34	55	3	1	–	–	–	–	–	2	–	–	–	Pregn-5-en-20-one, 3,21-bis[(trimethylsilyl)oxy]-, O-benzyloxime, (3β)-	57325-86-7	NI46209
582	328	180	148	111	87	436	117	**582**	100	99	17	6	6	6	5	5		12	–	–	–	1	8	–	–	2	–	–	–	–	Benzene, 1,1′-thiobis[2,3,4,5-tetrafluoro-6-iodo-	19638-37-0	LI09774
328	582	330	180	111	583	329	87	**582**	100	99	70	40	18	14	13	13		12	–	–	–	1	8	–	–	2	–	–	–	–	Benzene, 1,1′-thiobis[2,3,4,5-tetrafluoro-6-iodo-	19638-37-0	NI46210
582	328	180	583	111	329	87	436	**582**	100	39	19	14	9	6	6	5		12	–	–	–	1	8	–	–	2	–	–	–	–	Benzene, 1,1′-thiobis[2,3,4,5-tetrafluoro-6-iodo-	19638-37-0	NI46211
365	367	582	584	363	349	580	351	**582**	100	93	91	85	85	83	77	77		14	16	4	–	–	–	–	–	–	–	–	–	3	3-[o-Phenylenebis(dimethylarsino)]tungstentetracarbonyl		LI09775
554	552	69	582	78	277	555	580	**582**	100	62	39	26	24	22	20	16		18	5	1	–	–	8	–	–	–	–	–	–	1	Iridium, carbonyl(η^5-2,4-cyclopentadien-1-yl)(3,3′,4,4′,5,5′,6,6′-octafluoro[1,1′-biphenyl]-2,2′-diyl)-	51509-33-2	NI46212

MASS TO CHARGE RATIOS								M.W.	INTENSITIES								Parent	C	H	O	N	S	F	Cl	Br	I	Si	P	B	X	COMPOUND NAME	CAS Reg No	No
86	56	132	74	85	57	87	100	**582**	100	51	43	37	34	34	31	30	**1.00**	24	46	8	4	2	–	–	–	–	–	–	–	–	6,9,12,15,23,26,31,34-Octaoxa-1,3,18,20-tetraazabicyclo[18.8.8]hexatriaconta-2,19-dithione		MS10762
56	70	120	82	94	77	95	99	**582**	100	86	67	60	58	58	52	42	**0.30**	26	24	1	6	1	–	1	1	–	–	–	–	–	8-Bromo-6-(2-chlorophenyl)-1-[4-(2-phenoxyethyl)-1-piperazinyl]-4H-1,2,4 triazolo[4,3-a]thieno[3,2-f]-1,4-diazepine		MS10763
91	101	236	114	92	330	150	264	**582**	100	28	13	10	10	9	9	8	**0.24**	26	34	13	2	–	–	–	–	–	–	–	–	–	L-Serine, N-[benzyloxycarbonyl]-O-[3,4,6-tri-O-acetyl-2-(acetylamino)-2-deoxy-β-D-glucopyranosyl]-, methyl ester	25644-83-1	NI46213
43	498	91	106	396	393	456	354	**582**	100	19	14	12	10	10	9	9	**0.40**	28	22	14	–	–	–	–	–	–	–	–	–	–	2,5,8,9,11-Pentaacetoxy-2,3-dihydrospiro[4H-benzofuro[2,3-g]-1-benzopyran-4,2′(5′H)-furan]-5′-one		NI46214
73	253	75	451	343	107	147	361	**582**	100	99	71	54	40	37	36	30	**2.50**	30	58	5	–	–	–	–	–	–	3	–	–	–	5β-Pregnan-20-one, 17-hydroxy-3α,11β,21-tris(trimethylsiloxy)-	17563-00-7	NI46215
112	104	91	41	103	39	43	51	**582**	100	85	55	54	50	44	36	35	**0.00**	33	54	3	6	–	–	–	–	–	–	–	–	–	Triimidazo[1,5-a:1′,5′-c:1″,5″-e][1,3,5]triazine-1,5,9(2H,6H,10H)-trione, 2,6,10-tricyclohexylhexahydro-3,3,7,7,11,11-hexamethyl-	67969-52-2	NI46216
291	256	290	292	293	102	547	258	**582**	100	89	52	38	34	27	24	17	**13.00**	34	20	–	6	–	–	2	–	–	–	–	–	–	(1α,2α,3β,4β)-1,2-Bis(3-chloro-2-quinoxalinyl)-3,4-bis(4 -cyanophenyl)cyclobutane		MS10764
55	550	149	564	409	304	551	381	**582**	100	99	86	51	51	44	42	38	**4.90**	34	30	2	–	–	–	–	–	–	–	–	–	2	1′,1‴-Bis(α-hydroxybenzyl)biferrocene		MS10765
114	115	105	42	77	44	41	43	**582**	100	8	8	7	4	4	4	3	**1.08**	34	38	5	4	–	–	–	–	–	–	–	–	–	Pentanamide, N-(7-benzoyl-5,8-dioxo-3-phenyl-2-oxa-6,9-diazabicyclo[10.2.2]hexadeca-10,12,14,15-tetraen-4-yl)-2-(dimethylamino)-3-methyl-, [3R-[3R*,4S*(2S*,3R*),7S*]]-	21761-48-8	NI46217
366	367	211	426	253	199	368	351	**582**	100	30	30	26	19	13	11	10	**0.00**	36	54	4	–	1	–	–	–	–	–	–	–	–	3-Acetoxy-19-tosyloxycholestene		LI09776
514	515	461	383	582	448	516	475	**582**	100	45	38	24	18	18	17	17		37	50	2	–	–	–	–	–	–	2	–	–	–	Cannabidiol bis(phenyldimethylsilyl) ether		NI46218
390	391	486	582	487	392	44	346	**582**	100	49	21	19	15	12	12	11		38	34	4	2	–	–	–	–	–	–	–	–	–	N,N′-Dicycloheptyl-3,4:9,10-perylenebis(dicarboximide)	80525-30-0	MS10766
83	91	41	125	55	145	119	105	**582**	100	62	55	54	51	42	40	40	**38.00**	40	54	3	–	–	–	–	–	–	–	–	–	–	Capsanthone		LI09777
73	91	106	105	476	145	77	59	**582**	100	88	57	35	20	13	13	13	**4.50**	41	58	2	–	–	–	–	–	–	–	–	–	–	Hydroxyspirilloxanthin		MS10767
73	91	476	59	477	582	458	133	**582**	100	88	20	13	7	5	5	5		41	58	2	–	–	–	–	–	–	–	–	–	–	Hydroxyspirilloxanthin		LI09778
43	582	91	69	109	83	133	430	**582**	100	42	38	27	21	20	16	7		41	58	2	–	–	–	–	–	–	–	–	–	–	3′-O-Methyllutein		LI09779
73	69	91	476	105	41	106	582	**582**	100	27	25	12	11	11	10	9		41	58	2	–	–	–	–	–	–	–	–	–	–	Spheroidenone		MS10768
73	69	91	476	582	580	133	43	**582**	100	27	25	12	9	8	6	5		41	58	2	–	–	–	–	–	–	–	–	–	–	Spheroidenone		LI09780
194	165	166	360	332	195	283	254	**582**	100	37	28	21	15	15	11	7	**3.27**	42	30	3	–	–	–	–	–	–	–	–	–	–	1,3-Dioxan-4-one, 2,6-bis(diphenylmethylene)-5,5-diphenyl-	57438-11-6	NI46219
263	70	265	261	15	334	237	252	**583**	100	77	65	65	48	31	26	22	**0.00**	10	5	2	1	–	–	6	–	–	–	–	–	1	Mercury, (4,5,6,7,8,8-hexachloro-1,3,3a,4,7,7a-hexahydro-1,3-dioxo-4,7-methano-2H-isoindol-2-yl)methyl-	5902-79-4	NI46220
370	263	372	261	265	368	398	586	**583**	100	93	70	59	57	46	45	44	**7.00**	10	5	2	1	–	–	6	–	–	–	–	–	1	Mercury, (4,5,6,7,8,8-hexachloro-1,3,3a,4,7,7a-hexahydro-1,3-dioxo-4,7-methano-2H-isoindol-2-yl)methyl-	5902-79-4	NI46221
43	71	468	169	240	386	450	284	**583**	100	66	9	7	7	4	2	1	**0.00**	16	15	4	1	1	14	–	–	–	–	–	–	–	3-Methylbutyl N,S-bis(heptafluorobutyryl)cysteinate		NI46222
176	204	55	41	216	119	436	190	**583**	100	96	76	56	48	32	31	30	**0.00**	16	16	3	3	–	15	–	–	–	–	–	–	–	Spermidine triPFP		MS10769
73	186	132	378	174	147	103	75	**583**	100	25	25	24	23	22	22	22	**0.00**	23	57	6	1	–	–	–	–	–	5	–	–	–	Galactitol, 2-acetamido-2-deoxy-1,3,4,5,6-pentakis-O-(trimethylsilyl)-, D-	19127-01-6	NI46223
73	132	186	174	378	147	103	75	**583**	100	26	25	24	22	22	22	21	**0.00**	23	57	6	1	–	–	–	–	–	5	–	–	–	Galactitol, 2-acetamido-2-deoxy-1,3,4,5,6-pentakis-O-(trimethylsilyl)-, D-	19127-01-6	LI09781
73	96	132	103	186	147	174	75	**583**	100	26	19	16	15	14	13	12	**0.00**	23	57	6	1	–	–	–	–	–	5	–	–	–	Glucitol, 2-acetamido-2-deoxy-1,3,4,5,6-pentakis-O-(trimethylsilyl)-, D-	19127-02-7	NI46224
28	187	189	159	215	43	161	18	**583**	100	19	10	9	8	8	5	5	**0.00**	23	57	6	1	–	–	–	–	–	5	–	–	–	D-Mannitol, 2-(acetylamino)-2-deoxy-1,3,4,5,6-pentakis-O-(trimethylsilyl)	74464-41-8	NI46225
91	169	331	109	236	332	127	92	**583**	100	74	67	26	22	11	11	9	**0.30**	26	33	14	1	–	–	–	–	–	–	–	–	–	L-Alanine, N-[benzyloxycarbonyl]-3-[(2,3,4,6-tetra-O-acetyl-β-D-galactopyranosyl)oxy]-, methyl ester	14364-68-2	NI46226
101	114	115	86	43	232	117	102	**583**	100	59	41	14	12	12	11	7	**0.20**	26	49	13	1	–	–	–	–	–	–	–	–	–	1,5-Di-O-acetyl-7-O-[2-deoxy-2-(dimethylamino)-3,4,6-tri-O-methyl-α-D-glucopyranosyl]-2,3,4,6-tetra-O-methyl-L-glycero-D-manno-heptitol		NI46227
91	163	73	176	70	75	476	131	**583**	100	45	38	28	19	13	12	12	**0.25**	34	57	3	1	–	–	–	–	–	2	–	–	–	Pregnan-20-one, 3,6-bis[(trimethylsilyl)oxy]-, O-benzyloxime, (3α,5β,6α)	57326-07-5	NI46228
476	75	73	477	91	171	264	156	**583**	100	44	44	40	37	20	18	18	**0.50**	34	57	3	1	–	–	–	–	–	2	–	–	–	Pregnan-20-one, 3,16-bis[(trimethylsilyl)oxy]-, O-benzyloxime, (3β,5α,16α)-	57325-91-4	NI46229
73	91	492	75	476	186	493	440	**583**	100	99	83	73	63	39	38	36	**3.50**	34	57	3	1	–	–	–	–	–	2	–	–	–	Pregnan-20-one, 3,17-bis[(trimethylsilyl)oxy]-, O-benzyloxime, (3α,5α)-	57325-90-3	NI46230
73	91	492	75	479	186	493	440	**583**	100	99	91	75	71	42	39	31	**5.00**	34	57	3	1	–	–	–	–	–	2	–	–	–	Pregnan-20-one, 3,17-bis[(trimethylsilyl)oxy]-, O-benzyloxime, (3β,5α)-	57363-10-7	NI46231
91	73	75	358	251	477	492	77	**583**	100	82	77	48	35	32	26	25	**18.96**	34	57	3	1	–	–	–	–	–	2	–	–	–	Pregnan-20-one, 3,21-bis[(trimethylsilyl)oxy]-, O-benzyloxime, (3α,5β)-	57325-89-0	NI46232
221	42	55	57	40	91	69	105	**583**	100	90	72	70	64	58	53	52	**50.00**	40	55	3	–	–	–	–	–	–	–	–	–	–	Antheraxanthine		LI09782
91	221	105	82	69	119	81	67	**583**	100	79	79	79	79	67	64	64	**23.00**	40	55	3	–	–	–	–	–	–	–	–	–	–	Diadinoxanthine		LI09783
153	152	155	154	184	183	186	185	**584**	100	86	66	66	57	50	46	33	**0.00**	10	30	10	–	–	–	–	–	–	–	–	–	2	Niobium tantalum decamethoxide		MS10770
557	569	558	559	570	571	271	560	**584**	100	79	51	49	45	31	28	16	**8.99**	12	24	12	–	–	–	–	–	–	8	–	–	–	Tetramethyltetravinyloctasilsesquioxane	85233-78-9	NI46233
52	251	145	307	219	472	360	416	**584**	100	96	94	93	67	55	37	36	**1.80**	20	36	4	–	–	–	–	–	–	–	2	–	2	Tetracarbonyl[tellurobis(di-tert-butylphosphane)]chromium(0)		NI46234

MASS TO CHARGE RATIOS								M.W.	INTENSITIES								Parent	C	H	O	N	S	F	Cl	Br	I	Si	P	B	X	COMPOUND NAME	CAS Reg No	No
164	163	331	165	254	330	253	166	**584**	100	33	24	14	11	10	9	9	**0.00**	23	16	11	6	1	–	–	–	–	–	–	–	–	β-[7-(4-Hydroxybenzothiazolyl)]alanine		LI09784
313	57	325	41	29	314	326	327	**584**	100	99	55	37	35	35	24	16	**0.00**	24	56	8	–	–	–	–	–	–	4	–	–	–	1,3,5,7-Tetraethyl-1,3,5,7-tetrabutoxycyclotetrasiloxane	110991-13-4	NI46235
83	69	43	109	41	53	28	55	**584**	100	81	69	52	52	49	38	32	**0.03**	28	37	11	–	–	–	1	–	–	–	–	–	–	6β,12β,13,20-Tetraacetoxy-7α-chloro-4,9-dihydroxytigli-1-en-3-one		NI46236
369	45	69	255	370	107	105	55	**584**	100	41	30	24	23	23	21	21	**0.00**	28	38	6	–	–	6	–	–	–	–	–	–	–	Cholan-24-oic acid, 3,12-bis[(trifluoroacetyl)oxy]-, (3α,5β,12α)-	56438-17-6	NI46237
77	125	79	311	75	191	91	151	**584**	100	60	45	41	40	38	38	35	**35.00**	29	37	3	2	–	5	–	–	–	1	–	–	–	Androst-4-ene-3,17-dione, 11-[[dimethyl(pentafluorophenyl)silyl]oxy]-, bis(O-methyloxime), (11β)-	57691-90-4	MS10771
77	125	79	311	191	75	91	584	**584**	100	59	44	42	41	39	38	34		29	37	3	2	–	5	–	–	–	1	–	–	–	Androst-4-ene-3,17-dione, 11-[[dimethyl(pentafluorophenyl)silyl]oxy]-, bis(O-methyloxime), (11β)-	57691-90-4	NI46238
73	116	75	451	117	205	147	129	**584**	100	99	87	80	66	49	48	43	**2.00**	30	60	5	–	–	–	–	–	–	3	–	–	–	Pregnane-3,11,17,20,21-pentol, tris(trimethylsilyl) ether, (3α,5β,11β,20R)-	57325-69-6	NI46239
191	147	129	173	217	209	105	423	**584**	100	57	46	43	32	31	28	22	**0.00**	30	60	5	–	–	–	–	–	–	3	–	–	–	Prosta-5,13-dien-1-oic acid, 9,11,15-tris[(trimethylsilyl)oxy]-, methyl ester		LI09785
191	397	129	281	487	307	398	192	**584**	100	71	45	36	34	29	25	22	**0.00**	30	60	5	–	–	–	–	–	–	3	–	–	–	Prosta-11,15-dien-1-oic acid, 7,9,13-tris[(trimethylsilyl)oxy], methyl ester	56247-88-2	NI46240
191	147	129	173	217	103	207	307	**584**	100	54	47	43	31	29	24	22	**0.84**	30	60	5	–	–	–	–	–	–	3	–	–	–	Prosta-5,13-dien-1-oic acid, 9,11,15-tris[(trimethylsilyl)oxy]-, methyl ester, (5Z,9α,11α,13E,15S)-	52058-16-9	NI46241
191	423	237	333	494	513	584	569	**584**	100	48	48	36	33	18	5	5		30	60	5	–	–	–	–	–	–	3	–	–	–	Prosta-5,13-dien-1-oic acid, 9,11,15-tris[(trimethylsilyl)oxy]-, methyl ester, (5Z,9α,11α,13E,15S)-	52058-16-9	LI09786
73	191	147	173	217	129	75	74	**584**	100	29	14	12	12	11	11	8	**0.00**	30	60	5	–	–	–	–	–	–	3	–	–	–	Prosta-5,13-dien-1-oic acid, 9,11,15-tris[(trimethylsilyl)oxy]-, methyl ester, (5Z,9α,11α,13E,15S)-	52058-16-9	NI46242
165	586	584	353	585	587	322	315	**584**	100	66	65	42	24	23	15	13		31	37	6	–	–	–	–	1	–	–	–	–	–	1-(Bromomethyl)-2,3-bis(4,5-dimethoxy-2-methylphenyl)-2,3-dihydro-4,5-dimethoxy-7-methyl-1H-indene	94610-78-3	NI46243
299	286	110	584	226	300	210	287	**584**	100	95	55	50	45	41	20	19		33	36	6	4	–	–	–	–	–	–	–	–	–	21H-Biline-8,12-dipropanoic acid, 2,17-divinyl-1,10,19,22,23,24-hexahydro-3,7,13,18-tetramethyl-1,19-dioxo-	635-65-4	LI09787
286	44	227	299	108	300	211	584	**584**	100	93	51	46	44	28	27	22		33	36	6	4	–	–	–	–	–	–	–	–	–	21H-Biline-8,12-dipropanoic acid, 2,17-divinyl-1,10,19,22,23,24-hexahydro-3,7,13,18-tetramethyl-1,19-dioxo-	635-65-4	NI46244
507	525	447	465	387	343	567	325	**584**	100	64	46	13	9	9	8	4	**0.00**	33	44	9	–	–	–	–	–	–	–	–	–	–	Euphornin	80454-47-3	NI46245
105	43	398	44	366	367	77	41	**584**	100	75	70	56	40	36	35	34	**0.00**	34	49	3	–	–	–	–	1	–	–	–	–	–	5β,6β-Epoxy-7α-bromocholestane-3β-benzoate		NI46246
584	524	583	525	523	135	509	315	**584**	100	87	78	73	70	70	53	47		35	52	7	–	–	–	–	–	–	–	–	–	–	Methyl glabrate diacetate		LI09788
292	183	107	507	293	214	308	184	**584**	100	72	39	33	25	25	22	19	**13.89**	36	28	–	–	–	–	–	–	–	–	4	–	–	Dibenzo[c,g][1,2,5,6]tetraphosphocin, 5,6,11,12-tetrahydro-5,6,11,12-tetraphenyl-	38234-91-2	MS10772
538	539	288	274	228	189	584	249	**584**	100	40	7	7	6	5	2	2		36	56	6	–	–	–	–	–	–	–	–	–	–	Cyclamigen C. diacetate		LI09789
57	43	110	72	55	74	28	41	**584**	100	72	67	59	56	54	49	46	**0.12**	36	56	6	–	–	–	–	–	–	–	–	–	–	3-Deoxy-3-oxoingenol 5-hexadecanoate		NI46247
392	377	583	539	228	189	288	274	**584**	100	50	13	4	4	4	3	3	**1.00**	36	56	6	–	–	–	–	–	–	–	–	–	–	30β-Ethoxy-28,30-epoxyolean-12-en-3β,16α-diol diacetate		LI09790
584	301	270	540	585	239	509	465	**584**	100	50	38	33	32	32	6	6		37	28	7	–	–	–	–	–	–	–	–	–	–	1(3H)-Isobenzofuranone, 3,3'-(4-methoxy-1,3-phenylene)bis[3-(4-methoxyphenyl)-	55429-18-0	LI09791
584	301	270	540	585	239	525	271	**584**	100	50	40	33	32	32	13	13		37	28	7	–	–	–	–	–	–	–	–	–	–	1(3H)-Isobenzofuranone, 3,3'-(4-methoxy-1,3-phenylene)bis[3-(4-methoxyphenyl)-	55429-18-0	NI46248
135	584	553	107	292	478	122	121	**584**	100	59	50	29	12	10	8	8		39	44	1	4	–	–	–	–	–	–	–	–	–	Pycaanthinol		LI09792
91	109	105	106	145	83	478	55	**584**	100	87	52	51	37	37	26	26	**18.00**	40	56	3	–	–	–	–	–	–	–	–	–	–	Capsanthine		LI09793
221	91	181	504	352	287	566	133	**584**	100	53	39	37	35	31	27	26	**23.00**	40	56	3	–	–	–	–	–	–	–	–	–	–	β,ε-Carotene-3,3'-diol, 5,8-epoxy-5,8-dihydro-, (3S,3'R,5R,6'R,8S)-	27780-11-6	NI46249
221	91	504	352	181	584	492	133	**584**	100	63	53	42	39	34	27	27		40	56	3	–	–	–	–	–	–	–	–	–	–	β,ε-Carotene-3,3'-diol, 5,8-epoxy-5,8-dihydro-, (3S,3'R,5R,6'R,8R)-	512-29-8	NI46250
91	43	221	584	69	133	109	83	**584**	100	45	39	35	33	29	25	21		40	56	3	–	–	–	–	–	–	–	–	–	–	β,ε-Carotene-3,3'-diol, 5,6-epoxy-5,6-dihydro-, (3S,3'R,5R,6S,6'R)-	28368-08-3	NI46251
584	566	550	548	492	532	472	428	**584**	100	36	20	20	13	10	6	5		40	56	3	–	–	–	–	–	–	–	–	–	–	β,ε-Carotene-3,3',4-triol, (3S,3'R,6'R)-	68474-24-8	NI46252
205	165	221	584	181	218	271	424	**584**	100	61	45	34	27	24	20	19		40	56	3	–	–	–	–	–	–	–	–	–	–	Cryptochrome	73745-06-9	NI46253
221	181	584	504	352	585	287	222	**584**	100	56	51	48	32	22	22	22		40	56	3	–	–	–	–	–	–	–	–	–	–	Mutatoxanthin	31661-06-0	NI46254
73	91	106	105	40	77	81	79	**584**	100	50	26	18	9	7	6	6	**1.30**	41	60	2	–	–	–	–	–	–	–	–	–	–	Rhodovibrin		MS10773
73	91	478	69	43	133	460	109	**584**	100	50	5	5	4	3	2	2	**0.00**	41	60	2	–	–	–	–	–	–	–	–	–	–	Rhodovibrin		LI09794
68	262	569	539	147	247	119	570	**584**	100	94	71	67	42	35	35	34	**20.00**	44	56	–	–	–	–	–	–	–	–	–	–	–	Tetrakis(3,4,4,5-tetramethyl-2,5-cyclohexadien-1-ylidene)cyclobutane		NI46255
69	76	466	169	566	119	100	102	**585**	100	83	70	21	17	17	11	10	**9.01**	12	–	–	3	–	21	–	–	–	–	–	–	–	1,3,5-Triazine, 2,4,6-tris(heptafluoropropyl)-	915-76-4	NI46256
101	241	139	114	140	466	199	143	**585**	100	47	41	36	35	28	28	25	**0.50**	22	30	12	3	–	3	–	–	–	–	–	–	–	Asparagine, N-(2-acetamido-2-deoxy-β-D-glucopyranosyl)-N^2-(trifluoroacetyl)-, ethyl ester, 3',4',6'-triacetate, L-	16907-93-0	NI46257
585	586	73	587	570	571	588	103	**585**	100	46	44	28	28	13	10	9		22	43	4	5	1	–	–	–	–	4	–	–	–	Tetrakis(trimethylsilyl)-8,2'-thioanhydroguanosine		NI46258

MASS TO CHARGE RATIOS								M.W.	INTENSITIES								Parent	C	H	O	N	S	F	Cl	Br	I	Si	P	B	X	COMPOUND NAME	CAS Reg No	No
224	88	196	296	208	252	238	585	**585**	100	94	88	29	24	18	18	12		24	31	12	3	1	–	–	–	–	–	–	–	–	1,6-Dihydro-1-methyl-2-methylthio-5-((E)-2-methoxycarbonylvinyl)-4-(2,3,4,6-tetra-O-acetyl-β-D-glucopyranosylamino)-1H-pyrimidin-6-one		MS10774
109	210	182	238	194	296	224	280	**585**	100	93	60	20	20	17	13	10	**3.30**	24	31	12	3	1	–	–	–	–	–	–	–	–	1,6-Dihydro-2-methylthio-5-((E)-2-ethoxycarbonylvinyl)-4-(2,3,4,6-tetra-O acetyl-β-D-glucopyranaylamino)-1H-pyrimidin-6-one		MS10775
91	312	358	359	313	268	453	227	**585**	100	58	55	35	13	9	9	8	**0.20**	36	43	6	1	–	–	–	–	–	–	–	–	–	N-(2,2-diethoxyethyl)-1,2-bis(3-benzyloxy-4-methoxyphenyl)ethylamine		NI46259
43	55	57	155	69	487	474	71	**586**	100	49	35	34	34	30	30	27	**12.00**	19	22	2	–	–	16	–	–	–	–	–	–	–	1H,1H,9H-Hexadecafluorononyl decanoate		LI09795
43	28	55	31	57	155	69	41	**586**	100	99	49	48	35	34	34	34	**12.50**	19	22	2	–	–	16	–	–	–	–	–	–	–	1H,1H,9H-Hexadecafluorononyl decanoate		MS10776
73	586	147	571	587	103	588	75	**586**	100	99	74	48	40	39	31	28		22	42	5	4	1	–	–	–	–	4	–	–	–	Tetrakis(trimethylsilyl)-8,2′-thioanhydroxanthosine		NI46260
73	585	75	93	513	57	586	117	**586**	100	99	99	95	85	76	56	50		22	44	4	5	1	–	–	–	–	4	–	–	–	Tetrakis(trimethylsilyl)-8,3′-thioanhydroguanosine		NI46261
446	444	502	448	445	504	500	501	**586**	100	89	88	79	77	75	72	58	**22.78**	23	15	5	–	–	–	–	–	–	–	1	–	1	Tungsten, pentacarbonyl(triphenylphosphine)-	15444-65-2	NI46262
43	527	485	58	528	425	486	70	**586**	100	92	65	49	29	24	18	15	**4.00**	29	34	11	2	–	–	–	–	–	–	–	–	–	Dichotine, 19-hydroxy-11-methoxy-, triacetate (ester)	29484-57-9	NI46263
44	59	32	135	43	41	297	195	**586**	100	62	56	31	25	22	18	15	**0.00**	30	20	–	6	2	–	–	–	–	–	–	–	1	Nickel, [5,6,17,18-tetrahydro-6,18-diisothiocyanatotetrabenzo[b,f,j,n][1,5,9,13]tetraazacyclohexadecinato(2-)-N^5,N^{11},N^{17},N^{23}]-	72378-93-9	NI46264
502	504	457	459	487	43	503	393	**586**	100	71	55	48	37	36	35	33	**4.93**	30	28	8	–	–	–	2	–	–	–	–	–	–	1,1′-Diacetoxy-4,4′-dichloro-5,5′,8,8′-tetramethoxy-6,6′-dimethyl-2,2′-binaphthalene	89475-15-0	NI46265
220	316	255	173	425	295	406	317	**586**	100	82	77	65	53	52	43	38	**0.41**	30	62	5	–	–	–	–	–	–	3	–	–	–	Prost-5-en-1-oic acid, 9,11,15-tris[(trimethylsilyl)oxy]-, methyl ester, (5Z,9α,11α,15S)-	56051-52-6	NI46266
425	191	496	515	237	335	571	586	**586**	100	86	60	57	40	20	7	1		30	62	5	–	–	–	–	–	–	3	–	–	–	Prost-13-en-1-oic acid, 9,11,15-tris[(trimethylsilyl)oxy]-, methyl ester, (9α,11α,13E,15S)-	52058-15-8	LI09796
425	310	191	309	75	380	173	237	**586**	100	93	93	57	56	55	47	40	**0.17**	30	62	5	–	–	–	–	–	–	3	–	–	–	Prost-13-en-1-oic acid, 9,11,15-tris[(trimethylsilyl)oxy]-, methyl ester, (9α,11α,13E,15S)-	52058-15-8	NI46267
28	43	283	44	95	284	343	213	**586**	100	80	67	42	17	15	12	12	**2.10**	32	42	10	–	–	–	–	–	–	–	–	–	–	Khivarin		MS10777
151	586	281	253	317	271	311	299	**586**	100	12	12	9	8	6	5	5		32	42	10	–	–	–	–	–	–	–	–	–	–	Yuanhuadine		NI46268
314	243	315	215	77	271	230	231	**586**	100	36	29	17	9	7	6	5	**0.00**	33	23	2	4	–	–	–	1	–	–	–	–	–	2-(4,4′-Biphenyl)-6-oxo-8,10-diphenyl-4H,6H-pyrido[2,1-f][1,3,4]oxadiazino[2,3-c][1,2,4]triazin-11-ium bromide		MS10778
125	43	389	41	97	67	55	91	**586**	100	99	50	43	29	29	24	23	**0.50**	33	46	9	–	–	–	–	–	–	–	–	–	–	5β-Ergost-24-en-26-oic acid, 5,6β-epoxy-4β,18,22-trihydroxy-3-methoxy-1-oxo-, δ-lactone, diacetate, (20S,22R)-	21903-02-6	LI09797
125	43	389	168	41	97	67	55	**586**	100	96	48	46	45	30	30	23	**0.38**	33	46	9	–	–	–	–	–	–	–	–	–	–	5β-Ergost-24-en-26-oic acid, 5,6β-epoxy-4β,18,22-trihydroxy-3-methoxy-1-oxo-, δ-lactone, diacetate, (20S,22R)-	21903-02-6	NI46269
365	531	586	165	221	157	144	349	**586**	100	87	80	70	47	40	40	27		34	38	7	2	–	–	–	–	–	–	–	–	–	O-[3,4,5-Trimethoxycinnamoyl]vincamajine		LI09798
91	172	107	65	108	79	77	41	**586**	100	88	40	36	24	20	18	14	**0.00**	34	50	6	–	1	–	–	–	–	–	–	–	–	Spirostan-3,15-diol, 3-(4-methylbenzenesulphonate), (3β,5α,15β,25R)-	54965-90-1	NI46270
43	55	57	44	69	41	45	81	**586**	100	69	65	62	61	56	52	43	**40.00**	35	54	7	–	–	–	–	–	–	–	–	–	–	30-Norlupan-20-one, 3,21,28-tris(acetyloxy)-, (3β,21β)-	55401-96-2	NI46271
86	100	83	85	57	58	135	408	**586**	100	42	39	26	14	10	10	9	**1.00**	36	50	3	4	–	–	–	–	–	–	–	–	–	9-(Adamantan-1-oylhydrazono)-2,7-bis[2-(diethylamino)-ethoxy]-fluorene		MS10779
189	187	191	526	204	229	203	247	**586**	100	79	49	44	44	33	33	31	**2.20**	36	58	6	–	–	–	–	–	–	–	–	–	–	Gammacerane-2,3,22-triol, triacetate, (2α,3β,8α,9β,13α,14β,17α,18β,22α)-	43206-36-6	NI46272
266	43	57	73	55	41	60	69	**586**	100	95	82	79	74	68	63	53	**0.47**	36	58	6	–	–	–	–	–	–	–	–	–	–	Ingenol 3-palmitate		NI46273
43	526	262	189	202	135	203	187	**586**	100	98	96	76	68	65	62	58	**11.00**	36	58	6	–	–	–	–	–	–	–	–	–	–	Lupane-3,21,28-triol, triacetate, (3β,21β)-	55401-90-6	NI46274
390	98	586	391	488	587	489	530	**586**	100	92	72	57	30	27	25	16		38	38	4	2	–	–	–	–	–	–	–	–	–	N,N′-Bis(1-propylbutyl)-3,4:9,10-perylenebis(dicarboximide)	110590-82-4	MS10780
91	105	43	55	41	119	83	157	**586**	100	65	61	60	55	53	51	50	**18.00**	40	58	3	–	–	–	–	–	–	–	–	–	–	Capsanthol		LI09799
178	201	384	202	586	78	77	279	**586**	100	82	67	43	38	38	38	28		41	31	2	–	–	–	–	–	–	–	1	–	–	2-(Diphenyloxophosphinyl)-2,3,4,5-tetraphenyl-2,5-dihydrofuran		MS10781
586	587	201	383	202	385	279	384	**586**	100	57	55	24	21	17	13	10		41	31	2	–	–	–	–	–	–	–	1	–	–	cis-2-(Diphenyloxophosphinyl)-2,3,4,5-tetraphenyl-2,3-dihydrofuran		MS10782
201	586	178	77	279	587	202	384	**586**	100	98	86	82	51	48	45	41		41	31	2	–	–	–	–	–	–	–	1	–	–	trans-2-(Diphenyloxophosphinyl)-2,3,4,5-tetraphenyl-2,3-dihydrofuran		MS10783
69	100	119	61	587	440	51	70	**587**	100	81	74	61	53	49	47	45		17	4	5	1	–	15	–	–	–	–	–	–	–	5,6-Dihydroxyindole, tris(pentafluoropropionyl)-		NI46275
222	155	282	89	95	71	123	223	**587**	100	58	47	43	40	34	31	30	**0.01**	29	49	11	1	–	–	–	–	–	–	–	–	–	Diacetylpederin		LI09800
275	216	234	373	215	276	198	185	**587**	100	94	88	64	48	41	39	34	**24.00**	32	49	7	3	–	–	–	–	–	–	–	–	–	L-Proline, 1-[O-(1-oxohexyl)-N-[N-(1-oxohexyl)-L-valyl]-L-tyrosyl]-, methyl ester	56272-45-8	NI46276
203	587	107	133	324	119	189	588	**587**	100	80	69	55	46	45	43	38		39	57	3	1	–	–	–	–	–	–	–	–	–	Urs-12-en-28-amide, 3-(acetyloxy)-N-(4-methylphenyl)-, (3β)-	40575-30-2	NI46277
203	587	107	189	191	588	324	119	**587**	100	97	58	54	47	42	37	32		39	57	3	1	–	–	–	–	–	–	–	–	–	Olean-12-en-28-amide, 3-(acetyloxy)-N-(4-methylphenyl)-, (3β)-	40575-29-9	NI46278

MASS TO CHARGE RATIOS								M.W.	INTENSITIES								Parent	C	H	O	N	S	F	Cl	Br	I	Si	P	B	X	COMPOUND NAME	CAS Reg No	No
364	476	392	588	504	420	448		**588**	100	63	63	48	44	18	14			8	–	8	–	–	–	–	–	2	–	–	–	2	Di-μ-iodooctacarbonyldimanganese		MS10784
107	105	103	399	109	104	397	401	**588**	100	49	20	18	18	18	16	13	**5.92**	8	12	–	–	–	–	–	–	–	–	–	–	6	2,4,6,8,9,10-Hexaselenaadamantane, 1,3,5,7-tetramethyl-	25133-48-6	NI46279
166	84	70	167	277	41	128	139	**588**	100	42	39	19	18	18	17	10	**3.30**	24	34	6	4	–	6	–	–	–	–	–	–	–	N,N-Bis(trifluoroacetyl-L-prolyl)lysine, butyl ester		MS10785
166	84	277	167	41	128	70	139	**588**	100	73	41	23	22	21	18	11	**1.00**	24	34	6	4	–	6	–	–	–	–	–	–	–	N,N'-Bis(trifluoroacetyl-L-prolyl)-α-methylornithine butyl ester		MS10786
314	243	315	215	271	231	230	77	**588**	100	32	30	14	8	8	8	4	**0.00**	27	18	2	4	–	–	–	2	–	–	–	–	–	2-(4-Bromophenyl)-6-oxo-8,10-diphenyl-4H,6H-pyrido[2,1-f][1,3,4]oxadiazino[2,3-c][1,2,4]triazin-11-ium bromide		MS10787
43	169	109	331	91	127	151	365	**588**	100	64	55	26	26	21	16	5	**2.00**	29	32	13	–	–	–	–	–	–	–	–	–	–	Trichoside, tetraacetate		LI09801
43	169	109	331	91	127	151	258	**588**	100	65	55	26	26	21	16	6	**2.26**	29	32	13	–	–	–	–	–	–	–	–	–	–	Trichoside, tetraacetate	20633-70-9	NI46280
43	169	107	91	127	331	151	170	**588**	100	67	50	37	16	15	13	6	**0.04**	29	32	13	–	–	–	–	–	–	–	–	–	–	Trichoside, tetraacetate		MS10788
173	399	355	217	400	191	129	309	**588**	100	62	59	53	25	23	21	20	**0.00**	30	64	5	–	–	–	–	–	–	3	–	–	–	Prostan-1-oic acid, 7,9,13-tris[(trimethylsilyl)oxy]-, methyl ester, (9α)-	56282-42-9	NI46281
288	301	229	273	243	302	588	213	**588**	100	61	43	35	28	26	23	22		33	40	6	4	–	–	–	–	–	–	–	–	–	Phycocyanobilin		LI09802
95	108	125	110	122	82	93	124	**588**	100	97	70	69	46	31	23	21	**0.03**	33	40	6	4	–	–	–	–	–	–	–	–	–	D-Urobilin		LI09803
573	531	588	546	548	489	547	505	**588**	100	68	60	31	30	26	19	17		34	36	9	–	–	–	–	–	–	–	–	–	–	5,9,11-Triacetoxy-3,3-dimethyl-6,8-bis-(3-methylbut-2-enyl)-3H,12H-pyrano[3,2-a]xanthen-12-one		NI46282
189	136	204	191	279	187	347	498	**588**	100	78	62	46	31	17	15	13	**4.70**	36	68	2	–	–	–	–	–	–	2	–	–	–	21,21-Dimethyl-3,22-bis(trimethylsilyloxy)-29,30-dinorgammacerane, (3β,8α,9β,13α,14β,17α,18β,22α)-	75601-37-5	NI46283
240	121	90	588	133	468	343	241	**588**	100	97	79	64	53	47	47	44		39	40	5	–	–	–	–	–	–	–	–	–	–	2,6-Bis[2-hydroxy-3-(2-hydroxy-5-methylbenzyl)-5-methylbenzyl]-4-methylphenol		LI09804
588	589	121	95	590	91	136	574	**588**	100	55	38	36	34	28	22	20		40	48	2	2	–	–	–	–	–	–	–	–	–	2,5-Cyclohexadien-1-one, 4,4'-[(2,2',3,3',5,5',6,6'-octamethyl[1,1'-biphenyl]-4,4'-diyl)dinitrilo]bis[2,3,5,6-tetramethyl-	54245-95-3	NI46284
545	513	546	129	514	529	120	391	**589**	100	59	30	27	25	19	18	17	**0.00**	30	51	5	3	–	–	–	–	–	2	–	–	–	Pregna-1,4-dien-18-al, 3,20-bis(methoxyimino)-11,21-bis[(trimethylsilyl)oxy]-, O-methyloxime, (11β)-	69833-80-3	NI46285
171	170	169	168	172	589	356	115	**589**	100	38	17	16	15	6	5	5		31	31	5	3	2	–	–	–	–	–	–	–	–	1-Naphthalenesulphonic acid, 5-(dimethylamino)-, 4-[[[[5-(dimethylamino)1-naphthalenyl]sulphonyl]amino]methyl]phenyl ester	55836-92-5	NI46287
171	170	169	168	44	172	45	42	**589**	100	38	17	16	16	15	15	14	**5.50**	31	31	5	3	2	–	–	–	–	–	–	–	–	1-Naphthalenesulphonic acid, 5-(dimethylamino)-, 4-[[[[5-(dimethylamino)1-naphthalenyl]sulphonyl]amino]methyl]phenyl ester		NI46286
561	562	563	266	533	252	564	531	**590**	100	45	26	19	17	17	16	16	**0.00**	15	38	11	–	–	–	–	–	–	7	–	–	–	1-Methoxyperethylhomohexasilsesquioxane	73895-16-6	NI46288
433	489	461	444	544	388	589	570	**590**	100	71	68	9	7	7	3	3	**0.00**	16	30	8	–	–	4	–	–	–	–	2	–	1	(Tetrafluoroethylene)bis(triethoxyphosphino)dicarbonylruthenium		LI09805
73	147	57	41	281	74	415	327	**590**	100	33	21	21	12	9	8	7	**0.00**	19	54	7	–	–	–	–	–	–	7	–	–	–	3-Butoxy-1,1,1,7,7,7-hexamethyl-3,5,5-tris(trimethylsiloxy)tetrasiloxane		MS10789
333	364	350	319	334	446	365	386	**590**	100	99	65	64	55	32	31	22	**0.83**	20	23	8	2	–	9	–	–	–	–	–	–	–	5-Undecenedioic acid, 2,10-bis[(trifluoroacetyl)amino]-5-[[(trifluoroacetyl)oxy]methyl]-, dimethyl ester	56196-63-5	NI46289
73	315	357	299	217	147	243	75	**590**	100	40	39	33	24	19	13	11	**0.10**	20	51	8	–	–	–	–	–	–	5	1	–	–	D-erythro-2-Pentulose, 1,3,4-tris-O-(trimethylsilyl)-, 5-[bis(trimethylsilyl) phosphate]	55520-87-1	NI46290
73	315	357	299	217	243	358	147	**590**	100	81	74	52	45	24	20	17	**0.00**	20	51	8	–	–	–	–	–	–	5	1	–	–	D-threo-2-Pentulose, 1,3,4-tris-O-(trimethylsilyl)-, 5-[bis(trimethylsilyl) phosphate]	55520-85-9	NI46291
73	315	357	299	217	147	243	316	**590**	100	57	47	46	33	22	16	14	**0.00**	20	51	8	–	–	–	–	–	–	5	1	–	–	erythro-2-Pentulose, 1,3,4-tris-O-(trimethylsilyl)-, 5-[bis(trimethylsilyl) phosphate]	56192-89-3	NI46292
73	315	230	299	147	215	169	217	**590**	100	53	29	28	21	20	15	14	**0.00**	20	51	8	–	–	–	–	–	–	5	1	–	–	D-Ribofuranose, 1,2,3-tris-O-(trimethylsilyl)-, bis(trimethylsilyl) phosphate	69744-64-5	NI46293
73	217	299	315	147	75	343	169	**590**	100	54	38	23	23	16	14	14	**0.00**	20	51	8	–	–	–	–	–	–	5	1	–	–	D-Ribofuranose, 2,3,5-tris-O-(trimethylsilyl)-, bis(trimethylsilyl) phosphate	69688-46-6	NI46294
73	315	299	230	147	215	169	217	**590**	100	53	28	28	21	20	16	14	**0.50**	20	51	8	–	–	–	–	–	–	5	1	–	–	D-Ribofuranose, 2,3,5-tris-O-(trimethylsilyl)-, bis(trimethylsilyl) phosphate	69688-46-6	LI09806
73	315	299	230	147	215	316	169	**590**	100	51	26	24	23	15	13	13	**0.00**	20	51	8	–	–	–	–	–	–	5	1	–	–	D-Ribose, 2,3,4-tris-O-(trimethylsilyl)-, 5-[bis(trimethylsilyl) phosphate]	55520-86-0	NI46295
73	299	217	315	147	204	75	259	**590**	100	40	39	31	29	23	19	18	**0.00**	20	51	8	–	–	–	–	–	–	5	1	–	–	D-Xylopyranose, 2,3,4-tris-O-(trimethylsilyl)-, bis(trimethylsilyl) phosphate	69688-47-7	NI46296
298	174	199	280	175	44	237	45	**590**	100	68	64	61	54	39	34	31	**0.00**	24	19	7	–	–	9	–	–	–	–	–	–	–	Estra-1,3,5(10)-trien-17-one, 3,6,7-tris[(trifluoroacetyl)oxy]-, (6α,7α)-	56438-13-2	NI46297
105	164	488	246	149	530	575	470	**590**	100	8	5	5	5	4	1	1	**0.00**	30	38	12	–	–	–	–	–	–	–	–	–	–	1α-Benzoyloxy-6β,8β,9α,15-tetracetoxy-4β-hydroxydihydro-β-agarofuran		MS10790
180	73	231	410	179	217	181	147	**590**	100	63	33	30	26	22	18	18	**0.00**	30	54	4	–	–	–	–	–	–	4	–	–	–	Enterodiol		NI46298
547	84	70	279	56	160	590	182	**590**	100	24	19	17	17	13	10	8		33	42	6	4	–	–	–	–	–	–	–	–	–	2,6-Diacetyl-18-O-methylchaenorpin		DD01802

MASS TO CHARGE RATIOS								M.W.	INTENSITIES								Parent	C	H	O	N	S	F	Cl	Br	I	Si	P	B	X	COMPOUND NAME	CAS Reg No	No
547	278	590	121	250	548	84	100	**590**	100	67	56	52	50	39	34	30		33	42	6	4	–	–	–	–	–	–	–	–	–	4H-1,16-Etheno-5,15-(propaniminoethano)furo[3,4-l][1,5,10]triazacyclohexadecine-4,21-dione, 10,14-diacetyl-3,3a,6,7,8,9,10,11,12,13,14,15-dodecahydro-3-(4-methoxyphenyl)-	69721-59-1	NI46299
253	105	145	159	254	313	157	143	**590**	100	38	27	25	24	23	23	23	0.00	34	54	8	–	–	–	–	–	–	–	–	–	–	Cholestan-26-oic acid, 3,7,12-tris(acetyloxy)-, methyl ester, (3α,5β,7α,12α)-	60354-53-2	NI46300
589	590	575	174	295	145	144	190	**590**	100	76	26	24	18	13	13	5		36	34	6	2	–	–	–	–	–	–	–	–	–	(+)-Stephasubine		NI46301
590	517	591	518	531	503	575	443	**590**	100	50	32	22	19	19	14	13		36	38	4	4	–	–	–	–	–	–	–	–	–	21H,23H-Porphine-2,18-dipropionic acid, 7,12-divinyl-3,8,13,17-tetramethyl-, dimethyl ester	5522-66-7	NI46302
590	517	592	518	503	531	444	443	**590**	100	50	32	22	19	17	11	11		36	38	4	4	–	–	–	–	–	–	–	–	–	21H,23H-Porphine-2,18-dipropionic acid, 7,12-divinyl-3,8,13,17-tetramethyl-, dimethyl ester	5522-66-7	LI09807
349	175	355	590	575				**590**	100	78	61	53	5					37	38	5	2	–	–	–	–	–	–	–	–	–	DL-O-Methyltiliacorine		LI09808
349	335	590	175	575				**590**	100	61	57	56	7					37	38	5	2	–	–	–	–	–	–	–	–	–	O-Methyltiliacorinine		LI09809
309	55	43	69	41	99	57	32	**590**	100	66	50	45	44	32	32	32	0.00	38	70	4	–	–	–	–	–	–	–	–	–	–	9-Octadecenoic acid (Z)-, 1,2-ethanediyl ester	928-24-5	NI46303
197	590	591	198	289	303	274	592	**590**	100	44	25	19	14	12	10	8		39	38	–	2	–	–	–	–	–	2	–	–	–	Hydrazobenzene, 4-methyl-N,N'-bis(methyldiphenylsilyl)-	15951-44-7	NI46304
57	56	71	43	58	41	85	55	**590**	100	9	7	6	5	5	4	3	0.00	42	86	–	–	–	–	–	–	–	–	–	–	–	2,2,4,15,17,17-Hexamethyl-7,12-bis(3,5,5-trimethylhexyl)octadecane		AI02407
178	590	105	591	485	165	379	289	**590**	100	93	60	48	37	33	27	26		44	30	2	–	–	–	–	–	–	–	–	–	–	4,4'-Bis(2,3,4-triphenylcyclobut-2-en-1-on-4-yl)		MS10791
428	281	176	119	69	429	77	45	**591**	100	71	68	67	49	45	39	35	0.00	17	8	5	1	–	15	–	–	–	–	–	–	–	1,2-Benzenediol, 4-(2-aminoethyl)-, tris(2,2,3,3,3-pentafluoro-1-oxopropyl)-	52118-49-7	NI46305
428	176	119	281	429	69	77	78	**591**	100	37	37	25	19	12	9	7	0.00	17	8	5	1	–	15	–	–	–	–	–	–	–	1,2-Benzenediol, 4-(2-aminoethyl)-, tris(2,2,3,3,3-pentafluoro-1-oxopropyl)-	52118-49-7	MS10792
428	119	176	429	281	147	69	173	**591**	100	89	72	22	19	18	12	11	0.00	17	8	5	1	–	15	–	–	–	–	–	–	–	1,2-Benzenediol, 4-(2-aminoethyl)-, tris(2,2,3,3,3-pentafluoro-1-oxopropyl)-	52118-49-7	NI46306
165	428	266	129	164	100	149	121	**591**	100	76	40	40	33	28	24	23	0.00	17	8	5	1	–	15	–	–	–	–	–	–	–	Benzenemethanol, α-(aminomethyl)-4-hydroxy-, tris(2,2,3,3,3-pentafluoro 1-oxopropyl)-		NI46307
119	415	428	159	176	387	267	147	**591**	100	77	68	55	38	32	24	24	0.00	17	8	5	1	–	15	–	–	–	–	–	–	–	Benzenemethanol, α-(aminomethyl)-4-hydroxy-, tris(2,2,3,3,3-pentafluoro 1-oxopropyl)-		MS10793
176	119	428	415	387	105	69	159	**591**	100	72	65	42	23	21	20	15	1.60	17	8	5	1	–	15	–	–	–	–	–	–	–	2-(3-Hydroxyphenyl)-2-hydroxyethylamine, tris(pentafluoropropionyl)-		MS10794
428	165	429	468	552	503	117	109	**591**	100	38	18	3	1	1	1	1	0.00	17	8	5	1	–	15	–	–	–	–	–	–	–	2-(3-Hydroxyphenyl)-2-hydroxyethylamine, tris(pentafluoropropionyl)-		NI46308
428	176	429	281	265	119	430	387	**591**	100	19	18	17	4	4	3	3	0.00	17	8	5	1	–	15	–	–	–	–	–	–	–	Propanoic acid, pentafluoro-, 4-[2-[(2,2,3,3,3-pentafluoro-1-oxopropyl)amino]ethyl]-1,2-phenylene ester	59785-38-5	NI46310
428	119	176	69	281	429	77	65	**591**	100	77	63	37	30	21	19	16	0.00	17	8	5	1	–	15	–	–	–	–	–	–	–	Propanoic acid, pentafluoro-, 4-[2-[(2,2,3,3,3-pentafluoro-1-oxopropyl)amino]ethyl]-1,2-phenylene ester	59785-38-5	MS10795
428	281	176	119	69	429	77	45	**591**	100	71	68	67	49	45	39	35	0.00	17	8	5	1	–	15	–	–	–	–	–	–	–	Propanoic acid, pentafluoro-, 4-[2-[(2,2,3,3,3-pentafluoro-1-oxopropyl)amino]ethyl]-1,2-phenylene ester	59785-38-5	NI46309
324	325	102	223	73	326	179	220	**591**	100	29	14	13	13	11	6	6	0.00	29	53	4	1	–	–	–	–	–	4	–	–	–	Denopamine iso-M1, N,O,O,O-tetrakis(trimethylsilyl)-		NI46311
324	325	102	223	73	326	220	179	**591**	100	28	15	15	15	11	7	6	0.00	29	53	4	1	–	–	–	–	–	4	–	–	–	Denopamine-M1, N,O,O,O-tetrakis(trimethylsilyl)-		NI46312
86	84	534	154	506	421	69	576	**591**	100	69	26	23	19	16	16	14	3.50	30	49	7	5	–	–	–	–	–	–	–	–	–	Roseotoxin B	55466-29-0	NI46313
488	470	591	57	561	73	103	471	**591**	100	96	94	82	66	59	56	47		30	53	5	3	–	–	–	–	–	2	–	–	–	Pregn-4-en-18-al, 3,20-bis(methoxyimino)-11,21-bis[(trimethylsilyl)oxy]-, O-methyloxime, (11β)-	69833-79-0	NI46314
470	103	561	73	471	262	562	501	**591**	100	80	76	45	39	36	35	33	31.56	30	53	5	3	–	–	–	–	–	2	–	–	–	Pregn-4-en-18-al, 3,20-bis(methoxyimino)-11,21-bis[(trimethylsilyl)oxy]-, O-methyloxime, (11β,17α)-	69854-81-5	NI46315
519	520	521	591	75	517	593	86	**591**	100	53	23	22	14	9	7	7		30	57	3	1	–	–	–	–	–	4	–	–	–	Estriol, 2-amino-, tetrakis(trimethylsilyl)-		NI46316
519	520	521	518	75	591	522	517	**591**	100	70	31	12	12	10	10	10		30	57	3	1	–	–	–	–	–	4	–	–	–	Estriol, 4-amino-, tetrakis(trimethylsilyl)-		NI46317
222	202	55	250	110	58	194	166	**592**	100	90	90	75	51	41	40	39	6.36	10	–	10	–	–	–	–	–	–	–	–	–	3	Bis(pentacarbonylmanganese)mercury		MS10796
563	564	565	577	267	566	579	578	**592**	100	47	43	24	20	19	14	13	2.59	12	32	12	–	–	–	–	–	–	8	–	–	–	Tetramethyltetraethyloctasilsesquioxane	77699-55-9	NI46318
119	428	281	69	429	77	78	65	**592**	100	74	49	42	33	21	18	15	0.90	17	7	6	–	–	15	–	–	–	–	–	–	–	(3,4-Dihydroxyphenyl)ethanol tris(pentafluoropropionate)		MS10797
43	119	415	428	387	429	77	267	**592**	100	69	56	41	22	21	19	18	1.20	17	7	6	–	–	15	–	–	–	–	–	–	–	4-Hydroxyphenethylene glycol triPFP		MS10798
43	594	592	259	81	201	595	593	**592**	100	43	41	30	30	18	11	11		23	25	8	6	–	–	–	1	–	–	–	–	–	N-(2-acetoxycyclohexyl)-2-methoxy-5-acetamido-4-(2,4-dinitro-6-bromophenylazo)aniline		IC04603
60	99	258	229	300	315	592	564	**592**	100	57	12	11	6	1	0	0		28	32	14	–	–	–	–	–	–	–	–	–	–	14-O-(4-Hydroxysenecioyl)vernocistifolide-8-O-(2,3-epoxyisobutyrate)		MS10799
153	273	320	171	213	292	236	129	**592**	100	54	29	26	21	15	13	10	0.00	30	24	13	–	–	–	–	–	–	–	–	–	–	Eriocephaloside triacetate		NI46319
592	73	593	594	147	129	595	74	**592**	100	65	52	26	12	12	11	8		30	56	4	–	–	–	–	–	–	4	–	–	–	Estra-1,3,5(10)-triene, 2,3,16,17-tetrakis(trimethylsilyloxy)-, (16α,17β)-	51497-46-2	NI46320

MASS TO CHARGE RATIOS								M.W.	INTENSITIES								Parent	C	H	O	N	S	F	Cl	Br	I	Si	P	B	X	COMPOUND NAME	CAS Reg No	No
592	593	594	129	595	385	73	433	**592**	100	55	28	11	10	9	9	8		30	56	4	–	–	–	–	–	–	4	–	–	–	Estra-1,3,5(10)-triene, 2,3,16,17-tetrakis(trimethylsilyloxy)-, (16α,17β)-	51497-46-2	NI46321
592	593	594	595	503	577	596	73	**592**	100	57	30	10	5	4	3	3		30	56	4	–	–	–	–	–	–	4	–	–	–	Estra-1,3,5(10)-triene, 3,4,5,17-tetrakis(trimethylsilyloxy)-, (17β)-	74298-88-7	NI46322
592	593	594	129	595	433	399	385	**592**	100	58	28	15	11	10	10	10		30	56	4	–	–	–	–	–	–	4	–	–	–	Estra-1,3,5(10)-triene, 3,4,16,17-tetrakis(trimethylsilyloxy)-, (16α,17β)-		NI46323
444	413	412	322	414	445	229	323	**592**	100	95	57	48	40	39	31	27	**0.00**	30	56	4	–	–	–	–	–	–	4	–	–	–	Estra-1,3,5(10)-triene, 3,6,7,17-tetrakis(trimethylsilyloxy)-, (6β,7α,17β)-	74298-84-3	NI46324
502	503	412	295	242	322	504	296	**592**	100	50	30	29	27	26	23	22	**5.13**	30	56	4	–	–	–	–	–	–	4	–	–	–	Estra-1,3,5(10)-triene, 3,6,11,17-tetrakis(trimethylsilyloxy)-, (6α,11β,17β)-	74298-86-5	NI46325
502	503	322	504	230	412	309	229	**592**	100	46	30	21	16	15	15	14	**1.25**	30	56	4	–	–	–	–	–	–	4	–	–	–	Estra-1,3,5(10)-triene, 3,6,16,17-tetrakis(trimethylsilyloxy)-, (6α,16α,17β)-	57363-11-8	NI46326
73	502	503	229	504	322	230	147	**592**	100	99	48	27	23	21	20	20	**1.50**	30	56	4	–	–	–	–	–	–	4	–	–	–	Estra-1,3,5(10)-triene, 3,6,16,17-tetrakis(trimethylsilyloxy)-, (6α,16α,17β)-	57363-11-8	NI46327
309	322	592	57	502	129	412	71	**592**	100	62	54	46	44	41	37	36		30	56	4	–	–	–	–	–	–	4	–	–	–	Estra-1,3,5(10)-triene, 3,6,16,17-tetrakis(trimethylsilyloxy)-, (6β,16α,17β)-	69833-54-1	NI46328
191	73	192	502	147	190	193	592	**592**	100	66	25	23	17	15	13	11		30	56	4	–	–	–	–	–	–	4	–	–	–	Estra-1,3,5(10)-triene, tetrakis(trimethylsilyloxy)-, (15α,16α,17β)-	57305-24-5	NI46329
191	73	592	192	502	387	593	193	**592**	100	40	18	18	12	11	9	9		30	56	4	–	–	–	–	–	–	4	–	–	–	Estra-1,3,5(10)-triene, tetrakis(trimethylsilyloxy)-, (15α,16α,17β)-	57305-24-5	LI09810
191	192	502	193	347	503	386	283	**592**	100	21	14	9	7	6	6	6	**5.35**	30	56	4	–	–	–	–	–	–	4	–	–	–	Estra-1,3,5(10)-triene, tetrakis(trimethylsilyloxy)-, (15α,16α,17β)-	57305-24-5	NI46330
121	100	549	109	84	129	98	70	**592**	100	19	12	12	10	8	8	8	**1.00**	33	44	6	4	–	–	–	–	–	–	–	–	–	1,6,10,22-Tetraazatricyclo[9.7.6.1[12,16]]pentacosa-12,14,16(25)-triene-18,23-dione, 6,10-diacetyl-15-hydroxy-17-[(4-methoxyphenyl)methyl]-	69721-68-2	NI46331
591	592	365	349	351	183	160	206	**592**	100	78	59	40	39	32	31	20		36	36	6	2	–	–	–	–	–	–	–	–	–	(+)-2-Norcepharanthine		NI46332
591	592	577	561	296	578	191	204	**592**	100	94	62	38	28	25	17	13		36	36	6	2	–	–	–	–	–	–	–	–	–	(+)-Thalsivasine		NI46333
488	489	487	592	473	445	297	417	**592**	100	31	25	22	13	4	4	3		36	36	6	2	–	–	–	–	–	–	–	–	–	Warifteine	30996-86-2	LI09811
365	183	175	592	366	351	591	349	**592**	100	74	47	37	30	29	21	12		36	36	6	2	–	–	–	–	–	–	–	–	–	Yanangin		NI46334
77	340	105	130	183	102	252	51	**592**	100	60	33	29	24	21	18	16	**0.50**	37	28	4	4	–	–	–	–	–	–	–	–	–	3,5-Pyrazolidinedione, 4,4′-benzylidenebis[1,2-diphenyl-	2652-72-4	NI46335
311	43	55	57	99	82	309	69	**592**	100	63	58	51	50	47	38	38	**0.00**	38	72	4	–	–	–	–	–	–	–	–	–	–	Oleic acid, 2-hydroxyethyl ester stearate	26291-64-5	NI46336
99	311	55	43	57	41	69	98	**592**	100	85	78	75	57	56	42	34	**0.84**	38	72	4	–	–	–	–	–	–	–	–	–	–	Oleic acid, 2-hydroxyethyl ester stearate	26291-64-5	NI46337
57	264	322	71	55	69	43	28	**592**	100	94	89	85	76	72	72	72	**11.04**	39	76	3	–	–	–	–	–	–	–	–	–	–	Oleic acid, 3-(octadecyloxy)propyl ester	17367-41-8	NI46339
57	43	55	69	71	83	41	28	**592**	100	93	86	75	70	55	55	44	**5.04**	39	76	3	–	–	–	–	–	–	–	–	–	–	Oleic acid, 3-(octadecyloxy)propyl ester	17367-41-8	NI46338
592	593	590	594	595	575	412	361	**592**	100	41	24	18	9	2	2	2		40	48	4	–	–	–	–	–	–	–	–	–	–	β,β-Carotene-4,4′-dione, 7,7′,8,8′-tetradehydro-3,3′-dihydroxy-	31687-79-3	NI46340
313	57	43	71	55	83	69	97	**592**	100	89	78	51	46	43	42	38	**33.04**	40	80	2	–	–	–	–	–	–	–	–	–	–	Eicosanoic acid, eicosyl ester	22432-80-0	NI46341
285	57	43	71	55	83	97	69	**592**	100	73	56	46	45	44	41	40	**11.04**	40	80	2	–	–	–	–	–	–	–	–	–	–	Octadecanoic acid, docosyl ester	22413-03-2	NI46342
57	43	83	71	97	69	55	85	**592**	100	64	58	57	56	45	44	37	**0.10**	41	84	1	–	–	–	–	–	–	–	–	–	–	1-Hentetracontanol	40710-42-7	MS10800
73	91	43	518	133	69	592	412	**592**	100	30	4	2	2	2	1	1		42	56	2	–	–	–	–	–	–	–	–	–	–	2,2′-Dioxospirilloxanthin		LI09812
264	43	71	69	169	478	119	380	**593**	100	35	19	13	8	6	2	1	**0.00**	18	17	5	1	–	14	–	–	–	–	–	–	–	3-Methylbutyl N,O-bis(heptafluorobutyryl)hydroxyprolinate		NI46343
297	593	134	36	595	65	204	298	**593**	100	30	28	26	24	24	22	20		28	29	2	9	–	–	2	–	–	–	–	–	–	N-[4-(2-Chloroanilino)-triazin-2-ylaminobutyl]-N-ethyl-3-methyl-4-(4-nitrophenylazo)aniline		IC04604
536	508	451	423	593	537	452	424	**593**	100	72	57	55	48	43	30	30		30	51	7	5	–	–	–	–	–	–	–	–	–	Destruxin B	2503-26-6	NI46344
536	508	451	423	593	537	452	424	**593**	100	74	57	53	48	42	29	27		30	51	7	5	–	–	–	–	–	–	–	–	–	Destruxin B		LI09813
73	388	147	298	243	191	389	267	**593**	100	99	41	38	38	37	35	29	**3.25**	31	59	4	1	–	–	–	–	–	3	–	–	–	Pregn-4-en-3-one, 17,20,21-tris[(trimethylsilyl)oxy]-, O-methyloxime, (20R)-	35554-02-0	NI46345
73	388	147	298	267	243	191	75	**593**	100	99	48	38	35	26	24	20	**1.50**	31	59	4	1	–	–	–	–	–	3	–	–	–	Pregn-4-en-3-one, 17,20,21-tris[(trimethylsilyl)oxy]-, O-methyloxime, (20S)-	57325-74-3	NI46346
73	578	562	75	472	579	563	355	**593**	100	84	77	39	39	34	31	20	**5.10**	32	59	5	1	–	–	–	–	–	2	–	–	–	Methyl 3α,7β-dihydroxy-6-oxo-5β-cholan-24-oate, methyloximebis(trimethylsilyl)ether		NI46347
100	130	101	42	131	85	44	84	**593**	100	91	61	31	29	28	25	21	**4.40**	35	39	4	5	–	–	–	–	–	–	–	–	–	Integerrin		LI09814
100	130	101	42	131	85	84	44	**593**	100	11	7	4	3	3	3	3	**0.52**	35	39	4	5	–	–	–	–	–	–	–	–	–	Integerrin		MS10801
594	595	83	529	85	565	563	48	**594**	100	28	25	18	15	13	13	13		26	26	8	–	4	–	–	–	–	–	–	–	–	Tetramethyl 2,2′-(1,4-phenylene)bis[5-(ethylthio)-3,4-thiophenedicarboxylate]	110699-35-9	DD01803
594	83	529	85	565	563	48	59	**594**	100	25	18	15	13	13	13	10		26	26	8	–	4	–	–	–	–	–	–	–	–	Tetramethyl 2,2′-(1,4-phenylene)bis[5-(ethylthio)-3,4-thiophenedicarboxylate]	110699-35-9	DD01804
43	110	57	109	73	60	58	29	**594**	100	67	53	47	45	44	42	40	**0.00**	27	30	15	–	–	–	–	–	–	–	–	–	–	4H-1-Benzopyran-4-one, 2-[4-[(6-deoxy-α-L-mannopyranosyl)oxy]phenyl] 8-(β-D-glucopyranosyloxy)-5,7-dihydroxy-	57396-71-1	NI46348
169	109	331	205	163	247	222	510	**594**	100	78	50	50	44	28	17	14	**7.22**	27	30	15	–	–	–	–	–	–	–	–	–	–	Glucopyranose, 2,3,4,6-tetraacetate 1-(3,4-dihydroxycinnamate), diacetate, β-D-	18449-69-9	NI46349

MASS TO CHARGE RATIOS	M.W.	INTENSITIES	Parent	C	H	O	N	S	F	Cl	Br	I	Si	P	B	X	COMPOUND NAME	CAS Reg No	No
43 169 44 109 94 115 98 45	**594**	100 55 38 36 33 27 25 21	**0.00**	29	38	13	–	–	–	–	–	–	–	–	–	–	(+)-1-Abscisyl-β-D-glucopyranoside tetraacetate		NI46350
594 125 595 77 272 521 244 228	**594**	100 47 37 33 29 20 15 15		31	16	4	2	–	6	–	–	–	–	–	–	–	1H-Isoindole-1,3(2H)-dione, 5,5-(1,1,2,2,3,3-hexafluoro-1,3-propanediyl)bis[2-phenyl-	33734-35-9	NI46351
272 594 273 595 76 104 168 596	**594**	100 37 16 13 8 5 4 3		31	16	4	2	–	6	–	–	–	–	–	–	–	1H-Isoindole-1,3(2H)-dione, 5,5-(1,1,2,2,3,3-hexafluoro-1,3-propanediyl)bis[2-phenyl-	33734-35-9	NI46352
248 290 150 192 262 164 206	**594**	100 80 53 44 32 32 27	**0.00**	33	38	10	–	–	–	–	–	–	–	–	–	–	Anhydroacetylarthonia acid		LI09815
426 510 427 468 43 511 411 28	**594**	100 52 40 35 25 18 17 16	**0.70**	34	26	10	–	–	–	–	–	–	–	–	–	–	6,11-Dioxo-4,5,7,12-tetraacetoxy-2-(α-methylbenzyl)naphthacene		MS10802
426 43 510 427 468 511 411 42	**594**	100 63 34 28 23 12 12 10	**0.50**	34	26	10	–	–	–	–	–	–	–	–	–	–	5,12-Naphthacenedione, 1,6,10,11-tetrahydroxy-8-(α-methylbenzyl)-, tetraacetate	21127-34-4	MS10803
215 107 81 58 95 216 55 93	**594**	100 91 85 85 84 76 72 64	**28.00**	35	35	1	–	–	5	–	–	–	1	–	–	–	Cholestane, 3-[[dimethyl(pentafluorophenyl)silyl]oxy]-, (3β,5α)-	57691-87-9	NI46353
564 594 491 478 522 549	**594**	100 10 9 9 8 5		35	38	5	4	–	–	–	–	–	–	–	–	–	Purpurin 5, dimethyl ester		LI09816
55 149 105 100 83 292 77 79	**594**	100 75 75 64 64 53 47 44	**1.11**	35	46	8	–	–	–	–	–	–	–	–	–	–	16-Hydroxy-3-O-deca-2′,4′,6′-trienoylingenol-16-angelate		NI46354
439 440 131 219 154 77 362 144	**594**	100 58 52 46 41 40 29 29	**11.60**	36	30	3	–	–	–	–	–	–	3	–	–	–	Cyclotrisiloxane, hexaphenyl-	512-63-0	NI46355
191 381 594 593 367 382 174 146	**594**	100 90 85 59 45 27 15 11		36	38	6	2	–	–	–	–	–	–	–	–	–	Atherospermoline		LI09817
297 298 191 594 490 190 192 403	**594**	100 51 27 23 23 23 17 4		36	38	6	2	–	–	–	–	–	–	–	–	–	Dihydrowarifteine		LI09818
296 298 297 299 594 282 266 192	**594**	100 83 37 18 12 12 12 11		36	38	6	2	–	–	–	–	–	–	–	–	–	Hayatin		LI09819
298 297 296 299 191 162 190 594	**594**	100 26 25 21 17 17 16 12		36	38	6	2	–	–	–	–	–	–	–	–	–	Hayatin		MS10804
298 107 77 162 108 297 191 190	**594**	100 72 51 45 45 33 30 30	**15.00**	36	38	6	2	–	–	–	–	–	–	–	–	–	(–)-Isochondodendrine		MS10805
591 592 593 594 381 191 365 395	**594**	100 78 77 52 47 38 24 17		36	38	6	2	–	–	–	–	–	–	–	–	–	(–)-2′-Northalictine		NI46356
367 594 593 184 191 208 592 192	**594**	100 89 84 84 39 31 25 22		36	38	6	2	–	–	–	–	–	–	–	–	–	(+)-2′-Northaliphylline		NI46357
594 521 297 223 276 260	**594**	100 23 10 10 8 7		36	42	4	4	–	–	–	–	–	–	–	–	–	2,4-Diethyl-5,8-di-(2-methoxycarbonylethyl)-1,3,6,7-tetramethylporphin		LI09820
594 521 448 579 563 433 534 506	**594**	100 21 6 5 5 5 4 3		36	42	4	4	–	–	–	–	–	–	–	–	–	Mesoporphyrin-IX dimethyl ester		LI09821
326 276 105 275 426 566 454 436	**594**	100 82 82 81 62 46 46 46	**9.00**	36	50	7	–	–	–	–	–	–	–	–	–	–	2-Deoxy-α-ecdysone, 3-O-acetate 22-O-benzoate		NI46358
311 43 57 98 267 55 99 41	**594**	100 54 45 41 37 37 35 28	**0.34**	38	74	4	–	–	–	–	–	–	–	–	–	–	Stearic acid, ethylene ester	627-83-8	NI46359
311 43 57 267 98 310 99 55	**594**	100 37 35 34 34 30 26 25	**0.74**	38	74	4	–	–	–	–	–	–	–	–	–	–	Stearic acid, ethylene ester	627-83-8	NI46360
71 57 43 58 55 85 281 69	**594**	100 92 70 47 40 37 33 33	**1.04**	39	78	3	–	–	–	–	–	–	–	–	–	–	Stearic acid, 3-(octadecyloxy)propyl ester	17367-40-7	NI46361
71 57 281 325 43 85 83 55	**594**	100 84 65 54 53 45 40 40	**2.74**	39	78	3	–	–	–	–	–	–	–	–	–	–	Stearic acid, 3-(octadecyloxy)propyl ester	17367-40-7	NI46362
592 594 593 590 595 591 361 596	**594**	100 74 70 63 30 30 25 23		40	50	4	–	–	–	–	–	–	–	–	–	–	Asteric acid		LI09822
91 203 92 592 151 105 43 83	**594**	100 66 42 40 30 27 27 26	**6.66**	40	50	4	–	–	–	–	–	–	–	–	–	–	Asvacene		NI46363
594 91 595 233 232 234 95 596	**594**	100 85 45 42 22 16 15 10		40	54	2	2	–	–	–	–	–	–	–	–	–	6α-Nitro-5α-cholestano[3,2-b]-N-benzylindole		MS10806
43 57 55 41 69 83 56 71	**594**	100 95 74 72 59 51 49 46	**0.00**	40	82	2	–	–	–	–	–	–	–	–	–	–	Hexadecane, 1,1-bis(dodecyloxy)-	56554-64-4	NI46364
594 490 91 69 504 536 133 43	**594**	100 34 20 19 15 10 10 10		42	58	2	–	–	–	–	–	–	–	–	–	–	Rubixanthin acetate		LI09823
594 595 488 91 69 119 105 502	**594**	100 47 34 20 19 16 16 15		42	58	2	–	–	–	–	–	–	–	–	–	–	Rubixanthin acetate		MS10807
168 126 440 482 481 217 575 399	**595**	100 76 58 42 38 24 20 17	**1.50**	19	13	7	1	–	12	–	–	–	–	–	–	–	Acetic acid, trifluoro-, 4-[2-[(isopropyl)(trifluoroacetyl)amino]-1-[(trifluoroacetyl)oxy]ethyl]-1,2-phenylene ester	40629-67-2	NI46365
81 73 299 168 315 74 211 170	**595**	100 73 52 29 25 24 23 23	**0.50**	21	46	7	3	–	–	–	–	–	4	1	–	–	Cytidine, 2′-deoxy-N-(trimethylsilyl)-3′-O-(trimethylsilyl)-, 5′-[bis(trimethylsilyl) phosphate]	32645-72-0	NI46366
73 188 564 75 156 158 595 565	**595**	100 99 60 51 45 38 30 29		31	61	4	1	–	–	–	–	–	3	–	–	–	Pregnan-20-one, 3,11,17-tris[(trimethylsilyl)oxy]-, O-methyloxime, (3α,5β,11β)-	57325-77-6	NI46367
73 188 75 89 474 103 175 74	**595**	100 99 99 32 30 29 28 20	**11.00**	31	61	4	1	–	–	–	–	–	3	–	–	–	Pregnan-20-one, 3,11,21-tris[(trimethylsilyl)oxy]-, O-methyloxime, (3α,5α,11β)-	32206-69-2	NI46368
73 188 75 474 143 564 103 384	**595**	100 99 49 38 20 19 18 17	**13.00**	31	61	4	1	–	–	–	–	–	3	–	–	–	Pregnan-20-one, 3,11,21-tris[(trimethylsilyl)oxy]-, O-methyloxime, (3α,5β,11β)-	32206-68-1	NI46369
73 188 75 103 175 474 81 89	**595**	100 99 86 29 28 27 22 21	**6.00**	31	61	4	1	–	–	–	–	–	3	–	–	–	Pregnan-20-one, 3,11,21-tris[(trimethylsilyl)oxy]-, O-methyloxime, (3β,5α,11β)-	57325-78-7	NI46370
73 188 75 103 143 175 474 107	**595**	100 99 86 34 33 29 26 24	**6.00**	31	61	4	1	–	–	–	–	–	3	–	–	–	Pregnan-20-one, 3,11,21-tris[(trimethylsilyl)oxy]-, O-methyloxime, (3β,5β,11β)-	57325-79-8	NI46371
73 564 565 75 474 103 566 147	**595**	100 99 48 47 34 23 21 20	**20.00**	31	61	4	1	–	–	–	–	–	3	–	–	–	Pregnan-20-one, 3,17,21-tris[(trimethylsilyl)oxy]-, O-methyloxime, (3α,5β)-	56196-43-1	NI46372
105 106 124 77 509 271 148 95	**595**	100 74 43 27 15 14 10 7	**1.00**	32	37	10	1	–	–	–	–	–	–	–	–	–	1α-Benzoyloxy-6β-nicotinoyloxy-8β,9α-diacetoxy-4β-hyd roxydihydro-β agarofuran		MS10808
105 43 124 106 83 77 85 78	**595**	100 78 58 35 25 18 15 14	**7.50**	32	37	10	1	–	–	–	–	–	–	–	–	–	Maymyrsine	93772-06-6	NI46373

MASS TO CHARGE RATIOS								M.W.	INTENSITIES								Parent	C	H	O	N	S	F	Cl	Br	I	Si	P	B	X	COMPOUND NAME	CAS Reg No	No
74	56	428	260	456	372	102	158	**596**	100	87	45	44	37	33	33	32	**12.00**	12	–	12	–	–	–	–	–	–	–	–	–	3	Ruthenium, octacarbonyl(tetracarbonyliron)di-	55836-24-3	NI46374
569	581	570	571	582	583	572	584	**596**	100	66	50	45	34	26	17	9	**4.49**	13	24	12	–	–	–	–	–	–	8	–	–	–	Trimethylpentavinyloctasilsesquioxane		NI46375
131	187	188	243	242	189	133	153	**596**	100	84	15	12	12	8	8	7	**0.00**	16	36	4	–	4	–	–	–	–	–	2	–	1	Cadmium[II] bis(dibutyldithiophosphate)	43159-11-1	NI46376
101	102	129	55	73	45	56	511	**596**	100	94	85	73	27	24	21	20	**0.04**	17	13	4	–	–	–	9	–	–	–	–	–	–	Monohydrokelevan		MS10809
81	73	323	299	211	227	315	75	**596**	100	55	22	22	18	15	14	14	**1.80**	21	45	8	2	–	–	–	–	–	4	1	–	–	2(1H)-Pyrimidinone, 1-[2-deoxy-3-O-(trimethylsilyl)-β-D-erythro-pentofuranosyl]-4-(trimethylsiloxy)-, 5′-[bis(trimethylsilyl) phosphate]	32645-61-7	NI46377
81	73	323	299	211	227	315	75	**596**	100	55	22	22	18	15	14	14	**1.80**	21	45	8	2	–	–	–	–	–	4	1	–	–	5′-Uridylic acid, 2′-deoxy-1-(trimethylsilyl)-3′-O-(trimethylsilyl)-, bis(trimethylsilyl) ester	56211-37-1	NI46378
245	148	91	246	163	69	135	596	**596**	100	98	52	24	24	24	22	11		22	16	2	–	2	6	–	–	–	–	–	–	1	Bis[1,1,1-trifluoro-4-(4-methylphenyl)-4-thioxo-2-butanonato]palladium	53862-21-8	NI46379
186	121	56	39	95	81	94	65	**596**	100	99	79	27	15	10	9	8	**1.30**	24	20	4	–	–	–	–	–	–	–	–	–	4	Tetrakis[(π-cyclopentadienyl)(monocarbonyl)iron]		MS10810
91	150	330	210	168	101	278	108	**596**	100	39	33	20	19	17	15	13	**0.34**	27	36	13	2	–	–	–	–	–	–	–	–	–	L-Threonine, N-(benzyloxycarbonyl)-O-[3,4,6-tri-O-acetyl-2-(acetylamino)-2-deoxy-β-D-glucopyranosyl]-, methyl ester	23141-51-7	NI46380
335	454	377	436	412	362	334	394	**596**	100	98	56	46	46	44	30	29	**0.00**	30	28	13	–	–	–	–	–	–	–	–	–	–	1-Naphthacenecarboxylic acid, 5,7,10,12-tetrakis(acetyloxy)-2-ethyl-1,2,3,4,6,11-hexahydro-2-hydroxy-6,11-dioxo-, methyl ester	54725-40-5	NI46381
512	514	556	554	598	515	513	596	**596**	100	99	82	77	41	39	39	36		30	29	8	–	–	–	–	1	–	–	–	–	–	1′,4-Diacetoxy-3-bromo-2,3′-dimethyl-5,5′,8,8′-tetramethoxy-1,2′-binaphthalene	101459-36-3	NI46382
44	43	55	57	596	69	514	83	**596**	100	40	38	27	20	18	18	17		30	29	8	–	–	–	–	1	–	–	–	–	–	4,4′-Diacetoxy-3-bromo-5,5′8,8′-tetramethoxy-6,6′-dimethyl-1,1′-binaphthalene		NI46383
147	395	191	133	425	311	149	148	**596**	100	77	77	51	45	40	39	31	**1.98**	34	52	5	–	–	–	–	–	–	2	–	–	–	1-Pentanone, 1,1′-[5-[[3-(1-oxopentyl)-4-[(trimethylsilyl)oxy]phenyl]methyl]-2-[(trimethylsilyl)oxy]-1,3-phenylene]bis-	55724-90-8	NI46384
219	331	332	596	57	220	99	55	**596**	100	43	33	23	19	18	14	12		37	56	6	–	–	–	–	–	–	–	–	–	–	Bis(ethoxycarbonyl)bis-(3,5-di-tert-butyl-4-hydroxybenzyl)methane		IC04605
69	68	43	57	41	70	71	55	**596**	100	51	28	23	21	14	12	10	**0.00**	37	72	5	–	–	–	–	–	–	–	–	–	–	Hexadecaneperoxoic acid, 1,1-dimethyl-3-[(1-oxohexadecyl)oxy]propyl ester	56712-07-3	NI46385
444	428	426	408	596	578	580	564	**596**	100	10	10	2	1	1	1	1		40	52	4	–	–	–	–	–	–	–	–	–	–	Amarouciaxanthin B	92121-55-6	NI46386
580	596	564	581	488	594	363	578	**596**	100	72	68	47	45	43	42	41		40	52	4	–	–	–	–	–	–	–	–	–	–	Astaxanthin	472-61-7	NI46387
91	105	147	145	133	43	55	203	**596**	100	71	69	58	56	52	49	47	**6.50**	40	52	4	–	–	–	–	–	–	–	–	–	–	Astaxanthin	472-61-7	MS10811
580	596	581	488	363	594	362	594	**596**	100	72	69	48	46	44	43	39		40	52	4	–	–	–	–	–	–	–	–	–	–	Astaxanthin	472-61-7	LI09824
578	596	125	195	401	560	504	486	**596**	100	96	80	75	18	15	15	12		40	52	4	–	–	–	–	–	–	–	–	–	–	Mytiloxanthinone	83746-66-1	NI46388
578	596	560	471	401	195	504	486	**596**	100	94	16	8	8	8	6	3		40	52	4	–	–	–	–	–	–	–	–	–	–	Mytiloxanthinone	83746-66-1	NI46389
43	57	81	55	41	147	95	145	**596**	100	70	51	49	44	41	40	40	**0.00**	41	72	2	–	–	–	–	–	–	–	–	–	–	Cholest-5-en-3-ol (3β)-, tetradecanoate	1989-52-2	NI46390
73	91	490	43	596	133	109	69	**596**	100	16	3	3	1	1	1	1		42	60	2	–	–	–	–	–	–	–	–	–	–	Spirilloxanthin	34255-08-8	LI09825
73	91	106	105	74	490	77	43	**596**	100	16	8	6	6	3	3	3	**0.50**	42	60	2	–	–	–	–	–	–	–	–	–	–	Spirilloxanthin	34255-08-8	MS10812
29	265	70	27	28	26	263	261	**597**	100	65	55	53	18	18	11	8	**0.00**	11	7	2	1	–	–	6	–	–	–	–	–	1	Mercury, ethyl(4,5,6,7,8,8-hexachloro-1,3,3a,4,7,7a-hexahydro-1,3-dioxo-4,7-methano-2H-isoindol-2-yl)-	2597-93-5	NI46391
538	597	143	169	539	341	69	598	**597**	100	57	36	27	25	20	18	16		19	9	5	1	–	14	–	–	–	–	–	–	–	1H-Indole-3-acetic acid, 5-(2,2,3,3,4,4,4-heptafluoro-1-oxobutoxy)-1-(2,2,3,3,4,4,4-heptafluoro-1-oxobutoxy)-, methyl ester	55712-65-7	NI46392
434	432	354	435	433	508	384	460	**597**	100	95	24	22	22	12	10	8	**3.00**	23	21	8	1	–	–	–	2	–	–	–	–	–	Tetramethyl 5-dibromomethyl-10,11-dihydroazepino[1,2-a]-quinoline-7,8,9,10-tetracarboxylate		LI09826
115	101	43	117	335	246	214	73	**597**	100	88	77	72	58	32	31	29	**0.00**	26	47	14	1	–	–	–	–	–	–	–	–	–	1,5-Di-O-acetyl-7-O-(2-deoxy-2-acetamido-3,4,6-tri-O-methyl-α-D-glucopyranosyl)-2,3,4,6-tetra-O-methyl-L-glycero-D-manno-heptitol		NI46393
169	91	109	331	127	271	170	145	**597**	100	91	52	39	13	11	10	8	**0.10**	27	35	14	1	–	–	–	–	–	–	–	–	–	L-Threonine, N-(benzyloxycarbonyl)-O-(tetra-O-acetyl-β-D-glucopyranosyl)-, methyl ester	20638-79-3	NI46394
73	225	75	133	353	131	226	74	**597**	100	74	37	27	24	16	15	15	**2.20**	30	59	5	1	–	–	–	–	–	3	–	–	–	Prosta-5,13-dien-1-oic acid, 9-(methoxyimino)-11,15-bis[(trimethylsilyl)oxy]-, trimethylsilyl ester, (8ξ,12ξ)-	72377-56-1	NI46395
133	75	199	73	173	225	368	436	**597**	100	82	80	80	68	67	63	60	**5.00**	30	59	5	1	–	–	–	–	–	3	–	–	–	Prosta-5,13-dien-1-oic acid, 9-(methoxyimino)-11,15-bis[(trimethylsilyl)oxy]-, trimethylsilyl ester, (5Z,9E,11α,13E,15S)-	39003-24-2	NI46396
75	225	77	133	73	173	199	368	**597**	100	52	28	26	25	22	15	13	**3.00**	30	59	5	1	–	–	–	–	–	3	–	–	–	Prosta-5,13-dien-1-oic acid, 9-(methoxyimino)-11,15-bis[(trimethylsilyl)oxy]-, trimethylsilyl ester, (5Z,8β,9Z,11α,13E,15S)-	39003-25-3	NI46397
225	353	133	226	424	73	75	173	**597**	100	25	23	20	15	15	11	10	**3.00**	30	59	5	1	–	–	–	–	–	3	–	–	–	Prosta-5,13-dien-1-oic acid, 9-(methoxyimino)-11,15-bis[(trimethylsilyl)oxy]-, trimethylsilyl ester, (5Z,9Z,11α,13E,15S)-	39003-23-1	NI46398

MASS TO CHARGE RATIOS								M.W.	INTENSITIES								Parent	C	H	O	N	S	F	Cl	Br	I	Si	P	B	X	COMPOUND NAME	CAS Reg No	No
225	75	226	133	73	77	353	173	**597**	100	88	22	22	22	21	18	18	**7.00**	30	59	5	1	–	–	–	–	–	3	–	–	–	Prosta-5,11-dien-1-oic acid, 9-(methoxyimino)-11,15-bis[(trimethylsilyl)oxy]-, (5Z,8β,9E,11α,13E,15S)-	39003-26-4	NI46399
91	73	75	251	103	105	41	68	**597**	100	81	28	14	14	12	12	11	**3.00**	34	55	4	1	–	–	–	–	–	2	–	–	–	Pregnane-11,20-dione, 3,21-bis[(trimethylsilyl)oxy]-, 20-(O-benzyloxime), (3α,5β)-	57325-88-9	NI46400
135	246	79	93	136	452	55	41	**597**	100	33	15	14	13	11	10	8	**1.50**	35	55	5	3	–	–	–	–	–	–	–	–	–	Leucine, N-[N²,N⁶-bis(1-adamantylcarbonyl)-L-lysyl]-, methyl ester, L-	28417-15-4	NI46401
144	458	430	374	598	486	159	402	**598**	100	70	56	48	47	47	46	45		14	12	10	–	–	–	–	–	–	–	2	–	2	(μ-Tetramethyldiphosphine)bis(pentacarbonylmolybdenum)		MS10813
59	131	175	267	76	191	235	99	**598**	100	50	33	20	17	13	12	12	**1.08**	16	22	–	–	12	–	–	–	–	–	–	–	–	2,4,6,8,9-Pentathiatricyclo[3.3.1.1³,⁷]decane, 1,1'-dithiobis[3,5,7-trimethyl-	57274-53-0	NI46402
59	58	92	116	117	175	45	57	**598**	100	29	26	21	14	13	13	12	**0.49**	16	22	–	–	12	–	–	–	–	–	–	–	–	2,4,6,8,9-Pentathiatricyclo[3.3.1.1³,⁷]decane, 3,3'-dithiobis[1,5,7-trimethyl-	57274-52-9	NI46403
41	177	179	251	539	537	541	56	**598**	100	81	67	58	53	48	45	45	**0.00**	24	54	1	–	–	–	–	–	–	–	–	–	2	Distannoxane, hexabutyl-	56-35-9	NI46404
73	143	75	191	527	147	117	74	**598**	100	32	20	16	16	15	12	10	**0.16**	31	62	5	–	–	–	–	–	–	3	–	–	–	15-Methyl-PGF-2a, methyl ester, tris(trimethylsilyl)-	54142-23-3	NI46405
399	109	381	355	97	95	391	373	**598**	100	99	92	90	85	81	77	68	**0.00**	33	58	9	–	–	–	–	–	–	–	–	–	–	24ξ-(3'-O-Methyl-α-L-arabinofuranosyloxy)-5α-cholestane-3β,6α,8β,15α tetraol		NI46406
105	77	106	177	122	311	51	294	**598**	100	16	7	4	4	3	3	1	**0.00**	34	30	10	–	–	–	–	–	–	–	–	–	–	Glucitol, 1,2,5,6-tetrabenzoate, D-	20963-95-5	NI46407
44	598	569	599	570	43	270	269	**598**	100	59	34	31	13	11	10	10		34	30	10	–	–	–	–	–	–	–	–	–	–	Ustilaginoidin A, permethyl ether		NI46408
105	43	356	338	416	357	57	59	**598**	100	58	51	31	18	17	16	11	**0.00**	34	46	9	–	–	–	–	–	–	–	–	–	–	3,12,20-Triacetoxy-11-benzoyloxy-14-hydroxypregnane		MS10814
366	367	211	253	426	199	368	212	**598**	100	30	30	19	17	13	12	10	**0.00**	36	54	5	–	1	–	–	–	–	–	–	–	–	Cholest-5-ene-3,19-diol, 3-acetate 19-(4-methylbenzenesulphonate), (3β)-	21072-69-5	NI46409
366	211	253	367	212	199	353	226	**598**	100	71	34	31	28	27	21	20	**0.00**	36	54	5	–	1	–	–	–	–	–	–	–	–	Cholestene-3,19-diol, 19-acetate 3-(4-methylbenzenesulphonate)		LI09827
73	313	75	69	57	83	55	387	**598**	100	80	67	64	64	61	44	37	**0.00**	36	78	2	–	–	–	–	–	–	2	–	–	–	Tricosane-1,15-diol, bis(trimethylsilyl)-		NI46410
446	447	405	448	431	406	580	432	**598**	100	39	15	11	10	5	4	3	**1.81**	37	50	3	–	–	–	–	–	–	2	–	–	–	6α-Hydroxy-Δ¹-tetrahydrocannabinol bis(phenyldimethylsilyl) ether		NI46411
393	394	583	598	271	599	584	135	**598**	100	34	33	30	17	16	16	13		37	50	3	–	–	–	–	–	–	2	–	–	–	7-Hydroxy-Δ¹-tetrahydrocannabinol bis(phenyldimethylsilyl) ether		NI46412
43	463	369	309	464	123	69	121	**598**	100	95	73	47	35	31	30	29	**22.00**	37	58	6	–	–	–	–	–	–	–	–	–	–	(22R,23S)3β,22,23-Triacetoxy-24-methylene-5-α-lanost-8-ene		MS10815
580	582	598	565	490	488	506	562	**598**	100	52	45	20	20	15	8	7		40	54	4	–	–	–	–	–	–	–	–	–	–	Halocynthiaxanthin	81306-52-7	NI46413
73	91	492	598	133	43	474	109	**598**	100	85	23	6	6	6	4	4		41	58	3	–	–	–	–	–	–	–	–	–	–	2-Ketorhodovibrin		LI09828
73	600	494	568	536	462	431	508	**598**	100	18	16	8	7	6	5	3	**0.00**	42	62	2	–	–	–	–	–	–	–	–	–	–	ψ,ψ-Carotene, 3,4-didehydro-1,2,7',8'-tetrahydro-1,11-dimethoxy-		NI46414
564	101	310	240	146	425	216	261	**599**	100	81	69	66	65	42	41	37	**26.24**	–	–	–	7	–	–	9	–	–	–	6	–	–	3aH,6aH,9aH-1,3,4,6,7,9,9b-Heptaaza-2,3a,5,6a,8,9a-hexaphosphaphenalene, 2,2,3a,5,5,6a,8,8,9a-nonachloro-2,2,5,5,8,8-hexahydro-	33992-37-9	MS10816
167	164	197	193	407	405	107	136	**599**	100	46	44	42	40	40	40	38	**0.20**	19	16	7	3	3	–	3	–	–	–	–	–	–	2,2,2-Trichloroethyl 3-[(benzothiazol-2-yl)thiomethyl]-7-methoxycarbonylamino-3-cephem-4-carboxylate 1,1-dioxide	97609-93-3	NI46415
599	600	280	251	601	245	584	324	**599**	100	49	28	28	25	24	22	15		24	49	5	5	–	–	–	–	–	4	–	–	–	Guanosine, N,N-dimethyl-1-(trimethylsilyl)-2',3',5'-tris-O-(trimethylsilyl)-	55556-84-8	NI46416
88	196	224	310	208	526	252	264	**599**	100	78	75	25	22	19	19	13	**6.30**	25	33	12	3	1	–	–	–	–	–	–	–	–	1,6-Dihydro-1-methyl-2-methylthio-5-((E)-2-ethoxycarbonylvinyl)-4-(2,3,4,6-tetra-O-acetyl-β-D-glucopyranosylamino)-1H-pyrimidin-6-one		MS10817
73	75	133	355	74	130	426	43	**599**	100	28	15	11	9	8	7	7	**0.10**	30	61	5	1	–	–	–	–	–	3	–	–	–	Prost-13-en-1-oic acid, 9-(methoxyimino)-11,15-bis[(trimethylsilyl)oxy]-, trimethylsilyl ester, (8ξ,12ξ)-	72150-30-2	NI46417
73	75	355	133	426	130	225	74	**599**	100	30	20	19	18	13	9	9	**0.40**	30	61	5	1	–	–	–	–	–	3	–	–	–	Prost-13-en-1-oic acid, 9-(methoxyimino)-11,15-bis[(trimethylsilyl)oxy]-, trimethylsilyl ester, (8ξ,12ξ)-	72150-30-2	NI46418
75	355	426	130	133	225	427	356	**599**	100	93	84	55	48	37	30	28	**1.00**	30	61	5	1	–	–	–	–	–	3	–	–	–	Prost-13-en-1-oic acid, 9-(methoxyimino)-11,15-bis[(trimethylsilyl)oxy]-, trimethylsilyl ester, (8β,9E,11α,13E,15S)-	56085-38-2	NI46419
199	528	129	75	438	529	173	73	**599**	100	91	66	61	53	40	37	33	**3.00**	30	61	5	1	–	–	–	–	–	3	–	–	–	Prost-13-en-1-oic acid, 9-(methoxyimino)-11,15-bis[(trimethylsilyl)oxy]-, trimethylsilyl ester, (9E,11α,13E,15S)-	39062-24-3	NI46420
75	528	199	129	438	529	478	73	**599**	100	84	78	48	47	35	31	25	**2.00**	30	61	5	1	–	–	–	–	–	3	–	–	–	Prost-13-en-1-oic acid, 9-(methoxyimino)-11,15-bis[(trimethylsilyl)oxy]-, trimethylsilyl ester, (8β,9Z,11α,13E,15S)-	56144-51-5	NI46421
355	426	130	133	225	356	427	75	**599**	100	81	56	54	44	32	29	27	**0.50**	30	61	5	1	–	–	–	–	–	3	–	–	–	Prost-13-en-1-oic acid, 9-(methoxyimino)-11,15-bis[(trimethylsilyl)oxy]-, trimethylsilyl ester, (9Z,11α,13E,15S)-	39003-22-0	NI46422
598	599	195	584	205	600	191	356	**599**	100	95	65	50	48	35	35	25		32	41	10	1	–	–	–	–	–	–	–	–	–	7,8,14-Trimethoxy-13-methoxycarbonyl-15-(3,4,5-trimethoxybenzoyloxyalloberban)		NI46423
275	216	204	176	221	185	276	105	**599**	100	83	80	69	68	64	40	40	**1.20**	34	37	7	3	–	–	–	–	–	–	–	–	–	Proline, 1-[N-(N-benzoyl-L-valyl)-L-tyrosyl]-, methyl ester, benzoate (ester), L-	21026-94-8	NI46424
199	155	398	600	598	201	200	597	**600**	100	26	13	8	6	5	5	4		12	–	–	–	2	10	–	–	–	–	–	–	1	Mercury, bis[(pentafluorophenyl)thio]-	37612-94-5	NI46425

MASS TO CHARGE RATIOS	M.W.	INTENSITIES	Parent	C	H	O	N	S	F	Cl	Br	I	Si	P	B	X	COMPOUND NAME	CAS Reg No	No
57 41 385 441 397 369 43 442	600	100 47 31 29 14 12 12 11	0.00	22	52	9	–	–	–	–	–	–	5	–	–	–	1,3,5,7,9-Pentaethyl-3,5,9-tributoxybicyclo[5.3.1]pentasiloxane		NI46426
79 41 93 91 80 67 39 77	600	100 58 54 52 51 51 45 35	3.54	24	36	–	–	–	–	2	–	–	–	–	–	2	Rhodium, bis[5,6-bis(η^2-vinyl)cyclooctene]di-μ-chlorodi-, stereoisomer	74811-03-3	NI46427
79 93 91 41 133 67 39 77	600	100 81 80 73 66 56 54 45	3.52	24	36	–	–	–	–	2	–	–	–	–	–	2	Rhodium, [1,2-bis(η^2-vinyl)-4-vinylcyclohexane]di-μ-chlorodi-	74811-02-2	NI46428
331 169 193 540 253 569 600	600	100 93 37 12 7 3 2		27	36	15	–	–	–	–	–	–	–	–	–	–	5-Epiloganin pentaacetate		LI09829
169 331 109 193 600 569 540 253	600	100 82 25 17 1 1 1 1		27	36	15	–	–	–	–	–	–	–	–	–	–	Loganin pentaacetate	20586-11-2	LI09830
169 331 109 193 271 253 569 540	600	100 70 40 35 5 4 1 1	0.10	27	36	15	–	–	–	–	–	–	–	–	–	–	Loganin pentaacetate	20586-11-2	MS10818
73 191 289 103 261 75 147 290	600	100 24 22 15 13 11 9 9	0.50	27	52	7	–	–	–	–	–	–	4	–	–	–	Nivalenol, tetrakis(trimethylsilyl)-		NI46429
242 243 528 527 57 397 244 543	600	100 99 50 50 41 33 29 27	5.56	29	36	10	4	–	–	–	–	–	–	–	–	–	Riboflavin, 2′,3′,4′,5′-tetrapropanoate	7652-80-4	NI46430
242 243 527 397 528 244 543 256	600	100 99 50 34 33 30 27 27	6.00	29	36	10	4	–	–	–	–	–	–	–	–	–	Riboflavin, 2′,3′,4′,5′-tetrapropanoate	7652-80-4	NI46431
412 454 370 353 496 310 328 255	600	100 87 65 41 39 37 36 33	0.00	30	36	11	2	–	–	–	–	–	–	–	–	–	Acetamide, N-[2-[3,4-dihydroxy-α-[(N-methylacetamido)methyl]benzyl]-β,4,5-trihydroxyphenethyl]-N-methyl-, 3,4,4,5-tetraacetate	28215-67-0	NI46432
243 129 147 369 103 151 242 191	600	100 74 38 29 27 26 25 25	0.25	30	60	6	–	–	–	–	–	–	3	–	–	–	5-Octenoic acid, 8-[tetrahydro-4-[(trimethylsilyl)oxy]-5-[3-[(trimethylsilyl)oxy]-1-octenyl]-2-furanyl]-8-[(trimethylsilyl)oxy]-, methyl ester	35275-54-8	NI46433
73 75 349 55 111 43 41 324	600	100 28 27 22 19 18 18 17	0.00	30	60	6	–	–	–	–	–	–	3	–	–	–	Prost-13-en-1-oic acid, 6-oxo-9,11,15-tris[(trimethylsilyl)oxy]-, methyl ester, (9α, 11α, 13E, 15S)-	61799-72-2	NI46434
256 518 218 129 503 366 217 301	600	100 39 36 34 32 24 23 22	0.00	30	60	6	–	–	–	–	–	–	3	–	–	–	Thromboxane B2, methyl ester, tris(trimethylsilyl)-	63250-13-5	NI46435
256 510 218 129 511 366 217 301	600	100 58 36 34 31 24 23 22	0.00	30	60	6	–	–	–	–	–	–	3	–	–	–	Thromboxane B2, methyl ester, tris(trimethylsilyl)-	63250-13-5	NI46436
181 152 134 153 151 133 486 193	600	100 68 62 45 21 20 18 15	0.00	31	25	2	2	2	–	–	1	–	–	–	–	–	3-(4-Methoxyphenyl)-2-(4-phenylbenzoyl)-6-phenyl-2,3-dihydro-4H-thiazolo[2,3-b][1,3,4]thiadiazin-5-ium bromide		MS10819
600 601 358 602 343 89 179 359	600	100 46 36 19 16 16 14 12		32	40	4	4	–	–	–	–	–	2	–	–	–	1,2,4,5-Tetraaza-3,6-disilacyclohexane, 1,2,4,5-tetrakis(m-methoxyphenyl) 3,3,6,6-tetramethyl-	17082-89-2	NI46437
151 541 415 499 457 281 400 169	600	100 75 70 64 52 45 38 36	2.00	33	28	11	–	–	–	–	–	–	–	–	–	–	Sphagnorubin C acetate		NI46438
209 124 91 210 300 211 123 177	600	100 32 30 12 10 9 7 5	0.00	34	32	2	–	4	–	–	–	–	–	–	–	–	4-(4-Methylphenylthio)-3,6-bis(4-methoxyphenyl)-3-[2-(4-methylphenylthio)ethenyl]-3,4-dihydro-1,2-dithiin		MS10820
328 585 135 273 197 285 312 73	600	100 39 19 10 10 9 7 7	1.00	34	46	–	2	–	–	–	–	–	2	2	–	–	Silanamine, N,N′-[1,4-butanediylbis(diphenylphosphoranylidyne)]bis[1,1,1 trimethyl-	66055-17-2	NI46439
598 600 165 299 151 599 449 312	600	100 48 36 34 29 23 23 17		36	40	8	–	–	–	–	–	–	–	–	–	–	Tetrabenzo[a,d,g,j]cyclododecene, 5,10,15,20-tetrahydro-2,3,7,8,12,13,17,18-octamethoxy-	17873-58-4	NI46440
467 449 55 69 81 121 109 107	600	100 81 76 68 62 55 54 45	4.00	36	56	7	–	–	–	–	–	–	–	–	–	–	Methyl 3-methoxycarbonylacetylpolyporenate A		MS10821
323 107 95 383 201 267 480 540	600	100 80 74 31 28 7 6 4	0.00	36	56	7	–	–	–	–	–	–	–	–	–	–	Nepedinol 3,11,30-triacetate		NI46441
474 475 153 516 55 517 139 69	600	100 37 37 32 15 12 9 9	0.62	37	60	6	–	–	–	–	–	–	–	–	–	–	Bicyclo[3.1.1]hepta-1,3,5-triene-2,3,6-triol, 7-tetracosyl-, triacetate	55429-47-5	NI46442
109 83 55 43 41 91 127 57	600	100 52 46 46 42 41 35 33	5.00	40	56	4	–	–	–	–	–	–	–	–	–	–	Capsorubine		LI09831
502 600 582 503 352 584 601 520	600	100 52 47 43 43 28 22 21		40	56	4	–	–	–	–	–	–	–	–	–	–	Foliachrome		LI09832
43 91 105 221 181 582 600 502	600	100 30 20 11 9 6 5 5		40	56	4	–	–	–	–	–	–	–	–	–	–	Neochrome		LI09833
43 582 600 564 567	600	100 11 8 5 2		40	56	4	–	–	–	–	–	–	–	–	–	–	Neoxanthin		LI09834
91 57 55 221 157 145 119 105	600	100 95 92 91 91 82 82 82	25.00	40	56	4	–	–	–	–	–	–	–	–	–	–	Neoxanthin		LI09835
582 564 583 565 429 555 549 411	600	100 65 40 30 29 18 17 16	1.00	40	56	4	–	–	–	–	–	–	–	–	–	–	Siphonaxanthin	28526-44-5	NI46443
582 563 583 564 429 556 549 411	600	100 67 41 31 30 18 17 16	1.00	40	56	4	–	–	–	–	–	–	–	–	–	–	Siphonaxanthin		LI09836
91 221 119 105 69 81 121 95	600	100 79 79 77 70 63 57 57	35.00	40	56	4	–	–	–	–	–	–	–	–	–	–	Violaxanthine		LI09837
271 284 165 272 285 310 223 195	600	100 99 29 25 23 15 13 13	7.01	42	36	2	2	–	–	–	–	–	–	–	–	–	Benzene acetaldehyde, 4-methoxy-α,α-diphenyl-, [2-(4-methoxyphenyl)-2,2-diphenylethylidene]hydrazone	56666-84-3	NI46444
73 69 91 105 81 145 55 119	600	100 53 52 51 47 39 39 38	2.00	42	64	2	–	–	–	–	–	–	–	–	–	–	3,4,3′,4′-Tetrahydrospirilloxanthin		MS10822
73 91 69 133 109 83 536 494	600	100 59 53 23 22 9 3 3	0.00	42	64	2	–	–	–	–	–	–	–	–	–	–	3,4,3′,4′-Tetrahydrospirilloxanthin		LI09838
43 196 77 169 105 198 197 273	601	100 82 80 77 77 75 68 66	16.00	29	31	11	1	1	–	–	–	–	–	–	–	–	(4-Benzoylphenyl)aminocarbonylmethyl 2,3,4,6-tetra-O-acetyl-1-thio-α-D glucopyranoside	73364-62-2	NI46445
602 483 347 574 447 474 428 582	602	100 92 69 16 15 11 9 8		13	15	1	–	–	5	–	–	1	–	–	–	1	Iridium, carbonyliodo(pentafluoroethyl)[(1,2,3,4,5-η)-1,2,3,4,5-pentamethyl-2,4-cyclopentadien-1-yl]-	56784-35-1	MS10823
361 359 131 249 247 357 187 360	602	100 74 74 69 56 44 43 35	30.00	16	36	4	–	4	–	–	–	–	–	2	–	1	Tin[II] bis(dibutyldithiophosphate)	88478-80-2	NI46446
365 532 69 296 129 217 117 533	602	100 72 54 41 30 24 17 13	0.00	18	–	–	–	–	15	2	–	–	–	1	–	–	Tris(pentafluorophenyl) dichlorophosphorane		MS10824
277 279 202 139 460 278 155 205	602	100 35 32 23 16 13 13 11	3.00	24	14	5	–	–	–	2	–	–	–	–	–	2	10H-Phenoxarsine, 10,10′-oxybis[2-chloro-, 10,10′-dioxide	55429-19-1	NI46447

MASS TO CHARGE RATIOS								M.W.	INTENSITIES								Parent	C	H	O	N	S	F	Cl	Br	I	Si	P	B	X	COMPOUND NAME	CAS Reg No	No
134	185	183	193	102	157	155	135	**602**	100	97	95	55	44	43	40	38	**0.00**	25	20	2	2	2	–	–	2	–	–	–	–	–	2-(4-Bromobenzoyl)-3-(4-methoxyphenyl)-6-phenyl-2,3-dihydro-4h-thiazolo[2,3-b][1,3,4]thiadiazin-5-ium bromide		MS10825
245	147	129	191	173	246	204	243	**602**	100	45	37	36	24	20	20	18	**0.50**	30	62	6	–	–	–	–	–	–	3	–	–	–	2-Furanoctanoic acid, tetrahydro-η,4-bis[(trimethylsilyl)oxy]-5-[3-[(trimethylsilyl)oxy]-1-octenyl]-, methyl ester	35275-56-0	NI46448
106	170	171	169	168	42	154	602	**602**	100	99	68	44	40	26	19	18		32	34	4	4	2	–	–	–	–	–	–	–	–	N,N-Bis(1-dimethylaminonaphthalene-5-sulphonyl)-2-(4-aminophenyl)ethylamine		NI46449
106	170	171	169	168	602	107	172	**602**	100	99	68	42	40	20	14	12		32	34	4	4	2	–	–	–	–	–	–	–	–	1-Naphthalenesulphonamide, 5-(dimethylamino)-N-[4-[2-[[[5-(dimethylamino)-1-naphthalenyl]sulphonyl]amino]ethyl]phenyl]-	55836-94-7	NI46450
71	207	127	86	55	43	41	515	**602**	100	98	98	98	98	98	98	93	**1.23**	32	42	11	–	–	–	–	–	–	–	–	–	–	Anodendroside G, monoacetate	28905-47-7	NI46451
122	602	84	121	259	180	94	86	**602**	100	90	85	80	65	60	60	55		35	46	5	4	–	–	–	–	–	–	–	–	–	Tetrapyrrane A		LI09839
602	603	571	572	301	574	262	31	**602**	100	39	23	11	11	5	5	4		36	42	8	–	–	–	–	–	–	–	–	–	–	1,1′,6,6′,7,7′-Hexamethoxy-3,3′-dimethyl-5,5′-diisopropyl-2,2′-binaphthyl-8,8′-dialdehyde		LI09840
542	482	355	467	295	203	407	527	**602**	100	83	78	76	57	52	49	24	**5.20**	36	58	7	–	–	–	–	–	–	–	–	–	–	24S-Cycloartane-25-ol-3,16,24-triacetate		NI46452
451	225	43	524	449	109	69	95	**602**	100	67	66	61	60	51	48	46	**2.97**	36	58	7	–	–	–	–	–	–	–	–	–	–	Lanost-8(11)-ene-3,18,20,23-tetrol, 3,18,23-triacetate, (3β,20ξ)-	36871-80-4	NI46453
87	119	189	203	542	219	422	347	**602**	100	74	47	29	18	16	13	6	**1.30**	36	58	7	–	–	–	–	–	–	–	–	–	–	2α,3β,28-Triacetoxytaraxastan-30β-ol		NI46454
143	73	75	144	123	55	81	57	**602**	100	24	22	11	11	9	8	8	**0.00**	36	75	4	–	–	–	–	–	–	–	1	–	–	Phosphoric acid, dioctadecyl ester	3037-89-6	NI46455
559	131	560	253	385	386	561	89	**602**	100	55	44	40	33	22	18	12	**2.29**	37	70	2	–	–	–	–	–	–	2	–	–	–	3β,17β-Bis(tri-N-propylsilyloxy)androst-5-ene		NI46456
302	602	287	603	587	570	569	544	**602**	100	46	46	30	24	23	23	17		38	26	–	4	2	–	–	–	–	–	–	–	–	2,2′-Bis[3-cyano-4-(4-methylphenyl)-6-phenylpyridyl] disulphide		NI46457
43	368	602	584	538	510	492	472	**602**	100	3	1	1	1	1	1	1		40	58	4	–	–	–	–	–	–	–	–	–	–	5,6-Dihydro-β,β-carotene-3,3′,5,6-tetrol		NI46458
566	567	584	548	585	568	444	474	**602**	100	43	33	19	18	16	14	12	**1.00**	40	58	4	–	–	–	–	–	–	–	–	–	–	Siphonaxanthol	33474-08-7	NI46459
301	302	602	603	303	604	209	300	**602**	100	26	12	6	6	2	2	1		42	42	–	–	–	–	–	–	–	2	–	–	–	Disilane, hexa-p-tolyl-	15931-53-0	NI46460
135	203	272	262	330	398	341	409	**602**	100	21	5	5	3	2	2	1	**1.80**	43	70	1	–	–	–	–	–	–	–	–	–	–	Bombiprenone		LI09841
355	414	415	275	109	307	283	110	**603**	100	90	70	70	57	51	48	48	**29.03**	28	37	10	5	–	–	–	–	–	–	–	–	–	Glutaramic acid, 4-(carboxyamino)-N-[2-(1-carboxyimidazol-4-yl)-1-[(α-carboxyphenethyl)carbamoyl]ethyl]-, diethyl dimethyl ester	21178-22-3	NI46461
170	171	42	168	154	169	127	128	**603**	100	88	39	29	26	24	22	21	**12.50**	32	33	5	3	2	–	–	–	–	–	–	–	–	1-Naphthalenesulphonic acid, 5-(dimethylamino)-, 2-[2-[[[5-(dimethylamino)-1-naphthalenyl]sulphonyl]amino]ethyl]phenyl ester	55836-95-8	NI46462
170	171	168	169	172	263	203	107	**603**	100	87	29	25	14	12	11	11	**9.80**	32	33	5	3	2	–	–	–	–	–	–	–	–	1-Naphthalenesulphonic acid, 5-(dimethylamino)-, 2-[2-[[[5-(dimethylamino)-1-naphthalenyl]sulphonyl]amino]ethyl]phenyl ester	55836-95-8	NI46463
170	171	42	168	169	127	154	129	**603**	100	83	58	39	38	23	22	22	**7.90**	32	33	5	3	2	–	–	–	–	–	–	–	–	1-Naphthalenesulphonic acid, 5-(dimethylamino)-, 3-[2-[[[5-(dimethylamino)-1-naphthalenyl]sulphonyl]amino]ethyl]phenyl ester	55836-96-9	NI46465
170	171	168	169	603	172	167	203	**603**	100	82	42	32	18	14	11	8		32	33	5	3	2	–	–	–	–	–	–	–	–	1-Naphthalenesulphonic acid, 5-(dimethylamino)-, 3-[2-[[[5-(dimethylamino)-1-naphthalenyl]sulphonyl]amino]ethyl]phenyl ester	55836-96-9	NI46464
170	171	42	168	169	107	154	127	**603**	100	80	53	29	28	22	20	20	**6.20**	32	33	5	3	2	–	–	–	–	–	–	–	–	1-Naphthalenesulphonic acid, 5-(dimethylamino)-, 4-[2-[[[5-(dimethylamino)-1-naphthalenyl]sulphonyl]amino]ethyl]phenyl ester	13285-05-7	NI46467
170	171	168	169	172	263	167	603	**603**	100	84	30	28	12	11	8	7		32	33	5	3	2	–	–	–	–	–	–	–	–	1-Naphthalenesulphonic acid, 5-(dimethylamino)-, 4-[2-[[[5-(dimethylamino)-1-naphthalenyl]sulphonyl]amino]ethyl]phenyl ester	13285-05-7	NI46466
41	69	59	43	55	95	67	121	**603**	100	85	43	42	40	28	27	23	**0.00**	42	51	3	–	–	–	–	–	–	–	–	–	–	Allogrohiarubin acetate		NI46468
537	507	235	68	205	477	41	40	**604**	100	35	26	26	25	20	10	5	**0.10**	10	18	–	4	–	–	–	–	–	–	–	–	2	Imidazolyldimethylthallium dimer		LI09842
117	52	202	145	173	201	182	234	**604**	100	90	42	24	19	11	10	4	**3.80**	16	10	6	–	–	–	–	–	–	–	–	–	3	Bis(cyclopentadienyltricarbonylchromium)mercury		MS10826
117	52	202	145	173	201	182	604	**604**	100	90	42	24	19	11	10	4		16	10	6	–	–	–	–	–	–	–	–	–	3	Bis(cyclopentadienyltricarbonylchromium)mercury		LI09843
57	447	437	41	475	31	58	281	**604**	100	58	55	51	35	28	23	20	**0.00**	20	36	10	–	–	–	–	–	–	6	–	–	–	1,3,5,7,9,11-Hexavinyl-5,9-dibutoxytricyclo[5.5.1.1^{3,11}]hexasiloxane	110991-16-7	NI46469
73	315	299	182	204	147	357	75	**604**	100	49	36	33	25	23	17	14	**0.00**	21	53	8	–	–	–	–	–	–	5	1	–	–	D-arabino-Hexopyranose, 2-deoxy-1,3,4-tris-O-(trimethylsilyl)-, bis(trimethylsilyl) phosphate	55334-84-4	NI46470
73	315	299	182	204	357	147	316	**604**	100	60	47	34	31	24	22	15	**0.00**	21	53	8	–	–	–	–	–	–	5	1	–	–	D-arabino-Hexose, 2-deoxy-3,4,5-tris-O-(trimethylsilyl)-, 6-[bis(trimethylsilyl) phosphate]	55520-84-8	NI46471
376	69	267	377	41	55	97	321	**604**	100	80	60	42	32	28	27	20	**20.00**	25	21	7	–	–	9	–	–	–	–	–	–	–	Estra-1,3,5(10)-trien-17-one, 1-methyl-3,6,7-tris[(trifluoroacetyl)oxy]-, (6α,7α)-	56588-13-7	NI46472
73	265	217	75	204	69	147	41	**604**	100	48	26	23	22	22	15	15	**0.00**	30	56	10	2	–	–	–	–	–	–	–	–	–	1,3-Dimethyl(trimethylsilyloxy)secobarbital glucoranide, methyl ester		NI46473
170	171	604	168	154	169	64	155	**604**	100	32	28	19	14	13	13	12		31	28	7	2	2	–	–	–	–	–	–	–	–	1,2-Bis(1-dimethylaminonaphthalene-5-sulphonyloxy)(4-formyl)benzene		LI09844
105	122	77	78	441	91	95	94	**604**	100	99	90	75	72	70	62	62	**0.00**	35	40	9	–	–	–	–	–	–	–	–	–	–	12,21-Di-O-benzoylviminolone		LI09845
122	319	105	77	301	283	113	78	**604**	100	36	33	30	23	26	25	25	**0.00**	35	40	9	–	–	–	–	–	–	–	–	–	–	12,21-Di-O-benzoylviminolone	19308-45-3	NI46474
253	105	159	157	143	131	145	117	**604**	100	61	45	43	42	36	34	33	**0.00**	35	56	8	–	–	–	–	–	–	–	–	–	–	Cholestane-3,7,12,25-tetrol, tetraacetate, (3α,5β,7α,12α)-	60354-50-9	NI46475

MASS TO CHARGE RATIOS								M.W.	INTENSITIES								Parent	C	H	O	N	S	F	Cl	Br	I	Si	P	B	X	COMPOUND NAME	CAS Reg No	No
561	255	562	131	387	429	256	430	**604**	100	87	47	46	36	31	20	19	**2.34**	37	72	2	–	–	–	–	–	–	2	–	–	–	3α,17β-Bis(tripropylsilyloxy)-5α-androstane		NI46476
561	255	562	131	256	387	257	563	**604**	100	87	46	42	34	28	26	18	**0.00**	37	72	2	–	–	–	–	–	–	2	–	–	–	3α,17β-Bis(tripropylsilyloxy)-5β-androstane		NI46477
561	562	131	255	563	89	387	256	**604**	100	44	36	34	18	14	10	9	**0.98**	37	72	2	–	–	–	–	–	–	2	–	–	–	3β,17β-Bis(tripropylsilyloxy)-5α-androstane		NI46478
604	605	105	575	77	606	576	542	**604**	100	46	29	21	19	11	9	9		40	32	4	2	–	–	–	–	–	–	–	–	–	[(6-Anilino-2,7-dimethylpropyl)phenoxy]-2,3-dihydro-7H-dibenz(f,ij)isoquinoline		IC04606
69	137	205	263	604	399	467	273	**604**	100	25	8	5	3	3	2	2		43	72	1	–	–	–	–	–	–	–	–	–	–	6,10,14,18,22,26,30,34-Octamethylpentatriaconta-9,13,17,21,25,29,33-hepten-2-one		LI09846
57	43	71	85	55	41	69	83	**604**	100	92	61	44	37	32	30	25	**0.00**	43	88	–	–	–	–	–	–	–	–	–	–	–	12-Nonyl-12-decyltetracosane		AI02408
57	43	71	55	85	69	41	56	**604**	100	54	49	31	28	25	25	23	**0.00**	43	88	–	–	–	–	–	–	–	–	–	–	–	Tetracontane, 3,5,24-trimethyl-	55162-61-3	NI46479
57	71	43	85	55	83	69	56	**604**	100	64	48	42	29	22	22	16	**0.10**	43	88	–	–	–	–	–	–	–	–	–	–	–	Tritetracontane	7098-21-7	WI01599
604	605	105	104	77	527	395	575	**604**	100	50	10	8	8	7	7	5		44	32	1	2	–	–	–	–	–	–	–	–	–	2,4,5-Triphenyl-3-(3,4,5-triphenylpyrrol-2-yl)-3H-pyrrol-3-ol	65241-89-6	NI46480
604	527	605	528	499	500	104	77	**604**	100	85	50	38	32	14	12	12		44	32	1	2	–	–	–	–	–	–	–	–	–	1,3,4-Triphenyl-5-(3,4,5-triphenylpyrrol-2-yl)-6-oxa-2-azabicyclo[3.1.0]hex-3-ene	65241-88-5	NI46481
379	463	491	93	435	63			**605**	100	82	45	45	27	27			**11.73**	14	20	11	1	–	–	–	–	–	1	1	–	3	Iron, octacarbonyldihydro(trimethyl phosphite-P)[μ^3-[1,1,1-trimethylsilanaminato(2-)]]tri-, triangulo-	60370-89-0	NI46482
364	211	153	134	433	295	517	498	**605**	100	99	88	85	77	51	35	27	**19.00**	15	30	–	3	–	9	–	–	–	–	4	–	1	Tris(piperidinodifluorophosphino)trifluorophosphinonickel		NI46483
546	321	169	69	40	605	241	392	**605**	100	57	37	36	28	28	26	21		17	13	5	3	–	14	–	–	–	–	–	–	–	N(τ)-(2-hydroxyethyl)histidine N,O-bis(heptafluorobutyrate) methyl ester		NI46484
190	119	191	428	140	160	42	69	**605**	100	70	21	20	19	16	16	14	**0.00**	18	10	5	1	–	15	–	–	–	–	–	–	–	Benzenemethanol, 4-[tris(pentafluoropropionyl)oxy]-α-[(methylamino)methyl]-		MS10827
177	442	119	295	58	176	69	131	**605**	100	46	39	21	20	14	13	11	**6.80**	18	10	5	1	–	15	–	–	–	–	–	–	–	3-(3,4-Dihydroxyphenyl)propylamine tri(pentafluoropropionate)		MS10828
190	428	119	69	44	281	429	41	**605**	100	40	28	14	14	10	9	8	**0.00**	18	10	5	1	–	15	–	–	–	–	–	–	–	Epinine, tri(pentafluoropropionate)		MS10829
190	428	119	429	42	281	140	69	**605**	100	61	36	14	12	10	10	10	**0.00**	18	10	5	1	–	15	–	–	–	–	–	–	–	Epinine, tri(pentafluoropropionate)		MS10830
165	442	190	280	164	202	147	100	**605**	100	93	58	38	38	30	20	20	**0.00**	18	10	5	1	–	15	–	–	–	–	–	–	–	4-(Hydroxy-2-aminopropyl)phenol, tri(pentafluoropropionate)		NI46485
190	119	42	191	69	442	160	43	**605**	100	22	9	7	5	4	3	3	**0.00**	18	10	5	1	–	15	–	–	–	–	–	–	–	2-(3-Hydroxyphenyl)-2-hydroxy-N-methylethylamine tri(pentafluoropropionate)		MS10831
190	442	119	45	443	69	295	191	**605**	100	48	20	17	12	10	8	7	**0.00**	18	10	5	1	–	15	–	–	–	–	–	–	–	α-Methyldopamine, tri(pentafluoropropionate)		MS10832
190	442	119	443	191	415	149	147	**605**	100	48	17	10	6	3	2	2	**0.00**	18	10	5	1	–	15	–	–	–	–	–	–	–	Propanamide, N-[2-(3,4-dihydroxyphenyl)-isopropyl]-2,2,3,3,3-pentafluoro, bis(2,2,3,3,3-pentafluoro-1-oxopropyl)-	72347-75-2	NI46486
190	442	119	69	100	443	295	81	**605**	100	33	20	10	8	7	6	6	**0.00**	18	10	5	1	–	15	–	–	–	–	–	–	–	Propanoic acid, pentafluoro-, 4-[2-[(2,2,3,3,3-pentafluoro-1-oxopropyl)amino]propyl]-1,2-phenylene ester	74231-47-3	NI46487
498	497	101	88	127	158	159	572	**605**	100	90	75	75	60	50	48	34	**3.00**	37	67	5	1	–	–	–	–	–	–	–	–	–	N-(5α-cholestan-3β-yl)-2,3,4,6-tetra-O-methyl-β-D-glucosylamine		NI46488
91	136	348	513	604	350	84	258	**605**	100	32	25	20	18	17	16	15	**9.00**	39	43	5	1	–	–	–	–	–	–	–	–	–	(E)-2-(p-Benzyloxyhydrocinnamoyloxy)-4-(3-benzyloxy-4-methoxyphenyl)quinolizidine		LI09847
119	199	542	227	100	171	69	606	**606**	100	92	39	35	34	33	21	18		10	–	4	–	1	12	1	–	1	–	–	–	–	4-Chlorotetrafluorophenyl 5-iodo-3-oxa-perfluoropentanesulphonate	89847-83-6	NI46489
577	578	579	580	547	233	591	274	**606**	100	45	38	15	12	11	11	11	**1.19**	13	34	12	–	–	–	–	–	–	8	–	–	–	Trimethylpentaethyloctasilsesquioxane	77699-56-0	NI46490
135	136	164	108	178	237	54	49	**606**	100	44	42	27	19	9	8	8	**0.70**	14	17	6	6	–	–	6	–	–	–	1	–	–	N-(5′-Deoxyadenosyl)-di-O-(β,β,β-trichloroethyl)-phosphoramidate		MS10833
87	59	517	443	518	577	519	115	**606**	100	91	53	45	26	26	20	19	**0.99**	16	46	9	–	–	–	–	–	–	8	–	–	–	1,3,5,7,9,11,13,15-Octaethylbisyslo[7.7.1]octasiloxane	73113-21-0	NI46491
73	221	147	222	207	74	281	45	**606**	100	61	37	12	12	9	7	5	**0.00**	18	54	7	–	–	–	–	–	–	8	–	–	–	1,1,1,7,7,7-Hexamethyl-3,3,5,5-tetrakis(trimethylsiloxy)tetrasiloxane		MS10834
73	357	103	445	299	503	101	358	**606**	100	37	33	16	16	14	13	10	**0.27**	21	55	8	–	–	–	–	–	–	5	1	–	–	Phosphoric acid, bis[2,3-bis[(trimethylsilyl)oxy]propyl] trimethylsilyl ester	32046-28-9	NI46492
73	409	75	606	147	607	170	498	**606**	100	67	55	46	27	24	23	19		32	58	5	–	–	–	–	–	–	3	–	–	–	24-Norchola-20,22-dien-14-ol, 21,23-epoxy-3,12,23-tris[(trimethylsilyl)oxy]-, (3β,5β,12β,14β)-	40837-85-2	NI46493
73	409	75	606	147	170	498	183	**606**	100	68	55	47	28	24	19	14		32	58	5	–	–	–	–	–	–	3	–	–	–	24-Norchola-20,22-dien-14-ol, 21,23-epoxy-3,12,23-tris[(trimethylsilyl)oxy]-, (3β,5β,12β,14β)-	40837-85-2	MS10835
73	409	75	606	147	607	170	498	**606**	100	70	55	46	27	25	24	19		32	58	5	–	–	–	–	–	–	3	–	–	–	3,12,23-Tris(trimethylsilyloxy)digoxigenin		LI09848
447	429	385						**606**	100	59	32						**0.00**	33	50	10	–	–	–	–	–	–	–	–	–	–	Crustecdysone triacetate		LI09849
563	121	266	606	84	267	70	100	**606**	100	45	12	11	10	7	6	5		34	46	6	4	–	–	–	–	–	–	–	–	–	1,6,10,22-Tetraazatricyclo[9.7.6.1^{12,16}]pentacosa-12,14,16(25)-triene-18,23-dione, 6,10-diacetyl-15-methoxy-17-[(4-methoxyphenyl)methyl]-	69721-62-6	NI46494
606	607	73	57	43	341	608	69	**606**	100	47	35	27	25	24	20	18		35	66	4	–	–	–	–	–	–	2	–	–	–	cis-Dihydroubiquinone bis(trimethylsilylether)		LI09850
606	607	73	608	43	341	57	576	**606**	100	50	27	20	16	15	14	6		35	66	4	–	–	–	–	–	–	2	–	–	–	trans-Dihydroubiquinone bis(trimethylsilylether)		LI09851
606	547	607	574	548	461	608	459	**606**	100	67	55	36	31	28	19	19		36	38	5	4	–	–	–	–	–	–	–	–	–	Pheoporphyrin a5 methyl ester		LI09852
78	44	28	43	77	41	55	57	**606**	100	80	42	31	30	26	19	18	**1.92**	36	46	8	–	–	–	–	–	–	–	–	–	–	12-O-Tetradeca-2,4,6,8,10-pentaenoylphorbol-13-acetate	77550-17-5	NI46495

MASS TO CHARGE RATIOS								M.W.	INTENSITIES								Parent	C	H	O	N	S	F	Cl	Br	I	Si	P	B	X	COMPOUND NAME	CAS Reg No	No
502	503	501	606	504	607	486	312	**606**	100	35	35	17	8	7	7	4		37	38	6	2	–	–	–	–	–	–	–	–	–	Methyl warifteine	32728-54-4	LI09853
379	606	190	380	365	605	189	363	**606**	100	53	46	35	31	25	13	12		37	38	6	2	–	–	–	–	–	–	–	–	–	5-O-Methylyanangin		NI46496
197	289	319	274	606	198	275	607	**606**	100	55	50	42	32	20	19	17		39	38	1	2	–	–	–	–	–	2	–	–	–	N-Phenyl-N′-(p-anisyl)-N,N′-bis(diphenylmethylsilyl)hydrazine		LI09854
325	43	55	323	57	322	69	41	**606**	100	45	43	41	40	36	31	27	**0.00**	39	74	4	–	–	–	–	–	–	–	–	–	–	Oleic acid, 3-hydroxypropyl ester stearate	17367-45-2	NI46497
606	57	607	43	71	55	69	85	**606**	100	65	47	47	37	24	23	22		41	82	2	–	–	–	–	–	–	–	–	–	–	Hentetracontanoic acid	40710-29-0	NI46498
57	43	74	606	71	87	75	55	**606**	100	51	45	40	38	37	35	35		41	82	2	–	–	–	–	–	–	–	–	–	–	Octatriacontanoic acid, 34,36-dimethyl-, methyl ester	18082-12-7	NI46499
422	607	424	423	609	97	608	65	**607**	100	43	26	26	19	17	16	14		12	30	6	–	6	–	–	–	–	–	3	–	1	Chromium(III) tris(diethyldithiophosphate)	14177-95-8	NI46500
422	607	424	423	609	608	394	186	**607**	100	43	26	26	19	16	12	12		12	30	6	–	6	–	–	–	–	–	3	–	1	Chromium(III) tris(diethyldithiophosphate)	14177-95-8	NI46501
607	415	608	223	606	399	416	222	**607**	100	32	27	22	11	10	8	8		24	18	–	3	–	12	–	–	–	–	1	–	–	Tris(dimethylaminotetrafluorophenyl)phosphine		MS10836
73	402	504	403	147	505	243	191	**607**	100	99	42	32	25	18	17	17	**6.00**	31	57	5	1	–	–	–	–	–	3	–	–	–	Pregn-4-ene-3,11-dione, 17,20,21-tris[(trimethylsilyl)oxy]-, 3-(O-methyloxime), (20R)-	57325-72-1	NI46502
402	73	403	147	504	191	505	243	**607**	100	71	33	24	21	11	10	10	**3.00**	31	57	5	1	–	–	–	–	–	3	–	–	–	Pregn-4-ene-3,11-dione, 17,20,21-tris[(trimethylsilyl)oxy]-, 3-(O-methyloxime), (20S)-	57325-73-2	NI46503
412	533	443	103	73	534	75	413	**607**	100	99	98	69	52	48	48	42	**0.00**	31	57	5	1	–	–	–	–	–	3	–	–	–	Pregn-4-ene-3-one, 18,20-epoxy-11,20,21-tris[(trimethylsilyl)oxy]-, O-methyloxime, (11β)-	72101-52-1	NI46504
73	75	255	131	81	44	93	55	**607**	100	87	85	30	26	24	23	23	**1.00**	33	61	5	1	–	–	–	–	–	2	–	–	–	Glycine, N-[(3α,5β,12α)-24-oxo-3,12-bis[(trimethylsilyl)oxy]cholan-24-yl], methyl ester	57326-17-7	NI46505
581	582	583	593	594	584	595	553	**608**	100	50	38	31	16	13	12	8	**6.79**	14	24	12	–	–	–	–	–	–	8	–	–	–	Dimethylhexavinyloctasilsesquioxane	85248-20-0	NI46506
280	43	226	493	268	240	254	539	**608**	100	59	10	5	3	2	1	0	**0.20**	19	22	4	2	–	14	–	–	–	–	–	–	–	3-Methylbutyl N,N′-bis(heptafluorobutyryl)lysinate		NI46507
77	154	51	139	137	135	120	287	**608**	100	45	42	26	19	12	10	7	**1.17**	24	10	–	–	–	10	–	–	–	–	–	–	1	Stannane, bis(pentafluorophenyl)diphenyl-	1062-67-5	NI46508
217	147	204	127	132	305	143	133	**608**	100	79	67	58	57	48	39	36	**0.00**	26	56	8	–	–	–	–	–	–	4	–	–	–	6-(Octanoyl)glucuronide, tetrakis(trimethylsilyl)-		NI46509
73	217	75	77	147	57	45	47	**608**	100	82	77	73	43	42	40	27	**0.00**	26	56	8	–	–	–	–	–	–	4	–	–	–	2-Propylpentanoic acid glucuronide, tetrakis(trimethylsilyl)-		NI46510
496	494	493	495	490	28	498	492	**608**	100	77	66	56	52	52	39	36	**12.13**	30	24	4	–	–	–	–	–	–	–	2	–	1	Molybdenum, tetracarbonyl[1,2-ethanediylbis[diphenylphosphine]-P,P′]-	15444-66-3	NI46511
91	148	65	92	39	106	63	51	**608**	100	26	25	12	12	7	7	7	**2.31**	30	28	–	2	4	–	–	–	–	–	–	–	1	Zinc bis(dibenzyldithiocarbamate)		IC04607
429	368	430	326	325	308	428	369	**608**	100	46	26	23	20	20	19	15	**0.10**	30	40	13	–	–	–	–	–	–	–	–	–	–	4,6β-Dihydroxy-7β,9,12β,13,20-pentaacetoxytigli-1-en-3-one		LI09855
43	83	69	109	488	368	308	125	**608**	100	19	12	10	6	6	6	5	**0.00**	30	40	13	–	–	–	–	–	–	–	–	–	–	4,7α,12β,13,20-Pentaacetoxy-7,9-dihydroxy-7βH-tigli-1-en-3-one		NI46512
469	91	425	485	197	149	313	188	**608**	100	76	45	38	33	25	23	21	**0.00**	32	36	8	2	1	–	–	–	–	–	–	–	–	Dichotine, 19-(benzylthio)-11-methoxy-, acetate (ester)	29484-58-0	LI09856
469	91	43	470	425	485	197	313	**608**	100	75	49	47	44	38	32	28	**0.00**	32	36	8	2	1	–	–	–	–	–	–	–	–	Dichotine, 19-(benzylthio)-11-methoxy-, acetate (ester)	29484-58-0	NI46513
608	607	609	395	195	397	610	593	**608**	100	50	36	19	15	12	8	6		33	40	9	2	–	–	–	–	–	–	–	–	–	Isoreserpine		MS10837
195	608	251	607	397	381	414	609	**608**	100	52	38	32	25	25	20	19		33	40	9	2	–	–	–	–	–	–	–	–	–	Reserpine		MS10838
74	76	428	255	429	55	81	129	**608**	100	61	59	40	29	27	25	23	**0.00**	33	64	4	–	–	–	–	–	–	3	–	–	–	Cholan-24-oic acid, 3,7-bis[(trimethylsilyl)oxy]-, trimethylsilyl ester, (3α,5β,7α)-	55320-14-4	NI46514
341	313	327	382	342	340	359	606	**608**	100	31	30	23	23	22	17	15	**1.21**	34	56	9	–	–	–	–	–	–	–	–	–	–	2-Decarboxy-7-dehydroxy-3-demethoxy-3,7-dioxomonensin	92009-91-1	NI46515
313	296	608	593	297	327	311	281	**608**	100	75	39	36	29	23	22	22		35	28	10	–	–	–	–	–	–	–	–	–	–	Hinokiflavone pentamethyl ether		LI09857
608	521	580	549	419	592	493	433	**608**	100	20	12	9	5	4	4	4		36	40	5	4	–	–	–	–	–	–	–	–	–	2-Desvinyl-2-(3-oxoprop-1-enyl)isochlorin-e4, dimethyl ester		LI09858
55	389	311	91	43	83	203	69	**608**	100	96	86	86	86	79	71	71	**6.42**	36	48	8	–	–	–	–	–	–	–	–	–	–	12-O-Tetradeca-2,4,6,8-tetraenoylphorbol-13-acetate		NI46516
311	312	608	504	190	313	192	417	**608**	100	63	25	24	19	14	11	5		37	40	6	2	–	–	–	–	–	–	–	–	–	Methyl dihydrowarifteine		LI09859
191	381	608	367	192	174	609	382	**608**	100	74	51	41	41	24	23	23		37	40	6	2	–	–	–	–	–	–	–	–	–	Thalrugosine	33889-68-8	DD01805
298	312	608	607	299	204	311	174	**608**	100	92	91	50	24	23	19	19		37	40	6	2	–	–	–	–	–	–	–	–	–	Tubocuraran-12′-ol, 6,6′,7′-trimethoxy-2,2′-dimethyl-, (1β)-	31944-97-5	NI46517
325	43	57	100	55	113	98	41	**608**	100	72	67	41	40	38	35	32	**2.04**	39	76	4	–	–	–	–	–	–	–	–	–	–	Octadecanoic acid, 1,3-propanediyl ester	17367-44-1	NI46518
325	43	57	55	324	28	100	113	**608**	100	63	54	38	34	34	33	32	**0.64**	39	76	4	–	–	–	–	–	–	–	–	–	–	Octadecanoic acid, 1,3-propanediyl ester	17367-44-1	NI46519
441	259	442	260	443	363	181	261	**608**	100	89	43	21	15	8	7	6	**0.12**	43	36	–	–	–	–	–	–	–	2	–	–	–	Disilane, (diphenylmethyl)pentaphenyl-	74630-36-7	NI46520
530	362	234	76	451	155	293	169	**609**	100	38	7	6	5	5	4	4	**2.40**	–	–	–	3	–	–	–	6	–	–	3	–	–	1,3,5,2,4,6-Triazatriphosphorine, 2,2,4,4,6,6-hexabromo-2,2,4,4,6,6-hexahydro-	13701-85-4	NI46521
530	372	189	225.5	273	234	110	76	**609**	100	37	16	15	7	6	6	6	**2.40**	–	–	–	3	–	–	–	6	–	–	3	–	–	1,3,5,2,4,6-Triazatriphosphorine, 2,2,4,4,6,6-hexabromo-2,2,4,4,6,6-hexahydro-	13701-85-4	MS10839
281	124	251	139	201	59	452	186	**609**	100	56	41	40	39	16	13	9	**8.50**	5	5	1	1	–	–	–	–	3	–	–	–	2	Cyclopentadienylnitrosylcobaltgermaniumtriiodide		NI46522
73	578	488	75	579	147	103	489	**609**	100	99	74	69	47	41	35	30	**27.50**	31	59	5	1	–	–	–	–	–	3	–	–	–	Pregnane-11,20-dione, 3,17,21-tris[(trimethylsilyl)oxy]-, 20-(O-methyloxime), (3α,5α)-	57325-75-4	NI46523

MASS TO CHARGE RATIOS								M.W.	INTENSITIES								Parent	C	H	O	N	S	F	Cl	Br	I	Si	P	B	X	COMPOUND NAME	CAS Reg No	No
73	578	75	488	609	103	579	147	**609**	100	99	95	66	56	56	52	45		31	59	5	1	–	–	–	–	–	3	–	–	–	Pregnane-11,20-dione, 3,17,21-tris[(trimethylsilyl)oxy]-, 20-(O-methyloxime), (3α,5β)-	57325-76-5	NI46524
578	488	579	398	580	489	609	506	**609**	100	61	45	30	26	26	24	18		31	59	5	1	–	–	–	–	–	3	–	–	–	Pregnane-11,20-dione, 3,17,21-tris[(trimethylsilyl)oxy]-, 20-(O-methyloxime), (3α,5β)-	57325-76-5	NI46525
78	48	33	80	287	232	290	29	**610**	100	74	65	40	34	26	25	25	**0.00**	12	12	13	–	–	–	6	–	–	–	–	–	4	Oxohexakis(chloroacetato)tetraberyllium(II)		MS10840
41	57	115	73	44	64	76	150	**610**	100	65	62	28	26	12	10	8	**0.00**	18	36	–	2	4	–	–	–	1	–	–	–	1	Iodobis(N,N-di-isobutyldithiocarbamato)arsine	59196-55-3	NI46526
149	57	71	43	85	83	55	263	**610**	100	62	44	39	36	36	36	35	**0.00**	27	30	16	–	–	–	–	–	–	–	–	–	–	4H-1-Benzopyran-4-one, 2-(3,4-dihydroxyphenyl)-6,8-di-β-D-glucopyranosyl-5,7-dihydroxy-	29428-58-8	NI46527
105	149	164	246	202	488	306	550	**610**	100	42	38	16	16	4	4	3	**0.00**	33	38	11	–	–	–	–	–	–	–	–	–	–	1α,8β-Dibenzoyloxy-9α,15-diacetoxy-4β,6β-dihydroxydihydro-β-agarofuran		MS10841
450	610	195	399	608	212	579	395	**610**	100	19	17	12	10	5	3	3		33	42	9	2	–	–	–	–	–	–	–	–	–	Dihydroreserpine		MS10842
608	610	195	174	212	396	579	188	**610**	100	90	18	10	6	5	4	4		33	42	9	2	–	–	–	–	–	–	–	–	–	3,4-Secoreserpine		LI09860
77	403	366	269	488	262	281	311	**610**	100	23	22	17	15	13	8	4	**0.00**	35	30	10	–	–	–	–	–	–	–	–	–	–	Ethyl 2,3,4,5-tetra-o-benzoyl-l-xylonate		NI46528
610	611	300	612	313	241	225	311	**610**	100	40	31	27	26	22	18	16		35	38	6	4	–	–	–	–	–	–	–	–	–	Biliverdin dimethyl ester		LI09861
368	353	107	81	105	95	369	145	**610**	100	55	54	52	50	47	46	46	**45.00**	35	51	1	–	–	5	–	–	–	1	–	–	–	Cholest-5-ene, 3-[[dimethyl(pentafluorophenyl)silyl]oxy]-, (3β)-	57691-88-0	MS10843
368	107	353	81	105	95	93	55	**610**	100	57	56	54	51	50	46	46	**45.00**	35	51	1	–	–	5	–	–	–	1	–	–	–	Cholest-5-ene, 3-[[dimethyl(pentafluorophenyl)silyl]oxy]-, (3β)-	57691-88-0	NI46529
368	610	521	447	107	81	95	105	**610**	100	99	93	89	80	79	71	69		35	51	1	–	–	5	–	–	–	1	–	–	–	Cholest-5-ene, 3-[[dimethyl(pentafluorophenyl)silyl]oxy]-, (3β)-	57691-88-0	NI46530
57	305	303	301	43	71			**610**	100	77	58	45	45	37			**5.62**	36	30	–	–	–	–	–	–	–	–	–	–	2	Digermane, hexaphenyl-	2816-39-9	NI46531
305	303	301	304	149	151	307	306	**610**	100	76	58	37	35	26	24	20	**7.37**	36	30	–	–	–	–	–	–	–	–	–	–	2	Digermane, hexaphenyl-	2816-39-9	NI46532
610	338	394	268	611	424	302	393	**610**	100	96	72	46	45	45	30	28		36	34	2	–	–	–	–	–	–	–	–	–	2	1′,1‴-Bis-(α-methoxybenzyl)biferrocene		MS10844
610	592	537	579	551	549			**610**	100	26	25	24	24	17				36	42	5	4	–	–	–	–	–	–	–	–	–	2-Ethyl-4-(2-hydroxyethyl)-6,7-di-(2-methoxycarbonylethyl)-1,3,5,8-tetramethylporphin		LI09862
568	526	138	484	137	442	221	441	**610**	100	99	89	64	59	35	29	17	**5.00**	36	50	8	–	–	–	–	–	–	–	–	–	–	Tetraacetylstriatol		LI09863
44	319	123	610	611	515	95	395	**610**	100	90	84	68	32	29	26	20		39	24	5	–	–	2	–	–	–	–	–	–	–	Bis[4-(4-fluorobenzoyl)phenoxy]benzophenone		IC04608
542	451	489	543	452	490	610	503	**610**	100	96	85	49	39	38	33	29		39	54	2	–	–	–	–	–	–	2	–	–	–	Cannabidiol bis(benzyldimethylsilyl) ether		NI46533
390	391	500	392	610	501	373	346	**610**	100	46	15	12	10	10	10	9		40	38	4	2	–	–	–	–	–	–	–	–	–	N,N′-Dicyclooctyl-3,4:9,10-perylenebis(dicarboximide)	83054-85-7	MS10845
519	610	520	91	611	159	57	158	**610**	100	82	42	41	36	20	14	11		40	54	3	2	–	–	–	–	–	–	–	–	–	6α-Nitro-5α-cholestano[3,2-b]-5′-benzyloxyindole		MS10846
297	207	91	298	206	208	179	92	**610**	100	59	50	30	24	10	10	7	**4.00**	42	34	1	4	–	–	–	–	–	–	–	–	–	(2-Hydroxy-3-phenyl-1,2-dihydroquinoxalin-1-yl)phenyl(3-phenyl-1-benzyl-1,2-dihydroquinoxalin-2-yl)methane	91757-02-7	NI46534
203	610	263	530	223	394	611	518	**610**	100	62	55	48	32	22	20	20		42	58	3	–	–	–	–	–	–	–	–	–	–	Cryptoflavin acetate		NI46535
370	368	329	369	610	371	353	355	**610**	100	86	67	40	37	30	27	20		43	78	1	–	–	–	–	–	–	–	–	–	–	Cholest-5-ene, 3-(hexadecyloxy)-, (3β)-	22032-48-0	LI09864
369	370	368	329	610	371	95	353	**610**	100	25	22	17	9	8	8	7		43	78	1	–	–	–	–	–	–	–	–	–	–	Cholest-5-ene, 3-(hexadecyloxy)-, (3β)-	22032-48-0	NI46536
610	611	78	305	458	612	41	43	**610**	100	58	35	32	19	18	12	10		48	34	–	–	–	–	–	–	–	–	–	–	–	m-Octiphenyl	16716-11-3	NI46537
610	611	78	305	458	612	305.5	41	**610**	100	58	35	32	19	18	17	12		48	34	–	–	–	–	–	–	–	–	–	–	–	m-Octiphenyl	16716-11-3	AM00159
610	305	593	380	289	276	265	228	**610**	100	23	1	1	1	1	1	1		48	34	–	–	–	–	–	–	–	–	–	–	–	m-Octiphenyl	16716-11-3	MS10847
73	319	205	147	320	217	103	218	**611**	100	71	51	30	20	16	15	13	**0.00**	24	61	5	1	–	–	–	–	–	6	–	–	–	Galactose oxime, hexakis(trimethylsilyl)-,		NI46538
129	142	43	335	101	117	87	88	**611**	100	84	80	59	58	49	44	40	**0.00**	27	49	14	1	–	–	–	–	–	–	–	–	–	1,5-Di-O-acetyl-7-O-[2-deoxy-3,4,6-tri-O-methyl-2-(N-methylacetamido)-α-D-glucopyranosyl]-2,3,4,6-tetra-O-methyl-L-glycero-D-manno-heptitol		NI46539
75	225	73	77	147	173	423	199	**611**	100	50	30	27	23	22	14	12	**4.00**	31	61	5	1	–	–	–	–	–	3	–	–	–	Prosta-5,13-dien-1-oic acid, 9-(ethoxyimino)-11,15-bis[(trimethylsilyl)oxy]-, trimethylsilyl ester, (5E,8β,9Z,11α,13E,15S)-	56085-40-6	NI46540
225	75	73	147	77	226	353	173	**611**	100	90	27	24	21	20	18	17	**10.00**	31	61	5	1	–	–	–	–	–	3	–	–	–	Prosta-5,13-dien-1-oic acid, 9-(ethoxyimino)-11,15-bis[(trimethylsilyl)oxy]-, trimethylsilyl ester, (5Z,8β,11α,13E,15S)-	57377-99-8	NI46541
75	225	73	77	147	173	423	199	**611**	100	50	30	27	23	22	14	12	**4.00**	31	61	5	1	–	–	–	–	–	3	–	–	–	Prosta-5,13-dien-1-oic acid, 9-(ethoxyimino)-11,15-bis[(trimethylsilyl)oxy]-, trimethylsilyl ester, (5Z,8β,11α,13E,15S)-	57377-99-8	NI46542
225	147	353	226	73	75	566	173	**611**	100	25	24	20	14	12	10	10	**4.00**	31	61	5	1	–	–	–	–	–	3	–	–	–	Prosta-5,13-dien-1-oic acid, 9-(ethoxyimino)-11,15-bis[(trimethylsilyl)oxy]-, trimethylsilyl ester, (5Z,11α,13E,15S)-	57377-97-6	NI46544
225	423	173	75	147	199	73	368	**611**	100	94	88	88	85	79	77	66	**8.00**	31	61	5	1	–	–	–	–	–	3	–	–	–	Prosta-5,13-dien-1-oic acid, 9-(ethoxyimino)-11,15-bis[(trimethylsilyl)oxy]-, trimethylsilyl ester, (5Z,11α,13E,15S)-	57377-97-6	NI46543

MASS TO CHARGE RATIOS	M.W.	INTENSITIES	Parent	C	H	O	N	S	F	Cl	Br	I	Si	P	B	X	COMPOUND NAME	CAS Reg No	No
225 75 73 147 77 226 353 173	**611**	100 90 27 24 21 20 18 17	**10.00**	31	61	5	1	–	–	–	–	–	3	–	–	–	**Prosta-5,13-dien-1-oic acid, 9-(ethoxyimino)-11,15-bis[(trimethylsilyl)oxy]-, trimethylsilyl ester, (5Z,8β,9E,11α,13E,15S)-**	56085-39-3	**NI46545**
225 423 173 75 147 199 73 368	**611**	100 94 88 88 85 79 77 66	**8.00**	31	61	5	1	–	–	–	–	–	3	–	–	–	**Prosta-5,13-dien-1-oic acid, 9-(ethoxyimino)-11,15-bis[(trimethylsilyl)oxy]-, trimethylsilyl ester, (5Z,9E,11α,13E,15S)-**	56009-48-4	**NI46546**
225 147 353 226 73 75 566 173	**611**	100 25 24 20 14 12 10 10	**4.00**	31	61	5	1	–	–	–	–	–	3	–	–	–	**Prosta-5,13-dien-1-oic acid, 9-(ethoxyimino)-11,15-bis[(trimethylsilyl)oxy]-, trimethylsilyl ester, (5Z,9Z,11α,13E,15S)-**	56009-49-5	**NI46547**
496 556 465 434 612 584	**612**	100 86 52 49 24 24		10	18	8	–	–	3	1	–	–	–	2	–	1	**Chlorotrifluoroethylene-trans-dicarbonyl-cis-bis(trimethylphosphite)osmium**		**LI09865**
299 73 315 283 211 370 300 243	**612**	100 98 49 37 35 29 21 19	**0.00**	18	50	9	–	–	–	–	–	–	5	2	–	–	**Phosphoric acid, 1-[[(trimethylsilyl)oxy]methyl]-1,2-ethanediyl tetrakis(trimethylsilyl) ester**	36449-17-9	**NI46548**
73 299 357 283 211 315 300 75	**612**	100 74 46 37 21 18 18 18	**0.00**	18	50	9	–	–	–	–	–	–	5	2	–	–	**Phosphoric acid, 2-[(trimethylsilyl)oxy]-1,3-propanediyl tetrakis(trimethylsilyl) ester**	31038-36-5	**NI46549**
73 318 147 217 305 191 265 129	**612**	100 44 37 35 26 23 15 8	**0.10**	24	60	6	–	–	–	–	–	–	6	–	–	–	**Inositol, 1,2,3,4,5,6-hexakis-O-(trimethylsilyl)-, allo-**	29267-02-5	**NI46550**
73 147 217 305 191 129 318 265	**612**	100 54 51 35 32 15 14 13	**0.10**	24	60	6	–	–	–	–	–	–	6	–	–	–	**Inositol, 1,2,3,4,5,6-hexakis-O-(trimethylsilyl)-, cis-**	29412-27-9	**NI46551**
73 318 147 217 305 191 265 204	**612**	100 40 39 36 34 23 13 10	**0.20**	24	60	6	–	–	–	–	–	–	6	–	–	–	**Inositol, 1,2,3,4,5,6-hexakis-O-(trimethylsilyl)-, D-chiro-**	29412-25-7	**NI46552**
73 217 147 305 191 318 265 129	**612**	100 73 66 50 39 37 19 15	**0.10**	24	60	6	–	–	–	–	–	–	6	–	–	–	**Inositol, 1,2,3,4,5,6-hexakis-O-(trimethylsilyl)-, epi-**	29267-01-4	**NI46553**
73 318 305 217 147 191 265 204	**612**	100 44 43 41 31 20 10 8	**0.20**	24	60	6	–	–	–	–	–	–	6	–	–	–	**Inositol, 1,2,3,4,5,6-hexakis-O-(trimethylsilyl)-, muco-**	29412-26-8	**NI46554**
43 75 44 73 57 41 55 97	**612**	100 90 85 60 60 58 50 30	**0.00**	24	60	6	–	–	–	–	–	–	6	–	–	–	**Inositol, 1,2,3,4,5,6-hexakis-O-(trimethylsilyl)-, myo-**	2582-79-8	**NI46557**
73 147 217 305 191 318 74 129	**612**	100 27 23 16 12 9 8 6	**0.00**	24	60	6	–	–	–	–	–	–	6	–	–	–	**Inositol, 1,2,3,4,5,6-hexakis-O-(trimethylsilyl)-, myo-**	2582-79-8	**NI46555**
73 217 305 147 318 191 265 204	**612**	100 82 78 53 43 43 22 20	**1.60**	24	60	6	–	–	–	–	–	–	6	–	–	–	**Inositol, 1,2,3,4,5,6-hexakis-O-(trimethylsilyl)-, myo-**	2582-79-8	**NI46556**
73 318 147 217 191 432 305 265	**612**	100 52 42 37 19 15 14 13	**0.10**	24	60	6	–	–	–	–	–	–	6	–	–	–	**Inositol, 1,2,3,4,5,6-hexakis-O-(trimethylsilyl)-, neo-**	29307-70-8	**NI46558**
73 318 217 305 147 191 204 319	**612**	100 89 66 65 46 45 30 11	**0.40**	24	60	6	–	–	–	–	–	–	6	–	–	–	**Inositol, 1,2,3,4,5,6-hexakis-O-(trimethylsilyl)-, scyllo-**	14251-18-4	**NI46559**
581 273 77 317 612 582 125 340	**612**	100 70 49 45 40 40 37 23		31	41	3	2	–	5	–	–	–	1	–	–	–	**Pregn-4-ene-3,20-dione, 17-[[dimethyl(pentafluorophenyl)silyl]oxy]-, bis(O-methyloxime)**	57691-93-7	**MS10848**
581 273 77 317 612 582 125 340	**612**	100 68 49 44 41 41 36 22		31	41	3	2	–	5	–	–	–	1	–	–	–	**Pregn-4-ene-3,20-dione, 17-[[dimethyl(pentafluorophenyl)silyl]oxy]-, bis(O-methyloxime)**	57691-93-7	**NI46560**
43 69 95 55 57 425 41 81	**612**	100 55 52 49 44 43 43 40	**0.60**	32	53	3	–	–	–	–	–	1	–	–	–	–	**Lanostan-3β-ol, 11β,18-epoxy-19-iodo-, acetate**	28328-13-4	**NI46561**
43 425 407 395 485 443 612 567	**612**	100 43 30 20 18 10 1 1		32	53	3	–	–	–	–	–	1	–	–	–	–	**Lanostan-3β-ol, 11β,18-epoxy-19-iodo-, acetate**	28328-13-4	**LI09866**
423 73 513 191 333 307 129 424	**612**	100 88 60 54 46 44 36 36	**0.00**	32	64	5	–	–	–	–	–	–	3	–	–	–	**16,16-Dimethyl-f2a, methyl ester, tris(trimethylsilyl)-**	61306-61-4	**NI46562**
42 235 231 217 81 105 209 219	**612**	100 62 52 40 35 31 28 24	**19.71**	33	40	11	–	–	–	–	–	–	–	–	–	–	**Satratoxin H diacetate**		**NI46563**
215 107 95 81 58 216 55 93	**612**	100 90 86 85 80 73 72 65	**32.00**	35	53	1	–	–	5	–	–	–	1	–	–	–	**Cholestane, 3-[[dimethyl(pentafluorophenyl)silyl]oxy]-, (3β,5α)-**		**MS10849**
248 43 233 57 430 108 93 140	**612**	100 48 41 40 38 26 26 22	**0.30**	36	56	4	2	1	–	–	–	–	–	–	–	–	**N,N'-Dilauryl-4,4'-diaminodiphenylsulphone**		**IC04609**
91 92 209 253 181 73 65 193	**612**	100 50 37 33 25 18 17 11	**0.00**	37	44	6	–	–	–	–	–	–	1	–	–	–	**1-O-Trimethylsilyl-2,3,4,6-tetra-O-benzyl-α-D-glucopyranoside**	85982-87-2	**NI46564**
91 209 92 253 181 65 73 235	**612**	100 73 68 55 31 28 20 12	**0.00**	37	44	6	–	–	–	–	–	–	1	–	–	–	**1-O-Trimethylsilyl-2,3,4,6-tetra-O-benzyl-β-D-glucopyranoside**	85982-86-1	**NI46565**
135 612 613 107 136 134 506 478	**612**	100 77 34 26 24 20 18 16		40	44	2	4	–	–	–	–	–	–	–	–	–	**(+)-Pycnanthine**		**LI09867**
448 612 449 335 396 383 382 307	**612**	100 41 39 33 32 29 28 26		42	60	3	–	–	–	–	–	–	–	–	–	–	**ψ,ψ-Carotene, 20-(acetyloxy)-1,2-dihydro-1-hydroxy-**	56051-50-4	**NI46566**
204 73 75 217 205 79 103 147	**613**	100 96 51 40 25 25 24 21	**0.00**	24	63	5	1	–	–	–	–	–	6	–	–	–	**D-Glucitol, 2-deoxy-1,3,4,5,6-pentakis-O-(trimethylsilyl)-2-[(trimethylsilyl)amino]-**	56282-41-8	**NI46567**
171 217 156 202 172 117 115 28	**613**	100 80 70 28 26 13 13 12	**0.00**	24	63	5	1	–	–	–	–	–	6	–	–	–	**D-Mannitol, 2-deoxy-1,3,4,5,6-pentakis-O-(trimethylsilyl)-2-[(trimethylsilyl)amino]-**	74420-75-0	**NI46568**
542 199 75 452 129 543 478 568	**613**	100 93 68 53 53 47 40 31	**5.00**	31	63	5	1	–	–	–	–	–	3	–	–	–	**Prost-13-en-1-oic acid, 9-(ethoxyimino)-11,15-bis[(trimethylsilyl)oxy]-, trimethylsilyl ester, (8α,11α,13E,15S)-**	57377-98-7	**NI46569**
355 426 147 75 130 225 427 356	**613**	100 96 69 64 58 41 35 31	**3.00**	31	63	5	1	–	–	–	–	–	3	–	–	–	**Prost-13-en-1-oic acid, 9-(ethoxyimino)-11,15-bis[(trimethylsilyl)oxy]-, trimethylsilyl ester, (8β,9E,11α,13E,15S)-**	56085-41-7	**NI46570**
199 542 129 452 75 543 173 478	**613**	100 71 59 44 44 32 32 31	**3.00**	31	63	5	1	–	–	–	–	–	3	–	–	–	**Prost-13-en-1-oic acid, 9-(ethoxyimino)-11,15-bis[(trimethylsilyl)oxy]-, trimethylsilyl ester, (9E,11α,13E,15S)-**	56144-49-1	**NI46571**
542 199 75 452 129 543 478 568	**613**	100 93 68 53 53 47 40 31	**5.00**	31	63	5	1	–	–	–	–	–	3	–	–	–	**Prost-13-en-1-oic acid, 9-(ethoxyimino)-11,15-bis[(trimethylsilyl)oxy]-, trimethylsilyl ester, (8β,9Z,11α,13E,15S)-**	56085-42-8	**NI46572**
355 426 147 130 225 427 356 75	**613**	100 85 72 61 47 32 31 25	**1.00**	31	63	5	1	–	–	–	–	–	3	–	–	–	**Prost-13-en-1-oic acid, 9-(ethoxyimino)-11,15-bis[(trimethylsilyl)oxy]-, trimethylsilyl ester, (9Z,11α,13E,15S)-**	56009-50-8	**NI46573**

MASS TO CHARGE RATIOS								M.W.	INTENSITIES								Parent	C	H	O	N	S	F	Cl	Br	I	Si	P	B	X	COMPOUND NAME	CAS Reg No	No
240	613	585	317	425	323	346	163	**613**	100	24	24	24	23	22	18	8		35	34	1	1	–	–	–	–	–	–	1	–	1	Molybdenum, (α-tert-butylbenzylideneiminato)carbonyl-π-cyclopentadienyl(triphenylphosphine)-	34829-42-0	MS10850
272	458	191	169	91	459	170	273	**613**	100	82	50	48	45	40	32	23	**12.24**	36	43	4	3	1	–	–	–	–	–	–	–	–	Tubulosan, 10,11-dimethoxy-2′-[(4-methylphenyl)sulphonyl]-	2632-31-7	NI46574
162	120	613	180	614	108	121	119	**613**	100	72	42	39	19	18	17	17		39	67	4	1	–	–	–	–	–	–	–	–	–	L-Phenylalanine, N-(3-methoxy-1-oxo-4-octacosenyl)-, methyl ester	72087-99-1	NI46575
235	205	233	203	220	218	542	540	**614**	100	95	42	40	30	13	7	6	**0.00**	10	24	2	2	–	–	–	–	–	–	–	–	2	Acetoximatodimethyldithallium		LI09868
43	128	443	127	44	315	274	401	**614**	100	98	57	57	36	24	22	19	**0.00**	11	9	4	2	–	–	–	–	3	–	–	–	–	3,5-Diacetamido-2,4,6-triiodobenzoic acid	117-96-4	NI46576
307	113	68	310	256	288	106	132	**614**	100	23	10	9	6	5	4	3	**0.00**	12	–	–	–	4	18	–	–	–	–	–	–	–	2,2′-Bis[2,4,5-tris(trifluoromethyl)-1,3-dithiole]		LI09869
131	69	44	181	119	169	93	51	**614**	100	38	14	13	10	5	4	4	**0.00**	12	1	2	–	–	23	–	–	–	–	–	–	–	Dodecanoic acid, tricosafluoro-	307-55-1	NI46578
131	69	181	119	100	93	43	169	**614**	100	34	14	9	6	6	6	5	**0.00**	12	1	2	–	–	23	–	–	–	–	–	–	–	Dodecanoic acid, tricosafluoro-	307-55-1	NI46577
43	614	29	513	57	219	177	527	**614**	100	56	38	31	27	22	17	15		21	23	8	6	–	–	–	–	1	–	–	–	–	N-(2-Propionyloxypropyl)-2-methoxy-5-acetamido-4-(2,4-dinitro-6-iodophenydazo)aniline		IC04610
307	380	308	309	381	599	382	147	**614**	100	30	28	14	11	8	7	5	**0.00**	22	54	8	–	–	–	–	–	–	6	–	–	–	Butanedioic acid, tetrakis[(trimethylsilyl)oxy]-, bis(trimethylsilyl) ester	38166-10-8	NI46579
307	380	308	309	381	599	382	600	**614**	100	30	28	14	12	9	7	5	**0.00**	22	54	8	–	–	–	–	–	–	6	–	–	–	Butanedioic acid, tetrakis[(trimethylsilyl)oxy]-, bis(trimethylsilyl) ester	38166-10-8	LI09870
73	217	103	205	307	319	147	218	**614**	100	44	41	40	32	30	30	10	**0.00**	24	62	6	–	–	–	–	–	–	6	–	–	–	Galactitol, 1,2,3,4,5,6-hexakis-O-(trimethylsilyl)-	18919-39-6	NI46580
397	73	117	132	145	75	43	412	**614**	100	88	75	67	53	49	38	32	**0.00**	24	62	6	–	–	–	–	–	–	6	–	–	–	Galactitol, 1,2,3,4,5,6-hexakis-O-(trimethylsilyl)-	18919-39-6	NI46582
73	217	307	319	205	103	147	308	**614**	100	62	48	46	44	35	33	15	**0.00**	24	62	6	–	–	–	–	–	–	6	–	–	–	Galactitol, 1,2,3,4,5,6-hexakis-O-(trimethylsilyl)-	18919-39-6	NI46581
73	205	319	103	147	217	307	117	**614**	100	60	57	36	33	32	18	13	**0.00**	24	62	6	–	–	–	–	–	–	6	–	–	–	D-Glucitol, 1,2,3,4,5,6-hexakis-O-(trimethylsilyl)-	14199-80-5	NI46583
73	319	205	103	147	421	217	524	**614**	100	50	50	31	30	23	23	22	**0.00**	24	62	6	–	–	–	–	–	–	6	–	–	–	D-Glucitol, 1,2,3,4,5,6-hexakis-O-(trimethylsilyl)-	14199-80-5	NI46584
73	205	319	103	147	217	307	320	**614**	100	50	49	31	29	23	16	14	**0.00**	24	62	6	–	–	–	–	–	–	6	–	–	–	D-Glucitol, 1,2,3,4,5,6-hexakis-O-(trimethylsilyl)-	14199-80-5	NI46585
73	319	205	147	103	217	320	307	**614**	100	50	48	29	28	23	16	12	**0.00**	24	62	6	–	–	–	–	–	–	6	–	–	–	D-Mannitol, 1,2,3,4,5,6-hexakis-O-(trimethylsilyl)-	14317-07-8	NI46587
73	319	205	103	147	217	320	307	**614**	100	57	55	29	28	24	17	14	**0.00**	24	62	6	–	–	–	–	–	–	6	–	–	–	D-Mannitol, 1,2,3,4,5,6-hexakis-O-(trimethylsilyl)-	14317-07-8	NI46588
73	319	205	147	75	217	103	320	**614**	100	53	43	41	32	30	24	17	**0.00**	24	62	6	–	–	–	–	–	–	6	–	–	–	D-Mannitol, 1,2,3,4,5,6-hexakis-O-(trimethylsilyl)-	14317-07-8	NI46586
43	532	534	530	533	461	574	463	**614**	100	85	45	42	21	19	16	13	**0.00**	26	16	8	–	–	–	–	2	–	–	–	–	–	4,4′-Diacetoxy-3,3′-dibromo-7,7′-dimethyl-1,1′-binaphthalene-5,5′,8,8′-tetrone	101459-14-7	NI46589
167	105	614	253	107	149	91	77	**614**	100	37	21	16	16	9	9	8		26	22	14	4	–	–	–	–	–	–	–	–	–	Methyl 4,6-O-benzylidene-2,3-bis-O-(2,4-dinitrophenyl)-α-D-glucopyranoside		NI46590
88	101	45	71	115	75	111	187	**614**	100	54	30	25	24	24	19	17	**0.00**	27	50	15	–	–	–	–	–	–	–	–	–	–	L-Arabinose, O-2,3,4,6-tetra-O-methyl-β-D-galactopyranosyl-(1-6)-O-2,3,4-tri-O-methyl-β-D-galactopyranosyl-(1-3)-2,4,5-tri-O-methyl-	56248-59-0	NI46591
269	202	201	329	84	287	174	173	**614**	100	74	54	41	38	20	14	12	**7.06**	34	30	11	–	–	–	–	–	–	–	–	–	–	7H-Furo[3,2-g][1]benzopyran-7-one, 9-[[4-[2-(acetyloxy)-1,1-dimethyl-3-[(7-oxo-7H-furo[3,2-g][1]benzopyran-9-yl)oxy]propoxy]-3-methyl-2-butenyl]oxy]-	49586-44-9	MS10851
523	524	314	415	614	479	506	448	**614**	100	30	27	23	22	19	18	18		35	38	8	2	–	–	–	–	–	–	–	–	–	4,4′-Dimethyl-5,5′-bis(benzyloxycarbonyl)-3,3′-bis(methoxycarbonylethyl)2,2′-dipyrrolylmethane		LI09871
614	599	302	315	229	213	227		**614**	100	81	66	51	48	39	37			35	42	6	4	–	–	–	–	–	–	–	–	–	Didehydrophycocyanobilin dimethyl ester		LI09872
614	615	315	302	616	313	467	541	**614**	100	42	27	23	15	12	10	9		35	42	6	4	–	–	–	–	–	–	–	–	–	Glaucobilin IIIα dimethyl ester		NI46592
614	615	302	315	307	613	467	313	**614**	100	40	10	8	8	6	6	5		35	42	6	4	–	–	–	–	–	–	–	–	–	Glaucobilin IXα dimethyl ester		NI46593
614	315	302	467	313	541	316	583	**614**	100	15	15	7	7	5	5	4		35	42	6	4	–	–	–	–	–	–	–	–	–	Glaucobilin IXα dimethyl ester		LI09873
614	615	616	315	467	302	613	541	**614**	100	41	12	9	8	8	7	6		35	42	6	4	–	–	–	–	–	–	–	–	–	Glaucobilin IXα dimethyl ester		NI46594
614	615	616	498	360	255	244	229	**614**	100	41	12	12	9	9	9	7		35	42	6	4	–	–	–	–	–	–	–	–	–	Glaucobilin IXβ dimethyl ester		NI46595
244	229	360	213	287	199	215	214	**614**	100	62	50	30	27	25	20	20	**16.00**	35	42	6	4	–	–	–	–	–	–	–	–	–	Glaucobilin IXβ dimethyl ester		LI09874
614	615	616	302	613	467	315	541	**614**	100	35	9	6	5	5	5	4		35	42	6	4	–	–	–	–	–	–	–	–	–	Glaucobilin XIIIα dimethyl ester		NI46596
614	467	313	315	227	541	331	241	**614**	100	70	70	62	41	39	39	39		35	42	6	4	–	–	–	–	–	–	–	–	–	Mesobiliverdin dimethyl ester		LI09875
614	189	119	582	262	187	175	133	**614**	100	83	81	73	67	60	54	52		36	54	8	–	–	–	–	–	–	–	–	–	–	Dimethyl granulosate diacetate		NI46597
421	494	485	539	253	541	201	436	**614**	100	96	95	83	63	53	49	47	**0.00**	37	58	7	–	–	–	–	–	–	–	–	–	–	3β,6α,24,25-Tetrahydroxy-24S-cycloartan-16-one 3,6-diacetate 24,25-acetonide		MS10852
462	444	426	582	578	560	598	596	**614**	100	50	45	4	4	4	2	2	**1.00**	40	54	5	–	–	–	–	–	–	–	–	–	–	Amarouciaxanthin a		NI46598
614	494	442	582	495	443	550	583	**614**	100	54	48	26	24	19	15	13		42	62	3	–	–	–	–	–	–	–	–	–	–	ψ,ψ-Caroten-20-al, 1,1′,2,2′-tetrahydro-1,1′-dimethoxy-	56196-61-3	NI46599
91	614	615	599	600	616	220	233	**614**	100	99	48	43	21	12	11	10		45	46	–	2	–	–	–	–	–	–	–	–	–	5α-Androstano[3,2-b.17,16-b]N,N′-dibenzyldiindole		MS10853
360	303	402	43	69	361	528	343	**615**	100	93	50	27	21	17	16	16	**0.00**	20	15	5	1	–	14	–	–	–	–	–	–	–	N,O-Bis(heptafluorobutyryl)tyrosine propyl ester		MS10854
223	375	217	376	224	169	257	377	**615**	100	80	37	26	18	18	17	12	**0.00**	26	49	8	1	–	–	–	–	–	4	–	–	–	β-D-Glucopyranosiduronic acid, 4-(acetylamino)phenyl 2,3,4-tris-O-(trimethylsilyl)-, trimethylsilyl ester	72088-06-3	NI46600

MASS TO CHARGE RATIOS								M.W.	INTENSITIES								Parent	C	H	O	N	S	F	Cl	Br	I	Si	P	B	X	COMPOUND NAME	CAS Reg No	No
73	285	45	229	159	75	79	313	**615**	100	69	67	63	54	28	21	18	**0.03**	28	53	8	3	–	–	–	–	–	2	–	–	–	Tridecylamine, N-(2,4-dinitrophenyl)-2-methoxy-1-(methoxymethyl)-3,4-bis(trimethylsiloxy)-, erythro-	27089-69-6	NI46601
135	584	585	600	566	432	615	552	**615**	100	41	16	7	7	4	2	2		33	45	10	1	–	–	–	–	–	–	–	–	–	Secojesaconitine		MS10855
542	615	590	584	556	603	600	528	**615**	100	95	38	26	20	17	12	10		37	37	4	5	–	–	–	–	–	–	–	–	–	meso-Cyanoprotoporphyrin IX dimethyl ester		LI09876
476	364	280	336	392	220	308	448	**616**	100	75	63	61	59	54	51	44	**24.00**	12	1	12	–	–	–	–	–	–	–	–	–	4	Hydridododecarbonylrutheniumtricobalt		LI09877
134	470	546	237	273	547	544	135	**616**	100	81	49	48	35	33	22	18	**0.00**	20	30	–	–	–	–	4	–	–	–	–	–	2	Rhodium, di-μ-chlorodichlorobis[(1,2,3,4,5-η)-1,2,3,4,5-pentamethyl-2,4-cyclopentadien-1-yl]di-	12354-85-7	NI46602
475	57	41	476	477	487	587	56	**616**	100	89	41	41	28	23	16	16	**0.00**	20	48	10	–	–	–	–	–	–	6	–	–	–	1,3,5,7,9,11-Hexaethyl-5,9-dibutoxytricyclo[5.5.1.1^{3,11}]hexasiloxane	110991-17-8	NI46603
43	169	109	331	122	107	127	78	**616**	100	57	41	19	18	16	15	10	**0.00**	30	32	14	–	–	–	–	–	–	–	–	–	–	Salicyloylsalicin pentaacetate		MS10856
43	169	109	331	105	127	122	170	**616**	100	79	45	25	22	16	12	7	**0.10**	30	32	14	–	–	–	–	–	–	–	–	–	–	Salireposide pentaacetate		MS10857
43	91	169	109	331	127	244	271	**616**	100	86	85	85	46	38	15	9	**2.00**	30	32	14	–	–	–	–	–	–	–	–	–	–	Trichocarpin, pentaacetate	17019-77-1	LI09878
109	43	91	169	331	127	244	271	**616**	100	85	73	72	39	32	13	8	**2.22**	30	32	14	–	–	–	–	–	–	–	–	–	–	Trichocarpin, pentaacetate	17019-77-1	NI46604
43	169	91	109	127	245	271	211	**616**	100	67	53	46	13	5	3	2	**0.03**	30	32	14	–	–	–	–	–	–	–	–	–	–	Trichocarpin, pentaacetate	17019-77-1	MS10858
200	157	75	119	143	616	77	93	**616**	100	47	44	40	34	32	31	18		31	45	3	–	–	5	–	–	–	2	–	–	–	Spiro[androstane-17,4′-[1,3]dioxa[2]silacyclopentane], 3-[[dimethyl(pentafluorophenyl)silyl]oxy]-2′,2′,5′-trimethyl-, (3α,5β,5′S)-	69745-69-3	NI46605
43	207	55	150	179	178	400	616	**616**	100	99	79	65	62	59	58	57		32	40	12	–	–	–	–	–	–	–	–	–	–	Anodendroside E2, monoacetate	28905-46-6	NI46606
276	105	305	248	324	277	202	405	**616**	100	99	98	40	30	30	20	10	**0.00**	33	28	7	2	–	–	–	–	–	–	–	–	1	Pentacarbonyl[3-(dibenzoylamino)-1-(diethylamino)-2-methyl-3-phenyl-(E)-2-propenylidene]chromium(0)	123810-12-8	DD01806
155	254	60	239	197	57	255	56	**616**	100	69	49	45	44	43	43	40	**28.93**	33	44	11	–	–	–	–	–	–	–	–	–	–	1-O-Retinoyl β-D-methyl-2′,3′-4′-tri-O-acetylglucopyranuronate	101470-88-6	NI46607
493	315	302	494	180	316	229	614	**616**	100	63	61	59	40	35	32	30	**15.15**	35	44	6	4	–	–	–	–	–	–	–	–	–	Dihydroisomesobiliviolin IXα dimethyl ester		NI46608
493	494	492	346	614	315	461	618	**616**	100	41	29	18	14	12	11	10	**6.09**	35	44	6	4	–	–	–	–	–	–	–	–	–	Dihydromesobiliviolin IXα dimethyl ester		NI46609
493	346	492	495	618	461	333		**616**	100	38	16	14	11	10	10	9	**2.10**	35	44	6	4	–	–	–	–	–	–	–	–	–	Dihydromesobiliviolin XIIIα dimethyl ester		NI46610
492	493	494	614	346	315	616	478	**616**	100	63	26	22	17	14	13	12		35	44	6	4	–	–	–	–	–	–	–	–	–	Isomesobiliviolin IXα dimethyl ester		NI46611
492	493	494	346	616	418	614	495	**616**	100	72	24	17	9	9	7	7		35	44	6	4	–	–	–	–	–	–	–	–	–	Mesobiliviolin IIIα dimethyl ester		NI46612
492	493	614	494	346	315	302	616	**616**	100	59	30	29	19	18	18	16		35	44	6	4	–	–	–	–	–	–	–	–	–	Mesobiliviolin IXα dimethyl ester		NI46613
492	493	614	494	315	302	616	615	**616**	100	72	70	42	41	41	32	29		35	44	6	4	–	–	–	–	–	–	–	–	–	Mesobiliviolin XIIIα dimethyl ester		NI46614
273	201	496	601	556	293	481	423	**616**	100	85	69	46	42	39	35	31	**0.00**	37	60	7	–	–	–	–	–	–	–	–	–	–	Cyclocathogenin 3,6-di-acetate 24,25-acetonide		MS10859
197	212	181	119	598	368	163	106	**616**	100	90	87	76	44	40	39	39	**16.77**	40	56	5	–	–	–	–	–	–	–	–	–	–	β,β-Carotene, 6′,7′-didehydro-5,6-epoxy-5,5′,6,6′,7,8-hexahydro-3,3′,5′-trihydroxy-8-oxo-	7176-02-5	NI46615
43	221	105	91	197	119	598	580	**616**	100	56	47	47	35	15	13	13	**1.00**	40	56	5	–	–	–	–	–	–	–	–	–	–	Fucoxanthinol		LI09879
43	91	105	119	580	598	562	547	**616**	100	62	37	30	8	6	4	2	**2.00**	40	56	5	–	–	–	–	–	–	–	–	–	–	Fucoxanthinol		LI09880
91	105	197	221	119	580	598	368	**616**	100	63	48	35	27	12	10	9	**1.00**	40	56	5	–	–	–	–	–	–	–	–	–	–	Isofucoxanthinol		LI09881
253	454	69	81	544	400	178	274	**616**	100	69	57	54	53	40	34	31	**0.00**	41	57	2	–	–	–	1	–	–	–	–	–	–	Cholest-8(14)-ene, 7-chloro-3,15-dibenzyloxy-, (3β,5α,7α,15α)-	74420-79-4	NI46616
616	308	511	178	588	280	267	598	**616**	100	46	23	19	13	12	12	9		46	32	2	–	–	–	–	–	–	–	–	–	–	2,3-Dihydro-3,3′,4,4′,5,5′-hexaphenyl-2,2′-bis(cyclopentadienone)		LI09882
105	616	178	308	102	511	588		**616**	100	60	50	32	20	10	5			46	32	2	–	–	–	–	–	–	–	–	–	–	(E)-3,3′,4,4′,5,5′-Hexaphenylbis(3-cyclopentenylidene)-2,2′-dione		LI09883
616	410	204	178	333	382	305	105	**616**	100	22	18	12	11	6	6	6		46	32	2	–	–	–	–	–	–	–	–	–	–	3a,3b,6a,10b-Tetrahydro-1,2,5,6,6a-pentaphenylbenz[e]-as-indacene-3,4-dione		LI09884
602	603	119	280	617	69	455	454	**617**	100	34	24	15	10	9	6	6		19	10	5	1	–	15	–	–	–	–	–	–	–	Isosalsolinol tri(pentafluoropropionate)		NI46617
601	602	119	280	616	69	44	77	**617**	100	19	15	12	9	8	4	3	**1.20**	19	10	5	1	–	15	–	–	–	–	–	–	–	Isosalsolinol tri(pentafluoropropionate)		MS10860
602	603	119	455	471	335	470	280	**617**	100	36	26	18	16	16	15	15	**14.50**	19	10	5	1	–	15	–	–	–	–	–	–	–	Salsolinol tri(pentafluoropropionate)		NI46618
602	119	603	617	552	455	470	69	**617**	100	21	19	17	16	9	8	8		19	10	5	1	–	15	–	–	–	–	–	–	–	Salsolinol tri(pentafluoropropionate)		MS10861
170	171	168	277	169	172	617	167	**617**	100	53	17	15	12	8	4	3		33	35	5	3	2	–	–	–	–	–	–	–	–	1-Naphthalenesulphonic acid, 5-(dimethylamino)-, 4-[2-[[[5-(dimethylamino)-1-naphthalenyl]sulphonyl]amino]propyl]phenyl ester, (±)-	55836-93-6	NI46620
170	171	42	168	234	277	154	155	**617**	100	53	26	17	16	15	12	11	**3.30**	33	35	5	3	2	–	–	–	–	–	–	–	–	1-Naphthalenesulphonic acid, 5-(dimethylamino)-, 4-[2-[[[5-(dimethylamino)-1-naphthalenyl]sulphonyl]amino]propyl]phenyl ester, (±)-	55836-93-6	NI46619
279	294	590	478	534	618	506	562	**618**	100	99	20	17	15	10	10	1		10	9	5	–	–	–	–	–	–	–	–	–	4	4-Methyl-1,2,6-triarsatricyclo[2.2.1.0^{2,6}]heptanepentacarbonyltungsten		MS10862
323	380	260	209	618	419	561	505	**618**	100	80	60	10	8	6	3	2		17	27	5	–	–	–	–	–	–	–	–	–	2	Pentacarbonyl(tributylbismuthine)molybdenum	90188-76-4	NI46621

MASS TO CHARGE RATIOS								M.W.	INTENSITIES								Parent	C	H	O	N	S	F	Cl	Br	I	Si	P	B	X	COMPOUND NAME	CAS Reg No	No
446	28	77	91	115	78	92	59	**618**	100	99	86	70	56	53	45	43	**4.00**	18	6	10	–	–	–	–	–	–	–	–	–	4	μ-(Phenylacetylene)decacarbonyltetracobalt		MS10863
545	591	57	536	546	547	73	592	**618**	100	55	53	51	50	42	33	28	**0.00**	18	30	11	–	–	–	–	–	–	7	–	–	–	1-Butoxypervinylhomohexasilsesquioxane	94096-71-6	NI46622
59	185	149	145	277	117	41	62	**618**	100	38	22	22	16	16	15	13	**13.05**	22	34	–	–	10	–	–	–	–	–	–	–	–	2,4,6,8-Tetrathiatricyclo[3.3.1.1^{3,7}]decane, 1,1′-dithiobis[3,5,7,10,10-pentamethyl-	17525-43-8	NI46623
318	316	402	43	274	109	444	69	**618**	100	95	68	67	56	44	42	41	**0.00**	32	42	12	–	–	–	–	–	–	–	–	–	–	Card-20(22)-enolide, 11-(acetyloxy)-3-[(6-deoxy-3,4-O-methylenehexopyranos-2-ulos-1-yl)oxy]-5,14-dihydroxy-12-oxo-, (3β,5α,11α)-	56051-65-1	NI46624
264	575	236	617	618	100	576	265	**618**	100	48	45	43	35	32	22	22		34	42	7	4	–	–	–	–	–	–	–	–	–	4H-1,16-Etheno-5,15-(propaniminoethano)furo[3,4-l][1,5,10]triazacyclohexadecine-4,21-dione, 10,14-diacetyl-3-[4-(acetyloxy)phenyl]-3,3a,6,7,8,9,10,11,12,13,14,15-dodecahydro-	69755-30-2	NI46625
494	495	496	387	371	618	388	316	**618**	100	61	36	15	10	8	6	5		35	46	6	4	–	–	–	–	–	–	–	–	–	Mesobilirhodin dimethyl ester		LI09885
498	293	423	483	558	438	413	480	**618**	100	72	56	50	44	39	39	33	**0.20**	36	58	8	–	–	–	–	–	–	–	–	–	–	Cyclocanthogenin 3,6,16-triacetate		MS10864
265	119	105	95	93	107	190	189	**618**	100	65	58	56	55	51	51	45	**8.28**	39	54	6	–	–	–	–	–	–	–	–	–	–	Rubicoumaric acid	80489-66-3	NI46626
181	221	582	600	526	566	564	502	**618**	100	99	53	47	38	30	26	23	**15.00**	40	58	5	–	–	–	–	–	–	–	–	–	–	(3S,5R,6S,3′S,5′R,6′R)-6,7-Didehydro-5,6,5′,6′-tetrahydro-β,β-carotene-3,5,6,3′,5′-pentol		NI46627
600	618	181	221	582	526	564	502	**618**	100	55	36	31	22	6	4	3		40	58	5	–	–	–	–	–	–	–	–	–	–	(3S,5S,6S,3′S,5′R,6′R)-6,7-Didehydro-5,6,5′,6′-tetrahydro-β,β-carotene-3,5,6,3′,5′-pentol		NI46628
43	582	564	600	549	490	508	618	**618**	100	28	10	8	5	3	2	1		40	58	5	–	–	–	–	–	–	–	–	–	–	6,7-Dihydro-isofucoxanthinol		LI09886
43	91	105	119	221	181	502	582	**618**	100	55	54	53	51	46	32	23	**2.00**	40	58	5	–	–	–	–	–	–	–	–	–	–	Isofucoxanthol		LI09887
57	43	71	85	55	69	41	83	**618**	100	70	60	38	21	16	16	14	**0.05**	44	90	–	–	–	–	–	–	–	–	–	–	–	Tetratetracontane		AI02409
57	43	71	85	55	69	41	83	**618**	100	76	64	40	30	22	21	18	**0.28**	44	90	–	–	–	–	–	–	–	–	–	–	–	Tetratetracontane		AI02410
456	165	204	457	178	310	484	458	**619**	100	50	21	17	6	5	4	3	**0.00**	19	12	5	1	–	15	–	–	–	–	–	–	–	α-Methylsynephrine, tri(pentafluoropropionate)		NI46629
357	73	358	299	315	359	217	147	**619**	100	96	34	33	25	18	17	16	**0.10**	21	54	8	1	–	–	–	–	–	5	1	–	–	D-erythro-2-Pentulose, 1,3,4-tris-O-(trimethylsilyl)-, O-methyloxime, 5-[bis(trimethylsilyl) phosphate]	55319-73-8	NI46630
73	315	217	299	316	147	459	317	**619**	100	65	49	40	20	16	12	12	**0.00**	21	54	8	1	–	–	–	–	–	5	1	–	–	D-Ribose, 2,3,4-tris-O-(trimethylsilyl)-, O-methyloxime, 5-[bis(trimethylsilyl) phosphate]	55319-74-9	NI46631
315	73	217	299	316	147	459	317	**619**	100	99	75	61	30	24	19	19	**0.00**	21	54	8	1	–	–	–	–	–	5	1	–	–	D-Ribose, 2,3,4-tris-O-(trimethylsilyl)-, O-methyloxime, 5-[bis(trimethylsilyl) phosphate]	55319-74-9	NI46632
73	357	299	75	103	358	74	211	**619**	100	33	30	27	11	10	10	9	**0.02**	21	54	8	1	–	–	–	–	–	5	1	–	–	Serine, N-(trimethylsilyl)-, L-, trimethylsilyl ester, 2,3-bis(trimethylsiloxy)propyl trimethylsilyl phosphate	32204-75-4	NI46633
82	73	315	192	323	207	208	83	**619**	100	34	12	12	9	8	7	6	**1.00**	22	46	6	5	–	–	–	–	–	4	1	–	–	Adenosine, 2′-deoxy-N-(trimethylsilyl)-3′-O-(trimethylsilyl)-, 5′-[bis(trimethylsilyl) phosphate]	32645-70-8	NI46634
170	172	202	235	204	249	273	261	**619**	100	99	70	30	30	27	22	14	**10.00**	32	33	6	3	2	–	–	–	–	–	–	–	–	N,O-Didansyl octopamine		LI09888
170	171	42	168	235	154	169	127	**619**	100	77	38	27	25	22	21	21	**4.00**	32	33	6	3	2	–	–	–	–	–	–	–	–	N,O-Bis(1-dimethylaminonaphthalene-5-sulphonyl)-2-(4-hydroxyphenyl)-2 ethanolamine		NI46635
170	171	168	169	203	172	355	202	**619**	100	75	28	21	10	10	7	6	**5.57**	32	33	6	3	2	–	–	–	–	–	–	–	–	1-Naphthalenesulphonic acid, 5-(dimethylamino)-, 4-[2-[[[5-(dimethylamino)-1-naphthalenyl]sulphonyl]amino]-1-hydroxyethyl]phenyl ester	20289-21-8	NI46636
91	198	313	267	155	210	241	70	**619**	100	85	48	48	43	31	28	26	**0.50**	32	49	5	3	2	–	–	–	–	–	–	–	–	Benzenesulphonamide, 4-methyl-N-[3-[[(4-methylphenyl)sulphonyl]amino]propyl]-N-[3-(2-oxoazacyclotridec-1-yl)propyl]-	65545-59-7	NI46637
255	256	619	511	257	510	269	212	**619**	100	38	22	21	21	15	13	13		39	45	4	3	–	–	–	–	–	–	–	–	–	Tripyrrane B		LI09889
76	79	114	81	116	44	160	158	**620**	100	90	41	36	28	28	23	17	**0.00**	3	–	–	–	4	–	4	4	–	–	–	–	–	1,4,4,7-Tetrabromo-1,1,7,7,-tetrachloro-2,3,5,6-tetrathiaheptane		LI09890
591	592	593	594	281	605	595	282	**620**	100	46	33	14	8	8	7	5	**0.59**	14	36	12	–	–	–	–	–	–	8	–	–	–	Dimethylhexaethyloctasilsesquioxane	77699-57-1	NI46638
593	594	595	596	605	283	567	215	**620**	100	55	40	14	13	10	7	7	**6.09**	15	24	12	–	–	–	–	–	–	8	–	–	–	Methylheptavinyloctasilsesquioxane	87662-29-1	NI46639
591	531	87	503	592	59	593	532	**620**	100	89	71	58	57	52	44	42	**0.09**	16	44	10	–	–	–	–	–	–	8	–	–	–	1,3,5,7,9,11,13,15-Octaethyltricyclo[9.5.1.1^{3,9}]octasiloxane	73113-16-3	NI46640
251	307	572	460	193	516	404	544	**620**	100	91	29	18	18	10	7	6	**0.00**	20	26	4	–	–	–	–	–	–	–	2	–	2	Tetracarbonyl[tellurobis(di-tert-butylphosphane)]molybdenum(0)		NI46641
452	536	508	380	308	336	575	56	**620**	100	65	60	40	16	15	14	12	**7.00**	22	20	14	–	–	–	–	–	–	–	–	–	2	Iron, hexacarbonyl[μ-[(1,2,3,4-η:1,4-η)-1,2,3,4-tetrakis(ethoxycarbonyl)-1,3-butadiene-1,4-diyl]]di-, (Fe-Fe)	41551-08-0	MS10865
452	536	508	380	620	308	336	535	**620**	100	65	60	40	17	16	15	14		22	20	14	–	–	–	–	–	–	–	–	–	2	Iron, hexacarbonyl[μ-[(1,2,3,4-η:1,4-η)-1,2,3,4-tetrakis(ethoxycarbonyl)-1,3-butadiene-1,4-diyl]]di-, (Fe-Fe)	41551-08-0	NI46642
92	77	620	529	69	362	528	105	**620**	100	81	46	46	39	35	27	27		24	7	–	2	–	14	–	–	–	–	1	–	–	1-Phenyl-2-[bis(2-pentafluorophosphinyl)tetrafluorophenyl]hydrazine		MS10866

MASS TO CHARGE RATIOS								M.W.	INTENSITIES								Parent	C	H	O	N	S	F	Cl	Br	I	Si	P	B	X	COMPOUND NAME	CAS Reg No	No
146	622	620	91	607	605	542	208	**620**	100	92	82	70	63	58	37	36		24	25	7	6	1	–	–	1	–	–	–	–	–	N,N-Diethyl-2-methoxy-5-(4-tolysulphonylamino)-4-(2,4-dinitro-6-bromophenylazo)aniline		IC04611
43	331	169	109	111	83	115	112	**620**	100	40	39	21	20	16	12	11	**0.01**	26	36	17	–	–	–	–	–	–	–	–	–	–	6-Deoxy-1,2,3,2′,3′,4′,6′-β-maltose heptaacetate		MS10867
43	169	273	157	184	111	317	153	**620**	100	23	18	16	15	15	11	11	**0.01**	26	36	17	–	–	–	–	–	–	–	–	–	–	6′-Deoxy-1,2,3,6,2′,3′,4′-β-maltose heptaacetate		MS10868
73	317	169	171	318	41	217	75	**620**	100	68	55	19	18	18	17	15	**1.40**	26	45	9	–	–	–	1	–	–	3	–	–	–	D-Glucuronic acid, 2,3,4-tris-O-(trimethylsilyl)-, methyl ester, 2-(4-chlorophenoxy)-2-methylpropanoate	55373-82-5	NI46643
420	504	352	452	351	350	536	43	**620**	100	93	91	86	82	69	68	63	**30.00**	31	24	14	–	–	–	–	–	–	–	–	–	–	2,3,6,8,9-Penta-acetoxy-12-(acetoxymethyl)benzo[1,2-b:4,5-b′]bisbenzofuran		LI09891
149	77	105	150	122	485	106	620	**620**	100	77	72	25	25	20	20	15		31	26	6	4	–	–	2	–	–	–	–	–	–	N,N-Bis(benzoyloxyethyl)-3-methyl-4-(2,6-dichloro-4-nitrophenylazo)aniline		IC04612
299	83	211	139	217	170	157	155	**620**	100	95	29	24	15	12	11	10	**0.00**	34	52	10	–	–	–	–	–	–	–	–	–	–	Spathulasin diacetate	122899-05-2	DD01807
620	588	561						**620**	100	55	25							36	36	6	4	–	–	–	–	–	–	–	–	–	10(R)-Methyl phaophorbide		LI09892
363	620	337	253	69	132	378	159	**620**	100	79	73	58	38	34	32	24		36	49	1	–	–	5	–	–	–	1	–	–	–	3β,22E-Ergosta-5,7,22-triene, 3-[[dimethyl(pentafluorophenyl)silyl]oxy]-	57691-89-1	NI46644
204	206	205	177	379	178	148	176	**620**	100	17	17	16	13	13	12	10	**4.00**	37	36	7	2	–	–	–	–	–	–	–	–	–	Repanduline		LI09893
516	620	517	515	621	518	501	473	**620**	100	52	42	34	23	20	20	9		38	40	6	2	–	–	–	–	–	–	–	–	–	Dimethyl warifteine		LI09894
379	620	190	380	365	619	189	363	**620**	100	55	47	32	29	28	14	13		38	40	6	2	–	–	–	–	–	–	–	–	–	O,O-Dimethylyanangin		NI46645
620	605	590	576	295	288	310	302.5	**620**	100	74	43	37	34	27	24	17		38	50	–	4	–	–	–	–	–	–	–	–	1	α,γ-Dimethyl-α,γ-dihydro-octaethylporphinato-nickel		LI09895
357	339	283	603	358	604	340	265	**620**	100	96	76	71	26	24	24	22	**0.00**	39	72	5	–	–	–	–	–	–	–	–	–	–	1,2-Diolein		NI46646
603	339	357	604	340	621	605	283	**620**	100	70	48	43	16	10	10	9	**4.54**	39	72	5	–	–	–	–	–	–	–	–	–	–	1,2-Diolein		NI46647
43	105	119	91	602	584	566	510	**620**	100	50	43	43	33	30	7	6	**5.00**	40	60	5	–	–	–	–	–	–	–	–	–	–	6,7-Dihydro-isofucoxanthinol		LI09896
57	199	172	85	71	97	69	83	**620**	100	88	67	51	43	39	39	36	**6.14**	41	80	3	–	–	–	–	–	–	–	–	–	–	2-Furanhexanoic acid, 5-(24,26-dimethyloctacosyl)tetrahydro-, methyl ester	18082-14-9	NI46648
241	57	55	43	69	83	95	449	**620**	100	93	54	49	45	42	41	40	**1.01**	41	80	3	–	–	–	–	–	–	–	–	–	–	2-Furannonanoic acid, 5-(21,23-dimethylpentacosyl)tetrahydro-, methyl ester	18082-13-8	NI46649
241	57	43	55	69	449	83	214	**620**	100	96	51	49	41	38	38	37	**15.00**	41	80	3	–	–	–	–	–	–	–	–	–	–	2-Furannonanoic acid, 5-(21,23-dimethylpentacosyl)tetrahydro-, methyl ester	18082-13-8	NI46650
57	312	43	71	55	69	83	366	**620**	100	85	85	62	62	57	52	49	**13.23**	41	80	3	–	–	–	–	–	–	–	–	–	–	Hentetracontanoic acid, 18-oxo-	40710-25-6	NI46651
57	55	69	83	43	95	97	81	**620**	100	67	62	59	58	55	50	43	**5.94**	41	80	3	–	–	–	–	–	–	–	–	–	–	2H-Pyran-2-dodecanoic acid, 6-(17,19-dimethylheneicosyl)tetrahydro-, methyl ester	18082-15-0	NI46652
620	87	74	57	75	43	71	621	**620**	100	99	93	85	79	62	49	48		42	84	2	–	–	–	–	–	–	–	–	–	–	Hentetracontanoic acid, methyl ester	40710-33-6	NI46653
119	458	445	69	176	151	147	77	**621**	100	64	33	32	29	21	16	13	**5.10**	18	10	6	1	–	15	–	–	–	–	–	–	–	Normetadrenaline tri-PFP		MS10870
159	119	457	458	310	69	90	109	**621**	100	40	35	30	24	22	17	16	**2.40**	18	10	6	1	–	15	–	–	–	–	–	–	–	Normetadrenaline tri-PFP		MS10869
458	43	119	41	57	445	159	71	**621**	100	75	57	39	37	36	34	26	**4.26**	18	10	6	1	–	15	–	–	–	–	–	–	–	Normetadrenaline tri-PFP		MS10871
204	73	300	217	145	372	191	143	**621**	100	99	98	95	95	88	83	83	**0.00**	28	63	6	1	–	–	–	–	–	4	–	–	–	N-Decyl-D-glucuronamide, tetrakis(trimethylsilyl)-		NI46654
108	86	79	77	91	57	71	51	**621**	100	95	85	60	45	37	35	30	**0.00**	30	47	9	5	–	–	–	–	–	–	–	–	–	Glycine, N-[N-[N-[N-(N-carboxy-L-threonyl)-L-alanyl]-L-leucyl]-L-leucyl]-, N-benzyl methyl ester	27545-13-7	NI46655
324	325	102	223	326	73	209	250	**621**	100	28	13	13	11	10	4	4	**0.00**	30	55	5	1	–	–	–	–	–	4	–	–	–	Denopamine-M3 isomer 1, N,O,O,O-tetrakis(trimethylsilyl)-		NI46656
324	325	102	223	73	326	209	250	**621**	100	28	14	14	11	11	5	5	**0.00**	30	55	5	1	–	–	–	–	–	4	–	–	–	Denopamine-M3 isomer 2, N,O,O,O-tetrakis(trimethylsilyl)-		NI46657
324	325	223	102	326	73	209	250	**621**	100	28	13	13	11	11	4	4	**0.00**	30	55	5	1	–	–	–	–	–	4	–	–	–	Denopamine-M3 isomer 3, N,O,O,O-tetrakis(trimethylsilyl)-		NI46658
324	325	102	223	73	326	209	250	**621**	100	28	15	14	12	11	5	5	**0.00**	30	55	5	1	–	–	–	–	–	4	–	–	–	Denopamine-M3, N,O,O,O-tetrakis(trimethylsilyl)-		NI46659
73	186	285	157	258	75	247	132	**621**	100	99	55	46	43	38	35	33	**0.00**	30	71	4	1	–	–	–	–	–	4	–	–	–	1,3,4,5-Octadecanetetrol, 2-amino-, tetrakis(trimethylsilyl)-	72088-27-8	NI46660
311	183	55	57	58	43	71	41	**621**	100	64	64	53	49	45	33	27	**0.00**	32	64	8	1	–	–	–	–	–	–	1	–	–	3,5,9-Trioxa-4-phosphaheneicosan-1-aminium, 4-hydroxy-N,N,N-trimethyl-10-oxo-7-[(1-oxododecyl)oxy]-, 4-oxide, (R)-	18194-25-7	NI46661
74	148	308	98	73	154	468	306	**622**	100	80	75	60	58	45	34	30	**1.00**	12	4	–	–	–	–	–	6	–	–	–	–	–	1,1′-Biphenyl, 2,2′,4,4′,5,5′-hexabromo-	59080-40-9	NI46662
308	468	148	628	74	626	306	310	**622**	100	68	67	63	61	48	47	46	**1.23**	12	4	–	–	–	–	–	6	–	–	–	–	–	2,2′,4,4′,6,6′-Hexabromobiphenyl	59261-08-4	NI46663
74	308	148	628	98	73	154	626	**622**	100	79	68	52	51	44	38	35	**0.00**	12	4	–	–	–	–	–	6	–	–	–	–	–	3,3′,4,4′,5,5′-Hexabromobiphenyl	60044-26-0	NI46664
119	76	163	60	578	622	510	347	**622**	100	99	45	27	5	4	3	3		16	36	4	6	–	–	–	–	–	–	2	–	1	Bis[tris(dimethylamino)phosphine]tetracarbonyltungsten		MS10872
159	622	161	624	288	277	455	307	**622**	100	83	71	65	56	54	47	44		18	–	–	–	–	15	–	–	–	–	–	–	1	Tris(pentafluorophenyl)stibine		MS10873
119	445	69	458	295	459	77	297	**622**	100	48	44	27	21	19	16	14	**10.80**	18	9	7	–	–	15	–	–	–	–	–	–	–	(3-Hydroxy-4-methoxyphenyl)ethylene glycol tris(pentafluoropropionate)		MS10874
119	458	445	622	311	459	417	69	**622**	100	58	46	37	32	24	21	19		18	9	7	–	–	15	–	–	–	–	–	–	–	(4-Hydroxy-3-methoxyphenyl)ethylene glycol tris(pentafluoropropionate)	56728-12-2	MS10875
119	297	445	69	298	151	417	325	**622**	100	80	75	47	43	35	33	28	**7.00**	18	9	7	–	–	15	–	–	–	–	–	–	–	(4-Hydroxy-3-methoxyphenyl)ethylene glycol tris(pentafluoropropionate)	56728-12-2	NI46665

MASS TO CHARGE RATIOS								M.W.	INTENSITIES								Parent	C	H	O	N	S	F	Cl	Br	I	Si	P	B	X	COMPOUND NAME	CAS Reg No	No
244	347	533	332	257	242	519	418	**622**	100	49	37	33	32	21	18	15	**0.00**	25	58	6	2	–	–	–	–	–	5	–	–	–	D-erythro-3-Trimethylsiloxy-2-trimethylsiloxymethyl-L-lyxo-5,6-bis-trimethylsiloxy-7-trimethylsiloxymethyl-cis-anti-cis-perhydro-dipyrrolo[2,1-b:2′,3′-d]oxazole		LI09897
43	28	41	77	395	501	396	67	**622**	100	39	36	34	32	31	28	27	**6.60**	27	28	5	–	1	–	–	2	–	–	–	–	–	Phenol, 4,4′-(3H-2,1-benzoxathiol-3-ylidene)bis[2-bromo-3-methyl-6-isopropyl-, S,S-dioxide	76-59-5	NI46666
311	276	230	229	313	281	102	622	**622**	100	55	52	45	34	30	30	6		32	20	4	6	–	–	2	–	–	–	–	–	–	(1α,2α,3β,4β)-1,2-Bis(3-chloro-2-quinoxalinyl)-3,4-bis(4 -nitrophenyl)cyclobutane		MS10876
311	276	230	229	102	622	587	445	**622**	100	40	36	27	10	4	4	4		32	20	4	6	–	–	2	–	–	–	–	–	–	(1α,2α,3β,4β)-1,3-Bis(3-chloro-2-quinoxalinyl)-2,4-bis(4 -nitrophenyl)cyclobutane		MS10877
622	607	621	135	576	592	311	245	**622**	100	33	31	16	10	8	5	5		36	30	10	–	–	–	–	–	–	–	–	–	–	Amentoflavone hexamethyl ether		LI09898
622	621	135	592	311	132	245	607	**622**	100	38	26	18	14	14	11	8		36	30	10	–	–	–	–	–	–	–	–	–	–	Cupressuflavone hexamethyl ether		LI09899
190	605	365	379	606	174	620	591	**622**	100	98	98	97	95	67	31	29	**8.00**	37	38	7	2	–	–	–	–	–	–	–	–	–	(+)-Cepharanthine 2′-β-N-oxide		NI46667
206	622	621	174	381	204	176	191	**622**	100	69	43	35	33	33	32	25		37	38	7	2	–	–	–	–	–	–	–	–	–	Dihydrorepanduline		LI09900
622	621	191	607	593	311	190	381	**622**	100	80	30	18	14	11	9	8		37	38	7	2	–	–	–	–	–	–	–	–	–	Pseudoxandrine		NI46668
594	521	622	579	535	563	433		**622**	100	42	35	19	11	8	5			37	42	5	4	–	–	–	–	–	–	–	–	–	β-Formyl mesoporphyrin IX, dimethyl ester		LI09901
312	296	204	190	174	622	313	623	**622**	100	53	30	24	24	23	23	22		38	42	6	2	–	–	–	–	–	–	–	–	–	(–)-Curine dimethyl ether		MS10878
311	312	518	622	204	206	519	623	**622**	100	75	50	37	32	20	19	15		38	42	6	2	–	–	–	–	–	–	–	–	–	Dimethyl dihydrowarifteine		LI09902
622	295	312	310	591	297	160	67	**622**	100	99	65	65	40	35	35	35		38	42	6	2	–	–	–	–	–	–	–	–	–	(–)-Melanthioidine		LI09903
198	395	608	174	381	175	607	396	**622**	100	64	56	42	41	39	36	28	**0.00**	38	42	6	2	–	–	–	–	–	–	–	–	–	Oxycanthine-methyl ether		LI09904
339	55	114	338	57	43	127	340	**622**	100	51	46	40	32	31	29	25	**0.24**	40	78	4	–	–	–	–	–	–	–	–	–	–	Octadecanoic acid, 1-methyl-1,3-propanediyl ester	14251-40-2	NI46669
39	41	185	312	144	142	40	624	**624**	100	76	56	35	30	23	23	22		12	20	–	–	–	–	–	–	2	–	–	–	2	Rhodium, di-μ-iodotetrakis(η^3-2-propenyl)di-	12307-77-6	NI46670
482	118	77	451	55	540	105	624	**624**	100	50	47	43	42	38	28	20		23	24	7	2	–	–	–	–	–	–	–	–	1	(3-Cyclohexyl-5-ethoxy-1-methyl-2-oxo-5-phenyl-4-imidazolidinylidene)tungsten pentacarbonyl		DD01808
624	133	239	199	121	547	105	318	**624**	100	76	67	63	60	52	48	43		31	30	4	–	–	–	–	2	–	–	–	–	–	Bis-[3-(3′-bromo-2′-hydroxy-5′-methylbenzyl)-2-hydroxy-5-methylphenyl]methane		LI09905
626	133	239	199	121	547	628	624	**624**	100	76	67	63	60	52	50	50		31	30	4	–	–	–	–	2	–	–	–	–	–	Bis-[3-(3′-bromo-2′-hydroxy-5′-methylbenzyl)-2-hydroxy-5-methylphenyl]methane		LI09906
72	85	70	84	71	83	86	112	**624**	100	57	39	37	34	30	28	23	**9.86**	34	64	6	4	–	–	–	–	–	–	–	–	–	L-Valinamide, N-(1-oxodecyl)glycyl-N-[1-(2-methoxy-2-oxoethyl)-4-[(1-oxodecyl)amino]butyl]-, (S)-	33912-91-3	NI46671
43	72	85	284	70	84	142	44	**624**	100	79	45	37	31	30	29	29	**12.80**	34	64	6	4	–	–	–	–	–	–	–	–	–	L-Valinamide, N-(1-oxodecyl)glycyl-N-[1-(2-methoxy-2-oxoethyl)-4-[(1-oxodecyl)amino]butyl]-, (S)-	33912-91-3	NI46672
72	57	70	284	142	43	55	314	**624**	100	26	25	24	24	24	21	18	**10.50**	34	64	6	4	–	–	–	–	–	–	–	–	–	L-Valinamide, N-(1-oxodecyl)glycyl-N-[6-methoxy-6-oxo-4-[(1-oxodecyl)amino]hexyl]-, (S)-	33861-95-9	NI46673
72	70	71	314	102	341	73	69	**624**	100	24	22	18	15	14	14	13	**10.72**	34	64	6	4	–	–	–	–	–	–	–	–	–	L-Valinamide, N-(1-oxodecyl)glycyl-N-[6-methoxy-6-oxo-4-[(1-oxodecyl)amino]hexyl]-, (S)-	33861-95-9	NI46674
70	57	72	43	41	112	71	55	**624**	100	94	83	81	75	68	68	63	**16.30**	34	64	6	4	–	–	–	–	–	–	–	–	–	L-Valine, N-[N-[N^2,N^6-bis(1-oxodecyl)-L-lysyl]glycyl]-, methyl ester	33281-02-6	NI46675
70	57	72	43	41	112	71	55	**624**	100	93	83	81	75	68	68	63	**16.38**	34	64	6	4	–	–	–	–	–	–	–	–	–	L-Valine, N-[N-[1-oxo-3,6-bis[(1-oxodecyl)amino]hexyl]glycyl]-, methyl ester	34020-23-0	NI46676
121	195	181	444	311	624	154	135	**624**	100	92	84	50	40	38	36	22		36	32	10	–	–	–	–	–	–	–	–	–	–	Talbot flavone hexamethyl ether		LI09907
624	609	594	595	593	312	551		**624**	100	40	8	7	7	7	5			37	44	5	4	–	–	–	–	–	–	–	–	–	β-Methoxymesoporphyrin-IX-dimethyl ester		LI09908
518	339	624	329	386	445	388	403	**624**	100	99	82	80	53	41	41	21		37	52	8	–	–	–	–	–	–	–	–	–	–	Methyl 1-hexosyl-1,2-dihydro-3,4-didehydro-apo-8′-lycopeniate		LI09909
73	75	43	187	55	539	325	41	**624**	100	28	28	24	23	22	16	16	**0.60**	37	76	3	–	–	–	–	–	–	2	–	–	–	25-Trimethylsiloxyhentriacontane-14,16-dione, trimethylsilyl-		MS10879
182	57	43	71	268	326	142	240	**624**	100	78	72	54	52	48	41	31	**17.00**	38	76	4	2	–	–	–	–	–	–	–	–	–	Acetamide, 2,2′-[1,2-ethanediylbis(oxy)]bis[N,N-dioctyl-	67615-89-8	NI46677
43	57	55	41	73	98	60	71	**624**	100	88	73	62	58	57	49	43	**0.00**	39	76	5	–	–	–	–	–	–	–	–	–	–	Octadecanoic acid, 2-hydroxy-1,3-propanediyl ester	504-40-5	NI46678
43	57	41	55	267	341	71	69	**624**	100	74	55	54	45	36	34	33	**0.00**	39	76	5	–	–	–	–	–	–	–	–	–	–	Octadecanoic acid, 2-hydroxy-1,3-propanediyl ester	504-40-5	LI09910
73	92	39	91	106	105	74	119	**624**	100	37	34	30	17	9	5	4	**0.60**	42	56	4	–	–	–	–	–	–	–	–	–	–	2,2′-Dioxospirilloxanthin		MS10880
121	291	207	334	278	624	292	290	**624**	100	50	38	28	23	22	22	22		42	56	4	–	–	–	–	–	–	–	–	–	–	2,4,6,8,10,12-Tridecahexaenoic acid, 13-(4-methoxyphenyl)-, 2-decyl-3-methoxy-5-pentylphenyl ester	71176-02-8	NI46679
43	204	175	147	161	189	133	484	**624**	100	34	22	19	19	18	17	15	**13.90**	42	72	3	–	–	–	–	–	–	–	–	–	–	1,3,5-Tridodecanoylbenzene		NI46680
624	484	344	485	441	326	204	469	**624**	100	94	61	54	49	37	37	25		42	72	3	–	–	–	–	–	–	–	–	–	–	1,3,5-Tridodecanoylbenzene		NI46681
255	623	365	363					**624**	100	83	9	6					**0.00**	43	76	2	–	–	–	–	–	–	–	–	–	–	Cholest-5-en-3-ol (3β)-, hexadecanoate	601-34-3	NI46682
41	40	248	39	124	93	391	267	**625**	100	48	22	18	16	16	15	12	**0.01**	14	3	–	1	–	18	–	–	–	–	–	–	1	Acetonitrilo-tris(π-hexafluorobut-2-yne)molybdenum		MS10881

MASS TO CHARGE RATIOS								M.W.	INTENSITIES								Parent	C	H	O	N	S	F	Cl	Br	I	Si	P	B	X	COMPOUND NAME	CAS Reg No	No
73	298	75	147	300	259	420	205	**625**	100	39	37	29	24	21	20	19	**0.00**	25	55	9	1	–	–	–	–	–	4	–	–	–	D-Glycero-α-D-galacto-2-nonulopyranosidonic acid, methyl 5-(acetylamino)-3,5-dideoxy-4,7,8,9-tetrakis-O-(trimethylsilyl)-, methyl ester	53081-21-3	NI46683
629	627	631	625	628	488	632	566	**625**	100	99	39	38	35	12	10	6		27	18	2	1	–	–	–	3	–	–	–	–	–	3,6,8-Tribromo-1-(2-methoxycarbonylphenyl)-9-methyl-4-phenylcarbazole		LI09911
625	113	241	99	128	227	611	369	**625**	100	32	20	7	5	2	1	1		32	16	–	9	–	–	–	–	–	–	–	–	1	Nitrido-phthalocyaninotechnetium(V)		NI46684
305	548	151	625	320	228	471	243	**625**	100	63	60	49	30	30	20	20		36	31	–	1	–	–	–	–	–	–	–	–	2	Bis(triphenylgermyl)amine		LI09912
630	632	628	627	625	389	633	387	**626**	100	87	85	85	66	40	39	36	**0.00**	12	18	13	–	–	–	–	–	–	–	–	–	4	Oxohexakis(acetato)tetrazinc(II)		MS10882
627	628	464	319	293	629	481	318	**626**	100	16	8	3	2	1	1	1	**0.00**	16	13	5	4	–	15	–	–	–	–	–	–	–	Arginine methyl ester tris(pentafluoropropionate)		NI46685
73	299	211	297	387	315	147	300	**626**	100	82	39	30	25	25	21	20	**0.80**	18	48	10	–	–	–	–	–	–	5	2	–	–	3,5-Dioxa-4-phospha-2-silaoctan-8-oic acid, 7-[[bis[(trimethylsilyl)oxy]phosphinyl]oxy]-2,2-dimethyl-4-[(trimethylsilyl)oxy]-, trimethylsilyl ester, 4-oxide, (R)-	55334-89-9	NI46686
245	383	349	90	314	137	77	119	**626**	100	85	85	52	46	39	37	36	**3.00**	19	11	2	–	–	12	–	–	1	–	–	–	–	1,3-Dihydro-5-methyl-1-[1-phenyl-1-(trifluoromethyl)-2,2,2-trifluoroethoxy]-3,3-bis(trifluoromethyl)-1,2-benziodoxole		MS10883
468	353	432	469	261	273	293	460	**626**	100	79	55	45	45	35	34	29	**2.07**	27	26	12	6	–	–	–	–	–	–	–	–	–	L-Valine, N-[N,O-bis(2,4-dinitrophenyl)-L-tyrosyl]-, methyl ester	56051-46-8	NI46687
626	412	598	262	105	483	91	434	**626**	100	38	26	19	17	15	15	14		27	27	2	2	–	–	–	–	–	–	1	–	1	Tungsten, carbonyl(η^5-2,4-cyclopentadien-1-yl)[N-methyl-P,P-diphenyl-N(1-phenylethyl)phosphinous amide-P]nitrosyl-	75520-18-2	NI46688
101	291	117	88	161	87	129	43	**626**	100	64	63	59	24	22	19	18	**0.00**	27	46	16	–	–	–	–	–	–	–	–	–	–	1,5-Di-O-acetyl-6-O-[4,6-di-O-acetyl-2,3,7-tri-O-methyl-D-glycero-α-D-manno-heptopyranosyl]-2,3,4-tri-O-methyl-D-glucitol		NI46689
345	41	43	57	26	85	70	55	**626**	100	78	64	49	49	47	42	42	**0.20**	33	54	11	–	–	–	–	–	–	–	–	–	–	Ponasteroside A	20117-33-3	NI46690
345	327	145	109	309	127	428	363	**626**	100	40	35	34	25	25	18	15	**0.20**	33	54	11	–	–	–	–	–	–	–	–	–	–	Ponasteroside A		LI09913
626	563	627	282	266	121	105	59	**626**	100	60	37	22	11	11	7	5		34	26	8	–	2	–	–	–	–	–	–	–	–	Tetramethyl 2,2′-(1,3-phenylene)bis(5-phenyl-3,4-thiophenedicarboxylate)	110699-37-1	DD01809
596	460	597	461	121	595	133	230	**626**	100	94	43	34	34	26	26	21	**2.20**	36	34	10	–	–	–	–	–	–	–	–	–	–	Rhusflavanone hexamethyl ether		MS10884
626	595	559	567	565	521	566	564	**626**	100	39	15	4	4	4	3	3		36	42	6	4	–	–	–	–	–	–	–	–	–	2,4-Di-(2-hydroxyethyl)-6,7-di-(2-methoxycarbonylethyl)-1,3,5,8-tetramethylporphin		LI09914
518	339	624	329	338	388	340	445	**626**	100	99	82	80	68	52	52	41	**0.00**	37	54	8	–	–	–	–	–	–	–	–	–	–	8′-Apo-ψ,ψ-carotenoic acid, 1,2-dihydro-1-(1,2,3,4,5,6-hexahydroxyhexyl), methyl ester	56114-38-6	NI46691
611	626	596	582	298	291	313	305.5	**626**	100	46	34	32	15	11	9	8		38	50	–	4	–	–	–	–	–	–	–	–	1	α,γ-Dimethyl-α,γ-dihydro-octaethylporphinato-zinc		LI09915
312	122	121	176	313	177	284	57	**626**	100	54	49	48	42	42	38	38	**1.05**	39	62	6	–	–	–	–	–	–	–	–	–	–	3-O-Hexadecanoyl-5,20-di-O-isopropylidene-ingenol		NI46692
558	559	492	626	505	560	627	493	**626**	100	52	40	33	26	20	17	17		39	70	2	–	–	–	–	–	–	2	–	–	–	Cannabidiol bis(tripropylsilyl) ether		NI46693
626	627	44	76	104	628	77	105	**626**	100	53	34	31	31	13	8	7		40	18	8	–	–	–	–	–	–	–	–	–	–	2,2′:3′,2″:3″,2‴-Quaternaphthalene-1,1′,1″,1‴,4,4′,4″,4‴-octone		NI46694
85	423	86	57	43	67	55	41	**626**	100	8	6	6	6	5	4	4	**0.00**	40	66	5	–	–	–	–	–	–	–	–	–	–	Lup-20(29)-en-21-ol, 3,28-bis[(tetrahydro-2H-pyran-2-yl)oxy]-, (3β,21β)-	55401-93-9	NI46695
165	69	169	135	139	120	150	612	**627**	100	64	50	43	21	17	16	15	**0.00**	10	18	2	1	1	9	–	–	–	–	–	–	2	1-Butanesulphonamide, 1,1,2,2,3,3,4,4,4-nonafluoro-N,N-bis(trimethylstannyl)-	41006-34-2	MS10885
73	103	307	217	147	308	218	205	**627**	100	65	62	61	31	18	12	10	**0.00**	24	61	6	1	–	–	–	–	–	6	–	–	–	Fructose oxime hexa-TMS		MS10886
73	319	205	147	320	217	103	218	**627**	100	71	62	39	22	22	22	15	**0.00**	24	61	6	1	–	–	–	–	–	6	–	–	–	Galactose oxime hexa-TMS		MS10887
73	319	205	147	103	217	218	320	**627**	100	52	45	38	28	23	17	16	**0.00**	24	61	6	1	–	–	–	–	–	6	–	–	–	Glucose oxime hexa-TMS		MS10889
73	319	205	147	320	218	217	103	**627**	100	76	60	40	26	22	21	15	**0.00**	24	61	6	1	–	–	–	–	–	6	–	–	–	Glucose oxime hexa-TMS		MS10888
230	236	73	75	93	264	613	245	**627**	100	82	81	79	63	58	57	50	**0.00**	25	53	4	5	–	–	–	–	–	5	–	–	–	Adenosine penta-TMS		NI46696
127	254	247	501	628	245	499	120	**628**	100	99	19	18	14	14	13	13				–	–	–	–	–	–	4	–	–	–	1	Stannic iodide	7790-47-8	NI46697
205	235	284	220	299	269	221	314	**628**	100	40	13	6	4	2	2	1	**0.10**	6	18	4	–	2	–	–	–	–	–	–	–	2	Thallium, bis(μ-methanesulphinato)tetramethyldi-	26025-21-8	NI46698
76	44	114	70	336	72	64	323	**628**	100	91	58	52	48	33	32	26	**0.00**	10	16	–	2	4	–	–	–	1	–	–	–	1	Iodobis(pyrrolidyldithiocarbamato)bismuthine	59196-62-2	NI46699
59	69	131	119	100	169	181	28	**628**	100	28	19	9	7	6	4	4	**0.00**	13	3	2	–	–	23	–	–	–	–	–	–	–	Dodecanoic acid, tricosafluoro-, methyl ester	56554-52-0	NI46700
529	45	527	531	41	27	43	293	**628**	100	93	70	68	38	30	28	25	**3.00**	19	20	4	–	–	–	–	4	–	–	–	–	–	2,2-Bis[3,5-dibromo-4-(2-hydroxyethoxy)phenyl]propane		IC04613
200	391	312	470	549	228	284	256	**628**	100	93	52	38	34	34	22	15	**0.00**	20	8	4	–	–	–	–	4	–	–	–	–	–	5a,5b,11a,11b-Tetrabromo-5,6,11,12-tetraketo-5,5a,5b,6,11,11b,12-octahydrodibenzo[b,h]biphenylene		LI09916
529	348	530	448	215	429	265	257	**628**	100	15	12	12	8	6	4	4	**3.08**	20	28	8	–	–	–	–	–	–	–	–	–	1	Thorium, tetrakis(2,4-pentanedionato-O,O′)-	17499-48-8	NI46701
73	292	147	319	205	217	305	103	**628**	100	35	31	27	25	24	15	15	**0.00**	24	60	7	–	–	–	–	–	–	6	–	–	–	Galactonic acid, 2,3,4,5,6-pentakis-O-(trimethylsilyl)-, trimethylsilyl ester	55400-16-3	MS10890

MASS TO CHARGE RATIOS								M.W.	INTENSITIES								Parent	C	H	O	N	S	F	Cl	Br	I	Si	P	B	X	COMPOUND NAME	CAS Reg No	No
73	147	205	292	319	217	103	305	**628**	100	35	31	28	25	25	19	16	**0.00**	24	60	7	–	–	–	–	–	–	6	–	–	–	Galactonic acid, 2,3,4,5,6-pentakis-O-(trimethylsilyl)-, trimethylsilyl ester	55400-16-3	NI46702
73	333	292	147	319	205	217	305	**628**	100	45	43	39	31	30	22	20	**0.00**	24	60	7	–	–	–	–	–	–	6	–	–	–	D-Gluconic acid, 2,3,4,5,6-pentakis-O-(trimethylsilyl)-, trimethylsilyl ester	34290-52-3	MS10891
73	147	292	103	74	217	333	205	**628**	100	18	9	9	8	8	7	7	**0.00**	24	60	7	–	–	–	–	–	–	6	–	–	–	D-Gluconic acid, 2,3,4,5,6-pentakis-O-(trimethylsilyl)-, trimethylsilyl ester	34290-52-3	NI46703
73	333	147	205	292	217	319	103	**628**	100	41	39	28	26	22	20	19	**0.00**	24	60	7	–	–	–	–	–	–	6	–	–	–	D-Gluconic acid, 2,3,4,5,6-pentakis-O-(trimethylsilyl)-, trimethylsilyl ester	34290-52-3	NI46704
508	385	540	419	509	387	386	391	**628**	100	80	75	53	50	36	28	24	**0.40**	26	40	4	6	4	–	–	–	–	–	–	–	–	3,3′-(1,6-Hexanediyl)bis[1-(2-(N,N-dimethylaminothio)carbonylthioethyl)-6-methyl-2,4(1H,3H)-pyrimidinedione]		MS10892
242	243	528	527	397	244	543	57	**628**	100	99	75	75	33	30	27	27	**6.00**	31	40	10	4	–	–	–	–	–	–	–	–	–	Riboflavin tetrapropionate		LI09917
348	404	460	544	458	402	456	454	**628**	100	93	85	84	63	51	50	38	**9.30**	32	36	6	–	–	–	–	–	–	–	–	–	2	Iron, hexacarbonyl[μ-[(1,1′,2,2′-η:2,2′-η)[bi-1-cyclotridecen-7-yn-1-yl]-2,2′-diyl]]di-, (Fe-Fe)	39040-50-1	MS10893
57	43	41	71	405	403	55	391	**628**	100	80	80	69	67	57	51	49	**0.00**	32	60	4	–	–	–	–	–	–	–	–	–	1	Dibutyltin dilaurate		IC04614
137	281	569	527	179	485	400	401	**628**	100	98	96	92	92	87	86	83	**6.00**	34	28	12	–	–	–	–	–	–	–	–	–	–	Sphagnorubin B acetate		NI46705
95	109	123	107	97	253	121	235	**628**	100	78	75	70	70	69	65	47	**0.00**	34	60	10	–	–	–	–	–	–	–	–	–	–	29-(α-L-arabinofuranosyloxy)-5α-stigmastane-3β,6α,8,15α,16β-pentaol		NI46706
95	109	253	123	107	97	121	235	**628**	100	78	76	75	70	70	65	47	**0.00**	34	60	10	–	–	–	–	–	–	–	–	–	–	29-(α-L-arabinofuranosyloxy)-5α-stigmastane-3β,6α,8β,15α,16β-pentaol		NI46707
298	314	312	296	313	299	272	284	**628**	100	82	41	32	31	25	21	19	**1.00**	36	40	8	2	–	–	–	–	–	–	–	–	–	15,16-Dimethoxy-6-(15,16-dimethoxy-8-oxo-erythrinen-(6)-yl-(7)-oxy)-cis-erythrinan-7,8-dione		LI09918
298	314	296	312	313	272	299	284	**628**	100	73	28	27	21	21	18	18	**1.00**	36	40	8	2	–	–	–	–	–	–	–	–	–	15,16-Dimethoxy-6-(15,16-dimethoxy-8-oxo-erythrinen-(6)-yl-(7)-oxy)-trans-erythrinan-7,8-dione		LI09919
596	597	490	628	504	443	411	598	**628**	100	52	40	29	29	16	16	15		42	60	4	–	–	–	–	–	–	–	–	–	–	α-Carotene, 7,8-dihydro-3-hydroxy-3′,19-dimethoxy-8-oxo-, all-trans-	33474-09-8	NI46708
596	597	490	628	504	443	598	565	**628**	100	54	41	30	30	16	15	14		42	60	4	–	–	–	–	–	–	–	–	–	–	α-Carotene, 7,8-dihydro-3-hydroxy-3′,19-dimethoxy-8-oxo-, all-trans-		LI09920
73	75	115	147	378	74	173	191	**629**	100	13	12	10	10	9	9	8	**0.51**	31	63	6	1	–	–	–	–	–	3	–	–	–	6-Keto-PGF-1a mox, methyl ester, tris(trimethylsilyl)-	59403-25-7	NI46709
73	75	115	147	378	74	173	55	**629**	100	22	12	11	11	10	10	10	**0.90**	31	63	6	1	–	–	–	–	–	3	–	–	–	6-Keto-PGF-1a mox, methyl ester, tris(trimethylsilyl)-	59403-25-7	NI46710
73	115	147	378	75	173	191	55	**629**	100	20	17	16	16	13	11	11	**1.25**	31	63	6	1	–	–	–	–	–	3	–	–	–	6-Keto-PGF-1a, mox, methyl ester, tris(trimethylsilyl)-	59403-25-7	NI46711
629	108	91	123	415	150	630	479	**629**	100	99	80	50	40	40	30	25		36	43	7	3	–	–	–	–	–	–	–	–	–	Tripyrrane C		LI09921
102	101	129	55	237	73	56	45	**630**	100	84	78	59	22	21	20	18	**0.01**	17	12	4	–	–	–	10	–	–	–	–	–	–	Kelevan		MS10894
462	546	518	630	574	226	372	344	**630**	100	48	26	17	6	6	5	5		22	42	6	–	–	–	–	–	–	–	4	–	2	μ-Tetramethyldiphosphine-bis(triethylphosphinotricarbonylchromium)		MS10895
202	187	119	203	105	132	133	159	**630**	100	84	83	80	74	61	61	56	**0.80**	23	24	4	–	–	14	–	–	–	–	–	–	–	8,11-Bis(heptafluorobutyryl)culmorin		NI46712
88	101	45	71	89	75	187	111	**630**	100	67	46	45	37	21	16	16	**0.00**	28	54	15	–	–	–	–	–	–	–	–	–	–	Arabinitol, O-2,3,4,6-tetra-O-methyl-β-D-galactopyranosyl-(1-6)-O-2,3,4-tri-O-methyl-β-D-galactopyranosyl-(1-3)-1,2,4,5-tetra-O-methyl-, L	32581-26-3	NI46713
310	370	243	328	311	272	254	282	**630**	100	99	86	65	47	36	36	34	**1.50**	31	35	9	–	–	–	–	1	–	–	–	–	–	3-O-(4-bromobenzoyl)-neophorbal-13,20-diacetate		LI09922
208	210	211	209	315	317	212	316	**630**	100	82	48	35	30	29	17	14	**0.00**	33	42	–	–	6	–	–	–	–	–	–	–	–	7,10.18,21.29,32-Trisbenz-1,5,12,16,23,27-hexathiacyclotritriacontane		MS10896
128	96	95	41	177	55	71	210	**630**	100	55	55	55	50	50	45	44	**0.00**	33	54	–	6	3	–	–	–	–	–	–	–	–	Triimidazo[1,5-a:1′,5′-c:1″,5″-e][1,3,5]triazine-1,5,9(2H,6H,10H)-trithione, 2,6,10-tricyclohexylhexahydro-3,3,7,7,11,11-hexamethyl-	67969-54-4	NI46714
367	351	271	391	349	327	283	450	**630**	100	59	57	53	48	41	41	38	**20.00**	34	46	11	–	–	–	–	–	–	–	–	–	–	Δ^6-THC-C-4′-Glucuronide methyl ester triacetate		NI46715
451	409	630	343	221	231	547	299	**630**	100	66	46	27	27	20	19	15		34	46	11	–	–	–	–	–	–	–	–	–	–	Δ^6-THC-C-6′-Glucuronide methyl ester triacetate		NI46716
231	257	314	317	630	570			**630**	100	94	71	48	8	8				34	46	11	–	–	–	–	–	–	–	–	–	–	Δ^6-THC-O-Glucuronide methyl ester triacetate		NI46717
27	73	60	55	43	41	39	29	**630**	100	99	99	99	99	99	99	98	**0.07**	36	54	9	–	–	–	–	–	–	–	–	–	–	13-Acetoxy-4,9,20-trihydroxy-12-tetradecanoyloxy-1,5-tigliadien-3,7-dione		NI46718
83	43	57	69	310	292	55	125	**630**	100	53	47	33	31	29	29	27	**0.15**	37	58	8	–	–	–	–	–	–	–	–	–	–	4-O-Methyl-12-tetradecanoyl-phorbol-13-acetate		NI46719
159	587	131	588	201	589	89	160	**630**	100	82	60	39	24	22	15	14	**1.56**	39	74	2	–	–	–	–	–	–	2	–	–	–	3β,20α-Bis(tripropylsilyloxy)pregn-5-ene		NI46720
307	308	144	121	160	630	108	106	**630**	100	50	20	16	14	11	11	11		40	46	3	4	–	–	–	–	–	–	–	–	–	12′-Hydroxyisostrychnobiline		NI46721
415	119	387	416	69	77	160	57	**631**	100	44	28	15	12	8	7	6	**0.00**	19	8	6	1	–	15	–	–	–	–	–	–	–	α-MethylDOPA di-PFP PFP-azlactone		MS10897
105	239	77	211	111	110	210	272	**631**	100	66	38	15	15	13	11	7	**2.00**	31	29	10	5	–	–	–	–	–	–	–	–	–	Benzamide, N-[6-[2,3-di-O-acetyl-3-C-[(acetyloxy)methyl]-5-O-benzoyl-β-D-xylofuranosyl]-9H-purin-9-yl]-	75332-24-0	NI46722
76	87	116	59	64	72	44	336	**632**	100	89	74	74	57	54	51	33	**0.00**	10	20	–	2	4	–	–	–	1	–	–	–	1	Iodobis(N,N-diethyldithiocarbamato)bismuthine	59196-65-5	NI46723

MASS TO CHARGE RATIOS	M.W.	INTENSITIES	Parent	C	H	O	N	S	F	Cl	Br	I	Si	P	B	X	COMPOUND NAME	CAS Reg No	No
69 253 177 81 265 195 379 139	**632**	100 88 37 28 27 27 25 17	**0.60**	15	7	10	–	–	15	–	–	–	–	–	–	–	**Pentitol, pentakis(trifluoroacetate)**	55319-75-0	**NI46724**
605 606 551 289 607 275 552 553	**632**	100 58 30 26 25 25 18 17	**17.09**	16	24	12	–	–	–	–	–	–	8	–	–	–	**Pervinyloctasilsesquioxane**	69655-76-1	**NI46725**
603 604 605 606 559 41 57 56	**632**	100 52 38 14 13 12 11 11	**0.00**	18	44	11	–	–	–	–	–	–	7	–	–	–	**1-Butoxyperethylhomohexasilsesquioxane**	73895-19-9	**NI46726**
274 272 259 257 276 275 71 273	**632**	100 82 75 59 55 45 41 37	**2.20**	20	16	6	–	–	–	8	–	–	–	–	–	–	**19,20,21,22,23,24,25,26-Octachloro-2,3:11,12-dibenzo-1,4,7,10,13,16-hexaoxacyclooctadeca-2,11-diene**		**MS10898**
446 386 537 402 416 628 337 475	**632**	100 72 60 56 28 20 16 4	**0.00**	24	24	2	2	2	–	–	–	–	–	–	–	2	**Molybdenum, bis(benzenemethanethiolato)di-2,4-cyclopentadien-1-yldinitrosyldi-**	55955-45-8	**NI46727**
634 632 636 638 111 28 138 141	**632**	100 73 52 14 14 14 10 9		30	20	–	8	–	–	4	–	–	–	–	–	–	**Pyrimido[5,4-d]pyrimidine, 2,4,6,8-tetrakis(m-chloroanilino)-**	18710-95-7	**MS10899**
105 91 77 122 160 51 204 70	**632**	100 78 44 41 32 18 14 14	**0.00**	34	36	8	2	1	–	–	–	–	–	–	–	–	**2,3-Di-O-benzoyl-6-(N-carbobenzoxyprolineamide)-1-methylthio-α-DL-lyxo-hexopyranoside**		**NI46728**
206 131 542 207 147 137 205 116	**632**	100 98 50 33 32 31 26 25	**17.69**	36	68	3	–	–	–	–	–	–	3	–	–	–	**3β,5Z,7E-9,10-Secocholesta-5,7,10(19)-triene, 1,3,25-tris(trimethylsiloxy)**	55759-95-0	**NI46729**
131 132 527 221 632 429 355 133	**632**	100 11 7 7 6 6 6 5		36	68	3	–	–	–	–	–	–	3	–	–	–	**3β,5Z,7E-9,10-Secocholesta-5,7,10(19)-triene, 3,24,25-tris(trimethylsiloxy)-**		**NI46730**
118 147 117 119 145 208 219 121	**632**	100 62 49 37 24 20 17 15	**6.15**	36	68	3	–	–	–	–	–	–	3	–	–	–	**3β,5Z,7E-9,10-Secocholesta-5,7,10(19)-triene, 3,25,26-tris(trimethylsiloxy)-**	56009-11-1	**NI46731**
131 117 132 147 133 632 208 118	**632**	100 17 15 10 9 8 8 8		36	68	3	–	–	–	–	–	–	3	–	–	–	**3β,5Z,7E-9,10-Secocholesta-5,7,10(19)-triene,-3,21,25-tris(trimethylsiloxy)-**	56009-12-2	**NI46732**
335 168 377 321 336 590 548 632	**632**	100 60 37 34 31 30 23 10		38	36	7	2	–	–	–	–	–	–	–	–	–	**O,O-Diacetyltricordatine**		**NI46733**
589 159 590 131 283 591 201 160	**632**	100 81 50 29 26 20 11 11	**0.00**	39	76	2	–	–	–	–	–	–	2	–	–	–	**3α,20β-Bis(Tripropylsilyloxy)-5β-pregnane**		**NI46734**
589 131 159 590 591 89 201 283	**632**	100 66 49 47 19 15 12 8	**0.00**	39	76	2	–	–	–	–	–	–	2	–	–	–	**3β,20α-Bis(tripropylsilyloxy)-5α-pregnane**		**NI46735**
590 159 131 591 592 89 160 201	**632**	100 92 49 46 20 16 11 10	**0.00**	39	76	2	–	–	–	–	–	–	2	–	–	–	**3β,20β-Bis(tripropylsilyloxy)-5α-pregnane**		**NI46736**
262 203 147 205 187 371 632 468	**632**	100 81 68 20 16 9 8 7		40	56	6	–	–	–	–	–	–	–	–	–	–	**Urs-12-en-28-oic acid, 2-hydroxy-3-[[3-(4-hydroxyphenyl)-1-oxo-2-propenyl]oxy]-, methyl ester, (2α,3β)-**	75363-15-4	**NI46737**
632 633 105 541 603 119 423 555	**632**	100 52 22 18 16 14 13 9		46	36	1	2	–	–	–	–	–	–	–	–	–	**2-p-Tolyl-4,5-diphenyl-3-(3-p-tolyl-4,5-diphenylpyrrol-2-yl)-3H-pyrrol-3-ol**	65241-97-6	**NI46738**
632 555 633 513 541 616 589 104	**632**	100 70 52 31 27 15 10 10		46	36	1	2	–	–	–	–	–	–	–	–	–	**1-p-Tolyl-3,4-diphenyl-5-(3-p-tolyl-4,5-diphenylpyrrol-2-yl)-6-oxa-2-azabicyclo[3.1.0]hex-3-ene**	65241-96-5	**NI46739**
73 299 315 357 370 147 387 217	**633**	100 50 39 35 29 23 20 17	**0.10**	22	56	8	1	–	–	–	–	–	5	1	–	–	**D-arabino-Hexose, 2-deoxy-3,4,5-tris-O-(trimethylsilyl)-, O-methyloxime, 6-[bis(trimethylsilyl) phosphate]**	55401-79-1	**NI46740**
299 73 315 357 370 147 387 217	**633**	100 99 77 70 57 45 39 34	**0.00**	22	56	8	1	–	–	–	–	–	5	1	–	–	**D-arabino-Hexose, 2-deoxy-3,4,5-tris-O-(trimethylsilyl)-, O-methyloxime, 6-[bis(trimethylsilyl) phosphate]**	55401-79-1	**NI46741**
98 295 91 77 198 267 210 224	**633**	100 37 25 24 17 10 10 8	**0.00**	33	51	5	3	2	–	–	–	–	–	–	–	–	**Benzenesulphonamide, N,4-dimethyl-N-[3-[[(4-methylphenyl)sulphonyl][3(2-oxoazacyclotridec-1-yl)propyl]amino]propyl]-**	67370-88-1	**NI46742**
43 41 42 57 56 29 28 166	**633**	100 75 65 53 51 37 37 30	**3.00**	42	39	3	1	–	–	–	–	–	1	–	–	–	**3-(Diethylamino)-2-[4,4-diphenyl-3-(triphenylsiloxy)-3-butenoyl]-2-cyclobuten-1-one**	79139-25-6	**NI46743**
421 131 419 337 420 173 379 214	**634**	100 68 47 47 42 32 28 28	**13.00**	12	28	4	–	4	–	–	–	–	–	2	–	1	**Lead[II] bis(dipropyldithiophosphate)**	43159-15-5	**NI46744**
59 76 58 92 150 43 285 57	**634**	100 48 18 16 15 14 11 8	**1.25**	14	18	–	–	14	–	–	–	–	–	–	–	–	**2,4,6,8,9,10-Hexathiatricyclo[3.3.1.1^{3,7}]decane, 1,1′-dithiobis[3,5,7-trimethyl-**	57274-62-1	**NI46745**
605 606 607 608 609 288 533 575	**634**	100 53 46 16 8 7 7 6	**0.00**	15	38	12	–	–	–	–	–	–	8	–	–	–	**Methylheptaethyloctasilsesquioxane**	77626-18-7	**NI46746**
605 606 290 607 608 291 277 579	**634**	100 59 24 24 20 13 9 9	**3.99**	16	26	12	–	–	–	–	–	–	8	–	–	–	**Ethylheptavinyloctasilsesquioxane**	79235-35-1	**NI46747**
605 606 607 608 503 87 288 260	**634**	100 58 44 18 17 13 12 9	**0.00**	16	42	11	–	–	–	–	–	–	8	–	–	–	**1,3,5,7,9,11,13,15-Octaethyl-7,13-dihydrotetracyclo[9.5.1.1^{3,9}.1^{5,1}5]octasiloxane**	80177-44-2	**NI46748**
147 43 99 81 148 141 189 146	**634**	100 88 15 15 11 9 7 7	**0.01**	31	38	14	–	–	–	–	–	–	–	–	–	–	**Grandidentatin pentaacetate**		**MS10900**
45 254 634 272 83 619 121 317	**634**	100 45 37 35 33 29 22 9		32	30	8	2	2	–	–	–	–	–	–	–	–	**Tetramethyl 2,2′-(1,4-piperazinediyl)bis(5-phenyl-3,4-thiophenedicarboxylate)**	110699-41-7	**DD01810**
560 470 561 103 471 501 188 73	**634**	100 77 47 40 30 27 27 26	**4.41**	32	58	5	2	–	–	–	–	–	3	–	–	–	**Pregna-3,5-dien-18-al, 20-(methoxyimino)-3,11,21-tris[(trimethylsilyl)oxy], O-methyloxime, (11β,17α)-**	69688-41-1	**NI46749**
57 522 634 578 458 64 505 441	**634**	100 14 7 6 6 5 4 3		34	34	4	–	4	–	–	–	–	–	–	–	–	**2,2′-(1,4-Phenylene)bis[5-(tert-butylsulphonyl)-3-phenylthiophene]**	110699-60-0	**DD01811**
130 132 129 263 318 56 70 102	**634**	100 60 58 53 42 42 40 35	**0.32**	34	46	6	6	–	–	–	–	–	–	–	–	–	**(S,S)-Bistryptophyldiaza-18-crown-6**	93767-79-4	**NI46750**
75 73 131 55 43 199 212 243	**634**	100 92 65 65 64 63 60 16	**1.00**	34	58	7	–	–	–	–	–	–	2	–	–	–	**Colletotrichin di-TMS**		**NI46751**
591 121 592 537 84 85 155 100	**634**	100 63 43 20 20 16 14 13	**10.00**	35	46	7	4	–	–	–	–	–	–	–	–	–	**1,6,10,22-Tetraazatricyclo[9.7.6.1^{12,16}]pentacosa-12,14,16(25)-triene-18,23-dione, 6,10-diacetyl-15-(acetyloxy)-17-[(4-methoxyphenyl)methyl]-**	69721-69-3	**NI46752**

MASS TO CHARGE RATIOS								M.W.	INTENSITIES								Parent	C	H	O	N	S	F	Cl	Br	I	Si	P	B	X	COMPOUND NAME	CAS Reg No	No
43	185	57	69	55	210	41	95	**634**	100	80	59	53	45	41	35	34	**11.00**	36	58	5	–	2	–	–	–	–	–	–	–	–	Lanostane-7,11-dione, 3,18-bis(acetyloxy)-, cyclic 7-(1,2-ethanediyl mercaptole), (3β,20ξ)-	56298-07-8	NI46753
57	201	43	71	55	41	69	85	**634**	100	99	90	52	52	41	38	31	**0.70**	36	75	–	–	3	–	–	–	–	–	1	–	–	Trilauryl trithiophosphite		IC04615
201	69	71	83	70	97	85	433	**634**	100	93	79	66	62	45	44	42	**14.62**	36	75	–	–	3	–	–	–	–	–	1	–	–	Trilauryl trithiophosphite		IC04616
119	320	145	117	118	467	292	635	**635**	100	80	68	35	19	15	14	13		18	5	6	1	–	16	–	–	–	–	–	–	–	1,1,1,3,3,3-Hexafluoroisopropyl 5,6-dihydroxyindole 2-carboxylate, bis(pentafluoropropionyl)-		NI46754
431	459	119	69	432	460	147	93	**635**	100	65	54	18	13	10	10	9	**0.00**	19	12	6	1	–	15	–	–	–	–	–	–	–	β-Ethoxynoradrenaline tri-PFP		MS10901
190	119	458	140	160	42	69	147	**635**	100	27	26	18	12	11	8	5	**1.20**	19	12	6	1	–	15	–	–	–	–	–	–	–	Metadrenaline tri-PFP		MS10902
190	119	458	69	42	44	45	100	**635**	100	32	16	13	13	12	10	7	**1.00**	19	12	6	1	–	15	–	–	–	–	–	–	–	Metadrenaline tri-PFP		MS10903
73	217	204	143	75	147	218	205	**635**	100	80	73	38	37	33	26	22	**0.00**	24	46	6	1	–	–	–	1	–	4	–	–	–	N-(4-Bromophenyl)-D-glucuronamide, tetrakis(trimethylsilyl)-		NI46755
41	79	67	157	39	53	239	91	**636**	100	94	89	74	70	68	61	58	**0.00**	12	18	–	–	–	–	–	6	–	–	–	–	–	Cyclododecane, hexabromo-		NI46756
79	80	82	91	39	53	81	67	**636**	100	85	78	61	56	55	52	52	**0.00**	12	18	–	–	–	–	–	6	–	–	–	–	–	Cyclododecane, hexabromo-	25637-99-4	NI46757
607	608	609	610	611	291	290	276	**636**	100	54	53	22	12	11	10	6	**2.69**	16	28	12	–	–	–	–	–	–	8	–	–	–	Diethylhexavinyloctasilsesquioxane	79210-40-5	NI46758
587	267	59	414	119	147	636	549	**636**	100	63	53	44	31	21	18	18		18	7	8	–	–	15	–	–	–	–	–	–	–	Benzeneacetic acid, α,3,4-tris(2,2,3,3,3-pentafluoro-1-oxopropoxy)-, methyl ester	25637-99-4	MS10904
27	29	26	139	40	44	137	469	**636**	100	94	82	63	61	55	49	45	**29.41**	19	3	–	–	–	15	–	–	–	–	–	–	1	Stannane, methyltris(pentafluorophenyl)-	1062-71-1	NI46759
169	109	331	193	540	289			**636**	100	70	42	9	1	1			**0.00**	26	36	16	–	1	–	–	–	–	–	–	–	–	5-O-Methylsulphonyl-5-epiloganin tetra-acetate		LI09923
43	169	109	331	116	81	115	111	**636**	100	17	10	8	7	6	5	4	**0.01**	26	36	18	–	–	–	–	–	–	–	–	–	–	1,2,3,2′,3′,4′,6′-β-Maltose heptaacetate		MS10905
496	580	336	552	636	248	218	174	**636**	100	39	15	14	11	9	8	7		27	27	5	–	–	–	–	–	–	–	3	–	2	μ-Tetramethyldiphosphine-tricarbonyliron-triphenylphosphine-dicarbonyliron		MS10906
636	212	635	637	634	211	267	633	**636**	100	68	67	35	33	20	14	13		30	27	3	–	–	–	–	–	–	–	–	3	3	Triferrocenylboroxine		LI09924
91	92	66	77	39	251	130	120	**636**	100	80	47	38	37	30	30	30	**0.00**	30	32	6	6	2	–	–	–	–	–	–	–	–	1,2-Di(2′-carboxyphenylamino)ethane di-tosylhydrazide		LI09925
73	605	606	75	103	636	515	74	**636**	100	99	55	48	44	43	38	32		32	60	5	2	–	–	–	–	–	3	–	–	–	Pregn-4-ene-3,20-dione, 11,17,21-tris[(trimethylsilyl)oxy]-, bis(O-methyloxime), (11β)-	32221-26-4	NI46760
605	484	606	103	73	515	75	295	**636**	100	83	59	56	55	53	53	49	**5.13**	32	60	5	2	–	–	–	–	–	3	–	–	–	Pregn-4-ene-3,20-dione, 11,18,21-tris[(trimethylsilyl)oxy]-, bis(O-methyloxime), (11β)-	69833-64-3	NI46761
73	117	621	429	636	339	147	75	**636**	100	99	18	15	13	13	13	13		33	64	4	–	–	–	–	–	–	4	–	–	–	(3β,20S)-Pregna-5,11-diene-3,11,17,20-tetrakis(trimethylsiloxy)	57325-82-3	NI46762
235	290	496	468	234	178	552	56	**636**	100	89	76	73	66	45	43	41	**0.50**	34	20	6	–	–	–	–	–	–	–	–	–	2	Iron, hexacarbonyl[μ-[(1,2,3,4-η:1,4-η)-1,2,3,4-tetraphenyl-1,3-butadiene-1,4-diyl]]di-, (Fe-Fe)	33310-05-3	NI46763
161	100	107	84	43	44	268	70	**636**	100	37	33	28	26	23	22	19	**0.23**	35	48	7	4	–	–	–	–	–	–	–	–	–	1,5,9,13-Tetraazacycloheptadecan-6-one, 9,13-diacetyl-1-[3-[4-(acetyloxy)phenyl]-1-oxopropyl]-8-(4-methoxyphenyl)-, (S)-	42919-92-6	NI46764
105	344	183	259	345	452	421	343	**636**	100	70	70	59	47	45	38	35	**31.45**	36	36	–	–	–	–	–	–	–	6	–	–	–	Cyclohexasilane, 1,2,3,4,5,6-hexaphenyl-	60221-63-8	NI46765
135	621	636	151	622	311	132	590	**636**	100	91	66	60	56	27	26	23		37	32	10	–	–	–	–	–	–	–	–	–	–	Homocupressuflavone hexamethyl ether		NI46766
636	635	621	198	607	365	593	175	**636**	100	85	28	19	12	9	6	6		38	40	7	2	–	–	–	–	–	–	–	–	–	Pseudoxandrinine		NI46767
409	381	205	636	191	410	395	379	**636**	100	71	70	54	32	30	25	23		39	44	6	2	–	–	–	–	–	–	–	–	–	O-Ethyllauberine		LI09926
135	107	197	144	589	188	108	158	**636**	100	68	56	39	35	29	26	25	**6.00**	40	52	3	4	–	–	–	–	–	–	–	–	–	1,16-Cyclocorynan-17-ol, 2,7-dihydro-7-(20,21-dihydro-19-hydroxy-21,24-secoalstophyllan-21-yl)-, (2α,7α,19E)-	6514-07-4	NI46768
334	121	335	207	636	278	151	107	**636**	100	37	27	25	15	8	8	6		42	68	4	–	–	–	–	–	–	–	–	–	–	Benzenetridecanoic acid, 4-methoxy-, 2-decyl-3-methoxy-5-pentylphenyl ester	71142-40-0	NI46769
121	318	210	197	171	287	179	636	**636**	100	91	47	46	46	45	36	35		44	44	4	–	–	–	–	–	–	–	–	–	–	2′,3′,6′,7′-Tetrahydro-4′-(4-methoxyphenyl)-1′,2,5-tris[(E)-(4-methoxyphenyl)methylene]spiro[cyclopentan-1,5′(4′H)-indene]		MS10907
72	522	238	181	521	410	508	98	**637**	100	88	74	73	72	72	71	60	**46.90**	33	59	7	5	–	–	–	–	–	–	–	–	–	L-Valine, N-[N-[N-[N-[N-(3-hydroxy-1-oxododecyl)glycyl]-L-valyl]-D-Leucyl]-L-alanyl]-, ρ-lactone, (R)-	10409-85-5	NI46770
105	124	43	106	595	77	78	596	**637**	100	51	49	36	30	14	12	11	**1.90**	34	39	11	1	–	–	–	–	–	–	–	–	–	Acetylmaymyrsine	93772-07-7	NI46771
331	638	137	616	252	281	332	319	**638**	100	85	35	25	10	9	8	7		14	–	–	–	4	18	–	–	–	–	–	–	–	1,1,1,4,4,4-Hexafluoro-2,3-bis-[4,5-bis(trifluoromethyl)-1,3-dithiolanylidene]butane		LI09927
609	610	611	555	612	290	613	43	**638**	100	55	49	24	21	14	10	9	**1.49**	16	30	12	–	–	–	–	–	–	8	–	–	–	Triethylpentavinyloctasilsesquioxane		NI46772
326	129	425	327	69	130	43	638	**638**	100	29	24	16	10	8	7	6		22	16	4	2	–	14	–	–	–	–	–	–	–	Bis(heptafluorobutyryl)tryptophan propyl ester		MS10908

MASS TO CHARGE RATIOS								M.W.	INTENSITIES								Parent	C	H	O	N	S	F	Cl	Br	I	Si	P	B	X	COMPOUND NAME	CAS Reg No	No
87	43	567	565	640	638	88	487	638	100	24	10	10	7	7	5	3		24	27	10	6	–	–	–	1	–	–	–	–	–	N,N-bis(2-acetoxyethyl)-2-ethoxy-5-acetamido-4-(2,4-dinitro-6-bromophenylazo)aniline		IC04617
260	129	231	362	276	228	156	101	638	100	82	72	62	62	60	58	57	0.40	28	50	14	2	–	–	–	–	–	–	–	–	–	N-Glycolyl-muramic acid		LI09928
640	642	638	560	562	380	393		638	100	55	50	5	4	3	2			30	40	5	–	–	–	–	2	–	–	–	–	–	16,25-Dibromo-14-oxo-tetramethoxyturriane		LI09929
579	580	638	260	151	501	123	639	638	100	36	30	26	19	16	12	11		32	30	14	–	–	–	–	–	–	–	–	–	–	Secalonic acid		NI46773
562	472	563	296	530	350	382	473	638	100	56	44	36	35	29	27	25	0.00	32	62	5	2	–	–	–	–	–	3	–	–	–	Pregnan-18-al, 20-(methoxyimino)-3,11,21-tris[(trimethylsilyl)oxy]-, O-methyloxime, (3β,5α,11β)-	63503-03-7	NI46774
73	607	75	103	74	147	608	517	638	100	99	99	83	59	58	53	40	29.00	32	62	5	2	–	–	–	–	–	3	–	–	–	Pregnane-3,20-dione, 11,17,21-tris[(trimethylsilyl)oxy]-, bis(O-methyloxime), (5α,11β)-		NI46775
73	607	75	608	103	517	244	147	638	100	99	44	43	43	33	30	30	28.00	32	62	5	2	–	–	–	–	–	3	–	–	–	Pregnane-3,20-dione, 11,17,21-tris[(trimethylsilyl)oxy]-, bis(O-methyloxime), (5α,11β)-		NI46776
169	246	291	307	156	264	331	182	638	100	32	27	22	13	11	9	6	5.00	33	38	11	2	–	–	–	–	–	–	–	–	–	21(R)-(Tetraacetyl-β-D-glucopyranosyl)-hydroxysarpagan-17-al		MS10909
73	331	75	147	169	638	332	74	638	100	44	42	35	20	17	16	11		33	66	4	–	–	–	–	–	–	4	–	–	–	(3α,5β)-Pregn-20-ene, 3,17,20,21-tetrakis[(trimethylsilyl)oxy]-	42599-94-0	NI46777
73	117	75	147	118	74	341	169	638	100	99	27	25	12	12	11	11	6.00	33	66	4	–	–	–	–	–	–	4	–	–	–	(3α,5β,20S)-Pregn-11-ene, 3,11,17,20-tetrakis[(trimethylsilyl)oxy]-	57325-83-4	NI46778
73	251	75	117	341	147	129	143	638	100	99	99	90	75	65	53	43	0.50	33	66	4	–	–	–	–	–	–	4	–	–	–	(3β,11β,20S)-Pregn-5-ene, 3,11,17,20-tetrakis[(trimethylsilyl)oxy]-	32221-72-0	NI46779
73	251	341	117	75	147	143	129	638	100	99	95	86	84	50	46	40	1.50	33	66	4	–	–	–	–	–	–	4	–	–	–	Pregn-5-ene, 3β,11β,17,20β-tetrakis[(trimethylsilyl)oxy]-	32221-73-1	NI46780
535	445	536	446	355	217	265	537	638	100	77	53	30	24	24	22	21	0.00	33	66	4	–	–	–	–	–	–	4	–	–	–	Pregn-5-ene, 3β,16α,20α,21-tetrakis[(trimethylsilyl)oxy]-		LI09930
638	111	195	212	186	610	636	174	638	100	99	64	36	26	22	18	14		34	42	10	2	–	–	–	–	–	–	–	–	–	4-Formyl-3,4-secoreserpine		LI09931
253	368	458	75	369	343	254	226	638	100	54	26	26	24	24	20	16	0.00	34	66	5	–	–	–	–	–	–	3	–	–	–	Cholan-24-oic acid, 3,7,12-tris[(trimethylsilyl)oxy]-, methyl ester, (3α,5β,7α,12α)-	6818-43-5	LI09932
73	253	368	75	458	55	81	369	638	100	99	88	78	68	43	41	37	0.50	34	66	5	–	–	–	–	–	–	3	–	–	–	Cholan-24-oic acid, 3,7,12-tris[(trimethylsilyl)oxy]-, methyl ester, (3α,5β,7α,12α)-	6818-43-5	NI46781
253	73	368	369	75	458	254	55	638	100	86	69	39	31	30	24	24	0.00	34	66	5	–	–	–	–	–	–	3	–	–	–	Cholan-24-oic acid, 3,7,12-tris[(trimethylsilyl)oxy]-, methyl ester, (3α,5β,7α,12α)-	6818-43-5	NI46782
162	147	105	107	159	121	119	129	638	100	61	59	59	56	52	42	40	0.00	34	66	5	–	–	–	–	–	–	3	–	–	–	Methyl 23-hydroxychenodeoxycholate, tris(trimethylsilyl)-		NI46783
284	458	623	368	254	247	255	269	638	100	45	33	33	33	33	27	16	0.00	34	66	5	–	–	–	–	–	–	3	–	–	–	Methyl (22R)-3α,7α,12α-tris[(trimethylsilyl)oxy]-5β-cholan-24-oate	73502-77-9	NI46784
72	44	43	86	214	57	338	73	638	100	65	40	32	28	25	24	21	0.95	35	66	6	4	–	–	–	–	–	–	–	–	–	Leucine, N-[N-[N-(N-stearoyl-L-alanyl)-L-valyl]glycyl]-, methyl ester, L-	4560-73-0	NI46785
609	638	594	319	521	551	550		638	100	60	5	5	3	2	2			38	46	5	4	–	–	–	–	–	–	–	–	–	β-Ethoxymesoporphyrin-IX-dimethyl ester		LI09933
183	57	43	71	55	85	311	41	638	100	84	59	40	34	28	26	26	0.00	39	74	6	–	–	–	–	–	–	–	–	–	–	Dodecanoic acid, 1,2,3-propanetriyl ester	538-24-9	NI46786
439	440	441	117	201	183			638	100	32	4	4	1	1			0.00	39	74	6	–	–	–	–	–	–	–	–	–	–	Dodecanoic acid, 1,2,3-propanetriyl ester	538-24-9	NI46787
183	57	43	71	257	85	55	98	638	100	97	85	69	64	49	49	44	0.00	39	74	6	–	–	–	–	–	–	–	–	–	–	Dodecanoic acid, 1,2,3-propanetriyl ester	538-24-9	NI46788
390	391	124	514	392	373	167	123	638	100	44	12	11	11	10	10	10	0.00	42	42	4	2	–	–	–	–	–	–	–	–	–	N,N′-Dicyclononyl-3,4:9,10-perylenebis(dicarboximide)	110590-75-5	DD01812
311	221	220	91	312	92	222	193	638	100	48	22	21	19	8	6	5	1.00	44	38	1	4	–	–	–	–	–	–	–	–	–	[2-Hydroxy-3-(p-tolyl)-1,2-dihydroquinoxalin-1-yl]phenyl[1-benzyl-3-(p-tolyl)-1,2-dihydroquinoxalin-2-yl]methane	91736-73-1	NI46789
638	183	538	227	184	181	199	105	638	100	53	49	34	33	30	27	26		45	38	2	–	–	–	–	–	–	1	–	–	–	7-Silabicyclo[2.2.1]hept-5-ene-2-carboxylic acid, 1,4,5,6,7,7-hexaphenyl-, ethyl ester	56805-05-1	MS10910
69	81	107	135	160	203	189	638	638	100	77	62	34	26	8	8	5		46	70	1	–	–	–	–	–	–	–	–	–	–	2-Octaprenyl-phenol		LI09934
57	110	41	470	414	413	471	415	639	100	42	36	28	27	25	24	21	0.00	28	46	–	–	–	–	–	–	–	–	2	–	1	Platinum, bis[bis(tert-butyl)phenylphosphine]-	59765-06-9	NI46790
130	314	170	131	154	151	146	258	639	100	51	35	33	29	29	25	24	1.58	35	53	6	5	–	–	–	–	–	–	–	–	–	DL-Leucine, N-[N-[1-[N-(1-oxodecyl)glycyl]-DL-prolyl]-DL-tryptophyl]-, methyl ester	55822-84-9	NI46791
288	417	445	128	161	127	636	65	640	100	93	78	45	36	31	17	16	0.00	10	10	2	2	–	–	–	–	2	–	–	–	2	Molybdenum, bis(η^5-2,4-cyclopentadien-1-yl)di-μ-iododinitrosyldi-, (Mo-Mo)	55836-28-7	NI46792
605	613	615	606	614	607	616	617	640	100	89	68	56	44	42	24	18	4.19	14	21	12	–	–	–	1	–	–	8	–	–	–	Heptavinylchlorooctasilsesquioxane		NI46793
611	612	613	614	292	615	293	582	640	100	51	37	15	14	7	6	6	1.29	16	32	12	–	–	–	–	–	–	8	–	–	–	Tetraethyltetravinyloctasilsesquioxane	79210-42-7	NI46794
100	75	73	93	57	640	382	147	640	100	96	86	72	63	50	41	37		25	52	4	6	–	–	–	–	–	5	–	–	–	Pentakis(trimethylsilyl)-8,2′-aminoanhydroguanosine		NI46795
73	640	641	74	45	642	75	147	640	100	23	13	10	10	8	6	4		30	40	8	–	–	–	–	–	–	4	–	–	–	Benzo[1,2-b:4,5-b′]bisbenzofuran-6,12-dione, 2,3,8,9-tetrakis[(trimethylsilyl)oxy]-	51860-96-9	NI46796
439	43	95	55	57	350	151	83	640	100	90	57	50	45	40	40	40	0.10	30	40	15	–	–	–	–	–	–	–	–	–	–	Detigloyldihydroazadirachtin		NI46797
155	172	170	284	323	253	224	454	640	100	94	82	69	66	46	45	44	0.00	33	60	8	4	–	–	–	–	–	–	–	–	–	L-Leucine, N-[N-[N-[N-(1-oxodecyl)-L-α-aspartyl]-L-valyl]-L-alanyl]-, 1-butyl 4-methyl ester	56272-49-2	NI46798
357	267	73	129	537	147	358	75	640	100	97	84	53	38	32	30	27	0.25	33	68	4	–	–	–	–	–	–	4	–	–	–	(3α,5α,11β,20R)-Pregnane, 3,11,20,21-tetrakis[(trimethylsilyl)oxy]-	57363-12-9	NI46799
73	253	343	75	117	523	298	433	640	100	99	95	74	51	50	50	42	9.00	33	68	4	–	–	–	–	–	–	4	–	–	–	(3α,5β,11β,20R)-Pregnane, 3,11,17,20-tetrakis[(trimethylsilyl)oxy]-	57397-17-8	NI46800

MASS TO CHARGE RATIOS								M.W.	INTENSITIES								Parent	C	H	O	N	S	F	Cl	Br	I	Si	P	B	X	COMPOUND NAME	CAS Reg No	No
357	267	73	129	447	358	147	268	**640**	100	96	76	47	32	32	27	23	**0.06**	33	68	4	–	–	–	–	–	–	4	–	–	–	(3α,5β,11β,20R)-Pregnane, 3,11,20,21-tetrakis[(trimethylsilyl)oxy]-	57326-08-6	NI46801
357	267	73	129	358	447	147	537	**640**	100	90	79	48	32	29	27	26	**0.06**	33	68	4	–	–	–	–	–	–	4	–	–	–	(3α,5β,11β,20S)-Pregnane, 3,11,20,21-tetrakis[(trimethylsilyl)oxy]-	57326-09-7	NI46802
73	357	129	75	147	267	103	143	**640**	100	99	99	99	85	68	34	32	**0.00**	33	68	4	–	–	–	–	–	–	4	–	–	–	(3β,5α,11β,20R)-Pregnane, 3,11,20,21-tetrakis[(trimethylsilyl)oxy]-	57325-85-6	NI46803
73	147	75	255	435	243	191	107	**640**	100	99	99	71	64	55	45	39	**2.50**	33	68	4	–	–	–	–	–	–	4	–	–	–	(3β,5α,20R)-Pregnane, 3,17,20,21-tetrakis[(trimethylsilyl)oxy]-	57325-84-5	NI46804
91	105	160	47	70	43	523	77	**640**	100	57	37	29	27	25	22	20	**0.00**	34	44	8	2	1	–	–	–	–	–	–	–	–	2-O-Acetyl-3-O-benzoyl-6-(N-carbobenzoxyprolineamido)-1-N-hexylthio-α-DL-lyxo-hexopyranoside		NI46805
91	105	160	70	43	523	77	41	**640**	100	78	41	32	24	20	20	19	**0.00**	34	44	8	2	1	–	–	–	–	–	–	–	–	3-O-Acetyl-2-O-benzoyl-6-(N-carbobenzoxyprolineamido)-1-N-hexylthio-α-DL-lyxo-hexopyranoside		NI46806
368	167	96	329	40	70	94	353	**640**	100	67	64	61	48	47	46	45	**27.00**	34	55	1	–	–	7	–	–	–	1	–	–	–	Cholesterol 3,3,4,4,5,5,5-heptafluoropentyldimethylsilyl ester		NI46807
60	45	44	43	42	41	454	455	**640**	100	99	99	99	99	86	46	21	**1.59**	38	24	10	–	–	–	–	–	–	–	–	–	–	6,11,12,17-Tetraacetoxy-5,18-trinaphthylenedione	96722-24-6	NI46808
371	57	43	55	73	71	41	69	**640**	100	84	77	52	48	46	40	36	**0.00**	38	76	5	–	–	–	–	–	–	1	–	–	–	Hexadecanoic acid, 2-[(trimethylsilyl)oxy]-1,3-propanediyl ester	53212-95-6	NI46809
107	641	640	642	223	320	321	239	**641**	100	60	50	45	37	21	20	20		10	2	2	–	2	12	–	–	–	–	–	–	1	Platinum, bis(1,1,1,5,5,5-hexafluoro-4-thioxo-2-pentanonato-O,S)-	55450-75-4	NI46810
73	217	131	378	159	144	147	103	**641**	100	58	38	35	20	20	18	17	**0.00**	25	59	8	1	–	–	–	–	–	5	–	–	–	D-Glycero-β-D-galacto-2-nonulopyranosidonic acid, methyl 5-amino-3,5-dideoxy-4,7,8,9-tetrakis-O-(trimethylsilyl)-, trimethylsilyl ester	56247-31-5	NI46811
613	614	615	616	293	279	292	617	**642**	100	76	49	26	19	17	14	11	**1.09**	16	34	12	–	–	–	–	–	–	8	–	–	–	Pentaethyltrivinyloctasilsesquioxane	79210-43-8	NI46812
73	333	147	292	334	305	335	277	**642**	100	84	32	29	25	15	13	11	**0.00**	24	58	8	–	–	–	–	–	–	6	–	–	–	Galactaric acid, 2,3,4,5-tetrakis-O-(trimethylsilyl)-, bis(trimethylsilyl) ester	56272-61-8	NI46813
73	333	147	292	334	74	305	75	**642**	100	44	25	16	13	9	7	7	**0.00**	24	58	8	–	–	–	–	–	–	6	–	–	–	Glucaric acid, 2,3,4,5-tetrakis-O-(trimethylsilyl)-, bis(trimethylsilyl) ester	38165-96-7	NI46814
333	73	147	334	292	335	305	277	**642**	100	92	31	29	26	15	14	10	**0.00**	24	58	8	–	–	–	–	–	–	6	–	–	–	Glucaric acid, 2,3,4,5-tetrakis-O-(trimethylsilyl)-, bis(trimethylsilyl) ester	38165-96-7	NI46815
73	333	147	334	292	335	305	423	**642**	100	60	30	20	20	10	9	7	**0.00**	24	58	8	–	–	–	–	–	–	6	–	–	–	Bis(trimethylsilyl) 2,3,4,5-tetrakis(trimethylsilylol)adipate		LI09935
57	71	43	259	85	83	69	63	**642**	100	95	93	78	68	62	55	55	**0.00**	30	62	2	–	–	–	4	–	–	–	–	–	1	1,2-Bis(tetradecyloxy)ethane titanium tetrachloride complex		IC04618
189	147	43	305	146	190	148	477	**642**	100	86	57	41	14	13	10	7	**0.03**	32	34	14	–	–	–	–	–	–	–	–	–	–	Trichocarposide pentaacetate		MS10911
43	45	60	42	44	412	454	86	**642**	100	57	52	52	33	18	14	14	**0.00**	32	38	12	2	–	–	–	–	–	–	–	–	–	Acetamide, N-[2-[3,4-dihydroxy-α-[(N-methylacetamido)methyl]benzyl]-β,4,5-trihydroxyphenethyl]-N-methyl-, pentaacetate (ester)	28177-12-0	NI46816
73	75	147	191	129	74	173	217	**642**	100	25	19	17	12	9	8	7	**0.30**	32	66	5	–	–	–	–	–	–	4	–	–	–	Prosta-5,13-dien-1-oic acid, 9,11,15-tris[(trimethylsilyl)oxy]-, trimethylsilyl ester, (5Z,9α,11α,13E,15S)-	50669-95-9	NI46818
191	173	217	75	73	391	481	366	**642**	100	49	44	42	38	34	33	31	**2.00**	32	66	5	–	–	–	–	–	–	4	–	–	–	Prosta-5,13-dien-1-oic acid, 9,11,15-tris[(trimethylsilyl)oxy]-, trimethylsilyl ester, (5Z,9α,11α,13E,15S)-	50669-95-9	NI46817
73	75	191	147	129	217	74	173	**642**	100	24	16	16	16	9	9	7	**0.40**	32	66	5	–	–	–	–	–	–	4	–	–	–	Prosta-5,13-dien-1-oic acid, 9,11,15-tris[(trimethylsilyl)oxy]-, trimethylsilyl ester, (5Z,9β,11α,13E,15S)-	50669-96-0	NI46819
191	217	129	199	173	365	313	353	**642**	100	59	56	47	47	41	41	39	**0.80**	32	66	5	–	–	–	–	–	–	4	–	–	–	Prosta-5,13-dien-1-oic acid, 9,11,15-tris[(trimethylsilyl)oxy]-, trimethylsilyl ester, (5Z,9β,11α,13E,15S)-	50669-96-0	NI46820
642	606	608	644	569	533	535	571	**642**	100	46	40	35	10	10	8	5		36	39	5	4	–	–	1	–	–	–	–	–	–	Chloroethylporphyrin		LI09936
624	608	551	593	642				**642**	100	25	20	15	5					36	42	7	4	–	–	–	–	–	–	–	–	–	β-Hydroxy-2,4-di-(2-hydroxyethyl)-6,7-di-(2-methoxycarbonylethyl)-1,3,5,8-tetramethylporphin		LI09937
139	111	197	152	141	44	308	75	**642**	100	55	43	35	33	33	31	22	**6.94**	39	24	5	–	–	–	2	–	–	–	–	–	–	Bis[4-[4-(4-chloro-benzoyl)phenyl]phenyl] carbonate		IC04619
468	469	470	427	453	428	383	627	**642**	100	37	10	10	4	4	3	2	**0.00**	39	70	3	–	–	–	–	–	–	2	–	–	–	6α-Hydroxy-Δ^1-tetrahydrocannabinol bis(tripropylsilyl) ether		NI46821
455	456	642	457	643	627	458	628	**642**	100	40	12	12	6	4	4	3		39	70	3	–	–	–	–	–	–	2	–	–	–	7-Hydroxy-Δ^1-tetrahydrocannabinol bis(tripropylsilyl) ether		NI46822
146	105	277	79	77	147	350	103	**642**	100	36	35	18	18	13	10	10	**4.05**	40	48	–	4	–	–	–	–	–	–	–	–	1	Nickel, bis[N,N′-(1,2-dimethyl-1,2-ethanediylidene)bis[2,6-dimethylaniline]-N,N′]-	74779-91-2	NI46823
138	642	504	305	149	643	505	137	**642**	100	33	33	33	25	15	13	13		41	46	3	4	–	–	–	–	–	–	–	–	–	Vobtusine, anhydrode(methoxycarbonyl)-	26249-91-2	NI46824
390	111	642	112	129	391	123	368	**642**	100	95	89	78	77	63	55	41		42	46	4	2	–	–	–	–	–	–	–	–	–	N,N′-Bis(1-butylpentyl)-3,4:9,10-perylenebis(dicarboximide)	110613-98-4	MS10912
103	104	642	643	229	230	144	322	**642**	100	77	50	26	20	13	12	9		44	30	–	6	–	–	–	–	–	–	–	–	–	4,4′-Bis(2,4-diphenyltriazin-6-yl)stilbene		IC04620
424	219	409	57	425	161	203	41	**642**	100	76	66	42	32	24	23	19	**0.01**	44	66	3	–	–	–	–	–	–	–	–	–	–	2,6-Di-tert-butyl-4,4-bis(3,5-di-tert-butyl-4-hydroxy-benzyl)cyclohexa-2,5-dienone		IC04621
70	303	360	71	69	430	169	528	**643**	100	69	54	32	18	13	11	10	**0.10**	22	19	5	1	–	14	–	–	–	–	–	–	–	3-Methylbutyl N,O-bis(heptafluorobutyryl)tyrosinate		NI46825
73	324	147	103	280	245	643	74	**643**	100	18	15	13	12	12	10	10		25	53	5	5	–	–	–	–	–	5	–	–	–	9H-Purin-2-amine, N-(trimethylsilyl)-6-[(trimethylsilyl)oxy]-9-[2,3,5-tris-O-(trimethylsilyl)-β-D-ribofuranosyl]-	53294-38-5	NI46826

MASS TO CHARGE RATIOS								M.W.	INTENSITIES								Parent	C	H	O	N	S	F	Cl	Br	I	Si	P	B	X	COMPOUND NAME	CAS Reg No	No
324	245	643	73	103	217	644	230	**643**	100	80	76	73	47	42	40	38		25	53	5	5	–	–	–	–	–	5	–	–	–	9H-Purin-2-amine, N-(trimethylsilyl)-6-[(trimethylsilyl)oxy]-9-[2,3,5-tris-O-(trimethylsilyl)-β-D-ribofuranosyl]-	53294-38-5	NI46827
615	616	617	618	293	279	619	294	**644**	100	51	41	17	14	12	7	7	**0.89**	16	36	12	–	–	–	–	–	–	8	–	–	–	Hexaethyldivinyloctasilsesquioxane	79210-44-9	NI46828
197	453	198	256	289	410	384	356	**644**	100	39	31	15	8	3	2	2	**0.00**	21	24	–	3	3	9	–	–	–	–	–	–	1	Tris(1,1,1-trifluoro-4-mercapto-5-methylhex-3-en-2-onato)cobalt(III)		NI46829
324	322	326	164	647	649	645	243	**644**	100	74	70	56	24	20	18	14	**0.00**	24	12	–	2	–	–	–	4	–	–	–	–	–	9,9′-Bis(3,6-dibromocarbazole)	18628-03-0	NI46830
43	97	41	29	28	488	98	27	**644**	100	34	11	11	11	8	8	7	**0.30**	32	36	14	–	–	–	–	–	–	–	–	–	–	Rubratoxin B triacetate		MS10913
108	79	107	109	127	91	80	77	**644**	100	67	37	9	9	7	7	7	**0.23**	32	44	10	4	–	–	–	–	–	–	–	–	–	Bis(N-benzyloxycarbonylglycyl)diaza-18-crown-6	93788-19-3	NI46831
277	177	248	264	210	192	164	170	**644**	100	92	90	89	89	89	86	82	**22.00**	32	52	6	8	–	–	–	–	–	–	–	–	–	N-Decanoylhistidyl alanylhistidyl leucine methyl ester		LI09938
73	483	75	147	191	129	554	367	**644**	100	32	31	30	29	28	26	22	**0.20**	32	68	5	–	–	–	–	–	–	4	–	–	–	Prost-13-en-1-oic acid, 9,11,15-tris[(trimethylsilyl)oxy]-, trimethylsilyl ester, (9α,11α,13E,15S)-	50669-94-8	NI46832
368	191	367	483	173	554	129	217	**644**	100	80	57	53	51	44	39	38	**1.00**	32	68	5	–	–	–	–	–	–	4	–	–	–	Prost-13-en-1-oic acid, 9,11,15-tris[(trimethylsilyl)oxy]-, trimethylsilyl ester, (9α,11α,13E,15S)-	50669-94-8	NI46833
73	75	147	129	191	55	43	117	**644**	100	36	20	16	15	13	13	10	**0.10**	32	68	5	–	–	–	–	–	–	4	–	–	–	Prost-13-en-1-oic acid, 9,11,15-tris[(trimethylsilyl)oxy]-, trimethylsilyl ester, (9β,11α,13E,15S)-	55556-77-9	NI46834
644	571	616	613	629	483	585	512	**644**	100	40	15	14	10	10	9	9		38	40	4	6	–	–	–	–	–	–	–	–	–	Meso-cyano-4(2)-(2-cyanoethyl)-4(2)-desvinyl-3,4-(1,2)-dihydroprotoporphyrin IX dimethyl ester		LI09939
584	203	509	355	524	295	402	471	**644**	100	82	64	64	55	55	46	36	**3.00**	38	60	8	–	–	–	–	–	–	–	–	–	–	24S-Cycloartane-3,16,24,25-tetraacetate		NI46835
323	135	107	644	197	265	111	149	**644**	100	48	21	16	11	10	10	9		41	48	3	4	–	–	–	–	–	–	–	–	–	1,16-Cyclocorynan-17-oic acid, 2-alstophyllan-18-yl-19,20-didehydro-2,7-dihydro-, methyl ester, (2α,7α,16S,19E)-	66408-47-7	NI46836
73	103	217	91	147	307	57	75	**645**	100	55	35	30	27	26	24	21	**0.00**	28	59	6	1	–	–	–	–	–	5	–	–	–	Fructose benzyloxime penta-TMS		MS10914
73	91	147	205	75	217	207	77	**645**	100	75	23	19	19	16	14	10	**0.00**	28	59	6	1	–	–	–	–	–	5	–	–	–	Galactose benzyloxime penta-TMS		MS10916
73	205	319	147	217	91	103	320	**645**	100	62	58	36	30	30	20	17	**0.00**	28	59	6	1	–	–	–	–	–	5	–	–	–	Galactose benzyloxime penta-TMS		MS10915
73	147	75	217	205	204	201	77	**645**	100	38	38	34	31	28	26	18	**0.00**	28	59	6	1	–	–	–	–	–	5	–	–	–	Glucose benzyloxime penta-TMS		MS10917
73	91	147	75	205	217	103	117	**645**	100	88	28	21	20	18	16	11	**0.00**	28	59	6	1	–	–	–	–	–	5	–	–	–	Glucose benzyloxime penta-TMS		MS10918
73	91	147	103	75	129	217	205	**645**	100	73	22	21	18	17	16	11	**0.00**	28	59	6	1	–	–	–	–	–	5	–	–	–	Glucose benzyloxime penta-TMS		MS10919
100	69	81	153	85	95	183	143	**646**	100	95	60	30	30	16	12	9	**0.00**	16	9	10	–	–	15	–	–	–	–	–	–	–	D-arabino-Hexitol, 2-deoxy-, pentakis(trifluoroacetate)	26388-40-9	NI46837
69	100	81	113	95	305	191	153	**646**	100	80	68	17	17	15	15	13	**0.30**	16	9	10	–	–	15	–	–	–	–	–	–	–	L-Mannitol, 1-deoxy-, pentakis(trifluoroacetate)	26293-33-4	NI46838
617	618	619	620	294	621	280	264	**646**	100	48	41	16	11	8	7	5	**0.59**	16	38	12	–	–	–	–	–	–	8	–	–	–	Heptaethylvinyloctasilsesquioxane	79235-36-2	NI46839
133	121	43	559	334	41	81	108	**646**	100	39	39	30	29	29	24	23	**12.50**	22	24	4	2	–	14	–	–	–	–	–	–	–	Bis(heptafluorobutyryl)octahydrotryptophan propyl ester		MS10920
564	41	566	562	648	606	565	438	**646**	100	83	52	51	31	30	29	22	**14.71**	28	24	8	–	–	–	–	2	–	–	–	–	–	4,4′-Diacetoxy-3,3′-dibromo-5,5′,8,8′-tetramethoxy-1,1′-binaphthalene	101458-98-4	NI46840
28	183	113	199	325	446	243	506	**646**	100	11	11	6	5	3	3	2	**0.00**	34	46	12	–	–	–	–	–	–	–	–	–	–	Seneosterone-2,22-diacetate cyclic carbonate		LI09940
105	122	149	481	140	269	241	281	**646**	100	91	88	54	51	49	45	40	**17.00**	37	42	10	–	–	–	–	–	–	–	–	–	–	Gniditrin	55306-10-0	NI46841
323	135	324	646	107	149	111	109	**646**	100	51	24	19	19	17	15	11		41	50	3	4	–	–	–	–	–	–	–	–	–	1,16-Cyclocorynan-17-oic acid, 19,20-didehydro-2-(20,21-dihydroalstophyllan-18-yl)-2,7-dihydro-, methyl ester, (2α,7α,16S,19E)-	66408-48-8	NI46842
450	197	198	253	203	353	407	286	**647**	100	77	71	20	14	12	9	9	**0.00**	21	24	3	–	3	9	–	–	–	–	–	–	1	Tris(1,1,1-trifluoro-4-mercapto-5-methylhex-3-en-2-onato)iron(III)	56260-98-1	NI46843
616	135	617	647	632	648	646	633	**647**	100	40	36	5	4	2	2	2		34	49	11	1	–	–	–	–	–	–	–	–	–	Aljesaconitine A		MS10921
432	381	216	200	613	203	397	235	**648**	100	99	99	99	40	40	30	30	**0.00**	–	–	6	–	–	9	–	–	3	–	–	–	–	Tris[iodine trifluoride dioxide]	58735-48-1	NI46844
613	615	614	616	623	621	622	624	**648**	100	73	54	30	21	20	10	9	**0.59**	12	18	12	–	–	–	2	–	–	8	–	–	–	Hexavinyldichlorooctasilsesquioxane		NI46845
441	391	579	184	648	203	372	253	**648**	100	12	5	5	4	4	1	1		15	3	6	–	–	18	–	–	–	–	–	–	1	Tris(1,1,1,5,5,5-hexafluoro-2,4-pentanedionato)aluminium		LI09941
619	620	295	621	622	589	296	591	**648**	100	61	36	31	18	15	13	12	**0.69**	16	40	12	–	–	–	–	–	–	8	–	–	–	Perethyloctasilsesquiohane		NI46846
119	59	442	69	295	43	445	92	**648**	100	65	38	38	32	25	14	14	**7.20**	19	7	8	–	–	15	–	–	–	–	–	–	–	Methyl O,O′,O″-tris(pentafluoropropionoyl)catecholpyruvate		MS10922
73	183	318	147	199	319	191	74	**648**	100	49	48	37	24	16	14	10	**0.00**	23	57	9	–	–	–	–	–	–	5	1	–	–	L-chiro-Inositol, 1,2,3,5,6-pentakis-O-(trimethylsilyl)-, dimethyl phosphate	55334-86-6	NI46847
73	147	271	183	217	199	191	74	**648**	100	35	33	25	21	12	12	9	**0.20**	23	57	9	–	–	–	–	–	–	5	1	–	–	myo-Inositol, 1,3,4,5,6-pentakis-O-(trimethylsilyl)-, dimethyl phosphate	55569-48-7	NI46848

MASS TO CHARGE RATIOS	M.W.	INTENSITIES	Parent	C	H	O	N	S	F	Cl	Br	I	Si	P	B	X	COMPOUND NAME	CAS Reg No	No
73 217 204 75 147 185 253 45	**648**	100 53 37 36 22 18 13 12	**0.00**	27	52	10	2	–	–	–	–	–	3	–	–	–	α-D-Glucopyranosiduronic acid, 3-(5-ethylhexahydro-2,4,6-trioxo-5-pyrimidinyl)-1,1-dimethylpropyl 2,3,4-tris-O-(trimethylsilyl)-, methyl ester	55556-81-5	NI46849
73 253 75 217 147 185 317 45	**648**	100 40 37 23 23 20 17 13	**0.00**	27	52	10	2	–	–	–	–	–	3	–	–	–	β-D-Glucopyranosiduronic acid, 3-(5-ethylhexahydro-2,4,6-trioxo-5-pyrimidinyl)-1,1-dimethylpropyl 2,3,4-tris-O-(trimethylsilyl)-, methyl ester	55556-80-4	NI46850
334 447 429 385	**648**	100 70 57 19	**0.00**	35	52	11	–	–	–	–	–	–	–	–	–	–	Crustecdysone tetraacetate		LI09942
324 183 247 107 184 388 325 43	**648**	100 83 46 39 31 24 24 21	**0.21**	36	28	–	–	2	–	–	–	–	–	4	–	–	Tetrahydrotetraphenyldibenzo-1,2,5,6-tetraphosphocin disulphide		MS10923
445 209 159 193 41 71 43 105	**648**	100 99 99 95 84 81 81 67	**49.54**	36	40	11	–	–	–	–	–	–	–	–	–	–	Pentherin II	52748-45-5	NI46851
253 105 313 107 131 145 157 109	**648**	100 54 38 38 36 33 31 30	**0.00**	36	56	10	–	–	–	–	–	–	–	–	–	–	Cholestan-26-oic acid, 3,7,12,24-tetrakis(acetyloxy)-, methyl ester, (3α,5β,7α,12α)-	60354-51-0	NI46852
73 147 75 43 57 69 55 235	**648**	100 53 32 30 22 20 20 18	**11.00**	36	68	4	–	–	–	–	–	–	3	–	–	–	2β,3β,14α-Tris(trimethylsilyloxy)-5β-cholest-7-en-6-one		MS10924
105 151 281 241 269 648 253 271	**648**	100 96 19 18 16 15 14 12		37	44	10	–	–	–	–	–	–	–	–	–	–	Yuanhuacine	60195-70-2	NI46853
57 43 73 71 55 69 97 83	**648**	100 80 66 62 52 47 44 44	**8.00**	44	88	2	–	–	–	–	–	–	–	–	–	–	Docosyl docosanoate		MS10925
69 225 648 81 649 41 650 68	**648**	100 95 81 51 43 28 22 22		46	64	2	–	–	–	–	–	–	–	–	–	–	1,4-Naphthalenedione, 2-(3,7,11,15,19,23,27-heptamethyl-2,6,10,14,18,22,26-octacosaheptaenyl)-3-methyl-	13425-62-2	NI46854
486 119 455 415 69 487 387 77	**649**	100 63 36 27 21 16 12 11	**1.20**	19	10	7	1	–	15	–	–	–	–	–	–	–	N,O,O′-Tris(pentafluoropropionoyl)-L-DOPA methyl ester		MS10926
431 459 119 69 190 432 460 42	**649**	100 72 48 19 18 14 12 9	**0.00**	20	14	6	1	–	15	–	–	–	–	–	–	–	β-Ethoxyadrenaline tri-PFP		MS10927
73 102 548 147 75 549 117 74	**649**	100 99 70 43 42 33 31 27	**0.00**	26	59	6	3	–	–	–	–	–	5	–	–	–	Pentakis(trimethylsilyl) isotabtoxinine		NI46855
73 82 75 147 102 244 359 117	**649**	100 99 58 39 36 29 28 28	**0.00**	26	59	6	3	–	–	–	–	–	5	–	–	–	L-Threonine, N-[2-amino-4-(3-hydroxy-2-oxo-3-azetidinyl)-1-oxobutyl]-, pentakis(trimethylsilyl)-	56247-34-8	NI46856
170 171 168 169 203 172 202 174	**649**	100 71 29 25 15 9 8 7	**1.76**	33	35	7	3	2	–	–	–	–	–	–	–	–	1-Naphthalenesulphonic acid, 5-(dimethylamino)-, 4-[2-[[[5-(dimethylamino)-1-naphthalenyl]sulphonyl]amino]-1-hydroxyethyl]-2-methoxyphenyl ester	36600-91-6	NI46857
649 490 590 576 634 564 502 562	**649**	100 27 25 17 16 10 10 8		37	39	6	5	–	–	–	–	–	–	–	–	–	3-Cyano-3-desformyl rhodin-G7, trimethyl ester		LI09943
618 603 588 619 574 287 634 649	**649**	100 80 67 51 39 36 29 24		39	53	1	4	–	–	–	–	–	–	–	–	1	(α,γ-Dimethyl-α,γ-dihydrooctaethylporphinato)methoxyiron(III)		NI46858
119 149 486 455 69 59 57 77	**650**	100 61 46 41 37 32 22 19	**0.00**	19	9	8	–	–	15	–	–	–	–	–	–	–	Methyl O,O′,O″-tris(pentafluoropropionoyl)catechollactate		MS10928
182 146 142 98 110 128 170 187	**650**	100 51 51 51 45 44 42 41	**0.00**	27	38	18	–	–	–	–	–	–	–	–	–	–	α-D-Glucopyranose, 3-O-methyl-, 1,2,4-triacetate, anhydride with α-D-glucopyranose 1,2,4,6-tetraacetate	55887-84-8	NI46859
510 482 328 622 272 174 241 56	**650**	100 84 75 42 30 26 21 12	**6.00**	30	20	6	–	–	–	–	–	–	–	2	–	2	Di-μ-diphenyl-phosphino-bis(tricarbonyliron)		MS10929
538 281 458 510 354 269 255 539	**650**	100 48 38 22 15 14 13 11	**6.00**	30	24	4	–	–	–	–	–	–	–	–	–	3	cis-[1,2-Bis(diphenylarsino)ethane]chromium tetracarbonyl		LI09944
57 506 41 538 474 442 77 594	**650**	100 55 47 19 19 8 7 6	**0.00**	30	34	8	–	4	–	–	–	–	–	–	–	–	Tetramethyl 2,2′-(1,4-phenylene)bis[5-(tert-butylthio)-3,4-thiophenedicarboxylate]	110699-36-0	DD01813
73 103 75 74 439 147 529 619	**650**	100 99 95 64 60 57 44 43	**25.00**	32	58	6	2	–	–	–	–	–	3	–	–	–	Pregn-4-ene-3,11,20-trione, 6,17,21-tris[(trimethylsilyl)oxy]-, 3,20-bis(O-methyloxime), (6β)-	57326-06-4	NI46860
136 145 54 122 123 159 93 176	**650**	100 88 70 65 59 50 49 44	**0.00**	33	46	13	–	–	–	–	–	–	–	–	–	–	β-D-Glucopyranoside, [5-[2-(2,5-dihydro-2-oxo-3-furanyl)ethyl]decahydro 1,4a-dimethyl-6-oxo-1-naphthalenyl]methyl, tetraacetate, [1R-(1α,4aα,5α,8aβ)]-	72101-43-0	NI46861
43 169 331 109 303 135 81 55	**650**	100 93 41 40 38 20 19 14	**4.81**	33	46	13	–	–	–	–	–	–	–	–	–	–	Neoandrographolide norketone acetate		NI46862
28 496 234 468 290 552 56 178	**650**	100 79 71 68 58 52 46 30	**0.00**	35	22	6	–	–	–	–	–	–	–	–	–	2	Iron, hexacarbonyl[μ-[(1,2,3,4-η:1,5-η)-1,2,3,4-tetraphenyl-1,3-pentadiene 1,5-diyl]]di-, (Fe-Fe)	74421-53-7	NI46863
591 650 489	**650**	100 37 17		37	38	7	4	–	–	–	–	–	–	–	–	–	10(R)-Methoxy-methyl phaophorbide		LI09945
650 206 649 198 635 190 175 325	**650**	100 98 89 67 61 27 27 26		39	42	7	2	–	–	–	–	–	–	–	–	–	O,O-Dimethylpseudoxandrine		NI46864
650 651 325 352 652 616 633 36	**650**	100 53 15 13 12 11 10 9		42	22	6	2	–	–	–	–	–	–	–	–	–	1,1′,5,1″-Trianthrimide		IC04622
121 239 207 240 317 278 131 105	**650**	100 45 42 21 17 17 17 17	**12.00**	44	58	4	–	–	–	–	–	–	–	–	–	–	2,4,6,8,10,12,14-Pentadecaheptaenoic acid, 15-(4-methoxyphenyl)-, 2-decyl-3-methoxy-5-pentylphenyl ester	71246-59-8	NI46865
649 281 367 263	**650**	100 44 8 6	**0.00**	45	78	2	–	–	–	–	–	–	–	–	–	–	Cholest-5-en-3-ol, (3β)-, 9-octadecenoate, (Z)-	303-43-5	NI46866
655 653 657 656 532 530 659 651	**651**	100 68 66 25 20 20 16 16		22	9	3	1	–	–	–	4	–	–	–	–	–	Tetrabromo-2-(3-hydroxy-1,2-dihydro-quinol-2-ylidene)-2,3-dihydro-1H-benz[f]indene		IC04623

MASS TO CHARGE RATIOS	M.W.	INTENSITIES	Parent	C	H	O	N	S	F	Cl	Br	I	Si	P	B	X	COMPOUND NAME	CAS Reg No	No
460 99 130 112 594 129 366 472	651	100 74 43 38 37 35 32 31	13.57	35	65	6	5	–	–	–	–	–	–	–	–	–	Glycine, N-[N^5-[imino[(1-oxodecyl)amino]methyl]-N^2-(1-oxodecyl)-N-(1-oxohexyl)-L-ornithyl]-, methyl ester	55556-78-0	NI46867
651 652 586 52 587 117 653 534	651	100 61 54 35 31 23 22 14		47	35	–	–	–	–	–	–	–	–	–	–	1	Chromium, (η^5-2,4-cyclopentadien-1-yl)[(1',2',3',4',5',6'-η)-3',4',5',6'-tetraphenyl-1,1':2',1''-terphenyl]-	74811-14-6	NI46868
347 455 506 483 652 478 624 497	652	100 63 22 15 5 5 2 2		14	15	1	–	–	7	–	–	1	–	–	–	1	Iridium, carbonyl(heptafluoropropyl)iodo[(1,2,3,4,5-η)-1,2,3,4,5-pentamethyl-2,4-cyclopentadien-1-yl]-	56784-36-2	MS10930
73 147 281 267 415 207 461 327	652	100 64 47 40 37 21 21 19	0.00	16	48	10	–	–	–	–	–	–	9	–	–	–	Bis(pentamethylcyclotrisiloxy)hexamethyltrisiloxane	18142-95-5	NI46869
607 564 238 652 195 608 69 565	652	100 97 85 65 48 43 41 35		24	21	–	6	–	12	–	–	–	–	1	–	–	Tris(β-dimethylhydrazino)tetrafluorophenyl phosphine		MS10931
652 195 186 200 426 174 226 212	652	100 65 65 50 45 36 35 17		35	44	10	2	–	–	–	–	–	–	–	–	–	4-Acetyl-3,4-secoreserpine		LI09946
610 609 652 594 268 537 551 521	652	100 37 36 6 5 4 3 2		38	44	6	4	–	–	–	–	–	–	–	–	–	α-Acetoxy-2,8-diethyl-4,5-bis(2-methoxycarbonylethyl)-1,3,6,7-tetramethylporphin		LI09947
652 579 326 506 593 621 505 433	652	100 44 14 11 10 8 8 8		38	44	6	4	–	–	–	–	–	–	–	–	–	1-Ethyl-3,6,7-tris(2-methoxycarbonylethyl)2,4,5,8-tetramethylporphin		LI09948
213 652 651 174 426 461 190 637	652	100 62 31 27 17 16 13 12		39	44	7	2	–	–	–	–	–	–	–	–	–	Hernandezine		LI09949
125 165 527 529 654 291 652 326	652	100 31 23 22 21 18 17 16		40	32	–	–	–	–	4	–	–	–	–	–	–	4'-(4-Chlorophenyl)-1',2,5-tris[(E)-(4-chlorophenyl)methylene]-2',3',6',7'-tetrahydrospiro[cyclopentan-1,5'(4'H)-indene]		MS10932
43 91 133 69 109 83 592 652	652	100 82 51 32 30 9 6 4		44	60	4	–	–	–	–	–	–	–	–	–	–	Lutein diacetate		LI09950
121 221 291 362 213 263 290 105	652	100 99 83 50 50 39 34 33	33.00	44	60	4	–	–	–	–	–	–	–	–	–	–	2,4,6,8,10,12-Tridecahexaenoic acid, 13-(4-methoxyphenyl)-, 3-methoxy-2 (9-methyldecyl)-5-(4-methylpentyl)phenyl ester	71142-35-3	NI46870
652 512 344 204 469 484 513 326	652	100 60 54 48 44 40 36 31		44	76	3	–	–	–	–	–	–	–	–	–	–	1,3-Didodecanoyl-5-tetradecanoylbenzene		NI46871
43 204 189 175 147 161 133 162	652	100 60 37 36 27 27 21 21	16.43	44	76	3	–	–	–	–	–	–	–	–	–	–	1,3-Didodecanoyl-5-tetradecanoylbenzene		NI46872
234 43 69 203 57 202 216 190	652	100 39 38 37 31 30 30 28	0.00	44	76	3	–	–	–	–	–	–	–	–	–	–	3β-Myristoylolean-12-en-16β-ol		NI46873
203 28 204 69 43 57 234 216	652	100 20 20 16 15 14 14 12	0.00	44	76	3	–	–	–	–	–	–	–	–	–	–	3β-Myristoylolean-12-en-28-ol		NI46874
651 283 367 265	652	100 74 9 6	0.00	45	80	2	–	–	–	–	–	–	–	–	–	–	Cholest-5-en-3-ol, (3β)-, octadecanoate	35602-69-8	NI46875
205 652 437 409 558 653 559 331	652	100 73 70 70 53 37 30 30		46	36	4	–	–	–	–	–	–	–	–	–	–	Diphenyl 3,4-bis(diphenylmethylidene)-1,2-dimethylcyclobutane-1,2-dicarboxylate		MS10933
240 353 253 466 212 182 143 438	653	100 96 80 66 60 44 40 36	0.60	32	55	9	5	–	–	–	–	–	–	–	–	–	β-Alanine, N-[N-[N-[N-[1-(D-2-hydroxyvaleryl)-L-prolyl]-L-isoleucyl]-N-methyl-L-valyl]-N-methyl-L-alanyl]-, methyl ester, acetate	6686-63-1	NI46876
106 547 79 86 250 91 64 176	653	100 81 51 46 38 37 23 22	10.00	44	35	3	3	–	–	–	–	–	–	–	–	–	N,N',N''-Tris(4-methylphenyl)triptycene-1,8,16-tricarboxamide		NI46877
514 598 189 174 542 654 570 257	654	100 74 20 17 8 5 5 5		27	45	5	–	–	–	–	–	–	–	3	–	2	μ-Tetramethyldiphosphine-tricarbonyliron-trihexylphosphinedicarbonyl-iron		MS10934
73 130 157 103 147 75 217 205	654	100 97 88 24 23 21 19 18	0.60	29	54	7	2	–	–	–	–	–	4	–	–	–	2-O-(Indole-3-acetyl)-3,4,5,6-tetra-TMS-D-glucose O-methyloxime		NI46878
157 130 73 75 158 217 131 205	654	100 75 63 14 12 10 10 8	4.20	29	54	7	2	–	–	–	–	–	4	–	–	–	4-O-(Indole-3-acetyl)-2,3,5,6-tetra-TMS-D-glucose O-methyloxime		NI46879
73 130 157 290 147 217 75 160	654	100 92 55 36 28 24 21 18	0.80	29	54	7	2	–	–	–	–	–	4	–	–	–	6-O-(Indole-3-acetyl)-2,3,4,5-tetra-TMS-D-glucose O-methyloxime		NI46880
654 655 656 64 657 327 203 146	654	100 40 18 6 5 5 5 5		32	26	6	6	2	–	–	–	–	–	–	–	–	4,4'-Bis(4-phenyl-2H-1,2,3-triazol-2-yl)-2-methylsulpho-stilbene		IC04624
449 359 450 551 243 360 269 552	654	100 42 38 35 33 30 18 17	5.00	33	66	5	–	–	–	–	–	–	4	–	–	–	Pregnan-11-one, 3,17,20,21-tetrakis[(trimethylsilyl)oxy]-, (3α,5β,20R)-	56196-47-5	NI46882
73 449 359 147 551 450 243 75	654	100 54 28 25 21 21 21 19	2.90	33	66	5	–	–	–	–	–	–	4	–	–	–	Pregnan-11-one, 3,17,20,21-tetrakis[(trimethylsilyl)oxy]-, (3α,5β,20R)-	56196-47-5	NI46881
73 449 147 243 75 359 191 450	654	100 99 87 64 60 47 38 37	4.00	33	66	5	–	–	–	–	–	–	4	–	–	–	Pregnan-11-one, 3,17,20,21-tetrakis[(trimethylsilyl)oxy]-, (3α,5β,20R)-	56196-47-5	NI46883
73 147 449 75 359 269 450 74	654	100 99 95 95 51 45 40 40	1.00	33	66	5	–	–	–	–	–	–	4	–	–	–	Pregnan-11-one, 3,17,20,21-tetrakis[(trimethylsilyl)oxy]-, (3α,5β,20S)-	56196-46-4	NI46884
73 147 75 449 243 129 359 117	654	100 99 86 82 58 51 50 40	3.00	33	66	5	–	–	–	–	–	–	4	–	–	–	Pregnan-11-one, 3,17,20,21-tetrakis[(trimethylsilyl)oxy]-, (3α,5β,20S)-	56196-46-4	NI46885
449 450 359 269 551 451 243 360	654	100 38 37 22 17 15 13 12	2.30	33	66	5	–	–	–	–	–	–	4	–	–	–	Pregnan-11-one, 3,17,20,21-tetrakis[(trimethylsilyl)oxy]-, (3α,5β,20S)-	56196-46-4	NI46886
273 241 75 73 213 295 55 129	654	100 80 51 40 22 21 21 20	1.80	39	62	6	–	–	–	–	–	–	1	–	–	–	(2S,4S,6S,8R,9S)-4-(Dimethyl-tert-butylsiloxy)-2-[(5R,2E,6EZ,8E)-9-(5-methoxy-2-methoxycarbonyl-p-tolyl)-3,5-dimethyldeca-2,6,8-trienyl] 8,9-dimethyl-1,7-dioxaspiro[5.5]undecane		MS10935
149 75 165 97 83 57 73 69	654	100 54 33 31 31 30 29 27	4.30	40	86	2	–	–	–	–	–	–	2	–	–	–	3,38-Dioxa-2,39-disilatetracontane, 2,2,39,39-tetramethyl-	35177-13-0	NI46887
124 281 251 247 501 628 186 59	655	100 78 70 50 48 46 35 35	0.00	5	5	1	1	–	–	–	–	3	–	–	–	2	Cyclopentadienylnitrosylcobalttintri-iodide		NI46888
381 284 75 225 73 579 382 490	655	100 82 66 65 39 38 37 32	0.00	32	65	5	1	–	–	–	–	–	4	–	–	–	PGE-2, oxime, tetrakis(trimethylsilyl)-		NI46889
248 51 43 203 91 231 44 92	655	100 78 69 38 28 23 20 16	0.00	40	53	5	3	–	–	–	–	–	–	–	–	–	Carbamic acid, [α-[[[(20-oxo-5α-pregnan-3α-yl)carbamoyl]methyl]carbamoyl]phenethyl]-, benzyl ester, L-	16947-14-1	NI46890

MASS TO CHARGE RATIOS								M.W.	INTENSITIES								Parent	C	H	O	N	S	F	Cl	Br	I	Si	P	B	X	COMPOUND NAME	CAS Reg No	No
262	290	234	318	374	280	174	402	**656**	100	87	87	49	31	30	29	16	**0.00**	10	12	6	–	–	–	–	–	2	–	2	–	2	μ-Tetramethyldiphosphine-bis(iodotricarbonyliron)		MS10936
623	621	624	622	625	631	629	632	**656**	100	90	48	45	40	13	9	6	**0.59**	10	15	12	–	–	–	3	–	–	8	–	–	–	Pentavinyltrichlorooctasilsesquioxane		NI46891
314	173	428	271	542	511			**656**	100	38	23	13	5	5			**0.00**	27	33	8	–	–	9	–	–	–	–	–	–	–	Prostaglandin F2α tris(trifluoroacetyl) methyl ester		LI09951
334	599	614	498	135	73	512	197	**656**	100	28	18	17	14	14	12	7	**2.00**	31	54	–	4	–	–	–	–	–	4	2	–	–	1,3,2,4-Diazadiphosphetidine, 4-[bis(trimethylsilyl)amino]-2,2,4,4-tetrahydro-1-methyl-2,2,2-triphenyl-3-(trimethylsilyl)-4-[(trimethylsilyl)imino]-	57016-88-3	NI46892
621	619	617	620	622	586	128	623	**656**	100	76	63	48	46	21	18	16	**14.22**	32	16	–	8	–	–	2	–	–	–	–	–	1	Germanium phthalocyanine dichloride	19566-97-3	NI46893
282	656	309	310	354	325	218	169	**656**	100	69	51	32	17	11	9	9		33	36	14	–	–	–	–	–	–	–	–	–	–	Egonol tetra-O-acetylglucoside	77690-84-7	NI46894
312	476	164	121	477	463	313	181	**656**	100	50	22	20	16	10	10	10	**0.00**	37	36	11	–	–	–	–	–	–	–	–	–	–	3,8''-Biflavanone, 3'',4',4''',5,5'',7,7''-heptamethoxy-	18913-19-4	NI46895
501	502	656	255	544	231	213	105	**656**	100	32	24	24	22	19	18	14		37	53	2	–	–	–	–	–	1	–	–	–	–	Gorgost-7-en-3-ol, 4-iodobenzoate, (3β,5α)-	56362-46-0	NI46896
133	275	155	382	274	121	656	239	**656**	100	80	80	76	71	69	66	66		39	38	5	–	–	–	2	–	–	–	–	–	–	2,6-Bis[3-(3-chloro-2-hydroxy-5-methylbenzyl)-2-hydroxy-5-methylbenzyl]-4-methylphenol		LI09952
656	642	643	313	657	627	644	306	**656**	100	99	67	49	42	35	19	16		40	48	8	–	–	–	–	–	–	–	–	–	–	Pentacyclo[19.3.1.$1^{3,7}$.$1^{9,13}$.$1^{15,19}$]octacosa-1(25),3,5,7(28),9,11,13(27),15,17,19(26),21,23-dodecaene, 4,6,10,12,16,18,22,24-octamethoxy-2,8,14,20-tetramethyl-	55905-55-0	NI46897
231	134	77	121	426	149	657	69	**657**	100	95	88	56	47	47	45	40		20	12	2	–	2	6	–	–	–	–	–	–	1	Bis(1,1,1-trifluoro-4-phenyl-4-thioxo-2-butanonato)platinum	53776-99-1	NI46898
623	588	311.5	276.5	448	241.5	518	658	**658**	100	59	30	18	14	14	12	11		14	–	–	–	–	–	14	–	–	–	–	–	–	Perchlorobis(tol-4-yl)		LI09953
124	321	658	169	109	107	445	123	**658**	100	41	34	33	29	29	24	23		23	20	6	–	–	14	–	–	–	–	–	–	–	3,13-Bis(heptafluorobutyryl)sambucinol		NI46899
316	173	430	273	513	544			**658**	100	29	19	14	8	1			**0.00**	27	35	8	–	–	9	–	–	–	–	–	–	–	Prostaglandin F1α tris(trifluoroacetyl) methyl ester		LI09954
357	126	599	125	168	358	297	142	**658**	100	27	22	21	20	19	13	12	**0.00**	28	38	16	2	–	–	–	–	–	–	–	–	–	L-erythro-3-Acetoxy-2-(D-glycero-1,2-diacetoxyethyl)-1-acetyl-D-xylo-5,6 diacetoxy-7-(D-glycero-1,2-diacetoxyethyl)-cis-anti-cis-perhydro-dipyrrolo[2,1-b.2,3'd]oxazole		LI09955
88	101	75	71	45	219	187	111	**658**	100	62	37	27	27	24	18	17	**0.00**	29	54	16	–	–	–	–	–	–	–	–	–	–	D-Galactose, O-2,3,4,6-tetra-O-methyl-β-D-galactopyranosyl-(1-6)-O-2,3,4-tri-O-methyl-β-D-galactopyranosyl-(1-6)-2,3,4,5-tetra-O-methyl-	55975-29-6	NI46900
88	423	187	101	45	75	71	111	**658**	100	83	53	44	43	32	26	25	**0.00**	29	54	16	–	–	–	–	–	–	–	–	–	–	D-Galactose, O-2,3,4,6-tetra-O-methyl-β-D-glucopyranosyl-(1-2)-O-3,4,6-tri-O-methyl-β-D-glucopyranosyl-(1-4)-2,3,5,6-tetra-O-methyl-	56248-54-5	NI46901
187	101	45	111	88	71	219	155	**658**	100	54	36	33	30	25	17	15	**0.00**	29	54	16	–	–	–	–	–	–	–	–	–	–	α-D-Glucopyranoside, O-2,3,4,6-tetra-O-methyl-α-D-glucopyranosyl-(1-3) 1,4,6-tri-O-methyl-β-D-fructofuranosyl	38646-05-8	LI09956
187	101	45	111	88	71	219	75	**658**	100	54	36	33	30	25	17	15	**0.00**	29	54	16	–	–	–	–	–	–	–	–	–	–	α-D-Glucopyranoside, O-2,3,4,6-tetra-O-methyl-α-D-glucopyranosyl-(1-3) 1,4,6-tri-O-methyl-β-D-fructofuranosyl	38646-05-8	NI46902
101	219	88	187	45	71	111	75	**658**	100	99	98	84	64	52	44	42	**0.00**	29	54	16	–	–	–	–	–	–	–	–	–	–	α-D-Glucopyranoside, 1,3,4,6-tetra-O-methyl-β-D-fructofuranosyl 2,3,4-tri-O-methyl-6-O-(2,3,4,6-tetra-O-methyl-α-D-galactopyranosyl)-	34141-00-9	NI46903
219	190	658	235	191	659	45	205	**658**	100	34	28	26	26	22	18	15		29	54	16	–	–	–	–	–	–	–	–	–	–	α-D-Glucopyranoside, 1,3,4,6-tetra-O-methyl-β-D-fructofuranosyl 2,3,4-tri-O-methyl-6-O-(2,3,4,6-tetra-O-methyl-α-D-galactopyranosyl)-	34141-00-9	LI09957
88	101	75	71	45	187	219	111	**658**	100	45	42	24	18	13	11	10	**0.00**	29	54	16	–	–	–	–	–	–	–	–	–	–	D-Glucose, O-2,3,4,6-tetra-O-methyl-α-D-galactopyranosyl-(1-6)-O-2,3,4-tri-O-methyl-α-D-galactopyranosyl-(1-6)-2,3,4,5-tetra-O-methyl-	55956-04-2	NI46904
101	187	71	88	75	111	45	219	**658**	100	88	72	71	66	62	46	44	**0.00**	29	54	16	–	–	–	–	–	–	–	–	–	–	D-Glucose, O-2,3,4,6-tetra-O-methyl-β-D-galactopyranosyl-(1-3)-O-2,4,6-tri-O-methyl-β-D-galactopyranosyl-(1-4)-2,3,5,6-tetra-O-methyl-	55956-11-1	NI46905
88	187	101	45	219	75	111	71	**658**	100	37	37	34	31	31	25	20	**0.00**	29	54	16	–	–	–	–	–	–	–	–	–	–	D-Glucose, O-2,3,4,6-tetra-O-methyl-β-D-galactopyranosyl-(1-4)-O-2,3,6-tri-O-methyl-β-D-galactopyranosyl-(1-4)-2,3,5,6-tetra-O-methyl-	55975-32-1	NI46906
88	101	75	71	187	219	45	111	**658**	100	42	28	27	22	17	17	16	**0.00**	29	54	16	–	–	–	–	–	–	–	–	–	–	D-Glucose, O-2,3,4,6-tetra-O-methyl-β-D-galactopyranosyl-(1-6)-O-2,3,4-tri-O-methyl-β-D-galactopyranosyl-(1-4)-2,3,5,6-tetra-O-methyl-	55975-30-9	NI46907
88	187	75	101	45	111	71	219	**658**	100	49	38	36	29	24	23	18	**0.00**	29	54	16	–	–	–	–	–	–	–	–	–	–	D-Glucose, O-2,3,4,6-tetra-O-methyl-α-D-glucopyranosyl-(1-4)-O-2,3,6-tri-O-methyl-α-D-glucopyranosyl-(1-4)-2,3,5,6-tetra-O-methyl-	55956-09-7	NI46908
88	75	101	187	71	45	111	219	**658**	100	47	35	32	22	22	15	14	**0.00**	29	54	16	–	–	–	–	–	–	–	–	–	–	D-Glucose, O-2,3,4,6-tetra-O-methyl-α-D-glucopyranosyl-(1-4)-O-2,3,6-tri-O-methyl-α-D-glucopyranosyl-(1-6)-2,3,4,5-tetra-O-methyl-	55975-31-0	NI46909
88	101	187	75	45	111	219	89	**658**	100	38	34	32	23	17	15	12	**0.00**	29	54	16	–	–	–	–	–	–	–	–	–	–	D-Glucose, O-2,3,4,6-tetra-O-methyl-α-D-glucopyranosyl-(1-6)-O-2,3,4-tri-O-methyl-α-D-glucopyranosyl-(1-4)-2,3,5,6-tetra-O-methyl-	55956-06-4	NI46910
187	101	88	45	111	71	75	391	**658**	100	84	73	65	60	59	51	43	**0.00**	29	54	16	–	–	–	–	–	–	–	–	–	–	D-Glucose, O-2,3,4,6-tetra-O-methyl-β-D-glucopyranosyl-(1-2)-O-3,4,6-tri-O-methyl-β-D-glucopyranosyl-(1-2)-3,4,5,6-tetra-O-methyl-	56248-55-6	NI46911

MASS TO CHARGE RATIOS								M.W.	INTENSITIES								Parent	C	H	O	N	S	F	Cl	Br	I	Si	P	B	X	COMPOUND NAME	CAS Reg No	No
88	187	101	75	45	71	111	155	**658**	100	97	81	56	56	54	53	30	**0.00**	29	54	16	–	–	–	–	–	–	–	–	–	–	**D-Glucose, O-2,3,4,6-tetra-O-methyl-β-D-glucopyranosyl-(1-3)-O-2,4,6-tri-O-methyl-β-D-glucopyranosyl-(1-4)-2,3,5,6-tetra-O-methyl-**	55956-08-6	**NI46912**
88	71	187	45	75	145	101	111	**658**	100	50	49	46	42	35	31	29	**0.00**	29	54	16	–	–	–	–	–	–	–	–	–	–	**D-Glucose, O-2,3,4,6-tetra-O-methyl-β-D-glucopyranosyl-(1-4)-O-2,3,6-tri-O-methyl-β-D-glucopyranosyl-(1-3)-2,4,5,6-tetra-O-methyl-**	55956-10-0	**NI46913**
88	101	71	75	187	45	145	111	**658**	100	50	40	37	32	28	20	19	**0.00**	29	54	16	–	–	–	–	–	–	–	–	–	–	**D-Glucose, O-2,3,4,6-tetra-O-methyl-β-D-glucopyranosyl-(1-6)-O-2,3,4-tri-O-methyl-β-D-glucopyranosyl-(1-3)-2,4,5,6-tetra-O-methyl-**	55956-07-5	**NI46914**
88	75	101	71	279	187	45	219	**658**	100	46	34	25	17	15	15	12	**0.00**	29	54	16	–	–	–	–	–	–	–	–	–	–	**D-Glucose, O-2,3,4,6-tetra-O-methyl-β-D-glucopyranosyl-(1-6)-O-2,3,4-tri-O-methyl-β-D-glucopyranosyl-(1-6)-2,3,4,5-tetra-O-methyl-**	55956-05-3	**NI46915**
88	101	187	75	71	45	127	89	**658**	100	49	46	34	28	27	15	14	**0.00**	29	54	16	–	–	–	–	–	–	–	–	–	–	**D-Glucose, O-2,3,4,6-tetra-O-methyl-β-D-glucopyranosyl-(1-3)-2,4,6-tri-O-methyl-β-D-glucopyranosyl-(1-6)-2,3,4,5-tetra-O-methyl-**	55991-81-6	**NI46916**
88	101	75	71	187	45	111	89	**658**	100	67	60	57	54	30	28	19	**0.00**	29	54	16	–	–	–	–	–	–	–	–	–	–	**D-Glucose, O-2,3,4,6-tetra-O-methyl-β-D-glucopyranosyl-(1-3)-2,4,6-tri-O-methyl-β-D-glucopyranosyl-(1-6)-2,3,4,5-tetra-O-methyl-**	55991-81-6	**NI46917**
88	101	75	187	45	71	219	111	**658**	100	73	73	70	54	51	45	32	**0.00**	29	54	16	–	–	–	–	–	–	–	–	–	–	**D-Mannose, O-2,3,4,6-tetra-O-methyl-α-D-mannopyranosyl-(1-2)-O-3,4,6 tri-O-methyl-α-D-mannopyranosyl-(1-2)-3,4,5,6-tetra-O-methyl-**	56248-57-8	**NI46918**
88	187	45	219	101	75	111	71	**658**	100	42	39	37	31	31	21	21	**0.00**	29	54	16	–	–	–	–	–	–	–	–	–	–	**D-Mannose, O-2,3,4,6-tetra-O-methyl-β-D-mannopyranosyl-(1-4)-O-2,3,6 tri-O-methyl-β-D-mannopyranosyl-(1-4)-2,3,5,6-tetra-O-methyl-**	56248-56-7	**NI46919**
560	344	562	372	543	290	56	332	**658**	100	47	22	18	17	15	13	12	**0.00**	30	54	13	–	–	–	–	–	–	–	–	–	4	**Oxohexakis(isovalerato)tetraberyllium(II)**		**MS10937**
59	42	27	37	370	25	557	347	**658**	100	99	57	44	13	10	7	6	**0.00**	30	54	13	–	–	–	–	–	–	–	–	–	4	**Oxohexakis(pivalato)tetraberyllium(II)**		**MS10938**
455	345	457	58	90	56	57	370	**658**	100	34	25	21	16	13	12	11	**0.00**	30	54	13	–	–	–	–	–	–	–	–	–	4	**Oxohexakis(valerato)tetraberyllium(II)**		**MS10939**
129	147	301	211	369	104	151	243	**658**	100	68	61	37	33	33	27	24	**0.50**	32	66	6	–	–	–	–	–	–	4	–	–	–	**5-Octenoic acid, 8-[tetrahydro-4-[(trimethylsilyl)oxy]-5-[3-[(trimethylsilyl)oxy]-1-octenyl]-2-furanyl]-8-[(trimethylsilyl)oxy]-, trimethylsilyl ester**	35275-55-9	**NI46920**
273	201	538	293	643	405	598	480	**658**	100	89	67	67	45	41	28	27	**0.00**	39	62	8	–	–	–	–	–	–	–	–	–	–	**3,6,16-Triacetate 24,25-acetonide**		**MS10940**
207	334	385	155	658	325	208	121	**658**	100	50	20	20	16	15	15	15		42	55	4	–	–	–	1	–	–	–	–	–	–	**2,4,6,8,10,12-Tridecahexaenoic acid, 13-(3-chloro-4-methoxyphenyl)-, 2-decyl-3-methoxy-5-pentylphenyl ester**	71142-32-0	**NI46921**
44	42	197	109	91	118	105	69	**658**	100	78	76	69	69	66	65	64	**3.81**	42	58	6	–	–	–	–	–	–	–	–	–	–	**Fucoxanthin**	3351-86-8	**NI46922**
43	44	197	91	221	105	41	119	**658**	100	97	77	68	40	37	34	29	**6.00**	42	58	6	–	–	–	–	–	–	–	–	–	–	**Fucoxanthin**	3351-86-8	**LI09958**
43	197	91	105	119	221	640	622	**658**	100	52	49	25	18	15	8	3	**1.00**	42	58	6	–	–	–	–	–	–	–	–	–	–	**Isofucoxanthin**		**LI09959**
112	245	658	161	273	178	174	153	**658**	100	68	50	45	44	28	25	19		43	38	3	4	–	–	–	–	–	–	–	–	–	**(1S*,5S*)-1-(4-Methoxyphenyl)-5-morpholino-4-(2-oxo-1,3-diphenylcyclopent-4-en-1-yl)methyl-1,2,3-triazole**		**DD01814**
112	245	658	161	273	174	178	153	**658**	100	51	49	37	28	20	17	17		43	38	3	4	–	–	–	–	–	–	–	–	–	**(1S*,3R*)-1-(4-Methoxyphenyl)-5-morpholino-4-(2-oxo-1,3,4,5-tetraphenylcyclopent-4-en-1-yl)methyl-1,2,3-triazole**		**DD01815**
660	574	602	488	604	632	576	516	**660**	100	74	38	30	26	25	24	24		14	12	6	–	–	–	–	–	–	–	–	–	2	**Osmium, hexacarbonyl[μ-[(1,2,3,4-η:1,4-η)-1,2,3,4-tetramethyl-1,3-butadiene-1,4-diyl]]di-, (Os-Os)**	33310-11-1	**NI46923**
69	278	127	319	97	265	155	99	**660**	100	35	20	19	17	16	13	12	**0.00**	16	7	11	–	–	15	–	–	–	–	–	–	–	**D-Fructopyranose, pentakis(trifluoroacetate)**	56942-38-2	**MS10941**
69	319	265	252	97	177	81	253	**660**	100	30	27	24	22	18	14	12	**0.00**	16	7	11	–	–	15	–	–	–	–	–	–	–	**α-D-Galactofuranose, pentakis(trifluoroacetate)**	49560-99-8	**MS10942**
69	319	265	83	97	407	177	43	**660**	100	45	45	32	29	22	16	14	**0.00**	16	7	11	–	–	15	–	–	–	–	–	–	–	**α-D-Galactopyranose, pentakis(trifluoroacetate)**	49560-82-9	**MS10943**
603	660	604	661	546	528	437	415	**660**	100	60	18	12	10	7	5	5		20	45	–	–	–	–	–	–	–	–	–	–	5	**Penta-tert-butyl-cyclopentaarsane**	77256-92-9	**NI46924**
93	64	65	66	505	92	48	269	**660**	100	76	36	34	33	31	26	19	**2.00**	30	24	8	6	2	–	–	–	–	–	–	–	–	**1,4-Bis(2-nitro-4-anilinosulphonylanilino)benzene**		**IC04625**
303	147	129	304	191	173	213	243	**660**	100	33	33	25	20	18	17	13	**0.25**	32	68	6	–	–	–	–	–	–	4	–	–	–	**2-Furanoctanoic acid, tetrahydro-η,4-bis[(trimethylsilyl)oxy]-5-[3-[(trimethylsilyl)oxy]-1-octenyl]-, trimethylsilyl ester**	35275-57-1	**NI46925**
77	105	183	182	252	226	170	91	**660**	100	55	27	24	22	21	13	11	**3.00**	37	32	8	4	–	–	–	–	–	–	–	–	–	**Propanedioic acid, bis(3,5-dioxo-1,2-diphenyl-4-pyrazolidinyl)-, diethyl ester**	6139-71-5	**NI46926**
322	306	321	323	204	338	121	660	**660**	100	36	34	30	19	17	12	4		38	32	9	2	–	–	–	–	–	–	–	–	–	**Tetramethyloxybis-1-phenylmethyleneindolizine-2,3-dicarboxylate**		**LI09960**
322	660	181	197	336	263	180	170	**660**	100	88	87	79	72	70	43	42		41	48	4	4	–	–	–	–	–	–	–	–	–	**Villamin**		**LI09961**
73	186	285	258	157	132	276	247	**661**	100	88	52	33	30	23	21	20	**0.00**	32	71	5	1	–	–	–	–	–	4	–	–	–	**Acetamide, N-[1-[[(trimethylsilyl)oxy]methyl]-2,13,14-tris[(trimethylsilyl)oxy]-3-heptadecenyl]-**	56247-93-9	**NI46927**
631	135	633	601	661	617	646		**661**	100	27	11	8	6	6	5			35	51	11	1	–	–	–	–	–	–	–	–	–	**Aljesaconitine B**		**MS10944**
88	310	398	271	486	662	232	574	**662**	100	93	55	55	36	25	12	1		–	–	–	–	–	12	–	–	2	–	4	–	1	**Tetrakis(trifluorophosphine)iron diiodide**		**NI46928**
381	131	331	443	69	262	181	493	**662**	100	38	12	9	9	8	7	4	**1.00**	14	–	–	–	–	26	–	–	–	–	–	–	–	**Perfluoro-1,2-dicyclohexylethane**		**NI46929**

MASS TO CHARGE RATIOS	M.W.	INTENSITIES	Parent	C	H	O	N	S	F	Cl	Br	I	Si	P	B	X	COMPOUND NAME	CAS Reg No	No
79 91 93 41 80 67 105 77	**662**	100 67 59 47 45 45 39 39	**0.00**	14	20	–	–	–	–	–	6	–	–	–	–	–	**1,2-Bis(3,4-dibromocyclohexyl)-1,2-dibromoethane**		**IC04626**
83 280 55 67 41 43 520 82	**662**	100 91 83 48 42 41 29 26	**5.90**	23	28	4	2	–	14	–	–	–	–	–	–	–	**Bis(heptafluorobutyryl)-Nε-cyclohexyllysine propyl ester**		**MS10945**
323 338 69 45 51 50 44 163	**662**	100 77 45 43 39 29 28 27	**0.00**	28	27	8	–	–	9	–	–	–	–	–	–	–	**Pregn-4-ene-3,20-dione, 11β,17,21-trihydroxy-16-methylene-, tris(trifluoroacetate)**	33205-61-7	**NI46930**
121 180 138 120 205 122 181 124	**662**	100 52 29 28 20 19 7 6	**0.00**	28	33	10	–	–	7	–	–	–	–	–	–	–	**Toxin T2, 3-(heptafluorobutyryl)-**		**NI46931**
73 129 356 200 235 103 75 95	**662**	100 74 49 43 28 19 19 18	**0.00**	30	65	3	–	–	–	3	–	–	3	–	–	–	**Tetradecane, 3,5-dichloro-1-(1-chloroheptyl)-1,2,14-tris[(trimethylsilyl)oxy]-**	56248-45-4	**NI46932**
662 331 664 663 627 115 150 332	**662**	100 40 39 39 32 32 20 16		32	16	–	8	–	–	1	–	–	–	–	–	1	**Indium, chloro[29H,31H-phthalocyaninato(2–)-N[29],N[30],N[31],N[32]]-**	19631-19-7	**MS10946**
44 78 43 41 175 55 42 59	**662**	100 34 34 19 18 14 11 11	**1.50**	34	30	14	–	–	–	–	–	–	–	–	–	–	**Dimethyl 6,6′,9,9′-tetrahydroxy-1,1′-dimethyl-5,5′,10,10′-tetraoxo-8,8′-bis(naphtho[2,3-c]pyran)-3,3′-diacetate**		**NI46933**
662 663 64 45 73 618 48 103	**662**	100 36 36 20 16 14 14 13		34	30	14	–	–	–	–	–	–	–	–	–	–	**[8,8′-Bi-1H-naphtho[2,3-c]pyran]-3,3′-diacetic acid, 3,3′,4,4′-tetrahydro-9,9′,10,10′-tetrahydroxy-7,7′-dimethoxy-1,1′-dioxo-, dimethyl ester**	35483-50-2	**NI46934**
544 545 546 601 586 587 602 588	**662**	100 53 50 41 33 26 22 19	**1.20**	34	51	7	2	–	–	1	–	–	1	–	–	–	**3-O-(Dimethyl-tert-butyl)silyl-4,5-deoxymaytansinol**	75349-69-8	**NI46935**
662 628 627 626 613 589	**662**	100 15 15 13 10 9		36	40	4	4	–	–	2	–	–	–	–	–	–	**2,4-Bis(2-chloroethyl)-6,7-bis(2-methoxycarbonylethyl)-1,3,5,8-tetramethylporphin**		**LI09962**
107 619 100 84 70 113 98 129	**662**	100 31 31 29 23 17 14 12	**1.00**	36	46	8	4	–	–	–	–	–	–	–	–	–	**1,6,10,22-Tetraazatricyclo[9.7.6.1^{12,16}]pentacosa-12,14,16(25)-triene-18,23-dione, 6,10-diacetyl-15-(acetyloxy)-17-[[4-(acetyloxy)phenyl]methyl]**	69721-67-1	**NI46936**
634 633 636 632 635 631 637 662	**662**	100 63 54 49 39 39 21 20		37	44	1	4	–	–	–	–	–	–	–	–	1	**Ruthenium, carbonyl[2,3,7,8,12,13,17,18-octaethyl-21H,23H-porphinato(2–)-N[21],N[22],N[23],N[24]]-**	41636-35-5	**NI46937**
413 88 414 155 101 55 187 43	**662**	100 39 32 29 29 28 27 27	**2.61**	38	62	9	–	–	–	–	–	–	–	–	–	–	**D-Glucopyranoside, (3β,22α,25S)-22,25-epoxy-3-methoxyfurost-5-en-26-yl 2,3,4,6-tetra-O-methyl-**	56196-23-7	**NI46938**
170 171 277 168 169 172 203 167	**663**	100 51 21 20 16 8 7 5	**1.63**	34	37	7	3	2	–	–	–	–	–	–	–	–	**1-Naphthalenesulphonic acid, 5-(dimethylamino)-, 4-[2-[[[5-(dimethylamino)-1-naphthalenyl]sulphonyl]methylamino]-1-hydroxyethyl]-2-methoxyphenyl ester**	36600-92-7	**NI46939**
631 633 629 632 630 634 35 639	**664**	100 71 63 47 40 20 8 7	**0.39**	8	12	12	–	–	–	4	–	–	8	–	–	–	**Tetravinyltetrachlorooctasilsesquioxane**		**NI46940**
508 73 480 147 666 452 435 638	**664**	100 99 70 70 50 40 40 32	**0.00**	14	27	5	–	–	–	–	–	–	3	–	–	2	**Pentacarbonyltris(trimethylsilyl)stibinetungsten**		**NI46941**
45 69 339 51 44 163 55 161	**664**	100 97 81 80 79 70 67 60	**0.00**	28	29	8	–	–	9	–	–	–	–	–	–	–	**Pregnane-3,20-dione, 16-methylene-11,17,21-tris[(trifluoroacetyl)oxy]-, (5α,11α)-**	56438-16-5	**NI46942**
98 144 142 99 128 127 145 95	**664**	100 70 65 55 50 49 48 42	**0.00**	34	48	6	8	–	–	–	–	–	–	–	–	–	**N-Methylpyrrolidin-2-on-5-yloxo-N,N′-dimethylhistidyl-N,N′-dimethyltryptophyl-N-methyl-3-methoxyalanyl dimethylamide**		**NI46943**
100 107 84 622 268 110 242 70	**664**	100 76 56 46 44 36 33 32	**31.00**	36	48	8	4	–	–	–	–	–	–	–	–	–	**Acetamide, N-[3-(acetylamino)propyl]-N-[4-[15-(acetyloxy)-2-[[4-(acetyloxy)phenyl]methyl]-3,9-dioxo-4,8-diazabicyclo[10.3.1]hexadeca-1(16),12,14-trien-4-yl]butyl]-**	69721-70-6	**NI46944**
318 580 183 262 581 28 108 319	**664**	100 95 90 65 42 33 32 23	**4.32**	39	30	3	–	–	–	–	–	–	–	2	–	1	**Iron, tricarbonylbis(triphenylphosphine)-**	14741-34-5	**NI46945**
600 664 512 498 572 622 562 558	**664**	100 98 18 16 10 8 8 7		44	56	5	–	–	–	–	–	–	–	–	–	–	**2,3-Didehydrofritschiellaxanthin-diacetate**		**NI46946**
334 207 121 335 664 151 122 278	**664**	100 44 44 26 19 11 9 7		44	72	4	–	–	–	–	–	–	–	–	–	–	**Benzenepentadecanoic acid, 4-methoxy-, 2-decyl-3-methoxy-5-pentylphenyl ester**	71142-43-3	**NI46947**
362 221 121 363 664 222 292 151	**664**	100 50 30 26 17 8 7 5		44	72	4	–	–	–	–	–	–	–	–	–	–	**Benzenetridecanoic acid, 4-methoxy-, 3-methoxy-2-(9-methyldecyl)-5-(4-methylpentyl)phenyl ester**	71142-41-1	**NI46948**
218 203 189 664 408	**664**	100 22 16 2 2		46	80	2	–	–	–	–	–	–	–	–	–	–	**α-Amyrin palmitate**		**LI09963**
408 393 409 394 260 365 205 339	**664**	100 65 46 24 19 17 17 15	**5.00**	46	80	2	–	–	–	–	–	–	–	–	–	–	**Cycloartenyl palmitate**		**LI09964**
189 203 218 205 408 664	**664**	100 44 39 38 27 18		46	80	2	–	–	–	–	–	–	–	–	–	–	**Lupeol palmitate**		**LI09965**
217 215 417 416 446 415 216 214	**665**	100 80 70 67 64 64 64 64	**53.00**	3	9	–	1	–	–	–	–	–	–	–	–	3	**Tris(methylmercury)amine**		**LI09966**
148 149 574 70 147 135 91 287	**665**	100 12 10 9 7 7 7 5	**0.33**	39	47	5	5	–	–	–	–	–	–	–	–	–	**L-Prolinamide, N,N-dimethyl-L-phenylalanyl-N-(3-(1-isopropyl-5,8-dioxo 7-benzyl-2-oxa-6,9-diazabicyclo[10.2.2]hexadeca-10,12,14,15-tetraen 4-yl)-, [3R-(3R*,4S*,7S*)]-**	14051-13-9	**NI46949**
189 327 268 297 236 210 496 467	**666**	100 37 30 17 12 12 5 2	**0.00**	14	20	2	2	2	–	–	2	–	–	–	–	2	**Molybdenum, dibromobis(η^5-2,4-cyclopentadien-1-yl)bis[μ-(ethanethiolato)]dinitrosyldi-**	40671-84-9	**NI46950**

MASS TO CHARGE RATIOS								M.W.	INTENSITIES								Parent	C	H	O	N	S	F	Cl	Br	I	Si	P	B	X	COMPOUND NAME	CAS Reg No	No
305	333	277	249	334	250	306	278	666	100	57	43	36	12	9	5	4	0.00	16	10	6	–	–	–	–	–	–	–	–	–	2	(π-Cyclopentadienyl)tricarbonyltungsten dimer		LI09967
326	453	129	383	71	666	69	169	666	100	15	14	12	6	6	5	4		24	20	4	2	–	14	–	–	–	–	–	–	–	3-Methylbutyl N,N'-bis(heptafluorobutyryl)tryptophanate		NI46951
527	524	582	55	104	493	161	666	666	100	99	84	74	72	46	27	24		26	30	7	2	–	–	–	–	–	–	–	–	1	(1-Butyl-3-cyclohexyl-5-ethoxy-2-oxo-5-phenyl-4-imidazolidinylidene)tungsten pentacarbonyl		DD01816
55	526	70	105	148	160	582	468	666	100	61	61	43	26	22	20	18	0.00	26	30	7	2	–	–	–	–	–	–	–	–	1	(1-tert-Butyl-3-cyclohexyl-5-ethoxy-2-oxo-5-phenyl-4-imidazolidinylidene)tungsten pentacarbonyl		DD01817
169	109	331	666					666	100	60	45	1						34	38	12	2	–	–	–	–	–	–	–	–	–	N-Acetylvincoside lactam, tetraacetate		LI09968
73	75	43	341	359	81	57	129	666	100	41	39	36	25	24	23	22	0.00	37	74	4	–	–	–	–	–	–	3	–	–	–	Ergostan-5-ol, 3,6,12-tris[(trimethylsilyl)oxy]-, (3β,5α,6β,12α)-	56052-12-1	NI46952
117	73	217	191	119	43	430	558	666	100	95	86	55	34	27	25	24	0.00	37	74	4	–	–	–	–	–	–	3	–	–	–	Ergostan-5-ol, 3,6,12-tris[(trimethylsilyl)oxy]-, (3β,5α,6β,12β)-	56053-02-2	NI46953
666	593	607	606	608	534	635		666	100	12	8	8	7	6	4			38	42	7	4	–	–	–	–	–	–	–	–	–	2,4-Bis(2-acetoxyethyl)-β-hydroxy-6,7-bis(2-methoxycarbonylethyl)-1,3,5,8-tetramethylporphin		LI09969
650	576	666	619	605	589	635	507	666	100	42	21	21	17	14	4	4		38	42	7	4	–	–	–	–	–	–	–	–	–	2-Devinyl-2-(3-oxoprop-1-enyl)chlorin E6, trimethyl ester		LI09970
95	406	42	408	57	407	551	55	666	100	84	82	80	68	65	63	58	0.00	41	78	6	–	–	–	–	–	–	–	–	–	–	Hexadecanoic acid, 2-[(1-oxohexyl)oxy]-1,3-propanediyl ester	56588-23-9	NI46954
254	170	367	269	226	480	157	255	667	100	75	48	42	37	35	30	20	0.20	33	57	9	5	–	–	–	–	–	–	–	–	–	β-Alanine, N-[N-[N-[N-[1-(D-2-hydroxy-4-methylvaleryl)-L-prolyl]-L-isoleucyl]-N-methyl-L-valyl]-N-methyl-L-alanyl]-, methyl ester, acetate (ester)	30859-37-1	NI46955
254	170	367	268	226	480	157	452	667	100	77	50	46	37	36	29	15	0.15	33	57	9	5	–	–	–	–	–	–	–	–	–	Methyl O-acetyldestruxinate B		LI09971
105	77	14	106	16	139	164	18	667	100	18	11	7	4	2	2	1	0.00	42	25	6	3	–	–	–	–	–	–	–	–	–	Benzamide, N-[4-[[5-(benzoylamino)-9,10-dihydro-9,10-dioxo-1-anthracenyl]amino]-9,10-dihydro-9,10-dioxo-1-anthracenyl]-	128-89-2	NI46956
667	668	669	589	334	294	293	256	667	100	51	12	6	6	5	4	4		44	28	–	4	–	–	–	–	–	–	–	–	1	Manganese, [5,10,15,20-tetraphenyl-21H,23H-porphinato(2–)-N^{21},N^{22},N^{23},N^{24}]-	31004-82-7	NI46957
612	444	668	472	528	500	236	640	668	100	99	65	63	48	41	36	31		8	–	8	–	–	–	2	–	–	–	–	–	2	Di-μ-chloro(octacarbonyl)dirhenium		MS10947
43	458	584	500	542	374	416	626	668	100	56	46	42	39	28	17	16	3.00	32	28	16	–	–	–	–	–	–	–	–	–	–	2,3',4,5',6-Pentaacetoxy-4'-(3,5-diacetoxyphenoxy)diphenyl ether	81757-70-2	NI46958
43	169	109	331	127	163	107	170	668	100	87	56	22	18	13	13	9	0.00	34	36	14	–	–	–	–	–	–	–	–	–	–	Populoside hexaacetate		LI09972
149	71	83	167	113	279	668	637	668	100	38	33	29	7	2	1	1		36	48	10	2	–	–	–	–	–	–	–	–	–	Glaudelsine	78018-32-3	NI46959
509	581	653	668	510	147	582	75	668	100	47	39	38	37	32	26	25		37	60	5	–	–	–	–	–	–	3	–	–	–	1-Pentanone, 1,1'-[2-[(trimethylsilyl)oxy]-5-[[4-[(trimethylsilyl)oxy]-3-[1-[(trimethylsilyl)oxy]-1-pentenyl]phenyl]methyl]-1,3-phenylene]bis-	55724-91-9	NI46960
668	609	610	624	594	595	312	638	668	100	30	23	15	13	8	6	5		38	44	7	4	–	–	–	–	–	–	–	–	–	Mesoporphyrin 1X, dimethyl ester β-methyl carbonate		LI09973
153	123	70	607	166	207	236	668	668	100	80	74	73	64	60	50	43		38	52	10	–	–	–	–	–	–	–	–	–	–	4-Cyclohexene-1,3-dione, 4,4'-[[2,4,6-trihydroxy-5-(1-oxobutyl)-1,3-phenylene]bis(2-methylpropylidene)]bis[5-hydroxy-2,2,6,6-tetramethyl-	55822-95-2	NI46961
221	666	441	427	411	667	204	206	668	100	75	65	60	36	31	30	28	8.00	39	44	8	2	–	–	–	–	–	–	–	–	–	(–)-Hydroxythalidasine		NI46962
328	341	227	668	225	269	241	239	668	100	72	61	39	31	28	28	28		39	48	6	4	–	–	–	–	–	–	–	–	–	21H-Biline-8,12-dipropanoic acid, 2,17-diethenyl-1,10,19,22,23,24-hexahydro-3,7,13,18-tetramethyl-1,19-dioxo-, diisopropyl ester	50865-93-5	NI46963
668	669	670	334	666	590	294	293	668	100	51	14	11	8	6	6	5		44	28	–	4	–	–	–	–	–	–	–	–	1	Iron, [5,10,15,20-tetraphenyl-21H,23H-porphinato(2–)-N^{21},N^{22},N^{23},N^{24}]-	16591-56-3	NI46964
43	57	71	55	69	83	97	85	668	100	95	70	67	64	58	53	49	47.00	44	60	5	–	–	–	–	–	–	–	–	–	–	Antheraxanthine diacetate		LI09974
263	203	668	588	223	394	669	589	668	100	95	80	74	45	38	32	30		44	60	5	–	–	–	–	–	–	–	–	–	–	Mutatoxanthin diacetate		NI46965
285	383	367	369	401	368			668	100	96	60	23	22	19			0.00	45	80	3	–	–	–	–	–	–	–	–	–	–	Cholest-5-ene-3,7-diol, 7-octadecanoate, (3β,7α)-	58870-28-3	NI46966
367	285	423	385	383	405	669		668	100	23	10	10	4	2	1		0.00	45	80	3	–	–	–	–	–	–	–	–	–	–	Cholest-5-ene-3,7-diol, 7-octadecanoate, (3β,7α)-	58870-28-3	NI46967
367	302	402	385	384	383			668	100	20	16	6	6	6			0.00	45	80	3	–	–	–	–	–	–	–	–	–	–	Cholest-5-ene-3,7-diol, 7-octadecanoate, (3β,7α)-	58870-28-3	NI46968
217	668	130	333	230	117	106	319	668	100	61	56	52	50	48	31	26		48	36	–	4	–	–	–	–	–	–	–	–	–	Hexabenzo[b,f,j,p,t,x]-1,4,15,18-tetraazacyclooctacosine		NI46969
600	69	650	119	550	669	243	500	669	100	66	54	29	27	13	9	8		15	–	–	1	–	25	–	–	–	–	–	–	–	1-Azabicyclo[2.2.0]hexa-2,5-diene, 2,3,4,5,6-pentakis(pentafluoroethyl)-	30567-91-0	NI46970
600	69	650	119	550	93	455	500	669	100	92	48	38	25	18	10	9	6.00	15	–	–	1	–	25	–	–	–	–	–	–	–	1-Azatetracyclo[2.2.0.0^{2,6}.0^{3,5}]hexane, 2,3,4,5,6-pentakis(pentafluoroethyl)	30649-87-7	NI46971
600	650	69	119	669	550	255	462	669	100	46	32	11	10	9	6	5		15	–	–	1	–	25	–	–	–	–	–	–	–	Pyridine, pentakis(pentafluoroethyl)-	20017-53-2	NI46972
205	69	669	155	432	140	127	335	669	100	87	37	19	15	15	13	10		16	8	2	–	4	6	–	–	–	–	–	–	1	Bis[1,1,1-trifluoro-4-(2-thienyl)-4-thioxo-2-butanonato]platinum	55029-39-5	NI46973
173	171	223	502	225	504	500	330	669	100	99	4	3	3	2	2	2	0.00	28	22	2	3	–	–	–	3	–	–	–	–	–	1-(p-Bromoanilino)-2,7-bis(p-bromophenyl)-2,7-diazatetracyclo[4,4,2,0^{3,12}.0^{8,11}]dodecane-4,10-dione		MS10948
361	260	318	373	243	374	217	131	669	100	91	74	70	37	35	33	33	2.00	31	63	5	1	–	–	–	–	–	5	–	–	–	N-(Trimethylsilyl)-tetrakis-O-(trimethylsilyl)-N-((±)-1,2,3,4-tetrahydro-1 naphthyl)-D-galactosylamine		NI46974

MASS TO CHARGE RATIOS	M.W.	INTENSITIES	Parent	C	H	O	N	S	F	Cl	Br	I	Si	P	B	X	COMPOUND NAME	CAS Reg No	No
261 361 362 262 205 271 131 156	**669**	100 91 32 23 23 20 20 19	**0.90**	31	63	5	1	–	–	–	–	–	5	–	–	–	**N-(Trimethylsilyl)-tetrakis-O-(trimethylsilyl)-N-((±)-1,2,3,4-tetrahydro-1 naphthyl)-D-glucosylamine**		**NI46975**
73 464 91 147 191 243 465 374	**669**	100 99 99 52 50 41 38 32	**4.00**	37	63	4	1	–	–	–	–	–	3	–	–	–	**Pregn-4-en-3-one, 17,20,21-tris[(trimethylsilyl)oxy]-, O-(phenylmethyl)oxime, (20R)-**	57325-95-8	**NI46976**
655 656 657 658 639 320 659 623	**670**	100 70 52 22 15 15 13 13	**0.00**	10	30	15	–	–	–	–	–	–	10	–	–	–	**Permethyldecasilsesquioxane**	18106-15-5	**NI46977**
134 177 89 81 79 77 294 191	**670**	100 92 39 36 36 34 32 30	**0.00**	16	16	–	6	–	–	–	4	–	–	2	–	–	**5,5,10,10-Tetrabromo-1,6-dimethyl-3,8-diphenyl[1,3,2,4]diazadiphospheto[2,1-c:4,3-c']bis[1,2,4,3λ[5]]triazaphosphole**	79116-94-2	**NI46978**
43 58 259 301 60 231 77 135	**670**	100 23 18 14 11 8 7 6	**0.00**	23	24	8	6	–	–	–	–	1	–	1	–	–	**2',3'-O-Isopropylideneadenosine-5'-(1-acetyl-5-iodoindol-3-ylphosphate)**		**LI09975**
530 372 376 528 527 642 78 529	**670**	100 99 93 86 71 66 64 57	**12.90**	29	26	5	–	–	–	–	–	–	–	2	–	2	**Di-μ,μ'-[1-phenyl-3,4-dimethylphosphole](dicarbonyliron)(tricarbonylmolybdenum)**	77629-03-9	**NI46979**
323 409 599 311 540 598 329 427	**670**	100 79 61 51 49 35 27 25	**0.00**	36	62	11	–	–	–	–	–	–	–	–	–	–	**Monensin**	17090-79-8	**NI46980**
114 130 72 69 135 98 86 68	**670**	100 8 4 3 2 2 2 2	**0.21**	38	50	5	6	–	–	–	–	–	–	–	–	–	**Pentanamide, 2-(dimethylamino)-N-[1-(1H-indol-3-ylmethyl)-2-[3,3a,11,12,13,14,15,15a-octahydro-13-sec-butyl-12,15-dioxo-5,8-ethenopyrrolo[3,2-b][1,5,8]oxadiazacyclotetradecin-1(2H)-yl]-2-oxoethyl]**	38541-75-2	**NI46981**
334 207 335 155 308 278 670 336	**670**	100 40 26 11 10 10 9 9		42	67	4	–	–	–	1	–	–	–	–	–	–	**Benzenetridecanoic acid, 3-chloro-4-methoxy-, 2-decyl-3-methoxy-5-pentylphenyl ester**	71142-38-6	**NI46982**
69 219 131 264 502 119 100 414	**671**	100 56 44 15 7 7 5 4	**0.00**	12	–	–	1	–	27	–	–	–	–	–	–	–	**Tributylamine, heptacosafluoro-**	311-89-7	**NI46983**
69 131 219 100 119 169 264 70	**671**	100 30 29 11 8 3 2 1	**0.00**	12	–	–	1	–	27	–	–	–	–	–	–	–	**Tributylamine, heptacosafluoro-**	311-89-7	**IC04627**
69 131 219 100 119 114 264 181	**671**	100 33 23 15 14 8 5 5	**0.00**	12	–	–	1	–	27	–	–	–	–	–	–	–	**Tributylamine, heptacosafluoro-**	311-89-7	**NI46984**
73 160 217 147 205 103 307 191	**671**	100 40 24 24 23 23 22 15	**0.00**	26	65	7	1	–	–	–	–	–	6	–	–	–	**D-Glycero-D-gulo-heptose, 2,3,4,5,6,7-hexakis-O-(trimethylsilyl)-, O-methyloxime**	56196-15-7	**NI46985**
73 91 75 264 172 227 186 158	**671**	100 99 50 39 31 25 24 24	**1.50**	37	65	4	1	–	–	–	–	–	3	–	–	–	**Pregnan-20-one, 3,11,17-tris[(trimethylsilyl)oxy]-, O-benzyloxime, (3α,5β,11β)-**	57326-01-9	**NI46986**
73 264 91 75 474 564 251 103	**671**	100 99 88 63 38 36 36 34	**31.00**	37	65	4	1	–	–	–	–	–	3	–	–	–	**Pregnan-20-one, 3,11,21-tris[(trimethylsilyl)oxy]-, O-benzyloxime, (3α,5α,11β)-**	57325-98-1	**NI46987**
73 91 75 264 103 147 143 105	**671**	100 99 51 49 24 19 16 16	**3.68**	37	65	4	1	–	–	–	–	–	3	–	–	–	**Pregnan-20-one, 3,11,21-tris[(trimethylsilyl)oxy]-, O-benzyloxime, (3α,5β,11β)-**	57326-18-8	**NI46988**
91 73 75 264 103 92 107 74	**671**	100 53 27 17 8 8 5 5	**1.00**	37	65	4	1	–	–	–	–	–	3	–	–	–	**Pregnan-20-one, 3,11,21-tris[(trimethylsilyl)oxy]-, O-benzyloxime, (3β,5α,11β)-**	57325-99-2	**NI46989**
73 91 264 75 103 251 143 105	**671**	100 99 58 55 31 20 20 19	**6.00**	37	65	4	1	–	–	–	–	–	3	–	–	–	**Pregnan-20-one, 3,11,21-tris[(trimethylsilyl)oxy]-, O-benzyloxime, (3β,5β,11β)-**	57325-97-0	**NI46990**
73 91 75 564 103 580 105 147	**671**	100 99 61 43 43 36 31 29	**1.50**	37	65	4	1	–	–	–	–	–	3	–	–	–	**Pregnan-20-one, 3,17,21-tris[(trimethylsilyl)oxy]-, O-benzyloxime, (3α,5β)-**	57326-00-8	**NI46991**
73 564 91 580 565 75 581 103	**671**	100 78 70 64 40 39 34 27	**2.71**	37	65	4	1	–	–	–	–	–	3	–	–	–	**Pregnan-20-one, 3,17,21-tris[(trimethylsilyl)oxy]-, O-benzyloxime, (3β,5β)-**	57326-13-3	**NI46992**
105 91 77 459 122 294 216 195	**671**	100 85 71 53 17 6 5 4	**1.40**	40	33	9	1	–	–	–	–	–	–	–	–	–	**D-Arabinose, O-benzyloxime, 2,3,4,5-tetrabenzoate**	71641-38-8	**NI46993**
105 91 459 77 216 201 122 282	**671**	100 86 62 51 16 14 6 2	**0.20**	40	33	9	1	–	–	–	–	–	–	–	–	–	**D-threo-2-Pentulose, O-benzyloxime, 1,3,4,5-tetrabenzoate**	75422-14-9	**NI46994**
105 91 77 459 122 294 216 195	**671**	100 85 73 70 21 8 7 5	**1.30**	40	33	9	1	–	–	–	–	–	–	–	–	–	**D-Ribose, O-benzyloxime, 2,3,4,5-tetrabenzoate**	71641-37-7	**NI46995**
105 91 77 459 122 294 216 195	**671**	100 87 70 60 14 6 5 4	**0.80**	40	33	9	1	–	–	–	–	–	–	–	–	–	**D-Xylose, O-benzyloxime, 2,3,4,5-tetrabenzoate**	71641-36-6	**NI46996**
545 291 672 202 163 343 418 216	**672**	100 19 12 10 10 6 4 2		1	2	–	–	–	–	–	–	4	–	–	–	2	**Bis(diiodoarsino)methane**	51524-81-3	**NI46997**
639 641 643 637 640 35 638 642	**672**	100 89 47 42 40 26 24 21	**0.59**	6	9	12	–	–	–	5	–	–	8	–	–	–	**Trivinylpentachlorooctasilsesquioxane**		**NI46998**
323 217 212 186 275 267 236 324	**672**	100 45 40 38 35 35 35 33	**0.00**	10	30	10	–	–	–	–	–	–	–	–	–	2	**Ditantalum decamethoxide**		**MS10949**
657 493 315 285 269 283 147 299	**672**	100 41 20 20 10 9 9 8	**0.00**	13	36	–	–	–	–	–	–	–	–	–	–	4	**Stannane, methanetetrayltetrakis[trimethyl-**	56177-41-4	**NI46999**
473 199 107 274 155 398 395 200	**672**	100 90 73 41 40 31 28 19	**18.18**	18	–	–	–	3	15	–	–	–	–	–	–	1	**Arsenotrithious acid, tris(pentafluorophenyl) ester**	21459-35-8	**NI47000**
294 279 296 292 277 281 676 216	**672**	100 66 65 59 41 34 30 29	**6.10**	20	20	6	–	–	–	–	4	–	–	–	–	–	**20,21,24,25-Tetrabromo-2,3:11,12-dibenzo-1,4,7,10,13,16-hexaoxoacyclooctadeca-2,11-diene**		**MS10950**
457 211 672 261 407 388 387 242	**672**	100 13 9 8 5 1 1 1		30	18	6	–	–	9	–	–	–	–	–	–	1	**Tris(1,1,1-trifluoro-4-phenyl-2,4-butanedionato)aluminium**		**LI09976**
273 111 358 43 316 83 287 171	**672**	100 98 90 90 83 32 29 29	**7.30**	32	32	16	–	–	–	–	–	–	–	–	–	–	**Vogelin pentaacetate**		**MS10951**

MASS TO CHARGE RATIOS	M.W.	INTENSITIES	Parent	C	H	O	N	S	F	Cl	Br	I	Si	P	B	X	COMPOUND NAME	CAS Reg No	No
73 191 423 129 147 582 117 143	672	100 37 29 26 24 21 21 21	0.00	33	68	6	–	–	–	–	–	–	4	–	–	–	19-R-Hydroxy-PGF-2a, methyl ester, tetrakis(trimethylsilyl)-	67356-00-7	NI47001
143 44 115 120 36 185 154 326	672	100 75 32 30 24 21 18 15	8.00	34	20	3	4	–	–	4	–	–	–	–	–	–	2-Hydroxy-3-(naphth-1-ylaminocarbonyl)-1-[2,3,5,6-tetrachloro-4-(phenylaminocarbonyl)phenylazo]naphthalene		IC04628
108 79 107 141 229 114 109 91	672	100 63 40 17 10 10 9 9	0.00	34	48	10	4	–	–	–	–	–	–	–	–	–	(R,R)-Bis(N-benzyloxycarbonyl-α-alanyl)diaza-18-crown-6		MS10952
108 107 79 109 141 91 80 229	672	100 83 48 29 19 16 16 13	0.00	34	48	10	4	–	–	–	–	–	–	–	–	–	(S,S)-Bis(N-benzyloxycarbonyl-α-alanyl)diaza-18-crown-6		MS10953
108 107 79 141 229 156 109 90	672	100 86 86 31 27 23 18 16	0.00	34	48	10	4	–	–	–	–	–	–	–	–	–	Bis(N-benzyloxycarbonyl-β-alanyl)diaza-18-crown-6		MS10954
91 198 267 155 210 70 313 241	672	100 60 45 45 20 18 15 15	0.00	35	52	5	4	2	–	–	–	–	–	–	–	–	Benzenesulphonamide, N-(2-cyanoethyl)-4-methyl-N-[3-[[(4-methylphenyl)sulphonyl][3-(2-oxoazacyclotridec-1-yl)propyl]amino]propyl]-	67171-90-8	NI47002
234 219 307 295 203 75 235 159	672	100 49 48 46 28 19 16 12	0.80	39	64	7	–	–	–	–	–	–	1	–	–	–	(2S,4S,6S,8R,9S)-4-(Dimethyl-tert-butylsiloxy)-2-[(5R,2E,6EZ,8E)-9-(5-methoxy-2-methoxycarbonyl-p-tolyl)-3,5-dimethyl-6-hydroxydeca-2,8-dienyl]-8,9-dimethyl-1,7-dioxaspiro[5.5]undecane		MS10955
105 77 269 403 428 122 323 202	672	100 68 29 24 16 15 8 8	0.00	40	32	10	–	–	–	–	–	–	–	–	–	–	D-Arabinitol, pentabenzoate	71641-43-5	NI47003
105 77 269 403 428 331 323 202	672	100 30 14 8 6 4 4 3	0.00	40	32	10	–	–	–	–	–	–	–	–	–	–	Ribitol, pentabenzoate	71658-20-3	NI47004
105 77 269 403 428 331 323 202	672	100 61 28 25 12 7 7 7	0.00	40	32	10	–	–	–	–	–	–	–	–	–	–	Xylitol, pentabenzoate	36030-82-7	NI47005
43 57 310 155 328 85 41 69	672	100 75 65 59 52 38 35 34	1.69	40	64	8	–	–	–	–	–	–	–	–	–	–	Phorbol, 12,13-didecanoate		NI47006
197 672 657 673 182 183 181 403	672	100 50 30 23 23 19 19 17		42	48	4	4	–	–	–	–	–	–	–	–	–	Alstonisidine		LI09977
672 347 144 348 168 108 252 158	672	100 82 30 25 20 16 15 15		42	48	4	4	–	–	–	–	–	–	–	–	–	Isovoafoline		LI09978
672 458 347 144 168 158 108 162	672	100 85 43 30 25 25 25 22		42	48	4	4	–	–	–	–	–	–	–	–	–	Voafoline		LI09979
78 51 77 110 50 482 672 644	672	100 49 47 40 35 26 22 20		46	32	2	4	–	–	–	–	–	–	–	–	–	5,15-Diethyl-8,18-diphenyl-5,15-dihydrocarbazolo[3',2':5,6][1,4]oxazino[2,3-b]indolo[2,3-]phenoxazine		IC04629
466 673 209 228 416 259 604 278	673	100 26 21 20 15 13 5 4		15	3	6	–	–	18	–	–	–	–	–	–	1	Tris(1,1,1,5,5,5-hexafluoro-2,4-pentanedionate)chromium		LI09980
327 73 675 328 245 325 676 147	673	100 86 40 34 32 22 21 18	7.43	26	55	6	5	–	–	–	–	–	5	–	–	–	Guanosine, 7,8-dihydro-7-methyl-7,8-oxo, pentakis(trimethylsilyl)-	56273-02-0	NI47007
240 242 162 200 186 184 106 202	674	100 99 99 98 98 98 98 97	0.00	18	14	10	–	–	–	–	2	–	–	–	–	2	Copper, diaquabis[5-bromo-3-formyl-2-hydroxybenzeneacetato(2–)]di-	56615-46-4	NI47008
88 101 71 235 45 75 187 89	674	100 74 56 47 46 37 35 32	0.00	30	58	16	–	–	–	–	–	–	–	–	–	–	D-Galactitol, O-2,3,4,6-tetra-O-methyl-β-D-galactopyranosyl-(1-6)-O-2,3,4-tri-O-methyl-β-D-galactopyranosyl-(1-6)-1,2,3,4,5-penta-O-methyl-	32694-79-4	NI47009
187 101 45 88 111 89 71 391	674	100 99 89 83 58 52 48 40	0.00	30	58	16	–	–	–	–	–	–	–	–	–	–	D-Galactitol, O-2,3,4,6-tetra-O-methyl-β-D-glucopyranosyl-(1-2)-O-3,4,6-tri-O-methyl-β-D-glucopyranosyl-(1-4)-1,2,3,5,6-penta-O-methyl-	32581-23-0	NI47010
88 101 89 71 45 187 75 111	674	100 62 36 35 35 32 26 23	0.00	30	58	16	–	–	–	–	–	–	–	–	–	–	D-Glucitol, O-2,3,4,6-tetra-O-methyl-α-D-galactopyranosyl-(1-6)-O-2,3,4-tri-O-methyl-α-D-galactopyranosyl-(1-6)-1,2,3,4,5-penta-O-methyl-	32581-13-8	NI47011
101 187 71 45 111 75 235 89	674	100 90 76 75 67 53 40 40	0.00	30	58	16	–	–	–	–	–	–	–	–	–	–	D-Glucitol, O-2,3,4,6-tetra-O-methyl-β-D-galactopyranosyl-(1-3)-O-2,4,6-tri-O-methyl-β-D-galactopyranosyl-(1-4)-1,2,3,5,6-penta-O-methyl-	32831-67-7	NI47012
88 101 45 187 235 111 75 71	674	100 59 51 49 47 39 37 36	0.00	30	58	16	–	–	–	–	–	–	–	–	–	–	D-Glucitol, O-2,3,4,6-tetra-O-methyl-β-D-galactopyranosyl-(1-4)-O-2,3,6-tri-O-methyl-β-D-galactopyranosyl-(1-4)-1,2,3,5,6-penta-O-methyl-	32581-18-3	NI47013
88 101 71 45 187 89 75 111	674	100 67 43 41 40 32 32 31	0.00	30	58	16	–	–	–	–	–	–	–	–	–	–	D-Glucitol, O-2,3,4,6-tetra-O-methyl-β-D-galactopyranosyl-(1-6)-O-2,3,4-tri-O-methyl-β-D-galactopyranosyl-(1-4)-1,2,3,5,6-penta-O-methyl-	32581-15-0	NI47014
88 187 101 45 111 235 71 75	674	100 85 64 49 41 40 38 35	0.00	30	58	16	–	–	–	–	–	–	–	–	–	–	D-Glucitol, O-2,3,4,6-tetra-O-methyl-α-D-glucopyranosyl-(1-4)-O-2,3,6-tri-O-methyl-α-D-glucopyranosyl-(1-4)-1,2,3,5,6-penta-O-methyl-	32581-17-2	NI47015
88 187 101 45 75 71 89 111	674	100 57 52 42 32 32 28 27	0.00	30	58	16	–	–	–	–	–	–	–	–	–	–	D-Glucitol, O-2,3,4,6-tetra-O-methyl-α-D-glucopyranosyl-(1-4)-O-2,3,6-tri-O-methyl-α-D-glucopyranosyl-(1-6)-1,2,3,4,5-penta-O-methyl-	32694-80-7	NI47016
88 187 101 45 71 75 111 89	674	100 63 58 38 35 33 30 28	0.00	30	58	16	–	–	–	–	–	–	–	–	–	–	D-Glucitol, O-2,3,4,6-tetra-O-methyl-α-D-glucopyranosyl-(1-6)-O-2,3,4-tri-O-methyl-α-D-glucopyranosyl-(1-4)-1,2,3,5,6-penta-O-methyl-	32581-14-9	NI47017
187 101 45 111 71 88 75 89	674	100 75 58 54 48 46 39 36	0.00	30	58	16	–	–	–	–	–	–	–	–	–	–	D-Glucitol, O-2,3,4,6-tetra-O-methyl-β-D-glucopyranosyl-(1-3)-O-2,4,6-tri-O-methyl-β-D-glucopyranosyl-(1-3)-1,2,4,5,6-penta-O-methyl-	55658-19-0	NI47018
187 45 101 111 88 71 89 75	674	100 92 89 60 50 50 47 35	0.00	30	58	16	–	–	–	–	–	–	–	–	–	–	D-Glucitol, O-2,3,4,6-tetra-O-methyl-β-D-glucopyranosyl-(1-3)-O-2,4,6-tri-O-methyl-β-D-glucopyranosyl-(1-4)-1,2,3,5,6-penta-O-methyl-	32581-22-9	NI47019
88 187 101 45 111 75 71 89	674	100 56 51 27 26 26 25 22	0.00	30	58	16	–	–	–	–	–	–	–	–	–	–	D-Glucitol, O-2,3,4,6-tetra-O-methyl-β-D-glucopyranosyl-(1-3)-O-2,4,6-tri-O-methyl-β-D-glucopyranosyl-(1-6)-1,2,3,4,5-penta-O-methyl-	32581-21-8	NI47020
101 187 71 88 45 111 75 89	674	100 92 68 66 57 50 39 37	0.00	30	58	16	–	–	–	–	–	–	–	–	–	–	D-Glucitol, O-2,3,4,6-tetra-O-methyl-β-D-glucopyranosyl-(1-3)-O-2,4,6-tri-O-methyl-β-D-glucopyranosyl-(1-6)-1,2,3,4,5-penta-O-methyl-	32581-21-8	NI47021

MASS TO CHARGE RATIOS								M.W.	INTENSITIES								Parent	C	H	O	N	S	F	Cl	Br	I	Si	P	B	X	COMPOUND NAME	CAS Reg No	No
88	187	101	45	115	111	89	235	**674**	100	61	50	43	34	33	32	29	**0.00**	30	58	16	–	–	–	–	–	–	–	–	–	–	D-Glucitol, O-2,3,4,6-tetra-O-methyl-β-D-glucopyranosyl-(1-4)-O-2,3,6-tri-O-methyl-β-D-glucopyranosyl-(1-3)-1,2,4,5,6-penta-O-methyl-	32581-20-7	NI47022
88	101	187	45	75	71	89	111	**674**	100	49	37	34	30	28	24	19	**0.00**	30	58	16	–	–	–	–	–	–	–	–	–	–	D-Glucitol, O-2,3,4,6-tetra-O-methyl-β-D-glucopyranosyl-(1-6)-O-2,3,4-tri-O-methyl-β-D-glucopyranosyl-(1-3)-1,2,4,5,6-penta-O-methyl-	32581-16-1	NI47023
88	101	187	71	45	89	75	235	**674**	100	51	34	34	29	26	25	21	**0.00**	30	58	16	–	–	–	–	–	–	–	–	–	–	D-Glucitol, O-2,3,4,6-tetra-O-methyl-β-D-glucopyranosyl-(1-6)-O-2,3,4-tri-O-methyl-β-D-glucopyranosyl-(1-6)-1,2,3,4,5-penta-O-methyl-	32581-12-7	NI47024
101	187	235	88	45	71	89	111	**674**	100	97	93	88	87	71	61	57	**0.00**	30	58	16	–	–	–	–	–	–	–	–	–	–	D-Glucitol, O-2,3,4,6-tetra-O-methyl-β-D-glucopyranosyl-(1-2)-3,4,6-tri-O-methyl-β-D-glucopyranosyl-(1-2)-1,3,4,5,6-penta-O-methyl-	32581-24-1	NI47025
88	101	187	45	235	71	89	219	**674**	100	96	83	80	79	64	56	54	**0.00**	30	58	16	–	–	–	–	–	–	–	–	–	–	D-Mannitol, O-2,3,4,6-tetra-O-methyl-α-D-mannopyranosyl-(1-2)-O-3,4,6 tri-O-methyl-α-D-mannopyranosyl-(1-2)-1,3,4,5,6-penta-O-methyl-	32581-25-2	NI47026
88	45	101	187	235	115	75	219	**674**	100	49	42	38	35	29	28	27	**0.00**	30	58	16	–	–	–	–	–	–	–	–	–	–	D-Mannitol, O-2,3,4,6-tetra-O-methyl-β-D-mannopyranosyl-(1-4)-O-2,3,6 tri-O-methyl-β-D-mannopyranosyl-(1-3)-1,2,4,5,6-penta-O-methyl-	32581-19-4	NI47027
643	499	511	659	674	571	627	543	**674**	100	54	28	27	20	13	12	11		35	46	13	–	–	–	–	–	–	–	–	–	–	Chrysin 6-C-β-D-glucoside-8-C-α-L-arabinoside permethyl ether		NI47028
643	543	674	555	659	499	613	627	**674**	100	49	42	30	24	21	18	12		35	46	13	–	–	–	–	–	–	–	–	–	–	Chrysin 6-C-α-L-arabinoside-8-C-β-D-glucoside permethyl ether		NI47029
44	440	57	458	84	439	56	444	**674**	100	62	57	54	45	40	32	25	**0.00**	40	38	4	2	2	–	–	–	–	–	–	–	–	tert-Butyl 2-(tert-butoxycarbonyl)-3-[α-[2,5-diphenyl-4-(thiobenzoyl)-3-thienyl]benzylidene]carbazate	110315-33-8	DD01818
219	316	203	57	161	331	220	218	**674**	100	38	35	23	21	17	17	16	**0.01**	41	58	6	2	–	–	–	–	–	–	–	–	–	Diethyl 2,4-dicyano-2,4-di-(3,5-di-tert-butyl-4-hydroxybenzyl)glutarate		IC04630
55	177	119	556	176	269	415	428	**675**	100	83	83	82	69	57	48	44	**13.40**	23	20	5	1	–	15	–	–	–	–	–	–	–	3′,4′-Dihydroxy-8-phenyloctylamine, tris(pentafluoropropionyl)-		NI47030
202	272	77	440	316	76	338	237	**676**	100	99	32	22	19	18	11	9	**3.57**	16	10	–	–	–	–	2	–	–	–	–	–	2	Phenyl acetylide mercury chloride		NI47031
310	438	308	416	205	436	168	181	**676**	100	94	93	77	43	41	38	36	**9.40**	16	24	–	–	–	–	–	–	2	–	–	–	2	Bis[(1,5-cyclooctadiene)rhodium iodide]		LI09981
81	114	135	121	248	189	217	462	**676**	100	98	53	26	13	13	6	2	**2.00**	24	26	6	–	–	14	–	–	–	–	–	–	–	Methyl 10,11-bis(heptafluorobutyryloxy)-3,7,11-trimethyl-2,6-dodecadienoate		MS10956
81	114	135	121	189	248	249	617	**676**	100	98	51	22	17	12	8	7	**6.00**	24	26	6	–	–	14	–	–	–	–	–	–	–	Methyl 10,11-bis(heptafluorobutyryloxy)-3,7,11-trimethyl-3,6-dodecadienoate		MS10957
73	253	217	317	185	204	75	69	**676**	100	54	32	29	29	17	17	15	**0.10**	29	56	10	2	–	–	–	–	–	3	–	–	–	D-Glucopyranosiduronic acid, 3-(5-ethylhexahydro-1,3-dimethyl-2,4,6-trioxo-5-pyrimidinyl)-1-methylbutyl 2,3,4-tris-O-(trimethylsilyl)-, methyl ester	55556-79-1	NI47032
238	91	267	70	198	155	113	293	**676**	100	85	53	34	30	24	22	20	**0.30**	35	56	5	4	2	–	–	–	–	–	–	–	–	Benzenesulphonamide, N-(3-aminopropyl)-4-methyl-N-[3-[[(4-methylphenyl)sulphonyl][3-(2-oxoazacyclotridec-1-yl)propyl]amino]propyl]-	75422-06-9	NI47033
224	586	202	189	277	204	157	191	**676**	100	58	55	54	53	52	41	40	**5.30**	39	76	3	–	–	–	–	–	–	3	–	–	–	Germacerane, 2,3,22-tris[(trimethylsilyl)oxy]-, (2α,3β,8α,9β,13α,14β,17α,18β,22α)-	71899-28-0	NI47034
70	197	80	196	94	92	69	68	**676**	100	77	49	20	18	18	18	14	**0.35**	44	68	5	–	–	–	–	–	–	–	–	–	–	p-Benzoquinone, 2-(3-hydroxy-3,7,11,15,19,23,27-heptamethyl-6,10,14,18,22,26-octacosahexaenyl)-5,6-dimethoxy-3-methyl-	19721-60-9	NI47035
68	197	80	660	661	196	92	67	**676**	100	75	48	38	20	18	17	17	**4.00**	44	68	5	–	–	–	–	–	–	–	–	–	–	γ-Hydroxy-β,γ-dihydroubiquin-7-one		LI09982
207	265	121	208	334	266	151	187	**676**	100	66	27	19	11	11	11	10	**9.00**	46	60	4	–	–	–	–	–	–	–	–	–	–	2,4,6,8,10,12,14,16-Heptadecaoctaenoic acid, 17-(4-methoxyphenyl)-, 2-decyl-3-methoxy-5-pentylphenyl ester	69505-95-9	NI47036
118	218	106	339	676	91	130	232	**676**	100	50	46	32	30	30	28	27		48	44	–	4	–	–	–	–	–	–	–	–	–	5,6,7,8,25,26,27,28-Octahydrohexabenzo[b,f,j,p,t,x]-1,4,15,18-tetraazacyclooctacosine		NI47037
470	213	420	401	677	235	282	263	**677**	100	80	27	19	17	9	8	7		15	3	6	–	–	18	–	–	–	–	–	–	1	Tris-(1,1,1,5,5,5-hexafluoro-2,4-pentanedionato)iron		LI09983
472	42	119	514	590	415	473	220	**677**	100	69	32	31	19	18	17	12	**0.00**	21	14	7	1	–	15	–	–	–	–	–	–	–	N,O,O-Tris(pentafluoropropionyl)-D-3,4-dihydroxyphenylalanine isopropyl ester		MS10958
472	42	119	514	590	415	473	220	**677**	100	79	35	28	19	18	17	14	**0.00**	21	14	7	1	–	15	–	–	–	–	–	–	–	N,O,O-Tris(pentafluoropropionyl)-L-3,4-dihydroxyphenylalanine isopropyl ester		MS10959
431	459	159	176	43	119	181	219	**677**	100	92	37	35	35	27	17	16	**0.00**	22	18	6	1	–	15	–	–	–	–	–	–	–	β-Ethoxyisoprenalin, tris(pentafluoropropionate)-		MS10960
43	107	206	106	178	151	161	677	**677**	100	87	51	39	30	27	22	21		32	39	15	1	–	–	–	–	–	–	–	–	–	Evonine, O^6,O^9-dideacetyl-	35721-62-1	NI47038
339	211	55	57	43	71	285	69	**677**	100	86	62	57	48	45	38	33	**0.00**	36	72	8	1	–	–	–	–	–	–	1	–	–	3,5,9-Trioxa-4-phosphatricosan-1-aminium, 4-hydroxy-N,N,N-trimethyl-10-oxo-7-[(1-oxotetradecyl)oxy]-, hydroxide, inner salt, 4-oxide, (R)-	18194-24-6	NI47039
113	73	126	75	662	70	55	72	**677**	100	61	38	28	24	17	17	16	**0.70**	37	71	4	1	–	–	–	–	–	3	–	–	–	Pyrrolidine, 1-[3α,7α,12α-tris(trimethylsiloxy)-5β-cholan-24-oyl]-	33745-60-7	NI47040

MASS TO CHARGE RATIOS								M.W.	INTENSITIES								Parent	C	H	O	N	S	F	Cl	Br	I	Si	P	B	X	COMPOUND NAME	CAS Reg No	No
618	603	588	574	287	309	294	662	**677**	100	95	84	69	58	51	50	42	**30.00**	40	53	2	4	–	–	–	–	–	–	–	–	1	**Acetato(α,γ-dimethyl-α,γ-dihydrooctaethylporphinato)iron(III)**		**NI47041**
202	398	301	538	426	566	308	280	**678**	100	83	73	61	37	35	35	34	**20.00**	14	–	10	–	–	6	–	–	–	–	–	–	4	**μ-(Hexafluorobut-2-yne)decacarbonyltetracobalt**		**MS10961**
121	123	322	161	268	142	88	236	**678**	100	83	73	63	60	60	52	38	**0.00**	18	–	6	–	2	10	–	–	–	–	–	–	2	**Iron, hexacarbonylbis[μ-(pentafluorobenzenethiolato)]di-, (Fe-Fe)**	21747-16-0	**MS10962**
239	304	209	174	613	678	483	548	**678**	100	68	53	22	9	4	2	2		20	20	–	–	–	–	2	–	–	–	–	–	2	**Dichloro tetra(cyclopentadienyl)diytterbium**	12128-90-4	**NI47042**
139	57	137	29	41	135	621	55	**678**	100	88	74	66	57	45	39	30	**0.00**	22	9	–	–	–	15	–	–	–	–	–	–	1	**Stannane, butyltris(pentafluorophenyl)-**	1182-53-2	**NI47043**
169	331	60	44	45	43	109	127	**678**	100	86	85	85	84	83	44	38	**0.00**	28	38	19	–	–	–	–	–	–	–	–	–	–	**D-Galactose, 6-O-(2,3,5,6-tetra-O-acetyl-β-D-galactofuranosyl)-, 2,3,4,5-tetraacetate**	55649-74-6	**NI47044**
310	295	221	73	177	75	251	44	**678**	100	63	37	36	31	19	14	14	**0.00**	28	38	19	–	–	–	–	–	–	–	–	–	–	**D-Glucopyranose, 6-O-(2,3,4,6-tetra-O-acetyl-α-D-galactopyranosyl)-, tetraacetate**	23846-69-7	**NI47045**
169	109	331	97	127	99	115	619	**678**	100	42	34	25	18	17	14	13	**0.00**	28	38	19	–	–	–	–	–	–	–	–	–	–	**β-D-Glucopyranose, 4-O-(2,3,4,6-tetra-O-acetyl-α-D-glucopyranosyl)-, tetraacetate**	22352-19-8	**NI47046**
169	109	331	81	115	127	112	97	**678**	100	50	48	23	18	17	16	16	**0.00**	28	38	19	–	–	–	–	–	–	–	–	–	–	**β-D-Glucopyranose, 4-O-(2,3,4,6-tetra-O-acetyl-α-D-glucopyranosyl)-, tetraacetate**	22352-19-8	**LI09984**
169	43	109	211	127	331	97	170	**678**	100	75	69	53	17	11	11	9	**0.00**	28	38	19	–	–	–	–	–	–	–	–	–	–	**α-D-Glucopyranoside, 1,3,4,6-tetra-O-acetyl-β-D-fructofuranosyl, tetraacetate**	126-14-7	**NI47047**
291	305	247	309	277	519	299	377	**678**	100	98	90	89	87	65	65	58	**0.00**	33	42	15	–	–	–	–	–	–	–	–	–	–	**Deacetylazadirachtin**		**NI47048**
105	43	231	393	121	106	109	107	**678**	100	49	25	15	12	12	9	7	**0.00**	35	34	14	–	–	–	–	–	–	–	–	–	–	**Salicyloyl populin tetraacetate**		**MS10963**
105	43	169	77	393	106	121	109	**678**	100	45	12	12	10	10	9	9	**0.00**	35	34	14	–	–	–	–	–	–	–	–	–	–	**Salicyloyl tremuloidin tetraacetate**		**MS10964**
321	405	463	617	647	327	575	491	**678**	100	65	57	54	45	32	30	29	**0.00**	35	59	11	–	–	–	–	–	–	–	–	–	1	**3-De-O-methylmonensin sodium salt**		**NI47049**
314	71	153	133	286	121	155	365	**678**	100	95	92	92	90	86	85	84	**21.00**	37	42	12	–	–	–	–	–	–	–	–	–	–	**Pruninchalcone-6″-p-coumarate heptamethyl ether**		**NI47050**
588	501	620	678					**678**	100	30	15	14						38	38	8	4	–	–	–	–	–	–	–	–	–	**10(R)-Acetoxymethylphaophorbide**		**LI09985**
221	239	362	121	317	363	222	315	**678**	100	44	33	31	25	15	14	11	**4.00**	46	62	4	–	–	–	–	–	–	–	–	–	–	**2,4,6,8,10,12,14-Pentadecaheptaenoic acid, 15-(4-methoxyphenyl)-, 3-methoxy-2-(9-methyldecyl)-5-(4-methylpentyl)phenyl ester**	71394-73-5	**NI47051**
647	649	651	35	645	648	646	650	**680**	100	88	64	54	50	45	24	22	**1.59**	4	6	12	–	–	–	6	–	–	8	–	–	–	**Divinylhexachlorooctasilsesquioxane**		**NI47052**
162	93	143	680	255	100	446	502	**680**	100	98	90	55	48	48	43	42		10	–	6	–	–	12	–	–	–	–	4	–	2	**Diironhexacarbonyl-1,2,3,4-tetrakis(trifluoromethyl)cyclotetraphosphine**		**NI47053**
404	216	473	266	680	235.5	196	611	**680**	100	58	35	24	21	11	2	1		15	3	6	–	–	18	–	–	–	–	–	–	1	**Tris(1,1,1,5,5,5,-hexafluoro-2,4-pentanedionato)cobalt**		**LI09986**
87	59	517	503	531	429	518	504	**680**	100	81	63	53	53	34	33	27	**0.00**	18	52	10	–	–	–	–	–	–	9	–	–	–	**1,3,5,7,9,11,13,15,17-Nonaethylbicyclo[9.7.1]nonasiloxane**	73113-24-3	**NI47054**
680	681	467	320	133	91	665	105	**680**	100	30	25	20	17	15	14	14		27	26	4	–	–	14	–	–	–	–	–	–	–	**3,5-Androstadiene-3,17β-diol, di(heptafluorobutyrate)**		**MS10965**
680	682	646	684	575	577	634	636	**680**	100	77	45	21	14	13	11	10		36	26	6	4	–	–	2	–	–	–	–	–	–	**C.I. Pigment Violet 35**		**LI09987**
226	215	212	211	195	382	381	364	**680**	100	74	41	39	39	22	21	9	**0.00**	36	44	11	2	–	–	–	–	–	–	–	–	–	**Reserpine methacetate**		**MS10966**
298	594	58	593	299	609	595	564	**680**	100	38	27	22	18	17	16	9	**0.00**	37	42	6	2	–	–	2	–	–	–	–	–	–	**Tubocuraranium, 7′,12′-dihydroxy-6,6′-dimethoxy-2,2′,2′-trimethyl-, chloride, hydrochloride**	57-94-3	**NI47055**
94	434	66	65	107	586	228	482	**680**	100	68	68	64	56	52	48	47	**35.00**	42	32	9	–	–	–	–	–	–	–	–	–	–	**Distichol**		**MS10967**
680	681	664	665	682	586	666	508	**680**	100	71	52	39	24	10	8	8		44	28	1	4	–	–	–	–	–	–	–	–	1	**Oxo-5,10,15,20-tetraphenylporphinatochromium(IV)**	78833-34-8	**NI47056**
680	512	513	204	469	372	540	497	**680**	100	55	37	33	32	32	30	28		46	80	3	–	–	–	–	–	–	–	–	–	–	**1-Dodecanoyl-3,4-ditetradecanoylbenzene**		**NI47057**
43	204	189	175	147	161	680	162	**680**	100	27	19	16	12	11	10	9		46	80	3	–	–	–	–	–	–	–	–	–	–	**1-Dodecanoyl-3,5-ditetradecanoylbenzene**		**NI47058**
73	243	147	296	129	75	191	74	**681**	100	99	89	80	60	60	50	37	**3.50**	34	67	5	1	–	–	–	–	–	4	–	–	–	**Pregn-4-en-3-one, 11,17,20,21-tetrakis[(trimethylsilyl)oxy]-, O-methyloxime, (11β,20R)-**	57325-92-5	**NI47059**
73	296	147	75	243	129	74	191	**681**	100	99	95	62	60	39	38	36	**3.00**	34	67	5	1	–	–	–	–	–	4	–	–	–	**Pregn-4-en-3-one, 11,17,20,21-tetrakis[(trimethylsilyl)oxy]-, O-methyloxime, (11β,20S)-**	57325-93-6	**NI47060**
73	650	103	75	74	129	651	276	**681**	100	99	99	99	65	60	59	53	**9.00**	34	67	5	1	–	–	–	–	–	4	–	–	–	**Pregn-5-en-20-one, 3,11,17,21-tetrakis[(trimethylsilyl)oxy]-, O-methyloxime, (3β,11β)-**	57325-94-7	**NI47061**
91	155	518	444	197	526	125		**681**	100	52	40	22	13	11	6		**0.10**	35	47	5	5	2	–	–	–	–	–	–	–	–	**1-(4-Toluenesulphonyl)-3,5-dicyclohexyl-2-cyclohexylimino-4-(4-toluenesulphonylimino)-1,3,5-triazin-6-one**		**LI09988**
69	150	81	135	190	97	109	95	**681**	100	60	50	22	21	19	17	17	**6.00**	47	71	2	1	–	–	–	–	–	–	–	–	–	**Benzoic acid, 4-amino-3-(3,7,11,15,19,23,27,31-octamethyl-2,6,10,14,18,22,26,30-dotriacontaoctaenyl)-, (all-E)-**	33569-81-2	**NI47062**
655	656	657	682	683	658	684	605	**682**	100	70	48	44	24	20	19	17		20	26	12	–	–	–	–	–	–	8	–	–	–	**Heptavinylphenyloctasilsesquioxane**		**NI47063**

MASS TO CHARGE RATIOS								M.W.	INTENSITIES								Parent	C	H	O	N	S	F	Cl	Br	I	Si	P	B	X	COMPOUND NAME	CAS Reg No	No
69	45	162	51	44	91	166	50	**682**	100	81	80	54	51	34	32	24	**0.90**	25	27	8	4	–	9	–	–	–	–	–	–	–	L-Phenylalanine, N-[N-[N^2,N^5-bis(trifluoroacetyl)-L-ornithyl]-O-(trifluoroacetyl)-D-allothreonyl]-, methyl ester	56772-21-5	MS10968
682	683	593	684	594				**682**	100	60	31	25	17					38	54	2	4	–	–	–	–	–	3	–	–	–	Bis(trimethylsiloxy)etioporphyrin		LI09989
206	76	179	77	91	65	103	137	**682**	100	54	48	38	28	28	23	22	**0.00**	42	30	4	6	–	–	–	–	–	–	–	–	–	7,15-Bis(4-nitrophenyl)-6,14-diphenyl-6a,7,14a,15-tetrahydroquinoxalino[1′,2′:1,2]pyrazino[4,5-a]quinoxaline	91736-74-2	NI47064
43	91	119	105	664	604	590	544	**682**	100	67	42	42	39	37	12	4	**2.00**	44	58	6	–	–	–	–	–	–	–	–	–	–	4′-Anhydroisofucoxanthin acetate		LI09990
81	69	95	109	123	135	151	188	**682**	100	87	52	45	43	38	34	19	**4.50**	47	70	3	–	–	–	–	–	–	–	–	–	–	3-Octaprenyl-4-hydroxybenzoic acid		LI09991
683	543	613	508	438	648			**683**	100	51	47	27	9	4				19	–	–	–	–	–	13	–	–	–	–	–	–	Fluoren-9-yl, 1,2,3,4,5,6,7,8-octachloro-9-(pentachlorophenyl)-	32390-14-0	NI47065
683	543	613	306.5	271.5	508	236.5	341.5	**683**	100	51	47	30	30	27	18	12		19	–	–	–	–	–	13	–	–	–	–	–	–	Fluoren-9-yl, 1,2,3,4,5,6,7,8-octachloro-9-(pentachlorophenyl)-	32390-14-0	LI09992
73	169	315	351	147	299	75	74	**683**	100	26	24	23	19	18	15	12	**0.00**	24	54	8	3	–	–	–	–	–	5	1	–	–	5′-Cytidylic acid, N-(trimethylsilyl)-2′,3′-bis-O-(trimethylsilyl)-, bis(trimethylsilyl) ester	32653-16-0	NI47066
73	298	147	75	300	43	317	103	**683**	100	99	65	49	43	37	35	35	**0.09**	27	61	9	1	–	–	–	–	–	5	–	–	–	D-Glycero-D-galacto-2-nonulosonic acid, 5-(acetylamino)-3,5-dideoxy-4,6,7,8,9-pentakis-O-(trimethylsilyl)-, methyl ester	57397-00-9	NI47067
73	298	147	75	300	43	317	103	**683**	100	32	21	16	14	12	11	11	**0.00**	27	61	9	1	–	–	–	–	–	5	–	–	–	D-Glycero-D-galacto-2-nonulosonic acid, 5-(acetylamino)-3,5-dideoxy-4,6,7,8,9-pentakis-O-(trimethylsilyl)-, methyl ester	57397-00-9	NI47068
73	652	75	103	147	653	246	276	**683**	100	99	99	86	66	60	56	49	**27.00**	34	69	5	1	–	–	–	–	–	4	–	–	–	5α-Pregnan-20-one, 3α,11β,17,21-tetrakis(trimethylsiloxy)-, O-methyloxime	32221-30-0	NI47069
73	103	75	147	580	81	93	55	**683**	100	99	99	89	84	68	67	63	**20.00**	34	69	5	1	–	–	–	–	–	4	–	–	–	5β-Pregnan-20-one, 3α,11β,17,21-tetrakis(trimethylsiloxy)-, O-methyloxime	32206-70-5	NI47070
648	649	271	575	502	487	257	471	**683**	100	45	27	26	24	20	18	16	**11.80**	36	40	4	4	–	–	1	–	–	–	–	–	1	Iron, chloro[dimethyl 7,12-diethyl-3,8,13,17-tetramethyl-21H,23H-porphine-2,18-dipropanoato(2–)-N^{21},N^{22},N^{23},N^{24}]-, (sp-5-13)-	14126-91-1	NI47071
542	372	458	682	430	570	357	402	**684**	100	61	55	50	49	48	46	43	**0.00**	14	12	10	–	–	–	–	–	–	–	2	–	2	1-Pentacarbonylmolybdenum-2-pentacarbonyltungsten(tetramethyldiphosphine)		NI47072
73	169	315	299	147	75	230	316	**684**	100	63	50	28	19	19	13	12	**0.00**	24	53	9	2	–	–	–	–	–	5	1	–	–	2(1H)-Pyrimidinone, 1-[2,3-bis-O-(trimethylsilyl)-β-D-ribofuranosyl]-4-(trimethylsiloxy)-, 5′-[bis(trimethylsilyl) phosphate]	32645-59-3	NI47073
457	688	686	459	455	231	229	655	**684**	100	84	84	50	50	42	42	30	**28.00**	27	27	6	–	–	–	–	3	–	–	–	–	–	10,15-Dihydro-1,6,11-tribromo-2,3,7,8,12,13-hexamethoxy-5H-benzo[a,d,g]cyclononene		NI47074
334	135	197	183	105	73	320	654	**684**	100	20	17	10	10	10	5	4	**2.00**	33	58	–	4	–	–	–	–	–	4	2	–	–	1,3,2,4-Diazadiphosphetidine, 4-[bis(trimethylsilyl)amino]-2,2,4,4-tetrahydro-2,2,2-triphenyl-1-propyl-3-(trimethylsilyl)-4-[(trimethylsilyl)imino]-	67277-82-1	NI47075
343	441	423	325	407	344	439	342	**684**	100	80	42	39	38	37	28	27	**0.20**	37	64	11	–	–	–	–	–	–	–	–	–	–	Monensin, methyl ester	28636-21-7	NI47076
105	77	106	318	159	317	51	319	**684**	100	12	7	6	3	2	2	1	**0.00**	40	33	10	–	–	–	–	–	–	–	–	1	–	D-Glucitol, cyclic 3,4-(phenylboronate) 1,2,5,6-tetrabenzoate	74810-65-4	NI47077
207	348	135	278	273	155	137	121	**684**	100	40	27	20	16	13	13	13	**7.00**	44	57	4	–	–	–	1	–	–	–	–	–	–	2,4,6,8,10,12,14-Pentadecaheptaenoic acid, 15-(3-chloro-4-methoxyphenyl)-, 2-decyl-3-methoxy-5-pentylphenyl ester	71142-36-4	NI47078
133	327	207	169	192	551	328	193	**684**	100	47	34	18	17	14	13	9	**8.00**	52	60	–	–	–	–	–	–	–	–	–	–	–	2′,3′,6′,7′-Tetrahydro-4′-(2,4,6-trimethylphenyl)-1′,2,5-tris[(E)-(2,4,6-trimethylphenyl)methylene]spiro[cyclopentan-1,5′(4′H)-indene]		MS10969
57	536	480	684	444	500	592	386	**685**	100	94	42	35	27	25	19	15	**0.00**	16	36	–	–	–	–	2	–	–	–	2	–	2	Dichloro[tellurobis(di-tert-butylphosphane)]platinum(II)		NI47079
685	148	440	163	245	213	91	343	**685**	100	71	70	49	44	42	40	28		22	16	2	–	2	6	–	–	–	–	–	–	1	Bis[1,1,1-trifluoro-4-(4-methylphenyl)-4-thioxo-2-butanonato]platinum	55029-40-8	NI47080
73	91	75	103	147	594	105	578	**685**	100	99	35	31	28	18	18	17	**0.56**	37	63	5	1	–	–	–	–	–	3	–	–	–	Pregnane-11,20-dione, 3,17,21-tris[(trimethylsilyl)oxy]-, 20-[O-(phenylmethyl)oxime], (3α,5β)-	57325-96-9	NI47081
105	91	77	160	306	122	473	269	**685**	100	86	58	18	16	13	7	2	**0.10**	41	35	9	1	–	–	–	–	–	–	–	–	–	D-arabino-Hexose, 2-deoxy-, O-benzyloxime, 3,4,5,6-tetrabenzoate	71641-41-3	NI47082
105	91	77	473	122	281	230	308	**685**	100	82	72	45	19	6	6	5	**2.40**	41	35	9	1	–	–	–	–	–	–	–	–	–	L-Galactose, 6-deoxy-, O-benzyloxime, 2,3,4,5-tetrabenzoate	71641-39-9	NI47083
105	91	77	306	160	473	122	269	**685**	100	77	35	10	10	8	7	2	**0.20**	41	35	9	1	–	–	–	–	–	–	–	–	–	D-lyxo-Hexose, 2-deoxy-, O-benzyloxime, 3,4,5,6-tetrabenzoate	71641-42-4	NI47084
105	91	77	473	122	230	281	308	**685**	100	80	72	60	22	7	6	5	**1.60**	41	35	9	1	–	–	–	–	–	–	–	–	–	L-Mannose, 6-deoxy-, O-benzyloxime, 2,3,4,5-tetrabenzoate	71641-40-2	NI47085
559	531	456	404	329	432	658	357	**686**	100	80	60	5	5	4	3	3	**1.00**	2	4	–	–	–	–	–	–	4	–	–	–	2	Bis(diiodoarsino)ethane	28483-83-2	NI47086
69	169	335	119	131	219	285	100	**686**	100	87	36	32	13	11	8	7	**0.00**	12	–	3	–	–	26	–	–	–	–	–	–	–	Perfluoro-2,2,8-trimethyl-3,6,9-trioxadodecane		NI47087
204	73	103	205	217	147	337	206	**686**	100	61	22	20	19	17	9	9	**0.01**	27	66	8	–	–	–	–	–	–	6	–	–	–	2-O-Glycerol-α-D-galactopyranoside, hexakis(trimethylsilyl)-		NI47088

MASS TO CHARGE RATIOS								M.W.	INTENSITIES								Parent	C	H	O	N	S	F	Cl	Br	I	Si	P	B	X	COMPOUND NAME	CAS Reg No	No
544	546	77	236	55	180	105	318	**686**	100	94	90	71	66	46	40	30	**0.00**	28	26	7	2	–	–	–	–	–	–	–	–	1	**(3-Cyclohexyl-5-ethoxy-2-oxo-1,5-diphenyl-4-imidazolidinylidene)tungsten pentacarbonyl**		**DD01819**
175	143	88	111	101	129	75	115	**686**	100	66	30	26	26	15	14	11	**0.00**	30	54	17	–	–	–	–	–	–	–	–	–	–	**Methyl-2,3-di-O-methyl-4-O-[2-O-methyl-3,4-di-O-(2,3,4-tri-O-methyl-β-D-xylopyranosyl)-β-D-xylopyranosyl]-β-D-xylopyranoside**	78424-00-7	**NI47089**
175	143	101	235	88	111	75	115	**686**	100	87	29	24	22	20	17	14	**0.00**	30	54	17	–	–	–	–	–	–	–	–	–	–	**Per-O-methyl-β-D-xylotetraose**		**NI47090**
205	163	189	247	110	147	535	162	**686**	100	62	50	47	42	40	38	27	**0.00**	33	34	16	–	–	–	–	–	–	–	–	–	–	**Arbutin, p,3,4,6-tetraacetate 2-(3,4-dihydroxycinnamate) diacetate**	3757-39-9	**NI47091**
205	247	163	110	535	162	109	373	**686**	100	83	64	38	37	34	29	19	**0.00**	33	34	16	–	–	–	–	–	–	–	–	–	–	**β-D-Glucopyranoside, 4-(acetyloxy)phenyl, 2,3,4-triacetate 6-[3-[3,4-bis(acetyloxy)phenyl]-2-propenoate]**	3808-15-9	**NI47092**
686	492	246	461	687	477	655	151	**686**	100	73	47	46	43	35	29	25		39	42	11	–	–	–	–	–	–	–	–	–	–	**Flavan, 8-[2-(3,4-dimethoxyphenyl)-5,7-dimethoxy-2H-1-benzopyran-4-yl] 3,3′,4′,5,7-pentamethoxy-**	17941-23-0	**NI47093**
669	131	670	412	686	671	163	414	**686**	100	53	51	48	31	24	23	22		42	30	4	4	1	–	–	–	–	–	–	–	–	**Bis[1-amino-4-(4-toluidino)-anthraquinon-2-yl] sulphide**		**IC04631**
686	363	108	331	169	138	472	228	**686**	100	85	80	50	40	40	37	30		42	46	5	4	–	–	–	–	–	–	–	–	–	**Folicangine**		**LI09993**
221	362	363	247	686	325	292	222	**686**	100	75	40	30	20	15	15	15		44	59	4	–	–	–	1	–	–	–	–	–	–	**2,4,6,8,10,12-Tridecahexaenoic acid, 13-(3-chloro-4-methoxyphenyl)-, 3-methoxy-2-(9-methyldecyl)-5-(4-methylpentyl)phenyl ester**	71142-33-1	**NI47094**
105	390	391	686	567	165	437	285	**686**	100	92	25	22	19	14	11	11		49	34	4	–	–	–	–	–	–	–	–	–	–	**1-(9′-Fluorenylidene)-2,4,6-tribenzoyl-9-phenylnona-2,4,6,8-tetraene**		**LI09994**
375	295	223	217	376	296	257	169	**687**	100	85	74	37	32	23	20	20	**0.35**	29	57	8	1	–	–	–	–	–	5	–	–	–	**β-D-Glucopyranosiduronic acid, 4-[acetyl(trimethylsilyl)amino]phenyl 2,3,4-tris-O-(trimethylsilyl)-, trimethylsilyl ester**	55836-51-6	**NI47095**
371	180	152	165	77	229	210	107	**687**	100	80	65	58	50	43	31	31	**0.00**	39	30	–	3	–	6	–	–	–	–	–	3	–	**Tris(diphenylketiminoboron difluoride)**		**LI09995**
657	655	658	35	659	653	656	37	**688**	100	96	87	78	78	42	40	26	**0.59**	2	3	12	–	–	–	7	–	–	8	–	–	–	**Vinylheptachlorooctasilsesquioxane**		**NI47096**
268	39	178	77	65	230	51	688	**688**	100	48	39	30	24	22	20	19		34	15	1	–	–	10	–	–	–	–	–	–	1	**2,4-Bis(pentafluorophenyl)-3,5-diphenylcyclopentadienone-π-cyclopentadienylcobalt**		**MS10970**
363	688	252	168	139	630	108	344	**688**	100	89	14	13	13	12	12	11		42	48	5	4	–	–	–	–	–	–	–	–	–	**Isovoafolidene**		**LI09996**
474	688	363	168	139	252	138	108	**688**	100	70	35	15	13	12	12	12		42	48	5	4	–	–	–	–	–	–	–	–	–	**Voafolidine**		**LI09997**
441	443	439	165	445	163	442	440	**690**	100	97	97	97	93	93	92	92	**0.00**	6	18	–	–	–	–	–	–	–	–	–	–	6	**2,2,4,4,6,6-Hexamethyl-1,3,5-triselena-2,4,6-tristannacyclohexane**		**NI47097**
131	187	449	447	448	337	393	242	**690**	100	90	61	29	26	26	15	15	**7.00**	16	36	4	–	4	–	–	–	–	–	2	–	1	**Lead[II] bis(dibutyldithiophosphate)**	43159-18-8	**NI47098**
81	114	149	203	180	262	631	417	**690**	100	60	23	8	7	5	1	1	**1.00**	25	28	6	–	–	14	–	–	–	–	–	–	–	**Methyl 10,11-bis(heptafluorobutyryloxy)-3,7,11-trimethyl-2,6-tridecadienoate**		**MS10971**
81	114	203	180	262	631	690	417	**690**	100	73	12	11	9	4	3	2		25	28	6	–	–	14	–	–	–	–	–	–	–	**Methyl 10,11-bis(heptafluorobutyryloxy)-3,7,11-trimethyl-2,6-tridecadienoate**		**MS10972**
324	507	325	255	155	508	176	273	**690**	100	25	24	7	7	6	6	4	**0.00**	26	4	2	–	–	18	–	–	–	–	–	–	–	**1,1′-Biphenyl, 2,2′,3,3′,5,5′,6,6′-octafluoro-4,4′-bis[(pentafluorophenoxy)methyl]-**	38795-56-1	**NI47099**
181	182	161	176	93	81	69	31	**690**	100	13	6	5	3	3	3	3	**1.00**	26	4	2	–	–	18	–	–	–	–	–	–	–	**1,1′-Biphenyl, 2,2′,3,3′,5,5′,6,6′-octafluoro-4,4′-bis[(pentafluorophenyl)methoxy]-**	38795-54-9	**NI47100**
181	182	161	176	155	131	93	81	**690**	100	13	6	5	3	3	3	3	**1.00**	26	4	2	–	–	18	–	–	–	–	–	–	–	**4,4′-Bis(pentafluorobenzyloxy)octafluorobiphenyl**		**LI09998**
507	57	341	157	288	365	41	491	**690**	100	19	13	13	10	8	7	5	**0.00**	33	57	6	–	–	–	–	–	–	–	–	–	1	**Praseodymium, tris(2,2,6,6-tetramethyl-3,5-heptanedionato-O,O′)-, (OC-6 11)-**	15492-48-5	**NI47101**
315	167	285	690	524	345	301	641	**690**	100	69	50	47	42	41	28	13		38	42	12	–	–	–	–	–	–	–	–	–	–	**Octamethyldehydrodicatechin B4**		**LI09999**
91	181	253	153	151	431	539	601	**690**	100	17	7	6	6	2	1	0	**0.00**	38	43	7	–	–	–	–	1	–	–	–	–	–	**(1RS)-2-Bromo-1-ethoxyethyl 2,3,4,6-tetra-O-benzyl-β-D-glucopyranoside**		**MS10973**
547	160	100	84	279	70	56	268	**690**	100	50	46	41	33	32	26	25	**0.00**	38	50	8	4	–	–	–	–	–	–	–	–	–	**tert-Butyl 2,6-diacetyl-18-O-methylchaenorpin-30-carboxylate**		**DD01820**
109	125	110	77	218	423	405	462	**690**	100	95	85	80	57	2	2	1	**0.00**	38	62	7	–	1	–	–	–	–	1	–	–	–	**Prost-14-en-1-oic acid, 19-[[(tert-butyl)dimethylsilyl]oxy]-9-oxo-13-(phenylsulphinyl)-11-[(tetrahydro-2H-pyran-2-yl)oxy]-, methyl ester, (11α,13S,14E,19R)-(±)-**	68216-38-6	**NI47102**
352	321	365	222	690	364	353	322	**690**	100	99	74	33	31	28	25	25		42	50	5	4	–	–	–	–	–	–	–	–	–	**Ajmalan-16-carboxylic acid, 19,20-didehydro-1-[(20α)-20,21-dihydro-21-hydroxy-21-methyl-18-noralstophyllan-19-yl]-17-hydroxy-, methyl ester, (2α,17S,19E)-**	38899-91-1	**MS10974**
328	44	285	348	329	200	50	300	**690**	100	48	35	27	25	24	21	20	**0.00**	43	27	7	–	–	–	1	–	–	–	–	–	–	**6-Chloro-5,6-dihydro-6-[5,13a-dihydro-8,13a-dimethoxy-5-oxodinaphtho[1,2-b:2′,1′-d]furan-6-yl]-8-methoxy-5-oxodinaphtho[1,2-b:2′,1′-d]furan**	91913-19-8	**NI47103**
243	244	165	43	105	166	77	245	**690**	100	99	90	90	54	46	41	32	**0.09**	43	46	8	–	–	–	–	–	–	–	–	–	–	**12,13-Di-O-acetylphorbol-20-trityl ether**		**NI47104**

MASS TO CHARGE RATIOS								M.W.	INTENSITIES								Parent	C	H	O	N	S	F	Cl	Br	I	Si	P	B	X	COMPOUND NAME	CAS Reg No	No
344	301	314	302	345	287	259	271	**690**	100	36	23	23	22	15	15	13	**0.50**	44	50	–	8	–	–	–	–	–	–	–	–	–	Quadrigemine-A		NI47105
516	172	517	173	130	486	429	474	**690**	100	91	80	64	27	21	20	17	**12.00**	44	50	–	8	–	–	–	–	–	–	–	–	–	Quadrigemine-B		NI47106
376	503	392	517	350	265	606	663	**691**	100	65	54	48	32	32	23	8	**0.00**	17	19	1	1	–	–	–	–	2	–	–	–	1	Tungsten, (α-tert-butylbenzylideniminato)carbonyl-π-cyclopentadienyldiiodo-	35004-45-6	MS10975
478	691	479	480	310	352	131	436	**691**	100	52	30	26	26	22	22	20		18	42	6	–	6	–	–	–	–	–	3	–	1	Chromium(III) tris(di-N-propyldithiophosphate)	22466-20-2	NI47107
73	203	387	299	131	129	147	75	**691**	100	53	37	24	24	20	15	14	**0.80**	24	62	8	1	–	–	–	–	–	6	1	–	–	D-Glucose, 2-deoxy-3,4,5-tris-O-(trimethylsilyl)-2-[(trimethylsilyl)amino]-, 6-[bis(trimethylsilyl) phosphate]	55530-82-0	NI47108
73	676	217	147	677	372	678	458	**691**	100	49	43	40	31	22	18	16	**3.52**	27	61	6	3	–	–	–	–	–	6	–	–	–	1H-Pyrazole-5-carboxamide, N,1-bis(trimethylsilyl)-4-[(trimethylsilyl)oxy]-3-[2,3,5-tris-O-(trimethylsilyl)-β-D-ribofuranosyl]-	72360-96-4	NI47109
676	691	677	692	331	632	690	560	**691**	100	80	30	24	20	16	9	6		30	30	–	3	–	12	–	–	–	–	1	–	–	Tris(dimethylaminotetrafluorophenyl)phosphine		MS10976
128	100	470	111	109	485	236	163	**691**	100	87	35	33	25	17	16	15	**0.30**	34	46	10	3	–	–	1	–	–	–	–	–	–	Maytansine		NI47110
85	43	57	55	98	41	30	29	**691**	100	95	74	66	65	64	49	36	**0.00**	37	74	8	1	–	–	–	–	–	–	1	–	–	Hexadecanoic acid, 1-[[[(2-aminoethoxy)hydroxyphosphinyl]oxy]methyl]-1,2-ethanediyl ester	3026-45-7	NI47111
655	522	487	606	578	690	550	634	**692**	100	52	52	41	31	28	28	21	**0.00**	6	–	6	–	–	–	4	–	–	–	–	–	2	Hexacarbonyldiosmiumtetrachloride		LI10000
201	199	203	119	121	39	107	41	**692**	100	53	48	43	42	28	20	17	**0.00**	9	15	4	–	–	–	–	6	–	–	1	–	–	1-Propanol, 2,3-dibromo-, phosphate (3:1)	126-72-7	NI47112
39	41	119	137	57	219	217	27	**692**	100	95	71	64	50	26	24	24	**0.00**	9	15	4	–	–	–	–	6	–	–	1	–	–	1-Propanol, 2,3-dibromo-, phosphate (3:1)	126-72-7	IC04632
119	121	28	123	39	125	201	31	**692**	100	99	90	61	55	52	38	30	**0.00**	9	15	4	–	–	–	–	6	–	–	1	–	–	2-Propanol, 1,3-dibromo-, phosphate (3:1)	18713-51-4	NI47113
169	194	43	195	409	71	170	69	**692**	100	65	48	27	16	14	12	12	**0.00**	22	22	7	2	–	14	–	–	–	–	–	–	–	Butanoic acid, heptafluoro-, 1-[[hexahydro-1,3-dimethyl-5-(1-methylbutyl)-2,4,6-trioxo-5-pyrimidinyl]methyl]-1,2-ethanediyl ester	55538-88-0	NI47114
73	299	315	357	147	217	387	370	**692**	100	88	44	42	39	29	25	23	**0.00**	24	61	9	–	–	–	–	–	–	6	1	–	–	D-arabino-Hexonic acid, 2-deoxy-3,4,5-tris-O-(trimethylsilyl)-, trimethylsilyl ester, bis(trimethylsilyl) phosphate	55538-92-6	NI47115
73	315	147	299	230	589	316	74	**692**	100	54	33	31	22	21	13	12	**0.00**	24	61	9	–	–	–	–	–	–	6	1	–	–	β-D-Fructofuranose, 1,2,3,4-tetrakis-O-(trimethylsilyl)-, bis(trimethylsilyl) phosphate	28176-76-3	NI47116
73	217	445	299	147	75	446	328	**692**	100	56	43	30	25	22	17	17	**0.10**	24	61	9	–	–	–	–	–	–	6	1	–	–	β-D-Fructofuranose, 2,3,4,6-tetrakis-O-(trimethylsilyl)-, bis(trimethylsilyl) phosphate	55538-90-4	NI47117
73	315	299	147	589	230	590	316	**692**	100	42	24	24	22	15	11	11	**0.00**	24	61	9	–	–	–	–	–	–	6	1	–	–	D-Fructose, 1,3,4,5-tetrakis-O-(trimethylsilyl)-, 6-[bis(trimethylsilyl) phosphate]	55530-83-1	NI47118
73	204	147	299	443	442	75	315	**692**	100	22	22	20	14	14	13	11	**0.00**	24	61	9	–	–	–	–	–	–	6	1	–	–	D-Galactofuranose, 1,2,3,5-tetrakis-O-(trimethylsilyl)-, bis(trimethylsilyl) phosphate	69688-48-8	NI47119
73	387	299	217	315	147	388	75	**692**	100	29	29	29	21	21	11	11	**0.00**	24	61	9	–	–	–	–	–	–	6	1	–	–	D-Galactofuranose, 1,2,3,5-tetrakis-O-(trimethylsilyl)-, bis(trimethylsilyl) phosphate	69688-48-8	NI47120
299	73	217	204	147	315	129	75	**692**	100	99	70	64	46	37	21	19	**0.00**	24	61	9	–	–	–	–	–	–	6	1	–	–	α-D-Galactopyranose, 2,3,4,6-tetrakis-O-(trimethylsilyl)-, bis(trimethylsilyl) phosphate	55530-81-9	NI47121
73	299	217	204	147	315	129	75	**692**	100	77	54	49	35	29	16	15	**0.00**	24	61	9	–	–	–	–	–	–	6	1	–	–	α-D-Galactopyranose, 2,3,4,6-tetrakis-O-(trimethylsilyl)-, bis(trimethylsilyl) phosphate	55530-81-9	NI47122
73	204	387	299	129	147	357	217	**692**	100	42	39	34	26	23	20	18	**0.00**	24	61	9	–	–	–	–	–	–	6	1	–	–	D-Galactopyranose, 1,2,3,4-tetrakis-O-(trimethylsilyl)-, bis(trimethylsilyl) phosphate	69744-66-7	NI47123
73	204	299	387	129	217	147	191	**692**	100	35	32	28	25	18	15	13	**0.00**	24	61	9	–	–	–	–	–	–	6	1	–	–	D-Galactopyranose, 1,2,3,4-tetrakis-O-(trimethylsilyl)-, bis(trimethylsilyl) phosphate	69744-66-7	NI47124
73	443	299	204	442	315	387	147	**692**	100	39	39	37	35	35	24	21	**0.00**	24	61	9	–	–	–	–	–	–	6	1	–	–	D-Galactose, 2,3,4,5-tetrakis-O-(trimethylsilyl)-, 6-[bis(trimethylsilyl) phosphate]	55520-83-7	NI47125
73	299	147	217	204	75	315	133	**692**	100	48	45	39	31	26	20	12	**0.00**	24	61	9	–	–	–	–	–	–	6	1	–	–	α-D-Glucopyranose, 2,3,4,6-tetrakis-O-(trimethylsilyl)-, bis(trimethylsilyl) phosphate	55520-79-1	NI47127
299	73	147	217	204	75	315	133	**692**	100	99	94	81	66	54	43	25	**0.00**	24	61	9	–	–	–	–	–	–	6	1	–	–	α-D-Glucopyranose, 2,3,4,6-tetrakis-O-(trimethylsilyl)-, bis(trimethylsilyl) phosphate	55520-79-1	NI47126
73	204	387	299	129	147	388	205	**692**	100	41	31	23	19	18	11	10	**0.00**	24	61	9	–	–	–	–	–	–	6	1	–	–	D-Glucopyranose, 1,2,3,4-tetrakis-O-(trimethylsilyl)-, bis(trimethylsilyl) phosphate	28176-73-0	NI47128
73	204	387	147	299	129	315	388	**692**	100	38	29	20	18	16	12	10	**0.00**	24	61	9	–	–	–	–	–	–	6	1	–	–	D-Glucopyranose, 1,2,3,4-tetrakis-O-(trimethylsilyl)-, bis(trimethylsilyl) phosphate	28176-73-0	NI47129

MASS TO CHARGE RATIOS								M.W.	INTENSITIES								Parent	C	H	O	N	S	F	Cl	Br	I	Si	P	B	X	COMPOUND NAME	CAS Reg No	No
73	387	204	299	388	129	147	315	**692**	100	73	70	40	25	24	22	18	**0.00**	24	61	9	–	–	–	–	–	–	6	1	–	–	D-Glucose, 2,3,4,5-tetrakis-O-(trimethylsilyl)-, 6-[bis(trimethylsilyl) phosphate]	55520-80-4	NI47130
73	387	204	147	299	129	388	315	**692**	100	49	49	26	25	20	18	14	**0.00**	24	61	9	–	–	–	–	–	–	6	1	–	–	D-Glucose, 2,3,4,5-tetrakis-O-(trimethylsilyl)-, 6-[bis(trimethylsilyl) phosphate]	55520-80-4	NI47131
315	73	217	299	147	204	316	450	**692**	100	99	72	64	58	46	30	27	**0.00**	24	61	9	–	–	–	–	–	–	6	1	–	–	α-D-Mannopyranose, 2,3,4,6-tetrakis-O-(trimethylsilyl)-, bis(trimethylsilyl) phosphate	55520-81-5	NI47132
73	315	217	299	147	204	316	450	**692**	100	71	51	46	41	33	21	19	**0.00**	24	61	9	–	–	–	–	–	–	6	1	–	–	α-D-Mannopyranose, 2,3,4,6-tetrakis-O-(trimethylsilyl)-, bis(trimethylsilyl) phosphate	55520-81-5	NI47133
73	204	387	299	129	147	388	217	**692**	100	44	35	29	24	21	13	10	**0.00**	24	61	9	–	–	–	–	–	–	6	1	–	–	D-Mannopyranose, 1,2,3,4-tetrakis-O-(trimethylsilyl)-, bis(trimethylsilyl) phosphate	69744-65-6	NI47135
73	204	387	299	129	147	388	74	**692**	100	60	46	35	32	25	16	13	**0.00**	24	61	9	–	–	–	–	–	–	6	1	–	–	D-Mannopyranose, 1,2,3,4-tetrakis-O-(trimethylsilyl)-, bis(trimethylsilyl) phosphate	69744-65-6	NI47134
73	387	204	299	129	388	147	357	**692**	100	76	72	45	28	25	21	18	**0.00**	24	61	9	–	–	–	–	–	–	6	1	–	–	D-Mannose, 2,3,4,5-tetrakis-O-(trimethylsilyl)-, 6-[bis(trimethylsilyl) phosphate]	55520-82-6	NI47136
300	43	342	384	42	301	57	55	**692**	100	78	25	21	20	19	19	17	**0.00**	32	36	17	–	–	–	–	–	–	–	–	–	–	4H-1-Benzopyran-4-one, 5-(acetyloxy)-2-[3-(acetyloxy)-4-methoxyphenyl] 7-[[6-O-(6-deoxy-α-L-mannopyranosyl)-β-D-glucopyranosyl]oxy]-	57408-39-6	NI47137
43	205	163	162	247	81	99	141	**692**	100	92	89	64	34	32	27	23	**0.00**	33	40	16	–	–	–	–	–	–	–	–	–	–	Grandidentaside hexaacetate		LI10001
43	84	204	201	568	203	248	247	**692**	100	80	10	9	8	7	6	6	**0.00**	36	53	8	–	–	–	–	1	–	–	–	–	–	12α-Bromo-2α,3β,13β,23-tetrahydroxyoleanan-28-oic acid 13β,28-lactone,2α,3β,23-triacetate	89503-62-8	NI47138
617	321	661	405	618	491	575	463	**692**	100	56	43	39	36	30	22	20	**3.05**	36	61	11	–	–	–	–	–	–	–	–	–	1	Monensin, monosodium salt	22373-78-0	NI47139
617	661	618	321	405	662	491	463	**692**	100	45	40	39	20	17	17	16	**2.00**	36	61	11	–	–	–	–	–	–	–	–	–	1	Monensin, monosodium salt	22373-78-0	NI47140
603	307	647	604	391	477	308	449	**692**	100	63	40	38	35	24	22	20	**0.00**	36	61	11	–	–	–	–	–	–	–	–	–	1	Monensin, monosodium salt	22373-78-0	NI47141
115	86	87	99	100	158	101	238	**692**	100	99	80	53	50	31	28	27	**1.00**	37	68	6	6	–	–	–	–	–	–	–	–	–	Acetamide, N-[3-[acetyl[3-(acetylamino)propyl]amino]propyl]-N-[3-[acetyl[3-[acetyl[3-(2-oxoazacyclotridec-1-yl)propyl]amino]propyl]amino]propyl]-	75422-11-6	NI47142
100	649	70	169	84	114	155	126	**692**	100	62	54	46	35	29	25	25	**6.00**	37	68	6	6	–	–	–	–	–	–	–	–	–	13,17,21,25,29-Pentaacetyl-13,17,21,25,29-pentaaza-32-dotriacontanelactam	75422-13-8	NI47143
334	121	207	335	692	221	151	122	**692**	100	81	51	27	17	11	10	10		46	76	4	–	–	–	–	–	–	–	–	–	–	Benzeneheptadecanoic acid, 4-methoxy-, 2-decyl-3-methoxy-5-pentylphenyl ester	69506-03-2	NI47144
362	221	121	692	222	292	302	151	**692**	100	49	26	10	8	6	3	3		46	76	4	–	–	–	–	–	–	–	–	–	–	Benzenepentadecanoic acid, 4-methoxy-, 3-methoxy-2-(9-methyldecyl)-5-(4-methylpentyl)phenyl ester	63953-41-3	NI47145
299	197	254	360	263	378	399	693	**693**	100	99	96	52	48	18	14	6		21	24	3	–	3	9	–	–	–	–	–	–	1	Tris(1,1,1-trifluoro-4-mercapto-5-methylhex-3-en-2-onato)ruthenium(III)	56288-43-8	NI47146
503	531	694	638	664				**694**	100	25	20	16	12					12	19	8	–	–	6	1	–	–	–	2	–	1	1,1,1,4,4,4-Hexafluorobut-2-en-2-yl bis(trimethoxyphosphine) dicarbonyl osmium chloride		LI10002
665	666	318	667	619	304	319	635	**694**	100	65	46	29	29	13	13	13	**1.69**	18	46	13	–	–	–	–	–	–	8	–	–	–	1,3,5,7,9,11,13,15-Octaethyl-7,13-dimethoxytetracyclo[9.5.1.$1^{3,9}$.$1^{5,15}$]octasiloxane	72617-70-0	NI47147
73	81	135	114	93	107	99	121	**694**	100	76	36	28	22	15	12	10	**2.00**	25	29	5	–	–	15	–	–	–	–	–	–	–	Methyl 11-methoxy-3,7,11-trimethyl-10-(pentadecafluorooctangyloxy)-2,6 dodecadienoate		MS10977
347	278	348	297	309	247	675	279	**694**	100	21	10	7	5	5	3	3	**0.20**	26	2	–	–	–	20	–	–	–	–	–	–	–	Ethane, 1,1,2,2-tetrakis(pentafluorophenyl)-	5736-50-5	NI47148
347	278	297	309	247	675	328	279	**694**	100	21	7	5	5	3	3	3	**0.15**	26	2	–	–	–	20	–	–	–	–	–	–	–	Ethane, 1,1,2,2-tetrakis(pentafluorophenyl)-		LI10003
115	71	60	84	43	86	44	41	**694**	100	18	18	17	15	11	11	11	**2.00**	28	35	10	6	–	–	–	1	–	–	–	–	–	N,N-Bis(butyryloxyethyl)-2-ethoxy-4-acetamido-3-(6-bromo-2,4-dinitrophenylazo)aniline		IC04633
357	126	599	125	168	297	142	114	**694**	100	27	22	21	20	13	12	12	**0.00**	28	42	18	2	–	–	–	–	–	–	–	–	–	D-Glucose, 4-amino-4-deoxy-, 2,3,5,6-tetraacetate, dimer	56195-97-2	NI47149
301	173	303	302	431	73	211	217	**694**	100	46	26	26	23	15	14	10	**0.00**	33	78	5	–	–	–	–	–	–	5	–	–	–	1,9,10,12,13-Penta(trimethylsilyloxy)octadecane		LI10004
197	195	179	196	213	237	177	209	**694**	100	81	42	40	38	23	20	19	**0.00**	35	59	12	–	–	–	–	–	–	–	–	–	1	18-Hydroxy-3-de-O-methylmonensin sodium salt		NI47150
263	163	221	207	167	227	205	264	**694**	100	75	74	46	36	32	32	19	**0.00**	35	59	12	–	–	–	–	–	–	–	–	–	1	23-Hydroxy-3-de-O-methylmonensin sodium salt		NI47151
298	297	299	296	594	190	191	162	**694**	100	26	22	18	17	17	15	15	**0.00**	38	44	6	2	–	–	2	–	–	–	–	–	–	Tubocurarine chloride		MS10978
649	576	19	678	605	667	635	589	**694**	100	38	27	25	22	12	11	11	**2.00**	39	42	8	4	–	–	–	–	–	–	–	–	–	2-Deformyl-δ-formyl-2-(3-oxoprop-1-enyl)chlorin E6, trimethyl ester		LI10005
105	122	77	162	191	100	84	318	**694**	100	48	46	44	39	30	25	18	**0.50**	40	46	7	4	–	–	–	–	–	–	–	–	–	4H-1,16-Etheno-5,15-(propaniminoethano)furo[3,4-l][1,5,10]triazacyclohexadecine-4,21-dione, 10,14-diacetyl-22-benzoyl-3,3a,6,7,8,9,10,11,12,13,14,15-dodecahydro-3-(4-methoxyphenyl)-	69721-72-8	NI47152

MASS TO CHARGE RATIOS								M.W.	INTENSITIES								Parent	C	H	O	N	S	F	Cl	Br	I	Si	P	B	X	COMPOUND NAME	CAS Reg No	No
122	105	77	51	123	50	106	74	**694**	100	84	80	54	47	46	43	42	**0.00**	42	46	9	–	–	–	–	–	–	–	–	–	–	14β,17α-Pregn-5-ene-3β,8,12β,14,17,20α-hexol, 3,12,20-tribenzoate	15894-87-8	NI47153
390	391	392	542	543	373	346	345	**694**	100	49	14	12	9	9	7	7	**4.00**	46	50	4	2	–	–	–	–	–	–	–	–	–	N,N′-Dicycloundecyl-3,4:9,10-perylenebis(dicarboximide)	110590-83-5	MS10979
73	75	117	253	86	147	131	116	**695**	100	71	46	35	32	31	24	20	**0.00**	36	69	6	1	–	–	–	–	–	3	–	–	–	Glycine, N-[(3α,5β,7α,12α)-24-oxo-3,7,12-tris[(trimethylsilyl)oxy]cholan-24-yl]-, methyl ester	57326-16-6	NI47154
665	35	663	667	666	664	661	37	**696**	100	94	85	78	46	37	33	32	**0.59**	–	–	12	–	–	–	8	–	–	8	–	–	–	Perchlorooctasilsesquioxane		NI47155
69	97	231	159	131	627	209	181	**696**	100	47	12	9	5	4	4	4	**0.00**	12	–	4	–	1	24	–	–	–	–	–	–	–	Perfluoropinacol orthosulphate		LI10006
161	245	322	217	202	189	135	692	**696**	100	89	66	66	56	37	27	26		16	10	6	–	–	–	–	–	–	–	–	–	3	Bis[π-cyclopentadienyl(tricarbonyl)molybdenum]mercury		MS10980
277	696	227	276	258	425	697	444	**696**	100	79	59	39	34	24	21	21		24	–	–	–	–	20	–	–	–	1	–	–	–	Tetrakis(pentafluorophenyl)silane		MS10981
121	263	262	199	183	384	77	128	**696**	100	75	70	38	33	28	25	21	**0.00**	25	22	–	–	–	–	–	–	2	–	2	–	1	Nickel, diiodo[methylenebis[diphenylphosphine]-P,P′]-	50769-18-1	NI47156
640	638	584	637	634	324	556	248	**696**	100	69	67	66	66	60	57	54	**31.33**	30	24	4	–	–	–	–	–	–	–	–	–	3	cis-[1,2-Bis(diphenylarsino)ethane]tetracarbonylmolybdenum		LI10007
101	71	73	55	56	83	129	86	**696**	100	95	61	50	45	43	36	36	**0.00**	34	40	4	4	4	–	–	–	–	–	–	–	–	4,13-Ditosyl-8,9,17,18-dibenzo-1,7-dithia-4,10,13,16-tetraazacyclooctadeca-8,17-diene		MS10982
73	75	423	97	143	513	191	129	**696**	100	53	53	39	38	34	31	28	**0.00**	35	68	6	–	–	–	–	–	–	4	–	–	–	19-Hydroxy-16,16-dimethyl f2a, methyl ester, tetrakis(trimethylsilyl)-		NI47157
423	75	191	129	55	513	333	147	**696**	100	76	75	73	65	57	54	49	**0.00**	35	68	6	–	–	–	–	–	–	4	–	–	–	19-Hydroxy-16,16-dimethyl f2a, methyl ester, tetrakis(trimethylsilyl)-		NI47158
74	357	43	339	85	203	73	95	**696**	100	75	74	67	62	54	44	43	**0.70**	36	56	13	–	–	–	–	–	–	–	–	–	–	Thevebioside	33279-55-9	LI10008
74	357	339	203	356	358	161	246	**696**	100	75	66	53	28	23	22	21	**0.60**	36	56	13	–	–	–	–	–	–	–	–	–	–	Thevebioside	33279-55-9	NI47159
43	185	183	69	81	55	420	109	**696**	100	58	57	43	37	34	30	24	**9.09**	39	53	6	–	–	–	–	1	–	–	–	–	–	Lanost-9(11)-en-18-oic acid, 23-(acetyloxy)-3-[(4-bromobenzoyl)oxy]-20-hydroxy-, γ-lactone, (3β,20ξ)-	56259-26-8	NI47160
473	474	327	475	106	77	696	328	**696**	100	33	8	6	6	3	2	2		45	33	6	–	–	–	–	–	–	–	–	–	1	Aluminium, tris(1,3-diphenyl-1,3-propanedionato)-	14405-36-8	NI47161
363	361	365	322	243	241	364	162	**697**	100	54	50	28	22	22	19	17	**0.60**	28	15	1	1	–	–	–	4	–	–	–	–	–	Dispiro-(2,7-dibromofluorene)-9,4′-Δ²-2′-methyloxazoline-5′,9″,2″,7″-dibromofluorene		MS10983
482	697	286	412	236	267	344	173	**697**	100	49	18	14	9	7	4	1		30	18	6	–	–	9	–	–	–	–	–	–	1	Tris(1,1,1-trifluoro-4-phenyl-2,4-butanedionato)chromium		LI10009
202	73	361	217	147	203	103	362	**697**	100	91	71	61	32	29	24	23	**3.50**	31	59	7	1	–	–	–	–	–	5	–	–	–	1-O-(1-Trimethylsilylindole-3-acetyl)-2,3,4,5-tetra(trimethylsilyl)-β-D-glucopyranose		NI47162
80	82	127	126	96	140	110	195	**698**	100	99	62	51	22	19	19	17	**0.00**	22	36	4	8	2	–	–	2	–	–	–	–	–	3,3′-(1,6-Hexanediyl)bis[1,2-carbamimidoylthioethyl-6-methyl-2,4(1H,3H)pyrimidinedione] dihydrobromide		MS10984
77	51	139	137	244	135	50	120	**698**	100	69	49	40	32	25	15	13	**3.73**	24	5	–	–	–	15	–	–	–	–	–	–	1	Stannane, tris(pentafluorophenyl)phenyl-	1262-57-3	NI47163
272	270	77	167	169	371	180	349	**698**	100	92	43	41	38	32	19	18	**0.00**	26	20	–	2	–	–	–	4	–	–	–	2	–	Bis(diphenylketiminoboron dibromine)		LI10010
169	109	127	145	182	331	211	139	**698**	100	66	13	10	9	8	5	5	**4.00**	36	46	12	2	–	–	–	–	–	–	–	–	–	17-O-Acetyl-21-O-(tetra-O-acetyl-β-D-glucopyranosyl)ajmaline (tetraacetyl)acetylrauglucine		MS10985
130	91	87	129	127	148	101	117	**698**	100	93	73	59	44	30	30	26	**0.50**	36	46	12	2	–	–	–	–	–	–	–	–	–	Neoantimycin	22862-63-1	LI10011
136	87	146	101	219	117	164	246	**698**	100	73	29	29	27	25	19	10	**0.00**	36	46	12	2	–	–	–	–	–	–	–	–	–	Neoantimycin	22862-63-1	NI47164
136	91	87	566	548	135	133	538	**698**	100	92	73	60	59	59	44	34	**5.11**	36	46	12	2	–	–	–	–	–	–	–	–	–	Neoantimycin	22862-63-1	NI47165
121	105	162	77	191	100	84	134	**698**	100	87	70	39	36	30	24	20	**0.50**	40	50	7	4	–	–	–	–	–	–	–	–	–	Benzamide, N-[3-[10,14-diacetyl-3a,4,6,7,8,9,10,11,12,13,14,15-dodecahydro-15-(2-hydroxyethyl)-3-(4-methoxyphenyl)-4-oxo-1,16-ethenofuro[3,4-l][1,5,10]triazacyclohexadecin-5(3H)-yl]propyl]-	69721-76-2	NI47166
334	207	335	155	121	336	278	208	**698**	100	38	30	13	12	9	8	5	**4.00**	44	71	4	–	–	–	1	–	–	–	–	–	–	Benzenepentadecanoic acid, 3-chloro-4-methoxy-, 2-decyl-3-methoxy-5-pentylphenyl ester	71142-42-2	NI47167
362	221	363	155	698	292	222	121	**698**	100	52	31	14	8	8	7	7		44	71	4	–	–	–	1	–	–	–	–	–	–	Benzenetridecanoic acid, 3-chloro-4-methoxy-, 3-methoxy-2-(9-methyldecyl)-5-(4-methylpentyl)phenyl ester	71142-39-7	NI47168
390	391	698	530	586	460	392	531	**698**	100	63	36	35	20	17	17	15		46	54	4	2	–	–	–	–	–	–	–	–	–	N,N′-Bis(1-pentylhexyl)-3,4:9,10-perylenebis(dicarboximide)	110590-83-5	MS10986
475	680	670	682	477	672	698	701	**698**	100	74	55	44	44	43	24	12		46	82	4	–	–	–	–	–	–	–	–	–	–	1,18-Bis(cyclotetraadeca-2,14-dionyl)octadecane		LI10012
138	533	505	671	699	561	398	367	**699**	100	84	52	20	15	13	11	9		27	33	3	1	–	–	–	–	–	–	3	–	1	Rhenium, dicarbonyltris(dimethylphenylphosphine)(isocyanato)-, stereoisomer	33990-59-9	NI47169
615	614	616	532	533	462	531	617	**699**	100	87	86	49	46	38	36	25	**11.00**	32	58	–	–	–	–	–	–	–	–	2	–	1	(3,3-Dimethyl-1-butyne)[bis(dicyclohexylphosphino)ethane]platinum(0)		MS10987

MASS TO CHARGE RATIOS								M.W.	INTENSITIES								Parent	C	H	O	N	S	F	Cl	Br	I	Si	P	B	X	COMPOUND NAME	CAS Reg No	No
91	79	107	106	84	77	92	139	**699**	100	29	26	21	21	15	10	9	**0.22**	34	41	13	3	–	–	–	–	–	–	–	–	–	Glutamine, N-(2-acetamido-2-deoxy-β-D-glucopyranosyl)-N^2-carboxy-, dibenzyl ester, 3′,4′,6′-triacetate, L-	5123-47-7	NI47170
97	95	251	111	109	362	347	129	**700**	100	86	74	61	60	51	40	35	**0.00**	33	57	12	–	1	–	–	–	–	–	–	–	1	24ξ-(3′-O-Methyl-α-L-arabinofuranosyloxy)-5α-cholestane-3β,6α,8β,15α tetraol 5′-O-sodium sulphate		NI47171
616	617	308	55	134	135	117	41	**700**	100	43	28	18	13	11	11	11	**4.35**	39	66	3	–	–	–	–	–	–	–	2	–	1	Iron, tricarbonylbis(tricyclohexylphosphine)-	40697-14-1	NI47172
138	562	700	237	504	149	642	350	**700**	100	53	45	23	18	13	10	9		43	48	5	4	–	–	–	–	–	–	–	–	–	Owerreine		LI10013
562	138	700	504	563	701	530	149	**700**	100	58	54	31	30	27	19	18		43	48	5	4	–	–	–	–	–	–	–	–	–	Vobtusine, 2′,3′-didehydro-2′-deoxy-	25662-91-3	NI47173
562	138	700	504	530	149	305	642	**700**	100	77	70	39	25	25	23	20		43	48	5	4	–	–	–	–	–	–	–	–	–	Vobtusine, 2′,3′-didehydro-2′-deoxy-	25662-91-3	LI10014
43	197	91	368	105	119	682	664	**700**	100	31	31	24	15	13	5	2	**1.00**	44	60	7	–	–	–	–	–	–	–	–	–	–	Isofucoxanthin acetate		LI10015
28	93	503	63	109	74	32	124	**701**	100	12	8	8	6	5	5	4	**1.06**	16	29	13	1	–	–	–	–	–	1	2	–	3	Iron, heptacarbonyldihydrobis(trimethyl phosphite-P)[μ^3-[1,1,1-trimethylsilanaminato(2–)]]tri-, triangulo	60370-92-5	NI47174
486	221	701	436	177	417	290	271	**701**	100	26	17	7	6	5	4	3		30	18	6	–	–	9	–	–	–	–	–	–	1	Tris(1,1,1-trifluoro-4-phenyl-2,4-butanedionato)iron		LI10016
615	616	614	532	533	531	462	617	**701**	100	86	86	63	55	47	41	31	**3.00**	32	60	–	–	–	–	–	–	–	–	2	–	1	(2,3-Dimethyl-2-butene)[bis(dicyclohexylphosphino)ethane]platinum(0)		MS10988
615	616	614	532	533	611	617	531	**701**	100	90	87	39	36	34	30	29	**5.00**	32	60	–	–	–	–	–	–	–	–	2	–	1	(3,3-Dimethyl-1-butene)[bis(dicyclohexylphosphino)ethane]platinum(0)		MS10989
289	291	198	200	202	293	253	163	**702**	100	98	96	96	36	32	30	25	**0.00**	16	22	–	2	2	–	6	–	–	–	–	–	2	Tetrachlorodi-μ-chlorobis(dimethyl sulphide)bis(phenylimino)diniobium	91054-06-7	NI47175
336	334	517	57	519	320	337	335	**702**	100	52	41	15	9	9	8	7	**0.00**	33	57	6	–	–	–	–	–	–	–	–	–	1	Europium, tris(2,2,6,6-tetramethyl-3,5-heptanedionato-O,O′)-, (OC-6-11)-	15522-71-1	NI47176
702	703	320	671	351	121	639	105	**702**	100	44	24	16	12	8	3	2		40	30	8	–	2	–	–	–	–	–	–	–	–	Tetramethyl 2,2′-(4,4′-biphenylylene)bis(5-phenyl-3,4-thiophenedicarboxylate)	110699-38-2	DD01821
219	149	161	203	55	41	57	220	**702**	100	36	33	26	26	24	23	18	**0.05**	40	62	6	–	2	–	–	–	–	–	–	–	–	1,6-bis-[(2,6-di-tert-butyl-4-hydroxy)benzylthioacetoxy]hexane		IC04634
138	702	488	377	174	188	162	110	**702**	100	88	70	62	28	21	20	20		43	50	5	4	–	–	–	–	–	–	–	–	–	Desoxyvobtusine		LI10017
702	377	138	174	154	214	168	110	**702**	100	78	77	24	21	19	16	14		43	50	5	4	–	–	–	–	–	–	–	–	–	2′,3′-Dihydroanhydrovobtusine		LI10018
138	488	377	702	174	110	644	188	**702**	100	49	44	42	24	21	20	18		43	50	5	4	–	–	–	–	–	–	–	–	–	Goziline		LI10019
190	147	119	107	189	105	131	121	**702**	100	99	92	79	73	63	57	56	**1.80**	43	58	8	–	–	–	–	–	–	–	–	–	–	Rubicoumaric acid diacetate		NI47177
684	702	263	624	666	622	668	586	**702**	100	34	32	28	23	18	12	11		44	62	7	–	–	–	–	–	–	–	–	–	–	(3S,5R,6S,3′S,5′R,6′R)-6,7-Didehydro-5,6,5′,6′-tetrahydro-β,β-carotene-3,5,6,3′,5′-pentol, 3,3′-diacetate		NI47178
211	702	703	225	212	161	704	173	**702**	100	50	30	23	19	15	14	14		50	70	2	–	–	–	–	–	–	–	–	–	–	1,4-Naphthalenedione, 2-(3,7,11,15,19,23,27,31-octamethyl-2,6,10,14,18,22,26,30-dotriacontaoctaenyl)-, (all-E)-	29790-47-4	NI47179
57	71	43	85	55	69	83	41	**702**	100	63	63	42	31	28	26	22	**0.85**	50	102	–	–	–	–	–	–	–	–	–	–	–	11,20-Bis(decyltriacontane)		AI02411
57	43	71	85	55	41	69	83	**702**	100	73	60	39	31	27	25	23	**0.00**	50	102	–	–	–	–	–	–	–	–	–	–	–	11,20-Bis(decyltriacontane)		AI02412
57	43	71	85	55	69	83	41	**702**	100	73	61	39	30	25	22	22	**0.00**	50	102	–	–	–	–	–	–	–	–	–	–	–	11,20-Bis(decyltriacontane)		AI02413
57	43	71	85	55	41	69	83	**702**	100	86	59	41	38	33	32	29	**0.00**	50	102	–	–	–	–	–	–	–	–	–	–	–	13-Heptyl-13-decyltritriacontane		AI02414
57	43	71	85	55	83	69	41	**702**	100	77	67	48	29	27	26	23	**0.00**	50	102	–	–	–	–	–	–	–	–	–	–	–	17-Hexadecyltetratriacontane		AI02415
573	530	589	516	479	574	461	590	**703**	100	80	63	63	43	40	34	29	**15.00**	22	25	9	3	–	12	–	–	–	–	–	–	–	O,N-Tetra(trifluoroacetyl)[N-η-(5-amino-5-carboxypentanyl)]hydroxylysine methyl ester		LI10020
169	331	109	127	644	356	398	584	**703**	100	66	40	37	18	16	5	3	**0.60**	31	45	17	1	–	–	–	–	–	–	–	–	–	N-Cyclopentyl-2,2′,3,3′,4′,6,6′-hepta-O-acetyl-D-lactosylamine		NI47180
69	64	113	73	44	451	163	251	**704**	100	68	27	24	23	18	18	8	**5.50**	8	–	–	–	2	18	–	–	–	–	–	–	1	Bis(perfluoro-tert-butylmercapto)mercury		LI10021
209	57	323	380	267	647	704	591	**704**	100	80	70	60	60	20	18	4		17	27	5	–	–	–	–	–	–	–	–	–	2	Pentacarbonyl(tributylbismuthine)tungsten	90188-77-5	NI47181
165	278	219	332	184	445	504	671	**704**	100	90	53	47	18	13	4	3	**0.00**	20	35	–	5	–	8	–	–	–	–	5	–	1	Chromium, tetrakis[difluoro(dimethylamino)phosphine][methyliminobis(diphenylphosphine)]		MS10990
95	114	163	135	276	217	673	645	**704**	100	84	35	18	16	8	4	4	**1.00**	26	30	6	–	–	14	–	–	–	–	–	–	–	Methyl 7-ethyl-10,11-bis(heptafluorobutyryloxy)-3,11-dimethyl-2,6-tridecadienoate		MS10991
489	420	704	224	274	180	376		**704**	100	73	42	34	20	5	3			30	18	6	–	–	9	–	–	–	–	–	–	1	Tris(1,1,1-trifluoro-4-phenyl-2,4-butanedionato)cobalt		LI10022
592	55	240	676	271	209	326	564	**704**	100	72	34	30	26	26	13	10	**0.00**	32	20	8	–	–	–	–	–	–	–	2	–	2	Di-μ-diphenylphosphinobis(tetracarbonylmanganese)		MS10992
673	529	689	541	704	601	573	559	**704**	100	41	35	35	20	15	15	8		36	48	14	–	–	–	–	–	–	–	–	–	–	4H-1-Benzopyran-4-one, 5,7-dimethoxy-2-(4-methoxyphenyl)-6-(2,3,4,6-tetra-O-methyl-β-D-glucopyranosyl)-8-(tri-O-methylarabinosyl)-	75284-21-8	NI47182

MASS TO CHARGE RATIOS								M.W.	INTENSITIES								Parent	C	H	O	N	S	F	Cl	Br	I	Si	P	B	X	COMPOUND NAME	CAS Reg No	No
673	573	704	585	689	643	559	529	**704**	100	33	28	28	25	20	18	18		36	48	14	–	–	–	–	–	–	–	–	–	–	**4H-1-Benzopyran-4-one, 5,7-dimethoxy-2-(4-methoxyphenyl)-8-(2,3,4,6-tetra-O-methyl-β-D-glucopyranosyl)-6-(tri-O-methylarabinosyl)-**	75284-22-9	**NI47183**
673	529	689	704	541	626	573	657	**704**	100	31	25	23	23	14	14	10		36	48	14	–	–	–	–	–	–	–	–	–	–	**4H-1-Benzopyran-4-one, 5,7-dimethoxy-2-(4-methoxyphenyl)-6-(tetra-O-methyl-β-D-glucosyl)-8-(tri-O-methyl-α-L-arabinosyl)-**	71976-86-8	**NI47184**
122	136	180	182	181	704	135	124	**704**	100	62	50	45	45	30	23	20		43	52	5	4	–	–	–	–	–	–	–	–	–	**Voacamine**	3371-85-5	**NI47185**
122	180	181	136	182	194	511	524	**704**	100	75	68	52	31	20	18	11	**7.00**	43	52	5	4	–	–	–	–	–	–	–	–	–	**Voacamine**		**LI10023**
704	377	138	174	378	352	144	365	**704**	100	38	37	25	10	10	10	9		43	52	5	4	–	–	–	–	–	–	–	–	–	**Vobtusine, 2′-deoxy-2,3-dihydro-, (2′ϵ,3′ϵ)-**	56114-54-6	**NI47186**
221	362	265	363	187	121	292	222	**704**	100	97	79	39	21	21	20	18	**18.00**	48	64	4	–	–	–	–	–	–	–	–	–	–	**2,4,6,8,10,12,14,16-Heptadecaoctaenoic acid, 17-(4-methoxyphenyl)-, 3-methoxy-2-(9-methyldecyl)-5-(4-methylpentyl)phenyl ester**	69505-98-2	**NI47187**
142	114	470	109	111	485	236	472	**705**	100	57	21	17	14	11	10	9	**0.10**	35	48	10	3	–	–	1	–	–	–	–	–	–	**Maytanprine**	38997-09-0	**NI47188**
128	100	111	236	109	184	224	222	**705**	100	45	19	11	10	7	6	6	**0.00**	35	48	10	3	–	–	1	–	–	–	–	–	–	**9-O-Methylmaytansine**	64817-73-8	**NI47189**
170	30	43	705	171	71	706	268	**705**	100	99	95	75	56	42	41	27		43	67	5	3	–	–	–	–	–	–	–	–	–	**1-(1-Hydroxynaphth-2-ylcarboxylaminohexyl)-4-(N-isobutyryl)octadecyalmino-2,5-dioxopyrrole**		**IC04635**
566	444	416	538	482	398	370	510	**706**	100	95	70	68	61	60	53	43	**32.00**	12	1	12	–	–	–	–	–	–	–	–	–	4	**Hydridoosmiumtricobaltdodecacarbonyl**		**LI10024**
215	216	251	712	250	714	126	161	**706**	100	71	65	62	62	57	52	49	**8.24**	18	–	–	–	–	–	14	–	–	–	–	–	–	**Terphenyl, tetradecachloro-**	29629-84-3	**NI47190**
712	714	710	716	708	718	713	642	**706**	100	87	81	57	39	27	19	18	**7.37**	18	–	–	–	–	–	14	–	–	–	–	–	–	**Terphenyl, tetradecachloro-**	29629-84-3	**NI47191**
73	299	147	315	204	217	74	75	**706**	100	67	54	51	49	46	19	18	**0.00**	24	59	10	–	–	–	–	–	–	6	1	–	–	**D-Galactopyranuronic acid, 2,3,4-tris-O-(trimethylsilyl)-, trimethylsilyl ester, bis(trimethylsilyl) phosphate**	69745-86-4	**NI47192**
73	299	315	217	147	204	300	387	**706**	100	83	53	51	46	33	22	20	**0.00**	24	59	10	–	–	–	–	–	–	6	1	–	–	**D-Glucopyranuronic acid, 2,3,4-tris-O-(trimethylsilyl)-, trimethylsilyl ester, bis(trimethylsilyl) phosphate**	69727-23-7	**NI47193**
57	307	41	27	317	297	308	309	**706**	100	74	57	55	39	34	30	19	**0.00**	32	66	9	–	–	–	–	–	–	4	–	–	–	**1,3,5,7-Tetravinyl-1,1,3,5,7,7-hexabutoxytetrasiloxane**	110991-11-2	**NI47194**
706	663	664	191	649	353	437	381	**706**	100	58	50	30	19	9	5	5		41	42	9	2	–	–	–	–	–	–	–	–	–	**O,O-Diacetyloxandrine**		**NI47195**
706	705	663	191	664	691	621	381	**706**	100	55	34	34	28	13	8	4		41	42	9	2	–	–	–	–	–	–	–	–	–	**O,O-Diacetylpseudoxandrine**		**NI47196**
73	315	230	299	382	103	321	75	**707**	100	33	26	20	19	16	15	15	**0.00**	25	54	7	5	–	–	–	–	–	5	1	–	–	**3′-Adenylic acid, N-(trimethylsilyl)-2′,5′-bis-O-(trimethylsilyl)-, bis(trimethylsilyl) ester**	32653-17-1	**NI47197**
73	169	315	230	258	316	299	147	**707**	100	68	67	36	22	17	16	15	**1.30**	25	54	7	5	–	–	–	–	–	5	1	–	–	**5′-Adenylic acid, N-(trimethylsilyl)-2′,3′-bis-O-(trimethylsilyl)-, bis(trimethylsilyl) ester**	32653-14-8	**NI47198**
73	692	315	169	466	693	230	694	**707**	100	76	68	68	56	42	40	24	**14.00**	25	54	7	5	–	–	–	–	–	5	1	–	–	**5′-Adenylic acid, pentakis(trimethylsilyl)-**		**LI10025**
73	692	315	169	466	693	230	694	**707**	100	77	69	69	53	40	36	25	**12.50**	25	54	7	5	–	–	–	–	–	5	1	–	–	**5′-Adenylic acid, pentakis(trimethylsilyl)-**	56247-35-9	**NI47199**
73	169	315	230	258	147	192	316	**707**	100	82	80	58	36	29	26	25	**1.70**	25	54	7	5	–	–	–	–	–	5	1	–	–	**5′-Adenylic acid, pentakis(trimethylsilyl)-**	56145-14-3	**NI47200**
81	73	295	280	323	299	75	82	**707**	100	43	16	16	10	10	8	7	**3.50**	25	54	7	5	–	–	–	–	–	5	1	–	–	**9H-Purine, 9-[2-deoxy-3-O-(trimethylsilyl)-β-D-erythro-pentofuranosyl]-6-(trimethylsiloxy)-2-[(trimethylsilyl)amino]-, 5′-[bis(trimethylsilyl) phospate]**	32645-71-9	**NI47201**
707	708	709	706	711	710	712	354	**707**	100	90	83	72	26	26	11	9		32	16	–	8	–	–	–	–	–	–	–	–	1	**Platinum phthalocyanine**	14075-08-2	**NI47202**
679	680	681	59	115	682	325	683	**708**	100	61	52	30	29	22	13	10	**0.00**	18	48	12	–	–	–	–	–	–	9	–	–	–	**1,3,5,7,9,11,13,15,17-Nonaethyltetracyclo[9.5.1.1^{3,9}.3^{5,15}]nonasiloxane**	72617-68-6	**NI47203**
301	73	445	173	317	302	211	355	**708**	100	81	40	32	31	25	25	24	**0.00**	33	76	6	–	–	–	–	–	–	5	–	–	–	**Trimethylsilyl 9,10,12,13-tetra(trimethylsiloxy)stearate**		**LI10026**
243	244	165	105	77	51	166	122	**708**	100	99	99	99	97	54	52	51	**0.13**	47	48	6	–	–	–	–	–	–	–	–	–	–	**3-O-Methyl-5-O-benzoyl-20-trityl ingenol**		**NI47204**
43	204	175	147	189	161	133	109	**708**	100	27	24	18	17	14	14	13	**6.07**	48	84	3	–	–	–	–	–	–	–	–	–	–	**1,3,5-Tritetradecanoylbenzene**		**NI47205**
708	540	497	541	372	204	354	175	**708**	100	63	48	47	46	36	35	29		48	84	3	–	–	–	–	–	–	–	–	–	–	**1,3,5-Tritetradecanoylbenzene**		**NI47206**
708	397	396	398	412	602	381	382	**708**	100	99	99	70	54	51	48	44		50	52	–	4	–	–	–	–	–	–	–	–	–	**1H-Cyclopenta[b]quinoxaline, 2,2′-(3,7,12,16-tetramethyl-1,3,5,7,9,11,13,15,17-octadecanonaene-1,18-diyl)bis[1,1,3-trimethyl-, (all-E)-**	22453-08-3	**NI47207**
234	384	341	324	383	235	69	340	**709**	100	87	75	72	62	62	62	60	**0.88**	24	33	9	7	1	6	–	–	–	–	–	–	–	**Glycine, N-[N^5-[imino[(trifluoroacetyl)amino]methyl]-N^2-[1-[S-(2-methoxy-2-oxoethyl)-N-(trifluoroacetyl)-L-cysteinyl]-L-prolyl]-L-ornithyl]-, methyl ester**	72060-17-4	**NI47208**

MASS TO CHARGE RATIOS	M.W.	INTENSITIES	Parent	C	H	O	N	S	F	Cl	Br	I	Si	P	B	X	COMPOUND NAME	CAS Reg No	No
73 315 129 299 503 371 236 192	**709**	100 45 35 27 25 23 20 19	**5.00**	25	56	7	5	–	–	–	–	–	5	1	–	–	Phosphoric acid, 2-(trimethylsiloxy)-2-[2-(trimethylsiloxy)-1-[6-[(trimethylsilyl)amino]-9H-purin-9-yl]ethoxy]propyl bis(trimethylsilyl) ester	32645-63-9	NI47209
368 412 226 299 315 454 312 238	**709**	100 92 72 60 49 43 39 34	**0.10**	38	71	7	5	–	–	–	–	–	–	–	–	–	L-Lysine, N[6]-(1-oxodecyl)-N[2]-[N-[N-[N-(1-oxodecyl)-L-alanyl]glycyl]-L-leucyl]-, methyl ester	35146-64-6	NI47210
637 636 55 639 67 634 86 30	**710**	100 99 99 70 70 69 57 45	**0.00**	18	30	13	–	–	–	–	–	–	–	–	–	4	Oxohexakis(propionato)tetrazinc(II)		MS10993
129 185 259 157 158 217 111 147	**710**	100 98 96 95 60 48 45 40	**0.00**	33	42	17	–	–	–	–	–	–	–	–	–	–	Detigloylpyruvoyldihydroazadirachtin		NI47211
87 43 91 340 41 94 81 160	**710**	100 71 51 50 50 49 45 43	**0.00**	36	54	14	–	–	–	–	–	–	–	–	–	–	Card-20(22)-enolide, 3-[(2,6-dideoxy-4-O-β-D-glucopyranosyl-3-O-methyl β-D-ribo-hexopyranosyl)oxy]-5,14-dihydroxy-19-oxo-, (3β,5β)-	560-53-2	NI47212
87 43 340 91 41 95 113 160	**710**	100 72 50 50 50 48 44 43	**0.00**	36	54	14	–	–	–	–	–	–	–	–	–	–	Card-20(22)-enolide, 3-[(2,6-dideoxy-4-O-β-D-glycopyranosyl-3-O-methyl β-D-ribo-hexopyranosyl)oxy]-5,14-dihydroxy-19-oxo-, (3β,5β)-		LI10027
178 131 91 361 682 710 234 150	**710**	100 82 37 36 31 30 30 30		37	46	12	2	–	–	–	–	–	–	–	–	–	O-Anisamide, N-(15-benzyl-10-sec-butyl-3-isopropyl-7,13,13-trimethyl-2,5,9,12,14-pentaoxo-1,4,8,11-tetraoxacyclopentadec-6-yl)-3-formamido-	22862-49-3	NI47213
178 131 91 234 150 261 233 136	**710**	100 82 38 30 30 28 25 20	**3.00**	37	46	12	2	–	–	–	–	–	–	–	–	–	Neooxoantimycin methyl ether		LI10028
73 75 147 89 81 74 67 215	**710**	100 31 28 10 9 9 8 7	**0.00**	39	78	5	–	–	–	–	–	–	3	–	–	–	PGF-2a, methylester, tris(tert-butyldimethylsilyl)-	65844-25-9	NI47214
710 637 652 651 650 577	**710**	100 8 5 5 4 4		40	46	8	4	–	–	–	–	–	–	–	–	–	2,4-Bis(2-acetoxyethyl)-6,7-bis(2-methoxycarbonylethyl)-1,3,5,8-tetramethylporphyrin		LI10029
668 667 710 652 609 595 579 493	**710**	100 38 30 12 8 8 4 4		40	46	8	4	–	–	–	–	–	–	–	–	–	α-2-Ethyl-4,5,8-tris(2-methoxycarbonylethyl)-1,3,6,7-tetramethylporphin		LI10030
710 711 637 712 638 651 679 564	**710**	100 47 20 17 9 7 6 5		40	46	8	4	–	–	–	–	–	–	–	–	–	21H,23H-Porphine-2,7,12,17-tetrapropanoic acid, 3,8,13,18-tetramethyl-, tetramethyl ester	25767-20-8	NI47215
710 711 712 694 340 347 415 341	**710**	100 46 12 6 6 4 3 3		42	63	7	–	–	–	–	–	–	–	1	–	–	Tris(3,5-di-tert-butyl-4-hydroxyphenyl) phosphate		IC04636
710 711 57 712 41 340 55 29	**710**	100 48 47 13 8 6 4 4		42	63	7	–	–	–	–	–	–	–	1	–	–	Tris(3,5-di-tert-butyl-4-hydroxyphenyl) phosphate		IC04637
87 57 43 55 71 69 41 129	**710**	100 35 35 20 18 18 18 16	**6.00**	44	86	4	–	1	–	–	–	–	–	–	–	–	Distearyl thiodibutyrate		IC04638
87 43 85 386 339 57 71 69	**710**	100 77 22 21 21 18 11 10	**1.50**	44	86	4	–	1	–	–	–	–	–	–	–	–	Distearyl thiodibutyrate		MS10994
207 299 155 301 131 300 157 121	**710**	100 89 67 33 33 22 22 18	**2.00**	46	59	4	–	–	–	1	–	–	–	–	–	–	2,4,6,8,10,12,14,16-Heptadecaoctaenoic acid, 17-(3-chloro-4-methoxyphenyl)-, 2-decyl-3-methoxy-5-pentylphenyl ester	69505-91-5	NI47216
244 241 166 165 242 105 239 228	**710**	100 82 76 72 67 64 49 45	**22.67**	47	38	5	2	–	–	–	–	–	–	–	–	–	6H-Furo[2',3':4,5]oxazolo[3,2-a]pyrimidin-6-one, 2,3,3a,9a-tetrahydro-3-(triphenylmethoxy)-2-[(triphenylmethoxy)methyl]-, [2R-(2α,3β,3aβ,9aβ)]-	22860-36-2	NI47217
165 244 243 241 166 242 105 239	**710**	100 99 99 82 75 68 65 50	**23.00**	47	38	5	2	–	–	–	–	–	–	–	–	–	3',5'-Di-O-trityl-2,2'-anhydro-1-(β-D-arabinofuranosyl)uracil		NI47218
41 40 229 124 39 248 301 323	**711**	100 49 34 26 17 13 12 11	**1.60**	14	3	–	1	–	18	–	–	–	–	–	–	1	Acetonitrilotris(π-hexafluorobut-2-yne)tungsten		MS10995
169 109 331 81 85 127 98 115	**711**	100 43 29 28 23 20 13 11	**6.00**	32	41	17	1	–	–	–	–	–	–	–	–	–	D-Glucopyranosylamine, N-phenyl-4-O-(2,3,4,6-tetra-O-acetyl-β-D-galactopyranosyl)-, 2,3,6-triacetate	35573-83-2	NI47219
169 331 81 109 141 122 711 93	**711**	100 59 53 49 32 29 19 16		32	41	17	1	–	–	–	–	–	–	–	–	–	D-Glucopyranosylamine, N-phenyl-6-O-(2,3,4,6-tetra-O-acetyl-α-D-galactopyranosyl)-, 2,3,4-triacetate	35427-07-7	NI47220
169 109 331 81 85 127 98 115	**711**	100 43 29 28 23 20 13 11	**6.00**	32	41	17	1	–	–	–	–	–	–	–	–	–	N-Phenyllactosylamine acetate		LI10031
169 331 81 109 141 122 711 93	**711**	100 59 53 49 32 29 19 16		32	41	17	1	–	–	–	–	–	–	–	–	–	N-Phenylmelibiosylamine acetate		LI10032
618 603 588 574 287 309 294 711	**711**	100 95 84 53 41 24 22 11		44	55	1	4	–	–	–	–	–	–	–	–	1	(α,γ-Dimethyl-α,γ-dihydrooctaethylporphinato)phenoxyiron(III)		NI47221
712 107 498 483 93 121 81 215	**712**	100 72 63 63 61 60 60 48		29	34	4	–	–	14	–	–	–	–	–	–	–	3α,20α-Dihydroxy-5β-pregnane, diheptafluorobutyrate		MS10996
178 133 87 91 580 127 321 260	**712**	100 95 95 91 61 56 54 50	**5.13**	37	48	12	2	–	–	–	–	–	–	–	–	–	O-Anisamide, N-(15-benzyl-10-sec-butyl-14-hydroxy-3-isopropyl-7,13,13-trimethyl-2,5,9,12-tetraoxo-1,4,8,11-tetraoxacyclopentadec-6-yl)-3-formamido-	22862-64-2	NI47223
178 87 133 91 129 260 231 247	**712**	100 96 95 91 55 50 48 40	**0.80**	37	48	12	2	–	–	–	–	–	–	–	–	–	O-Anisamide, N-(15-benzyl-10-sec-butyl-14-hydroxy-3-isopropyl-7,13,13-trimethyl-2,5,9,12-tetraoxo-1,4,8,11-tetraoxacyclopentadec-6-yl)-3-formamido-	22862-64-2	LI10033
178 133 87 91 260 231 247 117	**712**	100 99 99 94 52 49 42 42	**0.65**	37	48	12	2	–	–	–	–	–	–	–	–	–	O-Anisamide, N-(15-benzyl-10-sec-butyl-14-hydroxy-3-isopropyl-7,13,13-trimethyl-2,5,9,12-tetraoxo-1,4,8,11-tetraoxacyclopentadec-6-yl)-3-formamido-	22862-64-2	NI47222
91 92 621 28 439 622 65 529	**712**	100 49 36 19 16 14 13 12	**0.00**	44	60	–	–	–	–	–	–	–	–	4	–	–	1,6,11,16-Tetraphosphacycloeicosane, 1,6,11,16-tetrakis(phenylmethyl)-	57978-17-3	NI47224

MASS TO CHARGE RATIOS								M.W.	INTENSITIES								Parent	C	H	O	N	S	F	Cl	Br	I	Si	P	B	X	COMPOUND NAME	CAS Reg No	No
221	362	363	292	222	155	135	273	**712**	100	63	27	20	20	20	20	17	**6.00**	46	61	4	–	–	–	1	–	–	–	–	–	–	**2,4,6,8,10,12,14-Pentadecaheptaenoic acid, 15-(3-chloro-4-methoxyphenyl)-, 3-methoxy-2-(9-methyldecyl)-5-(4-methylpentyl)phenyl ester**	71246-77-0	**NI47225**
712	713	714	167	367	343	267	467	**712**	100	61	19	18	12	12	11	10		56	40	–	–	–	–	–	–	–	–	–	–	–	**1,3,5,7-Cyclooctatetraene, 1,2,3,4,5,6,7,8-octaphenyl-**	2041-08-9	**NI47226**
602	630	714	574	658	458	686	245	**714**	100	80	64	64	60	60	54	52		12	12	8	–	–	–	–	–	–	–	2	–	2	**Di-μ-dimethylphosphinobis(tetracarbonyltungsten)**		**MS10997**
534	594	197	257	215	43	270	81	**714**	100	87	86	68	58	55	50	44	**1.00**	41	62	10	–	–	–	–	–	–	–	–	–	–	**Tetra-O-acetyl-O-isopropylidenegymnemagenin**		**LI10034**
491	714	492	105	715	77	61	267	**714**	100	56	56	29	25	17	17	10		45	33	6	–	–	–	–	–	–	–	–	–	1	**Scandium, tris(1,3-diphenyl-1,3-propanedionato)-**	15133-54-7	**NI47227**
276	364	187	275	452	118	188	59	**716**	100	82	69	66	63	57	52	46	**27.00**	–	1	–	–	–	20	–	–	–	–	7	–	2	**μ-Hydrido-μ-difluorophosphidohexakis(trifluorophosphine)dicobalt**		**NI47228**
43	326	284	238	41	146	145	270	**716**	100	86	52	25	24	20	16	15	**0.00**	20	22	6	2	2	14	–	–	–	–	–	–	–	**N,N′-Bis(heptafluorobutyryl)cystine, dipropyl ester**		**MS10998**
57	602	517	489	658	251	545	307	**716**	100	23	21	13	12	11	9	7	**0.00**	20	36	4	–	–	–	–	–	–	–	2	–	2	**Tetracarbonyl[tellurobis(di-tert-butylphosphane)]tungsten(0)**		**NI47229**
307	215	596	199	536	656	494	259	**716**	100	86	64	64	44	43	43	43	**2.90**	40	60	11	–	–	–	–	–	–	–	–	–	–	**3β,6α,16β,24,25-Pentaacetoxy-24R-cycloartan-11-one**		**MS10999**
716	717	610	626	718	611	624	627	**716**	100	59	46	29	25	23	20	13		46	76	2	–	–	–	–	–	–	2	–	–	–	**ψ,ψ-Carotene, 1,1′,2,2′-tetrahydro-1,1′-bis[(trimethylsilyl)oxy]-**	55622-63-4	**NI47230**
225	187	718	226	719	186	720	161	**716**	100	26	23	17	12	10	8	5	**0.43**	51	72	2	–	–	–	–	–	–	–	–	–	–	**1,4-Naphthalenedione, 2-methyl-3-(3,7,11,15,19,23,27,31-octamethyl-2,6,10,14,18,22,26,30-dotriacontaoctaenyl)-, (all-E)-**	523-38-6	**NI47231**
225	716	717	187	718	239	226	224	**716**	100	96	58	24	23	21	20	17		51	72	2	–	–	–	–	–	–	–	–	–	–	**1,4-Naphthalenedione, 2-methyl-3-(3,7,11,15,19,23,27,31-octamethyl-2,6,10,14,18,22,26,30-dotriacontaoctaenyl)-, (all-E)-**	523-38-6	**NI47232**
257	689	413	349	275	226	551	717	**717**	100	77	67	66	63	60	57	47		36	62	5	1	–	–	–	–	–	–	1	–	1	**Molybdenum, carbonyl(η^5-2,4-cyclopentadien-1-yl)nitrosyl[tris[5-methyl-2-isopropylcyclohexyl] phosphite-P]-, stereoisomer**	75442-65-8	**NI47233**
158	174	116	159	99	98	115	43	**717**	100	38	37	35	34	32	28	28	**2.60**	37	67	12	1	–	–	–	–	–	–	–	–	–	**Erythromycin, 12-deoxy-**	527-75-3	**NI47234**
372	391	484	187	568	428	512	327	**718**	100	60	54	53	49	49	43	43	**19.80**	10	–	10	–	–	–	–	–	–	–	–	–	3	**Rhenium, decacarbonyl-μ-zinciodi-**	33728-44-8	**NI47235**
69	41	44	81	58	53	340	165	**718**	100	97	82	73	72	69	66	66	**0.00**	32	48	8	4	1	–	2	–	–	–	–	–	–	**1,2-Pyridazinedicarboxylic acid, 4,4′-[thiobis(3-isopropenyltrimethylene)]bis[5-chloro-3,6-dihydro-, tetraethyl ester**	27959-98-4	**NI47236**
69	41	44	58	81	53	341	165	**718**	100	95	83	75	73	68	67	67	**0.00**	32	48	8	4	1	–	2	–	–	–	–	–	–	**1,2-Pyridazinedicarboxylic acid, 4,4′-[thiobis(3-isopropenyltrimethylene)]bis[5-chloro-3,6-dihydro-, tetraethyl ester**	27959-98-4	**LI10035**
138	504	393	505	305	338	149	139	**718**	100	72	29	28	27	22	21	21	**11.03**	43	50	6	4	–	–	–	–	–	–	–	–	–	**Vobtusine**	19772-79-3	**NI47237**
57	43	71	55	83	69	97	85	**718**	100	70	59	51	49	46	40	37	**0.29**	50	102	1	–	–	–	–	–	–	–	–	–	–	**1-Pentacontanol**	40710-43-8	**NI47238**
225	187	718	186	160	224	202	190	**718**	100	25	22	11	6	5	4	4		51	74	2	–	–	–	–	–	–	–	–	–	–	**1,4-Naphthalenedione, 2-methyl-3-(3,7,11,15,19,23,27,31-octamethyl-2,10,14,18,22,26,30-dotriacontaheptaenyl)-, (all-E)-**	21632-35-9	**NI47239**
225	718	719	187	226	720	721	161	**718**	100	46	26	26	16	15	6	5		51	74	2	–	–	–	–	–	–	–	–	–	–	**1,4-Naphthalenedione, 2-methyl-3-(3,7,11,15,19,23,27,31-octamethyl-2,10,14,18,22,26,30-dotriacontaheptaenyl)-, (all-E)-**	21632-35-9	**LI10036**
206	107	43	44	150	152	207	178	**719**	100	76	74	55	54	49	47	47	**0.00**	34	41	16	1	–	–	–	–	–	–	–	–	–	**Evonine, O²-deacetyl-**	36017-57-9	**NI47240**
43	107	206	719	151	178	161	106	**719**	100	91	47	34	31	25	23	19		34	41	16	1	–	–	–	–	–	–	–	–	–	**Evonine, O⁶-deacetyl-**	33540-08-8	**NI47241**
43	107	206	178	106	161	150	160	**719**	100	79	56	40	29	24	23	21	**8.53**	34	41	16	1	–	–	–	–	–	–	–	–	–	**Evonine, O⁶-deacetyl-, (all-ξ)-**	35721-65-4	**NI47242**
156	128	235	109	168	111	470	236	**719**	100	40	16	14	14	14	13	11	**0.10**	36	50	10	3	–	–	1	–	–	–	–	–	–	**Maytanbutine**	38997-10-3	**NI47243**
73	211	478	479	370	720	299	227	**720**	100	91	77	71	58	55	55	48		30	58	8	–	–	–	–	–	–	4	2	–	–	**Estradiol, bis[bis(trimethylsilyl) phosphate]**	33745-65-2	**NI47244**
377	57	367	423	387	378	368	379	**720**	100	97	86	63	47	36	30	22	**0.00**	30	60	10	–	–	–	–	–	–	5	–	–	–	**1,3,5,7,9-Pentavinyl-1,3,5,7,9-pentabutoxycyclopentasiloxane**		**NI47245**
169	97	43	81	109	91	331	131	**720**	100	95	84	79	53	47	25	25	**0.00**	34	40	17	–	–	–	–	–	–	–	–	–	–	**[1S-(1α,4aα,6α,7α,7aα)]-methyl 1-(tetraacetoxy-β-D-glucopyranosyloxy)-1,4a,5,6,7,7a-hexahydro-4a,7-dihydroxy-7-methyl-6-[(1-oxo-3-phenyl-2-propenyl)oxy]cyclopenta[c]pyran-4-carboxylate**		**NI47246**
670	259	233	185	217	195	449	235	**720**	100	98	95	85	77	77	65	60	**0.00**	35	44	16	–	–	–	–	–	–	–	–	–	–	**Azadirachtin**	11141-17-6	**NI47247**
673	573	704	585	689	559	643	529	**720**	100	37	33	22	16	14	11	11	**0.00**	36	48	15	–	–	–	–	–	–	–	–	–	–	**4H-1-Benzopyran-4-one, 5,7-dimethoxy-2-(4-methoxyphenyl)-8-[(2,3,4,6-tetra-O-methyl-β-D-glucopyranosyl)oxy]-6-[(tri-O-methylarabinosyl)oxy]-**	75311-69-2	**NI47248**

MASS TO CHARGE RATIOS	M.W.	INTENSITIES	Parent	C	H	O	N	S	F	Cl	Br	I	Si	P	B	X	COMPOUND NAME	CAS Reg No	No
204 73 170 720 185 117 429 75	720	100 85 48 38 38 25 15 15		38	68	7	–	–	–	–	–	–	3	–	–	–	Card-20(22)-enolide, 3-[[2,6-dideoxy-3,4-bis-O-(trimethylsilyl)-D-ribo-hexopyranosyl]oxy]-14-[(trimethylsilyl)oxy]-, (3β,5β)-	57289-27-7	NI47249
204 73 170 185 720 117 721 205	720	100 86 49 39 34 26 22 20		38	68	7	–	–	–	–	–	–	3	–	–	–	Card-20(22)-enolide, 3-[[2,6-dideoxy-3,4-bis-O-(trimethylsilyl)-ribo-hexopyranosyl]oxy]-14-[(trimethylsilyl)oxy]-, (3β,5β)-	56009-08-6	MS11000
204 73 170 185 716 117 717 205	720	100 86 50 40 35 25 22 20	0.00	38	68	7	–	–	–	–	–	–	3	–	–	–	Card-20(22)-enolide, 3-[[2,6-dideoxy-3,4-bis-O-(trimethylsilyl)-ribo-hexopyranosyl]oxy]-14-[(trimethylsilyl)oxy]-, (3β,5β)-	56009-08-6	NI47250
105 117 119 77 121 28 720 122	720	100 51 49 25 15 15 14 13		43	44	10	–	–	–	–	–	–	–	–	–	–	3,7,8-Tri-O-benzoyl-ingol-12-acetate		NI47251
362 221 363 121 720 292 222 151	720	100 41 28 28 10 6 6 4		48	80	4	–	–	–	–	–	–	–	–	–	–	Benzeneheptadecanoic acid, 4-methoxy-, 3-methoxy-2-(9-methyldecyl)-5-(4-methylpentyl)phenyl ester	69506-06-5	NI47252
57 71 43 479 85 55 83 69	720	100 71 65 54 52 38 37 37	1.24	48	96	3	–	–	–	–	–	–	–	–	–	–	Hexadecanoic acid, 2-(octadecyloxy)-, tetradecyl ester	56599-98-5	NI47253
315 73 217 299 147 316 459 357	721	100 99 56 36 31 28 24 20	0.00	25	64	9	1	–	–	–	–	–	6	1	–	–	D-Fructose, 1,3,4,5-tetrakis-O-(trimethylsilyl)-, O-methyloxime, 6-[bis(trimethylsilyl) phosphate]	55530-74-0	NI47254
73 299 387 315 247 147 300 357	721	100 59 34 24 24 24 20 18	0.00	25	64	9	1	–	–	–	–	–	6	1	–	–	D-Galactose, 2,3,4,5-tetrakis-O-(trimethylsilyl)-, O-methyloxime, 6-[bis(trimethylsilyl) phosphate]	55530-75-1	NI47255
299 73 387 147 315 247 300 357	721	100 99 58 41 40 40 34 30	0.00	25	64	9	1	–	–	–	–	–	6	1	–	–	D-Galactose, 2,3,4,5-tetrakis-O-(trimethylsilyl)-, O-methyloxime, 6-[bis(trimethylsilyl) phosphate]	55530-75-1	NI47256
387 73 299 388 147 315 160 357	721	100 99 80 43 43 42 42 41	0.00	25	64	9	1	–	–	–	–	–	6	1	–	–	D-Glucose, 2,3,4,5-tetrakis-O-(trimethylsilyl)-, O-methyloxime, 6-[bis(trimethylsilyl) phosphate]	55530-73-9	NI47257
73 387 299 388 147 357 315 160	721	100 52 42 23 23 22 22 22	0.00	25	64	9	1	–	–	–	–	–	6	1	–	–	D-Glucose, 2,3,4,5-tetrakis-O-(trimethylsilyl)-, O-methyloxime, 6-[bis(trimethylsilyl) phosphate]	55530-73-9	NI47258
73 387 299 357 315 147 388 160	721	100 67 40 26 26 23 20 17	0.00	25	64	9	1	–	–	–	–	–	6	1	–	–	D-Mannose, 2,3,4,5-tetrakis-O-(trimethylsilyl)-, O-methyloxime, 6-[bis(trimethylsilyl) phosphate]	55530-76-2	NI47260
387 73 299 357 315 147 388 160	721	100 99 61 39 39 34 30 25	0.00	25	64	9	1	–	–	–	–	–	6	1	–	–	D-Mannose, 2,3,4,5-tetrakis-O-(trimethylsilyl)-, O-methyloxime, 6-[bis(trimethylsilyl) phosphate]	55530-76-2	NI47259
375 101 139 143 150 153 115 168	721	100 78 72 53 42 38 35 32	0.00	30	43	19	1	–	–	–	–	–	–	–	–	–	Methyl 7-O-(2-acetamido-3,4,6-tri-O-acetyl-2-deoxy-α-D-glucopyranosyl) 2,3,4,6-tetra-O-acetyl-L-glycero-α-D-manno-heptopyranoside	92414-16-9	NI47261
174 115 331 271 229 169 212 91	721	100 78 76 64 62 54 52 43	2.80	32	39	16	3	–	–	–	–	–	–	–	–	–	α-D-Galactopyranoside, 2,3-bis(acetyloxy)-1-[(acetyloxy)(2-phenyl-2H-1,2,3-triazol-4-yl)methyl]propyl, 2,3,4,6-tetraacetate, [1S-[1R*(S*),2S*]]-	56298-40-9	NI47262
169 109 331 271 127 212 81 115	721	100 77 68 35 28 25 20 19	2.50	32	39	16	3	–	–	–	–	–	–	–	–	–	β-D-Glucopyranoside, 2,3-bis(acetyloxy)-1-[(acetyloxy)(2-phenyl-2H-1,2,3-triazol-4-yl)methyl]propyl, tetraacetate, [1S(R),2R]-	13977-37-2	NI47263
169 109 331 81 91 115 127 111	721	100 68 46 40 37 23 18 18	1.10	32	39	16	3	–	–	–	–	–	–	–	–	–	β-D-Glucopyranoside, 2,3,4-tris(acetyloxy)-1-(2-phenyl-2H-1,2,3-triazol-4-yl)butyl, tetraacetate	35405-82-4	NI47264
498 721 499 275 722 78 105 77	721	100 53 44 38 35 24 18 13		45	33	6	–	–	–	–	–	–	–	–	–	1	Chromium, tris(1,3-diphenyl-1,3-propanedionato-O,O')-	21679-35-6	NI47265
693 694 695 696 332 31 87 115	722	100 66 54 25 19 17 16 13	0.69	18	46	13	–	–	–	–	–	–	9	–	–	–	1-Hydroperethylhomooctasilsesquioxane	72617-64-2	NI47266
582 554 638 279 694 668 722 610	722	100 84 70 60 55 50 26 15		22	42	6	–	–	–	–	–	–	–	4	–	2	μ-Tetramethyldiphosphine-bis(triethylphosphinotricarbonylmolybdenum)		MS11001
722 147 69 105 169 119 467 55	722	100 99 98 92 84 80 70 70		29	28	5	–	–	14	–	–	–	–	–	–	–	20-Oxo-3,5-pregnadiene-3,21-diol-diheptafluorobutyrate		MS11002
253 151 185 559 125 111 109 141	722	100 67 60 50 50 50 45 43	0.00	35	46	16	–	–	–	–	–	–	–	–	–	–	Dihydroazadirachtin		NI47267
57 43 211 339 55 71 41 73	722	100 84 76 68 58 50 40 38	0.14	45	86	6	–	–	–	–	–	–	–	–	–	–	Tetradecanoic acid, 1,2,3-propanetriyl ester	555-45-3	NI47268
390 391 556 392 557 722 373 558	722	100 53 19 16 15 10 9 6		48	54	4	2	–	–	–	–	–	–	–	–	–	N,N'-Dicyclododecyl-3,4:9,10-perylenebis(dicarboximide)	80509-54-2	MS11003
390 391 556 392 557 722 373 558	722	100 53 19 16 15 10 9 6		48	54	4	2	–	–	–	–	–	–	–	–	–	N,N'-Dicyclododecyl-3,4:9,10-perylenebis(dicarboximide)	80509-54-2	DD01822
180 78 181 77 104 154 645 284	722	100 96 62 62 47 27 23 13	4.00	50	40	–	2	–	–	–	–	–	–	–	–	2	Diphenyl (α-phenylbenzylideneamino)-aluminium dimer		LI10037
616 630 722 361 707 631 617 433	722	100 74 68 51 46 44 42 42		51	54	–	4	–	–	–	–	–	–	–	–	–	Phenazine, 1,2-dihydro-2,2,4-trimethyl-3-[3,7,12,16-tetramethyl-18-(1,1,3-trimethyl-1H-cyclopenta[b]quinoxalin-2-yl)-1,3,5,7,9,11,13,15,17-octadecanonaenyl]-, (all-trans)-	38310-09-7	NI47269
540 666 357 723 482 326 708 510	723	100 57 35 18 7 6 3 3		33	57	6	–	–	–	–	–	–	–	–	–	1	Ytterbium, tris(2,2,6,6-tetramethyl-3,5-heptanedionato-O,O')-, (OC-6-11)-	15492-52-1	NI47270
57 357 355 356 127 354 359 41	723	100 67 61 49 42 38 32 27	0.00	33	57	6	–	–	–	–	–	–	–	–	–	1	Ytterbium, tris(2,2,6,6-tetramethyl-3,5-heptanedionato-O,O')-, (OC-6-11)-	15492-52-1	NI47271
107 108 387 28 219 39 63 77	724	100 88 77 60 54 46 44 40	20.99	18	12	4	–	–	–	–	–	3	–	–	–	1	Tris(p-iodophenoxo)oxovanadium	62449-97-2	NI47272

MASS TO CHARGE RATIOS								M.W.	INTENSITIES								Parent	C	H	O	N	S	F	Cl	Br	I	Si	P	B	X	COMPOUND NAME	CAS Reg No	No
309	89	724	69	133	212	310	161	**724**	100	43	33	21	19	18	11	10		20	10	2	–	2	6	–	2	–	–	–	–	1	**Bis[1,1,1-trifluoro-4-(4-bromophenyl)-4-thioxo-2-butanonato]palladium**	52375-08-3	**NI47273**
73	43	55	69	131	75	429	81	**724**	100	62	32	31	29	27	23	22	**0.00**	39	76	6	–	–	–	–	–	–	3	–	–	–	**Ergostane-5,25-diol, 3,6,12-tris[(trimethylsilyl)oxy]-, 25-acetate, (3β,5α,6β,12β)-**	56053-01-1	**NI47274**
481	449	482	397	656	605	477	450	**724**	100	48	46	40	26	19	16	15	**0.00**	40	68	11	–	–	–	–	–	–	–	–	–	–	**Nigericin**	28380-24-7	**NI47275**
85	57	96	86	83	67	43	41	**724**	100	11	6	6	6	5	5	4	**0.33**	46	76	6	–	–	–	–	–	–	–	–	–	–	**Lup-20(29)-en-21-ol, 3,28-bis[(tetrahydro-2H-pyran-2-yl)oxy]-, 3,3-dimethylbutanoate, (3β,21β)-**	55401-92-8	**NI47276**
169	331	725	109	127	149	136	191	**725**	100	56	42	36	16	15	15	14		33	43	17	1	–	–	–	–	–	–	–	–	–	**D-Glucopyranosylamine, N-(4-methylphenyl)-4-O-(2,3,4,6-tetra-O-acetyl-β-D-galactopyranosyl)-, 2,3,6-triacetate**	35427-10-2	**NI47277**
107	169	106	331	109	81	136	141	**725**	100	99	99	62	61	46	45	26	**22.36**	33	43	17	1	–	–	–	–	–	–	–	–	–	**D-Glucopyranosylamine, N-(4-methylphenyl)-6-O-(2,3,4,6-tetra-O-acetyl-α-D-galactopyranosyl)-, 2,3,4-triacetate**	35427-08-8	**NI47278**
169	109	331	139	127	136	107	725	**725**	100	59	36	36	26	24	24	23		33	43	17	1	–	–	–	–	–	–	–	–	–	**D-Glucopyranosylamine, N-(4-methylphenyl)-2-O-(2,3,4,6-tetra-O-acetyl-β-D-glucopyranosyl)-, 3,4,6-triacetate**	35427-13-5	**NI47279**
169	81	136	109	725	141	331	178	**725**	100	56	53	50	36	32	28	19		33	43	17	1	–	–	–	–	–	–	–	–	–	**D-Glucopyranosylamine, N-(4-methylphenyl)-3-O-(2,3,4,6-tetra-O-acetyl-β-D-glucopyranosyl)-, 2,4,6-triacetate**	35427-12-4	**NI47280**
208	502	681	473	457	223	488	443	**725**	100	57	32	24	23	22	21	18	**17.80**	36	57	6	6	–	–	–	–	–	–	–	–	1	**Ferriaspergillin**	77423-90-6	**NI47281**
502	725	223	443	473	683	696	166	**725**	100	99	67	57	54	47	23	16		36	57	6	6	–	–	–	–	–	–	–	–	1	**Ferrineoaspergillin**		**MS11004**
502	105	503	77	279	725	133	223	**725**	100	58	35	32	24	17	12	11		45	33	6	–	–	–	–	–	–	–	–	–	1	**Iron, tris(1,3-diphenyl-1,3-propanedionato-O,O′)-**	14405-49-3	**NI47282**
327	394	422	273	294	534	478	450	**726**	100	47	44	44	17	14	11	8	**1.90**	15	9	13	–	–	–	–	–	–	–	–	–	5	**7-Methyl-1,3,5-triarsa-2,4,9-trioxa-adamantan-As,As′-bis[pentacarbonylchromium(0)]**	80950-88-5	**NI47283**
290	336	91	107	96	93	183	119	**726**	100	48	30	27	27	20	19	19	**3.71**	32	38	12	8	–	–	–	–	–	–	–	–	–	**D,L-1,4-Bis-[1-Ethoxycarbonyl-2-(2,4-dinitrophenylhydrazone)cyclopentyl]butane**		**NI47284**
130	131	28	57	43	41	56	77	**726**	100	90	37	23	23	23	18	15	**0.00**	32	62	7	2	–	–	–	–	–	5	–	–	–	**D-Glucose, 2-O-[3-acetyl-1-(trimethylsilyl)-1H-indolyl]-3,4,5,6-tetrakis-O (trimethylsilyl)-, 1-(O-methyloxime)**	74421-39-9	**NI47285**
229	73	202	230	103	147	246	217	**726**	100	94	68	34	21	20	19	15	**1.50**	32	62	7	2	–	–	–	–	–	5	–	–	–	**D-Glucose, 2-O-[3-acetyl-1-(trimethylsilyl)-1H-indolyl]-3,4,5,6-tetrakis-O (trimethylsilyl)-, 1-(O-methyloxime)**	74421-39-9	**NI47286**
229	73	202	230	217	147	203	231	**726**	100	93	49	29	12	12	10	9	**1.50**	32	62	7	2	–	–	–	–	–	5	–	–	–	**D-Glucose, 4-O-[3-acetyl-1-(trimethylsilyl)-1H-indolyl]-2,3,5,6-tetrakis-O (trimethylsilyl)-, 1-(O-methyloxime)**	74421-40-2	**NI47287**
130	131	77	103	51	132	28	102	**726**	100	85	20	13	12	8	8	7	**0.00**	32	62	7	2	–	–	–	–	–	5	–	–	–	**D-Glucose, 4-O-[3-acetyl-1-(trimethylsilyl)-1H-indolyl]-2,3,5,6-tetrakis-O (trimethylsilyl)-, 1-(O-methyloxime)**	74421-40-2	**NI47288**
158	159	144	28	41	43	57	56	**726**	100	81	43	36	19	18	17	14	**0.00**	32	62	7	2	–	–	–	–	–	5	–	–	–	**D-Glucose, 6-O-[3-acetyl-1-(trimethylsilyl)-1H-indolyl]-2,3,4,5-tetrakis-O (trimethylsilyl)-, 1-(O-methyloxime)**	74421-41-3	**NI47289**
229	73	202	230	362	217	203	160	**726**	100	84	55	21	20	15	12	12	**1.70**	32	62	7	2	–	–	–	–	–	5	–	–	–	**D-Glucose, 6-O-[3-acetyl-1-(trimethylsilyl)-1H-indolyl]-2,3,4,5-tetrakis-O (trimethylsilyl)-, 1-(O-methyloxime)**	74421-41-3	**NI47290**
169	109	331	127	211	271	145	196	**726**	100	50	10	10	7	6	6	5	**0.30**	37	46	13	2	–	–	–	–	–	–	–	–	–	**1-N-Acetyl-17-O-acetyl-21-O-(tetraacetyl-β-D-glucopyranosyl)norajmaline (pentaacetyl)acetylnorrauglucine**		**MS11005**
253	143	343	456	159	147	254	546	**726**	100	59	35	31	30	28	24	23	**0.00**	37	74	6	–	–	–	–	–	–	4	–	–	–	**Methyl 23-hydroxycholate, tetrakis(trimethylsilyl)-**	111945-03-0	**NI47291**
336	105	191	162	121	305	100	337	**726**	100	41	38	35	35	24	24	23	**0.50**	41	50	8	4	–	–	–	–	–	–	–	–	–	**1,16-Ethenofuro[3,4-l][1,5,10]triazacyclohexadecine-15-acetic acid, 10,14-diacetyl-5-[3-(benzoylamino)propyl]-3,3a,4,5,6,7,8,9,10,11,12,13,14,15-tetradecahydro-3-(4-methoxyphenyl)-4-oxo-, methyl ester**	69721-75-1	**NI47292**
666	606	411	514	574	667	531	383	**726**	100	77	76	63	52	47	47	42	**30.00**	46	62	7	–	–	–	–	–	–	–	–	–	–	**α-Carotene, 7,8-dihydro-3,3′,19-trihydroxy-8-oxo-, triacetate, all-trans-**	33474-10-1	**LI10038**
666	606	411	514	574	531	485	471	**726**	100	76	75	62	51	46	43	42	**29.00**	46	62	7	–	–	–	–	–	–	–	–	–	–	**α-Carotene, 7,8-dihydro-3,3′,19-trihydroxy-8-oxo-, triacetate, all-trans-**	33474-10-1	**NI47293**
334	335	207	726	155	122	151	121	**726**	100	26	23	10	10	7	5	4		46	75	4	–	–	–	1	–	–	–	–	–	–	**Benzeneheptadecanoic acid, 3-chloro-4-methoxy-, 2-decyl-3-methoxy-5-pentylphenyl ester**	69505-99-3	**NI47294**
362	221	363	155	726	292	222	157	**726**	100	52	31	13	8	7	7	7		46	75	4	–	–	–	1	–	–	–	–	–	–	**Benzenepentadecanoic acid, 3-chloro-4-methoxy-, 3-methoxy-2-(9-methyldecyl)-5-(4-methylpentyl)phenyl ester**	63953-40-2	**NI47295**
280	279	560	277	616	554	672	558	**728**	100	70	59	50	47	47	45	44	**27.00**	34	20	6	–	–	–	–	–	–	–	–	–	2	**Ruthenium, hexacarbonyl[μ-(1,2,3,4-tetraphenyl-1,3-butadienylene)]di-,**	33310-08-6	**NI47296**
73	243	253	523	343	147	535	75	**728**	100	45	39	38	36	36	31	31	**3.98**	36	76	5	–	–	–	–	–	–	5	–	–	–	**3β,5α,11β,20S-Pregnane, 3,11,17,20,21-penta(trimethylsiloxy)-**	57326-11-1	**NI47297**
73	243	75	253	147	343	129	523	**728**	100	99	99	74	72	53	45	40	**3.00**	36	76	5	–	–	–	–	–	–	5	–	–	–	**5β-Pregnane, 3α,11β,17,20β,21-penta(trimethylsiloxy)-**	17563-09-6	**NI47298**
73	253	243	343	523	535	147	445	**728**	100	31	31	28	26	25	23	21	**3.20**	36	76	5	–	–	–	–	–	–	5	–	–	–	**5β-Pregnane, 3α,11β,17,20β,21-penta(trimethylsiloxy)-**	17563-09-6	**NI47299**

MASS TO CHARGE RATIOS								M.W.	INTENSITIES								Parent	C	H	O	N	S	F	Cl	Br	I	Si	P	B	X	COMPOUND NAME	CAS Reg No	No
343	253	523	243	433	524	535	344	**728**	100	92	79	59	36	35	28	27	**0.00**	36	76	5	–	–	–	–	–	–	5	–	–	–	5β-Pregnane, 3β,11β,17,20β,21-penta(trimethylsiloxy)-	32213-62-0	NI47300
108	79	107	169	257	370	156	114	**728**	100	57	35	22	21	16	12	11	**0.00**	38	56	10	4	–	–	–	–	–	–	–	–	–	(S,S)-Bis(N-benzyloxycarbonylvalyl)diaza-18-crown-6		MS11006
43	273	111	153	85	57	379	83	**728**	100	43	38	35	34	25	24	19	**0.00**	41	60	11	–	–	–	–	–	–	–	–	–	–	Cholesta-9(11),20(22)-dien-23-one, 6-(acetyloxy)-3-[(2,3,4-tri-O-acetyl-deoxy-β-D-galactopyranosyl)oxy]-, (3β,5α,6α)-	56312-53-9	NI47301
259	470	105	182	288	393	729	365	**728**	100	54	49	25	22	20	14	14	**0.00**	48	40	–	–	–	–	–	–	–	4	–	–	–	Octaphenylcyclotetrasilane		LI10039
603	604	602	521	522	520	453	605	**729**	100	91	89	59	52	43	34	33	**3.00**	34	64	–	–	–	–	–	–	–	–	2	–	1	[2-(2-Propyl)-3-methyl-1-butene][bis(dicyclohexylphosphino)ethane]platinum(0)		MS11007
69	357	548	454	381	330	287	332	**730**	100	55	50	34	33	27	27	20	**0.00**	12	–	13	–	–	18	–	–	–	–	–	–	4	Oxohexakis(triflouroacetato)tetraberyllium(II)		MS11008
365	198	366	69	110	129	296	730	**730**	100	88	67	61	39	29	21	20		24	–	–	–	–	20	–	–	–	–	2	–	–	Tetrakis(pentafluorophenyl) diphosphine		MS11009
730	728	726	729	732	731	290	727	**730**	100	70	45	34	30	26	23	15		24	–	–	–	2	16	–	–	–	–	–	–	1	10,10′-Spirobi[10H-phenothiagermanin], 1,1′,2,2′,3,3′,4,4′,6,6′,7,7′,8,8′,9,9′-hexadecafluoro-	19739-94-7	NI47302
688	140	123	108	260	633	730	163	**730**	100	82	76	69	56	41	40	40		25	20	9	–	–	14	–	–	–	–	–	–	–	3-Acetyl-7,15-bis(heptafluorobutyryl)deoxynivalenol		NI47303
444	321	135	145	730	377	445	163	**730**	100	49	47	44	42	39	37	36		25	20	9	–	–	14	–	–	–	–	–	–	–	15-Acetyl-3,7-bis(heptafluorobutyryl)deoxynivalenol		NI47304
57	385	441	386	41	397	442	56	**730**	100	83	56	47	44	25	23	23	**0.00**	30	70	10	–	–	–	–	–	–	5	–	–	–	1,3,5,7-Tetraethyl-3,5,7-tributoxy-1-ethyldibuthoxysiloxycyclotetrasiloxane		NI47305
730	699	731	700	639	299	519	357	**730**	100	94	34	30	27	27	26	25		40	42	13	–	–	–	–	–	–	–	–	–	–	Monoacetyl-octamethyl-procyanidine		LI10040
219	94	55	41	220	57	43	161	**730**	100	55	21	19	18	18	14	12	**0.06**	42	66	6	–	2	–	–	–	–	–	–	–	–	1,6-Bis(3,5-di-tert-butyl-4-hydroxyphenyl-methyl-thioethylcarboxy)hexane		IC04639
705	706	732	733	707	655	627	628	**732**	100	67	46	23	21	15	13	11		24	28	12	–	–	–	–	–	–	8	–	–	–	Hexavinyldiphenyloctasilsesquioxane	87911-37-3	NI47306
168	103	268	142	732	178	77	187	**732**	100	29	22	13	7	7	6	5		34	15	1	–	–	10	–	–	–	–	–	–	1	2,4-Bis(pentafluorophenyl)-3,5-diphenylcyclopentadienone-π-cyclopentadienylrhodium		MS11010
518	732	181	407	198	458	144	108	**732**	100	99	55	24	22	20	20	18		46	44	5	4	–	–	–	–	–	–	–	–	–	Acetovoafoline		LI10041
732	57	733	71	43	85	83	69	**732**	100	58	57	37	32	24	20	19		50	100	2	–	–	–	–	–	–	–	–	–	–	Pentacontanoic acid	40710-30-3	WI01600
73	204	217	412	484	75	147	143	**733**	100	78	55	45	30	25	22	19	**0.10**	36	79	6	1	–	–	–	–	–	4	–	–	–	N-Octadecyl-D-glucuronamide, tetrakis(trimethylsilyl)-		NI47307
158	43	98	41	115	71	159	83	**733**	100	65	35	30	28	26	25	24	**1.20**	37	67	13	1	–	–	–	–	–	–	–	–	–	Erythromycin	114-07-8	NI47308
115	83	127	43	126	31	223	99	**733**	100	84	79	77	58	54	38	33	**0.00**	37	67	13	1	–	–	–	–	–	–	–	–	–	Erythromycin, 12-deoxy-, N-oxide	21395-58-4	LI10042
115	127	83	223	43	99	125	113	**733**	100	91	73	49	48	36	35	31	**0.00**	37	67	13	1	–	–	–	–	–	–	–	–	–	Erythromycin, 12-deoxy-, N-oxide	21395-58-4	NI47309
43	57	239	55	367	71	313	41	**733**	100	99	85	82	72	61	58	54	**0.00**	40	80	8	1	–	–	–	–	–	–	1	–	–	3,5,9-Trioxa-4-phosphapentacosan-1-aminium, 4-hydroxy-N,N,N-trimethyl-10-oxo-7-[(1-oxohexadecyl)oxy]-, hydroxide, inner salt, 4 oxide	2644-64-6	NI47310
57	41	437	465	455	475	377	477	**734**	100	36	22	21	20	19	12	11	**0.00**	28	54	11	–	–	–	–	–	–	6	–	–	–	1,3,5,7,9,11-Hexavinyl-3,5,9,11-tetrabutoxybicyclo[5.5.1]hexasiloxane	110991-14-5	NI47311
331	43	169	109	165	115	103	121	**734**	100	52	50	40	30	25	23	20	**0.50**	35	42	17	–	–	–	–	–	–	–	–	–	–	Ligustroside pentaacetate		MS11011
153	195	137	283	488	530	590	237	**734**	100	40	39	36	28	27	22	21	**2.90**	37	34	16	–	–	–	–	–	–	–	–	–	–	Isosilychristin hexaacetate	57499-42-0	NI47312
499	341	515	500	325	516	395	355	**734**	100	44	37	29	13	11	10	10	**4.00**	37	50	15	–	–	–	–	–	–	–	–	–	–	6-C-Diglucosylapigenin		NI47313
499	341	515	325	467	355	311	397	**734**	100	89	45	28	14	12	11	8	**1.40**	37	50	15	–	–	–	–	–	–	–	–	–	–	Per-O-methylspinosin		NI47314
734	735	736	367	720	719	368	733	**734**	100	54	18	18	4	3	3	2		48	38	4	4	–	–	–	–	–	–	–	–	–	21H,23H-Porphine, 5,10,15,20-tetrakis(4-methoxyphenyl)-	22112-78-3	NI47315
239	43	57	55	41	71	69	83	**734**	100	82	80	70	46	45	44	40	**0.00**	48	94	4	–	–	–	–	–	–	–	–	–	–	Hexadecanoic acid, 1-tetradecyl-1,2-ethanediyl ester	56599-97-4	NI47316
267	57	43	71	55	69	268	41	**734**	100	48	38	29	28	23	21	20	**1.04**	48	94	4	–	–	–	–	–	–	–	–	–	–	Octadecanoic acid, 1-[(tetradecyloxy)carbonyl]pentadecyl ester	69688-49-9	NI47317
315	217	73	459	299	316	147	357	**735**	100	77	71	46	46	42	31	22	**0.40**	26	66	9	1	–	–	–	–	–	6	1	–	–	D-Fructose, 1,3,4,5-tetrakis-O-(trimethylsilyl)-, O-ethyloxime, 6-[bis(trimethylsilyl) phosphate]	55530-77-3	NI47318
201	244	288	69	121	202	68	289	**736**	100	81	75	65	64	62	62	61	**0.57**	31	43	11	4	1	3	–	–	–	–	–	–	–	L-Glutamic acid, N-[N-[N-[S-(2-methoxy-2-oxoethyl)-N-(trifluoroacetyl)-L-cysteinyl]-O-methyl-L-tyrosyl]-L-isoleucyl]-, dimethyl ester	69782-85-0	NI47319

MASS TO CHARGE RATIOS								M.W.	INTENSITIES								Parent	C	H	O	N	S	F	Cl	Br	I	Si	P	B	X	COMPOUND NAME	CAS Reg No	No
156	362	269	86	128	114	100	43	**736**	100	96	94	66	64	34	30	22	**0.00**	39	72	7	6	–	–	–	–	–	–	–	–	–	**Bis N,N′-(acetyl-N-methylvalyl-N-methylvalyl)-N,N′-dimethyl-α,α′-diaminodiisopropylacetone**		**LI10043**
185	369	111	167	143	69	125	43	**736**	100	73	70	51	44	36	33	33	**2.22**	40	64	12	–	–	–	–	–	–	–	–	–	–	**Nonactin**	6833-84-7	**NI47320**
736	447	721	644	431	368	629	630	**736**	100	99	90	86	71	68	60	59		52	56	–	4	–	–	–	–	–	–	–	–	–	**Phenazine, 2,2′-(3,7,12,16-tetramethyl-1,3,5,7,9,11,13,15,17-octadecanonaene-1,18-diyl)bis[3,4-dihydro-1,3,3-trimethyl-, (all-E)-**	28081-91-6	**NI47321**
137	121	107	109	135	131	190	105	**736**	100	39	38	37	33	19	17	17	**5.39**	52	80	2	–	–	–	–	–	–	–	–	–	–	**Phenol, 2-methoxy-6-(3,7,11,15,19,23,27,31,35-nonamethyl-2,6,10,14,18,22,26,30,34-hexatriacontanonaenyl)-**	10232-06-1	**NI47322**
137	121	109	107	135	175	177	131	**736**	100	39	37	37	33	32	20	19	**0.00**	52	80	2	–	–	–	–	–	–	–	–	–	–	**Phenol, 2-methoxy-6-(3,7,11,15,19,23,27,31,35-nonamethyl-2,6,10,14,18,22,26,30,34-hexatriacontanonaenyl)-**	10232-06-1	**NI47323**
709	710	711	340	712	679	680	341	**738**	100	64	51	25	24	20	17	17	**4.09**	18	46	14	–	–	–	–	–	–	9	–	–	–	**1-Oxyperethylhomooctasilsesquioxane**	72617-65-3	**NI47324**
73	259	75	147	389	129	171	81	**738**	100	40	34	25	22	18	17	17	**0.00**	34	78	7	–	–	–	–	–	–	5	–	–	–	**Octadecanoic acid, 9,10,12,13,18-pentakis(trimethylsiloxy)-, methyl ester**	21987-18-8	**NI47325**
105	122	77	202	200	183	185	51	**738**	100	68	62	42	42	36	35	26	**0.06**	41	39	8	–	–	–	–	1	–	–	–	–	–	**3,5-Di-O-benzoyl-ingenol-20-p-bromo-benzoate**		**NI47326**
379	411	185	113	153	85	81	226	**738**	100	92	92	47	38	11	11	6	**0.20**	41	70	11	–	–	–	–	–	–	–	–	–	–	**Nigericin methyl ester**		**LI10044**
57	43	55	83	69	71	98	41	**738**	100	83	69	59	59	52	50	43	**3.70**	46	90	4	–	1	–	–	–	–	–	–	–	–	**Di-Eicosyl thiodipropionate**		**IC04640**
221	299	362	264	363	313	302	239	**738**	100	95	84	39	37	29	29	26	**19.00**	48	63	4	–	–	–	1	–	–	–	–	–	–	**2,4,6,8,10,12,14,16-Heptadecaoctaenoic acid, 17-(3-chloro-4-methoxyphenyl)-, 3-methoxy-2-(9-methyldecyl)-5-(4-methylpentyl)phenyl ester**	69505-94-8	**NI47327**
711	665	712	666	31	667	713	341	**740**	100	70	61	39	37	36	20	18	**4.99**	20	52	14	–	–	–	–	–	–	8	–	–	–	**1,3,5,7,9,11,13,15-Octaethyl-5,7,13,15-tetramethoxytricyclo[9.5.1.1^{3,9}]octasiloxane**		**NI47328**
213	697	169	171	109	740	409	393	**740**	100	75	45	40	25	15	10	10		37	44	14	2	–	–	–	–	–	–	–	–	–	**Penta-acetylisovincoside**		**LI10045**
213	171	697	169	109	740	409	393	**740**	100	55	50	45	30	15	10	10		37	44	14	2	–	–	–	–	–	–	–	–	–	**Penta-acetylvincoside**		**LI10046**
105	162	121	84	191	100	77	83	**740**	100	64	53	43	40	40	36	33	**5.00**	42	52	8	4	–	–	–	–	–	–	–	–	–	**Benzamide, N-[3-[10,14-diacetyl-15-[2-(acetyloxy)ethyl]-3a,4,6,7,8,9,10,11,12,13,14,15-dodecahydro-3-(4-methoxyphenyl)-4-oxo-1,16-ethenofuro[3,4-l][1,5,10]triazacyclohexadecin-5(3H)-yl]propyl]-**	69721-77-3	**NI47329**
226	169	528	69	515	487	105	317	**741**	100	64	54	39	29	24	20	12	**0.00**	20	8	5	1	–	21	–	–	–	–	–	–	–	**(3-Hydroxyphenyl)ethanolamine, tris(heptafluorobutyrate)**		**NI47330**
169	528	515	240	69	487	226	317	**741**	100	84	78	58	53	37	28	25	**0.00**	20	8	5	1	–	21	–	–	–	–	–	–	–	**(4-Hydroxyphenyl)ethanolamine, tris(heptafluorobutyrate)**		**NI47331**
332	528	529	333	257	164	215	111	**741**	100	78	20	15	15	13	12	11	**0.00**	20	8	5	1	–	21	–	–	–	–	–	–	–	**(4-Hydroxyphenyl)ethanolamine, tris(heptafluorobutyrate)**		**NI47332**
169	69	528	515	226	487	317	28	**741**	100	68	56	52	39	25	19	18	**0.20**	20	8	5	1	–	21	–	–	–	–	–	–	–	**(4-Hydroxyphenyl)ethanolamine, tris(heptafluorobutyrate)**		**MS11012**
73	624	147	356	75	536	625	300	**741**	100	33	33	29	23	18	17	17	**0.00**	29	67	9	1	–	–	–	–	–	6	–	–	–	**D-Glycero-D-galacto-2-nonulosonic acid, 5-(acetylamino)-3,5-dideoxy-4,6,7,8,9-pentakis-O-(trimethylsilyl)-, trimethylsilyl ester**		**MS11013**
299	73	357	315	300	147	314	75	**742**	100	76	27	21	20	19	17	15	**0.00**	23	60	11	–	–	–	–	–	–	6	2	–	–	**D-erythro-2-Pentulose, 3,4-bis-O-(trimethylsilyl)-, bis[bis(trimethylsilyl) phosphate]**	55622-57-6	**NI47333**
296	575	742	573	444	740	571	738	**742**	100	71	54	52	43	39	39	29		24	–	–	–	–	20	–	–	–	–	–	–	1	**Tetrakis(pentafluorophenyl)germane**		**MS11014**
699	73	700	701	702	147	727	28	**742**	100	93	64	41	16	16	13	13	**1.31**	30	62	6	4	–	–	–	–	–	6	–	–	–	**Pyrimidine, 5,5′-(2-methylpropylidene)bis[2,4,6-tris[(trimethylsilyl)oxy]-**	61142-39-0	**NI47334**
217	363	273	191	231	204	232	245	**742**	100	96	91	80	48	43	41	33	**0.01**	30	70	9	–	–	–	–	–	–	6	–	–	–	**α-L-Mannopyranoside, 6-deoxy-2,3,4-tris-O-(trimethylsilyl)-**	56784-11-3	**MS11015**
629	72	387	630	412	98	70	68	**742**	100	39	35	34	24	23	18	14	**0.00**	36	66	13	–	–	–	–	–	–	–	–	–	4	**Oxohexakis(diethylacetato)tetraberyllium(II)**		**MS11016**
169	109	331	218	171	699	394	742	**742**	100	95	65	35	30	20	10	5		37	46	14	2	–	–	–	–	–	–	–	–	–	**Penta-acetyldihydrovincoside**		**LI10047**
121	105	162	191	100	84	77	70	**742**	100	94	75	63	50	39	35	24	**0.10**	42	54	8	4	–	–	–	–	–	–	–	–	–	**3,7,12-Triazabicyclo[13.3.1]nonadeca-1(19),15,17-triene-2-acetic acid, 3,7-diacetyl-12-[3-(benzoylamino)propyl]-16-methoxy-14-[(4-methoxyphenyl)methyl]-13-oxo-, methyl ester**	69721-74-0	**NI47335**
177	205	183	181	175	131	389	169	**743**	100	82	72	70	70	68	66	64	**7.20**	39	65	7	7	–	–	–	–	–	–	–	–	–	**L-Leucine, N-[N-[N-[N-[N-(2-pyridinylmethylene)-L-valyl]-L-isoleucyl]-L-alanyl]-L-leucyl]-L-leucyl]-, methyl ester**	42547-82-0	**NI47336**
224	268	423	156	226	494	237	395	**743**	100	64	60	48	40	34	34	27	**10.00**	41	69	7	5	–	–	–	–	–	–	–	–	–	**L-Phenylalanine, N-[N-[N-[N^{2},N^{5}-bis(1-oxodecyl)-L-ornithyl]-L-alanyl]-L alanyl]-, methyl ester**	35146-60-2	**LI10048**

MASS TO CHARGE RATIOS								M.W.	INTENSITIES								Parent	C	H	O	N	S	F	Cl	Br	I	Si	P	B	X	COMPOUND NAME	CAS Reg No	No
224	423	156	226	494	395	157	223	**743**	100	58	48	37	34	26	26	21	**10.99**	41	69	7	5	–	–	–	–	–	–	–	–	–	L-Phenylalanine, N-[N-[N-[N^2,N^5-bis(1-oxodecyl)-L-ornithyl]-L-alanyl]-L alanyl]-, methyl ester	35146-60-2	NI47337
661	703	662	704	663	705	697	664	**744**	100	72	61	47	45	39	17	17	**1.19**	24	40	12	–	–	–	–	–	–	8	–	–	–	Perallyloctasilsesquioxane	89038-78-8	NI47338
133	121	239	199	105	225	318	744	**744**	100	79	78	60	56	53	51	36		39	38	5	–	–	–	–	2	–	–	–	–	–	2,6-Bis-[3′-(2″-bromo-2″-hydroxy-4″-methylbenzyl)-2′-hydroxy-5′-methylbenzyl]-4-methylphenol		LI10049
133	238	120	239	201	199	222	105	**744**	100	78	78	70	61	61	57	57	**35.00**	39	38	5	–	–	–	–	2	–	–	–	–	–	2,6-Bis[3-(3-bromo-2-hydroxy-5-methylbenzyl)-2-hydroxy-5-methylbenzyl]-4-methylphenol		LI10050
310	434	189	124	744	59	651		**744**	100	18	13	6	5	3	1			41	35	6	–	–	–	–	–	–	–	2	–	1	Cyclopentadienyl-cobalt-bis(triphenylphosphite)		LI10051
180	151	491	684	744	449	685	492	**744**	100	90	69	48	42	26	23	20		41	44	13	–	–	–	–	–	–	–	–	–	–	(2R,3S)-2,3-trans-3-Acetoxy-6-[(2R,3S,4R)-2,3-trans-3,4-cis-3-acetoxy-3′,4′,7-trimethoxyflavan-4-yl]-3′,4′,7,8-tetramethoxyflavan		NI47339
684	624	685	152	625	153	151	299	**744**	100	54	42	40	34	28	28	22	**7.00**	41	44	13	–	–	–	–	–	–	–	–	–	–	Diacetyl-heptamethyl-procyanidine		LI10052
31	415	44	82	51	29	69	131	**746**	100	50	28	22	21	15	9	8	**0.21**	18	10	4	–	–	24	–	–	–	–	–	–	–	Bis(1,1,7-trihydroperfluoroheptyl)succinate		MS11017
31	415	51	69	130	416	48	101	**746**	100	50	21	10	9	4	4	3	**0.60**	18	10	4	–	–	24	–	–	–	–	–	–	–	Bis(1,1,7-trihydroperfluoroheptyl)succinate		LI10053
43	169	109	331	141	237	115	81	**746**	100	14	10	8	7	2	2	2	**0.01**	26	35	17	–	–	–	–	–	1	–	–	–	–	6-Deoxy-6-iodo-1,2,3,2′,3′,4′,6′-β-maltose heptaacetate		MS11018
43	169	243	237	399	619	109	141	**746**	100	16	15	15	13	10	10	7	**0.06**	26	35	17	–	–	–	–	–	1	–	–	–	–	6′-Deoxy-6′-iodo-1,2,3,6,2′,3′,4′-β-maltose heptaacetate		MS11019
475	57	476	477	56	487	385	441	**746**	100	99	42	29	28	27	17	14	**0.00**	28	66	11	–	–	–	–	–	–	6	–	–	–	1,3,5,7,9,11-Hexaethyl-3,5,9,11-tetrabutoxybicyclo[5.5.1]hexasiloxane	110991-15-6	NI47340
221	362	213	215	363	307	309	222	**746**	100	71	63	45	42	35	22	18	**2.50**	46	60	4	–	–	–	2	–	–	–	–	–	–	2,4,6,8,10,12,14-Pentadecaheptaenoic acid, 15-(3,5-dichloro-4-methoxyphenyl)-, 3-methoxy-2-(9-methyldecyl)-5-(4-methylpentyl)phenyl ester	71246-78-1	NI47341
57	43	312	71	54	69	83	59	**746**	100	76	60	56	51	50	49	46	**5.29**	50	98	3	–	–	–	–	–	–	–	–	–	–	Pentacontanoic acid, 18-oxo-	40710-26-7	NI47342
57	746	87	75	74	71	747	43	**746**	100	97	84	64	64	64	59	57		51	102	2	–	–	–	–	–	–	–	–	–	–	Pentacontanoic acid, methyl ester	40710-34-7	NI47343
605	545	419	702	307	391	353	606	**747**	100	92	66	59	57	49	49	38	**2.30**	40	68	11	–	–	–	–	–	–	–	–	–	1	Nigericin, monosodium salt	28643-80-3	NI47344
721	722	723	703	724	748	704	749	**748**	100	66	55	27	25	24	21	17		20	32	14	–	–	–	–	–	–	9	–	–	–	1-Ethoxypervinylhomooctasilsesquioxane		NI47345
57	535	41	545	591	536	27	637	**748**	100	93	67	63	49	45	44	42	**0.00**	26	48	12	–	–	–	–	–	–	7	–	–	–	1,3,5,7,9,11,13-Heptavinyl-3,11,13-tributoxytricyclo[5.5.1.3^{3,11}]heptasiloxane	110991-18-9	NI47346
195	283	347	318	303	332	326	150	**748**	100	75	64	27	24	21	20	19	**10.00**	34	28	4	–	–	10	–	–	–	–	–	–	1	Nickel(II) bis(3-pentafluorobenzoylcamphorate)		NI47347
184	594	383	593	240	402	370	352	**748**	100	95	65	55	40	31	28	21	**0.00**	34	41	9	2	–	–	–	–	1	–	–	–	–	Isoreserpine methiodide		MS11020
310	354	748	281	282	558	187	325	**748**	100	96	81	62	60	22	21	18		38	52	15	–	–	–	–	–	–	–	–	–	–	Egonol hepta-O-methylgentiobioside	77690-85-8	NI47348
85	43	373	359	299	358	57	169	**748**	100	77	73	50	50	46	46	32	**0.00**	39	56	14	–	–	–	–	–	–	–	–	–	–	Acetyl atractyloside		NI47349
216	190	189	69	191	95	57	81	**748**	100	80	38	38	33	27	27	23	**0.00**	46	75	4	–	–	3	–	–	–	–	–	–	–	3β-Myristoyl-16β-(trifluoroacetyl)olean-12-ene		NI47350
330	69	190	57	43	55	203	95	**748**	100	70	68	65	61	49	44	36	**0.00**	46	75	4	–	–	3	–	–	–	–	–	–	–	3β-Myristoyl-28-(trifluoroacetyl)olean-12-ene		NI47351
180	181	104	77	568	78	284	748	**748**	100	60	50	38	30	27	13	12		52	40	–	4	–	–	–	–	–	1	–	–	–	Silanetetramine, tetrakis(diphenylmethylene)-	23115-34-6	NI47352
151	109	123	137	191	179	135	111	**748**	100	53	50	38	36	35	33	31	**5.00**	53	80	2	–	–	–	–	–	–	–	–	–	–	2,5-Cyclohexadiene-1,4-dione, 2,3-dimethyl-5-(3,7,11,15,19,23,27,31,35-nonamethyl-2,6,10,14,18,22,26,30,34-hexatriacontanonaenyl)-, (all-E)-	4299-57-4	NI47354
189	748	69	749	203	750	81	217	**748**	100	63	35	29	29	27	24	19		53	80	2	–	–	–	–	–	–	–	–	–	–	2,5-Cyclohexadiene-1,4-dione, 2,3-dimethyl-5-(3,7,11,15,19,23,27,31,35-nonamethyl-2,6,10,14,18,22,26,30,34-hexatriacontanonaenyl)-, (all-E)-	4299-57-4	NI47353
348	201	749	150	321	95	320	91	**749**	100	87	79	79	59	55	53	51		28	28	4	–	–	14	–	–	–	–	–	–	1	Manganese(II) bis(3-heptafluorobutyrylcamphorate)	76059-35-3	NI47355
79	73	276	75	131	147	129	78	**749**	100	54	39	22	17	15	13	13	**0.00**	35	79	6	1	–	–	–	–	–	5	–	–	–	Acetamide, N-[2,9,10,14-tetrakis[(trimethylsilyl)oxy]-1-[[(trimethylsilyl)oxy]methyl]-3-heptadecenyl]-	56247-92-8	NI47356
73	223	361	217	103	147	222	191	**750**	100	90	85	38	33	33	32	31	**0.00**	32	66	10	–	–	–	–	–	–	5	–	–	–	8-O-acetylmioporoside, pentakis(trimethylsilyl)-		NI47357
720	735	721	736	722	737	723	738	**750**	100	68	66	52	35	32	18	12	**1.80**	33	58	8	–	–	–	–	–	–	6	–	–	–	1,2,5,6,8-Pentahydroxy-3-(hydroxymethyl)anthraquinone, hexakis(trimethylsilyl)-	91701-35-8	NI47358

MASS TO CHARGE RATIOS								M.W.	INTENSITIES								Parent	C	H	O	N	S	F	Cl	Br	I	Si	P	B	X	COMPOUND NAME	CAS Reg No	No
43	79	169	109	161	178	127	207	**750**	100	69	62	43	31	23	15	13	**0.00**	35	42	18	–	–	–	–	–	–	–	–	–	–	Methyl 1-(tetraacetoxy-β-D-glucopyranosyloxy)-1,4a,5,6,7,7a-hexahydro-4a,7-dihydroxy-6-[[3-(4-methoxyphenyl)-1-oxo-2-propenyl]oxy]-7-methylcyclopenta[c]pyran-4-carboxylate		NI47359
43	169	331	109	301	403	302	127	**750**	100	54	28	22	13	7	6	6	**5.50**	36	38	14	4	–	–	–	–	–	–	–	–	–	Galactopyranoside, D-erythro-2,3-dihydroxy-1-(1-phenyl-1H-pyrazolo[3,4-b]quinoxalin-3-yl)propyl, hexaacetate, β-D-	34948-23-7	NI47360
43	169	301	331	109	275	302	317	**750**	100	53	32	31	24	23	17	16	**4.38**	36	38	14	4	–	–	–	–	–	–	–	–	–	α-D-Glucopyranoside, 2,3-bis(acetyloxy)-3-(1-phenyl-1H-pyrazolo[3,4-b]quinoxalin-3-yl)propyl, 2,3,4,6-tetraacetate, (R*,S*)-	55658-20-3	NI47361
43	169	301	109	403	302	127	273	**750**	100	90	37	32	22	21	19	18	**3.58**	36	38	14	4	–	–	–	–	–	–	–	–	–	Glucopyranoside, D-erythro-2,3-dihydroxy-1-(1-phenyl-1H-pyrazolo[3,4-b]quinoxalin-3-yl)propyl, hexaacetate, α-D-	33796-99-5	NI47362
43	159	331	301	109	403	127	302	**750**	100	42	35	15	15	14	10	7	**3.46**	36	38	14	4	–	–	–	–	–	–	–	–	–	Glucopyranoside, D-erythro-2,3-dihydroxy-1-(1-phenyl-1H-pyrazolo[3,4-b]quinoxalin-3-yl)propyl, hexaacetate, β-D-	33797-00-1	NI47363
43	169	301	275	109	331	317	302	**750**	100	45	23	20	19	13	11	10	**2.27**	36	38	14	4	–	–	–	–	–	–	–	–	–	Glucopyranoside, D-erythro-2,3-dihydroxy-3-(1-phenyl-1H-pyrazolo[3,4-b]quinoxalin-3-yl)propyl, hexaacetate, α-D-	33797-01-2	NI47364
43	169	301	275	331	317	109	302	**750**	100	19	18	15	13	13	11	10	**0.08**	36	38	14	4	–	–	–	–	–	–	–	–	–	Glucopyranoside, D-erythro-2,3-dihydroxy-3-(1-phenyl-1H-pyrazolo[3,4-b]quinoxalin-3-yl)propyl, hexaacetate, β-D-	33909-98-7	NI47365
648	606	245	564	522	480	287	590	**750**	100	76	65	61	40	29	27	24	**0.01**	37	34	17	–	–	–	–	–	–	–	–	–	–	3,3',4',5,7-Pentaacetoxy-4-(2,4,6-triacetoxyphenyl)flavan		NI47366
57	43	367	183	71	55	239	495	**750**	100	87	68	68	55	55	44	43	**0.14**	47	90	6	–	–	–	–	–	–	–	–	–	–	Hexadecanoic acid, 2-[(1-oxododecyl)oxy]-1,3-propanediyl ester	56599-90-7	NI47367
467	57	183	468	71	43	85	55	**750**	100	48	38	35	31	30	21	16	**0.00**	49	98	4	–	–	–	–	–	–	–	–	–	–	Lauric acid, 2-(hexadecyloxy)-3-(octadecyloxy)propyl ester	10322-47-1	NI47368
390	391	570	571	750	392	751	572	**750**	100	71	34	27	22	22	12	10		50	58	4	2	–	–	–	–	–	–	–	–	–	N,N'-Dicyclotridecyl-3,4:9,10-perylenebis(dicarboximide)	110590-78-8	MS11021
69	151	81	41	189	55	191	93	**750**	100	83	64	28	26	25	24	24	**13.79**	53	82	2	–	–	–	–	–	–	–	–	–	–	2H-1-Benzopyran-6-ol, 3,4-dihydro-2,7,8-trimethyl-2-(4,8,12,16,20,24,28,32-octamethyl-3,7,11,15,19,23,27,31-tritriacontaoctaenyl)-, [R-(all-E)]-	4382-43-8	NI47369
723	724	725	726	347	727	348	665	**752**	100	64	52	23	19	11	11	9	**4.59**	19	48	14	–	–	–	–	–	–	9	–	–	–	1-Methoxyperethylhomooctasilsesquioxane	72617-66-4	NI47370
83	69	333	404	752	149	96	67	**752**	100	68	54	52	46	36	33	33		28	28	4	–	–	14	–	–	–	–	–	–	1	Nickel(II) bis(3-heptafluorobutyrylpinan-4-onate)	87306-50-1	NI47371
305	83	149	108	177	265	96	304	**752**	100	84	69	31	21	18	18	15	**2.00**	28	28	4	–	–	14	–	–	–	–	–	–	1	Nickel(II) bis(4-heptafluorobutyrylpinan-3-onate)		NI47372
752	724	754	726	753	725	696	709	**752**	100	79	46	37	35	26	24	16		28	28	4	–	–	14	–	–	–	–	–	–	1	Nickel(II) bis[(1R)-3-(heptafluorobutyryl)camphorate]	68457-34-1	NI47373
154	673	671	675	674	672	669	155	**752**	100	30	29	24	18	17	13	13	**0.70**	36	20	–	–	–	8	–	–	–	–	–	–	2	Dibenzo[b,e][1,4]digermanin, 1,2,3,4,6,7,8,9-octafluoro-5,10-dihydro-5,5,10,10-tetraphenyl-	19638-32-5	NI47374
171	131	172	81	425	564	173	426	**752**	100	80	77	71	68	62	45	30	**2.00**	39	76	6	–	–	–	–	–	–	4	–	–	–	Ecdysone 2β,3β,22,25-tetrakis(trimethylsilyl) ether		MS11022
77	93	91	124	119	183	105	252	**752**	100	47	46	43	34	31	20	15	**3.50**	45	32	6	6	–	–	–	–	–	–	–	–	–	[4,4':4',4''-Terpyrazolidine]-3,3',3'',5,5',5''-hexone, 1,1',1'',2,2',2''-hexaphenyl-	3812-22-4	NI47375
533	149	666	708	167	667	445	709	**752**	100	99	85	64	48	46	33	32	**4.40**	49	86	4	1	–	–	–	–	–	–	–	–	–	3β-[2-(14-Carboxytetradecyl)-2-ethyl-4,4-dimethyl-1,3-oxazolin-2-yl]cholest-5-ene-N-oxide		MS11023
683	753	613	543	473				**753**	100	61	28	19	11					19	–	–	–	–	–	15	–	–	–	–	–	–	Benzyl, α,2,3,4,5-pentachloro-α,6-bis(pentachlorophenyl)-	33517-72-5	NI47376
683	753	306.5	271.5	341.5	613	543	236.5	**753**	100	35	33	25	24	19	12	11		19	–	–	–	–	–	15	–	–	–	–	–	–	Benzyl, α,2,3,4,5-pentachloro-α,6-bis(pentachlorophenyl)-	33517-72-5	LI10054
683	753	306.5	341.5	613	271.5	543	473	**753**	100	61	36	29	28	25	19	11		19	–	–	–	–	–	15	–	–	–	–	–	–	Benzyl, α,2,3,4,5-pentachloro-α,6-bis(pentachlorophenyl)-	33517-72-5	LI10055
176	119	590	159	69	100	81	147	**753**	100	99	55	55	40	34	28	24	**0.00**	20	7	7	1	–	20	–	–	–	–	–	–	–	Noradrenaline tetra-PFP	55256-13-8	NI47377
159	119	109	442	69	589	90	176	**753**	100	46	19	18	18	16	12	9	**0.00**	20	7	7	1	–	20	–	–	–	–	–	–	–	Noradrenaline tetra-PFP	55256-13-8	MS11024
119	176	147	590	69	177	120	43	**753**	100	99	32	13	13	8	6	5	**0.00**	20	7	7	1	–	20	–	–	–	–	–	–	–	Noradrenaline tetra-PFP	55256-13-8	NI47378
726	754	348	697	151	727	755	320	**753**	100	80	51	40	33	30	27	27	**0.00**	28	28	4	–	–	14	–	–	–	–	–	–	1	Cobalt(II) bis(3-heptafluorobutyrylcamphorate)	83664-34-0	NI47379
119	69	147	44	590	100	45	50	**754**	100	31	16	12	11	9	9	7	**0.50**	20	6	8	–	–	20	–	–	–	–	–	–	–	(3,4-Dihydroxyphenyl)ethylene glycol tetrakis(pentafluoropropionate)		MS11025
362	221	155	754	121	222	292	157	**754**	100	67	19	11	11	10	7	7		48	79	4	–	–	–	1	–	–	–	–	–	–	Benzeneheptadecanoic acid, 3-chloro-4-methoxy-, 3-methoxy-2-(9-methyldecyl)-5-(4-methylpentyl)phenyl ester	69506-02-1	NI47380
390	391	530	586	531	392	284	754	**754**	100	55	39	27	15	15	14	13		50	62	4	2	–	–	–	–	–	–	–	–	–	N,N'-Bis(1-hexylheptyl)-3,4:9,10-perylenebis(dicarboximide)	110590-84-6	MS11026
390	391	530	586	531	392	284	754	**754**	100	55	39	27	15	15	14	13		50	62	4	2	–	–	–	–	–	–	–	–	–	N,N'-Bis(1-hexylheptyl)-3,4:9,10-perylenebis(dicarboximide)	110590-84-6	DD01823
105	754	755	77	377	121	300	677	**754**	100	63	33	25	10	5	3	1		52	34	2	–	2	–	–	–	–	–	–	–	–	2,2'-(1,4-Phenylene)bis(4-benzoyl-3,5-diphenylthiophene)	110718-10-0	DD01824
240	169	350	528	241	542	337	69	**755**	100	15	7	7	7	6	6	6	**0.00**	21	10	5	1	–	21	–	–	–	–	–	–	–	4-Hydroxy-N-methylphenylethanolamine, tris(hexafluorobutanoyl)-		NI47381

MASS TO CHARGE RATIOS								M.W.	INTENSITIES								Parent	C	H	O	N	S	F	Cl	Br	I	Si	P	B	X	COMPOUND NAME	CAS Reg No	No
176	263	160	290	289	148	188	200	**755**	100	91	79	62	62	58	55	34	**0.00**	40	49	6	7	1	–	–	–	–	–	–	–	–	L-Phenylalaninamide, N-[[4-(dimethylamino)phenyl]methylene]-L-tryptophyl-L-methionyl-L-α-aspartyl-, ethyl ester	40760-02-9	NI47382
756	294	588	700	672	644	291	616	**756**	100	65	57	48	45	42	41	39		22	12	6	–	–	–	–	–	–	–	–	–	2	Osmium, hexacarbonyl[μ-[(1,2,3,4-η:1,4-η)-1,4-diphenyl-1,3-butadiene-1,4 diyl]]di-, (Os-Os)	57174-04-6	NI47383
424	73	425	299	509	383	426	741	**756**	100	80	43	33	29	24	20	18	**0.60**	27	61	9	2	–	–	–	–	–	6	1	–	–	2,4(1H,3H)-Pyrimidinedione, 5-[2-O-[bis[(trimethylsilyl)oxy]phosphinyl]-3,5-bis-O-(trimethylsilyl)-β-D-ribofuranosyl]-1,3-bis(trimethylsilyl)-	56145-27-8	NI47384
108	79	107	271	384	183	109	442	**756**	100	96	87	23	21	21	19	19	**0.00**	40	60	10	4	–	–	–	–	–	–	–	–	–	(R,R)-Bis(N-benzyloxycarbonylleicyl)diaza-18-crown-6		MS11027
712	713	714	684	282	298	683	334	**756**	100	55	17	15	12	10	9	9	**8.50**	48	40	7	2	–	–	–	–	–	–	–	–	–	Bis[3-oxo-6″-diethylamino-spiro(phthalam-1,9″-xanth-2″-yl)] ether		IC04641
73	91	75	147	651	650	129	77	**757**	100	99	83	77	43	37	37	37	**3.50**	40	71	5	1	–	–	–	–	–	4	–	–	–	Pregn-5-en-20-one, 3,11,17,21-tetrakis[(trimethylsilyl)oxy]-, O-benzyloxime, (3β,11β)-	57326-03-1	NI47385
73	650	91	651	560	147	75	129	**757**	100	99	76	65	50	50	47	44	**1.50**	40	71	5	1	–	–	–	–	–	4	–	–	–	Pregn-5-en-20-one, 3,16,17,21-tetrakis[(trimethylsilyl)oxy]-, O-(phenylmethyl)oxime, (3β,16α)-	57326-04-2	NI47386
73	650	651	652	91	560	75	666	**757**	100	89	53	38	31	30	24	23	**2.49**	40	71	5	1	–	–	–	–	–	4	–	–	–	Pregn-5-en-20-one, 3,16,17,21-tetrakis[(trimethylsilyl)oxy]-, O-(phenylmethyl)oxime, (3β,16α)-	57326-04-2	NI47387
366	365	44	756	57	43	367	55	**757**	100	66	36	35	32	29	28	27	**22.00**	53	107	–	1	–	–	–	–	–	–	–	–	–	(2)-(Azacyclohexacosane)(cyclooctacosane)-catenane		MS11028
426	386	69	452	56	757	82	370	**757**	100	44	24	20	18	14	12	11		54	95	–	1	–	–	–	–	–	–	–	–	–	Di-5-α-cholestan-3-α-ylamine		LI10056
69	181	207	93	193	303	253	505	**758**	100	93	73	62	36	30	27	25	**3.00**	18	8	12	–	–	18	–	–	–	–	–	–	–	Hexitol, hexakis(trifluoroacetate)	55538-86-8	NI47388
197	257	215	578	188	43	638	270	**758**	100	87	86	84	80	77	59	43	**0.00**	42	62	12	–	–	–	–	–	–	–	–	–	–	Hexa-O-acetyl-gymnemagenin		LI10057
372	370	373	585	758	587	759	412	**758**	100	55	22	21	17	10	7	7		44	40	5	4	–	–	–	–	–	–	–	–	2	μ-Oxo-di[bis(2,3-dimethyl-8-quinolinolato)]dialuminium(III)		LI10058
535	536	758	759	105	224	223	77	**758**	100	70	63	23	19	11	11	11		45	33	6	–	–	–	–	–	–	–	–	–	1	Yttrium, tris(1,3-diphenyl-1,3-propanedionato)-	15688-50-3	NI47389
531	227	303	379	152	454	758		**758**	100	72	52	31	21	4	1			48	32	–	–	–	–	–	–	–	–	–	–	2	Bis-2,2′-biphenylylene[2′-(2,2′-biphenylylenearsino)-2-biphenylyl]arsenic		LI10059
73	91	75	147	103	105	143	652	**759**	100	99	62	45	39	30	26	25	**0.56**	40	73	5	1	–	–	–	–	–	4	–	–	–	Pregnan-20-one, 3,11,17,21-tetrakis[(trimethylsilyl)oxy]-, O-(phenylmethyl)oxime, (3α,5β,11β)-	57326-05-3	NI47390
142	198	395	381	175	608	607	174	**760**	100	94	70	51	45	44	40	40	**0.00**	39	41	6	2	–	–	–	–	1	–	–	–	–	Oxycanthine-methylether-dimethyliodide		LI10060
83	55	189	91	207	215	111	171	**760**	100	73	15	14	13	12	10	9	**0.00**	43	68	11	–	–	–	–	–	–	–	–	–	–	21-O-(4-O-Acetyl-3-O-angeloyl-β-D-fucopyranosyl)theasapogenol B		DD01825
362	221	363	292	222	151	189	137	**760**	100	53	29	8	7	7	5	5	**3.00**	46	74	4	–	–	–	2	–	–	–	–	–	–	Benzenepentadecanoic acid, 3,5-dichloro-4-methoxy-, 3-methoxy-2-(9-methyldecyl)-5-(4-methylpentyl)phenyl ester	71142-48-8	NI47391
57	88	71	101	89	43	760	85	**760**	100	66	63	59	59	56	52	41		52	104	2	–	–	–	–	–	–	–	–	–	–	Pentacontanoic acid, ethyl ester	40710-37-0	WI01601
57	760	88	101	89	71	761	43	**760**	100	99	70	68	67	63	57	57		52	104	2	–	–	–	–	–	–	–	–	–	–	Pentacontanoic acid, ethyl ester	40710-37-0	MS11029
760	57	761	88	101	89	71	43	**760**	100	81	57	56	55	54	51	46		52	104	2	–	–	–	–	–	–	–	–	–	–	Pentacontanoic acid, ethyl ester	40710-37-0	NI47392
165	470	254	428	284	221	120	344	**761**	100	91	81	66	54	54	53	52	**1.00**	27	63	–	1	–	–	–	–	–	–	–	–	3	Tris(tri-propylstannyl)amine		LI10061
43	93	178	206	129	57	132	106	**761**	100	98	58	42	27	24	23	20	**11.10**	36	43	17	1	–	–	–	–	–	–	–	–	–	Evonimine	41758-54-7	NI47393
43	107	178	206	133	106	160	134	**761**	100	95	72	58	25	25	23	23	**6.80**	36	43	17	1	–	–	–	–	–	–	–	–	–	Evonine	33458-64-9	NI47394
107	43	51	206	178	761	133	106	**761**	100	89	71	58	48	25	23	23		36	43	17	1	–	–	–	–	–	–	–	–	–	Evonine	33458-64-9	NI47395
57	27	605	606	607	633	623	634	**762**	100	58	56	30	26	24	19	14	**0.00**	24	42	13	–	–	–	–	–	–	8	–	–	–	1,3,5,7,9,11,13,15-Octavinyl-7,13-dibutoxytetracyclo[9.5.1.1^{3,9}.1^{5,15}]octasiloxane	110991-19-0	NI47396
603	57	547	604	41	605	548	559	**762**	100	76	74	52	41	37	33	28	**0.00**	26	62	12	–	–	–	–	–	–	7	–	–	–	1,3,5,7,9,11,13-Heptaethyl-5,9,13-tributoxytricyclo[5.5.1.1^{3,11}]heptasiloxane	110991-21-4	NI47397
602	559	685	643	625	661	702	670	**762**	100	66	56	44	44	25	12	11	**0.00**	37	46	17	–	–	–	–	–	–	–	–	–	–	O-Acetylazadirachtin		NI47398
203	527	262	187	528	495	467	263	**762**	100	93	81	81	37	30	30	22	**0.90**	43	70	11	–	–	–	–	–	–	–	–	–	–	Hepta-O-methylmedicagenic acid 3-O-β-D-glucopyranoside		NI47399
57	149	105	91	197	119	368	744	**762**	100	84	56	30	18	17	14	5	**1.00**	49	62	7	–	–	–	–	–	–	–	–	–	–	Isofucoxanthin benzoate		LI10062

MASS TO CHARGE RATIOS								M.W.	INTENSITIES								Parent	C	H	O	N	S	F	Cl	Br	I	Si	P	B	X	COMPOUND NAME	CAS Reg No	No
582	60	699	657	539	43	523	45	**763**	100	67	66	63	54	52	48	39	**0.00**	36	37	14	5	–	–	–	–	–	–	–	–	–	3H-Pyrazol-3-one, 2,4-dihydro-4-[[5-hydroxy-1-phenyl-3-[1,2,3-tris(acetyloxy)propyl]-1H-pyrazol-4-yl]imino]-2-phenyl-5-[1,2,3-tris(acetyloxy)propyl]-	35426-85-8	NI47400
659	657	660	658	661	662	605	607	**764**	100	70	50	39	33	13	10	9	**3.59**	14	22	12	–	–	–	–	2	–	8	–	–	–	β-Bromoethylhexavinylbromooctasilsesquioxane	88243-03-2	NI47401
73	318	315	299	147	387	319	316	**764**	100	72	59	43	41	23	22	17	**0.00**	27	69	9	–	–	–	–	–	–	7	1	–	–	D-myo-Inositol, 1,2,4,5,6-pentakis-O-(trimethylsilyl)-, bis(trimethylsilyl) phosphate	55568-91-7	NI47402
318	73	315	299	147	319	387	217	**764**	100	96	64	49	38	30	27	17	**0.00**	27	69	9	–	–	–	–	–	–	7	1	–	–	D-myo-Inositol, 1,2,4,5,6-pentakis-O-(trimethylsilyl)-, bis(trimethylsilyl) phosphate	55568-91-7	NI47403
73	315	299	147	318	75	316	191	**764**	100	39	36	28	26	12	10	10	**0.00**	27	69	9	–	–	–	–	–	–	7	1	–	–	L-myo-Inositol, 1,2,4,5,6-pentakis-O-(trimethylsilyl)-, bis(trimethylsilyl) phosphate	33800-38-3	NI47404
73	147	318	387	315	299	217	191	**764**	100	32	31	29	29	29	14	10	**0.00**	27	69	9	–	–	–	–	–	–	7	1	–	–	myo-Inositol, 1,3,4,5,6-pentakis-O-(trimethylsilyl)-, bis(trimethylsilyl) phosphate	33910-06-4	NI47405
73	318	315	147	299	387	319	191	**764**	100	40	33	31	29	13	12	10	**0.00**	27	69	9	–	–	–	–	–	–	7	1	–	–	myo-Inositol, 1,3,4,5,6-pentakis-O-(trimethylsilyl)-, bis(trimethylsilyl) phosphate	33910-06-4	NI47406
73	299	147	315	318	217	300	191	**764**	100	52	30	23	18	17	13	12	**0.00**	27	69	9	–	–	–	–	–	–	7	1	–	–	myo-Inositol, pentakis-O-(trimethylsilyl)-, bis(trimethylsilyl) phosphate	55518-06-4	NI47407
73	299	315	318	147	389	300	217	**764**	100	52	45	36	33	17	14	14	**0.10**	27	69	9	–	–	–	–	–	–	7	1	–	–	Inositol, 1,2,3,5,6-pentakis-O-(trimethylsilyl)-, bis(trimethylsilyl) phosphate, L-chiro-	33800-40-7	NI47408
529	545	530	371	339	546	385	355	**764**	100	38	30	28	21	12	8	8	**5.00**	38	52	16	–	–	–	–	–	–	–	–	–	–	6-C-Diglucosylluteolin		NI47409
764	370	326	298	297	187	342	574	**764**	100	69	61	46	31	29	21	15		39	56	15	–	–	–	–	–	–	–	–	–	–	2-(3,4-Dimethoxyphenyl)-5-(3-hydroxypropyl)-7-methoxybenzofuran hepta-O-methylgentiobioside		NI47410
146	73	147	170	217	75	159	131	**764**	100	90	43	37	36	24	22	17	**16.02**	39	68	9	–	–	–	–	–	–	3	–	–	–	Card-20(22)-enolide, 3-[[6-deoxy-3-O-methyl-2,4-bis-O-(trimethylsilyl)-α-L-glucopyranosyl]oxy]-19-oxo-14-[(trimethylsilyl)oxy]-, (3β,5β)-	56784-21-5	MS11030
146	73	147	217	170	75	159	131	**764**	100	90	40	36	36	23	22	17	**15.00**	39	68	9	–	–	–	–	–	–	3	–	–	–	Card-20(22)-enolide, 3-[[6-deoxy-3-O-methyl-2,4-bis-O-(trimethylsilyl)-α-L-glucopyranosyl]oxy]-19-oxo-14-[(trimethylsilyl)oxy]-, (3β,5β)-	56784-21-5	NI47411
146	73	147	170	217	75	159	131	**764**	100	90	43	36	34	23	22	17	**15.76**	39	68	9	–	–	–	–	–	–	3	–	–	–	24-Norchola-20,22-dien-19-al, 3-[[6-deoxy-3-O-methyl-2,4-bis(trimethylsilyl)-α-L-glucopyranosyl]oxy]-21,23-epoxy-14-hydroxy-23-[(trimethylsilyl)oxy]-, (3β,5β)-	40837-89-6	NI47412
73	43	58	57	69	113	29	41	**764**	100	78	75	72	67	63	62	58	**0.00**	41	64	13	–	–	–	–	–	–	–	–	–	–	Digitoxin	71-63-6	NI47413
357	203	95	113	339	131	73	69	**764**	100	98	97	96	94	94	93	92	**0.02**	41	64	13	–	–	–	–	–	–	–	–	–	–	Digitoxin	71-63-6	LI10063
357	303	95	113	339	131	73	69	**764**	100	99	97	96	92	91	90	90	**4.00**	41	64	13	–	–	–	–	–	–	–	–	–	–	Digitoxin	71-63-6	LI10064
91	108	94	255	122	269	357	301	**764**	100	79	50	24	24	22	21	21	**1.00**	44	52	8	4	–	–	–	–	–	–	–	–	–	Tetrapyrrane B		LI10065
205	69	81	167	105	199	67	131	**764**	100	70	57	33	32	31	31	30	**18.03**	53	80	3	–	–	–	–	–	–	–	–	–	–	2,5-Cyclohexadiene-1,4-dione, 5-methoxy-2-methyl-3-(3,7,11,15,19,23,27,31,35-nonamethyl-2,6,10,14,18,22,26,30,34-hexatriacontanonaenyl)-, (all-E)-	7200-28-4	NI47414
73	315	299	217	147	316	103	387	**766**	100	78	38	28	25	21	17	16	**0.00**	27	71	9	–	–	–	–	–	–	7	1	–	–	D-Glucitol, 2,3,4,5,6-pentakis-O-(trimethylsilyl)-, bis(trimethylsilyl) phosphate	55528-11-5	NI47415
73	387	299	315	147	217	388	357	**766**	100	48	36	29	27	18	17	17	**0.00**	27	71	9	–	–	–	–	–	–	7	1	–	–	Glucitol, 1,2,3,4,5-pentakis-O-(trimethylsilyl)-, bis(trimethylsilyl) phosphate, D-	33800-35-0	NI47416
299	73	147	387	315	205	388	217	**766**	100	99	97	82	79	43	36	36	**0.00**	27	71	9	–	–	–	–	–	–	7	1	–	–	Glucitol, 1,2,3,4,5-pentakis-O-(trimethylsilyl)-, bis(trimethylsilyl) phosphate, D-	33800-35-0	NI47417
73	387	299	315	147	217	388	357	**766**	100	43	33	30	28	19	16	15	**0.00**	27	71	9	–	–	–	–	–	–	7	1	–	–	Mannitol, 1,2,3,4,5-pentakis-O-(trimethylsilyl)-, bis(trimethylsilyl) phosphate, D-	33800-37-2	NI47418
73	387	315	299	217	147	388	357	**766**	100	71	51	44	31	29	25	20	**0.00**	27	71	9	–	–	–	–	–	–	7	1	–	–	Mannitol, 1,2,3,4,5-pentakis-O-(trimethylsilyl)-, bis(trimethylsilyl) phosphate, D-	33800-37-2	NI47419
387	73	315	299	217	147	388	357	**766**	100	99	71	62	44	41	35	28	**0.00**	27	71	9	–	–	–	–	–	–	7	1	–	–	Mannitol, 1,2,3,4,5-pentakis-O-(trimethylsilyl)-, bis(trimethylsilyl) phosphate, D-	33800-37-2	NI47420
99	71	69	57	55	43	42	41	**766**	100	96	96	96	96	96	96	96	**0.00**	47	90	7	–	–	–	–	–	–	–	–	–	–	1-(5-Hydroxy)-dodecanoyl-2,3-dihexadecanoyl glycerol		LI10066
69	81	95	93	121	109	107	55	**766**	100	78	37	31	26	26	25	24	**1.40**	55	90	1	–	–	–	–	–	–	–	–	–	–	Undecaprenol		LI10067
190	119	604	191	440	484	431	147	**767**	100	27	15	14	9	6	4	4	**0.00**	21	9	7	1	–	20	–	–	–	–	–	–	–	Propanamide, N-[2-(3,4-dihydroxyphenyl)-2-hydroxyisopropyl]-2,2,3,3,3-pentafluoro-, tris(2,2,3,3,3-pentafluoro-1-oxopropyl) deriv.	72347-74-1	NI47421

MASS TO CHARGE RATIOS								M.W.	INTENSITIES								Parent	C	H	O	N	S	F	Cl	Br	I	Si	P	B	X	COMPOUND NAME	CAS Reg No	No
190	119	100	69	81	61	70	51	767	100	34	22	19	16	9	8	8	0.00	21	9	7	1	–	20	–	–	–	–	–	–	–	Propanoic acid, pentafluoro-, 4-[2-[methyl(2,2,3,3,3-pentafluoro-1-oxopropyl)amino]-1-(2,2,3,3,3-pentafluoro-1-oxopropoxy)ethyl]-1,2-phenylene ester, (R)-	59785-42-1	NI47422
190	119	100	69	173	81	61	70	767	100	36	31	29	18	18	14	10	0.00	21	9	7	1	–	20	–	–	–	–	–	–	–	Propanoic acid, pentafluoro-, 4-[1-(2,2,3,3,3-pentafluoro-1-oxopropoxy)-2[(2,2,3,3,3-pentafluoro-1-oxopropyl)amino]propyl]-1,2-phenylene ester	74231-48-4	NI47423
70	694	86	130	170	72	328	314	767	100	61	60	52	40	38	29	29	8.00	40	61	8	7	–	–	–	–	–	–	–	–	–	DL-Alanine, N-[N-[1-[N-[N-[N-(1-oxopropyl)-DL-tryptophyl]-DL-leucyl-DL-valyl]-DL-prolyl]-DL-leucyl]-, methyl ester	55822-85-0	NI47425
314	70	413	306	86	526	483	225	767	100	91	70	59	57	51	47	30	1.00	40	61	8	7	–	–	–	–	–	–	–	–	–	DL-Alanine, N-[N-[1-[N-[N-[N-(1-oxopropyl)-DL-tryptophyl]-DL-leucyl-DL-valyl]-DL-prolyl]-DL-leucyl]-, methyl ester	55822-85-0	NI47424
314	70	413	306	86	526	483	225	767	100	91	70	60	57	53	48	30	1.00	40	61	8	7	–	–	–	–	–	–	–	–	–	Alanine, N-[N-[1-[N-[N-(N-propionyl-L-tryptophyl)-L-leucyl]-L-valyl]-L-prolyl]-L-leucyl]-, methyl ester, L-	31944-47-5	NI47426
114	568	372	327	299	439	764	271	768	100	86	71	69	68	63	52	43	0.00	10	–	10	–	–	–	–	–	–	–	–	–	3	Rhenium, μ-cadmiodecacarbonyldi-	33728-45-9	NI47427
73	172	437	217	303	257	75	159	768	100	48	42	27	26	25	24	18	0.30	31	72	8	2	–	–	–	–	–	6	–	–	–	1-Deoxy-1-L-theanino-D-fructopyranose hexa-TMS		MS11031
171	73	57	43	116	81	71	55	768	100	96	90	90	87	86	72	60	0.00	39	76	7	–	–	–	–	–	–	4	–	–	–	20-Hydroxyecdysone 2β,3β,22,25-tetrakis(trimethylsilyl) ether		MS11032
666	726	593	768	607				768	100	16	13	9	5					42	48	10	4	–	–	–	–	–	–	–	–	–	β-Acetoxy-2,4-di-(2-acetoxyethyl)-6,7-di-(2-methoxycarbonylethyl)-1,3,5,8-tetramethylporphin		LI10068
726	710	652	694	768	725	637	593	768	100	32	24	20	18	15	11	6		42	48	10	4	–	–	–	–	–	–	–	–	–	α-Acetoxy-2,4,5,8-tetra-(2-methoxycarbonylethyl)-1,3,6,7-tetramethylporphin		LI10069
57	768	769	507	384	770	221	701	768	100	87	47	34	26	24	23	20		54	72	3	–	–	–	–	–	–	–	–	–	–	1,3,5-Trimethyl-2,4,6-tris(3,5-di-tert-butyl-4-oxo-cyclohexa-2,5-dienylmethylene)benzene		MS11033
229	73	202	230	147	157	769	217	769	100	74	34	22	16	15	11	10		34	67	7	1	–	–	–	–	–	6	–	–	–	myo-Inositol, 2,3,4,5,6-pentakis(trimethylsilyl-, 1-DL-1-O-1-trimethylsilyl-(indole-3-acetyl)-		NI47428
229	73	202	230	147	769	770	203	769	100	85	48	22	18	15	11	10		34	67	7	1	–	–	–	–	–	6	–	–	–	myo-Inositol, 1,3,4,5,6-pentakis(trimethylsilyl)-, 2-O-1-trimethylsilyl-(indole-3-acetyl)-		NI47429
229	769	202	318	130	507	433	574	769	100	42	33	24	3	2	1	1		34	67	7	1	–	–	–	–	–	6	–	–	–	myo-Inositol, 1,3,4,5,6-pentakis(trimethylsilyl)-, 2-O-1-trimethylsilyl-(indole-3-acetyl)-		NI47430
229	202	318	769	130	679	507	433	769	100	30	24	5	4	2	1	1		34	67	7	1	–	–	–	–	–	6	–	–	–	myo-Inositol, 2,3,4,5,6-pentakis(trimethylsilyl)-, 1-O-1-trimethylsilyl-(indole-3-acetyl)-		NI47431
242	770	506	330	594	682	418	56	770	100	50	45	30	20	15	10	10		2	5	–	–	1	20	–	–	–	–	7	–	2	Iron, [μ-(difluorophosphino)][μ-(ethanethiolato)]hexakis(phosphorous trifluoride)di-, (Fe-Fe)	39733-63-6	MS11034
770	357	658	602	630	385	546	574	770	100	81	63	47	45	42	41	38		14	12	10	–	–	–	–	–	–	–	2	–	2	μ-Tetramethyldiphosphine-bis(pentacarbonyltungsten)		MS11035
131	243	43	57	41	55	71	56	770	100	72	68	67	52	47	43	41	0.00	32	68	4	–	4	–	–	–	–	–	2	–	1	Zinc bis(dioctyldithiophosphate)		IC04642
91	244	288	246	121	287	305	69	770	100	99	92	83	79	73	69	63	0.50	34	41	11	4	1	3	–	–	–	–	–	–	–	L-Glutamic acid, N-[N-[N-[S-(2-methoxy-2-oxoethyl)-N-(trifluoroacetyl)-L-cysteinyl]-O-methyl-L-tyrosyl]-L-phenylalanyl]-, dimethyl ester	69782-86-1	NI47432
667	770	771	668	595	772	564	418	770	100	97	56	56	31	30	27	27		42	58	6	4	–	–	–	–	–	2	–	–	–	2,4-Di-(1-hydroxyethyl)-6,7-di-(2-methoxycarbonylethyl)-1,3,5,8-tetramethylporphin, bis-TMS ether		LI10070
770	680	771	590	681	592	591	755	770	100	65	58	56	42	30	29	27		42	58	6	4	–	–	–	–	–	2	–	–	–	2,4-Di-(2-hydroxyethyl)-6,7-di-(2-methoxycarbonylethyl)-1,3,5,8-tetramethylporphin, bis-TMS ether		LI10071
770	680	771	590	681	592	591	755	770	100	65	59	57	41	30	29	28		42	58	6	4	–	–	–	–	–	2	–	–	–	21H,23H-Porphine-2,18-dipropanoic acid, 3,7,12,17-tetramethyl-8,13-bis[1-[(trimethylsilyl)oxy]ethyl]-, dimethyl ester	38574-18-4	NI47433
667	770	771	668	595	772	564	698	770	100	98	56	56	30	29	28	26		42	58	6	4	–	–	–	–	–	2	–	–	–	21H,23H-Porphine-2,18-dipropanoic acid, 3,7,12,17-tetramethyl-8,13-bis[2-[(trimethylsilyl)oxy]ethyl]-, dimethyl ester	38574-19-5	NI47434
151	165	770	347	241	209	227	619	770	100	54	39	28	25	25	23	20		49	54	8	–	–	–	–	–	–	–	–	–	–	(E)-1,3-bis[2-(3,4-dimethoxyphenyl)methyl-5-(3,4-dimethoxyphenyl)methylene-1-cyclopentenyl]propene		MS11036
57	589	770	153	560	251	683	385	770	100	18	11	11	10	3	2	2		51	94	4	–	–	–	–	–	–	–	–	–	–	20-Methyl-37-(1′-methyl-6′-cylohexylhexyl)-31-oxo-30,39,40-trioxatricyclo[33.3.1.1^{3,7}]tetracontane		LI10072
770	236	771	772	773	238	414	774	770	100	71	65	59	30	28	25	15		56	40	–	–	–	–	–	–	–	–	–	–	1	Nickel, bis[1,1′,1″,1‴-(η^4-1,3-cyclobutadiene-1,2,3,4-tetrayl)tetrakis(benzene)]-	61483-84-9	NI47435

MASS TO CHARGE RATIOS								M.W.	INTENSITIES								Parent	C	H	O	N	S	F	Cl	Br	I	Si	P	B	X	COMPOUND NAME	CAS Reg No	No
557	169	69	544	226	361	151	517	771	100	95	70	46	32	26	25	22	8.61	21	10	6	1	–	21	–	–	–	–	–	–	–	O,O,N-Tris(heptafluorobutyryl)normetanephrine		MS11037
771	773	772	774	698	769	700	699	771	100	56	48	24	16	9	9	8		40	44	8	4	–	–	–	–	–	–	–	–	1	Copper, [tetramethyl 3,8,13,18-tetramethyl-21H,23H-porphine-2,7,12,17-tetrapropanoato(2–)-N^{21},N^{22},N^{23},N^{24}]-	54111-68-1	NI47436
328	218	383	442	370	472	299	256	771	100	43	36	30	27	20	20	20	0.00	43	93	4	1	–	–	–	–	–	3	–	–	–	1,3,4-Tri-O-trimethylsilyl-N-(palmitayol)phytosphingosine		LI10073
284	282	286	328	330	250	249	71	772	100	95	84	82	81	67	52	43	0.00	18	–	–	–	–	–	15	–	–	–	1	–	–	Tris(pentachlorophenyl)phosphine		MS11038
43	71	354	169	284	401	386	772	772	100	76	16	5	5	1	1	1		24	30	6	2	2	14	–	–	–	–	–	–	–	Bis(3-methylbutyl) N,N′-bis(heptafluorobutyryl)cystinate		NI47437
534	772	770	532	267	737	386	519	772	100	55	20	15	15	12	12	11		36	47	2	4	–	–	–	–	–	–	–	–	1	Thallium, aquahydroxy[2,3,7,8,12,13,17,18-octaethyl-21H,23H-porphinato(2–)-N^{21},N^{22},N^{23},N^{24}]-	33339-93-4	NI47438
404	712	43	592	57	44	55	71	772	100	83	63	58	50	49	41	36	7.00	42	60	13	–	–	–	–	–	–	–	–	–	–	Hexa-O-acetyl-11-oxo-gymnemagenin		LI10074
774	772	776	775	777	773	779	277	772	100	70	61	52	34	34	26	24		44	24	–	4	–	–	4	–	–	–	–	–	1	Magnesium, [5,10,15,20-tetrakis(4-chlorophenyl)-21H,23H-porphinato(2–)-N^{21},N^{22},N^{23},N^{24}]-	39479-01-1	NI47439
57	55	43	41	91	69	71	83	772	100	77	77	67	59	58	50	40	3.00	48	40	6	2	1	–	–	–	–	–	–	–	–	Bis[3-oxo-6′-diethylamino-spiro(phthalam-1,9′-xanth-2′-yl)] sulphide		IC04643
560	262	108	121	317	125	307	107	774	100	88	85	82	75	62	48	47	0.00	27	24	10	–	–	14	–	–	–	–	–	–	–	7,8-Bis(heptafluorobutyryl)calonectrin		NI47440
474	317	501	427	227	500	185	560	774	100	99	92	89	83	68	55	51	0.00	27	24	10	–	–	14	–	–	–	–	–	–	–	3,8-Bis(heptafluorobutyryl)neosolaniol		NI47441
191	101	143	175	423	129	159	142	774	100	38	34	18	16	16	15	14	0.03	34	62	19	–	–	–	–	–	–	–	–	–	–	Permethyl β-D-xylopyranosyl-(1-3)-α-L-rhamnopyranosyl-(1-2)-α-D-galactopyranosyluronic acid-(1-4)-D-xylitol		NI47442
151	343	299	521	537	492	180	774	774	100	89	88	85	52	48	44	40		42	46	14	–	–	–	–	–	–	–	–	–	–	(2R,3S)-2,3-trans-3-Acetoxy-8-[(2R,3S,4S)-3,4-cis-2,3-trans-3-acetoxy-3′,4′,7,8-tetramethoxyflavan-4-yl]-3′,4′,5,7-tetramethoxyflavan		NI47443
521	151	299	537	180	714	774	715	774	100	99	76	58	51	34	32	19		42	46	14	–	–	–	–	–	–	–	–	–	–	(2R,3S)-2,3-trans-3-Acetoxy-6-[(2R,3S,4S)-2,3-trans-3,4-cis-3-acetoxy-3′,4′,7,8-tetramethoxyflavan-4-yl]-3′,4′,5,7-tetramethoxyflavan		NI47444
714	151	715	299	180	521	477	327	774	100	96	60	60	54	40	40	21	7.90	42	46	14	–	–	–	–	–	–	–	–	–	–	(2R,3S)-2,3-trans-3-Acetoxy-6-[(2R,3S,4S)-2,3-trans-3,4-trans-3-acetoxy-3′,4′,7,8-tetramethoxyflavan-4-yl]-3′,4′,5,7-tetramethoxyflavan		NI47445
151	343	299	714	180	521	654	492	774	100	67	63	58	48	32	25	25	10.60	42	46	14	–	–	–	–	–	–	–	–	–	–	(2R,3S)-2,3-trans-3-Acetoxy-8-[(2R,3S,4S)-2,3-trans-3,4-trans-3-acetoxy-3′,4′,7,8-tetramethoxyflavan-4-yl]-3′,4′,5,7-tetramethoxyflavan		NI47446
714	151	299	715	654	180	521	492	774	100	99	85	42	42	42	39	35	24.00	42	46	14	–	–	–	–	–	–	–	–	–	–	(2R,3S)-2,3-trans-3-Acetoxy-8-[(2R,3S,4S)-2,3-trans-3,4-trans-3-acetoxy-3′,4′,7,8-tetramethoxyflavan-4-yl]-3′,4′,5,7-tetramethoxyflavan		NI47447
151	774	180	193	714	167	655	654	774	100	29	22	21	19	16	10	10		42	46	14	–	–	–	–	–	–	–	–	–	–	(2R,3S)-2,3-trans-3-Acetoxy-5-[(2R,3S)-2,3-trans-3-acetoxy-3′,4′,7,8-tetramethoxyflavan-5-yl]-3′,4′,7,8-tetramethoxyflavan		NI47448
151	774	180	714	327	552	492	222	774	100	39	38	25	7	6	5	4		42	46	14	–	–	–	–	–	–	–	–	–	–	(2R,3S)-2,3-trans-3-Acetoxy-6-[(2R,3S)-2,3-trans-3-acetoxy-3′,4′,7,8-tetramethoxyflavan-5-yl]-3′,4′,5,7-tetramethoxyflavan		NI47449
151	774	180	714	343	332	331	715	774	100	81	65	48	46	46	45	40		42	46	14	–	–	–	–	–	–	–	–	–	–	(2R,3S)-2,3-trans-3-Acetoxy-6-[(2R,3S)-2,3-trans-3-acetoxy-3′,4′,7,8-tetramethoxyflavan-5-yl]-3′,4′,7,8-tetramethoxyflavan		NI47450
774	151	343	331	330	180	714	655	774	100	89	78	74	74	65	61	61		42	46	14	–	–	–	–	–	–	–	–	–	–	(R)-[(2R,3S)-2,3-trans-3-acetoxy-8-[(2R,3S)-2,3-trans-3-acetoxy-3′,4′,7,8-tetramethoxyflavan-5-yl]-3′,4′,5,7-tetramethoxyflavan]		NI47451
774	151	343	331	330	180	714	492	774	100	90	68	67	66	57	50	49		42	46	14	–	–	–	–	–	–	–	–	–	–	(S)-Æ(2R,3S)-2,3-trans-3-Acetoxy-8-[(2R,3S)-2,3-trans-3-acetoxy-3′,4′,7,8 tetramethoxyflavan-5-yl]-3′,4′,5,7-tetramethoxyflavanŒ		NI47452
714	654	151	299	503	246	492	165	774	100	94	90	88	83	76	75	68	21.00	42	46	14	–	–	–	–	–	–	–	–	–	–	Diacetyl-octamethyl-procyanidine		LI10075
684	774	223	263	756	624	714	666	774	100	93	54	44	26	20	19	18		47	70	7	–	–	–	–	–	–	1	–	–	–	(3S,5R,6S,3′S,5′R,6′R)-6,7-Didehydro-5,6,5′,6′-tetrahydro-β,β-carotene-3,5,6,3′,5′-pentol,3,3′-diacetate,5′-trimethylsilyl ether		NI47453
57	774	71	43	775	103	85	61	774	100	99	63	63	58	50	41	41		53	106	2	–	–	–	–	–	–	–	–	–	–	Pentacontanoic acid, propyl ester	40710-39-2	NI47454
732	57	733	71	43	85	83	18	774	100	58	57	37	32	24	20	20	0.00	53	106	2	–	–	–	–	–	–	–	–	–	–	Pentacontanoic acid, propyl ester	40710-39-2	WI01602
57	774	775	219	163	41	203	511	774	100	80	49	34	11	11	10	7		54	78	3	–	–	–	–	–	–	–	–	–	–	1,3,5-Trimethyl-2,4,6-tris[(3,5-di-tert-butyl-4-hydroxybenzyl)]benzene		IC04644
716	253	228	456	367	570	342	314	775	100	31	22	20	17	13	13	11	0.00	21	16	13	1	–	15	–	–	–	–	–	–	–	Methyl N-trifluoroacetyl-4,7,8,9-tetra-O-trifluoroacetylneuraminate methyl glycoside		NI47455
534	775	310	57	535	776	777	536	775	100	63	42	42	37	29	26	26		24	54	6	–	6	–	–	–	–	–	3	–	1	Chromium(III) tris(di-N-butyldithiophosphate)	90580-05-5	NI47456
534	775	310	535	776	777	536	131	775	100	63	42	37	29	26	26	21		24	54	6	–	6	–	–	–	–	–	3	–	1	Chromium(III) tris(di-N-butyldithiophosphate)	90580-05-5	NI47457
703	665	704	749	666	667	705	750	776	100	90	66	65	53	43	37	34	16.39	22	36	14	–	–	–	–	–	–	9	–	–	–	1-Butoxypervinylhomooctasilsesquioxane	94048-07-4	NI47458

MASS TO CHARGE RATIOS								M.W.	INTENSITIES								Parent	C	H	O	N	S	F	Cl	Br	I	Si	P	B	X	COMPOUND NAME	CAS Reg No	No
685	91	686	297	44	687	657	637	**776**	100	44	41	12	9	8	6	6	**0.73**	44	60	4	–	–	–	–	–	–	–	4	–	–	**1,6,11,16-Tetraphosphacycloeicosane, 1,6,11,16-tetrakisbenzyl-, 1,6,11,16 tetraoxide**	57978-16-2	**NI47459**
154	503	501	499	701	624	155	502	**778**	100	36	29	22	17	15	14	13	**2.93**	24	10	–	–	–	8	–	–	2	–	–	–	1	**Germane, diphenylbis(2,3,4,5-tetrafluoro-6-iodophenyl)-**	69688-61-5	**NI47460**
57	41	56	637	43	55	44	638	**778**	100	68	43	38	35	25	24	23	**0.00**	24	58	13	–	–	–	–	–	–	8	–	–	–	**1,3,5,7,9,11,13,15-Octaethyl-7,13-dibutoxytetracyclo[9.5.1.1^{3,9}.1^{5,15}]octasiloxane**	110991-20-3	**NI47461**
105	78	658	402	42	203	123	429	**778**	100	84	40	37	35	15	13	12	**0.00**	42	30	13	–	–	–	–	–	–	–	–	–	4	**Oxohexakis(benzoato)tetraberyllium(II)**		**MS11039**
105	666	57	667	77	778	722	588	**778**	100	73	55	37	29	8	5	2		48	42	2	–	4	–	–	–	–	–	–	–	–	**2,2′-(1,4-Phenylene)bis[4-benzoyl-5-(tert-butylthio)-3-phenylthiophene]**	110699-54-2	**DD01826**
57	43	211	71	239	523	55	98	**778**	100	74	62	62	45	40	40	38	**0.14**	49	94	6	–	–	–	–	–	–	–	–	–	–	**Hexadecanoic acid, 2-[(1-oxotetradecyl)oxy]-1,3-propanediyl ester**	56599-89-4	**NI47462**
55	57	43	28	41	71	69	98	**778**	100	97	97	66	65	54	45	42	**0.00**	49	94	6	–	–	–	–	–	–	–	–	–	–	**Octadecanoic acid, 2,3-bis[(1-oxotetradecyl)oxy]propyl ester**	56846-96-9	**NI47463**
523	57	524	71	43	239	85	55	**778**	100	53	40	35	34	32	25	17	**0.44**	51	102	4	–	–	–	–	–	–	–	–	–	–	**Hexadecanoic acid, 2,3-bis(hexadecyloxy)propyl ester**	10322-42-6	**NI47464**
390	391	584	585	778	392	779	586	**778**	100	65	32	28	24	23	14	11		52	62	4	2	–	–	–	–	–	–	–	–	–	**N,N′-Dicyclotetradecyl-3,4:9,10-perylenebis(dicarboximide)**	110590-79-9	**MS11040**
779	780	738	778	632	616	777	781	**779**	100	14	12	10	10	8	7	6		29	20	7	1	–	15	–	–	–	–	–	–	–	**Naltrexone tri-PFP**		**MS11041**
73	387	299	315	357	147	388	333	**780**	100	54	49	33	31	30	19	16	**0.00**	27	69	10	–	–	–	–	–	–	7	1	–	–	**Gluconic acid, 2,3,4,5-tetrakis-O-(trimethylsilyl)-, trimethylsilyl ester, bis(trimethylsilyl) phosphate, D-**	28176-77-4	**NI47465**
73	299	147	387	315	357	217	300	**780**	100	41	27	20	12	11	11	10	**0.00**	27	69	10	–	–	–	–	–	–	7	1	–	–	**Gluconic acid, 2,3,4,5-tetrakis-O-(trimethylsilyl)-, trimethylsilyl ester, bis(trimethylsilyl) phosphate, D-**	28176-77-4	**NI47466**
43	193	169	109	208	191	331	127	**780**	100	61	36	28	19	12	10	9	**0.00**	36	44	19	–	–	–	–	–	–	–	–	–	–	**Methyl 6-[[3-(3,4-dimethoxyphenyl)-1-oxo-2-propenyl]oxy]-1-(tetraacetoxy-β-D-glucopyranosyloxy)-1,4a,5,6,7,7a-hexahydro-4a,7 dihydroxy-7-methylcyclopenta[c]pyran-4-carboxylate**		**NI47467**
546	384	385	371	764	733	355	341	**780**	100	30	12	9	8	7	7	5	**0.00**	38	52	17	–	–	–	–	–	–	–	–	–	–	**4H-1-Benzopyran-4-one, 5,7-dimethoxy-2-[3-methoxy-4-[(2,3,4,6-tetra-O-methyl-β-D-glucopyranosyl)oxy]phenyl]-6-[(2,3,4,6-tetra-O-methyl-β-D-glucopyranosyl)oxy]-**	75299-52-4	**NI47468**
546	384	733	385	371	327	560	355	**780**	100	50	17	17	16	15	11	9	**0.00**	38	52	17	–	–	–	–	–	–	–	–	–	–	**4H-1-Benzopyran-4-one, 5,7-dimethoxy-2-[4-methoxy-3-[(2,3,4,6-tetra-O-methyl-β-D-glucopyranosyl)oxy]phenyl]-6-[(2,3,4,6-tetra-O-methyl-β-D-glucopyranosyl)oxy]-**	75299-53-5	**NI47469**
515	483	733	371	531	383	764	385	**780**	100	80	76	64	40	40	37	36	**0.00**	38	52	17	–	–	–	–	–	–	–	–	–	–	**4H-1-Benzopyran-4-one, 2-(3,4-dimethoxyphenyl)-5-methoxy-6,7-bis[(2,3,4,6-tetra-O-methyl-β-D-glucopyranosyl)oxy]-**	75299-51-3	**NI47470**
171	73	172	75	173	750	660	480	**781**	100	50	18	15	10	7	7	5	**4.87**	40	79	6	1	–	–	–	–	–	4	–	–	–	**Cholest-7-en-6-one, 14-hydroxy-2,3,22,25-tetrakis[(trimethylsilyl)oxy]-, O methyloxime, (2β,3β,5β)-**	55780-42-2	**NI47471**
171	73	81	131	172	147	75	472	**781**	100	50	26	25	20	15	15	12	**4.00**	40	79	6	1	–	–	–	–	–	4	–	–	–	**Cholest-7-en-6-one, 14-hydroxy-2,3,22,25-tetrakis[(trimethylsilyl)oxy]-, O methyloxime, (2β,3β,5β,22R)-**	53286-62-7	**NI47472**
458	299	348	323	142	103	531	416	**781**	100	34	29	25	18	18	17	16	**3.00**	44	91	4	1	–	–	–	–	–	3	–	–	–	**1,3,2′-Tris(trimethylsilyloxy)-N-(2-hydroxypalmitoyl)nonadecasphingadienine**		**NI47473**
411	720	486	413	728	488	409	782	**782**	100	94	94	94	87	87	85	52		30	24	4	–	–	–	–	–	–	–	–	–	3	**cis-[1,2-Bis(diphenylarsino)ethane]tungsten tetracarbonyl**		**LI10076**
50	59	118	268	69	327	62	210	**784**	100	97	81	78	21	20	19	14	**0.00**	24	5	10	–	–	5	–	–	–	–	–	–	4	**μ-(2,3,4,5,6-Pentafluorodiphenylacetylene)-decacarbonyltetracobalt**		**MS11042**
420	69	125	422	784	84	123	83	**784**	100	70	48	44	38	25	23	20	**0.00**	30	36	4	–	–	14	–	–	–	–	–	–	1	**Nickel(II) bis(heptafluorobutyrylcampholylmethanate)**		**NI47474**
591	592	593	288	273	561	589	563	**784**	100	56	16	14	10	10	9	9	**0.00**	45	53	1	6	–	–	1	–	–	–	–	–	1	**Dipyridine octaethylverdohemochrome monochloride**		**NI47475**
240	557	169	69	241	42	28	558	**785**	100	24	18	15	7	7	6	5	**1.30**	22	12	6	1	–	21	–	–	–	–	–	–	–	**O,O,N-Tris(heptafluorobutyryl)metanephrine**		**MS11043**
785	638	625	754	726	697			**785**	100	10	5	3	3	2				41	46	8	4	–	–	–	–	–	–	–	–	1	**Copper-2-devinyl-2′-(2-carbethoxycyclopropyl)-chlorin-E6 trimethyl ester**		**LI10077**
55	69	83	43	41	81	57	67	**785**	100	68	55	42	42	41	41	38	**0.00**	44	84	8	1	–	–	–	–	–	–	1	–	–	**Ethanaminium, 2-[[[2,3-bis[(1-oxo-9-octadecenyl)oxy]propoxy]hydroxyphosphinyl]oxy]-N,N,N-trimethyl , hydroxide, inner salt, (R)-**	56648-95-4	**NI47476**

MASS TO CHARGE RATIOS	M.W.	INTENSITIES	Parent	C	H	O	N	S	F	Cl	Br	I	Si	P	B	X	COMPOUND NAME	CAS Reg No	No
646 330 246 358 786 302 590 506	**786**	100 48 46 45 42 35 34 32		14	12	11	–	–	–	–	–	–	–	2	–	2	**Bis(pentacarbonyltungsten)(dimethylphosphinyl)oxide**		**NI47477**
69 119 667 717 767 169 629 529	**786**	100 85 35 23 22 11 6 6	**0.00**	18	–	–	–	–	30	–	–	–	–	–	–	–	**Hexakis(pentafluoroethyl)benzene**		**LI10078**
204 176 41 55 639 202 190 476	**786**	100 38 36 25 21 20 18 15	**0.00**	22	22	4	4	–	20	–	–	–	–	–	–	–	**Spermine tetra-PFP**		**MS11044**
73 75 315 299 262 129 120 149	**786**	100 18 17 13 11 11 11 9	**6.45**	35	67	8	2	–	–	–	–	–	4	1	–	–	**Pregna-1,4-diene-3,20-dione, 21-hydroxy-11β,17-bis(trimethylsiloxy)-, bis(O-methyloxime) bis(trimethylsilyl) phosphate**	33745-66-3	**NI47478**
219 94 220 55 41 203 57 161	**786**	100 25 18 14 14 13 13 12	**0.00**	46	74	6	–	2	–	–	–	–	–	–	–	–	**1,10-Bis(3,5-di-tert-butyl-4-hydroxyphenylmethyl-thioethylcarboxy)decane**		**IC04645**
83 185 144 283 685 74 786 587	**786**	100 46 29 22 14 12 10 6		52	98	4	–	–	–	–	–	–	–	–	–	–	**[6-(5′,23′-Dimethyl-28′-cyclohexyloctacosanyl)oxan-2-yl]-[6-(3′-methoxycarbonylpropyl)oxan-2-yl]-methane**		**LI10079**
786 534 326 502 253 754 755 294	**786**	100 98 85 65 65 55 40 40		53	102	3	–	–	–	–	–	–	–	–	–	–	**2-Dodecyl-6-(1′-methoxycarbonyl-7′,25′-dimethyl-30′-cyclohexyl-triacontyl)oxane**		**LI10080**
139 137 621 135 619 473 138 296	**788**	100 76 53 46 40 29 28 27	**16.00**	24	–	–	–	–	20	–	–	–	–	–	–	1	**Tetrakis(pentafluorophenyl)stannane**		**MS11045**
423 307 319 259 349 305 583 331	**788**	100 63 58 41 35 34 13 10	**0.00**	31	76	9	–	–	–	–	–	–	7	–	–	–	**Hepta-TMS-1-O-(α-6-arabopyranosyl)xylitol**		**LI10081**
73 15 316 299 75 168 317 147	**788**	100 92 31 30 24 19 16 14	**0.39**	35	69	8	2	–	–	–	–	–	4	1	–	–	**Pregn-4-ene-3,20-dione, 21-hydroxy-11β,17-bis(trimethylsiloxy)-, bis(O-methyloxime), bis(trimethylsilyl) phosphate**	33745-67-4	**NI47479**
478 261 788 168 199 695 142 103	**788**	100 53 44 32 6 4 3 0		41	35	6	–	–	–	–	–	–	–	2	–	1	**Cyclopentadienylrhodiumbis(triphenylphosphite)**	53556-34-6	**NI47480**
169 331 109 81 127 211 791 97	**789**	100 67 64 30 24 20 15 14	**14.00**	32	40	17	1	–	–	–	1	–	–	–	–	–	**D-Glucopyranosylamine, N-(4-bromophenyl)-4-O-(2,3,4,6-tetra-O-acetyl-β-D-galactopyranosyl)-, 2,3,6-triacetate**	35427-11-3	**NI47481**
169 331 109 81 141 127 155 115	**789**	100 59 53 49 28 24 19 18	**11.76**	32	40	17	1	–	–	–	1	–	–	–	–	–	**D-Glucopyranosylamine, N-(4-bromophenyl)-6-O-(2,3,4,6-tetra-O-acetyl-α-D-galactopyranosyl)-, 2,3,4-triacetate**	35427-09-9	**NI47483**
331 109 81 141 127 155 115 97	**789**	100 90 83 47 40 32 30 25	**20.00**	32	40	17	1	–	–	–	1	–	–	–	–	–	**D-Glucopyranosylamine, N-(4-bromophenyl)-6-O-(2,3,4,6-tetra-O-acetyl-α-D-galactopyranosyl)-, 2,3,4-triacetate**	35427-09-9	**NI47482**
152 224 140 330 441 153 745 457	**789**	100 98 53 42 26 23 21 12	**0.40**	35	51	19	1	–	–	–	–	–	–	–	–	–	**1-(N-Cyclopentylacetamido)-1-desoxy-2,2′,3,3′,4′,5,6,6′-octa-O-acetyl-D-lactitol**		**NI47484**
57 267 71 55 43 341 395 169	**789**	100 97 84 81 81 75 70 61	**0.00**	44	88	8	1	–	–	–	–	–	–	1	–	–	**3,5,9-Trioxa-4-phosphaheptacosan-1-aminium, 4-hydroxy-N,N,N-trimethyl-10-oxo-7-[(1-oxooctadecyl)oxy]-, hydroxide, inner salt, 4-oxide, (R)-**	816-94-4	**NI47485**
735 736 737 763 790 764 738 765	**790**	100 78 62 49 37 35 33 31		20	30	15	–	–	–	–	–	–	10	–	–	–	**Perethyldecasilsesquioxane**	71682-48-9	**NI47486**
43 169 331 109 155 81 92 91	**790**	100 15 14 8 6 6 4 4	**0.01**	33	42	20	–	1	–	–	–	–	–	–	–	–	**6-Tosyl-1,2,3,2′,3′,4′,6′-β-maltose heptaacetate**		**MS11046**
537 151 181 180 495 730 167 166	**790**	100 72 30 25 20 13 11 9	**2.50**	43	50	14	–	–	–	–	–	–	–	–	–	–	**(1S,2R)-2-Acetoxy-1-[(2R,3S)-2,3-trans-3-acetoxy-3′,4′,7,8-tetramethoxyflavan-6-yl]-1-(3,4-dimethoxyphenyl)-3-(2,3,4-trimethoxyphenyl)propane**		**NI47487**
537 670 315 730 181 790 151 165	**790**	100 75 71 70 69 65 46 37		43	50	14	–	–	–	–	–	–	–	–	–	–	**2H-1-Benzopyran-6-ethanol, 3-(acetyloxy)-β,2-bis(3,4-dimethoxyphenyl)-3,4-dihydro-4,7-dimethoxy-α-[(2,4,6-trimethoxyphenyl)methyl]-, acetate**	56890-00-7	**NI47488**
790 792 791 793 395 794 396 776	**790**	100 56 55 26 15 12 8 4		48	36	4	4	–	–	–	–	–	–	–	–	1	**Nickel, [5,10,15,20-tetrakis(4-methoxyphenyl)-21H,23H-porphinato(2–)-N^{21},N^{22},N^{23},N^{24}]-**	39828-57-4	**NI47489**
790 793 791 395 794 396 379 776	**790**	100 56 55 15 12 8 7 4		48	36	4	4	–	–	–	–	–	–	–	–	1	**Nickel, [5,10,15,20-tetrakis(4-methoxyphenyl)-21H,23H-porphinato(2–)-N^{21},N^{22},N^{23},N^{24}]-**	39828-57-4	**NI47490**
578 158 563 349 118 364 467 253	**792**	100 86 80 75 59 55 34 31	**0.00**	35	42	4	–	–	14	–	–	–	–	–	–	–	**25-Hydroxycholecalciferol bis(heptafluorobutyrate)**		**MS11047**
43 331 109 169 127 165 271 121	**792**	100 99 95 80 40 36 32 30	**0.26**	37	44	19	–	–	–	–	–	–	–	–	–	–	**Oleuropein hexaacetate**		**MS11048**
331 368 386 353 301 413 339 445	**792**	100 43 32 28 25 21 18 12	**4.30**	45	60	12	–	–	–	–	–	–	–	–	–	–	**Methyl 1-hexosyl-1,2-dihydro-3,4-didehydro-apo-8′-lycopeniate**		**LI10082**
684 792 342 538 288 790 343 273	**792**	100 65 59 55 49 48 43 37		50	48	9	–	–	–	–	–	–	–	–	–	–	**Distichol octamethyl ether**		**MS11049**
352 43 57 184 410 224 382 71	**792**	100 79 64 60 54 45 36 27	**14.00**	50	100	4	2	–	–	–	–	–	–	–	–	–	**Acetamide, 2,2′-[1,2-ethanediylbis(oxy)]bis[N,N-diundecyl-**	67615-90-1	**NI47491**
57 43 71 239 297 85 537 55	**792**	100 64 62 59 56 43 38 36	**0.00**	51	100	5	–	–	–	–	–	–	–	–	–	–	**Hexadecanoic acid, 1-[(hexadecyloxy)methyl]-1,2-ethanediyl ester**	1116-45-6	**NI47492**
57 43 241 71 267 509 85 225	**792**	100 65 63 60 49 44 42 36	**0.00**	51	100	5	–	–	–	–	–	–	–	–	–	–	**Octadecanoic acid, 1-[(dodecyloxy)methyl]-1,2-ethanediyl ester**	10322-32-4	**NI47493**

MASS TO CHARGE RATIOS								M.W.	INTENSITIES								Parent	C	H	O	N	S	F	Cl	Br	I	Si	P	B	X	COMPOUND NAME	CAS Reg No	No
231	189	69	232	81	190	68	55	**792**	100	43	25	16	15	6	5	5	**1.16**	55	84	3	–	–	–	–	–	–	–	–	–	–	2H-1-Benzopyran-6-ol, 3,4-dihydro-2,7,8-trimethyl-2-(4,8,12,16,20,24,28,32-octamethyl-3,7,11,15,19,23,27,31-tritriacontaoctaenyl)-, acetate	56282-30-5	NI47494
311	470	543	323	348	129	103	428	**793**	100	74	30	30	25	25	25	24	**2.00**	45	91	4	1	–	–	–	–	–	3	–	–	–	1,3,2′-Tris(trimethylsilyloxy)-N-(2-hydroxyheptadecenoyl)nonadecasphingadienine		NI47495
463	464	794	66	104	51	120	31	**794**	100	17	8	7	6	6	5	5		22	10	4	–	–	24	–	–	–	–	–	–	–	Bis(1,1,7-trihydroperfluoroheptyl)terephthalate		MS11050
765	766	767	709	768	710	29	41	**794**	100	69	57	30	25	20	19	17	**0.00**	22	54	14	–	–	–	–	–	–	9	–	–	–	1-Butoxyperethylhomooctasilsesquioxane	73895-20-2	NI47496
73	299	147	315	691	387	217	75	**794**	100	33	28	27	20	19	14	13	**0.00**	28	71	10	–	–	–	–	–	–	7	1	–	–	D-altro-2-Heptulose, 1,3,4,5,6-pentakis-O-(trimethylsilyl)-, bis(trimethylsilyl) phosphate	55470-93-4	NI47498
299	73	147	315	691	387	217	75	**794**	100	99	85	81	61	57	42	40	**0.00**	28	71	10	–	–	–	–	–	–	7	1	–	–	D-altro-2-Heptulose, 1,3,4,5,6-pentakis-O-(trimethylsilyl)-, bis(trimethylsilyl) phosphate	55470-93-4	NI47497
113	43	41	55	100	87	81	74	**794**	100	39	39	34	30	30	26	25	**0.00**	43	56	4	–	–	–	–	2	–	–	–	–	–	Cholest-8-ene-3,15-diol, 14-ethyl-, bis(4-bromobenzoate), (3β,5α,15α)-	74432-02-3	NI47499
152	73	28	43	41	245	55	69	**794**	100	82	78	74	64	50	44	40	**0.00**	44	62	11	2	–	–	–	–	–	–	–	–	–	Heneicomycin		MS11051
331	368	386	353	301	412	396	332	**794**	100	44	32	30	25	21	20	20	**0.00**	45	62	12	–	–	–	–	–	–	–	–	–	–	8′-Apo-ψ,ψ-carotenoic acid, 1,2-dihydro-1-[tetrakis(acetyloxy)dihydroxyhexyl]-, methyl ester	56193-57-8	NI47500
89	73	87	138	88	72	133	124	**794**	100	79	57	37	35	34	32	29	**0.00**	46	58	8	4	–	–	–	–	–	–	–	–	–	20′-Deoxyleurosidine		DD01827
119	77	138	82	124	83	93	136	**794**	100	71	56	42	20	18	16	14	**0.00**	46	58	8	4	–	–	–	–	–	–	–	–	–	20′-Deoxyvinblastine		DD01828
138	135	124	469	55	57	82	122	**794**	100	66	52	39	32	31	26	25	**0.00**	46	58	8	4	–	–	–	–	–	–	–	–	–	20′-Deoxyvincovaline		DD01829
89	87	73	88	138	72	84	71	**794**	100	57	36	28	27	21	20	19	**0.00**	46	58	8	4	–	–	–	–	–	–	–	–	–	20′-epi-20′-Deoxyvincovaline		DD01830
73	169	315	258	299	147	230	75	**795**	100	87	29	21	19	17	13	13	**8.51**	28	62	8	5	–	–	–	–	–	6	1	–	–	9H-Purine, 9-[2,3-bis-O-(trimethylsilyl)-β-D-ribofuranosyl]-6-(trimethylsiloxy)-2-[(trimethylsilyl)amino]-, 5′-[bis(trimethylsilyl) phospate]	32653-15-9	NI47501
472	335	323	313	545	348	204	142	**795**	100	19	19	18	14	14	9	8	**1.00**	45	93	4	1	–	–	–	–	–	3	–	–	–	1,3,2′-Tris(trimethylsilyloxy)-N-(2-hydroxyheptadecanoyl)nonadecasphingadienine		NI47502
767	768	769	770	369	771	87	44	**796**	100	71	60	29	16	14	13	11	**0.69**	20	52	14	–	–	–	–	–	–	10	–	–	–	1,3,5,7,9,11,13,15,17,19-Decaethylpentacyclo[13.5.1.1^{3,11}3.1^{5,1}1.1^{7,19}]decasiloxane	72617-72-2	NI47503
796	797	798	56	338	339	799	634	**796**	100	53	23	15	11	9	8	8		44	52	6	4	2	–	–	–	–	–	–	–	–	1,4-Bis(2,4,6-trimethyl-3-cyclohexylaminosulphonyl-anilino)anthraquinone		IC04646
108	79	107	404	448	405	109	114	**796**	100	86	60	56	20	15	13	12	**0.00**	44	52	10	4	–	–	–	–	–	–	–	–	–	(R,R)-Bis(N-benzyloxycarbonylphenylglycyl)diaza-18-crown-6		MS11052
627	365	457	625	455	363	626	456	**796**	100	89	87	76	72	69	56	46	**19.00**	48	36	4	–	–	–	–	–	–	–	–	–	1	Tetrakis(O-phenoxyphenyl)tin		LI10083
171	115	73	144	172	81	518	131	**797**	100	90	50	32	28	25	22	22	**4.00**	40	79	7	1	–	–	–	–	–	4	–	–	–	Cholest-7-en-6-one, 14,20-dihydroxy-2,3,22,25-tetrakis[(trimethylsilyl)oxy]-, O-methyloxime, (2β,3β,5β,14α)-		NI47504
171	73	172	518	75	536	173	168	**797**	100	43	29	28	19	18	18	15	**4.57**	40	79	7	1	–	–	–	–	–	4	–	–	–	Cholest-7-en-6-one, 14,20-dihydroxy-2,3,22,25-tetrakis[(trimethylsilyl)oxy]-, O-methyloxime, (2β,3β,5β,20ξ)-	55821-12-0	NI47505
171	115	73	144	172	81	518	131	**797**	100	90	50	32	28	25	22	22	**4.00**	40	79	7	1	–	–	–	–	–	4	–	–	–	Cholest-7-en-6-one, 14,20-dihydroxy-2,3,22,25-tetrakis[(trimethylsilyl)oxy]-, O-methyloxime, (2β,3β,5β,22R)-	53336-27-9	NI47506
86	198	311	170	199	312	84	100	**797**	100	94	54	24	21	20	20	19	**0.58**	42	79	9	5	–	–	–	–	–	–	–	–	–	Permethylated pepstatin		LI10084
769	770	771	723	724	725	370	772	**798**	100	73	60	48	33	29	27	26	**3.99**	21	54	15	–	–	–	–	–	–	9	–	–	–	1,3,5,7,9,11,13,15,17-Nonaethyl-7,13,17-trimethoxytetracyclo[9.5.1.1^{3,9}.3^{5,15}]nonasiloxane	72617-69-7	NI47507
798	799	725	708	726	709	652	579	**798**	100	58	51	47	24	21	16	11		43	54	9	4	–	–	–	–	–	1	–	–	–	21H,23H-Porphine-2,7,12,18-tetrapropanoic acid, 3,8,13,17-tetramethyl-7β(or 12β)-[(trimethylsilyl)oxy]-, tetramethyl ester	38684-68-3	NI47508
708	709	710	798	799	711	635	800	**798**	100	60	55	52	31	10	7	6		43	54	9	4	–	–	–	–	–	1	–	–	–	21H,23H-Porphine-2,7,18-tripropanoic acid, 8-(2-methoxy-2-oxoethyl)-3,13,17-trimethyl-12-[1-[(trimethylsilyl)oxy]ethyl]-, trimethyl ester	38574-17-3	NI47509

MASS TO CHARGE RATIOS								M.W.	INTENSITIES								Parent	C	H	O	N	S	F	Cl	Br	I	Si	P	B	X	COMPOUND NAME	CAS Reg No	No
798	799	725	708	726	709	652	579	**798**	100	58	51	47	24	21	16	11		43	54	9	4	–	–	–	–	–	1	–	–	–	21H,23H-Porphine-2,7,18-tripropanoic acid, 8-(2-methoxy-2-oxoethyl)-3,13,17-trimethyl-12-[1-[(trimethylsilyl)oxy]ethyl]-, trimethyl ester	38574-17-3	LI10085
356	218	411	470	398	357	471	299	**799**	100	49	39	35	31	29	24	24	**0.00**	45	97	4	1	–	–	–	–	–	3	–	–	–	Phytosphingosine, 1,3,4-tris[(trimethylsilyl)oxy]-N-stearoyl-		LI10086
408	392	799	798	407	57	409	43	**799**	100	95	50	42	35	33	32	31		55	109	1	1	–	–	–	–	–	–	–	–	–	(2)-(1-Acetylazacyclohexacosane)(cyclooctacosane)-catenane		MS11053
548	348	716	576	492	464	349	437	**800**	100	85	75	52	50	41	40	30	**10.00**	20	20	12	–	4	–	–	–	–	–	–	–	4	Manganese, tetrakis(tricarbonylethanethiolato)tetra-		MS11054
380	348	716	464	232	492	377	349	**800**	100	83	52	52	52	41	40	37	**12.00**	20	20	12	–	4	–	–	–	–	–	–	–	4	Manganese, tetrakis(tricarbonylethanethiolato)tetra-		LI10087
73	147	355	401	428	148	221	341	**800**	100	43	26	17	16	15	13	12	**0.00**	20	60	12	–	–	–	–	–	–	11	–	–	–	Bis(heptamethylcyclotetrasiloxy)hexamethyltrisiloxane	71449-67-7	NI47510
217	377	194	318	379	407	319	218	**800**	100	43	41	32	30	25	24	21	**0.00**	30	54	11	2	–	–	2	–	–	4	–	–	–	β-D-Glucopyranosiduronic acid, 2-[(dichloroacetyl)amino]-3-(4-nitrophenyl)-3-[(trimethylsilyl)oxy]propyl 2,3,4-tris-O-(trimethylsilyl)-, methyl ester, [R-(R*,R*)]-	40619-05-4	NI47511
83	55	57	111	207	189	311	211	**800**	100	48	20	12	10	8	7	7	**0.00**	46	72	11	–	–	–	–	–	–	–	–	–	–	21-O-(3,4-Di-O-angeloyl-β-D-fucopyranosyl)theasapogenol B		DD01831
399	145	459	446	398	315	187	117	**800**	100	29	24	20	20	19	19	19	**0.00**	47	46	5	4	–	–	–	–	–	–	–	–	2	Aluminium, μ-oxo-di[bis(2,3,4-trimethyl-8-quinolinolato)]di-		LI10088
57	71	43	85	55	83	97	56	**800**	100	67	63	44	34	32	26	19	**0.00**	48	97	–	–	–	–	–	–	1	–	–	–	–	Octatetracontane, 1-iodo-	40710-70-1	NI47512
83	199	158	88	297	800	685	601	**800**	100	39	32	15	11	6	5	4		53	100	4	–	–	–	–	–	–	–	–	–	–	[6-(5′,23′-Dimethyl-28′-cyclohexyloctacosanyl)oxan-2-yl]-[6-(3′-ethoxycarbonylpropyl)oxan-2-yl]-methane		LI10089
155	185	183	157	200	177	84	198	**801**	100	96	96	96	94	94	70	60	**0.10**	40	37	7	1	–	–	–	2	–	–	–	–	–	2-(4-Bromobenzoyloxy)-4-[3,4-(2-4-bromophenyl-2-oxo-ethyloxy)cinnamyloxy-4-methoxyphenyl]quinolizidine, (2S,4S,10R)-		NI47513
184	174	120	91	131	277	185	307	**801**	100	84	53	35	30	25	23	22	**10.00**	50	51	5	5	–	–	–	–	–	–	–	–	–	L-Phenylalanine, N-[N-[N-[N-[[4-(dimethylamino)-1-naphthalenyl]methylene]-L-phenylalanyl]-L-phenylalanyl]-L-phenylalanyl]-, methyl ester	57237-95-3	LI10090
185	175	121	92	132	278	302	198	**801**	100	85	54	35	30	25	23	21	**10.00**	50	51	5	5	–	–	–	–	–	–	–	–	–	L-Phenylalanine, N-[N-[N-[N-[[4-(dimethylamino)-1-naphthalenyl]methylene]-L-phenylalanyl]-L-phenylalanyl]-L-phenylalanyl]-, methyl ester	57237-95-3	NI47514
423	363	307	273	319	333	305	541	**802**	100	78	46	27	19	15	7	4	**0.00**	32	78	9	–	–	–	–	–	–	7	–	–	–	Xylitol, 1-O-(α-6-rhamnopyranosyl)-heptakis[(trimethylsilyl)oxy]-		LI10091
151	180	802	478	682	317	316	741	**802**	100	63	31	16	16	15	10	10		43	46	15	–	–	–	–	–	–	–	–	–	–	(2R,3S)-2,3-trans-3,7-Diacetoxy-6-[(2R,3S)-2,3-trans-3-acetoxy-3′,4′,7,8-tetramethoxyflavan-5-yl]-3′,4′,8-trimethoxyflavan		NI47515
151	802	180	301	742	358	317	165	**802**	100	72	53	34	32	29	29	29		43	46	15	–	–	–	–	–	–	–	–	–	–	(R)-Æ(2R,3S)-2,3-trans-3,7-diacetoxy-8-[(2R,3S)-2,3-trans-3-acetoxy-3′,4′,7,8-tetramethoxyflavan-5-yl]-3′,4′,5-trimethoxyflavanŒ		NI47516
28	64	36	432	43	48	41	38	**802**	100	93	80	52	50	43	35	35	**0.00**	58	42	2	–	1	–	–	–	–	–	–	–	–	Benzene, 1,1′,1″,1‴,1⁗,1′′′′′,1′′′′′′,1′′′′′′′-(sulphonyldi-1,3-cyclopentadiene 5,1,2,3,4-pentayl)octakis-	56728-01-9	NI47517
172	229	228	224	155	218	198	219	**803**	100	61	47	39	39	33	31	18	**0.00**	43	73	7	5	1	–	–	–	–	–	–	–	–	Glycine, N-[N-[N-[3-[(2-decanamidoethyl)thio]-N-decanoyl-L-alanyl]-3-phenyl-L-alanyl]-L-leucyl]-, methyl ester	19729-26-1	NI47518
195	454	400	370	442	429	230	374	**803**	100	92	88	80	74	67	65	59	**0.00**	45	73	11	1	–	–	–	–	–	–	–	–	–	D-Galacto-2,3,4,5,6-pentaacetoxyhexyl-3β-acetamido-5α-cholestane		NI47519
174	110	400	170	441	442	543	136	**803**	100	89	79	68	58	47	37	34	**1.50**	45	73	11	1	–	–	–	–	–	–	–	–	–	D-Gluco-2,3,4,5,6-pentaacetoxyhexyl-3β-acetamido-5α-cholestane		NI47520
635	43	71	70	266	504	492	334	**804**	100	63	58	54	52	50	44	43	**21.90**	21	17	5	4	–	21	–	–	–	–	–	–	–	Arginine, triheptafluorobutanoyl-, propyl ester		MS11055
237	279	195	153	152	109	154	169	**804**	100	75	73	65	52	51	38	36	**0.00**	36	36	21	–	–	–	–	–	–	–	–	–	–	Neocretanin, peracetyl-		MS11056
105	416	415	404	579	334	457	439	**804**	100	71	70	57	52	50	47	46	**0.00**	48	36	12	–	–	–	–	–	–	–	–	–	–	1,2,3,4,5,6-Hexa-O-benzoyl-L-xylo-hex-1-enitol	74894-95-4	NI47521
105	77	334	438	122	331	281	201	**804**	100	32	25	13	5	3	1	1	**0.00**	48	36	12	–	–	–	–	–	–	–	–	–	–	myo-Inositol, hexabenzoate	4099-90-5	NI47522
43	55	83	57	69	41	97	56	**804**	100	96	84	84	76	72	68	48	**21.00**	51	97	4	–	–	–	–	–	–	–	1	–	–	Phosphonic acid, O,O′-dioctadecyl-O″-(3,5-di-tert-butyl-4-hydroxybenzyl)-		IC04647
371	434	402	768	804	401	404	372	**804**	100	87	55	52	47	42	32	29		54	92	–	–	2	–	–	–	–	–	–	–	–	Cholestan-3-yl cholest-2-en-3-yl disulphide, (3α,5α,5′α)-		LI10092
105	91	77	593	122	294	195	297	**805**	100	79	49	27	15	5	3	2	**0.20**	48	39	11	1	–	–	–	–	–	–	–	–	–	D-Fructose, O-benzyloxime, 1,3,4,5,6-pentabenzoate	71641-44-6	NI47523

MASS TO CHARGE RATIOS	M.W.	INTENSITIES	Parent	C	H	O	N	S	F	Cl	Br	I	Si	P	B	X	COMPOUND NAME	CAS Reg No	No
105 91 77 593 122 350 269 195	**805**	100 78 64 36 20 5 4 4	**1.40**	48	39	11	1	–	–	–	–	–	–	–	–	–	**D-Galactose, O-benzyloxime, 2,3,4,5,6-pentabenzoate**	71641-46-8	**NI47524**
105 91 77 593 122 350 269 195	**805**	100 72 60 28 17 3 3 3	**0.40**	48	39	11	1	–	–	–	–	–	–	–	–	–	**D-Glucose, O-benzyloxime, 2,3,4,5,6-pentabenzoate**	71641-45-7	**NI47525**
105 91 77 593 122 350 269 195	**805**	100 62 45 22 13 3 2 2	**0.50**	48	39	11	1	–	–	–	–	–	–	–	–	–	**D-Mannose, O-benzyloxime, 2,3,4,5,6-pentabenzoate**	71641-47-9	**NI47526**
804 720 776 636 509 664 692 748	**806**	100 83 69 69 54 45 41 38	**0.00**	6	–	6	–	–	–	–	–	2	–	–	–	2	**Osmium, hexacarbonyldi-μ-iododi-, (Os-Os)**		**LI10093**
806 722 778 638 511 666 694 750	**806**	100 83 69 69 54 45 41 38		6	–	6	–	–	–	–	–	2	–	–	–	2	**Osmium, hexacarbonyldi-μ-iododi-, (Os-Os)**	22391-77-1	**NI47527**
625 651 778 722 750	**806**	100 46 37 4 1	**0.00**	19	22	3	–	–	–	–	–	2	–	2	–	1	**Osmium, diiodotricarbonylbis(dimethylphenylphosphine)-**		**LI10094**
275 405 239 745 304 487 680 615	**806**	100 62 46 36 29 14 9 8	**0.00**	30	34	4	–	–	–	–	–	–	–	–	–	2	**Tetracyclopentadienylbis(pentanonato)diytterbium**		**NI47528**
204 217 205 206 218 73 191 75	**806**	100 21 20 9 7 5 3 3	**0.00**	40	86	8	–	–	–	–	–	–	4	–	–	–	**Docosanoic acid, 13-[[2,3,4,6-tetrakis-O-(trimethylsilyl)-D-glucopyranosyl]oxy]-**	56728-07-5	**MS11057**
124 109 179 222 327 209 131 161	**806**	100 71 53 49 41 41 37 34	**0.40**	42	46	14	–	1	–	–	–	–	–	–	–	–	**4-Thiaheptane-1,7-diol, 2,6-bis(2-methoxyphenoxy)-3,5-bis(4-hydroxy-3-methoxyphenyl)-, tetraacetate**		**NI47529**
105 77 269 403 537 336 214 122	**806**	100 25 9 8 4 4 4 4	**0.00**	48	38	12	–	–	–	–	–	–	–	–	–	–	**D-Glucitol, hexabenzoate**	20869-38-9	**NI47530**
105 77 106 177 403 311 269 51	**806**	100 13 8 3 2 2 2 2	**0.00**	48	38	12	–	–	–	–	–	–	–	–	–	–	**L-Iditol, hexabenzoate**	21238-34-6	**NI47531**
105 77 269 403 336 537 214 294	**806**	100 40 17 14 9 8 8 7	**0.00**	48	38	12	–	–	–	–	–	–	–	–	–	–	**D-Mannitol, hexabenzoate**	7462-41-1	**NI47532**
551 523 579 267 239 211 550 578	**806**	100 97 80 78 63 63 60 55	**28.00**	51	98	6	–	–	–	–	–	–	–	–	–	–	**Hexadecanoic acid, 3-[(1-oxotetradecyl)oxy]-2-[(1-oxooctadecyl)oxy]propyl ester**		**LI10095**
551 239 57 550 43 313 367 71	**806**	100 83 71 54 54 49 41 41	**0.80**	51	98	6	–	–	–	–	–	–	–	–	–	–	**Hexadecanoic acid, 1,2,3-propanetriyl ester**	555-44-2	**NI47533**
43 57 55 367 551 71 41 73	**806**	100 99 99 97 95 76 71 66	**0.10**	51	98	6	–	–	–	–	–	–	–	–	–	–	**Hexadecanoic acid, 1,2,3-propanetriyl ester**	555-44-2	**NI47534**
57 43 71 55 183 41 85 73	**806**	100 81 55 52 42 36 35 32	**0.10**	51	98	6	–	–	–	–	–	–	–	–	–	–	**Octadecanoic acid, 2-[(1-oxododecyl)oxy]-1,3-propanediyl ester**	56183-45-0	**NI47535**
57 43 71 267 183 523 85 98	**806**	100 73 65 62 59 51 43 39	**0.10**	51	98	6	–	–	–	–	–	–	–	–	–	–	**Octadecanoic acid, 3-[(1-oxododecyl)oxy]-1,2-propanediyl ester**	35405-55-1	**NI47536**
551 523 579 267 239 211 550 578	**806**	100 97 79 78 65 64 61 55	**2.77**	51	98	6	–	–	–	–	–	–	–	–	–	–	**Octadecanoic acid, 1-[[(1-oxohexadecyl)oxy]methyl]-2-[(1-oxotetradecyl)oxy]ethyl ester**	55256-01-4	**NI47537**
57 43 71 211 41 55 85 98	**806**	100 99 56 49 45 44 39 33	**0.34**	51	98	6	–	–	–	–	–	–	–	–	–	–	**Octadecanoic acid, 3-[(1-oxohexadecyl)oxy]-2-[(1-oxotetradecyl)oxy]propyl ester**	56554-26-8	**NI47538**
371 370 404 401 355 389 806	**806**	100 88 80 49 41 39 1		54	94	–	–	2	–	–	–	–	–	–	–	–	**Bis(cholestan-3-yl) disulphide, (3α,3′α,5α,5′α)-**		**LI10096**
132 171 314 692 210 283 182 315	**807**	100 87 86 78 66 64 60 54	**36.00**	39	56	8	7	–	3	–	–	–	–	–	–	–	**L-Alanine, N-[N-[N-[N-[N-L-tryphtophanyl]-L-leucyl]-L-valyl]-L-prolyl]-L-leucyl-, methyl ester**		**LI10097**
325 484 129 323 103 348 142 557	**807**	100 68 57 25 25 21 20 16	**1.00**	46	93	4	1	–	–	–	–	–	3	–	–	–	**1,3,2′-Tris(trimethylsilyloxy)-N-(2-hydroxyoctadec-3-enoyl)nonadecasphingadienine**		**NI47539**
165 163 666 664 161 668 662 665	**808**	100 73 50 48 44 40 34 31	**8.01**	14	27	5	–	–	–	–	–	–	–	–	–	5	**Pentacarbonyltris(trimethylstannyl)stibinechromium**	31827-01-7	**NI47540**
73 147 217 358 129 103 361 169	**808**	100 27 19 17 15 13 10 10	**0.00**	31	72	7	4	–	–	–	–	–	7	–	–	–	**Vicine, heptakis(trimethylsilyl)-**		**NI47541**
73 103 307 217 147 308 574 14	**808**	100 64 53 43 25 16 11 10	**8.04**	33	76	7	2	–	–	–	–	–	7	–	–	–	**Deoxyfructosazine, heptakis(trimethylsilyl)-**	68361-17-1	**NI47542**
664 706 647 622 604 329 748 347	**808**	100 95 68 67 50 35 32 26	**0.01**	39	36	19	–	–	–	–	–	–	–	–	–	–	**3,3′,4′,5,5′,7-Hexaacetoxy-4-(2,4,6-triacetoxyphenyl)flavan**		**NI47543**
73 204 185 147 75 117 205 808	**808**	100 99 63 60 49 39 32 27		41	76	8	–	–	–	–	–	–	4	–	–	–	**Card-20(22)-enolide, 3-[[2,6-dideoxy-3,4-bis-O-(trimethylsilyl)-D-ribo-hexopyranosyl]oxy]-12,14-bis[(trimethylsilyl)oxy]-, (3β,5β,12β)-**	56009-07-5	**MS11058**
204 73 185 147 75 117 205 808	**808**	100 93 62 58 47 37 33 26		41	76	8	–	–	–	–	–	–	4	–	–	–	**Card-20(22)-enolide, 3-[[2,6-dideoxy-3,4-bis-O-(trimethylsilyl)-D-ribo-hexopyranosyl]oxy]-12,14-bis[(trimethylsilyl)oxy]-, (3β,5β,12β)-**	56009-07-5	**NI47544**
204 73 185 147 75 117 808 130	**808**	100 95 60 58 47 37 26 22		41	76	8	–	–	–	–	–	–	4	–	–	–	**Card-20(22)-enolide, 3-[[2,6-dideoxy-3,4-bis-O-(trimethylsilyl)-D-ribo-hexopyranosyl]oxy]-12,14-bis[(trimethylsilyl)oxy]-, (3β,5β,12β)-**	57304-67-3	**NI47545**
204 73 185 147 75 117 205 808	**808**	100 96 64 60 48 37 32 28		41	76	8	–	–	–	–	–	–	4	–	–	–	**Card-20(22)-enolide, 3-[[2,6-dideoxy-3,4-bis-O-(trimethylsilyl)-D-ribo-hexopyranosyl]oxy]-12,23-bis[(trimethylsilyl)oxy]-**		**LI10098**
605 423 155 422 536 606 537 156	**808**	100 96 67 65 57 49 35 33	**4.78**	43	80	8	6	–	–	–	–	–	–	–	–	–	**L-Alanine, N-[N-[N-[N-[N²,N⁵-bis(1-oxodecyl)-L-ornithyl]-L-leucyl]glycyl]-L-leucyl]-, methyl ester**	56009-16-6	**NI47546**
73 147 217 361 129 103 169 74	**809**	100 29 22 16 16 14 12 10	**0.00**	31	71	8	3	–	–	–	–	–	7	–	–	–	**Convicine, heptakis(trimethylsilyl)-**		**NI47547**
70 44 94 122 91 95 166 139	**809**	100 90 73 33 31 30 28 27	**6.00**	42	47	10	7	–	–	–	–	–	–	–	–	–	**Staphylomycin S1**		**LI10099**
486 323 327 348 559 111 444 335	**809**	100 23 21 16 15 14 10 9	**1.00**	46	95	4	1	–	–	–	–	–	3	–	–	–	**1,3,2′-Tris(trimethylsilyloxy)-N-(2-hydroxyoctadecanoyl)nonadecasphingadienine**		**NI47548**

MASS TO CHARGE RATIOS								M.W.	INTENSITIES								Parent	C	H	O	N	S	F	Cl	Br	I	Si	P	B	X	COMPOUND NAME	CAS Reg No	No
69	117	217	155	324	167	205	93	**810**	100	89	87	68	59	56	47	41	**2.60**	16	–	10	–	–	12	–	–	–	–	2	–	3	**Iron, [decarbonyl-1,2,3,4-tetrakis(trifluoromethyl)-1,2-diphosphacyclobut-3-enyl]tri-**		**NI47549**
781	782	783	784	376	785	751	753	**810**	100	73	61	31	17	16	12	9	**0.00**	20	50	15	–	–	–	–	–	–	10	–	–	–	**Perethyldecasilsesquioxane**		**NI47550**
603	529	367	441	691	323	279	543	**810**	100	29	20	15	13	5	5	3	**0.00**	30	75	15	–	–	–	–	–	–	–	–	–	5	**Aluminium ethoxide (pentamer)**		**LI10100**
70	91	205	128	94	177	122	106	**811**	100	60	54	36	32	28	28	28	**0.30**	42	49	10	7	–	–	–	–	–	–	–	–	–	**Staphylomycin S2**	33477-38-2	**NI47551**
70	91	205	138	44	94	122	106	**811**	100	60	58	39	37	35	29	29	**0.30**	42	49	10	7	–	–	–	–	–	–	–	–	–	**Staphylomycin S2**	33477-38-2	**LI10101**
70	91	205	138	44	94	177	106	**811**	100	59	57	38	35	34	30	29	**0.31**	42	49	10	7	–	–	–	–	–	–	–	–	–	**Staphylomycin S2**	33477-38-2	**NI47552**
733	189	372	530	327	575	268	248	**812**	100	24	22	21	19	13	11	10	**2.70**	–	–	–	4	–	–	–	8	–	–	4	–	–	**Tetraphosphonitrile, octabromo-**		**MS11059**
131	73	143	75	132	199	133	157	**812**	100	24	16	13	12	6	6	5	**0.00**	42	88	5	–	–	–	–	–	–	5	–	–	–	**Cholestan-3,7,12,23,25-pentol, (3α,5β,7α,12α,23ξ)-**		**MS11060**
131	73	132	75	129	133	147	321	**812**	100	31	14	11	9	8	6	4	**0.04**	42	88	5	–	–	–	–	–	–	5	–	–	–	**Cholestan-3,7,12,24,25-pentol, (3α,5β,7α,12α,24α)-**		**MS11061**
309	813	227	241	101	504	310	199	**813**	100	69	68	61	55	54	53	42		20	10	2	–	2	6	–	2	–	–	–	–	1	**Bis[1,1,1-trifluoro-4-(4-bromophenyl)-4-thioxo-2-butanonato]platinum**	53777-00-7	**NI47553**
24	32	42	25	30	84	75	48	**814**	100	95	66	65	63	33	29	25	**0.00**	12	6	13	–	–	–	12	–	–	–	–	–	4	**Beryllium (II), oxohexakis(dichloroacetato)tetra-**		**MS11062**
764	765	736	310	782	296	399	783	**814**	100	71	50	47	46	44	40	33	**21.78**	50	70	9	–	–	–	–	–	–	–	–	–	–	**Cholan-24-oic acid, 4-(23-carboxy-7,12-dioxo-24-nor-chol-3-en-3-yl)-3,7,12-trioxo-, dimethyl ester, (5β,5′β)-**	19974-59-5	**NI47554**
135	57	43	687	71	121	107	41	**814**	100	81	77	40	38	37	35	34	**1.15**	54	87	3	–	–	–	–	–	–	–	1	–	–	**Phosphorous acid, tris(4-dodecylphenyl) ester**		**IC04648**
183	185	258	259	84	416	414	218	**815**	100	99	88	44	31	18	18	14	**0.30**	41	39	7	1	–	–	–	2	–	–	–	–	–	**2-[4-(2-4-Bromophenyl-2-oxo-ethyloxy)cinnamyloxy]-4-[3,4-(2-4-bromophenyl-2-oxo-ethyloxy)cinnamyloxy-4-methoxyphenyl]quinolizidine, (2S,4S,10R)-**		**NI47555**
515	592	296	732	676	648	620	328	**816**	100	69	21	20	17	10	5	5	**3.00**	20	10	8	–	2	–	–	–	–	–	–	–	2	**Rhenium, octacarbonyldi-μ-benzenethiolatodi-**		**MS11063**
121	180	120	138	205	122	655	169	**816**	100	21	17	17	17	14	5	4	**0.00**	30	30	10	–	–	14	–	–	–	–	–	–	–	**3,4-Bis(heptafluorobutyryl)-HT-2 toxin**		**NI47556**
74	534	356	502	324	784	785	325	**816**	100	26	25	23	22	20	13	13	**2.60**	54	104	4	–	–	–	–	–	–	–	–	–	–	**Heptatetracontanoic acid, 47-cyclohexyl-18-(methoxycarbonyl)-24,42-dimethyl-, methyl ester**		**LI10102**
169	248	331	690	774	704	444	428	**817**	100	45	30	15	10	10	5	5	**1.00**	39	47	18	1	–	–	–	–	–	–	–	–	–	**Ipecoside, hexa-O-acetyl-**		**LI10103**
185	75	249	333	217	71	301	159	**818**	100	84	68	57	48	47	40	39	**0.00**	40	82	16	–	–	–	–	–	–	–	–	–	–	**Docosane, 1-(2,3-dimethoxypropoxy)-2,4,5,7,8,10,11,13,14,16,17,19,20-tridecamethoxy-**	56247-74-6	**NI47557**
83	55	71	271	171	111	205	223	**818**	100	90	20	17	16	11	8	3	**0.00**	45	70	13	–	–	–	–	–	–	–	–	–	–	**21-O-(4-O-Acetyl-3-O-angeloyl-β-D-fucopyranosyl)-22-O-acetylprotoaescigenin**		**DD01832**
69	818	135	819	150	300	204	120	**818**	100	33	33	20	20	8	8	8		56	82	4	–	–	–	–	–	–	–	–	–	–	**γ-Tocotrienol, 5-(γ-tocotrienyloxy)-**		**LI10104**
69	818	819	285	340	135	190	150	**818**	100	57	43	12	8	7	5	4		56	82	4	–	–	–	–	–	–	–	–	–	–	**γ-Tocotrienol, 5-(γ-tocotrienyloxy)-**		**LI10105**
120	148	91	189	295	175	267	472	**819**	100	60	28	22	20	18	12	11	**0.00**	51	57	5	5	–	–	–	–	–	–	–	–	–	**L-Phenylalanine, N-[N-[N-[N-[3-[4-(diethylamino)phenyl]-2-propenylidene]-L-phenylalanyl]-L-phenylalanyl]-L-phenylalanyl]-, ethyl ester**	42547-83-1	**NI47558**
120	148	91	189	279	295	131	175	**819**	100	58	28	22	20	19	19	18	**0.00**	51	57	5	5	–	–	–	–	–	–	–	–	–	**L-Phenylalanine, N-[N-[N-[N-[3-[4-(diethylamino)phenyl]-2-propenylidene]-L-phenylalanyl]-L-phenylalanyl]-L-phenylalanyl]-, ethyl ester**	42547-83-1	**LI10106**

MASS TO CHARGE RATIOS								M.W.	INTENSITIES								Parent	C	H	O	N	S	F	Cl	Br	I	Si	P	B	X	COMPOUND NAME	CAS Reg No	No
429	430	375	464	236	431	217	357	**821**	100	31	29	28	19	15	14	13	**0.28**	38	67	9	1	–	–	–	–	–	5	–	–	–	β-D-Glucopyranosiduronic acid, 7,8-didehydro-4,5-epoxy-17-methyl-6-[(trimethylsilyl)oxy]morphinan-3-yl 2,3,4-tris-O-(trimethylsilyl)-, trimethylsilyl ester, (5α,6α)-	52092-53-2	NI47559
409	379	600	491	218	570	493	386	**822**	100	27	10	7	4	2	2	2	**0.00**	34	30	2	2	4	–	–	–	–	–	–	–	2	Molybdenum, tetrakis(benzenethiolato)di-2,4-cyclopentadien-1-yldinitrosyldi-	55955-44-7	NI47560
149	75	165	97	57	83	71	103	**822**	100	49	41	30	30	29	24	23	**1.50**	52	110	2	–	–	–	–	–	–	2	–	–	–	3,50-Dioxa-2,51-disiladopentacontane, 2,2,51,51-tetramethyl-	35177-38-9	NI47561
73	387	299	315	147	388	357	471	**823**	100	63	37	29	27	25	22	21	**0.00**	29	74	10	1	–	–	–	–	–	7	1	–	–	D-altro-2-Heptulose, 1,3,4,5,6-pentakis-O-(trimethylsilyl)-, O-methyloxime, 7-[bis(trimethylsilyl) phosphate]	55470-94-5	NI47562
387	73	299	315	147	388	357	471	**823**	100	99	59	46	43	40	35	33	**0.00**	29	74	10	1	–	–	–	–	–	7	1	–	–	D-altro-2-Heptulose, 1,3,4,5,6-pentakis-O-(trimethylsilyl)-, O-methyloxime, 7-[bis(trimethylsilyl) phosphate]	55470-94-5	NI47563
70	205	118	290	134	177	122	93	**823**	100	22	20	16	16	15	15	15	**8.60**	43	49	10	7	–	–	–	–	–	–	–	–	–	Staphylomycin S	23152-29-6	NI47564
70	94	122	95	91	167	139	205	**823**	100	73	32	30	30	28	27	26	**0.00**	43	49	10	7	–	–	–	–	–	–	–	–	–	Staphylomycin S	23152-29-6	NI47565
70	205	134	290	58	44	177	94	**823**	100	24	18	17	17	17	16	16	**9.05**	43	49	10	7	–	–	–	–	–	–	–	–	–	Staphylomycin S	23152-29-6	NI47566
171	172	567	131	637	81	173	568	**824**	100	74	62	52	42	35	32	28	**6.00**	42	84	6	–	–	–	–	–	–	5	–	–	–	Ecdysone, 2,3,14,22,25-pentakis(trimethylsilyl)-, (2β,3β,14α)-		MS11064
108	79	107	109	91	566	304	80	**824**	100	83	63	13	9	9	9	9	**0.00**	46	56	10	4	–	–	–	–	–	–	–	–	–	(S,S)-Bis(N-benzyloxycarbonylphenylalanyl)diaza-18-crown-6		MS11065
601	603	602	600	599	604	605	598	**825**	100	94	80	76	74	64	47	40	**0.00**	32	57	–	–	–	–	–	–	1	–	2	–	1	Cis-(3,3-dimethyl-1-butynyl)iodo[bis(dicyclohexylphosphino)ethane]platinum(II)		MS11066
73	147	346	245	75	142	217	205	**826**	100	57	52	50	50	39	37	28	**0.00**	33	78	8	2	–	–	–	–	–	7	–	–	–	D-threo-3-Trimethylsiloxy-2-[D-glycero-1,2-bis(trimethylsiloxy)ethyl]-L-arabino-5,6-bis(trimethylsiloxy)-7-D-glycero-1,2-bis[(trimethylsiloxy)ethyl]perhydropyrrolo[2,1-b:2′,3′-d]oxazole		LI10107
43	41	71	155	57	232	73	55	**826**	100	94	93	86	86	85	67	66	**1.03**	50	82	9	–	–	–	–	–	–	–	–	–	–	Phorbol-12,13,20-tridecanoate		NI47567
384	218	439	498	426	312	299	528	**827**	100	56	40	36	36	23	23	20	**0.00**	47	101	4	1	–	–	–	–	–	3	–	–	–	Phytosphingosine, N-eicosanoyl-1,3,4-tri-O-trimethylsilyl-		LI10108
73	147	813	814	815	816	817	725	**828**	100	76	63	51	43	24	13	7	**0.00**	24	72	10	–	–	–	–	–	–	11	–	–	–	Trisiloxane, octakis(trimethylsilyloxy)-		LI10109
204	361	273	245	191	259	377	129	**828**	100	63	23	12	6	5	4	4	**1.20**	36	72	8	2	–	–	–	–	–	6	–	–	–	(2′S,3′S)-2-[2′,3′-Bis(trimethylsilyloxy)-4′-[tetrakis(trimethylsilyloxy)glucopyranosyl]butyl]quinoxaline		NI47568
149	167	284	279	150	215	230	328	**828**	100	85	76	67	53	50	38	35	**0.00**	48	60	12	–	–	–	–	–	–	–	–	–	–	Rosmanoyl-carnosate tetraacetate		MS11067
217	289	204	291	221	281	191	363	**830**	100	36	30	24	24	23	20	16	**0.00**	33	78	10	–	–	–	–	–	–	7	–	–	–	β-D-Glucopyranoside, 6-deoxy-2,3,4-tris-O-(trimethylsilyl)-α-L-mannopyranosyl 2,3,4,6-tetrakis-O-(trimethylsilyl)-	56784-15-7	MS11068
204	191	363	217	273	319	231	361	**830**	100	40	14	14	6	4	4	3	**0.02**	33	78	10	–	–	–	–	–	–	7	–	–	–	D-Glucose, 2-O-[6-deoxy-2,3,4-tris-O-(trimethylsilyl)-α-L-mannopyranosyl]-3,4,5,6-tetrakis-O-(trimethylsilyl)-	56784-14-6	MS11069
204	363	217	191	259	231	273	319	**830**	100	62	49	11	9	9	5	4	**0.04**	33	78	10	–	–	–	–	–	–	7	–	–	–	D-Glucose, 3-O-[6-deoxy-2,3,4-tris-O-(trimethylsilyl)-α-L-mannopyranosyl]-2,4,5,6-tetrakis-O-(trimethylsilyl)-	56784-13-5	MS11070
204	191	217	281	231	221	273	243	**830**	100	53	20	7	5	5	4	4	**0.02**	33	78	10	–	–	–	–	–	–	7	–	–	–	D-Glucose, 4-O-[6-deoxy-2,3,4-tris-O-(trimethylsilyl)-α-L-mannopyranosyl]-2,3,5,6-tetrakis-O-(trimethylsilyl)-	56784-12-4	MS11071
204	191	217	363	273	569	451	231	**830**	100	37	23	8	5	3	3	3	**0.02**	33	78	10	–	–	–	–	–	–	7	–	–	–	D-Glucose, 6-O-[6-deoxy-2,3,4-tris-O-(trimethylsilyl)-α-L-mannopyranosyl]-2,3,4,5-tetrakis-O-(trimethylsilyl)-	56784-10-2	MS11072
267	91	155	478	293	238	198	70	**830**	100	87	43	33	31	25	24	24	**0.10**	42	62	7	4	3	–	–	–	–	–	–	–	–	Benzenesulphonamide, 4-methyl-N-[3-[[(4-methylphenyl)sulphonyl]amino]propyl]-N-[3-[[(4-methylphenyl)sulphonyl][3-(2-oxoazacyclotridec-1-yl)propyl]amino]propyl]-	75422-07-0	NI47569

MASS TO CHARGE RATIOS								M.W.	INTENSITIES								Parent	C	H	O	N	S	F	Cl	Br	I	Si	P	B	X	COMPOUND NAME	CAS Reg No	No
43	266	492	663	832	478	280	239	**832**	100	17	10	8	4	3	3	2		23	21	5	4	–	21	–	–	–	–	–	–	–	3-Methylbutyl N,N′,N″-tris(heptafluorobutyryl)argininate		NI47570
189	43	147	109	169	120	70	95	**832**	100	98	97	75	70	47	42	34	**1.00**	42	40	18	–	–	–	–	–	–	–	–	–	–	Naringenin-7-O-(6″-O-p-coumaryl)-β-D-glucoside, hexaacetate	75659-57-3	NI47571
347	288	324	379	252	351	348	425	**834**	100	98	41	40	40	39	36	34	**4.00**	40	62	13	6	–	–	–	–	–	–	–	–	–	L-Proline, 1-[N-[N-[N^6-carboxy-N^2-(N-carboxy-L-valyl)-L-lysyl]-L-valyl]-L-tyrosyl]-, N,N^6-diethyl methyl ester, ethyl carbonate (ester)	21178-23-4	NI47572
347	288	379	324	351	252	348	425	**834**	100	94	40	40	39	39	37	34	**3.50**	40	62	13	6	–	–	–	–	–	–	–	–	–	L-Proline, 1-[N-[N-[N^6-carboxy-N^2-(N-carboxy-L-valyl)-L-lysyl]-L-valyl]-L-tyrosyl]-, N,N^6-diethyl methyl ester, ethyl carbonate (ester)		LI10110
57	43	71	367	55	41	579	69	**834**	100	87	55	54	52	40	38	38	**0.20**	53	102	6	–	–	–	–	–	–	–	–	–	–	Octadecanoic acid, 2-[(1-oxohexadecyl)oxy]-1-[[(1-oxohexadecyl)oxy]methyl]ethyl ester	2177-97-1	NI47574
57	43	239	71	579	267	55	313	**834**	100	85	67	62	61	55	51	44	**0.60**	53	102	6	–	–	–	–	–	–	–	–	–	–	Octadecanoic acid, 2-[(1-oxohexadecyl)oxy]-1-[[(1-oxohexadecyl)oxy]methyl]ethyl ester	2177-97-1	NI47575
57	43	55	71	367	41	85	69	**834**	100	85	65	58	55	40	38	38	**0.00**	53	102	6	–	–	–	–	–	–	–	–	–	–	Octadecanoic acid, 2-[(1-oxohexadecyl)oxy]-1-[[(1-oxohexadecyl)oxy]methyl]ethyl ester	2177-97-1	NI47573
57	43	71	55	41	551	69	85	**834**	100	93	49	44	40	39	30	27	**0.00**	53	102	6	–	–	–	–	–	–	–	–	–	–	Octadecanoic acid, 2-[(1-oxotetradecyl)oxy]-1,3-propanediyl ester	57396-99-3	NI47576
135	203	339	543	475	816	679	611	**834**	100	39	30	26	12	2	1	1	**0.17**	60	98	1	–	–	–	–	–	–	–	–	–	–	2,6,10,14,18,22,26,30,34,38,42,46-Octatetracontadodecaen-1-ol, 3,7,11,15,19,23,27,31,35,39,43,47-dodecamethyl-	17093-81-1	NI47577
73	186	147	145	79	75	258	157	**837**	100	46	31	29	27	22	18	17	**0.00**	38	87	7	1	–	–	–	–	–	6	–	–	–	Acetamide, N-[2,3,4,13,14-pentakis[(trimethylsilyl)oxy]-1-[[(trimethylsilyl)oxy]methyl]-3-heptadecenyl]-	56247-91-7	NI47578
70	44	94	122	95	91	139	41	**839**	100	60	46	37	30	26	25	24	**0.67**	43	49	11	7	–	–	–	–	–	–	–	–	–	Staphylomycin S3	33477-39-3	NI47579
70	94	122	95	139	91	205	177	**839**	100	44	34	26	24	24	20	18	**0.60**	43	49	11	7	–	–	–	–	–	–	–	–	–	Staphylomycin S3	33477-39-3	NI47580
70	44	94	122	95	91	139	205	**839**	100	63	48	38	33	26	25	24	**0.68**	43	49	11	7	–	–	–	–	–	–	–	–	–	Staphylomycin S3	33477-39-3	LI10111
576	400	488	312	254	235	664	645	**840**	100	85	79	69	59	53	46	25	**4.00**	2	6	–	2	–	22	–	–	–	–	8	–	2	Nickel, bis(difluorophosphinoamino)-1,2-bis[tris(trifluorophosphino)]di-		NI47581
169	331	109	43	163	271	121	283	**840**	100	97	38	24	18	13	13	10	**0.60**	41	44	19	–	–	–	–	–	–	–	–	–	–	Senburiside II methyl ester pentaacetate		MS11073
561	562	563	489	579	564	170	73	**840**	100	66	53	44	32	28	25	15	**0.00**	42	84	7	–	–	–	–	–	–	5	–	–	–	Ecdysone, 20-hydroxy-, 2,3,20,22,25-pentakis-O-(trimethylsilyl)-, (2β,3β)-		MS11074
799	800	408	436	801	44	392	43	**842**	100	60	21	19	18	9	7	6	**6.00**	56	110	2	2	–	–	–	–	–	–	–	–	–	(2)-(1-Acetylazacyclooctacosane)(1-acetylazacyclohexacosane)-catenane		MS11075
70	91	106	100	179	112	361	217	**843**	100	15	14	10	9	7	6	6	**0.03**	43	53	11	7	–	–	–	–	–	–	–	–	–	Staphylomycin S2, methyl ester	36261-13-9	NI47582
70	91	106	44	100	58	71	57	**843**	100	17	16	14	12	12	11	11	**0.05**	43	53	11	7	–	–	–	–	–	–	–	–	–	Staphylomycin S2, methyl ester	36261-13-9	NI47583
70	106	91	44	100	58	57	179	**843**	100	17	17	17	13	13	12	9	**0.05**	43	53	11	7	–	–	–	–	–	–	–	–	–	Staphylomycin S2, methyl ester	36261-13-9	LI10112
494	248	424	213	354	459	212	284	**844**	100	64	51	27	26	17	14	13	**0.00**	18	–	6	–	2	–	10	–	–	–	–	–	2	Cobalt, [[μ-bis(pentachlorophenyl)disulphide-S,S′:S,S′]]hexacarbonyldi-, (Co-Co)	35918-21-9	MS11076
73	299	315	147	75	230	300	316	**844**	100	83	81	31	30	29	24	22	**0.00**	27	70	12	–	–	–	–	–	–	7	2	–	–	β-D-Fructofuranose, 2,3,4-tris-O-(trimethylsilyl)-, 1,6-bis[bis(trimethylsilyl)]phosphate	55521-25-0	NI47584
73	315	299	230	316	147	445	215	**844**	100	95	46	39	24	23	21	16	**0.00**	27	70	12	–	–	–	–	–	–	7	2	–	–	D-Fructose, 3,4,5-tris-O-(trimethylsilyl)-, 1,6-bis[bis(trimethylsilyl)]phosphate	33800-33-8	NI47585
259	197	135	450	512	435	105	388	**844**	100	66	53	37	28	24	24	23	**8.00**	52	52	–	–	–	–	–	–	–	6	–	–	–	Cyclopentasilane, 1-methyl-2,2,3,3,4,4,5,5-octaphenyl-1-(trimethylsilyl)-		MS11077
619	354	226	285	304	732	521	550	**845**	100	70	66	53	42	22	20	19	**12.00**	7	21	–	5	–	16	–	–	–	–	8	–	2	Cobalt, bis[difluoro(dimethylamino)phosphine]tris[methyliminobis(difluorophosphine)]di-		MS11078
73	361	147	217	103	204	362	129	**846**	100	44	26	24	22	16	14	13	**0.00**	33	78	11	–	–	–	–	–	–	7	–	–	–	D-Turanose, heptakis(trimethylsilyl)-	60065-05-6	NI47586

MASS TO CHARGE RATIOS								M.W.	INTENSITIES								Parent	C	H	O	N	S	F	Cl	Br	I	Si	P	B	X	COMPOUND NAME	CAS Reg No	No
175	143	88	101	111	75	115	71	**846**	100	83	35	34	23	19	16	13	**0.00**	37	66	21	–	–	–	–	–	–	–	–	–	–	Per-O-methyl-β-D-xylopentaose	81685-75-8	NI47587
169	438	254	367	197	203	439	738	**846**	100	84	81	66	55	39	36	35	**15.00**	41	66	11	8	–	–	–	–	–	–	–	–	–	L-Leucine, N[N-[N-[N[N[N-[N-(N-carboxy-L-alanyl)-L-valyl]glycyl]leucyl]-L-alanyl]-L-valyl]glycyl]-, N-benzyl methyl ester	4249-33-6	LI10113
438	367	169	254	197	203	439	362	**846**	100	79	59	48	33	23	21	18	**0.21**	41	66	11	8	–	–	–	–	–	–	–	–	–	L-Leucine, N[N-[N-[N[N[N-[N-(N-carboxy-L-alanyl)-L-valyl]glycyl]leucyl]-L-alanyl]-L-valyl]glycyl]-, N-benzyl methyl ester	4249-33-6	NI47588
498	586	674	410	409	497	743	762	**850**	100	79	48	48	42	40	32	30	**22.00**	–	1	–	–	–	20	–	–	–	–	7	–	2	Cobalt, μ-hydrydo-μ-difluorophosphidotris(trifluorophosphine)-tris(trifluorophosphine)iridium-		NI47589
367	57	377	423	357	413	368	378	**850**	100	99	81	62	59	45	35	30	**0.00**	38	78	11	–	–	–	–	–	–	5	–	–	–	1,3,5,7,9-Pentavinyl-1,1,3,5,7,9,9-heptabutoxypentasiloxane		NI47590
43	310	328	370	612	588	570	388	**850**	100	50	25	6	2	2	2	2	**0.00**	52	82	9	–	–	–	–	–	–	–	–	–	–	9,12-Octadecadienoic acid (Z,Z)-, [9a-(acetyloxy)-1a,1b,4,4a,5,7a,7b,8,9,9a-decahydro-4a,7b-dihydroxy-1,1,6,8-tetramethyl-5-oxo-1H-cyclopropa[3,4]benz[1,2-e]azulen-3-yl]methyl ester, [1aR-(1aα,1bβ)]-	75311-71-6	NI47591
342	740	852	656	628	824	712	314	**852**	100	75	63	51	49	46	41	40		8	–	8	–	–	–	–	–	2	–	–	–	2	Rhenium di-μ-iodo-octacarbonyldi-		MS11079
194	267	375	195	220	376	268	165	**853**	100	60	51	46	35	16	15	15	**1.20**	39	71	10	1	–	–	–	–	–	5	–	–	–	Denopamine-glucuronide, pentakis(trimethylsilyl)-		NI47592
243	129	147	258	247	186	73	217	**853**	100	28	22	17	17	17	17	15	**0.10**	39	91	7	1	–	–	–	–	–	6	–	–	–	1,3,4,5,8,9-Hexakis(trimethylsilyloxy)-2-acetamido-9-methyloctadecane		NI47593
392	784	785	462	57	69	55	83	**853**	100	95	58	53	43	35	34	32	**31.00**	55	106	1	1	–	3	–	–	–	–	–	–	–	(2)-(1-(Trifluoroacetyl)azacyclohexacosane)(cyclooctacosane)-catenane		MS11080
70	122	94	84	95	134	91	98	**855**	100	27	26	24	23	18	16	15	**0.05**	44	53	11	7	–	–	–	–	–	–	–	–	–	Staphylomycin S, methyl ester	36261-12-8	NI47595
70	44	39	123	94	58	84	134	**855**	100	46	33	29	28	26	25	22	**0.08**	44	53	11	7	–	–	–	–	–	–	–	–	–	Staphylomycin S, methyl ester	36261-12-8	LI10114
70	39	44	122	94	84	58	95	**855**	100	32	31	29	28	25	25	24	**0.07**	44	53	11	7	–	–	–	–	–	–	–	–	–	Staphylomycin S, methyl ester	36261-12-8	NI47594
412	218	467	526	454	340	299	556	**855**	100	56	40	36	30	27	25	23	**0.00**	49	105	4	1	–	–	–	–	–	3	–	–	–	Phytosphingosine, N-(docosanoyl)-1,3,4-tri-O-(trimethylsilyl)-		LI10115
568	327	540	299	512	202	484	372	**856**	100	59	48	40	37	35	32	30	**0.00**	10	–	10	–	–	–	–	–	–	–	–	–	3	Rhenium, decacarbonyl-μ-mercuriodi-		NI47596
827	828	829	830	399	400	831	783	**856**	100	75	61	30	25	15	15	9	**2.59**	22	56	16	–	–	–	–	–	–	10	–	–	–	1,3,5,7,9,11,13,15,17,19-Decaethyl-9,17-dimethoxypentacyclo[13.5.1.1^{3,13}.1^{5,11}.1^{7,19}]decasiloxane	72617-71-1	NI47597
57	56	86	77	76	286	284	151	**856**	100	82	32	15	15	14	14	14	**0.00**	42	30	5	6	–	–	–	2	–	–	–	–	–	[2-Hydroxy-3-(4′-bromophenyl)-1,2-dihydroquinoxalin-1-yl](4-nitrophenyl)[1-(4-nitrobenzyl)-3-(4-bromophenyl)-1,2-dihydroquinoxalin-2-yl]methane	91736-71-9	NI47598
91	108	300	436	136	123	315	287	**856**	100	35	20	12	12	12	10	10	**0.00**	50	56	9	4	–	–	–	–	–	–	–	–	–	Dibenzyl 7-(2-carboxyethyl)-4,6-diethyl-2-[(2-methoxycarbonyl)ethyl]-1,3,5,8-tetramethyl-A-oxobilane-1′,8′-dicarboxylate		LI10116
91	121	108	122	180	107	148	120	**856**	100	10	10	8	7	7	5	5	**0.30**	50	56	9	4	–	–	–	–	–	–	–	–	–	Tetrapyrrane D		LI10117
91	108	389	197	180	162	480	120	**856**	100	40	10	9	8	8	7	6	**3.20**	50	56	9	4	–	–	–	–	–	–	–	–	–	Tetrapyrrane E		LI10118
93	206	178	43	161	134	132	28	**857**	100	67	61	57	23	15	15	15	**15.00**	41	47	19	1	–	–	–	–	–	–	–	–	–	Wilforgine	37239-47-7	NI47599
55	107	71	189	83	281	171	311	**858**	100	72	59	46	38	25	18	14	**0.00**	48	74	13	–	–	–	–	–	–	–	–	–	–	16-O-Acetyl-21-O-(3,4-di-O-angeloyl-β-D-fucopyranosyl)protoaescigenin		DD01833
43	635	798	297	199	606	585	683	**858**	100	17	12	10	10	8	4	2	**1.00**	55	102	6	–	–	–	–	–	–	–	–	–	–	[6-(1′-Cyclohexyl-6′,24′-dimethyl-7′-acetoxyoctacosyl)oxan-1-yl]-[6-(4′-methoxycarbonylbutyl)oxan-2-yl]methane		LI10119
73	147	563	475	564	565	401	476	**860**	100	35	20	20	10	9	9	9	**0.00**	20	60	14	–	–	–	–	–	–	12	–	–	–	1,5-Bis(heptamethylcyclotetrasiloxy)hexamethylcyclotetrasiloxane		NI47600
511	583	393	512	584	498	585	451	**860**	100	46	43	40	22	16	12	12	**0.00**	34	80	11	–	–	–	–	–	–	7	–	–	–	Melibioside, methyl heptakis-O-(trimethylsilyl)-		LI10120
579	57	55	43	41	69	71	604	**860**	100	87	79	78	55	54	51	49	**1.10**	55	104	6	–	–	–	–	–	–	–	–	–	–	9-Octadecenoic acid (Z)-, 3-[(1-oxohexadecyl)oxy]-2-[(1-oxooctadecyl)oxy]propyl ester	2190-28-5	NI47602

MASS TO CHARGE RATIOS								M.W.	INTENSITIES								Parent	C	H	O	N	S	F	Cl	Br	I	Si	P	B	X	COMPOUND NAME	CAS Reg No	No
57	43	55	579	71	69	83	41	**860**	100	83	82	73	61	61	56	52	**0.80**	55	104	6	–	–	–	–	–	–	–	–	–	–	9-Octadecenoic acid (Z)-, 3-[(1-oxohexadecyl)oxy]-2-[(1-oxooctadecyl)oxy]propyl ester	2190-28-5	NI47601
770	771	772	378	18	860	768	57	**861**	100	60	19	18	13	11	8	7	**7.00**	61	115	–	1	–	–	–	–	–	–	–	–	–	(2)-[N-(2-phenylethyl)azacyclohexacosane](cyclooctacosane)-catenane		MS11081
806	750	722	638	666	694	511	834	**862**	100	62	53	49	48	36	36	25	**19.00**	8	–	8	–	–	–	–	–	2	–	–	–	2	Osmium, octacarbonyldiiododi-, (Os-Os)	22587-71-9	NI47603
219	187	101	88	71	45	75	111	**862**	100	96	96	96	64	60	57	49	**0.00**	38	70	21	–	–	–	–	–	–	–	–	–	–	Stachyose, permethyl ether		LI10121
88	119	85	90	728	65	43	52	**862**	100	91	86	60	51	46	29	28	**0.00**	48	42	13	–	–	–	–	–	–	–	–	–	4	Beryllium (II), oxohexakis(phenylacetato)tetra-		MS11082
344	516	301	303	259	260	288	271	**862**	100	72	59	54	52	45	43	36	**5.00**	55	62	–	10	–	–	–	–	–	–	–	–	–	Psychotridine	52617-25-1	NI47604
43	57	239	355	117	98	71	55	**862**	100	99	82	79	76	64	58	48	**0.00**	55	106	6	–	–	–	–	–	–	–	–	–	–	Eicosanoic acid, 2-[(1-oxohexadecyl)oxy]-1-[[(1-oxohexadecyl)oxy]methyl]ethyl ester	56630-28-5	NI47605
239	43	355	57	117	98	71	171	**862**	100	96	95	91	88	62	53	48	**0.00**	55	106	6	–	–	–	–	–	–	–	–	–	–	Eicosanoic acid, 2-[(1-oxohexadecyl)oxy]-1-[[(1-oxohexadecyl)oxy]methyl]ethyl ester	56630-28-5	NI47606
57	43	267	579	55	71	41	395	**862**	100	90	71	66	64	62	46	45	**0.80**	55	106	6	–	–	–	–	–	–	–	–	–	–	Octadecanoic acid, 1-[[(1-oxohexadecyl)oxy]methyl]-1,2-ethanediyl ester	2177-99-3	NI47608
57	71	43	55	85	267	98	69	**862**	100	64	60	45	41	34	33	31	**0.00**	55	106	6	–	–	–	–	–	–	–	–	–	–	Octadecanoic acid, 1-[[(1-oxohexadecyl)oxy]methyl]-1,2-ethanediyl ester	2177-99-3	NI47607
57	43	55	395	71	69	41	85	**862**	100	87	72	66	57	40	40	38	**0.00**	55	106	6	–	–	–	–	–	–	–	–	–	–	Octadecanoic acid, 2-[(1-oxohexadecyl)oxy]-1,3-propanediyl ester	2190-24-1	NI47609
235	197	121	135	107	105	206	123	**862**	100	66	36	32	31	28	27	23	**0.66**	59	90	4	–	–	–	–	–	–	–	–	–	–	2,5-Cyclohexadiene-1,4-dione, 2-(3,7,11,15,19,23,27,31,35,39-decamethyl-2,6,10,14,18,22,26,30,34,38-tetracontadecaenyl)-5,6-dimethoxy-3-methyl-, (all-E)-	303-98-0	NI47610
205	235	197	175	121	107	109	135	**862**	100	90	57	24	22	22	19	18	**5.40**	59	90	4	–	–	–	–	–	–	–	–	–	–	2,5-Cyclohexadiene-1,4-dione, 2-(3,7-dimethyl-2,6-octadienyl)-5,6-dimethoxy-3-methyl-, (E)-		LI10122
455	57	427	437	465	456	428	58	**864**	100	99	69	64	59	45	34	32	**0.00**	36	72	12	–	–	–	–	–	–	6	–	–	–	1,3,5,7,9,11-Hexavinyl-1,3,5,7,9,11-hexabutoxycyclohexasiloxane		NI47611
169	109	229	187	331	228	127	213	**864**	100	68	49	41	23	21	18	14	**1.00**	40	48	21	–	–	–	–	–	–	–	–	–	–	β-D-Gentiobiosylcolumbianetin acetate		LI10123
151	83	94	73	43	633	55	691	**864**	100	90	87	80	73	67	67	59	**13.33**	41	60	16	–	–	–	–	–	–	2	–	–	–	O,O-Bis(trimethylsilyl)azadirachtin		NI47612
773	176	218	274	433	92	91	774	**864**	100	96	82	58	53	53	53	47	**0.00**	44	56	10	4	2	–	–	–	–	–	–	–	–	(S,S)-Bis(N-tosylphenylalanyl)diaza-18-crown-6		MS11083
666	746	133	121	866	199	318	108	**864**	100	98	90	88	80	67	63	63	**40.00**	47	46	6	–	–	–	–	2	–	–	–	–	–	Bis-[3-[3-(3-bromo-2-hydroxy-5-methylbenzyl)-2-hydroxy-5-methylbenzyl] 2-hydroxy-5-methylphenyl]methane		LI10125
666	746	133	121	864	199	318	108	**864**	100	98	90	88	80	67	63	63		47	46	6	–	–	–	–	2	–	–	–	–	–	Bis-[3-[3-(3-bromo-2-hydroxy-5-methylbenzyl)-2-hydroxy-5-methylbenzyl] 2-hydroxy-5-methylphenyl]methane		LI10124
204	217	83	397	191	361	73	305	**864**	100	13	7	3	3	2	2	2	**0.00**	47	92	6	–	–	–	–	–	–	4	–	–	–	δ(22)-Stigmastenyl-2,3,4,6-tetrakis(trimethylsilyl)-D-glucoside		NI47613
178	43	206	105	93	161	867	106	**867**	100	95	78	66	53	49	45	37		43	49	18	1	–	–	–	–	–	–	–	–	–	Wilforine	11088-09-8	NI47614
219	57	289	290	259	203	302	42	**867**	100	74	67	31	29	26	25	22	**8.00**	54	81	6	3	–	–	–	–	–	–	–	–	–	N,N′,N″-Tris(4-hydroxy-3,5-di-tert-butylphenylethylcarbonyl)triazine		IC04649
832	833	834	867	868	838	869	830	**867**	100	70	40	24	16	15	14	7		56	52	–	4	–	–	1	–	–	–	–	–	1	Chloro(5,10,15,20-tetramesityl)porphinatochromium(iii)	80584-29-8	NI47615
868	358	59	98	840	83	462	821	**868**	100	61	51	39	23	22	21	18		34	5	1	–	–	20	–	–	–	–	–	–	1	Tetrakis(pentafluorophenyl)cyclopentadienone-π-cyclopentadienylcobalt		MS11084
91	150	330	108	168	96	92	114	**869**	100	38	33	29	28	28	24	23	**0.60**	38	51	20	3	–	–	–	–	–	–	–	–	–	Alanine, 3-[[2-acetamido-4-O-(2-acetamido-2-deoxy-β-D-glucopyranosyl)-2-deoxy-β-D-glucopyranosyl]oxy]-N-carboxy-, N-benzyl methyl ester, pentaacetate (ester), L-	32887-58-4	NI47616
426	218	481	540	468	354	299	75	**869**	100	60	40	33	30	27	27	25	**0.00**	50	107	4	1	–	–	–	–	–	3	–	–	–	1,3,4-Tri-O-trimethylsilyl-N-(tricosanoyl)phytosphingosine		LI10126
91	108	122	422	450	228	107	92	**870**	100	28	25	21	16	16	16	16	**0.00**	51	58	9	4	–	–	–	–	–	–	–	–	–	Dibenzyl 4,6-diethyl-2,7-di-(2-methoxycarbonylethyl)-1,3,5,8-tetramethyl-A-oxobilane-1′,8′-dicarboxylate		LI10127
389	601	255	162	297	285	269	279	**870**	100	90	90	85	80	80	75	70	**1.00**	51	58	9	4	–	–	–	–	–	–	–	–	–	Tetrapyrrane C		LI10128

MASS TO CHARGE RATIOS	M.W.	INTENSITIES	Parent	C	H	O	N	S	F	Cl	Br	I	Si	P	B	X	COMPOUND NAME	CAS Reg No	No
331 169 91 109 332 127 139 317	**871**	100 79 74 23 15 14 11 10	**0.21**	38	49	22	1	–	–	–	–	–	–	–	–	–	**L-Alanine, N-[benzyloxycarbonyl]-3-[[2,3,4-tri-O-acetyl-6-O-(2,3,4,6-tetra-O-acetyl-α-D-galactopyranosyl)-β-D-glucopyranosyl]oxy]-, methyl ester**	25773-32-4	**NI47617**
169 91 109 331 127 98 271 115	**871**	100 60 38 35 15 14 13 13	**0.15**	38	49	22	1	–	–	–	–	–	–	–	–	–	**L-Serine, N-[benzyloxycarbonyl]-O-[2,3,6-tri-O-acetyl-4-O-(2,3,4,6-tetra-O-acetyl-α-D-glucopyranosyl)-β-D-glucopyranosyl]-, methyl ester**	26034-15-1	**NI47618**
91 340 160 131 133 81 146 145	**872**	100 99 99 94 87 87 86 86	**0.00**	42	64	19	–	–	–	–	–	–	–	–	–	–	**Card-20(22)-enolide, 3-[(O-β-D-glucopyranosyl-(1-6)-O-β-D-glucopyranosyl-(1-4)-2,6-dideoxy-3-O-methyl-β-D-ribo-hexopyranosyl)oxy]-5,14-dihydroxy-19-oxo-, (3β,5β)-**	33279-57-1	**NI47619**
340 160 91 131 133 145 81 322	**872**	100 98 98 94 87 85 85 78	**0.00**	42	64	19	–	–	–	–	–	–	–	–	–	–	**Card-20(22)-enolide, 3-[(O-β-D-glucopyranosyl-(1-6)-O-β-D-glucopyranosyl-(1-4)-2,6-dideoxy-3-O-methyl-β-D-ribo-hexopyranosyl)oxy]-5,14-dihydroxy-19-oxo-, (3β,5β)-**		**LI10129**
160 340 322 341 212 358 307 215	**872**	100 99 80 39 38 35 30 28	**0.00**	42	64	19	–	–	–	–	–	–	–	–	–	–	**Card-20(22)-enolide, 3-[(O-β-D-glucopyranosyl-(1-6)-O-β-D-glucopyranosyl-(1-4)-2,6-dideoxy-3-O-methyl-β-D-ribo-hexopyranosyl)oxy]-5,14-dihydroxy-19-oxo-, (3β,5β)-**	33279-57-1	**NI47620**
520 275 872 259 168 765 382 103	**872**	100 44 27 18 13 4 2 0		47	47	6	–	–	–	–	–	–	–	2	–	1	**Cyclopentadienylrhodiumbis(tri-p-tolylphosphite)**	53556-35-7	**NI47621**
57 71 872 85 43 873 83 97	**872**	100 68 62 50 49 41 37 31		60	120	2	–	–	–	–	–	–	–	–	–	–	**Hexacontanoic acid**	40710-31-4	**NI47622**
43 95 176 194 150 106 28 134	**873**	100 83 76 64 48 27 22 21	**10.00**	41	47	20	1	–	–	–	–	–	–	–	–	–	**Wilfortrine**	37239-48-8	**NI47623**
337 290 131 281 202 271 93 539	**876**	100 98 94 90 76 72 69 68	**26.80**	14	–	8	–	–	14	–	–	–	–	–	–	3	**Bis[heptafluoropropyl(tetracarbonyl)iron]mercury**		**LI10130**
227 709 375 208 225 226 708 707	**876**	100 71 65 61 48 42 34 31	**5.00**	24	–	–	–	–	20	–	–	–	–	–	–	1	**Tetrakis(pentafluorophenyl)plumbane**		**MS11085**
84 122 94 205 317 876 806 764	**876**	100 74 72 54 12 11 6 4		45	64	10	8	–	–	–	–	–	–	–	–	–	**Neoviridogrisein-MP**	81720-11-8	**NI47624**
525 57 535 526 536 545 591 527	**878**	100 97 94 52 44 33 32 29	**0.00**	34	66	13	–	–	–	–	–	–	7	–	–	–	**1,3,5,7,9,11,13-Heptaethenyl-3,5,7,11,13-pentabutoxybicyclo[7.5.1]heptasiloxane**		**NI47625**
246 316 314 439 348 416 337 386	**878**	100 40 39 27 10 4 3 2	**0.00**	38	38	2	2	4	–	–	–	–	–	–	–	2	**Molybdenum, tetrakis(benzenemethanethiolato)di-2,4-cyclopentadien-1-yldinitrosyldi-**	55955-43-6	**NI47626**
86 120 44 42 114 41 39 122	**878**	100 98 90 81 79 66 65 60	**20.00**	44	62	11	8	–	–	–	–	–	–	–	–	–	**Etamycin A**	299-20-7	**NI47628**
86 120 44 114 205 122 94 212	**878**	100 94 90 75 70 58 56 49	**20.00**	44	62	11	8	–	–	–	–	–	–	–	–	–	**Etamycin A**	299-20-7	**NI47627**
86 120 44 114 42 205 41 39	**878**	100 98 91 86 82 73 67 65	**20.00**	44	62	11	8	–	–	–	–	–	–	–	–	–	**Etamycin A**	299-20-7	**LI10131**
122 205 212 218 246 261 247 263	**878**	100 96 71 67 64 39 31 21	**13.30**	44	62	11	8	–	–	–	–	–	–	–	–	–	**Neoviridogrisein IV**		**MS11086**
194 297 375 225 250 298 376 217	**883**	100 76 54 49 34 24 17 15	**1.40**	40	73	11	1	–	–	–	–	–	5	–	–	–	**Denopamine-M2 glucuronide, pentakis(trimethylsilyl)-**		**NI47629**
43 176 194 150 105 106 28 134	**883**	100 77 63 54 48 39 23 21	**9.00**	43	49	19	1	–	–	–	–	–	–	–	–	–	**Evonimine, 8-(acetyloxy)-O^{2}-benzoyl-O^{2}-deacetyl-8-deoxo-26-hydroxy-, (8α)-**	37239-51-3	**NI47630**
440 218 495 554 299 482 75 368	**883**	100 66 43 36 33 30 30 27	**0.00**	51	109	4	1	–	–	–	–	–	3	–	–	–	**1,3,4-Tri-O-trimethylsilyl-N-(tetracosanoyl)phytosphingosine**		**LI10132**
855 856 857 413 858 414 859 87	**884**	100 79 68 39 37 24 20 17	**0.00**	22	56	16	–	–	–	–	–	–	11	–	–	–	**1-Hydroperethylhomodecasilsesquioxane**	80177-46-4	**NI47631**
117 145 169 359 331 333 135 294	**884**	100 82 60 45 35 31 29 27	**19.00**	27	17	9	–	–	21	–	–	–	–	–	–	–	**3,7,15-Tris(heptafluorobutyryl)deoxynivalenol**		**NI47632**
105 149 122 718 289 253 166 281	**884**	100 99 33 10 10 10 10 8	**3.00**	53	72	11	–	–	–	–	–	–	–	–	–	–	**Tanguticacine**	101910-66-1	**NI47633**
886 887 813 814 888 667 796 723	**886**	100 64 59 36 30 29 26 10		46	62	10	4	–	–	–	–	–	2	–	–	–	**21H,23H-Porphine-2,7,12,18-tetrapropanoic acid, 3,8,13,17-tetramethyl-β^{7},β^{12}-bis[(trimethylsilyl)oxy]-, tetramethyl ester**	38574-21-9	**NI47634**
886 887 813 814 888 667 796 723	**886**	100 63 58 36 31 28 25 10		46	62	10	4	–	–	–	–	–	2	–	–	–	**2,4-Di-[1-(trimethylsilyl)oxy]-2-methoxycarbonylethyl-1,3,5,8-tetramethylporphin**		**LI10133**
70 57 43 71 83 55 69 97	**886**	100 72 49 45 33 33 32 27	**0.64**	60	118	3	–	–	–	–	–	–	–	–	–	–	**Hexacontanoic acid, 18-oxo-**	40710-27-8	**NI47635**
57 886 75 71 887 87 43 74	**886**	100 82 65 63 56 56 52 49		61	122	2	–	–	–	–	–	–	–	–	–	–	**Hexacontanoic acid, methyl ester**	40710-35-8	**NI47636**

MASS TO CHARGE RATIOS	M.W.	INTENSITIES	Parent	C	H	O	N	S	F	Cl	Br	I	Si	P	B	X	COMPOUND NAME	CAS Reg No	No
663 887 719 607 691 635 747 212	**887**	100 67 60 40 27 27 17 10		15	15	10	1	–	–	–	–	–	–	–	–	4	Bis(pentacarbonyltungstendimethylarsinyl)methylamine		NI47637
204 420 638 173 217 521 191 330	**887**	100 54 53 50 44 37 35 34	**0.00**	35	81	11	1	–	–	–	–	–	7	–	–	–	D-Galactopyranose, 4-O-[2-(acetylamino)-2-deoxy-3,4,6-tris-O-(trimethylsilyl)-β-D-galactopyranosyl]-1,2,3,6-tetrakis-O-(trimethylsilyl)-	55956-13-3	NI47638
420 217 330 191 421 218 173 246	**887**	100 98 56 45 39 37 36 35	**0.00**	35	81	11	1	–	–	–	–	–	7	–	–	–	D-Galactose, 3-O-[2-(acetylamino)-2-deoxy-3,4,6-tris-O-(trimethylsilyl)-β D-galactopyranosyl]-2,4,5,6-tetrakis-O-(trimethylsilyl)-	55956-15-5	NI47639
204 217 552 191 173 638 218 205	**887**	100 65 43 42 41 26 23 22	**0.00**	35	81	11	1	–	–	–	–	–	7	–	–	–	D-Galactose, 6-O-[2-(acetylamino)-2-deoxy-3,4,6-tris-O-(trimethylsilyl)-β D-galactopyranosyl]-2,3,4,5-tetrakis-O-(trimethylsilyl)-	55956-14-4	NI47640
173 204 538 217 539 361 259 174	**887**	100 73 50 33 28 22 20 16	**0.00**	35	81	11	1	–	–	–	–	–	7	–	–	–	D-Glucose, 2-(acetylamino)-2-deoxy-6-O-[2,3,4,6-tetrakis-O-(trimethylsilyl)-α-D-glucopyranosyl]-3,4,5-tris-O-(trimethylsilyl)-	55956-16-6	NI47641
173 217 330 420 218 510 233 204	**887**	100 93 52 48 43 39 33 24	**1.56**	35	81	11	1	–	–	–	–	–	7	–	–	–	D-Mannose, 2-O-[2-(acetylamino)-2-deoxy-3,4,6-tris-O-(trimethylsilyl)-β-D-glucopyranosyl]-3,4,5,6-tetrakis-O-(trimethylsilyl)-	55956-12-2	NI47642
73 147 873 874 875 221 876 281	**888**	100 47 35 28 26 21 14 5	**0.00**	24	72	12	–	–	–	–	–	–	12	–	–	–	Octakis(trimethylsiloxy)cyclotetrasiloxane		LI10134
423 361 307 319 333 363 305 409	**890**	100 90 39 32 26 19 14 13	**0.00**	35	86	10	–	–	–	–	–	–	8	–	–	–	Octa(trimethylsilyl) 1-O-(β-glucopyranosyl)xylitol		LI10135
607 267 395 341 608 606 396 268	**890**	100 89 68 66 43 35 19 18	**0.17**	57	110	6	–	–	–	–	–	–	–	–	–	–	Octadecanoic acid, 1,2,3-propanetriyl ester	555-43-1	NI47643
395 57 607 267 451 55 71 43	**890**	100 51 46 46 40 40 39 37	**0.10**	57	110	6	–	–	–	–	–	–	–	–	–	–	Octadecanoic acid, 1,2,3-propanetriyl ester	555-43-1	NI47644
57 43 71 55 85 69 83 41	**890**	100 68 57 42 36 32 28 27	**0.00**	57	110	6	–	–	–	–	–	–	–	–	–	–	Octadecanoic acid, 1,2,3-propanetriyl ester	555-43-1	NI47645
69 73 222 481 223 890 135 120	**890**	100 67 28 18 18 9 9 7		59	90	4	–	–	–	–	–	–	1	–	–	–	5-(γ-Tocotrienyloxy)-γ-tocotrienol, trimethylsilyl ether		LI10136
44 97 134 69 41 78 106 151	**893**	100 99 96 86 82 61 50 46	**28.60**	47	72	10	5	–	–	–	–	–	–	–	–	1	Aluminium(mycobactinylpropyl)piperidine		IC04650
570 411 323 204 348 142 103 129	**893**	100 39 37 30 29 24 22 20	**1.00**	52	107	4	1	–	–	–	–	–	3	–	–	–	1,3,2'-Tris(trimethylsilyloxy)-N-(2-hydroxytetracosanoyl)nonadecasphingadienine		NI47646
415 455 288 39 65 161 128 262	**894**	100 96 96 78 67 59 56 39	**0.00**	10	10	2	2	–	–	–	–	4	–	–	–	2	Di-(π-cyclopentadienyl)diiododinitrosyldi-(μ-molybdenum iodide)		MS11087
65 288 128 414 445 127 66 63	**894**	100 78 78 73 61 45 40 40	**0.00**	10	10	2	2	–	–	–	–	4	–	–	–	2	Molybdenum, bis(η^5-2,4-cyclopentadien-1-yl)di-μ-iododiiododinitrosyldi-	12203-25-7	NI47647
173 73 75 204 147 103 117 174	**898**	100 96 72 50 44 39 38 34	**0.00**	35	74	13	4	–	–	–	–	–	5	–	–	–	L-Asparagine, N-[2-(acetylamino)-4-O-[2-(acetylamino)-2-deoxy-3,4,6-tris O-(trimethylsilyl)-β-D-glucopyranosyl]-2-deoxy-3,6-bis-O-(trimethylsilyl)-β-D-glucopyranosyl]-	56051-47-9	NI47648
814 760 900 646 702 758 730 674	**900**	100 85 82 67 64 57 54 54		12	3	12	–	–	–	–	–	–	–	–	–	3	Hydridorhenium tetracarbonyl trimer		MS11088
131 219 129 143 218 220 147 132	**900**	100 70 19 18 17 16 14 13	**0.00**	45	96	6	–	–	–	–	–	–	6	–	–	–	5β-Cholestane-3α,7α,12α,24,25,26-hexol, hexakis(trimethylsilyl)-		NI47649
43 295 279 239 331 323 240 859	**901**	100 5 4 4 3 3 3 2	**0.10**	43	51	20	1	–	–	–	–	–	–	–	–	–	4-[(1-Deoxycellobiit-1-yl)amino]benzophenone nonaacetate		NI47650
370 824 740 908 852 369 880 768	**908**	100 98 66 62 50 48 45 43		34	20	6	–	–	–	–	–	–	–	–	–	2	Osmium, hexacarbonyl[μ-[(1,2,3,4-η:1,4-η)-1,2,3,4-tetraphenyl-1,3-butadiene-1,4-diyl]]di-, (Os-Os)	33310-10-0	NI47651
73 147 318 461 103 217 191 129	**910**	100 33 21 17 15 14 13 12	**0.10**	33	83	11	–	–	–	–	–	–	8	1	–	–	D-myo-Inisitol, 1,2,4,5,6-pentakis-O-(trimethylsilyl)-, 2,3-bis[(trimethylsilyl)oxy]propyl trimethylsilyl phosphate, (R)-	55528-10-4	NI47652
86 44 42 114 41 39 120 94	**910**	100 99 84 61 58 58 54 53	**0.10**	45	66	12	8	–	–	–	–	–	–	–	–	–	Etamycin, methyl ester		LI10137
44 86 42 114 41 39 120 94	**910**	100 99 82 60 58 57 52 50	**0.10**	45	66	12	8	–	–	–	–	–	–	–	–	–	Glycine, N-[N-[N-[N-[cis-4-hydroxy-1-[N-[N-[(3-hydroxy-2-pyridinyl)carbonyl]-L-threonyl]-D-leucyl]-D-prolyl]-N-methylglycyl]-N,3-dimethyl-L-leucyl]-L-alanyl]-N-methyl-L-2-phenyl-, methyl ester	36261-15-1	NI47653

MASS TO CHARGE RATIOS								M.W.	INTENSITIES								Parent	C	H	O	N	S	F	Cl	Br	I	Si	P	B	X	COMPOUND NAME	CAS Reg No	No
44	86	114	120	94	180	122	118	**910**	100	99	58	50	48	40	40	40	**0.10**	45	66	12	8	–	–	–	–	–	–	–	–	–	Glycine, N-[N-[N-[N-[cis-4-hydroxy-1-[N-[N-[(3-hydroxy-2-pyridinyl)carbonyl]-L-threonyl]-D-leucyl]-D-prolyl]-N-methylglycyl]-N,3-dimethyl-L-leucyl]-L-alanyl]-N-methyl-L-2-phenyl-, methyl ester	36261-15-1	NI47654
259	652	470	182	575	105	729	288	**910**	100	65	62	51	28	28	26	20	**0.00**	60	50	–	–	–	–	–	–	–	5	–	–	–	Decaphenylcyclopentasilane		LI10138
375	267	220	252	376	180	217	169	**911**	100	68	46	37	33	32	31	21	**0.00**	41	77	10	1	–	–	–	–	–	6	–	–	–	Denopamine-M1 glucuronide, hexakis(trimethylsilyl)-		NI47655
800	772	660	716	632	576	744	912	**912**	100	78	59	54	52	50	43	37		12	–	12	–	–	–	–	–	–	–	–	–	3	Dodecacarbonyltriosmium		MS11089
168	141	145	91	358	305	207	299	**912**	100	66	42	24	23	21	18	15	**9.00**	34	5	1	–	–	20	–	–	–	–	–	–	1	Tetrakis(pentafluorophenyl)cyclopentadienone-π-cyclopentadienylrhodium		MS11090
227	229	152	77	456	306	151	154	**912**	100	73	73	72	68	57	52	33	**9.00**	36	30	–	–	–	–	–	–	–	–	–	–	6	Hexaphenylcyclohexaarsine		LI10139
561	193	73	147	562	143	75	81	**912**	100	80	46	45	35	35	35	30	**6.00**	45	92	7	–	–	–	–	–	–	6	–	–	–	20-Hydroxyecdysone 2β,3β,14α,20,22,25-hexakis(trimethylsilyl) ether		MS11091
150	227	907	304	531	454	603	680	**912**	100	80	21	10	5	5	4	2	**0.00**	48	40	–	–	–	–	–	–	–	–	–	–	4	Octaphenylcyclotetragermane		LI10140
523	424	275	522	310	154	352	293	**912**	100	85	74	72	72	68	66	64	**0.51**	49	80	10	6	–	–	–	–	–	–	–	–	–	L-Proline, 1-[O-(1-oxohexyl)-N-[N-[N^6-(1-oxohexyl)-N^2-[N-(1-oxohexyl)-L-valyl]-L-lysyl]-L-valyl]-L-tyrosyl]-, methyl ester	56272-43-6	NI47656
57	71	43	914	85	103	915	115	**914**	100	67	54	51	49	46	35	34		63	126	2	–	–	–	–	–	–	–	–	–	–	Hexacontanoic acid, propyl ester	40710-40-5	NI47657
73	299	315	147	387	318	300	28	**916**	100	75	36	26	23	18	18	16	**0.00**	30	78	12	–	–	–	–	–	–	8	2	–	–	Inositol, 1,2,4,5-tetrakis-O-(trimethylsilyl)-, bis[bis(trimethylsilyl) phosphate], myo-	33981-93-0	NI47658
327	294	394	273	422	502	474	586	**918**	100	65	57	57	33	29	25	20	**0.30**	20	9	18	–	–	–	–	–	–	–	–	–	6	7-Methyl-1,3,5-triarsa-2,4,9-trioxa-adamantan-As,As′,As″-tris[pentacarbonylchromium(0)]	80961-57-5	NI47659
361	73	217	362	147	204	103	815	**918**	100	91	36	32	28	26	23	17	**0.00**	36	86	11	–	–	–	–	–	–	8	–	–	–	D-Fructose, 3-O-[2,3,4,6-tetrakis-O-(trimethylsilyl)-α-D-glucopyranosyl]-1,4,5,6-tetrakis-O-(trimethylsilyl)-	39523-07-4	NI47660
217	73	204	361	147	218	129	103	**918**	100	99	73	54	30	22	22	18	**0.00**	36	86	11	–	–	–	–	–	–	8	–	–	–	D-Fructose, 6-O-[2,3,4,6-tetrakis-O-(trimethylsilyl)-α-D-glucopyranosyl]-1,3,4,5-tetrakis-O-(trimethylsilyl)-	56145-26-7	NI47661
583	451	361	584	452	585	453	362	**918**	100	96	60	57	43	35	22	21	**2.89**	36	86	11	–	–	–	–	–	–	8	–	–	–	D-Galactose, 6-O-[2,3,5,6-tetrakis-O-(trimethylsilyl)-β-D-galactofuranosyl]-2,3,4,5-tetrakis-O-(trimethylsilyl)-	55649-73-5	NI47662
204	73	217	191	205	361	147	129	**918**	100	41	21	21	20	12	12	9	**0.00**	36	86	11	–	–	–	–	–	–	8	–	–	–	D-Glucopyranose, 4-O-[2,3,4,6-tetrakis-O-(trimethylsilyl)-β-D-galactopyranosyl]-1,2,3,6-tetrakis-O-(trimethylsilyl)-	55529-68-5	NI47663
204	73	217	191	205	147	361	129	**918**	100	53	26	23	21	20	14	14	**0.00**	36	86	11	–	–	–	–	–	–	8	–	–	–	D-Glucopyranose, 4-O-[2,3,4,6-tetrakis-O-(trimethylsilyl)-β-D-galactopyranosyl]-1,2,3,6-tetrakis-O-(trimethylsilyl)-	55529-68-5	NI47664
361	73	362	217	147	363	271	103	**918**	100	61	31	30	16	15	14	14	**0.00**	36	86	11	–	–	–	–	–	–	8	–	–	–	α-D-Glucopyranoside, 1,3,4,6-tetrakis-O-(trimethylsilyl)-β-D-fructofuranosyl 2,3,4,6-tetrakis-O-(trimethylsilyl)-	19159-25-2	NI47665
204	73	205	361	191	217	147	206	**918**	100	32	20	17	17	15	10	8	**0.00**	36	86	11	–	–	–	–	–	–	8	–	–	–	D-Glucose, 4-O-[2,3,4,6-tetrakis-O-(trimethylsilyl)-β-D-glucopyranosyl]-2,3,5,6-tetrakis-O-(trimethylsilyl)-	56145-25-6	NI47666
361	362	569	583	451	363	570	584	**918**	100	71	47	45	45	37	25	23	**1.33**	36	86	11	–	–	–	–	–	–	8	–	–	–	D-Glucose, 6-O-[2,3,4,6-tetrakis-O-(trimethylsilyl)-α-D-glucopyranosyl]-2,3,4,5-tetrakis-O-(trimethylsilyl)-	55191-01-0	NI47667
361	362	451	363	569	435	539	452	**918**	100	68	41	36	22	22	20	18	**0.00**	36	86	11	–	–	–	–	–	–	8	–	–	–	D-Glucose, 6-O-[2,3,4,6-tetrakis-O-(trimethylsilyl)-β-D-glucopyranosyl]-2,3,4,5-tetrakis-O-(trimethylsilyl)-	55191-02-1	NI47668
204	73	191	217	205	147	361	103	**918**	100	70	28	24	21	19	12	12	**0.00**	36	86	11	–	–	–	–	–	–	8	–	–	–	Lactose, octakis(trimethylsilyl)-	42390-78-3	NI47669
204	73	191	217	205	361	147	103	**918**	100	95	45	24	21	21	18	16	**0.00**	36	86	11	–	–	–	–	–	–	8	–	–	–	Maltose, octakis(trimethylsilyl)-		NI47670
204	73	191	217	205	129	147	103	**918**	100	84	30	22	21	21	17	11	**0.00**	36	86	11	–	–	–	–	–	–	8	–	–	–	Melibiose, octakis(trimethylsilyl)-		NI47671
204	73	217	191	205	129	147	103	**918**	100	77	29	27	22	20	17	11	**0.00**	36	86	11	–	–	–	–	–	–	8	–	–	–	Melibiose, octakis(trimethylsilyl)-		NI47672
58	834	71	57	55	69	919	83	**919**	100	99	99	97	97	96	89	86		60	105	5	1	–	–	–	–	–	–	–	–	–	Azacyclohexacosan-14-one, 1-(28,29-dihydroxybicyclo[25.3.1]hentriaconta-1(31),27,29-trien-31-yl)-, diacetate (ester)	15462-29-0	NI47673

MASS TO CHARGE RATIOS	M.W.	INTENSITIES	Parent	C	H	O	N	S	F	Cl	Br	I	Si	P	B	X	COMPOUND NAME	CAS Reg No	No
57 91 922 92 923 865 423 39	**922**	100 62 58 41 40 26 23 21		64	90	4	–	–	–	–	–	–	–	–	–	–	**1,4-Bis[bis(4-hydroxy-3,5-di-tert-butylphenyl)methyl]benzene**		**IC04651**
262 183 108 263 261 107 184 185	**924**	100 62 38 22 15 15 13 10	**0.00**	54	45	–	–	–	–	1	–	–	–	3	–	1	**Rhodium, chlorotris(triphenylphosphine)-**	14694-95-2	**NI47674**
262 183 108 263 261 107 185 50	**924**	100 63 38 22 15 15 10 8	**0.00**	54	45	–	–	–	–	1	–	–	–	3	–	1	**Rhodium, chlorotris(triphenylphosphine)-**	14694-95-2	**LI10141**
73 75 147 217 191 361 204 129	**926**	100 77 75 63 47 45 23 19	**0.00**	38	82	12	–	–	–	–	–	–	7	–	–	–	**Pulchelloside II trimethylsilyl ether**		**MS11092**
73 75 147 217 191 129 207 103	**926**	100 47 43 25 18 18 16 16	**0.00**	38	82	12	–	–	–	–	–	–	7	–	–	–	**Pulchelloside I trimethylsilyl ether**		**MS11093**
229 202 130 73 230 157 147 203	**926**	100 49 18 17 15 13 10 9	**2.23**	44	74	8	2	–	–	–	–	–	6	–	–	–	**myo-Inositol, bis[N-(trimethylsilyl)indole-3-acetate]tetrakis-O-(trimethylsilyl) ether**		**NI47675**
869 767 854 665 461 563 402 106	**928**	100 62 50 46 40 33 29 21	**7.00**	36	84	12	–	–	–	–	–	–	–	–	–	4	**Lanthanum, tris(diisopropoxyaluminum)-**	20538-49-2	**NI47676**
83 313 55 153 111 213 189 171	**928**	100 40 35 28 25 24 15 15	**0.00**	51	76	15	–	–	–	–	–	–	–	–	–	–	**21-O-(4-O-Acetyl-3-O-angeloyl-β-D-fucopyranosyl)theasapogenol B tetraacetate**		**DD01834**
145 328 113 162 310 130 292 274	**930**	100 89 86 85 83 81 80 77	**0.00**	53	70	14	–	–	–	–	–	–	–	–	–	–	**Plocinine**	100508-98-3	**NI47677**
72 75 671 28 283 73 564 672	**931**	100 75 73 66 57 48 38 37	**3.40**	44	89	10	5	–	–	–	–	–	3	–	–	–	**Hydroxypepstatin A methyl ester, O-trimethylsilyl**		**NI47678**
944 942 946 948 940 152 784 624	**934**	100 85 80 50 50 35 28 28	**0.00**	12	–	–	–	–	–	–	10	–	–	–	–	–	**Decabromodiphenyl**		**IC04652**
223 73 83 705 95 224 75 55	**936**	100 70 60 20 20 19 19 19	**2.00**	44	68	16	–	–	–	–	–	–	3	–	–	–	**O,O,O-Tris(trimethylsilyl)azadirachtin**		**NI47679**
936 311 415 315 519 297 623 727	**936**	100 98 40 30 29 28 27 22		72	72	–	–	–	–	–	–	–	–	–	–	–	**Metacyclophane, (2 or 9)-**		**LI10142**
363 602 331 604 603 332 292 304	**940**	100 91 79 30 29 29 28 16	**2.48**	31	35	12	4	–	15	–	–	–	–	–	–	–	**5-Undecenedioic acid, 5-[[[6-methoxy-6-oxo-5-[(trifluoroacetyl)amino]-2-[(trifluoroacetyl)oxy]hexyl](trifluoroacetyl)amino]methyl]-2,10-bis[(trifluoroacetyl)amino]-, dimethyl ester**	59275-67-1	**NI47680**
297 252 375 298 250 217 299 644	**941**	100 42 41 41 26 17 13 12	**0.00**	42	79	11	1	–	–	–	–	–	6	–	–	–	**Denopamine-M3 glucuronide A, hexakis(trimethylsilyl)-**		**NI47681**
252 297 375 250 225 217 298 376	**941**	100 68 59 49 36 33 23 20	**1.60**	42	79	11	1	–	–	–	–	–	6	–	–	–	**Denopamine-M3 glucuronide B, hexakis(trimethylsilyl)-**		**NI47682**
282 309 944 310 169 354 325 326	**944**	100 65 51 51 49 21 15 9		45	52	22	–	–	–	–	–	–	–	–	–	–	**Egonol gentiobioside hepta-O-acetate**	77690-89-2	**NI47683**
946 843 947 948 844 949 874 771	**946**	100 93 75 43 32 17 16 16		48	74	8	4	–	–	–	–	–	4	–	–	–	**21H,23H-Porphine-2,18-dipropanoic acid, 7,12-bis[1,2-bis[(trimethylsilyl)oxy]ethyl]-3,8,13,17-tetramethyl-, dimethyl ester**	38574-20-8	**NI47684**
800 802 798 960 962 958 796 804	**950**	100 71 71 40 36 36 36 31	**0.00**	12	–	1	–	–	–	–	10	–	–	–	–	–	**Bis(pentabromophenyl) ether**	1163-19-5	**IC04653**
800 798 802 44 960 400 962 958	**950**	100 82 80 76 62 58 54 53	**0.00**	12	–	1	–	–	–	–	10	–	–	–	–	–	**Bis(pentabromophenyl) ether**	1163-19-5	**NI47685**
950 951 73 952 953 75 147 475	**950**	100 83 62 54 26 17 15 12		48	78	8	–	–	–	–	–	–	6	–	–	–	**Gossypol hexatrimethylsilyl ether**		**LI10143**
311 590 57 41 29 69 447 45	**952**	100 68 55 45 32 26 24 22	**0.00**	38	52	10	4	–	12	–	–	–	–	–	–	–	**5-Undecenedioic acid, 2,10-bis[butyl(trifluoroacetyl)amino]-5-[[[2-[butyl(trifluoroacetyl)amino]-2-carboxycyclopentyl](trifluoroacetyl)amino]methyl]-**	55282-80-9	**NI47686**
187 310 354 282 952 558 325 326	**952**	100 75 68 37 24 21 20 19		47	68	20	–	–	–	–	–	–	–	–	–	–	**Egonol gentiobioside, hepta-O-methyl-**		**NI47687**

MASS TO CHARGE RATIOS	M.W.	INTENSITIES	Parent	C	H	O	N	S	F	Cl	Br	I	Si	P	B	X	COMPOUND NAME	CAS Reg No	No
226 169 69 740 209 28 727 227	**953**	100 82 51 27 9 9 8 6	**0.00**	24	7	7	1	–	28	–	–	–	–	–	–	–	Norepinephrine, O,O,O,N-tetrakis(heptafluorobutyryl)-		MS11094
311 590 57 41 29 69 447 45	**954**	100 71 57 46 34 27 25 24	**4.00**	38	54	10	4	–	12	–	–	–	–	–	–	–	Merodesmosine, tetrakis(N-trifluoroacetyl)-, tributyl ester		LI10144
70 41 55 332 42 44 43 233	**955**	100 93 81 73 69 67 62 61	**0.24**	51	69	11	7	–	–	–	–	–	–	–	–	–	Alanine, N-[(3-methoxy-2-pyridinyl)carbonyl]-N,O-dimethyl-L-threonyl-N-methyl-D-α-aminobutyryl-L-prolyl-N-methyl-L-phenylalanyl-cis-4-methoxy-L-pipecoloyl-N-methyl-2-phenyl-, methyl ester	36261-14-0	NI47688
70 332 233 622 136 91 265 590	**955**	100 70 60 50 50 50 42 9	**0.10**	51	69	11	7	–	–	–	–	–	–	–	–	–	Alanine, N-[(3-methoxy-2-pyridinyl)carbonyl]-N,O-dimethyl-L-threonyl-N-methyl-D-α-aminobutyryl-L-prolyl-N-methyl-L-phenylalanyl-cis-4-methoxy-L-pipecoloyl-N-methyl-2-phenyl-, methyl ester	36261-14-0	NI47689
262 333 402 433 302 359	**957**	100 99 80 67 35 15	**0.00**	50	52	3	–	–	–	–	–	–	–	2	–	1	Acetylacetonato(8-methoxycyclo-oct-4-enyl)platinum, triphenylphosphine complex		LI10145
333 404 220 531 616 334 198 197	**958**	100 34 23 22 21 21 11 10	**0.34**	49	82	11	8	–	–	–	–	–	–	–	–	–	L-Leucine, N-methyl-N-[N-methyl-N-[N-methyl-N-[N-methyl-N-[N-methyl-N-[N-methyl-N-[N-methyl-N-[N-methyl-N-[(phenylmethoxy)carbonyl]-L-Ala]-L-Val]-Gly]-l-Leu]-L-Ala]-L-Val]Gly]-, methyl ester	56272-44-7	NI47690
960 370 298 614 169 341 282 658	**960**	100 42 29 20 19 18 16 14		46	56	22	–	–	–	–	–	–	–	–	–	–	2-(3,4-Dimethoxyphenyl)-5-(3-hydroxypropyl)-7-methoxybenzofuran hepta-O-acetylgentiobioside		NI47691
259 197 73 135 105 240 960 120	**960**	100 95 94 91 58 42 41 38		56	64	–	–	–	–	–	–	–	8	–	–	–	1,4-Dimethyl-2,2,3,3,5,5,6,6-octaphenyl-1,4-bis(trimethylsilyl)cyclohexasilane		MS11095
69 73 962 963 150 481 135 120	**962**	100 51 38 35 12 10 6 6		62	98	4	–	–	–	–	–	–	2	–	–	–	5-(γ-Tocotrienyl)-γ-tocotrienyl, bis(trimethylsilyl)-		LI10146
144 199 184 157 145 200 185 155	**964**	100 87 85 68 67 66 66 64	**0.15**	53	72	9	8	–	–	–	–	–	–	–	–	–	L-Lysine, N6-acetyl-N2-[N-[N-[N-(N2-acetyl-N,N,N2-trimethyl-L-asparaginyl)-N-methyl-L-phenylalanyl]-N-methyl-L-phenylalanyl]-N,1-dimethyl-L-tryptophyl]-N^2,N^6-dimethyl-, methyl ester	56818-04-3	MS11096
187 219 203 703 467 512 435 281	**966**	100 59 26 18 15 12 12 10	**0.70**	52	86	16	–	–	–	–	–	–	–	–	–	–	Deca-O-methylmedicagenic acid 3,28-di-O-β-D-glucopyranoside		NI47692
240 169 69 241 42 210 43 28	**967**	100 21 16 7 5 3 2 2	**0.00**	25	9	7	1	–	28	–	–	–	–	–	–	–	Epinephrine, O,O,O,N-tetrakis(heptafluorobutyryl)-		MS11097
83 55 57 69 189 355 215 153	**968**	100 63 30 26 22 11 11 8	**0.00**	54	80	15	–	–	–	–	–	–	–	–	–	–	21-O-(3,4-Di-O-angeloyl-β-D-fucopyranosyl)theasapogenol B tetraacetate		DD01835
977 262 714 636 634 975 712 978	**978**	100 98 81 80 63 60 60 53		54	44	–	–	–	–	–	–	–	–	3	–	1	Iridium, [2-(diphenylphosphino)phenyl-c,p]bis(triphenylphosphine)-, (SP-4 2)-	38316-81-3	MS11098
433 314 258 374 155 373 214 172	**984**	100 95 95 65 55 46 45 45	**0.00**	36	84	12	–	–	–	–	–	–	–	–	–	4	Gallium isopropoxide (tetramer)		LI10147
428 133 121 108 318 548 199 986	**984**	100 92 82 74 55 50 49 23	**11.00**	55	54	7	–	–	–	–	2	–	–	–	–	–	2,6-Bis[3-[3-(3-bromo-2-hydroxy-5-methylbenzyl)-2-hydroxy-5-methylbenzyl]-2-hydroxy-5-methylbenzyl]-4-methylphenol		LI10148

MASS TO CHARGE RATIOS								M.W.	INTENSITIES								Parent	C	H	O	N	S	F	Cl	Br	I	Si	P	B	X	COMPOUND NAME	CAS Reg No	No
428	133	121	108	318	548	199	984	**984**	100	92	82	74	55	50	49	23		55	54	7	–	–	–	–	2	–	–	–	–	–	2,6-Bis[3-[3-(3-bromo-2-hydroxy-5-methylbenzyl)-2-hydroxy-5-methylbenzyl]-2-hydroxy-5-methylbenzyl]-5-methylphenol		LI10149
83	156	313	55	111	213	171	184	**986**	100	67	52	45	43	38	21	6	**0.00**	53	78	17	–	–	–	–	–	–	–	–	–	–	21-O-(4-O-Acetyl-3-O-angeloyl-β-D-fucopyranosyl)-22-O-acetylprotoaescigenin tetraacetate		DD01836
447	348	376	475	377	349	448	476	**991**	100	99	96	82	42	40	36	35	**2.98**	52	93	11	7	–	–	–	–	–	–	–	–	–	L-Threonine, N-[N-[N-[1-[N-[N-[N-(3-hydroxy-1-oxoeicosyl)-L-threonyl]-L-valyl]-L-alanyl]-L-prolyl]-L-leucyl]-L-valyl]-, ψ-lactone	56272-51-6	NI47693
361	451	525	362	452	526	363	422	**992**	100	73	56	37	30	27	19	18	**0.00**	39	96	11	–	–	–	–	–	–	9	–	–	–	D-Galactitol, 6-O-[2,3,5,6-tetrakis-O-(trimethylsilyl)-β-D-galactofuranosyl]-1,2,3,4,5-pentakis-O-(trimethylsilyl)-	55649-72-4	NI47694
451	361	452	362	422	360	453	363	**992**	100	99	39	36	25	24	21	21	**0.00**	39	96	11	–	–	–	–	–	–	9	–	–	–	D-Galactitol, 1,2,3,4,6-pentakis-O-(trimethylsilyl)-, anhydride with 1,2,3,6-tetrakis-O-(trimethylsilyl)-β-D-galactofuranose	55823-10-4	NI47695
73	204	217	147	69	361	205	103	**992**	100	57	40	37	37	32	32	27	**0.00**	39	96	11	–	–	–	–	–	–	9	–	–	–	D-Glucitol, 3-O-(β-D-mannopyranosyl)-, nonakis-O-(trimethylsilyl)-		MS11099
361	319	525	345	307	451	305	363	**992**	100	85	50	38	31	25	25	19	**0.00**	39	96	11	–	–	–	–	–	–	9	–	–	–	D-Sorbitol, 2-O-(β-D-glucopyranosyl)-, nonakis-O-(trimethylsilyl)-		LI10150
361	273	345	451	319	525	331	363	**992**	100	88	84	68	57	30	30	23	**0.00**	39	96	11	–	–	–	–	–	–	9	–	–	–	D-Sorbitol, 4-O-(α-D-glucopyranosyl)-, nonakis-O-(trimethylsilyl)-		LI10151
307	319	361	305	345	363	331	525	**992**	100	92	46	21	11	9	9	8	**0.00**	39	96	11	–	–	–	–	–	–	9	–	–	–	D-Sorbitol, 6-O-(β-D-glucopyranosyl)-, nonakis-O-(trimethylsilyl)-		LI10152
73	146	218	77	147	75	148	74	**993**	100	75	45	35	30	25	15	15	**0.00**	24	35	4	1	–	–	–	–	4	3	–	–	–	Tris(trimethylsilyl)thyroxine		MS11100
69	235	197	81	28	95	135	93	**998**	100	99	94	91	40	38	34	32	**11.89**	69	106	4	–	–	–	–	–	–	–	–	–	–	Coenzyme Q12	24663-36-3	NI47696
57	71	85	44	43	97	28	83	**998**	100	74	51	51	42	41	38	36	**1.15**	69	138	2	–	–	–	–	–	–	–	–	–	–	Nonahexacontanoic acid	40710-32-5	NI47697
175	88	111	101	115	335	75	303	**1006**	100	26	22	21	12	11	11	8	**0.00**	44	78	25	–	–	–	–	–	–	–	–	–	–	Per-O-methyl-β-D-xylohexaose	81691-00-1	NI47698
169	109	212	271	115	127	331	81	**1009**	100	65	48	43	27	23	21	13	**0.00**	44	55	24	3	–	–	–	–	–	–	–	–	–	β-D-Glucopyranoside, 2,3-bis(acetyloxy)-1-[(acetyloxy)(2-phenyl-2H-1,2,3-triazol-4-yl)methyl]propyl 4-O-(2,3,4,6-tetra-O-acetyl-β-D-glucopyranosyl)-, triacetate (ester)	35405-85-7	NI47699
217	507	842	358	179	210	506	327	**1010**	100	34	33	32	24	20	17	13	**0.66**	30	–	6	–	–	20	–	–	–	–	2	–	2	Iron, bis(pentafluorophenyl)phosphidotricarbonyl-, dimer		MS11101
467	154	74	69	55	468	119	81	**1010**	100	44	37	37	34	28	28	27	**0.06**	37	39	8	–	–	21	–	–	–	–	–	–	–	Cholan-24-oic acid, 3,7,12-tris(2,2,3,3,4,4,4-heptafluoro-1-oxobutoxy)-, methyl ester, (3α,5β,7α,12α)-	57326-14-4	NI47700
57	43	71	55	69	83	85	41	**1012**	100	58	56	37	35	34	29	29	**0.29**	69	136	3	–	–	–	–	–	–	–	–	–	–	Nonahexacontanoic acid, 18-oxo-	40710-28-9	NI47701
57	71	28	43	44	85	55	83	**1012**	100	68	63	56	50	46	43	42	**4.12**	70	140	2	–	–	–	–	–	–	–	–	–	–	Nonahexacontanoic acid, methyl ester	40710-36-9	NI47702
169	109	81	107	106	149	127	98	**1013**	100	80	70	67	45	33	30	28	**10.00**	45	59	25	1	–	–	–	–	–	–	–	–	–	D-Glucopyranosylamine, O-2,3,4,6-tetra-O-acetyl-β-D-glucopyranosyl(1-3)-O-2,4,6-tri-O-acetyl-β-D-glucopyranosyl(1-3)-N-(4-methylphenyl)-, 2,4,6-triacetate	55331-35-6	NI47703
169	109	331	127	107	81	271	211	**1013**	100	47	33	14	14	13	12	12	**4.33**	45	59	25	1	–	–	–	–	–	–	–	–	–	D-Glucopyranosylamine, O-2,3,4,6-tetra-O-acetyl-β-D-glucopyranosyl(1-4)-O-2,3,6-tri-O-acetyl-β-D-glucopyranosyl(1-4)-N-(4-methylphenyl)-, 2,3,6-triacetate	55298-14-1	NI47704
91	181	92	105	106	77	240	253	**1014**	100	26	20	14	13	13	12	7	**0.00**	63	66	12	–	–	–	–	–	–	–	–	–	–	β-D-Glucopyranoside, benzyl 3,6-bis-O-benzyl-4-O-[2,3,4,6-tetrakis-O-benzyl-α-D-glucopyranosyl]-, acetate	56689-40-8	NI47705

MASS TO CHARGE RATIOS								M.W.	INTENSITIES								Parent	C	H	O	N	S	F	Cl	Br	I	Si	P	B	X	COMPOUND NAME	CAS Reg No	No
936	428.5	327	530	733	349.5	372	248	**1015**	100	40	24	21	20	16	13	13	**4.00**	–	–	–	5	–	–	–	10	–	–	5	–	–	**Decabromopentaphosphonitrile**		**MS11102**
79	67	42	91	93	41	55	81	**1022**	100	92	84	79	63	61	58	47	**0.00**	69	98	6	–	–	–	–	–	–	–	–	–	–	**Docosahexaenoic acid, 1,2,3-propanetriyl ester**	11094-59-0	**NI47706**
1019	346	164	182	601	327	255	582	**1024**	100	99	99	90	50	41	41	31	**0.00**	56	56	–	–	–	–	–	–	–	–	–	–	4	**Octa(p-tolyl)cyclotetragermane**		**LI10153**
262	401	358	301	458	501			**1025**	100	99	58	15	4	3			**0.00**	54	56	4	–	–	–	–	–	–	–	2	–	1	**Acetylacetonato-[8-(acetylacetonyl)cyclooct-4-enyl]platinum, triphenylphosphine**		**LI10154**
955	956	1032	957	1033	877	1034	878	**1032**	100	89	70	69	64	50	44	39		48	40	12	–	–	–	–	–	–	8	–	–	–	**Perphenyloctasilsesquioxane**	5256-79-1	**NI47707**
331	317	385	343	531	339	613	357	**1034**	100	17	6	3	2	2	1	1	**0.00**	38	51	25	–	–	–	–	–	1	–	–	–	–	**6^2-Deoxy-6^2-iodomaltosetriose decaacetate**		**MS11103**
83	43	55	311	211	111	205	453	**1034**	100	84	76	75	65	52	13	6	**0.00**	54	82	19	–	–	–	–	–	–	–	–	–	–	**22-O-Acetyl-21-O-(3,4-di-O-angeloyl-β-D-fucopyranosyl)protoaescigenin 3-O-β-D-glucuronopyranoside**		**DD01837**
57	71	43	85	103	83	1040	69	**1040**	100	65	55	44	33	33	32	29		72	144	2	–	–	–	–	–	–	–	–	–	–	**Nonahexacontanoic acid, propyl ester**	40710-41-6	**NI47708**
91	519	517	133	948	518	515	516	**1054**	100	68	56	39	35	33	30	24	**0.00**	60	78	1	–	–	–	–	–	–	–	–	–	2	**Distannoxane, hexakis(2-methyl-2-phenylpropyl)-**	13356-08-6	**NI47709**
41	57	43	55	71	69	83	97	**1058**	100	78	75	69	53	53	44	39	**0.00**	69	134	6	–	–	–	–	–	–	–	–	–	–	**Docosanoic acid, 1,2,3-propanetriyl ester**	18641-57-1	**NI47710**
898	870	618	842	814	730	758	702	**1066**	100	93	91	72	64	64	60	60	**57.00**	16	–	16	–	–	–	–	–	–	–	–	–	6	**Hexadecacarbonylhexarhodium**		**LI10155**
203	235	345	377	344	319	376	295	**1076**	100	77	68	66	66	64	62	60	**0.00**	48	88	24	2	–	–	–	–	–	–	–	–	–	**Nana-α-(2-3)-Gal-β-(1-4)-GlcNAc-β-(1-3)-Galol, methylated**		**MS11104**
169	43	331	109	44	193	127	139	**1082**	100	73	63	54	39	35	33	15	**0.00**	49	62	27	–	–	–	–	–	–	–	–	–	–	**Cantleyoside**		**LI10156**
44	149	148	101	36	150	18	38	**1083**	100	42	29	7	7	5	4	3	**0.00**	54	81	9	3	–	–	–	–	–	6	–	–	–	**myo-Inositol, tris-O-[3-acetyl-1-(trimethylsilyl)-1H-indolyl]tris-O-(trimethylsilyl)-**	74523-73-2	**NI47711**
229	73	202	157	130	230	147	217	**1083**	100	90	60	38	36	27	26	18	**1.00**	54	81	9	3	–	–	–	–	–	6	–	–	–	**myo-Inositol, tris-O-[3-acetyl-1-(trimethylsilyl)-1H-indolyl]tris-O-(trimethylsilyl)-**	74523-73-2	**NI47712**
109	169	211	341	331	619	115	455	**1088**	100	91	90	87	87	75	68	62	**0.00**	57	84	20	–	–	–	–	–	–	–	–	–	–	**Oleanolic acid diglycoside heptaacetate**		**NI47713**
648	545	736	824	560	1088	893	912	**1090**	100	86	76	50	22	16	12	11	**0.00**	–	–	–	–	–	24	–	–	–	–	8	–	2	**Octakis(trifluorophosphine)diiridium**		**NI47714**
259	470	182	105	393	288	365	575	**1092**	100	70	35	26	25	20	16	10	**0.00**	72	60	–	–	–	–	–	–	–	6	–	–	–	**Dodecaphenylcyclohexasilane**		**LI10157**

MASS TO CHARGE RATIOS								M.W.	INTENSITIES								Parent	C	H	O	N	S	F	Cl	Br	I	Si	P	B	X	COMPOUND NAME	CAS Reg No	No
151	1040	180	1041	684	714	491	744	**1100**	100	67	44	42	11	7	7	6	**0.00**	61	64	19	–	–	–	–	–	–	–	–	–	–	(2R,3S,4S)-2,3-trans-3,4-trans-3-Acetoxy-4-[(2R,3S)-2,3-trans-3-acetoxy-3′,4′,7,8-tetramethoxyflavan-6-yl]-6-[(2R,3S,4R)-2,3-trans-3,4-cis-3-acetoxy-3′,4′,7-trimethoxyflavan-4-yl]-3′,4′,7-trimethoxyflavan		NI47715
151	183	185	683	180	122	297	1009	**1100**	100	83	80	44	38	31	30	22	**1.30**	61	64	19	–	–	–	–	–	–	–	–	–	–	Tris(acetyloxy)decamethoxytrileucofisetinidin		MS11105
91	514	165	516	241	167	254	255	**1104**	100	11	10	10	9	9	8	8	**0.00**	80	70	–	–	–	–	–	–	–	–	–	–	1	Decabenzylgermanocene		NI47716
69	319	177	44	45	81	97	547	**1110**	100	90	43	40	36	33	28	24	**0.10**	28	14	19	–	–	24	–	–	–	–	–	–	–	D-Glucose, 4-O-[2,3,4,6-tetrakis-O-(trifluoroacetyl)-β-D-galactopyranosyl]-, 2,3,5,6-tetrakis(trifluoroacetate)	56890-07-4	MS11106
514	1028	734	811	500	486	888	972	**1140**	100	55	40	18	18	13	12	8	**3.00**	27	15	9	–	3	–	–	–	–	–	–	–	3	Rhenium, (phenylthio)tricarbonyl-, trimer		MS11107
304	753	150	377	531	1134	227	830	**1140**	100	71	71	15	14	12	11	6	**0.00**	60	50	–	–	–	–	–	–	–	–	–	–	5	Decaphenylcyclopentagermane		LI10158
1149	590	1150	1121	1093	545	151	107	**1149**	100	85	36	36	21	18	14	12		36	28	4	–	–	30	–	–	–	–	–	–	1	Manganese(II) bis(3-perfluorooctanoylcamphorate)	80148-55-6	NI47717
91	65	78	92	514	165	516	77	**1150**	100	14	11	8	7	6	5	5	**0.00**	80	70	–	–	–	–	–	–	–	–	–	–	1	Decabenzylstannocene		NI47718
1168	1056	1084	1140	330	972	316	1000	**1168**	100	75	64	32	21	20	20	17		12	–	16	–	–	–	–	–	–	–	–	–	4	Tetraosmiumtetroxidedodecacarbonyl		MS11108
105	579	77	231	335	122	214	201	**1174**	100	22	16	12	9	9	2	2	**0.00**	68	54	19	–	–	–	–	–	–	–	–	–	–	α-D-Glucopyranoside, 1,3,4,6-tetra-O-benzoyl-β-D-fructofuranosyl, 2,3,4,6-tetrabenzoate	2425-84-5	NI47719
105	77	122	579	231	335	214	201	**1174**	100	34	23	17	16	2	2	2	**0.00**	68	54	19	–	–	–	–	–	–	–	–	–	–	α-D-Glucopyranoside, 2,3,4,6-tetra-O-benzoyl-α-D-glucopyranosyl, tetrabenzoate	71658-21-4	NI47720
219	91	57	259	203	41	263	147	**1176**	100	79	68	56	43	43	27	23	**0.00**	73	108	12	–	–	–	–	–	–	–	–	–	–	Pentaerythritol tetra[β-(3,5-di-tert-butyl-4-hydroxyphenyl) propionate]		IC04654
1167	979	246	230	345	169	773	139	**1182**	100	23	21	19	18	15	10	10	**0.00**	6	18	–	12	–	15	–	–	–	–	9	–	3	1,3,5,2,4,6-Triazatriphosphorine, 2,2′,2″-(2,2,4,4,6,6-hexamethyl-1,3,5,2,4,6-triazatristannine-1,3,5-triyl)tris[2,4,4,6,6-pentafluoro-2,2,4,4,6,6-hexahydro-	52708-48-2	NI47721
43	57	282	71	520	85	548	254	**1184**	100	93	47	47	29	25	21	18	**0.30**	78	156	4	2	–	–	–	–	–	–	–	–	–	Acetamide, 2,2′-[1,2-ethanediylbis(oxy)]bis[N,N-dioctadecyl-	67615-91-2	NI47722
69	866	119	131	169	1166	771	471	**1185**	100	62	26	21	19	18	12	12	**1.86**	24	–	–	3	–	45	–	–	–	–	–	–	–	1,3,5-Triazine, 2,4,6-tris(pentadecafluoroheptyl)-	21674-38-4	NI47723
65	111	139	534	334	166	655	185	**1190**	100	68	24	23	18	18	17	16	**5.00**	34	70	18	–	–	–	–	–	–	–	6	–	3	Bis(η[5]-2,4-cyclopentadien-1-yl)hexakis[μ-(diethylphosphito-O″:P)]tin dicobalt		MS11109
100	130	210	380	352	507	366	493	**1207**	100	46	33	27	11	8	7	3	**0.00**	65	121	13	7	–	–	–	–	–	–	–	–	–	Permethylated surfactin		LI10159
673	914	1006	594	65	822	1208	763	**1208**	100	45	20	19	15	12	10	9		34	70	18	–	–	–	–	–	–	–	6	–	3	Bis(η[5]-2,4-cyclopentadien-1-ylcobalt)hexakis[μ-(diethylphosphito-O″:P)]barium		MS11110

MASS TO CHARGE RATIOS	M.W.	INTENSITIES	Parent	C	H	O	N	S	F	Cl	Br	I	Si	P	B	X	COMPOUND NAME	CAS Reg No	No
371 216 357 155 370 224 217 190	**1210**	100 57 49 46 44 32 29 29	**0.00**	62	98	12	8	2	–	–	–	–	–	–	–	–	**Decanoylcysteylphenylleucylglycine**		**NI47724**
530 1139 428.5 327 733 936 349.5 248	**1218**	100 78 20 18 14 11 11 8	**0.09**	–	–	–	6	–	–	–	12	–	–	6	–	–	**Dodecabromohexaphosphonitrile**		**MS11111**
73 147 75 346 245 142 217 205	**1222**	100 58 53 52 50 39 36 28	**0.00**	48	122	10	2	–	–	–	–	–	12	–	–	–	**D-Galactose, 4-[bis(trimethylsilyl)amino]-4-deoxy-2,3,5,6-tetrakis-O-(trimethylsilyl)-, dimer**	56247-39-3	**NI47725**
318 228 273 363 697 567 477 155	**1224**	100 90 89 63 37 31 23 14	**0.00**	36	90	18	–	–	–	–	–	–	–	–	–	6	**Gallium ethoxide (hexamer)**		**LI10160**
273 203 248 438 133 423 207 454	**1232**	100 70 38 26 22 6 4 3	**0.10**	63	92	24	–	–	–	–	–	–	–	–	–	–	**Raddeanin A**	89412-79-3	**NI47726**
91 514 516 241 255 165 254 167	**1238**	100 17 14 12 11 11 11 11	**0.00**	80	70	–	–	–	–	–	–	–	–	–	–	1	**Decabenzylplumbocene**		**NI47727**
73 217 147 103 361 204 191 129	**1244**	100 24 16 16 15 11 11 10	**0.00**	51	112	15	–	–	–	–	–	–	10	–	–	–	**Melittoside, decakis(trimethylsilyl)-**		**NI47728**
665 820 666 667 664 605 403 633	**1246**	100 75 71 68 54 46 45 43	**0.00**	67	114	17	4	–	–	–	–	–	–	–	–	–	**Mycoside C(B1)**		**MS11112**
618 603 588 574 287 309 294	**1252**	100 66 58 34 32 19 11	**0.00**	76	100	1	8	–	–	–	–	–	–	–	–	2	**μ-Oxobis(α,γ-dimethyl-α,γ-dihydrooctaethylporphyrinato)iron(III)**		**NI47729**
840 346 164 182 1274 255 419 491	**1280**	100 99 99 91 77 46 31 21	**0.00**	70	70	–	–	–	–	–	–	–	–	–	–	5	**Decakis(p-tolyl)cyclopentagermane**		**LI10161**
169 109 212 271 127 81 115 211	**1297**	100 63 42 28 23 21 16 12	**0.00**	56	71	32	3	–	–	–	–	–	–	–	–	–	**Cellotetrose N-phenylosotriazole acetate**		**LI10162**
169 109 98 331 118 107 149 97	**1301**	100 59 45 34 32 30 28 23	**0.59**	57	75	33	1	–	–	–	–	–	–	–	–	–	**N-p-Tolylcellotetraosylamine acetate**		**LI10163**
596 566 564 598 592 565 624 693	**1316**	100 99 99 88 76 75 67 65	**0.00**	72	124	17	4	–	–	–	–	–	–	–	–	–	**Permethylmycoside C(B1)**		**MS11113**
57 71 43 85 97 83 55 69	**1318**	100 72 58 53 41 41 38 35	**0.11**	94	190	–	–	–	–	–	–	–	–	–	–	–	**Tetranonacontane**		**AI02416**
43 331 301 169 109 317 275 285	**1326**	100 75 55 55 42 35 33 30	**0.11**	60	70	30	4	–	–	–	–	–	–	–	–	–	**Glucopyranoside, D-erythro-2,3-dihydroxy-3-(1-phenyl-1H-pyrazolo[3,4-b]quinoxalin-3-yl)propyl O-α-D-glucopyranosyl-(1-6)-O-α-D-glucopyranosyl-(1-6)-, dodecaacetate**	33910-03-1	**NI47730**
549 521 550 522 648 392 620 956	**1359**	100 94 41 41 32 30 28 25	**20.00**	72	129	15	9	–	–	–	–	–	–	–	–	–	**Fortuitine**		**LI10164**
304 150 980 227 753 1057 377 531	**1368**	100 64 41 22 16 12 12 10	**0.00**	72	60	–	–	–	–	–	–	–	–	–	–	6	**Dodecaphenylcyclohexagermane**		**LI10165**
73 204 361 217 205 147 103 451	**1370**	100 81 62 36 32 31 28 27	**0.00**	54	130	16	–	–	–	–	–	–	12	–	–	–	**Per-O-(trimethylsilyl)-(3-O-α-D-mannopyranosyl-4-O-β-D-glucopyranosyl-D-glucitol)**		**MS11114**

MASS TO CHARGE RATIOS								M.W.	INTENSITIES								Parent	C	H	O	N	S	F	Cl	Br	I	Si	P	B	X	COMPOUND NAME	CAS Reg No	No
621	456	293	279	202	143	748	546	**1384**	100	89	80	62	46	37	32	19	0.00	5	8	–	–	–	–	–	–	8	–	–	–	4	Tetrakis(diiodoarsinomethyl)methane	51524-82-4	NI47731
872	960	1048	784	1136	480	696	436	**1402**	100	98	95	91	80	59	54	49	0.00	–	–	–	–	–	27	–	–	–	–	10	–	3	Triiridium, μ^3-phosphidononakis(trifluorophosphino)-		NI47732
222	307	335	223	308	455	432	295	**1457**	100	25	22	16	13	11	9	7	0.00	76	111	14	15	–	–	–	–	–	–	–	–	–	(Pro-Pro-Gly-Phe-Ser-Pro-Phe-)-α,ω-bis[N-α-methyl-deguanido-5-(4′-spirocyclopentyl-1′-methyl-5′-oxo-2′-imidazolinylaminomethyl)]arginine		LI10166
595	600	599	596	598	601	628	602	**1460**	100	63	53	50	49	48	41	36	0.00	96	148	10	–	–	–	–	–	–	–	–	–	–	Ethyl-2-O-mycolyl-3-O-methyl-4-O-acetyl-6-deoxypyranoside		NI47733
1520	1352	968	1040	1464	1184	1240	1268	**1520**	100	77	71	67	20	20	8	6		36	20	12	–	4	–	–	–	–	–	–	–	4	(Phenylthio)rheniumtricarbonyl tetramer		MS11115
346	164	255	182	601	510	419	1092	**1536**	100	99	91	87	80	70	60	51	0.00	84	84	–	–	–	–	–	–	–	–	–	–	6	Dodeca(p-tolyl)cyclohexagermane		LI10167
1390	1416	1013	695	1039	708	1442	1374	**1558**	100	75	48	40	38	30	15	15	0.00	100	170	10	2	–	–	–	–	–	–	–	–	–	6,7,38,39-Tetraacetoxy-41,42-bis(14-oxo-1-azacyclohexacosyl)-tricyclo[$35.3.1.1^{4,8}$]dotetraconta-1(41),4,6,8(42),37,39-hexaene	83274-58-2	NI47734
636	662	757	800	618	842	678	1390	**1558**	100	48	13	12	6	4	4	3	0.00	100	170	10	2	–	–	–	–	–	–	–	–	–	5,6,37,38-Tetraacetoxy-1,42-diazapentacyclo[$40.25.25.2^{3,40}.0^{2,7}.0^{36,41}$]-tetranonaconta-2,4,6,36,38,40-hexaen-55,80-dione		NI47735
121	114	100	1009	170	507	161	142	**1580**	100	78	73	71	64	60	60	60	7.00	77	118	16	12	–	6	–	–	–	–	–	–	–	Acetylmethylleucylmethylalanyldimethyltrifluoroacetyllysylmethylvalyl-methylalanyldimethyltyrosylmethylvalyldimethyltyrosyldimethyltri-fluoroacetyllysylproline methyl ester		LI10168
169	109	212	331	81	271	127	115	**1585**	100	55	28	19	19	16	16	14	0.00	68	87	40	3	–	–	–	–	–	–	–	–	–	Cellopentose N-phenylosotriazole acetate		LI10169
235	197	157	189	246	361	638	606	**1961**	100	96	86	76	60	50	36	36	0.00	88	159	44	3	–	–	–	–	–	–	–	–	–	Fuc-α-(1-2)-Gal-β-(1-4)-GlcNAc-β-(1-6)-[Fuc-α-(1-2)-Gal-β-(1-4)-GlcNAc β-(1-3)]Gal-β-(1-4)GlcNAc-β-(1-3)-Galol, methyl ester		MS11116
757	553	756	655	552	451	801	437	**9999**	100	93	61	51	38	24	11	7	0.00	–	–	–	–	–	–	–	–	–	–	–	–	1	Aluminium isopropoxide polymer		MS11117
60	29	43	57	28	73	31	44	**9999**	100	62	58	57	52	52	46	45	0.00	–	–	–	–	–	–	–	–	–	–	–	–	1	Cellulose	9004-34-6	NI47736
45	46	44	47	61	48			**9999**	100	76	15	9	8	4				–	–	–	–	–	–	–	–	–	–	–	–	1	Methanethiol, homopolymer		NI47737
69	131	119	181	169	231	219	51	**9999**	100	37	36	25	22	12	9	8	0.00	–	–	–	–	–	–	–	–	–	–	–	–	1	Perfluorokerosene		MS11118
39	42	44	94	58	124	65	51	**9999**	100	96	95	80	58	52	40	40	0.00	–	–	–	–	–	–	–	–	–	–	–	–	1	DTA-Bis-phenol-diglycidyl ether polymer		LI10170
209	207	165	163	135	205	179	185	**9999**	100	72	66	50	48	46	44	40	0.00	–	–	–	–	–	–	–	–	–	–	–	–	1	Trimethyltin-acetate polymer		LI10171
185	165	183	163	187	181	161	155	**9999**	100	76	70	57	36	36	33	30	0.00	–	–	–	–	–	–	–	–	–	–	–	–	1	Trimethyltin-O-methyl-sulphinate polymer		LI10172